SCIENTIFIC AND TECHNICAL
Acronyms, Symbols, and Abbreviations

About the Authors

Uwe Erb is a professor in the Department of Materials Science and Metallurgy at the University of Toronto, Canada. He has published widely in the fields of materials science, metallurgy, and applied physics, and is author of several books and book chapters on grain boundaries and nanocrystalline materials as well as co-author of two dictionaries on abbreviations and acronyms and a database on engineering materials.

Harald Keller is a technical translator and writer with a broad background in mechanical engineering and materials testing. He is author of two dictionaries on abbreviations and acronyms as well as a database on engineering materials.

SCIENTIFIC AND TECHNICAL
Acronyms, Symbols, and Abbreviations

◆ ◆ ◆

Uwe Erb
Harald Keller

Ⓦ WILEY-INTERSCIENCE

JOHN WILEY & SONS, INC.

New York • Chichester • Weinheim • Brisbane • Singapore • Toronto

For ordering and customer service, call 1-800-CALL-WILEY.

Library of Congress Cataloging-in-Publication Data:
Library of Congress Cataloging-in-Publication Data is available.

ISBN 0-471-38802-5

Printed in the United States of America.

10 9 8 7 6 5 4 3 2 1

Contents

Introduction

This book contains approximately 200,000 entries, offering comprehensive, up-to-date coverage of abbreviations, acronyms, and symbols in more than 80 scientific and technical fields. Following the main section is the *Appendix*, including a comprehensive list of scientific and technical letter and nonletter signs and symbols.

It is very likely that no two authors would agree on which entries to include in a multidisciplinary compilation such as this. Nevertheless, it was the endeavor of the authors to cover the various scientific and technical fields in as much depth and breadth as possible and, at the same time, provide a certain balance between them. However, at times, this proved to be a challenge, since the number of acronyms and abbreviations varies greatly from field to field. For example, there are many more abbreviations and acronyms in fields such as computer science, communications and electronics than in architecture, mathematics, or mining.

As any reader may observe, *abbreviations* are used more frequently in scientific and technical writing than in the literature in general. Such "shortenings" or "contractions" are justified because, by the omission of letters or words, they save both time and space. In the absence of a single set of rules governing the formation and spelling of abbreviations, the following forms are most commonly encountered:

1. **Initials or first letters:** ED (Elastic Deformation), HW (Hot Water), MS (Multiple Sclerosis)
2. **First and last letters:** Bm (Beam), Dr (Doctor), Fd (Fjord), Ga (Georgia), Wd (Wood)
3. **Key letters:** Bldg (Building), Dwg (Drawing), Lgth (Length), Std (Standard), Stl (Steel)
4. **Syllables:** Biol (Biology), Can (Canada), Eq (Equation), Insul (Insulation), Lett (Letter).

There are also combinations of these forms: COSPEAR (COmmittee on Space Programs for EARth Observation), Naptalam (α-Napthylphthalamic Acid), Prepreg (Preimpregnated Material).

Abbreviations that are formed from the initial letters or syllables of words and are pronounced as such are called *acronyms*. These have become very common in the scientific and technical literature: LASER (Light Amplification by Stimulated Emission of Radiation), NASA (National Aeronautics and Space Administration), NATO (North Atlantic Treaty Organization), RADAR (Radio Detection and Ranging).

Letter symbols are conveniently used by scientists, engineers, and technologists for chemical elements, mathematical functions, physical quantities, units of measurement, electronic and electrical components, etc.: Au (Gold), f (Function), k (Boltzmann constant), kg (kilogram), M (Figure of merit), R (Resistor).

Some of the acronyms and abbreviations given in this book refer to proprietary devices, products, methods, or processes, e.g., the abbreviation "GBE" (Grain Boundary Engineering) is also a trademark of Integran Technologies Inc., Toronto, Ontario, Canada for a group of unique "grain-boundary engineered" materials with enhanced chemical, physical, and mechanical properties. The absence of a specific trademark designation or symbol in such cases should not be construed as affecting the ownership, scope, or validity of any trademark rights.

The authors greatly welcome any corrections and suggestions from readers, for these are an important source of information about the acceptance of this book, and an invaluable basis for the preparation of future editions.

U. Erb
H. Keller

Guide to This Book

This guide deals with the different kinds of entries and with their arrangement in this book. To use this compilation most effectively, it is recommended to read this section first.

1. Arrangement of Entries

In researching the literature, the authors have come across various styles used for listing acronyms, abbreviations, and symbols. In order to present the myriad of entries particular to this book, they have chosen a combination of these styles.

The abbreviations, acronyms and symbols in this dictionary are given in alphabetical order with key entries printed in boldface type. Uppercase spellings precede lowercase forms; entries containing spacings, superscripts, subscripts, ampersands, slashes, dashes, or numerals follow those without such characters. Abbreviations including numerals are listed under the corresponding letter, or letter sequence if the numeral is an integral part thereof, e.g., 3PDT (Triple-Pole Double Throw) follows "TPDT." Often, if the numeral is separated from the abbreviation by a space, dash, or slash, it is listed following the corresponding letter, e.g., SA 0 (Stuck-at-Zero Error) follows the entries under "SA." This is also true for most chemical and biochemical compounds, radioactive isotopes, military grades, etc. Abbreviations consisting of letters and subscripted or superscripted numerals, or letters and numerals, sometimes separated by dashes, are listed in alphabetical-numerical order following those without numerals, e.g., AP precedes AP-3 which, in turn, precedes AP-4. The following sequence of entries from letter "C" shows the general alphabetical order:

CB Canada Balsam

CB Large Cruiser [US Navy Symbol]

CB15 4'-(2-Methylbutyl)biphenyl-4-Carbonitrile

5CB Pentylcyanobiphenyl

23C14B 2,3-Dichloro-1,4-Benzoquinone

C/B Cold Water/Boiling Water (Immersion Ratio)

Cb Cerebellum [also cb] [Anatomy]

Cb-90 Columbium-90 [also ^{90}Cb, or Cb^{90}]

cB Coherent (Grain) Boundary (in Materials Science) [Symbol]

c-B Crystalline Boron

cb centibel [Unit]

c/b *c*-axis to *b*-axis ratio (of crystal lattice) (for monoclinic, triclinic and orthorhombic systems) [Symbol]

Greek letters are listed according to their English spellings, e.g., entries related to Γ (Gamma) follow the entries under "g," and entries related to Σ (Sigma) follow the "s" entries. Wherever possible equivalent or closely related explanations of the same abbreviation or acronym are considered one entry and separated by a semicolon, e.g.:

SEM Scanning Electron Micrograph; Scanning Electron Microscope; Scanning Electron Microscopy

In alphabetizing, many of the prefixes used in organic chemistry are disregarded, since they are not considered an essential part of the abbreviation; these include alpha (α), meta- (m-), N-, O-, tert-, sym-, trans-, as well as all numerals denoting structure, e.g., α-DPN (α-Diphosphopyridine Nucleotide) follows the entries under "DPN." However, the authors have found that, in some cases, prefixes of organic and biochemical compounds are included, e.g., BG (β-Galactosidase). Prefixes for chemical elements, inorganic compounds, metallurgical phases, etc., such as α (alpha) and γ (gamma), are usually considered an integral part of the abbreviation, consequently, related entries are found under A and G respectively, e.g., α-C (alpha carbon) follows the entries under "ac" and γ-Fe (gamma iron) follows those under "GFE."

In science and technology, it has become customary to omit periods after abbreviations and between the individual letters of abbreviations and acronyms for longer terms, e.g., MTS (Marine Technology Society) instead of M.T.S. and NIST (National Institute of Standards and Technology) instead of N.I.S.T. With some exceptions, the authors have adopted this style throughout this book.

2. Explanatory Information

For a multidisciplinary compilation such as this, the use of supplemental explanatory information is virtually essential. Depending on the type of information, it has been added either in parentheses or square brackets:

1. *Parentheses* are used to include information which, although not part of a particular abbreviation or acronym, will help the reader to categorize, or specify an entry, e.g., CAPS is given as "Current Advances in Plant Science (Database)," and "DICHILL" is explained as "Dilute-Chill (Process)." Parentheses are also used in cases in which an abbreviation or acronym forms part of the corresponding explanation, e.g., "SPCC" is given as "STS (Space Transportation System) Processing Control Center." Other instances in which parentheses are employed include certain chemical compounds, explanatory remarks, synonyms, additional spellings, plural forms, etc.

2. *Square brackets* are used to provide additional information on the particular scientific or technical field in which an abbreviation or acronym is most commonly found, as well as to specify affiliations, divisions, departments, and/or locations (city, state/province, country, etc.) of companies, institutes, laboratories, organizations, etc. English translations of foreign organizations are also given in square brackets. For

example, the explanation for "BVP (Boundary Value Problem)" is followed by the field label "[Mathematics]," which will help the reader to clearly categorize this entry; IBA (Industrial Biotechnology Association) is followed by "[US]," clearly specifying it as an organization based in the United States; and "DAGA" is given as "Deutsche Arbeitsgemeinschaft für Akustik" followed by the English translation "[German Acoustics Association]." Square brackets are also used for alternate or variant forms of abbreviations and acronyms, e.g., "Terr (Terrace) [also terr]" indicates that both forms "Terr" and "terr" are frequently used. The reader should note that the authors have included only those alternates and variants which they have actually found in the literature; there may well be others.

Field labels, if necessary, are used to identify the field of science or technology in which an abbreviation, acronym or symbol is most frequently found. While some are highly specialized, others could be applied only broadly owing to the overlapping between fields. Usually, if an entry is used in several fields, the field label is either omitted, or a more general field is specified, e.g., an abbreviation or acronym used in both particle and nuclear physics may be labeled "[Physics]."

Fields Covered in This Book

Following is an alphabetical list of the major scientific and technical fields covered in this dictionary:

Aeronautics and Aviation
Aerospace Science and Engineering
Agriculture
Anatomy
Anthropology
Archeology
Architecture
Astronomy
Astrophysics
Atomic Physics
Bacteriology
Biochemistry
Bioengineering
Biophysics
Biology
Botany and Horticulture
Building Construction
Ceramics
Chemistry and Chemical Engineering
Civil Engineering
Computer Science
Crystallography
Design Engineering
Electrical Engineering
Electronics
Engineering Education
Environmental Science and Engineering
Forestry
Geochemistry
Geodesy
Geography
Geology
Geophysics
Housing
Industrial Engineering
Immunology

Information Science and Technology
Manufacturing
Materials Science and Engineering
Mathematics
Mechanical Engineering
Medicine
Metallography
Metallurgy
Meteorology
Microbiology
Military Science and Engineering
Mineralogy
Mining
Molecular Biology
Nanotechnology
Navigation
Nuclear Science and Engineering
Nursing
Oceanography
Paleontology
Petrology
Pharmacy and Pharmacology
Photography
Physics
Physiology
Polymer Science and Engineering
Quality Control
Radio and Television
Statistics
Textile Technology
Thermodynamics
Toxicology
Transportation
Veterinary Medicine
Virology
Zoology

SCIENTIFIC AND TECHNICAL
Acronyms, Symbols, and Abbreviations

A   *a*-axis to *b*-axis ratio (of crystal lattice) [Symbol]
A   Abdomen; Abdominal
A   Abel [US Telephony Spelling Alphabet]
A   Aberdeen [Scotland]
A   Abies [Genus of Coniferous Trees Including the Firs]
A   Abiotic
A   Ablation; Ablator
A   Abort(ion)
A   Absolute
A   Absolute Temperature
A   Absorb(ent); Absorber
A   Absorbance [Symbol]
A   Absorbency [Symbol]
A   Absorption
A   Absorption Area [Symbol
A   Absorption Current [Symbol]
A   Absorptivity [Symbol]
A   Abstract(ion)
A   Acanthaster [Biology]
A   Acaracide
A   Accumulation; Accumulator
A   Acoustic(al); Acoustics
A   Accelerate; Acceleration; Accelerator
A   Accept(able); Acceptance; Acceptor
A   Acceptor
A   Access(ing)
A   Accessory
A   Accident(al)
A   Accommodate; Accommodation
A   Accommodation Amplitude (of Human Eye) [Symbol]
A   Accumulate; Accumulation
A   Accurate; Accuracy
A   Ace
A   Acer [Genus of Deciduous Trees Including the Maples]
A   Acetone

A   Acid(ic)
A   Acquire(d); Acquisition
A   Acre [Unit]
A   Actin [Biochemistry]
A   Actinic; Actinism
A   Actinomyces [Genus of Fungi]
A   Activate; Activation; Activator; Active; Activity
A   Active Length (of Deflected Springs) [Symbol]
A   (Optical) Activity [Symbol]
A   Activity (of Radionuclides) [Symbol]
A   Acuity; Acute
A   Adatom
A   Adapt(er); Adaption; Adaptive; Adaptivity
A   Addendum (of Gear Teeth) [Symbol]
A   Add(er); Addition
A   Addict(ed); Addiction
A   Address
A   Adelaide [Australia]
A   Adenine [Biochemistry]
A   Adenosine [Biochemistry]
A   Adhere; Adhesion
A   Adhesive
A   Adipose [Biology]
A   Adjust(ment)
A   Administer; Admininstration; Administrative
A   Admiral [also ADM, or Adm]
A   Admission
A   Adopt(ion)
A   Adrenaline [Biochemistry]
A   Adsorb(ent); Adsorption
A   Adult
A   Advance(d); Advancement
A   Advantage(ous)
A   Advice
A   Aedes [Genus of Mosquitos]
A   Aerial
A   Aerobacter [Genus of Bacteria]
A   Aerofoil
A   Aeronaut(ical); Aeronautics
A   Aerosol
A   Aerospace; Aerospatial
A   Aerospace Engineering (Technology) [Discipline Category Abbreviation]
A   Aesculus [Genus of Deciduous Trees Including the Buckeyes]
A   Affin(ity)
A   Afterburn(ed); Afterburner; Afterburning
A   Aftercool(ed); Aftercooler; Aftercooling
A   Afghanistan(i)
A   Africa(n)
A   Agave
A   Age(d); Ageing
A   Agency; Agent
A   Agglutinate; Agglutination

1

A   Agglutinin [Immunochemistry]
A   Aging
A   Agitation; Agitator
A   Agree(ment)
A   A Horizon [Horizon of Soil Eluviation Consisting of Several Subhorizons A_1, A_2, etc.] [Symbol]
A   Aid
A   Aileron
A   Aim
A   Air
A   Airfoil
A   Airbag
A   Airborne
A   Air-Cooled; Air Cooler; Air Cooling
A   Aircraft
A   Air-Launched (Vehicle) [USDOD Symbol]
A   Airman; Airwoman
A   Airplane
A   Airspeed
A   Airway(s)
A   Alabama [US]
A   (+)-Alanine [Biochemistry]
A   Alanyl [Biochemistry]
A   Alaska(n)
A   Albania(n)
A   Alberta [Canada]
A   Albumin [Biochemistry]
A   Albuquerque [New Mexico, US]
A   Alcaligenes [Microbiology]
A   Alcohol(ic)
A   Alcoholism
A   Alder
A   Aldose [Biochemistry]
A   Alexandria [Egypt]
A   Alfa [Phonetic Alphabet]
A   Algae
A   Algeria(n)
A   Algiers [Algeria]
A   Algorithm(ic)
A   Alkane
A   Alkene
A   Alkylation
A   Alkyne
A   Allele [Genetics]
A   Allergen; Allergic; Allergy
A   Alligator
A   Allium [Botany]
A   Allow(ance)
A   Alloy(ed); Alloying
A   Alnus [Genus of Deciduous Trees Including the Alders]
A   Aloe [Botany]
A   (Upper-case) Alpha [Greek Alphabet]
A   Alpine; Alps
A   Altitude

A   Alumina (or Aluminum Oxide) [Ceramics]
A   Aluminate
A   Aluminum
A   Aluminum Oxide (or Alumina) [Ceramics]
A   Aluminum Oxide Type (Grinding Wheel) Abrasive [Symbol]
A   Alveolus; Alveoli [Anatomy]
A   Amacrine
A   Amalgam
A   Amarillo [Texas, US]
A   A-Matrix
A   Amazon (River)
A   Amazonas [Brazilian State]
A   Amber
A   Ambulance; Ambulant
A   America(n)
A   6-Amino Purine
A   Amman [Jordan]
A   Ammeter
A   Ammonia
A   Ammonium
A   Amoeba
A   Amorph(ization); Amorphous
A   Amorphous Phase [Materials Science]
A   Amount
A   Ampere [Unit]
A   Amphibia(n)
A   Amplification; Amplifier
A   Amplifier Gain [Symbol]
A   Amplitude [Symbol]
A   Amplitude Factor (in Acoustics) [Symbol]
A   Amplitude Limiter
A   Amplitude Modulation [Symbol]
A   Amsterdam [Netherlands]
A   Amylase [Biochemistry]
A   Anabaena [Genus of Algae]
A   Analog(ous); Analogy
A   Analysis; Analytical; Analyze(r); Analyst
A   Anaphylactic; Anaphylaxis
A   Anatase [Mineral]
A   Anatomical; Anatomist; Anatomy
A   Anchorage [Alaska, US]
A   Ancylostoma [Genus of Roundworm]
A   Andean; Andes
A   Andorra
A   Andrew [UK Telephony Spelling Alphabet]
A   Androgen [Biochemistry]
A   Anemia
A   Anemometer; Anemometry
A   Anemone
A   Anerobe; Anerobic
A   Anesthesia; Anesthetic(s); Anesthetist
A   Angiosperm
A   Angle; Angular(ity)

A Angola(n)
A Angstrom (Unit)
A Aniline
A Animal
A Anisotropic; Anisotropy
A Ankara [Turkey]
A Annapolis [Maryland, US]
A Anneal(ed); Annealing
A Annealed Temper [Symbol]
A Annihilate; Annihilation
A Annual(ed)
A Annuities [Symbol]
A Anode; Anodic [General Abbreviation]
A Anode [Semiconductor Symbol]
A Anopheles [Genus of Mosquitoes]
A Answer(ing)
A Antagonist; Antagony
A Antarctic(a)
A Antenna
A Anthrax
A Antibody
A Anticline [Geology]
A Anticoagulant
A Antidote
A Antiferroelectric(ity)
A Antigen(ic)
A Antihistamine
A Antilles
A Antiparallel (Spin) [Symbol]
A Antiprism
A Antiseptic
A Antitoxin [Immunology]
A Antwerp [Belgium]
A Anus; Anal
A Aorta
A Aperture
A Apis [Genus of Insects Including the Bees]
A Apogee [Astronomy]
A Apoapsis [Astronomy]
A Apparatus
A Appendicitis
A Appendix
A Appetite
A Appliance
A Application; Applicator
A Approach
A Approval
A Approximate; Approximation
A Arabia(n)
A Arachnoid [Anatomy]
A Arc
A Arch
A Archeozoic [Geological Era]

A Archipelago
A Architect(ural); Architecture
A Archival; Archive
A Arctic
A Area; Areal
A Argentina; Argentinean
A Argon [Abbreviation; Symbol: Ar]
A Arithmetic
A Arizona [US]
A Arkansas [US]
A Arlington [Texas, US]
A Arm
A Armature
A Armature Acceleration [Controllers]
A Army
A Aromatization
A Arthritic; Arthritis
A Artillery
A Arrange(ment)
A Array
A Arrest
A Arrive; Arrival
A Arsonium [Chemistry]
A Arterial; Artery
A Arteriole [Anatomy]
A Articulate; Articulation
A Artificial
A Aruba [Netherlands Antillles]
A Asbestos
A Asbestos-Covered (Electric Wire)
A Ascaris [Genus of Roundworm]
A Ascend(ing)
A Aseptic; Asepsis
A Ash
A Asia(n)
A A-Size Sheet (of Paper) [8½" × 11"]
A Aspect Ratio [Symbol]
A Aspergillus [Genus of Fungi]
A Assemble(r); Assembly
A Assimilation
A Assist(ance); Assistant
A Associate; Association
A Assumption
A Astable
A Aster
A Asthma(tic)
A Astrological; Astrologist; Astrology
A Astrometric; Astrometry
A Astronaut(ics)
A Astronomer; Astronomic; Astronomy
A Asuncion [Paraguay]
A Asymmetric; Asymmetry
A Asynchronous

A   Athabasca (Bitumen) [Alberta Oil Sands, Canada]
A   Athens [Greece]
A   Atlanta [Georgia, US]
A   Atlantic (Ocean)
A   Atmosphere; Atmospheric
A   Atom(ic)
A   Atomic Weight [Symbol]
A   Atomize(r); Atomization; Atomizing
A   Atrophy [Medicine]
A   Attack
A   Attack Aircraft [US Navy Symbol]
A   Attempt Frequency [Symbol]
A   Attenuation [Symbol]
A   Attitude
A   Attract(ion)
A   Attribute
A   Atypical
A   Auckland [New Zealand]
A   Audible; Audibility
A   Audiogram
A   Audiological; Audiologist; Audiology
A   Auricle [Anatomy]
A   Aurora(l)
A   Austenite; Austenitic [Metallurgy]
A   Austin [Texas, US]
A   Australasia(n)
A   Australia(n)
A   Austria(n)
A   Author
A   Authority; Authorization
A   Auto-Immune; Auto-Immunity
A   Automata; Automaton
A   Automatic
A   Autotroph(ic)
A   Auxiliary
A   Auxin [Biochemistry]
A   Availability; Available
A   Availability Factor [Symbol]
A   Average
A   Aviation; Aviator
A   Axial(ity)
A   Axis
A   Axle
A   Axon [Anatomy]
A   Azide
A   Azimuth(al)
A   Azores [Portugal]
A   Blood Group "A" [Contains Red Blood Cells with Substance "A" on Surface]
A   Chemical Activity [Symbol]
A   Clearance (of Milled Bevel Gears) [Symbol]
A   Cross-Section(al Area) [Symbol]
A   Dorn Constant [Symbol]
A   Exchange Constant [Symbol]

A   Exchange Stiffness Constant (of Ferromagnetics) [Symbol]
A   Fraunhofer "A" Line [Extreme-Red Fraunhofer Line of Terrestrial Origin Having a Wavelength of 766.1 Nanometers] [Spectroscopy]
A   Grade Denoting Excellence
A   Groove Angle (of Welds) [Symbol]
A   Helmholtz Free Energy [Symbol]
A   Humid Tropical Climate [Symbol]
A   Influenza, Strain A [Symbol]
A   Load Ratio (in Fatigue) [Symbol]
A   Madelung Constant (in Solid-State Physics) [Symbol]
A   Mass Number [Symbol]
A   Maximum Work (of a Thermodynamic System) [Symbol]
A   Mean Areal Intercept (in Quantitative Metallography) [Symbol]
A   Medium-Alloy, Air-Hardening Type Cold-Work Tool Steels [AISI-SAE Symbol]
A   Metric Paper Size [Designated by the Letter "A" Followed by a Numeral, e.g., A1, A2, A3 etc.]
A   Molar Refractivity [Symbol]
A   Monopole Vector Potential [Symbol]
A   Optical Density [Symbol]
A   Planar Area of Intercepted Features, or Test Area (in Quantitative Metallography) [Symbol]
A   Pre-Exponential Constant [Symbol]
A   Principal Angle of Refraction (of a Prism) [Symbol]
A   Scattering Asymmetry [Symbol]
A   Sound Absorption [Symbol]
A   Star Group of Spectral Type A [Surface Temperature 10,000°C, or 18,000°F] [Letter Designation]
A   Strain-Energy Factor [Symbol]
A   Stress Ratio (i.e., Ratio of Alternating Stress Amplitude S_a to the Mean Stress S_m) (in Fatigue) [Symbol]
A   Surface Area [Symbol]
A   Tropical Rainforest Climates, Coolest Month above 64.4°F (18°C) [Classification Symbol]
A   (Electromagnetic) Vector Potential [Symbol]
A   Vertex (or Associated Polygonal Angle) [Symbol]
.A   Archive, or Library File [File Name Extension] [also .a]
-A   Amphibian [US Navy Suffix]
$\overline{A}$ Mean Areal Intercept (in Quantitative Metallography) [Symbol]
$\overline{A}$ Average Grain Area (in Metallography) [Symbol]
$\tilde{A}$ Adjoint (or Hermitian Conjugate) [also A*] [Symbol]
A*   Richardson Constant [Symbol]
A**   Effective Richardson Constant [Symbol]
A'   Aperture of an Internal Reflection Element [Symbol]
A'   Solid (Cross-Sectional) Area of (Magnetic) Core [Symbol]
A'   Transpose of Matrix "A" (in Mathematics) [Symbol]
A°   Absolute Temperature [Symbol]
A$^+$ Positive Acceptor (in Semiconductivity) [Symbol]
A$^-$ Anion [Symbol]
A$^-$ Negative Acceptor (in Semiconductivity) [Symbol]
A$_\perp$ Perpendicular Area (or Vertical Area) [Symbol]

A$_{||}$ Horizontal Area (or Parallel Area) [Symbol]

A$_0$ Original (Cross-sectional) Area [Symbol]

A$_0$ Proportionality Factor [Symbol]

A$_0$ Second Uppermost Soil Horizon Composed of Decomposing Organic Matter [Symbol]

A$_{00}$ Uppermost Soil Horizon Composed of Undecomposed Litter [Symbol]

A$_1$ Blood Supergroup A$_1$ [Symbol]

A$_1$ Face-Centered Cubic Crystal Structure [Symbol]

A$_1$ Michigan Influenza Strain [Symbol]

A$_1$ Uppermost Horizon of Soil Eluviation with High Humus Content [Symbol]

A$_2$ Asian Influenza Strain [Symbol]

A$_2$ Blood Supergroup A$_2$ [Symbol]

A$_2$ Body-Centered Cubic Crystal Structure [Symbol]

A$_2$ Middle Horizon of Soil Eluviation with Maximum Leaching [Symbol]

A$_3$ Bottom Horizon of Soil Eluviation–Transitional Zone [Symbol]

A$_3$ Hexagonal-Close Packed Crystal Structure [Symbol]

A$_4$ Cubic Diamond Crystal Structure [Symbol]

A$_5$ Body-Centered Tetragonal Crystal Structure [Symbol]

A$_6$ Face-Centered Tetragonal Crystal Structure [Symbol]

A$_7$ Rhombohedral Crystal Structure [Symbol]

A$_8$ Trigonal Crystal Structure [Symbol]

A$_A$ (Chemical) Activity of Component, or Species A [Symbol]

A$_A$ Areal Fraction [Symbol]

A$_A$ Atomic Weight of Anions [Symbol]

A$_b$ Absorptivity of Blackbody [Symbol]

A$_b$ Area of Burning Surface [Symbol]

A$_c$ Atomic Weight of Cations [Symbol]

A$_c$ (Total) Cross-sectional Area of Fiber-Reinforced Composite [Symbol]

A$_{CJ}$ Cottrell-Jaswon Interaction Parameter (in Physical Metallurgy) [Symbol]

A$_d$ Deformed Area (or Area of Deformation) [Symbol]

A$_e$ Exit Area (in Aerodynamics) [Symbol]

A$_F$ Fischer Interaction Parameter (in Physical Metallurgy) [Symbol]

A$_f$ Austentitic Finish Temperature (in Metallurgy) [Symbol]

A$_f$ Final (Cross-Sectional) Area [also A$_f$] [Symbol]

A$_f$ Fractured (Cross-Sectional) Area [Symbol]

A$_f$ (Total) Cross-Sectional Area of Fibers of Fiber-Reinforced Composite [also A$_f$] [Symbol]

A$_i$ (Chemical) Activity of Component, or Species "i" [Symbol]

A$_i$ Atomic Weight of Component, Element, or Species i [Symbol]

A$_i$ Initial (Cross-Sectional) Area [Symbol]

A$_{ij}$ Algebraic Complement, or Cofactor of a$_{ij}$ in a Determinant, or Matrix [Symbol]

A$_M$ Atomic Weight of Metal [Symbol]

A$_m$ (Total) Cross-Sectional Area of Matrix of Fiber-Reinforced Composite [Symbol]

A$_n$ Cross-Section of Electrical Conductor in Circular Mils (in American Wire Gauge System) [Symbol]

A$_o$ Formula (or Molecular) Weight of Oxide [Symbol]

A$_p$ Surface Area of Particle [Symbol]

A$_r$ Area Ratio of Soil Sampler, Sampling Spoon, or Sampling Tube [Symbol]

A$_r$ Relative Atomic Mass [Symbol]

A$_S$ Snoek Interaction Parameter (in Physical Metallurgy) [Symbol]

A$_s$ (Cross-Sectional) Area of the Sample [Symbol]

A$_s$ Austentitic Start Temperature (in Metallurgy) [Symbol]

A$_{SU}$ Suzuki Interaction Parameter (in Physical Metallurgy) [Symbol]

A$_t$ Throat Area (in Aerodynamics) [Symbol]

A$_{VCL}$ Closed-Loop Voltage Gain For Operational Amplifier [Semiconductor Symbol]

A$_{VOL}$ Open-Loop Voltage Gain For Operational Amplifier [Semiconductor Symbol]

A^T Transpose of Matrix "A" (in Mathematics) [Symbol]

A+ A Plus (or A Positive) (Terminal) [Symbol]

A0 Emission (or Transmission)–No Modulation [Symbol]

A0 Metric Paper Size [841 mm × 1189 mm]

A0A2 Emission (or Transmission)–Modulation Keyed, Carrier not Keyed [Symbol]

A1 Continuous Wave Telegraphy [Symbol]

A1 Metric Paper Size [594 mm × 841 mm]

A1 On-Off Keying Telregraphy–No Modulating Audio Frequency [Symbol]

A2 Continuous Signal of an Emission (or Transmission), Amplitude-Modulated by Low-Frequency Periodic Oscillation, Double Sideband [Symbol]

A2 Metric Paper Size [420 mm × 594 mm]

A2 Telegraphy by On-Off Keying of One, or Several Periodic Oscillations Low-Frequency Amplitude-Modulating the Emissions (or Transmissions), or by Keying of such Modulated Emissions (or Transmissions) [Symbol]

A2 Twofold (Crystallographic) Axis [Symbol]

A2H Continuous Signal of an Emission (or Transmission) Amplitude-Modulated by Periodic Oscillation, Single Sideband, Full Carrier [Symbol] [also A2h]

A3 Double Sideband Telephony, Amplitude Modulated, Full Carrier [Symbol]

A3 Metric Paper Size [297 mm × 420 mm]

A3A Single-Sideband Telephony–Amplitude Modulated, Reduced Carrier [Symbol] [also A3a]

A3B Telephony–Two Independent Sidebands, Amplitude Modulated, Carrier Reduced, or Suppressed [Symbol] [also A3b]

A3H Single-Sideband Telephony–Amplitude Modulated, Full Carrier [Symbol] [also A3h]

A3J Single-Sideband Telephony–Amplitude Modulated, Suppressed Carrier [Symbol] [A3j]

A4 Facsimile–Main-Carrier Modulation either by Frequency-Modulated Subcarrier, or Directly [Symbol]

A4 Metric Paper Size [210 mm × 297 mm]

A4A Facsimile–Subcarrier Frequency-Modulated by the Picture Signal and Amplitude-Modulating the Main Carrier, Single Sideband, Reduced Carrier [Symbol] [also A4a]

A4J Facsimile–Subcarrier Frequency-Modulated by the Picture Signal and Amplitude-Modulating the Main Carrier, Single Sideband, Suppressed Carrier [Symbol] [also A4j]

A5 Metric Paper Size [148 mm × 210 mm]

A5C Television–Amplitude Modulated, Vestigial Sideband, Picture Only [Symbol]

A6 Metric Paper Size [105 mm × 148 mm]

A7 Metric Paper Size [74 mm × 105 mm]

A7A Multichannel Voice-Frequency Telegraphy–Single Sideband, Reduced Carrier [Symbol] [also A7a]

A7B Multichannel Voice-Frequency Telegraphy–Independent Sideband, Reduced Carrier [Symbol] [also A7b]

A7J Multichannel Voice-Frequency Telegraphy–Single Sideband, Suppressed Carrier [Symbol] [also A7j]

A8 Metric Paper Size [52 mm × 74 mm]

A9 Metric Paper Size [37 mm × 52 mm]

A9B Speech and Multichannel Telegraphy Combinations–Independent Sideband, Reduced, or Suppressed Carrier [Symbol]

A10 Tenfold (Crystallographic) Axis [Symbol]

a   about

a   absolute

a   absorption coefficient; absorption parameter [Symbol]

a   absorptivity [Symbol]

a   acaracide

a   (linear) acceleration [Symbol]

a   account

a   acid

a   acre [Unit]

a   activity (e.g., of chemicals) [Symbol]

a   addendum (of gear pinion teeth, worm threads, etc.) [Symbol]

a   aerial

a   air

a   allele [Genetics]

a   allowance [Symbol]

a   alpha [English Equivalent]

a   ampere [Unit]

a   amplitude (of vibration) [Symbol]

a   amorphous

a   analog(ous)

a   anisotropy ratio [Symbol]

a   *(annus)* – year

a   anode

a   anorthic (crystal system) [Symbol]

a   *(ante)* – before

a   anterior

a   antiparallel (spin) [Symbol]

a   aperture diameter (in optics) [Symbol]

a   apothegm [(in mathematics) [Symbol]

a   append lines (command) [Edlin MS-DOS Line Editor]

a   aqua

a   are [Unit]

a   arithmetic mean (in statistics) [Symbol]

a   arterial; artery

a   asymmetric

a   asynchronous

a   atactic [Polymers]

a   atom(ic)

a   atom flux [Symbol]

a   atomic diameter [Symbol]

a   atom size [Symbol]

a   atto [SI Prefix for 10^{-18}]

a   autologous

a   autonomic

a   auxiliary switch (breaker) normally open [Controllers]

a   azimuth(al)

a   base (of exponential expressions, e.g., a^m, a^n etc.) [Symbol]

a   bond length (of atoms, or ions) [Symbol]

a   capillary constant [Symbol]

a   coefficient [Symbol]

a   constant [Symbol]

a   crack length (or crack size) (of surface crack) [Symbol]

a   crystal lattice length (or lattice parameter) along a-axis [Symbol]

a   depth [Symbol]

a   distance between crystal lattice sites [Symbol]

a   flaw size [Symbol]

a   half length of an internal crack [Symbol]

a   half thickness of conductor [Symbol]

a   initial term in arithmetic, or geometric progression [Symbol]

a   ionic diameter [Symbol]

a   ionization collision efficiency [Symbol]

a   (crystal) lattice parameter [Symbol]

a   length (of side) of a triangle, or polygon [Symbol]

a   level of significance (in statistics) [Symbol]

a   lever of resistance moment [Symbol]

a   linear absorption coefficient [Symbol]

a   linear acceleration [Symbol]

a   line segment (in geometry) [Symbol]

a   local necking extension (in tension testing) [Symbol]

a   molecular diameter [Symbol]

a   real part (of a complex number) (e.g., a is the real part and b the imaginary part of the complex number $a + bj$) [Symbol]

a   semimajor axis of an ellipse, or ellipsoid [Symbol]

a   semitransverse axis of a hyperbola [Symbol]

a   single sideband; reduced carrier [Symbol]

a   sonic velocity [Symbol]

a   thermal diffusivity [Symbol]

a   turns ratio (of transformer) [Symbol]

a   van der Waal's pressure constant [Symbol]

a   warmest month above 71.6° (22°C) [Subtype of Climate Region, e.g., in Ca, Caf, Caw, Csa etc.]

a   width [Symbol]

a   Williams-Comstock transition parameter (for magnetic media) [Symbol]

a   x-intercept of a curve, line, plane, or surface [Symbol]

a/   acting

a-   amorphous [Usually Followed by a Chemical Element, or Compound, e.g., a-Ni is Amorphous Nickel, or a-SiC Represents Amorphous Silicon Carbide, etc]
    average, or uniform (linear) acceleration [Symbol]

$a_{||}$ parallel to (or in-plane with) (crystal) unit cell axis "a" [Symbol] [also a||]

$a_\perp$ perpendicular (or normal) to (crystal) unit cell axis "a" [Symbol] [also a⊥]

[a]   specific optical rotation [Symbol]

[a]   surface concentration excess [Symbol]

$a_\pm$ mean ionic activity [Symbol]

a_0 Bohr radius ($5.2917725 \times 10^{-11}$ m) [Symbol]

a_0 equilibrium bond length (of ions) [Symbol]

a_0 equilibrium (crystal) lattice constant [Symbol]

a_0 flux line (crystal) lattice spacing [Symbol]

a_0 initial amplitude [Symbol]

a_0 initial term in a general (mathematical) expansion, or series [Symbol]

a_0 intermolecular distance [Symbol]

a_0 original crack size [Symbol]

a_0 original (cross-sectional) area [Symbol]

a_{AB} activity of electrolyte AB [Symbol]

a_{av} average acceleration [Symbol]

a_c critical crack size (or length) [Symbol]

a_c cubic-cell parameter [Symbol]

a_e effective crack size [Symbol]

a_f crack size at fracture [Symbol]

a_f final cross-sectional area [also a_f]

a_f (crystal) lattice parameter of film [Symbol]

a_{H+} hydrogen-ion activity [Symbol]

a_i acceleration on the i-th particle (e.g. atom) of a system [Symbol]

a_i activity of component i (in a solution) [Symbol]

a_i contrast coefficient (in statistics) [Symbol]

a_i term in a general (mathematical) expansion, or series [Symbol]

a_{ij} general term in a determinant, or matrix [Symbol]

a_j basis vector (of crystal system) [Symbol]

a_M (crystal) lattice parameter of matrix [Symbol]

a^n n-th power of a [Symbol]

a_n first Fourier coefficient (in Fourier Series) [Symbol]

a_n normal acceleration [Symbol]

a_n stationary orbit radius [Symbol]

a_{Na+} sodium-ion activity [Symbol]

a_P (crystal) lattice parameter of precipitate [Symbol]

a_p physical crack size [Symbol]

a_r radial acceleration [Symbol]

a_s (crystal) lattice parameter of substrate [Symbol]

a_s saturation activity [Symbol]

a_s thermal diffusivity of solid metal [Symbol]

a_T tangential (or linear) acceleration [Symbol]

a_t tangential acceleration [Symbol]

a_v coefficient of compressibility (for soil) [Symbol]

a_x x-coordinate of vector [Symbol]

a_y y-coordinate of vector [Symbol]

a_z z-coordinate of vector [Symbol]

α absorptance [Symbol]

α absorption coefficient; absorption factor [Symbol]

α absorptivity [Symbol]

α acceleration [Symbol]

α accommodation coefficient [Symbol]

α (lower-case) alpha [Greek Alphabet]

α alpha particle [Symbol]

α alpha phase (of a material) [Symbol]

α amplification factor [Symbol]

α (plane) angle [Symbol]

α angle of aberration [Symbol]

α angle of attack (of airfoils) [Symbol]

α angle of emergence (of Pulfrich refractometer) [Symbol]

α angle of incidence (or incidence angle) [Symbol]

α angle of inclination [Symbol]

α angle of obliquity (of soil, or rock solids) [Symbol]

α angle of opening (of a rotating circular sector) [Symbol]

α angle of optical rotation [Symbol]

α angle of repose [Symbol]

α angle of (molar magnetic) rotation [Symbol]

α angle of tilt (in centerless grinding) [Symbol]

α angle of wrap [Symbol]

α angular acceleration [Symbol]

α angular deflection (of shafts) [Symbol]

α angular displacement [Symbol]

α aperture angle [Symbol]

α area [Symbol]

α areal density [Symbol]

α asymptote angle [Symbol]

α attenuation (constant) [Symbol]

α Auger parameter (in physics) [Symbol]

α base transport factor (of transistor) [Semiconductor Symbol]

α bend(ing) angle [Symbol]

α bevel angle (of semiconductor wafers) [Symbol]

α clearance angle (or relief angle) (in machining) [Symbol]

α coefficient of linear thermal expansion [Symbol]

α coefficient of relative Auger sensitivity [Symbol]

α contact angle (of roller bearings) [Symbol]

α convergence angle (in electron microscopy) [Symbol]

α coupling constant [Symbol]

α damping constant (or factor) [Symbol]

α deformation stacking fault probability (in materials science) [Symbol]

α degree of (electrolytic) dissociation [Symbol]

α degree of hydration (of ceramics) [Symbol]

α deposition angle [Symbol]

a   dividing head angle (in machining) [Symbol]

a   draft angle (in forging) [Symbol]

a   Dundurs parameter (in physics) [Symbol]

a   edge angle (of milled bevel gears) [Symbol]

a   electric polarizability [Symbol]

a   (statistical) error of the first kind [Symbol]

a   exchange parameter [Symbol]

a   face angle of Vickers pyramidal diamond indenter [Symbol]

a   ferrite (solid solution) [Symbol]

a   (Sommerfeld) fine structure constant (7.29735×10^{-3}) [Symbol]

a   film thickness factor (for thrust bearings) [Symbol]

a   first-direction angle of a line, angle with the x-axis [Symbol]

a   fraction of association (or dissociation) [Symbol]

a   Gilbert's damping coefficient [Symbol]

a   grazing angle [Symbol]

a   half-angle (in optical microscopy) [Symbol]

a   heat-transfer coefficient [Symbol]

a   helix angle (of screw threads, spur gears, helical gears, etc.) [Symbol]

a   integration constant [Symbol]

a   isotope coefficient [Symbol]

a   lattice parameter–crystallographic angle between *b* and *c* axis of unit cell [Symbol]

a   lead angle (of worm gears) [Symbol]

a   molecular polarizability [Symbol]

a   nonlinearity coefficient (of a varistor) [Semiconductor Symbol]

a   optical rotation [Symbol]

a   phase [Symbol]

a   pitch angle (of cone clutches) [Symbol]

a   pitch cone angle (of milled bevel gears) [Symbol]

a   precipitated zone of reacted glass used for nuclear-waste end storage [Symbol]

a   prefix used with organic compounds to indicate the position of a substituent atom, or group

a   pressure angle (of cams, or splines) [Symbol]

a   radian [Symbol]

a   rake angle (in machining) [Symbol]

a   ratio of statistical segment lengths of A and B blocks (for block copolymers) [Symbol]

a   reciprocal of Young's modulus (i.e., $a = 1/E$) [Symbol]

a   recombination coefficient [Symbol]

a   reduced electromigration driving force [Symbol]

a   reduced pressure (i.e. ratio of pressure to critical pressure, p/p_c) [Symbol]

a   relative volatility in distillation [Symbol]

a   right ascension (in astronomy) [Symbol]

a   roughness exponent [Symbol]

a   scattering spin asymmetry ratio [Symbol]

a   Schmid stress resolution factor [Symbol]

a   segregation factor [Symbol]

a   semi-angle [Symbol]

a   separation factor [Symbol]

a   significance level (in statistics) [Symbol]

a   slope angle (in geometry) [Symbol]

a   smallest principal refractive index of a biaxial crystal [Symbol]

a   sound absorption coefficient [Symbol]

a   spectral absorbance [Symbol]

a   spectral parameter [Symbol]

a   spin wave function giving probability p = 1 for spin s = ½ and p = 0 for s = −½ [Symbol]

a   switch-up value (for Preisach-type hysteresis transducers) [Symbol]

a   take-off angle (of air or space vehicles) [Symbol]

a   temperature coefficient of resistance [Symbol]

a   thermal diffusivity [Symbol]

a   thermal expansion coefficient [Symbol]

a   tilt angle [Symbol]

a   Townsend primary ionization coefficient [Symbol]

a   transfer coefficient (in electroplating) [Symbol]

a   vaporization coefficient [Symbol]

   coefficient of linear expansion [Symbol]

[*a*]   specific rotation (of a chemical compound) [Symbol]

a′   coefficient of area (thermal) expansion [Symbol]

a′   hexagonal martensite (in metallurgy) [Symbol]

a″   coefficient of volume (thermal) expansion [Symbol]

a″   orthogonal martensite (in metallurgy) [Symbol]

$a_|$ linear thermal expansion coefficient along (composite) fiber axis [Symbol]

$a_\perp$ linear expansion coefficient across (composite) fiber axis [Symbol]

a_0 outermost precipitated salt layer of reacted glass used for nuclear-waste end storage [Symbol]

a_1 angle of incidence [Symbol]

a_1 precipitated glass layer of reacted nuclear glass used for nuclear-waste end storage [Symbol]

a^{-1} (spectral line) broadening factor [Symbol]

a_2 angle of reflection [Symbol]

a_c covalence (or covalency) [Symbol]

a_c Saffman-Taylor instability length [Symbol]

a_E eutectic of primary solid phase (in metallurgy) [Symbol]

a_E half-field-strength beam width (of antenna) [Symbol]

a_e absorptance [Symbol]

a_e effective rake angle (in machining) [Symbol]

a_e electronic polarizability [Symbol]

a_{EN} Esbjerg-Nørskov parameter [Symbol]

a_{eq} equilibrium composition of carbon in ferrite (in metallurgy) [Symbol]

a_H Hooge parameter (of diode) [Semiconductor Symbol]

a_I Ionicity [Symbol]

a_i angular aperture of illumination [Symbol]

a_i geometric (or shadowing) factor for species *i* (in ion spectroscopy scattering) [Symbol]

a_{ij} binary interaction coefficient of the *i-j* bond [Symbol]

a_j Stevens factor [Symbol]

a_l linear coefficient of thermal expansion [Symbol]

a_l polarizability [Symbol]

a_λ absorptivity of a body for wavelength λ [Symbol]

a_{lmn} Warren-Cowley short range order parameter (in solid-state physics) [Symbol]

a_n normal incidence sound absorption coefficient [Symbol]

a_{opt} optimum objective aperture angle (of electron microscope) [Symbol]

a_P primary solid phase in peritectic system (in metallurgy) [Symbol]

a_R rhombohedral angle (in crystallography) [Symbol]

a_{ss} supersaturated ferrite (in metallurgy) [Symbol]

a_t temperature coefficient of a material at temperature t [Symbol]

a_{th} coefficient of thermal expansion [Symbol]

a_v absorptance [Symbol]

a_v coefficient of (soil) compressibility (or compression) [Symbol]

a_v coefficient of volume thermal expansion [Symbol]

a_v velocity rake angle (in machining) [Symbol]

a_x nuclear Zeeman quenching factor [Symbol]

Å Angstrom (Unit) (= 10^{-10} meter) [Symbol]

$Å^{-1}$ reciprocal angstrom (or one per angstrom) [also 1/Å]

$Å^2$ square angstrom (= 10^{-20} square meter) [Symbol]

$Å^{-2}$ reciprocal square angstrom (or one per square angstrom) [also 1/Å²]

$Å^3$ cubic angstrom (= 10^{-30} cubic meter) [Symbol]

$Å^{-3}$ reciprocal cubic angstrom (or one per cubic angstrom) [also 1/Å³]

$Å^4$ angstrom to the fourth (= 10^{-40} meter to the fourth) [Symbol]

$Å^{-4}$ reciprocal angstrom to the fourth (power) (or one per angstrom to the fourth power) [also 1/Å⁴]

AA Absolute Alcohol

AA Absolute Altimeter

AA Absolute Altitude

AA Académie d' Agriculture [Academy of Agriculture, France]

AA Académie d' Armes [Military Academy, France]

AA Access Arm

AA Access Authorization

AA Acetic Acid

AA Acetic Anhydride

AA Acetoacetanilide

AA Acetylacetone

AA Acoustic Array

AA Acrylamide

AA Acrylic Acid

AA Activated Alumina

AA Activation Analysis [Nuclear Engineering]

AA Addis Ababa [Ethiopia]

AA Adenylic Acid

AA Adiabatic Approximation

AA Adipic Acid

AA Adjustable Angle

AA Adsorption Analysis

AA Advertising Association [UK]

AA Aerofoil Amplifier

AA Affirmative Action

AA Air Almanac [of US Naval Observatory]

AA Airborne Alert

AA Aircraft Alloy(s)

AA Airfoil Amplifier

AA Airship Association [US]

AA Air-to-Air [also A-A]

AA Alcoholics Anonymous

AA Algebraic Adder

AA Algorithm Analysis

AA All After

AA All Around

AA Allylalcohol

AA Alpha Alumina

AA Aluminum Alloy

AA Aluminum Association [Washington, DC, US]

AA Alumni Association

AA Alzheimer's Association [US]

AA Ambient Air

AA American Airlines [US]

AA American Airways [US]

AA American Assembly [New York City, US]

AA Aminoacetone

AA Amino Acid

AA Aminoacyl

AA 2-Amino Anthracene

AA Anchimeric Assistance [Organic Chemistry]

AA Angular Acceleration

AA Angular Aperture

AA Aniline Association [US]

AA Anion-Anion (Bond)

AA Anisylacetone

AA Annular Aperture

AA Anodized Aluminum

AA Antenna Array

AA Anthranilic Acid

AA Anti-Aircraft

AA Anticipated Answer [also aa]

AA Anticipatory Account

AA Antiproton Accumulator

AA Aortic Arch [Anatomy]

AA Appleton Anomaly

AA Approximate Absolute

AA Approximation Algorithm

AA Arachidic Acid

AA Arachidonic Acid

AA Arboricultural Association [UK]

AA Archeology Abroad [UK]

AA Architectural Acoustics

AA Architectural Association [UK]

AA Arithmetic Average [also aa]

AA Armature Accelerator

AA Arnold Arboretum [of Harvard University, Cambridge, Massachusetts, US]

AA Arsenic Analysis

AA Artificially Aged; Artificial Aging [Metallurgy]

AA As Annealed (Condition) [Metallurgy]

AA Ascorbic Acid

AA Asset Amount

AA Assigned Altitude

AA Associate Administrator

AA Associate in Accounting

AA Associate in (or of) Arts

AA Astro-Archeology

AA Atomic Absorption

AA Atomic Abundance

AA Auxiliary Anode [Metallurgy]

AA Atropic Acid

AA Attack Announcement

AA Attack Assessment

AA Audio Active

AA Audio Amplifier

AA Auger Amplitude [Physics]

AA Augusta Arsenal [Georgia, US]

AA Australian Academy (of Science)

AA Author Affiliation

AA Author's Alteration

AA Auto(matic) Acquisition

AA Auto(matic) Answer

AA Automobile Association [US]

AA Autoprobe Atomization

A&A Admission and Advancement [Institute of Electrical and Electronics Engineers Committee, US]

A&A Assembly and Association [UK Journal]

A+A Aluminium- und Automobil [Journal on Aluminum and Autombiles published by the Aluminum-Zentrale, Düsseldorf, Germany]

A+A Arts and Architecture [UK Publication]

A/A Air-to-Air [also A-A]

A/A Anti-Aircraft

A/A Austenite/Austenite (Boundaries) [Metallurgy]

A-A Air-to-Air [also A/A]

A-A Aluminum-Adhesive (Joint)

aA abampere [Unit]

aa (ana) – of each [also] [Medical Prescriptions]

A_1-A_2 Direct-Current Armature [Controllers]

AAA Acetoacetanilide

AAA Acetoacetic Anilide

AAA Acrylic Acid Anhydride

AAA Agricultural Adjustment Administration [US]

AAA Agricultural Aircraft Association [US]

AAA Alberta Association of Architects [Canada]

AAA Amateur Astronomers Association [US]

AAA American Academy of Allergy [US]

AAA American Accounting Association [US]

AAA American Airship Association [now Airship Association, US]

AAA American Angus Association [*Note:* Angus is a Breed of Cattle]

AAA American Antarctic Association [US]

AAA American Anthropological Association [US]

AAA American Arbitration Association [US]

AAA American Astronomers Association [US]

AAA American Automobile Association [US]

AAA Anguilla Anguilla Agglutinin [Biochemistry]

AAA Anti-Aircraft Artillery

AAA Appraisers Association of America [US]

AAA Architectural Aluminum Association [UK]

AAA Army Audit Agency [US]

AAA Aromatic Amino Acid

AAA Asian International Airports Association

AAA Association of Authors' Agents [UK]

AAA Astronomy and Astrophysics Abstracts (Database) [Germany]

AAA Australian Automobile Association

AAA Automated Amino-Acid Analysis

AAA Automobile Association of America

2AAA 2-Amino Acetic Acid [also 2AAa]

AAAA American-African Affairs Association [US]

AAAA American Association of Advertising Agencies [US]

AAAA Army Aviation Association of America [US]

AAAAW Anti-Aircraft Artillery Automatic Weapons

AAAB American Association of Architectural Bibliographers [US]

AAAC All Aluminum Alloy Conductor

AAAC American-Arab Affairs Council [US]

AAAC American Automatic Control Council [US]

AAAC Anti-Aircraft Artillery Command [US]

AAAC Archival Association of Atlantic Canada

AAAC Asociación de Astrónomos Autodidactas de Colombia [Association of Autodidactic Astronomers of Columbia]

AAAC Australian Aboriginal Affairs Council

AAAC Australian Army Aviation Corps

AAACC Association of American-Asian Chambers of Commerce

AAACE American Association of Agricultural College Editors [US]

AAACU Asian Association of Agricultural Colleges and Universities

AAAD Aromatic Amino Acid Decarboxylase [Biochemistry]

AAAD Association of Automotive Aftermarket Distributors [US]

AAAE American Association of Academic Editors [US]

AAAE American Association of Agricultural Engineers [US]

AAAE American Association of Airport Executives [US]

AAAF 2-(N-Acetoxyacetylamino)fluorene

AAAF Association Aéronautique et Astronautique de France [French Aeronautical and Astronautical Association]

AAAI Affiliated Advertising Agencies International [US]

AAAI American Academy of Allergy and Immunology [US]

AAAI American Association for Artificial Intelligence [US]

AAAI Association of Australian Aerospace Industries

AAAIA All-India Automotive and Ancillary Industries Association [India]

AAAID Arab Authority for Agricultural Investment and Development [Sudan]

AAAIP Advanced Army Aircraft Instrumentation Program [US]

AAAI Proc AAAI Proceedings [of American Association for Artificial Intelligence]

AAAL American Academy of Arts and Letters

AAALAC American Association for Accreditation of Laboratory Animal Care [US]

AAAM American Association of Aircraft Manufacturers

AAAM American Association for Automotive Medicine [US]

AAAOR Anti-Aircraft Artillery Operations Room

AAAP Asian-Australian Association of Animal Production Societies

AAAR American Association for Aerosol Research [US]

AAAS American Academy of Arts and Sciences [Boston, US]

AAAS American Association for the Advancement of Science [US]

AAASA Association for the Advancement of Agricultural Sciences in Africa [Ethiopia]

AAAS Obs AAAS Observer [Publication of the American Association for the Advancement of Science]

AAAS Rep AAAS Report [of American Association for the Advancement of Science]

A&A, Assem Assoc A&A – Assembly and Association [UK Journal]

AAAST African Association for the Advancement of Science and Technology

AAAT Anti-Aircraft Armored Truck

AAATP Asian Alliance of Appropriate Technology Practitioners [Philippines]

AAB Advertising Advisory Board

AAB Aircraft Accident Board

AAB Aircraft Anti-Collision Beacon

AAB Alliance Agricole Belge [Belgian Agricultural Alliance]

AAB All-to-All Broadcast

AAB American Association of Bioanalysts [US]

AAB 4-Aminoazobenzene

AAB Army Aviation Board [US Army]

AAB Association of Applied Biologists [Scotland]

AAb Autoantibody [also AAB] [Biochemistry]

A-AB Binary System of Component A and Intermediate Compound AB (Composed of Component A and B) [Symbol]

A₁A₂B Layered Heterostructure of Nonmetal Component B and Two Alternating Metal Components A_1 and A_2 [Symbol]

AABB American Association of Blood Banks [US]

AABC Associated Air Balance Council [US]

AABCP Advanced Airborne Command Post

AABD American Academy of Biological Dentistry [US]

AABDF Allied Association of Bleachers, Printers, Dyers and Finishers [UK]

AABE American Association of Blacks in Energy [US]

AABE Asian Association for Biology Education

AABFS Amphibious Assault Bulk Fuel System

AABG Alpha-Acetobromoglucose [Biochemistry]

AABGA American Association of Botanical Gardens and Arboretums [US]

AABI American Association of Bicycle Importers [US]

AABM Alpha-Acetobromoglucose [Biochemistry]

AABM Association of American Battery Manufacturers [US]

AABM Australian Association of British Manufacturers

AABNCP Advanced Airborne Command Post [US]

AABNF African Association for Biological Nitrogen-Fixation [Egypt]

AABP Acetylaminobiphenyl [Biochemistry]

AABP Aptitude Assessment Battery Program

AABP Australian Association of Business Publications

AABPA Amercian Association for Budget and Program Analysis [US]

AABR Australian Association of Bush Regenerators

AABS 2-(N-Acetylamino)butanoic Acid

AABSHIL Aircraft Anti-Collision Beacon System High-Intensity Light

AABZ Bis(acetylacetone)benzidine

AAC Abbreviated Address Calling

AAC a-N-Acetaminocinnamic Acid

AAC 2-Acetylamino trans-2-Butenoic Acid

AAC Aeronautical Advisory Council

AAC Aeronautical Approach Chart

AAC African Association of Cartography [Algeria]

AAC Afro-Asian Center [US]

AAC Agricultural Advisory Council [UK]

AAC Air Approach Control

AAC Air Carbon-Arc Cutting

AAC Airworthiness Advisory Circular [Australia]

AAC Alaskan Air Command [at Elmendorf Air Force Base, Alaska, US]

AAC Alkyl Amines Council [US]

AAC All Aluminum Conductor

AAC American Alpine Club [US]

AAC American Alumni Council [US]

AAC American Associoation of Colleges [US]

AAC Aminoalkyl Cellulose

AAC Annual Allowable Cut

AAC Anti-Aircraft Cannon

AAC Anti-Aircraft Command

AAC Antimicrobial Agents and Chemotherapy (Journal) [of American Society for Microbiology, US]

AAC Antiproton Accumulator Complex [of CERN–European Laboratory for Particle Physics Geneva, Switzerland]

AAC Architectural Anodizers Council [US]

AAC Argon Alternating Current

AAC Armagh Astronomy Center [Armagh, Northern Ireland]

AAC Army Air Corps [US]

AAC Arsenic Atmosphere Czochralski

AAC Associate and Advisory Committee

AAC Association of American Colleges [US]

AAC Australian Agricultural Council [now Agricultural and Resource Management Council of Australia and New Zealand]

AAC Automatic Amplitude Control

AAC Automatic Aperture Control

AAC Automatic Approach Control

AAC Automotive Advertisers Council [US]

AAC Average Annual Cost

AAC Aviation Administrative Communications

AAC Aviation Advisory Commission

AACA Aircraft Airworthiness Certification Agency

AACA Asociación Argentina de Control Automático [Argentinian Association for Automatic Control]

AACA Automotive Air Conditioning Association [now International Mobile Air Conditioning Association, US]

AACB Aeronautics and Astronautics Coordinating Board [of NASA]

AACB Australian Association of Clinical Biochemists

AACC Airport Associations Coordinating Council [Switzerland]

AACC American Association for Clinical Chemistry [US]

AACC American Association for Contamination Control [US]

AACC American Association of Cereal Chemists [US]

AACC American Automatic Control Council [US]

AACC Area Approach Control Center

AACC Association of Agricultural Computer Companies [US]

AACC Automatic Approach Control Complex

AACCA Associate of the Association of Certified and Corporate Accountants

AACCLA Association of American Chambers of Commerce in Latin America

AACCM Association of American Ceramic Component Manufacturers [US]

AACCM-C Association of American Ceramic Component Manufacturers–Consortium I [US] [also AACCMCI]

AACC Press American Association for Clinical Chemistry Press [Washington, DC, US]

AACD Antenna Adjustable Current Distribution

AACE African Association of Correspondence Education

AACE Airborne Alternate Command Echelon [of NATO]

AACE American Association of Cost Engineers [US]

AACE Trans AACE Transactions [of American Association of Cost Engineers]

AACFT Army Aircraft [also Aacft]

AACG American Association for Crystal Growth [US]

AACG Newsl AACG Newsletter [of American Association for Crystal Growth]

AACI Accredited Appraiser, Canadian Institute

AACI American Association of Ceramic Industries [US]

AACI Association of American Cancer Institutes [US]

AACIA American Association for Clinical Immunology and Allergy [US]

AACLI Australasian Advisory Committee on Land Information[now Australia New Zealand Land Information Council]

AACM Alpha Alumina Doped with Cupric Oxide (CuO) and Magnesia (MgO)

AACM ASEAN (Association of Southeast Asian Nations)–Australian Consultative Meeting

AACMCI Association of American Ceramic Component Manufacturers Consortium I

AACO Advanced and Applied Concept Office [US]

AACO Arab Air Carriers Organization [Lebanon]

AACO Assault Airlift Control Office

AACOBS Australian Advisory Council on Bibliographical Services

AACOM Army Area Communications

AACOMS Army Area Communications System [US]

AACP Advanced Airborne Command Post [US Air Force]

AACP American Association of Colleges of Pharmacy [US]

AACP American Association of Computer Professionals [US]

AACP Anglo-American Council for Productivity

AACPA Asian American Certified Public Accountants [US]

AACPP Association of Asbestos Cement Pipe Producers [US]

AACR American Association for Cancer Research [US]

AACR American Association of Clinical Research [US]

AACR Anglo-American Cataloguing Rules

AACRAO American Association of Collegiate Registrars and Admissions Officers [US]

AACS Advanced Automatic Compilation System

AACS Airborne Astrographic Camera System

AACS Airways and Air Communications Service [US Armed Forces]

AACS American Academy of Cosmetic Surgery [US]

AACS Army Alaska Communication System [US Air Force]

AACS Asynchronous Address Communications System

AACS Attitude Articulation Control System [of NASA]

AACSB American Assembly Collegiate Schools of Business

AACSCEDR Associate and Advisory Committee to the Special Committee on Electronic Data Retrieval [US]

AACSR Aluminum Alloy Conductor, Steel-Reinforced

AACSR/ACS Aluminum Alloy Conductor , Steel-Reinforced

AACT American Association of Candy Technologists [US]

AACT American Association of Clinical Toxicology [US]

AACT American Association of Commodity Traders [US]

AACTS African Centre for Technology Studies

AACV Assault Air Cushion Vehicle

AACVB Asian Association of Convention and Visitor Bureaus [Philippines]

AACW Active Adaptive Compliance Wrist

AAD Active Acoustic Device

AAD Administrative Applications Division

AAD Advanced Ammunitions Depot

AAD American Academy of Dermatology [US]

AAD Amino Acid Decarboxylase [Biochemistry]

AAD 3-Aminopyridine Adenine Dinucleotide [Biochemistry]

AAD Apparent Atomic Diameter

AAD Army Automation Directorate [US]

AAD Assigned Attitude Deviation

AAD Atlanta Army Depot [of US Army at Morris Army Air Field, Atlanta, Georgia, US]

AAD Australian Antarctic Division [of Department of the Environment]

Aad Alpha-Aminoadipic Acid [also AAD] [Biochemistry]

β-Aad Beta-Aminoadipic Acid [Biochemistry]

AADA Anti-Aircraft Defended Area

AADB Army Air Defense Board [US]

AADC Advanced Avionic Digital Computer

AADC Alberta Agriculture Development Corporation [Canada]

AADC All-Applications Digital Computer

AADC Anti-Aircraft Defense Commander

AADC Army Air Defense Command [US]

AADC Arnold Air Development Center [US Air Force]

AADC Automatic Aeromagnetic Digital Compensator

AADCP Army Air Defense Command Post [US]

AADE American Association of Dental Publishers [US]

AADE American Association of Diabetes Educators [US]

AADECA Asociación Argentina de Control Automático [Argentinian Association for Automatic Control] [also AA-DECA]

AADL Artillery Ammunition Development Laboratory [now Artillery Ammunition Rocket Development Laboratory]

AADP 3-Aminopyridine Adenine Dinucleotide Phosphate [Biochemistry]

AADS Advanced Army Defense System

AADS Advanced Automated Directional Scanning

AADS Airspeed and Direction Sensor

AADS American Academy of Dermatology and Syphilology [US]

AADS American Association of Dental Schools [US]

AADS Area Air Defense System

AADSF Advanced Automated Directional Solidification Furnace [Metallurgy]

AaDSV Aedes aegypti Densovirus

AA-dUTP 5-(3-Aminoallyl)-2'-Deoxyuridine 5'-Triphosphate [Biochemistry]

AADW Association of Artists and Designers in Wales [UK]

AAE Above Airport Elevation

AAE Acetoacetic Ester

AAE Aeronautical and Astronautical Engineering

AAE Aeroplane and Armament Establishment [UK]

AAE Affirmative Action Employer

AAE American Association of Engineers [US]

AAE Army Aviation Engineer

AAE Asociación Argentina de Electrotecnicos [Argentinian Association of Electrical Engineers]

AAE Association for Astronomy Education [UK]

AAE Australian Antarctic Expedition [1911 to 1914]

AAE Auto-Answer Equipment

AAE Automatic Answering Equipment

AAE Average Annual Earnings

AAEA African Adult Education Association

AAEA American Agricultural Editors Association [US]

AAEC Arnold Air Engineering Center [Tullahoma, Tennessee, US]

AAEC Australian Atomic Energy Commission

AAECP ASEAN (Assocociation of Southeast Asian Nations)–Australian Economic Cooperation Program

AAED Academic American Encyclopedia Database

AAED Bis(acetylacetone)ethylenediamine

AAEDC American Agricultural Economics Documentation Center [US]

AAEE Aircraft and Armament Experimental Establishment [of Ministry of Defense, UK]

AAEE American Academy of Environmental Engineers [US]

AAEE American Association of Economic Entomologists [US]

AAEE American Association of Electromyography and Electrodiagnosis [US]

AAEE Australian Association for Environmental Education

A&AEE Aircraft and Armament Experimental Establishment [of Ministry of Defense, UK]

AAeE Associate in Aeronautical Engineering

AAEER Aerophysics and Aerospace Engineering Research Report

AAEI American Association of Exporters and Importers [US]

AAEL American Association of Equipment Lessors [US]

AAELSS Active-Arm External Load Stabilization System

AAEM Acetoacetoxylethyl Methacrylate

AAEM American Academy of Environmental Medicine

AAEM Annals of Agricultural and Environmental Medicine [of Institute of Agricultural Medicine Poland]

AA/EOE Affirmative Action/Equal Opportunity Employer [also AAEOE]

AAES Advanced Aircraft Electrical System

AAES American Association of Engineering Societies [US]

AAES Australian Agricultural Economics Society

AAESDA Association of Architects, Engineers, Surveyors and Draftsmen of Australia

AAESWB Army Airborne Electronics and Special Warfare Board [US Army]

AA-EVP American Association–Electronic Voice Phenomena [US]

AAF Academie d'Agriculture de France [French Academy of Agriculture]

AAF Acetylaminofluorene

AAF Air-Assist Forming (of Plastics)

AAF American Advertising Federation [US]

AAF American Air Filter International [Louisville, Kentucky, US]

AAF American Air Force [US]

AAF American Architectural Foundation [US]

AAF Army Air Field [US Army]

AAF Army–Air Force

AAF Artificial Antiferromagnet(ic)

AAF Auxiliary Air Force

AAFA Architectural Aluminum Fabricators Association

AAFA Asthma and Allergy Foundation of America [Washington, DC, US]

AAFB Andrews Air Force Base [near Washington, DC, US]

AAFB Army Air Force Board [US]

AAFB Arnold Air Force Base [Tennessee, US]

AAFB Auxiliary Air Force Base

AAFC Anti-Aircraft Fire Control

AAFC Arab Aviation Finance Company

AAFC Army Air Force Center [US]

AAFCE Allied Air Forces Central Europe [of NATO]

AAFCO Association of American Feed Control Officials [US]

AAFCO Association of American Fertilizer Control Officials [US]

AAFE Advanced Applications Flight Experiment [of NASA]

AAFE Aero-Assist Flight Experiment

AAFEA Australian Airline Flight Engineers Association

AAFIF Automated Air Facility Information File [of Defense Mapping Agency Aerospace Center, US]

AAFM Association of American Feed Microscopists [US]

AAFMC Army Air Force Materiel Center

AAFNE Allied Air Forces Northern Europe [of NATO]

AAFPRS American Academy of Facial Plastic and Reconstructive Surgery [US]

AAFS American Academy of Forensic Sciences [US]

AAFS American Association of Foot Surgeons [US]

AAFS Atomic Absorption Flame Spectrophotometer

AAFSE Allied Air Forces Southern Europe [NATO]

AAFSS Advanced Aerial Fire Support System

AAFTS Army Air Force Technical School

AAFU All African Farmers Union

AAG Acquisition Advisory Group

AAG Aeromedical Airlift Group [US]

AAG Aeronautical Assignment Group [US]

AAG Air Adjutant General

AAG Airports Authority Group [Canada]

AAG Alpha-L-Acid Glycoprotein [Biochemistry]

AAG Anthropological Association of Greece

AAG Anti-Aircraft Gun

AAG Assistant Adjutant-General

AAG Association of American Geographers [US]

AAGP American Academy of General Practice [US]

AAGR Air-to-Air Gunnery Range

AAGR Average Annual Growth Rate

AAgr Associate in Agriculture

AAGS American Association for Geodetic Surveying [of American Congress on Surveying and Mapping, US]

AAGS Association of African Geological Surveys [France]

AAGW Air-to-Air Guided Weapon

α-AgZn Alpha Silver Zinc

AAH Advanced Attack Helicopter

AAH Anti-Armor Helicopter

AAHA American Animal Hospital Association [US]

AAHE American Association for Higher Education [US]

AAHL Australian Animal Health Laboratory

AAHM Association of Architectural Hardware Manufacturers

AAHO Afro-Asian Housing Organization [Egypt]

AAHQ Allied Air Headquarters [also AAHq]

AAHQS Australian Agricultural Health and Quarantine Service

AAHR Australian Association for Humane Research

AAHS American Aviation Historical Society [US]

AAI Acute Acquired Immunity

AAI African-American Institute

AAI Agricultural Ammonia Institute [US]

AAI Air Aid to Intercept

AAI Airlines Avionics Institute [US]

AAI American Association of Immunologists [of Federation of American Societies for Experimental Biology, US]

AAI American Association of Inventors [US]

AAI Angle of Approach Indicator

AAI Arab-American Institute

AAI Association of Advertisers in Ireland

AAI Association of Alabaster Importers and Wholesalers [UK]

AAI Azimuth Angle Increment

AAIB Air Accidents Investigation Branch [UK]

AAIC ASEAN (Association of Southeast Asian Nations) Aluminum Industry Club

AA/ICP Atomic Absorption/Inductively Coupled Plasma (Analysis)

AAIE Association for the Advancement of International Education [US]

AAIE Association of Applied Insect Ecologists [US]

AAIEE Associate of the American Institute of Electrical Engineers

AAIH American Academy of Industrial Hygiene [US]

AAII American Association for the Advancement of Invention and Innovation [US]

AAIM American Association of Industrial Management [US]

AAIMME Associate of the American Institute of Mining and Metallurgical Engineers

AAIMS An Analytical Information Management System

AAIN American Association of Industrial Nurses [US]

AAIN J American Association of Industrial Nurses Journal [US]

AAIPS American Association of Industrial Physicians and Surgeons [US]

AAIS African Association of Insect Scientists

AAITO Association of African Industrial Technology Organizations

AAJC American Association of Junior Colleges

AAJC Automatic Antijam Circuitry

AAL Above Aerodrome Level

AAL Absolute Assembly Language [Computer Language]

AAL Academia Argentina de Letras [Argentinian Academy of Letters, Buenos Aires, Argentina]

AAL Alternaria Alternata (Toxin)

AAL American Airlines [US]

AAL Ames Aeronautical Laboratory [US]

AAL Arctic Aeromedical Laboratory [US Air Force]

AAL Arctic Aerospace Laboratory [US Air Force]

AAL Association of Architectural Librarians [US]

AAL Association of Assistant Librarians [UK]

AAL ATM (Asynchronous Transfer Mode) Adaption Layer

AALA American Agricultural Law Society [US]

AALA American Association for Laboratory Accreditation [also A2LA, or A²LA]

AALA American Automotive Leasing Association [US]

AALAS American Association for Laboratory Animal Science [US]

AALC Advanced Airborne Launch Center [US]

AALC African American Labor Center

AALC Amphibious Assault and Landing Craft

AALC Amplified Automatic Level Control

a-AlCuRu Alpha Aluminum Copper Ruthenium (Compound)

a-AlCuY Aluminum Copper Yttrium Amorphous Alloy

AALD Association of Agricultural Librarians and Documentalists [Netherlands]

AALDEF Asian-American Legal Defense Education Fund

AALGCSU American Association of Land Grant Colleges and State Universities [US]

AALL American Association for Labor Legislation [US]

AALL American Association of Law Libraries [US]

a-AlLi Alpha Aluminum Lithium (Compound)

AALMG Anti-Aircraft Light Machine Gun

a-AlMnSi Alpha Aluminum Manganese Silicon (Compound)

a-AlNiCuCe Aluminum Nickel Copper Cerium Amorphous Alloy

a-AlNiCe Aluminum Nickel Cerium Amorphous Alloy

a-AlNiLa Aluminum Nickel Lanthanum Amorphous Alloy

a-AlNiTi Aluminum Nickel Titanium Amorphous Alloy

a-AlNiZr Aluminum Nickel Zirconium Amorphous Alloy

AALO Association of Academic Librarians of Ontario [Canada]

a-Al$_2$O$_3$ Amorphous Alumina (or Aluminum Oxide)

a-Al$_2$O$_3$ Alpha Alumina (or Alpha Aluminum Oxide)

a-Al(OH)$_3$ Gibbsite [Mineral]

AALS Acoustic Artillery Location System

AALS Association for Arid Lands Studies [US]

AALS Association of American Library Schools [US]

a-Al-LT-ET Aluminum Late-Transition-Metal Early-Transition-Metal Amorphous Alloy

AALT Alberta Association of Library Technicians [Canada]

AALUE Asymptotically Admissible Linear Unbiased Estimator

AAM Acrylamide

AAM Airline Administrative Message

AAM Air-to-Air Missile

AAM American Academy of Mechanics [US]

AAM American Academy of Microbiology [US]

AAM American Association of Museums [US]

AAM Amino Acid Mixture

AAM Anti-Aircraft Missile

AAM Automated Auger Microprobe

a-Am Alpha Americium [Symbol]

AAMA American Academy of Medical Acupuncture [US]

AAMA American Agricultural Marketing Association [US]

AAMA American Amusement Machine Association [US]

AAMA American Apparel Manufacturers Association [US]

AAMA American Architectural Manufacturers Association [US]

AAMA American Automobile Manufacturers Association [US]

AAMA Architectural Aluminum Manufacturers Association [US]

AAMA Asian American Manufacturers Association [US]

AAMA Automotive Accessories Manufacturers of America [US]

AAMC Alabama A&M (Agricultural and Mechanical) College [later Alabama Polytechnic Institute; now Auburn University]

AAMC Alcorn A&M (Agricultural and Mechanical) College [Lorman, Mississippi, US]

AAMC American Association of Medical Clinics [US]

AAMC Arkansas A&M (Agricultural and Mechanical) College [College Heights, US]

AAMC Army Air Materiel Command

AAMC Association of American Medical Colleges [US]

AAMCA Army Advanced Materiel Concepts Agency [US]

AAMD American Association on Mental Deficiency [US] [Defunct]

AAME Acetylarginine Methylester [Biochemistry]

AAME American Association of Microprocessor Engineers [US]

AAME Automated Multi-Media Exchange

AAMG Anti-Aircraft Machine Gun

AAMGAP Achieve as Many Goals as Possible

AAMI American Association of Machinery Importers [US]

AAMI Association for the Advancement of Medical Instrumentation [US]

AAM/ICOM International Council of Museums Committee of the American Association of Museums [US]

AAML Arctic Aeromedical Laboratory [US Air Force]

AAML Arctic Aerospace Medical Laboratory [US Air Force]

AAMOF As A Matter Of Fact

AAMOI As A Matter Of Interest

AAMRL American Association of Medical Record Librarians [US]

AAMS American Association of Meta-Science [US]

AAMTA American Apparel Machinery Trade Association [US]

AAMU Alabama A&M (Agricultural and Mechanical) University [Normal, US]

AAMVA American Association of Motor Vehicle Administrators [US]

AAMX Acetoacet-m-Xylidide

AAN Academy of Applied Nutrition [US]

AAN American Association of Nurserymen [US]

AAN Aminoacetonitrile

AANA American Anorexia Nervosa Association [US]

AANA American Association of Nurse Anesthetists [US]

AANC Aging Aircraft Nondestructive Inspection Development and Demonstration Center [of Sandia National Laboratories, Albuquerque, New Mexico, US]

A and M Agricultural and Mechanical (College) [also A&M]

AANHS Ayrshire Archeological and Natural Historical Society [UK]

AANP American Association of Naturopathic Physicians [US]

AANS American Agricultural News Service [US]

AAO Acetaldehyde Oxime

AAO Acetaldoxime

AAO Adjacent Arctic Ocean

AAO American Academy of Ophthalmology [US]

AAO American Academy of Optometry [US]

AAO American Association of Optometrists [US]

AAO Amino Acid Oxidase [Biochemistry]

AAO Anodic Aluminum Oxide

AAO Anti-Air Output

AAO Authorized Acquisition Officer

AAOA Acetoacet-o-Anisidine

AAOC Acetoacet-o-Chloroanilide

AAOC Anti-Aircraft Operations Center

AAODA Bis(acetylacetone)-o-Dianisidine

AAODC American Association of Oilwell Drilling Contractors [US]

AAOE Airborne Antarctic Ozone Experiment

AA of A Automobile Association of America [US]

AAOGAS American Association of Obstetricians, Gynecologists and Abdominal Surgeons [US]

AAOHN American Association of Occupational Health Nurses [US]

AAOM American Association of Occupational Medicine [US]

AAOO American Academy of Ophthalmology and Otolaryngology [US]

AAOP Antaircraft Observation Post

AAOR Anti-Aircraft Artillery Operations Room

AAOS American Academy of Orthopedic Surgeons [US]

AAOT Acetoacet-o-Toluidine

AAOU Asian Association of Open Universities

AAP Acridinylaminopropanol

AAP Adjunct Associate Professor

AAP Affirmative Action Plan

AAP Affirmative Action Program

AAP Aging Aircraft Program [of Federal Aviation Administration, US]

AAP Air at Atmospheric Pressure

AAP Airborne Acoustic Processor

AAP Anti-Alkaline Phosphatase [Biochemistry]

AAP Ambient Absolute Pressure

AAP American Academy of Pediatrics [US]

AAP American Academy of Periodontology [US]

AAP American Association of Physicians

AAP 4-Aminoantipyrine

AAP Apollo Applications Program [of NASA]

AAP Applications Access Point [Data Communications]

AAP Association of American Publishers [US]

AAP Association of Aviation Psychology [US]

AAP Associative Array Processor

AAP Attached Applications Processor

AAP Average Annual Precipitation

aap 4-Aminoantipyrine

9-AAP 9-Acridinylaminopropanol

9-AAP 9-Amino-Acridinepropanolol

AAPA American Alfalfa Processors Association [US]

AAPA American Association of Physical Anthropologists [US]

AAPA American Association of Port Authorities [US]

AAPA Association of Authorized Public Accountants [UK]

AAPA Australian Asphalt Pavement Association

AAPB American Association of Pathologists and Bacteriologists [US]

AAPC American Association of Professional Consultants [US]

AAPC Asociación Argentina para el Progreso de la Ciencia [Argentinian Association for the Progress of Science]

AAPC Australian Aluminum Production Commission

AAPCC American Association of Poison Control Centers [US]

AAPCO Association of American Pesticide Control Officials [US]

AAPD American Academy of Pediatric Dentistry [US]

AAPFCO Association of American Plant Food Control Officials [US]

AAPG American Association of Petroleum Geologists [US]

AAPICU American Association of Presidents of Independent Colleges and Universities [US]

AAPL Additional Programming Language

AAPL American Association of Professional Landmen [US]

AAPL An Array Processing Language [Computer Language]

AAPM American Association of Physicists in Medicine [US]

AAPMA Association of Australian Port and Marine Authorities (Inc.)

AAPP American Association of Police Polygraphists [US]

AAPP Auxiliary Airborne Power Plant

AAPPD Bis(Acetylacetone)-p-Phenylenediamine

AAPRCO American Association of Private Railroad Car Owners [US]

AAPRDTW Association for the Advancement of Policy, Research and Development in the Third World [US]

AAPS Alternative Automotive Power System

AAPS American Association for the Promotion of Science [US]

AAPS American Association of Pharmaceutical Scientists [US]

AAPS American Association of Phonetic Sciences [US]

AAPSD Alternative Automotive Power System Division [of US Environmental Protection Agency]

AAPSE American Association of Professors in Sanitary Engineering [US]

AAPS Newsl AAPS Newsletter [of American Association of Pharmaceutical Scientists]

AAPSP Advanced Automotive Power Systems Program [US]

AAPT Advanced Assembly and Packaging Technology

AAPT American Association of Physics Teachers [US]

AAPT Association of Asphalt Paving Technologists [US]

AAPTSR Australian Association for Predetermined Time Standards and Research [also AAPTS&R]

AAPU Auxiliary Airborne Power Unit

AAPWISM American Association of Public Welfare Information Systems Management [US]

AAQ Ambient Air Quality

AAR Against All Risks [also aar]

AAR Aircraft Accident Report

AAR Air-to-Air Refuelling

AAR Antigen-Antibody Reaction [Biochemistry]

AAR ASEAN (Association of Southeast Asian Nations) Association of Radiologists

AAR Association for Automated Reasoning [US]

AAR Association of American Railroads [US]

AAR Atlas of Australian Resources [of AUSLIG–Australian Surveying and Land Information Group]

AAR Automatic Alternative Routing

AAR Automotive Affiliated Representatives [US]

AAR Automotive Aftermarket Retailer [US]

AAR Average Annual Rainfall

a-Ar Amorphous Argon (Solid)

AARA Air-to-Air Refuelling Area

AARB Air-to-Air Refuelling Boom

AARB Australian Road Research Board

AARC Asociación de Agricultores del Rio Caulican [Agricultural Association of Rio Caulican, Mexico]

AARC Australian Applied Research Center

AARDL Artillery Ammunition Rocket Development Laboratory [Picatinny Arsenal, Dover, New Jersey US] [formerly AADL]

AARDS Australian Advertising Rate and Data Service

AARINENA Association of Agricultural Research Institutions in Near East North Africa

AARL Aeronautical and Astronautical Research Laboratory [US]

AARLMP African-American Resources and Library Manpower Project [Columbia University, New York State, US]

AARM Analytical Atomic Resolution Microscope

AARN Alberta Association of Registered Nurses [Canada]

AARNet Australian Academic and Research Network

AARPS Air Augmented Rocket Propulsion System

AARR Argonne Advanced Research Reactor [of Argonne National Laboratory, near Chicago, Illinois US] [1960 US Atomic Energy Commission Project] [also A^{2}R^2]

AARRO Afro-Asian Rural Reconstruction Organization [India]

AARS American Association of Railroad Superintendents [US]

AARS American Association of Railway Surgeons [US]

AARS Automatic Address Recognition System

AARSE African Association of Remote Sensing of the Environment

AARU Association of Arab Universities [Jordan]

AARV Armored Aerial Reconnaissance Vehicle

AAS *(Academiae Americanae Socius)* – American Academy of Arts and Science

AAS Academy of Applied Sciences [Boston, US]

AAS Academy of Arts and Science

AAS Adjusted Airspeed

AAS Advanced Antenna System

AAS Advanced Application Specialist

AAS Advanced Automation System

AAS Afghanistan Academy of Sciences [Kabul, Afghanistan]

AAS African Academy of Sciences

AAS Air Ambulance Service

AAS Airport Advisory Service

AAS Alabama Academy of Science [Birmingham, US]

AAS All-to-All Scatter

AAS Aluminum Ammonium Sulfate

AAS American Aeronautical Association [US]

AAS American Antiquarian Society [US]

AAS American Astronautical Society [US]

AAS American Astronomical Society [US]

AAS Ammoniacal Ammonium Sulfate (Leaching Process)

AAS Ammonium-Aluminum Silicate

AAS Ancient Astronaut Society [US]

AAS Anglesey Antiquarian Society [UK]

AAS Annunciator Alarm System

AAS Aquifer-Aquitard System

AAS Architectural Acoustics Society [US]

AAS Arithmetic Assignment Statement

AAS Arkansas Academy of Science [Little Rock, US]

AAS Army Air Service [US]

AAS Asociación Argentina de Sedimentologia [Argentinian Sedimentological Association]

AAS Associate in Applied Science

AAS Association of Architects and Surveyors [UK]

AAS Association of Asian Studies [US]

AAS Astronomy and Astrophysics Supplement (Series) [Published in France]

AAS Atomic Absorption Spectrometer; Atomic Absorption Spectrometry

AAS Atomic Absorption Spectrophotometer; Atomic Absorption Spectrophotometry

AAS Atomic Absorption Spectroscope; Atomic Absorption Spectroscopy

AAS Atomic Absorption Standard

AAS Auckland Astronomical Society [New Zealand]

AAS Australian Academy of Sciences [Sydney]

AAS Australian Acoustical Society

AAS Austrian Academy of Science [Vienna]

AAS Automated Accounting System

AAS Automatic Announcement Subsystem

AAS Monothioacetylacetone

a-As Amorphous Arsenic

a-As Alpha Arsenic [Symbol]

AAS2 Dithioacetylacetone

AASA Academy of Arts and Sciences of the Americas [US]

AASA American Association for the Study of Allergy [US]

AASA American Association of School Administrators [US]

AASB American Association of Small Business [US]

AASC Aerospace Application Studies Committee [of NATO]

AASC Airworthiness Authorities Steering Committee

AASC American Association of State Climatologists [US]

AASC Association for the Advancement of Science in Canada

AASCO Association of American Seed Control Officials [US]

AASCS American Association of Specialized Colleges and Schools [US]

AASCU American Association of State Colleges and Universities [US]

AAScW American Association of Scientific Workers [US]

AASDMCC American Association for Small Dredging and Marine Construction Companies [US]

AASDN Architectural and Archeological Society of Durham and Northumberland [UK]

AASE African Association of Science Editors

AASE Airborne Arctic Stratospheric Expedition [US – 1991-92]

AASE American Academy of Safety Education [US]

AASE American Academy of Sanitary Engineers [now American Academy of Environmental Engineers US]

AASE Australian Associated Stock Exchanges

AASERT Augmentation Award for Science and Engineering Research Training [of Air Force Office of Scientific Research]

AASF Advanced Air Striking Force [US]

AASG Association of American State Geologists [US]

AASH American Association for the Study of Headache [US]

AASH American Association of State Highways [US]

AAS Hist Ser AAS History Series [of American Astronautical Society]

AASHO American Association of State Highway Officials [US]

AASHTO American Association of State Highway and Transportation Officials [US]

AASI Association for the Advancement of Science in Israel

AASIR Advanced Atmospheric Sounder and Imaging Radiometer

AASIS American Association for the Study of Internal Secretions [US]

AASL Asian American Studies Library

AASLH American Association for State and Local History

AASM Association of American Steel Manufacturers [US]

AASME Associate of the American Society of Mechanical Engineers

AASMTC Association of American Steel Manufacturers Technical Committee [US]

AAS Newsl AAS Newsletter [of American Astronomical Society]

AASO Alberta Applied Science Office [Canada]

AASO American Association of Ship Owners [US]

AASP Advanced Automated Sample Processor

AASP American Association of Stratigraphic Palynologists [US]

AASP ASCII (American Standard Code for Information Interchange) Asynchronous Support Package

AASP Association of African Studies Programs

AASR Airports and Airways Surveillance Radar

AASRC American Association of Small Research Companies [US]

AASRI Arctic and Antarctic Scientific Research Institute

AASS *(Americanae Antiquarianae Societatis Socius)* – Fellow of the American Antiquarian Society

AASSA Association of American Schools in South America

AASU All Africa Students Union [Ghana]

AAS/UVR Atomic Absorption Spectroscopy/Ultraviolet Reflectance

AASW American Association of Scientific Workers [now United States Federation of Scientists and Scholars]

AAT Acetoacetoluidide

AAT Adenine/Aminopterin/Thymidine Medium [Biochemistry]

AAT Alpha-Antitrypsin (Enzyme) [Biochemistry]

AAT Aminoazotoluene

AAT Analytic Approximation Theory

AAT Anglo-Australian Telescope [Australia]

AAT Arme Automatique Transformable [French Machine Gun with Delayed Blowback]

AAT Association of Accounting Technicians [UK]

AAT Atomic Absorption Test

AAT Atomic Axial Tensor

AAT Australian Antarctic Territory

AAT Average Access Time

Aat Acetobacter aceti [Microbiology]

AATA All Africa Teachers Association

AATA Animal Transportation Association [UK]

AATB Army Aviation Test Board [US Army]

AATBN Aromatic Amine-Terminated Butadiene/Acrylonitrile [Biochemistry]

AATC Anti-Aircraft Training Center

AATC Army Aviation Test Command

AATC Automatic Air Traffic Control

AATCC American Association of Textile Chemists and Colorists [US]

AATCO Army Air Traffic Coordinating Office

AATD Australian Accounting and Taxation Database

AATD Aviation Applied Technology Directorate [US]

AATDC Association of American Teachers of the Diseases of Children [US]

AATech Associate in Automotive Technology

AATMS Advanced Air Traffic Management System

AATO Army Air Transport Organization [US]

AATO Association of Architectural Technologists of Ontario [Canada]

AATPO Association of African Trade Promotion Organizations [Morocco]

AATRI Army Air Traffic Regulations and Identification

AA-tRNA Aminoacyl Transfer Ribonucleic Acid [Biochemistry]

AATS American Association for Thoracic Surgery [US]

AATSE Australian Academy of Technological Science and Engineering

AATSR Advanced Along-Track Scanning Radiometer

AATT American Association for Textile Technology [US]

AATUF All-African Trade Union Federation

AAU Address Arithmetic Unit

AAU Association of African Universities [Ghana]

AAU Association of American Universities [US]

AAU Association of Atlantic Universities [Canada]

AAU Automatic Answering Unit

AAUA American Association of University Administrators [US]

AAUBO Association of Atlantic University Business Officers [Canada]

AA-UTP 5-(3-Aminoallyl)uridine 5'-Triphosphate [Biochemistry]

AAUG Association of Arab-American University Graduates

AAULC Association of Atlantic Universities Librarians Council [Canada]

AAUP American Association of University Professors [US]

AAUP Association of American University Presses [US]

AAUPF American Association of University Professors Foundation [US]

AAUS American Association of University Students [US]

AAUW American Association of University Women [US]

AAUWEF American Association of University Women Educational Foundation [US]

AAV Aboriginal Affairs Victoria (State) [Australia]

AAV Airborne Assault Vehicle

AAV Aviadenovirus

AAV Adeno-Associated Virus

AAV-2 Adeno-Associated Virus, Type 2

AAVCC Australian Agricultural and Veterinary Chemicals Council

AAVCS Automatic Aircraft Vectoring Control System

AAvTech Associate in Aviation Technology

AAVD Automatic Alternate Voice/Data

AAVIM American Association for Vocational Instructional Materials [US]

AAVS Aerospace Audiovisual Service

AAVSO American Association of Variable Star Observers [US]

AAVSO Bull AAVSO Bulletin [of American Association of Variable Star Observers]

AAVSO Circ AAVSO Circular [of American Association of Variable Star Observers]

AAVSO Monogr AAVSO Monographs [of American Association of Variable Star Observers]

AAVSO Rep AAVSO Reports [of American Association of Variable Star Observers]

AAVT Association of Audio-Video Technicians [US]

AAW Air-Acetylene Welding

AAW American Association of Woodturners [US]

AAW Anti-Air Warfare

AAW Automatic-Arc Welding

AAWC Australian Advisory War Council

AAWO American Automatic Weapons Association [US]

AAWH American Association for World Health [US]

AAWO Association of American Weather Observers [US]

AAWPI Association of American Wood Pulp Importers [US]

AAWS Anti-Armor Weapon System [of US Army]

AAWS-M Anti-Armor Weapon System–Medium [of US Army]

AAX Acetoacetxylidide

AAX Automated Attendant Exchange

AAZ Airport Approach Zone

AAZN American Association for Zoological Nomenclature [US]

AAZPA American Association of Zoological Parks and Aquariums [US]

AAZPA Newsl AAZPA Newsletter [of American Association of Zoological Parks and Aquariums]

AAZPA Reg Conf Proc AAZPA Regional Conference Proceedings [of American Association of Zoological Parks and Aquariums]

AB AB Battery

AB Able-Bodied Seaman [also ab]

AB Abrasive Belt

AB Abrasive Blaster; Abrasive Blasting

AB Abscisic Acid

AB Abstract

AB Acid Brine

AB Acoustic Bridge

AB Activated Bauxite

AB Adapter Booster

AB Address Buffer

AB Address Bus

AB Adiabatic Bend (Approximation)

AB Adjustable Bed (Press)

AB Aeronautical Board

AB Afterburner; Afterburning

AB Aharonov-Bohm (Effect) [Quantum Physics]

AB Air Balloon

AB Air Base

AB Air Bending [Metallurgy]

AB Air Bladder

AB Air Blanketing

AB Air Blast(ing)

AB Air-Blown; Air Blowing

AB Airborne [also A/B, or ab]

AB Air Brake

AB Air-Break

AB Alberta [Canada]

AB Alcian Blue

AB Alkyl Bromide

AB All Before

AB Amazon Basin [Brazil]

AB American and British (Standards)

AB Aminoazobenzene

AB Amino Borane

AB Analog Board

AB Anchor Bolt

AB Aneroid Barometer

AB Angle Bracket

AB Answer Back

AB Antibiotic

AB Antibody [also Ab]

AB Antibonding (Orbital) [Physics]

AB Antigua and Barbuda [West Indies]

AB Application Block

AB Application Brief

AB Applied Biology

AB Arc Bed

AB Arc Brazing

AB Arctic Bottom

AB *(Artium Baccalaureus)* – Bachelor of Arts

AB As Baked

AB As Batched

AB As Brazed

AB As Built

AB Ashby-Brown (Contrast) [Physics]

AB Astrobiological; Astrobiologist; Astrobiology

AB Asymmetric Bragg (Geometry) [X-Ray Analysis]

AB Asymptotic Behaviour

AB Atomic Beam

AB Atomic Bomb

AB Atomic Bond

AB Augis and Benett (Equation) [Materials Science]

AB Automated Bibliography

AB Automatic Balance

AB Automatic Blowdown

AB Automatic Brazing

AB Axial Bond

AB Azobenzene

AB Bachelor of Arts

AB Binary Alloy Composed of two Chemical Elements A and B [also A-B]

AB Binary Intermediate Compound of Component A and Component B With an A:B Ratio of 1:1 [Symbol]

AB Blood Group $\alpha AB\alpha$ [Contains Red Blood Cells with both Substance $\alpha A\alpha$ and $\alpha B\alpha$ on Surface]

AB Compound of Metal Component A and Nonmetal Component B [Symbol]

A/B Acid/Base (Ratio)

A/B Airborne [also AB, or ab]

A-B Allen-Bradley Company [US]

AB$_2$ Binary Alloy of Component A and Component B With an A:B Ratio of 1:2 [Symbol]

AB$_3$ Binary Alloy of Component A and Component B With an A:B Ratio of 1:3 [Symbol]

AB$_4$ Binary Alloy of Component A and Component B With an A:B Ratio of 1:4 [Symbol]

A$_2$B Binary Alloy of Component A and Component B With an A:B Ratio of 2:1 [Symbol]

A$_3$B Binary Alloy of Component A and Component B With an A:B Ratio of 3:1 [Symbol]

A$_x$B$_y$ Binary Ionic Compound With an A Ion to B Ion Ratio of x:y [Symbol]

A$_x$B$_{1-x}$ Binary Ionic Compound With an A Ion to B Ion Ratio of x:$(1-x)$ [Symbol]

Ab Alabamine [Chemistry]

Ab Albite [Mineral]

Ab Antibody [also AB]

Ab Asbestos

Aβ Feedback Factor (in Electronics) [Symbol]

Aβ Loop Gain [Symbol]

ab able-bodied (seaman) [also AB]

ab about

ab airborne [also AB]

ab- prefix used with cgs electromagnetic units (e.g., abampere, abvolt, etc.)

a/b a-axis to b-axis ratio (of crystal lattice) [Symbol]

(a,b) open interval between a and b (in differential calculus) [Symbol]

[a,b] closed interval between a and b (in differential calculus) [Symbol]

α-B Alpha Boron [Symbol]

ABA Abscisic Acid

ABA Académie des Beaux-Arts [Academy of Fine Arts, Institut de France, Paris]

ABA Acetoxybenzoic Acid

ABA 4-Acetyl Benzoyl Azide

ABA Active Brazing Alloy

ABA Agaricus Bisporus Agglutinin [Biochemistry]

ABA Air-Blown Asphalt

ABA Air Brake Association [US]

ABA American Bankers Association [US]

ABA American Bar Association [US]

ABA American Berkshire Association [US] [*Note:* Berkshire Refers to a Breed of Swine]

ABA American Birding Association [US]

ABA American Bison Association [US]

ABA American Booksellers Association [US]

ABA American Breed Association [US]

ABA American Buffalo Association [Name Changed to American Bison Association, US]

ABA American Burn Association [US]

ABA American Bus Association [US]

ABA American Business Association [US]

ABA Aminobenzoic Acid

ABA Aminobutyric Acid

ABA Antiquarian Booksellers Association [UK]

ABA ASEAN (Association of Southeast Asian Nations) Bankers Association [Singapore]

ABA Associate in Business Administration

ABA Association of British Archeologists

ABA Australian Booksellers Association

ABA Ayrshire Breeders Association [US] [*Note:* Ayrshire is a Breed of Cattle]

ABA Associação Brasiliera de Antropológica [Brazilian Association of Anthropology]

A$_1$BA$_2$ Layered Heterostructure of Metal Component A, Nonmetal Components B_1 and Metal Component A_2 [Symbol]

2ABA 2-Amino Benzoic Acid

3ABA 3-Amino Benzoic Acid

4ABA 4-Amino Benzoic Acid

ABAA Association of Business Advertising Agencies [UK]

ABAA Australian Business Aircraft Association

ABAAP 4-N-(4'-N,N-Dimethylaminobenzylidene) Aminoantipyrine [Biochemistry]

ABA Bank J ABA Banking Journal [of American Bankers Association]

ABAC A Basic Coursewriter

ABAC Active Brazing Alloy Compatible Ceramic

ABAC Association of Balloon and Airship Constructors [US]

ABAC Association of British Aero Clubs

ABAC Association of British Aviation Consultants

ABAC Association of Business and Administrative Computing [UK]

ABACUS AB Atomenergi Computerized User-Oriented Services [Sweden]

ABACUS Air Battle Analysis Center Utility System

ABACUS Association of Bibliographic Agencies of Britain, Australia, Canada and the United States

ABAF Alcian Blue Aldehyde Fuchsin

ABAFS American Board of Ambulatory Foot Surgery [US]

ABAG Association of Bay Area Governments [US]

ABAH Australian Bureau of Animal Health

ABAI American Boiler and Affiliated Industries [US]

abamp abampere [Unit]

abampere absolute ampere [Unit]

ABANA Artist-Blacksmiths Association of North America [US]

ABANK Annual State Databank [of Kentucky Economic Information Systems, US]

ABAR Advanced Battery Acquisition Radar

ABAR Alternate Battery Acquisition Radar

Å/bar Angstrom per bar [Unit]

ABARE Australian Bureau of Agricultural and Resource Economics

ABARES Australian Bureau of Agricultural and Resources Sciences

ABAS Association of Business Administration Studies [UK]

ABB American Board of Bioanalysts [US]

ABB Anweisung für den Bau von Betonfahrbahndecken [German Specification for the Construction of Concrete Surfacings, Germany]

ABB Asea Brown Boveri (Corporation) [Zurich, Switzerland]

ABB Automatic Back Bias

AB-B Binary System of Intermediate Compound *AB* (Composed of Components *A* and *B*) and Component *B* [Symbol]

AB$_1$B$_2$ Layered Heterostructure of Metal Component *A* and Two Alternating Nonmetal Components B_1 and B_2 [Symbol]

Abb Abbreviation [also abb]

abb abbreviate

ABBA American Bee Breeder Association [US]

ABBA American Brahman Breeders Association [US] [*Note:* Brahman is a Breed of Cattle]

ABBE Advisory Board on the Built Environment [now Building Research Board, US]

ABBET A Broad Based Environment for Test (Standards) [of Institute of Electrical and Electronics Engineers, US]

Abbey Newsl Abbey Newsletter [US]

ABBF Association of Brass and Bronze Founders [UK]

ABBIM American Brass and Bronze Ingot Manufacturers [US]

ABBM Automatic Baseband Monitor

ABBMM Association of British Brush Machinery Manufacturers

Abbr Abbreviation [also abbr]

abbr abbreviate(d)

ABBRA American Boat Builders and Repairers Association [US]

ABB Rev ABB Review [of Asea Brown Bovery, Switzerland]

Abbrev Abbreviation [also abbrev]

abbrev abbreviate(d)

Abbrn Abbreviation [also abbrn]

ABBBS Australian Bird and Bat Banding Scheme [of Australian Nature Conservation Agency]

ABC Abridged Building Classification

ABC Academia Brasileira de Ciencias [Brazilian Academy of Sciences, Rio de Janeiro]

ABC Acid-Base Catalysis

ABC Activity-Based Costing

ABC Acute Bacterial Conjunctivitis

ABC Adaptable Board Computer

ABC Advance Booking Charter

ABC Advanced Battery Consortium [US]

ABC Aerobic Bulk Cleanse

ABC African Bibliographic Center

ABC Airborne Control System

ABC Alfred Bader Collection (of Rare Chemicals)

ABC Already Been Converted

ABC Alum, Blood, Charcoal and Clay Process

ABC American Bibliographical Center [US]

ABC American Biotechnology Council

ABC American Blood Commission [US]

ABC American Botanical Council [US]

ABC American, British, and Canadian (Standards)

ABC American Broadcasting Company [US]

ABC American Bureau of Circulations [US]

ABC Annales de Biologie Clinique [Annals of Clinical Biology; published in France]

ABC Answer Back Code

ABC Antibody Binding Capacity (Unit)

ABC Approach by Concept

ABC Argentina–Brazil–Chile (Countries)

ABC Armored Bushed Cable

ABC Assisted by Computer

ABC Associated Builders and Contractors [US]

ABC Association of Biotechnology Companies [US and Canada]

ABC Association of Bituminous Contractors [US]

ABC Association of Boards of Certification [US]

ABC Association of British Climatologists [UK]

ABC Association of Building Centers [UK]

ABC Association of Business Centers [UK]

ABC Association of Business Communicators [UK]

ABC Asymptotic Boundary Condition

ABC Atanasoff-Berry Computer [First Vacuum-Tube Digital Electronic Computer Built by John Atanasoff and Clifford Berry in 1939]

ABC Atomic, Biological and/or Chemical (Weapons)

ABC Audit Bureau of Circulations [US]

ABC Australian Bird Count [of Australian Bird and Bat Banding Scheme]

ABC Australian Broadcasting Commission

ABC Australian Broadcasting Corporation

ABC Auto-Bill Calling

ABC Automated Bleed Compensation

ABC Automatic Background Control

ABC Automatic Bandwidth Control

ABC Automatic Bass Compensation [also abc]

ABC Automatic Bias Compensation

ABC Automatic Block Controller

ABC Automatic Boiler Control

ABC Automatic Brightness Control

$AB_{1-x}C_x$ Intermetallic Compound with an Atom Site Ratio $A{:}B{:}C$ of $1{:}(1\text{-}x){:}x$ [Symbol]

AB_2C_4 Spinel-Type Compound of Two Cations of Metallic Element A, Four Cations of Metallic Element B and Eight S (Sulfur) Anions [Ceramics]

$A_nB_mC_p$ Intermetallic Compound with n Sites of Atom A, m Sites of Atom B and p Sites of Atom C [Symbol]

$A_n(B_{m-x}C_x)$ Pseudobinary Intermetallic Compound with n Sites of Atom A, $m\text{-}x$ Sites of Atom B and x Sites of Atom C [Symbol]

a,b,c unit cell dimensions [Symbol]

a*,b*,c* reciprocal (crystal) lattice dimensions [Symbol]

(α,β,γ) Eulerian angles [Symbol]

ABCA American, British, Canadian and Australian (Standards)

ABCA American Building Contractors Association [US]

ABCA American Business Communications Association [US]

ABCA Army Bureau of Current Affairs [US]

ABCA Association of Biological Collections Appraisers [US]

ABCAS Active Beacon Collision Avoidance System

ABCAT Association of British Cutlers and Allied Trades [UK]

ABCB Air-Blast Circuit Breaker

ABCB Air-Break Circuit Breaker

ABCB American Bottlers of Carbonated Beverages [US]

ABCB Australian Broadcasting Control Board

ABCC American Board of Clinical Chemistry [US]

ABCC Association of British Chambers of Commerce [UK]

ABCC Association of British Correspondence Colleges [UK]

ABCC Atomic Bomb Casualty Commission [of United States Atomic Energy Commission]

ABCC Australian-British Chamber of Commerce

ABCCC Airborne Battlefield Command and Control Center [also ABC^3]

ABCCTC Advanced Base Combat Communication Training Center [US]

ABCD America–Britain–China–Dutch East Indies

ABCD Association of Better Computer Dealers [US]

ABCD Association for Bridge Construction and Design [US]

ABCD Atomic, Biological, Chemical and Damage Control

ABCI Advanced Business Communications Incorporated

ABCI Applied Biological Coatings and Implants [Dallas, Texas]

ABCI Automotive Booster Clubs International [US]

ABCL As-Built Configuration List

ABCM American Business Council of Malaysia [Malaysia]

ABCM Association of British Chemical Manufacturers

ABCM Association of Building Component Manufacturers [UK]

ABCMC Automotive Battery Charger Manufacturers Council [US]

ABCO Association of British Conference Organizers

A&B Comput A&B Computing [UK Journal]

ABCOP Association of Boards of Certification for Operational Personnel [now Association of Boards of Certification, US]

abcoul abcoulomb [Unit]

abcoulomb absolute coulomb [Unit]

ABCPO Association of British Professional Conference Organizers [UK]

ABCQ Association of Building Contractors of Quebec [Canada]

ABCPA Alberta Beef Cattle Performance Association [Canada]

ABCPF Association of British Columbia Professional Foresters [Canada]

ABCS Advisory Board for Cooperative Systems [US]

ABCS American Business Council of Singapore

ABCST American-British-Canadian Standards

ABCT American Board of Chelation Therapy [US]

ABCU Army Bureau of Current Affairs [US]

ABCYS Aluminum Trifluoride–Barium Difluoride–Calcium Fluoride–Yttrium Fluoride–Strontium Fluoride (Glass) [AlF_3–BaF_2–CaF_3–YF_3–SrF_2]

ABCYSNZ Aluminum Trifluoride–Barium Difluoride–Calcium Fluoride–Yttrium Fluoride–Strontium Fluoride–Sodium Fluoride–Zirconium Tetrafluoride (Glass) [AlF_3–BaF_2–CaF_3–YF_3–SrF_2–NaF–ZrF_4]

ABD Adhesive Bonding

ABD All But Dissertation (Status)

ABD Alloy Bulk Diffusion (Technique)

ABD American Board of Dermatology [US]

ABD American Bridge Division [of US Steel Corporation]

ABD Asian Development Bank

ABD Association Belge de Documentation [Belgian Association for Documentation]

ABD Atomic Beam Diffraction

ABD Average Business Day

ABDC After Bottom Dead Center

ABDEA Aminobenzyl Diethylamine

ABD-F (4-(Aminosulfonyl)-7-Fluoro-2,1,3-Benzoxadiazole)

ABDL Automatic Binary Data Link

ABDMA Aminobenzyl Dimethylamine

ABDO Association of British Dispensing Opticians [UK]

Abdom Abdomen; Abdominal [also abdom]

ABDP Association of British Directory Publishers [UK]

ABDR Aircraft Battle Damage Repair

ABE Acute Bacterial Endocarditis

ABE Advanced Bond Evaluator

ABE Aerodrome Beacon

ABE Alternate Molecular Beam Epitaxy

ABE Arithmetic Building Element

ABE Association of Business Executives [UK]

a-Be Alpha Beryllium [Symbol]

ABEC Aitken Bicentennial Exhibition Center [Canada]

ABEC Annular Bearing Engineers Committee [US]

ABEd Bachelor of Arts in Education

ABEI N-Ethyl-N-(4-Aminobutyl)-Isoluminol

ABEL Advanced Boolean Expression Language

ABEN Association of Biologists and Ecologists of Nicaragua

ABEND Abnormal End(ing) [also abend]

ABEPP American Board of Examiners in Professional Psychology [US]

ABES Aerospace Business Environment Simulator

ABES Association for Broadcast Engineering Standards [US]

ABES Australian Biological and Environmental Survey [Defunct]

ABES Automatic Burn-in and Environmental System

ABET Accreditation Board for Engineering and Technology [US]

ABETS Airborne Beacon Test Set

ab ex *(ab extra)* – from without

ABF America the Beautiful Fund [US]

ABF Arbetarnas Bildningsforbund [Workers Education Association, Sweden]

ABF Arbeitsgemeinschaft Betriebsfestigkeit [Association for Operational Strength of Materials Germany]

ABF Association des Bibliothécaires Français [Association of French Librarians]

ABF Automatic Ball Feeder

ABFA Azobisformamide

abfarad absolute farad [Unit]

ABFL Association of British Foam Laminators

ABFLO Association of Bedding and Furniture Law Officials [US]

ABFPRS American Board of Facial Plastic and Reconstructive Surgery [US]

ABFS Atomic Beam Frequency Standard

ABFS Auxiliary Building Filter System

ABG Abrasive Belt Grinding

ABG Anti-β-Galactosidase [Biochemistry]

ABG Aural Bearing Generator

ABGSM Association of British Generating Sets Manufacturers [UK]

ABGTS Auxiliary Building Gas Treatment System

ABGWIU Aluminum, Brick and Glass Workers International Union

ABH Actual Bodily Harm

ABH Average Busy Hour

A_3BH_n Interstitial Hydride of Intermetallic A15Type Compound With Atom Ratio of Metal A to Metal B to Hydrogen (H) is 3:1:n (n ≤ 3)

Abh Abhesion; Abhesive [also abh]

a-B:H Hydrogenated Amorphous Boron

AB(Hal)$_3$ Compound having Cations of Type A, and Type B respectively, and Halogen Anions (e.g. Chlorine, Fluorine, etc.) in Which the Ratio of the Cations of Type A to the Cations of Type B to the Halogen Anions is 1:1:3 [General Formula]

AB(Hal)$_4$ Compound having Cations of Type A, and Type B respectively, and Halogen Anions (e.g. Chlorine, Fluorine, etc.) in Which the Ratio of the Cations of Type A to the Cations of Type B to the Halogen Anions is 1:1:4 [General Formula]

AB$_2$(Hal)$_4$ Compound having Cations of Type A, and Type B respectively, and Halogen Anions (e.g. Chlorine, Fluorine, etc.) in Which the Ratio of the Cations of Type A to the Cations of Type B to the Halogen Anions is 1:2:4 [General Formula]

abhenry absolute henry [Unit]

Abhes Abhesion; Abhesive [also abhes]

ABHM Association of Builders' Hardware Manufacturers [UK]

ABHRS American Board of Hair Restoration Surgery [US]

ABHRS As Built Hardware Reporting System

ABI Abstracted Business Information

ABI Advance Boundary Information

ABI American Butter Institute [US]

ABI Application Binary Interface

ABI Association des Bibliothèques Internationales [Association of International Libraries, France]

ABI Association of British Insurers [UK]

ABI Automated Ball Indenter [Metallurgy]

α-**Bi** Alpha Bismuth [Symbol]

ABIA American Boardsailing Industries Association [US]

ABIH American Board of Industrial Hygienists [US]

ABI/INFORM Abstracted Business Information/Information Needs (Database)

ABIM Association of British Insecticide Manufacturers

ab init *(ab inito)* – from the beginning

AB in J Bachelor of Arts in Journalism

AB in SecEd Bachelor of Arts in Secondary Education

ABIOS Advanced BIOS (Basic Input/Output System)

ABIPC Abstract Bulletin of the Institute of Paper Chemistry [US]

ABIS Australian Biogeographic Information System [of Environmental Resources Information Network]

ABIST Automatic Built-In Self-Test

ABIST Autonomous Built-In Self Test

ABIX Australian Business Index [of Australian Information Network]

ABK Arrott-Belov-Kouvel (Plot)

α-**Bk** Alpha Berkelium [Symbol]

ABL Academia Brasileira de Letras [Brazilian Academy of Letters]

ABL Acid Brine Leaching

ABL Allegany Ballistics Laboratory [Cumberland, Maryland, US]

ABL Alpha-Acetyl-*γ*-Butyrolactone [Biochemistry]

ABL Architectural Block-Diagram Language

ABL Armoured Box Launcher

ABL Army Ballistics Laboratory [Aberdeen, Maryland, US]

ABL Army Biological Laboratories [Fort Detrick, Frederick, Maryland, US]

ABL Atlantic Basic Language

ABL Atlas BASIC (Computer) Language

ABL Atmospheric Boundary Layer [Meteorology]

ABL Atomic Beam Lamp

ABL Automated Biological Laboratory [of NASA]

ABL Automatic Bootstrap Loader

ABL Average Bond Length [Chemistry]

Abl Ablation; Ablative [also abl]

ABLB Alternate Binaural Loudness Balancing

ABLE Activity Balance Line Evaluation

ABLE Adaptive Battery Life Extender

ABLE Arctic Boundary Layer Experiment

ABLE Association for Biology Laboratory Education [Canada]

ABLE Atmospheric Boundary Layer Experiment [Meteorology]

ABLISS Association of British Library and Information Science Schools [UK]

ABLISS Association of British Library and Information Studies Schools [UK]

ABLO Australian Botanical Liaison Officer [of Australian Biological Resources Study]

ABLS American Bryological and Lichenological Society [US]

ABLS Bachelor of Arts in Library Science

ABLS Association of British Library Schools

ABLUE Asymptotically Best Linear Unbiased Estimator

ABM Ammonia-Beam Maser

ABM Anti-Ballistic Missile

ABM As Ball Milled (Condition)

ABM Associação Brasileira de Metales [Brazilian Association of Metals]

ABM Associação Brasileira de Metalurgia [now Associação Brasiliera de Metalurgia e Materiais]

ABM Associação Brasileira de Metalurgia e Materiais [Brazilian Metallurgy and MaterialsAssociation]

ABM Associate in Business Management [US]

ABM Association of Board Makers [UK]

ABM Asynchronous Balanced Mode

ABM Automated Batch Mixing

ABM Aviation Boatswain's Mate

ABMA American Boiler Manufacturers Association [US]

ABMA American Brush Manufacturers Association [US]

ABMA Army Ballistic Missile Agency [Redstone Arsenal, Huntsville, Alabama, US]

ABMAAI American Boiler Manufacturers Association and Affiliated Industries [now American Boiler Manufacturers Association, US]

ABMAC Abalone Management Advisory Committee [Australia]

ABMAC Association of British Manufacturers of Agricultural Chemicals

ABMD Air Ballistics Missile Division [of US Air Force]

ABMDA Advanced Ballistic Missile Defense Agency [of US Army]

ABMDA Associated Building Material Distributors of America [US]

ABME Annals of Biomedical Engineering [of Biomedical Engineering Society, US]

ABMEC Association of British Mining Equipment Companies [UK]

ABMEWS Anti-Ballistic Missile Early Warning System

ABMEX Association of British Mining Equipment Exporters [UK]

abmho absolute mho [Unit]

ABMIS Anti-Ballistic Missile Intercept System

ABMO Anti-Bonding Molecular Orbital

ABMR Atomic Beam Magnetic Resonance

ABMRC Association of British Market Research Companies [UK]

ABMRF Alcoholic Beverage Medical Research Foundation [Baltimore, Maryland, US]

ABMS American Bureau of Metal Statistics [US]

ABMS Anisotropic Bulk Magnetic Susceptibility

ABN 4-Acetyl Benzoylnitrene

ABN Acoustic Barkhausen Noise [Physics]

ABN Airborne [also Abn, or abn]

ABN Australian Bibliographic Network [of Australian Library Network]

ABN Azobisisobutyronitrile

Abn Airborne [also ABN, or abn]

a-**BN** Alpha Boron Nitride

a-BN Amorphous Boron Nitride

ABNO All But Not Only

ABNT Associação Brasiliera de Normas Technicas [Brazilian Standards Association]

ABO Alpha Barium Borate

ABO American Board of Otolaryngology [US]

ABO Antibonding (Molecular) Orbital

ABO Astable Blocking Oscillator

ABO$_2$ Compound of One Cation of Metallic Element A, One Cation of Metallic Element B and Two O (Oxygen) Anions [General Formula]

ABO$_3$ Perovskite-Type Compound of One Cation of Metallic Element A, One Cation of Metallic Element B and Three O (Oxygen) Anions [General Formula]

ABO$_4$ Compound of One Cation of Metallic Element A, One Cation of Metallic Element B and Four O (Oxygen) Anions [General Formula]

AB$_2$O$_4$ Spinel-Type Compound of One Cation of Metallic Element A, Two Cations of Metallic Element B and Four O (Oxygen) Anions [General Formula]

A$_2$B$_2$O$_6$ Pyrochlore-Type Compound of Two Cations of Metallic Element A, Two Cations of Metallic Element B and Six O (Oxygen) Anions [General Formula]

A$_2$B$_2$O$_7$ Compound of Two Cations of Metallic Element A, Two Cations of Metallic Element B and Seven O (Oxygen) Anions [General Formula]

A$_{n+1}$B$_n$O$_{3n+1}$ Ruddlesden-Popper-Type Compound in which the Ratio of the Cations of Metallic Element A, to the Cations of Metallic Element B and the Anions of Oxygen O is $(n+1) : n : (3n+1)$ [General Formula]

ABOA Australian Bibliography on Agriculture [of AUSINET]

ABOCF Association of British Organic and Compound Fertilizers [US]

abohm absolute ohm [Unit]

ABOI Association of British Oceanic Industries [UK]

ABOI Association of British Offshore Industries [UK]

ABOL Adviser Business-Oriented Language

ABOM American Board of Occupational Medicine [US]

A-Bomb Atomic Bomb [also A-bomb]

ABORT Abnormal Termination [also abort]

ABOS Advanced Bombardment System

ABP Abrasive Belt Polishing

ABP Actual Block Processor

ABP Adjustable Ball Pin

ABP Advanced Business Processor

ABP American Business Press [Name Changed to American Business Publishers, US]

ABP American Business Publishers [US]

ABP Aminobiphenyl

ABPA Aftermarket Body Parts Association [US]

ABPA American Book Producers Association [US]

ABPA Australian Book Publishers Association

ABPBC Association of Book Publishers of British Columbia [Canada]

ABPC Association of British Packing Contractors

ABPC Association of British Pewter Craftsmen [UK]

ABPI Association of the British Pharmaceutical Industry [UK]

ABPR Association of British Picture Restorers [UK]

ABPS American Board of Plastic Surgeons [US]

ABPS American Board of Podiatric Surgery [US]

ABPVM Association of British Plywood and Veneer Manufacturers [UK]

ABPy 2,2'-Azobis(pyridine) [also abpy]

ABQ Associação Brasileira de Quimica [Brazilian Chemical Association]

ABR Accomplishment-Based Renewal (Proposal)

ABR Acrylate-Butadiene Rubber

ABR Acrylic Ester-Butadiene Rubber

ABR Airborne Radar

ABR American Board of Radiology

ABR Atomic Beam Resonance

ABR Automatic Band Rate

ABR Automatic Baud Recognition

ABR Available Bit Rate

Abr Abrasive [also abr]

Abr Abridgement [also abr]

abr abridged

ABRACO Associação Brasileira de Corrosão [Brazilian Corrosion Association]

ABRAFE Associação Brasileira de Fabricantes de Ferroligas [Brazilian Association of Ferroalloy Manufacturers]

Abrasive Eng Soc Mag Abrasive Engineering Society Magazine [US]

ABRC Advisory Board for the Research Councils [UK]

ABRES Advanced Ballistic Reentry System

ABRFM Association of British Roofing Felt Manufacturers [UK]

ABRL Army Ballistics Research Laboratory [US Army]

ABRO Animal Breeding Research Organization [UK]

ABRS Australian Biological Resources Study [of Australian Nature Conservation Agency]

ABRS Automated Book Request System [of British Library]

ABRSAC Australian Biological Resources Study Advisory Council

Abrsv Abrasive [also abrsv]

ABS Absent Subscriber

ABS Absolute Value (Function) [also Abs]

ABS Acrylonitrile-Butadiene-Styrene (Resin)

ABS Address Book Synchronization

ABS Advanced Business System

ABS Air Bag System

ABS Air Bearing Surface [Magnetic Media]

ABS Air Break Switch

ABS Alkylbenzene Sulfonate

ABS American Biological Society [US]

ABS American Bladesmith Society [US]

ABS American Bureau of Shipping [US]

ABS American Business School [US]

ABS Animal Behavior Society [US]

ABS Anti-Block (Braking) System

ABS Antigen Binding Site [Biochemistry]

ABS Anti-Lock Brake System

ABS Association of British Sailmakers [UK]

ABS Association of British Spectroscopists [UK]

ABS Association of Broadcasting Staffs [UK]

ABS Association of Broadcasting Standards [US]

ABS Atom Beam Scattering

ABS Auckland Botanical Society [New Zealand]

ABS Australian Bat Society

ABS Australian Bureau of Statistics

ABS Automated Bond System

ABS Automatic Backwash Strainer

ABS Automatic Bibliographic Service

ABS Automatic Break System

ABS Auxiliary Building Sump

ABS Average Busy Season

AB(s) Surface of Compound *AB* (Consisting of Metal Component *A* and Nonmetal Component *B*) with *B* as the Surface Atom [Symbol]

Abs Absorb(er); Absorption [also abs]

Abs Abstract(ion) [also abs]

abs absolute(ly)

ABSA Aminobenzenesulfonic Acid

ABSAM Association of British Solid Fuel Appliance Manufacturers [UK]

ABSAP Airborne Search and Attack Plotter

ABSAP Aminobenzenesulfamidopyridine

ABSBH Average Busy Season Busy Hour

ABSC Alaska Biological Science Center [Anchorage, US]

Abs Corr Absorption Correction [also ABS CORR]

absiemens absolute siemens [Unit]

ABSLDR Absolute Loader [also Absldr, or absldr]

ABSM Alberta Bureau of Surveying and Mapping [Canada]

ABSM Association of British Sterilizer Manufacturers [UK]

ABSMA American Bleached Shellac Manufacturers Association [US]

absol absolute

Absorp Absorption [also absorp]

ABS-PA Acrylonitrile-Butadiene-Styrene/Polyamide (Blend) [also ABS/PA]

ABS-PC Acrylonitrile-Butadiene-Styrene/Polycarbonate (Blend) [also ABS/PC]

Abspn Absorption [also abspn]

ABS-PUR Acrylonitrile-Butadiene-Styrene/Polyurethane (Blend) [also ABS/PUR]

ABS-PVC Acrylonitrile-Butadiene-Styrene/Polyvinyl Chloride (Blend) [also ABS/PVC]

ABSR Association of British Science Writers

ABS-SAN Acrylonitrile-Butadiene-Styrene/Styrene Acrylonitrile (Blend) [also ABS/SAN]

ABS-SMA Acrylonitrile-Butadiene-Styrene/StyreneMaleic Anhydride (Blend) [also ABS/SMA]

ABST Alpha-Beta Solution Treatment [Metallurgy]

Abst Abstract [also abst]

abstat absolute electrostatic unit

ABSTI Advisory Board on Scientific and Technical Information [Canada]

Abstr Abstract [also abstr]

Abstr Brit Pat Abstracts of Specifications of British Patents [UK]

ABSW Air-Break Switch

ABSW Association of British Science Writers [UK]

ABT Abort [also abt]

ABT About [Amateur Radio]

ABT Air-Blast Transformer

ABT American Board of Toxicology [US]

ABT Association Belge de Télétravail [Belgian Teleworking Association]

ABT Association of Building Technicians [UK]

ABT Australian Broadcasting Tribunal

abT abtesla [Unit]

abt about

ABTA Allied Brewery Traders Association

ABTA Australian British Trade Association

abtesla absolute tesla [Unit]

ABTF Airborne Task Force

ABTICS Abstract and Book Title Index Card Service [UK]

a-BTO Amorphous $BaTiO_3$ (Barium Titanate)

ABTRD Advanced Battery Technology Research and Development (Program) [of US Department of Energy]

ABTS ASCII (American Standard Code for Information Interchange) Block Terminal Services

ABTS 2,2'-Azinobis(3-Ethyl-Benzthiazoline-6-Sulfonic Acid); 2,2'-Azino-di(3-Ethylbenzylthiazoline Sulfonic Acid-6)

ABTSS Airborne Transponder Subsystem

ABTT Association of British Theatre Technicians [UK]

ABU American Board of Urology [US]

ABU Asian Broadcasting Union [Japan]

ABU Asia-Pacific Broadcasting Union [Malaysia]

Abu *α*-Aminobutyric Acid

Abu *β*-Aminobutyric Acid

ABus Associate in Business

ABU Tech Rev ABU Technical Review [of Asian Broadcasting Union, Japan]

ABV Absolute Value

ABV Aquabirnavirus

abvolt absolute volt [Unit]

ABVP American Board of Veterinary Professionals [US]

Abwassertech Abfalltech Abwassertechnik mit Abfalltechnik [German Publication on Wastewater Technology and Waste Technology]

abwatt absolute watt [Unit]

abWb abweber [Unit]

ABWE Associated Business Writers of America [US]

abweber absolute weber [Unit]

ABWO Airborne Weather Operator

ABWR Advanced Boiling Water Reactor

ABWR American Beefalo World Registry [US]

ABX Ternary Alloy Composed of Three Chemical Elements A, B and X [also A-B-X]

ABX₂ Compound having Cations of Type A, Cations of Type B and Anions of Type X in Which the Ratio of the Cations of Type A to the Cations of Type B to the Anions of Type X is 1:1:2 [General Formula]

ABX₃ Perovskite Type Compound having Cations of Type A, Cations of Type B and Anions of Type X in Which the Ratio of the Cations of Type A to the Cations of Type B to the Anions of Type X is 1:1:3 (e.g., $BaTiO_3$, $SrZrO_3$, $SrSnO_3$, etc.) [General Formula]

AB₂X₄ Spinel Type Compound having Cations of Type A, Cations of Type B and Anions of Type X in Which the Ratio of the Cations of Type A to the Cations of Type B to the Anions of Type X is 1:2:4 (e.g., $FeAl_2O_4$, $MgAl_2O_4$, $MgFe_2O_4$, $MnFe_2O_4$, etc.) [General Formula]

A₂BX₂ Compound having Cations of Type A, Cations of Type B and Anions of Type X in Which the Ratio of the Cations of Type A to the Cations of Type B to the Anions of Type X is 2:1:2 [General Formula]

AₘBₙXₚ Compound of m Cations of Substance A, n Cations of Substance B and p Anions of Substance X [General Formula]

ABYA Association of Brokers and Yacht Agents [UK]

ABYC American Boat and Yacht Council [US]

ABZ Adsorbate (Layer) Brillouin Zone [Solid-State Physics]

ABZ Association of British Zoologists

AC Abbe Comparator

AC Abrasive Cutting

AC Absorption Coefficient

AC Absorption Control [Electronics]

AC Absorption Cooling

AC *Academia de Ciencias* – Spanish for "Academy of Sciences"

AC *Academia de Ciências* – Portuguese for "Academy of Sciences"

AC Accelerated Cement

AC Accepted Counts

AC Access Number

AC Accomodation Coefficient [Statistical Mechanics]

AC Accumulation; Accumulator

AC Acetal Polymer

AC Acetate

AC Acetyl

AC Acetone

AC Acetylated Cellulose

AC Acid-Catalyzed

AC Acid Chloride

AC Acidic Catalysis

AC Ackerman and Cox

AC Acoustic Concentration

AC Acoustic Coupler

AC Acrylic

AC Action Center [Meteorology]

AC Activated Carbon

AC Activated Complex (Theory) [Physical Chemistry]

AC Activating Center (of Brain)

AC Active Coil (of Springs) [Symbol]

AC Active Current

AC Activity Code

AC Adam's Catalyst [Chemical Engineering]

AC Adaptive Control

AC Address Counter

AC Adiabatic Calorimeter; Adiabatic Calorimetry

AC Adiabatic Cooling

AC Adjusted Charge [Explosives]

AC Adsorption Catalysis

AC Adsorption Chromatography

AC Advanced Ceramics

AC Advanced Composite

AC Advanced Certificate

AC Advertising Council [US]

AC Advice of Charge

AC Advisory Circular

AC Advisory Committee

AC Aerodynamic Center

AC Aerospace Corporation [US]

AC Aerial Crane

AC After Christ

AC Agribusiness Council [US]

AC Agriculture Canada

AC Aharonov-Casher (Effect) [Physics]

AC Air Canada

AC Air Carrier

AC Air Column

AC Air Commodore

AC Air Compression; Air Compressor

AC Air Conduction

AC Air Conditioner; Air Conditioning [also A/C]

AC Air-Cooled; Air Cooler; Air Cooling

AC Air Corps

AC Air Count

AC Aircraft [also A/C]

AC Aircraftman [also A/C]

AC Aircraftwoman [also A/C]

AC Air Cushion

AC Airworthiness Circular

AC Alabama College [Montevallo, US]

AC Alaska Coalition [US]

AC Aldol Condensation

AC Algebraic Code

AC Algonquin College [Canada]

AC Alicyclic Compound
AC Alkyl Chain
AC Alkyl Chloride
AC Allen-Cahn (Model) [Metallurgy]
AC Allyl Chloride
AC Alpine Club
AC Alternating Current [also ac]
AC Amacrine Cell
AC American and Canadian (Standards)
AC Amherst College [Massachusetts, US]
AC Ammonium Chloride
AC Ammonium Citrate
AC Analog Communications
AC Analog Computer
AC Analog Control(ler)
AC Analysis Chamber
AC Analysis Control
AC Analytical Chemist(ry)
AC Aneroid Chamber
AC Anisotropy Constant [Solid-State Physics]
AC Annual Cost
AC Annular Core (Reactor)
AC Anodized Coating
AC Answer Complete
AC *(Ante Christum)* – before the birth of Christ
AC Anti-Cancer (Drug)
AC Anticathode
AC Anticoagulant
AC Anti-Coincidence Counter; Anti-Coincidence Counting
AC Anti-Contamination
AC Anticyclone [Meteorology]
AC Applied Chemistry
AC Applied Climatology
AC Aqueous Corrosion
AC Arc Cast(ed); Arc Casting
AC Arc Chamber
AC Arc Cutting
AC Archeological Conservancy [US]
AC Arctic Circle
AC Argon Cooled; Argon Cooling [also ac]
AC Area Code
AC Armor-Clad
AC Armor-Clad Cable
AC Armored Cable
AC Armored Car
AC Army Corps
AC Arnold-Chiari (Syndrome) [Medicine]
AC Aromatic Compound
AC Art Conservation
AC As Cast (Condition)
AC Ascending Colon [also ac]
AC As Crystallized (Condition)
AC Assigned Code

AC Associate in Commerce
AC Astrochemical; Astrochemist(ry)
AC Astrochronological; Astrochronology
AC Astro-Compass
AC Astrocyte [Anatomy]
AC Asynchronous Channel
AC Asynchronous Communication
AC Asynchronous Counter
AC Atmospheric Corrosion
AC Atomic Clock
AC Atomic Core
AC Attachment Coefficient
AC Attenuation Coefficient
AC Auditory Canal [Anatomy]
AC Authority Code
AC Author's Correction
AC Auto-Calibration
AC Autocatalysis; Autocatalytic
AC Autocheck
AC Autoclave
AC Autocollimation; Autocollimator
AC Autoconvection [Meteorology]
AC Autocorrection; Autocorrector
AC Autocorrelation; Autocorrelator
AC Automatic Calibration
AC Automatic Calorimeter
AC Automatic Checkout
AC Automatic Computer [also ac]
AC Automatic Control(ler)
AC Automatic (Metal) Cutting
AC Auxiliary Carry (Flag) [Computers]
AC Avogadro's Constant
AC Awaiting Connection
AC Comptroller Office [of NASA]
A₂C 2-(Methoxyethoxy)ethyl 8-(cis-2-n-Octylcyclopropyl)octanoate
A/C Account [also a/c]
A/C Account Current [also a/c]
A/C Air Conditioner; Air Conditioning
A/C Aircraft [also AC]
A/C Aircraftsman; Aircraftswoman
A/C Anticoncrete
A/C Asbestos Cement
A-C Alternating Current [also a-c]
A-C Amorphous-Crystalline (Transition)
Ac Account [also ac]
Ac Acetate
Ac Acetic Acid
Ac Acetyl
Ac Acetylate(d)
Ac Acid [also ac]
Ac Actinium [Symbol]
Ac Altocumulus (Cloud) [also ac]
Ac⁻ Acetate Ion [Symbol]

Ac$_{cm}$ Temperature at Which Transformation of Austenite plus Cementite to Austenite is Completed during Heating of Hypereutectoid Steel [Symbol]

Ac$_0$ Temperature at Which a Magnetic Change Takes Place in Cementite (Curie Point of Cementite) [Symbol]

Ac$_1$ Temperature at Which Austenite Formation Begins During Heating of Iron and Steel [Symbol]

Ac$_2$ Temperature at Which Alpha Iron Changes to Beta Iron (Curie Temperature of Ferrite) [Symbol]

Ac$_3$ Temperature at Which Ferrite-to-Austenite Transformation is Completed During Heating of Iron and Steel [Symbol]

Ac$_4$ Temperature at Which the Transformation of Austenite to Delta Ferrite Occurs During Heating of Iron and Steel [Symbol]

Ac-222 Actinium-222 [also ^{222}Ac, or Ac^{222}]

Ac-223 Actinium-223 [also ^{223}Ac, or Ac^{223}]

Ac-224 Actinium-224 [also ^{224}Ac, or Ac^{224}]

Ac-225 Actinium-225 [also ^{225}Ac, or Ac^{225}]

Ac-226 Actinium-226 [also ^{226}Ac, or Ac^{226}]

Ac-227 Actinium-227 [also ^{227}Ac, Ac^{227}, or Ac]

Ac-228 Actinium-228 [also ^{228}Ac, Ac^{228}, or MsTh$_2$]

Ac-229 Actinium-229 [also ^{229}Ac, or Ac^{229}]

aC abcoulomb [Unit]

a-C amorphous carbon (or diamond like carbon)

ac account [also Ac]

ac acid [also Ac]

ac (a conto) – on account

ac acre(s) [Unit]

ac acute

ac alternating current [also AC]

ac (anni currentis) – this (current) year

ac (ante cibum) – before meals; before food [Medical Prescriptions]

a c (a condition) – upon request

a/c account [also A/C]

a/c account current [also A/C]

a/c amorphous-to-crystalline (transition)

a-c alternating current [also ac]

a-c amorphous-to-crystalline (transformation)

a-C Alpha Carbon [or C^a]

ACA Adjacent Channel Attenuation

ACA Advanced Ceramics Association [US]

ACA Advanced Control Algorithm

ACA Agricultural Computer Association

ACA Air Cargo Agent

ACA Air Compliance Advisor

ACA Aircrewman Association [UK]

ACA Alberta Construction Association [Canada]

ACA Allied Command Atlantic [of NATO]

ACA Amalgamated Construction Association (of British Columbia) [Canada]

ACA American Cartographic Association [US]

ACA American Chain Association [US]

ACA American Chiropractic Association [US]

ACA American Committee on Africa [US]

ACA American Communications Association [US]

ACA American Commuters Association [US]

ACA American Council for Aeronautics [US]

ACA American Crystallographic Association [US]

ACA Aminocephalosporanic Acid

ACA Ammoniacal Copper Arsenate

ACA Anelastic Creep and Aftereffect

ACA Anterior Cerebral Artery [Anatomy]

ACA Architectural Cladding Association [UK]

ACA Armament Control Agency

ACA Associate of the Institute of Chartered Accountants [UK]

ACA Association of Canadian Advertisers

ACA Association of Commuter Airlines [US]

ACA Association of Consultant Architects [UK]

ACA Asynchronous Communications Adapter

ACA Australasian Corrosion Association

ACA Australian Coal Association

ACA Australian Coastal Atlas

ACA Australian Consumer's Association

ACA Australian Council for Aeronautics

ACA Automatic Circuit Analyzer

ACA Automatic Communication Association

ACA Automatic Conference Arranger

ACA Automobil-Centrum Aachen [Automobile Centre at Aachen, Germany]

ACA Average Case Analysis

AC A Acetic Acid [also Ac A]

7-ACA 7-Aminocephalosporanic Acid

AcA Actinium A [also Po-215, ^{215}Po, or Po^{215}]

a-Ca Alpha Calcium [Symbol]

ACAA American Coal Ash Association [US]

ACAAC American College Admissions Advisory Center [US]

ACABQ Advisory Committee on Administratve and Budgetary Questions [of United Nations]

ACAC Acetylacetonate; Acetylacetonato; Acetylacetone [also AcAc, or acac]

ACAC Arab Civil Aviation Council

ACAC Asociación Colombiana para el Avance de la Ciencia [Colombian Association for the Advancement of Science]

ACAC Associate Committee on Automatic Control [of National Research Council of Canada]

ACAC Australian Chemicals Advisory Committee

ACACA Acetylacetic Acid [also AcAcA]

(ACAC)Cu(PMe$_3$) Acetylacetonate Copper (I) Trimethylphosphine Copper [also (acac)Cu(PMe$_3$)]

ACACEN N,N'-Bis(Acetylacetone)ethylenediamine

ACAD Air Containment Atmosphere Dilution

ACAD American Conference of Academic Deans [US]

ACAD AutoCAD (Computer-Aided Design) [Software Package]

Acad Academic; Academy [also acad]

ACADI Arab Center for Agricultural Documentation Information

ACADIA Association for Computer-Aided Design in Architecture

Acad Pol Sci Chim Academia Polonae Scientarium Chimica [Polish Academy of Chemical Sciences]

Acad Rep Fac Eng Tokyo Inst Polytech Academic Reports, Faculty of Engineering, Tokyo Institute of Polytechnics [Japan]

ACADS Association for Computer-Aided Design [Australia]

Acad Sin Academia Sinica [China]

Acad Serbe Sc Académie Serbe des Sciences [Serbian Academy of Sciences]

Acad Serbe Sc Arts Glas Cl Sci Tech Académie Serbe des Sciences et des Arts, Glas, Classe des Sciences Techniques [Serbian Academy of Arts and Sciences, Technical Sciences Class]

ACAE Australian Commission on Advanced Education

ACAH Acylcholine Acyl Hydrolase [Biochemistry]

ACAI Automatic Counting Accuracy Improvement

ACAIP Advisory Committee on Animal Import Priorities

ACAM American College for the Advancement of Medicine [Laguna, California, US]

ACAM Augmented Content Addressed Memory

A-CAM A-Cell Adhesion Molecule [Biochemistry]

ACAMAR Asociación Centroamericana de Armadores [Central American Association of Ship Owners Guatemala]

ACAN Advanced Contract Award Notices [of Open Bidding Services, Canada]

ACAN Army Command and Administrative Network [of US Army]

ACANZ Agricultural Council of Australia and New Zealand

ACAP Advanced Composite Airframe Program

ACAP Aeronautical Chart Automation Project [US]

ACAP Application Configuration Access Protocol

ACAP Automatic Circuit Analysis Program

A-CAP Environmental Grouping–Australia, Canada and Asia Pacific (incl. Japan, New Zealand and the United States of America)

ACAQ Advisory Committee on Air Quality

ACAR Aluminum Conductor Alloy Reinforced

ACAR Angular Correlation of Annihilation Radiation

ACARD Advisory Council for Applied Research and Development [UK]

ACARS ARINC (Aeronautical Radio Incorporated) Communications, Addressing and Reporting System

ACARTSOD African Center for Applied Research Training and Social Development

ACAS Advisory Conciliation and Arbitration Service [UK]

ACAS African Commission on Agricultural Statistics [Ghana]

ACAS Airborne Collision Avoidance System

ACAS Aircraft Collision Avoidance System

ACASP Air Conditioning Analytical Simulation Package [also A/CASP]

acaspOO' N-Acetyl-Aspartate [Biochemistry]

ACASPP American Committee to Advance the Study of Petroglyphs and Pictographs [US]

ACAST Advisory Committee on the Application of Science and Technology [of UNESCO]

ACAT Associate Committee on Air-Cushion Technology

ACAU Automatic Calling and Answering Unit

ACAV American Committee on Athropoid-Borne Viruses [US]

ACAVC Advisory Committee on Agricultural and Veterinary Chemicals [Australia]

ACAWU Aviation, Communication and Allied Workers Union [Venezuela]

ACB Access Method Control Block

ACB Adjusted Cost Base

ACB Air Circuit Breaker

ACB Alberta Cancer Board [Canada]

ACB American Council for the Blind [US]

ACB Application Control Block

ACB Army Classification Battery

ACB Association of Certification Bodies [UK]

ACB Association of Clinical Biochemists [UK]

ACB Association of Custom Brokers [US]

ACB Automatic Contrast and Brightness (Control)

AcB Actinium B [also Pb-211, ^{211}Pb, or Pb^{211}]

ACBFS Air-Cooled Blast-Furnace Slag

ACBLF Association Canadienne des Bibliothécaires de Langue Française [Canadian Association of French Language Librarians]

ACBM Advanced Cement-Based Materials [International Journal published in Affiliation with the American Concrete Institute] [also Adv Cem Mater, or Adv Cem Based Mater]

ACBM Association of Cartonboard Makers [UK]

ACBM Center for Science and Technology of Advanced Cement Based Materials [of National Science Foundation at Northwestern University, Evanston, Illinois, US]

ACBS Accrediting Commission for Business Schools

Ac-BSA Acetylated Bovine Serum Albumin [Biochemistry]

ACC Aboriginal Coordinating Council [Australia]

ACC Academia Chilena de Ciencias [Chilean Academy of Sciences]

ACC Academia de Ciencias de Cuba [Cuban Academy of Science, Havana]

ACC Accelerated Cooling [also AcC]

ACC Accumulator

ACC Acid Copper Chromate

ACC Adapter Common Card

ACC Additively Colored Crystal

ACC Administrative Committee on Coordination [of United Nations Environmental Program]

ACC Advanced Carbon/Carbon (Composite)

ACC Advanced Coating Center [Petten, Netherlands]

ACC Aft Cargo Carrier

ACC Agricultural Commodity Corporation [Guelph, Ontario, Canada]

ACC Air Control Center

ACC Air-Cooled Condenser

ACC Air Coordinating Committee

ACC Aldrich Chemical Company [US]

ACC Allied Command Channel [of NATO]

ACC Allied Control Council

ACC Amateur Computer Club [UK]

ACC Amazonian Cooperation Council [Brazil]

ACC American Carbon Committee [now American Carbon Society, US]

ACC American Chamber of Commerce [US]

ACC American Control Conference [US]

ACC American Copper Council [US]

ACC American Copyright Council [US]

ACC American Craft Council [US]

ACC 1-Amino Cyclopropane Carboxylic Acid; 1-Aminocyclopropane-1-Carboxylic Acid

ACC Amorphous Covalent Ceramic

ACC Analog Conversion Card

ACC Antarctic Circumpolar Current [Oceanography]

ACC Antenna Control Console

ACC Antigua Chamber of Commerce [Antigua Barbuda]

ACC Application Control Code

ACC Approved Capital Cost

ACC Area Control Center [Aeronautics]

ACC Arizona Corporation Commission [US]

ACC Army Chemical Center [Maryland, US]

ACC Arts and Communications Counsellor

ACC Assistant Chief Constable

ACC Association of Computer Clubs [UK]

ACC Association of Computer Consultants [now Association of Computer and CD-ROM Users US]

ACC Asynchronous Communications Controller

ACC Atlantic Capital Corporation [of Deutsche Bank, Germany]

ACC Atlantic Council of Canada

ACC Atomic Collision Cascade

ACC Australasian Corrosion Center [Australia]

ACC Australian Chamber of Commerce

ACC Automatic Carrier Control

ACC Automatic Chrominance Control

ACC Automatic Color Control

ACC Automatic Contrast Control

ACC Automatic Control Council [of Institution of Electrical Engineers, UK]

ACC Automotive Composites Consortium [of Chrysler Corporation, Ford Motor Company and General Motors Corporation]

AcC Accelerated Cooling [also ACC]

AcC Actinium C [also Bi-211, ^{211}Bi or Bi^{211}]

AcC′ Actinium C′ [also AcC$_1$, Po-211, ^{211}Po or Po^{211}]

AcC″ Actinium C″ [also AcC$_2$, Tl-207, ^{207}Tl or Tl^{207}]

Acc Acceleration; Accelerator [also acc]

Acc Acceptance [also acc]

Acc Account(ant) [also acc]

Acc Accumulation; Accumulator [also acc]

Acc Acinetobacter calcoaceticus [Microbiology]

acc accompanied; accompany

acc according

ACCA Accelerated Capital Cost Allowance

ACCA Air Chartered Carriers Association [UK]

ACCA Air Conditioning Contractors of America [US]

ACCA Air Courier Conference of America [US]

ACCA American Chamber of Commerce in Austria

ACCA Asociación Chilena de Control Automático [Chilean Association for Automatic Control]

ACCA Association of Certified and Corporate Accountants [UK]

ACCA Asynchronous Communications Control Attachment

ACCAD Advisory Committee for the World Climate Applications and Data Programs [of World Meteorological Organization]

ACCAP Autocoder-to-COBOL Conversion Aid Program

ACCAP Autocoder-to-COBOL Conversion and Program

ACCART Australian Council for Care of Animals in Research and Teaching

ACCAST Advisory Committee on Colonial Colleges of Arts, Science and Technology [now Colonial Colleges of Arts, Science and Technology, UK]

ACCB American Chamber of Commerce in Bolivia

ACCBD Advisory Committee on Conservation of Biological Diversity [Australia]

ACCC Ad Hoc Committee for Competitive Communication [US]

ACCC Association of Canadian Community Colleges

ACCCE Association of Consulting Chemists and Chemical Engineers [also ACC&CE, US]

Acc Chem Res Accounts of Chemical Research [Publication of the American Chemical Society] [also Acct Chem Res]

ACCCI American Coke and Coal Chemicals Institute [also AC&CCI, US]

ACCCIW Advisory Committee to the Canadian Center for Inland Waters

ACCD Accelerated Construction Completion Date

ACCD Approximately Coupled Cluster Doubles [Benzel-Dykstra]

accdg according

ACCE Advanced Composites Conference and Exposition [of ASM International/Engineering Society of Detroit, US]

ACCE Advisory Committee on Chemicals in the Environment [of Australian and New Zealand Environment and Conservation Council]

ACCE Aircrew-Cockpit Compatibility Evaluation

ACCE American Chamber of Commerce Executives [US]

ACCE American Council for Construction Education [US]

ACCEEO Advisory Committee on Environmental Employment Opportunities [Australia]

ACCEL Automated Circuit Card Etching Layout

Accel Acceleration [also accel]

ACCES Advanced Computing and Communications Educational System [Canada]

ACCESS Aachener Centrum für Erstarrung unter Schwerelosigkeit [Aachen Center for Zero-Gravity Solidification, Germany]

ACCESS Air Canada Cargo Enquiry and Service System

ACCESS Aircraft Communication Control and Electronic Signaling System

ACCESS Aircraft Communication Electronic Signaling System

ACCESS Architects Central Constructional Engineering Surveying Service [of Greater London Council UK]

ACCESS Architecture, Construction and Consulting Engineers Special Services [Australia]

ACCESS Argonne Code Center Exchange and Storage System [of US Atomic Energy Commission]

ACCESS Automated Catalog for Computer Equipment and Software Systems [of US Army]

ACCESS Automated City Clerk Enhanced Search System

ACCESS Automatic Computer-Controlled Electronic Scanning System

Access Accessory [also access]

ACCET Accrediting Council for Continuing Education and Training [US]

ACCFA Agricultural Credit and Cooperative Financing Administration [Philippines]

ACCG American Committee for Crystal Growth [now American Association for Crystal Growth, US]

ACCG American Conference on Crystal Growth [US]

ACCH Australian Center for Catchment Hydrology

ACCHAN Access Channel [also Acchan]

ACCHAN Allied Command Channel [of NATO]

ACCI American Council of Consumer Interests [US]

ACCI ASEAN (Association of Southeast Asian Nations) Chamber of Commerce and Industry

ACCI Association of Chambers of Commerce of Ireland

ACCI Australian Chamber of Commerce and Industry

Acciaio Inossid Acciaio Inossidabile [Italian Journal on Stainless Steel]

ACCIOS Azimuthal Closely Coupled, Infinite-Order Sudden (Method) [also ACC-IOS]

ACCIS Advisory Committee for Coordination of Information Systems [of United Nations]

ACCJ American Chamber of Commerce in Japan

ACCL Alberta Council of College Librarians [Canada]

ACCL American College of Computer Lawyers [US]

ACCL American Council of Commercial Laboratories [US]

Acclim Acclimations [Newsletter of the National Assessment of the Potential Consequences of Climate Variability and Change, US]

aCcm abcoulomb centimeter [Unit]

ACCMAIL Accessing the Internet via E-Mail

ACCN Academia Chilena de Ciencias Naturales [Chilean Academy of Natural Sciences]

ACCO Australian Council of Consumer Organisations

Acc of F Accuracy of Fire

ACCP American Chamber of Commerce of the Philippines

ACCP American College of Chest Physicians [US]

ACCP Atlantic Climate Change Program [of National Oceanic and Atmospheric Administration, US]

ACCRA American Chamber of Commerce Researchers Association [US]

accrd accrued

Accrd Int Accrued Interest [also accrd int]

ACC-ROC American Chamber of Commerce in the Republic of China [Taiwan]

ACCS Association of Casualty and Surety Companies

ACCS Automatic Calling Card Service

ACCS Automatic Checkout and Control System

ACCSA Adiabatic Capture Centrifugal Sudden Approximation

ACCSA Allied Communications Computer Security Agency [of NATO]

ACCT Alliance for Coal and Competitive Transportation

ACCT American Chamber of Commerce in Thailand

Acct Account(ant) [also acct]

Acct Chem Res Accounts of Chemical Research [Publication of the American Chemical Society] [also Acc Chem Res]

Acctg Accounting [also acctg]

ACCTI American Chamber of Commerce for Trade with Italy [US]

Accts Accounts [also accts]

ACCTVS Association of Closed-Circuit Television Surveyors [UK]

ACCU Advisory Committee of CERN Users [of CERN– European Laboratory for Particle Physics Geneva, Switzerland]

ACCU Advisory Committee on Computer Usage [US]

ACCU Association of Computer and CD-ROM Users [US]

Accum Accumulation [also accum]

Accumr Accumulator [also accumr]

Acc V Accelerating Voltage [also Acc Volt, acc volt, or acc v]

ACCW Alternating Current, Continuous Wave

ACCWP Acquisition, Cataloguing and Circulation Working Party [UK]

ACD Acetaldehyde

ACD Acid Citrate Dextrose [Biochemistry]

ACD Active Core Diameter

ACD Advanced Conceptual Design

ACD Alarm, Control and Display

ACD American College of Dentists [US]

ACD Annealed Cold-Drawn [Metallurgy]

ACD Anode-to-Cathode Distance

ACD Antenna Control Display

ACD Anticoagulant Citrate-Dextrose [Biochemistry]

ACD Anti-Coherency Dislocation [Materials Science]

ACD Anti-Contamination Device

ACD Armor Chemical Division [Chicago, Illinois, US]

ACD Associated Construction Distributors (International) [US]

ACD Association of Curriculum Development [UK]

ACD Atomic Charge Density

ACD Automatic Call Distribution; Automatic Call Distributor [Telecommunication]

AcD Actinium D [also AcPb]

ACDA American Component Dealers Association [US]

ACDA Arms Control and Disarmament Agency [US]

ACDB Airport Characteristics Databank

AC-DC Alternating Current/Direct Current [also AC/DC, ACDC, ac-dc, dc/dc or acdc]

ACD-ESS Automatic Call Distributor–Electronic Switching System

ACDI Agricultural Cooperative Development International [US]

ACDI Asynchronous Communications Device Interface

ACDM Association of Chairmen of Departments of Mechanics [US]

ACDO Air Carrier District Office [US]

ACDPI American Cultured Dairy Products Institute [US]

ACDRI Advisory Committee on the Development of Research for Industry [of Council for Scientific and Industrial Research, South Africa]

a-CdSnAs₂ Amorphous Cadmium Tin Arsenide

ACE Academia de Ciencias Económicas [Academy of Economic Sciences, Argentina]

ACE Acceptance Checkout Equipment

ACE Access Control Entry [Computers]

ACE Achieving the Competitive Edge [Organizational Analysis of Battelle Memorial Institute Columbus, Ohio]

ACE Advanced Certificate in Education

ACE Advanced Computing Environment

ACE Advanced Control Experiment

ACE Adverse Channel Enhancements [Computers]

ACE Advisory Center for Education [UK]

ACE Aerosol Characterization Experiment [of National Oceanic and Atmospheric Administration–Pacific Marine Environmental Laboratory, US]

ACE Aerosol Chemistry Experiment [of International Global Atmospheric Chemistry (Program)]

ACE Agricultural Communicators in Education [US]

ACE Air Combat Evaluator

ACE Air-Conditioning Engineer(ing)

ACE Air Cushion Equipment

ACE Airspace Coordination Element

ACE Alberta Caravan Exhibition (Foundation) [Canada]

ACE Allied Command Europe [of NATO]

ACE Alternative Chemical or Electron Impact Ionization

ACE Altimeter Control Equipment

ACE Ambush Communication Equipment

ACE American Council on Education [US]

ACE American Council on the Environment [US]

ACE Analog Contrast Enhancement

ACE Angiotensin-Converting Enzyme [Biochemistry]

ACE Animated Computer Education

ACE Anode Current Efficiency

ACE Arctic Cloud Experiment

ACE Area Control Error

ACE Army Corps of Engineers [US]

ACE Assemblée Consultative Européenne [European Consultative Assembly, France]

ACE Association for Computer Educators [US]

ACE Association for Continuing Education

ACE Association for Cooperation in Engineering

ACE Association for the Conservation of Energy [UK]

ACE Association of Conference Executives [UK]

ACE Association of Conservation Engineers [US]

ACE Association of Consulting Engineers [UK]

ACE Association of Cost Engineers [UK]

ACE Attitude Control Electronics

ACE Audio Connecting Equipment

ACE Automated Cable Expertise

ACE Automated Claims Environment [of United Service Automobile Association, US]

ACE Automated Cost Estimate

ACE Automatic Calling Equipment

ACE Automatic Checkout Equipment

ACE Automatic Circuit Exchange

ACE Automatic Clutter Eliminator

ACE Automatic Compensation of Emissivity

ACE Automatic Computer Evaluation

ACE Automatic Computing Engine

ACE Automatic Correction of Emissivity

ACE Automation and Control Engineering Exposition [Canada]

ACE Automation Conference and Exhibition

ACE Auxiliary Control Element

ACE Aviation and Computer Enthusiasts [US]

ACE Aviation Construction Engineer(ing)

ACE-1 Southern Hemisphere (Marine) Aerosol Characterization Experiment [of National Oceanic and Atmospheric Administration–Pacific Marine Environmental Laboratory, US]

ACE-2 North Atlantic Regional Aerosol Characterization Experiment [of National Oceanic and Atmospheric Administration–Pacific Marine Environmental Laboratory, US]

Ace Acetone [also ace]

α-Ce Alpha Cerium [Symbol]

ACEA American Cotton Exporters Association [US]

ACEA Association of Consulting Engineers of Australia

ACEA Association of Cost and Executive Accountants [UK]

ACEAA Advisory Committee on Electrical Appliances and Accessories

ACEAF Alternating-Current Electric-Arc Furnace [also AC EAF]

ACE-Asia Aerosol Characterization Experiment–Asia [International Global Atmospheric Chemistry (Program); from 2000 to 2001]

ACEC Advisory Council on Energy Conservation [UK]

ACEC American Consulting Engineers Council [Washington, DC, US]

ACEC Association of Consulting Engineers of Canada

ACEC Ateliers de Constructions Electriques de Charleroi S.A. [Belgian Electrical Equipment Manufacturer; located at Charleroi]

ACE-Cl 1-Chloroethyl Chloroformate

ACEC Rev ACEC Review [of Ateliers de Constructions Electriques de Charleroi S.A., Belgium]

ACEC/RMF American Consulting Engineers Council/Research and Management Foundation [Washington, DC, US]

ACEE Aircraft Energy Efficiency

ACEEE American Council for an Energy Efficient Economy [US]

ACEF Asian Cultural Exchange Foundation

ACEF Association of Commodity Exchange Firms [US]

ACEFQNZ Academia de Ciencias Exactas, Fisico Quimicas y Naturales de Zaragoza [Zaragoza Academy of Exact, Physicochemical and Natural Sciences, Spain]

ACEI Association of Consulting Engineers of Ireland

ACEID Asian Center for Educational Innovation Development

ACEJMC Accrediting Council on Education in Journalism and Mass Communications [US]

ACEK Association of Consulting Engineers of Kenya

ACEL Aerospace Crew Equipment Laboratory [US Navy]

ACEL Alternating Current Electroluminescent

ACELSCO Associated Civil Engineers and Land Surveyors of Santa Clara County [US]

ACEM Association of Consulting Engineers of Malaysia

ACEM Association of Consulting Engineers of Manitoba [Canada]

AcEm Actinium Emanation (or Actinon) [also Ac-Em, or An]

ACEN Association of Consulting Engineers of Norway

a Cen Alpha Centauri [Astronomy]

ACEO Assistant Chief Education Officer

ACEP Advisory Committee on Export Policy [US]

ACEP American College of Emergency Physicians [US]

ACEQ Advisory Council on Environmental Quality [US]

ACER Advisory Committee on Environmental Resources

ACER Australian Council for Educational Research

ACerS American Ceramic Society [US]

ACES (2-Acetamido)-2-Aminoethanesulfonic Acid; N-(2-Acetamido)-2-Aminoethanesulfonic Acid

ACES Acoustic Containerless Experiment System

ACES Aerosol Containment Evacuation System

ACES Air Collection and Enrichment System

ACES Air Collection Engine System

ACES 2-[(2-Amino-2-Oxoethyl)amino]ethanesulfonic Acid

ACES Annual Cycle Energy System

ACES Applied Computational Electromagnetics Society [US]

ACES Arab Center for Energy Studies

ACES Arc Cluster Evaporation Source

ACES Association of Consulting Engineers of Saskatchewan [Canada]

ACES Automated Code Evaluation System

ACES Automatic Checkout and Evaluation System [US Air Force]

ACESA Arizona Council of Engineering and Scientific Associations [US]

ACESA Australian Commonwealth Engineering Standards Association

ACE-S/C Acceptance Checkout Equipment–Spacecraft

ACET Advisory Committee for Electronics and Telecommunications

ACET Association of Consulting Engineers of Tanzania

ACET Automatic Cancellation of Extended Targets

Acet Acetone [also acet]

Acet Acetylene [also acet]

Acet A Acetic Acid [also Acet Ac, acet ac, or acet a]

ACETM Automatic Compensation of Emissivity for Temperature Measurement

ACETM Automatic Compensation of Emissivity for Temperature Monitoring

Acetoacetyl-S-ACP Acetoacetyl-S-Acyl Carrier Protein [also acetoacetyl-S-ACP] [Biochemistry]

Acetyl CoA Acetyl Coenzyme A [also acetyl CoA] [Biochemistry]

Acetyl-S-ACP Acetyl-S-Acyl Carrier Protein [also acetylS-ACP] [Biochemistry]

ACF Access Control Facility

ACF Access Control Field

ACF Access Cost Factor

ACF A (Al_2O_3), C (CaO), F (FeO + MgO) (Diagram) [Petrology]

ACF A (Al_2O_3 + Fe_2O_3 (Na_2O + K_2O)), C (CaO), F (FeO + MgO + MnO) (Diagram) [Petrology]

ACF Activated Carbon Fiber

ACF Active Color Formation

ACF Administration for Children and Families [of US Department of Health and Human Services]

ACF Adsorbing Colloid Flotation

ACF Advanced Ceramics Facility [of Max Planck Institut, Germany]

ACF Advanced Communication(s) Function

ACF Agricultural Cooperative Federation [UK]

ACF Air Cargo Forwarder

ACF Air Combat Fighter [US Air Force]

ACF Alternate Communications Facility

ACF American Chestnut Foundation [US]

ACF Analytical Characterization Facility [of Stevens Institute of Technology, Hoboken, New Jersey, US]

ACF Anisotropic Conductive Film

ACF Area Computing Facility

ACF ARM (Atmospheric Radiation Measurement) Central Facility [of NASA Langley Research Center, Hampton, Virginia]

ACF ASEAN (Association of Southeast Asian Nations) Constructors Federation

ACF Association of Consulting Foresters [US]

ACF Australian Conservation Foundation

ACF Australian Cotton Foundation

ACF Autocorrelation Function

ACF Automobile Club of France

a-C:F Fluorinated Amorphous Carbon

a-Cf Alpha Californium [Symbol]

ACFA Army Cadet Force Association [UK]

ACFC Aircraftman (or Aircraftwoman), First Class [also Acfc]

ACFE Association of Colleges for Further Education [UK]

ACFF Air Cargo Fast Flow

ACFG Automatic Continuous Function Generation

ACFM Actual Cubic Feet per Minute [also acfm]

ACFM Advisory Committee on Fishery Management [of International Council for the Exploration of the Sea, Denmark]

ACFM Association of Cereal Food Manufacturers [UK]

ACFO Association of Car Fleet Operators [UK]

ACFOA Australian Council for Overseas Aid

ACFSTA Association of Chinese Food Scientists and Technologists in America [now Chinese American Food Service, US]

Acft Aircraft [also acft]

ACF/TCAM Advanced Communication Function/Telecommunications Access Method

ACFVS Advanced Cold Fusion Vacuum System

ACFVS Automated Cold Fusion Vacuum System

ACF/VTAM Advanced Communication Function/Virtual Telecommunications Access Method

ACG Adjacent Charging Group

ACG Air Cargo Glider

ACG Alternating-Current Generator

ACG Angiocardiography

ACG Automatic Code Generator

AcG Accelerator Globulin [Biochemistry]

ACGB Aircraft Corporation of Great Britain

ACGC Asian Coordination Group Chemistry

AcGEPC 1-O-Alkyl-2-Acetyl-syn-Glycero-3-Phosphocholine [Biochemistry]

ACGIH American Conference of Governmental and Industrial Hygienists [US]

ACGIH TLV American Conference of Governmental Industrial Hygienists' Threshold Limit Value

ACGM Atikokan Conceptual Geosphere Model [Canada]

ACGR Associate Committee on Geotechnical Research [Canada]

ACGRA Australian Cotton Growers Research Association

ACGSC Army Command General Staff College [Fort Leavenworth, Kansas, US]

ACH Acetone Cyan(o)hydrin (Process)

ACH Acetylcholine [also Ach] [Biochemistry]

ACH Active Core Height

ACH Adrenal Cortical Hormone [Biochemistry]

ACH Air Changes per Hour [also ach]

ACH Aluminum Chlorohydrate (Process)

ACH American College of Health

ACH Association of Computers and Humanities

ACH Attempts per Circuit per Hour

ACH Automated Clearinghouse

ACh Acetylcholine [also Ach] [Biochemistry]

AcH Acetaldehyde

a-C:H Hydrogenated Amorphous Carbon

a-ch alpha-chain specific (immunoglobulin)

ACHA American Catholic History Society [US]

ACHA American College of Hospital Administrators [US]

ACHA American Council of Highway Advertisers [US]

ACHA American Cutlery Manufacturers Association [US]

Achats Entret Achats et Entretien [French Publication on Purchasing and Maintenance]

ACHE Acetylcholinesterase [also AchE] [Biochemistry]

ACHE Association for Continuing Higher Education [US]

AChE Acetylcholinesterase [also ACHE] [Biochemistry]

ACHEE ASEAN (Association of Southeast Asian Nations) Council of Higher Education Environment

ACHEMA Air-Cooled Heat Exchanger Manufacturers Association [US]

ACHEMA Ausstellungs-Tagung für Chemisches Apparatewesen [International Meeting on Chemical Engineering and Biotechnology; DECHEMA–Deutsche Gesellschaft für Chemisches Apparatewesen, Germany]

ACHI Application Channel Interface

a-C:H:N Hydrogenated and Nitrogenated Amorphous Carbon

ACHP Advisory Council on Historic Preservation

AChR Acetylcholine Receptor [Biochemistry]

a Chr *(ante Christum)* – before Christ

a Chr n *(ante Christum natum)* – before Christ

ACHTR Advisory Committee for Humid Tropics Research [of UNESCO]

AcHz Acetic Acid Hydrazide

ACI Adjacent Channel Interference

ACI Adjusted Calving Interval

ACI African Culture Institute

ACI Air Combat Intelligence

ACI Alloy Casting Institute [US]

ACI American Concrete Institute [US]

ACI Army Control Instruction

ACI Association of Construction Inspectors

ACI Association for Conservation Information [US]

ACI Atlantic Canada Institute

ACI Atlantic Coal Institute [Canada]

ACI Austenitic Cast Iron

ACI Australian Chemical Institute

ACI Automatic Card Identification

ACI Automatic Component Insertion

ACIA Asociación Colombiana de Ingenieros Agronomos [Colombian Association of Agricultural Engineers]

ACIA Associated Cooperage Industries of America [US]

ACIA Asynchronous Communications Interface Adapter

ACIAR Australian Center for International Agricultural Research

ACIAS Automated Calibration Interval Analysis System

ACIASAO American Council of Industrial Arts and State Association Officials [now Council of Technology Education Association, US]

ACIATE American Council of Industrial Arts Teacher Education [now Council of Technology Teacher Education, US]

ACIB Air Cushion Ice-Breaking Bow

ACIC Aeronautical Chart and Information Center [of US Air Force]

ACIC Alberta Council for International Cooperation [Canada]

ACIC American Committee for International Conservation [US]

ACIC Associate of the Canadian Institute of Chemistry

ACIC Australian Chemical Industry Council

ACICE Authority for the Coordination of Inland Transport in Central Europe

ACID Advisory Committee on Industrial Development [Zimbabwe]

ACID Association of Canadian Industrial Designers

ACID Association of Colleges for Implementing Diploma and Higher Education Programs [UK]

ACID Automatic Classification and Interpretation of Data

1,2,4-Acid 4-Amino-3-Hydroxy-1-Naphthalenesulfonic Acid [also 1,2,4-acid]

1,2,4-Acid 1-Amino-2-Naphthol-4-Sulfonic Acid [also 1,2,4-acid]

1,8,2,4-Acid 1-Amino-8-Naphthol-2,4-Disulfonic Acid [also 1,8,2,4-acid]

Aciers Spec Aciers Speciaux [French Journal on Special Steels]

ACIFIC Aspartame Committee of the International Food Information Council [US]

ACIGY Advisory Council of the International Geophysical Year

ACIIW American Council of the International Institute of Welding [US]

ACIJ African-Caribbean Research Institute

ACIL American Council of Independent Laboratories [US]

ACIL Newsl ACIL Newsletter [of American Council of Independent Laboratories]

ACIM Accident Cost Indicator Model

ACIM Axis-Crossing Interval Meter

ACIMA Aktiengesellschaft für die Chemische Industrie [Chemical Industrial Corporation, Switzerland]

ACI Mater J ACI Materials Journal [of American Concrete Institute]

ACIMET Acid Methane (Process)

ACIP Advisory Committee on Immunization Practice [US]

ACIP American Council of International Personnel [US]

ACIP Assembly Configuration and Integration Panel

ACIR Aviation Crash Injury Research

ACIRL Australian Coal Industry Research Laboratories

ACIS Alternating-Current Impedance Spectroscopy

ACIS American Committee for Interoperable Systems

ACIS Asia Credit Information Service, Limited [Taiwan]

ACIS Associate of the Canadian Institute of Surveying (and Mapping)

ACIS Association for Computing and Information Sciences

ACIS Automated Client Information Service

ACI SP American Concrete Institute Special Publication [US] [also ACI Spec Publ]

ACI Struct J ACI Structural Journal [of American Concrete Institute]

ACISUD Authority for the Coordination of Inland Transport in Southern Europe

ACITS Advisory Committee on Information Technology Standardization [Europe]

ACIUCN Australian Committee for the International Union for the Conservation of Nature (and Natural Resources)

ACIUGG American Committee of the International Union of Geodesy and Geophysics

ACJE Alternating-Current Josephson Effect

ACK Acknowledgment (Character)

ACK Aveston, Cooper and Kelly (Crack Growth Theory)

ACK0 Acknowledge Even Blocks

ACK1 Acknowledge Odd Blocks

AcK Actinium K [also Fr-223, ^{223}Fr, or Fr^{223}]

Ack Acknowledge(ment) [also ack]

Ackt Acknowledgment [also ackt]

ACL Academia Chilena de la Lengua [Chilean Academy of Letters]

ACL Academia des Ciências de Lisboa [Academy of Sciences at Lisbon, Portugal]

ACL Access Control List [Internet]

ACL Advanced CMOS (Complimentary MetalOxide Semiconductor) Logic

ACL Aeronautical Computers Laboratory [US Navy]

ACL Allis-Chalmers/Lepol Kiln

ACL Allowable Cabin Load

ACL Alpha-Chapman Layer

ACL Alternate Concentration Limits

ACL Altimeter Check Location

ACL Analytical Chemical Laboratory

ACL Analytical Chemistry Laboratory [of Argonne National Laboratory, Illinois, US]

ACL Anterior Cruciate Ligament

ACL Antigen-Carrier Lipid [Biochemistry]

ACL Application Control Language

ACL Applied Color Label

ACL Association for Computational Linguistics [US]

ACL Atlantic-Container Lines

ACL Atlas Commercial Language

ACL Atlas Computer Laboratories [UK]

ACL Audit Command Language

ACL Automated Coagulation Laboratory

ACL Automatic Circuit Layout

Acl Acetal [also acl]

ACLAM American College of Laboratory Animal Medicine [US]

ACLANT Allied Command Atlantic [of NATO]

ACLD Above Clouds

ACLG Air-Cushion Landing Gear

ACLI American Council of Life Insurance [US]

ACLM Advisory Council on Labour and Manpower (of Quebec) [Canada]

ACLM Association of Contact Lens Manufacturers [UK]

ACLO Agena-Class Lunar Orbiter [of NASA]

ACLP Above Core Load Pad

ACLS Air-Cushion Landing System

ACLS All-Weather Carrier Landing System

ACLS American Council of Learned Societies [US]

ACLS Automated Control Landing System

ACM Active Countermeasures

ACM Address Calculation Machine

ACM Address-Code Multiplex

ACM Advanced Composite Material

ACM Advanced Cruise Missile

ACM Advanced Cure Monitor

ACM Air Chief Marshall

ACM Air Combat Maneuver

ACM Air Composition Monitor

ACM Alarm Control Module

ACM Alterable Control Memory

ACM Alternating-Current Motor

ACM Andean Common Market

ACM Arab Common Market

ACM Area Contamination Monitor

ACM Artificial Channel Method

ACM Asbestos Containing Material

ACM Asbestos-Covered Metal

ACM Associated Colleges of the Midwest [US]

ACM Association for College Management [UK]

ACM Association for Computing Machinery [US]

ACM Association of Crane Makers

ACM Associative Communication Multiplexer

ACM Atmospheric Corrosion Monitor

ACM Audio Compression Manager [Computers]

ACM Australian Chamber of Manufacturers

ACM Automatic Clutter Mapping

ACM Avalanche Current Multiplication

ACM-2 N-[2-(7'-Azidocoumarin-4'-Carbonamide)-ethoxycarboxymethyl]maleimide

ACM-5 N-[2-(7'-Azidocoumarin-4'-Carbonamide)-pentoxycarboxymethyl]maleimide

ACM-5M N-[2-(7'-Azidocoumarin-4'-Acetomide)-pentoxycarboxymethyl]maleimide

Acm Acetamidine

A/cm^2 Ampere per square centimeter [also A cm^{-2}, or amp(s)/cm^2]

A$^+$/cm^2 Positive acceptors per square centimeter [also A$^+$ cm^{-2}]

A$^-$/cm^2 Negative acceptors per square centimeter [also A$^-$/cm^{-2}]

a-Cm Alpha Curium [Symbol]

ACMA Agricultural Cooperative Managers Association [UK]

ACMA American Cast Metals Association [US]

ACMA Associate of the Institute of Cost and Management Accountants [UK]

ACMAC Association for Computing Machinery Accreditation Committee [US]

ACMAD African Centre of Meteorological Applications for Development

ACMC Automotive Chemical Manufacturers Council [US]

ACM Conf Proc ACM Conference Proceedings [of Association for Computing Machinery, US]

ACME Acme Thread [also Acme]

ACME Advanced Computer for Medical Research Environment

ACME Association of Consulting Management Engineers [US]

ACME-C Acme Thread, Centralizing [also Acme-C]

ACME-G Acme Thread, General Purpose [also Acme-G]

ACMET Advisory Council on Middle East Trade [UK]

ACM/GAMM Association for Computing Machinery/German Association for Applied Mathematics and Mechanics

ACMI Air Combat Maneuvering Instrumentation

ACMI Art and Craft Materials Institute [US]

ACML Anti Common Market League [UK]

ACM/IEEE Association for Computing Machinery/Institute of Electrical and Electronics Engineers [US]

ACM/IEEE-CS Association for Computing Machinery/Institute of Electrical and Electronics Engineers–Computer Society [US]

A/cm^2·K^2 Amper per square centimeter square kelvin [also A cm^{-2} K^{-2}]

ACMLA Association of Canadian Map Libraries and Archives

ACMLC American Composite Manufacturing Learning Center [Florida, US]

ACMMM Annual Conference on Magnetism and Magnetic Materials

ACMO Afloat Communications Management Office [of Naval Ship Engineering Center, US Navy]

ACMOS Advanced Complimentary Metal-Oxide Semiconductor

ACMP Advisory Committee on Marine Pollution [of International Council for the Exploration of the Sea, Denmark]

ACMPPOH 4-Acetyl-3-Methyl-1-Phenyl 2-Pyrazoline-5-One

ACM Press Association of Computing Machinery Press [US]

ACMR Air Combat Maneuvering Range

ACMRI Air Combat Maneuvering Range Instrumentation

ACMRR Advisory Committee on Marine Resource Research [of Food and Agricultural Organization]

ACMS Advanced Configuration Management System

ACMS Application Control and Management System

ACMS Application Control Management System

ACMSC Association for Computing Machinery Standards Committee [US]

ACMST Association for Computers in Mathematics and Science Teaching [US]

ACM Trans Comp Syst ACM Transactions on Computer Systems [of Association for Computing Machinery, US]

ACM Trans Database Syst ACM Transactions on Database Systems [of Association for Computing Machinery, US]

ACM Trans Graph ACM Transactions on Graphics [of Association for Computing Machinery, US]

ACM Trans Inf Syst ACM Transactions on Information Systems [of Association for Computing Machinery, US]

ACM Trans Math Softw ACM Transactions on Mathematical Software [of Association for Computing Machinery, US]

ACM Trans Off Inf Syst ACM Transactions on Office Information Systems [of Association for Computing Machinery, US]

ACM Trans Program Lang Syst ACM Transactions on Programming Languages and Systems [of Association for Computing Machinery, US]

ACMU American Canoe Manufacturers Union [US]

ACN Accession Number

ACN Acyl Cyamide

ACN Air Consignment Note

ACN Aircraft Classification Number

ACN American College of Nutrition [US]

ACN Artificial Cloud Nucleation

ACN Asbestos Cloth Neck

ACN Azobiscyclohexanecarbonitrile

a-C:N Nitrogenated Amorphous Carbon

a-C$_3$N$_4$ Alpha Carbon Nitride

Acn Acetonitrile [also acn]

Acn Acrylonitrile [also acn]

ACNA Arctic Institute of North America [Canada]

ACNAM Associations Council of the National Association of Manufacturers Associations [US]

ACNAS Advanced Cableship Navigation Aid System

ACNBC Associate Committee on the National Building Code [of National Research Council of Canada]

ACNet Alternating-Current Network Analysis Program [also Acnet]

ACNM American College of Nurse-Midwives [US]

ACNM American Cordage and Netting Manufacturers [US]

ACNOCOMM Assistant Chief of Naval Operations for Communications and Cryptology

ACNT Australian Council of National Trusts

ACNW American College of Nurse-Midwives [US]

ACNY Advertising Club of New York [US]

ACO Abrasive Cut-Off

ACO Accounting Control Office

ACO Architectural Conservancy of Ontario [Canada]

ACO Army Corps Ordnance [UK]

ACO ASEAN (Association of Southeast Asian Nations) Cooperative Organization

ACO Association of Conservation Officers [UK]

ACO Automatic Call Origination [also AUTOCALL, or autocall]

AcO Acetoxy-

Ac$_2$O Acetic Oxide (or Acetic Acid Anhydride)

Aco Aerosol Custom Pharmaceuticals Limited

a-Co Amorphous Cobalt

a-Co Alpha Cobalt [Symbol]

ACOA Atlantic Canada Opportunities Agency

a-CoBSi Cobalt Boron Silicon Amorphous Alloy

ACOC Area Communications Operations Center

a-CoCrB Cobalt Chromium Bron Amorphous Alloy

ACODEX Asociación Colombiana de Exportadores [Colombian Association of Exporters]

a-CoFeNiBSi Cobalt Iron Nickel Boron Silicon Amorphous Alloy

a-CoFeSiB Cobalt Iron Silicon Boron Amorphous Alloy

ACOG American College of Obstetricians and Gynecologists [US]

AcOH Acetic Acid [also ACOH]

ACOIN Australian Core Inventory of Chemical Substances

ACOLF Advisory Committee on Live Fish

ACOLI Advanced Circuit Order and Layout Information

ACOLS Australasian Conference on Optics, Lasers and Spectroscopy

ACOM Automatic Coding System

ACOM Associate Committee on Meteorites [Canada]

ACOM Association for Convention Operations Management [US]

ACOM Awards Committee [of Institute of Electrical and Electronics Engineers, US]

ACOMPLIS A Computerized London Information Service [UK]

a-CoNbB Cobalt-Niobium-Boron Amorphous Alloy

a cond *(a condition)* – upon request

a-CoNiP Cobalt Nickel Phosphorus Amorphous Alloy

ACONS Automated Compressor On-Line System

a-CoP Cobalt Phosphorus Amorphous Alloy

a-CoPd Cobalt Palladium Amorphous Alloy

ACOPP Abbreviated COBOL Preprocessor

ACOPS Advisory Committee on Pollution of the Sea [UK]

ACORBAT Association for Cooperation in Banana Research in the Caribbean and Tropical America

ACORC Artificial Cells and Organs Research Center [of McGill University, Canada]

ACORD Advisory Council on Research and Development (for Fuel and Power) [UK]

ACORN A Classification of Residential Neighbourhoods [Geographic Information System]

ACORN Associative Content Retrieval Network

ACOS Acid Catalyzed Organosolv Saccharification

ACOS Advisory Committee on Safety

ACOS Assistant Chief of Staff

ACOS Automatic Check-Out System

ACOS Automatic Control System

a-CoSiB Cobalt Silicon Boron Amorphous Alloy

a-CoSiBFe Cobalt Silicon Boron Iron Amorphous Alloy

a-CoSiBFeNb Cobalt Silicon Boron Iron Niobium Amorphous Alloy

ACOSS Australian Council of Social Services

a-CoTi Cobalt Titanium Amorphous Alloy

ACOUFAT Acoustic Fatigue

Acous Acoustic(s) [also Acoust, acoust, or acous]

Acoust Aust Acoustics Australia [Journal of the Australian Acoustical Society]

Acoust Lett Acoustics Letters [UK]

Acoust Phys Acoustical Physics [Russian Translation Journal jointly published by the American Institute of Physics (US) and Maik Nauka (Russia)]

Acox Acetoxime

ACP Aboriginal Cadetship Program [Australia]

ACP Accelerated Carburizing Process

ACP Acid Phosphatase [Biochemistry]

ACP Activated Carbon Fiber, Pitch-Based [Designation]

ACP Acyl Carrier Protein [Biochemistry]

ACP Advanced Computational Processor

ACP Advanced Computing (Division) [of Fermilab, Batavia, Illinois, US]

ACP Advanced Cooperative Project [NASA]

ACP Aerospace Computer Program [US Air Force]

ACP African, Caribbean and Pacific (Countries)

ACP Agricultural Conservation Program [US]

ACP Airborne Collision Prevention

ACP Airlift Command Post

ACP Airline Control Program

ACP Albany College of Pharmacy [New York State, US]

ACP Allied Communications Publication

ACP Alternating-Current Polarography

ACP American College of Physicians [US]

ACP Amorphous Calcium Phosphate

ACP Ancillary Control Processor

ACP Ancillary Control Program

ACP Anodal Closing Picture

ACP Anticoincidence Point

ACP Arc-Cast Powder

ACP Arithmetic and Control Processor

ACP Arithmetic Control Processor

ACP As Cold-Pressed (Condition)

ACP Asian Commercial Paper

ACP Aspen Center for Physics [Colorado, US]

ACP Associated Construction Publications [US]

ACP Associate of the College of Preceptors [UK]

ACP Association of Canadian Publishers

ACP Association of Computer Professionals [US]

ACP Atmospheric Chemistry Program

ACP Auxillary Control Process

ACP Azimuth Change Pulse

AcP Acid Phosphatase [Biochemistry]

Acp Aminocaproic Acid

εAcp ε-Aminocaproic Acid

ACPA Activated Carbon Producers Association [Belgium]

ACPA Adaptive Controlled Phased Array

ACPA American Concrete Pavement Association [US]

ACPA American Concrete Pipe Association [US]

ACPA American Concrete Pumping Association [US]

ACPA American Council on Public Affairs [US]

ACPA Association of Computer Programmers and Analysts [US]

ACPAC AIDAB (Australian International Development Assistance Bureau) Center for Pacific Development and Training [of Department of Foreign Affairs and Trade]

ACPAR Angular Correlation of Positron Annihilation Radiation

AcPb Actinium-Lead [also AcD]

ACPC Advisory Committee on Procurement and Contracts [of European Commission]

ACPC Advisory Committee on Public Contracts

ACPC 1-Amino Cyclopentanecarboxylic Acid

ACPC Association of Coffee Producing Countries

ACPD Airborne Collision Prevention Device

ACPD Alternating-Current Potential Drop

ACPD 1-Aminocyclopentane-1,3-Dicarboxylic Acid

ACPDS Advisory Committee on Personal Dosimetry Services [of National Science Foundation, US]

ACPE American Council on Pharmaceutical Education [US]

ACP-EEC (Convention between) African, Caribbean and Pacific Countries and the European Economic Community

ACPF Acoustic Containerless Processing Facility

ACPF Asociación del Congreso Panamericano de Ferrocarriles [Pan-American Railway Congress Association, Argentina]

ACPF Average Coupled Pair Functional (Method)

ACPI Advanced Configuration and Power Interface

ACPI Automated Component Placement and Insertion

ACPM Acoustic Chamber Position Module

ACPM Attitude Control Propulsion Motor

ACPMAPS Advanced Computing Division Massively Parallel Supercomputer [of Fermilab, Batavia, Illinois, US]

ACPMME Association of Concentrated and Powdered Milk Manufacturers of the EEC (European Economic Community) [France]

ACPPA American Concrete Pressure Pipe Association [US]

ACPR Advanced Core Performance Reactor

ACPR American College of Puerto Rico [Bayamón]

ACPR Annular Core Pulse Reactor

ACPSM Australasian College for Physical Scientists in Medicine [Australia]

Acpt Acceptance [also acpt]

ACPTC Association of College Professors of Textile and Clothing [US]

ACP/TF Airline Control Program/Transaction Processing Facility

ACPU Auxiliary Computer Power Unit

3AcPy 3-Acetyl Pyridine

ACQ Ammoniacal Copper Quat

ACQM Arrangement-Channel Quantum Mechanics

ACR Accelerated Cost Recovery

ACR Access Control Register

ACR Accounts of Chemical Research [Publication of the American Chemical Society, US]

ACR Active Cavity Radiometer

ACR Active Radar

ACR Advanced Capabilities Radar

ACR Advanced Cracking Reactor

ACR Aerodrome Control Radar [also acr]

ACR Aircraft Control Room

ACR Airfield Control Radar [also acr]

ACR Alaskan Communications Region [US Air Force]

ACR Alberta Chamber of Resources [Canada]

ACR Alkalipolyacrylate

ACR Allowed Cell Rate

ACR Alternate CPU (Central Processing Unit) Recovery

ACR American College of Radiology [US]

ACR American Computer Referral (Database) [US]

ACR Antenna Coupling Regulator

ACR Anti-Carmout Ribs (Removal Bit)

ACR Applied Computer Research

ACR Approach Control Radar [also acr]

ACR Ashmore/Cartier Reef (Region) [Australia]

ACR Association for Clinical Research [UK]

ACR Audio Cassette Recorder

ACR Automatic Call Recording

ACR Automatic Compression Regulator

ACRA Aluminum Can Recycling Association [UK]

ACRA Association of College Registrars and Administrators [UK]

ACRA Association of Company Registration Agents [UK]

ACRA Australian Cultivar Registration Authority

ACRBC Acceptance, Checkout, Retest and Backout Criteria

ACRC Association of Commercial Records Centers [US]

ACRC Assured Crew Return Capability [NASA]

ACRD Australian Centennial Roads Development

Acrd Acridine [Biochemistry]

ACRE Advanced Center for Research in Electronics [of Indian Institute of Technology, Bombay]

ACRE Automatic Call-Recording Equipment

ACRE Automatic Checkout and Readiness (Equipment)

acre-ft acre-foot [Unit]

acre-ft/d acre-foot per day [Unit]

acre-in acre-inch [Unit]

ACRES Airborne Communications Relay Station [US Air Force]

ACRES Australian Centre for Remote Sensing [of AUSLIG–Australian Surveying and Land Information Group]

ACRF Atmospheric Cloud Radiative Forcing

ACRH Argonne Cancer Research Hospital [Chicago, Illinois, US]

ACRI Air Conditioning and Refrigeration Institute [US]

ACRI American Cocoa Research Institute [US]

ACRIM Active Cavity Radiometer Irradiance Monitor

ACRIT Advisory Committee for Research on Information Transfer

ACRL Aluminum Casting Research Laboratory [of Worcester Polytechnic Institute, Massachusetts [US]

ACRL Association of College and Reference Libraries [US]

ACRL Association of College and Research Libraries [of American Library Association, US]

Acrl Acrylic Acid

ACRM American College of Radio Marketing [US]

ACRM Atmospheric Corrosion Rate Monitor

ACRMA Air Conditioning and Refrigerating Manufacturers Association [US]

ACROSS Automated Cargo Release and Operations Service System

ACRPI Association for Clinical Research in the Pharmaceutical Industry [UK]

ACRR Annular Core Research Reactor [of Sandia National Laboratories, Albuquerque, New Mexico, US]

ACRS Advisory Committee on Reactor Safeguards [of US Atomic Energy Commission]

ACRS Advisory Committee on Reactor Standards

ACRSO Atlantic Center for Remote Sensing of the Oceans

ACRSP Association for Canadian Registered Safety Professionals

ACRT Accelerated Crucible Rotation Technique [Metallurgy]

ACRT Analysis Control Routine

ACRV Assured Crew Return Vehicle [of NASA]

ACRW American Council of Railroad Women [US]

ACS Absolute Contamination Standard

ACS Absorption Cross Section [Radar]

ACS Academy of Cosmetic Surgery [US]

ACS Access

ACS Access Control Set

ACS Access Control System

ACS Accumulator Switch

ACS Acrylonitrile Chlorinated Polyethylene Styrene (Terpolymer)

ACS Active Control System

ACS Adaptive Control System

ACS Administrative Control System

ACS Adrenocortico Steroid [Biochemistry]

ACS Advanced Ceramics Society [US]

ACS Advanced Communications Service [US]

ACS Advanced Computer Series

ACS Advanced Computer Services

ACS Advanced Computer System

ACS Aeronautical Communications Subsystem

ACS Agena Control System [NASA]

ACS Agricultural Cooperative Service [of US Department of Agriculture]

ACS Air Catering Service

ACS Aircraft Control Surveillance

ACS Alarm and Control System

ACS Alaska Communication System [US]

ACS Allied Chiefs of Staff

ACS Alternating Current Synchronous

ACS Altitude Control System

ACS Aluminium-Clad Steel

ACS American Canal Society [US]

ACS American Cancer Society [US]

ACS American Carbon Society [US]

ACS American Cartographic Society [US]

ACS American Ceramic Society [US]

ACS American Cetacean Society [US]

ACS American Chemical Society [US]

ACS American Chestnut Society [US]

ACS American College of Surgeons [US]

ACS American Conifer Society [US]

ACS American Copyright Society [US]

ACS Anodal Closing Sound

ACS Antarctic Conservation Strategy [of International Union for Conservation of Nature (and Natural Resources)]

ACS Anticurl System

ACS Antireticular Cytotic Serum [Biochemistry]

ACS Application Customizer Service

ACS Applied Computer Science

ACS Assembly Control System

ACS Assistant Chief of Staff

ACS Assistant Chief Statistician

ACS Associate in Commercial Science

ACS Association of Commonwealth Students

ACS Asynchronous Communication Server

ACS Attachment Cross-Section

ACS Attitude Command System

ACS Attitude Control System

ACS Austempered Cast Steel

ACS Australian Ceramic Society

ACS Australian Chamber of Shipping

ACS Australian Computer Society

ACS Australian Construction Services

ACS Australian Customs Service

ACS Auto-Calibration System

ACS Autocollimating Spectrograph

ACS Automated Communications Set

ACS Automatic Checkout System

ACS Automatic Color System

ACS Automatic Control System

ACS Automobile Club of Switzerland

ACS Auxiliary Code Storage

ACS Auxiliary Core Storage

ACS Auxiliary Cooling System

Acs Access [also acs]

ACSA Alberta Cities Safety Association [Canada]

ACSA Allied Communications Security Agency [of NATO]

ACSA American Cotton Shippers Association [US]

ACSA Association of Collegiate Schools of Architecture [US]

ACSA Australian Collieries Staff Association

ACSAD Arab Center for the Studies of Arid Zones and Dry Lands [Syria]

ACSB Amplitude Compandered Sideband

ACSC Activation Component Service Center [of General Electric Aircraft Engines, US]

ACSC Aircraftman (or Aircraftwoman), Second Class [also Acsc]

ACS/CAS American Chemical Society/Chemical Abstracts Service [US]

ACSCE Army Chief of Staff for Communications Electronics [US]

ACS-DoCT American Chemical Society–Division of Chemical Technicians [US]

ACSF Attack Carrier Striking Force

ACSI Advanced Chemical Systems Inc [US]

ACSI Automotive Cooling Systems Institute [US]

ACSID Adjunct Cervical Spine Immobilization Device

ACSIL Admiralty Center for Scientific Information and Liaison [UK]

ACSIRO Australian Commonwealth Scientific and Industrial Research Organization

ACSL Advanced Continuous Simulation Language

ACSL Altocumulus Standing Lenticular [Meteorology]

ACSL American Computer Science League [US]

ACSL Antisymmetric Coincidence Site Lattice [Materials Science]

ACSLG Artificially Controlled Superlateral Growth (Process)

ACSM American College of Sports Medicine [US]

ACSM American Congress on Surveying and Mapping [Falls Church, Virginia, US]

ACSM Associate of the Cambourne School of Mines [US]

ACSM-ASPRS American Congress on Surveying and Mapping/American Society for Photogrammetry and Remote Sensing

ACSM Bull ACSM Bulletin [of American Congress on Surveying and Mapping]

ACSN Appalachian Community Service Network [US]

AcSoc Acoustical Society of America [US]

ACS-OLSA American Chemical Society–Office of Local Section Activities [US]

ACSP Advanced Control Signal Processor [NASA]

ACSP Advisory Council on Scientific Policy [UK]

A/cs Pay Accounts Payable [also a/cs pay]

ACS Polym Mat Sci Eng ACS Polymeric Materials Science and Engineering [of American Chemical Society, US]

ACS Polym Preprints ACS Polymer Preprints [of American Chemical Society, US]

ACS-PRF American Chemical Society–Petroleum Research Fund [also ACS/PRF, US]

AC-SPV Alternating-Current Surface Photo-Voltage [also ac-SPV]

ACSR Aluminum Cable Steel-Reinforced [also acsr]

ACSR Aluminum-Clad, Steel-Reinforced [also acsr]

ACSR Aluminum Conductor, Steel-Reinforced [also acsr]

ACSR/ACS Aluminum Conductor, Aluminium-Clad, Steel-Reinforced [also acsr/acs]

ACSR/AW Aluminum Conductor, Steel-Reinforced using Aluminum-Clad Steel Wire [also acsr/aw]

ACSR/AZ Aluminum Conductor, Steel-Reinforced using Aluminum-Coated Steel Wire [also acsr/az]

ACS-RD American Chemical Society–Rubber Division [US]

A/cs Rec Accounts Receivable [also a/cs rec]

ACSR/GA Aluminum Conductor, Steel-Reinforced using Class A Zinc-Coated Steel Wire [also acsr/ga]

ACSR/GB Aluminum Conductor, Steel-Reinforced using Class B Zinc-Coated Steel Wire [also acsr/gb]

ACSR/GC Aluminum Conductor, Steel-Reinforced using Class C Zinc-Coated Steel Wire [also acsr/gc]

ACSR/HS Aluminum Conductor, Steel-Reinforced, High Strength [also acsr/hs]

ACSR/SD Aluminum Conductor, Steel-Reinforced, Self Damping [also acsr/sd]

ACSS Air Combat and Surveillance System

ACSS Analog Computer Subsystem

ACSS Association of Canadian Security Services

ACSS Association of Colleges and Secondary Schools [US]

ACSS Automatic Continuous Spectrum Stabilization

ACSS Automatic Control Society of Slovenia

ACSSB American Certified Shake and Shingle Bureau [US]

ACSSB Amplitude-Companded Single Sideband

ACSSM Associate Committee on Soil and Snow Mechanics [Canada]

ACS Symp Ser ACS Symposium Series [of American Chemical Society, US]

ACST Access Time

ACST Advisory Committee on Science and Technology [of Economic and Social Council– Organization of American States]

Acst Acoustics [also acst]

AC STM Alternating-Current Scanning Tunneling Microscope [also AC-STM]

ACSTTO Association of Certified Survey Technicians and Technologists of Ontario [Canada]

ACSU Association for Computer Science Undergraduates [US]

ACSW Academy of Certified Social Workers

ACSYS Accounting Computer System

ACSYS Arctic Climate System Study [World Climate Research Program of the World Meteorological Organization]

ACT Activated Complex Theory [Physical Chemistry]

ACT Acute Catarrhal Laryngitis

ACT Adjustable Cable Temple

ACT Adrenocorticotropic; Adrenocorticotropin [Biochemistry]

ACT Advance Corporation Tax

ACT Advanced Combustion Technologies (Program) [of Energy Technology Branch, Natural Resources Canada]

ACT Advanced Communications Technology

ACT Advanced Composites Technology (Research and Technology Program) [of NASA]

ACT Advances in Characterization and Testing

ACT Advisory Council on Technology

ACT Air Control Team

ACT Air-Cooled Triode [also act]

ACT Air Cushion Trailer

ACT Algebraic Compiler and Translator

ACT Alternative Cleaning Technology

ACT Alternative Community Telephone [Australia]

ACT Alternative Control Technology

ACT American College of Toxicology [US]

ACT American College Test

ACT American College Testing Form [US]

ACT American Council for Turfgrass [US]

ACT Analog Circuit Technique

ACT Applied Circuit Technology

ACT Applied Computer Technique

ACT Area Correlation Tracker [US Air Force]

ACT Asphalt Core Tomography

ACT Association of Commercial Television [US]

ACT Association of Communications Technicians [US]

ACT Association of Cycle Traders [UK]

ACT Association of Cytogenetic Technologists [US]

ACT Asymmetric Crystal Tomography

ACT Asymmetric Crystal Topography

ACT Atom Concentrator Tube

ACT Australian Capital Territory

ACT Automatic Code Translation

ACT Automatic Control Theory

Act Action [also act]

Act Activation [also act]

Act Activity [also act]

Act Actuation [also act]

act active

act actual

ACTA Advanced Civilian Technology Agency [US]

ACTA Alternative Carrier Telecommunications Association [US]

ACTA American Cement Trade Alliance [US]

ACTA Arms Control Treaties and Agreements [US]

ACTA Australian City Transit Authority

Acta Acad Aboensis Acta Academiae Aboensis [Finland]

Acta Acad Sci Pol Acta Academiae Scientiarum Polonae [of Polish Academy of Sciences, Warsaw]

Acta Acoust Acta Acoustica [PR China]

Acta Anatom Acta Anatomica [International Journal]

Acta Astron Acta Astronomica [Poland]

Acta Astronaut Acta Astronautica [UK]

Acta Astron Sin (China) Acta Astronomica Sinica (China) [of Chinese Astronomical Society, PR China]

Acta Astrophys Sin (China) Acta Astrophysica Sinica (China) [PR China]

Acta Autom Sin (China) Acta Automatica Sinica (China) [PR China]

Acta Biomed Ateneo Parm Acta Biomedica Ateneo Parmense [Italy]

Acta Bioquím Clin Latinoam Acta Bioquímica Clinica Latinoamericana [Publication of the Latin-American Confederation of Clinical Biochemistry, Colombia]

ACTAC Association of Community Technical Aid Centers [UK]

Acta Chim Acad Sci Hung Acta Chimica Academiae Scientarium Hungaricae [of Hungarian Academy of Sciences, Budapest]

Acta Chim Hung Acta Chimica Hungaricae [of Hungarian Academy of Sciences, Budapest]

Acta Chim Scand Acta Chemica Scandinavica [Denmark]

Acta Chim Scand A Acta Chemica Scandinavica A [Denmark]

Acta Chim Scand B Acta Chemica Scandinavica B [Denmark]

Acta Chim Sin (China) Acta Chimica Sinica (China) [of Chinese Chemical Society, PR China]

Acta Chim Sin (Engl Ed) Acta Chimica Sinica (English Edition) [Netherlands]

Acta Cienc Indica Math Acta Ciencia Indica, Mathematics [India]

Acta Cienc Indica Phys Acta Ciencia Indica, Physics [India]

Acta Crystallogr Acta Crystallographica [Published for the International Union of Crystallography in Denmark]

Acta Crystallogr A Acta Crystallographica, Section A [Published for the International Union of Crystallography in Denmark]

Acta Crystallogr A, Found Crystallogr Acta Crystallographica, Section A, Foundations of Crystallography [Published for the International Union of Crystallography in Denmark]

Acta Crystallogr B Acta Crystallographica, Section B [Published for the International Union of Crystallography in Denmark]

Acta Crystallogr B, Struct Sci Acta Crystallographica, Section B, Structural Science [Published for the International Union of Crystallography in Denmark]

Acta Crystallogr C Acta Crystallographica, Section C [Published for the International Union of Crystallography in Denmark]

Acta Crystallogr C, Struct Commun Acta Crystallographica, Section C, Crystal Structure Communications [Published for the International Union of Crystallography in Denmark]

Acta Crystallogr D Acta Crystallographica, Section D, Biological Crystallography [Published for the International Union of Crystallography in Denmark]

Acta Crystallogr D, Biol Crystallogr Acta Crystallographica, Section D, Biological Crystallography [Published for the International Union of Crystallography in Denmark]

Acta Crystallogr Sect A Acta Crystallographica, Section A [Published for the International Union of Crystallography in Denmark]

Acta Crystallogr Sect B Acta Crystallographica, Section B [Published for the International Union of Crystallography in Denmark]

Acta Crystallogr Sect C Acta Crystallographica, Section C [Published for the International Union of Crystallography in Denmark]

Acta Crystallogr Sect D Acta Crystallographica, Section D, Biological Crystallography [Published for the International Union of Crystallography in Denmark]

Acta Cybernet Acta Cybernetica [Hungary]

Acta Dermato-Venereol Acta Dermato-Venereologica [US]

Acta Electron Sin Acta Electronica Sinica [PR China]

Acta Genet Med Gemellol Acta Geneticae Medicae et Gemellologiae [US]

Acta Geod Geophys Montan Hung Acta Geodaetica, Geophysica et Montanistica Hungarica [Hungary]

Acta Geophys Pol Acta Geophysica Polonica [Poland]

Acta Geophys Sin (China) Acta Geophysica Sinica (China) [PR China]

Acta Geophys Sin (USA) Acta Geophysica Sinica (USA) [English translation of: *Acta Geophysica Sinica (China)*; published in the US]

Acta Haematol Acta Haematologica [International Journal]

Acta Hort Acta Horticulturae [of International Society for Horticultural Science, Netherlands]

Acta Inform Acta Informatica [Germany]

Acta Manil Acta Manilana [Manila, Philippines]

Acta Mater Acta Materialia [formerly Acta Metall; later Acta Metall Mater]

Acta Mat Compos Sin Acta Materiae Compositae Sinica [PR China]

Acta Math Acta Mathematica [Sweden]

Acta Math Appl Sin Acta Mathematica Applicatae Sinica [PR China]

Acta Math Sin Acta Mathematica Sinica [PR China]

Acta Mech Acta Mechanica [Germany]

Acta Mech Sin Acta Mechanica Sinica [PR China]

Acta Mech Solida Sin Acta Mechanica Solida Sinica [PR China]

Acta Med Scand Acta Medica Scandinavica

Acta Metl Acta Metallurgica [also Acta Metall; later Acta Metall Mater; now Acta Mater]

Acta Metall Acta Metallurgica [also Acta Met; later Acta Metall Mater; now Acta Mater]

Acta Metall Mater Acta Metallurgica et Materialia [also Acta Met Mater; formerly Acta Metall; now Acta Mater]

Acta Metall Sin Acta Metallurgica Sinica [PR China]

Acta Metall Sin (China) Acta Metallurgica Sinica (China) [PR China]

Acta Metall Sin (Engl Ed) A Acta Metallurgica Sinica (English Edition) Series A [Published in the US]

Acta Metall Sin (Engl Ed) B Acta Metallurgica Sinica (English Edition) Series B [Published in the US]

Acta Meteorol Sin Acta Meteorologica Sinica [PR China]

Acta Met Mater Acta Metallurgica et Materialia [also Acta Metall Mater; formerly Acta Met; now Acta Mater]

Acta Mex Cienc Tecnol Acta Mexicana de Ciencia y Tecnologia [Mexico]

Acta Mineral Petrogr Acta Mineralogica Petrographica [Hungary]

Acta Odontol Scand Acta Odontologica Scandinavica

Acta Oncol Acta Oncologica [Sweden]

Acta Opt Sin Acta Optica Sinica [PR China]

Acta Opt Sin (China) Acta Optica Sinica (China) [PR China]
Acta Orthop Scand Acta Orthopaedica Scandinavica

Acta Paed Scand Acta Paediatrica Scandinavica

Acta Pathol Microbiol Scand Acta Pathologica Microbiologica Scandinavia

Acta Pharm Fenn Acta Pharmaceutica Fennicae [Finland]

Acta Pharm Hung Acta Pharmaceutica Hungarica [Hungary]

Acta Pharm Jugosl Acta Pharmaceutica Jugoslavia [Yugoslavia]

Acta Pharm Nord Acta Pharmaceutica Nordica [Scandinavia]

Acta Pharm Technol Acta Pharmaceutica Technologica [US]

Acta Phys Acad Sci Hung Acta Physica Academiae Scientarium Hungarica [of Hungarian Academy of Sciences]

Acta Phys Chim Sin Acta Physico-Chimica Sinica [of Chinese Chemical Society, PR China]

Acta Phys Hung Acta Physica Hungarica [Hungary]

Acta Physiochim USSR Acta Physiochimica USSR

Acta Physiol Pharmacol Latinoam Acta Physiologica et Pharmacologica Latinoamericana [Argentina]

Acta Physiol Scand Acta Physiologica Scandinavica

Acta Phys Pol Acta Physica Polonica [Poland]

Acta Phys Pol A Acta Physica Polonica A [Poland]

Acta Phys Pol B Acta Physica Polonica B [Poland]

Acta Phys Sin Acta Physica Sinica [PR China]

Acta Phys Sin (China) Acta Physica Sinica (China) [PR China]

Acta Phys Slovacia Acta Physica Slovacia [Slovakia]

Acta Phys Temp Humilis Sin Acta Physica Temperaturae Humilis Sinica [PR China]

Acta Phys Univ Comen Acta Physica Universitatis Comenianae [Czechoslovakia]

Acta Phyto Acad Scient Hung Acta Phytobiologica Academiae Scientiarum Hungarica [of Hungarian Academy of Sciences]

Acta Politec Mex Acta Politecnica Mexicana [Mexico]

Acta Pol Pharm Acta Poloniae Pharmaceutica [Poland]

Acta Polym Acta Polymerica [Germany]

Acta Polytech Česke Vys Uc Tech Pr I Acta Polytechnica Česke Vysoke Uceni Technicke v Prace I [Czechoslovakia]

Acta Polytech Česke Vys Uc Tech Pr II Acta Polytechnica Česke Vysoke Uceni Technicke v Prace II [Czechoslovakia]

Acta Polytech Česke Vys Uc Tech Pr III Acta Polytechnica Česke Vysoke Uceni Technicke v Prace III [Czechoslovakia]

Acta Polytech Scand Acta Polytechnica Scandinavica [Finland]

Acta Polytech Scand, Appl Phys Acta Polytechnica Scandinavica, Applied Physics [Finland]

Acta Polytech Scand, Appl Phys Ser Acta Polytechnica Scandinavica, Applied Physics Series [Finland]

Acta Polytech Scand Chem Technol Metall Ser Acta Polytechnica Scandinavica, Chemical Technology and Metallurgy Series [Finland]

Acta Polytech Scand, Electr Eng Ser Acta Polytechnica Scandinavica, Electrical Engineering Series [Finland]

Acta Polytech Scand, Math Comp Sci Ser Acta Polytechnica Scandinavica, Mathematics and Computer Science Series [Finland]

Acta Polytech Scand, Mech Eng Ser Acta Polytechnica Scandinavica, Mechanical Engineering Series [Finland]

Acta Psych Scand Acta Psychiatrica Scandinavica

Acta Radiol Acta Radiologica [Sweden]

Acta Radiol Suppl Acta Radiologica, Supplement

Acta Sci Math Acta Scientiarum Mathematicarum [of University of Szeged, Hungary]

Acta Sci Nat Univ Pekin Acta Scientiarum Naturalium Universitatis Pekinensis [of Peking University, PR China]

Acta Sci Nat Univ Sunyatseni Acta Scientiarum Naturalium Universitatis Sunyatseni [of Sunyatseni University, PR China]

Acta Seismol Sin Acta Seismologica Sinica [PR China]

Acta Stereol Acta Stereologica [of International Society for Stereology, Germany]

Acta Tech Acad Sci Hung Acta Technica Academiae Scientiarum Hungaricae [of Hungarian Academy of Sciences]

Acta Tech ČSAV Acta Technica ČSAV [Prague, Czechoslovakia]

Acta Univ Agr Brno Acta Universitatis Agriculturae Brno [of Brno University School of Agriculture, Czech Republic]

Acta Univ Carolin Math Phys Acta Universitatis Carolinae, Mathematica et Physica [of Carolinean University]

Acta Univ Szeged Acta Phys Chem Acta Universitatis Szegediensis, Acta Physica et Chimica [of Szeged University, Hungary]

Acta Univ Wratislav Mat Fiz Astron Acta Universitatis Wratislaviensis, Matematyka Fizyka Astronomica [of Wratislav University, Poland]

Acta Vitaminol Enzymol Acta Vitaminologica et Enzymologica [US]

Acta Zool Fennica Acta Zoologica Fennica [Finland]

ACTB Acute Caseous Tuberculosis

Actd Acetamide

ACT DELP Australian Capital Territory Department of Environment, Land and Planning

ACTE Agkistrodon Contortrix Thrombin-Like Enzyme [Biochemistry]

ACTE Association of Commercial and Technical Employees

ACTEC American Coalition on Trade Expansion with Canada [US]

ACTF Altitude Control Test Facility

ACTFEL Alternating-Current Thin Film Electroluminescence; Alternating-Current Thin Film Electroluminescent [also AC-TFEL]

Actg Acting [also actg]

Actg Actuating [also actg]

ACTS AIDS Clinical Trials Group

ACTH Adrenal Cortical Steroid Hormone; Adrenocorticotropic Hormone [Biochemistry]

ACTILIS Australian Capital Territory Integrated Land Information System [Defunct]

ACTION Accepting Challenge Through Interaction with Others and Nature

ACTIS A Computerized Tribology Information System

ACTIS Advanced Computed Tomography Imaging System

ACTIS AIDS Clinical Trials Information Service

ACTIS Auckland Commercial and Technical Information Service [New Zealand]

ACTL Advanced Construction Technology Laboratory [of National Research Council of Canada]

Actld Acetanilide

ACTLIC Australian Capital Territory Land Information Council

ACTM American Cordage and Twine Manufacturers [now American Cordage and Netting Manufacturers, US]

ACTM Ashridge Center for Transport Management [UK]

ACTMAP Australian Capital Territory Graphics Land Database System [of Australian Capital Territory Department of Environment, Land and Planning]

ACTO Association of Chief Technical Officers [UK]

ACTO Automatic Computing Transfer Oscillator

Acto Acetone

ACTOL Air-Cushion Takeoff and Landing

ACTORS Atlantic Canada Thin-Film Organic Semiconductors

ACTP Advanced Composite Thermoplastics

ACTP Advanced Computer Technology Project [UK]

Act Passive Electron Compon Active and Passive Electronic Components [Journal published in the UK]

ACTPCS Australian Capital Territory Parks and Conservation Service [also ACTP&CS]

Actr Actuator [also actr]

ACTRAM Advisory Committee on the Safe Transport of Radioactive Material

ACTRAN Analog Computer Translator

ACTRAN Autocoder-to-COBOL Translating Service

ACTRAN Autocoder-to-COBOL Translation

ACTRUS Automatically-Controlled Turbine Run-up System

ACTS Acoustic Control and Telemetry System

ACTS Advanced Communications Technology Satellite [of NASA]

ACTS Advanced Computed Tomography System

ACTS African Center for Technology Studies

ACTS All-Channel Television System

ACTS Association of Competitive Telecommunications Suppliers

ACTS Automated Computer Time Service

ACTS Automatic Coin Toll Service

ACTS Automatic Computer Telex Services

ACTSU Association of Computer Time Sharing Users [US]

ACTT Advanced Communication and Timekeeping Technology

ACTT Association of Cinematograph, Television and Allied Technicians [UK]

ACTU Australian Council of Trade Unions

Actual Chim L'Actualité Chimique [French Chemical Newsletter; published by of Société Française de Chimie]

Actual Combust Energ Actualité, Combustible, Energie [French News Bulletin on Combustible Materials and Energy]

Actual Inf Cient Tec Actualidades de la Información Cientifica y Tecnica [Cuban News Bulletin on Scientific and Technical Information]

ACTW Amalgamated Clothing and Textile Workers

Acty Activity [also acty]

Acty Actuary [also acty]

ACU Acknowledgement Unit

ACU Adaptive Control Unit

ACU Address Control Unit

ACU Adenine–Cytosine–Uridine (Trinucleotide) [Biochemistry]

ACU Alarm Control Unit

ACU Antenna Control Unit

ACU Arithmetic Control Unit

ACU Asian Currency Unit

ACU Association of Commonwealth Universities [UK]

ACU Association of Computer Users [US]

ACU Australian Catholic University

ACU Automatic Call(ing) Unit

ACU Auto-Cycle Union [UK]

AcU Actinium-Uranium; Actonouranium [also U-235, ^{235}U, or U^{235}]

Acu Altocumulus (Cloud)

α-Cu Alpha Copper

ACUA Association of College and University Auditors [US]

α-CuAl Alpha Copper Aluminum

ACUG Atex (Computer) Commercial Users Group [US]

ACUHO-I Association of College and University Housing Officers–International [US]

ACUI Automatic Calling Unit Interface

ACU-I Association of College Unions–International [US]

ACUIB Association of Canadian University Information Bureaus

ACUMG Association of College and University Museums and Galleries [US]

ACUNS Academic Council on the United Nations System

ACUP Association of Canadian University Presses

ACUP Association of College and University Printers [US]

ACUPS Association of Calendered UPVC (Unplasticized Polyvinyl Chloride) Suppliers [UK]

ACUS Administrative Conference of the United States [US]

ACUTA Association of College and University Telecommunications Administrators [US]

ACUTE Accountants Computer Users Technical Exchange [US]

a-CuTi Copper Titanium Amorphous Alloy

a-CuZr Copper Zirconium Amorphous Alloy

a-CuY Copper Yttrium Amorphous Alloy

ACV Air-Cushion Vehicle

ACV Alarm Check Valve

ACV Alternating-Current Voltammetry

ACV Armored Command Vehicle

ACVEN Advisory Committee on Vehicle Emissions and Noise

ACVJ 9-([E]-2Carbamoyl-2-Cyanovinyl)julolidine [Biochemistry]

ACVL Association of Cinema and Video Laboratories [US]

ACVT Air Cushion Vehicle Technician

ACW Aircraft Control (and) Warning (Service)

ACW Aircraftwoman

ACW As Cold-Worked (Condition)

AC&W Air Control and Warning (Service)

AC&W Aircraft Control and Warning (Service)

ACWA Amalgamated Clothing Workers of America

ACWA American Clean Water Association [US]

ACWCRP Australian Committee for the World Climate Research Programme

ACWIS American Committee for the Weizmann Institute of Science [US]

ACWL Army Chemical Warfare Laboratory [US Army]

ACWP Actual Cost of Work Performed

ACWRON Aircraft Control and Warning Squadron

AC&WS Air Control and Warning Station

AC&WS Aircraft Control and Warning Station

AcX Actinium X [also Ra-223, ^{223}Ra, or Ra^{223}]

Acyc Anticyclone; Anticyclonic [also acyc] [Meteorology]

ACZ Academia de Ciencias de Zaragoza [Zaragoza Academy of Sciences, Spain]

ACZAA Alternating Magnetic Current Zeeman Atomic Absorption [also AC-ZAA]

AD Abi Dhabi [United Arab Emirates]

AD Absorbed Dose

AD Access Door

AD Account Directory

AD Acid Deposition

AD Acoustic Dispersion

AD Acoustic Domain

AD Active Deposit

AD Active Duty

AD Address/Data

AD Address Decoding

AD Administrative Decision

AD Administrative Domain

AD Advanced Design

AD Advanced Development

AD Aerodrome [also ad]

AD Aerodynamic(s)

AD Affine Deformation [Geology]

AD Aircraft Division

AD Air Defense

AD Air Density

AD Air Director

AD Air Dose

AD Air-Dried; Air Drying

AD Air Drill

AD Airworthiness Directive

AD Algerian Dinar [Currency]

AD Algorithm Design

AD Alloy(ed) Diffused; Alloy Diffusion [Transistors]

AD Alzheimer's Disease

AD Amylodextrin [Biochemistry]

AD Ammunition Depot

AD Ammunition Dump

AD Ampere Demand Meter

AD Amplitude Detection

AD Amplitude Distortion

AD Anaerobic Digestion

AD Analog Data

AD Analog Device

AD Analog Divider

AD Analog to Digital

AD Analytical Division [of American Chemical Society, US]

AD Anderson-Darling (Test) [Physics]

AD Andorra [ISO Code]

AD Angular Distribution [Nuclear Engineering]

AD Anisotropy Dispersion

AD *(Anno Domini)* – in the year of our lord

AD Anode

AD Anodal Deviation

AD Anodic Dissolution

AD Anomalous Dispersion

AD Antidiuretic(s)

AD Antidote

AD Antiphase Domain [Solid-State Physics]

AD Anxiety Disorder

AD Aperture Door

AD Aprotic Dipolar [Chemistry]

AD Arbeitsgemeinschaft Druckbehälter e.V. [Pressure Vessel Association, Germany]

AD Arbor Day [US]

AD Arc Discharge

AD Area Drain [Civil Engineering]

AD Ascending Difference

AD As Deposited

AD As Determined

AD As Drawn

AD As Dried

AD Assembly District

AD Assignment Date

AD Assistant Director

AD Associate Director

AD ASTIA (Armed Services Technical Information Agency) Document

AD Astrodynamic(s)

AD Asymmetric Device

AD Asymmetric Dihydroxylation

AD Atmospheric Drag

AD Atomic Diameter

AD Auger Deexitation [Physics]

AD Automatic Detection

AD Automatic Diffractometer

AD Automatic Dryer

AD Avalanche Diode

AD Average Density

AD Average Deviation [Statistics]

AD Awaiting Disconnection

AD Axial Direction

AD Axiomatic Development [Physics]

AD Azeotropic Distillation

AD Destroyer Tender [US Navy Symbol]

A/D Altitude/Depth

A/D Analog-to-Digital [also A-D]

Ad Adductor (Muscle) [Anatomy]

Ad Advertisement [also ad]

ad adsorb(ed)

ad advertise(ment) [also ad]

ad after date

a d (*a dato*) – from date of issue

.ad Andorra [Country Code/Domain Name]

A$ Australian Dollar [Currency]

ADA 2-Acetamido-2-Iminodiacetic Acid; N-(2-Acetamido)-2-Iminodiacetic Acid

ADA Acetonedicarboxylic Acid

ADA 4-Acetyl-4'-Biphenoyl-Azide

ADA Action Data Automation

ADA Adenosine Deaminase [Biochemistry]

ADA Adjustable Derivative Action [Controllers]

ADA Advisory Area [Aeronautics]

ADA Agricultural Development Association [UK]

ADA Airborne Data Automation

ADA Air Defense Area [US Army]

ADA Air Defense Artillery

ADA Aluminum Development Association [UK]

ADA American Dairy Association [US]

ADA American Dehydration Association [US]

ADA American Dehydrators Association [now American Alfalfa Processors Association, US]

ADA American Dental Association [US]

ADA American Diabetes Association [US]

ADA American Dietetic Association [US]

ADA Americans with Disabilities Act [US]

ADA Ammonium Dihydrogen Arsenate [Biochemistry]

ADA Anti-Dumping Authority [Australia]

ADA Association of Drainage Authorities [UK]

ADA Atomic Development Administration [US]

ADA Automatic Data Acquisitions

ADA Automatic Distillation Analyzer

ADA Automatic Distributing Apparatus

ADAA Anxiety Disorders Association of America [US]

ADAAC Affiliated Distributed Active Archive Center

ADAAG Americans with Disabilities Act Accessibility Guidelines

ADAB Australian Development Assistance Bureau

ADABAS Adaptable DAta BAse System

ADAC Allgemeiner Deutscher Automobil-Club [German Automobile Club]

ADAC Analog-Digital-Analog Converter

ADAC Automated Direct Analog Computer

ADACC Advanced Damage Control Console

ADACC Automatic Data Acquisition and Computer Complex [US Air Force]

ADADL Ada-Based Design and Documentation Language

ADAF Advanced Design Array Radar

α-D-AH α-D-Aldohexose [Biochemistry]

Ada Lett Ada Letters [Association of Computing Machinery, US] [Note: *Ada* is a High-Level Computer Programming Language]

ADALERT Advance Alert [also Adalert, or adalert]

ADALT Advanced Radar Altimeter

ADAM A Data Management System

ADAM Advanced Data Acquisition Module

ADAM Advanced Data Management

ADAM Advanced Design and Manufacturing

ADAM Advanced Direct-Landing Apollo Mission [of NASA]

ADAM Affordable Design And Manufacturing [Joint NASA-Industry Program, US]

ADAM Air Deflection and Modulation

ADAM Angular Distribution of Auger Measurements [Physics]

ADAM 9-Anthryldiazomethane

ADAM Association of Distributors of Advertising Material [Scotland]

ADAM Associometrics Data Management System

ADAM Australian Data Archive for Meteorology

ADAM Automatic Distance and Angle Measurement

ADAM Automatic Document Abstracting Method

ADAMAS Aquatic Database Management System [US]

ADAMHA Alcohol, Drug Abuse and Mental Health Administration [of US Department of Health and Human Services–Public Health Service]

ADAP Angular Distribution of Annihilation Photons [Physics]

ADAPS Anthropometric Design Assessment Program System

ADAPS Automatic Display and Plotting System

ADAPSO Association of Data Processing Service Organizations [Canada/US]

ADAPT Adaptation of APT (Automatically Programmed Tools)

ADAPTICOM Adaptive Communication

Adapt&Lrng Sys Adaptive and Learning Systems [Division of Computer Society] [of Institute of Electrical and Electronics Engineers, US]

ADAPTS Air-Delivered Antipollution Transfer System

ADAR Advanced Design Array Radar

ADAR Automatic Data Acquisition Routine

ADAS African Demonstration Center on Sampling Agricultural Surveys [of Food and Agricultural Organization]

ADAS Agricultural Development Advisory Service [UK]

ADAS Angular-Dependent Auger Spectroscopy

ADAS Automatic Data Acquisition System

ADAS Automatic Data Analysis System

ADAS Automatic Dialing Alarm System

ADAS Auxiliary Data Annotation Set

ADASS Conference (on) Astronomical Analysis Software and Systems

ADAT Automatic Data Accumulation and Transfer

adatom adsorbed atom

ADATS Air Defense Antitank System

ADAU Auxiliary Data Acquisition Unit

ADAW Arbeitsgemeinschaft Deutscher Aussteller für Werkstoffprüfung [Association of German Exhibitors for Materials Testing; of Verband Deutscher Maschinen und Anlagenbau, Germany]

A-Day Army Day

ADB Activated Diffusion Bonding [Metallurgy]

ADB Adhesive Bonding

ADB Administrative Data Base

ADB African Development Bank [of United Nations]

ADB Apple Desktop Bus

ADB Arizona Development Board [US]

ADB Asian Development Bank [of United Nations]

ADB Atlantic Development Board [Canada]

ADB Authority Database

ADB Gesellschaft Produktionstechnik (im VDI) [Production Engineering Society of the VDI (Association of German Engineers)]

ADBIS Administrative Data Base Information System

ADBMS Available Database Management System

ADBO Authority Database Operation

ADBS Advanced Database System

ADBS Association des Documentalistes et des Bibliothécaires Spécialisés [Association of Documentalists and Special Librarians, France]

ADC Adaptive Data Compression (Protocol)

ADC Add with Carry [Data Communications]

ADC Advanced Device Center

ADC Advanced Digital Control

ADC Advise Duration and Charge

ADC Aerodrome Control [also adc]

ADC Aerospace Defense Command [at Ent Air Force Base, Colorado Springs, US]

ADC Affiliated Data Center

ADC Agricultural Development Council [US]

ADC Aide-de-Camp [Military]

ADC AIDS Dementia Complex

ADC Airborne Digital Computer

ADC Air Data Computer

ADC Air Defense Command [of US Air Force]

ADC Air Development Center

ADC Air Diffusion Council [US]

ADC Air-Dried Coal

ADC Algebraic Diagrammatic Construction (Method)

ADC Allyl Diglycol Carbonate

ADC Analog-to-Digital Conversion

ADC Analog-to-Digital Converter [also A/DC]

ADC Anodal Duration Contraction

ADC Antenna Dish Control

ADC Application Development Center

ADC (International Conference on the) Applications of Diamond Coatings (and Related Materials)

ADC Applied Diamond Conference

ADC Ardeer Double Cartridge Test

ADC Army Dental Corps [US]

ADC Arsenic Development Committee [France]

ADC Association of Diving Contractors [US]

ADC Astronomical Data Center [at Goddard Space Flight Center, Greenbelt, Maryland, US]

ADC Atlantic Development Council [Canada]

ADC Automatic Data Collection (System)

ADC Automatic Drive Control

ADC Aviation Development Council [US]

ADC Azodicarbonamide

A/DC Analog-to-Digital Converter [also ADC]

adc aide-de-camp [Military]

ADCA Aerospace Department Chairmen's Association [US]

ADCA Aminodecephalosporanic Acid

7-ADCA 7-Aminodecephalosporanic Acid

ADCAD Airways Data Collection and Distribution

AdCADD Advanced Computer-Aided Design and Drafting

ADCC Antibody-Dependent Cellular Cytotoxicity [Biochemistry]

ADCC Air Defense Control Center

ADCC Asynchronous Data Communications Channel

ADCCP Advanced Data Communications Control Procedure

ADCE Attitude Determination and Control Electronics

ADCI American Die Casting Institute [US]

ADCIS Analog-Digital CMOS (Complimentary Metal-Oxide Semiconductor) IC Integrated (Circuit)

ADCIS Association for the Development of Computer-Based Instructional Systems [US]

ADCLS Advanced Data Collection and Location System

ADCN Aeronautical Data Communication Network

ADCOM Administrative Committee [of Institute of Electrical and Electronics Engineers Committee, US]

Ad Com Advisory Committee [of Institute of Electrical and Electronics Engineers, US]

ADCON Address Constant [also Adcon, or adcon]

ADCON Analog-Digital Converter

ADCOS Australian Development Cooperation Scholarships

ADCP Acoustic Doppler Current Profiler

ADCS Advanced Defense Communications Satellite

ADCS Attitude Determination and Control Subsystem

ADCSP Advanced Defense Communications Satellite Program [US]

ADCU Alarm Display and Control Unit

ADCU Association of Data Communications Users [US]

ADCUS Advise Customs

ADCVR Analog-to-Digital Converter

ADD Acoustic Deterrent Device

ADD Aerospace Digital Development

ADD Agency for Defense Development [of the South Korean Government at Taejon]

ADD Atomic Density Distribution (Function)

ADD Attention Deficit Disorder

ADD Automatic Diagnostic Display

ADD Automatic Document Detection [Word Processing]

ADD The International Arctic-Environment Data Directory

AD&D Accidental Death and Dismemberment

Add Addendum; Addenda [also add]

Add Addition [also add]

Add Additive [also add]

Add Address [also add]

ADDA (International Conference on Powder Metallurgy in) Aerospace, Defense and Demanding Applications [of Metal Powder Industries Federation, US]

ADDA Australian Database Development Association

AD/DA Analog-to-Digital/Digital-to-Analog (Converter)

ADDAM Air Deployable Datum Marker Buoy

ADDAR Automatic Digital Data Acquisition and Recording

ADDAS Automatic Digital Data Assembly System

ADDB Antarctic Digital Database

ADDC Ammonium N,N-Diethyl Dithiocarbamate

ADDC Association of Desk and Derrick Clubs [US]

ADDCL Add Clause [Computer Programming]

ADDDS Automatic Direct Distance Dialling System

ADDER Automatic Digital Data Error Recorder

Addit Addition [also addit]

ADDK Agency for Defense Development of Korea [Taejon, South Korea]

addl additional [also add'l]

ADDN Automated Defense Data Network

Addn Addition [also addn]

addnl additional

ADDPEP Aerodynamic Deployable Decelerator Performance Evaluation Program

Add Poly Additives for Polymers [Journal published in the UK]

ADDR Adder [also Addr, or addr]

ADDR Address (Function) [also Addr]

ADDR Address Register [Computers]

Addr Address [also addr]

ADDROUT Address Out

Addrs Address [also addrs]

ADDS Advanced Detector Development System

ADDS Applied Digital Data Systems

ADDS Automatic Data Digitizing System

ADE Address Error

ADE (Thermal) Adsorption-Desorption Equilibrium

ADE Advanced Data Entry

ADE Aerodynamics Engineer(ing)

ADE Air Defense Emergency

ADE Air Density Explorer

ADE Alberta Department of Energy [Canada]

ADE Antisite Defect Energy [Physics]

ADE Armament Design Establishment [UK]

ADE Association for Documentary Editing [US]

ADE Attracted Disk Electrometer

ADE Audible Doppler Enhancer [Radar]

ADE Automated Debugging Environment

ADE Automated Design Engineering

ADE Automatic Drafting Equipment

ADEA Association Belge pour le Développement Pacifique de l'Energie Atomique [Belgian Association for the Peaceful Development of Atomic Energy]

ADEC Advanced Development and Engineering Center

ADEC Alaska Department of Environmental Conservation [US]

ADECP Association of Directors of European Centers for Plastics [UK]

ADED Association of Driver Educators for the Disabled [US]

ADEE Automated Design and Engineering for Electronics

ADEF N,N-Diethyl Phosphoramidic Acid Diethyl Ester

ADEKS Advanced Design Electronic Key System

ADELE *(Automatische Datenerfassung durch Lochkarten-Eingabegeräte)* – German for "Automatic Data Acquisition by (Punched) Card Input Units"

ADELT Automatically Deployable Emergency Locator Transmitter

ADEM Adaptively Data Equalized Modem

ADEM Alabama Department of Environmental Management [US]

ADEM Automated Digital Electron Microscope; Automatic Digital Electron Microscope

ADEOM Advanced Earth Observing Mission

ADEOS Advanced Earth Observing Satellite [of National Space Development Agency of Japan]

ADEOS-1 First Advanced Earth Observing Satellite [of National Space Development Agency of Japan]

ADEPT Automated Direct Entry Packaging Technique

ADEPT Automatic Data Extractor and Plotting Table

ADES Angle-Dispersed Electron Spectroscopy

ADES Association of Directors of Education in Scotland

ADES Automatic Digital Encoding System

ADESH Atomistic Defect Simulation Handler (Software)

ADEU Automatic Data Entry Unit

ADF Acid Detergent Fiber

ADF Acoustic Depth Finding

ADF Adapter Description File

ADF Aerial Direction Finding

ADF Aerodynamic Force

ADF African Development Foundation

ADF African Development Fund

ADF Airborne Direction Finder

ADF Annular Dark Field (Image)

ADF Aqueous Dissolution and Fluorination (Process)

ADF Asian Development Fund

ADF Astrophysics Data Facility [of NASA Goddard Space Flight Center, Greenbelt, Maryland, US]

ADF Automatically Defined Function

ADF Automatic Direction Finder

ADF Automatic Document Feed

ADF Australian Defense Forces

ADF Auxiliary Detonating Fuse

.ADF Adapter Description File [IBM File Name Extension]

ADFA Alaska Department of Fish and Game [US]

ADFA Australian Defense Force Academy [Canberra]

ADFA Australian Dried Food Association

adfilm adsorbed film

ad fin *(ad finem)* – to the end

ADFRF Ames-Dryden Flight Research Facility [at Edwards Air Force Base, US]

ADFSC Automatic Data Field Systems Command [US Army]

ADG Australian (Code for the Transportation of) Dangerous Goods (by Road and Rail)

ADGA American Dairy Goat Association [US]

ADGB Air Defense of Great Britain

ADGB Allgemeiner Deutscher Gewerkschaftsbund [German Federation of Trade Unions]

ADGE Air Defense Ground Environment

ADGE AutoCAD Developers Group Europe

Adgez Rasplav Paika Mater Adgeziya Rasplavov i Paika Materialov [Journal on Materials; published in Kiev, Ukraine]

ADGF N,N-Diheptyl Phosphoramidic Acid Diethyl Ester

ADH Activated Diffusion Heating

ADH Aerodynamic Heating

ADH Alcohol Dehydrogenase [Biochemistry]

ADH Antidirected Hamiltonian

ADH Antidiuretic Hormone [Biochemistry]

ADH United Arab Emirates Dirham [Currency]

Adh Adhesion; Adhesive [also adh]

ADHA American Dental Hygienists Association [US]

ADHD Attention Deficit Hyperactivity Disorder [also AD-HD]

Adhes Adhesion; Adhesive [also adhes]

Adhes Abstr Adhesive Abstracts

Adhes Age Adhesives Age [US Publication]

Adhes Soc Newsl Adhesion Society Newsletter [US]

Adhes Trends Adhesive Trends [Publication of the Adhesives Manufacturers Association, US]

ad h l *(ad hunc locum)* – at the place

ad hoc *(ad hoc)* – for this

ADI Acceptable Daily Intake

ADI Acoustic Data Interface

ADI Address Incomplete

ADI Air Distribution Institute [US]

ADI Alternating Direction Implicit

ADI Altitude Direction Indicator

ADI American Documentation Institute [now American Society for Information Science, US]

ADI Attitude Direction Indicator

ADI Austempered Ductile Iron

ADI AutoCad/AutoDesk Device Interface (Driver)

ADI Autodesk Device Interface

ADI Automatic Direction Indicator

ADIC Aktueel Dokumentie en InformatieCentrum [Current Documentation and Information Center, Netherlands]

ADIC Analog-to-Digital Converter

ADICEP Association of Directors of European Centers for Plastics [UK]

ADIDAC Argonne Digital Data Acquisition Computer [of Argonne National Laboratory, Illinois, US]

ADIDS Aeronautical Digital Information Display System

ADIE Automatic Dialing and Indicating Equipment

ADIF Amplitude-Dependent Internal Friction

ADIF Antarctic Directory Interchange Format

ADIJ Association pour le Développement de l'Informatique Juridique [Association for the Development of Legal Informatics, France]

ADINA Automatic Dynamic Incremental Nonlinear (Finite-Element) Analysis (Code)

ad inf *(ad infinitum)* – to infinity

ad init *(ad initium)* – at the beginning

ad int *(ad interim)* – in the interim; in the meantime

ADINTELCEN Advanced Intelligence Center

ADIOS Analog-to-Digital Input/Output System

ADIOS Automatic Diagnosis Input/Output System

ADIOS Automatic Digital Input/Output System

Adip Adipic Acid

ADIRS Air Data Inertial Reference System

ADIS Advanced Data-Acquisition Imaging and Storage (System)

ADIS Air Defense Integrated System

ADIS Association for the Development of Instructional Systems [US]

ADIS Automated Display and Inspection System

ADIS Automatic Data Interchange System

ADIS Groupe Armées, DGA (Délégation Generale pour l'Armement), Industrie sur l'Interopérabilité des Simulations [Army DGA (French Armaments Directorate) Industry Work Group on Simulations Interoperability, France]

ADISP Aeronautical Data Interchange System Panel

ADIT Analog-to-Digital Integration Translator

ADITEC Aachener Demonstrationslabor für Integrierte Produktionstechnik [Aachen Demonstration Laboratory for Integrated Production Engineering, Germany]

ADIZ Air Defense Identification Zone

Adj Adjective [also adj]

Adj Adjoining [also adj]

Adj Adjunct [also adj]

Adj Adjust(ment) [also adj]

Adj Adjutant [also adj]

adj adjacent

adj adjoint (in mathematics)

adj adjourned [also adj]

AdjA Adjunct in Arts

adj A adjoint of square matrix aAa [Symbol]

ADJM Meter Adjustment [Automobiles]

ADJMT Meter Transaction [Automobiles]

Adjmt Adjustment [also adjmt]

Adj Prof Adjunct Professor

Adjt Adjutant [also adjt]

ADJUTD UTD (Utilization to Date) Adjustment [Automobiles]

ADJUTDC UTD (Utilization to Date) Change [Automobiles]

ADL Activity (or Activities) of Daily Living

ADL Address Data Latch

ADL Air-Dry Loss

ADL Applications Development Language

ADL Arthur D. Little Company [US]

ADL Automatic Data Link

ad l *(ad libitum)* – freely; without limit, or restraint

ADLAT Adaptive Lattice Filter

adlayer adsorbed layer

ADLC Adaptive Lossless Data Compression

ADLC Advanced Data Link Controller

ADLC Angle-Dependent Line of Centers (Model)

ADLC Asynchronous Data Link Control

ADLIB Adaptive Library Management System

ad lib *(ad libitum)* – freely; without limit, or restraint; at one's pleasure [also ad libit]

ADLIPS Automated Data Link Plotting System

ad loc *(ad locum)* – at the place

ADLRI Arthur D. Little Research Institute [UK]

ADLS Automatic Drag-Limiting System

ADLs Activities of Daily Living [Medicine]

ADM Academy of Dental Materials [US]

ADM Activity Data Method

ADM Activity Description Memorandum

ADM Adaptable Data Manager

ADM Adaptive Delta Modulation

ADM Address Modification

ADM Admiral(ty) [also Adm]

ADM Advanced Design Model

ADM Advanced Development Model

ADM Air Decoy Missile

ADM Air Deployable Marker

ADM American Down Association [US]

ADM Analog Data Module

ADM Association for Domestic Management [UK]

ADM Association of Drum Manufacturers [UK]

ADM Asymmetric Dimer Model

ADM Asynchronous Disconnected Mode

ADM Atomic Demolition Munition

ADM Automatic Drafting Machine

AdM Adductor Muscle [Anatomy]

Adm Administration; Administrative; Administrator [also adm]

Adm Admiral(ty) [also ADM]

A/dm² Amperes) per square decimeter [also A dm^{-2}]

ADMA Abu Dhabi Marine Areas [United Arab Emirates]

ADMA Advanced Direct Memory Access (Controller)

ADMA Alkyldimethyl Amine

ADMA Amusement Device Manufacturers Association [now American Amusement Machine Association, US]

ADMA 4-(9'-Anthracenyl)-N,N-Dimethylaniline

ADMA Asymmetric Dimethylarginine [Biochemistry]

ADMA Aviation Distributors and Manufacturers Association [US]

ADMACS Apple Document Management And Control System

ADM/ADACC Advanced Development Model/Advanced Damage Control Console

ADMATCH Address Matching Software [Geographic Information System]

ADMCS Allyldimethylchlorosilane

ADMD Administration (or Administrative) Management Domain [Electronic Mail]

ADMET Acyclic Diene Metathesis

ADMH Asymmetric Dimethylhydrazine

ADMI American Dry Milk Insitute [US]

Admin Administration; Administrative; Administrator [also admin]

ADMIRAL Automatic and Dynamic Monitor with Immediate Relocation, Allocation and Loading

ADMIRE Adaptive Decision Maker in an Information Retrieval Environment

ADMIRE Automated Diagnostic Maintenance Information Retrieval System

ADMIS Automated Data Management Information System

Adm Manage Administrative Management [Journal published in the US]

Adm of F Admiral of the Fleet [US Navy]

ADMR Absorption Detected Magnetic Resonance

Admr Administrator [also admr]

ADMRO Angular-Dependent Magnetoresistance Oscillation

ADMS Accessibility Data Management System

ADMS Allyldimethylsilyl

ADMS Antarctic Data Management System

ADMS Assistant Director of Medical Services [US Military]

ADMS Asynchronous Data Multiplexer Synchronizer

ADMS Automatic Digital Message Switch(ing)

ADMSC Automatic Digital Message Switching Center

ADMT Air-Dried Metric Ton

Admy Admiralty [also admy]

ADN 4-Acetyl-4'-Biphenoyl-Nitrene

ADN Adiponitrile

ADN Allgemeiner Deutscher Nachrichtendienst [Former East German News Service]

.ADN Add in Utility [Lotus 1-2-3 File Name Extension] [also .adn]

ADNAC Air Defense of the North American Continent

ADNOC Abu Dhabi National Oil Company [United Arab Emirates]

ADO Active Data Objects

ADO Advanced Development Objective [of US Air Force]

ADO Air Defense Officer

ADO Application Development Operation

ADO Autodyne Oscillator

ADO Automatic Dial-Out

ADO Averaged Dipole Orientation

Ado Adenosine [Biochemistry]

ADOC Air Defense Operations Center

Adoc Adamentyloxycarbonyl-

ADOGA American Dehydrated Onion and Garlic Association [US]

AdoHcy S-(5-Adenosyl)-L-Homocysteine [Biochemistry]

ADOIT Automatically Directed Outgoing Intertoll Trunk

ADOM Acid Deposition and Oxidants Model

ADONIS Automated Document Delivery over Networked Information Service

ADONIS Automatic Digital On-Line Instrument(ation) System

ADONIS Automatic Document On-Line Information System

ADOPE Automatic Decisions Optimizing Predicted Estimates

ADOS Advanced Disk Operating System

ADP Acceptance Data Package

ADP Adapter [also Adp, or adp]

ADP Adenosine Diphosphate; Adenosine-5'-Diphosphate [Biochemistry]

ADP Administrative Data Processing

ADP Advanced Data Processing

ADP Advanced Display Processor

ADP Advanced Ducted Prop

ADP Airborne Data Processor

ADP Air Defense Position

ADP Airport Development Program

ADP Amateur Data Processor

ADP 4-Aminodiphenyl

ADP Ammonium Dihydrogen Phosphate

ADP Association of Database Producers [UK]

ADP Atmospheric Dew Point

ADP Atomic Displacement Parameter

ADP Automated Data Processing; Automatic Data Processing

ADP Automatic Die Positioner

ADP Auxiliary Docking Port [of NASA]

Adp Adapter [also adp]

ADPA Acetone Diphenylamine

ADPA American Defense Preparedness Association [US]

ADPA 4-Aminodiphenylamine

ADPA 9,10-Anthracenedipropanoic Acid

ADPase Adenosine 5'-Triphosphatase [Biochemistry]

ADP-β-S Adenosine 5'-O-(2-Thiodiphosphate) [Biochemistry]

ADPC Agricultural Development Planning Center [of Association of Southeast Asian Nations]

ADPC Automatic Data Processing Center

ADPCM Adaptive Delta Pulse Code Modulation

ADPCM Adaptive (or Adapted) Differential Pulse Code Modulation [also adpcm]

ADPCM Association for Data Processing and Computer Management [US]

ADPE Automatic Data Processing Equipment

ADP/EP Automatic Data Processing/Extramural Program (Coordinating Committee)

ADPE/S Automatic Data Processing Equipment/System

ADPESO Automatic Data Processing Equipment Selection Office

ADPG Adenosine-5'-Diphosphoglucose [Biochemistry]

ADP-Glu Adenosine 5'-Diphosphoglucose [Biochemistry]

ADPI American Dairy Products Institute [US]

ADP-KDP Ammonium Dihydrogen Phosphate–Potassium Dihydrogen Phosphate Crystal)

ADPLAN Advancement Planning

ADPP Automatic Data Processing Program

ADPR Adenosine 5'-Diphosphoribose [Biochemistry]

ADP-Rib Adenosine 5'-Diphosphoribose [Biochemistry]

ADPRP Adenosine Diphosphate Ribose Phosphate [Biochemistry]

ADPRP 2-Monophosphoadenosine 5'-Diphosphoribose [Biochemistry]

ADPS Acceptance Data Package System

ADPS Automatic Data Processing System

ADPS Automatic Display and Plotting System

ADP-β-S Adenosine-5'-[β-Thio]diphosphate [Biochemistry]

ADPSO Association of Data Processing Service Organizations

ADPSO Automatic Data Processing Selection Office

Adpt Adapter [also Adptr, adptr, or adpt]

Ad-Q Advertising Quotient

ADR Address

ADR Adsorption/Desorption/Recovery

ADR Adverse Drug Reaction

ADR Advisory Route [Aeronautics]

ADR Aircraft Direction Room

ADR Alternative Dispute Resolution

ADR American Depository Receipt

ADR Applied Data Research

ADR Asset Depreciation Rate

ADR Audit Discrepancy Report

ADR Automatic Direction Finder

ADR Aviation Design Research [US Navy]

ADRA Adreßbuchausschuß der Deutschen Wirtschaft [German Trade Directory Committee]

ADRA Animal Disease Research Association [UK]

ADRA Automotive Dismantlers and Recyclers Association [US]

ADRAC Automatic Digital Recording and Control

ADRDA Alzheimer's Disease and Related Dementias Association [US]

ADRDE Air Defense Research and Development Establishment [UK]

ADRES Army Data Retrieval System [US Army]

ADRF Adiabatic Demagnetization (in the) Rotating Frame

ADRG Arc Digital Raster Graphics

ADRI Animal Disease Research Institute

ADRIQ Association des Directeurs de Recherches Industrielles de Québec [Association of Industrial Research Directors of Quebec, Canada]

ADRM Airdrome [also Adrm, or adrm]

ADRMP Auto-Dialed Recorded Message Player

ADRMP Automated Dialer and Recorded Message Player

ADRRP Activated Dissociation Reduction Reaction Process

ADRS Address [also Adrs]

ADRS Address Register

ADRS Analog-to-Digital Data Recording System

ADRS Automatic Document Request Service [UK]

Adrs Address [also adrs]

ADRT Analog Data Recorder Transcriber

ADS Accessory Drive System

ADS Accurately-Defined System

ADS Acid Deposition System (Database)

ADS Activity Data Sheet

ADS Address [also Ads, or ads]

ADS Address Data Strobe

ADS Administration of Designed Services

ADS Administrative Data System

ADS Advanced Data System

ADS Advanced Debugging System

ADS Advanced Digital System

ADS Advanced Dressing Station

ADS Advanced Dryer System

ADS Airborne Data System [US National Center for Atmospheric Research Airborne Infrared Lidar System]

ADS Aircraft Development Service [of Federal Aviation Administration, US]

ADS Air Defense Sector

ADS Air Defense Ship

ADS Air Defense System

ADS Air Development Service [of Federal Aviation Administration, US]

ADS Alloy Descaling Salt

ADS Antidiuretic Substance [Biochemistry]

ADS Application Development Solutions [of AT&T Corporation, US]

ADS Application Development System

ADS Articulated Diving Suit

ADS Association of Diesel Specialists [US]

ADS Astrophysics Data System [of NASA]

ADS Asymptotic Diffuse Scattering

ADS Atmospheric Detritiation System

ADS Automated Design System

ADS Automated Drafting System

ADS Automatic Data Set

ADS Automatic Dependent Surveillance

ADS Automatic Depressurization System

ADS Automatic Distribution System

ADS Automatic Duplicating System

ADS Automatic Door Seal

Ads Address [also ads]

Ads Adsorb(ed); Adsorption [also ads]

ADSA American Dairy Science Association [US]

ADSA American Dental Society of Anaesthesia [US]

ADSA Atomic Defense Support Agency

ADSAF Automated Data System(s) for the (US) Army in the Field [also Adsaf]

ADSAS Air-Derived Separation Assurance System

ADSATIS Australian Defense Science and Technology Information System

ADSC Address Status Changed

ADSC Adobe Document Structuring Conventions

ADSC Association of Drilled Shaft Contractors [US]

ADSC Automatic Data Service Center

ADSC Automatic Digital Switching Center

ADSCOM Advanced Shipboard Communication

ADSEL Address Selective [UK Aviation]

ADSF Advanced Directional Solidification Furnace

ADSF Automated Directional Solidification Furnace

ads/Bov/Ho/Hu Antibody Adsorbed with Bovine, Horse and Human Serum Proteins

ads/HIgG/RS Antibody Adsorbed with Human Immunoglobulin G and Rat Serum Proteins

ads/HIgG Antibody Adsorbed with Human Immunoglobulin G

ads/HS Antibody Adsorbed with Human Serum Proteins

ADSI Active Directory Service Interface

ADSI Administrative Design Service Information

ADSI Agricultural Development Service, Inc. [Philippines]

ADSI Analog Display Services Interface

ad-Si Adsorbed Silicon (Atom)

ADSIA Allied Data System Interoperability Agency [of NATO]

ADSID Air-Delivered Seismic Intrusion Detector [also Adsid]

ADSL Asymmetric Digital Subscriber Line

ADSM Adstar Distributed Storage Manager

ADSM Air Defense Suppression Missile

ADSM Association for the Development of Synthetic Materials [US]

ads/MS Antibody Adsorbed with Mouse Serum Proteins

Adsorp Adsorption [also adsorp]

Adsorp Sci Technol Adsorption Science and Technology [Journal published in the US]

ADSP Advanced Digitial Signal Processor

ADSP Apple Data-Stream Protocol

ADSR Attack, Decay, Sustain and Release

ADSRI Animal and Dairy Science Research Institute [South Africa]

ads/RS Antibody Adsorbed with Rat Serum Proteins

ADSS A Diesel Supply Set

ADSS Aircraft Damage Sensing System

ADSS Australian Defense Scientific Service

ADST Atlantic Daylight Saving Time

ADSTAR Automatic Document Storage and Retrieval

ADS-TP Administrative Data System–Teleprocessing

ADSUP Automatic Data Systems Uniform Practices

AdSV Adsorptive Stripping Voltammetry

ADT (Thermal) Adsorption-Desorption Transient

ADT Abstract Data Type

ADT Adenosine Triphosphate [Biochemistry]

ADT Adjustable Digital Thermometer

ADT Admission, Discharge and Transfer

ADT Aeronautical Data Terminal

ADT Alaska Daylight Time

ADT Arc-Discharge Tube

ADT Alloyed Diffused Transistor

ADT American District Telegraph

ADT Anomalous Diffraction Theory

ADT Application Data Types

ADT Application-Dedicated Terminal

ADT Articulated Dump Truck

ADT Asynchronous Data Transceiver

ADT Atlantic Daylight Time

ADT Automatic Data Translator

ADT Automatic Data Transmission

ADT Automatic Deflection-Temperature Tester

ADT Autonomous Data Transfer

ADT Average Daily Traffic

A$/t Australian Dollars per ton

ADTA American Dental Trade Association [US]

ADTC Anglo-Dutch Trade Council [Netherlands]

ADTD Association of Data Terminal Distributors [US]

ADTECH Advanced Decoy Technology

ADTS Approved Departure Times

ADTS Automated Data and Telecommunications Service [of General Services Administration, US]

ADTS Automatic Data Test System

ADTSEA American Driver and Traffic Safety Education Association [US]

ADTT Accelerator Driven Transmutation Technology [Physics]

ADTT Average Daily Truck Traffic

ADTU Auxiliary Data Translator Unit

ADU Accumulation Distribution Unit

ADU Adelaide University [Australia]

ADU Ammonium Diuranate (Process)

ADU Automatic Data Unit

ADU Automatic Dialing Unit

ADV Acid Demand Value

ADV Advice [Radio Engineering]

ADV Arbeitsgemeinschaft für Datenverarbeitung [Data Processing Association, Austria]

Adv Advance(s) [also adv]

Adv Advection [also adv]

Adv Adverb [also adv]

Adv Adversus [also adv]

Adv Advertisement [also adv]

Adv Advice [also adv]

Adv Advocate [also Adv, or adv]

adv advanced

ad v *(ad valorem)* – according to value, or according to the value of [also ad val]

Adv Alloys Met Advanced Alloys and Metals [Journal published in Spain]

AD-VAM Advanced Development of a Vulnerability Assessment Model [of Department of National Defense, Canada]

Advan Advance [also advan]

Adv Appl Probab Advances in Applied Probability [Journal published in the UK]

Adv Appl Microbiol Advances in Applied Microbiology [Journal published in the US]

Adv Astron Sci Advances in Astronautical Sciences [of American Astronautical Society]

Adv Battery Technol Advanced Battery Technology [Journal published in the US]

Adv Biochem Eng Biotechnol Advances in Biochemical Engineering and Biotechnology [International Journal]

Adv C Advanced Certificate

Adv Cancer Res Advances in Cancer Research

Adv Carbohyd Chem Biochem Advances in Carbohydrate Chemistry and Biochemistry [Journal published in the US]

Adv Catal Advanced Catalysis [Journal published in the UK]

Adv Cell Neurobiol Advances in Cellular Neurobiology [Journal]

Adv Cem Res Advances in Cement Research [Journal published in the US]

Adv Cem Mater Advanced Cement-Based Materials [International Journal published in Affiliation with the American Concrete Institute] [also Adv Cem Based Mater, or ACBM]

Adv Ceram Advances in Ceramics [Journal published in the UK]

Adv Ceram Mater Advanced Ceramic Materials [of American Ceramic Society, US] [also Adv Ceram Mat]

Adv Ceram Rep Advanced Ceramics Report [UK]

Adv Chem Eng Advances in Chemical Engineering [US Journal]

Adv Chem Ser Advances in Chemistry Series [US]

Adv Chromatogr Advances in Chromatography [Published in the US]

Adv Chromatogr (Houston) Advances in Chromatography (Houston) [Published in Houston, Texas, US]

Adv Chgs Advance Charges

Adv Colloid Interface Sci Advances in Colloid and Interface Science [Journal published in the Netherlands]

Adv Compos Advanced Composites [US Journal]

Adv Compos Bull Advanced Composites Bulletin [UK]

Adv Conv Pkg Technol Advancing Converting and Packaging Technologies [of US Technical Association of the Pulp and Paper Industry]

Adv Cryog Eng Advances in Cryogenic Engineering [Publication of the Cryogenic Engineering Conference]

Adv Cryog Eng Mater Advances in Cryogenic Engineering Materials [Publication of the International Cryogenic Materials Conference]

Adv Dent Res Advances in Dental Research [Journal published in the UK]

ADV DN Advance Down [also AdvDn]

Adv Drying Advances in Drying [Journal published in the US]

Adv Earth Plant Sci Advances in Earth and Planetary Sciences [Published in Japan]

Adv Electron Phys Advances in Electronics and Physics [Journal published in the US]

Adv Eng Softw Advances in Engineering Software [Journal published in the UK]

Adver Advertising [also adver]

Adv Exp Med Biol Advances in Experimental Medical Biology [US]

Adv Frt Advance Freight [also adv frt]

Adv Graph Mag Advanced Graphics Magazine [UK]

Adv Heat Transfer Advances in Heat Transfer [Journal published in the US]

Adv Inorg Chem Radiochem Advances in Inorganic Chemistry and Radiochemistry

Adv Instrum Advances in Instrumentation [of Instrument Society of America]

ADVISOR Advanced Integrated Safety and Optimizing Computer

ADV LFT Advance Left [also AdvLft]

Adv LTA Rev Advanced Lighter-Than-Air Review [Publication of the Airship Association, US]

ADVM Adaptive Delta Modulation Voice Modem

Adv M Advanced Master

Adv Macromol Chem Advanced Macromolecular Chemistry [International Journal]

Adv Manage J Advanced Management Journal [of Society for Advancement of Management, US]

Adv Manuf Process Advanced Manufacturing Processes [Journal published in the US]

Adv Manuf Technol Advanced Manufacturing Technology [Journal published in Ireland]

ADVMAT Environmental Effects on Advanced Materials Symposium [of NACE International]

Adv Mater Advanced Materials [US Journal]

Adv Mater (Germany) Advanced Materials (Germany) [Journal published in Germany]

Adv Mater Manuf Process Advanced Materials and Manufacturing Processes [Journal published in the US]

Adv Mater Process Advanced Materials and Processes [of ASM International, US]

Adv Mater Process inc Met Prog Advanced Materials and Processes incorporating *Metal Progress* [of ASM International, US]

Adv Mater Technol Int Advanced Materials Technology International [Journal published in the UK]

Adv Mater Technol: Monitor Advances in Materials Technology: Monitor [Journal published in Austria]

Adv Met Technol Advanced Metals Technology [UK]

Adv Microb Ecol Advances in Microbial Ecology [Journal of the International Committee on Microbial Ecology, US]

Adv Microb Physiol Advances in Microbial Physiology [International Journal]

Adv Model Simul Advances in Modelling and Simulation [Journal published in France]

Adv Myocardiol Advances in Myocardiology [Journal]

Adv Neurol Advances in Neurology

Adv Nutr The Advancement of Nutrition [of International Academy of Preventive Medicine]

Adv Perform Mater Advances in the Performance of Materials [Journal published in the US]

Adv Phys Advances in Physics [Journal published in UK]

Adv Phys Org Chem Advances in Physics–Organic Chemistry [Journal published in the UK]

Adv Pmt Advance Payment [also adv pmt]

Adv Polym Sci Advances in Polymer Science [Journal published in Germany]

Adv Polym Technol Advances in Polymer Technology [Journal published in the US]

Adv Printing Sci Technol Advances in Printing Science and Technology [Journal published in the UK]

Adv Prot Chem Advances in Protein Chemistry [Journal]

ADV RGT Advance Right [also AdvRgt]

Adv Robot Advanced Robotics [Journal published in the Netherlands]

Adv Sol Energy Advances in Solar Energy [International Journal]

Adv Space Res Advances in Space Research [Journal of the Committee on Space Research, France]

Adv Space Res Advances in Space Research [Journal published in the UK]

Adv Spec Electrometall Advances in Special Electrometallurgy [Translation of: *Problemy Spetsial'noi Elektrometallurgii (USSR)*; published in the UK]

Advsr Advisor [also advsr]

Advt Advertisement [also advt]

Advtg Advertising [also advtg]

AdvToCol Advance to Column

AdvToLn Advance to Line

ADV UP Advance Up [also AdvUp]

Adv Water Resour Advances in Water Resources [Journal published in the UK]

Adv X-Ray Anal Advances in X-Ray Analysis [Journal published in the US]

ADW Aerial Distribution Wire

ADW Air Defense Warning

AdW Akademie der Wissenschaften [Academy of Science, Germany]

ADWA Atlantic Deeper Waterways Association [US]

ADX Asymmetric Data Exchange

ADX Automatic Data Exchange

ADXPS Angular Dependent X-Ray Photoelectron Spectroscopy

ADXRD Angle-Dispersive X-Ray Diffraction

α-Dy Alpha Dysprosium [Symbol]

a-DyCo Dysprosium Cobalt Amorphous Alloy

a-DyFe Dysprosium Iron Amorphous Alloy

a-DyFeB Dysprosium Iron Boron Amorphous Alloy

a-DyGdNi Dysprosium Gadolinium Nickel Amorphous Alloy

a-DyNi Dysprosium-Nickel Amorphous Alloy

AE Above, or Equal
AE Accelerating Electrode
AE Accelerator Electrode
AE Accidental Error
AE Acid Equivalent [also ae]
AE Acoustic Emission
AE Acousto-Elastic(ity) [also A-E]
AE Acousto-Electric(al) [also A-E]
AE Acousto-Electron(ics) [also A-E]
AE Acrylic Emulsion
AE Activation Energy
AE Active Aerial
AE Adiabatic Efficiency
AE Adult Education
AE Advanced ECL (Emitter-Coupled Logic)
AE Aerial
AE Aeroelasticity
AE Aeroelectron(ics)
AE Aero-Engine [also: A-E]
AE Aeronautical Engineer(ing)
AE Aerosol Dispenser
AE Aerospace Engineer(ing)
AE After Effect
AE Age Equation
AE Agricultural Economist; Agricultural Economy
AE Agricultural Engineer(ing)
AE Airborne Electronics
AE Air-Equivalent [Nuclear Engineering]
AE Air Escape
AE Airways Engineer
AE Algebraic Equation
AE Alkaline Earth
AE All Electron
AE Alpha Emission
AE Aminoethyl
AE Ammunition Ship [US Navy Symbol]
AE Angioneurotic Edema
AE Anion Exchang(er)
AE Anisotropy Effect
AE Antienzyme
AE Apoenzyme
AE Application Engineer(ing)
AE Application/Expert
AE Architect/Engineer [also A/E]
AE Architectural Engineer(ing)
AE Arithmetic Element
AE Arithmetic Expression
AE Army Engineer
AE Arrhenius Equation [Physical Chemistry]
AE As-Extruded (Condition)
AE Associate in Education
AE Associate in Engineering
AE Asymmetric Epoxidation

AE Atmospheric Evaporation
AE Atmospheric Explorer
AE Atomic Emission
AE Atomic Environment
AE Attenuation Equalizer
AE Atomic Energy
AE Audio Engineer(ing)
AE Auger Effect [Physics]
AE Auger Electron [Physics]
AE Auger Emission [Physics]
AE Auroral Electron
AE Auto-Emission
AE Automatic Exposure
AE Automobile Engineer(ing); Automotive Engineer(ing)
AE Autumnal Equinox [Astronomy]
AE Auxiliary Electrode [Physical Chemistry]
AE Auxiliary Equation
AE Aviation Engineer(ing)
AE United Arab Emirates [ISO Code] [also ae]
A(E) Auger Line Shape [Symbol]
A&E Accident and Emergency
A&E Azimuth and Elevation
A/E Absorption/Emission Ratio
A/E Absorptivity/Emissivity Ratio [Solar Radiation]
A/E Architect/Engineer [also AE]
A-E Acousto-Elastic(ity) [also AE]
A-E Acousto-Electric(al) [also AE]
A-E Acousto-Electron(ics) [also AE]
A-E Aero-Engine [also AE]
Ae Aedes [Genus of Mosquitoes]
Ae Aerial [also ae]
Ae Aeronautics
A/e Atom-to-Electron (Ratio)
Ae$_{cm}$ Temperature at Which Transformation of Austenite to Austenite plus Cementite Occurs in Hypereutectoid Steels under Equilibrium Conditions [Symbol]
Ae$_1$ Temperature at Which Transformation of Ferrite plus Cementite to Austenite plus Cementite Occurs under Equilibrium Conditions in Hypereutectoid Steels; or Temperature at Which Transformation of Ferrite plus Cementite to Ferrite plus Austenite Occurs under Equilibrium Conditions in Hypoeutectoid Steels [Symbol]
Ae$_3$ Temperature at Which the Transformation of Austenite plus Ferrite to Austenite Begins in Iron or Steel under Equilibrium Conditions [Symbol]
Ae$_4$ Temperature at Which Transformation of Austenite to Delta Ferrite Occurs During Cooling of Iron or Steel at Equilibrium [Symbol]
ae (*aetatis*) – of age; aged
.ae United Arab Emirates [Country Code/Domain Name]
AEA Acoustic Emission Analysis; Acoustic Emission Analyzer
AEA Aerospace Education Association [US]
AEA Agricultural Education Association [UK]
AEA Agricultural Engineers Association [UK]

AEA Aircraft Electronics Association [US]

AEA Air-Entraining (or Entrainment) Agent

AEA All-Electric Aeroplane

AEA Alabama Education Association [US]

AEA Alpha Energy Analysis

AEA Aluminum Extruders Association [UK]

AEA American Economic Association [US]

AEA American Electrology Association [US]

AEA American Electronics Association [US]

AEA American Engineering Association [US]

AEA American Evaluation Association [US]

AEA Association for Environmental Archeology [UK]

AEA Association of Engineers and Architects [US]

AEA Association of Engineers and Associates

AEA Association of European Airlines [Belgium]

AEA Atomic Energy Act [US]

AEA Atomic Energy Authority [Harwell, Oxfordshire, UK]

AEA Attestation d'Etudes Approfondies [French Certificate of Advanced Studies]

AEA Automotive Electric Association

AEACA Acetyl-Epsilon-Aminocapronic Acid

AEACII Atomic Energy Advisory Committee on Industrial Information

AEAI Association Européenne des Assureurs Industriels [European Association of Industrial Insurers, Belgium]

AEAI Association of Engineers and Architects in Israel

AEAP Atmospheric Effects of Aviation Project [of NASA]

AEAPS Auger Electron Appearance Potential Spectroscopy [also AEPS]

AEAPT [N-(2-Aminoethyl) 3-Aminopropyl]trimethoxysilane

AEA-T AEA Technology

AEB American Egg Board [US]

AEB Analog Expansion Bus

AEB Area Electricity Board [UK]

AEB Atomic Energy Board [South Africa]

AEB Atomic Energy Bureau

AEB Auxiliary Equipment Building

AEBIG ASLIB (Association of Special Libraries) Economics and Business Information Group [US]

AEBSF 4-(2-Aminoethyl)benzenesulfonyl Fluoride

AEC Acousto-Electric Current

AEC Active Error Compensation

AEC Adult Education Center

AEC Agricultural Extension Center [Canada]

AEC Airship Experimental Center [US Navy]

AEC Alberta Environmental Center [Canada]

AEC Alkaline Earth Cuprate

AEC Aluminum Extruders Council [US]

AEC American Engineering Council [US]

AEC Aminoethyl Carbazole

AEC 3-Amino-9-Ethyl Carbazole

AEC Aminoethyl Cellulose

AEC Aminoethyl Cysteine

AEC Anelastic Creep

AEC Anisotropic Elastic Constant

AEC Applied Electrochemistry

AEC Architectural, Engineering and Construction; Architecture, Engineering and Construction

AEC Army Education Corps [US]

AEC Army Electronics Command [US]

AEC Army Engineer Center [US]

AEC Army Environmental Center [US]

AEC Associated European Capital Corporation [New York, US]

AEC Association Européenne de Ceramique [European Ceramics Association, France]

AEC Association of Electronic Cottagers [US]

AEC Atmospheric Emission Camera

AEC Atomic Energy Commission [later Energy Research and Development Administration; now US Department of Energy]

AEC Atomic Energy Commission [Egypt]

AEC Atomic Energy Commission [of United Nations]

AEC Atomic Energy Corporation (of South Africa) [Pretoria, South Africa]

AEC Atomic Energy Council [Taiwan]

AEC Australian Education Council

AEC Australian Environment Council [now Australian and New Zealand Environment and Conservation Council]

AEC Automatic Exposure Control

AECA Asociación Espanola de Control Automático [Spanish Association for Automatic Control]

AECA Atomic Energy Control Act [Canada]

AECB Association for Environmentally Conscious Building [UK]

AECB Association for the Export of Canadian Books

AECB Atomic Energy Control Board [Canada]

AECC Arkansas Electric Cooperative Corporation [US]

AECC Asociación Espanola de Control de Calidad [Spanish Society for Quality Control]

AECC Automobile Emissions Control by Catalysts [of Conseil Européen des Fédérations de l'Industrie Chimique, Belgium]

AéCF Aéro-Club de France [Flying Club of France]

AECG African Elephant Conservation Group

AECI African Explosives and Chemical Industries

AECI Association of Electrical Contractors, Ireland

AECIP Atlantic Energy Conservation Investment Program [Canada]

AECL Advanced Emitter-Coupled Logic

AECL Atomic Energy of Canada Limited

AECL-CANDU Atomic Energy of Canada Limited– Canada Deuterium Uranium (Reactor) [Canada]

AECL-CFFTP Atomic Energy of Canada Limited– Canadian Fusion Fuels Technology Project [Canada]

AECL-CRNL Atomic Energy of Canada Limited– Chalk River Nuclear Laboratory [Canada]

AECL-IRUS Atomic Energy of Canada Limited– Instrusion-Resistant Underground Structure [Chalk River Ontario, Canada]

AECL Rep Atomic Energy of Canada Limited Report [Chalk River, Ontario, Canada]

AECL-WNRE Atomic Energy of Canada Limited– Whiteshell Nuclear Research Establishment [Manitoba Canada]

AECM (International Symposium on) Acoustic Emission from Composite Materials

AECM Albert Einstein College of Medicine [New York City, US]

AECM Association of European Candle Manufacturers [France]

AECM Atomic Energy Commission Manual

AECMA Association Européenne des Constructeurs de Matériel Aerospatial [European Association of Aerospace Equipment Manufacturers, France]

AECPR Atomic Energy Commission Procurement Regulations

AECRC Atomic Energy of Canada Research Company [of Natural Resources Canada]

AECS Australia-Europe Container Service

AeCS Aero Club of Switzerland

AECSA Atomic Energy Corporation of South Africa [Pretoria]

AECT Association for Educational Communications and Technology [US]

AED Academy for Educational Development [US]

AED Advanced Electronic Design

AED Aerodynamic Equivalent Diameter

AED ALGOL Extended for Design

AED Alternative Energy Division [of Canada Center for Mineral and Energy Technology, Ottawa]

AED Associated Equipment Distributors [US]

AED Association of Electronic Distributors [US]

AED Association of Engineering Distributors [UK]

AED Atomic Emission Detection; Atomic Emission Detector

AED Auger Electron Diffraction [Physics]

AED Automated Engineering Design

AED Automatic Exposure Device

AED United Arab Emirates Dirham [Currency]

AEd Associate in Education

AEDANS N-Iodoacetyl-N'-(5-Sulfo-1-Naphthyl)ethylene-diamine

AEDC American Economic Development Council [US]

AEDC Arnold Engineering Development Center [of US Air Force at Tullahoma, Tennessee, US]

AEDC Atlantic Engineering Design Competition

AEDE Annual Effective Dose Equivalent

AEDF Asian Economic Development Fund

AEDOT Advanced Energy Design and Operation Technologies

AEDP Aboriginal Employment and Development Program [Australia]

AEDP Association for Educational Data Processing [US]

AEDS Association for Educational Data Systems [now International Association for Computing in Education, US]

AEDS Atomic Energy Detection System

AEDS J AEDS Journal [of Association for Educational Data Systems, US]

AEDS Monit AEDS Monitor [of Association for Educational Data Systems, US]

AEDU Admiralty Experimental Diving Unit [UK]

AEE Acousto-Electric Effect

AEE Airborne Evaluation Equipment

AEE Alliance for Environmental Education [US]

AEE Asociación Electrotecnica Espanola [Spanish Electrotechnical Association]

AEE Association for Environmental Education [Australia]

AEE Association of Electrical Engineers [Finland]

AEE Association of Energy Engineers [US]

AEE Associazione Elettrotecnica ed Elettronica [Association for Electrotechnics and Electronics, Italy]

AEE Atomic Energy Establishment [UK]

AEE Autoelectronic Emission

AeE Aeronautical Engineer

AEEC Airlines Electronic Engineering Committee [US]

AEEC Animal Experimentation Ethics Committee [Australia]

AEEC Association of European Express Carriers

AEED3A N-(Acetoxyethyl)ethylenediamine- N,N',N'-Triacetate [also aeed3a]

AEEE Army Equipment Engineering Establishment

AEEF Association Européenne des Exploitations Frigorifiques [European Association of Refrigeration Enterprises, Belgium]

AEEI Associazione Elettrotecnica ed Elettronica Italiana [Italian Association of Electrotechnics and Electronics]

AEEL Aeronautical Electronics and Electrical Laboratory [US Navy]

AEEMS Automatic Electric Energy Management System

AEEN Agence Européenne pour l'Energie Nucléaire [European Agency for Nuclear Energy]

AeEng Aeronautical Engineer

AEEP Association of Environmental Engineering Professors [US]

AEEP Association of European Engineering Periodicals

AEES Auger Electron Emission Spectroscopy

AEESCA Association for Engineering Education in South-Central Asia

AEESEA Association for Engineering Education in Southeast Asia

AEET Atomic Energy Establishment at Trombay [India]

AEF Acousto-Electric Field

AEF Aerospace Education Foundation [US]

AEF Afrique Equatoriale Française [French Equatorial Africa]

AEF Airfields Environment Federation [UK]

AEF Allied Expeditionary Force

AEF American Expeditionary Force(s)

AEF Analyzing Electric Field

AEF Aviation Engineer Force

AEFA Association of European Federations of Agro-Engineers

AEF Natl Newsl AEF National Newsletter [of Aerospace Education Foundation, US]

AEFPG Association of Engineering Firms Practicing in the Geosciences [of Association of Soil and Foundation Engineers, US]

AEG Acoustic Emission Group [Los Angeles, California, US]

AEG Active Element Group

AEG Allgemeine Elektrizitätsgesellschaft AG [Major German Manufacturer and Supplier of Electrical and Electronic Products]

AEG Association of Engineering Geologists [US]

AEG Association of Exploration Geochemists [Canada]

AEG Association of Exploration Geophysicists [India]

AEG Axial-Emission Gauge

AEGC Association Européenne de Génie Civil [European Association of Civil Engineers]

AEGE ASEAN (Association of Southeast Asian Nations) Experts Group on the Environment

AEGGF European Agricultural Guidance and Guarantee Fund [of European Community]

AEGIS Agricultural Ecological and Geographical Information System

AEGIS Assessment of Effectiveness of Geologic Isolation

AEGIS Australian Electronic Government Information Service

AEGIS Australian Environmental Geographic Information System [Defunct]

AEG Newsl AEG Newsletter [of Association of Engineering Geologists, US]

AEGPL Association Européenne des Gaz de Pétrole Liquéfiés [European Liquefied Petroleum Gas Association, France]

AEGRAFLEX Association Européenne des Graveurs et des Flexographes [European Association of Engravers and Flexographers, Germany]

AEGS Association of European Geological Studies

AEH Anhydroenneaheptitol [Biochemistry]

AEHA Allergy and Environmental Health Association [US]

AEHC Alberta Environmental Health Center [Canada]

AEI Acclimatization Experience Institute [now Institute for Earth Education, US]

AEI Acoustic Emission Inspection

AEI Aerial Exposure Index

AEI Alternate Energy Institute

AEI Alternate Enterprises Initiative Program [Canada]

AEI Armor Effectiveness Investigation

AEI Associated Electrical Industries [UK]

AEI Association of Electrical Industries [UK]

AEI Associazione Elettrotecnica Italiana [Italian Association for Electrotechnics]

AEI Authorized Engineering Information [of National Electrical Manufacturers Association, US]

AEI Automatic Error Interrogation

AEI Australian Education Index [of Australian Council for Educational Research]

AEI Average Efficiency Index

AEIA Asociación Espanola de Informatica y Automatica [Spanish Association of Informatics and Automation]

AEIC Association of Edison Illuminating Companies [US]

α-EIC Equal-Interface-Composition Contour in the Alpha Single Phase Region [Metallurgy]

AEICP Association of Entertainment Industry Computer Professionals [US]

AEIDA b-Aminoethyliminodiacetate [also aeida]

AEIS Aeronautical En-Route Information Service

AEIST Associação dos Estudantes do Instituto Superior Tecnico [Association of Students of the Higher Technical Institute, Lisbon, Portugal]

AEJI Association of European Jute Industries

AEJMC Association for Education in Journalism and Mass Communications [US]

AEL Acceptable Exposure Limit

AEL Acute Edematous Laryngitis

AEL Aeronautical Engineering Laboratory [US Naval Base, Phila, Pennsylvania]

AEL American Export Lines [US]

AEL Atomic Energy Level

AEL Audit Entry Language

AELS Atomic Energy Level Scheme

AEM Acoustic Emission Monitor

AEM Airborne Electromagnetics

AEM Air-Equivalent Material

AEM Alkaline Earth Metal

AEM Analytical Electron Microscope; Analytical Electron Microscopy

AEM Analyzer Energy Modulation

AEM Applications Explorer Mission [of NASA Earth Radiation Budget Satellite]

AEM Applied and Environmental Microbiology [Journal of the American Society for Microbiology]

AEM ASEAN (Association of Southeast Asian Nations) Economic Market

AEM ASEAN (Association of Southeast Asian Nations) Economic Ministers

AEM Assemblability Evaluation Method

AEM Association of Electronic Manufacturers [US]

AEM Auger Electron Microscope; Auger Electron Microscopy

AEM-2 Applications Explorer Mission, Second Mission (or Mission B) [of NASA Earth Radiation Budget Satellite] [also AEM-B]

AEMA Asphalt Emulsion Manufacturers Association [US]

AEMA Auger Electron Microanalysis

AEMAS N-2-3-4-(2,5-Dioxo-1-Pyrrolidinyl)oxyl-4 Oxobutylamino-3-Oxopropyldithioethyl-2-Propenamide

AEMB Alliance for Engineering in Medicine and Biology [Washington, DC, US]

AEM-B Applications Explorer Mission, Mission B (or Second Mission) [of NASA Earth Radiation Budget Satellite] [also AEM-2]

AEMBA Advanced Executive Master of Business Administration

AEMC American Engineering Model Society [US]

AEMC Atomic Energy and Minerals Center

AEMCC Air and Expedited Motor Carriers Conference [US]

AEME Association of Export Marketing Executives [UK]

AEM-EELS Analytical Electron Microscopy–Electron Energy Loss Spectroscopy

AEM-EDS Analytical Electron Microscopy–Energy Dispersive Spectroscopy

AEM/HREM Analytical Electron Microscope/High-Resolution Electron Microscope; Analytical Electron Microscopy/High-Resolution Electron Microscopy

AEMR Advanced Engineering Materials Research

AEMRP Advanced Engineering Materials Research Profile [of Synergistic Technologies Inc., Research Triangle Park, North Carolina, US]

AEMS Airborne Electromagnetic Survey

AEMS American Engineering Model Society [US]

AEMS Association of Electronic Manufacturers

AE/MS Acoustic Emission/Microseismic

AEMSA Army Electronics Material Support Agency

AEMSAT Association of European Manufacturers of Self-Adhesive Tapes [France]

AEMSM Association of European Metal Sink Manufacturers [Scotland]

AEMT Association of Electrical Machinery Trades [UK]

AEMTM Association of European Machine Tool Merchants [UK]

AEN Advanced Electronics Network [UK]

AEN Agence de l'Energie Nucléaire [Nuclear Energy Agency; of Organization for Economic Cooperation and Development, France]

AEN Attenuation Equivalent Nettiness

AENA American Ephemeris and Nautical Almanac [of US Naval Observatory]

AEng Associate in Engineering

AEO Acoustoelectric Oscillator

AEO Air Engineer Officer

AEO ALCOM Education Outreach [Newsletter of Advanced Liquid Crystalline Optical Materials Science and Technology Centers, National Science Foundation, US]

AEO Alkali(ne)-Earth Oxide

AEO Association of Exhibition Organizers [UK]

AEOC Advanced Epithelial Ovarian Cancer

AEOC Aminoethylhomocysteine [Biochemistry]

AEOG Air Ejection Off Gas

AENOR Asociación Española de Normalisación y Certificación [Spanish Standards and Certification Association]

AEOS Australian Environmental On-Line Service [Succeeded by Environment Australia On-Line Service]

AEP Advertising Exchange Program

AEP Agence Européenne de Productivité [European Agency for Productivity, Organization for Economic Cooperation and Development, France]

AEP Airports Economy Panel [of International Civil Aviation Organization]

AEP Alkaline Earth Plumbate

AEP American Electric Power

AEP 1-(2-Aminoethyl)piperazine

AEP Associação Eurocast-Portugal [Eurocast Association-Portugal]

AEP Austrian Energy Forum

AEP Averaged Evoked Potential

AEPCO Association of Economic Poison Control Officials [US]

AEPD Aminoethylpropanediol

AEPECS Auger Electron-Photoelectron Coincidence Spectroscopy

AEPG Army Electronic Proving Ground [US]

AEPI Atmospheric Emissions Photo Imaging

AEPP Azimuthal Equidistant Polar Projection

AEPPF Albert Einstein Peace Price Foundation

AEPS Advanced Electronic Publishing System

AEPS American Electric Power System

AEPS Auger Electron Appearance Potential Spectroscopy [also AEAPS]

AEPSC Atomic Energy Plant Safety Committee

AEPV Alkyl Ethers of Phenylvinyl

aeq *(aequales)* – equal [also aequal]

AER Advanced Electric Reactor

AER Aft Engine Room

AER Alpha Epsilon Rho [of University of South Carolina, Columbia, US]

AER American Export Register [US Trade Publication]

AER Anion Exchange Resin

AER Approach End Runway

AER Assembly of European Regions

AER Association for Education by Radio [US]

AER Atmospheric and Environmental Research

AER Average Evoked Response

*α***-Er** Alpha Erbium [Symbol]

AERA American Educational Research Association [US]

AERA Automated En-Route ATC (Air Traffic Control)

AERA Automotive Engine Rebuilders Association [US]

AERC Agricultural Economics Research Council [Canada]

AERC Applied Electrostatics Research Center [Canada]

AERCB Alberta Energy Resources Conservation Board [Canada]

AERCG African Environmental Research and Consulting Group [US]

AERD Agricultural Engineering Research and Development

AERD Atomic Energy Research Department [of United States Atomic Energy Commission]

AERDC Agricultural Extension and Rural Development Center [UK]

AERDL Army Electronics Research and Development Laboratory [US]

AERE Association of Environmental and Resource Economists [US]

AERE Atomic Energy Research Establishment [Harwell, Oxfordshire, UK]

AERE Atomic Energy Research Establishment [Pinawa, Manitoba, Canada]

AERE-R Atomic Energy Research Establishment Report [UK] [also AERE Rep, or AERE Rpt]

AERG Advanced Environmental Research Group

AERI Agricultural Economics Research Institute [UK]

AERI Atmospheric Emitted Radiance Interferometer

AERI Atomic Energy Research Institute [of Kinki University, Higashi-Osaka, Japan]

AERIC Applied Economic Research and Information Center [Canada]

AERIS Airways Environmental Radar Information System

AERL Atomic Energy Research Laboratory [of Musashi Institute of Technology, Kawasaki-shi, Japan]

AERNO Aeronautical Equipment Reference Number

AERO Air Education and Recreation Organization [UK]

AERO Alternate Energy Resources Organization [US]

AERO Association of Electronic Reserve Officers [US]

AERO Code Word for Aeronautical Weather Reports

Aero Aeronautic(s); Aeronautical [also aero]

AEROCE Atmosphere-Ocean Chemistry Experiment

Aerod Aerodynamic(s) [also Aerodyn, aerodyn, or aerod]

AeroE Aerospace Engineer [also Aero Eng, or AeroEng]

Aero Eng Aerospace Engineer(ing)

AEROF Aerological Officer

Aerojet Rev Aerojet Review [US]

Aerojet Technol Aerojet Technology [Journal published in the US]

AEROMAT Advanced Aerospace Materials and Processes Conference and Exposition [of ASM International, US] [also AeroMat]

Aeroméxico Aerolíneas México [Mexican Air Lines]

Aeron Aeronautic(s); Aeronautical [also Aeronaut, aeronaut, or aeron]

Aéronaut Astronaut Aéronautique et l'Astronautique [French Journal on Aeronautics and Astronautics]

Aeronaut J Aeronautical Journal [of Royal Aeronautical Society, UK]

Aeronaut Meridiana Aeronautica Meridiana [South African Publication]

Aeronaut Quart Aeronautical Quarterly [UK Journal]

Aeronaut Res Lab Rep Aeronautical Research Laboratories Report [Australia]

Aeronaut Satel News Aeronautical Satellite News [of International Maritime Satellite Organization, UK]

AEROS Aerometric and Emissions Reporting System [of US Environmental Protection Agency]

AEROS Aeronomy Satellites [Series of NASA-Launched German Satellites] [also Aeros]

AEROSAT Aeronautical Satellite System

AEROSAT Aeronautical Services Satellite [also Aerosat]

Aerosol Rep Aerosol Report [Germany]

Aerosol Sci Technol Aerosol Science and Technology [Journal of the American Association for Aerosol Research]

Aerosp Aerospace [also aerosp]

Aerosp Aerospace [Journal of the Royal Aeronautical Society, UK]

AEROSPACECOM Aerospace Communications

Aerosp Am Aerospace America [US Journal]

Aerosp Compos Mater Aerospace Composites and Materials [Journal published in the UK]

Aerosp Def Rev Aerospace and Defence Review [UK]

Aerosp Def Sci Aerospace and Defense Science [Journal published in Malaysia]

Aerosp Eng Aerospace Engineering [Journal of the Society of Automotive Engineers, US]

Aerosp Mater Aerospace Materials [Journal published in the UK] [also Aerosp Mat]

AERP Aquatic Effects Research Program [US]

AERS Atlantic Estuarine Research Society [US]

AERSG African Elephant and Rhino Specialists Group [Kenya]

AES Abrasive Engineering Society [US]

AES Aerodrome Emergency Service

AES Aeronautical Earth Station

AES Aerospace and Electronic Systems (Society) [of Institute of Electrical and Electronics Engineers, US]

AES Aerospace Electrical Society [US]

AES Agricultural Economics Society [UK]

AES Agricultural Employment Services [Canada]

AES Agricultural Experiment Station [of US Department of Agriculture]

AES Airways Engineering Society [US]

AES Alternate Energy Systems

AES Amateur Entomologists Society

AES American Electrochemical Society [US]

AES American Electronic Society [US]

AES American Electroplaters' Society [now American Electroplaters and Surface Finishers Society, US]

AES American Entomological Society [US]

AES American Ethnological Society [US]

AES American Eugenics Society [now Society for the Study of Social Biology, US]

AES Apollo Experiment Support [of NASA]

AES Apollo Extension System [of NASA]

AES Applied Electrostatics

AES Arctic Environmental Strategy [Canada]

AES Artificial Earth Satellite [of NASA]

AES Asian Environment(al) Society

AES Atlantic Economic Society

AES Atmospheric Environment Service [Canada]

AES Atmospheric Exchange System

AES Atomic Emission Spectrometry; Atomic Emission Spectroscopy

AES Audio Engineering Society [US]

AES Auger Electron Spectrometer; Auger Electron Spectrometry

AES Auger Electron Spectroscopy

AES Auger Electron Spectrum

AES Auger Emission Spectrometer; Auger Emission Spectrometry

AES Auger Emission Spectroscopy

AES Australian Entomological Society

AES Autoelectronic Selector

AES Automated Environmental Station

AES Automatic Emission Spectroscopy

AES Automatic Evacuation System

AES Automatic External Standardization

A&ES Architectural and Engineering Series

a-Es Alpha Einsteinium [Symbol]

AESA 2-Amino Ethanesulfonic Acid

AESA Association of Environmental Scientists and Administrators

AESA ATM (Asynchronous Transfer Mode) End System Address

AESAU Association of Eastern and Southern African Universities

AESC Aerospace and Electronic Systems Society [US]

AESC American Engineering Standards Committee [US]

AESC Automatic Electronic Switching Center

AESD Advanced Energy Systems Division [of Westinghouse Electric Corporation, US]

AES/DFO Atmospheric Environment Service/Department of Fisheries and Oceans [Canada]

AESE Airborne Electrical Support Equipment

AESE Association of Earth Science Editors [US]

AES-E Auger Electron Spectroscopy with Electron Excitation

AESF American Electroplaters and Surface Finishers Society [US]

AESF Association for Electroplating and Surface Finishing

AESF/EPA American Electroplaters and Surface Finishers Society/Environmental Protection Agency (Joint Conference)

AESF SUR/FIN American Electroplaters and Surface Finishers Society Annual Conference and Exhibit(ion) [now Conference and Exhibition for the Surface Finishing Industry]

AESFS American Electroplaters and Surface Finishers Society [US]

AESG (International) Asian Elephant Specialist Group

AESG Axial-Emission Suppression Gauge

AES-I Auger Electron Spectroscopy with Ion Excitation

AESIS Australian Earth Sciences Information System [of Australian Information Network]

AESIS-AC Australian Earth Sciences Information System Advisory Committee

AESJ Atomic Energy Society of Japan

AES-LEED Auger Electron Spectroscopy–Low-Energy Electron Diffraction

AESMC Automotive Exhaust Systems Manufacturers Council [US]

AESMG Axial-Emission Self-Modulating Gauge

AESOP An Evolutionary System for On-Line Processing

AESOP An Experimental Structure for On-Line Planning

AESOP Artificial Earth Satellite Observation Program

AESOP Association for Energy Systems, Operations and Programming

AESOP Australian Environmental Statistics Project

AESOP Automated Environmental Station for Overtemperature Probing

AESOPP An Estimator of Physical Properties (Databank) [of Japanese Union of Scientists and Engineers]

AESP Association of Energy Services Professionals [US]

AES-P Auger Electron Spectroscopy with Photon Excitation

AESR Acoustic Electron Spin Resonance

AESS Aerospace and Electronic Systems Society [of Institute of Electrical and Electronics Engineers, US]

AESS Appraisal, Evaluation and Sectoral Studies

AES/SAM Auger Electron Spectroscopy/Scanning Auger Microscopy

AESTE Association for Exchange of Students for Technical Experience [UK]

AES/WMO Atmospheric Environment Service/World Meteorological Organization

AET Acoustic Emission Technology

AET Acoustic Emission Test(ing)

AET Action in Education Team [of ASM International, US]

AET Advanced Engine Technology

AET Alternative Energy Technology

AET Aminoethylisothiourea Dihydrobromide

AET 2-Aminoethylisothiouronium Bromide Hydrobromide

AET Associate in Electrical Technology

AET Atomic Environment Type

AET Automobile Engineering Technology

aet *(aetatis)* – of age; at the age of; aged

AETA American Embryo Transfer Association [US]

aetat *(aetatis)* – of age; at the age of; aged

AETB Alumina-Enhanced Thermal Barrier

AETE Aerospace Engineering Test Establishment

AETM Analytical Electron Transmission Microscopy

AETR Advanced Engineering Test Reactor [US]

AETC Advanced Epithermal Thorium Reactor [US]

AETS Association for the Education of Teachers in Science [US]

AETT Association of Educational and Training Technology [UK]

AETU All Ethiopian Trade Union

AEU Amalgamated Engineering Union [UK]

AEU Asia Electronics Union [Japan]

AEU (Japan) AEU (Japan) [Publication of the Asia Electronics Union, Japan]

AEUSA Amalgamated Engineering Union of South Africa

AEV Acousto-Electric Voltage

AEV Aerothermodynamic Elastic Vehicle

AEV Avian Encephalomyelitis Virus

AEW Airborne Early Warning

AEWB Army Electronic Warfare Board [US Army]

AEWC Airborne Early Warning and Control [also AEW&C]

AEWES Army Engineer Waterways Experiment Station [at Vicksburg, Mississippi, US]

AEWIS Army Electronic Warfare Information System [US Army]

AEWTU Airborne Early Warning Training Unit
AF Abrasive Finishing
AF Académie Française [French Academy; Division of the Institut de France, Paris]
AF Accumulation Factor [Mathematics]
AF Acid-Fast (Stain)
AF Acoustic Filter
AF Active Filter
AF Adaptive Filter
AF Address Field
AF Adsorption Fluorometry
AF Advantage Factor [Nuclear Reactor]
AF Aerofoil
AF Afghanistan [ISO Code]
AF After Firing
AF Agglomerate-Free (Powder)
AF Agriservices Foundation [US]
AF Air Field
AF Air Filter
AF Air Flow
AF Air Force
AF Air France
AF Air Furnace [Metallurgy]
AF Aldehyde Fuchsin
AF Algebraic Function
AF Alkaline-Free
AF Allied Forces
AF Alternating Fields
AF Alumina Fiber
AF Aluminum Foil
AF 2-Aminofluorene
AF Amniotic Fluid [Medicine]
AF Amplification Factor [Electronics]
AF Amplitude Factor
AF Anadromous Fish
AF Analog Filter
AF Angle Frame
AF Anglo-French [also A-Fr]
AF Angular Frequency
AF Anharmonic Force
AF Anisotropy Field [Solid-State Physics]
AF Annealing Furnace
AF Anodic Film
AF Antarafacial (Process)
AF Antiferroelectric(ity)
AF Antiferromagnet(ic); Antiferromagnetism
AF Antifoam(ing)
AF Antifouling
AF Antifriction
AF Antilles Françaises [French Antilles]
AF Antireflection Film
AF Apparatus Function
AF Application Function
AF Approximation Function

AF Aqua Fortis (or Nitric Acid)
AF Aramid Fiber
AF Arc Furnace [Metallurgy]
AF Arithmetic Function
AF Armed Forces
AF Armored Force
AF Asbestos Fiber
AF Aseptically Filled (Solution)
AF As-Fabricated (Condition)
AF As-Fired (Condition)
AF Ash Fusibility [Geology]
AF Atomic Fluorescence
AF Atomic Forum
AF Atomic Fraction
AF Atrial Fibrillation [Medicine]
AF Audio Frequency [also af]
AF Autofining
AF Auto-Focus; Automatic Focus
AF Automatic Following
AF Auxiliary Feedwater
AF Auxiliary (Carry) Flag [Computers]
AF Availability Factor
AF Axial Flow
AF_0 Antiferromagnetic Phase 0 [Solid-State Physics]
AF_1 Antiferromagnetic Phase 1 (or Transverse Spin Density Wave Phase) [Solid-State Physics]
AF_2 Antiferromagnetic Phase 2 (or Longitudinal Spin Density Wave Phase) [Solid-State Physics]
A&F August and February [Business Finance]
A/F Across Flat
A/F Air-to-Fuel (Ratio)
A/F Austenite/Ferrite (Boundaries) [Metallurgy]
Af Afghani [Currency]
Af Africa(n)
Af 2-Aminophenol
Af Tropical Rainforest Climate, No Dry Season [Classification Symbol]
aF abfarad [Unit]
af (anni futuri) – of next year
a-F Alpha Fluorine [Symbol]
AFA Absorption-Filtration-Adsorption Process
AFA Accelerated File Access
AFA Access Flooring Association [UK]
AFA Afghani [Currency]
AFA Air Force Academy [at Colorado Springs, Colorado, US]
AFA Air Force Association [US]
AFA Air Freight Association (of America) [US]
AFA American Fan Association [US]
AFA American Flock Association [US]
AFA American Forestry Association [US]
AFA American Foundrymen's Association [US]
AFA Army Flight Activity
AFA Asociación Física Argentina [Argentinian Association of Physics]

AFA Association Française d'Astronomie [French Astronomy Association]

AFA Average Field Approximation

AFAA American Fire Alarm Association [US]

AFAA ASEAN (Association of Southeast Asian Nations) Federation of Automotive Associations

AFAC Air Force Armament Center [Eglin, Florida, US]

AF/ACTS Air Force Advanced Computed Tomography System [US]

AFADS Advanced Forward Air Defense System

AFAFC Air Force Accounting and Finance Center [US]

AFAIK As Far As I Know

AFAL Air Force Armstrong Laboratory [at Brooks Air Force Base, Texas, US]

AFAL Air Force Avionics Laboratory [at Wright Patterson Air Force Base, Ohio, US]

AFAMM ASEAN (Association of Southeast Asian Nations) Federation of Agricultural Machinery Manufacturers

AFAN Ammonium Phosphate and Ammonium Nitrates

AFAP Atlantic Fisheries Adjustment Program [of Department of Fisheries and Oceans, Canada]

AFAPL Air Force Aero-Propulsion Laboratory [US]

AFAR Advanced Field Array Radar

AFARB Association of Friends of the Achievement of Richard Bellman [France]

AFAS Abteilung für Angewandte Systemanalyse [Applied Systems Analysis Division, Kernforschungsanlage Karlsruhe, Germany]

AFAS Association Française pour l'Avancement des Sciences [French Association for the Advancement of Sciences]

AFASE Association for Applied Solar Energy [US]

AFA-SEF Air Force Association–Space Education and Foundation [US]

AFAIUI As Far As I Understand It

AFaxA American Facsimile Association [also AFAXA, US]

AFB Acoustic Feedback

AFB Air Force Base

AFB American Foulbrood [Bee Disease]

AFB American Foundation for the Blind [US]

AFB (*Amtliches Fernsprechbuch*) – German for "Official Telephone Directory"

AFB Antifriction Bearing

AFB Atmospheric Fluidized Bed

AFB Automatic Fallback [Facsimile]

AFBA Armed Forces Broadcasters Association [US]

AFBA Asociación Farmaceutica y Bioquímica Argentina [Argentinian Association for Pharmaceutics and Biochemistry]

AFBC Advanced Fluidized Bed Combustion

AFBC Atmospheric Fluidized-Bed Combustion

AFBD Association of Futures Brokers and Dealers

AFBDC Air Force Base Development Center [Eglin, Florida, US]

AFBF American Farm Bureau Federation [US]

AFBMA Antifriction Bearing Manufacturers Association [US]

AFBMA-ANSI Antifriction Bearing Manufacturers Association/American National Standards Institute (Standard) [US]

AFBMD Air Force Ballistic Missile Division [Inglewood, California, US]

AFBRF American Farm Bureau Research Foundation [US]

AFBS Acoustic Feedback System

AFC Acupuncture Foundation of Canada

AFC African Forestry Commission

AFC Air Flow Controller

AFC Air Force Cross

AFC Air/Fuel Concentration

AFC Airman, First Class [also Afc]

AFC Airplane Flight Control

AFC Alkaline Fuel Cell

AFC Alumni Fund Committee

AFC American Forest Council [US]

AFC (Latin-)American Forestry Commission [of Food and Agricultural Organization]

AFC Amplitude Frequency Characteristic

AFC Aquatic Federation of Canada

AFC Area Forecast Center

AFC Area Frequency Coordinator

AFC Arkansas Forestry Commission [US]

AFC Atomic Fluid Cells

AFC Australian Fisheries Council

AFC Australian Forestry Council

AFC Automatic Flatness Control [Metallurgy]

AFC Automatic Focus Control

AFC Automatic Font Change

AFC Automatic Fraction Collector

AFC Automatic Frequency Control [also afc]

AFC Axial-Flow Compressor

AFCA Alberta Floor Covering Association [Canada]

AFCA Anadromous Fish Conservation Act [US]

AFCA Armed Forces Chemical Association [US]

AFCAC African Civil Aviation Commission [Senegal]

AFCAL Association Française de Calcul [French Computing Association]

AFCALTI Association Française de Calcul et de Traitement de l'Information [French Association for Computing and Information Processing]

AFCAM Association of Fluorocarbon Consumers and Manufacturers [Australia]

AFCAPS Association of Flexible PVC (Polyvinyl Chloride) Suppliers [UK]

AFCAS Automatic Flight Control and Augmentation System

AFCASI Associate Fellow of the Canadian Aeronautics and Space Institute

AFCC Available Frame Capacity Count

AFCCE Association of Federal Communications Consulting Engineers [US]

AFCE Automatic Flight Control Equipment

AFCEA Armed Forces Communications and Electronics Association [US]

AFCEE Air Force Centers for Environmental Excellence [Various Centers throughout the US, e.g., at Massachusetts Military Reserve, Brooks Air Force Base, etc.]

AFCENT Allied Forces Central Europe [of NATO]

AFCET Association Française pour la Cybernetique Economique et Technique [French Association for Cybernetics and Technology, France]

AFCET Interf AFCET Interfaces [French Journal on Cybernetics and Related Topics; published by Association Française pour la Cybernetique Economique et Technique]

AFCFP Arab Federation of Chemical Fertilizer Producers [Kuwait]

AFCG American Fine China Guild [US]

AFCI Association of Faculty Clubs International

AFCIA Armed Forces Civilian Instructors Association [US]

AFCMA Aluminum Foil Container Manufacturers Association [UK]

AFCMA Australian Fiberboard Container Manufacturers Association

AFCO Aluminum Foil Conference

AFCO Australian Federation of Consumer Organisations

AFCO (1,3-Butadiene)tricarbonyl Iron

AFCOMMSTA Air Force Communications Station [US]

AFCR Air Force Contractor Report [US]

AFCRC Air Force Cambridge Research Center [Bedford, Massachusetts, US]

AFCRL Air Force Cambridge Research Laboratories [of Office for Aerospace Research, US]

AFCS Adaptive Flight Control System

AFCS Air Force Communications Service [at Scott Air Force Base, Illinois, US]

AFCS Air Force Communications System [US]

AFCS Automatic Flight Control System

AFCS Avionic Flight Control System

AFCSS Air Force Communications Support System [US]

AFCT Air Force Construction Troops

AFCU Association of Federal Computer Users [US]

AFCV Average Formal Copper Valence

AFD Accelerated Freeze Drying

AFD Aft Flight Deck

AFD Amplitude-Frequency Distortion

AFD Automatic File Distribution

AFD Automatic Forging Design

AFD Axial Flux Density

AFD Axial Flux Difference

AFDA Axial Flux Difference Alarm

AFDAC Association Française pour la Documentation Automatique en Chimie [French Association for Automatic Documentation in Chemistry]

AFDASTA Air Force Data Station [US]

AFDB African Development Bank [also AfDB]

AFDBS Association Française des Documentalistes et des Bibliothécaires Spécialisés [French Association of Documentalists and Special Librarians] [also ADBS, France]

AFDC Aid to Families with Dependent Children [US]

AFDC Air Force Defense Command [of US Air Force]

AFDDA Air Force Director of Data Automation [US]

AFDEC Association of Franchised Distributors of Electronic Components [UK]

AFDIN African Development Information Network

AFD Labs Anglo-French Drug Company Limited

AFDO Association of Food and Drug Officials [US]

AFDOUS Association of Food and Drug Officials of the United States

AFDP Australian Fast Delivery Processor

AFDP Automatic Flight Data Processing

AFDREW Air Force Director of Reconnaissance and Electronic Warfare [US]

AFDW Active Framework for Data Warehousing

AFE Aeroassist Flight Experiment

AFE American Flight Echocardiograph

AFE Analog Front End

AFE Antiferroelectric(s)

AFE$_{In}$ Incommensurate Antiferroelectric

AFE$_O$ Orthorhombic Antiferroelectric

a-Fe Amorphous Iron

α-Fe Alpha Iron [Symbol]

AFEA American Farm Economic Association [US]

AFEA Aviation Facilities Energy Association [US]

AFEAS Alternative Fluorocarbons Environmental Acceptability Study [of US Department of Energy]

AFEB Armed Forces Epidemiological Board

a-FeB Iron Boron Amorphous Alloy

a-FeBSi Iron Boron Silicon Amorphous Alloy

a-FeBSiC Iron Boron Silicon Carbon Amorphous Alloy

a-FeC Iron Carbon Amorphous Alloy

AFECI Association des Fabricants Européens de Chauffe-Bains et Chauffe-Eau Instantanes et de Chaudières Murales au Gaz [Association of European Manufacturers of Instantaneous Gas Water Heaters and Wall-Hung Boilers, Belgium]

AFECOGAZ Association des Fabricants Européens d'Appareils de Controle [European Control Equipment Manufacturers Association, Belgium]

a-FeCo Alpha Iron Cobalt

a-FeCoSiB Iron Cobalt Silicon Boron Amorphous Alloy

a-FeCrB Iron Chromium Boron Amorphous Alloy

a-FeCrCo Iron Chromium Cobalt Amorphous Alloy

a-FeCuNdSiB Iron Copper Neodymium Silicon Boron Amorphous Alloy

AFEDEF Association des Fabricants Européens d'Equipement Ferroviaires [Association of European Railway Equipment Manufacturers, France]

AFEDES Association Française pour l'Etude et le Développement des Applications de l'Energie Solaire [French Association for Study and Development of Solar Energy Applications]

AFEE Association Française pour l'Etude des Eaux [French Association for the Study of Waters]

AFEEC ASEAN (Association of Southeast Asian Nations) Federation of Electrical Engineering Contractors

AFE-FE Antiferroelectric-Ferroelectric (Phase Boundary)

a-FeHf Iron Hafnium Amorphous Alloy

AFEI Americans for Energy Independence [US]

AFEI Arab Federation of Engineering Industries

AFEI Association for Enterprise Integration [US]

AFEJ Asian Forum of Environmental Journalists

AFELIS Air Force Engineering and Logistics Information System [US]

AF-ELS Alkaline-Free Electroless Solution

a-FeNiB Iron Nickel Boron Amorphous Alloy

AFEO ASEAN (Association of Southeast Asian Nations) Federation of Engineering Organizations [Indonesia]

α-Fe$_2$O$_3$ Alpha Iron(III) Oxide (or Hematite)

α-FeO(OH) Alpha Iron Oxide Hydroxide (or Goethite) [also α-FeOOH]

γ-FeO(OH) Gamma Iron Oxide Hydroxide (or Lepidocrocite) [also γ-FeOOH]

AFERA Association des Fabricants Européens des Rubans Autoadhesifs [European Self-Adhesive Tape Manufacturers Association]

AFES Association Française pour l'Etude du Sol [French Association for Soil Research]

α-FeSi Alpha Iron Silicon

α-Fe(Si) Silicon-Doped Alpha Iron

a-FeSc Iron Scandium Amorphous Alloy

AFESD Arab Fund for Economic and Social Development

a-FeSi Iron Silicon Amorphous Alloy

a-FeSiB Iron Silicon Boron Amorphous Alloy

AFETAC Air Force Environmental Tactical Applications Center [US]

AFETR Air Force Eastern Test Range [now Eastern Space and Missile Center, US]

a-FeZr Amorphous Iron-Zirconium (Compound)

AFF Aberration Free Focus

AFF Above Finished Floor

AFF Aerial Firefighter

AFF American Farm Foundation [US]

AFF Army Field Forces [US]

AFF Autofocus Function

AFF Automatic Fast Feed

AFF Automatic Frequency Follower

AFF Axial-Flow Fan

AF-F Antiferromagnetic-to-Ferromagnetic (Transition)

Aff Affairs [also aff]

Aff Affirmative [also aff]

AF FCC Antiferromagnetic Face-Centered Cubic (Lattice)

AFFDL Air Force Flight Dynamics Laboratory [US]

AFFF Aqueous Film-Forming Foam

AFFI Arab Federation of Food Industries [Iraq]

Affil Affiliation [also affil]

AFFIRM Association for Federal Information Resources Management [US]

AFFM Australian Financial Futures Market

AF-FM Antiferromagnetic-Ferromagnetic (State) [also AF/FM]

AFFMA ASEAN (Association of Southeast Asian Nations) Federation of Furniture Manufacturers Associations

AFFP Air-Filed Flight Plan

AFFPI ASEAN (Association of Southeast Asian Nations) Federation of Food Processing Industries

Afft Affidavit [also afft]

AFFTC Air Force Flight Test Center [Edwards Air Force Base, California, US]

AFG Afghani [Currency]

AFG Analog Function Generator

AFG Arbitrary Function Generator

Afg Afghan(istan)

AFGC American Forage and Grassland Council [US]

AFGE American Federation of Government Employees [US]

aFGF Acidic Fibroblast Growth Factor [Biochemistry]

AFGL Air Force Geophysical Laboratory [US]

AFGM American Federation of Grain Millers [US]

AFGP Antifreeze Glycoprotein

AFGWC Air Force Global Weather Center (or Central) [US]

AFHB ASEAN (Association of Southeast Asian Nations) Food Handling Bureau

AFHF Air Force Historical Foundation [US]

AFHF American Food Health Foundation [US]

AFHP Advanced Flash Hydropyrolysis

AFHRL Air Force Human Resources Laboratory [US]

AFHRL Australian Fish Health Reference Laboratory

AFI Address Format Identifier

AFI Airways Flight Inspector

AFI American Film Institute [US]

AFI American Forest Institute [now American Forest Council, US]

AFI Armed Forces Institute [of US Department of Defense]

AFI Assistant Flying Instructor

AFI Association for Futures Investment

AFI Association of Food Industries [US]

AFI Authority and Format Identifier

AFI Automatic Fault Isolation

AF-I Simple Antiferromagnetic (Phase)

AFIA American Feed Industry Association [US]

AFIA American Foreign Insurance Association [US]

AFIA Association Française Interprofessionelle des Agrumes [French Interprofessional Association for Tropical Fruit]

AFIC Australian Fishing Industry Council

AFICCS Air Force Interim Command and Control System [US]

AFICE Association for International Cotton Emblem [Belgium]

AFICTIC Association Française des Ingénieurs, Chemistes et Techniciens des Industries du Cuir [French Association of Engineers, Chemists and Technicians in the Leather Industry]

AFID Alkali Flame Ionization Detector

AFID American Federation of Information and Data Processing [now American Federation of Information Processing Societies]

AFIDA Agricultural Foreign Investment Disclosure Act [US]

AFIDA Asociación de Ferias Internacionales de America [Association of International Trade Fairs of America, Peru]

AFII Association for Font Information Interchange [US]

AFIL Air Filed (Flight Plan)

AFIP Armed Forces Institute of Pathology [Washington, DC, US]

AFIPS American Federation of Information Processing Societies [US]

AFIRM Automated Fingerprint Image Reporting and Match

AFIRMS Air Force Integrated Readiness Measurement System

AFIRO Association Française d'Informatique et de Recherche Operationnelle [French Association of Informatics and Operations Research]

AFIS American Forces Information Service [US]

AFIS Automated Fingerprint Identification System

AFIS Automatic Flight Inspection System

AFIT Air Force Institute of Technology [at Wright Patterson Air Force Base, Dayton, Ohio, US]

AFIT American Fabricating Institute of Technology [US]

AFITI Association Française des Ingénieurs et Techniciens de l'Aeronautique [French Association of Aeronautical Engineers and Technicians]

AFJA Association Française des Journalistes Agricoles [French Association of Agricultural Journalists]

AFJWG Australia/France Joint Working Group (on the Environment)

AFK Phosphoric Acid Monoalkyl Ester

AFL Above Field Level

AFL Abstract Family of Languages

AFL Aeroflot (Airlines) [Russia]

AFL Alberta Federation of Labour [Canada]

AFL American Federation of Labor [also AF of L]

AFL Association Française Laitière [French Dairy Association]

AFL Automatic Fault Location

AFL Axial-Flow Laser

Afl Anabaena flos-aquae [Microbiology]

Afl Netherlands Antillean Florin [Currency]

AFLA Asian Federation of Library Associations

AFLA Automotive Fleet and Leasing Association [US]

AFLC Air Force Logistics Command [at Wright-Patterson Air Force Base, Dayton, Ohio, US]

AFL-CIO American Federation of Labor and Congress of Industrial Organizations [US]

AFLCM Air Force Logistics Command Manual [US]

AFLCON Air Force Logistics Command Operations Network [US]

AFLM Accredited Farm and Land Manager

aflt afloat

AFM Abrasive Flow Machining

AFM Academy of Feline Medicine [US]

AFM Acoustic Flowmeter

AFM Adobe Font Metrics (File)

AFM A (Al_2O_3), F (FeO), M (MgO) (Diagram) [Petrology]

AFM Air Force Manual [US]

AFM Air Force Medal [US]

AFM Air Force Museum [at Wright-Patterson Air Force Base, Dayton, Ohio, US]

AFM Air Freight Motor Carriers Conference [now Air and Expedited Motor Carriers Conference, US]

AFM Analysis and Forecasting Mode

AFM Antiferromagnet(ic); Antiferromagnetics; Antiferromagnetism

AFM Antifriction Metal

AFM Applied Fracture Mechanics

AFM Association of Facilities Managers

AFM Atmospheric Field Measurement (Facility) [of Argonne National Laboratory, Illinois, US]

AFM Atomic Force Microscope; Atomic Force Microscopy

AFM Automatic Flight Management

.AFM Adobe Font Metrics [File Name Extension]

AFMA American Feed Manufacturers Association [US]

AFMA American Fiber Manufacturers Association [US]

AFMA American Furniture Manufacturers Association [US]

AFMA ASEAN (Association of Southeast Asian Nations) Federation of Mining Associations

AFMA Association of Food Marketing Agencies (in Asia and the Pacific) [Thailand]

AFMA Australian Fisheries Management Authority

AFMAG Audio-Frequency Magnetic; Audio-Frequency Magnetics

AFMC Aeronautical Frequency Management Committee [UK]

AFMC Air Force Materiel Command [of US Air Force at Wright-Patterson Air Force Base, Ohio]

AFMC Armed Forces Marketing Council [US]

AFMC Asian Fluid Mechanics Committee [Japan]

AFMC Automotive Filter Manufacturers Council [US]

AFMDC Air Force Machinability Data Center [US]

AFMDC Air Force Missile Development Center [US]

AFMED Allied Forces Mediterranean

AFMFIC Associated Factory Mutual Fire Insurance Companies [US]

AFM-FM Antiferromagnetic-Ferromagnetic (State) [also AFM/FM]

AFMI Association of Food Marketing Institutions [now Association of Food Marketing Agencies (in Asia and the Pacific), Thailand]

AFML Air Force Materials Laboratory [of WrightPatterson Air Force Base, Dayton, Ohio, US]

AFML TR Air Force Materials Laboratory Technical Report [also AFML-TR, or AFML Tech Rep]

AFMM Association of Fish Meal Manufacturers [UK]

AFMMC Aluminum-Fiber Metal-Matrix Composite

AFM/MFM Atomic-Force Microscope/Magnetic Force Microscope; Atomic-Force Microscopy/Magnetic-Force Microscopy

AFMMP Advanced Field Monitoring Methods Program [of US Environmental Protection Agency]

AFMR Antiferromagnetic Resonance

AFMS Advanced Flight Management System

AFMS American Federation of Mineralogical Societies [US]

AFM/STM Atomic Force Microscope/Scanning Tunneling Microscope

AFMTC Air Force Missile Test Center [Patrick Air Force Base, Cocoa, Florida, US]

AFN Alaska Federation of Natives [US]

AFN All-Figure Numbering

AFN American Forces Network

AFN Australian Fiducial Network

AFNC American Food and Nutrition Center [Cutten, California, US]

AFNE Allied Forces Northern Europe [of NATO]

AFNETF Air Force Nuclear Engineering Test Facility

AFNETSTA Air Force Networks Station [US]

AF NN Antiferromagnetic Nearest Neighbor (Exchange)

AFNOR Association Française de Normalisation [French Association for Standards]

AFNORTH Allied Forces Northern Europe [of NATO]

AFO Accounting and Finance Office

AFO Advanced File Organization

AFO Airport Fire Officer

AFO Association of Field Ornithologists [US]

AFO Axial Flux Offset

AFOCEL Association Foret-Cellulose [Forest/Cellulose Association, France]

AF of L American Federation of Labor [also AFL]

AFOMJ Association of Fat and Oil Manufacturers of Japan

AFORA Advanced Forest Automation

AFOS Agriculture, Forestry and Other-Human Activities Subgroup [of Intergovernmental Panel on Climate Change/Response Strategies Working Group]

AFOS Automated Forcasting Operational System

AFOS Automation of Field Operations and Services

AFOSR Air Force Office of Scientific Research [at Bolling Air Force Base, Washington, DC, US]

AFOSR Ann Rep Air Force Office of Scientific Research Annual Report [US] [also AFOSR Annu Rep]

AFOSR/AOARD Air Force Office of Scientific Research/ Asian Office of Aerospace Research and Development [Tokyo, Japan]

AFOSR TR Air Force Office of Scientific Research Technical Report [also AFOSR-TR, or AFOSR Tech Rep]

AFOSR URI Air Force Office of Scientific Research– University of Rhode Island (Program) [also AFOSR-URI]

AFOTC Air Force Operational Test Center [US]

AFP Adiabatic Fast Passage (Technique)

AFP Advanced Function Presentation

AFP Agence France Presse [French News / Press Agency]

AFP Air Force Program

AFP Alpha-Fetoprotein (or α-Fetoprotein) [Biochemistry]

AFP Alternative Future Project [Norway]

AFP Appletalk Filing Protocol [Apple-Macintosh Computers]

AFP Armed Force Police [US]

AFP Association for Finishing Processes [of the Society of Manufacturing Engineers, US]

AFP Association of Family Physicians [US]

AFP Attached FORTRAN Processor

AFP Australian Federal Police

AFP Automatic Flow Process

AFPA Alabama Forest Products Association [US]

AFPA American Forest and Paper Association [also AF&PA]

AFPA Australian Fire Protection Association

AFPA Automatic Flow Process Analysis

AFPAM Automatic Flight Planning and Monitoring

AFPAV Airfield Pavement [also AF Pav]

AFPC Armed Forces Policy Council [US]

AFPC Automatic Frequency and Phase Control

AF/PC Automatic Frequency/Phase Controlled (Loop)

AFPD American Federation of Physicians and Dentists

AFPD Australian Federal Police Database

AFPI American Forest Products Industries, Inc. [US]

AFPI Association Française des Professionels de l'Internet [Association of French Internet Professionals]

AFPL Air Force Packaging Laboratory [US]

AFPL Air Force Phillips Laboratory [US]

AFPMCW Arab Federation of Petroleum, Mines and Chemical Workers

AFPRO Air Force Plant Representative Office [US]

AFPS Armed Forces Policy Council [of US Department of Defense]

AFP/SME Association for Finishing Processes of the Society of Manufacturing Engineers [US]

AFPT Auxiliary Feed Pump Turbine

AFPT Axial-Flow Propeller Turbine

AFPTRC Air Force Personnel and Training Research Center [San Antonio, US]

AfPU African Postal Union

AFQT Armed Forces Qualification Test [US]

AFR Admiralty Fuel Research [UK]

AFR Advanced Fault Recognition

AFR After Flame Ratio [Aerospace]

AFR Air Force Regulation

AFR Air Force Reserve [US]

AFR Air Force Route

AFR Alternating Frequency Rejection

AFR Amplitude Frequency Response

AFR Antiferrorotational

AFR Application Function Routine

AFR Audio-Frequency Range

AFR Automatic Field/Format Recognition

AFR Awaiting Forward Release

AFR Away from Reactor [Spent Fuel Storage]

AFr Anglo-French [also AF]

Afr Africa(n)

AFRA American Farm Research Association [US]

AFRAA African Airlines Association [Kenya]

AFRAeS Associate Fellow of the Royal Aeronautical Society
AFRAL Association Française de Reglage Automatique
[French Association for Automatic Regulation]

AFRASEC Afro-Asian Organization for Economic
Cooperation [Egypt]

AFRATC African Air Traffic Conference

AFRC Agricultural and Food Research Council [UK]

AFRC Air Force Records Center [San Antonio, Texas, US]

AFRC American Food Research Council [US]

AFRC Armed Forces Reserve Center

AFRCE Air Force Regional Civil Engineer [US]

AFRes Air Force Reserves [US]

AFRICA Environ Wildl AFRICA: Environment and Wildlife
[Journal]

AFRL Air Force Research Laboratory [US]

AFROTC Air Force Reserve Officers' Training Corps [US]

AFRP Aramid Fiber Reinforced Plastic

AFRPL Air Force Rocket Propulsion Laboratory [US]

AFRRI Armed Forces Radiobiology Research Institute [US]

AFRS Armed Forces Radio Service

Afrs Affairs [also afrs]

AFRSI Advanced Flexible Reusable Surface Insulation

Afr Stud African Studies [International Journal]

AFRTC Air Force Reserve Training Center [US]

AFRTS Armed Forces Radio and Television Service [US]

AFS Acoustic Frequency Standards

AFS Aerodrome Fire Service

AFS Aeronautical Fixed (Telecommunication) Service

AFS Air Force Specialty

AFS Air Force Station

AFS Air Force Supply

AFS American Fisheries Society [US]

AFS American Foundrymen's Society [US]

AFS Analog Flight Simulator [at NASA Langley Research
Center, Hampton, Virginia, US]

AFS Andrews File System

AFS Army Field Services [Fort Monroe, Virginia, US]

AFS Army Fire Service

AFS Asian Fisheries Societies

AFS Atomic Fluorescence Spectrometry

AFS Atomic Fluorescence Spectroscopy

AFS Atomic Force Spectroscopy

AFS Audio-Frequency Shift

AFS Australian Fishing Service

AFS Australian Forestry School [Canberra, Australian
Capital Territory]

AFS Auxiliary Fire Service

AFSA American Fire Sprinkler Association [US]

AFSA Application for Federal Student Aid [US]

AFSAB Air Force Science Advisory Board [US]

AFSARI Automation for Storage and Retrieval of
Information

AFSAT African Satellite

AFSATCOM Air Force Satellite Communications

AFSAU Association of Faculties of Science at African
Universities

AFSB American Federation of Small Business [US]

AFSC Air Force Space Command [US]

AFSC Air Force Systems Command [at Andrews Air Force
Base, Maryland, US]

AFSC Alaska Fisheries Science Center [of National Oceanic
and Atmospheric Administration, US]

AFSC Armed Forces Staff College [Norfolk, Virginia, US]

AFSC Australian Food Standards Committee

AFSC Automatic Focus Control

AFSCA Amalgamated Flying Saucer Clubs of America [US]

AFSCF Air Force Satellite Control Facility [US]

AFSCM Air Force Systems Command Manual [US]

AFSCME American Federation of State, County and
Municipal Employees [US]

AFSCN Air Force Satellite Control Network [US]

AFSE Allied Forces Southern Europe [of NATO]

AFSHP Association of Federal Safety and Health
Professionals [US]

AFS Int Cast Met J American Foundrymen's Society
International Cast Metals Journal

AFSK Audio Frequency-Shift Keying

AFSK Auto Frequency-Shift Keying

AF/SLR Auto-Focus/Single Lens Reflex

AFSM Air Force Systems Command [of US Air Force]

AFSM Association of Field Service Managers

AFSM Augmented Finite-State Machine

AFSMAS Association Francophone de Spectrométrie de
Masse de Solides [French-Speaking Association of Solid
Mass Spectrometry, France]

AFSMI Association of Field Service Managers, International
AFSOUTH Allied Forces Southern Europe [of NATO]

AFSR Argonne Fast Source Reactor [of Argonne National
Laboratory, Illinois, US]

AFSS Air Force Security Service [US]

AFSSRN Asian Fisheries Social Science Research Network

AFST Auxiliary Feedwater Storage Tank

AFS Trans American Foundrymen's Society Transactions
[US]

AFSWC Air Force Special Weapons Center [Kirkland Air
Force Base, New Mexico, US]

AFSWP Armed Forces Special Weapons Project [Changed to
DASA]

AFT Adaptive Ferroelectric Transformer

AFT Adiabatic Flame Temperature [Physical Chemistry]

AFT American Federation of Teachers [US]

AFT Analog Facility Terminal

AFT Audio-Frequency Transformer [also aft]

AFT Automated Funds Transfer

AFT Automatic Fine Tuning

AFT Automatic Focus Tracer

AFT Axial-Flow Turbine

Aft Afternoon [also aft]

A/ft² Ampere per square foot [Unit]

AFTA Aft Frame Tilt Actuator

AFTA Australian Fishing Tackle Association

AFTAC Air Force Technical Applications Center [US]

AFTAC American Fiber, Textile, Apparel Coalition [US]

AFTE American Federation of Technical Engineers [US]

AFTE Arab Federation for Technical Education [Iraq]

AFTE Association of Firearm and Tool Mark Examiners [US]

AFTE Autogenic Feedback Training Experiment

AFTEC Air Force Test and Evaluation Center [US]

AFTER Air Force Thermionic Engineering and Research

AFTEX ASEAN (Association of Southeast Asian) Federation of Textile Industries

AFTI Advanced Fighter Technology Integration [US Air Force]

AFTN Aeronautical Fixed Telecommunications Network

Aftn Afternoon [also aftn]

AFTO Association of Flight Training Organizations [UK]

AFTP Anonymous File Transfer Protocol

AFTP Association Française des Techniciens du Pétrole [French Association of Petroleum Technicians]

AFTRA American Federation of Television and Radio Artists [US]

AFTRCC Aerospace Flight Test Radio Coordinating Council

AFTS Aeronautical Fixed Telecommunications Service [UK]

AFU Archives for UFO (Unidentified Flying Object) Research [Sweden]

AFUA ARMS Inc. (Computer) Firms Users Association [US]

AFUD American Foundation for Urologic Disease [Baltimore, Maryland, US]

AFUH Association Française pour l'Urbanisme et l'Habitation [French Association for Town Planning and Habitation]

AFULE Australian Federated Union of Locomotive Engineers

AFUS Armed Forces of the United States (of America)

AFV Armored Fighting Vehicle

AFVA Armored Fighting Vehicle Association [US]

AFVMC Association of Faculties of Veterinary Medicine in Canada

AFVPA Advertising Film and Video Producers Association [UK]

AFWA Australian Federation for the Welfare of Animals

AFWAL Air Force Wright Aeronautical Laboratories [at Wright-Patterson Air Force Base, Dayton, Ohio, US]

AFWAL TR Air Force Wright Aeronautical Laboratories Technical Report [of Wright-Patterson Air Force Base, Dayton, Ohio, US] [also AFWAL-TR, or AFWAL Tech Rep]

AFWETS Air Force's Weapon Effectiveness Testing System [US]

AFWL Air Force Weapons Laboratory [US]

AFWL-MD Air Force Weapons Laboratory–Materials Directorate [US]

AFWL TR Air Force Weapons Laboratory Technical Report [US] [also AFWL-TR, or AFWL Tech Rep]

AFWTR Air Force Western Test Range [US]

AFZ Australian Fishing Zone

AFZA Australian Fishing Zone Authority

AFZIS Australian Fisheries Zone Information Service

AG Abrikosov-Gorkov (Theory) [Physics]

AG Acoustic Grating

AG Acute Glomerulonephritis

AG Adjutant General

AG Adrenal Gland

AG Advanced General-Purpose

AG Afterglow

AG Agent General

AG Air Gunner

AG Air-to-Ground [also A/G, a/g, A-G, or a-g]

AG Aktiengesellschaft [Business Form of Organization in German-Speaking Countries; Equivalent to Corporations in North America, and Public Limited Companies in the UK]

AG Algebraic Geometry

AG All-Glass

AG Alumina Gel

AG Aluminum Gallate

AG American Gage

AG Amplification Gas

AG Amplitude Grating

AG Anhydroglucose [Biochemistry]

AG Antenna Gain

AG Antigen [Immunology]

AG Antigua and Barbuda [ISO Code]

AG Application Generator

AG *(Arbeitsgruppe)* – German for "Task Group" or "Working Group"

AG Armor Grating

AG Arresting Gear

AG As-Grown (Film)

AG Assault Gun

AG Astrogeodetic; Astrogeodesy

AG Astrogeological; Astrogeologist; Astrogeology

AG Astronomische Gesellschaft [Astronomical Society, Germany]

AG Attitude Gyro

AG Attorney General

AG Autogenous Groove (Weld)

AG Autogiro

AG Autoleather Guild [US]

AG-4 Antigen of Herpes Virus [Immunology]

A/G Air-to-Ground [also AG, A-G, a-g, or a/g]

A/G Antigas

A-G Air-Glass (Interface)

A-G Air-to-Ground [also AG, A/G, a/g, or a-g]

Ag Agarose [Biochemistry]

Ag *(Agentum)* – Silver

Ag Antigen [Biochemistry]

Ag August

A(g) (Metallic) Element A in Gas Phase [Symbol]

Ag⁺ Silver Ion [Symbol]

Ag246 Morpholino-Ethyl-2 Methyl-4-Phenyl-6-Pyridazone-3-Chlorohydrate

Ag-104 Silver-104 [also ^{104}Ag or Ag^{104}]

Ag-105 Silver-105 [also ^{105}Ag or Ag^{105}]

Ag-106 Silver-106 [also ^{106}Ag or Ag^{106}]

Ag-107 Silver-107 [also ^{107}Ag or Ag^{107}]

Ag-108 Silver-108 [also ^{108}Ag or Ag^{108}]

Ag-109 Silver-109 [also ^{109}Ag or Ag^{109}]

Ag-110 Silver-110 [also ^{110}Ag or Ag^{110}]

Ag-111 Silver-111 [also ^{111}Ag or Ag^{111}]

Ag-112 Silver-112 [also ^{112}Ag or Ag^{112}]

Ag-113 Silver-114 [also ^{113}Ag or Ag^{113}]

Ag-114 Silver-114 [also ^{114}Ag or Ag^{114}]

Ag-115 Silver-115 [also ^{115}Ag or Ag^{115}]

ag attogram [1 ag = 10^{-18} g] [Unit]

.ag Antigua and Barbuda [Country Code/Domain Name]

AGA Aberdeen Granite Association [UK]

AGA Abrasive Grain Association [US]

AGA Advanced Graphics Adapter

AGA Alternating Gradient Accelerator

AGA American Galvanizers Association [US]

AGA American Gas Association [US]

AGA American Gastroenterological Association [US]

AGA American Genetic Association [US]

AGA American Gynecological Association [US]

AGA Asociación General de Agricultores [General Association of Agriculturalists, Guatemala]

AGA Associated Geographers of America [US]

AGA Australian Gas Association

AgA Silver Azide

α-Ga Alpha Gallium [Symbol]

AGAA Attitude Gyro Accelerometer Assembly

AGAAA Australian Government Acronyms and Abbreviations [Publication of the Department of Employment, Education and Training]

a-GaAs Amorphous Gallium Arsenide

Ag-ABA Silver-Based Active Brazing Alloy

AgAc Silver Acetate

Ag(ACAC) Silver Acetylacetonate [also Ag(AcAc), or Ag(acac)]

AGACS Automatic Ground-to-Air Communications Systems

AGA Gas Energy Rev AGA Gas Energy Review [of American Gas Association]

AGAL Australian Government Analytical Laboratories

AgAl Silver Aluminum (Spin Glass)

Ag-Al Silver-Aluminum (Alloy System)

Ag-Al₂O₃ Silver-Aluminum Oxide (System)

AGA Mon AGA Monthly [of American Gas Association]

AGAMP Automatic Gain Adjusting Amplifier

a-GaN Amorphous Gallium Nitride

AGANI Apollo Guidance and Navigation Information

AGARD Advisory Group for Aerospace Research and Development [of NATO]

AGARD (NATO) Advisory Group for Aerospace Research and Development (NATO) [Publication of NATO]

AGARD Rep AGARD Reports [of NATO Advisory Group for Aerospace Research and Development]

Ag-As Silver-Arsenic (Alloy System)

AGASP Arctic Gas and Aerosol Sampling Program

Ag-Au Silver-Gold (Alloy System)

AGB Advanced Geometry Blade

AGB American Glycerin Bomb

AGB Association of Governing Boards (of Universities and Colleges) [US]

AGB Asymptotic Giant Branch

AGB Austenite Grain Boundary [Metallurgy]

AGB Automatic Grid Bias [Electronics]

AGBM Association of Grey Board Makers [UK]

AgBr Silver Bromide

AGC Abort Guidance Computer

AGC Adjutant General's Corps [of US Army]

AGC Advanced Graphics Control

AGC Aerojet General Corporation [US]

AGC Agriculture Canada

AGC Agricultural Genetics Company

AGC Air Group Commander

AGC Alaska Game Commission [US]

AGC Apollo Guidance Computer [of NASA]

AGC Arkansas Geological Commission [US]

AGC Associated General Contractors (of America) [Washington, DC, US]

AGC Atlantic Geoscience Center [Nova Scotia, Canada]

AGC Automatic Ga(u)ge Control

AGC Automatic Gain Control [also agc]

AGC Automatic Gain Correction

AGC Automatic Generation Control

AGCA Association of General Contractors of America [US]

AGCA Automatic Ground-Controlled Approach

AGCAS Association for Graduate Careers Advisory Services [UK]

AgCan Agriculture Canada

AGCC Air-Ground Communications Channel

AGCC American Guernsey Cattle Club [US]

AGCC Arkansas Geological and Conservation Commission [US]

AGCD Association of Green Crop Driers [UK]

AgCd Silver Cadmium (Spin Glass)

Ag-Cd Silver-Cadmium (Alloy System)

Ag/CdO Cadmium Oxide Modified Silver

Ag-CdO Silver-Cadmium Oxide (System)

AGCL Automatic Ground-Controlled Landing

AgCl Silver Chloride

AgCl-Al₂O₃ Silver Chloride-Aluminum Oxide (System)

AGCM Atmospheric General Circulation Model

AgCN Silver Cyanide

Ag/Co Silver/Cobalt (Multilayer)

AGCR Advanced Gas-Cooled Reactor

AGCRC Australian Geodynamics Cooperative Research Center [of Commonwealth Scientific and Industrial Research Organization]

AGCRSP Army Gas-Cooled Reactor Systems Program [US]

AGCS Air-to-Ground Communications System

Ag-Cu Silver-Copper (Alloy System)

Agcy Agency [also agcy]

AGD Adjutant General's Department [of US Army]

AGD Advanced Graduate Certificate

AGD *Arbeitsgruppe für Graphische Datenverarbeitung* – Working Group on Graphic Data Processing, Germany]

AGD Australian Geodetic Datum [Mapping]

AGD Axial Gear Differential

AGD66 Australian Geodetic Datum 1966 [Mapping]

AGD84 Australian Geodetic Datum 1984 [Mapping]

a-**Gd** Alpha Gadolinium [Symbol]

AGDB Australian Geographic Data Base [of AUSLIG– Australian Surveying and Land Information Group]

AGDC Alaska Geospatial Data Clearinghouse [of US Geological Survey]

a-GdCo Gadolinium Cobalt Amorphous Alloy

a-GdCoB Gadolinium Cobalt Boron Amorphous Alloy

AGDEX Agricultural Index [UK]

a-GdFe Gadolinium Iron Amorphous Alloy

AGDIC Astro Guidance Digital Computer

a-GdNi Gadolinium-Nickel Amorphous Alloy

AGDS Airport Graphics Display System

AGDS American Gage Design Standard

AGDT Aging Date

a-GdTbCo Gadolinium Terbium Cobalt Amorphous Alloy

AGE Admiralty Gunnery Establishment [UK]

AGE Aerospace Ground Equipment

AGE Allyl Glycidyl Ether

AGE Applied General Equilibrium (Model)

AGE Asian Center for Geotechnical Engineering [Taiwan]

AGE Associate in General Education

AGE Association of German Engineers [also Verein Deutscher Ingenieure, Germany]

AGE Automatic Ground Equipment

AGE Automatic Guiding Electronics

AgE Agricultural Engineer [also AgEng]

a-Ge Amorphous Germanium

Age Ageing Age and Ageing [Journal published by Oxford University Press, UK]

AGECON American Bibliography of Agricultural Economics (Database) [US]

AGED Advisory Group on Electronic Devices [US]

a-Ge:H Hydrogenated Amorphous Germanium

AgEng Agricultural Engineer [also AgE]

a-Ge$_3$N$_4$:H Hydrogenated Amorphous Germanium Nitride

a-GeN:H Hydrogenated Amorphous Germanium Nitrogen (Alloy)

AGEP Advisory Group on Electronic Parts [US]

Ag:Er Erbium-Doped Silver [also Ag:Er^{3+}]

a-GeSe Amorphous Germanium Selenide

AGET Advisory Group on Electron Tubes [US]

a-GeTe Amorphous Germanium Telluride

AGF Arbeitsgemeinschaft der Großforschungsanlagen [Association of Large Research Centers, Bonn, Germany]

AGF Arbeitsgemeinschaft Getreideforschung [Wheat Research Association, Germany]

AGF Army Ground Forces

AgF Silver Fluoride

AGFC Alabama Game and Fish Commission [US]

AGFD Agricultural and Food Chemistry Division [of American Chemical Society, US]

Ag/Fe Silver-Iron (Thin Film)

AG-FIZ Arbeitsgemeinschaft der Fachinformationszentren [Association of Technical Information Centers, Germany]

AGFM Alternating Gradient Field Magnetometer

AGFM Alternating Gradient Force Magnetometer

Ag(fod) (Heptafluorodimethyloctanedionato)silver

AGFUND Arab Gulf Program for United Nations Development Organisations

Agg Aggregate [also agg]

Agg Aggregation [also agg]

Ag-Ga Silver-Gallium (Alloy System)

AgGaS$_2$ Silver Gallium Disulfide (Semiconductor)

AgGaSe$_2$ Silver Gallium Diselenide (Semiconductor)

AgGaS$_2$-OPO Silver Gallium Disulfide–Optic Parametric Oscillator

Agglut Agglutinating; Agglutination [also agglut]

Aggr Aggregate [also aggr]

AGH Akademia Gorniczo-Hutnicza [Academy of Mining and Metallurgy, Krakow, Poland]

AGHCF Alpha Gamma Hot Cell Facility [of Argonne National Laboratory, near Chicago, US] [1962 US Atomic Energy Commission Project]

Ag-Hg Silver-Mercury (Alloy System)

AGHS Active Gas Handling System [of Joint European Torus]

AGHS Australian Garden History Society

AGI Alliance Graphique Internationale [International Graphics Alliance, Switzerland]

AGI American Geological Institute [US]

AGI Association for Geographical Information

AGI Associazione Genetica Italiana [Italian Genetics Association]

AgI Silver Iodide

AGIC Australian Goldmining Industry Council

AGID Association of Geoscientists for International Development [Thailand]

AG-IDL Arbeitsgemeinschaft Information und Dokumentation–Literaturversorgung [Association for Information and Documentation–Literature Supply, Germany]

AGIF Adipogenesis Inhibitory Factor [Biochemistry]

AGIFORS Airlines Group of the International Federation of Operational Research Societies

AGIL Airborne General Illumination Light

AGILE Autonetics General Information Learning Equipment

Ag-In Silver-Indium (Alloy System)

AGIS Attorney General's Information Service [Australia]

Agitn Agitation [also agitn]

AGK Arbeitsgemeinschaft Korrosion [Association for Corrosion, Germany]

AGK1 Erster Katalog der Astronomischen Gesellschaft [First Catalog of the Astronomical Society, Germany]

AGK2 Zweiter Katalog der Astronomischen Gesellschaft [Second Catalog of the Astronomical Society, Germany]

AGL Above Ground Level

AGL Airborne Gun-Laying (Radar) [also agl]

AGL Allied Geophysical Laboratories [at University of Houston, Texas, US]

AGL Argon Glow Lamp

AGL Association of German Librarians [Germany]

AGL Atomic Gas Laser

AGLAE Accélérateur Grand Louvre d'Analyse Elémentaire [Louvre Tandem Particle Accelerator, France]

A-Glass High-Alkali Fiberglass [also A-glass]

Ag-Li Silver-Lithium (Alloy System)

AGLINE Agricultural On-Line (Database) [US]

AGLINET Agricultural Libraries Information Network [US]

AGLR Airborne Gun-Laying Radar [also aglr]

AGLT Airborne Gun-Laying Turret [also aglt]

AGM Air-to-Ground Missile

AGM Alternating Gradient Magnetometer

AGM Annual General Meeting

AGM Auxiliary General Missile

AGMA American Gear Manufacturers Association [US]

AGMAP Agricultural Market Assistance Program

AgMg Silver Magnesium (Spin Glass)

Ag-Mg Silver-Magnesium (Alloy System)

Ag-Mn Silver-Manganese (Alloy System)

AgMn Silver Manganide (Spin Glass)

AGMR Advanced Gas-Cooled Maritime Reactor

AGN Active Galactic Nuclei [Astronomy]

AGN Acute Glomerulonephritis

AGN Aerojet General Nucleonics (Reactor) [US]

AGN Again [Amateur Radio]

AGN Alberta Government Network [Canada]

AGN Average Group Number (Parameter)

Ag-NaCl Silver-Sodium Chloride (System)

Ag/NaCl Silver on Sodium Chloride (Substrate)

AGNCS Aluminum Gooseneck Clamp Strap

AGNES Algorithm for Generating Structural Surrogates of English Text

Ag(Ni) Nickel-Doped Silver

Ag-Ni Silver-Nickel (Alloy System)

Ag-NiO Silver-Nickel Oxide (System)

AGNIS Apollo Guidance and Navigation Industrial Support [US Army]

AGNIS Azimuth Guidance Nose-in-Stand

AGNPS Agricultural Nonpoint Source Pollution Model

AGNS Allied General Nuclear Services

AGO Alkali-Graphitic-Oxide

AGO Aluminum Gallium Oxide

AgOCN Silver Cyanate

AGOR Auxiliary General Oceanographic Research

Agora Inform Chang World Agora, Informatics in a Changing World [Publication of the Intergovernmental Bureau for Informatics, Italy]

AGOS Air-Ground Operations Section

AGOSS Automated Ground Operations Scheduling System

AGP α_1-Acid Glycoprotein [Biochemistry]

AGP Accelerated Graphics Port

AGP Advanced Graphics Port

AGP Antisymmetrized Geminal Power

AGP Applied Geophysics

AGP Arbeitsgemeinschaft Plasmaphysik [Plasma Physics Association, Germany]

AGP Association of Glucose Producers

AGPAT Agricultural Patents [also AG/PAT]

AGPB Association Générale des Producteurs de Blé et Autres Céréales [General Association of Producers of Wheat and Other Cereals, France]

Ag(Pb) Lead-Doped Silver

Ag-Pb Silver-Lead (Alloy System)

Ag-Pd Silver-Palladium (Alloy System)

AGPL Association Générale des Producteurs de Lin [General Association of Flax Producers, France]

AGPLAN Arbeitsgemeinschaft für Planungsrechnung [Budgetary Accounting Association, Germany]

AGPM Associated Glass and Pottery Manufacturers [US]

AGPM Association Générale des Producteurs de Mais [General Association of Maize Producers, France]

AGPO Association Générale des Producteurs d'Oléagineux [General Association of Producers of Oleaginous Products, France]

AGPO Australian Government Publication Office

AGPS Australian Government Publishing Service [Canberra, Australian Capital Territory]

Ag-Pt Silver-Platinum (Alloy System)

AGQ Association of Geologists of Quebec [Canada]

AgQRE Silver Quasi-Reversible Electrode

AGR Advanced Gas-Cooled Reactor

AGR Annual Growth Rate

AGR Association of Graduate Recruiters [UK]

AGR Australian Gold Refineries

Agr Agricultural; Agriculture [also agr]

a-Gr Amorphous Graphite

AGRA Army Group Royal Artillery [UK]

AGRAS Antiglare-Antireflective-Antistatic

AGRE Army Group Royal Engineer [UK]

AGRE Atlantic Gas Research Exchange

AGREE Advisory Group on Reliability of Electronic Equipment [US]

AGREP Agricultural Research Projects (Database) [of Commission of the European Communities]

Ag-Rep Agricultural Representative [also ag-rep]

Agri Agriculture [also Agri, or agri]

AGRI C Agriculture Canada

Agric Agricultural; Agriculture [also agric]

Agric Biol Chem Agricultural Biology and Chemistry [International Journal]

Agric Eng Agricultural Engineering [of American Society of Agricultural Engineers, US]

Agric Eng Mag Agricultural Engineering Magazine [of American Society of Agricultural Engineers, US]

Agric Food Chem Agriculture and Food Chemistry [Journal]

AGRICOLA Agricultural On-Line Access (Database) [also Agricola] [of US Department of Agriculture]

Agricola Gesellschaft beim Deutschen Museum zur Förderung der Geschichte und der Technik e.V. [Association for the Advancement of History and Technology–German Federal Museum]

AGRINET Agricultural Network [Canada]

AGRINET Inter-American System for Agricultural Information

AGRINTER Inter-American Information System for the Agricultural Sciences

AGRIS Agricultural Information System [of Food and Agricultural Organization]

AGRIS International Information System for Agricultural Sciences and Technology [of Food and Agricultural Organization]

Agritechnica Internationale Fachausstellung für Agrartechnik [International Exhibition for Agricultural Machines and Tractors, Frankfurt/Main, Germany]

agrl agricultural

AGRO Division of Agrochemicals [of American Chemical Society]

Agron Agronomical; Agronomist; Agronomy [also agron]

AGRR American Goat Record and Registry [US]

AGRU Acid Gas Removal Unit

Ag-Ru Silver-Ruthenium (Alloy System)

AGS Abort Guidance System

AGS Adreno-Genital Syndrome

AGS Advanced Graphics System

AGS Airborne Radar Gunsight [also ags]

AGS Aircraft General Standards

AGS Air Gunnery School [UK]

AGS Alaska Geological Society [US]

AGS Alternating Gradient Synchrotron [of Brookhaven National Laboratory, Upton, New York, US]

AGS American Geographical Society [US]

AGS American Geriatrics Society [US]

AGS American Goat Society [US]

AGS Annulus Gas System

AGS Appalachian Geolocial Society [US]

AGS Arizona Geological Survey [US]

AGS Associate in General Studies

AGS Association of Graduate Schools [of Association of American Universities, US]

AGS Austin Geological Society [Texas, US]

AGS Automated Graphic System

AGS Automatic Gain Stabilization

AGS Silver Gallium Sulfide

Ag(S) Sulfur-Doped Silver

Ag₂S Silver Sulfide (Ceramics)

Ag-S Silver-Sulfur (Alloy System)

AGSAN Astronomical Guidance System for Air Navigation

Ag-Sb Silver-Antimony (Alloy System)

AGSC Advanced Graduate Specialist Certificate

AgSCN Silver Thiocyanate

Ag₂Se Silver Selenide (Semiconductor)

AGSG Alliance of Genetic Support Groups [US]

Ag/Si Silver-on-Silicon

Ag-Si Silver-Silicon (Alloy System)

Ag-Si-Al Silver-Silicon-Aluminum (Coating)

AgSn Silver Stannide (Spin Glass)

Ag(Sn) Tin-Doped Silver

Ag-Sn Silver-Tin (Alloy System)

AGSO Australian Geological Survey Organization

AGSP Atlas General Survey Program

AGSRO Association of Government Supervisors and Radio Officers [UK]

AGSS Attitude Ground Support System

AGT Abnormality of Glucose Tolerance

AGT Advanced Gas Turbine

AGT Advanced Ground Transport

AGT Alberta Government Telephones [Canada]

AGT Algorithmic Graph Theory

AGT Association of Geology Teachers [US]

AGT Audiographic Tele-Conference

AGT Automotive Gas Turbine

AGT Aviation Gas Turbine

Agt Agent [also agt]

Agt Agreement [also agt]

AGTA Airport Ground Transportation Association [US]

AGTD Allison Gas Turbine Division [of General Motors in Indianapolis, US]

AgTCNQ Silver Tetracyanoquinodimethane [also Ag(TCNQ)ₓ]

AGTERA Advanced Gas Turbine for Engineering Research Association [Japan]

AGU Address-Generation Unit

AGU American Geophysical Union [Washington, DC, US]

AGU Anhydroglucose Unit [Biochemistry]

AGU Arabian Gulf University

AGU Arbeitsgemeinschaft für Umweltfragen e.V. [Association for Environmental Issues, Germany]

AGU J American Geophysical Union Journal

AGV Aniline Gentian Violet

AGV Automated Guided Vehicle; Automatic Guided Vehicle

AGV Avion a Grande Vitesse [French Hypersonic Transport Aircraft]

AGVS Automatic (or Automated) Guided Vehicle System

AGVS Automatic Guided Vehicle Systems [Division of the Materials Handling Institute, US]

AGW Actual Gross Weight

AGW Advanced Graphic Workstation

AGW Allowable Gross Weight

AGW Arbeitsgemeinschaft der technisch-wissenschaftlichen Vereine des Werkstoffwesens [Federation of German Materials Associations]

Ag-W Silver-Tungsten (Alloy System)

AGWSE Association of Ground Water Scientists and Engineers [of National Ground Water Association, US]

AGWT American Ground Water Trust [Dublin, Ohio, US]

Ag/YBCO Silver/Yttrium Barium Copper Oxide (Thin Film)

AGZ Actual Ground Zero

AgZn Silver Zinc (Spin Glass)

Ag-Zn Silver-Zinc (Alloy System)

Ag-ZnO Silver-Zinc Oxide (System)

Ag-ZrAg Zirconium-Silver Reinforced Silver

AH Abrasion Hardness

AH Absolute Humidity

AH Acceptor Handshake

AH Acoustic Hologram; Acoustic Holography

AH Advanced Helicopter

AH Afterheat

AH Age-Hardened; Age Hardening [Metallurgy]

AH Air Hammer

AH Air Hardened; Air Hardening [Metallurgy]

AH Air Heater; Air Heating

AH Airy-Heiskanen (System) [Geodesy]

AH Aldohexose [Biochemistry]

AH Alkali(ne) Halide

AH Allylic Hydrogen

AH Alpha Helix [Molecular Biology]

AH Ambegaokar-Halperin (Phase Slip Theory) [Physics]

AH Ambient Humidity

AH American Hybrid

AH Ampere-Hour [also Ah, or ah]

AH Amplifier Head

AH Analog Hybrid

AH Anhydrous Hydrogen

AH *(Anno Hegira)* – in the year of the Hegira

AH Antenna Housing

AH Antihemophilic (Factor) [Biochemistry]

AH Antihistamine [Biochemistry]

AH Antihydrogen

AH Aqueous Humor (of Eye)

AH Arc Heating [Metallurgy]

AH Army Helicopter

AH Artificial Heart

AH Artificial Horizon

AH Aspiration Hazard

AH Atmospheric Humidity

AH Atomic Heat

AH Atomic Hydrogen

AH Available Head [Water Turbines]

AH Available Hours

AH Axon Hillock [Anatomy]

AH Hospital Ship [US Navy Symbol]

β-D-AH β-D-Aldohexose [Biochemistry]

AH₃ Alumina Trihydrate (or Aluminum Hydroxide) [$Al_2O_3 \cdot 3H_2O$] [Ceramics]

γ-AH₃ Gamma Aluminum Hydroxide [$Al_2O_3 \cdot 3H_2O$] [Ceramics]

A/H Air over Hydraulic

Ah Ampere-hour [also ah, A-h, or AH]

aH abhenry [Unit]

α-H Alpha Hydrogen [also H^{α}]

Å/h Angstrom(s) per hour [also Å h^{-1}, Å/hr, or Å hr^{-1}]

AHA Age-Hardenable Alloy(s)

AHA Airco Hydrogen-Atmosphere (System)

AHA Alpha Hydroxy Acid

AHA American Hardboard Association [US]

AHA American Heart Association [US]

AHA American Hereford Association [US]

AHA American Historical Society [Washington, DC, US]

AHA American Hospital Association [US]

AHA Autoimmune Hemolytic Anemia

AHAM Association of Home Appliance Manufacturers [US]

AHAM American Home Appliance Manufacturers (Association) [US]

AHAMS Advanced Heavy Assault Missile System

AHASCES Association of the Higher Academic Staff of Colleges of Education in Scotland

AHB Anhydrous Hydrogen Bromide

AHB Austrian Historical Bibliography (Database) [of Institut für Maschinelle Dokumentation, Austria]

AHBA 3-Amino-4-Hydroxybenzoic Acid

AHC Advanced Health Center [US]

AHC Alberta Housing Corporation [Canada]

AHC American Horticultural Society [US]

AHC Aminohexyl Cellulose

AHC Anhydrous Hydrogen Chloride

AHC Animal Health Committee

AHC Appalachian Hardwood Club [now Appalachian Hardwood Manufacturers, Inc, US]

AHC Arkansas History Commission [US]

AHC Aromatic Hydrocarbon

AHC Association for the History of Computing [UK]

AHC Australian Heritage Commission

AHC Australian Horticultural Corporation

AHC Award Hunters Club [Radio]

AHC Act Australian Heritage Commision Act

AHCAS Ad Hoc Committee for American Silver [US]

AHCAT Advanced Hot Cell Analytical Project [of US Department of Energy]

AHCC American-Hellenic Chamber of Commerce [Greece]

AHCM Academy of Hazard Control Management [US]

AHCPR Agency for Health Care Policy and Research [of US Department of Health and Human Services]

AHCS Advanced Hybrid Computer System

AHCT Advanced High-Speed Complementary Technology

AHCT Ascending Horizon Crossing Time [Aeronautics]

AHCTL N-Acetylhomocysteinethiolactone [Biochemistry]

AHD Adiabatic Heat Drop

AHD Alaska-Hawaii Daylight (Time) [Greenwich Mean Time +11:00]

AHD Animal Health Division [Now Part of Animal and Plant Health Inspection Service, US Department of Agriculture]

AHD Artificial Heart Device

AHD Australian Height Datum [Mapping]

AHDGA American Hot Dip Galvanizers Association [US]

AHDRI Animal Husbandry and Dairy Research Institute [South Africa]

AHDS Association of Heads of Surveying

AHE Acute Health Effects

AHE Associate in Home Economics

a-**He** Alpha Helium [Symbol]

AHEA American Home Economics Association [US]

AHEAD Advanced High-Efficiency Alloy Drumstick

AHEM Association of Hydraulic Equipment Manufacturers [UK]

Ahep Apollo Heat Europe [European Industrial Furnace Manufacturer]

AHERA Asbestos Hazard Emergency Response Act [US]

AHEX Association of Host Exhibitors

AHF American Health Foundation [Valhalla, New York, US]

AHF Anhydrous Hydrogen Fluoride

AHF Antihemophilic Factor [Biochemistry]

AHF Architectural History Foundation [US]

AHF Army Historical Foundation [US]

AHF Asian Hospital Federation

a-**Hf** Alpha Hafnium [Symbol]

AHFMR Alberta Heritage Foundation for Medical Research [Edmonton, Canada]

AHFR Argonne High Flux Reactor [of Argonne National Laboratory, Illinois, US]

AHG Acute Hemorrhagic Glomerulonephritis

AHG Antihemophilic Globulin [Biochemistry]

AHGP Australian Heritage Grants Program

a-**HgS** Red Mercury Sulfide (or Cinnabar)

a-**HgS:Co²⁺** Cobalt-Doped Alpha Mercury Sulfide (Crystal)

AHHA American Holistic Health Association [US]

A-HHFE Antisymmetric Heavy-Hole Free-Exciton [Solid-State Physics]

AHI American Heart Institute [US]

AHI Anhydrous Hydrogen Iodide

AHI Animal Health Institute [US]

AHI Asian Health Institute

AHIB Alpha-Hydroxoisobutyric Acid

AHIDC Australian Housing Industry Development Council

AHIP Advanced Helicopter Improvement Program [of US Air Force]

AHIT Autohomologous Immunotherapy

AHITI Animal Health and Industry Training Institute [Kenya]

AHJ American Heart Journal [US]

Ah/kg Ampere-hour(s) per pound [also A-h/kg]

Ah/lb Ampere-hour(s) per pound [also A-h/lb]

AHLI American Home Lighting Association [now American Lighting Institute, US]

AHM Ampere-Hour Meter [also ahm]

AHM Antiferromagnetic Heisenberg Model

AHM Atomic Hydrogen Maser

AHMA American Hardware Manufacturers Association [US]

AHMA American Holistic Medical Association [US]

AHMI Appalachian Hardwood Manufacturers, Inc. [US]

AHMSA Altos Hornos de Mexico S.A. [Major Mexican Steelmaker]

AHNA American Holistic Nurses Association [US]

a-**Ho** Alpha Holmium [Symbol]

AHOA Acethydroxamic Acid [also Ahoa]

AHONDA Ad Hoc Committee on New Directions of the Research and Technical Services Division [of American Library Association, US]

A Horizon A Horizon [Horizon of Soil Eluviation Consisting of Several Subhorizons A_1, A_2, etc.] [also A horizon]

AHP Air Horsepower

AHP Analytic Hierarchy Process [Mathematics]

AHP Army Heliport [of US Army]

AHP Ash Handling Plant

AHP As Hot-Pressed (Condition) [Metallurgy]

AHP Atomic Habit Plane [Crystallography]

AHPA American Honey Producers Association [US]

AHPCRC Army High-Performance Computing Research Center [of University of Minnesota, Minneapolis, US]

AHPL A Hardware Programming Language

AHPW Alberta Housing and Public Works [Canada]

AHQ Allied Headquarters

AHR Acceptable Hazard Rate

AHR Adsorptive Heat Recovery (Process)

AHR American Heritage Radio [US]

AHR Aqueous Homogeneous Reactor

AHRC Australian Housing Research Center

AHRP Australian Heritage Research Program [of Australian Heritage Commission]

AHRS Altitude and Heading Reference System

AHS Agricultural History Society [US]

AHS Airborne Hardware Simulator

AHS Alabama Historical Society [US]

AHS Alaska-Hawaii Standard (Time) [Greenwich Mean Time +10:00]

AHS American Hearing Society [US]

AHS American Helicopter Society [US]

AHS American Horticultural Society [US]

AHS Arizona Historical Society [US]

AHS Atlanta Historical Society [Georgia, US]
AHS Australian Herpitological Society [New South Wales]
AHS Automated Handling System
AHS Automated Highway System
AHSA American Home Satellite Association [US]
AHSA Art, Historical and Scientific Association [Canada]
AHSD Authority Health and Safety Division [UK]
AHSE Assembly, Handling and Shipping Equipment
AHSR Air-Height Surveillance Radar
AHSR American High Speed Rail
AHT Acoustic(ally) Homing Torpedo
AHT Aminoheptyltriazole
AHT Animal Health Trust [UK]
AHT As Heat-Treated (Condition) [Metallurgy]
AHT Average Holding Time [Telecommunications]
AHTD Association of High Tech Distributors [US]
AHTS Anchor Handling, Tug and Supply
AHU Air Handling Unit
AHU Antihalation Undercoat
AHV Ateline Herpes Virus
AHV-1 Ateline Herpes Virus, Type 1
AHV-2 Ateline Herpes Virus, Type 2
AHW As Hot-Worked (Condition)
AHW Atomic Hydrogen Welding
AHWC Ad Hoc Working Group
A²/Hz Square Ampere per Hertz [also A² Hz⁻¹]
AI Acoustic Image; Acoustic Imaging
AI Acoustic Impedance
AI Acoustic Interferometer; Acoustic Interferometry
AI Acquired Immunity
AI Acquisition Institute [US]
AI Active Immunity
AI Active Ingredient [also ai]
AI Acupressure Institute [Berkeley, California, US]
AI Acute Infection
AI Adapter Interface
AI Address Incomplete
AI Admiralty Islands
AI After-Image
AI Agglomeritic Ice
AI Agricultural Index
AI Airbag Injury
AI Airborne Intercept(ion)
AI Air-Insulated
AI Air Intake
AI Air Intensifier
AI Air Interception (Radar)
AI Aleutian Islands
AI Alloys Index [Metals Abstracts Index]
AI Altitude Indication; Altitude Indicator
AI Amplifier Input
AI Analog Indicator
AI Analog Input

AI Analog Interface
AI Anguilla [ISO Code]
AI Annealed Iron
AI Annoyance Index
AI Anti-Icing
AI Anti-Interference
AI Application(s) Interface
AI Application Invariant
AI Area of Intersection
AI Artificial Insemination
AI Artificial Intelligence
AI Asbestos Institute [Canada]
AI Astronomical Institute
AI As Is (Basis)
AI Asphalt Institute [US]
AI Atomic Imaging
AI Atomics International
AI Atropisomer
AI Attitude Indicator
AI Audio Interface
AI Auger Ionization [Physics]
AI Autoignition
AI Auto-Immune; Auto-Immunity
AI Auto-Indexing
AI Auto-Ionization [Physics]
AI Automated Inspection
AI Automated Instrument(ation)
AI Automatic Input
AI Automation Institute [US]
AI Impregnated Asbestos Type (Conductor Insulation) [Symbol]
A&I Abstracting and Indexing
A-I Smectic-A to Isotropic (Phase Transition) [also AI]
AIA Abrasive Industries Association [UK]
AIA Aerial-Lift Industries Association [US]
AIA Aerospace Industries Association (of America) [US]
AIA Affinity Isolated Antibody [Biochemistry]
AIA Aircraft Industries Association [US]
AIA Alberta Institute of Agrology [Canada]
AIA Aluminum Industries Abstracts [Joint Database of ASM International in the US and the Institute of Materials in the UK]
AIA American Insurance Association [US]
AIA American Institute of Accountants [US]
AIA American Institute of Aeronautics [US]
AIA American Institute of Architects [US]
AIA American Insurance Association [US]
AIA American Inventors Association [US]
AIA Anthracite Industry Association [US]
AIA Apiary Inspectors of America [US]
AIA Applications Integration Architecture [Computers]
AIA Archeological Institute of America [New York City, US]
AIA Asbestos Information Association
AIA Asbestos International Association [UK]

AIA Asociación de Ingenieros Agronomos [Association of Agricultural Engineers, Uruguay]

AIA Associate of the Institute of Actuaries

AIA Association for Industrial Archeology [UK]

AIA Association of International Accountants [UK]

AIA Auto International Association [US]

AIA Automated (or Automatic) Image Analysis

AIA Automobile Importers Association [US]

AIA Automobile Industries Association [Canada]

AIA Aviation Industry Association [New Zealand]

AIAA Aerospace Industries Association of America [US]

AIAA American Industrial Arts Association [now International Technology Education Association, US]

AIAA American Institute of Aeronautics and Astronautics [Washington, DC, US]

AIAA Area of Intense Air Activity

AIAA Association Interprofessionnelle de l'Aviation Agricole [Interprofessional Association of Agricultural Aviation, France]

AIAA Association of International Advertising Agencies [US]

AIAA J American Institute of Aeronautics and Astronautics Journal [also AIAA J, US]

AIAA Proc American Institute of Aeronautics and Astronautics Proceedings

AIAA Stud J AIAA Student Journal [of American Institute of Aeronautics and Astronautics]

AIAC Aerospace Industries Association of Canada

AIAC Artificial Intelligence Advisory Committee

AIAC Automotive Industries Association of Canada

AIACS Association of International Air Courier Services [now Association of International Courier and Express Services, UK]

AIADA American International Automobile Dealers Association [US]

AIAEE Association for International Agricultural and Extension Education [US]

AIAF American Institute of Architect Foundations [now American Architectural Foundation, US]

AIAF Association de l'Industrie et de l'Agriculture Françaises [French Association of Industry and Agriculture]

AIAG Aluminium Industrie AG (Aktiengesellschaft) [Major Swiss Aluminum Manufacturer]

AIAG Automotive Industries Action Group [US]

AIAI Artificial Intelligence Applications Institute [UK]

AIA Newsl AIA Newsletter [of Aerospace Industries Association of America]

AIA/NM Asbestos Information Association/North America [US]

AIAR American Institute for Archeological Research [US]

AIAS American Institute of Architecture Students [US]

AIES American Institute of Ayurvedic Sciences [US]

AIAS Australian Institute of Aboriginal Studies

AIAS Australian Institute of Agricultural Science

AIASA American Industrial Arts Students Association [US]

AIAU Atomic Institute of the Austrian University

AIB Address and Instruction Bus

AIB American Institute of Baking [Manhattan, Kansas, US]

AIB Analog Input/Output Board

AIB Associate of the Institute of Bankers [UK]

AIB Association for Independent Businesses [UK]

AIB Atlantic Institute of Biotechnology

AIB Automatic Intercept Bureau

AIB Avian Infectious Bronchitis

AIBA Agricultural Information Bank Asia

AIBA American Industrial Bankers Association [US]

AIBA Association of International Border Agencies [US]

AIBC Architectural Institute of British Columbia [Canada]

AIBD American Institute of Building Design [US]

AIBD Asia-Pacific Institute for Broadcasting Development [Malaysia]

AIBD Association of International Bond Dealers

AIBDA Asociación Interamericana de Bibliothecarias y Documentalistas Agricoles [Inter-American Association of Agricultural Librarians and Documentalists, Costa Rica]

AIBF N,N-Diethyl Phosphoramidic Acid Mono(2-Methylpropyl) Ester

AIBN Azobisisobutyronitrile

AIBNRM American Institute of Bolt, Nut and Rivet Manufacturers [now Industrial Fasteners Institute, US]

AIBS American Institute of Biological Sciences [US]

AIBT American Institute of Business Technologies [US]

AIBTC Association for Information Brokerage and Technological Consultancy [Germany]

AIBV Avian Infectious Bronchitis Virus

AIC Academia de la Investigación Cientifica [Academy of Scientific Research, Mexico City]

AIC Action Industrial Computers

AIC Advanced Industrial Concepts (Program) [of US Department of Energy]

AIC Aeronautical Information Circular

AIC Africa Information Center

AIC Agricultural Improvement Council [UK]

AIC Agricultural Institute of Canada

AIC Aid India Consortium

AIC Aircraft Industry Conference [US Army]

AIC AIX(Advanced Interactive Executive) Windows Interface Composer

AIC American Institute of Chemists [US]

AIC American Institute of Conservation (of Historic and Artistic Works) [US]

AIC American Institute of Constructors [US]

AIC American Institute of Cooperation [US]

AIC American International College [Springfield, Massachusetts, US]

AIC Ammunition Identification Code

AIC Analytical Image Carrier

AIC Appraisal Institute of Canada

AIC Arab Information Center

AIC Architectural Institute of Canada

AIC Arctic Institute of Canada [of National Research Council of Canada]

AIC Asbestos Information Center [UK]

AIC Associate of the Institute of Chemistry

AIC Association Internationale de la Couleur [International Color Association, France]

AIC Astronomical Institute of Czechoslovakia

AIC Automatic Initiation Circuit

AIC Automatic Intercept Center

AIC Automatic Intersection Control

AIC Automobile Importers of Canada

AIC Automotive Information Council [US]

AICA 4-/5-Amino 3H-Imidazole-5/-4-Carboxamide

AICA 5-Aminoimidazole-4-Carboxamide Hydrochloride

AICA Association Internationale pour Calcul Analogique [International Association for Analog Computing]

AICA Associazione Italiana per il Calculo Automatica [Italian Computing Association]

AICAE American Indian Council of Architects and Engineers [US]

AICAR 5-Aminoimidazole-4-Carboxamide-1-Ribofuranosyl [Biochemistry]

AICB Association Internationale Contre le Bruit [International Association for Noise Abatement, Switzerland]

AICBM Anti-Intercontinental Ballistic Missile

AICBM Air Intercontinental Ballistic Missile

AICC Army Intelligence Corps Command [at Fort Holabird, Maryland, US]

AICC Association of Independent Corrugated Converters [US]

AICC Association of Independent Crop Consultants [UK]

AICCF Association Internationale du Congrès de Chemins de Fer [International Association of Railway Congresses, Belgium]

AICCP Association of the Institute for Certification of Computer Professionals [US]

AICD Advanced Industrial Concepts Division [of (Battelle) Pacific Northwest National Laboratories, Richland, Washington, US]

AICE American Institute of Chemical Engineers [US]

AICE American Institute of Consulting Engineers [US]

AICE American Institute of Crop Ecology [US]

AICE Associate of the Institute of Civil Engineers

AICES Association of International Courier and Express Services [UK]

AIChE American Institute of Chemical Engineers [also AICHE, US]

AIChE J AIChE Journal [of American Institute of Chemical Engineers, US]

AIChEMI American Institute of Chemical Engineers Modular Instruction (Series) [Published by AIChE]

AIChEMI Modul Instr Ser AIChEMI Modular Instruction Series [of American Institute of Chemical Engineers]

AIChE Symp AIChE Symposium [of American Institute of Chemical Engineers, US]

AIChE Symp Ser AIChE Symposium Series [of American Institute of Chemical Engineers, US]

AIChExtra American Institute of Chemical Engineers Extra [Published by AIChE]

AICJ American Institute of Constructors Journal [also AIC J, US]

AICMA Association Internationale des Constructeurs de Matériel Aerospatial [International Association of Aerospace Manufacturers, France]

AICMR Association Internationale des Constructeurs de Matériel Roulant [International Association of Rolling Stock Builders, France]

AICOM Artificial Intelligence Communications [Publication of the European Coordinating Committee for Artificial Intelligence, France]

AI Commun Artificial Intelligence Communications

AICP American Institute of Certified Planners [US]

AICP Association of Independent Commercial Producers [US]

AICPA American Institute of Certified Public Accountants [US]

AICRC Association of Independent Clinical Research Contractors [UK]

AICS Associate of the Institute of Charted Ship Brokers

AICS Association of Independent Colleges and Schools [US]

AICS Association of Independent Computer Specialists [UK]

AICS Australian Inventory of Chemical Substances

AICU All India Cooperative Union

AICU Association of International Colleges and Universities [US]

AID Agency for International Development [of International Development Cooperation Agency, US]

AID Aircraft Installation Delay

AID Algebraic Interpretive Dialog

AID Alkali-Flame Ionization Detector

AID Alkali Ion Diode

AID American Institute of Interior Design [US]

AID Analytical Instrumentation Division [of Siemens Industrial Automation, Inc.]

AID Architects Information Directory

AID Argon Ionization Detector

AID Army Intelligence Department [US]

AID Artificial Insemination by Donor

AID Associated Independent Distributors [now Independent Distributors Association, US]

AID Attention Identification; Attention Identifier

AID Auger Induced Desorption [Physics]

AID Augmented Index and Digest

AID Auto-Immune Disease

AID Auto-Interactive Design

AID Automated Inspection Device

AID Automated Interaction Detector

AID Automatic Identification [also A/ID]

AID Automatic Imaging Device

AID Automatic Industrial Drilling

AID Automatic Information Distribution

AID Automatic Insertion Dip

AID Automatic Interrogation Distortion

A/ID Automatic Identification [also AID]

AIDA Analysis of Interconnected Decision Areas

AIDA Associazione Italiana per la Documentazione Avanzata [Italian Association for Advanced Documentation]

AIDA Attention-Interest-Desire-Action (Model) [Marketing]

AIDA Automatic Instrumented Diving Assembly

AIDA Automatic Intelligent Defect Analysis

AIDAB Australian International Development Assistance Bureau [of Department of Foreign Affairs and Trade]

AIDAS Advanced Instrumentation and Data Analysis System

AIDATS Army In-Flight Data Transmission System

AIDC Aerospace Industry Development Center [US]

AIDC American Industrial Development Council [now American Economic Development Council, US]

AIDC Arkansas Industrial Development Commission [US]

AIDC Asian Industrial Development Council

AIDC Australian Industry Development Corporation

AIDC Automatic Image Density Control

AIDD American Institute for Design and Drafting [US]

AIDD Auckland Industrial Development Division [New Zealand]

AIDE Adapted Identification Decision Equipment

AIDE Airborne Insertion Display Equipment

AIDE Aircraft Installation Diagnostic Equipment

AIDE Army Integrated Decision Equipment [US]

AIDE Artificial Intelligence Development Environment

AIDE Automated Image Device Evaluation

AIDE Automated Integrated Design Engineering

AIDE Automatic Identification Decision Equipment

AIDEA Alaska Industrial Development and Export Authority [US]

AIDEC Association for European Industrial Development and Economic Cooperation [Netherlands]

AIDF African Industrial Development Fund

AIDGAP Aids to Identification of Difficult Groups of Animals and Plants [of Field Studies Council, UK]

AIDI Associazione Italiana di Illuminazione [Italian Association of Illumination]

AIDI Associazione Italiana per la Documentazione e l'Informazione [Italian Association for Documentation and Information]

AIDJEX Arctic Ice Dynamics Joint Experiment

AIDL Auckland Industrial Development Laboratory [New Zealand]

AIDO Arab Industrial Development Organization [Iraq]

AIDS Abstract Information Digest Service [US]

AIDS Acoustic Intelligence Data System

AIDS Acquired Immunodeficiency Syndrome; Acquired Immue Deficiency Syndrome [also Aids]

AIDS Advanced Interactive Debugging System

AIDS Aerospace Intelligence Data System

AIDS Agricultural Information Development Scheme [Thailand]

AIDS Airborne Integrated Data System

AIDS Aircraft Integrated Data System [of Air Force Systems Command, US]

AIDS Air Force Intelligence Data-Handling System

AIDS American Institute for Decision Sciences [Atlanta, Georgia, US]

AIDS Automated Information Dissemination System

AIDS Automated Integrated Debugging System

AIDS Automated Intelligence and Data System [of US Air Force]

AIDS Automatic Integrated Debugging System

AIDS Automation Instrument Data Service [UK]

AIDSCOM Army Information Data Systems Command [US]

AIE Association des Ingénieurs Electriciens [Association of Electrical Engineers, Belgium]

AIE Association Internationale des Entreprises d'Equipement Electrique [International Association of Electrical Contractors, France]

AIE Australian Institute of Energy [Wahroonga, New South Wales]

AIE Automated Insurance Environment [of United Service Automobile Association, US]

AIEA Agence Internationale de l'Energie Atomique [International Agency for Atomic Energy, France]

AIEC Automotive Industry in Expanding Countries (Conference)

AIECA Associated Independent Electrical Contractors of America [now Independent Electrical Contractors, US]

AIED American Institute for Economic Development [US]

AIED Angle-Integrated Energy Distribution

AIEd Associate in Industrial Education

AI EDAM Artificial Intelligence for Engineering Design, Analysis and Manufacturing [Published in the UK]

AIEE American Institute of Electrical Engineers [now Institute of Electrical and Electronics Engineers, US]

AIEE Associate of the Institution of Electrical Engineers [UK]

AIEH Australian Institute of Environmental Health

AIENDF Atomics International Evaluation Nuclear Data File

AIEO American Indian Environmental Office [US]

AIEP Association of Independent Electricity Producers [UK]

AIEST Association Internationale d'Experts Scientifiques du Tourisme [International Association of Scientific Experts in Tourism, Switzerland]

AIET Average Instruction Execution Time

AIETA Airborne Infrared Equipment for Target Analysis

AIF Affiliated Inventors Foundation [US]

AIF Arbeitsgemeinschaft Industrieller Forschungsvereinigungen e.V. [Federation of Industrial Research Associations, Berlin, Germany] [also AiF]

AIF Arc-Image Force

AIF Arc-Image Furnace

AIF Association Internationale Futuribles [Futuribles International, France]

AIF Atomic Industrial Forum; Atomic Industries Forum

AIF Attitude Instrument Flying

AIF Audio Interchange Format

AIFA Agricultural Insecticide and Fungicide Association [now National Agricultural Chemicals Association, US]

AIFRB American Institute of Fishery Research Biologists [US]

AIFS African Improved Farming Scheme

AIFS American Institute for Foreign Study [US]

AIFSSF American Institute for Foreign Study Scholarship Foundation [US]

AIG Accident Investigation Group

AIG Address Indicating Group

AIG Arbeitsgemeinschaft Instandhaltung Gebäudetechnik [Association for Maintenance and Technical Building Equipment, of Verband Deutscher Maschinen und Anlagenbau, Germany]

AIG Artificial Intelligence Group [of Massachusetts Institute of Technology, Cambridge, US]

AIG Association Internationale de Géodésie [International Association of Geodesy, France]

AIG Australian Institute of Geoscientists

AIGA American Institute of Graphic Arts [US]

AIGE Association for Individually Guided Education

AIGMF All India Glass Manufacturers

AIGS Auxiliary Inerting Gas Subsystem

AIH American Institute of Hydrology [US]

AIH Anderson Impurity Hamiltonian [Materials Science]

AIH Artificial Insemination by Husband

AIH Australian Institute of Horticulture

AIHA American Industrial Hygiene Association [US]

AIHAJ American Industrial Hygiene Association Journal [also AIHA J]

AIHC American Industrial Health Conference [US]

AIHENP Symposium on New Computing Techniques in Physics Research: Artificial Intelligence, Software Engineering, Expert Systems, Neural Nets, Genetic Algorithms, Computer Algebra, Automatic Calculations

AIHP American Institute of the History of Pharmacy [US]

AIHSA American Insurers Highway Safety Alliance [US]

AIHTTR African Institute for Higher Technical Training and Research

AIHX Auxiliary Intermediate Heat Exchanger

AII American Interprofessional Institute [US]

AIIA Australian Information Industry Association

AIIA Australian Institute of International Affairs

AIIC Army Imagery Intelligence Corps

AIIC Association Internationale des Interprètes de Conference [International Association of Conference Interpreters, Switzerland]

AIIE American Institute of Industrial Engineers [now Institute of Industrial Engineers, US]

AIIF Amplitude-Independent Internal Friction

AIII Association of International Industrial Irradiation [France]

AIIM Association for Information and Image Management [US]

AIIM Association of Independent Investment Managers [UK]

AIIMS All-India Institute of Medical Sciences [Delhi]

AIIP Association of Independent Information Professionals [US]

AIIPH All India Institute of Hygiene and Public Health

AIIP Newsl AIIP Newsletter [of Association of Independent Information Professionals, US]

AIIR Air Intelligence Information Report

AIIS American Institute for Imported Steel [US]

AIJ Architectural Institute of Japan

AIKR Artificial Intelligence Knowledge Representation

AIL Achromatic Interferometric Lithography

AIL Airborne Instruments Laboratory

AIL Air In-Leakage

AIL Amber Indicating Lamp

AIL Amorphous Intergranular Layer(s) [Metallurgy]

AIL Array Interconnection Logic

AIL Artificial Intelligence Laboratory [of Stanford University, Stanford, California, US]

AIL Association of International Libraries [Switzerland]

Ail Aileron [Aeronautics]

AILA American Institute of Landscape Architects [now American Society of Landscape Architects, US]

AILA Asociación Industrial Latinoamericana [Latin American Industrial Association]

AILAS Automatic Instrument Landing Approach System

Ailn Aileron

AILS Advanced Integrated Landing System

AILS Automatic Instrument Landing System

AIM Abridged Index Medicus [of National Library of Medicine, US]

AIM Abstracts of Instructional Materials (Database) [now Abstracts of Instructional Materials and Research Materials (Database)] [of Ohio State University, Columbus, US]

AIM Academy for Interscience Methodology [US]

AIM Access Isolation Mechanism

AIM Acquisition Interface Module

AIM Adaptive Injection Molding

AIM Adsorption Isothermal Measurement

AIM Advanced Industrial Material

AIM Advanced Industrial Materials (Program) [of US Department of Energy–Office of Industrial Technology]

AIM Advanced Information Manager

AIM Aerial Independent Model

AIM Age of Intelligent Machines

AIM Agroecosystem Integrated Management

AIM Air Induction Melt(ing) [Metallurgy]

AIM Air Intercept Missile

AIM Air-Isolated Monolithic

AIM Airmen's (or Airwomen's) Information Manual

AIM Alarm Indication Monitor

AIM American Innerspring Manufacturers [US]

AIM American Institute of Management [US]

AIM American Interactive Media [US]

AIM Analytical Ion Microscope; Analytical Ion Microscopy

AIM Argon Induction Melting [Metallurgy]

AIM Artificial Intelligence in Medicine

AIM Associated Information Managers [US]

AIM Association Européenne des Industries de Produits de Marque [European Association of Industries of Branded Products, Belgium]

AIM Association for Information Management [US]

AIM Association for Innovation and Management [UK]

AIM Association for International Marketing [UK]

AIM Association of Information Managers [US]

AIM Association of Innerspring Manufacturers [Name Changed to American Innerspring Manufacturers, US]

AIM Associative Index Method

AIM Associazione Italiana di Métallurgica [Italian Association for Metallurgy]

AIM Atomic Imaging and Manipulation

AIM Atoms-in-Molecules (Method)

AIM Attitude and Integration Management (System) [of NASA]

AIM Australasian Institute of Metals [Sydney]

AIM Australian Institute of Management

AIM Automated Industrial Monitoring

AIM Automated Internal Management

AIM Automatic Identification Manufacturers [UK]

AIM Automatic Identification Manufacturers (International) [US]

AIM Avalanche-Induced Migration

AIM Awaiting Incoming Message

AIMA American Incense Manufacturers Association [US]

AIMA Association Internationale des Musées d'Agriculture [International Association of Agricultural Museums, Denmark]

AIMA Association of Incorporated Managers and Administrators [UK]

AIMACO Air Materiel Computer

AI Mag Artificial Intelligence Magazine [of American Association for Artificial Intelligence]

AIM/ARM Abstracts of Instructional Materials and Research Materials (Database) [of Ohio State University, Columbia, US]

AIMBE American Institute of Medical and Biological Engineering [US]

AIMC Academic Information Management Center [US]

AIMCAL Association of Industrial Metallizers, Coaters and Laminators [US]

AIME American Institute of Mechanical Engineers [New York City, US]

AIME American Institute of Mining, Metallurgical and Petroleum Engineers [Warrendale, Pennsylvania, US] [also AIMMPE]

AIME Applied Innovative Management Engineering

AIMechE Associate of the Institution of Mechanical Engineers [UK]

AIME Conf Proc American Institute of Mining, Metallurgical and Petroleum Engineers Conference Proceedings

AIMED Association of Independent Mailing Equipment Dealers [US]

AIMES Automated Inventory Management Evaluation System

AIMEX Australia's International Mining and Exploration Exhibition

AIMF Australasian Institute of Metal Finishing

AIMILO Army/Industrial Material Information Liaison Office [US]

AIMIS Advanced Integrated Modular Instrument System

AIMLC Association of Island Marine Laboratories of the Caribbean [now Association of Marine Laboratories of the Caribbean, Puerto Rico]

AIMLO Auto-Instructional Media for Library Orientation

AIMM Associate of the Institution of Mining and Metallurgy [UK]

AIMM Australasian Institute of Mining and Metallurgy

AIMM Australian Indonesian Ministerial Meeting

AIMMAT AIM (Australian Institute of Management) Management and Training (Database)

AIM&ME American Institute of Mining and Metallurgical Engineers [US]

AIMMPE American Institute of Mining, Metallurgical and Petroleum Engineers [also American Institute of Mining, Metallurgical and Petroleum Engineers, US]

AIMO Audibly-Instructed Manufacturing Operation

AIMP Ab initio Model Potential (Method)

AIMP American Institute of Medical Prevention [US]

AIMP Anchored Interplanetary Monitoring Platform

AIMPE Australian Institute of Marine and Power Engineers

AIMPR Asociación de Ingenieros Mecanicos de Puerto Rico [Association of Mechanical Engineers of Puerto Rico]

AIMR Airborne Imaging Microwave Radiometer

AIMR Association for the Improvement of the Mississippi River [US]

AIM/R Association of Manufacturers Representatives [also AIMR, US]

AIMRA Agricultural and Industrial Manufacturers Representatives Association [US]

AIMS Advanced Integrated-Manufacturing System

AIMS Advanced Intercontinental Missile System

AIMS American Institute of Merchant Shipping [US]

AIMS Associated Institutions for Materials Sciences

AIMS Association of Independent Motor Stores [UK]

AIMS Australian Institute of Marine Science

AIMS Author Index Manufacturing System

AIMS Automated Industrial Management System

AIMS Automated Information (and) Management System

AIMS Automated Inventory Management System

AIMSC Australian Industry Marine Science Council

AIMTECH Association for Integrated Manufacturing Technology [US]

AIM-TWX Abridged Index Medicus–Teletypewriter Exchange Network [of National Library of Medicine, US]

AIM UK Automatic Identification Manufacturers (of the United Kingdom) [also AIM UK]

AIM USA Automatic Identification Manufacturers, Inc., USA

AIN Advanced Intelligent Network [of Bell Atlantic, US]

AIN American Institute of Nutrition [of Federation of American Societies for Experimental Biology, US]

AIN Arab Institute for Navigation

A/in Ampere(s) per inch [also amp-in]

A/in² Ampere(s) per square inch [Unit]

AINA Arctic Institute of North America [of University of Calgary, Alberta, Canada]

AINBN Association for the Introduction of New Biological Nomenclature [Belgium]

AINSE Australian Institute of Nuclear Science and Engineering [Lucas Heights, New South Wales, Australia]

AInstP Associate of the Institute of Physics

AIO Action Information Organization

AIO Analog Input/Output Board

AIO Arecibo Ionospheric Observatory [of Cornell University's National Astronomy and Ionospheric Center; located near Arecibo, Puerto Rico]

AIO Artificial Internal Organ

AIOA American Iron Ore Association [US]

AIOE Australian Institute of Energy

AIOEC Association of Iron Ore Exporting Countries

AIOP Automatic Identification of Outward Dialing

AIOPI Association of Information Officers in the Pharmaceutical Industry [UK]

AIOU Analog Input/Output Unit

AIP Acoustical Information Processing

AIP Acute Infectious Polyneuritis

AIP Adiabatic Ionization Potential

AIP Aeronautical Information Publication

AIP Agreement-in-Principle

AIP Air Independent Propulsion (System)

AIP Airport Improvement Program

AIP Algorithm Intercomparison Project [Part of Global Precipitation Climatology Project]

AIP Allied Intelligence Publication

AIP Alphanumeric Impact Printer

AIP Aluminum Isopropoxide

AIP Aluminum Isopropylate

AIP American Institute of Physics [New York City, US]

AIP American Institute of Planners [US]

AIP Approval-in-Principle

AIP Arc Ion Plating (Process)

AIP Articles in Physics [Publication of American Institute of Physics, US]

AIP ASEAN (Association of Southeast Asian Nations) Industrial Project

AIP Association Internationale de Photobiologie [International Photobiology Association, Switzerland]

AIP Astrophysikalisches Institut Potsdam [Potsdam Institute of Astrophysics, Germany]

AIP Australian Institute of Packaging

AIP Australian Institute of Petroleum (Ltd.)

AIP Australian Institute of Physics [Sydney]

AIP Autoignition Point

AIPAAP 4-N-(Acetylisopropylidene)-Aminoantipyrine [Biochemistry]

AIPC Association Internationale des Palais de Congrès [International Association of Congress Centers]

AIPCEE Association des Industries du Poisson de la Communauté Economique Européenne [Association of the Fishery Industries of the European Economic Community, Belgium]

AIP Conf Proc AIP Conference Proceedings [of American Institute of Physics, US]

AIPE American Institute of Plant Engineers [US]

AIPEA Association Internationale Pour l'Etude des Argiles [International Association for the Study of Clays, France]

AIPE Facil Manage Oper Eng AIPE Facilities Management: Operations and Engineering [Journal of the American Institute of Plant Engineers]

AIPG American Institute of Professional Geologists [US]

AIP Hist Phys Newsl AIP History of Physics Newsletter [of American Institute of Physics, US]

AIPLA American Intellectual Property Law Association [US]

AIPO African Intellectual Property Organization

AIPP American Institute of Pollution Prevention [US]

AIPPA Association Internationale pour la Protection de la Propriété Industrielle [International Association or the Protection of Industrial Property, Switzerland]

AIP-PAPS American Institute of Physics–Physics Auxiliary Publication Service [US]

AIP Press American Institute of Physics Press [US]

AIPR Advances in Computer-Assisted Recognition (Meeting) [of SPIE–International Society for Optical Engineering]

AIPR American Institute of Pacific Relations [US]

AIPS Advanced Information Processing System

AIPS Astronomical Image Processing System

AIPT Association for International Practical Training [US]

AIQS Associate of the Institute of Quantity Surveyors [UK]

AIR Aerosol Ionic Redistribution [Atomic Absorption Spectrometry]

AIR Aerospace Information Report

AIR Air-Arming Impact Rocket

AIR Airborne Intercept Radar

AIR Air Injection Reaction; Air Injection Reactor

AIR Air Intercept Rocket

AIR Airline Industrial Relations (Conference) [US]

AIR Alliance of Independent Retailers [UK]

AIR All-India Radio

AIR American Institute of Refrigeration [US]

AIR American Institute of Research [US]

AIR Association for Institutional Research

AIR Atmospheric Instrumentation Research, Inc. [US]

AIR Automatic Infrared Rangefinder

AI/R Artificial Intelligence and Robotics

AIRA American Independent Refiners Association [US]

AIRAC Aeronautical Information Regulation and Control

AIRAC All-Industry Research Advisory Council

AIRAC Australian Ionizing Radiation Advisory Council

AIRAD Airmen (or Airwomen) Advisory

AIRAH Australian Institute of Refrigeration, Air Conditioning and Heating

AIRC Association of Independent Radio Companies [UK]

Aircrft Aircraft [also Aircft, or aircft]

AIRCOM Aerospace Communications Complex

AIRCOM Air Command

AIRCON Automated Information and Reservation Computer-Oriented Network

Air Cond Air Conditioning

Air Cosm Mon Air and Cosmos Monthly [Journal published in France]

Aircr Aircraft [also aircr]

Aircr Eng Aircraft Engineering [Journal published in the UK]

Aircr Eng Aerosp Technol Aircraft Engineering and Aerospace Technology [Journal published in the UK]

AIRD Asian Institute for Rural Development

AIRDF Artificial Intelligence Research and Development Fund [of Industry and Science Canada]

AIRE Australian Institution of Radio Engineers

AIREW Airborne Infrared Early Warning

Air Force Space Dig Air Force and Space Digest [of Air Force Association, US]

AIR HP Air Horsepower [also Air HP, or air hp]

AIRI Applied Industrial Research Institute [of Milwaukee School of Engineering, Wisconsin, US]

AIRI Association of Independent Research Institutes [US]

AIRIA All India Rubber Industries Association [India]

AIRL Aeronautical Icing Research Laboratory [US]

AIRLO Air Liaison Officer

AIRMARC Aircraft-Marine Radio Corporation [also Airmarc]

AIRMIC Association of Insurance and Risk Managers in Industry and Commerce [UK]

AIRMICS Army Institute for Research in Management Information and Computer Science [US]

AIRO Association Internationale de Recherche Operationelle [International Association of Operational Research, France]

AIRPASS Airborne Interception Radar and Pilot's Attack Sight System

AIRPLAN Air Pollution Action Network

Airpt Airport [also Airpt, or airpt]

Airpt Forum Airport Forum [Journal published in Germany]

AIRR Adjusted Internal Rate-of-Return

AIRS Advanced Inertial Reference Sphere

AIRS Alliance of Information and Referral Systems [Indianapolis, Indiana, US]

AIRS American Information Retrieval Service [US]

AIRS Artificial Intelligence and Robotics Society

AIRS Atmospheric Infrared Sounder

AIRS Automatic Image Retrieval System

AIRS Automatic Information Retrieval System

AIRSIM Airways/Airport Simulation Model

AIRSO Association of Industrial Road Safety Officers [UK]

AIRSS ABRES (Advanced Ballistic Reentry System) Instrument Range Safety System

AIRTO Association of Independent Research and Technology Organizations [UK]

AIRXRS American Industrial Radium and X-Ray Society

AIS Academio Internacia de la Scienco [International Academy of Sciences, Germany]

AIS Accounting Information System

AIS Address Information System [Geographic Information System]

AIS Advanced Information System

AIS Advanced Instructional System

AIS Aeronautical Information Service [of Civil Aviation Authority, UK]

AIS Airbag Injury Study [by Transport Canada]

AIS Airborne Imaging Spectrometer

AIS Airborne Instrumentation System

AIS Alarm Indication Signal

AIS Alarm Inhibit Signal

AIS Altitude Indication System

AIS American Institute of Stress [US]

AIS American Interplanetary Society [Now Part of American Institute of Aeronautics and Astronautics, US]

AIS Analog Input System

AIS Applied Information Systems, Inc. [US]

AIS Association Internationale de la Savonnerie (et de la Detergence) [International Association of the Soap and Detergent Industry, Belgium]

AIS Audio Integrating System

AIS Australian Information Service

AIS Automated Identification System [of National Crime Information Center, US]

AIS Automated Information Security

AIS Automated Information System

AIS Automated Injection System

AIS Automatic Intercept System

AIS Automatic Intercity Station

AIS Avionics Intermediate Shop

AIS AXAF (Advanced X-Ray Astrophysics Facility) Information System [of NASA]

AISA American Institute for Shipper Associations [US]

AISA Association of International Schools in Africa

AISAG Aeronautical Information Service Automation Group [of International Civil Aviation Organization]

AISAP Aeronautical Information Service Automation Specialist Panel [of International Civil Aviation Organization]

AISB Association of Imaging Service Bureaus

AISC American Institute of Steel Construction [US]

AISC Association of Independent Software Companies

AISE American Iron and Steel Engineers [US]

AISE Association of Iron and Steel Engineers [US]

AISES American Indian Science and Engineering Society [US]

AISES Newsl AISES Newsletter [of American Indian Science and Engineering Society]

AISG Aeronautical Information Services Group

AISI American Iron and Steel Institute [US]

AISI/DOE American Iron and Steel Institute–Department of Energy (Joint Program) [US]

AISIF ASEAN (Association of Southeast Asian Nations) Iron and Steel Industry Federation

AISI-ISS American Iron and Steel Institute–Iron and Steel Society [US]

AISI/SAE American Iron and Steel Institute–Society of Automotive Engineers (Specification) [also AISI-SAE]

AISIST Australian Institute of Spatial Information Sciences and Technology

AISM American Institute of Sports Massage [US]

AISP Association of Information Systems Professionals [US]

AIS–PL/I Applied Information Systems Inc.–Programming Language One

AIST Agency of Industrial Science and Technology [of Ministry of International Trade and Industry, Japan]

AIST Alberta Institute of Science and Technology [Calgary, Canada]

AIST Automatic Information(al) Station

AIST/MITI Agency of Industrial Science and Technology/Ministry of International Trade and Industry [Ibaraki, Japan]

AISU Arab Iron and Steel Union

AIT Acute Infectious Tracheitis

AIT Advanced Individual Training

AIT Advanced Information Technology

AIT Advanced Instruction Technique

AIT Advanced Intelligent Tape (Drive)

AIT Aeritalia [Italian State Airline]

AIT Aichi Institute of Technology [Japan]

AIT Aircraft Instrument Technician

AIT Akita Research Institute of Advanced Technology [Japan]

AIT American Institute of Technology [US]

AIT Architect-in-Training

AIT Artificial Intelligence Technology

AIT Asian Institute of Technology [Bangkok, Thailand]

AIT Auto-Ignition Temperature

AI&T Assembly, Integration and Testing

AITC Acylisothiocyanate

AITC Allylisothiocyanate

AITC American Institute of Timber Construction [US]

AITC Association Internationale des Traducteurs de Conference [Association of Conference Translators, Switzerland]

AITC Australian Industry and Technology Commission

AITC 2,3,4-Tri-O-Acetyl-γ-D-Arabinopyranosylisothiocyanate

AITE Aircraft Integrated Test Equipment

AITE Automatic Intercity Telephone Exchange

AITES Association Internationale des Travaux en Souterrain [International Tunneling Association, France]

AITI Aero-Industries Technical Institute [US]

AITI Associazione Italiano Truduttori ed Interpreti [Italian Association of Translators and Interpreters]

AITIA American Institute of Technical Illustrators Association [US]

AITIPE Asociación de Investigación Tecnica de la Industria Papelera Espanola [Association for Technical Studies of the Spanish Paper Industry, Spain]

AITIT Association Internationale de la Teinture et de l'Impression Textiles [International Association of Textile Dyers and Printers, UK]

AITP Aerospace (or Aeronautical) Industry Technology Program [of NASA]

AITRP AIDS (Acquired Immunodeficiency Syndrome) International Training and Research Program

AITS Action Item Tracking System

AITS Automatic Integrated Telephone System

AI2S Automated Integrated Inspection System [also AI²S]

AITT Association of Industrial Truck Trainers [UK]

AIU Advanced Instrumentation Unit [of National Physics Laboratory, University of Illinois at Urbana-Champaign, US]

AIU Alarm Interface Unit

AIUI As I Understand It

AIUM American Institute of Ultrasound in Medicine [US]

AIUPS Angle-Integrated Ultraviolet Photoelectron Spectroscopy

AIV Air Inlet Valve

AIV Aluminum-Intensive Vehicle

AIV Institut für Beratung und Entwicklung in der "Automatischen Informations-Verarbeitung" [Institute for Consultation and Development in Automatic Information Processing, Darmstadt, Germany]

AIW Auroral Intrasonic Wave

AIW Average Industrial Wage

AIWM American Institute of Weights and Measures

AIWRS Arctic International Wildlife Range Society

AIX Advanced Interactive Executive [IBM System]

AJ Adhesive Joint

AJ Aerojet [US]

AJ Air Jet

AJ Analog Junction

AJ Anti-Jamming [also aj]

AJ Arc Jet (Engine)

AJ Area Junction

AJ Assembly Jig

AJAC Automobile Journalists Association of Canada

AJAI Anti-Jamming Anti-Interference

AJD Anti-Jam Display

AJCC American Jersey Cattle Club [US]

AJE Arc-Jet Engine

AJES Australian Journal of Earth Sciences

AJHG American Journal of Human Genetics [of University of Chicago Press, US]

AJM Abrasive Jet Machining

AJM Air Jet Milled; Air Jet Milling

AJM American Journal of Medicine

AJM Analog Junction Module

AJM SiO$_2$ Air Jet Milled Quartz

AJN American Journal of Nephrology [US]

AJN American Journal of Nursing [US]

AJNR American Journal of Nuclear Research [also Am J Nucl Res]

Ajour Ind-Tek Ajour Industrii-Teknikk [Norwegian Publication on Modern Industrial Technology]

AJP American Journal of Physics [of American Institute of Physics, US]

AJP American Journal of Physiology [of American Physiological Society, US]

AJPH American Journal of Public Health

AJR American Journal of Resonance [also Am J Reson]

AJRCCM American Journal of Respiratory and Critical Care Medicine [of American Lung Association, US]

AJRCMB American Journal of Respiratory Cell and Molecular Biology [of American Lung Association, US]

AJS American Journal of Science [of Yale University, New Haven, Connecticut, US]

AJSE Arabian Journal for Science and Engineering

AJSM Association of Jute Spinners and Manufacturers [UK]

AK Adenylate Kinase [Biochemistry]

AK Aggregation Kinetics

AK Air Knife

AK Akademik Kurchatov R/V [Kurchatov Academy, Russian Federation]

AK Alaska [US]

AK Alpha Keratin [Biochemistry]

AK Aluminum-Killed (Steel)

AK Annealing Kinetics [Metallurgy]

AK Cargo Ship [US Navy Symbol]

A/K² Ampere per Square Kelvin [also A K^{-2}]

Ak *(Arbeitskreis)* – German for "Committee", "Division, or "Study Group" [also AK]

AKA All Kidding Aside [Communications]

AKA Arbeitskreis Physik und Abrüstung [Committee "Physics and Disarmament" of the German Physical Society]

aka also known as

Akad Nauk SSR, Ural Fil Tr Inst Khim Akademiya Nauk SSR, Ural'skii Filial, Trudy Instituta Khimii [Chemistry Journal of the Academy of Sciences of the USSR]

Akad Nauk Ukr SSR, Metallofiz Akademiya Nauk Ukrainskai SSR, Metallofizika [Metal Physics Journal of the Academy of Sciences of the Ukrainian Soviet Socialist Republic]

Akad Wiss Lit Mainz Math-Natur Kl Akademie der Wissenschaften und Literatur Mainz, Mathematisch-Naturwissenschaftliche Klasse IIb [Academy of Sciences and Literature, Mayence, Mathematical Sciences Class, Germany]

AKC American Kennel Club [US]

AKCLIS Australian Key Center in Land Information Studies [Queensland]

A/K²·cm² Ampere per square kelvin square centimeter [also A K^{-2} cm^{-2}]

AKDQ Aluminum-Killed Drawing Quality (Mild Steel)

AKE Arbeitskreis Energie [Committee "Energy Matters" of German Physical Society]

AKF A (Al$_2$O$_3$ + Fe$_2$O$_3$ + (CaO + Na$_2$O)), K (K$_2$O), F (FeO + MgO + MnO) (Diagram) [Petrology]

AKF Arbeitskreis Festkörperphysik [Condensed Matter Division of German Physical Society]

AKFM Association of Knitted Fabrics Manufacturers [US]

***a*-KG** Alpha Ketoglutarate [Biochemistry]

AKK Atomkraftkonsortiet Krangede [Nuclear Power Study Group at Krangede, Sweden]

AKM Apogee Kick Motor [Aerospace]

AKO Arbeitskreis Optionen für die Zukunft [Committee "Future Options" of German Physical Society]

AKR Address Key Register

AKR Auroral Kilometric Radiation

AKRO Acknowledge Receipt of Order

Akron Bus Econ Rev Akron Business and Economic Review [US]

AKS Aluminum-Potassium-Silicon (Compound)

AKS Aluminum Potassium Sulfate

AKS Tungsten Aluminum-Potassium-Silicon Doped Tungsten [also AKS tungsten]

Aktual Probl Inf Dok Aktualne Problemy Informacji i Dokumentacji [Polish Journal on Current Problems in Information and Documentation]

Akust Mol Kwant Akustyka Molecularni i Kwantowa [Journal on Molecular Acoustics; published by the Polish Academy of Sciences]

Akust Zh Akusticheskii Zhurnal [Russian Journal on Acoustics]

AKWIC Author and Key Word in Context

AKZ Angolan Kwanza [Currency]

AL Above-Cloud Layer

AL Abscission Layer [Botany]

AL Academic Library

AL Acceptance Level

AL Acceptance Limit

AL Acceptor Level [Solid-State Physics]

AL Aclinic Line

AL Acoustic Lens

AL Acoustic Load

AL Action Level

AL Activated Lime

AL Acute Lethality

AL Additional Listing

AL Adsorboluminescence; Adsorboluminescent

AL Aeronomy Laboratory [of National Oceanic and Atmospheric Administration, US]

AL Age Limit

AL Air League [UK]

AL Airlift

AL Airline

AL Air Lock [also A/L]

AL Alabama [US]

AL Albania [ISO Code]

AL Alcohol

AL Allegheny Ludlum Corporation [Pittsburgh, Pennsylvania, US]

AL Amber-Glow Lamp

AL Ambient Light

AL Ames Laboratory [of US Department of Energy at Iowa State University]

AL Amplitude-Limited; Amplitude Limiter

AL Analog Link

AL Analysis Library

AL Analytical Laboratory

AL Analytical Limit(s)

AL Antilymphocytic (Serum) [Immunology]

AL Approach and Landing [also A/L]

AL Approach Light

AL Arc Lamp

AL Argon Laser

AL Armstrong Laboratory [of US Air Force at Brooks Air Force Base, Texas, US]

AL A Robot-Programming Language

AL Arrhenius Laboratory [Stockholm, Sweden]

AL Artificial Line

AL Assembly Language

AL Associated Laboratories [US]

AL Associated Liquids

AL Astronomical League [US]

AL Attenuation Length

AL Auroral Luminosity

AL Autograph Letter

AL Auto-Lag

AL Avian Leukosis [Poultry Disease]

AL Avionics Laboratory [US Air Force]

AL Axial(ly) Load(ed); Axial Loading

A/L Air Lock [also AL]

A/L Approach and Landing [also AL]

Al Aluminate

Al Aluminum [Symbol]

Al Allyl

Al^{3+} Aluminum Ion [also Al^{+++}] [Symbol]

Al-24 Aluminum-24 [also ^{24}Al, or Al^{24}]

Al-25 Aluminum-25 [also ^{25}Al, or Al^{25}]

Al-26 Aluminum-26 [also ^{26}Al, or Al^{26}]

Al-27 Aluminum-27 [also ^{27}Al, Al^{27}, or Al]

Al-28 Aluminum-28 [also ^{28}Al, or Al^{28}]

Al-29 Aluminum-29 [also ^{29}Al, or Al^{29}]

.al Albania [Country Code/Domain Name]

ALA Ada Language

ALA Alighting Area

ALA American Laminators Association [US]

ALA American Landrace Association [US] [*Note:* Landrace Refers to a Breed of Swine]

ALA American Laryngological Association [US]

ALA American Library Association [US]

ALA American Logistics Association [US]

ALA American Lung Association [US]

ALA 5-Aminolevulinic Acid; λ-Aminolevulinic Acid [Biochemistry]

ALA Asia and Latin America

ALA As Large As

ALA Associate of the Library Association [UK]

ALA Automobile Legal Association [US]

Ala Alabama [US]

Ala (+)-Alanine; Alanyl [Biochemistry]

ala Alaninate [Biochemistry]

β-Ala β-Alanine [Biochemistry]

a-La Alpha Lanthanum [Symbol]

Al/a-Al$_2$O$_3$ Alpha Alumina Reinforced Aluminum (Composite)

ALABOL Algorithmic and Business-Oriented Language

ALAC Australian Land Adminstrators Conference

Al(Ac)$_3$ Aluminum Acetate

Al(ACAC)$_3$ Aluminum Acetylacetonate [also Al(AcAc)$_3$, or Al(acac)$_3$]

ALACE Autonomous Lagrangian Circulation Explorer (float)

ALA Committee Committee for Aid to Developing Countries in Asia and Latin America

ALAE Asociación Latinoamericana de Entomologia [Latin American Association of Entomology]

ALAE Association of Licensed Aircraft Engineers [UK]

ALAE Atmospheric Lyman Alpha Experiment [of NASA]

ALAF Asociación Latinoamericana de Ferrocarriles [Latin American Association for Rolling Stock]

Al-Ag Aluminum-Silver (Alloy System)

Ala-Gln Alanyl-Glutamine [Biochemistry]

Ala-Gly Alanylglycine [Biochemistry]

ALAIH Asociación Latinoamericana de Ictiologos y Herpetologos [Latin American Association of Ichthyology and Herpetology]

ALA/ISAD American Library Association–Information Science and Automation Division [US]

Al/AlAs Aluminum on Aluminum Arsenide (Substrate)

ALALC Asociación Latinoamericana de Libre Comercio [Latin American Free Trade Association]

Al/Al$_4$C$_3$ Aluminum Carbide Reinforced Aluminum (Composite)

Al/Al$_2$Cu Aluminum Copper Reinforced Aluminum (Composite)

Al/AlN Aluminum Nitride Carbide Reinforced Aluminum(Composite)

Al/AlN MMC Aluminum Nitride Reinforced Aluminum Metal-Matrix Composite

Al/Al$_3$Ni Aluminum Nickelide Reinforced Aluminum (Composite)

Al-Al$_2$O$_3$ Aluminum-Alumina (System)

Al/Al$_2$O$_3$ MMC Alumina Reinforced Aluminum Metal-Matrix Composite

Al/Al$_3$Ti Aluminum Titanide Particulate Reinforced Aluminum (Composite)

ALAM Association of Lightweight Aggregate Manufacturers [UK]

ALAMAR Asociación Latinoamericana de Armadores [Latin American Association of Ship Owners, Uruguay]

Alan Alanine [Biochemistry]

ALAP As Low As Practical

ALAP Associative Linear Array Processor

ALAPCO Association of Local Air Pollution Control Officials [US]

ALARA As Low As Reasonably Achievable [Radiation Control]

ALARM Air-Launched Advanced Ramjet Missile

ALARM Automatic Light Aircraft Readiness Monitor

ALARR Air-Launched Air Recoverable Rocket

ALARS Association of Light Alloy Refiners and Smelters [UK]

ALART Army Low-Speed Air Research Tasks

ALAS Aminolevulinic Acid Synthetase [Biochemistry]

ALAS Asynchronous Lookahead Simulator

ALAS Automated Literature Alerting System

AlAs Aluminum Arsenide (Semiconductor)

Alas Alaska(n) [US]

α-La$_2$S$_3$ Alpha Lanthanum Sesquisulfide

AlAs DBR Aluminum Arsenide Distributed Bragg Reflector [Physics]

AlAs/GaAs SL Aluminum Arsenide/Gallium Arsenide Superlattice

AlAsN Aluminum Arsenide Nitride (Semiconductor)

AlAsSb Aluminum Arsenide Antimonide (Semiconductor)

ALATEP As Low as Technically and Economically Practicable

ALB Anti-Lock Brake

Alb Albania(n)

Alb Albany [New York, US]

Alb Albedo [Physics]

Alb Albite [Mineral]

Alb Albumin [Biochemistry]

alb *(albus)* – white [Medical Prescriptions]

Alba Alberta [Canada]

Al-BaM Aluminum-Substituted M-Type Barium Ferrite

Alban Albania(n)

AlBBe Aluminum Boron Beryllide

Al-Be Aluminum-Beryllium (Alloy System)

ALBEDOS Airborne Laser Base Enhanced Detection and Observation System

ALBENE Alcohol-Benzene (Process)

ALBI Air-Launched Ballistic Intercept

ALBIS Australian Library-Based Information System

ALBM Air-Launched Ballistic Missile

AlBN Aluminum Boronitride

ALBO Automatic Line Build-Out

AlBO Aluminoborate

Al-Bronze Aluminum Bronze [also Al bronze]

ALBRSO Association of London Borough Road Safety Officers [UK]

Al:BSCCO Aluminum Substituted Bismuth Strontium Calcium Copper Oxide (Superconductor)

AlBSi Aluminum Borosilicide (Compound)

AlBu$_3$ Tri-i-Butyl Aluminum

Al(BuO)$_3$ Aluminum(III) Butoxide (or Aluminum(III) Butylate)

ALBW Antiland Base Warfare

ALC Acetyl-L-Carnitine [Biochemistry]

ALC Adaptive Logic Circuit

ALC Agricultural Land Commission [UK]

ALC Airlock Chamber

ALC Alberta Livestock Cooperative [Canada]

ALC Approximate Lethal Concentration

ALC Arithmetic and Logic Circuit

ALC Arkansas Library Commission [US]

ALC Aromatic Liquid Crystal

ALC Assembly Language Coding

ALC Automated (or Automatic) Liquid Chromatography

ALC Automatic Level Control [also alc]

ALC Automatic Light Control

ALC Automatic Load Control [also alc]

ALC Automatic Locking Circuit

Alc Alcohol(ic) [also alc]

Alc Ethyl Alcohol

ALCA American Leather Chemists Association [US]

ALCA Associated Landscape Contractors of America

Alcan Alaska-Canada (Highway)

Alcan Aluminum Company of Canada

ALCAPP Automatic List Classification and Profile Production

ALCC Airborne Launch Control Center

ALCC Association of London Computer Clubs [UK]

ALCE Air Lift Control Element

ALCELL Alcohol-Cellulose (Process)

Al-Ce-LN Aluminum–Cerium–Lanthanide Metal (Alloy)

ALCH Approach Light Contact Height

ALCHEMI Atom Location by Channeling-Enhanced Microanalysis (Technique) [Transmission Electron Microscopy]

ALCI ARBED (Aciéries Réunies de Burbach-Eich-Dudelange) Lance Coal Injection [Metallurgy]

ALCL Association of London Chief Librarians [UK]

ALCM Air-Launched Cruise Missile

ALCS Authors Lending and Copyright Society [UK]

ALCS Authors Licensing and Collecting Society [UK]

ALCM Air-Launched Cruise Missile

AlCo Aluminum Cobalt (Compound)

Al-Co Aluminum-Cobalt (Alloy System)

Alcoa Aluminum Company of America [US]

Alcoa Technol Rep Alcoa Technology Report [Published by Alcoa Corporation, Pittsburgh, Pennsylvania, US]

AlCoCuSi Aluminum-Cobalt-Copper-Silicon (Coating)

ALCOM Advanced Liquid Crystalline Optical Materials

ALCOM Algebraic Compiler

ALCOM Science and Technology Centers for Advanced Liquid Crystalline Optical Materials [of National Science Foundation, US]

AlCoMn Aluminum Cobalt Manganese (Compound)

ALCOR ARPA (Advanced Research Project Agency) Lincoln Coherent Observable Radar [of US Department of Defense]

AlCoRe Aluminum Cobalt Rhenium (Compound)

ALCORSS Australian Liaison Committee on Remote Sensing by Satellite

AlCoTi Aluminum Cobalt Titanium (Compound)

ALCP Area Local Control Panel

Al-Cr Aluminum-Chromium (Alloy System)

AlCrMg Aluminum-Chromium-Magnesium (Coating)

AlCrSi Aluminum-Chromium-Silicon (Coating)

ALCS Aluminum Locator-Nose Clamp Strap

ALCU Arithmetic Logic and Control Unit

ALCU Asynchronous Line Control Unit

AlCu Aluminum Copper (Spin Glass)

AlCu Altocumulus (Cloud)

Al-Cu Aluminum-Copper (Alloy System)

Al-Cu/Al$_2$O$_3$ MMC Alumina Reinforced Aluminum-Copper Metal-Matrix Composite

Al-CuAl$_2$ Copper Dialuminide Reinforced Aluminum (Composite)

Al-Cu/B$_4$C Boron Carbide Reinforced Aluminum Copper (Composite)

AlCuFe Aluminum Copper Iron (Compound)

AlCuLi Aluminum-Copper-Lithium (Coating)

Al-Cu MMC Aluminum-Copper Metal-Matrix Composite

AlCuRu Aluminum Copper Ruthenium (Compound)

Al-Cu/SiC Silicon Carbide Reinforced AluminumCopper (Composite)

%alc/vol percent of alcohol per volume

ALD Algerian Dinar [Currency]

ALD Allegheny Lock and Dam [US]

ALD Analog Line Driver

ALD Asynchronous Line Driver

ALD Automated Logic Diagram

ALD Automatic Location Device

ALD Automatic Louver Damper

Ald Aldehyde [also ald]

ALDA Agricultural Land Development Assistance [Canada]

ALDA Association for Librarianship and Documentation in Agriculture [Germany]

ALDEP Automated Layout Design Program

ALDEV African Land Development Board [Kenya]

ALDOC Arab League Documentation Information Center

ALDP Automatic Language Data Processing

ALDS Apollo Launch Data System [of NASA]

ALE Address Latch Enable

ALE Annual Loss Expectancy

ALE Application Linking and Embedding [Computers]

ALE Arbitrary Lagrangian Eulerian [Mathematics]

ALE Arid-Lands Ecology (Project) [of Pacific Northwest Laboratory, Richland, Washington, US]

ALE Atmospheric Lifetime Experiment

ALE Atomic Layer Epitaxy

ALE Atomic Layer Etch(ing)

ALE Al$_2$O$_3$ Atomic Layer Epitaxy Aluminum Oxide

ALEAS Asociacion Latinoamericana de Educación Agricola [Latin American Association for Agricultural Education]

ALEC Arid Lands Environment Center [Australia]

ALE CdTe Atomic Layer Epitaxy Cadmium Telluride

ALECSO Arab League Educational, Cultural and Scientific Organization

ALE EL Atomic Layer Epitaxy/Electroluminescent (Device) [also ALE/EL, or ALE-EL]

ALEM Association of Loading and Elevation Equipment Manufacturers [UK]

ALEMS Apollo Lunar Excursion Module Sensors [NASA]

ALEPH A Low Energy PHysics (Experiment) [of CERN–European Laboratory for Particle Physics, Geneva, Switzerland]

ALEPH Automatic Library Expendable Program, Hebrew University [Israel]

ALERFA Alert Phase [Aeronautics]

ALERT Acute Launch Emergency Reliability Tip

ALERT Assistance for Liquid Electronic Reliability Testing

ALERT Automated Law Enforcement Response Team

ALERT Automated Linguistic Extraction and Retrieval Technique

ALERT Automatic Logical Equipment Readiness Tester

ALES American Labour Education Service [US]

ALESA American League for Exports and Security Assistance [US]

ALESCO Arab League Educational, Scientific and Cultural Organizations [Tunesia]

AlEt$_3$ Triethyl Aluminum (or Aluminum Triethyl)

Al(EtO)$_3$ Aluminum(III) Ethoxide (or Aluminum(III) Ethylate)

ALEX Analytical Laboratory Equipment Conference and Exposition

ALEX Automatic Linear Earing Die Measuring System

ALE ZnS(Mn) Atomic Layer Epitaxy Manganese-Doped Zinc Sulfide (Film)

ALF Absorption Limiting Frequency

ALF Alien Life Form

ALF Application Library File

ALF Automatic Letter Facer

ALF Automatic Line Feed

ALF Auxiliary Landing Field

ALFA Air-Lubricated Free Attitude

Alfaatih Univ Bull Fac Eng Alfaatih University Bulletin of the Faculty of Engineering [Libya]

ALFCE Allied Land Forces Central Europe

Al-Fe Aluminum-Iron (Alloy System)

ALFED Aluminum Federation [UK]

ALFENE Alfa Olefene (Process)

ALFTRAN ALGOL-to-FORTRAN Translator

ALG Advanced Landing Ground

ALG Antilymphocyte Globulin; Antilymphocytic Globulin [Biochemistry]

ALG Atomic Layer Growth

Alg Algebra(ic) [also alg]

Alg Algeria(n)

Alg Algiers [Algeria]

ALGA Australian Local Government Association

Al-Ga Aluminum-Gallium (Alloy System)

AlGaAs Aluminum Gallium Arsenide (Semiconductor)

AlGaAs/GaAs Aluminum Gallium Arsenide/Gallium Arsenide (Superlattice)

AlGaAs/GaAs SL Aluminum Gallium Arsenide/Gallium Arsenide Superlattice

AlGaAs PIN PD Aluminum Gallium Arsenide PositiveIntrinsic-Negative Photodetector

AlGaAsSb Aluminum Gallium Arsenide Antimonide (Semiconductor)

AlGaInN Aluminum Gallium Indium Nitride (Semiconductor)

AlGaInAsSb Aluminum Gallium Indium Arsenide Nitride (Semiconductor)

AlGaN Aluminum Gallium Nitride (Semiconductor)

AlGaP Aluminum Gallium Phosphide (Semiconductor)

AlGaSb Aluminum Gallium Antimonide (Semiconductor)

Al-Ge Aluminum-Germanium (Alloy System)

ALGM Air-Launched Guided Missile

ALGN Align [also Algn]

ALGO Algebraic Compiler

ALGOL Algebraic Oriented Language [also Algol]

ALGOL Algorithmic (Oriented) Language [Computer Language] [also Algol]

Al/Gr MMC Graphite-Reinforced Aluminum Metal-Matrix Composite

ALGTB Australian Local Government Training Board

ALGU Association of Land Grant Universities [US]

ALH Allan Hills (Iron Meteorite)

Al(HFAC)$_3$ Aluminum Hexafluoroacetylacetonate [also Al(hfac)$_3$]

ALHFAM Association for Living Historical Farms and Agricultural Museums [US]

ALHT Apollo Lunar Hand Tool [of NASA]

ALI Alarm Inhibit [Data Processing]

ALI Alberta Laser Institute [Canada]

ALI Allowable Limit of Intake

ALI American Ladder Institute [US]

ALI American Lighting Institute [US]

ALI Asynchronous Line Interface

ALI Automated Logic Implementation

ALI Automotive Lift Institute [US]

α-Li Alpha Lithium [Symbol]

ALIBI Adaptive Location of Internetworked Bases of Information

ALIC Australian Land Information Council [now Australian and New Zealand Land Information Council]

ALICE Adiabatic Low-Energy Injection and Capture Experiment [also Alice] [Physics]

ALICE Airborne Line-Scan Image Processor [also Alice]

ALICE A Large Ion Collider Experiment [of CERN–European Laboratory for Particle Physics, Geneva, Switzerland] [also Alice]

ALICE Alaska Integrated Communications Exchange [also Alice]

Align Alignment [also align]

ALIM Air-Launched Intercept Missile

ALIMG Australian Land Information Management Group

ALIN Alcoa Laboratories Information Network [US]

A Line A Line [Exreme-Red Fraunhofer Line of Terrestrial Origin Having Wavelength of 766.1 Nanometers [also A line] [Spectroscopy]

AlInAs Aluminum Indium Arsenide (Semiconductor)

AlInGaP Aluminum Indium Gallium Phosphide (Semiconductor)

ALINK Active Link [HyperText Markup Language]

AlInP Aluminum Indium Phosphide (Semiconductor)

ALIP Annular Linear Induction Pump

ALIS Advanced Life Information System

ALIS Artificial and Logical Intelligence System

ALIS Automated Library Information System

ALISA American League for International Security Assistance [now American League for Exports and Security Assistance, US]

ALISA Australian Library and Information Science Abstracts

ALISE Association for Library and Information Science Education [US]

ALIT Association of Laser Inspection Technologies

ALIT Automatic Line Insulation Tester

ALIWEB Archie Like Indexing in the Web [Internet]

ALK Aaronson, Laird and Kinsman (Theory) [Metallurgy]

ALK Alkyl [also Alk]

Alk Alkali(ne) [also alk]

Alk Alkoxide [also alk]

AlKα Aluminum K-Alpha (Radiation) [also AlK$_\alpha$]

ALKAR Alkylation of Aromatics (Process)

AlKβ Aluminum K-Beta (Radiation) [also AlK$_\beta$]

Al/Kr Aluminum/Krypton (Interface)

Alky Alkalinity [also alky]

ALL Accelerated Learning of Logic

ALL Acute Lymphoblastic Leukemia

ALL Albanian Lek [Currency]

ALL Application Language Liberator

ALL Artificial Language Laboratory [Michigan State University, East Lansing, US]

ALL Association for Active Learning [UK]

ALL Astrological Lodge of London [UK]

ALL Atomic Layer-by-Layer (Technique)

.ALL Printer Information File [WordPerfect File Name Extension] [also .all]

All β-D-Allose [Biochemistry]

ALLA Allied Long Lines Agency [of NATO]

ALLA Automated Laboratory Liquor Analyzer

ALLBD Alternate Layer Langmuir-Blodgett Deposition [Physical Chemistry]

ALLC Association for Literary and Linguistic Computing [UK]

ALL-CAPS All Capital Letters [also all-caps]

ALLC J ALLC Journal [of Association for Literary and Linguistic Computing, UK]

Allegheny Ludlum Horiz Allegheny Ludlum Horizons [Published by Allegheny Ludlum Corporation, Pittsburgh, Pennsylvania, US]

Allg Papier-Rundschau Allgemeine Papier Rundschau [German Publication on the Paper Industry]

Allg Vliesstoff-Rep Allgemeiner Vliesstoff-Report [German Report on Nonwovens]

Al-Li Aluminum-Lithium (Alloy System)

Al-Li (International) Aluminum-Lithium Conference

AlLiCu Aluminum Lithium Copper (Compound)

Al/Li MMC Aluminum/Lithium Metal-Matrix Composite

ALL-MBE Atomic Layer-by-Layer Molecular Beam Epitaxy (Technique) [Physics]

Al-Ln Aluminum–Lanthanide Metal (Alloy)

Al-Ln-TM Aluminum–Lanthanide Metal–Transition Metal (Alloy)

Alloc Allocation [also alloc]

Allow Allowance [also allow]

Alloy Dig Alloy Digest [US]

ALLS Apollo Lunar Logistic Support [of NASA]

Al-LT-ET Aluminum (Based) Late-Transition-Metal Early-Transition-Metal (Alloy)

Al-LTM Aluminum (Based) Late-Transition-Metal (Glass)

Al-LTM-Ln Aluminum (Based) Late-Transition-Metal (Glass)

Al-LTM-ETM Aluminum (Based) Late-Transition-Metal Early-Transition-Metal (Glass)

Allum Leghe Alluminio e Leghe [Italian Journal on Aluminum and Aluminum Alloys]

Allum Mag Alluminio Magazine [Italian Aluminum Magazine]

ALM Alarm [also Alm, or alm]

ALM Applied Laboratory Method

ALM (Novell) AppWare Loadable Modules [Software]

ALM Assembler Language for MULTICS (Multiplexed Information and Computing Service)

ALM Asynchronous Line Module

ALM Asynchronous Line Multiplexer

ALM Audiolingual Method

ALM Aviation Lean Mixture

Alm Alarm [also alm]

ALMA Alphanumeric Language for Music Analysis

ALMA Alusuisse Maleic Anhydride (Process)

ALMA American Loudspeaker Manufacturers Association [US]

ALMA Analytical Laboratory Managers Association [US]

Alma Allylmethylamine

Al-27 MAS NMR Aluminum-27 Magic Angle Spinning Nuclear Magnetic Resonance [also ^{27}Al MAS NMR, Al-27 MAS-NMR, or ^{27}Al MAS-NMR]

ALMC Army Logistic Management Center [US]

ALMC Australian Labour Ministers Conference

AlMe$_3$ Trimethyl Aluminum [also Al(Me)$_3$]

Al-Mg Aluminum-Magnesium (Alloy System)

Al MgF$_2$ Aluminum with Magnesium Fluoride (Overcoating)

Al-MgO Aluminum-Magnesium Oxide (or Magnesia) (System)

AFML Air Force Materials Laboratory [US]

ALMM Aviation Lean Mixture Method

Al MMC Aluminum Metal-Matrix Composite

AlMn Aluminum Manganese (Compound)

Al-Mn Aluminum-Manganese (Alloy System)

Al-Mn-LN Aluminum–Manganese–Lanthanide Metal (Alloy)

AlMnSi Aluminum Manganese Silicon (Compound)

ALMO African Livestock Marketing Organization [Kenya]

ALMR Advanced Liquid Metal (Nuclear) Reactor

ALMS Aircraft Landing Measurement System

ALMS Analytic Language Manipulation System

ALMS Auxiliary Liquid Metal System

ALN Arid Lands Newsletter [of International Arid Lands Consortium]

ALN Asynchronous Learning Network

AlN Aluminum Nitride (Semiconductor)

2H-AlN Hexagonal Aluminum Nitride [Ramsdell Notation: 2 Refers to Number of Aluminum and Nitrogen Bilayers Necessary to Produce a Unit Cell, and H refers to the Hexagonal Symmetry]

ALNA Alpha Linolenic Acid (or Omega-3-Superunsaturated Fatty Acid)

AlN-Al$_2$O$_3$ Aluminum Nitride-Aluminum Oxide (System)

AlNbCr Aluminum-Niobium-Chromium (Compound)

AlN/BN Aluminum Nitride/Boron Nitride (Superlattice)

AlNd Aluminum Neodymium (Compound)

Al-Nd Aluminum-Neodymium (Alloy System)

Al/n-GaAs Aluminum on N-Type Gallium Arsenide (Substrate)

AlN:H Hydrogenated Aluminum Nitride

Al-Ni Aluminum-Nickel (Alloy System)

Al$_3$Ni Aluminum Nickelide

Al$_2$(NMe$_2$)$_6$ Hexakis(dimethylamido)dialuminum [General Formula]

Al-27 NMR Aluminum-27 Nuclear Magnetic Resonance [also ^{27}Al NMR]

AlNO Aluminonitrate

AlN/Si Aluminum Nitride on Silicon (Substrate)

AlN/SiC Aluminum Nitride/Silicon Carbide (Superlattice)

AlN/TiN Aluminum Nitride/Titanium Nitride (Superlattice)

AlN/ZrN Aluminum Nitride/Zirconium Nitride (Superlattice)

ALO Accredited Leasing Officer

ALO Analytical Laboratory Operations [of US Department of Energy]

Al$_2$O$_3$ Aluminum Oxide (or Alumina)

α-Al$_2$O$_3$ Alpha Aluminum Oxide (or Alpha Alumina)

β-Al$_2$O$_3$ Beta Aluminum Oxide (or Beta Alumina)

δ-Al$_2$O$_3$ Delta Aluminum Oxide (or Delta Alumina)

η-Al$_2$O$_3$ Eta Aluminum Oxide (or Eta Alumina)

γ-Al$_2$O$_3$ Gamma Aluminum Oxide (or Gamma Alumina, or Gamal)

θ-Al$_2$O$_3$ Theta Aluminum Oxide (or Theta Alumina)

Al$_2$O$_3$(f) Aluminum Oxide (or Alumina) Fiber [also Al$_2$O$_3$f]

Al$_2$O$_3$(p) Aluminum Oxide (or Alumina) Particulate [also Al$_2$O$_3$p]

Al$_2$O$_3$(w) Aluminum Oxide (or Alumina) Whiskers [also Al$_2$O$_3$w]

ALOA Associated Locksmiths of America [US]

Al$_2$O$_3$/Al Alumina Reinforced Aluminum (Composite)

Al$_2$O$_3$/Al Aluminum on Alumina (Substrate)

Al$_2$O$_3$(f)/Al Alumina Fiber Reinforced Aluminum (Composite) [also Al$_2$O$_3$p/Al]

Al$_2$O$_3$(p)/Al Alumina Particulate Fiber Reinforced Aluminum (Composite) [also Al$_2$O$_3$f/Al]

Al$_2$O$_3$/Al-Mg Alumina Reinforced Aluminum-Magnesium (Composite)

Al$_2$O$_3$/Al-Mg Aluminum-Magnesium on Alumina (Substrate)

Al(OBu)$_2$ Aluminum(II) sec-Butoxide

Al(OBu)$_3$ Aluminum(III) Butoxide (or Aluminum(III) Butylate)

Al(OBu^s)$_3$ Aluminum Tri-sec-Butoxide (or Aluminum sec-Butoxide) [also Al(OBu^{sec})$_3$, or Al(sec-OBu)$_3$]

Al(OBu^s)$_2$(CH$_3$COO) Aluminum Di-sec-Butoxide Ethylacetoacetate [also Al(sec-OBu)$_2$(CH$_3$COO), or Al(OBu^s)$_2$(CH$_3$COO)]

AlOC Aluminum Oxycarbide (Ceramic)

Al$_2$O$_3$-Cr$_2$O$_3$ Aluminum Oxide–Chromic Oxide (System)

Al$_2$O$_3$/Ep Alumina Reinforced Epoxy (Composite) [also Al$_2$O$_3$/E]

Al$_2$O$_3$:Er Erbium-Doped Aluminum Oxide

Al(OEt)$_3$ Aluminum(III) Ethoxide (or Aluminum(III) Ethylate)

Al$_2$O$_3$(f) Aluminum Oxide (or Alumina) Fiber [also Al$_2$O$_3$f]

Al$_2$O$_3$–Fe$_2$O$_3$ Aluminum Oxide-Ferric Oxide (System)

Al$_2$O$_3$(f)/Al Alumina Fiber Reinforced Aluminum (Composite) [also Al$_2$O$_3$p/Al]

ALOFT Airborne Light Optical Fiber Technology

ALOG Applied Laser Optics Group

AlOH Allyl Alcohol

Al(O-i-Pr)$_3$ Aluminum(III) Isopropoxide [also Al(OPr^i)$_3$]

Al$_2$O$_3$/Mg Alumina Reinforced Magnesium (Composite)

Al$_2$O$_3$/Mg Magnesium-on-Alumina (Substrate)

Al$_2$O$_3$(p)/Mg Alumina Particulate Reinforced Magnesium (Composite) [also Al$_2$O$_3$p/Mg]

Al$_2$O$_3$:Mg Magnesium-Doped Aluminum Oxide

ALON Aluminum Oxide/Aluminum Nitride

ALON (γ-)Aluminum Oxynitride

Al$_2$O$_3$/Nb Alumina Reinforced Niobium (Composite)

Al$_2$O$_3$(p) Aluminum Oxide (or Alumina) Particulate [also Al$_2$O$_3$p]

Al$_2$O$_3$(p)/Al Alumina Particulate Fiber Reinforced Aluminum (Composite) [also Al$_2$O$_3$f/Al]

Al$_2$O$_3$(p)/Mg Alumina Particulate Reinforced Magnesium (Composite) [also Al$_2$O$_3$p/Mg]

Al(OPr^i)$_3$ Aluminum(III) Isopropoxide [also Al(O-iPr)$_3$]

ALOR Advanced Lunar Orbital Rendezvous

Al(OR)$_3$ Aluminum (III) Alkoxide [General Formula; R represents an Alkyl Group]

ALOS Advanced Land Observing (or Observation) Satellite [of National Space Development Agency of Japan]

ALOS Artificially-Made Layered Oxide Superconductor

Al$_2$O$_3$-SiO$_2$ Aluminum Oxide-Silicon Dioxide (System)

ALOT Airborne Lightweight Optical Tracker

Al$_2$O$_3$(Tl) Thallium-Doped Aluminum Oxide

Al$_2$O$_3$(w) Aluminum Oxide (or Alumina) Whiskers [also Al$_2$O$_3$w]

ALOTS Airborne Lightweight Optical Tracking System

Al$_2$O$_3$-YSTZ Alumina Reinforced Yttria-Stabilized Tetragonal Zirconia (Composite)

ALP Acute Lobular Pneumonia [Medicine]

ALP Alkaline Phosphatase [Biochemistry]

ALP Arithmetic Logic Processor

ALP Articulated Loading Platform

ALP Assembly Language Preprocessor; Assembly Language Processing

ALP Assembly Language Program

ALP Associated Laguerre Polynomial

ALP Atomic Layer Processing

ALP Automated Language Processing

ALP Automated Learning Process

AlP Alkaline Phosphatase [Biochemistry]

AlP Aluminum Phosphide (Semiconductor)

Al/p-Si Aluminum on P-Type Silicon

ALPA Alaskan Long Period Array

ALPAC Automated Language Processing Advisory Committee [of National Academy of Sciences, US]

Al-Pb Aluminum-Lead (Alloy System)

ALPC Adaptive Linear Protective Coding

Al-Pd Aluminum-Palladium (Alloy System)

AlPdMn Aluminum Palladium Manganese (Compound)

Alpeth Aluminum Cable Jacket with Sprayed Polyethylene Coating [also alpeth] [Electric Cables]

ALPHA Automated Literature Processing, Handling and Analysis

Alphanum Alphanumeric(s) [also alphanum]

ALPHGR Average Linear Planar Heat Generation Rate

ALPID Analysis of Large Plastic Incremental Deformation (Computer Simulation)

ALPIDP Analysis of Large Plastic Incremental Deformation, Powder Metals

ALPIDT Analysis of Large Plastic Incremental Deformation, Nonisothermal

ALPL Association of Librarians in Public Libraries [Germany]

AlPN Aluminum Phosphide Nitride

ALPO Association of Lunar and Planetary Observers [US]

AlPO Aluminophosphate

ALPR Argonne Low Power Reactor [of Argonne National Laboratory, near Illinois, US; Started up in 1958] [also CP-9]

AlPr Aluminum Praseodymium (Compound)

ALPS Advanced Linear Programming System

ALPS Advanced Liquid Propulsion System

ALPS Alyeska Oil Pipeline Service Company [Alaska, US]

ALPS Assembly Line Planning System

ALPS Associated (or Associative) Logic Parallel System

ALPS Automated Library Processing Service

ALPSP Association of Learned and Professional Society Publishers [UK]

ALPURCOMS All-Purpose Communications System

ALQ₃ Tris(8-Hydroxyquinoline) [also Alq₃]

ALQDS All Quadrants

ALR Active Line Rotation

ALR Adiabatic Lapse Rate [Meteorology]

ALR Advanced Logic Research, Inc. [US]

ALR Agricultural Land Reserve [Canada]

ALR Alarm Reset

ALR Allergenic Effects

ALRC Anti-Locust Research Center [London, UK]

AlRe Aluminum Rhenium (Compound)

ALRI Airborne Long-Range Input

ALRI Airborne Long-Range Intercept

ALRL Aluminum Research Laboratory [of Alcoa Inc., US]

ALRO Atomic Long-Range Order(ing) [Solid-State Physics]

ALROS American Laryngological, Rhinological and Otological Society [US]

ALRR Ames Laboratory Research Reactor [of US Department of Energy at Iowa State University]

ALRT Advanced Light Rapid Transit

ALRT Automated Light Rapid Transit

ALRU Automatic Line Record Update

AlRu Aluminum Ruthenium (Compound)

ALRUTS Automated Line Replaceable Unit Tracking System

ALS Advanced Launch System

ALS Advanced Life Support

ALS Advanced Light Source [of Berkeley National Laboratory, California, US]

ALS Advanced Logistics System

ALS Advanced Low-Power Schottky (Logic)

ALS Agricultural Land Service [US]

ALS Airborne Lidar System

ALS Aircraft Landing System

ALS Alberta Land Surveyors [Canada]

ALS Alternating Layer Spin [Physics]

ALS American Littoral Society [US]

ALS American Lunar Society [US]

ALS Ammonium Lignosulfonate

ALS Amyotrophic Lateral Sclerosis

ALS Analytical Laboratory System

ALS Antilymphocyte Globulin; Antilymphocytic Serum [Biochemistry]

ALS Anti-Mouse Lymphocyte Serum [Biochemistry]

ALS Approach Light System

ALS Arithmetic and Logic Section

ALS Augmented Landing Site

ALS Australian Littoral Society

ALS Automated Library System

ALS Automatic Landing System

AlS Aluminum Sulfide (Semiconductor)

Al-S Aluminum-Sulfur (Alloy System)

ALSA Area Library Service Authority

ALSARM Advanced Life Support Automated Robotic Mechanism

ALSB Aqueous Lithium Salt Blanket

AlSb Aluminum Antimonide (Semiconductor)

AlSb/GaSb Aluminum Antimonide/Gallium Antimonide (Semiconductor)

ALSC American Lumber Standards Committee [US]

ALSCP Appalachian Land Stabilization and Conservation Program [US]

ALSEP Apollo Lunar Surface Experiment Package [of NASA]

Al SG Spheroidal Graphite Aluminum (Cast Iron)

Al SG Iron Spheroidal Graphite Aluminum (Cast) Iron [also Al SG iron]

Al-Si Aluminum-Silicon (Alloy System)

Al-Si/B₄C Boron Carbide Reinforced AluminumSilicon (Composite)

Al/SiC Silicon Carbide Reinforced Aluminum (Composite)

Al/SiC MMC Silicon Carbide Reinforced Aluminum Metal-Matrix Composite

Al/SiC(p) Silicon Carbide Particulate Reinforced Aluminum-Matrix Composite [also Al/SiCp, or Al/SiCₚ]

AlSiCr Aluminum-Silicon-Chromium (Coating)

Al/SiCw Silicon Carbide Whiskers Reinforced Aluminum-Matrix Composite [also Al/SiCw]

Al-Si MMC Aluminum-Silicon Metal-Matrix Composite

Al SiO Aluminum with Silicon Monoxide (Overcoating)

Al/SiO₂ Silicon Dioxide Reinforced Aluminum (Composite)

Al-Si/SiC Silicon Carbide Reinforced Aluminum Silicon (Composite)

Alsk Alaska

Als-Lor Alsace-Lorraine [Historic Region in Northeastern France]

AlSO Aluminosulfate

ALSOCO Aluminum Solder Company [US]

ALSPEI Association of Land Surveyors of Prince Edward Island [Canada]

ALSS Advanced Location Strike System

ALSS Airborne Location and Strike System

ALSS Apollo Logistic Support System [of NASA]

ALSTG Altimeter Setting

ALSTTL Advanced Low-Power Schottky Transistor-Transistor Logic [also ALS TTL]

ALT Accelerated-Life Test(er); Accelerated Life Testing

ALT Acoustic Leak Test(ing)

ALT Activated Liquid Thromboplastin [Biochemistry]

ALT Airborne Laser Tracker

ALT Alanine Aminotransferase [Biochemistry]

ALT Alternate (Key) [Computers]

ALT Approach and Landing Test (Program) [NASA, Space Shuttle Program]

Alt Alteration [also alt]

Alt Alternate; Alternating; Alternation; Alternative; Alternator [also alt]

Alt Altimeter

Alt Altitude [also alt]

Alt D-Altrose [Biochemistry]

ALTA Active Long-Term Archive

ALTA Alberta Library Trustees Association [Canada]

ALTA Association of Local Transport Airlines

Alta Alberta [Canada]

ALTAC Algebraic Transistorized Automatic Computer

ALTAC Algebraic Translator and Compiler

Alta Freq Alta Frequenza [Journal of Associazione Elettrotecnica ed Elettronica Italiana, Italy]

Alta Freq, Riv Elettron Alta Frequenza – Rivista di Elettronica [Electronics Review; published by Associazione Elettrotecnica ed Elettronica Italiana, Italy]

ALTAIR ARPA (Advanced Research Projects Agency) Long-Range Tracking and Instrumentation Radar [of US Department of Defense]

ALTAN Alternate Alerting Network [US Air Force]

ALTARE Automatic Logic Testing and Recording Equipment

ALTECH Alterra Laser Technologies Project [of University of Alberta, Canada]

Alter Alteration [also alter]

Alter Alternating; Alternative [also alter]

ALTERM Australia's Long-Term Ecological Research and Monitoring (Program)

Altern Alternative [also altern]

Altern Med Dig Alternative Medicine Digest [US]

Altern Med Rev Alternative Medicine Review [International Journal]

Altern Sources Energy Alternative Sources of Energy [Journal published in the US]

ALT/GPT Alanine Aminotransferase/Glutamic-Pyruvic Transaminase [Biochemistry]

Al-Ti Aluminum-Titanium (Alloy System)

Al/TiB$_2$ Titanium Diboride Reinforced Aluminum

AlTiC Alumina Reinforced Titanium Carbide

Al/TiC Titanium Carbide Reinforced Aluminum (Composite)

Al/TiC(p) Titanium Carbide Particulate Reinforced Aluminum (Composite) [also Al/TiCp, or Al/TiC$_p$]

Al/TiN Titanium Nitride Reinforced Aluminum (Composite)

Al-TM-LN Aluminum–Transition Metal–Lanthanide Metal (Alloy)

Altm Altimeter [also altm]

Al(TMHD)$_3$ Aluminum Tris(2,2,6,6-Tetramethyl-3,5-Heptanedionate) [also Al(tmhd)$_3$]

Al-TM-Ln Aluminum–Transition-Metal–Lanthanide Metal (Alloy)

Al-TM-RE Aluminum–Transition-Metal–Rare Earth (Ternary Alloy)

Al-TM$_1$-TM$_2$-RE Aluminum–Transition-Metal 1–Transition Metal 2–Rare Earth (Quaternary Alloy)

Altn Alternation; Alternative [also altn]

ALTO Association of Local Television Operators [UK]

Alto Cu Alto-Cumulus (Cloud) [also alto cu]

ALTPLAN Accelerated Life Test Plan (Development Software)

ALTRACO Allied Transportation and Trading Corporation

ALTRAN Algebraic Translator

ALTRAN Assembly Language Translator

ALTRV Altitude Reservation

ALTS Advanced Lunar Transportation System

ALTS Automated Land Titles System [Database of the States of New South Wales and Victoria in Australia]

ALTS Automated Library Technical Services [US]

ALTS Automatic Line Test Set

Al-TZP Alumina-Stabilized Tetragonal Zirconia Polycrystal

ALU Aboriginal Liaison Unit [Australia]

ALU Advanced Levitation Unit

ALU Arithmetic Logic Unit

ALU Asynchronous Line Unit

Alu Arthrobacter luteus [Microbiology]

α-Lu Alpha Lutetium [Symbol]

ALUE Admissible Linear Unbiased Estimator

Alum Aluminum [also alum]

Alum Appl Aluminium Applications [Journal published in the UK]

Alum Assoc Rep Aluminum Association Report [of Aluminum Association, US]

Alum Dev Dig Aluminum Developments Digests [US]

Alum Ind Aluminium Industry [Journal published in the UK]

Alum Kur Aluminium Kurier [German Aluminum Courier]

Alum Mon Aluminium Monthly [UK]

Alum Remodeling News Aluminum Remodeling News [US]

Alum Rep Aluminum Report [of Aluminum Association, US]

Alum Rev Aluminium Review [South Africa]

Alum Situat Aluminum Situation [of Aluminum Association, US]

Alum Stat Rev Aluminum Statistical Review [of Aluminum Association, US]

Alum Today Aluminium Today [UK]

Alum Use Aluminium in Use [Australia]

Alurama Int Alurama International [Published in Sweden]

ALUSELECT Materials Database for Aluminum Alloys

ALV Airlock Valve

ALV Autonomous Land Vehicle

ALV Avian Leukosis Virus

ALVIN Antenna Lobe for Variable Ionospheric Nimbus

ALW Advanced Laboratory Workstation

Alw Acinetobacter Iwoffii [Microbiology]

ALWIN Algorithmic Wiswesser Notation

ALWIND Along (with) Wind Direction

AlwN Acinetobacter Iwoffii N [Microbiology]

ALWR Advanced Light Water Reactor

Al-Y Aluminum-Yttrium (Alloy System)

Aly Alloy [also aly]

Al-Zn Aluminum-Zinc (Alloy System)

Al-Zr Aluminum-Zirconium (Alloy System)

AM Aberration Microscopy

AM Abrasive Machining

AM Absolute Magnetometer

AM Absolute Magnitude

AM Académie de Marine [Marine Academy, France]

AM Académie de Médecine [Academy of Medicine, Paris, France]

AM Academy of Medicine [Canada]

AM Access Mechanism [Computers]

AM Access Method

AM Accounting Machine

AM Accumulator Metal

AM Acoustic Material

AM Acoustic Microscope; Acoustic Microscopy

AM Acrylamide

AM Actinomycosis [Medicine]

AM Active Matrix

AM Acute Myocarditis

AM Address Mark

AM Address Memory

AM Address Mode

AM Address Modifier

AM Advanced Material

AM Advance Mailer

AM Aeromechanic(s)

AM Aerometeorograph(y)

AM Aeroméxico [Mexican Air Lines]

AM Aeronava de Mexico [Mexican Air Carrier]

AM Aircraft Mechanic

AM Airlock Module

AM Air Management

AM Air Marshall

AM (Optical) Air Mass (Number)

AM Air-Melted; Air Melting [Metallurgy]

AM Air Ministry [UK]

AM Alkali Metal

AM Alternating Magnetic Field

AM American Motors (Corporation) [US]

AM Ammeter

AM Amorphous Material

AM Amorphous Metal

AM Amplifier [also Am, or am]

AM Amplitude [also Am, or am]

AM Amplitude Modulation [also am, or a-m]

AM Analog Monolithic

AM Anderson Model [Physics]

AM Angular Momentum

AM *(Anno Mundi)* – in the year of the world

AM Anodized Magnesium

AM Anomalous Mode

AM *(Ante Meridiem)* – before noon [also am]

AM Anterior-Median (Eyes) [Spiders]

AM Antimatter [Physics]

AM Appalachian Mountains [US]

AM Apparent Magnitude

AM Applied Mathematics

AM Applied Mechanics

AM Applied Meteorology

AM Applied Metrology

AM Aquatic Microbiology

AM Arachnoid Membrane [Anatomy]

AM Archeological Metallurgist; Archeological Metallurgy; Archeometallurgist; Archeometallurgy

AM Arc-Melted; Arc Melting [Metallurgy]

AM Arithmetic Mean

AM Armenia [ISO Code]

AM Army Manual [UK]

AM *(Artium Magister)* – Master of Arts

AM Asia Minor

AM As Milled (Condition)

AM Associate Member

AM Associative Memory

AM Astable Multivibrator

AM Astrometric; Astrometry

AM Asymmetric Membrane

AM Asynchronous Modem

AM Asynchronous Motor

AM Atomic Mass

AM Atomic Model

AM Atomistic Modelling

AM Attritor Mill(ing)

AM Auger Mining

AM Austenitic Martensite [Metallurgy]

AM Australian Museum [Sydney, Australia]

AM Automated Mapping

AM Auxiliary Memory

AM Aviation Medicine

AM Aviation Mixture

AM Master of Arts

AM0 (Optical) Air Mass 0 [also AM 0]

AM1 (Optical) Air Mass 1 [also AM 1]

AM1.5 (Optical) Air Mass 1.5 [also AM 1.5]

AM2 (Optical) Air Mass 2 [also AM 2]

AM1 Austin Model 1 (Semiempirical Quantum Chemical Method)

A&M Agricultural and Mechanical [also A and M]

A-M Austenite-to-Martensite (Transformation) [Metallurgy] [also A→M]

Am Alabamine

Am America(n)

Am Americium [Symbol]

Am Ammonium [Symbol]

Am Amyl [also am]

Am Amplifier [also AM, or am]

Am Amplitude [also AM, or am]

Am Tropical Rainforest Climate, Short Dry Season [Classification Symbol]

Am-237 Americium-237 [also ^{237}Am, or Am^{237}]

Am-238 Americium-238 [also ^{238}Am, or Am^{238}]

Am-239 Americium-239 [also ^{239}Am, or Am^{239}]

Am-240 Americium-240 [also ^{240}Am, or Am^{240}]

Am-241 Americium-241 [also ^{241}Am, Am^{241}, or Am]

Am-242 Americium-242 [also ^{242}Am, or Am^{242}]

Am-243 Americium-243 [also ^{243}Am, or Am^{243}]

Am-244 Americium-244 [also ^{244}Am, or Am^{244}]

A/m Ampere(-turn) per meter [also A m^{-1}]

A/m² Ampere per square meter [also A m^{-2}]

A·m² Ampere-meters squared [also Am², A·m², or amp-m²]

am ambient

am amorphous (state) [Chemistry]

am ampere-minute [Unit]

am *(ante meridiem)* – before noon [also AM]

a/m above mentioned

a-m amplitude modulation [also AM, or am]

.am Armenia [Country Code/Domain Name]

a/m³ alpha-decay events per cubic meter [also α^{-3}]

AMA Academy of Model Aeronautics [of National Aeronautics Association, US]

AMA Acoustical Materials Association [US]

AMA Acrylate Maleic Anhydride (Terpolymer)

AMA Actual Mechanical Advantage

AMA Adhesives Manufacturers Association [US]

AMA Aero Medical Association [US]

AMA Against Medical Advice

AMA Agricultural Marketing Administration [US]

AMA Air Material Area [US Air Force]

AMA Alberta Medical Association [Canada]

AMA Alberta Motor Association [Canada]

AMA Amarillo, Texas [Meteorological Station Designator]

AMA American Management Association [US]

AMA American Manufacturers Association [US]

AMA American Maritime Association [US]

AMA American Marketing Association [US]

AMA American Medical Association [US]

AMA American Meteor Association [US]

AMA American Military Attaché

AMA American Motorcycle Association [US]

AMA American Municipal Association [US]

AMA Amorphous Magnetic Alloy

AMA Anti-Markovnikov Addition [Organic Chemistry]

AMA Arbeitsgemeinschaft Meßwert-Aufnahme [Measurement Recording Association, Germany]

AMA Architectural Metal Association [Canada]

AMA Area Minimum Altitude

AMA Asbestos Mining Association [Canada]

AMA Association of Manufacturers and Allied Electrical Industries [UK]

AMA Associative Memory Address

AMA Associative Memory Array

AMA Auckland Mathematical Association [New Zealand]

AMA Automatic Memory Allocation

AMA Automatic Message Accounting [Telecommunications]

AMA Automobile Manufacturers Association [US]

AMAA Agencia de Medio Ambiente de Asturias [Asturias Environmental Agency, Spain]

AMAB Arctic Marine Advisory Board [Canada]

AMAC 5-Aminobarbituric-N,N-Diacetic Acid [also Amac]

AMAC Automatic Material Completion

AMAC Ayurvedic Medical Association of Canada

A-MAC Low-Bandwidth Multiplexed Analog Components

AmAc Ammonium Acetate

Am Acad Environ Eng Dipl American Academy of Environmental Engineers Diplomate [US Publication]

AMACUS Automated Microfilm Aperture Card Updating System [of US Army]

AMAD Activity Median Aerodynamic Diameter

AMAD Aircraft-Mounted Accessory Drive

AMADS Airframe-Mounted Auxiliary Drive System

AMA/I American Management Association International [US]

AMAIS Agricultural Materials Analysis Information Service [UK]

AMAL Aviation Medical Acceleration Laboratories [of US Navy]

Amal Amalgam(ation) [also amal]

Am Alc Amyl Alcohol [also am alc]

Amalg Amalgamation [also amalg]

AMAN Automatic Material Number

AMANDDA Automated Messaging and Directory Assistance

AMA/NET American Medical Association Network [US]

Am Anthrop The American Anthropologist [US Journal]

Am Anthrop Mem American Anthropological Memoirs [US]

Am Antiq American Antiquity [US Journal]

AMAP Arctic Monitoring and Assessment Programme

AMAQ 1-Amino-2-Methylanthraquinone

AMARS Air Mobile Aircraft Refueling System

Am Assoc Industr Nurses J American Association of Industrial Nurses Journal [US]

Amat Radio Amateur Radio [US Magazine]

AMAU Asynchronous Message Assembly Unit

AMAUS Aero Medical Association of the United States

AMARV Advanced Maneuvering Reentry Vehicle

AMAS Advanced Midcourse Active System

AMAS Automatic Material Status

AMAS Maleimidoacetic Acid N-Hydroxysuccinimide Ester

AMASE Astrophysics Multi-Spectral Archive Search Engine [of NASA Astrophysics Data Facility, Goddard Space Flight Center, US]

AMATYC American Mathematical Association of Two-Year Colleges [US]

Amaz Amazon (River)

Amaz Amazonas [Brazilian State]

AMB AIDS Malignancy Bank

AMB Airways Modernization Board [Now Part of Federal Aviation Administration, US]

AMB Anti-Motor Boat

AMB Asbestos Millboard

AMB (Internationale) Ausstellung der Metallbearbeitenden Industrie [International Exhibition for Metalworking, Stuttgart, Germany]

AMB Auto-Manual Bridge (Control)

A/MB Anti-Motor Boat

Amb Ambassador [also amb]

amb amber

amb ambient

AMBA American Malting Barley Association [US]

AMBA Association of Military Banks of America [US]

AMBCO Association of Major City Building Officials [US]

AMBCS Alaskan Meteor Burst Communication System

AMBD Automatic Multiple Blade Damper

AMBE Accelerated Molecular Beam Epitaxy

Am/Be Americium/Beryllium (Neutron Source)

Am Bee J American Bee Journal [US]

Am Book Publ Rec American Book Publishing Record [US Publication]

AMBES Association of Metropolitan Borough Engineers and Surveyors [UK]

AMBIM Associate Member of the British Institute of Management

AMBIT Algebraic Manipulation by Identity Translation [Computer Language]

AmBr Ammonium Bromide

AMC Acceptable Means of Compliance

AMC Account Manager Code

AMC Acetylmethylcarbon

AMC Aerodynamic Mean Chord

AMC Aerospace Medical Center [at Brooks Air Force Base, San Antonio, Texas, US]

AMC AIDS Malignancy Consortium

AMC Airborne Management Computer

AMC Aircraft Manufacturers Council

AMC Air Materiel Center [of US Navy]

AMC Air Materiel Command [US Air Force]

AMC Albany Medical College [New York State, US]

AMC Alberta Microelectronics Center [Canada]

AMC Aluminum Manganese Copper

AMC American Mining Congress [US]

AMC American Motors Corporation [Now part of Chrysler Corporation, US]

AMC American Movers Conference [US]

AMC Amorphous-to-Crystalline (Transformation) [also AM-C]

AMC 7-Amino-4-Methylcoumarin

AMC Antarctic Minerals Convention

AMC Army Materiel Command [of US Army at Washington, DC]

AMC Army Medical Center [US]

AMC Army Medical Corps [US]

AMC Army Missile Command [at Redstone Arsenal in Huntsville, Alabama, US]

AMC Arthritis Medical Center [Fort Lauderdale, Florida, US]

AMC Association of Management Consultants

AMC Atlantic Marine Center [of National Oceanic and Atmospheric Administration at Norfolk, Virginia, US]

AMC Australian Manufacturing Council

AMC Australian Maritime College

AMC Automated Meter Calibration

AMC Automatic Melt-Rate Control [Metallurgy]

AMC Automatic Message Counting

AMC Automatic Mixture Control

AMC Automatic Modulation Control

AMC Automatic Monitoring Circuit

AMC Auto Meter Correct

AMC Autonomous Multiplexer Channel

AMC Avionics Maintenance Conference [US]

AMCA Advanced Materiel Concept Agency [US]

AMCA Air Movement and Control Association [US]

AMCA Air Moving and Conditioning Association [Name Changed to Air Movement and Control Association, US]

AMCA Amateur Motor-Cycle Association [UK]

AMCA American Mosquito Control Association [US]

AMCA 7-Amino-4-Methylcoumarin-3-Acetic Acid

AMCA Asociación Mexicana de Control Automático [Mexican Association for Automatic Control]

AMCAP Advanced Microwave Circuit Analysis Program

Am Cart American Cartographer [Publication of the American Congress on Surveying and Mapping]

AMCB Asian Mass Communication Bulletin [of Asian Mass Communications Research and Information Center, Singapore]

AMCBMC Air Materiel Command Ballistic Missile Center [US Air Force]

AMCCW Association of Manufacturers of Chilled Car Wheels [US]

AMCEC Allied Military Communications Electronics Committee

AMCEE Association for Media-Based Continuing Education for Engineers [US]

AMCEN African Ministerial Conference on the Environment

Am Ceram Soc American Ceramic Society [US]

Am Ceram Soc Bull American Ceramic Society Bulletin [US]

AMCF Alkali Metal Cleaning Facility

AMCHA 6-(Aminomethyl)cyclohexanecarboxylic Acid

AMCHAM American Chamber of Commerce [also AmCham]

Am Chem Soc American Chemical Society [US]

AMC J American Mining Congress Journal [also AMCJ]

AmCl Ammonium Chloride

AMCM Academia Mexicana de Ciencia de Materiales [Mexican Academy of Materials Science]

AMCO Association of Management and Consulting Organizations [Ireland]

AMCOM Automated Machine Control for Optimized Mining

Am Concr Inst American Concrete Institute [US]

AMCORD Australian Model Code for Residential Development

AMCP Abnormal Milk Control Program

AMCP Allied Military Communications Panel

Am Cryst Assoc American Crystallographic Association [US]

AMCS Airborne Missile Control System

AMCS Aircraft Maintenance Control System

AMCS Association for Marine Catering and Supply [UK]

AMCS Association for Mexican Cave Studies [US]

AMCS Association of Military Colleges and Schools [US]

AMCTB Associated Motor Carriers Tariff Bureau [US]

AMD Acid Mine Drainage

AMD Active Matrix Display

AMD Advanced Microcomputer Device

AMD Advanced Micro Devices, Inc. [US]

AMD Aerospace Medical Division

AMD Ambulance Manufacturers Division [of National Truck Equipment Association, US]

AMD Antarctic Master Directory

AMD Applied Mechanics Division [of American Society of Mechanical Engineers International]

AMD Arbeitskreis Mittlere Datentechnik [Association for Medium-Size Data Systems Technology, Germany]

AMD Associated Manufacturing and Design [Alexandria, Virginia]

AMD Atomized Melt Deposition [Metallurgy]

AMD Automated Multiple Development System

AMDA Airlines Medical Directors Association

AMDA Associated Minicomputers Dealers of America [US]

AMDA Association of Medical Doctors in Asia

AMDA Automated Multilayer Deposition Accessory

AMDAA Accounting Machine Dealers Association of America [later AMMDA; now AMDA]

AMDAC Amdahl Diagnostic Resistance Center [US]

AMDB Agricultural Machinery Development Board [UK]

AMDCB Applied Moment Double-Cantilever Beam (Method) [Mechanics]

AMDCL Association of Metropolitan District Chief Librarians [UK]

AMDE Association of Medical Doctors in Europe

AMDEA Association of Domestic Electrical Appliances [UK]

AMDEC Agricultural Marketing Development Executive Committee [UK]

AMDECL Association of Metropolitan District Education Librarians [UK]

AMDEL Australian Mineral Development Laboratories [also Amdel]

Amdel Bull Amdel Bulletin [of Australian Mineral Development Laboratories]

Am Dent Assoc American Dental Association

AMDG *(Ad Majorem Dei Gloriam)* – to the greater glory of God

AMDM Association of Microbiological Diagnostic Manufacturers [US]

AMDMA American Metal Detector Manufacturers Association [US]

AMDS Australian MARC (Machine-Readable Catalogue) Distribution Service

AMDSB Amplitude Modulation, Double Sideband [also AM DSB, or AM-DSB]

Amdt Amendment [also amdt]

Am Dyestuff Rep American Dyestuff Reporter [US Publication]

AME Advanced Master of Education

AME Advanced Microstructural Evaluation for Materials (Program) [of University of Toronto, McMaster University and Ontario Hydro Technologies, Canada]

AME Advanced Modelling Extension [AutoCAD Software]

AME Amplitude Modulation Equivalent

AME Angle-Measuring Equipment

AME Association of Manufacturing Excellence [US]

AME Association of Membership [US]

AME Authorized Medical Examiner

AME Automatic Microfiche Editor

AMEA Association of Machinery and Equipment Appraisers [US]

AMEC Advanced Materials Engineering Center [Canada]

AMEC Aerospace Metals Engineering Committee [of Society of Automotive Engineers, US]

AMEC Association of Mining and Exploration Companies (Incorporated)

AMEC Australian Minerals and Energy Council

AMECD Association for Measurement and Evaluation in Counseling and Development [US]

AME/COTAR Angle Measuring Equipment/Correlation Tracking and Ranging

AMED Association for Management Education and Development [UK]

AM Ed Advanced Master of Education

AMEDAS Automated Meteorological Data Acquisition System

AMEDD Army Medical Department [US]

AMEDPC Automotive Manufacturers Electronic Data Processing Council [US]

AMEDS Army Medical Service

AMEE Admiralty Marine Engineering Establishment [UK]

AMEE Agriculture, Materials, Earth and Engineering Sciences

AMEE Association of Managerial Electrical Executives [UK]

AMEEF Australian Minerals and Energy Environmental Foundation

AMEG Association for Measurement and Evaluation in Guidance

AMEIC Associate Member of the Engineering Institute of Canada

AMEL Aircraft Multi-Engine Land

AMEM Association of Marine Engine Manufacturers [US]

AMEM Association of Miniature Engine Manufacturers [UK]

AMEME Association of Mining, Electrical and Mechanical Engineers [UK]

AMEO 3-Aminopropyl Triethoxy Silane

Amer America(n)

Amer Std American Standard

AMES Aerospace Materials Engineering Committee [of Society of Automotive Engineers, US]

AMES Aged Materials Evaluation and Study Network [of Institute for Advanced Materials, Netherlands]

AMES Air Ministry Experimental Station [UK]

AMES Applied Mechanics and Engineering Sciences [of University of California at San Diego, US]

AMES Association for Media Education in Scotland

AMES Association of Marine Engineering Schools [UK]

Ames Lab Ames Laboratory [of US Department of Energy at Iowa State University]

AMET Accumulated Mission Elapsed Time

AMETA Army Management Engineering Training Agency [US]

Am Ethnol American Ethnologist [US Journal]

AMETS Artillery Meteorological System

AMEU Association of Municipal Electrical Utilities

AMEX American Express [also Amex]

AMEX American Stock Exchange [also Amex, US]

AMEX Amine Extraction (Process)

AMEX Australian Monsoon Experiment

Amex American Express [also AMEX]

AMF ACE (Allied Command Europe) Mobile Force [of NATO]

AMF Advanced Manufacturing Facility

AMF Air Mail Facility

AMF Arc-Melting Furnace [Metallurgy]

AMF Arab Monetary Fund

AMF Assistant Manager of Facility Transition

AMF Astronaut Memorial Foundation [US]

AMF Australian Mineral Foundation

AMF Automatic Mirror Furnace

AMF Automatic Mirror Facility

AmF Ammonium Fluoride

AMF(A) ACE (Allied Command Europe) Mobile Force (Air) [of NATO]

Am Fam Phys American Family Physician [US Journal]

AMFAX Amplitude Modulation–Facsmilie [also AM FAX, or AM-FAX]

AMFC Adaptive Model-Following Control

AMFEP Association of Microbial Food Enzyme Producers [Belgium]

AMFI Aviation Maintenance Foundation International [US]

AMFI Ind News AMFI Industry News [of Aviation Maintenance Foundation International, US]

AMFIS American Microfilm Information Society [US]

AMFIS Automatic Microfilm Information Society

AMFIS Automatic Microfilm Information System

AMF(L) ACE (Allied Command Europe) Mobile Force (Land) [of NATO]

AM/FM Amplitude Modulation/Frequency Modulation

AM/FM Automated Mapping/Facilities Management [Geographic Information System]

Am Forests American Forests [Publication of the American Forestry Association, US]

AMFS American Federation of Mineralogical Societies

AMFS Newsl AMFS Newsletter [of American Federation of Mineralogical Societies]

AMFWA Association of Midwest Fish and Wildlife Agencies [US]

AMG Abrasives Marketing Group

AMG Allied Military Government

AMG American Military Government

AMG Association of Malayan Geographers

AMG Australian Map Grid

AMG Avalanche Microwave Generator

a-MgCuY Magnesium Copper Yttrium Amorphous Alloy

a-Mg-LT-ET Magnesium Late-Transition-Metal–Early-Transition-Metal Amorphous Alloy

AMGOT Allied Military Government of Occupied Territory

AMGRA American Milk Goat Record Association [US]

Am Health America's Health [US Journal]

Am Heart J American Heart Journal [US]

AMHF American Military History Foundation [now American Military Institute, US]

AMHIC Automatic Merchanizing Health Industry Council

AMHS American Material Handling Society [now International Material Management Society, US]

AMHS Automated Message-Handling System

AMHSA Automated Materials Handling Systems Association [UK]

Am-HTDO Amorphous Hydrous Titanium Dioxide [also am-HTDO]

AMI Acoustic Micro-Imaging (Technique)

AMI Active Microwave Instrument

AMI Acute Myocardial Infarction

AMI Advanced Microwave Instrument

AMI Advertising and Market Intelligence (Database) [of New York Times Information Service, US]

AMI Agricultural Mechanization Institute

AMI Aircraft Maintenance Inspector

AMI Air Movement Institute [US]

AMI Alkali-Metal Impurity

AMI Alliance of Metalworking Industries [US]

AMI Alternate Mark Inversion

AMI Alternative Mark Inversion (Signal)

AMI American Megatrends, Inc. [US]

AMI American Military Institute [Washington, DC, US]

AMI American Mushroom Institute [Kennett Square, Pennsylvania, US]

AMI Analytical Methods and Instrumentation [International Journal]

AMI Applied Microelectronics Institute

AMI Association for Multi-Image [US]

AMI Australian Medical Index [Part of AUSTHealth Database]

AmI Ammonium Iodide

AMIA American Metal Importers Association [US]

AMIA Australian Map Industry Association

AMIAS Automatic Metallurgical Image Analysis System

Amiben 3-Amino-2,5-Dichlorobenzoic Acid [also amiben]

AMIC Aerospace Materials Information Center [US]

AMIC American Marine Insurance Clearinghouse [US]

AMIC Asian Mass Communications Research and Information Center [Singapore]

AMIC Asociación Mexicana de Ingenieros en Corrosion [Mexican Association of Corrosion Engineers]

AMIC Australian Mining Industry Council

AMICA Automated Modules for Industrial Control Analysis

AMICE Associate Member of the Institute of Civil Engineers

AMICOM Army Missile Command [US]

Amicus J Amicus Journal [Publication of the Natural Resources Defense Council, US]

AMIDS Airport Management and Information Display System

AMIE Advanced Manufacturing in Electronics

AMIEE Associate Member of the Institute of Electrical Engineers

AMILAT American Military Attaché

AMIME Asociación Mexicana de Ingenieros Mecanicos y Electricistas [Mexican Association of Mechanical and Electrical Engineers]

AMIME Associate Member of the Institute of Mechanical Engineers

AMIMechE Associate Member of the Institution of Mechanical Engineers

AMIMinE Associate Member of the Institution of Mining Engineers

AMIN Advertising and Marketing International Network [US]

AMIN Arab Medical Information Network

A min Ampere-minute [also A-min, or A·min]

Å/min Angstrom(s) per minute [also Å min^{-1}]

AMINA Association Mondiale des Inventeurs [World Association of Inventors, Belgium]

Am Ind American Indian(s)

Am Ind Hyg Assoc J American Industrial Hygiene Association Journal

AM in Ed Master of Arts in Education

Am Ink Maker American Ink Maker [US Publication]

Amino-PAH Amino Polycyclic Aromatic Hydrocarbon [also amino-PAH]

AMInstCE Associate Member of the Institute of Civil Engineers

Am Inst Min Eng American Institute of Mining, Metallurgical and Petroleum Engineers

Am Inst Phys American Institute of Physics [US]

Am Inst Phys Conf Proc American Institute of Physics Conference Proceedings [US]

AMIP Atmospheric Model Intercomparison Project

AMIRA Australian Mining Industry Research Association

AMIS Agricultural Management Information System [of European Economic Community]

AMIS Aircraft Movement Information Service

AMIS Air Movement Identification Service

AMIS Assembly Management Information System

AMIS Australian Municipal Information System

AMIS Automated Management Information System

AMISB Amplitude Modulation–Independent Sideband [also AM ISB, or AM-ISB]

AMISC Army Missile Command [US]

Am J Archeol American Journal of Archeology [of Archeological Institute of America, New York City]

Am J Bot American Journal of Botany [of Botanical Society of America]

Am J Cardiol American Journal of Cardiology

Am J Clin Nutr American Journal of Clinical Nutrition

Am J Dent American Journal of Dentistry

Am J Dis Children American Journal of Diseases of Children [of American Medical Association, US]

Am J EEG Technol American Journal of EEG Technology [of American Society of Electroencephalographic Technologists]

Am J Epidemol American Journal of Epidemology

Am Jewel Manuf American Jewelry Manufacturer [US Journal]

Am J Gastroenterol American Journal of Gastroenterology

Am J Hosp Pharm American Journal of Hospital Pharmacy

Am J Hum Genet American Journal of Human Genetics [of University of Chicago Press, US]

Am J Ind Med American Journal of Industrial Medicine

Am J Infect Control American Journal of Infection Control [US]

Am J Math American Journal of Mathematics [of Johns Hopkins Press, US]

Am J Math Manage Sci American Journal of Mathematical and Management Sciences

Am J Med American Journal of Medicine

Am J Nat Med American Journal of Natural Medicine

Am J Nephrol American Journal of Nephrology

Am J Nucl Res American Journal of Nuclear Research [also AJNR]

Am J Nursing American Journal of Nursing [of American Nurses Association, US]

Am J Obstet Gynecol American Journal of Obstetrics and Gynecology

Am J Optom Physiol Opt American Journal of Optometry and Physiological Optics

Am J Orthodont Dentof Orthoped American Journal of Orthodontics and Dentafacial Orthopedics

Am J Orthopsych American Journal of Orthopsychiatry

Am J Pharmacol American Journal of Pharmacology

Am J Phys American Journal of Physics [of American Institute of Physics, US]

Am J Phys Anthrop American Journal of Physical Anthropology

Am J Physiol American Journal of Physiology [of American Physiological Society, US]

Am J Psych American Journal of Psychiatry

Am J Psychol American Journal of Psychology

Am J Psychother American Journal of Psychotherapy

Am J Publ Health American Journal of Public Health

Am J Reson American Journal of Resonance [also AJR]

Am J Resp Cell Mol Biol American Journal of Respiratory Cell and Molecular Biology [of American Lung Association, US]

Am J Resp Crit Care Med American Journal of Respiratory and Critical Care Medicine [of American Lung Association, US]

Am J Sci American Journal of Science [of Yale University, New Haven, Connecticut, US]

Am J Vet Res American Journal of Veterinary Research [of American Veterinary Medical Association, US]

AMK Anti-Misting Kerosene

Am Kennel Gaz American Kennel Gazette [of American Kennel Club]

A-m²/kg Ampere-meters squared per kilogram [also Am²/kg, A·m²/kg, Am² kg^{-1}, or amp-m²/kg]

AML Actual Measured Loss

AML Admiralty Materials Laboratory [UK]

AML Advanced Manipulator Language

AML Advanced Manufacturing Laboratory [of Alcoa Corporation, US]

AML Advanced Materials Laboratory

AML Advanced Materials Laboratory [of Sandia National Laboratory, Los Alamos National Laboratory and the University of New Mexico; located at Albuquerque, US]

AML Advanced Materials Laboratory [of Texas A&M University, College Station, US]

AML Advanced Materials Laboratory [of Lawrence Berkeley Laboratory, California, US]

AML Aeromedical Laboratory [of US Air Force at Wright Air Development Center, Dayton, Ohio, US]

AML Aeronautical Materials Laboratory [US]

AML A Manufacturing Language [Programming Language]

AML Amplitude-Modulated Link

AML Applied Mathematics Laboratory [of Department of Scientific and Industrial Research, UK]

AML ARC/INFO Macro Language [*Note:* ARC/INFO is a Proprietary GIS (Geographic Information System) Software]

AML Arizona Materials Laboratory [Tucson, US]

AML Array Machine Language

AML Association of Lightermen and Barge Owners [UK]

AML Automatic Modulation Limiting

AML Aviation Materials Laboratory [US Army]

Am Lab American Laboratory [US Journal]

AMLC Association of Marine Laboratories of the Caribbean [Puerto Rico]

AMLC Asynchronous Multiline Controller

AMLCC Asynchronous Multiline Communications Coupler

AMLCD Active Matrix (Addressed) Liquid Crystal Display [also AM-LCD]

Am:LiYF₄ Americium-Doped Lithium Yttrium Fluoride [also Am³⁺:LiYF₄]

Am Logger Lumberman American Logger and Lumberman [US Journal]

AMLRDC Australian Meat and Livestock Research and Development Corporation

AMLS Master of Arts in Library Science

AMM Additional Memory Module

AMM Ammeter [also Amm, or amm]

AMM Analog Monitor Module

AMM Analog Multimeter

AMM Anomalous Magnetic Moment

AMM Anti-Missile Missile

AMM ASEAN (Association of Southeast Asian Nations) Ministerial Meeting

AMM Associated Maintenance Module

AMM Aviation Machinist's Mate

AMM Aviation Mixture Method

Amm Ammeter [also amm]

Amm Ammonium

Amm Ammunition [also amm]

A/mm² Ampere(s) per square millimeter [Unit]

AMMA Acrylonitrile-Methylmethacrylate

AMMA Advanced Memory Management Architecture

AMMA Art Metalware Manufacturers Association [UK]

Am Mach American Machinist [US Journal]

Am Mach Autom Manuf American Machinist and Automated Manufacturing [US Journal]

Am Mach Spec Rep American Machinist, Special Report [US Publication]

Am Math Mon American Mathematical Monthly [of Mathematical Association of America, US]

Am Math Soc American Mathematical Society [US]

AMMCS Army Missile and Munitions Center and School [at Redstone Arsenal, Huntsville Alabama, US]

AMMDA Accounting Machine/Minicomputer Dealers Association of America [now Associated Minicomputers Dealers of America, US]

Am Met Mark American Metal Market [US Publication]

Am Met Mark, Metalwork News Ed American Metal Market, Metalworking News Edition [US]

AMMI American Merchant Marine Institute [US]

Am Microsc Soc American Microscopical Society [US]

Am Mineral The American Mineralogist [Journal of the Mineralogical Society of America]

AMMIP Aviation Materiel Management Improvement Program [US Army]

AMMLA American Merchant Marine Library Association [US]

AMMM Association of Microbiological Media Manufacturers [now Association of Microbiological Diagnostic Manufacturers, US]

AMMO 3-(Trimethoxysilyl) Propanamine

Ammo Ammunition [also ammo]

AMMOS Advanced Multi-Mission Operations System [of NASA Planetary Data System]

AmmP Ammonium Picrate

AMMPEC Advanced Materials and Manufacturing Processes for Economic Competitiveness (Alliance Inc.)

AMMPTA (4-Amino-2-Methyl-5-Pyrimidinylmethylthio) Acetate

AMMR Airborne Multichannel Microwave Radiometer

AMMRC Army Materials and Mechanics Research Center [now Army Materials Technology Laboratory, Watertown, Massachusetts, US]

AMMRC Connect Army Materials and Mechanics Research Center Connections [now AMTL Connect]

AMMRC TR Army Materials and Mechanics Research Center Technical Report [also AMMRC-TR, or AMMRC Tech Rep] [now AMTL TR]

AMMS Advanced Microwave Moisture Sounder

AMN 1-(Aminomethyl)naphthalene

AM&N Arkansas Agricultural, Mechanical and Normal College [Pine Bluff, US]

a-**Mn** Alpha Manganese [Symbol]

AMNC Agricultural, Mechanical and Normal College [Pine Bluff, Arkansas, US]

AM-NC Amorphous-to-Nanocrystalline (Transformation) [also AM→NC]

AMNH American Museum of Natural History [of Smithsonian Institution, Washington, DC, US]

am-Ni Amorphous Nickel

AMNIP Adaptive Man-Machine Nonarithmetical Information Processing

AMNIP Adaptive Man-Machine Nonnumeric Information Processing

AMNIPS Adaptive Man-Machine Nonarithmetical Information Processing System

AMNIPS Adaptive Man-Machine Nonnumeric Information Processing System

Am Nit Ammonium Nitrate [also am nit]

a-**Mn₃O₄** Alpha Manganese Tetroxide (or Hausmannite)

Am Nucl Soc American Nuclear Society [US]

AMNS Arctic Marine Navigation Systems [of Canadian Coast Guard]

AMO Advanced Materials for Optics and Electronics [formerly Journal of Molecular Electronics]

AMO Aeronautics Material Officer

AMO Air Ministry Order [UK]

AMO Alternate Molecular Orbital

AMO American Standard Microscope Objective Thread [Symbol]

AMO Antibonding Molecular Orbital

AMO Antisymmetrized Molecular Orbital

AMO Atomic, Molecular and Optical (Physics)

AmO Amyl Oxide

AmOD Ammonium Deuteroxide

AmOH Ammonium Hydroxide

AmOH Amyl Alcohol

AMOA Atmospheric Monitor Oxygen Analyzer

Amoco American Oil Company [US]

(AMOC)2O Di-tert-Amyldicarbonate

a-MoGe Molybdenum Germanium Amorphous Alloy

AMONO Agricultural Machinery Operation and Management Office [PR China]

AMOP Association of Mail Order Publishers [UK]

amor amorphous [also amorph]

Amort Amortization [also amort]

AMOS Acoustic Meteorological Observing System

AMOS Acoustic Meteorological Oceanographic Survey

AMOS Alpha Microsystems Operating Systems [US]

AMOS American Maritime Officers Service [US]

AMOS Associative Memory Organizing System

AMOS Australian Meteorological and Oceanographic Society

AMOS Automatic Meteorological Observing Station

AMOS Automatic Meteorological Observing System

AMOSA Association of Aviation Maintenance Organizations in South Africa

a-Mo₃Si Amorphous Molybdenum Silicide

AMP Acetaldehyde Monoperacetate

AMP Adaptation Mathematical Processor

AMP Adenosine Monophosphate [Biochemistry]

AMP Advanced Manned Penetrator

AMP Aerospace Medical Panel [of NATO Advisory Group for Aerospace Research and Development]

AMP Alliances for Minorities Participation (Program) [of National Science Foundation, US]

AMP Aluminum Metaphosphate

AMP American Melting Point

AMP 2-Amino-2-Methyl-1-Propanol

AMP Ammonium Molybdatophosphate

AMP Applied Mathematical Panel

AMP ARGOS (Atmospheric Research Geostationary Orbit Satellite) Meteorological Platform

AMP Associative Memory Processor

AMP Assured Mail Program

AMP Augmented Materials Production

AMP N-Bis(hydroxymethyl)-2-Amino-2-Methyl-1-Propanol

AM&P Advanced Materials and Processes [Journal of ASM International, US]

A5MP Adenosine-5-Monophosphoric Acid [Biochemistry]

2'-AMP Adenosine 2'-Monophosphate [Biochemistry]

3'-AMP Adenosine 3'-Monophosphate [Biochemistry]

3',5'-AMP Adenosine 3',5'-Cyclic Monophosphate [Biochemistry]

5'-AMP Adenosine 5'-Monophosphate [Biochemistry]

bis-AMP N,N-bis(Hyroxylethyl)-2-Amino-2-Methyl-1-Propanol

Amp Amplifier [also amp]

amp 2-Amino-6-Methyl Pyridine [also AMP]

amp amperage

amp ampere [Unit]

amp *(ampulla)* – Ampoule [Medical Prescriptions]

AMPA American Manganese Producers Association [US]

AMPA American Medical Publishers Association [US]

AMPA DL-α-Amino-3-Hydroxy-5-Methylisoxazole-4-Propionic Acid

AMPA Auger Microprobe Analysis

AMPAC Advanced Material Processing and Analysis Center [of University of Central Florida, Orlando, US]

Am Papermaker American Papermaker [US Journal]

AMPAS Academy of Motion Picture Arts and Sciences [US]

AMPC Automatic Message Processing Center

amp(s)/cm² ampere(s) per square centimeter [also A/cm², or A cm^{-2}]

AMPCO Association of Major Power Consumers in Ontario [Canada]

AMP-CP α,β-Methyleneadenosine 5'-Diphosphate (or Adenosine-5'-[Trihydrogenmethylene-bis(phosphonic Acid)]) [Biochemistry]

AMP-CPP α,β-Methyleneadenosine 5'-Triphosphate (or Adenosine-5'-[Hydrogen{(hydroxy(phosphonooxy)-phosphinyl)methyl}phosphonic Acid)]) [Biochemistry]

AMPD 2-Amino-2-Methyl-1,3-Propanediol

AMPERE Atomes et Molecules par Etudes Radio-Electriques [Atoms and Molecules for Radioelectric Studies, Switzerland]

Amph Amphibian; Amphibious [also amph]

Am Phil Soc American Philosophical Society [US]

amp hr Ampere-hour [also amp-hr]

Amph Reptile Conserv Amphibian and Reptile Conservation [International Journal]

AMPI Associated Milk Producers Information

AMPIC Atomic and Molecular Processes Information Center

amp/in ampere(s) per inch [also A/in]

AMPL Advanced Materials and Processing Laboratory [of University of Alberta, Edmonton, Canada]

AMPL Advanced Microprocessor Programming Language

AMPL Advanced Microprocessor Prototyping Laboratory [US]

Ampl Amplification; Amplifier [also ampl]

Ampl Amplitude [also ampl]

AMPLG Amplidyne Generator

Amplr Amplifier [also amplr]

AMPLMG Amplidyne Motor Generator

AMP-lyase Adenylosuccinate Lyase [Biochemistry]

amp-m² ampere-meters squared [also Am², or Am²]

AM/PM Amplitude Modulation/Pulse Modulation (Conversion)

amp min ampere-minute [also amp-min]

AMP-NH2 Adenosine-5'-Monophosphoroamidate [Biochemistry]

Am Potato J American Potato Journal [US]

AMP-PNP 5-Adenylylimidodiphosphate

AMPNS Asymmetric Multiple Position Neutron Source

AMPP Advanced Materials and Processing Program [Program of the US Office of Science and Technology Policy involving the Departments of Agriculture, Commerce, Defense, Energy, Health and Human Services, Transportation and the Interior, the Environmental Protection Agency, NASA and the National Science Foundation]

AMPP Association of Makers of Packaging Papers [UK]

AMP-PCP β,γ-Methyladenosine 5'Triphosphate (or Adenosine-5'-[β,γ-Methylene]triphosphate) [Biochemistry]

AMP-PNP 5'-Adenylic Acid Monoanhydride with Imidodiphosphoric Acid [Biochemistry]

AMPR Advanced Microwave Precipitation Radiometer

AMPR Aeronautical Manufacturers Planning Report

Am Printer American Printer [US Journal]

Am Prof Constr American Professional Constructor [Journal of the American Institute of Constructors]

AMPS Acrylamido-2-Methyl 1-Propanesulfonic Acid; 2-Acrylamido-2-Methylpropanesulfonic Acid

AMPS Advanced Mobile Phone Service

AMPS Army Mine Planting Service

AMPS Association of Management and Professional Staffs [UK]

AMPS Automatic Message Processing System

AMP-S Adenosine-5'-Thiomonophosphate [Biochemistry]

AmPS Amsterdam Pulse Stretcher (Project) [Netherlands] [also AMPS]

amps amperes

amp sec ampere-second [also amp-sec]

amp/sec ampere(s) per second [also A/sec, or A/s]

AMPSO Aminomethylhydroxypropanesulfonic Acid

AMPSO 3-[(1,1-Dimethyl-2-Hydroxyethyl)amino]-2-Hydroxy- 1-Propanesulfonic Acid; 3(N-(α, α-Dimethyl Hydroxyethyl)amino)-2-Hydroxypropane Sulfonic Acid

AMPSS Advanced Manned Penetrator Strike System

AMPSS Advanced Manned Precision Strike System

Am Psychol American Psychologist [Journal published in the US]

α-MPT α-Methyl-p-Tyrosine [Biochemistry]

AMPTE Active Magnetospheric Particle Tracer Experiment

amp-turn ampere-turn [Unit]

AMPW Association of Makers of Printings and Writings [UK]

AMPY 2-Aminomethyl Pyridine

AMQ 8-Aminoquinoline

AMQUA American Quaternary Association [US]

AMR Académie Militaire Royal [Royal Military Academy, Brussels, Belgium]

AMR Active Magnetic Regenerator

AMR Advance Material Request

AMR Airport Movement Radar

AMR Anisotropic Magnetoresistance

AMR Anomalous Angular Magnetoresistance

AMR Applied Materials Research

AMR Atlantic Missile Range [US Air Force]

AMR Automated Management Report

AMR Automated Meter Reading

AMR Automatic Message Registering

AMR Automatic Message Routing

AMR Automatic Meter Reading

AMR Autonomous Mobile Robot

AMR Autonomous Mobile Rocket

AMRA American Medical Record Association [US]

AMRA Army Materials Research Agency [US]

AMRA Automotive Manufacturers Racing Association [UK]

AMRA Automatic Meter Reading Association [US]

AMRAAM Advanced Medium-Range Air-to-Air Missile

AMRAC Anti-Missile Research Advisory Council [US]

AMRA CR Army Materials Research Agency–Contract Report [US]

AMRAD ARPA (Advanced Research Projects Agency) Measurements Radar

AMRC Advanced Materials Research Center [of University of Florida, Gainesville, US]

AMRC Army Mathematics Research Center [of University of Wisconsin, US]

AMRC Association of Municipal Recycling Coordinators [Canada]

AMRC Australian Materials Research Committee

AMRC Automotive Market Research Council [US]

AMRCO Association of Motor Racing Circuit Owners [UK]

AMREF African Medical Research Foundation

Am Rev Resp Dis American Review of Respiratory Diseases

Am Rev Tubercul American Review of Tuberculosis [also Am Rev TB]

AMRF Automated Manufacturing Research Facility [of National Institute for Standards and Technology, Gaithersburg, Maryland, US]

AMR/GMR Anisotropic Magnetoresistance/Giant Magnetoresistance (Structure)

AMRI Advanced Materials Research Institute [of University of New Orleans, Louisiana, US]

AMRIP Australian Marine Research in Progress [Publication]

AMRIR Advanced Medium Resolution Imaging Radiometer

AMRI-UNO Advanced Materials Research Institute–University of New Orleans [Louisiana, US]

AMRL Aerospace Medical Research Laboratory [US]

AMRL Army Medical Research Laboratories

AMR-MRAM Anisotropic Magnetoresistance–Magnetoresistive Random Access Memory

AMRNL Army Medical Research and Nutrition Laboratory [US]

AMRR (US) Army Materials Research Reactor [Formerly located at Watertown, Massachusetts, now moved to Aberdeen, Maryland]

AMRS Australian MARC (Machine-Readable Catalogue) Record Service

A-MRS Australian Materials Research Society

AMRU Aerospace Medical Research Unit [of McGill University, Montreal, Canada]

AMRV Atmospheric Maneuvering Reentry Vehicle

AMS Academy of Medicine of Singapore

AMS Accelerator Mass Spectrometry

AMS Access Method Services

AMS Acoustic Material Signature

AMS Acoustic Monitor System

AMS Administrative Management Society [US]

AMS Advanced Manufacturing System

AMS Advanced Memory System

AMS Advanced Meteorological Satellite

AMS Advanced Microcomputer System

AMS Advanced Microsystems, Inc.

AMS Administrative Management Society [US]

AMS Aeronautical Material Specification

AMS Aeronautical Mobile Service

AMS Aerospace Materials Society [US]

AMS Aerospace Material Specification [of Society of Automotive Engineers, US]

AMS Agricultural Manpower Society [UK]

AMS Agricultural Marketing Service [of US Department of Agriculture]

AMS Air Mail Service

AMS Air Mass

AMS Alarm Monitoring System

AMS Alma Mater (Student) Society

AMS Alpha Magnetic Spectrometer [of NASA and US Department of Energy]

AMS Alternate Mission Scientist

AMS American Magnolia Society [now The Magnolia Society, US]

AMS American Mathematical Society [US]

AMS American Meteorological Society [US]

AMS American Meteor Society [US]

AMS American Microchemical Society [US]

AMS American Microscopical Society [US]

AMS American Miscellaneous Society [US]

AMS Ammonium Sulfamate [also AmS]

AMS Amorphous Material Structure

AMS Applied Materials Science

AMS Apollo Mission Simulator [of NASA]

AMS Army Map Service [Corps of Engineers, US]

AMS Army Medical Service

AMS Association of Metal Sprayers [US]

AMS Asymmetric Multiprocessing System

AMS Atmospheric Monitor System

AMS Atomic Mass Scale

AMS Attitude Measurement Sensor

AMS Australian Mathematical Society

AMS Automated Manifest System

AMS Automated Manufacturing Show (and Exhibition) [Canada]

AMS Automated (or Automatic) Meteorological Station

AMS Automatic Music Search

AMS Australian Mathematical Society

AmS Ammonium Sulfamate

AMS Astronomy, Mathematics, Statistics [of University of California at Berkeley–Physical Science Libraries, US]

AMSA Advanced Manned Strategic Aircraft

AMSA American Meat Science Association [US]

AMSA American Medical Society of Alcoholism [of National Council on Alcoholism, US]

AMSA American Medical Student Association [US]

AMSA American Metal Stamping Association [US]

AMSA Aminomethanesulfonic Acid

AMSA Association of Medical Schools in Africa

AMSA Association of Metropolitan Sewerage Agencies [US]

AMSA Australian Marine Sciences Association

AMSA Australian Maritime Safety Authority

m-AMSA 4-(9-Acridinylamino)-N-(Methanesulfonyl)-m-Anisidine

AMSAM Anti-Missile Surface-to-Air Missile

AMsartacn 9-Amino-1,4,7,11,14,19-Hexaazatricyclo-[7.7.4.2^{4,14}]docosane

AMSAT (Radio) Amateur Satellite Corporation [US]

AMSC American Marine Standards Committee [US]

AMSC American Superconductor Corporation [US]

AMSC Army Mathematics Steering Committee [US]

AMSC Australian Materials Science Committee

AMSC Automatic Message Switching Center

AMSCE Army Map Service, Corps of Engineers [US]

Am Sci American Scientist [US Journal]

AMSCO American Sterilizer Company [Erie, Pennsylvania, US]

AMSE Association of Muslim Scientists and Engineers [US]

AMSEF Anti-Mine Sweeping Explosive Float

Am Semicond Dev Technol American Semiconductor Devices and Technology [US Publication]

AMSES Advanced Materials Science and Engineering Society [now Materials Research Society of Japan]

AMSIS Accepted Manuscript Status Inquiry System [of American Institute of Physics, US]

AMSL Above Mean Sea Level [also amsl]

AMS-M Advanced Microcomputer System with Multibus Architecture

AMS Newsl AMS Newsletter [of American Meteorological Society, US]

AMSO Association of Market Survey Organizations [UK]

AMSOC American Miscellaneous Society [US]

Am Soc Agric Eng American Society of Agricultural Engineers [US]

Am Soc Civil Eng American Society of Civil Engineers [US]

Am Soc Comp American Society for Composites [US]

Am Soc Met American Society of Metals [now ASM International]

Am Soc Met Q American Society of Metals Quarterly [also Am Soc Met Quart, ASM Q, or ASM Quart]

AMSP Allied Military Security Publications

AMSR Advanced Microwave Scanning Radiometer

AMSRL Army Materials Structural Research Laboratory [of US Army at Adelphi, Maryland]

AMSS Aeronautical Mobile Satellite Service

AMSSB Amplitude Modulation, Single-Sideband [also AM SSB, or AM-SSB]

AMSSP Aeronautical Mobile Satellite Service Panel [of International Civil Aviation Organization]

AMST Advanced Medium STOL (Short Takeoff and Landing) Transport

AMST Association of Maximum Service Telecasters [US]

Amst Amsterdam [Netherlands]

AMSTAC Australian Marine Science and Technologies Advisory Council

AMSTAC Australian Marine Science and Technologies Committee

Am Stat American Statistician [Journal of the American Statistical Society, US]

AMSTEC Alabama Microelectronics Science and Technology Center [at Auburn University, Normal, Alabama, US]

AMSTP Association of Makers of Soft Tissue Papers [UK]

AMSU Advanced Microwave Sounding Unit

AMSU Auto-Manual Switching Unit

AMSU-A Advanced Microwave Sounding Unit–A

AMSU-B Advanced Microwave Sounding Unit–B

AMSUS Association of Military Surgeons of the United States

AMT Accelerated Mission Test(ing)

AMT (Disk) Access Motion Time

AMT Advanced Manufacturing Technology (Initiative) [Canada]

AMT Advanced Materials Technology

AMT Advanced Materials Technology, Inc. [US]

AMT Advanced Metals Technology

AMT Alternative Minimum Tax

AMT 2-Amino-5,6-Dihydro-6-Methyl-4H-1,3-Thiazine

AMT Ammonium Metatungstate

AMT Amplitude Mode Theory

AMT Associate in Mechanical Technology

AMT Associate in Medical Technology

AMT Association of Medical Technologists [UK]

AMT Audiofrequency Magnetotellurics

AMT Audio Magnetotellurics

AMT Automated Manufacturing Technology

AMT Master of Arts in Teaching

Amt Amount [also amt]

AMTA Antenna Measurement Techniques Association [US]

AMTAP Advanced Manufacturing Technologies Application Program

AMTAP Advanced Manufacturing Technology Awareness Program [of Industry, Science and Technology Canada]

AMTB Acute Miliary Tuberculosis

AMTB Anti-Motor Torpedo Boat

AMTC Automated Manufacturing Training Center [Canada]

AMTCL Association for Machine Translation and Computational Linguistics [US]

AMTCS Amyltrichlorosilane

AMTD Automatic Magnetic Tape Dissemination [of Defense Documentation Center, US]

AMTDA Agricultural Machinery and Tractor Dealers Association [UK]

AMTDA American Machine Tool Distributors Association [US]

AMTE Admiralty Marine Technology Establishment

AMTEC Alkali Metal Thermoelectric Converter

AMTEC Association for Media and Technology in Education in Canada

AMTeC Advanced Manufacturing Technology Center [UK]

AMTEX Air-Mass Transportation Experiment [Japan]

AMTEX American Textile Industries

Am:ThO$_2$ Americium-Doped Thoria [also Am^{3+}:ThO$_2$]

AMTI Airborne Moving Target Indicator

AMTI Area Moving Target Indicator

AMTI Automatic Moving Target Indicator

AMTIDE Aircraft Multipurpose Test Inspection and Diagnostic Equipment

AMTL Army Materials Technology Laboratory [Watertown, Massachusetts, US] [formerly AMMRC]

AMTL Connect Army Materials Technology Laboratory Connections [formerly AMMRC Connect]

AMTL TR Army Materials Technology Laboratory Technical Report [also AMTL-TR, or AMTL Tech Rep] [formerly AMMRC TR]

AMT n.v. Advanced Materials Technology n.v. [Herk-de-Stad, Belgium]

Amtrac Amphibious Tractor [also amtrac]

Amtrak National Railroad Passenger Corporation [US]

AMTRAN Automatic Mathematical Translator

AMTRI Advanced Manufacturing Technology Research Institute [UK]

AMTRT Alberta Ministry of Technology, Research and Telecommunications [Canada]

AMTS Advanced Moisture and Temperature Sounder

AMTS Automated Materials Testing System

Amts-Mittbl Bundesanst Mater forsch-prüf Amts- und Mitteilungsblatt der Bundesanstalt für Materialforschung und- prüfung [Official Gazette and Communication of the Federal Institute for Materials Research and Testing, Germany]

AMTU Advanced Materials Technology Unit [Canada] [Defunct]

AM TV Amplitude Modulation–Television [also AM-TV]

AMTZ 2-Amino-5-Methyl-1,3,4-Thiadiazole [also amtz]

AMU African Mathematical Union [Congo]

AMU American Malacological Union [US]

AMU Associated Midwestern Universities [US]

AMU Association of Minicomputer Users [US]

AMU Astronaut Maneuvering Unit [of NASA]

AMU Atomic Mass Unit (1.66 × 10^{-24} g) [also amu]

AMUMeV Atomic Mass Unit Mega-Electronvolt

AMV Advanced Marine Vehicle

AMV Alfalfa Mosaic Virus

AMV Ammonium Metatungstate

AMV Astable Multivibrator

Am Vac Soc American Vacuum Society [US]

Am Vac Soc Ser American Vacuum Society Series [Periodical Series]

AMVER Atlantic Merchant Vessel Alert

AMVER Automated Merchant Vessel Reporting

Am Vet Med Assoc American Veterinary Medicine Association [US]

AMVETS American Veterans of World War II, Korea and Vietnam [Washington, DC, US]

AMVTA Association of Motor Vehicle Training Agents [Scotland]

AMWA American Medical Women's Association [US]

AMWA American Medical Writers Association [US]

AMWA Association of Metropolitan Water Agencies [US]

AMWG AIDS Malignancies Work(ing) Group

AMX Automatic Message Exchange

A$_2$MX$_4$ Compound of an Organic Cation, Divalent Metal Cation and a Halogen [Ceramics]

Am Zool American Zoologist [Journal of the American Society of Zoologists, US]

AN Absorbed Nitrogen

AN Abstract Number

AN Accession Number

AN Account Number

AN Acetonitrile

AN Acetyl Number

AN Acorn Nut

AN Acrylonitrile

AN Adiabatic Nuclei (Approximation)

AN Advanced Navigator

AN Aerial Navigation

AN Aeronaut(ical); Aeronautics

AN Air Force–Navy

AN Air Navigation [also an]

AN Aitken Nuclei [Meteorology]

AN *Academiya Nauk* – Russian for "Academy of Sciences"

AN Algebraic Number

AN Alphanumeric(s) [also A/N]

AN Aluminum Naphthenate (Polymer)

AN Aluminum Nitrate

AN Amino Nitrogen

AN Ammonium Nitrate

AN Analog Input, or Output

AN Anderson-Newns (Model)

AN Anorexia Nervosa

AN Antineutron

AN Antinode; Antinodal

AN Apparent Noon [Astronomy]

AN Army-Navy [US]

AN Arrival Note

AN Asaro-Needleman (Model) [Metallurgy]

AN Ascending Node [Astronomy]

AN As-Nitrided (Condition)

AN Associate in Nursing

AN Associative Neuron [Anatomy]

AN Astronavigation(al)

AN Atomic Number

AN Auditory Nerve [Anatomy]

AN Auger Neutralization [Physics]

AN Australian National (Railways)

AN Automotive News [Publication]

AN Netherland Antilles [ISO Code]

A/N Alphanumeric(s) [also AN]

A/N As Needed

AN- Nicaragua [Civil Aircraft Marking]

AN(I) Ammonium Nitrate, Phase I [also AN-I]

AN(II) Ammonium Nitrate, Phase II [also AN-II]

An Acetone

An Actinide (Series)

An Actinon [also Rn-219, ^{219}Rn, or Rn^{219}]

An Anatase [Mineral]

An Anorthite [Mineral]

An Antimony [Abbreviation; Symbol: Sb]

An Anus [also an]

an (anno) – in the year

.an Netherland Antilles [Country Code/Domain Name]

a-N Alpha Nitrogen [Symbol]

ANA Allyl Nadic Anhydride

ANA American Nature Association [US]

ANA American Nautical Almanac [of United States Naval Observatory]

ANA American Neurological Association [US]

ANA American Nurses Association [US]

ANA 3-Aminonocardicinic Acid

ANA 8-Anilino Naphthalene-1-Sulfonic Acid

ANA Antinuclear Antibody [Biochemistry]

ANA Army Navy Aeronautical

ANA Article Number(ing) Association [UK]

ANA Assigned Night Answer

ANA Associate of the National Academy of Design [US]

ANA Association of National Advertisers [US]

ANA Association of Naval Aviation [US]

ANA Association of Nordic Aeroclubs [Norway]

ANA Australian Nuclear Association

ANA Automatic Network Analyzer

ANA Automatic Number Analysis

ANA Automatic Number Announcer

a-Na Alpha Sodium [Symbol]

ANAB Anorexia Nervosa and Bulimia Association

ANA-C Antinuclear Antibody, Centromere Pattern [Biochemistry]

An Acad Bras Cienc Anais da Academia Brasileira de Ciencias [Annals of the Brazilian Academy of Sciences]

ANACHEM Association of Analytical Chemists [US]

ANACOM Analog Computer

ANADP Association of North American Directory Publishers [US]

ANAE Académie Nationale d'Air et de l'Espace [National Aeronautics and Space Academy, Toulouse, France]

ANAFC Australian National Average Fuel Consumption

ANA-H Antinuclear Antibody, Homogeneous Pattern [Biochemistry]

ANAHL Australian National Animal Health Laboratory

ANAIC Asian Network of Analytical Inorganic Chemistry

Anal Analogy; Analogous [also anal]

Anal Analysis; Analytic(al); Analyst [also anal]

Anal Abstr Analytical Abstracts [of Royal Society of Chemistry, UK]

Anal Biochem Analytical Biochemistry [Journal published in the US]

AnalChem Analytical Chemist(ry [also Anal Chem]

Anal Chem Analytical Chemistry [Journal of the American Chemical Society, US]

Anal Chem Analytical Chemistry [Journal of the Chinese Chemical Society, PR China]

Anal Chem Analytische Chemie [German Journal on Analytical Chemistry]

Anal Chim Acta Analytica Chimica Acta [Journal on Analytical Chemistry published in the Netherlands]

Anal Commun Analytical Communications [Journal of the Royal Society of Chemistry, UK]

Analele Stiinţ Univ 'AlI Cuza' Iaşi (Ser Noua) IB Fiz Analele Stiinţifice ale Universitatu 'AlI Cuza' din Iaşi (Serie Noua) Sectiuneas IB Fizica [Scientific Annals of the All Cuza University of Iasi (New Series), Section IB Physics, Romania]

Analele Univ Bucur Ser Fiz Analele Universitatii Bucaresti, Fizica [Annals of the University of Bucharest, Physical Series, Romania]

Analele Univ Bucur Ser Mat Analele Universitatii Bucuresti, Seria Matematica [Annals of the University of Bucharest, Mathematical Series, Romania]

Analele Univ Galati, Mec Constr Masini Analele Universitatii din Galati, Mecanica si Constructii de Masini [Annals of the University of Galati, Mechanics and Mechanical Engineering, Romania]

Analele Univ Galati, Metal Analele Universitatii din Galati, Metalurgie [Annals of the University of Galati, Metallurgy, Romania]

Anal Instrum Analytical Instrumentation [Journal published in the US]

ANALIT Analysis of Automatic Line Insulation Tests

Anal Lab Analytical Laboratory [Published in PR China]

Anal Lett Analytical Letters [US]

ANALOG Comput ANALOG Computing [US]

Anal Proc Analytical Proceedings [of Royal Society of Chemistry, UK]

Anal Proc R Soc Chem Analytical Proceedings of the Royal Society of Chemistry [UK]

Anal Sci Analytical Sciences [Journal published in Japan]

Analyst The Analyst [of Royal Society of Chemistry, UK]

analyt analytical

Analyt Abs Analytical Abstracts [of Royal Society of Chemistry, UK]

Analyt Chem Analytical Chemist(ry)

Analyt Chem Analytical Chemistry [Journal of the American Chemical Society, US]

Analytica International Trade Fair for Biochemical and Instrumental Analysis, Diagnostics and Laboratory Technology [Germany]

ANAMET Anaerobic Methane (Process)

ANA-N Antinuclear Antibody, Nucleolar Pattern [Biochemistry]

ANAO Australian National Audit Office

ANAPROP Anomalous Propagation

ANARE Australian National Antarctic Research Expedition(s)

ANAREMAGIC ANARE Mapping and Geographic Information Committee

ANARESAT Australian National Antarctic Research Expeditions (Communications) Satellite

ANA-S Antinuclear Antibody, Speckled Pattern [Biochemistry]

Anat Anatomical; Anatomist; Anatomy [also anat]

ANATRAN Analog Translator

ANB Air Navigation Bureau

ANB Alpha-Naphthyl Butyrate

ANB Aluminum-n-Butoxide

ANB Australian National Bibliography [of National Library of Australia]

ANBFM Adaptive Narrow-Band FM (Frequency Modulation) Modem

ANBG Australian National Botanic Gardens [also Canberra Botanic Gardens]

ANBLS Association of New Brunswick Land Surveyors [Canada]

ANBS Air Navigation and Bombing School

ANBS Asian Network of Biological Sciences

ANC Academia Nacional de Córdoba [National Academy of Cordoba, Argentina]

ANC Acid-Neutralizing Capacity

ANC Active Noise Control

ANC Aerojet Nuclear Company [US]

ANC Air Navigation Commission [of International Civil Aviation Organization]

ANC Air Navigation Conference

ANC All-Number Calling

ANC Alphanumeric Code

ANC Argentine Naval Commission [New York, US]

ANC Army-Navy Civil Aeronautics Committee [US]

ANC Army Nurse Corps

ANC Association of Noise Consultants [UK]

ANC Automatic Nutation Control

anc ancient

ANCA Allied Naval Communications Agency [of NATO]

ANCA Australian National Coastal Authority

ANCA Australian National Council on AIDS

ANCA Australian Nature Conservation Agency [Succeeded by Biodiversity Group–Environment Australia]

ANCAR Australian National Committee for Antarctic Research

ANCAT Abatement of Nuisances Caused by Air Transport [of European Civil Aviation Conference]

ANCC Algerian National Chamber of Commerce

ANCIAWPRC Argentinian National Committee of the International Association on Water Pollution Research and Control

ANCIAWPRC Australian National Committee of the International Association on Water Pollution Research and Control

ANCIAWPRC Austrian National Committee of the International Association on Water Pollution Research and Control

ANCIGBP Australian National Committee, International Geosphere Biosphere Program

ANCOVA Analysis of Covariance [also Ancova]

ANCS Advanced Numeric Crash System

ANCS American Numerical Control Society [US]

ANCU Airborne Navigation Computer Unit

ANC UNESCO Austalian National Commission for UNESCO [of United Nations]

AND Academics for Nuclear Disarmament

AND Air Force-Navy Design

AND Alphanumeric Display

AND Army-Navy Design

And Andromeda [Astronomy]

α-Nd Alpha Neodymium [Symbol]

ANDA Association Nationale pour le Développement Agricole [National Association for Agricultural Development, France]

ANDB Air Navigation Development Board [US]

a-NdCo Neodymium Cobalt Amorphous Alloy

a-NdDyFeB Neodymium Dysprosium Iron Boron Amorphous Alloy

ANDF Architecture-Neutral Distribution Format [Computers]

a-NdFe Neodymium Iron Amorphous Alloy

a-NdFeB Neodymium Iron Boron Amorphous Alloy

a-NdFeC Neodymium Iron Carbon Amorphous Alloy

ANDIP American National Dictionary for Information Processing [US]

And Is Andaman Islands

A-nDNA-A Anti-native DNA (Deoxyribonucleic Acid) Antibody [Biochemistry]

AND/NOR And Not-Or (Gate)

AND/OR And/Or (Gate)

Andr Andromeda [Astronomy]

ANDREE Association for Nuclear Development and Research in Electrical Engineering

Andrew Seybold's Outlook Prof Comput Andrew Seybold's Outlook on Professional Computing [Published in the US]

ANDSA Amino Naphthol Disulfonic Acid

andz anodize

ANE Aeronautical and Navigational Electronics

ANE Angioneurotic Edema

ANEC American Nuclear Energy Council [US]

ANEDA Association Nationale d'Etudes pour la Documentation Automatique [National Association for Studies in Automatic Documentation, France]

ANEF Australian Noise Exposure Forecast

ANEN African Network of Environmental NGO's (Nongovernmental Organizations)

ANEN Australian NGO (Nongovernmental Organizations) Environment Network

[9]aneN3 1,4,7-Triazacyclononane

[10]aneN3 1,4,7-Triazacyclodecane

[11]aneN3 1,4,8-Triazacycloundecane

[12]aneN3 1,5,9-Triazacyclododecane

1,4,7-[11]aneN3 1,4,7-Triazacycloundecane

1,4,7-[13]aneN3 1,4,7-Triazacyclotridecane

[15]aneN4 1,4,8,12-Tetraazacyclopentadecane

[15]aneN5 1,4,7,10-13-Pentaazacyclopentadecane

[16]aneN4 1,5,9,13-Tetraazacyclohexadecane

[16]aneN5 1,4,7,10,13-Pentaazacyclohexadecane

[17B]aneN5 1,4,7,11,14-Pentaazacycloheptadecane

[17C]aneN5 1,4,7,10,13-Pentaazacycloheptadecane

[18A]aneN5 1,4,7,11,15-Pentaazacyclooctadecane

[18B]aneN5 1,4,8,11,15-Pentaazacyclooctadecane

[18C]aneN5 1,4,7,10,13-Pentaazacyclooctadecane

[19]aneN5 1,4,8,12,16-Pentaazacyclononadecane

[19B]aneN5 1,7,10,13-Pentaazacyclononadecane

[20]aneN5 1,5,9,13,17-Pentaazacycloeicosane

[9]aneNO2 1,4-Dioxa-7-Azacyclononane

[9]aneN2O 1-Oxa-4,7-Diazacyclononane

[9]aneN2S 1-Thia-4,7-Diazacyclononane

[15]aneN3S2 1,4-Dithia-7,10,13-Triazacyclopentadecane

[9]aneS3 1,4,7-Trithiacyclononane

[12]aneS3 1,5,9-Trithiacyclododecane

[13]aneS4 1,4,7,10-Tetrathiacyclotridecane

[14]aneS4 1,4,8,11-Tetrathiacyclotetradecane

[15]aneS5 1,4,7,10,13-Pentathiacyclopentadecane

[18]aneS6 1,4,7,10,13,16-Hexathiacyclooctadecane

[24]aneS6 1,5,9,13,17,21-Hexathiacyclotetracosane

Anes Anesthesia; Anesthetic(s); Anesthetist [also Anesth]

ANF Academia Nacional de Farmacia [National Academy of Pharmacy, Brazil]

ANF Advanced Nuclear Fuels Corporation [US]

ANF American National Form

ANF Anti-Nuclear Factor

ANF Arkansas National Forest [US]

ANF Atlantic Nuclear Force

ANF Atrial Natriuretic Factor [Biochemistry]

ANFCE Allied Naval Force Central Europe

An Fis A, Fenom Interacciónes Anales de Fisica, Serie A, Fenomenos e Interacciónes [Annals of Physics, Series A, Phenomena and Interactions, Spain]

An Fis B, Apl Métodos Instrum Anales de Fisica, Serie B, Aplicaciones, Métodos e Instrumentos [Annals of Physics, Series B, Applications, Methods and Instrumentation, Spain]

An Fis Ser A Anales de Fisica, Serie A [Annals of Physics, Series A, Spain]

An Fis Ser B Anales de Fisica, Serie B [Annals of Physics, Series B, Spain]

ANFO Ammonium Nitrate and Fuel Oil Mixture (Explosive) [also AN-FO]

ANG Air National Guard [of US Air Force]

ANG Netherlands Antillean Guilder [Currency]

Ang Angola(n)

ANGB Air National Guard Base

ANG-CE Air National Guard–Civil Engineering [US]

Angew Chem Angewandte Chemie [German Journal on Applied Chemistry]

Angew Chem Int Angewandte Chemie International [Applied Chemistry International; Journal published in Germany]

Angew Chem Int Ed Angewandte Chemie, International Edition [Applied Chemistry, International Edition; Journal published in Germany]

Angew Chem Int Ed Engl Angewandte Chemie, International Edition Englisch [Applied Chemistry, International Edition, English; Journal published in Germany]

Angew Inform Angewandte Informatik [German Journal on Applied Computer Science]

Angew Makromol Chem Angewandte Makromolekulare Chemie [German Journal on Applied Macromolecular Chemistry]

ANGIS Australian National Geoscience Information Service [of Australian Geological Survey Organization]

Angl Angle [also angl]

Anglo-Saxon Chron Anglo-Saxon Chronicle [UK]

ANGOLO Asian NGO (Nongovernmental Organization) Coalition for Agrarian Reform and Rural Development

ANGOP Angolan Government News Agency

ANGUS A Navigable General-Purpose Underwater Surveyor [also Angus]

ANH Australian National Herbarium [of Commonwealth Scientific and Industrial Research Organization]

anh anhydrous

ANHC Animal Natural Health Center [Eugene, Oregon, US]

ANHS American Natural Hygiene Society [US]

ANHSIR Australian National Herbarium Specimen Information Register (Database) [of Commonwealth Scientific and Industrial Research Organization]

ANHSO Association of National Health Service Officers [UK]

An-HTDO Anatase-Type Hydrous Titanium Dioxide

anhyd anhydrous

ANI Advanced Network Integration

ANI Annular Isotropic

ANI Any [Amateur Radio]

ANI Automatic Number Identification [Telecommunications]

.ANI Animated Cursor [Microsoft File Name Extension] [also .ani]

a-Ni Amorphous Nickel

ANIA Associazione Nazionale Italiana per l'Automazione [Italian National Association for Automation]

ANIB Australian News and Information Bureau

ANIC Australian National Insect Collection

ANIE Associazione Nazionale Industrie Elettrotecniche [National Association of the Electrotechnical Industry, Italy]

ANIEL Asociación Nacional de Industrias Electronicas [National Association of Electronics Industries, Spain]

ANIELB Association of Northern Ireland Education and Library Boards [UK]

Anil Aniline [also anil]

a-NiLa Nickel Lanthanum Amorphous Alloy

ANILOL Aniline-Alcohol Mixture [also anilol]

ANIM Association of Nuclear Instrument Manufacturers [US]

ANIN Associated Northern Ireland Newspapers

ANIP Army-Navy Instrumentation Program [US]

ANIP Army-Navy Integrated Presentation

ANIPAC National Association for the Plastics Industry [US]

ANIS (*Allgemeines Nichtnumerisches Info-System*) – German for "Common Nonnumerical Information System"

ANIS Annular Isotropic Source

a-NiTa Nickel Tantalum Amorphous Alloy

a-NiZr Nickel Zirconium Amorphous Alloy

α-NiZr Alpha Nickel Zirconium

ANL Accademia Nazionale dei Lincei [National Academy, Rome, Italy]

ANL American National (Standard) Label

ANL Annoyance Noise Level

ANL Argonne National Laboratory [of US Department of Energy at University of Chicago, Illinois]

ANL Associação Naval de Lisboa [Naval Association of Lisbon, Portugal]

ANL Association Nationale de Logiciel [National Software Association, France]

ANL Australian National Line

ANL Automatic Noise Leveling

ANL Automatic Noise Limiter [also anl]

anl anneal(ed)

ANL-CAT Argonne National Laboratory–Collaborative Access Team [US]

ANL-E Argonne National Laboratory–East [of US Department of Energy near Chicago, Illinois]

ANLEMC Argonne National Laboratory Electron Microscopy Center [also ANL-EMC, US]

ANLG Analog [Communications]

Anlg Annealing [also anlg]

anlg analog

ANLG GND Analog Ground [Communications]

ANLOR Angle Order

ANL-W Argonne National Laboratory–West [of US Department of Energy near Idaho Falls, US]

ANM AFTN (Aeronautical Fixed Telecommunications Network) Notification Message

ANM N-(4-Anilino-1-Naphthyl)-Maleimide; N-(1-Anilinonaphthyl) Maleimide

ANM Annals of Nutrition and Metabolism [US]

ANM Australian Newsprint Mills Limited

α/nm² alphas per square nanometer [Unit] [also α nm^{-2}]

ANMA Australian National Maritime Association, Inc.

ANMB Army and Navy Munitions Board

ANMC American National Metric Council [US]

An Mec Electr Anales de Mecánica y Electricidad [Annals of Mechanics and Electricity, Spain]

ANML Amsterdam-Nijmegen Magnet Laboratory [Netherlands]

ANMM Australian National Maritime Museum

ANMSA Aminonaphthol Monosulfonic Acid

ANN Adaptive Neural Network

ANN Answer, No-Charge

ANN Artificial Neural Network

.ANN Annotations [IBM File Name Extension] [also .ann]

Ann Annealing [also ann]

Ann Annals [also ann]

Ann Annuity [also ann]

Ann Annunciator [also ann]

ann anneal(d)

ann annual

ANNA Army-Navy-NASA-Air (Satellite)

Ann AAG Annals of the Association of American Geographers

Ann Acad Med Sing Annals of the Academy of Medicine of Singapore

Ann Acad Sci Fenn A Annales Academiae Scientiarum Fennicae, Series A [Annals of the Finnish Academy of Sciences, Series A]

Ann Acad Sci Fenn A, Math Annales Academiae Scientiarum Fennicae, Series A, Mathematica [Annals of the Finnish Academy of Sciences, Series A, Mathematics]

Ann Acad Sci Fenn AI Annales Academiae Scientiarum Fennicae, Series AI [Annals of the Finnish Academy of Sciences, Series AI]

Ann Acad Sci Fenn AII Annales Academiae Scientiarum Fennicae, Series AII [Annals of the Finnish Academy of Sciences, Series AII]

Ann Acad Sci Fenn AII, Chem Annales Academiae Scientiarum Fennicae, Series AII, Chemica [Annals of the Finnish Academy of Sciences, Series AII, Chemistry]

Ann Acad Sci Fenn AVI Annales Academiae Scientiarum Fennicae, Series AVI [Annals of the Finnish Academy of Sciences, Series AVI]

Ann Acad Sci Fenn AVI, Phys Annales Academiae Scientiarum Fennicae, Series AVI, Physica [Annals of the Finnish Academy of Sciences, Series AVI, Physics]

Ann Agric Environ Med Annals of Agricultural and Environmental Medicine [of Institute of Agricultural Medicine]

Ann Allerg Annals of Allergy

Ann Appl Biol Annals of Applied Biology [of Association of Applied Biologists, UK]

Ann Biol Clin Annales de Biologie Clinique [Annals of Clinical Biology published in France]

Ann Biomed Eng Annals of Biomedical Engineering [of Biomedical Engineering Society, US]

Ann Braz Ent Soc Annals of the Brazilian Entomological Society

Ann Chim Annales des Chimie [Annals of Chemistry, France]

Ann Chim Fr Annales des Chimie, France [Annals of Chemistry published in France]

Ann Chim Phys Annales des Chimie et des Physique [Annals of Chemistry and Physics, France]

Ann Chim, Sci Matér Annales des Chimie, Science des Matériaux [Annals of Chemistry, Materials Science, France]

Ann Clin Lab Sci Annals of Clinical and Laboratory Science

Ann Compos Annales des Composites [Annals of Composites, France]

Ann Conf Annual Conference [also Annu Conf]

Ann Conf Australas Inst Met Annual Conference of the Australasian Institute of Metals [also Ann Conf AIM, Annu Conf Australas Inst Met, or Annu Conf AIM]

Ann Conf Chin Soc Mater Sci Annual Conference of the Chinese Society for Materials Science [also Annu Conf Chin Soc Mater Sci]

Ann Emer Med Annals of Emergency Medicine [of American College of Emergency Physicians, US]

Ann Ent Soc Am Annals of the Entomological Society of America

Ann Fac Sci Toulouse Math Annales de la Faculté des Sciences de Toulouse, Mathematiques [Annals of the Faculty of Science of Toulouse, Mathematics, France]

Ann Fond Louis Broglie Annales de la Fondation Louis de Broglie [Annals of the Louis de Broglie Foundation, France]

Ann Fr Microtech Chronom Annales Françaises des Microtechniques et de Chronometrie [French Annals of Microtechnology and Chronometry, France]

Ann Geogr Annales de Géographie [Annals of Geography, France]

Ann Geophys Annales Geophysicae [Annals of Geophysics; published by the European Geophysical Society]

Ann Geophys (CNRS) Annales de Geophysique (Conseil National de la Recherche Scientifique) [French Annals of Geophysics published by CNRS]

Ann Geophys Atmos Hydrospheres Space Sci Annales Geophysicae: Atmospheres, Hydrospheres and Space Sciences [Annals of Geophysics: Atmospheres, Hydrospheres and Space Science; published by the European Geophysical Society]

Ann Glaciol Annals of Glaciology [of International Glaciological Society] [also Ann Glac]

Ann Hist Comput Annals of the History of Computing [American Federation of Information Processing Societies, US]

Ann Human Biol Annals of Human Biology [of Society for the Study of Human Biology, UK]

ANNI Axial Next Nearest Ising (Model) [Solid-State Physics]

Ann ICRP Annals of the International Commission on Radiological Protection [UK]

Ann IGY Annals of the International Geophysical Year

Ann Int Med Annals of Internal Medicine

Ann Inst Henri Poincaré Annales de l'Institut Henri Poincaré [Annals of Henri Poincare Institute, France]

Ann Inst Henri Poincaré Phys Theor Annales de l'Institut Henri Poincaré Physique Theorique [Annals of Henri Poincare Institute of Theoretical Physics, France]

Ann Inst Pasteur Annales de l'Institut Pasteur [Annals of the Pasteur Institute, Paris, France]

Ann Isr Phys Soc Annals of the Israel Physical Society [Israel]

Ann Ist Speriment Selvicol Annali dell' Istituto Sperimentale per la Selvicoltura [Annals of the Silvicultural Research Institute, Italy]

Ann Ist Super Sanita Annali dell'Istituto Superiore di Sanita [Annals of the Higher Institute of Sanita, Italy]

Ann Mat Annali di Matematica [Italian Annals of Mathematics]

Ann Math Annals of Mathematics [of Princeton University Press, Princeton, New Jersey, US]

Ann Math Stud Annals of Mathematics Studies [of Princeton University Press, US]

Ann Math Stat Annals of Mathematical Statistics [of Institute of Mathematical Statistics, US]

Ann Mines Annales des Mines [Annals of Mining, France]

Ann Mtg Annual Meeting [also Annu Mtg]

Ann Mtg Am Ceram Soc Annual Meeting of the American Ceramic Society [also Annu Mtg Am Ceram Soc, or Annu Mtg ACS]

Ann Mtg JSME Annual Meeting of the Japanese Society of Mechanical Engineering [also Annu Mtg JSME]

Ann Neurol Annals of Neurology

ANNNI Anisotropic Next-Nearest-Neighbour Ising (Model) [Solid-State Physics]

ANNNI Axial Next-Nearest-Neighbour Ising (Model) [Solid-State Physics]

Ann Nucl Energy Annals of Nuclear Energy [UK]

Ann Nutr Metabol Annals of Nutrition and Metabolism [US]

Ann NYAS Annals of the New York Academy of Sciences [US] [also Ann NY Acad Sci]

Ann Occup Hyg Annals of Occupational Hygiene [US]

Ann Oper Res Annals of Operations Research [Switzerland]

Ann Paléont Annales de Paléontologie [Annals of Paleontology; published in France]

Ann Pharmacol Annals of Pharmacology

Ann Phys Annals of Physics

Ann Phys (Germany) Annalen der Physik (Germany) [Annals of Physics (Germany)]

Ann Phys (France) Annales de Physique (France) [Annals of Physics (France)]

Ann Phys (USA) Annals of Physics (USA) [Annals of Physics (USA)]

Ann Probab Annals of Probability [of Institute of Mathematical Statistics, US]

Ann Proc Assoc Sci Tech Soc South Afr Annual Proceedings of the Associated Scientific and Technical Societies of South Africa

Ann Pure Appl Log Annals of Pure and Applied Logic [Netherlands]

Ann Quim Annales de Quimica [Annals of Chemistry]

Ann Quim Int Ed Annales de Quimica, International Edition [Journal]

Ann Radioelectricité Annales de Radioelectricité [Annals of Radioelectricity, France]

Ann Rech Sylvicoles Annales Recherches Sylvicoles [Annals of Sylvicultural Research, France]

Ann Rep Ceram Eng Res Lab Annual Report of Ceramic Engineering Research Laboratory [Japan]

Ann Rep Eng Res Inst Fac Eng Univ Tokyo Annual Report of the Engineering Research Institute, Faculty of Engineering, University of Tokyo [Japan]

Ann Rep Eng Res Inst, Tokyo Univ Annual Report of the Engineering Research Institute, Tokyo University [Japan]

Ann Rep Inst Phys, Acad Sin Annual Report of the Institute of Physics, Academia Sinica [Taiwan]

Ann Rep Jpn Soc Technol Plast Annual Report of the Japan Society for Technology of Plasticity

Ann Rep Res Inst Catal, Hokkaido Univ Annual Report of the Research Institute for Catalysis, Hokkaido University [Japan]

Ann Rep, Res React Inst, Kyoto Univ Annual Reports, Research Reactor Institute, Kyoto University [Japan]

Ann Rev Biochem Annual Review of Biochemistry [US]

Ann Rev Biophys Biophys Chem Annual Review of Biophysics and Biophysical Chemistry [of Biophysical Society, US]

Ann Rev Ecol Syst Annual Review of Ecological Systems [US]

Ann Rev Fluid Mech Annual Review of Fluid Mechanics [US]

Ann Rev Mater Sci Annual Review of Materials Science [US]

Ann Rev Microbiol Annual Review of Microbiology [US]

Ann Rev Nuc Part Sci Annual Review of Nuclear Particle Physics [US]

Ann Rev Phys Chem Annual Review of Physical Chemistry [US]

Ann Rev Plant Physiol Annual Review of Plant Physiology [US]

Ann Rheum Diseases Annals of the Rheumatic Diseases

Ann Sim Symp Proc Annual Simulation Symposium Proceedings [of Society for Computer Simulation International, US]

Ann Soc Fr Chim Annuaire de la Société Française de Chimie [Annual Report of the French Chemical Society]

Ann Soc Pol Math Annales de la Société Polonaise de Mathématique [Annales of the Polish Mathematical Society]

Ann Surg Annals of Surgery

Ann Univ Sofia Fac Phys Annuaire de l'Université de Sofia Faculté de Physique [Annual Report of the University of Sofia, Faculty of Physics, Bulgaria]

Ann Sci Annals of Science [UK]

Ann Soc Log Eng Annals of the Society of Logistics Engineers [US]

Ann Soc Sci Brux Annales de la Société Scientifique de Bruxelles [Annals of the Scientific Society of Brussels, Belgium]

Ann Soc Sci Brux I, Sci Math Astron Phys Annales de la Société Scientifique de Bruxelles, Serie I (Sciences Mathematiques, Astronomiques et Physiques) [Annals of the Scientific Society of Brussels, Series I (Mathematical, Astronomical and Physical Sciences, Belgium]

Ann Stat Annals of Statistics [of Institute of Mathematical Statistics, US]

Ann Surg Annals of Surgery [US]

Ann Telecommun Annales des Télécommunications [Annals of Telecommunications; published by Centre National d'Etudes des Télécommunications, France]

Ann Thoracic Surg Annals of Thoracic Surgery

Ann Tokyo Astron Obs Annals of the Tokyo Astronomical Observatory [Japan]

annu annual

Annu Conf Annual Conference [also Ann Conf]

Annu Conf Australas Inst Met Annual Conference of the Australasian Institute of Metals [also Annu Conf AIM, Ann Conf Australas Inst Met, or Ann Conf AIM]

Annu Conf Chin Soc Mater Sci Annual Conference of the Chinese Society for Materials Science [also Ann Conf Chin Soc Mater Sci]

Annu HITEMP Rev Annual HITEMP (High-Temperature Engine Materials Technology Program) Review [of NASA–Lewis Research Center, Cleveland, Ohio, US] [also Annu (NASA) HITEMP Rev]

Annu Mtg Annual Meeting [also Ann Mtg]

Annu Mtg Am Ceram Soc Annual Meeting of the American Ceramic Society [also Ann Mtg Am Ceram Soc, or Annu Mtg ACS]

Annu Mtg JSME Annual Meeting of the Japanese Society of Mechanical Engineering [also Ann Mtg JSME]

Annu (NASA) HITEMP Rev Annual HITEMP (HighTemperature Engine Materials Technology Program) Review [of NASA–Lewis Research Center, Cleveland, Ohio, US] [also Annu HITEMP Rev]

Ann Univ Turku AI, Astron–Chem–Phys–Math Annales Universitasis Turkuensis, Series AI, Astronomica–Chemica– Physica–Mathematica [Annals of the University of Turku, Series AI, Astronomy–Chemistry– Physics– Mathematics, Finland]

Ann Univ Turku Ser AI Annales Universitasis Turkuensis, Series AI [Annals of the University of Turku, Series AI, Finland]

Annu Proc Assoc Sci Tech Soc South Afr Annual Proceedings of the Associated Scientific and Technical Societies of South Africa

Annu Rep Ceram Eng Res Lab Annual Report of Ceramic Engineering Research Laboratory [Japan]

Annu Rep Eng Res Inst Fac Eng Univ Tokyo Annual Report of the Engineering Research Institute, Faculty of Engineering, University of Tokyo [Japan]

Annu Rep Eng Res Inst, Tokyo Univ Annual Report of the Engineering Research Institute, Tokyo University [Japan]

Annu Rep Inst Phys, Acad Sin Annual Report of the Institute of Physics, Academia Sinica [Taiwan]

Annu Rep Jpn Soc Technol Plast Annual Report of the Japan Society for Technology of Plasticity

Annu Rep Res Inst Catal, Hokkaido Univ Annual Report of the Research Institute for Catalysis, Hokkaido University [Japan]

Annu Rep, Res React Inst, Kyoto Univ Annual Reports, Research Reactor Institute, Kyoto University [Japan]

Annu Rep USNM Annual Reports of the United States National Museum

Annu Rev Anthropol Annual Review of Anthropology [US]

Annu Rev Astron Astrophys Annual Review of Astronomy and Astrophysics [US]

Annu Rev Biochem Annual Review of Biochemistry [US]

Annu Rev Biophys Biophys Chem Annual Review of Biophysics and Biophysical Chemistry [of Biophysical Society, US]

Annu Rev Cell Biol Annual Review of Cell Biology [US]

Annu Rev Earth Planet Sci Annual Review of Earth and Planetary Sciences [US]

Annu Rev Ecol Syst Annual Review of Ecological Systems [US]

Annu Rev Ent Annual Review of Entomology [of Entomological Society of America, US]

Annu Rev Fluid Mech Annual Review of Fluid Mechanics [US]

Annu Rev Genet Annual Review of Genetics [US]

Annu Rev Mater Sci Annual Review of Materials Science [US]

Annu Rev Microbiol Annual Review of Microbiology [US]

Annu Rev Nutr Annual Reviews of Nutrition [US]

Annu Rev Phys Chem Annual Review of Physical Chemistry [US]

Annu Rev Plant Physiol Annual Review of Plant Physiology [US]

Annu Rev Psychol Annual Review of Psychology

Annu Sim Symp Proc Annual Simulation Symposium Proceedings [of Society for Computer Simulation International, US]

Annu Soc Fr Chim Annuaire de la Société Française de Chimie [Annual Report of the French Chemical Society]

Annu Univ Sofia Fac Phys Annuaire de l'Université de Sofia Faculté de Physique [Annual Report of the University of Sofia, Faculty of Physics, Bulgaria]

ANO Air Navigation Order

ANO Alphanumeric Output

ANO Atomic Natural Orbital

ANO Approximate Natural Orbital

ANO Average Natural Orbital

ANOCOVA Analysis of Covariance [also ANOCOVAR]

Anod Anodizing [also anod]

anom anomalous

ANON Another Node On the interNet [Internet Site on Computers, Geosciences and Mathematics]

anon anonymous(ly)

ANOP Australian National Opinion Polls

ANORTAL Anorthosite Aluminum (Process)

ANOVA Analysis of Variance [also ANOVAR]

ANP Aircraft Nuclear Propulsion (Program)

ANP Air Navigation Plan

ANP Allied Navigation Publication

ANP 2-Amino-4-Nitrophenol

ANP Atrial Natriuretic Peptide [Biochemistry]

α-Np Alpha Neptunium [Symbol]

ANPA American Newspaper Publishers Association [US]

ANPBC Association of Naturopathic Physicians of British Columbia [Canada]

ANPC Australian National Parks Council

ANPC Australian Network for Plant Conservation

ANPD Aircraft Nuclear Propulsion Department

ANPD/GE Aircraft Nuclear Propulsion Department, General Electric [US]

ANP-KLH Atrial Natriuretic Peptide–Keyhole Limpet Hemocyanin (Immunogen) [Biochemistry]

ANPO Aircraft Nuclear Propulsion Office [of United States Atomic Energy Commission]

ANPO Association of National Park Officers [UK]

ANPO 2-(α-Naphthyl)-5-Phenyl Oxazole

ANPOD Antenna Positioning Device

ANPP Army Nuclear Power Program [US]

ANPP Association Nationale pour la Protection des Plantes [French National Association for Plant Protection]

ANPP 4-Azido-2-Nitrophenyl Phosphate

ANPR Advance(d) Notice of Proposed Rulemaking [also ANPRM]

ANPS American Nail Producers Society [US]

ANPT Aeronautical National Form Taper Pipe Thread [Symbol]

ANPWS Australian National Parks and Wildlife Service [of Australian Nature Conservation Agency]

ANQ Solution Annealed and Quenched (Condition)

ANR Adiabatic Nuclear Rotation (Method) [Physics]

ANR Advanced Non-Rigid (Airship)

ANR Alphanumeric Replacement

ANR Association of Neutron Radiographers [US]

ANR Automatic Nine-Barrel Repeater

ANR Awaiting Number Received

ANRAC Aids to Navigation Radio Control(led) [also anrac]

ANRC American National Red Cross [Washington, DC, US]

ANRC Animal Nutrition Research Council [US]

ANRC Australian National Railways Commission

ANRC Australian National Research Council

ANRIC Annual National Information Retrieval Colloquium [US]

Anritsu Tech Bull Anritsu Technical Bulletin [Japan]

ANRPC Association of Natural Rubber Producing Countries [Malaysia]

ANRS Automatic Noise Reduction System

ANRT Association Nationale de la Recherche Technique [National Association for Technical Research, France]

ANS Académie National de Sciences [National Academy of Sciences, France]

ANS Academy of Natural Sciences [Philadelphia, US]

ANS Acyl Nucleophilic Substitution [Biochemistry]

ANS Advanced Neutron Source

ANS Air Navigation School

ANS American National Standard [US]

ANS American Nuclear Society [US]

ANS Ammonium Nickel(II) Sulfate

ANS Anilinenaphthalene Sulfonate

ANS 8-Anilino-1-Naphthalenesulfonic Acid

ANS Answer [also Ans, or ans]

ANS Aquatic Nuisance Species

ANS Artificial Neural System

ANS Astronautical Netherlands Satellite

ANS Astronomical Netherlands Satellite

ANS ATM (Asynchronous Transfer Mode) Network Switches

ANS Australian Nuclear Society

ANS Autonomic Nervous System

2,6-ANS 2-Anilinonaphthalene-6-Sulfonic Acid

bis-ANS 4,4'-Dianilino-1,1'-Binaphthyl-5,5'-Disulfonic Acid

Ans Anisole

Ans Answer [also ans]

ANSA Advanced Network System Architecture

ANSA 6-Amino-2-Naphthalenesulfonic Acid

ANSA 8-Anilino-1-Naphthalenesulfonic Acid

ANSA Automatic New Structure Alert

ANSAC American Natural Soda Ash Corporation [US]

ANSAM Antinuclear Surface-to-Air Missile

ANSB Army-Navy Safety Board [Washington, DC, US]

ANSC American National Standards Committee [now National Information Standards Organization, US]

ANSC Anisotropic Nanometer-Scale Composite

ANSCAD Associate of Nova Scotia College of Art and Design [Canada]

ANS COBOL American National Standard Common Business-Oriented Language [also ANSI COBOL]

ANSCR Alphanumeric System for Classification of Recordings

ANS/ENS American Nuclear Society/European Nuclear Society (International Conference)

ANSFMN Accademia Nazionale dei Scienze Fisiche, Matematiche e Naturali [National Academy of Physical, Mathematical and Natural Sciences, Italy]

ANSI American National Standards Institute [US]

ANSI Area of Natural and Scientific Interest

ANSI/AFBMA American National Standards Institute/ Antifriction Bearing Manufacturers Association [US]

ANSI/AGMA American National Standards Institute/American Gear Manufacturers Association [US]

ANSI/ASQC American National Standards Institute/ American Society for Quality Control [US]

ANSI/ASTM American National Standards Institute/ American Society for Testing and Materials [US]

ANSI/AWS American National Standards Institute/American Welding Society [US]

ANSI C++ American National Standards Institute– C-Plus-Plus [Object-Oriented Programming Language]

ANSI COBOL American National Standards Institute Common Business-Oriented Language [also ANS COBOL]

ANSI FORTRAN American National Standards Institute Formula Translator (Programming Language)

ANSI/IEEE American National Standards Institute/ Institute of Electrical and Electronics Engineers [US]

ANSI/ISO American National StandardsInstitute/ International Standards Organization (Standard)

ANSI/NEMA American National Standards Institute/ National Electrical Manufacturers Association [US]

ANSI/NSF American National Standards Institute/National Science Foundation (Standard) [US]

ANSL American National Standard Label

ANSLS Association of Nova Scotia Land Surveyors [Canada]

ANS-NH₄ 8-Anilino-1-Naphthalenesulfonic Acid, Ammonium Salt

An Soc Ent Bras Anais da Sociedade Entomologica do Brasil [Annals of the Brazilian Entomological Society]

ANSP Academy of Natural Sciences of Philadelphia [US]

ANSP Association of Navy Safety Professionals [US]

ANSS American Nature Study Society [US]

ANSSSR Academiya Nauk Sojus Sowjetskich Sozialistitscheskich Respublik [Academy of Sciences of the USSR, Moskow] [also AN SSSR]

ANSTC Australian Nuclear Science and Technology Council [of Department of Primary Industries and Energy]

ANSTI African Network for Scientific and Technical Institutions [of UNESCO in Kenya]

ANSTI Australian Nuclear Science and Technology Information

An Stiinț Univ 'All Cuza' Iaşi (Ser Noua) IB Fiz Analele Stiințifice ale Universitatu 'All Cuza' din Iaşi (Serie Noua) Sectiuneas IB Fizica [Scientific Annales of the All Cuza University of Iasi (New Series), Section IB Physics, Romania]

ANSTO Australian Nuclear Science and Technology Organization

ANSVIP American National Standard Vocabulary for Information Processing [US]

ANSW Antinuclear Submarine Warfare

ANSWER American Nuclear Society's Worldwide Electronic Resource

ANT Acetonitrile

ANT Administración Nacional de Telecomunicaciones [National Telecommunications Administration, Paraguay]

ANT Articulated Navigation Testbed

ANT Australian National Territory (Region)

Ant Antenna [also ant]

Ant Antlia [Astronomy]

Ant Antonym [also ant]

ant anterior

ANTAC Air Navigation Tactical Control System

Antarct Antarctic(a) [also antarc]

Antarct O Antarctic Ocean [also Antarc O]

ANTARES A Natural Tool to Achieve the Required Engineering Shape (Computer Software)

ANTC Antichaff Circuit

ANTEC Annual Technical Conference [of Society of Petroleum Engineers of the American Institute of Mining, Metallurgical and Petroleum Engineers, US]

ANTGWDPEC Association of National Trade Groups of Wood and Derived Products in the EEC (European Economic Community) Countries [Denmark]

Anth Anthology [also anth]

Anthrop Anthropological; Anthropologist; Anthropology [also anthrop]

Anthropom Anthropometric; Anthropometry

Anthrop Q Anthropological Quarterly [International Journal]

ANTI Anticoincidence

Anti-A-CAM Anti-A-Cell Adhesion Molecule [Biochemistry]

anti-*a*-Sm-1 Monoclonal Anti-*α*-Smooth Muscle Actin [Biochemistry]

anti-*a*-Sr-1 Monoclonal Anti-*α*-Sarcomeric Actin [Biochemistry]

Anti-BrdU (Monoclonal) Antibody to 5-Bromo-2'-Deoxyuridine [Biochemistry]

Anti-Corros Methods Mater Anti-Corrosion Methods and Materials [Journal published in the UK]

Anti-D Ig Anti-D Immunoglobulin [also anti-D Ig]

ANTI-DNA Anti-Deoxyribonucleic Acid [also Anti-DNA] [Biochemistry]

ANTI-DNP Anti-Deoxyribonucloprotein [also Anti-DNP] [Biochemistry]

Antifr Antifreeze [also antifr]

Anti-HNK Anti-Human NK (Cell) [Biochemistry]

antilog antilogarithm [Mathematics]

Antimicrob Agents Chemother Antimicrobial Agents and Chemotherapy (Journal) [of American Society for Microbiology, US]

Anti-N-CAM Anti-Neural Cell Adhesion Molecule [Biochemistry]

Antiobiot Antiobiotiki [Russian Journal on Antibiotics]

Antiq Antiquarian; Antiquary [also antiq]

anti-Rh Anti-Rhesus [also anti Rh] [Immunology]

ANTI-Sm Anti-Smith [Antibody Determination to Smith Antigen] [Immunology]

ANTIVOX Voice-Actuated Transmitter Keyer Inhibitor

AN/TN Amino Nitrogen/Total Nitrogen (Ratio)

AN-TNT Ammonium Nitrate and Trinitrotoluene Mixture

Antropol Antropología [Brazilian Journal on Anhropology]

ANTS Airborne Night Television System

ANTS ARPA (Advanced Research Projects Agency) Network Terminal System [of US Department of Defense]

ANTU *α*-Naphthylthiourea [also Antu]

ANU Airplane Nose Up

ANU Andong National University [Andong, South Korea]

ANU Australian National University [Canberra, Australia]

ANUG Atex (Computer System) Newspaper Users Group [US]

ANUGA Allgemeine Nahrungs– und Genußmittel-Ausstellung [World Food Fair and Exhibition, Germany]

ANUL Australian National University Library [Canberra, Australia]

An Univ Bucur Ser Fiz Analele Universitatii Bucuresti, Fizica [Annals of the University of Bucharest, Physical Series [Romania]

An Univ Bucur Ser Mat Analele Universitatii Bucuresti, Seria Matematica [Annals of the University of Bucharest, Mathematical Series [Romania]

An Univ Galati, Mec Constr Masin Analele Universitatii din Galati, Mecanica si Constructii de Masini [Annals of the University of Galati, Mechanics and Mechanical Engineering, Romania]

An Univ Galati, Metal Analele Universitatii din Galati, Metalurgie [Annals of the University of Galati, Metallurgy, Romania]

ANUZ Australian National University Zoology Department [Canberra]

ANV Adiabatic Nuclear Vibration (Method) [Physics]

ANVIS Aviator's Night Vision Imaging System

ANVM Association of Night Vision Manufacturers [US]

ANWC Australian National Wildlife Collection [of Commonwealth Scientific and Industrial Research Organization]

ANWR Arctic National Wildlife Range [Alaska]

ANX Automotive Network Exchange [Joint Program of Chrysler, Ford and General Motors]

ANYL Division of Analytical Chemistry [of American Chemical Society, US]

Anyst Analyst [also anyst]

ANZ Air New Zealand

ANZAAS Australian and New Zealand Association for the Advancement of Science [also Anzaas]

ANZAC Australian and New Zealand Army Corps [also Anzac]

ANZEC Australian and New Zealand Environment Council [now Australian and New Zealand Environment and Conservation Council]

ANZECC Australian and New Zealand Environment and Conservation Council

ANZFAC Australian and New Zealand Fisheries and Aquaculture Council

ANZFAS Australian and New Zealand Federation of Animal Societies

ANZLIC Australia and New Zealand Land Information Council

ANZMEC Australian and New Zealand Minerals and Energy Council

Anz Österr Akad Wiss Math-Nat wiss Kl Anzeiger der Österreichischen Akademie der Wissenschaften, Mathematisch-Naturwissenschaftliche Klasse [Gazette of the Austrian Academy of Sciences, Mathematical Sciences Class] [also Anz Oesterr Akad Wiss Math-Nat wiss Kl]

ANZSES Australian New Zealand Scientific Exploration Society

ANZSRLO Australian and New Zealand Scientific Research Liaison Office

ANZUK Australia-New Zealand-United Kingdom

ANZUS Australia-New Zealand-United States

AO Abnormal Occurrence

AO Absorber (or Absorption) Oil

AO Access Opening

AO Acousto-Optic(s) [also A-O]

AO Aircraft Operation; Aircraft Operator

AO Airport Operation; Airport Operator

AO Alkali(ne) Oxide

AO Aluminum Oxide

AO Always On (State)

AO 2-Amino-2-Methyl 3-Butanone Oxime

AO Amplifier Output

AO Analog Output

AO Angola [ISO Code]

AO Announcement of Opportunity

AO Anodic Oxidation

AO Answer Only

AO Antiferromagnetic Order [Solid-State Physics]

AO Anti-Operation

AO Antioxidant

AO Antisense Oligonucleotide [Biochemistry]

AO Applied Optics [Journal of the Optical Society of America]

AO Approving Official

AO Arctic Ocean

AO Armagh Observatory [Armagh, Northern Ireland]

AO Army Order

AO Assembly Order

AO Atlantic Ocean

AO Atomic Orbital

AO Atomic Oxygen

AO Audio Operator

AO Automated Office

AO Automated Operator

AO Auto-Oscillation

AO Fixed-Wing Observation Aircraft [US Army Symbol]

AO Oiler (Ship) [US Navy Symbol]

AO1 Antiferromagnetic Order 1

AO2 Antiferromagnetic Order 2

A&O April and October [Business Finance]

A/O account of [also a/o]

A-O Acousto-Optic(s) [also AO]

A-O And-Or (Circuit)

.ao Angola [Country Code/Domain Name]

a/o account of [also A/O]

a/o atomic percent [Symbol]

aΩ abohm [Unit]

$(a\Omega)^{-1}$ abmho [Unit]

α-O Alpha Oxygen [Symbol]

AOA Abort Once Around

AOA Administration on Aging [of US Department of Health and Human Services]

AOA Aerodrome Owners Association [UK]

AOA Airborne Optical Adjunct (System)

AOA Airport Operation Area

AOA American Oceanology Association [now American Oceanic Organization, US]

AOA American Ophthalmological Association [US]

AOA American Optometric Association [US]

AOA American Ordnance Association [US]

AOA American Orthopsychiatric Association [US]

AOA American Osteopathic Association [US]

AOA (Airplane) Angle of Attack

AOA Army Ordnance Association [Washington, DC, US]

AOA At, or Above

AΩA Alpha Omega Alpha (Student Society)

AOAA American Organization of Analytical Chemists [US]

AOAA Aminooxyacetic Acid

AOAC Association of Official Agricultural Chemists (of North America) [US]

AOAC Association of Official Analytical Chemists [now AOAC International, US]

AOAC Automobile Owners' Action Council [US]

AOAC Int'l Association of Official Analytical Chemists International [formerly AOAC, US]

AOAD Arab Organization for Agricultural Development [Sudan]

AOARD Asian Office of Aerospace Research and Development [of Air Force Office of Scientific Research in Tokyo, Japan]

AOB Address Operation Block

AOB Air-O-Brinell (Hardness Tester)

AOB Alcohol on Breath

AOB Analytical Operations Branch

AOB Any Other Business

AOB At or Below

AOBD Acousto-Optic Beam Deflector

AOC Aerodrome Obstruction Chart

AOC Aircraft Operations Center [of National Oceanic and Atmospheric Administration at MacDill Air Force Base, Florida, US]

AOC Airline Operational Control

AOC Air Officer Commanding

AOC Air Operators Certificate [UK]

AOC Airport Operators Council

AOC Alberta Opportunity Company [Canada]

AOC American Orchid Society [US]

AOC Applied Organometallic Chemistry [International Journal]

AOC Arc-Oxygen (or Oxygen-Arc) Cutting

AOC Army Ordnance Corps

AOC Associated Overseas Countries [of European Community]

AOC Association of Old Crows [Note: "Old Crows" are Electronic Warfare Specialists]

AOC Atlantic Ocean Circulation

AOC Automatic Output Control

AOC Automatic Overload Circuit

AOC Automatic Overload Control

AOC Awaiting Outgoing Continuity

Aoc tert-Amyloxycarbonyl-

AOCARS African Organization for Cartography and Remote Sensing

AOCB Any Other Competent Business

AOCE Apple Open Collaboration Environment [Computers]

AOCI Airport Operators Council International [US]

aΩcm abohm centimeter [Unit]

AOCR Advanced Optical Character Reader

AOCS Airline Operational Control Society [US]

AOCS Alpha Omega Computer System

AOCS American Oil Chemists Society [US]

AOCS Attitude and Orbit Control System

AOCU Arithmetic Output Control Unit

AOCU Associative Output Control Unit

AOD Above Ordnance Datum

AOD Acousto-Optic Device

AOD Advanced Ordnance Deport

AOD Angle of Deflection

AOD Anniston Ordnance Depot [Anniston, Alabama, US]

AOD Argon-Oxygen Decarburized; Argon-Oxygen Decarburization [Metallurgy]

AOD Arithmetic Output Data

AOD Army Ordnance Department

AOD Army Ordnance Depot

AOD Arsenal Operations Division

AODC Acridine Orange Direct Count [Biochemistry]

AODC Association of Offshore Diving Contractors [UK]

AODC Australian Oceanographic Data Center

AO/DI Always On/Dynamic ISDN (Integrated Services Digital Network)

AOD/VAR Argon-Oxygen Decarburization/Vacuum Arc Remelting [Metallurgy]

AOE Airport of Entry

AOE Application Operating Environment [of AT&T Corporation]

AOE Association of Overseas Educators [US]

AOE Atomic-Oxygen Exposure

AOEC Airways Operations Evaluation Center [US]

AOET Atomic-Oxygen Exposure Testing

AOETF Atomic Orbital Electron Translation Factor [also AO-ETF]

AOEW Airplane Operating Empty Weight

AOF Advanced Operating Facility

AOF Afrique Occidentale Française [French West Africa]

AOF American Optometric Foundation [US]

AOF Angle of Friction

AOF Australian Orchid Foundation

A of F Admiral of the Fleet [US Navy]

AOFA Atlantic Offshore Fishermen's Association [US]

AOFAS American Orthopedic Foot and Ankle Society [US]

AOG Aircraft on (the) Ground

AOG Augmented Off Gas

AOI Acousto-Optical Interaction

AOI Advance Ordering Information

AOI Ammonium Oxalate, Insoluble

AOI And-Or Invert (Gate) [also A-O-I]

AOI Arab Organization for Industrialization

AoI Association of Illustrators [UK]

AOI Automated Operator Interface

AOI Automated (or Automatic) Optical Inspection

A-O-I And-Or-Invert (Gate) [also AOI]

AOIM Automated Orientation Imaging Microscopy

AOIPS Atmospheric and Oceanographic Information Processing System

AOIS Australian Overseas Information Service (of Department of Foreign Affairs and Trade)

AOJ Association of Orthodox Jewish Scientists [US]

AOK Angolan Kwanza [Currency]

AOL Absent over Leave

AOL Airborne Oceanographic Lidar

AOL America On-Line

AOL Application-Oriented (Computer) Language

AOL Atlantic Oceanographic Laboratory [of National Oceanic and Atmospheric Administration, US]

AOLO Advanced Orbital Launch Operation

AOLS Association of Ontario Land Surveyors [Canada]

AOM Angular Overlap Model

AOMA American Occupational Medical Association [US]

AOMC Army Ordnance Missile Command

AOML Atlantic Oceanographic and Meteorological Laboratories [of National Oceanic and Atmospheric Administration, US]

AOMR Arab Organization for Mineral Resources

AON Assignment Order Number

AONB Area of Outstanding Natural Beauty [of Countryside Commission, UK]

AONM Australian Obligated Nuclear Material

AONM Australian Origin Nuclear Material

AONS Air Observer's Navigation School

AOO American Oceanic Organization [US]

AOO Applied Optics Online [Publication of the Optical Society of America]

AOP 21-Acetoxy Pregnenolone

AOP Advanced Oxidation Process

AOP Aerodrome Operations

AOP Air Observation Post

AOPA Aircraft Owners and Pilots Association [US]

AOPA/ASF Aircraft Owners and Pilots Association/Air Safety Foundation [Frederick, Maryland, US]

AOPL Association of Oil Pipelines [US]

AOPU Asian-Oceanic Postal Union [now Asian-Pacific Postal Union, Philippines]

AOQ Average Outgoing Quality

AOQL Average Outgoing Quality Level

AOQL Average Outgoing Quality Limit

AOR Angle of Reflection

AOR Army Ordnance Regulations

AOR Atlantic Ocean Region
AOR Atomic-Oxygen Resistant
AOR Axis of Rotation
AORB Aviation Operational Research Branch
AORC Association of Official Racing Chemists [US]
AORE Army Operational Research Establishment [UK]
AORF Anisotropic Optical Response Function
AORG Army Operational Research Group [UK]
AOS Acquisition of Signal
AOS Add Or Subtract
AOS Advanced Operating Service
AOS Advanced Operating System
AOS Agricultural Organization Society [UK]
AOS Air Observer's School
AOS Alberta Oil Sands [Canada]
AOS Algebraic Operating System
AOS Alpha-Olefin-Sulfonate
AOS American Oriental Society [US]
AOS Angle of Sight
AOS Apparent Opening Size [Geotextiles]
AOS Army Ordnance Service
AOS Automated Office System
AOS Axis of Symmetry
AOS Azimuth Orientation System
AOSA Association of Official Seed Analysts [US]
AOSC Association of Oilwell Servicing Contractors [US]
AOSCA Association of Official Seed Certifying Agencies [US]
AOSERP Alberta Oil Sands Environmental Research Program [Canada]
AOSI Alberta Oil Sands Index [of AOSIC, Canada]
AOSIC Alberta Oil Sands Information Center [Canada]
AOSIS Alliance of Small Island States
AOSO Advanced Orbiting Solar Observatory
AOSP Automatic Operating and Scheduling Program
AOSS (Airplane) Angle of Sideslip
AOSSM American Orthopedic Society for Sports Medicine [US]
AOSTRA Alberta Oil Sands Technology and Research Authority [Canada]
AOSTRA/APMC Alberta Oil Sands Technology and Research Authority/Alberta Petroleum Marketing Commission [Canada]
AOSTRA J Res AOSTRA Journal of Research [of Alberta Oil Sands Technology and Research Authority, Canada]
AOS/VS Advanced Operating System/Virtual Storage
AOT Aerosol Diiso-Octyl Sulfosuccinate
AOT Alignment Optical Telescope
AOT Arizona Office of Tourism [US]
AOT Bis(2-Ethylhexyl) Sodium Sulfosuccinate
AOTA American Occupational Therapy Association [US]
AOTF Acousto-Optic Tunable Filter
AOTS Aviation Ordnance Test Station [Chincoteague, Virginia, US]
AOTT Automatic Outgoing Trunk Test

AOTV Aero-Assisted Orbital Transfer Vehicle
AOU American Ornithologists Union [US]
AOU Apparent Oxygen Utilization
AOU Arithmetic Output Unit
AOU Associative Output Unit
AOU Automated Offset Unit
AOU Azimuth Orientation Unit
AOW Alabama Ordnance Works [Childersburg, Alabama, US]
AOX Absorbable Organic Halogen
AOX Adsorbable Organic Halogen
AP Abel-Pensky (Flash Point Test)
AP Absolute Pressure
AP Absorption Plant
AP Academic Press, Inc. [US]
AP Academy of Pharmacy
AP Accelerator Physics
AP Access Panel
AP Access Permit(tee)
AP Access Point
AP Access Primitives
AP Accident Prevention
AP Accounts Payable [also A/P]
AP Acetophenone
AP Acid-Proof
AP Acoustical Physics [Russian Translation Journal jointly published by the American Institute of Physics and MAIK Nauka]
AP Acoustical Plaster
AP Acoustical Properties
AP Acoustic Phonon [Solid-State Physics]
AP Action Plan
AP Action Potential [Biology]
AP Acute Pericarditis
AP Adaptin [Biochemistry]
AP Address Programming
AP Adenosine Phosphoric Acid [Biochemistry]
AP Adjunct Processor
AP Administrative Processor
AP Admission Port
AP Adsorption Pump
AP Advanced Placement (Credit) [College Entrance]
AP Advance Payment
AP Aerial Photo(grapher); Aerial Photography
AP Aerial Prospecting; Aerial Prospector
AP Aerophysical; Aerophysicist; Aerophysics
AP After Peak
AP After Perpendicular
AP Agricultural Pilot
AP Agricultural Powder
AP Airborne Prospecting
AP Air Pollution
AP Airport
AP Air Preheater

AP Air Pressure
AP Air Publication
AP Air Pump
AP Aldopentose [Biochemistry]
AP Alkaline Permanganate
AP Alkaline Phosphatase [Biochemistry]
AP Alloy Phase
AP Alloy Processing
AP All-Pass Filter
AP Alpha Particle
AP Alpha Protein [Biochemistry]
AP Aluminum Phosphate
AP Aluminum Powder
AP Ambient Pressure
AP American Patent
AP 2-Aminophenol
AP Aminophosphonobutyric Acid
AP Aminopromazine
AP Aminopropyltriethoxysilane
AP 2-Aminopyridine
AP Ammonium Perchlorate
AP Ammonium Picrate
AP Amorphous Polymer
AP Amylopectin [Biochemistry]
AP Anharmonic Potential
AP Aniline Point [Chemical Engineering]
AP Anionic Polymerization
AP Annealed Pipe
AP Annealed Powder
AP Annealing Point
AP Anomalous Propagation
AP Antarctica Project [US]
AP Antennas and Propagation (Society) [of Institute of Electrical and Electronics Engineers, US]
AP Anterior-Posterior (Projection) [X-Ray Technology]
AP Antimony Powder
AP Antiparallel (Spin) [Physics]
AP Antiparticle [Physics]
AP Antiperiodic [also ap]
AP Antipersonnel [also A/P]
AP Antiphase [Solid-State Physics]
AP Antiproton
AP Antipyrine [Pharmacology]
AP Appearance Potential [Mass Spectrometry]
AP Application Package
AP Application Processor
AP Application Program
AP Applications Processor
AP Applied Pharmacology
AP Applied Physics
AP Applied Probability
AP Archimedes Principle
AP Areal Precipitation
AP Argentinean Peso [Currency]

AP Argument Programming
AP Arithmetic Processor
AP Arithmetic Progression
AP Armor-Piercing (Bullet)
AP Aromatic Polymer
AP Array Processor
AP Aseptic Processing
AP As Plated (Condition)
AP As Polymerized (Condition)
AP As Prepared
AP As Pressed (Condition)
AP Assembler Programming
AP Assistant Professor
AP Assembly of Parties
AP Assembly Protein [Biochemistry]
AP Associated Press [US]
AP Associate Professor
AP Associative Processor
AP Assumed Position
AP Astrophotographer; Astrophotographical; Astrophotography
AP Astrophysical; Astrophysicist; Astrophysics
AP Asynchronous Protocol
AP Atmospheric Physics
AP Atomic Part(s)
AP Atomic Physicist; Atomic Physics
AP Atomic Pile
AP Atomic Polarization [Physical Chemistry]
AP Atomic Property
AP Atom Probe
AP Attached Processor [also A/P]
AP Auger Parameter [Physics]
AP Auroral Precipitation
AP Author's Proof
AP Automatic Pilot
AP Automatic Plating
AP Automatic Polisher; Automatic Polishing
AP Automatic Press
AP Automatic Process
AP Automatic Program(ming)
AP Autopilot
AP Autoprotolysis
AP Axial Plasma
AP Azimuthal Projection
AP Transport (Ship) [US Navy Symbol]
AP600 Advanced Passive 600 Megawatt (Reactor) [US]
2AP 2-Amino Phenol
3AP 3-Amino Phenol
4AP 4-Amino Phenol
AP- Pakistan [Civil Aircraft Marking]
AP-3 DL-2-Amino-3-Phosphonopropionic Acid
D-AP-3 D-2-Amino-3-Phosphonopropionic Acid
L-AP-3 L-2-Amino-3-Phosphonopropionic Acid
AP-4 DL-2-Amino-4-Phosphonobutyric Acid

D-AP-4 D-2-Amino-4-Phosphonobutyric Acid

L-AP-4 L-2-Amino-4-Phosphonobutyric Acid

AP-5 DL-2-Amino-5-Phosphonovaleric Acid

D-AP-5 D-2-Amino-5-Phosphonovaleric Acid

L-AP-5 L-2-Amino-5-Phosphonovaleric Acid

AP-6 DL-2-Amino-6-Phosphonohexanoic Acid

D-AP-6 D-2-Amino-6-Phosphonohexanoic Acid

AP-7 DL-2-Amino-7-Phosphonoheptanoic Acid

D-AP-7 D-2-Amino-7-Phosphonoheptanoic Acid

L-AP-7 L-2-Amino-7-Phosphonoheptanoic Acid

AP-8 DL-2-Amino-8-Phosphonooctanoic Acid

5-AP 5-Amino-1,10-Phenanthroline [Biochemistry]

4-AP 4-Aminopyridine

A&P Atlantic and Pacific

A/P Account Paid [also a/p]

A/P Account Purchase

A/P Accounts Payable [also AP]

A/P Amorphous (Material) on Polycrystalline (Material)

A/P Antipersonnel [also AP]

A/P Attached Processor [also AP]

A/P Authority for Purchase

A-P Antiferromagnetic-Paramagnetic (Change)

A-5'-P Adenosine 5'-Monophosphate [Biochemistry]

Ap Aperture [also ap]

Ap Apothecaries' (Measure or Weight) [Pharmacology]

Ap April

A(p) Aperture Function [Symbol]

aP Pearson Symbol for Primitive (Simple) Space Lattice in Triclinic (Anorthic) Crystal System [Symbol Followed by the Number of Atoms per Unit Cell, e.g. aP8, aP12, etc.]

ap additional premium

ap *(anni preteriti)* – of last year

ap antiperiodic

ap antipyrine [Pharmacology]

ap apothecaries' (measure or weight) [Pharmacology]

a/p account paid [also A/P]

APA Abrus Precatorius Agglutinin [Biochemistry]

APA Accidents in Private Aviation

APA Accredited Public Accountant

APA Acetone Producers Association [Belgium]

APA Acetophenone Acetone

APA (4-Acetylphenoxy)carbonyl Azide

APA Acrylamide Producers Association [US]

APA Adaptive Packet Assembly

APA Administrative Procedures Act [US]

APA Advertising Photographers of America US]

APA Agricultural Precast Association [US]

APA Agricultural Publishers Association [US]

APA Airline Pilots Association [US]

APA Alaska Power Authority [US]

APA Alberta Publishers Association [Canada]

APA All Points Addressable

APA Aluminum Prefabricators Association UK]

APA Amalgamated Printers Association [US]

APA American Pain Association [US]

APA American Parquet Association [US]

APA American Pharmaceutical Association [US]

APA American Philosophical Association [US]

APA American Photoengravers Association [US]

APA American Pilots Association [US]

APA American Planning Association [US]

APA American Plywood Association [US]

APA American Podiatric Association [US]

APA American Polygraph Association [US]

APA American Poultry Association [US]

APA American Press Association [US]

APA American Proctological Association [US]

APA American Psychiatric Association [US]

APA American Psychological Association [US]

APA American Pulpwood Association [US]

APA American Pyrotechnics Association [US]

APA 6-Aminopenicillanic Acid

APA Aminopropionic Acid

APA Architectural Precast Association [US]

APA Arithmetic Processing Accelerator

APA Army Parachute Association [UK]

APA Associate in Public Administration

APA Association of Practising Accountants [UK]

APA Atlantic Publishers Association [Canada]

APA Attack Transport Vessel, Military Sea Transportation Service [US Navy Symbol]

APA Atom Probe Analysis

APA Automobile Protection Association [Canada]

APA Axial Pressure Angle

2APA 2-Amino Propanoic Acid [also 2Apa]

3APA 3-Amino Propanoic Acid [also 3Apa]

3APA 3-Amino Propionic Acid

A[5']P6[5']A P^1, P^6-Di(adenosine-5'-) Hexaphosphate [Biochemistry]

A[5']P5[5']A P^1, P^5-Di(adenosine-5'-) Pentaphosphate [Biochemistry]

A[5']P3[5']A P^1, P^3-Di(adenosine-5'-) Triphosphate [Biochemistry]

A[5']P4[5']A P^1, P^4-Di(adenosine-5'-) Tetraphosphate [Biochemistry]

6-APA 6-Aminopenicillanic Acid

ApA Adenylyladenosine [Biochemistry]

Apa Acetobacter pasteurianus [Microbiology]

α-Pa Alpha Protactinium [Symbol]

APAA ASEAN (Association of Southeast Asian Nations) Port Authorities Association

APAA Asian Patent Attorneys Association [Japan]

APAA Automotive Parts and Accessories Association [US]

APAAP Alkaline Phosphatase-Anti-Alkaline Phosphatase [Biochemistry]

APABC (Institute of) Accredited Public Accountants of British Columbia [Canada]

APAC Alkaline Permanganate Ammonium Citrate

APAC Asia-Pacific (Countries)

APAC Conf Asia-Pacific Conference

APAC Asociación Peruana para el Avance de la Ciencia [Peruvian Association of the Advancement of Science]

APACE Aldermaston Project for the Application of Computers to Engineering [of UK Atomic Energy Authority]

APACHE Accelerator for Physics and Chemistry of Heavy Elements

APACS Airborne Position and Altitude Camera System

APACS Association of Payment Clearing Services [UK]

APACUE Atlantic Provinces Association for Continuing University Education [Canada]

APAD 3-Acetylpyridine Adenine Dinucleotide [Biochemistry]

APADS Automatic Programmer and Data System

APAE 2-(3-Aminopropyl)amino Ethanol

APAE Association of Public Address Engineers [UK]

APAFO Advanced Particles and Fields Observer

APAG American Photographic Artisans Guild [US]

APAG Association Européenne des Producteurs d'Acides Gras [European Association of Fatty Acid Producing Companies] [of Conseil Européen des Fédérations de l'Industrie Chimique, Belgium]

APAG Atlantic Policy Advisory Group [of NATO]

APAIS Australian Public Affairs Information Service [of National Library of Australia]

APAIS-Health APAIS Health (Database) [Australia]

APAL Array Processor Assembly Language

ApaL Acetobacter pasteurianus L [Microbiology]

APALA Asian/Pacific American Libraries Association

APAM Adler Planetarium and Astronomical Museum [Chicago, Illinois]

APAM Array Processor Access Method

APAMM Associação Portuguesa dos Armadores de Marinha Mercante [Portuguese Association of Merchant Ship Owners]

APANAC Asociacion Panamena para el Avance de la Ciencia [Panamanian Association of the Advancement of Science]

APANS 4-[(2-Arsonophenyl)azo]-3-Hydroxy Naphthalene-2,7-Disulfonic Acid

APAO p-Aminophenylarsine Oxide

APAP N-Acetyl-p-Aminophenol

ApApC Adenylyladenylylcytidine [Biochemistry]

APAR Active Phased Array Radar

APAR Authorized Program Analysis Report [of IBM Corporation, US]

APAR Automatic Processing and Recording

APAR Automatic Programming and Recording

APARE East Asia-North Pacific Region

APAREN Address Parity Enable

APAS Automatic Performance Analysis System

APASE Association for the Promotion and Advancement of Science Education [Canada]

APAT Armor-Piercing, Antitank

APATI Abbreviated Precision Approach Path Indicator

APATS Automatic Programmer and Test System; Automatic Programming and Testing System

APB Academic Planning Board

APB Advanced Pouring Box [Metallurgy]

APB Agricultural Products Board [Canada]

APB Air Portable Bridge

APB All Points Bulletin [Police]

APB Antiphase Boundary [Solid-State Physics]

APB As-Purchased Basis

APB Associated Press Broadcasters [US]

APB p-Azidophenacyl Bromide

α-**Pb** Alpha Lead

APBA American Power Boat Association [US]

APBA Atlantic Provinces Booksellers Association [Canada]

APBF Accredited Poultry Breeders Federation [UK]

α-**PbF$_2$** Alpha Lead Fluoride

APBT Advanced Polybutylene Terephthalate

aPBT Articulated Poly(p-Phenylene-Benzobisthiazole)

APC Accident Prevention Council

APC Acetylsalicylic Acid, Phenacetin and Caffeine

APC Acoustic Propagation Constant

APC Activated Protein C [Biochemistry]

APC Active Path Corrosion

APC Adaptive Payload Carrier

APC Adaptive Predictive Coding

APC Adaptive Processing Control

APC Adenoidal-Pharyngeal-Conjunctival (Viruses)

APC Adsorption-Partition Chromatography

APC Advanced PEEK (Polyetheretherketone) Composite

APC Advanced Personal Computer

APC Advanced Pocket Computer

APC Advanced Polymer Composite

APC Advanced Processing Computer

APC Advanced Processor Card

APC Advanced Professional Certificate

APC Advanced Professional Computer

APC Advanced Programming Course

APC Aerobic Plate Counts

APC Aeronautical Planning Chart

APC Aeronautical Public Correspondence

APC African Pacific and Caribbean

APC African Pacific and Caribbean Council

APC Air Pollution Control

APC Air Purification Control

APC Alberta Press Council [Canada]

APC Alloy Phases Committee [of Minerals, Metals and Materials Society of AIME, US]

APC All-Purpose Computer

APC Alternative Press Center [Canada]

APC American Plastics Council [US]

APC American Power Conference [US]

APC American Power Conversion

APC Ammonium Perchlorate

APC Amplitude Phase Conversion

APC Antigen-Presenting Cell [Biochemistry]

APC Antiphlogistic Corticoid [Biochemistry]

APC Approach Control [also apc]

APC Area Position Control

APC Argon Purge Cart

APC Arkansas Polytechnic College [Russellville, US]

APC Armored Personnel Carrier

APC Armor Piercing Capped

APC Army Petroleum Center [US]

APC Aromatic Polymer Composite

APC Array Processor Controller

APC Artificial Pinning Center [Solid-State Physics]

APC Aspirin–Phenacetin–Caffeine (Tablets)

APC Assistant Planning Controller

APC Associative Processor Control

APC Atlantic Press Council

APC Atomic Power Constructions [Builder of Dungeness B Power Station, UK]

APC Atomic Power Converter

APC Automatic Particle Counter

APC Automatic Performance Control

APC Automatic Phase Control [also apc]

APC Automatic Picture Control

APC Automatic Position Control

APC Automatic Pressure Control

APC Automatic Process Control

APC Automatic Program Control

APC Autoplot Controller

APC Auxiliary Power Converter

APC Available Page Count

APC Average Power Control

APC Average Propensity to Consume

ApC Adenylylcytidine [Biochemistry]

APCA Air Pollution Control Association [now Air and Waste Management Association, US]

APCA American Petroleum Credit Association[US]

APCA American Pollution Control Association [now Air and Waste Management Association, US]

APCA Architectural Precast Cladding Association [UK]

APCAC Asia-Pacific Council of American Chambers of Commerce [Japan]

APCAS Asia and the Pacific Commission on Agricultural Statistics [of Food and Agricultural Organization in the Philippines]

APC-BC Armor Piercing Capped, Ballistic Cap

APCC Acetylsalicylic Acid, Phenacetin, Caffeine and Codeine (Phosphate)

APCC Advanced Physical Coal Cleaning

APCC Agricultural Planning and Coordinating Committee [of Ministry of Economic Affairs, Taiwan]

APCC Air Pollution Control Commission [US]

APCC American Potash and Chemical Corporation [US]

APCC Asian and Pacific Coconut Community [Indonesia]

APCC Association of Professional Computer Consultants [UK]

APCCC Asian Pacific Corrosion Conference and Congress

APCChE Asian Pacific Confederation of Chemical Engineering [Australia]

APCD Air Pollution Control District [San Luis Obispo and Santa Barbara Counties, California, US]

APCEF Advanced Power Conversion Experimental Facility [of US Army]

APCEM Asia-Pacific Conference on Electron Microscopy

APCF Angular Position Correlation Function

APCHE Armor Piercing, Capped, High Explosive

APCHE Automatic Programmed Checkout Equipment

APCI Actual Process Capability Index

APCI Armor Piercing Capped Incendiary

APCI Atmospheric Pressure Chemical Ionization

APCI/MS Atmospheric Pressure Chemical Ionization/Mass Spectrometry

APCI-T Armor Piercing Capped Incendiary with Tracer

APCL Association of Professional Color Laboratories [US]

APCL Atmospheric Physics and Chemistry Laboratory [of National Oceanic and Atmospheric Administration in Boulder, Colorado, US]

APC-LC Armor Piercing Capped, Long Case

APCM Adaptive Pulse Code Modulation

APCM Adhesive Prepregs for Composite Manufacturing [US]

APCO Associated Public-Safety Communications Officers [US]

APCO Association of Pleasure Craft Operators [UK]

ApCpC Adenylylcytidylylcytidine [Biochemistry]

APCR Air Pollution Control Regulation

APCRC Australian Petroleum Cooperative Research Center [of Commonwealth Scientific and Industrial Research Organization]

APCS Associative Processor Computer System

APCS Automated Project Control System

APCS Automatic Program Control System

APCT (International Meeting on) Advanced Processing and Characterization Technologies

APCT Association of Polytechnic and College Teachers [UK]

APC-T Armor Piercing Capped with Tracer

APCTT Asian and Pacific Center for Transfer of Technology [India]

APCU Auxiliary Power Converter Unit

APCUG Association of Personal Computer User Groups [US]

APCVD Atmospheric-Pressure Chemical Vapor Deposition [also AP CVD, or Atm-CVD]

APC with C Acetylsalicylic Acid, Phenacetin and Caffeine with Codeine (Phosphate)

APD Aerospace Power Division [US Air Force]

APD Air Particulate Detector

APD Aldus Printer Description (File)

APD Alloy Phase Diagram [Metallurgy]

APD Amplitude-Phase Diagram

APD Amplitude Probability Distribution

APD Angular Position Digitizer

APD Antarctic Plant Database [British Antarctic Survey]

APD Antiphase Domain [Solid-State Physics]
APD Argon-Potassium Dating [Geology]
APD Army Pay Department
APD Atomic Products Division
APD Automated Powder Diffraction (Database) [of Joint Committee on Powder Diffraction Standards]
APD Automatic Power Down
APD Avalanche Photodetector
APD Avalanche Photodiode
APD Avalanche Photodiode Detector
APD Azimuthal Photoelectron Diffraction
APDA Appliance Parts Distributors Association [US]
APDA Atomic Power Development Associates [US]
APDB Antiphase Domain Boundary [Solid-State Physics]
APDC Ammonium Pyrrolidine Dithiocarbamate
APDC Apple and Pear Development Council [UK]
APDC Asian and Pacific Development Center [Malaysia]
APDC 1-Pyrrolidinecarbodithioic Acid, Ammonium Salt
a-PdCuP Palladium Copper Phosphorus Amorphous Alloy
a-PdCuSi Palladium Copper Silicon Amorphous Alloy
APDF Africa Project Development Facility
APDF Asian Pacific Dental Federation
APDF Association of Professional Design Firms [US]
APDG Antisymmetrized Product of Delocalized Geminals
APDI Asian Pacific Development Institute
APDIC Alloy Phase Diagram International Commission (Program on Binary Alloy Phases)
APDMS Axial Power Distribution Monitoring System
a-PdNiP Palladium Nickel Phosphorus Amorphous Alloy
APDS Acoustic Phonon Density of States [Solid-State Physics]
APDS Allyl Propyl Disulfide
APDS Armour Piercing Discarding Sabot
a-PdSi Palladium Silicon Amorphous Alloy
APDSMS Advanced Point Defense Surface Missile System [US Navy]
APDTC Ammonium Pyrrolidinedithiocarbamate
APDTC 1-Pyrrolidinecarbodithioic Acid, Ammonium Salt
4-APDTC 1-Aminophenazone Dithiocarbamic Acid [also 4-Apdtc]
APDU Association of Public Data Users [US]
APE Abbreviated Plain English (Language)
APE Accumulated Photon Echo [Physics]
APE Aminopropyl Epoxy
APE Antenna Positioning Electronics
APE Association of Professional Engineers [Canada]
APE Asynchronous Processing Element
APE Atomic Property Expression
APEA Advances in Engineering Plasticity and Its Application (Meeting)
APEA Association of Petroleum and Explosives Administrators [UK]
APEA Association of Professional Engineers of Alberta [Canada]
APEA Australian Petroleum Exploration Association

APEBC Association of Professional Engineers of British Columbia [Canada]
APEC All-Purpose Electronic Computer
APEC Asia-Pacific Economic Cooperation (Organization)
APEC Atlantic Provinces Economic Council [Canada]
APEC Automated Procedures for Engineering Consultants [US]
APEC Automotive Products Emissions Committee [US]
APEC Automotive Products Export Council [US]
APECA American Package Express Carriers Association [US]
APECMS Atmospheric Pressure Electron Capture Mass Spectrometry
APECS Auger Photoelectron Coincidence Spectroscopy
APED Atomic Power Equipment Department [US]
APEE Aminophenolethylether
APEF Association des Pays Exportateurs de Minéral de Fer [Association of Iron Ore Exporting Countries, Switzerland]
APEG Alkaline Polyethyleneglycol
APEGGA Association of Professional Engineers, Geologists and Geophysicists of Alberta [Canada]
APEGGNWT Association of Professional Engineers, Geologists and Geophysicists of the Northwest Territories [Canada]
APEL Aeronautical Photographic Experimental Laboratory [US]
APELL Awareness and Preparedness for Emergencies at Local Levels [of United Nations Environmental Program]
APEM Association of Professional Energy Managers [US]
APEM Association of Professional Engineers of Manitoba [Canada]
APEN Association of Professional Engineers of Newfoundland [Canada]
APEO Association of Professional Engineers of Ontario [Canada]
APEP 3-Amino-1-Phenyl-5-Ethyl Phenanthridinium
APEPEI Association of the Professional Engineers of Prince Edward Island [Canada]
APER Association of Publishers Educational Representatives [UK]
APERU Applied Plant Ecology Research Unit
Apers Antipersonnel
APES Adiabatic Potential Energy Surface
APES American Petroleum Equipment Suppliers [now Petroleum Equipment Suppliers Association, US]
APES Association of Professional Engineers of Saskatchewan [Canada]
APEX Advanced Productivity Exposition [US]
APEX Advance Purchase Excursion (Fare)
APEX Antiproton Experiment [of Fermilab, Batavia, Illinois, US]
APEX Assembler and Process Executive
APEX ATLAS (Argonne Tandem Linear Accelerator System) Positron Experiment [of US Department of Energy at Argonne National Laboratory, Illinois]

APEYT Association of Professional Engineers of the Yukon Territory [Canada]

APF Ablative-Pit Forming

APF Accurate Position Finder [also apf]

APF Adaptive Positive Feedback

APF Advanced Printer Function

APF Animal Protein Factor [Biochemistry]

APF Asian Packaging Federation [Japan]

APF Asphalt Plank Floor

APF Association de Paléontologie Française [French Paleontological Association]

APF Association of Professional Foresters [UK]

APF Atomic Packing Factor

APF Authorized Program Facility

APF Automatic Program Finder

APF Autopilot Flight Director

APFA American Pipe Fittings Association [US]

APFC American Plant Food Council [US]

APFC American Printed Fabrics Council [US]

APFC Asian Pacific Forestry Commission

APFC Automatic Profile and Flatness Control [Metallurgy]

APFCS Automatic Power Factor Control System

APFIM Atom Probe Field-Ion Microscope; Atom Probe Field-Ion Microscopy [also AP-FIM]

APFS Air Pressure Flow System

AP-FS-DS Armor-Piercing Fin-Stabilized, Discarding Sabot (Arrow)

APG Aberdeen Proving Ground [of US Army in Maryland]

APG Advanced Plasma Gun

APG Air Proving Ground [of US Air Force]

APG American Pewter Guild [US]

APG Aminopropyl Glass

APG Association of Polish Geomorphologists

APG Automatic Priority Group

APG Automatic Pressure Gelation

APG Automatic Program Generation; Automatic Program Generator

APG β-Azidophenylglyoxal Hydrate

APG Azimuth Pulse Generator

ApG Adenylylguanosine [Biochemistry]

APGA American Personnel Guidance Association [US]

APGA American Public Gas Association [US]

APGA Aminophenyl Glyoxalic Acid

APGC Air Proving Ground Center

APGE Association of Petroleum Geochemical Explorationists [US]

APGGQ Association of Professional Geologists and Geophysicists of Quebec [Canada]

ApGpU Adenylylguanylyluridine [Biochemistry]

APGM Autonomous Precision Guided Munition

APGS Association of Professional Geological Scientists [now American Institute of Professional Geologists , US]

APH Alkylated Phenol

APHA American Printing History Association [US]

APHA American Public Health Association [US]

a-phase amorphous phase [Materials Science]

APHAZ Aircraft Proximity Hazards

APHE Armor-Piercing. High Explosive

APHEBC Armor-Piercing, High Explosive, Ballistic Cap

APHEDA Australian People for Health, Education and Development Abroad

APHC Aminophenol Hydrochloride

APHI Association of Public Health Inspectors [US]

APHIS Animal and Plant Health Inspection Service [of US Department of Agriculture]

APHV Armor-Piercing, Hypervelocity

API Absolute Position Indicator

API Addition-Reaction Polyimide

API Aerial Photo Interpretation

API Air-Position Indicator

API Alabama Polytechnic Institute [formerly Alabama A&M (Agricultural and Mechanical) College; now Auburn University]

API Alternative Press Institute [of Alternative Press Center, Canada]

API American Paper Institute [US]

API American Petroleum Institute [US]

API American Potash Institute [US]

API Antecedent Precipitation Index [Meteorology]

API Application Program(ming) Interface [Internet]

API Arab Planning Institute

API Armor Piercing Incendiary

API Associated Photographers International [US]

API Association of Pharmaceutical Importers [UK]

API Atmospheric Pressure Ionization

API Automatic Priority Interrupt

APIA Asociación Peruana de Ingenieros Agronomos [Peruvian Association of Agricultural Engineers]

APIA Association for the Promotion of Industry–Agriculture [France]

APIC Advanced Programmable Interrupt Controller

APIC Apollo Parts Information Center [of NASA]

APIC Association of Power and Industrial Consultants [UK]

APICS American Production and Inventory Control Society [US]

APICS Atlantic Provinces Inter-University Council on the Sciences [Canada]

APICS E&R American Production and Inventory Control Society Educational and Research (Foundation) [US]

APID Applications Process Identification

API Fiber Sum API (Monthly) Fiber Summary [of American Paper Institute, US]

API Ind Fact Sheet API Industry Fact Sheet [of American Paper Institute, US]

APIL Axial Power Imbalance Limit

APILIT American Petroleum Institute Literature (Database) [of American Petroleum Institute, US]

APIMS Atmospheric Pressure Ionization Mass Spectrometry

APIO Array Processor Input/Output

APIN Association for Programmed Instruction in the Netherlands

API Newsprint Rep API Newsprint Report [of American Paper Institute, US] [also API Newsprint Rept]

APIP Alberta Petroleum Incentive Program [Canada]

APIPAT American Petroleum Institute Patents (Database) [of American Petroleum Institute, US]

APIRG African Regional Planning and Implementation Regional Planning Group [of International Civil Aviation Organization]

APIS Advanced Passenger Information System

APIS Army Photographic Interpretation Section [US]

API Statistical Sum API (Monthly) Statistical Summary [of American Paper Institute, US]

API Std American Petroleum Institute Standard [US]

API-T Armor Piercing Incendiary with Tracer [also APIT]

API Woodpulp Data API Woodpulp Data [of American Paper Institute, US]

APJT (Electric) Arc Plasma Jet-Treated; (Electric) Arc Plasma Jet Treatment

APK Amplitude Phase Keyed; Amplitude Phase Keying

APKD Adult Polycystic Kidney Disease

APL Aero-Propulsion Laboratory [of US Air Force]

APL Airplane [also Apl, or apl]

APL Alaskan Pipeline

APL Algorithmic/Procedural Language

APL Algorithmic Programming Language

APL All-Pass Loop [Electronics]

APL Alternate Pollutant Limits

APL Anneal and Pickle Line [Metallurgy]

APL Applied Physics Laboratory

APL Applied Physics Laboratory [of Johns Hopkins University, Laurel, Maryland, US]

APL Applied Physics Letters [Publication of American Institute of Physics, US]

APL A Programming Language

APL Association of Photographic Laboratories [UK]

APL Association of Programmed Learning

APL Associative Programming Language

APL Authorized Possession Limits

APL Automatic Phase Lock

APL Average Picture Level

Apl Airplane [also apl]

APLA American Patent Law Association [US]

APLA Aminophenyl Lactic Acid [Biochemistry]

APLA Arctic Production and Loading Atoll

APLA Atlantic Provinces Library Association [Canada]

APLE Association of Public Lighting Engineers [US]

APLG Antisymmetrized Product of Localized Geminals

APLHGR Average Planning Heat Generation Rate

Apl Mat Aplikace Matematiky [Czech(oslovak) Journal on Applied Mathematics]

APLN Armor-Piercing, Long Nose

APLS American Plant Life Society [US]

APL Tech Dig Applied Physics Laboratory Technical Digest [of Johns Hopkins University, Laurel, Maryland, US]

APL-UW Applied Physics Laboratory of the University of Washington [Seattle, US]

APM Accident Prevention Manual

APM Advanced Power Management [IBM Operating System 2]

APM Advanced Process Modeling

APM Affected Pedigree Member

APM Air Particulate Monitor

APM Aluminum Powder Metal

APM Ammonia–Hydrogen Perioxide–Water (Cleaning Solution)

APM Ammonium Paramolybdate

APM Ammonium Phosphomolybdate

APM Amplitude and Phase Modulation

APM Analog Panel Meter

APM Antenna Positioning Mechanism

APM Aspartame

APM Association of Project Managers [UK]

APM Associative Principle for Multiplication

APM Asynchronous Packet Manager

APM Atomic Parameter Matrix

APM Atom-Probe Microanalysis

APM Atom-Probe Microscopy

APM Australian Paper Manufacturers

APM Autoclavable Powdered Medium

APM Automated People Mover

APM Automatic Page Makeup

APM Automatic Predictive Maintenance

APM Azinphos Methyl

Apm α-Aminopimelic Acid

α-Pm Alpha Promethium [Symbol]

APMA American Pharmaceutical Manufacturers Association [US]

APMA American Podiatric Medical Association [US]

APMA Atom Probe Microanalysis

APMA Automotive Parts Manufacturers Association [Canada]

APMB Armor-Piercing, Monoblock

APMC Alberta Petroleum Marketing Commission [Canada]

APME Aminophenol Methylether

APME Association of Plastic Manufacturers in Europe [Belgium]

APMHC Association of Professional Material Handling Consultants [US]

APMI American Powder Metallurgy Institute [now APMI International]

APMI Area Precipitation Measurement Indicator

APMI Association of Printing Machinery Importers [UK]

APMI Int'l American Powder Metallurgy Institute International

APMI/MPIF American Powder Metallurgy Institute/Metal Powder Industries Federation [US]

APMO Association of Private Market Operators [UK]

AP-MO-CVD Atmospheric-Pressure Metallorganic Chemical Vapor Deposition [also AP MO CVD, or AP-MOCVD]

APMS Advanced Power Management System

APMS Aquatic Plant Management Society [US]

APMSF (4-Amidinophenyl)methanesulfonyl Fluoride

p-APMSF (p-Amidinophenyl)methanesulfonyl Fluoride

APN [(4-Acetylphenoxy)carbonyl]nitrene

APN Asia-Pacific Network

APN Augmented Proportional Navigation

Apn Antipyrine [also apn]

APNB Association of Professional Engineers of New Brunswick [Canada]

APNI Australian Plant Name Index (Database)

APNIC Asia-Pacific Network Information Center [Internet]

APNIC Automatic Programming National Information Center [UK]

APNO Atomic Pair Natural Orbital

APO After Payout

APO Amorphous Polyolefin

APO Anti-Parallel Ordering [Solid-State Physics]

APO Apocrine [Biochemistry]

APO Apolipoprotein [Biochemistry]

APO Army Post Office

APO Asian Productivity Organization [Japan]

APO Association of Professional Organizers [now National Association of Professional Organizers, US]

APO Astrophysical Observatory

APO Australian Patents Office

APO tris(1-Aziridinyl) Phosphine Oxide

Apo Apoenzyme [Biochemistry]

α-Po Alpha Pollonium [Symbol]

APOA Arctic Petroleum Operators Association [Canada]

APO A-1 Apolipoprotein A-1 [Biochemistry]

APOAC Asia Pacific Office Automation Control Council

APO B Apolipoprotein B [Biochemistry]

APO cyst Aprocrine Cystic Metaplasia

APO E Apolipoprotein E [Biochemistry]

APOL Alkaline Pressure Oxidation Leaching (Process)

apo-LDL Apolipoprotein B, Low-Density Lipoprotein [Biochemistry]

APOLLO Article Procurement with On-Line Local Ordering

APOMA American Precision Optics Manufacturers Association [US]

APOP Authenticated Post Office Protocol

APOS Advanced Polar Orbiting Satellite

APOTA Automatic Positioning Telemetering Antenna

APP Acoustic Phase Plate

APP Acquisition Pollution Prevention

APP Adjustable-Pitch Propeller

APP Ammonium Polyphosphate

APP Amorphous Polypropylene

APP Amyloid β/A4 Protein Precursor [Biochemistry]

APP Applet [Applications of Hot Java Internet Browser]

APP Approach Control (Office)

APP Approach Control Service

APP Arctic Pilot Project

APP Associative Parallel Processor

APP Atactic Polypropylene

APP Automated Process Planning

APP Auxiliary Power Plant

APP n-(3-Aminopropyl)pyrrole

.APP Application [R:BASE File Name Extension]

App Apparatus [also app]

App Appendix [also app]

App Application [also app]

App Appointment [also App, or app]

App Apprentice(ship) [also app]

a-PP Atactic Polypropylene

app apparent(ly)

app appended

app applied

app appointed

APPA African Petroleum Producers Association

APPA Aircraft Preparation and Paint Application

APPA Alberta Professional Photographers Association [Canada]

APPA American Public Power Association [US]

APPA American Pulp and Paper Association [US]

APPA Association pour la Prévention de la Pollution Atmospherique [Association for the Prevention of Atmospheric Pollution, France]

A[5']PP[5']A P^1, P^2-Di(adenosine-5'-) Pyrophosphate [Biochemistry]

Appar Apparatus [also appar]

appar apparent(ly)

AP&PB Association of Print and Packaging Buyers [UK]

APPC Advanced Program-to-Program Communication(s)

APPC Approach Control

APPCLU Advanced Program-to-Program Communications Logical Unit

appd approved [also app'd]

APPDA Atlantic Provinces Power Development Act [Canada]

APPE Association of Petrochemicals Producers in Europe [Belgium]

APPECS Adaptive Pattern Perceiving Electronic Computer System

APPEN Asia-Pacific People's Environment Network [Malaysia]

APPG Adjacent Phase Pulse Generator

APPH Auger Peak-to-Peak Height [Spectroscopy]

APPI Advanced Peer-to-Peer Internetworking

APPI Advanced Planning Procurement Information

APPITA Australian (and New Zealand) Pulp and Paper Industry Technical Association

APPL Aircraft Precision Position Location (Equipment)

Appl Appliance [also appl]

Appl Application [also appl]

appl applied

Appl Acoust Applied Acoustics [Journal published in the UK]

Appl Anal Applicable Analysis [Journal published in the UK]

Appl Artif Intell Applied Artificial Intelligence [Journal published in the US]

Appl Biochem Biotechnol Applied Biochemistry and Biotechnology [Journal published in the US]

Appl Catal Applied Catalysis [Journal published in the Netherlands]

Appl Clay Sci Applied Clay Science [Journal published in the Netherlands]

Appl Compos Mater Applied Composite Materials [US Journal] [also Appl Comp Mat]

appld applied [also appl'd]

APPLE Apple (Computer) Puget Sound Program Library Exchange [US]

APPLE Associative Processor Programming Language Evaluation

Apple Assem Line Apple Assembly Line [US Publication]

Appl Electron Tech Applied Electronic Technique [Journal published in PR China]

Appl Energy Applied Energy [Journal published in the UK]

Appl Environ Microbiol Applied Environmental Microbiology [International Journal]

Applenet Apple Computer Network [US]

Appl Eng Applied Engineering [also appl eng]

Appl Eng Agric Applied Engineering in Agriculture [Journal published in the Netherlands]

Appl Environ Microbiol Applied and Environmental Microbiology [Journal of the American Society for Microbiology]

APPLETS Applications [Internet] [also Applets, or applets]

Appl Geochem Applied Geochemistry

Appl Geochem Applied Geochemistry [Journal of the International Association of Geochemistry and Cosmochemistry, Canada]

applicat application [Apothecary]

Appl Ind Hyg Applied Industrial Hygiene [Journal published in the US]

Appl Magn Reson Applied Magnetic Resonance [Journal published in Germany]

Appl Math Applied Mathematics

Appl Math Comput Applied Mathematics and Computation [Journal published in the US]

Appl Math Mech Applied Mathematics and Mechanics [Translation of: *Prikladnaya Matematika i Mekhanika (USSR)*; published in the US]

Appl Math Model Applied Mathematical Modelling [Journal published in the UK]

Appl Math Optim Applied Mathematics and Optimization [Journal published in the US]

Appl Mech Applied Mechanics

Appl Mech Applied Mechanics [Journal published in Romania]

Appl Mech Rev Applied Mechanics Reviews [of American Society of Mechanical Engineers International]

Appl Metrol Applied Metrology

Appl Microbiol Applied Microbiology [Discipline]

Appl Microbiol Applied Microbiology [Journal]

Appl Microbiol Biotechnol Applied Microbiology and Biotechnology [Journal published in Germany]

Appl Microgravity Technol Applied Microgravity Technology [Journal published in Germany]

appln Application [also appln]

Appl Numer Math Applied Numerical Mathematics [Journal of the International Association for Mathematics and Computer in Simulation]

Appl Ocean Res Applied Ocean Research [Journal published in the UK]

Appl Opt Applied Optics

Appl Opt Applied Optics [Journal of the Optical Society of America]

Appl Phys Applied Physics

Appl Phys Applied Physics [Journal published in Germany]

Appl Phys A Applied Physics A [Journal published in Germany]

Appl Phys A, Solids Surf Applied Physics A, Solids and Surfaces [Journal published in Germany]

Appl Phys B Applied Physics B [Journal published in Germany]

Appl Phys B, Photophys Laser Chem Applied Physics B, Photophysics and Laser Chemistry [Journal published in Germany]

Appl Phys Commun Applied Physics Communications [US]

Appl Phys Lett Applied Physics Letters [of American Institute of Physics, US]

Appl Polym Symp Applied Polymer Symposium [Periodical Series]

Appl Psychol Monogr Applied Psychology Monographs [US]

Appl Radiat Isot Applied Radiation and Isotopes [Journal published in the UK]

Appls Appliances [also appls]

Appl Sci Applied Science

Appl Sci Res Applied Scientific Research [Journal published in the Netherlands]

Appl Sci Technol Index Applied Science and Technology Index [Published in the US]

Appl Sol Energy Applied Solar Energy [Translation of: *Geliotekhnika*; published in the US]

Appl Spectrosc Applied Spectroscopy [Journal of the Society for Applied Spectroscopy, US]

Appl Spectrosc Rev Applied Spectroscopy Reviews [US]

Appl Spectry Applied Spectroscopy [of Society for Applied Spectroscopy, US]

Appl Supercond Applied Superconductivity [A Section of the International Journal "Solid State Electronics"]

Appl Surf Sci Applied Surface Science [Published in the Netherlands]

Appl Virol Applied Virology [US Journal]

APPM Atomic Parts per Minute [also appm]

APPM Australian Pulp and Paper Mills

AppME Applied Mechanics Engineer

APPMI American Peanut Product Manufacturers, Inc. [US]

APPMSA American Pulp and Paper Mills Superintendents Association [US]

APPN Advanced Peer-to-Peer Networking(Protocol)

App[NH]p 5'-Adenylylimidodiphosphate

APPPC Asia and Pacific Plant Protection Commission [Thailand]

APPR Army Package Power Reactor

appr approximately

Appro Approval [also appro]

approp appropriate

approx approximate(ly)

APPS Adenosine 3'-Phosphate-5'-Phosphosulfate [Biochemistry]

APPS Annual Procurement Plan and Strategy [of Department of Supplies and Services, Canada]

APPS Atmospheric Pressure Plasma Spraying

Appls Applications [also apps]

appm atomic part(s) per million [Unit]

Appshp Apprenticeship [also appshp]

Appt Appoint(ment) [also appt]

apptd appointed [also appt'd]

APPU Asian-Pacific Postal Union [Philippines]

APPU Australian Primary Producers Union

Appval Approval [also appval]

Appx Appendix [also appx]

APQ Available Page Queue

APQC American Productivity and Quality Center [US]

APR Acoustic Paramagnetic Resonance

APR Active Prominence Region [Astronomy]

APR Advanced Production Release

APR Airborne Profile Recorder

APR Air-Purifying Respirator

APR Alternate Path Reentry

APR Alternate Path Retry

APR American Petroleum Re-Refiners [US]

APR Annual(ized) Percentage Rate

APR Annulus Pressure Responsive

APR ARF Power Reactor [US]

APR Automatic Pressure Relief

Apr April

α-Pr Alpha Praseodymium [Symbol]

APRA Association of Public Relations Associations [UK]

APRA Automotive Parts Rebuilders Association [US]

APRACA Asian and Pacific Regional Agricultural Credit Association

APRC Automotive Public Relations Council [US]

APRD Atmosphere Particulate Radioactivity Detector

APRDL Advanced Products Research and Development Laboratory [of Motorola Corporation at Austin, Texas, US]

APRES American Peanut Research and Education Society [US]

APRF Army Pulsed Radiation Facility [US]

a-PrFeB Praseodymium Iron Boron Amorphous Alloy

APRFR Army Pulsed Radiation Facility Reactor [US]

APRI Absolute Rod Position Indication [Nuclear Engineering]

APRI Aqua Planing Risk Indicator

APRIL Aqua Planing Risk Indicator for Landing

APRIL Automatically Programmed Remote Indication Logged

APRIS Alcoa Picturephone Remote Information System [US]

APRL Army Prosthetic Research Laboratory [US]

APRL Automated Primate Research Laboratory

APRM Automatic Position Reference Monitor

APRM Average Power Range Monitor

APRN Armor-Piercing, Round Nose

APRO Aerial Phenomena Research Organization

APRO Airline Public Relations Organization [UK]

APRP Adaptive Pattern Recognition Processing

APRR Association for Planning and Regional Reconstruction [UK]

APRR Association of Petroleum Re-Refiners [US]

APRS American Park and Recreation Society [US]

APRS Association for the Preservation of Rural Scotland

APRS Association for the Protection of Rural Scotland

APRS Association of Professional Recording Services [UK]

APRS Automatic Position Reference System

APRT Adenine Phosphoribosyltransferase [Biochemistry]

aprxly approximately

APS Abstract Preparation Sheet

APS Acquisition Processor Subsystem

APS Adenosine-5'-Phosphosulfate [Biochemistry]

APS Advanced Packaging System

APS Advanced Photon Source [of US Department of Energy at Argonne National Laboratory, Illinois; Established in 1993]

APS Advanced Photo System

APS Advanced Printing Service

APS Advanced Programming Software

APS Aerodynamic Particle Sizer

APS Aircraft Prepared for Service

APS Air Plasma Sprayed; Air Plasma Spraying

APS Allegheny Power Systems [US]

APS Alphanumeric Photocomposer System

APS Alternate Payload Specialist

APS American Philatelic Society, Inc. [US]

APS American Philosophical Society [US]

APS American Physical Society [US]

APS American Physiological Society [US]

APS American Phytopathological Society [US]

APS American Polar Society [US]

APS American Pomological Society [US]

APS American Purchasing Society [US]

APS Bis(4-Aminophenyl)sulfone

APS Aminopolystyrene (Resin)

APS Aminopropyl Siloxane

APS Aminopropyltriethoxysilane

APS Angular Power Spectrum

APS Antennas and Propagation Society [of Institute of Electrical and Electronics Engineers, US]

APS Appearance Potential Spectroscopy

APS Appearance Potential Spectrum

APS Application Process Subsystem

APS Armor-Piercing Sabot

APS Array Processor Software

APS Assembly Programming System

APS Association de la Presse Suisse [Swiss Press Association]

APS Association of Pacific Systematists

APS Association of Plastics Societies [US]

APS Asynchronous Protocol Specification

APS Atmospheric Plasma Spraying

APS Atmospheric Pressure Sensor

APS Atom Probe Spectroscopy

APS Attached Processor for Speech

APS Attached Processor System

APS Attended Pay Station

APS Australian Public Service

APS Automated Patent System [US]

APS Automatic Page Search

APS Automatic Patching System

APS Auxiliary Power System

APS Auxiliary Program Storage

APS Average Particle Size [also aps]

APS Average Propensity to Save

Aps Apus [Astronomy]

aPS Porous Amorphous Silicon

a-PS Atactic Polystyrene [also aPS]

APSA Advanced Product Supply Approach

APSA Automatic Particle Size Analyzer

APS/AAPT (Joint) American Physical Society/American Association of Physics Teachers (Meeting) [US]

APS/AIP (Joint) American Physical Society/American Institute of Physics (Project) [US]

APSC Arkansas Public Services Commission [US]

APSC Asian Pacific Society for Cardiology

APSC Austin Peay State College [Clarksville, Tennessee, US]

APSCE Association of Professional Staffs for College Education [Ireland]

ApSci Applied Science; Applied Scientist

APSD Advanced Photon Source Division [of Argonne National Laboratory, Illinois, US]

APSDEP Asian and Pacific Skill Development Program [of International Labour Organization in Pakistan]

APSDIN Asian and Pacific Skill Development Information Network [Asian and Pacific Skill Development Program, International Labour Organization]

APSE Ada (Language) Programming Support Environment

APSF Australian Pacific Science Foundation

APSG Antisymmetrized Product of Strongly Orthogonal Geminals

APSG Australian Plant Specialists Group [of International Union for the Conservation of Nature (and Natural Resources)]

APSH 2-Acetylpyridine-p-Toluenesulfonyl Hydrazonate

APSH-H 2-Acetylpyridine-p-Toluenesulfonyl Hydrazone

APSK Amplitude Phase Shift Keying

APSK Audio Phase Shift Keying

APS LINAC Advanced Photon Source Linear Accelerator [of Argonne National Laboratory, Illinois, US] [also APS Linac]

APSM Academy of Product Safety Management [US]

APSM Association for Physical and System Mathematics [US]

APS News American Physical Society News [US Monthly Publication]

APSP African Primary Science Programme

APSP Array Processor Subroutine Package

APSR Antiproton Storage Ring [Particle Physics]

APSR Axial Power Shaping Rod

APSRU Agricultural Production Systems Research Unit

APSS Atmospheric Plasma Spray System

APSS Automated Program Support System

APSTM Apertured Photon Scanning Tunneling Microscopy

APT Accelerator Production of Tritium [Physics]

APT Address Pass Through

APT Advanced Parallel Technology

APT Advanced Passenger Train

APT Advanced Passenger Transport

APT Advanced Processing Technology

APT Airman (or Airwoman) Proficiency Test

APT Airport [also Apt, or apt]

APT 3-Aminopropyl-Isothiuronium

APT (3-Aminopropyl)trimethoxysilane

APT Ammonium Paratungstate

APT Apartment [also Apt, or apt]

APT Apparent Solar Time [Astronomy]

APT Applied Technology Council [US]

APT Applied Probability Trust [of University of Sheffield, UK]

APT Arbitrage Pricing Theory

APT Armor-Piercing with Tracer [also AP-T]

APT Asia Pacific Telecommunity

APT Association for Preservation Technology

APT Atomic Polar Tensor

APT Attached Proton Test

APT Augmented Programming Training

APT Automatic Picture Taking

APT Automatic Picture Transmission

APT Automatic Position Telemetering

APT Automatic Process Testing

APT Automatic(ally) Programmed Tool; Automatic Program(ming) Tool

APT Automation Planning and Technology

APT Autonomous Picture Transmission (System)

AP&T Automation Products and Technology [Canada]

AP-T Armor Piercing with Tracer [also APT]

Apt Airport [also apt]

Apt Apartment [also apt]

Apt Appoint(ment) [also apt]

APTA American Physical Therapy Association [US]

APTA American Public Transit Association [US]

APTA Atlantic Provinces Trucking Association [Canada]

APTC Army Physical Training Corps [US]

APTC Atlantic Provinces Transport Commission [Canada]

APTI Advanced Processing Technology Institute [US]

APTI Arab Petroleum Training Institute [Iraq]

APTI Association of Principals of Technical Institutions [UK]

APTIC African Pyrethrum Technical Information Center

APTIC Air Pollution Technical Information Center [US]

APTMS Advanced Payload Transfer Guidance System

APTP N-(4-Azidophenylthio)phthalimide

APTR Advanced Pressure Tube Reactor [US]

APTS Automatic Picture Transmission Subsystem

APTS Automatic Picture Transmission System

APTT Activated Partial Thromboplastin Time (Test) [Medicine]

APTU African Postal and Telecommunications Union [Congo]

APU Acid Purification Unit

APU Alaska Pacific University [Anchorage, US]

APU Analog Processing Unit

APU Arab Postal Union

APU Arithmetic Processing Unit

APU Asynchronous Processing Unit

APU Audio Playback Unit

APU Automatic Power Up

APU Auxiliary Power Unit

APU Auxiliary Processing Unit

ApU Adenylyl(3'-5'-Uridine) [Biochemistry]

α-**Pu** Alpha Plutonium [Symbol]

APUCOT Automated Piezoelectric Ultrasonic Composite Oscillator Technique

APUG AutoPrep (Computer) Users Group [US]

APUHF Approximately Projected Unrestricted Hartree-Fock [Quantum Mechanics]

APUHS Automatic Program Unit, High Speed

APULS Automatic Program Unit, Low Speed

APUMP Approximately Projected Unrestricted Moeller-Plesset [Physics]

ApUpG Adenylyluridylylguanosine [Biochemistry]

ApUpU Adenylyluridylyluridine [Biochemistry]

α-**P-32 UTP** *α*-Phosphorus-32 Uridine 5'-Triphosphate [also *α*-^{32}P-UTP] [Biochemistry]

APV All-Passenger Vehicle

APV All-Purpose Van

APV All-Purpose Vehicle

APV 2-Amino-5-Phosphonovaleric Acid

APV Arbeitsgemeinschaft für Pharmazeutische Verfahrenstechnik e.V. [Association for Pharmaceutical Process Engineering, Germany]

APV Avipoxvirus

APVD Accelerated Physical Vapor Deposition

apvd approved [also apv'd]

APW Apparent Polar Wander

APW Augmented Plane Wave (Calculation)

APW Average Piece Weight

APWA American Public Works Association [US]

APWC Association of Professional Writing Consultants [US]

APWI Airborne Proximity Warning Indicator

APWO Association of Public Works Officials

APWP Apparent Polar Wander Path

APWR Advanced Pressurized Water Reactor

APWSS Asian Pacific Weed Science Society [Philippines]

APX Appendix [also Apx]

APXS Alpha-Proton X-Ray Spectrometer [of NASA]

2APy 2-Amino Pyridine

3APy 3-Amino Pyridine

4APy 4-Amino Pyridine

APYS Auger Electron Partial Yield Spectroscopy

a-PZT Amorphous Lead Zirconate Titanate

AQ Achievement Quotient

AQ Acoustic Quadrupole

AQ Air Quality

AQ Air-Quenched; Air Quenching

AQ Aminoquinoline

AQ Antarctica [ISO Code]

AQ 9,10-Anthraquinone

AQ Any Quality

AQ Any Quantity

AQ As Quenched (Condition)

AQ Autoquote

aq *(aqua)* – water

aq aqueous (state)

aq aquatic

.aq Antarctica [Country Code/Domain Name]

AQA Air Quality Act [US]

AQA Analytical Quality Assurance

AQA Approved Quality Assurance

aq ad *(aqua ad)* – water up to [Medical Prescriptions]

AQAM Air-Quality Assessment Model

AQAP Allied Quality Assurance Publications [Canada]

AQAP Applied Quality Assurance Publication(Requirements) [of NATO]

AQC Air Quality Compliance

aq dest *(aqua destillata)* – distilled water

9AQDS 9,10-Anthraquinone-1,8-Disulfonic Acid

9AQ5DS 9,10-Anthraquinone-1,5-Disulfonic Acid

AQE Airman (or Airwoman) Qualification Examination

Aqil Aquila [Astronomy]

Aq Insects Aquatic Insects [Journal published in the Netherlands]

AQIS Australian Quarantine and Inspection Service

AQL Acceptable Quality Level

Aql Aquila [Astronomy]

AQMC Association of Quality Management Consultants [UK]

AQMD Air Quality Management District [South Coast of California, US]

AQMG Assistant Quartermaster General

AQO Aminoquinoline Oxide

AQP Aquaporin

AQP Association for Quality and Participation [US]

AQP1 Aquaporin-1
AQR Air Quality Regulations [US]
Aqr Aquarius [Astronomy]
Aq Reg Aqua Regia [also aq reg]
AQS Aircraft-Quality Steel
AQS Air Quality Standard
9AQ26S 9,10-Anthraquinone-2,6-Disulfonic Acid
9AQ27S 9,10-Anthraquinone-2,7-Disulfonic Acid
9AQ1S 9,10-Anthraquinone-1-Sulfonic Acid
9AQ2S 9,10-Anthraquinone-2-Sulfonic Acid
Aq Soln(s) Aqueous Solution(s) [also aq soln(s)]
AQT Acceptable Quality Test
AQTX Aquatic Toxicity
Aqua Abstr Aqualine Abstracts [US] [also Aqualine Abstr]
AQUAMIN Secretariat on the Assessment of Aquatic Effects of Mining in Canada
AQUIS Air Quality Utility Information System [of Argonne National Laboratory, Illinois, US]
AQUREB Aquatic Research Board [Australia]
AQW Asymmetric Quantum Well [Electronics]
AR Abbe Refractometer
AR Aberration Resistance
AR Abrasion Resistance
AR Access Road
AR Accounts Receivable [also A/R]
AR Accumulator Register [Computers]
AR Acetal Resin
AR Achiral Radical
AR Acid Rain
AR Acid Reaction
AR Acid Resistance; Acid-Resistant; Acid-Resisting
AR Acknowledgement of Receipt
AR Acoustic Radiometer
AR Acoustic Reflex [Medicine]
AR Acoustic Resonance
AR Acoustic Resonator
AR Acoustic Ringing
AR Acquisition Radar
AR Acree-Rosenheim (Test) [Biochemistry]
AR Acrylic Resin
AR Actinide Research
AR Adaptive Robot
AR Addition Reaction
AR Address Register
AR Administration Room
AR Advice of Receipt (Form)
AR Aerial Refueling
AR Aircraft Rocket
AR Air Register
AR Air-Release (Valve)
AR Air Resistance
AR Air Retarder
AR Alkali Resistance; Alkali-Resistant; Alkali-Resisting
AR Alpha Ray

AR Amazon River
AR Ament and Rado (Procedure) [Physics]
AR Ammonia Recovery
AR Amphiregulin
AR Amplitude Resonance
AR Analog Radio System
AR Analytical (Grade) Reagent
AR Anchor Ring
AR Angular Resolution
AR Annihilation Radiation
AR Annual Report
AR Annual Return
AR Annual Review
AR Annual Reviews [US]
AR Annual Ring [Wood]
AR Anti-Reflection; Antireflective
AR Antiresonance
AR Applied Research
AR Applied Rheology
AR Aqua Regia
AR Arc Rectifier
AR Area Ratio
AR Area Rule [Aeronautics]
AR Argentina [ISO Code]
AR Argyll-Robertson (Pupil) [Medicine]
AR Arithmetic Register
AR Arkansas [US]
AR Army Regulation
AR (Lightning) Arrester
AR Artificial Rainfall
AR Aspect Ratio
AR As Received (Basis)
AR As Required
AR As Rolled (Condition) [Metallurgy]
AR Assembly and Repair
AR Associative Register
AR Astronomy Research
AR Attention Routine
AR Atmospheric Radiation
AR Atmospheric Research
AR Atomic Reactor
AR Atomic Resolution
AR Audio Range
AR Automated Reasoning
AR Automatic Ranging
AR Automatic Regulation; Automatic Regulator
AR Automatic Rifle
AR Autoradiograph(y)
AR Autoranging
AR Autoregression
AR Auto-Regulator
AR Auto-Reverse
AR Aviation Radionavigation [of Federal Communications Commission, US]

AR Avionic Requirements

AR Axis of Rotation

AR Awaiting Reply

AR Repair Ship [US Navy Symbol]

A&R Air and Rail [also a&r]

A&R Automation and Robotics

A/R Accounts Receivable [also AR]

A/R All Risks [also a/r]

A/R Amplitude-to-Radius (Ratio)

%AR Percent Area Reduction

Ar Arabia(n)

Ar Arabic

Ar Aragonite [Mineral]

Ar Argon [Symbol]

Ar Arrival

Ar Aryl (Radical) [Chemistry]

Ar Aromatic Group (or Ring) [Chemistry]

A(r) Space-Vector-Dependent Plane Wave [Symbol]

Ar^+ Argon Ion

Ar_{cm} Temperature at Which Transformation of Austenite to Austenite plus Cementite Occurs during Cooling of Hypereutectoid Steel [Symbol]

Ar_1 Temperature at Which Austenite to Ferrite, or Austenite to Ferrite plus Cementite Transformation is Completed during Cooling of Iron and Steel [Symbol]

Ar_3 Temperature at Which Austenite-to-Ferrite Transformation Starts during Cooling of Iron and Steel [Symbol]

Ar_4 Temperature at Which Transformation of Delta Ferrite to Austenite Occurs during Cooling of Iron and Steels [Symbol]

Ar-35 Argon-35 [also ^{35}Ar, or Ar^{35}]

Ar-36 Argon-36 [also ^{36}Ar, or Ar^{36}]

Ar-37 Argon-37 [also ^{37}Ar, or Ar^{37}]

Ar-38 Argon-38 [also ^{38}Ar, or Ar^{38}]

Ar-39 Argon-39 [also ^{39}Ar, or Ar^{39}]

Ar-40 Argon-40 [also ^{40}Ar, Ar^{40}, or Ar]

Ar-41 Argon-41 [also ^{41}Ar, or Ar^{41}]

Ar-42 Argon-42 [also ^{42}Ar, or Ar^{42}]

ar arrive(d)

a(r) antibonding orbital [Symbol]

.ar Argentina [Country Code/Domain Name]

ARA Aerial Rocket Artillery

ARA Aerial Ropeway Association [UK]

ARA Agricultural Research Administration [US]

ARA Aircraft Replaceable Assemblies

ARA Aircraft Research Association [UK]

ARA Aluminum Recycling Association [US]

ARA Amateur Rocket Association [US]

ARA American Railway Assiciation [US]

ARA American Rental Association [US]

ARA American Retreaders Association [US]

ARA American Rheumatism Association [US]

ARA American Royal Association [US]

ARA Angular Rate Assembly

ARA AppleTalk Remote Access [Apple Computer, Inc., US]

ARA Arab Roads Association [Egypt]

ARA Aromatic Apolar [Chemistry]

ARA Asian Recycling Association [Philippines]

ARA Associate of the Royal Academy [UK]

ARA Associates for Radio Astronomy

ARA Atikokan Research Area [Canada]

ARA Australian Robot Association

ARA Automotive Recyclers Association [US]

ARA Automotive Retailers Association[Canada]

Ara-A Adenine Arabinose [also ARA-A] [Biochemistry]

Ara-A Arabinofuranosyl Adenine [Biochemistry]

Ara-A 9-β-D-Arabinofuranosyladenine [also ara-A] [Biochemistry]

Ara-AMP 9-β-D-Arabinofuranosyladenine 5'-Monophosphate [Biochemistry]

Ara-ATP 9-β-D-Arabinofuranosyladenine 5'-Triphosphate [Biochemistry]

Arab Arabia(n)

Arab Arabic

ARABC Automotive Retailers Association of British Columbia [Canada]

Arab Gulf J Sci Res Arab Gulf Journal of Scientific Research [Sauda Arabia]

Arab J Sci Eng Arabian Journal for Science and Engineering [Saudi Arabia]

Arab Metall News Arab Metallurgical News [Algeria]

Arab Min J Arab Mining Journal [Jordan]

ARABSAT Arab Countries Regional Communications Satellite

Arab Wildl Online Arabian Wildlife Online

ARAC Aerospace Research Applications Center [of Indiana University, Bloomington, US]

Ara-C Cytosine β-D-Arabinofuranoside [Biochemistry]

Ara-C Cytosine Arabinoside [also ARA-C] [Biochemistry]

Arachnol Arachnological; Arachnologist; Arachnology

Arachnol Arachnology [International Journal]

Ara-CMP Cytosine β-D-Arabinofuranoside 5'-Monophosphate [Biochemistry]

Ara-CTP Cytosine β-D-Arabinofuranoside 5'-Triphosphate [Biochemistry]

ARAD Airborne Radar and Doppler

ARADCOM Army Air Defense Command [of US Army in Illinois]

ARAEM Atomic Resolution Analytical Electrical Microscope; Atomic Resolution Analytical Electrical Microscopy

ARAEN Reference Speech Power for the Measurement of AEN (Attenuation Equivalent Nettiness)

ARAES Angle-Resolved Auger Electron Spectroscopy

ARAF Air Reserve Augmentation Flights

ARAG Antireflective-Antiglare

ARAI Automotive Research Association of India

ARAL Automatic Record Analysis Language

ARALL Aramid-Aluminum Laminate [also Arall]

ARAM Association of Railroad Advertising and Marketing [US]

ARAM Association of Records Administrators and Managers [US]

ARAM Automated Remediation Assessment Methodology

ARAMCO Arabian-American Oil Company [Dharan, Saudi Arabia] [also Aramco]

ARA Memo ARA Memorandum [of Aircraft Research Association, US]

ARAMIS Antarctic Remote Sensing Aerial Photography and Mapping Information System

ARAN Association for the Reduction of Aircraft Noise [UK]

ARAP Aeronautical Research Associates of Princeton [of Princeton University, New Jersey, US]

ARAR Applicable or Relevant and Appropriate Requirement(s)

ARA Rep ARA Report [of Aircraft Research Association, US]

ARAS Antireflective-Antistatic

ARAS Atomic Resonance Absorption Spectrophotometry

ARAT Avion de Recherche Atmospherique [French Atmospheric Research Plane]

ARB Academie Royale de Belgique [Royal Belgian Academy, Brussels]

ARB Air Registration Board [UK]

ARB Air Research Bureau

ARB Air Resources Board [California, US]

ARB Airworthiness Requirements Board [UK]

ARB All Routes Busy [Telecommunications]

ARB American Research Bureau [US]

ARB Repair Ship for Battle Damage [US Navy Symbol]

Arb Arbitration [also arb]

arb arbitrary

ARBA American Road Builders Association [now American Road and Transportation Builders Association, US]

ARBBA American Railway Bridge and Building Association [US]

ARBD Allgemeiner Radio-Bund Deutschlands [General Radio Association of Germany]

ARBEC ASEAN Review of Biodiversity and Environmental Conservation [of Association of Southeast Asian Nations]

ARBED Aciéries Réunies de Burbach-Eich-Dudelange [United Steel Works of Burbach, Eich and Dudelange, Luxembourg]

Arbido Bull Arbido Bulletin [Switzerland]

Arbido Rev Arbido Revue [Switzerland]

ARBIT Arbeitsgemeinschaft der Bitumenindustrie [Bitumen Industry Association, Germany]

ARBOR Argonne Boiling Water Reactor [of Argonne National Laboratory, Illinois, US]

ArBr Aryl Bromide

ARBS Angle Rate Bombing System

Arb U Arbitrary Unit [also arb u, AU, or au]

ARC Academic Research Cooperation (Program) [Germany]

ARC Academy of Roofing Contractors [US]

ARC Accelerating Rate Calorimeter

ARC Action Resource Center

ARC Addiction Research Center [of National Institute on Drug Abuse, US]

ARC Additive Reference Carrier [Television]

ARC Advanced Reentry Concept

ARC Advisory Research Council [Queen's University, Kingston, Canada]

ARC Aeronautical Research Council [UK]

ARC Aggregation of Red-Blood Cells

ARC Agricultural Relations Council [US]

ARC Agricultural Research Center [Athens, Georgia, US]

ARC Agricultural Research Council [UK]

ARC Africa Resource Center

ARC AIDS-Related Complex

ARC Aiken Relay Calculator

ARC Aircraft Radio Corporation

ARC Airworthiness Requirements Committee

ARC Albany Research Center [of US Department of Energy, located at Albany, Oregon]

ARC Alberta Research Council [Canada]

ARC Aldol Ring Closure [Organic Chemistry]

ARC Almaden Research Center [of IBM Corporation at San Jose, California, US]

ARC Alpha-Ray Counter

ARC Alternate Route Cancel

ARC Altitude Rate Command

ARC American Red Cross [US]

ARC American Research Center [Egypt]

ARC American Research Corporation

ARC Ames Research Center [of NASA at Mountain View, California, US]

ARC Amplitude and Rise-Time Compensation

ARC Animal Resources Center

ARC Annual Report on Carcinogens [of National Institutes of Health, Bethesda, Maryland, US]

ARC Antireflection (or Antireflective) Coating

ARC Arab Research Center

ARC (Circular) Arc (Function) [Programming] [also Arc]

ARC Architects Registration Council [UK]

ARC Argonne Reactor Computation [Argonne National Laboratory, Illinois, US]

ARC Asian Research Center

ARC Asian Resource Center

ARC Association of Railway Communicators [US]

ARC Atlantic Reference Center [Canada]

ARC Atlantic Research Corporation [Alexandria, Virginia, US]

ARC Atlantic Richfield Company [US]

ARC Attached Resource Computer

ARC Attended Resource Computer

ARC Augmentation Research Center [US]

ARC Australian Research Council

ARC Automatic Relay Calculator

ARC Automatic Remote Control

ARC Auxiliary Roll Control

ARC Average Response Computer

ARC Awards and Recognition Committee [of Institute of Electrical and Electronics Engineers, US]

.ARC Archive [File Name Extension] [also .arc]

ARCA Advanced RISC (Reduced Instruction Set Computing) Computing Architecture

ARCA Air-Conditioning and Refrigeration Contractors of America

ARCA Asbestos Removal Contractors Association [UK]

ARCAS Automatic Radar Chain Acquisition System

ARCC Airworthiness Requirements Coordinating Committee

ARCC Animal Research and Conservation Center [now Wildlife Conservation International, US]

ARCCA Agricultural Research Council of Central Africa

ARCCLMA Aromatic Red Cedar Closet Lining Manufacturers Association [US]

ARCCOH Asian Regional Coordinating Committee on Hydrology

arccos arccosine [also arc cos, or $\cos^{-1}$]

arccosec arccosecant [also arc cosec, $\csc^{-1}$, arccsc, arc csc, or $\csc^{-1}$]

arccot arccotangent [also arc cot, $\cot^{-1}$, arcctn, arc ctn, or ctn^{-1}]

arccsc arccosecant [also arc csc, or $\csc^{-1}$]

arcctn arccotangent [also arc ctn, ctn^{-1}, arccot, arc cot, or $\cot^{-1}$]

ARCE Amphibious River Crossing Equipment

ARCEDEM African Regional Center for Engineering Design and Manufacturing [Nigeria]

ARCFCP Alliance for Responsible CFC (Chlorofluorocarbon) Policy [US]

ARCFS Angle-Resolved Constant Final State [Physics]

ARCG American Research Committee on Grounding [US]

ARCH Articulated Computing Hierarchy [UK]

ARCH Australian Architecture Database

Arch Archipelago [also arch]

Arch Architect; Architectural; Architecture [also arch]

Arch Archival; Archive [also arch]

arch archiac

Ar/CH$_4$ Argon/Methane (Gas Mixture)

Arch Acoust Archives of Acoustics [Warsaw, Poland]

Arch Appl Mech Archive of Applied Mechanics [Published in Germany]

Arch Autom Telemech Archiwum Automatyki i Telemechanika [Archives of Automation and Telemechanics, Warsaw, Poland]

Arch Biochem Biophys Archives of Biochemistry and Biophysics [US]

Arch Budowy Masz Archiwum Budowy Maszyn [Archives of Mechanical Engineering and Machine Design, Warsaw, Poland]

Arch Combust Archivum Combustionis [Archives of Combustion, Warsaw, Poland]

ArchD Doctor of Architecture

Arch Dermatol Archives of Dermatology [of American Medical Association]

Arch Devel Biol Archives of Developmental Biology

ArchE Architectural Engineer [also Arch Eng]

ARCHEDDA Architectures for Heterogeneous European Distributed Databases

Arch Eisenb tech Archiv für Eisenbahntechnik [Archive for Railway Engineering, Germany]

Arch Eisenhüttenwes Archiv für das Eisenhüttenwesen [Archive for Ferrous Metallurgy published by Verein der Deutschen Eisenhüttenleute, Düsseldorf, Germany]

Arch Electron Übertrag tech Archiv für Elektronik und Übertragungstechnik [Archive for Electronics and Transmission Engineering]

Arch Elektrotech Archiv für Elektrotechnik [Archive for Electrical Engineering, Germany]

Arch Elektrotech Archiwum Elektrotechniki [Archives of Electrical Engineering, Poland]

Arch Elektrotech (Germany) Archiv für Elektrotechnik (Germany) [Archive for Electrical Engineering (Germany)]

Arch Elektrotech (Poland) Archiwum Elektrotechniki (Poland) [Archives of Electrical Engineering (Poland)]

ArchEng Architectural Engineer [also Arch Eng, or ArchE]

Arch Eng Architectural Engineer(ing) [also Arch Engr]

Arch Environ Health Archives of Environmental Health [US]

Archeo Archeological; Archeologist; Archeology [also archeo]

Archeoastron Archeoastronomical; Archeoastronomer; Archeoastronomy [also archeoastron]

Archeol Archeological; Archeologist; Archeology [also archeol]

Archeol Archeology [Magazine of the Archeological Institute of America, New York City, US]

Archeomater Archeomaterials [US Journal] [also Archeomat]

Archeomet Archeometallurgical; Archeometallurgist; Archeometallurgy [also Archeometall, archeomet, or archeometall]

Archeotechnol Archeotechnological; Archeotechnologist; Archeotechnology [also archeotechnol]

Arch Gen Psych Archives of General Psychiatry

Arch Hutn Archiwum Hutnictwa [Archives of Metallurgy, Warsaw, Poland]

Arch Hydrotech Archiwum Hydrotechniki [Archives of Hydraulic Engineering, Warsaw, Poland]

Arch Ind Health Archives of Industrial Health [of American Medical Association, US]

Arch Insect Biochem Physiol Archives of Insect Biochemistry and Physiology

Arch Int Pharmacodyn Therap Archives Internationales de Pharmacodynamie et de Thérapie [International Archives of Pharmacodynamics and Therapy, France]

Arch Int Med Archives of Internal Medicine [of American Medical Association, US]

Archip Archipelago [also archip]

Archit Architect; Architectural; Architecture [also archit]

Archit J Architect's Journal [UK]

Archit Met Architectural Metals [Journal published in the US]

Archit Q Architectural Quarterly [of National Concrete Masonry Association, US]

Arch Math Archiv der Mathematik [Archive of Mathematics; published by the Mathematische Forschungsinstitut, Germany]

Arch Math Log Archive for Mathematical Logic [Germany]

Arch Mech Archives of Mechanics [Warsaw, Poland]

Arch Metall Archives of Metallurgy [Warsaw, Poland]

Arch Microbiol Archives of Microbiology [Published in Germany]

Arch Nauki Mater Archiwum Nauki o Materialach [Archive of Materials Science, Warsaw, Poland]

Arch Neurol Archives of Neurology [US Journal]

Arch Neurol Psych Archives of Neurology and Psychiatry [of American Medical Association, US]

Arch Ochr Śr Archiwum Ochrony Środowiska [Warsaw, Poland]

Arch Ophthal Archives of Ophthalmology [of American Medical Association, US]

Arch Oral Biol Archives of Oral Biology

Arch Otolaryngol Archives of Otolaryngology [of American Medical Association, US]

Arch Pathol Archives of Pathology [of American Medical Association, US]

Arch Post- Fernmeldewes Archiv für das Post- und Fernmeldewesen [Archive for Post and Telecommunications, Germany]

Arch Psychol Archives of Psychology

Arch Ration Mech Anal Archive for Rational Mechanics and Analysis [Germany]

Arch Sci Archives des Sciences [Archives of Science, Switzerland]

Arch Sex Behav Archives of Sexual Behavior [US Journal]

Arch Surg Archives of Surgery [of American Medical Association, US]

Arch Toxicol Archives of Toxicology

Archv Archive [also archv]

ARCI American Railway Car Institute [US]

ARCIS African Regional Center for Information Science

ARCIS Angle-Resolved Constant Initial-Energy Spectra [Physics]

ArCl Aryl Chloride

ARCN Agricultural Research Council of Norway

ARCnet Attached Resource Computer Network

ARCO Atlantic Richfield Company [US]

ARCOM Association of Researchers in Construction Management

ArCom Awards and Recognition Committee [of Institute of Electrical and Electronics Engineers Standards Association, US]

ARCOMSAT Arab League Communications Satellite

ArCOOH Aryl Group [General Formula]

arcosh inverse hyperbolic cosine [also $\cosh^{-1}$]

arcoth inverse hyperbolic cotangent [also $\coth^{-1}$]

ARCP Accredited Roofing Contractors Program [of Academy of Roofing Contractors, US]

ARCP Aerodrome Reference Code Panel [of International Civil Aviation Organization]

ARCRL Agricultural Research Council Radiobiological Laboratory [UK]

ARCS Advanced Reconfigurable Computer System

ARCS Advanced Remote Controlled Submarine

ARCS Airline Request Communication System

ARCS Air Resupply and Communication Service

ARCS Alternate Remedial Contracting Strategy

ARCS Associate of the Royal College of Science [UK]

ARCS Atmospheric Radiation and Clouds Station

ARCS Automated Ring Code Search

ARCS Autonomous Remote(ly) Controlled Submersible

arcsch inverse hyperbolic cosecant [also csch^{-1}]

arcsec arcsecant [also arc sec, or $\sec^{-1}$]

arcsin arcsine [also arc sin, or $\sin^{-1}$]

ARCSS Arctic System Sciences Program [of National Science Foundation, US]

ARCST Associate of the Royal College of Science and Technology [UK]

ARCSyM Arctic Regional Climate System Model

ARCT African Regional Center for Technology [Senegal]

arct arctic

arctan arctangent [also arc tan, or $\tan^{-1}$]

AR Ctd Anti-Reflection Coated (Lens) [also AR ctd]

ARCTIC Alaskan Resources, Commodities, Trading and Investment Corporation [US]

Arct O Arctic Ocean

ARCUK Architects Registration Council of the United Kingdom

ARCV American Research Corporation of Virginia [Radford, US]

ARC/W Arc Weld [also Arc/W]

ARD Academy of Rheumatoid Diseases [US]

ARD Acid Rock Drainage

ARD Acoustic Research Division [of Bell Laboratories, Murray Hill, New Jersey, US]

ARD Acute Respiratory Disease [Medicine]

ARD Advanced Reactor Division [of Westinghouse Company, US]

ARD Airborne Respirable Dust

ARD Answering Recording and Dialing

ARD Arbeitsgemeinschaft der öffentlich-rechtlichen Rundfunkanstalten in Deutschland [Association of German Public Broadcasting Stations]

ARD Armament Research Department [UK]

ARD Association of Research Directors [US]

ARD Automated Retrieval of Data

ARDA Agriculture and Rural Development Act [Canada]

ARDA American Railway Development Association [US]

ARDA Analog Recording Dynamic Analyzer

ARDA Atomic Research and Development Authority [US]

ARDB Aeronautical Research and Development Board [India]

ARDC Adult Respiratory Distress Syndrome

ARDC Agricultural and Rural Development Corporation [Burma]

ARDC Air Research and Development Command [US Air Force]

ARDC American Research and Development Corporation [US]

ARDC Armament Research and Development Center

ARDC/BMD Air Research and Development Command, Ballistic Missile Division [Palo Alto, California, US]

ARDE Armament Research and Development Establishment [now Royal Armament Research and Development Establishment, UK]

ARDEC Armament Research and Development Establishment [now Canadian Armament Research and Development Establishment]

ARDEC Army Research, Development and Engineering Center [at Picatinny Arsenal, New Jersey, US]

ARDF Airborne Radio Direction Finding

ARDI Aviation Research Development Institute

ARDIC Association pour la Recherche et le Développement en Informatique Chimique [Association for Research and Development in Chemical Informatics, France]

AR-DIET Angle-Resolved Desorption Induced by Electronic Transition [also AR DIET]

ARDIS Army Research and Development Information System [US Army]

ARDS Advanced Remote Display Station

ARDS Artillery Regimental Data System [Canada]

ARDSA Agricultural and Rural Development Subsidiary Agreement [Canada]

ARDS/ADM Artillery Regimental Data Systems/Advanced Development Model

ARE Activated Reactive Evaporation

ARE Admiralty Research Establishment

ARE Airline Revenue Enhancement

ARE Arabian Republic of Egypt

ARE Armament Research Establishment [Fort Halstead, Kent, UK]

ARE Association for Research and Enlightenment [US]

ARE Asymptotic Relative Efficiency

ARE Automated Responsive Environment

AREA American Railway Engineering Association [US]

AREA American Recreational Equipment Association [US]

AREA Army Reactor Area [US]

AREA Association for Research and Environmental Aid, Ltd.

AREA Association for Rational Environmental Alternatives [US]

AREA Australian Resources and Environment Assessment (Project) [of Environment Australia]

AREA Bull AREA Bulletin [of American Railway Engineering Association]

AREAL Atmospheric Research and Exposure Assessment Laboratory [US]

AREC Agricultural Research and Educational Center [Lebanon]

ARED Aperture Relay Experiment Definition

AREDC Angle-Resolved Energy Distribution Curve [Physics]

AREG Apparatus Repair-Strategy Evaluation Guidelines

AREG Antarctic Research Evaluation Group

ARELEM Arithmetic Element Program

ARENTS ARPA (Advanced Research Projects Agency) Experimental Test Satellite [US]

AREP Applied Rangeland Ecology Program [Northern Territories, Australia]

AREP Automated Reliability Estimation Program

AREP Average Relativistic (Core) Potential [Physics]

ARE/RA Armaments Research Establishment, Royal Arsenal [Woolwich, UK]

ARERE Admiralty Research Establishment Rotating Arm [UK]

ARES Advanced Railroad Electronics System

ARES Advanced Rocket Engine/Storable (Liquid Fuel Study) [of Aerojet General, US]

ARESE ARM (Atmospheric Radiation Measurement) Enhanced Short-Wave Experiment [of NASA Langley Research Center, Hampton, Virginia, US]

AREST Analytical Repository Source Term

AREST-CT Analytical Repository Source Term with Chemical Transport

ARET Accelerated Reduction and Elimination of Toxics (Initiative) [Canada]

ARET Angle-Averaged Relative Energy Transfer [Physics]

a-RE-TM Amorphous Rare Earth-Transition Metal (Alloy)

ARETO Arab Republic of Egypt Telecommunications Organization

ARETS Alberta Renewable Energy Test Site [Canada]

ARF Advertising Research Foundation [US]

ARF Amplitude Reflection Factor

ARF Antianemic Factor [Biochemistry]

ARF Anti-Resonance Frequency

ARF Aquatic Research Facility [Canada]

ARF Archeological Research Facility [of University of California at Berkeley, US]

ARF Armor Research Foundation [US]

ARF Arthritis and Rheumatism Foundation [US]

ARF Assembly and Refurbishment Facility

ARF Automatic Report Feature

ArF Argon Fluoride (Excimer Laser)

ArF Aryl Fluoride

ARFA Allied Radio Frequency Agency [of NATO]

ARFAA Atlantic Region Freight Assistance Act [Canada]

ARFC Automatic Radar Frequency Control

ARFF Airport Rescue Fire Fighting

ARFOR Area Forecast

ARFP Automatic Radiographic Film Processing

ARG Air Reserve Group

ARG Allergy Research Group [US]

ARG American Resources Group [US]

ARG Analytical Reference Glass

ARG Argument [also Arg] [Computers]

Arg Argentina; Argentine; Argentinian

Arg (+)-Arginine; Arginyl [Biochemistry]

Arg Argo [Astronomy]

Arg Argument [also arg]

arg argument [Mathematics]

ARGC Australian Research Grants Committee

ARGE Arbeitsgemeinschaft der Verbände der Europäischen Schloß- und Beschlagindustrie [European Federation of Associations of Lock and Builders Hardware Manufacturers, Germany]

Argent Argentina; Argentine; Argentinian

Arge Pham Arbeitsgemeinschaft physikalische Analytik und Meßtechnik [Association for Physical Analysis and Measurement Technology, Gießen, Germany]

ArGeZ Arbeitsgemeinschaft Zulieferindustrie [Industrial Subcontractor's Association, Germany]

AR-Glass Alkali-Resistant Fiberglass [also AR-glass]

ARGMA Army Rocket and Guided Missile Agency [Huntsville, Alabama, US]

L-Arg-MCA L-Arginine-4-Methylcoumaryl-7-Amide [Biochemistry]

ARGONAUT Argonne Nuclear Assembly for University Training [of Argonne National Laboratory, US; Established in 1957] [also Argonaut, or CP-11]

ARGOS Atmospheric Research Geostationary Orbit Satellite [also Argos]

ARGOS Argos Data Collection and Location System [France]

ARGS Australian Research Grants Scheme

ARGUS Automatic Routine Generating and Updating System

Ar/H$_2$ Argon/Hydrogen (Atmosphere)

Ar/He$_2$ Argon/Helium (Atmosphere)

ArHe Argon-Helium (Mixture)

ARI Abrasion-Resistance Index

ARI Academic Research Infrastructure (Program) [of National Science Foundation, US]

ARI Aeronautical Research Institute [of Tokyo University, Japan]

ARI Agricultural Research Institute [Edmonton, Alberta, Canada]

ARI Airborne Radio Instrument

ARI Air-Conditioning and Refrigeration Institute [US]

ARI Aldose Reductase Inhibitor [Biochemistry]

ARI Aluminium Research Institute [US]

ARI American Refractories Institute [US]

ARI Annual Recurrence Interval

ARI Aquatic Research Institute [US]

ARI Atomic Resolution Imaging

ARI Australian Road Index (Database) [of Australian Road Research Board]

ARI *(Autofahrer-Rundfunk-Information)* – German for "Road Traffic Information Service" [*Note:* A Radio Broadcast in Central Europe]

ArI Aryl Iodide

Ari Aries [Astronomy]

ARIA Advanced Range Instrumentation (or Instrumented) Aircraft

ARIA Apollo/Range Instrumentation Aircraft [also A/RIA]

ARIBA Associate of the Royal Institute of British Architects

ARIC Acid Rain Information Center

ARIC Associate of the Royal Institute of Chemistry

ARICS Associate of the Royal Institute of Chartered Surveyors

Arid ID Arid Integrated Demonstration [also arid ID]

Arid Lands Newsl The Arid Lands Newsletter [Publication of the Office of Arid Lands Studies; of University of Arizona, Tucson, US]

ARIDO Association of Registered Interior Designers of Ontario [Canada]

ARIES Advanced Radar Information Evaluation System

ARIES Angle-Resolved Ion Electron Spectroscopy

ArIM Argon Ion Milling

ARIMA Autoregressive Integrated Moving Average

ARINC Aeronautical Radio Inc. [US]

ARIP Automatic Rocket Impact Predictor

ARIPE Angle-Resolved Inverse Photoemission (Spectroscopy)

ARIPS Angle-Resolved Inverse Photoemission Spectroscopy [also ARIPES]

ARIPS Angle-Resolved Inverse Photoemission Spectrum

ARIS Active Rack Isolation System

ARIS Advanced Range Instrumentation Ship [of NASA]

ARIS Apollo Range Instrumentation Ship [of NASA]

ARIS Atomic Resolution Imaging System

ARIS Australian Resources Information System

ARIST Annual Review of Information on Science and Technology

ARISTOTELES Applications and Research Involving Space Technologies Observing the Earth's Field from a Low-Earth-Orbiting Satellite [NASA]

ARISTOTLE Annual Review of Information and Symposium on the Technology of Training, Learning and Education [US]

ARIS-UHV Atomic Resolution Imaging System–Ultrahigh Vacuum

Arith Arithmetic(s) [also arith]

Ariz Arizona [US]

Ariz Hwys Arizona Highways [US Publication]

.ARJ Compressed File [Jung File Name Extension]

ARK Arkansas, Kansas [Meteorological Station Designator]

Ark Arkansas [US]

Ark Fys Semin Trondheim Arkiv for det Fysiske Seminar i Trondheim [Archive of Physical Seminars in Trondheim, Norway]

Ark Mat Arkiv för Matematik [Archive for Mathematics; published by the Royal Swedish Academy of Sciences]

ARL Acceptable Reliability Level [Industrial Engineering]

ARL Adjusted Ring Length

ARL Admiralty Research Laboratory [UK]

ARL Aeronautical Research Laboratory [of Defense Science and Technology Organization, Melbourne, Australia]

ARL Aerospace Research Laboratory [UK]

ARL Aircraft Rocket Launcher

ARL Air Resources Laboratory [of National Oceanic and Atmospheric Administration, US]

ARL Akademie für Raumforschung und Landesplanung [Academy for Regional Research and Planning, Hannover, Germany]

ARL Alden Research Laboratory [US]

ARL Aluminum Research Laboratories [US]

ARL Amman Ritson Limit

ARL Applied Research Laboratories [US]

ARL Applied Research Laboratories S.A. [Switzerland]

ARL Arctic Research Laboratory [US]

ARL Arizona Research Laboratory [of University of Arizona, Tucson, US]

ARL Army Radiation Laboratory [US Army]

ARL Army Research Laboratories [Fort Monmouth, New Jersey, US]

ARL Army Research Laboratories [Watertown, Massachusetts, US]

ARL Association of Research Libraries [US]

ARL Atlantic Research Laboratory [of National Research Council of Canada]

ARL Auditory Research Laboratory [of Princeton University, New Jersey, US]

ARL Australian Radiation Laboratory

ARL Automation Research Laboratory [Kyoto University, Japan]

ARL Average Run Length

ARL Repair Ship for Landing Craft [US Navy Symbol]

ARL FRD Air Resources Laboratory–Forecast Research Development [of National Oceanic and Atmospheric Administration, US]

ARLIS Arctic Research Laboratory Ice Station [US Navy]

ARLL Advanced Run Length Limited

ARM Accumulator Read-in Module [Computers]

ARM Advanced RISC (Reduced Instruction Set Computing) Machine

ARM Anhysteretic Remanent Magnetization

ARM Annotated Reference Manual

ARM Anti-Radar Missile

ARM Anti-Radiation Missile

ARM Area Radiation Monitor

ARM Articulated Remote Manual

ARM Associate in Risk Management

ARM Association of Railway Museums [US]

ARM Association of Researchers in Medicine [UK]

ARM Association of Rooflight Manufacturers [UK]

ARM Association of Rotational Molders [US]

ARM Asynchronous Response Mode

ARM Atmospheric Radiation Measurement Experiment [of NASA]

ARM Atmospheric Radiation Measurement Program [of US Department of Energy]

ARM Atmosphere Radiation Monitor

ARM Atomic Resolution Microscope; Atomic Resolution Microscopy

ARM Automated Route Management

ARM Automatic Reel Mounting

ARM Automatic Robotic Module

ARM Availability, Reliability and Maintainability

Arm Armature [also arm]

Arm Armenia(n)

Arm Armo(u)r [also arm]

ARMA Aluminum Radiator Manufacturers Association [UK]

ARMA American Records Managers and Administrators [US]

ARMA American Rock Mechanics Association [US]

ARMA Asphalt Roofing Manufacturers Association [US]

ARMA Association of Records Managers and Administrators [US]

ARMA Autoregression Moving Average; Autoregressive Moving Average

Armada Int Armada International [Technical Journal published in Switzerland]

ARMAN Artificial Methods Analyst

ARMAR Airborne Rain Mapping Radar

ARMA Rec Manage Q ARMA Records Management Quarterly [of Association of Records Managers and Administrators, US]

ARMAX Autoregressive Moving Average Exogenous

ARMC Automotive Research and Management Consultants

ARMCANZ Agricultural and Resource Management Council of Australia and New Zealand

ARMCAS Arctic Radiation, Microphysics, Clouds and Aerosols Study

ARMCO American Rolling Milling Company [Middletown, Ohio, US] [also Armco]

armd armored [also arm'd]

Armen Armenia(n)

ARMF Advanced Reactivity Measurement Facility [of Idaho National Engineering Laboratory, Idaho Falls, US]

ArMgBr Arylmagnesium Bromide

ARMI Analytical Reference Materials Incorporated [US]

ARMM Association of Reproduction Materials Manufacturers [US]

ARMMS Automated Reliability and Maintainability Measurement System

Arm-Pl Armor Plate

ARMRC Atlantic Regional Magnetic Resonance Center [of Dalhousie University, Halifax, Canada]

ARMS Advanced Receiver Model System

ARMS Aerial Radiological Measurements and Surveys; Aerial Radiological Measuring Surveys

ARMS Amateur Radio Mobile Society [US]

ARMS Atmospheric Roving Manipulator System [Oceanography]

ARMS Automated Record Management System

Arm SSR Armenian Soviet Socialist Republic [USSR]

Armt Armament [also armt]

Armyanskii Khim Zh Armyanskii Khimicheskii Zhurnal [Armenian Journal of Chemistry]

Army Mag Army Magazine [of Association of the United States Army]

ARN Atmospheric Radio Noise

Ar/N$_2$ Argon/(Molecular) Nitrogen) (Gas Mixture)

aRNA Activator Ribonucleic Acid [also a-RNA, or ARNA] [Biochemistry]

ARNAB African Research Network on Agricultural Byproducts

ARNEWS Acid Rain National Early Warning System(Project) [Canada]

ARNG Air National Guard [US]

ARNGB Air National Guard Base [US]

ARNIS Advanced Rocket Nozzle Inspection System

ARNM African Regional Network on Microbiology

ARNMD Association for Research in Nervous and Mental Diseases [US]

ARNO Association of Royal Navy Officers [UK]

ARNOT Aera Notice

ARNS Arkansas River Navigation System [US]

ARO After Receipt of Order

ARO Agricultural Research Organization

ARO Airborne Range Only [also aro]

ARO Air Radio Officer

ARO All Rods Out

ARO Army Research Office [of US Department of Defense in Research Triangle Park, North Carolina, US]

ARO Asian Regional Office [of International Confederation of Free Trade Unions]

ARO ATS (Air Traffice Service) Reporting Office [of Federal Aviation Administration, US]

ARO Automatic Recovery Option

Ar/O$_2$ Argon/Oxygen (Gas Mixture)

ARO/AFOSR Army Research Office/Air Force Office of Scientific Research [US]

AROD Airborne Range and Orbit Determination

ARODS Airborne Radar Orbital Determination System

AROM Alterable Read-Only Memory

AROM Associative Read-Only Memory

Arom Aromatic(s) [also arom]

arom aromatic [Apothecary]

AROS Alterable Read-Only Operating System

AROS Automated Regional Operations System

AROS Autoregressive, One-Lag, Single-Site (Model)

AROSAT Aromatics Saturation (Process)

AROT Australian Rare or Threatened (Plants) [of Environment Australia–Endangered Species Unit]

AROU Aviation Repair and Overhaul Unit

ARP Acid Rain Program [of US Environmental Protection Agency]

ARP Addictions Rehabilitation Program [of Canadian Armed Forces]

ARP Address Resolution Protocol [Internet]

ARP Adjunct Research Professor

ARP Advanced Radiator Program [of NASA]

ARP Advanced Reentry Program

ARP Advanced Research Project

ARP Aeronautical Recommended Practice

ARP Aerospace Recommended Practice

ARP Airborne Radar Platform

ARP Airport Reference Point

ARP Air Raid Precautions

ARP Anderson-Rayleigh-Polya (Localization) [Physics]

ARP Angle-Resolved Photoemission (Spectroscopy)

ARP Annihilation Radiation of Positrons [Particle Physics]

ARP Aramid-Reinforced Plastics

ARP Argentinean Peso [Currency]

ARP Aromatic Polar [Chemistry]

ARP Assistant Research Professor

ARP Atmospheric Refractivity Profile

ARP Aviation Regulatory Proposal

ARP Azimuth Reset Pulse

ARPA Advanced Research Projects Agency [of US Department of Defense]

ARPA Archeological Resources Protection Act [US]

ARPA Automated Radar Plotting Aid

ARPA/NASA Advanced Research Projects Agency/National Aeronautics and Space Administration (Contract)

ARPAnet Advanced Research Project Agency Network [of US Department of Defense] [Replaced by NSFnet]

ARPA/NIST Advanced Research Projects Agency/National Institute of Standards and Technology (Project)

ARPA/ONR Advanced Research Projects Agency/Office of Naval Research (Project) [US Department of Defense]

ARPAT ARPA (Advanced Research Projects Agency) Terminal

ARPE American Registry of Professional Entomologists [US]

ARPEFS Angle-Resolved Photoelectron Extended Fine Structure

ARPEFS Angle-Resolved Photoemission Extended Fine Structure

ARPES Angle-Resolved Photoelectron Spectroscopy

ARPES Angle-Resolved Photoemission Spectroscopy

ARPES Angle-Resolved Photoemission Spectrum

ARPG Asphalt Rubber Producers Group [US]

ARPHOS 1-Diphenylphosphino-2-Diphenylarsinoethane [also arphos]

ARPI Analog Rod Position Indicator

ARPI Automotive Refrigeration Products Institute [US]

ARPL Adjust Requested Privilege Level

ARPMA Aluminum Rolled Products Manufacturers Association [UK]

ARPR Academia Republicii Populare Romine [Academy of People's Republic of Romania, Bucharest]

ARPS Aerospace Research Pilot School [US Air Force]

ARPS Angle-Resolved Photoelectron Spectroscopy

ARPS Associate of the Royal Photographic Society [UK]

ARPS Association of Recycled Paper Suppliers [UK]

ARPS Australian Radiation Protection Society

ARQ Automatic Repeat Request

ARQ Automatic Request for Correction

ARQ Automatic Request for Repetition

ARQ Automatic Response Query

Arq Univ Bahia Arquivos de Universidade de Bahia [Archives of the University of Bahia, Brazil]

ARQX Air Quality Experimental Studies Division [of Atmospheric Environment Service, Canada]

ARR Academia Republicii Romania [Academy of the Republic of Romania]

ARR Address Record Register

ARR Adjustable Reset Response

ARR Alberta Resources Railway [Canada]

ARR Alligator Rivers Region (Region) [Australia]

ARR Anti-Reciprocity Relation

ARR Anti-Repeat Relay

ARR Argonne Research Reactor [of Argonne National Laboratory, Illinois, US]

ARR Association for Radiation Research [UK]

ARR Audio Response System

ARR Automatic Rerouting

ARR Automatic Retransmission Request

Arr Arrangement [also arr]

Arr Arrest(or) [also arr]

Arr Arrival [also arr]

arr arranged

arr arrive(d)

ARRA American Rod and Reel Association [US]

ARRA Asphalt Recycling and Reclaiming Association [US]

ARRB Australian Road Research Board

ARRDO Australian Railways Research and Development Organization

ARRE Alarm Receiving and Reporting System

ARRI Aboriginal Rural Resource Initiative [Australia]

ARRI Automation and Robotics Research Institute [US]

ARRIS All Round Information System for Standard Production

ARRL Aeronautical Radio and Radar Laboratory

ARRL American Radio Relay League [US]

ARRP Animal Research Review Panel

ARRR Aerotest Research and Radiography Reactor [of Aerotest Operations Inc., San Ramon, Calfiornia, US]

ARRRI Alligator Rivers Region Research Institute [of Office of the Supervising Scientists, Australia]

ARRS Aerospace Rescue and Recovery Service [US Air Force]

ARRS American Roentgen Ray Society [US]

ARRS American Registered Respiratory Therapist

ARRT American Registry of Radiologic Technologists [US]

ARRU Alligator Rivers Research Unit [Australia]

ARS Active Repeater Satellite

ARS Active Response System [Automobiles]

ARS Advanced Reconnaissance Satellite

ARS Advanced Record System [of General Services Administration, US]

ARS Advanced Recovery System (Process)

ARS Aeronautical Research Scientist

ARS Agricultural Research Service [of US Department of Agriculture]

ARS Air Rescue Service

ARS Alpha-Ray Spectrometer; Alpha-Ray Spectrometry

ARS Alpha-Ray Spectrum

ARS American Radium Society [US]

ARS American Rhinologic Society [US]

ARS American Rocket Society [Now Part of American Institute of Aeronautics and Astronautics, US]

ARS American Rose Society [US]

ARS Amplified Response Spectrum

ARS Archive and Retrieval Subsystem

ARS Asbestos Roof Shingle

ARS Asian Research Service

ARS As Rapidly Solidified (Condition) [Metallurgy]

ARS Atmospheric Radiance Study

ARS Automatic Recovery System

ARS Repair Ship, Salvage Vessel [US Navy Symbol]

ARSA Aeronautical Repair Station Association [US]

ARSA Associate of the Royal Scottish Academy

ARSA Associate of the Royal Society of Arts

ARSAP Agricultural Requisites Scheme for Asia and the Pacific

ARSB Automated Repair Service Bureau

ARSB Anchored Radio Sono Buoy [also arsb]

ARSBB Association of Railway Superintendents of Bridges and Buildings [now American Railway Bridge and Building Association, US]

ARSC African Remote Sensing Council

ARSC Automatic Resolution Selection Control

ARSD Repair Ship, Salvage Lifting Vessel [US Navy Symbol]

ARSDP Asian Regional Skill Development Program [now Asian and Pacific Skill Development Program; of International Labour Organization in Pakistan]

arsech inverse hyperbolic secant [also sech^{-1}]

ARSES Angle-Resolved Secondary Electron Spectroscopy

ARSFC Australian Recreation and Sport Fishing Confederation

arsinh inverse hyperbolic sine [also $\sinh^{-1}$]

ARSL Agricultural Research Service Laboratory [Fargo, North Dakota, US]

ARSLBB Académie Royal des Sciences, des Lettres et des Beaux-Arts de Belgique [Royal Belgian Academy of Sciences, Letters and Fine Arts]

ARSM Associate of the Royal School of Mines [UK]

ARSME All-Round Shape Memory Effect [Metallurgy]

Arsn Arsenal [also arsn]

ARSO African Regional Standardization Organization [Kenya]

ArSO$_2$OR Alkyl Ester of Sulfonic Acid

ARSP Aerospace Research Satellite Program [US Air Force]

ARSP Aerospace Research Support Program [US Air Force]

ARSR Academia Republicii Socialiste Romania [Academy of the Socialist Republic of Romania, Iasi]

ARSR Air-Route Surveillance Radar

ARSS Applied Resolved Shear Stress [Mechanics]

ARST Repair Ship, Salvage Tender [US Navy Symbol]

ART Accredited Record Technician

ART Additional Reference Transmission

ART Additive Reference Transmission [Television]

ART Admissible Rank Test
ART Advanced Reactor Technology
ART Advanced Recording Technology
ART Advanced Refractory Technologies, Inc. [Buffalo, US]
ART Advanced Research and Technology
ART Airborne Radiation Thermometer
ART Aircraft Radio Technician
ART Aircraft Reactor Test
ART Alarm Reporting Telephone
ART Algebraic Reconstruction Technique
ART Army Research Task
ART Artificial Resynthesis Technology
ART Asphalt Residuum Treating (Process)
ART Automated Request Transmission
ART Automated Rotor Test System
ART Automatic Range Tracker; Automatic Range Tracking
ART Automatic Reporting Telephone
ART Autonomous Robotics Technology
ART Average Response Time
ART Average Retrieval Time
Art Artery [also art]
Art Article [also art]
Art Artillery [also art]
A(r,t) Space-Vector and Time-Dependent Plane Wave [Symbol]
art artificial
ARTA American Reuseable Textile Association [US]
ARTA Apple Real-Time Architecture
artanh inverse hyperbolic tangent [also tanh^{-1}]
ARTBA American Road and Transportation Builders Association [US]
ARTC Aircraft Research and Testing Committee
ARTC Air-Route Traffic Control
ARTCC Air-Route Traffic Control Center [US]
ARTCD Aboriginal Recruitment Training and Career Development [Australia]
ARTDO Asian Regional Trade Development Organization
ARTE Admiralty Reactor Test Establishment [UK]
ARTEM Atomic Resolution Transmission Electron Microscope; Atomic Resolution Transmission Electron Microscopy
ARTEMIS Automatic Retrieval of Test through European Multipurpose Information Services
Arterioscl Thomb Vasc Biol Arteriosclerosis, Thrombosis and Vascular Biology [of American Heart Association, US]
ARTES Advance Research in the Telecommunications Systems Program [Canada]
ARTF Australian Road Transport Federation
artfl artificial
ARTG Azimuth Range and Timing Group
Arthrop Arthropathic; Arthropathy [Medicine]
ARTI Arab Regional Telecommunications Institute
ARTIC A Real-Time Interface Coprocessor
ARTIC Associometrics Remote Terminal Inquiry Control System
Artic Articulated Lorry [UK] [also artic]

artif artificial
Artif Intell Artificial Intelligence [Journal published in the Netherlands]
Artif Intell Eng Artificial Intelligence in Engineering [Journal published in the UK]
Artif Intell Eng Des Anal Manuf Artificial Intelligence for Engineering Design, Analysis and Manufacturing [Journal published in the UK]
Artif Intell Rev Artificial Intelligence Review [UK]
Artif Life Artificial Life [of Massachusetts Institute of Technology Press, Cambridge, US]
Artill Artillery [also artill]
ARTOC Army Tactical Operations Center [US]
ARTP Association of Reclaimed Textile Processors [UK]
ARTRAC Advanced Range Testing, Reporting and Control
ARTRAC Advanced Real-Time Range Control
Artron Artificial Neutron [also atron]
ARTS Advanced Radar Terminal System
ARTS Advanced Radar Traffic (Control) System
ARTS Advanced Radio Telephone Service
ARTS American Radio Telephone System
ARTS Army Research Task Summary
ARTS Asynchronous Remote Takeover Server
ARTS Automated Radar Terminal System
ARTS Automated Resource Tracking System
.ARTS Cultual Groups [Internet Domain Name] [also .arts]
Arts D Doctor of Arts
ARTSM Association of Road Traffic Sign Makers [UK]
Arts Manuf Arts et Manufactures [French Publication on Arts and Manufacture]
ARTT Asynchronous Remote Takeover Terminal
ARTU Automatic Range Tracking Unit
Arty Artillery [also arty]
Ärztl Prax Ärztliche Praxis [German Journal on Medical Practice]
ARU Acoustical Resistance Unit (or Acoustic Ohm)
ARU Address Recognition Unit
ARU American Railway Union
ARU Arithmetic Unit
ARU Audio Response Unit
ARUBIS Angle-Resolved Ultraviolet Bremsstrahlung Isochromate Spectroscopy
ARUPS Angle-Resolved Ultraviolet Photoelectron Spectroscopy
ARUPS Angle-Resolved Ultraviolet Photoelectron Spectrum
ARUPS Angle-Resolved Ultraviolet Photoemission Spectroscopy [also ARUPES]
ARUPS Angle-Resolved Ultraviolet Photoemission Spectrum
ARV Accepted Reference Value
ARV Aeroballistic Reentry Vehicle
ARV AIDS-Associated Retrovirus [Immunochemistry]
ARV Armored Recovery Vehicle
ARVE Repair Ship for Aircraft Engines [US Navy Symbol]
ARVS Autonomous Rescue Vehicle System

ARVS Autonomous Research Vehicle System

ARW Air-Conditioning and Refrigeration Wholesalers [US]

ARW Air Reserve Wing

Ar-WCl₆ Argon-Tungsten Hexachloride (Gas Mixture)

Ar-WF₆ Argon-Tungsten Hexafluoride (Gas Mixture)

ArX Aryl Halide [Formula]

ARXES Angle-Resolved X-Ray Emission Spectroscopy

ARXPD Angle-Resolved X-Ray Photoelectron Diffraction

ARXPS Angle-Resolved X-Ray Photoelectron Spectroscopy

ARY Arab Republic of Yemen

AS Able Seaman [Canadian Forces]

AS Abrasive Slurry

AS Absorption Spectrophotometer

AS Absorption Spectroscopy

AS Absorption Spectrum

AS Account of Sales [also A/S]

AS Academia Sinica [China]

AS Académie des Sciences [Academy of Sciences, Paris, France]

AS Academy of Sciences

AS Acetanisidine

AS Acid Sludge [Chemical Engineering]

AS Acoustic Scattering

AS Acoustic Spectrometer; Acoustic Spectrometry

AS Acoustic Spectrum

AS Acrylonitrile Styrene Copolymer

AS Actinium Series

AS Active Satellite

AS Active Sun [Astronomy]

AS Actuator Sensor

AS Adaptive System

AS Adhesion Science

AS Adhesion Society [US]

AS Adiabatic Stretch (Approximation) [Physics]

AS Administrative Support

AS Advanced Schottky (Logic)

AS Advanced System

AS Aeronautical Standards

AS Aerospace

AS Aerospace Science

AS Aerospace Standards

AS Aerostatic(s)

AS After Sight

AS Aircraft Scientist

AS Air Screw

AS Air Separation; Air Separator

AS Air Service [UK]

AS Air Setting

AS Air Speed

AS Air-Spaced (Cable) [also as]

AS Air Station

AS Alabama Section [of Materials Research Society, US]

AS Algebraic(al) Sum

AS Alkyl Shift

AS Alloy Steel

AS Alloy System

AS Alkali(ne) Sulfide

AS Alongside

AS Altimeter Setting

AS Altitude Sensor

AS Aluminate Substrate

AS Aluminosilicate (Glass)

AS Aluminum-Oxide Special-Grain (Grinding Wheel)

AS Aluminum Silicate

AS Aluminum Sulfate

AS American Samoa

AS Aminosalicylic Acid

AS Ammeter Switch

AS Ammonia Synthesis

AS Amorphous Semiconductor

AS Amorphous Solid

AS Amplitude Separator

AS Amplitude Spectrum

AS Analytical Separation

AS Anaphylactic Shock

AS Angle Strain

AS Anglo-Saxon [also A-S]

AS Ankylosing Spondylitis

AS Annealed Steel

AS Anthranilate Synthetase [Biochemistry]

AS Antislip

AS Antistatic

AS Anti-Stokes (Spectral Lines)

AS Antistructure

AS Antisubmarine [also A/S, or as]

AS Antisymmetrization

AS Aortic Stenosis

AS Aperture Size

AS Aperture Stop

AS Application-Specific

AS Applications Software [also ASW]

AS Application Study

AS Application System

AS Applied Science

AS Applied Spectroscopy

AS Aprotic Solvent

AS Aqueous Solution

AS Arabian Sea

AS Aral Sea [Kazakhstan/Uzbekistan]

AS Arakawa-Schubert [Physics]

AS Arc Source

AS Arc Spectrum

AS Area Surveillance

AS Armature Shunt [Controllers]

AS Arsenosilicate

AS Arteriosclerosis

AS Articulation Score

AS Ascent Stage [also A/S] [of NASA Apollo Program]
AS Ash Sediment
AS Associate in Science
AS As Soldered (Condition)
AS As Sprayed (Condition)
AS As Spun (Condition)
AS As Synthesized (Condition)
AS Asymmetric(al)
AS Atomic Science; Atomic Scientist
AS Atomic Size
AS Atomic Spectrometer; Atomic Spectrometry
AS Atomic Spectroscope; Atomic Spectroscopy
AS Atomic Spectrum
AS Audio Signal
AS Auger Spectroscopy
AS Auger Spectrum [Physics]
AS Austenitic Steel
AS Australian Sceptics
AS Australian Standard
AS Austrian Schilling [Currency]
AS Automated Storage
AS Auto(matic) Sampler; Automatic Sampling
AS Automatic Sprinkler
AS Automatic Switching
AS Automatic System
AS Automation Society [US]
AS Automation System
AS Autonomous System [Internet]
AS Auxiliary Storage
AS Avicultural Society [UK]
AS Avogadro-Siegbahn (Constant) [Physics]
AS Avoidance System
AS Axis of Symmetry
AS Azonal Soil [Geology]
AS Silver Selenide
AS Surface-Treated Fiber [Designation]
AS Submarine Tender [US Navy Symbol]
AS$_3$ Aluminum Silicate [$Al_2O_3 \cdot 3SiO_2$]
AS3 1,1,1-Tris(diphenylarsinomethyl)ethane
5-AS 5-Aminosalicylic Acid [also 5AS]
2-AS 2-(9-Anthroyloxy)stearic Acid
12-AS 12-(9-Anthroyloxy)stearic Acid
9-AS 9-(9-Anthrylcarboxy)stearic Acid
12-AS 12-(9-Anthrylcarboxy)stearic Acid
A$_3$S$_2$ Trialuminum Disilicate [$3Al_2O_3 \cdot 2SiO_2$]
A/S Account of Sales [also A/S]
A/S Anti-Submarine
A/S Ascent Stage [of NASA Apollo Program]
A-S Abrikosov-Suhl (Resonance) [Solid-State Physics]
A-S Anglo-Saxon [also AS]
A-S^2 Studievereniging A-Eskwadraat [Research Association "A-S-Squared," of University of Utrecht, Netherlands]
As Altostratus (Cloud) [also as]

As Ampere-second [also A·s]
As Arsenic [Symbol]
As Asia(n)
As^{3-} Arsenide [also As^{---}] [Symbol]
As^{5+} Pentavalent Arsenic Ion [Symbol]
As$_4$ Molecular Arsenic [Symbol]
As-70 Arsenic-70 [also ^{70}As, or As^{70}]
As-71 Arsenic-71 [also ^{71}As, or As^{71}]
As-72 Arsenic-72 [also ^{72}As, or As^{72}]
As-73 Arsenic-73 [also ^{73}As, or As^{73}]
As-74 Arsenic-74 [also ^{74}As, or As^{74}]
As-75 Arsenic-75 [also ^{75}As, As^{75}, or As]
As-76 Arsenic-76 [also ^{76}As, or As^{76}]
As-77 Arsenic-77 [also ^{77}As, or As^{77}]
As-78 Arsenic-78 [also ^{78}As, or As^{78}]
As-79 Arsenic-79 [also ^{79}As, or As^{79}]
A(s) Atoms of Metal Component A as Surface Atoms [Symbol]
A(s) Input Elements (in Automatic Control) [Symbol]
aS absiemens [Unit]
a-S amorphous sulfur
as attosecond [10^{-18} second]
a/s after sight
.as American Samoa [Country Code/Domain Name]
αS alpha sulfur [Symbol]
Å/s Ångstrom(s) per second [also Ås^{-1}, Å/sec, or Å sec^{-1}]
Å²/s Square Ångstrom(s) per second [also Å² s^{-1}, Å²/sec, or Å² sec^{-1}]
ASA Abort Sensor Assembly [of NASA, Apollo Program]
ASA Accessible Surface Area (Model)
ASA Acetylsalicylic Acid [Pharmacology]
ASA Acoustical Society of America [US]
ASA Acrylate-Styrene-Acrylonitrile
ASA Acrylic-Styrene-Acrylonitrile (Terpolymer)
ASA Acrylonitrile-Styrene-Acrylate (Terpolymer)
ASA Acrylonitrile-Styrene-Acrylonitrile
ASA Active Surface Area
ASA Air Security Agency
ASA Air Services Agreement
ASA Aluminum Stockholders Association [UK]
ASA American Snowmobile Association [US]
ASA American Society of Agronomy [US]
ASA American Society of Anesthesiologists [US]
ASA American Society of Appraisers [US]
ASA American Soybean Association [US]
ASA American Standards Association [now American National Standards Institute, US]
ASA American Statistical Association [US]
ASA American Student Association [US]
ASA American Subcontractors Association [US]
ASA American Supply Association [US]
ASA American Surgical Association [US]
ASA Aminosalicylic Acid
ASA Aniline Sulfonic Acid

ASA Antarctic Shipping Association

ASA Antistatic Additive

ASA Applied Systems Analysis

ASA Army Security Agency [US]

ASA Army Signal Association [US]

ASA Asian Students Association [Hong Kong]

ASA Association des Statisticiens Agricoles [Association of Agricultural Statisticians, France]

ASA Association of Sealant Applicators [UK]

ASA Association of Southeast Asia [Now part of Association of Southeast Asian Nations]

ASA Association of Subscription Agents [UK]

ASA Astronomical Society of Australia

ASA Atlantic Salmon Association [US]

ASA ATM (Apollo Telescope Mount) Solar Array [of NASA]

ASA Atomic-Scale Architecture

ASA Atomic Scientists Association [UK]

ASA Atomic Security Agency [US]

ASA Atomic-Sphere Approximation [Physics]

ASA Audiological Society of Australia

ASA Australian Society of Accountants

ASA Automotive Service Association [US]

ASA Average Sphere Approximation

ASA Avicultural Society of America [US]

ASA Azide-Styphnate-Aluminum [Explosive]

5-ASA 5-Amino Salicylic Acid

A3SA Aniline-3-Sulfonic Acid

A4SA Aniline-4-Sulfonic Acid

Asa βCarboxyaspartic Acid

ASAAWE Association of South Asian Archeologists in Western Europe [UK]

ASAB Association for the Study of Animal Behaviour [UK]

ASAC Accounting Standards Authority of Canada

ASAC Antarctic Sciencific Advisory Committee

ASAC Asian Standards Advisory Committee

ASAC Award Symposium for Analytical Chemistry [of American Chemical Society, US]

ASACS Airborne Surveillance and Control System

ASACUSA Atomic Spectroscopy And Collisions Using Slow Antiprotons (Experiment) [of CERN–European Laboratory for Particle Physics, Geneva, Switzerland]

ASADA Atomic and Space Development Authority [US]

ASAE Advanced School of Automobile Engineering

ASAE American Society of Agricultural Engineers [US]

ASAE American Society of Association Executives [US]

ASAE Association Suisse pour l'Aménagement des Eaux [Swiss Association for Water Management]

ASAES Army Small Arms Experimental Station

ASAF Assistant Secretary of the Air Force [US]

ASAI Adjunct Switch Application Interface

ASAI American School of Aircraft Instruments [US]

ASAIHL Association of Southeast Asian Institutions of Higher Learning [Thailand]

ASAIO American Society of Artificial Internal Organs [US]

ASAIO Trans American Society of Artificial Internal Organs Transactions [US]

ASAL Arid and Semi-Arid Lands

ASALM Advanced Strategic Air-Launched Missile

ASAM Alkali-Sulfite Anthraquinone Methanol (Process)

ASAM American Society for Abrasive Methods [now Abrasive Engineering Society, US]

ASAME Automated Strain Analysis and Measurement Environment

ASAO Association of Show and Agricultural Organizers [UK]

ASAP Accelerated Surface Area and Porosimetry

ASAP Advanced Systems Applications and Products

ASAP Aerospace Supplier Accreditation Program

ASAP Agricultural Safety Audit Program [Canada]

ASAP American Society of Access Professionals [US]

ASAP American Society of Aerospace Pilots [US]

ASAP American Society of Animal Production [US]

ASAP Analog System Assembly Pack

ASAP Antisubmarine Attack Plotter

ASAP Applied Systems and Personnel

ASAP Army Scientific Advisory Panel [US]

ASAP As Soon As Possible [also asap]

ASAP Australian Society of Animal Production

ASAP Automated Shipboard Aerological Program [Canada]

ASAP Automated Statistical Analysis Program

ASAP Automatic Switching And Processing

AS3AP ANSI (American National Standards Institute) SQL (Structured Query Language) Standard Scalable and Portable [also AS^{3}AP]

ASA/PC Acrylonitrile-Styrene-Acrylate/Polycarbonate (Blend)

ASAQS Association of South African Quantity Surveyors

A-SAR Advanced Synthetic Aperture Radar

ASARC Army Systems Acquisition Review Council [US]

ASARCO American Smelting and Refining Company [also Asarco]

ASARCO American Smelting and Refining Company Process

ASAS American Society of Agricultural Sciences [US]

ASAS American Society of Animal Science [US]

ASASC Aviation Safety Authorities Steering Committee

ASASP Axially-Scattering Aerosol Spectrometer Probe

ASAT Antisatellite

ASAT Arbeitsgemeinschaft Satellitenträgersystem [Association for Satellite Vehicle Systems, Germany]

ASAXS Anomalous Small-Angle X-Ray Scattering

ASB Academy of Sciences of Belarus [Minsk]

ASB Adiabatic Shear Band [Metallurgy]

ASB Agricultural Stabilization Board [Canada]

ASB Air Safety Board

ASB Alternating Sideband (Method)

ASB Alternatives to Slash and Burn (Global Project)

ASB Aluminum sec-Butoxide

ASB Ammonium Sulfobetaine

ASB Amt für Strassenbau [Federal Road Construction Office, Switzerland]

ASB Anti-Sideband

ASB Arbeitsgemeinschaft für wirtschaftliche Betriebsführung und soziale Betriebsgestaltung [Association for Economic Plant Management and Social Plant Structuring, Germany]

ASB Association of Shell Boilermakers [UK]

ASB Association of Southeastern Biologists [US]

ASB Australian Space Board

ASB-1 Ammonium Sulfobetaine-1

ASB-2 Ammonium Sulfobetaine-2

ASB-3 Ammonium Sulfobetaine-3

ASB-4 Ammonium Sulfobetaine-4

Asb Asbestos [also asb]

asb apostilb [Unit]

αSb Alpha Antimony [Symbol]

ASBA American Shorthorn Breeders Association [US] [*Note:* Shorthorn is a Breed of Cattle]

ASBA American Small Business Association [US]

ASBA American Southdown Breeders Association [US] [*Note:* Southdown is a Breed of Sheep]

ASBA Australian Small Business Association

ASBC Advanced Standard Buried Collector

ASBC American Safety Belt Council [US]

ASBC American Seat Belt Council [US]

ASBC American Society of Biological Chemists [now American Society for Biochemistry and Molecular Biology; of Federation of American Societies for Experimental Biology, US]

ASBC American Society of Biophysics and Cosmology [US]

ASBC American Society of Brewing Chemists [US]

ASBC American Standard Building Code [US]

ASBC Newsl ASBC Newsletter [of American Society of Brewing Chemists]

ASBD Advanced Sea-Based Deterrent

ASBE American Society of Bakery Engineers [US]

ASBE American Society of Body Engineers [US]

ASBHCA Average Season Busy Hour Call Attempt

ASBHCC Average Season Busy Hour Call Completion

ASBI Advisory Service for the Building Industry [UK]

A/S/Bi Alumina/Silicon Carbide/Binder [Ceramics]

ASBMB American Society for Biochemistry and Molecular Biology [US]

ASBO Association of School Business Officials [US]

ASBOG (National) Association of State Boards of Geology [US]

ASBPA American Shore and Beach Preservation Association [US]

ASBPE American Society of Business Press Editors [US]

ASBR Academically Separated Budgeted Research

ASBS Australian Systematic Botany Society

a-SBT Amorphous Strontium Bismuth Tantalate

ASBU Arab States Broadcasting Union

ASC Absorbed Specimen Current

ASC Accrediated Standards Committee [of Institute of Electrical and Electronics Engineers, US]

ASC 4-Acetylaminobenzenesulfonyl Chloride

ASC Acetylsulfanilyl Chloride

ASC Adams State College [Alamoso, Colorado, US]

ASC Adelphi Suffolk College [Garden City, New York, US]

ASC Adhesives and Sealants Council [US]

ASC Administrative Services Council [US]

ASC Adsorbent-Copper, Silver and (Hexavalent) Chromium (Impregnated)

ASC Advanced Safety Certificate

ASC Advanced Scientific Computer; Advanced Scientific Computing

ASC Advertising Standards Council

ASC Aeronautical Systems Center [at Wright-Patterson Air Force Base, Dayton, Ohio, US]

ASC Aeronautical Systems Command [of US Air Force Systems Command]

ASC Agricultural Stabilization and Cultivation

ASC AIN (Advanced Intelligent Network) Switch Capabilities

ASC Airport Security Council [US]

ASC Air-Powered Swing Clamp

ASC Airman, Second Class [also Asc]

ASC Alabama State College [Montgomery, US]

ASC Albany State College [Albany, Georgia, US]

ASC Alberta Safety Council [Canada]

ASC Alberta Securities Commission

ASC Albuquerque Seismological Center [of National Oceanic and Atmospheric Administration in New Mexico, US]

ASC American Sandblasting Contractors [US]

ASC American Satellite Corporation [US]

ASC American Society for Composites [US]

ASC American Society for Cybernetics [US]

ASC American Society of Cartographers [US]

ASC American Specialty Contractors [US]

ASC American Standard Code

ASC American Systematic Collections [US]

ASC Antislip Control

ASC Applied Superconductivity Conference

ASC Applied Surface Chemistry

ASC Arizona State College [Flagstaff, US]

ASC Arkansas State College [State College, US]

ASC Army Service Corps [US]

ASC Army Signal Corps [US]

ASC Ascent [also Asc]

ASC ASCII (Function) [Computer Programming]

ASC ASEAN (Association of Southeast Asian Nations) Standing Committee

ASC Assistant Sector Controller

ASC Associated Schools of Construction [US]

ASC Associate in Science [also ASc]

ASC Association for Student Counselling [UK]

ASC Association of Systematic Collections [US]

ASC Associative Structure Computer

ASC Asthma Society of Canada

ASC Atlantic Systems Conference

ASC Atomic-Scale Composite

ASC Australian Seismological Center

ASC Automatic Selectivity Control

ASC Automatic Sensitivity Control

ASC Automatic Sequence Control(ler)

ASC Automatic Setting Control

ASC Automatic Slip Control

ASC Automatic Spark Control

ASC Automatic Stability Control

ASC Automatic Switching Center

ASC Automatic System Controller

ASC Automotive Sales Council [US]

ASC Auxiliary Switch, (Normally) Closed

ASC AXAF (Advanced X-Ray Astrophysics Facility) Science Center [of NASA]

.ASC ASCII Text [File Name Extension] [also .asc]

ASc Associate in Science [also ASC]

Asc Ascent [also asc]

asc ascend(ing)

αSc Alpha Scandium [Symbol]

ASCA Advanced Satellite for Cosmology and Astrophysics [of NASA]

ASCA American Society for Conservation Archeology [US]

ASCA American Society of Consulting Arborists [US]

ASCA Architectural Spray Coaters Association [US]

ASCA Association of Science Cooperation in Asia

ASCA Automated Spin Chemistry Analyzer

ASCA Automatic Science Citation Alerting

ASCA Automatic Subject Citation Alert

ASCAC Anti-Submarine Classification Analysis Center [US Navy]

AscaDF Advanced Satellite for Cosmology and Astrophysics Data Facility [of NASA Astrophysics Data Facility, Goddard Space Flight Center, Greenbelt, Maryland, US]

ASCAP Aeronautical Satellite Communications Processor

ASCAP American Society of Composers, Authors and Publishers [US]

A-SCAT Advanced Scatterometer

ASCATS Apollo Simulation Checkout and Training System [of NASA]

ASCB American Society for Cell Biology [US]

ASCB Avionics Standard Communications Bus

ASCC Aeronautical Satellite Communications Center [US]

ASCC Air Standardization Coordinating Committee [US]

ASCC American Society for Concrete Construction [US]

ASCC American Society for the Control of Cancer [now American Cancer Society, US]

ASCC Army Strategic Communications Command [Washington, DC, US]

ASCC Australian Soil Conservation Council

ASCC Automatic Sequence Controlled Calculator

ASCCO Aluminum Strontium Calcium Copper Oxide (Suoerconductor)

ASCD American Society of Computer Dealers [US]

ASCD Association for Supervision and Curriculum Development [US]

ASCE Airlock Signal Conditioning Electronics

ASCE American Society of Civil Engineers [New York City, US]

ASCE Proc ASCE Proceedings [American Society of Civil Engineers, US]

ASCEA American Society of Civil Engineers and Architects [US]

ASCEND Advanced System for Communications and Education in National Development

ASCENT Assembly System for Central Processor

ASCEP Australasian Society for Clinical Experimental Pharmacologists

ASCET American Society of Certified Engineering Technicians [US]

ASCF Analog Switched Capacitor Filter

ASCG Automatic Solution Crystal Growth

ASCGP Administrative Special Canadian Grains Program [Canada]

ASCH American Society for Clinical Hypnosis [US]

ASch Austrian Schilling [Currency]

ASCI American Society for Clinical Investigation [US]

ASCII American Standard Code for Information Interchange

ASCII-8 American Standard Code for Information Interchange, Eight-Bit Version

ASCIO Asian Oceanic Computing Industry Organization

as-CIPed As Cold Isostatically Pressed [Powder Metallurgy]

ASCIS Australian Schools Cataloguing Information System

ASCLD American Society of Crime Laboratory Directors [US]

ASCM Advanced Semiconductor Manufacturing Conference [of Institute of Electrical and Electronics Engineers and Semiconductor Equipment and Materials Institute]

ASCM Aluminum-Silicon-Copper-Magnesium (Alloy)

ASCM Antiship Cruise Missile

ASCMA American Sprocket Chain Manufacturers Association [now American Chain Association, US]

ASC Newsl ASC Newsletter [of Association of Systematic Collections, US]

ASCO American Society for Clinical Oncology [US]

ASCO American Society of Composites [US]

ASCO Arab Satellite Communications Organization [Saudi Arabia]

ASCO Automatic Sustainer Cutoff

ASCOMACE Association des Constructeurs de Machines à Coudre [Association of Sewing Machine Manufacturers; of European Community, Belgium]

ASCOPE ASEAN (Association of Southeast Asian Nations) Council on Petroleum [Indonesia]

ASCOT Atmospheric Studies in Complex Training

ASCP American Society of Clinical Pathologists [US]

ASCP American Society of Consulting Planners [US]

ASCP Automatic System Checkout Program

ASCPP Advanced School on Cosmology and Particle Physics

ASCR Academy of Sciences of the Czech Republic [Prague] [also AS-CR]

ASCR Asymmetric Silicon Controlled Rectifier

ASCRS American Society of Colon and Rectal Surgeons [US]

ASCRT Association for the Study of Canadian Radio and Television

ASCS Agricultural Stabilization and Conservation Service [of US Department of Agriculture]

ASCS Aluminum Straight Clamp Strap

ASCS American School of Classical Studies [Locations at Athens, Greece and Rome, Italy]

ASCS Area Surveillance Control System

ASCS Automatic Stabilization and Control System

ASCSA American School and Community Safety Association [US]

ASCT Address Space Control Task

ASC+T Automatic Stability Control and Traction Control

ASCTA Association of Short-Circuit Testing Authorities

ASCU Association of Small Computer Users [US]

ASCUE Association of Small Computer Users in Education [US]

ASCVD Arteriosclerotic Cardiovascular Disease

ASCWSA Association of Scientific Workers of South Africa

ASD Accelerated Standards Development

ASD Access Storage Device

ASD Adjustable-Speed Drive

ASD Administrative Services Division [of Canada Center for Mineral and Energy Technology; of Natural Resources Canada, Ottawa, Canada]

ASD Adverse State Detector

ASD Aeronautical Systems Division [US Air Force]

ASD Aerospace Systems Division [US Air Force]

ASD Allowable-Stress Design

ASD American Society of Dowsers [US]

ASD Ammunition Sub-Depot

ASD Analysis and Simulation Division [of US Air Force Geophysical Laboratory]

ASD Anti-Structure Defect [Materials Science]

ASD Apparent Solar Day [Astronomy]

ASD Applied Science Division

ASD Asociación Salvadoreana de Demógrafia [Salvadoran Demographic Association]

ASD Associated Surplus Dealers [US]

ASD Association of Steel Distributors [US]

ASD Association Suisse de Documentation[Swiss Documentation Association]

ASD Atmospheric Sciences Department

ASD Atmospheric Sciences Division [of NASA Langley Research Center, Hampton, Virginia, US]

ASD Auger-Stimulated Desorption [Physics]

ASD Automated Systems Division [of Panametrics Inc., US]

ASD Automatic Synchronized Discriminator

ASD Avionic System Division [of Society of Automotive Engineers International, US]

asd as-deposited

ASDA Accelerate–Stop Distance Available

ASDA American Sleep Disorder Association [US]

ASDAR Aircraft-to-Satellite Data Relay

AsDB Asian Development Bank

ASDC American Society of Dentistry for Children [US]

ASDC Automatic Sample Data Collection

ASDE Airport Surface Detection Equipment [also asde]

ASDE American Society of Danish Engineers [US and Canada]

as-dep as-deposited

ASDG Aircraft Storage and Disposition Group [US Air Force]

ASDI Association for Science Documentation Information [of Tokyo Institute of Technology, Japan]

ASDI Australian Spatial Data Infrastructure

ASDI Automatic Selective Dissemination of Information

ASDIC Allied Submarine Devices Investigating Committee [UK]

ASDIC Antisubmarine Depth Indicating Control

ASDIC Antisubmarine Detection Investigation Committee [also Asdic]

ASDIC Armed Services Document Intelligence Center

ASDIRS Army Study Documentation and Information Retrieval System [US]

ASDL Advanced Systems Development Laboratory

ASDL Automated Ship Data Library

ASDMA Architectural and Specialist Door Manufacturers Association [UK]

ASD/NA National Aerospace Plane Joint Program Office [of NASA at Langley Research Center, Hampton, Virginia, US]

ASDR Airport Surface Detection Radar

ASDSO Association of State Dam Safety Officials [US]

ASDSRS Automatic Spectrum Display and Signal Recognition System

ASDSVN Army Switched Data and Secure Voice Network

ASE Academia de Studii Economice [Academy for Economic Studies, Bucharest, Romania]

ASE Acute Simple Endocarditis

ASE Admiralty Signal Establishment [UK]

ASE Agence Spatiale Européenne [European Space Agency, France]

ASE Airborne Search Equipment

ASE Airborne Support Equipment

ASE Airport Surface-Detection Equipment [also ase]

ASE Albany Society of Engineers [US]

ASE Alberta Stock Exchange [Canada]

ASE Alkylsulfonic Acid Ester

ASE Alliance-to-Save Energy [US]

ASE Allowable Steering Error

ASE Alternative Sources of Energy [Milaca, Minnesota, US]

ASE Altimetry System Error

ASE Amalgamated Society of Engineers [UK]

ASE American Society for Educators [US]

ASE American Society of Echocardiography [US]

ASE American Stock Exchange [New York City, US]

ASE Amplified Spontaneous Emission (Spectroscopy)

ASE Analytical Systems Engineering

ASE Applied Science and Engineering

ASE Association of Science Education [UK]

ASE Association of Scientists and Engineers [of Naval Sea Systems Command, US]

ASE Association of Space Explorers [US]

ASE Association Suisse des Electriciens [Swiss Association of Electricians]

ASE Astronomical Society of Edinburgh [Scotland]

ASE Astronomical Society of Egypt

ASE AutoCAD SQL (Structured Query Language) Extension (Software)

ASE Automatic Spectroscopic Ellipsometry

ASE Automatic Support Equipment

ASE Automation Systems Engineering

ASE (National Institute for) Automotive Service Excellence [US]

a-Se Amorphous Selenium

αSe Alpha Selenium [Symbol]

ASEA Agricultural Show Exhibitors' Association [UK]

ASEA American Solar Energy Association [US]

Asea J Asea Journal [Published by Asea AB, Vasteras, Sweden]

ASEAN Association of Southeast Asian Nations [Includes Indonesia, Malaysia, Philippines, Singapore and Thailand]

ASEANCUPS Association of Southeast Asian Nation Countries Union in Polymer Science

ASEANIP Association of Southeast Asian Nations Institute of Physics

ASEB Aeronautics and Space Engineering Board [US]

ASEB American Society for Experimental Biology [US]

ASEC Albert Schweitzer Ecological Center [Switzerland]

ASEC American Standard Elevator Code [US]

ASEC Applied Solar Energy Corporation [US]

ASEC Auburn Science and Engineering Center [of University of Akron, Ohio, US]

Å/sec Angstrom(s) per second [also Å/s, or Å s^{-1}]

Å²/sec Square Ångstrom(s) per second [also Å/s]

ASECH Acetylselenocholine [also ASECh] [Biochemistry]

ASED Bis(anisaldehyde)ethylenediamine

ASED Atom Superposition and Electron Delocalization; Atom Superposition and Electron Delocalizer

ASEDA Aboriginal Studies Electronic Data Archive [of Australian National University, Canberra]

ASED-MO Atom Superposition and Electron Delocalization (or Delocalizer) Molecular Orbital (Theory)

ASEE Advanced Semiconductor Equipment Exposition

ASEE American Society for Engineering Education [Washington, DC, US]

ASEE American Society for Environmental Education [US]

ASEE Association of Supervising Electrical Engineers [UK]

ASEE Association of Supervisory and Executive Engineers [UK]

ASEE-EED American Society for Engineering Education–Engineering Education Division [US]

ASEE-ELD American Society for Engineering Education–Engineering Libraries Division [US]

ASEE J Eng Educ American Society for Engineering Education–Journal of Engineering Education

ASEE/ONT American Society for Engineering Education/Office of Naval Technology (Fellowship) [US]

ASEERE Assistant Secretary for Energy Efficiency and Renewable Energy [of US Department of Energy]

ASEE Rep American Society for Engineering Education Report

ASEG Australian Society of Exploration Geophysicists

ASEH American Society for Environmental History [US]

ASEL Aircraft Single Engine Land

A.SEL Auto-Select

ASEM Active Source Electromagnetic Correction

ASEM American Society for Engineering Management [US]

ASEM Analytical Scanning Electron Microscopy

ASEM Australian Society of Electron Microscopy

ASE Mag ASE Magazine [of Alternative Sources of Energy, Milaca, Minnesota, US]

ASEMC Autobody Supply and Equipment Manufacturers Council [US]

ASEMRC Allied Signal Engineered Materials Research Center [US]

ASEP American Society for Electroplated Plastics [US]

ASEP American Society for Experimental Pathology [US]

ASEP ASEAN (Association of Southeast Asian Nations) Subregion Environment Program

ASES Active Solar Energy System

ASES American Solar Energy Society [US]

ASES Automated Software Evaluation System

ASESA Armed Services Electronic Standards Agency [US]

ASESS Aerospace Environment Simulation System

ASET Aeronautical Services Earth Terminal

ASET American Society for Engineering Technology [US]

ASET American Society of EEG (Electroencephalographic) Technologists [US]

ASET Association of Super-Advanced Electronics Technologies [of Ministry of International Trade and Industry, Japan]

ASE-USA Association of Space Explorers–United States of America

ASEV American Society for Enology and Viticulture [US]

ASF Active Streaming Format [Computers]

ASF Air Safety Foundation [US]

ASF Alaska SAR (Synthetic Aperture Radar) Facility

ASF American Scandinavian Foundation [US]

ASF Ampere(s) per Square Foot [Unit]

ASF Arab Sugar Federation [Sudan]

ASF Army Service Forces [US]

ASF Army Special Forces [US]

ASF Artificially Structured Films

ASF Artificial Superlattice Film [Solid-State Physics]

ASF Atlantic Salmon Federation [Canada]

ASF Atmospheric Stabilization Framework

ASF Atomic Scattering Factor [Physics]

ASF Australian Science Foundation

ASF Austrian Science Foundation

ASF Automatic Sheet Feeder

ASF Australian Speleological Federation

ASF Austrian Association of Sciences

ASF Auxiliary Supporting Feature

ASF Axial Shielding Factor [Nuclear Engineering]

ASFA Aquatic Sciences and Fisheries Abstracts [of United Nations Food and Agricultural Organization]

ASFB Australian Society for Fish Biology

ASFCEW Association of Sea Fisheries Committees in England and Wales [UK]

ASFD American Society of Furniture Designers [US]

ASFDO Antisubmarine Fixed Defense Office [US]

ASFE Association of Soil and Foundation Engineers [US]

ASFE/AEFPG Association of Soil and Foundation Engineers/Association of Engineering Firms Practicing in the Geosciences [US]

ASFG Atmospheric Sound-Focusing Gain

ASFI Association of Suppliers of the Furniture Industry [UK]

ASFIR Active Swept-Frequency Interferometer Radar

ASFIS Aquatic Sciences and Fisheries Information System [of United Nations]

ASFMRA American Society of Farm Managers and Rural Appraisers [US]

AS4P Antisymmetric Four-Point Loading [Bending]

ASFPCM Association of Structural Fire Protection Contractors [UK]

ASFPM Association of State Floodplain Managers [US]

ASFSE American Swiss Foundation for Scientific Exchange

ASFTS Airborne Systems Functional Test Stand

ASFV African Swine Fever Virus

ASFW Arbeitsgruppe Schweizerischer Fachverbände Werkstofftechnik [Working Group of Swiss Materials Associations]

ASG Advanced Study Group

ASG Aeronautical Standards Group [US]

ASG African Seabird Group [South Africa]

ASG Aluminosilicate Glass

ASG Antiferromagnetic Spin Glass

ASG Apparent Specific Gravity

ASG Association Suisse des Sables et Graviers [Swiss Association for Sand and Gravel]

ASG Australasian Seabird Group [New Zealand]

ASGAP Association of Societies for Growing Australian Plants

ASGB Aeronautical Society of Great Britain

ASGC Australian Standard Geographical Classification [of Australian Bureau of Statistics]

ASGCA American Society of Golf Course Architects [US]

ASGE American Society of Gas Engineers [US]

AsGe Arsenic Germanium (Spin Glass)

ASGLS Advanced Space-Ground Link Subsystem

ASGM Aperiodic Stepwise Growth Model [Metallurgy]

Asgmt Assignment

ASGRP Atlantic Salmon Genetics Research Program [of Atlantic Salmon Federation, Canada]

ASGS American Scientific Glassblowers Society [US]

ASGV Apple Stem Grooving Virus

ASH Air-Sparged Hydroclone

ASH (Schweizerische) Aktionsgemeinschaft Sparsamer Heizen [Swiss Action Group for Economic Heating]

ASH Assault Support Helicopter

ASH Assistant Secretary for Health [US]

ASH Australian Society of Herpitologists (Inc.)

As-H Arsenic-Hydrogen (System)

Ash Airship [also ash]

ASHA American Social Health Association [US]

ASHA American Speech and Hearing Association [US]

ASHAE American Society of Heating and Air Conditioning Engineers [now American Society of Heating, Refrigeration and Air Conditioning Engineers]

ASHC Aeronautics and Space Historical Center [France]

ASHD Arteriosclerotic Heart Disease

ASHE American Society for Hospital Engineering [US]

ASHG American Society of Human Genetics [US]

ASHI American Society of Home Inspectors [US]

as-HIPed As Hot Isostatically Pressed [Powder Metallurgy]

ASHOE Airborne Southern Hemisphere Ozone Expedition

ASHP American Society for Hospital Pharmacists [US]

ASHRAE American Society of Heating, Refrigeration and Air Conditioning Engineers [Formerly Subdivided into ASHAE and ASHVE]

ASHRAE J ASHRAE Journal [of American Society of Heating, Refrigerating and Air-Conditioning Engineers]

ASHRAE Trans ASHRAE Transactions [of American Society of Heating, Refrigerating and Air-Conditioning Engineers]

ASHS American Society for Horticultural Science [US]

ASHS Newsl ASHS Newsletter [of American Society for Horticultural Science, US]

Ashton-Tate Q Ashton-Tate Quarterly [US]

ASHVE American Society of Heating and Ventilating Engineers [formerly also ASH&VE; now ASHRAE]

ASHW Antisymmetric Shear Horizontal Wave

ASI Advanced Sciences, Inc. [US]

ASI Advanced Scientific Instrument

ASI Advanced Silicon Imaging

ASI Advanced Studies Institute [of NATO]

ASI Advanced Systems Institute [Vancouver, British Columbia, Canada]

ASI Advanced Systems Institute [of National Resource Information Center, Australia]

ASI Aeronautical Society of India

ASI African Scientific Institute

ASI Agenzia Spaziale Italiana [Italian Space Agency]

ASI Airspeed Indicator

ASI Altimeter Setting Indicator

ASI American Society of Interpreters [US]

ASI American Society of Inventors [US]

ASI American Specifications Institute [US]

ASI American Standards Institute [US]

ASI American Statistical Index (Database) [of Cranbrook Institute of Science,Detroit Michigan, US]

ASI Ampere(s) per Square Inch [Unit]

ASI Architects and Surveyors Institute [UK]

ASI Articulated Subject Index

ASI Astronomical Society of India

ASI Asynchronous Series Interface

ASI Atmospheric Structure Instrument [of NASA]

ASI Atomic-Scale Imaging

ASI Australian Science Index [Publication of Commonwealth Scientific and Industrial Research Organization]

ASI Australian Standards Institute

ASI Austrian Standards Institute

ASI Aviation Safety Inspector

ASI Axial Shape Index

AS-I Actuator-Sensor-Interface

a-Si Amorphous Silicon

ASIA Automotive Service Industry Association [US]

ASIAC Australian Surveying Industry Advisory Committee

ASIAG Association Interprofessionnelle de l'Aviation Agricole [Interprofessional Association of Agricultural Aviation, France]

Asia M Asia Minor

Asian Autotech Rep Asian Autotech Report

Asian-Pac Popul J Asian-Pacific Population Journal [of United Nations Economic and Social Commission for Asia and the Pacific]

Asian Persp Asian Perspective [Journal]

Asian Sources Electron Asian Sources Electronics [Published in Hong Kong]

Asian Surv Asian Survey [Publication]

AS IAP Academy of Sciences–Institute of Atmospheric Physics [Russia]

Asia-Pac J Oper Res Asia-Pacific Journal of Operational Research [Singapore]

ASIB Accademia delle Scienze dell' Istituto di Bologna [Academy of Science of the Institute of Bologna, Italy]

ASIC American Society of Irrigation Consultants [US]

ASIC Application-Specific Integrated Circuit

ASIC Asian Students Information Center

ASIC Association Suisse des Ingénieurs-Conseils [Swiss Association of Consulting Engineers]

ASIC Australian Standard Industry Classification

a-SiC Amorphous Silicon Carbide

αSiC Alpha Silicon Carbide

a-SiC:B Boronated Amorphous Silicon Carbide

ASIC CMOS Application-Specific Integrated Circuit Complementary Metal-Oxide Semiconductor

a-SiC:H Hydrogenated Amorphous Silicon Carbide

a-SiC/Si Amorphous Silicon Carbide/Silicon (Heterojunction Diode)

ASID Address Space Identifier

ASID American Society for Industrial Designers [US]

ASID American Society of Interior Designers [US]

ASID Analytical Scanning Imaging Device

ASIDIC Association of Scientific Information and Dissemination Centers [US]

ASIDP American Society of Information and Data Processing [US]

ASID Rep ASID Report [of American Society of Interior Design, US]

ASIF Aldosterone Secretion Inhibiting Factor [Biochemistry]

a-Si:F Fluorinated Amorphous Silicon

a-SiGe Amorphous Silicon Germanide

a-Si/Ge Amorphous Silicon/Germanium (Heterostructure)

a-SiGe:H Hydrogenated Amorphous Silicon Germanide

a-Si:H Hydrogenated Amorphous Silicon

αSi:H Hydrogenated Alpha-Silicon

a-Si:(H,F) Hydrogenated and Fluorinated Amorphous Silicon

a-Si:H TFT Hydrogenated Amorphous Silicon Thin-Film Transistor

ASIH American Society of Ichthyologists and Herpetologists [US]

ASII American Science Information Institute [US]

ASIM Advanced Simulation

ASIM Applied Simulation

ASI/MET Atmospheric Structure Instrument/Meteorology Experiment [of NASA]

ASIN Agricultural Service Information Network [of National Agricultural Library, US]

a-Si$_3$N$_4$ Amorphous Silicon Nitride

a-SiN Amorphous Silicon-Nitrogen (Compound)

αSi$_3$N$_4$ Alpha Silicon Nitride

ASI Newsl ASI Newsletter [of American Society of Inventors, US]

a-Si$_3$N$_4$:H Hydrogenated Amorphous Silicon Nitride

a-SiN:H Hydrogenated Amorphous Silicon-Nitrogen (Compound)

a-SiO$_2$ Amorphous Silica (or Amorphous Silicon Dioxide)

a-SiO Amorphous Silicon Monoxide

a-SiO$_2$:H Hydrogenated Amorphous Silica (or Silicon Dioxide)

a-SiO:H Hydrogenated Amorphous Silicon Monoxide

a-Si:O Oxygenated Amorphous Silicon

αSiO$_2$ Alpha Silica (or Alpha Silicon Dioxide, or Alpha Quartz)

ASIP Aircraft Structural Integrity Program [US Air Force]

ASIPT Advanced School of ITEP (Institute of Theoretical and Experimental Physics) on Particle Theory [Moscow, Russia]

ASIRC Aquatic Sciences Information Retrieval Center [US]

ASIS Abort Sensing and Instrumentation System

ASIS American Society for Industrial Security [US]

ASIS American Society for Information Science [US]

ASIS Aromatic Solvent Induced Shift

AS/ISES American Section of the International Solar Energy Society [US]

a-SiSn Amorphous Silicon Stannide

a-SiSn:H Hydrogenated Amorphous Silicon Stannide

ASIST Advanced Scientific Instruments Symbolic Translator

ASIT Adaptive Surface Interface Terminal

ASIT Advanced Security and Identification Technology

ASIT Association Suisse d'Inspection Technique [Swiss Technical Inspection Association]

ASIWPCA Association of State and Interstate Water Pollution Control Administrators [US]

A Size Metric Paper Size [Designated by the Letter "A" Followed by a Numeral, e.g., A1, A2, A3, etc.] [also A size]

ASJ All-Service Jacket

ASJ Astronomical Society of Japan

ASK Access to Sources of Knowledge

ASK Agricultural Society of Kenya

ASK Alcohol Server Knowledge (Program) [US]

ASK Amplitude Shift Keying

ASK Anomalous State of Knowledge

ASKA Automatic System for Kinematic Analysis

AsKα Arsenic K-Alpha (Radiation) [also AsK$_\alpha$]

AsKβ Arsenic K-Beta (Radiation) [also AsK$_\beta$]

AskERIC Ask the Educational Resources Information Center [Internet]

As/kg Ampere-second per kilogram [also As kg^{-1}]

ASKI Arbeitskreis Schweizerische Kunststoff-Industrie [Swiss Association of the Plastics Industry, Switzerland]

ASKS Automatic Station Keeping System

ASKT American Society of Knitting Technologists [US]

ASL Above Sea Level

ASL Accademia di Scienze e Lettere [Academy of Science and Letters, Milan, Italy]

ASL Acting Sub-Lieutenant (or Second Lieutenant) [Canadian Forces]

ASL Adaptive Speed Levelling

ASL Aeronautical Structures Laboratory [US Navy]

ASL Albuquerque Seismological Laboratory [of US Geological Survey in New Mexico]

ASL American Sign Language

ASL Analytical Sciences Laboratory [of University of Idaho, US]

ASL Anti-Stokes (Spectral) Line

ASL Association for Symbolic Logic [US]

ASL Association Suisse de Logistique [Swiss Logistics Association]

ASL Atmospheric Sciences Laboratory [US]

ASL Automated Symbols Library

ASL Available Space List

ASL Average Signal Level

ASL Average Staffing Level

AsL Arsenic L (Radiation)

ASLA American Society of Landscape Architects [US]

ASLAP Atomic Safety and Licensing Appeal Panel [US]

ASLAS Australian Society for Laboratory Animal Science

ASLB Atomic Safety and Licensing Board [US]

ASLBM Air-to-Ship Launched Ballistic Missile

ASLBP Atomic Safety and Licensing Board Panel [US]

ASLC Association of Street Lighting Contractors [UK]

ASLE American Society of Lubrication Engineers [now Society of Tribologists and Lubrication Engineers, US]

ASLEEP Automated Scanning Low-Energy Electron Probe

ASLEF Amalgamated Society of Locomotive Engineers and Firemen [also ASLE&F, UK]

ASLE Prepr American Society of Lubrication Engineers Preprint [also ASLE Prpt]

ASLE Trans American Society of Lubrication Engineers Transactions [US]

ASLG Australian Surveying and Land Group

ASLHA American Speech, Language and Hearing Association [US]

ASLIB Association for Information Management [UK]

ASLIB Association of Special Libraries [US]

ASLIB Association of Special Libraries and Information Bureau [US]

ASLIB Inf ASLIB Information [of Association of Information Management, UK]

ASLIB Proc ASLIB Proceedings [of Association of Information Management, UK]

ASLIC Association of Special Libraries and Information Centers [India]

ASLM American Society of Laser Medicine

ASLM Apple (Macintosh) Shared Library Manager

ASLMS American Society of Laser Medicine and Surgery [US]

ASLO American Society of Limnology and Oceanography [US]

ASLO Australian Scientific Liaison Office

ASLP Association of Special Libraries in the Philippines

ASLRA American Short Line Railroad Association [US]

ASLS American Society of Laser Surgery

ASLT Advanced Solid Logic Technology

ASLt Acting Sub-Lieutenant (or Second Lieutenant) [Canadian Forces]

ASLUC All Scrap with Lump Coal (Process) [of Sumitomo Metal Industries, Japan]

ASM Abrasive Slurry Machining

ASM Acoustic Surface Measurement

ASM Advanced Surface-to-Air Missile

ASM Airfield Surface Movement (Indicator)

ASM Airline Station Manager

ASM Airport Surface Movement (Indicator)

ASM Airspace Management

ASM Air-to-Ship Missile

ASM Air-to-Surface Missile

ASM Alaska State Museum [Juneau, US]

ASM Algorithmic State Machine

ASM Algorithmic State Matrix

ASM All-Sky Monitor

ASM Amalgamated Steel Mills [Malaysia]

ASM American Samoa [ISO Code]

ASM American Society for Materials [formerly American Society for Metals; now ASM International, US]

ASM American Society for Metals [later American Society for Materials; now ASM International, US]

ASM American Society for Microbiology [US]

ASM American Society of Mammologists [US]

ASM Analog Subscriber Module

ASM Analyze System Manual

ASM Angular Second Momentum

ASM Anion Selective Membrane

ASM Apollo Service Module [of NASA]

ASM Apollo Systems Manual [of NASA]

ASM Application-Specific Module

ASM Application Support and Management

ASM Argon Secondary Metallurgy

ASM Arizona State Museum [Tucson, US]

ASM Artificially Structured Material(s)

ASM Assembler [also Asm, or asm]

ASM Association for Systems Management [US]

ASM Association Patronale Suisse de l'Industrie des Machines [Association of Swiss Engineering Employers]

ASM Asynchronous State Machine

ASM Audiosonometer; Audiosonometry

ASM Auxiliary Storage Manager

ASM Auxiliary Storage Medium

ASM Available Seat Miles

.ASM Assembler Source Language [File Name Extension] [also .asm]

AS&M Analytical Services and Materials, Inc. [US]

Asm Assembler [also asm]

As m Ampere-second meter [also As-m, or As·m]

As/m^2 Ampere-second per square meter [also As m^{-2}]

As/m^3 Ampere-second per cubic meter [also As m^{-3}]

αSm Alpha Samarium [Symbol]

$\alpha\Sigma\mu$ Alpha Sigma Mu (Metallurgy Student Society)

ASMA Aerospace Medical Association [US]

ASMA Antarctic Specially Managed Area

as-MAed As Mechanically Alloyed (Condition)

ASM-ATAC ASM International–Advisory Technical Awareness Council [US]

ASMB Acoustical Standards Management Board [of American National Standards Institute, US]

ASMC Automatic Systems Management and Control

ASMC Aviation Surface Material Command [US Army]

ASM CDTF ASM International Chapter Development Task Force [US]

ASM CMS ASM International Chapter Management Seminar [US]

ASMD Anti-Ship Missile Defense

ASMD Association of Science Museum Directors [US]

ASME American Society of Mechanical Engineers [now ASME International]

ASMEA American Society of Mechanical Engineers, Auxiliary [US]

ASME-AMD American Society of Mechanical Engineers–Applied Mechanics Division [also ASME AMD]

ASME-AMD Ice Mech American Society of Mechanical Engineers–Applied Mechanics Division Ice Mechanics [Publication]

ASME EC American Society of Mechanical Engineers Publication EC [US]

ASMED Association Suisse des Fournisseurs de Produits Médicaux [Association of Swiss Medical Suppliers]

ASME-ETCE American Society of Mechanical Engineers–Energy-Sources Technology Conference and Exhibition [US]

ASME FACT American Society of Mechanical Engineers Publication FACT [US]

ASME FED American Society of Mechanical Engineers Publication FED [US]

ASME HTD American Society of Mechanical Engineers Publication HTD [US]

ASMEIGTI American Society of Mechanical Engineers International Gas Turbine Institute [US]

ASME Int'l American Society of Mechanical Engineers International [formerly ASME]

ASME Proc American Society of Mechanical Engineers Proceedings

ASME Publ EC American Society of Mechanical Engineers Publication EC [US]

ASME Publ FACT American Society of Mechanical Engineers Publication FACT [US]

ASME Publ FED American Society of Mechanical Engineers Publication FED [US]

ASME Publ HTD American Society of Mechanical Engineers Publication HTD [US]

ASMER Association for the Study of Man-Environment Relations [US]

ASM/EMAA ASM International/Engineering Materials Achievement Award [US]

ASM/ESD ASM International/Engineering Society of Detroit [US]

ASME/STLE Conf American Society of Mechanical Engineers/Society of Tribologists and Lubrication Engineers Conference [US]

ASME Trans, J Tribol American Society of Mechanical Engineers Transactions, Journal of Tribology [US]

ASMFC Atlantic States Marine Fisheries Commission [US]

ASMFE Association of Soil Mechanics and Foundation Engineering

ASM-FEMS ASM International–Federation of European Materials Societies (Liaison Committee)

ASMFER ASM International–Foundation for Education and Research [US]

ASM GPAC ASM International Government and Public Affairs Committee [US]

ASM HTS ASM International Heat Treating Society [US]

ASMI Airfield Surface Movement Indicator; Airport Surface Movement Indicator

ASMI American Society for Materials International [formerly ASM, US]

ASMI Association of Singapore Marine Industries

ASMIC Australian Surveying and Mapping Industries Council

ASM/IIM ASM International/Indian Institute of Materials (Joint Project)

ASM/IMS ASM International/International Metallographic Society (Program)

ASM Int'l American Society for Materials International [formerly ASM; also ASMI]

ASMM American Society of Machine Manufacturers [US]

ASMMA American Supply and Machinery Manufacturers Association [US]

ASM Mat.DB ASM International Materials Database

ASM-MEI ASM International Materials Engineering Institute [US]

ASM-MSD ASM International Materials Science Division [US]

ASM/NIST ASM International/National Institute of Standards and Technology (Joint Project)

ASMO Arab Organization for Standardization in Metrology [Jordan]

ASMOD Adaptive Spline Modelling of Observation Data (Algorithm)

ASMOS (*Automatische Senderegie Mono oder Stereo*) – German for "Automatic Program Continuity Control–Mono, or Stereo"

ASMP Asymmetric Multprocessing

ASM Q American Society of Metals Quarterly [also ASM Quart, Am Soc Met Q, or Am Soc Met Quart]

ASMR Advanced Short-Takeoff-and-Landing, Medium Range

ASMR Australian Society for Medical Research

ASMS Advanced Surface Missile System

ASMS American Society for Mass Spectroscopy [US]

ASMS Automatic Search Music System

ASM-SED ASM International Surface Engineering Division [US]

ASMT Air-Space Multiple-Twin (Cable)

ASMT American Society of Medical Technologists [US]

Asmt Assortment [also asmt]

ASM TDB ASM International Technical Divisions Board [US]

ASM/TMS ASM International/The Minerals, Metals and Materials Society (Meeting) [US]

ASM Trans Q American Society of Metals Transactions Quarterly [also ASM Trans Quart]

ASM TSD ASM International Thermal Spray Division [US]

ASM TSS ASM International Thermal Spray Society [US]

As m/V/m Ampere-second meter per Volt per meter [also As m/Vm^{-1}, As m^2/V, or As m^2 V^{-1}]

ASN Abstract Syntax Notation [Internet]

ASN Advanced Silicon Nitride

ASN American Society of Naturalists [US]

ASN Army Serial Number [US]

ASN Atlantic Satellite Network

ASN Autonomous System Number

ASN Average Sample Number

ASN.1 Abstract Syntax Notation One [Internet]

Asn Arthrobacter species NCM [Microbiology]

Asn (+)-Asparagine; Asparaginyl [Biochemistry]

αSn Alpha Tin [Symbol]

ASNA American SMR (Special Mobile Radio) Network Association [US]

ASNDT American Society for Nondestructive Testing [now American Society for Nondestructive Testing, US]

ASNE American Society of Naval Engineers [Alexandria, Virginia, US]

ASNE American Swedish News Exchange

ASNT American Society for Nondestructive Testing [US]

ASO American Society of Ocularists [US]

ASO Anti-Submarine Operations [US Navy]

ASO Applied Science Office

ASO Area of Safe Operation

ASO Australian Space Office

ASO Australian Survey Office

ASO Automated Systems Operations

ASO Auxiliary Switch, (Normally) Open

ASO Silver Selenium Oxide

As$_2$O$_3^-$ Arsenite [Symbol]

ASOC Adaptive Self-Organizing Computer Solutions

ASOC Antarctica and Southern Oceans Coalition [US]

ASODDS ASWEPS (Anti-Submarine Warfare Environmental Prediction System) Submarine Oceanographic Digital Data System [of US Navy]

ASOE Antarctic and Southern Ocean Environment

ASOE ASEAN (Association of Southeast Asian Nations) Senior Officials on the Environment

As(OEt)$_3$ Arsenic Triethoxide

ASOP Automatic Scheduling and Operating Program

ASOP Automatic Structural Optimization Program

ASOPS American Society of Oral and Plastic Surgeons [US]

ASOR American School of Oriental Research [Jerusalem, Israel]

ASOS Airport Surface Observing System

ASOS (International Colloquium on) Atomic Spectra and Oscillator Strengths

ASOS Automatic Storm Observation Service

ASOS Automated Surface Observing System [of National Weather Service, US]

ASOVAC Asociación Venezolana para el Avance de la Ciencia [Venezuelan Association for the Advancement of Science]

ASOVII Asociación Venezolana de Ingenieros Industriales [Venezuelan Association for Industrial Engineers]

ASP Active Server Page [Internet]

ASP Adaptive Signal Processing

ASP Adjusted Selling Price

ASP Advanced Signal Processor

ASP Aerodynamic Source of Power

ASP Airborne SAR (Synthetic Aperture Radar) Processor

ASP Airborne Signal Processor

ASP Alcoa Smelting Process

ASP Alloy Steel Plant

ASP Aluminosilicophosphate

ASP Aluminum Silicate Pigment

ASP Air-Speeded Post

ASP American Selling Price

ASP American Society for Photobiology [US]

ASP American Society of Parasitologists [US]

ASP American Society of Perfumers [US]

ASP American Society of Pharmacology [US]

ASP American Society of Photogrammetry [now American Society of Photogrammetry and Remote Sensing, US]

ASP American Society of Photographers [US]

ASP Ammunition Supply Point

ASP Analog Signal Processing [also asp]

ASP Analytical Statistics Printer

ASP Antenna Signal Processing

ASP Anti-Segregation Process

ASP Applied Science Publications [London, UK]

ASP Army Supply Point

ASP Asea-Stora Process

ASP Aspheric (Lens)

ASP Association of Service Providers [UK]

ASP Association of Shareware Professionals [US]

ASP Associative Storing Processor

ASP Associative Structures Package

ASP Astronomical Society of the Pacific [San Francisco, California, US]

ASP Asymmetric Multiprocessing System

ASP Attached Support Processor

ASP Attitude Sensor Package

ASP Audio Signal Processing

ASP Auger Sputtering Profile [Physics]

ASP Authorized Service Provider

ASP Automatic Schedule Procedure

ASP Automatic Servo Plotter

ASP Automatic Synthesis Program

ASP Average Selling Price

.ASP Active Server Page [File Name Extension] [also .asp]

A/SP Auto/Steel Partnership [US Consortium]

AsP Arsenic Phosphide

Asp Asparagine Acid [Biochemistry]

Asp (+)-Aspartic Acid; Aspartyl [Biochemistry]

ASPA American Society for Personnel Administration [US]

ASPA American Sod Producers Association[US]

ASPA Antarctic Specially Protected Area

ASPA Association of South Pacific Airlines [Fiji]

ASPA Australian Sugar Producers Association

Aspa Aspartic Acid [Biochemistry]

ASPAC Asian and Pacific Council

ASPA-I American Society for Personnel Administration International [US]

Aspartame Aspartylphenylalanine Methylester [also aspartame]

ASPB Asian Student Press Bureau

ASpB Arbeitsgemeinschaft der Spezialbibliotheken [Association of Special Libraries, Germany]

ASPC Air-Space Paper-Core (Cable)

ASPC American Sheep Producers Council [US]

ASPCA American Society for the Prevention of Cruelty to Animals

ASPD American Society of Podiatric Dermatology [US]

AsPdSe Arsenic Palladium Selenide

ASPE Advances in Solar Physics Euroconference

ASPE American Society of Plastic Engineers [US]

ASPE American Society of Plumbing Engineers [US]

ASPE American Society of Professional Ecologists

ASPE American Society of Professional Estimators [US]

ASPEA Association Suisse pour l'Energie Atomique [Swiss Association for Atomic Energy]

ASPEC Association of Sorbitol Producers in the European Community [Belgium]

ASPEC Automated Solid-Phase Extraction on Disposable Columns

ASPEI Association of South-Pacific Environmental Institutions

ASPEICP Associated Schools Project in Education for International Cooperation and Peace [of UNESCO]

ASPEN Advanced System for Process Engineering

ASPEN Advanced System for Project Engineering

ASPEN Arctic Ship(ping) Probability Evaluation Network [Canada]

ASPEN Asian Physics Education Network

ASPENPLAN Asian and Pacific Energy Planning Network [Malaysia]

ASPEP Association of Scientists and Professional Engineering Personnel [US]

ASPER Assembly Program for Peripheral Process

ASPER Assembly Program for Peripheral Processors

Asph Asphalt [also asph]

ASPI Advanced SCSI (Small Computer Systems Interface) Programming Interface

ASPI American Society for Performance Improvement [US]

ASPI Asynchronous/Synchronous Programmable Interface

ASPIC Author's Standard Prepress Interface Code

ASPIL American Standard Practice for Industrial Lighting

ASPIP Arab Society for the Protection of Industrial Property

ASPIRE Aluminum Self-Propagating Interfacial Reaction [Metallurgy]

ASPJ Advanced Self-Protective Jammer

ASPJ Airborne Self-Protective Jammer

ASPM Association of Sanitary Protection Manufacturers [UK]

ASPN American Society of Precision Nailmakers [US]

Asp(NH2) Asparagine [Biochemistry]

ASPO American Society of Planning Officials [US]

ASPO Apollo Spacecraft Project Office [of NASA]

ASPOE American Society of Petroleum Operations Engineers [US]

ASPP Alloy-Steel Protective Plating

ASPP American Society of Picture Professionals [US]

ASPP American Society of Plant Physiologists [US]

ASPP Auto/Steel Partnership Program [US]

ASPPD Bis(anisaldehyde)-p-Phenylenediamine

Asp-Phe Aspartic Acid-Phenylalanine [Biochemistry]

ASPP Newsl ASPP Newsletter [of American Society of Plant Physiologists]

ASPPR Arctic Shipping Pollution Prevention Regulations [Canada]

ASPPRC Advanced Steel Processing and Products Research Center [of Colorado School of Mines, Golden, US]

ASPR American Society for Psychical Research Inc. [US]

ASPR Armed Services Procurement Regulations [US]

ASPR Average Specific Polymerization Rate

ASPRC Advanced Steel Processing and Products Research Center [of Colorado School of Mines, Golden, US]

ASPRI American Society for Psychical Research, Inc. [US]

ASPRS American Society for Photogrammetry and Remote Sensing [US]

ASPRS American Society of Plastic and Reconstructive Surgeons [US]

ASPS Advanced Signal Processing System

ASPS African Succulant Plant Society

ASPS Argon-Shrouded Plasma Spraying

ASPT Adaptive Signal Processing Testbed

ASPT American Society of Plant Taxonomists [US]

ASPT Newsl ASPT Newsletter [of American Society of Plant Taxonomists, US]

as-Q As-Quenched (Condition)

ASQC American Society for Quality Control [US]

ASQDE American Society of Questioned Document Examiners [US]

ASR Academy of Sciences of Russia

ASR Accumulator Shift Right

ASR Active Status Register

ASR Address Shift Register

ASR Address Space Register

ASR Aerospace Sponsored Research

ASR Airborne Search Radar [also asr]

ASR Airborne Surveillance Radar

ASR Airport Surveillance Radar [also asr]

ASR Air Search Radar

ASR Air-Sea Rescue

ASR Alberta Sulfur Research (Process)

ASR Alkali-Silica Reaction; Alkali-Silica Reactivity [Concrete]

ASR Alkylene Sulfide Rubber

ASR Altimeter Setting Region

ASR Answer, Send and Receive

ASR Antiproton Storage Ring [Physics]

ASR Aqueous Solution Reaction

ASR Arithmetic Shift Right (Function)

ASR Atomic Size Ratio

ASR Automatic Send and Receive; Automatic Send-Receive

ASR Automatic Speech Recognition

ASR Automatic Sprinkler Riser

ASR Automobile Shredder Residue

ASR Available Supply Rate

αSr Alpha Strontium [Symbol]

ASRA American Society of Refrigerating Engineers [US]

ASRA Automatic Stereophonic Recording Amplifier

ASRAAM Advanced Short-Range Air-to-Air Missile [also AMRAM]

ASRC Alabama Space and Rocket Center [Huntsville, US]

ASRC Atmospheric Sciences Research Center [of State University of New York at Albany, US]

ASRCT Applied Scientific Research Corporation of Thailand

ASRD Astronomy, Space and Radio Division [of Science and Engineering Research Council , UK]

ASRE Admiralty Signal and Radar Establishment [UK]

ASRE Admiralty Signal Research Establishment [UK]

ASRE American Society of Refrigeration Engineers [now American Society of Heating, Refrigeration and Air Conditioning Engineers]

ASRET Affiliation of Societies Representing Engineers and Technicians [South Africa]

ASRF American Seed Research Foundation [US]

ASR Gel Alkali Silica Reaction Gel [also ASR gel]

ASRI Academy of Sciences Research Institute [Accra, Ghana]

ASRL Aeroelastic and Structures Research Laboratory [of Massachusetts Institute of Technology, Cambridge, US]

ASRLO Australian Scientific Research Liaison Office

ASRM Advanced Shuttle Rocket Motor [of NASA]

ASRM Advanced Solid Rocket Motor

ASRM American Association of Range Management [US]

Asrn Assurance [also asrn]

ASRO Atomic Short-Range Order(ing) [Solid-State Physics]

ASROC Antisubmarine Rocket [also Asroc]

ASRPA Army Signal Radio Propagation Agency [US]

ASRR Academy of the Socialist Republic of Romania

ASRRC Australian Special Rural Research Council [now Rural Industry Research and Development Corporation]

ASRS Amalgamated Society of Railway Servants

ASRS Automated Storage and Retrieval System [also AS/RS]

ASRS Automated Support Requirements System

ASRS Automatic (or Automated) Seat Reservation System

ASRS Automatic Storage and Retrieval System [also AS/RS]

ASRS Aviation Safety Reporting System [of Eurocontrol]

as-RS as-rapidly solidified

ASRT Aperture Synthesis Radio Telescope

ASRWPM Association of Semi-Rotary Wing Pump Manufacturers

ASS Acetic Acid–Salt Spray (Test)

ASS Acoustical Society of Scandinavia

ASS Aerospace Support System [of Aerospace and Electronic Systems Society, US]

ASS Aerospace Surveillance System

ASS African Sleeping Sickness

ASS Air Stage Service

ASS Air Supply System

ASS American Statistical Society [US]

ASS Analog Switching Subsystem

ASS Applied Surface Science (Journal)

ASS Assembler [also Ass, or ass]

ASS Associate in Secreterial Science

ASS Associate in Secreterial Studies

ASS Austenitic Stainless Steel

ASS Automatic Selector Switch

AsS Arsenosilicate

Ass Assembler [also ass]

Ass Assistant [also ass]

Ass Assurance [also ass]

αSS Ferritic Stainless Steel

ASSA Association for the Study of Snow and Avalanches [France]

ASSA Astronomical Society of South(ern) Africa

ASSA Austrian Solar and Space Agency

ASSAP Association for the Scientific Study of Anomalous Phenomena [UK]

ASSASSIN Agricultural System for Storage and Subsequent Selection of Information

ASSBT American Society of Sugar Beet Technologists [US]

αSSC Alpha(-Phase) Sintered Silicon Carbide

ASSD All-Solid-State Device

ASSE American Society of Safety Engineers [US]

ASSE American Society of Sanitary Engineers [US]

ASSE American Society of Swedish Engineers [US]

AS-SEED Application-Specific Self-Electrooptic Effect Device

Assem Assembly [also assem]

Assem Assoc Assembly and Association [UK Journal] [also A&A]

Assem Autom Assembly Automation [UK Journal]

ASSESS Analytical Studies of Surface Effects of Submerged Submarines

ASSET Aerothermodynamic (Elastic) Structural Systems Environment(al) Test

ASSET American Society of Scientific and Engineering Translators [US]

ASSET Association of Supervisory Staffs, Executives and Technicians [UK]

ASSG American Association for the Study of Goiter [US]

AsSG Arsenosilicate Glass

ASSGP Association Suisse des Fabricants de Spécialités Grand Public [Swiss Association of Manufacturers of Non-Prescription Medicine]

αSSiC Alpha(-Phase) Sintered Silicon Carbide

ASSIGN Association of Industrial Graphics and Nameplate Manufacturers [UK]

AssignCrt Associate Text File with Cathode-Ray Tube [Pascal Function]

ASSILEC Association de l'Industrie Laitière de la Communauté Européenne [European Community Dairy Trade Association, France]

Assim Assimilation [also assim]

ASSL Advanced Solid-State Laser

ASSLC Advanced Solid-State Laser Conference

ASSM Association of Shopfront Section Manufacturers [UK]

ASSMC Aligned Short-Fiber Sheet Molding Compound

ASSMC Association of State Supervisors of Mathematics [US]

Assn Association [also assn]

Assoc Associate; Association [also assoc]

assocd associated [also assoc'd]

Assocn Association [also assocn]

Assoc Off Anal Chem J Association of Official Analytical Chemists Journal [US]

Assoc Prof Associate Professor

Assoc Sc Associate in Science

ASSP Acoustics, Speech and Signal Processing Society [of Institute of Electrical and Electronic Engineers, US]

ASSP Aerosol Scattering Spectrometer Probe

ASSP Atomic and Solid-State Physics

ASSP Transport Submarine [US Navy Symbol]

ASSP Mag ASSP Magazine [of IEEE Acoustics, Speech and Signal Processing Society, US]

ASSR Airborne Sea and Swell Recorder

ASSR Airport Surface Surveillance Radar

ASSR Autonomous Soviet Socialist Republic

ASSS Aerospace Systems Safety Society

ASSS African Soil Science Society

ASSS Air Special Staff School [US Air Force]

ASSS Australian Society of Soil Science

ASSSDE Association of State Supervisors of Safety and Driver Education

ASST Advanced Supersonic Transport

ASST American Society for Steel Treating [later American Society for Metals; now ASM International, US]

Asst Assessment [also asst]

Asst Assistant [also asst]

asstd assorted [also asst'd]

Asst Prof Assistant Professor

Asst Res Prof Assistant Research Professor

ASSU Air Support Signal Unit [UK]

Assy Assembly [also assy]

Assyr Assyria(n)

AST Above-Ground Storage Tank

AST ACE (Arctic Cloud Experiment) Science Team

AST Activation Strain Tensor

AST Adjustable Spatula Temple

AST Advanced Simulation Technology

AST Advanced Supersonic Technology

AST Advanced Supersonic Transport

AST Advanced Surface Technology

AST Aerospace Technologist; Aerospace Technology

AST Alanine Serum Transaminase [Biochemistry]

AST Anti-Sidetone [Telecommunications]

AST Apparent Solar Time [Astronomy]

AST Applied Statistical Thermodynamics

AST ARM (Atmospheric Radiation Measurement) Science Team [of NASA]

AST ARESE (ARM Enhanced Shortwave Experiment) Science Team [of NASA]

AST Army Satellite Tracking Center [US]

AST Aspartate Aminotransferase [Biochemistry]

AST Associates of Science and Technology

AST Astronomical Society of Tasmania

AST Astro-Star Tracker

AST Atlantic Standard Time [Greenwich Mean Time +04:00]

AST Atomic Site Tunneling

AST Automatic Shop Tester

AST Auxiliary Segment Table

ASTA Aerial Surveillance and Target Acquisition

ASTA American Seed Trade Association [US]

ASTA American Surgical Trade Association [US]

ASTA Association of Short-Circuit Testing Authorities

ASTA Auckland Science Teachers Association [New Zealand]

AStA Allgemeiner Studenten-Ausschuss [General Committee of Students, Germany]

AS3AP ANSI (American National Standards Institute) SQL (Structured Query Language) Standard Scalable and Portable [also AS³AP]

ASTAS Antiradar Surveillance and Target Acquisition System

ASTAS Australian Sponsored Training and Scholarships

ASTC Advanced Semiconductor Technology Center [of IBM Corporation, US]

ASTC Airport Surface Traffic Control

ASTC Association of Science and Technology Centers [US]

ASTD American Society for Training and Development [US]

ASTDS Auger Spectroscopy Thermal Desorption Spectra

ASTE American Society of Test Engineers [US]

ASTE American Society of Tool Engineers [US]

ASTE Association for Software Testing and Evaluation [US]

ASTE Bull ASTE Bulletin [of American Society of Test Engineers, US]

ASTEC Advanced Solar Turboelectric Conversion

ASTEC Advanced Systems Technology

ASTEC Antisubmarine Technical Evaluation Center [US]

ASTEC Australian Science and Technology Council

ASTEM Analytical Scanning Transmission Electron Microscopy

ASTER Advanced Spaceborne Thermal Emission and Reflection (Radiometer)

ASTER Antisubmarine Terrier Missile

ASTE Rev ASTE Review [of Association for Software Testing and Evaluation, US]

ASTEX Atlantic Stratocumulus Transition Experiment

ASTeX Applied Science and Technology, Inc. [Woburn, Massachusetts, US]

ASTF Aero-Propulsion Systems Test Facility [US Air Force]

ASTG Aerospace Test Group

AST/GOT Aspartate Aminotransferase/Glutamic-Oxalacetic Transaminase [Biochemistry]

ASTHO Association of State and Territorial Health Offices [US]

ASTI Applied Science and Technology Index

ASTI Area of Special Tourist Interest

ASTI Association for Science, Technology and Innovation [US]

ASTIA Armed Services Technical Information Agency [now Defense Documentation Center, US]

ASTI Newsl ASTI Newsletter [of Association for Science, Technology and Innovation, US]

ASTIC Arctic Science and Technology Information Center

ASTIIT American Society for the Technion Israel Institute of Technology [US]

ASTINFO Asian Scientific and Technological Information Network [Thailand]

ASTINFO Scientific and Technical Information in Asia [of UNESCO]

ASTIS Arctic Science and Technology Information Service [of Arctic Institute of North America, University of Calgary, Alberta, Canada]

ASTIS Australian Science and Technology Information Service

ASTL American Society of Transportation and Logistics [US]

ASTM American Society for Testing and Materials [US]

ASTM American Society for Tropical Medicine [US]

ASTM American Standard Test Method

ASTM COP American Society for Testing and Materials Committee on Publication [US]

ASTM COS American Society for Testing and Materials Committee on Standards [US]

ASTM COT American Society for Testing and Materials Committee on Terminology [US]

ASTME American Society of Tool and Manufacturing Engineers [now Society of Manufacturing Engineers, US]

ASTMH American Society of Tropical Medicine and Hygiene [US]

ASTM ISR American Society for Testing and Materials Institute for Standards Research [US]

ASTM J Compos Technol Res ASTM Journal of Composites Technology and Research [of American Society for Testing and Materials] [ASTM J Comp Technol Res]

ASTM J Test Eval ASTM Journal of Testing and Evaluation [of American Society for Testing and Materials]

ASTM-NBS-NSF American Society for Testing and Materials–National Bureau of Standards–National Science Foundation (Symposium)

ASTM-NIST-NSF American Society for Testing and Materials–National Institute of Standards and Technology–National Science Foundation (Symposium)

ASTM/NRWG American Society for Testing and Materials/Neutron Radiography Working Group [US]

ASTMS Association of Scientific, Technical and Managerial Staffs [UK]

ASTM Spec Techn Publ American Society of Testing and Materials Special Technical Publication [also ASTM STP, or ASTM-STP]

ASTM Stand News American Society of Testing and Materials Standardization News

ASTM STP American Society for Testing and Materials Special Technical Publication [also ASTM-STP, or ASTM Spec Tech Publ]

ASTO Arkansas State Tourist Office [US]

ASTO Association of Sea Training Organizations [UK]

as tol as tolerated [Medicine]

ASTOR Anti-Ship Torpedo

ASTOVL Advanced Short Takeoff and Vertical Landing (Aircraft)

ASTP Advanced Space Transportation Program [of NASA]

ASTP Advanced Systems and Technology Program

ASTP Apollo-Soyuz Test Project [of NASA]

ASTP Army Specialized Training Program

Astr Astronomer; Astronomical; Astronomy [also astr]

ASTRA Application of Space Techniques Relating to Aviation

ASTRA Association in Scotland to Research into Astronautics

ASTRA Automatic Scheduling with Time-Integrated Resource Allocation

ASTRAC Arizona Statistical Repetitive Analog Computer

ASTRAL Analog Schematic Translator to Algebraic Language

ASTRAL Assurance and Stabilization Trends for Reliability by Analysis of Lots

ASTREC Atomic Strike Recording System [US Air Force]

αStrep Alpha Streptococcus Group (or Streptococcus viridans Group) [Microbiology]

ASTRO Advanced Spacecraft Transport Reusable Orbiter

ASTRO Aerodynamic Spacecraft Two-Stage Reusable Orbiter

ASTRO Airspace Travel Research Organization [US]

ASTRO Arctic Stratospheric Ozone

ASTRO (International) Association of State Trading Organizations (of Developing Countries)

ASTRO Ultraviolet Astronomy Observatory [of NASA]

AST/RO Antarctic Submillimeter Telescope and Remote Observatory [of University of Chicago, US]

Astrobiol Astrobiologist; Astrobiology [also astrobiol]

Astrochem Astrochemical; Astrochemist(ry) [also astrochem]

Astrochron Astrochronological; Astrochronologist; Astrochronology [also astrochron]

Astrodyn Astrodynamic(s) [also astrodyn]

Astrofiz Issled-Izv Spets Astrofiz Obs Astrofizicheskie Issledovaniya–Izvestiya Spetsial'noi Astrofizicheskoi Observatorii [Russian Journal on Astrophysical Research–Special Investigations of the Astrophysical Observatory]

Astrogeod Astrogeodesy; Astrogeodetic [also astrogeod]

Astrogeol Astrogeological; Astrogeologist; Astrogeology [also astrogeol]

Astrol Astrologer; Astrological; Astrology [also astrol]

Astrom Astrometric; Astrometry [also astrom]

Astron Astronomer; Astronomical; Astronomy [also astron]

Astronaut Astronautics [also astronaut]

Astron Astrophys Astronomy and Astrophysics [Journal published in Germany]

Astron Astrophys Suppl Ser Astronomy and Astrophysics Supplement Series [Published in France]

Astronav Astronavigation(al) [also astronav]

Astron Circ (Acad Sin) Astronomical Circulars (Academia Sinica) [PR China]

Astron Geophys Astronomy and Geophysics [Magazine of the Royal Astronomical Society, UK]

Astron Her Astronomical Herald [of Astronomical Society of Japan]

Astron J Astronomical Journal [of American Astronomical Society, US]

Astron Letters Astronomy Letters [Russian Translation Journal jointly published by the American Institute of Physics (US) and Maik Nauka (Russia)]

Astron Nachr Astronomische Nachrichten [Astronomical News; published in Germany]

Astron Now Astronomy Now [UK]

Astronomer The Astronomer [UK]

Astron Rep Astronomy Reports [Russian Translation Journal jointly published by the American Institute of Physics (US) and MAIK Nauka (Russia)]

Astron Tidsskr Astronomisk Tidsskrift [Swedish Astronomical Journal]

Astron Vestn Astronomicheskii Vestnik [Russian Journal on Solar System Research]

Astron Zh Astronomicheskii Zhurnal [Russian Astronomical Journal]

Astrophys Astrophysical; Astrophysicist; Astrophysics [also astrophys]

Astrophys J Astrophysical Journal [of American Astronomical Society, US]

Astrophys J Lett Astrophysical Journal Letters [of American Astronomical Society, US]

Astrophys J, Lett Ed Astrophysical Journal, Letters to the Editor [of American Astronomical Society, US]

Astrophys J Suppl Ser Astrophysical Journal Supplement Series [of American Astronomical Society, US]

Astrophys Lett Commun Astrophysical Letters and Communications [UK]

Astrophys Space Sci Astrophysics and Space Science [Journal published in the Netherlands]

ASTROS Advanced Star and Target Reference, Optical Sensor

ASTSSA Associated Scientific and Technical Societies of South Africa

ASTSWMO Association of State and Territorial Solid Waste Management Officials [US]

ASTT Aeronautics and Space Transportation Technology

ASTT American Society of Traffic and Transportation [now American Society of Transportation and Logistics, US]

ASTTL Advanced Schottky Transistor-Transistor Logic [also AS-TTL, or AS TTL]

ASU Acknowledgement Signal Unit

ASU Analytical Services Unit

ASU Appalachian State University [Boone, North Carolina, US]

ASU Apparatus Slide-In Unit

ASU Arizona State University [Tempe, US]

ASU Arkansas State University [State University, US]

ASU Automatic Switching Unit

AS/U Advanced Server for Unix

Asu αAminosuberic Acid; Aminosuberyl

ASUG American Software Users Group [US]

ASU-HREM Arizona State University–High-Resolution Electron Microscopy (Facility)

ASUN Associated Students of the University of Nevada [US]

ASU-NSF Arizona State University–National Science Foundation (Center)

ASUPA Aluminum Sulfate Producers Association [Belgium]

ASUS Arts and Science Undergraduate Society

ASUSSR Academy of Sciences of the Union of Soviet Socialist Republics [Moskow] [also AS USSR]

ASUT Association Suisse d'Usagers de Télécommunications [Swiss Association of Telecommunications Users] [also asut]

ASUW Antisurface Unit Warfare

ASV Absorptive Stripping Voltametry

ASV Active Service

ASV Airborne-Radar to locate Surface Vessels

ASV Aircraft Searching Apparatus

ASV Air(craft)-to-Surface Vessel (Radar)

ASV Aluminum-Structured Vehicle

ASV Angle Stop Valve

ASV Anodic Stripping Voltage

ASV Anodic Stripping Voltametry

ASV Automatic Self-Verification

ASV Avian Sarcoma Virus

As/V Ampere-second per Volt [also A·s/V, or As V^{-1}]

ASV/CSV Anodic Stripping Voltage/Cathodic Stripping Voltage

ASVIP American Standard Vocabulary for Information Processing

As/V·m Ampere-second per Volt Meter [also As V^{-1}m^{-1}]

ASVP Application System Verification and Transfer Program

ASVS Advanced Space Vision System

ASVS Automatic Signature Verification System

ASVW Antisymmetric Shear Vertical Wave

ASW Acoustic Surface Wave [also asw]

ASW American Steel and Wire

ASW Antisubmarine Warfare [also A/SW, or asw]

ASW Antisubmarine Weapon

ASW Applications Software [also AS]

ASW Arc-Sprayed Wire

ASW Attenuation Sound Wave

ASW Augmented Spherical Wave (Method)

ASW Automatic Socket Weld

ASW Auxiliary Switch

ASWA American Steel Warehouse Association [now Steel Shipping Container Institute, US]

ASWCR Airborne Surveillance, Warning and Control Radar

ASWE Admiralty Surface Weapon Establishment [UK]

ASWEPS Anti-Submarine Warfare Environmental Prediction System; Anti-Submarine Weapons Environmental Prediction System [of US Navy]

ASWG American Steel (and) Wire Gage [also AS&WG]

ASWM Association of State Wetland Managers [US]

ASWORG Antisubmarine Warfare Operations Research Group [US Navy]

ASWS Ammonium Sulfide Water Stripper

ASW/SCCS Antisubmarine Warfare/Ship Command and Control System

ASWSPO Antisubmarine Warfare Systems Project Office [US]

Asx Asparagine [Biochemistry]

Asx Aspartic Acid [Biochemistry]

ASXRED American Society for X-Ray and Electron Diffraction [US]

ASXRT American Society of X-Ray Technicians [US]

ASY Active Sun Year [Astronomy]

ASYM Association of Synthetic Yarn Manufacturers [US]

Asym Asymmetric(al); Asymmetry [also asym]

asym asymmetrical [Chemistry]

Async Asynchronization; Asynchronize; Asynchronous [also Asynch, async, or asynch]

ASZ American Society of Zoologists [US]

ASZD American Society for Zero Defects [now American Society for Performance Improvement, US]

AT Abbe Theory [Physics]

AT Absolute Temperature

AT Absorption Tower

AT Accelerating Tube

AT Acceleration Transducer

AT Acceptance Tag

AT Acceptance Test

AT Access Tandem [Telecommunications]

AT Acid Treated; Acid Treatment

AT Acoustical Tile

AT Acoustic Test(ing)

AT Acoustic Tomography

AT Active Transport [Medicine]

AT Acute Toxicity

AT Address Translation

AT Adenine/Thymidine Medium [Biochemistry]

AT Adhesion Technology

AT Adipose Tissue [Anatomy]

AT Advanced Technology

AT Aerofoil Theory

AT Aerospace Technologist; Aerospace Technology

AT Affine Transformation [Mathematics]

AT Aging Test

AT Agricultural Technologist; Agricultural Technology

AT Agyromeritic Term; Agyroscopic Term

AT Airfoil Theory

AT Air Temperature [also at]

AT Air Terminal

AT Airtight(ness)

AT Air Transport(ation)

AT Alignment Telescope

AT Alcoxotechnology

AT Almeida-Thouless (Line) [Physics]

AT Aluminothermic; Aluminothermy [Metallurgy]

AT Aluminum Titanate

AT Ambient Temperature

AT American Terms [Commerce]

AT Americium Titanate

AT Aminotriazole

AT Ampere-Turn [also at]

AT Analog Transmission

AT Annealed and Tempered; Annealing and Tempering [Metallurgy]

AT Annealing Twin [Metallurgy]

AT Antarctic Treaty

AT Antenna [Joint Army-Navy]

AT Anti-Tank

AT Antitetanus (Serum) [Immunology]

AT Antithrombin [Biochemistry]

AT Anti-Torpedo

AT Antitoxin [Immunology]

AT Applied Technology

AT Approximation Theory

AT Architectural Technician

AT Architectural Technologist; Architectural Technology

AT Arc Transmitter

AT Array Theory

AT Assay Ton [Mining]

AT As Tempered (Condition)

AT Astronomical Telescope

AT Asymmetric(al) Tilt

AT Asynchronous Transmission

AT Atomic Theory

AT Atomic Time

AT Attention

AT Audit Trail

AT Aufbereitungs-Technik [German Journal on Processing Technology]

AT Auger Transition [Physics]

AT Auricular Therapist; Auricular Therapy

AT Ausformed-and-Tempered; Ausforming and Tempering [Metallurgy]

AT Australia Telescope [of Commonwealth Scientific and Industrial Research Organization, Australia]

AT Austria [ISO Code]

AT Austenitic Transformation [Metallurgy]

AT Automated Tool

AT Automatic Ticketing

AT Automatic Translation

AT Automatic Transmission

AT Automotive Technician

AT Automotive Technologist; Automotive Technology

AT Auto-Transductor

AT Auto-Transformer

AT Auto-Throttle

AT Avalanche Threshold [Electronics]

AT Azores Time [Greenwich Mean Time +02:00]

AT Tape-Armoured Aerial Cable

AT-III Antithrombin III [Biochemistry]

A(T) Temperature-Dependent Absorptivity [Symbol]

A&T Agricultural and Technological; Agriculture and Technology

A-T Adenine-Thymine [DNA Research]

A/T Antitank

At Ampere-turn(s) [also A·t]

At Astatine [Symbol]

At Atom(ic) [also at]

At$_2$ Molecular Astatine [Symbol]

At-203 Astatine-203 [also ^{203}At, or At^{203}]

At-204 Astatine-203 [also ^{204}At, or At^{204}]

At-205 Astatine-203 [also ^{205}At, or At^{205}]

At-206 Astatine-203 [also ^{206}At, or At^{206}]

At-207 Astatine-203 [also ^{207}At, or At^{207}]

At-208 Astatine-203 [also ^{208}At, or At^{208}]

At-209 Astatine-203 [also ^{209}At, or At^{209}]

At-211 Astatine-203 [also ^{211}At, At^{211}, or At]

At-212 Astatine-203 [also ^{212}At, or At^{212}]

At-214 Astatine-203 [also ^{214}At, or At^{214}]

At-215 Astatine-203 [also ^{215}At, or At^{215}]

At-216 Astatine-203 [also ^{216}At, or At^{216}]

At-217 Astatine-203 [also ^{217}At, or At^{217}]

At-219 Astatine-203 [also ^{219}At, or At^{219}]

A(t) Modulated Signal [Symbol]

at (technical) atmosphere [Unit]

at% atomic percent [also AT%]

.at Austria [Country Code/Domain Name]

α(T) temperature-dependent absorptivity [Symbol]

α(t) time-dependent crystallized volume fraction (in metallurgy) [Symbol]

ATA Abstracts on Tropical Agriculture [of Royal Tropical Institute, Netherlands]

ATA Academic Travel Abroad [US]

ATA Actual Time of Arrival

ATA Advanced Tactical Aircraft

ATA Advanced Tactical Attack

ATA Advanced Technology Bus Attachment

ATA Advanced Transit Association [US]

ATA Aeronautical Telecommunications Agency

ATA African Technical Association

ATA Air-to-Air

ATA Air Training Command

ATA Air Transport Association [US]

ATA Alberta Trucking Association [Canada]

ATA Alkyl Trifluoro Acetic Acid Ester

ATA Aloe Technology Association [US]

ATA American Telemarketing Association [US]

ATA American Teleport Association [US]

ATA American Topical Association [US]

ATA American Transit Association [US]

ATA American Translators Association [US]

ATA American Transport Association [US]

ATA American Tree Association [US]

ATA American Trucking Association [US]

ATA American Tube Association [US]

ATA 3-Amino-1H-1,2,4-Triazole

ATA Anthranilamide

ATA Arcnet Trade Association [US]

ATA Associate Technical Aide

ATA Association of Technical Artists [US]

ATA Asynchronous Terminal Adapter

ATA Atlantic Treaty Association [of NATO]

ATA Aurintricarboxylic Acid

ATA Automatic Trouble Analysis

ATA Automotive Technicians Association [US]

ata (technical) atmosphere, absolute

ATAA Air Transport Association of America [US]

ATAA Air Transport Auxiliary Association [also ATA Assn, UK]

ATAB Area-Array Tape-Automated Bonding

ATAC Advisory Technical Awareness Council [of ASM International, US]

ATAC Air Transport Advisory Council [US]

ATAC Air Transport Association of Canada

ATAC Army Tank Automotive Center [US Army]

ATAC Australian Transport Advisory Council

AT/AC Alignment Telescope/Autocollimator

ATAE Associated Telephone Answering Exchanges [now Association of Telemessaging Services International, US]

ATAE Automotive Trade Association Executives [US]

ATAF Allied Tactical Air Force [of NATO]

ATAGR Air-to-Air Gunnery Range

ATALA Association pour le Traitement Automatique de Langue [Association for Automatic Language Processing, France]

ATAM Association of Teaching Aids in Mathematics [UK]

ATAMS Advanced Tactical Attack/Manned System

ATAPI Advanced Technology Attachment Packet Interface

ATAR Antitank Aircraft Rocket

ATARC Advanced Technology Alcoa Reduction Cell

ATARS Aircraft Traffic Advisory Resolution System

ATAS Academy of Television Arts and Science [US]

ATAS Advanced Technology Alert System

ATAS Association of Telephone Answering Services [US]

ATB Access Type Base

ATB Access Type Bits

ATB Acetylene-Terminated Bisphenol

ATB Address Translation Buffer

ATB Advanced Technology Bomber

ATB Aerated Test Burner (Number)

ATB Air Transport Board

ATB All Trunks Busy

ATB Aluminum tert-Butoxide

ATB Asphalt Tile Base

ATB Association for Tropical Biology [US]

ATB Asymmetrical Tilt (Grain) Boundary [Materials Science]

ATB Institute für Agrartechnik Bornim e.V. [Bornim Institute for Agricultural Engineering, Potsdam-Bornim, Germany]

a-Tb Alpha Terbium [Symbol]

ATBA Automated Ticket and Boarding (Pass)

ATBA Automatic Test Break and Access

ATBCB Architectural and Transportation Barriers Compliance Board [US]

a-TbCo Terbium-Cobalt Amorphous Alloy

a-TbDyFe Terbium-Dysprosium-Iron Amorphous Alloy

a-TbFe Terbium-Iron Amorphous Alloy

a-TbFeCo Terbium-Iron-Cobalt Amorphous Alloy

a-TbGdCo Terbium-Gadolinium-Cobalt Amorphous Alloy

a-TbGdFe Terbium-Gadolinium-Iron Amorphous Alloy

ATBM Advanced Tactical Ballistic Missile

ATBM Anti-Tactical Ballistic Missile

ATBM Antitank Ballistic Missile

ATBM Articulated Total Body Model

ATB Metall ATB Metallurgie [Metallurgical Journal; published in Brussels, Belgium]

ATBOA Australian Tuna Boat Owners Association

ATC Absolute Time Command

ATC Acoustic Tile Ceiling

ATC Adiabatic Toroidal Compressor [of Princeton University, New Jersey, US]

ATC Advanced Technology Center

ATC Aerial Tuning Capacitor

ATC Aerothermochemistry

ATC Aircraft Technical Committee

ATC Airman, Third Class [also Atc]

ATC Air-Route Traffic Control(ler)

ATC Air Temperature Control(ler)

ATC Air Traffic Control(ler) [also atc]

ATC Air Training Command [at Randolph Air Force Base, Texas, US]

ATC Air Training Corps [US]

ATC Air Transport Command

ATC Air Transport Committee [of International Civil Aviation Organization]

ATC Alcoa Technology (or Technical) Center [of Alcoa Corporation at Alcoa Center, Pennsylvania, US]

ATC Alloy-Tin Couple

ATC American Technical Ceramics [US]

ATC Antenna Tuning Capacitor

ATC Antithymocytic

ATC Applied Technology Center [of Milwaukee School of Engineering, Wisconsin, US]

ATC Applied Technology Council [US]

ATC Armament Training Camp

ATC Armament Training Center

ATC Armored Troop Carrier

ATC Army Topographic Command [US]

ATC Army Training Center

ATC Ashton-Tate Corporation [US]

ATC Assembly Test Chip [Electronics]

ATC Association of Translation Companies [UK]

ATC Australian Telecommunications Commission

ATC Australian Tourist Commission

ATC Australian Trade Commission

ATC Atom Count

ATC Authorization to Copy [Computer Software]

ATC Authorized Training Center

ATC Automatic Temperature Compensation

ATC Automatic Threshold Control

ATC Automatic Through Center

ATC Automatic Timing Control

ATC Automatic Toggle Clamp

ATC Automatic Tool Change

ATC Automatic Train Control

ATC Automatic Tuning Control [also atc]

ATC Automation Training Center [US]

ATC Ethyl Trichloro Silane

a-tC amorphous-tetrahedral carbon

ATCA Advanced Tanker/Cargo Aircraft

ATCA Air Traffic Control Association [US]

ATCA Allied Tactical Communications Agency [of NATO]

ATCA Australian Tuna Canners Association

ATCA Automatic Tuned Circuit Adjustment

ATCAA Automatic Tuned Circuit Adjustment Amplitude

ATCAC Air-Traffic Control Advisory Committee

ATCAP Air Traffic Control Automatic Procedure

ATCAP Air-Traffic Control Automation Panel

ATCAS African Training Center for Agricultural Statistics [of Food and Agricultural Organization]

ATCBI Air Transport Control Beacon Interrogator

ATCC Air-Route Traffic Control Center; Air-Traffic Control Center

ATCC American-Type Culture Collection [Rockville, Maryland, US]

ATCCC Advanced Technical Command Control Capability

ATCCS Air-Traffic Control Communication System

ATCDE Association of Teachers in Colleges and Departments of Education [UK]

ATCE Automatic Test and Checkout Equipment

ATCER Appropriate Technology Center–Environment [Kenya]

ATCF Axial Tube Component Feeder

ATCH Attached [also Atch, or atch]

ATCM Airborne Toxic Control Measures

ATCM Antarctic Treaty Consultative Meeting

at/cm² atoms per square centimeter [Unit]

at/cm³ atoms per cubic centimeter [Unit]

at/cm²s atoms per square centimeter-second [also at/cm²-s] [Unit]

ATCO Air-Traffic Controllers Organization [US]

ATCO Air-Traffic Control Officer

ATCO Association of Transport Coordinating Officers [UK]

ATCOS Atmospheric Composition Satellite [of NASA]

ATCP Antarctic Treaty Consultative Party

ATCP Asociación Mexicana de Tecnicos de las Industrias de la Celulosa y del Papel [Mexican Technical Association of the Cellulose and Paper Industries]

ATCPA Air Taxi and Commercial Pilots Association [US]

ATCPM Advanced Technology Center for Precision Manufacturing [of University of Connecticut, Storrs, US]

ATCRBS Air-Traffic Control Radar Beacon System

ATCS Active Thermal Control System

ATCS Advanced Train Control System

ATCS Air-Traffic Control Service

ATCS Air-Traffic Control Specialist

ATC-SEM Advanced Technology Center for Surface Engineered Materials [of Stevens Institute of Technology, Hoboken, US] [also ATC SEM]

ATCSS Air-Traffic Control Signaling System

ATCT Air-Traffic Control Tower

ATCV Australian Trust for Conservation Volunteers

ATD Acceptance and Takeover Date

ATD Actual Time of Departure

ATD Admission, Transfer and Discharge System

ATD Advanced Technical Device

ATD Advanced Technology Development

ATD Aerospace Technology Division

ATD Aerothermodynamic(s)

ATD Alkali Thermoionization Detector

ATD Along Track Distance [Aeronautics]

ATD Alpha Track Detector

ATD American Truck Dealers [US]

ATD Association of Tar Distillers [UK]

ATD Asynchronous Time Dilation

ATD Atmosphere Technology Division [of National Center for Atmospheric Research, US]

ATD Automatic Threat Detection (System)

ATDA Appropriate Technology Development Association [India]

ATDA Augmented Target Docking Adapter [of NASA]

At Data Nucl Data Tables Atomic Data and Nuclear Data Tables [US]

ATDC Advanced Technology Development Center

ATDC After Top Dead Center

ATDM Asynchronous Time-Division Multiplexing

ATDMA Advanced Time-Division Multiple Access

ATDMA Automotive Tool and Die Manufacturers Association [now Detroit Tooling Association, US]

ATDP Attention Dial Pulse

ATDS Airborne Tactical Data System

ATDS Automatic Telemetry Decommutation System

ATDT Attention Dial Tone

ATE Advanced Technology Engine

ATE Aerothermoelastic(ity)

ATE Air Traffic Engineer

ATE Air-Turbo Exchanger

ATE Altitude Transmitting Equipment

ATE Aluminum Triethyl

ATE Artificial Traffic Equipment

ATE Automated Test Equipment

ATE Automatic Telephone Exchange

ATE Automatic Test Equipment

ATE Automotive Test Equipment

a-Te Amorphous Tellurium

ATEA American Technical Education Association [US]

ATEA Army Transportation Engineering Agency [US]

ATEC Agence Transequatoriale de Communications [Transequitorial Agency for Communications, Central Africa]

ATEC Army Test and Evaluation Command

ATEC Automated Technical Control

ATEC Automatic Test Equipment Complex

ATEC Aviation Technician Education Council [US]

ATech Associate in Technology

ATEE N-Acetyl-L-Tyrosine Ethyl Ester (Monohydrate) [Biochemistry]

*a*TeGe Alpha Tellurium Germanide

ATEGG Advanced Turbine Engine Gas Generator

ATEM Analytical Transmission Electron Microscopy

ATEN Association Technique pour (la Production et l'Utilisation de) l'Energie Nucléaire [Technical Association for Nuclear Energy, France]

At Energ Atomnaya Energiya [Soviet Journal of Atomic Energy]

At Energy Rev Spec Issue Atomic Energy Review, Special Issue [of International Atomic Energy Agency, Austria]

*a*TeO$_2$ Alpha Tellurium Dioxide

ATEP Asian Trade Expansion Program

ATERC Australian Telecommunications and Electronics Research Board

ATES Airborne Trial and Evaluation Section [of Canadian Airborne Center]

*a*TeSn Alpha Tellurium Stannide

ATEWS Advanced Tactical Early Warning System

ATEX Atlantic Tropospheric Experiment

ATF Accelerator Test Facility [of Brookhaven National Laboratory, Upton, New York, US]

ATF Acceptance Test Facility

ATF Acid-Treated Florisil

ATF Advanced Tactical Fighter(plane)

ATF Advanced Technology Fighter(plane)

ATF Advanced Toroidal Facility

ATF Advanced Turbofan

ATF African Tick Fever

ATF Aft Turbofan

ATF (Bureau of) Alcohol, Tobacco and Firearms Bureau [US]

ATF Alternative Transportation Fuels (Program) [of Natural Resources Canada]

ATF American Typecasting Fellowship [US]

ATF Associated Theoretical Function

ATF Association Technique de la Sidérurgie Française [Technical Association of the French Iron and Steel Industry]

ATF Attention To File

ATF Australian Teachers Foundation

ATF Automatic Test Formatter

ATF Automatic Track Finding

ATF Automatic Transmission Fluid

ATF Aviation Turbine Fuel

ATFA Association for Testing and Failure Analysis [US]

ATFAC American Turpentine Farmers Association Cooperative [US]

ATFM Air Traffic Flow Management

ATFMU Air Traffic Flow Management Unit

ATFOS Alignment and Test Facility for Optical Systems

ATFR Automatic Terrain-Following Radar

ATFSC Amorphous Thin-Film Solar Cell

ATFT Additive Thin Film Technology

ATG Advanced Technology Group

ATG Air-to-Ground

ATG Air Transport Group [of Department of National Defense, Canada]

ATG Air Turbine Generator

ATG American Traders Group [US]

ATG Asaro-Tiller-Grinfield (Instability)

ATG Automatic Test Generator

A/TG Antitank Gun

ATGAR Antitank Guided Air Rocket

ATGB Asymmetric(al) Tilt Grain Boundary [Materials Science]

ATGC Asian Textiles and Garments Council

ATGM Antitank Guided Missile

AT-GPIB Advanced Technology/General-Purpose Interface Bus

ATGSB Advanced Test for Graduate Studies in Business

ATH Alumina Trihydrate

ATH Attention Hang-Up [Modem]

*a*Th Alpha Thorium [Symbol]

ATHENA AnTiHydrogEN Apparatus (Experiment) [of CERN–European Laboratory for Particle Physics, Geneva, Switzerland [also Athena]

ATI Above-Threshold Ionization

ATI Access to Information

ATI Aerial Tuning Inductance [also ati]

ATI Air Technical Intelligence [US Air Force]

ATI American Training International [Los Angeles, US]

ATI Amorphous Technologies International [Laguna Niguel, California, US]

ATI Antenna Tuning Inductance [also ati]

ATI Appropriate Technology International

ATI Area-Time Integral

ATI Armored Transportation Institute [US]

ATI Army Training Instruction

ATI Associate of the Textile Institute [UK]

ATI Association of Technical Institutions [UK]

ATI Atmospheric Temperature Inversion

ATI Auckland Technical Institute [New Zealand]

ATI Australian Tourism Index

a-Ti Amorphous Titanium

*a*Ti Alpha Titanium [Symbol]

ATIA Access to Information Act [Canada]

ATIA Australian Tourism Industry Association

αTi₃Al Alpha Titanium Aluminide

a-TiBeSi Titanium Beryllium Silicon Amorphous Alloy

a-TiBeZr Titanium Beryllium Zirconium Amorphous Alloy

ATIBT Association Technique Internationale des Bois Tropicaux [International Technical Tropical Timber Association, France]

ATIC Aerospace Technical Intelligence Center

ATIC Air Technical Intelligence Center [Name Changed to Aerospace Technical Intelligence Center]

ATICCA Australian Tertiary Institutions Consulting Companies Association

ATIF Advanced Technology Innovation Fund [of Industry Canada]

ATIF Australian Timber Importers Federation

ATIG Alternative Technology Information Group [UK]

αTiN Alpha Titanium Nitride

ATIO Association of Translators and Interpreters of Ontario [Canada]

ATiO₃ Perovskite-Type Ceramics [General Formula; A = Barium, Calcium, Strontium]

ATIP Aluminum (III) Isopropoxide

ATIP Analog Tune in Progress

ATIP Association Technique de l'Industrie Papetière [Technical Association of the Paper Industry, France]

ATIPCA Asociación de Tecnicos de la Industria Papelera y Celulosica Argentina [Argentinian Technical Association of the Cellulose and Paper Industries]

ATIS AIDS/HIV Treatment Information Service

ATIS Air Traffic Information Service

ATIS Automatic Terminal Information Service

ATIS Automatic Traffic Information System

ATITA Air Transport Industry Training Association [UK]

ATJ Automatic Through Junction

ATJS Advanced Tactical Jamming System

ATK Antriebstechnisches Kolloquium [Colloquium on Power Transmission Engineering, Germany]

ATK Available Tonne/Kilometer

ATK Aviation Turbine Kerosine

ATL Active Template Library

ATL Aeronautical Turbine Laboratory [of US Navy]

ATL American Tariff League [US]

ATL Analog-Threshold Logic

ATL Appliance Testing Laboratory

ATL Anti-Thrust Law

ATL Artificial Transmission Line

ATL Attitude Timeline

ATL Automated Tape Library

ATL Automatic Tape Layer

Atl Atlantic

αTl Alpha Thallium [Symbol]

A2LA American Association for Laboratory Accreditation [also AALA, or A²LA]

ATLAS Abbreviated Test Language for All Systems

ATLAS Abbreviated Test Language for Avionic Systems

ATLAS Adaptive Test and Logic Analysis System

ATLAS Antitank Laser Assisted System

ATLAS Argonne Tandem Linear Accelerator System [of Argonne National Laboratory, Illinois, US]

ATLAS Association of Atlas Group [France]

ATLAS Atmospheric Laboratory for Applications and Science

ATLAS A Toroidal LHC (Large Hadron Collider) ApparatuS (Experiment) [of CERN–European Laboratory for Particle Physics, Geneva, Switzerland]

ATLAS Automated Tape Lay-Up System

ATLAS Automated Telephone Line Address System

ATLAS Automatic Tabulating, Listing and Sorting (System)

ATLAS Automatic Test Language Specification

ATLB Air Transport Licensing Board [UK]

ATLC Advanced Technology Learning Center [of Texas Technical University, Lubbock, US]

ATLID Atmospheric Lidar [also Adlid, or adlid]

ATLIS Army Technical Libraries and Information Systems [US]

ATLIS Automatic Tracking Laser Illumination System

Atl Mon The Atlantic Monthly [US Journal]

ATLO Assembly, Test, Launch and Operations (Phase) [of NASA Mars Pathfinder Mission]

Atl O Atlantic Ocean

AT LORAN Air-Transportable LORAN (Long-Range Navigational System) Set [also AT Loran, or at loran]

ATLS Australian Transport Literature Information System [of Australian Information Network]

ATLSS Advanced Technology for Large Structural Systems

ATM Adaptive Triangular Mesh [Geographic Information System]

ATM Adobe Typeface Manager

ATM Advanced Technology Management

ATM Advanced Technology Materials [Danbury, Connecticut, US]

ATM Advanced Test Module

ATM Airborne Terrain Mapper [of NASA]

ATM Airborne Thematic Mapper

ATM Air Turbine Motor

ATM Aluminum Trimethyl

ATM Amateur Telescope Maker

ATM Apollo Telescope Mount [of NASA]

ATM Archiv für Technisches Messen [Archive for Metrology, Germany]

ATM Association of Teachers of Mathematics [UK]

ATM Asynchronous Transfer Mode (Technology)

ATM Ataxia-Telangiectasia Mutant (Family) [Genetics]

ATM Automated Teller Machine

ATM Auxiliary Tape Memory

ATM Axilrod-Teller-Moto (Theory)

Atm Atmosphere; Atmospheric(s) [also atm]

At/m Ampere-turn(s) per meter [also A·t/m, or At m⁻¹]

aTm Autologous Tumor Vaccine [Immunology]

atm (standard) atmosphere [Unit]

at/m² atoms per square meter [Unit]

at/m³ atoms per cubic meter [Unit]

αTm Alpha Thulium [Symbol]

ATMA Adhesive Tape Manufacturers Association [UK]

ATMA American Textile Machinery Association [US]

ATMA American Textile Manufacturers Association [US]

ATMAM Analytical Testing Methodologies for Design with Advanced Materials

ATMC Airspace and Traffic Management Center

ATMC Automotive Training Managers Council [US]

atm cm³ (standard) atmosphere cubic centimeter

atm·cm³/s atmosphere cubic centimeters per second [also atm·cm³/sec, or atm·cm³ s⁻¹]

atm·cm³/sec atmosphere cubic centimeters per second [also atm·cm³/s, or atm·cm³ s⁻¹]

Atm-CVD Atmospheric-Pressure Chemical Vapor Deposition [also APCVD, or AP CVD]

ATME Automatic Transmission Measuring Equipment

ATMFORUM Asynchronous Transfer Mode Forum

ATMG Airspace and Traffic Management Group

atm He/div atmospheric helium per division [Unit]

ATMI Advanced Technology Materials, Inc. [US]

ATMI American Textile Manufacturers Institute [US]

ATMOS Atmosphären- und Ozeanforschungs-Satellit [Atmospheric and Ocean Research Satellite, Germany]

ATMOS Atmospheric Trace Molecules Observed by Spectroscopy [of NASA Jet Propulsion Laboratory, Pasadena, California, US]

ATMOS Atmospheric and Ocean Research Satellite [Germany]

ATMOS Atmospheric Trace Molecule Spectroscopy

Atmos Atmosphere; Atmospheric [also atmos]

atmos atmosphere [Unit]

Atmos Chem Atmospheric Chemist(ry)

Atmos Environ Atmospheric Environment [UK Journal]

Atmos Environ A, Gen Top Atmospheric Environment, Part A, General Topics [UK Journal]

Atmos Environ B, Urban Atmos Atmospheric Environment, Part B, Urban Atmosphere [UK Journal]

Atmos Ocean Opt Atmospheric and Oceanic Optics [US Journal]

Atmos Phys Atmospheric Physicist; Atmospheric Physics

Atmos Ocean Atmosphere and Ocean [US Journal]

Atmos Res Atmospheric Research [Dutch Journal]

Atmos Technol Atmospheric Technologist; Atmospheric Technology

Atmos Technol Atmospheric Technology [Journal of the National Center for Atmospheric Research, US]

ATMR Advanced Technology Medium-Range

ATMRT Advanced Technology Medium-Range Transport

ATMS Advanced Text Management System

ATMS Association of Telephone Messaging Suppliers [US]

AMTS Automated Transportation Management System

ATMS Automatic Transmission Measuring System

at/m²s atoms per square meter second [also at m⁻² s⁻¹, or atoms/m²s]

ATN Aeronautical Telecommunications Network [of International Civil Aviation Organization]

ATN Arctangent (Function) [Computer Programming]

ATN Attention [Data Processing]

ATN Augmented Transition Network

ATN Augmented Transmission Network

atn arctangent [Symbol]

ATNF Australia Telescope National Facility [of Commonwealth Scientific and Industrial Research Organization, Australia]

At No Atomic Number [also at no]

ATO Abort to Orbit

ATO Accumulated Time Off

ATO African Timber Organization [Gabon]

ATO Agricultural Trade Office [US]

ATO Air Transfer Order

ATO Antarctic Terminal Operations [of Antarctic Shipping Association]

ATO Antimony-Doped Tin Oxide

ATO Asia Tele-Tech Organization

ATO Assisted Takeoff

ATO Australian Taxation Office

ATO Automatic Trunk Office

ATOA Air Transport Operators Association [UK]

ATOC Acoustic Thermometry of Ocean Climate (Study) [US]

A to D Analog to Digital

ATOF Australian Transport Officers Federation

ATOL Assisted Takeoff and Landing

ATOL Atochem Polymerization Process

ATOLL Acceptance, Test, or Launch Language

ATOM Against Tests on Mururoa

ATOM Apollo Telescope Orientation Mount [of NASA]

ATOM Automatic Transmission of Mail

ATOMDEF Atomic Defense

ATOMIC Automatic Test of Monolithic Integrated Circuits

Atomkernenerg/Kerntech Atomkernenergie/Kerntechnik [German Journal on Nuclear Energy/Nuclear Engineering]

Atom Phys Atomic Physicist; Atomic Physics

ATOMS Automated Technical Order Maintenance Sequences

atoms/cm² atoms per square centimeter [Unit]

atoms/cm³ atoms per cubic centimeter [Unit]

atoms/cm²s atoms per square centimeter-second [Unit]

atoms/m² atoms per square meter [Unit]

atoms/m³ atoms per cubic meter [Unit]

Atomwirtsch–Atomtech Atomwirtschaft–Atomtechnik [German Journal on Nuclear Power Industry and Nuclear Technology; published by Kerntechnische Gesellschaft]

A/torr Ampere(s) per torr [Unit]

ATP Accelerated Training Program

ATP Acceptance Test Procedure

ATP Accepted Test Procedure

ATP Adenosine Triphosphate; Adenosine-5'-Triphosphate [Biochemistry]

ATP Advanced Technology Park

ATP Advanced Technology Program [of US Department of Commerce]

ATP Advanced Technology Program [of National Institute for Standards and Technology, US]

ATP Advanced Technology Program [of American Society of Mechanical Engineers, US]

ATP Advanced Technology Project [of Defense Advanced Research Projects Agency, US]

ATP Advanced Turboprop (Airliner)

ATP Airline Transport Pilot (Certificate)

ATP Air Travel Plan

ATP Allied Technical Publication

ATP Aluminothermic Process

ATP Angstrom Technology Partnership [Ibaraki, Japan]

ATP Army Training Plan

ATP Arsenical Tough Pitch (Copper)

ATP Association of Technical Professionals [US]

ATP Asynchronous Transaction Processing

ATP Auger-Type Process [Physics]

ATP Austrian Patent [also AtP]

ATP Automated Test Plan

ATP Automated Tow Placement

ATP Automatic Train Protection

ATP O,O-Dinonyl-N-Phenylnaphthylamidothiophosphate

ATP Silver Titanium Orthophosphate (Ceramics) [$AgTi_2(PO_4)_3$]

5'-ATP Adenosine-5'-Triphosphate [Biochemistry]

A&TP Assembly and Test Pit [of High-Temperature Sodium Facility, US]

AtP Austrian Patent

at% atomic percent [also AT%]

ATPase Adenosine Triphosphatase; Adenosine 5'-Triphosphatase [Biochemistry]

ATPC Association of Tin Producing Countries [Malaysia]

ATPC Australian Timber Producers Council

at pct atomic percent [also at%]

ATP Cu Arsenical Tough Pitch Copper

ATPG Automatic Test Pattern Generation; Automatic Test Pattern Generator

ATPG Automatic Test Program Generator

ATP-γS Adenosine 5'-O-(3-Thiotriphosphate) [Biochemistry]

At Phys Atomic Physicist; Atomic Physics

ATPI American Textbook Publishers Institute [US]

ATPI American Transfer Printing Institute [now International Transfer Printing Institute, US]

ATPISO Acetyl-Terminated Polyimidesulfone

ATPL Airline Transport Pilots License

at ppb atomic parts per billion [Unit]

at ppm atomic parts per million [Unit]

at ppt atomic parts per trillion [Unit]

ATP-Ribose 2'-Monophosphoadenosine 5'-Diphosphatase [Biochemistry]

ATPS AppleTalk Printing Services

ATPS Automatic Tile Processing Navigation System

ATP-γS Adenosine-5'-[γThio]triphosphate [Biochemistry]

ATR Advanced Terminal Reactor

ATR Advanced Test Reactor

ATR Aircraft Transmitter Receiver [also atr]

ATR Aircraft Trouble Report

ATR Air Traffic Regulations

ATR Air Training Resources [of University of Toronto, Canada]

ATR Air Transport Racking

ATR Air Transport Radio

ATR Air Transport Rating

ATR Air Turbo Rocket

ATR Aluminothermic Reaction

ATR Answering Time Recorder

ATR Anti-Transmit-Receive

ATR Attenuated Total Reflectance (Spectroscopy)

ATR Attenuated Total Reflection

ATR Attribute [also Atr]

ATR Automatic Target Recognition

ATR Automatic Terminal Recognition

ATR Automatic Traffic Recorder

ATR Average Transfer Rate

Atr Attribute [also atr]

ATRA Advanced Transit Association

ATRA Automatic Transmission Rebuilders Association [US]

atRA all-trans-Retinoic Acid [Biochemistry]

ATRAN Automatic Terrain Recognition and Navigation [also Atran]

ATRAX Air Transportable Communications Complex

ATRC Advanced Technology Research Center [of Agency for Defense Development, Taejon, South Korea]

ATRC Agricultural Tools Research Center

ATRC Alberta Telecommunications Research Center [Canada]

ATRC Antitracking Control

ATRCE Advanced Test Reactor Critical Experiment

ATRCW African Training and Research Center for Women

ATREG African Technical Regional Environment Group

ATREM Average Time Remaining

ATREP Air Traffic Control Representative

ATREX Astrophysics Transient Explorer

ATR FTIR Attenuated Total Reflectance Fourier Transform Infrared Spectroscopy [also ATR-FTIR]

ATR FTIRS Attenuated Total Reflectance Fourier Transform Infrared Spectroscopy [also ATR-FTIRS]

ATRI Australian Turfgrass Research Institute

ATRIB Average Transfer Rate of Information Bits

ATRID Automatic Target Recognition Identification and Detection

ATRIP (International Association for the) Advancement of Teaching and Research in Intellectual Property [US]

ATRL Advanced Technology Research Laboratory [of Nippon Steel Corporation, Japan]

ATRP Air Transport Regulation Panel [of International Civil Aviation Organization]

ATRS Attenuated Total Reflection Spectroscopy

ATRS Automated Temporary Roof Support

ATRT Anti-Transmit-Receive Tube

ATS Absolute Timed Sequence

ATS Academic Text Service [of Stanford University, California, US]

ATS Acetylene-Terminated Sulfone

ATS Acquisition and Tracking System

ATS Action Tracking System

ATS Administrative Terminal System

ATS Advanced Technology Satellite

ATS Advanced Test System

ATS Agence Télégraphique Suisse [Swiss News Agency]

ATS Aided Tracking System [also ats]

ATS Air Tactical School [of US Air Force]

ATS Air Traffic Service [of Federal Aviation Administration, US]

ATS Air Traffic Surveillance

ATS Air Traffic System

ATS Alarm Termination Subsystem

ATS Alkali-Tin-Silicate (Glass)

ATS Alliance for Traffic Safety [US]

ATS Alloy Thermo-Sorter

ATS American Technical Society [US]

ATS American Television Society [US]

ATS American Therapeutic Society [US]

ATS Aminopropyl Triethoxysilane

ATS Ammonium Thiosulfate (Process)

ATS Analog Test System

ATS Antarctic Treaty System

ATS Antitetanus Serum

ATS Apple Terminal Services

ATS Applications Technology Satellite [Series of NASA Satellites]

ATS Applied Test Systems, Inc. [US]

ATS Armament Training Station

ATS Army Transport Service

ATS Association Technique de la Sidérurgie Française [Technical Association of the French Iron and Steel Industry]

ATS Astronomical Time Switch

ATS Atmospheric Transmission Spectrum

ATS Audio Test Set

ATS Austrian Schilling [Currency]

ATS Automated Telemetry System

ATS Automated Tilting System

ATS Automated Tracking System

ATS Automatic Telephone System

ATS Automatic (Money) Transfer Service [US]

ATS Automatic Transport System

ATS Automatic Trunk Synchronizer

ATS Auto-Transformer-Starter

ATS Auxiliary Territorial Service [US Military]

ATS Auxiliary Tug Service

ATS-I First Applications Technology Satellite [of NASA] [also ATS-1]

ATSA American Traffic Services Association[US]

ATSA Aminotoluene Sulfonic Acid

ATSAR Advanced Technology Synthetic Aperture Radar

ATSC Acetone-Thiosemicarbazone

ATSC Advanced Television Systems Committee [US]

ATSC Australian Tree Seed Center [of Commonwealth Scientific and Industrial Research Organization]

ATSCM Antarctic Treaty Special Consultative Meeting

ATSD Airborne Traffic Situation Display

ATSDR Agency for Toxic Substances and Disease Registry [of US Department of Health and Human Services]

ATSI Association of Telemessaging Services International [US]

ATSIC Aboriginal and Torres Strait Islander Commission [Australia]

ATSIhealth Australian and Torres Strait Islander Health Database [Part of AUSThealth]

ATSIT Automatic Techniques for the Selection and Identification of Targets

ATSO Association of Trading Standards Officers [UK]

ATSORA Air Traffic Services Outside Regulated Airspace [UK]

ATSP Air Transport Statistical Program [of International Civil Aviation Authority, Canada]

ATSP Association of Technical and Supervisory Professional [US]

At Spectrosc Atomic Spectroscopy [Journal published in the US]

ATSPM Air Traffic Service Planning Manual

ATSR Along-Track Scanning Radiometer

ATSR Argonne Thermal Source Reactor [of Argonne National Laboratory, Illinois, US]

ATSR/MS Along-Track Scanning Radiometer and Microwave Sounder

ATSS Acquisition Tracking Subsystem

ATSS Automatic Telephone Switching System

ATSSA American Traffic Safety Services Association [US]

At Strom Atom und Strom [German Publication on Nuclear Energy and Electric Power Generation; published by Vereinigung Deutscher Elektrizitätswerke]

ATSU Air Traffic Service Unit

ATSU Association of Time Sharing Users [US]

ATT Advanced Transport Technology

ATT Attenuation Level (Units)

ATT Automatic Testing Technology

ATT Automatic Toll Ticketing

ATT Avalanche Transit Time (Diode)

AT&T American Telegraph and Telephone Company [US]

Att Attaché

Att Attachment [also att]

Att Attention [also att]

Att Attenuation; Attenuator [also att]

Att Attitude [also att]

Att Attorney [also att]

att attached

ATTA American Tin Trade Association [US]

ATTAC Advanced Technologies Testing Aircraft

ATTAP Advanced Turbine Technology Application Program [US]

ATTB Advanced Technology Transit Bus

ATTC American Towing Tank Conference [US]

ATTC ASEAN (Association of Southeast Asian) Timber Technology Center

ATTC Automatic Transmission Test and Control

ATTCOM AT&T Communications [US]

ATTCS Automatic Takeoff Thrust Control System

ATTD Avalanche Transit-Time Device

ATTDOC Attitude Documentation

Atten Attenuation; Attenuator [also atten]

Att Gen Attorney General

ATTI Arizona Transportation and Traffic Institute [US]

ATTI Association of Teachers in Technical Institutions [UK]

ATTI AT&T International

Atti Accad Ligure Sci Lett Atti della Accademia Ligure di Scienze e Lettere [Italian Scientific Journal]

Atti Accad Naz Lincei, Ottava Atti della Accademia Nazionale dei Lincei, Ottava [Journal on Physical, Mathematical and Natural Sciences; published by Accademia Nazionale dei Lincei, Rome, Italy]

Atti Accad Naz Lincei, Rend Cl Sci Fis Mat Nat Atti della Accademia Nazionale dei Lincei, Rendiconti, Classe di Scienze Fisiche, Matematiche e Naturali [Journal on Physical, Mathematical and Natural Sciences; published by Accademia Nazionale dei Lincei, Rome, Italy]

Atti Accad Sci Ist Bologna Cl Sci Fis Rend XIII [Atti della Accademia delle Scienze dell' Istituto di Bologna, Classe di Scienze Fisiche, Rendiconti, Serie XIII [Proceedings of the Academy of Sciences of the Bologna Institute, Physical Sciences Class, Series XIII, Bologna, Italy]

Atti Fond Giorgio Ronchi Atti della Fondazione Giorgio Rochi [Proceedings of the Giorgio Ronchi Foundation, Italy]

ATTIS AT&T Information System [of AT&T Company, US]

Atti Semin Mat Fis Univ Módena Atti del Seminario Matematico e Fisico dell' Universita di Módena [Proceedings of the Physical and Mathematical Seminars of the University of Modena, Italy]

Atti Soc Pelorit Sci Fis Math Nat Atti della Società Peloritana di Scienze Fisiche, Matematiche e Naturali [Proceedings of the Peloritan Society of Physical, Mathematical and Natural Sciences, Italy]

ATTITB Air Transport and Travel Industry Training Board

ATTIX AT&T Interexchange Carrier

Attm Attachment [also attm]

ATTMCA Association of Tile, Terrazzo, Marble Contractors and Affiliates [now National Tile Contractors Association, US]

ATTN Attention [Communications]

Attn Attention [also attn]

atto- SI prefix representing 10^{-18}

ATTP Advanced Transport Technology Program

ATTRIB (File) Attribute (Command) [also attrib]

Attrib Attribute [also attrib]

ATTS Automatic Tape Time Select (Process)

ATTS Automatic Telemetry Tracking System

ATTT Association pour le Traitement Thermique [Heat Treatment Association, France]

ATTT AT&T Technologies [US]

AT&T Tech J AT&T Technical Journal [of AT&T Company, US]

AT&T Technol AT&T Technology [Publication of AT&T Corporation, US]

ATTW Association of Teachers of Technical Writing [US]

Atty Attorney [also atty]

Atty Gen Attorney General

ATU Address Translation Unit

ATU Aerial Tuning Unit

ATU African Telecommunications Union

ATU Amalgamated Transit Union

ATU Antenna Tuning Unit

ATU Antithrombin Unit [Biochemistry]

ATU Arab Telecommunications Union [Iraq]

ATU Association Internationale des Universités [International University Association, France]

ATU Atlantic Telephone System

ATU Auckland Technical University [New Zealand]

ATU Automatic Tracking Unit

ATU Autonomous Transfer Unit

At U Atomic Unit [also u, AU, or au]

atu (technical) atmosphere, underpressure[Unit]

atü atmosphere, excess pressure [obsolete] [Unit]

ATUA Air Transport Users Committee [UK]

ATUC African Trade Union Confederation

ATUC ASEAN (Association of Southeast Asian Nations) Trade Union Council

A turn Ampere turn [Symbol]

ATURS Automatic Traffic Usage Recording System

ATV Abwassertechnische Vereinigung [Association for Wastewater Technology, Germany]

ATV Air-to-Vessel (Radar) [also atv]

ATV All-Terrain Vehicle

ATV Automatic Threshold Variation

ATVB Arteriosclerosis, Thrombosis and Vascular Biology [of American Heart Association, US]

ATV J ATV Journal [German Journal on Wastewater Technology; published by Abwassertechnische Vereinigung]

at vol atomic volume

ATW Accelerator Transmutation of Waste

ATW Aircraft Tail-Warning (Radar) [also atw]

ATW Aluminothermic Welding

ATW Automatic Tube (Butt) Weld

ATW Auto-Tracing White Balance (of Video Equipment)

AT/W Atomic Hydrogen Weld

ATWA Association of Third World Affairs

ATWE Association of Third World Economists

ATWS Anticipated Transient Without Scram

At Wt Atomic Weight [also at wt]
at wt% atomic weight percent
ATX Advanced Technology Extended
ATX Automatic Telex Exchange
ATX II Anemonia sulcata Toxin II
A2YSC Associated Two-Year Schools in Construction [US]
ATZ Aerodrome Traffic Zone
ATZ Airport Traffic Zone
ATZ Attention Restore [Modem]
AU Aberdeen University [Scotland]
AU Absorbance Unit
AU Acadia University [Canada]
AU Acousto-Ultrasonic(s)
AU Adelphi University [Garden City, New York, US]
AU Agra University [India]
AU Agricultural Union
AU Air University [at Maxwell Air Force Base, Alabama, US]
AU Ajou University [Suwon, South Korea]
AU Akita University [Akita, Japan]
AU Albuminuria [Medicine]
AU Alfaatih University [Libya]
AU Alfred University [New York, US]
AU Alpha Uranium
AU American University [Washington, DC, US]
AU Amidinourea [also au]
AU Amplifier Unit
AU Angstrom Unit [also au, ÅU, or åu]
AU Applied Ultrasonics
AU Arbitrary Unit [also au, Arb U, or arb u]
AU Arithmetic Unit
AU Aston University [Birmingham, UK]
AU Astronomical Unit
AU Atlanta University [Georgia, US]
AU Atomic Unit [also au, At U, or at u]
AU Auburn University [Normal, Alabama, US]
AU Australia [ISO Code]
AU Axial(ly) Unload(ed); Axial Unloading
AU Untreated Fiber [Designation]
.AU Audio [File Name Extension] [also .au]
A-U Adenine–Uridine [Biochemistry]
Au Audio [also au]
Au *(Aurum)* – Gold [Symbol]
Au Author [also au]
Au$^+$ Monovalent Gold Ion [Symbol]
Au^{3+} Trivalent Gold Ion [also Au^{+++}] [Symbol]
Au-191 Gold-191 [also ^{191}Au, or Au^{191}]
Au-192 Gold-192 [also ^{192}Au, or Au^{192}]
Au-193 Gold-193 [also ^{193}Au, or Au^{193}]
Au-194 Gold-194 [also ^{194}Au, or Au^{194}]
Au-195 Gold-195 [also ^{195}Au, or Au^{195}]
Au-196 Gold-196 [also ^{196}Au, or Au^{196}]
Au-197 Gold-197 [also ^{197}Au, Au^{197}, or Au]
Au-198 Gold-198 [also ^{198}Au, or Au^{198}]

Au-199 Gold-199 [also ^{199}Au, or Au^{199}]
Au-200 Gold-200 [also ^{200}Au, or Au^{200}]
Au-201 Gold-201 [also ^{201}Au, or Au^{201}]
Au-203 Gold-203 [also ^{203}Au, or Au^{203}]
au atomic unit [also AU, At U, or at u]
au^3 atomic unit cubed
au^2 atomic unit squared
au^{-1} one per atomic unit [also 1/au]
au^{-3} one per atomic unit cubed [also 1/au^3]
au^{-2} one per atomic unit squared [also 1/au^2]
.au Australia [Country Code/Domain Name]
αU Alpha Uranium [Symbol]
ÅU Ångstrom Unit [also AU, au, or åu]
AUA Agricultural Unit of Account
AUA American Underground-Space Association [US]
AUA American Urological Society [US]
AUA Anisotropic United Atom (Model)
AUA Argonne Universities Association [Illinois, US]
AUA Association of University Architects [US]
Au-ABA Gold-Based Active Brazing Alloy
Au-Ag Gold-Silver (Alloy System)
Au-Al Gold-Aluminum (Alloy System)
AUBC Association of Universities of the British Commonwealth [UK]
Au-Be Gold-Beryllium (Alloy System)
AUBLS American Uniform Boiler Law Society [US]
AuBr Gold(I) Bromide (or Aurous Bromide)
AUBRCC Australian Uniform Building Regulations Coordinating Council
AUBTW Amalgamated Union of Building Trade Workers [UK]
AUC Alberta Universities Commission[Canada]
AUC Algoma University College [Canada]
AUC Ammonium Uranyl Carbonate (Process)
AUC *(Anno Urbis Conditae)* – in the year of the founding of the city
AUC Atlanta University Center [Atlanta, Georgia, US]
AUC Auckland University College [New Zealand]
AUC Australian Universities Commission
AUC Authentication Center [Telecommunications]
AUCBE Advisory Unit for Computer-Based Education [UK]
AUCBM Arab Union for Cement and Building Materials [Syria]
AUCC Association of Universities and Colleges of Canada [Ottawa, Ontario]
AUCCP Arab Union for Cement and Cement Products [now Arab Union for Cement and Building Materials, Syria]
AuCd Gold Cadmium (Spin Glass)
Au-Cd Gold-Cadmium (Alloy System)
Au-CdS Gold-Cadmium Sulfide (Material)
AuCl Gold(I) Chloride (or Aurous Chloride)
AuCN Gold(I) Cyanide (or Aurous Cyanide)
Au-Co Gold-Cobalt (Alloy System)
Au/Co Gold/Cobalt (Multilayer)
Au-Cr Gold-Chromium (Alloy System)

AUCS Atlantic University Computer Study

AuCu Gold Copper (Spin Glass)

Au-Cu Gold-Copper (Alloy System)

AUD Australian Dollar [Currency]

Aud Audibility [also aud]

Aud Audio [also aud]

Aud Audit(or) [also aud]

AUDACIOUS Automatic Direct Access to Information with On-Line UDC System [of American Institute of Physics, US]

AUDAR Autodyne Detecting and Ranging

AUDI Arab Urban Development Institute [Saudi Arabia]

Audiol Audiological; Audiologist; Audiology [also audiol]

Audiol Akust Audiologische Akustik [German Journal on Audiological Acoustics]

Audiol Neurootol Audiology and Neurootology [International Journal]

Audio Vis Audio Visual [UK Journal]

Audio-Vis Commun Rev Audio-Visual Communication Review [US]

AUDIT Automated Data Input Terminal

AUDIT Automatic Unattended Detection Inspection Transmitter

AUDM Advanced Unified Defect Model

AUDREY Audio Reply

Au:Er Erbium-Doped Gold [also Au:Er^{3+}]

AUEW Amalgamated Union of Engineering Workers [UK]

AuFe Gold Iron (Spin Glass)

Au-Fe Gold-Iron (Alloy System)

Aufbereit-Tech Aufbereitungstechnik [German Journal on Processing Technology]

AUFIS Automated Ultrasonic Flaw Inspection System

AUFS American Universities Field Staff [US]

AUFSC Association of University Forestry Schools of Canada

AUFW Amalgamated Union of Foundry Workers [UK]

AUFWPA Association of University Fisheries and Wildlife Program Administrators [US]

AUG Add-On's and Upgrades

AUG Amdahl (Computer) Users Group [US]

Aug August

Au-Ga Gold-Gallium (Alloy System)

AUGC All-University Gerontological Center [of Fordham University, New York City, US]

Au-Ge Gold-Germanium (Alloy System)

AUGER Auger Electron Analysis

Au-Gr Gold-Graphite (System)

Au-H Gold-Hydrogen (Complex)

AUHF Annihilated Unrestricted Hartree-Fock (Method) [Statistical Mechanics]

Au-Hg Gold-Mercury (Alloy System)

AUI Acadia University Institute [Wolfville, Nova Scotia, Canada]

AUI Associated Universities, Inc. [of Brookhaven National Laboratory, Upton, New York State, US]

AUI Attachment Unit Interface

AUI Autonomous Unit Interface

AuI Gold(I) Iodide (or Aurous Iodide)

AUID Association of University Interior Design [US]

Au-In Gold-Indium (Alloy System)

Au-KCl Gold-Potassium Chloride (System)

AUKML Association of United Kingdom Media Librarians

AUKOI Association of United Kingdom Oil Independence

AUL Above Upper Limit

AUL Air University Library [at Maxwell Air Force Base, Alabama, US]

AUL Assistant University Librarian; Associate University Librarian

AuLα Gold L-Alpha (Radiation) [also AuL$_α$]

Au-LiF Gold-Lithium Fluoride (System)

AUM Air-to-Underwater Missile

AUMA Ausstellungs-und Messeausschuß der deutschen Wirtschaft e.V. [German Committee for Exhibitions and Trade Fairs]

AuMg Gold Magnesium (Spin Glass)

Au-Mg Gold-Magnesium (Alloy System)

Au-MgO Gold-Magnesium Oxide (System)

Au/Mn Gold/Manganese (Interface)

AUMS Aberdeen University Marine Studies Ltd. [Scotland]

AuNα Gold N-Alpha (Radiation) [also AuN$_α$]

Au-NaCl Gold-Sodium Chloride (System)

Au/n-GaAs Gold on N-Type Gallium Arsenide (Substrate)

Au-Ni Gold-Nickel (Alloy System)

Au/Ni Gold/Nickel (Multilayer)

Au/n-InP Gold on N-Type Indium Phosphide (Substrate)

Au/n-Si Gold on N-Type Silicon (Substrate)

Au/n-Si(Ct) Gold on N-Type Silicon Control Sample

Au/n-Si(H) Gold on Hydrogenated N-Type Silicon

Au/n-Si(H)(SB) Gold on Hydrogenated N-Type Silicon (Schottky Barrier)

AUNT Automatic Universal Translator

αU$_3$O$_8$ Alpha Triuranium Octoxide

AuOH Gold(I) Hydroxide (or Aurous Hydroxide)

AUP Acceptable Use Policy [Internet]

AUP African Union of Physics [Nigeria]

AUP Arab Union of Producers

AuP Australian Patent [also AUP]

Au-Pb Gold-Lead (Alloy System)

Au-Pd Gold-Palladium (Alloy System)

Au/Pd/Ti Gold/Palladium/Titanium (Contact)

Au/Pd/Zn Gold/Zinc/Gold (Layer)

AUPI Air University Periodical Index [Published at Maxwell Air Force Base, Alabama, US]

AUPS American University Press Services [now Association of American University Presses, US]

Au-Pt Gold-Platinum (Alloy System)

AUPTDE Arab Union of Producers, Transporters and Distributors of Electricity

AUR Association of University Radiologists [US]

Aur Auriga [Astronomy]

AURA Association of Universities for Research in Astronomy [Baltimore, Maryland, US]

AURA/STScI Association of Universities for Research in Astronomy/Space Telescope Science Institute [US]

aur dext *(auris dexter)* – in (or to) the right ear

Auri Auriga [Astronomy]

AURIG American University Research Incentive Grant [US]

AURIGA An Ultracryogenic Gravitational Waves Detector (Experiment) [of Instituto Nazionale de Fisico Nucleare, Italy]

AURIO Auroral Imaging Observatory

AURIS Aberdeen University Research and Industrial Services [Scotland]

AURISA Australasian Urban and Regional Information Systems Association, Inc.

aur laev *(auris laevus)* – in (or to) the left ear

AUROPA Automatische Ultraschall-Radoberflächen-Prüfanlage [Automatic Ultrasonic Wheel Surface Testing Facility, Hamburg and Munich, Germany]

AURP American Universities Research Program [US]

AURRP Association of University Related Research Parks [US]

aur sinist *(auris sinister)* – in (or to) the left ear

Au-Ru Gold-Ruthenium (Alloy System)

AUS Army of the United States (of America)

AUS Austrian Schilling [Currency]

AuS Gold(I,III) Sulfide (or Auroauric Sulfide)

Aus Australasia(n)

Aus Australia(n)

Aus Austria(n)

au/s arbitrary unit(s) per second [also au s^{-1}, au/sec, or au sec^{-1}]

au/s atomic unit(s) per second [also au s^{-1}, au/sec, or au sec^{-1}]

AUSA Association of the United States Army [US]

Au-Sa Gold-Sapphire (System)

AusAID Australian Agency for International Development

AUSAT Australian Satellite Users Association

a-USb Uranium-Antimony Amorphous Alloy

a-USbMn Uranium-Antimony-Manganese Amorphous Alloy

AUSBC ASEAN-US (Association of Southeast Asian Nations–United States) Business Council [Thailand]

AUSD African Union for Scientific Development

AUSDEC Australasian Spatial Data Exchange Center

au/sec arbitrary unit(s) per second [also au sec^{-1}, au/s, or au s^{-1}]

au/sec atomic unit(s) per second [also au sec^{-1}, au/s, or au s^{-1}]

AuSi Gold-Silicon (Compound)

Au-Si Gold-Silicon (Alloy System)

AusIMM Australian Institute of Mining and Metallurgy [now Australasian Institute of Mining and Metallurgy] [Parkville, Victoria, Australia]

AusIMM Bull Proc Australasian Institute of Mining and Metallurgy Bulletin and Proceedings

AusIMM Proc Australasian Institute of Mining and Metallurgy Proceedings

AUSINET The Australian Information Network [of Commonwealth Scientific and Industrial Research Organization]

AUSLIG Australian Surveying and Land Information Group

AUSMARC Australian MARC (Machine-Readable Catalogue)

AUSMIISL Association of United States Members of the International Institute of Space Law

Au-Sn Gold-Tin (Alloy System)

AUSPI Auburn University Space Power Institute [Alabama, US]

AUSPIC Australian Government Photographic Service [of Department of Administrative Services]

AUSS Advanced Unmanned Search System

AUSS Association of University Summer Sessions [US]

AUSS Austenitic Stainless Steel

AUSS Automatic Ultrasonic Scanning System

AUSSAT Australian Satellite Network

AUSSI Alternative Uses of Steel Slags in Industry (Team) [of Michigan Technological University, Houghton, US]

Aust Australia(n)

Aust Austria(n)

Aust Asian Pac Electr World Australian, Asian and Pacific Electrical World [Australian Journal]

AUSTCERAM International Ceramic Conference in Australia [also Austceram]

Aust Commun Netw Australian Communications Networks

Aust Comput J Australian Computer Journal [of Australian Computer Society]

Aust Comput Sci Commun Australian Computer Science Communications

Aust Electron Eng Australian Electronics Engineering [Journal]

AUSTEMEX Australian Environment Management Export Corporation, Ltd.

Aust Forestry Australian Forestry [Journal of the Institute of Foresters of Australia]

AUSThealth Australian Health Database

Aust J Appl Sci Australian Journal of Applied Science

Aust J Audiol Australian Journal of Audiology [of Audiological Society of Australia]

Aust J Biotechnol Australian Journal of Biotechnology

Aust J Chem Australian Journal of Chemistry [of Commonwealth Scientific and Industrial Research Organization, Australia]

Aust J Earth Sci Australian Journal of Earth Sciences

Aust J Geod Photogramm Surv Australian Journal of Geodesy, Photogrammetry and Surveying [of University of New South Wales, Kensington, Australia]

Aust J Phys Australian Journal of Physics [of the Commonwealth Scientific and Industrial Research Organization, Australia]

Aust J Phys Astrophys Suppl Australian Journal of Physics, Astrophysical Supplement [of Commonwealth Scientific and Industrial Research Organization, Australia]

Aust J Plant Physiol Australian Journal of Plant Physiology [of Commonwealth Scientific and Industrial Research Organization, Australia]

Aust J Psychol Australian Journal of Psychology

Aust J Soil Res Australian Journal of Soil Research

Austl Australasia(n)

Austl Australia(n)

Aust Meteorol Meg Australian Meteorological Magazine [of Bureau of Meteorology]

Aust Min Australian Mining [Journal]

Aust NZ J Med Australia and New Zealand Journal of Medicine

AUSTPAC Australian Packet Switching Service

Aust Pat Australian Patent [also Austr Pat]

Aust Phys Australian Physicist [Journal of the Australian Institute of Physics]

Aust Pkg Australian Packaging [Journal]

Austr Australia(n)

AUSTRADE Australian Trade Commission

AUSTRIB International Tribology Conference in Australia [of Institution of Engineers]

Austral Australia(n)

Australas Australaisia(n)

Australas Corros Eng Australasian Corrosion Engineering [Journal]

Austral Asian Pac Electr World Australian, Asian and Pacific Electrical World [Australian Journal]

Australas Phys Eng Sci Med Australasian Physical and Engineering Sciences in Medicine [Journal]

Austral Inst Met Australasian Institute of Metals [Sydney]

Austria Mach Steel Austria Machinery and Steel [Journal]

Aust Road Res Australian Road Research [Journal of the Australian Road Research Board]

Aust Road Res Prog Australian Road Research in Progress [of Australian Road Research Board]

Austr Pat Australian Patent [also Aust Pat]

Aust Telecommun Res Australian Telecommunication Research [of Telecommunication Society of Australia]

AUSTUDY Australian Government Scheme of Educational Allowances

Aust Weld J Australian Welding Journal

Aust Weld Res Australian Welding Research [Journal of the Australian Welding Research Association]

Aust Road Res Australian Road Research [Journal of the Australian Road Research Board]

Aust Road Res Prog Australian Road Research in Progress [Journal of the Australian Road Research Board]

AUSTROM Australian Periodical Abstracts on CD-ROM

Aust Telecommun Res Australian Telecommunication Research [Journal of the Telecommunication Society of Australia]

Aust Weld J Australian Welding Journal

Aust Weld Res Australian Welding Research [Journal of the Australian Welding Research Association]

AUSUDIAP Association of United States University Directors of International Agricultural Programs

Ausz Eur Patentanmeld I Auszüge aus den Europäischen Patentanmeldungen, Teil I [Abstracts of European Patent Applications, Part I, Germany]

Ausz Eur Patentanmeld II Auszüge aus den Europäischen Patentanmeldungen, Teil II [Abstracts of European Patent Applications, Part II, Germany]

AUT Acousto-Ultrasonic Testing

AUT Advanced User Terminal

AUT Aristiole University of Thessaloniki [Greece]

AUT Association of University Teachers [UK]

Aut Author [also aut]

Aut Automatic [also aut]

Aut Automobile [also aut]

AuTe Gold Telluride (or Calaverite)

AUTEC Atlantic Undersea Test and Evaluation Center [US]

AUTECS Automated Eddy Current (Inspection) System

Auth Author(ized); Authorization; Authority [also auth]

auth authentic

Author Authorization [also author]

AUTHR Authorization Response [Telecommunications]

Au/Ti Gold/Titanium (Contact)

AUTO Automatic Line Number Generation and Incrementation (Command) [Programming]

Auto Automatic; Automation [also auto]

Auto Automaton [also auto]

Auto Automobile; Automotive [also auto]

AUTOABSTRACT Automatic Abstract [also autoabstract]

AUTOALARM Automatic Alarm [also Autoalarm, or autoalarm]

Autobiog Autobiographer; Autobiographical; Autobiography [also autobiog]

AUTOCALL Automatic Calling; Automatic Call Origination [also Autocall, or autocall]

AUTOCOM Automotive Composites Conference and Exhibition

AUTOCONT Auto(matic) Continue [also Autocont]

AUTO CV Automatic Check Valve

AUTODIN Automatic Digital (Information) Network [of US Department of Defense]

AUTODOC Automated Documentation

AUTOEXEC Automatic Execution (File) [also autoexec]

AUTOFACT Automated and Integrated Factory (Conference and Exposition) [of the Computer and Automated Systems Association of the Society of Manufacturing Engineers, International] [also Autofact]

AUTO ID Automatic Identification (System) [also Auto ID]

Autoign Autoignition [also autoign]

Autoign Temp Autoignition Temperature [also autoign temp]

AUTO-LF Auto(matic) Line-Feed [also Auto-LF]

Autom Automatic; Automation [also autom]

AUTOMAP Automatic Machining Program

AUTOMAST Automatic Mathematical Analysis and Symbolic Translation

Automat Weld (USSR) Automatic Welding (USSR) [Published in the UK]

Autom Control Automation and Control [Published in New Zealand]

Autom Control Comput Sci Automatic Control and Computer Sciences [Translation of: *Avtomatika i Vychislitell'naya Tekhnika (USSR)*; published in the US]

Autom Data Process Inf Bull Automatic Data Processing Information Bulletin [of International Social Security Association, Brussels, Belgium]

Autom Doc Math Linguist Automatic Documentation and Mathematical Linguistics [Translation of: *Nauchno-Tekhnicheskaya Informatsiya, Seriya 2 (USSR)*; published in the US]

automechanika International Trade Fair for Motor Car Workshop and Service Station Equipment, Spare Parts and Accessories, Frankfurt/Main, Germany]

Autom Electr Power Syst Automation of Electric Power Systems [Chinese Journal]

AUTOMEX Automatic Message Exchange Service [also Automex]

Autom Instrum Automatica e Instrumentación [Journal on Automation and Instrumentation published by CETISA Co., Barcelona, Spain]

Autom Integrata Automazione Integrata [Italian Journal on Integrated Automation]

Autom Mach Automatic Machining [US Journal]

Autom Manuf Strategy Automated Manufacturing Strategy [Journal published in the Netherlands]

Automob Automobile [also automob]

Automob Abs Automobile Abstracts [of Motor Industry Research Association, UK]

Automob Ind Automobil-Industrie [German Journal on the Automotive Indystry]

Automob Q Automobile Quarterly (Magazine) [US]

automot automotive

Automot Des Dev Automotive Design and Development [Journal published in the US]

Automot Eng Automotive Engineering [Journal of the Society of Automotive Engineers, US]

Automot Eng (UK) Automotive Engineer (UK) [Journal published in the UK]

Automot Ind Automotive Industries [US Journal]

Automot News Automotive News [US Journal]

Automot Outlook Automotive Outlook (Journal)

Automot Technol Int Automotive Technology International (Journal)

Autom, Prod, Inf Ind Automatique, Productique, Informatique Industrielle [French Journal on Industrial Automation, Production and Information Science; published by Association Française pour le Cybernetique, Economique et Technique]

Autom Prod Technol Automation Products and Technology [Canadian Journal]

Autom Remote Control Automation and Remote Control [Translation of: *Avtomatika i Telemekha-nicheska (USSR)*; published in the US]

Autom Strum Automazione e Strumentazione [Italian Journal on Automation and Instrumentation published by Associazione Nazionale Italiana per l'Automazione]

Autom Syst Automation Systems [Canada]

Autom tech Prax Automatisierungstechnische Praxis [German Journal on Practical Automation Technology]

auton autonomous

AUTONET Automatic Network Display

Autophon Bull Autophon Bulletin [Switzerland]

AUTOPIC Automatic Personal Identification Code

Autopilot Automatic Pilot [also autopilot]

Autopoll Automatic Polling [also autopoll]

AUTOPROG Automatic Programming [also Autoprog]

AutoProj Automotive Pollution Prevention Project [Joint Project of Chrysler Corporation, Ford Motor Company and General Motors Corporation]

AUTOPROMPT Automatic Programming of Machine Tools [also Autoprompt]

AUTOPSY Automatic Operating System

AUTO-QC Automatic Quality Control

AUTO RECL Automatic Reclosing [also Auto Recl]

AUTOSATE Automated Data Systems Analysis Technique

AUTOSCRIPT Automated Systems for Composing, Revising, Illustrating and Phototypesetting

AUTOSDI Automatic Selective Dissemination of Information

AUTOSEVCOM Automatic Secure Voice Communications [also AUTOSEVOCOM]

AUTOSPOT Automatic System for Positioning (of) Tools [also Autospot]

AUTOSTRAD Automated System for Transportation Data

AUTOSTRT Automatic Starter [also Autostrt]

AUTOSTRTG Automatic Starting [also Autostrtg]

AUTOTESTCON Automobile Testing Conference [US]

AUTO TR Auto-Transformer

AUTO-TRIP Automatic Transportation Research Investigation Program

AUTOVERIFIER Automatic Verifier [also Autoverifier]

AUTOVON Automatic Voice Network [also Autovon] [of US Department of Defense]

AUTRAN Automatic Target Recognition Analysis

AUTRAN Automatic Utility Translator

AUTS Association of University Teachers of Scotland [also AUT(S)]

AUU Association of Urban Universities [US]

AUUA American UNIVAC (Computer) Users Association [US]

AUV Armored Utility Vehicle

AUV Autonomous Underwater Vehicle

AUVS Association for Unmanned Vehicle Systems [US]

AUW Air-to-Underwater

AUW All-Up Weight [Aeronautics]

AUW Anti-Underwater Warfare

AUW Wausau, Utah [Meteorological Station Designator]

AUWE Admiralty Underwater Weapons Establishment [UK]

aux auxiliary

Aux AAF Auxiliary Army Air Field [of US Army]

Aux AHP Auxiliary Army Heliport [of US Army]

AUXI-ATOME Societé Auxiliare pour l'Energie Atomique [Auxiliary Society for Atomic Energy, France]

auxil auxiliary

Auxs Auxiliaries [also auxs]

Au/YBCO Gold/Yttrium Barium Copper Oxide (Thin Film)

AuZn Gold Zinc (Spin Glass)

Au-Zn Gold-Zinc (Alloy System)

Au/Zn/Au Gold/Zinc/Gold (Layer)

AuZn/p-GaAs Gold Zinc/P-Type Gallium Arsenide (Contact)

AV Abrikosov Vortex [Physics]

AV Absolute Viscosity

AV Absolute Vorticity

AV Accelerating Voltage

AV Active Voltage

AV Added Value

AV Adenovirus

AV Aerial Videography

AV Air Valve

AV Air Volume

AV Angular Velocity

AV Antivenom; Antivenin

AV Anti-Virus

AV Arbeitsvorbereitung [Journal for Production Planning and Control, Germany]

AV Arbovirus

AV Arctic Vessel

AV Armored Vehicle

AV Arteriovenous

AV Asbestos-Covered Varnish-Treated (Electric Wire)

AV Astrovirus

AV Athropoid-Borne Virus

AV Atrioventricular (Node) [also A-V] [Anatomy]

AV Audio-Visual [also A-V, or A/V]

AV Authenticity Verification

AV Authorized Version

AV Availability

AV Average Value

AV Average Variability

A&V Artery and Vein; Arterial and Venous

A/V *(Ad Valorem)* – according to value

A/V Ampere per Volt [also A V^{-1}]

A/V Area-to-Volume (Ratio)

A/V Armored Vehicle

A/V Audio/Video

A-V Atrioventricular (Node) [also AV] [Anatomy]

A-V Audio-Visual [also A/V, or AV]

Av Avenue [also av]

Av Average [also av]

Av Aviation; Aviator [also av]

aV abvolt [Unit]

av (arithmetic) average

av avoirdupois [Unit]

a v *(a vista)* – at sight

a/v *(ad valorem)* – according to the value

Å/V Angstrom(s) per Volt [Unit]

AVA Absolute Virtual Address

AVA Academy of Visual Arts [UK]

AVA Accelerating Voltage Alteration

AVA Adventitious Viral Agent

AVA Aerodynamische Versuchsanstalt [Research Institute for Aerodynamics, Germany]

AVA Alberta Veterinary Association [Canada]

AVA American Video Association [US]

AVA American Vocational Association [US]

AVA ASEAN (Association of Southeast Asian Nations) Valuers Association [Malaysia]

AVA Asbestos and Varnished Cambric, Asbestos Braid Type (Conductor Insulation) [Symbol]

AVA Audio Visual Association [UK]

AVA Audio Visual Authoring

AVA Australian Veterinary Association

AVA Automated Vision Association [of Robotics Industries Association, US]

AVA Azimuth versus Amplitude

Ava Anabaena variabilis [Microbiology]

Ava I Anabaena variabilis I [Microbiology]

Ava II Anabaena variabilis II [Microbiology]

AVACS Average Number of Active Systems (per Grain) (on Stress-Strain Curve) [Mechanical Testing]

AVAD Automatic Voice Alerting Device

avail available

AVA/RIA Automated Vision Association of Robotics Industries Association [US]

AVAS Automated Vacancy Announcement System

AVAS Automatic VFR (Visual Flight Rules) Advisory Service

AVASI Abbreviated Visual Approach Slope Indicator

AVB Asbestos and Varnished Cambric, Cotton Braid Type (Conductor Insulation) [Symbol]

AVBL Armored Vehicle Bridge Launcher

AVC Advanced Biomedical Capsule [Aerospace]

AVC American Veterans Committee [Washington, DC, US]

AVC American Veterinary Congress [Washington, DC, US]

AVC American Viscose Company [US]

AVC Association of Vitamin Chemists [US]

AVC Audio Visual Connection

AVC Automatic Valve Control

AVC Automatic Vent Control

AVC Automatic Voltage Compression

AVC Automatic Voltage Control

AVC Automatic Volume Contractor

AVC Automatic Volume Control

AVC Axial Vector Current

AVCA Agriculture and Veterinary Chemicals Association

AVCAA Agriculture and Veterinary Chemicals Association of Australia

AVCASA Agricultural and Veterinary Chemicals Association of South Africa

AVCC Australian Vice-Chancellors' Committee [also AV-CC]

AVCCIOS Azimuthal and Vibrational Closely Coupled, Infinite-Order Sudden (Method) [also AVCC-IOS]

AVCF Angular Velocity Correlation Function

AVCG Automatic Vapor Crystal Growth

AvCIR Aviation Crash Injury Research (Division) [of Flight Safety Foundation, US] [also AvCir]

AVCOM Aviation Materiel Command [US]

AVCS Advanced Vidicon Camera System

AVCS Assistant Vice Chief of Staff

AVCU Agriculture and Veterinary Chemicals Unit

AVD Alternate Voice/Data

AVD Alternating Voice and Data

AVD Aluminum Vacuum Degassing [Metallurgy]

AVD Automatic Voice/Data

AvD Automobilverein von Deutschland [German Automobile Association]

AVDA American Venereal Disease Association [US]

AVDC Association for Vaccine-Damaged Children [Canada]

avdp avoirdupois [Unit]

AVE Aerospace Vehicle Electronics

AVE Automatic Voltammetric Electrode

AVE Automatic Volume Expander; Automatic Volume Expansion

Ave Avenue [also ave]

AVEA American Veterinary Exhibitors Association [US]

AVEC Amplitude Vibration Exciter Control

AVEG AIDS Vaccine Evaluation Group

AVEM Association of Vacuum Equipment Manufacturers [US]

AVERE Association Européenne des Vehicules Electriques Routièrs [European Electric Road Vehicle Association, Belgium]

AVERT Association of Volunteer Emergency Radio Teams [US]

Avesta Stainless Bull Avesta Stainless Bulletin [Published in Italy]

AVEU AIDS Vaccine Evaluation Unit

AVF Academie Vétérinaire de France [French Academy of Veterinary Science]

AVF Azimuthally Varying Field

AVG L-α-(2-Aminoethoxyvinyl)glycine [Biochemistry]

Avg Average [also avg]

Avgas Aviation Gasoline [also avgas]

AVGN Additive Voice Gaussian Noise

AVGP Armored Vehicle–General Purpose

AvH Alexander von Humboldt-Stiftung [Alexander von Humboldt Foundation, Germany]

Avh Nor Vidensk-Akad Oslo I Avhandlingar utgitt av det Norske Videnskaps-Akademi i Oslo I [Transactions of the Norwegian Academy of Sciences in Oslo, I]

Avh Nor Vidensk-Akad Oslo II Avhandlingar utgitt av det Norske Videnskaps-Akademi i Oslo II [Transactions of the Norwegian Academy of Sciences in Oslo, II]

Avh Nor Vidensk-Akad Oslo I, Mat-Natvidensk Kl Avhandlingar utgitt av det Norske Videnskaps-Akademi i Oslo, I, Matematisk-Naturvidenskaplige Klasse [Transactions of the Norwegian Academy of Sciences in Oslo I, Mathematical and Natural Sciences Class]

AVHRR Advanced Very-High-Resolution Radiometer (Satellite System)

AVHRR GAC AVHRR Global Area Coverage

AVHRR LAC AVHRR Local Area Coverage

AVI Airborne Vehicle Identification

AVI Audio Video Interleave(d); Audio Visual Interleave(d)

AVI Automated Visual Inspection

.AVI Audio Visual Interleave (File) [Microsoft File Name Extension]

Aviat Aviation [also Avi, avi, or aviat]

Aviat Educ News Bull Aviation Education News Bulletin [of Aviation Distributors and Manufacturers Association, US]

Aviat Mech Bull Aviation Mechanics Bulletin [of Flight Safety Foundation, US]

Aviat Space Environ Med Aviation, Space and Environmental Medicine

Aviat Week Space Technol Aviation Week and Space Technology [US]

AVID Advanced Visual Information Display

AVIEM Asociación Venezolana de Ingenieros Electricos y Mecanicos [Venezuelan As-sociation of Electrical and Mechanical Engineers, Caracas, Venezuela]

Avion Avionics [also avion]

Avionics Aviation Electronics [also avionics]

Avion News Avionics News [of Aircraft Electronics Association, US]

Avion Newsl Avionics Newsletter [of National Avionics Society, US]

AVIOS American Voice Input/Output Society [US]

AVIOS Conf Proc AVIOS Conference Proceedings [of American Voice Input/Output Society]

AVIOS J AVIOS Journal [of American Voice Input/Output Society]

AVIP Association of Viewdata Information Providers [UK]

AVIRIS Airborne Visual and Infrared Imaging Spectrometer [of NASA Jet Propulsion Laboratory, Pasadena, California, US]

AVIS Association of Voluntary/Independent Schools [UK]

AVIS Automatic Video Inspection System

AVISA Asociación Venezolana de Ingenieria Sanitaria y Ambiental [Venezuelan Chapter of the Inter-American Association of Sanitary Engineers]

AVL Adel'son-Vel'skii and Landis Trees

AVL Asbestos and Varnished Cambric, Lead Sheath Type (Conductor Insulation) [Symbol]

AVL Automatic Vehicle Location

AVL Available [Computers]

5AVL 5-Aminovaleric Acid [also 5Avl]

Av L Average Length [also av l]

AVLB Armored Vehicle Launch Bridge

AVLC Automatic Vehicle Location and Control System

AVLINE Audiovisual On-Line (Database) [US]

AVLIS Atomic Vapor Laser Isotope Separation [Physics]

AVLS Automatic Vehicle Locating System

AVLSI Analog Very Large-Scale Integration

AVM Airborne Vibration Monitoring

AVM Air Velocity Meter

AVM Air Vice-Marshall

AVM Arc-Vacuum Melted; Arc-Vacuum Melting [Metallurgy]

AVM Automatic Vehicle Monitoring

A/V/m Ampere per Volt per meter [also $AV^{-1}m^{-1}$]

AVMA American Veterinary Medical Association [US]

AVMA Audio-Visual Management Association [US]

AVMARC Audiovisual MARC (Machine-Readable Catalog)

AVMC Association for Vertical Market Computing [US]

AVMC Audio-Visual Media Center [of University of California at Berkeley, US]

AvMed Aviation Medicine [also Avmed, or avmed]

Av Mol Wt Average Molecular Weight [also av mol wt]

AVMR Association of Visual Merchandise Representatives [US]

AVMRI Arctic Vessel and Marine Research Institute

AVMS Annulus Vacuum Maintenance System

Avn Aviation [also avn]

AVNIR Advanced Visible and Near-Infrared Radiometer

AVNL Automatic Video Noise Limiting

AVO Advanced Video Option

AVOID Airfield Vehicle Obstacle Indication Device

avoir avoirdupois [Unit]

avoir oz avoirdupois ounce [Unit]

AVOLO Automatic Voice Link Observation

α-VOPO$_4$ Alpha I Vanadium Phosphate

α-VOPO$_4$ Alpha II Vanadium Phosphate

AVOSS Added Value Operating Support System

AVP Advanced VLSI (Very-Large-Scale Integration) Packaging

AVP Attached Virtual Processor

AVP Avionics Panel [of NATO Advisory Group for Aerospace Research and Development]

AVP Seaplane Tender [US Navy Symbol]

AVPC Association of Vice-Principals of Colleges [UK]

AVPO Axial Vapor-Phase Oxidation

AVPU Alert–Verbal–Pain–Unresponsive [Patient Responsiveness Assessment]

AVR Acidification/Volatilization/Reneutralization (Process)

AVR Automatic Voice Recognition

AVR Automatic Voltage Regulator

AVR Automatic Volume Recognition

AVR Automatic Volume Regulation

AVRA Audio-Visual Research Foundation [US]

AVRAC Agricultural and Veterinary Research Advisory Committee [East Africa]

AVRC AIDS Vaccine Research Committee

AVRDC Asian Vegetable Research and Development Center [Taiwan]

AVRS Automated Vehicle Roading System

AVS Advanced Vacuum System

AVS Advanced Visualization System

AVS Aerospace Vehicle Simulation

AVS American Vacuum Society [US]

AVS American Veterinary Society [US]

AVS Application Visualization System

AVS Applications Visual System

AVS Association of Veterinary Students [UK and Ireland]

AVS Automated Vision Association [US]

AVSAT Aviation Satellite [also AvSat]

AVSCOM Army Aviation Systems Command [US]

AVSE Armored Vehicle Survivability Enhancement

AVSF Advanced Vertical Strike Fighter

AVSL Association of Visual Science Librarians [US]

AVSS Audio-Video Support System

AVS Ser American Vacuum Society Series [Periodical Series]

AVT All Volatile Treatment [Boiler Feedwater]

AVT Applied Voice Technology

AVTA Automatic Vocal Transaction Analysis

Avtag Aviation Turbine Gasoline [also avtag]

Avtom Prom-st Avtomobilnaya Promyshlennost' [Latvian Journal on Automotive Engineering]

Avtom Svarka Avtomaticheskaya Svarka [Ukrainian Journal on Automation in Welding]

Avtom Telemekh Avtomatika i Telemekhanika [Russian Journal on Automation and Remote Control]

Avtom Vychisl Tekh Avtomatika i Vychislitel'naya Tekhnika [Latvian Journal on Automatic Control and Computer Sciences]

AVTR Airborne Video-Cassette Tape Recorder

Avtur Aviation Turbine (Kerosine) [also avtur]

Av W Average Width [also av w]

AW Above Water

AW Acid-Washed; Acid Washing

AW Acoustic Wave

AW Acoustic Waveguide

AW Adler-Weisberger (Rule)

AW Aircraft Warning

AW Aircraftwoman

AW Air Warning

AW Air Woman

AW Airworthiness

AW Annealed Wire

AW Arc-Welded; Arc Welding

AW Arming Wire

AW Aruba [ISO Code] [also aw]

AW Assembly Work

AW As Welded (Condition)

AW Atomic Weight [also aw]

AW Automated Warehouse

AW Automatic Weapon

AW Automatic Welding

AW Available Water

AW Aviation Writer

A/W Actual Weight

A/W Air/Water (Ratio)

A/W Alcohol-in-Water (Solution)

A/W Ampere(s) per Watt [also A W^{-1}]

Aw Tropical Savanna Climate with Wet and Dry Seasons [Classification Symbol]

aW abwatt [Unit]

aw atomic weight [also AW]

.aw Aruba [Country Code/Domain Name]

a/w normalized crack length (i.e., ratio of crack length "a" to sample width "w") [Symbol]

αW Alpha Tungsten [Symbol]

AWA Alliance of Women in Architecture [US]

AWA All-Weather Aircraft

AWA Aluminum Window Association [UK]

AWA American Whitewater Affiliation [Phoenicia, New York, US]

AWA American Wilderness Alliance [now American Wildlands, US]

AWA American Wire Association [US]

AWA Animal Welfare Act [US]

AWA Argonne Wakefield Accelerator [of Argonne National Laboratory, Illinois, US]

AWA Association of Women in Architecture[US]

AWA Aviation/Space Writers Association [US]

AWAC Airborne Warning and Control

AWAC Animal Welfare Advisory Council

AWACS Airborne Warning and Control System [also Awacs, US Air Force]

AWAD Assembly Work Authorization Document

AWADS All-Weather Aerial Delivery System

AWAE Automotive Wholesalers Association Executives [US]

AWAR Area-Weighted Average Resolution

AWARES Automated Warehousing and Retrieval System

AWARS Airborne Weather and Reconnaissance System

AWASTA African Women's Association for Scientific and Technical Advancement

AWAT Area Weighted Average T-Number

AWAXS Anomalous Wide-Angle X-Ray Scattering

AWB Agricultural Wages Board [UK]

AWB Air Waybill

AWB Asian Wetland Bureau

AWB Australian Wheat Board

AWB Australian Wool Board

AWB Automatic White Balance (of Video Equipment)

A/Wb Ampere(s) per Weber [also A Wb^{-1}]

AWBM As-Welded Base Metal

AWC Active Wire Concentrator

AWC Air War College [at Maxwell Air Force Base, Alabama, US]

AWC American Waterfowl Council [US]

AWC American Wood Council [US]

AWC American Wool Council [of American Sheep Producers Council, US]

AWC Army War College [Carlisle, Pennsylvania, US]

AWC Association for Women in Computing [US]

AWC Automatic Width Control

AWC Auto White-Balance Control (of Video Equipment)

αW$_2$C Alpha Ditungsten Carbide [Symbol]

AWCEA American Wood Chip Export Association [US]

AWCI American Wire Cloth Institute [US]

AWCI Association of the Wall and Ceiling Industries–International [US]

AWCLS All-Weather Carrier Landing System

AWCMA American Window Covering Manufacturers Association [US]

AWCS Air Weapons Control System

AWD Acoustic Warning Device

AWD Acoustic Wave Device

AWD Action World Development

AWD All-Wheel Drive

AWD Association of Welding Distributors [UK]

AWDA Automotive Warehouse Distributors Association [US]

AW-DMCS Acid-Washed Dimethylchlorosilane

AWDS Ash Water Dense Slurry

AWDS Ash Water Dense Suspension

AWE Advanced Wave Effects

AWEA American Wind Energy Association [US]

AWEBB Association of Wholesalers and Electrical Bulk Buyers [UK]

AWEC Albury-Wodonga Environment Center [Australia]

AWES Association of West European Shipbuilders [UK]

AWEU Assembly of the Western European Union [also A-WEU]

AWF African Wildlife Foundation [US]

AWF Ausschuß für Wirtschaftliche Fertigung [Committee for Economic Manufacturing, Germany]

AWG Acoustic Waveguide

AWG American Wire Ga(u)ge

AWG Arbitrary Wave Generator

AWG Art Workers Guild [UK]

AWG Asian Women's Institute [Pakistan]

AWG Association for Women Geoscientists [US]

AWGN Additive White Gaussian Noise

AWHG Australian and World Heritage Group

AWI Alfred-Wegener-Institute (für Polar- und Meeresforschung) [Alfred Wegener Institute (for Polar and Marine Research), Bremerhaven, Germany]

AWI American Welding Institute [Knoxville, Tennessee, US]

AWI Architectural Woodwork Institute [US]

AWIC Animal Welfare Information Center [US]

AWIC Association for Women in Computing

AWID Automated Window Inspection Device

AWIM Association for Women in Mathematics

AWIPS Advanced Weather Interactive Processing System [of National Weather Service, US]

AWIPS-90 Advanced Weather Interactive Processing System for the Nineties [of National Weather Service, US]

AWIS Association of Women in Science [Washington, DC, US]

AWIS Aviation Weather Information Service

AWJ Abrasive Water Jet

AWJC American Water Jet Conference [US]

AWK Aho-Weinberger-Kernighan [also Awk] [Unix Computer Language]

AWL Absent with Leave [also awl]

AWL American Wildlands [formerly American Wilderness Alliance, US]

AWLREM Association of Webbing Load Restraint Equipment Manufacturers [UK]

AWLS All-Weather Landing System

AWM Active Walker Model

AWM Air-Wall Material

AWM Appliance Wiring Materials

AWM Association for Women in Mathematics [US]

AWM Automated Weaving Machine

AWMA Air and Waste Management Association [also A&WMA, US]

AWMA American Walnut Manufacturers Association [now Fine Hardwood Veneer Association/American Walnut Manufacturers Association, US]

AWMS Australasian Wildlife Management Society

AWN Automatic Weather Network [of US Air Force]

AWO Airborne Weather Operator

AWO All-Weather Operation

AWO American Waterways Organization [US]

a-WO₃ Amorphous Tungsten Trioxide

AWOA American West Overseas Association [US]

AWOC All-Weather Operations Committee

AWOL Absent Without (Official) Leave [also awol]

AWOP All-Weather Operations Panel [of International Civil Aviation Organization]

AWOS Airport Weather Observing System

AWOS Automated Weather Observing System

AWP Allied Weather Publication

AWP Annual Work Plan

AWP Antarctic Wilderness Park

AWP Average Weighted Pressure

AWPA American Wire Producers Association [US]

AWPA American Wood Preservers Association [US]

AWPB American Wood Preservers Bureau [US]

AWPI American Wood Preservers Institute [US]

AWRA American Water Resources Association [US]

AWRA Australian Welding Research Association

AWRAC Australian Water Research Advisory Council

AWRAP Australian Wool Research and Promotion Organization

AWRC Australian Water Resources Council

AWRC Australian Wool Realization Council

AWRE Atomic Weapons Research Establishment [UK]

AWRF Associated Wire Rope Fabricators [US]

AWRNCO Aircraft Warning Company [US Marines]

AWS Acoustic Wave Simulation

AWS Advanced Workstations and Systems (Group) [of IBM Corporation, US]

AWS Agricultural Wholesale Society [UK]

AWS AIDS Associated Wasting Syndrome [Medicine]

AWS Aircraft Warning Service

AWS Air Warning System

AWS Air Weapon System

AWS Air Weather Service

AWS American War Standards [US]

AWS American Welding Society [US]

AWS Association of Women Scientists

AWS Automatic Weather Station

AWS/AMS American Welding Society/Aerospace Material Specification (Classification)

AWS/ANSI American Welding Society/American National Standards Institute (Welding Code)

AWSC American Waterways Shipyard Conference [US]

AWSF Alpha Waste Storage Facility [Nuclear Engineering]

AWS FMC American Welding Society Filler Metal Comparison Chart

AWSR Adler-Weisberger Sum Rule

AWSS Association of Women Soil Scientists [US]

AWST Association of Women Science Teachers [UK]

AWT Abstract Window Toolkit [Java]

AWT Abwassertechnik [German Journal on Wastewater Technology] [also awt]

AWT Actual Work Time

AWT Advanced Wastewater Treatment

AWT Arbeitsgemeinschaft Wärmebehandlung und Werkstofftechnik [Heat Treating and Materials Technology Association, Germany]

AWT Automatic Weld Tube

AWTAC American Welding Technology Application Center [now American Welding Institute, US]

AWTAO Association of Water Transportation Accounting Officers [US]

AWTE Association for World Travel Exchange [US]

AWTP Advanced Waste Treatment Process

AWTS Atlantic Wind Test Site [Prince Edward Island, Canada]

AWU Agricultural Workers Union [South Africa]

AWU Associated Western Universities (Inc.) [Salt Lake City, Utah, US]

AWU Association for the World University [US]

AWU Atomic Weight Unit [also awu]

AWU NW Associated Western Universities (Inc.)–Northwest [US]

AWV Ausschuß für Wirtschaftliche Verwaltung [Committee for Economic Administration, Germany]

AWVS American Women's Voluntary Service [US Military]

AWW Alert Weather Watch

AWWA American Water Works Association [US]

AWWA Australian Water and Wastewater Association

AWWA J AWWA Journal [of American Water Works Association]

AWWM Association of Wholesalers of Woollen Merchants [UK]

AWWRF American Water Works Research Foundation [US]

AWY Airway [also Awy, or awy]

AX Activity of Component "X" (in Thermodynamics) [Symbol]

AX Architecture Extended

AX Attack Experimental

AX Automatic Exchange

AX Automatic Transfer

AX Automatic Transmission

AX Molecule Composed of Metal Atom(s) A and Halide Atom(s) X [Symbol]

AX Rocksalt Type Compound with Equal Number of Cations of Substance A and Anions of Substance X [Symbol]

AX$_2$ Fluorite Type Compound Having Cations of Type A and Anions of Type X in Which the Cation to Anion Ratio is 2:1 (e.g., CaF_2, UO_2, ThO_2, etc.) [Symbol]

A$_m$X$_p$ Compound of m Cations of Substance A and p Anions of Substance X [Ceramics]

Ax Annex

Ax Axiom [also ax]

A(x) Projected Area of (Hardness) Indenter [Symbol]

ax axial

AXAA Australian X-Ray Analytical Association

AXAF Advanced X-Ray Astrophysics Facility [of NASA]

AXAF-I Advanced X-Ray Astrophysics Facility–Imaging [of NASA]

AXAFS Atomic X-Ray Absorption Fine Structure

AXBT Airborne Expendible Bathythermograph

AXCP Airborne Expendible Current Profiler

AXD Auxiliary Drum

AXFMR Automatic Transformer

AXION Workshop on Axions [Particle Physics]

AXP Axial Pitch

AXREM Association of X-Ray Equipment Manufacturers [also AxrEM]

AXS Anomalous X-Ray Scattering

AY Academic Year

A-y Ampere-year [also A-yr]

αY Alpha Yttrium [Symbol]

αYb Alpha Ytterbium [Symbol]

AYC American Yorkshire Club [US]

AYD American Yarn Distributors [US]

AYEI Association of Yugoslav Electrical Industry

a-YIG Amorphous Yttrium-Iron Garnet

A-y/kg Ampere-year(s) per kilogram [also A-yr/kg]

a-YNi Yttrium Nickel Amorphous Alloy

αY$_2$O$_3$ Alpha Yttria (or Alpha Yttrium Oxide)

AYRS Amateur Yacht Research Society [UK]

AYSA American Yarn Spinners Association [US]

AYSM Association of Yugoslav Societies for Microbiology

AZ Absolute Zero (i.e., $-273.16°C$, $-459.69°F$, or 0 Kelvin)

AZ Academy of Zoology [India]

AZ Alloyed Zone

AZ Aluminum Zirconate

AZ Amorphous Zone

AZ Arizona [US]

AZ Auroral Zone [Geophysics]

AZ Azerbaijan [ISO Code]

AZ Azimuth [also Az, or az]

Az Azobenzene

Az Azores

az azure

.az Azerbaijan [Country Code/Domain Name]

AZA American Zinc Association [US]

azacapten 8-Methyl 6,10,19-Trithia-1,3,13,16-Tetraazabicycl[6.6.6]eicosane

azaMEsar 8-Methyl-1,3,6,10,13,16,19-Heptaazabicyclo-[6.6.6]eicosane

AZAS Adjustable Zero Adjustable Span

azasartacn 1,4,7,9,11,14,19-Heptaazatricyclo-[7.7.4.2(4,14)]docosane

AZBN Azobisisiobutyronitrile

AZDN Azodiisobutyronitrile

AZE Anomalous Zeeman Effect [Spectroscopy]

AZEL Azimuth Elevation

Azerbayd Azerbaydzhan

Azerbayd SSR Azerbaydzhan Soviet Socialist Republic [USSR]

AZG American Zinc Gauge

AZGS Arizona Geological Survey [also AzGS, US]

Azido-FDA Azidofluorescein Diacetate

Azim Azimuth(al) [also Azm, Azim, or azim]

αZnS Alpha Zinc Sulfide (or Wurtzite)

AZO Aluminum-Doped Zinc Oxide

AZON Azimuth Only [also azon]

AZQ 2,5-bis[1-Aziridinyl]-3,6-bis-carboethoxyamino-1,4-Benzoquinone

AZQ Diaziquone

a-Zr Amorphous Zirconium [Symbol]

αZr Alpha Zirconium [Symbol]

AZC Ammonium Citrato(oxo)zirconate (IV) Dihydrate

AZRAN Azimuth and Range

Azrd Aziridine

a-ZrO$_2$ Amorphous Zirconia

αZr(O) Oxygen-Doped Alpha Zirconium

AZS Alumina-Zirconia-Silica (Ceramics)

AZS Automatic Zero Set

AzSSR Azerbaidzhan Soviet Socialist Republic [USSR]

AZT Alumina-Zirconia-Titania (Ceramics)

AZT Azidodeoxythymidine; 3'-Azido-3'-Deoxythymidine [Biochemistry]

AZT Azidothymidine [Anti-AIDS Drug]

Azt Azetidine [Biochemistry]

AZT-BSA Azidothymidine–Bovine Serum Albumin (Immunogen)

AZU Akademija Znanosti i Umjetnosti [Academy of Sciences and Fine Arts, Yugoslavia]

Azx Azoxy

B

B Additional Designation of Emission (or Transmission) Type–Two Independent Sidebands [Symbol]

B Adsorption Coefficient [Symbol]

B Area of Base of a Solid [Symbol]

B Asparagine [Biochemistry]

B Aspartic Acid [Biochemistry]

B Bachelor

B Bacillus

B Background

B Backlash (of Gears) [Symbol]

B Bacteria(l); Bacterium

B Bactericidal; Bactericide [also b]

B Badger

B Baghdad [Iraq]

B Bahamas

B Bahía [Brazilian State and City]

B Bahrain

B *(Baie)* – French for "Bay"

B Bain Distortion [Symbol]

B Bainite; Bainitic [Metallurgy]

B Baku [Azerbaidzhan]

B Balance; Balancing

B Balantidium [Genus of Protozoans]

B Bale

B Balboa [Currency of Panama]

B Ball(ing)

B Ballast Tube

B Ballistic(s)

B Balloon

B Baltimore [Maryland, US]

B Band(ing)

B Bandwidth

B Bangalore [India]

B Bangkok [Thailand]

B Bangladesh

B Bank(ing)

B Bar

B Barbados

B Barcelona [Spain]

B Bar Size [Symbol]

B Baryon [Particle Physics]

B Basal

B Base

B Base [Semiconductor Symbol]

B Basel (or Basle) [Switzerland]

B Basic

B Basis

B Bass

B Bat

B Batch

B Battery

B Baud

B Baumé (Gravity) [also Bé]

B Bauxite

B Bavaria(n) [Germany]

B Bay

B Bayerite [Mineral]

B Beacon

B Beam(ing)

B Bear

B Bearing

B Beat(ing)

B Beaver

B Behavio(u)r

B Beirut [Lebanon]

B Bel [Unit]

B Belfast [Northern Ireland]

B Belgian; Belgium

B Belgrade [Yugoslavia]

B Belize

B Bell

B Belugou (Coefficient)

B Bend(ing); Bender

B Benin

B Benzene

B Bergen [Norway]

B Bering (Standard Time)

B Berkeley [California, US]

B Berlin [Germany]

B Bern(e) [Switzerland]

B (Upper-case) Beta [Greek Alphabet]

B Betula [Genus of Deciduous Trees Including the Birches]

B B Horizon [Horizon of Soil Eluviation Consisting of Several Subhorizons B_1, B_2, etc.] [Symbol]

B Bias-Belted (Tire)

B Biceps [Anatomy]

B Bicuspid [Dentistry]

B Bilateral Hole System (of Fits) [Symbol]

B Bile

B Binary

B   Binocular(s)

B   Biological; Biologist; Biology

B   Biomedical Engineering (Technology) [Discipline Category Abbreviation]

B   Biotin [Biochemistry]

B   Bipolar(ity)

B   Birmingham [Alabama, US]

B   Birmingham [UK]

B   Birth

B   Bissau [Guinea-Bissau]

B   Bit

B   Bit Cell Length [Magnetic Media]

B   Black [On Pencils]

B   Blackbody

B   Bladder

B   Blade

B   Blank(ing)

B   Blastoma [Medicine]

B   Blastomyces [Genus of Fungi]

B   Blastula [Embryology]

B   Blight

B   Blind

B   Bloch Constant (in Solid-State Physics) [Symbol]

B   Block(ing)

B   Blood

B   Blood Group "B" [Contains Red Blood Cells with Substance "B" on Surface]

B   Blue

B   Blue Primary (Color) [Symbol]

B   Body

B   Body Length (of Drills) [Symbol]

B   Bogota [Columbia]

B   Boil(er); Boiling

B   Boise [Idaho, US]

B   Bold(type)

B   Bolívar [Currency of Venezuela]

B   Bolivia(n)

B   Bologna [Italy]

B   Bomb; Bombing

B   Bombay [India]

B   Bomber [US Air Force Symbol]

B   Bombyx [Genus of Insects Including the Silkworm]

B   Bond(ing)

B   Bond Number (in Fluid Mechanics) [Symbol]

B   Bond Strength [Symbol]

B   Bone

B   Bonn [Germany]

B   Book

B   Boom

B   Boost(er)

B   Bordeaux [France]

B   Border

B   Borneo

B   Boron [Symbol]

B   Boston [Massachusetts, US]

B   Botrytis [Genus of Fungi]

B   Botswana

B   Bottle

B   Bottom(ing)

B   Bottoming Chamfer (of Thread Taps) [Symbol]

B   Bound(ing)

B   Boundary

B   Bovine

B   Brace

B   Braid(ing)

B   Brain

B   Branch(ing)

B   Brasilia [Brazil]

B   Brass

B   Bravo [Phonetic Alphabet]

B   Brazil(ian)

B   Brazing

B   Breadth

B   Break

B   Breath(ing)

B   Breed(er); Breeding

B   Bremen [Germany]

B   Brewster [Unit]

B   Bridgeport [Barbados]

B   Bright(ness)

B   Brightness (or Luminance) [Symbol]

B   Brisance

B   Brisbane [Australia]

B   Bristol [UK]

B   Britain; British

B   British Thermal Unit

B   Brittle(ness)

B   Brno [Czech Republic]

B   Broadcast(er); Broadcasting

B   (Spectral Line) Broadening

B   Broken

B   Bromination

B   Bronchi(al); Bronchiole; Bronchitis

B   Broth

B   Brother

B   Brucella [Genus of Bacteria]

B   Brush(ing)

B   Brussels [Belgium]

B   B-Size Sheet (of Paper) [11" × 17"]

B   Bubble; Bubbling

B   Bucharest [Romania]

B   Bud

B   Budapest [Hungary]

B   Buff(ing); Buffer

B   Buffalo [New York State, US]

B   Bulb

B   Bulgaria(n)

B Bulge; Bulging

B Bulk(ing)

B Bulk Compliance [Symbol]

B Bulk Modulus (of Volume Elasticity)[Symbol]

B Bundle(d)

B Buoy(ancy); Buoyant

B (Total) Burgers Vector (in Crystallography) [Symbol]

B Burma; Burmese

B Burn(er); Burning

B Bursa(e) [Anatomy]

B Burst(er); Bursting

B Burundi

B Bus

B Butadiene

B Butane

B Butanol

B Butoxide

B Butyl(ate)

B Byte

B Cubic Boron Nitride Type (Grinding Wheel) Abrasive [Symbol]

B Damping Force [Symbol]

B Dedendum (of Gear Teeth) [Symbol]

B Deep-Level Entropy Factor [Deep-Level Transient Sprectroscopy]

B (Crystallographic) Dislocation Density Function [Symbol]

B Dorn Constant [Symbol]

B Drag Coefficient [Symbol]

B Dry Climates [Classification Symbol]

B Flux Line Density [Symbol]

B Formability [Symbol]

B Fraunhofer "B" Line [Red Fraunhofer Line of Terrestrial Origin Having a Wavelength of 686.7 Nanometers] [Spectroscopy]

B Grade Denoting Above-Average but not Outstanding Work

B Hall-Petch Slope (in Metallurgy) [Symbol]

B Height of Barometric Pressure [Symbol]

B Hildebrandt's Constant (in Mining) [Symbol]

B Influenza, Strain B [Symbol]

B Liquidity Index (or Water-Plasticity Ratio) [Symbol]

B Luminance (or Brightness) [Symbol]

B Magnetic Flux (Density) [Symbol]

B Magnetic Flux Vector [Symbol]

B Magnetic Hyperfine Field [Symbol]

B Magnetic Induction [Symbol]

B Magnetization [Symbol]

B Maryland Influenza Strain [Immunology]

B Multiple Launched (Vehicle) [US Department of Defense Symbol]

B Normal (Magnetic) Induction [Symbol]

B Overall Breadth of Compressive Flange of (Reinforced Concrete) Beam [Symbol]

B Phase [Symbol]

B Phonon and Electron Drag Coefficient [Symbol]

B Pitch Line Circumference (of Thrust Bearings) [Symbol]

B Planck Function [Symbol]

B Resinoid Type (Grinding Wheel) Bond

B Rotational Constant (in Physical Chemistry) [Symbol]

B Seventh Tone in the Scale of C Major [Acoustics]

B Specimen Thickness [Symbol]

B Star Group of Spectral Type B [Surface Temperature 13,900°C, or 25,000°F] [Letter Designation]

B Strain Displacement Matrix [Symbol]

B Susceptance [Symbol]

B Telegraph Speed in Bauds [Symbol]

B Thickness [Symbol]

B Vertex (or Associated Polygonal Angle) [Symbol]

B- Taiwan [Civil Aircraft Marking]

B- Boeing (Aircraft) [Followed by a Numeral, e.g., B-737 stands for Boeing 737]

-B Special Armament (Aircraft) [US Air Force Suffix]

.B Static Binary File [File Name Extension] [also .b]

B$_{||}$ Magnetic Flux Parallel to Magnetic Field [Symbol] [also B||]

B$_\perp$ Magnetic Flux Perpendicular to Magnetic Field [Symbol] [also B⊥]

B° Baumé Degree [Symbol]

B+ B Plus (of Anode Supply Terminal) [Symbol]

B− B Minus (of Anode Supply Terminal) [Symbol]

B$_1$ Magnetoelastic Coupling (in Solid-State Physics) [Symbol]

B$_1$, B$_2$ Permissible Average and Local Bond Stresses [Symbol]

B$_1$ Uppermost Horizon of Soil Illuviation–Transitional Zone [Symbol]

B$_2$ Middle Horizon of Soil Illuviation with Maximum Deposition [Symbol]

B$_3$ Bottom Horizon of Soil Illuviation–Transitional Zone [Symbol]

B^{3+} Boron Ion [also B^{+++}] [Symbol]

B^{3-} Boride [Symbol]

B$_a$ Anisotropic Magnetic Field [Symbol]

B$_b$ Biased Induction [Symbol]

B$_C$ Capacitive Susceptance [Symbol]

B$_c$ Core Flux Density (of Transformer) [Symbol]

B$_c$ Critical Magnetic Flux Density (for Superconductivity) [Symbol]

B$_d$ Flux Density [Symbol]

B$_d$ Remanent Induction [Symbol]

B$_d$ Total External Permeance [Symbol]

B$_\Delta$ Incremental Induction [Symbol]

B$_{dm}$ Magnetic Remanence [Symbol]

B$_f$ Bainitic Finish Temperature (in Metallurgy) [Symbol]

B$_g$ Gravity Bond Number [Symbol]

B$_{hf}$ Magnetic Hyperfine Field [also B$_{hf}$] [Symbol]

B$_{hkl}$ X-Ray Diffraction Peak Width at Half Maximum [Symbol]

B$_i$ Internal Field [Symbol]

B_i Intrinsic Induction [Symbol]

$(B_i)_P$ Saturation Induction [Symbol]

B_{is} Intrinsic Saturation Induction [Symbol]

B_k Permeability Bond Number [Symbol]

B_L Inductive Susceptance [Symbol]

B_M Mercury-Vapor Boiler [Symbol]

B_m Amplitude of Field Change [Symbol]

B_m Maximum Adsorption Coefficient [Symbol]

B_m Maximum Induction [Symbol]

B_m Saturation Flux Density (of Transformer) [Symbol]

B_{max} Maximum Induction [Symbol]

B_N Net Thickness (of Fracture-Testing Specimen) [Symbol]

B_n Debye Parameter of Atom n [Symbol]

B_p Change in Applied Field [Symbol]

B_p Columnar Pin Density (in Physics) [Symbol]

B_Φ (Tangent) Magnetic Vector [Symbol]

B_r Remanent Magnetization (or Magnetic Remanence) [Symbol]

B_r Residual Induction [Symbol]

$B_r{'}$ Apparent Residual Induction [Symbol]

B_{rs} Magnetic Retentivity [Symbol]

B_S Steam Boiler [Symbol]

B_s Bainitic Start Temperature (in Metallurgy) [Symbol]

B_s (Magnetic) Saturation Induction (or Saturation Flux Density) [Symbol]

B_z Magnetic Field in z Direction [Symbol]

/B   Build All Option [Turbo Pascal]

B-8   Boron-8 [also ^{8}B, or B^8]

B-10   Boron-10 [also ^{10}B, or B^{10}]

B-11   Boron-11 [also ^{11}B, or B^{11}]

B-12   Boron-12 [^{12}B, or B^{12}]

b   additional designation of emission (or transmission) type–two independent sidebands [Symbol]

b   auxiliary switch (breaker) normally closed [Controllers]

b   bactericide [also B]

b   bale

b   bandwidth

b   bar [Unit]

b   barn [Unit]

b   base

b   base [Semiconductor Symbol]

b   base (length) of a plane polygon [Symbol]

b   Basquin's exponent (or fatigue strength exponent) [Symbol]

b   bath

b   battery

b   bay

b   bed plate

b   bel [Unit]

b   bench

b   beta [English Equilvalent]

b   binary

b   bicuspid [Dentistry]

b   bit

b   blackbody

b   blend

b   blue

b   boiling

b   (atomic) bond order [Symbol]

b   born

b   bottom

b   bottom-type quark (in particle physics) [Symbol]

b   brass

b   breadth [Symbol]

b   brick

b   bright

b   broad

b   broad [Spectroscopy]

b   buoyancy; buoyant

b   Burgers vector (in materials science) [Symbol]

b   burner

b   coefficient [Symbol]

b   constant [Symbol]

b   crystal lattice length along b-axis [Symbol]

b   damping constant (or mechanical resistance) [Symbol]

b   dedendum (of gear pinion teeth, worm threads, etc.) [Symbol]

b   face width (of gears) [Symbol]

b   half-width (in x-ray diffraction) [Symbol]

b   imaginary part (of a complex number) (e.g., a is the real part and b the imaginary part of the complex number $a + bj$) [Symbol]

b   length (of side) of a triangle, or polygon [Symbol]

b   line segment (in geometry) [Symbol]

b   molality [Symbol]

b   phase [Symbol]

b   remaining ligament (in fracture testing) [Symbol]

b   sample optical path length [Symbol]

b   sample path length (of spectrophotometer) [Symbol]

b   semi-minor axis of an ellipse, or ellipsoid [Symbol]

b   semitransverse axis of a hyperbola [Symbol]

b   susceptance (in alternating-current circuit) [Symbol]

b   van der Waals volume constant [Symbol]

b   warmest month below 71.6°F (22°C) [Subtype of Climate Region, e.g., in Cb, Cbf, Cbw, Db, Dbf, etc.]

b   width [Symbol]

b   y-intercept of a curve, line, plane, or surface [Symbol]

$b_{||}$ parallel to (or in-plane with) (crystal) unit cell axis "b" [Symbol]

$b_\perp$ perpendicular (or normal) to (crystal) unit cell axis "b" [Symbol]

b_0 original uncracked ligament (in fracture testing) [Symbol]

b_1 Fraunhofer line for magnesium having wavelength of 517.8 nanometers [Spectroscopy]

b_2 Fraunhofer line for magnesium having wavelength of 517.3 nanometers [Spectroscopy]

b_c Burgers vector component perpendicular to crack plane (in materials science) [Symbol]

b_e edge component of Burgers vector (in materials science) [Symbol]

b_F Frank-type partial Burgers vector (in materials science) [Symbol]

b_{fs} common-source small-signal forward transfer susceptance [Semiconductor Symbol]

b_i molality [Symbol]

b_{is} common-source small-signal input susceptance [Semiconductor Symbol]

b_j reciprocal basis vector (of crystal structures) [Symbol]

b_{LC} (Burgers vector of a) Lomer-Cottrell dislocation (in materials science) [Symbol]

b_n scattering length (in neutron diffraction) [Symbol]

b_n second Fourier coefficient (in Fourier Series) [Symbol]

b_{os} common-source small-signal output susceptance [Semiconductor Symbol]

b_p (Burgers vector of a) partial dislocation (in materials science) [Symbol]

b_{rs} common-source small-signal reverse transfer susceptance [Semiconductor Symbol]

b_S Shockley-type partial Burgers vector (in materials science) [Symbol]

b_s crystal lattice length along b-axis [Symbol]

b_s screw component of Burgers vector (in materials science) [Symbol]

β absorption coefficient [Symbol]

β (plane) angle [Symbol]

β angle of obliquity (of soil, or rock solids) [Symbol]

β angle of refraction (or refraction angle) [Symbol]

β anisotropy constant [Symbol]

β (lower-case) beta [Greek Alphabet]

β beta factor (in classical physics) [Symbol]

β beta particle [Symbol]

β beta phase (of a material) [Symbol]

β bifurcation parameter (in solidification theory) [Symbol]

β Bohr magneton [Symbol]

β brightness [Symbol]

β celestial latitude [Symbol]

β coefficient [Symbol]

β coefficient of volumetric thermal expansion [Symbol]

β common emitter current gain [Semiconductor Symbol]

β (coefficient of) compressibility [Symbol]

β consumer's risk [Symbol]

β (x-ray) diffraction breadth [Symbol]

β dislocation density (in materials science) [Symbol]

β Dundurs parameter (in cracking theory) [Symbol]

β electron moment (9.27×10^{-24} A·m²) [Symbol]

β (grain boundary, or interfacial) enrichment factor (in materials science) [Symbol]

β (statistical) error of the second kind [Symbol]

β fluid bulk modulus [Symbol]

β growth exponent [Symbol]

β (magnetic) hysteretic angle [Symbol]

β imaginary part of refractive index n [Symbol]

β integral (peak) breadth (in x-ray diffraction) [Symbol]

β integration constant [Symbol]

β inverse temperature [Symbol]

β Kohlrausch-Williams-Watts Exponent (in materials science) [Symbol]

β lattice parameter–crystallographic angle between a and c axis of unit cell [Symbol]

β linear density [Symbol]

β load angle (of roller bearings) [Symbol]

β Mach angle [Symbol]

β modulation index (i.e. ratio of frequency swing to modulation frequency) [Symbol]

β overall (amplifier) gain [Symbol]

β phase [Symbol]

β phase (or wavelength) constant [Symbol]

β phase factor (in solid-state physics) [Symbol]

β pitch cone angle (of bevel gear pinions) [Symbol]

β prefix used with aromatic compounds to denote the side-chain attachment of a chemical unit

β prefix used to indicate the position of a substituent atom, or radical

β prefix used to indicate the second position in a naphthalene ring

β radiance factor [Symbol]

β reaction zone of glass used for nuclear-waste end storage [Symbol]

β reduced volume (i.e. ratio of volume to critical volume, V/V_c) [Symbol]

β relaxation time (i.e., mean life of an energy quantum) [Symbol]

β second-direction angle of a line angle with the y-axis [Symbol]

β second largest principal refractive index of a biaxial crystal [Symbol]

β specific heat constant [Symbol]

β spin wave function giving probability p = 1 for spin s = $-\frac{1}{2}$ and p = 0 for s = $\frac{1}{2}$ [Symbol]

β Tafel Constant (or Tafel Slope) (in electrochemistry) [Symbol]

β thermal emission (of electrons and holes) [Symbol]

β wave number [Symbol]

β wedge angle (n maching) [Symbol]

β^* critical wave number [Symbol]

β' beta prime phase (of a material) [Symbol]

β'' beta double-prime phase (of a material) [Symbol]

β^+ positive beta particle (or positron) [Symbol]

β^- negative beta particle (or negatron, or electron) [Symbol]

β_0 major depletion zone of reacted glass used for nuclear-waste end storage [Symbol]

β_1 gradient zone of reacted glass used for nuclear-waste end storage [Symbol]

β_2 diffusion zone of reacted glass used for nuclear-waste end storage [Symbol]

β_2 solidification shrinkage [Symbol]

β_a anodic Tafel constant (in electrochemistry) [Symbol]

β_C junction capacitance [Symbol]

β_c cathodic Tafel constant (in electrochemistry) [Symbol]

β_E eutectic of second solid phase (in metallurgy) [Symbol]

β_{hkl} integral peak breadth for (hkl) lattice plane set (in x-ray diffraction) [Symbol]

β_i enrichment ratio (of fractured grain boundaries, e.g., β_P is the enrichment ratio for phosphorus) (in materials science)

β_L inductance of two-dimensional array unit cell [Symbol]

β_n overall formation constant [Symbol]

β_P second solid phase in peritectic system (in metallurgy) [Symbol]

β_σ Bauschinger stress parameter [Symbol]

BA   Bachelor of Arts

BA   Background Activity

BA   Ball Bearing, Single-Row, Angular-Contact, Non-Separable, Nominal Contact Angle from Above 22° to 32° [Symbol]

BA   Bank Angle [also B/A]

BA   Barium Aluminate

BA   Basal Area

BA   Batch-Annealed; Batch Annealing [Metallurgy]

BA   Bayard-Alpert (Gauge)

BA   Beam Aerial

BA   Beam Attenuation ; Beam Attenuator

BA   Bearing Alloy

BA   Bend Allowance

BA   Bell Atlantic (Company) [US]

BA   Benicia Arsenal [California, US]

BA   Benzaldehyde

BA   Benzanthracene

BA   Benzla Acetone

BA   Benzoic Acid

BA N^6-Benzyladenine [Biochemistry]

BA   Benzylamine

BA   6-Benzylaminopurine

BA   Benzalacetone

BA   Binary Adder; Binary Addition

BA   Biological Abstracts [of Biosciences Information Service, US]

BA   Biological Activity

BA   Bipolar Amplifier

BA   Blast Atomizer

BA   Blocking Antibody [Immunology]

BA   Body Axis [Aerospace]

BA   Boiler Availability

BA   Bond Angle [Physical Chemistry]

BA   Booksellers Association [UK]

BA   Boolean Algebra

BA   Booster Assembly [of NASA]

BA   Boric Acid

BA   Born Approximation [Solid-State Physics]

BA   Bosnia-Herzegovina [ISO Code]

BA   Boundary Angle

BA   Box Annealing [Metallurgy]

BA   Bragg Angle [Solid-State Physics]

BA   Braking Action

BA   Breathing Air

BA   Breathing Apparatus

BA   Brewster Angle

BA   Bright Annealing [Metallurgy]

BA   British Academy

BA   British Admiralty

BA   British Airways

BA   British Aluminium (Company)

BA   British Association (Screw Thread)

BA   Bromoacetone

BA   Bromoalcohol

BA   Bromoalkane

BA   Bronze Age [Archeology]

BA   Buenos Aires [Argentinean City and Province]

BA   Buffer Amplifier [also B/A]

BA   Budget Allocation

BA   Budget Authority

BA   Bundesanstalt für Arbeit [Federal Labor Ministry, Germany]

BA   Bureau of Aeronautics [Washington, DC, US]

BA   Burning Area

BA   Bus Architecture [Computers]

BA   Business Administration; Business Administrator

BA   Butyl Acetate

BA   Butylamine

BA   Butyric Acid

BA_6 Barium Hexaaluminate [$BaO \cdot 6Al_2O_3,$]

BA_4 Barium Tetraaluminate [$BaO \cdot 4Al_2O_3$]

B_3A Tribarium Aluminate [$3BaO \cdot Al_2O_3$]

B/A   Bank Angle [also BA]

B/A   Barometric Altimeter

B/A   Boron/Aluminum

B/A   Buffer Amplifier [also BA]

B-A   Bayard-Alpert (Gauge) [also B/A]

Ba   Bacillus anthracis [Microbiology]

Ba   Baddeleyite [Mineral]

Ba   Balboa [Currency of Panama]

Ba   Barium [Symbol]

Ba^{2+} Barium Ion [also Ba^{++}] [Symbol]

Ba II   Barium II [Symbol]

Ba-129   Barium-129 [also ^{129}Ba, or Ba^{129}]

Ba-130   Barium-130 [also ^{130}Ba, or Ba^{130}]

Ba-131   Barium-131 [also ^{131}Ba, or Ba^{131}]

Ba-132   Barium-132 [also ^{132}Ba, or Ba^{132}]

Ba-133   Barium-133 [also ^{133}Ba, or Ba^{133}]

Ba-134   Barium-134 [also ^{134}Ba, or Ba^{134}]

Ba-135   Barium-135 [also ^{135}Ba, or Ba^{135}]

Ba-136   Barium-136 [also ^{136}Ba, or Ba^{136}]

Ba-137   Barium-137 [also ^{137}Ba, or Ba^{137}]

Ba-138   Barium-138 [also ^{138}Ba, or Ba^{138}]

Ba-139 Barium-139 [also ^{139}Ba, or Ba^{139}]

Ba-140 Barium-140 [also ^{140}Ba, or Ba^{140}]

Ba-141 Barium-141 [also ^{141}Ba, or Ba^{141}]

Ba-142 Barium-142 [also ^{142}Ba, or Ba^{142}]

Ba-143 Barium-143 [also ^{143}Ba, or Ba^{143}]

Ba-144 Barium-144 [also ^{144}Ba, or Ba^{144}]

Ba-213 Ba$_2$YCu$_3$O$_{7-x}$ [A Barium Yttrium Copper Oxide Superconductor]

Ba-2121 Ba$_2$YCu$_2$WO$_y$ [A Barium Yttrium Copper Tungsten Oxide Superconductor]

Ba-4112 Ba$_4$YCuW$_2$WO$_{6-y}$ [A Barium Yttrium Copper Tungsten Oxide Superconductor]

Ba-8314 Ba$_8$Y$_3$CuW$_4$WO$_{24-z}$ [A Barium Yttrium Copper Tungsten Oxide Superconductor]

.ba Bosnia-Herzegovina [Country Code/Domain Name]

BAA Bachelor of Applied Arts

BAA Ball Bearing, Double-Row, Angular-Contact, Non-Separable, Contact Angle Vertex Inside

Bearing, Two-Piece Outer Ring [Symbol]

BAA Nα-Benzoyl-L-Arginine Amide [Biochemistry]

BAA Biodiversity Analysis for Africa

BAA British Accounting Association

BAA British Acetylene Association

BAA British Agrochemicals Association

BAA British Airports Authority

BAA British Archeological Association

BAA British Astronomical Association

BAA Broad Agency Announcement [of US Department of Defense]

BAA Broadband Antenna Amplifier

BAA N-n-Butylacetanilide

BAA Butyraniline Aldehyde

BA$_1$A$_2$ Layered Heterostructure of Nonmetal Component B, Metal Component A_1 and Metal Component A_2 [Symbol]

BAAA British Association of Accountants and Auditors

BAAC British Association of Aviation Consultants

BAAC British Aviation Archeological Council

Ba(Ac)$_2$ Barium Acetate [also Ba(ac)$_2$]

Ba(ACAC)$_2$ Barium Acetylacetonate [also Ba(AcAc)$_2$, or Ba(acac)$_2$]

BAACMIR British Aerospace Air Combat Maneuvering Instrumentation Range

BAAEMS British Association of Airport Equipment Manufacturers and Services

Ba[Al((OBu^{sec})$_3$]$_2$ Barium (II) Aluminum sec-Butoxide [also Ba[Al((OBu^s)$_3$]$_2$]

BAAP 4-N-(Benzylidene)-Aminoantipyrine [Biochemistry]

BAAR Board for Aviation Accident Research [US Army]

BAArch Bachelor of Arts in Architecture

BAAS Bristol and Avon Archeological Society [UK]

BAAS British Association for the Advancement of Science

BAB 2-Benzoylamino Butanoic Acid

BAB Biotechnology and Applied Biochemistry [Journal of the International Union of Biochemistry and Molecular Biology, UK]

BAB Bundesautobahn [German Express Highway]

B$_1$AB$_2$ Layered Heterostructure of Nonmetal Component B_1, Metal Component A_1 and NonmetalComponent B_2 [Symbol]

Bab Babbitt [also bab]

Bab Babylon(ian)

BABA Balloons Bassee Altitude

BABA α-N-Benzoylamino Butanoic Acid

BABA Biacetyl-bis(anil)

BABA British Air Boat Association

BABA British Anaerobic and Biomass Association (Limited)

BABA British Artist Blacksmiths Association

BABM 2-Benzoylamino Butanoic Acid Methyl Ester

BABS Background Analysis Center

BABS Barometric Altitude Control

BABS Beam Approach Beacon System

BABS Blind Approach Beacon System

BABS British Aluminum Building Service

BABS British Association for Brazing and Soldering

BABS Binary Asymmetric Channel

BABS Bird Association of California [US]

BABS N,N'-bis(acryloyl)-Cystamine [Biochemistry]

BABS 1,3-bis(2-Aminoethyl)cyclohexane

BABS Block Advisory Committee [India]

BABS Blood Alcohol Concentration

BABS British Airways Corporation

BABS British Association of Chemists

BABS British Atlantic Committee

BABT British Approval Board for Telecommunications

Babyl Babylonia(n)

BAC Back Association of Canada

BAC α-N-Benzoylaminocrotonic Acid

BAC Biological Activated Carbon

BAC Biotinamidocaproate

BAC N,N'-bis(Acryloyl)cystamine [Biochemistry]

BAC Boeing Aircraft Company [US]

BAC Booster Assembly Contractor [NASA]

BAC Buffer Access Card

BAC Business Archives Council [UK]

BAc Benzoic Acid

Bac (*Baccalauréat*) – French for "Baccalaureate"

BACA Baltic Air Charter Association [UK]

Bac A Baccalaureate in Art, Communications, Language, Literature and Philosophy [France][now Bac L]

BACAH British Association of Consultants in Agriculture and Horticulture

BACAIC Boeing Airplane Company Algebraic Interpretive Computing System

BACAN British Association for the Control of Aircraft Noise

Bac B Baccalaureate in Business, Current Affairs and Economics [France] [now Bac ES]

Bac C Baccalaureate in Science [France] [Now included in Bac S]

BAcc Bachelor of Accountancy

Bac D Baccalaureate in Mathematics [France] [Now included in Bac S]

BACE Basic Automatic Checkout Equipment

BACE British Association of Consulting Engineers

BACE British Association of Corrosion Engineers

BACE Bureau of Agricultural Chemistry and Engineering [US]

Bac E&F Baccalaureate in Technology [France] [Now included in Bac S]

Bac ES Baccalaureate in Business, Current Affairs and Economics [France] [formerly Bac B]

BACFI British Association of Canned Food Importers [UK]

BACFOX Bacterial Film Oxidation (Process)

BACG British Association for Crystal Growth

BACH N-(+)-Biotinyl-6-Aminocaproic Acid Hydrazide

B-Acid 1-Amino-8-Naphthol-3,5-Disulfonic Acid [also B-acid]

BACIE British Association for Commercial and Industrial Education

BACIU Bricklayers and Allied Craftsmen International Union

BACK Boublik-Alder-Chen-Kreglewski (Equation of State)

Back EMF Back Electromotive Force [also back emf]

Bac L Baccalaureate in Art, Communications, Language, Literature and Philosophy [France] [formerly Bac A]

BACM α-N-Benzoylaminocrotonic Acid Methyl Ester

BACM Best Available Control Measures

BACM British Association of Colliery Management

BACM Business and Commercial Management Group [of Metal Trades Industry Association, Australia]

BACMA British Aromatic Compound Manufacturers Association

BACMI British Aggregate Construction Materials Industries

BACMM British Association of Clothing Machinery Manufacturers

BACO 1,4-Diazabicyclo[2.2.2]octane

BACON British Association for Control of Aircraft Noise

Bacor Baddeleyite-Corundum-Silica (Refractory) [also bacor]

BACP Bandwidth Allocation Control Protocol

BACR British Association for Cancer Research

BACS Banks Automated Clearing System

BACS Bibliographic Access and Control System

BACS Body Axis Coordinate System [Aerospace]

BACS British Association for Chemical Specialties

Bac S Baccalaureate in Science, Mathematics and/or Technology [France] [Formerly subdivided into Bac C, Bac D and Bac E&F]

Bac SMS Baccalaureate in Medicine and Social Sciences [France]

Bac STAE Baccalaureate for Technicians in Agriculture and the Environment [France]

Bac STI Baccalaureate for Technicians in Mechanical, Electrical and Civil Engineering Industries [France]

Bac STL Baccalaureate for Technicians in Industrial Laboratories in Physical and Life Sciences [France]

Bac STPA Baccalaureate for Technicians in Food Production [France]

Bac STT Baccalaureate for Technicians in Commerce and Business [France]

BAC-SulfoNHS Biotinamidocaproate-N-Hydroxysulfosuccinimide Ester

BACT Best Available Control Technology

BACT British Association of Conference Towns

Bact Bacteriological; Bacteriologist; Bacteriology [also bact]

Bact Bacterium; Bacteria

BACTE British Association for Commercial and Technical Education

Bacter Bacteriological; Bacteriologist; Bacteriology [also Bacteriol, bacteriol, or bacter]

Bact Proc Bacterial Proceedings [of American Society of Microbiology, US]

BAD Base Ammunition Depot

Ba$ Bahamian Dollar [Currency of the Bahamas]

BADA British Audio Dealers Association [UK]

BADAS Binary Automatic Data Annotation System

BADC Binary Asymmetric Dependent Channel

BADEA Banque Arabique pour le Développement Economique en Afrique [Arab Bank for Economic Development in Africa]

BADF Bond Angle Distribution Function [Physical Chemistry]

BADGE Base Air Defense Ground Environment

BADGE Base Air Defense Ground Experiment

BADIC Biological Analysis Detection Instrumentation and Control

BAdm Bachelor of Administration [also Badmin]

BAdmEng Bachelor of Administrative Engineering

BAdmin Bachelor of Administration [also Badm]

Ba(DPM)$_2$ Barium (II) Dipivaloylmethanoate (or Bis(dipivaloylmethanato)barium (II)) [also Ba(dpm)$_2$]

BAE Bachelor of Aeronautical Engineering

BAE Bachelor of Agricultural Engineering

BAE Bachelor of Architectural Engineering

BAE Bachelor of Arts in Education

BAE Beacon Antenna Equipment

BAE Belfast Association of Engineers [Northern Ireland]

BAE Bovine Aortic Endothelial (Cell)

BAE British Association Electrolysists [UK]

BAE Bureau of Agricultural Economics [Australia]

BAE Bureau of American Ethnology [of Smithsonian Institution, Washington, DC, US]

BAe British Aerospace

BAEA British Agricultural Exhibitors Association

BAEA British Atomic Energy Authority

BAeA British Aerobatics Association

BAEB Bituminous and Aggregate Equipment Bureau [US]

BAEC Bangladesh Atomic Energy Center [Dhaka, Bangladesh]

BAEC British Agricultural Export Committee

BAEC British Agricultural Export Council

BAEC British Amateur Electronics Club

BAED Bachelor of Arts in Environmental Design

BAEd Bachelor of Arts in Education

BAEE Benelux Association of Energy Economists [Europe]

BAEE Nα-Benzoyl-L-Arginine Ethyl Ester [Biochemistry]

BAeE Bachelor of Aeronautical Engineering

BAER Brainstem Auditory Evoked Response

BAERE British Atomic Energy Research Establishment

BAF B-Cell Activating Factor [Biochemistry]

BAF Bioaccumulation Factor

BAF Bonded Acetate Fiber

BAF British Abrasive Federation

Baf Baffle [also baf]

BAFB Bolling Air Force Base [of US Air Force]

BaFe Barium Ferrite

Ba(FOD)₂ Bis(1,1,1,2,2,3,3-Heptafluoro-7,7-Dimethyl-4,6-Octanedionato)barium [also Ba(fod)₂]

Ba(FOD)₂ Bis(6,6,7,7,8,8,8-Heptafluoro-2,2-Dimethyl-3,5-Octanedionato)barium [also Ba(fod)₂]

BAFM British Association for Forensic Medicine

Bafög Bundesausbildungsförderungsgesetz [Federal Law on Grants for Higher Education, Germany] [also BaföG]

BAFPA British Association of Fencing and Pallet Agents

BAFRA British Aluminum Foil Rollers Association

BA Freiburg Bergakademie Freiburg [Freiburg Academy of Mining, Germany]

BAFS British Association of Forensic Science

BAFSM British Association of Feed Supplement Manufacturers

BAFT Bankers Association for Foreign Trade [US]

BAG Bayard-Alpert Gauge

BAG Beryllium Aluminum Gallate

BAG Bibliographic and Grouping System

BAG British Aviation Group

BAg Bachelor of Agriculture

BAg Silver Brazing Alloy

β-AgAl Beta Silver Aluminum

BAGC Biased Automatic Gain Control

BAGCC British Association of Golf Course Constructors

BAGCD British Association of Green Crop Driers

β-AgCd Beta Silver Cadmium

BAGD Baltimore Academy of General Dentistry [US]

BAgE Bachelor of Agricultural Engineering

BAGMA British Agricultural and Garden Machinery Association

BAGO Beryllium Aluminum Gallium Oxide

BAGS Bullpup All-Weather Guidance System

β-AgZn Beta Silver Zinc

BAH Biologische Anstalt Helgoland [Heligoland Biological Institute, Germany]

Bah Bahamas

BAHC Biosphere Aspects of the Hydrological Cycle (Program) [of International Geosphere/Biosphere Programme]

Ba(HFA)₂ Barium Hexafluoroacetylacetonate [also Ba(hfa)₂]

Ba(HFAC)₂ Barium Hexafluoroacetylacetonate [also Ba(hfac)₂]

BAHPA British Agricultural and Horticultural Plastics Association

BAHS British Agricultural History Society

BAI Bank Administrative Institute [Rolling Meadows, Illinois, US]

BAI Barometric Altitude Indicator

BAI Beratender Ausschuß der Industriephysiker [Advisory Committee "Physicists in Industry" of the German Physical Society]

BAI Baryon-Antibaryon Interactions [Particle Physics]

BAIB Beta-Aminoisobutyric Acid

Ba i-Bu Barium Isobutyrate

BAIC Bureau of Agricultural and Industrial Chemistry [US]

BAID Bachelor of Interior Design

BAIE British Association of Industrial Editors

BA in Ed Bachelor of Arts in Education

BA in J Bachelor of Arts in Journalism

BA in LS Bachelor of Arts in Library Science

BA in Nurs Bachelor of Arts in Nursing

BAINS Basic Advanced Integrated Navigation System

Ba-i-Pr Barium Isopropoxide

BAIR Berkeley Artificial Intelligence Research [of University of California at Berkeley, US]

BAIR Breathing Air [also B Air]

BAISYS Blast Furnace Artificial Intelligence System

BAIT Bacterial Automated Identification Technique

BAJ Boettinger-Aziz-Jackson (Solidification Model) [Metallurgy]

BA(Jour) Bachelor of Arts in Journalism

BAK Baboon Kidney

BAK Basler Arbeitsgruppe für Konjunkturforschung [Basle Working Group for Economic Research, Switzerland]

BAK (Microsoft) Binary Adaptation Kit

.BAK Backup [File Name Extension] [also .bak]

BaK Barium K (Absorption Edge) [X-Ray Analysis]

BAL Balboa [Currency of Panama]

BAL Basic Assembler (or Assembly) Language

BAL Bright Anneal(ing) Line

BAL Bristol Aerospace Limited [Winnipeg, Manitoba, Canada]

BAL British Anti-Lewisite

BAL Business Application Language

BAL 2,3-Dimercapto-1-Propanol

B/Al Boron Reinforced Aluminum (Composite)

B_f/Al Boron Fiber Reinforced Aluminum (Composite)

Bal Balance [also bal]

Bal Ballistic(s) [also bal]

Bal Brevibacterium Albidum [Microbiology]

BaLα Barium L-Alpha (Radiation) [also BaL_α]

β-Ala β-Alanine [Biochemistry]

BaLβ₁ Barium L-Beta 1 (Radiation) [also BaL_β1]

BaLβ₂ Barium L-Beta 2 (Radiation) [also BaL_β2]

BALC Borehole Acoustics and Logging Consortium [of Massachusetts Institute of Technology Earth Resources Laboratory, US]

BALCO Bahrain Saudi Aluminum Marketing Company [also Balco]

β-**AlCu** Beta Aluminum Copper

BALGOL Burroughs Algebraic Compiler

BALGOL Burroughs Algorithmic Language

BALI British Association of Landscape Industries

BALIS Bayerisches Landwirtschaftliches Informations-system (Datenbank) [Bavarian Agricultural Information System (Databank), Germany]

Bal Is Balearic Islands

Ball Ballast [also ball]

BALLAST Balanced Loading via Automatic Stability and Trim

Ball Bear J Ball Bearing Journal [Australia]

BALLOTS Bibliographic Automation of Large Library Operations using a Time-Sharing System [now Research Libraries Information Network, US]

BALLUTE Balloon Parachute

BalMi Ballistic Missile [also BALMI]

BalMort Ballistic Mortar

BALOP Balopticon

β-**Al$_2$O$_3$** Beta Aluminum Oxide (or Beta Alumina)

β'-**Al$_2$O$_3$** Beta Prime Aluminum Oxide

β''-**Al$_2$O$_3$** Beta Double Prime Aluminum Oxide

BALPA British Airline Pilots Association

β-**AlPr** Beta Aluminum Praseodymium

BalPend Ballistic Pendulum

BALS Balancing Set

BALS Blind Approach Landing System

BAlSi Aluminum-Silicon Brazing Alloy

Baltimore Bull Educ Baltimore Bulletin of Education [US]

BALUN Balanced-to-Unbalanced (Transformer)

BAM Bachelor of Applied Mathematics

BAM Ballistic Missile

BAM Balloon Anisotropy Measurements

BAM Basic Access Method

BAM Baykal-Amur Mainline (Railroad) [Siberia]

BAM Bituminous Aggregate Mixture

BAM Block Access Method

BAM Bovine Adrenal Medulla

BAM Boyan Action Module

BAM Brewster-Angle Microscope; Brewster-Angle Microscopy

BAM Broadcasting Amplitude Modulation

BAM Bundesanstalt für Materialforschung und –prüfung [Federal Institute for Materials Research and Testing, Berlin, Germany]

BaM M-Type Barium Ferrite

β-**Am** Beta Americium [Symbol]

BAMBI Ballistic Missile Boost Intercept

BAMBP 4-Butyl-2-(α-Methylbenzyl)phenol

BAMC Brooke Army Medical Center [San Antonio, Texas, US]

BAME Nα-Benzoyl-L-Arginine Methyl Ester [Biochemistry]

BAME British Aerosol Manufacturers Association

BamH Bacillus amyloliquefaciens H [Microbiology]

BAMIRAC Ballistic Missile Radiation Analysis Center [US]

BAMM Balloon Altitude Mosaic Measurement

BAMN Basic Acrylic Monomer Manufacturers Association [US]

BAMO Bureau of Aeronautics Material Officer

BAMPA British Aeromedical Practitioners Association [also BAPA]

BAMPITC 4-(tert-Butoxycarbonyl-Aminomethyl) Phenylisothiocyanate

BAMS Bachelor of Arts, Master of Science

BAMS Brazilian Association of Mechanical Sciences

BAMS Bulletin of the American Mathematical Society [US]

BAMT Boric Acid Mixing Tank

BAMTM British Association of Machine Tool Merchants

BAN Best Asymptotically Normal

BAN British Approved Name

BAN α-Bromo-2'-Acetonaphthone

BAN Budget Allocation Notice

BAN Bulgarskata Akademiya na Naukite [Bulgarian Academy of Sciences, Sofia]

Ban Bacillus aneurinolyticus [Microbiology]

Ban I Bacillus aneurinolyticus I [Microbiology]

Ban II Bacillus aneurinolyticus II [Microbiology]

BANA Nα-Benzoyl-DL-Arginine-β-Naphthylamide [Biochemistry]

BANC British Association of Nature Conservatists

BANCS Bell Administrative Network Communications System

BAND Bandolier [also Band, or band]

B and S Brown and Sharpe (Wire Gauge) [also: B&S]

B and W Black and White [also b and w, or B&W]

Bangladesh J Sci Ind Res Bangladesh Journal of Scientific and Industrial Research [of Bangladesh Council of Scientific and Industrial Research]

BangT Bangalore Torpedo

BANI Nα-Benzoyl-DL-Arginine-4-Nitroanilide [Biochemistry]

BANI Benzoyl Arginine Nitroanilide Hydrochloride [Biochemistry]

BANK Banking [also Bank, or bank]

Bank Syst Technol Bank Systems + Technology [Journal published in the US]

Bank Technol Banking Technology [Journal published in the US]

BANS Basic Air Navigation School

BANS Bright Alphanumeric Subsystem

BANS British Association of Numismatic Societies

B anthracis Bacillus anthracis [Microbiology]

BANZARE British, Australian and New Zealand Antarctic Expedition [1929 to 1931]

Bányász Kohász Lapok (Kohász) Bányászati és Kohászati Lapok (Kohászati) [Hungarian Journal]

Bànyàsz Kohàsz Lapok (Öntöde) Banyaszati és Kohaszati Lapok (Ontode) [Hungarian Journal]

BAO Barium Aluminum Oxide

BAO 2,5-Bis(4-Aminophenyl)-1,3,4-Oxadiazole

BAO Boulder Atmospheric Observatory [Boulder, Colorado, US]

BaO Barium Oxide

Ba(OAc)$_2$ Barium(II) Acetate

BAODA British Association of Operating Department Assistants

Ba(OEt)$_2$ Barium(II) Ethoxide

Ba(OPr^i)$_2$ Barium (II) Isopropoxide

Ba(OR)$_2$ Barium Alkoxide [General Formula]

BaPr^i Barium Isopropoxide

BA ohm British Association ohm [UK]

BAOR British Army of the Rhine

BAP Band Amplitude Product

BAP Basic Assembler Program

BAP Beam-Assisted Processing

BAP Benzalacetophenone

BAP Benzopyrene

BAP Benzyl-P-Aminophenol

BAP 6-Benzylaminopurine

BAP Biologically Active Peptides [Biochemistry]

BAP 1,4-Bis(acryloyl)piperazine

BAP Branch Arm Piping

BAP British Association for Psychopharmacology

6-BAP 6-Benzylamino Purine

BaP Benzopyrene

BAPA British Aeromedical Practitioners Association [also BAMPA]

BAPA British Amateur Press Association

DL-BAPA Nα-Benzoyl-DL-Arginine-p-Nitroanilide [also DL-Bapa, or DL-bapa] [Biochemistry]

L-BAPA Nα-Benzoyl-L-Arginine-p-Nitroanilide [also L-Bapa, or L-bapa] [Biochemistry]

BAPABA Nα-Benzoyl-DL-Arginine-4-Aminobenzoic Acid [Biochemistry]

BAPC British Aircraft Preservation Council

BAPC British Association of Printers and Copyshops

BAPC British Aviation Preservation Council

BAPD Binary Alloy Phase Diagrams [Publication of ASM International, US]

BAPE Branch Arm Piping Enclosure

BAPEX British Association of Paper Exporters

BAPI Business Application Programming Interface

BAPL Bettis Atomic Power Laboratory [US]

BAPLA British Association of Picture Libraries and Agencies

BAPMoN Background Air Pollution Monitoring Network [of Global Environmental Monitoring System]

BAPNA Nα-Benzoyl-DL-Arginine-p-Nitroanilide [Biochemistry]

BAPNA Benzoyl Arginine Nitroanilide Hydrochloride [Biochemistry]

D-BAPNA Nα-Benzoyl-D-Arginine-p-Nitroanilide [Biochemistry]

L-BAPNA Nα-Benzoyl-L-Arginine-p-Nitroanilide [Biochemistry]

BAPP Bis(aminophenoxy Phenylpropane)

BAppArts Bachelor of Applied Science [also BApplArts]

BAPPENAS Baden Perencanaan Pembagunan Nasional [National Development Planning Agency, Indonesia]

BApplArts Bachelor of Applied Science [also BappArts]

BApplSc Bachelor of Applied Science [also BAppS, or BAppSc]

BApplStat Bachelor of Applied Statistics [also BAppStat]

BAppSc Bachelor of Applied Science [also BAppS, or BApplSc]

BAppStat Bachelor of Applied Statistics [also BApplStat]

BAPS Baseline Air Pollution Station

BAPS Berg Analysis and Prediction System

BAPS Branch Arm Piping Shielding

BAPS Bulletin of the American Physical Society

BAPT Bundesamt für Telekommunikation und Post [Federal Post and Telecommunications Office, Germany] [now RegPT]

BAPTA Bearing and Power Transfer Assembly

BAPTA 1,2-bis(2-Aminophenoxy)ethane-N,N,N',N'-Tetraacetic Acid

BAPTA 1,2-bis(o-Aminophenoxy)ethanetetraacetic Acid

BAPTA-AM 1,2-bis(2-Aminophenoxy)ethane-N,N,N',N'-Tetraacetic Acid tetrakis-(Acetoxymethyl)ester

BAQ Bundesanstalt für Qualitätsforschung [Federal Institute for Quality Research, Germany]

BAR Bar Address Register

BAR Base Address Register

BAR Basic Axisymmetric Remelting [Metallurgy]

BAR Broadcast Advertisers Report [US]

BAR Browning Automatic Rifle

BAR Buffer Address Register

BAR Bureau of Aeronautics Representative [US]

BAr Bachelor of Architecture

Bar Barometer; Barometric [also bar]

bar barrel [Unit]

BARA Board of Airline Representatives

BARB Ballast Aerating Retrieval Boom

BARC Bangladesh Agricultural Research Council

BARC Bhabha Atomic Research Center [Bombay, India]

BARC British Automobile Racing Club

BArch Bachelor of Architecture

BArchE Bachelor of Architectural Engineering [also BarchEng]

BARD Bangladesh Academy for Rural Development

BARE Biased-Activated Reactive Evaporation

BAREMA British Anaesthetic and Respiratory Equipment Manufacturers Association [UK]

BARI Bangladesh Agricultural Research Institute

BARIF Banjarbaru Research Institute for Food

BARITT Barrier Injection Transit Time (Diode)

BARMA Boiler and Radiator Manufacturers Association

Barom Barometer [also Boro, baro, or barom]

BARR Bureau of Aeronautics Resident Representative [US]

BARS Backup Attitude Reference System

BARS Ballistic Analysis Research System

BARS Baseline Accounting and Reporting System

BARS Biostatic Auroral Radar System

BARSTUR Barking Sands Tactical Underwater Test Range [Hawaii, US]

BART Baronet [also Bart]

BART Bay Area Rapid Transit (Project) [San Francisco, US]

BART British Army Rifle Team

BArt Bachelor of Arts

Bart Baronet [also BART]

BARTD Bay Area Rapid Transit District [San Francisco, US]

Bar3D Draw a Three-Dimensional Bar [Pascal Function]

BARTS Bell Atlantic Regional Timesharing

BARV Beach Armored Recovery Vehicle

BARZREX Bartok Archives Z-Symbol Rhythm Extraction

BAS Bachelor in Agricultural Science

BAS Bachelor of Applied Science

BAS Ball Bearing, Single-Row, Angular-Contact, Separable Inner Ring, Nominal Contact Angle from Above 22° to 32° [Symbol]

BAS Bangladesh Academy of Sciences [Dacca, Bangladesh]

BAS Barium Aluminosilicate (or Celsian) [$BaO \cdot Al_2O_3 \cdot 2SiO_2$]

BAS Basic Activity Subset

BAS Basic Airspeed

BAS Basic Auto-Oxidation Scheme

BAS Belarus Academy of Sciences

BAS Belgian Academy of Sciences [Brussels, Belgium]

BAS Bessemer Aerial Steel

BAS Bibliotheks-Automatisierungssystem [Library Automation System, Germany]

BAS Blind Approach System

BAS Border Agricultural Society [South Africa]

BAS Bosnian Academy of Sciences [Sarajevo]

BAS Brazilian Academy of Sciences

BAS British Acoustical Society [UK]

BAS British Antarctic Survey [of Natural Environment Research Council, UK]

BAS Bulgarian Academy of Sciences [Sofia, Bulgaria]

BAS Bulletin of Atomic Scientists [US]

BAS Bureau of Analysed Samples Limited [of BCIRA– International Center for Cast Metals Technology, UK]

BAS Byelorussian Academy of Sciences [Minsk, Russian Federation]

.BAS BASIC (Computer) Language [File Name Extension] [also .bas]

BAS₂ Barium Aluminate Silicate [$BaO \cdot Al_2O_3 \cdot 2SiO_2$]

B-AS Britian-Australia Society [UK]

BA(s) Surface of Compound *AB* (Consisting of Metal Component *A* and Nonmetal Component *B*) with *A* as the Surface Atom [Symbol]

BaS Barium Sulfide

bas basic

β-As Beta Arsenic [Symbol]

BASA British Adhesives and Sealants Association

BASA British Air Survey Association

BASA British Architectural Students Association

BASA British Association of Seed Analysts

BASA British Automatic Sprinkler Association

BASc Bachelor of Applied Science

BASCELT British Association of State Colleges of English Language Teaching

BASCOM BASIC Compiler

BASD Bell Aerospace Systems Division

BASE Beaufort Arctic Storms Experiment

BASE Beta-Alumina Solid Electrolyte

BaSe Barium Selenide (Semiconductor)

BASEC British Approval Service for Electric Cables

BASEEFA British Approval Service for Electrical Equipment in Flammable Atmospheres

BASEX Extended BASIC [Advanced BASIC Programming Language]

BASF Badische Anilin– und Sodafabriken AG [German Chemical Manufacturer headquartered in Ludwigshafen]

BASF Badische Anilin– und Sodafabriken Process [Acetylene Production Process developed by BASF]

BASH Booksellers Association Service House [UK]

BASH Bourne Again Shell [Unix Operating System]

BASI Bureau of Air Safety Investigation [Australia]

BASIC Basel Information Center for Chemistry [Switzerland]

BASIC Basic Algebraic Symbolic Interpretative Compiler

BASIC Basic Automatic Stored Instruction Computer

BASIC Battle Area Surveillance and Integrated Communications

BASIC Beginner's Algebraic Symbolic Interpretive Compiler

BASIC Beginner's All-Purpose Symbolic Instruction Code [also Basic]

BASIC Biological Abstracts Subjects in Context [of Biosciences Information Service, US]

BASICPAC BASIC Processor and Computer

BASICS Battle Area Surveillance and Integrated Communication System

Basin Res Basin Research [Journal]

BASIS Bank Automated Service Information System

BASIS Battelle Automatic Research Information System [of Battelle Memorial Institute, US]

BASIS British Airways Staff Information System

BASIS Bulletin of the American Society for Information Science [US]

BASIS Burroughs Applied Statistical Inquiry System

BASIS Business and Safety Integration Survey [of US National Safety Council]

BASIS-E Bibliothekarisch-Analytisches System zur Informationsspeicherung und -erschliessung [Analytical Library System for Information Storage and Retrieval, Germany]

BASJE Bolivian Air Shower Joint Experiment

BASM Bachelor of Arts, Master of Science

BASM Built-in Assembler [Computers]

BASMA Boot and Shoe Manufacturers Association [UK]

BAS NET Basic Network [also Bas Net]

BAS-NET Belarus Academy of Sciences Network

BaSO Bariosulfate

BASRA British Amateur Scientific Research Association

BASRM British Association of Synthetic Rubber Manufacturers

BaSrTiO Barium Strontium Titanium Oxide (Semiconductor)

BASS Backup Avionics System Software

BAST Boric Acid Storage Tank

BAST Bundesanstalt für Strassenwesen [Federal Institute for Road Traffic and Transportation, Germany] [also BASt]

BASW Bell Alarm Switch

BASYS Basic System

BAT Barium Aluminum Titanate

BAT Basic Assurance Test

BAT Battalion Anti-Tank

BAT Best Available Technology

BAT Best Available Treatment Technology

BAT Biological Abstracts on Tape [of Biosciences Information Service, US]

BAT Bipolar Axon Terminal

BAT Block Address Translation

BAT Bond–Assembly–Test

BAT British Anarctic Territory

BAT Bundesangestelltentarif [Statutory Salary Scale, Germany]

BAT Bureau of Apprenticeship and Training [of Employment and Training Administration, US]

.BAT Batch Processing [File Name Extension] [also .bat]

BA/T Battalion Anti-Tank

Bat Batch [also bat]

Bat Battery [also bat]

BATA British Air Transport Association

Bat Chg Battery Charger

BATCLM British Association of Trade Computer Label Manufacturers

BATEA Best Available Technology Economically Available

BATF Bureau of Alcohol, Tobacco and Firearms [US]

Bath Bathroom [also bath]

Ba(THD)$_2$ Barium Bis(tetramethylheptanedionate) [also Ba(thd)$_2$]

BaTi(OR)$_6$ Barium Titanium Alkoxide [General Formula]

BA TiSULC Batch-Annealed Titanium Stabilized ULC (Ultralow Carbon) (Steel)

BA TiSULC Steel Batch-Annealed Titanium Stabilized ULC (Ultralow Carbon) Steel [also BA TiSULC steel]

BATMA Bookbinding and Allied Trades Management Association [UK]

Ba(TMHD)$_2$ Barium Bis(2,2,6,6-Tetramethyl-3,5-Heptanedionate) [also Ba(thmd)$_2$]

BATNAP Bat National Action Plan (Database) [Australia]

BATNEEC Best Available Techniques Not Entailing Excessive Costs [European Community]

BaTO Bariotungstate

BATP Boric Acid Transfer Pump

BATR Bullet At Target Range

BATRE Battle Reconnaissance

BATS Basic Additional Teleprocessing Support

BATS Bay Area Tracking System [San Francisco, US]

BATS Bit Access Test System

Batt Battalion [also batt]

Batt Batten [also batt]

Batt Battery [also batt]

Battn Battalion [also battn]

Batty Battery [also batty]

BATU Brotherhood of Asian Trade Unionists [Philippines]

BATY Barium Difluoride–Aluminum Trifluoride–Thorium Fluoride–Yttrium Fluoride (Glass) [BaF$_2$–AlF$_3$–ThF$_4$–YF$_3$]

BAU Bangladesh Agricultural University [Mymensingh]

BAu Gold Brazing Alloy

BAUA Business Aircraft Users Association [UK]

BauA Bundesanstalt für Arbeitsschutz und Arbeitsmedizin [Federal Institute for Industrial Safety and Medicine, Germany]

β-AuCd Beta Gold Cadmium

BAUD Baudot Code

Bauen Stahl Bauen in Stahl [Swiss Publication on Building with Steel]

BAUFO Bauforschungsprojekte und -berichte (Datenbank) [Database on Civil Engineering Research Projects and Reports; of Fraunhofer Informationszentrum Raum und Bau, Germany]

β-AuGa Beta Gold Gallium

BAUMA Internationale Fachmesse für Baumaschinen und Baustoffmaschinen [International Trade Fair for Building and Building Material Machinery and Equipment, Munich, Germany]

BAURES Bangladesh Agricultural University Research System

Bautec Internationale Baufachmesse [International Trade Fair for Construction Engineering and Technology, Germany]

β-AuZn Beta Gold Zinc

BAV Bovine Adenovirus

Bav Bavaria(n) [Germany]

BAVIP British Association of Viewdata Information Providers

BAW Boltzmann Active Walker (Model) [Physics]

BAW Bond-Order Alternation Wave

BAW Bulk Acoustic Wave

BAW Bundesanstalt für Wasserbau [Federal Institute for Hydraulic Engineering, Germany]

BAWE British Association of Women Entrepreneurs

BAWI Bundesamt für Wirtschaft [Federal Office for Trade and Commerce, Germany]

BaWO Bariotungstate

BAXD N,N'-Bis(4-Aminophenyl)-1,3-Xylene Diamine

BAX(s) Reactant AX (e.g. a Metal Halide) Adsorbed on Surface B [Symbol]

Bay Bayerite [Mineral]

BAY CAND DC Bayonette Candelabra, Double Contact

Bayer Rep Bayer Reports [Published by Bayer AG, Leverkusen, Germany]

BAYS British Association of Young Scientists

BB Backboard

BB Ball Bearing [also bb]

BB Barbados [ISO Code]

BB Barium Bismuthate

BB Barrage Balloon

BB Barrel Bridge [Horology]

BB Baseband

BB Battleship [US Navy Symbol]

BB Beattie-Bridgman (Equation of State) [Thermodynamics]

BB Bending Beam

BB Berthelot Bomb

BB Best Best (Wire)

BB Big Bang [Astronomy]

BB Blackbody

BB Black Box

BB Black Brant (Sounding Rocket)

BB Bloch-Boltzmann (Theory) [Physics]

BB Block Brazing

BB Blood Bank

BB Body Burden [Toxicity]

BB Bordetella Bacteria

BB Bottom Blowing; Bottom-Blown [Metallurgy]

BB Bottom Boundary

BB Bragg-Brentano (Geometry) [Physics]

BB Brass Boot

BB Breadboard

BB Bright Blue (Fluorescence)

BB Broadband

BB Bromobenzene

BB 2-Bromobutane

BB Brush Box

BB Bucket Brigade [Electronics]

BB Building Block

BB Bulb Barometer

BB Bulk Burning

BB Bulletin Board

BB Bunch Block

BB Bunsen Burner

BB Buoyancy Bomb

BB Bus Bar

BB Double Black [On Pencils]

B&B Bell and Bell

B_r/B_s Residual Induction to Saturation Induction (i.e., Magnetic Squareness) [Symbol]

B-B Bright-Bright (Fringe)

B-B Butane-Butene Mixture

25B14B 2,5-Dibromo-1,4-Benzoquinone

β-B Beta Boron [Symbol]

BBA Bachelor of Business Administration

BBA Bear Biology Association [now International Association for Bear Research and Management, Canada]

BBA o-Benzoylbenzoic Acid

BBA Better Blowing Agents (Project) [of Dow Corporation]

BBA Biologische Bundesanstalt für Land- und Forstwirtschaft [Federal Biological Institute for Agriculture and Forestry, Germany]

BBA British Bankers Association

BBA British Business Administration

BBA British Businessmen's Association

BBA Broad-Beam Absorption

BBA Bromobenzoic Acid

BBA p-Bromo-sec-Butylbenzene

BBA p-sec-Butylbenzoic Acid

B_1B_2A Layered Heterostructure of Nonmetal Components B_1 and B_2 and Metal Component A [Symbol]

BBAC British Balloon and Airship Club

BBAO N-Benzylidene-tert-Butylamine-N-Oxide

BBB Better Business Bureau

BBB Bisbenzimidazobenzophenanthroline [Biochemistry]

BB/B Body Bound Bolts

BBBA British Boat Builders Association

BBBS Bird and Bat Banding Scheme [of Australian Nature Conservancy Agency]

BBC Bachelor of Building Construction

BBC Beam-to-Beam Correlation

BBC Before Business Clearance

BBC Breeding Bird Census [US]

BBC British Broadcasting Corporation

BBC Broadband Conducted; Broadband Conduction

BBC Bromobenzylcyanide

BBC Brown Boveri Company

BBC Building Better Cities

BB15C5 t-Butylbenzo-15-Crown-5 [also BB15Crown5]

BBCr Bis(biphenylchromium)

BBC-Nachr BBC-Nachrichten [Newsletter published by Brown, Boveri & Cie AG, Germany]

BBC-Tech BBC-Technik [Brown Boveri Technology Reports; published in Zurich, Switzerland]

BBD Barbados Dollar [Currency]

BBD Beam-Blanking Device

BBD 7-Benzylamino-4-Nitrobenz-2-Oxa-1,3-Diazole

BBD 2,5-Bis(4-Biphenylyl)-1,3,4-Oxadiazole

BBD Bubble Bath Detector

BBD Bucket-Brigade Device [Electronics]

BBDC Before Bottom Dead Center

BBE Black Body Emission

β-Be Beta Beryllium [Symbol]

BBEA Brewery and Bottling Engineers Association

BBER Bureau for Business and Economic Research [of University of Oklahoma, US]

BBF Baseband Frequency

BBF Benzofluoranthene

BBF Bond Breaking Factor

BBF (International) Brotherhood of Boilermakers, Iron Shipbuilders, Blacksmiths, Forgers and Helpers [UK]

BBFA British Binders and Finishers Association

BBG Board of Broadcasting Governors

BBG Brooklyn Botanical Garden [New York, US]

BBGKY Bogoliubov-Born-Green-Kirkwood-Yvon [Physics]

BBGS British Business Graduates Society

BBHIR Broad Bandpass Hemispheric Infrared Radiometer

BBHSR Broad Bandpass Hemispheric Solar Radiometer

BBI Baryon-Baryon Interactions [Particle Physics]

BBI Black Body Instrument

BBI British Bottlers Institute

BBIG Broad Beam Ion Gun

BBIM Brass and Bronze Ingot Manufacturing

β-Bi_2O_3 Beta Bismuth Trioxide

BBIR Broadband Infrared

BBIRA British Baking Industries Research Association

β-**Bk** Beta Berkelium [Symbol]

BBKA British Beekeepers Association

BBL Basic Business Language

BBL Benzimidazobenzophenanthroline [Biochemistry]

BBL Bird Banding Laboratory [of US Geological Survey/Biological Resources Division and Canadian Wildlife Service]

bbl barrel [Unit]

bbl/d barrels per day [also bbl d^{-1}, BPD, or bpd]

BBM Beam-Blanking Method

BBM Beam Brightness Modulation

BBM Break Before Make

BBM Bureau of Broadcast Measurement

BBMA British Bath Manufacturers Association

BBMA British Battery Manufacturers Association

BBM-CCS Beam Brightness Modulation/Computer-Controlled Data Acquisition and Processing System

BBMRA British Brush Manufacturers Research Association

BBN Bolt, Beranek and Newman [Internet]

BBN Borabicyclononane

BBN Broad Band Noise

9-BBN 9-Borabicyclo[3.3.1]nonane

β-**BN** Beta Boron Nitride

BBNA Black Bolt and Nut Association [UK]

BBNO Bismuth Barium Niobium Oxide

BBO Barium Bismuth Oxide (Superconductor)

BBO Beta Barium Borate

BBO 2,5-Bis(4-Biphenylyl)oxazole; 2,5-Di(4-Biphenylyl)-oxazole

BBOD 2,5-bis[(1,1'-Biphenyl)-4-yl]1,3,4-Oxadiazole

BBOT 2,5-bis(5-tert-Butyl-2-Benzoxazolyl)thiophene; 2,5-bis-(5'-tert-Butylbenzoxazolyl-2')thiophene

BBP Benzyl Butyl Phthalate; Butyl Benzyl Phthalate

BB&PA British Box and Packaging Association

BBPT Baseband Pulse Transmission

BBQ Benzimidazobenzisoquinolinone; 7H-Benzimidazo-[2,1-a]benz[de]isoquinoline

B14BQ Monobromo-1,4-Benzoquinone

BBR Bend-Bend-Roll

BBR Bending Beam Rheometer

BBR Blackbody Radiation

BBR Broadband Radiated

BBRG Ball Bearing

B-Brom-9-BBN 9-Bromo-9-Borabicyclo[3.3.1]nonane

BBRR Brookhaven Beam Research Reactor [of Brookhaven National Laboratory, Upton, New York, US]

BBRU Bituminous Binder Research Unit [South Africa]

BBS Balanced Biaxial Stretching

BBS Ball-Bearing Steel

BBS Bermuda Biological Station

BBS Breeding Bird Survey [US]

BBS British Biophysical Society

BBS British Bryological Society

BBS Bulletin Board Service

BBS Bulletin Board Software

BBS Bulletin Board System

BBSA British Blind and Shutter Association

BBSR Bermuda Biological Station for Research

BBSS Balloon Bourne Sounding System

BBT Black Body Temperature

BBT Broadband Transmission

Bbt Bombardment [also bbt]

BBTA British Bureau of Television Advertising

BBTV Banana Bunchy Top Virus

BBU Baseband Unit

B(BuO)₃ Boron Butoxide

BBW Bohr-Breit-Wigner (Theory) [Nuclear Physics]

BBWV-1 Broad Bean Wilt Virus, Type 1

BC Baby Cyclotron

BC Bachelor of Chemistry

BC Bachelor of Commerce

BC Back Connected; Back Connection

BC Back Course

BC Bacterial Conjugation

BC Bacterial Culture

BC Ball Bearing, Single-Row, Radial-Contact, Non-Filling Slot Assembly [Symbol]

BC Bar Code

BC Bare Copper [also bc]

BC Barium Carbonate

BC Barium Citrate

BC Barium Cuprate

BC Barometric Condenser

BC Barrier Coating

BC Barrier Crack

BC Baryta Crown (Glass)

BC Base-Catalyzed

BC Base Circle

BC Basic Catalysis

BC Basic Circle

BC Basic Control

BC Batchelor College [Australia]

BC Bathyconductograph(y)

BC Battery Charger

BC Bayonet Cap [Electron Tube]

BC Beam Current

BC Bedford College [UK]
BC Before Christ
BC Belt Conveyor
BC Belted Cable
BC Benzyl Chloride
BC Bernoulli's Constant
BC Bessemer Converter [Metallurgy]
BC Between Centers
BC Bibliographic Classification
BC Bicrystal
BC Binary Code
BC Binary Comparator
BC Binary Counter
BC Binary Cycle
BC Bioceramic(s)
BC Biochemical; Biochemist(ry)
BC Biocompatibility; Biocompatible
BC Bioconversion
BC Biological Chemist(ry)
BC Biological Control
BC Biological Conversion
BC Biological Corrosion
BC Biological Council [UK]
BC Bituminous Coal
BC Blastocyst [Embryology]
BC Bliss Classification
BC Block Copolymer
BC Blocking Capacitor
BC Blome Process
BC Blood Cell
BC Blood Count
BC Blue Cross [US]
BC Blumlein Circuit
BC Blunting Configuration
BC Bogert-Cook (Synthesis)
BC Bolt Circle
BC Boltzmann Constant
BC Bomb Calorimeter
BC Bond-Centered; Bond Centering [Chemistry]
BC Bond Charge
BC Bonded Single Cotton (Electric Wire)
BC Bone Conduction
BC Boron Carbide
BC Boston College [Massachusetts, US]
BC Bottom Chord [Civil Engineering]
BC Bottom Contour
BC Boundary Condition
BC Bowman's Capsule [Urology]
BC Branched Chain
BC Brayton Cycle [Thermodynamics]
BC Break Control
BC Breaking Current
BC Breakpoint on Code (Command) [Pascal Programming]

BC Bridge Crane
BC British Columbia [Canada]
BC British Commonwealth
BC Broadcast Control
BC Bronchiogenic Carcinoma [Lung Cancer]
BC Brooklyn College [of City University of New York, US]
BC Brown Coal
BC Bubble Chamber
BC Buccal Cavity [Anatomy]
BC Budd-Chiari (Syndrome)
BC Budget Costs
BC Building Code
BC Bunsen Cell
BC Burgers Circuit [Materials Science]
BC Burn Center
BC Bus Controller
BC Bus Coupler
BC Bureau of Census [of US]
BC Buried Channel [Electronics]
BC Burnhampthorpe College [Canada]
BC Burning Chamber
BC Business Class
BC_2 Barium Cuprate [$BaCu_2O_2$, or $BaO \cdot Cu_2O$]
B_4C Boron Carbide (Ceramics)
B12C4 Benzo-12-Crown-4 [also B12-Crown-4, or B12-crown-4]
B13C4 Benzo-13-Crown-4 [also B13-Crown-4, or B13-crown-4]
B15C5 Benzo-15-Crown-5 [also B15-Crown-5, or B15-crown-5]
B18C6 Benzo-18-Crown-6 [also B18-Crown-6, or B18-crown-6]
$B_{||}c$ Magnetic Flux Parallel (or In-Plane) to c-Axis (of Superconductor) [Symbol]
$B_{\perp}c$ Magnetic Flux Perpendicular (or Normal) to c-Axis (of Superconductor) [Symbol]
B/C Bench Check
B/C Benefit-to-Cost (Ratio)
B-C Battery-Capacitor
b/c b-axis to c-axis ratio (of crystal lattice) [Symbol]
β-C Beta Carbon [or C^{β}]
BCA Barium Calcium Aluminate
BCA Barium Chromium Aluminate
BCA Barium Cobalt Aluminate
BCA Basilar Cerebral Artery [Anatomy]
BCA Battery Control Area
BCA Beale Cypher Association [US]
BCA Benefit-Cost Analysis
BCA Benzene Carboxylic Acid
BCA N-Benzyl Cyclopropyl Amine
BCA Bicinchoninic Acid
BCA Binary Collision Approximation [Physics]
BCA Bliss Classification Association [UK]
BCA Box Culvert Association [UK]

BCA Brazilian Corrosion Association

BCA British Carton Association

BCA British Cement Association

BCA British College of Accountancy

BCA Bureau of Consular Affairs [of US Department of State]

BCA Bureau of Coordination of Arabization [Morocco]

BCA Business Council of Australia

BCA 4,4'-Dicarboxy-2,2'-Biquinoline

BC$_4$A$_2$ Barium Tetracalcium Dialuminate [BaO·4CaO·2Al$_2$O$_3$]

β-Ca Beta Calcium [Symbol]

BCAA Branched-Chain Amino Acid

BCAA Bristol Center for the Advancement of Architecture [UK]

BCAB British Control Advisory Bureau

BCABP Bureau of Competitive Assessment and Business Policy [US]

BCAC Boeing Commercial Airplanes Corporation [US]

BCAC British Columbia Aviation Council [Canada]

BCAC British Conference on Automation and Computation

BCACA British Columbia Arts and Crafts Association [Canada]

BCAE Ballarat College of Advanced Education [Ballarat, Victoria, Australia]

BCAE Bendigo College of Advanced Education [Bendigo, Victoria, Australia]

BCAG Boeing Commercial Airplanes Group [Seattle, US]

BCALT British Columbia Association of Library Technicians [Canada]

BCAR British Civil Airworthiness Requirements

BCAS Beacon-Based Collision Avoidance System [also B-CAS]

BCAS Beacon Collision Avoidance System

BCAS British Columbia Agricultural Society [Canada]

BCAS British Compressed Air Society

BCAS Business Center for Academic Societies [Tokyo, Japan]

BCASJ Business Center for Academic Societies, Japan [Tokyo]

BCAST Belize College of Arts, Science and Technology [Belize City]

B-cat Branhamell catarrahalis (Bacterium) [Microbiology]

BCAVM British Catalogue of Audiovisual Materials

BCB Benzocyclobutene

BCB British Consultants Bureau [UK]

BCB Broadcast Band

BCB Bromocresol Blue

BCB Brown Cardboard

BCB Butyl-1-Cyclohexyl Barbituric Acid

B$_4$C/B Boron Carbide Coated Boron

BCBA Beef Cattle Breeders Association [Israel]

BCC Baltimore Chamber of Commerce [US]

BCC Benzylchlorocarbonate

BCC Bioaccumulative Chemicals of Concern

BCC Biological Council of Canada

BCC Blind Carbon Copy

BCC Block Character Check

BCC Block Check Character

BCC Blocked Calls Cleared

BCC Body-Centered Cubic (Crystal) [also bcc]

BCC Boiler Construction Code [of American Society of Mechanical Engineers, US]

BCC British Ceramic Confederation

BCC British Chamber of Commerce

BCC British Cleaning Council

BCC British Color Council

BCC British Communications Corporation

BCC British Copyright Council

BCC British Cryogenics Council

BCC Bus and Coach Council [UK]

Bcc Blind Carbon Copy [also bcc]

BCCA British Columbia Cancer Agency [Canada]

BCCA British Columbia Construction Association [Canada]

BCCC British-Chilean Chamber of Commerce

BCCD Bulk Charge-Coupled Device

BCCD Buried-Channel Charge-Coupled Device

BCCE Biennial Conference on Chemical Education

BCCF British Calcium Carbonate Federation

BCCF British Cast Concrete Federation

BCCF British Columbia Conservation Foundation [Canada]

BCCG British Chamber of Commerce, Germany

BCCJ British Chamber of Commerce, Japan

BCCL Birkbeck College Computation Laboratory [UK]

BCCRC British Columbia Cancer Research Center [Canada]

BCCT Break Control Command Transducers

BCCT British Chamber of Commerce, Turkey

BCC(T) Transformed to Body-Centered Cubic (from Face-Centered Cubic) [also bcc(t)] [Metallurgy]

BCD Barrels per Calendar Day [also bcd]

BCD Base-Catalyzed Destruction

BCD Baseline Change Proposal

BCD Battelle Columbus Division [Columbus, Ohio, US]

BCD Baylor College of Dentistry [Dallas, Texas, US]

BCD Binary-Coded Decimal [also bcd]

BCD Borderline between Comfort and Discomfort (Glare)

BCD Burst Cartridge Detection

BCDA Barge and Canal Development Association [UK]

BCD/BIN Binary-Coded Decimal to Binary (Converter)

BCDC (San Francisco) Bay Conservation and Development Commission [US]

BCDC N-Benzyl-Cinchonidinium Chloride

BCDC Binary-Coded Decimal Counter

BCDC British Columbia Development Corporation [Canada]

BCD/DEC Binary-Coded Decimal to Decimal (Decoder)

BCDIC Binary-Coded Decimal Interchange Code

BCDMOS Bipolar CMOS (Complimentary Metal-Oxide Semiconductor)–DMOS (Double-Diffused Metal-Oxide Semiconductor)

BCD/Q Binary-Coded Decimal/Quaternary

BCDS Bulk Chemical Delivery System

BCDSFU British Columbia Deep Sea Fishermen's Union [Canada]

BCDTA British Chemical and Dyestuff Traders Association

BCE Bachelor of Chemical Engineering

BCE Bachelor of Civil Engineering

BCE Backup Control Electronics

BCE Before the Common Era

BCE Bell Canada Enterprises

BCE Biochemical Engineer(ing)

BCE Bus Control Element

β-Ce Beta Cerium [Symbol]

BCEC Battle Creek Engineers Club [US]

BCECA British Chemical Engineering Contractors Association

BCECEC British Conference and Exhibition Centers Export Council

BCECC British and Central-European Chamber of Commerce

BCECF 2',7'-Bis(carboxyethyl)-4/5-Carboxyfluorescein; 2',7'-Bis(carboxyethyl)carboxy Fluorescein

BCECF-AM 2',7'-Bis(carboxyethyl)-5,6-Carboxy Fluorescein Pentaacetoyl Methyl Ester

BCEF N,N-Bis-(2-Cyanoethyl) Formamide

BCEF British Coal Exporters Federation

BCEMA British Combustion Equipment Manufacturers Association

BCEN N,N'-Bis(β-Carbamoyethyl)ethylenediamine

β Cen Beta Centauri [Astronomy]

BCerE Bachelor of Ceramic Engineering

BCET Biochemical Engineering Technologist; Biochemical Engineering Technology

BCF Baseline Cosine Fitting

BCF B-Cell Stimulating Factor [Biochemistry]

BCF Billion Conductor Feet

BCF Billion Cubic Feet [also Bcf]

BCF Blood Cells Foundation [US]

BCF Bolivar Coastal Field [Venezuela]

BCF British Concrete Federation

BCF Bromochlorofluoromethane

BCF Bulked Continuous Fiber

BCF Bureau of Commercial Fisheries [now National Marine Fisheries Service, US]

BCF Burton-Cabrera-Frank (Crystal Growth Theory) [Metallurgy]

BCF Byte Cipher Feedback

β-Cf Beta Californium [Symbol]

BCFA British Columbia Federation of Agriculture [Canada]

BCFBA British Columbia Food Brokers Association [Canada]

BCFCA British Columbia Floor Covering Association [Canada]

BCF/CD Billion Cubic Feet per Calendar Day [also bcf/cd]

BCF/d Billion Cubic Feet per Day [also Bcf/day, or Bcf per day]

BCFGA British Columbia Fruit Growers Association [Canada]

BCFL British Columbia Federation of Labour [Canada]

BCFMCA British Columbia Frequency Modulation Communications Association [Canada]

BCFP British Columbia Forest Products [Canada]

BCFS British Columbia Forestry Service [Canada]

BCFSK Binary-Coded Frequency-Shift Keyed (or Keying)

BCF/yr Billion Cubic Feet per Year [also Bcf/yr, BCF/Y, or Bcf/y]

BCG Bacille Calmette-Guérin (Vaccine)

BCG Ballistocardiogram; Ballistocardiography

BCG Bromocresol Green

BCGA British Commercial Gas Association

BCGA British Compressed Gas Association

BCGA British Cotton Growers Association

BCGF B-Cell Growth Factor [Biochemistry]

BCGF I B-Cell Growth Factor I [Biochemistry]

BCGF II B-Cell Growth Factor II [Biochemistry]

BCGLO British Commonwealth Geographical Liaison Office

BCGMA British Commercial Glasshouse Manufacturers Association

BCGO Barium Cerium Gadolinium Oxide

BCH 2-Amino-2-Norbornanecarboxylic Acid

BCH Banco Central de Honduras [Central Bank of Honduras]

BCH Binary Coded Hexadecimal

BCH Binary-Coded Hollerith

BCH Bits per Circuit per Hour

BCH Block Control Header

BCH Blocked Calls Held

BCH Bose-Chaudhuri-Hocquenghem (Computer Code)

BCH Boston City Hospital [Massachusetts, US]

BCH British Columbia Hydro [also BC Hydro, Canada]

BCH Bromocyclohexane

BCh Bachelor of Chemistry

BCh Bachelor of Chirurgery (ie., Surgery)

Bch Bunch [also bch]

BCHA British Columbia Health Association[Canada]

B-CHANNEL Bearer Channel [also B-Channel]

BCHA-PPV Polydicholestanoxy-p-Phenylenevinylene

BCHC Battle Creek Health Center [Michigan, US]

BCHCM Board of Certified Hazard Control Management [US]

BCHD Bicyclo[2.2.1]hepta-2,5-Diene

BChE Bachelor of Chemical Engineering

BChem Bachelor of Chemistry

BCHPA British Columbia Hydro and Power Authority [Canada]

BCHS Bronx County Historical Society [US]

BCHTN N,N'-Bis(β-Carbamoylethyl)-2-Hydroxytrimethylenediamine

BC Hydro British Columbia Hydro [also BCH, Canada]

BCI Bat Conservation International

BCI Battery Council International [US]

BCI Binary-Coded Information

BCI Bit Count Integrity

BCI Bituminous Coal Institute [US]

BCI Broadcast Interference

BCIA British Clothing Industry Association

BCIA British Columbia Institute of Agrologists [Canada]

BCIAWPRC Belgian Committee of the International Association on Water Pollution Research and Control

BCIAWPRC Brazilian Committee of the International Association on Water Pollution Research and Control

BCILA British Columbia Independent Loggers Association [Canada]

BCIM Bromochloroiodomethane

BCIP Belgian Center for Information Processing

BCIP Bromochloroindolyl Phosphate; 5-Bromo-4-Chloro-3-Indolyl Phosphate

BCIP/NBT 5-Bromo-4-Chloro-3-Indolyl Phosphate/Nitro Blue Tetrazolium

BCIPPA British Cast Iron Pressure Pipe Association

BCIQ Bureau of Commodity Inspection and Quarantine [of Ministry of Economic Affairs, Taiwan]

BCIRA British Cast Iron Research Association [Name changed to International Center for Cast Metals Technology]

BCIRA International Center for Cast Metals Technology [formerly British Cast Iron Research Association]

BCIRA/BAS International Center for Cast Metals Technology/Bureau of Analysed Samples Limited [UK]

BCIRA J British Cast Iron Research Association Journal [now International Center for Cast Metals Technology Journal]

BCIS Biodiversity Conservation Information System [US]

BCIS Building Cost Information Service [UK]

BCISC British Chemical Industrial Safety Council

BCIT British Columbia Institute of Technology [Burnaby, Canada]

BCJ British Corrosion Journal

BCK Backward [Communications]

BCK Biochemical Kinetics

BCKA Branched Chain A-Keto Acid

BCKCD Back-Order Code

BCKGND Background [also Bckgnd]

BCL Bare-Coded Label

BCL Barrel Calorimeter

BCL Base Contour Line; Basic Contour Line

BCL Batch Command Language

BCL Battelle Columbus Laboratories [Ohio, US]

BCL Beta-Chapman Layer

BCL Binary Cutter Location

BCL Biological Computer Laboratory [US]

BCL Broadcast Listener

BCL Burroughs Common Language

Bcl Bacillus caldolyticus [Microbiology]

BCLA British Columbia Library Association [Canada],

BCLDI British Clayware Land Drainage Association

BCLMA British Columbia Lumber Manufacturers Association [Canada]

BCLN British Columbia Library Network [Canada]

BCLS British Columbia Lifeboat Society[Canada]

B-3-Cl-SEDI Bis-3-Chlorosalicylaldehyde Ethylenediimine

BCL Bank Cubic Meter [also bcm]

BCL Beyond Capability of Maintenance

BCL Binary Coded Matrix

BCL British Catalogue of Music

BCL Bromochloromethane

BCM Back Course Marker

BCM Baylor College of Medicine [Houston, Texas, US]

BCM Binary Collision Model

BCM Blood-Clotting Mechanism Effects (of Hazardous Materials)

BCM Bromochloromethane

BCM Brooklyn College of Medicine [New York, US]

β-Cm Beta Curium [Symbol]

BCMA BEAMA (British Electrical and Allied Manufacturers Association) Capacitor Manufacturers Association [UK]

BCMA British Chain Manufacturers Association

BCMA British Chipboard Manufacturers Association

BCMA British Color Makers Association

BCMA British Columbia Medical Association [Canada]

BCMAF British Columbia Ministry of Agriculture and Food [Canada]

BCMC British Cable Makers Confederation

BCMCD Brevard County Mosquito Control District [Florida, US]

BCME Bischloromethylether

BCMEA British Columbia Maritime Employees Association [Canada]

BCMEN N,N'-Bis(β-Carbamoylethyl)-1,2-Propylenediamine

BCMF British Ceramic Manufacturers Federation

BCMH British Columbia Ministry of Health [Canada]

BCML Burroughs Current Mode Logic

BC-MOSFET Buried-Channel Metal-Oxide Semiconductor Field-Effect Transistor

BCMS Boron Concentration Measuring System

BCMTA British Columbia Motor Transport Association [Canada]

BCN Backbone Concentrate Node

BCN Beacon [also Bcn, or bcn]

BCN Biblioteca de Congreso de la Nacion [Library of National Congress, Argentina]

BCN Biomedical Communications Network [US]

BCN Boron-Carbon-Nitrogen (Thin Film Technology)

Bcn Beacon [also bcn]

β-CN Beta Carbon Nitride [also β-C_3N_4]

BCNC N-Benzyl-Cinchoninium Chloride

BCNA British Columbia Naturopathic Association [Canada]

BCNI Business Council on National Issues

BCNM Bastyr College of Naturopathic Medicine [Bothell, Washington, US]

BCNO Bismuth Calcium Niobium Oxide

BCNU N,N'-Bis(2-Chloroethyl) N-Nitroso Urea

BCNU British Columbia Nurses Union [Canada]

BCNU Carmustine [Chemotherapeutic Cancer Drug]

BCNU I'll Be Seeing You [Amateur Radio and Internet]

BCO Bachelor of the College of Ophthalmologists

BCO Barium Cerium Oxide (Ceramics)

BCO Barium Copper Oxide (Superconductor)

BCO Battelle Columbus Operations [Ohio, US]
BCO Battery Cutoff
BCO Beryllium Chromium Oxide
BCO Bill in Care of
BCO Binary-Coded Octal
BCO Black Copper Oxide
BCO Bridge Cutoff
BCo Cobalt Brazing Alloy
β-Co Beta Cobalt [Symbol]
BCOA Bituminous Coal Operators Association [US]
BCofC Bradford Chamber of Commerce [UK]
B coli Bacillus coli (Bacteria) [Microbiology]
B coli Balantidium coli [Genus of Protozoans]
BCOM Burroughs Computer Output to Microfilm
BComm Bachelor of Commerce [also BCom]
BCompS Bachelor of Computer Science
BComSc Bachelor of Commercial Science
BCORA British Colliery Owners Research Association
BCotIRA British Cotton Industry Research Association
BCP Bachelor of City Planning
BCP Base Condemnation Percent
BCP Baseline Change Proposal
BCP Benchmark Control Point
BCP Best Commercial Practice
BCP Best Current Practices [Internet]
BCP Better Cities Program
BCP Bisynchronous Communications Processor
BCP Bromochloropropene
BCP Bromocresol Purple
BCP Bromocyclopentane
BCP Bulk Copy Program
BCP Bureau of Competition Policy [of Industry Canada]
BCP Bureau of Consumer Protection [US]
BCP Burroughs Control Program
BCP Butyl Carbitol Piperonylate
BCP Byte Control Protocol
BCPA British Commonwealth Pacific Airlines
BCPA British Concrete Pumping Association
BCPA British Copyright Protection Association
BCPB Bromochlorophenol Blue
BCPC British Columbia Press Council [Canada]
BCPC British Crop Protection Council
BCPC sec-Butyl-N-(3-Chlorophenyl)carbamate
BCPE 1,1-Bis(4-Chlorophenyl)ethanol
BCPIT British Council for the Promotion of International Trade
BCPL Basic Combined Programming Language
BCPL Basic Computer Programming Language
BCPM Bis-Methylcyclopentadienyl Manganese
BCPMA British Chemical Plant Manufacturers Association
BCPO British Commonwealth Producers Organization
BCPS Basic Call Processing System
BCPS Beam Candle Power Seconds
BCPSM Board of Certified Product Safety Management [US]

BCR Bar Chart Report
BCR Bar-Code Reader
BCR Benefit-to-Cost Ratio
BCR Bibliographic Center for Research [US]
BCR Billing, Collecting, Remitting
BCR Bituminous Coal Research [now Bituminous Coal Research National Laboratory, US]
BCR British Columbia Railways [Canada]
BCR Brush Contact Resistance
BCR Buffer Control Register
BCR Bureau of Commercial Research [UK]
BCR Byte Count Register
BCRA British Cave Research Association
BCRA British Ceramic Research Association
BCRA British Coke Research Association
BCRC British Columbia Research Corporation [Canada]
BCRC British Columbia Research Council [Canada]
BCRD Basic Consolidated Requirements Document
BCRF British Columbia Registered Forester [Canada]
BCRG Brine Cavity Research Group [now Solution Mining Research Institute, US]
BCRI Biological and Chemical Research Institute [Rydalmere, Australia]
BCRLM Bending-Corrected Rotating Linear Model
BCRNL BCR (Bituminous Coal Research) National Laboratory [US]
BCRS Bell Canada Relay Service
BCRT Binary Code Range Time
BCRT Bright Cathode-Ray Tube
BCRU British Committee on Radiological Units
BCRUM British Committee on Radiation Units and Measurements
BCS Bachelor of Commercial Science
BCS Bachelor of Computer Science
BCS Backscatter Cross Section
BCS Bar Code Sorter
BCS Bardeen-Cooper-Schrieffer (Theory of Superconductivity)
BCS Barium Calcium Silicate
BCS Basic Combined Subset
BCS Basic Control System (Satellite)
BCS Batch Control Software
BCS Batch Control System
BCS Best Cast Steel
BCS Bilby-Cottrell-Swinden (Model) [Metallurgy]
BCS Binary Communications–Synchronous [also BISYNC]
BCS Biochemical Science(s)
BCS Biomedical Computing Society [US]
BCS Block Check Sequence
BCS Block Control Signal
BCS Boeing Computer Services [US]
BCS Boston Computer Society [US]
BCS Bridge Control System
BCS British Cartographic Society
BCS British Ceramic Society

BCS British Composites Society

BCS British Computer Society

BCS Broadcasting Station

BCS Buffer-Rod–Couplant–Sample (Configuration) [Ultrasonic Testing]

BCS Business Communication System

BCS Business Consulting Service

BCS Business Customer Service

BCSA British Constructional Steelwork Association

BCSA British Cutlery and Silverware Association

BCSC British Columbia Safety Council [Canada]

BCSC British Computer Services Corporation

BCSCO Bismuth Calcium Strontium Copper Oxide (Superconductor)

BCSD Business Council for Sustainable Development

BCSIR Bangladesh Council of Scientific and Industrial Research [Dacca, Bangladesh]

BCSMT British Columbia Society of Medical Technologists [Canada]

BCS Newsl BCS Newsletter [of British Computer Society]

BCSO British Commonwealth Scientific Office [UK]

BCSP Board of Certified Safety Professionals [US]

BCSR Boeing Computer Services Richland, Inc. [Washington State, US]

BCSS British Cactus and Succulent Society

Bcst Broadcast [also bcst]

BCT Belfast College of Technology [Northern Ireland]

BCT Best Conventional Technology

BCT Between Commands Testing

BCT Body-Centered Tetragonal (Crystal) [also bct]

BCT Boettinger, Coriell and Trivedi (Model) [Metallurgy]

BCT Building Conservation Trust [UK]

BCT Bus Configuration Table

BCT Bushing Current Transformer

BCTA British Canadian Trade Association

BCTC British Carpet Technical Center

BCTC British Ceramic Tile Council

BCTC Building and Construction Trades Council [US]

BCTD Building and Construction Trades Department

BCTDC British Columbia Trade Development Corporation [Canada]

BC Tel British Columbia Telephone Company [Canada]

BCTMP Bleached Chemithermomechanical Pulp

BCTN N,N'-Bis(β-Carbamoylethyl)trimethylenediamine

BCTP British Continental Trade Press [UK]

BCTV Beet Curley Top Virus

BCU Bacterial Chemistry Unit [of Centers for Disease Control and Prevention, Atlanta, Georgia, US]

BCU Barium Chromium Uranate

BCU Bayamon Central University [Puerto Rico]

BCU Binary Counting Unit

BCU Block Control Unit

BCU British Commonwealth Union

BCU Buffer Control Unit

BCU Bus Control Unit

BCu Copper-Base Brazing Alloy

β-CuAlNi Beta Copper Aluminum Nickel

BCUC British Coal Utilities Commission

β-CuGa Beta Copper Gallium

β-CuIn Beta Copper Indium

BCuP Copper-Phosphorus Brazing Alloy

BCURA British Coal Utilization Research Association [now Coal Utilization Research Laboratories, UK]

BCuZn Copper-Zinc Brazing Alloy

β-CuSn Beta Bronze (Phase)

β-CuZn Beta Brass (Phase)

β'-CuZn Beta Prime Brass (Phase)

BCV Barge-Carrying Vessel

BCVA British Columbia Veterinary Association [Canada]

BCW Barium Copper Tungstate

BCW Base Curb Weight [Motor Vehicle]

BCW Buffer Control Word

BCW Burst Codewords

BCWA British Clock and Watch Association

BCWF British Columbia Wildlife Federation [Canada]

BCWL Basic Carbonate White Lead

BCWMA British Clock and Watch Manufacturers Association

BCWO Barium Copper Tungsten Oxide

BCWP Budget(ed) Costs for Work Performed

BCWS Budget(ed) Costs for Work Scheduled

BCWSG British Columbia Wood Specialties Group [Canada]

BCWWA British Columbia Water and Waste Association [Canada]

BD Back Diode

BD Backup Drive

BD Backward

BD Bahraini Dinar [Currency of Bahrain]

BD Balanced Draft

BD Ball Bearing, Double-Row, Angular-Contact, Filling Slot Assembly, Contact Angle Vertex Inside Bearing [Symbol]

BD Band Direction

BD Bangladesh [ISO Code]

BD Bank Draft [also B/D]

BD Barometric Damper

BD Barrel Distortion

BD Barrels per Day [Unit]

BD Base Detonating

BD Base Down (Light Bulb)

BD Basic Dimension

BD Beam Direction

BD Beams Division [of Fermilab, Batavia, Illinois, US]

BD Benzofurane Derivative

BD Beta Decay

BD Bile Duct [Anatomy]

BD Bills Discounted [also B/D]

BD Binary Decoder

BD Binary Digit

BD Binary Discrete

BD Binary Divide

BD Binary-to Decimal [also B/D]
BD Binomial Distribution
BD Biotechnology Department
BD Biotechnology Division
BD Block Diagram
BD Blocker Deflector
BD Blocking Device
BD Blowing Dust
BD Boiler Drum
BD Boltzmann-Drude (Conductance) [Physics]
BD Bomb Disposal
BD Bonner Durchmusterung [Star Catalog of the Bonn Observatory, Germany]
BD Bottlenecked Decay
BD Bottom Down
BD Bound
BD Brans-Dicke (Prediction) [Astrophysics]
BD Breakdown
BD Bright Dipped; Bright Dipping [Metallurgy]
BD Brought Down [also B/D]
BD Brush Discharge
BD Bulk Density
BD Bursting Disk
BD Business Design(er)
BD Butadiene
B$_2$D$_6$ Deuterated Diborane
B/D Bank Draft [also BD]
B/D Barrels per Day [Unit]
B/D Bills Discounted [also BD]
B/D Binary-to-Decimal [also BD]
B/D Brittle-to-Ductile (Transition) [also B-D]
B/D Brought Down [also BD]
B/D Bulldozer
B-D Bright-Dark (Fringe)
B-D Brittle-to-Ductile (Transition) [also B/D]
Bd Band [also bd]
Bd Baud [also bd]
Bd Board [also bd]
Bd Bond [also bd]
bd *(bis in die)* – twice daily [Medical Prescriptions]
bd bound
b/d brought down [also B/D]
B$ Belize Dollar [Currency of Belize]
B$ Brunei Dollar [Currency of Brunei]
BDA Backup Drive Amplifier
BDA Bermuda (Island) [NASA Space Tracking and Data Network]
BDA BIOS (Basic Input/Output System) Data Area
BDA Blast Danger Area
BDA Bomb Damage Assessment
BDA Booster-Distributor Amplifier
BDA Brick Development Association [UK]
BDA Bright-Dipped and Anodized (Aluminum)
BDA British Dental Association

BDA British Diabetes Association
BDA British Drilling Association
BDA Bulb Distributors Association [UK]
BDA Bund Deutscher Architekten [German Federation of Architects]
BDA Bundesvereinigung der Deutschen Arbeitgeberverbände [Federation of German Employers Associations]
BDA Butendiol Diacetate
14Bda 1,4-Butanediamine
bda 1,3-Butanedionate
BDAA Bio-Dynamic Agricultural Association [UK]
BDAB Benzyl-2,5-Diacetoxybenzoate
Bda$ Bermuda Dollar [Currency]
BDAC Biological Diversity Advisory Committee
BDAM Basic Data Access Method
BDAM Basic Direct Access Method
BDAT Best Demonstrated Available Technology
BDB Bundesverband Deutscher Banken [German Bankers Federation]
BDC Backup Domain Controller [Data Communications]
BDC Battelle Development Corporation [US]
BDC Bi-Directional Converter
BDC Binary Decimal Counter
BDC Boltzmann-Drude Conductance [Physics]
BDC Bonded Double Cotton (Electric Wire)
BDC Book Development Council [UK]
BDC Bottom Dead Center
BDC Bureau of Domestic Commerce
BDC Business Development Bank of Canada
BDCB Buffered Data/Control Bus
BDCC Beta Decay Coupling Constant
BDCE 1-Bromo-1,2-Dichloroethane
BDCF Baseline Data Collection Facility
BDC-OH 4,4'-Bis(dimethylamino)diphenylcarbinol
BDCR Baseline Document Change Request
BDCS Butyldimethylchlorosilane
BDCT Bradford Durfee College of Technology [Fall River, Massachusetts, US]
BDD Baseline Definition Document
BDD Binary-to-Decimal Decoder
BDD Bureau of Dangerous Drugs [of Health Canada]
BD$ Barbados Dollar [Currency]
BDDA Benzene-1,2-Dioxydiacetate
BDDG Bundesverband des Deutschen Dental-Medizinischen Grosshandels e.V. [Federation of German Dental Wholesalers]
BDDT Banque de Données en Toxicologie [Database on Toxicology; of Institut National de la Santé et de la Recherche Médicale, France]
BDE Baseband Distribution Equipment
BDE Batch Data Exchange
BDE N,N'-Bis(4-Aminophenyl) N,N'-Dimethylethylene Diamine
BDE Bond Dissociation Energy
BDE Borland Database Engine
BDE Brigade [also Bde, or bde]

BDE 1,4-Butanediol Diglycidyl Ether

BDEA tert-Butyl Diethanolamine

BDEC Bendigo and District Environment Council [Victoria State, Australia]

B-D EFC Bright-Dark Extreme Fringe Contrast

BD ELIM Band Elimination [also Bd Elim]

BDES Batch Data Exchange Service

BDes Bachelor of Design

BDF Backward Differentiation Formula (Technique)

BDF Banque de France [Bank of France]

BDF Baseband Distribution Frame

BDF Base-Detonating Fuse

BDF B-Cell Differentiation Factor [Biochemistry]

BDF Benzene-Derived Fiber

BDF Binary Difference Field

BDFA British Dairy Farmers Association

bd ft board foot [also bd-ft, bdft, fmb, or ft bd]

BDG Butyl Diglycol

BDGA Bundesverband des Deutschen Gross– und Aussenhandels e.V. [Federation of German Wholesale and Foreign Trade]

BDGH Binding Head [also BDG H, or Bdg H]

Bdgt Budget [also bdgt]

BDH Bearing, Distance and Heading

BDH British Drug Houses Limited

BDHI Bearing, Distance and Heading Indicator

BDHI Bearing Distance Homing Indicator

BDHT Blowdown Heat Transfer Program

BDI Bearing Deviation Indicator [also bdi]

BDI Bundesverband der Deutschen Industrie [Federation of German Industry]

BDI Bureau of Dairy Industry [US]

BDIA Base Diameter [also bdia]

BDIA Bund Deutscher Innenarchitekten [Federation of German Interior Designers]

BDIAC Battelle Defender Information Analysis Center [of Battelle Memorial Institute, US]

BDIC Binary-Coded Decimal Interchange Code

BDIR Bus Direction

B-1 DIV B-1 Division [of Rockwell International, US]

BDL Below Detection Limit

BDL Boltzmann Distribution Law [Statistical Mechanics]

BDL Boundary Diffraction Lattice [Metallurgy]

Bdl Bundle [also bdl]

BDLC Burroughs Data Link Control

Bdle Bundle [also bdle]

BDLI Bundesverband der Deutschen Luft–, Raumfahrt– und Ausrüstungsindustrie [Federation of German Aeronautics, Aerospace and Equipment Industries]

BDLS Bidirectional Loop Switching

BDM Biodiversity Data Management (Project) [of Global Environment Fund]

BDM Bockris-Devanathan-Mueller (Model)

BDM Bomber Defense Missile

BDMA Benzyldimethylamine

BDMA British Direct Marketing Association

BDMA British Disinfectant Manufacturers Association

BDMA Butylene Glycol Dimethacrylate

BDMAc N-[(4-Ethenylphenyl)methyl]-N-Methyl Acetamide

BDM15C5 Benzodimethyl-15-Crown-5 [also BDM15CROWN5]

BDMF N-[(4-Ethenylphenyl)methyl]-N-Methyl Formamide

BDMPAB N,N-Bis(3,5-Dimethylpyrazol-1-ylmethyl) aminobenzene [also bdmpab]

BDMS Brewer Data Management System

BDMS Butyl Dimethylsilyl Ether

BDN Bank Draft Number

BDN Bell Data Network

BDN Blocked Deoxynucleoside [Biochemistry]

BDNA Benzyldinonylame

BDNF Brain-Derived Neurotrophic Factor [Biochemistry]

BD Nickel Beryllia-Dispersed Nickel [also BD nickel]

BDOS Basic Disk Operating System

BDOS Batch Disk Operating System

BDOS Bulk Density of States [Solid-State Physics]

BDP Bellman's Dynamic Programming

BDP Bioenergy Development Program [of Panel on Energy Research and Development, Canada]

BDP Bond Directionality Principle

BDP Bonded Double Paper (Electric Wire)

BDP Breakdown Potential [Solid-State Physics]

BDP Business Data Processing

BDPA α,β-Bisdiphenylene-β-Phenylallyl

BDPA 1,3-Bisdiphenylene-2-Phenylallyl

BDPA Black Data Processing Associates [US]

BDPA British Disposable Products Association

BDPCP Trans-1,2-Bis(diphenylphosphinoxy)cyclopentane

BDPEC Bureau of Disease Prevention and Environmental Control [US]

BDPOP 2,4-Bis(diphenylphosphinoxy)pentane

BDPP Barrier Development Program Plan

BDPP 2,4-Bis(diphenylphosphino)pentane

BDPP Bis[2-(Diphenylphosphino)ethylphenylphosphine [also bdpp]

BDPP 1-Bromo-1,2-Diphenylpropane

BDPP Butyl Diphenyl Phosphate

BDPS Beam Deflection Phase Shifter

BDPSK Binary Differential Phase-Shift Keying

BDPU Bubble, Drop and Particle Unit

BDR Bell Doesn't Ring

BDR Bi-Duplexed Redundancy

BDR Binary Dump Routine

BDR Bus Device Request

Bdrm Bedroom [also bdrm]

Bdr Binder [also bdr]

Bdry Boundary [also bdry]

bdry boundary (or frontier) [Mathematics] [also Fr, or fr]

BDS Benzenediazothioether

BDS Bomb Disposal Squad [now Explosives Ordnance Disposal Team]

BDS Bonded Double Silk (Electric Wire)

BDS Bulk Data Switching

BDS British Display Society

BDS Bundesverband Deutscher Stahlhandel [Federation of German Steel Industries]

Bds Boards [also bds]

Bds Bonds [also bds]

Bds$ Barbados Dollar [Currency]

BDSA Benzene Disulfonic Acid

BDSA Business and Defense Services Administration [Washington, DC, US]

BDSC Boride-Dispersion-Strengthened Copper

BDSL Broadband ADSL (Asymmetric Digital Subscriber Line)

BDT Bangladesh Taka [ISO Currency Code]

BDT Basic Document Terminal

BDT Benzenedithiol

BDT Binary Deck to Binary Tape

BDT Block Data Transfer

BDT Brittle-Ductile Transition [Metallurgy]

BDT Bulk Data Transfer

BDT Bureau de Développement des Télécommunications [Telecommunications Development Bureau of International Telecommunications Union]

BDT Business Design Technology

BdT Bureau de Terminologie [of European Commision]

BDTA Benzophenone Tetracarboxylic Dianhydride; 3,3',4,4'-Benzophenone Tetracarboxylic Dianhydride

BDTA British Dental Trades Association

BDTA meso-2,3-Butanediaminetetraacetate [also bdta]

BDTC Sodium N,N-Dibutyl-Dithiocarbamate

BDTH Base Down to Horizontal (Light Bulb)

BDTP Bisdiphenyltriazinylpyridine

2,4-BDTP 2,4-Bis(5,6-Diphenyl-1,2,4-triazin-3-yl)pyridine

BDTS Buffered Data Transmission Simulator

BDTS Bulk Data Transfer Subsystem

BDTT Brittle-(to-)Ductile Transition Temperature [Metallurgy]

BDTTD Bermuda Department of Tourism and Trade Development

BDU Baseband Distribution Unit

BDU Basic Device Unit

BDU Basic Display Unit

BDU Biodiversity Unit [of Environment Australia]

BDU Bomb Disposal Unit

BDU Bundesverband Deutscher Unternehmensberater [Federation of German Business Consultants]

BDÜ Bundesverband der Dolmetscher und Übersetzer [Federation of German Interpreters and Translators]

BDV Bovine Diarrhea Virus

BDV Breakdown Voltage [Solid-State Physics]

BDW Bundesverband der Deutschen Wirtschaft e.V. [Federation of German Trade and Commerce]

BDW Buried Distribution Wire

BDWT Battelle Drop-Weight-Tear (Test)

BDWTT Battelle Drop-Weight-Tear Test

Bdy Body [also bdy]

Bdy Boundary [also bdy]

β-Dy Beta Dysprosium [Symbol]

BDYFLP Body Flap

BDZ Bund(esverband) Deutscher Zahnärzte [German Dentist Federation]

BE Bachelor of Economics

BE Bachelor of Education

BE Bachelor of Engineering

BE Back End

BE Backscattered Electron

BE Bacterial Endocarditis

BE Bakery Engineer(ing)

BE Baking Enamel

BE Balance Equation [Mathematics]

BE Ball Bearing, Double-Row, Angular-Contact, Filling Slot Assembly, Contact Angle Vertex Outside Bearing [Symbol]

BE Baltic Exchange [London, UK]

BE Band Elimination

BE Barkhausen Emission

BE Barnett Effect [Physics]

BE Base Ejection

BE Base Exchange

BE Basic Electrode

BE Battelle Europe

BE Baumann Exhaust

BE Bauschinger Effect [Metallurgy]

BE Beacon Electronic (Device)

BE Beam Expander; Beam Expansion

BE Becquerel Effect [Physics]

BE Beige

BE Belgium [ISO Code]

BE Bell End

BE Bellows Extension [Photomicrography]

BE Below, or Equal

BE Benchmark Experiment

BE Best Estimate Model

BE Bill of Exchange [also B/E]

BE Binding Energy [also be]

BE Binomial Expansion

BE Bioelectric(ity)

BE Bioengineer(ing)

BE Biological Engineer(ing)

BE Biomedical Engineer(ing)

BE Biphenyl Ether

BE Blended Elemental (Powder)

BE Board of Education

BE Body Engineer(ing)

BE Boiler Efficiency

BE Boltzmann Equation [Physics]

BE Bond and Etch (Method) [Electronics]

BE Bond Energy [Physical Chemistry]

BE Bonding Electrons [Physical Chemistry]

BE Booster Engine

BE Boron Equivalent [Metallurgy]

BE Bose-Einstein (Statistics) [Statistical Mechanics]

BE Bottom Edge

BE Boundary Element

BE Bound Energy

BE Bound Exciton [Solid-State Physics]

BE Bragg Effect [Physics]

BE Braking Ellipses [Aerospace]

BE Breaker End

BE Break-Even

BE Brewer Engel (Theory)

BE Burst-Error

BE Business Economics

BE 2-Butoxy Ethanol

B/E Bill of Entry

B/E Bill of Exchange [also b/e]

B/E Boron-Reinforced Epoxy (Composite) [also B/Ep]

Be Benzene

Be Beryllium [Symbol]

Be Bessel (Function) [Symbol]

Be$_f$ Beryllium Fiber [also Be(f)]

Be^{2+} Beryllium Ion [also Be^{++}] [Symbol]

Be-7 Beryllium-7 [also ^{7}Be, or Be^7]

Be-8 Beryllium-8 [also ^{8}Be, or Be^8]

Be-9 Beryllium-9 [also ^{9}Be, Be^9, or Be]

Be-10 Beryllium-10 [also ^{10}Be, or Be^{10}]

Bé Baumé (Gravity) Unit [also B]

be beige

.be Belgium [Country Code/Domain Name]

BEA Baugesellschaft für Elektrische Anlagen AG [Manufacturer of Electric Equipment for Industrial Applications; headquartered in Düsseldorf, Germany]

BEA N-Benzylethanolamine

BEA Biological Environment Affectors [US]

BEA Boundary Element Analysis (System)

BEA British Egg Association

BEA British Electrical Authority [now Central Electrical Authority]

BEA British Engineers Association

BEA British European Airways

BEA Broadcast Education Association [US]

BEA Bureau d'Etudes Automatismes [Research Bureau for Control Systems, France]

BEA Bureau of Economic Analysis [of US Department of Commerce]

BEA Business Education Association [US]

BEAB British Electrical Approvals Board

BEAB British Electrotechnical Approvals Board

BEAC Banque des Etats de l'Afrique Centrale [Bank of the Central African States]

BEAC Boeing Engineering Analog Computer

BEAC British European Airways Corporation

BEAC British Export Advisory Council

Be(Ac)$_2$ Beryllium Acetate

BEACON British European Airways Computer Network

BEAIRA British Electrical and Allied Industries Research Association

Be-Al Beryllium-Aluminum (Alloy)

BEAM Brewer Earth Atmosphere Measurements

BEAM Building Equipment, Accessories and Materials Program [Canada]

BEAM Burroughs Electronic Accounting Machine

BEAMA British Electrical and Allied Manufacturers Association

BEAMA TDA BEAMA Transmission and Distribution Association [UK]

BEAMOS Beam-Accessed Metal-Oxide Semiconductor

BEAMOS Beam-Addressable (or Addressed) Metal-Oxide Semiconductor [also Beamos]

BEAMS British Emergency Air Medical Service

BEAP Boundary Element Application Package

BEAR Berkeley Elites Automated Retrieval [of University of California at Berkeley, US]

BEAR Biological Effects of Atomic Radiation (Committee) [of National Academy of Sciences, US]

Bear Bearing [also bear]

BEARP Beaufort Environmental Assessment Review Panel [Canada]

BEAS Boundary Element Analysis System

BEAST Basic Experimental Automatic Syntactic Translator

BEATRIX Breeder Experimental Matrix; Breeder Materials Exchange Matrix [Nuclear Engineering]

Beauty International Conference on B-Physics

BEB Beach Erosion Board [US]

BEBA Bureau of Economic and Business Affairs [of US Department of State]

BEBC Big European Bubble Chamber [of CERN–European Laboratory for Particle Physics, Geneva, Switzerland]

BEBO Bond Energy Bond Order [Physical Chemistry]

Be-Bronze Beryllium Bronze [also Be bronze]

BEC Background Equivalent Concentration

BEC Beginning of the Equilibrium Cycle

BEC Belgian Trade Franc [Belgium and Luxembourg]

BEC Bengal Engineering College [Howrah, India]

BEC Bioenergy Council

BEC Biquadratic Exchange Coupling [Physics]

BEC Book Exporters Council [of Singapore Book Publishers Association]

BEC (Recent Advances in) Bose-Einstein Condensation (Conference) [Physics]

BEC British Electricity Commissioners

BEC British Engineers Club

BEC Building Employers Association [UK]

BEC Burst-Error Correction

BECA British Exhibition Contractors Association

BECC Burst-Error Correction Code

BEChemEng Bachelor of Engineering in Chemical Engineering [also BSChemE, or BEChE]

Beck Isol tech Beck Isoliertechnik [Publication on Electrical Insulation and Insuylators; published by Beck Elektro-Isoliersysteme, Hamburg, Germany]

BECM British Electrical Conduit Manufacturers

BECMA British Electroceramic Manufacturers Association

BECN Backward Explicit Congestion Notification

BECO Booster Engine Cutoff

BECS Basic Error Control System

BECS Building Energy Conservation Sector

BECTIS Bell College Technical Information Service [UK]

BECTO British Electric Cable Testing Organization

BECU Billion European Currency Units

BeCu Beryllium Copper

Be-Cu Beryllium-Copper (Alloy)

BED Bachelor of Environmental Design

BED Bermuda Dollar [Currency]

BED Bridge Element Delay

BED Building Equipment Division [of US Department of Energy]

BED Business Equipment Digest [UK]

BEd Bachelor of Education

Bed Bedroom [also bed]

BEDA Bachelor of Environmental Design in Architecture

BEDA British Electrical Development Association

BEDA Bureau of European Designers Associations [UK]

BED3A N-Benzylethylenediamine-N,N',N'-Triacetate [also bed3a]

BEDC Berlin Economic Development Corporation [Germany]

BEDC Building Economic Development Committee [UK]

BEDes Bachelor of Environmental Design

BEDO Burst Extended Data Out

B2EDP Di-n-Butylethane-1,2-Diphosphonate

BEDS Bachelor of Environmental Design Studies

Beds Bedforshire [UK]

Beds Bedrooms [also beds]

BEDSe-TSeF Bis(ethylenediselena)tetraselenafulvalene

BEDT Bis(ethylenedithio)-

BEDT-TTF Bis(ethylenedithio)tetrathiafulvalene

BEDT-TTN Bis(ethylenedithio)tetrathionaphthalene

BEE Bachelor of Electrical Engineering

BEE Biology, Energy and Ecology

BE(E) Binding Energy of Electron [also BE(e)]

BEEA British Educational Equipment Association

BEEC Binary Error-Erasure Channel

BEEC Buildings Energy Efficiency Conferation [UK]

BEEE Bachelor of Electrical and Electronics Engineering

BEEF Business and Engineering Enriched FORTRAN

BEEM Ballistic Electron Emission Microscopy

BEEN Bureau d'Etude de l'Energie Nucléaire [Bureau for the Study of Nuclear Energy, Belgium]

BEER Back-Emission Electron Radiography

BEES Ballistic Electron Emission Spectroscopy

BeEt₂ Diethylberyllium

BEF Band Elimination Factor

BEF Band Elimination Filter

BEF Blunt End Forward

BEF Bovine Embryonic Fluid

BEF Bovine Ephemeral Fever

BEF British Expeditionary Force(s)

BEF Buffered Emitter Follower

Be(f) Beryllium Fiber [also Be$_f$]

bef before

BEFAP Bell Laboratories FORTRAN Assembly Program

BEG Begin(ning) [Computers]

BEG Bundesverband des Elektro-Grosshandels e.V. [Federation of Electrical Wholesalers, Germany]

BEG Blume-Emery-Griffiths (Model) [Physics]

BEG Bureau of Economic Geology [Texas, US]

Beg Begin(ning)

beg beginning [Woven Fabrics]

BEH Bragg-Effect Hologram

BEHA British Export Houses Association

Behav Behavior(al) [also behav]

Behav Biol Behavioral Biology [Journal published in the US]

Behav Brain Sci Behavioral and Brain Sciences [US Journal]

Behav Inf Technol Behaviour and Information Technology [Journal published in the UK]

Behav Res Methods Instrum Behaviour Research Methods and Instrumentation [Published by Psychonomic Society, Inc., US]

Behav Res Methods Instrum Comput Behaviour Research Methods, Instruments and Computers [Published by Psychonomic Society, Inc., US]

Behav Sci Behavioral Science [Publication of the University of Louisville, Kentucky, US]

Behav Sci Res Behavior Science Research [International Journal]

BEHP Bis(2-Ethylhexyl)phthalate

BEH-PPV Polybutoxy Ethyl Hexyloxy-p-Phenylenevinylene

BEI Backscattered Electron Image; Backscattered Electron Imaging

BEI Backscattering of Energetic Ions

BEI Biological Exposure Index

BEI Bridgeport Engineering Institute [Connecticut, US]

BEI British Education Index [of British Library, UK]

Beil Beilstein's Handbuch der Organischen Chemie [A Handbook of Organic Chemistry compiled by the German Chemist F.P. Beilstein]

BEIP Biomedical Engineering and Instrumentation Program [of National Center for Research Resources, US]

BEIR Biological Effects of Ionizing Radiation

Beitr Atmos phys Beiträge zur Atmosphärenphysik [Contributions to Atmospheric Physics–German Journal]

Beitr Mineral Petrol Beiträge zur Mineralogie und Petrologie [Contributions to Mineralogy and Petrology–German Journal]

Beitr Plasmaphys Beiträge zur Plasmaphysik [Contributions to Plasma Physics–German Journal]

BEK Bleached Eucalyptus Kraft

BEK Butyl Ethyl Ketone

BEL Belgian (Financial) Franc [Currency of Belgium and Luxembourg]

BEL Bell (Character)

Bel Belgian; Belgium

BELF Bundesministerium für Ernährung, Landwirtschaft und Forsten [Federal Ministry for Food, Agriculture and Forestry, Germany]

Belg Belgian; Belgium

Belg J Oper Res Stat Comput Sci Belgian Journal of Operations Research, Statistics and Computer Science [Published in Brussels, Belgium]

BELINDIS Belgian Information Dissemination Service

BELLCORE Bell Communications Research

BELLREL Bell Laboratories Library Real-Time Loan [US]

Bell Syst Tech J Bell System Technical Journal [of AT&T Company, US]

BEM Bachelor of Engineering of Mining

BEM Band Edge Movement

BEM Bioelectromagnetic(s)

BEM Boundary Element Method

BEMA Bakery Equipment Manufacturers Association [US]

BEMA Bristol Engineering Manufacturers Association [UK]

BEMA British Essence Manufacturers Association

BEMA Business Equipment Manufacturers Association [now Computer and Business Equipment Manufacturers Association, US]

BEMAC British Exports Marketing Advisory Committee

BEMAS British Educational Management and Administration Society

BEMB Bituminous Equipment Manufacturers Bureau [now Bituminous and Aggregate Equipment Bureau, US]

BEMB British Egg Marketing Board

BEMEKO Belgische Meßtechnische Konföderation [Belgian Measurement Confederation]

BEMI Bioelectromagnetics Institute [US]

BEMP t-Butylimino-Diethylamino-Dimethyl-Perhydrodiazaphosphorine

BEMS Bioelectromagnetics Society [US]

BEMSA British Eastern Merchant Shipping Association

BEM SIG Bioelectromagnetics Special Interest Group [US]

BEN Bias(ed) Enhanced Nucleation (Technique) [Metallurgy]

BEN 7-Bromo-1-Ethylnaphthalene

BEN Bureau d'Etudes Nucléaires [Bureau for Nuclear Studies, Belgium]

BE(N) Binding Energy of Neutron [also BE(n)]

Ben Benzimidazole

BENACEN 3,8-Dimethyl-1,10-Diphenyl 4,7-Diazadecane-1,10-Dione

BENELUX Belgium, Netherlands and Luxembourg [also Benelux]

BENETUG Benelux Tandem Users Group

BEng Bachelor of Engineering

Beng Bengal

BEngMgt Bachelor of Engineering and Management

BEngSc Bachelor of Engineering and Science

BENHS British Entomological and Natural History Society

BeNi Beryllium Nickel

Be-Ni Beryllium-Nickel (Alloy)

BENILITE Beneficiation of Ilmenite (Process)

BenMet Benelux Métallurgie

BENSAT Benzene Saturation (Process) [Chemical Engineering]

Benz Benzene [also benz]

BeO Beryllium Oxide

BEOL Back-End-of-the-Line (Processing)

BeO(w) Beryllium Oxide (or Beryllia) Whisker [also BeO$_f$]

BEP Bachelor of Engineering Physics

BEP Back-End Processor

BEP Back Error Propagation

BEP Base Exchange Process

BEP Belgian Patent [also BeP]

BEP Bilateral Exchange Program [of Natural Sciences and Engineering Research Council, Canada]

BEP Brevet d'Etudes Professionelles [French Certificate of Vocational Studies]

BEP Bureau of Engraving and Printing [US]

BEP Business Emergency Plan

BE(P) Binding Energy of Proton [also BE(p)]

B/Ep Boron Reinforced Epoxy (Composite) [also B/E]

BeP Belgian Patent [also BEP]

BEPA British Egg Products Association

BEPA British European Potato Association

BEPAC The Building Energy Performance Analysis Club

BEPC Beijing Electron Positron Collider [PR China]

BEPC British Electrical Power Convention

BE(PE) Binding Energy of Photoelectron

BEPO British Experimental Pile Operation [of Atomic Energy Research Establishment at Harwell, UK]

BEPQ Bureau of Entomology and Plant Quarantine [US]

BEPS Building Energy Performance Standards

BER Basal Energy Requirement

BER Basic Encoding Rules [Internet]

BER Berliner Forschungsreaktor [Research Reactor at Hahn-Meitner Institute, Berlin, Germany]

BER Biomedical and Environmental Research

BER Bit Error Rate

BER British Experimental Rocket

BER Bureau of Educational Research [US]

Ber Bermuda

BERA Biomass Energy Research Association [US]

BERA British Educational Research Association

Ber Abwassertech Ver Berichte der Abwassertechnischen Vereinigung [Reports of the German Association for Wastewater Technology]

Ber Bunsenges Berichte der Bunsengesellschaft [Reports of the German Bunsen Society]

Ber Bunsenges Phys Chem Berichte der Bunsengesellschaft für Physikalische Chemie [Reports of the German Bunsen Society for Physical Chemistry]

BER CONT Bit Error Rate Continuous [also BER Cont]

Ber Dtsch Chem Ges Berichte der Deutschen Chemischen Gesellschaft [also Ber deut chem Ges]

BERGDB Berggrundsgeologisk Databas [Bedrock Geology Database, of Swedish Geological Survey]

Berg Hüttenmänn Monatsh Berg- und Hüttenmännische Monatshefte [German Monthly on Mining and Metallurgy published in Vienna, Austria]

Berks Berkshire [UK]

BERM Basic Encyclopedic Redundancy Media

BERM Bit Error Rate Monitor

BERNET Berlin Computer Network [Germany]

BERP British Experimental Rotor Program

BERPM Basic Exchange Rate Planning Model

BERT Basic Energy Reduction Technology

BERT Bit Error Rate Test(er); Bit Error Testing

Berth Berthing [also berth]

BERTS Bit Error Rate Test Set

BES Bachelor of Engineering Science

BES Bachelor of Environmental Studies

BES Basic Energy Sciences

BES (Office of) Basic Energy Sciences [of US Department of Energy]

BES Basic Executive System

BES Beijing Electron Synchrotron [PR China]

BES Bioelectrochemical Society [Germany]

BES Biological Engineering Society [UK]

BES Biomedical Engineering Society [US]

BES N,N-bis(2-Hydroxyethyl)-2-Aminoethanesulfonic Acid

BES Black Enamel Slate

BES Bose-Einstein Statistics [Statistical Mechanics]

BES Boundary Element Solver

BES Brazilian Entomological Society

BES British Ecological Society

BES Business Expansion Scheme

BeS Beryllium Sulfide

β-Es Beta Einsteinium [Symbol]

BESA British Electrical Systems Association

BESA British Engineering Standards Association

BESA Building Energy System Analysis

BESc Bachelor of Engineering Sciences

BES/DMS (Office of) Basic Energy Sciences/Division of Materials Science [of US Department of Energy] [also BES-DMS]

Bes Gr *(Besoldungsgrad)* – German for "Pay Grade"

Beskontaktn Elektr Mash Beskontaktnye Elektricheskie Mashiny [Latvian Publication on Electrical Machinery]

BESMA British Electrostatic Manufacturers Association

BES-MRC (Office of) Basic Energy Sciences/Materials Science Centers [of US Department of Energy] [also BES/MRC]

BESO British Executive Service Overseas

BeSO Beryllium Sulfate

BES-OER (Office of) Basic Energy Sciences/Office of Exploratory Research [of US Department of Energy] [also BES/OER]

BESOI Bond and Etch(ed Back) Silicon-on-Insulator (Method)

BESOM Brookhaven Energy System Optimization Model [of Brookhaven National Laboratory, Upton, New York, US]

BESP Building Energy Standards Program

BESRL Behavioral Science Research Laboratory [US Army]

BESS Base Ejection Smoke Shell

BESS Binary Electromagnetic Signal Signature

Bess Bessemer [Metallurgy]

BESSRC Basic Energy Science Synchrotron Radiation Center [of Argonne National Laboratory, Illinois, US]

BESSRC-CAT Basic Energy Science Synchrotron Radiation Center–Collaborative Access Team [of Argonne National Laboratory, the Northern Illinois University, Oak Ridge National Laboratory and the University of Tennessee, US]

BESSY Berliner Elektronen-Strahlungssynchrotron [Berlin Electron Radiation Synchrotron, Germany]

BESSY Bestellsystem [German Teleordering System]

BEST Bahamas Environment, Science and Technology Commission

BEST Ballastable Earthmoving Sectionized Tractor

BEST Ballistic Electron Schottky-Gate Transistor

BEST Basic Engineering Series and Tools [McGraw-Hill Texbook Series]

BEST Basic Extraction Sludge Treatment

BEST Battery Energy Storage Test

BEST Bio-Energetic Synchronization Technique [Chiropractic]

BEST Biomonitoring Environmental Status and Trends (Program) [of US Geological Survey/Biological Resources Division]

BEST Bioprocessing Expert System Tool

BEST Boehler Electroslag Topping [Metallurgy]

BEST Business EDP (Electronic Data Processing) Systems Technique

BEST Business Electronic Systems Technique

BEST Business Equipment Software Technique

BEST 2-(Trimethylsilyl) Methanethiol

B-Ester Hydroxy Acetic Acid Butyl Ester [also B-ester]

BET Best Estimate of Trajectory

BET Biological Engineering Technologist; Biological Engineering Technology

BET Biomedical Equipment Technician

BET Blended Elemental Technique

BET Brunauer, Emmett, and Teller (Method) [Physical Chemistry]

bet between

BETA Battlefield Exploitation and Target Acquisition

BETA Bureau for Education Technology and Administration [US]

BETA Business Equipment Trade Association [US, and UK]

BETEC Building Environment and Thermal Envelope Council [US]

BETECH (International Conference on) Boundary Element Technology

Be(TFAC)₂ Beryllium Trifluoroacetylacetonate [also Be(tfac)₂]

BethForge Bethlehem Forge Division [of Bethlehem Steel Corporation, Pennsylvania, US]

BETM Broadband Electromagnetic Testing Method

BETO Burlington Environmental Technology Office [of Wastewater Technology Center, Canada]

Beton Stahlbetonbau Beton- und Stahlbetonbau [German Publication on Concrete and Reinforced Concrete Construction]

BETP Buildings Energy Technologies Program [of Energy Technology Branch, Natural Resources Canada]

BETRA British Export Trade Research Organization

BET SA Brunauer, Emmett, and Teller Surface Area [Physical Chemistry]

BETSCE Brilliant Eyes Ten-K Sorption Cooler Experiment [of NASA]

BETT Buildings Energy Technology Transfer

betw between

BEU Basic Encoding Unit

BEU Bengal Engineering College [India]

BeV Billion Electron Volts [also Bev, or bev]

Bev Bevel(led) [also bev]

Bev-Bd Bevel Board

BEWA British Effluent and Water Association

Bewag Berliner Kraft- und Licht AG [Energy Supply Company located in Berlin, Germany]

BEX Broadband Exchange

BExA British Exporters Association

BF Bachelor of Forestry

BF Backface

BF Back Feed

BF Bad Flag [Computers]

BF Bag Filter

BF Balance Forward [also B/F]

BF Ball Bearing, Double-Row, Radial-Contact, Filling Slot Assembly [Symbol]

BF Balsam Fir

BF Band Filter

BF Banque de France [Bank of France]

BF Barium Ferrite

BF Barrage Fire

BF Base File

BF Base Fuse

BF Basic Fiber

BF Basic Frequency

BF Basic Function

BF Batch File

BF Beat Frequency

BF Before Firing

BF Belgian Franc

BF Bessel Function [Mathematics]

BF Bilby-Frank (Equation) [Physics]

BF Black Forest [Germany]

BF Blast Furnace

BF Blind Flying

BF Blocking Factor [Computers]

BF Board Foot [Lumber]

BF Boiler Feed

BF Boldface [also bf]

BF Boolean Function

BF Both Faces

BF Bottom Face

BF Box, Filled [Programming]

BF Bragg-Fresnel (Optics)

BF Brake Fluid

BF Branching Filter

BF Breakdown Field

BF Brightfield (Micrograph) [Microscopy]

BF British Forces

BF Brody-Fleming (Equation)

BF Brought Forward [also B/F]

BF Bubble Forming

BF Buckminsterfullerene [Chemistry]

BF Buffer

BF Burkina Faso [ISO Code]

BF Butterworth Filter

BF Resinoid Reinforced Type (Grinding Wheel) Bond

BF Boron Difluoride Ion [Symbol]

BF Boron Trifluoride Ion [Symbol]

BF Boron Tetrafluoride Ion [Symbol]

BF₆ Barium Hexaferrite [$BaO \cdot 6Fe_2O_3$]

B₂F Dibarium Ferrite [$2BaO \cdot Fe_2O_3$]

B₃F Tribarium Ferrite [$3BaO \cdot Fe_2O_3$]

B₅F Pentabarium Ferrite [$5BaO \cdot Fe_2O_3$]

B₂F₃ Dibarium Triferrite [$2BaO \cdot 3Fe_2O_3$]

B&F Bell and Flange

B&F Bound and Free [also B+F] [Biochemistry]

B/F Background/Foreground

B/F Brought Forward [also b/f]

Bf Belgian Franc [Currency]

B(f) Boron Fiber

bf boldface [also BF]

β-F Beta Fluorine [Symbol]

BFA Bachelor of Fine Arts

BFA Barium Iron Aluminate

BFA Batch File Access

BFA Binary Full-Adder

BFA Blended Friable Abrasive

BFA Bonded Friable Aluminum Oxide

BFA British Fabric Association

BFA British Foundry Association

BFA Bundesstelle für Außenhandelsinformation [Federal Office for Foreign Trade Information, Germany]

B-F-A Body Fluid Alcalizer

Bfafi Bundesforschungsanstalt für Fischerei [Federal Fisheries Research Institute, Germany]

BFAi Bundesstelle für Außenhandelsinformation [Federal Department for Foreign Trade Information, Germany] [also BFAI]

BfArM Bundesinstitut für Arzneimittel und Medizinprodukte [Federal Institute for Drugs and Medical Products, Germany]

BFAS Basic File Access System

BFB Bromofluorobenzene

B1B Boundary-Corrected First Born (Approximation) [Physics]

BFBS British Forces Broadcasting Service [Germany]

BFC Backup Flight Control (System)

BFCMA British Flue and Chimney Manufacturers Association

BFCO Band Filter Cutoff

BFCO (2-Methyl-1,3-Butadiene)tricarbonyl Iron

BFD Back Focal Distance

BFD Basic Floppy Disk

BF/DF Brightfield/Darkfield (Microscopy)

BFDK Before Dark

BFE Beam Forming Electrode

BFE Berufsgenossenschaft der Feinmechanik und Elektrotechnik [Association for Precision Mechanics and Electrical Engineering, Cologne, Germany]

BFE Bromotrifluoroethylene

BFE Bundesforschungsanstalt für Ernährung [Federal Nutrition Research Institute, Germany]

β-Fe Beta Iron [Symbol]

BFEC British Food Export Council

BFER Base Field Effect Register

β-FeSi$_2$ Beta Iron Disilicide

BFET Bipolar-Mode Field-Effect Transistor

BF$_3$-Ether Complex Boron Fluoride Etherate

BF$_3$·Et$_2$O Boron Trifluoride Ethyl Ether

BFF Binary File Format

BFF Budget Furniture Forum [of National Office Products Association, US]

BFF Bureau of Flora and Fauna [Defunct]

BFFN Body Frame Fixed Nuclei (Method) [also BF-FN]

BFG Binary Frequency Generator

BFG Blast-Furnace Gas

BFG Board of Fisheries and Game [Connecticut, US]

BFG Büro für Geophysik Lorenz [Lorenz Bureau for Geophysics, Berlin, Germany]

BfG Bank für Gemeinwirtschaft [Bank of Commerce, Germany]

BfG Bundesanstalt für Gewässerkunde [Federal Hydrological Institute, Germany]

BFG Citizen BFG Citizen [Published by BF Goderich, US]

bFGF Basic Fibroblast Growth Factor [Biochemistry]

BFH Bundesanstalt für Forst- und Holzwirtschaft [Federal Institute for Forestry and Timber Industry, Germany]

BFHA Benzoyl Phenyl Hydroxylamine

BFI Betriebsforschungsinstitut [Plant Research Institute; of Verein Deutscher Eisenhüttenleute, Düsseldorf, Germany]

BFI Bismuth Formic Iodide

BFI Brightfield Illuminated; Brightfield Illumination [Microscopy]

BFI Brightfield Image [Microscopy]

BFIA British Forging Industries Association

BFICC British Facsimile Industry Consultative Committee

BFIO Büro Führungskräfte zu internationalen Organisationen [Bureau for International Organization Executives; of Zentralstelle für Arbeitsvermittlung, Germany]

BFL Back Focal Length

BFL Buffered FET (Field-Effect Transistor) Logic

BFL Busy Flash [Telecommunications]

BFMA British Farm Mechanization Association

BFMA British Floorcovering Manufacturers Association

BFMA Business Forms Management Association [US]

BFMC British Friction Materials Council

BF$_3$-MEA Boron Trifluoride Monoethylamine

BF$_3$-MeOH Boron Trifluoride Methanol

BFMF British Furniture Manufacturers Federation

BFMI Benjamin Franklin Memorial Foundation [US]

BFMIRA British Food Manufacturing Industries Research Association

BFMP British Federation of Master Printers

BFMSA British Firework Manufacturers Safety Association

BFN Beam Forming Network

BFN British Forces Network

BFN Bundesamt für Naturschutz [Federal Nature Conservancy Office, German]

BFN Bye For Now [Internet]

BFO Back Flashover

BFO Barium Ferrite [BaFe$_x$O$_y$]

BFO Beat-Frequency Oscillator

BFO Beryllium Iron Oxide

BFO Bragg-Fresnel Optics

BFO/SSB Beat-Frequency Oscillator/Single Sideband

BFP Back Focal Plane

BFP Biologic False Positive (Reaction) [Medicine]

BFP Boiler Feed Pump

BFP Bureau of Fire Prevention

BFPA British Fiberboard Packaging Association

BFPA British Fluid Power Association

BFPC British Farm Produce Council

BFPDDA Binary Floating Point Digital Differential Analyzer

BFPMS British Federation for Printing Machinery and Supplies

BFPO Bis(dimethylamino)fluorophosphine Oxide

BFPO British Forces Post Office

BFPR Basic Fluid Power Research Program [US]

BFPSA British Fire Protection Systems Association

BFR Bridged Frequency Ringing

BFr Belgian Franc [Currency]

Bfr Buffer [also bfr]

BFRC British Flat Roofing Council

BFRL Building and Fire Research Laboratory [of National Institute of Standards and Technology, US]

BFS Bachelor of Foreign Service

BFS Backface Strain Gauge (Method)

BFS Backup Flight System

BFS Barium Fluosilicate

BFS Barium Iron Scandate

BFS Beam-Foil Spectroscopy

BFS Bozzolo-Ferrante-Smith (Method)

BFS Brute Force Supply

BFS Bundesamt für Strahlenschutz [Federal Radiation Protection Office, Germany]

BFS Bundesanstalt für Flugsicherung [Federal Institute for Air Traffic Control, Germany]

BFSA British Fire Services Association

bFSH Bovine Follicle-Stimulating Hormone

BFSK Binary Frequency Shift Keying

BFSO Barium Iron Scandium Oxide

BFT Bachelor of Foreign Trade

BFT Barium Iron Tantalate

BFT Barium Iron Titanate

BFT Betonwerk + Fertigteil-Technik [German Journal for Concrete and Precast Concrete Engineering]

BFT Binary File Transfer

BFTA British Fur Traders Association

BFTB Better Fabrics Testing Bureau [US]

BF-TEM Brightfield Transmission Electron Microscope; Brightfield Transmission Electron Microscopy [also BFTEM, or BF TEM]

BFTO Barium Iron Titanium Oxide (Magnet)

BFTP Batch File Transfer Protocol [Internet]

BFU Barium Iron Uranate [Ba_2FeUO_6]

BFUW British Federation of University Women

BF&VCA British Fruit and Vegetable Canners Association

BFVCC Body Frame Vibrational Close-Coupling (Method) [also BF-VCC]

BFW Boiler Feed Water

BG Back Gear

BG Background

BG Ball Bearing, Double-Row, Angular-Contact, Non-Filling Slot Assembly, Contact Angle Vertex Outside Bearing [Symbol]

BG Ballistic Galvanometer

BG Barring Gear

BG Bearing

BG β-Galactosidase [Biochemistry]

BG β-Glucuronidase [Biochemistry]

BG Better Grade (Lens)

BG Bevel Gear

BG Beyer-Graff (Star Chart)

BG Biguanide [Biochemistry]

BG Billing Group

BG Biodiversity Group [of Environment Australia]

BG Biogenesis

BG Bioglass

BG Birmingham Gauge

BG Blasting Gelatin

BG Bloch-Grüneisen (Model) [Solid-State Physics]

BG Blue-Green (Algae)

BG Board of Governors

BG Boehmite Gel

BG Bonded Goods

BG Bordet-Gengou (Medium) (for Whooping Cough)

BG Born-Green (Equation) [Physics]

BG Botanic(al) Garden

BG Bottom Grille

BG Bourdon Gauge

BG Bowling Green [Ohio, US]

BG Bragg-Gray (Principle) [Physics]

BG Brigadier General

BG British Gas Plc [UK]

BG British (Standard) Gauge

BG British Guiana

BG Bulgaria [ISO Code]

BG Bunsen-Gesellschaft [Bunsen Society, Germany]

BG Butyl Glycol

B(G) Magnetic Field [Symbol]

B/G Blue/Green (Laser)

Bg Bag [also bg]

Bg Bearing [also bg] [Navigation]

Bg Beauveria globulifera [Chinch Bug Fungus]

B$_n$(g) Nonmetal Component (of Compound AB Consisting of Metal Component A and Nonmetal Component B) in Gas Phase as a Multi-Atom Molecule [Symbol]

.bg Bulgaria [Country Code/Domain Name]

BGA Ball Grid Array

BGA Barre Granite Association [US]

BGA Blue-Green Algae

BGA Brilliant Green Agar

BGA British Gear Association

BGA British Gliding Association

BGA Butyl Glycol Acetate

β-Ga Beta Gallium [Symbol]

BGABG β-Galactosidase Anti-β-Galactosidase (Complex) [Biochemistry]

B-GAP Branched Glycidyl Azide Polymer

BGAS Bristol and Gloucestershire Archeological Society [UK]

BGAV Blue Green Algae Virus

BG/BARC Board of Governor's Budget and Accounts Review Committee [INTELSAT]

BGC Bay State Gas Company

BGC Binary-to-Gray Conversion

BGC Biogeochemical; Biogeochemist(ry)

BGC British Gas Corporation [UK]

BGC British Gas Council [UK]

BGC British Glaciological Society [now International Glaciological Society, UK]

BG Chemie Berufsgenossenschaft der chemischen Industrie [Association of the Chemical Industry, Germany]

BGCI Botanic Gardens Conservation International

BGCOLOR Background Color [HyperText Markup Language]

BGCS Botanic Gardens Conservation Secretariat

BGD Billion Gallons per Day [also bgd]

β-Gd Beta Gadolinium [Symbol]

BGDE Brigade [also Bgde, or bgde]

BGE Bachelor of Geological Engineering

BGE Branch if Greater, or Equal

BGE Butyl Glycidyl Ether

BGen Brigadier General

BGenEd Bachelor of General Education

BGFMA Bridge Grid Flooring Manufacturers Association [US]

BGFO Beryllium Gallium Iron Oxide

BGG Bovine Gamma Globulin

BGH Boschgotthardthütte [Iron and Steel Works located at Siegen, Germany]

BGH Bovine Growth Hormone

BGH Bundesgerichtshof [Federal Supreme Court, Germany]

BGHIP Benzo(g,h,i)perylene

bght bought

BGI BioGeo (Biogeosciences) Images [International Conference]

BGI Boeing Georgia Inc. [Macon, Georgia, US]

BGI Borland Graphic Interface

BGI Bureau Gravimetrique International [International Gravimetric Bureau; of BRGM (Bureau de Recherches Géologiques et Minières), France]

.BGI Borland Graphic Interface [File Name Extension]

BGIRA British Glass Industry Research Association

BGK Bhatnagar-Gross-Krook (Model) [Kinetic Equation]

BGL Bulgarian Lev [Currency]

Bgl Bacillus globigii [Microbiology]

Bgl I Bacillus globigii, Type I [Microbiology]

Bgl II Bacillus globigii, Type II [Microbiology]

BGMA British Gear Manufacturers Association

BGMC British Glass Manufacturers Confederation

BGMV Bean Golden Mosaic Virus

BGN Board on Geographic Names [US]

BGO Beryllium Gallium Oxide

BGO Bismuth Germanate (Semiconductor); Bismuth Germanium Oxide (Semiconductor)

BGOD Blue Grass Ordnance Depot [Richmond, Kentucky, US]

BGP Becker, Green and Pearson (Equation) [Materials Science]

BGP Border Gateway Protocol [Internet]

BGPC Bunsen-Gesellschaft für Physikalische Chemie [Bunsen Society for Physical Chemistry, Germany]

BG/PC Board of Governor's Planning Committee [of International Telecommunications Satellite Organization, US]

B-G Phos B Vitamins–Glycerophosphates

BGR Bombing and Gunnery Range

BGR Bundesanstalt für Geowissenschaften und Rohstoffe [Federal Institute for Geosciences and Natural Resources, Germany]

BGRAF Basic Graphics Software

BGRIMM Beijing General Research Institute of Mining and Metallurgy [PR China]

BGRR Brookhaven's Graphite Research Reactor [of Brookhaven National Laboratory, Upton, New York, US]

BGRV Boost Glide Reentry Vehicle

BGS Bachelor of General Studies

BGS Beta Gamma Signal

BGS Bombing and Gunnery School

BGS Bosson, Gutmann and Simmons (Equation)

BGS Brazilian Geochemical Society

BGS British Geological Survey

BGS British Geotechnical Society

BGS British Grassland Society

BGS Bundesgrenzschutz [Federal Border Patrol, Germany]

Bgs Bags [also bgs]

BGSU Bowling Green State University [Bowling Green, Ohio, US]

BGT Branch if Greater Than

BGU Ben-Gurion University [Beer Sheva, Israel]

BgVV Bundesinstitut für gesundheitlichen Verbraucherschutz [Federal Consumer Health Institute, Germany]

BGWF British Granite and Whinstone Federation

BGWP Battersea Gas Washing Process

BGY Billion Gallons per Year [also bgy]

BGY Born-Green-Yvon (Equation) [Physics]

BH Bahrain [ISO Code]

BH Bake Hardenability; Bake Hardenable; Bake Harden(ed); Bake Hardening

BH Ball Bearing, Single-Row, Radial-Contact, Non-Separable Counterbore Assembly [Symbol]

BH Bardeen-Herring (Source) [Materials Science]

BH Barium Hafnate

BH Barrier Height

BH Benzhydrol

BH Benzylic Hydrogen

BH Binary-to-Hexadecimal [also B-H]

BH Black Hole

BH Blankholder [Metallurgy]

BH Blasthole

BH Block Handler; Block Handling

BH Block Heater; Block Heating

BH Boiler House

BH Born-Haber (Process) [Solid-State Physics]

BH Born-Huang (Theory) [Physics]

BH Brazed-on-Hub

BH Brinell Hardness

BH British Honduras

BH Bromohydrin

BH Brooks-Herring (Approach) [Metallurgy]

BH Brush Holder

BH Burden and Hunt (Microsegregation Model) [Metallurgy]

BH Buried Heterostructure

BH Busy Hour

BH Butterworth and Heineman Publishers [US]

(BH)$_m$ Maximum (Magnetic) Energy Product [also (BH)$_{max}$] [Symbol]

BH$_3$ Borohydride

B$_d$H$_d$ (Magnetic) Energy Product [Symbol]

(B$_d$H$_d$)$_m$ Maximum (Magnetic) Energy Product [also (B$_d$H$_d$)$_{max}$] [Symbol]

B$_2$H$_6$ Diborane

B$_5$H$_9$ Pentaborane

B$_{10}$H$_{14}$ Decaborane

B$_{20}$H$_{26}$ Dodecaborane

B/H Permeance Coefficient (i.e., Ratio of Magnetic Induction (B) to Magnetizing Field (H))

B-H Binary-to-Hexadecimal [also BH]

.bh Bahrain [Country Code/Domain Name]

β-H Beta Hydrogen [or H$^\beta$]

BHA Ball Bearing, Double-Row, Radial-Contact, Non-Separable, Two-Piece Outer Ring [Symbol]

BHA Barium Hexa-Aluminate

BHA Base Helix Angle

BHA Benzyl Hydroxamic Acid

BHA Bottomhole Assembly

BHA British Hardmetal Association

BHA Bus History Association [Canada]

BHA Butylated Hydroxyanisole

BHA 2-(or 3-)-tert-Butyl-4-Hydroxyanisole

BHAB British Helicopter Advisory Board

BHB Bifurcated Hydrogen Bond

BHB Di-tert-Butyl Hydroxybenzene

BHBMA British Hacksaw and Bandsaw Manufacturers Association

BHBT 4-(5,6-Dimethoxybenzothiazol-2-yl)benzoic Acid Hydrazide

BHC Benzene Hexachloride

BHC Binary-to-Hexadecimal Conversion

α-BHC α-Benzene Hexachloride

β-BHC β-Benzene Hexachloride

γ-BHC γ-Benzene Hexachloride (or Lindane)

δ-BHC δ-Benzene Hexachloride

BHCA Busy Hour Call Attempt(s)

BHCC British Hellenic Chamber of Commerce

BHCDA Bureau of Health Care Delivery and Assistance [of Health Resources and Services Administration, US]

BHCS Business and Home Computer Show [US]

BHD Bahraini Dinar [Currency]

Bhd Bulkhead [also bhd]

BHDA Benzyl Heptadecyl Amine

β-He Beta Helium [Symbol]

BHEDC British Hospital Equipment Display Center [UK]

BHEL Bharat Heavy Electricals Ltd. [New Delhi, India]

BHET Bishydroxyethyl Terephthalate

BHF Barium Hexa-Ferrite

BHF Blankholder Force [Metallurgy]

BHF British Hardware Federation

BHF Buffered Hydrofluoric Acid; Buffered Hydrogen Fluoride (Solution)

β-Hf Beta Hafnium [Symbol]

BHFBR Brookhaven High-Flux Beam Reactor [of Brookhaven National Laboratory, Upton, New York, US]

BHFP Bottomhole Flowing Pressure

BHGA British Hang Gliding Association

β-HgS Black Mercury Sulfide (or Metacinnabar)

BHHBP Bishydroxyhexoxy Biphenyl; 4,4'-Bis(6-Hydroxyhexoxy)biphenyl

BHHMA British Hardware and Housewares Manufacturers Association

BHI Bald Head Island [North Carolina, US]

BHI Bechtel Hanford Inc. [US]

BHI Branch if Higher

BHI British Horological Institute [Upton, UK]

BHI Bureau Hydrogaphique International [International Hydrographical Bureau, Monaco] [now Organisation Hydrographique International (International Hydrographical Organization)]

BHJT Bipolar Heterojunction Transistor

BHL Busy Hour Load

BH&L John Bell, Hills and Lucas Limited [Pharmaceutical Manufacturer]

BHLS Below/Hook Lifters Section [of the Materials Handling Institute, US]

BHM Berg- und Hüttenmännische Monatshefte [Monthly Series on Mining and Metallurgy; published in Vienna, Austria]

BHMA Builders Hardware Manufacturers Association [US]

BHMF 1,1'-Bis(hydroxymethyl)ferrocene

BHMF 2,5-Bis(hydroxymethyl)furan

BHMORD Bureau of Health Maintenance Organization and Resources Development [of Health Resources and Services Administration, US]

BHMT Bis(hexamethylene)triamine

BHMT Bishydroxymethyl Terephthalate

BHMT Amine Bishydroxymethyl Terephthalate Amine

BHN Brinell Hardness Number [also Bhn]

BH$_3$·NH$_3$ Borane-Ammonia Complex

BHO Barium Hafnate; Barium Hafnium Oxide

BHOD Black Hills Ordnance Depot [South Dakota, US]

B Horizon B Horizon [Horizon of Soil Eluviation Consisting of Several Subhorizons B$_1$, B$_2$, etc.] [also B horizon]

BHOS Branch if Higher, or Same

BHP Boiler Horsepower [also Bhp, or bhp]

BHP Bottom Hole Pressure

BHP Brake Horsepower [also Bhp, or bhp]

BHP British Horsepower

BHP Broken Hill Proprietary (Company) [Australia]

BHP Bulk Handling Plant

BHP Bureau of Health Professionals [of Health Resources and Services Administration, US]

BHP sec-Butyl Hydrogen Phthalate

BHPC Broken Hill Proprietary Company [Australia]

BHPhr Brake Horsepower-Hour [also BHP-hr, bhphr, or bhp-hr]

BHP J Broken Hill Proprietary Journal [Australia]

BHP Tech Bull Broken Hill Proprietary Technical Bulletin [Australia]

BHR Biotechnology and Human Research

BHR Block Handling Routine

BHRA British Hydromechanics Research Association

BHRA News BHRA News [of British Hydromechanics Research Association]

BHRD Bureau of HMO (Health Maintenance Organization) and Resources Development [of Health Resources and Services Administration, US]

BHS Binding Head, Steel (Screw)

BHS British Herpetological Society

BHS British Holstein (Cattle) Society

BHS British Horological Society

BHS British Horse Society

BHS British Hydrological Society

BHS Business History Society [US]

BHSA British Heavy Steel Association

BHSC Black Hills State College [Spearfish, South Dakota, US]

BHSL Basic Hytran Simulation Language

BHSMA British Hay and Straw Merchants Association

BHT Baht [Currency of Thailand]

BHT Blowdown Heat Transfer

BHT Bottom Hole Temperature

BHT (Braunkohlen-Hochtemperaturkoks) – German for "High-Temperature Carbonized Coke"

BHT Butylated Hydroxytoluene

BHT Butylated Hydroxytoluol

Bht Baht [Currency of Thailand]

BHTA British Herbal Trade Association

BHT-BHA Butylated Hydroxytoluene–Butylated Hydroxyanisole

BH$_3$·THF Boron Trihydride Tetrahydrofuran

BHU Banaras Hindu University [Varanasi, India]

Bhu Bhutan

BHW Boiling Heavy Water (Nuclear Reactor)

BHWR Boiling Heavy Water Reactor

BHyg Bachelor of Hygiene

BI Backward Indicator

BI Baffin Island [Canada]

BI Balearic Islands [Spain]

BI Bank Indicator [Aerospace]

BI Bar Iron

BI Base Ignition; Base Initiating

BI Basic Index

BI Basicity Index

BI Batch Input

BI Bedford Institute [Canada]

BI Beijer Institute [of Royal Swedish Academy of Sciences]

BI Billing Instructions

BI Binary Input

BI Bioindication

BI Bismuth Institute [Belgium]

BI Black Iron

BI Blanking Input

BI Board of Investigation

BI Borland International

BI Break Interrupt

BI Bridge Indentation

BI British Isles

BI Brookings Institution [Washington, DC, US]

BI Burundi [ISO Code]

BI Bus Interconnect

B&I Base and Increment

Bi Biot [Unit]

Bi Biot Number [Thermal-Shock Resistance] [Symbol]

Bi Bismuth [Symbol]

Bi^{5+} Bismuth Ion [Symbol]

Bi-198 Bismuth-198 [also ^{198}Bi, or Bi^{198}]

Bi-199 Bismuth-199 [also ^{199}Bi, or Bi^{199}]

Bi-200 Bismuth-200 [also ^{200}Bi, or Bi^{200}]

Bi-201 Bismuth-201 [also ^{201}Bi, or Bi^{201}]

Bi-202 Bismuth-202 [also ^{202}Bi, or Bi^{202}]

Bi-203 Bismuth-203 [also ^{203}Bi, or Bi^{203}]

Bi-204 Bismuth-204 [also ^{204}Bi, or Bi^{204}]

Bi-205 Bismuth-205 [also ^{205}Bi, or Bi^{205}]

Bi-206 Bismuth-206 [also ^{206}Bi, or Bi^{206}]

Bi-207 Bismuth-207 [also ^{207}Bi, or Bi^{207}]

Bi-209 Bismuth-209 [also ^{209}Bi, Bi^{209}, or Bi]

Bi-210 Bismuth-210 [also ^{210}Bi, Bi^{210}, or RaE]

Bi-211 Bismuth-211 [also ^{211}Bi, Bi^{211}, or AcC]

Bi-212 Bismuth-212 [also ^{212}Bi, Bi^{212}, or ThC]

Bi-213 Bismuth-213 [also ^{213}Bi, or Bi^{213}]

Bi-214 Bismuth-214 [also ^{214}Bi, Bi^{214}, or RaC]

Bi-215 Bismuth-215 [also ^{215}Bi, or Bi^{215}]

Bi-221 Bi$_2$Sr$_2$CuO$_6$ [A Bismuth Strontium Copper Oxide Superconductor]

Bi-1112 BiSrCaCu$_4$O$_x$ [A Bismuth Strontium Copper Oxide Superconductor]

Bi-2201 Bi$_2$Sr$_2$Ca$_0$CuO$_x$ [A Bismuth Strontium Copper Oxide Superconductor]

Bi-2212 Bi$_2$Sr$_2$CaCu$_3$O$_x$ [A Bismuth Strontium Calcium Copper Oxide Superconductor]

Bi-2223 Bi$_2$Sr$_2$Ca$_2$Cu$_3$O$_x$ [A Bismuth Strontium Calcium Copper Oxide Superconductor]

Bi-2234 Bi$_2$Sr$_2$Ca$_3$Cu$_4$O$_x$ [A Bismuth Strontium Calcium Copper Oxide Superconductor]

Bi-2245 Bi$_2$Sr$_2$(Ca,Dy)$_4$Cu$_5$O$_x$ [A Bismuth Strontium Calcium Dysprosium Copper Oxide Superconductor]

Bi-2278 Bi$_2$Sr$_2$Ca$_7$Cu$_8$O$_x$ [A Bismuth Strontium Calcium Copper Oxide Superconductor]

Bi-4334 Bi$_4$Sr$_3$Ca$_3$Cu$_4$O$_x$ [A Bismuth Strontium Calcium Copper Oxide Superconductor]

Bi-4336 Bi$_4$Sr$_3$Ca$_3$Cu$_6$O$_x$ [A Bismuth Strontium Calcium Copper Oxide Superconductor]

Bi-4441 Bi$_3$(Sb$_{1-x}$Pb$_x$)Sr$_4$Ca$_4$CuO$_x$ [A Bismuth (Antimony Lead) Strontium Calcium Copper Oxide Superconductor]

Bi-11905 $Bi_{11}(Sr, Ca)_9Cu_5O_x$ [A Bismuth Strontium Calcium Copper Oxide Superconductor]

.bi Burundi [Country Code/Domain Name]

BIA Berufsgenossenschaftliches Institute für Arbeitssicherheit [Institute of the Association for Occupational Safety, Germany]

BIA Bicycle Institute of America [US]

BIA Bio-Industry Association [UK]

BIA Boating Industry Association

BIA Boost, Insertion and Abort

BIA Brick Institute of America [US]

BIA British Ironfounders Association

BIA Bureau of Indian Affairs [of US Department of the Interior]

BIA Buses International Association [US]

BIA Buyers Information Advisory

BIAA British Industrial Advertising Association

BIAA Bureau of Inter-American Affairs [US]

BIAC Bio-Instrumentation Advisory Council [of American Institute of Biological Sciences, US]

BIAC British Institute of Agricultural Consultants

BIAC Business (and) Industry Advisory Committe [of Organization for Economic Cooperation and Development, France]

Bi(Ac)$_2$ Bismuth (II) Acetate [also Bi(ac)$_2$]

Bi(Ac)$_3$ Bismuth (III) Acetate [also Bi(ac)$_3$]

BIADS Beam Injection Assessment of Defects in Semiconductors

BIAM Beijing Institute of Aeronautical Materials [PR China]

BIAMPA Banque d'Information Automatisée sur les Médicaments Principes Actifs [Automated Information Databank on Active Ingredients of Drugs] [also BIAM PA, France]

BIAMS Banque d'Information Automatisée sur les Médicaments Spécialités [Automated Information Databank on Special Pharmaceutical Products] [also BIAM S, France]

BIAN (International Conference on) Bianisotropics

BIAS Battlefield Illumination Airborne System

BIAS Biomedical Instrumentation Advisory Service [UK]

BIAS Bremer Institut für Angewandte Strahlentechnologie [Bremen Institute of Applied Beam Technology, Germany]

BIAS British Industrial Archeology Society

BIAS Brooklyn Institute of Arts and Sciences [New York, US]

BIAT British Institute of Architectural Technicians

BIATA British Independent Air Transport Association

BIB Baby Incendiary Bomb

BIB Backward Indicator Bit

BIB Balanced Incomplete Block (Design) [Statistics]

BIB Burn-In-Board

.BIB Bibliography [File Name Extension] [also .bib]

BIBA Bremer Institut für Betriebstechnik und angewandte Arbeitswissenschaft [Bremen Institute for Plant Technology and Applied Industrial Science, Bremen, Germany]

BIBB Bundesinstitut für Berufsbildung [Federal Institute for Professional Training, Germany]

BIBD Balanced Incomplete Block Design [Statistics]

BIBBES Bibliographic Book Entry System

BIBDES Bibliographic Data Entry System [of British Library Automated Information Service, UK]

Bibl Bibliographic; Bibliographer; Bibliography [also Bibliog, bibliog, or bibl]

BIBLIOS Book Inventory Building Library Information Oriented System [US]

BIBM Bureau International du Béton Manufacture [International Bureau for Precast Concrete, Belgium]

BIBO Bounded Input, Bounded Output

BiBr Bismuth Monobromide

BIBRA British Industrial Biological Research Association [now Bibra Toxicology International, UK] [also Bibra]

BIBRA Bull BIBRA Bulletin [of BIBRA Toxicology International, UK]

BIBUQ 4,4'''-Bis(2-Butyloctyloxy)-p-Quaterphenyl

BIC Beef Industry Council [of National Livestock and Meat Board, US]

BIC Biodeterioration Center [of University of Aston in Birmingham, UK]

BIC Biospecific Interaction (Affinity) Chromatography

BIC Bipolar Integrated Circuit

BIC Book Importers Council [of Singapore Book Publishers Association]

BIC British Importers Confederation

BIC Building Information Center [UK]

BIC Bureau International des Containers [International Container Bureau, France]

BIC Bureau of International Commerce

BIC Business Information Center

BIC Bus Interface Controller

BICAMS Barnett Institute of Chemical Analysis and Materials Science [US]

B-ICC Britain-Israel Chamber of Commerce [UK]

Bi-Cd Bismuth-Cadmium (Alloy System)

BICEMA British Internal Combustion Engine Manufacturers Association

BICEPT Book Indexing with Context and Entry Point from Text

BICERA British Internal Combustion Engine Research Association

BICERI British Internal Combustion Engine Research Institute

BICI Bureau International du Commerce et de l'Industrie [International Bureau of Commerce and Industry, France]

BiCl$_3$-GIC Bismuth Chloride Graphite Bi-Intercalation Compound

BICINE N,N-bis(2-Hydroxyethyl)glycine [also Bicine, or bicine]

BICMA British Industrial Ceramic Manufacturers Association

BICMOS Bipolar Complementary Metal-Oxide Semiconductor [also BiCMOS, or Bi-CMOS]

BICROS Binaural Contralateral Routing-of-Signal (Process)

BICS British Institute of Cleaning Sciences

BICS Building Industry Consulting Service

BiCSCO Bismuth Calcium Strontium Copper Oxide (Superconductor)

BICT Bipolar Integrated Circuit Technology

BICTA British Investment Casters Technical Association

BICTA British Investment Casting Trade Association

BiCu-1212 $(Bi,Cu)Sr_2YCu_2O_{7-\delta}$ [A Bismuth Copper Strontium Yttrium Copper Oxide Superconductor]

BID Bachelor of Industrial Design

BID Bachelor of Interior Design

BID Background Information Document

BID Blocker Initial Design; Blocker Initial-Guess Design; Blocker Initial-Guess Device

BID Brinell Indentation Diameter

bid *(bis in die)* – twice a day [Medical Prescriptions] [also BID]

BIDAC Biological Diversity Advisory Committee

BIDAP Bibliographic Data Processing Program

BIDCO Built-in Digital Circuit Observer

BiDi Bi-Directional

BIDMC Bismuth-Dimethyl Dithiocarbamate

BIDS Burroughs Information Display System

Bi:DyIG Bismuth Substituted Dysprosium Iron Garnet (Ferrite) [also Bi-DyIG]

BIE Bachelor of Industrial Engineering

BIE Boundary Integral Element (Method)

BIE Boundary Integral Equation

BIE British Institute of Engineers

BIE Bureau International des Expositions [International Bureau of Exhibitions, France]

BIE Bureau of Industry Economics

BIEAP Burrard Inlet Environmental Action Program [of Environment Canada]

BIEE British Institute of Electrical Engineers

BIEE British Institute of Energy Economics

BIECA Belgium International Express Courier Association

BIEM Boundary Integral Equation Method

BIET British Institute of Engineering Technology

BiEt₃ Triethyl Bismuth

BIF Banded Iron-Formation

BIF Basic in Flow

BIF Beef Improvement Federation [US]

BIF Berliner Institut für Weiterbildung von Führungskräften der Wirtschaft e.V. [Berlin Institute for Continuing Education of Executives in Trade and Commerce, Germany]

BIF Best Image Field

BIF Boiler and Industrial Furnaces

BIF British Industries Fair

BIF Burundi Franc [Currency]

.BIF Binary Initialization File [File Name Extension]

Bif Bildfernsprechen [Picture Telephony Service, Germany]

BIFA British International Freight Association

BIFAC British Isles Federation of Agricultural Cooperatives

BIFCA British Industrial Furnace Construction Association

BIFET Bipolar (Compatible) Field-Effect Transistor

BIFEX Baltic International Freight Futures Exchange

BIFF Binary Interchange File Format

BIFF British Inustrial Fasteners Federation

BIFMA Business and Institutional Furniture Manufacturers Association [US]

BIFOA Betriebswirtschaftliches Institut für Organisation und Automation [Institute for Organization and Automation in Business Management, Germany]

BIFU Banking, Insurance and Finance Union [UK]

BIG Bibliography and Index of Geology

BIG Bio-Informatics Group [of Australian National University, Canberra]

BIG British Inatitute of Graphologists

Big Biguanide [Biochemistry]

BIGA Bundesamt für Industrie, Gewerbe und Arbeit [Federal Department for Industry, Business and Labour, Switzerland]

BIGAS Bituminous Gas (Process)

BIGCA British Institute of Golf Course Architects

BIGCAP N,N'-Octanoylimino)bis(trimethylene)bis-(D-Gluconamide) [Biochemistry]

BIGCHAP Bis(gluconamidopropyl)cholamide; N,N-Bis(3-D-Gluconamidopropyl)cholamide [Biochemistry]

BIGENA Bibliography and Index of Geology Exclusive of North America

BIGFET Bipolar Insulated-Gate Field-Effect Transistor

bigit binary digit [also bit]

BIHFS British Institute of Hardwood Flooring Specialists

Bi-HTSC Bismuth High-Temperature Superconductor

BIIA British Institute of Industrial Art

BIIBA British Insurance and Investment Brokers Association

BIIC British Insurers International Committee

BIIL Basic Impulse Isolation Level

BIIM 2,2'-Biimidazole

BIIR Bromobutyl Rubber

BIIT Bureau International d'Information pour les Télécommunications [International Information Bureau for Telecommunications, Switzerland]

BIKAS Bibliothekskatalogsystem [Automated Library Cataloguing System, Germany]

BIKE Biotechnologischer Informations-Knoten für Europa [Biotechnological Information Network for Europe] [of Gesellschaft für Biotechnologische Forschung, Germany]

BIL Basic Impulse Level

BIL Basic Insulation Level

BIL Block Input Length

BIL Built-In Logic

BIL Buried Injection Logic

BILA Battelle Institute Learning Automation [of Battelle Memorial Institute, US]

BILA Bureau of International Labor Affairs [of US Department of Labor]

BILBO Built-In Logic Block Observability; Built-In Logic Block Observation; Built-In Logic Block Observer

BILD Binaural Intelligibility Level Difference

BILD Board for Industrial Leadership and Development [Canada]

BILD Business Industrial Leadership Development

Bild Ton Bild und Ton [German Publication on Television Engineering and Cinematography]

BILE Balanced Indicator Logical Element

BILE Beam-Induced Light Emission

BILGE Binary Load Generation

billi billion

BIM Beginning of Information Marker

BIM Benzimidazole

BIM Best in Match

BIM Blade Integrity Monitor

BIM Branch "If" Multiplexer

BIM British Institute of Management

BIM Bus Interface Module

BIMA British Interlining Manufacturers Association

BiMα Bismuth M-Alpha (Radiation) [also BiM$_\alpha$]

BIMAC Bistable Magnetic Core [also BIMAG]

BIMC Blessed Information-Memory-Concentration (Test) [Alzheimer's Disease]

BIMCAM British Industrial Measuring and Control Apparatus Manufacturers Association

BIMCO Baltic and International Maritime Council [Denmark]

BIMDA Benzimidazole

BIMOS Bipolar Metal-Oxide Semiconductor [also BiMOS]

BIMR Barton Institute for Medical Research [of University of Arkansas, US]

BIMRAB BuWeps-Industry Materiel Reliability Advisory Board [US Navy]

BImSchV Bundes-Immissionsschutzgesetz [Federal Air Pollution Laws, Germany]

BIN Basic Identification Number

BIN Binary [also bin]

BIN Bureau International de Normalisation [International Standards Bureau]

BIN Bureau of Information on Nickel [UK]

BIN Business Information Number

.BIN Binary [File Name Extension] [also .bin]

bin binary [also BIN]

BINAC Binary Automatic Computer

BINAP Binaphthyl; [1,1'-Binaphthalene]-2,2'-Diylbis-(diphenylphosphine); 2,2'-Bis(diphenylphosphino)-1,1'-Binaphthyl

BCD/BIN Binary to Binary-Coded Decimal (Converter)

BINC Building Industries National Council [UK]

BIND Berkeley Internet Name Domain [of University of California at Berkeley, US]

BInd Bachelor of Industry

BIndD Bachelor of Industrial Design [also BIndDes]

BIndEd Bachelor of Industrial Education

BINDT British Institute of Nondestructive Testing [Northampton, UK] [also BI NDT, or BI-NDT]

BINE Bürger Information Neue Energietechniken [Citizens' Information about New Energy Technologies (Project), of Fachinformationszentrum Karlsruhe, Germany]

BINHEX Binary Hexadecimal

BIN-HEX Binary-to-Hexadecimal (Conversion) [also Bin-Hex]

BINOBJ Binary File to Object File (Conversion) [Turbo Pascal]

BINOCT Binary-Octal

BIN-OCT Binary-to-Octal (Conversion) [also Bin-Oct]

BINOMEXP Binomial Expansion

BINOVC Break in Overcast

BINR Basic Intrinsic Noise Ratio

BINS Boreal Institute for Northern Studies [Canada]

BInstNDT British Institute of Nondestructive Testing

BINSW Bicycle Institute of New South Wales [US]

BIntArch Bachelor of Interior Architecture

BIntDes Bachelor of Interior Design

BIO Barium Indium Oxide

BIO Bedford Institute of Oceanography [Dartmouth, Nova Scotia, Canada]

BIO Bio Industry Association [UK]

BIO Biomedical Information Processing Organization [US]

BIO Black Iron Oxide

BIO Buffered Input/Output

BIOA Bureau of International Organization Affairs [of US Department of State]

BIOAL Bioastronautics Laboratory

BIOALRT Bioastronautics Laboratory Research Tool <19

Bioastron Bioastronautics [also bioastron]

Biodiv Biodiversity [also biodiv]

BiOBr Bismuth Oxybromide

Biochem Biochemical; Biochemist(ry) [also Bioch, bioch, or biochem]

Biochem Biochemistry [Journal of the American Chemical Society, US]

Biochem Biophys Res Comm Biochemical and Biophysical Research Communications [US Publication]

Biochem Cell Biol Biochemistry and Cell Biology [Publication of the National Research Council of Canada]

Biochem Educ Biochemical Education [Publication of the International Union of Biochemistry, US]

Biochem Eng J Biochemical Engineering Journal [Switzerland]

Biochem Int Biochemistry International [of International Union of Biochemistry, US]

Biochem J Biochemical Journal [of Biochemical Society, UK]

Biochem J Biochemistry Journal [of American Chemical Society, US]

Biochem Med Biochemical Medicine [Journal]

Biochem Mol Med Biochemical and Molecular Medicine [US Publication]

Biochem (Mosc) Biochemistry (Moscow) [English translation of *Biokhimiya* published by the Moscow State University, Russia]

Biochem Pharmacol Biochemical Pharmacology [US Journal]

Biochem Physiol Pflanzen Biochemie und Physiologie der Pflanzen [German Publication on Plant Biochemistry and Physiology]

Biochem Soc Trans Biochemical Society Transactions [UK]

Biochem (USSR) Biochemistry (USSR)

Biochem Z Biochemische Zeitschrift [German Biochemical Journal]

Biochim Biophys Acta Biochimica et Biophysica Acta [International Journal]

BiOCl Bismuth Oxychloride

BIOCLIM Bioclimatic Prediction and Modeling System [Software]

Bioconj Chem Bioconjugate Chemistry [Journal of the American Chemical Society, US]

BIOD Bell Integrated Optical Device

Bio-7-dATP N^6-(N-Biotinyl-6-Aminohexyl)-2'-Deoxyadenosine 5'-Triphosphate [Biochemistry]

Bio-11-dCTP 5-[N-(N-Biotinyl-ε-Aminocaproyl)-3-Aminoallyl]-2'-Deoxycytidine 5'-Triphosphate [Biochemistry]

Bio-11-dUTP 5-[N-(N-Biotinyl-ε-Aminocaproyl)-3-Aminoallyl]-2'-Deoxyuridine 5'-Triphosphate; 5-[N-(N-Biotinyl-ε-Aminocaproyl)-γ-Aminobutyryl)-3-Aminoallyl]-2'-Deoxyuridine 5'-Triphosphate [Biochemistry]

Bioelectrochem Bioenerg Bioelectrochemistry and Bioenergetics [Journal published in Switzerland]

BioE Bioengineer [also Bio Eng, or BioEng]

Bioeng Bioengineer(ing) [also Bioeng]

Bio Eng Bioengineer [also BioEng, or BioE]

BiOF Bismuth Oxyfluoride

Biofeedback Self-Regul Biofeedback and Self-Regulation [US Publication]

BIOFOR/BIOQUAL Implementation of Biological Processes for the Forest Industry and Environmental Quality [Canada]

Biog Biographer; Biographical; Biography [also biog]

Biogeochem Biogeochemical; Biogeochemist(ry) [also biogeochem]

Biogeogr Biogeographical; Biogeographer; Biogeography [also biogeogr]

Biohaz Biohazard(ous) [also biohaz]

Bioinform Bioinformatics [Journal published by Oxford University Press, UK]

Bioinorg Chem Bioinorganic Chemistry [US Journal]

BiOI Bismuth Oxyiodide

Biokhim Biokhimiya [Russian Biochemistry Journal published by Moscow State University]

BIOL Division of Biological Chemistry [of American Chemical Society]

Biol Biologic(al); Biologist; Biology [also biol]

Biol Biologicals [US Journal]

Biol Abstr Biological Abstracts [of Biosciences Information Service, US]

Biol Cybern Biological Cybernetics [Journal published in Germany]

BiolE Biological Engineer [also Biol Eng, or BiolEng]

Biol Eng Biological Engineer(ing)

Biol Gr Biological Grade [Chemistry]

Biol Int Biology International [Publication of the International Union of Biological Sciences, France]

Biol Neonate Biology of the Neonate [International Journal]

Biol Psychiat Biological Psychiatry [Journal]

Biol Reprod Biology of Reproduction [Journal of the Society for the Study of Reproduction, US]

Biol Sci Biological Science; Biological Scientist

Biol Signals Receptors

Biological Signals and Receptors [International Journal]

Biol Trace Elem Res Biological Trace Element Research [US Journal]

Biol Wastes Biological Wastes [UK Publication]

BIOMAIL Bedford Institute of Oceanography Marine Advisory and Industrial Liaison Office [Canada]

BIOMASS Biological Investigations of Marine Antarctic Systems and Stocks

Biomat Biomaterial [also Biomater]

Biomater Biomaterials [Journal published in the Netherlands]

Biomater Artif Cells Artif Organs Biomaterials, Artificial Cells and Artificial Organs [Journal published in the US]

Biomater Med Devices Artif Organs Biomaterials, Medical Devices and Artificial Organs [Journal published in the US]

BIOMED International Conference on Computers in Biomedicine

Biomed Biomedical; Biomedicine [also biomed]

Biomed Eng Biomedical Engineering [Translation of: *Meditsinskaya Tekhnika (USSR)*; published in the US]

Biomed Instrum Technol Biomedical Instrumentation and Technology [Publication of the National Society of Biomedical Equipment Technicians, US]

Biomed Mass Spect Biomedical Mass Spectrometry [Journal]

Biomed Mater Biomedical Materials [Journal published in the UK] [also Biomed Mat]

Biomed Mater Eng Biomedical Materials and Engineering [International Journal] [also Biomed Mat Eng]

Biomed Polym Biomedical Polymers [Journal published in the UK]

Biomed Res Biomedical Research [Journal published in the US]

Biomed Tech Biomedizinische Technik [Germany]

Biomed Technol Today Biomedical Technology Today [Publication of the Association for the Advancement of Medical Instrumentation, US]

Biometr Biometric(s) [also biometr]

Biometr Bull Biometric Bulletin [of Biometric Society, US]

BioMEMS Biomicroelectromechanical System

Bi_2O_3-MnO_2 Bismuth Trioxide-Manganese Dioxide (System)

BIONESS Biological (Zooplankton) Net Sampling System

BIONET Biodiversity Action Network [US]

Bionic Biological and Electronic [also bionic]

Bionics Biological Electronics [also bionics]

Bioorg Bioorganic [also bioorg]

Bioorg Bioorganism [also bioorg]

Bioorg Chem Bioorganic Chemistry [Published in the US]

Bioorg Med Chem Bioorganic and Medicinal Chemistry [Published in the US]

Bioorg Med Chem Lett Bioorganic and Medicinal Chemistry Letters [Published in the US]

BIOP Buffer Input/Output Processor

Biophys Biophysical; Biophysicist; Biophysics [also biophys]

Biophys J Biophysical Journal [of Biophysical Society, US]

Biopolym Biopolymers [International Journal]

Bi(OPr^i)$_3$ Bismuth (III) Isopropoxide

Bioprocess Eng Bioprocess Engineering [Published in Germany]

Bioprocessing Technol Bioprocessing Technology [Published in the US]

BIOR Business Input-Output Rerun

BIORED Biological Resources Development

BIOS Baffin Island Oil Spill Project [Canada]

BIOS Basic Input/Output (Operating) System

BIOS Biological Investigation of Space (Project) [of NASA]

BIOS Biological Satellite

BIOS British Institute for Organ Studies

BIOS British Intelligence Objectives Subcommittee

BiOS International Biomedical Optics Symposium

Biosci Bioscience; Bioscientist [also biosci]

Biosci Bioscience [Journal of the American Institute of Biological Sciences]

Biosci Biotech Biochem Bioscience, Biotechnology and Biochemistry [Journal]

Biosens Bioelectron Biosensors and Bioelectronics [Journal published in the UK]

BiOS Europe European Biomedical Optics Week

BIOSIM Simulation Modelling in Bioengineering

BIOSIS Biosciences Information Service [US] [also Biosis]

Biosis Data Biosciences Information Service Data

Biosph Biosphere [also biosph]

Biostat Biostatistical; Biostatistician; Biostatistics [also biostat]

BIOSTYR Biological Polystyrene (Process) [also Biostyr]

BioSync (Structural) Biology Synchrotron Users Organization [US]

BIOT British Indian Ocean Territory

BIOT Division of Biochemical Technology [of American Chemical Society, US]

BIOTA Biological Institute of Tropical America [US]

Biotech Biotechnological; Biotechnology; Biotechnology [also biotech]

BIOTECHNICA International Trade Fair for Biotechnology

BIOTechnol The BIOTechnologist [Publication of the American Chemical Society–Division of Biochemical Technology]

Biotechnol Biotechnological; Biotechnology; Biotechnology [also biotechnol]

Biotechnol Appl Biochem Biotechnology and Applied Biochemistry [Journal of the International Union of Biochemistry and Molecular Biology, UK]

Biotechnol Bioeng Biotechnology and Bioengineering [Journal published in the US]

Biotechnol Prog Biotechnology Progress [Joint Publication of the American Institute of Chemical Engineers and the American Chemical Society]

Biotechnol Lett Biotechnology Letters [US]

Biotechnol Tech Biotechnology Techniques [Journal]

BIOTEX Biotechnological Experiment

BIOTOX Biological and Toxin Research

BIOTRAC Biosphere Transport and Consequences (Model)

Biotrop Biotropica [International Journal]

Bio-11-UTP 5-[N-(N-Biotinyle-Aminocaproyl)-3-Aminoallyl] uridine 5'-Triphosphate [Biochemistry]

Bio-X-X-NHS 6-(Biotinamidocaproylamido)caproic Acid N-Hydroxysuccinimide Ester [Biochemistry]

BIP Bacterial Intravenous Protein [Biochemistry]

BIP Balanced-in-Plane (Contour)

BIP Banco Industrial del Peru [Industrial Bank of Peru]

BIP BASIC Interpreter Package

BIP Binary Image Processor

BIP Bismuth Iodoform Paraffin

BIP Books in Print (Database)

BIP British Institute of Plastics

BIPAAP 4-N-(Benzoylisopropylidene)-Aminoantipyrene

BIPAC Bibliographic Procedures and Control Committee [of Research Libraries Group, US]

Bi-Pb Bismuth-Lead (Alloy System)

BiPb-1212 Bismuth Lead Strontium Calcium Copper Oxide (Superconductor) [(BiPb)Sr$_2$CaCu$_2$O$_y$]

BiPb-2223 Bismuth Lead Strontium Calcium Copper Oxide Superconductor [(BiPb)$_2$Sr$_2$Ca$_2$Cu$_3$O$_y$]

BiPb-3221 Bismuth Lead Strontium Calcium Copper Oxide (Superconductor) [(BiPb)$_3$Sr$_2$Ca$_2$CuO$_{10}$]

BIPCO Built-in Place Component

BIPD Bi-Parting Doors

BIPE Bureau d'Informations et de Prévisions Economiques [Bureau for Economic Information and Forecasts]

BIPEA British Independent Plastic Extruders Association,

Bi(Ph)$_3$ Triphenylbismuth [also Bi(ph)$_3$]

BIPK n-Butyl Isopropyl Ketone

BIPM N-[4-(2-Benzimidazolyl)-Phenyl]-Maleimide; N-(p-(2-Benzimidazolyl)-Phenyl]-Maleimide

BIPM Bureau International des Poids et Mésures [International Bureau of Weights and Measures, France]

BIPP British Institute of Professional Photography

BIP Plast Rev BIP Plastics Review [of British Institute of Plastics]

BIPS Billion Instructions per Second

BiPSCCO Bismuth Lead Strontium Calcium Copper Oxide (Superconductor) [also BIPSCCO, or BPSCCO]

Bi-PST Polysynthetically Twinned (Crystal) [also bi-PST]

BiPtTe Bismuth Platinum Telluride

BIPY Bipyramidal [also BI-PY, Bi-Py, bi-py, Bipy, or bipy]

BIPY 2,2'-Bipyridine; 2,2'-Bipyridyl [also bipy]

BIPYAM Bis(2-Pyridyl)amine

BIPYO 2,2'-Bipyridine N,N'-Dioxide [also bipyo]

Biq 2,2'-Biquinoline [also biq]

BIQUIN 2,2'-Biquinoline

BIR B-50 Immunoreactivity

BIR Bio-Imaging Research, Inc. [US]

BIR Bioinorganic Reaction

BIR Bird Impact-Resistant (Aircraft Windshield)

BIR British Institute of Radiology [London, UK]

BIR Bureau International de la Récupération [International Reclamation Bureau, Switzerland]

BIR Bureau of Industry Research

BIR Bureau of Intelligence and Research [of US Department of State]

BIR Bureau of Internal Revenue [US]

BIR Bureau of International Recycling [London, UK]

BIRC Barnes Infrared Camera

BIRC Bio-Integral Resource Center [US]

BIRD Banque International pour la Reconstruction et le Développement [International Bank for Reconstruction and Development] [of United Nations]

BIRD Bilinear Rotation Decoupling

BIRD Birdseye Intelligent Reconnaissance Demonstrator

BIRDIE Battery Integration and Radar Display Equipment

BIRE British Institute of Radio Engineers

BIRES Broadband Isotropic Real-Time Electric Field Sensor

BIRF Brewing Industry Research Foundation

BIRL Basic Industrial (or Industry) Research Laboratory [of Northwestern University, Evanston, Illinois, US]

BIRM Bio-Inorganic Reaction Mechanism

BIRS Basic Indexing and Retrieval System

BIRS Basic Information Retrieval System

BIRS British Institute of Recorded Sound

BIS Bank for International Settlement [of United Nations]

BIS Basic Interchange System

BIS Biodiversity Information System

BIS Bio-Geographic Information System

BIS Biological Information Service [US]

BIS Board of Inspection and Survey [of US Navy]

BIS Boiler Inspection System

BIS Bremsstrahlung Isochromat Spectroscopy

BIS British and Irish Skeptics [Ireland]

BIS British Ichthyological Society

BIS British Imperial System

BIS British Information Service

BIS British Interplanetary Society

BIS Brought into Service

BIS Business Information System

BIS N,N'-Methylenebis(2-Propenamide)

BiS Bismuth Monosulfide

BISA Bibliographic Information on Southeast Asia [University of Sydney, Australia]

BISAC Book Industry Systems Advisory Committee [of Book Industry Study Group, US]

BISAM Basic Indexed Sequential Access Method

bis-AMP Bisaminomethylpropanol

bis-AMP N,N-bis(Hydroxyethyl)-2-Amino-2-Methyl-1-Propanol

bis-ANS 4,4'-Dianilino-1,1'-Binaphthyl-5,5'-Disulfonic Acid

BiSb Bismuth Antimonide (Semiconductor)

Bi-Sb Bismuth-Antimony (Alloy System)

BISC Back Intersystem Crossing (Emission Spectroscopy)

BISC Berkeley Integrated Sensor Center [of University of California at Berkeley, US]

BISC British Intelligence Subcommittee

BISF Business and Industry Service Center [Canada]

BISCC British Iron and Steel Consumer Council

BiSCCAO Bismuth Strontium Calcium Copper Aluminum Oxide (Superconductor) [also BISCCAO, or BSCCCAO]

BiSCCO Bismuth Strontium Calcium Copper Oxide (Superconductor) [also BISCCO, or BSCCO]

BiSCO Bismuth Strontium Copper Oxide (Superconductor) [also BISCO, or BSCO]

BISCOM Business Information System–Communications

BISCUS Business Information System–Customer Service

BISDIEN 1,4,7,13,16,19-Hexaaza-10,22-Dioxacyclotetracosane

B-ISDN Broadband Integrated Services Digital Network [also BISDN]

BISF British Iron and Steel Federation

BISFA British Industrial and Scientific Film Association

BISFA Bureau International pour la Standardisation de la Rayonne et des Fibres Synthétiques [International Bureau for the Standardization of Man-Made Fibers, Switzerland]

BISG Book Industry Study Group [US]

BisGMA Bisphenolglycidyldimethacrylate

Bis-GMA Bisphenol-A Bis(2-Hydroxypropyl)methacrylate; Bisphenol-A/Glycidyl Methacrylate (Mixture)

bis-homo(Tris) tris(3-Hydroxypropyl)aminomethane

BISHOP Bernice P. Bishop Museum [Honolulu, Hawaii, US]

BISITS British Industrial and Scientific International Translation Service

BISITS British Iron and Steel Industry Translation Service

BISL British Information Service Library

BISMRA Bureau of Inter-Industrial Statistics and Multiple Regression Analysis [US]

bis-MSB 1,4-Bis(2-Methylstyryl)benzene; p-bis(o-Methylstyryl)benzene

bis-NAD Bis(carbonylmethyl)nicotinamide Adenine Dinucleotide [Biochemistry]

Bi-Sn Bismuth-Tin (Alloy System)

BISNET Bank Information System Network [US]

BISNY Binary Synchronous Communication

BISPA British Independent Steel Producers Association

BISPAD Bissilylated Phenylaminodiol

BiSPCO Bismuth Strontium Lead Copper Oxide (Superconductor) [also BISPCO, or BSPCO]

BISPICTN Bis(2-Picolinyl)trimethylenediamine [also bispictn]

BISRA British Iron and Steel Research Association

BISRA Rep British Iron and Steel Research Association Report

BISS Base and Installation Security System

BISSC Baking Industry Sanitation Standards Committee [US]

BIST Beijing Institute of Science and Technology [PR China]

BIST Built-In Self Test

BISTREN Trioxaoctaazabicyclopentatriacontane

BIS-TRIS Bis(2-Hydroxyethyl)iminotris(hydroxymethyl) methane; Bisiminotrismethane [also bis-tris]

BIS-TRIS 2-Bis(2-Hydroxyethyl)-Amino-2-(hydroxymethyl)-1,3-Propanediol

BIS-TRIS 2,2-Bis(hydroxymethyl)-2,2',2"-Nitrilotriethanol [also bis-tris]

BiSYNC Binary Synchronous Communication (Protocol) [also BISYNC, bisync, or BSC]

Bisync Binary Synchronous; Bisynchronous [also bisync]

BISYNCH Binary Synchronous Communication [also bisynch]

Bisynch Binary Synchronous; Bisynchronous [also bisync]

BIT Bachelor of Industrial Technology

BIT Backscatter Imaging Tomography

BIT Barium Indium Tantalate

BIT Beijing Institute of Technology [PR China]

BIT Bihar Institute of Technology [Sindri, India]

BIT Bipolar Integrated (Circuit) Technology

BIT Boric-Acid Injection Tank

BIT Boron Injection Tank

BIT Brighton Information Technology Conference

BIT Built-In Test

BIT Bureau International du Temps [International Time Bureau, Paris, France]

BIT Bureau International du Travail [International Labour Office; of United Nations]

bit binary digit [also Bit, or BIT]

BITA British Industrial Truck Association

BITBLT Bit (Binary Digit) Block Transfer

Bit Bytes Rev Bits and Bytes Review [US]

BITE Backward Interworking Telephony Event

BITE Built-In Test Equipment

BITEL Bildschirm-Telefon [German Videotext Telephone]

BITER British Institute for Traffic Education and Research

BITL Bureau International Technique de l'ABS [International Technical Bureau of ABS (Acrylonitrile-Butadiene-Styrene) Producers, of Conseil Européen des Fédérations de l'Industrie Chimique, Belgium]

Bitblt Bit Block(-Level) Transfer [also biltblt]

BITM Bureau International Technique du Methanol [International Technical Bureau of Methanol Producers, of Conseil Européen des Fédérations de l'Industrie Chimique, Belgium]

Bi(TMHD)₃ Bismuth Tris(2,2,6,6-Tetramethyl-3,5-Heptanedionate) [also Bi(thmd)₃]

bit·m/s bit-meter per second [also bit/sec, or bit·m s⁻¹]

BITN Bilateral Iterative Network

BITNET Because It's Time Network [also Bitnet]

BITNIC Bitnet Network Information Center

BITPI Bureau International Technique des Polyésters Insaturés [International Technical Bureau of Producers of Unsaturated Polyesters, of Conseil Européen des Fédérations de l'Industrie Chimique, Belgium]

BITS Baseline Information Tracking System

BITS Binary Information Transfer System

BITS Birla Institute of Technology and Science [Pilani, India]

bit(s)/s bit(s) per second [also bit/sec, or bit s⁻¹]

BITU Bustamante Industrial Trade Union [Jamaica]

BIU Bar-Ilan University [Ramat-Gan, Israel]

BIU Basic Income Unit

BIU Basic Information Unit

BIU British Import Union

BIU Bus Interface Unit

BIV Best Image Voltage

BIVAR Bivariant Function Generator

BIW Body-in-White (Cost) [or B-I-W]

BIW Business Intelligence Warehouse [of Systems, Applications and Products Company, US]

BIX Binary Information Exchange

BIX Byte Information Exchange (Bulletin Board System)

BIXE Bombardment Induced X-Ray Emission

Bi:YIG Bismuth Substituted Yttrium Iron Garnet [also Bi-YIG]

BIZ Barium Indium Zirconate [$3BaO·In_2O_3·ZrO_2$, or $Ba_3In_2ZrO_8$]

BIZYbT Barium Fluoride–Indium Fluoride–Zinc Fluoride–Ytterbium Fluoride–Thorium Fluoride (Glass)

BIZYT Barium Fluoride–Indium Fluoride–Zinc Fluoride–Yttrium Fluoride–Thorium Fluoride (Glass)

BJ Bachelor of Journalism

BJ Bainbridge-Jordan (Mass Spectrometer)

BJ Ball Bearing, Double-Row, Angular-Contact, Non-Filling Slot Assembly, Contact Angle Vertex Inside Bearing [Symbol]

BJ Bar Joist

BJ Bence Jones (Protein) [Biochemistry]

BJ Benin [ISO Code]

BJ Brass Jacket

BJ Bubble Jet

B_r/J_s (Magnetic) Remanence-to-Saturation (Ratio)

BJA Burlap and Jute Association [US]

BJCEB British Joint Communications and Electronics Board

BJCEM British Joint Committee for Electron Microscopy

BJCG British Joint Corrosion Group

BJF Batch Job Foreground

BJH Barrett, Joyner and Halenda (Gas Adsorption Method) [Physical Chemistry]

BJM Between Job Monitor

BJMA Bangladesh Jute Mills Association

BJNDT British Journal of Nondestructive Testing [also Br J NDT, or Brit J NDT]

BJP Bubble Jet Printer

BJSM British Joint Services Mission [US]

BJT Band Jahn-Teller (Effect) [Physics]

BJT Base Junction Transistor

BJT Bipolar Junction Transistor

BJT/LED Bipolar Junction Transistor/Light-Emitting Diode (Interconnect)

BJTRA British Jute Trades Research Association

BJTU Beteiligungskapital für jungeTechnologie-unternehmen (Modellversuch) [Share Capital for New High Tech Companies (Model); of Bundesministerium für Forschung und Technologie, Germany]

BK Backward [also bk]

BK Ball Bearing, Double-Row, Radial-Contact, Non-Filling Slot Assembly [Symbol]

BK Bar Knob

BK Below Knee

BK Black [also Bk, or bk]

BK Break [Amateur Radio]

BK Break-In Keying

BK Burridge-Knopoff (Model)

.BK Backup (File) [WordPerfect File Name Extension]

.BK1 Temporary Backup (File) for Document One [WordPerfect File Name Extension]

.BK2 Temporary Backup (File) for Document Two [WordPerfect File Name Extension]

.BK! (Original) Backup (File) [WordPerfect File Name Extension]

B&K Bruel and Kjaer (Company)

Bk Bank [also bk]

Bk Barkometer [Chemical Engineering]

Bk Berkelium [Symbol]

Bk Block [also bk]

Bk Book [also bk]

Bk Brake [also bk]

Bk-243 Berkelium-243 [also ^{243}Bk, or Bk^{243}]

Bk-244 Berkelium-244 [also ^{244}Bk, or Bk^{244}]

Bk-245 Berkelium-245 [also ^{245}Bk, or Bk^{245}]

Bk-246 Berkelium-246 [also ^{246}Bk, or Bk^{246}]

Bk-247 Berkelium-247 [also ^{247}Bk, or Bk^{247}]

Bk-248 Berkelium-248 [also ^{248}Bk, or Bk^{248}]

Bk-249 Berkelium-249 [also ^{249}Bk, or Bk^{249}]

Bk-250 Berkelium-250 [also ^{250}Bk, or Bk^{250}]

bk black

BKA Bundeskriminalamt [Federal Criminal Police Office, Germany]

BKBO Barium Potassium Bismuth Oxide

BKCEC British Knitting and Clothing Export Council

Bkcy Bankruptcy [also bkcy]

BKD Backscattered Kikuchi Diffraction

BKDP Backscattered Kikuchi Diffraction Pattern [Crystallography]

BKDP Book/Disk Pack

BKDP3.5 Book/Disk Pack with 3.5" Disk(ette)(s)

BKDP5.25 Book/Disk Pack with 5.25" Disk(ette)(s)

BKF Benzo[k]fluoranthene

Bkg Banking [also bkg]

Bkgd Background [also Bkgrd, bkgrd, or bkgd]

Bkkpr Bookkeeper [also bkkpr]

BKME Bleached Kraft Mill Effluent

BKP Bleached (or Bleachable) Kraft Pulp

BkPdr Black Powder

Bkpg Bookkeeping [also bkpg

BKPM Bleached Kraft Pulp Mill

Bkpr Bookkeeper [also bkpr]

Bkpt Bankrupt [also bkpt]

Bkr Breaker [also bkr]

Bkrpt Bankrupt [also bkrpt]

Bks Barracks [also bks]

Bks Books [also bks]

BKSP Backspace [also Bksp]

BKSTS British Kinematograph, Sound and Television Society

BKT Berezinskii-Kosterlitz-Thouless (Transition) [Physics]

Bkt Basket [also bkt]

Bkt Bracket [also bkt]

BKZ Bernstein, Kearsley and Zapas (Model)

BL *(Baccalaureus Legum)* – Bachelor of Laws

BL Backlit

BL Bale [also Bl, or bl]

BL Ball Bearing, Single-Row, Radial-Contact, Filling Slot Assembly [Symbol]

BL Barrier Layer [Electronics]

BL Baseline [also B/L]

BL Base Load

BL Base-Loaded (Shell)

BL Beam Lead [Electronics]

BL Bed Load [Geology]

BL Bell

BL Bend Line

BL Between Layers

BL Bibliographic Level

BL Bilayer

BL Bioluminescence; Bioluminescent

BL Bit Line

BL Black [also bl]

BL Bloch Line [Solid-State Physics]

BL Block Length

BL Block Loading

BL Blotting Grade [Immunology]

BL Blue Light

BL Blumlein Line

BL Bond Length [Physical Chemistry]

BL Bottom Layer

BL Boundary Layer

BL Boundary Light

BL Boundary Lubrication
BL Boyle's Law
BL Bragg's Law [Solid-State Physics]
BL Bravais (Crystal) Lattice
BL Breadth-Length
BL Breaking Load
BL Break Line
BL Breech Loading
BL British Leyland [UK]
BL British Library [London, UK]
BL Brönsted-Lowry (Theory) [Chemistry]
BL Buffer Layer
BL Building Line
BL Burkitt's Lymphoma
BL Burnside Laboratory [of DuPont Company at Penns Grove, New Jersey, US]
BL Busy Lamp
BL Büttiger-Landauer (Time)
B&L Bausch and Lomb [Manufacturer of Laboratory Equipment]
B/L Baseline [also BL]
B/L Bill of Lading
Bl Bladder [also bl]
Bl Blank(ing) [also bl]
Bl Blasting [also bl]
Bl Block [also bl]
Bl Blotting [also bl]
bl bale
bl barrel [Unit]
bl black
bl block
bl blue
BLA Bachelor of Landscape Architecture
BLA Ball Bearing, Single-Row, Radial-Contact, Spherical Outside Surface, Filling Slot Assembly [Symbol]
BLA Binary Logical Association
BLA Blocking Acknowledgement Signal
BLA British Library Association
BLA British Lift Association
BLA British Light Aviation Center
BLA Bureau de Liaison des Syndicats Européens (CEE) des Produits Aromatiques [Liaison Bureau of the European and EEC Unions of Aromatic Products, Belgium]
β-La Beta Lanthanum [Symbol]
BLAC British Light Aviation Center
BLADE Basic Level Automation of Data through Electronics
BLADES Bell Laboratories Automatic Design System [also BLADS, US]
BLAISE British Library Automated Information Service
BLAM Ballistically-Launched Aerodynamic Missile
BLAS Basic Linear Algebra Subroutines (or Subprograms)
β-La₂S₃ Beta Lanthanum Sesquisulfide
BLandArch Bachelor of Landscape Architecture

BLAST Basic Local Alignment Search Tools [of National Center for Biological Information, US]
BLAST Blocked Asynchronous Transmission
BLAST Boolean Logic and State Transfer
B/L att Bill of Lading Attached
BLAVA British Laboratory Animals Veterinary Association
BLB Brillouin-Levy-Berthier (Theorem) [Solid-State Physics]
BLBA British List Brokers Association
BLBM Binary Linear Bubble Model (of Grain Growth) [Metallurgy]
BLBPEN N,N'-Bis(2,6-Lutidinyl)-N,N'-Bis(2-Picolinyl) ethylenediamine [also blbpen]
BLBS British Library Bibliographic Services
BLBSD British Library Bibliographic Services Division
BLC Barrier-Layer Capacitance
BLC Base-Loaded Capped (Shell)
BLC Beef Liver Catalase [Biochemistry]
BLC Board Level Computer
BLC Boundary-Layer Concept
BLC Boundary-Layer Control
BLC Breech-Loading, Converted
BLC British Lighting Council
BLCC Building Life-Cycle Cost
BLD Beam-Lead Device
BLD Binary Load Dump
BLD Blood Effects (of Hazardous Materials)
.BLD BASIC BLoad Graphics (File Name Extension)
Bldg Building [also bldg]
BldgE Building Engineer [also Bldg Eng, or BldgEng]
Bldg Mat Building Material [also bldg mat]
Bldr Builder [also bldr]
BLDSC British Library Document Supply Center
BLE Best Linear Estimator
BLE Bombardment-Induced Light Emission
BLE Branch if Less, or Equal
BLE Brotherhood of Locomotive Engineers [Now part of United Transportation Union]
BLEM Beijing Laboratory of Electron Microscopy [of Chinese Academy of Sciences]
BLEND Birmingham Loughborough Electronic Network Development [UK]
BLER Block Error
BLERT Block Error Rate Test(ing)
BLES Best Linear Estimators
BLESSED Bell Little Electronic Symbolic System for the Electrodata
BLEU Belgium-Luxembourg Economic Union
BLEU Blind Landing Experimental Unit [of Royal Aerospace Establishment, UK]
BLEVE Boiling Liquid Expanding Vapor Explosion
BLEW Burum, Linder, Ernst Windowless (Analysis) [Physics]
BLF Band Limiting Filter
BLF British Laminated Plastics Fabricators Association

BLF British Leather Federation

BLF British Lubricants Federation

BLF Bulk-Loss Function

BLFE Brotherhood of Locomotive Firemen and Enginemen [Now part of United Transportation Union]

BLG Benzyl-L-Glutamate

BLG Breech Loading Gun

BlG Blasting Gelatin

BLHP Bradley Lake Hydroelectric Project [of Alaska Power Authority, US]

BLI Basic Learning Institute

BLI Birdlife International

β-Li Beta Lithium [Symbol]

BLIC Bureau de Liaison de l'Industrie Caoutchouc [Liaison Bureau of the Rubber Industries of the European Community, Belgium]

BLIE Best Linear Invariant Estimator

BLINE Baseline [also Bline]

B Line B Line [Fraunhofer Line of Terrestrial Origin Having a Wavelength of 686.7 Nanometers] [also B line] [Spectroscopy]

b_1 line b_1 line [Fraunhofer line for magnesium having a wavelength of 517.8 nanometers] [Spectroscopy]

b_2 line b_2 line [Fraunhofer line for magnesium having a wavelength of 517.3 nanometers] [Spectroscopy]

B-LINK Birmingham Library and Information Network [UK]

BLIP Background-Limited Incident Power

BLIP Background Limited Infrared Photoconductor

BLIP Block Diagram Interpreter Program

BLIS Baffle/Liner Interface Seal

BLIS Bell Laboratories Interpretative System

BLISS Baby Life Support System

BLISS Basic Language for the Implementation of System Software

BLISS Butylene Isomerization System (Process)

BLitt Bachelor of Letters

BLJ British Library Journal [of British Library, London, UK]

BLK Block [Computers]

BLK Bund–Länder-Kommission für Bildungsplanung und Forschungsförderung [Federal/State Commission for Education Planning and Research Promotion, Germany]

Blk Blank [also blk]

Blk Block [also blk]

Blk Bulk [also blk]

blk black

Blk Car Bulk Carrier [also Blk Car, or blk car]

Blkg Blocking [also blkg]

blksh blackish

BLL Below Lower Limit

BLL Bobadilla, Lacaze and Lesoult (Microsegregation Model) [Metallurgy]

BLLD British Library Lending Division

Blldzr Bulldozer [also blldzr]

BLM Bachelor of Land Management

BLM Background Luminance Monitor

BLM Basic Language Machine

BLM Bilayer Lipid Membrane

BLM Bureau of Land Management [of US Department of the Interior]

BLMA British Ladder Manufacturers Association

BLMA British Leathergoods Manufacturers Association

BLMC Buried Logic Macrocell

BLMF British Lawnmower Manufacturers Association

BLMRF British Lawn Mower Racing Association

BLMRA British Leather Manufacturers Research Association

BLMRA J BLMRA Journal [of British Leather Manufacturers Research Association]

BLMRCP Bureau of Labor-Management Relations and Cooperative Programs [of US Department of Labor]

BLMSS Bayfield Laboratory for Marine Science and Surveys

BLN Barium Lanthanum Niobate

Bln Balloon [also bln]

BLNO Barium Lanthanum Niobium Oxide

BLO Below Clouds [Meteorology]

BLO Blocking Signal

BLO γ-Butyrolactone

Blo Blower [also blo]

BLOB Binary Large Object

BLOC Booth Library On-Line Circulation [of Eastern Illinois State University, US]

BLODI Block Diagram

Blood Cells Mol Disease Blood Cells, Molecules and Disease [of Blood Cells Foundation, US]

Blood Purif Blood Purification [International Journal]

BLOS Branch if Lower, or Same

Blot Immunoblotting

BLOW Booster-Lift-Off Weight

BLP Ball Lock Pin

BLP Behlen Laboratory of Physics [of University of Nebraska, US]

BLP Biotechnology Loan Program [of Western Economic Diversification Canada]

BLP Bypass Label Processor

Bl&P Blind Loaded and Plugged

BLPC Blocking Layer Photocell

BLR Baseline Restorer

BLR Breech Loading Rifle

Blr Boiler [also blr]

BLRD British Library Reference Division

BLR&DD British Library Research and Development Division

BLS Bachelor of Library Science

BLS Base Load Station

BLS Bottom Left Side

BLS Branch Line Society

BLS Brillouin Light Scattering

BLS Bureau of Labor Statistics [of US Department of Labor]

BLSGMA British Lamp Blown Scientific Glassware Manufacturers Association

blsh bluish

BLSN Blowing Snow

Blstg Blasting [also blstg]

Blstg Pwd Blasting Powder

BLT Bachelor of Laboratory Technology

BLT Basic Language Translator

BLT Borrowed Light

BL&T Blind Loaded and Traced; Blind Loaded with Tracer

BLTC Bottom Loading Transfer Cask

BLTI Better Lawn and Turf Institute [now Lawn Institute, US]

BLTP Bogoliubov Laboratory of Theoretical Physics [of Joint Institute for Nuclear Research, Dubna, Russia]

BLU Basic Link Unit

BLU Basic Logic Unit

BLU Bipolar Line Unit

BLUE Best Linear Unbiased Estimator

BLUES Best Linear Unbiased Estimators

Bluo-Gal 5-Bromo-3-Indolyl β-D-Galactoside [Biochemistry]

Bluo-GlcA 5-Bromo-3-Indolyl β-D-Glucuronide [Biochemistry]

BLV Bleed(er) Valve

BLV Bovine Leukemia Virus

Blvd Boulevard [also blvd]

BLW Below [also blw]

BLWA British Laboratory Ware Association

Blwr Blower [also blwr]

BLX Benelux Economic Union

BLZ Bayerisches Laserzentrum [Bavarian Laser Center, Erlangen, Germany]

Blzd Blizzard [also blzd]

BM Babbitt Metal

BM Bachelor of Medicine

BM Back Marker

BM Bacterial Meningitis

BM Bainitic Martensite [Metallurgy]

BM Ball Bearing, Single-Row, Radial-Contact, Separable Assembly [Symbol]

BM Ballistic Material

BM Ballistic Missile

BM Ball Mill(ing); Ball Milled

BM Band Model [Physics]

BM Barium Manganate

BM Baroclinic Model [Meteorology]

BM Barotropic Model [Meteorology]

BM Barrels per Month

BM Basal Metabolism

BM Base Material

BM Base Metal

BM Basic Motion

BM Beam [also Bm, or bm]

BM Bearing Metal

BM Bellows Metering Valve

BM Bench Mark

BM Bending Magnet

BM Bend(ing) Moment

BM Bénélux Métallurgie [Joint Metallurgy Organization of Belgium, Netherlands and Luxembourg]

BM Bermuda [ISO Code]

BM Bhagavantam's Method

BM Bill of Materials [also B/M]

BM Bimolecular Mechanism

BM Binary Multiply

BM Binocular Microscope

BM Biological Material

BM Biological Microscopy

BM Biomass

BM Biomaterial

BM Biomathematical; Biomathematics

BM Biomechanic(al); Biomechanics

BM Biomedical; Biomedicine

BM Biomembrane

BM Biomolecular; Biomolecule

BM Bipolar Membrane

BM Bistable Multivibrator

BM Board Measure [also bm] [Lumber]

BM Body Mass

BM Body-Mounted (Device)

BM Bohr Magneton [Physics]

BM Bone Marrow

BM Born-Mayer (Potential) [Solid-State Physics]

BM Boundary Marker

BM Bound Magnon (Theory) [Solid-State Physics]

BM Bowel Movement

BM Bowles-Mackenzie (Theory of Martensitic Transformation) [Metallurgy]

BM Boyle-Mariotte (Law)

BM Branch Manager

BM Breakpoint on Memory (Command) [Pascal Programming]

BM Breech Mechanism

BM Brigade Major

BM British Museum [London, UK]

BM Bromomethane

BM Brownian Motion (or Movement)

BM Bubble Memory

BM Buffer Module

BM Building Material

BM Bulk Modulus

BM Bureau of Meteorology [Melbourne, Australia]

BM Bureau of Mines [US]

BM Business Machine

BM Business Manager

BM Byte Machine

2B5M14 2-Bromo-5-Methyl-1,4-Benzoquinone

B/M Bill of Materials [also BM]

Bm Beam [also bm]

Bm Bombyx mori [Silkworm]

.bm Bermuda [Country Code/Domain Name]

BMA Bachelor of Management Arts

BMA Bank Marketing Association [US]

BMA Barium Manganese Aluminate

BMA Bayesian Multivariate Analysis

BMA BEAMA (British Electrical and Allied Manufacturers Association) Metering Association [UK]

BMA Benzylmalonic Acid

BMA Bicycle Manufacturers Association (of America) [US]

BMA Biomedical Marketing Association [US]

BMA Blausäure-Methan-Ammoniak-Prozess [Hydrocyanic-Acid Methane Ammonia Process]

BMA Body-Mounted Accelerometer

BMA Bornylmethacrylate

BMA British Manufacturers Association

BMA British Medical Association

BMA Bundesministerium für Arbeit und Sozialordnung [Federal Ministry of Labor and Social Services, Germany]

BMA Butyl Methacrylate

BMAA β-N-Methylamino-L-Alanine [Biochemistry]

BMAA British Marine Aquarists Association

BMAA British Microlight Aircraft Association

BMAG Body-Mounted Attitude Gyro

BM-AGA Bureau of Mines–American Gas Association (Coal Test)

BMAP Background Measurement and Analysis Program

BMAP Buffer Map

BMAR Ballistic Missile Acquisition Radar

BMarE Bachelor of Marine Engineering

BMAS Barium Magnesium Aluminosilicate

BMAT Bill of Materials (Generator Software)

BMath Bachelor of Mathematics

BMAW Bare Metal-Arc Welding

BMB Baltic Marine Biologists [Sweden]

BMB Base Maintenance Building

BMB British Metrication Board

BMB 4-Bromomethyl-7-Methoxy-2-Oxo-2H-Benzopyran

BMBF Bundesministerium für Bildung, Wissenschaft, Forschung und Technologie [Federal Ministry for Education, Science, Research and Technology, Germany]

BMBL Biosafety in Microbiological and Biomedical Laboratories

BMBT Bismethoxybenzylidenebitoluidine; n,n'-Bis(p-Methoxybenzylidine)-α,α'-Bi-p-Toluidine

BMBW Bundesministerium für Bildung und Wissenschaft [Federal Ministry for Education and Science, Germany] [Now Part of BMBF, Germany]

BMC Ballester-Malinet-Castauer (Perchlorination)

BMC Bimolecular Collision

BMC Biomedical Cement

BMC Block Multiplexer Channel

BMC Brittle Matrix Composite

BMC 4-Bromomethyl-7-Methoxy-Cumarin

BMC Bubble Memory Controller

BMC Bulk Media Conversion

BMC Bulk Molding Compound

BMC Burst Multiplexer Channel

BMC Bureau of Management Consulting [of Department of Supply and Services, Ottawa, Ontario, Canada]

BMCC British Metal Castings Council

BMCI Bureau of Mines Correlation Index

BMCM Bohr-Mottelson Collective Model [Physics]

BmCPV Bombyx mori Cypovirus

BmCPV-1 Bombyx mori Cypovirus, Type 1

BMCS Baan Manufacturing Control System

BMCS Bureau of Motor Carrier Safety [US]

BMD Badische Maschinenfabrik Durlach GmbH [Machinery Manufacturer headquartered in Karlsruhe, Germany]

BMD Bahamian Dollar [Currency]

BMD Ballistic Missile Defense (Program) [of US Department of Defense]

BMD Ballistic Missile Division [of Air Research and Development Command, Palo Alto, California, US]

BMD Ballistic Missile Division [of US Air Force at Inglewood, California, US]

BMD Benchmark Monitor Display System

BMD Bermuda Dollar [Currency]

BMD Biomathematics Division [of US Army Biological Laboratory]

BMD Biomechanical Device

BMD Brittle Materials Design

BMD Bubble Memory Device

BMD Building Materials Division [of National Institute of Standards and Technology, Gaithersburg, Maryland, US]

BMDC Bismuth Dimethyl Dithiocarbamate

BMDH Basic Minimum Descent Height

BMDMCS Bromomethyl-Dimethylchlorosilane

BMDMS (Bromomethyl)dimethylsilyl

BMDO Ballistic Missile Defense Organization [formerly Strategic Defense Initiative Organization, US]

BMDP BioMeDical (Computer) Programs

BMDS Benzene-m-Disulfonate

BmDSV Bombyx mori Densovirus

BME Bachelor of Mechanical Engineering

BME Bachelor of Mining Engineering

BME Basal Medium Eagle [Biochemistry]

BME Bench Maintenance Equipment

BME Biomass Energy

BME Biomechanical Engineer(ing)

BME Biomedical Engineer(ing)

BME Bulk Magnetoelastic Energy

BME Bundesverband Materialwirtschaft, Einkauf und Logistik e.V. [Federation for Materials Management, Purchasing and Logistics, Germany]

BMEA British Marine Equipment Association

BMEC Ball Bearing Manufacturers Engineers Committee [US]

BMEC British Marine Equipment Council

BMEF British Mechanical Engineering Federation

B megaterium Bacillus megaterium [Microbiology]

B(MeO)₃ Boron Methoxide

BMEP Brake Mean Effective Pressure

BMES Biomedical Engineering Society [US]

BMES Bull BMES Bulletin [of Biomedical Engineering Society, US]

BMet Bachelor of Metallurgy

B MET A Birmingham Metallurgical Association [also BMETA, UK]

BMetE Bachelor of Metallurgical Engineering [also BMetEng]

BMETO Ballistic Missiles European Task Organization

B-3-MetSEDI Bis-3-Methylsalicylaldehyde Ethylenediimine

BMEU Biomedical Engineering Unit [of Queen's University, Canada]

BMEWS Ballistic Missile Early Warning System

BMF Biomagnification Factor

BMF British Microcomputer Federation

BMF British Motorcyclists Federation

BMF Builders Merchants Federation

BMF Bulk Mail Facility

BMFF British Man-Made Fibers Federation

BMFT Bundesministerium für Forschung und Technologie [Federal Ministry for Research and Technology, Germany] [Now Part of Bundesministerium für Bildung, Wissenschaft, Forschung und Technologie, Germany] [now BMBF]

BMFT J BMFT Journal [of Bundesministerium für Forschung und Technologie, Germany]

BMFT-MATFO Bundesministerium für Forschung und Technologie–Materialforschung [Materials Research Program of the German Federal Minstry of Research and Technology]

BMG Browning Machine Gun

BMG Bundesministerium für Gesundheit [Federal Ministry of Health, Germany]

BMG Business Machine Group

BMg Magnesium-Base Brazing Alloy

BMGE Bovine Mammary Gland Epithelial (Cells)

BMGT Division of Business Development and Management [of American Chemical Society, US]

BMgtE Bachelor of Management Engineering

BMgtSc Bachelor of Management Science

BMH Baptist Memorial Hospital [Memphis, Tennessee, US]

BMH Born-Mayer-Huggins (Potential) [Solid-State Physics]

BMHC Bell and Howell/Mamiya Company [US]

BMHF British Materials Handling Federation

BMI Basic Measuring Instrument

BMI Battelle Memorial Institute [Columbus, Ohio, US]

BMI Bibliography Master Index

BMI Biomedical Implant

BMI Bismaleimide

BMI Body Mass Index

BMI Book Manufacturers Institute [US]

BMI Branch if Minus

BMI Building Maintenance Information [UK]

BMI Bundesministerium des Inneren [Federal Ministry of the Interior, Germany]

BMIC Bachelor of Microbiology

BMIC (Intel) BusMaster Interface Controller

BMIF British Marine Industries Federation

BMILS Bottom-Mounted Impact Location System

BMinE Bachelor of Mining Engineering

BMIS Bank Management Information System

BMIS Battle Management Information System [of Department of National Defense, Canada]

BMIU Bricklayers and Masons Independent Union [Canada]

BMJ British Medical Journal [of British Medical Association]

BMJ Bundesministerium für Justiz [Federal Ministry of Justice, Germany]

Bmkr Boilermaker [also bmkr]

BML Bed-Material Load [Geology]

BML British Museum Library [of British Library]

BML Bulk Material Length

BML Bundesministerium für Ernährung, Landwirtschaft und Forsten [Federal Ministry of Nutrition, Agriculture and Forestry, Germany]

BMLA Biological Material License Agreement(s) [of National Institute of Allergy and Infectious Diseases, US]

BMLA British Maritime Law Association

BMLA British Medical Laser Association

BML&BS British Matchbox, Label and Booklet Society

BMLSc Bachelor of Medical Laboratory Science

BMLSS British Marine Life Study Society

BMM Biomedical Material

BMM Biomolecular Material

BMMA British Meat Manufacturers Association

BMMA British Micrographic Manufacturers Association

BMMC Basic Monthly Maintenance Charge

BMMD Body Mass Measuring Device

BMMG British Microcomputer Manufacturers Group

BMMPIC Basic Metals and Mineral Processing Industry Council [Australia]

BMN Barium Magnesium Niobate

β-Mn Beta Manganese [Symbol]

BMN-BZ Barium Magnesium Niobate–Barium Zirconate

BMNH Natural History Museum, London [also BM(NH)]

β-Mn$_2$O$_3$ Beta Manganese(III) Oxide

β-Mn$_3$O$_4$ Beta Manganese Tetroxide

β-MnS Beta Manganese Sulfide

BMO Ballistic Missile Office

BMO Bloch Molecular Orbit [Physical Chemistry]

BMO Bonding Molecular Orbital [Physical Chemistry]

BMO British Meteorological Office

BMOC Ballistic Missile Orientation Course

BMOM Base Maintenance and Operations Model

B mori Bombyx mori [Silkworm]

BMOT Bayonne Military Ocean Terminal [of US Army in New Jersey]

B-3-MoxSEDI Bis-3-Methoxysalicylaldehyde Ethylenediimine

B-3-MoxSPHDI Bis-3-Methoxysalicylaldehyde Phenylenediimine

B-3-NO2-SEDI Bis-3-Nitrosalicylaldehyde Ethylenediimine

BMP Batch Message Processing

BMP Best Management Practices (Program) [of Department of National Defense, Canada]

BMP Biomedical Polymer

BMP Bitmap [Microsoft Windows]

BMP Bit Mapped; Bit Mapping

BMP Bone Morphogenetic Protein [Biochemistry]

BMP Bound Magnetic Polaron [Solid-State Physics]

BMP Bricklayers, Masons and Plasterers Union

BMP Brownian Motion Process

BMP (National Council of) Building Material Producers [UK]

.BMP Bitmap [File Name Extension] [also .bmp]

BMPCA British Metallurgical Plant Constructors Association

BMPPOH 4-Benzoyl-3-Methyl-1-Phenyl-2-Pyrazolin-5-One

BMPRI Basic Metals Processing Research Institute [of University of Pittsburgh, Pennsylvania, US]

BMPS β-Maleimidopropionic Acid N-Hydroxysuccinimide Ester

BMR Basal Metabolic (or Metabolism) Rate

BMR Bipolar Magnetic Region

BMR Brookhaven's Medical Reactor [of Brookhaven National Laboratory, Upton, New York, US]

BMR Bureau of Mineral Resources [now Australian Geological Survey Organization, Canberra, Australia]

Bmr Bomber [also bmr]

BMRBS Bundesministerium für Raumordnung, Bauwesen und Städtebau [Federal Ministry for Environmental Planning, Housing and Urban Development]

BMRC British Manufacture and Research Company

BMRC British Medical Research Council

BMRC Bureau of Meteorology Research Center [Australia]

BMRGG Bureau of Mineral Resources, Geology and Geophysics [now Australian Geological Survey Organization, Canberra, Australia]

BMR J Aust Geol Geophys BMR Journal of Australian Geology and Geophysics [of Bureau of Mineral Resources, Canberra, Australia]

BMS Bachelor of Marine Science

BMS Background Measurement Satellite

BMS Ballistic Missile Ship [US Navy]

BMS Basic Mapping Support

BMS Basic Medical Sciences

BMS Battle Management System

BMS Biomedical System

BMS Biometric Society [US]

BMS Biomimetic System

BMS Birmingham Metallurgical Society [UK]

BMS Black Mercury Sulfide

BMS Boeing Materials Specification

BMS Boranemethylsulfide

BMS Boston Museum of Science [Massachusetts, US]

BMS British Meteor Society

BMS British Mycological Society

BMS Bureau of Medicine and Surgery [of US Navy]

BMS Burner Management System

BMS Burstein-Moss Shift [Physics]

Bm$ Bermuda Dollar [Currency]

BMSA Benzene Monosulfonic Acid

BMSGMA British Maize Starch and Glucose Manufacturers

BMSOA British Motor Ship Owners Association

BMSP British Modal Speaking Position [Telecommunications]

BMT Bachelor of Medical Technology

BMT Barium Magnesium Tantalate

BMT Barium Manganese Tantalate

BMT Base Metal Tailings

BMT Basic Motion-Time Study

BMT Baumaschine + Bautechnik [German Journal on Construction Machinery and Construction Engineering]

BMT Biomechanical Transducer

BMT British Maritime Technology Limited

BMT British Mean Time

BMTA British Mining Tools Association

BMTA British Motor Trade Association

BMTC Bureau of Meterology Training Center [Australia]

BMTD Ballistic Missile Terminal Defense

BMTFA British Malleable Tube Fittings Association

BMTI Block Mode Terminal Interface

BMTS Ballistic Missile Test System

BMTS Bureau of Mines Test Station [Bruceton, Pennsylvania, US]

BMTT Buffered Magnetic Tape Transfer; Buffered Magnetic Tape Transport

BMU Barrel Muon Chamber [Physics]

BMU Bundesministerium für Umwelt, Naturschutz und Reaktorsicherheit [Federal Ministry of the Environment, Nature Conservancy and Reactor Safety, Germany]

BMU Bus Monitor Unit

BMUS British Medical Ultasound Society

BMUX Buffered Communication Unit Multiplexer

BMV Brome Mosaic Virus

BMV Bundesministerium für Verkehr [Federal Ministry of Transportation, Germany]

BMVA British Machine Vision Association

BMVg Bundesministerium für Verteidigung [Federal Ministry of Defense, Germany]

BMW Bayrische Motorenwerke [German Automobile and Engine Manufacturer headquartered in Munich, Germany]

BMW Beamwidth [also Bmw]

BMWE Brotherhood of Maintenance of Way Employees

BMWi Bundesministerium für Wirtschaft [Federal Ministry of Trade and Commerce, Germany]

BM-WM Base Metal–Weld Metal (Combination)

BMWZ Bundesministerium für wirtschaftliche Zusammenarbeit und Entwicklung [Federal Ministry for Economic Cooperation and Development, Germany]

BMZ Baumusterzentrale [Industrial Design Center, Austria]

BN Bachelor of Nursing

BN Baker-Nutting (Orientation Relationship) [Crystallography] [also B-N]

BN Ball Bearing, Single-Row, Angular-Contact, Non-Separable, Nominal Contact Angle from Above 10° to 22° [Symbol]

BN Ballistic Number

BN Barkhausen Noise [Physics]

BN Baryon Number [Physics]

BN Becklin-Neugebauer (Object) [Astronomy]

BN Benzonitrile

BN Bernoulli-Navier (Hypothesis) [Mechanics]

BN Bibliothèque Nationale [National Library, Paris, France]

BN Binary Number

BN Bischler-Napieralski (Synthesis)

BN (Data) Block Number

BN Bomber Navigator

BN Borazon (Compound)

BN Boron Nitride

BN Branch Name

BN Bromonaphthalene

BN Bromonitrate

BN Brunei Darussalam [ISO Code]

BN Bureau de Normalisation [Bureau of Standards]

BN Bureau Number

BN Burlington Northern (Railroad) [US]

B/N Boron-to-Nitrogen (Ratio)

Bn Battalion [also bn]

bn billion

bn meso-2,3-butanediamine

.bn Brunei Darussalam [Country Code/Domain Name]

β-N Beta Nitrogen [Symbol]

BNA Barkhausen Noise Amplitude [Physics]

BNA Base-Neutral/Acid (Compound)

BNA Basle Nomina Anatomica

BNA Boeing Network Architecture

BNA British Naturalists Association

BNA British Naval Attaché

BNA British North America

BNA Bureau of National Affairs, Inc. [US]

B/N/A Base/Neutral and Acid-Extractable (Compound)

β-Na Beta Sodium [Symbol]

BNAA British North America Act

BNAC British North American Committee

BNAG Bibliography of North American Geology

BNAH 1-Benzyl-1,4-Dihydronicotinamide

BN/AlN Boron Nitride/Aluminum Nitride (Superlattice)

BN/Al$_2$O$_3$ Alumina Fiber Reinforced Boron Nitride (Composite)

BNAS Bolivian National Academy of Sciences

BNB Benzene-Nitrogen-Benzene

BNB 1,1-Bis-(p-Chlorophenyl)-2-Nitrobutane

BNB British National Bibliography [of British Library Bibliographic Services Division]

BNB Bromonitrobenzene

BNB Bulk Nucleate Boiling

BNB Bullet Nose Bushing

BNB 2,4,6-Tri-tert-Butylnitrosobenzene

BNBC British National Book Center

BNC Baby 'N' Connector

BNC Bayonet Norm Connector

BNC Bayonet Neil-Concelman (Connector)

BNC Bayonet Nut Connector

BNC Bayonet Nut Coupling

BNC British National Committee

BNCC British National Committee for Chemistry

Bnchbd Benchboard [also bnchbd]

BNCI Banque Nationale pour le Commerce et l'Industrie [National Bank for Commerce and Industry, Belgium]

BNCIA Bayonet Nut Connector Interface Adapter

BNCIAWPRC Belgian National Committee of the International Association on Water Pollution Research and Control

BNCIAWPRC Brazilian National Committee of the International Association on Water Pollution Research and Control

BNCM British National Committee on Materials

BNCOR British National Committee for Oceanographic Research

BNCS Board on Nuclear Codes and Standards

BNCS British Numerical Control Society

BNCSAA British National Committee on Surface-Active Agents

BNCSR British National Committee on Space Research

BNCT Boron-Neutron Capture Therapy

BNC-T Bayonet Nut Connector Tee

BNCWPC British National Committee for World Petroleum Congresses

BND Benzoylated, Naphthoylated DEAE (Diethylaminoethyl)

BND Brunei Dollar [Currency of Brunei]

BND Bullet Nose Dowel

BND Bundesnachrichtendienst [Federal Intelligence Service, Germany]

β-Nd Beta Neodymium [Symbol]

BNDD Bureau of Narcotics and Dangerous Drugs [US]

BNDO Bureau National des Données Oceanique [National Bureau for Oceanic Data, France]

BNDS Banque Nationale de Développement du Sénégal [National Development Bank of Senegal]

BNE Branch if Not Equal

BNEA British Naval Equipment Association

BNEC British National Export Council

BNEC British Nuclear Energy Conference

BNES British Nuclear Energy Society

BNF Backus-Naur Form; Backus-Normal Form

BNF Bomb Nose Fuse

BNF British Nuclear Forum

BNF British Nuclear Fuels plc

BNF British Nutrition Foundation

BNFA British Narrow Fabrics Association

BNF ABS British Nonferrous Metals Abstracts [of British Nonferrous Metals Technology Cente] [also BNF Abs]

BNFL British Nuclear Fuels Limited

BNFMF British Nonferrous Metals Federation

BNFMRA British Nonferrous Metals Research Association

BNFMS Bureau of Nonferrous Metals Statistics [UK]

BNFMTC British Nonferrous Metals Technology Center

BNFP Barnwell Nuclear Fuel Plant [US]

BNFSA British Nonferrous Metals Smelters Association

BNG Branch No Group

BNGL British National Gridiron League

Bnh Burnish(ing) [also bnh]

BNHA Badlands Natural History Association [US]

BNHS (+)-Biotin-N-Hydroxysuccinimidester

BNI Beijing Neurosurgical Institute [PR China]

BNI Bernhard-Nocht-Institut für Tropenmedizin [Bernhard Nocht Institute for Tropic Medicine, Hamburg, Germany]

BNi Nickel-Base Brazing Alloy

B:Ni₃Al Boron-Doped Trinickel Aluminide

β-NiAl Beta Nickel Aluminide

β-NiGa Beta Nickel Gallium

BNIST Bureau National de l'Information Scientifique et Technique [National Bureau for Scientific and Technical Information, France]

BNITA British Nautical Instrument Trade Association

β-NiZn Beta Nickel Zinc

BNL Bell Northern Laboratories [Ottawa, Canada]

BNL Berkeley Nuclear Laboratory [California, US]

BNL Brookhaven National Laboratory [of US Department of Energy at Upton, Long Island, New York]

BNL/ATF Brookhaven National Laboratory/Accelerator Test Facility [US]

BNM Bureau National de Métrologie [National Bureau of Metrology, France]

BNMA British Nonwovens Manufacturers Association

BNMC Bethesda Naval Medical Center [Maryland, US]

B-11 NMR Boron-11 Nuclear Magnetic Resonance [also ^{11}B NMR]

BNN Barium Sodium Niobate

BNO Backus Normal Form

BNO Barium Niobium Oxide

BNO Black Nickel Oxide

BNO British Naval Officer

BNOA β-Naphthoxyacetic Acid

BNOC British National Oil Corporation

bNOS Brain-Derived Nitric Oxide Synthase [Biochemistry]

BNP (+)-Biotin-4-Nitrophenylester

BNP Brain Natriuretic Peptide [Biochemistry]

BNP Bureau of Naval Personnel [of US Navy]

β-Np Beta Neptunium [Symbol]

BNPA Beta-Nitropropionic Acid

BNPF Beginning Negative Positive Finish

BNPP Bis-(p-Nitrophenyl) Phosphate

BNPS-Skatole 3-Bromo-3-Methyl-2-(2-Nitrophenylmercapto)-3H-Indole

BNR Bell-Northern Research [of Northern Telecom, Ottawa, Canada]

BNR Bonded Negative Resistance (Diode)

BNRL British Nuclear Reactors Limited

BNRS British National Radio School

BNS Bachelor of Natural Sciences

BNS Bachelor of Naval Science

BNS Backbone Network Service

BNS Ball Bearing, Single-Row, Angular-Contact, Separable Outer Ring, Nominal Contact Angle from Above 10° to 22° [Symbol]

BNS Bermuda News Service

BNS Bill Number Screening

BNS Binary Number(ing) System

BNS Boston Naval Shipyard [US Navy]

BNSC British National Space Center

BNSF Belgian National Science Foundation

BNSI Barbados National Standards Institution

BN/SiC Silicon Carbide Fiber Reinforced Boron Nitride (Composite)

BNT Ball Bearing, Single-Row, Angular-Contact, Separable Inner Ring, Nominal Contact Angle from Above 10° to 22° [Symbol]

BNT (Salt) Bath Nitriding

BNT Boreal Northern Titles [University of Alberta, Edmonton, Canada]

BN/TiB₂ Titanium Diboride Reinforced Boron Nitride (Composite) [also BN-TiB₂]

BNTVA British Nuclear Test Veterans Association

BNU N-Butyl N-Nitroso Urea

BNW Battelle-Northwest [Richland, Washington, US]

BNW Bureau of Naval Weapons [also BuWEPS, BUWEPS, or BuWeps] [of US Navy]

BNWL Battelle Northwest Laboratories [Richland, Washington, US]

BNX British Nuclear Export Executive

BNZS Bipolar with N Zeros Substitution

B-NZTC Britain-New Zealand Trade Council [UK]

BO Back Order

BO Bail-Out [Aeronautics]

BO Beat Oscillator

BO Bend-Over

BO 2-Benzoxazolone

BO Beryllium Oxide

BO Binary Output

BO Binary-to-Octal [also B-O]

BO Biochemical Oxidation

BO Black Oil

BO Blackout

BO Blanking Oscillator

BO Blocking Oscillator

BO Blown Out

BO Blow-Off [Aerospace]

BO Blow-Out

BO Body Odo(u)r

BO Bolivia [ISO Code]

BO Bonding Orbital [Physical Chemistry]

BO Bond Order [Physical Chemistry]

BO Born-Oppenheimer (Approximation) [Physics]

BO Box Office

BO Branch Office [also B/O, or bo]

BO Breakout [also B/O]

BO Bridging Oxygen

BO Bright Object

BO Broker's Order [also bo]

BO Burn-Off

BO Burnout [also B/O]

BO Buyer's Option

B/O Bad Order

B/O Booster/Orbiter

B/O Branch Office [also BO, or bo]

B/O Breakout [also BO]

B/O Brought Over [also b/o]

B/O Burnout [also BO]

B-O Binary-to-Octal [also BO]

Bo Boehmite [Mineral]

Bo Boltzmann Number [Symbol]

bo branch office [also BO, or B/O]

.bo Bolivia [Country Code/Domain Name]

b/o brought over [also B/O]

β-O Beta Oxygen [Symbol]

BOA Benzyl iso-Octyl Adipate

BOA Born-Oppenheimer Approximation

BOA British Optical Association

BOA o-Butoxy-Acetanilide

BOAA β-N-Oxalylamino-L-Alanine [Biochemistry]

BOAA Born-Oppenheimer Adiabatic Approximation [Physics]

BOAC British Ordnance Ammunition Corps

BOAC British Overseas Airways Corporation

BOAM Bell Owned and Maintained (Equipment)

BOAW Bond Order Alternation Wave [Physics]

BOB Best on Best

BOB BLAST (Basic Local Alignment Search Tools) Output Browser [also BoB]

BOB Break-Out Box [Data Communications] [also BoB]

Bob Bobbin [also bob]

BOBMA British Oil Burners Manufacturers Association

BOC Base Operations Contract

BOC Basic Operator Console

BOC Beginning of Cycle

BOC Bell Operating Company [US]

BOC Bevatron Orbit Code

BOC Binary-to-Octal Conversion

BOC Bio-Organic Chemistry

BOC Black Oil Conversion (Process)

BOC Block Oriented Computer

BOC Blowout Coil

BOC Bond-Order Conservation [Physical Chemistry]

BOC Bottom of Conduit

BOC British Ornithologists Club

BOC Brominated Organic Compound

BOC Butoxycarbonyl

(BOC)20 (Butoxycarbonyl)20 [also (Boc)20]

(BOC)2NH (Butoxycarbonyl)2NH [also (Boc)2NH]

Boc 1-Butyloxycarbonyl

BOCA Building Officials and Code Administrators (International) [US]

BOCA Building Officials Conference of America [US]

BOCBA British Overseas and Commonwealth Banks Association

BOC-DAKLI N-tert-Butoxycarbonyl Dynorphin A Analog κ Ligand [Biochemistry]

BOC-L-DOPA N-tert-Butoxycarbonyl-L-3,4-Dihydroxyphenylalanine [Biochemistry]

BOCES Board of Cooperative Educational Services [US]

BOC-MP Bond-Order Conservation–Morse Potential [also BOCMP] [Physical Chemistry]

BOCOL Basic Operating Consumer-Oriented Language

BOC-ON N-(tert-Butoxycarbonyloxyimino)-2-Phenylacetonitrile

BOC-ONP tert-Butoxycarbonyl-4-Nitrophenol

BOC-OSU N-tert-Butyloxycarbonyl-Succinimide

BOCS Bendix Optimum Configuration Satellite

BOC-S Carbonothioic Acid O-(1,1-Dimethylethyl) S-(4,6-Dimethyl-2-Pyrimidinyl)-Ester

BOD Barrels of Oil per Day [also bod, bo/d, BO D^{-1}, or bo/d^{-1}]

BOD Base Ordnance Depot

BOD Beneficial Occupancy Date

BOD Biochemical Oxygen Demand

BOD Biological Oxygen Demand

BOD Birmingham Ordnance District [Alabama, US]

BOD Bistable Optical Device

BOD Boston Ordnance District [Massachusetts, US]

BOD Bottom of Duct

BOD Breakover Diode

BO/D Barrels of Oil per Day [also bo/d, BOD, bod, BO D^{-1}, or bo/d^{-1}]

BoD Board of Directors [of Institute of Electrical and Electronics Engineers, US]

BOD$_5$ Biological Oxygen Demand over 5-Day Period [Wastewater Treatment]

BOD/COD Biological Oxygen Demand/Chemical Oxygen Demand

Body Eng Body Engineering [US Journal]

Body Eng J Body Engineering Journal [of American Society of Body Engineers]

BOE Beginning of Extent

BOE Break of Entry

BOE Brick-on-Edge (Course)

BOE Buffered Oxide Etch (Solution)

BOE Bureau of Explosives [now Hazardous Materials System (Bureau of Explosives), US]

BOEAF Basic-Oxygen Electric-Arc Furnace [Metallurgy]

BOF Basic-Oxygen Furnace [Metallurgy]

BOF Beginning of File [also bof]

BOF Birds Of a Feather [Discussion Group]

BOF British Organic Farmers

BOFA Bibliography of Agriculture [of National Agricultural Library, US]

B of B Back of Board

BOF/BB Basic Oxygen Furnace/Bottom Blown [Metallurgy]

B of C Bank of Canada

B of E Bank of England

B of E Board of Education

BOFGA Biodynamic Farming and Gardening Association [US]

BOFS Black Oil Finish Slate

BOFT Board on Foreign Trade [of Ministry of External Affairs, Taiwan]

BOF/TB Basic Oxygen Furnace/Top Blown [Metallurgy]

BOG Bismuth Orthogermanate

BOG International Bogoliubov Conference (on Problems of Theoretical and Mathematical Physics) [Russia]

BOH Breakoff Height

Bohem Bohemia(n)

BOHS British Occupational Hygiene Society

BOI Beginning of Information

BOI Branch Output Interrupt

BOI Break of Integrity

BOIESA Bureau of Oceans and International Environmental and Scientific Affairs [of US Department of State]

BOL Beginning of Life (Fuel Condition) [Nuclear Engineering]

Bol Bolivia(n)

BOLD Bibliographic On-line Library Display

Bol Fac Ing Univ Repub Boletin de la Facultad de Ingeneria Universidad de la Republica [Bulletin of the Faculty of Engineering, University of the Republic, Montevideo, Uruguay]

Bol Inform Boletin Informativo [Information Bulletin; published in Montevideo, Uruguay]

Bol Inst Tonantzintla Boletin del Instituto de Tonantzintla [Bulletin of the Tonantzintla Institute; published by Instituto Nacional de Astrofisica, Optica y Electronica, Puebla, Mexico]

Boll Geofis Teor Appl Bolletino di Geofisica Teorica ed Applicata [Bulletin of Theoretical and Applied Geophysics; published by Osservatorio Geofisico Sperimentale, Trieste, Italy]

Boll Oceanol Teor Appl Bolletino di Oceanologia Teorica ed Applicata [Bulletin of Theoretical and Applied Oceanology; published by Osservatorio Geofisico Sperimentale, Trieste, Italy]

Boll Riv Bolletino delle Riviste [Italian Review Bulletin]

Boll Soc Ital Biol Sperim Bolletino–Socièta Italiana Biologia Sperimentale [Bulletin of the Italian Society of Experimental Biology]

Boll Tec Finsider Bolletino Tecnico Finsider [Italian Technical Bulletin]

Boll Unione Mat Ital Bollettino Unione Matematica Italiana [Bulletin of the Italian Mathematical Union]

BOLMADA British Ophthalmic Lens Manufacturers and Distributors Association

Bols Bolster [also bols]

Bol SMF Boletin de la Sociedade Mexicana de Fisica [Bulletin of the Mexican Physical Society]

Bol Soc Esp Ceram Vidrio Boletin Sociedad Espanola de Cerámica y Vidrio [Bulletin of the Spanish Society of Ceramics and Glass]

Bol Soc Mat Mex Boletin Sociedad Matemática Mexicana [Bulletin of the Mexican Mathematical Society]

Bol Soc Mat São Paulo Boletim Sociedade de Matemática de São Paulo [Bulletin of the Mathematical Society of São Paulo, Brazil]

Bol Museu Nac Boletim do Museu Nacional [Bulletin of the National Museum, Rio de Janeiro, Brazil]

Bol Soc Port Mat Boletim Sociedade Portuguesa de Matemática [Bulletin of the Portuguese Society for Mathematics]

BOLT Beam of Light Transistor

BOLT British Columbia Organization of Library Technicians [Canada]

Bol Tec Boletin Tecnico [Brazilian Technical Bulletin]

Bol Tec, Niquel Boletin Técnico, Niquel [Cuban Technical Bulletin on Nickel]

BOM Basic Operating Monitor

BOM Beginning of Message

BOM Beginning of Month

BOM Bill of Materials

BOM Binary Order of Magnitude

BOM Bureau of Meterology [Melbourne, Australia] [also BoM]

BoM Bureau of Mines [of US Department of the Interior, Washington, DC] [also BOM, USBoM, or USBOM]

BOMA British Overseas Mining Association

BOMA Building Owners and Managers Association

BOMA Business Owners Management Association

BOMAEC Bureau of Mines/Atomic Energy Commission Incinerator [US]

BOMC Blessed Orientation-Memory-Concentration (Test) [Alzheimer's Disease]

B(OMe)$_3$ Trimethyl Borate

BOMEN Bulletin Officiel du Ministère de l'Education Nationale [Official Bulletin of the Minister of National Education, France]

BOMEX Barbados Oceanographic and Meteorological Experiment

BOM IC Bureau of Mines Information Circular [US]

BOMMA British Ophthalmic Mass Manufacturers Association

BOMP Base Organization and Maintenance Processor

BOMS Bureau of Overseas Medical Service [UK]

BON Beta-Oxynaphthoic Acid

BONAC Broadcasting Organizations of Non-Aligned Countries [Formerly in Yugoslavia] [obsolete]

BOND Bandwidth On Demand

BONUS Boiling-Water Nuclear Superheat Reactor [of Puerto Rico Water Resources Authority at Punta Higuera]

BONUS Boiling-Water-Reactor (Nuclear) Superheat Reactor Project

BONUS-CX Boiling-Water Nuclear Superheat Critical Experiment

BOO Bond-Orientational Order [Physics]

Boo Boötes [Astronomy]

BOOG British Osborne Owners Group

BOOGIE British Osborne Owners Group Information Exchange

BOOL Boolean (Algebra) [also Bool]

BooNE Booster Neutrino Experiment [at Fermilab, Batavia, Illinois, US] [also BooNe]

BOOST Booster [also Boost, or boost]

BOOT Bootstrap [also Boot, or boot]

Boot Boötes [Astronomy]

BOOTP Bootstrap Protocol [Internet]

BOP Balance of Payments

BOP Balance of Plant

BOP Balance of Power

BOP Baseline Operations Plan

BOP Basic Operating Program

BOP Basic Operator Panel

BOP Basic Oxygen Process [Metallurgy]

BOP Bead-on-Plate (Weld)

BOP Bend-Over Point

BOP Binary Output Program

BOP Bipolar Amplifier

BOP Bit-Oriented Protocol

BOP Bitumen Oxidizing Plant

BOP Benzotriazolyl-N-Oxy-tris (dimethylamino)phosphonium

BOP Blowout Preventer [Oil Drilling]

BOP Bolivian Peso [Currency]

BOP Burlington Ordnance Plant [New Jersey, US]

BOP Business Opportunities Program [of Ontario Ministry of Economic Development, Trade and Tourism, Canada]

BOPACE Boeing Plastic Analysis Capability for Engines

BOP-Cl Bis(2-Oxo-3-Oxazolidinyl)phosphinic Chloride

BOPD Barrels of Oil per Day [also bopd]

BOPET Biaxially-Oriented Polyethylene Terephthalate

BOP-LD Basic Oxygen Process–Linz-Donawitz (Steelmaking Converter) [also BOP/LD] [Metallurgy]

BOPOB Bis[2-(5-P-Biphenylyloxazoyl)]benzene

BOPP Biaxially Oriented Polypropylene

BOPS Billion Operations per Second

Bopress Boiler Pressure [also bopress]

BOQ Bachelor Officers' Quarters [US Navy]

BOR Bureau of Reclamation [of US Department of the Interior] [also BoR]

BOR Bus Out Register

B-OR Burgers Orientation Relationship [also BOR] [Crystallography]

Bor Borneo

Bor Borough [also bor]

BORAD British Oxygen Research and Development Association

BORAM Block-Oriented Random-Access Memory [also Boram, or boram]

BORAPP Borland Business Application Products Forum [of Borland International, US]

BORAX Boiling Reactor Experiment [of Idaho National Engineering Laboratory, US]

BORAX I Boiling Reactor Experiment, Safety [1953]

BORAX II Boiling Reactor Experiment, Pressurized [1954]

BORAX III Boiling Reactor Experiment, Power [1955]

BORAX IV Boiling Reactor Experiment, Fuel Test [1956]

BORAX V Boiling Reactor Experiment [1959]

BORE Beryllium Oxide Reactor Experiment [US]

BOREAS Boreal Ecosystem-Atmosphere Study [of Saskatchewan Research Council, Canada]

BOREQ Broadcast Request

BORI Bhandarkar Oriental Research Iinstitute [Poona, India]

BORIS Book Order, Register and Invoicing System

Boro Borough [also boro]

BORPQU Borland Pro Quattro [Computers]

BORQU Borland Quattro [Computers]

BORSCHT Battery, Overvoltage, Ringing, Supervision, Coding, Hybrid and Test Access; Battery Overvoltage Protection Ringing Signaling Coding Hybrid Testing

Borsic Silicon Carbide Coated Boron [also borsic]

BOS Background Operating System

BOS Back of Slip

BOS Backup Operating System

BOS Basic Operating System

BOS Basic Oxygen Steel(making) [Metallurgy]

BOS Batch Operating System

BOS Bell Operating System

BOS Book Order System

BOS Brewer Ozone Spectrophotometer

BOS Bright Object Sensor

BOS Bundle of Sticks

BOS Business Office Supervisor

BOS Business Operating Software

B1S First-Stage Binary Structure [Symbol]

BOSAC Bofors Sulfuric Acid Concentrator (Process)

BOSC Beryllium Oxide Strengthened Copper

BOSCA British Oil Spill Control Association

Bosch Tech Ber Bosch Technische Berichte [Bosch Technical Reports; published by Robert Bosch GmbH, Stuttgart, Germany]

BOSE 1,2-Bis(n-Octylsulfinyl)ethane

Bosei Kanri (Rust Prev Control) Bosei Kanri (Rust Prevention Control) [Japanese Publication]

Boshoku Gijutsu (Corros Eng) Boshoku Gijutsu (Corrosion Engineering) [Japanese Publication]

BOSFET Bidirectional Output Switch Field-Effect Transistor

BOSM Bis(n-Octylsulfinyl)methane

Bosn Bosnia(n)

Bo's'n Boatswain [also bo's'n]

BOSOR Buckling of Shell of Revolution

BOSS Basic Operating System Software

BOSS Batch Operating Software System

BOSS Behaviour of Off-Shore Structures Conference

BOSS Benthic Organic Seston Sampler

BOSS Bioastronautic Orbiting Space Station

BOSS Boeing Operational Supervisory System

BOSS Book of SEMI (Semiconductor Equipment and Materials Institute) Standards

BOSS BRE (Building Research Establishment) On-Line Search System

BOSS Brinell Optical Scanning System

BOSS Business Opportunities Sourcing System

BOSS Business-Oriented Software System

BOSSF British Office Systems and Stationery Federation

BOST Basic Offshore Survival Training

BOSTI Buffalo Organization for Social and Technological Innovation [US]

BOSVA British Offshore Support Vessels Association

BOT Basic Offshore Training

BOT Beginning of Table

BOT Beginning of Tape [also BoT, or bot]

BOT Beginning of Task

BOT Benzoxazole-2-Thione

BOT Board of Trade [UK]

BOT Botswana [NASA Space Tracking and Data Network]

BOT Build, Operate, Transfer (Project) [Building Construction]

B1T Basic One Ton [Unit]

Bot Botanical; Botanist; Botany [also bot]

Bot Bottle [also bot]

Bot Bottom [also bot]

BOTAC Board of Trade Advisory Committee [UK]

BOTB British Overseas Trade Board

Bot Gaz Botanical Gazette [Published by University of Chicago Press, US]

Bot Helv Botanica Helvetica [Publication of the Swiss Botanical Society]

BOTMA British Office Technology Manufacturers Alliance

Bot Rev Botanical Review [International Journal]

BOU British Ornithologists Union

Bound Boundary [also bound]

Bound Lay Meteor Boundary-Layer Meteorology [Journal published in the Netherlands]

BOV Brown Oil of Vitriol

Bov Bovine [Immunochemistry]

BOVC Base of Overcast

BOW Badger Ordnance Works [Baraboo, Wisconsin, US]

BOW Bill of Work

BOX Buried Oxide [also BOx, or Box]

BP Backplane [alsp bp]

BP Back Pressure

BP Back Projection

BP Bacteriophage

BP Ballistic Pendulum

BP Ball Plunger

BP Band-Pass (Filter)

BP Barber Pole

BP Barrel Plating [Metallurgy]

BP Basal Plane [Crystallography]

BP Baseline Program

BP Base Paper

BP Base Peak

BP Base Plate

BP Base Point

BP Base Pointer

BP Base-Promoted (Halogenation)

BP Basic Protection (Coverall)

BP Batch Process(ing); Batch Processor

BP Beam Processing

BP Before Present

BP Bend Plane

BP Benzoyl Peroxide

BP Bessemer (Steelmaking) Process [Metallurgy]

BP Beta Particle

BP Between Perpendiculars

BP Bijvoet Pair

BP Bills Payable [also bp]

BP Binding Post

BP Binding Protein [Biochemistry]

BP Biodegradable Polymer

BP Biological Physicist; Biological Physics

BP Biological Processing

BP Biophysical; Biophysicist; Biophysics

BP Biopolymer

BP Bipolar(ity)

BP Bit Processor

BP Black Phosphorus

BP Blueprint [also B/P]

BP Block Parity

BP Blood Plasma

BP Blood Pressure

BP (Cholesteric) Blue Phase(s) (of Liquid Crystals)

BP Body Protection; Body Protector

BP Boilerplate

BP Boiling Point [also bp]

BP Bolivian Peso [Currency]

BP Bolza Problem

BP Bonded Single Paper (Electric Wire)

BP Bonded Phase

BP Bond Percolation [Fluid/Porous Media]

BP Bordered Pit (in Wood)

BP Bordoni Peak [Metallurgy]

BP Boron Phosphide (Semiconductor)

BP Boson Peak [Physics]

BP Bottled Gas

BP Bottom Part

BP Bragg-Pierce (Law) [Physics]

BP Brake Power [also bp]

BP Branch Point

BP Brake Power [also bp]

BP Breakpoint

BP British Patent

BP British Petroleum

BP British Pharmacopoeia [also bp]

BP 1-Bromopropane

BP Bronchopenumonia

BP Buffered Printing

BP Bulk Plasmon [Solid-State Physics]

BP Bundespost [(German) Federal Post Office]

BP Buried P-Layer (Field-Effect Transistor)

BP Burnable Poison [Nuclear Engineering]

BP Bypass

BP Byproduct

BP I (Cholesteric) Blue Phase I (of Liquid Crystals) [also BPI]

BP II (Cholesteric) Blue Phase II (of Liquid Crystals) [also BPII]

BP III (Cholesteric) Blue Phase(s) (of Liquid Crystals) [also BPIII]

B&P Budgetary and Planning

B/P Baryon-to-Photon (Ratio)

B/P Bills Payable [also b/p]

B/P Blueprint [also BP]

B-P Bullet-Proof

B$ Bolivian Peso [Currency]

Bp Bacillus proteus [Microbiology]

Bp Bacillus popilliae [Microbiology]

Bp Birthplace [also bp]

bp base pair [Biochemistry]

bp below par

bp bills payable

BPA Bachelor of Professional Arts

BPA Bachelor of Public Administration

BPA Bandpass Amplifier

BPA Basic Probability Assignment

BPA Bauhinia Purpurea Agglutinin [Immunology]

BPA N-Benzyl-9-(2-Tetrahydropyranyl)adenine [Biochemistry]

BPA Biological Photographic Association [US]

BPA Bioprocessing Acid

BPA Biotin-Propranolol Analog [Biochemistry]

BPA Bisphenol A

BPA 1,2-Bis(4-Pyridyl)ethane

BPA Bonneville Power Administration [Oregon, US]

BPA Book Packagers Association

BPA British Parachute Association

BPA British Philatelic Association [UK]

BPA British Photographic Association

BPA British Ploughing Association

BPA British Professional Association

BPA British Pyrotechnists Association

BPA α-Bromopropionic Acid

BPA Bureau of Public Affairs [of US Department of State]

β-Pa Beta Protactinium [Symbol]

B/PAA Business/Professional Advertising Association [US]

B-PABA N-(+)-Biotinyl-p-Amino Benzoic Acid

BPAD BSC (Binary Synchronous Communications Protocol) Packet Assembly/Disassembly

BPADCy Bisphenol A Dicyanate

BP Add British Pharmacopoeia Addendum

BPAE N,N-Bis(pyrazol-1-ylmethyl)aminoethane [also bpae]

BPAI Board of Patent Appeals and Interferences [of United States Patent and Trademark Office]

BPAM Basic Partitioned Access Method

BPA-PC Bisphenol A–Polycarbonate

B-PAS Benzoyl p-Aminosalicylic Acid

B Pay Bills Payable [also b pay]

BPB Bank Post Bill

BPB BIOS (Basic Input/Output) Parameter Block

BPB Black Powder Bag

BPB Bromophenol Blue

BPB Bubble Pressure Barrier

BPBD Butyl Phenylbiphenylyloxadiazole

BPBF British Paper Box Federation

β-PbF$_2$ Beta Lead Fluoride

BPBIF British Paper and Board Industry Federation

BPBIRA British Paper and Board Industry Research Association

BPBO Barium-Lead-Bismuth Oxide (Superconductor)

BPBZIM 1-Benzyl-2-Phenylbenzimidazole [also bpbzim]

BPC Back Pressure Control

BPC Basic Peripheral Channel

BPC Bilateral Private Circuit

BPC Binding Post Chamber

BPC Biomass Production Chamber

BPC Biophysical Chemist(ry)

BPC Bonded-Phase Chromatography

BPC Bond Percolation Cluster [Fluid/Porous Media]

BPC Boost Protective Cover [of NASA]

BPC British Pharmaceutical Codex

BPC British Productivity Council

BPC N-Butyl Pyridinium Chloride

BPC Bypass Capacitor

BPCA British Pest Control Association

BPCC 2,2'-Bipyridinium Chlorochromate

BPCC British Printing and Publishing Communications Corporation

BPCD Barrels (of Oil) per Calendar Day

BPCF British Precast Concrete Federation

BPCVMRA British Paint, Color and Varnish Manufacturers Association

BPD Barrels (of Oil) per Day [also bpd, bbl d^{-1}, or bbl/d]

BPD Baseline Program Documentation

BPD Beam Positioning Drive

BPD Bushing Potential Device

BPDA Bibliographic Pattern Discovery Algorithm

BPDA Biphenyltetracarboxylic Dianhydride; Bisphenyldianhydride; 3,3',4,4'-Biphenyl Tetracarboxylic Acid Dianhydride

β-PdAl Beta Palladium Aluminide

BPDA-PDA 3,3',4,4'-Biphenyl Tetracarboxylic Acid Dianhydride-p-Phenylene Diamine [also BPDA/PDA]

BPDA-PDA Poly(Bisphenyldianhydride-p-Phenylenediamine) (Polyimide) [also BPDA/PDA]

BPDA-PPD Poly(Bisphenyldianhydride-p-Phenylenediamine)-p-Phenylenediamine [also BPDA/PPD]

BPDO 2,2'-Bipyridine N,N'-Dioxide

BPDTO 1,8-Di-2-Pyridyl-3,6-Dithiaoctane [also bpdto]

BPDU Bridge Protocol Data Unit

BPDZ 3,3'-Bipyridazine [also bpdz]

BPE Bioprocess Engineer(ing)

BPE trans-1,2-Bis(4-Pyridyl)ethylene [also bpe]

BPE Boiling Point Elevation [Chemistry]

BPE 4-(Bromomethyl)phenylacetic Acid Phenacyl Ester

BPE 1-Bromo-1-Phenylethane

BPEA Belfast Port Employers Association [Northern Ireland]

BPEA 9,10-Bis(phenylethynyl)anthracene

BPEC Building Products Executives Conference

BPEG British Photographic Enterprise Group

BPEM Best Practice Environmental Management (in Mining) [Australia]

BPEN 5-12-Bis(phenylethynyl)naphthacene

BPEN N,N'-Bis(2-Picolinyl)ethylenediamine [also bpen]

BPEO Best Practical Environmental Option

BPetE Bachelor of Petroleum Engineering

BPF Bandpass Filter

BPF Basal Plane Facet [Crystallography]

BPF Bezirksdirektion für Post– und Fernmeldewesen [District Office for Post and Telecommunications, Germany]

BPF Blueprint Files

BPF Bottom Pressure Fluctuation

BPF British Philatelic Federation

BPF British Plastics Federation

BPF British Plastics Foundation

BPF British Ports Federation

BPF British Poultry Meat Federation

BPF British Property Association

BPF Byproduct Fuel

BPFMA British Plumbing Fittings Manufacturers Association

BPF News BPF News [of British Plastics Foundation]

BPGMA British Pressure Gauge Manufacturers Association

BPGO Barium Praseodymium Gadolinium Oxide

BPH Bachelor of Public Health

BPH Barrels (of Oil) per Hour [also bph]

BPH Benign Prostatic Hyperplasia; Benign Prostatic Hypertrophy

BPH Botanico Periodicum Huntianum [Listing of Standard Botanical Journal Abbreviations]

BPH 2,2'-Methylene-Bis(4-Methyl-6-tert-Butyl)phenol

BPh Bachelor of Philosophy [also BPhil]

B(Ph) Tetraphenylboron Ion

BPHA N-Benzoyl-N-Phenyl Hydroxylamine

N-BPHA N-Benzyl-N-Phenyl Hydroxylamine

BPharm Bachelor of Pharmacy

BPHE Bureau of Public Health Education [of University of Michigan, US]

BPhil Bachelor of Philosophy [also BPh]

BPhys Bachelor of Physics

BPI Belorusski Politekhnicheskii Institut [Belarus Polytechnic Institute, Minsk, Belarus]

BPI Bits per Inch [also bpi]

BPI Boost Phase Intercept

BPI British Phonographic Industry

BPI Bytes per Inch [also bpi]

BPIA British Photographic Importers Association

BPIC Butylperoxy Isopropyl Carbonate; Terbutyl Peroxoy Isopropyl Carbonate

BPICS British Production and Inventory Control Society

BPID Book Physical Inventory Difference

BPIDP Book Publishing Industry Development Program [Canada]

BPIF British Printing Industries Federation

BPKT Basic Programming Knowledge Test

BPL Beta-Propiolactone [Biochemistry]

BPL Binary Program Load

BPL Bone Phosphate of Lime

BPL Boston Public Library [Massachusetts, US]

BPL Branch if Plus

BPL British Physical Laboratories

BPL Burroughs Program Loader

Bpl Birthplace [also bpl]

BP LDD SAGFET Buried p-Layer Lightly-Doped-Drain Self-Aligned-Gate Field-Effect Transistor

BPM Balanced Processing Monitor

BPM Barrels per Month [also bpm]

BPM Batch Processing Monitor

BPM Beam Position Monitor

BPM Beats per Minute

BPM Best Practical Means

BPM 4,4'-Bipyrimidine [also bpm]

BPM Bits per Millimeter [also BPMM, bpm, or bpmm]

β-Pm Beta Promethium [Symbol]

BPMA Biotechnology Purchasing Management Association [US]

BPMA British Photographic Manufacturers Association

BPMA British Pottery Manufacturers Association

BPMA British Premium Manufacturers Association

BPMA British Printing Manufacturers Association

BPMA British Pump Manufacturers Association

BPMA Bureau of Politico-Military Affairs [of US Department of State]

BPMC Butylphenylmethylcarbamate

BPMD Battelle Project Management Division [US]

BPMEN N,N'-Bis(2-Picolinyl)-N,N'-Dimethyl Ethylenediamine [also bpmen]

BPMF British Postgraduate Medical Federation

BPMF British Pottery Manufacturers Federation

BPMF British Powder Metal Federation

BPMM Bits per Millimeter [also BPM, bpm, or bpmm]

BPMMA British Paper Machinery Makers Association

BPMP 2,6-Bis[Bis(2-Pyridylmethyl)aminomethyl]-4-Methyl Phenol [also bpmp]

BPMS Blood Pressure Measuring System

BPMU N,N'-Bis(pentamethylene)urea

BPN Biological Polymer Network

BPN Breakdown Pulse Noise [Electronics]

BPN British Pendulum Number [Surface Friction Testing]

BPNL Battelle Pacific Northwest Laboratories [Richland, Washington, US]

BPO Bartlesville Project Office [for National Oil Program at Bartlesville, Oklahoma, US]

BPO Battelle Portland Operations [Oregon, US]

BPO Before Payout

BPO Benzoyl Peroxide

BPO 2-(4-Biphenylyl)-5-Phenyloxazole

BPO Borophosphate

BPO British Post Office [London, UK]

BPO Building Prices Order

BPO Bypass Orifice

β-Po Beta Polonium [Symbol]

BPOC Biphenylisopropyloxycarbonyl

BPP 2,3-Bis(2-Pyridyl)pyrazine [also bpp]

BPP Black Powder Pellet

BPP Bloembergen-Purcell-Pound (Nuclear-Magnetic Resonance Relaxation)

BPP Bulk Paid Publication

BPPA Book Publishers Professional Association

BPPA British Precision Pilots Association

BPPCF Billion Particles per Cubic Foot [also bppcf]

BPPF Basic Program Preparation Facility

BPPFA Bis(diphenylphosphino)ferrocenylethyl Dimethylamine

BPPM Butyldiphenylphosphinomethylpyrrolidine Carboxylate

BPPMA British Power Press Manufacturers Association

BPPZ 2,5-Bis(2-Pyridyl)pyrazine [also bppz]

BPQ 2,3-Bis(2-Pyridyl)quinoxaline [also bpq]

BPR Bachelor of Public Relations

BPR Ball-to-Powder (Weight) Ratio [Ball Milling]

BPR Bromophenol Red

BPR Bureau of Public Roads [US]

BPR Bypass Ratio

BPR Byproduct Recycle (Process)

β-Pr Beta Praseodymium [Symbol]

BPRA Book Publishers Representatives Association [UK]

BPRA Burnable Poison Rod Assembly [Nuclear Engineering]

BPRC Byrd Polar Research Center [of Ohio State University, Columbus, US]

Bprf Bulletproof [also bprf]

BPRI British Polarographic Research Institute

BPROGA Borland Programming Languages Forum A–Pascal and Basic [of Borland International, US]

BPROGB Borland Programming Languages Forum B–C and Prolog [of Borland International, US]

BPRS British Polarographic Research Society

BPRTHM Bureau of Public Roads Transport Highway Mobilization [Canada]

BPS Bachelor of Professional Studies

BPS Back-Pressure Switch

BPS Basic Programming Support

BPS Basic Programming System

BPS Batch Processing System

BPS Battersea Power Station [London, UK]

BPS Belarussian (or Byelorussian) Physical Society

BPS Belgian Physical Society

BPS Bell Production Services [Multimedia]

BPS Biophysical Society [US]

BPS Bits per Second [also bps]

BPS Boron Phosphate Silicate

BPS Borophosphosilicate

BPS British Pharmacological Society

BPS British Photobiology Society

BPS British Phycological Society

BPS British Printing Society

BPS British Pterodological Society

BPS Bulk Processing Stream

BPS Bulk Processing System

BPS Bytes per Second [also Bps]

BPSA British Plastics Stockholders Association

BPSA Business Products Standards Association [US]

BPSCCO Bismuth Lead Strontium Calcium Copper Oxide (Superconductor) [also BiPSCCO, or BIPSCCO]

BPSCCO-1212 $(BiPb)Sr_2CaCu_2O_y$ [A Bismuth Lead Strontium Calcium Copper Oxide Superconductor]

BPSCCO-2223 $(BiPb)_2Sr_2Ca_2Cu_3O_y$ [A Bismuth Lead Strontium Calcium Copper Oxide Superconductor]

BPSCCO-3221 $(BiPb)_3Sr_2Ca_2CuO_{10}$ [A Bismuth Lead Strontium Calcium Copper Oxide Superconductor]

BPSD Barrels per Stream Day [also bpsd]

BPSG Boron Phosphate Silicate Glass; Borophosphosilicate Glass

BPSI Bits per Square Inch [also bpsi]

B:p-Si Boron-Doped P-Type Silicon

BPSK Binary Phase-Shift Keyed (or Keying); Biphase Shift Keyed (or Keying)

BPSS Basic Packet Switched System

BPSSCCO Bismuth Lead Antimony Strontium Calcium Copper Oxide (Superconductor)

BPST 2-(2'-Benzothiazolyl)-5-Styryl-3-(4'-Phthalhydrazidyl) tetrazolium

BPT Back-Pressure Turbine

BPT Baseband Pulse Transmission

BPT Best Practicable Technology; Best Practical Technology; Best Practice Technology

BPT Bipolar Technology

BPT Bipolar Transistor

BPT Blade Passage Tones

BPT Body Point

BPT Borderline Pumping Temperature

B Pt Boiling Point [also BPt, bpt, or b pt]

BPTA British Polyolefin Textiles Association

BPTCA Best Practicable Technology Currently Available

BPTI Bovine Pancreas Trypsin Inhibitor [Biochemistry]

BPTN N,N'-Bis(2-Picolinyl)trimethylenediamine [also bptn]

BPTS 1N-Benzenesulfonyl-4N-Phenylthiosemicarbazidate

BPTS Bipolar Transistor Switch

BPTS-H 1N-Benzenesulfonyl-4N-Phenylthiosemicarbazide

BPTZ 3,6-Bis(2-Pyridyl)-1,2,4,5-Tetrazine [also bptz]

BPU Beijing Polytechnic University [PR China]

β-Pu Beta Plutonium [Symbol]

BPV Bipropellant Valve

BPV Bovine Varvovirus

B&PV Boiler and Pressure Vessel [also B+PV]

BPVC Boiler and Pressure Vessel Code [of American Society of Mechanical Engineers, US]

BPVC Boiler and Pressure Vessel Committee [of American Society of Mechanical Engineers, US]

BPW (Federation of) Business and Professional Women [UK]

BPWH British Petroleum Western Hemisphere (Test Program)

BPWR Burnable Poison Water Reactor

BPWS Banked Position Withdrawal Sequence

BPX Burning Plasma Experiment

BPY 4,4'-Bipyridine [also Bpy, or bpy]

BPym 2,2'-Bipyrimidine [also bpym]

BPyz 2,2'-Bipyrazine [also bpyz]

BPZ4 Tetrakis(1-Pyrazolyl)borate

BPz 2,2'-Bipyrazine [also bpz]

BQ Benzoquinone

BQ Brine Quenched; Brine Quenching [Metallurgy]

BQ 2,6-Di-tert-Butylquinone

12BQ 1,2-Benzoquinone

Bq Becquerel [Unit]

BQA British Quality Association

BQE Bi-Quadratic Equation

Bq/kg Becquerel per kilogram [also Bq kg^{-1}]

Bq/L Becquerel per liter [also Bq/l, or Bq L^{-1}]

Bq/m³ Becquerel per cubic meter [also Bq m^{-3}]

Bq/mmol Becquerel per millimole [Unit]

BQRTT Bipolar Quantum Resonant Tunneling Transistor

B-Quark Beauty (or Bottom) Quark [also b-quark] [Particle Physics]

BQUIN 2,2'-Biquinoline [also bquin]

BQL Basic Query Language

BQL Batch Query Language

BR Background Radiation

BR Back Reflection (Method) [Crystallography]

BR Bacteriorhodopsin [Biochemistry]

BR Bad Register [Computers]

BR Ballast Resistor

BR Band-Rejection (Filter)

BR Bank Rate

BR Baryon Resonance [Particle Physics]

BR Basal Rate [Medicine]

BR Base Register

BR Basic Research(er)

BR Batch Reactor

BR Baton Rouge [Louisiana, US]

BR Bayrischer Rundfunk [Bavarian State Broadcasting Station, Germany]

BR Beacon Receiver

BR Beam Rocking

BR Beckmann Reaction [Chemistry]

BR Beevers-Ross (Site)

BR Belgian Reactor [at Mol, Belgium]

BR Bellows Regulating Valve

BR Bend Radius

BR Berliner Rundfunk [Broadcasting Station of Berlin, Germany]

BR Beta Ray

BR Biased Reception

BR Bills Receivable

BR Biological Reactivation

BR Bioreactor

BR Biorheological; Biorheology

BR Biradical

BR Bit Rate

BR Blue Record

BR Boil Resistance; Boil-Resistant

BR Bottom Register

BR Bottoms Recycle [Nuclear Engineering]

BR Boundary Representation [Geographic Information System]

BR Brake [Controllers]

BR Brake Relay

BR Brazil [ISO Code]

BR Breakpoint on Register (Command) [Pascal Programming]

BR Breath(ing) Rate

BR Breeder Reactor

BR Breit-Rabi (Formula) [Physics]

BR Briefing Room

BR British Rail(ways) [UK]

BR Brittle Rupture

BR Brown [also br]

BR Bruggeman-Rayleigh Formula [Physics]

BR Bureau of Reclamation [of US Department of the Interior]

BR Burlington Resources Inc. [US]

BR Butadiene Rubber

BR Butyl Rubber

BR-1 Belgian Reactor 1 [at Mol, Belgium]

BR-2 Belgian Reactor 2 [at Mol, Belgium]

B&R Budget(ing) and Reporting (Classification)

B/R Bills Receivable [also b/r]

Br Birr [Currency of Ethiopia]

Br Brake [also br]

Br Branch [also br]

Br Brand [also br]

Br Brass [also br]

Br Braze [also br]

Br Brinkmann Number (of Viscous Flow) [Symbol]

Br Briquet(te) [also br]

Br Britain; British

Br Bromine [Symbol]

Br Bronze [also br]

Br Brother [also br]

Br Brucella [Genus of Bacteria]

Br Brucite [Mineral]

Br Brush [also br]

Br⁻ Bromine Ion [Symbol]

Br_2 Molecular Bromine [Symbol]

Br-74 Bromine-74 [also ^{74}Br, or Br^{74}]

Br-75 Bromine-75 [also ^{75}Br, or Br^{75}]

Br-76 Bromine-76 [also ^{76}Br, or Br^{76}]

Br-77 Bromine-77 [also ^{77}Br, or Br^{77}]

Br-78 Bromine-78 [also ^{78}Br, or Br^{78}]

Br-79 Bromine-79 [also ^{79}Br, or Br^{79}]

Br-80 Bromine-80 [also ^{80}Br, or Br^{80}]

Br-81 Bromine-81 [also ^{81}Br, or Br^{81}]

Br-82 Bromine-82 [also ^{82}Br, or Br^{82}]

Br-83 Bromine-83 [also ^{83}Br, or Br^{83}]

Br-84 Bromine-84 [also ^{84}Br, or Br^{84}]

Br-85 Bromine-85 [also ^{85}Br, or Br^{85}]

Br-87 Bromine-87 [also ^{87}Br, or Br^{87}]

Br-88 Bromine-88 [also ^{88}Br, or Br^{88}]

Br-89 Bromine-89 [also ^{89}Br, or Br^{89}]

B(r) Magnetic Flux Density (or Magnetic Induction)

bR bacteriorhodopsin [Biochemistry]

br braze(d)

br brown

.br Brazil [Country Code/Domain Name]

b(r) normalized bonding orbital [Physical Chemistry]

BRA Base Rate Area [Telecommunications]

BRA Basic Rate Access

BRA Bee Research Association [UK]

BRA Beta-Resorcylic Acid [Biochemistry]

BRA Bioreactor Analysis

BRA British Refrigeration Association

BRA British Robot Association

BRAB Building Research Advisory Board [now Building Research Board, US]

BRAC Bangladesh Rural Advancement Committee

BRAC Brotherhood of Railway and Airline Clerks [US]

BrAc 2-Bromoacetic Acid

Br-AcAc Bromo-Acetylacetonate

BRadPA British Radiation Protection Association

BRAF Braking Action Fair

BRAG Braking Action Good

BRAGS Bioelectrical Repair and Growth Society [US]

BRAHMS Broad RAnge Hadron Magnetic Spectrometers (Experiment) [at the Relativistic Heavy Ion Collider of Brookhaven National Laboratory, Upton, New York, US]

Brain Behav Evol Brain, Behavior and Evolution [International Journal]

Brain Behav Immun Brain, Behavior and Immunity [US Journal]

Brain Cognit Brain and Cognition [US Journal]

Brain Lang Brain and Language [US Journal]

Brain Res Brain Research [International Journal]

Brain Res, Brain Res Protoc Brain Research, Brain Research Protocols [International Journal]

Brain Res, Brain Res Rev Brain Research, Brain Research Reviews [International Journal]

Brain Res, Cogn Brain Res Brain Research, Cognitive Brain Research [International Journal]

Brain Res, Devel Brain Res Brain Research, Developmental Brain Research [International Journal]

Brain Res, Mol Brain Res Brain Research, Molecular Brain Research [International Journal]

BRAM Bit Rate Agile Modem

BRAMA British Rubber and Resin Manufacturers Association

BRAN Braking Action Nil

BRANA Bumper Recycling Association of North America [US]

BRANE Bombing Radar Navigation Equipment

BRAP Braking Action Poor

BRAS Ballistic Rocket Air Suppression

BRASS Bottom Reflection Active Sonar System

BRASTACS Bradford Scientific, Technical and Commercial Service [UK]

Br Astron Assoc Circ British Astronomical Association Circular

BRAVO Business Risk and Value of Operation in Space

Braz Brazil(ian)

Braz Sold Brazing and Soldering [UK Journal]

BRB Base Rate Boundary [Telecommunications]

BRB Benefits Review Board

BRB Be Right Back [Internet]

BRB British Railway Board

BRB Building Research Board [US]

BRBMA Ball and Roller Bearing Manufacturers Association [US]

BRBO Barium Rubidium Bismuth Oxide (Superconductor)

BrBu 2-Bromobutanoic Acid

Br Busin British Business [Publication of the UK Department of Trade and Industry]

BrBz Bromobenzene

BRC Below Regulatory Concern

BRC Biological Records Center [UK]

BRC Branch Conditional

BRC Brazilian Cruzeiro [Currency]

BRC Brazilian Research Council

BRC Breeder Reactor Corporation [US]

BRC British Radio Communication

Br-cADP 8-Bromo-Cyclic Adenosine Diphosphate [Biochemistry]

BRCBED Beam-Rocking Method Convergent Beam Electron Diffraction

Br Ceram Proc British Ceramic Proceedings [of Institute of Ceramics]

Br Ceram Trans British Ceramic Transactions [of Institute of Ceramics]

Br Ceram Trans J British Ceramic Transactions and Journal [of Institute of Ceramics]

BrCl Bromine Chloride

BRCMA British Radio Cabinet Manufacturers Association

Br Comm British Commonwealth [also Brit Comm]

Br Corros J British Corrosion Journal [of Institute of Materials]

BRCS Basic Reference Coordinate System

BRD Binary Rate Divider

BRD Biological Resources Division [of United States Geological Survey]

BRD Bioreactor Design

BRD Brunei Dollar [Currency of Brunei]

BRD (*Bundesrepublik Deutschland*) – German for "Federal Republic of Germany"

BR$ Brunei Dollar [Currency of Brunei]

Brd Board [also brd]

Brd Braid [also brd]

BRDA Boxboard Research and Development Association [US]

BRDB British Rubber Development Board

BRDC British Research and Development Corporation

Br Dent J British Dental Journal

BRDF Bidirectional Reflectance Distribution Function

BRDG Bituminous Research and Development Group

Brdg Bridge [also brdg]

Br-DMEQ 3-Bromomethyl-6,7-Dimethoxy-1-Methyl-2-(1H)-Quinoxalinone

BRDP Bit Rate Distance Product

BRDU 5-Bromo-2'-Deoxyuridine [also Br-dU] [Biochemistry]

Br-dU Bromodeoxyuridine [also BrdU] [Biochemistry]

BRE Beta-Ray Electroscope

BRE Building Research Establishment [UK]

BRE Bureau of Research and Engineering

Br E Afr British East Africa(n)

BREAM Beaufort Region Environmental Assessment and Monitoring Program [Canada]

B Rec Bills Receivable [also b rec]

BRECOM Broadcast Radio Emergency Communication [US Air Force]

BREG British Rivet Export Group

BREL Boeing Radiation Effects Laboratory [US]

BREMA British Radio (and Electronic) Equipment Manufacturers Association

BREM-SAT Bremen Research Satellite [of Zentrum für Angewandte Raumfahrttechnologie und Mikrogravitation, University of Bremen, Germany]

BREN Bare Reactor Experiment, Nevada [of US Department of Energy]

Br Engine Tech Rep British Engine Technical Reports

Brennst Wärme Kraft Brennstoff–Wärme–Kraft [German Journal on Fuels, Heat and Power; published by Verein Deutscher Ingenieure]

B-REP Boundary Representation [also B-rep]

BResMA British Resin Manufacturers Association

BRET Breeder Reprocessing Engineering Test

Brev Brevet [Military]

BRF Band-Rejection Filter

BRF Bartol Research Foundation [of Franklin Institute at Swarthmore, Pennsylvania, US]

BRF Bass Research Foundation [US]

BRF Bell Rings Faintly

BRF Bereich Rundfunk und Fernsehen (im Ministerium für Post– und Fernmeldewesen) [Department of Radio and Television (of the Federal Ministry for Post and Telecommunications), Germany]

BRF Biochemical Research Foundation [US]

BRF British Road Federation

BRFFI Biochemical Research Foundation of the Franklin Institute [US]

BRG Baud Rate Generator

BRG Beacon Reply Group

BRG Bearing [also Brg, or brg]

BRG Budget Review Group

BRGC Binary Reflected Gray Code

Br Geol British Geologist [Journal of the Institution of Geologists, UK]

BRGHD Bridgehead [also Brghd, or brghd]

BRGM Bureau de Recherches Géologiques et Minières [Bureau for Geological and Mining Research, Orléans, France] [French Geological Survey]

BRGS Baton Rouge Geological Society [Louisiana, US]

Br Gui British Guiana

BRGW Brake Release Gross Weight

BRH Bureau of Radiological Health [of US Department of Health and Human Services]

Br Heart J British Heart Journal

BRI Bartol Research Institute [of University of Delaware, Newark, US]

BRI Basic Rate Interface

BRI Basin Research Institute [of Louisiana State University, US]

BRI Behavioral Research Institute [US]

BRI Biological Research Institute [of University of California at Los Angeles, US]

BRI Biosystematics Research Institute [Ottawa, Canada]

BRI Biotechnology Research Institute [of National Research Council of Canada]

BRI Book Review Index

BRI Bose Research Institute [Calcutta, India]

BRI Brace Research Institute [Canada]

BRI Brain Response Interface

BRI Bread Research Institute [Australia]

BRI Building Research Institute

BrI Bromine Iodide

BRIAN Barrier Reef Image Analysis [Software]

BRIC Bureau de Recherche pour l'Innovation et la Convergence [Research Bureau for Innovation and Convergence, France]

BRICS Bureau of Research Information Control Systems [of United States Office of Education]

Bridges Mon Rev Bridges Monthly Review [Publication of the International Center for Trade and Sustainable Development]

Bridges Weekly Trade News Dig Bridges Weekly Trade News Digest [Publication of the International Center for Trade and Sustainable Development]

BRIE Berkeley Roundtable on the International Economy [of University of California at Berkeley, US]

Brig Brigade; Brigadier [also brig]

Brig Gen Brigadier General

BRIM Biosphere Reserve Integrated Monitoring

BRIMAFEX British Manufacturers of Malleable Tube Fittings Export Group

BRIMARC Brighton MARC (Machine-Readable Catalogue) Program [UK]

BRIMT Beijing Research Institute of Material Technology [PR China]

BRINCO British Newfoundland Corporation [Canada] [also Brinco]

BRINDEX (Association of) British Independent Oil Exploration Companies

BRIT Bureau Robotique, Informatique et Telematique [Bureau for Robotics, Informatics and Telematics, France]

Brit Britain; British

BRITCHAM British Chamber of Commerce

Brit Ceram Proc British Ceramic Proceedings [of Institute of Ceramics]

Brit Ceram Trans British Ceramic Transactions [of Institute of Ceramics]

Brit Ceram Trans J British Ceramic Transactions and Journal [of Institute of Ceramics]

BrCl Bromine Chloride

BRCMA British Radio Cabinet Manufacturers Association

Brit Comm British Commonwealth [also Brit Comm]

Brit Corros J British Corrosion Journal [of Institute of Materials]

BRITE Basic Research in Industrial Technology for Europe (Program) [also Brite]

BRITE Bright Radar Indicator Tower Equipment

BRITE/EURAM Basic Research in Industrial Technology for Europe/European Research on Advanced Materials (Program) [of European Community] [also Brite/EuRam]

BRITEQ Bureau de Recherche sur l'Industrie de la Tourbe dans l'Est du Québec [Research Bureau of the Peat Industry in Eastern Quebec [Canada]

Brit Foundryman British Foundryman [UK Publication]

Brit Heart J British Heart Journal

Brit J Appl Phys British Journal of Applied Physics

Brit J Biomed Sci British Journal of Biomedical Science

Brit J Cancer British Journal of Cancer

Brit J Clin Pharmacol British Journal of Clinical Pharmacology

Brit J Dermatol British Journal of Dermatology

Brit J Educ Psychol British Journal of Educational Psychology

Brit J Educ Technol British Journal of Educational Technology [of Council for Educational Technology for the United Kingdom]

Brit J Exp Pathol British Journal of Experimental Pathology

Brit J Haematol British Journal of Haematology

Brit J Hist Sci British Journal for the History of Science [of British Society for the History of Science]

Brit J Hosp Med British Journal of Hospital Medicine

Brit J Ind Med British Journal of Industrial Medicine

Brit J Non-Destr Test British Journal of Non-Destructive Testing [of British Institute of Non-Destructive Testing] [also Br J Non-Destr Test, Br J NDT, or Brit J NDT]

Brit J Nutr British Journal of Nutrition

Brit J Pharmacol British Journal of Pharmacology

Brit J Philos Sci British Journal for the Philosophy of Science

Brit J Phot British Journal of Photography

Brit J Psych British Journal of Psychiatry

Brit J Psychol British Journal of Psychology

Brit J Radiol British Journal of Radiology [of British Institute of Radiology]

Brit J Rheumatol British Journal of Rheumatology [Published by Oxford University Press, UK]

Brit J Surgery British Journal of Surgery

Brit Libr J British Library Journal [of British Library, London, UK]

Brit Med Bull British Medical Bulletin

Brit Med J British Medical Journal [of British Medical Association]

BritP British Patent [also Brit Pat]

Brit Plast Rubb British Plastics and Rubber [UK Publication]

Brit Polym J British Polymer Journal [Publication of the Society of Chemical Industry]

Brit Printer British Printer [UK Publication]

Brit Steel British Steel [UK Publication]

Brit Steelmaker British Steelmaker [UK Publication]

Brit Telecom J British Telecom Journal [Published by Post Office Central Headquarters, London, UK]

Brit Telecommun Eng British Telecommunications Engineering [Journal of the Institution of British Telecommunications Engineers]

Brit Telecom Technol J British Telecom Technology Journal [of British Telecom Research Laboratories]

Brit Telecom World British Telecom World [Publication of the British Telecom Center]

Brit Weld J British Welding Journal

BRIX Building Research Information Exchange [US]

Br J Appl Phys British Journal of Applied Physics

Br J Biomed Sci British Journal of Biomedical Science

Br J Cancer British Journal of Cancer

Br J Clin Pharmacol British Journal of Clinical Pharmacology

Br J Dermatol British Journal of Dermatology

Br J Educ Psychol British Journal of Educational Psychology

Br J Educ Technol British Journal of Educational Technology [of Council for Educational Technology for the United Kingdom]

Br J Exp Pathol British Journal of Experimental Pathology

Br J Haematol British Journal of Haematology

Br J Hist Sci British Journal for the History of Science [of British Society for the History of Science]

Br J Hosp Med British Journal of Hospital Medicine

Br J Ind Med British Journal of Industrial Medicine

Br J Non-Destr Test British Journal of Non-Destructive Testing [of British Institute of Non-Destructive Testing] [also Brit J Non-Destr Test, Br J NDT, or Brit J NDT]

Br J Nutr British Journal of Nutrition

Br J Pharmacol British Journal of Pharmacology

Br J Philos Sci British Journal for the Philosophy of Science

Br J Phot British Journal of Photography

Br J Psych British Journal of Psychiatry

Br J Psychol British Journal of Psychology

Br J Radiol British Journal of Radiology [of British Institute of Radiology]

Br J Rheumatol British Journal of Rheumatology [Published by Oxford University Press, UK]

Br J Surgery British Journal of Surgery

Brk Break [also brk]

Brk Brick [also brk]

BrKα Bromine K-Alpha (Radiation) [also BrK$_\alpha$]

BRKHIC Breaks in Higher Overcast

Brkg Brokerage [also brkg]

Brkr Breaker [also brkr]

Brkt Bracket [also brkt]

BRL Balance Return Loss

BRL Ballistic Research Laboratory [of US Army at Aberdeen, Maryland]

BRL Basic Research Laboratory

BRL Basic Research Laboratory [of Nippon Telegraph and Telephone Corporation, Japan]

BRL Bromanil [also brl]

BRL Bureau of Research and Laboratories [Philippines]

BRL Butterwick Research Laboratories [UK]

BRLESC Ballistics Research Laboratories Electronic Scientific Computer

BRLI Bjorksten Research Laboratories, Inc. [US]

Br Libr J British Library Journal [of British Library, London, UK]

Br Lig Broad Ligament [Anatomy] [also br lig]

BRLP Burlap [also Brlp, or brlp]

BRLS Barrier Ready Light System

BRL-TR Ballistics Research Laboratory Technical Report [also BRL TR, or BRL Tech Rep]

BRM Barometer [also Brm, or brm]

BRM Binary Relationship Model

BRM Biological Research Module

BRMA Business Records Manufacturers Association [US]

BRMA British Reinforcement Manufacturers Association

BRMA British Rubber Manufacturers Association

BRMA Rev BRMA Review [of British Rubber Manufacturers Association]

BRMCA British Ready Mixed Concrete Association

BRMD Bureau of Radiation and Medical Devices [of Health Canada]

Br Med Bull British Medical Bulletin

Br Med J British Medical Journal [of British Medical Association]

BRMUG Biomedical Research Macintosh Users Group

brn brown

brnsh brownish

BrO Bromine Oxide

Bro Broach [also bro]

Bro Brother [also bro]

Broadcast Technol Broadcast Technology [Canadian Publication]

BROM Bipolar Read-Only Memory

Bromo-PADAP 2-(5-Bromo-2-Pyridylazo)-5-Diethylaminophenol

Brookhaven Bull Brookhaven Bulletin [of Brookhaven National Laboratory, Upton, New York, US]

BROOM Ballistic Recovery of Orbiting Man

BroP Bromo-tris(Dimethylamino)phosphonium Hexafluorophosphate

Bros Brothers [also bros]

Brown Boveri Mitt Brown Boveri Mitteilungen [Brown Boveri Communications; published in Zurich, Switzerland]

Brown Boveri Rev Brown Boveri Review [Published in Zurich, Switzerland]

Brown Boveri Tech Brown Boveri Technik [Brown Boveri Technology Report; published in Zurich, Switzerland]

Browser Browsing On-Line with Selective Retrieval [also browser]

BRP Backfile Registration Project

BRP Back-Reflection Photography

BRP Bathroom Privileges

BRP Beacon Ranging Pulse

BRP Bureau de Recherches de Pétrole [Petroleum Research Bureau, France]

BrP British Patent [also BrPat]

Br Plast Rubb British Plastics and Rubber [UK Publication]

BRPM Breath(ing) Rate per Minute

Br Polym J British Polymer Journal [Publication of the Society of Chemical Industry]

Br Printer British Printer [UK Publication]

5-Br-PSAA (5-Bromo-2-Pyridylazo)-(N-Propyl-3-Sulfopropylamino)aniline

BRPRA British Rubber Producers Research Association

2BrPy 2-Bromo Pyridine

BRR Battelle Research Reactor [US]

BRR Brookhaven Research Reactor [of Brookhaven National Laboratory, Upton, New York, US]

BRR Bureau of Rural Resources

BRRA British Rayon Research Association

BRRA British Refractories Research Association

BRRAMA British Rubber and Resin Adhesive Manufacturers Association

BRRI Bangladesh Rice Research Institute

BRRL British Road Research Laboratory

BRRS Banana River Repeater Station

BRRU Business Regulation Review Unit

BRS Banard's Runaway Star [Astronomy]

BRS Beta-Ray Spectrometer; Beta-Ray Spectrometry

BRS Beta-Ray Spectrum

BRS Bibliographic Retrieval Services [US]

BRS Block Received Signal

BRS B-Mode Receiving Station

BRS Boron Recycle System

BRS Bottom Right Side

BRS Break Request Signal

BRS British Road Service Federation

BRS Building Research Station [UK]

BRS Bureau of Railroad Safety [US]

BRS Bureau of Resource Sciences

BRS Burma Research Society

BRS Business Recovery Services [of IBM Corporation, US]

BrS Brown and Sharpe (Gage)

Brs Brass [also brs]

BRSCC British Racing and Sports Car Club

BRSL Biochemical Research Service Laboratory [of University of Kansas, Lawrence, US]

Br Steel British Steel [UK Publication]

Br Steelmaker British Steelmaker [UK Publication]

Br Std British Standard

BRSV Bovine Respiratory Syncytial Virus

BRT Binary Run Tape

BRT Brotherhood of Railway Trainmen [Now part of United Transportation Union]

Brt Bright(ness) [also brt]

BRTA British Road Tar Association

BR&TC Better Roads and Transportation Council [US]

Br Telecom J British Telecom Journal [Published by Post Office Central Headquarters, London, UK]

Br Telecommun Eng British Telecommunications Engineering [Journal of the Institution of British Telecommunications Engineers]

Br Telecom Technol J British Telecom Technology Journal [of British Telecom Research Laboratories]

Br Telecom World British Telecom World [Publication of the British Telecom Center]

BRU Basic Resolution Unit

BRU Branch Unconditional

BRU Brown University [Providence, Rhode Island, US]

Bruel & Kjaer Tech Rev Bruel and Kjaer Technical Review [Publication of Bruel & Kjaer, Naerum, Denmark]

BRUFMA British Rigid Urethane Foam Manufacturers Association

BRUNET Brown University Network [US]

BRUSYS Bruesseler-System [Programming Language for Numerical Control Applications]

Br$_2$-VGCF Brominated Vapor-Grown Carbon Fiber

BRVMA British Radio Valve Manufacturers Association

BRW Bar, Rod and Wire

Br W Afr British West Africa(n)

Br Weld J British Welding Journal

BRWM Board on Radioactive Waste Management

Brz Bronze [also brz]

Brzg Brazing [also brzg]

BS Bachelor of Science [also Bsc]

BS Backing Store

BS Backscatter(ed); Backscattering

BS Backscattering Spectrometry; Backscattering Spectroscopy

BS Backspace (Character)

BS Backup System

BS Backward Signalling

BS Bahamas [ISO Code]

BS Bakable Space

BS Balance Sheet [also bs]

BS Ball Bearing, Double-Row, Self-Aligning, Raceway of Outer Ring Spherical [Symbol]

BS Band Screen

BS Band Setting

BS Bandeiraea Simplicifolia [Biochemistry]

BS Band Spectrum

BS Band Structure [Solid-State Physics]

BS Bardhan-Sengupta (Synthesis)

BS Barium Silicate [BaO·SiO$_2$, or BaSiO$_3$]

BS Baryon Spectroscopy [Particle Physics]

BS Base Stock [Paper]

BS Basic Sediment [Petroleum Engineering]

BS Batch Still

BS Beam Splitter; Beam Splitting

BS Beam Spread

BS Beam Stop

BS Beaufort Scale [Meteorology]

BS Below Slab

BS Bend(ing) Strength

BS Bend(ing) Stress

BS Benedict's Solution

BS Benzenesulfonate

BS Bessemer Steel
BS Bethe-Salpeter (Equation) [Particle Physics]
BS Bile Salt [Biochemistry]
BS Bill of Sale [also bs]
BS Binary Star
BS Binary Subtract(er)
BS Biochemical Society [UK]
BS Biological Shield
BS Biological Science(s)
BS Biological Society [US]
BS Biometric Society [US]
BS Biophysical Society [US]
BS Biosatellite
BS Bioscience; Bioscientist
BS Biosynthesis
BS Bismuth Silicate
BS Bismuth Sulfite
BS Black Sea
BS Blind Spot
BS Block Specification
BS Blood Serum
BS Blowing Snow
BS Blue Shield (Medical Insurance) [US]
BS Body Stress
BS Bomb Sight
BS Bonded Single Silk (Electric Wire)
BS Bond Strength
BS Booster Stage
BS Born Series [Physics]
BS Borosilicate (Glass)
BS Both Sides
BS Bottom Sediment [Petroleum Engineering]
BS Boundary Surface
BS Bowel Sounds
BS Bragg Spectrometer
BS Brake Skid
BS Break Seal
BS Break(ing) Strength
BS Breaking Stress
BS Breath Sounds
BS Bremsstrahlung
BS Bremsstrahlung Spectroscopy
BS British Standard
BS British Steel plc [Gwent, UK]
BS Broadcasting Satellite
BS Broadcast(ing) Station
BS Brown-Séquard (Syndrome)
BS Buffer Stage
BS Building Steel
BS Bureau of Standards
BS Bursting Strength
BS Steppe Climate [Classification Symbol]
BS$_2$ Barium Disilicate [$BaO \cdot 2SiO_2$]

B$_2$S Dibarium Silicate [$2BaO \cdot SiO_2$]
BS-I Bandeiraea Simplicifolia I [Biochemistry]
BS-II Bandeiraea Simplicifolia II [Biochemistry]
B1S First-Stage Binary Structure [Symbol]
B2S Second-Stage Binary Structure [Symbol]
B&S Beams and Stringers
B&S Bell and Spigot (Pipe Joint)
B&S Brown and Sharpe (Wire) Gauge [also: B and S]
B/S Bill of Sale [also b/s]
B/S Bits per Second [also b/s]
Bs Bacillus subtilis [Microbiology]
Bs Bismuth [Abbreviation; Symbol Bi]
Bs Brosyl [Chemistry]
Bs Dry Climate, Semiarid [Symbol]
B(s) Feedback Signal [Automatic Control Symbol]
.bs Bahamas [Country Code/Domain Name]
β-S Beta Sulfur [Symbol]
BSA Bachelor of Science in Agriculture
BSA Bandeiraea Simplicifolia Agglutinin [Biochemistry]
BSA Bank Stationers Association [now Financial Stationers Association, US]
BSA Bearing Specialists Association [US]
BSA Benzene Sulfonic Acid
BSA Bibliographical Society of America [US]
BSA Bimetal Steel-Aluminum
BSA Biological Signal Analysis
BSA Biophysical Society of America
BSA N,O-Bis(Trimethylsilyl)acetamide
BSA N-Bis(Trimethylsilyl)acetamine
BSA Blood Serum Albumin
BSA Board of Scientific Advisors [US]
BSA Body Surface Area
BSA Borie, Sass and Andreassen (Model) [Metallurgy]
BSA Boston Society of Architects [Massachusetts, US]
BSA Botanical Society of America [US]
BSA Bovine Serum Albumin [Biochemistry]
BSA British Sign Association
BSA British Society of Audiology
BSA British South Africa
BSA British Speleological Association
BSA Broadside (Antenna) Array
BSA Building Societies Association
BSA Business Software Alliance
BSA Bund Schweizer Architekten [Federation of Swiss Architects]
BSA Business Software Association [US]
BSA N-[(4-Ethenylphenyl)methyl]-N-Methyl Methanesulfonamide
BSA I Bandeiraea Simplicifolia Agglutinin I [Biochemistry]
BSA II Bandeiraea Simplicifolia Agglutinin II [Biochemistry]
BSAA Bachelor of Science in Applied Arts
BSAA British South American Airways

BSAC Berkeley Sensor and Actuator Center [of University of California at Berkeley, US]

BSACI British Society for Allergy and Clinical Immunology

BSAD Bachelor of Science in Architectural Design

BSAdv Bachelor of Science in Advertising

BSAE Bachelor of Science in Aeronautical Engineering

BSAE Bachelor of Science in Agricultural Engineering

BSAE Bachelor of Science in Architectural Engineering

BSAeE Bachelor of Science in Aeronautical Engineering

BSAEM British Society of Allergy and Environmental Medicine

BSAF British Sulfate and Ammonia Federation

BSAg Bachelor of Science in Agriculture

BSAgE Bachelor of Science in Agricultural Engineering [also BSAgrEng]

BSAgr Bachelor of Science in Agriculture

BsaH Bacillus stearothermophilus [Microbiology]

BSAL Block Structured Assembly Language

BSAM Basic Sequential Access Method

BSAP British Society of Animal Production

BSA/PBS Bovine Serum Albumin in Phosphate Buffered Saline [Biochemistry]

BSAR Babcock and Wilcox Safety Analysis Report

BSArch Bachelor of Science in Architecture

BSArchE Bachelor of Science in Architectural Engineering [also BSArchEng]

BSArchTech Bachelor of Science in Architectural Technology

BSAS Bachelor of Science in Architectural Studies

BSATA Ballast, Sand and Allied Trades Association

BSA-TBS Bovine Serum Albumin–Tris-Buffered Saline [Biochemistry]

BSAVA British Small Animal Veterinary Association

BSB Bayrische Staatsbibliothek [Bavarian State Library, Munich]

BSB British Standard Beam

BSBA Bachelor of Science in Business Administration

BSBC Bachelor of Science in Building Construction

BSBC British Social Biology Council

BSBI Botanical Society of the British Isles [UK]

BSBldgSci Bachelor of Science in Building Science [also BSBldgSc]

BSBL Bond Strength–Bond Length

BSBMA British Structural Bearings Manufacturers Association

BSBP British Standard Bulb Plate

BSBSPA British Sugar Beet Seed Producers Association

BSBus Bachelor of Science in Business

BSBusMgt Bachelor of Science in Business Management

BSC Bachelor of Science in Commerce

BSC Base Site Control

BSC Basic Message Switching Center

BSC Benzenesulfonyl Chloride

BSC Bermidji State College [Minnesota, US]

BSC Bethlehem Steel Corporation [Bethlehem, Pennsylvania, US]

BSC Bibliographical Society of Canada

BSC Bile Salts–Cascara (Compound) [Pharmacology]

BSC Binary Symmetric Channel

BSC Binary Synchronous Communication

BSC Biological Stain Commission [US]

BSC Binary Synchronous Communications (Protocol) [also BISYNC, or bisync]

BSC N,O-Bis(trimethylsilyl)carbamate

BSC Bisynchronous Communication

BSC Bloomsburg State College [Pennsylvania, US]

BSC Bluefield State College [West Virginia, US]

BSC Board of Scientific Counselors [US]

BSC Borosilicate Crown (Glass)

BSC Boston State College [Massachusetts, US]

BSC Bowie State College [Maryland, US]

BSC Branch, or Skip on Condition

BSC Brake Skid Control(ler)

BSC British Safety Council

BSC British Seeds Council

BSC British Shippers Council

BSC British Society of Cinematographers

BSC British Society of Commerce

BSC British Standard Channel

BSC British Steel Corporation

BSC Building Systems Council [of National Association of Home Builders, US]

BSC Business Service Center

.BSC Boyan Script [File Name Extension] [also .bsc]

B/SC Brake Skid Control

BSc Bachelor of Science [also BS]

β-Sc Beta Scandium [Symbol]

BSCA Binary Synchronous Communications Adapter

BSCA British Sulfate of Copper Association

BSCA Building Service Contractors Association (International) [US]

BScA Bachelor of Science in Agriculture

BSCAC Beaufort Sea Community Advisory Committee [Canada]

BScAgr Bachelor of Science in Agriculture

BScAgrE Bachelor of Science in Agricultural Engineering [also BScAgrEng]

BSCAI Building Service Contractors Association International

BScArch Bachelor of Science in Architecture

BSCB British Society of Cell Biology

BSCBA Brown Swiss Cattle Breeders Association [US]

BSC/BISPA British Steel Corporation/British Independent Steel Producers Association (Project)

BScBldgDes Bachelor of Science in Building Design

BSCC British Society of Clinical Cytology

BSCC British Soviet Chamber of Commerce [defunct]

BSCC British Swedish Chamber of Commerce

BSCC British Swiss Chamber of Commerce

BSCC British Synchronous Clock Conference

BScC Bachelor of Commercial Studies

BSCCAO Bismuth Strontium Calcium Copper Aluminum Oxide (Superconductor) [also BiSCCAO, or BISCCCAO]

BScCE Bachelor of Science in Civil Engineering

BScCE Bachelor of Science in Construction Engineering

BSCCGO Bismuth Strontium Calcium Copper Gallium Oxide (Superconductor)

BSCCO Barium Strontium Calcium Copper Oxide (Superconductor)

BSCCO Bismuth Strontium Calcium Copper Oxide (Superconductor) [also BiSCCO, or BISCCO]

BSCCO-2201 $Bi_2Sr_2Ca_0CuO_x$ [A Bismuth Strontium Copper Oxide Superconductor]

BSCCO-2212 $Bi_2Sr_2CaCu_3O_x$ [A Bismuth Strontium Calcium Copper Oxide Superconductor]

BSCCO-2223 $Bi_2Sr_2Ca_2Cu_3O_x$ [A Bismuth Strontium Calcium Copper Oxide Superconductor]

BSCCO-2234 $Bi_2Sr_2Ca_3Cu_4O_x$ [A Bismuth Strontium Calcium Copper Oxide Superconductor]

BSCCO-2245 $Bi_2Sr_2(Ca,Dy)_4Cu_5O_x$ [A Bismuth Strontium Calcium Dysprosium Copper Oxide Superconductor]

BSCCO-2278 $Bi_2Sr_2Ca_7Cu_8O_x$ [A Bismuth Strontium Calcium Copper Oxide Superconductor]

BSCCO-4334 $Bi_4Sr_3Ca_3Cu_4O_x$ [A Bismuth Strontium Calcium Copper Oxide Superconductor]

BSCCO-4336 $Bi_4Sr_3Ca_3Cu_6O_x$ [A Bismuth Strontium Calcium Copper Oxide Superconductor]

BSCCO-11905 $Bi_{11}(Sr, Ca)_9Cu_5O_x$ [A Bismuth Strontium Calcium Copper Oxide Superconductor]

BScCom Bachelor of Commercial Science

BScCS Bachelor of Science in Computer Science

BScD Bachelor of Science in Dentistry [also BScDent]

BScDA Bachelor of Science in Data Analysis

BScD(AN) Bachelor of Science in Dentistry (Anesthesia) [also BScDent(AN), BScD(Anes)]

BScDent Bachelor of Science in Dentistry [also BScD]

BScDent(AN) Bachelor of Science in Dentistry (Anesthesia) [also BScD(AN), or BScD(Anes)]

BSCE Bachelor of Science in Civil Engineering

BSCE Bird Strike Committee Europe [Denmark]

BSCE Boston Society of Civil Engineers [Massachusetts, US]

BScE Bachelor of Science in Engineering [also BScEng, or BSc(Eng)]

BScEcon Bachelor of Science in Economics

BscEE Bachelor of Science in Electrical Engineering [also BSc(EE)]

BScEng Bachelor of Science in Engineering [also BScE, or BSc(Eng)]

BSCer Bachelor of Science in Ceramics

BSCerE Bachelor of Science in Ceramic Engineering [also BSCerEng]

BSCerSci Bachelor of Science in Ceramic Science

BSCh Bachelor of Science in Chemistry

BSChE Bachelor of Science in Chemical Engineering

BSCF Barium Tin Chlorine Fluoride

BScF Bachelor of Science in Forestry

BScFE Bachelor of Science in Forestry Engineering

BSCFO Barium Strontium Calcium Iron Oxide (Superconductor)

BSCFL Binary Synchronous Frame Level

BSCh Bachelor of Science in Chemistry

BSChemEng Bachelor of Science in Chemical Engineering [also BSChemE]

BSc(Hons) Bachelor of Science (Honours)

BSc in Agr Bachelor of Science in Agriculture

BSc in CE Bachelor of Science in Civil Engineering

BSc in Dent Bachelor of Science in Dentistry

BSc in ME Bachelor of Science in Mechanical Engineering

BSc in Med Bachelor of Science in Medicine

BSc in Nurs Bachelor of Science in Nursing

BSc in Opt Bachelor of Science in Optometry

BSc in Pharm Bachelor of Science in Pharmacy [also BS in Phar]

BSc in Rest Mgt Bachelor of Science in Restaurant Management

BSCL Bell System Common Language

BSCM Bisynchronous Communications Macro

BScME Bachelor of Science in Mechanical Engineering [also BScMechE]

BScMed Bachelor of Science in Medicine

BScMet Bachelor of Science in Metallurgy

BScMetE Bachelor of Science in Metallurgical Engineering [also BScMetEng]

BScMnlE Bachelor of Science in Mineral Engineering [also BScMnlEng]

BScMnlProcEng Bachelor of Science in Mineral Processing and Engineering

BSCN Bit Scan

BScN Bachelor of Science in Nursing [also BScNurs]

BSC-NAHB Building Systems Council of the National Association of Home Builders [US]

BScNurs Bachelor of Science in Nursing [also BscN]

BSCO Barium Strontium Copper Oxide (Superconductor)

BSCO Bismuth Strontium Copper Oxide (Superconductor)

BSCO-221 $Bi_2Sr_2CuO_6$ [A Bismuth Strontium Copper Oxide (Superconductor)

BSCom Bachelor of Science in Commerce [alsoBSComm]

BSConEng Bachelor of Science in Construction Engineering [also BSConE]

BSConTech Bachelor of Science in Construction Technology

BScOpt Bachelor of Science in Optometry

BSCP Biological Sciences Communication Project [US]

BScPharm Bachelor of Science in Pharmacy [also BScPhar and BScPhm]

BSCRA British Steel Castings Research Association

BSCRP Bachelor of Science in City and Regional Planning

BSCS Bachelor of Science in Computer Science

BSCS Biological Sciences Curriculum Study [of American Institute of Biological Sciences]

BSC/SS Binary Synchronous Communications /Start Stop

BScTech Bachelor of Science and Technology

BScTech Bachelor of Technical Science

BSD Bachelor of Science in Design

BSD Backscattering Detector

BSD Bahamian Dollar [Currency]

BSD Ballistic Systems Division [US Army]

BSD Barrels per Stream Day [also bsd]

BSD Base Supply Depot

BSD Basic Science Division

BSD Basic Science Division [of National Renewable Energy Laboratory, Golden, Colorado, US]

BSD Beam Scan Driver

BSD Berkeley Software Distribution [of University of California at Berkeley, US]

BSD Berkeley Standard Distribution [of University of California at Berkeley, US]

BSD Bibliographic Services Division [of British Library]

BSD Bulk Storage Device

BSDA British Sheep Dairying Association

BSDA British Soft Drink Association

BSDB British Society of Developmental Biology

BSDC Binary Symmetric Dependent Channel

BSDC British Standard Data Code

BSDCO Bismuth Strontium Dysprosium Copper Oxide (Superconductor)

BSDes Bachelor of Science in Design

BSDHyg Bachelor of Science in Dental Hygiene

BSDL Boresight Datum Line

BSDL Boundary Scan Description Language

BSDP Bibliographic Service Development Program [of Council of Library Resources, US]

BSDP Booster Stage Discharge Pressure

BSDR British Society of Dental Research

BSDS Bulk Slurry Distribution System

BSE Bachelor of Science in Education

BSE Bachelor of Science in Engineering

BSE Bachelor of Science in Sanitary Engineering

BSE Backscattered Electrons

BSE Bonn European School of Economics e.V. [Germany]

BSE Booster Systems Engineer

BSE Bovine Spongiform Encephalopathy

BSE Bridge and Structural Engineer

BSE British Shipbuilding Exports

BSE Broadcast Satellite Experiment [Japan]

BSE Building Services Engineer

β-Se Beta Selenium [Symbol]

BSEA British Standard Equal Angle

BSEA British Steel Export Association

BSEc Bachelor of Science in Economics

B/sec Bits per Second [Unit]

BSED Bachelor of Science in Environmental Design

BSEd Bachelor of Science in Education

BSEDI Bis-Salicylaldehyde Ethylenediimine

BSEE Bachelor of Science in Electrical Engineering [also BSEEngr]

BSEET Bachelor of Science in Electrical Engineering Technology

b-SEI Biased Secondary Electron Imaging [also B-SEI]

BSELCH Buffered Selector Channel

BSElE Bachelor of Science in Electronic Engineering

BSEM Bachelor of Science in Engineering of Mines

BSEng Bachelor of Science in Engineering

BSEP Bachelor of Science in Engineering Physics

BSERC British Science and Engineering Research Council

BSES Bachelor of Science in Engineering Sciences

BSET Bachelor of Science in Engineering Technology

BSF Bachelor of Science in Forestry

BSF Back Surface Field

BSF Band-Stop Filter

BSF B-Cell Stimulatory Factor [Biochemistry]

BSF Binational Science Foundation [Israel-US]

BSF Bipolar Shape Formation

BSF N,N-Bis(trimethylsilyl)formamide

BSF Bit Scan Forward

BSF British Scrap Federation

BSF British Shipping Federation

BSF British Slag Federation

BSF British Society of Flavourists

BSF British Standard Fine (Screw Thread)

BSF British Stone Federation

BSF Bulk Shielding Facility [of Oak Ridge National Laboratory, Tennessee, US]

BSFA British Science Fiction Association

BSFA British Steel Founders Association

BSFM Bachelor of Science in Forest Management

BSFMA British Spectacle Frame Manufacturers Association

BSFMgt Bachelor of Science in Fisheries Management

BSFO Barium Scandium Iron Oxide

BSFor Bachelor of Science in Forestry

BSFS Bachelor of Science in Foreign Service

BSFT Bachelor of Science in Fuel Technology

BSFuelSc Bachelor of Science in Fuel Science

BSG Borosilicate Glass

BSG British Standard Gauge

B&SG Brown and Sharpe Gauge

BSGE Bachelor of Science in General Engineering

BSGenEd Bachelor of Science in General Education

BSGeog Bachelor of Science in Geography

BSGeolE Bachelor of Science in Geological Engineering

BSGL Branch System General License

BSGMgt Bachelor of Science in Game Management

BSGph Bachelor of Science in Geophysics

BSGT British Society of Glass Technology

BSH Benzenesulfonylhydrazide

BSH Bleached Sulfate Hardwood (Pulp)

BSH Bundesamt für Seeschiffahrt und Hydrographie [Federal Office for Ocean Shipping and Hydrography]

BSh Tropical and Subtropical Steppe Climate [Classification Symbol]

bsh bushel [Unit]

BSHA Bachelor of Science in Hospital Administration

BSHE Bachelor of Science in Home Economics [also BSHEc]

BSHEd Bachelor of Science in Health Education

BSHM British Society for the History of Mathematics

BSHP British Society for the History of Pharmacy

BSHS British Society for the History of Science

BSH/S Bleached Sulfate Hardwood/Softwood (Pulp Blends)

BSHyg Bachelor of Science in Hygiene

B/s/Hz Bits per second per Hertz

BSI Bandeiraea Simplicifolia Isolectin [Biochemistry]

BSI Basic Shipping Instructions

BSI Biochemical Society of Israel

BSI Bit Sequence Independence

BSI Boeing Services International, Inc.

BSI Botanical Society of Israel

BSI British Society for Immunology

BSI British Standards Institution

BSI Building Stone Institute [US]

BSI Building Systems Institute [US]

BSI Bundesamt für Sicherheit in der Informationstechnik [Federal Office for Security in Information Technology]

BSIA British Security Industry Association

BSIA British Starch Industry Association

BsiA Benzenesulfinic Acid

BSI-A$_4$ Bandeiraea Simplicifolia Isolectin A$_4$ [Biochemistry]

BSI-AB$_3$ Bandeiraea Simplicifolia Isolectin AB$_3$ [Biochemistry]

BSI-A$_2$B$_2$ Bandeiraea Simplicifolia Isolectin A$_2$B$_2$ [Biochemistry]

BSI-A$_3$B Bandeiraea Simplicifolia Isolectin A$_3$B [Biochemistry]

β-SIALON Beta Silicon Aluminum Oxynitride(Ceramic) [also β-SiAlON, or β-Sialon]

BSI-B$_4$ Bandeiraea Simplicifolia Isolectin B$_4$ [Biochemistry]

BSIC Binary Symmetric Independent Channel

β-SiC Beta Silicon Carbide

β-SiC$_w$ Beta Silicon Carbide Whiskers [also β-SiC(w)]

β-SiC$_w$/Al Beta Silicon Carbide Whisker ReinforcedAluminum (Composite) [also β-SiC(w)/Al]

BSIE Bachelor of Science in Industrial Education

BSIE Bachelor of Science in Industrial Engineering

BSIE Banking Systems Information Exchange [US]

β-Si$_3$N$_4$ Beta Silicon Nitride

BS in Acc Bachelor of Science in Accountancy

BS in AE Bachelor of Science in Aeronautical Engineering

BS in Ae Bachelor of Science in Aeronautics

BS in Agr Bachelor of Science in Agriculture

BS in Agr Ed Bachelor of Science in Agricultural Education

BS in Agric Bachelor of Science in Agriculture

BS in App Arts Bachelor of Science in Applied Arts

BS in Arch Bachelor of Science in Architecture

BS in Arch Engr Bachelor of Science in Architectural Engineering

BS in BA Bachelor of Science in Business Administration

BS in BMS Bachelor of Science in Basic Medical Sciences

BS in Bus Ed Bachelor of Science in Business Education

BS in C Bachelor of Science in Commerce

BS in C and BA Bachelor of Science in Commercial and Business Administration

BS in C and E Bachelor of Science in Commerce and Economics

BS in Cart Bachelor of Science in Cartography

BS in CE Bachelor of Science in Civil Engineering

BS in Cer Engr Bachelor of Science in Ceramic Engineering

BS in Cer Tech Bachelor of Science in Ceramic Technology

BS in ChE Bachelor of Science in Chemical Engineering

BS in Chem Tech Bachelor of Science in Chemical Technology

BS in Com Bachelor of Science in Commerce

BS in Com Ed Bachelor of Science in Commercial Eduation

BS in DH Bachelor of Science in Dental Hygiene

BS in Dent Bachelor of Science in Dentistry

BS in Diet Bachelor of Science in Dietetics

BS in E Bachelor of Science in Engineering

BS in EE Bachelor of Science in Electrical Engineering

BS in E Law Bachelor of Science in Engineering Law

BS in EM Bachelor of Science in Engineering of Mines

BS in EMath Bachelor of Science in Engineering Mathematics

BS in EP Bachelor of Science in Engineering Physics [also BS in Engr Phys]

BSI News BSI News [of British Standards Institution]

BS in Fin Bachelor of Science in Finance

BS in GS Bachelor of Science in General Studies

BS in Gen Bus Bachelor of Science in General Business

BS in Glass Tech Bachelor of Science in Glass Technology

BS in GphE Bachelor of Science in in Geophysical Engineering

BS in H and PE Bachelor of Science in Health and Physical Education

BS in H and RA Bachelor of Science in Hotel and Restaurant Administration

BS in Home Ec Bachelor of Science in Home Economics

BS in Home Ec Ed Bachelor of Science in Home Economics Education

BS in IA Bachelor of Science in Industrial Arts

BS IndEd Bachelor of Science in Industrial Education

BS Ind Eng Bachelor of Science in Industrial Engineering [also BS Ind Engr]

BS Ind Mgt Bachelor of Science in Industrial Management

BS in Ind Chem Bachelor of Science in Industrial Chemistry

BS in J Bachelor of Science in Journalism

BS in LS Bachelor of Science in Library Science

BS in Lab Tech Bachelor of Science in Laboratory Technology

BS in MA Bachelor of Science in Mechanical Arts

BS in ME Bachelor of Science in Mechanical Engineering

BS in Mech Arts Bachelor of Science in Mechanical Arts

BS in Med Bachelor of Science in Medicine

BS in Med Rec Bachelor of Science in Medical Records

BS in Med Rec Lib Bachelor of Science in Medical Records Librarianship

BS in Med Sc Bachelor of Science in Medical Science

BS in Med Tech Bachelor of Science in Medical Technology

BS in Mgt Engr Bachelor of Science in Management Engineering

BS in Mktg Bachelor of Science in Marketing

BS in MS Bachelor of Science in Military Science

BS in MT Bachelor of Science in Medical Technology

BS in N Bachelor of Science in Nursing

BS in N Ed Bachelor of Science in Nursing Education

BS in NS Bachelor of Science in Natural Science

BS in Nurs Bachelor of Science in Nursing

BS in Occ Ther Bachelor of Science in Occupational Therapy

BS in Opt Bachelor of Science in Optics

BS in Opt Bachelor of Science in Optometry

BS in PA Bachelor of Science in Public Administration

BS in PE Bachelor of Science in Petroleum Engineering

BS in PetE Bachelor of Science in Petroleum Engineering

BS in Ph Bachelor of Science in Pharmacy [also BS in Pharm]

BS in PHN Bachelor of Science in Public Health Nursing

BS in PrGe Bachelor of Science in Professional Geology

BS in PrMet Bachelor of Science in Professional Meteorology

BS in RT Bachelor of Science in Radiological Technology

BS in SE Bachelor of Science in Sanitary Engineering

BS in SS Bachelor of Science in Social Science [also BS in SSc]

BS in SecSci Bachelor of Science in Secreterial Science

BS in TE Bachelor of Science in Textile Engineering [also BS in Textile Eng]

BS in VocAg Bachelor of Science in Vocational Agriculture

BS in VocEd Bachelor of Science in Vocational Education

β-Si$_3$N$_4$(w) Beta Silicon Nitride Whiskers [also β-Si$_3$N$_{4(w)}$]

β-Si$_3$N$_4$(w)/Al Beta Silicon Nitride Whisker Reinforced Aluminum (Composite [also β-Si$_3$N$_{4(w)}$/Al]

BSIntArch Bachelor of Science in Interior Architecture

β-SiO$_2$ Beta Silica (or Beta Silicon Dioxide or Beta Quartz)

BSIR Bachelor of Science in Industrial Relations

BSIR Bibliography of Scientific and Industrial Reports [of US Department of Commerce]

BSIR Board for Scientific and Industrial Research [Israel]

BSIRA British Scientific Instrument Research Association

BSIS Bleached Sulfite Softwood (Pulp)

BSI Sales Bull BSI Sales Bulletin [of British Standards Institution]

BSIT Bachelor of Science in Industrial Technology

BSIT Bipolar-Mode Static Induction Transistor

BSJ Bachelor of Science in Journalism

BSk Middle Latitude Steppe Climate [Classification Symbol]

Bskt Basket [also bskt]

BSL Bachelor of Science in Law

BSL Baseline [also Bsl]

BSL Bit Serial Link

BSL Botanical Society of London [UK]

BSLA Bachelor of Science in Landscape Architecture [also BSLArch]

BS Lab Rel Bachelor of Science in Labor Relations

BSLM Bachelor of Science in Landscape Management

BSLS Bachelor of Science in Library Science

BSM Bachelor of Science in Medicine

BSM Basic Storage Module

BSM Bit-Slice Microprocessor

BSM Booster Separation Motor

BSM Brushless Servo Motor

BSM Bureau of Standards and Measures [now National Institute for Standards and Materials, US]

Bsm Bacillus stearothermophilus NUB36 [Microbiology]

β-Sm Beta Samarium [Symbol]

BSMA British Secondary Metals Association

BSMA Building Societies Members Association[UK]

BSMatEng Bachelor of Science in Materials Engineering [also BSMatE]

BSMatScEng Bachelor of Science in Materials Science and Engineering [also BSMatSciEng orBSMatSE]

BSME Bachelor of Science in Mechanical Engineering

BSME Bachelor of Science in Mining Engineering

BSMechEng Bachelor of Science in Mechanical Engineering [also BSMechE]

BSMed Bachelor of Science in Medicine

BSMedTech Bachelor of Science in Medical Technology

BSMet Bachelor of Science in Metallurgy

BSMetE Bachelor of Science in Metallurgical Engineering [also BSMetEng]

BSMGP British Society of Master Glass Painters

BSMin Bachelor of Science in Mineralogy

BSMnlEng Bachelor of Science in Mineral Engineering

BSMnlProcEng Bachelor of Science in Mineral Processing and Engineering

BSMP Bit-Slice Microprocessor

B/SMPL Bits per Sample [also B/smpl]

BSMPT British Standard Metric Pipe Thread

BSMS Backscatter Mössbauer Spectrometer

BSMSP Bernoulli Society for Mathematical Statistics and Probability [Netherlands]

BSMT Bachelor of Science in Medical Technology

BSMT Barium Strontium Magnesium Tantalate

Bsmt Basement [also bsmt]

BSMV Barley Stripe Mosaic Virus

BSN Bachelor of Science in Nursing

BSN Back-End Storage Network

BSN Backward Sequence Number

BSN Barium Sodium Niobate

BSN Bismuth Strontium Niobate

BSN Bisegmental Auditory Neuron

β-Sn Beta Tin (or White Tin) [Symbol]

BSNA Bachelor of Science in Nursing Administration

BSNDT British Society for Nondestructive Testing

BSNH Boston Society of Natural History [Massachusetts, US]

BSNO Barium Sodium Niobium Oxide

BSNO Bismuth Strontium Niobium Oxide

BSNS Buffalo Society of Natural Sciences [US]

Bsns Business [also bsns]

BSNSC Bifacial Silicon Nitride Solar Cell

BSO Backside Selective Oxidation

BSO Beryllium Silicon Oxide (or Phenacite)

BSO Blue Stellar Object

BSO Bonded Silicon Oxide

BSO Broadside On Impact of Projectile

BSO Methyl-p-Vinylbenzyl Sulfoxide

BSO2 Methyl-p-Vinylbenzylsulfone

BSOCOES Bis[2-(Succinimido-Oxycarbonyloxy)ethyl]sulfone

BSOIW (International Association of) Bridge,Structural and Ornamental Iron Workers

B-Sol Quantitative Analysis of Bismuth Strontium Calcium Copper Oxide by Solid-Phase Reaction Method

BSOpt Bachelor of Science in Optometry

BSOrnHort Bachelor of Science in Ornamental Horticulture

BSOT Bachelor of Science in Occupational Therapy

BSOT Back-Twinned Second Order Twin (System) [Metallurgy]

BSoUP British Society of Underwater Photographers

BSP Bachelor of Science in Pharmacy

BSP Backspace

BSP Bell System Practice

BSP Biodiversity Support Program [US]

BSP Bioseparation Process

BSP Boltzmann Superposition Principle [Physics]

BSP Botanical Society of Pennsylvania [US]

BSP Bond-Selective Photochemistry

BSP Boron Superphosphate

BSP Boundary Scattering of Phonons [Solid-State Physics]

BSP British Society of Parasitology

BSP British Standard Pipe Thread

BSP British Steel plc [Gwent, UK]

BSP Broken Symmetry Phase [Materials Science]

BSP Bromosulphalein

BSP Building Systems Program [of US Department of Energy]

BSP Burroughs Scientific Processor

BSP Bulk Synchronous Parallelism

BSPA Bachelor of Science in Public Administation

BSPB British Society of Plant Breeders

BSPCO Bismuth Strontium Lead Copper Oxide (Superconductor) [also BiSPCO, or BISPCO]

BSPH Bachelor of Science in Public Health

BSPharm Bachelor of Science in Pharmacy [also BSPhar]

BSPhDI Bis-Salicyclaldehyde Phenylenediimine

BSPHN Bachelor of Science in Public Health Nursing

BSPlmSc Bachelor of Science in Polymer Science

BSPMC Brake System Parts Manufacturers Council [US]

BSPP British Society of Plant Pathology

BSPS British Society for the Philosophy of Science

BSPT Bachelor of Science in Physical Therapy

BSPT 2-(2'-Benzothiazolyl)-5-Styryl-3-(4'Phthalhydrazidyl) tetrazolium (Chloride)

BSPT Bound-State Perturbation Theory [Physics]

BSQ Bias-Sputtered Quartz

BSR Basic System Release

BSR Bend Stress Relaxation (Test)

BSR Bit Scan Reverse

BSR Bit Status Register

BSR Blip-Scan Ratio [Electronics]

BSR Board of Standards Review [of American National Standards Institute, US]

BSR Body-Section Radiography

BSR British Society of Rheology

BSR Buffered Send/Receive

BSR Bulk Shielding Reactor [of Oak Ridge National Laboratory, Tennessee, US]

b/sr barn(s) per steradian [also b sr^{-1}]

β-Sr Beta Strontium [Symbol]

BSRA British Ship(building) Research Association

BSRA British Sound Recording Association

BSRA British Sugar Refiners Association

BSRadio-TV Bachelor of Science in Radio and Television

BSRAE British Society for Research in Agricultural Engineering

BS-RAM Burst Static Random-Access Memory

BSRC Battelle Seattle Research Center [US]

BSRF Borderland Sciences Research Foundation [US]

BSRFS Bell System Reference Frequency Standard

b/sr/fu barn(s) per steradian per formula unit [also b sr^{-1} fu$_{-1}$]

BSRIA Building Services Research and Information Association [UK]

BSRIP Basic Scientific Research Incentive Program

BSRL Broadcasting Science Research Laboratories [of Nippon Hoso Kyokai, Tokyo, Japan]

BSRM Booster Solid Rocket Motor [of NASA]

BSRN Baseline Surface Radiation Network

BSRP Biological Sciences Research Paper

BSRS Bell System Repair Specification

BSRT Bachelor of Science in Radiological Technology

BSS Bacharach Smoke Scale

BSS Bachelor of Secretarial Science

BSS Bachelor of Social Science

BSS Backup System Services

BSS Balanced Saline Solution

BSS Balanced Salt Solution

BSS Band of Secondary Slip [Crystallography]

BSS Basic Synchronized Subset

BSS Binary Solid Solution [Materials Science]

BSS Bis(trimethylsilyl)sulfate

BSS Bleached Sulfate Softwood (Pulp)

BSS Bloch-Siegert Shift [Physics]
BSS Block Started by Symbol
BSS Botanical Society of Scotland
BSS Brillouin Scattering Spectroscopy
BSS British Satellite Service
BSS British Standard Screen
BSS British Standard Section
BSS British Standard Specification
BSS British Standards Society
BSS Broadcasting Satellite Service
BSS Bulk Storage System
BSS Bundles of Slip Steps [Crystallography]
BSS Business Services Section [of Fermilab, Batavia, Illinois, US]
BSSA Bachelor of Science in Secretarial Administration
BSSA British Stainless Steel Association
BSSC Bit Synchronizer/Signal Control
β-SSC Beta(-Phase) Sintered Silicon Carbide
BSSE Bachelor of Science in Secondary Education
BSSE Basis Set Superposition Error
BS-SEM Backscattered Scanning Electron Micrograph; Backscattered Scanning Electron Microscopy [also SEM-BS]
BSSG Biological Sciences Study Group
BSSG British Society of Scientific Glassblowers
β-SSiC Beta(-Phase) Sintered Silicon Carbide
BSSM British Society for Strain Measurement
BSSMA Business Systems and Security Marketing Association [US]
BSSR Byelorussian Soviet Socialist Republic [USSR]
BSSRS British Society for Social Responsibility inScience
BSSRS Bureau of Safety and Supply Radio Services [US]
BSSS Bachelor of Science in Secretarial Studies
BSSS Bachelor of Science in Social Science
BSSS British Society of Soil Science
BSSSA British Surgical Support Suppliers Association
BST Baeyer's Strain Theory
BST Barium Scandium Tantalate
BST Barium Stannate Titanate
BST Barium Strontium Titanate
BST Beam-Switching Tube
BST Bering Standard Time
BST Beta Solution Treatment
BST Binary Synchronous Transmission
BST Blast Saturation Temperature
BST Block the SPADE Terminal Command
BST Blood Serological Test
BST Bosonic String Theory [Physics]
BST Bovine Somatotropin
BST British Summer Time
BST British Standard Tee
BST British Standard Thread
BST British Standard Time [Greenwich Mean Time]
BST British Summer Time

BST Bulk Supply Tariff [UK]
BST Business Systems Technology
B/St Bill of Sight
bst booster amplifier
BSTA British Surgical Trades Association
β-STA Beta Solution (Heat) Treated and Aged (Titanium Alloy)
BSTAN Beta Solution Treated and Annealed; Beta Solution Treatment and Anneal(ing)
B/START Behavioral Science Track Award for Rapid Transition
BstB Bacillus stearothermophilus B [Microbiology]
Bstd Bastard (File)
BstE Bacillus stearothermophilus ET [Microbiology]
BSTex Bachelor of Science in Textiles
BSTFA Bistrifluoroacetamide
BSTFA N,O-bis(Trimethylsilyl)trifluoroacetamide
BST&IE Bachelor of Science in Trade and Industrial Education
BSTJ Bell Systems Technical Journal
BstN Bacillus stearothermophilus N [Microbiology]
BSTO Barium Strontium Titanium Oxide (or Barium Strontium Titanate)
BSTOA Beta-Solution-Treated and Overaged; Beta Solution Treatment and Anneal(ing)
BSTP British Society of Toxicological Pathologists
BSTR Booster [also Bstr, or bstr]
BSTrans Bachelor of Science in Transportation
β-Strep Beta Streptococcus Group (or Hemolytic Streptococci Group)
BSTSA British Surface Treatment Suppliers Association
BstU Bacillus stearothermophilus U [Microbiology]
BStW Berliner Stahlwerke AG [Steelmaker located in Berlin, Germany]
BstX Bacillus stearothermophilus X [Microbiology]
BSU Ball State University [Muncie, Indiana, US]
BSU Barium Scandium Uranate
BSU Baseband Switching Unit
BSU Basic Sounding Unit
BSU Basic Structural Unit
BSU Beam Steering Unit [Electronics]
BSU N,N'-Bis(trimethylsilyl)urea
BSU Boise State University [Idaho, US]
BSU Bowie State University [Texas, US]
BSU Business Service Unit
BSU Byelorussian State University [Minsk]
Bsu Bacillus subtilis (Bacteria) [Microbiology]
Bsu36 I Bacillus subtilis 36 I [Microbiology]
BSUA British Standard Unequal Angle
B subtilis Bacillus subtilis (Bacteria) [Microbiology]
BSUG Bedford Systems Users Group [US]
BSUP Bachelor of Science in Urban Planning
BSV Bushy Stunt Virus
.BSV BASIC Save Graphics [File Name Extension]
BSW Botanical Society of Washington [US]

BSW Bottom Sediment and Water; Bottom Settlings and Water [Petroleum Engineering]

BSW British Standard Whitworth (Screw) Thread

BS&W Basic Sediment and Water

BS&W Bottom Sediment and Water [Petroleum Engineering]

BSWG British Standard Wire Gauge

BSWL Basic White Lead

BSWM Bureau of Solid Waste Management [US]

BSXF Burst Sync Failure [SPADE]

BSY Busy [also Bsy, or bsy]

BSYNC Binary Synchronous Communications (Protocol) [also bsynch]

BSYPH Baikal School for Young Researchers on the Astrophysics and Microworld Physics [Irkutsk City, near Lake Baikal, Russia]

BT Bacillus Thuringiensis [also Bt] [Microbiology]

BT Backtracker

BT Bainitic Transformation [Metallurgy]

BT Ball Bearing, Single-Row, Angular-Contact, Separable Inner Ring, Nominal Contact Angle from Above 22° to 32° [Symbol]

BT Band Theory (of Solids) [Solid-State Physics]

BT Barium Titanate

BT Barred Trunk

BT Basal Temperature [Medicine]

BT Bathythermograph(y)

BT Beacon Transmitter

BT Beam Tetrode

BT Beginning of Tape

BT Bellini-Tosi (Direction Finding System)

BT Bend Test

BT 2-Benzothiazolone

BT Benzotriazole [also Bt]

BT Bernoulli Theorem

BT Bhatia-Thornton (Partial Structure Factor) [Liquid Alloys]

BT Bhutan [ISO Code]

BT Bias Temperature

BT Biotechnological; Biotechnologist; Biotechnology

BT Bipolar Transmission

BT Bismaleimide-Triazine

BT Bismuth Titanate

BT Bite-Type (Tube Fitting)

BT 2,2'-Bithiophene

BT Bit Test

BT Blalock-Taussig (Operation)

BT Block Terminal

BT Blood Test

BT Blue Tetrazolium

BT Blue Tongue (Virus)

BT Body Temperature

BT Boltzmann Transformation [Physics]

BT Bomb Technician

BT Booster Transformer

BT Bourdon Tube

BT Branch Tee

BT Brevet de Technicien [French Technician's Certificate]

BT Bridge Transition

BT Brightness Temperature

BT British Telecom

BT Brittleness Temperature [Metallurgy]

BT Broadcast Technician

BT Broadcast Technologist; Broadcast Technology

BT Broad(er) Term

BT Bromotoluene

BT Bubble Tray [Chemical Engineering]

BT Building Technician

BT Building Technology

BT Bulb Turbine

BT (Engine) Burn Time

BT Bus Tie

BT Busy Tone

BT Long Break [Morse Telegraphy]

BT Tape-Armoured Buried Cable

BT^{2-} Blue Tetrazolium (Salt) Ion

BT$_2$ Barium Dititanate [BaO·2TiO$_2$]

BT$_3$ Barium Trititanate [BaO·3TiO$_2$]

BT$_4$ Barium Tetratitanate [BaO·4TiO$_2$]

BT$_5$ Barium Pentatitanate [BaO·5TiO$_2$]

B$_2$T$_9$ Dibarium Nonatitanate [2BaO·9SiO$_2$]

o-BT o-Bromotoluene

B(T) Planck Function at Temperature T [Physics]

Bt Bacillus thuringiensis [also BT] [Microbiology]

Bt Baht [Currency of Thailand]

Bt Baronet [also bt]

Bt Benzotriazole [also BT]

Bt Boat [also bt]

Bt1 1-Benzotriazole

Bt2 2-Benzotriazole

B-t Boat-tailed (Bullet)

B/t Magnetic Flux Density to Sample Thickness (Ratio)

b(T) reduced magnetization [Symbol]

bt bent

bt berth terms [Commerce]

bt bought

.bt Bhutan [Country Code/Domain Name]

BTA Barium Tetraaluminate

BTA Basic Trading Areas [of Rand McNally, US]

BTA 1H-Benzotriazole

BTA Benzoyltrifluoroacetone

BTA Bicyclo[2.2.2]oct-7-En-2,3,5,6-Tetracarbonic Dianhydride

BTA *(Biologisch-technische(r) Assistent(in))* – German for "Laboratory Technician"

BTA Biotechnische Abfallverwertung GmbH (Prozess) [German Biotechnological Waste Utilization Process]

BTA Book Trade Association [South Africa]

BTA Boring-Trepanning Association

BTA British Tourist Authority

BTA British Travel Association

BTA British Tugowners Association

BTAB Barrier Technical Advisory Board

BTAB Bumped Tape-Automated Bonding

BTAM Basic Tape Access Method

BTAM Basic Telecommunications Access Method [of IBM Corporation, US]

BTAM Basic Teleprocessing Access Method

BTAM Basic Terminal Access Method

BTAM Batch Terminal Access Method

BTAO Bureau of Technical Assistance Operations [of United Nations]

BTB Behind-the-Bars

BTB Belgische Transportarbeidersbond [Belgian Transport Workers Union]

BTB Branch Target Buffer

BTB Bromothymol Blue

BTB Bus Tie Breaker

β-Tb Beta Terbium [Symbol]

BT-BASIC Backtracker-BASIC

BTBC 1,1'-Bis[6-(Trifluoromethyl)-1H-Benzotriazolyl] carbonate

BTC Belgian Trade Commission

BTC Benzotrichloride

BTC Binary Time Code

BTC Binary-Tree Computer

BTC Bit Test and Complement

BTC Blue Tetrazolium

BTC Block Transfer Controller

BTC British Telecom Center [London, UK]

BTC British Textile Council

BTC British Transport Commission [London, UK]

BTC Bulawayo Technical College [Zimbabwe]

BTC Business and Technology Center [UK]

BTC Bus Tie Contractor

BTC Butyrylthiocholine

BTCA Benzenetricarboxylic Acid

BTCA Bedlington Terrier Club of America [US]

BTCC Benzothiazolocarbon Cyanine

BTCC Board of Transportation Commissioners of Canada

BTCD Business and Technology Coordination Division [of Canada Center for Mineral and Energy Technology; of Natural Resources Canada, Ottawa]

BTCE Bureau of Transport and Communications Economics

BTCh Bachelor of Textile Chemistry

BTCM Bromotrichloromethane

BTCMPI British Technical Council of the Motor and Petroleum Industries

BTCN 3-Bromo-1,1,1-Trichlorononane

β-TCP β-Tricalcium Phosphate

BTCS Benzyltrichlorosilane

BTCV British Trust for Conservation Volunteers

BTD Binary-to-Decimal

BTD Bomb Testing Device

BTD Brightness Temperature Difference

BTDA 3,3',4,4'-Benzophenonetetracarboxylic Dianhydride

BTDA/ODA 3,3',4,4'-Benzophenonetetracarboxylic Dianhydride/ 4,4'-Oxydianiline

BTDC Before Top Dead Center

BTDE 3,3',4,4'-Benzophenonetetracarboxylic Acid

BTDF Bellini-Tosi Direction Finding (System) [also BT-DF]

BTDL Back Transient Diode Logic

BTDL Basic Transient Diode Logic

BTDU Benzylthiodihydrouracil

BTE Bachelor of Textile Engineering

BTE Baldwin, Tate and Emery Theory

BTE Battery Terminal Equipment

BTE Bench Test Equipment

BTE Bidirectional Transceiver Element

BTE Boltzmann Transport Equation [Statistical Mechanics]

BTE Brake Thermal Efficiency

BTE Bureau of Transport Economics [now Bureau of Transport and Communications Economics]

BTE Business Terminal Equipment

BTEAC Benzyltriethylammonium Chloride

BTech Bachelor of Technology

BTEC Brucellosis and Tuberculosis Eradication Campaign

BTEE N-Benzoyl-L-Tyrosine Ethyl Ester [Biochemistry]

β-TeGe Beta Tellurium Germanide

BTEMA British Tanning Extract Manufacturers Association

β-TeO$_2$ Beta Tellurium Dioxide

BTES Bromotriethoxysilane

β-TeSn Beta Tellurium Stannide

BTEU Bis(trinitroethyl)urea

BTEX Benzene, Toluene, Ethylbenzene and Xylene

BTF Bomb Tail Fuse

BTF British Turkey Federation

BTF Bulk Transfer Facility

BTFA Benzoyltrifluoroacetone

BTFA Bis(trifluoroacetamide)

BTFI Basic Technologies for Future Industries

Btfly Vlv Butterfly Valve

BTFM Bromotrifluoromethane

bt fwd brought forward

BTG Barium Titanium Germanate

BTG British Technology Group

BTH Basic Transmission Header

β-Th Beta Thorium [Symbol]

BThU British Thermal Unit [also BTU, or Btu]

BTI Bangladesh Textile Institute [Dacca]

BTI Battelle Technology International [US]

BTI Bibra Toxicology International [formerly BIBRA, UK]

BTI Bis(trifluoroacetyloxy)phenyl Iodine

BTI Bridged Tap Isolator

BTI British Technology Index

BTI British Telecommunications International

β-Ti Beta Titanium [Symbol]

B-TIP Battelle Technical Inputs to Planning [US]

BTIPR Boyce Thompson Institute for Plant Research

B-TiSULC Boron and Titanium Stabilized ULC (Ultralow Carbon) (Steel)

BTK Blonder-Tinkham-Klapwijk (Theory of Superconductivity)

BTL Backplane Transceiver Logic

BTL Balanced Transformerless (Circuit)

BTL Beginning of Tape Level

BTL Bell Telephone Laboratories [US]

BTL Between Layers

β-Tl Beta Thallium [Symbol]

BTLE Between-the-Lines Entry

Btlg Bottling [also btlg]

BTLIA British Turf and Land Irrigation Association

BTM Basic Time-Sharing Monitor

BTM Batch Time-Sharing Monitor

BTM Bell Telephone Manufacturing Company

BTM Benzylthiomethyl [also Btm]

BTM Bis(trimethylsilyl)methane

BTM Bromotrifluoromethane

BTM Business and Technical Management

Btm Benzylthiomethyl

Btm Bottom [also btm]

BTMA Benzyltrimethylammonium

BTMA Best Technical Means Available

BTMA British Textile Machinery Association

BTMA British Timber Merchants Association

BTMA British Turned-Parts Manufacturers Association

BTMA British Typewriter Manufacturers Association

BTMA Busy-Tone Multiple Access

BTMABr3 Benzyltrimethylammonium Tribromide

BTMAC Benzyltrimethylammonium Chloride

BTMA-ICl2 Benzyltrimethylammonium Dichloroiodide

BTM Bull BTM (Business and Technical Management) Bulletin [UK]

BTMF Block Type Manipulation Facility

BTMS Body Temperature Measuring System

BTMSA Bis(trimethylsilyl)acetylene [also btmsa]

BTMSBD 1,4-Bis(trimethylsilyl)-1,3-Butadiyne

BTMU N'-[(4-Ethenylphenyl)methyl]-N,N,N'-Trimethylurea

BTN Barium Titanium Niobate

BTN Billing Telephone Number

BTn Baccalauréat de Technicien [French Technician's Baccalaureate]

BTNES Bis(trinitroethyl)succinate

BTO Barium Titanyl Oxalate

BTO Bismuth Titanium Oxide

BTO Blocking Tube Oscillator [also bto]

BTO Bombing through Overcast

BTO Bomb Thermocouple Oxygen

BTO British Trust for Ornithology

BTO Brussels Trade Organization [Belgium]

BTOA Binary To ASCII (American Standard Code for Information Interchange) [also BtoA]

B to B Back to Back

B304SS Borated AISI-SAE 304 Stainless Steel

BTOG British Transport Officers Guild

BTO/STO Barium Titanate/Strontium Titanate (Ceramics)

BTP Batch Transfer Program

BTP Botswana Pula [Currency]

BTP Branch Technical Position

BTP Bromine Trifluoride Process

BTPA British Tractor Pullers Association

BTPPC Benzyl Triphenyl Phosphonium Chloride

BTR Behind-the-Tape Reader Board

BTR Better [Amateur Radio]

BTR Bit Test and Reset

BTR Bit Time Recovery

BTR Brittle Temperature Range [Metallurgy]

BTR Broadcast and Television Receivers (Group) [of Institute of Electrical and Electronics Engineers, US]

BTR (Engine) Burn Time Remaining

BTR Bus Transfer

BTRA Bombay Textile Research Association [India]

BTRA British Truck Racing Association

BTRL British Telecom Research Laboratories [Ipswich, Suffolk, UK]

Btry Battery [also btry]

BTS Barium Titanate Stannate

BTS Barium Titanium Silicate

BTS Base Transceiver Station

BTS Batch Terminal Simulation

BTS Bellini-Tosi System (of Radio Direction Finding)

BTS Bias Temperature Stress [Electrical Engineering]

BTS Bisynchronous Terminal Support

BTS Bit Test and Set

BTS Brazilian Thorium Sludge

BTS Brevet de Technicien Supérieur [French Advanced Technician's Certificate]

BTS British Tarantula Society

BTS British Telecommunications Systems

BTS British Toxicology Society

BTS British Tunneling Society

BTS Broadcast Technology Society [of Institute of Electrical and Electronics Engineers, US]

BTS Business Telecommunications Services [Australia]

B2S Second-Stage Binary Structure [Symbol]

BTSC Broadcast Television Systems Committee

BTSD Basic Training for Skill Development

BTSP Bootstrap [also Btsp, or btsp]

BTSS Basic Time Sharing System

BTSS Braille Time-Sharing System

Bt St Billet Steel

BTT Barium Titanium Tantalate

BTT Business Transfer Tax [Canada]

BTTG British Textile Technology Group

BTTN Butanetriol Trinitrate

BTTP British Towing Tank Panel

BTTRI Beijing Television Technology Research Institute [PR China]

BTU Basic Transmission Unit

BTU Board of Trade Unit [= Kilowatt per hour]

BTU Bus Terminal Unit

BTU tert-Butylthiourea

Btu British thermal Unit [also BTU, or BthU]

Btu$_{IT}$ International Table British thermal unit

Btu$_{mean}$ Mean British thermal unit

Btu$_{60/61}$ Sixty Degrees Fahrenheit British thermal unit

BTUC Botswana Trade Union Center

Btu/cu ft British thermal units per cubic foot [alsoBtu/ft³]

Btu/day·ft² British thermal unit per day per square foot [also Btu/day/ft², or Btu/day/sq ft]

Btu/°F British Thermal Unit per degree Fahrenheit [Unit]

Btu/°F·lb British Thermal Unit per degree Fahrenheit pound [also Btu/°F/lb]

Btu/ft² British thermal units per square foot [alsoBtu/sq ft]

Btu/ft³ British thermal units per cubic foot [alsoBtu/cu ft]

Btu/ft²·h British thermal units per square foot hour [also Btu/ft²/hr, Btu/sq ft/h, Btu/sq ft/hr]

Btu/ft·h·°F British thermal units per foot hour degree Fahrenheit [also Btu/ft/h/°F, Btu/ft·hr·°F,Btu/ft/hr/°F]

Btu/ft²·h·°F British thermal units per square foot hour degree Fahrenheit [also Btu/ft²/h/°F,Btu/ft²·hr·°F, or Btu/ft²/hr/°F]

Btu/ft²·h·°F·in British thermal units per square foot hour degree Fahrenheit per inch [also Btu/ft²/h/°F/in, Btu/ft²·hr·°F·in or Btu/ft²/hr/°F/in]

Btu-ft/h·ft²·°F British thermal unit feet per hour square foot degree Fahrenheit [also Btu-ft/h/ft²/°F]

Btu/ft²·h·°R⁴ British thermal units per square foot hour degree Rankine to the fourth [also Btu/ft²/h/°R⁴]

Btu/ft²·hr British thermal units per square foot hour [also Btu/ft²/hr]

Btu/ft²·hr·°F British thermal units per square foot hour degree Fahrenheit [also Btu/ft/hr/°F]

Btu-ft/hr·ft²·°F British thermal unit feet per hour square foot degree Fahrenheit [also Btu-in/hr/ft²/°F]

Btu/ft²·hr·°R⁴ British thermal units per square foot hour degree Rankine to the fourth [also Btu/ft²/hr/°R⁴]

Btu/ft²·min British thermal units per square foot minute [also Btu/ft²/min]

Btu/ft²·s British thermal units per square foot second [also Btu/ft²/s, Btu/ft²·sec or Btu/ft²/sec]

Btu/ft²·sec British thermal units per square foot second [also Btu/ft²/sec, Btu/ft²·s or Btu/ft²/s]

Btu-ft/sec·ft²·°F British thermal unit feet per second square foot degree Fahrenheit [also Btu-in/sec/ft²/°F]

Btu-ft/s·ft²·°F British thermal unit feet per second square foot degree Fahrenheit [also Btu-in/s/ft²/°F]

Btu/h British thermal units per hour [also Btu/hr]

Btu/h·ft² British thermal units per hour square foot [also Btu/h/ft²]

Btu/h·ft·°F British thermal units per hour square foot degree Fahrenheit [also Btu/h/ft/°F, Btu/hr·ft·°F, or Btu/hr/ft/°F]

Btu/h·ft²·°R British thermal units per hour square foot degree Rankine [also Btu/h/ft/°R, Btu/hr·ft²·°F, or Btu/hr/ft²/°F]

Btu/hr British thermal units per hour [also Btu/h]

Btu/hr·ft·°F British thermal units per hour foot degree Fahrenheit [also Btu/hr/ft/°F, Btu/h·ft·°F, or Btu/h/ft/°F)]

Btu/hr·ft² British thermal units per hour square foot [also Btu/hr/ft²]

Btu/hr·ft²·°F British thermal unit per hour square foot degree Fahrenheit [also Btu/hr/ft²·°F]

Btu/hr·ft²·°R British thermal units per hour square foot degree Rankine [also Btu/hr/ft²/°R,Btu/h·ft²·°F, or Btu/h/ft²/°F]

Btu/in British thermal units per inch [Unit]

Btu/in² British thermal units per square inch [also Btu/sq in]

Btu-in/h·ft²·°F British thermal unit-inches per hour square foot degree Fahrenheit [also Btu-in/h/ft²/°F, Btu-in/hr·ft²·°F, or Btu in/hr/t²/°F]

Btu-in/hr·ft²·°F British thermal unit-inches per hour square foot degree Fahrenheit [also Btu-in/hr/ft²/°F, Btu-in/h·ft²·°F, or Btu in/h/ft²/°F]

Btu-in/s·ft²·°F British thermal unit inches per second square foot degree Fahrenheit [also Btu-in/s/ft²/°F, BTU-in/sec·ft²·°F, or Btu-in/sec/ft²/°F]

Btu-in/sec·ft²·°F British thermal unit inches per second square foot degree Fahrenheit [also Btu -in/sec/ft²/°F, BTU-in/s·ft²·°F, or Btu-in/sec/ft²/°F]

Btu/in²·min·°F British thermal unit per square inch minute degree Fahrenheit [also Btu/in²/min/°F]

Btu/in²·s British thermal units per square inch second [also Btu/in²/s, Btu/in²·sec or Btu/in²/sec]

Btu/in²·sec British thermal units per square inch second [also Btu/in²/s, Btu/in²·sec or Btu/in²/sec]

Btu-in/s·ft²·°F British thermal unit inches per second square foot degree Fahrenheit [also Btu-in/s/ft²/°F, Btu-in/sec·ft²·°F, or Btu-in/sec/ft²/°F]

Btu-in/sec·ft²·°F British thermal unit inches per second square foot degree Fahrenheit [also Btu-in/sec/ft²/°F, Btu-in/s·ft²·°F, or Btu-in/s/ft²/°F]

Btu/kw-hr British thermal units per kilowatt-hour [Unit]

Btu/lb British thermal units per pound [Unit]

Btu/lb·°F British thermal units per pound degree Fahrenheit [also Btu/lb-°F, or Btu/lb/°F]

Btu/lb$_m$·°F British thermal units per pound degree Fahrenheit [also Btu/lbm-°F, or Btu/lb$_m$/°F]

Btu/min British thermal units per minute [also Btu/min]

Btu/s British thermal units per second [also Btu/sec]

Btu/s·in²·°R British thermal units per second square inch degree Rankine [also Btu/s/in²/°R,Btu/sec·in²·°R, or Btu/sec/in²/°R]

Btu-sec British thermal unit-second [Unit]

Btu/sec British thermal units per second [also Btu/s]

Btu/sec·ft²·°F British thermal units per second square foot degree Fahrenheit [also Btu/sec/ft²/°F]

Btu/sec·in²·°R British thermal units per second square inch degree Rankine [also Btu/sec/in²/°R,Btu/s·in²·°R, or Btu/s/in²/°R]

Btu/s·ft²·°F British thermal units per second square foot degree Fahrenheit [also Btu/s/ft²/°F]

Btu/sq ft British thermal units per square foot [also Btu/ft²]

Btu/sq ft/h/°F British thermal units per square foot per hour per degree Fahrenheit [also Btu/ft²/h/°F, Btu/ft²·h·°F, Btu/sq ft·h·°F]

Btu/sq in British thermal units per Square Inch [also Btu/in²]

Btu/ton British thermal units per ton [Unit]

BTV Barium Titanium Vanadate [$Ba_2TiV_2O_8$]

BTV Bluetongue Virus

BTV-1 Bluetongue Virus, Type 1

BTW Behind-the-Wheel

BTW By The Way [Internet]

BTWCVER Biological and Toxin Weapons Convention Verification

BTX Batrachotoxin

BTX Benzene, Toluene and/or Xylene (Hydrocarbons)

BTX Bildschirmtext [also Btx; German Viewdata System]

B2X Binary To Hexadecimal [Restructured Extended Executor (Language)]

BTYGZ Barium Fluoride–Thorium Fluoride–Yttrium Fluoride–Gallium Fluoride–Zinc Fluoride (Glass)

BU Backup [also B/U]

BU Baseband Unit

BU Base Up (Light Bulb)

BU Bath University [UK]

BU Basic User

BU Baylor University [Waco, Texas, US]

BU Beijing University [PR China]

BU Beneficial Use

BU Beta Uranium

BU Binding Unit

BU Binghamton University [New York State, US]

BU Bond University [Australia]

BU Boston University [Massachusetts, US]

BU Bottom Up

BU Bradley University [Preoria, Illinois, US]

BU Branch Unit

BU Brandeis University [Waltham, Massachusetts, US]

BU Brandon University [Manitoba,Canada]

BU Bristol University [UK]

BU Brock University [St. Catharines, Ontario,Canada]

BU Brown University [Providence, Rhode Island, US]

BU Brunel University [Uxbridge, Middlesex,UK]

BU Bucknell University [Lewisburg, Pennsylvania, US]

BU Build-Up

BU Butler University [Indianapolis, Indiana, US]

B&U Buildings and Utilities

B/U Backup [also BU]

Bu Bouguer Number [Symbol]

Bu Bureau [also bu]

Bu Butanoic Acid

Bu Butyl (Alcohol)

Bu Butyryl

Bu Buzzer [also bu]

Bu₂ Dibutyl

Bu₃ Tributyl

Bu₄ Tetrabutyl

Bu₆ Hexabutyl

t-Bu tert-Butyl

bu bushel [Unit]

β-U Beta Uranium [Symbol]

BUA Bureau of Aeronautics [US Navy]

BuA Butylacrylate

BUAA Beijing University of Aeronautics and Astronautics [PR China]

n-BuAc n-Butyl Acetate

BUAER Bureau of Aeronautics [also BuAER orBuAer] [Now Part of BuWEPS of US Navy]

BUBL Bulletin Board for Libraries

BuBr Butyl Bromide

n-BuBz n-Butylbenzene

BUC Backup Computer

BUC Buckhorn, California [NASA Ground Spacecraft (or Space Flight) Tracking and Data Network]

BuCAIM Buffered Citric Acid Indicator Mixture

Bucks Buckinghamshire [UK]

Bucky ball (60-Carbon-Atom-)Buckminsterfullerene Molecule

BuCl Butyl Chloride

t-BuCl tert-Butyl Chloride

BUCLASP Buckling of Laminated Stiffened Plates

t-Bu₂CuLi Lithium Di(tert-Butyl)copper

BUD Biological Up-Down (Probe)

1,4-(Bu)dab 1,4-Di-tert-Butyl-1,4-Diazabutadiene [also1,4-(Bu)DAP]

Budavox Telecommun Rev Budavox Telecommunication Review [Hungary]

BUDC Before Upper Dead Center

BUdR 5-Bromo-2'-Deoxyuridine [also Br-dU] [Biochemistry]

BUDWSR Brown University Display for Working Set References [US]

BUE Built-Up Edge [Machining]

BUEC Back-Up Emergency Communications

BUEMG Bristol University Electron Microscopy Group [UK]

BUET Bangladesh University of Engineering and Technology [Dacca]

BuF Butyl Fluoride

t-BuF tert-Butyl Fluoride

Buf Buffer [also buf]

BUFC British Universities Film Council

BUFCA British Urethane Foam Contractors Association

Buff Buffer [also buff]

BUFORA British Unidentified Flying Objects Research Association

BUFVC British Universities Film and Video Council

BUG Brooklyn Union Group [US]

BuGDN Butyleneglycoldinitrate

Bugas Butane Gas [also bugas]

Bu₃Ge Tributylgermane

BUGS Brown University Graphic System [US]

BUI Buildup Index [Forest Fire Danger]

BuI Butyl Iodide

BUIC Back-Up Interceptor Control

BUILD Australian Building Database

BUILD Base for Uniform Language Definition

Build Building [also build]

Build Res Establ Dig Building Research Establishment Digest [UK]

Build Res Pract Building Research and Practice [Journal of the International Council for Building Research, Studies and Documentation, Netherlands]

Build Sci Building Science [Journal of the National Institute of Building Sciences, US]

Build Serv Building Services [Publication of the Chartered Institution of Building Services Engineers, UK]

Build Serv Eng Res Technol Building Service Engineering Research and Technology [Publication of the Chartered Institution of Building Services Engineers, UK]

Build Serv Environ Eng Building Services and Environmental Engineer [Journal published in the UK]

Build Steel Building with Steel [UK Journal]

BUIRA British Universities Industrial Relations Association

BUIS Barrier Up Indicator System

BUIST Beijing University of Iron and Steel Technology [PR China]

BUK Burmese Kyat [Currency of Burma]

Bulg Bulgaria(n)

Bulg J Phys Bulgarian Journal of Physics [of Bulgarian Academy of Sciences]

BuLi Butyllithium

n-BuLi n-Butyllithium

s-BuLi sec-Butyllithium

t-BuLi tert-Butyllithium

Bul Inst Politeh Bucur, Autom-Calc Buletinul Institutului Politehnic Bucuresti, Automatica-Calculatoare [Bulletin of the Bucharest Polytechnic Institute, Automation and Computation, Romania]

Bul Inst Politeh Bucur, Chim Buletinul Institutului Politehnic Bucuresti, Chimie [Bulletin of the Bucharest Polytechnic Institute, Chemistry, Romania]

Bul Inst Politeh Bucur, Constr Masini Buletinul Institutului Politehnic Bucuresti, Constructii de Masini [Bulletin of the Bucharest Polytechnic Institute, Mechanical Engineering, Romania]

Bul Inst Politeh Bucur, Electroteh Buletinul Institutului Politehnic Bucuresti, Electrotehnica [Bulletin of the Bucharest Polytechnic Institute, Electrical Engineering, Romania]

Bul Inst Politeh Bucur, Electron Buletinul Institutului Politehnic Bucuresti, Electronica [Bulletin of the Bucharest Polytechnic Institute, Electronics, Romania]

Bul Inst Politeh Bucur, Energ Buletinul Institutului Politehnic Bucuresti, Seria Energetica [Bulletin of the Bucharest Polytechnic Institute, Energetics Series, Romania]

Bul Inst Politeh Bucur, Mec Buletinul Institutului Politehnic Bucuresti, Mecanica [Bulletin of the Bucharest Polytechnic Institute, Mechanics, Romania]

Bul Inst Politeh Bucur, Metal Buletinul Institutului Politehnic Bucuresti, Metalurgie [Bulletin of the Bucharest Polytechnic Institute, Metallurgy, Romania]

Bul Inst Politeh Bucur, Transpt-Aeronave Buletinul Institutului Politehnic, Transporturi-Aeronave [Bulletin of the Bucharest Polytechnic Institute, Air Transportation and Aeronautics, Romania]

Bul Inst Politeh 'Gheorghe Gheorghiu Dej', Chim Buletinul Institutului Politehnic, 'Gheorghe Gheorghiu-Dej', Chimie [Bulletin of the 'Gheorghe Gheorghiu-Dej' Polytechnic Institute, Chemistry, Romania]

Bul Inst Politeh 'Gheorghe Gheorghiu Dej', Chim-Metal Buletinul Institutului Politehnic, 'Gheorghe Gheorghiu-Dej', Chimie-Metalurgie [Bulletin of the 'Gheorghe Gheorghiu-Dej' Polytechnic Institute, Chemistry–Metallurgy, Romania]

Bul Inst Politeh 'Gheorghe Gheorghiu Dej', Constr Masini Buletinul Institutului Politehnic, 'Gheorghe Gheorghiu-Dej', Constructii de Masini [Bulletin of the 'Gheorghe Gheorghiu-Dej' Polytechnic Institute, Mechanical Engineering, Romania]

Bul Inst Politeh 'Gheorghe Gheorghiu Dej', Electroteh Buletinul Institutului Politehnic, 'Gheorghe Gheorghiu-Dej', Electrotehnica [Bulletin of the 'Gheorghe Gheorghiu-Dej' Polytechnic Institute, Electrical Engineering, Romania]

Bul Inst Politeh 'Gheorghe Gheorghiu Dej', Mec Buletinul Institutului Politehnic, 'Gheorghe Gheorghiu-Dej', Mecanica [Bulletin of the 'Gheorghe Gheorghiu-Dej' Polytechnic Institute, Mechanics, Romania]

Bul Inst Politeh 'Gheorghe Gheorghiu Dej', Metal Buletinul Institutului Politehnic, 'Gheorghe Gheorghiu-Dej', Metalurgie [Bulletin of the 'Gheorghe Gheorghiu-Dej' Polytechnic Institute, Metallurgy, Romania]

Bul Inst Politeh 'Gheorghe Gheorghiu Dej', Transpt-Aeronave Buletinul Institutului Politehnic, 'Gheorghe Gheorghiu-Dej', Transporturi-Aeronave [Bulletin of the 'Gheorghe Gheorghiu-Dej' Polytechnic Institute, Air Transport and Aeronautics, Romania]

Bul Inst Politeh Iaşi I Institutului Politehnic din Iaşi, Sectia I [Bulletin of the Iasi Polytechnic Institute, Chemistry, Romania]

Bul Inst Politeh Iaşi II Buletinul Institutului Politehnic din Iaşi, Sectia II [Bulletin of the Iasi Polytechnic Institute, Section II, Romania]

Bul Inst Politeh Iaşi III Buletinul Institutului Politehnic din Iaşi, Sectia III [Bulletin of the Iasi Polytechnic Institute, Section III, Romania]

Bul Inst Politeh Iaşi III, Electroteh Electron Autom Buletinul Institutului Politehnic din Iaşi, Sectia III, Electrotehnica, Electronica, Automatizari [Bulletin of the Iasi Polytechnic Institute, Section III, Electrical Engineering, Electronics, Automation, Romania]

Bul Inst Politeh Iaşi III, Electroteh Energ Electron Autom Buletinul Institutului Politehnic din Iaşi, Sectia III, Electrotehnica, Energetica, Electronica, Automatizari [Bulletin of the Iasi Polytechnic Institute, Section III, Electrical Engineering, Energetics, Electronics, Automation, Romania]

Bul Inst Stud Proiect Energ Buletinul Institutului de Studii si Proiectari Energetice [Bulletin of the Institute of Energetic Project Studies, Romania]

Bulk Handl Bulk Handling [Journal published in the US]

Bulk Solids Handl Bulk Solids Handling [Journal published in Switzerland]

Bulk Syst Int Bulk Systems International [Journal published in the US]

Bull Bullet [also bull]

Bull Bulletin [also bull]

Bull AAAS Bulletin of the American Academy of Arts and Sciences [US]

Bull AAUP Bulletin of the American Association of University Professors [US]

Bull Acad Pol Sci Bulletin de l'Academie Polonaise des Sciences [Bulletin of the Polish Academy of Sciences]

Bull Acad Sci DPR Korea Bulletin of the Academy of Sciences of the DPR (Democratic People's Republic of) Korea [South Korea]

Bull Acad Sci USSR, Chem Ser Bulletin of the Academy of Sciences of the USSR, Chemical Series [Translation of: *Izvestiya Akademii Nauk SSSR, Khimicheskaya (USSR)*; published in the US]

Bull Acad Sci USSR, Phys Ser Bulletin of the Academy of Sciences of the USSR, Physical Series [Translation of: *Izvestiya Akademii Nauk SSSR, Fizicheskaya (USSR)*; published in the US]

Bull Acad Serbe Sci Arts, Cl Sci Tech Bulletin de l'Academie Serbe des Sciences et des Arts, Classe des Sciences Techniques [Bulletin of the Serbian Academy of Arts and Sciences, Technical Sciences Class]

Bull Aichi Inst Technol A Bulletin of Aichi Institute of Technology, Part A [Japan]

Bull Aichi Inst Technol B Bulletin of Aichi Institute of Technology, Part B [Japan]

Bull Alloy Phase Diagr Bulletin of Alloy Phase Diagrams [of ASM International, US]

Bull Am Assoc Var Star Obs Bulletin of the American Association of Variable Star Observers [US]

Bull Am Anthrop Assoc Bulletin of the American Anthropological Association [US]

Bull Am Astron Soc Bulletin of the American Astronomical Society [US]

Bull Am Ceram Soc Bulletin of the American Ceramic Society [US]

Bull Am Math Soc Bulletin of the American Mathematical Society [US]

Bull Am Met Soc Bulletin of the American Meteorological Society [also Bull Am Meteorol Soc]

Bull Am Soc Inf Sci Bulletin of the American Society for Information Science

Bull Anal Lab Managers Assn Bulletin of the Analytical Laboratory Managers Association [US]

Bull Annu Soc Suisse Chronom Lab Suisse Rech Horlog Bulletin Annuel de la Société Suisse de Chronométrie et du Laboratoire Suisse de Recherche Horlogères [Annual Bulletin of the Swiss Society of Chronometry and the Swiss Laboratory of Horological Research]

Bull APD Bulletin of Alloy Phase Diagrams [of ASM International, US]

Bull Astron Inst Czech Bulletin of the Astronomical Institutes of Czechoslovakia [of Czechoslovak Academy of Sciences]

Bull Astron Soc India Bulletin of the Astronomical Society of India [Published by the Tata Institute of Fundamental Research, Bombay, India]

Bull At Sci Bulletin of the Atomic Scientists [US]

Bull Aust Math Soc Bulletin of the Australian Mathematical Society [Australia]

Bull Bismuth Inst Bulletin of the Bismuth Institute [Belgium]

Bull Bur Miner Resour Geol Geophys Bulletin–Bureau of Mineral Resources, Geology and Geophysics [Australia]

Bull Calcutta Math Soc Bulletin of the Calcutta Mathematical Society [India]

Bull Cancer Bulletin du Cancer [French Bulletin on Cancer]

Bull Can Pet Geol Bulletin of Canadian Petroleum Geology

Bull Cercle Etud Mét Bulletin du Cercle d'Etudes des Métaux [Bulletin of the Study Circle for Metals, St. Etienne, France]

Bull Chem Soc Japan Bulletin of the Chemical Society of Japan

Bull Cl Sci Acad R Belg Bulletin de la Classe des Sciences, Academie Royale de Belgique [Bulletin of the Science Class, Royal Academy of Belgium]

Bull Coll Eng Hosei Univ Bulletin of the College of Engineering, Hosei University [Japan]

Bull Coll Sci Univ Ryukyus Bulletin of the College of Science, University of Ryukyus [Okinawa, Japan]

Bull Crime Astrophys Obs Bulletin of the Crimean Astrophysical Observatory [Translation of: *Izvestiya Krymskoi Astroficheskoi Observatorii (USSR)*; published in the US]

Bull Daido Inst Technol Bulletin of Daido Institute of Technology [Japan]

Bull Dir Etud Rech A Bulletin de la Direction des Etudes et Recherches, Série A [Bulletin of the Directorate of Studies and Research, Series A; published by Electricité de France, Direction des Etudes et Recherches, Paris, France]

Bull Dir Etud Rech B Bulletin de la Direction des Etudes et Recherches, Série B [Bulletin of the Directorate of Studies and Research, Series B; published by Electricité de France, Direction des Etudes et Recherches, Paris, France]

Bull Dir Etud Rech C Bulletin de la Direction des Etudes et Recherches, Série C [Bulletin of the Directorate of Studies and Research, Series C; published by Electricité de France, Direction des Etudes et Recherches, Paris, France]

Bull Earthq Res Inst Univ Tokyo Bulletin of the Earthquake Research Institute, University of Tokyo [Japan]

Bull Electrochem Bulletin of Electrochemistry [India]

Bull Electron Microsc Soc India Bulletin of the Electron Microscopy Society of India

Bull Electrotech Lab Bulletin of the Electrotechnical Laboratory [Published by Agency of Industrial Science and Technology, Ministry of International Trade and Industry, Ibaraki, Japan]

Bull Ent Soc Am Bulletin of the Entomological Society of America

Bull Ent Soc Can Bulletin of the Entomological Society of Canada

Bull Environ Contam Toxicol Bulletin of Environmental Contamination and Toxicology

Bull Eur Assoc Theor Comput Sci Bulletin of the European Association for Theoretical Computer Science [Published by the Department of Computer Science, Technical University of Graz, Graz, Austria]

Bull Fac Eng, Hiroshima Univ Bulletin of the Faculty of Engineering, Hiroshima University [Japan]

Bull Fac Eng, Hokkaido Univ Bulletin of the Faculty of Engineering, Hokkaido University [Japan]

Bull Fac Eng, Ibaraki Univ Bulletin of the Faculty of Engineering, Ibaraki University [Japan]

Bull Fac Eng, Kyushu Tokai Univ Bulletin of the Faculty of Engineering, Kyushu Tokai University [Kumamoto, Japan]

Bull Fac Eng, Miyazaki Univ Bulletin of the Faculty of Engineering, Miyazaki University [Japan]

Bull Fac Eng, Tokushima Univ Bulletin of the Faculty of Engineering, Tokushima University [Japan]

Bull Fac Eng, Toyama Univ Bulletin of the Faculty of Engineering, Toyama University [Japan]

Bull Fac Eng, Univ Ryukyus Bulletin of the Faculty of Engineering, University of Ryukyus [Japan]

Bull Fac Eng, Yokohama Natl Univ Bulletin of the Faculty of Engineering, Yokohama National University [Japan]

Bull Forestry Forest Prod Res Inst Bulletin of the Forestry and Forest Products Research Institute [Ibaraki, Japan]

Bull Geod Bulletin Géodésique [Bulletin of Geodesy; published by Association Internationale de Géodésie, France]

Bull Geophys Bulletin of Geophysics [Published by National Central University, Taiwan]

Bull Gov Ind Res Inst Bulletin of the Government Industrial Research Institute [Osaka, Japan]

Bull Gov Ind Res Inst Osaka Bulletin of the Government Industrial Research Institute, Osaka [Japan]

Bull GRC Bulletin of the Geothermal Resources Council [US]

Bull Hist Med Bulletin of the History of Medicine

Bull Hyg Bulletin of Hygiene

Bull IAE Bulletin of the International Association of Engineering

Bull IASS Bulletin of the International Association for Shell and Spatial Structures [Spain]

Bull Indian Natl Sci Acad Bulletin of the Indian National Science Academy [New Delhi, India]

Bull Indian Vac Soc Bulletin of the Indian Vacuum Society

Bull Ind Res Inst, Hiroshima Prefect, West Bulletin of the Industrial Research Institute, Hiroshima Prefecture, West [Japan]

Bull Inf Cent Donnees Stellaires Bulletin d'Information du Centre de Données Stellaires [Information Bulletin of the Stellar Data Center, Observatoire de Strasbourg, France]

Bull Inf Cybern Bulletin of Informatics and Cybernetics [Published by the Research Association of Statistical Sciences, Kyushu University, Fukuoka, Japan]

Bull Inst At Energy, Kyoto Univ Bulletin of the Institute of Atomic Energy, Kyoto University [Japan]

Bull Inst Chem Res, Kyoto Univ Bulletin of the Institute for Chemical Research, Kyoto University [Japan]

Bull Inst Corros Sci Technol Bulletin of the Institute of Corrosion Science and Technology [UK]

Bull Inst Eng (India) Bulletin of the Institution of Engineers (India) [Calcutta]

Bull Int Assoc Eng Bulletin of the International Association of Engineering

Bull Int ISSA Prev Occup Risks Electr Bulletin of the International Section of the ISSA for the Prevention of Occupational Risks due to Electricity [Published by Berufsgenossenschaft der Feinmechanik und Elektrotechnik, Cologne, Germany]

Bull Int Peat Soc Bulletin of the International Peat Society [Finland]

Bull Isr Phys Soc Bulletin of the Israel Physical Society [of Hebrew University, Jerusalem]

Bull Isr Soc Spec Libr Inf Cent Bulletin of the Israel Society of Special Libraries and Information Centers [Tel Aviv]

Bull Jpn Inst Met Bulletin of the Japan Institute of Metals [Sendai, Japan]

Bull Jpn Soc Mech Eng Bulletin of the Japan Society of Mechanical Engineers [Tokyo, Japan]

Bull Jpn Soc Prec Eng Bulletin of the Japan Society of Precision Engineering [Tokyo, Japan] [also Bull Jpn Soc Precis Eng]

Bull Korean Inst Met Bulletin of the Korean Institute of Metals [Seoul, South Korea]

Bull Kurume Inst Technol Bulletin of Kurume Institute of Technology [Japan]

Bull Kyushu Inst Technol Bulletin of Kyushu Institute of Technology [Kitakyushu, Japan]

Bull Kyushu Inst Technol (Math Nat Sci) Bulletin of Kyushu Institute of Technology (Mathematics, Natural Sciences) [Kitakyushu, Japan]

Bull Kyushu Inst Technol (Sci Technol) Bulletin of Kyushu Institute of Technology (Science and Technology) [Kitakyushu, Japan]

Bull Liaison Rech Inf Autom Bulletin de Liaison de la Recherche en Informatique et en Automatique [Liaison Bulletin for Research in Computer Science and Control; published by Institut National de Recherche en Informatique et en Automatique, France]

Bull Mar Eng Soc Jpn Bulletin of Marine Engineering Society of Japan

Bull Mater Sci Bulletin of Materials Science [of Materials Research Society of India and Indian Academy of Sciences, Bangalore] [also Bull Mat Sci]

Bull Mech Eng Lab Bulletin of Mechanical Engineering Laboratory [Tokyo, Japan]

Bull Met Mus Bulletin of the Metals Museum [Japan]

Bull Minéral Bulletin de Minéralogie [French Bulletin on Mineralogy]

Bull Nagoya Inst Technol Bulletin of Nagoya Institute of Technology [Japan]

Bull NASSP Bulletin of the National Association of Secondary School Principals [US]

Bull NRLM Bulletin of the National Research Laboratory of Metrology [Ibaraki, Japan] [also Bull Natl Res Lab Metrol]

Bull Offic Prop Ind Bulletin Officiel de la Propriété Industrielle [Official Bulletin of Industrial Property, France]

Bull Okayama Univ Sci A Bulletin of the Okayama University of Science A [Japan]

Bull Okayama Univ Sci A, Nat Sci Bulletin of the Okayama University of Science A, Natural Science [Japan]

Bull Okayama Univ Sci B Bulletin of the Okayama University of Science B [Japan]

Bull Okayama Univ Sci B, Hum Sci Bulletin of the Okayama University of Science A, Human Science [Japan]

Bull PAIS Bulletin of the Public Affairs Information Service [US]

Bull Pol Acad Sci, Chem Bulletin of the Polish Academy of Sciences, Chemistry [Warsaw, Poland]

Bull Pol Acad Sci, Tech Sci Bulletin of the Polish Academy of Sciences, Technical Sciences [Warsaw, Poland]

Bull Primary Tungsten Assoc Bulletin of the Primary Tungsten Association [UK]

Bull Proc Australas Inst Min Metall Bulletin and Proceedings of the Australasian Institute of Mining and Metallurgy [Australia]

Bull Prf Bullet-Proof [also Bullprf, bull prf, or bullprf]

Bull Res Inst Electron Shizuoka Univ Bulletin of the Research Institute of Electronics, Shizuoka University [Japan]

Bull Res Inst Miner Dressing Metall Bulletin of the Research Institute of Mineral Dressing and Metallurgy [Katahira, Sendai, Japan]

Bull Res Inst Sci Meas, Tohoku Univ Bulletin of the Research Institute for Scientific Measurements, Tohoku University [Sendai, Japan]

Bull Res Lab Nucl React Bulletin of the Research Laboratory for Nuclear Reactors [of Tokyo Institute of Technology, Japan]

Bull Res Lab Precis Mach Electron Bulletin of the Research Laboratory of Precision Machinery and Electronics [of Tokyo Institute of Technology, Japan]

Bull SCGM Bulletin de la Société Canadienne de Génie Mécanique [Bulletin of the Canadian Society for Mechanical Engineering]

Bull Schweiz Elektrotech Bulletin der Schweizerischen Vereinigung für Elektrotechniker [Bulletin of the Swiss Association of Electrical Engineers]

Bull Schweiz Elektrotech Ver Verb Schweiz Elektr werke Bulletin des Schweizerischen Elektrotechnischen Vereins und des Verbandes Schweizerischer Elektrizitätswerke [Bulletin of the Swiss Electrotechnical Association and the Association of Swiss Electric Power Companies]

Bull Schweiz Ges Mikrotech Bulletin der Schweizerischen Gesellschaft für Mikrotechnik [Bulletin of the Swiss Society of Microtechnology]

Bull Sci A Bulletin Scientifique A [Scientific Bulletin, Series A, Yugoslavia]

Bull Sci Assoc Ing Electr Inst Electrotech Montefiore Bulletin Scientifique de l'Association des Ingénieurs Electriciens sortis de l'Institut Electrotechnique Montefiore [Scientific Bulletin of the Association of Electrical Engineers; published by the Montefiore Institute of Electrical Engineering, Liege, Belgium]

Bull Sci B Bulletin Scientifique B [Scientific Bulletin, Series B, Yugoslavia]

Bull Sci Eng Res Lab, Waseda Univ Bulletin of the Science and Engineering Research Laboratory, Waseda University [Japan]

Bull Sci Instrum Soc Bulletin of the Scientific Instrument Society [UK]

Bull Sci Math Bulletin des Sciences Mathématiques [Bulletin of Mathematical Sciences; published in France]

Bull Seismol Soc Am Bulletin of the Seismological Society of America

Bull SME Bulletin of the Society of Manufacturing Engineers [US]

Bull Soc Chim Belg Bulletin des Sociétés Chimiques Belges [Bulletin of the Belgian Chemical Societies]

Bull Soc Chim Fr Bulletin de la Société Chimique de France [Bulletin of the French Chemical Society]

Bull Soc Chim Fr I Bulletin de la Société Chimique de France, Partie I [Bulletin of the French Chemical Society, Part I]

Bull Soc Chim Fr II Bulletin de la Société Chimique de France, Partie II [Bulletin of the French Chemical Society, Part II]

Bull Soc DV-Xα Bulletin of the Society for Discrete Variational X-Alpha (Method) [also Bull Soc DVX$_\alpha$]

Bull Soc Fr Céram Bulletin de la Société Française de Céramique [Bulletin of the French Ceramic Society]

Bull Soc Fr Mineral Crystallogr Bulletin de la Société Française de Minéralogie et de Cristallographie [Bulletin of the French Society for Mineralogy and Crystallography]

Bull Soc Fr Photogramnm Teledetect Bulletin de la Société Française de Photogrammetrie et des Télédetection [French Society of Photogrammetry and Remote Sensing]

Bull Soc Manuf Eng Bulletin of the Society of Manufacturing Engineers [US]

Bull Soc Math Belg Bulletin de la Société Mathématique de Belgique [Bulletin of the Mathematical Society of Belgium]

Bull Soc Math Fr Bulletin de la Société Mathématique de France [Bulletin of the Mathematical Society of France]

Bull Soc Math Grèce Bulletin Société Mathématique de Grèce [Bulletin of the Mathematical Society of Greece]

Bull Soc R Sci Liège Bulletin de la Société Royale des Sciences de Liège [Bulletin of the Royal Scientific Society of Liege, Belgium]

Bull Soc Sci Bretagne Bulletin de la Société Scientifique de Bretagne [Bulletin of the Scientific Society of Brittany, University of Rennes, France]

Bull Spec Astrophys Obs-North Caucasus Bulletin of the Special Astrophysical Observatory – North Caucasus [Translation of: *Astrofizicheskie Issledovaniya–Izvestiya Spetsial'noi Astrofizicheskoi Observatorii (USSR)*; published in the US]

Bull SSA Bulletin of the Seismological Society of America

Bull Taiwan Forestry Res Inst Bulletin of the Taiwan Forestry Research Institute

Bull Tech Univ Istanb Bulletin of the Technical University of Istanbul [Turkey]

Bull Tokyo Gakugei Univ I Bulletin of the Tokyo Gakugei University, Series I [Japan]

Bull Tokyo Gakugei Univ II Bulletin of the Tokyo Gakugei University, Series II [Japan]

Bull Tokyo Gakugei Univ III Bulletin of the Tokyo Gakugei University, Series III [Japan]

Bull Tokyo Gakugei Univ IV Bulletin of the Tokyo Gakugei University, Series IV [Japan]

Bull Tokyo Gakugei Univ IV, Math Nat Sci Bulletin of the Tokyo Gakugei University, Series IV, Mathematics and Natural Sciences [Japan]

Bull Univ ElectroComm Bulletin of the University of ElectroCommunications [of University of ElectroCommunications, Tokyo, Japan]

Bull Univ Osaka Prefect, A Bulletin of the University of Osaka Prefecture, Series A [Japan]

Bull Univ Osaka Prefect, B Bulletin of University of Osaka Prefecture, Series B [Japan]

Bull USNM Bulletin of the United States National Museum

Bull Volcanol Bulletin of Volcanology [of International Association of Volcanology and Chemistry of the Earth's Interior, Germany]

Bull WHO Bulletin of the World Health Organization [of United Nations]

Bull Vyskum Ustavu Papieru Celulozy Bulletin Vyskumneho Ustavu Papieru a Celuloy [Bulletin of the Research Institute for Paper and Cellulose, Bratislava, Slovakia]

Bull Yamagata Univ (Eng) Bulletin of the Yamagata University (Engineering) [Japan]

Bull Yamagata Univ (Nat Sci) Bulletin of the Yamagata University (Natural Science) [Japan]

Bul Ştiinţ Inst Politeh Cluj-Napoca, Chim–Metal Buletinul Ştiinţific al Institutului Politechnic Cluj-Napoca, Seria Chimie–Metalurgie [Scientific Bulletin of the Polytechnic Institute of ClujNapoca, Series Chemistry–Metallurgy, Romania]

Bul Ştiinţ Teh Inst Politeh 'Traian Vuia' Timiş Ser Chim Buletinul Ştiinţific si Tehnic al Institutului Politechnic 'Traian Vuia' Timişiora, Seria Chimie [Scientific and Technical Bulletin of the Traian Vuia Polytechnic Institute of Timisiora, Romania]

Bul Ştiinţ Teh Inst Politeh 'Traian Vuia' Timiş Ser Elektroteh Buletinul Ştiinţific si Tehnic al Institutului Politechnic 'Traian Vuia' Timişiora, Seria Electrotehnica [Scientific and Technical Bulletin of the Traian Vuia Polytechnic Institute of Timisiora, Romania]

BULPAC Bulgarian Packet Switched Network

BULTEX Bulgarian Videotex Network

Bul Univ Braşov A Buletinul Universitatii din Braşov, Seria A [Bulletin of the University of Brasov, Series A, Romania]

Bul Univ Braşov A, Mec Apl Electroteh Electron Constr Mas Tehnol Prelucr Met

Buletinul Universitatii din Braşov, Seria A, Mecanica Aplicata, Electrotehnica si Electronica, Constructia de Masini si Tehnologia Prelucrarii Metalelor [Bulletin of the University of Brasov, Series A, Applied Mechanics, Electrical Engineering and Electronics, Mechanical Engineering and Metal Processing Technology, Romania]

Bul Univ Brasov B Buletinul Universitatii din Brasov Seria B [Bulletin of the University of Brasov, Series A, Romania]

Bul Univ Galati, Fasc V, Tehnol Constr Masini, Metal Buletinul Universitatii din Galati, Fascicula V, Tehnologii in Constructia de Masini, Metalurgie [Bulletin of the University of Galati, Faculty V, Mechanical Engineering Technology, Metallurgy, Romania]

BUM Back-Up Mode

BUM Break-Up Missile

BUMC Boston University Medical Center [Massachusetts, US]

BUMED Bureau of Medicine and Surgery [US Navy]

t-BUMEOC 1-(3,5-Di-tert-Butylphenyl)-1-Methylethoxycarbonyl

n-BuMgBr n-Butylmagnesium Bromide

sec-BuMgBr sec-Butylmagnesium Bromide

n-BuMgCl n-Butylmagnesium Chloride

BuMines (United States) Bureau of Mines [also BUMINES, BOM, or BoM]

BUMP Bottom Up Modular Programming

BUMS Boston University School of Medicine [Massachusetts, US]

BUN Biomass Users' Network

BUN Blood Urea Nitrogen

BUNA Butadien-Natrium-Prozess [Butadiene Sodium Process for Synthetic Rubber Manufacture]

Buna Butadiene Rubber made with Sodium(Na) Catalyst

Buna A Acrylonitrile-Butadiene Copolymer made with Sodium (Na) Catalyst

BUNAC British Universities North America Club[US]

Buna N Acrylonitrile-Butadiene Rubber made with Sodium (Na) Catalyst

Buna S Styrene-Butadiene Copolymer made with Sodium (Na) Catalyst

BuNH$_2$ tert-Butylamine

Bunny ball Osmylated (OsO$_4$) (60-Carbon-Atom) Buckminsterfullerene Molecule

BuO Butoxy; Butoxide; Butylate

Bu$_2$O Butyl Oxide (or Butyl Ether)

β-U$_3$O$_8$ Beta Triuranium Octoxide

BuOCl t-Butylhypochlorite

BuOH Butanol (or Butyl Alcohol)

t-BuOH tert-Butanol (or tert-Butyl Alcohol)

BuOK Potassium Butoxide

t-BuOK Potassium tert-Butoxide

t-BuOOOH tert-Butyl Hydroperoxide [also t-BuO-O-OH]

BUORD Bureau of Ordnance [also BuORD, or BuOrd] [Now Part of BuWEPS of US Navy]

BuOSO₃H t-Butyl Hydrogen Sulfate

BUOU Backup Optical Unit

BUP Baratron Upgrade Program

BUP Bottom-Up Parsing

BUP British United Press

BUPA British United Provident Association

t-Bu₄Pc tert-Tetrabutyl Phthalocyanine

BUPCR Bath University Program of Catalogue Research [UK]

BUPERS Bureau of Naval Personnel [US Navy]

bupy butylpyridine

t-bupy 4-tert-butylpyridine

BUR Back-Up Rate

BUR Back-Up Register

BUR Burmese Kyat [Currency of Burma]

Bur Bureau [also bur]

Bur Burma; Burmese

Bur Burundi

BURA British Urban Regeneration Association

BURB Back-Up Roll Bending

Bur Educ Res Monogr Bureau of Educational Research Monographs [US]

Bur Fr Burundi Franc [Currency of Burundi]

BURISA British Urban and Regional Information Systems Association

BurMines Bureau of Mines [US]

BurMinesTS Bureau of Mines Test Station [Bruceton, Pennsylvania, US]

BURO Bureau Universitaire de Recherche Operationnelle [University Bureau for Operational Research, France]

Bur Std (United States) Bureau of Standards [also Bur St]

Buryat ASSR Buryat Autonomous Soviet Socialist Republic

BUS Broadcast and Unknown Server

BUS Broken-Up Structure

Bus Business [also bus]

bus bushel [Unit]

BUSAK Bus Acknowledgement

BUSANDA Bureau of Supply and Accounts [of US Navy]

Bus Commun Rev Business Communications Review [US]

Bus Comput Commun Business Computing and Communications [Journal published in the UK]

BUSEN Bus Enable

Bus Environ Business and the Environment [US Monthly News Magazine]

Bus Equip Dig Business Equipment Digest [UK]

Bus Forms Labels Syst Business Forms, Labels and Systems [Published in the US]

Bus Forms Syst Business Forms and Systems [Published in the US]

Bush Bushing [also bush]

BuShips (United States) Bureau of Ships [formerly also BUSHIPS; now NSEC]

Busin Business [also busin]

Bus Jpn Business in Japan [Publication] [also Busin Jpn, Bus Jap, or Busin Jap]

BUSM Boston University School of Medicine [Massachusetts, US]

Bus Mgr Business Manager [also BusMgr]

Bus Monitor Rubb Business Monitor: Rubber [Published in the UK]

Bus Monitor Synth Business Monitor: Synthetic Resins [Published in the UK]

BuSn n-Butyltin

Bu₃Sn Tributyltin

BUSRQ Bus Request

Bus Softw Business Software [Journal published in the US]

Bus Softw Rev Business Software Review [US]

Bus Sol Business Solutions [Journal published in the UK]

Bu Std (United States) Bureau of Standards [also BuStd]

Bus Syst Equip Business Systems and Equipment [UK Publication]

BUST Beijing University of Science of Technology [PR China]

Bus Tech Manage Business and Technical Management [UK Journal]

Bus Today Business Today [Published in the UK]

Bus Transp Bus Transportation (Magazine) [US]

BUT Broadband Unbalanced Transformer

But Button [also but]

But Butyl

But Alc Butyl Alcohol [also but alc]

BUTAMER Butadiene Isomerization (Process)

BUTESOM Butene Isomerization (Process)

BUTEX Dibutoxy Diethyl Ether

BUTSC Butyraldehyde Thiosemicarbazone

Butt Buttress (Screw Thread)

BuTTN Butanetriol Trinitrate

ButylPBD 2-(4Biphenylyl)5-(4-tert-Butylphenyl)-1,3,4-Oxadiazole

n-Butyryl-S-ACP n-Butyryl-S-Acyl Carrier Protein [Biochemistry]

BUV Backscatter Ultraviolet

BUWAL Bundesamt für Umwelt, Wald und Landschaft [Federal Department forEnvironment, Forestry and Countryside, Switzerland]

BuWEPS Bureau of Naval Weapons [also BUWEPS, BuWeps, or BNW] [of US Navy]

Buz Buzzer [also buz]

n-BuZr n-Butylzirconium

BV Back View

BV Balanced Force [Controllers]

BV Ball Valve

BV Baudot-Verdan (Telegraph)

BV Bauer-Vogel (Oxide Conversion Process)

BV Bed Volume

BV Berne Virus

BV Biological Value [Biochemistry]

BV Blood Vessel

BV Bochumer Verein [Vacuum Degassing Process]
BV Book Value
BV Boundary Value
BV Bouvet Island [ISO Code]
BV Bovine Virus
BV Breakdown Voltage [Electronics]
BV Brookfield Viscometer
BV Bureau Veritas [France]
BV Burgers Vector [Crystallography]
BV Busy Verification
BV Butterfly Valve
BV_{dg} Reverse Drain to Gate Voltage [Semiconductor Symbol]
B/V Bleed(er) Valve
BVA Bachelor of Vocational Agriculture
BVA Billing Validation Application [Telecommunications]
BVA British Veterinary Association
BVA British Videogram Association
BVAMA British Valve and Actuators Manufacturers Association
BVC Black Varnish Cambric (Insulation)
BVC British Vacuum Council
BVCA British Venture Capital Association
BVD Belgische Vereniging voor Dokumentation [Belgian Documentation Association]
BVD Bovine Virus Diarrhea
BVD Brand-Verhütungs-Dienst (für Industrie und Gewerbe) [Fire Prevention Service (for Industry and Trade), Switzerland]
BVDU (E)-5-(2-Bromovinyl)2'-Deoxyuridine [Biochemistry]
BVE Bachelor of Vocational Education
BVE Bivariate Exponential Distribution
BVE Butyl Vinyl Ether
BVF Belgische Vereniging voor Fotogrammetrie [Belgian Society for Photogrammetry]
BVH Base Video Handler
BVHA British Veterinary Hospitals Association
BVI British Virgin Islands
BVM Bureau of Veterinary Medicine [of US Food and Drug Administration]
BvM Bond voor Materialienkennis [Society for Materials, Netherlands]
BVMA British Valve Manufacturers Association
BVO Brominated Vegetable Oil
β-VOPO$_4$ Beta Vanadium Phosphate
BVP Back Vertex Power
BVP Boundary Value Problem [Mathematics]
BVP Bulb-Venturi Pump
BVRA British Veterinary Radiology Association
BVRLA British Vehicle Renting and Leasing Association
BVS Bachelor of Veterinary Science
BVS Bakable Vacuum System
BVS Bibliotheksverbundsystem [German Library Network]
BVS Bond Valence Sum
BVSc Bachelor of Veterinary Science

BVSV Bimetal Switching Valve
BVT Barium Vanadium Tantalate
Bvt Brevet [Military]
BVU Brightness Value Unit
BVU (E)-5-(2-Bromovinyl)uridine [Biochemistry]
BVV Ball Bearing, Double-Row, Angular-Contact, Separable, Contact Angle Vertex Outside Bearing, Two-Piece Inner Ring [Symbol]
BVW Backward Volume Wave
BVZS British Veterinary Zoological Association
BW Backing Wind
BW Backward Wave
BW Bacteriological Warfare
BW Bandwidth [also Bw, or bw]
BW Barbed Wire
BW Barium Tungstate
BW Bauwirtschaft [German Journal for the Construction Industry]
BW Berliner Welle [Berlin Wave] [A German Radio Station]
BW Biological Warfare
BW Biological Wastewater
BW Biological Weathering
BW Black and White [also B&W, or B/W]
BW Bloch Wall [Solid-State Physics]
BW Bloch Wave [Physics]
BW Boiling Water [also bw]
BW Both Ways
BW Botswana [ISO Code]
BW Bow Wave
BW Bragg-Williams (Ordering Theory) [also B-W] [Solid-State Physics]
BW Braided Wire
BW Brain Wave
BW Breit-Wigner (Theory) [Nuclear Physics]
BW Bridge Wire
BW Buried Wire
BW Burnished Ware
BW Desert Climate [Classification Symbol]
B&W Babcock and Wilcox Company [US]
B&W Black and White [also BW, or B/W]
B-W Bragg-Williams (Ordering Theory) [also BW] [Solid-State Physics]
Bw Bundeswehr [Federal Armed Forces, Germany]
Bw Dry Climate, Arid [Symbol]
B(w) Boron Whisker
.bw Botswana [Country Code/Domain Name]
β-W Beta Tungsten [Symbol]
BWA Backward Wave Amplifier
BWA Bandwidth Allocation
BWA Betriebswirtschafts-Akademie [Academy of Industrial Engineering, Germany]
BWA Bragg-Williams Approximation [Solid-State Physics]
BWA Bridge-Deck Waterproofing Association [UK]
BWA British Waterfowl Association

BWA British Waterways Association

BWA British Waterworks Association

BWA British West Africa

BWA British Wildlife Appeal [of Royal Society for Nature Conservation, UK]

BWAHDA British Warm Air Hand Drier Association

BWAS Bristol and Warwickshire Archeological Society [UK]

BWB Bundesamt für Wehrtechnik und Beschaffung [Federal Office for Military Engineering and Procurement, Germany]

BWC Backward Wave Converter

BWC Board Wood Cellulose

BWC Buffer Word Counter

β-WC Beta Tungsten Carbide

BWCC British Weed Control Council

BWCMG British Watch and Clockmakers Guild

BW&Co Burroughs Wellcome and Company Limited [Pharmaceutical Manufacturer]

BWCS British White Cattle Society

BWD Barrels of Water per Day [also bwd, BW/D, bw/d, BW D^1, or bw d^1]

BWDA Bicycle Wholesale Distributors Association [US]

BWDD Biaxial Wedge Disclination Dipole [Metallurgy]

BWE Bachelor of Welding Engineering

BWE Bucket-Wheel Excavator

BWEA British Wind Energy Association

BWEN Brailsford-Wynblatt Encounter Modified Model of Microstructural Coarsening) [Metallurgy]

BWEN Brillouin-Wigner (Perturbation Theory with) Epstein-Nesbert (Partitioning) [Physics]

BWF Bloch Wave Function [Physics]

BWF Breit-Wigner-Fano (Lineshape) [Physics]

BWF British Whiting Federation

BWF British Woodworking Federation

BWG Biphenyl Working Group [US]

BWG Birmingham Wire Gauge

BWG Bragg-Williams-Gorsky (Model) [also B-W-G] [Solid-State Physics]

BWG British Imperial Wire Gauge

BWH Barrels of Water per Hour [also bwh,BW/H, or bw/h]

BWh Tropical and Subtropical Desert Climate [Classification Symbol]

BWHC Babcock and Wilcox Hanford Company [US]

BWI Bahamas and West Indies Airlines

BWI Betriebswirtschaftliches Institut [Institute for Industrial Engineering; of Verein Deutscher Eisenhüttenleute, Germany]

BWI Boating Writers International [US]

BWI British West Indies

BWID Buried (Nuclear) Waste Integrated Demonstration [of US Department of Energy]

BWIG British Water Industries Group

BWIP Basalt Waste Isolation Project [of US Department of Energy]

BWK Brennstoff–Wärme–Kraft [German Journal on Fuels, Heat and Power; published by Verein Deutscher Ingenieure]

Bwk Middle Latitude Desert Climate [Classification Symbol]

BWM Backward-Wave Magnetron

BWM Barrels of Water per Minute [also bwm]

BWM Block-Write Mode

BWMA British Woodwork Manufacturers Association

BWMB British Wool Marketing Board

BWNMA British Wire Netting Manufacturers Association

BWO Backward-Wave Oscillator

BWO Backwave Oscillator

BWO Barium Tungsten Oxide

BWO Battelle Washington Office [US]

BWOC By Weight of Cement

BWP Babcock and Wilcox Protec Inc. [US]

BWP Botswana Pula [Currency]

BWP Brown Wrapping Paper

BWPA Backward-Wave Power Amplifier

BWPA British Wastepaper Association

BWPA British Women Pilots Association

BWPA British Wood Preserving (and Dampproofing) Association

BWPA British Wood Pulp Association

BWPD Barrels of Water per Day [also bwpd]

BWPUC British Wastepaper Utilization Council

BWR Bandwidth Ratio

BWR Boiling-Water Reactor

BWRA British Welding Research Association

BWRE Biological Warfare Research Establishment [UK]

BWRRA British Wire Rod Rollers Association

BWSC Bureau of Waste Site Cleanup [of Massachusetts Department of Environmental Protection, US]

BWST Borated Water Storage Tank

BWT Backward-Wave Tube

BWT Bilateral Wilms' Tumor [Medicine]

BWT Biological Wastewater Treatment

BWT Bloch Wave Theory [Physics]

BWTA British Wood Turners Association

BWTSDS Base Wire and Telephone System Development Schedule

BWV Back Water Valve

BWW Biological Warfare Weapons

BWW Buses Worldwide [UK]

BWWEA British Woven Wire Export Association

BWX Babcock and Wilcox Inc. [US]

BWXT Babcock and Wilcox Technologies Inc. [Lynchburg, Virginia, US]

BWXT Forum Babcock and Wilcox Technologies Forum [Quarterly newsletter published in Lynchburg, Virginia, US]

BX Base Exchange

BX Bicrystal

BX Branch Exchange

BX Flexible Armored Cable

BX₃ Boron Halide [Formula]

B2X Binary-To-Hexadecimal [Restructured Extended Executor (Language)]

Bx Biopsy

Bx Box [also bx]

Bx$_a$ Acute Bisectrix (of a Biaxial Crystal) [Symbol]

Bx$_o$ Obtuse Bisectrix (of a Biaxial Crystal) [Symbol]

B(x) Dislocation Distribution Function (in Materials Science)[Symbol]

β(x) dislocation distribution (in materials science) [Symbol]

BXB British Crossbar

Bxs Boxes [also bxs]

BY Ball Bearing, Single-Row, Angular-Contact, Two-Piece Outer Ring [Symbol]

BY Belarus [ISO Code]

BY Budget Year

BY Byte Pointer

BY Molecule Composed of Nonmetal Atom(s) *B* and Hydride Atom(s) *Y* [Symbol]

.by Belarus [Country Code/Domain Name]

β-Y Beta Yttrium [Symbol]

BYAS Bypass Ammonia Synthesis (Process) [Chemical Engineering]

β-Yb Beta Ytterbium [Symbol]

BYCO Barium Yttrium Copper Oxide (Superconductor)

BYCO-213 Ba$_2$YCu$_3$O$_{7-x}$ [A Barium Yttrium Copper Oxide Superconductor]

BYCOAg Barium Yttrium Copper Oxide–Silver (Composite)

BYCWO Barium Yttrium Copper Tungsten Oxide (Superconductor)

BYCWO-2121 Ba$_2$YCu$_2$WO$_y$ [A Barium Yttrium Copper Tungsten Oxide Superconductor]

BYCWO-4112 Ba$_4$YCuW$_2$WO$_{6-y}$ [A Barium Yttrium Copper Tungsten Oxide Superconductor]

BYCWO-8314 Ba$_8$Y$_3$CuW$_4$WO$_{24-z}$ [A Barium Yttrium Copper Tungsten Oxide Superconductor]

BYD Bureau of Yards and Docks [US]

byd beyond

BYDV Barley Yellow Dwarf Virus

BYF Barium Yttrium Fluoride (Crystal)

BYMUX Byte-Multiplexer Channel

BYMV Barley Yellow Mosaic Virus

BYN Barium Ytterbium Niobate

BYO Beryllium Yttrium Oxide

ByoA Trans-2-Butynoic Acid

BYWO Barium Yttrium Tungsten Oxide

Byp Bypass [also byp]

Byr Buyer [also byr]

BYTE PTR Byte Pointer [also Byte Ptr]

BYU Barium Yttrium Uranate

BYU Brigham Young University [Provo, Utah, US]

BYV Beet Yellows Virus

Byz Byzantine

BZ Ball Bearing, Single-Row, Angular-Contact, Two-Piece Inner Ring [Symbol]

BZ Barium Zirconate

BZ Belize [ISO Code]

BZ Benzidine

BZ Boundary Zone

BZ Brillouin Zone [also Bz] [Solid-State Physics]

BZ Quinuclidinyl Benzilate

Bz Benzene [also bz]

Bz Benzoate

Bz Benzoyl

Bz Benzyl

Bz Brillouin Zone [also BZ] [Solid-State Physics]

Bz Bronze

bz benzene [also BZ]

.bz Belize [Country Code/Domain Name]

BZA British Zeolite Association

BzAc Benzoylacetonate; Benzoylacetone [also Bzac]

BzAc Benzyl Acetate

BzAl Aluminum Bronze

BzAspOO' N-Benzoyl Aspartate [also bzaspOO'] [Biochemistry]

BZBA 1,3-Diphenylpropane-1,3-Dione

BZD Belize Dollar [Currency of Belize]

BZD Benzidine

Bzd Benzamide

Bz1,3,5d$_3$ 1,3,5-Trideuteriobenzene

Bzde Benzidine

Bzdl Benzimidazole

BzDMTdA N6-Benzoyl-5'-O-(4,4-Dimethoxytrityl)-2'-Deoxyadenosine [Biochemistry]

BzDMTdC N4-Benzoyl-5'-O-(4,4-Dimethoxytrityl)-2'-Deoxycytidine [Biochemistry]

BzEt$_2$ Diethylbenzene

BzEt$_3$ Triethylbenzene

BzgA Bundeszentrale für gesundheitliche Aufklärung [Federal Center for Health Education, Germany]

BzH Benzaldehyde

Bzh Benzyhydryl

Bzhoc Benzhydryloxycarbonyl

BIMID Benzimidazole

BZK Benzalkonium Chloride

Bzl Benzoyl

Bzl Benzyl

BzLArgMCA N-(α)-Benzoy-L-Arginine-4-Methylcoumaryl-7-Amide [Biochemistry]

BzLAspOO' N-Benzoyloxycarbonyl Aspartate [also bzlaspOO'] [Biochemistry]

BzlMgCl Benzylmagnesium Chloride

Bzm Benzamidine

Bzma Benzylmethylamine

BzMe$_3$ Trimethylbenzene

Bz5MeDMTdC N4-Benzoyl-5'-O-(4,4-Dimethoxytrityl)-5-Methyl-2'-Deoxyadenosine [Biochemistry]

β-ZnS Beta Zinc Sulfide (or Sphalerite) (Structure)

BZO Barium Zirconium Oxide

BZO Barium Zirconyl Oxalate

BzOH Benzoic Acid

BzOH Benzyl Alcohol

β-Zr Beta Zirconium [Symbol]

BZS Britain Zimbabwe Society [UK]

BZS Bundesamt für Zivilschutz [Federal Office for Civil Defense, Germany]

Bzs Benzylmercaptan

BZT Barium Zinc Tantalate

Bztfac Benzoyltrifluoroacetonate [also bztfac]

Bztrz 1H-Benzotriazole [also bztrz]

C   Additional Designation of Emission (or Transmission) Type–Vestigial Sideband [Symbol] [also c]

C   American Standard Steel Channel [AISI/AISC Designation]

C   Base-Centered Space Lattice in Monoclinic, Orthorhombic and Hexagonal Crystal Systems [Herman Mauguin Symbol]

C   Basic Load Rating (of a Ball, or Roller Bearing) [Symbol]

C   Bearing Capacity [Symbol]

C   Cache

C   Cactus

C   Caecilioides [Genus of Snails]

C   Caffeine

C   Cairo [Egypt]

C   Calcification; Calcify

C   Calcium Oxide [Ceramics]

C   Calculate; Calculation; Calculator

C   Calcutta [India]

C   Calgary [Alberta, Canada]

C   Calibrate; Calibration; Calibrator

C   Calibration Constant [Symbol]

C   California [US]

C   Caliper

C   Call(er); Calling

C   Calm [also C]

C   Calories; Calorific

C   Calorimetric; Calorimetry

C   Cambrian [Geological Period]

C   Camel

C   Cameroon

C   Can

C   Canner(y)

C   Canning

C   Canada; Canadian

C   Canal

C   Canberra [Australia]

C   Cancel(lation)

C   Cancer(ous)

C   Candida [Genus of Fungi]

C   Candidate

C   Candle [Unit]

C   Canine [Zoology]

C   Canine [Dentistry]

C   Canis [Genus of Mammals Comprising Coyotes, Dogs, Foxes, Wolves, etc.]

C   Cannabis

C   Canton [PR China]

C   Cap

C   Capability; Capable

C   Capacitance [Symbol]

C   Capacitor [Symbol]

C   Capacity [also c]

C   Cape

C   Capillary, Capillarity

C   Capillary Number (in Fluid Flow) [Symbol]

C   Capped; Capper; Capping

C   Capsicum [Botany]

C   Capsule

C   Car

C   Caracas [Venezuela]

C   Caramel

C   Carapace

C   Carbamite

C   Carbide

C   Carbon [Symbol]

C   Carbonaceous

C   Carbon Black

C   Carboniferous [Geological Period]

C   Carbonize(r); Carbonization

C   Carboxy-

C   Carboxylate [also c]

C   Carburize(r); Carburization

C   Card

C   Carder; Carding

C   Cardia(c)

C   Cardiff [UK]

C   Cardiological; Cardiologist; Cardiology

C   Care

C   Cargo

C   Cargo/Transport (Aircraft) [US Air Force and US Army Symbol]

C   Caribbean

C   Carnivora [Order of Mammals Including Bears, Cats and Dogs]

C   Carnivore; Carnivorous

C   Carp

C   Carpal; Carpus [Anatomy]

C   Carrier

C   Cartilage

C   Cartoid [Zoology]

C   Cartridge

C   Carya [Genus of Deciduous Trees Including the Hickories]
C   Casablanca [Morocco]
C   Cascade
C   Case
C   Cast(ing); Caster
C   Castanea [Genus of Deciduous Trees Including the Chestnuts and Chinkapins]
C   Cat
C   Catalysis; Catalyst; Catalytic; Catalyzer
C   Catechol
C   Category
C   Cathode; Cathodic
C   Catholic
C   Caucasia(n)
C   Caucasian; Caucasoid [Anthropology]
C   Caucasus (Mountains)
C   Causal(ity)
C   Cave
C   Cavity
C   c-axis to a-axis ratio (of crystal lattice) (for hexagonal, rhombohedral and tetragonal systems) [Symbols]
C   c-axis to b-axis ratio (of crystal lattice) (for monoclinic, triclinic and orthorhombic systems) [Symbol]
C   Cecum (or Caecum) [Anatomy]
C   Cedrus [Genus of Trees Referring to the True Cedars of Lebanon, Northern Africa and the Himalayas]
C   Ceiling
C   Ceiling (Exposure) Limit [Toxicology]
C   Celebes [Indonesia]
C   Cell(ular)
C   Cellulose
C   Celsius
C   Cement(itious)
C   Cemented; Cementing; Cementation
C   Cementite [Metallurgy]
C   Cenozoic [Geological Era]
C   Cental [Unit]
C   Center
C   Center Distance (of Gearings) [Symbol]
C   Central(ization)
C   Centigrade [also c]
C   Centimeter [also c]
C   Centrifugal; Centrifuge
C   Centripetal
C   Ceramic(s)
C   Cerebellar; Cerebellum [Anatomy]
C   Cerebral; Cerebrum [Anatomy]
C   Ceres [Astronomy]
C   Certificate; Certify
C   Cervical; Cervix [Anatomy]
C   Cetacea [Order of Mammals Including Dolphins, Porpoises and Whales]
C   Ceylon [now Sri Lanka]

C   Chad
C   Chain
C   Chains [Heterogeneous Connection of Atoms] [Symbol]
C   Chair
C   Chamber
C   Chameleon
C   Chance
C   Change(r) Changing
C   Channel
C   Chapter [also c]
C   Character; Characteristic(s)
C   Charge
C   Charge Conjugation (Symmetry) [Particle Physics]
C   Charlie [Phonetic Alphabet]
C   Charlotte [North Carolina, US]
C   Charlottetown [Prince Edward Island, Canada]
C   Chart
C   Chelate(d); Chelation
C   Chemical; Chemist(ry)
C   Chemisorption; Chemisorptive
C   Chesapeake [Virginia]
C   Chicago [Illinois, US]
C   Chile(an)
C   Chip(ping)
C   Chiral(ity) [Chemistry]
C   Chiropodist; Chiropody
C   Chiropractic; Chiropractor
C   Chiroptera [Order of Mammals Including the Bats]
C   Chlamydia [Microbiology]
C   Chlamydomonas [Genus of Algae]
C   Chlorella [Genus of Algae]
C   Chlorination
C   Chlorite
C   Chloroform
C   Chlorogonium [Genus of Algae]
C   Chloroplast [Botony]
C   Chop(per); Chopping
C   Chord
C   Chorea [Medicine]
C   Chorion(ic)
C   C Horizon (Composed of Weathered Soil Parent Material) [Symbol]
C   Chromatograph(y)
C   Chromosome
C   Chronic
C   Chronological; Chronology
C   Chungking (or Chongqing) [PR China]
C   Cincinnati [Ohio, US]
C   Cinnamon
C   Circle; Circular [also c]
C   Circle of Confusion (in Optics) [Symbol]
C   Circuit [also c]
C   Circulate; Circulating; Circulation
C   Circumference [Symbol]

C   Circumference; Circumferential
C   Cis (Configuration) [Chemistry]
C   Citrus
C   City
C   Cladophora [Genus of Algae]
C   Class
C   Classifier; Classification; Classify
C   Clean(er); Cleaning
C   Cleanliness
C   Clear(ing)
C   Clearance
C   Cleveland [Ohio, US]
C   Client
C   Clitoris
C   Climatic; Climate
C   Climatological; Climatologist; Climatology
C   Clinic(al); Clinician
C   Cloaca [Zoology]
C   Clock
C   Closure
C   Clot(ting)
C   Cloth(ing)
C   Cloud(y)
C   Coagulate; Coagulation
C   Coal
C   Coat(er); Coating
C   Cocaine
C   Coccal; Coccus [Microbiology]
C   Coccyx [Anatomy]
C   Cochlea(r)
C   Cocoa
C   Cocconeis [Genus of Algae]
C   Cod
C   Code
C   Coefficient
C   Coelastrum [Genus of Algae]
C   Coffee
C   Coffin Type (Vehicle Launch) [USDOD Symbol]
C   Coherence; Coherent
C   Cohesion; Cohesive
C   Coil(ing)
C   Coiler
C   Cold Lake (Bitumen) [Alberta Oil Sands]
C   Colon [Anatomy]
C   Coke
C   Collagen [Biochemistry]
C   Collapse
C   Collar
C   Collect(ion); Collector
C   Collector [Semiconductor Symbol]
C   College
C   Collegiate
C   Collimation; Collimator

C   Collision
C   Colloid(al)
C   Cologne [Germany]
C   Colombia(n)
C   Colombo [Sri Lanka]
C   Colon
C   Colón [Currency of Costa Rica]
C   Colonial; Colony [Geosciences]
C   Colony [Biosciences]
C   Colorado [US]
C   Color(ation); Coloring
C   Columbia [South Carolina]
C   Columbus [Ohio, US]
C   Column(ar)
C   Coma
C   Combat
C   Combination; Combinatory; Combine
C   Combustibility; Combustible
C   Comet
C   Command(ing)
C   Commensal(ism) [Ecology]
C   Commensurate (Phase) [Materials Science]
C   Comment
C   Commerce; Commercial
C   Commission(er)
C   Committee
C   Commonwealth (International)
C   Communicability; Communicable
C   Communicate; Communication; Communicator
C   Commutative (Number)
C   Commutator
C   Compact(ion); Compactor
C   Comparator; Compare; Comparison
C   Compatibility; Compatible
C   Complain(t)
C   Complement(arity); Complementary
C   Complementation
C   Complete; Completion
C   Complex(ity)
C   Complex Cubic [Crystallography]
C   Compliance; Compliant
C   Compliance (in Mechanics) [Symbol]
C   Complication
C   Component
C   Compose; Composition
C   Composite
C   Composite Strength [Symbol]
C   Compost(er); Composting
C   Compound(ing)
C   Compress(ed); Compression; Compressor
C   Computation; Computer; Computing
C   Concentrate; Concentration; Concentrator
C   Concentration Rate [Symbol]

C   Concept(ion)
C   Concord [New Hampshire]
C   Concrete
C   Condensation; Condense(r)
C   Condition(al)
C   Conductance (of a Fluid) [Symbol]
C   Conduction; Conductive; Conductor
C   Conduit
C   Cone; Conical
C   Cone Distance (of Bevel Gears) [Symbol]
C   Conference; Conferencing
C   Confidence; Confidential
C   Configuration
C   Confirm(ation)
C   Conform(ation)
C   Congest(ion)
C   (Congius) – gallon [Medical Prescription]
C   Congo
C   Congress(ional)
C   Conjugation
C   Conjunction
C   Conjunctiva [Anatomy]
C   Connecticut [US]
C   Connect(ion); Connector
C   Connectivity
C   Consent
C   Conservation(ist); Conserve
C   Consolidate(d); Consolidation
C   Constant
C   (Arbitrary) Constant of Integration [Symbol]
C   Constituent
C   Constitution(al)
C   Constrict(ion)
C   Consult(ant); Consultation
C   Consumption
C   Contact [also c]
C   Contain(er)
C   Contaminate; Contaminant; Contamination
C   Content
C   Contention
C   Continuation; Continuity; Continuous
C   Continuous Exposure [Toxicology]
C   Continuum
C   Contour(ing)
C   Contract(or)
C   Contractile; Contraction
C   Contracture
C   Contradict(ion)
C   Contrast
C   Control
C   Controller
C   Control Dependency [Digital Logic]
C   Conus

C   Convalescence; Convalescent
C   Convection
C   Convention(al)
C   Convergence; Convergent
C   Convolute; Convolution
C   Conversion; Convert(er)
C   Coolant; Cool(er); Cooling
C   Coordinate; Coordination
C   Copenhagen [Denmark]
C   Copier; Copy(ing)
C   Copyright [also c]
C   Coral
C   Cord(age)
C   Córdoba [Currency of Nicaragua]
C   Core [also c]
C   Corium
C   Cornea (of the Eye)
C   Corner [or c]
C   Corona
C   Coronary
C   Coroner
C   Corps
C   Correct(ed); Correction
C   Correlation Function [Symbol]
C   Corrosion; Corrosive
C   Cortex
C   Cortisone [Biochemistry]
C   Corundum
C   Corynebacterium [Microbiology]
C   Cost(ing)
C   Cotton
C   Coulomb [Unit]
C   Coulombic
C   Coulombmeter
C   Council
C   Count(ing); Counter
C   Coupe
C   Couple(d); Coupling
C   Coupling Constant [Symbol]
C   Course
C   Court
C   Cover(age)
C   Covering
C   C Programming Language
C   Crane
C   Cranial; Cranium
C   Crash(ing)
C   Crate
C   Cream
C   Creep(ing); Creepage
C   Crest
C   Cretaceous [Geological Period]
C   Crete; Cretan

C　Critic(al); Criticality
C　Critical Angle (or Total Reflection Angle) [Symbol]
C　Crop [Ornithology]
C　Crop(per); Cropping
C　Cross(wise)
C　Crosswind
C　Croup [Medicine]
C　Crown
C　Crush(er); Crushing
C　Cryostat(ic)
C　Crystal(line); Crystallinity
C　Crystallographer; Crystallographic; Crystallography
C　Crystal Phase
C　C-Size Sheet (of Paper) [17" × 22"]
C　Cuba(n)
C　Cube; Cubic
C　Cubicle
C　Cubic Slip (in Metallurgy) [Symbol]
C　Cucumber
C　Cucumis [Genus of Plants Including the Cucumbers]
C　Culex [Genus of Mosquitoes]
C　Culcoides [Species of Midges]
C　Cultural; Culture [Anthropology]
C　Culture [Biology]
C　Cumulative
C　Cupola
C　Cupressus [Genus of Trees Including the True Cypresses]
C　Cure
C　Curettage [Medicine]
C　Curie [Unit]
C　Curie (-Weiss) Constant [Symbol]
C　Current
C　Curvature; Curve(d)
C　Cuspid [Dentistry]
C　Cuticle [Biology]
C　Cut(ter); Cutting
C　Cyan
C　Cyanamide
C　Cybernetic(s)
C　Cycle [also c]
C　Cycle Ratio (in Fatigue) [Symbol]
C　Cyclization
C　Cyclotella [Genus of Algae]
C　Cylinder; Cylindrical
C　Cymbella [Genus of Algae]
C　Cyprus
C　Cyst(ic)
C　(-)-Cysteine; Cysteinyl [Biochemistry]
C　Cytidine [Biochemistry]
C　Cytoplasm
C　Cytosine [Biochemistry]
C　Czech
C　Czechoslovakia(n)

C　Decentration (of Decentered Lenses) [Symbol]
C　Dynamic Load Rating (of Roller Bearing) [Symbol]
C　Electrostatic Capacity [Symbol]
C　Feed Factor [Symbol]
C　First Tone in the Scale of C Major [Acoustics]
C　Fraunhofer "C" Line [Strong-Red Fraunhofer Line for Hydrogen Having a Wavelength of 766.1 Nanometers [also Hα line] [Spectroscopy]
C　Giant Magnetoresistance Curvature [Symbol]
C　Grade denoting Average Work
C　Heat Capacity [Symbol]
C　Humid Warm (or Mesothermal) Forest Climates, Coldest Month above 32°F (0°C), but below 64.4°F (18°C)/Warmest Month above 50°F (10°C) [Classification Symbol]
C　(One) Hundred
C　Influenza, Strain C [Symbol]
C　Lay Approximately Circular Relative to Center of Nominal Surface [Symbol]
C　Molar Heat (of Solid) [Symbol]
C　Number of Components (in a Mixture) [Symbol]
C　One Hundred [Roman Numeral]
C　Pitch Cone Radius (of Milled Bevel Gears) [Symbol]
C　Pre-exponential Constant [Symbol]
C　Root-Mean-Square Velocity of Molecules [Symbol]
C　Silicon Carbide Type (Grinding Wheel) Abrasive [Symbol]
C　Single-Cotton-Covered (Electric Wire)
C　Specific Dynamic Capacity (of Roller Bearings) [Symbol]
C　Specific Heat Capacity [Symbol]
C　Speed (or Velocity) of Sound [Symbol]
C　(Average) Speed of Propagation [Symbol]
C　Spring Index [Symbol]
C　Stiffness Tensor [Symbol]
C　Storage Compliance [Symbol]
C　Thermal Conductance [Symbol]
C　Vertex, or Associated Polygonal Angle [Symbol]
C　Warm Temperate Rainy Climate [Symbol]
.C　C Source Code [File Name Extension]
.C　C++ (Plus-Plus) Source Code File [File Name Extension]
-C　Carrier Conversion of Noncarrier Type Aircraft [US Navy Suffix]
$\bar{C}$　Average Composition [Symbol]
$\bar{C}$　Carbide [Symbol]
$\bar{C}$　Mean Velocity (of Gas Molecules, Particles, etc.) [Symbol]
$\bar{C}$　One Hundred Thousand [Roman Numeral]
C*　Characteristic Velocity [Symbol]
C*　Chroma [Symbol]
C*　Critical Concentration [Symbol]
C*　Critical Crack Size [Symbol]
C*　Effective Composite Strength [Symbol]
C*　Effective Stiffness Tensor [Symbol]
C*　Magnetic Propagation Vector [Symbol]
°C　Degrees Celsius (or Centigrade) [Unit]
(°C)⁻¹　One per Degree Celsius [Unit]

C_∞ Bulk Concentration [Symbol]

C^+ Carbenium Ion (i.e., Positively Charged Carbon Atom) [Symbol]

C^+ Cation [Symbol]

C^+ Concentration of Positive Ions (in a Solution) [Symbol]

C^- Carbeniate Ion (i.e., Negatively Charged Carbon Atom) [Symbol]

C^- Concentration of Negative Ions (in a Solution) [Symbol]

C_L Center Line [Symbol]

C^0 Continuous Function with Noncontinuous Derivatives [Symbol]

C_0 Bulk Concentration [Symbol]

C_0 Capacitance of Free Space (Vacuum) [Symbol]

C_0 (Unconfined, or Uniaxial) Compressive Strength (of Soil, or Rock) [Symbol]

C_0 Initial Compliance [Symbol]

C_0 Initial Composition [Symbol]

C_0 Initial Concentration (of Impurities, Solutes, etc.) in the Liquid Phase [Symbol]

C_0 Self-Capacitance [Symbol]

C_0 Static Load Rating (of Roller Bearing) [Symbol]

C^1 Continuous Function with Continuous Derivatives [Symbol]

C^{-1} Inverse Capacitance (Matrix) [Symbol]

C_1 Capacitance of Electric Circuit 1 [Symbol]

C^2 Mean Square Velocity (for a System of Particles, Molecules, etc.) [Symbol]

C_2 Ethyl [Symbol]

C_2^{2-} Acetylide [Symbol]

C^3 Cleaved Coupled Cavity [also CCC]

C^3 Command, Control and Communication [also CCC]

C^3 Computer, Communications and Components [also CCC]

C_3 Methyl [Symbol]

C^{4+} Carbon Ion [Symbol]

C_4 Butyl [Symbol]

C_5 Pentyl [Symbol]

C_6 Hexyl [Symbol]

C_7 Heptyl [Symbol]

C_8 Octyl [Symbol]

C_8 Octanoates [Symbol]

C_9 Nonyl [Symbol]

C_{10} Decyl [Symbol]

C_{10} Decanoates [Symbol]

C_{12} Dodecyl [Symbol]

C_{12} Dodecanoates (or Laurates) [Symbol]

C_{16} Hexadecyl [Symbol]

C_{18} Octadecanoates (or Stearates) [Symbol]

C_{18} Octadecyl [Symbol]

C_{44} Buckminsterfullerene Molecule with 44 Carbon Atoms [Symbol]

C_{50} Buckminsterfullerene Molecule with 50 Carbon Atoms [Symbol]

C_{60} Buckminsterfullerene Molecule with 60 Carbon Atoms [Symbol]

C_{70} Buckminsterfullerene Molecule with 70 Carbon Atoms [Symbol]

C_A Alloy Concentration [Symbol]

C_A Capacitance of Amorphous Layer [Symbol]

C_A Concentration of Substance A [Symbol]

C_a Modified Capillary Number (in Fluid Flow) [Symbol]

C_a Total Adhesion (of Soil, or Rock) [Symbol]

C^a Alpha Carbon [or α-C]

C_a Carbon Concentration of Ferrite in Steel [Symbol]

C_a Composition of Alpha Phase (of an Alloy) [Symbol]

C^*_{ab} Chroma [now C*]

C_{ac} Interelectrode Capacitance between Anode and Cathode (of a Triode) [Semiconductor Symbol]

C_{aE} Composition of Alpha Phase (of an Alloy) at the Eutectic Temperature T_E [Symbol]

C_{ag} Anode-to-Grid Capacitance [Semiconductor Symbol]

C_b Bending Rigidity [Symbol]

C_b Bulk Concentration (of Impurities, Solutes, etc.) [Symbol]

C^β Beta Carbon [or β-C]

C_β Composition of Beta Phase (of an Alloy) [Symbol]

$C_{\beta E}$ Composition of Beta Phase (of an Alloy) at the Eutectic Temperature T_E [Symbol]

C_C Chromatic Aberration Coefficient [Symbol]

C_c Compression Index (of Soil, or Rock) [Symbol]

C_{cb} Collector-Base Interterminal Capacitance [Semiconductor Symbol]

C_{cd} Catalogue on CD-ROM [of Elsevier Science, US]

C_{ce} Collector-Emitter Interterminal Capacitance [Semiconductor Symbol]

C_{crit} Critical Capacity (of a Capacitor) [Symbol]

C_D Capacitance of Diode [Symbol]

C_D Drag Coefficient [Symbol]

C_D Mass Discharge Coefficient [Symbol]

C_d Coefficient of Discharge [Symbol]

C_d Diamond Modification of Carbon [Symbol]

C_{Di} Induced Drag Coefficient [Symbol]

C_{ds} Drain-Source Capacitance [Semiconductor Symbol]

C_{du} Drain-Substrate Capacitance [Semiconductor Symbol]

C_{DW} Drag Wave Coefficient [Symbol]

C_E Eutectic Concentration (or Composition) [Symbol]

C_e Eutectoid Concentration (or Composition) [Symbol]

C_e Unloading Compliance

C_{eb} Emitter-Base Interterminal Capacitance [Semiconductor Symbol]

C_{eq} Equivalent Capacity (of a Capacitor) [Symbol]

C_{EXT} External Capacitance (or Capacitor) [Symbol]

C_F Force Coefficient [Symbol]

C_f Compliance of (Test) Load Frame [Symbol]

C_f Skin Friction Coefficient [Symbol]

C_{fb} Flatband Capacitance [also C_{fb}] [Semiconductor Symbol]

C_G Gate Capacitance (in Electronics) [Symbol]

C_g Graphite Modification of Carbon [Symbol]

C_g Specific Heat of Glass [Symbol]

C_g Total Surface Cloudiness [Symbol]

C^γ Gamma Carbon [or γ-C]

C_γ Carbon Concentration of Austenite in Steel [Symbol]

C_{ga} Interelectrode Capacitance between Grid and Anode [Semiconductor Symbol]

C_{gc} Interelectrode Capacitance between Grid and Cathode [Semiconductor Symbol]

C_H Hydrogen Concentration [Symbol]

C_I Interstitial Concentration [Symbol]

C_i Catalogue on the Internet [of Elsevier Science, US]

C_i Complex Magnetostatic Mode [Symbol]

C_i Concentration of Element, Species, Constituent, or Component i (in Weight Percent) [Symbol]

C_i Impurity Concentration [Symbol]

C_i Input Capacitance [Semiconductor Symbol]

C_i Interface Layer Capacitance [Semiconductor Symbol]

C_i' Concentration of Element, Species, Constituent, or Component i (in Atomic Percent) [Symbol]

C_i^B Bulk Concentration of Species i [Symbol]

C_i^ϕ Equilibrium Grain Boundary Mole Fractional Monolayer Segregation Level of Species "i" [Symbol]

C_{ibo} Common-Base Open-Circuit Input Capacitance [Semiconductor Symbol]

C_{ibs} Common-Base Short-Circuit Input Capacitance [Semiconductor Symbol]

C_{ieo} Common-Emitter Open-Circuit Capacitance [Semiconductor Symbol]

C_{ies} Common-Emitter Short-Circuit Input Capacitance [Semiconductor Symbol]

C_{ij} Voigt (Elastic Stiffness) Matrix [Symbol]

C_{ijkl} Elastic Stiffness Constants [Symbol]

C_{INT} Internal Capacitance (or Capacitor) [Symbol]

C_{iss} Common-Source Short-Circuit Input Capacitance [Semiconductor Symbol]

C_L Composition of Liquid (Phase) [Symbol]

C_L Lift Coefficient [Symbol]

C_L Load Capacitance [Symbol]

C_L^* Solute Content of the Liquid at the Solid/Liquid Interface [Symbol]

C_l Composition of Liquid [Symbol]

C_l Concentration (of Impurities, Solutes, etc.) in Liquid (Phase) [Symbol]

C_M Matrix Concentration (of Composites) [Symbol]

C_m Mutual Capacitance (of Two Electric Circuits) [Symbol]

C_m Signal Alternating-Current Capacitance [Semiconductor Symbol]

C_{max} Maximum Capacitance [Symbol]

C_{MG} Monkman-Grant Constant (in Metallurgy) [Symbol]

C_{min} Minimum Capacitance [Symbol]

C_{MM} Magnetic-Magnetic Correlation Function [Symbol]

C_{mp} Molar Heat Capacity at Constant Pressure [Symbol]

C_{mv} Molar Heat Capacity at Constant Volume [Symbol]

C_n Neutralizing (or Balancing) Capacitor [Semiconductor Symbol]

C_n Refractive Index Structure Function Parameter [Symbol]

$C_{n,r}$ Binomial Coefficient Preferably Designated as $\binom{n}{r} = [n(n-1)\cdots(n-r+1)]/[1-2-\cdots-r]$ [also $_nC_r$, or C_r^n] [Symbol]

$C_{n,r}$ Number of Combinations of n Distinct Objects taken r at a Time without Repetition [also $_nC_r$, or C_r^n]

C_o (Composite) Fiber Orientation Factor

C_o Output Capacitance [Semiconductor Symbol]

C_{obo} Common-Base Open-Circuit Output Capacitance [Semiconductor Symbol]

C_{obs} Common-Base Short-Circuit Output Capacitance [Semiconductor Symbol]

C_{oeo} Common-Emitter open-circuit Output Capacitance [Semiconductor Symbol]

C_{oes} Common-Emitter Short-Circuit Output Capacitance [Semiconductor Symbol]

C_{oss} Common-Source Short-Circuit Output Capacitance [Semiconductor Symbol]

C_{ox} Oxide Capacitance per Unit Area [Semiconductor Symbol]

C_P Perod Surface [Symbol]

C_p (Statistical) Capability Index (of Manufacturing Processing) [Symbol]

C_p Equivalent Parallel Capacitance (of a Dielectric) [Symbol]

C_p Molar Heat (or Molar Heat Capacity) at Constant Pressure [Symbol]

C_p Most Probable Velocity of Molecules [Symbol]

C_p Specific Heat at Constant Pressure [Symbol]

C_p° Molar Heat at Standard Pressure [Symbol]

C_{pd} Power Dissipation Capacitance (per Logic Gate) [Semiconductor Symbol]

C_{pk} (Statistical) Capability Index (versus Specifications) [Symbol]

C_{pix} Pixel Capacitance (in Electronics) [Symbol]

C_{pmin} (Cavitation) Coefficient of Minimum Pressure [Symbol]

C_r Relative Consistency (of Soil) [Symbol]

C_{rbs} Common-Base Short-Circuit Reverse Transfer Capacitance [Semiconductor Symbol]

C_{rcs} Common-Collector Short-Circuit Reverse Transfer Capacitance [Semiconductor Symbol]

C_{res} Common-Emitter Short-Circuit Reverse Transfer Capacitance [Semiconductor Symbol]

C_{rss} Common-Source Short-Circuit Reverse Transfer Capacitance [Semiconductor Symbol]

C_s Compliance of (Test) Specimen [Symbol]

C_s Composition of Solid [Symbol]

C_s Concentration (of Impurities, Solutes, etc.) in Solid (Phase) [Symbol]

C_s Spherical Aberration Coefficient [Symbol]

C_s Surface Concentration [Symbol]

C_s^* Solute Content of the Solid at the Solid/Liquid Interface [Symbol]

C_{sat} Saturation Capacity (of a Capacitor) [Symbol]

C_{SM} Structural-Magnetic Correlation Function [Symbol]

C_{SS} Structural-Structural Correlation Function [Symbol]

C_T Capacitance of Trap [Semiconductor Symbol]

C_T (Total) Equivalent Capacitance [Semiconductor Symbol]

C_t Capacity at Time t [Symbol]

C_t Depletion-Layer Capacitance [Semiconductor Symbol]

C_t Torsional Rigidity [Symbol]

C_t Transverse Velocity of Sound [Symbol]

C_{tc} Collector Depletion-Layer Capacitance [Semiconductor Symbol]

C_{te} Emitter Depletion-Layer Capacitance [Semiconductor Symbol]

C_u Coefficient of (Soil) Uniformity [Symbol]

C_u Unit Cost [Symbol]

C^*_{uv} Chroma [now C^*]

C_v Coefficient of Variation [Symbol]

C_v Vacancy Concentration [Symbol]

C_v Capacitance of Vacuum [Symbol]

C_v Flow Coefficient [Symbol]

C_v Molar Heat (or Molar Heat Capacity) at Constant Volume [Symbol]

C_{va} Specific Heat of Translational Degrees of Freedom [Symbol]

C_{vi}^{st} Total Heat of Vibration (in a Gas) [Symbol]

C_{vi}^{t} Effective Total Heat of Vibration (in a Gas) at an Instant [Symbol]

C^x Continuity (or Smoothness) of Function (or Curve) (in Mathematics) [Symbol]

C_x (Atomic) Concentration of Element x [Symbol]

C+ C Plus (High-Level Programming Language)

C+ C Plus (i.e., Positive Terminal of C Battery)

C++ C Plus Plus [Object-Oriented Programming Language]

C– C Minus (i.e., Negative Terminal of C Battery)

C1 Chief Petty Officer, First Class [Military]

C2 Chief Petty Officer, Second Class [Military]

C3 Carbon Cycle in Plant Varieties, e.g., Barley, Rice, Potatoes and Wheat [Symbol]

C4 Carbon Cycle in Plant Varieties, e.g., Maize, Sorghum and Sugar Cane] [Symbol]

C4 Center for Clouds, Chemistry and Climate [at Scripps Institute of Oceanography, University of California at San Diego, US]

C4 Controlled Collapse (Micro) Chip Connection

C6 Trichloroacetic Acid Hexyl Ester

C7 Trichloroacetic Acid Heptyl Ester

C8 Trichloroacetic Acid Octyl Ester

C9 Trichloroacetic Acid Nonyl Ester

C10 Trichloroacetic Acid Decyl Ester

C11 Trichloroacetic Acid Undecyl Ester

C12 Trichloroacetic Acid Lauryl Ester

C14 Trichloroacetic Acid Myristyl Ester

C16 Trichloroacetic Acid Palmityl Ester

C18 n-Alkyl Thiol

C18 Trichloroacetic Acid Stearyl Ester

C21 n-Heneicosanoic Acid

C22 Dococanoic Acid (or Behenic Acid)

9C3 9-Crown-3 [1,4,7-Trioxacyclononane] [also 9-Crown-3, or 9-crown-3]

12C4 12-Crown-4 [1,4,7,10-Tetraoxacyclododecane] [also 12-Crown-4, or 12-crown-4]

15C5 15-Crown-5 [1,4,7,10,13-Pentaoxacyclopentadecane] [also 15-Crown-5, or 15-crown-5]

16C4 16-Crown-4 [1,5,9,13-Tetraoxacyclohexadecane] [also 16-Crown-4, or 16-crown-4]

18C6 18-Crown-6 [1,4,7,10,13,16-Hexaoxacyclooctadecane] [also 18-Crown-6, or 18-crown-6]

21C7 21-Crown-7 [1,4,7,10,13,16,19-Heptaoxacycloheneicosane] [also 21-Crown-7, or 21-crown-7]

24C8 24-Crown-8 [1,4,7,10,13,16,19,22-Octaoxacyclotetracocane] [also 24-Crown-8, or 24-crown-8]

C-10 Carbon-10 [also ^{10}C, or C^{10}]

C-11 Carbon-11 [also ^{11}C, or C^{11}]

C-12 Carbon-12 [also ^{12}C, C^{12}, or C]

C-13 Carbon-13 [also ^{13}C, or C^{13}]

C-14 Carbon-14 [also ^{14}C, or C^{14}]

C-15 Carbon-15 [also ^{15}C, or C^{15}]

C-50 Buckminsterfullerene Molecule with 50 Carbon Atoms [Symbol]

C-60 Buckminsterfullerene Molecule with 60 Carbon Atoms [Symbol]

C-70 Buckminsterfullerene Molecule with 70 Carbon Atoms [Symbol]

3C- 3C Polytype [Ramsdell Notation: 3 Refers to Number of Layers Necessary to Produce a Unit Cell, and C Refers to the Cubic Symmetry, e.g., 3C-SiC]

4H-C 4H Polytype of Diamond [Ramsdell Notation; "4" Denotes Layer Periodicity; "H" Denotes Hexagonal Structural Symmetry]

6H-C 6H Polytype of Diamond [Ramsdell Notation; "6" Denotes Layer Periodicity; "H" Denotes Hexagonal Structural Symmetry]

8H-C 8H Polytype of Diamond [Ramsdell Notation; "8" Denotes Layer Periodicity; "H" Denotes Hexagonal Structural Symmetry]

10H-C 10H Polytype of Diamond [Ramsdell Notation; "10" Denotes Layer Periodicity; "H" Denotes Hexagonal Structural Symmetry]

15R-C 15R Polytype of Diamond [Ramsdell Notation; "15" Denotes Layer Periodicity; "R" Denotes Rhombohedral Structural Symmetry]

21R-C 21R Polytype of Diamond [Ramsdell Notation; "21" Denotes Layer Periodicity; "R" Denotes Rhombohedral Structural Symmetry]

ℭ C-Fraktur (in Mathematics) [Symbol]

c acceptance number (in acceptance sampling) [Symbol]

c additional designation of emission (or transmission) type–vestigial sideband [Symbol] [also C]

c average speed [Symbol]

c calm [also C]

c calorie [Unit]

c cancel

c candle

c canine [Dentistry]

c capacity [also C]

c   carat [Unit]
c   carry
c   cathode
c   cell constant [Symbol]
c   celerity (or phase velocity) [Symbol]
c   cent
c   center distance [Symbol]
c   centi- [SI Prefix for 10^{-2}]
c   centigrade [also C]
c   centimeter [also C]
c   chains [heterogeneous connection of atoms] [Symbol]
c   chapter [also C]
c   charmed quark [Symbol]
c   chip
c   chord (of a circle) [Symbol]
c   (circa) – approximately; about
c   circle; circular [also C]
c   circle of confusion [Symbol]
c   circuit [also C]
c   cirrus [Meteorology]
c   clear
c   clearance (of gears, etc.) [Symbol]
c   clockwise
c   coating
c   coefficient [Symbol]
c   cognate
c   coherent (defect) [Metallurgy]
c   cohesion (of soil) [Symbol]
c   cold
c   collector [Semiconductor Symbol]
c   color
c   composite
c   compressibility [Symbol]
c   compressor
c   concentration [Symbol]
c   conditional
c   constant [Symbol]
c   contact [also C]
c   continental air mass [Symbol]
c   copy lines (command) [Edlin MS-DOS Line Editor]
c   copyright [also C]
c   core [also C]
c   corner [also C]
c   cost(ing)
c   crystal lattice length along c-axis [Symbol]
c   crystalline (phase, state, etc.)
c   cubic (crystal system) [Symbol]
c   (cum) – with [Medical Prescriptions]
c   curie [Unit]
c   current
c   cycle [also C]
c   cycles per second [Unit]
c   cylindrical

c   damping constant [Symbol]
c   depth of cut (in machining) [Symbol]
c   distance from neutral axis to extreme fiber [Symbol]
c   distance from neutral axis of cross section to column side under compression [Symbol]
c   effective exhaust velocity [Symbol]
c   fatigue ductility exponent [Symbol]
c   heat capacity [Symbol]
c   length of a surface crack [Symbol]
c   length (of side) of a triangle [Symbol]
c   less than four months over 50°F (10°C) [Subtype of Climate Region, e.g., Cc, Dc, Dcf, Dcw, etc.]
c   molar concentration [Symbol]
c   number of defects (in a sample) (in statistics) [Symbol]
c   one half the length of an internal crack [Symbol]
c   semiaxis of an ellipsoid [Symbol]
c   specific heat [Symbol]
c   speed (or velocity) of light (in vacuum) (0.2998×10^9 m/s) [Symbol]
c   spring constant [Symbol]
c   permissible compression stress for (reinforced concrete) column bar [Symbol]
c   permissible stress in reinforced concrete in direct compression [Symbol]
c   phase velocity (of waves) [Symbol]
c   power of continuum (in mathematics) [Symbol]
c   transfinite cardinal number of the set of all real numbers [Symbol]
c   unit cell edge length [Symbol]
c   z-intercept of a plane, or surface [Symbol]
.c   C source code file [File Name Extension]
c-   crystalline [Usually Followed by a Chemical Element, or Compound, e.g., c-Ni is crystalline nickel, or c-SiC represents crystalline silicon carbide]
$c_{||}$ parallel to (or in-plane with) c-axis of superconductor [also c||] [Symbol]
$c_{||}$ parallel to (or in-plane with) (crystal) unit cell axis "c" [also c||] [Symbol]
$c_{||}$ texture with c-axis of film material parallel to substrate surface (for epitaxial growth) [also c||] [Symbol]
$c_{\perp}$ perpendicular (or normal) to c-axis of superconductor [also c⊥] [Symbol]
$c_{\perp}$ perpendicular (or normal) to (crystal) unit cell axis "c" [also c⊥] [Symbol]
$c_{\perp}$ texture with c-axis of film material perpendicular to substrate surface (for epitaxial growth) [also c⊥] [Symbol]
$\bar{c}$ average (or relative) cost [Symbol]
$\bar{c}$ average molecular speed [Symbol]
$\bar{c}$ average number of defects (in a sample) (in statistics) [Symbol]
$\bar{c}$ (cum) – with [Medical Prescriptions]
$\bar{c}$ mean particle velocity [Symbol]
$\bar{c}^2$ mean (or average) square velocity (in kinetic molecular theory) [Symbol]
c*   characteristic exhaust velocity [Symbol]

c*   concentration of electrolyte [Symbol]

c′   true average number of defects per sample (in statistics) [Symbol]

c_0 (initial) bulk concentration (of diffusion species) [Symbol]

c_0 concentration of solvent [Symbol]

c_0 speed of light (in vacuum) [Symbol]

c_1 first radiation constant (3.741775×10^{-16} W·m²) [Symbol]

c_1 speed (or velocity) of light in medium 1 [Symbol]

c_1 Williams-Landel-Ferry constant 1 [Symbol]

c_2 second radiation constant (1.438769×10^{-2} m·K) [Symbol]

c_2 Williams-Landel-Ferry constant 2 [Symbol]

c_A concentration of species A [Symbol]

c_a unit adhesion (of soil, or rock) [Symbol]

$c_{A,b}$ bulk concentration of species A [Symbol]

$c_{A,e}$ concentration of species A at Electrode [Symbol]

c_{av} average concentration [Symbol]

c_c critical damping coefficient [Symbol]

c_d atomic defect concentration [Symbol]

c_e concentration at electrode [Symbol]

c_e equivalent damping constant [Symbol]

c_g group velocity (of waves) [Symbol]

c_l concentration of component l [Symbol]

c_i concentration of macromolecules of size range i (e.g. in a polymer) [Symbol]

c_i concentration of species i (e.g., in a solution) [Symbol]

c_i interstitial concentration [Symbol]

c_{mp} molar heat capacity at constant pressure [Symbol]

c_{mv} molar heat capacity at constant volume [Symbol]

c_n speed (or velocity) of light in a medium with refractive index n [Symbol]

c_P constant-pressure molar heat capacity [Symbol]

c_p specific heat (or specific heat capacity) at constant pressure [Symbol]

c_R Rayleigh wave speed [Symbol]

c_s saturation concentration (of ions in solution) [Symbol]

c_s substrate crystal lattice length along c axis [Symbol]

c_s surface concentration (of diffusion species) [Symbol]

c_T constant-temperature molar heat capacity [Symbol]

c_v coefficient of (soil) consolidation [Symbol]

c_v specific heat (or specific heat capacity) at constant volume [Symbol]

c_v vancancy concentration [Symbol]

c_x concentration of diffusion species in x direction [Symbol]

/c   cpar maxalloc (maximum allocation) space option [MS-DOS Linker]

ℂ   Complex Number [Symbol]

©   Copyright [Symbol]

¢   cent [Unit of Money]

¢   Colón [Currency of Costa Rica and El Salvador]

¢   Cedi [Currency of Ghana]

χ angle [Symbol]

χ chemical scale [Symbol]

χ (lower-case) chi [Greek Alphabet]

χ chi function (or enthalpy function) [Symbol]

χ chi phase (of a material) [Symbol]

χ compressibility [Symbol]

χ crack blunting [Symbol]

χ electrical susceptibility [Symbol]

χ electron affinity [Symbol]

χ homogeneity level [Symbol]

χ Flory parameter [Symbol]

χ nonlinearity [Symbol]

χ oscillation [Symbol]

χ shear moduli ratio [Symbol]

χ' real (or in-phase) component of a-c susceptibility [Symbol]

χ'' imaginary (or out-of-phase) component of a-c susceptibility [Symbol]

$\chi_{||}$ magnetic susceptibility in horizontal, or parallel direction [Symbol]

$\chi_\perp$ magnetic susceptibility in vertical, or perpendicular direction [Symbol]

χ_0 chi meson (in particle physics) [Symbol]

χ_0 free electron susceptibility [Symbol]

χ_0 initial (magnetic) susceptibility [Symbol]

χ_0 Pauli spin susceptibility (in physics) [Symbol]

χ_0 temperature-independent (magnetic) susceptibility [Symbol]

χ_1 first harmonic [Symbol]

χ^2 chi square distribution (a measure of goodness of fit) [Symbol]

χ_2 second harmonic [Symbol]

χ_3 third harmonic [Symbol]

χ_4 fourth harmonic [Symbol]

χ_5 fifth harmonic [Symbol]

χ_{ac} alternating-current (magnetic) susceptibility [Symbol]

χ_{cw} Curie-Weiss susceptibility [Symbol]

χ_d differential susceptibility [Symbol]

χ_K complex magnetooptic polar Kerr angle [Symbol]

χ_L^2 lower fractile of chi square (statistical) distribution [Symbol]

χ_l linear (magnetic) susceptibility [Symbol]

χ_m magnetic susceptibility [SI Symbol]

χ_m mass susceptibility [Symbol]

χ_m' magnetic susceptibility [CGS-EMU Symbol]

χ_{mol} molar susceptibility [Symbol]

χ_q roughness parameter [Symbol]

χ_ρ mass susceptibility [Symbol]

χ_s electron affinity [Semiconductor Symbol]

χ_u^2 upper fractile of chi square (statistical) distribution [Symbol]

CA α-Bromobenzylcyanide

CA Cabibbo Angle [Particle Physics]

CA Cadmium Association [UK]

CA Calcium Aluminate

CA California [US]
CA Calixarene
CA Calorimetric Analysis
CA Cam-Actuated (Press)
CA Canada [ISO Code]
CA Canadian Army [now Canadian Armed Forces]
CA Canonical Assembly [Statistical Mechanics]
CA Capacity Assignment
CA Capillary Action
CA Capillary Attraction
CA Capital Asset
CA Carbohydrate Analysis
CA Carbon Acid
CA Carbon Arc
CA Carrier Aircraft
CA Cartoid Artery [Zoology]
CA Casting Axis [Metallurgy]
CA Celestial Axis
CA Cellular Automaton; Cellular Automata [Computers]
CA Cellulose Acetate
CA Center of Area
CA Central America(n)
CA Centrifugal Accelerator
CA Centrifugal Atomization; Centrifugal Atomizer
CA Certified Accountant
CA Chaney Adapter
CA Channel Adapter
CA Chartered Accountant
CA Chelating Agent
CA Chemical Abstracts [of American Chemical Society, US]
CA Chemical Agent
CA Chemical Analysis
CA Chief Accountant
CA Chloroalkane
CA Chromatic Aberration
CA Chromic Acid
CA Chronological Age
CA Cinnamic Acid
CA Circuit Analysis
CA Circular Aperture
CA Citric Acid
CA Civil Aircraft
CA Civil Aviation
CA Clamp Assembly
CA Clean Air
CA Clear Aperture
CA Clinical Analysis
CA Close-Annealed; Close Annealing [also C/A] [Metallurgy]
CA Coarse Aggregate
CA Coast Artillery
CA Cobalt-Base Alloy
CA Cold Air
CA Collision Activation

CA Collision Avoidance
CA Color Analyzer
CA Column Approximation [Electron Microscopy]
CA Commercial Agent
CA Commonwealth of Australia
CA Communications Adapter
CA Commutative Algebra
CA Company Address
CA Compressed Air
CA Compression Axis
CA Computer Age
CA Computer Algorithm
CA Computer Analysis
CA Computer Application
CA Computer Architecture
CA Computers and Automation
CA Conditional Authorization
CA Cone Angle
CA Conflict Alert
CA Conformational Analysis [Physical Chemistry]
CA Conservation Area
CA Constant Amplitude (Test)
CA Construcciónes Aeronauticas [Spanish Aeronautics Company]
CA Construction Analysis
CA Construction Authorization
CA Consular Agent
CA Consulting Architect
CA Consumers Association [UK]
CA Contact Allergen; Contact Allergy
CA Contingency Abort [Computers]
CA Continue-Any Mode
CA Continuous(ly) Annealed; Continuous Annealing [Metallurgy]
CA Contract Administrator
CA Contract Authorization
CA Contract Award
CA Contraction Allowance
CA Contraction of Area
CA Contrast Analysis
CA Control Accelerometer
CA Control Area
CA Controlled Atmosphere
CA Controller of Accounts
CA Conventional Alloy
CA Coriolis Acceleration
CA Coronary Artery
CA Corrective Action [also C/A]
CA Cortisone Acetate [Biochemistry]
CA Cost Account(ing)
CA Cottrell Atmosphere
CA Crab Apple [US Computer Users Group]
CA Cresyl Acetate
CA Critical Angle

CA Critical Assembly [Nuclear Engineering]

CA Cross-Reacting Antigen [Immunology]

CA Crotonic Acid

CA Croup-Associated Virus

CA Cruising Altitude

CA Cryonics Association [US]

CA Cryptoanalysis; Cryptoanalyst

CA Crystal Axis

CA Current Account

CA Cycloaddition

CA Cycloalkanes

CA Heavy Cruiser [US Navy Symbol]

CA$_2$ Calcium Dialuminate [$CaO \cdot 2Al_2O_3$]

CA$_6$ Calcium Hexaaluminate [$CaO \cdot 6Al_2O_3$]

C$_3$A Tricalcium Aluminate [$3CaO \cdot Al_2O_3$]

C$_3$A$_5$ Tricalcium Pentaaluminate [$3CaO \cdot 5Al_2O_3$]

C$_5$A$_3$ Pentacalcium Trialuminate [$5CaO \cdot 3Al_2O_3$]

C$_{12}$A$_7$ Dodecacalcium Heptaaluminate [$12CaO \cdot 7Al_2O_3$]

C$_m$A$_n$ Valence Compound Having m Cations and n Anions

CA65 California List of Known Cancer, or Reproductive-Effect Causing Chemicals [US]

C/A Calcium-to-Aluminum (Ratio) [Ceramics]

C/A Capital Account [also c/a]

C/A Close-Annealed; Close Annealing [also CA] [Metallurgy]

C/A Corrective Action [also CA]

C/A Counter-Attack

C/A Credit Account [also c/a]

C/A Current Account [also c/a]

C-A Crystalline-(to-)Amorphous (Transition)

Ca Cable

Ca Calcium [Symbol]

Ca Cancel [also ca]

Ca Candle [also ca]

Ca Capillary Number [Symbol]

Ca Cathode

Ca Humid Subtropical Climate, Warm Summer [Classification Symbol]

Ca^{2+} Calcium Ion [also Ca^{++}] [Symbol]

Ca-39 Calcium-39 [also ^{39}Ca, or Ca^{39}]

Ca-40 Calcium-40 [also ^{40}Ca, or Ca^{40}]

Ca-41 Calcium-41 [also ^{41}Ca, or Ca^{41}]

Ca-42 Calcium-42 [also ^{42}Ca, or Ca^{42}]

Ca-43 Calcium-43 [also ^{43}Ca, or Ca^{43}]

Ca-44 Calcium-44 [also ^{44}Ca, or Ca^{44}]

Ca-45 Calcium-45 [also ^{45}Ca, or Ca^{45}]

Ca-46 Calcium-46 [also ^{46}Ca, or Ca^{46}]

Ca-47 Calcium-47 [also ^{47}Ca, or Ca^{47}]

Ca-48 Calcium-48 [also ^{48}Ca, or Ca^{48}]

Ca-49 Calcium-49 [also ^{49}Ca, or Ca^{49}]

ca centare (or centiare) [Unit]

ca *(circa)* – approximately; about

.ca Canada [Country Code/Domain Name]

c/a c-axis to a-axis ratio (of crystal lattice) (for hexagonal, rhombohedral and tetragonal Systems) [Symbols]

c-a crystalline-to-amorphous (transformation)

CAA Canadian Acoustical Association

CAA Canadian Arctic Archipelago

CAA Canadian Automobile Association

CAA Caragana Arborescens Agglutinin [Immunology]

CAA Carboxyarsenazo-

CAA Center for American Archeology [US]

CAA Central African Airways

CAA Chinese Association of Automation [PR China]

CAA Chloroacetic Acid

CAA Chromic Acid Anodizing

CAA Chromic Anodized Aluminum (Alloy)

CAA Cicer Arietinum Agglutinin [Immunology]

CAA Civil Aeronautics Administration [later Federal Aviation Agency; now Federal Aviation Administration, US]

CAA Civil Aviation Authority [UK]

CAA Clean Air Act [UK]

CAA Clean Air Act [of US Environmental Protection Agency]

CAA Commonwealth Association of Architects [UK]

CAA Community Aid Abroad

CAA Computer-Aided Analysis

CAA Computer-Assisted Accounting

CAA Computer-Assisted Analysis

CAA Concepts Analysis Agency [of US Army]

CAA Contaminant Analysis Automation

CAA Cost Accountants Association [UK]

CAA Criticality Alarm Annunciator [Nuclear Engineering]

CAA Cuprous Ammonium Acetate (Process)

CAA1 Clean Air Act 1: Hazardous Air Pollutants List [of US Environmental Protection Agency]

CAA2 Clean Air Act 2: Ozone Depletion Chemicals List [of US Environmental Protection Agency]

CAAA Canadian Academic Accounting Association

CAAA Canadian Association of Advertising Agencies

CAAA Clean Air Act Amendments [US Environmental Protection Agency]

CAAAL Classified Abstract Archive of the Alcohol Literature

CAAB Canadian Advertising Advisory Board

CAAC Civil Aviation Administration of China [PR China]

Ca(Ac)$_2$ Calcium (II) Acetate [also Ca(ac)$_2$]

Ca(ACAC)$_2$ Calcium (II) Acetylacetonate [also Ca(AcAc)$_2$, or Ca(acac)$_2$]

CAACE Comité des Associations d'Armateurs des Communautés Européennes [Organization of Shipowners of the European Communities]

CAAD Computer-Aided Armor Design/Analysis (Project) [of Southwest Research Institute, San Antonio, Texas, US]

CAAD Computer-Aided Automobile Design

CAAE Canadian Association for Adult Education

CAAE Certificat d'Aptitude à l'Administration des Entreprises [Certificate of Aptitude for Business Administration, France]

CAAE Chinese Academy of Atomic Energy [PR China]

CAADRP Civil Aircraft Airworthiness Data Recording Program

CAAL Consortium for Automated Analytical Laboratories

Ca/Al Calcium-to-Aluminum (Ratio)

CAAM Center of Advanced Aerospace Materials [of Pohang University of Science and Technology, South Korea]

CAAMP Canadian Acid Aerosol Measurement Program [of Environment Canada]

CAAOT Credit Amount Available at any One Time [also caaot]

CAAP Certified Advertising Agency Practitioner

CAARC Commonwealth Advisory Aeronautical Research Council [UK]

CAAS Canadian Association of Administrative Studies

CAAS Ceylon Association for the Advancement of Science [now Sri Lanka Association for the Advancement of Science]

CAAS Chinese Academy of Agricultural Science

CAAS Civil Aviation Authority of Singapore

CAAS Computer-Aided Approach Sequencing

CAAS Computer-Assisted Acquisition System

CAAS Connecticut Academy of Arts and Sciences [US]

CaAs Calcium Arsenide (Semiconductor)

CAASA Commercial Aviation Association of Southern Africa [also CAA/SA, South Africa]

CAAST Canadian Alliance Against Software Theft

CAAT Canadian Academic Aptitude Test

CAAT College of Applied Arts and Technology

CAAT Council Against Arms Trade

CAAT Computer-Assisted Audit Techniques

CAATCM Canadian Association of Acupuncture and Traditional Chinese Medicine

CAATS Canadian Automated Air Traffic System

CAAV Central Association of Agricultural Valuers [UK]

CAB Cable-Television Advertising Bureau [US]

CAB Canadian Accreditation Board [of Canadian Council of Professional Engineers]

CAB Canadian Association of Broadcasters

CAB Canadian Automated Building

CAB Captive Air Bubble

CAB Captured Air Bubble (Ship)

CAB Cation Antiphase Boundary [Solid-State Physics]

CAB Cellulose Acetate Butyrate

CAB Cellulose Acetobutyrate

CAB Centro Atómico Bariloche [Bariloche Nuclear Center, Argentina]

CAB Civil Aeronautics Board [US] [Dissolved]

CAB Commonwealth Agricultural Bureau [now Commonwealth Agricultural Bureau International, UK]

CAB Communications Adapter Board

CAB Community Advisory Board

CAB Consumers' Advisory Board [US]

CAB Conventional Acoustic Beamformer

CAB Corporate Advisory Board

CAB Cost Audit Board

CAB Critical Air Blast

CAB Current Awareness Bulletin [US]

.CAB Cabinet [Microsoft File Name Extension] [also .cab]

Cab Cabin [also cab]

Cab Cabinet [also cab]

Cab Cabriolet [also cab]

CABA Canadian Automated Buildings Association

CABA Compressed Air Breathing Apparatus

Cabal Calcium Oxide, Boric Oxide, Aluminum Oxide (Glass) [also cabal]

CABATM Civil Aeronautics Board Air Transport Mobilization [US]

CABC Canadian Airborne Center

CABC Chelation Association of British Columbia [Canada]

CABC Craftsmen's Association of British Columbia [Canada]

CAB Curr Aware Bull CAB Current Awareness Bulletin [US]

CABE Canadian Association of Business Economics

CABEI Central American Bank for Economic Integration

CABET Canadian Association of Business Economics Teachers

CABG Coronary Artery Bypass Graft

CABI Commonwealth Agricultural Bureau International [formerly Commonwealth Agri-cultural Bureau, UK]

Cable Satell Eur Cable and Satellite Europe [Published in the UK]

Cable Telev Eng Cable Television Engineering [Journal of the Society of Cable Television Engineers, UK]

CABLIS Current Awareness Bulletin for Librarians and Information Scientists [of British Library]

CABMA Canadian Association of British Manufacturers and Agencies

CABMB Calcium Oxide-Aluminum Oxide-Barium Oxide-Magnesium Oxide-Boron Oxide (Glass Ceramic)

CABO Council of American Building Officials [US]

Cabot Dig Cabot Digest [Published by Cabot Corporation, Kokomo, Indiana US]

CaBP Calcium Binding Protein [Biochemistry]

CABRA Copper and Brass Research Association

CABS Computer-Aided Batch Searching

CABS Computerized Annotated Bibliographic System [of University of Alberta, Edmonton, Canada]

CABS Current Awareness in Biological Sciences

CABSA Copper and Brass Service Center Association [also CBSA, US]

CA-B-TiSULC Continuous(ly) Annealed Boron and Titanium Stabilized ULC (Ultralow Carbon) (Steel)

CAC California Avocado Commission [US]

CAC Canada-ASEAN (Association of Southeast Asian Nations) Center (Program)

CAC Canadian Armored Corps

CAC Carbon-Arc Cutting

CAC Center for Advanced Computation [of University of Illinois, US]

CAC Center for Analytical Chemistry [of National Institute for Standards and Technology, US]

CAC Center for Atmospheric Chemistry [Canada]

CAC Central Arizona College [Coolidge, US]

CAC Cis-Anti-Cis (Geometry) [Chemistry]

CAC Citizens Advisory Committee

CAC Civil Aircraft

CAC Civil Applications Committee [US]

CAC Clean Air Council [UK]

CAC Climate Analysis Center [of National Oceanic and Atmospheric Administration, US]

CAC Coal Association of Canada

CAC Coast Artillery Corps

CAC Codex Alimentarius Commission [Joint United Nations Commission of the Food and Agricultural Organization and the World Health Organization]

CAC Commonwealth Aircraft Corporation [Australia]

CAC Community Advisory Committee

CAC Computer-Aided Classification

CAC Consumers Association of Canada

CAC Containment Atmosphere Control

CAC Crossed Aldol Condensation

CaC_2 Calcium Carbide (Structure)

CACA Canadian Agricultural Chemicals Association

CACAC Civil Aircraft Control Advisory Committee [UK]

CACAS Civil Aviation Council of the Arab States

CACB Canadian Association of Convention Bureaus

CACB Canadian Association of Customs Brokers

CACB Center Aisle Connector Bracket

CACB Compressed-Air Circuit Breaker

CACC Central Australian Conservation Council

CACC Civil Aviation Communications Centre [UK]

CACC Communications and Configuration Console

CACCI Confederation of Asian-Pacific Chambers of Commerce and Industry [Taiwan]

CACD Computer-Aided Circuit Design

CACDP California Association of County Data Processors [US]

CACDS Commonwealth Advisory Committee on Defense Science

CACDT Committee on Alternative Chemical Demilitarization Technologies [US]

CACE Computer-Aided Cost Engineering

CACEQ Citizens' Advisory Committee on Environmental Quality [US]

CACFOA Chief and Assistant Chief Fire Officers Association [UK]

CACGP Commission of Atmospheric Chemistry and Global Pollution

CACHE Computer-Controlled Automated Cargo Handling Envelope

CACI Chicago Association of Commerce and Industry [Illinois, US]

C-Acid 3-Amino-1,5-Naphthalenedisulfonic Acid; 7-Hydroxy-1,5-Naphthalenedisulfonic Acid; 2-Naphthylamine-4,8-Disulfonic Acid [also C-acid]

Ca_3Cit_2 Calcium Citrate

$C_4ACl_2H_{10}$ Calcium Chloroaluminate (or Friedel Salt)

CACM Central American Common Market

CACM Communications of the Association for Computing Machinery [US]

CACMF Central American Common Market Fund

CA/CN Cellulose Acetate/Cellulose Nitrate (Mixture)

CACOM Central American Common Market

CACON Cargo Container

CA CON Chemical Abstracts Condensates [of American Chemical Society Chemical Abstract Service, US]

$CaCO_3$-$SrCO_3$ Calcium Carbonate-Strontium Carbonate (System)

CACP Cartridge-Actuated Compaction Press

CACP Cis-Diammine Dichloro Platinum

CACQ Chartered Accountants Corporation of Quebec [Canada]

CACRS Canadian Advisory Council on Remote Sensing

CACS Computer-Aided Communications System

CACS Content Addressable Computer System

CACSUSS Canadian Advisory Committee on Scientific Utilization of Space Stations

CACT Canadian Advisory Committee on Terminology

CACT Center for Advanced Ceramic Technology [of New York State College of Ceramics at Alfred University, US]

CACT Center for Advanced Combustion Technology

CACT Newsl Center for Advanced Ceramic Technology Newsletter [Publication of the New York State College of Ceramics, Alfred University, US]

CACTS Canadian Air Cushion Technology Society

CACUL Canadian Association of College and University Libraries

CACW Central American Confederation of Workers [Costa Rica]

CACW Core Auxiliary Cooling Water [Nuclear Engineering]

CAC&W Continental Aircraft Control and Warning

CACWS Core Auxiliary Cooling Water System [Nuclear Engineering]

CAD Cabling Diagram

CAD Canadian Dollar [Currency]

CAD Cartridge-Activated Device

CAD Cash Against Disbursement

CAD Cash Against Documents [also cad]

CAD Cathodic Arc Deposition

CAD Central Ammunition Depot

CAD Character Assemble Disassemble

CAD Charged Area Development

CAD Coincidence Axis Direction, or Coincident Axial Direction (Grain Boundary Model) [Materials Science]

CAD Collisionally-Activated Dissociation [Mass Spectroscopy]

CAD Commutative–Associative–Distributive (Properties) [Mathematics]

CAD Compensated Avalanche Diode

CAD Computer Access Device

CAD Computer-Aided Design

CAD Computer Aided Design [UK Publication]

CAD Computer-Aided Detection

CAD Computer-Aided Diagnosis

CAD Computer-Aided Dispatch

CAD Computer-Aided Drafting

CAD Computer Applications Digest [UK]

CAD Computer-Assisted Design

CAD Computer-Assisted Diagnosis

CAD Concept Assessment Document

CAD Containment Atmosphere Dilution

CAD Coronary Artery Disease

CADA Cellulose Acetate Diethylaminoacetate

CADA Computer-Aided Design Analysis

CADAM Computer-Aided Design and Manufacturing

CADAM Computer-Graphics Augmented Design and Manufacturing (System)

CADAPSO Canadian Association of Data Processing Service Organizations; Canadian Association of Data and Professional Service Organizations

CADAR Computer-Aided Design, Analysis and Reliability

CADAS Coventry and District Archeological Society [UK]

CADAT Computer Aided Design and Testing

CADC Cambridge Automatic Digital Computer

CADC Canadian Astronomy Data Center [of National Research Council of Canada at Dominion Astrophysical Observatory, Victoria, British Columbia]

CADC Central Air Data Computer [of NASA]

CADC Computer-Aided Design Center [of Department of Trade and Industry, UK]

CAD/CADD Computer-Aided Design/Computer-Aided Design and Drafting

CAD/CAE Computer-Aided Design/Computer-Aided Engineering

CAD/CAM Computer-Aided Design/Computer-Aided Manufacturing

CAD/CAM/CAE Computer-Aided Design/Computer-Aided Manufacturing/Computer-Aided Engineering

CAD/CAM Dig CAD/CAM Digest [UK]

CadCam Int CadCam International [Published in the UK]

CAD/CAM Robot CAD/CAM and Robotics [Canadian Journal]

CAD/CAM Syst CAD/CAM Systems [Canadian Journal]

CAD/CAM Technol CAD/CAM Technology [US Journal]

CAD/CAT Computer-Aided Design/Computer-Aided Testing

CADCOMP International Conference on Computer-Aided Design in Composite Material Technology [also Cadcomp]

CADD Coding and Analysis of Drillhole Data

CADD Computer-Aided Design and Drafting; Computer-Aided Drafting and Design; Computer-Assisted Drafting and Design

CADD Computer-Aided Design Drawing

CADD/CAM/CAE Computer-Aided Drafting and Design/Computer-Aided Manufacturing/Computer-Aided Engineering

CADDE Canadian Association of Deans and Directors of Education

CADDET Centre for Analysis and Dissemination of Demonstrated Energy Technologies

CADDIF Computer-Aided Design Data Interchange Format

CADDS Computervision Automated Design and Drafting System

CADE Canadian Association of Drilling Engineers

CADE Client/Server Application Development Environment [Internet]

CADE Coalition Against Dangerous Exports

CADE Computer-Aided Design Engineering

CADE Computer-Aided Design Evaluation

CADE Computer-Assisted Data Entry

CADE Controller/Attitude-Direct Electronics

CADEF Centro Argentino de Estudios Forestales [Argentinian Center of Forestry Studies]

CADEP Computer-Aided Design of Electronic Products

CADES Computer-Aided Development and Evaluation System

CADET Computer-Aided Design Experimental Translator

CADETS Classroom-Aided Dynamic Educational Time-Sharing System

CADF Commutated-Antenna Direction Finder

CADF CRT (Cathode-Ray Tube) Automatic Direction Finding

CADFISS Computation and Data Flow Integrated Subsystem

CADI Computer-Aided Diagnostic Information

CADI Computer-Aided Dimensional Inspection

CADIA Centro Argentino de Ingenieros Agronomos [Argentinian Center of Agricultural Engineers]

CADIC Chemical Analysis Detection Instrumentation Control

CADIC Computer-Aided Design of Integrated Circuits

CADIF Computer-Aided Design Instructional Facility [of Cornell University, Ithaca, New York, US]

CADIN Canadian Integration North

CADIN Continental Air Defense Integration North

CADIS Computer-Aided Design of Information Systems

CADIZ Canadian Air Defense Identification Zone

CADL Cadkey Advanced Design Language

CADL Computer-Aided Design Language

CADM Clustered Airfield Defeat Submunition

CAdm Certified Administrator

CADMAT Computer-Aided Design, Manufacture and Test; Computer-Aided Design, Manufacturing and Testing

Cadmium Res Dig Cadmium Research Digest [of International Lead Zinc Research Organization, US]

CADO Central Air Documents Office [later Armed Services Technical Information Agency; now Defense Documentation Center, US]

cADP Cyclic Adenosine Diphosphate [also c-ADP] [Biochemistry]

CAD/PM Coincident Axial Direction/Planar Matching

Ca(DPM)$_2$ Calcium (II) Dipivaloylmethanoate (or Bis(dipivaloylmethanato)calcium (II)) [also Ca(dpm)$_2$]

CADPO Communications and Data Processing Operation

CADPR Center for Advanced Deformation Processing Research [of Carnegie-Mellon University, Pittsburgh, Pennsylvania, US]

CADRA Computer-Aided Design and Drafting System

CADRE Complete ADR Environment

CADRE Current Awareness and Document Retrieval for Engineers

CADS Chemical Agent Detection System

CADS Command and Data Simulator

CADS Computer-Aided Design Software

CADS Computer-Aided Design System

CADS Computer Analysis and Design System

CADS Containment Atmosphere Dilution System [Nuclear Engineering]

CADS Counterfeiting, Altering, Duplicating and Simulating

CADS Currency Authenticating and Denominating System

CADSI Communications and Data Systems Integration

CADSI Computer-Aided Design and Software Incorporated [Canada]

CADSS Combined Analog-Digital Systems Simulator

CADT Cathodic Arc Deposition Technique

CADU Control and Display Unit

CADW Civil Air Defense Warning (System)

CAE Canadian Academy of Engineering

CAE Canadian Association of Exhibitors

CAE Canadian Aviation Electronics

CAE Carbon Alcohol Extract

CAE Cathodic Arc Evaporation

CAE Center of Advanced Electroplating [US]

CAE Central African Empire [now Central African Republic]

CAE Client Application Enabler [Internet]

CAE College of Advanced Education

CAE Common Applications Environment [Computers]

CAE Compare Alpha Equal

CAE Computer-Aided Education

CAE Computer-Aided Engineering

CAE Computer-Aided Engineering [US Publication]

CAE Computer-Assisted Engineering

CAE Computer-Assisted Enrolment

CAE Constant Analyzer Energy [X-Ray Photoelectron Spectroscopy]

CAE Customer Application Engineering

Cae Caelum [Astronomy]

CAEAL Canadian Association of Environmental Analytical Laboratories

CAEC Canadian Architectural, Engineering and Construction Show (and Exposition)

CAEC Central American Energy Commission [Guatemala]

CAE/CAD Computer-Aided Engineering/Computer-Aided Design

CAE/CAM Computer-Aided Engineering/Computer-Aided Manufacturing

CAE Centre Computer-Aided Engineering Centre [UK]

CAED Canadian Association of Equipment Distributors

CAED CANMET (Canada Center for Mineral and Energy Technology)/Alternative Energy Division [of Natural Resources Canada]

CAEDA Compressed Air Equipment Distributors Association

CAEE Committee on Aircraft Engine Emissions [of International Civil Aviation Organization]

CAEF Comité des Associations Européennes de Fonderie [Committee of European Foundry Associations, France]

CAEFMS Canadian Agricultural Economics and Farm Management Society

CA-EG Citric Acid–Ethylene Glycol (Polymeric Mixture)

CAE J CAE Journal [Journal on Computer-Aided Engineering; published in Germany]

CAE J Computer-Aided Engineering Journal [of Institution of Electrical Engineers, UK]

CAEM Canadian Association of Exposition Managers

CAEM Center for Advanced Engineering Materials [of Hong Kong University of Science and Technology]

CAEM Controlled Atmosphere Electron Microscopy

CAeM Commission for Aeronautical Meteorology

CAEME Center for Computer Applications in Electromagnetic Education [of National Science Foundation International and Institute of Electrical and Electronics Engineers, US]

CAEN Computer-Aided Engineering Network [of University of Michigan, US]

CAEP Committee on Aviation Environmental Protection

CAER Center for Applied Energy Research [of University of Kentucky at Lexington, US]

CAER Community Awareness and Emergency Response

CAES Canadian Agricultural Economics Society

CAES Compressed-Air Energy Storage

CAES Connecticut Agricultural Experiment Station [US]

CAES Current Associated with Elastic Strain [Corrosion Fatigue Testing]

CAESA Canada-Alberta Environmentally Sustainable Agriculture (Agreement)

CAESAR Center for Advanced European Studies and Research [Bonn, Germany]

CAESS Customer Application Engineering Support System

CAEU Council of Arab Economic Unity [Jordan]

CAEX Computer-Aided Exploration

CAF Canadian Advertising Foundation

CAF Canadian Armed Forces

CAF Chemical Analysis Facility

CAF Cleared as Filed

CAF Cobalt Aluminum Ferrite

CAF Computer-Automated Fixturing

CAF Content-Addressable File

CAF Continuous Annealing Furnace [Metallurgy]

CAF Corrected Airflow

CAF Cost and Freight

CAF Cost, Assurance and Freight

CAF Council on Alternate Fuels [US]

CAF Critical Absorption Frequency

C_4AF Tetracalcium Aluminoferrate [$4CaO·Al_2O_3·Fe_2O_3$]

C Af Central Africa(n)

CaF Calcium-Free

CaF_2 Calcium Fluoride (Structure)

Caf Caffine [also caf]

Caf Moderate Continental Forest Climate, Mild Winters [Classification Symbol]

$CaF_2-Al_2O_3$ Calcium Fluoride-Aluminum Oxide (Composite)

CAFB Chemically Active Fluidized Bed (Process)

CAFC Canadian Association of Fire Chiefs

CAFCG Constant Amplitude Fatigue Crack Growth [Mechanics]

CAFD Computer-Aided Filter Design

CAFD Contact Analog Flight Display

CaF_2:Dy Dysprosium-Activated Calcium Fluoride (Thermoluminescence Dosimeter)

CAFE Cellular Automaton–Finite Element (Solidification Model) [also CA-FE] [Metallurgy]

CAFE Computer-Aided Film Editor

CAFE Corporate Average Fuel Economy

CAFEA Commission on Asian and Far Eastern Affairs

CAFEE Critical Assembly Fuel Element Exchange [Nuclear Engineering]

CaF_2:Er Erbium-Doped Calcium Fluoride

CaF_2:Eu Europium-Doped Calcium Fluoride

CAFF Conservation of Arctic Flora and Fauna

CAFGE Colombian Association of Flower Growers and Exporters

CAFM Computer-Aided Facilities Management

CAFM Computer-Assisted Facilities Management

C-AFM Contact Atom Force Microscope

C-AFM Conventional Atom Force Microscope

CaF_2:Mn Manganese-Activated Calcium Fluoride (Thermoluminescence Dosimeter)

CaF_2:Nd Neodymium-Doped Calcium Fluoride

$Ca(FOD)_2$ Calcium Bis(1,1,1,2,2,3,3-Heptafluoro-7,7-Dimethyl-4,6-Octanedionate) [also $Ca(fod)_2$]

CAFPRS Canadian Academy of Facial Plastic and Reconstructive Surgery

C Afr Central Africa(n)

CAFRG Canadian Advanced Fluids Research Group

CAFS Canadian Association for Future Studies

CAFS Chinese American Food Service [US]

CAFS Content-Adressable File Store

CAFSAC Canadian Atlantic Fisheries Scientific Advisory Committee

CaF_2/Si Calcium Fluoride on Silicon (Substrate)

CAFTA Central American Free Trade Association

CaF_2:YF$_3$ Yttrium Fluoride Doped Calcium Fluoride

CAG Canadian Air Group [of Canadian Armed Forces]

CAG Canadian Association of Geographers

CAG Carcinogen Assessment Group [of US Environmental Protection Agency]

CAG Cartographic Applications Group [of NASA Jet Propulsion Laboratory, Pasadena, California, US]

CAG Center of Advanced Electroplating [Denmark]

CAG Citizens Advisory Group [US]

CAG Civil Aviation Group

CAG Column Address Generator

CAG Computer-Aided Graphics

CAG Computer Applications Group [of US Air Force]

CAG Computer Applications Group [of Association of Special Libraries and Information Bureau, US]

CAG Cooperative Automation Group [of British Library Bibliographic Services, UK]

$CaGa_2S_4$:Ce Cerium-Doped Calcium Gallium Sulfide (Phosphor)

CAGC Center for Advanced Geophysical Computing [of Massachusetts Institute of Technology Earth Resources Laboratory, Cambridge, US]

CAGC Clutter Automatic Gain Control

CAGC Coded Automatic Gain Control

CAGC Continuous Access Guided Communication(s)

CAGD Computer-Aided Geometric Design

CAGD Computer-Aided Geometric Design [Published in the Netherlands]

CAGE Compiler and Assembler by General Electric

CAGE Computer-Aided Genetic Engineering

CAGE Computerized Aerospace Ground Equipment

CAGE/GEM Computer-Aided Genetic Engineering/Genetic(ally) Engineering Machine

CAGEX CERES/ARM (Clouds and the Earth's Radiant Energy System/Atmospheric Radiation Measurement)/GEWEX (Global Energy and Water Cycle Experiment) [NASA Experiment]

CAGI Compressed Air and Gas Institute [US]

CAGR Compound Annual Growth Rate

CAGS Canadian Arctic Gas Study

CAGS Canadian Association of General Surgery

CAGS Canadian Association of Graduate Schools

CAGS Certificate of Advanced Graduate Studies

CAGS Chinese Academy of Geological Sciences

CAH Canadian Association of Homeopathists

CAH Certified Auricular Hygienist

CAH_{10} Calcium Aluminate Decahydrate [$CaO \cdot Al_2O_3 \cdot 10H_2O$]

C_2AH_8 Dicalcium Aluminate Octahydrate [$2CaO \cdot Al_2O_3 \cdot 8H_2O$]

C_2AH_{13} Dicalcium Aluminate Hydrate [$2CaO \cdot Al_2O_3 \cdot 13H_2O$]

C_3AH_6 Tricalcium Aluminate Hexahydrate [$3CaO \cdot Al_2O_3 \cdot 6H_2O$]

C_4AH_{13} Tetracalcium Aluminate Hydrate [$4CaO \cdot Al_2O_3 \cdot 13H_2O$]

Cah Anal Donnees Cahiers de l'Analyse des Donnees [French Publication on Data Analysis]

Cah Cent Etud Rech Oper Cahiers du Centre d'Etudes de Recherche Operationelle [Publication of the Operations Research Center, Free University of Brussels, Belgium]

CAHE Core Auxiliary Heat Exchanger [Nuclear Engineering]

$Ca(HFAC)_2$ Calcium Hexafluoroacetylacetonate [also $Ca(hfac)_2$]

Cah Groupe Fr Rhéol Cahiers de Groupe Français de Rhéologie [Publication of the French Rheology Group]

Cah Hist Mond Cahiers de Histoire Mondiale [French Publication on World History]

CAHI Canadian Association of Home Inspectors

Cah Ind Metall Electr Cahiers des Industries Métallurgiques et Electriques [French Publication on the Metallurgical and Electrical Industries]

CAHS Canadian Aviation Historical Society

CAHS Center for Applied Health Studies [UK]

CAHT Computer Aids for Human Translation [of Carnegie-Mellon University, Pittsburgh, Pennsylvania, US]

CAHTS Computer-Aided Heat-Treating System [also CA-HTS]

CAI Cabin Air Intake

CAI Canadian Acupressure Institute [Victoria, British Columbia]

CAI Canadian Aeronautical Institute

CAI Centro de Automaticación Industrial [Center for Industrial Automation, Havana, Cuba]

CAI Close Approach Indicator

CAI Color Alteration Index

CAI Comité Arctique International [International Arctic Committee, Norway]

CAI Common Air Interference [Telecommunications]

CAI Compression-After-Impact (Test)

CAI Compressive-Strength After Impact

CAI Computer-Administered Instruction

CAI Computer-Aided Instruction

CAI Computer Analog Input

CAI Computer-Assisted Instruction

CAI Confederation of Aerial Industries Limited [UK]

CAI Confederation of Australian Industry

CAI Core Area Initiative

CAIA Computer-Aided Image Analysis

CAIBE Chemically-Assisted Ion-Beam Etching

CAIC Caribbean Association of Industry and Commerce [Barbados]

CAIC Computer-Assisted Indexing and Classification

CAI/CAL Computer-Aided Instruction/Computer-Aided Learning

CAICISS Coaxial Impact Collision Ion-Scattering Spectroscopy

CAIL Computer-Assisted Instruction and Learning

CAIM Citric Acid Indicator Mixture

CAIMAF Canadian Advanced Industrial Materials Forum

CAIMAW Canadian Association of Industrial, Mechanical and Allied Workers

CAIN Cataloging and Indexing System [of National Agricultural Library, US]

CAINS Carrier Aircraft Inertial Navigation System

CAINT Computer-Assisted Interrogation

CAI/O Computer Analog Input/Output

CAI/OP Computer Analog Input/Output

CAIP Canadian Artificial Intelligence Products

CAIP Civil Aircraft Inspection Procedures

CAIP Computer-Assisted Indexing Program

CAIR Conference on Artificial Intelligence and Robotics

CAIR Confidential Aviation Incident Reporting (Program) [Australia]

CAIRN Collaborative Advanced Interagency Research Network

CAIRS Canadian Institute for Radiation Safety

CAIRS Computer-Assisted Information Retrieval System

CAIS Canadian Association for Information Science

CAIS Catalog of Automated Information Systems [US]

CAIS Central Abstracting and Indexing Service

CAIS Central Artificial Insemination Station

CAIS Communication and Information Systems Division [of Microelectronic Education Program, UK]

CAIS Computer-Aided Instruction System

CAIS Computer-Assisted Information Retrieval System

CAISF Chemical Abstracts Integrated Subject File [of Chemical Abstracts Service, US]

CAISIM Computer-Assisted Industrial Simulation [US Army]

CAISU Canadian Alumni for the International Space University

CAIT Coalition for the Advancement of Industrial Technology [now Council on Research and Technology, US]

CAITS Center for Alternative Industrial and Technological Systems [UK]

CAJ Caulked Joint

CAK Command Access Key

CAK Command Acknowledgment

CAK Cycle Activating Kinase [Biochemistry]

CaKα Calcium K-Alpha (Radiation) [also CaK$_\alpha$]

Ca/kg Calcium per Kilogram [Unit]

CAL Carbon-Arc Lamp

CAL Center for Applied Linguistics [US]

CAL Certificate of Advanced Librarianship

CAL Certificate in Applied Linguistics

CAL China Air Lines [Taiwan]

CAL Clean Air Legislation

CAL Client Access License

CAL Cocoa Association of London [UK]

CAL Column-Address Latch

CAL Common Assembler Language

CAL Composition Analysis Laboratory [of Colorado State University, Fort Collins, US]

CAL Computer-Aided Learning

CAL Computer Animation Language

CAL Computer-Assisted Learning

CAL Confined Area Landing

CAL Continuous Annealing Line [Metallurgy]

CAL Conservation Analytical Laboratory [of Smithsonian Institution, Washington, DC, US]

CAL Continuous Annealing Line [Metallurgy]

CAL Conversational Algebraic Language

CAL Cornell Aeronautical Laboratory [of Cornell University, Ithaca, New York, US]

CAL Cray Assembly Language

C/Al Carbon (Fiber) Reinforced Aluminum (Composite) [also C-Al]

Cal Calendar [also cal]

Cal Caliber [also cal]

Cal Calibrate; Calibration; Calibrator [also cal]

Cal California [US]

Cal Calomel [Mineral]

Cal Calorie [i.e., Kilocalorie]

cal caliber [length of cannon]

cal calorie (i.e. gram-calorie) [Unit]

cal_{15} fifteen-degrees calorie [Unit]

cal_{IT} International Table calorie [Unit]

cal_{th} thermalchemical calorie [Unit]

CaLα Calcium L-Alpha Radiation [also CaL$_\alpha$]

CALAS Canadian Association for Laboratory Animal Science

CalArts California Institute of the Arts [Los Angeles, US] [also Calarts, or CIA]

CaLaSOAP Calcium Lanthanum Silicate Oxyapatite

CALC Cargo Acceptance and Load Control

CALC Customer Access Line Charge

CaLC Calcia-Containing Lanthanum Chromite

cal/°C calorie(s) per degree Celsius (or centigrade) [also cal °C^{-1}]

Calc Calculation; Calculator [also calc]

calc calculate(d)

calcd calculated [also calc'd]

Calc fluor Calcium Fluoride (Tissue Salt)

Calcif Tissue Int Calcified Tissue International [Journal]

Calcif Tissue Res Calcified Tissue Research [Journal]

cal·cm/·sK·cm² calorie centimeter(s) per second kelvin centimeter squared [Unit]

cal/cm calorie(s) per centimeter [also cal cm^{-1}]

cal/cm² calorie(s) per square centimeter [also cal cm^{-2}]

cal/cm·°C·s calorie(s) per centimeter degree Celsius second [also cal/cm/°C/s, or cal cm^{-1} °C^{-1} s^{-1}]

cal/cm·°C·sec calorie(s) per centimeter degree Celsius second [also cal/cm/°C/sec, or cal cm^{-1} °C^{-1} sec^{-1}]

cal/cm²·h calorie(s) per square centimeter hour [also cal/cm²/h, or cal cm^{-2} h^{-1}]

cal/°C·mol calorie(s) per degree Celsius per mole [also cal/°C/mol, or cal °C^{-1} mol^{-1}]

cal/cm²·s calorie(s) per square centimeter second [also cal cm^{-2} s^{-1}]

cal/cm·s·°C calorie(s) per centimeter second degree Celsius [also cal/cm/s/°C, or cal cm^{-1} s^{-1} °C^{-1}]

cal/cm·sec·°C calorie(s) per centimeter second degree Celsius [also cal/cm/sec/°C, or cal cm^{-1} sec^{-1} °C^{-1}]

calcg calculating

Calcn Calculation [also calcn]

Calc phos Calcium Phosphate (Tissue Salt)

Calc sulph Calcium Sulphate (Tissue Salt)

Calc Tissue Int Calcified Tissue International [Journal]

CALDA Canadian Airline Dispatchers Association

CALEOT Center of Applications of Laser and Electro-Optic Technologies [US]

CALFAB Computer-Aided Layout and Fabrication

cal/g calorie(s) per gram [also cal g^{-1}]

cal/g·°C calorie(s) per gram per degree Celsius [also cal g^{-1} °C^{-1}]

cal/g·K calorie(s) per gram kelvin [also cal/g-K, or cal g^{-1} K^{-1}]

CALGRI Commonwealth Agency Liaison Group on Regional Initiatives

cal/g-sec calorie(s) per gram second [also cal g^{-1} sec^{-1}, cal/g-s, or cal g^{-1} sec^{-1}]

cal/h·cm² calorie(s) per hour square centimeter [also cal/h/cm², or cal h^{-1} cm^{-2}]

cal/hr·cm² calorie(s) per hour square centimeter [also cal/hr/cm², or cal hr^{-1} cm^{-2}]

CALI Computer-Assisted Language Instruction

CALI Computer-Assisted Learning and Instruction

CALICO Computer-Assisted Language Learning and Instruction Consortium [US]

CALICO J CALICO Journal [of Computer-Assisted Language Learning and Instruction Consortium, US]

Calif California [US]

CALINDO Canadian Alumni in Indonesia

Cal Inst Tech California Institute of Technology [Pasadena, US]

CalIT Calorie, International Steam Table [Unit]

Cal J Educ Res California Journal of Educational Research [US] [also Calif J Educ Res]

CALL Computer-Assisted Language Learning

CAL-LAB Calibration Laboratory [also Cal-Lab]

cal/lb/day calories per pound per day [Animal Feed]

CALLICON Computer-Assisted Language Learning and Instruction Consortium [Brigham Young University, Provo, Utah, US]

CALLS California Academic Libraries List of Serials [of California Libraries Authority for Systems and Services, US]

CALM Catenary Anchor Leg Mooring

CALM Collected Algorithms for Learning Machines

CALM Common Assembly Language for Microprocessors

CALM Computer Archive of Language Materials [of Stanford University, California, US]

CALM Computer-Assisted Library Mechanization

CALM (Department of) Conservation and Land Management [States of New South Wales and Western Australia]

CALM Council of Academic Libraries of Manitoba [Canada]

cal/M calories per mole [also cal M^{-1}]

CalMed California Institute of Medicine [Los Angeles, US] [also CIM]

cal/min calorie(s) per minute [also cal min^{-1}]

c-AlN Cubic Aluminum Nitride

cal/mol calorie(s) per mole [also cal mol^{-1}]

cal/mol·°C calorie(s) per mole degree Celsius [also cal/mol-°C, or cal mol^{-1} °C^{-1}]

cal/mol·K calorie(s) per mole kelvin [also cal/mol-K, or cal mol^{-1} K^{-1}]

C-Al$_2$O$_3$ Carbon (Fiber) Reinforced Alumina (Composite)

χ-Al$_2$O$_3$ Chi Alumina (or Chi Aluminum Oxide)

CALOGSIM Computer-Assisted Logistics Simulation [US Army]

Calorimetry Therm Anal Calorimetry and Thermal Analysis [Published in Japan]

cal·Ω/sK² calorie-ohm(s) per second kelvin squared [Unit]

CALPA Canadian Airline Pilots Association

CALPHAD Calculation of Phase Diagrams (Method)

CALPHAD Computer Coupling of Phase Diagrams and Thermochemistry [Published in the UK] [also Calphad]

CALPHAD: Comput Coupling Phase Diagr Thermochem CALPHAD: Computer Coupling of Phase Diagrams and Thermochemistry [Published in the UK]

Cal Poly California Polytechnic State University [San Luis Obispo, US] [also Cal-Poly]

CALR Computer-Assisted Legal Retrieval

CALRAB California Raisin Advisory Board [US]

CALS Cloud and Aerosol Lidar System

CALS College of Agriculture and Life Sciences

CALS Computer-Aided Acquisition in Logistic Support

CALS Computer-Aided Logistics Support

cals calories [Unit]

cal/s calorie(s) per second [also cal s^{-1}]

cal/s·cm² calorie(s) per second square centimeter [also cal/s/cm², or cal s^{-1} cm^{-2}]

cal/s-cm-K calorie(s) per second centimeter kelvin [also cal/s/cm/K, or cal s^{-1} cm^{-1} K^{-1}]

cal/sec calorie(s) per second [also cal sec^{-1}]

cal/sec·cm² calorie(s) per second square centimeter [also cal/sec/cm², or cal sec^{-1} cm^{-2}]

Cal State California State University [US]

Caltech California Institute of Technology [Pasadena, US] [also CalTech, or CIT]

Caltrans California Department of Transportation [also CALTRANS]

CALURA Corporation and Labour Union Returns Act [Canada]

CALUS Center for Advanced Land Use Studies [UK]

Calutron California University Cyclotron [also calutron]

Cal/Val Calibration/Validation

CAM Canada Air Mail

CAM Canadian Association of Movers

CAM Career Asset Manager

CAM Carrier Aircraft Modification

CAM Cascade Access Method

CAM Catchment Area Model

CAM c-Axis Modulated [Materials Science]

CAM Cell Adhesion Molecule [Biochemistry]

CAM Cellular Automata Machine

CAM Cellulose Acetate Methacrylate

CAM Cement Aggregate Mixture

CAM Center for Advanced Materials [Lawrence Berkeley Laboratory, California, US]

CAM Center for Applied Metallography [US]

CAM Center for Aquatic Microbiology [US]

CAM Central Address Memory

CAM Central Australian Museum [Alice Springs]

CAM Centre d'Aéronautique Météorologique [Center for Meteorological Aeronautics, France]

CAM Centro Armamento Marinha [Center for Naval Armament, Brazil]

CAM Civil Aeronautics Manual

CAM Civil Aircraft Marking

CAM Checkout and Maintenance

CAM Cluster Activation Method [Metallurgy]

CAM Coherent Anomaly Method

CAM Commercial Air Movement

CAM Common Access Method

CAM Communications Access Method

CAM Computer Access Matrix

CAM Computer-Addressable Memory; Computer-Addressed Memory

CAM Computer-Address Matrix

CAM Computer-Aided Manufacture; Computer-Aided Manufacturing

CAM Computer-Aided Mapping

CAM Computer-Aided Modeling

CAM Computer Annunciation Matrix

CAM Computer-Assisted Makeup

CAM Computer-Assisted Manufacture; Computer-Assisted Manufacturing

CAM Computer-Assisted Microscopy

CAM Conservatoire des Arts et Métiers [Industrial Museum, Paris, France]

CAM Containment Atmospheric Monitoring

CAM Content-Addressable Memory

CAM Continuous Air Monitor

CAM Corrosion-Allowance Material

CAM Coupled Angular Momentum (Representation)

CAM Cryogenic Attritor Milling

CAM Custom Application Module

CAM Cybernetic Anthropomorphous Machine

Cam Camelopardalis [Astronomy]

Cam Camera [also cam]

CAMA Canadian Appliance Manufacturers Association

CAMA Canadian Automatic Merchandising Association

CAMA Centralized Automatic Message Accounting

CAMA Civil Aviation Medical Association [US]

CAMA Coated Abrasives Manufacturers Association

Cam Sci Abstr Cambridge Scientific Abstracts [US Database] [also Camb Sci Abstr, or CSA]

CAMAC Center for Applied Mathematics and Advanced Computation

CAMAC Computer-Automated Measurement and Control

CAMAL Continuous Airborne Missile-Launched and Low-Level (System)

CAMA-ONI Centralized Automatic Message Accounting–Operator Number Identification

CAMAR Common-Aperture Malfunction Array Radar

CAMAR Common-Aperture Multifunction Array Radar

CAMARCO Canadian Marine Coasting Advisory [also Camarco]

CAMAS Confederation of African Medical Associations and Societies [Nigeria]

CAMBA China-Australia Migratory Birds Agreement

Camborne Sch Mines Camborne School of Mines (Journal) [Redruth, Cornwall, UK]

Camb Sci Abstr Cambridge Scientific Abstracts [US Database] [also Cam Sci Abstr, or CSA]

Camb Tracts Math Math Phys Cambridge Tracts in Mathematics and Mathematical Physics [Publication of Cambridge University Press, UK] [also Cam Tracts Math Math Phys]

Cambs Cambridgeshire [UK]

CAMC Canadian Army Medical Corps

CAMC Colorado A&M (Agricultural and Mechanical) College [Fort Collins, US]

CAMCA Canadian-American Motor Carriers Association

CAMCOM Canadian International Conference on Composites

CAMCOS Computer-Assisted Maintenance Planning and Control System

CAMD Center for Advanced Microstructures and Devices [of Louisiana State University, Baton Rouge, US]

CAMD Computer-Aided Molecular Design

CAMD Council of Australian Museum Directors

CAMDEC Ceramics Advanced Manufacturing Development Engineering Center [of Oak Ridge National Laboratory, Tennessee, US]

CAMDF Canadian Agricultural Market Development Fund [of Agriculture Canada]

CAMEA Comité des Applications Militaires de l'Energie Atomique [Committee on Military Applications of Atomic Energy, France]

CAMEL Collapsible Airborne Military Equipment Lifter

CAMEL Component and Material Evaluation Loop

CAMELSPIN Cross-Relaxation Appropriation for Mini-Molecules Emulated by Locked Spins

CAMEO Computer-Assisted Management and Emergency Operations

CAMEO Creative Audio and Music Electronics Organization [US]

Ca(MeO)$_2$ Calcium Methoxide

C Amer Central America(n)

Camer Cameroon

CAMESA Canadian Military Electronic Standards Agency

CAMET Center for the Advancement of Mathematical Education in Technology [UK]

CAMEX Convection and Moisture Experiment (Mission) [of NASA]

CAMF Canadian Association of Metal Finishers

Camf Camouflage [also camf]

CAMFORGE Computer-Aided Microstructural Control During Forging and Cooling (Software)

Ca-Mg Calcium-Magnesium (Alloy System)

CAMI Canadian Conference on Computer Applications in the Mineral Industry

CAMI Civil Aviation Medical Institute [US]

CAMI Coated Abrasives Manufacturers Institute [US]

CAMI Computer Applications in the Mineral/Mining Industry

CAM-I Computer-Aided Manufacturing International [also CAMI, US]

CAMIFA Campaign for Independent Financial Advice [UK]

CAMIS Computer-Assisted Make-Up and Imaging System

CAMM Canadian Association of Medical Microbiologists

CAMM Computer-Aided Manufacturing Management

CAMM Computer-Aided Modeling and Manufacturing

CAMM Computer-Aided Modeling Machine

CAMM Council of American Maritime Museums [US]

CAMMD Canadian Association of Manufacturers of Medical Devices

CAMMS Computer-Assisted Material Management System

CAMN Committee of Australian Museum Directors

CAMP Cabin Air Manifold Pressure

CAMP Capital Asset Management Process

CAMP Center for Advanced Materials Processing [of Clarkson University, Potsdam, New York, US]

CAMP Central Access Monitor Program

CAMP Compiler for Automatic Machine Programming

CAMP Computer-Aided Mineral Processing

CAMP Computer-Aided Modeling Program

CAMP Computer-Assisted Mathematics Program

CAMP Computer-Assisted Menu Planning

CAMP Computer-Assisted Movie Production

CAMP Continuous Air Monitoring Program

CAMP Controls and Monitoring Processor

CAMP Cooperative African Microform Project

cAMP Cyclic Adenosine Monophosphate; Cyclic Adenosine-3',5'-Monophosphate [also c-AMP] [Biochemistry]

CAMPUS Computer-Aided Material Preselection by Uniform Standards

CAM-PC Cellular Automata Machine for Personal Computers

CAMR Center for Advanced Materials Research

CAMR Center for Advanced Materials Research [of Brown University, Providence, Rhode Island, US]

CAMR Center for Advanced Materials Research [of Oregon State University, Corvallis, US]

CAMR Center for Advanced Materials Research [of Seoul National University, South Korea]

CAMR Center for Applied Microbiology Research [UK]

Camr Camera [also camr]

CAMRAS Computer-Assisted Mapping and Records Activities System

CAMRDC Central African Mineral Resources Development Center [Congo]

CAMROC Cambridge Radio Observatory Committee [US]

CAMRT Canadian Association of Medical Radiation Technologists

CAMS Cambridge Accelerator for Materials Science [UK]

CAMS Canadian Association of Management Consultants

CAMS Certificate of Advanced Management Studies

CAMS Collisional Activation Mass Spectrum

CAMS Cybernetic Anthropomorphous Machine System

CAMSE (Conference and Exhibition on) Computer Applications to Materials and Molecular Science and Engineering

CAMSE (International Conference and Exhibition on) Computer Applications to Materials Science and Engineering

CAMSIM Computer-Assisted Maintenance Simulation [US Army]

Cam Tracts Math Math Phys Cambridge Tracts in Mathematics and Mathematical Physics [Publication of Cambridge University Press, UK] [also Camb Tracts Math Math Phys]

CAMU Corrective Action Management Unit

CAMUS Common Aperture Multisensor System

CAMVAP Canadian Motor Vehicle Arbitration Plan

CAN Calcium Ammonium Nitrate (Process)

CAN Canadian Aquaculture Network

CAN Cancel (Character) [Data Communications]

CAN Ceric Ammonium Nitrate; Cerium(IV) Ammonium Nitrate

CAN Certification Analysis Network

CAN Committee on Aircraft Noise [of International Civil Aviation Organization]

CAN Conservation Action Network [of World Wildlife Fund]

CAN Controller Area Network (Technology)

CAN Copper-Aluminum-Nickel (Alloy)

CAN Correlation Air Navigation

CAN Customer Access Network

Can Canada; Canadian

Can Cancellation [also can]

Can Cancer [also can]

Can Canister [also can]

can cancel(ed)

CANAC Computer Assisted National ATC (Air Traffic Control) Center [Belgium]

CANACERO Mexican National Chamber of Commerce for the Iron and Steel Industry

Can Acoust Canadian Acoustics [Journal of the Canadian Acoustical Association]

Canad Canada; Canadian

Can Aeron Space J Canadian Aeronautics and Space Journal [of Canadian Aeronautics and Space Institute] [also Can Aeronaut Space J]

Can Agric Eng Canadian Agricultural Engineering [Journal]

CANAIRDEF Canadian Air Force Defense Command

CANAIRDIV Canadian Air Force Division

CANAIRHEQ Canadian Air Force Headquarters

CANAL Command Analysis

CANAPES Canadian Acoustic Parabolic Equation System

CANARI Caribbean Natural Resources Institute

CANARIE Canadian Network for the Advancement of Research, Industry and Education

Can Artif Intell Canadian Artificial Intelligence [Publication of the Canadian Society for Computational Studies of Intelligence]

CANAS Canadian Naval Air Station

CANASA Canadian Alarm and Security Association

CANAVHED Canadian Naval Headquarters

CANBIOCON Canadian Biotechnology Conference (and Exhibition)

Can Bus Canadian Business [Canadian Journal]

Canc Cancel(lation) [also canc]

Canc Cancer [Astronomy]

CANCAM Canadian Congress of Applied Mechanics [also CAN/CAM]

CANCAPS Canadian Research Asia Pacific Security [of York University, Ontario, Canada]

CANCASS Canadian Command Active Sonobuoy System

CAN/CAT Canadian Cataloging Subsystem

CANCEE Canadian National Committee for Earthquake Engineering

Can Ceram Q Canadian Ceramics Quarterly

Cancer Epidemiol Cancer Epidemiology [US Journal]

Cancer Lett Cancer Letters [US Journal]

Cancer Nursing Cancer Nursing (Journal)

Cancer Res Cancer Research (Journal)

Cancer Treat Rep Cancer Treatment Report

Can Chem News Canadian Chemical News [of Chemical Institute of Canada]

Can Chem Process Canadian Chemical Processing [Canadian Journal]

Cancl Cancel [also cancl]

CANCOLD Canadian Commission on Large Dams

CANCOM Canadian International Conference on Composite Structures and Materials

CANCOM Canadian Satellite Communications

Can Copper Canadian Copper [Journal]

CAND Computer-Aided Network Design

Cand Candelabra [also cand]

Cand Candidate [also cand]

Can Datasyst Canadian Datasystems [Canadian Journal]

Can$ Canadian Dollar [Currency]

CANDE Command and Edit Language

C and F Cost and Freight [also C&F]

C and I Cost and Insurance [also C&I]

Can$/gal Canadian Dollar(s) per Gallon [Unit]

Can$/kW Canadian Dollar(s) per Kilowatt (of Electricity) [Unit]

CANDIS Canadian Disarmament Information Service

cand med *(candidatus medicinae)* – Candidate in Medicine [Final-Year Medical Student]

CANDOC Canadian Documentation [also CAN/DOC] [of Canada Institute for Scientific and Technical Information]

cand phil *(candidatus philosophiae)* – Candidate in Philosophy [Final-Year Arts Student]

cand sci *(candidatus scientiae)* – Candidate in Science [Final-Year Science Student]

CANDU Canada Deuterium Uranium (Reactor)

CANDU-BHW Canada Deuterium Uranium–Boiling Heavy Water [also CANDU BHW]

CANDU-BHWR Canada Deuterium Uranium–Boiling Heavy Water Reactor [also CANDU BHWR]

CANDU-BLW Canada Deuterium Uranium–Boiling Light Water [also CANDU BLW]

CANDU-BLWR Canada Deuterium Uranium–Boiling Light Water Reactor [also CANDU BLWR]

CANDU-PHW Canada Deuterium Uranium–Pressurized Heavy Water [also CANDU PHW]

CANDU-PHWR Canada Deuterium Uranium–Pressurized Heavy Water Reactor [also CANDU PHWR]

C and W Chip and Wire [Electronics]

CANEL Connecticut Advanced Nuclear Engineering Laboratory [US]

CANEL Connecticut Aircraft Nuclear Experiment [US]

Can Electr Eng J Canadian Electrical Engineering Journal [of Canadian Society for Electrical Engineering]

Can Electron Eng Canadian Electronics Engineering [Canadian Journal]

Can Ent The Canadian Entomologist [Journal of the Entomological Society of Canada]

CA* net Canadian Research Network

CANEWS Canadian Naval Electronic Warfare System

CANFARMS Canadian Farm Management Data System

Can Forestry Serv Dept Publ Canadian Forestry Service, Departmental Publications

Can Forestry Serv Inf Rep Dig Canadian Forestry Service, Information Reports Digest

CANFUT Canadian Future Study Group

Can Geogr Canadian Geographic (Magazine)

Can Geotech J Canadian Geotechnical Journal [of Canadian Geotechnical Society]

Can Inst Food Sci Technol J Canadian Institute of Food Science and Technology Journal

Can Is Canary Islands

Can J Appl Physiol Canadian Journal of Applied Physiology

Can J Bot Canadian Journal of Botany [of National Research Council of Canada]

Can J Chem Canadian Journal of Chemistry [of National Research Council of Canada]

Can J Chem Eng Canadian Journal of Chemical Engineering

Can J Chin Med Canadian Journal of Chinese Medicine

Can J Civ Eng Canadian Journal of Civil Engineering [of Canadian Society for Civil Engineering]

Can J Commun Canadian Journal of Communication [of Canadian Communication Association]

Can J Earth Sci Canadian Journal of Earth Sciences [of National Research Council of Canada]

Can J Electr Comput Eng Canadian Journal of Electrical and Computer Engineering [of Canadian Society for Electrical Engineering]

Can J Fisheries Aq Sci Canadian Journal of Fisheries and Aquatic Sciences [of Public Works and Government Services Canada]

Can J Forest Res Canadian Journal of Forest Research [of National Research Council of Canada]

Can J Herb Canadian Journal of Herbalism

Can J Inf Sci Canadian Journal of Information Science [of Canadian Association for Information Science]

Can J Math Canadian Journal of Mathematics [of University of Toronto, Canada]

Can J Microbiol Canadian Journal of Microbiology [of National Research Council of Canada]

Can J Phys Canadian Journal of Physics [of National Research Council of Canada]

Can J Physiol Pharmacol Canadian Journal of Physiology and Pharmacology

Can J Psychol Canadian Journal of Psychology

Can J Remote Sens Canadian Journal of Remote Sensing [of Canadian Aeronautics and Space Institute]

Can J Spectrosc Canadian Journal of Spectroscopy

Can J Zool Canadian Journal of Zoology [of National Research Council of Canada]

Can Lab Canadian Laboratory [Canadian Journal]

CanLaunch Canadian Capability for Launch Site Study

CAN/LAW Canadian Computer-Assisted Legal Research System

Can Libr J Canadian Library Journal [of Canadian Library Association]

Can Mach Metalwork Canadian Machinery and Metalworking [Canadian Journal]

Canmaking Cann Int Canmaking and Canning International [Journal published in the UK]

CANMAP Canadian Marketing Assistance Program

CANMARC Canadian Machine-Readable Cataloguing [of National Library of Canada] [also CAN/MARC]

CAN-MATE Canadian Manufacturing Advanced Technology Exchange

Can Med Assoc J Canadian Medical Association Journal [also CMAJ]

CANMET Canada Center for Mineral and Energy Technology [of Natural Resources Canada, Ottawa, Ontario] [also Canmet]

CANMET/AED CANMET/Alternative Energy Division [of Natural Resources Canada]

Can Metall Q Canadian Metallurgical Quarterly

CANMET/ASD CANMET/Administrative Services Division [of Natural Resources Canada]

CANMET/BTCD CANMET/Business and Technology Coordination Division [of Natural Resources Canada]

CANMET/CCRL CANMET/Combustion and Carbonization Research Laboratory [of Natural Resources Canada]

CANMET/CRL CANMET/Coal Research Laboratories [of Natural Resources Canada]

CANMET/CTMD CANMET/Technology Marketing Division [of Natural Resources Canada]

CANMET/ED CANMET/Explosives Division [of Natural Resources Canada]

CANMET/EDRL CANMET/Energy Diversification Research Laboratory [of Natural Resources Canada]

CANMET/EED CANMET/Energy Efficiency Division [of Natural Resources Canada]

CANMET/EMR CANMET/(Department of) Energy, Mines and Resources [now CANMET/NRCan]

CANMET/ERL CANMET/Energy Research Laboratories [of Natural Resources Canada]

CANMET ETC CANMET Energy Technology Center [of Natural Resources Canada]

CANMET/ETSD CANMET/Engineering and Technical Services Division [of Natural Resources Canada]

CANMET/MRL CANMET/Mining Research Laboratories [of Natural Resources Canada]

CANMET/MSL CANMET/Mineral Sciences Laboratories [of Natural Resources Canada]

CANMET/MTB CANMET/Mineral Technology Branch [of Natural Resources Canada]

CANMET/MTL CANMET/Metals Technologies Laboratories [of Natural Resources Canada]

CANMET/NRCan CANMET/Natural Resources Canada

CANMET/RPO CANMET/Research Program Office [of Natural Resources Canada]

CANMET/TMD CANMET/Technology Marketing Division [of Natural Resources Canada]

CANMET WRC CANMET Western Research Center [of Natural Resources Canada in Devon, Alberta]

CAN-MEX WKSP Canadian-Mexican Comparative Perspectives on Security Workshop

Can Mineral Canadian Mineralogist [Journal published by the University of Manitoba, Winnipeg]

Can Min J Canadian Mining Journal

Can Min Metall Bull Canadian Mining and Metallurgical Bulletin [of Canadian Institute of Mining and Metallurgy]

Can Nurse The Canadian Nurse [Journal]

Can Occup Safety Mag Canadian Occupational Safety Magazine

CAN/OLE Canadian On-Line Enquiry [of Canada Institute for Scientific and Technical Information]

CANOPUS Canadian Auroral Network for the Open Program Unified Study

CANOZE Canadian Ozone Experiment

CANP Civil Air Notification Procedure

Can Phys Canadian Physics [Canadian Journal]

Can Pkg Canadian Packaging [Canadian Journal]

Can Plast Canadian Plastics [Canadian Journal]

CANPOTEX Canadian Potash Export Association

Can Printer Publ Canadian Printer and Publisher [Canadian Journal]

CANQUA Canadian Quaternary Association

Can Res Canadian Research [Canadian Journal]

CANRIS Coastal and Marine Resources Information System

CANS Computer-Assisted Network Scheduling System

CANSARP Canadian Search and Rescue Planning (Program) [of Transport Canada]

CAN/SDI Canadian Service for the Selective Dissemination of Information [of Canada Institute for Scientific and Technical Information]

CANSG Civil Aviation Navigational Services Group

CANSIA Canadian Solar Industries Association

CAN/SIM Canadian Socio-Economic Information Management System

CanSIS Canadian Soil Information System [also CAN/SIS]

CAN/SND Canadian Service for Scientific Numeric Databases

CANSPECS CANMET's (Canada Center for Mineral and Energy Technology) Service Program for the Evaluation of Codes and Standards

Cant Cantonese

Cantab *(Cantabrigiensis)* – from Cambridge University [UK]

CAN/TAP Canadian Technical Awareness Program

CANTASS Canadian Towed Array Sonar System

CANTAT Canadian Transatlantic Telephone (Cable)[Canada/UK]

CANTi Copper-Aluminum-Nickel-Titanium (Shape-Memory Alloy)

CANTiM Copper-Aluminum-Nickel-Titanium-Manganese (Shape-Memory Alloy)

CANTO Caribbean Association of National Telecommunications Organizations

CANTRAN Cancel in Transmission; Cancel Transmission

CAN-UK Canada–United Kingdom

CANUNET Canadian University Computer Network

CANUSE Canada-United States Eastern Interconnection

CANUSE Canada-United States Eastern Power Complex

CANUTEC Canadian Transport Emergency Center [Ottawa, Ontario]

CANWEC Canadian National Committee of the World Energy Conference

Can Weld Fabr Canadian Welder and Fabricator [Canadian Journal]

CANZ Canada, Australia and New Zealand (Group) [of G-77–Group of 77 (Nations)]

CAO Chief Administrative Officer

CAO Circuit Allocation Order

CAO Communications Authorization Order

CAO Completed as Ordered

CAO Computer-Aided Optimization

CAO Coniferyl Alcohol Oxidase [Biochemistry]

CAO Crimean Astrophysical Observatory [Ukraine]

CaO Calcium Oxide

CaO-Al$_2$O$_3$ Calcium Oxide-Aluminum Oxide (System)

CAOC Constant Axial Offset Control

CaO-CaF$_2$ Calcium Oxide-Calcium Fluoride (Slag)

CaO-CaF$_2$-NaF$_2$ Calcium Oxide-Calcium Fluoride-Sodium Fluoride (Slag)

CAOCI Commercially Available Organic Chemicals Index [of Chemical Notation Association, UK]

CAOD Computer-Aided Optical Design

CAODC Canadian Association of Oil-Well Drilling Contractors

Ca(OEt)$_2$ Calcium (II) Ethoxide (or Calcium (II) Ethylate)

CaO-Fe$_2$O$_3$ Calcium Oxide-Ferric Oxide (System)

Ca(OMe)$_2$ Calcium (II) Methoxide (or Calcium (II) Methylate)

CAORB Civil Aviation Operational Research Branch [UK]

CAORC Council of American Overseas Research Centers [US]

CAORE Canadian Army Operation Research Establishment

CAOS Completely Automatic Operating System

CAOS Computer-Assisted Organic Synthesis

CaO-SiO$_2$ Calcium Oxide-Silicon Dioxide (System)

CaO-ZrO$_2$ Calcium Oxide-Zirconium Dioxide (System)

CAP Campus Advisory Panel

CAP Canadian Assistance Plan

CAP Canadian Association of Palynologists

CAP Canadian Association of Pathologists

CAP Canadian Association of Physicists
CAP Canadian Association of Principals
CAP Canadian Astronaut Program
CAP Canadian Patent [also CaP]
CAP Capacity Assurance Plan(s)
CAP Carburizing Atmosphere Process
CAP Card Assembly Program
CAP Carrierless Amplitude and Phase-Modulation
CAP Cataloguing in Advanced Publication [of British Library Bibliographic Services Division, UK]
CAP CCMS (Checkout, Control and Monitor System) Application Programs
CAP Cellulose Acetate Phthalate
CAP Cellulose Acetate Propionate
CAP Census of Australian (Vascular) Plants [Now Part of Australian Plant Name Index (Database)]
CAP Center for Academic Publications [Japan]
CAP Center for Aerosol Processes [of University of Cincinnati, Ohio, US]
CAP Central Arbitration Point
CAP Central Arizona Project [US]
CAP Certificat d'Aptitude Professionnelle [Certificate of Professional Aptitude, France]
CAP Chloramphenicol
CAP Chloroaluminum Phthalocyanine
CAP Circuit Access Point
CAP Civil Air Patrol [US]
CAP Civil Air Publication
CAP Civil Aviation Publication
CAP Cleaner Air Package
CAP Cloud-Atmospheric Parameter
CAP College of American Pathologists [US]
CAP Combat Air Patrol
CAP Committee on Applications of Physics [of American Physical Society, US]
CAP Common Agricultural Policy [European Community]
CAP Common Avionics Processor
CAP Commonwealth Association of Planners [Canada]
CAP Communication Application Platform
CAP Community-Acquired Pneumonia
CAP Competitive Ablation and Polymerization (Mechanism)
CAP Competitive Access Provider
CAP Compliance Audit Program
CAP Computer Access Panel
CAP Computer-Aided Planning
CAP Computer-Aided Production
CAP Computer-Aided Programming
CAP Computer-Aided Publishing
CAP Computer-Aided Purchasing
CAP Computer Analysts and Programmers [US]
CAP Computer Application Program
CAP Computer-Assisted Processing
CAP Computer-Assisted Production
CAP Computer-Assisted Programming

CAP Computerized Area Pricing
CAP Consolidated by (or under) Atmospheric Pressure; Consolidation by (or under) Atmospheric Pressure [Metallurgy]
CAP Consulting Assistance Program
CAP Consumers Association of Penang [Malaysia]
CAP Contact Approach
CAP Continuous Airworthiness Panel [of International Civil Aviation Organization]
CAP Continuous Audit Program
CAP Contractor Acquired Property
CAP Contracts Administrative Procedure
CAP Controlled Atmosphere Packaging
CAP Cost Account Package
CAP Council on Advanced Programming
CAP Council to Advance Programming
CAP Crew Activity Plan
CAP Cryotron Associative Processor
CAP Customer Access Panel
.CAP Capture [File Name Extension]
Ca/P Calcium-to-Phosphorus (Ratio)
Cap Capacitance; Capacitor [also cap]
Cap Capacity [also cap]
Cap Capillary [also cap]
Cap Capitalization [also cap]
Cap Capital Letter [also cap]
Cap Capricornus [Astronomy]
cap capitalize
cap (caput) – chapter
CAPA Canadian Aircraft Preservation Association
CAPA Canadian Association of Purchasing Agents
CAPA Commission on Asian and Pacific Affairs
CAPA Commonwealth Association of Polytechnics in Africa [Kenya]
CAPA Computer-Aided Plant Analysis
CAPA Confederation of Asian and Pacific Accountants
CA-PA Calixarene-Polyaniline (Mixture)
CAPAC Canadian Association of Primary Air Carriers
Capacity Manage Rev Capacity Management Review [US]
CAPAL Computer and Photographically Assisted Learning
CAPAM Commonwealth Association for Public Administration and Management [Canada]
CAPB Catchment Areas Protection Board
cAPB Conservative Antiphase Boundary [Solid-State Physics]
CAPC Canadian Army Pay Corps
CAPC Canadian Association of Professional Conservators
CAPC Civil Aviation Planning Committee
CAPCA Canadian Advanced Protein Crystallization Apparatus
CAPCC Computer Application Process Control Committee
CAPCOM Capsule Communications [Aerospace]
CAPCOM Capsule Communicator [of NASA]
Cap Cost Capital Cost [also cap cost]
CAPD Computer-Aided Process Design

CAPD Computing to Assist Persons with Disabilities [of Johns Hopkins University, US]

CAPD Continuous Ambulatory Peritoneal Dialysis

CAPDAC Computer-Aided Piping Design and Construction

CAPD LAB Computer-Aided Process Design Laboratory [of Queen's University, Canada]

CAPDM Canadian Association of Physical Distribution Management

CAPE Children's Alliance for Protection of the Environment

CAPE Coalition of Aerospace Employees

CAPE Communication Automatic Processing Equipment

CAPE Computer-Aided Plant Engineering

CAPE Computer-Assisted Patient Evaluation

CAPE Computer Assisted Planning Experiment

CAPE Computer Assisted Program Evaluation

CAPE Computer-Assisted Project Execution

CAPE Concurrent Art-to-Product Environment

CAPE Convective Available Potential Energy

CAPE Coordinated Assessment and Program Planning for Education

CaPE Convection and Precipitation/Electrification Experiment

CAPER Canadian Association of Publishers Educational Representatives

CAPERS Canadian Acoustic Pulse Echo Repeater System

CAPERT Computer-Assisted Program Evaluation Review Technique

CAPERTSIM Computer-Assisted Program Evaluation Review Technique Simulation

CAPH Calcium Aluminate Phosphate Hydrate (Gel) [also C-A-P-H]

CAPHE Consortium for the Advancement of Private Higher Education [US]

CAPIC Canadian Association for Production and Inventory Control

CAPL Canadian Association of Petroleum Landmen

CAPL Canadian Association of Public Libraries

CAPL Continuous Annealing and Processing Line [Metallurgy]

CAP-Li₂ Dilithium Carbamoylphosphate

CAPM Capital Asset Pricing Model

CAPM Computer-Aided Patient Management

CAPM Computer-Aided Production Management

CAPM Containment Atmosphere Particulate Monitor

CAPM CPU (Central Processing Unit) Access Port Monitor

CAPMON Canadian Air and Precipitation Monitoring Network [also CAPMoN]

CAPMON Canadian Atmospheric Precipitation Monitoring Network [also CAPMoN]

Cap'n Captain

CAPO Canadian Army Post Office

CaPO Calciophosphate

CAPOSS Capacity Planning and Operation Sequence System

CAPOSS-E Capacity Planning and Operation Sequence System Extended Program

CAPP Commission on Air Pollution Prevention [Germany]

CAPP Computer-Aided Process Planning

CAPP Computer-Aided Production Planning

CAPP Computer-Aided Pulse Plating

CAPP Content Addressable Parallel Processor

CAPPA Crusher and Portable Plant Association [US]

CAPPI Constant-Altitude Plan Position Indicator

CAPPS Computer-Aided Part Programming System

CAPPS/EDGES Computer-Aided Part Programming System/Expert DMIS (Directory Management Information System) Graphical Editor and Simulator

CAPR Catalogue of Programs

CAPR Colegio Americana de Puerto Rico [American College of Puerto Rico, Bayamón]

CAPRI Captive Reset Ignitor

CAPRI Card and Printer Remote Interface

CAPRI Center for Applied Polymer Research [of Case Western Reserve University, Cleveland, Ohio, US]

CAPRI Civil Aircraft Protection Against Ice

CAPRI Code(d) Address Private Radio Intercom

CAPRI Computer-Aided Personal Reference Index System [of United Kingdom Atomic Energy Authority]

CAPRI Computerized Advance Personnel Requirements and Inventory

CAPRI Computerized Area Pricing

CAP/RPSP College of American Pathologists/Reference Preparation for Serum Proteins [US]

CAPS Call Attempts per Second

CAPS Cassette Programming System

CAPS Cassini Plasma Spectrometer [of NASA]

CAPS Center for Analysis and Prediction of Storms [of University of Oklahoma, Norman, US]

CAPS Cell Atmosphere Processing System

CAPS Center for Advanced Purchasing Studies

CAPS Certificate of Advanced Professional Studies

CAPS Civil Aviation Purchasing Service [of International Civil Aviation Organization]

CAPS Common Attitude Pointing System

CAPS Computer-Aided Polymer Selection

CAPS Computer-Aided Product Selection

CAPS Computer-Aided Programming Software Simulator

CAPS Computer-Assisted Problem Solving

CAPS Control and Auxiliary Power Supply System

CAPS Courtauld's All-Purpose Simulator

CAPS Crew Altitude Protection System

CAPS Current Advances in Plant Science (Database) [UK]

CAPS Current Associated with Plastic Strain [Corrosion Fatigue Testing]

CAPS 3-(Cyclohexylamino)-1-Propanesulfonic Acid

Caps Capital Letters [also caps]

Caps Capsule [also caps]

caps *(capsula)* – capsule [Medical Prescriptions]

Cap Scr Cap Screw

CAPSIM Captive Simulation

CAPSIN Civil Aviation Packet Switching Integrated Network

CAPSO 3-(Cyclohexylamino)-2-Hydroxy-1-Propanesulfonic Acid

CAPST Capacitor Start

Ca-PSZ Calcia-Partially Stabilized Zirconia

CAPT Center for Advanced Pyrometallurgical Technology [of University of Utah, Salt Lake City, US]

Capt Captain [also CAPT]

CAPTAIN Character and Pattern Telephone Access Information Network [Japan]

CAPTEX Cross-Appalachian Tracer Experiment

Capt(N) Captain (Navy)

CAPTOR Encapsulated Torpedo

Capt(R) Captain (Regular)

Cap'y Capacity [also cap'y]

CAQ Computer-Aided Quality Assurance

CAQC Computer-Aided Quality Control

CAQ/SPC Computer-Aided Quality Control/Statistical Process Control

CAR Canadian Association of Radiologists

CAR Canadian Automotive Report

CAR Carcinogenic Effects (of Hazardous Materials) [Toxicology]

CAR Center for Accounting Research [University of Southern California, Los Angeles, US]

CAR Center for Atmospheric Research [Canada]

CAR Central African Republic

CAR Central Apparatus Room

CAR Certification Approval Request

CAR Channel Address Register

CAR Channel Assignment Record

CAR Civil Air Regulations [US]

CAR Cloud Absorption Radiometer

CAR Collision Avoidance Radar

CAR Combined Autothermal Reforming (Process)

CAR Community Antenna Relay

CAR Comprehensive, Adequate and Representative [Forest Reserve System]

CAR Computer-Aided Recording

CAR Computer-Aided Repair

CAR Computer-Aided Retrieval

CAR Computer-Assisted Recognition

CAR Computer-Assisted Research

CAR Computer-Assisted Retrieval

CAR Conditioned Avoidance Response

CAR Configuration (and) Acceptance Review

CAR Containment Air Recirculation

CAR Containment Air Removal

CAR Contractors' All Risks (Insurance)

CAR Control Advisory Release

CAR Corrective Action Request

CAR Critical Angle Refractometer

C_{60}/Ar Fullerene/Argon (Mixture)

Car Carcinogen(ic) [also car]

Car Carcinoma [also car]

Car Cargo [also car]

Car Caribbean (Region)

Car Carina [Astronomy]

car carat [Unit]

CARA Cargo and Rescue Aircraft

CARA Combat Aircrew Rescue Aircraft

CARAC Civil Aviation Radio Advisory Committee [UK]

CARAFE Cloud, Aerosol and Radiation Arctic Field Experiment

CARAM Content-Addressable Random-Access Memory

CARAS Canadian Academy of Recording Arts and Sciences

CARAT Cargo Agents Reservation, Airwaybill Insurance and Tracking System

CARB California Air Resources Board [US]

CARB Center for Advanced Research in Biotechnology [US]

CARB Compact Aligning Roller Bearing

CARB Division of Carbohydrate Chemistry [of American Chemical Society, US]

Carb Carbonate; Carbonation [also carb]

Carb Carburetor [also carb]

Carb Carburization [also carb]

carb carbenicillin [Biochemistry]

Carbide Tool Carbide and Tool [Publication of ASM International, US]

Carbide Tool J Carbide and Tool Journal [of Society of Carbide and Tool Engineers, US]

CARBINE Computer-Automated Real-Time Betting Information Network

Carbohyd Carbohydrate [also carbohyd]

Carbohyd Chem Carbohydrate Chemistry [Journal]

Carbohyd Polymers Carbohydrate Polymers [Journal published in the UK]

Carbohyd Res Carbohydrate Research [Journal published in the Netherlands]

CARC Canadian Agricultural Research Council

CARC Canadian Agri-Food Research Council

CARC Canadian Arctic Resources Committee

CARC Canadian Audio Research Consortium

CARC Central America Resource Center

CARC Chemical Agent Resistant Coating

CARC Clark Atlanta Research Center [of Clark Atlanta University, Atlanta, Georgia, US]

CARCA Computer-Aided Rocking Curve Analysis

CARCAE Caribbean Regional Council for Adult Education [Barbados]

CARCS Canadian Amphibian and Reptile Conservation Society

CARD Canadian Advertising Rates and Data

CARD Cartographic Representation of Data [Geographic Information System]

CARD Center for Agricultural and Rural Development [of Iowa State University, Ames, US]

CARD Channel Allocation and Routing Data

CARD Coastal and Arctic Research Division [of National Oceanic and Atmospheric Administration– Pacific Marine Environmental Laboratory, US]

CARD Compact Automatic Retrieval Device

CARD Compact Automatic Retrieval Display

CARD Constraints and Restrictions Document

CARDA Canadian Agricultural Rehabilitation and Development Association

CARDCODER Card Automatic Code System

CARDE Canadian Armament Research and Development Establishment [of Defense Research Board, Canada]

CARDI Caribbean Agricultural Research Development Institute

Cardiol Cardiological; Cardiologist; Cardiology [also cardiol]

Cardiol Cardiology [International Journal]

CARDO Center for Architectural Research and Development Overseas [UK]

CARDS Comprehensive Aerological Reference Data Set

Cards Cardiganshire [UK]

Card Treat Rep Cardiac Treatment Reports

CARE Center for Applied Research in Education [UK]

CARE Center for Aviation Research and Education [now National Association of State Aviation Officials, US]

CARE (Consortium for) Ceramic Applications in Reciprocating Engines [UK]

CARE Computer-Aided Reliability Estimation

CARE Computer-Assisted Remediation and Education Project [Canada]

CARE Cooperative for American Relief Everywhere, Inc. [US]

CARE Cooperative for American Remittances to Everywhere, Inc. [now Cooperative for American Relief Everywhere, Inc.]

Ca-RE Calcium Rare-Earth (Compound)

CAREIRS Conservation and Renewable Energy Inquiry and Referral Service [US]

CARES Civil Air Rescue Emergency Services

CARESS Career Retrieval Search System [of University of Pittsburgh, Pennsylvania, US]

CARETS Central Atlantic Regional Ecological Test Site

CARF Canadian Advertising Research Foundation

CARF Canadian Amateur Radio Federation

CARF Central Altitude Reservation Facility

CARF Consumer Affairs and Regulatory Functions [of Office of the Assistant Secretary, US]

CARI Canadian Association of Recycling Industries

CARI Council of Air-Conditioning and Refrigeration Industry [US]

Carib Caribbean

CARIC Computerized Automation and Robotics Information Center [of Machine Vision Association, US]

CARICOM Caribbean Community and Common Market

CARID Customer Acceptance Review Item Disposition

Caries Res Caries Research [International Journal]

CARIFTA Caribbean Free Trade Agreement

CARIFTA Caribbean Free Trade Area [now Caribbean Common Market]

CARIFTA Caribbean Free Trade Association

CARIPOL Caribbean Pollution Research and Monitoring Program (International)

CARIRI Caribbean Industrial Research Institute

CARIS Computer-Aided Resource Information System

CARIS Computerized Agricultural Research Information System [of Food and Agricultural Organization]

CARIS Current Agricultural Research Information System

CARITAG Caribbean Trade Advisory Group [UK]

CARL Canadian Academic Research Libraries

CARL Canadian Association of Research Libraries

CARL Chemical Algorithm for Reticulation Linearization

CARL Colorado Alliance of Research Libraries [US]

CARLAM Société Carolorégienne de Laminage SA [Belgian Rolling Mill]

Carlsberg Res Comm Carlsberg Research Communications

CARM Canadian Association of Risk Managers

Carmarths Carmarthenshire [UK]

CARN Conditional Analysis for Random Networks

CARNACS Conventional Arms Reconnaissance [Canada]

Carnegie Corp NY Q Carnegie Corporation of New York Quarterly [US]

Carnegie Inst Wash Yrb Carnegie Institute of Washington Yearbook [US]

CARNET Canadian Aging Research Network

CAROL Circulation and Retrieval On-Line [of James Cook University, Australia]

CAROL Computer-Oriented Language

CAROM Carbide Aromatics Extraction (Process)

CAROT Centralized Automatic Reporting on Trunks

CARP Call Accounting Reconciliation Process

CARP Committee for Acoustic Emission in Reinforced Plastics [of American Society for Nondestructive Testing, US]

CARP Computed Air Release Point

CARP Computer-Assisted Regional Planning [Geographic Information System]

Carp Carpenter; Carpentry [also carp]

CARPAS Comision Asesora Regional de Pesca para el Atlantico Sudoccidental [Regional Fisheries Advisory Commission for the Southwest Atlantic, Italy]

Carptr Carpenter [also carptr]

Carptry Carpentry [also carptry]

CARR Customer Acceptance Readiness Review

Carr Carriage; Carrier [also carr]

Carr Pd Carriage Paid [also carr pd]

CARRS Close-In Automatic Route Restoral System [of North American Aerospace Defense Command]

CARS Center for Advanced Radiation Studies [of University of Chicago, US]

CARS Coherent anti-Stokes Raman Scattering

CARS Coherent anti-Stokes Raman Spectroscopy

CARS Common Antenna Relay System

CARS Community Aerodrome Radio System

CARS Community Antenna Relay Service

CARS Computer-Aided Readability System

CARS Computer-Aided Routing System

CARS Computer-Assisted Reference Service [of University of Arizona, Tucson, US]

CARS Computerized Applications and Reference System (Software) [of Auto/Steel Partnership Program, US]

CARS Computerized Automatic Rating System

CARS Computerized Automotive Reporting Service

CARS Continuous Alarm Reporting Service

CARSP Canadian Association of Road Safety Professionals

CARSS Center for Atmospheric and Remote Sounding Studies

CART Center for Advanced Resource Technology [Canada]

CART Central Automatic Reliability Test

CART Central Automated Replenishment Technique

CART Classification and Regression Tree

CART Cloud and Atmospheric Radiation Testbed (Facility)

CART Community Alliance for Responsible Transport

CART Complete Automatic Reliability Testing

CART Computerized Automatic Rating Technique

Cart Cartographer; Cartographic; Cartography [also cart]

Cart Carton [also cart]

Cart Cartridge [also cart]

Cart Art Common Cartoid Artery [also cart art]

CarTech Carpenter Technology [of Carpenter Steel, Reading, Pennsylvania, US]

CARTG Canadian Amateur Radio Teletype Group

Carth Carthage; Carthaginian

Cart Inf Cartographic Information [Publication of the North American Cartographic Information Society]

CARTIS Computer-Aided Real-Time Inspection System

CARTS Capacitors and Resistors Technology Symposium

CARTS-Europe European Capacitors and Resistors Technology Symposium

CAS Cable Activity System

CAS Calcium Aluminosilicate (or Anorthite)

CAS Calculated Air Speed

CAS Calibrated Air Speed

CAS California Academy of Sciences [San Francisco, US]

CAS California Avocado Society [US]

CAS Call Accounting Subsystem

CAS Canadian Anaesthetists Society

CAS Canadian Army Staff

CAS Canadian Astronomical Society

CAS Center for Advanced Study [India]

CAS Center for Agricultural Stragegy [UK]

CAS Center for Alcohol Studies [of Rutgers State University of New Jersey, Piscataway, US]

CAS Center for Auto Safety [Washington, DC, US]

CAS Center of Atmospheric Studies [US]

CAS Central Accounting System

CAS Central Alarm Station

CAS Central Alarm System

CAS Central Amplifier Station

CAS CERN Accelerator School [of CERN–European Laboratory for Particle Physics, Geneva, Switzerland]

CAS Certificate of Advanced Studies

CAS Channel-Associated Signaling

CAS Chemical Abstracts Service [of American Chemical Society, US] [also Chem Abstr Serv]

CAS Chester Archeological Society [UK]

CAS Chicago Academy of Sciences [Illinois, US]

CAS Chinese Academy of Sciences [PR China]

CAS Chinese Association for Standardization [PR China]

CAS Chinese Astronomical Society [PR China]

CAS Circuits and Systems (Society) [of Institute of Electrical and Electronics Engineers, US]

CAS Clean Air Society

CAS Clean Air System

CAS Close Air Support

CAS College of Arts and Sciences

CAS Collision Avoidance System [of Federal Aviation Administration, US]

CAS Column Address Select

CAS Column Address Strobe

CAS Command Augmentation System

CAS Commission for Atmospheric Science [of World Meteorological Organization]

CAS Commission for Atmospheric Sciences [of National Academy of Sciences, US]

CAS Communications Application Specification

CAS Complete Active Space

CAS Composition Adjustment by Sealed Argon Bubbling (Steelmaking) [Metallurgy]

CAS Computer-Aided Service; Computer-Aided Servicing

CAS Computer-Aided Styling

CAS Compressed Air System

CAS Computer-Aided Simulation

CAS Computer-Aided Strategy and Sales Controlling

CAS Computer-Aided System

CAS Computer-Assisted Steelmaking

CAS Computerized Acquisition System

CAS Condition Assessments Survey

CAS Connecticut Academy of Sciences [Hartford, US]

CAS Control Augmentation System

CAS Control Automation System

CAS Controlled Airspace

CAS Cooperative Applications Satellite [Canada]

CAS Cornwall Archeological Society [UK]

CAS Cuban Academy of Sciences [Havana, Cuba]

CAS Current Awareness Service

CAS Customer Administrative Service

CAS Customs Agency Service [US]

CAS Czech Academy of Sciences [Prague, Czech Republic]

CAS Czechoslovak Academy of Sciences [Prague, Czechoslovakia] [now Czech Academy of Sciences]

CAS$_2$ Calcium Aluminum Disilicate (or Anorthite) [$CaO \cdot Al_2O_3 \cdot 2SiO_2$]

C$_2$AS Dicalcium Aluminum Silicate [$2CaO \cdot Al_2O_3 \cdot SiO_2$]

C As Central Asia(n)

CaS Calcium Sulfide

Cas Cassiopeia [Astronomy]

c-As Crystalline Arsenic

CASA Canadian Advertising and Sales Association

CASA Canadian Alarm and Security Association

CASA Canadian Automatic Sprinkler Association

CASA Car Audio Specialists Association [US]

CASA Computer-Aided Structural Analysis

CASA Computer and Automated Systems Association [of Society of Manufacturing Engineers, US]

CASAA Canadian Association of Student Awards Administrators

CAS-ACS Chemical Abstracts Service of the American Chemical Society [US]

CASAFA (Interunion) Commission on the Application of Science to Agriculture, Forestry and Aquaculture [Canada]

CASALS Congress on Advances in Spectroscopy and Laboratory Sciences

CASANZ Clean Air Society of Australia and New Zealand

CASAO Chartered Accountant Students Association of Ontario [Canada]

CASARA Civil Air Search and Rescue Association

CASAS Commonwealth Association for Scientific and Agricultural Societies

CASA/SME Computer and Automated Systems Association of the Society of Manufacturing Engineers [US]

CASAW Canadian Association of Smelters and Allied Workers

CASB Canadian Aviation Safety Board

CASB Cost Accounting Standards Board [US]

CASC Canadian Army Service Corps

CASC Canadian Association for Studies in Cooperation

CASC Canadian Automobile Sport Club

CASC Central Administrative Support Center [of National Oceanic and Atmospheric Administration, US]

CASC Certified Alfalfa Seed Council [US]

CASCC Canadian Agricultural Services Coordinating Committee

CASD CANMET (Canada Center for Mineral and Energy Technology)/Administrative Services Division [of Natural Resources Canada]

CASD Chemical and Atmospheric Sciences Directorate [of Air Force Office of Scientific Research, US]

CASD Computer-Aided System Design

CASDAC Computer-Aided Ship Design and Construction

CAS DDS Chemical Abstracts Service–Document Delivery Service [of American Chemical Society, US]

CASE Campaign for State Education [UK]

CASE Citizens Association for Sound Energy

CASE Committee on Academic Science and Engineering [of Federal Council for Science and Technology, US]

CASE Common Application Service Element (Protocol)

CASE Computer-Aided Software Engineering

CASE Computer-Aided System Evaluation

CASE Computer-Aided Systems Engineering

CASE Computer Automated Support Equipment

CASE Computing and Automated Systems Association [of Society of Mechanical Engineers, US]

CASE Confined Access Support Equipment

CASE Consolidated Aerospace Supplier Evaluation

CASE Cooperative Awards in Science and Engineering [UK]

CASE Coordinating Agency for Suppliers Evaluation [US]

CASE Council for the Advancement and Support of Education [Canada]

CASE Counselling Assistance to Small Enterprises [Canada]

CASE Curriculum Alignment Services for Educators [of Educational Products Information Exchange (Institute), US]

CASEA Canadian Association for the Study of Educational Administration

CASearch Chemical Abstracts Search [of American Chemical Society, US]

CASEC Confederation of Associations of Specialist Engineering Contractors [UK]

CASE CCR Common Application Service Element (Protocol) Committment, Concurrency and Recovery

CASEE Canadian Army Signals Engineering Establishment

CASES Capabilities Assessment Expert System

CASE/SME Computing and Automated Systems Association of the Society of Mechanical Engineers [US]

CaS:Eu^{2+} Europium-Doped Calcium Sulfide

CaS:Eu^{2+}:Sm^{2+} Europium and Samarium-Doped Calcium Sulfide

CAS Files Chemical Abstracts Service Files [of American Chemical Society, US]

CASG Collaborative Antiviral Study Group

CASH Computer-Assisted Subject Headings Program [of University of California at San Diego, US]

C$_2$ASH$_8$ Dicalcium Aluminate Silicate Octahydrate (or Straetlingite) [$2CaO \cdot Al_2O_3 \cdot SiO_2 \cdot 8H_2O$]

CASI Canadian Aeronautics and Space Institute [Ottawa, Canada]

CASI Center for AeroSpace Information [of NASA at Hanover, Maryland, US]

CASI Compact Airborne Spectral Imager; Compact Airborne Spectrographic Imager

CASI Convenient Automotive Services Institute [US]

Ca/Si Calcium/Silicon (Ratio)

CASIA Chemical Abstracts Subject Index Alert [of American Chemical Society, US]

CaSiAlON Calcium Silicon Aluminum Oxynitride

CASING Crosslinking by Activec (or Activated) Activated Species of Inert Gases

CASIS Computerized Alloy Steel Information System

CASI TRS Center for AeroSpace Information Technical Reports Server [of NASA]

CASL Crosstalk Application Scripting Language

CASLE Commonwealth Association of Surveying and Land Economy [UK]

Čas Lék Česk Časopis Lékařů Českých [Czech Medical Journal]

CASLIS Canadian Association of Special Libraries and Information Services

CASMT Central Association of Science and Mathematics Teachers [US]

Cas Nut Castle Nut

CASO Cancellation Addendum Sales Order

CaSO Calciosulfate

CAS-OB CAS (Composition Adjustment by Sealed Argon Bubbling) with Oxygen Blowing (Steelmaking) [Metallurgy]

CaSO$_4$:Dy Dysprosium Doped (or Activated) Calcium Sulfate

CaSO$_4$:Mn Manganese Doped (or Activated) Calcium Sulfate

CaSO$_4$:Tm Thulium Doped (or Activated) Calcium Sulfate

CASP Canadian Airspace Systems Plan

CASP Canadian Atlantic Storms Program

CASP CDS (Central Data System) Application Support Program

CASP Central American Society of Pharmacology [Panama]

CASPA Certificate of Advanced Study in Public Administration

CASPA Computer-Aided Sculptured Pre-APT (System) [*Note:* APT stands for Automatic Programmed Tool]

CASPAR Cambridge Analog Simulator for Predicting Atomic Reactions

CASPAR Cushion Aerodynamics System Parametric Assessment Research

CASPPR Canadian Arctic Shipping Pollution Prevention Regulations

CAS RN Chemical Abstracts Services Registry Number [also CASRN]

CASS California Agricultural Statistics Service [US]

CASS CITE (Cargo Integration Test Equipment) Augmentation Support System

CASS Commanded Active Sonobuoy System

CASS Common Address Space Section

CASS Computer-Assisted Search Service

CASS Conférence des Académies Scientifiques Suisses [Conference of Swiss Scientific Academies]

CASS Consolidated Automatic Support System

CASS Copper-Accelerated Salt Spray (Test)

CASS Copper Acetic Salt Spray (Test)

CASSA Coarse Analog Sun Sensor Assembly

CASSANDRA Chromatogram Automatic Soaking Scanning and Digital Recording Apparatus

CASSCF Complete Active Space Self-Consistent Field [Physics]

CASSI Chemical Abstracts Service Source Index [of American Chemical Society, US]

CAS-SiC Silicon Carbide (Fiber) Reinforced Calcium Aluminosilicate (Glass Ceramic) [also CAS/SiC]

CASSIS Classified and Search Support Information System

CASSI KWOC CASSI (Chemical Abstracts Service Source Index) Keyword Out-of-Context [of American Chemical Society, US]

CASSIS Classification and Search Support Information System (Database) [of Patent and Trademark Depository Libraries, US]

CASSM Context Addressed Segment Sequential Memory

CaS:Sm^{2+} Samarium-Doped Calcium Sulfide

CAST Canadian Air-Sea Transportable Brigade Group

CAST Canadian Air-Sea Transportable Combat Group

CAST Capillary Action Shaping Technique

CAST Chemical Abstract Searching Terminal

CAST Clearinghouse Announcements in Science and Technology

CAST College of Arts, Science and Technology [Kingston, Jamaica]

CAST Computer-Aided Solidification Technique

CAST Computer-Assisted Surgical Techniques

CAST Computerized Automatic Systems Tester

CAST Convention/Conference for the Advancement of Science Teaching

CAST Cooperation in Applied Science and Technology (with the Newly Independent States of the Former Soviet Union) (Program) [of US National Research Council]

CAST Council for Agricultural Science and Technology [US]

CAST CRC (Cooperative Research Center) for Alloy Solidification Technology [of University of Queensland, St. Lucia, Australia]

Cast Casting [also cast]

CAST ASIA Conference on the Application of Science and Technology to the Development of Asia

Cast Dig Casting Digest [of ASM International, US]

Cast Eng/Foundry World Casting Engineering/Foundry World [Published in the US]

CASTI Centers for the Analysis of Scientific and Technical Information [US]

CASTME Commonwealth Association of Science, Technology and Mathematics Educators [UK]

Cast Met Cast Metals [Journal published in the UK]

CASTOR Corps Airborne Stand-Off Radar

Cast Plant Technol Casting Plant and Technology [Journal published in Germany]

CASTS Czechoslovak Associations of Scientific Technical Societies [now Czech and Slovak Associations of Scientific-Technical Societies]

Cast World Casting World [Published in the US]

CASU Carrier Aircraft Service Unit [US Navy]

CASU CAST (College of Arts, Science and Technology) Academic Staff Union [Jamaica]

CASW Council for the Advancement of Science Writing [Somerville, Massachusetts, US]

CASWS Canadian Acoustic Surveillance Work Station

CASWS Close Air Support Weapon System

CAT Calcium Aluminum Tantalate

CAT Capacity-Activated Transducer

CAT Carburetor Air Temperature

CAT Catechol

CAT Center for Adhesive Technology [UK]

CAT Center for Advanced Technologies

CAT Center for Alternative Technology

CAT Central Alaska Time [Greenwich Mean Time +10:00]

CAT Certificat d'Aptitude Technique [Certificate of Technical Aptitude, France]

CAT Character Asignment Table

CAT Chemical Addition Tank

CAT Chemical Annealing Treatment [Metallurgy]

CAT 2-Chloro-4,6-Bis(ethylamino)-1,3,5-Triazine

CAT Choline Acetyltransferase [Biochemistry]

CAT Cis-Anti-Trans (Configuration) [Chemistry]

CAT Civil Air Transport

CAT Clean Air Today [of Clean Air Council, UK]

CAT Clear Air Transport

CAT Clear-Air Turbulence

CAT Collaborative Access Team [US]

CAT College of Advanced Technology

CAT College of Agricultural Technology

CAT Commercial Air Transport

CAT Compile and Test

CAT Composition Analysis by Thickness Fringe

CAT Computer Adaptive Test

CAT Computer-Aided Teaching

CAT Computer-Aided Test(ing)

CAT Computer-Aided Tomography

CAT Computer-Aided Training

CAT Computer-Aided Transcription

CAT Computer-Aided Translation

CAT Computer-Aided Typesetting

CAT Computer-Assisted Teaching

CAT Computer-Assisted Testing

CAT Computer-Assisted Tomography

CAT Computer-Assisted Training

CAT Computer-Assisted Treatment

CAT Computer Automated Test

CAT Computer Average(d) Transient(s)

CAT Computer(ized) Axial Tomography; Computer(ized) Assisted Tomography

CAT Computer of Average Transients

CAT Conditionally Accepted Tag [Computers]

CAT Constant Analyzer Transmission

CAT Control and Analysis Tool

CAT Controlled Attenuator Timer

CAT Cooled-Anode Transmitting Valve

CAT Crack Arrest Temperature [Mechanics]

CAT Cumulated Aging Times [Metallurgy]

CAT Cumulative Abbreviated Trouble (File)

CAT International Trade Fair for Computer-Aided Engineering, Design and Production with User Conference [Stuttgart, Germany]

.CAT Catalog [File Name Extension] [also .cat]

Cat Catalog(ue); Catalog(u)ing [also cat]

Cat Catalysis; Catalyst [also cat]

Cat Catapult [also cat]

Cat Category [also cat]

CATA Canadian Advanced Technology Association

CATA Canadian Air Transportation Administration

CATA Cobble Automatic Thermoanalyzer

CATA Commissão de Avaliação Toxicologica dos Aditivos Alimentares [Committee for Toxicologic Evaluation of Food Additives, Portugal]

CATAC Creative, Analytical, Technical, Alternative, Competent (Calculating)

CATACARB Catalyzed Removal of Carbon Dioxide (Process)

CATADIENE Catalytic Butadiene (Process)

Catal Catalog(ue) [also catal]

Catal Catalysis; Catalyst [also catal]

Catal Lett Catalysis Letters [US Journal]

Catal Rev Sci Eng Catalysis Reviews, Science and Engineering [US]

Catal Soc Newsl Catalysis Society Newsletter [US]

Catal Spec Period Rep Catalysis, Specialists Periodical Reports

Catal Today Catalysis Today [Journal]

CATC Canadian Air Transport Commission

CATC Caribbean Appropriate Technology Center

CATC Circular-Arc-Toothed Cylindrical

CATC Commonwealth Air Transport Commission [UK]

CATC Commonwealth Air Transport Council [UK]

CATCA Canadian Air Traffic Control Association

CATCALL Completely Automated Technique for Cataloguing and Acquisition of Literature for Libraries

CATCC Canadian Association of Textile Colorists and Chemists

CATCH Computer-Aided Testing and Checking

CATCH Computer-Aided Tolerance Charting

CATCH Computer Analysis of Thermochemical Data

Cat Classif Q Cataloging and Classification Quarterly [US]

CATCON Catalyzed Condensation (Process)

Cat-CVD Catalytic Chemical Vapor Deposition (Method) [also CAT-CVD, or cat-CVD]

CATD Computer-Aided Tool Design

CATD Center for Advanced Technology Development [US]

CATE Center for Advanced Technology Education [of Ryerson Polytechnic University, Toronto, Canada]

CATE Ceramic Applications in Turbine Engines [US Department of Energy Program]

CATE Computer-Assisted Training in Endoscopy

CATE Coordinator of the Air Transport in Europe

CATED Centre d'Assistance Technique et de Documentation [Center for Technical Assistence and Documentation, France]

Categ Category

CATER Collection and Analysis of Terminal Records (Service) [US]

CATF Center for Alternative Transportation Fuels [Vancouver, Canada]

CATF Rev Center for Alternative Transportation Fuels Newsletter [Canada]

CATH Class–Architecture–Topology–Homologous Superfamilies [Protein Structure Analysis]

Cath Cathartic; Catheter [also cath]

Cath Cathedral [also cath]

Cath Cathode [also cath]

CATHALAC Centro de Agua para los Tropicos Humidos do Latinoamerica e los Caribicos [Water Center for the Humid Tropics of Latin America and the Caribbean, Panama]

Ca(THD)₂ Calcium Bis(2,2,6,6-Tetramethyl-3,5-Heptanedionate) [also Ca(thd)₂]

CATI Caribbean Aviation Training Institute

CATI Computer-Assisted Telephone Interviewing (Facility)

CATI Computer-Assisted Television Instruction

CATIE Centro Agronomica Tropical Investigacion y Ensenanza [Tropical Agriculture Research and Training Center, Costa Rica]

CaTiO₃ Calcium Titanate (or Perovskite) (Structure)

CATIS Computer-Aided Tactical Information System

Catlg Catalog(ue) [also catlg]

CaTl Calcium Thallide

CATLAS Centralized Automatic Trouble-Locating and Analysis System

CATLINE Catalog On-Line [of National Library of Medicine, Bethesda, Maryland, US]

CATM Canadian Achievement Test in Mathematics

Ca(TMHD)₂ Calcium Bis(2,2,6,6-Tetramethyl-3,5Heptanedionate) [also Ca(tmhd)₂]

CATNIP Computer-Assisted Technique for Numerical Index Preparation

Cat No Catalog(ue) Number [also cat no]

CATO Civil Air Traffic Operations

CATO Compiler for Automatic Teaching Operation

CATOFIN Catalytic Olefin (Process)

CATOX Catalytic Oxidation (Process)

CATP Controlled Atmosphere Temperature Pressure

CATRA Cutlery and Allied Trades Research Association

CATS Center for Advanced Television Studies [US and UK]

CATS Central Automated Transit System

CATS Computer-Aided Teaching System

CATS Computer-Aided Testing System

CATS Computer-Assisted Trading System

CATS Computer-Assisted Training System

CATS Computer Automated Test System

CATS Corrective Action Tracking System

CAT Scan Computer(ized) Axial Tomography Scan [also CAT scan]

CATSE Capacity of the Air Transport System

CATSI Compact Atmospheric Sounding Interferometer

CATSS Catalogue Support System [of University of Toronto, Canada]

CA TiSULC Continuously-Annealed Titanium Stabilized ULC (Ultralow Carbon) (Steel)

CA TiSULC Steel Continuously-Annealed Titanium Stabilized ULC (Ultralow Carbon) Steel [also CA TiSULC steel]

CATT Campaign Against Arms Trade [UK]

CATT Centralized Automatic Toll Ticketing

CATT Controlled Avalanche Transit-Time Triode

CATT Conveyorized Automatic Tube Tester

CATTCM Canadian Achievement Test in Technical and Commercial Mathematics

CATU Ceramic and Allied Trades Union [UK]

CATV Cable Television

CATV Common Antenna Television

CATV Community Antenna Television (System)

CATY Computer-Aided Telephone Interviewing

CAU Clark Atlanta University [Atlanta, Georgia, US]

CAU Command Acquisition Unit

CAU Command Activation Unit

CAU Command/Arithmetic Unit

CAU Compare Alpha Unequal

CAU Controlled Access Unit

CAU Controller Adapter Unit

CAU CPU (Central Processing Unit) Access Unit

CAU Crypto Ancillary Unit

CAU Customer Acquisition Unit

CAUBO Canadian Association of University Business Officers

Cauc Caucasia(n); Caucasus

CAUCE Canadian Association for University Continuing Education

CAUDO Canadian Association for University Development Officers

CAUFN Caution Adviced Until Further Notice

CAUML Computers and Automation Universal Mailing List

CAURA Canadian Association of University Research Administrators

CAUS Citicens Against UFO (Unidentified Flying Object) Secrecy [US]

CAUS Color Association of the United States

CAUS Newsl CAUS Newsletter [of Color Association of the United States]

CAUSE College and University Systems Exchange [US]

caust caustic

CAUT Canadian Association of University Teachers

Caut Caution [also caut]

CAV Cavity (Resonance)

CAV Chicken Anemia Virus

CAV Component Analog Video

CAV Composite Analog Video

CAV Constant Angular Velocity

CAV Constant Arc Voltage

Cav Cavalry [also cav]

Cav Cavity [also cav]

CAVE Computer Augmented Video Education

CAVE Computer Automatic Virtual Environment

CAVLP Chronoampèrométrie à Variation Linéaire de Potentiel [Chronoamperometry for Linear Variation of Potential]

CAV-OK Ceiling and Visibility Okay [Aeronautics]

CAVORT Coherent Acceleration and Velocity Observations in Real Time

CAVP Census of Australian Vascular Plants [Now part of Australian Plant Name Index (Database)]

CAVP Complex-Arithmetic Vector Processor

CAVS Census of Australian Vertebrate Species

CAVU Ceiling and Visibility Unlimited [Aeronautics]

ČAVU Česká Akademie Věd a Uměni [Czech Academy of Arts and Sciences]

CAVY Cavalry [also Cavy, or cavy]

CAW Canadian Autoworkers Union

CAW Carbon-Arc Welding

CAW Channel Address Word

CAW Current Acid Waste

Caw Subtropical Climate, Winter Drought and Summer Rain [Classification Symbol]

CAWC Central Advisory Water Committee [UK]

CAW-G Gas Carbon-Arc Welding

CAWIS Canadian Association for Women in Science

CaWO Calciotungstate

CAWP Cost Accounting Work Plan

CAWS Common Aviation Weather System

CAW-S Shielded Carbon-Arc Welding

CAWSS Council of Australian Weed Science Societies

CAW-T Twin Carbon-Arc Welding

CAX Community Automatic Exchange

CAX Computer-Aided Experimentation

CayI$ Cayman Island Dollar [Currency]

Ca:YIG Calcium-Substituted Yttrium Iron Garnet [also Ca^{2+}:YIG]

Cay Is Cayman Islands

CAZRI Central Arid Zone Research Institute [India]

CAZS Center for Arid Zone Studies [UK]

CB Canada Balsam

CB Cantilever Beam

CB Capacitor Bank

CB Cape Breton [Nova Scotia, Canada]

CB Carbon Black

CB Cargo Bay

CB Carrier-Based

CB Cartridge Brass

CB Cataclysmic Binary

CB Catch Basin

CB Cathode Bias

CB Cell Biologist; Cell Biology

CB Cell Block (of Grains) [Metallurgy]

CB Cell Body

CB Cell Boundary [Metallurgy]

CB (+)-Cellobiose

CB Cellulose Butyrate

CB Center of Buoyancy

CB Central Battery

CB Certification Body

CB Cesium Beam

CB Check Bit

CB Chemical Bond(ing)

CB Chesapeake Bay [Maryland/Virginia, US]

CB Chief Biologist

CB *(Chirurgiae Baccalaureus)* – Bachelor of Surgery

CB Chlorobenzene

CB Chlorobromomethane

CB Circuit Board

CB Circuit Breaker

CB Citizens' Band (Radio)

CB Clear Back

CB Clear Bit (Function)

CB Clinical Biochemist(ry)

CB Coal Bed

CB Coal Beneficiation

CB Coated Back [Carbonless Paper]

CB Coherent Backscattering

CB Coin Box

CB Coincidence (Grain) Boundary [Materials Science]

CB Cold-Blooded (Animal)

CB Cold Box

CB Colliding Beam

CB Commensurate (Grain) Boundary [Materials Science]

CB Common Base (Circuit)

CB Common Battery

CB Common Branch

CB Communications Buffer

CB Communications Bus

CB Compact Beam

CB Component Board

CB Computing Board

CB Conduction Band [Solid-State Physics]

CB Confined to Barracks [Armed Forces]

CB Congo Basin [Africa]

CB Connecting Box

CB Connector Bracket

CB Constant Bandwidth

CB Construction Ball

CB Construction Battalion

CB Containment Building [Nuclear Reactors]

CB Continuous Blowdown [Metallurgy]

CB Contract Bolt

CB Control Bus

CB Convergent Beam

CB Coupled Biquad

CB Covalent Bond

CB Cross Beta

CB Crystal Ball (Scintillation Counter)

CB Cumulonimbus (Cloud) [also Cb]

CB Currency Bond

CB Cyanobiphenyl

CB Cyanogen Bromide

CB Large Cruiser [US Navy Symbol]

CB15 4'-(2-Methylbutyl)biphenyl-4-Carbonitrile

5CB Pentylcyanobiphenyl

8CB 4-n-Octylcyanobiphenyl

12CB Dodecylcyanodiphenyl

23C14B 2,3-Dichloro-1,4-Benzoquinone

25C14B 2,5-Dichloro-1,4-Benzoquinone

26C14B 2,6-Dichloro-1,4-Benzoquinone

C/B Cold Water/Boiling Water (Immersion Ratio)

C/B Coloring/Bleaching (Process)

C/B Cost-Benefit (Ratio)

Cb Cerebellum [also cb] [Anatomy]

Cb Columbium [= Niobium]

Cb Cumulonimbus (Cloud) [also CB]

Cb Marine West Coast (of North America), Cool Summer [Classification Symbol] [also Cc]

Cb-90 Columbium-90 [also ^{90}Cb, or Cb^{90}]

Cb-91 Columbium-91 [also ^{91}Cb, or Cb^{91}]

Cb-92 Columbium-92 [also ^{92}Cb, or Cb^{92}]

Cb-93 Columbium-93 [also ^{93}Cb, or Cb^{93}]

Cb-94 Columbium-94 [also ^{94}Cb, or Cb^{94}]

Cb-95 Columbium-95 [also ^{95}Cb, or Cb^{95}]

Cb-96 Columbium-96 [also ^{96}Cb, or Cb^{96}]

Cb-97 Columbium-97 [also ^{97}Cb, or Cb^{97}]

Cb-99 Columbium-99 [also ^{99}Cb, or Cb^{99}]

cB Coherent (Grain) Boundary (in Materials Science) [Symbol]

c-B Crystalline Boron

cb centibel [Unit]

c/b c-axis to b-axis ratio (of crystal lattice) (for monoclinic, triclinic and orthorhombic systems) [Symbol]

CBA Canadian Bankers Association

CBA Canadian Booksellers Association

CBA Canadian Botanical Association

CBA Carboxybenzaldehyde

CBA C-Band Transponder Antenna

CBA Chemical Bond Approach

CBA Chlorobenzoic Acid

CBA Cold Bed Adsorption (Process)

CBA College of Business Administration

CBA Colliding Beam Accelerator

CBA Commonwealth Broadcasting Association [UK]

CBA Computer-Based Automation

CBA Confederation of British Associations [UK]

CBA Conjugated Bile Acid

CBA Contact-Breaking Ammeter

CBA Cost-Benefit Analysis

CBA Coulomb-Born Approximation [Physics]

CBA Council for British Archeology

CBA Current Biotechnology Abstracts [of Royal Society of Chemistry, UK]

C/BA Computer/Bus Architecture [Institute of Electrical and Electronics Engineers–Computer Society Committee, US]

CBAA Canadian Business Aircraft Association

CBABG Commonwealth Bureau of Animal Breeding and Genetics

CBAC Chemical-Biological Activities [of American Chemical Society, US]

CB/ac Circuit Breaker (alternating current only)

12CB-ad_2 Deuterated Dodecylcyanodiphenyl

CbAc Carbamoyl Acetic Acid

CBAE Commonwealth Board of Architectural Education [of Commonwealth Association of Architects, UK]

CBAL Counterbalance [also Cbal]

C BAND SCATT C-Band Scatterometer [also C Band Scatt]

CBARC Conference Board of Associated Research Councils

cbar centibar [Unit]

CBAS Chemical Bond Approach Study

C-BASIC Commercial BASIC [Computer Language]

CBAST Concentrated Boric Acid Storage Tank

CBAT Central Bureau for Astronomical Telegrams [US]

CB/ATDS Carrier-Based Airborne Tactical Data System

CBBB Council of Better Business Bureaus [US]

CBC Canadian Bibliographic Center [Ottawa]

CBC Canadian Brigade Corps

CBC Canadian Broadcasting Corporation

CBC Cannabichromen

CBC Can't Be Called

CBC Carbobenzoxy Chloride

CBC Center for Biological Chemistry [of University of Nebraska, US]

CBC Central Bank of China [Taiwan]

CBC Central Bureau of Compensation [Belgium]

CBC Chain Block Character

CBC Chemically-Bonded Ceramic(s)

CBC Christmas Bird Count [US]

CBC Cipher Block Chaining

CBC Circulating Bed Combustor

CBC Columbia Basin College [US]

CBC Complete Blood Count

CBC Computer Brokers of Canada

CBC Confederation of Building Contractors [UK]

CBC Conference Board of Canada

CBC Contact-Breaking Clock

CBC Continuity Bar Connector

CbC Columbium Carbide

CBCA Cyclobutanecarboxylic Acid

CBCC Chemical-Biological Coordination Center [of National Research Council, US]

CBCCO Chromium Barium Calcium Copper Oxide (Superconductor)

CBCES Chesapeake Bay Center for Environmental Studies [of Smithsonian Institution; located near Annapolis, Maryland, US]

CbCN Columbium Carbonitride

CBCNO Cerium Barium Copper Niobium Oxide (Superconductor)

CBCPW Conductor-Backed Coplanar Waveguide

CBCR Channel Byte Count Register

CBCRL Cape Breton Coal Research Laboratories [Nova Scotia, Canada]

CBCS Commonwealth Bureau of Census and Statistics [Australia]

C-B₄C-SiC Carbon–Boron Carbide–Silicon Carbide (Composite)

CBCSM Council of British Ceramic Sanitaryware Manufacturers

CBCT Council for British Cotton Textiles [UK]

CBCT Customer-Bank Communications Terminal; Customer Banking Communications Terminal

CBD Call Box Discrimination

CBD Cannabidiol

CBD Cash before Delivery [also cbd]

CBD Cellulose Binding Domain

CBD Central Business District

CBD Chemical Bath Deposition

CBD Chronic Beryllium Disease

CBD Commerce Business Daily

CBD Common Bile Duct [Anatomy]

CBD Configuration Block Diagram

CBD Constant Bit Density

CBD Continuous Blow-Down [Metallurgy]

CBD Convention on Biological Diversity

CBD Convergent Beam Diffraction

C-BD C-Band [3900 to 6200 megahertz]

CBDA Cannabidiolcarbonic Acid

CBDB Conference Board Database [US]

CBDC Cape Breton Development Corporation [Nova Scotia, Canada]

CB/DC Circuit Breaker/Direct Current Only [also CB/dc]

CBDCA Cyclobutane-1,1-Dicarboxylate [also cbdca]

CBDCA 1,3-Cyclobutanedicarboxylic Acid

CBDI Control Red Bank Demand Indicator

CBDICE Chemical/Biological Defense Industrial Center of Expertise [Canada]

CBDN Canadian Bacterial Diseases Network

CBDP Convergent Beam Diffraction Pattern

Cb(DPM)₅ Columbium Dipivaloylmethanoate [also Cb(dpm)₅]

CBDS Circuit Board Design System

CBDS Connectionless Broadband Data Service

CBDST Commonwealth Bureau of Dairy Science and Technology [UK]

CBDT Can't Break Dial Tone

CBDT Computer-Based Documentation and Training

CBE Centralized Branch Exchange

CBE Chemical Beam Epitaxy

CBE Circuit Breaker for Equipment

CBE Citizens for a Better Environment [US]

CBE Cleveland Board of Education [Ohio, US]

CBE Commander of the Order of the British Empire

CBE Communities for a Better Environment [San Francisco, California, US]

CBE Companion of the Order of the British Empire

CBE Competency-Based Education

CBE Computer-Based Education

CBE Congresso Brasileira de Entomológica [Brazilian Entomological Congress]

CBE Connector Bracket, Experiment

CBE Council of Biology Editors [US]

C&BE Computer and Business Equipment

CBEA Commonwealth Banana Exportes Association [Saint Lucia]

CBEC Canadian Book Exchange Center

CBEC Concentration-Based Exemption Criteria [of US Environmental Protection Agency]

CBED Convergent-Beam Electron Diffraction

CBEDP Convergent-Beam Electron-Diffraction Pattern

CBEL Computational Bioscience and Engineering Laboratory [of Division of Computer Research and Technology, National Institutes of Health, Bethesda, Maryland, US]

CBEMA Canadian Business Equipment Manufacturers Association

CBEMA Computer and Business Equipment Manufacturers Association [US]

CBEP Community-Based Environmental Protection (Project) [US Federal/State Project]

CBERS Chinese-Brazilian Earth Resources Satellite

CBET Certified Biomedical Equipment Technician

CBF Carbon Baking Furnace

CBF Correlated Basis Functions

Cbf Moderate Marine Forest Climate, Mild Winter [Classification Symbol]

CBFC Copper and Brass Fabricators Council [US]

CBFCO Cyclobutadiene Iron Tricarbonyl

CBFS Critical Brittle Fracture Stress

CBG Canberra Botanic Gardens [also Australian National Botanic Gardens]

CBG Cannabigerol

CBG Corticosteroid-Binding Globulin [Biochemistry]

CBG Craniofacial Biology Group [of International Association for Dental Research, US]

CBGA Cannabigerol Acid

CBGA Carpathian Balkan Geological Association [Poland]

CBGA Ceramic Ball Grid Array

CBGC Central Bank Governors' Committee

CBH Can't be Heard

CbH Columbium Hydride

CBHA p-Chlorobenzohydroxamic Acid Ester

CBHL Council on Botanical and Horticultural Libraries [US]

CBI Canadian Back Institute

CBI Canadian Business Index

CBI Cape Breton Island [Nova Scotia, Canada]

CBI Center for Biological Informatics [of US Geological Survey/Biological Resources Division at Denver, Colorado]

CBI Charles Babbage Institute [US]

CBI Chesapeake Bay Institute [US]

CBI China–Burma–India

CBI Common Batch Identification

CBI Compound Batch Identification

CBI Computer-Based Instruction

CBI Computer-Based Instrumentation

CBI Confederation of British Industry [UK]

CBI Confidential Business Information

CBI Cooperative Business International [US]

CB&I Chicago Bridge and Iron [US]

CBIC Canadian Biodiversity Informatics Consortium

CBIC Canadian Book Information Center

CBIC Complementary Bipolar Integrated Circuit

CBIE Canadian Bureau of International Education

CBIL Common and Bulk Items List

CBIN Caribbean Basin Information Network

CBIS Computer-Based Instructional System

CBIT Cape Breton Institute of Technology [Nova Scotia, Canada]

Cbk Checkbook [also cbk]

CBKP Convergent Beam Kikuchi Pattern

CBKR Cross-Bridge Kelvin Resistance

CBL Canadian Bronze Company Ltd.

CBL Cement Bond Log

CBL Chesapeake Biological Laboratory [of National Resources Institute, US]

CBL Cloud Boundary Layer

CBL Commercial Bill of Lading

CBL Computer-Based Learning

.CBL COBOL Source Code [File Name Extension]

CBLF Compressible Boundary Layer Flow

CBLM Cluster Bethe Lattice Method (or Model) [Materials Science]

CBM Calcium Bismuth Manganate

CBM Carbamoyl

CBM Carbon Budget Monitor

CBM Certified Ballast Manufacturers Association [US]

CBM Chlorobromomethane

CBM Coal Bed Methane (Region)

CBM Colliding Beam Mode

CBM Commonwealth Bureau of Meteorology [Australia]

CBM Conduction Band Minimum [Solid-State Physics]

CBM Confidence Building Material

CBM Confidence Building Measure

CBM Constant-Boiling Mixture

CBM Containerized Batch Mixer

CBM Conventional Ballistic Missile

CBM Conventional Buoy Mooring

CBM Convergent-Beam Microdiffraction

CBM Convergent-Beam Microdiffractometer; Convergent-Beam Microdiffractometry

CBM Creep Brittle Material

CBMA Canadian Battery Manufacturers Association

CBMA Canadian Brush Manufacturers Association

CBMA Canadian Business Manufacturers Association

CBMA Carbamylmethyl Amine [also Cbma]

CBMA Certified Ballast Manufacturers Association

CBMC Canadian Book Manufacturing Council

CBM-CFS Carbon Budget Monitor–Canadian Forest Sector

CBME Chemical, Bio and Materials Engineering (Department) [of Arizona State University, Tempe, US]

CBMIS Computer-Based Management Information System

CBMM Companhia Brasileira de Metalurgia e Mineração [Brazilian Metallurgical and Mining Company located at Araxá, Minas Gerais]

CBMM Council of Building Materials Manufacturers [US]

CBMO Calcium Bismuth Manganese Oxide

Cb/MoSi$_2$ Columbium-Fiber-Reinforced Molybdenum Silicide (Composite)

CBMPE Council of British Manufacturers of Petroleum Equipment

CBMS Chill Block Melt Spinning [Metallurgy]

CBMS Computer-Based Mail System

CBMS Computer-Based Message Service

CBMS Computer-Based Message System

CBMS Conference Board of the Mathematical Sciences [US]

CBMU Canadian Board of Marine Underwriters

CBMU Current Bit Motor Unit

CBMUA Canadian Boiler and Machinery Underwriters Association

CBN Cannabinol

CBN Cesium Bismuth Niobate

CBN Cubic Boron Nitride [also c-BN, or cBN]

CBNA Cannabinolcarbonic Acid

CBNB Chemical Business Newsbase

CBN/B$_4$C Cubic Boron Nitride/Boron Carbide (Film) [also c-BN/B$_4$C]

CBNC Canadian Business Networks Coalition

CBNM Central Bureau for Nuclear Measurements

CBNO Calcium Barium Niobium Oxide (Superconductor)

CBNY Chemical Bank of New York [US]

CBO Conference of Baltic Oceanographers [Germany]

CBO Congressional Budget Office [US]

CbO Columbium(II) Oxide

Cbo Carbobenzoxy-

CBOA Canadian Building Officials Association

CBOE Chicago Board Options Exchange [US]

CBOM Canadian Board of Occupational Medicine

CBOM Current Break-Off and Memory

C-Bomb Cobalt Bomb [also C-bomb]

CBOOA Cyanobenzylidene Octyloxyaniline

Cbore Counterbore [also cbore]

CBOSS Count Back Order and Sample Select

CBOT Chicago Board of Trade [US]

CBOTA Cape Breton Offshore Trade Association [Nova Scotia, Canada]

CBP Canadian Business Press

CBP Chesapeake Bay Program [of US Environmental Protection Agency]

CBP Coffee Bean Parchment

CBP Compact Beam Production

CBP Connector Bracket Power

CBP Construction Ball Pad

CBP Convergent Beam Pattern

CBPAH Council for British Plastics in Agriculture and Horticulture

CBPC Canadian Book Publishers Council

CBPE Centro Brasileiro de Pesquisas Educacionais [Brazilian Center for Educational Research]

CBPF Centro Brasileiro de Pesquisas Fisicas [Brazilian Center for Research in Physics, Rio de Janeiro]

CBPI Canadian Business Periodicals Index

CBPqE Centro Brasileiro de Pesquisas Educacionais [Brazilian Center for Educational Research]

CBPqF Centro Brasileiro de Pesquisas Fisicas [Brazilian Center for Research in Physics, Rio de Janeiro]

CBPT CLIRA (Closed Loop In-Reactor Assembly) Backup Plug Tool [Nuclear Engineering]

C14BQ Monochloro-1,4-Benzoquinone

CBR California Bearing Ratio

CBR Case-Based Reasoning

CBR Cavity-Backed Radiator

CBR Center for Biotechnology Research [Beijing, PR China]

CBR Center for Business Research [UK]

CBR Chemical, Bacteriological and Radiological (Methods)

CBR Chemical, Biological and Radiological (Warfare)

CBR Chlorinated Butyl Rubber

CBR Cloud Base Recorder

CBR Commercial Breeder Reactor

CBR Community Bureau of Reference [US]

CBR Constant Bit Rate

CBRA Chemical, Biological and Radiological Agency [US]

CBRC Community Business Resource Center [Canada]

CBRET Canadian Board for Registration of EEG (Electroencephalography) Technologists

CBRI Central Building Research Institute [India]

CBRP Chimpanzee Breeding and Research Program

CBRS Canadian Bond Rating Service

CBRS Chemical, Biological, and Radiological Section [of US Military]

CBR SRM Community Bureau of Reference–Standard Reference Material [US]

CBRTGW Canadian Brotherhood of Railway Transport and General Workers

CBS Canadian Biochemical Society

CBS Canadian Blood Services

CBS Canadian Business Service

CBS Call Box Station

CBS Central Battery Signalling

CBS Central Battery System

CBS Chinese Biochemical Society

CBS Columbia Broadcasting System, Inc. [US]

CBS Commission for Basic Systems [of World Meteorological Organization]

CBS Common Battery Signaling

CBS Common Battery System

CBS Commonwealth Bureau of Soils

CBS Complete Basis Set

CBS Connector Bracket Signal

CBS N-Cyclohexyl-2-Benzothiazole Sulfenamide

CBSA Cargo Bay Stowage Assembly

CBSA Center for Business Systems Analysis [UK]

CBSA Copper and Brass Service Center Association [also CABSA, US]

CB/SA Circuit Breaker/Surge Arrester

CBSC Canada Business Service Center

CBSE Center for Biomolecular Science and Engineering [of Naval Research Laboratory, Washington, DC, US]

CBSE Commonwealth Bureau of Survey Education [UK]

CBSE Converted Backscattered Electron Secondary Electron

CBSI Chartered Building Societies Institute [UK]

CBSI Convergent Beam Shadow Image

CBSR Coupled Breeding Superheating Reactor [US]

CBT Chemical and Biological Treatment

CBT Chicago Board of Trade [US]

CBT Cincinnati Board of Trade [US]

CBT Committee for Better Transit [US]

CBT Computer-Based Terminal

CBT Computer-Based Training

CBT Core Block Table

CBTA Canadian Business Telecommunications Alliance

CBTA Cape Breton Transit Authority [Canada]

CBTCD CANMET (Canada Center for Mineral and Energy Technology)/Business and Technology Coordination Division [of Natural Resources Canada]

CbTe Columbium Telluride (Semiconductor),

CBTP 4-Carboxybutyl Triphenyl Phosphonium

CBU Coefficient of Beam Utilization

CBV Cabin Bleed Valve

CBV Contact-Breaking Voltmeter

CBV Coxsackie B4 Virus

CBW Chemical and Biological Warfare

CBW Chemical and Biological Weapon(s)

CBW Constant Bandwidth

CBW Convert Byte to Word

Cbw Moderate Marine Forest Climate, Dry Season in Winter [Classification Symbol]

CBWA Canadian Bottled Water Association

CBWA Copper and Brass Warehouse Association [now Copper and Brass Service Center, US]

CBX Cam Box

CBX C-Band Transponder

CBX Centralized Branch Exchange

CBX Computer Branch Exchange

CBX Computer-Controlled Branch Exchange

CBX Computerized (Private) Branch Exchange

Cbyd Cubic yard [Unit]

Cbyd/hr Cubic yard per hour [also Cbyd/h]

CBZ Carbobenz(yl)oxy- [also Cbz]

N-CBZ N-Carbobenzoxy- [also N-Cbz]

CC Cadmium Council [US]

CC Calcium Carbonate

CC Calculator

CC Call Check
CC Calorimetry Conference
CC Canada Council
CC Canaries Current [Oceanography]
CC Canterbury College [New Zealand]
CC Cap Core
CC Cape Canaveral [Florida, US]
CC Capital Cost
CC Capsule Communicator
CC Carbocation
CC Carbonaceous Chondrite [Geology]
CC Carbon-Carbon (Composite) [also C-C, or C/C]
CC Carbon Chemist(ry)
CC Carbon Copy [also cc]
CC Carbon Cycle [Ecology]
CC Carbon Fiber/Calcium Phosphate (Composite)
CC Carbon-Fiber-Reinforced Carbon (Composite) [also C-C, or C/C]
CC Carleton College [Northfield, Minnesota, US]
CC Carnegie Commission [Washington, DC, US]
CC Carnot Cycle [Thermodynamics]
CC Carriage Control
CC Carrier Concentration [Solid-State Physics]
CC Carrier Current
CC Carrying Capacity
CC Carrying Cost
CC Carson City [Nevada, US]
CC Cascade Control
CC Catalyst Carrier
CC Catalytic Cracking [Chemical Engineering]
CC Catecholamine Club [US]
CC Category Code
CC Cathode Copper
CC Cation-Cation (Bond)
CC Cell Carcinoma [Medicine]
CC Centers Curve
CC Centiseconds [also cc]
CC Central Computer
CC Central Control
CC Centrifugal(ly) Cast; Centrifugal Casting [Metallurgy]
CC Centrifugal Circulator
CC Ceramic Coating
CC Cerebral Cortex [also cc]
CC Cerenkov Counter
CC Certificat de Capacité [French Certificate of Capacity]
CC Chamber of Commerce
CC Channel Command
CC Channel Controller [also cc]
CC Channel Coordinator
CC Chapman Conference (on Magnetic Helicity in Plasmas) [of American Geophysical Union, US]
CC Charge Conjugation [Particle Physics]
CC Charge Conveyor
CC Chartered Cartographer

CC Chemical Ceramics
CC Chemical Corps [of US Army]
CC Chemistry Consortium [US]
CC Chemometrics in Analytical Chemistry
CC Chief Complaint [Medicine]
CC Chief Constable
CC Chief Counsel [NASA Kennedy Space Center Directorate, Florida, US]
CC Chip Carrier
CC Chiral Center [Organic Chemistry]
CC Chlorocarbon
CC Chondrocyte [Anatomy]
CC Chronocoulometer; Chronocoulometry
CC Chrysler Canada (Corporation)
CC Chrysler Corporation [US]
CC Circuit Closing
CC Circumnavigators Club [US]
CC Cirrocumulus (Cloud) [also Cc]
CC Cis-Cis (Configuration) [Chemistry]
CC City College [of City University of New York, US]
CC Claisen Condensation [Organic Chemistry]
CC Classification Code
CC Clean Credit [Business Finance]
CC Climbing Crane
CC Clinical Center
CC (Warren Grant Magnuson) Clinical Center [of National Institutes of Health, Bethesda, Maryland, US]
CC Clinical Computing
CC Clock Correction
CC Close-Coupled; Close Coupling
CC Closed Cell
CC Closed Circuit
CC Closed Cup (Flash Point Test)
CC Closing Coil
CC Cloud Chamber
CC Cluster Center [Metallurgy]
CC Cluster Controller
CC Coal Carbonization (Process)
CC Coal Chemist(ry)
CC Coal Cleaning
CC Coal Combustion
CC Coal Conversion
CC Coarse Control
CC Coarse-Grained Crystalline; Coarse-Grained Crystallite
CC Cocos (Keeling) Islands [ISO Code]
CC Code Controller
CC Coe College [Cedar Rapids, Iowa, US]
CC Cogeneration Council [now Cogeneration and Independent Power Coalition of America, US]
CC Coiled Coil (Filament)
CC Coin Collect(ion)
CC Colby College [Waterville, Maine, US]
CC Cold Cathode
CC Collect Call

CC Collision Cascade
CC Collodion Cotton
CC Colloid Chemist(ry)
CC Colorado College [Colorado Springs, US]
CC Colorado Springs [US]
CC Color Code
CC Color Compensation
CC Color Correction
CC Column Chromatograph(y)
CC Combined Carbon
CC Combined Cycle
CC Combustion Chamber
CC Command Chain
CC Common Carrier
CC Common Collector
CC Common Control
CC Communication Center
CC Communication Control(ler)
CC Communications Canada
CC Comparison Circuit
CC Complete Combustion
CC Complexant Concentrate
CC Compression Cable
CC Compressor Contactor
CC Computational Chemistry
CC Computer Center
CC Computer Communications
CC Computer Community
CC Computer Control
CC Computer Conference
CC Concentric Cable
CC Concept Code
CC Concord College [Athens, West Virginia, US]
CC Condenser Coil
CC Condition Code
CC Conduction Current
CC Conductive Carbon
CC Cone Cell (of the Eye)
CC Cone Clutch
CC Configuration Control
CC Connecticut College [New London, US]
CC Constant Current
CC Construction Cost
CC Consular Corps
CC Consulting Chemist
CC Contact Closure
CC Continuous(ly) Cast; Continuous Casting [Metallurgy]
CC Continuous Coarsening [Metallurgy]
CC Continuous(ly) Cast; Continuous Cooling [Metallurgy]
CC Continuous Current [also cc]
CC Contracts Canada [of Public Works and Government Services Canada]
CC Control Center
CC Control Character

CC Control Code
CC Control Computer
CC Control Console
CC Control Counter
CC Convalescent Care
CC Convolutional Code
CC Cooling Coil
CC Coordinating Committee
CC Coordination Catalysis
CC Cornell College [Mount Vernon, Iowa, US]
CC Correlation Coefficient [Statistics]
CC Cost Center
CC Cotton Count
CC Cotton-Covered (Cable)
CC Council on Competitiveness [US]
CC Counter Clockwise [also cc]
CC Countercurrent
CC Country Code
CC Countryside Commission [UK]
CC County Council [UK]
CC Coupled Channel
CC Coupling Constant
CC Crevice Corrosion
CC Crew Certified [also cc]
CC Crew Compartment
CC Cross-Correlation
CC Cross Couple
CC Cryochemist(ry)
CC Cryptocenter [of NATO]
CC Crystal Characterization
CC Crystal Chemist(ry)
CC Crystal Class
CC Crystal-Controlled; Crystal Controller
CC Cubic Capacity
CC Current Carrier
CC Current Commutator
CC Current Complaint [Medicine]
CC Current Contact
CC Current Control(ler)
CC Cursor Control
CC Curvilinear Coordinates
CC Cyclic Code
CC Cyclic Compound
CC Cycloconverter
CC Cyclone Collector
CC- Chile [Civil Aircraft Marking]
C15C5 Cyclohexano-15-Crown-5 [also C15Crown-5, or C15-crown-5]
C18C6 Cyclohexano-18-Crown-6 [also C18Crown-6, or C15-crown-6]
C&C Casgrain and Charbonneau Limited [Pharmaceutical Manufacturer]
C&C Command and Control
C&C Communication and Cognition

C&C Computers and Communications

C/C Clean Credit [Business Finance]

C/C Carbon/Carbon (Composite) [also CC, cc, C-C, or c-c]

C/C Carbon-Fiber-Reinforced Carbon (Composite)

C/C Combustion Chamber

C_{60}/C_{70} Buckminsterfullerene Mixture of Molecules with 60 Carbon Atoms and Molecules with 70 Carbon Atoms

C-C Carbon-Carbon (Composite) [also CC, or C/C]

C-C Carbon-Fiber-Reinforced Carbon (Composite) [also CC, or C/C]

C-C Center to Center [also c-c]

Cc Carbon Copy [Internet]

Cc Cirrocumulus (Cloud) [also CC]

c-C Crystalline Carbon

Cc Cumulocirrus (Cloud)

Cc Marine West Coast (of North America), Cool Summer [Classification Symbol] [also Cb]

cc carbon copy [also CC]

cc chapters

cc complex conjugate [Mathematics]

cc copies

cc creep in tension and compression [Mechanical Testing]

cc cubic centimeter [also cu cm, or cm^3]

c/c center to center

CCA Cam Clamp Assembly

CCA Canadian Cartographers Association

CCA Canadian Cat Association

CCA Canadian Center of Architecture

CCA Canadian Chemical Association

CCA Canadian Chiropractic Association

CCA Canadian Communication Association

CCA Canadian Construction Association

CCA Canadian Consumers Association

CCA Capital Cost Allowance

CCA Carbohydrate-Containing Antibiotics

CCA Carrier-Controlled Approach (System)

CCA Cement and Concrete Association [UK]

CCA Central Computer Accounting

CCA Central Computer Agency [UK]

CCA Channel-to-Channel Adapter

CCA Chemical Coaters Association [US]

CCA Chemical Communications Association [US]

CCA Chemical Corps Association [US]

CCA Chromated Copper Arsenate

CCA Cluster-Cluster Aggregation (Model) [Metallurgy]

CCA Coagulation Control Abnormal

CCA Coastal Conservation Association [US]

CCA Cogeneration Council of America [now Cogeneration and Independent Power Coalition of America, US]

CCA College Chemists Association [UK]

CCA Common Communication Adapter

CCA Communications Carrier Assembly

CCA Computer and Control Abstracts [of Institution of Electrical Engineers, UK]

CCA Consumer and Corporate Affairs [Canada]

CCA Continuing Crown Allocation

CCA Contract Change Authorization

CCA Coolant Control Assembly

CCA Cooperative Communicators Association [US]

CCA Copper Conductors Association [UK]

CCA Council of Chemical Associations [now Council of Chemical Association Executives, US]

CCA Coupled Cluster Approach [Physics]

CCA Crop Condition Assessment [US]

CCA Crime Control Act [US]

CCA Cyclic Creep Acceleration [Mechanics]

CCAA Cement and Concrete Association of Australia

CCAAP Central Committee for the Architectural Advisory Panels [UK]

CCAB Canadian Circulation Audit Board

CCAB Consumer and Corporate Affairs Canada

CCABC Cancer Control Agency of British Columbia [Canada]

CCAC California College of Arts and Crafts [Oakland, US]

CCAC Canadian Council on Animal Care

CCAC Colorado Center for Advanced Ceramics [of Colorado School of Mines, Golden, US]

CCAC Conference on Composites and Advanced Ceramics [US]

CCAC Coordinating Committee on Agricultural Chemicals

CCACD Canadian Center for Arms Control and Disarmament

CCAE Canberra College of Advanced Education [Australia]

CCAE Center for Computer Assisted Engineering [of Colorado State University, Fort Collins, US]

CCAE (Joint) Congressional Committee of Atomic Energy [US]

CCAE Council of Chemical Association Executives [US]

CCAFS Cape Canaveral Air Force Station [Florida, US]

CCAHC Central Council for Agricultural and Horticultural Cooperation [UK]

CCAI Canadian Center for Advanced Instrumentation

CC-AI Communication and Cognition–Artificial Intelligence [also CCAI, Belgium]

CCALI Center for Computer-Assisted Legal Instruction [Minneapolis, Minnestoa, US]

CCAM Canadian Congress on Applied Mechanics

CCAM Common Central American Market

CCAM Conversational Communication Access Method

CCAMLR Commission for the Conservation of the Antarctic Marine Living Resources [Australia]

CCAMLR Convention for the Conservation of the Antarctic Marine Living Resources [Australia]

CCAP Communications Control Application Program

CCAP Culture Center for Algae and Protozoa [UK]

C-CAP Coastal Change Analysis Program

C-CAP Coastwatch Change Analysis Program [of National Oceanic and Atmospheric Administration, US]

CCAQ Consultative Committee on Administrative Questions [of United Nations]

CCAR Colorado Center for Astrodynamics Research [of University of Colorado at Boulder, US]

C/CARES Composite Ceramics Analysis and Reliability of Structures (Computer Code)

CCARM Centre Canadien d'Automatisation et de Robotique Minière [Canadian Center for Automation and Robotics in Mining]

CCARR Coordinating Committee for the Alligators River Region [Australia]

CCAS Cape Canaveral Air Station [Florida, US]

CCAS Carrier-Controlled Approach System

CCAS Convention for the Conservation of Antarctic Seals

CCASM Canadian Council of the American Society for Metals

CCAST Center for Condensed Matter and Radiation Physics [of World Laboratory, Beijing, PR China]

CCAT Centralia College of Agricultural Technology [Canada]

CCAT Computer-Controlled Automatic Test Equipment

CCAT Conglutinating Complement Absorption Test [Biochemistry]

CCATS Command, Communication and Telemetry System

CCB Calcium Channel Blocker [Medicine]

CCB Change Control Board

CCB Channel Command Block

CCB Channel Control Block

CCB Character Control Block

CCB Circuit Concentration Bay

CCB Close Control Bombing [also ccb]

CCB Coin Collecting Box

CCB College of Cape Breton [Nova Scotia, Canada]

CCB Combined Communications Board

CCB Command Control Block

CCB Common Carrier Bureau

CCB Communications Control Block

CCB Composites and Ceramics Branch [of Naval Research Laboratory, Washington, DC, US]

CCB Configuration Change Board

CCB Configuration Control Board

CCB Continuing Calibration Blank

CCB Continuum Configurational Bias

CCB Controlled-Collapse Bonding

CCB Convertible Circuit Breaker

CCBA Coupled Channel Born Approximation [Physics]

CCBC Council of Community Blood Centers [US]

CCBD Configuration Control Board Directive

CCBDA Canadian Copper and Brass Development Association

CCB-MC Continuum Configurational Bias–Monte Carlo (Method)

CCBN Center for Conservation Biology Network [Rice University, Houston, Texas, US]

CCBS Clear Channel Broadcasting System

CCBS Conventional Core-Barrel Sampling

CCC Calculators Collector Club [UK]

CCC California Coastal Commission [US]

CCC California Conservation Council [US]

CCC Canadian Chamber of Commerce

CCC Canadian Climate Center [of University of British Columbia, Vancouver]

CCC Canadian Commercial Corporation

CCC Canadian Committee on Cataloguing

CCC Canadian Computer Complex

CCC Canadian Computer Conference

CCC Canister Centerline Cooling

CCC Canola Council of Canada

CCC Capricorn Conservation Council

CCC Carbon-Carbon Composite

CCC Car Care Council [US]

CCC Caribbean Conservation Corporation [US]

CCC Carrier-Current Communication [also ccc]

CCC Central Communications Controller

CCC Central Computer Complex

CCC Ceramic Chip Carrier

CCC Change and Configuration Control

CCC Chemical Coal Cleaning

CCC Chemicals Consultative Committee [of Australian and New Zealand Environment and Conservation Council]

CCC Chief of the Chemical Corps [US Army]

CCC Chlorocholine Chloride

CCC Cis-Cis-Cis (Geometry) [Chemistry]

CCC City Communications Center [UK]

CCC Civilian Conservation Corps [US] [Abolished]

CCC Cleaved Coupled Cavity

CCC Cleveland Chamber of Commerce [Ohio, US]

CCC Cleveland Chiropractic College [Kansas City, Missouri, US]

CCC Cleveland Convention Center [Ohio, US]

CCC Command, Control and Communication

CCC Commodity Credit Corporation [of US Department of Agriculture]

CCC Communications Control Console

CCC Communities of Common Concern

CCC Comparative Capital Cost

CCC Complex Control Center

CCC Computer Communication Console

CCC Computer Control Complex

CCC Computing Coordinating Committee

CCC Conductive Carbon Cement

CCC Consolidated Contractors International Company Limited [Riyadh, Saudi Arabia]

CCC Consumer Credit Code

CCC Contaminant Control Cartridge

CCC Continuing Care Coordinator

CCC Continuous Cooling Compression (Test)

CCC Controller Checkout Console

CCC Convert Character Code

CCC Copy Control Character

CCC Copyright Clearance Center [Salem, Massachusetts, US]

CCC Corporate Computer Center

CCC Corporate Conservation Council

CCC Council on Clinical Cardiology [of American Heart Association, US]

CCC Countercurrent Chromatography

CCC Cross-Channel Cable

CCC Crossed Claisen Condensation [Chemistry]

CCC Customs Charges Collectable

CCC Customs Cooperation Council [Belgium]

CC&C Command, Control and Communications [also CCC, or C^3]

ccc core-core-core (transition) [Auger Electron Spectroscopy]

CCCA Canadian Corrugated Case Association

CCCA Center for Chemical Characterization and Analysis [of Texas A&M University, College Station, US]

CCCA Comprehensive Crime Control Act [US]

CCCB Cámara de Comercio Colombo Británica [Colombia-Britain Chamber of Commerce]

CCCB CRAF (Comet Rendevous/Asteroid Fly-By) Cassini Common Buy [of NASA]

CCCC Computerized Conferencing and Communications Center [of New Jersey Institute of Technology, US]

CCCC Coordinating Council for Computers in Construction [US]

CCCC Council for Car Care Centers [US]

CCCCD Conductively-Connected Charge-Coupled Device [also C^4D]

CC-CCR Continuous Casting and Cold-Charge Rolling [Metallurgy]

CCCE California Council of Civil Engineers [US]

CCCE Canadian Chemical Conference and Exhibition

CCCE Commission for Certification of Consulting Engineers [US]

CCCEC Commission for Certification of Consulting Engineers in Colorado [US]

CCCE&LS California Council of Civil Engineers and Land Surveyors [US]

CCCESD Council of Chairmen of Canadian Earth Science Departments

CCCG Combined-Cycle Coal Gasification

CCCI Command, Control, Communication and Intelligence [also C^3I]

CCCI Correlation-Consistant Configuration Interaction

CCCI Cyprus Chamber of Commerce and Industry

CCCL Complementary Constant-Current Logic [also C^3L]

CCCM Center for Cement Composite Materials [US]

CCCN Customs Cooperation Council Nomenclature

CCCNO Cerium Calcium Copper Niobium Oxide (Superconductor)

CCCO Calcium Cerium Copper Oxide (Superconductor)

CCCO Committee on Climatic Changes and the Oceans [France]

CCCP Carbon-Carbon Connectivity Plot [Nuclear Magnetic Resonance]

CCCP Carbonyl Cyanide-3-Chlorophenylhydrazone

CCCP- Soviet Union [Civil Aircraft Marking]

CCCRAM Continuously Charge-Coupled Random-Access Memory [also C^3RAM]

CCCRL CANMET (Canada Center for Mineral and Energy Technology)/Combustion and Carbonization Research Laboratory [of Natural Resources Canada]

CCCS Canadian Cooperative Credit Society

CCCS Command, Control and Communications System

CCCS Core Component Cleaning System [Nuclear Engineering]

CCCS Core Component Conditioning Station [Nuclear Engineering]

CCCS Current Controlled Current Source

CCCSA Conservation Center and Council of South Australia

CCCT Canadian Center for Creative Technology

CCCT Compound Cyclic Corrosion Test

CCD Calcite Compensation Depth [Geology]

CCD Capacitor-Charged Device

CCD Census Collection District

CCD Charge-Coupled Device

CCD Checkout Command Decoder

CCD Circumscribing Circle Diameter

CCD Coarse Control Damper

CCD Cold Cathode Discharge

CCD Commonwealth Committee on Defense [now Commonwealth Defense Conference, India]

CCD Computer-Controlled Display

CCD Constant Change Display

CCD Continuous-Countercurrent Decantation (Process)

CCD Contract Completion Date

CCD Controlled Current Distribution

CCD Cooled Camera Detector

CCD Core Current Driver

CCD Counter-Clockwise Diamond (Test) [Mechanical Testing]

CCD Countercurrent Decantation

CCD Countercurrent Digestion [Ore Leaching]

CCD Coupled-Cluster Doubles

CCDA Commercial Chemical Development Association [US]

CCDBP Continuous(ly) Cast Direct Billet Process [Metallurgy]

CCDC Construction Contract Development Association

CCDC Current Carrier Density Collapse

CCDCG Clinical Chemistry Data Communication Group

CCDF Complementary Cumulative Distribution Function

CCDLERDSWA Commission for Controlling the Desert Locust in the Eastern Region of its Distribution in Southwest Asia [Italy]

CCDLNE Commission for Controlling the Desert Locust in the Near East [of Food and Agricultural Organization in Saudi Arabia]

CCDLNWA Commission for Controlling the Desert Locust in Northwest Africa [of Food and Agricultural Organization in Algeria]

CCDM Change and Configuration Control/Development and Maintenance

CCDM Communications, Command and Data Management

CCDN Corporate Consolidated Data Network [of IBM Corporation, US]

CCDP Canadian Continental Drilling Program

CCDP Climate Change Detection Project [of World Meteorological Organization]

CCDR Contractor Critical Design Review

CCDS Center(s) for Commercial Development of Space

CCDS Center for the Commercial Development of Space Power [of NASA; located at Auburn University, Auburn, Alabama, US]

CCDS Countercurrent Drum Separation

CCDSO Command and Control Defense Systems Office

CCD TV Charge-Coupled Device Television (Camera)

CCE Carboline Carboxylate Ethyl

CCE Carbon Chloroform Extract

CCE Caribbean Conservation Association [Barbados]

CCE Cathode Current Efficiency

CCE Certified Cost Engineer

CCE Chief Construction Engineer

CCE Coal Conversion Engineer(ing)

CCE College of Chemical Engineers [of Institution of Engineers of Australia, Barton, Australia]

CCE Collisional Charge Transfer

CCE Comisión Colombiana de Commercio [Colombian Trade Commission]

CCE Commission des Communautés Européennes [Commission of the European Communities, Brussels, Belgium]

CCE Communication Control Equipment

CCE Computer Command Engineer

CCE Constant-Current Electrolysis

CCEA Commonwealth Council for Education Administration [UK/Australia]

CCEBI Center for Continuing Education in the Building Industry [UK]

CCEC China Civil Engineering and Construction Corporation [Taiwan]

CCECA Consultative Committee on Electronics for Civil Aviation [US]

CCECE Canadian Conference on Electrical and Computer Engineering

CCEE Canadian Conference on Engineering Education

CCEE Committee of Concerned Electrical Engineers

C2E2 Collaborating Center on Energy and Environment [of United Nations Environment Program] [also c2e2]

CCEL Collidge Center for Environmental Leadership [US]

CCEN Comisió Chilena Energía Nuclear [Chilean Nuclear Energy Commission, Santiago]

CCEO Caribbean Council of Engineering Organizations

CCEPC Canadian Civil Engineering Planning Committee

CCES Calder Center for Environmental Sciences [of Fordham University; located in Armonk, New York State, US]

CCES Canadian Congress of Engineering Students

CCES Common Control Echo Suppressor

CCES Congress of Canadian Engineering Schools

CCES Congress of Canadian Engineering Students

CCET Center for Computers in Education and Training [UK]

CCETT Canadian Council of Engineering Technicians and Technologists

CCETT Centre Commun d'Etudes de Télévision et de Télécommunications [Center for Television and Telecommunication Studies, France]

CCEU Council on the Continuing Education Units [US]

CCF Cavity Correlation Function

CCF Central Competitive Field

CCF Central Communications Facility [US Air Force]

CCF Central Computing Facility [of NASA]

CCF Central Control Function

CCF Cephalin Cholesterol Flocculation [Biochemistry]

CCF Climate Control Factor

CCF Close-Coupling Formulation

CCF Color-Correction Filter

CCF Common Communications Format

CCF Communications Control Field

CCF Configuration Control Function

CCF Converter Compressor Facility [Metallurgy]

CCF Cooperative Commonwealth Federation (of Canada)

CCF Cross-Correlation Function [Communications]

CCF Cumulative Cost per Foot

CCFA Caribbean Cane Farmers Association [Jamaica]

CCFA Combined Cadet Force Association [UK]

CCFF Canadian Cystic Fibrosis Foundation

CCFF Cape Canaveral Forecast Facility [Florida, US]

CCFL Countercurrent Flow Limiting

CCFM Centre Canadien de Fusion Magnétique [Canadian Center for Magnetic Fusion, Varennes, Quebec, Canada]

CCFM Cryogenic Continuous-Film Memory

CCFM/TdeV Centre Canadien de Fusion Magnétique/ Tokamak de Varennes [Canadian Center for Magnetic Fusion/ Tokamak de Varennes, Quebec, Canada]

CCFO Charge-Charge Flux Overlap

CCFP Canadian College of Family Physicians

CCFP Closed Cup Flash Point

CCFR Constant Current Flux Reset (Test Method)

CCFR Coordinating Committee on Fast Reactors [of European Community]

CCFT Cold Cathode Fluorescent Tube

CCFT Controlled Current Feedback Transformer

CCG Calibration Coordination Group [of US Department of Defense]

CCG Canada Communications Group

CCG Canadian Coast Guard

CCG Catalytic Coal Gasification (Process)

CCG Community Constituency Group

CCG Computer Communications Group

CCG Computer-Controlled Goniometer

CCG Constant-Current Generator

CCG Creep Crack Growth [Mechanics]

cc/g cubic centimeter per gram [Unit]

CCGA Ceramic Column Grid Array

CCGC Canadian Coast Guard College

CCGCR Closed Cycle Gas Cooled Reactor

CCGCS Containment Combustion Gas Control System [Nuclear Engineering]

CCGE Cold Cathode Gauge Experiment

CCGMA Cabrera-Celli-Goodman-Manson Approximation [Physics]

CCG-P Canada Communications Group–Publishing

CCGR Creep Crack Growth Rate [Mechanics]

CCGS Canadian Center for Global Security

CCGS Canadian Coast Guard Service

CCGS Canadian Coast Guard Ship

CCGS Closed Curve Graphite-Like Carbon

CCGSE Concentric-Circle Grating Surface-Emitting (Laser)

CCGT Closed-Cycle Gas Turbine

CC-GTA Constant-Voltage Gas Tungsten-Arc (Welding)

CC-GTAW Constant-Voltage Gas Tungsten-Arc Welding

CCGTM Commonwealth Consultative Group on Technology Management

CCH Calcium Chloride Hexahydrate

CCH Canadian College of Homeopathy

CCH Channel-Check Handler

CCH Computerized Criminal History

CCH Connections per Circuit per Hour

CCH Cyclohexylidene Cyclohexane

CCH Industrial Calcium Hypochlorite

CC-HCR Continuous Casting and Hot-Charge Rolling [Metallurgy]

CC-HDR Continuous Casting and Hot-Direct Rolling [Metallurgy]

CCHE Carnegie Commission on Higher Education [US]

CChem Certified Chemist [UK]

CCHP Center for Corporate Health Promotion [Reston, Virginia, US]

CCHS Capital City Historical Society [Olympia, Washington, US]

CCHS Cylinder-Cylinder-Head Sector

CCHSSA Cyprus Computer Hardware and Software Suppliers Association

CC-HTSC (International Workshop of) Crystal Chemistry of High-Temperature Superconductors

CCHX Component Cooling Heat Exchanger

CCI Cadmium Council, Inc. [US]

CCI Canada Composting Inc. [Newmarket, Ontario, Canada]

CCI Canadian Carpet Institute

CCI Canadian Conservation Institute [of Heritage Canada, Ottawa, Canada]

CCI Canadian Copyright Institute

CCI Center for Communication [US]

CCI Chamber of Commerce and Industries [Namibia]

CCI Chamber of Commerce International

CCI Chamber of Commerce of Ireland

CCI Chambre de Commerce Internationale [International Chamber of Commerce, France]

CCI Co-Channel Interface

CCI Comité Consultatif International [International Consultative Committee]

CCI Common Client Interface

CCI Computer Composition International [US]

CCI Consultative Committee International

CCI Continuous Computation of Impact Points

CCI Coordination Chemistry Institute [PR China]

CCI Cotton Council International [US]

CCI Crevice Corrosion Index

CCIA Cellular Communications Industry Association [now Cellular Telecommunications Industry Association, US]

CCIA Chemical Corps Intelligence Agency

CCIA Computer and Communications Industry Association [US]

CCIA Console Computer Interface Adapter

CCIBP Canadian Committee for the International Biological Program

CCIC Cabled Conductor in Conduit

CCIC Canadian Council for International Cooperation

CCIC Chapter Chairmen's Invitational Conference [of ASM International, US]

CCIC Contracts Canada Information Center [of Public Works and Government Services Canada]

CCIC CRAF (Comet Rendevous/Asteroid Fly-By)/Cassini Integrated Circuits [NASA]

CCICED China Council for International Cooperation on Environment and Development

CCICED Newsl China Council for International Cooperation on Environment and Development Newsletter

CCID Community Colleges for International Development [US]

CCIF Comité Consultatif International Téléphonique [International Telephone Consultative Committee] [Now Part of Comité Consultatif International Télégraphique et Téléphonique]

CCIF Consultative Committee on International Frequencies

CCIGC Capillary Column Inverse Gas Chromatography

CCIIW Canadian Council of the International Institute of Welding

CCIL Continuously Computed Impact Line

CCILMB Committee for Coordination of Investigations of the Lower Mekong Basin [Thailand]

CCIM Certified Commercial Industrial Member

CCIM Cold Crucible Induction Melter; Cold Crucible Induction Melting [Metallurgy]

CCIM Command Computer Input Multiplexer

CCIP Chambre de Commerce et d'Industrie du Paris [Paris Chamber of Commerce and Industry, France]

CCIP Continuously Computed Impact Point

CCIR Comité Consultatif International des Radiocommunications [International Radio Consultative Committee; of International Telecommunications Union, United Nations, Geneva, Switzerland]

CCIRN Coordinating Committee for Intercontinental Research Networks

CCIS China Credit Information Service Limited [Taipei, Taiwan]

CCIS Command (and) Control Information System

CCIS Command, Control and Information System

CCIS Common Control Interoffice Signaling

CCIS Common Channel Interoffice Signaling

CCIS/ADP Command, Control and Information Systems and Automatic Data Processing Committee [of NATO]

CCIT Central China Institute of Technology [PR China]

CCIT Comité Consultatif International Télégraphique [International Telegraph Consultative Committee] [Later Part of Comité Consultatif International Télégraphique et Téléphonique]

CCITT Comité Consultatif International Télégraphique et Téléphonique [International Telegraph and Telephone Consultative Committee] [now International Telecommunication Union–Telecommunication Standards Section]

CCITT Consultative Committee for International Telegraphy and Telephony [now International Telecommunication Union–Telecommunication Standards Section]

CCITT CALS Comité Consultatif International Télégraphique et Téléphonique (International Telegraph and Telephone Consultative Committee)/Computer-Aided Logistics Support

CC-IUMRS Chinese Committee of the International Union of Materials Research Societies

CCIW Canada Center for Inland Waters

CCK Cholecystokinin [Biochemistry]

CCKF Continuity Checktone Failure [SPADE]

CCK-8-KLH Chloecystokinin–8-Keyhole Limpet Hemocyanin (Immunogen)

cckw counterclockwise [also ccw]

CCL Capacitor-Coupled Logic

CCL Caribbean Congress of Labour

CCL Catalytic Coal Liquids (Process)

CCL Centenary College of Louisiana [US]

CCL Center for Computer Law [Manhattan Beach, California, US]

CCL Central Coatings Laboratory [of Metcut Research Inc., Cincinnati, Ohio, US]

CCL Chemists Club Library

CCL Closed Circuit Loop

CCL Closed COS/MOS (Complementary Symmetry Metal-Oxide Semiconductor) Logic

CCL Coating and Chemical Laboratory

CCL Cold-Cathode Lamp

CCL Command Control Language

CCL Commonality Candidate List

CCL Common-Carrier Line

CCL Common Command Language

CCL Common Control Language

CCL Communications Control Language

CCL Configuration Control Logic

CCL Connection Control Language

CCL Continuous Coarsened Lamellae [Metallurgy]

CCL Core Current Layer

CCL Critical Current Limitation (of Superconductors)

CCL Cursor Control Language

CCL Cyber Control Language

CCl$_4$ Carbon Tetrachloride

C-Cl Carbon-Chlorine (Bond)

CCLA Centenary College of Los Angeles [US]

cc/lb cubic centimeter per pound [also cm^3/lb]

CCl$_2$F$_2$/H$_2$ Dichlorodifluoromethane/Hydrogen (Gas Mixture)

CCl$_4$/H$_2$ Carbon Tetrachloride/Hydrogen (Gas Mixture)

CCLI Charlotte Chemical Laboratories, Inc. [North Carolina, US]

CCLN Council Computerized Library Networks [US]

CCLOW Canadian Congress for Learning Opportunities for Women

CCLS Canadian Council of Land Surveyors

CCLS Central Committee on Lumber Standards [now American Lumber Standards Committee, US]

CCLSD Cooperative Centre for Local Sustainable Development

cclw counterclockwise

CCM California College of Medicine [of University of California at Irvine, US]

CCM Call Count Meter

CCM Canada Center for Mapping [of Natural Resources Canada]

CCM Canadian Committee on MARC (Machine Readable Catalogue)

CCM Carboline-3-Carboxylic Acid Methyl Ester

CCM Caribbean Common Market [formerly CARIFTA]

CCM Cation Chelating Mechanism

CCM Center for Composite Materials [of University of Delaware, Newark, US]

CCM Center for Composite Materials [of University of New Mexico, Albuquerque, US]

CCM Center for Composites Manufacturing (Science and Engineering) [US]

CCM Cerium Calcium Manganate

CCM Certified Crane Manufacturer

CCM Chain Crossing Model

CCM Channeling Contrast Microscopy

CCM Chicago Cyclotron Magnet [of Fermilab, Batavia, Illinois, US]

CCM Chromatographie Analytiques sur Couche Mince [Analytical Chromatography on Thin Films]

CCM Chromium-Cobalt-Molybdenum (Alloy)

CCM Cobalt-Chromium-Molybdenum (Alloy)

CCM Cold Crucible Melter [Metallurgy]

CCM Comité Consultatif pour la Masse et les Grandeurs Apparentées [Consultative Committee for Apparent Mass and Size]

CCM Communications Control Module

CCM Communications Controller Multichannel

CCM Community Climate Model [of National Center for Atmospheric Research, US]

CCM Computer-Coupled Machine

CCM Concentric Cylinder Model (for Composites) [Materials Science]

CCM Continuous Cold-Rolling Mill [Metallurgy]

CCM Continuous Core Measurements [Nuclear Engineering]

CCM Controlled-Carrier Modulation [also ccm]

CCM Council of Communication Management [US]

CCM Counter Countermeasures

CCM Coupled-Cluster Method

CCM Crew/Cargo Module

CCM Cross-Correlation Matrix

CCM Cross-Correlation Method

CCM Crucible Ceramic Mold

β-CCM β-Carboline-3-Carboxylic Acid Methyl Ester

C/cm² Coulomb per square centimeter [also C cm^{-2}]

°C/cm Degree Celsius (or Centigrade) per centimeter [also °C cm^{-1}]

ccm cubic centimeter [also cu cm, or cm^3]

CCMA Canadian Council of Management Associations

CCMA Certified Color Manufacturers Association [US]

CCMA Converging Computing Methodologies in Astrophysics (Conference)

CCMA Corrugated Case Materials Association [UK]

CCMA Cyprus Clothing Manufacturers Association

CCMAI Crystal Chemical Model of Atomic Interactions

CCMB Completion of Calls Meeting Busy

CCMC Canadian Construction Materials Center

CCMC Carbido Carbonyl Metal Cluster

CCMC Comité des Constructeurs d'Automobiles du Marché Commun [Committee of Common Market Automobile Constructors]

CCMC Commonwealth Cable Management Committee

CCMD Continuous Current Monitoring Device

CCME Canadian Council of Ministers of Education

CCME Canadian Council of Ministers of the Environment

CCMET Canadian Center for Mineral and Energy Technology [Ottawa]

CCMI Center for Construction Market Information [UK]

cc/min cubic centimeter(s) per minute [cm^3/min, or cm^3 min^{-1}]

cc/min/g cubic centimeter(s) per minute per gram [also cm^3/min/g, or cm^3 min^{-1} g^{-1}]

CCMIS Cartridge Case Measurement Inspection System

CCMMI Congress of the Council of Mining and Metallurgical Institutions

CCMMI Council of Commonwealth, Mining and Metallurgical Institutions [now Council of Mining and Metallurgical Institutions, UK]

CCMO Cerium Calcium Manganese Oxide

cc/mol cubic centimeter per mole [also cm^3/mol, or cm^3 mol^{-1}]

CCMRG Commonwealth Consultative Group on Mineral Resources and Geology

CCMS Center for Composite Materials and Structures [at Virginia Polytechnic Institute and State University, Blacksburg, US]

CCMS Center for Condensed Matter Sciences [of National Taiwan University, Taipei, Taiwan]

CCMS Checkout, Control and Monitor Subsystem [NASA Launch Processing System] [also ccms]

CCMS Checkout, Control and Monitor System

CCMS Committee on Challenges of Modern Society [of NATO]

CCMSC Caribbean Common Market Standards Council [Guyana]

CCMTA Canadian Council of Motor Transport Administrators

CCMTC Cape Canaveral Missile Test Center [of NASA]

CCN Catalog Card Number [of Library of Congress, US]

CCN Cloud Condensation Nuclei

CCN Comité de Ciencia Nuclear [Nuclear Science Committee, Spain]

CCN Computer Communication Network

CCN Contract Change Negotiation

CCN Contract Change Notice

cCnA cis-Cinnamic Acid

CCNC Conservation Council of North Carolina [US]

CCNDT Canadian Council for Nondestructive Testing

CC-NDT Can't Call–No Dial Tone

CCNG Computer Communications Network Group [of University of Waterloo, Ontario, Canada]

CCNG Consortium on the Conversion of Natural Gas [Canada]

CCNIM Computer-Controlled Nuclear Instrument Module

CCNM Canadian College of Naturopathic Medicine [Toronto, Ontario]

CCNR Canadian Coalition for Nuclear Responsibility

CCNR Central Commission for Navigation on the Rhine [France]

CCNR Current-Controlled Negative (Differential) Resistance

CCNT Conservation Commission of the Northern Territory [Australia]

CCNU N-(2-Chloroethyl)-N'-Cyclohexyl-N-Nitrosa Urea

CCNY Carnegie Corporation of New York [US]

CCO Calcium Cerium Oxide

CCO Calcium Cobalt Oxide

CCO Calcium Copper Oxide (Superconductor)

CCO Chief Conservation Officer [of International Union for the Conservation of Nature (and Natural Resources)]

CCO Conservation Council of Ontario [Canada]

CCO Crystal-Controlled Oscillator

CCO Current-Controlled Oscillator

CCOA Chinese Cereals and Oils Association [PR China]

CCOFI California Cooperative Oceanic Fisheries Investigation [of National Oceanic and Atmospheric Administration, US]

CCOH Corrosive Contaminants, Oxygen and Humidity

CCOHS Canadian Center for Occupational Health and Safety [Hamilton, Ontario]

CCOHTA Canadian Coordinating Office for Health Technology Assessment

CCOL Coordinating Committee on the Ozone Layer [of United Nations Environmental Program]

CCOP Chlorine-Catalyzed Oxidative Pyrolysis (Process)

CCOP Constant-Control Oil Pressure

CCOP Coordinating Committee for Offshore Prospecting [of UNESCO]

CCOP/SOPAC Committee for Coordination of Joint Prospecting for Mineral Resources in South Pacific Offshore Areas [of UNESCO in Fiji]

C-CORE Center for Cold Ocean Resources Engineering [of Memorial University of Newfoundland, Canada]

CCOT Cycling Clutch Orifice Tube

CCP Call Control Processing

CCP Center Console Panel

CCP Central Canada Potash

CCP Central Collecting Point

CCP Central Control Position

CCP Centrifugal Charging Pump

CCP Certificate in Computer Programming

CCP Certified Computer Programmer

CCP Channel Control Processor

CCP Charge Capacitance Probe

CCP China Clay Producers Trade Association [US]

CCP Chlorocyclopropane

CCP Cobalt Chromium Platinum (Alloy)

CCP Combined Cycle Plant

CCP Command Console Processor

CCP Commercial Change Proposal

CCP Committee on Commodity Problems [of United Nations]

CCP Communication Control Package

CCP Communication Control Panel

CCP Communication Control Processor

CCP Communication Control Program

CCP Composite Correction Program

CCP Computer Control Panel

CCP Confederación Cientifica Panamericana [Pan-American Scientific Union, Argentina]

CCP Conference on Computational Physics

CCP Configuration Change Point

CCP Configuration Control Panel

CCP Configuration Control Phase

CCP Console Command Processor

CCP Console Control Package

CCP Contingency Control Panel

CCP Continuous Cooling Precipitation [Metallurgy]

CCP Contract Change Proposal

CCP Contract Configuration Process

CCP Core Component Pot [Nuclear Engineering]

CCP Cost Control Program

CCP Critical Compression Pressure

CCP Cross Connection Point

CCP Cubic Close-Packed (Crystal) [also ccp]

C&CP Corrosion and Cathodic Protection [Institute of Electrical and Electronics Engineers Industry Applications Society, US]

C-13 CP Carbon-13 Cross-Polarization [also ^{13}C CP]

CCPA California Canning Peach Association [US]

CCPA Canadian Chemical Producers Association

CCPA Cemented Carbide Producers Association [US]

CCPA 2-Chloro-N^6-Cyclopentyladenosine [Biochemistry]

CCPA Choline Chloride Producers Association [of Conseil Européen des Fédérations de l'Industrie Chimique, Belgium]

CCPA Constant Centrifugal Potential Approximation

CCPAB California Cling Peach Advisory Board [US]

CCPC Civil Communications Planning Committee [of NATO]

CCPD Continuous Cyclic Peritoneal Dialysis

CCPD Council for Cultural Planning and Development [Taiwan]

CCPD Coupling Capacitor Potential Device

CCPE Canadian Council of Professional Engineers

CCPEOF Consulting Committee of the Professional Electro-Engineers Organization of Finland

CCPES Canadian Council of Professional Engineers and Scientists

CCPIT China Council for the Promotion of International Trade [PR China]

C-CPM Center for Chemical Process Metallurgy [of University of Toronto, Canada]

C-13 CP MAS NMR Carbon-13 Cross-Polarization (with) Magic-Angle Spinning Nuclear Magnetic Resonance [also ^{13}C CP MAS NMR]

CCPMO Consultative Council of Professional Management Organizations [UK]

CCPO Cerium Copper Phosphorus Oxide

CCPO Conference on Charged Particle Optics

CCPPA Coupled Cluster Polarization Propagator Approximation [Physics]

CCPR Codex Committee on Pesticide Residues

CCPS Cell Census Plus System [Immunochemistry]

CCPS Center for Chemical Process Safety [of American Institute of Chemical Engineers, US]

CCPS Consultative Committee for Postal Studies [of Universal Postal Union]

CCR Canadian Copper Refinery [of Noranda Inc.]

CCR Ceiling Cavity Ratio

CCR Center for Catalogue Research [of Bath University, UK]

CCR Center for Ceramic Research [of Rutgers University, Piscataway, New Jersey, US]

CCR Channel Command Register

CCR Charge Control Ring [also C^{2}R]

CCR Chemical Cartridge Respirator

CCR Chemical Compounds Registry [of Chemical Abstract Service–American Chemical Society, US]

CCR Closed-Cycle Refrigerator

CCR Cold-Charge Rolling [Metallurgy]

CCR Condition Code Register

CCR Committment, Concurrency and Recovery [Common Application Service Elements (Protocol)]

CCR Computer and Computation Research [US]

CCR Computer Cassette Recorder

CCR Condition Code Register

CCR Continuous Catalyst Regeneration (Process)

CCR Contractor Change Request

CCR Control Center Rack

CCR Control Contactor

CCR Conventional Controlled Rolling [Metallurgy]

CCR Coriell Cell Repositories [of University of Medicine and Dentistry of New Jersey, Newark, US]

CCR Council for Chemical Research [US]

CCR Council on Concrete Research [of American Concrete Institute, US]

CCR Creep Crack Growth [Mechanics]

CCR Crevice-Corrosion Resistance

CCR Critical Compression Ratio

CCR Crossed Cannizzaro Reaction [Organic Chemistry]

CCR Cyclic Catalytic Reforming (Process)

CCR Cyclic Creep Retardation [Mechanics]

CC&R Canadian CAD/CAM (Computer-Aided Design/Computer-Aided Manufacturing) and Robotics Exhibition and Conference

CCRA Canada Customs and Revenue Agency

CCRA Cape Canaveral Reference Atmosphere [formerly Cape Kennedy Reference Atmosphere]

CCRC Continuing Care Retirement Community

CCRC Core Component Receiving Container [Nuclear Engineering]

CCRD Canadian Council on Rural Development

CCRD Consumer and Commercial Relations Department [Canada]

CCR&D Canadian Corporate Research and Development (Database/Directory)

CCRE Canadian Council for Research in Engineering

CCREM Canadian Council of Resource and Environment Ministers

CCRF Consolidated Communication Recording Facility

CCRG Canadian Classification Research Group

CCRH Center for Computer Research in Humanities [University of Colorado at Boulder, US]

CCRL Cement and Concrete Reference Laboratory [US]

CCRL Ceramic Composites Research Laboratory [of University of Michigan, Ann Arbor, US]

CCRL Combustion and Carbonization Research Laboratory [of Canada Center for Mineral and Energy Technology, Natural Resources Canada]

CCRL Countercurrent Reaction Launder (Process)

CCRM Constant Control Relay Module

CCRM Crew Commander Remote Monitor

CCRMP Canadian Certified Reference Materials Project

CCROS Card Capacitor Read-Only Storage

CCRP Continuously Computed Release Point

CCRS Canada Center for Remote Sensing

CCRT Center for Conservation Research and Technology [of University of Maryland, Baltimore, US]

CCRVDF Codex Committee on Residues of Veterinary Drugs in Food

CCS Canadian Cancer Society

CCS Canadian Ceramic Society

CCS Canadian Computer Show

CCS Canadian Control Status

CCS Cancer Control Society [US]

CCS Cartesian Coordinate System

CCS Center Core Signal

CCS Center for Coastal Studies [US]

CCS Center for Computer Studies [of Ngee Ann Polytechnic, Singapore]

CCS Central Certificate Service [now Depository Trust Company, US]

CCS Central Computer Station

CCS Central Control Section

CCS Central Control Station

CCS Change Control System

CCS Chinese Chemical Society [PR China]

CCS Collective Call Signal

CCS Color Calibration System

CCS Colo(u)rvision Constant Speed

CCS Combined Chiefs of Staff

CCS Command and Communication System

CCS Command and Control Subsystem

CCS Command, Control and Subordinate Systems

CCS Commercial Communications Satellite [Japan]

CCS Commitment Control System

CCS Committee of Concerned Scientists

CCS Common-Channel Signalling

CCS Common Command Set

CCS Common Communications Services

CCS Common Communications Support

CCS Communication Control System

CCS Complex Control Set

CCS Computer-Controlled (Data Acquisition and Processing) System

CCS Computer Core Segment

CCS Confederation of Construction Specialists

CCS Console Communication System

CCS Containment Cooling System [Nuclear Engineering]

CCS Contamination Control System

CCS Continuous(ly) Cast Steel

CCS Continuous Color Sequence

CCS Continuous Commercial Service

CCS Continuous Composite Servo

CCS Controlled Combustion System

CCS Conversational Compiling System

CCS Cost Compilation Sheet

CCS Countryside Commission for Scotland

CCS Current Control Source

CCS Custom Computer System

CCS Cyber (Computer) Credit System

CCS Czechoslovak Chemical Society

CCS Hundred Call Seconds [Telecommunications]

.CCS Color Calibration System [File Name Extension]

CC&S Central Computer and Sequencer [of NASA]

CC&S Central Computer and Sequencing

CCSA Canadian Committee on Sugar Analysis

CCSA Common Carrier Special Application

CCSA Common-Channel Signalling Arrangement

CCSA Common-Control Switching Arrangement

CCSA Conservation Council of South Australia

CCSB CRAF (Comet Rendevous/Asteroid Fly-By)/Cassini Science Buy [of NASA]

CCS Bull CCS Bulletin [of Chinese Chemical Society, PR China]

CCSC Central Connecticut State College [New Britain, US]

CCSC Coordinating Committee for Satellite Communications [Switzerland]

CCSCC Certificate of the Canadian Society of Clinical Chemists

CCSD Cellular Circuit-Switched Data

CCSD Command Communication Service Designator

CCSD Coupled Cluster (Method Including All) Single and Double (Excitations) [Physics]

CCSDS Consultative Committee for Space Data Systems [NASA Planetary Data System]

cc/sec cubic centimeter(s) per second [cm^3/s, cm^3/sec, or $cm^3\ s^{-1}$]

CCSEM Computer-Controlled Scanning Electron Microscope; Computer-Controlled Scan-ning Electron Microscopy

CCSERC Conservation Council of the South-East Region and Canberra [Australia]

CCSF City College of San Francisco [US]

CCSG Computer Components and Systems Group

CCSL Comparative Current Sinking Logic

CCSL Compatible Current Sinking Logic

CCSL Constrained Coincidence Site Lattice [Materials Science]

CCSL/DSC Constrained Coincidence Site Lattice/Displacement Shift Complete [Materials Science]

CCSM Center for Compound Semiconductor Microelectronics [of University of Illinois at Urbana-Champain, US]

CCSM Composite-Chopped Strand Mat

CC-SMA Constant-Voltage Shielded Metal-Arc (Welding)

CC-SMAW Constant-Voltage Shielded Metal-Arc Welding

CCSN Community College Satellite Network [of Instructional Telecommunication Consortium, US]

CCSP Center for Client and Supplier Promotion [of Public Works and Government Services Canada]

CCSP Cottrell College Science Program [of Research Corporation, Tucson, Arizona, US]

CCSR Copper Cable Steel Reinforced (Cable)

CCSS Canada Center for Space Sciences

CCSS Common-Channel Signalling System

CCSS Continuous Cast Stainless Steel

CCSSO Council of Chief State School Officials [US]

CCSSRRT Continuous(ly) Cast Stainless Steel Round Robin Test [Metallurgy]

CCST Caribbean Council for Science and Technology

CCST Center for Catalytic Science andTechnology [of University of Delaware, Newark, US]

CCST Center for Computer Sciences and Technology [of National Institute of Standards and Technology, Gaithersburg, Maryland, US]

CCST Coordinating Committee on Science and Technology

CCSTG Carnegie Commission on Science, Technology and Government [US]

CCSW Component Cooling Service Water

CCSW Copper-Clad Steel Wire [also ccsw]

CCT California Consolidated Technologies [US]

CCT Canada (or Canadian) Center for Toxicology

CCT Center-Crack Tension (Test Specimen)

CCT Chamber of Coal Traders [UK]

CCT China Coast Time [Greenwich Mean Time −8:00]

CCT Chocolate-Coated Tablet(s) [Pharmacology]

CCT Cis-Cis-Trans (Geometry) [Chemistry]

CCT Clarkson College of Technology [Potsdam, New York, US]

CCT Common Customs Tariff [European Community]

CCT Communications Control Team

CCT Complete Calls To

CCT Component Check Test

CCT Computer-Compatible Tape

CCT Conductivity-Controlled Transistor

CCT Constant Current Transformer

CCT Continuity Check Transceiver

CCT Continuous Cooling Transformation [Metallurgy]

CCT Correlated Color Temperature

CCT Coupler Cut Through

CCT Creosote Coal Tar

CCT Critical Crevice-Corrosion Temperature

CCT Crystal-Controlled Transmitter

Cct Circuit [also cct]

CCTA Canadian Cable Television Association

CCTA Central Computer and Telecommunications Agency [UK]

CCTC Canada-China Trade Council

CCTEAH Coordinating Committee of the Trans-East African Highway [now Trans-East African Highway Authority, Ethiopia]

CCTest Component Check Test

CCTF Canadian CCSDS (Consultative Committee on Space Data Systems) Testbed Facility (Project)

CCTF Cylindrical Core Test Facility [Nuclear Engineering]

CCTG Configuration Control Task Group

CCTI Composite Can and Tube Institute [US]

CCTL Casing Cooling Tank Level

CCTL Collection-Coupled Transistor Logic

CCTL Core Component Test Loop [of Argonne National Laboratory, Illinois, US]

CCTMA Closed Circuit Television Manufacturers Association [US]

CCTP California Competitive Technology Program [US]

CCTP Clean Coal Technology Program

CCTR Center for Cell and Tissue Research [UK]

CCTR Culham Conceptual Tokamak Reactor [UK]

CCTS Coordinating Committee for Telecommunication by Satellite

CCTT Canadian Council of Technicians and Technologists

CCTU Committee of Corporate Telecommunications Users [US]

CCTV Closed-Circuit Television [also cctv]

CCTVA Canadian Cable Television Association

CCTV/CCD Closed-Circuit Television/Charge-Coupled Device (Camera)

CCTWT Coupled Cavity Traveling Wave Tube

CCU Camera Control Unit

CCU Canadian Commission for UNESCO

CCU Cardiac Care Unit

CCU Caribbean Consumers Union [Antigua-Barbuda]

CCU Central Control Unit

CCU Comité Consultatif des Unités [Consultative Committee on Units, France]

CCU Common Control Unit

CCU Communication and Consultation Unit [of Murray-Darling Basin Commission, Australia]

CCU Communications Carrier Umbilical

CCU Communications Control Unit

CCU Computer Control Unit

CCU Confederation of Canadian Unions

CCU Contaminant Collection Unit

CCU Coronary Care Unit

CCU Coupling Control Unit

CCU Crew Communications Unit

CCU Crewman Communications Umbilical

CCU Critical Care Unit

CCUAP Computerized Cable Upkeep Administration Program

CCUBC Canadian Council of University Biology Chairs

CCULP Canada/China University Linkage Program [of Association of Universities and Colleges of Canada]

CCUR Center for Crop Utilization [of Iowa State University, Ames, US]

CCURR Canadian Council on Urban and Regional Research

CCV Chamber Coolant Valve

CCV Conservation Council of Victoria [Australia]

CCV Continuing Calibration Verification

CCV Control Configured Vehicle

ccv core-core-valence (transition) [Auger Electron Spectroscopy]

CCVA Chamber Coolant Valve Actuator

CCVJ 9-([E]-2-Carboxy-2-Cyanovinyl)julolidine

CCVD Combustion Chemical Vapor Deposition (Process)

CCVS COBOL Compiler Validation System

CCVS Current-Controlled Voltage Source

CCW Channel Command Word

CCW Channel Control Word

CCW Circulation Controlled Wing

CCW Commutatively Compound-Wound

CCW Component Cooling Water

ccw counterclockwise [also cckw]

CCWA Conservation Council of Western Australia

CCWBAD Counterclockwise Bottom Angular Down [Metallurgy]

CCWBAU Counterclockwise Bottom Angular Up [Metallurgy]

CCWBH Counterclockwise Bottom Horizontal [Metallurgy]

CCWCP Coordination Committee World Climate Program

CCWDB Counterclockwise Down Blast [Metallurgy]

CCWE Canadian Committee on Women in Engineering

CCWE Constituent Concentration in a Waste Extract

CCWEST Canadian Conference of Women in Engineering, Science and Technology

CCWM Commutatively Compound-Wound Motor

CCWO Cryptocenter Watch Officer [of NATO]

CCWP Close-Coupling Wave Packet (Method)

CCWS Component Cooling Water System

CCWTAD Counterclockwise Top Angular Down [Metallurgy]

CCWTAU Counterclockwise Top Angular Up [Metallurgy]

CCWTH Counterclockwise Top Horizontal [Metallurgy]

CCWUB Counterclockwise Up Blast [Metallurgy]

CCZ California Coastal Zone

CCZ Coastal Confluence Zone

CCZCC California Coastal Zone Conservation Commission [US]

CD Cable Duct

CD Caderock Division [of Naval Surface Warfare Center, Annapolis, Maryland, US]

CD Cage Dipole

CD Canine Distemper

CD Canonical Distribution [Statistical Mechanics]

CD Capacitor Diode

CD Capacitor Discharge

CD Carbon Dating

CD Carried Down

CD Carrier Detect

CD Cash Discount [also cd]

CD Casting Densification (Process) [Metallurgy]

CD Catalytic Dehydrogenation

CD Cathode of Diode

CD Cathodoluminescence; Cathodoluminescent

CD Cauchy Distribution [Statistics]

CD Ceiling Diffuser

CD Cellular Dislocation [Materials Science]

CD Census Division

CD Center Director [NASA Kennedy Space Center Directorate, Florida, US]

CD Center Distance

CD Central Difference

CD Cerenkov Detector

CD Certificate of Deposit [also C/D]

CD Change Directory (Command) [also cd]

CD Charge Density

CD Check Digit

CD Chemical Deposition

CD Chemistry Division

CD Chemical Diffusion

CD Chiral Discrimination

CD Chlorinated Dioxin

CD Choleski Decomposition

CD Circuit Description

CD Circular Dichroism

CD Circulator Drive [Nuclear Power Stations]

CD Civil Defense
CD Classical Dynamics
CD Climb Dislocation [Materials Science]
CD Clock Driver
CD Clockwise Diamond (Test) [Mechanical Testing]
CD Cluster of Differentiation [Biochemistry]
CD Coal Division
CD Coal Division [of Canadian Institute of Mining and Metallurgy]
CD Coal Division [of Natural Resources Canada]
CD Coal Division [of Society of Mining Engineers, US]
CD Coal Dust
CD Coast Defense (Radar) [UK]
CD Codirectional
CD Coefficient of Drag
CD Coherency Dislocation [Materials Science]
CD Coherent Detector
CD Cold Drawing; Cold-Drawn
CD Cole-Davidson (Formula) [Solid-State Physics]
CD Collection District
CD Collision Detection
CD Color Display
CD Command
CD Common Digitizer
CD Communicable Disease
CD Compact Disk
CD Compensation Depth [Oceanography]
CD Composites Division [of ASM International, US]
CD Compression Direction
CD Computer Design
CD Computer Design [Journal published in the US]
CD Computing Division [of Fermilab, Batavia, Illinois, US]
CD Conceptual Design
CD Congressional District [US]
CD Conjugated Diene
CD Constant Deviation
CD Constrained Deconvolution
CD Continental Divide [Rocky Mountains]
CD Continental Drift [Geology]
CD Continuous Decomposition
CD Continuous Distribution [Statistics]
CD Continuous Duty
CD Contract Demand
CD Contracting Definition
CD Control(-Rod) Drive [Nuclear Engineering]
CD Convolution Difference
CD Convolution Distance
CD Corbino Disk
CD Core Diameter
CD Core Dump
CD Corona Discharge
CD Corps Diplomatique [Diplomatic Corps]
CD Corrosion Data
CD Countdown [also C/D]

CD Coxsackie Disease [Medicine]
CD Crack Direction
CD Crack Displacement
CD Critical Damping
CD Critical Dimension
CD Critical Dose
CD Cross Direction
CD Crystal Defect
CD Crystal Detector
CD Crystal Diffraction
CD Crystal Driver
CD Crystal Dynamics
CD Current Density [also cd]
CD Current Distribution
CD Customs Declaration
CD Cyclic Deformation
CD Cyclodextrin [Biochemistry]
CD Cystic Duct [Anatomy]
CD International Conference on Circular Dichroism
CD_4 Deuterated Methane (or Methane-d_4)
C_8D_{18} Deuterated Octane (or Octane-d_{18})
CD4 Cluster of Differentiation 4 (Receptor) [Biochemistry]
CD-4 Compatible Discrete Four-Channel Sound
α-CD α-Cyclodextrin [Biochemistry]
β-CD β-Cyclodextrin [Biochemistry]
γ-CD γ-Cyclodextrin [Biochemistry]
C&D Charman and Darlington (Hot Top) [Metallurgy]
C&D Collection and Delivery
C&D Control and Display
C/D Bearing-Journal Diametral Clearance to Journal Diameter [Symbol]
C/D Certificate of Deposit [also CD]
C/D Control Data
C/D Countdown [also CD]
C-D Carbon-Deuterium (Vibration)
C-D Channel-Die (Compression)
C-D Convergence-Divergence (Principle)
C-D Converging-Diverging (Nozzle)
C1D Conducting along One Dimension
C2D Conducting along Two Dimensions
C$ Canadian Dollar [Currency]
C$ Córdoba [Currency of Nicaragua]
Cd Cadmium [Symbol]
Cd Card [also cd]
Cd Cedi [Currency of Ghana]
Cd Cord [also cd]
Cd Corynaebacterium diphtheriae [Microbiology]
Cd^{2+} Cadmium Ion [also Cd^{++}] [Symbol]
Cd II Cadmium II [Symbol]
Cd-106 Cadmium-106 [also ^{106}Cd, or Cd^{106}]
Cd-107 Cadmium-107 [also ^{107}Cd, or Cd^{107}]
Cd-108 Cadmium-108 [also ^{108}Cd, or Cd^{108}]
Cd-109 Cadmium-109 [also ^{109}Cd, or Cd^{109}]

Cd-110 Cadmium-110 [also ^{110}Cd, or Cd^{110}]

Cd-111 Cadmium-111 [also ^{111}Cd, or Cd^{111}]

Cd-112 Cadmium-112 [also ^{112}Cd, or Cd^{112}]

Cd-113 Cadmium-113 [also ^{113}Cd, or Cd^{113}]

Cd-114 Cadmium-114 [also ^{114}Cd, or Cd^{114}]

Cd-115 Cadmium-115 [also ^{115}Cd, or Cd^{115}]

Cd-116 Cadmium-116 [also ^{116}Cd, or Cd^{116}]

Cd-117 Cadmium-117 [also ^{117}Cd, or Cd^{117}]

cD Coherent Dislocation (in Materials Science) [Symbol]

cd candela [Unit]

cd cash discount [alaso CD]

cd cord [Unit]

c/d carried down

$\chi(\Delta)$ polarization dependence of spin susceptibility [Symbol]

CDA Canadian Dental Association

CDA Canadian Department of Agriculture

CDA Canadian Diabetes Association

CDA Canadian Drilling Association

CDA Capital Dividend Account

CDA Central Dredging Association [UK]

CDA Charge Density Analysis

CDA Charge Distribution Analysis

CDA Chemists Defense Association [UK]

CDA Civil Defense Act [US]

CDA Civil Defense Administration [later Office of Civil Defense; then Defense Civil Preparedness Agency; now Federal Emergency Management Agency, US]

CDA Clean Dry Air

CDA Coin Detection and Announcement

CDA Command and Data Acquisition

CDA Commercial Development Association [US]

CDA Completely Denatured Alcohol

CDA Compound Document Architecture

CDA Computer Dealers Association [US]

CDA Comprehensive Development Area

CDA Conference of Defense Association

CDA Containment Depressurization Activation [Nuclear Reactors]

CDA Containment Depressurization Alarm [Nuclear Reactors]

CDA Copier Dealers Association [US]

CDA Copper Development Association [US]

CDA Core Disruptive Accident [Nuclear Engineering]

CDA Core Dump Analysis

CDA Corrosion Data Acquisition

CDA Cosmic Dust Analyzer

CDA Critical Design Audit

CDA Cross-Disciplinary Activities

CDA Cumulative Deduction Amout

CDA Cylindrical Deflector Analyzer

C/DA Computer/Design Automation [Institute of Electrical and Electronics Engineers–Computer Society Committee, US]

CdA 2-Chloro-2'-Deoxyadenosine [Biochemistry]

Cda Canada

CDAA Chlorodiallylacetamide

Cd(Ac)$_2$ Cadmium(II) Acetate

Cd-Ag Cadmium-Silver (Alloy System)

CDAP 1-Cyano-4-Dimethylaminopyridinium Tetrafluoroborate

CDAPSO Canadian Data Processing Service Organization

CDAS Central Data Acquisition System

CDAS Command and Data Acquisition Station

CDAS Computerized Document Access System

CDATA Census Data [of Australian Bureau of Statistics] [Usually followed by a date, e.g., CDATA91 for Census Data of 1991]

CDB Caribbean Development Bank [Barbados]

CDB Center for Drugs and Biologics [US]

CDB Central Data Bank

CDB Central Data Buffer

CDB Coal Database [UK]

CDB Command Data Base

CDB Common Data Bus

CDB Commonwealth Development Bank

CDB Companhia Dinamitos do Brasil [Brazilian Dynamite Company]

CDB Computerized Database

CDB Conjugated Double Bond

CDB Current Data Bit

CDBAB California Dried Bean Advisory Board [US]

CDBFR Common Data Buffer

CDBG Community Development Block Grant (Program) [US]

CDBI N,N'-Carbonyldibenzimidazole

Cd-Bi Cadmium-Bismuth (Alloy System)

CDBL Convert to Double-Precision Number (Function) [Programming]

CDBN Column-Digit Binary Network

CDBS Central Database System

CDBS Commonwealth Digital Boundaries (Database) System

CDC Call Directing Character

CDC Call Directing Code

CDC Canadian Dairy Commission

CDC Canadian Development Corporation

CDC Centers for Disease Control [now Centers for Disease Control and Prevention]

CDC Centers for Disease Control and Prevention [Atlanta, Georgia, US]

CDC Characteristic Distortion Compensation

CDC Chicago Development Corporation [Illinois, US]

CDC Civil Defense Corps [UK]

CDC Climate Diagnostics Center [of National Oceanic and Atmospheric Administration, US]

CDC Code Directing Character

CDC Cold Dynamic Compaction

CDC Combat Developments Command [of US Army at Fort Belvoir, Virginia, US]

CDC Command and Data-Handling Console

CDC Commonwealth Defense Conference [India]

CDC Commonwealth Development Corporation

CDC Committee for Development and Cooperation

CDC Communications and Data Center

CDC Composite Development Center [US]

CDC Computer Display Channel

CDC Configuration Data Control

CDC Confined Detonating Cord

CDC Connecticut Development Commission [US]

CDC Consensus Development Conference [of National Institutes of Health, Bethesda, Maryland, US]

CDC Conservation Data Center [US]

CDC Conservation for Development Center

CDC Construction Design Criteria

CDC Control Data Corporation

CDC Copper Data Center [of Copper Development Association, US]

CDC Copper Diethyldithiocarbamate

CDC Corporate Data Center

CDC Countdown Clock

CDC Cryogenic Data Center [of National Institute for Standards and Technology, US]

CDC Cycloheptaamylose Dansyl Chloride

CDCC Caribbean Development Cooperation Committee

CDCCP Control Data Communications Control Procedure

CD_3CDO Deuterated Acetaldehyde

CDCE Central Data Conversion Equipment

CD-CHROM Chromatography Database on CD-ROM (Compact-Disk Read-Only Memory)

Cd_3Cit_2 Cadmium Citrate

CDC-OCCE Commonwealth Defense Conference–Operation Clothing and Combat Equipment [India]

CDCQ N-Carboxymethyl-6-(2,2-Dicyanovinyl) 1,2,3,4-Tetrahydroquinoline

CDCR Center for Documentation and Communication Research [of Case Western Reserve University, Cleveland, Ohio, US]

CDCR Conceptual Design and Cost Review

CDCT Centro de Documentação Científica e Tecnica [Center for Scientific and Technical Documentation, Portugal]

CDCTM Centro de Documentación Científica y Técnica de Mexico [Mexican Center for Scientific and Technical Documentation]

CDCU Communications Digital Control Unit

CdCuSb Cadmium Copper Antimonide

CDCVR Code Converter

CDD Chemical Deformation Density

CDD Chlorinated Dibenzo-p-Dioxin

CDD Common Data Dictionary

CDD Conference on Dual Distribution

CDD Contrast-Detail-Dose (Diagram)

CDD Control of Diarrheal Diseases Program [of World Health Organization]

CDD 1,5,9-Cyclododecatriene

CDDA Conseil Departemental du Développement Agricole [Departmental Council for Agricultural Development, France]

CD-DA Compact Disk–Digital Audio

CDDC Copper Dimethyldithiocarbamate

CDDD Center for Diagnostics and Drug Recovery [US]

CDDIS Crustal Dynamics Data Information System

CDDL Conference of Directors of Danube Lines [Hungary]

CDDP Canadian Deep Drilling Project

CdDMC Cadmium Dimethyl Dithiocarbamate

CDDT Countdown Demonstration Test

CDDT Coupled Diffusional/Displacive Transformation (Model) [Metallurgy]

Cd5MC Cadmium Pentamethylene Dithiocarbamate

CDE Cadkey Dynamic Extension (Module)

CDE Canadian Depletion Expense

CDE Canadian Development Expense

CDE Card Data Entry

CDE Certificate in Data Education

CDE Certified Diabetes Educator

CDE Common Desktop Environment

CDE Complex Data Entry (Generator)

CDE Control and Display Equipment

CDE Customer Design Engineering

CDEC Center for Design of Educational Computing [of Carnegie Mellon University, Pittsburgh, Pennsylvania, US]

CDEC 2-Chloroallyl-N,N-Diethyldithiocarbamate

CDECA Certified Decorators Association [also CdecA]

CDEE Chemical Defense Experimental Establishment [of Ministry of Defense, UK]

CDEP Climate Dynamics and Experimental Prediction

CDEP Community Development Employment Projects

CDES Computer Data Entry System

CDF Cable Distribution Frame

CDF Carrier Distribution Frame

CDF Centered Dark Field (Technique) [Microscopy]

CDF Central Data Facility

CDF Charge Density Fluctuation

CDF Chlorodibenzofuran

CDF Circuit Design Fabrication

CDF Class Determination and Findings

CDF Coal-Derived Fuel

CDF Collider Detector at Fermilab (Experiment) [of Fermilab, Batavia, Illinois, US]

CDF Combination Distribution Frame

CDF Combined Die Forging

CDF Combined Distribution Frame

CDF Common Data Format [also cdf]

CDF Communications Data Field

CDF Community Development Fund

CDF Computer Dealers Forum [of National Office Products Association, US]

CDF Confined Detonating Fuse

CDF Continuous Drossing Furnace [Metallurgy]

CDF Cool-Down Facility

CDF Cooperative Development Foundation [Canada]

CDF Cumulative Damage Function

CDF Cumulative Distribution Function [also cdf]

.CDF Comma Delimited Format [File Name Extension]

.CDF Common Data Form

CdF Charbonnages de France [Coal Mine Union of France]

CdFeSe Cadmium Iron Selenide (Semiconductor)

CdFeTe Cadmium Iron Telluride (Semiconductor)

CDF/HREM Centered Dark Field/High-Resolution Electron Microscopy

CDFI Center(ed) Darkfield Image

CDFIN Center for Designing Foods to Improve Nutrition [of Iowa State University, Ames, US]

CDFR Commercial Demonstration Fast Reactor

cd ft cord foot [Unit]

cd/ft² candela per square foot [also cd ft^{-2}, or cd/sq ft]

CDF&TDS Circuit Design and Fabrication and Test Data Systems

CDFU Cordova District Fishermen United [US]

CDG Capacitor Diode Gate

CDG Christian-Doppler-Gesellschaft [Christian Doppler Society, Austria]

CDH Cable Distribution Head

CDH Ceramide Dihexoside [Biochemistry]

CDH Command and Data Handling [also C&DH]

CDH Congenital Dislocation of the Hip

CDH Constant Delta Height

C&DH Command and Data Handling [also CDH]

CdHgTe Cadmium Mercury Telluride (Semiconductor)

CDHP Cis-2,6-Dimethyl-3,6-Dihydro-2H-Pyrane

CDHS California Department of Health Services [US]

CDHS CERN, Dortmund, Heidelberg and Saclay (Neutrino Experiment) [of CERN–European Laboratory for Particle Physics, Geneva, Switzerland]

CDI Capacitor Discharge Ignition

CDI Carbodiimide

CDI N,N'-Carbonyl Diimidazole

CDI Center for the Development of Industry [Belgium]

CDI Cobalt Development Institute

CDI Collector Diffusion Isolation (Transistor)

CDI Common-Rail Direct Injection

CDI Community Development Index

CDI Compact Disk–Interactive [also CD-I]

CDI Comprehensive Dissertation Index [of University Microfilms International, US]

CDI Continuous Deionization

CDI Control Data Institute [US]

CDI Control Deviation Indicator

CDI Control Direction Indicator

CDI (Conference on) Corrosion-Deformation Interactions

CDI Course Deviation Indicator

CDI Cutting Die Institute [US]

CD-I Compact Disk–Interactive [also CDI]

CDIA Canadian Investment Abroad (Conference)

CDIAC Carbon Dioxide Information Analysis Center [of Oak Ridge National Center, Tennessee, US]

CDIB Collector, Diffusion, Isolation, Bipolar

CDIC Canadian Development Investment Corporation

CDIF Component Development and Integration Facility

CDIF Crystal Data Identification File [also CRYSTDAT; of National Institute of Standards and Technology, US]

CDII Canadian Dental Implants Institute

cd/in² candela(s) per square inch [also cd in^{-2}, or cd/sq in]

CDIS Community Data Information System

C/DIS Computer/Distributed Interactive Simulation [Institute of Electrical and Electronics Engineers–Computer Society Committee, US]

CDITC Commonwealth Department of Industry, Technology and Commerce

CDIU COARE (Coupled Ocean-Atmosphere Response Experiment) Data Information Unit

CDJM Canadian Journal of Mathematics

CDK Channel Data Check

CDL Call Description Language

CDL Commercial Driving (or Drivers) License

CDL Common Data-Form Language [also cdl]

CDL Common Display Logic

CDL Computer Description Language

CDL Computer Design Language

CDL Computer Development Laboratory

CDL Core-Diode Logic

.CDL Common Data-Form Language [File Name Extension] [also .cdl]

CDLA Centro Demografica Latinoamericana [Latin-American Demographic Center, San José, Costa Rica]

CDLA Cluster-Diffusion Limited Aggregation (Model) [Physics]

CDLA Computer Dealers and Lessors Association [US]

CDLC Cellular Data Link Control

CDLE Colorado Department of Employment and Labor [US]

CDLI Clinical and Diagnostic Laboratory Immunology [Journal of the American Society for Microbiology]

CDLRD Confirming Design Layout Report Date

CDLS Canadian Defense Liaison Staff

C-DLTS Capacitance-Based Deep Level Transient Spectroscopy

cd/lx candela(s) per lux [also cd lx^{-1}]

cd/lx·m candela(s) per lux meter [also cd/(lx·m), cd/lx/m, or cd lx^{-1} m^{-1}]

cd/lx·m² candela(s) per lux square meter [also cd/(lx·m²), cd/lx/m², or cd lx^{-1} m^{-2}]

CDM Central Data Management

CDM Centre de Documentation de la Mécanique [Documentation Center for Mechanics, France]

CDM Chlorodimeform

CDM Chromosome Diagnostic Medium

CDM Code-Division Multiplex(ing)

CDM Color-Division Multiplex(ing)

CDM Common Data Model

CDM Communications/Data Manager

CDM Companded Delta Modulation

CDM Continuous Diffusion Model

CDM Continuum (Creep) Damage Mechanics (Model)

CDM Critical Dimension Metrology

cdm cubic decimeter [also dm³, or cu dm]

cd/m² candela(s) per square meter [also cd m^{-2}]

CDMA Cartridge Direct Memory Access

CDMA Code-Division Multiple Access [Telecommunications]

CDMI N-Cyclododecylmaleimide

cd/m²·lx candela(s) per square meter lux [also cd/(m²·lx), cd/m²/lx, or cd m^{-2} lx^{-1}]

CdMnS Cadmium Manganese Sulfide (Semiconductor)

CdMnSe Cadmium Manganese Selenide (Semiconductor)

CdMnTe Cadmium Manganese Telluride (Semiconductor)

CdMnTe:In Indium-Doped Cadmium Manganese Telluride (Semiconductor)

CDMS Command and Data Management Subsystem

CDMS Command and Data Management System

CDMS COMRADE (Computer-Aided Design Environment) Data Management System

CDMS Cryogenic Dark Matter Search (Experiment) [of Fermilab, Batavia, Illinois, US]

CDMSCS Committee for the Development and Management of Fisheries in the South China Sea [of Indo-Pacific Fishery Commission, Thailand]

CDMTCS Center for Discrete Mathematics and Theoretical Computer Science [of Rutgers State University of New Jersey, Piscataway, US]

CDN Canadian Dollar [Currency]

CDN CDR (Critical Design Review) Discrepancy Notice

Cdn Canadian

cDNA Chloroplast Deoxyribonucleic Acid [Biochemistry]

cDNA Complementary Deoxyribonucleic Acid [also c-DNA] [Biochemistry]

CDNet CD Network

Cd-Ni Cadmium-Nickel (Alloy System)

C-13 DNP Carbon-13 Dynamic Nuclear Polarization (Spectroscopy) [also ^{13}C DNP]

C-13 DNP-DPMAS Carbon-13 Dynamic Nuclear Polarization Direct-Polarization Magic-Angle Spinning (Spectroscopy) [also ^{13}C DNP-DPMAS]

C-13 DNP-MAS Carbon-13 Dynamic Nuclear Polarization Magic-Angle Spinning (Spectroscopy) [also ^{13}C DNP-MAS]

C-13 DNP-NMR Carbon-13 Dynamic Nuclear Polarization Nuclear Magnetic Resonance (Spectroscopy) [also ^{13}C DNP-NMR]

CDNR CDR (Critical Design Review) Discrepancy Notice Record

CDNSWC Caderock Division, Naval Surface Warfare Center [of US Navy at Annapolis, Maryland]

CDNT 4-Chloro-3,5-Dinitro Trifluorotoluene

CDO Civil Defense Organization [Canada]

CDO Community Dial Office

CDO Completely Disordered (State) [Solid-State Physics]

CDO P-Benzoquinone Dioxime

CdO Cadmium Oxide

CDOA Car Department Officers Association [US]

CdO-Fe₂O₃ Cadmium Oxide-Ferric Oxide (System)

CDOIPS Central Dispatching Organization of the Interconnected Power Systems [Czech Republic]

CDOM Center for Development of Materials

CDOS Chain Density of States

CDOS Concurrent Disk Operating System

CDOS Customer Data and Operations System

CdO-SiO₂ Cadmium Oxide-Silicon Dioxide (System)

CDP Center for Design Planning [US]

CDP Central Data Processor

CDP Centralized Data Processing

CDP Certificate in Data Processing

CDP Certified Data Processor

CDP Certified Digital Provider

CDP Checkout Data Processor

CDP Circular Diffraction Pattern`

CDP Command Data Processor

CDP Committee of Directors of Polytechnics

CDP Communications Data Processor

CDP Compact Disk Player

CDP Compressor Discharge Pressure

CDP Computerized Data Processing

CDP Conceptual Design Plan

CDP Correlated Data Processor

CDP Cresyl Diphenyl Phosphate

CDP Critical Decision Point

CDP Cytidine 5'-Diphosphate [Biochemistry]

5'-CDP Cytidine-5'-Diphosphate [Biochemistry]

CDPA Cellular Digital Packet Data [Telecommunications]

CDPA Copyright Designs and Patent Act [US]

Cd-Pb Cadmium-Lead (Alloy System)

CDPC Canadian Database Promotion Center

CDPC Central Data Processing Computer

CDPC Commercial Data Processing Center

CDPD Cellular Digital Packet Data

CDPF Canadian Data Processing Facility

CDPHE Colorado Department of Public Health and Environment [US]

CDPI Corrected Discretized Path Integral

CDPIR Crash Data Position Indicator Recorder

CDPIS Command Data Processing and Instrumentation System

Cd Pl Cadmium Plate

CDPM Colored Digital Panel Meter

Cd5MC Cadmium Pentamethylene Dithiocarbamate

CDPR Card to Card and/or Printer

CDPR Colorado Department of Public Relations [US]

CDPR Customer Dial Pulse Receiver

CDPS Computing and Data Processing Society

CD-PROM Compact Disk–Programmable Read-Only Memory [also CD PROM]

CDQ Coke Dry Quenching

CDQR Constant Displacement and Quick Return

CDQR Critical Design and Qualification Review

CDQW Coupled Double Quantum Well [Solid-State Physics]

CDR Call Data Recording

CDR Call Detail Recording

CDR Card Reader

CDR Centre de Données Recherche [Center of Research Data, France]

CDR Circular Depolarization Ratio

CDR Cleaning Decontamination Request

CDR Command Destruct Receiver

CDR Commander [also Cdr]

CDR Compact-Disk Reader

CDR Compact-Disk–Rewritable [also CD-R]

CDR Composite Damage Risk

CDR Conceptual Design Report

CDR Conceptual Design Requirement

CDR Conceptual Design Review

CDR Contract(ual) Data Requirements

CDR Contract Documentation Requirements

CDR Controlled Decomposition/Reformation (Process)

CDR CorelDraw (Software)

CDR Council on Documentation Research [US]

CDR Critical Design Review

CDR Current Directional Relay

CD-R Compact Disk–Rewritable [also CDR]

Cdr Commander [also CDR]

CDRA Card to Random Access

CDRA Canadian Drilling Research Association

CDRA Committee of Directors of Research Associations [UK]

CDRAP Collaborative Diesel Research Advisory Panel

CDRB Canadian Defense Research Board

CDRD Chemical Defense Research Department

CDRE Chemical Defense Research Establishment [UK]

CDRE Convolution Difference Resolution Enhancement

CDRH Center for Devices and Radiological Health [of Food and Drug Administration, US]

CDRI Chihuahuan Desert Research Institute [US]

CDRI Configuration-Dependent Reactive Incorporation (Model)

Cdrill Counterdrill [also cdrill]

CDRL Contract(ual) Data Requirements List

CD-ROM Compact Disk–Read-Only Memory [also CD ROM]

CD-ROM Libr CD-ROM Library [Journal published in the US]

CD-ROMSPAG CD-ROM Standards and Practices Action Group [UK]

CDRR Call Detail Recording and Reporting

CDRR Contract Documentation Requirements Records

CDRS Call Detail Reporting System

CDRS Computerized Data Retrieval System

CD-RTOS Compact Disk–Real-Time Operating System [also CD RTOS]

CD RW Compact Disk–Rewritable [also CDR-W]

CDRX Critical External Damping Resistance [Electrical Engineering]

CDS Cataloging Distribution Service

CDS Catalog of Data Sources

CDS Central Data Subsystem [of NASA Launch Processing System]

CDS Central Data System

CDS Centre de Documentation Sidérurgique [Documentation Center of the Iron and Steel Industry, France]

CDS Centre de Données Stellaires [Center for Stellar Data, Strasbourg, France]

CDS Charged Droplet Scrubber

CDS Chief of Defense Staff [of Canadian Forces]

CDS Closeout Door System

CDS Coding Sequence [of US National Center for Biotechnology Information–Genetic Codes Databank]

CDS Cold-Drawn Steel

CDS Collision Detector System

CDS Command and Data Simulator [also C&DS]

CDS Command Data System

CDS Communications and Data Subsystem

CDS Compatible Duplex System

CDS Component Disassembly Station

CDS Comprehensive Display System

CDS Computer Data System

CDS Computer Duplex System

CDS Computerized Documentation System [of UNESCO]

CDS Conceptual Design Study

CDS Configuration Data Set

CDS Container Delivery System

CDS Continuous Detonation System

CDS Control Data System

CDS Crystal Diffraction Spectrometer; Crystal Diffraction Spectrometry

CDS Single-Cotton, Double-Silk Covered (Electric Wire)

C&DS Command and Data Simulator [also CDS]

C/DS Cache/Disk System

CdS Cadmium Sulfide (Semiconductor)

CDSA Comprehensive Digestive and Stool Analysis

CdSb Cadmium Antimonide (Semiconductor)

Cd-Sb Cadmium-Antimony (Alloy System)

CDSC Communications Distribution and Switching Center

CDSCC Canberra Deep Space Communications Complex

$CdS:Co^{2+}$ Cobalt-Doped Cadmium Sulfide

CDSE Computer-Driven Simulation Environment

CdSe Cadmium Selenide (Semiconductor)

CdSe:In Indium-Doped Cadmium Selenide (Semiconductor)

CDSF Commercially Developed Space Facility

CDSF COMRADE (Computer-Aided Design Environment) Data Storage Facility

$CdSiO_x$ Cadmium Silicates (e.g., $CdSiO_3$) [also CdSiOx]

CDSL Conference on Data System Langauges [US]

Cd-Sn Cadmium-Tin (Alloy System)

CdSO Cadmiosulfate

cd/sq ft candela(s) per square foot [also cd/ft^2, or $cd \ ft^{-2}$]

cd/sq in candela(s) per square inch [also cd/in^2, or $cd \ in^{-2}$]

cd sr candela steradian [Unit]

cd/sr candela(s) per steradian [Unit]

CDSS Computer Digital Switching System

C&DSS Communication and Data Subsystem

CdSSe Cadmium Sulfide Selenide

CDST Central Daylight Saving Time

CDST Centre de Documentation Scientifique et Technique [Center for Scientific and Technical Documentation; of Conseil National de la Recherche Scientifique, France]

CdS$_x$Te$_{1-x}$ Cadmium Sulfur Telluride (Semiconductor)

CdS/ZnS Cadmium Sulfide/Zinc Sulfide (Heterojunction)

CDT 1,1'-Carbonyl-Di(1,2,4-Triazole)

CDT Central Daylight Time

CDT College of Dental Technologists

CDT Color Detection Tube

CDT Command Descriptor Table

CDT Compressed Data Tape

CDT Compressor Discharge Temperature

CDT Conductivity, Depth and Temperature

CDT Configuration Data Table

CDT Control Data Terminal

CDT Countdown Time

CDT Cyclododecatriene

.CDT Corel Draw Template [File Name Extension]

CD-1,2,4-T 1,1'-Carbonyldi-1,2,4-Triazole

C$/t Canadian Dollars per ton [Unit]

CDTA 1,2-Cyclohexanediaminetetraacetic Acid; 1,2-Cyclohexanediamine Tetraacetate; trans-1,2-Diaminocyclohexane-N,N,N',N'-Tetraacetic Acid [also cdta]

CdTe Cadmium Telluride (Semiconductor)

CdTe:Cl Chlorine-Doped Cadmium Telluride (Semiconductor)

CdTe/GaAs Cadmium Telluride on Gallium Arsenide

CdTe-HgTe Cadmium Telluride-Mercury Telluride (System)

CdTe/Si Cadmium Telluride on Silicon (Substrate)

CdTe:V Vanadium-Doped Cadmium Telluride (Semiconductor)

CDTF Chapter Development Task Force [of ASM International, US]

CDTI Centro para el Desarrollo Tecnológico Industrial [Technological and Industrial Development Center, Spain]

CDTI Cockpit Display of Traffic Information

CDTL Common Data Translation Language

CDTN Centro de Desenvolvimento da Tecnologia Nuclear [Nuclear Technology Development Center, Brazil]

CDTO College of Dental Technologists of Ontario

CDTP Card to Tape

CDTS Centralized Digital Telecommunications System

CDTTM California Department of Transportation Test Methods [US]

CDU Cartridge Disk Unit

CDU Central Display Unit

CDU Cesium Demonstration Unit

CDU Color Developing Unit

CDU Command Detector Unit

CDU Computer Display Unit

CDU Control Detector Unit

CDU Control (and) Display Unit

CDU Coolant Distribution Unit

CDU Coupling Data Unit

CDU Coupling Display Unit [of NASA, Apollo Project]

CDU CRT (Cathode-Ray Tube) Display Unit

CDV Canine Distemper Virus

CDV Check Digit Verification

CDV Compact Disk Video

CDV Component Digital Video

CDVO cis-2,4-Dimethyl-trans-3-Vinyl Oxetane

CDVTPR Centre de Documentation du Verre Textile et des Plastiques Renforcés [Documentation Center for Textile Glass and Reinforced Plastics, France]

CDW Capacitative Discharge Welding; Capacitor Discharge Welding

CDW Charge Density Wave [Solid-State Physics]

CDW Chaotic Dynamics Workbench (Software) [of American Institute of Physics, US]

CDW Civil Defense Warning

CDW Command Data Word

CDW Computer Data Word

CDW Continuum Distorted Wave (Approximation) [Physics]

CDWC Charge Density Wave Conductor [Physics]

CD-WO Compact Disk Write Once

CdWO Cadmiotungstate

CDWR California Department of Water Resources [US]

CDX Control Differential Transmitter

.CDX Compound Index [FoxPro File Name Extension]

CdX Cadmium Chalcogenide [X = Sulfur, Selenium, or Tellurium]

Cd-Zn Cadmium-Zinc (Alloy System)

CdZnTe Cadmium Zinc Telluride (Semiconductor)

CdZnTe/Si Cadmium Zinc Telluride on Silicon (Substrate)

CE Cable Equalizer

CE California Encephalitis

CE Calomel Electrode

CE Capillary Electrophoresis

CE Carbamazepine-10,11-Epoxide

CE Carbon Electrode

CE Carbon Equivalence; Carbon Equivalent [Metallurgy]

CE Carcinoembryonic(s)

CE Card Error

CE Carnot Efficiency

CE Carpenter-Evans (Process)

CE Cathode Emission

CE Cation Exchange(r)

CE Cause and Effect

CE Caustic Embrittlement

CE Cavitation Effect

CE Celestial Equator

CE CENELEC Mark (of Electronic Components) [European Union]

CE Center of Excellence

CE Central Europe(an)

CE *(Centre d'Etude)* – French for "Study Center," or "Research Center"

CE Ceramic Engineer(ing)

CE Change Evaluation

CE Channel End

CE Chapman-Enskog (Theory) [Statistical Mechanics]

CE Charge Exchange [Physics]

CE Chelate Effect

CE Chemical Element

CE Chemical Energy

CE Chemical Engine

CE Chemical Engineer(ing)

CE Chemical Equilibrium

CE Chief Engineer

CE 1-Chloroethane

CE Chronic Endocarditis

CE Chronoellipsometry

CE Civil Engineer(ing)

CE Cloudiness Element

CE Coal Equivalent

CE Coal Extraction

CE Coanda Effect

CE Coastal Engineer(ing)

CE Coefficient of Entry

CE Coextruded; Coextrusion

CE Cold Emission

CE Cold Extrusion

CE College of Engineering

CE Collision Energy

CE Combustion Engine

CE Combustion Engineer(ing)

CE Combustion Engineering, Inc. [US]

CE Common Emitter

CE Common Era

CE Commonwealth Edison [Electric Utility Company]

CE Communauté Européenne [European Community]

CE Communication Electronic (Equipment)

CE Communication Equipment

CE Communications–Electronics

CE Commutator End

CE Composition Exploding

CE Compressor Efficiency

CE Compton Effect [Quantum Effect]

CE Compton Electron [Quantum Effect]

CE Compulsory Expenditure

CE Computational Enumeration

CE Computed Endoscopy

CE Computer Engineer(ing)

CE Computer Equipment

CE Concurrent Engineering

CE Conditional Exemption

CE Conducted Emission

CE Conjugative Effect

CE Conseil de l'Europe [Council of Europe, Strasbourg, France]

CE (Office of) Conservation and Renewable Energy [of US Department of Energy]

CE Conservation Engineer(ing)

CE Consulting Engineer

CE Consumable Electrode

CE Consumer Electronics

CE Control Electrode

CE Control Electronics

CE Control Element

CE Control Engineer(ing)

CE Control Equipment

CE Controlled Environment

CE Controlled Expansion (Alloy)

CE Conversion Efficiency

CE Coriolis Effect

CE Corps of Engineers [US Army]

CE Corrosion Engineer(ing)

CE Cost Element

CE Cost Engineer(ing)

CE Coulomb Excitation

CE Council of Engineering

CE Council of Europe [Strasbourg, France]

CE Counter Electrode

CE Creep Embrittlement

CE Cross-Elasticity

CE Cryogenic Engineer(ing)

CE Crystal Engineering

CE Cubic Equation

CE Current Efficiency

CE Current Expendable

CE Custom Engineering

CE Customer Engineer

CE N-Methyl-N,2,4,6-Tetranitro Benzenamine

C_1E_3 Triethylene Glycol Monomethyl Ether

C_2E_2 Diethylene Glycol Monoethyl Ether

C_5E_2 Diethylene Glycol Monopentyl Ether

C_6E_3 Triethylene Glycol Monohexyl Ether

C_6E_5 Pentaethylene Glycol Monohexyl Ether

C_7E_3 Triethylene Glycol Monoheptyl Ether

C_7E_4 Tetraethylene Glycol Monoheptyl Ether

C_8E_3 Triethylene Glycol Monooctyl Ether

C_8E_4 Tetraethylene Glycol Monooctyl Ether

C_8E_5 Pentaethylene Glycol Monooctyl Ether

$C_{10}E_3$ Triethylene Glycol Monodecyl Ether

$C_{10}E_4$ Tetraethylene Glycol Monodecyl Ether

$C_{10}E_5$ Pentaethylene Glycol Monodecyl Ether

$C_{10}E_6$ Hexaethylene Glycol Monodecyl Ether

$C_{10}E_7$ Heptaethylene Glycol Monodecyl Ether

$C_{10}E_8$ Octaethylene Glycol Monodecyl Ether

$C_{12}E_3$ Triethylene Glycol Monolauryl Ether

$C_{12}E_4$ Tetraethylene Glycol Monolauryl Ether

$C_{12}E_5$ Pentaethylene Glycol Monolauryl Ether

$C_{12}E_6$ Hexaethylene Glycol Monolauryl Ether

$C_{12}E_7$ Heptaethylene Glycol Monolauryl Ether

$C_{12}E_8$ Octaethylene Glycol Monolauryl Ether

$C_{12}E_9$ Nonaethylene Glycol Monolauryl Ether (or Polidocanol)

$C_{12}E_{10}$ Decaethylene Glycol Monolauryl Ether

$C_{12}E_{23}$ Polyoxyethylene(23) Monolauryl Ether

$C_{13}E_{10}$ Decaethylene Glycol Tridecyl Ether

$C_{14}E_3$ Triethylene Glycol Monomyristyl Ether

$C_{14}E_4$ Tetraethylene Glycol Monomyristyl Ether

$C_{14}E_5$ Pentaethylene Glycol Monomyristyl Ether

$C_{14}E_6$ Hexaethylene Glycol Monomyristyl Ether

$C_{14}E_7$ Heptaethylene Glycol Monomyristyl Ether

$C_{14}E_8$ Octaethylene Glycol Monomyristyl Ether

$C_{16}E_2$ Diethylene Glycol Monocetyl Ether

$C_{16}E_3$ Triethylene Glycol Monocetyl Ether

$C_{16}E_4$ Tetraethylene Glycol Monocetyl Ether

$C_{16}E_5$ Pentaethylene Glycol Monocetyl Ether

$C_{16}E_6$ Hexaethylene Glycol Monocetyl Ether

$C_{16}E_7$ Heptaethylene Glycol Monocetyl Ether

$C_{16}E_8$ Octaethylene Glycol Monocetyl Ether

$C_{16}E_{10}$ Decaethylene Glycol Monocetyl Ether

$C_{16}E_{20}$ Icosaethylene Glycol Monocetyl Ether

$C_{18-1}E_2$ Diethylene Glycol Monooleyl Ether

$C_{18}E_2$ Diethylene Glycol Monostearyl Ether

$C_{18}E_3$ Triethylene Glycol Monostearyl Ether

$C_{18}E_4$ Tetraethylene Glycol Monostearyl Ether

$C_{18}E_5$ Pentaethylene Glycol Monostearyl Ether

$C_{18}E_6$ Hexaethylene Glycol Monostearyl Ether

$C_{18}E_7$ Heptaethylene Glycol Monostearyl Ether

$C_{18}E_8$ Octaethylene Glycol Monostearyl Ether

$C_{18}E_{10}$ Decaethylene Glycol Monostearyl Ether

$C_{18}E_{20}$ Icosaethylene Glycol Monostearyl Ether

$C_{18}E_{21}$ Polyoxyethylene(21) Monostearyl Ether

$C_{18}E_{100}$ Polyoxyethylene(100) Monostearyl Ether

$C_{18-1}E_{10}$ Decaethylene Glycol Monooleyl Ether

$C_{18-1}E_{20}$ Icosaethylene Glycol Monooleyl Ether

C/E Calculation/Experiment

Ce Cerium [Symbol]

Ce^{3+} Trivalent Cerium Ion [also Ce^{+++}] [Symbol]

Ce^{4+} Tetravalent Cerium Ion [Symbol]

Ce-134 Cerium-134 [also ^{134}Ce, or Ce^{134}]

Ce-136 Cerium-136 [also ^{136}Ce, or Ce^{136}]

Ce-138 Cerium-138 [also ^{138}Ce, or Ce^{138}]

Ce-139 Cerium-140 [also ^{139}Ce, or Ce^{139}]

Ce-140 Cerium-140 [also ^{140}Ce, or Ce^{140}]

Ce-141 Cerium-141 [also ^{141}Ce, or Ce^{141}]

Ce-142 Cerium-142 [also ^{142}Ce, or Ce^{142}]

Ce-143 Cerium-143 [also ^{143}Ce, or Ce^{143}]

Ce-144 Cerium-144 [also ^{144}Ce, or Ce^{144}]

Ce-146 Cerium-146 [also ^{146}Ce, or Ce^{146}]

$\chi_q(E_F)$ response function corresponding to antiferromagnetic ordering wave vector q [Symbol]

CEA Cambridge Electron Accelerator [US]

CEA Canadian Economics Association

CEA Canadian Education Association

CEA Canadian Electrical Association

CEA Canadian Export Association

CEA Carcinoembryonic Antigen [Immunology]

CEA Carnuntum Excavations [Austria]

CEA Cement Employers Association [US]

CEA Central Electrical Authority [UK]

CEA Centre d'Etudes Atomiques [Atomic Research Center, France]

CEA Certified Environmental Auditor

CEA Chemical Engineering Abstracts [of Royal Society of Chemistry, UK]

CEA Cinematograph Exhibitors Association [UK]

CEA Circle of Equal Altitude

CEA Circular Error Average

CEA Coal Exporters Association (of the United States)

CEA Combustion Engineering Association [UK]

CEA Comite Electrotécnico Argentina [Argentinian Electrotechnical Committee]

CEA Commissariat à l'Energie Atomique [Atomic Energy Commission, France]

CEA Commodity Exchange Authority [US]

CEA Communications-Electronics Agency [US]

CEA Competitive Equipment Analysis

CEA Completion Engineering Association [US]

CEA Confédération Européenne del'Agriculture [European Confederation of Agriculture, Switzerland]

CEA Conservation Education Association [US]

CEA Construction Equipment Advertisers

CEA Consulting Engineers Association [US]

CEA Consulting Engineers of Alberta [Canada]

CEA Control Electronics Assembly [for NASA Manned Maneuvering Unit]

CEA Control Element Assembly

CEA Cooperative Editorial Association [now Cooperative Communicators Association, US]

CEA Core Element Assembly [Nuclear Engineering]

CEA Corrosion Engineering Association [of Republic of China, Taiwan]

CEA Cost Effective Analysis

CEA Council for Energy Awareness [US]

CEA Council of Economic Advisers (to the President) [US]

CEA Cyclohexyl Ethyl Amine

2-CEA N-(2-Chloroethyl)-N-Ethyl-2-Methyl Benzenemethanamide

CEAC Central European Analysis Commission

CEAC Combined Environmental Acoustic Chamber

CEAC Committee for European Airspace Coordination [of NATO]

CEAC Consulting Engineers Association of California [US]

CEAC Control Element Assembly Computer

CEAC Cost and Economic Analysis Center [of US Army]

CEAC County Engineers Association of California [US]

Ce(Ac)$_3$ Cerium (III) Acetate

Ce(ACAC)$_3$ Cerium(III) Acetylacetonate [also Ce(AcAc)$_3$, or Ce(acac)$_3$]

CEA-CENG Commissariat à l'Energie Atomiques–Centre d'Etudes Nucléaires de Grenoble [Commissariat for Atomic Energy–Grenoble Nuclear Research Center, France]

CEA-CEN Valrho Commissariat à l'Energie Atomique–Centre d'Etudes Nucléaires de Valrho [Commissariat for Atomic Energy–Valrho Nuclear Research Center, Bagnols-sur-Cèze, France]

CEA-CNRS Centre d'Etudes Atomiques–Centre National de la Recherche Scientifique [Nuclear Research Center–National Center for Scientific Research, Saclay, France]

CEADI Colored Electronic Altitude Director Indicator

CEAEN Centre d'Etudes pour les Applications de l'Energie Nucléaire [Research Center for Nuclear Energy Applications, France]

CE/AF Charge Exchange Antiferromagnetic (Ordering) [also CE AF] [Solid-State Physics]

CEA Grenoble Centre d'Etudes Atomiques de Grenoble [Grenoble Nuclear Research Center, France]

CEAL Canadian Explosive Atmospheres Laboratory [of Canada Center for Mineral and Energy Technology/Mining Research Laboratories]

CEAL Civil Engineering Automation Library (System)

CeAl Cerium Aluminide (Compound)

CEAM Center for Exposure Assessment Modeling [of US Environmental Protection Agency]

CEAM Center of Excellence for Advanced Materials [of University of California at San Diego, US]

CEAM Concerted European Action on Magnets [Part of EURAM Program]

CEAN Centre d'Etude pour les Applications de l'Energie Nucléaire [Center for the Study of Nuclear Energy, Belgium]

CeAN Cerium(IV) Ammonium Nitrate

CEANAR Commission on Education in Agriculture and Natural Resources [of National Research Council, US]

CEA Notes Inf CEA Notes d'Information [French Publication on Atomic Energy; published by Commissariat à l'Energie Atomique]

CEAO Communauté Economique de l'Afrique Occidentale [West African Economic Community]

CEAP Canada Export Award Program [of Department of Foreign Affairs and International Trade]

CEAP Canadian Energy Audit Program

CEA PRC Construction Equipment Advertisers and Public Relations Council [US]

CEAR Committee on the Effects of Atomic Radiation [of United Nations]

CEARAM Commission for the Exploration of Archeological Remains of Asia Minor [Austria]

CEARC Canadian Environmental Assessment Research Council

CEARC Computer Education and Applied Research Center [US]

CEAREX Coordinated Eastern Arctic Experiment

CEAS Center of European Agricultural Studies

CEAS College of Engineering and Applied Science [of University of Wisconsin at Milwaukee, US]

CEA Saclay Centre d'Etudes Atomiques de Saclay [Saclay Nuclear Research Center, France]

CEASI Concerted European Action on Structural Intermetallics (Program)

CEAT Coordination Européene des Amis de la Terre [European Organization of Friends of the Earth]

CEAU Continuing Education Achievement Unit

CEB Central Electricity Board [UK]

CEB Council for Environmental Balance [now National Council for Environmental Balance, US]

CeB$_6$ Cerium Hexaboride

CEBA Chemical Engineering and Biotechnology Abstracts [also Chem Eng Biotech Abstr]

CEBAF Continuous Electron Beam Accelerator Facility [of Jefferson Laboratotry, Newport News, Virginia, US]

CEBC Consulting Engineers of British Columbia [Canada]

CEBEDEAU Centre Belge d'Etude et de Documentation des Eaux [Belgian Center for Water Studies and Documentation]

CEBELCOR Centre Belge d'Etude de la Corrosion [Belgian Center for Corrosion Studies]

CeBIT Welt-Centrum Büro-Information-Telekommunikation [World Center Office–Information– Telecommunications, Hanover, Germany]

CEBM Corona, Eddy Current, Beta Ray, Microwave

CEC California Energy Commission [US]

CEC Canada Employment Center

CEC Canadian Electrical Code

CEC Canadian Engineering Competition

CEC Carboxy(l) Ethyl Cysteine [Biochemistry]

CEC Caribbean Employers Confederation [Trinidad]

CEC Cation Exchange Capacity

CEC Center for Economic Cooperation [of Food and Agricultural Organization]

CEC Center for Marine Conservation [US]

CEC Central Electronic Complex

CEC Central European Command

CEC Central European Country

CEC Ceramic Education Council [US]

CEC Character-Erase Character

CEC Charlotte Engineers Club [North Carolina, US]

CEC Chemical Electronics Company

CEC Chemical Evaluation Committee [for US National Toxicology Program]

CEC Chiba Engineering College [Japan]

CEC Chick Embryo Cell

CEC China Electronics Corporation

CEC Civil Engineer Corps [of US Navy]

CEC Clarence Environment Center [Australia]

CEC Clothing Export Council (of Great Britain)

CEC Code-Extension Character

CEC Colorado Engineering Council [US]

CEC Combination Emission Control

CEC Commission for Environmental Cooperation [Canada]

CEC Commission of the European Communities [Brussels, Belgium]

CEC Commission on Education and Communication [of International Union for Conservation of Nature and Natural Resources]

CEC Committee for Environmental Conservation [of European Commission]

CEC Commodity Exchange Commission [US]

CEC Commonwealth Economic Committee [UK] [Dissolved]

CEC Commonwealth Education Cooperation

CEC Commonwealth Engineering Conference

CEC Commonwealth Engineers Council [UK]

CEC Community Environmental Council [US]

CEC Confédération Internationale des Cadres [International Confederation of Executive and Managerial Staffs, France]

CEC Congruently Evaporating Composition

CEC Consolidated Electrodynamics Corporation

CEC Consulting Engineers Council [UK and US]

CEC Control Encoder Coupler

CEC Cooperative Engagement Capability [US Navy]

CEC Coordinating European Council (for the Development of Performance Tests for Lubricants and Engine Fuel) [of European Union]

CEC Corporate Executive Committee

CEC Council for Education in the Commonwealth [UK]

CEC Council for Environmental Conservation [UK]

CEC Council on Environmental Cooperation

CEC Critical Electrolyte Concentration

CEC Crop End Control [Metallurgy]

CEC Cryogenic Engineering Conference

CECA Canadian Elevator Contractors Association

CECA Communauté Europénne du Charbon et de l'Acier [European Community for Steel and Coal]

CECA Consumers Energy Council of America [US]

CECAF Committee for Eastern Central Atlantic Fisheries [Senegal]

CECC California Educational Computing Consortium [US]

CECC CENELEC Electronic Components Committee [Belgium]

CECC Commonwealth Economic Consultative Committee [UK]

CECC Consulting Engineers Council of Colorado [US]

CEC/CAL Consulting Engineers Council of California [US]

CECD Coulson Electrolytic Conduction Detector

CECE Combined Electrolysis Catalytic Exchange

CECE Committee for European Construction Equipment [UK]

CECEC Centre of Economic Computation and Economic Cybernetics, Bucharest, Romania]

CECED Conseil Européen de la Construction Electrodomestique [European Committee of Manufacturers of Domestic Electrical Equipment, UK]

CEC/ICMC Cryogenic Engineering Conference and International Cryogenic Materials Conference [also CEC-ICMC]

CECIF Chambre Européenne pour le Développement du Commerce, de l'Industrie et des Finances [European Chamber for the Development of Trade, Industry and Finances, Belgium]

CECIMO Comité Européen de Coopération des Industries de la Machine-Outil [European Committee for Cooperation of Machine Tool Industries, Belgium]

CEC/JRC Commission of the European Communities/Joint Research Center [Petten, Netherlands] [also CEC-JRC]

CECL Cascade-Emitter Coupled Logic

CECM Centre d'Etudes de Chimie Métallurgique [Metallurgical Chemistry Research Center, Vitry,-sur-Seine, France]

CECM Council of Eastern Caribbean Manufacturers

CECM/CNRS Centre d'Etudes de Chimie Métallurgique/Conseil National de la Recherche Scientifique [Metallurgical Chemistry Research Center/National Scientific Research Council Vitry-sur-Seine, France] [also CECM-CNRS]

CEC/MINN Consulting Engineers Council of Minnesota [US]

CEC/NYS Consulting Engineers Council of New York State [US]

CECO Consulting Engineers Council of Oregon [US]

CECO Cost Estimate Change Order

CEC-OC Canada Employment Center–On Campus

CECOM Communications-Electronics Command [US Army]

CECON Center for Conservation Studies

CECOPHIL Council of Engineering Consultants of the Philippines

CECP Compatibility Engineering Change Proposal

Ce(Cp)$_3$ Tris(cyclopentadienyl)cerium

CECPP Canadian Electronic Chart Pilot Project

CECRI Central Electrochemical Research Institute [Karaikudi, India] [also CE-CRI]

CECS Congress of the European Ceramic Society

CEC/TEX Consulting Engineers Council of Texas [US]

CECTK Committee for Electrochemical Thermodynamics and Kinetics [Belgium]

CECU Consulting Engineers Council of Utah [US]

CECUA Confederation of European Computer Users Associations [Belgium]

CECUA Conference of European Computer User Association

CED CANMET (Canada Center for Mineral and Energy Technology)/ Explosives Division [of Natural Resources Canada]

CED Canning Electronic Division [Berlin, Germany]

CED Capacitance Electronic Disk

CED Carbon-Equivalent-Difference [Metallurgy]

CED Centro Elettronico di Documentazio [Electronic Documentation Center, Italy]

CED Chemical Engineering Department

CED Civil Engineer Directorate [of US Air Force]

CED Cohesive Energy Density

CED Committee for Economic Development [US]

CED Communauté Européenne de Défense [European Defense Community]

CED Conditional Exponential Distribution

CED Constant Energy Difference

CED Cupriethylenediamine

CED Cyanoethyl Diisopropyl

CEDA Canadian Electrical Distributors Association

CEDA Catering Equipment Distributors Association (of Great Britain)

CEDA Center for Economic Development of Australia

CEDA Central Dredging Association [Netherlands]

CEDA Committee for Economic Development of Australia

CEDAC Cause-and-Effect Diagrams with Addition of Cards

CEDAR Central-European Environmental Data Request (Facility) [Austria]

CEDAR Coupling, Energetics and Dynamics of Atmospheric Regions (Program) [US]

CEDAT Center for Educational Development and Training

CEDB Civil Engineering Database

CEDC Canadian Engineering Data Center [of ASM International, US]

CEDC Canadian Engineering Design Competition

CEDC Cyclic Error Detection Code

CEDE Committed Effective Dose Equivalent

CEDEFOP Centre Européen pour le Développement de la Formation Professionelle [European Center for the Development of Vocational Training]

CEDeS Center for Engineering and Design Systems [of University of Washington, US]

CEDIC Comité Européen des Ingenieurs-Conseils du Marche Commun [Committee of the Consulting Engineers of the Common Market; of European Community, Belgium]

CEDIP Canadian Exploration and Development Incentive Program

CEDM Control Element Drive Motor

CEDO Consulting and Engineering Design Organization

CEDOCAR Centre de Documentation de l'Armement [Ordnance Documentation Center, France]

CEDOCOS Centre de Documentation sur les Combustibles Solides [Documentation Center on Solid Fuels, Belgium]

CEDPA California Educational Data Processing Association [US]

CEDR Comité Européen de Droit Rural [European Committee for Agrarian Law, Belgium]

CEDRL CANMET (Canada Center for Mineral and Energy Technology)/Energy Diversification Research Laboratory [of Natural Resources Canada]

CEDU Community Economic Development Unit [of Industry Canada]

CEE Canadian Exploration Expense

CEE Center for Environmental Education [Canada]

CEE College Entrance Examination [US]

CEE Commission Economique pour l'Europe [Economic Commission for Europe; of United Nations]

CEE (International) Commission on Rules for the Approval of Electrical Equipment

CEE Competitive Entrance Examination [Kenya]

CEE Comprehensive Environmental Evaluation

CEE Council for Environmental Education [UK]

CEEA Communauté Européenne de l'Energie Atomique [European Community for Atomic Energy; of EURATOM]

CEEA N-(2-Cyanoethyl)-N-Ethylamine

CEEAS Centre Européen d'Etudes de l'Acide Sulfurique [European Center for Studies of Sulfuric Acid; of Conseil Européen des Fédérations de l'Industrie Chimique, Belgium]

CEEB College Entrance Examination Board [US]

CEEC Comité Européenne des Economistes de Construction [European Committee of Construction Economists, France]

CEEC Committee of Engineers for Environmental Control

CEEC Committee of European Economic Cooperation [France]

CEEC Controlled Environmental Exposure Chamber

CEED CANMET (Canada Center for Mineral and Energy Technology)/Energy Efficiency Division [of Natural Resources Canada]

CEED Center for Economic and Environmental Development [UK]

CEED Community Economics and Ecological Development (Council) [US]

CEEFAX United Kingdom Teletext Service

CEEK Cargo Element Extension Kit

CEELS Characteristic Electron Energy-Loss Spectroscopy

CEEM Center for Electronics and Electrooptical Materials [of University of Buffalo, US]

CEEMA Committee on Engineering Education in Middle Africa

CEEMT N-(2-Cyanoethyl)-N-Ethyl-m-Toluidine

CEEN Centre d'Etude de l'Energie Nucléaire [Center for Nuclear Energy Studies, Belgium]

CEEOC Commonwealth Equal Employment Opportunity Council

CEEP Centre Européen de l'Entreprise Publique [European Center of Public Enterprises, Belgium]

CEER Chemical Economy and Engineering Review [Japan]

CEERI Central Electronics Engineering Research Institute [New Delhi, India]

CEES Center for Energy and Environmental Studies [of Princeton University, New Jersey, US]

CEES Center for Environmental Engineering and Science [of New Jersey Institute of Technology, Newark, US]

CEES Committee on Earth and Environmental Sciences

CEESA Conference of European Engineering Students Associations

CEEST Center for Environmental Engineering and Science Technology [of University of Massachusetts at Lowell, US]

CEET Clean and Efficient Energy Technologies (Program) [of European Commission]

CEF Cable Entrance Facility

CEF Canadian Expeditionary Force

CEF Carrier Elimination Filter

CEF Center for Economic Forecasting [UK]

CEF Construction Employers Federation [UK]

CEF Critical Experiments Facility [Physics]

CEF Crystal(line) Electric Field

CEFAC Civil Engineering Field Activities Center [UK]

CEFBC Continuous Elution Flat-Bed Chromatography

CEFIC Conseil Européen des Fédérations de l'Industrie Chimique [Council of European Chemical Manufacturers Federations, Belgium]

CEFIGRE International Training Center for Water Resources Management

CEFNO Carbethoxyformonitrile

CEFPI Council on Educational Facility Planning International

CEFRACOR Centre Française de la Corrosion [French Center for Corrosion]

CEG Center of Exploration Geophysics [Osmania University, Hyderabad, India]

CEG Certified Engineering Geologist

CEG Chemical Engineering Group [London, UK]

CEG Collège d'Enseignement Général [College of General Education, France]

CEG Combined Environment Groups

CEG Cryogenic Engineering Conference [US]

CEGA Conference of Governmental Statisticians of the Americas [of Inter-American Statistical Institute]

CEGABA Carboxyethyl-γ-Aminobutyric Acid

CEGB Central Electricity Generating Board [London, UK]

CEGB Res CEGB Research [Publication of the Central Electricity Generating Board, London, UK]

CEGEP Collège d'Enseignement Général et Professionel [College for General and Professional Education, Quebec, Canada]

CEGIS Construction Engineering Geographic Information System

CEGS Canadian Exploration Geophysical Society

CEH Center for Environmental Health [of Centers for Disease Control and Prevention, Atlanta, Georgia, US]

CEH Chemical Economics Handbook [Stanford Research Institute, California, US]

CEH Conférence Européenne des Horaires des Trains de Voyageurs [European Passenger Train Timetable Conference, Switzerland]

CEHR Committee on Education and Human Resources [Federal Coordinating Council for Science, Engineering and Technology, US]

CEHU College of Engineering, Hosei University [Japan]

CEI Canadian Education Index

CEI Center for Environmental Interpretation [UK]

CEI Chemical Exposure Index

CEI Cloudtop Entrainment Instability

CEI Commission Electrotechnique Internationale [International Electrotechnical Commission, Switzerland]

CEI Committee for Environmental Information [US]

CEI Communications-Electronics Instructions

CEI Complete Elliptic Integral

CEI Compliance Evaluation Inspection

CEI Computer Extended Instruction

CEI Conducted Electromagnetic Interference

CEI Configuration End Item

CEI Continuing Education Institute [US]

CEI Contract End Item

CEI Council of Engineering Institutions [UK]

CEI Cycle Engineers Institute [UK]

CEIAC Coastal Engineering Information Analysis Center [US]

CEI-BOIS Confédération Européenne des Industries du Bois [European Confederation of Woodworking Industries, Belgium]

CEIC Canada Employment and Immigration Commission

CEI/CSTI Council of Engineering Institutions and Council of Science and Technology Institutes [UK]

CEIE Conformational Equilibrium Isotope Effect

CEI-Europe Continuing Education Institute–Europe

CEIF Council of European Industrial Federations

CEIF Cyprus Employers and Industrialists Federation

Ceil Ceiling [also ceil]

CEIM Conseil International pour l'Exploration de la Mer [International Council for the Exploration of the Sea, Denmark]

CEIP Canadian Exploration Incentive Program

CEIP Carnegie Endowment for International Peace [US]

CEIP Center for Environmental Intern Programs (Fund) [US]

CEIP Communications-Electronics Implementation Plan

CEIRD Confirming Engineering Information Report Date

CEIS Caribbean Energy Information System

CEIS Center for Environmental Information Science [Tokyo, Japan]

CEIS Center for Environmental Information and Statistics [of US Environmental Protection Agency]

CEIS Center for European Industrial Studies [UK]

CEISMC Center for Education Integrating Science, Mathematics and Computing [of Georgia Institute of Technology, Atlanta, US]

CEIT Centro de Estudios e Investigaciones Técnicas (de Guipúzcoa) [Guipuzcoa Technical Research Center, San Sebastian, Spain]

CEIT Crew Equipment Integration Test

CEIT Crew Equipment Interface Test

CEIU Canada Employment and Immigration Union

CEL Carbon Equilibrium Loop

CEL Carbon-Equivalent, Liquidus [Metallurgy]

CEL Central European (Pipe-)Line

CEL Chemical Engineering Laboratory

CEL Commission on Environmental Law

CEL Committee on Engineering Laws [US]

CEL Contrast Enhanced Lithography

CEL (Two Photon) Correlated Emission Laser

CEL Critical Experiment Laboratory [of Babcock & Wilcox Inc., Mount Athos, Virginia, US] [Now a Nuclear Historic Landmark]

CEL Cryogenic Engineering Laboratory [of National Institute for Standards and Technology, US]

CELA Canadian Environmental Law Association

CeLα Cerium L-Alpha (Radiation) [also CeL$_\alpha$]

CELADE Centro Latinoamericano de Demografía [Latin American Demographic Center, Chile]

celest celestial

Celest Mech Celestial Mechanics [US Journal]

Celest Mech Dyn Astron Celestial Mechanics and Dynamical Astronomy [Journal published in the Netherlands]

Ce:LiGdF$_4$ Cerium-Doped Lithium Gadolinium Fluoride (Laser) [also Ce^{3+}:LiGdF$_4$]

CELIMAC Comité Européen de Liaison des Industries de la Machine à Coudre [European Liaison Committee for the Sewing Machine Industries, Germany]

Ce:LiYbF$_4$ Cerium-Doped Lithium Ytterbium Fluoride (Laser) [also Ce^{3+}:LiYbF$_4$]

Ce:LiYF$_4$ Cerium-Doped Lithium Yttrium Fluoride (Laser) [also Ce^{3+}:LiYF$_4$]

CELL Division of Cellulose, Paper and Textile [of American Chemical Society, US]

Cell Cellulose [also cell]

cell cellular

Cell Biol Int Cell Biology International [of International Federation of Cell Biology, Canada]

Cell Biol Int Rep Cell Biology International Reports [of International Federation of Cell Biology, Canada]

Cell Immunol Cellular Immunology [US Journal]

Cell Mol Biol Cellular and Molecular Biology

Cell Mol Biol Lett Cellular and Molecular Biology Letters

Cell Mol Biol Res Cellular and Molecular Biology Research [International Journal incorporating Cellular and Molecular Biology]

Cello Cellophane [also cello]

Cell Physiol Biochem Cellular Physiology and Biochemistry [International Journal]

Cell Polym Cellular Polymers [Journal published in the UK]

Cells Mater Cells and Materials [International Journal]

Cell Tissue Res Cell Tissue Research [Journal published in the US]

Cellulosa Carta Cellulosa e Carta [Italian Publication on Cellulose and Paper; published by Ente Nazionale per la Cellulosa e per la Carta]

Cellulose Chem Technol Cellulose Chemistry and Technology [Romanian Journal]

Celuloza Hirtie Celuloza si Hirtie [Romanian Cellulose Chemistry and Technology Journal]

CELP Code Excited Linear Prediction

CELRF Canadian Environmental Law Research Foundation

CELS Characteristic Energy-Loss Spectroscopy

CELS Corning Engineering Laboratory Services [US]

Cels Celsius

CELSS Controlled Ecological Life Support System

CELT Charge-Energy-Limited Tunneling (Model) [Physics]

CELT Coherent Emitter Location Testbed

CELTE Constructeurs Européens de Locomotives Thermiques et Electriques [European Manufacturers of Thermal and Electric Locomotives, France]

CELTIC Charge-Energy-Limited Tunneling Conduction (Model) [Physics]

CELU Confederation of Ethiopian Labor Unions

Ce:LuPO$_4$ Cerium-Doped Lutetium Phosphate (Crystal) [also Ce^{3+}:LuPO$_4$]

CEM Calcium Europium Manganate

CEM Center for Electron Microscopy

CEM Center for Electron Microscopy [of Materials Research Laboratory, University of Illinois, US]

CEM Center for Entrepreneurial Management [US]

CEM Center for Environmental Mechanics [of Commonwealth Scientific and Industrial Research Organization, Australia]

CEM Cercle d'Etudes des Métaux [Study Circle on Metals, France]

CEM Channel Electron Multiplier

CEM Channeltron Electron Multiplier

CEM Communications–Electronics–Meteorology

CEM Computer Education for Management

CEM Computerized Electron Microscopy

CEM Contrast Enhancing Material

CEM Conventional Electron Microscope; Conventional Electron Microscopy

CEM Conversion Electron Moessbauer (Spectroscopy)

CEM Core-Excitation Model [Physics]

CEM Corrected Effective Medium

CEM Counter Electromotive Cell

Cem Cement [also cem]

Cem Cementite [Metallurgy]

CEMA Canadian Egg Marketing Association

CEMA Canadian Electrical Manufacturers' Association

CEMA Catering Equipment Manufacturers Association [UK]

CEMA Center d'Etude des Matériaux Avancés [Advanced Materials Research Center; of Université de Rennes, France]

CEMA Channel Electron Multiplier Array

CEMA Cleaning Equipment Manufacturers Association [US]

CEMA Comité Européen des Groupements de Constructeurs du Machinisme Agricole [European Committee of Associations of Manufacturers of Agricultural Machinery, France]

CEMA Converting Equipment Manufacturers Association [US]

CEMA Conveyor Equipment Manufacturers Association [US]

CEMA N-(2-Cyanoethyl)-N-Methylaniline

Cem A Cement Asbestos

CEMAA Special Commission for the Amazonian Environment

Cem AB Cement Asbestos Board

CEMAFON Comité Européen des Matériels et Produits pour la Fonderie [European Committee of Foundry Materials and Products, France]

CEMAGREF Centre National du Machinisme Agricole [Center for Agricultural Machinery and Equipment, France]

CEMAID Center of Excellence in Molecular and Interfacial Dynamics (Network) [Canada]

CEMAP Cotton Export Market Acreage Program

CEMATEX European Committee of Textile Machinery Manufacturers [Switzerland]

CEMBS Committee for European Marine Biological Symposia [UK]

CEMBUREAU Bureau Européen du Ciment [European Cement Association, France]

CEMC Canadian Engineering Manpower Council

CEMC Curriculum Evaluation and Management Center [UK]

Cem Concr Aggr J Cement, Concrete and Aggregates Journal [of American Society for Testing and Materials]

Cem Concr Compos Cement and Concrete Composites [Journal published in the US]

Cem Concr Res Cement and Concrete Research [Published in the US]

CEMD Center for Electrophotonic Materials and Devices [of McMaster University, Hamilton, Canada]

CEMD Ceramic and Electronic Materials Division [of National Science Foundation, US]

CEMES Centre d' Elaboration des Matériaux et d'Etudes Structurales [Center for Materials Elaboration and Structural Studies, Toulouse France]

CEMES-CNRS Centre d' Elaboration des Matériaux et d'Etudes Structurales–Centre National de Recherche Scientifique [Center for Materials Elaboration and Structural Studies of the National Center for Scientific Research, Toulouse France]

CEMF Counter Electromotive Force [also cemf]

Cem Fl Cement Floor

CEMI Canadian Engineering Manpower Inventory

CEMI Commission Européenne de Marketing Industriel [European Committee for Industrial Marketing, UK]

CEMIRT Civil Engineering Maintenance, Inspection, Repair and Training Team [US Air Force]

CEMM Center for Electron Microscopy and Microanalysis

CeMM Cerium Mischmetall

Cem Mort Cement Mortar

CEMO Calcium Europium Manganese Oxide

CEMON Customer Engineering Monitor

CEMP Canadian Ecosystem Monitoring Program

CEMP Center for Environmental Management and Planning [Scotland]

CEMP Coastal Environment Management Plans

Cem Plas Cement Plaster

Cem PC Cement Plaster Ceiling

CEMR CANMET (Canada Center for Mineral and Energy Technology)/(Department of) Energy, Mines and Resources [now CANMET/NRCan]

CEMR Center for Energy and Mineral Resources [of Texas A&M University, College Station, US]

CEMREL Central Midwest Regional Educational Laboratory [US]

CEMS Center for Electrochemical and Materials Science [of University of St. Andrews, UK]

CEMS Central Electronic Management System

CEMS Chinese Electron Microscopy Society [PR China]

CEMS (Department of) Chemical Engineering and Materials Science

CEMS College of Earth and Mineral Sciences [at Pennsylvania State University, University Park, US]

CEMS Continuous Emission Monitoring System

CEMS Conversion Electron Moessbauer Scattering

CEMS Conversion Electron Moessbauer Spectroscopy

CEMS Conversion Electron Moessbauer Spectrum

CE&MS Chemical Engineering and Materials Science

CE-MS Capillary Electrophoresis/Mass Spectroscopy

CEMT Conference Européenne des Ministres des Transport [European Conference of Ministers of Transport]

CEMUL Center of Mechanics and Materials of Technical University of Lisbon [Portugal]

CEN Centre d'Etudes Nucléaires [Nuclear Research Center, France]

CEN Centre d'Etudes Nucléaires [Nuclear Research Center at Mol, Belgium]

CEN Comité Européen de Normalisation [European Standards Committee, Belgium]

Cen Centaurus [Astronomy]

Cen Century [also cen]

Cen X-3 Centaurus X-3 [Astronomy]

cen central

Cen Afr Central Africa(n) [also Cent Afr]

CENB Consulting Engineers of New Brunswick [Canada]

CENBG Centre d'Etudes Nucléaires de Bordeaux-Gradignan [Bordeaux-Gradignan Nuclear Research Center, France]

CENC Centre d'Etudes Nucléaires de Cadarache [Cadarache Nuclear Research Center, France] [also CEN Cadarache]

CENCAL Central California Oceanographic Cooperative [US]

CENCER CEN (Comité Européen de Normalisation) Certification Body

CENCER CEN (Comité Européen de Normalisation) Certification Branch

CENDEC Centralized-Decentralized (Software Program)

CENDIT Center for Development of International Technology

CENELEC Comité Européen de Normalisation Electrotechnique [European Committee for Electrotechnical Standardization] [also CENEL, Belgium]

CE-NET The Concurrent Engineering NETwork of Excellence

CENFAM Centro Nazionale per la Fisica della Atmosfera e la Meteorologia [National Center of Atmospheric Physics and Meteorology, Italy]

CENFAR Centre d'Etudes Nucléaires de Fontenay-au-Roses [Fontenay-au-Roses Nuclear Research Center, France] [also CEN Fontenay-au-Roses]

CENG Centre d'Etudes Nucléaires de Grenoble [Grenoble Nuclear Research Center, France]

CEng Certified Engineer [UK]

CEng Chemical Engineer

CEN Grenoble Centre d'Etudes Nucléaires de Grenoble [Grenoble Nuclear Research Center, France]

CENID Centro Nacional de Información y Documentación [National Center of Information and Documentation, Chile]

CENIM Centro Nacional de Investigaciónes Metalurgicas [National Center for Metallurgical Research, Spain]

CeNiSn Cerium-Nickel-Tin (Antiferromagnet)

CENNA Convention on Early Notification of a Nuclear Accident

CENR Committee on the Environment and Natural Resources [of National Science and Technology Council, US]

CENRTC Capital Equipment Not Related to Construction

CENS Centre d'Etudes Nucléaires de Saclay [Saclay Nuclear Research Center, France]

CENS China Economic News Service [Taiwan]

CENSA Council of European and Japanese National Shipowners Associations [UK]

CEN Saclay Centre d'Etudes Nucléaires de Saclay [Saclay Nuclear Research Center, France]

CenSIS Center of Spatial Information Systems [US]

Cent Centigrade [also cent]

Cent Centrifugal; Centrifuge [also cent]

Cent Century [also cent]

cent centimeter [Unit]

cent central

cent *(centum)* – hundred

CENTA Combined Edible Nut Trade Association [UK]

CENTAG Central Army Group [of NATO]

Cent Afr Central Africa(n) [also Cen Afr]

Cent Afr J Med Central African Journal of Medicine

Cent Am Central America(n) [also Cent Amer]

CENTAS Central Army Group

CEN/TC CEN (Comité Européen de Normalisation)/Technical Commission

CENTCON Centralized Control Facility

Cent Doc Sider, Circ Inf Tech Centre de Documentation Sidérurgique, Circulaire d'Informations Techniques [Documentation Center of the Iron and Steelmaking Industry, Technical Information Circular; published in Paris, France]

centihg centimeter(s) of mercury [Unit]

centl central [also cent'l]

centl centrifugal [also cent'l]

CENTO Central European Treaty Organization [of NATO]

Centr Centralite [Explosive]

CEN Valrho Centre d'Etudes Nucléaires de Valrho [Valrho Nuclear Research Center, Bagnols-sur-Cèze, France] [also CENV]

CEO Catalyzed Electrochemical Oxidation

CEO Center for Earth Observation

CEO Chief Education Officer

CEO Chief Executive Officer

CEO Cleaved-Edge Overgrowth

CEO Comprehensive Electronic Office

CEO Consulting Engineers of Ontario [Canada]

CeO$_x$ Cerium Oxides [CeO$_2$, or Ce$_2$O$_3$] [also CeOx]

CEOA Central Europe Operating Agency [of NATO]

CEOC Colloque Européene des Organisations de Contrôle [European Colloquium of Inspection Organizations]

CEOC Confédération Européenne d'Organismes de Contrôle [European Confedera-tion of Inspection Organizations]

CEOCOR Comité d'Etude de la Corrosion et de la Protection des Canalisation [Committee for the Study of Pipe Corrosion and Protection, Belgium]

CeO$_2$:Ga^{3+} Trivalent Tantalum-Doped Cerium Dioxide

Ce(O-i-Pr) Cerium(IV) Isopropoxide

CEONet Canadian Earth Observation Network

CeOPr^i Cerium(IV) Isopropoxide

CEOS Committee on Earth Observation Satellites [Canada]

CeO$_2$/Si Cerium Dioxide on Silicon (Substrate)

CEOS-IDN Committee on Earth Observation Satellites–International Directory Network [Canada]

CEOT Conference on Emerging Optoelectronic Technologies [of SPIE–International Society for Optical Engineering, US]

CeO$_2$:Ta^{5+} Pentavalent Tantalum-Doped Cerium Dioxide

CeOx Cerium Oxides [i.e., CeO$_2$, or Ce$_2$O$_3$] [also CeO$_x$]

CeO$_2$:Y$_2$O$_3$ Yttria-Doped Cerium Dioxide

CeO$_2$-ZrO$_2$ Cerium Dioxide-Zirconium Dioxide (System)

CEP Catalytic Extraction Process(ing)

CEP Cathode Electrodeposited Paint

CEP Central East Pacific (Ocean Region)

CEP Central European Pipeline

CEP Certificat d'Education Professionelle [French Certificate of Vocational Education]

CEP Certified Environmental Professional

CEP Chemical Engineering Progress [Publication of the American Institute of Chemical Engineers, US]

CEP Chief Education Planner

CEP Circle of Equal Probability; Circular Error of Probability; Circular Error Probability [Aerospace]

CEP Citizens' Energy Project

CEP Civil Emergency Planning [of NATO]

CEP Civil Engineering Package

CEP Clinical Experimental Pharmacologist

CEP Command Executive Procedures

CEP Commission and the European Parliament

CEP Committee for Environmental Protection

CEP Community Education Program

CEP Compact Effective (Core) Potentials [Physics]

CEP Computer Entry Punch

CEP Concentrated Employment Program [US]

CEP Constraint Exploiter Process

CEP Corporation de l'Ecole Polytechnique [Polytechnic Institute Corporation, Montreal, Quebec, Canada]

CEP Cotton Equalization Program

CEP Cylinder Escape Probability

C/Ep Carbon (Graphite) Fiber-Reinforced Epoxy (Composite)

Cep Cepheus [Astronomy]

CEPA California Environmental Protection Agency [US]

CEPA Canadian Environmental Protection Act

CEPA Civil Engineering Programming Applications

CEPA Commonwealth Environment Protection Agency

CEPA (National Society for) Computer Applications in Engineering, Planning and Architecture [US]

CEPA Council of Europe Parliamentary Assembly

CEPA Coupled Electron-Pair Approximation [Physics]

CEPAC Center for Process Analytical Chemistry [of University of Washington, Seattle, US]

CEPAC Confédération Européenne de l'Industrie des Pates, Papiers et Cartons [European Confederation of Pulp, Paper and Board Industries, Belgium]

CEPACS Cargo Entry Processing and Collection System

CEPACS Customs Entry Processing and Control System

CEPANCRM Contract Employment Program for Aboriginal People in Natural and Cultural Resource Management [Australia]

CEPA Newsl CEPA Newsletter [of National Society for Computer Applications in Engineering, Planning and Architecture, US]

CEPA-PNO Coupled Electron-Pair Approximation using Pair Natural Orbitals [Physics]

CEPC Canada Civil Engineering Planning Committee

CEPCEO Comité d'Etude des Producteurs de Charbon d'Europe Occidentale [Association of the Coal Producers of the European Community, Belgium]

CEPD Council for Economic Planning and Development [Taiwan]

CEPE Central Experimental and Proving Establishment [Canada]

CEPE Comité Européen des Associations des Fabricants de Peinture, d'Encres, d'Imprimerie et de Couleurs [European Committee of Paint, Printing Ink and Artists Colors Manufacturers Associations, Belgium]

CEPEA N-(2-Cyanoethyl)-N-Phenyl Ethanolamine

CEPER Combined Engineering Plant Exchange Record

CEPEX Central Equatorial Pacific Experiment

CEPEX Controlled Ecosystem Pollution Experiment

CEPH Centre d'Etude du Polymorphisme Humain [Center for Human Polymorphism Research, France]

CEPhAG Centre d'Etudes des Phénomès Aléatoires et Géophysiques [Research Center for Hazardous Geophysical Phenomena, France]

CEPLA Commission on Environment Policy, Law and Administration [of International Union for the Conservation of Nature (and Natural Resources)]

CEPO Central European Pipeline Office [of NATO]

CEPOD Catalyzed Electrochemical Plutonium Oxide Dissolution

CEPP Center for Emergency Preparedness and Provisioning [Los Angeles, US]

CEPP Chemical Emergency Preparedness Program [US]

CEPP Créations, Editions et Productions Publicitaires [France]

CEPPC Central European Pipeline Policy Committee [of NATO]

CEPR Centre for Economic Policy Research [of Australian National University, Canberra]

CEPR Center for Energy Policy and Research [of New York Institute of Technology, US]

CEPRC Chemical Emergency Planning and Response Commission [US]

CEPS Central Europe Pipeline System [of NATO]

CEPS Command Module Electrical Power System

CEPS Cornish Engines Preservation Society [UK]

CEPT Conférence Européene des Administrations des Postes et des Télécommunications [Conference of European Postal and Telecommunications Administrations, Switzerland]

CEPT Conference of European Postal and Telecommunications Administrations [Switzerland]

CEPT Council of European Post and Telegraph

CEQ Committee on Environmental Quality [of Federal Council for Science and Technology, US]

CEQ Corporation of Engineers of Quebec [Canada]

CEQ Council on Environmental Quality [US]

CER Cation Exchange Resin

CER Center for Estuarine Research [Canada]

CER Ceramic Electrolytic Reactor

CER Cerotate (Crystal)

CER Chlorination with Energy Recovery (Process)

CER Civil Engineering Report

CER Classical Electron Radius

CER Closer Economic Relations

CER Combat Engineer Regiment

CER Complete Engineering Release

CER Compression-Expansion-Recompression

CER Constant Extension Rate

CER Consultative Environmental Review

CER Contactless Electroreflectance

CER Coordinated Experimental Research

CER Cost Estimating Relationship

CER Critical Experiment Reactor

CER Cumulative Excess Returns

Cer Ceramic(s) [also cer]

CERA Canadian Electronic Representatives Association

CERA Civil Engineering Research Association [now Construction Industry Research and Information Association, UK]

CERA Concurrent Engineering Research Association (Journal)

CERAB Ceramic Abstracts On-Line [of American Ceramic Society, US]

CERAD Consortium to Establish a Registry for Alzheimer's Disease [US]

Ceram Ceramic(s) [also ceram]

Ceram Bull Ceramics Bulletin [US]

Ceram Eng Sci Proc Ceramic Engineering and Science Proceedings [of American Ceramic Society, US]

Ceramet Ceramic/Metal (Combination) [also ceramet]

Ceram Forum Int Ceramic Forum International [Ceramic Forum International–Reports of the German Ceramic Society] [also CFI, or cfi]

Ceram Ind Int Ceramic Industries International [Journal published in the UK]

Ceram Ind J Ceramic Industries Journal [US]

Ceram Int Ceramics International [Published in the US]

Ceram Int News Ceramics International News

CERAMITEC Internationale Fachmesse Maschinen, Geräte, Anlagen und Rohstoffe für die gesamte keramische Industrie [International Trade Fair–Machinery, Equipment and Raw Materials for the Ceramic Industry, Munich, Germany]

Ceram Jpn Ceramics Japan [Journal published in Japan]

Ceram Sci Eng Proc Ceramic Science and Engineering Proceedings

Ceram Technol Newsl Ceramic Technology Newsletter [US]

Ceram Trans Ceramic Transactions [of American Ceramic Society, US]

Ceram Trans Ceram Powder Sci Ceramic Transactions, Ceramic Powder Science [of American Ceramic Society, US]

CERAP Center Radar Approach Control

Cerberus Electron Cerberus Electronics [Publication of Cerberus Ltd., Mannedorf, Switzerland]

CERC Civil Engineering Research Council [UK]

CERC Coastal Engineering Research Center [US Army]

CERC Coastal Engineering Research Council [US]

CERC Columbia Environmental Research Center [of US Geological Survey/Biological Resources Division at Columbia, Missouri]

CERC Composites Education and Research Center [of Georgia Institute of Technology, Atlanta, US]

CERCA Centre de Recherche pour Combustibles Atomiques [Research Center for Nuclear Fuels, France]

Cercet Metal Cercetari Metalurgice [Romanian Metallurgical Sciences Journal]

CERCHAR Centre d'Etudes et de Recherches des Charbonnages [Center for the Study and Research of Coal Mines, France]

CERCLA Comprehensive Environmental Response, Compensation and Liability Act [US]

CERCLA/SARA Comprehensive Environment Response, Compensation and Liability Act/Superfund Amendment and Reauthorization Act [US]

CERCLIS Comprehensive Environmental Responsibility, Compensation and Liability System [US]

CERDC Canadian Ethanol Research and Development Center

CerDIP Ceramic Dual In-Line Package [also Cerdip] [Electronics]

CERE Consortium for Environmental Risk Evaluation [US]

CerE Ceramic Engineer [also Cer Eng, or CerEng]

Cereal Chem Cereal Chemistry [Journal of the American Association of Cereal Chemists, US]

Cerebrovasc Disease Cerebrovascular Diseases [International Journal]

Cer Eng Ceramic Engineer(ing)

CERES California Environmental Resources Evaluation System [US]

CERES Clouds and Earth's Radiant Energy System (Project) [of NASA Langley Research Center, Hampton, Virginia, US]

CERES Coalition for Environmentally Responsible Economies [US]

CERES Controlled Environment Research Laboratory [of Commonwealth Scientific and Industrial Research Organization, Australia]

CERESIS Centro Regional de Sismología para America del Sud [Regional Center for Seismology for South America, Peru]

CERF Civil Engineering Research Foundation [US]

CERF Coastal Education and Research Foundation [Charlottesville, Virginia, US]

CERF Critical Experiment Reactor Facility

CERG Chemical Ecology Research Group

CERG Concrete-Encased Ring Ground

CERI Canadian Energy Research Institute

CERI Center for Educational Research Innovation [of Organization for Economic Cooperation and Development, France]

CERI Central Electrochemical Research Institute [Karaikudi, India]

CERI Clean Energy Research Institute [US]

CERL Cambridge Electronics Research Laboratory

CERL Canadian Explosives Research Laboratory [of CANMET/Mining Research Laboratories, Canada]

CERL Central Electricity Research Laboratories [US]

CERL Ceramic Engineering Research Laboratory [Japan]

CERM Centre d'Etudes et Recherches sur les Matériaux [Materials Research and Study Center; of Centre d'Etudes Nucléaire de Grenoble, France]

Cermet Ceramic/Metal (Combination); Ceramic-to-Metal [also cermet]

CERN Centre Européen pour la Recherche Nucléaire [European Laboratory for Particle Physics, Geneva, Switzerland]

CERN Conseil Européen pour la Recherche Nucléaire [European Organization for Nu-clear Research]

CERN Cour CERN Courier [of CERN–European Laboratory for Particle Physics, Geneva, Switzerland]

CERN/LEAR CERN Low-Energy Anti-Photon Ring [of CERN–European Laboratory for Particle Physics, Geneva, Switzerland]

CERNOX Ceramic Nitric Oxide (Process)

CERN SPS CERN Super Proton Synchrotron [of CERN–European Laboratory for Particle Physics, Geneva, Switzerland] [also CERN-SPS]

CERN-TH CERN Theroetical Physics Division [of CERN–European Laboratory for Particle Physics, Geneva, Switzerland]

CERN-TH News CERN Theroetical Physics Division News [Newsletter of CERN–European Laboratory for Particle Physics, Geneva, Switzerland]

CERO Centre d'Etudes de Recherche Operationnelle [Operations Research Center, Brussels, Belgium]

CERP Confédération Européenne des Relation Publiques [European Confederation of Public Relations, Brussels, Belgium]

Cer Proc Ceramic Processing [also cer proc]

CERQUAD Ceramic Quad Flat Pack [also Cerquad] [Electronics]

CERR Center for Earth Resources Research [of Memorial University of Newfoundland, St. Johns, Canada]

Cer Res Ceramic Research(er)

CERS Certified Energy Reduction Specialist

CERS Commission pour l'Encouragement de la Recherche Scientifique [Commission for the Encouragement of Scientific Research, Berne, Switzerland]

CERS Cornell Electron Storage Ring [of Cornell University, Ithaca, New York, US]

CERT Centre d'Etudes et de Recherches de Toulouse [Research and Study Center at Toulouse; of Office National d'Etudes et de Recherches Aérospatiales, France]

CERT Character Error Rate Test

CERT Combined Environmental Reliability Test(ing)

CERT Computer Emergency Response Team [Internet]

CERT Constant Extension Rate Tensile (Test)

CERT Constant Extension Rate Test(er); Constant Extension Rate Testing

Cert Certificate; Certification [also cert]

cert certified; certify

CERTAN Computer Equipment Resources and Technology Acquisition for NIH (National Institutes of Health) [US]

CertEd Certificate of Education [UK]

Cert Eng Certified Engineer

Cert Eng Tech Mag Certified Engineering Technician Magazine [of American Society of Certified Engineering Technicians]

CERTICO Certification Committee [of American National Standards Institute, US]

Certif Certificate; Certification [also cert]

certif certified; certify

Certif Eng Certificated Engineer [Publication of the Institution of Certificated Mechanical and Electrical Engineers, South Africa]

CertPA Certificate in Public Administration

Cert Pros Certified Prosthodontist

CeRu$_2$Se$_2$ Cerium Ruthenium Selenide

CERVED Centri Elettronica Reteconnessi Valutazione Elaborazione Dati [Electronic Value-Added Data Network Center, Italy]

CES Capillary Electrophoresis System

CES Capstone Engineering Society [US]

CES Center for Economic Studies [of United States Census Bureau]

CES Center for Energy Studies [of University of Texas at Austin, US]

CES Center for Environmental Studies [of Arizona State University, Tempe, US]

CES Center for Environmental Studies [of Smithsonian Institution on Chesapeake Bay near Annapolis, Maryland, US]

CES Center for Epidemiological Studies [US]

CES Centre d'Etudes de Saclay [Saclay Research Center, Gif-sur-Yvette, France]

CES Centre Européen des Silicones [European Silicone Center; of Conseil Européen des Fédérations de l'Industrie Chimique, Belgium]

CES Civil Engineering Section

CES Cleveland Engineering Society [Ohio, US]

CES Coast Earth Station (Satellite)

CES Committee on Earth Sciences

CES Committee on Economic Studies [of International Iron and Steel Institute]

CES Commonwealth Employment Service

CES Community Energy Systems (Program) [of Natural Resources Canada]

CES Conference of European Statisticians

CES Conjugated Estrogens [Biochemistry]

CES Constant Elasticity of Substitution

CES Consumer Electronics Society [of Institute of Electrical and Electronics Engineers, US]

CES Control Electronics Section

CES Control Electronics System

CES Coordinate Evaluation System

CES Crew Escape System

CES Critical Experiment Station [of Westinghouse Reactor Evaluation Center, US]

CES Cyanoethyl Sucrose [Biochemistry]

CESA Canadian Engineering Standards Association

CESA Committee of EEC (European Economic Community) Shipbuilders Associations [Belgium]

CESA Concurrent Education Students Association [Canada]

CESAC Communications-Electronics Scheme Accounting and Control [US Air Force]

CE Saclay Centre d'Etudes de Saclay [Saclay Research Center, Gif-sur-Yvette, France]

CESAR Canadian Expedition to Study the Alpha Ridge

CESAR CERN Electron Storage and Accumulation Ring [of CERN–European Laboratory for Particle Physics, Geneva, Switzerland]

CESAR Combustion Engineering Safety Analysis Report

CESARS Chemical Evaluation, Search and Retrieval System [US]

CESBRA Companhia Estanifera do Brasil [Stanniferous Products Company of Brazil]

CESC Consulting Engineers of South Carolina [US]

CESCE Comité Européen des Services des Conseillers [European Committee of Consultant Services, Belgium]

CESD Composite External Symbol Directory

CES-D Center for Epidemiological Studies–Depression Scale [US]

CESE Comparative Education Society in Europe [Belgium]

CESEMI Computer Evaluation of Scanning Electron Microscope Images

CESF College for Environmental Science and Forestry [of State University of New York, US]

CESFS Constant Energy Synchronomy Fluorescence Spectrometry

CESI Centro Elettrotecnico Sperimentale Italiano [Italian Center for Experimental Electrical Engineering]

CESI Council for Elementary Science International [US]

CeSi$_2$ Cesium Disilicide

Cesk Cas Fyz A Ceskoslovensky Casopis pro Fyziku, Sekce A [Physics Journal, Part A; published by the Czechoslovak Academy of Sciences]

Cesk Cas Fyz B Ceskoslovensky Casopis pro Fyziku, Sekce B [Physics Journal, Part B; published by the Czechoslovak Academy of Sciences]

Cesk Farm Ceskoslovensky Farmacie [Pharmaceutical Journal; published by the Czechoslovak Academy of Sciences]

Cesk Inf Teor Praxe Ceskoslovenska Informatika, Teorie a Praxe [Czechoslovak Journal on Computer Science, Theory and Practice]

CESNEF Centro di Studi Nucleari Enrico Fermi [Enrico Fermi Center for Nuclear Studies, Italy]

CESO Canadian Executive Service Organization

CESO Canadian Executive Service Overseas

CESP Commission on Environmental Strategy and Planning

CESPA Canadian Engineering Student Publications Association

CESQG Conditionally Exempt Small Quantity Generator

CESQT Conditionally Exempt Small Quantity Treatment

CESR Conduction Electron Spin Resonance

CESSE Council of Engineering and Scientific Society Executives [US]

CESSIM Centre d'Etudes des Supports d'Information Medicale [Study Center for the Support of Medical Information, France]

CES-SPEC Centre d'Etudes de Saclay–Service de Physique de l'Etat Condensé [Saclay Research Center–Division for Condensed State Physics, France]

CESSS Council of Engineering and Scientific Society Secretaries

CEST Center for Exploitable Science and Technology

CEST Computer Evaluation of Surface Topography

CESW Conditionally Exempt Specified Wastestreams

CeSZ Cerium-Substituted Zirconia

2.5CeSZ 2.5 mol% Ceric Oxide-Substituted Zirconia

5CeSZ 5 mol% Ceric Oxide-Substituted Zirconia

CET Cement Evaluation Tool

CET Center for Educational Technology [US]

CET Center for Engineering Tribology [of Northwestern University, Evanston, US]

CET Central European Time [Greenwich Mean Time –01:00]

CET Centre d'Etudes de Télécommunications [Center for Telecommunication Studies, France]

CET Certified Electronics Technician

CET Certified Engineering Technician

CET Certified Engineering Technologist

CET Certified Environmental Trainer

CET Chemical Engineering Technologist; Chemical Engineering Technology

CET 2-Chloro-4,6-Bis(ethylamino)-1,3,5-Triazine

CET Civil Engineering Technician

CET Civil Engineering Technologist; Civil Engineering Technology

CET Collège d'Enseignement Technique [College of Technical Education, France]

CET College of Engineering Technology

CET Columnar-(to-)Equiaxed Transition [Metallurgy]

CET Commission on Education and Training [of International Union for the Conservation of Nature and Natural Resources]

CET Common External Tariff

CET Corrected Effective Temperature

CET Correlated Electron Transfer

CET Council for Educational Technology [London, UK]

CET Counter-Field Electron Transfer

CET Critical Experiment Tank [Nuclear Engineering]

CET Cross-Linkable Epoxy Thermoplastic (Technology)

CET Cumulative Elapsed Time

Cet Cetus [Astronomy]

CETA Community Economic Transformation Agreements [Canada]

CETA Comprehensive Employment and Training Act [US]

CETATS Cetyltrimethylammonium Tosylate

CETC CANMET (Canada Center for Mineral and Energy Technology) Energy Technology Center [of Natural Resources Canada]

CETD Calculated Estimated Time of Departure

CETDC China External Trade Development Council [Taiwan]

CETECH Environmental Technology for Canadian Industry Trade Show

CETEDOC Centre de Traitement Electronique de Documents [Center for Electronic Document Processing, Belgium]

Ce(TFAC)$_3$ Cerium(III) Trifluoroacetylacetonate [also Ce(tfac)$_3$]

CETI Communication with Extraterrestrial Intelligences

CETIM Centre Technique des Industries Mécaniques [Technical Center of the Machinery Industry; of Centre de Documentation de la Mécanique, France]

CETIM Inf CETIM Informations [of Centre Technique des Industries Mécaniques, France]

CETIS Centre Européen de Traitement de l'Information Scientifique [European Processing Center for Scientific Information, France]

CETISA Compania Espanola de Editoriales Tecnologicas Internacionales S.A. [Spanish Company for International Technological Editorials, Barcelona, Spain]

CETME Centro de Estudios Técnicos de Materiales Especiales [Technical Center for the Study of Special Materials, Spain]

Ce(TMHD)$_4$ Cerium Tetra(2,2,6,6-Tetramethyl-3,5-Heptanedionate) [also Ce(tmhd)$_4$]

CETO Calculated Estimated Time of Overflight

CETP Canada Employment Training Program

CETP Centre d'Etudes des Environnements Terrestre et Planétaire [Research Center for Terrestrial and Planetary Environments, France]

CETR Center of Explosives Technology Research [of New Mexico Institute of Mining and Metallurgy, Socorro, US]

CETR Consolidated Edison Thorium Reactor [US]

CETRA China External Trade Research Association [Taiwan]

CETS Conférence Européenne pour Télécommunication par Satellites [European Conference for Telecommunication via Satellites, France]

CETS Conference on European Telecommunications Satellites

CETS Control Element Test Stand

CETSD CANMET (Canada Center for Mineral and Energy Technology)/Engineering and Technical Services Division [of Natural Resources Canada]

CETUK Council for Educational Technology for the United Kingdom [London, UK]

Ce-TZP Ceria-Stabilized Tetragonal Zirconia Polycrystal [also CeTZP]

Ce-TZP/Al₂O₃ Ceria-Stabilized Tetragonal Zirconia Polycrystal Alumina (Composite)

CEU Carbonate-Evaporite Universal

CEU Central Executive Committee

CEU Channel Extension Unit

CEU Commission of the European Union

CEU Communications Expansion Unit

CEU Continuing Education Units [US]

CEU Construction Engineering Unit

CEU Council of the European Union

CEU Coupler Electronics Unit

CEU Customs Excise Union

CEV Citrus Exocortis Viroid

CEV Combat Engineer Vehicle

CEV Corona Extinction Voltage

CEVAR Consumable-Electrode Vacuum-Arc Remelting [Metallurgy]

CE Valrho Centre d'Etudes Valrho [Valrho Research Center, Bagnols-sur-Cèze, France]

CEVE Chloroethyl Vinyl Ether

CEVM Consumable Electrode Vacuum Melting [Metallurgy]

CEW Chemical Engineering World [Journal published in the US]

CEW Coextrusion Welding

CEW Chem Eng World CEW Chemical Engineering World [Journal published in the US]

CEWS Cargo Element Work Stand

CEX Canadian Environmental Exposition

CEX Central Excitation

CEX Controlled Experiment

CEX Crystal Extraction (Experiment) [of Fermilab, Batavia, Illinois, US]

CEY Consulting Engineers of the Yukon (Territory) [Canada]

Ce:YAG Cerium-Activated Yttrium-Aluminum Garnet (Crystal) [also Ce³⁺:YAG]

Ce:YIG Cerium-Activated Yttrium-Iron Garnet (Crystal) [also Ce³⁺:YIG]

Ce:ZBLAN Cerium-Doped Zirconium Tetrafluoride–Barium Difluoride–Lanthanum Trifluoride–Aluminum Trifluoride–Sodium Fluoride (Laser) [also Ce³⁺:ZBLAN]

CEZUS (Société) Compagnie Européenne du Zirconium [European Zirconium Company, Paris, France]

CF Calcium Ferrite

CF Canadian Forces

CF Canadian Formulary [Grade of Chemicals]

CF Canadian-French [ISO Code]

CF Canonic(al) Form

CF Can't Find

CF Capacity Factor [Electrical Power]

CF Carbofuchsin

CF Carbon Fiber

CF 5-/6-Carboxyfluorescein

CF Card Format

CF Carried Forward [also cf]

CF Carrier Frequency [also cf]

CF Carry-over Factor [Structure Analysis]

CF Cathode Follower

CF Cathodic Film

CF Cationized Ferritin

CF Center Fire

CF Center Forward

CF Center Frequency [also C/F]

CF Center of Figure

CF Center of Flotation [Oil Drilling]

CF Central African Republic [ISO Code]

CF Central Facility

CF Central Field (Approximation) [Physics]

CF Central File

CF Central Force

CF Centrifugal Fan

CF Centrifugal Force

CF Centripetal Force

CF Ceramic Fiber

CF Cerium-Free

CF Character Figure

CF Characteristic Frequency

CF Characteristic Function (in Statistics) [also cf]

CF Characterization Factor [Chemical Engineering]

CF Charcoal Filter

CF Chebyshev Filter [Electronics]

CF Chemical Family

CF Chemical Formula

CF Chromatic Free (Optical System)

CF Chromium Ferrite

CF Chronic Fatigue [Medicine]

CF Circuit Finder

CF Circulating Flow

CF Citrovorum Factor

CF Clear Flag (Function)

CF Coated Front [Carbonless Paper]

CF Cobalt Ferrite

CF Cohesive Force

CF Cold Finger [or cf]

CF Cold Finish(ed); Cold Finishing [Metallurgy]

CF Cold Front [Meteorology]
CF Cold Fusion
CF Collège de France [of University of Paris, France]
CF Collision Frequency [Physics]
CF Colloidal Fuel
CF Color Filter
CF Column Flotation
CF Column Fractionation
CF Combined Function
CF Combustion Flame
CF Commonwealth Foundation [UK]
CF Communauté Française (d'Afrique) [French Community (of Africa)]
CF Communications Facility
CF Complement-Fixation; Complement Fixing [Immunology]
CF Complex Fault [Metallurgy]
CF Composite Fermion [Physics]
CF Compound Fracture [Medicine]
CF Compressible Flow
CF Concept Formulation
CF Condenser Filter
CF Conductance Fluctuation(s)
CF Conducting Furnace (Black)
CF Conduction Fluctuation
CF Confinement Factor
CF ConFlat Flange
CF Conservation Foundation [US]
CF Constant Fraction
CF Construction Forces
CF Consumption-Fuel [Automobiles]
CF Contemporary Force
CF Context Free
CF Continued Fraction
CF Continuous Fiber
CF Continuous Flow
CF Continuous Form
CF Continuous Function
CF Continuous Furnace [Metallurgy]
CF Control Footing
CF Controlled Fusion [Physics]
CF Conversion Factor
CF Copper Ferrate
CF Copy Furnished
CF Coriolis Force
CF Corner Fracture
CF Corn Flour
CF Corrected Field
CF Correlation Factor
CF Correlation Function
CF Corrosion Fatigue
CF Cosanti Foundation [US]
CF Cost and Freight [also C&F, c&f, or cf]
CF Couette Flow

CF Coulomb Force
CF Council of Fellows
CF Counterfire
CF Counterflow
CF Count Forward
CF Crack Front [Mechanics]
CF Cracking Furnace
CF Creeping Flow
CF Cresolformaldehyde
CF Crest Factor
CF Critical Field [Electronics]
CF Critical Flow
CF Crude Fiber
CF Cryogenic Focusing
CF Crystal Field (Parameter) [Physics]
CF Crystalline Film
CF Cubic Foot [also cf, cu ft, or ft^3]
CF Cumulative Frequency
CF Curvature of Field
CF Cycles to Failure
CF Cyclone Furnace
CF Cylinder Function
CF Cystic Fibrosis
CF Field Forcing (Increasing on Variable Voltage) [Controllers]
CF- Canada [Civil Aircraft Marking]
CF_2 Carbon Difluoride
CF_3 Carbon Trifluoride
CF_4 Carbon Tetrafluoride
$(CF)_x$ Carbon Fluoride [General Formula]
$(CF_x)_n$ Substoichiometric Graphite Fluoride (or Fluorinated Coke) [General Formula; x = 0.3 to 1.1]
C_2F Dicalcium Ferrite [$2CaO \cdot Fe_2O_3$]
C_4F Tetracalcium Ferrite [$4CaO \cdot Fe_2O_3$]
C&F Cost and Freight [also c&f, CF and cf]
C/F Carried Forward [also c/f]
C/F Center Frequency [also CF]
C-F Carbon-Fluorine (Bond)
C-F Corrugated-Flat (Fiberboard)
Cf Californium [Symbol]
Cf Humid Warm Climate, No Dry Season [Classification Symbol]
Cf-244 Californium-244 [also ^{244}Cf, or Cf^{244}]
Cf-246 Californium-246 [also ^{246}Cf, or Cf^{246}]
Cf-249 Californium-249 [also ^{249}Cf, or Cf^{249}]
Cf-252 Californium-249 [also ^{252}Cf, or Cf^{252}]
C-f Capacitance-Frequency [also C-f]
C(f) C-Glass Fiber [Symbol]
cF Pearson symbol for face-centered space lattice in cubic crystal system (this symbol is followed by the number of atoms per unit cell, e.g. cF4, cF24, etc.)
cf (confer) – compare
cf cost and freight [also CF, C&F, or c&f]
cf cubic feet [also cu ft, or ft^3]

cf crest factor (for periodically alternating quantities) [Symbol]

C4 Controlled Collapse (Micro) Chip Connection

CFA California Fertilizer Association [US]

CFA Canadian Federation of Agriculture

CFA Canadian Federation of Aromatherapists

CFA Canadian Field Artillery

CFA Canadian Forces Attache

CFA Canadian Forestry Association

CFA Canadian Foundry Association

CFA Caribbean Federation of Aeroclubs [Netherlands Antilles]

CFA Carrier Frequency Alarm

CFA Cascade-Failure Analysis

CFA Cat Fanciers Association, Inc. [US]

CFA Centre de Formation d'Apprentis [Center for Training Apprentices, France]

CFA Chinese Foundrymen's Association [Taiwan]

CFA Cloud Fractional Absorptance

CFA Code of Federal Regulations

CFA College of Fine Arts [of Carnegie-Mellon University, Pittsburgh, Pennsylvania, US]

CFA Committee of Food Aid (Policies and Programs)

CFA Commonwealth Forestry Association [UK]

CFA Communauté Financière Africaine [African Financial Community]

CFA Communauté Française d'Afrique [French Community of Africa] [dissolved]

CFA Complete Freund's Adjuvant [Immunology]

CFA Component Failure Analysis

CFA Computer Family Architecture

CFA Consumer Federation of America [US]

CFA Continuous Flow Analysis

CFA Contract Flooring Association

CFA Council of Iron Founders Associations

CFA Council on Fertilizer Application [US]

CFA Country Fire Authority

CFA Core Flood Alarm

CFA Cross(ed)-Field Amplifier

CFA Cross Flow Analyzer [Aerospace]

CFAA Canadian Fire Alarm Association

CFAC Canadian Forces Air Defense Command [also CFADC]

CFAC Commercial Fishing Advisory Council

CFAD(s) Contoured Frequency by Altitude Diagram(s)

CFAE Contractor-Furnished Aircraft Equipment

CFAES Carbon Furnace Atomic Emission Spectrometry

CFA Fr Communauté Financière Africaine Franc [Currency of Benin, Cameroon, Central African Republic, Chad, Comoros, Congo, Gabon, Ivory Coast, Niger, Reunion, Senegal, Togo and Upper Volta/Burkina Faso]

CF/Al Carbon-Fiber Aluminum Composite

CFANS Canadian Forces Air Navigation School

cf ante *(confer ante)* – compare above

CFAO Canadian Forces Administrative Order

CFAP Copenhagen Frequency Allocation Plan

CFAR Constant False-Alarm Rate

CFARS Centers for AIDS (Acquired Immunodeficiency Syndrome) Research [US]

CFAT Canadian Federation of Aroma Therapists

CFAT Carnegie Foundation for the Advancement of Teaching [US]

CFAV Canadian Forces Arctic Vessel

CFAW Canadian Food and Allied Workers

CFB Canadian Forces Base

CFB Charcoal Filter Breather

CFB Chemische Fabrik Budenheim (Process)

CFB Cipher Feedback

CFB Circulating Fluid(ized Bed) (Combustion)

CFB Coated Front and Back [Carbonless Paper]

CFB Commonwealth Forestry Bureau [Australia]

CFB Computing Facilities Branch [of Division of Computer Research and Technology, US]

CFB Current Feedback

CFBA Canadian Food Brokers Association

CFBC Circulating Fluidized-Bed Combustion

CFBL Commonwealth Fishing Boat Licence

CF Black Conducting Furnace Black [also CF black]

CFBR Commercial Fast Breeder Reactor

CFBS Canadian Federation of Biological Sciences

CFBS Canadian Federation of Biological Societies

CFC Canadian Forces College

CFC Capillary Filtration Coefficient

CFC Carbon-Fiber Coating

CFC Carbon-Fiber Composite

CFC Carbon-Fiber-Reinforced Carbon

CFC Carbon Fuel Company [Charleston, West Virginia, US]

CFC Caribbean Food Corporation

CFC Central(ized) Fire Control

CFC Channel Flow Control

CFC Chlorinated Fluorocarbon

CFC Chlorofluorocarbon

CFC Coin and Fee Checking (Relay Set)

CFC Colony Forming Cells [Microbiology]

CFC Consolidated Freight Classification

CFC Continuous Flow Calorimetry

CFC Controlled Furnace Cooled; Controlled Furnace Cooling [Metallurgy]

CFC Crossed-Film Cryotron

CFC-11 Trichlorofluoromethane (or Freon-11)

CFC-12 Dichlorodifluoromethane (or Freon-12)

CFC-21 Dichloromonofluoromethane (or Freon-21)

CFC-112 1,2-Difluoro-1,1,2,2-Tetrachloroethane (or Freon-112)

CFC-114 Dichlorotetrafluorethane (or Freon-114)

CFCA Classical Franck-Condon Approximation [Physical Chemistry]

CFCC Canadian Forces Communications Command

CFCC Continuous Fiber(-Reinforced) Ceramic Composite

CFCE Centre Française du Commerce Extérieur [French Center of Foreign Trade]

CFCF Central Flow Control Facility

CFCG Corrosion Fatigue Crack Growth

CF_4/CHF_3 Carbon Tetrafluoride/Trifluoromethane (Gas Mixture)

CFCM Continuous Flow Cell Culture [Microbiology]

CFCMC Ceramic-Fiber Ceramic Matrix Composite

CFCO trans-1,3-Pentadiene Iron Tricarbonyl

CFCS Caribbean Food Crops Society [Puerto Rico]

CF-CVC Combustion Flame–Chemical Vapor Condensation

CFD Call For Discussion

CFD Compact Floppy Disk

CFD Computational Fluid Dynamics

CFD Constant Fraction Discriminator

CFD Continuous Flow Diffusion

CFD Converter, Frequency to DC Voltage

CFD Counterflow Diode

CFD Cubic Feet per Day [also cfd]

CFD Cusped Field Device

CFDA Carboxyfluorescein Diacetate

CFDA Catalog of Federal Domestic Assistance [US]

CFDA Current File Disk Address

5-CFDA 5-Carboxyfluorescein Diacetate

6-CFDA 6-Carboxyfluorescein Diacetate

CFDC Canadian Film Development Corporation

CFDD Compact Floppy Disk Drive

CFDE Call Failure Detection Equipment

CFDM Companded Frequency-Division-Multiplex

CFDM/FM Companded Frequency-Division-Multiplex/Frequency Modulation

CFDP Caribbean Fisheries Development Project [US]

CFDRC Computational Fluid Dynamics Research Corporation [Huntsville, Alabama, US]

CFE Canadian Forces Europe

CFE Cauchy Functional Equation

CFE Central Fighter Establishment [UK]

CFE Certified Food Executive

CFE Chief Financial Executive

CFE Chlorotrifluoroethylene

CFE Cold Field Emission

CFE Comisión Federal de Electricidad [Federal Electricity Commission, Mexico]

CFE Continuous Flow Electrophoresis

CFE Continuous Fuel Economizer

CFE Contractor-Furnished Equipment

CFE Conventional (Armed) Forces (in) Europe (Treaty)

CFE Polychlorotrifluoroethylene (Resin)

CFED Corporate Facilities Engineering Department [of Dow Corning Corporation, US]

CFEE Canadian Foundation for Economic Education

CFELS Center for Free-Electron Laser Strudies [of University of California at Santa Barbara, US]

CFEM Conference on Fossil Energy Materials [US]

CFEMC Canadian Forces Europe Medical Center

CFEN Certificat de Fin d'Etudes Normales [French Certificate of Completion of Normal School Studies]

CFER Collector Field Effect Register

C-FER Center for Frontier Engineering Research [Edmonton, Alberta, Canada] [also CFER]

CFES Canadian Federation of Engineering Students

CFES Center for Energy Systems [US]

CFES Continuous Flow Electrophoresis System

CFESA Commercial Food Equipment Service Association [US]

CFES-CCPE Canadian Federation of Engineering Students/Canadian Council of Professional Engineers (Agreement)

CFE-SEM Cold Field Emission (High-Resolution) Scanning Electron Microscope; Cold Field Emission (High-Resolution) Scanning Electron Microscopy [also CFE SEM]

CFEST Coupled Fluid Energy and Solute Transport

CFF Cat Fanciers Federation [US]

CFF (Commonwealth) Commission For the Future

CFF Conventional Fast Firing [Powder Metallurgy]

CFF Corrected Fuel Flow

CFF Critical Flicker Frequency [Optics]

CFF Critical Fusion Frequency

CFF Current Fault File

CFF Cystic Fibrosis Foundation [US]

CFFA Chemical Fabrics and Film Association [US]

CF-FAB Continuous Flow Fast Atom Bombardment

CFFC Combined Forced and Free Convection

CFFED Canadaian Federation of Farm Equipment Dealers

C-FFOX Carbon-Ferrailles Fusion Oxydantes [Sollac Process]

CFFR Cushman Foundation for Forominiferal Research [US]

CFFTP Canadian Fusion Fuels Technology Project

CFG Compact-Flake-Graphite (Iron) [Metallurgy]

CFG Context-Free Grammar

.CFG Configuration [File Name Extension] [also .cfg]

CFG Iron Compact-Flake-Graphite Iron [also CFG iron]

C_xF GIC Carbon Fluoride Graphite Intercalation Compound [x = 4, etc.]

CFH Cubic Feet per Hour [also cfh, or cu ft/hr]

CFHC Chlorofluorohydrocarbon

CFHQ Canadian Forces Headquarters

CFHT Canada-France-Hawaii Telescope (Corporation) [Kamuela, Hawaii, US]

CFI California Fig Institute [US]

CFI Call for Instruction

CFI Campaign for Industry [UK]

CFI Canada Foundation for Innovation

CFI Canadian Fertilizer Institute

CFI Card Format Identifier

CFI Ceramic Forum International–Berichte der Deutschen Keramischen Gesellschaft [Ceramic Forum International–Reports of the German Ceramic Society] [also cfi]

CFI Chief Flying Instructor

CFI Clothing and Footware Institute

CFI Committee on Foreign Investment

CFI Commonwealth Forestry Institute [UK]

CFI Cost, Freight and Insurance [also cfi]

CFI Cremer Forschungsinstitut [Cremer Research Institute, Roedental, Germany]

CFIA Component Failure Impact Analysis

CFIA Core Flood Isolation Valve Assembly [Nuclear Reactors]

CFIAM Canadian Forces Institute of Aviation Medicine

CFIB Canadian Federation of Independent Business

CFIBC Council of Forest Industries of British Columbia [Canada]

CFIC Canadian Forestry Industries Council

CFIC Commercial Fishing Industries Council [Canada]

CFIDS Chronic Fatigue Immune Dysfunction Syndrome

CFIEI Canadian Farm and Industrial Equipment Institute

CFIEM Canadian Forces Institute of Environmental Medicine

CFIF Continuous Flow Isoelectric Focusing

CFIS California Fiscal Information System [US]

CFIUS Committee on Foreign Investment in the United States [of US Department of Treasury]

CFIX Communications Futures Index [Database on New Communications Technologies]

CFL Calibrated Focal Length

CFL Call Failure Signal

CFL Canadian Federation of Labour

CFL Carbon-Filament Lamp

CFL Clear Flight Level

CFL Context-Free Language

CF&L (Department of) Conservation, Forests and Land [of State of Victoria, Australia]

CFLA Canadian Forces Leadership Academy [Canada]

CFLS Canadian Forces Language School

CFM Cathode Follower Mixer

CFM Centre du Fauga-Mauzac [Fauga-Mauzac Center, ONERA-CERT (Office National d'Etudes et de Recherches Aérospatiales–Centre d'Etudes et de Recherches de Toulouse), France]

CFM Cerium-Free Mischmetall

CFM Center Frequency Modulation

CFM Centro de Fisica Molecular [Molecular Physics Center; of Universidade Tecnico de Lisboa, Lisbon, Portugal]

CFM Chemical Force Microscope; Chemical Force Microscopy

CFM Chlorofluoromethane

CFM Companded Frequency Modulation; Companding and Frequency Modulation

CFM Confirm [Computers]

CFM Continuous Film Memory

CFM Continuous Flow Manufacturing

CFM Continuous Focus Microscope

CFM Cubic Feet per Minute [also cfm, cf/min, ft³/min, or cu ft/min]

CFM Customer-Furnished Material

CFMA Chair Frame Manufacturers Association [UK]

CFMA Construction Financial Management Association [US]

CFMA Cyprus Foodwear Manufacturers Association

CFMC Caribbean Fishery Management Council

CFMEU Construction, Forestry, Mining and Energy Union

CFMF Crip Flow Management Facility

CFMR Center for Fundamental Materials Research [of Michigan State University, East Lansing, US]

CFMS Chained File Management System

CFMS Computer(-Based) Financial Management System

CFMT Concentrator and Feed Makeup Tank [Nuclear Engineering]

CFMU Centralized Flow Management Unit

CFN Canadian Forces Network

CFN Centro de Fisica Nuclear [Nuclear Physics Center; of University of Lisbon, Portugal]

CFNBS Committee on the Formation of the National Biological Survey [US]

CFNEC Cairns and Far North Environment Council [Australia]

CFNI Caribbean Food Nutrition Institute

CFO Calcium Iron Oxide

CFO Carrier Frequency Oscillator

CFO Central Forecast Office

CFO Chief Financial Officer

CFO Cobalt Iron Oxide

CFO Critical Flashover

CF$_4$/O$_2$ Carbon Tetrafluoride (or Tetrafluoromethane)/Oxygen (Gas Mixture)

Cfo Clostridium formicoaceticum [Microbiology]

CFOCF (Commonwealth) Commission For Our Common Future

CF$_3$OF Trifluoromethyl Hypofluorite

CFOMR Center for Fiber-Optic Materials Research [of Rutgers University, Piscataway, New Jersey, US]

CFOP Club des Fibres Optiques et Plastiques [Plastic and Optic Fibers Club, France]

CFP Call for Papers

CFP Canadian Forces Publication

CFP Canadian Forest Products

CFP Common Functioning Principle

CFP Communauté Financière Pacifique [Pacific Financial Community]

CFP Communauté Française du Pacifique [French Pacific Community] [dissolved]

CFP Compagnie Française des Pétroles [French Petroleum Company]

CFP Conceptual Flight Profile

CFP Crystal Field Parameter [Physics]

CFPA Canadian Fluid Power Association

CFPA Canadian Food Processors Association

CFPA Chlorophenyltrifluoromethyl Phenoxyacetate

CFPARU Canadian Forces Personnel Applied Research Unit

CFPD Canadian Forces Publication Depot

CFP Fr Communauté Financière Pacifique Franc [Currency of French Polynesia, New Caledonia and Tahiti]

CFPHT Constant Fraction of Pulse Height Trigger

CFPIM Certified Fellow in Production and Inventory Management

CFPMO Canadian Forces Project Management Office

CFPO Canadian Forces Post Office

cf post *(confere post)* – compare after

CFPP Canadian Federation of Pulp and Paper Workers

CFPP Cold Filter Plugging Point

CFPT Canadian Forces Parachute Team

CFQMC Correlation Function Quantum Monte Carlo [Physics]

CFR Carbon Fiber Reinforced; Carbon Fiber Reinforcement

CFR Carbon-Film Resistor

CFR Catastrophic Failure Rate

CFR Center for Field Research

CFR Center for Field Research [Watertown, Massachusetts, US]

CFR Center for Futures Research [US]

CFR Centre de Faibles Radioactivités [Center for Weak Radioactivity, France]

CFR Code of Federal Regulations [of Occupational Safety and Health Administration, US]

CFR Commercial Fast-Breeder Reactor

CFR Commercial Fast Reactor [UK]

CFR Concrete-Faced Rockfill (Dam)

CFR Contact Flight Rules

CFR Continuous Flow Reactor

CFR Cooperative Fuel Research (Engine)

CFR Coordinating Fuel Research

CFR Council on Foreign Relations [New York State, US]

CFR Cumulative Failure Rate

CFR Crash Fire Rescue

CFR Crash, Firefighting and Rescue Services [of Transport Canada]

cfr *(confer)* – compare

CFRA Canadian Forces Reorganization Act

CFRAl Carbon-Fiber-Reinforced Aluminum

CFRC Canadian Forces Recruiting Center

CFRC Carbon-Fiber-Reinforced Carbon

CFRC Carbon-Fiber-Reinforced Cement

CFRC Carbon-Fiber-Reinforced Composite

CFRC Committee on Fiber-Reinforced Concrete [US]

CFRC Continuous Fiber-Reinforced Composite

CFRC Cooperative Fuel Research Council [US]

CFRCC Carbon-Fiber-Reinforced Cement Composite

CFRE Circulating-Fuel Reactor Experiment [of Oak Ridge National Laboratory, Tennessee, US]

CFRI Central Fuel Research Institute [India]

C$_2$F$_6$ RIE Hexafluoroethane Reactive Ion Etching

CFRM (International) Conference on Fusion Reactor Materials

CFR MMC Carbon-Fiber-Reinforced Metal-Matrix Composite

CFRP Carbon-Fiber-Reinforced Polymer

CFRP Carbon-Fiber-Reinforced Plastic(s)

CFRP Comité Français de la Recherche sur la Pollution de l'Eau [French Committee for Research on Water Pollution]

CFRTI Centre Français de Renseignements Techniques Industriels [French Center for Technical and Industrial Information]

CFRTP Carbon-Fiber Reinforced Thermoplastic(s)

CFRUO Committee on the Future Role of Universities in Ontario [Canada]

CFS Cadmium Iron Selenide (Semiconductor)

CFS Calcium Fluosilicate

CFS Call Failure Signal

CFS Call for Service Signal

CFS Canadian Forces Ship

CFS Canadian Forces Station

CFS Canadian Forces Supplementary [of Department of National Defense]

CFS Canadian Forestry Service [of Environment Canada]

CFS Canadian Forest Sector

CFS Carrier Frequency Shift

CFS Center for Future Studies

CFS Center of Forensic Sciences

CFS Central Flying School [UK]

CFS Chronic Fatigue Syndrome

CFS Constant Final-Energy Spectra

CFS Constant Final State

CFS Constant Final-State Spectroscopy

CFS Container Freight Station

CFS Continuous Form Stacker

CFS Council of Fleet Specialists [US]

CFS Crystal Field Stabilization [Physics]

CFS Cubic Feet per Second [also cusec, cfs, ft^3/s, or ft^3/sec]

CFS Custom Font System

CFSA California Flyers School of Aeronautics [US]

CFSA Canadian Fire Safety Association

CFSAN Center for Food Safety and Applied Nutrition [of US Food and Drug Administration]

CFSBA Cold Finished Steel Bar Institute [US]

CFSBE Commission Fédérale Suisse de Bourses pour Etudiants Etrangers [Swiss Federal Commission for Foreign Student Scholorships]

CF-SCAN Canadian Forces–Second Career Assistance Network

CFSE Crystal Field Stabilization Energy [Physics]

CFSEA Canadian Food Service Executives Association

CFSG Cometary Feasibility Study Group [of European Space Research Organization]

CFSK Coherent Frequency Shift Keying

CFSO Crystal Field Surface Orbital [Physics]

CF-SO Crystal Field–Spin Orbit (Split Level) [Physics]

CFSO-BEBO Crystal Field Surface Orbital–Bond Energy Bond Order (Method) [Physics]

CFSPES Constant Final-State Photoelectron Spectroscopy

CFSPL Canadian Forces Special Project Laboratory

CFSRS Canadian Forces Supplementary Radio Systems

CFSS Canadian Forces Supply System

CFSS Constant Final-State Spectroscopy

CFSSB Central Flight Status Selection Board

CFSTI Clearinghouse for Federal Scientific and Technical Information [of US Department of Commerce]

CFSTR Continuous Flow Stirred Tank Reactor [also CSTR]

CFT Cadmium Iron Telluride (Semiconductor)

CFT Calcium Iron Titanate

CFT (–)-2β-Carbomethoxy-3β-(4-Fluorophenyl)tropane

CFT Charge Flow Transistor

CFT Common Facilities Test

CFT Complement-Fixation Test [Immunology]

CFT Concentration Functional Theory

CFT Constant Fraction Trigger

CFT Continuous Flow Test

CFT Controlled Finish(ing) Temperature [Metallurgy]

CFT Core Flood Tank

CFT Cray FORTRAN

CFT Cubic Feet per Ton [also cft, ft^3/t, or cu ft/t]

C$_3$FT Tricalcium Ferrotitanate [$3CaO \cdot Fe_2O_3 \cdot TiO_2$]

C$_4$FT$_2$ Tetracalcium Ferrotitanate [$4CaO \cdot Fe_2O_3 \cdot 2TiO_2$]

C-f/T Capacitance-Frequency versus Temperature (Diagram) [also C-f/T]

CFTA Canadian Film and Television Association

CFTA Committee of Foundry Technical Associations

CFTC Commodity Futures Trading Commission [US]

CFTC Commonwealth Fund for Technical Cooperation

CFTC Confédération Française des Travailleurs Chrétiens [French Confederation of Christian Workers]

CFTD Constant Fraction Timing Discriminator

CFTDC Canadian Forces Training Development Center

CFT/DQ/T Controlled Finish Temperature Rolling/Direct Quenching/Tempering (Process) [Metallurgy]

CFTMA Caster and Floor Truck Manufacturers Association [US]

C-13 FT-NMR Carbon-13 Fourier Transform Nuclear Magnetic Resonance (Spectra) [also ^{13}C FT-NMR]

CFTR Cystic Fibrosis Transmembrane Conductance Regulator

CFTRI Central Food Technological Research Institute [India]

CFTS Canadian Forces Training System

CFU Colony Forming Unit [Microbiology]

CFU Current File User

CFUBMS Closed Field Unbalanced Magnetron Sputter Ion Plating

CFUR Continuously Fed Unstirred Reactor

CFU-SA Colony Forming Unit-Stimulating Activity [Microbiology]

CFUW Canadian Federation of University Women

CFVD Constant Frequency, Variable Dot (System)

CFVS Cold Fusion Vacuum System

CFWA Canadian Fruit Wholesalers Association

CFWD Canadian Foundation for World Development

CFWMS Coherent Four-Wave Mixing Spectroscopy

CFWTL Chronic Freshwater Toxicity Level

CFY Company Fiscal Year

CFY Current Fiscal Year

D-CFZ 3-Chloro-2-Hydroxy-N,N,N-Trimethyl-1-Propanaminium Chloride

CFZF Canadian Float Zone Furnace [of Dalhousie University, Halifax, Canada]

CFZF Commercial Float Zone Furnace

CG Cardiogram

CG Cardiograph(er); Cardiography

CG Carry Generate [Combinational Logic]

CG Cascade Generator

CG Cathode Glow

CG Cell Growth

CG Center of Gravity [also Cg, or cg]

CG Centre de Géostatistique [Geostatistics Center, of Ecole Nationale Supérieure des Mines de Paris, France]

CG Chain Grate

CG Channel Gas

CG Chorionic Gonadotropin [Biochemistry]

CG Circuit Group

CG Clebsch-Gordon (Coefficient) [Quantum Mechanics]

CG Closed Grain (Woods)

CG Cloud-to-Ground

CG Coarse Grain(ed)

CG Coal Gas(ification)

CG Coal Gasifier

CG Coast Guard

CG Cobalt Gallate

CG Code Generation

CG Coincidence Gate [Electronics]

CG Colloidal Gel

CG Color Group

CG Columnar Grain [Metallurgy]

CG Commanding General

CG Commercial Grade

CG Commissary-General

CG Common Ground

CG Compacted Graphite [Metallurgy]

CG Complex Geometry

CG Compressed Gas

CG Computational Geometry

CG Computer Graphics

CG Condensable Gas

CG Conference Grant

CG Congo [ISO Code]

CG Conjugate Gradient (Method) [also cg] [Micromagnetics]

CG Consul General

CG Crack Growth

CG Crown Glass

CG Cryogenic Grade (Material)

CG Crystal Growth

CG Cubic Gauche (Structure) [also cg]

CG Cyanoguanidine [Biochemistry]

CG Cyclone Gasifier

CG Guided-Missile Cruiser [US Navy Symbol]

C/G Carbide/Graphite (Ratio)

C-G Cytosine-Guanidine [Biochemistry]

Cg Grid Capacitance [Electronics]

C/g Coulomb per gram [also C g^{-1}]

cg centigram [also cgm]

cg Conjugate Gradient (Method) [also CG] [Micromagnetics]

CGA Cadmium Germanium Arsenide (Semiconductor)

CGA Canadian Gas Association

CGA Carrier Group Alarm

CGA Certified General Accountant

CGA Coal-Gas Atmosphere

CGA Coal-Gold Agglomeration (Process)

CGA Coast Guard Academy [New London, Connecticut, US]

CGA Color Graphics Adapter

CGA Color Guild Associates [US]

CGA Compressed Gas Association [US]

CGA Configurable Gate Array

CGA Contrast Gate Amplifier

CGA Cyprus Geographical Association

CGAA Canadian Graphic Arts Association

CGAA Certified General Accountants of Alberta [Canada]

C:GaAs Carbon-Doped Gallium Arsenide

CGA/EGA Color Graphics Adapter/Enhanced Graphics Adapter

CGAHi Color Graphics Adapter, High Mode

¢/gal cent(s) per gallon [also ¢/gall]

CG Alloy Compacted Graphite (Ferro-)Alloy [also CG alloy]

CGAMEEC Committee of Glutamic Acid Manufacturers of the European Economic Community [France]

c-GaN Cubic Gallium Nitride

CGAS Coast Guard Air Station

CGAU Cabin Gas Analysis Unit

CGB Ceramics and Graphite Branch [US Air Force]

CGB Coast Guard Bulk Hazardous Materials (List) [of US Department of Transportation]

CGB Coincidence Grain Boundary [Materials Science]

CGB Commensurate Grain Boundary [Materials Science]

CGB Commonwealth Geographical Bureau [Australia]

CGB Convert Gray to Binary

CGBAPS Cape Grim Baseline Atmospheric Pollution Station [Australia]

CGBD Coincidence Grain Boundary Distribution [Materials Science]

CGBD Consultative Group on Biological Diversity [at Rockefeller Brothers Fund, US]

CGBM Cooperative Grain Boundary Migration [Materials Science]

CGBOC City and Guilds Boiler Operators Certificate [of City and Guilds of the London Institute, UK]

CGBS Cooperative Grain Boundary Sliding [Materials Science]

CGC Calavo Growers Council [US]

CGC Canadian Geoscience Council

CGC Canadian Grain Commission

CGC Canadian Gypsum Council

CGC Capillary Gas Chromatography

CGC Center for Global Communications [of International University of Japan]

CGC Clebsch-Gordon (3-j, 6-j and 9-j) Coefficients [Quantum Mechanics]

CGC Coaxial Gas Capacitor

CGC Cold Gas Compression

CGC Command Guidance Computer

CGC Compact Glass Column

CGC Coulomb-Gap Conduction (Model)

CGCB Committee of Governors of Central Banks [European Community]

CGCC Circuit Group Congestion

CGCC Clock Generator Controller

CGCC Chinese General Chamber of Commerce [Hong Kong]

CGCC Coal Gasification, Combined Cycle (Process)

CGCC Compact Glass Column Chromatography

CGCP Canadian Global Change Program

CGCRI Central Glass and Ceramics Research Institute [Calcutta, India]

CGCRI/CSIR Central Glass and Ceramics Research Institute/Council for Scientific and Industrial Research [Calcutta, India]

CGD Cathode Glow Discharge

CGD Cell Growth Determination [Microbiology]

CGD Constituent Gaussian Distribution

CGD Cosmic Gas Dynamics

CGDE Contact-Glow Discharge Electrolysis

cGDP Cyclic Guanosine Diphosphate [also cGDP] [Biochemistry]

CGE Canadian General Electric (Company)

CGE Capillary Gel Electrophoresis

CGE Center for Global Education [UK]

CGE Cold Gas Efficiency

CGE College of General Education

CGE Commission Geographical Education [of International Geographical Union, Canada]

Cge Carriage [also cge]

c-Ge (Single) Crystalline Germanium

CGEC Canadian Government Exposition Center

CGED Caribbean Group for Cooperation in Economic Development [US]

CGEL Cover Gas Evaluation Loop

CGER Center for Global Environmental Research

CGES Center for Global Energy Studies [UK]

CGF Carrier Gas Fusion

CGF Continuous Glass Fiber

CGF Crystal Growth Furnace

Cgf Grid-Filament Capacitance [Electronics]

CGG Compagnie Générale de Géophysique [A French Geophysical Supply Company]

CGG Continuous Galvanizing Grade (of Slab Zinc)

CGH Cape of Good Hope [South Africa]

CGH Colorado General Hospital [US]

CGH Computer-Generated Hologram; Computer-Generated Holograph

CGHAZ Coarse-Grain(ed) Heat-Affected Zone [Metallurgy]

CGI Common Gateway Interface [Internet]

CGI Computer-Generated Image(ry)

CGI Computer Graphics Interface

CGI Ground Control Interceptor

CGIA Center for Geographical Information and Analysis

CGIAR Consultative Group on International Agricultural Research [US]

CGIB Consumer Goods Industries Bureau [Japan]

CGIC Ceramics and Graphite Information Center [US Air Force]

CGIC Compressed-Gas-Insulated Cable

CGIL Center for Genetic Improvement of Livestock [of University of Guelph, Ontario, Canada]

CGIL Confederazione Generale Italiana del Lavoro [Italian General Confederation of Labor]

CGIP Channel Gas Inlet Pressure

CG Iron Compacted Graphite (Cast) Iron [also CG iron]

CGIS Canadian Geographical Information System

CGIS Canadian Geographic Information Service [of Environment Canada]

CGIS COMSAT (Communications Satellite) General Integrated System

CGIS Continental Geographic Information System

CGIT Compressed-Gas-Insulated Transmission Line

Cgk Grid-Cathode Capacitance [Electronics]

CGL Charge Generation Layer

CGL Circling Guidance Lights

CGL Complex Ginzburg-Landau (Equation) [Solid-State Physics]

CGL Comprehensive General Liability

CGL Computer-Generated Letter

CGL Continuous Galvanizing Line [Metallurgy]

C-Glass (Acid and) Chemical-Resistant Fiberglass [also C-glass]

CGLBI Canadian Geophysical Long Baseline Interferometry

CGLF Convention of the Great Lakes Fisheries

CGLI City and Guilds of the London Institute [UK]

CGLO Commonwealth Geological Liaison Office

CGM (International Conference on Joining of) Ceramics, Glass and Metal

CGM Character Generation Module

CGM Coarse-Grained Material

CGM Commission for the Geological Map of the World

CGM Committee on Guided Missiles [US]

CGM Computer Graphics Metafile

CGM Conjugate Gradient Method [Micromagnetics]

CGM Continuous Growth Model [Metallurgy]

.CGM Computer Graphics Metafile [File Name Extension] [also .cgm]

.CGM Graph [Lotus 1-2-3 File Name Extension]

cgm centigram [also cg, or ctgm]

CGMA Compressed Gas Manufacturers Association [US]

CGMID Character Generation Module Identifier

CG/MOI Center of Gravity/Moment of Inertia

CGMP Current Good Manufacturing Practices [also cGMP]

cGMP Cyclic Guanosine Monophosphate [Biochemistry]

3',5'-cGMP Cyclic-Guanosine-3',5'-Monophosphate [Biochemistry]

CGMW Commission for the Geological Map of the World [France]

CGN Coast Guard Noxious Liquid Substances (List) [of US Department of Transportation]

CGO Cobalt Gallium Oxide

CGO Compton Gamma-Ray Observatory

CGO Conventional Grain-Oriented (Product) [Metallurgy]

CGOFC Cryogenic Grade Oxygen-Free Copper

CGOP Channel Gas Outlet Pressure

CGOS Computer Graphics Operating System

CGP Central Graphics Processor

CGP Central Grounding Point

CGP Coalition for Government Procurement [US]

CGP College of General Practitioners [UK]

CGP Color Graphics Printer

CGP Computer General Processing

Cgp Grid-Plate Capacitance [Electronics]

CGPA Canadian Gas Processors Association

CGPC Canadian Government Publishing Center

CGPC Cellular General-Purpose Computer

CGPM Conference Générale des Poids et Mésures [General Conference on Weights and Measures, France]

CGPO Canadian Government Printing Office

CGPS Canadian Government Printing Service

CGPSA Canadian Gas Processors and Suppliers Association

CGR Coast Guard Reserve [US]

CGR Compound Growth Rate

CGR Cooperative Grain Rotation [Materials Science]

CGR Cosmic Gamma Ray(s)

CGR Crack Growth Rate

CGRA China and Glass Retailers Association [UK]

CGRA Consortium Général des Recherches Aeronautiques [General Consortium for Aeronautical Research, France]

CGRAM Clock-Generated Random-Access Memory

CGRD Center for Grinding Research and Development [of University of Connecticut, Storrs, US]

CGRI Canadian Gas Research Institute

CGRM Containment Gaseous Radiation Monitor [Nuclear Engineering]

CGRP Calcitonin Gene-Related Peptide [Biochemistry]

CGRP Circuit Group

CGRP-II β-Calcitonin Gene-Related Peptide [Biochemistry]

CGRP-KLH Calcitonin Gene Related Peptide–Keyhole Limpet Hemocyanin (Immunogen)

CGS Calcium Gallium Sulfide

CGS Canadian Geographical Society

CGS Canadian Geotechnical Society

CGS Carolina Geological Society [at Duke University, Durham, North Carolina, US]

CGS Centimeter-Gram-Second (System) [also cgs]

CGS Central Gunnery School

CGS Certificate of Graduate Studies
CGS Chief of (the) General Staff
CGS Circuit Group Congestion Signal
CGS Coast Guard Station
CGS College of General Studies
CGS Colorado Geological Survey [US]
CGS Common Gateway Services [also cgs]
CGS Computer Graphics Service [US]
CGS Computer Graphics Society [Japan]
CGS Consistent Global State
CGS Control Guidance Subsystem
CGS Copper Gallium Selenide
CGSA Cellular Geographic Service Area
CGSB Canadian General Standards Board
CGSB Canadian Government Specifications Board
CGSB Canadian Government Standards Board
CGSC Chemistry Grants Selection Committee [of Natural Sciences and Engineering Research Council, Canada]
CGSE CGS (Centimeter-Gram-Second) Electrostatic System [also cgse]
CGS-EMU Centimeter-Gram-Second/Electromagnetic Unit (System of Units) [also cgs-emu]
CGSET Circuit Group Set
CGSL (Centimeter-Gramm-Second-Ladungseinheit) – German for "Centimeter-Gram-Second-Coulomb" (System)] [also cgsl]
CGSM CGS (Centimeter-Gram-Second) Electromagnetic System [also cgsm]
CGSS Cryogenic Gas Storage Subsystem
CGSU CGS (Centimeter-Gram-Second) Unit [also cgsu]
CGSUC Claremont Graduate School and University Center [California, US]
CGT Capital Gains Tax
CGT Confederación General del Trabajo [General Confederation of Labour, Argentina]
CGT Confédération Générale du Travail [General Confederation of Labour, France and Algeria]
CGT Corrected Gross Thrust
CGT Current Gate Tube
CGTF Contracted Gaussian-Type Function [Physics]
CGTO Contracted Gaussian-Type Orbital [Physics]
CGTP Confederación General dos Trabajadores [General Confederation of Peruvian Workers]
CGU Canadian Geophysical Union
ČGU Česky Geologicky Ustav [Czech Geological Survey]
CGU Compu/Graphics Users Association [US]
CGWB Canadian Government Wheat Board
CGWP Certified Ground Water Professional
Cgy Calgary [Alberta, Canada]
CH Cahn-Hilliard (Equation) [Metallurgy]
CH Calcium Hydroxide
CH Calder Hall (Nuclear Reactor) [UK]
CH Can't Hear
CH Cape Horn [Chile]
CH Carbohydrate

CH Card Hopper
CH Case-Hardened; Case Hardening [Metallurgy]
CH Catalytic Hydrogenation (Process)
CH Ceiling Height
CH Central Heating
CH Cerebral Hemisphere [also ch]
CH Certified Herbologist
CH Chain Home (Radar)
CH Chapman and Hall Publishers [New York City, US]
CH Chlorohydrin
CH Clearinghouse
CH Coal Hook
CH Coal Hydrogenation
CH Coastal Harbour
CH Cold-Work Hardened; Cold-Work Hardening [Metallurgy]
CH Conductor Head
CH Confederatio Helvetica [Swiss Confederation]
CH Contact-Handled; Contact Handling [Nuclear Engineering]
CH Contention Handling
CH Control Heading
CH Convex Hull [Mathematics]
CH Cord and Hook (Installation)
CH Core Height
CH Corey-House (Synthesis)
CH Courthouse
CH Crack Healing
CH Critical Height
CH Cross-Hatch(ed); Cross Hatching
CH Cross-Head [Mechanical Testing]
CH Cushion Height [Automobiles]
CH Custom House
CH Cyclohexane
CH Cyclohexanol
CH High Cloud Type Code
CH Switzerland [ISO Code]
CH_x Hydrogenated Carbon [Symbol]
CH_4 Methane
C_mH_n Hydrocarbon
C_nH_{2n} Alkene [General Formula]
C_nH_{2n+2} Alkane [General Formula]
C_nH_{2n-2} Cycloalkene [General Formula]
$C_{60}H_2$ Fullerene Dihydride [Hydrocarbon Derivative of Buckminsterfullerene C_{60})]
$C_{70}H_2$ Fullerene Dihydride [Hydrocarbon Derivative of Buckminsterfullerene C_{70})]
1,9-C70H2 1,9-Fullerene Dihydride [Hydrocarbon Derivative of Buckminsterfullerene C_{70})]
7,8-C70H2 7,8-Fullerene Dihydride [Hydrocarbon Derivative of Buckminsterfullerene C_{70})]
C(H) Magnetic Field-Dependent Heat Capacity
C/H Carbon/Hydrogen Ratio
C-H Carbon-Hydrogen
C-H Conventional-Hydrothermal (Technique) [Materials Science]

Ch Chain [also ch]

Ch Channel [also ch]

Ch Chapter [also ch]

Ch Chelon [Chemistry]

Ch Chief [also ch]

Ch Child(ren) [also ch]

Ch China; Chinese

Ch Choke [also ch]

Ch Cholesteric Phase [Physical Chemistry]

Ch Chopper [also ch]

C11/h11 Loose Running Fit; Basic Shaft System [ISO Symbol]

°C/h Degree(s) Celsius (or Centigrade) per hour [also °C/hr, or $°C\ h^{-1}$]

ch candle hour [Unit]

ch chain [Unit]

ch *(cheval vapeur)* – French for "horsepower"

ch chi [English Equivlalent]

ch hyperbolic cosine [Symbol]

.ch Switzerland [Country Code/Domain Name]

χ(H) Field-Dependent (Magnetic) Susceptibility [Symbol]

CHA Calcium Hydroxyapatite

CHA Canadian Historical Association

CHA Canadian Hydrographers Association

CHA Caprinohydroxamic Acid

CHA Child Health Association [US]

CHA Cold Work Hardened, Aged [Metallurgy]

CHA Commercial Horticultural Association

CHA Concentric Hemispherical Analyzer

CHA Correlation Height Analysis

CHA Cross Hatch Angle [Magnetic Media]

CHA Cyclohexanol Acetate

CHA Cyclohexylamine

Cha Chameleon [Astronomy]

Cha Cyclohexanamine

CHABA Committee on Hearing and Bioacoustics [US]

ChAc Choline Acetyltransferase [Biochemistry]

CHAD Code for Handling Angular Data

CHAFC Committee of Heads of Australian Fauna Collections

CHAG Chain Arrester Gear

CHAG Compact High-Performance Aerial Gun

CHAH Committee of Heads of Australian Herbaria

CHAIS Consumer Hazards Analytical Information Service [of Laboratory of the Government Chemists, UK]

CHAL Division of Chemistry and the Law [of American Chemical Society, US]

Ch-Alc 2-Chloroethanol

Cham Chamfer [also cham]

CHAMMP Computer Hardware, Advanced Mathematics and Model Physics Program

CHAMP Character Manipulation Procedure

CHAMP Communications Handler for Automatic Multiple Programs

CHAMP Community Health Air Monitoring Program

CHAMPS Computerized History and Maintenance Planning System

Chan Channel [also chan]

Chanc Chancellor; Chancery

CHANCOM Channel Committee [of NATO]

Chang Changing [also chang]

Change Mag Change Magazine [Publication on Environmental Changes]

CHANHI Upper Channel Corresponding to Half-Amplitude Point of a Distribution

Chan Is Channel Islands

CHANLO Lower Channel Corresponding to Half Amplitude Point of a Distribution

CHAOS Canadian High Acceptance Orbit Spectrometer (Experiment) [of Tri-Universities Meson Facility, University of British Columbia, Canada]

CHAP Challenge-Handshake Authentication Protocol

Chap Chapter [also chap]

Chap Bull Chapter Bulletin [of Experimental Aircraft Association, US]

CHAPI Compact Helicopter Approach Path Indicator

CHAPS 3-[(3-Cholamidopropyl)dimethylammonio]-1-Propanesulfonate

CHAPS Clearinghouse Automated Payments System [US]

Chaps Chapters [also chaps]

CHAPSO 3-[(3-Cholamidopropyl)dimethylammonio]-2-Hydroxy-1-Propanesulfonate

CH_4/Ar Methane/Argon (Gas Mixture)

Char Character(istic) [also char]

CHARA Center for High Angular Resolution Astronomy [US]

CHARIBDIS Chalk River Bibliographic Data Information System [of Atomic Energy of Canada Limited, Canada]

CHARLEN Character String Length (i.e. Number of Characters in String)

CHARM CAA (Civil Aeronautics Administration) High-Altitude Remote Monitoring

CHARM CAA (Civil Aviation Authority) High-Altitude Remote Monitoring

CHARM Combined Hadonic Response Measurement [Particle Physics]

CHARPOS Character String Position

CHART Computerized Hierarchy and Relationship Table

Chart Charter(ed) [also chart]

chart powder (wrapped) [Medical Prescriptions]

Chart Mech Eng Chartered Mechanical Engineer [Journal published in the UK]

Chart Quant Surv Chartered Quantity Surveyors [of Royal Institute of Chartered Surveyors, UK]

CHARTS Change Handling and Routing System

Chart Surv Wkly Chartered Surveyors Weekly [of Royal Institute of Chartered Surveyors, UK]

CHAS Division of Chemical Health and Safety [of American Chemical Society, US]

CHASM Coventry Health and Safety Movement [UK]

CHAT CLIRA (Closed Loop In-Reactor Assembly) Holddown Assembly Tool [Nuclear Engineering]

CHAT Computer Harmonized Application Tailored

CHAT Conversational Hypertext Access Technology [Internet]

CHb Cerebrellar Hemangioblastoma

ChB *(Chirurgiae Baccalaureus)* – Bachelor of Surgery

CHBA Canadian Home Builders Association [formerly Housing and Urban Development Asso ciation of Canada]

CHBADCB 1,2-Bis(3,5-Dichloro-2-Hydroxybenzamido)-4,5-Dichlorobenzene

CHBA-Et 1,2-Bis(3,5-Dichloro-2-Hydroxybenzamido)ethane

CHBE Cyclohexyl tert-Butyl Ether

C-HBT C-Doped Heterojunction Bipolar Transistor

CHC Carbohydrate Chemistry

CHC Chemical Hydrogen Cracking

CHC Child Health Clinic

CHC Chlorinated Hydrocarbon

CHC Chlorohydrocarbon

CHC Choke Coil

CHC Clean Harbors Cooperatives [US]

CHC Clearinghouse for Copyright

CHC Colorado Heritage Center [Denver, US]

CHC Cyclohexanecarboxylic Acid

CHC Cyclohexylamine Carbonate

CH₃CCl₃ Methyl Chloroform

CH₄-CD₄ Methane-Deuterated Methane (Gas Mixture)

CHCF Component Handling and Cleaning Facility

CHCK Channel Check

°C/h/cm² Degree(s) Celsius (or Centigrade) per hour per square centimeter [also °C/hr/cm², or °C h⁻¹ cm⁻²]

ChCMV Chrysanthemum Chlorotic Mottle Viroid

CHCP Change (Current) Code Page [also chcp]

CHCS Cabin Humidity Control Subsystem

CHCT Conjugated Honeycomb Chained Trimer

CHCU Channel Control Unit

CHD Carbohydrate Derivative

CHD Chlordane

CHD Coronary Heart Disease

Ch$ Chilean Peso [Currency of Chile]

Chd Chord [also chd]

12CHDA 1,2-Cyclohexanediamine

Ch D'Aff Chargé d'Affaires

CHDB Compatible High Density Bipolar

CHDC Canadian Housing Design Council

CHDC Child and Human Development Council [of National Institutes of Health, Bethesda, Maryland, US]

CHDIR Change Directory (Command) [also ChDir, or chdir]

CHDL Computer Hardware Definition Language

CHDM Cyclohexanedimethanol

CHDN Dicyanoazocyclohexane

CHDPM Bis(dicyclohexylphosphinyl)methane

CHDTA 1,2-Cyclohexanediamine-N,N,N',N'-Tetraacetic Acid

CHE Cahn-Hilliard Equation [Metallurgy]

CHE Cascade Heat Exchanger

CHE Center for the History of Entomology [of Pennsylvania State University, University Park, US]

CHE Channel End

CHE Channel Hot Electron

CHE Cholinesterase [also ChE] [Biochemistry]

CHE Chronic Health Effects

CHE Committee on the Human Environment

CHE Conference on the Human Environment

CHE Cyclohexylcarboxylic Acid Methyl Ester

CHE Cylindrical Hydraulic Engineer

ChE Chemical Engineer [also Chem Eng, or ChemEng]

ChE Cholinesterase [Biochemistry]

CHEC Cascade Holistic Economic Consultants [US]

CHEC Channel Evaluation and Call

CHEC Commonwealth Human Ecology Council [UK]

CHED Division of Chemical Education [of American Chemical Society, US]

CHED Newsl Division of Chemical Education Newsletter [of American Chemical Society, US]

CheE Chemical Engineer [also ChemE]

CHEF Chelation-Enhanced Fluorescence

CHEF Chemistry of High Elevation Fog

CHEK Cargo Handling Equipment Kit

CHEM Centre des Hautes Etudes Militaires [Center for Higher Military Studies]

Chem Chemical; Chemist(ry) [also chem]

Chem Abstr Chemical Abstracts [of American Chemical Society, US]

Chem Abstr Serv Chemical Abstracts Services [of American Chemical Society, US] [also CAS]

Chem Age India Chemical Age of India [Journal published in India]

Chem Anal Chemical Analysis

Chem Anal Chemia Analytyczna [Polish Journal on Analytical Chemistry]

Chem Anlagen Verf Chemie Anlagen und Verfahren [German Publication on Chemical Equipment and Processes]

Chem Ber Chemische Berichte [Chemical Reports; published in Germany and US]

Chem Biochem Eng Q Chemical and Biochemical Engineering Quarterly [US]

Chem Biol Chemistry and Biology [International Journal]

Chem Br Chemistry in Britain [UK Publication] [also Chem Brit]

Chem Bus Chemical Business [Journal published in the US]

Chem Crystal Chemical Crystallographer; Chemical Crystallography

Chem Commun Chemical Communications [Journal of the Royal Society of Chemistry, UK] [also Chem Commun (Cambridge)]

CHEMDEX Chemical Index [also Chemdex]

ChemE Chemical Engineer [also Chem Eng, or ChemEng]

Chem Econ Eng Rev Chemical Economy and Engineering Review [Japan]

Chem Eng Chemical Engineer(ing)

Chem Eng Chemical Engineer [Journal of the Institution of Chemical Engineers, UK]

Chem Eng Chemical Engineering [Journal published in the US]

Chem Eng Abstr Chemical Engineering Abstracts [of Royal Society of Chemistry, UK]

Chem Eng Aust Chemical Engineering in Australia [Publication of the Institution of Engineers of Australia]

Chem Eng Biotech Abstr Chemical Engineering and Biotechnology Abstracts [also CEBA]

Chem Eng Commun Chemical Engineering Communications [Switzerland]

Chem Eng Educ Chemical Engineering Education [Publication of the American Society for Engineering Education, US]

Chem Eng J Chemical Engineering Journal [Switzerland]

Chem Eng J Biochem Eng J Chemical Engineering Journal and Biochemical Engineering Journal [Switzerland]

Chem Eng Mach Chemical Engineering and Machinery [PR China]

Chem Eng News Chemical and Engineering News [Publication of the American Chemical Society, US]

Chem Eng (NY) Chemical Engineering (New York) [US]

Chem Eng Process Chemical Engineering and Processing [Journal published in Switzerland]

Chem Eng Prog Chemical Engineering Progress [Publication of the American Institute of Chemical Engineers]

Chem Eng Res Bull Chemical Engineering Research Bulletin [US]

Chem Eng Res Des Chemical Engineering Research and Design [Publication of the Institution of Chemical Engineers, UK]

Chem Eng Sci Chemical Engineering Science [Published in US and UK]

Chem Eng Technol Chemical Engineering and Technology [Journal published in Germany]

Chem Eng (UK) Chemical Engineer (UK) [Journal of the Institution of Chemical Engineers]

Chem Eng World Chemical Engineering World [Journal published in the US]

Chem Erde Chemie der Erde [German Journal on Geochemistry]

Chem-Eur J Chemistry–European Journal

Chem Express Chemistry Express [Publication of the Kinki Chemical Society, Osaka, Japan]

CHEMFET Chemical(ly Selective Field-Effect Transistor; Chemically Sensitive Field-Effect Transistor [also ChemFET, or chemFET]

Chem Geol Chemical Geologist; Chemical Geology

Chem Geol Chemical Geology [Journal published in the Netherlands]

Chem Geol (Isot Geosci Sect) Chemical Geology (Isotope Geoscience Section) [Journal published in the Netherlands]

Chem Hazards Ind Chemical Hazards in Industry [Publication of the Royal Society of Chemistry, UK] [also Chem Haz Ind]

Chem Health Safety Chemical Health and Safety [Journal of the American Chemical Society, US]

Chemico-Biol Interact Chemico-Biological Interactions [International Journal]

Chemiefasern Textilind Chemiefasern Textilindustrie [German Journal on Synthetic Fibers and the Textile Industry]

Chem Ind Chemistry and Industry [Publication of the Society of Chemical Industry, UK]

Chem Ind Chemische Industrie [German Journal on the Chemical Industry; published by Verband der Chemischen Industrie]

Chem Ind (Duesseldorf) Chemische Industrie [German Journal on the Chemical Industry; published in Duesseldorf]

Chem Ind Int Chemische Industrie International [German International Journal on the Chemical Industry; published by Verband der Chemischen Industrie, Germany]

Chem Ind (London) Chemistry and Industry (London) [Journal published in the UK]

Chem Ind Mag Chemistry and Industry Magazine [UK]

Chem Inf Bull Chemical Information Bulletin [of American Chemical Society, US]

Chem Ing Tech Chemie-Ingenieur-Technik [German Publication on Chemical Engineering and Technology]

Chem Int Chemistry International [Journal of the International Union of Pure and Applied Chemistry, UK]

Chemist The Chemist [Journal of the American Institute of Chemists, US]

Chem J Chemie Journal [German Chemistry Journal; published by Verband der Chemischen Industrie]

Chem Lett Chemistry Letters [of Chemical Society of Japan]

CHEMLINE Chemical Information On-Line [US]

Chem Listy Chemické Listy [Czech Chemical Publication]

Chem Mark Reporter Chemical Marketing Reporter [US]

Chem Mater Chemistry of Materials [Publication of the American Chemical Society, US]

Chem Met Chemical Metallurgist; Chemical Metallurgy

Chem Mineral Chemical Mineralogist; Chemical Mineralogy

CHEMNAME Chemical Names (Database) [of American Chemical Society, US]

Chem News Lett Chemical News Letter [of Chinese Chemical Society, PR China]

Chemo Chemoterapy [also chemo]

Chem Oceanogr Chemical Oceanographer; Chemical Oceanography

Chemometr Intell Lab Syst Chemometrics and Intelligent Laboratory Systems [Journal published in the Netherlands]

Chemother Chemotherapy [International Journal]

Chem Pet Eng Chemical and Petroleum Engineering [Journal published in the US]

Chem Pharm Bull Chemical and Pharmaceutical Bulletin [Journal published in the US]

Chem Phys Chemical Physicist; Chemical Physics

Chem Phys Chemical Physics [Journal published in the Netherlands]

Chem Phys Lett Chemical Physics Letters [Journal published in the Netherlands]

Chem Proc Chemical Processing [also chem proc]

Chem Process Chemical Processing [Journal published in the US]

Chem Prod Chemische Produktion [German Publication on Chemical Production]

Chem Prop Chemical Properties [also chem prop]

Chem Prum Chemicky Prumsyl [Czech Chemical Publication]

CHEMRAWN Chemical Research Applied to World Needs Committee [of the United States National Committee for International Union of Pure and Applied Chemistry]

Chem React Eng Technol Chemical Reaction Engineering and Technology [Journal published in the US]

Chem Res Chemical Research [also chem res]

Chem Res Toxicol Chemical Reseach in Toxicology [Publication of the American Chemical Society, US]

Chem Rev Chemical Reviews [of American Chemical Society, US]

Chem Sci Chemical Science; Chemical Scientist

Chem Scr Chemica Scripta [Published for the Royal Swedish Academy of Sciences by Cambridge University Press, UK]

CHEMSDI Chemical Abstracts Selective Dissemination of Information [of American Chemical Society, US]

Chem Senses Chemical Senses [Journal published by Oxford University Press, UK]

ChemSoc Chemical Society [also Chem Soc, UK]

Chem Soc Jpn Chemical Society of Japan [also CSJ]

Chem Soc Jpn Bull Chemical Society of Japan Bulletin [also CSJ Bull]

Chem Soc Rev Chemical Society Reviews [of Royal Society of Chemistry, UK]

Chem Solids Chemistry of Solids [Journal published in the US]

Chemsphere Am Chemsphere Americas [Journal published in the US]

CHEMTECH Chemical Technology (Meeting)

Chem Tech Chemical Technologist; Chemical Technology

Chem Tech Chemie Technik [German Publication on Chemistry and Technology]

Chem Tech Chemische Technik [German Publication on Chemical Technology]

Chem Tech (Heidelberg) Chemie Technik (Heidelberg) [German Publication on Chemistry and Technology; published in Heidelberg]

Chem Tech (Leipzig) Chemische Technik (Leipzig) [German Publication on Chemical Technology; published in Leipzig]

Chem Technol Fuels Oils Chemistry and Technology of Fuels and Oils [Journal published in the US]

CHEMTREC Chemical Transportation Emergency Center [Division of Chemical Manufacturers Association in Washington, DC, US]

Chem Vlakna Chemicke Vlakna [Slovak Chemical Publication]

Chem Warf Chemical Warfare [also ChemWarf]

Chem Week Chemical Week [US Publication]

Chemy Chemistry [also chemy]

Chem Z Chemische Zeitung [German Chemical Journal] [also Chem Ztg]

Chem Zvesti Chemicke Zvesti [Slovak Chemical Publication]

Ch Eng Chief Engineer

CHEOPS Chemical Information Systems Operators [now European Association of Scientific Information Dissemination Centers, UK]

CHEP Computing in High Energy Physics (Conference)

CHERUB Chemical Engineering Reference User Bibliography

CHES 2-(N-Cyclohexylamino)ethanesulfonic Acid

Ches Cheshire [UK]

CHESNAVFAC Chesapeake Naval Division Facilities Engineering Command [US Navy]

CHESS Canadian Health Education Specialists Society

CHESS Chemical Shift-Selective (Imaging)

CHESS Community Health and Environmental Surveillance System

CHESS Cornell High Energy Synchrotron Source [of Cornell University, Ithaca, New York, US]

CHEVMA Canadian Heat Exchanger and Vessel Manufacturers Association

CHF Congestive Heart Failure

CHF Coupled Hartree-Fock [Physics]

CHF Critical Heat Flux

CHF Swiss Franc [Currency of Switzerland and Liechtenstein]

Chf Chief [also chf]

Chf Chloroform [also chf]

CHFA Consumer's Health Forum of Australia

CHF$_3$/Ar Trifluoromethane/Argon (Mixture)

CHF$_3$/CO$_2$ Trifluoromethane/Carbon Dioxide (Mixture)

CHFIE Cordell Hull Foundation for International Education [US]

CHFN Change Finger [Unix Operating System]

CHF$_3$/O$_2$ Trifluoromethane/Oxygen (Mixture)

CHFPT Coupled Hartree-Fock Perturbation Theory [Physics]

Ch Fwd Charges Forward [also ch fwd]

Chg Change [also chg]

Chg Charge [also chg]

Chge Charge [also chge]

Chgr Charger [also chgr]

CHGP Charging Pump

CHGRP Change Group

CH$_4$/H$_2$ Methane/Hydrogen (Mixture)

CH$_4$/H$_2$/Ar Methane/Hydrogen/Argon (Mixture)

CH$_4$/He Methane/Helium (Mixture)

CHI CH2M Hill Hanford, Inc. [US]

CHI Computer/Human Interaction

CHI Computer/Human Interface

CHIA Canadian Hovercraft Industries Association

CHIC Canadian Housing Information Center [of Canadian Mortgage and Housing Corporation]

CHIC Cyclohexyl Isocyanate

CHICA Community and Hospital Infection Control Association [Canada]

Chicago Purch Chicago Purchaser [Published in the US]

CHIEF Controlled Handling of Internal Executive Functions

Chief Eng Chief Engineer

CHIF Channel Interface

Chil Chile; Chilean

CHIL Current-Hogging Injection Logic

CHILD Cognitive Hybrid Intelligent Learning Device

CHILD Computer Having Intelligent Learning and Development

Child Devel Child Development (Journal)

CHILL CCITT High-Level (Computer) Language [of Consultative Committee for International Telegraphy and Telephony]

ChIM Chemical Inventory Management System (Software)

Chim Chron Chimika Chronika [Chemical Chronicles; published in Athens, Greece]

ChIME Chemical Industry for Minorities in Engineering [US]

Chim Ind Chimica e l'Industria [Italian Journal on Chemistry and Industry]

Chim Ind (Milan) Chimica e l'Industria (Milan) [Journal on Chemistry and Industry; published in Milan, Italy]

Chimp Chimpanzee [also chimp]

Chin China; Chinese

China Pulp Paper China Pulp and Paper [Journal published in PR China]

China Steel Tech Rep China Steel Technical Report [Published in Taiwan, Republic of China]

Chin Astron Astrophys Chinese Astronomy and Astrophysics [Selected translations from: *Acta Astrophysica Sinica, Acta Astronomica Sinica* and *Chinese Journal of Space Science*; published in the UK]

Chin J Biomed Eng Chinese Journal of Biomedical Engineering [of Chinese Society of Biomedical Engineering, PR China]

Chin J Chromotogr Chinese Journal of Chromotography [of Chinese Chemical Society, PR China]

Chin J Comput Chinese Journal of Computers [PR China]

Chin J Infrared Res Chinese Journal of Infrared Research [PR China]

Chin J Lasers Chinese Journal of Lasers [PR China]

Chin J Low Temp Phys Chinese Journal of Low Temperature Physics [PR China]

Chin J Mater Sci Chinese Journal of Materials Science [Taiwan, Republic of China]

Chin J Mech Eng Chinese Journal of Mechanical Engineering [PR China]

Chin J Nucl Phys Chinese Journal of Nuclear Physics [PR China]

Chin J Phys Chinese Journal of Physics [of Physical Society of the Republic of China, Taiwan]

Chin J Polym Sci Chinese Journal of Polymer Science [Published in the Netherlands]

Chin J Sci Instrum Chinese Journal of Scientific Instruments [PR China]

Chin J Semicond Chinese Journal of Semiconductors [PR China]

Chin J Space Sci Chinese Journal of Space Science [PR China]

Chin Pat Chinese Patent [also Chn Pat]

Chin Phys Chinese Physics [Selected Translations from Chinese Journals on Physics and Astronomy; published by the American Institute of Physics in the US]

Chin Phys Lasers Chinese Physics-Lasers [Translation of: *Chinese Journal of Lasers*; published in the US]

Chin Phys Lett Chinese Physics Letters [PR China]

Chin Sci Bull Chinese Science Bulletin [PR China]

Chin Soc Mater Sci Chinese Society of Materials Science

CHINT Charge Injection Transistor

CHIP Canadian Home Insulation Program

CHIP Chemical Hazard Information Profile

CHIP Chip Hermetically in Plastic [Electronics]

CHIP Cold and Hot Isostatic Pressing (with Intermediate Sintering) [Powder Metallurgy]

C-HIP Conventional Hot Isostatic Pressing [Powder Metallurgy]

CHiPR Center for High-Pressure Research [of Carnegie Institute of Washington, DC, US]

CHIPS Clearinghouse Interbank Payment System [New York, US]

Chips Tips Chips and Tips [Published in the US]

CHIRAPHOS 2,3-Bis(diphenylphosphino)butane [also Chiraphos]

CHIRP Confidential Human Factors Incident Reporting System [UK]

Chittagong Univ Stud I Chittagong University Studies, Part I [Bangladesh]

Chittagong Univ Stud II Chittagong University Studies, Part II [Bangladesh]

Chittagong Univ Stud II, Sci Chittagong University Studies, Part II, Science [Bangladesh]

Ch J Chief Justice

CHK Check [also Chk, or chk]

.CHK CHKDSK (Check Disk) [File Name Extension]

Chklst Checklist [also chklst]

CHKDSK Check Disk (Command) [also chkdsk]

Chkpt Checkpoint [also chkpt]

CH$_4$/Kr Methane-Krypton (Gas Mixture)

Chkr Checker [also chkr]

CHL Certified Hardware List

CHL Channel

CHL Chlamydia [Microbiology]

CHL Chloranil

CHL Cracked HAZ (Heat-Affected Zone) Length [Metallurgy]

CHL Current-Hogging Logic

Chl Channel [also chl]

Chl Chloral

Chl Chloroform [also chl]

Chl Chlorophyll

Chl a Chlorophyll a [Biochemistry]

Chl b Chlorophyll b [Biochemistry]

Chlf Chloroform [also chlf]

CHLOREP Chlorine Emergency Plan (or Procedure) [of The Chlorine Institute, US]

3-Chloro-CCP 3-Chloro-Carbonylcyanidephenylhydrazone

Chloro-IPC Isopropyl-N-(3-Carbophenyl) Carbamate [also chloro-IPC]

CHLW Commercial High-Level Waste

CHM Canadian Historical Museum [Montreal]

CHM Chemical Machining

CHM Chemical Milling

CHM Clearing House Mechanism

CHM Cleveland Health Museum [Ohio, US]

CHM Cold Hearth Melting (Process) [Metallurgy]

CHM Council on Health Manpower [of American Medical Association, US]

ChM *(Chirurgiae Magister)* – Master of Surgery

Chm Chairman [also chm]

Chmbr Chamber [also chmbr]

CHMM Certified Hazardous Materials Manager

Chmn Chairman [also chmn]

CHMOD Change Mode

CHMOS Complementary High-Density Metal-Oxide Semiconductor

CHMR Center for Hazardous Materials Research [US]

CHMSL Center, High-Mounted Stop Light

CHMT Components, Hybrids and Manufacturing Technology

CHMTS Components, Hybrids and Manufacturing Technology Society [of Institute of Electrical and Electronics Engineers, US]

CHN Carbon, Hydrogen and Nitrogen

CHN Chakravarty, Halperin and Nelson (Approach) [Quantum Mechanics]

CH$_4$/N$_2$ Methane/(Molecular) Nitrogen (Gas Mixture)

C/H/N Carbon/Hydrogen/Nitrogen (Mixture)

Chn Channel [also chn]

Chn Chinese

CHN-HN Chakravarty, Halperin and Nelson–Hasenfratz and Niedermeyer (Formula) [Quantum Mechanics]

Chnl Channel [also chnl]

CHNO Carbon, Hydrogen, Nitrogen and Oxygen

C/H/N/O Carbon/Hydrogen/Nitrogen/Oxygen

Chn Pat Chinese Patent [also Chin Pat]

ChNPP Chernobyl Nuclear Power Plant [Ukraine]

CHO Carbohydrate

CHO Carbon, Hydrogen and Oxygen

CHO Chinese Hamster Ovary [Biochemistry]

CH$_4$/O$_2$ Methane/Oxygen (Mixture)

C$_2$H$_3$O Acetate Ion [Symbol]

C$_m$(H$_2$O)$_n$ Carbohydrate [Formula]

CH$_4$/O$_2$/Ar Methane/Oxygen/Argon (Mixture)

CH$_4$/O$_2$/H$_2$ Methane/Oxygen/Hydrogen (Mixture)

CHOC Center for History of Chemistry [University of Philadelphia, Pennsylvania, US]

CHOC Consumer Health Organization of Canada [Toronto]

CHOGM Commonwealth Heads of Government Meeting

CHOICES Computerized Heuristic Occupational Information and Career Exploration System [Canada]

CHOL Common High-Order Language

CHORD Chief of Office of Research and Development

CHORI Chief of Office of Research and Inventions [US]

C Horizon C Horizon (Composed of Weathered Soil Parent Material) [also C horizon]

CHORTLE Carbon-Hydrogen Correlations from One-Dimensional Polarization Transfer Spectra by Least-Squares Analysis

CHOWN Change Owner

CHP Channel Processor

CHP Chemical Hygiene Plan

CHP Combined Heat and Power (Plant)

CHP Continental Horsepower

CHP Corning High Purity (Vitreous Silica)

CHP Cryohydric Point [Physical Chemistry]

CHP Cumene Hydroperoxide

CHP Cyclohexyl Isopropyl Methylamine

CHP N-Cyclohexyl 2-Pyrrolidinone

CHP Swiss Patent [also CH P, CP, or Ch P]

CHp 2-Cyclohepten-1-One

ChP Chilean Peso [Currency of Chile]

Chp Chapter [also chp]

CHPA Combined Heat and Power Association [UK]

CHPAC Center for High-Performance Adhesives and Composites [of Virginia Polytechnic Institute and State University, Blacksburg, US]

CHPAE Critical Human Performance and Evaluation

CHPB Canadian Health Protection Branch [of Health Canada]

CHPC Center for High-Performance Computing [of University of Texas, US]

CHP/DH Combined Heat and Power/District Heating

Chpers Chairperson [also chpers]

CHPG Combined Heat and Power Generation

Ch Ppd Charges Prepaid [also ch ppd]

CHPR Center for High-Pressure Research [of State University of New York at Stony Brook, US]

CHPS Characters per Second

ChPT Chiral Perturbation Theory [Physics]

CHQ 2-Chlorohydroquinone

CHQ 5,7-Dichloro-8-Hydroxyquinoline

Chq Cheque [also chq]

CHR Center for Human Radiobiology [of Argonne National Laboratory, Illinois, US]

CHR Character (Function) [also Chr]

CHR Cold Hearth Refinery [Metallurgy]

CHR Cooper-Harper Rating

.CHR Character [File Name Extension] [also .chr]

Chr Chair [also chr]

Chr Character [also chr]

°C/hr Degree(s) Celsius (or Centigrade) per hour [also °C/h, or °C h^{-1}]

c-hr candle-hour [Unit]

Chrg Charge [also chrg]

CHREM Center for High-Resolution Electron Microscopy [at Arizona State University, Tempe, US]

CHRGN Character Generator

CHRIS Chemical Hazard Response Information System [of US Coast Guard]

Christ Sci Monit Christian Science Monitor [US]

CHRM Center for Holistic Resource Management [US]

Chromium Rev Chromium Review [South Africa]

Chron Chronological; Chronology [also chron]

Chron Hort Chronica Horticulturae [Publication of the International Society for Horticultural Science]

Chronic Diseases Can Chronic Diseases in Canada [Journal published by Health Canada]

CHRP Common Hardware Reference Platform

CHRS Canadian Heritage Rivers System

CHRS Center for Hospitality Research and Service [US]

CHRS Containment Heat Removal System [Nuclear Engineering]

CHR$ Character String (Function)

CHS Calder Hall (Nuclear) Reactor [UK]

CHS California Historical Society [US]

CHS Canadian Hemophilia Society

CHS Canadian Hydrographic Service

CHS Case Hardening Steel

CHS Central Heading System

CHS Change Sign (Function)

CHS Charged Hard-Sphere (Reference System)

CHS Chicago Historical Society [US]

CHS Citizens for Highway Safety [US]

CHS Colorado Historical Society [US]

CHS Connecticut Historical Society [US]

CHS Cross-Head Speed [Mechanical Testing]

CHS Cyclohexane Sulfonate

CHS Cylinder Head Sector

$C_6H_5S^-$ Thiophenoxide Ion [Formula]

CH(s) Hydrogenated Surface Carbon Radical [Symbol]

Chs Chapters [also chs]

Chs Characters [also chs]

CHSA Crane and Hoist Services Association

CHSB Construction Health and Safety Branch [of Ministry of Labour, Canada]

CHSC Cambridge Higher School Certificate [UK]

CHT Call Hold and Trace

CHT Certified Holistic Therapist

CHT Charactron Tube

CHT Chemical Heat Treatment

CHT Conductive Heat Transfer

CHT Continuous Heating Transformation

CHT Convective Heat Transfer

CHT Cycloheptatriene

Cht Chart [also cht]

CHTA Contract Heat Treatment Association [UK]

CHTM Center for High Tech Materials [of University of New Mexico, Albuquerque, US]

CHTP Certified Healing Touch Practitioner

CH-TRU Contract-Handled Transuranic (Waste) [also CH TRU]

CHTSC Cyclohexanone Thiosemicarbazone

CHU Celsius Heat Unit [also Chu, or chu]

CHU Centigrade Heat Unit [also Chu, or chu]

CHU Channel Unit [SPADE]

CHU_{mean} Mean Centigrade Heat Unit [also Chu_{mean}]

CHUG Canadian Honeywell Users Group

Chunnel Channel Tunnel [Railroad Tunnel Between France and Great Britain] [also chunnel]

Chute Parachute [also chute]

CHW Cladding Hull (Nuclear) Waste

CHWE Closed Hot Wall Epitaxy

Chwm Chairwoman [also chwm]

CHX Cabin Heat Exchanger

CHx 2-Cyclohexenone

chxn cis-1,2-Cyclohexanediamine

1,3-CHXN cis-1,3-Cyclohexanediamine

Chxo Cyclohexanone

CHY Chymotrypsinogen [Biochemistry]

°C/Hz Degree(s) Celsius per Hertz [also °C Hz^{-1}]

CI Call Indicator

CI Canary Islands

CI Canvey Island (Liquefied Natural Gas Terminal) [UK]

CI Caroline Islands

CI Carry Input [Combinational Logic]

CI Card Input

CI Caroline Institute [Stockholm, Sweden]

CI Cascade Impactor

CI Cast Iron

CI Cavitation Inception

CI Cayman Islands

CI Cement Industry

CI Center Island

CI Center of Inertia

CI Certificate of Insurance

CI Certification Inspection

CI Chain Index

CI Channel Islands [UK]

CI Charge Independence [Physics]

CI Chemical Inspectorate [of Ministry of Defense, UK]

CI Chemical Ionization

CI Chief of Information [of US Army]

CI Chlorine Institute [US]

CI Christmas Island

CI Chronic Infection

CI Circuit Interrupter

CI Cirrus (Cloud) [also Ci]

CI Class Interval [Statistics]

CI Cold-Iron (Soldered Joint) [also ci]

CI Collective Index (of Chemical Abstracts) [American Chemical Society, US]

CI Colloidal Ion

CI Colloidal Iron

CI Color Index

CI Combat Information (Network)
CI Combat Intelligence
CI Combustion Institute [US]
CI Comfort Index
CI Communications Interface
CI Competitive Inhibitor
CI Compexity Index (in Quantitative Metallography) [Symbol]
CI Component Interface
CI Composites Institute [US]
CI Compressor Inlet
CI Computational Intelligence
CI Computer Interconnect(ion)
CI Conditional Instability [Meteorology]
CI Confidence Interval [Statistics]
CI Configurational Isomer
CI Configuration Interaction [Physical Chemistry]
CI Configuration Inspection
CI Configuration Item
CI Conformational Isomer
CI Conservation International [Washington, DC, US]
CI Consular Invoice
CI Consumers Interpol [Malaysia]
CI Containment Integrity [Nuclear Engineering]
CI Continuity Index (for Intermetallics) [Materials Science]
CI Continuous Interlock
CI Control Interval
CI Cordage Institute [US]
CI Cost and Insurance [also ci, C&I, or c&i]
CI Côte d'Ivoire/Ivory Coast [ISO Code]
CI Cotton Incorporated [US]
CI Counter-Intelligence
CI Course Indicator
CI Course Invalid
CI Crew Interface
CI Christmas Island
CI Critical Inclination
CI Critical Item
CI Cropping Index [Agriculture]
CI Customer Information
CI Crystal Impedance
CI Cubic Inch [also ci]
CI Cumulative Index
C&I Control and Indication
C&I Control and Instrumentation
C&I Cost and Insurance [also c&i, CI, or ci]
C/I Carrier-to-Interface (Ratio)
C/I Carrier-to-Interference (Ratio)
C/I Certificate of Insurance [also c/i]
C/I Cost, Insurance [also c/i]
C³I Command, Control, Communications and Intelligence [also CCCI]
Ci Cirrus Cloud [also CI]
Ci Curie [Unit]
C.i. One Gallon [Apothecary]

cI Pearson symbol for body-centered space lattice in cubic crystal system (this symbol is followed by the number of atoms per unit cell, e.g. cI2, cI40, etc.)
ci Catalogue on Internet [of Elsevier Science Limited]
.ci Côte d'Ivoire/Ivory Coast [Country Code/ Domain Name]
CIA California Institute of the Arts [Los Angeles, US] [also CalArts]
CIA Central Intelligence Agency [US]
CIA Ceramics International Association [US]
CIA Certified Internal Auditor
CIA Chemical Industries Association [UK]
CIA Chemical Institute of Australia
CIA Chemiluminescence Immunoassay
CIA Cobb Institute of Archeology [US]
CIA Collision-Induced (Infrared) Absorption
CIA Commonwealth Industries Association [UK]
CIA Communications Industry Association
CIA Communications Interrupt Analysis
CIA Computer Industry Association [now Computer and Communications Industry Association, US]
CIA Computer Interface Adapter
CIA Construction Industries Association [now Construction Industry Manufacturers Association, US]
CIA Containment Isolation A [Nuclear Reactors]
CIA Control(ler) Interface Assembly
CIA Current Instruction Address
Cia (Companhia) – Portuguese for "Company"
Cia (Compañia) – Spanish for "Company"
CIAA Canadian Industrial Arts Association
CIAB Coal Industry Advisory Board
CIAC Ceramics Information Analysis Center [of Purdue University, West Lafayette, Indiana, US]
CIAC Computer Incident Advisory Capability
CIAC Construction Industry Advisory Council [US]
CIAD Center of Industrial Automation and Development [Cuba]
CIAG Construction Industry Institute Action Group
CIAJ Communications Industry Association of Japan
CIAM Computerized Integrated and Automated Manufacturing
CIAP Climatic Impact Assessment Program [of US Department of Transportation]
CIAP Climatic Implications of Atmospheric Pollution
CIAP Comite Interamericana del Alianza para el Progreso [Interamerican Committee of the Alliance for Progress]
CIAP Community-Based Industrial Adjustment Program
CIAPA Chemical Industries Accident Prevention Association
CIAPY Centro de Investigación Agricolas de la Peninsula de Yucatan [Agricultural Research Center of the Peninsula of Yucatan, Mexico]
CIAQR Center for Indoor Air Quality Research [Canada]
CIAR Canadian Institute for Advanced Research
CIArb Chartered Institute of Arbitrators [UK]
CIAS CCRS (Canada Center for Remote Sensing) Image Analysis System

CIAT Centro Internacional de Agricultural Tropical [International Center for Tropical Agriculture at Cali, Columbia]

CIATF Comité International des Associations Techniques de Fonderie [International Committee of Foundry Technical Associations, Switzerland]

CIATM Conferencia Interamericana de Tecnologica Materiales [Interamerican Conference for Materials Technology, Chile]

CIB Centralized Intercept Bureau

CIB Centro de Investigaciónes Basicas [Center for Basic Research, Mexico]

CIB Change Impact Board

CIB Change Implementation Board

CIB Channel Interface Bus

CIB Chartered Institute of Building [Ascot, Berkshire, UK]

CIB COBOL Information Bulletin [US]

CIB Community Inquiry Bureau

CIB Conseil International du Bâtiment pour la Recherche, l'Etude et la Documentation [International Council for Building Research, Studies and Documentation, Netherlands]

CIB Containment Isolation B [Nuclear Reactors]

CIB Counterfeiting Intelligence Bureau [of International Chamber of Commerce, UK]

CIBADS Canadian Integrated Biological Agent Detection System

Ciba-Geigy J Ciba-Geigy Journal [Published by Ciba-Geigy Ltd., Basle, Switzerland]

Ciba-Geigy Tech Notes Ciba-Geigy Technical Notes [Published by Ciba-Geigy Ltd., Cambridge, UK]

CIBC Commonwealth Institute of Biological Control [Trinidad]

CIBCA Computer Interconnect Bus Control Adapter

CIBD Centre d'Information des Banques de Données [Data Bank Information Center, France]

CIBER Cellular Intercarrier Billing Exchange Record [Telecommunications]

CIBMBE Combined Ion Beam and Molecular Beam Epitaxy

CIBO Council of Industrial Boiler Owners [US]

CIBS Chartered Institution of Building Services [now CIBSE]

CIBSE Chartered Institution of Building Services Engineers [London, UK]

CIC Cambridge International College [UK]

CIC Canada Immigration Center

CIC Carbon Intercalation Compound

CIC Carrier Identification Code [Telecommunications]

CIC Centro de Información de Comunicaciónes [Communication Information Center, Havana, Cuba]

CIC Chemical Industries Council [US]

CIC Chemical Institute of Canada

CIC China Institute of Communications [PR China]

CIC Climate Impacts Center [of Macquarie University, Australia]

CIC Cogeneration Coalition

CIC Combat Information Center

CIC Command Input Buffer

CIC Command Interface Control

CIC Comisión de Investigaciones Cientificas [Scientific Research Commission, Argentina]

CIC Committee on Institutional Cooperation [US]

CIC Common Interface Circuit

CIC Communication Intelligence Channel

CIC Community Information Center

CIC Computer Intelligence Corps [US Army]

CIC Conseil International de la Chasse et de la Conservation du Gibier [International Council for Game and Wildlife Conservation, France]

CIC Construction Industry Commission [Canada]

CIC Construction Industry Council [UK]

CIC Consumer Information Center [Pueblo, Colorado, US]

CIC Control and Information Center

CIC Coordinated Information Center [of Reynolds Electrical and Engineering Company, Las Vegas, Nevada, US]

CIC Copper/Invar/Copper (Material)

CIC Council of Independent Colleges [US]

CIC Counter-Intelligence Corps [US]

CIC Cover-Integrated Cell

CIC Crew Interface Coordinator

CICA Canadian Industrial Communicators Assembly

CICA Canadian Institute of Chartered Accountants [Toronto, Ontario, Canada]

CICA Chemical and Industrial Consultants Association [UK]

CICA Chicago Industrial Communications Association [US]

CICA Competition in Contracting Act [US]

CICA Confederation of International Contractor Associations [France]

CICA Construction Industry Computing Association [UK]

CICA Convention on International Civil Aviation

CICAD Inter-Paliamentary Commission for Environment and Development [Australia]

CICAT Center for International Cooperation and Appropriate Technology

CICC Cargo Integration Control Center

CICC Custom Integrated Circuit Conference [of Institute of Electrical and Electronics Engineers, US]

CICCA Committee on International Cooperation of Cotton Associations

CIC-Can Cambridge International College of Canada

CICD Computer-Integrated Circuit Design

CICE Council for International Congresses of Entomology [UK]

CICERO Control Information (System) Concepts (based on) Encapsulated Real-Time Objects (Experiment) [of CERN–European Laboratory for Particle Physics, Geneva, Switzerland]

CICG Centre d'Informatique et Computation de Grenoble [Center for Informatics and Computation of the University of Grenoble, France]

CICHE Consortium for International Cooperation in Higher Education [US]

CICI Confederation of Information Communication Industries [UK]

CICIG Commissione Italiana del Comitato Internazionale di Geofisica [Italian Commission of the International Committee of Geophysics]

CICIN Conference on Interlibrary Communications and Information Networks [of American Library Association, US]

CICIREPATO Committee for International Cooperation in Information Retrieval among Examining Patent Offices

CICM Committee on Intersociety Cooperation in Materials [US]

CICP Communications Interrupt Control Program

CICPBA Comisión de Investigaciones Cientificas de la Provincia de Buenos Aires [Scientific Research Commission of the Province of Buienos Aires, Argentina]

CICRED Comité International de Coopération dans les Recherches Nationales en Démographie [Committee for International Cooperation in National Research in Demography, France]

CICRIS Cooperative Industrial and Commercial Reference and Information Service [UK]

CICS Canadian Industrial Computer Society

CICS Chemical Inventory Control System [of US Environmental Protection Agency]

CICS Committee for Index Cards for Standards

CICS Customer Information Control System

CICSA Canadian Indepent Computer Services Association

CICS/VS Customer Information Control System/Virtual Storage

CICU Communication Interface Control Unit

CICU Computer Interface Conditioning Unit

CICU Computer Interface Control Unit

CICU Coronary Intensive Care Unit

CICT Commission on International Commodity Trade

CICYT Comisión Interministerial de Ciencia y Technológia [Interministerial Commission of Science and Technology, Spain] [also CI-CYT]

CID Cayman Islands Dollar [Currency]

CID Center for Infectious Diseases [of Centers for Disease Control and Prevention, Atlanta, Georgia, US]

CID Center for Information and Documentation [of European Atomic Energy Community]

CID *(Centro de Investigación y Desarrollo)* – Spanish for "Research and Development Center"

CID Charge-Injection Device

CID Chemical Inorganic Deposition

CID Circular Intensity Difference

CID Collision-Induced Decomposition

CID Collision-Induced Desorption

CID Collision-Induced Dissociation

CID Combined Immunological Deficiency

CID Commerce and Industry Department [Hong Kong]

CID Commercial Item Description

CID Committee for Instructional Development [US]

CID Communication Identification

CID Communication Identifier

CID Component Identification

CID Computer Interface Device

CID Computerized Index of Delivery Data

CID Configuration/Installation/Distribution

CID Criminal Investigation Department [of Scotland Yard, London, UK]

CID Cubic Inch Displacement [Automobiles]

CID Current-Image Diffraction (Test)

CIDA Canadian International Development Agency

CIDA Centre International de Documentation Arachnologique [International Center for Arachnological Documentation, France]

CIDA Channel Indirect Data Addressing

CIDA Current Input Differential Amplifier

CIDA/NSERC Canadian International Development Agency/ Natural Sciences and Engineering Research Council (Joint Project) [Canada]

CIDAS Climatological Ice Data Archiving System

CIDAS Conversational Interactive Digital/Analog Simulator

CIDB Canadian International Development Board

CIDB Chemie Information und Dokumentation, Berlin [Berlin Center for Chemical Information and Documentation, Germany]

CIDC Commercial and Industrial Development Corporation

CIDC Computer-Integrated Design and Construction

CIDE Center for International Development and Environment [US]

CIDE Commission Intersyndicale de Deshydrateurs Européens [European Dehydrators Association, France]

CIDEP Chemically Induced Dynamic (Magnetic) Electron Polarization

CIDF Control Interval Definition Field

CIDF Cultural Industries Development Fund [of Business Development Bank of Canada]

CIDI Collision-Induced Dissociative Ionization

CIDIE Committee of International Development Institutions on the Environment

CIDIN Common ICAO (International Civil Aviation Organization) Data Interchange Network [of United Nations]

CIDL Configuration Item Data List

CIDNP Chemically-Induced Dynamic Nuclear-Spin Polarization

CIDNP-COSY Chemically-Induced Dynamic Nuclear-Spin Polarization–Correlation Spectroscopy

cIDP Cyclic Inosine Diphosphate [also c-IDP] [Biochemistry]

CIDR Classless Inter-Domain Routing [Internet]

CIDR Critical Intermediate Design Review

CIDS CEPEX (Central Equatorial Pacific Experiment) Integrated Data System

CIDS Chemical Information and Data System [US Army]

CIDS Communication, Information and Documentation System

CIDS Concrete Island Drilling System

CIDST Committee for Information and Documentation on Science and Technology [UK]

CIE Center for Integrated Electronics [of Rensselaer Polytechnic Institute, Troy, New York, US]

CIE Center for Interfacial Engineering [National Science Foundation Engineering Research Center at the University of Minnesota, Minneapolis, US]

CIE Center for International Economics

CIE Center for International Education

CIE Certificate in Electronics

CIE Chinese Institute of Engineers [Taiwan]

CIE Commission Internationale de l'Eclairage [International Commission on Illumination, Austria]

CIE Common Ion Effect [Chemistry]

CIE Commonwealth Institute of Entomology [UK]

CIE Communications Interface Equipment

CIE Compression-Ignition Engine

CIE Computer-Integrated Engineering

CIE Computer-Integrated Enterprise (Program)

CIE Córas Iompair Eireann [Irish Transport System]

CIE Council for International Education [UK]

CIE Counter Immunoelectrophoresis

Cie *(Compagnie)* – French for "Company"

CIEA Committee on International Education in Agricultural Sciences [Netherlands]

CIEBAA Comité International pour l'Etude des Bauxites, d'Alumine et d'Aluminium [International Committee for the Study of Bauxites, Alumina and Aluminum]

CIECD Council for International Economic Cooperation and Development

CIEDS Computer-Integrated Electrical Design Series

CIEE Council on International Educational Exchange [US]

CIEEM Center for Integrated Electronics and Electronics Engineering [of Rensselaer Polytechnic Institute, Troy, US]

CIEF Continuous Isoelectric Focusing

CIEH Charge Iterated Extended Hueckel [Physical Chemistry]

CIEI Center for International Environment Information [now World Environment Center, US]

CIE/ISO CIE (Commission Internationale de l'Eclairage)/ International Standards Organization (Specification)

CIE J CIE Journal [of Commission Internationale de l'Eclairage, Austria]

CIEL Center for International Environmental Law [Washington, DC, US]

CIELAB CIE (Commission Internationale de l'Eclairage) L* a* b* Scale Color Difference

Ciel Espace Ciel et Espace [French Publication on Sky and Space; published by Association Française d'Astronomie]

Ciel Terre Ciel et Terre [Belgian Publication on Sky and Earth; published by Société Belge d'Astronomie, de Météorologie et de Physique du Globe]

CIELUV CIE (Commission Internationale de l'Eclairage) L* u* v* Scale Color Difference

CIEMAT Centro de Investigaciones Energeticas Medioambientales y Technológicas [Research Center for Environmental Energetics and Technologies, Spain]

CIEN Commision Interamericana de Energía Nuclear [Inter-American Commission on Nuclear Energy; of Organization of American States]

Cienc Tec Fis Mat Ciencias Tecnicas, Fisicas y Matematicas [Cuban Journal on Technical Sciences, Physics and Mathematics; published by Academia de Ciencias de Cuba]

Cienc Tierra Espac Ciencias de la Tierra y del Espacio [Cuban Journal on Earth and Space Sciences; published by Academia de Ciencias de Cuba]

CIEP Canadian Institute for Economic Policy

CIER Comisión de Integración Electrica Regional [Regional Commission for Electrical Integration, Uruguay]

CIER Conseil International des Economies Régionales [International Council for Local Development, France]

CIES Comparative and International Education Society [US]

CIES Council for the International Exchange of Scholars [Washington, DC, US]

CIESIN Center for International Earth Sciences Information Network

CIESIN Consortium for International Earth Science Information Networks

CIETA Centre International d'Etude des Textiles Anciens [International Center for the Study of Ancient Textiles, France]

CIE/TC CIE (Commission Internationale de l'Eclairage) Technical Committee

CIE-USA Chinese Institute of Engineers–United States of America

CIF Canadian Institute of Forestry

CIF Captive Installation Function

CIF Central Index File

CIF Central Information File

CIF Central Instrumentation Facility

CIF Central Integration Facility

CIF Chemical Industry Federation [Finland]

CIF Cold-Insoluble Fibrinogen [Biochemistry]

CIF Common Interchange Format

CIF Common Intermediate Format

CIF Computer Integrated Factory

CIF Cork Industry Federation [UK]

CIF Cost, Insurance and Freight [also cif]

CIF Crystallographic Information File

CIF Customer Information Feed

CIFA Comité International de Recherche et d'Etude de Facteurs de l'Ambience [International Committee for Research and Study of Environmental Factors]

CIFAR Cooperative Institute for Arctic Research [of University of Alaska at Fairbanks, US]

Cifax Enciphered Facsimile Communication

CIFC Center for Interfirm Comparison [UK]

CIFC Cost, Insurance, Freight and Commission [also cifc, CIF&C, or cif&c]

CIFCI Cost, Insurance, Freight, Commission and Interest [also cifci, CIFC&I, cif&i]

CIFE Conference for Independent Further Education [UK]

CIFE Cost, Insurance, Freight and Exchange [also cife]

CIFEG Centre International pour la Formation et les Echanges Géologiques [International Center for Training and Exchanges in the Geosciences, France]

CIFFA Canadian International Freight Forwarders Association

CIFI Cost, Insurance, Freight and Interest [also cifi, CIF&I, or cif&i]

CIFKU Conditional Instability of the First Kind Upside-Down [Meteorology]

CIFM Consorzio Interuniversitario di Fisica della Materia [Inter-University Consortium on Matter Physics, Trieste, Italy]

CIFOR Center for International Forestry Research

CIFRI Central Inland Fisheries Research Institute [India]

CIFRR Common Instrument Flight Rules Room

CIFS Committee on the International Freedom of Scientists [of American Physical Society, US]

CIFS Common Internet File System

CIFST Canadian Institute of Food Science and Technology

CIFT Canadian Institute of Fisheries Technology [of Technical University of Nova Scotia, Halifax, Canada]

CIFW Cost, Insurance, Freight plus War Risk [also cifw]

CIG Cable Integrity Group

CIG Ceiling [also Cig, or cig]

CIG Coal Iron Gasification (Process)

CIG Cold-Walled Induction Guide (Forming)

CIG Comité International de Géophysique [International Geophysics Committee]

CIG Communications and Interface Group

CIG Compressed Industrial Gases (Company)

CIG Computer Image Generation

CIG Conference Interpreters Group [UK]

CIG Cryogenic-in-Ground

Ci/g Curie per gram [also Ci g^{-1}]

CIGAS Cambridge Intercollegiate Graduate Application Scheme [UK]

CIGGT Canadian Institute for Guided Ground Transport [Kingston, Ontario, Canada]

CIGI Canadian International Grain Institute

CIGL Centre d'Informatique Général de Liège [Center for General Informatics at Liege, Belgium]

CIGM Chemically Induced Grain Boundary Migration [Materials Science]

CIGM Chief Inspector for Gun Mounting [UK]

CIGR Commission Internationale du Génie Rural [International Commission of Agricultural Engineers, France]

CIGRE Conférence Internationale des Grands Réseaux Electriques à Haute Tension [International Conference on Large High-Voltage Electric Systems, France]

CIGS Centro Interdipartimentale Grandi Strumenti [Interdepartmental Center for Large Instruments; of University of Modena, Italy]

CIGS Chief of the Imperial General Staff

CIGS Copper Indium Gallium Diselenide (Thin Film)

CIGTF Central Inertial Guidance Test Facility

CIH Certified Industrial Hygienist

CIH Commonwealth Institute of Health [of Sydney University, Australia]

CIHSI Chemically Induced Hyperthermal Surface Ionization

CII Call Identity Index

CII Center for Innovation in Industry [UK]

CII Chartered Insurance Institute [UK]

CII Commercial, Industrial and/or Institutional (Facility)

CII Confederation of the Irish Industry

CII Construction Industry Institute

CII Containerization and Intermodal Institute [US]

CII Council for Industrial Innovation

CIIA Canadian Information Industry Association

CIIA Canadian Institute of International Affairs

CIIA Commission Internationale des Industries Agricoles et Alimentaires [International Commission for Food Industries, France]

CIIA Newsl CIIA Newsletter [of Canadian Information Industry Association]

CIIC Canadian Industrial Innovation Center

CIIC Centre d'Information de l'Industrie des Chaux et Ciments [Information Center of the Lime and Cement Industries, France]

CIIFO CIRRUS I IFO (Intensive Field Observations) [First ISCCP (International Satellite Cloud Climatology Project) Regional Experiment, Phase I (1984-1989)]

CIIG Construction Industry Information Group [UK]

CIIME Committee for International Investment and Multinational Enterprises [of Organization for Economic Cooperation and Development, France]

CIIMS Canadian Information and Image Management Society

CIIR Canadian Institute of International Relations

CIIR Chlorinated Isobutylene-Isoprene Rubber; Chlorobutyl Rubber

CIIT Canadian Institute of Industrial Technology [of National Research Council of Canada]

CIIT Chemical Industry Institute of Toxicology [US]

CIJ Cour Internationale de Justice [International Court of Justice, Netherlands]

CIL Cadet Instructor List

CIL Cadmium Isotope Lamp

CIL Call Identification Line

CIL Call Information Logging

CIL Carbon-in-Leach (Process) [also C-I-L]

CIL Common Interface Language

CIL Computer-Independent Language

CIL Critical Items List

CIL Current Injection Logic

C-I-L Carbon-in-Leach (Process) [also CIL]

Ci/L Curie per Liter [also Ci/l, or Ci L^{-1}]

CILA Centro Interamericano de Libros Academicos [Inter-American Center of Free Academies, Mexico]

CILA Chartered Institute of Loss Adjusters [UK]

CILC Confédération Internationale du Lin et du Chanvre [International Linen and Hemp Confederation, France]

CILE Call Information Logging Equipment

CILER Cooperative Institute for Limnology and Ecosystems Research [of University of Michigan, Ann Arbor, US]

CILOP Conversion in lieu of Procurement

CILP Current Index to Legal Periodicals (Database)

CILRT Containment Integrated Leak Rate Test [Nuclear Engineering]

CILS Collision-Induced Light Scattering

CILS Compressive Interlaminar Shear Strength

CILSS Comité Permanent Inter-Etats de Lutte Contre la Secheresse dans le Sahel [Permanent Interstate Committee for Drought Control in the Sahelian Zone, Burkina Faso]

CIM Canadian Institute of Management

CIM Canadian Institute of Marketing

CIM Canadian Institute of Mining, Metallurgy and Petroleum [also CIMM]

CIM Center for Irradiation of Materials [of Alabama A&M University, Normal, US]

CIM Certificate in Information Management

CIM Certificate in Management

CIM Certified Industrial Manager

CIM Charge Imaging Matrix

CIM Chartered Institute of Marketing [UK]

CIM Chinese Society for Measurement

CIM Common Information Model

CIM Communications Improvement Memorandum

CIM Communications Interface Monitor

CIM Compression Injection Molding

CIM CompuServe Information Manager

CIM Computer Input (from) Microfilm

CIM Computer Input Multiplexer

CIM Computer-Integrated Manufacturing

CIM Constituent Interchange Model

CIM Continuous Image Microfilm

CIM Custom Injection Molding

CIMA Chartered Institute of Management Accountants [UK]

CIMA (International Congress on) Computational Intelligence: Methods and Applications

CIMA Construction Industry Manufacturers Association [US]

CIMAC Computer-Integrated Manufacturing Control

CIMAC Congrès International des Machines à Combustion [International Congress on Combustion Engines, France]

CIMAC Conseil International des Machines à Combustion [International Council on Combustion Engines, France]

CIMAS Cooperative Institute for Marine and Atmospheric Studies [of University of Miami, Florida, US]

CIMAV Centro de Investigaciónes en Materiales Avanzados [Research Center for Advanced Materials, Chihuahua, Mexico]

CIMB Construction Industry Management Board [US]

Cim Bétons Plâtres Chaux Ciments Bétons Plâtres Chaux [French Journal on Cement, Concrete, Gypsum and Lime]

CIM Bull Canadian Mining and Metallurgical Bulletin [of Canadian Institute of Mining, Metallurgy and Petroleum]

CIMCO Card Image Correction

CIMD Combined Ion and Molecular Deposition

CIM&D Canadian Integrated Manufacturing and Design Show and Conference [also CIM/D, or CIMD]

CIMD/CAEC Canadian Integrated Manufacturing and Design Show and Conference/Canadian Architectural, Engineering and Construction Exposition

CIM CAPC Canadian Institute of Mining, Metallurgy and Petroleum/Canadian Association of Professional Conservators

CIM CD Canadian Institute of Mining, Metallurgy and Petroleum/Coal Division

CIM CMP Canadian Institute of Mining, Metallurgy and Petroleum/Canadian Mineral Processors

CIMD Combined Ion and Molecular Deposition

CIM-D (Canadian) Computer-Integrated Manufacturing/Design (Show)

CIME Chartered Institute of Marine Engineers [US]

CIME Chilean Institute of Mining Engineers

CIME Computer-Integrated Manufacturing and Engineering

CIME Computer-Integrated Mechanical Engineering (Show and Exhibition)

CIMechE Companion of the Institute of Mechanical Engineers

CIME-DESIGN Computer-Integrated Manufacturing and Engineering/Design Show [alsoCIME-Design]

CIM GD Canadian Institute of Mining, Metallurgy and Petroleum/Geology Division

CIMH Caribbean Institute for Meteorology and Hydrology

CIMH Comité International pour la Métrologie Historique [International Committee for Historical Metrology, France]

CIM HRC Canadian Institute of Mining, Metallurgy and Petroleum/Human Resources Committee

CIM IMD Canadian Institute of Mining, Metallurgy and Petroleum/Industrial Mineral Division

CIMM Canadian Institute of Mining, Metallurgy and Petroleum [also CIM]

CIMM Computer-Integrated Manufacturing and Management

Ci/mM Curie per millimole [also Ci/mmol, Ci mmol^{-1}, or Ci mM^{-1}]

CIM MEC Canadian Institute of Mining, Metallurgy and Petroleum/Mineral Economics Committee

CIM MED Canadian Institute of Mining, Metallurgy and Petroleum/Maintenance Engineering Division

CIM MMD Canadian Institute of Mining, Metallurgy and Petroleum/Metal Mining Division

Ci/mmol Curie per millimole [also Ci/mM, Ci mmol^{-1}, or Ci mM^{-1}]

CIMMS Cooperative Institute for Mesoscale Meteorological Studies [of University of Oklahoma, Norman, US]

CIM MS Canadian Institute of Mining, Metallurgy and Petroleum/Metallurgical Society

CIMMYT Centro Internacional de Mejoramiento de Maiz y Trigo [International Center for Corn and Wheat Improvement, Mexico]

CIMP Computer-Integrated Materials Processing

CIMP Cornell Injection Molding Program [of Cornell University, Ithaca, New York, US]

cIMP Cyclic-Inosine-3',5'-Monophosphate [also c-IMP] [Biochemistry]

CIMPC Computer-Integrated Manufacturing Programmable Controller

CIM PS Canadian Institute of Mining, Metallurgy and Petroleum/Petroleum Society

CIMR Center for International Media Research

CIM Rep CIM Reporter [Publication of the Canadian Institute of Mining, Metallurgy and Petroleum]

CIM Rev CIM (Computer-Integrated Manufacturing) Review [US]

CIM RMSCC Canadian Institute of Mining, Metallurgy and Petroleum/Rock Mechanics and Strata Control Committee

CIMRS Cooperative Institute for Marine Resources Studies [of Oregon State University, Corvallis, US]

CIMS Center for Inorganic Membrane Studies [of Worchester Polytechnic Institute, Massachusetts, US]

CIMS Chemical Ionization Mass Spectrometry; Chemical Ionization Mass Spectroscopy

CIMS Computer-Integrated Manufacturing System

CIMS Coordination and Interference Management System [of International Telecommunications Satellite Organization]

CIMSA Chloroiodomethanesulfonic Acid

CIMSE Computerized Instruction in Materials Science and Engineering

CIMSS Cooperative Institute for Meteorological Satellite Studies [of University of Wisconsin at Madison, US]

CIMT Center for Instructional Media and Technology

CIMTC Construction and Industrial Machinery Technical Committee [of Society of Automotive Engineers, US]

CIMTEC World Ceramics Congress (with HighTemperature Superconductor Symposium)

CIMTECH (National) Center for Information Media and Technology [at Hatfield Polytechnic, Hatfield, Hertfordshire, UK] [also Cimtech]

CIM Technol CIM (Computer-Integrated Manufacturing) Technology [Publication of the Computer and Automated Systems Association of the Society of Manufacturing Engineering, US]

CIMV Cauliflower Mosaic Virus

CIMVAR CIMplicity Value Added Remarketer [*Note:* CIM stands for Computer-Integrated Manufacturing]

CIN Carrier Input

CIN Change Identification Number

CIN Chemical Industry Notes (Database) [US]

CIN Communication Identification Navigation

Cin 8-Cinnolinol

CINAHL Nursing and Allied Health (Database)

CINC Commander-in-Chief [also C-in-C, or C in C]

CINCAF Commander-in-Chief of Allied Forces

CINCC Coal Industry National Consultative Council [UK]

CINCEUR Commander-in-Chief (of the US Forces) in Europe

CINCHAN Commander-in-Chief Channel [of NATO]

CINCNORAD Commander-in-Chief North American Aerospace Defense Command

CINDA Computer Index for Neutron Data [of United Kingdom Atomic Energy Authority]

CINDAS Center for Information and Numerical Data Analysis (and Synthesis) [of Purdue University, West Lafayette, Indiana, US]

CINDY Code for Internal Dosimetry

CINF Division of Chemical Information [of American Chemical Society, US]

CINF Bull Chemical Information Bulletin [of American Chemical Society, US]

CINF News Division of Chemical Information News [Newsletter of American Chemical Society, US]

CINS CENTO (Central European Treaty Organization) Institute of Nuclear Science [of NATO]

CINT Convert, Round to Next Integer (Function) [Programming]

CINTERFOR Centro Interamericano de Investigación y Documentación Sobre Formación Profesional [Inter-American Center for Research and Documentation on Vocational Training, Uruguay]

CINVESTAV Centro de Investigación y de Estudios Avanzados [Center for Research and Advanced Studies, Mexico]

CINVESTAV-IPN Centro de Investigación y de Estudios Avanzados del Instituto Politecnico Nacional [Center for Advanced Research and Studies of the National Polytechnic Instuitute, Mexico]

CIO Carrier Insertion Oscillator

CIO Central Input/Output (Multiplexer)

CIO Chief Information Officer

CIO Coal-in-Oil (Suspension)

CIO Confirming Informal Order

CIO Congress of Industrial Organizations [US]

CIOB Chartered Institute of Building [UK]

CIOCS Communications Input/Output Control System

CIOM Communications Input/Output Multiplexer

CIOMS Council for International Organizations of Medical Sciences [Switzerland]

CIOP Communications Input/Output Processor

CIOPORA Communauté Internationale des Obtenteurs de Plantes Ornamentales et Fruitière à Reproduction Asexuée [International Community of Breeders of Asexually Reproduced Fruit Trees and Ornamental Varieties, Switzerland]

CIOR Confederation Interallieé des Officiers de Reserve [Inter-Allied Confederation of Reserve Officers; of NATO]

CIOS Combined Intelligence Objectives Subcommittee

CIOS Conseil International de l'Organisation Scientifique [International Council for Scientific Management, Malaysia]

CIOU Custom Input/Output Unit

CIP Cahn-Ingold-Prelog System [Stereochemistry]

CIP Canadian Institute of Planning

CIP Carbon-in-Pulp [also C-I-P]

CIP Cascade Improvement Program

CIP Cassette In Place

CIP Cast-in-Place (Concrete)

CIP Cast Iron Pipe

CIP Cataloging in Publication

CIP Cell-Proliferation Inhibiting Protein [Biochemistry]

CIP Centro Internacional de la Papa [International Potato Center at Lima, Peru]

CIP Changchun Institute of Physics [Changchun, PR China]

CIP Circulator Installed Power [Nuclear Power Station]

CIP Clean(ing)-in-Place

CIP Cold Isostatic Press(ed) [also CIPed]

CIP Cold Isostatic Pressing [also CIPing]

CIP Command Interface Port

CIP Commercial Instruction Processor

CIP Common Image Part

CIP Common Indexing Protocol

CIP Common-Integrated Processor

CIP Communications Interrupt Program

CIP Compatible Independent Peripherals

CIP Composite Interlayer Bonding

CIP Computer Image Processing

CIP Computers in Physics [Publication of American Institute of Physics, US]

CIP Conduction-in-Plane (Magnetoresistance) [Solid-State Physics]

CIP Continuous Induction Process (Furnace) [Metallurgy]

CIP Council of Iron Producers

CIP Carriage and Insurance Paid [also cip]

CIP Current Injection Probe

CIP (Electric) Current In-Plane (Configuration) [Solid-State Physics]

C/IP Construction/Inspection Procedure

C-I-P Carbon-in-Pulp [also CIP]

CIPA Canadian Industrial Preparedness Association

CIPA Chartered Institute of Patent Agents [UK]

CIPA Comité International de Photogrammetrie Architecturale [International Committee of Architectural Photogrammetry, France]

CIPA Comité International des Plastiques en Agriculture [International Committee of Plastics in Agriculture, France]

CIPAC Collaborative International Pesticides Analytical Council [Netherlands]

CIPAC Copyright, Inventions and Patent Association of Canada

CIPAD Consejo Internaciónal de Procesamiento Automático de Datos [International Council for Automatic Data Processing, Madrid, Spain]

CIPASH Committee on International Programs in Atmospheric Sciences and Hydrology [of National Academy of Sciences/National Research Council, US]

CIPB Carnegie Institute of Plant Biology [Stanford, California, US]

CIPC Canadian Institute on Pollution Control

CIPC Chloro-Isopropyl-N-Phenylcarbamate

CIPCA Cogeneration and Independent Power Coalition of America [US]

CIP'd Cold Isostatic(ally) Pressed [also CIP, CIPed, or CIPped]

CIPE Center for International Private Enterprise [US]

CIPEC Canadian Industry Program for Energy Conservation [of Natural Resources Canada]

CIPEC Conseil Intergouvernemental des Pays Exportateurs de Cuivre [Intergovernmental Council of Copper Exporting Countries, France]

CIPed Cold Isostatic(ally) Pressed [also CIP, or CIP'd]

CIPG Coal Industry Tripartite Group [UK]

CIPH Canadian Institute of Plumbing and Heating

5-CIPHEN 5-Chloro-1,10-Phenanthroline

CIPHEX Canadian Instiitute of Plumbing, Heating and Cooling Exposition

CIPHI Canadian Institute of Public Health Inspectors

CIP/HIP Cold Isostatic Pressing/Hot Isostatic Pressing [also CIP-HIP]

CIPHONY Cipher and Telephone Equipment

CIPing Cold Isostatic Pressing [also CIP]

CIPM Comité International des Poids et Mésures [International Committee of Weights and Measures, France]

CIPM Council for International Progress in Management

CIPMP Commission Internationale pour la Protection de la Moselle contre la Pollution [International Commission for the Protection of the Moselle against Pollution, France]

CIP-MR Magnetoresistance with (Electric) Current in (Layer) Plane [Solid-State Physics]

CIPO Bis(2-Carboisopentyloxy-3,5,6-Trichlorophenyl)oxalate

CIPO Computer Integrated Plant Organization

CIPOM Computers, Information Processing and Office Machines (Committee) [of Canadian Standards Association]

CIPP Comprehensive Integrated Planning Process

CIPped Cold Isostatic(ally) Pressed [also CIP, CIPed, or CIP'd]

CIPPOCS Cast-In-Place, Push-Out Cylinders

CIPQ Centre International de Promotion de la Qualité [International Center for Promotion of Quality, France]

CIPR Commission Internationale de Protection contre les Radiations [International Commission of Radiological Protection, UK]

CIPRA Cast Iron Pipe Reseach Association [now Ductile Iron Pipe Research Association, US]

CIPRA Commission Internationale pour la Protection des Régions Alpines [International Commission for the Protection of Alpine Regions, Liechtenstein]

CIPS Canadian Information Processing Society

CIPS Continuous Ice Particle Sampler

CIPSI Configuration Interaction (Treatment) by Perturbation Selected Iterations

CIPS Rev CIPS Review [of Canadian Information Processing Society]

CIPW Cross, Iddings, Pirsson, and Washington (Classification) [Petrology]

CIR Canada-India Reactor [Bombay, India]

CIR Cargo Integration Review

CIR Carrier-to-Interference Ratio

CIR Center for Industrial Research [Norway]

CIR Center for International Research

CIR Characteristic Instants of Restitution

CIR Color Infrared

CIR Commission for Industrial Relations [UK]

CIR Committed Information Rate
CIR Configuration Inspection Report
CIR Control and Interrupt Register
CIR Cost Information Report
CIR Critical Isotope Reactor
CIR Current Instruction Register
CIr Certified Iridologist
Cir Circinus [Astronomy]
Cir Circle; Circular [also cir]
Cir Circuit [also cir]
Cir Circulate; Circulation [also cir]
Cir Circumference [also cir]
cir (circa) – about; approximately
cir circular
CIRA Cast Iron Research Association [UK]
CIRA Committee on Industrial Research Assistance [Industrial Research Assistance Program, Canada]
CIRA Committee on International Reference Atmosphere
CIRA (International Symposium on) Computational Intelligence in Robotics and Automation
CIRA Cooperative Institute for Research in the Atmosphere [of Colorado State University, Fort Collins, US]
CIRA COSPAR (Committee on Space Research) International Reference Atmosphere
CIRAC Canadian Institute for Research in Atmospheric Chemistry [of York University, Toronto, Canada]
CIRB Canadian Industrial Renewal Board [of Department of Regional and Industrial Expansion]
CIRC Center for Industrial Research and Consulting
CIRC Centralized Information Reference and Control
CIRC Centre International de Recherche sur le Cancer [International Agency for Research on Cancer; of World Health Organization; located at Lyon, France]
CIRC Circular Reference [Computers]
CIRC Cross-Interleaved Reed-Solomon Code
CIRC Cylindrical Internal Reflection Cell
Circ Circle; Circular;Circulation [also circ]
Circ Circuit [also circ]
Circ Circular(s) [also circ]
Circ Circumference [also circ]
circ (circa) – about; approximately
CIRCA Computerized Information Retrieval and Current Awareness
CIRCAL Circuit Analysis
Circ Brkr Circuit Breaker [also circ brkr]
Circ Electrotech Lab Circulars of the Electrotechnical Laboratory [Ibaraki, Japan]
Circuit Des Circuit Design [Journal published in the US]
Circuits Manuf Circuits Manufacturing [Journal published in the US]
Circuits Syst Signal Process Circuits, Systems and Signal Processing [Journal published in the US]
Circul Circulation [Journal of the American Heart Association, US]
Circul Res Circulation Research [Journal of the American Heart Association, US]

Circum Circumference [also circum]
CIRES Communication Instructions for Reporting Enemy Sightings
CIRES Cooperative Institute for Research in Environmental Sciences [of University of Colorado, Boulder, US]
CIRF Corn Industries Research Foundation [US]
CIRGA Critical Isotope Reactor, General Atomics [US]
CIRHS Critical Items and Residual Hazards List
CIRI Canadian Industrial Risk Insurers
CIRIA Construction Industry Research and Information Association [UK]
CIRIA Newsl CIRIA Newsletter [of Construction Industry Research and Information Association, UK]
CIRIS Complete Integrated Reference Instrumentation System
CIRK Computer Information Retrieval from Keyboards
CIRL Canadian Institute of Resources Law
CIRL Cotton Insects Research Laboratory [of US Department of Agriculture in Brownsville, Texas]
CIRLA Cloud Information Reference Archive and Library
CIRM Centre International de Rencontres Mathematiques [International Center for Mathematical Meetings, France]
CIRM Comité International de Radio Maritime [International Committee on Maritime Radio, UK]
CIRMES Center for Integrated Resource Management and Environmental Science
cir mil circular mil [Unit]
cirnav circumnavigate
CIRP Canadian Industrial Renewal Program
CIRQUE Conference on Industry and Resources, Queen's University Engineering [Kingston, Ontario, Canada]
CIRRA Canadian Industrial Relations Research Association
CIRRUS I Cirrus IFO (Intensive Field Observations), Phase I (Wisconsin 1986) [First ISCCP (International Satellite Cloud Climatology Project) Regional Experiment]
CIRRUS II Cirrus IFO (Intensive Field Observations), Phase II (Kansas 1991) [First ISCCP (International Satellite Cloud Climatology Project) Regional Experiment]
CIRS Composite Infrared Spectrometer
CIRT Computer and Information Resources and Technology Center [of University of New Mexico, Albuquerque, US]
CIS Cadmium Indium Sulfide (Semiconductor)
CIS Cadmium Iron Selenide (Semiconductor)
CIS Canadian Institute of Surveying [now Canadian Institute of Surveying and Mapping]
CIS Cancer Information Service [US]
CIS Card Information Structure
CIS Cargo Information System
CIS Central Integration Site
CIS Centre International d'Informations de Sécurité et d'Hygiène du Travail [International Information Center for Industrial Safety and Health; of International Labour Organization]
CIS Certificate in Information Science
CIS Chain-Initiating Step [Chemistry]
CIS Change Impact Summary

CIS Channel and Isolation Supervision

CIS Character Instruction Set

CIS Characteristic Isochromat Spectroscopy

CIS Chemical Information Service [UK]

CIS Chemical Information System

CIS Client Information System

CIS Closed Ion Source

CIS Colloque International de Salzbourg [International Salzburg Colloquium–Automobiles and Materials]

CIS Commercial Instruction Set

CIS Commonwealth of Independent States [Formed in 1991 by 11 Former Soviet States]

CIS Communication and Instrumentation System

CIS Communication Information System

CIS Communication Interface System

CIS Component Identification Sheet

CIS CompuServe Information Service

CIS Computer Imaging System

CIS Computer Information System

CIS Constant Initial Energy Spectra

CIS Constant Initial State

CIS Constant Initial State(-Energy) Spectroscopy

CIS Constant Injection System

CIS Contact Image Sensor

CIS Contact to Inner Solution

CIS Containment Isolation System [Nuclear Reactors]

CIS Continuous Injection System

CIS Contractor's Information Submittal

CIS Cooperative Independent Surveillance

CIS Copper Indium Selenide (Solar Cell)

CIS Course Information System

CIS Cranbrook Institute of Science [Detroit, Michigan, US]

CIS Cryogenic Interferometer Spectrometer

CIS Crystal Ion Slicing (Technique)

CIS Cue Indexing System

CIS Current Index to Statistics (Database)

CIS Current Information Selection

CIS Customer Information System

CI$ CompuServe Information $ervice

cis- prefix denoting a stereoisomer that has atoms, or atom groups attached to the same side of a molecular chain [Chemistry]

CISA Canadian Industrial Safety Association

CISA Canadian Intelligence and Security Association

CISA Casting Industry Suppliers Association [US]

CISAM Compressed Index Sequential Access Method

CISC Canadian Institute for Steel Construction

CISC Complex Instruction Set Computer; Complex Instruction Set Computing [also cisc]

CISC Construction Industry Stabilization Committee [US] [Defunct]

CISCA Ceilings and Interior Systems Construction Association [US]

CISCO Compass Integrated System Compiler

CISCO Computer Information System for Center Operations

CISD Configuration Interaction (Including All) Single and Double Excitations

CISE Computer and Information Science and Engineering (Program) [of National Science Foundation, US]

CISE Consortium for International Studies Education [US]

CISEM Center for Interface Science and Engineering of Materials [of (South) Korea Science and Engineering Foundation]

CISEP Cellulose Industry Standards Enforcement Program [US]

CISET Committee on International Science, Engineering and Technology

CIS Extended Current Index to Statistics Extended (Database)

CISFFEL Conférence Internationale pour le Soudage et la Fusion aux Faisceau d'Electrons et Laser [International Conference on Welding and Melting by Electron and Leaser Beams, France]

CISHEC Chemical Industries Association's Safety and Health Council [UK]

CISIR Ceylon Institute of Scientific and Industrial Research [now Sri Lanka Institute of Scientific and Industrial Research]

CISK Conditional Instability of the Second Kind [Meteorology]

CISL Confederazione Italiana Sindacati Lavoratori [Italian Confederation of Labor Unions]

CISM Canadian Institute of Surveying and Mapping

CISM Centre International des Sciences Mécaniques [International Center for Mechanical Sciences, France]

CISM J CISM Journal [of Canadian Institute of Surveying and Mapping]

CISP Canadian Institute of Surveying and Photogrammetry

CISPI Cast Iron Soil Pipe Institute [US]

CISPR Comité International Spécial des Perturbations Radioélectriques [International Special Committee on Radio Interferences, UK]

CISR Center for Information Systems Research [of Massachusetts Institute of Technology, Cambridge, US]

CISR Center for International Systems Research [of United States Department of State]

CISRC Computer and Information Science Research Center [of Ohio State University, Columbus, US]

CISRI Central Iron and Steel Research Institute [Beijing, PR China]

CISS Canadian Institute of Strategic Studies

CISS Centaur Integrated Support Structure

CISS Center for International Strategic Studies [of York University, Toronto, Canada]

CISS Constant Initial State Spectroscopy

CISS Corporate Information Systems and Services

CiSt Cirrostratus (Cloud)

CISTI Canada Institute for Scientific and Technical Information [of National Research Council of Canada]

CISTI HIA Canada Institute for Scientific and Technical Information–Herzberg Institute of Astrophysics [of National Research Council of Canada]

CISTOD Confederation of International Scientific and Technological Organizations for Development [France]

CIT Cadmium Iron Telluride (Semiconductor)

CIT California Institute of Technology [Pasadena, US] [also Caltech, or CalTech]

CIT Call-in-Time

CIT Canadian Import Tribunal

CIT Canberra Institute of Technology [Australia]

CIT Carnegie Institute of Technology [of Carnegie-Mellon University, Pittsburgh, Pennsylvania, US]

CIT Case Institute of Technology [Now Part of Case Western Reserve University, Cleveland, Ohio, US]

CIT Caulfield Institute of Technology [now Monash University–Caulfield Campus, Australia]

CIT Center for Information Technology [of National Institutes of Health, Bethesda, Maryland, US]

CIT Center for Innovative Technology [Virginia, US]

CIT Cesium Iodide Thallide

CIT Chartered Institute of Transport [UK]

CIT Chiba Institute of Technology [Chiba, Japan]

CIT (Committee on) Civilian Industrial Technology [US]

CIT Collection Implanted Technology

CIT Comité International des Transports Ferroviares [International Rail Transport Committee, Switzerland]

CIT Commensurate-Incommensurate Transition [Materials Science]

CIT Compact Ignition Tokamak [US]

CIT Compact Ignition Torus [of Princeton University, New Jersey, US]

CIT Compressor Inlet Temperature

CIT Computer-Integrated Telephony

CIT Computer-Integrated Testing

CIT Computer Interface Technology

CIT Court of International Trade [US]

CIT Crack-Induced Tension (Method) [Mechanics]

CIT Cranfield Institute of Technology [Cranfield, Bedford, UK]

CIT Critical Item Tag

C&IT Communications and Information Technology

Cit Citation [also cit]

Cit Citizen [also cit]

Cit Citrate; Citric Acid

Cit Citrulline [Biochemistry]

cit cited

Cit A Citric Acid [also cit a]

CITAB Computer Instruction and Training Assistance for the Blind

CITB Construction Industry Training Board [UK]

CITC Canadian Institute of Timber Construction

CITC Canadian International Trade Classification

CITC Committee of International Technical Cooperation [Taiwan]

CITC Construction Industry Training Center [US]

CITCA Committee of Inquiry into Technological Change in Australia

CITE Capsule-Integrated Test Equipment [Aerospace]

CITE Cargo-Integrated Test Equipment; Cargo Integration Test Equipment

CITE Compression, Ignition and Turbine Engine

CITE Contractor-Independent Technical Effort

CITE Council of Institute of Telecommunications Engineers [US]

CITE Current Information Transfer in English

CITEP Commonwealth Industrial Training Experience Program

CITEPA Centre Interprofessionel Technique d'Etudes de la Pollution Atmosphérique [Interprofessional Technical Center for the Study of Atmospheric Pollution]

CITES (Washington) Convention on International Trade in Endangered Species (of Wild Fauna and Flora)

CIT/GAL California Institute of Technology/Guggenheim Aeronautical Laboratory [Pasadena, US]

CIT/HDL California Institute of Technology/Hydrodynamics Laboratory [Pasadena, US]

CIT/JPL California Institute of Technology/Jet Propulsion Laboratory [Pasadena, US]

CITHA Confederation of International Trading Houses Associations [Netherlands]

CITL Canadian Industrial Transportation League

CITM Charge Injection Transistor Memory

CITO Communications and Information Technology in Ontario [of Ministry of Economic Development, Trade and Tourism, Canada]

CITR Canadian Institute for Telecommunications Research

CITREX Citric Acid Extraction (Process)

CITS Commission Internationale Technique de Sucrerie [International Commission of Sugar Technology, Belgium]

CITS Computer-Integrated Testing System

CITS Current Imaging Tunneling Spectroscopy

CITT Canadian Institute of Traffic and Transportation

CITTA Confédération Internationale des Fabricants de Tapis et de Tissus [International Confederation of Manufacturers of Carpets and Furnishing Fabrics, France]

CITU Confederation of Independent Trade Unions [Luxembourg]

City Transp City Transport [Journal published in the UK]

CIU Cable Interface Unit

CIU Channel Interface Unit

CIU Chlorella International Union [US]

CIU Clinical Information Utility

CIU Color Intensity Unit

CIU Communications Interface Unit

CIU Computer Interface Unit

CIU Control Indicator Unit

CIU Controller Interface Unit

CIU Coupler Interface Unit

CIUR Commission Internationale des Unités et des Mésures Radiologiques [International Commission on Radiological Units and Measures]

CIV Center Island Vessel

CIV Chilo Iridescent Virus

CIV City Imperial Volunteers

CIV Containment Isolation Valve [Nuclear Reactors]

CIV Corona Inception Voltage

Civ Civil(ian) [also civ]

CIVA Charge Induced Voltage Alternation

CivE Civil Engineer [also Civ Eng, or CivEng]

Civ Eng Civil Engineer(ing)

Civ Eng Civil Engineering [Journal published in the UK]

Civ Eng Jpn Civil Engineering in Japan [Journal published in Japan]

CIVR Computer and Interactive Voice Response

CIVT Cargo Interface Verification Test

CIVT Center for Interventional Vascular Therapy [of Stanford University Hospital, California, US]

CIVVA Current-Induced Voltage-Variation Analysis

CIW Carnegie Institution of Washington [Washington, DC, US]

CIW Current In (Domain) Wall (Direction) [Solid-State Physics]

CIW MR Current In (Domain) Wall Magnetoresistance [Solid-State Physics]

CIWS Central Instrument Warning System

CIWS Close-In Weapons System

CIX Commercial Internet Exchange

CIX Compulink Information Exchange

Ci/yr Curie per year [also Ci yr^{-1}]

CJ Chief Justice

CJ Chuck Jaw

CJ Cold Junction

CJ Computer Journal [of British Computer Society]

CJ Contiguous Junction

CJ Cottrell-Jawson (Parameter) [Physics]

c.j. One gallon [Apothecary]

CJA (Brotherhood of) Carpenters and Joiners of America

CJA Commonwealth Journalists Association [UK]

CJCC Commonwealth Joint Communications Committee [UK]

CJChE Canadian Journal of Chemical Engineering

CJCP Corrected Coupled J Cross-Polarization

CJCS Conventional Joint Control System

CJCSM Chinese Joint Committee of Societies of Materials

CJF Connecticut Joint Federation [US]

CJIS Canadian Journal of Information Science

CJJ Coupled Josephson Junction [Electronics]

CJLI Command Job Language Interpreter

CJM Canadian Journal of Mathematics

CJP Canadian Journal of Physics

CJP Communication Jamming Processor

CJS Canadian Jobs Strategy

CJS Canadian Jobs Studies

CJTC Canada-Japan Trade Council

CK Calcia-Killed (Steel) [Metallurgy]

CK Casein Kinase [Biochemistry]

CK Cement Kiln

CK Check [also Ck, or ck]

CK Chemical Kinetics [Physical Chemistry]

CK Commonwealth of Kentucky [US]

CK Competitive Kinetics

CK Cook Islands [ISO Code]

CK Coster-Kronig (Transition) [Physics]

CK Creatine Kinase [Biochemistry]

CK Cyanogen Chloride [Symbol]

CK Cytokeratin [Biochemistry]

CK I Casein Kinase I [also CK 1] [Biochemistry]

CK II Casein Kinase II [also CK 2] [Biochemistry]

Ck Cask [also ck]

Ck Check [also ck]

c(κ) compositional scattering vector (in materials science) [Symbol]

χ(k) photoelectron wave vector dependent oscillation [Symbol]

χ(κ) EXAFS (Extended X-Ray Absorption Fine Structure) Interference Function [Symbol]

CKα Carbon K-Alpha (Radiation) [also Ck$_\alpha$]

CKAFS Cape Kennedy Air Force Station [now Cape Canaveral Air Force Station]

CKAO Coalition to Keep Alaska Oil [US]

CK-BB Creatine Kinase, Isoenzyme BB [Biochemistry]

CKC Canadian Kennel Club

CKD Completely Knocked Down

CKD Count-Key Data Device

CKDIG Check Digit

C/kg Coulomb per kilogram [also C kg^{-1}]

C/kg·s Coulomb per kilogram second [also C/(kg·s), C/kg/s, C/kg^{-1}·s^{-1}, or C kg^{-1} s^{-1}]

CKI Cocos/Keeling Island (Region) [Australia]

CKM Cabibbo-Kobayashi-Maskawa (Matrix) [Physics]

CK-MB Creatine Kinase, Isoenzyme MB [Biochemistry]

CK-MB Creatine Kinase, Isoenzyme MM [Biochemistry]

CKMTA Cape Kennedy Missile Test Annex [of NASA in Florida, US]

CKO Checking Operator

CKO Closs-Kapstein-Oosterhoff (Radical Pairs Model)

Ckpt Checkpoint

CKR Czech(oslovakian) Koruna [Currency]

CKRA Cape Kennedy Reference Atmosphere [now Cape Canaveral Reference Atmosphere]

CKS Cell Kinetics Society [US]

Cks Checks [also cks]

cks centistokes [Unit]

CKSNI Cape Kennedy Space Network [of NASA]

Ckt Circuit [also ckt]

ckw counterclockwise

¢/kW cents per kilowatthour (of electricity)

¢/kWh cents per kilowatthour (of electricity)

CL Cable Line

CL Cable Link

CL Cadmium Lamp

CL Caldwell-Luc (Operation) [Medicine]

CL Camera Length [Microscopy]

CL Caprolactam
CL Caprolactone
CL Carload [also C/L, or cl]
CL Cathodoluminescence; Cathodoluminescent
CL Cavendish Laboratory [of Cambridge University, UK]
CL Ceiling Level
CL Ceiling Limit
CL Center Line [also cl]
CL Centerline Lights
CL Cerro Largo [Uruguay]
CL Chapman's Law [Physics]
CL Chemiluminescence; Chemiluminescent
CL Chemical Laser
CL Chile [ISO Code]
CL Cholestryl Linoleate (Liquid Crystal)
CL Chronic Laryngitis
CL Classic
CL Clinical Laboratory
CL Closed Loop [also C/L]
CL Coefficient of Lift [Aerospace]
CL Coherence Length [Physics]
CL Coil Loading [Communications]
CL Cold Layer
CL College Library
CL Collimator Lens
CL Command Language
CL Communications Language
CL Compatibility List
CL Compound Layer
CL Compound Leaf [Botany]
CL Compressed Length (e.g., of Springs) [Symbol]
CL Compressibility Limit
CL Computational Linguistics
CL Computer Language
CL Concentric Louver
CL Condenser Lamp
CL Condenser Lens
CL Conference Location
CL Confidence Level [Statistics]
CL Confidence Limit(s) [Statistics]
CL Connecting Line
CL Conservation Line
CL Contact Light
CL Contact Line
CL Continuous Loading
CL Contour Line
CL Contrast Liquid
CL Control Language
CL Control Law
CL Control Leader
CL Control Limit
CL Control Logic
CL Conversion Loss

CL Coordinated Ligand [Chemistry]
CL Cord and Loop (Installation)
CL Core Loss
CL Corpus Luteum [Medicine]
CL Course Line
CL Crack Layer (Model)
CL Crystal Lattice
CL Crystalloluminescence; Crystalloluminescent
CL Culham Laboratory [Oxon, UK]
CL Current Layer
CL Current Logic
CL Current Loop
CL Cutter Location [Numerical Control]
CL Cutting Lubricant
CL Cyclic Loading
CL Cylindrical Lens
CL Light Cruiser [US Navy Symbol]
CL Low-Cloud-Type Code
CL_c Control Limit for "c" Control Chart (in Statistics) [Symbol]
CL_{Rm} Control Limit for Moving-Average (Rm) Control Chart (in Statistics) [Symbol]
CL_u Control Limit for "u" Control Chart (in Statistics) [Symbol]
C/L Carload [also CL, or cl]
C/L Checklist
C/L Closed Loop [also CL]
C-L Face Orientation (i.e., Circumferential Direction–Forging Direction) [Forging]
Cl Claim [also cl]
Cl Class; Classification [also cl]
Cl Clause [also cl]
Cl Clavicle [Anatomy]
Cl Clay
Cl Clearance [also cl]
Cl Clerk [also cl]
Cl Close(d); Closing [also cl]
Cl Clostridium [Microbiology]
Cl Cluster [also cl]
Cl Clutch [also cl]
Cl Ceiling Average [Symbol]
Cl Chlorine [Symbol]
Cl^- Chlorine Ion [Symbol]
Cl_2 Molecular Chlorine [Symbol]
Cl-33 Chlorine-33 [also ^{33}Cl, or Cl^{33}]
Cl-34 Chlorine-34 [also ^{34}Cl, or Cl^{34}]
Cl-35 Chlorine-35 [also ^{35}Cl, or Cl^{35}]
Cl-36 Chlorine-36 [also ^{36}Cl, or Cl^{36}]
Cl-37 Chlorine-37 [also ^{37}Cl, or Cl^{37}]
Cl-38 Chlorine-38 [also ^{38}Cl, or Cl^{38}]
Cl-39 Chlorine-39 [also ^{39}Cl, or Cl^{39}]
$C(\lambda)$ Wavelength-Dependent Concentration [Symbol]
cL centiliter [also cl]
cl carload [also CL, or C/L]
cl *(citato loco)* – as mentioned above

cl   clear

cl   close(ly)

cl   closure [Topology]

.cl   Chile [Country Code/Domain Name]

CLA   Canadian Labour Association

CLA   Canadian Library Association

CLA   Canadian Lumbermen's Association

CLA   Catch Limit Algorithm

CLA   Center for Laser Applications [of University of Tennessee Space Institute, Tullahoma, US]

CLA   Centerline Average

CLA   Certified Laboratory Assistant

CLA   Clear and Add

CLA   Communication Line Adapter

CLA   Communication Link Analyzer

CLA   Computer Law Association [US]

CLA   Computer Lessors Association [US]

CLA   Conjugated Linoleic Acid

CLA   Copyright Licensing Agency [UK]

Cla   Caryophanon latum L [Microbiology]

CLAA   Antiaircraft Cruiser [US Navy Symbol]

CLAC   Certificate in Library and Archives Conservation

CLAC   Christian Labour Association of Canada

CLAC   Closed Loop Approach Control

ClAc   Chloroacetic Acid

Cl-AcAc   Chloro-Acetylacetonate [also Cl-acac]

CLACC   Confederación Latinoamericano de Bioquímica Clinica [Latin American Confederation of Clinical Biochemistry, Colombia]

Clad   Alclad (Alloy)

CLADES   Latin American Center for Economic and Social Affairs [of United Nations]

CLAES   Cryogenic Limb Array Etalon Spectrometer

CLAF   Centro Latinoamericano de Física [Latin American Center for Physics, Brazil]

CLAIM   Center for Library and Information Management [UK]

CLAIRA   Chalk, Lime and Allied Industries Research Association [now Welwyn Hall Research Association, UK]

CLAM   Chemical Low-Altitude Missile

CLAMP   Center for Laser-Aided Material Processing [of University of Illinois at Urbana-Champaign, US]

CLAN   Clean Air Society in the Netherlands

C&L Appl   C&L Applications [UK Journal on Computers and Libraries]

CLAR   Channel Local Address Register

Cl_2/Ar (Molecular) Chlorine/Argon (Gas Mixture)

Clar   Clarendon [UK]

CLARA   Cornell Learning and Recognizing Automaton [of Cornell University, Ithaca, US]

CLARB   Council of Landscape Architectural Registration Boards [US]

CLARCS   Copyright Licensing Agency Rapid Clearance Service [UK]

CLARET   Cloud Lidar and Radar Exploratory Tests

CLAS   Clean Laminated Air-Station

CLAS   Computerized Library Acquisitions System

CLAS   Computer Library Applications Service

CLAS   Counter-Gravity Low-Pressure Air-Melt(ed) Sand Casting [Metallurgy]

CLASP   Closed Line Assembly for Single Particles [Nuclear Fusion]

CLASP   Composite Launch and Spacecraft Program

CLASP   Computer Language for Aeronautics and Space Programming [of NASA]

CLASS   California Libraries Authority for Systems and Services [US]

CLASS   Client Access to Systems and Services

CLASS   Closed Loop Accounting for Store Sales

CLASS   Composite Laminate Analysis Systems

CLASS   Computer-Based Laboratory for Automated School Systems

CLASS   Concrete Lintel Association [UK]

CLASS   Conservation Learning Activities for Science and Social Studies [of National Wildlife Federation, US]

CLASS   Cooperative Library Agency for Systems and Services

CLASS   Cross-Chain Loran Atmospheric Sounding System

CLASS   Current Literature Alerting Search Service [US]

CLASS   Custom Local-Area Signaling Services [Telecommunications]

Class   Classification [also class]

class   classic(al)

CLASSMATE   Computer Language to Aid and Simulate Scientific, Mathematical and Technical Education

Class Quantum Gravity   Classical and Quantum Gravity [Journal of Institute of Physics, UK]

CLAT   Aircraft Latitude [Loran Navigation]

CLAT   Communications Line Adapter for Teletype

CLAUSPOL   Claus Polyethylene Glycol (Process)

CLAVR   Clouds from AVHRR (Advanced Very-High-Resolution Radiometer)

CLAW   Clustered Atomic Warhead

CLAYCOP   Clay-Supported Cupric Nitrate

CLAYFEN   Clay-Supported Ferric Nitrate

Clay Miner   Clay Minerals [Publication of the Mineralogical Society, UK]

Clay Res   Clay Research [Journal published in the US]

Clay Sci   Clay Science [Journal published in the UK]

Clays Clay Miner   Clays and Clay Minerals [Journal published in the US]

CLB   Central Logic Bus

CLB   Cerium Lanthanum Boride (Kondo Alloy)

CLB   Community Land Bank

CLB   Cone Locator Bushing

CLBA   Canadian Log Builder Association

CLBA   Canadian Long Baseline Array

CLBDA   Cyprus Land and Building Developers Association

CLBM   Continuous Linear Bubble Model (of Grain Growth) [Metallurgy]

ClBNH   Iminochloroborane

ClBz   Chlorobenzene

CLC Canadian Labour Code
CLC Canadian Labour Congress
CLC Caribbean Labour Congress
CLC Central Land Council
CLC Central Logic Control
CLC Change Letter Control
CLC Cholesteric Liquid Crystal
CLC Circle of Least Confusion (in Optics)
CLC Classical Liquid Crystal
CLC Clear Carry Flag [Computers]
CLC Column Liquid Chromatography
CLC Combinational Logic Circuit
CLC Commercial Letter of Credit
CLC Commonwealth Liaison Committee [UK]
CLC Communications Line Control
CLC Communications Link Controller
CLC Constant Light Compensating
CLC Containment Leakage Control [Nuclear Reactors]
CLC Course Line Computer
CLC Cost of Living Council [US]
Cl/C Chlorine-to-Carbon (Ratio) [Chlorocarbon Compounds]
CLCA Comité de Liaison de la Construction Automobile [Liaison Committee for the Motor Vehicle Industry, of European Economic Community]
8-Cl-cAMP 8-Chloro-Cyclic Adenosine-3',5'-Monophosphate [Biochemistry]
Cl4CAT Tetrachloro-Catechol
CLCC Ceramic Leaded Chip Carrier [Electronics]
CLCCR Comité de Liaison de la Construction de Carrosseries et de Remorques [Liaison Committee of the Body and Trailer Building Industry, Germany]
CLCD Color Liquid Crystal Display
CLCGM Closed Loop Cover Gas Monitor
CLCIS Closed Loop Cover and Instrumentation System
CLCM Cooperating Libraries of Central Maryland [US]
CLCO Calcium Lanthanum Copper Oxide (Superconductor)
Cl Coll Clinch Collar
CLCPE California Legislative Council on Professional Engineers [US]
CLCR Controlled Letter Contract Reduction
CLCS Central Laboratory for Control Systems [of Bulgarian Academy of Sciences, Sofia]
CLCS Checkout and Launch Control System
CLCS Chinese Language Computer Society
CLCS Closed-Loop Control System
CLCS Consequence Limiting Control System
CLCS Current Logic, Current Switching
CLCV Cold Leg Check Valve
CLD Called Line
CLD Capacitance Level Detector
CLD Caprolactam Disulfide
CLD Central Library and Documentation Branch [of International Labour Organization]
CLD Chlorine Leak Detector
CLD Closed Loop Drive

CLD Cloud [also Cld, or cld]
CLD Constant Level Discriminator
CLD Course Line Deviation
CLD Crystal Lattice Defect
CLD Clear Direction Flag [Computers]
CLD Cleared Customs
Cld Cloud [also CLD, or cld]
cld cleared
cld cooled
CLDAS Clinical Laboratory Data Acquisition System
CLDB (Interlab) Cell Line Database
CLDCC Ceramic Leaded Chip Carrier [Electronics]
CLDI Course Line Deviation Indicator
CLDPE Chlorinated Low-Density Polyethylene
CLDS Canada Land Data System [of Environment Canada]
CLDS Closed Loop Drive System
CLDS Closed Loop Dynamic Stability
CLDS Clouds [also Clds, or clds]
CLDST Closed Loop Dynamic Stability Test
CLE Capillary Liquid Epitaxy
CLE Coefficient of Linear Expansion
CLE Communications Line Expander
CLE Continuum Linear Elasticity (Approach) [Materials Science]
CLE Croisot-Loire Enterprises [France]
CLEA Canadian Library Exhibitors Association
CLEA Conference on Laser Engineering and Applications [of Institute of Electrical and Electronic Engineers, US]
CLEAN Case Based Learning (Project) [Europe]
CLEAN Commonwealth Law Enforcement Assistance Network
CLEAR Campaign for Lead-Free Air [UK]
CLEAR Compiler, Executive Program, Assembler Routines
CLEAR Components Life Evaluation and Reliability
CLEAR Copper Leach Electrolysis and Regeneration (Process)
CLEC Chiral Ligand Exchange Chromatography
CLEC Crosslinked Enzymes Microcrystal(s)
CLECAT Comité de Liaison Européen des Commissionnaires et Auxiliaires de Transport [European Liaison Committee of Forwarders, Belgium]
CLEFT Cleavage of Lateral Epitaxial Film for Transfer
CLEHA Conference of Local Environmental Health Administrators
CLEM Cargo Lunar Excursion Module [of NASA]
CLEM Closed Loop Ex-Vessel Machine [Nuclear Engineering]
CLEM Composite for the Lunar Excursion Module
CLEO Clear Language for Expressing Orders
CLEO Conference (and Trade Show) on Lasers and Electrooptics
CLEO/IQEC Conference on Lasers and Electrooptics/International Quantum Electronics Conference [also CLEO-IQEC]

CLEO/QELS Conference on Lasers and Electrooptics/Quantum Electronics and Laser Science [also CLEO-QELS]

CLEP Cavendish Laboratory of Experimental Physics [of University of Cambridge, UK]

CLEPA Comite de Liaison de la Construction d'Equippements et de Piece d'Automobiles [Liaison Committee of Manufacturers of Motor Vehicle Parts and Equipment, UK]

CLES Certified Laboratory Equipment Specialist

CLETS California Law Enforcement Telecommunications System [US]

CLF Cadmium Lanthanum Fluoride

CLF Canadian Liver Foundation

CLF Capacitive Loss Factor

CLF Cellular Ligand Field (Model)

CLF Clear Forward

CLF Closed Line Field

CLF Cutter Location File [Numerical Control]

ClF Chlorine Monofluoride

CLFB Canadian Livestock Feed Board

clfd classified [also clf'd]

CLFMI Chain Link Fence Manufacturers Institute [US]

CLG Calling Line

CLG Ceiling [also Clg, or clg]

CLG Controlled Lead Grade (of Slab Zinc)

CLG Cooperative Libraries Group [now Cooperative Automation Group, UK]

CLG Crosswind Landing Gear

Clg Catalog(ue) [also clg]

Clg Ceiling [also clg]

CL GB Clean Grain Boundary (i.e., Without Impurities) [Materials Science]

Clg No Catalog(ue) Number [also clg no]

CLGP Cannon-Launched Guided Projectile

CLGW Cement, Lime and Gypsum Workers (International Union)

CLHIA Canadian Life and Health Insurance Association

CLI Calling Line Identification

CLI Call-Level Interface

CLI Canadian Land Inventory

CLI Canadian Lifeboat Institution

CLI Cathodoluminescence Imaging

CLI Chemical Liquid Infiltration

CLI Clear Interrupt Flag

CLI Client Library Interface

CLI Coin Level Indicator

CLI Command Language Interpreter

CLI Command Line Interface

CLI Continuity Limit Index [Materials Science]

CLI Control Language Interpreter

CLI Cutter Location Information [Numerical Control]

CLIA Clinical Laboratory Improvement Act [US]

CLIB C Library [Computers]

CLIC Canadian Law Information Council

CLIC Command Language for Interrogating Computers [of Royal Radar Establishment, UK]

CLIC Compact Linear Collider Study [of CERN–European Laboratory for Particle Physics, Geneva, Switzerland]

CLIC Computer Liability Insurance Coverage

CLIC Conversational Language for Interactive Computing

CLID Calling Line Identification

CLIF Commonwealth Land Information Forum [now Commonwealth Spatial Data Committee]

CLIFFORD International Conference on Clifford Algebras (and their Applications in Mathematical Physics)

CLIM Cable Line Inspection Mechanism

CLIM Cellular Logic In Memory

Clim Climate; Climatic [also clim]

CLIMAP Climate Long-Range Investigation Mapping and Prediction

Climatol Climatological; Climatologist; Climatology [also climatol]

Clim Change Climatic Change [Journal published in the Netherlands]

CLIMEX Information Exchange System on Country Activities on Climate Change

Clim Imp Netw Newsl Climate-Related Impacts Network Newsletter [Publication of the Environmental and Societal Impacts Group of the National Center for Atmospheric Research, US]

Clim Res Climate Research [Journal]

Clin Clinic(al); Clinician [also clin]

CLINAC Clinical Linear Accelerator (Device)

Clin Cancer Res Clinical Cancer Research [Journal of the American Association for Cancer Research]

Clin Chem Clinical Chemistry [Journal of the American Association for Clinical Chemistry]

Clin Chim Acta Clinica Chimica Acta [Journal published in the US]

Clin Diagn Lab Immunol Clinical and Diagnostic Laboratory Immunology [Journal of the American Society for Microbiology]

C Line Strong-Red Fraunhofer Line for Hydrogen Having a Wavelength of 766.1 Nanometers [also Hα line] [Spectroscopy]

Clin Immunol Immunopathol Clinical Immunology and Immunopathology [US Journal]

Clin Med Clinical Medicine [Journal]

Clin Microbiol Rev Clinical Microbiology Reviews [of American Society for Microbiology]

Clin Mycol Clinical Mycology; Clinical Mycologist

Clin Neuropharmacol Clinical Neuropharmacology [International Journal]

Clin Obstetr Gynecol Clinical Obstetrics and Gynecology [International Journal]

Clin Phys Physiol Meas Clinical Physics and Physiological Measurement [Publication of the Institute of Physical Sciences in Medicine, UK]

Clin Practice Guidel Clinical Practice Guideline–Quick Reference Guide for Clinicians [of Agency for Health Care Policy and Research, Department of Health and Human Services, US]

Clin Rheum Disease Clinics in Rheumatic Diseases [Journal]

Clin Sci Clinical Science [UK Journal of the Biochemical Society and the Medical Research Society]

Clin Therapeut Clinical Therapeutics [Journal]

Clin Vis Sci Clinical Vision Sciences [Journal published in the UK]

CLIO Conversation Language Input/Output

CLIP Cellular Logic Image Processor

CLIP Closed-Loop Incremental Positioner

CLIP Combined Laser Instrumentation Package

CLIP Compiler Language for Information Processing

CLIP Computer Language for Information Processing

CLIP Contributions to Laboratory Investigations Program

CLIRA Closed Loop In-Reactor Assembly

CLIS Certificate of Library and Information Science

CLISG Commonwealth Land Information Support Group

CLIST Command List [Computers]

CLIV Cold Leg Isolation Valve

CLIVAR Climate Variability and Predictability Program [of World Climate Research Program of World Meteorological Organization]

CLJA Closed Loop Jumper Assembly

Clk Clerk [also clk]

Clk Clock [also clk]

Clkg Caulking [also clkg]

CLK GEN Clock Generator [also Clk Gen]

CLK IN Clock In [Data Transmission]

CLK OSC Clock Oscillator [also Clk Osc]

CLK OUT Clock Out [Data Transmission]

CLL Central Light Loss

CLL Chronic Lymphocytic Leukemia

CLLCC Ceramic Leadless Chip Carrier [Electronics]

CLM Care Logic Module

CLM Coal-Liquid Mixture

CLM Communications Line Multiplexer

CLM Constrained Local Moment (Model) [Physics]

C/LM Computer/Local and Metropolitan Area Networks [of Institute of Electrical and Electronics Engineers Committee, US]

CLMA Clinical Laboratory Management Association [US]

CLMC Central Logistics Management Center

C/L min Carload, Minimum Weight

CLML Chicago Linear Music Language

CLMPC Canadian Labour Market and Productivity Centre [Ottawa, Canada]

C LNB C-Band Low-Noise Block (Downconverter)

Clnc Clearance [also clnc]

Clng Cleaning [also clng]

CLNGG Calcium Lithium Niobium Gallium Garnet

CLNP Connectionless Network Protocol

Clnr Cleaner [also clnr]

CLO Cornell Laboratory of Ornithology [of Cornell Univerrsity, Ithaca, New York, US]

ClO Chlorine Monoxide

ClO Chlorosyl (Radical) [Symbol]

ClO⁻ Hypochlorite (Ion) [Symbol]

ClO₂ Chloryl (Radical) [Symbol]

ClO Chlorite (Ion) [Symbol]

ClO Chlorate (Ion) [Symbol]

ClO Perchlorate (Ion) [Symbol]

Clo Closet [also clo]

Clo Clothing [also clo]

CLOAX Corrugated Laminated Coaxial Cable

CLOB Core Load Overlay Builder

Cloc Collocation [also cloc]

CLOG Chrysler Lemon Owners Group [Canada]

CLONG Aircraft Longitude [Loran Navigation]

CLOS Common LISP (List Processing) Object System [Object-Oriented Programming]

Clos Closure [also clos]

CLP Carnegie Laboratory of Physics [Dundee, Scotland]

CLP Cell Loss Priority

CLP Chilean Peso [Currency of Chile]

CLP City of London Polytechnic [UK]

CLP Communication Line Processor

CLP Competition Law and Policy (Committee) [of Organization for Economic Cooperation and Development]

CLP Cone Locator Pin

CLP Constraint Logic Programming

CLP Contract Laboratory Program [of US Environmental Protection Agency]

CLP Current Line Pointer

.CLP Clipboard [Windows File Name Extension]

Clp Clamp [also clp]

5-Cl-PADAP 2-(5-Chloro-2-Pyridylazo)-5-Diethylaminophenol

CLPE Cross-Linked Polyethylene

ClPh Chlorophenyl

5-ClPHEN 5-Chloro-1,10-Phenanthroline

Clpyz Chloropyrazine [also Clpyz]

CLR Calcium-Rust-Lime

CLR Center for Laser Research [of Noble Research Center, Oklahoma State University, Stillwater, US]

CLR Central Logic Rack

CLR Chemical Laboratory Report

CLR Clean Liquid Radwater

CLR Clear Screen

CLR Combined Line and Recording (Traffic) [also clr]

CLR Computer Language Recorder

CLR Computer Language Research

CLR Constant Load Rupture

CLR Coordinating Lubricants Research

CLR Cost Limit Review

CLR Council of Library Resources [US]

CLR Crack-Length Ratio

CLR Current-Limiting Reactor

CLR Current-Limiting Resistor

Clr Clearance [also clr]

clr clear

CLRA Construction Labour Relations Association [Canada]

CLRB Canada Labour Relations Board

CLRB Cost Limit Review Board

CLRC Circuit Layout Record Card

CLRC Cooperative League of the Republic of China [Taiwan]

ClrEol Clear (All Characters to) End of Line [Pascal Function]

CLRI Central Leather Research Institute [UK]

CLRI Central Leather Research Institute [Madras, India]

CLRS Clear and Smooth

ClrScr Clear Screen [Pascal Function]

CLRU Cambridge Language Research Unit [US]

CLRV Canadian Light Rail Vehicle

CLS Calcium Lanthanum Sulfide

CLS Canada Land Surveyor

CLS Canadian Lumber Standard

CLS Cask Loading Station

CLS Cathodoluminescence Spectroscopy

CLS Characteristic Loss Spectroscopy

CLS Cislunar Space

CLS Classical Least-Squares

CLS Clear and Subtract

CLS Clear Screen (Statement) [also cls]

CLS Clear To Send

CLS Closed-Loop Stripper

CLS Closed-Loop System

CLS Cloud Lidar System

CLS Coherent-Light System

CLS Coincidence (Crystal) Lattice Site [Materials Science]

CLS College of Letters and Science [of University of Wisconsin, US]

CLSb Commission of Life Sciences [of National Research Council, US]

CLS Common Language System

CLS Communications Line Switch

CLS Comparative Systems Laboratory [US]

CLS Computerized Library System

CLS Concept Learning System

CLS Condenser Lens Supply

CLS Constant Level Speech

CLS Constant Light Signal

CLS Containment Leakage System [Nuclear Reactors]

CLS Contingency Landing Site(s)

CLS Control Launch Subsystem

CLS Core Level Spectroscopy

CLS Cosmetic Laser Surgery

Cls Claims [also cls]

CLSA Canadian Laboratory Suppliers Association

CLSA Closed-Loop Stripping Analysis

clsd closed

CLSE Center for Laser Science and Engineering [of University of Iowa, Iowa City, US]

CLSEG Confined Lateral Selective Epitaxial Growth

Clsg Closing [also clsg]

CLSI Computer Library Services, Inc. [US]

CLSM Confocal Laser Scan(ning) Microscope; Confocal Laser Scan(ning) Microscopy

CLSMDA Closed Loop System Meltdown Accident [Nuclear Engineering]

CLSO Contingency Landing Support Officer

CLSP Community Landcare Support Program

CLSR Computer Law Service Reporter

CLSS Communication Link Subsystem

CLT Central-Limit Theorem [Statistics]

CLT Certificate of Laboratory Technology

CLT Clinical Laboratory Technologist

CLT Closed Loop Test

CLT Communications Line Terminal

CLT Computer Language Translator

CLT Constant Load Tensile

CLT Continuous Lumber Tester

Clt Collect [also clt]

CLTP Connectionless Transport Protocol

CLTS Clear Task Switch Flag [Computer]

CLTV Constant Linear Time Velocity

CLU Central Logic Unit

CLU Circuit Line-Up

CLU Continuous Loading/Unloading [Mechanical Testing]

CLU Creusot-Loire-Uddeholm (Process) [Metallurgy]

CLUE Comprehensive Loss Underwriting Exchange

CLUI Command Line User Interface

CLUT Color Look-Up Table

CLUT Computer Logic Unit Tester

CLUW Coalition of Labor Union Women [US]

CLV Carnation Latent Virus

CLV Constant Linear Velocity

Clv Clevis [also clv]

CLW College of Librarianship, Wales [UK]

clw clockwise

CLWS Coal-Lime-Water Slurry

CLX Clear X (Function)

CM Cache Memory

CM Calculating Machine

CM Calibrated Magnification; Calibration Magnification

CM Cameroon [ISO Code]

CM Campaign

CM Cannon-Muskegon Corporation [Muskegon, Michigan, US]

CM Capacitor Motor

CM Capacity Management

CM Carbon(aceous) Material

CM Carboxymethyl (Cellulose)

CM Cardiac Muscle

CM Cards per Minute

CM Cargo Management

CM Carnegie Museum [Pittsburgh, Pennsylvania, US]

CM Carpometacarpal (Joints)

CM Catalytic Material

CM Celestial Meridian
CM Celestial Mechanics
CM Cell Membrane
CM Cellular Matrix
CM Cellulosic Material
CM Cementitious Material
CM Center Matched (Joints)
CM Center of Mass [also cm]
CM Centimetric
CM Central Memory
CM Centrifugal Moment
CM Centro Morgardshammar AB [Swedish Rolling Technology Manufacturer located at Smedjebacken]
CM Ceramic Magnet
CM Ceramic Material
CM Certified Manager
CM Certified Master
CM Chemical Manufacturing
CM Chemical Metallurgist; Chemical Metallurgy
CM Chemical Meterorologist; Chemical Meterorology
CM Chemical Microscopy
CM Chemical Milling
CM Chemical Mortar
CM Chief of Maintenance [UK]
CM Chief Microbiologist
CM Chiral Medium
CM Chiral Molecule
CM Chloromethane
CM Chronic Myocarditis
CM Circular Measure
CM Circular Mil
CM City Manager
CM Classical Mechanics
CM Class Mark [Statistics]
CM Class Master
CM Clausius-Mosotti (Equation)
CM Clinical Medicine
CM Clinical Modification
CM Cluster Management; Cluster Manager [Computer Multiproccessing]
CM Coal Measures [Geology]
CM Coal Mine; Coal Mining
CM Coincidence Method
CM Cold Mix [Civil Engineering]
CM Color Monitor
CM Command Module [NASA Apollo Program]
CM Common Meter
CM Common Mode
CM Commonwealth of Massachusetts [US]
CM Communications Management; Communications Manager
CM Communications Medium
CM Communications Multiplexer
CM Commutating Machine
CM Commutator Motor

CM Composite Material
CM Composition Modulation
CM Compressibility Modulus
CM Computational Mathematics
CM Computational Mechanics
CM Computational Model
CM Computer Management
CM Computer Management [Journal published in the UK]
CM Computer Model(ling)
CM Computer Module
CM Conceptual Modelling
CM Condensed Matter
CM Condenser Microphone
CM Configuration Management [also cm]
CM Configuration Mixing
CM Confocal Microscopy
CM Connection Machine [Computers]
CM Construction and Machinery
CM Construction Management
CM Consumable-Electrode Remelted; Consumable-Electrode Remelting [Metallurgy]
CM Consumables Management [also cm]
CM Contact Mode
CM Continuity Message
CM Continuum Mechanics
CM Controlled Manual
CM Controlled Mine-Field
CM Controlling Magnet
CM Control Mark
CM Control Mechanism
CM Control Memory
CM Control Monitor [also cm]
CM Core Memory
CM Corrective Maintenance [Computers]
CM Correlation Method
CM Corresponding Member
CM Corrosive Material
CM Coulombmeter
CM Council of Ministers
CM Countermeasure
CM County Municipality
CM Coupled Modes
CM Coupling Model
CM Court Martial
CM Cowell Method [Aerospace]
CM Crew Module [also cm]
CM Critical Mass [Nuclear Sciences]
CM Critical Material(s)
CM Cross-Modulated; Cross Modulation
CM Cryogenically Milled; Cryogenic Milling; Cryo-Milled; Cryo-Milling
CM Cubic Measure
CM Culture Medium [Microbiology]
CM Cumulative [also Cm, or cm]

CM Cutting Machine [Mining]

CM Cyclically Magnetized (Condition)

CM Cylinder Mount

CM Cylindrical Mirror (Analyzer)

CM Middle Cloud Type Code

2C5M14 2-Chloro-5-Methyl-1,4-Benzoquinone

C&M Control and Monitoring

C/M Ceramic-to-Metal (Joint)

C/M Communications Multiplexer

C/M Control Monitor

Cm Cement

Cm Cotesia melanoscela [Biology]

Cm Coulomb-meter [also C-m, or C·m]

Cm Curium [Symbol]

Cm-238 Curium-238 [also ^{238}Cu, or Cu^{238}]

Cm-239 Curium-239 [also ^{239}Cu, or Cu^{239}]

Cm-240 Curium-240 [also ^{240}Cu, or Cu^{240}]

Cm-241 Curium-241 [also ^{241}Cu, or Cu^{241}]

Cm-242 Curium-242 [also ^{242}Cu, or Cu^{242}]

Cm-243 Curium-243 [also ^{243}Cu, or Cu^{243}]

Cm-244 Curium-244 [also ^{244}Cu, or Cu^{244}]

Cm-245 Curium-245 [also ^{245}Cu, or Cu^{245}]

C/m^2 Coulomb(s) per square meter [also C m^{-2}]

C/m^3 Coulomb(s) per cubic meter [also C m^{-3}]

cM centimorgan [Unit of Genetic Map Distance]

cm centimeter [Unit]

cm cumulative

c.m. circular mil [also c m]

cm^{-1} one per centimeter [also 1/cm]

cm^{-2} one per square centimeter [also 1/cm^2]

cm^{-3} one per cubic centimeter [also 1/cm^3]

cm^2 square centimeter [also sq cm]

cm^3 cubic centimeter [also cu cm]

cm^4 centimeter to the fourth power [Unit]

c/m^2 candles per square meter [also c cm^{-2}]

CMA Cable Makers' Association [UK]

CMA Cadmium Manganese Arsenide (Semiconductor)

CMA Calcium Magnesium Acetate

CMA Calcium Methanearsonate

CMA Calcium Monoaluminate

CMA Canadian Manufacturers Association

CMA Canadian Marketing Association

CMA Canadian Medical Association

CMA Canadian Metric Association

CMA Canadian Mineral Analysts

CMA Canadian Motorcycle Association

CMA Canadian Museums Association

CMA Carboxymethylamylose

CMA Case Makers Association [UK]

CMA Cast Metals Association [US]

CMA Castor Manufacturers Association [UK]

CMA Census Metropolitan Areas

CMA Center for Management in Agriculture [UK]

CMA Center for Mathematical Analysis [of Australian National University, Canberra]

CMA Central Mapping Authority [now Land Information Center, Bathurst, New South Wales, Australia]

CMA Certificate in Management Accounting

CMA Certified Management Accountant

CMA Certified Medical Assistant

CMA Chartered Mechanical Engineer

CMA Chemical Manufacturers Association [US]

CMA Chemical Microanalysis

CMA Chinese Meteorological Administration

CMA N-(3-Chloro-4-Methylphenyl)-2-Methyl Pentanamide

CMA Chronic Menopausal Arthritis

CMA Circular Mil Area

CMA Classified Mail Address

CMA Closure Manufacturers Association [US]

CMA Colleges of Mid-America [US]

CMA Commonwealth Medical Association [US]

CMA Communication Managers Association [UK]

CMA Communications Market Association [US]

CMA Compañía Mexicana de Aviación [Mexican Aviation Company]

CMA Composites Manufacturing Association [of Society of Manufacturing Engineers, US]

CMA Composition(ally)-Modulated Alloy (Method)

CMA Compositionally Multilayered Alloy

CMA Computer-Aided (X-Ray) Microanalyzer

CMA Computer Management Association [US]

CMA Computer Monitor Adapter

CMA Concert Multi-Thread Architecture [Computers]

CMA Configuration Management Accounting

CMA Consulting Management Engineer

CMA Contact-Making Ammeter

CMA Contract Management Assistance

CMA Contract Manufacturers Association [US]

CMA Copper-Magnesium-Aluminum (Alloy)

CMA Crane Manufacturers Association

CMA Cryptomelane-type Manganic Acid

CMA Cylindrical Mirror Analyzer

CMa Canis Major [Astronomy]

CMAA Coalbed Methane Association of Alabama [US]

CMAA Construction Management Association of America [US]

CMAA Crane Manufacturers Association Act [US]

CMAA Crane Manufacturers Association of America [US]

CMAC Centralized Media Access Control

C-MAC Direct-Broadcast Television Multiplexed Analog Components

CMACL Composite Mode Adjective Checklist

CMACS Central Monitor and Control System

CM-AFM Contact-Mode Atomic Force Microscope; Contact-Mode Atomic Force Microscopy

(C,M)$_4$AH$_{13}$ Calcium Magnesium Aluminate Hydrate [4(CaO,MgO)·Al$_2$O$_3$·13H$_2$O]

CMAHK Chinese Manufacturers Association of Hong Kong

CMAJ Canadian Medical Association Journal [also Can Med Assoc J]

CMA-Na Carboxymethylamylose, Sodium Salt

CMANY Communications Managers Association of New York [US]

CMAO Contract Management Assistance Officer

CMAP (Digitized) Census Field Maps (on Compact Disk) [of Australian Bureau of Statistics]

CMAP Central Memory Access Priority

CMAP Climate Modeling, Analysis and Prediction Program

CMAR Control Memory Access Register

CMARC Current Machine-Readable Catalog [of Library of Congress, US]

C-13 MAS Carbon-13 Magic-Angle Spinning [also ^{13}C MAS]

C-13 MAS NMR Carbon-13 Magic-Angle Spinning Nuclear Magnetic Resonance [also ^{13}C MAS NMR]

CMA/SME Composites Manufacturing Association of the Society of Manufacturing Engineers [also CMA-SME, US]

CMAT Compatible Material(s)

CMB Carbolic Methylene Blue

CMB Center for Mechanistic Biology [of Argonne National Laboratory, Illinois, US]

CMB Center for Molecular Biotechnology [of University of Washington, Seattle, US]

CMB Climate Monitoring Bulletin

CMB Concrete Median Barrier

CMB Continuous Moving Bed

CMB Corporate Management Branch

CMB Cosmic Microwave Background

CMBA Corporate Master of Business Administration

CMBA Crossed Molecular Beam Apparatus

CmBCO Curium Barium Copper Oxide (Superconductor)

CMBES Canadian Medical and Biological Engineering Society

CMBG Canadian Mechanized Brigade Group

CMBL Clapp Marine Biological Laboratory [of US Bureau of Fisheries at Woods Hole, Massachusetts, US]

CMC Cable Maintenance Center

CMC Canada Meat Council

CMC Canadian Manpower Center

CMC Canadian Meteorological Center

CMC Canadian Microelectronics Corporation

CMC Canadian Military College

CMC Carbon-Matrix Composite

CMC Carboxymethyl Cellulose [also CM Cellulose]

CMC Cellular Mobile Carrier [Telecommunications]

CMC Center for Marine Conservation [US]

CMC Center for Materials Chemistry [of University of Texas at Austin, US]

CMC Ceramic Matrix Composite

CMC Certified Management Consultant

CMC Code for Magnetic Characters

CMC Collective Measures Committee [of United Nations]

CMC Command Module Computer [NASA Apollo Program]

CMC Common Mail Calls

CMC Common Mezzanine Card [of Institute of Electrical and Electronics Engineers, US]

CMC Common Messaging Calls

CMC Communications Channel

CMC Communication Management Configuration

CMC Communications Mode Control

CMC Complement Carry Flag [Computers]

CMC Compliant Metal/Ceramic

CMC Computer-Mediated Communications [Internet]

CMC Computer Musicians Cooperative [US]

CMC Conservation Monitoring Center [now World Conservation Monitoring Center]

CMC Constant Mean Curvature

CMC Construction Metrication Council [US]

CMC Contact-Making Clock

CMC Container Marketing Committee [of American Iron and Steel Institute, US]

CMC Coordinating Manual Control

CMC Copper-Molybdenum-Copper

CMC Cornell Medical Center [US]

CMC Crew Module Computer

CMC Critical Micelle Concentration [also cmc] [Physical Chemistry]

CMC Crucible Materials Corporation [US]

CMC 1-Cyclohexyl-3-(2-Morpholinoethyl)carbodiimide Metho-p-Toluenesulfonate

CMCA Canadian Masonry Contractors Association

CMCA Character Mode Communications Adapter

CMCA Cornell Medical College Association [US]

CMCA Cruise Missile Carrier Aircraft

CMCC Canadian Memorial Chiropractic College [Toronto, Ontario]

CMCC Canadian Museum Construction Corporation

CMCD Centro de Missão de Coleta de Dados [Data Collection Mission Center in Cachoeira Paulista; of Brazilian National Institute for Space Research]

CM Cellulose Carboxymethyl Cellulose [also CM cellulose, or CMC]

CMCI 1-Cyclohexyl-3-(N-2-Morpholinoethyl)carbodiimide

cm/cm centimeters per centimeter [also cm cm^{-1}]

CMCP Canadian Museum of Contemporary Photography

CMCP Cruise Missile Conversion Project

CMCR Center for Mass Communication Research [UK]

CMCR Continuous Melting, Casting and Rolling [Metallurgy]

CMCS Corporate Materials Catalog System

CMCSA Canadian Manufacturers of Chemical Specialties Association

CMCSC Citadel Military College of South Carolina [Charleston, US]

CMCTL Current Mode Compementary Transistor Logic

CMD Carboxymuconolactone Decarboxylase [Biochemistry]

CMD Center for Management Development [of Memorial University of Newfoundland, St. John's Canada]

CMD Center for Materials Data [of ASM International, US]

CMD Circuit Mode Data

CMD Cold-Mix Design [Civil Engineering]

CMD Communications Management Device

CMD Condensed Matter Division [of European Physical Society]

CMD Core Memory Driver

CMD Corporate Management Division

CMD Command [also Cmd, or cmd]

.CMD Command [File Name Extension] [also .cmd]

CM-D Dextran Carboxymethyl Ether

Cmd Command [also cmd]

CMDA Cornish Mining Development Association [UK]

CMDAC Current Mode Digital-to-Analog Converter

CMD DCDR Command Decoder [also Cmd Dcdr]

Cmdg Commanding [also cmdg]

CMDL Climate Monitoring and Diagnostics Laboratory [of National Oceanic and Atmospheric Administration, US]

CMDMCS Chloromethyl Dimethyl Chlorosilane

CMDNJ College of Medicine and Dentistry of New Jersey [Newark, US]

CmdO Commissioned Officer [also Cmd O]

CMDR Center for Molecular Design and Recognition [of University of South Florida, Tampa, US]

CMDR Command Reject

Cmdr Commander

Cmdre Commodore

CMDS Centralized Message Data System

CMDS Consortium for Materials Development in Space [of University of Alabama at Huntsville, US]

cm/dyn centimeter(s) per dyne [Unit]

CME Canadian Military Engineers

CME Center for Manufacturing Engineering

CME Center for Molecular Electronics [of Carnegie-Mellon University, Pittsburgh, Pennsylvania, US]

CME Central Memory Extension

CME Centrifuge Moisture Equivalent (of Soil)

CME Chartered Mechanical Engineers [of Institution of Mechanical Engineers, UK]

CME Chemically-Modified Electrode

CME Chicago Mercantile Exchange [US]

CME Chloromethyl Ether

CME Clay-Modified Electrode

CME Common Minimum Evaluation

CME Comprehensive Maintenance Evaluation

CME Computer Measurement and Evaluation

CME Concurrent Machine Environment

CME Continuing Medical Education

CME Council of Medical Education [US]

CME Council of Ministers of Education [Canada]

CME Council on Medical Education [of American Medical Association, US]

CME Cresyl Methylether

CME Crucible Melt Extraction [Metallurgy]

C-Me Carbon-Metal (Bond)

CMEA Council of Mutual Economic Aid; Council of Mutual Economic Assistance [also COMECON, or Comecon]

CMEC Center for Micro-Engineered Ceramics [of National Science Foundation at University of New Mexico, Albuquerque, US]

CMEC Council of Ministers of Education, Canada

CMEC Crucible Melt Extraction [Metallurgy]

CME-CDI Cyclohexyl-2-(4-Methyl-Morpholino) ethylcarbodiimide Tosylate

CMEH Council on Medical Education and Hospitals [of American Medical Association, US]

CMEM International Conference on Computational Methods and Experimental Measurements

CMEQ Corporation of Master Electricians of Quebec [Canada]

CMER College of Mineral and Energy Resources [of West Virginia University, Morgantown, US]

CMER College of Mines and Earth Resources [of University of Idaho, Moscow, US]

CMER Conseil Mondial de l'Environnement et des Ressources [World Environment and Resources Council, Belgium]

CMERI Central Mechanical Engineering Research Institute [Durgapur, India]

CMES Chinese Mechanical Engineering Society

CMES Conversion Electron Moessbauer Spectroscopy

CMES-PEI Chinese Mechanical Engineering Society Power Engineering Institution

$cm^{-2} \cdot eV^{-1}$ one per square centimeter per electron volt [also $1/cm^2 \cdot eV$, or $1/cm^2/eV$]

$cm^{-3} \cdot eV^{-1}$ one per cubic centimeter per electron volt [also $1/cm^3 \cdot eV$, or $1/cm^3/eV$]

CMF Calcium Magnesium Fluoride

CMF Cartesian Mapping Function

CMF Center for Materials Fabrication [of Battelle Columbus Laboratory, Ohio, US]

CMF Center for Metals Fabrication [now Center for Materials Fabrication]

CMF Central Maintenance Facility

CMF Coal Merchants Federation [UK]

CMF Coherent Memory Filter

CMF Common Mode Failure

CMF Composition(ally) Modulated (Thin) Film

CMF Coriolis Mass Flowmeter

CMF Creative Music Format

CMF Cross-Flow Microfiltration

CMF Cross Modulation Factor [Communications]

CMF Cymomotive Force [also cmf]

cmf circular mil-foot [also CMF]

CMFA Common-Mode-Failure Analysis

CMfgE Certified Manufacturing Engineer

CMfgT Certified Manufacturing Technologist

CMFI Conversion, Memory and Fault Indication

CMFLPD Core Maximum Fraction of Limiting Power Density [Nuclear Engineering]

CMFRI Central Marine Fisheries Research Institute [India]

CMG Center for Marine Geology [of Dalhousie University, Halifax, Nova Scotia, Canada]

CMG Color Marketing Group [US]

CMG Commission for Marine Geology [Germany]

CMG Computer Management Group [of General Electric Company, Bridgeport, Connecticut, US]

CMG Computer Measurement Group [US]

CMG Computer Modeling Group

CMG Control Moment Gyro(scope) [Aerospace]

C/mg Columb per milligram [also C mg^{-1}]

cmg centimeter gram [Unit]

cm²/g square centimeter(s) per gram [also cm^2 g^{-1}]

cm^3/g cubic centimeter(s) per gram [also cm^3 g^{-1}]

CMGA Canadian Mushroom Growers Association

CMGA Newsl Canadian Mushroom Growers Association Newsletter

cm^{-1} GPa^{-1} one per centimeter per gigapascal [also 1/cm/GPa, or cm^{-1}/GPa]

CMG Proc CMG Proceedings [of Computer Measurement Group, US]

CMG Trans CMG Transactions [of Computer Measurements Group, US]

cm/h centimeter(s) per hour [also cm/hr, or cm h^{-1}]

cm^3/h cubic centimeter(s) per hour [also cm^3/hr, or cm^3 h^{-1}]

CMHA Canadian Manufactured Housing Association

CMHA Canadian Mental Health Association

CMHC Canada Mortgage and Housing Corporation

CMHC Community Mental Health Center

CMHEC Carboxymethyl Hydroxyethyl Cellulose

cm Hg centimeters of mercury [also cmHg]

CMHI Canadian Manufactured Housing Institute

cm/hr centimeter(s) per hour [also cm/h, or cm h^{-1}]

cm^3/hr cubic centimeter(s) per hour [also cm^3/h, or cm^3 h^{-1}]

CMHS Canadian Military Historical Society

CMHSA Coal Mine Health and Safety Act [US]

cm Hz$^{1/2}$ W^{-1} centimeter square root of hertz per watt [also cm $\sqrt{Hz}$ W^{-1}]

CMI Canadian Motor Industries

CMI Can Manufacturers Institute [New York City, US]

CMI Caribbean Meteorological Institute [Barbados]

CMI Casting Metals Institute

CMI CEA (Core Element Assembly) Motion Inhibit [Nuclear Engineering]

CMI Cell-Mediated Immunity [Immunology]

CMI Cell Multiplication Inhibition

CMI Christian Michelsen Institute [Norway]

CMI Coating Measurement Instrument

CMI Code Mark Inversion

CMI Colorado Mining Association [US]

CMI Comité Maritime International [International Maritime Committee, Belgium]

CMI Commonwealth Mycological Institute [UK]

CMI Computational Mechanics Institute [UK]

CMI Computer-Managed Instruction

CMI Construction Management Institute

CMI Cordage Manufacturers Association [UK]

CMI Cultured Marble Institute [US]

CMi Canis Minor [Astronomy]

CMIA Coal Mining Institute of America [US]

CMIA Cyprus Metal Industry Association

CMIARD Collaborative Minority Institution Alcohol Research Development (Grant) [of National Institute on Alcohol Abuse and Alcoholism, Bethesda, Maryland, US]

CMIEE Canadian Mining and Industrial Equipment Exposition

CMIG Canadian MAP (Manufacturing Automation Protocol) Information Group

CMIG Canadian MAP (Manufacturing Automation Protocol) Interest Group

CMIG Collaborative Mucosal Immunity Groups [Immunology]

cmil circular mil [Unit]

cmils circular mils

CMIM Center for Measurement and Information in Medicine [UK]

°C/min Degrees Celsius (or Centigrade) per Minute [also °C min^{-1}]

CMIP Common Management Information Protocol

C-MIRB Committee on Megaproject Industrial and Regional Benefits

CMIS Common Management Information Services

CMIS Common Management Information System

CMIS Complementary Metal-Insulator Semiconductor

CMIST Consultative Group on Marine Industries Science and Technology [Australia]

CMIT Center for Materials for Information Technology [of University of Alabama, Tuscaloosa, US]

CMITP Canada Manpower Industrial Training Program

CMJ Canadian Mining Journal

CMJ Construtora Mendes Junior SA [Brazilian Power Plant Manufacturing Company]

CMJ Croatian Medical Journal

cm^3/kg cubic centimeter per kilogram [also cm^3 kg^{-1}]

CML Cell-Mediated Lympholysis

CML Chemical Markup Language

CML Common Mode Logic

CML Compound Middle Lamella [Wood Science]

CML Computational Microgravity Laboratory [of NASA Lewis Research Center, Cleveland, Ohio, US]

CML Computer-Managed Learning

CML Conceptual Modelling Language

CML Coupled Map Lattice

CML Critical Mass Laboratory [of Pacific Northwest National Laboratories, Richland, Washington, US]

CML Current-Mode Logic

cm^3/L cubic centimeter(s) per liter [also cm^3/l]

cml chemical

cml commercial

cm^3/lb cubic centimeter per pound [also cc/lb]

CMLC Chemical Corps [US]

CMLCENCOM Chemical Corps Engineering Command [US]

CMLE Carboxymuconate Lactonizing Enzyme [Biochemistry]

Cm:LiYF$_4$ Curium-Doped Lithium Yttrium Fluoride [also Cm^{3+}:LiYF$_4$]

CMM Capability Maturity Model

CMM Center for Mechanics of Materials and Instabilities [of Michigan Technological University, Houghton, US]

CMM Center for Microscopy and Microanalysis [of University of Queensland, St. Lucia, Australia]

CMM Centrum für Molekulare Medizin [Molecular Medicine Center, Berlin, Germany]

CMM Centre de Météorologie Marine [Marine Meteorology Center, France]

CMM Ceramic Matrix Metal

CMM Commission for Marine Meteorology [of World Meteorological Organization]

CMM Common Memory Manager

CMM Communications Multiplexer Module

CMM Component Maintenance and Mockup

CMM Computerized Modular Monitoring

CMM Computer Main Memory

CMM Concentration Module Main

CMM Condition Monitored Maintenance

CMM Coordinate Measuring Machine

CMM Core Mechanical Mockup

C/MM Computer/Microprocessors and Microcomputers [of Institute of Electrical and Electronics Engineers, US]

cm^3/m^3 cubic centimeter(s) per cubic meter [also cm^3 m^{-3}]

CMMA Concrete Mixer Manufacturers Association

CMMC Center for Molecular and Microstructure of Composites [of Case Western Reserve University, Cleveland, US]

CMMC Ceramic and Metal Matrix Composite

CMMC 7-[(Chlorocarbonyl)methoxy]-4-Methylcoumarin

CMMC Continuously-Reinforced Metal-Matrix Composite

cm^3/m^2·d·bar cubic centimeter(s) per square meter per day per bar [also cm^3/(m^2·d·bar), cm^3/m^2/d/bar, or cm^3 m^{-2} d^{-1} bar^{-1}]

CMME Chloromethyl Methyl Ether

CMMF Component Maintenance and Mockup Facility

CMMI Council of Mining and Metallurgical Institutions [UK]

cm/min centimeter(s) per minute [also cm min^{-1}]

cm^2/min square centimeter(s) per minute [also cm^2 min^{-1}]

cm^3/min cubic centimeter(s) per minute [also cm^3 min^{-1}]

cm^3/min/g cubic centimeter(s) per minute per gram [also cm^3 min^{-1} g^{-1}]

CMMMC Common-Matrix Micromacrocomposite [also CM^{3}C]

cm^3/mol cubic centimeter(s) per mole [also cm^3 mol^{-1}, or cc/mol]

CMMP Canada Manpower Mobility Program

CMMP Carnegie Multi-Mini Processor

CMMP Commodity Management Master Plan

CMMP Condensed Matter and Materials Physics (Conference) [UK]

CMMS Computerized Maintenance Management Software

CMMS Cation Micro-Membrane Suppressor

CMMS Chloromethylmethylsulfide

CMMS Computerized Maintenance Management System

cm/μs centimeter(s) per microsecond [also cm μs^{-1}]

cm/μsec centimeter(s) per microsecond [also cm μsec^{-1}]

CMMU Cache/Memory Management Unit [of Motorola Inc., US]

CMN Canadian Museum of Nature

CMN Cerium Magnesium Nitrate; Cerous Magnesium Nitrate

CMN 1-Chloromethylnaphthalene

CMN Control Motion Noise [Aerospace]

Cmn Commission [also cmn]

Cmn Crewman

Cmnd Command [also Cmnd]

CMNH Chicago Museum of Natural History [Illinois, US]

CMNH Cleveland Museum of Natural History [Ohio, US]

CMNOS Complementary Metal-OxideSemiconductor

CMNT 4-Chloro-3-Nitrobenzotrifluoride

cm^{3}NTP/g Cubic Centimeter Normal Temperature and Pressure per Gram [also cm^{3}NTP g^{-1}]

CMO Carboxymethyl Oxime

CMO Caribbean Meteorological Organization [Trinidad and Tobago]

CMO Center for Micro- and Optoelectronics [of Lawrence Livermore National Laboratory, California, US]

CMO Centruum voor Material-Onderzoeg [Materials Research Center; of University of Twente, Enschede, Netherlands]

CMO Committee Management Office [of National Institutes of Health, Bethesda, Maryland, US]

CMO Common Mode Operation

CMO Configuration Management Office

cm^{-1}·Ω^{-1} one per centimeter per ohm [also 1/cm·Ω, or 1/cm-Ω]

CMOD Crack Mouth Opening Displacement

cm^2/Ω eq square centimeter per ohm equivalent [also cm^2 Ω^{-1} eq^{-1}, cm^2/ohm eq]

cm^2/ohm eq square centimeter per ohm equivalent [also cm^2/Ω eq, cm^2 Ω^{-1} eq^{-1}]

C/mol Coulomb per mole [also C mol^{-1}]

CMOMR Center for Microelectronic and Optical Materials Research [of University of Nebraska, Lincoln, US]

CMOP Canadian Market Opportunities Program

CMOS Canadian Meteorological and Oceanographic Society

CMOS Complementary Metal-Oxide Semiconductor [also C/MOS, or C-MOS]

CMOS Complementary Metal-Oxide Silicon [also C/MOS]

CMOS Custom Metal-Oxide Semiconductor [also C/MOS]

CMOS IC Complementary Metal-Oxide Semiconductor Integrated Circuit

CMOSL Complementary Metal-Oxide Semiconductor Logic

CMOS LSI Complementary Metal-Oxide Semiconductor/Large-Scale Integration (Subsystem) [also CMOS/LSI, or CMOS-LSI]

CMOS/LSTTL Complementary Metal-Oxide Semiconductor/Large-Scale Transistor-Transistor Logic [also CMOS-LSTTL]

CMOS PLD Complementary Metal-Oxide Semiconductor Programmable Logic Device [also CMOS-PLD]

CMOS-RAM Complementary Metal-Oxide Semiconductor–Random Access Memory [also CMOS-RAM]

CMOS/SIMOX Complementary Metal-Oxide Semiconductor–Separation by Implanted Oxygen (Circuit) [also CMOS-SIMOX]

CMOS/SOI Complementary Metal-Oxide Semiconductor/Silicon-on-Insulator (Process)

CMOS/SOS Complementary Metal-Oxide Semiconductor/Silicon-on-Sapphire (Process)

CMOS SRAM Complementary Metal-Oxide Semiconductor Static Random-Access Memory

CMOS UART Complementary Metal-Oxide Semiconductor Universal Asynchronous Receiver Transmitter

CMOU Canadian Marine Officers Union

CMOV Conditional Move

CMP Canadian Mineral Processors [of Canadian Institute of Mining and Metallurgy]

CMP Capacity-Coupled Microwave Plasma

CMP Center for Materials Production [of Electric Power Research Institute, Palo Alto, California, US]

CMP Center for Metal Production

CMP Central Monitoring Position

CMP Ceramic Matrix Polymer

CMP Ceramic Mold Process

CMP Chemical Mechanical Planarization

CMP Chemical Mechanical Polishing

CMP Circolo Matematico di Palermo [Palermo Mathematical Circle, Italy]

CMP Coextrusion of Metal and Plastic

CMP Compagnie du Chemin de Fer Métropolitain de Paris [Paris Underground Railway Company, France]

CMP Compare [also Cmp, or cmp] [Computers]

CMP Compounded Mobilized Planes (Theory)

CMP Compression Amplifier

CMP Computer [also Cmp]

CMP Condensed Matter Physics

CMP Configuration Management Plan

CMP Controlled Materials Production

CMP Cooperative Marketing Partner

CMP Cooperative Marketing Program

CMP Cooperative Market Participant

CMP Cytidine Monophosphate; Cytidine 5'-Monophosphate [Biochemistry]

2'-CMP Cytidine-2'-Monophosphate [Biochemistry]

3'-CMP Cytidine-3'-Monophosphate [Biochemistry]

5'-CMP Cytidine-5'-Monophosphate [Biochemistry]

Cmp Campaign

cmp compare

CMPA Cyprus Master-Printers Association

CMPAA Certified Milk Producers Association of America [US]

CMP-CIM Canadian Mineral Processors–Canadian Institute of Mining and Metallurgy

CMPDA Canadian Motion Picture Distributors Association

CMPE Center for Metallurgical Process Engineering [of University of British Columbia, Vancouver, Canada]

CMPE Contractors and Mechanical Plant Engineers [UK]

CMPF Core Maximum Power Fraction [Nuclear Engineering]

CMPO Carbomoyl Methylphosphine Oxide

Cmplx Complex [also cmplx]

CMPM Committee on Microwave Processing of Materials [of National Research Council, US]

CMPM Computer-Managed Parts Manufacture

CMPM Cubic Meters per Minute [also cmpm, m^3/min, or cu m/min]

CMP-NAN Cytidine 5'-Monophospho-N-Acetylneuraminic Acid [Biochemistry]

CMPO Carbomoylmethylphospine Oxide

CMPP 2-(4-Chloro-2-Methylphenoxy)propanoic Acid

CMPP Chloromethylphenoxypropionic Acid

CMPP 2-(4-Chloro-O-Tolyloxy)propionic Acid

CMPR Compare [also Cmpr, or cmpr]

Cmprt Compartment [also cmprt]

CMPS Compare (Word) String

cmps centimeter per second [Unit]

CMPSS Conference on Molecular Processes on Solid Surfaces

Cmpt Component [also cmpt]

Cmpt Computer [also cmpt]

cmptd computed [also cmpt'd]

Cmptr Computer [also cmptr]

Cmpter Acty Computer Activity

CMPX Computerized Engineering Index [also COMPENDEX]

CMQW Coupled Multiple Quantum Well [Solid-State Physics]

CMR Carbon-13 (Nuclear) Magnetic Resonance (Spectroscopy)

CMR Catalytic Membrane Reactor

CMR CBX (Computerized Branch Exchange) Management Reporter

CMR Center for Materials Research

CMR Center for Materials Research [of Stanford University, California, US]

CMR Center for Materials Research [of Ohio State University, Columbus, US]

CMR Center Materials Representative

CMR Centralized Mail Remittance

CMR Chemical Marketing Register

CMR Collège Militaire Royal [Royal Military College, Saint-Jean-sur-Richelieu, Quebec, Canada]

CMR Collège Militaire Royal [Royal Military College, St. Cyr, France]

CMR Colossal Magnetoresistance

CMR Committee on Manpower Resources [US]

CMR Common-Mode Rejection

CMR Communications Moon Relay

CMR Communications Monitoring Report

CMR Competitive Market Research

CMR Contact Microradiography

CMR Continuous Maximum Rating

CMR Contributing Margin Report

CMR Council on Marine Resources [US]

CMRA Center for Materials Research and Analysis [of University of Nebraska at Lincoln, US]

CMRA Chemical Marketing Research Association [US]

CMRB California Melon Research Board [US]

CMRB Contractor Material Review Board

CMRC Caribbean Marine Research Center [at West Palm Beach, Florida, US]

CMRC Coal Mining Research Center [Canada]

CMRC Corporate Manufacturing Research Center [of Motorola, Inc., Schaumburg, Illinois, US]

CMRCI Contracted Multireference Configuration Interaction (Method)

CMREF Committee on Marine Research, Education and Facilities [US]

CMRI Carnegie Mellon Research Institute [Pittsburgh, Pennsylvania, US]

CMRL CANMET (Canada Center for Mineral and Energy Technology)/Mining Research Laboratories [of Natural Resources Canada]

CMRL Construction Materials Reference Laboratory [US]

CMRR Center for Magnetic Recording Research [of University of California at San Diego, US]

CMRR Common-Mode Rejection Ratio

CMRS Cellular Mobile Radio (Telecommunications) Service

CMRS Central Mining Research Station

C-MRS Chinese Materials Research Society

CMRSD Condensed Matter and Radiation Sciences Division [of US Naval Research Laboratory]

CMRTO College of Medical Radiation Technologists of Ontario [Canada]

CMRSJ Collège Militaire Royal de Saint-Jean [Royal Military College of Saint Jean, Quebec, Canada]

CMRT Center for Materials Research and Technology [Florida State University, Tallahassee, US]

CMS Cadmium Manganese Sulfide

CMS Calcium Magnesiosilicate (or Monticellite)

CMS Calcium Magnesium Sulfide

CMS Calcutta Mathematical Society [India]

CMS California Macadamia Society [US]

CMS Cambridge Materials Selector [Materials Database of Cambridge University, UK]

CMS Cambridge Monitor(ing) System (Computer Program) [of IBM Corporation, US]

CMS Canadian Micrographics Society [now Canadian Information and Image Management Society]

CMS Canadian Mine Services

CMS Carbon Molecular Sieve

CMS Caribbean Meteorological Service [now Caribbean Meteorological Organization]

CMS Cavity Monitoring System

CMS Center for Management Systems [US]

CMS Center for Mass Spectrometry

CMS Center for Marine Studies [now Maritime Research Department of Webb Institute of Naval Architecture, US]

CMS Center for Materials Science

CMS Center for Materials Science [of National Institute for Standards and Technology, Gaithersburg, Maryland, US]

CMS Center for Materials Science [of Los Alamos National Laboratory, Albuquerque, New Mexico, US]

CMS Center for Materials Simulation [of University of Connecticut, Storrs, US]

CMS Center for Measurement Standards [of Industrial Technology Research Institute, Hsinchu, Taiwan]

CMS Centre de Météorologie Spatiale [Center for Space Meteorology, France]

CMS Centrifuge Melt Spinning

CMS Certificate in Museum Studies

CMS Certified Metrication Specialist

CMS Change Management System

CMS Chapter Management Seminar [of ASM International, US]

CMS Chemical Management System

CMS Chemical Metallizing System (Process)

CMS Chief Master Sergeant

CMS Chinese Mathematical Society

CMS Chlorinated Polymethyl Styrene

CMS Chloromethylated Polystyrene

CMS Chromatographic Mode Sequencing

CMS Chrome-Managnese-Silicon (Alloy)

CMS Circuit Maintenance System

CMS Civil and Mechanical Systems

CMS Clay Minerals Society [US]

CMS Club Micro Son [France]

CMS Code Management System

CMS Coincidence Moessbauer Spectroscopy

CMS Cold Melt Steel

CMS College of Marine Studies [of University of Delaware, Lewes, US]

CMS Combustion and Melting System

CMS Common Modular Simulator

CMS Communications Management System

CMS Compact Muon Solenoid (Large Hadron Collider Experiment) [of CERN–European Laboratory for Particle Physics, Geneva, Switzerland]

CMS Compiler Monitor(ing) System

CMS Computerized Manufacturing System

CMS Computer Management System

CMS Condensed Matter Science

CMS Conference Management System [of The Minerals, Metals and Materials Society of the American Institute of Mining, Metallurgical and Petroleum Engineers]

CMS Configuration Management System

CMS Conservation Materials and Services

CMS (International Convention on the) Conservation of Migratory Species of Wild Animals

CMS Continuous Mining System

CMS Conversation(al) Monitor System

CMS Coordinate Measurement System

CMS Corrective Measure Study

CMS Council of Medical Staffs [US]

CMS Current-Mode Switching

CMS Cylinder Management System

CMS Czechoslovak Medical Society

CM&S Communications Maintenance and Storage

C&MS Consumer and Marketing Service

CMS$_2$ Calcium Magnesium Disilicate [$CaO \cdot MgO \cdot 2SiO_2$]

C$_2$MS$_2$ Dicalcium Magnesium Disilicate [$2CaO \cdot MgO \cdot 2SiO^2$]

C$_3$MS$_2$ Tricalcium Magnesium Disilicate [$3CaO \cdot MgO \cdot 2SiO_2$]

C$_3$MS$_6$ Tricalcium Magnesium Hexasilicate (or Merwinite) [$3CaO \cdot MgO \cdot 6SiO_2$]

CM(s) Methylated Surface Carbon Radical [Symbol]

CM*(s) Methylene (—CH_2—) Radical Bound to Surface Carbon [Symbol]

cm/s centimeter(s) per second [also cm/sec, or cm s^{-1}]

cm/s^2 centimeter(s) per second squared [also cm/sec^2, or cm s^{-2}]

cm^2/s square centimeter(s) per second [also cm^2/sec, or cm^2 s^{-1}]

cm^{-2}s^{-1} one per square centimeter per second [also 1/cm$^2 \cdot$s, or 1/cm^2/s]

cm^3/s cubic centimeter(s) per second [also cm^3/sec, or cm^3 s^{-1}]

cm^{-3}s^{-1} one per cubic centimeter per second [also 1/cm$^3 \cdot$s, or 1/cm^3/s]

CMSA Canning Machinery and Supplies Association [now Food Processing Machinery and Suppliers Association, US]

CMSA Chamber of Mines of South Africa

CMSA Coal Mine Safety Act [US]

CMSA Consolidated Metropolitan Statistical Areas

CMSB Certified Metrication Specialists Board [of United States Metric Association]

CMSC Canadian Materials Science Conference

CMSC Central Missouri State College [Warrensburg, US]

CMSC Composite Materials and Structures Center [of Michigan State University, East Lansing, US]

CMSC Cornell Materials Science Center [of Cornell University, Ithaca, New York, US]

CMSCI Council of Mechanical Specialty Contracting Industries [now American Specialty Contractors, US]

CMSE Center for Materials Science and Engineering

CMSE Center for Materials Science and Engineering [of University of Texas at Austin, US]

CMSE-MIT Center for Materials Science and Engineering–Massachusetts Institute of Technology [Cambridge, US]

CMSe Cadmium Manganese Selenide

cm/sec centimeter(s) per second [also cm/s, or cm s^{-1}]

cm/sec^2 centimeter(s) per second squared [also cm/s^2, or cm s^{-2}]

cm^{-2} sec^{-1} one per square centimeter per second [also cm^{-2} sec^{-1}, 1/cm$^2 \cdot$s, or 1/cm^2/s]

cm^2/sec square centimeter(s) per second [also cm^2/s, or cm^2 s^{-1}]

cm^{2}sec^{-1} square centimeter per second [also cm^2 sec^{-1}]

cm^3/sec cubic centimeter(s) per second [also cm^3/s, or cm^3 s^{-1}]

CMSER Commission on Marine Science, Engineering and Resources [US]

CMSG Canadian Merchant Service Guild

CMSI Checkout/Control and Monitor Subsystem Interface

CMSL CANMET (Canada Center for Mineral and Energy Technology)/Mineral Sciences Laboratories [of Natural Resources Canada]

CMS-NIST Center for Materials Science–National Institute of Standards and Technology [US]

cm^{-1}sr^{-1} one per centimeter per steradian [also 1/cm·sr, or 1/cm/sr]

CMSS Center on Materials for Space Structures [of Case Western Reserve University, Cleveland, Ohio, US]

CMSS Configuration Management Support System

CMSS Controlled Mechanical Storage Systems

CMST Characterization, Monitoring, Sensor Technology

CMSWAN Chamber of Mines of South West Africa/Namibia

CMT Cadmium Mercury Telluride (Semiconductor)

CMT Calcium Manganese Tantalate

CMT Cassette Magnetic Tape

CMT Center for Management Technology [US]

CMT Center for Materials Tribology [of Vanderbilt University, Nashville, Tennessee, US]

CMT Center for Microelectronics Technologies [of Sandia National Laboratory, Albuquerque, New Mexico, US]

CMT Code Matching Technique

CMT Computational Methods and Testing for Engineering Integrity (International Conference)

CMT Computer-Managed Training

CMT Consejo Mexicana de Turismo [Mexican Tourist Council]

CMT Conversational Mode Terminal

CMT Corrugating Medium-Flat Test

CMT Council for Mineral Technology [Randburg, South Africa]

cm^{-1} T^{-1} one per centimeter per Tesla [also cm^{-1} T^{-1}]

CMTA Canadian Marine Transportation Administration [of Transport Canada]

CMTA Constant Momentum Transfer Averaging

CMTC Cadmium Mercury Thiocyanate

CMTC Composite Manufacturing Technology Center [of Alcoa Laboratories, US]

CMTC Coupled Monostable Trigger Circuit

CMTC Cyano-Methyltrimethylene Carbonate

CMTDA Canadian Machine Tool Distributors Association

Cm:ThO$_2$ Curium-Doped Thoria [also Cm^{3+}:ThO$_2$]

CMTe Cadmium Manganese Telluride

CMTI Celestial Moving Target Indicator

CMTL Canadian Mining Technology Laboratory [of CANMET/Mining Research Laboratories, Natural Resources Canada]

CMTM Communications and Telemetry

CMTMDS 1,3-Bis(chloromethyl)-1,1,3,3-Tetramethyl Disilazane

CMTO Calcium Manganese Tantalum Oxide

CMTP Canadian Manpower Training Program

CMTP Continuum Mechanics of Textured Polycrystals (Method)

CMTS Canadian Machine Tool Show (and Exhibition)

CMTS Centralized Maintenance Test System

CMTS Computerized Maintenance Test System

CMTT Commonwealth Military Training Team

CMU Cambridge Management Unit [UK]

CMU Carnegie Mellon University [Pittsburgh, Pennsylvania, US]

CMU Central Michigan University [Mount Pleasant, US]

CMU Chiang Mai University [Thailand]

CMU Chlorophenyldimethylurea

CMU Computer Memory Unit

CMU Concrete Masonry Unit

CMU Control Maintenance Unit

CMU-NSF Carnegie-Mellon University/National Science Foundation (Joint Program) [US]

CMU-TR Carnegie-Mellon University Technical Report [also CMU TR, or CMU Tech Rep] [Published in Pittsburgh, Pennsylvania, US]

CMV Common-Mode Voltage

CMV Contact-Making Voltmeter

CMV Cucumber Mosaic Virus

CMV Cytomegalovirus

CmV Cotesia melanoscela Virus

cm/V centimeter(s) per Volt [also cm V^{-1}]

CMVC Configuration Management Version Control

CMVDR Center of Marine Vessel Design and Research [of Technical University of Nova Scotia, Halifax, Canada]

CMVM Contact-Making Voltmeter

CMVPCB California Motor Vehicle Pollution Control Board [US]

cm²/V·s square centimeter per volt-second [also cm²/Vs, cm² $V^{-1}s^{-1}$, cm²/V·sec, or cm²/V-sec]

CMVSA Canadian Motor Vehicle Safety Act

CMVSA Commercial Motor Vehicle Safety Act [US]

CMVSS Canadian Motor Vehicle Safety Standard

CMVTSS Canadian Motor Vehicle Tire Safety Standard

CMW Calcium Magnesium Tungstate (or Calcium Magnesium Wolframate)

CMW Canadian Manufacturing Week [Journal published in Canada]

CMW Centrimetric Wave

CMW Compartmented Mode Workstation

cm²/W square centimeter per watt [also cm² W^{-1}]

CMX Character Multiplexer

CMX Concentration Module Extension

CMXD Circular Magnetic X-Ray Dichroism

CMY Cyan/Magenta/Yellow

CMYK Cyan/Magenta/Yellow/Black

CMZP Calcium Magnesium Zirconium Phosphate (Ceramics)

CN Cadmium Niobate

CN Calcineurin [Biochemistry]

CN Canadian National Railways

CN Carbon Nanotube

CN Carbon Number [Chemistry]

CN Cascade Nozzle

CN Cathode Nickel

CN Caudate Nucleus [Anatomy]

CN Celestial Navigation

CN Celonavigation

CN Cellulose Nitrate

CN Center Notch

CN Central Nodes

CN Cetane Number

CN Change Notice

CN Chemical Name

CN China [ISO Code]

CN Chloronaphthalene

CN Chromium Niobate

CN Circular Note [also C/N]

CN Citation Number

CN Class Name

CN Close Nipple [Pipe Fittings]

CN Cold Neutron [Solid-State Physics]

CN Common Neighbor (Analysis Method) [Materials Science]

CN Communication Network

CN Company Name

CN Computer Network

CN Complete Name

CN Concentric Neutral

CN Condensation Nucleus (or Nuclei) [Meteorology]

CN Conseil National [National Council, Switzerland]

CN Consignment Note

CN Contract Number

CN Control Node

CN Coordination Number

CN Corporate Name

CN Cosmic Neutrino

CN Country Name

CN Crab Nebula [Astronomy]

CN Cranial Nerve [Anatomy]

CN Credit Note [also C/N]

CN Cubic Number

CN Cyanide

CN Cyano-

CN Cyanogen

CN Cycle Number [Mechanics]

CN- Morocco [Civil Aircraft Marking]

CN⁻ Cyanide (Ion) [Symbol]

CN$_x$ Nitrogenated Carbon [Symbol]

CN⁻ Cyanamide [Symbol]

C$_3$N$_4$ Carbon Nitride (Thin Film)

4C1N 4-Chloro-1-Naphthol [also 4-C1N]

C/N Carbon-to-Nitrogen (Ratio)

C/N Carrier-to-Noise (Ratio)

C/N Charge Nurse

C/N Circular Note [also CN]

C/N Credit Note [also CN]

Cn Chain [also cn]

Cn Corn [also cn]

Cn Neutralizing Capacitor [Symbol]

cn cosine of the amplitude

.cn China [Country Code/Domain Name]

CNA Cadmium Neodymium Aluminate

CNA Calcium Neodymium Aluminate

CNA Canadian Nuclear Association

CNA Canadian Nurses Association

CNA Center for Naval Analyses [US Navy]

CNA Centre Nucléaire des Ardennes [Nuclear Center of the Ardennes, Belgium]

CNA Certified NetWare Administrator

CNA Certified Network Administrator

CNA Certified Novell Administration [of Novell Inc., US]

CNA Chemical Notation Association [UK]

CNA Committee for Nautical Archeology [UK]

CNA Communications Network Architecture

CNA Concentrated Nitric Acid (Process)

CNA Copper-Nickel Alloy

CNA Cosmic Noise Absorption

CNA Cyanamide

CNAA Committee on Nucleation and Atmospheric Aerosols [US]

CNAA Council for National Academic Awards [UK]

CNAC CCIS (Common Channel Interoffice Signalling) Network Administration Center

CNAC China National Aviation Corporation

CNAc 2-Cyano Acetic Acid

CNACAC Cyano-Acetylacetonate [also CN-AcAC, or CN-acac]

3-CNACAC 3-Cyanopentane-2,4-Dione

CNAct 2-Cyanoacetamide

CNAD Conference on National Armament Directors [of NATO]

CNAd Cyanic Acid

CNAM Conservatoire National des Arts et Métiers [National Conservatory of Arts and Crafts, Paris, France]

CNAPS Coprocessing Node Architecture for Parallel Systems

CNAS Chemical Nomenclature Advisory Service [of Laboratory of the Government Chemists, UK]

CNAS Civil Navigation Aids System

CN-ATC Cyanide Amenable to Chlorination

CNB Centrale Nucléaire Belge [Belgian Nuclear Power Station]

CNB Chevron-Notched-Beam (Test) [Mechanics]

CNB Customs Nomenclature of Brussels

CNB Cyclical Neutral Balance

$C_{58}NB$ Fullerene Nitroboride [Derivative of Buckminsterfullerene C_{58}]

CNBMDA Canadian National Building Materials Distribution Association

CNBR The Computer Network of Building Researchers

CNBr Cyanogen Bromide

CNC Canadian National Committee

CNC Carbon-Nitrogen Cycle [Nuclear Physics]

CNC Centre National de la Cinématographie [National Cinematography Center, France]

CNC Chief of Naval Communications [of US Department of the Navy]

CNC (Agricultural and Technical) College of North Carolina [Greensboro, US]

CNC Computer(ized) Numerical Control

CNC Condensation Nuclei Counter [Meteorology]

CNC Configurable Network Computing

CN-CA Cellulose Nitrate–Cellulose Acetate

CNCC Customer Network Control Center

CNC/DNC Computer Numerical Control/Digital Numerical Control

CNCE Council for Non-Collegiate Continuing Education

CNCIAPS Canadian National Committee for the International Association on the Properties of Steam [also CNC/IAPS]

CNCIAWPRC Canadian National Committee of the International Association on Water Pollution Research and Control [also CNC/IAWPRC]

CNCIAWPRC Chilean National Committee of the International Association on Water Pollution Research and Control

CNCIAWPRC Cyprus National Committee of the International Association on Water Pollution Research and Control

CNCIAWPRC Czech(oslovakian) National Committee of the International Association on Water Pollution Research and Control

CNCIFAC Canadian National Committee for the International Federation of Automation Control [also CNC/IFAC]

CNCIUTAM Canadian National Committee for the International Union of Theoretical and Applied Mechanics [also CNC/IUTAM]

CNCl Cyanogen Chloride

CNCO Calcium Neodymium Copper Oxide (Superconductor)

CNCP Canadian National/Canadian Pacific (Telecommunications)

CN/CPT Canadian National/Canadian Pacific Telecommunications

CNCRM Canadian National Committee on Rock Mechanics

CNCS Central Navigation Control School

Cnct Connect [also cnct]

Cnctn Connection [also cnctn]

Cnctr Connector [also cnctr]

cnctrc concentric

CND Campaign for Nuclear Disarmament [UK]

CND Cluster of N-Defects [Physics]

Cnd Condition [also cnd]

Cnd Conduit [also cnd]

Cn3D Three-Dimensional Macromolecular Structure Viewer [for Molecular Modelling Database of National Center for Biological Information, US]

CNDA Canadian National Distribution Authority

Cndct Conductivity [also cndct]

Cndctn Conduction [also cndctn]

Cndctr Conductor [also cndctr]

CNDE Center for Nondestructive Evaluation [of Johns Hopkins University, Baltimore, Maryland, US]

cndl conditional

CNDL EOP Conditional End of Page [also Cndl EOP]

CNDO Complete Neglect of Differential Overlap (Approximation) [Physics]

CNDP Centre National de Documentation Pédagogique [National Center for Educational Documentation, France]

CNDP Communication Network Design Program

CNDP Continuing Numerical Data Project

Cnds Condensate [also cnds]

CNDST Centre National de Documentation Scientifique et Technique [National Center of Scientific and Technical Documentation, France]

CNE Canadian National Exhibition [Toronto, Canada]

CNE Certified NetWare Engineer

CNE Charge Neutrality Equation [Physics]

CNE Communications Network Emulator

CNE Compare Numerical Equal

CNEA Comisión Nacional de Energía Atómica [National Atomic Energy Commission, Buenos Aires, Argentina]

CNea Cyanoethylamine

CN-ECO 3-Cyano-7-Ethoxy Coumarin

CNEEL Centre National d'Etude et d'Expérimentation des Eoliennes [Windmill Study and Experimentation Center, France]

CNEN Centro Nacional de Energia Nuclear [National Nuclear Energy Center, Sao Paulo, Brazil]

CNEN Comisión Nacional de Energía Nuclear [National Nuclear Energy Commission, Mexico]

CNEN Comissão Nacional de Energia Nuclear [National Nuclear Energy Commission, Brazil]

CNEN Comitato Nazionale per Energia Nucleare [National Nuclear Energy Committee, Italy]

CNERT Center for Natural Resources, Energy and Transport [of United Nations]

CNES Centre Nationale d'Etudes Spatiales [National Center for Space Studies, France]

CNESER Conseil National de l'Enseignement Supérieur et de la Recherche [National Council of Higher Education and Research, France]

CNET Centre National d'Etudes des Télécommunications [National Telecommunications Research Center, Meylan, France]

CNET Communications Network

CNEt Cyanoethyl

CNEUPEN Commission Nationale pour l'Etude de l'Utilisation Pacifique de l'Energie Nucléaire [National Commission for the Study of the Peaceful Uses of Nuclear Energy, Belgium]

CNF Canadian Nature Federation

CNF Controlled Nuclear Fusion

CNF Cyanogen Fluoride

.CNF Configuration [File Name Extension] [also .cnf]

CNFRL Columbia National Fisheries Research Laboratory [US]

CNFSFACM Committee of the National Ferrous Scrap Federations and Associations of the Common Market [Belgium]

CNFWMP Canadian Nuclear Fuel Waste Management Program

CNG Calling (Tone) [Telecommunications]

CNG Compressed Natural Gas

CNG Consolidated Natural Gas

CNGA California Natural Gas Association [US]

CN_xH_y Hydrogenated Carbon Nitride

Cn Hd Corn Head (of Combines) [Agriculture]

CNHM Chicago Natural History Museum [Illinois, US]

CNHS Cherokee National Historical Society [US]

CNI Center for Nuclear Information [Brazil]

CNI Centre Nucléaire Interescaut [Interescaut Nuclear Center, Belgium]

CNI Certified Novell Instructor [of Novell Inc., US]

CNI Changed Number Interception

CNI Coalition for Networked Information [Internet]

CNI Cognitive Neuroscience Institute [of Massachusetts Institute of Technology, Cambridge, US]

CNI Communication, Navigation and Identification

CNI Consiglio Nazionale Ingeneri [National Engineering Council, Italy]

CNI Consolidated National Intervenors

CNI Cyanogen Iodide

CNIC Centre National d'Information Chimique [National Center for Chemical Information, France]

CNIC Consorcio Nacional de Industriales del Caucho [National Consortium of Rubber Industries, Spain]

CNIDR Clearinghouse for Network Information and Discovery and Retrieval [Internet]

CNIE Comisión Nacional de Investigaciónes Espaciales [National Commission on Space Research, Argentina] [now Comisión Nacional de Actividades Espaciales]

CNIE Committee for the National Institutes of the Environment [US]

CNIL Conferazione Nazionale Italiana de Lavoro [Italian National Confederation of Labor]

CNIL Cumulative Net Investment Loss

CNIM Centro Nacional de Investigaciónes Metalurgicas [National Metallurgical Research Center, Madrid, Spain]

CNIPA Committee of National Institutes of Patent Agents [UK]

CNIRI Chugoku National Industrial Research Institute [of Agency of Industrial Science and Technology, Japan]

CNJA Comité Nationale des Jeunesses Agriculteurs [National Committee of Young Agriculturalists, France]

C(n)K Potassium Alkanecarboxylates

CNL Cancellation Message

CNL Circuit Net Loss

CNL Commonwealth National Library [Canberra, Australia]

CNL Constant Net Loss

CNLA Council on National Library Associations [now Council of National Library and Information Associations]

CNLIA Council of National Library and Information Associations [US]

CNLS Center for Non-Linear Studies [UK]

CNM Centro Nacional de Microelectrónica [National Microelectronics Center, Spain]

CNM Centro Nuclear de Mexico [Mexican Nuclear Center, Salazar Estado de Mexico]

CNM Certified Nurse-Widwife

CNM Classical Nucleation Model [Metallurgy]

CNM Collective Nuclear Model

CNM Concast Nozzle Manipulator

C²/Nm² Coulumb squared per Newton-meter squared [also C²/N·m², or $C^2 N^{-1} m^{-2}$]

CNMA Canadian Naturopathic Medicine Association

CNMA Communications Network for Manufacturing Applications [Europe]

CNMa Cyanomethylamine

CNMB Central Nuclear Measurements Bureau

CNM-CSIC Centro Nacional de Microelectrónica–Consejo Superior de Investigaciónes Cientificas [National Microelectronics Center of the National (Scientific) Research Council at Bellaterra, Spain]

CNMDA Canada/Newfoundland Mineral Development Agreement

CNMI Communications Network Management Interface

CNMP Conference on National Materials Policy [of Federation of Materials Societies, US]

CNMR Carbon(-13) Nuclear Magnetic Resonance [also C-NMR]

CNMR Catalytic Nonpermselective Membrane Reactor

C-NMR Carbon Nuclear Magnetic Resonance [also C NMR, or C-NMR]

C-13 NMR Carbon-13 Nuclear Magnetic Resonance [also ^{13}C NMR, or ^{13}C NMR]

C-15 NMR Carbon-15 Nuclear Magnetic Resonance [also ^{15}C NMR, or ^{15}C-NMR]

^{15}C NMR Spectrosc Carbon-15 Nuclear Magnetic Resonance Spectroscopy [US Journal]

CNMT Consejo Nacional Mexicana de Turismo [Mexican National Tourism Council]

CNN Cable News Network [US]

CNN Composite Network Node

CNO Cadmium-Niobium Oxide

CNO Chief of Naval Operations [of US Department of the Navy]

CNO Cyanate [Formula]

CNOGEDC China National Oil and Gas Exploration Development Corporation

CNOM Center for Nonlinear Optical Materials [of Stanford University, California, US]

CNOOC China National Offshore Oil Corporation

CNOP Conditional Non-Operational; Conditional No Operation

CNOUS Centre National des Oeuvres Universitaires et Scolaires [National Center of University and Academic Affairs, France]

CNP Celestial North Pole [Astronomy]

CNP Communications Network Procedure

CNP Communications Network Processor

CNP Council for National Parks [UK]

CNP C-Type Natriuretic Peptide [Biochemistry]

CNPase Cyclic Nucleotide Phosphodiesterase [Biochemistry]

CNPC Chinese National Petroleum Corporation [PR China]

CN-PEG Mono-Methoxypolyethylene Glycol activated with Cyanuric Chloride

CNPF Conseil National des Personnels Françaises [National Council for French Employees, France]

CNPFT Centro Nacional de Pesquisa de Fruteiras de Clima Temperado [National Research Center of Temperate Weather Fruit Trees, Brazil]

CNPI Comisión Nacional de Productividas Industrial [National Commission of Industrial Productivity, Spain]

CNPP Centre National de Prévention et de Protection [National Center for Prevention and Protection, France]

CNPPA Commission on National Parks and Protected Areas [of International Union for Conservation of Nature (and Natural Resources)]

CNPPSDP Cooperative National Plant Pest Survey and Detection Program [US]

CNPq Conselho Nacional de Pesquisas [National (Scientific) Research Council, Brazil] [now Conselho Nacional de Desenvolvimento Científico e Tecnológico]

3CNPy 3-Cyanopyridine

C(n)PyBr Alkylpyridinium Bromide

CNR Canadian National Railways

CNR Carrier-to-Noise Ratio

CNR Center for Neutron Research [of National Institute of Science and Technology, Gaithersburg, Maryland, US]

CNR Centro Nazionale delle Ricerche [National Research Center, Italy]

CNR Composite Noise Rating

CNR (Department of) Conservation and Natural Resources [of State of Victoria, Australia]

CNR Consiglio Nazionale delle Ricerche [National Research Council, Italy]

C(n,r) Number of r Subsets of a Set of n Elements (in Mathematics) [Symbol]

CNRC Canadian National Research Council

CNRF Cold Neutron Radiation Facility [of Japan Atomic Energy Research Institute, Tokai, Japan]

CNRF Cold Neutron Research Facility [of National Institute for Standards and Technology–Materials Science and Engineering Laboratory, Gaithersburg, Maryland, US]

CNR-GNSM Consiglio Nazionale delle Ricerche, Gruppo Nazionale di Struttura della Materia [National Research Council, National Group on the Structure of Matter, Italy]

CNRI Caribbean Natural Resources Institute

CNRL Communications and Navigation Research Laboratory [US]

CNRM Centre National de Recherche Métallurgiques [National Center for Metallurgical Research, Liege, Belgium]

CNRM Centre National de Recherche Météorologique [National Center for Meteorological Research, France]

CNR-MADESS Consiglio Nazionale delle Ricerche–Progretto Finalizzato Materiali e Dispositio per l'Elettronica allo Stato Solido [National Research Council–Final Progress Report on Materials and Depositions for Solid-State Electronics, Italy]

CNRN Comitato Nazionale per le Ricerche Nucleare [National Committee for Nuclear Research] [now Comitato Nazionale per Energia Nucleare, Italy]

CNRS Centre National de la Recherche Scientifique [National Scientific Research Center, France]

CNRS Conseil National de la Recherche Scientifique [National Scientific Research Council, France]

CNRS Bordeaux Centre National de la Recherche Scientifique à Bordeaux [National Center for Scientific Research at Bordeaux, France]

CNRS Grenoble Centre National de la Recherche Scientifique à Grenoble [National Center for Scientific Research at Grenoble, France]

CNRS-INPL Conseil National de la Recherche Scientifique–Institut National Polytechnique de Lorraine [National Scientific Research Council–National Polytechnic Institute of Lorraine, Nancy, France]

CNRS-L2M Conseil National de la Recherche Scientifique–Laboratoire de Microstructures et de Microélectronique [National Scientific Research Council–Laboratory for Microstructures and Microelectronics, Bagneux, France]

CNRSM Centro Nazionale per la Ricerca e lo Sviluppo dei Materiali [National Center for Research and Development of Materials, Mesagne, Italy]

CNRS-ONERA Conseil National de la Recherche Scientifique–Office National d'Etudes et de Recherches Aérospatiales [CNRS National Office for Aerospace Studies and Research, France] [also CNRS/ONERA]

CNRS Res CNRS (Conseil National de la Recherche Scientifique) Research [Published in US]

CNRS Strasbourg. Centre National de la Recherche Scientifique à Strasbourg [National Center for Scientific Research at Strasbourg, France]

CNRS Toulouse Centre National de la Recherche Scientifique à Toulouse [National Center for Scientific Research at Toulouse, France]

CNS Canadian Navigation Society

CNS Canadian Nuclear Society

CNS Canaveral National Seashore [Florida, US]

CNS Cardiff Naturalists Society [UK]

CNS Center for Northern Studies [Canada]

CNS Central Navigation School

CNS Central Nervous System

CNS Coherent Neutron Scattering

CNS College of Natural Sciences [of Seoul National University, South Korea]

CNS Commodity News Service [US]

CNS Communications Navigation and Surveillance System

CNS Communications Network Service

CNS Communications Network Simulation

CNS Consejo Nacional Sindical [National Trade Union Council, Guatemala]

C NS Count, Non-Summing (Mode)

CNSA Centro Nacional de Sanidad Ambiental [National Center for Environmental Health, Spain]

CNSC Carrying Nuclear-Strike Cruiser

CNSF Cornell National Supercomputer Facility [of Cornell University, Ithaca, New York, US]

CNSG Consolidated Nuclear Steam Generator [of Babcock and Wilcox Company, US]

CNSH Calcium Disodium Silicate Hydrate (or Pectolite) [$CaO \cdot Na_2O \cdot SiO_2 \cdot H_2O$, or $CaNa_2SiO_4 \cdot H_2O$]

CNSI Canadian Newspaper Services International

Cnsl Console [also cnsl]

Cnsl Counsel [also cnsl]

cnsld consolidated [also cnsl'd]

CNSM Council on Nutritional Sciences and Metabolism [of American Diabetes Association, US]

CNSN Canadian National Seismograph Network

CNSR Canadian Network for Space Research

CNSR Chevron-Notched Short Rod (Fracture Toughness Test) [Mechanics]

CNSS Center for National Space Study

CNSS Core Nodal Switching Subsystem [Internet]

CNT Canadian National Telecommunications

CNT Celestial Navigation Trainer

CNT Centre National de Tourisme [National Tourist Center, France]

CNT Classical Nucleation Theory [Metallurgy]

CNT Confederación Nacional del Trabajo [National Confederation of Labour, Spain]

CNT Convención Nacional de Trabajadores [National Convention of Workers, Uruguay]

CNT Cyanotoluene

.CNT Contents [File Name Extension] [also .cnt]

CN-T Total Cyanide

Cnt Count(er) [also cnt]

CNTA Council for Nordic Teachers Associations [Denmark]

CNTB Choking, Nose, Tear and Blister Gases

CNTBT Comprehensive Nuclear Test Ban Treaty

Cn3D Three-Dimensional Macromolecular Structure Viewer [for Molecular Modelling Database of National Center for Biological Information, US]

CNTF Ciliary Neurotrophic Factor [Biochemistry]

Cntl Control [also cntl]

Cntor Contractor [also cntor]

CNTP Committee for a National Trade Policy

Cntr Container [also cntr]

Cntr Contract [also cntr]

Cntr Counter [also cntr]

Cntrl Control(ler) [also cntrl]

cntrl central

Cntrls Controllers [also cntrls]

Cntrls Controls [also cntrls]

CNTS Centre National des Techniques Spatiales [National Space Technology Center, Algeria]

CNTS Confédération Nationale des Travailleurs Sénégalaises [National Confederation of Senegalese Workers]

cnts/min counts per minute [also cts/min, CPM, or cpm]

cnts/sec counts per second [also cts/sec, CPS, or cps]

CNTT Convegno Nazionale Trattamenti Termici [National Heat Treatment Convention, Italy]

CNTU Confederation of National Trade Unions [Canada]

CNU Christopher Newport University [Newport News, Virginia, US]

CNU Changwon National University [Changwon, South Korea]

CNU Chungbuk National University [Chongju, South Korea]

CNU Chungnam National University [Taejon, South Korea]

CNU Compare Numeric Unequal

CNUCE Centro Nazionale Universitario di Calcalo Elettronica [National University Center for Electronic Calculation, Italy]

CNUJ Committee for Nordic Universities of Journalism [Sweden]

C-Number Commutative Number [also C-number]

CNV Contingent Negative Variation

CNV Conventional (Memory)

CNV Convert [also Cnv, or cnv] [Computers]

CNVF Corpo Nazionale-Vigili del Fuoco [Italian National Fire Service]

Cnvr Conveyor [also cnvr]

CNVT Convert [also Cnvt, or cnvt]

Cnvtr Converter [also cnvtr]

C&NW Chicago and North Western (Railway System) [US]

CNWDI Critical Nuclear Weapons Design Information

CNWRA Center for Nuclear Waste Regulatory Analyses [of Southwest Research Institute, San Antonio, Texas, US]

CNX Certified Network Expert

CNY Chinese Yuan [Currency of PR China]

CO Carbon Monoxide

CO Carbonyl (Radical) [Symbol]

CO Cargo Operations [NASA Kennedy Space Center Directorate, Florida, US]

CO Carry Output [Combinational Logic]

CO Cash Order [also C/O, Co, or c/o]

CO Center of Oscillation

CO Central Office [Telecommunications]

CO Chairman of Operations

CO Change Order

CO Change-Over

CO Charge-Ordered; Charge Ordering [Solid-State Physics]

CO Chemical Oceanographer; Chemical Oceanography

CO Chief of Ordnance

CO Chief Operator

CO Circuit Order

CO Cleanout

CO Close-Open Operation

CO Coble Creep [Metallurgy]

CO Coke Oven

CO College of Ophthalmologists

CO Colonial Office

CO Color

CO Colombia [ISO Code]

CO Colorado [US]

CO Combined Operations

CO Commanding Officer

CO Command Output

CO Commonwealth Observatory

CO Communications Office(r)

CO Compressor Outlet

CO Computer Operations; Computer Operator

CO Condensate Overflow

CO Conscientious Objector

CO Conservation Officer

CO Consumption-Oil [Automobiles]

CO Contracting Officer

CO Convert Out

CO Crack Opening [Mechanics]

CO Crude Oil

CO Crystal(-Controlled) Oscillator

CO Cubo-Octahedron

CO Cutoff [also C/O]

CO Cut Out

CO_2 Carbon Dioxide (Laser)

CO^- Carbonate (Ion) [Symbol]

C_2O^- Oxalate (Ion) [Symbol]

$C_{60}O$ Fullerene Monoepoxide [Oxygen Derivative of Buckminsterfullerene C_{60}]

C/O Carbon-to-Oxygen (Ratio)

C/O Carried Over [also CO, co, or c/o]

C/O Cash Order [also CO, co, or c/o]

C/O Changeout

C/O Checkout

C/O Contamination/Overpressure

C/O Cutoff [also CO]

C-O Condenser-Objective

C=O Carbonyl Group [Symbol]

Co Cobalt [Symbol]

Co Coenzyme [Biochemistry]

Co Company [also co]

Co County [also co]

Co Cordierite [Mineral]

Co Cyclooctadiene

Co^{2+} Divalent Cobalt Ion [also Co^{++}] [Symbol]

Co^{3+} Trivalent Cobalt Ion [also Co^{+++}] [Symbol]

Co I Coenzyme I (or Diphosphopyridine Nucleotide) [Biochemistry]

Co II Coenzyme II (or Diphosphopyridine Nucleotide) [Biochemistry]

Co-55 Cobalt-55 [also ^{55}Co, or Co^{55}]

Co-56 Cobalt-56 [also ^{56}Co, or Co^{56}]

Co-57 Cobalt-57 [also ^{57}Co, or Co^{57}]

Co-58 Cobalt-58 [also ^{58}Co, or Co^{58}]

Co-59 Cobalt-59 [also ^{59}Co, Co^{59}, or Co]

Co-60 Cobalt-60 [also ^{60}Co, or Co^{60}]

Co-61 Cobalt-61 [also ^{61}Co, or Co^{61}]

Co-62 Cobalt-62 [also ^{62}Co, or Co^{62}]

co carried over [also c/o]

.co Colombia [Country Code/Domain Name]

c/o care of [also %]

c/o carried over [also C/O, CO, or co]

c/o cash order [also co, Co, or C/O]

COA Center Operations Area

COA Certificate of Analysis [also CoA]

COA Certificate of Approval

COA Certificate of Authenticity

COA Certified Office Administrator

COA College of Aeronautics [UK]

COA Committee on Accreditation [US]

COA Conversion of Acetyl

COA Council of Agriculture [Taiwan]

CoA Coenzyme A [Biochemistry]

COAC Commission on Agriculture [of Food and Agricultural Organization]

$Co(Ac)_2$ Cobalt(ous) Acetate

$Co(ACAC)_2$ Cobalt(II) Acetylacetonate [also $Co(AcAc)_2$, or $Co(acac)_2$]

$Co(ACAC)_3$ Cobalt(III) Acetylacetonate [also $Co(AcAc)_3$, or $Co(acac)_3$]

COACH Canadian Organization for Advancement of Computers in Health

COADS Comprehensive Oceanic and Atmospheric Data Set [of National Oceanic and Atmospheric Administration, US]

CO/AF Charge-Ordered Antiferromagnetic (Phase) [also CO AF] [Solid-State Physics]

Co-Ag Cobalt-Silver (Alloy System)

CoAl Cobalt Aluminide (Coating)

CoAl Cobalt Aluminum (Spin Glass)

Co-Al Cobalt-Aluminum (Alloy System)

Coal J Coal Journal [US]

Coal Min Coal Mining [Journal published in the UK]

$Co-Al_2O_3$ Cobalt-Bonded Aluminum Oxide (Cermet)

Coal Prep Coal Preparation [Journal published in the UK]

COAL PRO Coal Research Project

COAM Customer Owned and Maintained (Equipment)

Coam Coaming [also coam]

COAMS Canadian Ocean Acoustic Measurement System

COARE Coupled Ocean-Atmosphere Response Experiment

COAS Course Optical Alignment Sight

COAS Crew(man) Optical Alignment Sight

CoASH Coenzyme A with Sulfhydryl Functional Group [Biochemistry]

CoAsS Cobalt Arsenic Sulfide (or Cobaltite)

COAST Card On A Stick (Module)

COAT Coherent Optical Array Technique

COAT Corrected Outside Air Temperature

Coat Coating [also coat]

COATS Canadian Over-the-Counter Automated Trading System

Coat Technol Coating Technology [Journal published in the US]

Co-Au Cobalt-Gold (Alloy System)

Co/Au Cobalt on Gold (Substrate)

Coax Coaxial (Cable) [also coax]

COAXIS Centralized Oceanographic and Acoustic Cross-Referenced Information System

COB Central Obrera Boliviano [Bolivian Central Labour Organization]

COB Chip On Board

COB Close of Business

COB Communications Office Building

COB Company of Biologists Limited [UK]

COB Cut-Off Bias [Electronics]

.COB COBOL Source Code [File Name Extension]

CoB Cobaltic Boride

CoB Coenzyme B [Biochemistry]

CoB_{12} Coenzyme B_{12} [Biochemistry]

Co-B Cobalt-Boron (Alloy System)

Cobalt TPP 5,10,15,20-Tetraphenyl-21H,23H-Porphine Cobalt(II) [also cobalt TPP]

COBASE Collaboration in Basic Science and Engineering (Program) [Joint US Program of the National Academy of Science and National Research Council]

COBE Cosmic Background Explorer (Project) [of NASA Astrophysics Data Facility, Goddard Space Flight Center, Greenbelt, Maryland, US]

CoBe Cobalt Beryllide

Co-Be Cobalt-Beryllium (Alloy System)

COBESTCO Computer-Based Estimating Technique for Contractors

COBH Chinonoxime Benzoylhydrazone

COBI Coded Biphase

COBIS Computer-Based Instruction System

COBLIB COBOL Library

COBLOC CODAP (Control Data Assembly Program) Language Block-Oriented Compiler

COBLOS Computer-Based Loans System

COBOL Common Business-Oriented Language [also Cobol]

CO-Bomb Cobalt Bomb [also Co-bomb]

CO-BR Butadiene Rubber based on Cobalt Catalyst

COBRA Consolidated Omnibus Budget Reconciliation Act [US]

COBRA Cosmic Background Radiation Anisotropy [University of Chicago, Illinois, US]

COBSA Computer Service Bureaus Association [UK]

COBSEA Coordinating Body on the Seas of East Africa

COBSEA Coordinating Body on the Seas of East Asia

COBSTRAN Composite Blade Structural Analyser (Computer Code)

COC Carbon Oxide Chemisorption

COC Chain of Custody

COC Chamber of Commerce

COC Chapter Operations Committee [of ASM International, US]

COC Chlorinated Organic Compound

COC Cleveland Open Cup (Flash-Point Tester)

COC Close-Open-Close

COC Coal Operators Conference [of Canadian Institute of Mining and Metallurgy]

COC Coded Optical Character

COC Combat Operations Center

COC Compiler Object Code

COC Copper Oxychloride

COC Crystal Observation Chamber

COC Cube-on-Corner (Texture) [also CoC] [Metallurgy]

COC Cycloolefin Copolymer

Coc Cholesteryloxycarbonyl

COCA Council of Ontario Contractors Associations [Canada]

COCAM Canadian-Ontario Center for Advanced Manufacturing

COCAST Colonial Colleges of Arts, Science and Technology [UK]

COCATRAM Comision Centroamericana de Transporte Maritimo [Central American Commission of Maritime Transport, El Salvador]

COCC Character Oriented Communications Controller

COCD Canadian Ownership and Control Development (Program)

COCESNA Central American Corporation for Air Navigation Services

COCF Center for Our Common Future

COCF Conditional Two-Point Orientation Correlation Function [Crystallography]

COCH$_3$ Acetyl Group [Formula]

CO$_2$-CH$_4$ Carbon Dioxide–Methane (Gas System)

CO$_2$-C$_2$H$_2$ Carbon Dioxide–Acetylene (Gas System)

CO$_2$-C$_3$H$_8$ Carbon Dioxide–Propane (Gas System)

COCI Consortium on Chemical Information [UK]

CoCl$_3$-GIC Cobalt Trifluoride Graphite Intercalation Compound

CoCl$_3$-K-GBC Cobalt Trifluoride-Potassium Graphite BiIntercalation Compound

COCO Committee Code

COCO (Word) Count and Concordance Generator on Atlas [UK]

CO/CO$_2$ Carbon Monoxide/Carbon Dioxide (Ratio)

Co-CoBe Cobalt Beryllide Reinforced Cobalt (Composite)

COCONOSY Combined Correlated and Nuclear Overhauser Enhancement Spectroscopy

COCODE Compressed Coherency Detection

COCOM Coordinating Committee

CoCom Coordinating Committee for Multilateral Export Controls [also COCOM, or Cocom]

Cocomo Cost Constructive Model

COCORP Consortium for Continental Reflection Profiling (Program) [of Cornell University, Ithaca, New York, US]

COCOT Coin-Operated Customer-Owned Telephone

Co(Cp)$_2$ Bis(cyclopentadienyl)cobalt(II) (or Cobaltocene)

CoCp(CO)$_2$ Cyclopentadienylcobalt Dicarbonyl

Co-Cr Cobalt-Chromium (Alloy System)

CoCrAl Cobalt-Chromium-Aluminum (Coating)

CoCrAlY Cobalt-Chromium-Aluminum-Yttrium (Coating)

COCRIL Council of City Research and Information Librariers [UK]

CoCrPt Cobalt Chromium Platinum (Alloy)

CoCrPtTa Cobalt Chromium Platinum Tantalum (Alloy)

CoCrTa Cobalt Chromium Tantalum (Alloy)

COCS Common Occupational Classification System

COCS Copper Oxychloride Sulfate

Co-Cu Cobalt-Copper (Alloy System)

COD Carrier Onboard Delivery

COD Carrier-Operated Device

COD Cash on Delivery [also cod]

COD Catastrophic Optical Damage

COD Center Operations Directorate [of NASA Johnson Space Center, Houston, Texas, US]

COD Charleston Ordnance Depot [South Carolina, US]

COD Chemical Organic Deposition

COD Chemical Oxygen Demand

COD Chicago Ordnance Depot [Illinois, US]

COD Cincinnati Ordnance Depot [Ohio, US]

COD Clean-Out Door

COD Cleveland Ordnance Depot [Ohio, US]

COD Coefficient of Diffusion

COD Coefficient of Draft

COD Collect on Delivery [also cod]

COD Crack Opening Displacement [Mechanics]

COD Crack Opening Distance [Mechanics]

COD Critical Opening Displacement [Mechanics]

COD Crystallographic Orientation Distribution

COD Cyclooctadiene

1,5-COD 1,5-Cyclooctadiene

.COD Code List [File Name Extension] [also .cod]

C1D Conducting along One Dimension

CODA Conservation Options and Decision Analysis (Software)

CODAC Coordination of Operating Data by Automatic Computer

CODAG Combined Diesel and Gas

CODAN Carrier-Operated Device Antinoise (Muting System)

CODAN Coded Weather Analysis [of US Navy]

CODAP Control Data Assembly Program

CODAR Coastal Ocean Dynamics Application Radar

CODAR Coherent Display Analyzing and Recording

CODAR Correlation, Detection, and Ranging [also Codar, or codar]

CODAR Correlation Display Analyzing and Recording

CODAS Control and Data Acquisition System

CODAS Customer-Oriented Data Acquisition System

CoDAS Condensed Matter Direct Alerting Service [of Institute of Physics Publishing, UK]

CODASYL Conference on Data System Languages [Lexington, Massachusetts, US] [also Codasyl]

CODATA Committee on Data for Science and Technology [of International Council of Scientific Unions]

CODATA Bull CODATA Bulletin [of Committee on Data for Science and Technology]

CODATU Conference on the Development and Improvement of Urban Transport

CODAZR Committee on Desert and Arid Zones Research [US]

CODC Canadian Oceanographic Data Center

[(1,5-COD)CuCl]₂ Bis(1,5-Cyclooctadiene) Copper (I) Chloride

CODE Canadian Organization for Development through Education

CODE Center of Disciplinary Expertise [Department of Fisheries and Oceans, Canada]

CODE Client-Server Open Development Environment [Internet]

CODE COBOL Program Development

CODEC Coder/Decoder (Equipment); Coding/Decoding (Equipment) [also Codec, or codec]

CODEC Compression/Decompression

CODECA Corporation for Economic Development in the Caribbean

CODED Computer Design of Electronic Devices

CODELACS Computer Developments Limited Automatic Coding System

CODEMIN Empresa de Desenvolvinmento de Recursos Minerais SA [Brazilian Mineral Resources Development Enterprise]

CODES Computer Design and Education System

CODES Computer Design and Evaluation System

CODEVER Code Verification

CODEX Codex Alimentarius Commission

CODF Crystallite Orientation Distribution Function [Crystallography]

CODIA Colegio Dominicano de Ingenieros, Arquitectos y Agrimensores [Dominican College of Engineers, Architects and Surveyors, Puerto Rico]

CODIAC Centralized Operation Deterministic Interface Access Control [also Codiac]

CODIC Color Difference Computer

CODIC Computer-Directed Communications

Co(dien)₃ Cobalt(III) (Diethylenetriamine)

CODIL Content Dependent Information Language

CODIL Control Diagram Language

CODIPHASE Coherent Digital Phased Array System

CODIS Controlled Digital Simulator

CODIT Computer Direct to Telegraph

CODMAC Committee on Data Management and Archiving and Computation [NASA Planetary Data System]

CODOC Cooperative Documents Project [of Ontario University Libraries Cooperation System, Canada]

CODOG Combined Diesel or Gas

CODORAC Coded Doppler Radar Command

CODSIA Council of Defense and Space Industry Associations [US]

CODSTRAN Composite Durability Structural Analysis (Computer Code)

CODZR Committee on Desert and Arid Zone Research

COE Cab over Engine

COE Central Office Equipment

COE Coefficient of Elasticity

COE Coefficient of Expansion

COE Commission on Ecology [of International Union for Conservation of Nature (and Natural Resources)

COE Committee on Education [of American Physical Society, US]

COE Conservation of Energy

COE Cooperative Office Education

COE Corps of Engineers [of US Army]

COE Cost of Electricity

COE Cube-on-Edge (Texture) [also CoE] [Metallurgy]

COEC Council Operations and Exercises Committee [of NATO]

COED Char Oil Energy Development (Process)

COED Coal-Oil-Energy Development (Process)

COED Computer-Operated Electronic Display

Co-ed Co-Education [also Coed, coed, or co-ed]

COEES Central Office Equipment Engineering System

Coef Coefficient [also Coeff, coeff, or coef]

COEM Commercial Original Equipment Manufacturer

CoEnCo Council for Environmental Conservation [now The Environment Council]

Co(en)₃ Cobalt(III) (1,2-Diaminoethane)

Co(en)₂Cl₂ Cobalt(II) (1,2-Diaminoethane) Dichloride

CO₂-EOR Carbon Dioxide Enhanced Oil Recovery

COER Central Office Equipment Report

COESA Committee on Extension to the Standard Atmosphere [US]

COF Canadian Order of Foresters

COF Carry-Over Factor [Structure Analysis]

COF Cause of Failure

COF Coefficient of Friction

COF Coffeyville, Kansas [Meteorological Station Designator]

COF Confusion Signal

COF Construction of Facilities

COF Correct Operation Factor

COF Curvature of Field

COF Cut-Off Frequency [Electronics]

COF Cube-on-Face (Texture) [also CoF] [Metallurgy]

COFAB Committee on Finance and Banking [of Association of Southeast Asian Nations]

COFAF Committee on Food, Agriculture and Forestry [of Association of Southeast Asian Nations]

COFC Container on Flat Car

C of C Chamber of Commerce

COFDM Companded Frequency-Division Multiplex(ing)

CoFe Cobalt Iron (Alloy)

Co-Fe Cobalt-Iron (Alloy System)

COFEB Confederation of European Bath Manufacturers [Scotland]

COFEGES Conceil Fédérale des Groupes Economiques de Sénégal [Federal Council of Economique Groups of Senegal]

CoFeNi Cobalt Iron Nickel (Alloy)

CoFeSi Cobalt Iron Silicon (Alloy)

CoFeSiB Cobalt Manganese Silicon Boron (Alloy)

COFF Common Object File Format [Unix Operating System]

C of F Construction of Facilities

C of F Cost of Facilities

C of F Cost of Freight [also CofF]

COFFI Communications Frequency and Facility Information System [of International Civil Aviation Organization]

COFFS Coffs Harbour, New South Wales Forestry Commission Herbarium [Australia]

C of G Center of Gravity

COFI Checkout and Fault Isolation (on Board)

COFI Committee of Fisheries [of Food and Agricultural Organization]

COFI Council of (British Columbian) Forest Industries [Vancouver, Canada]

COFIL Core File

CoFIQ Coalition des Facultés d'Ingénièrie du Québec [Coalition of Engineering Faculties of Quebec, Canada]

COFIS Canadian On-Line Financial Information Service

COFO Committee on Forestry [of United Nations]

COFOG Classification of the Functions of Government

C of ORD Chief of Ordnance [also COFORD, or C of Ord]

COFR Certificate of Flight Readiness

C of R Center of Resistance

C of R Center of Rotation

C of R Commencement of Rifling

COFS Closed-Open-Face Sandwich

C of S Chief of Staff

C of U Council of Universities

COFRDA Canada-Ontario Forest Resource Development Agreement

COFW Certificate of Flight Worthiness

COG Canberra Ornithologists Group [Australia]

COG CANDU (Canada Deuterium Uranium Reactor) Owners Group [Canada]

COG Center of Gravity

COG Center of Gyration

COG Central Office Ground

COG Chip-on-Glass

COG Coke-Oven Gas

COG Commentary Graphics

COG Commercial Operations Group [of Battelle, US]

COG Composite Group

COG Computer Operations Group

COG Council of Australian Governments

cog cognate

CoGa Cobalt Gallium

Co/GaAs Cobalt on Gallium Arsenide (Substrate)

COGAG Combined Gas and Gas

CoGa/GaAs Cobalt Gallium on Gallium Arsenide (Substrate)

COGAS Coal Gasification (Process)

COGB Certified Official Government Business

COGD Circulator Outlet Gas Duct

COGECA Comité Général de la Coopération Agricole de la Communauté Européenne [General Committee of Agricultural Cooperation in the European Community, Belgium]

COGENE Committee on Genetic Engineering [of International Council of Scientific Unions, France]

COGENT Compiler and Generalized Translator

COGENT Compiler, Generator and Translator

COGEO Coordinate(d) Geometry [Computers]

COGEODATA Committee on the Storage, Automatic Processing and Retrieval of Geological Data [of Geological Survey of Canada]

CoGeV Cobalt-Germanium-Vanadium (Structure)

COGIS Comprehensive Oil and Gas Information System [of University of Michigan, US]

CoGIS Cooperative Geographic Information System [Canberra, Australian Capital Territory]

COGLA Canada Oil and Gas Lands Administration

Cogn Cognition; Cognitive [also cogn]

Cogn Psychol Cognitive Psychology [US Journal]

Cogn Sci Cognitive Science [Journal published in the US]

COGO Coordinate(d) Geometry (Programming Language); Coordinate(d) Geometry-Oriented Program

COGPE Canadian Oil and Gas Property Expense

COGR Council on Governmental Relations [US]

COGS Computer-Oriented Geological Society [US]

COGS Concordance Generation System [of University of Toronto, Canada]

COGS Continuous Orbital Guidance Sensor

COGS Continuous Orbital Guidance System

COGSME Composite Group of SME (Society of Manufacturing Engineers) [US]

COH Carboxyhemoglobin [also COHb] [Biochemistry]

COH Coefficient of Haze [Meteorology]

CO/H$_2$ Carbon Monoxide/(Molecular) Hydrogen (Mixture)

CO$_2$H Carboxy (or Carboxyl) (Radical) [Symbol]

COHA Canadian Occupational Health Association

CO/H$_2$/CH$_4$ Carbon Monoxide/(Molecular) Hydrogen/Methane (Mixture)

COHb Carboxyhemoglobin [also COH] [Biochemistry]

COHMEX Cooperative Huntsville Meteorological Experiment [US]

COHN Certified Occupational Health Nurse

COHO Coherent Oscillator [also Coho, or coho] [Radar]

COHSE Confederation of Health Service Employees [UK]

COHSL Coherent Superlattice [Materials Science]

COI Central Office of Information [UK]

COI Coast Orbital Insertion

COI Communication Operation Instruction

CO-I Co-Investigator [also Co-I]

COIA Canadian Ocean Industries Association

COIA Conservative Orthopedics International Association [UK]

COIC Canadian Oceanographic Identification Center

COIC Careers and Occupational Information Center [UK]

COIC Citibank Overseas Investment Corporation

COID Council of Industrial Design [UK]

COIL Concurrent Oil (Process)

COIM Checkout Interpreter (Software) Module

COIME Committee on Industry, Minerals and Energy [of Association of South East Asian Nations]

COIN Coal Oxygen Injection Process [Metallurgy]

COIN COBOL Indexing and Maintenance Package

COIN Coin-Phone Operational and Information Network (System)

COIN Committee on Information Needs [US]

COIN Complete Operating Information

COIN Computer and Information

COIN Computerized Ontario Investment Network [Canada]

COIN Console Interrupt [Data Processing]

COIN Counter Insurgency

Coinc Coincidence [also coinc]

COINIM Center for Information on Standardization and Metrology [of Polish Committee for Standardization, Measures and Quality Control, Poland]

COINS Computer and Information Sciences

COINS Counter-Insurgency

COINS Communications Oriented Language

COINS Computerized Information System

COINS Computer-Oriented Language

COINS Control in Information Systems

COIP Classification of Outlays of Industry by Purpose

CoKα Cobalt K-Alpha (Radiation) [also CoK$_\alpha$]

CoKβ Cobalt K-Beta (Radiation) [also CoK$_\beta$]

Coke Chem USSR Coke and Chemistry USSR [Journal]

COL Chain Overseas Low (Radar Network) [UK]

COL Character Outline Limit

COL Checkout Language

COL Cholestane

COL Computer-Oriented Language

COL Cost of Living

CoL Cobalt Ligand

Col Collagen [Biochemistry]

Col Collect(ion); Collector [also col]

Col College [also col]

Col Collegiate [also col]

Col Collision [also col]

Col Colombia(n)

Col Colonel [also COL]

Col Colonial; Colonist; Colony [also col]

Col Color(less) [also col]

Col Colorado [US]

Col Colorimeter; Colorimetry [also col]

Col Columba [Astronomy]

Col Columbia [South Carolina, US]

Col Columbia [River in Western North America]

Col Column [also col]

Col I Type I Collagen [Biochemistry]

Col II Type II Collagen [Biochemistry]

col collect(ed)

col color(ed)

COLA (International) Conference on Laser Ablation

COLA Cooperation in Library Automation

COLA Cost-of-Living Adjustment; Cost-of-Living Allowance

CoLα Cobalt L-Alpha Radiation [also CoL$_\alpha$]

COL-AMCHAM Colombian-American Chamber of Commerce [Colombia]

COLASL Compiler/Los Alamos Scientific Laboratory [New Mexico, US]

COLC Circle of Least Confusion (in Optics)

Colcrete Colloidal Concrete [also colcrete]

COLD Chronic Oibstructive Lung Disease

COLD Computer Output to Laser Disk

COLDCO Considerably Older Less Dramatic (Super-) Conductors [Solid-State Physics]

COL DEF (Text) Column Definition [also Col Def]

Cold Harbor Symp Quant Biol Cold Harbor Symposium on Quantitative Biology [US]

Cold Reg Sci Technol Cold Regions Science and Technology [Journal published in the US]

Colgrout Colloidal Grout [also colgrout]

Colgunite Collidal Gunite [also colgunite]

COLIDAR Coherent Light Detection and Ranging [also Colidar, or colidar]

COLIN Centrox (Inc.) On-Line Interactive Network [US]

COLINE Comité Legislatif d'Information Ecologique [Legislative Committee for Ecological Information] [also Coline]

COLINET College Libraries Information Network

COLINGO Compile On-Line and Go

COLIPA Comité de Liaison des Associations Européennes de l'Industrie de la Parfumerie, des Produits Cosmetiques et de Toilette [European Federation of Perfume, Cosmetics and Toiletries Industries, Belgium]

COLIWASA Containerized Liquid Waste Sampler; Composite Liquid Waste Sampler [also COLI/WASA]

COLL Division of Colloid and Surface Chemistry [of American Chemical Society, US]

Coll Collator [also coll]

Coll Colleague [also coll]

Coll Collect(ion); Collective; Collector [also coll]

Coll College [also coll]

Coll Collegiate [also coll]

Coll Collision [also coll]

Coll Colloid(al) [also coll]

coll collateral [also collat]

coll colloquial [also colloq]

Collab Collaboration; Collaborator [also collab]

collat collateral [also coll]

Collecn Collection [also collecn]

Collect Collection [also collect]

Collect Math Collectanea Mathematica [Mathematical Journal; published by the University of Barcelona, Spain]

Collagen Rel Res Collagen Release Research [Journal published in the US]

Coll Czech Chem Commun Collection of Czechoslovak Chemical Communications [of Czechoslovak Academy of Sciences]

COLLEGE College Scholarship Service Financial Aid Form [US]

Coll J USSR Colloid Journal of the USSR [Translation of: *Kolloidnyi Zhurnal*; published in the US]

Coll Microcomput Collegiate Microcomputer [Journal published in the US]

Colloid J USSR Colloid Journal of the USSR [Translation of: *Kolloidnyi Zhurnal*; published in the US] [also Coll J USSR]

Colloid Polym Sci Colloid and Polymer Science [Journal published in Germany] [also Coll Polym Sci]

Colloids Surf Colloids and Surfaces [Journal published in the Netherlands] [also Coll Surf]

Colloq Colloquium [also colloq]

colloq colloquial [also coll]

Colloq Math Colloquium Mathematicum [Colloquium of Mathematics published in Wroclaw, Poland] [also Coll Math]

Colloq Phys Colloque de Physique [French Colloquium of Physics] [also Coll Phys]

Colloq Publ Colloquium Publication [of the American Mathematical Society, US]

colly *(collyrium)* – Eye Lotion, or Eye Water [also collyr]

Colm Column [also colm]

Colo Colorado [US]

COLOC Correlation Spectroscopy for Long-Range Couplings

COL OFF (Text) Column Mode Off [also Col Off]

colog cologarithm [Symbol]

Colom Colombia(n)

COL ON (Text) Column Mode On [also Col On]

Colorado Sch Mines Q Colorado School of Mines Quarterly [US]

Color Res Appl Color Research and Application [Journal published in the US]

colo(u)rl colo(u)rless

COLP Center for Oceans Law and Policy [US]

Col-P Collagen Fragment P [Biochemistry]

Col$ Colombian Peso [Currency]

COLR Circuit Order Layout Record

COLREGS Convention on the International Regulations for Preventing Collisions

COLRS CLASS On-Line Reference Service [US]

Col-Sergt Colonel-Sergeant [also Col-Sgt]

COLSS Core Operating Limits Supervisory System

COLT Communication Line Terminator

COLT Computerized On-Line Testing

COLT Computer-Oriented Language Translator

COLT Control Language Translator

COLT Council on Library-Media Technical Assistants [US]

Co Ltd Company Limited [Commerce]

COLUMA Comite de Lutte Contre les Mauvaises Herbes [Weed Control Committee]

COM Cassette Operating Monitor

COM Center of Mass

COM Chief of Maintenance

COM Committee on Minorities [of American Physical Society, US]

COM Communication Port (Device)

COM Communications Society [of Institute of Electrical and Electronics Engineers, US]

COM Compander [also Com, or com]

COM Component Object Model [Computers]

COM Compressed Multisound

COM Computer Operations Manager

COM Computer Output Microfiche [also com]

COM Computer Output Microfilm(er); Computer Output Microform; Computer (on, or to) Microfiche; Computer (on, or to) Microfilm

COM Conference of Metallurgists [of Canadian Institute of Mining and Metallurgy, Canada]

COM Confocal Optical Microscope; Confocal Optical Microscopy

.COM Command [File Name Extension] [also .com]

.COM Commercial Business [Internet Domain Name]

COM1 First Serial Port [Computers]

COM2 Second Serial Port [Computers]

COM3 Third Serial Port [Computers]

COM4 Fourth Serial Port [Computers]

Com Coma Berenices [Astronomy]

Com Combine [Agriculture]

Com Command [also com]

Com Commander

Com Commerce; Commercial [also com]

Com Commission(er) [also com]

Com Committee [also com]

Com Commodore

Com Common(ality) [also com]

Com Commonwealth

Com Communication(s) [also com]

Com Community [also com]

com common

COMA Canadian Oilfield Manufacturers Association

COMA Coke Oven Manufacturers Association [UK]

COMA Computer Operations Management Association [US]

COMAC Continuous Multiple-Access Collator

COMAD Compound Optoelectronic Materials and Devices (Meeting)

COMAL Command Algorithmic Language

COMAL Common Algorithmic Language [Computer Language]

COMANSEC Computation and Analysis Section [of Defense Research Board, Canada]

COMAPS (House of) Commons Automated Publishing System [of the Canadian Parliament]

COMAR Committee on Man and Radiation [of Institute of Electrical and Electronics Engineers, US]

COMAR Computer Aerial Reconnaissance

COMARIS Coastal Marine Information System

COMARO Composite Magic-Angle Rotation [Physics]

COMAS Combined Orbital Maneuvering and Abort System

COMAS Computerized Maintenance and Administration Support

COMAS Concentration-Modulated Absorption Spectroscopy

COMAT Committee on Materials [US]

COMAT Compatibility of Materials

COMAT Computer-Assisted Training

COMB Console-Oriented Model Building

Comb Combination; Combining [also comb]

Comb Combustion [also comb]

COMBATEX Combat Exercises

combd combined [also comb'd]

Comb Explos Shock Waves Combustion, Explosion and Shock Waves [Translation of: *Fizika Goreniya i Vzryva (USSR)*; published in the US]

Comb Flame Combustion and Flame [Journal published in the US]

Combn Combination [also combn]

COMBO Combination Support Chip

COMBOIS Communauté d'Intérêt Européenne des Grossistes de Machine à Bois [Community of Interest of Leading European Woodworking and Wood-Processing Machinery Resellers, Germany]

Comb Sci Technol Combustion Science and Technology [Journal published in the US]

Combstn Combustion [also combstn]

Combust Combustion [also combust]

Combust Explos Shock Waves Combustion, Explosion and Shock Waves [Translation of: *Fizika Goreniya i Vzryva (USSR)*; published in the US]

Combust Flame Combustion and Flame [Journal published in the US]

Combust Sci Technol Combustion Science and Technology [Journal published in the US]

Combust Theory Model Combustion Theory and Modeling [Journal of the Institute of Physics, UK]

COMBZ Combat Zone [also CombZ]

COMCANLANT Command of the Canadian Atlantic

Com Cap Commonwealth Capital [also Com cap]

Com Cart Art Common Cartoid Artery [also com cart art]

COMCAT Commercial Catch Database [US]

COMCEN Communications Center

COMCIAM Commonwealth Climate Impacts Assessment and Management (Program)

COMCM Communication Countermeasures

COMCO Computerized Operation and Maintenance Concept

Comd Command [also comd]

COMDA Canada/Ontario Mineral Development Agreement

COMDA Canadian Office Machinery Dealers Association

COMDEX Computer Dealers Exposition [US]

Comdg Commanding [or comdg]

COMDP Cooperative Overseas Market Development Program

Comdr Commander [also Cdr]

Comdt Commandant

COMEC Communications Security

COMECON Council for Mutual Economic Aid; Council for Mutual Economic Assistance [also Comecon, or CMEA]

COMED Combined Moving Map and Electronic Display

COMEINDORS Composite Mechanized Information and Documentation Retrieval System

COMEL Coordinating Committee for Common Market Associations of Manufacturers of Rotating Electrical Machinery [UK]

COMEPP Cornell Manufacturing Engineering and Productivity Program [of Cornell University, Ithaca, New York, US]

COMER College of Mineral and Energy Resources

COMESA Committee on Meteorological Effects of Stratospheric Aircraft [UK]

COMET Commercial Experiment Transporter [of NASA]

COMET Computer Message Transmission

COMET Computer-Operated Management Evaluation Technique

COMET Cornell Macintosh Terminal Emulator

COMET Council on Middle East Trade [UK]

COMETEC-GAZ Comité d'Etudes Economiques de l'Industrie du Gaz [Economic Research Committee of the Gas Industry, Belgium]

COMETS Communications and Broadcasting Engineering Test Satellite [of National Space Development Agency of Japan]

COMETS Comprehensive On-Line Manufacturing and Engineering Tracking System

COMETT (Action Program of the) Community in Education and Training for Technology [European Union] [also Comett]

COMEX Commodity Exchange [US]

COMEX Computer and Office Machinery Exhibition

COMEXT Community External Trade Statistics [of European Economic Community, Belgium]

COMFET Conductively-Modulated Field-Effect Transistor; Conductivity-Modulated Field-Effect Transistor

COMFOR Commercial Wire Center Forecast Program

Co-Mg Cobalt-Magnesium (Alloy System)

Co:MgF$_2$ Cobalt-Doped Magnesium Fluoride (Laser)

COMIBOL Corporación de Minera de Bolivia [Bolivian Mining Corporation]

COMICT Comisión Interministerial para Ciencia y Tecnológica [Interministerial Commission for Science and Technology]

COMIDES (Interstate) Committee of Ministers on Desertification

Com Il Art Common Iliac Artery [also com il art]

Com Il Vein Common Iliac Vein [also com il vein]

Cominière Société Commerciale et Minière du Congo [Congo (State) Mining Company]

COMINT Communications Intelligence

COMIT Compiler, Massachusetts Institute of Technology [Cambridge, US]

COMIT Computing System, Massachusetts Institute of Technology [Cambridge, US]

COMITEXTIL Coordinating Committee for the Textile Industries [of European Economic Community, Belgium]

COML Commercial Language

coml commercial [also com'l]

COMLA Commonwealth Library Association [Jameica]

COMLAB Commerce Laboratory

COMLO Compass Locator

COMLOGNET Combat Logistics Network

COMLOGNET Communications Logistics Network [also Com Log Net, or ComLogNet]

COMM Commercial Mission

COMM Communications Society [of Institute of Electrical and Electronics Engineers, US]

COMM Commutator [also Comm]

COM-M Common Mode [also com-m]

Comm Command [also comm]

Comm Commander

Comm Commentary [also comm]

Comm Commerce; Commercial [also comm]

Comm Commission(er) [also comm]

Comm Committee [also comm]

Comm Communication(s) [also comm]

Comm Commutator [also comm]

comm commercial

comm common

COMMANDS Computer-Operated Marketing, Mailing and News Distribution System

COMMAD Conference on Optoelectronic and Microelectronic Materials and Devices

COMMAG Combined Sound and Picture–Magnetic Sound Record

COMMCEN Communications Center

COMMECH Computational Mechanics

COMMEN Compiler Oriented for Multiprogramming and Multiprocessing Environments

COMMEND Computer-Aided Mechanical Engineering Design System

Comment Math Helv Commentarii Mathematici Helvetici [Comments on Mathematics; published by Swiss Mathematical Society]

Comment Phys-Math Commentationes Physico-Matematicae [Comments on Mathematical Physics; published by Finnish Scientific Society, Helsinki]

Comments Astrophys, Comments Mod Phys A Comments on Astrophysics, Comments on Modern Physics: Part A [Canada]

Comments Astrophys, Comments Mod Phys B Comments on Astrophysics, Comments on Modern Physics: Part B [UK]

Comments Astrophys, Comments Mod Phys C Comments on Astrophysics, Comments on Modern Physics: Part C [UK]

Comments At Mol Phys Comments on Atomic and Molecular Physics [UK]

Comments Condens Matter Phys Comments on Condensed Matter Physics [UK]

Comments Mol Cell Biophys, Comments Mod Biol A Comments on Molecular and Cellular Biophysics, Comments on Modern Biology: Part A [UK]

Comments Mol Cell Biophys,Comments Mod Biol B Comments on Molecular and Cellular Biophysics, Comments on Modern Biology: Part B [UK]

Comments Nucl Part Phys Comments on Nuclear and Particle Physics [UK]

Comments Plasma Phys Control Fusion Comments on Plasma Physics and Controlled Fusion [UK]

ComMET Commercialization Model for Environmental Technologies

Comm Eur Commun Rep Commission of the European Communities Report [Luxembourg]

comml commercial [also comm'l]

COMM MUX Communications Multiplexer

Commn Commission [also commn]

Commod Commodity [also commod]

Commod Markets Devel Countries Commodity Markets and the Developing Countries [Publication of the World Bank]

Commodore Comput Int Commodore Computing International [Journal published in the UK]

Commod Res Q Rep, Copper Commodity Research Quarterly Report, Copper [Journal published in the US]

Commod Res Q Rep, Tin Commodity Research Quarterly Report, Tin [Journal published in the US]

COMMP (International Conference on) Computer-Assisted Materials Design and Process Simulation

COMM PRINT Commercial Printing [also Comm Print, or comm print]

COMMPUTE Computer-Oriented Music Materials Processed for User Transformation, or Exchange [of State University of New York, US]

COMMS Central Office Maintenance Management System

COMMS Communications [also Comms, or comms]

COMMS Communications Interface

COMM-STOR Communications Storage Unit

COMM SYST Communication System [also Comm Syst, or comm syst]

Commun Communication [also commun]

Commun ACM Communications of the Association for Computing Machinery [US]

Commun Appl Numer Methods Communications in Applied Numerical Methods [Journal published in the UK]

Commun Broadcast Communication and Broadcasting [Journal published in the UK]

Commun Dublin Inst Adv Stud A Communications of the Dublin Institute for Advanced Studies, Series A [Ireland]

Commun Dublin Inst Adv Stud B Communications of the Dublin Institute for Advanced Studies, Series B [Ireland]

Commun Eng Int Communications Engineering International [Journal published in the UK]

Commun Inf Communication Information [Publication of the Département d'Information et Communication, Université Laval, Sainte Foy, Quebec, Canada]

Commun Int Communications International [Journal published in the UK]

Commun Law Communications and the Law [Journal published in the UK]

Commun Manage Communications Management [Journal published in the UK]

Commun Math Phys Communications in Mathematical Physics [Journal published in Germany]

Commun News Communications News [US]

Commun Now Communications Now [Journal published in the UK]

Commun Outlook Communication Outlook [Publication of the Artificial Language Laboratory, Michigan State University, US]

Commun Pure Appl Math Communications on Pure and Applied Mathematics [US Publication]

Commun R Soc Edinb Communications of the Royal Society of Edinburgh [UK]

Commun R Soc Edinb, Phys, Sci Communications of the Royal Society of Edinburgh, Physical Sciences [UK]

Commun Sém Math Communications du Séminaire Mathématique [Communications of the Mathematical Seminary, University of Lund, Sweden]

Commun Stat–Simul Comput Communications in Statistics– Simulation and Computation [US]

Commun Stat–Theory Methods Communications in Statistics–Theory and Methods [US]

Commun Syst Worldw Communications Systems Worldwide [Published in the US]

Commun Theor Phys Communications in Theoretical Physics [Published by Huazhong University of Science and Technology, Wuhan, PR China]

Commut Commutation [also commut]

Commut Transm Commutation and Transmission [Journal published in France]

COM(n) Communication (Statement)

Co-Mn Cobalt-Manganese (Alloy System)

Comn Communication [also comn]

COMNAP Council of Managers of National Antarctic Programs

COM/NAV Communication and Navigation Aids

COMNET Communications Network

COMNET Computer Network

COMNET International Network of Centers for Documentation and Communication Research and Policies [France]

CoMnSi Cobalt Manganese Silicon (Alloy)

CoMnSiB Cobalt Manganese Silicon Boron (Alloy)

Co-Mo Cobalt-Molybdenum (Alloy System)

COMOPT Combined Sound and Picture–Optical Sound Record

COMP Charlotte Ordnance Missile Plant [North Carolina, US]

COMP Computer Society [of Institute of Electrical and Electronics Engineers, US]

COMP Division of Computers in Chemistry [of American Chemical Society, US]

Comp Comparison [also comp]

Comp Compass [also comp]

Comp Compatible; Compatibility [also comp]

Comp Compendium [also comp]

Comp Compensate; Compensation [also comp]

Comp Compilation; Compiled; Compiler [also comp]

Comp Complex [noun; also comp]

Comp Component [also comp]

Comp Composite [also comp]

Comp Composition; Compositor [also comp]

Comp Compound [also comp]

Comp Compress(ion) [also comp]

Comp Computation(al); Computer [also comp]

comp comparative

comp complete

comp complex [verb]

COMPAC Commonwealth Pacific Cable

COMPAC Computer Output Microfilm Package

COMPAC Computer Program for Automatic Control

COMPACT Compatible Algebraic Compiler and Translator

COMPACT Computerization of World Facts [of Stanford Research Institute, California, US]

COMPACT Computer Planning and Control Technique

COMPACT Computer-Programmed Automatic Checkout and Test System

Comp Anat Comparative Anatomy [also comp anat]

Compander Compressor-Expander [also compander]

compar comparative

COMPARE Computerized Performance and Analysis Response Evaluator

COMPARE Console for Optical Measurement and Precise Analysis of Radiation from Electronics

COMPAS Committee on Physics and Society [of American Institute of Physics, US]

COMPAS Computer-Controlled Automatic Continuity System

COMPAS Computer Oriented Metering, Planning and Advisory System

COMPASS COmmon Muon Proton Apparatus for Structure and Spectroscopy (Experiment) [of CERN–European Laboratory for Particle Physics, Geneva, Switzerland]

COMPASS Compiler-Assembler [also Compass, or compass]

COMPASS Comprehensive Assembly System

COMPASS Computer-Assisted Classification and Assignment System

COMPAY Computerized Payroll System

Comp Biochem Physiol Compendium of Biochemical Physiology [US]

COMPCON Computer Convention [of Institute of Electrical and Electronics Engineers, US]

COMPDUPE Compare and Duplicate Disk(ette) (Command) [also compdupe]

Compd Compound [also compd]

COMPEL Compute Parallel

COMPEL (International Journal for) Computation and Mathematics in Electrical and Electronic Engineering [Ireland]

COMPEN Compensator [also Compen]

COMPENDEX Computer(ized) Engineering Index [also Compendex, or CMPX]

COMPENDEX Corporate Engineering Data Exchange

COMPINT Computer-Integrated Technologies (Conference)

COMPINT International Conference on Computer-Integrated Technologies [of Institute of Electrical anmd Electronics Engineers, US]

Compl Complement [also compl]

Compl Completion [also compl]

compl complete(d)

Complex Syst Complex Systems [Journal published in the US]

Comp Mech Computational Mechanics [also comp mech]

Compn Composition [also compl]

COMP News Division of Computers in Chemistry News [of American Chemical Society, US]

Compo Composition [also compo]

Compo Board Composition Board [also compo board]

Compo Image Composition Image [also compo image]

Compole Commutating Pole [also compole]

Compo Math Compositio Mathematica [Mathematical Journal of Wiskundig Genootschap te Amsterdam (Mathematical Society of Amsterdam)]

Compon Component [also compon]

COMPOOL Common Data Pool

COMPOOL Communications Pool

Compos Composite [also compos]

Compos Adhes Newsl Composites and Adhesives Newsletter [US]

COMPOSE Computerized Production Operating System Extension

Compos Eng Composites Engineering [Journal published in the US]

Compos A Composites Part A [Published in the Netherlands]

Compos A: Appl Sci Manuf Composites Part A: Applied Science and Manufacturing [Published in the Netherlands; incorporates *Composites* and *Composites Manufacturing*]

Compos B Composites Part B [Published in the Netherlands]

Compos B: Eng Composites Part B: Engineering [Published in the Netherlands]

Compos Manuf Composites in Manufacturing [Journal of the Society of Manufacturing Engineers, US]

Compos Mater Sci Composite Materials Science [Journal published in the UK]

Compos Plast Renf Fibres Verre Text Composites Plastiques Renforcés par Fibres de Verre Textile [French Journal on Glass-Textile Fiber-Reinforced Plastic Composites; published by Centre de Documentation du Verre Textile et des Plastiques Renforcés]

Compos Polym Composite Polymers [Publication of the Association of Directors of European Centers for Plastics, UK]

Compos Sci Technol Composites Science and Technology [Journal published in the UK]

Compos Struct Composite Structures [Journal published in the US]

Compos Technol Res Composites Technology and Research [Publication of the American Society for Testing and Materials]

Compos Technol Rev Composites Technology Review [US]

COMPOW Committee on Professional Opportunities for Women

Comp Prog Computer Program(ming); Computer Programmer [also comp prog]

Compr Compress(ed); Compressor [also compr]

Compr Air Compressed Air [Journal published in the US]

Compreg Compregnate(d) [also compreg]

Compreg Compregnated Wood [also compreg]

Comp Rend Acad Bulg Sci Comptes Rendus, Academie Bulgare des Sciences [Proceedings of the Bulgarian Academy of Sciences]

Comp Rend Acad Sci Comptes Rendus de l'Academie des Sciences [Proceedings of the Academy of Sciences, France]

Comp Rend Acad Sci I, Math Comptes Rendus de l'Academie des Sciences, Série I: Mathematics [Proceedings of the Academy of Sciences, Series I: Mathematics, France]

Comp Rend Acad Sci II, Méc Phys Chim Sci Terre Univers Comptes Rendus de l'Academie des Sciences, Série II: Mécanique, Physique, Chimie–Sciences de la Terre et de l'Univers [Proceedings of the Academy of Sciences, Series II: Mechanics, Physics and Chemistry–Earth and Cosmic Sciences, France]

Comp Rend Acad Sci Sér I Comptes Rendus de l'Academie des Sciences, Série I [Proceedings of the Academy of Sciences, Series I, France]

Comp Rend Acad Sci Sér II Comptes Rendus de l'Academie des Sciences, Série II [Proceedings of the Academy of Sciences, Series II, France]

Comp Rend Acad Sci Sér III Comptes Rendus de l'Academie des Sciences, Série III [Proceedings of the Academy of Sciences, Series III, France]

Comp Rend Acad Sci Sér A Comptes Rendus de l'Academie des Sciences, Série A [Proceedings of the Academy of Sciences, Series A, France]

Comp Rend Acad Sci Sér B Comptes Rendus de l'Academie des Sciences, Série B [Proceedings of the Academy of Sciences, Series B, France]

Comp Rend Acad Sci Sér C Comptes Rendus de l'Academie des Sciences, Série C [Proceedings of the Academy of Sciences, Series C, France]

Comp Rend Acad Sci Sér Gén Vie Sci Comptes Rendus de l'Academie des Sciences, Série Générale, la Vie des Sciences [Proceedings of the Academy of Sciences, Series General Life Sciences, France]

Comp Rend Hebd Séances Acad Sci I Comptes Rendus Hebdomadaire des Séances de l'Academie des Sciences, Série I [Proceedings of the Weekly Sessions of the Academy of Sciences, Series I, France]

Comp Rend Hebd Séances Acad Sci II Comptes Rendus Hebdomadaire des Séances de l'Academie des Sciences, Série II [Proceedings of the Weekly Sessions of the Academy of Sciences, Series II, France]

Comp Rend Hebd Séances Acad Sci III Comptes Rendus Hebdomadaire des Séances de l'Academie des Sciences, Série III [Proceedings of the Weekly Sessions of the Academy of Sciences, Series III, France]

Comp Rend Hebd Séances Acad Sci, Sér A Comptes Rendus Hebdomadaire des Séances de l'Academie des Sciences, Série A [Proceedings of the Weekly Sessions of the Academy of Sciences, Series A, France]

Comp Rend Hebd Séances Acad Sci, Sér B Comptes Rendus Hebdomadaire des Séances de l'Academie des Sciences, Série B [Proceedings of the Weekly Sessions of the Academy of Sciences, Series B, France]

Comp Rend Hebd Séances Acad Sci, Ser C, Sci Chem Comptes Rendus Hebdomadaire des Séances de l'Academie des Sciences, Serie C: Sciences Chimique [Proceedings of the Weekly Sessions of the Academy of Sciences, Series C: Chemical Sciences, France]

COMPRESS Computer Research, Systems and Software

COMPROC Command Processor

COMPROG Computer Program

Comprsn Compression [also cmprsn]

COMPS Commercial Materials Processing Support Facility

Comps Computers [also comps]

COMPSAC Computer Software and Applications Conference

Comp Sci Computer Science; Computer Scientist

Comp Sim Computer Simulation [also comp sim]

COMPSO Computer Software and Peripherals Show

COMPSTAT European Meeting on Computational Statistics [of International Association for Statistical Computing, Netherlands]

Compt Compartment [also compt]

Compt Comptroller

COMPTEL Competitive Telecommunications Association [US]

Compunications Computers and Communications [also compunications]

Comput Computation(al); Computing [also comput]

Comput Account Computers in Accounting [Journal published in the US]

Comput Aided Des Computer Aided Design [Journal published in the UK]

Comput-Aided Eng Computer-Aided Engineering [Journal published in the US]

Comput-Aided Eng J Computer-Aided Engineering Journal [of Institution of Electrical Engineers, UK]

Comput-Aided Geom Des Computer-Aided Geometric Design [Journal published in the Netherlands]

Comput Appl Eng Educ Computer Applications in Computer Engineering [Publication of Institute of Electrical and Electronics Engineers, US]

Comput Appl Power Computer Applications in Power [Publication of Institute of Electrical and Electronics Engineers, US]

Comput Graph Appl Computer Graphics and Applications [Publication of Institute of Electrical and Electronic Engineers–Computer Society, US]

Comput Archit News Computer Architecture News [of Association for Computing Machinery, US]

Comput Artif Intell Computers and Artificial Intelligence [Publication of the Slovak Academy of Sciences]

Comput Bank Computers in Banking [Journal published in the US]

Comput Biol Med Computers in Biology and Medicine [Journal published in the UK]

Comput Biomed Res Computers and Biomedical Research [Journal published in the US]

Comput Bull Computer Bulletin [of British Computer Society]

Comput Bus Computing for Business [Journal published in the US]

Comput Chem Computers and Chemistry [Journal published in the US]

Comput Chem Eng Computers and Chemical Engineering [Journal published in the UK]

Comput Commun Computer Communications [Journal published in the UK]

Comput Commun Decis Computer and Communications Decisions [Journal published in the US]

Comput Commun Rev Computer Communication Reviews [of Association for Computing Machinery, US]

Comput Compos Computers and Composition [Publication of Michigan Technological University, Houghton, US]

Comput Control Eng J Computing and Control Engineering Journal [of Institution of Electrical Engineers, UK]

Comput Coupling Phase Diagr Thermochem Computer Coupling of Phase Diagrams and Thermochemistry [Journal published in the UK]

Comput Data Computer Data [Canadian Journal]

Comput Decis Computer Decisions [Journal published in the US]

Comput Des Computer Design [Journal published in the US]

Comput Dig Computer Digest [Issued with *RadioElectronics*] [Journal published in the UK]

Comput Disp Rev Computer Display Review [Journal published in the US]

Comput Econ Rep Computer Economics Report [US]

Comput Educ Computer Education [Publication of the National Computer Center, Manchester, UK]

Comput Educ J Computers in Education Journal [of American Society for Engineering Education]

Comput Electr Eng Computers and Electrical Engineering [Journal published in the US]

Comput Electron Computers and Electronics [US Magazine]

Comput Electron Agric Computers and Electronics in Agriculture [Journal published in the Netherlands]

Comput Eng Computing for Engineers [of CERN– European Laboratory for Particle Physics, Geneva, Switzerland]

Comput Enhanc Spectrosc Computer Enhanced Spectroscopy [Journal published in the UK]

Comput Environ Urban Syst Computers, Environment and Urban Systems [Journal published in the UK]

Comput Equip Rev Computer Equipment Review [US]

Comput Fluids Computers and Fluids [Journal published in the UK]

Comput Fraud Secur Bull Computer Fraud and Security Bulletin [UK]

Comput Geosci Computers and Geosciences [Publication of the International Association for Mathematical Geology, US]

Comput Geosci Computers and Geosciences [Journal published in the UK]

Comput Geotech Computers and Geotechnics [Journal published in the UK]

Comput Graph Computer Graphics [Publication of the Association for Computing Machinery, US]

Comput Graph Computers and Graphics [Journal published in the UK]

Comput Graph Art Computer Graphics and Art [Journal published in the US]

Comput Graph Forum Computer Graphics Forum [Publication of EUROGRAPHICS]

Comput Graph Today Computer Graphics Today [Journal published in the US]

Comput Graph World Computer Graphics World [Journal published in the US]

Comput Hum Behav Computers in Human Behaviour [Journal published in the US]

Comput Hum Serv Computers in Human Services [Journal published in the US]

Comput Ind Computers in Industry [Journal published in the Netherlands]

Comput Ind Eng Computers and Industrial Engineering [Journal published in the UK]

Comput-Integr Manuf Syst Computer-Integrated Manufacturing Systems [Journal published in the UK]

Comput Intell Computational Intelligence [Publication of the National Research Council of Canada]

Comput J Computer Journal [of British Computer Society]

Comput Lang Computer Language [Journal published in the US]

Comput Lang Computer Languages [Journal published in the UK]

Comput Law Computers and Law [Journal published in the UK]

Comput/Law J Computer/Law Journal [Published by Center for Computer/Law, Manhattan Beach, California, US]

Comput Law Pract Computer Law and Practice [UK]

Comput Law Secur Rep Computer Law and Security Report [UK]

Comput Libr Computers and Libraries [Journal published in the UK]

Comput Libr Computers in Libraries [Journal published in the US]

Comput Life Sci Educ Computers in Life Science Education [Publication of the National Resource for Computers in Life Science Education, University of Washington, Seattle, US]

Comput Linguist Computational Linguistics [Publication of the Association for Computational Linguistics, US]

Comput Mag Computation, The Magazine [UK]

Comput Manuf Computerised Manufacturing [Publication of the Institution of Production Engineering, UK]

Comput Math Appl Computers and Mathematics with Applications [UK]

Comput Mater Sci Computational Materials Science [Journal published in the Netherlands] [also Comput Mat Sci]

Comput Mech Computational Mechanics [Germany]

Comput Mech Eng Computers in Mechanical Engineering [of American Society of Mechanical Engineers, US]

Comput Meth Appl Mech Eng Computer Methods in Applied Mechanics and Engineering [Journal published in the Netherlands]

Comput Meth Programs Biomed Computer Methods and Programs in Biomedicine [Journal published in the Netherlands]

Comput Music J Computer Music Journal [Published by MIT Press, Cambridge, US]

Comput Netw ISDN Syst Computer Networks and ISDN (Integrated Services Digital Network) System [Journal published in the Netherlands]

Comput News Computer News [Published in the US]

Comput Newsp Computing, The Newspaper [UK]

Comput Oper Res Computers and Operations Research [Journal published in the UK]

Comput Peripher Rev Computer Peripherals Review [US]

Comput Pers Computer Personnel [Publication of the Association for Computing Machinery, US]

Comput Phys Computers in Physics [Publication of the American Institute of Physics, US]

Comput Phys Commun Computer Physics Communications [Published in the Netherlands]

Comput Phys Rep Computer Physics Report [Netherlands]

Comput Polym Sci Computational Polymer Science [International Journal]

Comput Process Chin Orient Lang Computer Processing of Chinese and Oriental Languages [of Chinese Language Computer Society, US]

Comput Recht Computers und Recht [German Publication on Computers and Law]

Comput Reseller Mon Computer Reseller Monthly [US]

Comput Rev Computer Reviews [of Association for Computing Machinery, US]

Comput Sch Computers in the Schools [Journal published in the US]

Comput Sci Econ Manage Computer Science in Economics and Management [Journal published in the Netherlands]

Comput Sci Inf Computer Science and Informatics [Journal of the Computer Society of India]

Comput Secur Computers and Security [Journal published in the UK]

Comput Secur Audit Controls Computer Security, Auditing and Controls [Published in the US] [also COMSAC]

Comput Secur J Computer Security Journal [of Computer Security Institute, US]

Comput Seismol Computational Seismology [Translation of: *Vychislitel'naya Seismologiya (USSR)*; published in the US]

Comput Soc Computers and Society [Publication of the Association for Computing Machinery, US]

Comput Speech Lang Computer Speech and Language [Journal published in the UK]

Comput Stand Interfaces Computer Standards and Interfaces [Journal published in the Netherlands]

Comput Stat Data Anal Computational Statistics and Data Analysis [Journal published in the Netherlands]

Comput Stat Q Computational Statistics Quarterly [Journal published in Germany]

Comput Struct Computers and Structures [Journal published in the US]

Comput Surv Computer Survey [Journal published in the US]

Comput Surv Computing Surveys [Publication of the Association for Computing Machinery, US]

Comput Syst Computer Systems [Journal published in the UK]

Comput Syst Computing Systems [Publication of University of California Press, Berkeley, US]

Comput Syst Eur Computer Systems Europe [Journal published in the UK]

Comput Syst Sci Eng Computer Systems Science and Engineering [Journal published in the UK]

Comput Teach Computer Teacher [Journal of the International Council for Computers in Education, US]

Comput Tech Computing Techniques [Journal published in the UK]

Comput Termin Rev Computer Terminals Review [US]

Comput Transl Computers and Translation [Journal published in the Netherlands]

Comput Vis Graph Image Process Computer Vision, Graphics and Image Processing [Journal Pubished in the US]

Comput Wkly Computer Weekly [Journal published in the US]

COMR Communication Research

Comr Commissioner

COMRADE Computer-Aided Design Environment

COMR&DSAT Communication Research and Development Satellite

COMS Canadian Ocean Mapping System [of Department of Fisheries and Oceans]

COM-SAC Computer Security, Auditing and Controls [US Publication]

COMSAT Commercial Satellite Corporation [US]

COMSAT Communications Satellite [also Comsat]

COMSAT Communications Satellite Corporation [Washington, DC, US] [also Comsat]

COMSAT Tech Rev COMSAT Technical Review [of Communications Satellite Corporation, US]

COMSCAM Combat Scenario Assessment; Combat Scenario Model

COMSEC Communications Security

COMSEC Communications Security Association [US]

COMSL Communication System Simulation Language

COMSOAL Computer Method of Sequencing Operations for Assembly Lines

COMSOC Communications Society [of Institute of Electrical and Electronics Engineers, US]

COMSOC Communications Spacecraft Operation Center

COMSPEC Command Processor Specification (Environment)

COMST Conference Organizers in Medicine, Science and Technology

COMSTARS International Communications Satellites [also Comstars]

COMSYL Communications System Language

COMSYS Communication System

COMT Catechol-O-Methyltransferase [Biochemistry]

COM/T&A Communications/Transmission and Access Systems [Institute of Electrical and Electronics Engineers Committee, US]

COMTASS Compact Towed Sonar System

COMTEC Computer and Technology Trade Fair [Germany]

COMTEC Computer Micrographics and Technology Group [US]

COMTRAN Commercial Translator

COMZ Communications Zone

CON Console Device [Computer Programming]

CO/N₂ Carbon Monoxide/Nitrogen (Gas Mixture)

Con Conclusion [also con]

Con Connect(ion) [also con]

Con Console [also con]

Con Consul(ar); Consulate [also con]

Con Contact [also con]

Con Contactor [also con]

Con Contrast [also con]

Con Control(ler) [also con]

con consolidated

con contra

Con A Concanavalia Ensiformis; Concanavalin A [Biochemistry]

CONABIO Comisión Nacional para el Conocimiento y Uso de la Biodiversidad [National Commission on Conservation and Use of Biodiversity, Mexico]

CONAC Continental Air Command [of US Air Force]

CONACYT Consejo Nacional de Ciencia y Tecnologia [National Council for Science and Technology, Mexico]

CONAD Continental Air Defense Command [of US Air Force]

CONAE Comisión Nacional de Actividades Espaciales [National Commission on Space Activities, Argentina]

CONAF Corporación Nacional Foerstal y de Protección de Recursos Naturales Renovables [National Corporation for Forestry and the Protection of Renewable Natural Resources, Chile]

CONAES Committee on Nuclear and Alternative Energy Systems [US]

CONARC Continental Army Command [Fort Monroe, Virginia, US]

CONAT Concrete Articulated Tower

Co(Nb) Niobium-Doped Cobalt

Co-Nb Cobalt-Niobium (Alloy System)

CONBAT Converted Battalion Antitank

Conc Concentrate; Concentration [also conc]

Conc Concentric(ity) [also conc]

Conc Concrete [also conc]

conc concentrate(d)

Concast AG Continuous Casting Standards AG [Zurich, Switzerland]

Concast Stand News Concast Standard News [Published by Concast, Zurich, Switzerland]

Concast Technol News Concast Technology News [Published by Concast, Zurich, Switzerland]

Concat Concatenate [Pascal Function]

CONCAWE Conservation of Clean Air and Water, Western Europe [now The Oil Companies' International Study Group for Conservation of Clean Air and Water–Europe]

Conc B Concrete Block

Conc C Concrete Ceiling

concd concentrated [also conc'd]

Conc F Concrete Floor

Concg Concentrating [also concg]

Conch Conchologica; Conchologist; Conchology [also conch]

Concln Conclusion [also concln]

Concn Concentration [also concn]

CONCOM Council of Nature Conservation Ministers [of Australian and New Zealand Environment and Conservation Council]

Concord Concordance [also concord]

Concr Concrete [also concr]

Concr Int Des Constr Concrete International: Design and Construction [Publication of the American Concrete Institute, US]

Cond Condenser; Condensing [also cond]

Cond Condition(er) [also cond]

Cond Conduct(ion); Conductivity; Conductor [also cond]

Cond Mat News Condensed Materials News [US]

Condn Condition [also condn]

Condo Condominium [also condo]

Condr Conditioner [also condn]

CONDUIT Computers at Oregon State University, North Carolina Educational Computing Services, Dartmouth College, Iowa State University and University of Texas at Austin [US Computer Network]

Condy Conductivity [also condy]

CONECS Connectorized Exchange Cable Splicing

Con Ed Consolidated Edison (Company) [US]

CONELRAD Control of Electromagnetic Radiations [also Conelrad]

Conf Conference [also conf]

Conf Confidential(ity) [also conf]

Conf Confirm(ed); Confirmation [also conf]

conf *(confer)* – compare

Confed Confederation [also conded]

CONFIDAL Conjugate Filter Data Link

Conf Exhib Int Conferences and Exhibitions International [Journal published in the UK]

CONFIG Configure; Configuration [Programming] [also config]

Config Configuration [also config]

CONFIGHD Configure Hard Drive [also confighd]

CONFLEX Conditional Reflex

Conformal Geom Dyn Conformal Geometry and Dynamics [Journal of the American Mathematical Society, US]

Conformer Conformational Isomer [also conformer]

CONFRAC Continued Fractions: From Analytic Number Theory to Constructive Approximation (Conference)

Cong Congress(ional) [also cong]

CONHAN Contextural Harmonic Analysis

CO/NH₃ Carbon Monoxide/Ammonia (Mixture)

$[Co(NH_3)_6]X_3$ Hexaamminecobalt (III) [X = Halogen (Cl, Br, etc.), such as in Hexaamminecobalt (III) Chloride]

CoNi Cobalt Nickel (Thin Film)

Co-Ni Cobalt-Nickel (Alloy System)

CONICET Consejo Nacional de Investigaciónes Cientificas y Técnicas [National Council for Scientific and Technical Research, Argentina]

CONICET-UNR Consejo Nacional de Investigaciónes Cientificas y Técnicas–Universidad Nacional de Rosario [National Council for Scientific and Technical Research–National University of Rosario, Argentina]

CONICIT Consejo Nacional de Investigaciónes Cientificas y Tecnológicas [National Council for Scientific and Technological Research, Costa Rica and Venezuela]

CONICYT Comisión Nacional de Investigaciónes Cientificas y Técnicas [National Scientific and Technical Research Commission, Argentina]

CoNiCrAlY Cobalt Nickel Chromium Aluminum Yttrium (Coating)

CONIO Console Input/Output

CoNiO Cobalt Nickel Oxygen (Alloy)

CONIT Connector for Networked Information Transfer

Conj Conjugate; Conjugation [also conj]

Conj Conjunction [also conj]

Conjg Conjugate [also conjg]

CONLIS Committee on National Library Information Systems [US]

CONMAH Configuration Change, Mixing, Activation and Heating (Model) [Shock Compression]

CONMAT National Program for High-Performance Construction Materials and Systems [US]

Conn Connect(ion); Connective; Connector [also conn]

Conn Connecticut [US]

Conn Dia Connection Diagram [also conn dia]

Connect Connection; Connective [also connect]

Connect Technol Connection Technology [Journal published in the US]

Connect Tissue Res Connective Tissue Research [Journal]

Conn Tpk Connecticut Turnpike [US]

CO/NO Current Operator/Next Operator

Conoco Continental Oil Company [US]

CONP Cyclooctylamine Nitropyridine

CONRAD Committee on Radiology

CONRAD Contraceptive Research and Development Program [Agency for International Development, US]

Conrail Consolidated Rail Corporation [US]

CONROLL Continuous Thin Slab Casting and Rolling Technology [Metallurgy]

CONS Carrier-Operated Noise Suppression

CONS Connection-Oriented Network Service

Cons Console [also cons]

Cons Constable

Cons Consul(ar) [also cons]

Cons Consulate [also cons]

cons consolidate(d)

cons consult

CONSAL Congress of Southeast Asian Librarians [Philippines]

Consciousn Cogn Consciousness and Cognition [Journal published in the US]

Consens Devel Conf Summ Consensus Development Conference Summaries [Publication of the National Library of Medicine–National Institutes of Health, Bethesda, Maryland, US]

Consens Statem Consensus Statement [Publication of National Institutes of Health–Office of Medical Applications of Research, Bethesda, Maryland, US]

CONSER Conversion of Serials Project [Canada and US]

Conserv Conservation [also conserv]

Conserv Biol Conservation Biology [Journal]

Conserv Ecol Conservation Ecology [Journal]

Conserv Recycling Conservation and Recycling [Journal published in the UK]

Consg Consulting [also consg]

Conslt Consultant [also conslt]

ConSoc Conservation Society [UK]

consol consolidated

CONSORT Conversation System with On-Line Remote Terminals

Consorzio INFM Consorzio Interuniversitario di Fisica della Materia [Inter-University Consortium on Matter Physics, Italy]

Consr Consumer [also consr]

Const Constitution [also const]

Const Construction [also const]

Const Constancy; Constant [also const]

const constant [Mathematics]

constg consisting

Constr Construct(ion) [also constr]

Constr Build Mater Construction and Building Materials [Journal published in the UK]

Constr Comput Construction Computing [Publication of the Chartered Institute of Building, UK]

Constr Dim Mag Construction Dimensions Magazine [of Association of the Wall and Ceiling Industries International, US]

Constr Maş Construcţia de Maşini [Romanian Journal on Machine Construction]

Constrn Construction [also const]

Constr Surv Construction Surveyor [of Construction Surveyors Institute, UK]

CONSUL Control Subroutine Language

Consult Eng Consulting Engineer [Journal published in the UK]

Consum Rep Consumers Reports [of the Consumers Union of the United States, Inc.]

CONT Continue Program Execution (Command)

Cont Contact [also cont]

Cont Container [also cont]

Cont Content(s) [also cont]

Cont Continent(al) [also cont]

Cont Contingency [also cont]

Cont Continuation; Continue(d); Continuity [also cont]

Cont Contract(or) [also cont]

Cont Contraction [also cont]

Cont Control(ler) [also cont]

cont contain(ing)

cont continue(d)

cont continuous

cont controlled

Contact EMF Contact Electromotive Force [also contact emf]

Contact PD Contact Potential Difference [also contact pd]

Contam Contaminate; Contamination [also contam]

contd continued [also cont'd]

contemp contemporary

Contemp Math Contemporary Mathematics [Journal of the American Mathematical Society, US]

Contemp Phys Contemporary Physics [Journal published in the UK]

Cont Eur Continental Europe

Contg Containing [also contg]

Contg Contracting [also contg]

Cont HP Continental Horsepower [also cont hp]

contig contiguous

CONTIMELT Continuous Melting and Refining (Process) [Metallurgy]

Contin Continuum [also contin]

contin continuous

Contin Mech Thermodyn Continuum Mechanics and Thermodynamics [Journal published in Germany]

contl Continental [also cont'l]

Contn Continuation [also contn]

CONTONE Continuous Tone

Contr Container [also contr]

Contr Contract(or) [also contr]

Contr Contraction [also contr]

Contr Control(ler) [also contr]

Contrail(s) Condensation Trail(s) [also contrail(s)]

CONTRAN Control Translator

CONTRANS Conceptual Thought, Random Net Simulation

Cont Rel Newsl Controlled Release Newsletter [of Controlled Release Society, US]

CONTREQS Contingency Transportation Requirements System

Contrg Contracting [also contrg]

Contrib Contribution; Contributor [also contrib]

Contrib Atmos Phys Contributions to Atmospheric Physics [Journal published in Germany]

Contrib Geophys Inst Slovak Acad Sci Contributions of the Geophysical Institute of the Slovak Academy of Sciences

Contrib Geophys Inst Slovak Acad Sci Ser Meteorol Contributions of the Geophysical Institute of the Slovak Academy of Sciences, Series of Meteorology

Contrib Inst Low Temp Sci A Contributions from the Institute of Low Temperature Sciences, Series A [of Hokkaido University, Sapporo, Japan]

Contrib Inst Low Temp Sci B Contributions from the Institute of Low Temperature Sciences, Series B [of Hokkaido University, Sapporo, Japan]

Contrib Mineral Petrol Contributions to Mineralogy and Petrology [Journal published in Germany]

Contrib Plasma Phys Contributions to Plasma Physics [Journal published in Germany]

Contr O Contracting Officer

CONTROL International Trade Fair for Quality Assurance and Control [Germany]

Control Controller [also control]

Control Cibern Autom Control Cibernetica y Automatización [Cuban Journal on Cybernetic Control and Automation; published by Centro de Automatización Industrial]

Control Comput Control and Computers [Canadian Journal]

Control Cybern Control and Cybernetics [Journal published in Poland]

Control Eng Control Engineering [Journal published in US]

Control Eng Pract Control Engineering Practice [Journal]

Control Instrum Control and Instrumentation [Journal published in UK]

Cont Shelf Res Continental Shelf Research [Journal published in UK]

Cont Sys Control Systems [also cont sys]

Control Syst Mag Control Systems Magazine [of Institute of Electrical and Electronics Engineers, US]

CONUS Contiguous United States

CONUS Continental United States Complex [US Communication System]

CONUS Continental United States

Conv Conversion; Converter; Converting [also conv]

Conv Mag Converting Magazine [US]

Convn Convention [also convn]

convt Convert

Conv't Convertible [Automobiles]

COO Carboxylate

COO Chairman of Operations

COO Chief Operating Officer

COO Chief Operations Officer

CoO Cobalt(II) Oxide (or Cobaltous Oxide)

CoO Cost of Ownership

CoO$_x$ Cobalt Oxides (i.e., CoO, Co$_2$O$_3$, and Co$_3$O$_4$) [also CoOx]

CoOEP 2,3,7,8,12,13,17,18-Octaethylporphine Cobalt(II)

COOEt Carbethoxy Group [Formula]

CoO-Fe$_2$O$_3$ Cobaltous Oxide-Ferric Oxide (System)

COOH Carboxy (or Carboxyl) (Radical) [Symbol]

COOH Carboxylic Acid

COOKI Coordinated Keysort Index

COOL Checkout-Oriented Language

COOL Control-Oriented Language

COOP Crystal Orbital Overlapping Population [Physics]

Coop Cooperation; Cooperative [also coop]

Co-op Cooperation; Cooperative [also co-op]

Coop M Cooperative Measures [also coop m]

Coop Res Monogr Cooperative Research Monographs [US Office of Edication, Washington, DC]

Coord Coordinate; Coordination; Coordinator [also coord]

Coord Bd Coordinating Board [also coord bd]

Coordn Coordination [also coordn]

COORS Communications Outage Restoral Section

COOSRA Canadian Offshore Oilspill Research Association

CoOx Cobalt Oxides (i.e., CoO, Co$_2$O$_3$, and Co$_3$O$_4$) [also CoO$_x$]

COP Central Order Processing (System)

COP Certificate of Performance [also CoP]

COP Change Over Point

COP Circle of Position

COP Circulator Outlet Pressure [Nuclear Power Stations]

COP Citizens on Patrol (Program)

COP Coastal Ocean Program [of National Oceanic and Atmospheric Administration, US]

COP Code of Practice

COP Coefficient of Performance [also cop]

COP Coherent Optical Processing

COP Colombian Peso [Currency]

COP Committee on Publication [of American Society for Standards and Materials, US]

COP Common On-Line Package

COP Common Operational Concept

COP Communication Output Printer

COP Computer Operating Properly (Reset Timer)

COP Computer Optimization Package

COP Conference of the Parties

COP Conservation of Parity [Quantum Mechanics]

COP Contingency Operations Plan

COP Continuous Optimization Program

COP Copolyester Thermoplastic Elastomer

COP Cornhusker Ordnance Plant [Nebraska, US]

COP Crystal Originated Particle(s)

COP-I First Session of the Conference of the Parties (to the Framework Convention on Climate Change)

CoP Cobalt Phosphide

CoP Colombian Peso [Currency]

Co-P Cobalt-Phosphorus (Alloy System)

Co-P Co-Pilot [also co-p]

Cop Copper

Cop Copyright [also cop]

COPA Canadian Office Products Association

COPA Copolyamide (Elastomer)

COPA Council on Post-Secondary Accreditation [US]

COPAC Committee for the Promotion of Aid to Cooperatives [Italy]

COPAC Community Patent Appeal Court [of European Economic Community, Belgium]

COPAC Continuous Operation Production Allocation and Control

COPAFS Council of Professional Associations of Federal Statistics [US]

COPAG Collision Prevention Advisory Group [of Federal Aviation Administration, US]

COPAL Cocoa Producers Alliance

COPAN Command Post Alerting Network

COPANIT Comisión Panameña de Normas Industriales y Técnicas [Panamenian Commission for Industrial and Technical Standards]

COPANT Comisión Panamericana de Normas Técnicas [Pan-American Standards Commission, Argentina]

COPAR Cooperative Preservation of Architectural Records [US]

CoPc Cobalt Phthalocyanine

COPD Chronic Obstructive Pulmonary Disease

Co-Pd Cobalt-Palladium (Alloy System)

COPE Chronic Obstructive Pulmonary Emphysema

COPE Claus Oxygen-Based Process Expansion (Process)

COPE Coastal Observation Program, Engineering [Australia]

COPE Communications-Oriented Processing Equipment

COPE Console Operator Proficiency Examination

COPE Consumer-Optimized Product Engineering

COPE Copolyester

COPEP Committee on Public Engineering Policy [of National Academy of Engineering, US]

COPES Conceptually-Oriented Program in Elementary Science

COPES Consumer-Optimized Product Engineering System

COPESCAL Comision de Pesca Continental para America Latina [Commission for Inland Fisheries of Latin America, Italy]

COP/FCCC Conference of the Parties to the Framework Convention on Climate Change

COPI Computer-Oriented Programmed Instruction

COPI Cooperative Projects with Industry

Co-PI Co-Principal Investigator

COPICS Communications-Oriented Production Information and Control System

COPIS Communication-Oriented Production Information System

COPISA CO (Carbon Monoxide) Pressure Induced Selective Adsorption (Process)

COPL Center for Optics, Photonics and Lasers [of Laval University, Sainte Foy, Quebec, Canada]

COPOL Council on Polytechnic Librarians [UK]

Copol Copolarization [also Co-Pol, co-pol, or copol]

COPOLCO Consumer Policy Committee [of International Standards Organization]

Copper Stud Copper Studies [Journal published in the US]

Copper Top Copper Topics [Publication of the Copper Development Association, US]

Copper Top (New York) Copper Topics (New York) [Journal published in the US]

Copper TPP 5,10,15,20-Tetraphenyl-21H,23H-Porphine Copper(II) [also copper TPP]

Cop Pl Copper Plate

COPPS Committee on Power Plant Siting [US]

COPR Center for Overseas Pest Research [UK]

Copr Copyright [also copr]

Coprod Coproduct [also Co-Prod, co-prod, or coprod]

COPS Calculator-Oriented Processor System

COPS Canadian Occupational Projection(s) System

COPS Central Order Processing System

COPS Command Operations

COPS Component Placement System

COPS Computer Oracle and Password System

COPSA COCO Pressure Swing Adsorption (Process)

COPSS Committee of Presidents of Statistical Societies [US]

COPSTOC Cornell Program for the Study of the Continents [of Cornell University, Ithaca, New York, US]

CoPt Cobalt Platinum (Alloy)

Co/Pt Cobalt on Platinum (Substrate)

Co-Pt Cobalt-Platinum (Alloy System)

CoPtCr Cobalt Platinum Chromium (Thin Film)

COPUL Council of Prairie University Libraries

COPUOS Committee on the Peaceful Uses of Outer Space [of United Nations]

COPUS Canadian Organization of Part-Time University Students

CoPy Cobalt Permalloy

Copy Copyright [also copy]

COQ Certificate of Qualification

CoQ Coenzyme Q [Biochemistry]

CoQ_0 Coenzyme Q_0 [Biochemistry]

CoQ_1 Coenzyme Q_1 (or Ubiquinone-5) [Biochemistry]

CoQ_2 Coenzyme Q_2 (or Ubiquinone-10) [Biochemistry]

CoQ_6 Coenzyme Q_6 (or Ubiquinone-30) [Biochemistry]

CoQ_7 Coenzyme Q_7 (or Ubiquinone-35) [Biochemistry]

CoQ_9 Coenzyme Q_9 (or Ubiquinone-45) [Biochemistry]

CoQ_{10} Coenzyme Q_{10} (or Ubiquinone-50) [Biochemistry]

COR Canadian Ownership Regulations

COR Carrier-Operated Relay

COR Coefficient of Reduction

COR Coefficient of Restitution

COR Cold Optics Radiometer

COR Common Object Runtime

COR Communications Operations Report

COR Conditioned Orientation Reflex

COR Contracting Officer's Representative

COR Contract Organization Review

COR Cordoba [Currency of Nicaragua]

COR Corrosive Effects (of Hazardous Materials)

Co_5R Cobalt Rare Earth (Magnet) (e.g., Cobalt Samarium, Co_5Sm]

Cor Corner [also cor]

Cor Coronagraph [also cor] [Astronomy]

Cor Coroner

Cor Corpus [also cor]

Cor Correction [also cor]

Cor Correlation [also cor]

Cor Correspond(ence); Correspondent; Corresponding [also cor]

Cor Corrosion; Corrosive [also cor]

Cor Corundum [Mineral]

cor corrected [also corr]

CORA Canadian Ocean Research Associates (Incorporated)

CORA Coherent Radar Array

CORA Conditioned Reflex Analog

CORA Conditioned Response Analog (Machine)

CORAD Correlation Radar [also Corad, or corad]

CORAE Chief of Operational Research and Analysis Establishment

CORAL Class-Oriented Ring Associated Language

CORAL Command Radio Link

CORAL Common Real-Time Applications Language [Computer Language]

CORAL Computer On-Line Real-Time Applications Language

CORAL Correlated Radio Link

CORAPRAN Cobelda Radar Automatic Preflight Analysis

CORBA Common Object Request Broker Architecture [also Corba] [Computers]

CORC Cornell Computing Language [of Cornell University, Ithaca, New York, US]

CORDE Cooperative Research, Development and Education (Project) [of Foundation for Education with Production, Botswana]

CORDIC Coordinate Rotation Digital Computer

CORDIS Cold Reflex Discharge Ion Source

CORDIS Community Research and Development Information Service [Europe]

CORDPO Correlated Data Printout

CORDPO-SORD Correlated Data Printout–Separation of Radar Data

CORDS Coherent-On Receive Doppler System

CORDS Coordination of Record and Database System

CORE Canadian Offshore Resources Exposition

CORE Coherent-On-Receive

CORE Cold Ocean Resources Engineering

CORE Committee on Research Expenditure [of Council for Scientific and Industrial Research, Pretoria, South Africa]

CORE Computer-Operated Readiness Equipment

CORE Common Operational Research Equipment

CORE Computer-Oriented Reporting Efficiency

Co$_5$RE Cobalt Rare Earth (Magnet) (e.g., Cobalt Samarium, Co$_5$Sm]

CORELAP Computerized Relationship Layout Planning

CoREN Corporation for Research and Enterprise Network

COREP Combined Overload Repair Control

CORF Committee on Radio Frequencies

CORG Combat Operations Research Group [Fort Monroe, Virginia, US]

CORGI Confederation of Registration Gas Installers [UK]

CoRh Cobalt Rhodium

CORINE Coordinated Information on the European Environment; Coordinated Information System on the State of the Environment and Natural Resources [of European Economic Community]

CORINE Coordination of Information on the Environment (Program) [of Swedish Space Corporation]

CORKS Computer-Oriented Record Keeping System

Corn Cornwall [UK]

CORNAP Cornell Network Analysis Program [also Cornap] [of Cornell University, Ithaca, New York, US]

Cornw Cornwall [UK]

CORNET Corporation Network [also Cornet]

CORODIM Correlation of the Recognition of Degradation with Intelligibility Measurements

Corp Corporal

Corp Corporate; Corporation [also corp]

Corp Comput Corporate Computing [Journal published in the UK]

Corpn Corporation [also corpn]

Corr Correction [also corr]

Corr Correlation; Correlator [also corr]

Corr Correspond(ence); Correspondent; Corresponding [also corr]

Corr Corrosion [also corr]

Corr Corrugating; Corrugation [also corr]

corr corrected [also cor]

corr corrugated

CORREGATE Correctable Gate [Electronics]

Corr Med Corrugating Medium [also corr med]

Corrn Corrosion [also corrn]

CORROS Expert System for Prediction of Corrosion Behaviour [of Märkische Fachhochschule Isarlohn, Germany]

Corros Corrosion [also corros]

Corros Abstr Corrosion Abstracts [of NACE International, US]

Corrosão Proteção Materiais Corrosão e Protecção de Materiais [Portuguese Journal on Corrosion and Corrosion Protection of Materials]

Corros Australas Corrosion Australasia [Publication of the Australasian Corrosion Association]

Corros Coatings, S Afr Corrosion and Coatings, South Africa [Journal published in Cape Town, South Africa]

Corros Control Corrosion Control [Canadian Journal]

Corros Eng Corrosion Engineering [Journal of the Japan Society of Corrosion Engineering]

Corrosion-NACE Corrosion–National Association of Corrosion Engineers International [US]

Corros Inf Anal Corrosion Information and Analysis [Journal published in the US]

Corros J Sci Eng Corrosion–The Journal of Science and Engineering [of NACE International, US]

Corros Maint Corrosion and Maintenance [Indian Publication]

Corros Prev Control Corrosion Prevention and Control [Journal published in the UK]

Corros Prev Inhib Dig Corrosion Prevention/Inhibition Digest [of ASM International, US]

Corros Prot Corrosion y Proteccion [Spanish Journal on Corrosion and Corrosion Protection]

Corros Rev Corrosion Reviews [UK]

Corros Sci Corrosion Science; Corrosion Scientist

Corros Sci Corrosion Science [Journal of the Institute of Corrosion, UK]

Corros Technol Corrosion Technology [International Journal]

CORS Canadian Operational Research Society

CORSA Cosmic Radiation Satellite [Japan]

CORSAIR Computer-Oriented Reference System for Automatic Information Retrieval

Cor Sec Corresponding Secretary

CortCO Consortium of Retail Teaching Companies [UK]

CORTECH Council on Research and Technology [US]

CORTEX Communications-Oriented Real-Time Executive

CORTS Canada-Ontario Rideau-Trent-Severn [Canada]

CORTS Convert Range Telemetry Systems

Co/Ru Cobalt on Ruthenium (Substrate)

COS Calculator On Substrate

COS Carbon Oxysulfide (or Carbonyl Sulfide)

COS Carry-on Oxygen System

COS Cash on Shipment [also cos]

COS Cassette Operating System

COS Centered Optical System

COS Change of Subscribers

COS Chief of Staff

COS Class of Service

COS Cloud Observing System [Atlantic Stratocumulus Transition Experiment]

COS Code Operated Switch

COS Committee on Standards [of American Society for Testing and Materials, US]

COS Communications Operating System

COS Communications-Oriented Software

COS Compact Operating System

COS Compatible Operating System

COS Computer Operating System

COS Concurrent Operating System

COS Console Operating System

COS Contactor Starting

COS Cooper Ornithological Society [US]

COS Core Operating System

COS Corporation for Open Systems International [US]

COS Cosine Function [Programming]

COS Cosmic Rays and Trapped Radiation Committee [of European Space Research Organization, France]

COS Cray Operating System

CoS Cobaltous Sulfide

Co-S Cobalt-Sulfur (Alloy System)

Cos Companies [also cos]

Cos Counties [also cos]

cos cosine [Symbol]

cos^{-1} anticosine (or inverse cosine) [also arccos]

COSA Canadian Oil Scouts Association

COSA Carwash Owners and Suppliers Association [US]

COSAC Computing Systems for Air Cargo

COSAG Combination of Steam and Gas

Co(SALEN)$_2$ Bis(salicylidene)ethylenediamineocobalt (II)

COSAM Cobol Shared Access Method

COSAM Conservation of Strategic Aerospace Materials

COSAM Co-Site Analytical Model

COSA NOSTRA Computer-Oriented System and Newly Organized Storage-to-Retrieval Apparatus

COSAP Cooperative On-Line Serials Acquisition Project

COSAR Compression Scanning Array Radar

COSATI Committee on Scientific and Technical Information [of Federal Council for Science and Technology, US]

COSB Canadian Organization of Small Business

CoSb Cobalt(II) Antimonide

COSBA Computer Services and Bureau Association [UK]

COSBE Committee for Small Business Exports [US]

COSC Cambridge Overseas School Certificate [Lesotho]

COSCL Common Operating System Control Language [CODASYL Committee]

CosCom Coastal Command [of Royal Singapore Navy]

cos/cos^2 cosine/cosine squared (pulse)

COSE Combined Office Standard Environment

COSE Common Open Software/Systems Environment

COSE Commonwealth Open Systems Environment

CoSe Cobalt(II) Selenide (or Cobaltous Selenide)

COSEC Coordinating Secretariat of National Unions of Students [Netherlands]

COSEC Culham On-Line Single Experimental Console

cosec cosecant [also csc]

cosec^2 cosecant squared (reflector)

COSEPUP Committee on Science, Engineering and Public Policy [US]

cos γ magnetic power factor [Symbol]

C-O-S-H Carbon-Oxygen-Sulfur-Hydrogen (Gases)

cosh hyperbolic cosine [Symbol]

cosh^{-1} inverse hyperbolic cosine [also arcosh]

COSI Closeout System Installation

COSI Committee on Scientific Information [US]

COSIRES International Conference on Computer Simulation on Radiation Effects in Solids

CoSi Cobalt Monosilicide (Compound)

Co/Si Cobalt on Silicon (Substrate)

Co-Si Cobalt-Silicon (Alloy System)

CoSiB Cobalt Silicon Boron (Alloy)

CoSiBFeNb Cobalt-Silicon-Boron-Iron-Niobium (Alloy)

COSIE Commission on Software Issues in the Eighties

COSINE Committee on Computer Science in Electrical Engineering Education [of Association for Computing Machinery, US]

COSIP College Science Improvement Program [of National Science Foundation, US]

COSIPA Companhia Siderurgica Paul [Brazil] [Metallurgy]

CoSi$_2$/p-Si Cobalt Disilicide on P-Type Silicon (Substrate)

COSIST Committee on Strengthening of Infrastructure in Science and Technology [of Center for Advanced Study, India]

CoSi$_2$/Si Cobalt Disilicide on Silicon (Substrate)

COSLANE Constant Optimum Separation Lane

Cosm Cosmic; Cosmos [also cosm]

COSMAR Committee on Surface Mining and Reclamation

COSMAT Committee on the Survey of Materials Science and Engineering [US]

COSMIC Common System for Main Interconnecting

COSMIC Computer-Oriented Structure Models from Inadequate Derived Coupling Data

COSMIC Computer Software Management and Information Center [of University of Georgia, Athens, US]

Cosmog Cosmogonical; Cosmogonist; Cosmogony [also cosmog]

Cosmog Cosmographer; Cosmographical; Cosmography [also cosmog]

Cosmol Cosmological; Cosmologist; Cosmology [also cosmol]

CosmoLEP Cosmic-Ray Experiment (to Detect Atmospheric Muons) by Low Energy Physics [of CERN–European Laboratory for Particle Physics, Geneva, Switzerland] [also CosmoLep]

COSMON Component Open/Short Monitor

COSMOS Coastal Multidisciplinary Oceanic System

COSMOS Complementary Symmetry Metal-Oxide Semiconductor [also CMOS, C-MOS, or C/MOS]

COSMOS Computer-Oriented System for Management Order Synthesis

COSMOS Computer System for Mainframe Operations

COSMOS Console-Oriented Statistical Matrix Operator System

COSMOS Consortium of Selected Manufacturers Open Systems

COSMOS Cornell Simulator of Manufacturing Operations [of Cornell University, Ithaca, New York, US]

COSMOS COsmologically Significant Mass Oscillation Search (Experiment) [of Fermilab, Batavia, Illinois, US]

COS/MOS Complementary-Symmetry Metal-Oxide Semiconductor

Cosm Res Cosmic Research(er)

Cosm Res Cosmic Research [Translation of: *Kosmicheske Issledovaniya (USSR)*; published in the US]

Co-Sn Cobalt-Tin (Alloy System)

CoSn Cobalt Stannide (Structure)

CoSO Cobaltosulfate

COSORB CO (Carbon Monoxide) Adsorption (Process)

COSOS Conference on Self-Operating Systems [US]

COSP Canadian Oil Substitution Program

COSP Central Office Signaling Panel

COSPA Canada's Ocean Strategy Project–The Atlantic (Study)

COSPAR Committee on Space Research [also Cospar] [of International Council of Scientific Unions]

COSPAS COSPAR Satellite

COSPAS/SARSAT COSPAR Satellite/Search and Rescue Satellite

COSPEAR Committee on Space Programs for Earth Observation [of NASA]

COSPUP Committee on Science and Public Policy [of National Academy of Sciences, US]

COSR Committee on Space Research [US]

COSRIMS Committee on Research in the Mathematical Sciences [of National Academy of Science, US]

COSRO Conical Scan On Receive Only

COSS Common Object Services Specification

COSS Computer-Optimized Storehouse System

COSS Correlation with Shift Scaling

COSS Correlation (with) Shift Slicing

COSSA Commonwealth Office of Space Science and Applications; CSIRO Office for Space Science Applications [of Commonwealth Scientific and Industrial Research Organization, Australia]

COSSAC Chief of Staff to Supreme Allied Commander

COST Committee on Science and Technology [of Association of Southeast Asian Nations–Australian Economic Cooperation Program]

COST Cooperation in (the Field of) Scientific and Technological Research [Joint of Program of the European Community and the European Free Trade Association]

cos θ power factor [Symbol]

COSTA Cost Accounting (Code)

COSTADE Cost Optimization Software for Transport Aircraft Design Evaluation [of NASA]

COSTAR Computer-STored Ambulatory Record [of Massachusetts General Hospital, Boston, US]

COSTAR Conversational On-Line Storage and Retrieval

COSTAR Corrective Optics Space Telescope Axial Replacement [of Hubble Space Telescope]

CostE Cost Engineer [also Cost Eng, or CostEng]

Cost Eng Cost Engineer(ing)

Cost Eng Cost Engineering [Journal of the American Association of Cost Engineers, US]

COSTDC Committee on Science and Technology in Developing Countries [US]

COSTHA Council on the Safe Transportation of Hazardous Articles [US]

COSTI Center on Scientific and Technical Information [Israel]

COSTI Committee on Scientific and Technical Information [US]

COSTPRO Canadian Organization for the Simplification of Trade Procedures

Costr Met Costruzioni Mettalliche [Italian Journal on Metal Construction]

COSWA Conference on Science and World Affairs

COSY Compiler System

COSY Compressed Symbolic Source Language

COSY Correction System

COSY Correlated Spectroscopy; Correlation Spectroscopy

COSY-LR Correlation Spectroscopy (with) Long-Range (Coupling) [also COSY LR]

COSY MAS NMR Correlation Spectroscopy Magnetic-Angle Spinning Nuclear Magnetic Resonance [also COSY-MASNMR]

COSY-NMR Correlation Spectroscopy Nuclear Magnetic Resonance [also COSY NMR]

COT Center for Office Technology [US]

COT Committee on Terminology [of American Society for Testing and Materials, US]

COT Continuity Signal

COT Create Occurrence Table

COT Customer-Oriented Terminal

COT Cyclooctatetraene [also cot]

Co T True Course [also CoT] [Navigation]

Cot Cotter [also cot]

Cot Cotton [also cot]

cot cotangent [also ctn]

cot^{-1} inverse cotangent (or anticotangent)

COTA Certified Occupational Therapy Assistant

COTAC Committee on Transport and Communications [of Association of South East Asian Nations]

COTAR Correlated Orientation Tracking and Ranging; Correlation Tracking and Ranging [also Cotar, or cotar]

COTAR-AME COTAR Angle Measuring Equipment

COTAR-DAS COTAR Data Acquisition System

COTAR-DME COTAR Distance Measuring Equipment

COTAT Correlation Tracking and Triangulation [also Cotat, or cotat]

CoTaZr Cobalt Tantalum Zirconium (Alloy)

COTC Canadian Officers Training Corps

COTC Canadian Overseas Telecommunications Corporation

COTCO Committee on Terminology Technical Committee Operations [of American Society for Testing and Materials, US]

CoTe Cobalt(II) Telluride

CO_2TEA Carbon Dioxide Transversely Excited Atmosphere (Laser) [also CO_2(TEA)]

COTF Canadian Occupational Therapy Foundation

COTH Cadmium Oxalate Trihydrate

coth hyperbolic cotangent [also ctnh]

$coth^{-1}$ inverse hyperbolic cotangent [also $ctnh^{-1}$]

COTI Coordinator of Operational Technical Investigations [of Canadian Armed Forces]

CoTi Cobalt Titanium (Spin Glass)

Co-Ti Cobalt-Titanium (Alloy System)

COTM Customer Owned and Telephone Company Maintained

Co/TM Cobalt/Transition Metal (Multilayer)

$Co(tn)_3$ Cobalt(III) (1,3-Diaminopropane)

CoTPP 5,10,15,20-Cobalt Tetraphenylporphyrin

COTR Contracting Officer's Technical Representation

COTR Cyclooctatriene

COTRAN COBOL-to-COBOL Translator

COTS Commercial Off-the-Shelf [Computers]

COTSAC Crown of Thorns Starfish Advisory Committee [Australia]

COTSARC Crown of Thorns Starfish Advisory Review Committee [Australia]

COTSREC Crown of Thorns Starfish Research Committee [Australia]

COTT Committee on Trade and Tourism [of Association of South East Asian Nations]

Cot Web Cotton Webbing

COU Council of Ontario Universities [Canada]

COUD-I Collectors of Unusual Data–International [US]

Coul Coulomb [also coul]

Coun Council [also coun]

CouncilNET Australian Local/Commonwealth Government Network

COUPLE Communications-Oriented User Programming Language

COUR Courier (Font) [also Cour]

Courr Norm Courrier de la Normalisation [French Publication on Standardization]

COUSA Confederation of Ontario University Staff Associations [Canada]

COV Coefficient of Variability; Coefficient of Variation [Statistics]

COV Coefficient of Viscosity

COV Covariance [also cov] [Statistics]

COV Cross-Over Value

COV Cut-Off Valve

COV Cut-Off Voltage [Electronics]

Cov Covariance [also cov] [Statistics]

Cov Cover [also cov]

COVA Computer Code Validation

COVAR Covariance [also Covar] [Statistics]

Covar Cobalt-Vanadium-Rhodium (Alloy System)

COVENIN Comisión Venezolana de Normas Industriales [Venezuelan Standards Commission]

COVER Cut-Off Velocity and Range [Aerospace]

covers coversed sine

COVET Cooperation Venture in the Education of Teachers [Canada]

CoVis (Learning through) Collaborative Visualization (Project) [of Northwestern University, Evanston, Illinois, US]

COVOA Canadian Offshore Vessel Operators Association

covrs coversed sine [also cvrs]

COW Cactus Ordnance Depot [Dumas, Texas, US]

COW Cherokee Ordnance Works [Danville, Pennsylvania, US]

COW Committee of the Whole

COW Coventry Ordnance Works [UK]

COW Crude Oil Washing

Co-W Cobalt-Tungsten (Alloy System)

COWAR (Coordinating) Committee on Water Research [France]

Cu/WC_p Tungsten-Carbide Particulate Reinforced Cobalt (Composite) [also Cu/WC(p)]

Cowl Cowling [also cowl] [Aerospace]

COWLIS Coastal Ocean Water Level Information System [also Cowlis]

CoWO Cobaltotungstate

COWPS Council on Wage and Price Stability

COWSIP Country Towns Water Supply Improvement Program

CoX Binary Cobalt Alloy [*X* Represents One, or More Metalloids)]

Cox Coxswain

COXE Combined Operations Experimental Establishment

Coy 5-Hydroxy-2-(Hydroxymethyl) 4-(4H)-Pyranone

Co-YBCO Cobalt-Doped Yttrium Barium Copper Oxide (Superconductor)

COZAC Conservation Zone Advisory Committee

COZI Communications Zone Indicator

CoZr Cobalt Zorconium (Alloy)

Co-Zn Cobalt-Zinc (Alloy System)

Co-Zr Cobalt-Zirconium (Alloy System)

CoZrCr Cobalt Zirconium Chromium (Alloy)

CoZrMo Cobalt Zirconium Molybdenum (Alloy)

CoZrNb Cobalt Zirconium Niobium (Alloy)

CoZrRe Cobalt Zirconium Rhenium (Alloy)

CoZrTa Cobalt Zirconium Tantalum (Alloy)

CP Calcium Phosphate

CP Calendar Process

CP Call Processor

CP Call to Several (Radio) Stations

CP Calorific Power

CP Canadian Pacific

CP Canadian Press

CP Candlepower [also cp]

CP Carbon Paper

CP Carbon Potential [Metallurgy]

CP Carbon Process

CP Card Punch

CP Cardinal Points

CP Cardiopulmonary

CP Cargo Program Office [of NASA Kennedy Space Center, Florida, US]

CP Cargo Projects (or Program) Office [of NASA Kennedy Space Center, Florida, US]

CP Car-Parrinello (Method) [Physics]

CP Carretera Panamericana [Pan-America Highway]

CP Carr-Parrinello (Molecular-Dynamics Method) [Physics]

CP Carriage Paid [also cp]

CP Carry Propagate [Combinational Logic]

CP Cars of the Past [US]

CP Cartesian Product [Mathematics]

CP Casing Pressure [Oil Drilling]

CP Catalytic Process

CP Cathodic Protection [Metallurgy]

CP Cathodophosphorescence; Cathodophosphorescent

CP Cationic Polymerization

CP Celestial Pole [Astronomy]

CP Cell Plate [Biology]

CP Cellulose Phosphate

CP Cellulose Propanoate

CP Cellulose Propionate

CP Cement Points

CP Center of Percussion

CP Center of Pressure [Aerospace]

CP Center Point

CP Center Punch

CP Central Peak [Astronomy]

CP Central Plane

CP Central Point (Software)

CP Central Processing; Central Processor

CP Centrifugal Pump

CP Ceramic Processing

CP Ceramics Program

CP Cerebral Palsy

CP Certified Planner

CP Cesspool

CP Chain Printer

CP Chain Procedure

CP Chamber Pressure [Aerospace]

CP Change Proposal

CP Channel Program

CP Characteristic Polynomial

CP Character Printer

CP Charge-Conjugation Parity [Particle Physics]

CP Charge-Conjugation Space-Inversion (Invariance) [Particle Physics]

CP Charged Particle

CP Charge Parity [Particle Physics]

CP Charge Pump(ing)

CP Check Pilot

CP Chemically Pure [also cp]

CP Chemical Physicist; Chemical Physics

CP Chemical Pneumonitis

CP Chemical Potential

CP Chemical Power

CP Chemical Precipitation

CP Chemical Preparation

CP Chemical Process(ing)

CP Chemical Properties

CP Chemical Purity

CP Chevrel Phase [Solid-State Physics]

CP Chicago Pile [Series of US Atomic Reactors; mainly at Argonne National Laboratory near Chicago, Illinois] [See also CP-1 through CP-11]

CP Chief Programmer

CP Chlorinated Paraffin

CP Chlorinated Polyolefin

CP Chloropentene

CP Chlorophenol

CP Chloroplast [Botany]

CP Chloropurine [Pharmacology]

CP Chromatographically-Purified

CP Chromoprotein [Biochemistry]

CP Chronopotentiometry

CP Circle of Position

CP Circuit Package
CP Circularly Polarized; Circular Polarization
CP Circulating Pump
CP Circular Pendulum
CP Circular Pitch (of Gears)
CP Clamping Pin
CP Classical Physics; Classical Physicist
CP Clinical Pathologist; Clinical Pathology
CP Clinical Pharmacologist; Clinical Pharmacology
CP Clock Pulse
CP Closed Pore
CP Close-Packed (Crystal) [also cp]
CP Cloud Physicist; Cloud Physics
CP Cloud Point [Chemical Engineering]
CP Coal Processing
CP Code Page [also cp]
CP Coefficient of Performance
CP Coherent Potential
CP Cold Plate [also C/P]
CP Cold Punch(ing)
CP Collecting Power
CP Collective Paramagnetism
CP Collective Pinning
CP College of the Pacific [Stockton, California, US]
CP Collision Probability
CP Colloidal Particle
CP Colombian Peso [Currency]
CP Color Photography
CP Color Plate
CP Color Plotter; Color Plotting
CP Color Print(er); Color Printing
CP Combustion Powder
CP Command Post
CP Command Privilege
CP Command Processor
CP Command Pulse [Electronics]
CP Commensurate Phase [Physics]
CP Commercially Pure [also cp]
CP Commercial Purity
CP Common Point
CP Common-Pointed (Shell)
CP Common Process
CP Commonwealth of Pennsylvania [US]
CP Communication Package
CP Communications Processor
CP Compacted Powder [Metallurgy]
CP Compressor Power
CP Compressor Pump
CP Compton Process [Physics]
CP Compton Profile [Physics]
CP Computational Physics
CP Computer [also Cp or cp]
CP Computer Port

CP Computer Processing
CP Computer Professional
CP Computer Program(ming); Computer Programmer
CP Concrete-Piercing (Projectile) [also C/P]
CP Condensation Pump
CP Conditional Probability [Statistics]
CP Conducting Polymer; Conductive Polymer
CP Cone Pedicel [Biology]
CP Conference Publication
CP Conical Pendulum
CP Conic Projection [Cartography]
CP Conference Paper
CP Conjugated Protein [Biochemistry]
CP Connection Pending
CP Connection Point
CP Connection Principle
CP Conservation Park
CP Console Processor
CP Constant Photocurrent
CP Constant Potential
CP Constant Power
CP Constant Pressure
CP Construction Permit
CP Consulting Planner
CP Contact Potential
CP Contemporary Physics
CP Contingency Planning
CP Continuous Path
CP Continuous-Path Programming
CP Continuous Phase
CP Continuous Pickling [Metallurgy]
CP Continuous Polymerization (Process)
CP Continuous Precipitation [Metallurgy]
CP Continuum Physics
CP Contracting Party
CP Controllable Pitch
CP Controllable-Pitch Propeller
CP Controlled Product
CP Control Panel
CP Control Point
CP Control Processor
CP Control Program
CP Conventional Purity
CP Convertiplane [Aerospace]
CP Cooling Pond [Chemical Engineering]
CP Cooper Pair [Solid-State Physics]
CP Coordination Polyhedron [Chemistry]
CP Coordination Polymer(ization)
CP Co-Pilot
CP Copolymer(ization)
CP Copper Powder
CP Co-Precipitated; Co-Precipitation
CP Copy Protected; Copy Protection

CP Coriolis Perturbation

CP Correspondence Principle [Quantum Mechanics]

CP Corrosion-Protected; Corrosion Protection

CP Coulomb Potential

CP Counterpoise

CP Counterpressure

CP Coupling [also Cp or cp]

CP Coventry Polytechnic [UK]

CP Covering Power

CP Cracking Pressure

CP Crack Plane [Mechanics]

CP Crack Propagation [Mechanics]

CP Cratering Process

CP Crazing Process

CP Critical Path

CP Critical Point [also cp]

CP Critical Pressure

CP Cross-Ply [Composites Engineering]

CP Cross-Polarization; Cross-Polarized; Cross-Polarizer

CP Cross-Product [Mathematics]

CP Cryopump(ed); Cryopumping

CP Crystalline Polymer

CP Crystal Physicist; Crystal Physics

CP Crystal Pulling

CP Cubic Parabola

CP Curie Point [Solid-State Physics]

CP Current Paper

CP Current Pointer

CP Cushion Pressure [Automobiles]

CP Customized Processor

CP Cyclic Polymer

CP Cyclopentadiene

CP Cyclopropane

CP 6-Cyclopurine

CP Cytidine Phosphate [Biochemistry]

CP Cytoplasm; Cytoplast [Biology]

CP- Bolivia [Civil Aircraft Marking]

CP-1 Chicago Pile 1 [First US Atomic Pile/Reactor in 1942]

CP-2 Chicago Pile 2 [Second US Atomic Pile/Reactor for Graphite Research in 1943]

CP-3 Chicago Pile 3 [Third US Heavy-Water Moderated Atomic Pile/Reactor in 1944]

CP-3' Chicago Pile 3' [US Heavy Water-Enriched University Research Reactor in 1950]

CP-4 Chicago Pile 4 [Experimental Breeder Reactor of US Atomic Energy Commission in 1951] [also EBR-1 or EBR-I]

CP-5 Chicago Pile 5 [US Heavy Water-Moderated Uranium-235 Research Reactor in 1954]

CP-6 Chicago Pile 6 [US Savannah River Basic Design Reactor in 1954]

CP-7 Chicago Pile 7 [US Experimental Boiling-Water Reactor in 1956]

CP-8 Chicago Pile 8 [Experimental Breeder Reactor II of US Atomic Energy Commission in 1964] [also EBR-2 or EBR-II]

CP-9 Chicago Pile 9 [ALPR (Argonne Low Power Reactor) in 1958] [of Argonne National Laboratory, near Chicago, US] [also ALPR]

CP-10 Chicago Pile 10 [US Isotope Producing Reactor, 1953 Design]

CP-11 Chicago Pile 11 [ARGONAUT (Argonne Nuclear Assembly for University Training) in 1957] [of Argonne National Laboratory, near Chicago, Illinois, US] [also Argonaut]

6-CP 6-Chloropurine [Pharmacology]

C-5'-P Cytidine 5'-Monophosphate [Biochemistry]

C£ Cyprus Pound [Currency]

C/P Carriage Paid [also c/p]

C/P Concrete-Piercing (Projectile)

C/P Cold Plate [also CP]

C-P Cambrian-Precambrian (Boundary) [Geology]

Cp Cap [also cp]

Cp Casseopeium [now lutetium (Lu)]

Cp Cathode of Pentode

Cp Coupling [also cp]

Cp Coupon [also cp]

Cp Cyclopentadienyl

Cp Cytidylyl [Biochemistry]

Cp^N endo-2-(Cyclopentadienylmethyl)norborn-5-ene

cP Continental Polar Air [Meteorology]

cP centipoise [Unit]

cP coherent perfect (crystal) cattice [Symbol]

cP Pearson symbol for primitive (simple) space lattice in cubic crystal system (this symbol is followed by the number of atoms per unit cell, e.g. cP7, cP20, etc.)

c-P Crystalline Phosphorus

cp candlepower [Unit]

cp carriage paid [also cp]

cp centipoise [Unit]

cp chemically pure

cp commercially pure

cp compare

cp coupling

cp creep in tension, plastic strain in compression [Mechanical Testing]

cp critical point (in 3-D space)

$\chi(p)$ wave aberration function [Symbol]

CPA California Pharmaceutical Association [US]

CPA California Pistachio Association [US]

CPA Canadian Pacific Airlines [also CP Air]

CPA Canadian Paraplegic Association

CPA Canadian Petroleum Association

CPA Canadian Pharmaceutical Association

CPA Canadian Phytopathological Association

CPA Canadian Plywood Association

CPA Canadian Postmasters Association

CPA Canadian Psychological Association

CPA Certificate in Preservation Administration

CPA Certified Professional Aromatherapist

CPA Center for Particle and Astrophysics [of University of California at Berkeley, US]

CPA Center for Photographic Arts [Chicago, Illinois, US]

CPA Certified Public Accountant

CPA Chemical Process Assessment

CPA Chemical Propulsion Abstracts [of Chemical Propulsion Information Agency, US]

CPA Chlorobenzene Producers Association [US]

CPA p-Chlorophenoxyacetic Acid

CPA Civilian Production Administration [US]

CPA Closest Point of Approach

CPA Coherent Potential Approximation [Physics]

CPA Cold Plasma Analyzer

CPA Color-Phase Alternation [also cpa]

CPA Commonwealth Pharmaceutical Association [UK]

CPA Commutative Principle of Addition

CPA Compressed-Pulse Altimeter

CPA Computer Architecture

CPA Computer Performance Analysis

CPA Computer Press Association [US]

CPA Computer Process Automation

CPA Concrete Pipe Association [US]

CPA Concurrent Photon Amplification

CPA Condensed Phosphoric Acid

CPA Constant Phase Angle

CPA Construction Plant-hire Association [UK]

CPA Contingency Planning Aid

CPA Contractors Plant Association [UK]

CPA Control of Pollution Act [UK]

CPA Control Program Assist

CPA Copolar Attenuation

CPA Cracking Pressure Adjusted Valve

CPA Craftmen and Potters Association [UK]

CPA Critical Path Analysis

CPA Cross Program Auditor

CPA Cytidylyladenosine [also CpA] [Biochemistry]

4-CPA 4-Chlorophenoxyacetic Acid

C/PA Computer/Portable Applications [of Institute of Electrical and Electronics Institute, US]

CpA Cytidylyl Adenosine [also CPA] [Biochemistry]

CPAA Charged-Particle Activation Analysis

CPAA Cycle Parts and Accessories Association [US]

CPAC Center for Process Analytical Chemistry [US]

CPAC Collaborative Pesticides Analytical Committee

CPACS Center for Peace and Conflict Studies [of Sydney University, Australia]

CPAE Cold Plasma Analyzer Experiment

CPAF Cost Plus Award Fee

CP Air Canadian Pacific Airlines [also CPA]

CPAL Containment Person Air Lock

CP Al Commercially Pure Aluminum [also CP-Al]

C-PAM Cationic Polyacrylamide

CPAP Continuous Positive Airway Pressure [Medicine]

cPAPS Cyclic 3'-Phosphoadenosine-5'-Phosphosulfate [also c-PAPS] [Biochemistry]

CPAR Construction Procurement Advancement Research (Program) [of US Army]

CPAR Cooperative Pollution Abatement Research

CPAS S-4-Chlorophenyl-2,4,5-Trichlorophenylazosulfide

CPAS Construction Program Administration System

CPAV Central Point (Software) Anti-Virus

CPAWS Computer-Planning and Aircraft Weighing Scales

CPB Cardiopulmonary Bypass

CPB Channel Program Block

CPB Colombo Planning Bureau [Sri Lanka]

CPB Competitive Protein Binding

CPB Contractors Pump Bureau [US]

CPB Corporation for Public Broadcasting [Washington, DC, US]

CPBA 3-Chloroperoxybenzoic Acid

CPBF Campaign for Press and Broadcasting Freedom [UK]

CPBMA Canadian Paper Box Manufacturers Association

CpBr Cetylpyridinium Bromide

CPBS Chlorophenylbenzenesulfonate

CPBX Computerized Private Branch Exchange

CPC Calgary Petroleum Club [Alberta, Canada]

CPC California Pistachio Commission [US]

CPC Calling Party Category

CPC Calling Party Control

CPC Canada Ports Corporation

CPC Canada Post Corporation

CPC Canadian Pallet Council

CPC Canadian Publishers Council

CPC Carbohydrate Protein Conjugates (of Proteases) [Biochemistry]

CPC Card-Programmed Calculator

CPC Center for Plant Conservation [of International Union for Conservation of Nature (and Natural Resources)]

CPC Central Planning Center

CPC Central Product Classification [US]

CPC Cerium Pyridinium Chloride

CPC Channel Program Command

CPC Characteristic Properties Code

CPC Chemical Protective Clothing

CPC Chinese Petroleum Corporation [Taiwan]

CPC Circular Paper Chromatography

CPC Clock Pulse(d) Control

CPC Coil Planet Centrifuge

CPC Collective Pitch Control [Helicopters]

CPC Committee for Programs and Coordination [of United Nations]

CPC Common Peripheral Channel

CPC Common-Pointed Capped (Shell)

CPC Commonwealth Palaeontological Collection

CPC Compound Parabolic Concentrator

CPC Computerized Plasma Control

CPC Computer Planning Consultant

CPC Computer Power Center

CPC Computer Process Control

CPC Computer Program Change

CPC Computer Program Component

CPC Computer Programming Concepts

CPC Condensation Particle Counter

CPC Constant Point Calculation

CPC Continuous Path Control

CPC Controlled-Potential Coulometry

CPC Core Protection Computer

CPC Cycle Program Control

CPC Cycle Program Counter

CPC Cyclic Pitch Control [Helicopters]

CPC Cyclopropylcarbinyl

CPC Cyclopropanecarboxylic Acid

CpC Cytidylyl Cytidine [Biochemistry]

CPCA Canadian Pest Control Association

CPCA Canadian Portland Cement Association

CPCA Country Pest Controllers Association [UK]

CPCA 5'-(N-Cyclopropyl)carboxamidoadenosine [Biochemistry]

CPCB Crew Procedures Control Board

CPC(BPN') Carbohydrate Protein Conjugates of Subtilisin BPN' [Biochemistry]

7-CPC(BPN') Carbohydrate Protein Conjugates of Subtilisin BPN' [Biochemistry]

CPCCI Conférence Permanente des Chambres de Commerce et d'Industrie [Permanent Conference of Chambers of Commerce and Industry] [of European Economic Community, Belgium]

CPC(CT) Carbohydrate Protein Conjugates of α-Chrymotrypsin [Biochemistry]

7-CPC(CT) 7-Carbohydrate Protein Conjugates of α-Chrymotrypsin [Biochemistry]

8-CPC(CT) 8-Carbohydrate Protein Conjugates of α-Chrymotrypsin [Biochemistry]

9-CPC(CT) 9-Carbohydrate Protein Conjugates of α-Chrymotrypsin [Biochemistry]

CPCDPD Chemical Protective Clothing Degradation and Permeation Database

CPCEI Computer Program Contract End Item

CpCl Cetylpyridinium Chloride

CPCEMR Circum-Pacific Council for Energy and Mineral Resources [US]

CPCFA Council of Pollution Control Financing Agencies

CPCGN Canadian Permanent Committee on Geographical Names

CPCH Calling Party Cannot Hear

CPCH Cyanophenylcyclohexane

CPCI Canadian Prestressed Concrete Institute

CPCI Computer Program Change Instruction

CPCI Computer Program Configuration Item

CPCI Coupled-Perturbed Configuration Interaction [Physics]

CPCI CPU (Central Processing Unit) Power Calibration Instrument

CPCL Computer Program Change Library

CPCL Computer Program Control Library

CPCO Calcium Praseodymium Copper Oxide (Superconductor)

(Cp)$_2$Co Bis(cyclopentadienyl)cobalt(III)

(Cp)$_2$CoOH Bis(cyclopentadienyl)cobalt(III) Hydroxide

CPCR Computer Program Change Request

CPCR Crew Procedures Change Request [of NASA]

CPCRA Community Programs for Clinical Research on AIDS

CPCRR Cyclopropylcarbinyl Radical Rearrangement

CPCS Check Processing Control System

CPCT Center for Photoinduced Charge Transfer [at University of Rochester, New York, US]

CPC(Th) Carbohydrate Protein Conjugates of Thermolysin [Biochemistry]

7-CPC(Th) 7-Carbohydrate Protein Conjugates of Thermolysin [Biochemistry]

CPC(Try) Carbohydrate Protein Conjugates of Trypsin [Biochemistry]

CP Cu Commercially Pure Copper [also CP-Cu]

CPCUG Capital Personal Computer User Group [US]

CPD Canadian Paleomagnetic Database

CPD Carboxypeptidase [Biochemistry]

CPD Cards per Day

CPD Central Pulse Distributor

CPD Charged Particle Detector

CPD Charterer Pays Duties

CPD Citrate Phosphate Dextrose [Biochemistry]

CPD Community Planning and Development [of Organization of American States]

CPD Consolidated Programming Document

CPD Contact Potential Difference (Method)

CPD Continuing Professional Development

CPD Contractor's Preliminary Design

CPD Converter, Pulse to Direct-Current Voltage

CPD Crew Passive Dosimeter

CPD Crew Procedures Division [of NASA]

CPD Critical Point Drier; Critical Point Drying

CPD Cumulative Population Doublings

CPD Cumulative Probability Distribution [Statistics]

CPD Charterer Pays Duties [also cpd]

Cpd Capreomycidine (or (S,S)-α-(2-Iminohexahydro-4-Pyrimidyl)glycine) [Biochemistry]

Cpd Compound [also cpd]

cpd charterer pays duties [also CPD]

CPDA Clay Pipe Development Association [UK]

CPDA Cyclopentane Tetracarboxylic Acid Dianhydride

CPD-A Citrate-Phosphate-Dextrose Solution with Adenine [Biochemistry]

CPDAMS Computer Program Development and Management System

CPDC China Petrochemical Development Corporation [Taipei, Taiwan]

CPDC Committee on the Participation of Developing Countries

CPDCA 1,3-Cyclopentanedicarboxylic Acid

CPDD cis-Platinum-Diammino-Dichloride (or cis-Diammine Dichloroplatinum)

CPDD Command Post Digital Display

CPDD Conceptual Program Design Description

CPDDS Computer Program Detail Design Specification

CPDL Canadian Patents and Developments Limited [of Department of Regional and Industrial Expansion]

CPDLC Chiral Polymer-Doped Liquid Crystal

CP-DPO Cargo Projects–Deployable Payloads Office [of NASA Kennedy Space Center, Florida, US]

CPDR Contractor's Preliminary Design Review

CPDS Computer Program Design Specification

CPDS Computer Program Development Specification

CPDS Crew Procedures Documentation System

Cpds Compounds [also cpds]

CPDU Continuous Process Development Unit [of Canada Center for Mineral and Energy Technology, Natural Resources Canada]

CPE Canadian Painters, Etchers and Engravers

CPE Carbon Paste Electrode

CPE Center for Packaging Education [US]

CPE Center for Petroleum Engineering [of University of New South Wales, Sydney, Australia]

CPE Center for Professional Education

CPE Central Pennsylvania Experiment

CPE Central Planned Economy

CPE Central Processing Element

CPE Central Programmer and Evaluator

CPE Certificate of Primary Education [now Kenya Certificate of Primary Education]

CPE Charged Particle Equilibrium [Particle Physics]

CPE Chief Program Engineer

CPE Chlorinated Polyethylene

CPE Circular Probable Error

CPE Computer Performance Evaluation

CPE Computer Premises Equipment

CPE Constant Phase Element

CPE Contact Potential Emitter

CPE Continuous Particle Electrophoresis

CPE Contractor Performance Evaluation

CPE Control Processing Electronics

CPE Conventional Polyethylene

CPE Correlated Particles Expansion

CPE Crew Procedure Engineer

CPE Cross-Roll Piercing Elongation (Rolling Process) [Metallurgy]

CPE Customer Premise Equipment [Telecommunications]

CPE Customer Provided Equipment

CPE Custom Power Engineering

CPE Cylindrical Photolithographic Etching

CPE 2-Cyclopentenone

.CPE Cover Page [File Name Extension] [also .cpe]

CPEA College of Physical Education Association

CPEA Concentrated Phosphate Export Association [US]

CPEA Cyprus Professional Engineers Association

CPEB Center for Psychotherapy and Emotional Bodywork [Canada]

CPEBS Central Processing Element Bit Slice

CPEC Cranfield Product Engineering Center [UK]

CPED Continuous Particle Electrophoresis Device

CPED Convergent Probe Electron Diffraction

C/PEEK Carbon (Graphite) Fiber-Reinforced Polyetheretherketone (Composite)

CPEI Computer Program End Item

CPEM Conference on Precision Electromagnetic Measurements [US]

CPEM Conventional Photoelectron Microscope

CPEQ Corporation of Professional Engineers of Quebec [Canada]

CPES Crew Procedures Evaluation Simulator

C-PET Crystallized Polyethylene Terephthalate

CP-ENDOR Circular-Polarized Electron Nuclear Double Resonance

CPEUG Computer Performance Evaluation Users Group [of National Institute for Standards and Technology, US]

CPF Calcium Phosphate Free

CPF Canadian Patrol Frigate

CPF Cargo Processing Facility

CPF Central Processing Facility

CPF Circular Polarized Fluorescence

CPF Complete Power Failure

CPF Control Program Facility

CPF Corporation Pharmaceutique Française Limited [French Pharmaceutical Manufacturer]

CPF Cost per Flight

CpF Cetylpyridinium Fluoride

CP-FEO Cargo Facilities and GSE (Ground Support Equipment) Projects Office [of NASA Kennedy Space Center, Florida, US]

CPFF Cost plus Fixed Fee

CPFMS COMRADE (Computer-Aided Design Environment) Permanent File Management System

CPFP Canadian Patrol Frigate Program

CPFP Commission des Publications Française de Physique [Commission for French Publications in Physics]

CPFR Calling Party Forced Release

CPFSK Continuous Phase Frequency Shift Keying

CPFT Customer Premises Facility Terminal

C-PFZ Carbon Precipitation-Free Zone [Metallurgy]

CPG Certified Professional Geologist

CPG Change Planning Group

CPG Chemical Paleogenetics

CPG Clean Power Generation (Process)

CPG Clock Pulse Generator

CPG COBOL Program Generator

CPG Collaborative Project Grants (Program) [of Natural Sciences and Engineering Research Council, Canada]

CPG College Publishers Group

CPG Controlled Pore Glass

CPG Critical Pressure Gauge

CpG Cytidylyl Guanosine [Biochemistry]

CPGA Ceramic Pin Grid Array [Electronics]

C:p-GaAs Carbon-Doped P-Type Gallium Arsenide

CPGE Center for Petroleum and Geosystems Engineering [of University of Texas, US]

CPGR Commission on Plant Genetic Resources [of Food and Agricultural Organization]

CPH Certificate in Public Health

CPH Characters per Hour [also cph]

CPH Close-Packed Hexagonal (Crystal) [also cph]

CPH Coal Preheating [Chemical Engineering]

CPH 2,2'-Methylene-bis-(4-Methyl-6-Cyclohexylphenol)

CPH Cyclopentadiene

CPHA Canadian Ports and Harbours Association

CPHA Canadian Public Health Association

CPhA Canadian Pharmaceutical Association

c-phase crystalline phase [Materials Science]

CPHF Coupled Perturbed Hartree-Fock (Method) [Physics]

CPHK City Polytechnic of Hong Kong [Kowloon]

CPhil Candidate in Philosophy [also C Phil]

CPHS Containment Pressure High Signal

CPhys Certified Physicist [UK]

CPI Cable Pair Identification

CPI Call Progress Indicator

CPI Canadian Plastics Institute

CPI Canine Parainfluenza (Virus)

CPI Carbon Predominance Index

CPI Carbon Preference Index

CPI Central Patents Index [UK]

CPI Certificate in Planning Information

CPI Characters per Inch [also cpi]

CPI Chemical Process(ing) Industries

CPI Chittagong Polytechnic Institute [Bangladesh]

CPI Clock Per Instruction [Computers]

CPI Code Page Information [also cpi]

CPI Commission Permanente Internationale Européenne des Gaz Industriels et du Carbure de Calcium [Permanent International European Commission on Industrial Gases and Calcium Carbide, France]

CPI Common Programming Interface

CPI Computer-PBX (Private-Branch Exchange) Interface

CPI Computer Power Index

CPI Computer Prescribed Instruction

CPI Condensation-Reaction Polyimide

CPI Confederation of the Photographic Industry [UK]

CPI Conference Papers Index

CPI Conference Proceedings Index

CPI Consumer Price Index

CPI Continuous Process Improvement

CPI Control Position Indicator

CPI Cost Performance Index

CPI Cotton Producers Institute [now Cotton Inc., US]

CPI Council for Professional Interest

CPI Council of the Printing Industries [Canada]

CPI Crash Position Indicator

CPI Crop Protection Institute [US]

CPI Cross Pointer Indicator

CPI Current Physics Index [of American Institute of Physics, US]

.CPI Code Page Information [MS-DOS File Name Extension] [also .cpi]

CPIA Canadian Photovoltaic Industries Association

CPIA Cathodic Protection Industry Association [US]

CPIA Chemical Propulsion Information Agency [of Johns Hopkins University, Baltimore, Maryland, US]

CPIA Chlorinated Paraffins Industry Association [US]

CPIA Close-Pair Interstitial Atoms [Materials Science]

CPIB Chlorophenoxyisobutyrate; p-Chlorophenoxy Isobutyric Acid

CPIB Ester Ethyl p-Chlorophenoxyisobutyric Acid [also CPIB ester]

CPIC Canadian Police Information Center

CPIC Canadian Professional Information Center [Mississauga, Ontario]

CPIC Chemical Process Industries of Canada

CPIC Crop Protection Institute of Canada

CPI-C Common Programming Interface for Communications

CPICOR Clean Power from Integrated Coal/Ore Reduction

CPID Color Printer and Imaging Division [of Tektronic, Inc., US]

CPID Computer Program Integrated Document

CPID Current Pulse Interface Demarcation

CPIF Cost Plus Incentive Fee

CPILS Correlation-Protected Instrument Landing System

CPIM Certified Production Inventory Manager

CPIMA Canadian Printing Ink Manufacturers Association

CPIN Computer Program Identification Number

CPIO Copy In and Out [Unix Operating System]

CPIP Computer Pneumatic Input Panel

CPIP Computer Program Implementation Process

26CPIP 2,6-Dichlorophenol Indophenol

CPIRA Copying Product and Inked Ribbon Association [US]

CPITC Coumarin Phenyl Isothiocyanate

CPIV Comité Permanent des Industries du Verre [Permanent Committee of the Glass Industries; of European Economic Community, Belgium]

CPK Correy-Pauling-Koltun (Model) [Chemistry]

CPK Creatine Phosphokinase [Biochemistry]

CPL Canadian Plastics

CPL Capability Password Level

CPL CAST (Computer-Aided Solidification Technique) Programming Language

CPL Certified Professional Landman

CPL Characters per Line [also cpl]

CPL Chicago Public Library [Illinois, US]

CPL Circular Polarized Luminescence (Spectroscopy)

CPL Close-Packed Plane [Crystallography]

CPL Combined Programming Language

CPL Commercial Pilot License

CPL   Common Programming Language
CPL   Computer Program Library
CPL   Continuous Pickling Line [Metallurgy]
CPL   Conversational Programming Language
CPL   Core Performance Log
CPL   Critical Path Length
CPL   Current Privilege Level
.CPL   Control Panel [File Name Extension] [also .cpl]
CP&L   Carolina Power and Light Company [US]
Cpl   Compliance [also cpl]
Cpl   Corporal [also CPL]
Cpl   Couple [also cpl]
CPLD   Complex Programmable Logic Device
CPLEAR   Measurements of CP (Charge-Parity) and T (Time) Violation in the Low-Energy Anti-Photon Ring (Experiment) [of CERN–European Laboratory for Particle Physics, Geneva, Switzerland]
CPLEE   Charged Particle Lunar Environment Experiment
Cplg   Coupling [also cplg]
CPLI   Canadian Professional Logistics Institute
CPLQ   Corporation of Professional Librarians of Quebec [Canada]
cplt   complete
CPM   Call Protocol Message
CPM   Carbide Powdered Metal
CPM   (Punched) Cards per Minute [also cpm]
CPM   Central Processor Module
CPM   Certificate in Public Management
CPM   Cesium Phosphomolybdate
CPM   Characters per Minute [also Cpm or cpm]
CPM   Chemical Process Metallurgy
CPM   Chemical Process Modelling
CPM   Colliding-Pulse Mode-Locked (Dye Laser)
CPM   Commutative Principle of Multiplication
CPM   Complex Permeability Measurement
CPM   Computer Performance Management
CPM   Computer Program Module
CPM   Connection Point Manager
CPM   Constant Photocurrent Method
CPM   Continuous Phase Modulation
CPM   Continuum Potts Model [Physics]
CPM   Control Path Method
CPM   Conversational Program Module
CPM   Counts per Minute [also cpm, cnts/min, or cts/min]
CPM   Critical Path Method
CPM   Cross-Product Matrix [Mathematics]
CPM   Crucible Particle Metallurgy (Process) [also Cpm]
CPM   Current Physics Microform [of American Institute of Physics, US]
CPM   Current Processor Mode
CPM   Cycles per Minute [also cpm]
CP/M   Control Program for Microcomputers (or Microprocessors)
CP/M   Control Program/Monitor

cpm   copies per minute
cpm   counts per minute [also CPM, cnts/min, or cts/min]
cpm   Crucible Particle Metallurgy [also CPM]
CPMA   Canadian Podiatric Medical Association
CPMA   Central Processor Memory Address
CPMA   Computer Peripherals Manufacturers Association [US]
CPMAR   Cross-Polarization Magic Angle Rotation [also CP MAR, or CP/MAR]
CPMAS   Cross-Polarization (with) Magic Angle Spinning [also CP MAS, or CP/MAS]
CPMB   Concrete Plant Manufacturers Bureau [US]
CPMC   Chlorophenyl-Methyl-Carbamate
CPMC   Columbia Presbyterian Medical Center [US]
CPMCSCF   Coupled Perturbed Multiconfiguration Self-Consistent Field [Physics]
CPMES   Committee on Physical, Mathmatical and Engineering Sciences [of Federal Coordinating Council for Science, Engineering and Technology, US]
CPMET   Coupled-Pair Many-Electron Theory [Physics]
CPMG   Carr, Purcell, Meiboom, Gill (Nuclear Magnetic Resonance Pulse Sequence)
CPMLS   Center for Petroleum and Mineral Law Studies [UK]
CPMP   Crew Procedures Management Plan
CPMR   Conference of Peripheral Maritime Regions [of European Economic Community, Belgium]
CPMS   Cable Pressure Monitoring System
CPMS   Canadian Pest Management Society
CPMS   Canadian Project Management Society
CPMS   Center for Policy and Management Studies
CPMS   Components, Packaging and Manufacturing Society [US]
CPMT   Center for Powder Metallurgical Technology [US]
CPMT   Components, Packaging and Manufacturing Technology (Society) [of Institute of Electrical and Electronics Engineers, US]
CPMTC   Canadian Pest Management Training Center
CPN   Cesium Lead Niobate
CPN   Community Psychiatric Nurse
Cpn   Coupon [also cpn]
CPNF   Coated Particle Nuclear Fuel
CP Ni   Commercially Pure Nickel [also CP-Ni]
CP-NiAl   Commercial Purity Nickel Aluminide
CPO   Canada Post Office
CPO   Canadian Passport Office
CPO   Catalytic Partial Oxidation
CPO   Chief Petty Officer
CPO   Chlorinated Polyolefin
CPO   Code Practice Oscillator
CPO   Command Pulse Output [Electronics]
CPO   Commonwealth Producers Organization [UK]
CPO   Compulsory Purchase Order
CPO   Concurrent Peripheral Operations
CPO   Cyclic Phosphine Oxide
CPODA   Contention Priority-Oriented Demand Assignment
CPOL   Communications Procedure-Oriented Language

CPP Calcium Pyrophosphate

CPP Card Punching Printer

CPP Career Planning and Placement

CPP Cast Polypropylene

CPP Center for Plutonium Production [France]

CPP Chemical Power Plant

CPP Chlorinated Polypropylene

CPP Coal Preparation Plant

CPP Coherent Plume Pattern

CPP Computer Professional for Peace

CPP Conductive Plastic Potentiometer

CPP Containment Pressure Protection

CPP Controllable-Pitch Propeller

CPP Core Polarization Potential [Physics]

CPP Critical Pitting Potential [Corrosion Science]

CPP Curie-Point Pyrolysis

CPP Current Papers in Physics (Service) [of Information Services Physics, Electrical and Electronics, and Computers and Control, UK]

CPP (Electric) Current Perpendicular to (Layer) Plane (Configuration)

CPP Cyclopentenophrenanthrene

.CPP C++ (Plus-Plus) File [File Name Extension] [also .cpp]

CPPA Canadian Pulp and Paper Association

CPPA Coal Preparation Plant Association [US]

CPPA Newsprint Stat CPPA Newsprint Statistics [Publication of the Canadian Pulp and Paper Association]

CPPA Newsprint Data CPPA Newsprint Data [Publication of the Canadian Pulp and Paper Association]

CPPA Tech Sect Proc CPPA Technical Section Proceedings [of Canadian Pulp and Paper Association]

CPPA Trans Tech Sect CPPA Transactions of the Technical Section [of Canadian Pulp and Paper Association]

CPPC Caribbean Plant Protection Commission [Trinidad and Tobago]

CP-PCO Cargo Projects–Program Control Office [of NASA Kennedy Space Center, Florida, US]

CPPD Calcium Pyrophosphate Deposition Disease

CPPD Career Planning and Placement Division [of American Physical Society, US]

CPPD Collaborative Program for Professional Development

CP Pd Commercially Pure Palladium [also CP-Pd]

CPPF Canadian Project Preparation Facility

CPP-GMR Current Perpendicular-to-Plane Giant Magnetoresistance

CPPI Consultative Panel on Public Information [of United Nations]

CPPMA Canadian Public Personnel Management Association

CPP-MR Magnetoresistance with (Electric) Current Perpendicular to (Layer) Plane

CPPO Bis(2-Carbopentyloxy-3,5,6-Trichlorophenyl) Oxalate

CPPO Certified Public Purchasing Officer

CPPP Computerized Production Process Planning

CPPR Center for Protein and Polymer Research [Philadelphia, Pennsylvania, US]

CPPS Comisión Permanente de Pacifico Sud [Permanent South Pacific Commission, Colombia]

CPPS Critical Path Planning and Scheduling

C/PPS Carbon (Graphite) Fiber-Reinforced Polyphenylene Sulfide (Composite)

CPPSS Culham Plasma Physics Summer School [Abingdon, Oxon, UK]

CP Pt Commercially Pure Platinum [also CP-Pt]

CPPU N-(2-Chloro-4-Pyridyl)-N'-Phenylurea [also 4-CPPU]

CPR Cam Plate Output

CPR Canadian Pacific Railways [also CP Rail]

CPR Cardiopulmonary Resuscitation

CPR Card Punch and Reader

CPR Casting-Pressing-Rolling [Compact (Metal) Strip Production Process]

CPR Center for Population Research [US]

CPR Center of Polish Research

CPR Center for Polymer Research [of University of Texas at Austin, US]

CPR Central Premonitions Registry [US]

CPR Chlorophenol Red

CPR Coal-Pile Runoff (System)

CPR Committee of Permanent Representatives [of European Community]

CPR Committee on Petroleum Reserves [of American Petroleum Institute, US]

CPR Committee on Polar Research [of National Academy of Science/National Research Council, US]

CPR Component Pilot Rework

CPR Continuous Progress Indicator

CPR Correspondence Pattern Recognition

CPR Corrosion Penetration Rate

CPR Crack Propagation Rate [Mechanics]

CPR Critical Power Ratio

CPR Critical Problem Report

CPR Cyclonene-Pyrethrin-Rotenone

CP-R Control Program–Real-Time

Cpr Copper

CP Rail Canadian Pacific Railways [also CPR]

CPRC Canadian Police Research Center

CPRD Committee on Prosthetics Research and Development [of National Research Council, US]

CPRDC Coordinated Program of Research in Distributed Computing [of Science and Engineering Research Council, UK]

CPRE Council for the Protection of Rural England

CPREA Canadian Peace Research and Education Association

CPRF Canadian Psychiatric Research Foundation

CPRG Computer Personnel Research Group [now Special Interest Group for Computer Personnel Research of the Association for Computing Machinery, US]

CPRI Canadian Peace Research Institute

CPRI Central Psi Research Institute [US]

CPRL Ceramics Processing Research Laboratory [of Massachusetts Institute of Technology, Cambridge, US]

CPRL Chemical and Physical Research Laboratories [Australia]

CPR-MD Coalition of Physicians for Responsible Medical Democracy

CPRS Canadian Public Relations Society

CPRS Centralized Personnel Record System

CPRW Council for the Protection of Rural Wales [UK]

CPS Calcium Poly(styrene Sulfonate)

CPS Calling Processing Subsystem

CPS Cambridge Philosophical Society [of Cambridge University, UK]

CPS Canadian Parks Service

CPS Canadian Physiological Society

CPS (Punched) Card Programming System

CPS Cards per Second

CPS Carnivorous Plant Society [UK]

CPS Cathode-Potential-Stabilized (Emitron)

CPS Central Point Software, Inc. [Beaverton, Oregon, US]

CPS Central Processing (or Processor) System

CPS Certified Plastic Surgeon

CPS Certified Productivity Specialist

CPS C-Frame Profile Scanner; C-Frame Profile Scanning

CPS Chain-Propagating Step [Chemistry]

CPS Characters per Second [also cps]

CPS Chinese Petroleum Society

CPS Circuit Package Schematic

CPS Coils per Slot

CPS College of Pharmaceutical Sciences [New York City, US]

CPS College Press Service [US]

CPS Columns per Second

CPS Combined Planning Staff

CPS Commission on the Patent System [US]

CPS Committee for Production Sharing [US]

CPS Communications Processor System

CPS Compendium of Pharmaceuticals and Specialties [of Canadian Pharmaceutical Association]

CPS Conservation Protection Service [Canada]

CPS Contour Plotting System

CPS Controlled Path System

CPS Control Programs Support

CPS Conversational Programming System

CPS Conversion Program System

CPS Cooperative Processes Software

CPS Counts per Second [also cps, cnts/sec, or cts/sec]

CPS Critical Path Scheduling

CPS Current Population Survey [of US Department of Commerce]

CPS Curved Position Sensitive (Detector)

CPS Customer Premises System

CPS Cycles per Second [also cps]

.CPS Central Point Signature [File Name Extension]

cPS Porous Crystalline Silicon

cps centipoise [Unit]

cps critical points (in 3-D space)

CPSA Civil and Public Services Association [UK]

CPSA Consumer Product Safety Act [US]

CPSC Consumer Product Safety Commission [US]

CPSC Consumer Protection and Safety Commission [US]

CPSCI Central Personnel Security Clearance Index

CPSD Counting Position Sensitive Detector

CPSE Common Payload Support Equipment

Cpse Counterpoise [also cpse]

CPSG Common Power Supply Group

CPSK Coherent Phase-Shift Keying

CPSK Conitinuous-Phase-Shift Keying

CPSM Critical Path Scheduling Method

CP Sn Commercially Pure Tin [also CP-Sn]

CP-SPO Cargo Projects–Spacelab Projects Office [of NASA Kennedy Space Center, Florida, US]

CPSR Computer Professionals for Social Responsibility [Palo Alto, California, US]

CPSR Controlled Process Serum Replacement [Immunology]

CPSS Certificate in Public Service Studies

CPSS Certified Professional Soil Scientist

CPSS Circular Polarization Selective Surface

CPSS Cold Plate Support Structure

CPSS Committee of Presidents of Statistical Societies

CPSS Common Programming Support System

CPSS Computer Power Support System

CPSS Critical Phase System Software

CPSSRB Canada Public Service Staff Relations Board

CPST Commission on Professionals in Science and Technology [US]

CPSU California Polytechnic State University [San Luis Obispo, US] [also Cal Poly]

cps/V cycles per second per volt [Unit]

CPT Canadian Pacific Telecommunications

CPT 1-Carbamic Acid Ethylester-4-Phenyl-2-(1H)-Pyrimidinethionate

CPT Cargo Processing Technician

CPT Carr-Purcell Technique [also CP-T] [Physics]

CPT Casting Plant and Technology [Journal published in Germany]

CPT Certified Personal Trainer

CPT Charge-Conjugation Space-Inversion Time-Reversal (Theorem) [Particle Physics]

CPT Charge, Parity, Time (Theory) [Particle Physics]

CPT Chief Programmer Team

CPT Cluster Perturbation Theory [Physics]

CPT Color Picture Tube

CPT Command Pass Through

CPT Computer Program Tape

CPT Cone Penetrometer

CPT Confederación Paraguaya de Trabajadores [Confederation of Paraguayan Workers]

CPT Congress of Physical Therapy

CPT Control Power Transformer

CPT Covariant Perturbation Theory [Physics]

CPT Critical Path Technique

CPT Critical Pitting Temperature [Corrosion Science]

Cpt Captain [also CPT]

CPTA Canadian Paper Trade Association

CPTA Computer Programming and Testing Activity

CPTA Cyclopentane-trans-1,2-Diamine

CP-TAPF Carr-Purcell Time-Averaged Precession Frequency [Physics]

CP-TAPF Cross-Polarization using Time-Averaged Precession Frequency [also CPTAPF]

CPTB Clay Products Technical Bureau [UK]

CPT/Cast Plant Technol CPT/Casting Plant and Technology [Journal published in Germany]

CPTE Computer Program Test and Evaluation

CPTEC Centro de Previsão de Tempo e Estudos Climáticos [Center for Weather Forecast and Climatic tudies; of Brazilian National Institute of Space Research]

CPTEO 3-Chloropropyl Triethoxysilane

CPT-H 1-Carbamic Acid Ethylester-4-Phenyl-2-(1H)-Pyramidinethione

CP Ti Commercially Pure Titanium [also CP-Ti]

CPTL Chemical Process Technology Center [of General Electric Research and Development Center, Schenctady, New York, US]

CPTMO 3-Chloropropyl Trimethoxysilane

CPTP Civil Pilot Training Program [US]

Cptr Computer [also cptr]

CPTS Collidine-P-Toluenesulfonate

CPTU Continuous Process Development Unit

CPTU Council of Progressive Trade Unions [Trinidad]

CPU Canadian Paperworkers Union

CPU Capacitive Pick-Up

CPU Central Processing (or Processor) Unit [also cpu]

CPU Collective Protection Unit

CPU Commonwealth Press Union [UK]

CPU Communications Processor Unit

CPU Computer Peripheral Unit

CPU Computer Printer Unit

CPU Computer Processor Unit

CPU Control Process Unit

CpU Cytidylyl Uridine [Biochemistry]

CPUID Central Processing Unit Identification Number

Cpunch Counterpunch [also cpunch]

CPUOS Committee on the Peaceful Uses of Outer Space [of United Nations]

CPUS Coalition for Peaceful Uses of Space

CPUSAC Crafted with Pride in the USA Council

CPUSOF Committee on the Peaceful Uses of the Seabed and the Ocean Floor (Beyond the Limits of National Jurisdiction) [of United Nations]

CPV Canine Parvovirus

CPV Cotia Pox Virus

CPV Cypovirus

CPV-1 Cypovirus, Type 1

CPVA Chemisch-Physikalische Versuchsanstalt [Chemical and Physical Research Institute, Berlin, Germany]

CPVC Chlorinated Polyvinyl Chloride

CPW Commercial Processing Workload

CPW Commercial Projected Window

CPW Coplanar Waveguide

CPW Current Perpendicular to (Domain) Wall (Direction) [Solid-State Physics]

CPW MR Current Perpendicular to (Domain) Wall Magnetoresistance [Solid-State Physics]

CPX Charged Pigment Xerography

CPX Clear Prefix (Function)

2CPy 2-Chloropyridine

3CPy 3-Chloropyridine

Cpy Copy [also cpy]

CPZ Chlorpromazine

CQ (General) Call to All (Radio) Stations

CQ Charge of Quarters

CQ Charlie Quebec [Radiotelephony]

CQ Chloroquine

CQ Clinical Quality

CQ Commercial Quality

CQ Controlled Quality

C&Q Compressed and Quenched; Compression and Quench(ing) [Metallurgy]

C(q) Charge Structure Factor [Symbol]

$\chi(q)$ Lindhard-type linear response function [Symbol]

$\chi(q)$ spin susceptibility (in physics) [Symbol]

CQA Construction Quality Assurance

CQAK Commercial Quality Aluminum-Killed (Steel)

CQC Certificate in Quality Control

CQC Citizens for a Quieter City [US]

CQC Construction Quality Control

CQD Coated Quantum Dot

CQD Come–Quick–Danger [also cqd] [Navigation]

CQD Critical Qualification Design

CQD Customary Quick Dispatch

CQDR Critical Qualification Design Review

CQE Certified Quality Engineer

CQE Cognizant Quality Engineer

CQEFP Center for Quality Engineering and Prevention

CQES Center for Quantized Electronic Structures [of University of California at Santa Barbara, US]

CQFP Ceramic Quad Flat Pack [Electronics]

CQMA Cost/Quality Management Assessment

CQMS Circuit Quality Monitoring System

CQP Ceramic Quad Pack [Electronics]

CQ Radio Amat J CQ (Charlie Quebec) Radio Amateur's Journal [US]

CQRS Commercial Quality Rimmed Steel

CQS Certified Quality System

CQS Common Queue Space

CQS Constant-Q Spectrometer

CQU Central Queensland University [Australia]

C-Quark Charmed Quark [also c-quark] [Particle Physics]

CR Calcium Ruthenate

CR Call Reader
CR Call Request
CR Cannizzaro Reaction [Organic Chemistry]
CR Canola Rape [Agriculture]
CR Capacitance-Resistance (Law)
CR Carbon(aceous) Residue
CR Carbon Resistor
CR Card Reader
CR Carriage Return (Character) [Data Communications]
CR Carrier Recovery
CR Carry Register [Computers]
CR Catalytic Reforming
CR Cathode Ray [also cr]
CR Cavity Resonance
CR Cedar Rapids [Iowa, US]
CR Ceiling Register
CR Cellular Respiration
CR (Centre de Recherche) – French for "Research Center"
CR Ceramic Research
CR Cerenkov Radiation
CR Certification Requirement
CR Certified Reflexologist
CR Ceylon Rupee [Currency of Sri Lanka]
CR Chain-Radar (System)
CR Chain Reaction
CR Change Request
CR Channel Request
CR Characteristic Radiation
CR Charge Ratio
CR Chart Recorder
CR Chemically Reactive; Chemical Reaction; Chemical Reactivity; Chemical Reactor
CR Chemical Recorder
CR Chemical Report
CR Chemical Resistance; Chemical-Resistant
CR Chicago Reactor [of Argonne National Laboratory, near Chicago, Illinois, US]
CR Chiral Radical
CR Chiral Reagent
CR Chlorinated Rubber
CR Chloroprene Rubber
CR Churchill River [Labrador, Canada]
CR Clamp Rest
CR Clean Room
CR Clemmensen Reduction (or Reaction) [Chemistry]
CR Cobalt Rhenate
CR Coherent Radiation
CR Cold-Reduced; Cold Reduction [also cr] [Metallurgy]
CR Cold-Rolled; Cold Rolling [also cr] [Metallurgy]
CR Collector Ring
CR Colorado River [US]
CR Columbia River [US/Canada]
CR Command Register
CR Communications Register

CR Compact Rail
CR Company's Risk [also C/R]
CR Complement Receptor [Biochemistry]
CR Complete Round
CR Complex Root
CR Composites Removal
CR Composites Research
CR Compression Ratio
CR Comptes Rendus [Extensive Series of Proceedings of the Académie des Sciences, Paris, France]
CR Computed Radiography
CR Computer Research
CR Computer Run
CR Computing Reviews [Published in the US]
CR Conditioned Reflex [Psychology]
CR Conditioned Reinforcer
CR Conditioned Response
CR Conference Report
CR Configuration Review
CR Conflict Resolution
CR Congressional Record
CR Conservation Reserve
CR Constant Rate
CR Contact Resistance
CR Containment Rupture
CR Continuously Reinforced; Continuous Reinforcement
CR Contract(or) Report
CR Controlled Rectifier
CR Controlled Release
CR Controlled Rolled; Controlled Rolling [Metallurgy]
CR Control Ratio
CR Control Relay
CR Control Rod [Nuclear Reactors]
CR Control Room
CR Control Routine
CR Converter Reactor
CR Cooling Rate
CR Copper Rhenate
CR Corona Resistance
CR Corrosion Rate
CR Corrosion Research(er)
CR Corrosion Resistance; Corrosion-Resistant
CR Cosmic Radiation
CR Cosmic Ray(s)
CR Costa Rica(n)
CR Costa Rica [ISO Code]
CR Cost Reduction
CR Counterrotating; Counterrotation
CR Cramer's Rule [Mathematics]
CR Creep Rate [Mechanics]
CR Creep Resistance; Creep-Resistant [Mechanics]
CR Creep Rupture [Mechanics]
CR Critical Region
CR Critical Requirements

CR Cruzeiro [Currency of Brazil]

CR Crystallographic Relationship

CR Crystallographic Research

CR Crystal Rectifier

CR *(currentis)* – this year, month, etc.

CR Current Rate

CR Current Relay

CR Curtius Rearrangement (or Reaction) [Chemistry]

CR Cycloreversion

CR Cyclotron Resonance

CR Czech Republic

C$_0$(R) Height-Height Correlation Function [Symbol]

C/R Chamfer, or Radius

C/R Commutation Rate

C/R Company's Risk [also CR]

C-R Edge Orientation (i.e., Circumferential Direction–Radial Direction) [Forging]

Cr Chromite [Mineral]

Cr Chromium [Symbol]

Cr Credit(or) [also cr]

Cr Creek [also cr]

Cr Crew [also cr]

Cr Cristobalite [Mineral]

Cr Crown [also cr]

Cr Crystal(line) [also cr]

Cr^{2+} Divalent Chromium Ion [also Cr^{++}] [Symbol]

Cr^{3+} Trivalent Chromium Ion [also Cr^{+++}] [Symbol]

Cr^{6+} Hexavalent Chromium Ion [Symbol]

Cr-48 Chromium-48 [also ^{48}Cr or Cr^{48}]

Cr-49 Chromium-49 [also ^{49}Cr or Cr^{49}]

Cr-50 Chromium-50 [also ^{50}Cr or Cr^{50}]

Cr-51 Chromium-51 [also ^{51}Cr or Cr^{51}]

Cr-52 Chromium-52 [also ^{52}Cr or Cr^{52}]

Cr-53 Chromium-53 [also ^{53}Cr or Cr^{53}]

Cr-54 Chromium-54 [also ^{54}Cr or Cr^{54}]

C(r) Correlation Integral [Symbol]

.cr Costa Rica [Country Code/Domain Name]

χ/ρ magnetic susceptibility [Symbol]

CRA Calcium Reserve Assembly [Poultry Science]

CRA California Redwood Association [US]

CRA Canadian Residential Appraiser

CRA Catalog Recovery Area

CRA Chemical Reaction Alignment (Method)

CRA Chemical Recovery Association [UK]

CRA Chicago Research Association [US]

CRA Clear Rear Access

CRA Cold-Rolled and Annealed; Cold Rolling and Annealing [Metallurgy]

CRA Colorado River Association [US]

CRA Community Radio Association [UK]

CRA Complete Round of Ammunition

CRA Composite Research Aircraft

CRA Comprehensive Regional Assesment

CRA Concrete Repair Association [UK]

CRA Control Rod Assembly [Nuclear Engineering]

CRA Conzinc Rio Tinto Australia Limited

CRA Coordinated Regional Assessment

CRA Corn Refiners Association [US]

CRA Cosmic-Ray Age

CrA Corona Australis [Astronomy]

CRAB Cosmic Radiation Laboratory [of Institute of Physical and Chemical Research, Saitama, Japan]

CRAC Careers Research and Advisory Center [UK]

Cr(Ac)$_3$ Chromium Acetate

Cr(ACAC)$_2$ Chromium(II) Acetylacetonate [also Cr(AcAc)$_2$, or Cr(acac)$_2$]

Cr(ACAC)$_3$ Chromium(III) Acetylacetonate [also Cr(AcAc)$_3$, or Cr(acac)$_3$]

CR Acad Bulg Sci Comptes Rendus, Academie Bulgare des Sciences [Proceedings of the Bulgarian Academy of Sciences]

CR Acad Sci Comptes Rendus des Séances de l'Academie des Sciences [Proceedings of the Academy of Sciences, France]

CR Acad Sci I, Math Comptes Rendus de l'Academie des Sciences, Série I: Mathematics [Proceedings of the Academy of Sciences, Series I: Mathematics, France]

CR Acad Sci II, Méc Phys Chim Sci Terre Univers Comptes Rendus de l'Academie des Sciences, Série II: Mécanique, Physique, Chimie–Sciences de la Terre et de l'Univers [Proceedings of the Academy of Sciences, Series II: Mechanics, Physics and Chemistry– Earth and Cosmic Sciences, France]

CR Acad Sci Paris Comptes Rendus des Séances de l'Academie des Sciences de Paris [Proceedings of the Academy of Sciences of Paris, France]

CR Acad Sci Sér I Comptes Rendus de l'Academie des Sciences, Série I [Proceedings of the Academy of Sciences, Series I, France]

CR Acad Sci Sér II Comptes Rendus de l'Academie des Sciences, Série II [Proceedings of the Academy of Sciences, Series II, France]

CR Acad Sci Sér III Comptes Rendus de l'Academie des Sciences, Série III [Proceedings of the Academy of Sciences, Series III, France]

CR Acad Sci Sér A Comptes Rendus de l'Academie des Sciences, Série A [Proceedings of the Academy of Sciences, Series A, France]

CR Acad Sci Sér B Comptes Rendus de l'Academie des Sciences, Série B [Proceedings of the Academy of Sciences, Series B, France]

CR Acad Sci Sér C Comptes Rendus de l'Academie des Sciences, Série C [Proceedings of the Academy of Sciences, Series C, France]

CR Acad Sci Sér Gén Vie Sci Comptes Rendus de l'Academie des Sciences, Série Générale, la Vie des Sciences [Proceedings of the Academy of Sciences, Series: General Life Sciences, France]

CRAD Chief of Research and Development [of National Defense Headquarters, Ottawa, Canada]

CRAD Committee for Research into Apparatus for the Disabled

CRADA Cooperative Research and Development Agreement(s Program) [US]

CRAF Civilian Reserve Air Fleet

CRAF Comet Rendevous/Asteroid Fly-By

CRAFT Changing Radio Automatic Frequency Transmission

CRAFT Computerized Relative Allocation of Facilities Technique

CRAFT Cooperative Research Action for Technology [of European Community]

CRAG Composite Research Advisory Group [UK]

CRAGS Chemistry Records and Grading System

Cr-Al Chromium-Aluminum (Alloy System)

Cr:Al₂O₃ Chromium-Doped Aluminum Oxide [also Cr-Al₂O₃ or Cr³⁺:Al₂O₃] [Al₂O₃ in Ruby Variety]

Cr-Al₂O₃ Chromium-Bonded Aluminum Oxide (Cermet)

CrAlY Chromium-Aluminum-Yttrium (Coating)

CRAM Card Random Access Memory

CRAM Computerized Reliability Analysis Method

CRAM Conditional Relaxation Analysis Method

CRAM Core and Random Access Manager

CRAM Cyberspatial Reality Advancement Movement

CRAMPS Combined Rotation and Multiple-Pulse Spectroscopy

CRAMRA Convention on the Regulation of Antarctic Mineral Resource Activities

CRAN Cross Scan

CRANZ Coal Research Association of New Zealand

CRAP California Rivers Assessment Project

Crapo Chauffage avec Régulation Automatique de la Pause par Ordinateur [A French Automated Furnace Heating System]

CRAQ Controlled Rolling, Air Quenching (Steel)

CRAR Committee for the Recovery of Archeological Remains [US]

CRAR Control ROM (Read-Only Memory) Address Register

CRAS Cost Reduction Alternative Study

CrAs Chromium Arsenide (or Chromic Arsenide)

CRB Change Review Board

CRB Community Reference Bureau (Database) [of European Atomic Energy Community]

CRB Customer Records and Billing System

CrB Chromium Boride (or Chromic Boride)

CrB Corona Borealis [Astronomy]

CrB₂/C Chromium Diboride/Carbon (Multilayer)

CRBDI Control Red Bank Demand Indicator

CRBL Charles River Breeding Laboratories [US]

CRBR Clinch River Breeder Reactor [of Tennessee Valley Authority, US]

CRBR Controlled Recirculation Boiling-Water Reactor

CRBR-CX Clinch River Breeder Reactor–Critical Assembly [of Tennessee Valley Authority, US]

CRBRP Clinch River Breeder Reactor Plant [of Tennessee Valley Authority, US]

CRBRP Clinch River Breeder Reactor Project [of Tennessee Valley Authority, US]

Cr(Bz)₂ Bis(benzene) Chromium

CRC CAD (Computer-Aided Design) Resource Center [Scarborough, Ontario, Canada]

CRC Camera-Ready Copy [also crc]

CRC Career Resource Center

CRC Carriage Return Character

CRC Carriage Return Contact

CRC Chemical Referral Center [US]

CRC Chemical Rubber Company [US]

CRC Civil Rights Commission [US]

CRC Clinical Research Center [UK]

CRC Communication Relay Center

CRC Communications Regulatory Commission [US]

CRC Communications Research Center [of Department of Communications, Canada]

CRC Composing Reducing Camera

CRC Computer Robot Control

CRC Confederation of Roofing Contractors [UK]

CRC Contractor Recommended Code

CRC Control and Reporting Center

CRC Cooperative Research Center

CRC Coordinating Research Council [US]

CRC Copy Research Council

CRC Corporate Research Center

CRC Corrosion Research Center [of University of Minnesota, Minneapolis, US]

CRC Costa Rican Colon [Currency]

CRC Cost Reduction Curve

CRC Counter-Rota(tion) Cutter [Microanalysis]

CRC Critical Reactor Component

CRC Cumulative Results Criterion

CRC Cyclic Redundancy Check

CR¢ Colón [Currency of Costa Rica]

CRCA Canadian Research Center for Anthropology

CRCA Canadian Roofing Contractors Association

CRCA Cold-Rolled, Close-Annealed; Cold Rolling and Close Annealing [Metallurgy]

CRCA Component Refurbishment Chemical Analysis

Crcb Crucible [also crcb]

CRCC Cyclic Redundancy Check Character

Cr₃C₂/C Chromium Carbide/Carbon (Multilayer)

CRC Crit Rev Biomed Eng CRC Critical Reviews in Biomedical Engineering [Published by CRC Press Inc., US]

CRC Crit Rev Mat Sci CRC Critical Reviews in Materials Science [Published by CRC Press Inc., US] [also CRC Crit Rev Mater Sci]

CRC Crit Rev Solid State Mat Sci CRC Critical Reviews in Solid State Materials Science [Published by CRC Press Inc., US] [also CRC Crit Rev Solid State Mater Sci]

CRC Food Sci Nutr CRC Food Science and Nutrition (Journal) [Published by CRC Press, Inc., US]

CRCGR Cyclic Redundancy Check Generator

CRCE Chicago Rice and Cotton Exchange [US]

Cr Chf Crew Chief [also CR CHF]

CRCIA Columbia River Comprehensive Impact Assessement [US]

CRCMC Continuously Reinforced Ceramic Matrix Composite

Crcmf Circumference [also crcmf]

CRCO Control Route Charges Office

Cr-Co Chromium-Cobalt (Alloy System)

CRCP Continuously-Reinforced Concrete Pavement

Cr(Cp)$_2$ Bis(cyclopentadienyl)chromium

CRCPD Conference of Radiation Control Program Directors [US]

Cr-Cr$_2$O$_3$ Chromium-Bonded Chromic Oxide (Cermet)

CRCS Canadian Red Cross Society

CRCSHM Cooperative Research Center for Southern Hemisphere Meteorology

CRCTA Composite Reactor Components Test Activity

CRCVD Constant Rate Chemical Vapor Deposition

CRD Capacitor-Resistor Diode

CRD Capital Regional District

CRD Card [Amateur Radio]

CRD Card Reader [Computers]

CRD Cathode Ray (Tube) Display

CRD Change Request Disposition

CRD Chronic Respiratory Disease

CRD CODATA (Committee on Data for Science and Technology) Referral Database

CRD Collaborative Research and Development (Grant Program) [of Natural Sciences and Engineering Research Council, Canada]

CRD Comité pour la Recherche et le Développement en Matière d'Energie [Committee on Energy Research and Development; of Organization for Econonic Cooperation and Development, France]

CRD Constant Ringing Drop

CRD Control Rod Drive [Nuclear Engineering]

CRD Cooperative Research and Development

CRD Corporate Research and Development (Center) [of General Electric in Schenectady, US]

.CRD Cardfile [File Name Extension] [also .crd]

.CRD Chord (Music) [File Name Extension]

Cr$ Cruzeiro [Currency of Brazil]

CRDA Control Rod Drive Assembly [Nuclear Reactors]

CRDA Control Rod Drop Accident [Nuclear Reactors]

CRDA Cooperative Research and Development Agreement

CRDB Caribbean Regional Development Bank

CRDC Cotton Research and Development Corporation

CRDEWPA Center for Research Documentation and Experimentation on Water Pollution Accidents [France]

CRDF Canadian Radio-Direction Finder

CRDF Cathode-Ray Direction Finder; Cathode-Ray Direction Finding

CR/DIR Change Request Directive

CRDL Chemical Research and Development Laboratories [US Army]

CRDM Control Rod Drive Mechanism [Nuclear Reactors]

CRDM Control Rod Drive Motor [Nuclear Reactors]

CRDME Committee for Research into Dental Materials and Equipment

CRDP Computer Resources Development Plan

CRDQ Controlled Rolling, Direct Quenching (Steel)

CRDS Chemical Reactions Documentation Service [UK]

CRDSD Current Research and Development in Scientific Documentation

CRE Centro Ricerche Energetiche [Energy Research Center, Italy]

CRE Chemical Reaction Engineering

CRE Coal Research Establishment [of National Coal Board, UK]

CRE Commander Royal Engineers

CRE Constant Rate-of-Extension (Tensile Testing Machine)

CRE Controlled Residual Element

CRE Current Ring End

CREA Committee on the Relation of Electricity to Agriculture [US]

CREAM Canadian Research on Exposure Assessment Modelling

CREAM Comprehensive Risk Evaluation and Management

CREaMRC Contaminant Risk Evaluation and Management Risk Center [of US Department of Energy]

Creat Creative; Creativity [also creat]

Creat Comp Creative Computing [US Magazine]

CREATE Chalk River Experiment to Assess Tritium Emission [of Chalk River Nuclear Laboratories, Ontario, Canada]

Creat Innov Netw Creative and Innovation Network [Publication of the University of Manchester, UK]

Creat Innov Yearb Creative and Innovation Yearbook [Publication of the University of Manchester, UK]

CREB Conservation and Renewable Energy Board

CREC Centro Ricerche Energetiche Casaccia [Casaccia Energy Research Center, Rome, Italy] [also CRE Casaccia]

CREDA Conservation and Renewable Energy Demonstration Agreement

CREDAS Centro Regional de Datos Satelitales [Regional Satellite Data Center, Argentina]

Cr(EDA)$_3$Cl$_3$ Tris(ethylenediamine)chromium Trichloride

CREDO Central Reliability Data Organization

CREDOC Centre de Recherches et de Documentation sur la Consommation [Research and Documentation Center on Consumption, France]

CREE Cathode-Ray Excited Emission Spectroscopy

CREF Centro Ricerche Energia Frascati [Frascati Energy Research Center, Rome, Italy]

CREF College Retirement Equities Fund [US]

CREF Concrete Research and Education Foundation [US]

CREFAL Centro Regional para el Desarrollo del Educación Fundamental en América Latina [Regional Center for the Development of Fundamental Education in Latin America]

CRE Frascati Centro Ricerche Energia Frascati [Frascati Energy Research Center, Rome, Italy] [also CREF]

CREM Cast-Products with Refined-Grain-Structure by Electromagnetic Stirring [Metallurgy]

CREM.GP Centre de Recherche en Electrochimie Minerale et Genie des Procedes [Research Center for Mineral Electrochemistry and Process Engineering of Institut National Polytechnique de Grenoble, France]

CREN Computer Research Education Network

CREN Corporation for Research and Educational Networking [US]

CREO Conservation and Renewable Energy Office

CREOL Center for Research in Electro-Optics and Lasers [of University of Florida, Gainesville, US]

CREPM Centre de Recherche en Electronique et Photonique Moléculaires [Molecular Electronics and Photonics Research Center, University of Mons, Belgium]

CREPS Compagnie de Recherches et d'Exploitation de Pétrole au Sahara [Company for Research and Exploitation of Sahara Petroleum, France]

CREQ Center for Research and Environmental Quality

CRES Center for Resource and Environmental Studies [of Australian National University, Canberra]

CRES Certified Radiologic Equipment Specialist

CRES Chemical Reaction and Equilibrium Software

CRES Chinese Rare Earth Society [PR China]

CRES Corrosion-Resistant Steel

Cres Crescent [also cres]

CRESEX Cresol Extraction (Process) [Chemical Engineering]

CRESP Coding Region Expression Selection Plasmid [Genetics]

CRESS Center for Research in Experimental Space Science [Canada]

CRESS Central Regulatory Electronic Stenographic System

CRESS Combined Reentry Effort for Small Systems

CRESS Computerized Reader Enquiry Service System

CRESS Computer Reader Enquiry Service System

CRESST Cryogenic Rare Event Search with Superconducting Thermometers (Experiment) [of Laboratori Nazionali del Gran Sasso, Italy]

CREST Committee on Reactor Safety and Technology [of European Nuclear Energy Agency, France]

CRESTech Center for Research in Earth and Space Technology [of Ministry of Economic Development, Trade and Tourism, North York, Ontario, Canada]

CRESTS Courtauld's Rapid Extracting, Sorting and Tabulating System

CRESUF Centre de Recherche en Economie Spatiale de l'Université de Fribourg [Research Center in Spatial Economics at the University of Fribourg, Switzerland]

Cret Res Cretaceous Research [International Journal]

CRETC Combined Radiation Effects Test Chamber

CREVS Control Room Emergency Ventilation System

CREW Consortium of Regional Electrical Wholesalers [UK]

CREWEX Consortium for Research in Elastic Wave Exploration [of University of Calgary, Alberta, Canada]

CRF Cable Retransmission Facility

CRF Capital Recovery Factor

CRF Cave Research Foundation [US]

CRF Central Retransmission Facility

CRF Channel Replacement Furnace (Black)

CRF Clean Room Filled (Facility)

CRF Cloud Radiative Forcing

CRF Coincidence Rangefinder

CRF Compressor Research Facility

CRF Computer Retailing Forum [now Computer Dealers Forum, US]

CRF Conservation and Research Foundation [US]

CRF Control Relay Forward

CRF Correspondence Routing Form

CRF Corticotropin-Releasing Factor [Biochemistry]

CRF Cross-Reference Facility

CRF Cross-Reference File

CRF Cryptographic Repair Facility [US Military]

CRFA Canadian Restaurant and Food Services Association

CRF Black Channel Replacement Furnace Black [also CRF black]

CRF-BP Corticotropin-Releasing Factor–Binding Protein [Biochemistry]

Cr-Fe Chromium-Iron (Alloy System)

CRFI Custom Roll Forming Institute [US]

CrFol Crown Folio [Book size of 9.5" × 15"]

CRF-R Corticotropin-Releasing Factor Receptor [Biochemistry]

CRF-R2β Corticotropin-Releasing Factor Receptor Type-2β [Biochemistry]

CRFSA Canadian Restaurant and Food Services Association

CRG Catalytic Rich Gas (Process) [Chemical Engineering]

CRG Change Review Group

CRG Cineradiography

CRG Classification Research Group [UK]

CRG Clear Register (Function)

CRG Collaborative Research Grants (Program) [of NATO]

CRG Composites Research Group

CRG Cooperative Republic of Guyana

CRG Correspondence Review Group

Crg Carriage [also crg]

CrGa Chromium Gallium (Compound)

Cr-Ga Chromium-Gallium (System)

Crge Carriage [also crge]

CRGR Coalition for Responsible Genetic Research [US]

CRH Caliber-Radius-Head

CRH Channel Reconfiguration Hardware

CRH Cold-Rolled, Hard (Condition) [Metallurgy]

CRH Constant Rate of Heating

CRH Corticotropin-Releasing Hormone [Biochemistry]

CRH Council on Rural Health [of American Medical Association, US]

CrH Chromium Hydride

CRHA Canadian Railroad Historical Association

CRHA Canadian Retail Hardware Association

CR Hebd Séances Acad Sci I Comptes Rendus Hebdomadaire des Séances de l'Academie des Sciences, Série I [Proceedings of the Weekly Sessions of the Academy of Sciences, Series I, France]

CR Hebd Séances Acad Sci II Comptes Rendus Hebdomadaire des Séances de l'Academie des Sciences, Série II [Proceedings of the Weekly Sessions of the Academy of Sciences, Series II, France]

CR Hebd Séances Acad Sci III Comptes Rendus Hebdomadaire des Séances de l'Academie des Sciences, Série III [Proceedings of the Weekly Sessions of the Academy of Sciences, Series III, France]

CR Hebd Séances Acad Sci, Sér A Comptes Rendus Hebdomadaire des Séances de l'Academie des Sciences, Série A: Sciences Chimique [Proceedings of the Weekly Sessions of the Academy of Sciences, Series A, France]

CR Hebd Séances Acad Sci, Sér B Comptes Rendus Hebdomadaire des Séances de l'Academie des Sciences, Série B [Proceedings of the Weekly Sessions of the Academy of Sciences, Series B, France]

CR Hebd Séances Acad Sci, Sér C Comptes Rendus Hebdomadaire des Séances de l'Academie des Sciences, Série C [Proceedings of the Weekly Sessions of the Academy of Sciences, Series C, France]

CRHH Cold-Rolled, Half-Hard (Condition) [Metallurgy]

CR-hi Channel Request, High Priority

CRHSI Center for Research in the Hospitality Service Industry [now Center for Hospitality Research and Service, US]

CRI Canadian Research Institute

CRI Cancer Research Institute [of Sapporo Medical College, Japan]

CRI Carbohydrate Research Institute [Canada]

CRI Caribbean Research Institute

CRI Cement Research Institute

CRI Central Research Institute

CRI Central Research Institute [of the Electric Power Industry of Japan]

CRI Central Research Institute [of Fukuoka University, Japan]

CRI Centre de Recherches en d'Irradiations [Center for Irradiation Research, France]

CRI Ceramic Research Institute [Calcutta, India]

CRI Cold Rolled Iron

CRI Colloid Research Institute [Kitakyushu, Japan]

CRI Color Rendering Index

CRI Color Reproduction Indices

CRI Committee for Reciprocity Information

CRI Concentrated Rust Inhibitor

CRI Cray Research Inc. [US]

CRIB Computerized Resources Information Bank [of United States Geological Survey]

CRID Centre de Recherche et de l'Information pour le Dévéloppement [Center for Developmental Research and Information, France]

CRID Centro di Riferimento Italiano DIANE [Italian Reference Center for the Direct Information Access Network for Europe, EURONET–European Public Data Network]

CRIE Crossed Radioimmunoelectrophoresis [Biochemistry]

CRIEPI Central Research Institute of Electric Power Industry [Kanagawa, Japan]

CRIES Comprehensive Resource Inventory and Evaluation System

CRIF Centre de Recherche Scientifique et Technique de l'Industrie des Fabrications Métalliques [Scientific and Technical Research Center of the Metalworking Industry, Belgium]

CRIFC Columbia River Inter-Tribal Fish Commission [State of Oregon, US]

CRIFO Civilian Research, Interplanetary Flying Objects

CRIM Centre de Recherche Informatique de Montréal [Montreal Center for Information Research, Quebec, Canada]

CRIM Clinical Research Institute of Montreal [Quebec, Canada]

CRIM Composite Reaction Injection Molding

CRIME Censorship Records and Information, Middle East [US Military]

CRIMM Changsha Research Institute of Mining and Metallurgy [PR China]

CRIP Controlled Retracting Injection Point

CRIQ Centre de Recherche Industrielle du Québec [Quebec Center for Industrial Research, Canada]

CRIS Calibration Recall and Information System

CRIS Command Retrieval Information System

CRIS Cooperative Research Information System

CRIS Creep Isostatic Pressing [Metallurgy]

CRIS Current Research Information System [of US Department of Agriculture]

CRISC (Combined) Complex and Reduced Instruction Set Computer

CRISP Car Radio Industry Specialists Association [UK]

CRISP Center for Remote Imaging, Sensing and Processing [Singapore]

CRISP Computer Resources Integrated Support Plan

CRISP Computer Retrieval of Information on Scientific Projects

CRISP Cooperative Research Information System in Physics (Workshop)

CRISPE Computerized Retrieval Information Service on Precision Engineering [of Cranfield Institute of Technology, UK]

CRISTAL Contract Regarding a Supplement to Tanker Liability for Oil Pollution

Crit Critic; Critical(ity); Criticism [also crit]

CRITIC Chalk River In-Reactor Tritium Instrumented Capsule [of Chalk River Nuclear Laboratory, Ontario, Canada]

CRITICOMM Critical Intelligence Communications System [US Air Force]

Crit Rev Environ Cont Critical Reviews in Environmental Control [US]

Crit Rev Microbiol Critical Reviews in Microbiology [US]

Crit Rev Surf Chem Critical Reviews in Surface Chemistry [US]

CRJE Conversational Remote Job Entry

Crk Crank [also crk]

Cr-K EXAFS Chromium-Potassium Extended X-Ray Absorption Fine Structure

CRKA Community Right to Know Act [US]

CrKα Chromium K-Alpha (Radiation) [also CrK$_\alpha$]

Crkc Crankcase [also crkc]

Cr-K XAFS Chromium-Potassium X-Ray Absorption Fine Structure

CRL Cambridge Research Laboratories [of US Air Force]

CRL Casting Research Laboratory [of Waseda University, Tokyo, Japan]

CRL Center for Research in Librarianship [of University of Toronto, Canada]

CRL Center for Research Libraries [US]

CRL Central Research Laboratory

CRL Central Research Laboratory [of Hitachi Ltd., Japan]

CRL Central Research Laboratory [of Texas Instruments, Dallas, US]

CRL Ceramic Research Laboratory [of Nagoya Institute of Technology, Japan]

CRL Certified Reporting Limit

CRL Chalk River Laboratories [of Atomic Energy of Canada Limited, Ontario, Canada]

CRL Chemical Research Laboratory [UK]

CRL Coal Research Laboratories [of Canada Center for Mineral and Energy Technology, Natural Resources Canada]

CRL Coherent Radiation Laboratory [US]

CRL Coincident Reciprocal (Crystal) Lattice [Crystallography]

CRL Communications Research Laboratory [Canada]

CRL Communications Research Laboratory [of Ministry of Post and Telecommunications, Tokyo, Japan]

CRL Constant Rate-of-Load(ing) (Tensile Test)

CRL Construction Robotics Laboratory [US]

CRL Corporate Research Laboratories [of Eastman Kodak Company, Rochester, US]

CRL Cosmic Radiation Laboratory [of Institute of Physical and Chemical Research, Saitama, Japan]

CRLA Canadian Railway Labour Association

CRLA College and Research Libraries Association

CR/LF Carriage Return/Line Feed

Cr:LiCAF Chromium-Doped Lithium-Calcium Aluminum Fluoride (Laser Material)

Cr:LiSAF Chromium-Doped Lithium-Strontium Aluminum Fluoride (Laser Material)

Cr:LiSGAF Chromium-Doped Lithium-Strontium Gallium-Aluminum Fluoride (Laser Material)

CR-lo Channel Request, Low Priority

CRLR Chemical and Radiological Laboratories [US Army]

CRM Canadian Risk Managers

CRM Centre de Recherches Métallurgiques [Metallurgical Research Center, Liège, Belgium]

CRM Centre de Recherche Minérales [Minerals Research Center, Quebec, Canada]

CRM Certified Record Manager

CRM Certified Reference Material

CRM Chemical Release Module

CRM Chemical Remanent Magnetization

CRM Cloud Resolving Model

CRM Collision Risk Model

CRM Comprehensive Resource Management

CRM Concurrent Resource Manager [Computers]

CRM Confusion Reflector Material

CRM Containment Radiation Monitor [Nuclear Engineering]

CRM Continuous Random Mat [Composites Engineering]

CRM Control and Reproducibility Monitor

CRM Coordinated Resource Management

CRM Cooperative Research Monographs [Series of Publications by the US Office of Education]

CRM Core Restraint Mechanisms [Nuclear Engineering]

CRM Corrosion-Resistant Material

CRM Counter-Radar Measures

CRM Counter-Radar Missile

CRM Count Rate Meter [Nuclear Engineering]

CRM Crew Resource Management

CRM Cross-Reacting Material

CRM Cyber Record Manager

CRMA Canadian Research Managers Association

CRMA Canadian Rock Mechanics Association

CRMA Commercial Refrigerator Manufacturers Association [US]

CRM/BAM Cyber Record Manager/Basic Access Method

CRMCC Centre de Recherche sur les Méchanismes de la Croissance Cristalline [Research Center for Crystal Growth Mechanisms, Marseilles, France] [also CRMC2]

CRMCC-CNRS Centre de Recherche sur les Méchanismes de la Croissance Cristalline–Conseil National de la Recherche Scientifique [Research Center for Crystal Growth Mechanisms of the French National Scientific Research Council located at Marseilles, France] [also CRMC2-CNRS]

CRMD Centre de Recherche sur la Matière Divisée [Research Center for Divided Matter, of Université d'Orléans, France]

CRMD Computer Resources Management Data

CRMD-CNRS Centre de Recherche sur la Matière Divisée–Conseil National de la Recherche Scientifique [Research Center for Divided Matter of the French National Scientific Research Council at the Université d'Orléans, France]

Cr/Me Chromium/Metal (Tape)

CR-med Channel Request, Medium Priority

CRMG Calgary Rock-Mechanics Group [Canada]

CRML Coalition for Responsible Mining Law [US]

CRMMC Carbon (Fiber) Reinforced Metal Matrix Composite

CRMMC Continuously Reinforced Metal Matrix Composite

CRM Metall Rep CRM Metallurgical Reports [Published by Centre de Recherches Métallurgiques Liège, Belgium]

CrMn Chromium-Manganese (Compound)

Cr-Mo Chromium-Molybdenum (Alloy System)

Cr Moly Chrome Molybdenum (Steel)

CRMP Coastal Resources Management Project [of the United States of America and the Association of Southeast Asian Nations]

CRMP Coordinated Resource Management Plan [US]

CRMR Continuous-Reading Meter Relay

CR-MRP Central Region–Mineral Resources Program [of US Geological Survey]

CRMS Continuous Repetitive Measurement of Spectra

CRN Cellular Radio Network

CRN Centre de Recherche Nucléaire [Nuclear Research Center, Strasbourg, France]

CRN Conflict Resolution Network

CRN Continuous Random Network

CRN Contract Revision Number

CrN Chromic Nitride

CRNA Certified Registered Nurse Anesthetist

CRNA Certified Registered Nurses Association

Cr-Nb Chromium-Niobium (Alloy System)

Cr-Ni Chromium-Nickel (Alloy System)

Cr-NiAl Nickel Aluminide Reinforced Chromium (Composite)

Cr:NiO Chromium-Doped Nickel Oxide [also Cr^{3+}:NiO]

CRNL Chalk River Nuclear Laboratories [of Atomic Energy of Canada Limited, Ontario]

CRNL-CRITIC Chalk River Nuclear Laboratories/Chalk River In-Reactor Tritium Instrumented Capsule [of Atomic Energy of Canada Limited]

CRO Calcium Ruthenium Oxide (or Calcium Ruthenate)

CRO Cathode-Ray Oscillograph; Cathode-Ray Oscilloscope [also cro]

CRO Centre de Recherche Océanographique [Oceanographic Research Center, France]

CRO Control Room Operator

CRO Copy Receipt Office

CRO Cosmic Ray Observatory

CrO Chromous Oxide

CrO_2 Chromyl (Radical) [Symbol]

CrO^- Chromate (Ion) [Symbol]

Cr_2O^- Dichromate (Ion) [Symbol]

Croat Croatia(n)

Croat Chim Acta Croatica Chimica Acta [Croatia]

Croat Med J Croatian Medical Journal

Cr-ODA Chromium Oxydianiline (Complex)

CR-OFC Corrosion-Resistant Oxygen-Free Copper

$CrO-Fe_2O_3$ Chromous Oxide-Ferric Oxide (System)

CROM Capacitive Read-Only Memory

CROM Control Read-Only Memory

CROP Cation Ring Opening Polymerization

$CrOPr^i$ Chromium Isopropoxide

CROS Capacitor Read-Only Storage

CROS Contra-Lateral Routing-of-Signal (Process)

CROSS Computerized Rearrangement of Subject Specialties [of Biosciences Information Service, US]

CROSSBOW Computerized Retrieval of Organic Structures Based on Wiswesser

Cross Pol Cross Polarization [also cross pol]

Crotonyl-S-ACP Crotonyl-S-Acyl Carrier Protein [Biochemistry]

CROW Counter-Rotating Optical Wave

CRP Capacity Requirements Planning

CRP Cape Roberts (Antarctic Drilling) Project [of BGR–Bundesanstalt für Geowissenschaften und Rohstoffe, Germany]

CRP Card Reader Punch

CRP Cast-Ribbon Process [Solar Cells]

CRP Chain-Reaction Polymerization

CRP Community Relations Plan

CRP (North Sea) Community Research Project [of Natural Environmental Research Council, UK]

CRP Compact Rail Production

CRP Configuration Requirements Processing

CRP Constant Rate of Penetration

CRP Control and Reporting Post

CRP Counterrotating Propeller

CRP Controlled Reliability Program

CRP Cooperative Research Project (Program) [of Research Institute of Electrical Communications, Tohoku University, Japan]

CRP Copper-Rich Precipitate

CRP Corrosion-Resistant Pump

CRP Cosmic-Ray Physics

CRP C-Reactive Protein [Biochemistry]

CrP Chromic Phosphide

Crp Creep [also Crp, or crp]

CRPA Cotton Research and Promotion Act [US]

CRPA C-Reactive Protein Antiserum [Biochemistry]

CRPC Canadian Research and Publication Center

CRPC Center for Research on Parallel Computation [of Rice University, Houston, Texas, US]

CRPE Centre de Recherches en Physiques de l'Environnement [Environmental Physics Research Institute, France]

Cr-PMDA Chromium Pyromellitic Dianhydride (Composite)

CRPL Central Radio Propagation Laboratory [now Institute for Telecommunication Sciences and Aeronomy, US]

CRPL Cosmic Ray Physics Laboratory [US]

Cr Pl Chromium Plate

CRPM Communication Registered Publication Memorandum

CRPMC Continuously Reinforced Polymer-Matrix Composite

$Cr(PMCp)_2$ Bis(pentamethylcyclopentadienyl)chromium

CRPO CANMET (Canada Center for Mineral and Energy Technology)/Research Program Office [of Natural Resources Canada]

CRPO Central Radio Propagation Office [US]

CRPO Continuous Rating Permitting Overload

CRPP Correlated Radical Pair Polarization

CRPPH Committee on Radiation Protection and Public Health [of Organization for Econo mic Cooperation and Development/Nuclear Energy Agency]

CRPQF Commisão Reguladora dos Productos Quimicos e Farmaceuticos [Regulatory Committee for Chemical and Pharmaceutical Products, Portugal]

Crpt Carport [also crpt]

CRPV Cottontail Rabbit Papilloma Virus

CRQ Conseil de Recherche du Québec [Research Council of Quebec, Canada]

CRQ Console Reply Queuing

CRR Cable Routing Rotation

CRR Cathode Ray Recording

CRR Center for Radiation Research [of National Institute of Standards and Technology, Gaithersburg, Maryland, US]

CRR Center for Renewable Resources

CRR Churchill Research Range [US Air Force]

CRR Computer Run Report

CRR Constant Relative Resolution

CRR Constant Ringing Relay

CRR Critical Requirements Review

CRR Customer Response Representative

CRREL Cold Regions Research and Engineering Laboratory [of US Army at Hanover, New Hampshire]

CRREL Monogr Cold Regions Research and Engineering Laboratory Monographs [of US Army]

CRREL Spec Rep Cold Regions Research and Engineering Laboratory Special Report [of US Army]

CRRI Central Rice Research Institute [India]

CRRI Colorado Resources Research Institute [of Colorado State University, Fort Collins, US]

CRRL Cosmic Ray Research Laboratory [of Nagoya University, Japan]

CRRP Community Rainforest Reforestation Program [Australia]

CRS Canadian Rocket Society

CRS Center for Radio Science [Canada]

CRS Center for Remote Sensing [UK]

CRS Center for Resource Studies [Canada]

CRS Centralized Results System

CRS Centre de la Recherche Scientifique [Scientific Research Center, France]

CRS Cerium Ruthenium Selenide

CRS Character Recognition System

CRS Chemical Registry System [of American Chemical Society, US]

CRS Chinese Rare-Earth Society [PR China]

CRS Chinese Restaurant Syndrome

CRS Clamp Rest Screw

CRS Climax Research Services [US]

CRS Coating Removal System

CRS Coasting Richer System

CRS Cold-Rolled Section

CRS Cold-Rolled Steel

CRS Command Retrieval System

CRS Commercially Rapidly Solidified [Powder Metallurgy]

CRS Common Rail System

CRS Computerized Retrieval Service

CRS Computer Recognition System

CRS Computer Reservation System (Network)

CRS Congressional Research Service [of Library of Congress, Washington, DC, US]

CRS Containment Rupture Signal

CRS Controlled Release Society [US]

CRS Conversion Resource (File)

CRS Coordinate Representation Sudden

CRS CO_2 Reduction Subsystem

CRS Corrosion-Resistant Steel

CRS Cosmic-Ray Scattering

CRS Crystal Spectrometer

CrS Chromous Sulfide

Crs Course [also crs]

Crs Cresol

Crs Cross [also crs]

Cr-S Chromium-Sulfur (System)

CRSA Cold-Rolled Sections Association [UK]

CRSA Control Rod Scram Accumulator [Nuclear Engineering]

Cr-Sb Chromium-Antimony (Alloy System)

CrsBl Cresol Blue

CRSC Center for Research on Scientific Communication [of Johns Hopkins University, Baltimore, Maryland, US]

Crse Course [also crse]

CrSe Chromium(II) Selenide

Crsfd Crossfeed

CRSI Ceramic Reusable Surface Insulation

CRSI Concrete Reinforcing Steel Institute [US]

CrSi Chromium Silicide

Cr-SiO$_2$ Chromium-Bonded Silicon Dioxide (Cermet)

CRSL Corporate Research Science Laboratory [of Exxon Research and Engineering, Annandale, New Jersey, US]

Cr-Sn Chromium-Tin (Alloy System)

CRSO Common Rolling Stock Organization

CRSOA County Road Safety Officers Association [UK]

CRSP Canadian Registered Safety Professional

CRSR Center for Radiophysics and Space Research [of Cornell University, Ithaca, New York, US]

CRSR Coherent Resonant Stokes Rotation

CRSS Critical Resolved Shear Stress [Mechanics]

CRST Cold-Rolled Steel

Crsvr Crossover

CRT Cathode-Ray Tube [also crt]

CRT Center for Research on Transportation [Canada]

CRT Center for Rural Transport [UK]

CRT Charactron Tube

CRT Circuit Requirement Table

CRT Clean Room Technology

CRT Constant Rate-of-Traverse (Tensile Testing Machine)

CRT Controlled Release Technology

CRT Cosmic Ray Telescope

CRT Crystal Research and Technology

Crt Crater [Astronomy]

C(r,t) Concentration on Site r at Time t [Symbol]

crt *(currentis)* – this year, month, etc.

CRTA Cost Recoverable Technical Assistance (Program) [Canada]

Cr-Ta Chromium-Tantalum (Alloy System)

Cr Tan Lthr Chromium Tanned Leather

CRTC Canadian Radio-Television and Telecommunications Commission

CRTC Cathode-Ray Tube Controller

CRTC Clay Roofing Tile Council [UK]

CRTD Cathode-Ray Tube Drive

CRTEE Canadian Roundtable on Environment and Energy

Crtg Cartridge [also crtg]

Cr-Ti Chromium-Titanium (Alloy System)

CRTK Community Right-to-Know (Laws) [US]

Cr(TMHD)₃ Chromium Tris(2,2,6,6-Tetramethylheptane-dionate) [also Cr(tmhd)₃]

Cr:Tm:YAG Chromium- and Thulium-Doped Yttrium Aluminum Garnet [also $Cr^{3+}:Tm^{3+}:YAG$]

CRTOS Cathode-Ray Tube Operating System

CR&TP Committee on Research and Technical Planning [of American Society for Testing and Materials, US]

CRTPB Canadian Radio Technical Planning Board

CRTS Constant Return to Scale

CRTS Controllable Radar Target Simulator

CRTT Certified Respiratory Therapy Technician

CRTU Combined Receiving and Transmitting Unit

CRTV Composite Reentry Test Vehicle

CRTVS Central Radio and Television School

CRU Card Reader Unit

CRU Centre de Recherches d'Ugine [Ugine Research Center, France]

CRU Combined Rotating Unit

CRU Commodities Research Unit [UK]

CRU Conformational Repeat Unit

CRU Cooperative Research Unit

CRU Cryogenic Unit

Cru Crux [Astronomy]

CRUC Cooperative Research Unit Center [US]

Cruc Crucible [also cruc]

CRUD Chalk River Unidentified Deposit [Ontario, Canada]

CRUD Create, Retrieve, Update, Delete

Cruis Cruising [also cruis]

CRUS Center for Research on User Studies [of University of Sheffield, UK]

CRU-SEM Cryogenic Unit Scanning-Electron Microscope

CRV Carnation Ringspot Virus

CRV Centre de Recherches de Voreppe [of Pechiney Corporation, France]

CRV Contact Resistance Variation

Cr-V Chromium-Vanadium (Alloy System)

Crv Corvus [Astronomy]

Crv Curve [also crv]

CRVA Canadian Recreational Vehicle Association

Cr Van Chrome Vanadium (Steel)

Crv Dwg Curve Drawing

CRVM Cathode-Ray Voltmeter

CRW Cladding Removal (Nuclear) Waste

CRW Community Radio Watch

CRW Control Read/Write

CRW Correlated Random Walkers

CRWC Connecticut River Watershed Council [US]

CRWI Coalition for Responsible Waste Incineration [US]

CRWM Committee on Radioactive Waste Management [of National Academy of Science, National Research Council, US]

CRWMP Civilian Radioactive Waste Management Program [of US Department of Energy at Lawrence Livermore National Laboratory, Livermore, California, US]

CRWO Coding Room Watch Officer [US Navy]

CRWPC Canadian Radio Wave Propagation Committee

Cry Crystal(line) [also cry]

Cr:YAG Chromium-Doped Yttrium Aluminum Garnet (Laser) [also $Cr^{3+}:YAG$]

Cryo Cryogenic(s) [also cryog]

Cryobiol Cryobiologist; Cryobiologist

Cryobiol Cryobiology [Journal published in the US]

Cryog Cryogenic(s) [also cryog]

Cryog Mater Ser Cryogenic Materials Series [Publication of the International Cryogenic Materials Conference, US]

Cryo TEM Cryo(genic) Transmission Electron Microscopy [also Cryo-TEM, cryo TEM, or cryo-TEM]

Cryomilling Cryogenic Ball Milling [also cryomilling]

crypt cryptand[2.2.2] [Chemistry]

Cryst Crystal(line); Crystallize [also cryst]

Crystal Crystallographer; Crystallography [also Crystallogr, crystallogr, or crystal]

Crystallogr Rep Crystallography Reports [Russian Translation Journal jointly published by the American Institute of Physics (US) and Maik Nauka (Russia)]

crystd crystallized [also cryst'd]

CRYSTDAT Crystal Data Identification File [of National Institute for Standards and Technology, Gaithersburg, Maryland, US] [also CDIF]

Crystg Crystallizing [also crystg]

CRYSTIN Inorganic Crystal Structure Database [of Canada Institute for Scientific and Technical Information, Canada] [also ICSD]

Cryst Latt Def Crystal Lattice Defects [Journal published in the US]

Cryst Latt Def Amorph Mater Crystal Lattice Defects and Amorphous Materials [Journal published in the US]

CRYSTMET NRC (National Research Council) Metals Crystallographic Data File [of Canada Institute for Scientific and Technical Information]

Crystn Crystallization [also crystn]

CRYSTOR Cambridge Crystallographic Database [of Canada Institute for Scientific and Technical Information]

Cryst Prop Prep Crystal Properties and Preparation [Journal published in the Switzerland]

Cryst Rep Crystallography Reports [Russian Translation Journal jointly published by the American Institute of Physics (US) and Maik Nauka (Russia)]

Cryst Res Technol Crystal Research and Technology [Journal published in the US]

Crypto Cryptographer; Cryptographic; Cryptography [also crypto]

CRYPTONET Crypto-Communications Network [also Cryptonet]

CRYSYS Cryospheric System

CS Cableship

CS Cadmium Stannate

CS Calahan and Snider [Physics]

CS Calcium Silicate (or Wollastonite)

CS Calls per Second [also CPS or cps]

CS Calogero-Sutherland (Model)

CS Canadian Shield
CS Capital Stock [also cs]
CS Capsule Suspension
CS Carbon and Sulfur (Analyzer)
CS Carbon Monosulfide
CS Carbon Steel
CS Card Stack(er)
CS Caribbean Sea
CS Carrier System
CS Cascade Stack
CS Casein
CS Cassette
CS Cast Steel
CS Catalysis Society [US]
CS Cathode Spot
CS Cathode Sputtering
CS Celestial Sphere
CS Cell Size
CS Center of Suspension
CS Center Section
CS Centers Surface
CS Centrifugal Separator
CS Centro-Symmetry
CS Cerebrospinal
CS Cesarean Section
CS Chain Site [also cs] [Chemistry]
CS Change Status
CS Channel Status
CS Characteristic Spectrum
CS Chartered Ship Broker
CS Chartered Surveyor
CS Checkout Station
CS Check Sorter
CS Check Surface [Numerical Control]
CS Chemical Science(s)
CS Chemical Sensor
CS Chemical Society [UK]
CS Chemical Spill
CS Chemical System
CS Chemisorption; Chemisorptive
CS Chemosynthesis; Chemosynthetic [Biochemistry]
CS Chief of Staff
CS Chief Scientist
CS Chip Select
CS Chondroitin Sulfate [Biochemistry]
CS Christian Science; Christian Scientist
CS Circuit Switching
CS Circulator Speed [Nuclear Power Stations]
CS Circumsporozoite [Biology]
CS Cirrostratus (Cloud) [also Cs]
CS Citrate Synthase [Biochemistry]
CS Citrate Synthetase [Biochemistry]
CS Civil Service

CS Clamp Strap
CS Class of Service
CS Clear to Send
CS Closed System [Thermodynamics]
CS Cloud Fractional Coverage from Satellite
CS Coal Store
CS Coblentz Society (for Spectroscopy) [US]
CS Code Segment
CS Coding Specification
CS Coincidence (Crystal Lattice) Site [Materials Science]
CS Cold Stage [Microscopy]
CS Cold Storage
CS Colepterists Society [US]
CS College of Science
CS Collimator Stop
CS Collision Sequence
CS Colloidal Stability
CS Colloid Science
CS Colorado Springs [US]
CS Color Science
CS Color Specification
CS Color Symmetry
CS Color System
CS Combustion Science
CS Combustion Spectrum
CS Combustion Synthesis (Process) [Metallurgy]
CS Commercial Standard
CS Commercial System
CS Common Set
CS Common Shell
CS Common Source [Electronics]
CS Commonwealth Secretariat [UK]
CS Communications Services
CS Communications Simulator
CS Communication(s) Satellite
CS Communication Station
CS Communication Status
CS Communication System
CS Community Service
CS Complex Structure
CS Compositional Separation
CS Composition Service
CS Compression Spring
CS Compression Shock
CS Compression Stroke
CS Compressive Strength
CS Compressive Stress
CS Compton Scattering [Physics]
CS Computer Science
CS Computer Simulation
CS Computer Society [of Institute of Electrical and Electronics Engineers, US]
CS Computer System
CS Concentric Spheres

CS Concrete Slab

CS Concrete Society [UK]

CS Conditioned Stimulus

CS Condensed Solid

CS Conducted Susceptibility

CS Conformation Substrate

CS Connecting Segment

CS Conservation Science

CS Constitutional Supercooling [Metallurgy]

CS Consumables Status [also cs]

CS Contact Section

CS Containment Spray [Nuclear Engineering]

CS Continental Shelf

CS Continue-Specific Mode

CS Continuous Spectrum

CS Continuous System

CS Controlled Substance

CS Control Section

CS Control Segment

CS Control Set

CS Control Stick

CS Control Store

CS Control Switch

CS Control System

CS Control Systems (Society) [of Institute of Electrical and Electronics Engineers, US]

CS Conventional System

CS Conversion Start

CS Conveyors Section [of Materials Handling Institute, US]

CS Cooling Stage

CS Cooling System

CS Coordinate System

CS Coqblin-Schrieffer (Model) [Physics]

CS Core Sample [Geology]

CS Core Segment

CS Core Shift

CS Core Storage

CS Coriell-Sekerka (Rapid Solidification Model) [Metallurgy]

CS Corporate Source

CS Corrected Speed

CS Corrosion Scientist; Corrosion Science

CS Corticosterone [Biochemistry]

CS Cottrell-Stokes (Law) [Physics]

CS Countermeasure Simulator

CS County Seat

CS Coupled States [Physics]

CS Crew Station [also cs]

CS Critical Speed

CS Cross [Pipe Fittings]

CS Cross Section(al)

CS Cross-Slip (Model) [Metallurgy]

CS Cruising Speed

CS Crystalline Solid

CS Crystal(line) Symmetry

CS Crystallographic Shear

CS Crystal Spectrometer; Crystal Spectrometry

CS Currency Sign

CS Current Strength

CS Cutting Speed

CS Cycle Shift

CS D-Cycloserine [Biochemistry]

CS Cyclostationary

CS Cyclosynchrotron

CS Czechoslovakia [ISO Code]

CS Single-Cotton, Single-Silk (Electric Wire)

CS Spacelab Operations Directorate [of NASA]

CS STS (Space Transportation System) Cargo Operations [NASA Kennedy Space Center Directorate, Florida, US]

CS- Portugal [Civil Aircraft Marking]

CS_2 Carbon Disulfide (Gas)

CS_3 Thiocarbonate

C_2S Dicalcium Silicate [$2CaO \cdot SiO_2$]

C_3S Tricalcium Silicate [$3CaO \cdot SiO_2$]

C_3S_2 Tricalcium Disilicate [$3CaO \cdot 2SiO_2$]

C_4S Chemical Shift–Specific Slice Selection (Method) [Physics]

CS1 Global Coordinate System (X*, Y*, Z*) [Crystallographic Transformation]

CS2 Crystallographic Coordinate System (X, Y, Z) [Crystallographic Transformation]

CS3 Local Coordinate System (n, m, q) [Crystallographic Transformation]

C&S Colloid and Surface (Science Symposium)

C&S Computers and Systems

C/S Calcium Oxide-to-Silicon Dioxide (Ratio)

C/S Call Signal

C/S Cases

C/S Chief of Staff

C/S Client/Server [Internet]

C/S Counts per Second

C/S Cycles per Second [also c/s]

Cs Cesium [Symbol]

Cs Cirrostratus (Cloud)

Cs Humid Warm Mediterranean or Subtropical Climate, Dry Summer [Classification Symbol]

Cs^+ Cesium Ion [Symbol]

Cs-125 Cesium-125 [also ^{125}Cs or Cs^{125}]

Cs-127 Cesium-127 [also ^{127}Cs or Cs^{127}]

Cs-128 Cesium-128 [also ^{128}Cs or Cs^{128}]

Cs-129 Cesium-129 [also ^{129}Cs or Cs^{129}]

Cs-130 Cesium-130 [also ^{130}Cs or Cs^{130}]

Cs-131 Cesium-131 [also ^{131}Cs or Cs^{131}]

Cs-132 Cesium-132 [also ^{132}Cs or Cs^{132}]

Cs-133 Cesium-133 [also ^{133}Cs, Cs^{133} or Cs]

Cs-134 Cesium-134 [also ^{134}Cs or Cs^{134}]

Cs-135 Cesium-135 [also ^{135}Cs or Cs^{135}]

Cs-136 Cesium-136 [also ^{136}Cs or Cs^{136}]

Cs-137 Cesium-137 [also ^{137}Cs or Cs^{137}]

Cs-138 Cesium-138 [also ^{138}Cs or Cs^{138}]

Cs-139 Cesium-139 [also ^{139}Cs or Cs^{139}]

Cs-140 Cesium-140 [also ^{140}Cs or Cs^{140}]

Cs-141 Cesium-141 [also ^{141}Cs or Cs^{141}]

Cs-142 Cesium-142 [also ^{142}Cs or Cs^{142}]

Cs-143 Cesium-143 [also ^{143}Cs or Cs^{143}]

Cs-144 Cesium-144 [also ^{144}Cs or Cs^{144}]

C(s) Controlled Variable [Symbol]

C*(s) Surface Carbon Radical [Symbol]

°C/s Degrees Celsius (or Centigrade) per Second [also °C/sec, °C s^{-1} or °C sec $^{\circ}$1]

cS centistoke [Unit]

cs case

cs chain site [also CS] [Chemistry]

c/s cases

c/s cycles per second [also C/S]

CSA Calendaring and Scheduling API (Application Program Interface)

CSA Called Subscriber Answer

CSA Cambridge Scientific Abstracts [US Database] [also Cam Sci Abstr, or Camb Sci Abstr]

CSA Camphor Sulfonic Acid

CSA Canadian Science Association

CSA Canadian Shipowners Association

CSA Canadian Space Agency

CSA Canadian Standards Association

CSA Carbon and Sulfur Analyzer

CSA Caribbean Shipping Association

CSA Cast Section Angle

CSA Celestial Sensor Assembly [of NASA Mars Global Surveyor]

CSA Center for Sustainable Agriculture [US]

CSA Centrifugal Sudden (Coupled States) Approximation [Physics]

CSA Chemical Shift Anistropy

CSA Chemical Sources Association [US]

CSA Chemical Structure Association [UK]

CSA Chief of Staff of the Army [US Department of Defense]

CSA Chiral Solvating Agent

CSA Chondroitin Sulfonic Acid [Biochemistry]

CSA Cold Starting Aid

CSA Colorado Safety Association [US]

CSA Common Service Area

CSA Common System Area

CSA Communications Satellite Act [US]

CSA Community Service Activities [of American Federation of Labor and Congress of Industrial Organizations, US]

CSA Community Service Administration

CSA Computer Science Association

CSA Computer System Architecture

CSA Computer Systems Association [US]

CSA Computing Services Association [UK]

CSA Confederate States of America

CSA Conference on Superconductivity and Applications

CSA Continental Shelf Act [UK]

CSA Contract Services Association (of America) [US]

CSA Controlled Separation Amplifier

CSA Council of Supervisory Associations [US]

CSA Cross-Sectional Area

CSA Cryogenic Society of America [US]

CSA Current Science Association [India]

CSA Cyclic Strain Attenuator

CSA Cytisus Sessilifolius Agglutinin [Biochemistry]

CSA Czech(oslovakian) Airlines

Csa Mediterranean or Dry Summer Subtropical Climate, Warmest Month above 71.6°F (22°C), Mild Climate with Summer Drought and Winter Rain [Classification Symbol]

CSAA Canadian Society of Applied Art

CSAB Center for the Study of Anorexia and Bulimia [New York City, US]

CSAC Council on Superconductivity for American Competitiveness [US]

CsAc Cesium Acetate

CSACIS Center for the Study of Arms Control and International Security [UK]

CSACPS Canadian Society of Aesthetic Cosmetic Plastic Surgery

CSAE Canadian Society of Agricultural Engineers

CSAF Chief of Staff of the Air Force [US Department of Defense]

CSAGI Comité Spécial de l'Année Geophysique Internationale [Special Committee of the International Geophysical Year]

CSAI Compressive Strength After Impact

CSAM Canadian Society of Aviation Medicine

CSAM Circular Sequential Access Memory

CSAM C-Mode Scanning Acoustic Microscope; C-Mode Scanning Acoustic Microscopy [also C-SAM]

CSAM Conventional Scanning Acoustic Microscopy

CSAM Crinkled Single Aluminized Mylar

CSAMT Controlled Source Audiofrequency Magnetotellurics

CSAMT Controlled Source Audio-Magnetotelluric Tensor

CSAO Construction Safety Association of Ontario [Canada]

CSAO Council of Safety Associations of Ontario [Canada]

CSAP Canadian Society of Animal Production

CSAR Canadian Space Agency Rocket

CSAR Canadian Synthetic Aperture Radar

CSAR Channel System Address Register

CSAR Communications Satellite Advanced Research

CSAR Computational Sequence Analysis Resource

CSAR Control Store Address Register

CSAR-1 Canadian Space Agency Rocket One

CSAR-CSCI Canadian Synthetic Aperture Radar–Computer Software Configuration Item

CSAS Classical Scattering Aerosol Spectrometer

CSAS Command and Stability Augmentation System

CSAS Computerized Status Accounting System

CSAS Cove Standby Actuation Signal [US]

CsAs Cesium Arsenide (Semiconductor)

CSASI Canadian Society of Air Safety Investigators

CSA-SSRP Canadian Space Agency–Space Science Radioastronomy Program

CSASTS Czech and Slovak Associations of Scientific-Technical Societies [formerly CASTS]

CSAT Combined Systems Acceptance Test

CSAUDP Canadian Space Agency User Development Program

CSB Called Subscriber Busy

CSB Canada Service Bureau

CSB Carrier and Sideband

CSB Coarse Slip Band [Metallurgy]

CSB Coherent Surface Bremsstrahlung

CSB Companies Services Branch [of Ministry of Consumer and Commercial Relations, Canada]

CSB Concentrate Storage Building

CSB Concrete Splash Block

CSB Consumer Sounding-Board

CSB Customer Services Branch [of Division of Computer Research and Technology, National Institutes of Health, Bethesda, Maryland, US]

Csb Mediterranean or Dry Summer Subtropical Climate, Warmest Month below 71.6°F (22°C) [Classification Symbol]

CSBA Calcium Silicate Brick Association [UK]

CSBCO Calcium Strontium Barium Copper Oxide (Superconductor)

CSBE Chinese Society of Biomedical Engineering [PR China]

CSBH Channeled-Substrate Buried Heterostructure (Laser)

CSBISSS Commission on Soil Biology of the International Society of Soil Science [Netherlands]

CSBMCB Canadian Society for Biochemistry and Molecular and Cellular Biology

CsBr Cesium Bromide

CSC Cadmium-Sulfide Cell

CSC Calcium Strontium Cuprate

CSC California Science Center [US]

CSC California State College [Bakersfield, US]

CSC Cambridge School Certificate

CSC Camphor–Sparteine–Creosote

CSC Canada Safety Council

CSC Canadian Society for Chemistry

CSC Canadian Spectroscopy Conference

CSC Canadian Symposium on Catalysis

CSC Cartridge Short Case

CSC Cartridge Storage Case

CSC Cascade Synchronous Speed

CSC Catalytic Surface Conversion

CSC Center for Scientific Computing [Finland]

CSC Center for Superconductivity [of University of Illinois at Urbana-Champaign, US]

CSC Central State College [Wilberforce, Ohio, US]

CSC Central State College [Edmond, Oklahoma, US]

CSC Central Switching Center

CSC Centrifugal Shot Casting

CSC CERN School of Computing [of CERN–European Laboratory for Particle Physics, Geneva, Switzerland]

CSC Chadron State College [Nebraska, US]

CSC Cheyney State College [Pennsylvania, US]

CSC Chico State College [California, US]

CSC Chief Sector Control

CSC China Steel Corporation [Kaohsiung, Taiwan]

CSC Circuit Switching Center

CSC Cis-Syn-Cis (Configuration) [Chemistry]

CSC Civil Service Commission [US] [Abolished]

CSC Clarion State College [Pennsylvania, US]

CSC Coastal Services Center [of National Oceanic and Atmospheric Administration at Charleston, South Carolina, US]

CSC Colloid and Surface Chemistry

CSC Colorado State College [Greeley, US]

CSC Color Sub-Carrier

CSC Commercial Space Centers [of NASA– Office of Life and Microgravity Sciences and Applications]

CSC Commercial Synchronous Communication(s)

CSC Common Signalling Channel

CSC Commonwealth Science Council

CSC Commonwealth Scientific Committee

CSC Commonwealth Scientific Corporation [US]

CSC Communications Satellite Corporation

CSC Communications Systems Center

CSC Computer Science Corporation [at NASA Ames Research Center, Mountain View, California, US]

CSC Computer-Services Companies

CSC Computer Society of Canada

CSC Computer Support Coordinator

CSC Conical Shaped Charge [also csc]

CSC Conserved Successive Collision (Model) [Physics]

CSC Constant Structure Creep [Mechanics]

CSC Construction Specifications Canada

CSC Contingency Support Center [of Cape Canaveral Air Force Station, Florida, US]

CSC Core Standby Cooling [Nuclear Reactors]

CSC Course and Speed Computer

CSC Cube-Surface Coil

CSC Customer Support Center

CsC$_n$ Graphitide [General Formula; $n = 8, 24, 36$, etc.]

CsC$_8$ Stage 1 Cesium Graphitide

CsC$_{24}$ Stage 2 Cesium Graphitide

CsC$_{35}$ Stage 3 Cesium Graphitide

CsC$_{56}$ Stage 4 Cesium Graphitide

csc cosecant [also cosec]

csc^{-1} inverse cosecant [also arccsc]

CSCB Contractor's Summary Cost Breakdown

CSCC Canadian Society of Clinical Chemists

CSCC Canadian Steel Construction Council

CSCC Certificate, Society of Clinical Chemists

CSCC Chloride Stress Corrosion Cracking

CSCC Commonwealth-State Consultative Committee

CSCC Cumulative Sum Control Chart

CSCCO Chromium Strontium Calcium Copper Oxide (Superconductor)

CSCD Committee on Sugar Cane Diseases [of Hawaiian Sugar Planters Association, US]

CSCD Common Signalling Channel Demodulator

CSCE Canadian Society for Chemical Engineering

CSCE Canadian Society for Civil Engineering

CSCE Coffee, Sugar and Cocoa Exchange [New York, US]

CSCE Committee on Security and Cooperation in Europe

CSCE Communications Subsystem Checkout Equipment

CSCE Conference on Security and Cooperation in Europe [of NATO]

CSCE Connecticut Society of Civil Engineers [US]

CSCE/FACT Conference on Security and Cooperation in Europe/Fact-Finding and Dispute Resolution

CSCF California State College at Fullerton [US]

CSCF Complex Self-Consistent Field [Physics]

CSCFN Canadian Science Committee on Food and Nutrition

CSCH California State College at Hayward [US]

csch hyperbolic cosecant

$csch^{-1}$ inverse hyperbolic cosecant [also arcsch]

CSChE Canadian Society for Chemical Engineering

CSchEC Canadian Schools of Engineering Conference

CSCIDC Chlorosulfonylcarbonimidoyl Dichloride

CSCI Computer Software Configuration Item

CsCl Cesium Chloride (Structure)

CSCLA California State College at Los Angeles [US]

CSCM Common Signalling Channel Modem

CsCN Cesium Cyanide

CSCNO Cerium Strontium Copper Niobium Oxide (Superconductor)

CSCO Calcium Strontium Copper Oxide (Superconductor)

CSCO Cerium Strontium Cobalt Oxide

CSCP Centro di Studio per la Chimica dei Plasmi [Plasma Chemistry Research Center, Bari, Italy]

CSCP Chinese Society for Corrosion Protection [of Beijing University of Iron and Steel Technology, PR China]

CSCP-CNR Centro di Studio per la Chimica dei Plasmi–Consiglio Nazionale delle Ricerche [Plasma Chemistry Research Center of the National Research Council at Bari, Italy]

CSCR Central Society for Clinical Research [US]

CSCR Complementary Silicon-Controlled Rectifier

CSCRF Computer System for Crop Response to Fertilizers [of Food and Agricultural Organization]

CSCS Common Signalling Channel Synchronizer

CSCS Core Standby Cooling System [Nuclear Reactors]

CSCSAT Commercial Synchronous Communication Satellite

C/SCSC Cost/Schedule Control Systems Criteria

CSCSI Canadian Society for Computational Studies of Intelligence

CSCT Canadian Society for Chemical Technology

CSCW Computer Supported Cooperative Work

C_2S-CZ Dicalcium Silicate–Calcium Zirconate

CSD Cambridge Structural Database

CSD Census Subdivision

CSD Centrifugal(ly) Spray Deposited; Centrifugal Spray Deposition

CSD Charge State Distribution [Solid-State Physics]

CSD Chartered Society of Designers [UK]

CSD Chemical Systems Division [of NASA]

CSD Circuit-Switched Data

CSD Civil Service Department [UK]

CSD Cluster Size Distribution [Physics]

CSD Cold (Reactor) Shutdown

CSD Commission on Sustainable Development [of International Union for the Conservation of Nature and Natural Resources]

CSD Communication Systems Division

CSD Composite Structures Division [of Alcoa Corporation, US]

CSD Computer Science Department

CSD Computer Science Division

CSD Computer Services Department

CSD Conservation and Survey Division [Nebraska, US]

CSD Constant Speed Drive

CSD Constant-Stress-Difference (Test)

CSD Controlled-Slip Differentials [Metallurgy]

CSD Controlled Spontaneous-Emission Diode

CSD Controlled Spray Deposition

CSD Control System Development

CSD Corrective Service Diskette

CSD Crack Sliding Displacement [Mechanics]

CSD Crew Systems Division

CSD Critical Solvent De-Ashing (Process)

CSD Crystallite Size Distribution [Crystallography]

CSDA Canadian Soft Drink Association [US]

CSDA Concrete Sawing and Drilling Association [US]

CSDB Continuous Seam Diffusion Bonding [Metallurgy]

CSDC Circuit Switched Digital Capability

CSDC Circuit Switched Digital Circuitry

CSDC Commonwealth Spatial Data Committee

CSDD Conceptual System Design Description

CSDD Control Systems Development Division [of NASA Johnson Space Center, Houston, Texas, US]

CSDD Cornell Scale for Depression in Dementia [of Cornell University, Ithaca, New York, US]

CSDF Central Source Data File

CSDF Cold Storage and Distribution Federation

CSDF Core Segment Development Facility

CSDL Charles Stark Draper Laboratory [of Massachusetts Instutite of Technology, Cambridge, US]

CSDL Current Switching Diode Logic

CSDM Continuous Slope Delta Modulation

CSDN Circuit Switched Data Network

CSDR Control Store Data Register

CSDS Circuit Switched Data Service

CSDS Community Surface Drainage Scheme [of Victoria State, Australia]

CSDT Computer Software Data Tapes

CSDW Centrifugal-Sudden Distorted-Wave (Method) [Solid-State Physics]

CSDW Commensurate Spin-Density Wave [also C SDW, or C-SDW] [Solid-State Physics]

CSDW-P Commensurate Spin-Density Wave– Paramagnetic (Transition) [Solid-State Physics]

CSE Canadian Security Establishment

CSE Center for Science and Environment [India]

CSE Center for Software Engineering [UK]

CSE Center for the Study of the Environment

CSE Certificate in Stationary Engineering

CSE Certificate of Secondary Education [UK]

CSE Certified System(s) Engineer

CSE Chemisorptive Electron Emission

CSE Coefficient of Self-Excitation [Explosives]

CSE Colorado Society of Engineers [US]

CSE Commission des Substances Explosives [Explosives Commission, France]

CSE Commission on Science Education

CSE Common Stock Equivalent

CSE Common Support Equipment

CSE Communications Security Establishment [Canada]

CSE Communications Systems Engineer

CSE Computerized Shrinkage Evaluation

CSE Computer Science and Engineering

CSE Computer Support Equipment

CSE Configuration Switching Equipment

CSE Consortium for Superconducting Electronics [US]

CSE Containment Steam Explosion

CSE Containment Systems Experiment [US]

CSE Control System(s) Engineer(ing)

CSE Copper Sulfate Electrode

CSE Core Storage Element

CSE Council of Stock Exchange [US]

C/SE Computer/Software Engineering [of Institute of Electrical and Electronics Engineers, US]

CSEA California State Electronics Association [US]

C-SEAL Combat System Evaluation and Analysis Laboratory [of Defense Research Establishment Atlantic, Canada] [also CSEAL]

°C/sec Degrees Celsius (or Centigrade) per second [also °C/s, °C s^{-1}, or °C sec^{-1}]

CSECE Canadian Society for Electrical and Computer Engineering

CSECT Control Section [Computers]

CSED Coordinated Ship Electronics Design

CS/EDS Computer Society/Electrodischarge Machining (Event) [of Institute of Electrical and Electronics Engineers, US]

CSEE Canadian Society for Electrical Engineering

CSEE Certificate, Society of Clinical Chemists

CSEE Chinese Society of Electrical and Electronics Engineers [PR China]

CSEE Committee of Stock Exchanges in the European Community [Belgium]

CSEF Current Switch Emitter Follower

CSEG Canadian Society of Exploration Geophysicists

CSEG Current Value of Control Segment Register [also CSeg] [Computers]

CSEHMA Cyprus Solar and Electric Heaters Manufacturers Association

CSEL Charge-Stripping Energy Loss

CSELT Centro Studi e Laboratori Telecommunicazione [Research Center and Telecommunications Laboratory, Torino, Italy]

CSELT Tech Rep CSELT Technical Reports [of Centro Studi e Laboratori Telecomunicazioni, Italy]

CSEM Centre Séismologique Européoméditerranéen [Europe-Mediterranean Seismological Center, Strasbourg, France]

CSEM Centre Suisse d'Electronique et de Microtechnique [Swiss Center for Electronics and Microtechnology]

CSEM Committee of Societies for Electron Microscopy

CSEM Computerized Scanning Electron Microscope

CSEM Conventional Scanning Electron Microscope

CSEN Conseil Supérieur de l'Education Nationale [High Council of National Education, France]

CSEP Center for the Study of Ethics in the Professions [of Illinois Institute of Technology, Chicago, US]

CSEP Ceramic Science and Engineering Program [of Pennsylvania State University, University Park, US]

CSEPA Central Station Electrical Protection Association [US]

CSEPP Chemical Stockpile Emergency Preparedness Program [US]

CSERB Computer Systems and Electronics Requirements Board [UK]

CSES Center for the Study of Earth from Space [of University of Colorado, US]

CSE Solution CSE Solution [Publication of the Center for the Study of the Environment]

CSEU Confederation of Shipbuilding and Engineering Unions [UK]

CSF Canadian Schizophrenia Foundation

CSF Canadian Standard Freeness (Unit) [Papermaking]

CSF Casio Science Foundation [Japan]

CSF Catalytic Seed Fund

CSF Cat Scratch Fever

CSF Central Supply Facility

CSF Central Switching Facility

CSF Cerebrospinal Fever

CSF Cerebrospinal Fluid

CSF Classical Signal Field (Method)

CSF Colony-Stimulating Factor [Biochemistry]

CSF Compagnie Générale de Télégraphie sans Fil [French Wireless Telegraphy Company]

CSF Complex Stacking Fault [Materials Science]

CSF Configuration State Function

CSF Consol Synthetic Fuel (Process)

CSF Constant Scattering Factor Averaging

CSF Containment Support Fixture

CSF Continuous Streamflow

CSF Contract Stationers Forum [of National Office Products Association, US]

CSF Controlled Surface Flow

CSF Cost Sensitivity Factor

CSF Council on Synthetic Fuels [now Council on Alternate Fuels, US]

CSF Cumulative Size-Selection Function

CSF Customer Support Facility

CsF Cesium Fluoride

CSFA Canadian Scientific Film Association

CSFC Canadian Salt Fish Corporation

CSFE Canadian Society of Forestry Engineers

CS FET Common Source Field-Effector Transistor [also CS-FET or CSFET]

CSFGR Canadian Society for Fifth Generation Research

CSFI Communication Subsystem For Interconnection

CSFM Commercial Spent Fuel Management [Nuclear Engineering]

CSFP Commonwealth Scholarship and Fellowship Plan

CSFPCPM Chambre Syndicale des Fabricants de Papier à Cigarettes et Autres Papiers Minces [Union of Producers of Cigarette Papers and Other Lightweight Papers, France]

CSFPN Commission on Soil Fertility and Plant Nutrition [India]

CSFR Czechoslovak Federal Republic [formerly CSSR] [Now Separated into Czech Republic and Slovakia]

CSFS Canadian Society of Forensic Science(s)

CSFS Colorado State Forest Service [of Colorado State University, Fort Collins, US]

CSFS Computer-Simulated Fracture Surface

CSFS Continuous Streamflow Simulation

CSFTI Committee on Southern Forest Tree Improvement [US]

CSG Canada Systems Group

CSG Cetacean Specialist Group [of the Species Survival Commission of the International Union for the Conservation of Nature and Natural Resources]

CSG Combat Service Group

CSG Constant Size Grain(s) [Metallurgy]

CSG Constructive Solid Geometry

CSG (Lotus) Consulting Services Group

CSG Correlated Spin Glass [Solid-State Physics]

CSG Council of State Governments [US]

Csg Casing [also csg]

CSGA Canadian Seed Growers Association

CSGA Canadian Society of Graphic Arts

CSGCC Commission on Soil Genesis, Classification and Cartography [Netherlands]

CSGWPP Comprehensive State Ground Water Protection Program [US]

CSH Calcium Silicate Hydrate

CSH Called Subscriber Held

CSH Chicago State Hospital [Illinois, US]

CSH Convective Stratiform Heating [Meteorology]

CSH Convector Superheater

CSH Cyclic Strain Hardening [Metallurgy]

.CSH C-Shell Script [File Name Extension] [also .csh]

C$_2$SH Dicalcium Silicate Hydrate (or Hillebrandite) [$2CaO \cdot SiO_2 \cdot H_2O$]

C$_5$S$_6$H$_x$ Pentacalcium Hexasilicate Hydrate (or Tobermorite) [$5CaO \cdot 6SiO_2 \cdot xH_2O$]

C$_9$S$_6$H$_{11}$ Nonacalcium Hexasilicate Hydrate (or Jennite) [$9CaO \cdot 6SiO_2 \cdot 11H_2O$]

C-S-H Calcium–Silicate–Hydrate (Ceramics)

CsH Cesium Hydride

CSHC California State Highway Commission [US]

CSHCS Center for the Studies of Human Communities in Space [now Space Settlement Studies Program, US]

CSHE Center for the Study of Higher Education [of University of California at Berkeley, US]

CSHF Closed-Shell Hartree-Fock (Theory) [Physics]

CSHL Cold Spring Harbor Laboratory [New York State, US]

CSHO Compliance Safety and Health Office

CSHP Canadian Society of Hospital Pharmacists

C-SHRP Canadian Strategic Highway Research Program

CSHS California School of Herbal Studies [US]

CSI Campbell Scientific, Inc. [US]

CSI Canadian Securities Institute

CSI Chartered Surveyors Institution [UK]

CSI Chemical Substances Index [of American Chemical Society, US]

CSI Chlorosulfonyl Isocyanate

CSI Coalition of Service Industries [US]

CSI Coelliptic Sequence Initiation [also csi]

CSI College of Staten Island [of City University of New York, US]

CSI Colloquium Spectroscopicum Internationale [International Spectroscopy Colloquium]

CSI Command Sequence Introducer

CSI Command String Interpreter

CSI Common Sense Initiative [of US Environmental Protection Agency]

CSI Compliance Sampling Inspection

CSI CompuServe Inc. [Now Part of America On-Line]

CSI Computer Security Institute [Northborough, Massachusetts, US]

CSI Computer Society of India

CSI Construction Specifications Institute [US]

CSI Construction Surveyors Institute [UK]

CSI Consumer Satisfaction Index

CSI Control Sequence Introducer

CSI Control Servo Input [also csi]

CSI Coral Sea Islands

CSI Cosmetic Surgery Institute [US]

CSI Crew Software Interface [also csi]

CSI Crystal Structure Imaging

CSI Current Source Inverter

CSi$_2$ Carbon Disilicide

C-Si Carbon-Silicon (Bond)

CsI Cesium Iodide

c-Si (Single) Crystalline Silicon

CSIA Canadian Solar Industries Association

CSIC Cervical Spine Immobilization Device

CSIC Chlorosulfonyl Isocyanate

CSIC Computer System Interface Circuits

CSIC Consejo Superior de Investigaciónes Científicas [National (Scientific) Research Council, Madrid, Spain]

CSIC Customer-Specified Integrated Circuit

C-SiC Carbon-Fiber-Reinforced Silicon Carbide (Composite) [also C/SiC or Cf/SiC]

CSICC Canadian Steel Industries Construction Council

CSIC-ITMA Consejo Superior de Investigaciónes Científicas–Instituto Tecnológico de Materiales de Asturias [National (Scientific) Research Council–Asturias Institute of Materials Technology; located at Oviedo, Spain]

CSICOP Committee for the Scientific Investigation of Claims of the Paranormal [US]

CSID Call Subscriber Identification

CSID Compound Semiconductor Industry Technology [of Elsevier Advanced Technology, Oxford, UK]

CSIE Canadian Society for Industrial Engineering

CSIET Council on Standards for International Educational Travel [US]

C-Si-Ge Carbon-Silicon-Germanium (Material)

c-Si:H Hydrogenated Crystalline Silicon

CSII Continuous Subcutaneous Insulin Infusion [Medicine]

CSIM Combustion Synthesis of Inorganic Materials

CSIM Compact Short-Channel IGFET (Insulated Gate Field-Effect Transistor) Model

CSIMM Central South Institute of Mining and Metallurgy [Changsha, PR China]

CSIMP Center for Studies in Income Maintenance Policy [of New York University, US]

CSIN Chemical Substances Information Network (Database)

CSIO Central Scientific Instruments Organization [Chandigarh, India]

c-SiO$_2$ Crystalline Silica

CSIO Commun CSIO Communications [of Central Scientific Instruments Organization, India]

CS:IP Code Segment:Instruction Pointer [Computers]

CSIPR Comité Spéciale et Internationale sur les Parasites Radiotélégraphiques [International Special Committee on Radiotelegraphic Atmospherics; of International Electrotechnical Commission, Switzerland]

CSIR Computer Systems Integration Review

CSIR Commonwealth Scientific and Industrial Research [now Commonwealth Scientific and Industrial Research Organization]

CSIR Council for Scientific and Industrial Research [New Delhi, India]

CSIR Council for Scientific and Industrial Research [Pretoria, South Africa]

CSIRA Canadian Steel Industry Research Association

CSIRAC Commonwealth Scientific and Industrial Research Automatic Computer [Australia]

CSIRO Commonwealth Scientific and Industrial Research Organization [Melbourne, Australia]

CSIRO DAR CSIRO Division of Atmospheric Research [of Commonwealth Scientific and Industrial Research Organization]

CSIRO DBCE CSIRO Division of Building, Construction and Engineering [of Commonwealth Scientific and Industrial Research Organization]

CSIRO DCR CSIRO Division of Computing Research [of Commonwealth Scientific and Industrial Research Organization, Australia]

CSIRO DCWT CSIRO Division of Chemical and Wood Technology [of Commonwealth Scientific and Industrial Research Organization, Australia]

CSIRO DIT CSIRO Divison of Information Technology [of Commonwealth Scientific and Industrial Research Organization]

CSIRO Div Chem Wood Technol Tech Papers CSIRO Division of Chemical and Wood Technology, Technical Papers [of Commonwealth Scientific and Industrial Research Organization, Australia]

CSIRO Div Chem Wood Technol Res Rev CSIRO Division of Chemical and Wood Technology, Research Review [of Commonwealth Scientific and Industrial Research Organization, Australia]

CSIRO DO CSIRO Division of Oceanography [of Commonwealth Scientific and Industrial Research Organization, Australia]

CSIRO DPR CSIRO Division of Petroleum Resources [Commonwealth Scientific and Industrial Research Organization, Australia]

CSIRO DS CSIRO Division of Soils [Commonwealth Scientific and Industrial Research Organization, Australia]

CSIRO DTCP CSIRO Division of Tropical Crops and Pastures [of Commonwealth Scientific and Industrial Research Organization, Australia]

CSIRO DWE CSIRO Division of Wildlife and Ecology [of Commonwealth Scientific and Industrial Research Organization, Australia]

CSIRO DWR CSIRO Division of Water Resources [of Commonwealth Scientific and Industrial Research Organization]

CSIRO ENTO CSIRO Division of Entomology [of Commonwealth Scientific and Industrial Research Organization]

CSIRONET Commonwealth Scientific and Industrial Research Organization Network [Australia]

CSIRONET News CSIRONET News [of CSIRONET, Australia]

CSIRO OSSA CSIRO Office of Space Science and Applications [of Commonwealth Scientific and Industrial Research Organization, Australia]

CSIRO TSC CSIRO Tree Seed Center (Canberra) [of Commonwealth Scientific and Industrial Research Organization, Australia]

CSIS Canadian Security Intelligence Service

CSIS Centre for Spatial Information Systems [of Commonwealth Scientific and Industrial Research Organization]

CSIS Centralized Storm Information System [of US National Weather Service]

CSIS Core Spray Injection System [Nuclear Reactors]

CSISRS Cross-Section Information Storage and Retrieval System [of National Neutron Cross-Section Center, US]

CSIST Chung Shan Institute of Science and Technology [Lung-Tan, Taiwan]

CsI(CO₃) Carbonate-Doped Cesium Iodide

CsI(Tl) Thallium-Doped Cesium Iodide

CSIU Core Segment Interface Unit [Computers]

CSJ Carbon Society of Japan

CSJ Ceramic Society of Japan

CSJ Chemical Society of Japan [also Chem Soc Jpn]

CSJ Crystallographic Society of Japan

CSJ Bull Chemical Society of Japan Bulletin [also Chem Soc Jpn Bull]

CSK Czechoslovakian Koruna [Currency]

Csk Countersink(ing); Countersunk [also csk]

Csk-O Countersink other Side

CSL Canada Steamship Lines

CSL Canadian Sleep Laboratory

CSL Chemical Services Laboratory [US]

CSL Climate Simulation Laboratory [of National Center for Atmospheric Research, Boulder, Colorado, US]

CSL Closest Specification Limit

CSL Code Selection Language

CSL Coincidence (or Coincident) Site Lattice [Materials Science]

CSL Command Signal Limiter

CSL Comparative Systems Laboratory [US]

CSL Computer Sensitive Language

CSL Computer-Simulated Language

CSL Computer Status Lights

CSL Conceptual Schema Language

CSL Connecticut State Library [Hartford, US]

CSL Constant Scattering Length

CSL Continuous Sheet Lamination

CSL Control and Simulation Language

CSL Coordinated Science Laboratory [of University of Illinois at Urbana-Champaign, US]

CSL Corrosion Science Laboratory [of Academia Sinica, Shenyang, PR China]

CSL Crew Systems Laboratory [of NASA]

CSL Current-Sinking Logic

CSL Current Switch Logic

CSL Cytisus Scoparius Lectin [Biochemistry]

CSLATP Canadian Society of Landscape Architects and Town Planners

CSLB Coincidence Site Lattice Boundary [Materials Science]

CSLC Canadian Softwood Lumber Committee

CSL-DSC Coincidence Site Lattice–Displacement Shift Complete (Model) [also CSL/DSC] [Materials Science]

CSLE Coincidence Site Lattice Enhanced (Sample) [Materials Science]

CSL GB Coincidence Site Lattice Grain Boundary [Materials Science]

CSLIP Compressed Serial Line Interface Protocol [Internet]

CSLM Confocal Scanning Laser Microscope

CSLMA Canadian Softwood Lumber Manufacturers Association

CSLO Canadian Scientific Liaison Office

CSLP Canada Student Loans Plan

CSLP Center for Short Lived Phenomena [US]

CSLT Canadian Society of Laboratory Technologists

Cslty Casualty [also cslty]

CSM Calcium Samarium Manganate

CSM Call Supervision Module

CSM Camborne School of Mines [Redruth, Cornwall, UK]

CSM Canadian Society of Microbiologists

CSM Center for Sensor Materials [of National Science Foundation at Michigan State University, East Lansing, US]

CSM Center for the Study of Man [of Smithsonian Institution, Washington, DC, US]

CSM Centro Sviluppo Materiali SpA [Materials Development Center, Italy]

CSM Cerebrospinal Meningitis

CSM Cerium Strontium Manganate

CSM Chemical Surety Material(s)

CSM Chinese Society for Measurement

CSM Chinese Society for Metals

CSM Chopped Strand Mat [Composites Engineering]

CSM Climate System Monitoring (Project) [of World Meteorological Organization/World Climate Program]

CSM Code Set Map (File)

CSM Colorado School of Mines [Golden, US]

CSM Colorado State Museum [Denver, US]

CSM Combined Symbol Matching

CSM Command Service Module [of NASA Apollo Program]

CSM Commission for Synoptic Meteorology

CSM Commission on Soil Mineralogy [France]

CSM Common Support Module [also csm]

CSM Communications Services Manager

CSM Communications System Monitoring

CSM Complex Shear Modulus

CSM Computerized Scanning Electron Microscopy

CSM Computer Sedimentation Model

CSM Computer Status Matrix [also csm]

CSM Conceptual Site Model(ling)

CSM Constant System Monitor

CSM Continuous Sheet Memory

CSM Continuous Slowing-Down Models

CSM Continuous Steelmaking Mitsubishi

CSM Continuous Stiffness Measurement

CSM Continuum Spectrum Method

CSM Corn Meal–Soy Flour–Milk (Blend)

CSM Czech Metallurgical Society

CSMA Camborne School of Mines Association [UK]

CSMA Carrier-Sense Multiple-Access

CSMA Chain Saw Manufacturing Association [now Portable Power Equipment Manufacturers Association, US]

CSMA Chemical Specialties Manufacturers Association [US]

CSMA Communications Systems Management Association

CSMA/CA Carrier Sense Multiple Access with Collision Avoidance [also CSMA-CA]

CSMA/CD Carrier-Sense Multiple-Access with Collision Detect(ion) [also CSMA-CD]

CSMCS Continuous Space Monte Carlo Simulation

CSME Canadian Society for Mechanical Engineering

CSME Chinese Society of Mechanical Engineers [PR China]

CSME Communications System Monitoring Equipment

CSME-HTI Chinese Society of Mechanical Engineers–Heat Treating Institution

CSME Trans CSME Transactions [of Canadian Society for Mechanical Engineering]

Csmith Coppersmith [also csmith]

CSML Continuous Self Mode Locking

CSMMFRA Cotton, Silk and Man-Made Fibers Research Association [UK]

CSMO Calcium Samarium Manganese Oxide

CSMO Cerium Strontium Manganese Oxide

CSMP Continuous System Modelling Program

CSM Q Colorado School of Mines Quarterly [US]

CSMS Centre Sciences des Matériaux et des Structures [Science Center for Materials and Structures; of Ecole des Mines de Saint-Etienne, France]

CSMS Charge-Separation Mass Spectrometry

CSMS Chinese Society of Materials Science [Hsinchu, Taiwan]

CSMS Customer Support Management System

CSMT Center for Space Microelectronic Technology [at NASA Jet Propulsion Laboratory, Pasadena, California]

CSN Cable Satellite Network

CSN Card-Select Number

CSN Circuit Switching Network

CSN Common Services Network

CSN Conductive Solids Nebulizer

CSNA Classification Society of North America [US]

CSNC Chemical Societies of the Nordic Countries [Sweden]

CSNDT Canadian Society for Nondestructive Testing

CSNDT J CSNDT Journal [of Canadian Society for Nondestructive Testing]

CSNET Computer Science NETwork [also CSnet]

CSNG Convert to Single-Precision Number (Function) [Programming]

CSNI Committee on the Safety of Nuclear Installations [US]

CSNI Communications System Network Interoperability (Software)

CSNMT Czech Society for New Materials and Technology

c-SnO Crystalline Stannous Oxide (or Crystalline Tin (II) Oxide)

CSNS Canadian Science News Service

CSNSM Centre de Spectrométrie Nucléaire et de Spectrométrie de Masse [Nuclear and Mass Spectrometry Center; of IN2P3–National Institute of Nuclear and Particle Physics, Orsay, France]

CSNSM-CNRS Centre de Spectrométrie Nucléaire et de Spectrométrie de Masse–Conseil National de la Recherche Scientifique [Nuclear and Mass Spectrometry Center–National Scientific Research Council, France]

CSO Cadmium Stannate [$CdSnO_3$]

CSO Central Services Organization

CSO Central Statistical Office [UK]

CSO Centralized Service Observation

CSO Chained Sequential Operation

CSO Chief Signal Officer

CSO Coastal States Organization [US]

CSO Color Separation Overlay

CSO Combined Sewer Overflow

CSO Committee of Senior Officials

CSO Community Service Order

CSO Complementary Symmetry Operation

CSO Computer Systems Officer

CSO Crossed Second Order

CSOC Consolidated Space Operations Center [of NASA]

CsOH Cesium Hydroxide

CSOHS Canadian Society for Occupational Health and Safety

CSOM Confocal Scanning Optical Microscope

CSOP Crew Systems Operating Procedures

CSOV Constrained Space Orbital Variation

CSP Canadian Space Program

CSP Cast Steel Plate

CSP Celestial South Pole [Astronomy]

CSP Center for Space Power [of NASA Texas Engineering Experiment Station]

CSP Certified Safety Professional

CSP Certified Systems Professional

CSP Channel Substrate Planar [Electronics]

CSP Chemical Spray Pyrolysis

CSP (Micro) Chip Scale Package; (Micro) Chip Scale Packaging

CSP Chlorosulfonated Polyethylene

CSP Collaborative Special Projects (Grant) [of Natural Sciences and Engineering Research Council, Canada]

CSP Coder Sequential Pulse

CSP Combustion Spray Pyrolysis

CSP Commercial Subroutine Package

CSP Committee on Scientific Programs [of National Academy of Sciences, US]

CSP Communicating Sequential Processes

CSP Communications Security Publication [US Military]

CSP Compact (Metal) Strip Production (Process) [of Schloemann Siemag AG, Germany]

CSP Completely Self-Protecting (Transformer)

CSP CompuCom Speed Protocol

CSP Computer Support Program

CSP Conservation Science Program [of World Wildlife Fund]

CSP Constant-Speed Propeller

CSP Continuous Sampling Plan

CSP Continuous Strip Production (Process) Metallurgy]

CSP Control Signal Processor

CSP Control Switching Point

CSP Cooperative Software Program

CSP Council for Scientific Policy

CSP Crack Shape Parameter [Mechanics]

CSP Cross-Slip Pinning (Model) [Metallurgy]

CSP Cross System Product

CSP Crystallographic Shear Plane

CSP Current Set Point

C/SP Communications/Symbiont Processor

CsP Cesium Phosphide (Semiconductor)

CsPb Cesium-Lead (Compound)

CSPC California State Polytechnic College [San Luis Obispo, US]

CSPCA Canadian Society for the Prevention of Cruelty to Animals

CSPDN Circuit Switched Public Data Network

CSPE Chlorosulfonated Polyethylene

CsPFO Cesium Perfluorooctanoate

CSPG Canadian Society of Petroleum Geologists

CSPG Chondroitin Sulfate Proteoglycan [Biochemistry]

CSPG Collaborative Special Projects Grant [of Natural Sciences and Engineering Research Council, Canada]

CSPI Center for Science in the Public Interest [Washington, DC, US]

CSPI Corrugated Steel Pipe Institute

CSPO Communications Satellite Project Office

CSPODP Canada South Pacific Ocean Development Project

CSPP Chemically-Sensitive Paper Tape (Detector)

CS Press Computer Society Press [of Institute of Electrical and Electronics Engineers, US]

CSPS Canadian Society of Plastic Surgeons

CSPU Core Segment Processing Unit [Computers]

CSR Canadian Signals Regiment

CSR Center for Scientific Review [of National Research Council, US] [formerly Division of Research Grants]

CSR Center for Superconductivity Research [of University of Maryland, College Park, US]

CSR Certification Status Report

CSR Cesium Recovery

CSR Channel Select Register

CSR Check Signal Return

CSR Chinese Standardized (Natural) Rubber

CSR Chiral Shift Reagents [Chemistry]

CSR Committee on Space Research [France]

CSR Common Services Rack

CSR Compressive Strain at Rupture [Mechanics]

CSR Continuous Speech Recognition

CSR Continuous Stirred Reactor [Chemical Engineering]

CSR Contractile Strain Ratio [Mechanics]

CSR Control (and) Status Register

CSR Controlled Silicon Rectifier

CSR Control Status Register

CSR Council for Scientific Research [Baghdad, Iraq]

CSR Crack-Sensitivity Ratio [Mechanics]

CSR Crack Subsceptibility Region [Mechanics]

CSR Crew Station Review

CSR Cursor [also Csr]

CSR Customer Service Representative

CSR Czechoslovak Republic [later CSSR; now CSFR]

CSRA Canadian Semiotics Research Association

CSRB Canadian Security Review Board

CSRC CALS Shared Resource Center [US]

CSRF Canadian Synchrotron Radiation Facility [at Synchrotron Radiation Center, University of Wisconsin at Madison, Stoughton, US]

CSRF Computing Sector Resource Facility [of (Ontario) Ministry of Economic Development, Trade and Tourism, Canada]

CSRG Chemical Sciences Research Group

CSRG Computer Systems Research Group [of University of Toronto, Canada]

CSRLIN Current Line Position of Cursor (Variable) [Programming]

CSRO Chemical Short-Range Order(ing) [Solid-State Physics]

CSRO Compositional Short-Range Order(ing) [Solid-State Physics]

CSRP Chemical Sciences Research Paper

CSRP Computers and Software Review Panel

CSRS Canadian Symposium on Remote Sensing

CSRS Coherent Stokes-Raman Spectroscopy

CSRS Coherent Stokes-Raman Spectrum

CSRS Cooperative State Research Service [of US Department of Agriculture]

CSRT Canadian Society of Radiological Technicians

CS RVE Concentric Spheres Representative Volume Element (Model) [also CS-RVE] [Mechanics]

CSS Canada Standard Size

CSS Canadian Space Station

CSS Cascading Style Sheets [Internet]

CSS Cask Support Structure [Nuclear Engineering]

CSS Cast Semi-Steel

CSS Central Security Service [US]

CSS Certificate in Special Studies

CSS Character Start Stop

CSS Chiral Spin State [Particle Physics]

CSS Circuits and Systems Society [of Institute of Electrical and Electronics Engineers, US]

CSS Close-Spaced Sublimation (Deposition Method) [Materials Science]

CSS Coarse Sun Sensor

CSS Cognitive Science Society [US]

CSS Colorado Scientific Society [US]

CSS Command Substitution System

CSS Communications Subsystem

CSS Comprehensive Support Software

CSS Computer Sharing Services [US]

CSS Computer Software Specialist

CSS Computer Special Systems (Group) [of Digital Equipment of Canada]

CSS Computer Subsystem [also css]

CSS Computer System Simulator

CSS Contact Start-Stop (Disk Head)

CSS Containment Spray System [Nuclear Engineering]

CSS Contaminated Surface Soil

CSS Continuous-Cast Stainless Steel

CSS Continuous System Simulator (Language)

CSS Control Stick Steering [also css]

CSS Control Systems Society [of Institute of Electrical and Electronics Engineers, US]

CSS Conversational Software System

CSS Cordless Switchboard Section

CSS Core Segment Simulator [also css] [Computers]

CSS Core Support Structure [Nuclear Reactors]

CSS Critical Shear Stress [Mechanics]

CSS Customer Support Center

CSS Customer Switching System

CSS County Surveyors Society [UK]

CSS Cyclic Stress-Strain (Curve) [Mechanics]

C2S Congrès Scientifique Services [Saint Cloud, France]

CSSA Canadian Space Station Agency

CSSA Cleaning Support Services Association [UK]

CSSA Computer Society of South Africa

CSSA Concrete Society of Southern Africa

CSSA Crop Science Society of America [US]

CSSB Cedar Shake and Shingle Bureau [US]

CSSB Companded Single-Sideband

CSSB Compatible Single-Sideband

CSSB Cesium Antimonide (Semiconductor)

CSSB/AM Companded Single-Sideband/Amplitude Modulation

CSSBI Canadian Sheet Steel Building Institute

CSSC Canadian Steel Service Center

CSSC Constant Stress-Strain Cycle [Fatigue Testing]

CSSC Construction Software Standards Council [now National Construction Software Associa tion, US]

CSSC Cyclic Stress-Strain Curve [Mechanics]

CSSCI Canadian Steel Service Center Institute

CSSE Canadian Society of Safety Engineers

CSSE Conference of State Sanitary Engineers [US]

CSSE Control System Simulation Equipment

CSSe Carbon Selenosulfide

CSSG Chemical Sciences Study Group

CSSG Customer Service and Support Group

CSSL Coincidence Site Superlattice [Materials Science]

CSSL Continuous System Simulation Language

CSSM Center for Strategic Standardization Management [US]

CSSM Client-Server Systems Management [Internet]

CSSM Computer Stream Sedimentation Model

CSSM Czechoslovak Scientific Society for Metals

CSSME Center for Studies in Science and Mathematics Education [UK]

CSSP Canadian Shellfish Sanitation Program [of Environment Canada]

CSSP Canadian Space Station Program [of Canadian Space Agency]

CSSP Certified Security and Safety Professional

CSSP Council on Scientific Society Presidents [US]

CSSR Czechoslovak Socialist Republic [Now Divided into Czech Republic and Slovakia]

CSSRA Canadian Shipbuilding and Ship Repairing Association

CSSS Canadian Society of Soil Science

CSSS Center for Solid-State Science [of Arizona State University, Tempe, US]

CS/SS Card Service/Socket Service

CSST Center for Solder Science and Technology [of Sandia National Laboratories, US]

CSST Committee on Science, Space and Technology [US]

CSST Computer System Science Training

CSST Coordination Committee for Science and Technology

CSSW Chemically-Sustained Shock Waves

CST Calcium Strontium Titanate

CST Canadian Scholarship Trust (Foundation)

CST Carrier Power Supply, Transistorized

CST Center for Science and Technology

CST Central Standard Time [Greenwich Mean Time +06:00] [also cst]

CST Certified Shiatsu Therapist

CST Certified Survey Technician

CST Certified Survey Technologist

CST Channel Status Table

CST Cis-Syn-Trans (Configuration) [Chemistry]

CST Clean Steel Technology

CST Code Segment Table [Computers]

CST College of Science and Technology [of Nihon University, Tokyo, Japan]

CST Combined Systems Test

CST Companhia Siderurgia de Tubarão [Brazilian Steel Manufacturer]

CST Complex Systems Theory

CST Composite Science and Technology

CST Constitutional Solution Treatment [Metallurgy]

CST Contract Supplemental Tooling

CST Coordinated Storm Tracking

CST Crew Station Trainer

CST Crew Systems Trainer [NASA's One-G Trainer]

C/ST Combined Station/Tower

Cst Coelostat [also cst]

cST centistoke [Unit]

cst constant

CSTA Canadian Society of Technical Agriculturalists

CSTA Canadian Surface Transportation Administration

CSTA Coal and Slurry Technology Association [US]

CSTA Computer-Supported Telephony Applications

CSTA Crew Software Training Aid [of NASA]

CSTAR Classified Scientific and Technical Aerospace Reports [of NASA]

CSTD Center for Science and Technology for Development [of United Nations]

CSTD Chemical Sciences and Technology Division [of Los Alamos National Laboratory, New Mexico, US]

CSTe Carbon Tellurosulfide

CSTEC Canadian Steel Trade and Employment Congress

CSTEI Center for Scientific, Technical and Economic Information [Sofia, Bulgaria]

CSTG Ceramic Science and Technology Group [of Los Alamos National Laboratory, New Mexico, US]

Cstg Casting [also cstg]

CSTHA Canadian Science and Technology Historical Association

Cs(THF)$_{1.75}$C$_{24}$ Cesium Tetrahydrofuran Graphitide

CSTEM Cross-Sectional Transmission Electron Microscopy

CSTEX Cost Extension [also CST EX or Cst Ex]

CSTF Continuous Stirred Tank Fermenter

CSTI Committee on Scientific and Technological Information [US]

CSTI Council of Science and Technology Institutes [UK]

CSTL Computer System Terminal Log

CSTO Calcium Strontium Titanium Oxide

CSTP Conceptual Site Treatment Plan [Nuclear Engineering]

CSTPA Council on Soil Testing and Plant Analysts [US]

CSTPC Cassette Magnetic Tape Controller

CSTPC Cost Price [also Cost Pc]

CSTR Center for Speech Technology Research [UK]

CSTR Committee on Solar Terrestrial Research [US]

CSTR Constant Stirred Tank Reactor

CSTR Continuous(-Flow) Stirred Tank Reactor [also CFSTR]

Cstr Canister [also cstr]

CSTS Combined System Test Stand

CSTS Computer Systems Technical Support

CSTS Cryogenic Storage and Transfer System

CSTS Czech Scientific and Technical Society [Czech Republic]

CSTT Contaminated Sediment Treatment Technology

CSTTP Center for Science, Trade and Technology Policy [of George Mason University, Fairfax, Virginia, US]

CSTUN Cost per Unit [also CST UN or Cst Un]

CSU Cache Store Unit

CSU California State University [US]

CSU Central State University [Edmond, Oklahoma, US]

CSU Central Switching Unit

CSU Channel Service Unit

CSU Channel Switching Unit

CSU Charles Sturt University [Australia]

CSU Chicago State University [Illinois, US]

CSU Check Switching Unit

CSU Circuit Switching Unit

CSU Cleveland State University [Ohio, US]

CSU College of Southern Utah [Cedar City, US]

CSU Colorado State University [Fort Collins, US]

CSU Combined Shaft Unit

CSU Common Services Unit

CSU Continuous Speech Understanding

CSU Crude Steel Unit

CSU Customer Service Unit

CSU Customer Setup

CSUA Computer Science Undergraduate Association [US]

CSUC California State University at Chico [US]

CSUCA Confederación Universitaria Centroamericana [Confederation of Central American Universities, Puerto Rico]

CSUDH California State University at Dominguez Hills [Carson, US]

CSUF California State University at Fresno [US]

CSUF California State University at Fullerton [US]

CSUFC Colorado State University at Fort Collins [US]

CSUH California State University at Hayward [US]

CSULA California State University at Los Angeles [US]

CSULB California State University at Long Beach [US]

CSUN California State University at Northridge [US]

CSUOT Central South University of Technology [Changsha, PR China]

CSU/PO Colorado State University Project Office [US]

CSUR College of Science, University of Ryukyus [Japan]

CSURF Colorado State University Research Foundation [US]

CSUS California State University at Sacramento [US]

CSUSA Chief of Staff, United States Army

CSUSAF Chief of Staff, United States Air Force

CSUT Central-South University of Technology [Changsha, Hunan, PR China]

CSV Cathodic Stripping Voltage

CSV Cathodic Stripping Voltammetry

CSV Chrysanthemum Stunt Viroid

CSV Circuit-Switched Voice

CSV Comma-Separated Value

CSV Comma-Separated Variable

CSV Common Services Verbs (Interface)

CSV Corona Start(ing) Voltage

CsV Campoletis sonorensis Virus

CSVD Close-Spaced Vacuum Diode

CSVS Canadian Space Vision System [of Canadian Space Agency]

CSVT Close-Spaced Vapor Transport (Deposition Method) [Materials Science]

CSW Caustic Slurry Waste

CSW Certified Social Worker

CSW Channel Status Word

CSW Continuous Strip Welding

CSW Control Switch

CSW Crude Steel Weight

CSW Custom Switch Module

CSWA Canadian Science Writers Association

CSWG Chemical Selection Working Group [of National Cancer Institute, Bethesda, Maryland, US]

CSWP Committee on the Status of Women in Physics [of American Physical Society, US]

CSWPRC Chinese Society on Water Pollution Research and Control [PR China]

CsX Cesium Halide [General Formula; e.g., CsBr, CsCl, CsF, or CsI]

CSZ Calcia-Stabilized Zirconia
CSZ Ceria-Stabilized Zirconia
CsZn Cesium Zinc (Compound)
CSZP Calcium Strontium Zirconium Phosphate
CT Cabibbo Theory [Particle Physics]
CT Cable Test
CT Cadmium Telluride (Semiconductor)
CT Calcination Temperature
CT Calcitonin [Biochemistry]
CT Calcium Titanate (Ceramics)
CT Caloric Theory
CT Canonical Transformation
CT Cape Town [South Africa]
CT Carbon Tetrachloride
CT Carrier Telegraph (Channel)
CT Carrier Transmission
CT Cartilaginous Tissue [Anatomy]
CT Cassette Tape
CT Cauchy Theorem [Mathematics]
CT Cell-Culture Tested
CT Cellulose Triacetate
CT Center Tap
CT Center Thickness (of Lenses)
CT Central Tablelands [New South Wales, Australia]
CT Central Time [also ct]
CT Cephalothorax [Zoology]
CT Ceramic Technologist; Ceramic Technology
CT Certified Teacher
CT Charge Transfer
CT Chemical Technologist; Chemical Technology
CT Chemical Thermodynamics
CT Chemical Transport
CT Chemotherapy
CT Chief of Transportation
CT Chisel Toughness
CT Chiyoda Thorougbed (Process)
CT Chlorotoluene
CT Chronic Toxicity
CT Chymotrypsin; α-Chrymotrypsin [Biochemistry]
CT Ciguatoxin [Biochemistry]
CT Circuit Theory
CT Cis-Trans (Configuration) [Chemistry]
CT Civilian Transport
CT Classical Trajectory (Method) [Physics]
CT Cleaning Technician
CT CLEM (Cargo Lunar Excursion Module) Transporter
CT Clinical Trial
CT Coal Tar
CT Coastal Telegrapher
CT Coating Technologist; Coating Technology
CT Code Telegram
CT Coding Theory
CT Coherence Transfer

CT Coherent Twin (Boundary) [Materials Science]
CT Coherent Tunneling [Electronics]
CT Cold Trap
CT College of Technology
CT Collision Time [Physics]
CT Collision Theory
CT Colloidal Thorium
CT Color Temperature [Statistical Mechanics]
CT Color Transmission
CT Combinatorial Theory [Mathematics]
CT Commercial Translator
CT Commercial Traveller
CT Communications Technician
CT Communications Technologist; Communications Technology
CT Communications Terminal
CT Compact Tension (Test Specimen) [also C-T] [Mechanics]
CT Compact Toroid (Fuelling Gun) [at Tokamak de Varennes, Quebec, Canada]
CT Compact-Type (Test Specimen) [also C-T] [Mechanics]
CT Complex Transformation
CT Composites Technology
CT Compound Turbine
CT Compressed Tablet(s) [Pharmacology]
CT Compression Test [Mechanics]
CT Computed Tomography; Computerized (Axial) Tomography
CT Computer Technologist; Computer Technology
CT Concrete Technician
CT Condensation Trail
CT Condensing Turbine
CT Conference Terms
CT Conference Title
CT Conjugation Tube (of Algae)
CT Connecticut [US]
CT Connective Tissue [Anatomy]
CT (The) Conservation Trust
CT Constant Teeming (Rate) [Metallurgy]
CT Continuous Transformation [Mathematics]
CT Continuous Time
CT Continuum Theory [Physics]
CT Control (Sample) [Symbol]
CT Controlled Temperature
CT Controlled Term
CT Control Tag
CT Control Term
CT Control Theory
CT Control Tower [Aeronautics]
CT Control Transformer
CT Conversion Time [Computers]
CT Convolution Theorem [Mathematics]
CT Cooling Tower
CT Cordless Telephone; Cordless Telephony
CT Core Transformer

CT Coronary Thrombosis [Medicine]

CT Corrosion-Protected Cable

CT α-Corticotropin [Biochemistry]

CT Counter Tube

CT Counting Time

CT Coupled Thickness

CT Crack Tip [Mechanics]

CT Crawler Transporter

CT Creep Test(ing)

CT Critical Temperature [Physical Chemistry]

CT Crosstalk [also XT]

CT Cryogenic Trap(ping)

CT Current Tracer; Current Tracing

CT Current Transformer

CT Curtis Tuirbine

CT Cytogenetic Technologist; Cytogenetic Technology

α-CT α-Chrymotrypsin [Biochemistry]

C_3T_2 Tricalcium Dititanate (Ceramics) [$3CaO \cdot 2TiO_2$]

C_4T_3 Tetracalcium Trititanate (Ceramics) [$4CaO \cdot 3TiO_2$]

CT-2 Digital Cordless Telephony, Second Generation

CT-3 Digital Cordless Telephony, Third Generation

C&T Chips and Technologies

C&T Communication(s) and Tracking

C/T Carrier-to-Noise Temperature (Ratio)

C/T Controlled Target

C-T Capacitance-Temperature (Characteristic)

C-T Compact Tension (Test Specimen) [also CT] [Mechanics]

C-T Compact-Type (Test Specimen) [also CT] [Mechanics]

C-T Concentration-Time (Curve)

C-T Cretaceous-Tertiary (Boundary) [also K-T] [Geology]

C(T) Temperature-Dependent Specific Heat

$C_v(T)$ Molar Heat (or Molar Heat Capacity) as a Function of Temperature

Ct Canton [also ct]

Ct Conduit [also ct]

Ct Connecticut [US]

Ct Count(er); Counting [also ct]

Ct Count(y) [also ct]

Ct Court [also ct]

C(t) Autocovariance Function [Symbol]

C(t) Time-Dependent Concentration [Symbol]

C(t) Time-Dependent Correlation Function [Symbol]

C(t) Transient Capacitance [Symbol]

$C_s(t)$ Time-Dependent Surface Concentration [Symbol]

ct carat [Unit]

ct cent [Currency]

ct *(centum)* – one hundred

ct conference terms

ct continental tropical air [Symbol]

ct *(cum tempore)* – 15 minutes after the time announced

$c_i(t)$ concentration of species I as a function of time [Symbol]

c't Magazin für Computertechnik [German Journal for Computer Technology]

χ-T (magnetic) susceptibility versus temperature (curve) [Symbol]

$\chi(T)$ temperature dependent (magnetic) susceptibility [Symbol]

$\chi_l(T)$ temperature dependent linear (magnetic) susceptibility [Symbol]

CTA Cable Television Association [UK]

CTA Calculated Time of Arrival

CTA California Teachers Association [US]

CTA California Trucking Association [US]

CTA Canadian Transit Association

CTA Canadian Trucking Association

CTA Canadian Tuberculosis Association

CTA Cellulose Triacetate

CTA Centre Technique de Cooperation Agricole et Rural [Technical Center for Agricultural and Rural Cooperation, Netherlands]

CTA Centre Technologique en Aérospatiale [Aerospace Technology Center, Quebec, Canada]

CTA Cetyltrimethylammonium

CTA Channel Tunnel Association [UK]

CTA Citraconic Anhydride

CTA Commercial Trailer Association [UK]

CTA Compatibility Test Area

CTA Computer Technology Associates [at NASA Goddard Space Flight Center, Greenbelt, Maryland, US]

CTA Computer Traders Association [UK]

CTA Concrete Technicians Association [US]

CTA Connecticut Transportation Authority [US]

CTA Constant-Temperature Anemometer

CTA Constant-Time Anemometer

CTA Control Area [Aeronautics]

CTA Controlled Airspace

CTA Controlled Thrust Assembly

CTA Cotton Textile Arrangement [US] [Defunct]

CTA Cyanuric Triazide

CTA Cystine Trypticase Agar [Biochemistry]

CTAB Canadian Technology Accreditation Board

CTAB Cetyltrimethylammonium Bromide

CTAB Commerce Technical Advisory Board [US]

C8TAB n-Octyltrimethylammonium Bromide

C10TAB n-Decyltrimethylammonium Bromide

C14TAB n-Tetradecyltrimethylammonium Bromide

CTABr Cetyltrimethylammonium Bromide

CTAC Cetyltrimethylammonium Chloride [also CTACl]

CTACl Cetyltrimethylammonium Chloride [also CTAC]

CTA3,5ClBz Cetyltrimethylammonium 3,5-Dichlorobenzoate

CTACN Cetyltrimethylammonium Cyanide

CTAF Common Traffic Advisory Frequency [US]

CTAHE Ceramic Technology for Advanced Heat Engines

CTAHEP Ceramic Technology for Advanced Heat Engine Program [of US Department of Energy]

CTAHS Cetyltrimethylammonium Hydrogen Sulfate
CTAK Cipher Text Auto Key
CTAL Confederación de Trabajadores de América Latina [Confederation of Workers of Latin America]
CTAOH Cetyltrimethylammonium Hydroxide
CTAS Canadian Thermal Analysis Society
CTAT Cetyl Trimethyl Ammonium Tosylate
CTAX Cetyltrimethylammonium Halide
CTB California Test Bureau [Los Angeles, US]
CTB Cipher Type Byte
CTB Code Table Buffer
CTB Coherent Twin (Grain) Boundary [Materials Science]
CTB Comprehensive Test Ban
CTB Concentrator Terminal Buffer
CTBL Cloud-Topped Boundary Layer
CT-BL Combined Transport Bill of Lading
CTBM Chief Testboard Man
CTBN Carboxyl-Terminated Butadiene Acrylonitrile
CTBS Canadian Test of Basic Skills
CTBT Comprehensive Test Ban Treaty
CTBUH Council on Tall Buildings and Urban Habitat [US]
CTC Camera, Timing and Control
CTC Canadian Transport Commission
CTC Carbon Tetrachloride
CTC Center for Technology Commercialization [US]
CTC Centralized (Railway) Traffic Control
CTC Central Training Council [UK]
CTC Central Trust of China [Taiwan]
CTC Channel-to-Channel
CTC Chargeable Time Clock
CTC Charge Transfer Complex [Chemistry]
CTC Chief Test Conductor
CTC Chlorotetracycline
CTC Chonju Technical College [South Korea]
CTC Cis-Trans-Cis (Geometry) [Chemistry]
CTC City Technology Center [UK]
CTC Civilian Technology Corporation [US]
CTC Client-To-Client (Protocol) [Internet]
CTC Colombian Trade Commission
CTC Communications Transistor Corporation [US]
CTC Compact Transpiration Cooling [Aerospace]
CTC Computer Training Center
CTC Concurrent Technologies Corporation [Johnstown, Pennsylvania, US]
CTC Conditional Transfer of Control
CTC Confederación dos Trabajadores Colombiana [Confederation of Colombian Workers]
CTC Consumer's Transport Council
CTC Control Technology Center
CTC Counter/Timer Circuit
CTC Cultural Training Center
CTCA Canadian Telecommunications Carrier Association
CTCA Ceramic Tile Contractors Association
CTCA Channel and Traffic Control Agency

CTCA Channel-to-Channel Adapter
CTCAC Ceramic Technology Center for Advanced Ceramics
CTCC Combustion Turbine Combined Cycle
Ctch Catechol
CTCI 1-Cyclohexyl-3-(3'-Trimethylaminopropyl) carbodiimide Iodide
CTCL Columnar Transcrystalline Layer
CTCL Community and Technical College Libraries
C$_{60}$-TCNQ Buckminsterfullerene-60 Tetracyanoquinodimethane
CTCP Client-To-Client Protocol [Internet]
CTCP Clinical Toxicology of Commercial Products [Database of US Environmental Agency]
CTCS Consolidated Telemetry Checkout System
C&TCS Communications and Tracking Checkout System
CTCU Canadian Textile and Chemical Union
CTD Charged Tape Detection
CTD Charge-Transfer Device
CTD Chemical Technology Division
CTD Chemical Technology Division [of Argonne National Laboratory, Illinois, US]
CTD Chemical Technology Division [of Oak Ridge National Laboratory, Tennessee, US]
CTD Chemical Thermodynamics
CTD Chemical Transport and Deposition
CTD Coated (Lens)
CTD Combination Thermal Drive
CTD Commercial Technology Division [of NASA]
CTD Communications Technology Division [of NASA Lewis Research Center, Cleveland, Ohio, US]
CTD Conductivity, Temperature and Density
CTD Conductivity, Temperature and Depth
CTD Continuity Tone Detector
CTD Corporate Technical Development
CTD Cross Track Distance [Aeronautics]
CTD Cumulative Trauma Disorder [Medicine]
C2D Character to Decimal [Restructured Extended Executor (Language)]
C2D Conducting along Two Dimensions
ctd coated
CTDA Ceramic Tile Distributors Association [US]
CTDAS Canadian Trade Document Alignment System
CT-DEC Centre Technique de l'Industrie du Décolletage [Technical Center for the Free-Cutting Industries, France]
CTDH Command and Telemetry Data Handling
CTDS Code Translation Data System
C/TDS Count/Time Data System
CTE Cable Television Engineer
CTE Central Timing Equipment
CTE Centro de Tecnologias Especiais [Center for Special Technologies; of Brazilian National Institute of Space Research]
CTE Channel Translating Equipment
CTE Charge-Transfer Efficiency
CTE Coefficient of Thermal Expansion

CTE Computer Telex Exchange
CTE Confederación de Trabajadores del Ecuador [Confederation of Ecuadorean Workers]
Cte Carbon Tetrachloride
CTEA Canadian Telephone Employees Association
CTEA Channel Transmission and Engineering Activation
CTEA Council of Technology Education Association [US]
CTEA Cyanotriethylammonium Tetrafluoroborate
CTEC Commonwealth Tertiary Education Commission
c-Te₃Cl₂ Crystalline Tellurium Chloride
CTEF Coherence Transfer Echo Filtering [Physics]
CTEF Comité Technique Européenne pour le Fluor [European Technical Committee for Fluorine; of Conseil Européen des Fédérations de l'Industrie Chimique, Belgium]
CTEI Cloud-Top Entrainment Instability
CTEM Conventional Transmission Electron Microscope; Conventional Transmission Electron Microscopy
CTEM Cross-Section Transmission Electron Microscopy
CTEM-LACBED Conventional Transmission Electron Microscopy/Large-Angle Convergent-Beam Electron Diffraction
CTEM/STEM Conventional Transmission Electron Microscopy/Scanning Transmission Electron Microscopy
CTEN Count Enable (Input) [Digital Counter]
CTEP Center for Transport Engineering Practice [UK]
CTES Chlorotriethoxysilane
CTES Coiled Tubing Engineering Services [US]
C2E2 Collaborating Center on Energy and Environment [of United Nations Environment Program] [also c2e2]
CTF Cask Tilting Fixture [Nuclear Reactors]
CTF Chlorine Trifluoride
CTF Cleanroom Technology Forum [US]
CTF Coal Tar Fuel
CTF Coffee Trade Federation [UK]
CTF Colloquium on Thin Films [Solid-State Physics]
CTF Colorado Tick Fever
CTF Commander Task Force
CTF Common Translation Factor
CTF Contrast Transfer Function
CTF Controlled Temperature Furnace (Atomizer)
CTF Controlled Thermonuclear Fission
CTF Cottonseed Flour
CTF Counting Test Facility
Ctf Certificate [also ctf]
CTFCB 1-Chloro-1,2,2-Trifluorocyclobutane
CTFE Chlorotrifluoroethylene (Fluoropolymer)
CTFM Continuous-Transmission, Frequency-Modulated
CTFMP Constrained Thin-Film Melt Polymerization
CTFV Colorado Tick Fever Virus
CTG Cardiotocograph(y)
CTG Communications Task Group [of CODASYL–Conference on Data System Languages]
CTG Control Techniques Guidelines
Ctg Cartage [also ctg]

Ctg Cartridge [also ctg]
Ctg Crating [also ctg]
CTGC California Table Grape Commission [US]
Ctge Cartage [also ctge]
Ctge Cartridge [also ctge]
ctgm centigram [also cgm or cg]
CTGN Centre des Technologies du Gaz Naturel [Center for Natural Gas Technologies, Quebec, Canada]
CTH Chalmers Tekniska Högskola [Chalmers University of Technology, Gothenburg, Sweden]
cth, dn concentration detection limit (of odorants) [Symbol]
cth, rn concentration recognition threshold (of odorants) [Symbol]
CTI Canadian Telecommunications Industry
CTI Canadian Textile Institute
CTI Canadian Trade Index
CTI Centralized Ticket Investigation
CTI Centre de Traitement de l'Information [Information Processing Center, Belgium]
CTI Cleveland Trainrail International [Ohio, US]
CTI Color Television Inc. [US]
CTI Color Transmission Index
CTI Comparative Track(ing) Index (of Insulators)
CTI Computerized Tomographic Imaging
CTI Computer-Telephony Integration
CTI Configuration and Tuning Interface
CTI Computer to Telephony Integration
CTI Cooling Tower Institute [US]
CTI Critical Technologies Institute [now Science and Technology Policy Institute, US]
CTI Critical Transportation Item
CTI Current Technology Index
CTIA Canadian Textile Importers Association
CTIA Cellular Telecommunications Industry Association [US]
CTIA Cellular Telephone Industry Association [US]
CTIAC Concrete Technology Information Analysis Center [US]
CTIC Canadian Translators and Interpreters Council
CTIF Comité Technique International de Prévention et d'Extinction du Feu [International Technical Committee for the Prevention and Extinction of Fire, Switzerland]
CTIO Cerro Tololo Inter-American Observatory [La Serena, Chile]
CTIOA Ceramic Tile Institute of America [US]
CTIO/ESO Cerro Tololo Inter-American Observatory/European Southern Observatory (Workshop)
CTIS Computerized Transport Information System
CTIS Crawler Transporter Intercom System
CTL CAGE (Compiler and Assembler by General Electric) Test Language
CTL Canoga Test Laboratory [US]
CTL Cassette Tape Loader
CTL Charge Transport Layer
CTL Checkout Test Language
CTL Cincinnati Testing Laboratories [Ohio, US]
CTL Coaxial Transmission Line

CTL Columbus Testing Laboratory [Ohio, US]

CTL Commercial Tape Laying Machine

CTL Complementary Transistor Logic

CTL Constant Time Locus

CTL Constructive Total Loss

CTL Control [also Ctl]

CTL Core Transistor Logic

CTL Cytotoxic T Lymphocytes [Biochemistry]

Ctl Control [also ctl]

ctl central

CTLM Commercial Tape Laying Machine

CTLZ Control Zone [also Ctl Z]

CTM Chemithermomechanical; Chemithermomechanics

CTM Combustion Theory and Modeling [Journal of Institute of Physics, UK]

CTM Communications Terminal Module

CTM Complete Treatment Module

CTM Confederación de Trabajadores Mexicana [Confederation of Mexican Workers]

CTM Continuity Transceiver Module

CTM Contract Technical Manager

CTM Costs to Manufacture

CTM Counter Timer Multimeter

CTM Crystalline Transitional Material

CTMA Canadian Tooling Manufacturers Association

CTMA n-Cetyltrimethylammoniumchloride

CTMA Civil Engineering Test Equipment Manufacturers Association [UK]

CTMA Cutting Tool Manufacturers Association [now United States Cutting Tool Institute]

CTMC Canadian Tobacco Manufacturers Council

CTMC Communication Terminal Module Controller

CTMD CANMET (Canada Center for Mineral and Energy Technology)/Technology Marketing Division [of Natural Resources Canada]

CTMF Ceramic Tile Marketing Federation [US]

CTMP Chemithermomechanical Pulp [Papermaking]

CTMS Carrier Transmission Maintenance System

CTMS Computer-Controlled Test Management System

CTMT Combined Thermomechanical Treatment [Metallurgy]

Ctmt Containment [also ctmt]

CTMTN Cyclotrimethylenetrinitramine [Explosive]

CTN Cable Termination Network

CTN Certification Test Network

CTN Close-to-Net (Rolling Mill) [Metallurgy]

Ctn Carton [also ctn]

ctn cotangent [also cot]

ctn^{-1} inverse hyperbolic cotangent

CTNB Charge Transfer No-Bond

CTNE Compania Teléfonica Nacional de Espania [National Telephone Company of Spain]

ctnh hyperbolic cotangent [also coth]

$ctnh^{-1}$ inverse hyperbolic cotangent [also $coth^{-1}$]

C13-NMR Carbon-13 Nuclear Magnetic Resonance

C13-NMR Carbon-13 Nuclear Magnetic Resonance Database [of Information System Karlsruhe, Germany]

CTNR Controlled Thermonuclear Reaction

CTO Charge Transforming Operator

CTO Chief Technical Officer

CTO Chief Technology Officer

CTO Close Tolerance Orifice

CTO Combined Transport Operator

CTO Commonwealth Telecommunications Organization [UK]

CTO Corporate Telecommunications Operation

CTO Cut-Off

CTOA Crack Tip Opening Angle [Mechanics]

C to C Center to Center [also c to c]

C-to-C Computer-to-Computer

CTOD Crack Tip Opening Displacement [Mechanics]

$CTOD_c$ Critical Crack Tip Opening Displacement [Mechanics]

C to F Center to Face [also c to f]

CTOL Conventional Takeoff and Landing

CTON Character to Number (Conversion)

CTOS Cassette Operating System

CTOS Cassette Tape Operating System

CTOS Computerized Tomography Operating System

CTOS Convergent Technologies Operating System

CTP Calcium Titanium Orthophosphate [$CaTi_4(PO_4)_6$]

CTP Carboxyl-Terminated Polybutadiene

CTP Center for Theoretical Physics

CTP Central Transfer Point

CTP Ceramic Technology Project [of US Department of Energy]

CTP Character Table Pointer

CTP Character Translation Pointer

CTP Charge Transforming Parameter

CTP Chemical Treatment Pond

CTP Coal Tar Pitch

CTP Columbus Telescope Project [US]

CTP Command Translator and Programmer

CTP Communications Timing Procedure

CTP Confederación de Trabajadores del Peru [Confederation of Peruvian Workers]

CTP Construction Test Procedure

CTP Continuous Tender Panel

CTP Controlled Temperature Profile

CTP Coordinated Test Plan

CTP Crystal Transition Point

CTP Cyclic Time Processor

CTP Cyclohexylthiophthalimide

CTP Cytidine 5'-Triphosphate [Biochemistry]

5'-CTP Cytidine-5'-Triphosphate [Biochemistry]

$\chi(T,P)$ Temperature and Pressure Dependent (Magnetic) Susceptibility

CTPA Coax(ial)-to-Twisted-Pair Adapter

CTPB Carboxyl-Terminated Polybutadiene Binder

CTPD Crew Training and Procedures Division [of NASA Johnson Space Center, Houston, Texas, US]

CTPR Confederación del Trabajo de Puerto Rico [Puerto Rico Federation of Labour]

CTPS sym-Collidinium (Toluol-p-Sulfonate)

CTPV Coal Tar Pitch Volatiles

CTR California Tumor Registry [US]

CTR Carbothermal Reduction (Process) [Metallurgy]

CTR Center for Transportation Research [of Argonne National Laboratory, Illinois, US]

CTR Central Tumor Registry

CTR Ceramic Turbocharger Rotor

CTR Certified Test Requirement

CTR Certified Test Results

CTR Chemisch-Technische Reichsanstalt [Institute for Chemical Engineering Research, Germany] [now Chemisch-Technische Versuchsanstalt]

CTR Civil Tilt Rotor

CTR Collective Television Reception

CTR Computerized Tumor Registry

CTR Computing-Tabulating-Recording

CTR Contract Technical Representative

CTR Control Zone [Aeronautics]

CTR Controlled Temperature Resistor

CTR Controlled Thermonuclear Reaction; Controlled Thermonuclear Reactor

CTR Cosmetic Transdermal (Hair) Reconstruction [Medicine]

CTR Council for Tobacco Research [US]

CTR Counter [also Ctr or ctr]

CTR Crack-Thickness Ratio [Mechanics]

CTR Crystal Truncation Rod

CTR Current Transfer Ratio

Ctr Center(ing) [also ctr]

Ctr Contour [also ctr]

Ctr Control [also ctr]

Ctr Counter [also ctr]

CTRA Coal Tar Research Association [UK]

CTRAP Customer Trouble Report Analysis System

CTRCO Calculating-Tabulating-Recording Company [now IBM (International Business Machines) Corporation, US]

CTR DIV x Counter with x States [e.g., CTR DIV 16 is a Counter with Sixteen States] [Logic Symbol]

Ct Rend Acad Sci Paris Comptes Rendus des Séances de l'Academie des Sciences de Paris [Proceedings of the Academy of Sciences of Paris, France]

Ct Rend Acad Sci Sér A Comptes Rendus de l'Academie des Sciences, Série A [Proceedings of the Academy of Sciences, Series A, France]

Ct Rend Acad Sci Sér B Comptes Rendus de l'Academie des Sciences, Série B [Proceedings of the Academy of Sciences, Series B, France]

Ct Rend Acad Sci Sér C Comptes Rendus de l'Academie des Sciences, Série C [Proceedings of the Academy of Sciences, Series C, France]

CTRF Canadian Transport Research Forum

CTRIPS Charge Transfer Reaction Inverse Photoemission

CTRL Control [also Crtl, or ctrl] [Computers]

CTRL Central Technical Research Laboratory [Nagoya and Yokohama, Japan]

CTR/N Carbothermal Reduction and Nitration (Process) [Metallurgy]

CTRS Component Test Requirements Specification

Ctrs Contrast [also ctrs]

CTR-USA Council for Tobacco Research–USA

CTRW Continuous Time Random Walk

CTS Cable Terminal System

CTS Cable Transmission System

CTS Cable Turning System

CTS Call to Stations

CTS Canadian Technology Satellite

CTS Carpal Tunnel Syndrome [also Repetitive Strain Injury]

CTS Carrier Test Switch

CTS Central Target Simulation

CTS Centre de Télécommunications Spatiales [Space Telecommunications Center, France]

CTS Cesium Time Standard

CTS Chain-Terminating Step [Chemistry]

CTS Charge-Transfer Spectrum

CTS Charge-Transfer State

CTS Chemically-Treated Steel

CTS Clear to Send

CTS Committee on the Teaching of Science [of International Council of Scientific Unions]

CTS Common Test Subroutine

CTS Communications and Tracking Subsystem

CTS Communications and Tracking System

CTS Communications Technology Satellite [US/Canada]

CTS Communications Terminal Synchronous

CTS Compact-Tension Specimen [Mechanics]

CTS Complex-Tension Sample

CTS Computer-Controlled Testing System

CTS Computerized Training System

CTS Computer Telewriter System [US]

CTS Computer Test Sequence

CTS Computer Test Set

CTS Computer Typesetting

CTS Contract Technical Services

CTS Contralateral Threshold Shift

CTS Contrast Transfer Function

CTS Controlled Thermal Severity Test

CTS Convention on the Territorial Sea [of United Nations]

CTS Conversational Terminal System

CTS Conversational Time Sharing

CTS Coordinate Transformation System

CTS Counter-Timer System

CTS Courier Transfer Station [US Military]

CTS Correction-to-Scaling (Exponent)

CTS Curved Track Simulator

CTS Customer Telephone System

C2S Congrès Scientifique Services [Saint Cloud, France]

cts cents [also ¢]

CTSA Crucible and Tool Steel Association [UK]

CTSC Cleveland Technical Societies Council [Ohio, US]

CT Scan Computed (Axial) Tomography

CTSD Computer Test Sequences Document

CTSD Crack Tip Sliding Displacement [Mechanics]

CTSE Commodity Classification for Transport Statistics in Europe

CTSF Crack-Tip Stress Field [Mechanics]

CTSI Central Terminal Signaling Interface

CTSIBV Centre Technique et Scientifique de l'Industrie Belge du Verre [Scientific and Technical Center of the Belgian Glass Industry]

cts/lb cents per pound [also ¢/lb]

cts/min counts per minute [also cnts/min, CPM, or cpm]

cts/sec counts per second [also cnts/min, CPS, or cps]

CTS/RTS Clear-to-Send/Request-to-Send

CTSS Compatible Time-Sharing System

CTSS Conversational Time Sharing System

C&TSS Communication and Tracking Subsystem

CTST Classical Transition State Theory [Physics]

CTST Conventional Transition State Theory [Physics]

CTST Critical Trade Skills Training [Canada]

cts/t oz cents per troy ounce [also ¢/t oz]

CTT Cable Trouble Ticket

CTT Capital Transfer Tax [UK]

CTT Central Trunk Terminal

CTT Character Translation Table

CTT Chlorotris(1-Methylethanolato)titanium

CTT Cis-Trans-Trans (Geometry) [Chemistry]

CTT Cloud Top Temperature

CTT College of Trade and Technology

CTT Consejo Técnica de Telecommunicación [Technical Council for Telecommunications, Madrid, Spain]

CTT Core-Type Transformer

CTT Cross-Type Telescope

C/TT Computer/Test Technology [of Institute of Electrical and Electronics Engineers, US]

CTTB Checkout Techniques Test Bed

CTTC Canadian Trade and Tariffs Committee

CTTE Council of Technology Teacher Education [US]

CTTF Continuous Thermal Treatment Facility

CTTL Charge Transfer to Ligand

CTTL Complementary Transistor-Transistor Logic

CTTM Charge Transfer to Metal

CTTMA Canadian Truck and Trailer Manufacturers Association

CTTS Charge Transfer to Solvent

CTTY Change Terminal (Command) [also ctty]

CTU Caribbean Telecommunications Union

CTU Cartridge Tape Unit

CTU Cassette Tape Unit

CTU Centigrade Thermal Unit [also Ctu, or ctu]

CTU Central Terminal Unit

CTU Central Timing Unit

CTU Channel Testing Unit

CTU Communication Terminal Unit

CTU Compatibility Test Unit

CTU Czech Technical University [Prague, Czech Republic]

CTUC Committee on Tunneling and Underground Construction [US]

CTUC Commonwealth Trade Union Council [UK]

Ctu/ft² Centigrade thermal unit(s) per square foot [also Ctu/sq ft]

Ctu/ft·h Centigrade thermal unit(s) per foot hour [also Ctu/(ft·h)]

Ctu/ft·h·°C Centigrade thermal unit(s) per foot hour degree centigrade [also Ctu/(ft·h·°C)]

Ctu/ft²·h·°C Centigrade thermal unit(s) per square foot hour degree centigrade [alsoCtu/ft²·h·°C, or Ctu/sq ft·h·°C]

Ctu/in² Centigrade thermal unit(s) per square inch [also Ctu/sq in]

Ctu/lb Centigrade thermal unit(s) per pound [also Ctu/lb]

Ctu/lb·°C Centigrade thermal unit(s) per pound degree centigrade [also Ctu/(lb·°C)]

Ctu/SCF Centigrade thermal unit(s) per Standard Cubic Foot

Ctu/SCF·°C Centigrade thermal unit(s) per Standard Cubic Foot per Degree Centigrade (or Celsius) [also Ctu/(SCF·°C)]

Ctu/sq ft Centigrade thermal unit(s) per square foot [also Ctu/ft²]

Ctu/sq ft·h·°C Centigrade thermal unit(s) per square foot hour degree centigrade [also Ctu/ft²·h·°C, or Ctu/(ft²·h·°C)]

Ctu/sq in Centigrade thermal unit(s) per square inch [also Ctu/in²]

CTV Canadian Television

CTV Commercial Television

CTV Control Test Vehicle

CTV Cultural Television

CTV Chemisch-Technische Versuchsanstalt [Institute for Chemical Engineering and Testing, Berlin, Germany]

Ctwt Counterweight [also ctwt]

CTX Committed to Excellence

CTX Continuously Variable Transaxle

C2X Character To Hexadecimal [Restructured Extended Executor (Language)]

CTXCO Centrex Central Office

CTXCU Centrex Customer

CTXSN Centrex System Number

CU Cairo University [Egypt]

CU Cambridge University [UK]

CU Carleton University [Ottawa, Ontario, Canada]

CU Charles University [Prague, Czech Republic]

CU Chiba University [Chiba, Japan]

CU Chicago University [Illinois, US]

CU Chittagong University [Chittagong, Bangladesh]

CU Chongqing University [Chongqing, PR China]

CU Chubu University [Aichi, Japan]

CU Chulalongkorn University [Bangkok, Thailand]

CU Chungkung University [Tainam, Taiwan]

CU Cirrouncinus (Cloud) [also Cu]

CU Clark University [Worcester, Massachusetts, US]

CU Clemson University [South Carolina, US]

CU Close-Up

CU Coefficient of Utilization [Illumination Engineering]

CU Colgate University [Hamilton, New York, US]

CU Columbia University [New York City, US]

CU Common User

CU Communications Unit [of International Union for the Conservation of Nature and Natural Resources]

CU Community User

CU Common Update

CU Computer Usage; Computer User

CU Concordia University [Montreal, Quebec, Canada]

CU Construction Unit

CU Consumers Union (of the United States)

CU Control Unit [also cu] [Computers]

CU Cooling Unit

CU Cooperativity Unit

CU Cornell University [Ithaca, New York, US]

CU Crosstalk Unit

CU Crude Unit [Oil Refining]

CU (Piezoelectric) Crystal Unit

CU Cuba [ISO Code]

CU See You [Amateur Radio and Internet]

CU- Cuba [Civil Aircraft Marking]

Cu Cirrouncinus (Cloud) [also CU]

Cu Cumulus (Cloud)

Cu *(Cuprum)* – Copper [Symbol]

Cu Customer [also cu]

Cu$^+$ Monovalent Copper Ion [Symbol]

Cu^{2+} Divalent Copper Ion [Symbol]

Cu-7 Copper-7 (Intrauterine Device)

Cu-58 Copper-58 [also ^{58}Cu, or Cu^{58}]

Cu-60 Copper-60 [also ^{60}Cu, or Cu^{60}]

Cu-61 Copper-61 [also ^{61}Cu, or Cu^{61}]

Cu-62 Copper-62 [also ^{62}Cu, or Cu^{62}]

Cu-63 Copper-63 [also ^{63}Cu, Cu^{63}, or Cu]

Cu-64 Copper-64 [also ^{64}Cu, or Cu^{64}]

Cu-65 Copper-65 [also ^{65}Cu, or Cu^{65}]

Cu-66 Copper-66 [also ^{66}Cu, or Cu^{66}]

Cu-67 Copper-67 [also ^{67}Cu, or Cu^{67}]

cu cubic

CUA Catholic University of America [Washington, DC, US]

CUA Circuit Unit Assembly

CUA Commercial Utilization Area

CUA Common User Access [IBM Corporation]

CUA Compugraphics Users Association

CUA Computer Users Association

CUA Confederation of University Administrators [UK]

Cu(Ac)$_2$ Copper(II) Acetate (or Cupric Acetate) [also Cu(ac)$_2$]

Cu(ACAC)$_2$ Copper(II) Acetylacetonate [also Cu(AcAc)$_2$, or Cu(acac)$_2$]

Cu-Ag Copper-Silver (Alloy System)

CuAl Copper Aluminum (Alloy)

CuAl Copper Monoaluminide (Compound)

Cu-Al Copper-Aluminum (Alloy System)

CuAlMn Copper Aluminum Manganese (Shape Memory Alloy)

CuAlNi Copper Aluminum Nickel (Shape Memory Alloy)

Cu-Al$_2$O$_3$ Alumina Dispersion-Strenghened Copper

Cu/Al$_2$O$_3$ Copper Reinforced Alumina (or Aluminum Oxide)

Cuam Cuprammonium

CUAP Catholic University of America Press [Washington, DC, US] [also CUA Press]

CUAS Computer Utilization Accounting System

Cu-As Copper-Arsenic (Alloy System)

CuAu Copper Gold (Spin Glass)

CuAu I Copper Gold I (Superlattice)

CuAu II Copper Gold II (Superlattice)

Cu-Au Copper-Gold (Alloy System)

CUB Commonality Usage Board

CUB Council for UHF (Ultrahigh Frequency) Broadcasting [US]

CUB Cursor Backward

Cub Cuba(n)

cub cubic

CUBE Concertation Unit for Biotechnology in Europe [also Cube]

CUBE Cooperating Users of Burroughs Equipment [US]

CuBe Copper Beryllium (Compound)

Cu-Be Copper-Beryllium (Alloy System)

Cu-BeO Beryllia Dispersion Strenghened Copper

Cu-Bi Copper-Bismuth (Alloy System)

CUBIC Cosmic Unresolved X-Ray Background Instrument [of Penn State University, University Park, US]

CUBOL Computer Usage Business-Oriented Language

Cub$ Cuban Peso [Currency of Cuba]

CuBr Copper(I) Bromide (or Cuprous Bromide)

CUC Claremont University Center [Californian University Center including Pomona College, Claremont Graduate School, Scripps College, Claremont KcKenna College, Harvey Mudd College and Pitzer College]

CUC Coal Utilization Council [UK]

CUC Computer Usage Control

CUC Computer Users Committee

CUC Cooperative Union of Canada

Cu-Ca Copper-Calcium (Alloy System)

CUCC Concordia University Computer Centre [Montreal, Quebec, Canada]

CUCCA Canadian University and College Counselling Association

CUCCA Columbia University Center for Computing Activities [New York City, US]

Cu-Cd Copper-Cadmium (Alloy System)

CuCl Copper(I) Chloride (or Cuprous Chloride)

CuCl$_2$ Copper(II) Chloride (or Cupric Chloride)

CuCl$_2$-GIC Copper(II) Chloride (or Cupric Chloride) Graphite Intercalation Compound

cu cm cubic centimeter [also cm^3]

CuCN Copper(I) Cyanide (or Cuprous Cyanide)

Cu-CO Copper-Cobalt (Alloy System)

CUCPS Columbia University College of Physicians and Surgeons [New York City, US]

Cu-Cr Copper-Chromium (Alloy System)

CuCTez Copper Chlorotetrazole

CUD Could [Amateur Radio]

CUD Craft Union Department [of American Federation of Labor and Congress of Industrial Organizations, US]

CUD Cursor Down

CUDAT Common User Data

CuDBC Copper Dibutyl Dithiocarbamate

CuDIP Copper Diisopropyl Dithiophosphate

Cu-DLP Deoxidized, Low-Phosphorus Copper

cu dm cubic decimeter [also cdm, or dm^3]

Cu(DMAE)$_2$ Copper(II) Dimethylaminoethoxide [also Cu(dmae)$_2$]

CuDMC Copper Dimethyl Dithiocarbamate

CUDN Common User Data Network

CUDOS Continuously-Updated Dynamic Optimizing System

Cu(DPM)$_2$ Copper (II) Dipivaloylmethanoate (or Bis(dipivaloylmethanato)copper (II)) [also Cu(dpm)$_2$]

CUDS Cumulative Data Statistics

Cu:Dy Dysprosium-Doped Copper [also Cu:Dy^{3+}]

CUE Common Usage Equipment

CUE Computer Updating Equipment

CUE Control Unit End [Computers]

CUE Cooperating Users Exchange

CUE Custom Updates and Extras

CUEA Coastal Upwelling Ecosystem Analysis

CUEBS Commission on Undergraduate Education in the Biological Sciences [of American Institute of Biological Sciences, US]

CUED Cambridge University Engineering Department [UK]

CUED-TR Cambridge University Engineering Department Technical Report [also CUED TR, or CUED Tech Rep]

CuEDTA Copper Ethylenediaminetetraacetate [also CuEdta]

CUEFA Centre Universitaire de Formation des Adultes [University Center for Adult Education, Grenoble, France]

Cuen Cupriethylenediamine [also cuen]

CUEP Central Unit on Environmental Pollution [UK]

Cu:Er Erbium-Doped Copper [also Cu:Er^{3+}]

CUES Center for Urban Environmental Studies [now National Center for Urban Environmental Studies, US]

CUES Computer Utility Educational System

Cu(EtO)$_2$ Copper(II) Ethoxide

CUETS Credit Union Electronic Transaction Service [Canada]

CUEW Canadian Union of Educational Workers

CUF Cross Utilization File

CUF Cumuliform (Cloud)

CUF Cursor Forward

CuF Cuprous Fluoride

CUFA Concordia University Faculty Association [Montreal, Quebec, Canada]

CUFC Consortium of University Film Centers [US]

Cu(Fe) Iron-Doped Copper

Cu-Fe Copper-Iron (Alloy System)

CuFeAl Copper-Iron-Aluminum (Compound)

Cu(FOD)$_2$ Copper(II) Bis(1,1,1,2,2,3,3-Heptafluoro-7,7-Dimethyl-4,6-Octanedionate) [also Cu(fod)$_2$]

CUFOS Center for UFO (Unidentified Flying Object) Studies [US]

CUFS Consumers United for Food Safety [Seattle, US]

CUFT Center for the Utilization of Federal Technology [US]

cu ft cubic foot [also cu-ft, Cf, cf, or ft^3]

cu ft/h cubic feet per hour [also cu-ft/h, cu ft/hr, cu-ft/hr, ft^3/h, or ft^3/hr]

cu ft/lb cubic feet per pound [also cu-ft/lb, or ft^3/lb]

cu ft/min cubic feet per minute [also cu-ft/min, ft^3/min, CFM, cfm, or cf/min]

CUG China University of Geosciences [Wuhan, PR China]

CUG Closed User Group

CUG CP/M (Control Program for Microcomputer) User Group [US]

CUG Crosfield Users Group [US]

CuGa Copper Gallium (Compound)

Cu-Ga Copper-Gallium (Alloy System)

CuGaS$_2$ Copper Gallium Selenide (Semiconductor)

CuGaSe$_2$ Copper Gallium Selenide (Semiconductor)

Cu-Ge Copper-Germanium (Alloy System)

CUH Chaminade University of Honolulu [Hawaii, US]

CUH Chinese University of Hong Kong

Cu(HFAC) Copper(I) Hexafluoroacetylacetonate [also Cu(hfac)]

Cu(HFAC)$_2$ Copper(II) Hexafluoroacetylacetonate [also Cu(hfac)$_2$]

Cu(HFAC)BTMSA Copper (I)1,1,1,5,5,5-Hexafluoro-2,4-Pentanedione Bis(trimethylsilyl)acetylene [also Cu(hfac)btmsa]

Cu(HFAC)COD Copper (I) 1,1,1,5,5,5-Hexafluoro-2,4-Pentanedione 1,5-Cyclooctadiene [also Cu(hfac)COD]

Cu(HFAC)-1,5-COD Copper (I) 1,1,1,5,5,5-Hexafluoro-2,4-Pentanedione 1,5-Cyclooctadiene [alsoCu(hfac)-1,5-COD]

Cu(HFAC)PMe$_3$ Copper (I) 1,1,1,5,5,5-Hexafluoro-2,4-Pentanedione Trimethylphosphine [also Cu(hfac)PMe$_3$]

Cu(HFAC)TMVS Copper (I) 1,1,1,5,5,5-Hexafluoro-2,4-Pentanedione Trimethylvinylsilane [also Cu(hfac)tmvs]

Cu(HFAC)VTMS Copper (I) 1,1,1,5,5,5-Hexafluoro-2,4-Pentanedione Vinyltrimethylsilane [also Cu(hfac)vtms]

Cu-Hg Copper-Mercury (Alloy System)

CUHK Chinese University of Hong Kong

CU-HKSM Columbia University, Henry Krumb School of Mines [New York City, US]

CUI Character-Oriented User Interface

CUI Common User Interface

CUI Consistent User Interface

CuI Copper(I) Iodide (or Cuprous Iodide)

Cu(I) i-Bu Copper (I) Isobutyrate

Cu(II) i-Bu Copper (II) Isobutyrate

CUICAC Canadian University-Industry Council on Advanced Ceramics

CUIIPS Center Unit for International and Investment Policies Studies [Canada]

CUIL Common Usage Item List

CuIn Copper Indium (Spin Glass)

Cu-In Copper-Indium (Alloy System)

cu in cubic inch [also cu-in, CI, ci, or in³]

CuInGaSe₂ Copper Indium Gallium Selenide (Semiconductor)

cu in/lb cubic inch(es) per pound [also in³/lb]

cu in/min cubic inch(es) per minute [also cu-in/min, or in³/min]

cu in/rad cubic inch(es) per radian [also in³/lrad]

CuInS₂ Copper Indium Sulfide (Semiconductor)

CuInSe₂ Copper Indium Selenide (Semiconductor)

CUJT Complementary Unijunction Transistor

CuK Copper K (Absorption Edge) [X-Ray Analysis]

CuKα Copper K-Alpha (Radiation) [also CuK$_\alpha$]

CuKα₁ Copper K-Alpha 1 (Radiation) [also CuK$_{\alpha1}$]

CuKβ Copper K-Beta (Radiation) [also CuK$_\beta$]

cu km cubic kilometer [also km³]

CUL Cambridge University Library [UK]

CUL Canadian Underwriters Laboratory

CuL Copper Ligand

Cu-La Copper-Lanthanum (System)

CuLα Copper L-Alpha (Radiation) [also CuL$_\alpha$]

Cu L/F Copper-Based Leadframe

CULO Cornell University Laboratory of Ornithology [now Cornell Laboratory of Ornithology, Ithaca, New York, US]

CULP Computer Usage List Processor

CuLPCN Copper Leucophalocyanine

CULT Chinese University Language Translation [Hong Kong]

CUM Cumulative Effects (of Hazardous Materials)

cum cumulative

cu m cubic meter [also cu-m, or m³]

cu μ cubic micron [also cu mu, cu-mu, or μm³]

CUMA Canadian Underwater Mine Apparatus

CUMA Canadian Urethane Manufacturers Association

CUMARC Cumulated Machine-Readable Catalogue

Cumb Cumberland [UK]

CUMC Cornell University Medical Center [New York City, US]

cu m/d cubic meter(s) per day [also cu-m/d, or m³/d]

cum div with dividend

Cu(MeO)₂ Copper(II) Methoxide

Cum Freq Cumulative Frequency

Cu-Mg Copper-Magnesium (Alloy System)

Cu-MgO Magnesium Oxide Dispersion Strengthened Copper

CUMM Council of Underground Machinery Manufacturers

cu mm cubic millimeter [also cu-mm, or mm³]

cu m/min cubic meters per minute [also cu-m/min, CMPM, cmpm, or m³/min]

CuMn Copper Manganese (Spin Glass)

Cu-Mn Copper-Manganese (Alloy System)

CuMnAu Copper Manganese Gold (Spin Glass)

Cu-MnO Magnesium Oxide Dispersion Strengthened Copper

Cu-Mo Copper-Molybdenum (Alloy System)

Cu/Mo$_p$ Molybdenum Particulate Reinforced Copper (Composite)

CUMOX Cumene Oxidation (Process)

Cum Pref Cumulative Preference

CUMREC College and University Machine Records Conference [US]

Cu-MTS Copper Muffin Tin Sphere [Physics]

cu mu cubic micron [also cu-mu, cu μ, or μm³]

Cu-Nb Copper-Niobium (Alloy System)

Cu-Ni Copper-Nickel (Alloy System)

Cu-NiO Nickel Oxide Dispersion Strengthened Copper

cu nm cubic nanometer [also nm³]

Cu-63 NMR Copper-63 Nuclear Magnetic Resonance [also ⁶³Cu NMR]

CUNY City University of New York [US]

CUNY-BC City University of New York–Brooklyn College [US]

CUNY-CC City University of New York–City College [US]

CUNY-CSI City University of New York–College of Staten Island [US]

CUNY-HC City University of New York–Hunter College [US]

CUNY-MC City University of New York–Manhattan College [US]

CUNY-QC City University of New York–Queens College [Flushing, US]

CUNY-YC City University of New York–York College [US]

CUO Cambridge University Observatory [UK]

CuO Copper(II) Oxide (or Cupric Oxide)

Cu(OAc)₂ Copper (II) Acetate

Cu₄(OBuᵗ)₈ Copper (II) tert-Butoxide (or Copper (II) tert-Butylate)

CUOE Canadian Union of Operating Engineers and General Workers

CuOEP 2,3,7,8,12,13,17,18-Octaethylporphine Copper(II)

Cu-OF Oxygen-Free Copper

Cu-OFHC Oxygen-Free, High-Conductivity Copper

CuOH Cuprous Hydroxide

Cu(OMe)₂ Copper(II) Methoxide (or Copper (II) Methylate)

Cu(OMe)₂(en)₂ Copper(II) Methoxy bis(1,2-Diaminoethane)

Cu-O-X Copper-Oxygen-Metal (System)

CUP Cambridge University Press [London, UK]

CUP Canadian University Press

CUP Coefficient d'Utilisation Pratique [Modified Trauzl Test Value]

CUP Columbia University Press [New York, US]

CUP Commonality Usage Proposal

CUP Communications User Program

CUP Cornell University Press [Ithaca, New York, US]

CUP Cuban Peso [Currency]

CUP Cursor Position

CuP Copper Porphyrine

CuP Cuban Peso [Currency]

Cu-Pb Copper-Lead (Alloy System)

CuPc Copper Phthalocyanine (Organic Semiconductor)

Cu/Pd Copper/Palladium (Superlattice)

Cu-Pd Copper-Palladium (Alloy System)

CUPE Canadian Union of Public Employees

Cu/PI Copper on Polyimide (Substrate)

CUPID Completely Universal Processor Input/Output Design

CUPID Cornell University Program for Integrated Devices [US]

CUPM Committee on the Undergraduate Program in Mathematics [of Mathematical Association of America, US]

Cu-PMDA Copper Pyromellitic Dianhydride (Composite)

Cu-PMMA Copper Polymethylmethacrylate (Composite)

Cu Pl Copper Plate

CU Press Cambridge University Press [London, UK]

CUPR Center for Urban Policy Research [US]

CUPR Catholic University of Puerto Rico [Ponce]

CUPREX Copper Extraction (Process)

CuPt Copper Platinum (Compound)

Cu-Pt Copper-Platinum (Alloy Sysytem)

CUPTE Canadian Union of Professional and Technical Employees

CUPW Canadian Union of Postal Workers

CUR Complex Utility Routine Current

CUR Cost per Unit Requirement

CUR Council on Undergraduate Research [US]

.CUR Cursor [File Name Extension] [also .cur]

Cur Currency [also cur]

Cur Current [also cur]

cur current(ly)

CURE Citizens United to Reduce Emissions [of Formaldehyde Poisoning Association, US]

CURE Color Uniformity Recognition Equipment

CURE Computer User Research and Evaluation Laboratory [Canada]

CURE Council for Unified Research and Education

CURES Computer Utilization Reporting System

CURI College-University Resource Institute [US]

CURL Coal Utilization Research Laboratories [UK]

CURP Certificate in Urban and Regional Planning

Curr Current [also curr]

Curr Adv Mater Process Current Advances in Materials and Processes [Published in Japan]

Curr Anthrop Current Anthropology [International Journal] [also Curr Anthropol]

Curr Ass Current Assets

Curr Biol Current Biology [International Journal]

Curr Cont Agr Biol Envir Sci Current Contents: Agriculture, Biology and Environmental Sciences [Publication of the Institute for Scientific Information, US]

Curr Cont Eng Technol Appl Sci Current Contents: Engineering, Technology and Applied Sciences [Publication of the Institute for Scientific Information, US]

Curr Cont Life Sci Current Contents: Life Sciences [Publication of the Institute for Scientific Information, US]

Curr Cont Phys Chem Earth Sci Current Contents: Physical, Chemical and Earth Sciences [Publication of the Institute for Scientific Information, US]

Curr Cont SCI Current Contents: Science Citation Index [Publication of the Institute for Scientific Information, US] [also Current Cont Sci Cit Ind]

Curr Cont SCISEARCH Current Contents: Science Citation Index Database [Publication of the Institute for Scientific Information, US]

Curr Eye Res Current Eye Research [Journal published by Oxford University Press, UK]

Curr Genet Current Genetics [International Journal]

Curr Ind Rep Current Industrial Reports [US]

Curr Liabs Current Liabilities

Curr Med Res Opin Current Medical Research Opinion

Curr Opin Biotechnol Current Opinion in Biotechnology [International Journal]

Curr Opin Cardiol Current Opinion in Cardiology [International Journal]

Curr Opin Cell Biol Current Opinions in Cell Biology [International Journal]

Curr Opin Chem Biol Current Opinion in Chemical Biology [International Journal]

Curr Opin Genet Devel Current Opinion in Genetics and Development [International Journal]

Curr Opin Hematol Current Opinion in Hematology [International Journal]

Curr Opin Immunol Current Opinion in Immunology [International Journal]

Curr Opin Nephrol Hypertens Current Opinion in Nephrology and Hypertension [International Journal]

Curr Opin Neurobiol Current Opinion in Neurobiology [International Journal]

Curr Opin Neurol Current Opinion in Neurology [International Journal]

Curr Opin Obstetr Gynecol Current Opinion in Obstetrics and Gynecology [International Journal]

Curr Opin Pediatr Current Opinion in Pediatrics [International Journal]

Curr Opin Pulmon Med Current Opinion in Pulmonary Medicine [International Journal]

Curr Opin Rheumatol Current Opinion in Rheumatology [International Journal]

Curr Opin Struct Biol Current Opinion in Structural Biology [International Journal]

Curr Phys Index Current Physics Index [of American Institute of Physics, US]

Curr Sci Current Science [of Current Science Association, Raman Research Institute, Bangalore, India]

Curr Sci ((India) Current Science (India) [of Current Science Association, Raman Research Institute, Bangalore]

Curr Serials Rec'd Current Serials Received [Published in the UK]

Curr Therap Res Current Therapeutic Research [International Journal]

CURS Center for Urban and Regional Studies [US]

CURSOR CURrent Set Of Records

CURTS Common User Radio Transmission System

CURV Cable-Controlled Underwater Recovery Vehicle; Cable-Controlled Underwater Research Vehicle

CUS Clean-Up System

CUS Common User System

CUS Coordinatively Unsaturated Site

CuS Copper(II) Sulfide (or Cupric Sulfide)

Cus Customer [also cus]

Cu-Sb Copper-Antimony (Alloy System)

CUSBL Cumulus Under Stratocumulus Boundary Layer

CUSCL Customer Class [also Cus Cl]

CuSCN Copper(I) Thiocyanate (or Cuprous Thiocyanate)

CuSe Copper Selenide

CUSEA Council for University Students of East Africa

CUSEC Canada-United States Environment(al) Council

CUSEC Cubic Foot per Second [also cusec, CFS, or cfs]

CUSF Correlative Unsymmetrized Self-Consistent Field (Method) [Physics]

CUSI Configurable Unified Search Engine [Internet]

Cu-Si Copper-Silicon (Alloy System)

Cu-SiO$_2$ Silicon Dioxide Dispersion Strengthened Copper

CUSIP Committee on Uniform Securities Identification Procedures [US]

CUSMAP Conterminous United States Mineral Assessment Program

CuSn Copper-Tin (or Bronze)

Cu-Sn Copper-Tin (Alloy System)

CUSO Canadian Universities Service Overseas

CuSO Cuprosulfate

CUSP Central Unit for Scientific Photography [of Royal Aeronautical Establishment, UK]

CUSP Commonly Used System Programs

CUSPAD Columbia University Scale for Psychopathology in Alzheimer's Disease [US]

CUSPH Columbia University School of Public Health [New York City, US]

CUSRPG Canada-United States Regional Planning Group [of NATO]

CUSS Computerized Uniterm Search System [US]

CUSS Cooperative Union Serials System [of Ontario University Libraries Cooperation System, Canada]

CUSSCO College, University and School Safety Council of Ontario [Canada]

CUST Chinese University of Science and Technology [Hefei, PR China]

Cust Customer [also cust]

CUSTARD Canada, United States, Taiwan, Australia, Russia and Danmark

CuSum Cumulative Sum (Control Charts) [also CUSUM] [Industrial Engineering]

CUT Chalmers University of Technology [Gothenburg, Sweden]

CUT Circuit Under Test

CUT Colegio Universitario del Turabo [Turabo University College, Caguas, Puerto Rico]

CUT Control Unit Terminal

CUT Control Unit Tester

CUT Continuous-Use Temperature

cut cutaneous [Anatomy]

CUTA Canadian Urban Transit Association

CUTCD Council on Uniform Traffic Control Devices [US]

CuTCNQ Copper Tetracyanoquinodimethane (Organic Semiconductor) [also Cu(TCNQ)]

Cu(TDF)$_2$ Copper (II) 1,1,1,2,2,3,3,7,7,8,8,9,9,9-Tetradecafluorononanel-4,6-Dione [also Cu(tdf)$_2$]

CUTE Clarkston University Terminal Emulator

CUTE Common-Use Terminal Equipment

Cu-Te Copper-Tellurium (Alloy System)

Cu(TFAC)$_2$ Copper (II) Trifluoroacetylacetonate) [also Cu(tfac)$_2$]

Cu(THD)$_2$ Copper (II) 2,2,6,6-Tetramethyl-3,5Heptanedionate [also Cu(thd)$_2$]

Cu-Ti Copper-Titanium (Alloy System)

Cu-Ti/Al$_2$O$_3$ Alumina Reinforced Copper-Titanium (Composite)

Cu-TiC Copper-Titanium Carbide (System)

Cu-Tl Copper-Thallium (Alloy System)

Cu(TMHD)$_2$ Copper Bis(2,2,6,6-Tetramethyl-3,5-Heptanedionate) [also Cu(tmhd)$_2$]

CuTPP 5,10,15,20-Copper Tetraphenylporphyrin

CUTS Canadian University Travel Service

CUTS Cassette User Tape System

CUTS Coalition for Urban Transport Sanity

CUTS Computer-Utilized Turning System

Cutting Tool Eng Cutting Tool Engineering [Journal published in the US]

CUU Cursor Up

Cu-UO$_2$ Uranium Dioxide Dispersion Strengthened Copper (System)

CUW Committee on Undersea Warfare [of US Department of Defense]

Cu-W Copper-Tungsten (Alloy System)

Cu/Wp Tungsten Particulate Reinforced Copper (Composite) [also Cu/W$_p$]

cu yd cubic yard [also cu-yd, or yd^3]

CUZ City University Zurich [Switzerland]

CuZn Copper-Zinc (or Brass) [also Cu-Zn]

Cu-Zn Copper-Zinc (Alloy System)

CuZnAl Copper Zinc Aluminum (Shape Memory Alloy)

Cu-Zn-SOD Copper-Zinc Superoxidase Dismutase [Biochemistry]

Cu-Zr Copper-Zirconium (Alloy System)

CV Aircraft Carrier [US Navy Symbol]

CV *(Caballo de Vapor)* – Spanish for "Horsepower"

CV Calibrated Volume

CV Calicivirus

CV Calorific Value [also cv]

CV Capacitance versus Voltage (Curve)

CV Cape Verde

CV Cape Verde [ISO Code]

CV Capillary Viscometer

CV Cardiovascular
CV Cation Vacancy (Concentration)
CV Ceiling Value
CV Center of Vision
CV Center of Volume
CV Central Vacuole [Plant Cytology]
CV Cepheid Variables [Astronomy]
CV Cerebrovascular
CV Cervical Vertebra [Anatomy]
CV Challenge Virus
CV Characteristic Value [Mathematics]
CV Charpy V-Notch [Mechanical Testing]
CV Check Valve [also cv]
CV *(Cheval Vapeur)* – French for "Horsepower"
CV Chief Value
CV Cited Volume
CV Coefficient of Variance; Coefficient of Variation [also cv] [Statistics]
CV Color Vision
CV Column Volume [also cv]
CV Combining Volumes
CV Commonwealth of Virginia [US]
CV Communicating Vessels
CV Computational Vision
CV Computerized Vision
CV Computer Vision
CV Condenser Vacuum
CV Condensing Vacuole
CV Constant-Arc Voltage
CV Constant Value
CV Constant Velocity (Joint) [Automobiles]
CV Constant Viscosity
CV Constant Voltage
CV Constant Volume
CV Consumer Video
CV Containment Vessel [Nuclear Reactors]
CV Continuously Variable
CV Continuous Variable [Computers]
CV Continuous Voltage
CV Continuous Vulcanization
CV Contractile Vacuole [Biology]
CV Controlled Valence
CV Control Valve
CV Conventional Vehicle
CV Coronary Vein
CV Correspondence Variant [Crystallography]
CV Costovertebral (Joints) [Medicine]
CV Counter Voltage
CV Coxsackie Virus
CV Crest Value
CV Cross-Viscosity
CV Curriculum Vitae
CV Cyclic Voltammetry; Cyclic Voltammograph [Physical Chemistry]

CV Cyclovoltammetry
CV Deployable Payloads Operation [NASA Kennedy Space Center Cargo Operations Directorate, Florida, US]
CV Expendable Vehicles Operations [NASA Kennedy Space Center Directorate, Florida, US]
CV Single-Cotton-Covered Varnished (Electric Wire)
CV% Coefficient of Variation Unevenness (in Textiles) [Symbol]
CV% Percent Coefficient of Variation [Statistics]
$C(V_g)$ Capacitance between Front Gate and Electron Gas [Semiconductor Symbol]
C/V Capacitance versus Voltage (Curve)
C/V Coulomb per Volt [also C V^{-1}]
C-V Capacitance-Voltage (Characteristic)
C-V Current-Voltage (Relationship)
cV Coherent Vacancy (in Materials Science) [Symbol]
CVA Attack Aircraft Carrier [US Navy Symbol]
CVA California Vehicle Act [US]
CVA Canadian Vocational Association
CVA Cerebrovascular Accident
CVA Certified Veterinary Assistant
CVA Characteristic Vector Analysis [Mathematics]
CVA Commonwealth Veterinary Association
CVA Computerized Vibration Analysis
CV(A) Coefficient of Variation of Area [also CV[A], or CV(a)]
CVAA Cold Vapor Atomic Absorption Spectroscopy
CVAD Converter, Voltage, AC to DC
CVAS Configuration Verification Accounting System
CVB Battle Aircraft Carrier [US Navy Symbol]
CVBC Camára Venezolana Británica de Comercio [Venezuela-Britain Chamber of Commerce]
CVBE Chemical Vapor Beam Epitaxy
CVBS Composite Video Broadcast Signal
CVBU Computervision Business Unit
CVC Carrier Virtual Circuit
CVC Chemical Vapor Cleaning
CVC Chemical Vapor Composite
CVC Chemical Vapor Condensation (Process)
CVC Conserved Vector Current (Theory) [Particle Physics]
CVC Constant Voltage Conditioner
CVC Continuously Variable Crown (Rolling Technology) [Metallurgy]
CVC Current Voltage Characteristics
CVCA Canadian Venture Capital Association
CV/CC Constant Voltage/Constant Current (Power Supply)
CVCM Collected Volatile Condensable Material(s)
CVCP Committee of Vice-Chancellors and Principals [of Copyright Licensing Agency, UK]
CVCS Chemical and Volume Control System
CVD Cash versus Documents
CVD Cathode Voltage Drop
CVD Chemical(ly) Vapor-Deposited (Coating) [Often Followed by a Chemical Symbol, or Formula, e.g., CVD Al Represents Chemically Vapor Deposited Aluminum, or CVD SiC Stands for Chemically Vapor Deposited Silicon Carbide]

CVD Chemical Vapor Deposition

CVD Communication Valve Development [UK]

CVD Compact Video Disk

CVD Convert 8-Byte String to Double-Precision Number (Function) [Programming]

CVD Countervailing Duty

CVD Coupled Vibration-Dissociation

CVD Current-Voltage Diagram

CVDA Converter, Voltage Discrete Alternating Current

CVD/CVI Chemical(ly) Vapor Deposited/Chemical(ly) Vapor Infiltrated; Chemical Vapor Deposition/Chemical Vapor Infiltration

CV-DORS Computervision Developers Open Resource Software

CVD/PVD Chemical(ly) Vapor Deposited/Physical(ly) Vapor Deposited; Chemical Vapor Deposition/Physical Vapor Deposition

CVDV Coupled Vibration-Dissociation-Vibration

CVE Cluster Valence Electron

CVE Coefficient of Volume(tric) Expansion

CVE Complete Vehicle Erector

CVE Constant-Voltage Electrolysis

CVE Critical Voltage Effect

CVE Escort Aircraft Carrier [US Navy Symbol]

C Veh Combat Vehicle

C Verd Is Cape Verde Islands

CV Esc Caboverdianos Escudo [Currency of Cape Verde Islands] [also CV esc]

CVF Circular Variable Filter

CVF (Microsoft) Compressed Volume File

CVF Controlled Visual Flight

C-V-f Capacitance-Voltage-Frequency (Characteristic) [also C-V-f]

CVFC Concord Video and Film Council [UK]

CVFR Controlled Visual Flight Rules

CVGA Color Video Graphics Array

CVGB Cable Vault Ground Bar

CV-GMA Constant-Voltage Gas Metal-Arc (Welding)

CV-GMAW Constant-Voltage Gas Metal-Arc Welding

CVI Cape Verde Islands

CVI Certified Vendor Information

CVI Chemical Vapor Infiltrated; Chemical Vapor Infiltration

CVI Communication, Navigation and Identification

CVI Continuous Variation of Incidence

CVI Control Volume Integral

CVI Convert 2-Byte String to Integer (Function) [Programming]

CVI Counterflow Virtual Impactor

CVIA Computer Virus Industry Association [US]

CVIC Conditional Variable Incremental Computer

CVIM Configurable Vision Input Module

CVIS Computerized Vocational Information System

CVL Light Aircraft Carrier [US Navy Symbol]

CV(L) Coefficient of Variation of Length [also CV[L], or CV(l)]

CVLF Consolidated Very-Low Frequency (Program)

CVM Center for Veterinary Medicine [US]

CVM Cluster Variation Method [Metallurgy]

CVM College of Veterinary Medicine

CVM COBOL Virtual Machine

CVM College of Veterinary Medicine

CVM Control Valve Module [also cvm]

CVM Consumable(-Electrode) Vacuum-Melted; Consumable(-Electrode) Vacuum Melting [Metallurgy]

C/V·m Coulomb(s) per volt-meter [also C/(V·m), C/V/m, or C V^{-1}m^{-1}]

CVMA Canadian Vehicle Manufacturers Association

CVMA Canadian Veterinary Medical Association

CVMO Cluster Valence Molecular Orbital

CVN Attack Aircraft Carrier, Nuclear-Powered US Navy Symbol]

CVN Chevron-Notched (Test Specimen) [Mechanical Testing]

CVN Charpy V-Notch (Impact Test) [Mechanical Testing]

CVn Canes Venatici [Astronomy]

CVP Central Valley (Hydroelectric) Project [of US Bureau of Reclamation in California]

CVP Central Venous Pressure

CVP Chimeric Virus Particle [Biology]

CVP Congreso Veterinario Panamericana [Pan-American Veterinary Congress]

CVP Containment Vacuum Pump [Nuclear Reactors]

CVP Corporación Venezolana de Petróleos [Petroleum Corporation of Venezuela]

CVPI Constant Volume Pressure Increase

CVPO Chemical Vapor Phase Oxidation

CVP Resinotes CVP (Cray Valley Publishers) Resinotes [Published in the UK]

CVPT Canonical Van Vleck Perturbation Theory [Physics]

CVPWA Central Valley (Hydroelectric) Project Water Association [California, US]

CVR Change Verification Record

CVR Chemical Vapor Reaction (Process)

CVR Cockpit Voice Recorder

CVR Configuration Verification Review

CVR Continuous Vertical Retort

CVR Continuous Video Recorder

CVR Contrast Variation Ratio

CVR Controlled Visual Rules

CVRD Converter, Variable Resistance, to DC Voltage

cvrs coversed sine [also covrs]

CVS CAD (Computer-Aided Design) for VLSI (Very-Large-Scale-Integtrated) Systems

CVS Caisson Vessel System

CVS Cambridge Veterinary School [UK]

CVS Cardiovascular System

CVS Chemical Vapor Synthesis

CVS Chorionic Villi Sampling [Medicine]

CVS Chinese Vacuum Society

CVS City Vehicle (or Vehicular) System

CVS Composite Variability Study

CVS Computer-Controlled Vehicle System

CVS Constant Volume Sampling

CVS Continuous Velocity Survey [Geology]

CVS Convert 4-Byte String to Single-Precision Number (Function) [Programming]

CVS Cyclic Voltametric Stripping

CVS Support Aircraft Carrier [US Navy Symbol]

CVSA Canadian Vehicle Safety Alliance

CVSA Commercial Vehicle Safety Alliance [US]

CVSB Compatible Vestigial Sideband

CVSD Continuously Variable Slope Delta Modulation [also CVSDM]

CVS-4-HS Constant Volume Sampling–Four-High Stand–Horizontal Work-Roll Stabilization (Cold Rolling Mill) [Metallurgy]

CVSG Channel Verification Signal Generator

CVT Canonical Variational (Transition State) Theory [Physics]

CVT Capacitance-Voltage-Temperature (Method)

CVT Chemical Vapor Transport

CVT Committee on Vacuum Techniques [US]

CVT Communications Vector Table

CVT Concept Verification Testing

CVT Constant-Voltage Transformer

CVT Continuously Variable Transmission

CVT Convert(ible) [also Cvt, or cvt] [Computers]

CVT Crystal Violet Tetrazolium

CVT Customer Verification Test

C-V/T Capacitance-Voltage versus Temperature (Diagram)

cv(TM) vacancy concentration at melting point (in materials science) [Symbol]

CVT-MOCVD Chemical Vapor Transport–Metal(lo)organic Chemical Vapor Deposition

CVTR Carolinas Virginia Tube Reactor [at Parr, South Carolina, US] [Shutdown]

CVTST Canonical Variational Transition-State Theory [Physics]

CVU Constant Voltage Unit

CVUVRS Cyclic Voltammetry with Ultraviolet-Visible Reflectance Spectroscopy

CV(V) Coefficient of Variation of Volume [also CV[V], CV(v), or CV(v)]

cvv core-valence-valence (transition) [Auger Electron Spectroscopy]

CVW Code View for Windows

CVW Consumable-Electrode Vacuum Melting [Metallurgy]

CV(X) Coefficient of Variation of X [Symbol]

CW Call Waiting

CW Capillary Water [Hydrology]

CW Carnauba Wax

CW Carrier Wave [also C/W]

CW Cell Wall [Plant Cytology]

CW Center Wavelength

CW Chemical War(fare)

CW Chemical Waste

CW Chemical Weapon

CW Chemical Weathering

CW Clean Water

CW Clockwise [also cw]

CW Cockcroft-Walton (Accelerator) [Physics]

CW Code Word

CW Cold Water

CW Cold Welded; Cold Welding

CW Cold Work(ed); Cold Working [Metallurgy]

CW Combined Water [Geochemistry]

CW Combining Weight [Chemistry]

CW Combustion Wave

CW Combustion Wire

CW Command Word [also cw]

CW Composite Wave

CW Compound Winding; Compound-Wound

CW Compression Wave

CW Compton Wavelength [Quantum Mechanics]

CW Computer World [Published in the US]

CW Concentration Wave

CW Configuration Wave-Function

CW Constant Wattage

CW Continuous Wave [also cw]

CW Continuous Wetting

CW Control Wheel

CW Control Word

CW Cooling Water

CW Coupled Wave

CW Crosswind

CW Curie-Weiss (Law) [Solid-State Physics]

CW C-Washer

CW Cyclone Washer

CW Cylindrical Wave

%CW Percent of Cold Work [Metallurgy]

C&W Caution and Warning [also C/W]

C&W Chip and Wire [Electronics]

C/W Carrier Wave [also CW]

C/W Cast and Wrought [Metallurgy]

C/W Caution and Warning [also C&W]

C/W Commercial Weight [also c/w]

C-W Continuous Wave [also c-w]

Cw Humid Warm Climate, Dry Winter [Symbol]

cw clockwise [also CW]

CWA Canada Water (Database) [of Environment Canada]

CWA Canadian Waferboard Association [now The Waferboard Association, Canada]

CWA Canadian Warehousing Association

CWA Caution and Warning Annunciator

CWA Chemical Warfare Agent

CWA Civil Works Aministration

CWA Clean Water Act [US]

CWA Clean Work(ing) Area

CWA Cockcroft-Walton Accelerator [Physics]

CWA Communication Workers of America

CWA Conference Work Area

CWA Constant-Wattage Autotransformer

CWA Construction Writers Association [US]

CWA Controlled Work Area

CWA Control Word Address

CWA Country Women's Association

CWAA Cotton Warehouse Association of America [US]

CWAC Canadian Women's Army Corps

CWAI Confederation of Western Australian Industry

CWAP Clean Water Action Project [US]

CWAR Continuous-Wave Acquisition Radar

CWARC Canadian Workplace Automation Research Center

CWAS Contractor Weighted Average Share

CWB Canadian Welding Bureau

CWB Canadian Wheat Board

CWB Ceramic Wiring Board

CWB Communications and Electrical Workers of Canada

CWB Communications Work Committee [Canada]

CWB Compound Water Cyclone

CWBAD Clockwise Bottom Angular Down [Metallurgy]

CWBAU Clockwise Bottom Angular Up [Metallurgy]

CWC Canadian Wood Council

CWC Center for Wireless Communications [Singapore]

CWC Chemical Weapons Convention

CWCE Central Washington College of Education [Ellensburg, Washington State, US]

CWCG Copper Wire Counterpoise Ground

CW CO$_2$ Continuous-Wave Carbon Dioxide (Laser) [also CW-CO$_2$]

CW COMP Chemical Weapons Compendium

CWD Change Working Directory [Internet]

CWD Colonial Waterbird Database [US]

CWD Convert Word to Double Word

CWD Creosoted Wood Duct

CWDB Clockwise Down Blast [Metallurgy]

CWDC Canadian Wood Development Council

CWDE Center for World Development Education [UK]

CWDI Canadian Welding Development Institute [now Welding Institute of Canada]

CWDMA Canadian Window and Door Manufacturer's Association

CWDS Clean (Aircraft) Wing Detection System

CWE Caution and Warning Electronics

CWEA Caution and Warning Electronic Assembly

CWED Cold Weld Evaluation Device

CWEI Canadian Wood Energy Institute

CW-ENDOR Continuous Wave–Electron Nuclear Double Resonance

CWEU Caution and Warning Electronics Unit

CWF Canadian Wildlife Federation

CWF Coal-Water Fuel Technology

CWF Construction Workers Federation [San Marino]

CWFM Continuous Wave/Frequency Modulation [also CW/FM]

CWFSP Caution and Warning/Fire Suppression Panel

CWG Carburetted Water Gas

CWG Closed Waveguide

CWG Committee for Women in Geophysics

CWG Constant Wear Garment

CWG Continuous-Wave Generator

CWI Call Waiting Indicator

CWI Certified Welding Inspector

CWI China Welding Institute [PR China]

CWI Clean World International [UK]

CWIF Continuous Wave Intermediate Frequency

CWIP Construction Work In-Progress

CWIS Campus Wide Information Service [Internet]

CWIS Campus Wide Information System [Internet]

CWIS Community Wide Information Service [Internet]

CWIS Community Wide Information System [Internet]

CWIT Cooling Water Inlet Temperature

CWJR Center for Welding and Joining Research [of Colorado School of Mines, Golden, US]

CWL Chemical Warfare Laboratories

CWL Continuous-Wave Laser

CWL Continuous Working Level

CWLM Caution and Warning Limit Module

CWLR Chemical Warfare Laboratories Reports

CWLS Canadian Well Logging Society

CWM Chemical Waste Management

CWM College of William and Mary [Williamsburg, Virginia, US]

CWM Compound-Wound Motor

CWM Connolly-Williams Method [Phase Transformation Theory]

CWMC Canadian Waste Management Conference

CWME Canadian Waste Materials Exchange

Cwmn Crewwoman [also cwmn]

CWMS Commercial Waste Management Statement

CWMTU Cold-Weather Material Test Unit

CW-Nd:YAG Continuous-Wave Neodymium-Doped Yttrium Aluminum Garnet (Laser) [also CW Nd:YAG]

CW-NMR Continuous-Wave Nuclear Magnetic Resonance

CWO Capital Work Order

CWO Carrier Wave Oscillator

CWO Cash with(out) Order [also cwo]

CWO Chief Warrant Officer

CWO Continuous Wave Oscillator

CWO Custom Work Order

CWP California Water Project [US]

CWP Center for Wave Phenomena [US]

CWP Central West Pacific (Ocean Region)

CWP Circulating Water Pump

CWP Coalworker's Pneumoconiosis

CWP Cold-Work Peak [Metallurgy]

CWP Communicating Word Processor

CWP Contractor Work Plan
CWP Cooling Water System
CWP Current Word Pointer
CWPCA California Water Pollution Control Association [US]
CWPCA Canadian Wood Pallet and Container Association
CWQA Canadian Water Quality Association
CWR Center of Welding Research [of Colorado School of Mines, Golden, US]
CWR Continuous-Wave Reflectometer
CWR Continuous Welded Rail
CWRA Canadian Water Resources Association
CWRA Clean Water Restoration Act [US]
CWRC CANMET (Canada Center for Mineral and Energy Technology) Western Research Center [of Natural Resources Canada in Devon, Alberta]
CWRC Central Water and Power Commission [India]
CWRC Chemical Warfare Review Council
CWRE Chemical Warfare Royal Engineers [UK]
CWRT Center for Waste Reduction Technologies [US]
CWRU Case Western Reserve University [Cleveland, Ohio, US]
CWS Canadian Welding Society [now Welding Institute of Canada]
CWS Canadian Wildlife Service
CWS Caucus for Women in Statistics [US]
CWS Caution and Warning Status [also cws]
CWS Caution and Warning System [of NASA Apollo Program]
CWS Cell Wall Skeleton [Plant Cytology]
CWS Center Wireless Station
CWS Central Warning System
CWS Chemical Warfare Service [US Army]
CWS Colonial Waterbird Society [Canada]
CWS Complex Waveform Simulator
CWS Continuous-Wave Spectrometer
CWS Cooling Water System
CWS Cooperative Wholesale Society
CWS Copper Weld Steel
CWSA Chemical Warfare Service, Army [US]
CWSC Central Washington State College [Ellensburg, Washington, US]
CWSF Canada Wide Science Fair
CWSF Coal-Water Slurry Fuel
CWSF Customer Waste Solidification Facility
CWSR Cold-Worked Stress Relieved (Condition) [Metallurgy]
CWSRA Chester White Swine Record Association [US]
CWSU Caution and Warning Status Unit [also cwsu]
CWSU Condensate Water Servicing Unit [also cwsu]
CWT Carrier-Wave Telegraphy
CWT Character Width Table
CWT Coded Wire Tagging
CWT Cooperative Wind Tunnel
CWT Critical Water Temperature
cwt hundredweight [Unit]

CWTAD Clockwise Top Angular Down [Metallurgy]
cwtad hundredweight, air dry [Unit]
CWTAU Clockwise Top Angular Up (Metallurgy)
CWTC Chemical Waste Transportation Council [US]
CWTG Computer World Trade Group [UK]
CWTH Clockwise Top Horizontal [Metallurgy]
CWTI Chemical Waste Transportation Institute [US]
cwtn hundredweight net [Unit]
CWU Central Washington University [US]
CWU Communication Workers Union [Trinidad]
CWUB Clockwise Up Blast [Metallurgy]
CWV Continuous Wave Video
CWWA Canadian Water Well Association
CWWIA Cyprus Woodworking Industry Association
Cwy Clearway [also cwy] [Aeronautics]
CX Central Exchange
CX Character Transfer
CX Christmas Island [ISO Code]
CX Color Exterior (Film) [also cx]
CX Compatible Expansion (System)
CX Composite
CX Composite Signalling
CX Control Transmitter
CX Cord [Telephony]
CX Critical Experiment [Nuclear Engineering]
CX- Uruguay [Civil Aircraft Marking]
CX-1 First Critical Experiment [Nuclear Experiment Carried Out by Babcock and Wilcox Inc. for the US Atomic Energy Commission in 1957]
C2X Character-to-Hexadecimal [Restructured Extended Executor (Language)]
C(x) Distance Dependent Concentration
C(x) Local Stiffness Tensor [Symbol]
c(x) point cluster [Symbol]
CXA Central Exchange Area
CXAES Continuous X-Ray-Induced Auger Electron Spectroscopy
CXC Carribbean Examination Council [Jamaica]
cXDP Cyclic Xanthosine Diphosphate [also cXDP] [Biochemistry]
CX-HLS Cargo Experimental, Heavy Logistic Support (Aircraft) [US Air Force Symbol]
CXL Center for X-Ray Lithography [of University of Wisconsin at Madison, US]
CXR Carrier [Communications]
CXRO Center for X-Ray Optics [of Lawrence Berkeley Laboratory, California, US]
CXT Common External Tariff
C(x,t) Distance- and Time-Dependent Concentration
CXY Cyclohexylidene
CY Calendar Year [Astronomy]
CY Carry (Flag) [Computers]
CY Common Year [Astronomy]
CY Container Yard
CY Cosmic Year [Astronomy]

CY Cyclohexyl

CY Cyclophoshamide

CY Cyan

CY Cyprus [ISO Code]

Cy Copy [also cy]

Cy Country [also cy]

Cy Currency [also cy]

Cy Cycle [also cy]

Cy Cyclohexane

Cy Cylinder [also cy]

cy cubic yard [also cu yd, or yd³]

.cy Cyprus [Country Code/Domain Name]

CYA Canadian Yachting Association

CYAM Cyanide Ammonia (Process)

CYAP O,O-Dimethyl-O-(p-Cyanophenyl)phosphorothioate

Cybern Cybernetic(s) [also Cybern, or cybern]

Cybern Comput Technol Cybernetics and Computing Technology [Translation of: *Kibernetika i Vychislitel'naya Tekhnika (USSR)*; published in the US]

Cybern Syst Cybernetics and Systems [Journal published in the US]

Cyborg Cybernetic Organism [also cyborg]

Cyc Cycle; Cyclic [also cyc]

Cyc Cyclone; Cyclonic [also cyc]

Cyc Cyclopedia [also cyc]

CYCH Cyclohexyl

Cycl Cyclopedia [also cycl]

cycl cyclic

CYCLAM 1,4,8,11-Tetraazacyclotetradecane [also cyclam]

CYCLAR Cyclization of Light Hydrocarbons to Aromatics (Process)

CYCLEN 1,4,7,10-Tetraazacyclododecane [also cyclen]

CYCLOPS Cyclically Ordered Phase Sequence

CYCLOPS 1,1-Difluoro-4,5,11,12-Tetramethyl-1-Borata-3,6,10,13-Tetraaza-2,14-Dioxocyclotetradeca-3,5,10,12-Tetraene [also cyclops]

CYCLYD Cyclic Alkyd [also cyclid]

CYCPHOS 1,2-Bis(diphenylphosphino)-1-Cyclohexylethane [also cycphos]

Cyd Cytidine [Biochemistry]

CyDTA Cyclohexenediaminetetraacetic Acid

CYFO Calcium Yttrium Iron Oxide (Powder)

CYFRONET Academic Computer Center Network [Poland]

Cyg Cygnus [Astronomy]

Cyg A Cygnus A [Astronomy]

Cyg X-1 Cygnus X-1 [Astronomy]

Cyg X-3 Cygnus X-3 [Astronomy]

CYI Council for Yukon Indians [Canada]

Cyl Cylinder [also cyl]

Cyl L Cylinder Lock

CYMET Cyprus Metallurgical (Process)

CYMEX Cymene Extraction (Process)

CYMK Cyan/Yellow/Magenta/Black

CYMV Commelina Yellow Mottle Virus

Cyn Cyanide

CYO Constant Yield Oil

CyOx Tetrahydro-3,5-Dinitro-1,3,5,2H-Oxadiazine

CYP Cape York Peninsula [Queensland, Australia]

CYP Commonwealth Youth Program

CYP p-Cyanophenyl Ethyl Phenylphosphoric Acid

CYP Cyprus Pound [Currency]

CYPLUS Cape York Peninsula Land Use Study [Australia]

Cys Cysteine; Cysteinyl [Biochemistry]

CYSA Cape York Space Agency [Queensland, Australia]

Cys-Cys (–)-Cystine [Biochemistry]

Cys-Gly Cysteinylglycine [Biochemistry]

Cysh Cysteine [Biochemistry]

Cyt Cytosine [Biochemistry]

Cyt Cytochrome [Biochemistry]

Cyt a$_3$ Cytochrome a$_3$; Cytochrome Oxidase [Biochemistry]

Cyt b Cytochrome b [Biochemistry]

Cyt b$_2$ Cytochrome b$_2$; L-Lactic Dehydrogenase, Type IV [Biochemistry]

Cyt c Cytochrome c [Biochemistry]

CYTD Calendar Year to Date

Cyt f Cytochrome b$_2$; L-Lactic Dehydrogenase, Type IV [Biochemistry]

Cyt h Cytochrome h [Biochemistry]

Cytochem Cytochemical; Cytochemist(ry) [also cytochem]

Cytogenet Cell Genet Cytogenetics and Cell Genetics [International Journal]

Cytokine Rev Commun Cytokine Reviews and Communications [International Journal]

Cytol Cytological; Cytologist; Cytology [also cytol]

CZ Calcium Zirconate

CZ Canal Zone [Panama]

CZ Cavitating Zone; Cavitation Zone

CZ Cerium Zirconate

CZ Closure Zone [Metallurgy]

CZ Coastal Zone

CZ Combat Zone

CZ Common Zone

CZ Commutating Zone

CZ Conservation Zone

CZ Control Zone [Aeronautics]

CZ Cubic Zirconia

CZ Czech Republic [ISO Code]

CZ Czochralski (Crystal) [also Cz]

CZ Czochralski Zone

Cz Czech

Cz Czechoslovakia(n)

Cz Czochralski (Crystal) [also CZ]

C(z) Concentration at Depth z [Symbol]

.cz Czech Republic [Country Code/Domain Name]

CZARIS Czech Agricultural Research Information System [of Institute of Agricultural and Food Information, Ministry of Agriculture, Czech Republic]

CZBA Canal Zone Biological Area [of Smithsonian Institution; located at Barro Colorado Island in Gatun Lake, Panama Canal Zone]

CZC Chromated Zinc Chlorate

CZCS Coastal Zone Color Scanner; Coastal Zone Color Scanning [Remote Sensing]

CZ/CZ Czochralski/Czochralski (Wafer)

CZD Combined Zone Dispersion (Process)

CZE Capillary Zone Electrophoresis

CZE Compare Zone Equal

Czech Czechoslovakia(n)

Czech Heavy Ind Czech(oslovak) Heavy Industry [Published in Prague, Czech Republic]

Czech J Phys Czechoslovak Journal of Physics [Published in US]

Czech J Phys A Czechoslovak Journal of Physics, Section A [Published by the Czechoslovak Academy of Sciences]

Czech J Phys B Czechoslovak Journal of Physics, Section B [Published by the Czechoslovak Academy of Sciences]

CZ-GaAs Czochralski-Grown Gallium Arsenide (Wafer) [also Cz-GaAs]

CZ-Ge Czochralski-Grown Germanium (Wafer) [also Cz-Ge]

CZ-InSb Czochralski-Grown Indium Antimonide (Wafer) [also Cz-InSb]

CZM Coastal Zone Management

CZMA Coastal Zone Management Act [US]

CZMS Coastal Zone Management Subgroup [of Intergovernmental Panel on Climate Change]

CZN Calcium Zinc Nitride

CZP Calcium Zirconium Phosphate (Ceramics)

CZP Cubic Zirconia Polycrystal [also c-ZrO$_2$]

CZR Central Zone Remelting [Metallurgy]

CZR Continuous Zone Refiner; Continuous Zone Refining [Metallurgy]

c-ZrO$_2$ Crystalline Zirconia

c-ZrO$_2$ Cubic Zirconia (Polycrystal) [also C-ZrO$_2$, or CZP]

CZS Calcium Zirconium Silicate

CZ-Si Czochralski-Grown Silicon (Wafer) [also CZ Si, Cz-Si, or Cz-Si]

Cz-Sl Czecho-Slovakia

CZT Cadmium Zinc Telluride (Semiconductor)

CZT Calcium Zirconium Titanate

CZU Compare Zone Unequal

D Absorbed Dose (of Radiation) [Symbol]
D Additional Designation of Type of Emission (or Transmission)–Amplitude Modulated Pulses [Symbol] [also d]
D Anisotropy Constant (in Solid-State Physics) [Symbol]
D Asparagine Acid
D (+)-Aspartic Acid; Aspartyl
D Average Deviation (of the Mean) [Symbol]
D Bond Dissociation Energy [Symbol]
D Charge Density [Symbol]
D Cold Snow Forest Climate [Symbol]
D Crystallite Size (in Crystallography) [Symbol]
D Dacca [Bangladesh]
D Daily
D Dairy
D Dakar [Senegal]
D Dalasi [Currency of Gambia]
D Dallas [Texas, US]
D Dalton [Unit]
D Damage
D Damascus [Syria]
D Danger(ous)
D Danish; Denmark
D Danube (River)
D Daphnia [Genus of Water Fleas]
D Darcy [Unit]
D Dark(ness)
D Darmstadt [Germany]
D Data
D Data-Input (Line) [Symbol]
D Date [also D]
D Daughter [Biology and Physics]
D Davit
D Day
D Dayton [Ohio, US]
D Dead Time [Symbol]
D Death

D Debye (Unit) [Symbol]
D Deca- [SI Prefix]
D Decagonal (Phase) [Materials Science]
D Decay
D Deceleration; Decelerator
D December
D Decimal
D Decision
D Deck
D Declaration
D Declination
D Decode(r)
D Decodon [Ecology]
D Decollation; Decollator
D Decompose; Decomposition
D Decompress(ion)
D Decoy (Vehicle) [USDOD Symbol]
D Decrease
D Decyl
D Dee [Nuclear Engineering]
D Default
D Defect(ive)
D Defend; Defense
D Deficiency; Deficient
D Definition
D Deflection
D Deflector
D Deform(ation)
D Degeneracy; Degenerate
D Degrade; Degradation
D Degree
D Degree of Order (in Metallurgy) [Symbol]
D Dehydration
D Delaware [US]
D Delay
D Delete; Deletion
D Delhi [India]
D Deliver(y)
D Delta
D Delta [Phonetic Alphabet]
D Demagnetization
D Demagnetization Factor [Symbol]
D Demagnification
D Demand
D Demodulation; Demodulator
D Demographic; Demography
D Demolish; Demolition
D Demonstrate; Demonstration; Demonstrator
D Dendrite; Dendritic [Crystallography]
D Dendrite (or Dendron) [Anatomy]
D Denmark
D Dense; Density
D Density [Symbol]

D   Dental; Dentist(ry)
D   Dentin [Dentistry]
D   Denture; Denturism; Denturist
D   Denver [Colorado, US]
D   Depart(ure)
D   Dependence; Dependent
D   Deplete(d); Depletion
D   Deposit(ed); Deposition
D   Depot
D   Depreciation
D   Depress(ion)
D   Depth
D   Derange(ment)
D   Derail(ment)
D   Derivation; Derivative; Derive(d)
D   Dermacentor [Genus of Ticks]
D   Dermatological; Dermatologist; Dermatology
D   Dermis [Anatomy]
D   Derrick
D   Descend(ing)
D   Description
D   Desensitization; Desensitize(r)
D   Designate(d); Designation
D   Destination
D   Destroy; Destruction; Destructive
D   Desynchronization
D   Desynchronized Sleep
D   Detach(ment)
D   Detail
D   Detect(ion); Detectivity; Detector
D   Detergent
D   Deteriorate; Deterioration
D   Determinant
D   Determination; Determine
D   Detonate; Detonation; Detonator
D   Detritus [Geology]
D   Detroit [Michigan, US]
D   Deuterium [also ^{2}H, or H-2]
D   Develop(ment)
D   Deviation
D   Device
D   Devonian [Geological Period]
D   Dew
D   Dextran
D   Dextrorotatory [Chemistry]
D   D Horizon [Composed of Weathered Soil Parent
    Material] [Symbol]
D   Di- [Chemistry]
D   Diabetes; Diabetic
D   Diagnosis; Diagnostic(s)
D   Diagonal
D   Diagonal-Bias (Tire)
D   Diagram
D   Dial(ing)

D   Dialyser; Dialysis
D   (External, or Outside) Diameter [Symbol]
D   Diamond
D   Diamond (Crystal System) [Symbol]
D   Diamond Type (Grinding Wheel) Abrasive [Symbol]
D   Diaphragm
D   Diastole [Medicine]
D   Diatom [Botany]
D   Diatomite [Geology]
D   Dirofilaria [Genus of Filarial Worms]
D   Dielectric(s)
D   Dielectric Displacement [Symbol]
D   Diet(itian)
D   Differ(ence); Differential
D   Differential Operator [e.g., $D_x y$] [Symbol]
D   Difficult(y)
D   Diffract(ion)
D   Diffractometer; Diffractometer
D   Diffuse; Diffusion; Diffuser
D   Diffuse Line [Spectroscopy]
D   Diffusion Coefficient [Symbol]
D   Diffusivity [Symbol]
D   Digest(ion); Digestive
D   Digitalis [Pharmacology]
D   Digit(izer); Digital
D   Dilate; Dilatation
D   Diluent; Dilute; Dilution
D   Dimension(al)
D   Dimer(ization)
D   Diode
D   Diopter [also d]
D   Depeptidase [Biochemistry]
D   Diphyllobothrium [Genus of Tapeworms]
D   Diphtheria [Medicine]
D   Diploma
D   Dipole
D   Direct
D   Direction(al)
D   Directorate
D   Directional Quantity [Symbol]
D   Directive; Director
D   Director (Aircraft) [USDOD Symbol]
D   Discharge
D   Discontinuity; Discontinuous
D   Discount
D   Discover(y); Discoverer
D   Discriminate; Discrimination; Discriminator
D   Disease(d)
D   Disinfect(ant); Disinfection
D   Disintegrate; Disintegration
D   Disintegrator
D   Disk
D   Dislocate(d); Dislocation [Medicine]

D Dislocation [Materials Science]
D Dislocation Density (in Materials Science) [Symbol]
D Disorder(ed); Disordering
D Dispense(d); Dispenser
D Disperse(d); Dispersion
D Displace(ment)
D Displacement Field (in Electrostatics) [Symbol]
D Display
D Display Command [Programming]
D Disposal; Dispose
D Dissipate(d); Dissipation
D Dissipated Energy
D Dissipation Factor [Symbol]
D Dissociate(d); Dissociation
D Dissolution; Dissolve(d)
D Distance; Distant
D Distil(lation)
D Distort(ion)
D Distortion Factor [Symbol]
D Distribute; Distribution; Distributor
D District
D Disturb(ance)
D Divert(er)
D Diverticulitis [Medicine]
D Diverticulum [also d]
D Dividend
D Divide(r); Dividing
D Division [also d]
D Djibouti [Africa]
D D-Load (for Concrete Pipe) [Symbol]
D Dock(ing)
D Doctor(ate)
D Document(ation)
D Domain
D Don (River) [Russia]
D Dong [Currency of Vietnam]
D Donor
D Dopant; Dope; Doping
D Dormant
D Dorsal; Dorsum
D Dortmund [Germany]
D Dosage; Dose
D Dosimeter; Dosimetry
D Dot
D Double
D Doublet
D Down
D Down Contactor, or Relay
D Drachma [Currency of Greece]
D Draft
D Drag
D Drain(age); Draining
D Drain [Semiconductor Symbol]

D Drawing
D Dream(ing)
D Dresden [Germany]
D Drift
D Drill
D Drip
D Drive(r); Driving
D Drive Gear [Symbol]
D Drizzle
D Drogue
D Drop(let)
D Drug(ist)
D Drum
D Dry(ing)
D D-Size Sheet (of Paper) [22" × 34"]
D Dublin [Ireland]
D Duct(ing)
D Ductwork
D Ductile; Ductility
D Dump
D Dunnite [Explosive]
D Duodenum [Anatomy]
D Duplex
D Duplex Structure [Metallurgy]
D Durability; Durable
D Duration
D Durham [North Carolina, US]
D Düsseldorf [Germany]
D Dust(ing)
D Dutch
D Dyad [Mathematics]
D Dynamic(s)
D Electric Displacement [Symbol]
D Electric Flux Density [Symbol]
D Equivalent (Particle) Diameter (or Equivalent Size) [Symbol]
D (Photographic) Film Density [Symbol]
D Five Hundred [Roman Numeral]
D Fraunhofer "D" Line [Yellow Fraunhofer Line for Sodium Having a Wavelength of 589.3 Nanometers] [Spectroscopy]
D Grade for Below-Average Work
D Group-Delay Dispersion [Symbol]
D High-Carbon, High-Chromium Type Cold-Work Tool Steels [AISI-SAE Symbol]
D Humid Cold (or Microthermal) Climates, Coldest Month below 32°F (0°C), Warmest Month above 50°F (10°C) [Classification Symbol]
D Interdiffusion Coefficient [Symbol]
D Lidar Off-Zenith Pointing [Symbol]
D Linear Density [Symbol]
D Local Anisotropy Field [Symbol]
D Magnetic Declination [Symbol]
D Major Diameter (of Internal Threads) [Symbol]

D   Optical Density [Symbol]

D   Pitch Diameter (of Gears, Splines, etc.) [Symbol]

D   Restoring Torque [Symbol]

D   Second Tone in the Scale of C Major [Acoustics]

D   Sensitivity (or Figure of Merit) (of a Galvanometer) [Symbol]

D   Sound Energy Density [Symbol]

D   Spacing of Grain-Boundary Dislocations (in Materials Science) [Symbol]

D   Spin-Wave Stiffness (in Solid-State Physics) [Symbol]

D   Strain-Rate Tensor [Symbol]

D   Tensile Compliance [Symbol]

D   Tortuosity (of Interfaces) (in Materials Science) [Symbol]

D   Transmission Density (in Optics) [Symbol]

D-   Germany [Civil Aircraft Marking]

D-   Prefix Denoting the Right-Handed Enantiomer of an Optical Isomer

-D   Drone Control (Aircraft) [US Navy Suffix]

/D   Default Option [MS-DOS Shell]

$\bar{D}$ Average (or Mean) Particle Diameter (in Quantitative Metallography) [Symbol]

$\bar{D}$ Interdiffusion Coefficient [Symbol]

$\bar{D}$ Size-Effect Correction [Symbol]

D*   Complex Tensile Compliance [Symbol]

D*   Detectivity [Symbol]

D*   Electrical Flow Density [Symbol]

D*   Specific Detectivity [Symbol]

D^+   D-Plus Particle [Symbol]

D^+   Positive Donor [Semiconductor Symbol]

D^-   Deuteride [Symbol]

D^-   Negative Donor [Semiconductor Symbol]

D^0   Donor (Ground State) [Semiconductor Symbol]

D_0   Crystallite Size at Time $t = 0$ (in Crystallography) [Symbol]

D_0   Preexponential Factor for Diffusivity [Symbol]

D_0   Surface Charge Density of Plate Capacitor in a Vacuum [Semiconductor Symbol]

D_1   Fraunhofer "D_1" Line [Fraunhofer Line for Sodium Having a Wavelength of 589.3 Nanometers] [Spectroscopy]

D_2   Molecular Deuterium [Symbol]

D_2   Fraunhofer "D_2" Line [Fraunhofer Line for Sodium Having a Wavelength of 588.9 Nanometers] Spectroscopy]

D^3   Defense, Description and Designation [also DDD]

D_3   Drive Gear (First Three Gears Selected) [Symbol]

D_3   Fraunhofer "D_3" Line [Fraunhofer Line for Sodium Having a Wavelength of 587.6 Nanometers] [Spectroscopy]

D_{10}   Effective (Soil Particle) Diameter (or Size) [Symbol]

D_{10}   (Soil) Particle Diameter which Corresponds to 10% Finer on Cumulative Particle-Size Distribution Curve [Symbol]

D_{60}   (Soil) Particle Diameter which Corresponds to 60% Finer on Cumulative Particle-Size Distribution Curve [Symbol]

D_a   Aerodynamic Diameter [Symbol]

D_a   Cumulative Dose [Symbol]

D_{av}   Weighted-Average Diffusivity [Symbol]

D_B   Demagnetizing Coefficient [Symbol]

D_b   Diffusion Coefficient of Grain Boundary (in Materials Science) [Symbol]

D_b   Mobility Equivalent Diameter [Symbol]

D_c   Cluster Diffusion Constant [Symbol]

D_c   Critical Diameter [Symbol]

D_c   Critical Single Domain Particle Size (of Magnetic Material) [Symbol]

D_c   Dissipation Factor [Symbol]

D_{cm}   Deflection Sensitivity [Symbol]

D_D   Deuterium Diffusion Coefficient (or Deuterium Diffusivity) [Symbol]

D_d   Displacement Dose (in Radiology) [Symbol]

D_d   Drill Diameter [Symbol]

D_d   Relative Density (of Soil) [Symbol]

D_e   Effective (Soil Particle) Diameter (or Size) [Symbol]

D_e   Knudsen Diffusion Coefficient [Symbol]

D_e   Maximum External Diameter of Soil Sampling Spoon [Symbol]

D_e   Mean Droplet Diameter (in Meteorology) [Symbol]

D_{eff}   Effective Diffusion Coefficient [Symbol]

D_F   Degree of Freedom [Symbol]

D_F   Fractal Dimension [also D_f] [Symbol]

D_f   Depth of Field [Symbol]

D_f   Fatigue Ductility [Symbol]

D_{gm}   Geometric Mean Diameter (of a Liquid Drop) [Symbol]

D_H   Hydrogen Diffusion Coefficient (or Hydrogen Diffusivity) [Symbol]

D_{hkl}   Crystallite Size as per Scherrer Relationship (in Crystallography) [Symbol]

D_i   Devaluation Enthalpy [Symbol]

D_i   Diameter of i-th Liquid Particle [Symbol]

D_i   Diffusion Coefficient (or Diffusivity) of the i-th Species [Symbol]

D_i   Induced Drag [Symbol]

D_i   Inside Diameter (of a Pipe) [Symbol]

D_i   Internal Transmission Density [Symbol]

D_i   Interstitial Diffusion Coefficient (in Metallurgy) [Symbol]

D_i   Maximum Internal Diameter of Soil Sampling Spoon [Symbol]

D_i^*   Self-Diffusion Coefficient [Symbol]

D_{IN}   Data-Input Line [Symbol]

D_{IT}   Interface Trap Density (in Materials Science) [Symbol]

D_L   (Crystal) Lattice Diffusivity [Symbol]

D_L   Liquid Diffusivity (or Liquid Diffusion Coefficient) [Symbol]

D_L   Profile Fractal Dimension [Symbol]

D_m   Diameter of Milling Cutter [Symbol]

D_m   Hydraulic Motor Displacement [Symbol]

D_m   Magnetic Dissipation Factor [Symbol]

D_m   Mean Droplet Diameter (in Meteorology) [Symbol]

D_m   Measured Density [Symbol]

D_{max}   Maximum Diameter [Symbol]

D$_{max}$ Maximum (or Largest) Dimension [Symbol]

D$_{max}$ Maximum Photographic Density [Symbol]

D$_{min}$ Minimum Diameter [Symbol]

D$_{min}$ Minimum (or Smallest) Dimension [Symbol]

D$_{min}$ Minimum Photographic Density [Symbol]

D$_n$ Characteristic Diameter of Ice Crystal Distribution [Symbol]

D$_o$ Outside Diameter (of a Pipe) [Symbol]

D$_{OUT}$ Data-Output Line [Symbol]

D$_p$ Particle Diameter [Symbol]

$\overline{D}_{pq}$ Mean (Liquid) Droplet Diameter [Symbol]

D$_{qe}$ Jominy Distance (in Metallurgy) [Symbol]

D$_R$ Reflection Density (in Optics) [Symbol]

D$_R$ Rosin-Rammler Diameter (in Rosin-Rammler (Liquid) Drop Size Distribution) [Symbol]

D$_\rho$ Reflectance Density [Symbol]

D$_S$ Self-Diffusivity (or Self-Diffusion Coefficient) [also D$_s$] [Symbol]

D$_S$ Surface Fractal Dimension [Symbol]

D$_s$ Surface Diffusion Coefficient [Symbol]

D$_T$ Transmission Density [also D$_t$] [Symbol]

D$_t$ Diameter of Workpiece in Turning [Symbol]

D$_V$ Vacancy Diffusion Constant (in Materials Science) [Symbol]

D$_v$ Volume Diffusion Coefficient [Symbol]

D$_x$ Density Calculated from X-Ray Measurements [Symbol]

D$_{xf}$ Mean (Liquid) Droplet Diameter [Symbol]

D0 Day Zero

D0 D-Zero Detector [of Fermilab, Batavia, Illinois, US]

D3.5 3.5" Disk(ette)

D5.25 5.25" Disk(ette)

D10 Decagonal [Crystallographic Symmetry Group Symbol]

D12 Dodecagonal [Crystallographic Symmetry Group Symbol]

D-1 Erste Deutsche Spacelab Mission [First German Spacelab Mission]

D-2 Zweite Deutsche Spacelab Mission [Second German Spacelab Mission]

0D Zero-Dimensional [also 0-D]

1D One-Dimensional [also 1-D]

2D Two Days

2D Two-Dimensional [also 2-D]

2D2 Two-Dimensional Two-Grain Junction [Metallurgy]

2D3 Two-Dimensional Three-Grain Junction [Metallurgy]

3D Three-Dimensional [also 3-D]

3D2 Three-Dimensional Two-Grain Junction [Metallurgy]

3D3 Three-Dimensional Three-Grain Junction [Metallurgy]

3D3a Three-Dimensional Three-Grain Junction, axisymmetric [Metallurgy]

3D4 Three-Dimensional Four-Grain Junction [Metallurgy]

3D4a Three-Dimensional Four-Grain Junction, axisymmetric [Metallurgy]

4D Four-Dimensional [also 4-D]

5D Five-Dimensional [also 5-D]

6D Six-Dimensional [also 6-D]

7D Seven Digit (Number)

2,4-D 2,4-Dichlorophenoxyacetic Acid

d additional designation of type of emission (or transmission)–amplitude modulated pulses [Symbol] [also D]

d average deviation (of a single measurement) [Symbol]

d average grain diameter (in materials science) [Symbol]

d common difference (in arithmetic progressions) [Symbol]

d dalton [Unit]

d dam

d damping factor [Symbol]

d dark

d date [also D]

d daughter [Radioactive Elements]

d day [Symbol]

d decay rate (of sound) [Symbol]

d deceased

d deci- [SI Prefix]

d decompose(s)

d deficit

d deflection (of beams) [Symbol]

d degree [also deg, or °]

d delete

d delete line(s) (command) [Edlin MS-DOS Line Editor]

d delta [English Equilvalent]

d *(denarius)* – penny [UK Symbol]

d dendrite-arm spacing (in metallurgy) [Symbol]

d (relative) density [Symbol]

d depart

d deposit [also D]

d depth [Symbol]

d depth of cut (in machining) [Symbol]

d depth of thread [Symbol]

d derivative (of) [Symbol]

d desorption rate [Symbol]

d deuterated

d deuterium [also D, ^{2}H, or H-2]

d deuteron [Physics]

d deviation (in statistics) [Symbol]

d dextrorotatory [Chemistry]

d diagonal [Symbol]

d (internal, or inside) diameter [Symbol]

d died

d difference [Symbol]

d differential

d differential coefficient [Symbol]

d differential operator [e.g., dy/dx] [Symbol]

d digital

d dimensional

d diopter [also D]

d diverticulum [also D]

d distance [Symbol]

d distorted-coherent (defect) [Physics]

d   division [also D]

d   d-orbital [Symbol]

d   dose

d   double

d   doublet [Spectroscopy]

d   down(-type) quark (in particle physics) [Symbol]

d   *(drachma)* – a drachm; a dram [Apothecary]

d   drain

d   drain [Semiconductor Symbol]

d   drive letter [Computers]

d   duty cycle [Radar]

d   dyne [Unit]

d   fractal dimension [Symbol]

d   grain size (in crystallography) [Symbol]

d   interferometer sensitivity [Symbol]

d   interplanar spacing (in crystallography) [Symbol]

d   (core) lamination thickness [Symbol]

d   less than four months over 50°F (10°C), coldest month below –36.4°F (–38°C) [Subtype of Climate Region, e.g., in Cd, Ddw, etc.]

d   linear clear lens aperture [Symbol]

d   long diagonal of Knoop hardness impression [Symbol]

d   major diameter (of external threads) [Symbol]

d   mean diagonal of Vickers hardness indentation [Symbol]

d   number of defectives (in a sample) (in statistics) [Symbol]

d   penny size (i.e., length of a nail) [Symbol]

d   piezoelectric coefficient [Symbol]

d   pitch diameter (of gear pinions, worms, etc.) [Symbol]

d   phase factor (in acoustics) [Symbol]

d   resolution (of optical devices) [Symbol]

d   superlattice period (in solid-state physics) [Symbol]

d   thickness [Symbol]

d   zone size [Symbol]

d-   prefix denoting a substance as being dextrorotatory

-d   ending of chemical compounds containing deuterium (e.g., ammonia-d)

/d   data group allocation option [MS-DOS Linker] average grain diameter (in crystallography) [Symbol]

d^* subgrain size (in crystallography) [Symbol]

d_0 initial diameter [Symbol]

d_0 initial grain size (in crystallography) [Symbol]

d_0 initial interplanar spacing (d-spacing) of strain-free material [Symbol]

d_0 original diameter [Symbol]

d^{20} specific gravity with respect to water at 20°C [Symbol]

d_c critical density [Symbol]

d_c critical grain size (in crystallography) [Symbol]

d_c critical (crystal lattice) spacing [Symbol]

d_e Eden cluster (fractal) [Symbol]

d_e effective path length (or thickness) (in internal reflection spectroscopy) [Symbol]

d_f (reinforcing) fiber diameter [Symbol]

d_f final diameter [also d_f] [Symbol]

d_f fractal dimension [Symbol]

d_g mean grain diameter (in crystallography) [Symbol]

d_g density of gas [Symbol]

d_{hkl} interplanar spacing between two parallel crystallographic planes of Miller indices h, k and l [Symbol]

d_i drop diameter (in sampling size interval) [Symbol]

d_l density of liquid [Symbol]

d_m number of divisions on main scale of vernier [Symbol]

d_o optical spectral dimension [Symbol]

d_p depth of penetration (in internal reflection spectroscopy) [Symbol]

d_s density of solid [Symbol]

d_s length of secondary scale of vernier [Symbol]

d_t total depth per stroke (in broaching) [Symbol]

d_v vapor density [Symbol]

Δ   Apparent Free (Edge-to-Edge) Distance in a Lamellar Structure (in Quantitative Metallography) [Symbol]

Δ   Band Separation [Symbol]

Δ   Change in Quantity [Symbol]

Δ   Degree of Flattening (of Faradaic Current) [Symbol]

Δ   (Upper-case) Delta [Greek Alphabet]

Δ   Delta Connection (in Electrical Engineering) [Symbol]

Δ   Determinant [Symbol]

Δ   Deviator Stress (for Soil, or Rock) [Symbol]

Δ   Dielectric Constant [Symbol]

Δ   Diffusion Coefficient [Symbol]

Δ   Dilation Energy [Symbol]

Δ   Distance (in Astronomy) [Symbol]

Δ   Double Bond (in Chemistry) [Symbol]

Δ   Energy Gap (in Bulk Superconductor) [Symbol]

Δ   Finite Difference, or Change [Symbol]

Δ   Forward Difference in Interpolation Theory (e.g., $\Delta x_i = x_i + 1 - x_i$) [Symbol]

Δ   Laplacian (or Laplace Operator) [also ∇^2, or div grad]

Δ   Logarithmic Decrement [Symbol]

Δ   Ellipsometric Parameter [Symbol]

Δ   Energy Barrier [Symbol]

Δ   Increment (e.g., Δx) [Symbol]

Δ   Interaction Strength [Symbol]

Δ   Maximum Retardation (of Interferometer) [Symbol]

Δ   Order Parameter Value at Absolute Zero [Symbol]

Δ   (Relative) Phase Retardation [Symbol]

Δ   Polarization [Symbol]

Δ   Phase Shift [Symbol]

Δ   Quadrupole Splitting Parameter (in Spectroscopy) [Symbol]

Δ   Range [Symbol]

Δ   Sleet (in Meteorology) [Symbol]

Δ   Snoek Peak Damping (in Metallurgy) [Symbol]

Δ   Superconducting Gap [Symbol]

Δ   Supersaturation [Symbol]

Δ   Triangle (in Plane Geometry) [Symbol]

Δ^{**} Enthalpy Thickness [Symbol]

Δ_0 Peak-Shift Parameter [Symbol]

Δ_0 True Free (Edge-to-Edge) Distance in a Lamellar Structure (in Quantitative Metallography) [Symbol]

Δ_c Charge Excitation Gap [Symbol]

Δ_χ Incremental Tolerance [Symbol]

Δ_s Spin (in Physics) [Symbol]

Δ_x Increment of x [Symbol]

δ angle [Symbol]

δ angle between crystallographic directions (in cubic system) [Symbol]

δ angle of compound rest (for milling bevel gears) [Symbol]

δ angle of external (or wall) friction (of soil) [Symbol]

δ angle of set (of aeronautical control surfaces) [Symbol]

δ boundary layer thickness [Symbol]

δ chemical chift (in nuclear magnetic resonance spectroscopy) [Symbol]

δ central difference in interpolation theory [Symbol]

δ crack-(tip) opening displacement [Symbol]

δ damping coefficient [Symbol]

δ declination (in astronomy) [Symbol]

δ decrement [Symbol]

δ (axial) deflection (of beam) [Symbol]

δ deformation [Symbol]

δ (lower-case) delta [Greek Alphabet]

δ delta phase (of a material) [Symbol]

δ density [Symbol]

δ depth [Symbol]

δ deviation (in statistics) [Symbol]

δ diametral interference (of a shrink fit) [Symbol]

δ displacement [Symbol]

δ dissipation constant [Symbol]

δ domain wall width (in solid-state physics) [Symbol]

δ effective (magnetic) recording depth [Symbol]

δ (total) elongation [Symbol]

δ form factor [Symbol]

δ (electron) hole density [Symbol]

δ incommensuratality parameter (of superlattices) [Symbol]

δ increment [Symbol]

δ indentation depth (in hardness testing) [Symbol]

δ ineffective length (of a reinforcing fiber) [Symbol]

δ isomer shift [Symbol]

δ (crystal) lattice mismatch [Symbol]

δ least difference of practical importance [Symbol]

δ Lorentzian line-shape parameter [Symbol]

δ (dielectric) loss angle [Symbol]

δ optical path difference (or optical retardation) [Symbol]

δ phase (defect) angle [Symbol]

δ positive number usually dependent on another, ε, used in limit process arguments [Symbol]

δ reduced distance between adjacent principal points (of lens combination) [Symbol]

δ resolving power [Symbol]

δ secondary emission ratio [Symbol]

δ skin depth [Symbol]

δ slip band width (in crystallography) [Symbol]

δ slow field to fast field diffusion constant [Symbol]

δ speed fluctuation coefficient (of flywheels) [Symbol]

δ static spring deflection [Symbol]

δ taper (of thrust bearings) [Symbol]

δ thickness [Symbol]

δ total interference (of a fit) [Symbol]

δ variation (e.g., δx is the variation of x) [Symbol]

δ very small increment [Symbol]

δ_I mode I displacement (in mechanical testing) [Symbol]

δ_{II} mode II displacement (in mechanical testing) [Symbol]

δ_0 exchange length (of magnetic materials) [Symbol]

δ_c critical crack-tip opening displacement [Symbol]

δ_c deflection per (spring) coil [Symbol]

δ_{ic} installed deflection per spring coil [Symbol]

δ_{ij} Kronecker delta (tensor) [Symbol]

δ_{IS} Isomer shift [Symbol]

δ_{mc} maximum deflection per coil [Symbol]

δ_s (total) spring deflection [Symbol]

δ_T tetragonal (crystal lattice) deformation [Symbol]

δ_z^{ind} induced (crystal lattice) deformation (due to external field) [Symbol]

δ_z^s spontaneous (crystal lattice) deformation (due to external field) [Symbol]

$\$$ Dollar

DA Danger Area

DA Data Acquisition

DA Data Administrator

DA Data Analysis

DA Data Area

DA Data Array

DA Data Available

DA Daughter Atom [Physics]

DA Days after Acceptance [also D/A]

DA Deactivated for Acidic Compounds [Gas Chromatography]

DA Dealers Alliance [US]

DA Debye Approximation [Physics]

DA Decarburization Anneal(ing)

DA Decimal Add(er)

DA Decimal-to-Analog

DA Decision Altitude

DA Decontaminating Agent

DA Decylamine

DA Defect Analysis

DA Define Address

DA Define Area

DA Degree of Anisotropy

DA Delay(ed) Action
DA Delay Amplifier
DA Demand Assignment
DA Dendrite Arm [Metallurgy]
DA Dental Alloy
DA Dental Amalgam
DA Departmental Approval
DA Department of Agriculture
DA Department of Astronomy
DA Department of the Army [US]
DA Deployed (or Deployment) Assembly
DA Deposit Account [also D/A]
DA Depth Analysis
DA Design Allowables
DA Design Analysis
DA Design Automation
DA Destination Address
DA Detroit Arsenal [Michigan, US]
DA Diacetamide
DA Diacetylene
DA Dielectric Absorption
DA Dianhydride
DA Diels-Alder (Reaction) [Chemistry]
DA Differential Amplifier
DA Differential Analysis; Differential Analyzer
DA Diffraction Analysis
DA Diffusional Activity
DA Diffusion-Alloyed; Diffusion Alloying [Metallurgy]
DA Diffusion Annealing [Metallurgy]
DA Digit Absorbing
DA Digital Amplifier
DA Digital-to-Analog
DA Dimensional Analysis
DA Dinar, Algerian [Currency of Algeria]
DA Diode Array
DA Dipole Absorption
DA Direct Access
DA Direct Address
DA Direct Ammonolysis [Chemistry]
DA Direction Action
DA Directional Antenna
DA Directory Assistance
DA Disaccommodation
DA Discharge Afloat
DA Discrete Address
DA Disk Access
DA Disk Address(ing)
DA Display Adapter
DA Display ASCII (Command) [Pascal Programming]
DA Dissociated Ammonia
DA Dissolved Acetylene
DA Distortion Analyzer
DA Distribution Assembly

DA District Attorney [US]
DA Divisional Artillery
DA Doctor of Administration
DA Doctor of Archeology
DA Doctor of Arts
DA Document Analyst
DA Documentary Acceptance
DA Documents against Acceptance
DA Dodecanedioic Acid
DA Doesn't (or Don't) Answer
DA Dopamine [Biochemistry]
DA Dorsal Aorta [Anatomy]
DA Double-Acting (Machine)
DA Double Aged; Double Aging [Metallurgy]
DA Double Amplitude
DA Double Armored (Cable)
DA Drainage Area [Hydrology]
DA Drive by Air-Side
DA Drug Addict(ion)
DA Duplex Anneal(ing) [Metallurgy]
DA Dynamic Alignment
DA Dynamic Allocation
DA Dynamic Analysis; Dynamic Analyzer
DA 10,12-Pentacosadiynoic Acid
6F-DA 6F-Dianhydride [also 6FDA]
$A Australian Dollar [Currency]
D/A Days after Acceptance [also DA]
D/A Deposit Account [also DA]
D/A Digital-to-Analog [also D-A]
D/A Discharge Afloat [also DA]
D/A Documents Against Acceptance [also d/a]
D/A Documents Attached
D-A Decimal-to-Analog
D-A Digital-to-Analog [also D/A]
D-A Document Against Acceptance
D-A Donor-Acceptor (Pair) [Solid-State Physics]
Da Dalton [Unit]
Da Damkoehler Number (of Fluid Flow) [Symbol]
Da Darcy Number (of Fluid Flow) [Symbol]
Da Humid Continental Climate, Warm Summer [Classification Symbol]
Da_1 Darcy Number 1 (of Fluid Flow) [Symbol]
Da_2 Darcy Number 2 (of Fluid Flow) [Symbol]
Da I Damköhler Number I (of Fluid Flow) [Symbol]
Da II Damköhler Number II (of Fluid Flow) [Symbol]
Da III Damköhler Number III (of Fluid Flow) [Symbol]
Da IV Damköhler Number IV (of Fluid Flow) [Symbol]
Da V Damköhler Number V (of Fluid Flow) [Symbol]
D/a Days after acceptance
dA 2'-deoxyribosyladenine [Biochemistry]
dA deoxyadenosine [Biochemistry]
dA surface element [Symbol]
da day [Unit]
da deca- [SI Prefix]

d/a documents against acceptance

ΔA Absorbance Change [Symbol]

ΔA Surface Area Change [Symbol]

Δa crack extension [Symbol]

Δa_B pseudo-crack advance [Symbol]

Δa_e difference (or change) in effective crack size [Symbol]

Δa_p difference (or change) in physical crack size [Symbol]

$\Delta \alpha$ absorption coefficient change [Symbol]

$\Delta \alpha$ linear change in thermal expansion [Symbol]

$\Delta \alpha$ ultrasonic attenuation [Symbol]

$\Delta \alpha_{OH^-}$ hydroxide absorption change [Symbol]

DAA Danish Astronomical Society

DAA Data-Access Arrangement

DAA Decimal Adjust for Addition

DAA Department of Agriculture, Adelaide [South Australia]

DAA Diacetone Acrylamide

DAA Diacetone Alcohol

DAA Direct-Access Arrangement

DAA Director of Army Aviation

DAA Dominion Auto Association [Canada]

DAA Durene Association of America [US]

$\Delta A/A$ Relative Change in Area (or Area Change per Unit Area) [Symbol]

DAAB Diazoaminobenzene

DAAB 4-Dimethylaminoazobenzene

DAAC Distributed Active Archive Center

DAAD Deutscher Akademischer Austauschdienst [German Academic Exchange Service]

DAAG Deputy Assistant-Adjutant-General

DAAIS Danger Area Activity Information Service

DAAM Diacetone Acrylamide

DAAP Dialkylalkylphosphonate

DAAP Dialkylaminopyridine

DAAP Diamylamylphosphonate

DAAS Drilling Activity Analysis System [US]

DAB Data Acquisition Bus

DAB Defense Acquisition Board [US]

DAB Delayed Action Bomb

DAB Diaminobenzene

DAB 3,3'-Diaminobenzidine

DAB 1,4-Diaminobutane

DAB Dictionary of American Biography

DAB Digital Audio Broadcast(ing)

DAB 4-N,N-Dimethylaminoazobenzene

DAB Display Arrangement Bits

DAB Display Assignment Bits

DAB Display Attention Bits

DAB Dual-Polarization and Backscatter

L-DAB L-2,4-Diamino Butanoic Acid [also L-Dab]

DABA 4,4'-Diaminobenzanilide

DABAL Data Bank Language

DABC 3,5-Diaminobenzene Chloride

DABCC Diploma of the American Board of Clinical Chemists

DABCO 1,4-Diazabicyclo[2.2.2]octane

DABCYL 4-(4-Dimethylaminophenylazo)benzoyl

DAB·4HCl 3,3'-Diaminobenzidine Tetrahydrochloride

DABIA N-(4-Dimethylaminoazobenzene-4')iodoacetamide

DABITC 4-(N,N-Dimethylamino)azobenzene-4'-Isothiocyanate [also DAB-ITC]

DABL Daisy Behavioral Language

DABO Diploma of the American Board of Otolaryngology

DABP Di-2-Amyl-2-Butylphosphonate

DABP 4,4'-Dithiobis(phenyl Azide

DABPath Diploma of the American Board of Pathology

DABPathAnal Diploma of the American Board of Pathological Anatomy

DABPH Diploma of the American Board of Public Health

DABR Diploma of the American Board of Radiology

DABS D Absolute [Forth Computer Language]

DABS [4-{(4-Dimethylaminophenyl)azo}benzene]sulfonyl

DABS Discrete(ly) Address(ed) Beacon System

DABSA 2,5-Diaminobenzenesulfonic Acid

DABS-Cl 4-[(4-Dimethylaminophenyl)azo]benzenesulfonyl Chloride

DABSYL 4-(Dimethylamino)azobenzene-4'-Sulfonyl (Chloride)

DABTC N,N-Dimethylamino Azobenzene Thiocarbonyl

DABTT Bis(p-Dimethylaminobenzaldehyde)-triethylenetetramine

DAB-UL Dual-Polarization and Backscatter–Unattended Lidar

DAC Acetic Acid, α,α-Dicyanoethyl Ester

DAC Danish CAA (Civil Aviation Authority)

DAC Data Acquisition [also Dac, or dac]

DAC Data Acquisition and Control (System)

DAC Data Acquisition Camera

DAC Data Acquisition Center

DAC Data Analysis Computer

DAC Data Authentication Code

DAC Delay, Attenuate and Compare (Circuit)

DAC Delayed Atomization Cuvette

DAC Demand Assignment Controller

DAC Derived Air Concentration

DAC Design Augmented by Computer

DAC Design Automation Conference

DAC Development Assistance Committee [of Organization for Economic Cooperation and Development, France]

DAC Diallyl Chlorendate

DAC 1,2-Diaminocyclohexane

DAC Diamond Anvil Cell

DAC Digital Amplitude Curve

DAC Digital Analysis Converter

DAC Digital Arithmetic Center

DAC Digital-to-Analog Conversion; Digital-to-Analog Converter [also D/AC]

DAC Display Analysis Console

DAC Distance Amplitude Compensation

DAC Distance Amplitude Correction

DAC Division of Academic Centers [of Miami University, Oxford, Ohio, US]

DAC DME (Distance-Measuring Equipment) for Area Coverage

DAC Dual Attachment Concentrator

DA/C Data Acquisition

D/AC Digital-to-Analog Converter; Digital-to-Analog Conversion [also DAC]

DAc Doctor of Acupuncture

Dac Data Acquisition [also DAC, or dac]

DAcAc Diacetylacetone

DACB Data Acquisition and Control Buffer

DACBU Data Acquisition and Control Buffer Unit

DACBU Data Acquisition Control and Buffer Unit

DACC Direct-Access Communications Channel

DACE Data Acquisition and Control Executive

DACE Data Administration Center Equipment

DACH 1,2-Diaminocyclohexane [also dach]

DACH 1,4-Diazacycloheptane [also dach]

Dach+Wand International Trade Fair for Roof, Wall and Insulation Techniques, Hannover, Germany]

D-Acid 4-Hydroxy-1,6-Naphthalenedisulfonic Acid [also D-acid]

δ-Acid 7-Amino-2-Naphthalenesulfonic Acid; 4-Hydroxy-1,5-Naphthalenedisulfonic Acid; 2-Naphthylamine-7-Sulfonic Acid [also δ-acid]

DACM 7-Dimethylamino-4-Methyl-3-Coumarinyl) maleimide;N-(7-Dimethylamino-4-Methylcoumarinyl) maleimide

DACM-3 N-(7-Dimethylamino-4-Methyl-3-Coumarinyl) maleimide

DAC Newsl Division of Analytical Chemistry Newsletter [of American Chemical Society, US]

DACO 1,5-Diazacyclooctane [also daco]

DACOM Data Communications

DACOM Datascope Computer Output Microfilmer

DACOM DIAL (Differential-Absorption LIDAR (Light Detection and Ranging) and CO_2 Measurement (Project) [of NASA]

DACON Data Controller

DACON Digital-to-Analog Converter

DACOR Data Correction

DACOR Data Correlator

DACOS Data Communication Operating System

DACQ Data Acquisition

DACS Danger Area Crossing Service

DACS Data Acquisition (and) Control System

DACS Data and Analysis Center for Software

DACS Designers and Artists Copyright Society [UK]

DACS Digital Access and Cross-Connect System

DACS Digital Acquisition and Control System

DACS Directorate of Aerospace Combat Systems

DACT Disposable Absorption Collection Trunk

DACU Digitizing and Control Unit

DACUM Development of a Curriculum

DACVR Digital-to-Analog Converter

DAD Desktop Application Director

DAD Digital-Assisted Dispatching

DAD Digital Audio Disk

DAD Diode Array Detector

DAD Direct-Access Device

DAD Discharged Area Development

DAD Divisional Ammunition Dump

DAD Documents Against Discretion (of Collecting Bank)

DAD Drum and Display

DA&D Data Acquisition and Distribution

DADA Designers and Art Directors Association [also D&ADA, UK]

DADA Diisopropylammonium Dichloroacetate

D-ADA Deuterated Ammonium Dihydrogen Arsenate

DADB Data Analysis Database

DADC Digital Air Data Computer

DADD 1,10-Diamino-4,7-Diazadecane

DADE Data Acquisition and Decommutation Equipment

DADEE Dynamic Analog Differential Equation Equalizer

DADI Direct Analysis of Daughter Ions

DADLE [D-Alanine, D-Leucine] Enkephalinamide [Biochemistry]

DAdm Doctor of Administration

DADN 1,9-Diamino-3,7-Diazanonane

da/dN fatigue crack growth rate (i.e. crack extension per cycle of fatigue, or loading) [Symbol]

DADNPh Diazodinitrophenol

dADO 2'-Deoxyadenosine [Biochemistry]

D-ADP Deuterated Ammonium Dihydrogen Phosphate

dADP 2'-Deoxyadenosine Diphosphate [also d-ADP] [Biochemistry]

DADPM 4,4'-Diaminodiphenylmethane

DADPS 4,4'-Diaminodiphenylsulfone

DADPU Diaminodiphenylurea

DADPTU Diaminodiphenylthiourea

DADS Data Acquisition and Distribution System [of NASA]

DADS Data Acquisition Display System

DADS Data Acquisition Display Subsystem

DADS Data Archive Distribution System

DADS Diaminodiphenylsulfone

DADS Digital Air Data System

DADS Digitally-Assisted Dispatch System

DADS Dual Air Density Satellite

DADS Dynamic Analysis and Design System

DADSM Direct-Access Device Space Management

DADT Diaminobenzene Dithiol

DADT Durability and Damage Tolerance

dA/dt real generation rate (of solar cells)

dA/dt time rate of change in cross-sectional area [Symbol]

da/dt crack growth rate (i.e., crack length change per unit time) [Symbol]

d*a*/dt time rate of change of (linear) thermal expansion coefficient [Symbol]

DADTA Durability and Damage Tolerance Assessment

DADT-M Diaminobenzene-Dithiol-Containing Transition Metal Polymer

DAE Data Acquisition Equipment

DAE Department of Aerospace Engineering

DAE Department of Atomic Energy [Government of India]

DAE Diacetic Acid

DAE Distaloy AE [Proprietary Iron-Nickel-Copper-Molybdenum Alloy]

DAE Distributed Application Environment

DAE Distribution of Adsorption Energies

DAE Dynamic Asian Economy

DAEC Diethyl Aminoethyl Cellulose

DAEd Doctor of Arts in Education

DAEMON Data Adaptive Evaluator and Monitor [also Daemon]

DAEP Directorate of Aircraft Equipment Production [UK]

DAEP Division of Advanced Energy Projects [of US Department of Energy]

DAEP Division of Atomic Energy Production [UK]

DAEP II Diazaetioporphyrin(II)

DAEP IV Diazaetioporphyrin(IV)

DAER Department of Aeronautical Engineering Research [UK]

DAES Department of Applied Earth Sciences

DAEs Dynamic Asian Economies

DAF Data Acquisition Facility

DAF Data Analysis Facility

DAF Delayed Action Fuse

DAF Delivered At Frontier [also]

DAF Denmark-America Foundation [Denmark]

DAF Department of Agriculture and Fisheries [New South Wales]

DAF Department of the Air Force [of US Department of Defense]

DAF Destination Address Field

DAF Diacetylferrocene

DAF 4,5-Diazafluorene

DAF Dilution Attenuation Factor

DAF Direct-Access File

DAF Direct-Arc Furnace

DAF Discard at Failure

DAF Dissolved-Air Flotation [Chemical Engineering]

DAF Distributed Acquisition Facility

DAF Dry and Ash-Free [also daf]

DAF Dynamitron Accelerator Facility [of Argonne National Laboratory, Illinois, US]

Daf Continental Forest Climate, Warm Summer [Classification Symbol]

daf delivered at frontier [also DAF]

DAFA Data Accounting Flow Acceptance

DAFC Digital Automatic Frequency Control

DAFCS Digital Automatic Flight Control System

DAFF Deutsche Akademie für Film und Fernsehen [German Film and Television Academy, Berlin]

DAFM Direct-Access File Manager

DAFM Discard-at-Failure Maintenance

DAFS Deepwater Actively Frozen Seabed

DAFS Department of Agriculture and Fisheries of Scotland

DAFS Department of Agriculture and Forest Service

DAFS Diffraction Anomalous Fine Structure [Physics]

DAFT Data Acquisition Frequency Table

DAFT Digital/Analog Function Table

DAG Deputy Adjutant General

DAG Deputy Assistant General

DAG Deutsche Angestellten-Gewerkschaft [German Employees Union]

DAG Deutsche Automobil-Gesellschaft mbH [German Automobile Manufacturer located at Braunschweig]

DAG Development Assistance Group

DAG Directed Acyclic Graph [Computers]

DAG Distance Amplitude Gate

DAG Dysprosium-Aluminum Garnet

dag dekagram [Unit]

DAGA Deutsche Arbeitsgemeinschaft für Akustik [German Acoustics Association]

DAGAS Dangerous Goods Advisory Service [of Laboratory of the Government Chemistry, UK]

DAGC Delayed Automatic Gain Control

DAGF N,N,N',N'-Tetraethyl Phosphorodiamidioc Acid Heptyl Ester

DAGK Deutsche Arbeitsgemeinschaft für Kybernetik [German Cybernetics Association]

DAGM Deutsche Arbeitsgemeinschaft für Mustererkennung [German Association for Pattern Recognition]

DAGO D-Alanine-N-Methyl-Phenylalanine, Glycinol]-Enkephalin [Biochemistry]

DAgr Doctor of Agriculture

DAGV Differential Automatic Gas Volumetry

DAH Data Acquisition Hardware

DAH Department of Archives and History

DAH Disordered Action of the Heart

Dah Dahomey

DAHC Drain Avalanche Hot Carrier [Electronics]

DAHD 3,4-Diacetyl-2,4-Hexadiene-2,5-Diol [also dahd]

DAHE Department of the Arts, Heritage and Environment [now Department of the Environment, Australia]

DAHP N,N,N'N'-Tetraethyl Phosphorodiamidic Acid Heptyl Ester [also DAHPA]

DAHQ Di-tert-Amylhydroquinone

DAHS Derbyshire Agricultural and Horticultural Society [UK]

DAI Deutsches Archäologisches Institut [German Archeological Institute]

DAI Direct Action Impact

DAI Direct Aqueous Injection [Gas Chromatography]

DAI Dissertation Abstracts International

DAI Distributed Artificial Intelligence

1,3-DAI 1,3-Diaxial Interaction [Chemistry]

DAIB Diacetoxyiodobenzene

DAIDS Division of AIDS (Acquired Immunodeficiency Syndrome) [of National Institute of Allergy and Infectious Diseases, US]

DAI/GC Direct Aqueous Injection/Gas Chromatography

DAIM Digital Acceleration Integration Module

Dainichi-Nippon Cables Rev Dainichi-Nippon Cables Review [Japan]

DAIP Department of Atomic and Interface Physics

DAIP Diallyl Isophthalate

DAIR Direct Altitude and Identity Readout [of Federal Aviation Administration, US]

DAIR Driver Aid, Information and Routing

DAIR Dynamic Allocation Interface Routine

DAIRS Dial Access Information Retrieval System

Dairy Counc Dig Dairy Council Digest [of National Dairy Council, US]

Dairy Ind Int Dairy Industry International (Journal)

Dairy Prod J Dairy Products Journal [of American Cultured Diary Products Institute]

Dairy Res Dig Dairy Research Digest [of Diary Research Inc., US]

Dairy Sci Technol Dairy Science and Technology (Journal)

DAIS Data Avionics Information System

DAIS Defense Automatic Integrated Switch(ing System)

DAIS Digital Avionics Information System

DAIS District Agricultural Improvement Station [PR China]

DAIS Doctor of Arts in Information Science

DAISY Dairy Information System [UK]

DAISY Data Acquisition and Interpretation System

DAISY Decision-Aided Information System

DAISY Digital Acoustic Imaging System

DAISY Double-Precision Automatic Interpretative System

DAISY Düsseldorfer Analyse- und Iterationssystem [Dusseldorf Analysis and Iteration System]

DAIV Data Area Initializer and Verifier

DAIWR Department of the Army, Institute for Water Resources [also DA IWR, US]

DAJFT Dual-Axis Joint-Fourier Transform (Correlator)

DAK Decision Acknowledgement

DAK Deny All Knowledge

DAK Deutsche Angestellten-Krankenkasse [German Employees' Health Insurance]

DAK Dynamic Air Knife

Dak Dakota [US]

DAKLI Dynorphin A Analog κ Ligand [Biochemistry]

BOC-DAKLI N-tert-Butoxycarbonyl Dynorphin A Analog Kappa Ligand [Biochemistry]

DAL (Apple Computer) Data Access Language

DAL Data-Access Line [Telecommunications]

DAL Data Acquisition List

DAL Data Accession List

DAL Data Address List

DAL Data-Aided Loop

DAL Delta Air Lines [US]

DAL Diallyl Phthalate

DAL Direct Address Line

DAL Disk Access Lockout

daL decaliter [also dal]

D-Ala(P) D(+)-1-Aminoethylphosphonic Acid

DALK Data Link Controller

dalm dekalumen [Unit]

δ-Al$_2$O$_3$ Delta Alumina (or Delta Aluminum Oxide)

DALR Dry Adiabatic Lapse Rate [Meteorology]

DALS Digital Approach and Landing System

Dalton Trans Dalton Transactions [of Royal Society of Chemistry, UK]

DAM Bis(diphenylarsino)methane

DAM Data Acquisition and Monitoring

DAM Data Addressed Memory

DAM Data Association Manager

DAM Data Association Message

DAM Delayed Action Mine

DAM Dental Amalgam Mercury

DAM Department of Astronomy and Meteorology

DAM Descriptor Attribute Matrix

DAM Diallyl Maleate

DAM Diaminomesitylene

DAM 1,2-Diamino-2-Methylpropane

DAM Digital-to-Analog Multiplier

DAM Direct-Access Memory

DAM Direct-Access Method

DAM Discrete Atom Method [Materials Science]

DAM Doctor of Applied Management

DAM Double Aluminized Mylar

DAM Driver Amplifier Module

DAM Dual Absorption Model

Dam Damage [also dam]

dam dekameter [Unit]

dam^2 square dekameter [Unit]

dam^3 cubic dekameter [Unit]

DAMA (Particle) DArk MAtter (Searches with Low Activity Scintillators) [of Laboratori Nazionali del Gran Sasso, Italy] [also Dama]

DAMA Demand-Assigned Multiple Access

DAMA Demand-Assignment Multiple Access

DAMA Dialkyl Methylamine [also Dama]

DAMA Diallylmethacrylamide

DAMA-10 N-Decyl-N-Methyl-1-Decanamine

DAMC Desert Arthritis Medical Clinic [Desert Hot Springs, California, US]

DAMC 3-(4-Isothiocyanatophenyl)-7-Diethylamino-4-Methylcoumarin

DAM CON Damage Control [also Dam Con]

DAMDA Dairy Appliance Manufacturers and Distributors Association [UK]

DAME Data Acquisition and Monitoring Equipment

DAME Department of Aeronautical and Mechanical Engineering

DAMF Methylphosphonic Acid Bis(3-Methylbutyl)ester [also DAMFK]

DAMGO [D-Alanine-N-Methyl-Phenylalanine, Glycinol]-Enkephalin [Biochemistry]

DAMIT Data Analysis Massachusetts Institute of Technology [US]

DAMN Diaminomaleonitrile

DAMOP Division of Atomic, Molecular and Optical Physics [of American Physical Society, US]

DAMP Database of Atomic and Molecular Physics [of Queen's University, Belfast, Northern Ireland]

DAMP Diisoamylmethylphosphonate

DAMP Downrange Anti-Missile Measuring Project

DAMP Downrange Anti-Missile Program [US Army]

dAMP 2'-Deoxyadenosine 5'-Monophosphate [also d-AMP] [Biochemistry]

DAMPA Diisoamylmethylphosphonate

DAMPS Data Acquisition Multiprogramming System

DAMPS Digital Advanced Mobile Phone Service

DAMS Defense Against Missile Systems

DAMS Dental Amalgam Mercury Syndrome

DAMS Differential Amplification Magnetic Sensor

DAMS Direct-Access Management System

DAMSU Digital Auto-Manual Switching Unit

DAMTP 4,6-Diamino-2-Methylthiopyrimidine

DAN Data Analysis

DAN 2,3-Diaminonaphthalene

DAN 4-(N,N-Dimethylamino)-3-Acetamidonitrobenzene

DAN Distribution Analysis

DAN Doklady Akademii Nauk [Proceedings of the Academy of Sciences (of the USSR)]

Dan Danish; Denmark

daN decanewton [Unit]

DAN Azerb SSR Doklady Akademii Nauk Azerbaidzhanskoi SSR [Doklady, Proceedings of the Academy of Sciences of the Azerbayjanian SSR]

DAN BSSR Doklady Akademii Nauk BSSR [Doklady, Proceedings of the Academy of Sciences of the Byelorussian SSR]

DAN Belorusskoi SSR Doklady Akademii Nauk Belorusskoi SSR [Doklady, Proceedings of the Academy of Sciences of the Byelorussian SSR]

DANC Decontaminating Agent, Noncorrosive

DANCED Danish Cooperation for Environment and Development [Denmark]

d and a dry and abandoned [Oil and Gas Industry]

D and C Dila(ta)tion and Curettage [Induced Abortion]

D and E Dila(ta)tion and Evacuation [Induced Abortion]

DANDOK Danish Committee for Scientific and Technical Information and Documentation

D and V Diarrhea and Vomiting [also D&V]

DANES Diffraction Anomalous Near Edge Structure

Danfoss J Danfoss Journal [Published by Danfoss Manufacturing Co., Nordborg, Denmark]

DANIDA Danish International Development Agency

DanP Danish Patent

DANRIC Department of Agriculture and Natural Resources Information Council [Philippines]

DANS 5-(Dimethylamino)naphthalen-1-ylsulfonyl

DANS Dimethylaminonaphthalenesulfonyl Chloride

DANS-BBA N-Dansyl-3-Aminobenzeneboronic Acid

DAN SSSR Doklady Akademii Nauk SSSR [Proceedings of the Academy of Sciences of the USSR]

DANSYL 5-Dimethylamino-1-Naphthalenesulfone; 5-Dimethylaminonaphthalene-1-Sulfonyl [also Dansyl, or dansyl]

DAN Tadzh SSR Doklady Akademii Nauk Tadzhikshoi SSR [Doklady, Proceedings of the Academy of Sciences of the Tadzhik SSR]

DANTE Delays Alternating with Nutations for Tailored Excitation [Nuclear Magnetic Resonance]

DANTEC Danish Center for Technical Information [Skovlunde, Denmark]

DANTEC Inf DANTEC Information [Denmark]

DANTHRON 1,8-Dihydroxyanthraquinone [also Danthron, or danthron]

DAN USSR Doklady Akademii Nauk USSR [Proceedings of the Academy of Sciences (of the USSR)]

DAN Uzb SSR Doklady Akademii Nauk Uzbekskoi SSR [Doklady, Proceedings of the Academy of Sciences of the Uzbek SSR]

DAO Data Access Object

DAO Data Assimilation Office [of NASA Goddard Space Flight Center, Greenbelt Maryland, US]

DAO Dioctadecyladipate

DAO Dominion Astrophysical Observatory [of National Research Council in Victoria, British Columbia, Canada]

DAOF N,N,N',N'-Tetraethylphosphorodiamidic Acid Octyl Ester

DAOMeTSC Diacetyl-O-Methyl Monoxime Thiosemicarbazone

DAOS N-Ethyl-N-(2-Hydroxy-3-Sulfopropyl)-3,5-Dimethoxyaniline

DAOTSC Diacetyl Monoxime Thiosemicarbazone

DAP Data-Access Protocol

DAP Data Acquisition Plan

DAP Data Acquisition Processor

DAP Data-Analysis Package

DAP Death Associated Protein

DAP Deformation of Aligned Phases

DAP Department of Applied Physics

DAP Developer Assistance Program

DAP Diabetes Associated Peptide

DAP 2,6-Diacetylpyridine

DAP Dial-A-Program

DAP Dialkylphosphate

DAP Diallyl Phthalate (Thermoset)

DAP Diaminophenol

DAP Diaminopimelate; Diaminopimelic Acid

DAP 1,2-Diaminopropane

DAP Diammonium Phosphate

DAP Diammonium Phosphate Plant

DAP Diamylphthalate

DAP Dibasic Ammonium Phosphate

DAP Digital Access Point

DAP Digital Assembly Program [of Engineers and Architects Institute, US]

DAP Digital Autopilot

DAP 1,4-Dihydrazinophthalazine

DAP Dihydroxyacetone Phosphate

DAP Directionally Attached Piezoelectric (Element)

DAP Directory Access Protocol [Internet]

DAP Distributed Array of Processors

DAP Distributed Array Processor

DAP Division of Air Pollution [of US Department of Energy]

DAP Division of Applied Physics [of Commonwealth Scientific and Industrial Research Organization, Sydney, Australia]

DAP Division of Astrophysics [of American Physical Society, US]

DAP Donor-Acceptor Pair [Solid-State Physics]

13-DAP 1,3-Diaminopropane

β-DAP β-Dimethylaminopropiophenone

DAPA Diaminodipropylamine

DAPA Di-n-Amylphosphoric Acid

DAPCA Development and Production Costs for Aircrafts

DAPD 2,6-Diacetylpyridine Dioximate [also dapd]

DAPD Directorate of Aircraft Production Development [UK]

DAPDA Dimethylaminophenylpentadienal

DAPEX Dialkylphosphoric Acid Extraction (Process) [Chemical Engineering]

DAPHC Diaminophenol Hydrochloride

DAPI 4,6-Diamidino-2-Phenylindole; 4'-6-Diamidino-2-Phenylindole (Dihydrochloride)

DAPI Diamine Polyimide

DAPI Diaminophenylindane

DAPIE Developers Application Programming Interface Extensions

DAPIS Danish Agricultural Products Information Service

DAPK Death Associated Protein Kinase

DAPMAT Data Analysis Package for Materials Testing

DAPN Diamylmethylpropanol

DAPN 1-([2-(5-Dimethylamino)naphthalene-1-Sulfonylaminoethyl]amino)-3-(1-Naphthaleneoxy)-2-Propanol

DAPO Digital Advance Production Order

DAPP Data Acquisition and Processing Program

DAPR Digital Automatic Pattern Recognition

DAPS Department of Applied Physical Science

DAPS Direct-Access Programming System

DAPS Disappearance Potential Spectroscopy

DAPS Distributed Application Processing System

DAPSC 2,6-Diacetylpyridine Bis(semicarbazone)

DApSc Doctor of Applied Science

DAPSONE 4,4'-Diaminodiphenyl Sulfone; p-Aminophenyl Sulfone [also DDS, Dapsone, or dapsone]

DAPT 2,4-Diamino-5-Phenyl Thiazole

DAPU Data Acquisition and Processing Unit

DAQ Data Acquisition

DAQ International Data Acquisition Workshop

DAQC Data Acquisition Card

DAQMG Deputy Assistant Quartermaster-General

DAR Daily Activity Report

DAR Damage Assessment Routine

DAR Data Access Register

DAR Data Aided Receiver

DAR Defense Acquisition Radar

DAR Department of Applied Research

DAR Destination Access Register

DAR Deutscher Akkreditierungsrat [German Accreditation Council; of German Calibration Service]

DAR Deviation Approval Request

DAR Diacetylresorcinol

DAR Diels-Alder Reaction [Organic Chemistry]

DAR Differential Absorption Ratio

DAR Digital AIDS (Aircraft Integrated Data System) Recorder

DAR Digital Autopilot Requirements

DAR Directorate of Atomic Research

DAR Division of Atmospheric Research [of Commonwealth Scientific and Industrial Research Organization, Australia]

DAR Drawing Analysis Record

DARA Department of Agriculture and Rural Affairs [of Victoria State, Australia]

DARA Deutsche Agentur für Raumfahrt-Angelegenheiten [German Space Agency]

DARC Deutscher Amateur-Radio-Club [German Radio Amateur Club]

DARC Digital Automated Rocking Curve (System)

DARC Direct-Access Radar Channel

DARC Documentation and Automation of Research on Correlations

DArch Doctor of Architecture

DARD Dynamic-Annealing Related Diffusion [Metyallurgy]

DARE Data Rescue

DARE Data Retrieval System

DARE Document Abstract Retrieval Equipment

DARE Documentation Automated Retrieval Equipment

DARE Doppler and Range Evaluation

DARE Doppler Automatic Reduction Equipment

DAREOD Damaged Airfield Reconnaissance and Explosive Ordnance Disposal

DARES Data Analysis and Reduction System

DARI Database Application Remote Interface [of IBM Corporation, US]

DARLI Digital Angular Readout by Laser Interferometry

DARME Directorate of Armament Engineering [Canada]

DARMS Digital Alternate Representation of Musical Symbols

DARPA Defense Advanced Research Projects Agency [of US Department of Defense]

DARPA/AFOSR Defense Advanced Research Projects Agency/Air Force Office of Scientific Research [of US Department of Defense]

DARPA ATP Defense Advanced Research Projects Agency–Advanced Technology Project [of US Department of Defense] [also DARPA-ATP]

DARPA/ESTO Defense Advanced Research Projects Agency/Electronic Systems Technology Office [of US Department of Defense]

DARPA/ONR Defense Advanced Research Projects Agency/Office of Naval Research [of US Department of Defense] [also DARPA-ONR]

DARPA/OSP Defense Advanced Research Projects Agency/Office of Scientific Projects [of US Department of Defense] [also DARPA-OSP]

DARPA URI Defense Advanced Research Projects Agency/University Research Initiative (Program) [of US Department of Defense]

DARS Data Acquisition and Reduction System

DARS Digital Adaptive Recording System

DARS Digital Attitude and Rate System

DARS Digital Attitude Reference System

DARSS Diode Array Rapid-Scan Spectrometer

DART Daily Automatic Rescheduling Technique

DART Data Accumulation and Retrieval of Time (System)

DART Data Acquisition and Recording Terminal

DART Data Acquisition in Real Time

DART Data Analysis Recording Tape

DART Defect Area Revisit Technology [Wafer Inspection]

DART Defense Assistance Response Team

DART Dependable and Reliable Transit

DART Development Advanced Rate Technique

DART Digital Audio Reconstruction Technology

DART Digital Automatic Readout Tracker

DART Direct-Access Radio Transceiver

DART Director and Response Tester

DART Dual Axis Rate Transducer

DART Dynamic Acoustic Response Trigger

DARTS Digital Automated Radar Tracking System

DART Disaster Assistance Response Team [of Department of National Denfense, Canada]

DARTS Digital Azimuth Range Tracking System

DAS Acetyl Sulfide (or Acetic Thioanhydride)

DAS Danish Academy of Sciences [Copenhagen]

DAS Danish Automation Society

DAS Data Access Security

DAS Data Acquisition System

DAS Data Acquisition Subsystem

DAS Data Analysis Software

DAS Data Analysis System

DAS Data Automation System

DAS Data Auxiliary Set

DAS Datapac Access Software

DAS Datatron Assembly System

DAS De-Ashing Solvent

DAS Decay-Associated Spectrum

DAS Decimal Adjust for Subtraction

DAS Delivered Alongside Ship

DAS Dendrite-Arm Spacing [Metallurgy]

DAS Department of Administrative Services

DAS Department of Applied Science

DAS Deputy Assistant Secretary [US]

DAS Derbyshire Archeological Society [UK]

DAS Deutscher Ausschuß für Stahlbeton [German Reinforced Concrete Association]

DAS Device Access Software

DAS Devon Archeological Society [UK]

DAS Diacetoxyscirpenol

DAS Diacetylsulfide

DAS Diaminostilbene

DAS 4,4'-Diaminostilbene-2,2'-Disulfonic Acid

DAS Differential Absorption and Scattering

DAS Digital Address System

DAS Digital Analog Simulator

DAS Digital Attenuator System

DAS Digital Avionics System

DAS Dimer Adatom Stacking-Fault (Model) [Surface Science]

DAS 9,10-Dimethoxy 2-Anthracenesulfonic Acid

DAS Direct-Access Storage

DAS Direction of Armament Supply [UK]

DAS Directory Assistance System

DAS Discount Access Services

DAS Discrete Algebraic Structures

DAS Disturbance Analysis System

DAS Division of Administrative Services [of Alaska Department of Environmental Conservation, US]

DAS Division of Applied Science

DAS Division of Applied Science [of Harvard University, Cambridge, Massachusetts, US]

DAS Documentation Accountability Sheet

DAS Double Antibody Sandwich (Enzyme-Linked Immunosorbent Assay)

DAS Double-Arm Spectrometer

DAS Dry, As-Molded

DAS Dual-Attached Station

DAS Dutch Academy of Sciences

DAS Dynamic-Angle Spinning [Nuclear Magnetic Resonance Technique]

DAS 1,2-Phenylenebis(dimethylarsine)

DASA Daimler Benz Aerospace [Ulm, Germany]

DASA Data Acquisition and Signal Analysis (System)

DASA Defense Atomic Support Agency [of US Department of Defense]

DASA Deutsche Aerospace [Germany]

DASA Domestic Appliance Service Association [UK]

DASA Dual Aerosurface Servo Amplifier

DASC Defense Automotive Supply Center

DASC Direct Air Support Center

DASC Division of Advanced Scientific Computing [of National Science Foundation, US]

DASc Doctor of Applied Science

DASCOTAR Data Acquisition System Correlating Tracking and Ranging

DASD Direct-Access Storage Device

DASDL Data and Structure Definition Language

DASDR Direct-Access Storage Dump Restore

DASET Department of the Arts, Sport, the Environment and Territories [now Department of the Environment, Australia]

DASETT Department of the Arts, Sport, the Environment, Tourism and Territories [now Department of the Environment, Australia]

DASF Direct-Access Storage Facility

DASH Design Aid Schematic Helpmate

DASH Direct Access Storage Handler

DASH Drone Anti-Submarine Helicopter

DASH Dual Access Storage Handling

DASHER Dynamic Analysis of Shells of Revolution

DASL Data-Access System Language

DASM Direct-Access Storage Media

DASM Directional Acoustic Sensor Module

DASP Double-Arm Spectrometer

DASp Deutscher Arbeitskreis für Angewandte Spektroskopie [German Association for Applied Spectroscopy]

DASS Demand-Assignment Signaling and Switching (Device)

DASS Diesel Air Start System

DASS Digital Access Signalling System

DAST Diethylaminosulfur Trifluoride

DAST Diploma of Advanced Studies in Teaching

DAST Division for Advanced Systems Technology

DASt Deutscher Ausschuß für Stahlbau [German Steel Construction Committee]

DAT Designation–Acquisition–Track

DAT Deutsch-Atlantische Telegraphengesellschaft [German-Atlantic Telegraphy Company]

DAT Diaminotoluene

DAT Differential Aptitude Test

DAT Diffused-Alloy Transistor

DAT Digital Archival Tape

DAT Digital Audio Tape

DAT Dipole Array Telescope

DAT Director of Advanced Technology

DAT Disconnect Actuating Tools

DAT Disk Access Time

DAT Disk Allocation Table

DAT Disk Array Technology

DAT Dynamic Address Translation; Dynamic Address Translator

.DAT Data [File Name Extension] [also .dat]

Dat Datum [also dat]

DATA Design and Technology Association [UK]

DATA Development and Technical Assistance

DATA 1,2-Diaminocyclohexanetetraacetic Acid [also Data]

DATA Draftsmen and Allied Technicians Associations [UK]

Database Netw J Database and Network Journal [UK]

DATAC Data Analog Computer

DATAC Digital Automatic Tester and Classifier

DATAC Digital Autonomous Terminal Access Communication

DATACOL Data Collection System

DATACOM Data Communication(s) [also Datacom or DC]

DATACOM Data Communication Network [US Air Force]

DATACOM Data Communication Service [US]

DATACOM Data Communication System

Data Commun Data Communications [Journal published in the US]

Data Commun Process Data Communication and Processing [Journal published in Japan]

DATAGEN Data File Generator

Data Knowl Eng Data and Knowledge Engineering [Journal published in the Netherlands]

DATALINK Digital Circuit-Switched Data Network [of Telecom Canada]

Data Manage Data Management [Publication of the Data Processing Management Association, US]

DATAN Data Analysis

DATANET Data Network [also Datanet]

DATAPAC Packet-Switched Data Network [of Telecom Canada] [also Datapac]

Data Process Data Processor [Publication of International Business Machines Corporation, US]

Data Process Pract Data Processing Practitioner [Publication of the Institute of Data Processing, UK]

DATAR Digital Automatic Tracking and Ranging [also Datar, or datar]

DATAR Digital Auto-Transducer and Recorder

DATAROM Data Read-Only Memory

DATAROUTE Digital Data Transmission Service [of Telecom Canada] [also Dataroute]

DATAS Data Link and Transponder Analysis System

Data Train Data Training [Published in the US]

DATC S-2,3-Dichloroallyl N,N-Diisopropyl Thiocarbamate

DATCO Duty Air Traffic Control Officer

DATCOM Data Support Command [US Army]

DATD N,N'-Diallyltartardiamine

DATD N,N'-Diallyl Tartaric Acid Diamide

DATDC Data Analysis and Technique Development Center

DATE Dynamics, Acoustics and Thermal Environment

DATEL Data Telecommunication [also Datel or datel]

DATEX Data Exchange Service [of German Federal Post Office] [also Datex or datex]

DATEX Data Text [also Datex]

DATEX-L Data Exchange Service–Circuit Switching [also Datex-L] [Germany]

DATEX-P Data Exchange Service–Packet Switching [also Datex-P] [Germany]

DAtF Deutsches Atomforum [German Nuclear Forum]

DATFUS Data Fusion

DATICO Digital Automatic Tape Intelligence Checkout

DATNB 1,3-Diamino-2,4,6-Trinitrobenzene

DATP 4,6-Diamino-1,2-Dihydro-2-Thiopyrimidine

DATP p-Methoxyphenyl-N,N'-Bis(ethoxyphenyl) diamidothiophosphate

dATP 2'-Deoxyadenosine 5'-Triphosphate [also d-ATP] [Biochemistry]

DATR Design Approval Test Report

DATRAN Data Transmission

DATRI Division of AIDS Treatment Research Initiatives [of National Institute of Allergy and Infectious Diseases, US]

DATRIX Direct Access to Reference Information

DATRIX DIRECT Computerized Data Search and Retrieval System for Dissertations [of University Microfilms International, US]

DATS Danish Academy of Technical Sciences

DATS Despun Antenna Test Satellite

DATS Digital Avionics Transmission System

DATS Dynamic Accuracy Test Set

DATSA Depot Automatic Test System for Avionics

DATTS Data Acquision, Telecommand and Tracking Station

DAU Data Access Unit

DAU Data Acquisition Unit

DAU Data Adapter Unit

DAU Digital Adapter Unit

DAU Disposable Adsorbent Unit

Dau Daughter [also dau]

DAUG Danger Area Users Group

DAV Data above Voice

DAV Data Available

DAV Data Valid [General-Purpose Interface Bus Control Line]

DAV Deutsche Arbeitsgemeinschaft Vakuum [German Vacuum Association]

DAV Digital Audio-Video

DAVC Delayed Automatic Volume Control

DAVI Dynamic Antiresonant Vibration Isolator

DAVIC Digital Audio-Visual Council [US]

DAVID Data Above Video System

DAVIE Digital Alphanumeric Video Insertion Equipment

DAVINS Direct Analysis of Very Intricate NMR (Nuclear Magnetic Resonance) Spectra

DAV-L Data Available-Low

DAVSS Doppler Acoustic Vortex Sensing Equipment

DAW Data Address Word

DAW Deutsche Akademie der Wissenschaften [German Academy of Sciences, Berlin]

DAW Dry Active Waste

DAWA Department of Agriculture, Western Australia

DAWNS Design of Aircraft Wing Structures

DAX Data Acquisition and Control

DAX Deutscher Aktienindex [German Stock Index]

DAZD Double Anode Zener Diode

DB Base Diameter (of Involute Splines) [Symbol]

DB Dangling Bond [Solid-State Physics]

DB Databank; Database

DB Data Buffer(ing)

DB Data Bus

DB Data Byte

DB Daw and Baskes (Potential) [Materials Science]

DB Day Book

DB Daytona Beach [Florida, US]

DB Deactivated for Basic Compounds [Gas Chromatography]

DB Deadband

DB Dead Beat

DB Debug Byte (Values) [Operation Code]

DB Decimal-to-Binary [also D-B]

DB Decision Bar

DB Deformation Band(ing) (Model) [Metallurgy]

DB Delayed Boiling

DB Delayed Broadcast

DB Density Bottle

DB Department of Biology

DB Depth Bomb

DB Design Baseline

DB Deutsche Bibliothek [German Library, Frankfurt]

DB Deutsche Bundesbahn [German Federal Railways]

DB Deutsche Bundesbank [German Federal Bank]

DB Deutsche Bundespost [German Federal Post, Telephone and Telegraph] [now Deutsche Telekom]

DB Device Bay

DB Di-n-Butyramide

DB 4-(2,4-Dichlorophenoxy)butanoic Acid

DB Dichlorophenoxybutyric Acid

DB Die-Bonded; Die Bonding

DB Dielectric Breakdown

DB Diffuse Background (Structure Analysis)

DB Diffused Base (Technique) [Electronics]

DB Diffusion Battery

DB Diffusion Bond(ing) [Metallurgy]

DB Dip Brazing [Metallurgy]

DB Directed Beam

DB Display Bytes (Command) [Pascal Programming]

DB Distribution Box

DB Dive Bomber

DB Divergent Beam

DB Diving Bell

DB Domain Boundary [Solid-State Physics]

DB Doppler-Broadening [Spectroscopy]

DB Double Barreled

DB Double Barrier [Physics]

DB Double Beam

DB Double-Biased (Relay)

DB Double Bond

DB Double Bottom

DB Double Braid (Electric Wire)

DB Double-Break

DB Drill Box (Application) [Agriculture]

DB Drilling Barge

DB Driving Band

DB Dry Bulb

DB Dumbbell

DB Dynamic Braking

DB Dynamic Breaking

DB2 Database 2 [of IBM Corporation]

DB6 Hexylphenyl Cyanobenzoyloxy Benzoate

2,4-DB 4-(2,4-Dichlorophenoxy)butanoic Acid

2,4-DB 2,4-Dichlorophenoxybutyric Acid

4-(2,4-DB) 4-(2,4-Dichlorophenoxy)butanoic Acid

DB÷ Double Divide (Function) [Computers]

$B Belize Dollar [Currency of Belize]

D/B Documentary Bill

D-B Dark-Bright (Fringe)

D-B Decimal-to-Binary [also DB]

Db Humid Continental Climate, Cool Summer [Classification Symbol]

dB decibel [also db]

dB de Broglie (Wavelength)

dB distorted-coherent (grain) boundary (in materials science) [Symbol]

dB$_A$ decibel referred to A curve [Unit]

dB$_B$ decibel referred to B curve [Unit]

ΔB Change in Magnetic Induction [Symbol]

ΔB Magnetic Excursion Range [Symbol]

DBA Database Administration; Database Administrator

DBA Design Basis Accident

DBA Design Business Association [UK]

DBA 0,0'-Diallyl Bisphenol A

DBA Dibenzalacetone

DBA Dibenz[a,h]anthracene

DBA Dibenzylamine

DBA Dibenzylideneacetone

DBA 9,10-Dibromo Anthracene

DBA Dibutyl Amine

DBA 4,4'-Dichlorobenzilic Acid

DBA Doctor of Business Administration

DBA Doing Business As [also dba]

DBA Dolichos Biflorus Agglutinin [Biochemistry]

dB(A) decibel based on 40 phon equal loudness contour [Unit]

dBa decibel, adjusted [also dBA] [Unit]

dBa0 dBA at zero transmission level [Unit]

dba doing business as [also DBA]

DBAA Dibenzylacetic Acid

DBAAM Disk Buffer Area Access Method

DBACS Database Administrator Control System

dBae decibel(s) of acoustic emission amplitude [Unit]

DBAF Database Access Facility

dBa(F1A) dBa measured by set with F1A line weighting [Unit]

dBa(H1A) dBa measured by set with H1A receiver weighting [Unit]

DBAlCl Di-i-Butylaluminum Chloride

DBAlH Di-i-Butylaluminum Hydride

DBAM Database Access Method

DBAM Database Access Module

DBAO Digital Block AND-OR Gate

DBAPA Dibutyl Allylphosphonic Acid

DBAR Doppler Broadening of the Annihilation Radiation (Method) [Physics]

DBAWG Database Administration Working Group [of CODASYL–Conference on Data System Languages]

DBB Detector, Back Bias

DBB Detector, Balanced Bias

DBB Deutscher Beamtenbund [German Public Employees Union]

DBB Dibromobenzene

DBB Dibromobutane

dB(B) decibel based on 70 phon equal loudness contour [Unit]

DBBP Dibenzoylbiphenyl

DBBP Dibromobiphenyl

DBBP Dibutyl Butyl Phosphate

DBBP Dibutyl Butyl Phosphonate

DBBP Dibutylphosphonic Acid Dibutyl Ester

DBBW Daw, Baskes, Bisson and Wolfer (Potential) [Materials Science]

DBC Databank COMECON [also DB COMECON]

DBC Database Computer

DBC Data Bridging Capacity

DBC Data Bus Control(ler)

DBC Data Bus Coupler

DBC Decimal-to-Binary Conversion

DBC Deformation Banding Constraint [Metallurgy]

DBC Dense Barium Crown (Glass)

DBC Denver Botanic Gardens [Colorado, US]

DBC Department of Biological Chemistry

DBC Device Bay Controller

DBC Diameter Bolt Circle

DBC 1,2-Dibromo-3-Chloropropane

DBC Dibutyl Carbinol

DBC Digital-to-Binary Conversion; Digital-to-Binary Converter

DBC Dynamic-Breaking Contactor

DBC Dysprosium Barium Cuprate

DB14C4 Dibenzo-14-Crown-4

DB15C5 Dibenzo-15-Crown-5

DB18C5 Dibenzo-18-Crown-5

DB18C6 Dibenzo-18-Crown-6

DB21C7 Dibenzo-21-Crown-7

DB24C8 Dibenzo-24-Crown-8

DB27C9 Dibenzo-27-Crown-9

DB30C10 Dibenzo-30-Crown-10

dBC C-scale sound level in decibels [Unit]

dB(C) decibel based on 100 phon equal loudness contour [Unit]

dB/c decibel attenuation per 100 feet [Unit]

DBCAT 3,5-Di-tert-Butylcatechol [also DBCat]

DBCB Database Control Block

DBCE Division of Building, Construction and Engineering [of Commonwealth Scientific and Industrial Research Organization, Australia]

DBCI Decibel(s) with Respect to a Circular Polarized Antenna

DBCL Database Command Language

dB/cm decibel(s) per centimeter [also dB cm^{-1}] [Unit]

DB$_5$CN Pentylphenyl Cyanobenzoyloxy Benzoate

DBCO Dysprosium Barium Copper Oxide (Superconductor) [also DyBCO]

DBCO-123 DyBa$_2$Cu$_3$O$_{7-x}$ [A Dysprosium Barium Copper Oxide Superconductor]

DB COMECON Databank Council for Mutual Economic Assistance [also DB Comecon]

DBCP 1,2-Dibromo-3-Chloropropane

DBCP 1,3-Dibromocyclopentane

DBCP Drifting Buoy Cooperation Panel

DBCS Database Control System

DBCS Delivery Bar Code Sorter

DBCS Double-Byte Character Set

DBD Database Definition

DBD Database Description

DBD Database Diagnostics

DBD Dibromodulcite

DBD Double Backscattering Diffractometer

DBD Double-Base Diode

DBD Ductile-Brittle-Ductile (Transition) [Materials Science]

dBd antenna gain in decibels referenced to that of a dipole [Unit]

DBDA Design Basis Depressurization Accident

DBDA Dibenzyldodecylamine

DBDB18C6 Di-t-Butyldibenzo-18-Crown-6

DB/DC Database/Data Communications

DBDECMP Dibutyl-N,N'-Diethylcarbamylmethylene-phosphonate

DBDECP Dibutyl-N,N-Diethylcarbamylphosphonate

dB/dH permeability (i.e., change of magnetic induction with magnetic field strength) [Symbol]

ΔB/ΔH Incremental Magnetic Permeability

DBDL Database Definition Language

DBDM15C5 Dibenzodimethyl-15-Crown-5

DBDM18C6 Dibenzodimethyl-18-Crown-6

dB/dP change of uulk modulus with pressure [Symbol]

DBDPO Decabromodiphenyloxide

dB/dt time rate of change of magnetic induction [Symbol]

dB/dx magnetic field gradient (i.e., change in magnetic field per unit distance) [Symbol]

DBE Dame Commander of the Order of the British Empire

DBE Databank Eurocontrol

DBE Data Bus Element

DBE Data Bus Enable

DBE Department of Biomedical Engineering

DBE Design-Basis Earthquake

DBE Design Basis Event

DBE Detroit Board of Education [Michigan, US]

DBE Development Bank of Ethiopia

DBE 1,2-Dibromoethane

DBE Dibutyl Ether

DBEC Danish Building Export Council

DBED Dibenzylethylene Diamine

DBEDDA N,N'-Dibenzylethylenediamine-N,N'-Diacetate [also dbedda]

DBEM Dual Boundary Element Method

DBEP Di(butoxyethyl)phosphate

DBEPA 2-(Dibutylphosphonato)diethylsuccinate

DBER Division of Biomedical and Environmental Research [of Economic and Regional Development Agreements, US]

DBF Database File

DBF Delayed Brittle Fracture [Metallurgy]

DBF Demodulator Band Filter

DBF Design Basis Fault

DBF Dibenzofuran

DBF Dynamic Beam Focus

DBF Dynamic Beam Forming

DBF Phosphoric Acid Dibutyl Ester

.DBF Database File [File Name Extension] [also .dbf]

Dbf Continental Forest Climate, Cool Summer [Classification Symbol]

dBf decibels referred to one femtowatt [Unit]

DBFAS Digital Beam-Focusing Array Signal

DBFN Data Bus File Number

DBFN Data Base File Number

DBFS Deep Bed Farming Society [US]

DBFS Dull Black Finish Slate

DBG Database Generator

DBG Data Bus Group

DBG Desert Botanical Gardens [Phoenix, Arizona, US]

DBGM Deutsches Bundesgebrauchsmuster [German Registered Pattern, or Design]

DBGMP Data Bus Generation and Maintenance Package

DBH Bis(4-Chlorophenyl)methanone

DBH Diameter at Breast Height (of Tree)

DBH Dopamine β-Hydroxylase [Biochemistry]

DBHI Dopamine β-Hydroxylase Inhibitor [Biochemistry]

DBHPI Dibutylhexylphosphonate

DBHT Dibutylhydroxytoluene

D-β-Hydroxybutyryl-S-ACP D-β-Hydroxybutyryl-S-Acyl Carrier Protein [Biochemistry]

DBI Database Index

DBI Database Integrity

DBI Data Bus Interface

DBI Deutsches Bibliotheksinstitut [German Library Institute, Berlin]

DBI Diazepam Binding Inhibitor [Biochemistry]

DBI Differential Bearing Indicator

DBI Double-Byte Interleaved

dBi antenna gain in decibels referenced to that of an isotrope [Unit]

DBIA Data Bus Interface Adapter

DBIA Data Bus Isolation Amplifier

DBIA Design-Build Institute of America [US]

DBIN Data Bus In

DBIOC Database Input/Output Control

DBIS Dun and Bradstreet Information Services [US]

DBIU Data Bus Interface Unit

DBIU-L Data Bus Interface Unit-Launch

DBIV Deutscher Braunkohlen-Industrie-Verein [German Association of the Brown Coal Industry

dBJ decibel(s), Jerrold [Unit]

DBK Deutsche Bibliotheks-Konferenz [German Library Conference, Germany]

dB/K decibel(s) per degree Kelvin [Unit]

dBk decibel(s) referred to one kilowatt [also dbK] [Unit]

dB/kfci decibel(s) per kilo flux change(s) per inch [Unit]

dB/km decibel(s) per kilometer [also dB km^{-1}] [Unit]

DBL Database Locking

dbl double [also dble]

Dblr Doubler [also dblr]

DBLTG Database Language Task Group [of CODASYL–Conference on Data System Languages]

Dbl Und Double Underline [also dbl und]

DBM Database Management; Database Manager

DBM Database Model

DBM Data Buffer Module

DBM Data Bus Monitor

DBM Dead-Burnt Magnesia

DBM Dense Branding Morphology

DBM Deutsches Bergbau-Museum [German Mining Museum, Bochum]

DBM Dibenzoylmethane

DBM Dibromomannitol [Biochemistry]

DBM Dibutyl Maleate

DBM Dielectric Breakdown Model

DBM Direct Bombardment Mode

DBM Direct Branch Mode

dBm decibel(s) referred to one milliwatt [also dbm] [Unit]

dBm0 dBm (decibel(s) referred to one milliwatt) at zero transmission level [Unit]

dB/m decibel(s) attenuation per meter [Unit]

dbm 1,3-Diphenyl 1,3-Propanedione, ion(1-)

DBMAPA Dibutyl-1-Propynylphosphonate

DBMC Data Bus Monitor/Controller

DBMC Di-tert-Butyl-M-Cresol

DBMC Di-tert-Butyl-Methylphenol

dB/mi decibel(s) per mile [Unit]

DBMIB Dibromomethyl Isopropyl Benzoquinone

dBm/m² dbm (decibel(s) referred to one milliwatt) per square meter [Unit]

dBm/m²/Mhz dBm (decibel(s) referred to one milliwatt) per square meter per megahertz [Unit]

dBm0p dBm (decibel(s) referred to one milliwatt) at zero transmission level, psophometrically weighted [Unit]

dBm0ps dBm (decibel(s) referred to one milliwatt) at zero transmission level, psophometri-cally weighted for sound program [Unit]

dBm(PSOPH) dBm (decibel(s) referred to one milliwatt) measured by a set with psophometric weighting [Unit]

DBMS Database Management Software

DBMS Database Management System

DBMSPSM Database Management System Problem Specification Model

dBμV decibel(s) above one microvolt [Unit]

dBμV/MHz decibel(s) above one microvolt per megahertz [Unit]

DBN Data Bus Network

DBN DEC (Digital Equipment Corporation) Business Network

DBN 1,5-Diazabicyclo[4.3.0]non-5-ene

DBN 2,6-Dichlorobenzonitrile

DBO 1,4-Diazabicyclo[2.2.2]octane

DBO Drop Build-Out Capacitor

DBoA Delayed Breeder, or Alternative

DBOBPA Dibutyl 3-Oxobutyl-1-Phosphonate

DBOF Octylphosphonic Acid Dibutyl Ester

DBOMP Database Organization and Maintenance Processor

DB$_8$ONO$_2$ Octyloxyphenyl Nitrobenzoyloxy Benzoate

DBOP Dibutyloctylphosphonate

DBOS Disk-Based Operating System

DBQW Double-Barrier Quantum Well (Heterostructure) [Solid-State Physics]

DBP Database Processor

DBP Design Baseline Program

DBP Deutsche Bundespost [German Federal Post Office] [Now Deutsche Telekom]

DBP Deutsches Bundespatent [German (Federal) Patent]

DBP Diastolic Blood Pressure [Medicine]

DBP Dibasic Barium Phosphate

DBP 2,6-Di-tert-Butyl-4-Methylphenol

DBP Di-n-Butylphospate

DBP Dibutyl Phthalate

DBP 2,4'-Dichlorobenzophenone; 4,4'-Dichlorobenzophenone

DBP Directed-Beam Processing

DBP Division of Biological Physics [of American Physical Society, US]

DBP 5-Phenyl-5H-Dibenzophosphole

2,4'-DBP 2,4'-Dichlorobenzophenone

4,4'-DBP 4,4'-Dichlorobenzophenone

dBp decibel(s) above one picowatt [Unit]

dBp decibel(s) signal-to-noise ratio psophometrically weighted [Unit]

DBPA 1,10-Decanediylbis(phosphonic Acid)

DBPA Di-n-Butyl Phosphoric Acid

DBP angem Deutsches Bundespatent angemeldet [German (Federal) Patent Pending]

DBPC 2,6-Di-tert-Butyl-p-Cresol [also dbpc]

DBPCB Database Program Communication Block [also DB/PCB]

DBPhP Dibutylphenylphosphonate

DBPP Dibutylphenylphosphate

DBPPA Dibutylpropylphosphonic Acid

DBPrP Dibutylpropylphosphonate

DBPTelekom Deutsche Bundespost Telekommunikation [German Post and Telecommunications Office]

dBQ decibel(s) relative to a reference voltage, measured with quasi-peak noise meter [also dBq] [Unit]

DBR Database Retrieval

DBR Directorate of Biosciences Research [of Defense Research Board, Canada]

DBR Distributed Bragg Reflector (Mirror) [Physics]

DBR Division of Building Research [of National Research Council of Canada]

DBR DOS (Disk-Operating System) Boot Record

DBR Double Remainder (Function)

DBr Deuterium Bromide

dBr decibel(s) relative to reference [also dBR] [Unit]

dBrap decibel(s) above reference acoustical power [also dBRAP] [Unit]

DBRN Data Bank Release Notice

dBrn decibel(s) above reference noise of one micromicrowatt at one kilohertz [also dBRN, or dbRN] [Unit]

dBrnC decibel(s) above reference noise, C-message weighted [also dBRNC] [Unit]

dBrnC0 decibel(s) above reference noise, C-message weighted, at zero transmission level [also dBRNC0] [Unit]

DBRT Double-Barrier Resonant Tunneling [Physics]

DBRTS Double-Barrier Resonant Tunneling Structures

DBS Database Server [Internet]

DBS Database Supplier

DBS Database System

DBS Data Bridging Service

DBS Dibromosalicil

DBS Dibutyl Sebacate

DBS Dielectric Breakdown Strength

DBS Direct Broadcast(ing) by (or from) Satellite

DBS Direct Broadcast Satellite

DBS Direct Broadcast(ing Satellite) System

DBS Division of Biological Sciences [of National Research Council of Canada]

DBS Division of Biological Standards [of National Institutes of Health, Bethesda, Maryland, US]

DBS Doppler Beam-Sharpening

DBS Doppler Broadening Spectroscopy

DBS Dodecylbenzene Sulfonate

DBS Dominion Bureau of Statistics [now StatCan]

DBS Drill/Bolt/Screening Head

dB/s decibel(s) per second [also dB/sec, or dB s^{-1}] [Unit]

DBSC N,N'-Dibutylmonoselenocarbamate [also dbsc]

DBSD Double Backscattering Diffractometer

dB/sec decibel(s) per second [also dB/s, or dB s^{-1}] [Unit]

DBSH Benzene-1,3-Disulfohydrazide

DBSn Dibutyltin

DBSnO Dibutyltin Oxide

DBSnS Dibutyltin Sulfide

DBSO Dibenzylsulfoxide

DBSO Di-n-Butylsulfoxide

DB Sound Eng Mag Decibel, The Sound Engineering Magazine [US]

DBSP Double-Based Solid Propellant

DB-SPF Diffusion Bonding–Superplastic Forming (Technology) [Metallurgy]

DBST Double British Summer Time

DBT Data Byte Transfer

DBT Design Basis Tornado

DBT Dibenzothiophene

DBT Di-n-Butyltin

DBT Diffused-Base Transistor

DBT Dogbone Tension (Test Specimen) [Mechanics]

DBT Dry-Bulb Teperature

DBT Dry-Bulb Thermometer

DBT Ductile-(to-)Brittle Transition [Materials Science]

DBTC Data Byte Transfer Control

DBTCP Dibutyltetrachlorophthalate

DBTDA Di-n-Butyltin Diacetate

DBTG Database Task Group [of CODASYL–Conference on Data System Languages]

DBTG Database Technology Group

DBTM18C6 Dibenzotetramethyl-18-Crown-6

DBTO Di-(1-Benzotriazolyl)oxalate

DBTO Di-n-Butyltin Oxide

DBTS Di-n-Butyltin Sulfide

DBTT Ductile-to-Brittle Transition Temperature [Materials Science]

DBTT Ductile-to-Brittle Transition Transformation [Materials Science]

DBTTeF Dibutyl(ene)tetratellurafulvalene

DBTU Dibutyl Thiourea

DBU Decay of Buildup

DBU 1,8-Diazabicyclo[5.4.0]undec-7-ene

DBU Digital Buffer Unit

Dbu α,γ-Diaminobutyric Acid

DBuPh Dibutyl Phthalate

DBUR Data Bank Update Request

DBUT Database Update Time

DBV Deutscher Beton-Verein [German Concrete Association]

DBV Deutscher Bibliotheksverband [German Library Association]

DBV Digital Broadcast Video

DBV Doppler Broadening Velocity

dBV decibel(s) relative to one volt [also dBv, or dbv] [Unit]

dBVg decibel(s) voltage gain [Unit] [also dBvg]

DBV-OSI Deutscher Bibliotheksverband–Open System Interconnect [A Distributed Library System Jointly Developed by the German Library Association and Fachinformationszentrum Karlsruhe]

DBW Data Bus Wire

DBW Design Bandwidth

dBW decibel(s) relative to one watt [also dBw, or dbw] [Unit]

DBWP Disk-Based Word Processor

DBWP Double-Braid, Weatherproof (Cable)

DBX Depth Bomb Explosive

DBX Digital Branch Exchange

DBX Double Multiply (Function)

dBx decibel(s) above reference coupling [Unit]

DBXRD Divergent-Beam X-Ray Diffraction

DBZ Deuterated Benzene [also dBZ, or d-BZ]

DBZM Dibenzoylmethane

DBZP 2,6-Dibenzylphenol

DC Daimler-Chrysler AG [Merger of German Daimler AG and US Chrysler Corporation]

DC Damage Control

DC Danube Commission [Hungary]

DC Dark Current

DC Dartmouth College [Hanover, New Hampshire, US]

DC Data Cartridge

DC Data Cassette

DC Data Center

DC Data Channel

DC Data Check

DC Data Classifier

DC Data Code

DC Data-Collecting; Data Collection

DC Data Communication

DC Data Communications [also DATACOM]

DC Data Control

DC Data Conversion; Data Converter

DC Data Coordinator

DC Dawn Chorus

DC Dead Center

DC Debris Control

DC Decay Constant (or Decay Coefficient) [Physics]

DC Decimal Classification

DC Decimal Code

DC Decompression Chamber

DC Decorators Club [US]

DC Defect Center [Crystallography]

DC Defect Chemistry

DC Defect Control

DC Define Constant [Computers]

DC Define Contrast

DC Deflection Coil

DC Delay Cable

DC Delayed Coking (Process) [Chemical Engineering]

DC Department of Ceramics

DC Department of Chemistry

DC Department of Conservation

DC Depolarization Current

DC Deposited Carbon

DC Depth Charge

DC Depth of Cut [Machining]

DC Derivative Control

DC Descending Chromatography

DC Descending Colon [also dc]

DC Descriptor Code

DC Design Center

DC Design Change

DC Design Contractor

DC Design Curve

DC Desk Check(ing)

DC Detail Condition

DC Detective Constable

DC Developed Country

DC Developing Country

DC Development Center

DC Device Control (Character) [Data Communications]

DC Device Coordinate

DC Dial Comparator

DC Diamond-Cubic (Crystal Structure) [also dc]

DC Dickinson College [Carlisle, Pennsylvania, US]

DC Die Cast(ing) [Metallurgy]

DC Dieckmann Condensation [Chemical Engineering]

DC Dielectric Constant

DC Diesel Cycle

DC Differential Calorimetry

DC Differential Coating

DC Differential Current

DC Diffraction Contrast

DC Diffusion Coating

DC Diffusion Constant [Solid-State Physics]

DC Digital Code

DC Digital Communication(s)

DC Digital Comparator

DC Digital Computer

DC Digital Control(ler)

DC Digital Counter

DC Dimensional Coordination

DC N-(Dimethylaminoethyl)carbamate

DC Diode Cathode

DC Diphenylarsine Cyanide

DC Direct-Chill (Casting) [also dc]

DC Direct Connection

DC Direct Cost

DC Direct Coupling

DC Direct Current [also dc, D-C, or d-c]

DC Directional Coupler

DC Direction Cycle

DC Directory Clearinghouse [US]

DC Discontinuous Coarsening [Metallurgy]

DC Disk Cartridge

DC Disk Clutch

DC Disk Controller

DC Disk Cylinder

DC Dispatcher Console

DC Displacement Cascade

DC Displacement Chromatography

DC Display Code

DC Display Console

DC Display Coupler

DC Distillation Column

DC Distorted Communications [also dc]

DC Distributed Computer; Distributed Computing

DC Distributed Control

DC Distribution Chromatography

DC District of Columbia [US]

DC Doctor of Chiropractic

DC Dodge City [Kansas, US]

DC Donnan Coefficient [Physical Chemistry]

DC Dot Cycle [also dc] [Telecommunications]

DC Double Channel

DC Double Column [also D/C]

DC Double-Cone

DC Double Contact

DC Double Cotton Covered (Magnet Wire)

DC Double-Crucible

DC Double Current [also dc]

DC Drag Coefficient

DC Drawing Change

DC Drift Constant

DC Drive Coil

DC Driver Control

DC Driving Cycle

DC Drought Code

DC Dry Cell

DC Dry Cleaning (Process)

DC Drying Control

DC Dual Cursor

DC Dual-Cycle

DC Dublin Core [Metadata Elements for the World Wide Web]

DC Dust Cap

DC Dust Cloud

DC Dust Collection; Dust Collector

DC Duty Cycle

DC Dynamic Crowdion

DC- McDonnell Douglas Airplane [Followed by a Numeral, e.g., DC-8, DC-10, etc.]

DC_1 First Punch On

DC_2 Second Punch On

DC_3 Tape Reader On

DC1 Device Control One (Transmit On) [Data Communications]

DC2 Device Control Two [Data Communications]

DC3 Device Control Three (Transmit Off) [Data Communications]

DC4 Device Control Four [Data Communications]

D2C Decimal To Character [Restructured Extended Executor (Language)]

D&C Dila(ta)tion and Curettage [Induced Abortion]

D&C Display and Control

D/C Displays/Controls

D/C Double Column [also DC]

D/C Down-Converter

D/C Drift Correction

Dc Subarctic Climates [also Dd] [Classification Symbol]

D(c) Diffusivity Function of Concentration [Symbol]

dC Deoxycytidine [Biocemistry]

dc diamond-cubic (crystal structure) [also DC]

d/c double column

ΔC Composition Difference (or Chanhe in Composition)

ΔC_{mix} Specific Heat of Mixing [Symbol]

ΔC_0 Total Magnitude of Transient Capacitance [Symbol]

Δc_p excess specific heat [Symbol]

DCA Data Communications Administrator

DCA Decade Counting Assembly

DCA Defense Communications Agency [now Defense Information Systems Agency, US]

DCA Defense Contract Audit

DCA Deoxycorticosterone Acetate [Biochemistry]

DCA Design Change Authorization

DCA Device Cluster Adapter [Computers]

DCA Dichloroacetic Acid

DCA Dichloroaniline

DCA 4,4'-Dichlorodiphenylacetic Acid

DCA 9,10-Dicyanoanthracene

DCA Digital Command Assembly

DCA Digital Communications Associates

DCA Digital Computer Association

DCA Diploma in Computer Application

DCA Direct Chip Attach

DCA Direct Colorimetric Analysis

DCA Direct Configurational Averaging

DCA Direct-Current Arc

DCA Directorate of Civil Aviation [UK]

DCA Distance of Closest Approach

DCA Distributed Communications Architecture

DCA Distribution Control Assembly

DCA Document Content Architecture

DCA Double Chain Approximation

DCA Driver Control Area

DCA Dynamic Contact Analyzer

DCAA Defense Contract Audit Agency [US]

DCAA Devon County Agricultural Association [UK]

DCAA Dichloroacetic Acid

DCAA Dynamic Contact Angle Analyzer

DCAD 2,4-Dichlorobenzaldehyde

DC/AC Direct Current/Alternating Current [also D-C/A-C, dc/ac, or d-c/a-c]

DCA/DIA Document Content Architecture/Document Interchange Architecture

DCAES Direct-Current Arc Emission Spectroscopy

DCAF 2',4'-Bis[di(carboxymethyl)aminomethyl]fluorescein; 2,4-Bis(N,N'-Di(carbomethyl)aminomethyl)fluorescein [Biochemistry]

DCAF Distributed Console Access Facility [of IBM Corporation, US]

DCAM Data Collection Access Method

DCAM Digital Camera

DCAM Direct Chip Attach Module

$Can Canadian Dollar [also Can$]

$Can/bu Canadian Dollar per Bushel [Unit]

$Can/gal Canadian Dollar per Gallon [Unit]

$Can/g Canadian Dollar per Gram [Unit]

$Can/kW Canadian Dollar per Kilowatt [Unit]

$Can/kWh Canadian Dollar per Kilowatthour [Unit]

$Can/L Canadian Dollar per Liter [Unit]

$Can/lb Canadian Dollar per Pound [Unit]

$Can/ton Canadian Dollar per Ton [Unit]

$Can/yr Canadian Dollar per year [Unit]

DCAOC Defense Communications Agency Operations Center [US]

DCAP Degraded Composite Analysis Program

DCAR Design Corrective Action Report

DCAS Data Collection and Analysis System [of NASA]

DCAS Defense Contract Administration Service [US]

DCAS Deputy Commander Aerospace System

DCAS Digital Control and Automation System

DCASA Dyers and Colorists Association of South Africa

DCAT Drug, Chemical and Allied Trade Association [US]

DCB Data Control Block

DCB Data Control Bus

DCB Defense Communications Board [US]

DCB Define Control Block

DCB 1,3-Di(N-Carbazolyl)propane

DCB Department of Chemistry and Biochemistry

DCB Device Control Block

DCB Dichlorobenzene

DCB Dichlorobenzidine

DCB Dichloroborazine

DCB 1,4-Dichlorobutane

DCB Dicyanobenzene

DCB Disk Coprocessor Board [of Novell Inc., US]

DCB Division of Cancer Biology [of National Cancer Institute, Bethesda, Maryland, US]

DCB Double-Cantilever Beam (Method) [Mechanical Testing]

DCB Drawout Circuit Breaker

dcb 1,4-Diamino-cis-2-Butene

1,4-DCB 1,4-Dichlorobenzene

DCBA Dichlorobenzaldehyde

DCBA Dichlorobenzamide

DCBA 2,4-Dichlorobenzoic Acid

DCBC 2,4-Dichlorobenzyl Chloride

DCBD Define Control Block Dummy

DCBDC Division of Cancer Biology, Diagnosis and Center [of National Cancer Institute, Bethesda, Maryland, US]

DCBP Decachlorobiphenyl

DCBP 4,4'-Dichlorodibenzophenone

DCBRE Defense Chemical, Biological and Radiation Establishment [of Defense Research Board, Canada]

DCBRL Defense Chemical, Biological and Radiation Laboratories [now Defense Chemical, Biological and Radiation Establishment, Canada]

DCBS N,N-Dicyclohexyl-2-Benzothiazole Sulfenamide

DCBS Devon Cattle Breeders Society [UK]

DCBS Dimer-Centered Basis Set

DCBTA 1,4-Diamino-cis-2-Butene-N,N,N',N'-Tetraacetate [also dcbta]

DCBTF Dichlorobenzotrifluoride

DCBWR Dual-Cycle Boiling-Water Reactor

DCC Data Channel Converter

DCC Data Circuit Concentration; Data Circuit Concentrator

DCC Data Communication Channel

DCC Data Communications Controller

DCC Data Computation Complex [at NASA Johnson Space Center, Houston, Texas, US]

DCC Data Country Code

DCC Defense Construction Canada

DCC Department of Culture and Communications [Canada]

DCC Deputy Chief Constable

DCC Destination Code Cancelled

DCC Detroit Cultural Center [Michigan, US]

DCC Device Cluster Controller

DCC Device Control Character

DCC N,N'-Dicyclohexylcarbodiimide

DCC Digital Command Control

DCC Digital Communications and Control

DCC Digital Compact Cassette

DCC Digital Cross Correct

DCC Digital Cross Current

DCC Dipolar Coupling Constant

DCC Direct Cable/Client Connection [Internet]

DCC Direct-Chill Continuous Casting (Process) [Metallurgy]

DCC Direct Coagulation Casting [Metallurgy]

DCC Direct Computer Control

DCC Direct Contact Condensation; Direct-Contact Condenser

DCC Direct Control Channel

DCC Direct-Current Comparator

DCC Disoriented Chiral Condensate

DCC Display Channel Complex

DCC Display Combination Code

DCC District Communications Center

DCC Document Control Center

DCC Double Cotton Covered (Electric Wire)

DCC Droplet Counter-Current Chromatograph

DC14C4 Dicyclohexano-14-Crown-4

DC18C6 Dicyclohexano-18-Crown-6

DC21C7 Dicyclohexano-21-Crown-7

DC24C8 Dicyclohexano-24-Crown-8

DC30C10 Dicyclohexano-30-Crown-10

DCCA 2-Chloro-N^6-Cyclopentyl-1-Deazaadenosine [Biochemistry]

DCCA Design Change Cost Analysis

DCCA Drying Control Chemical Additive

DCCC Data Communication Control Character

DCCC Double Current Cable Code

DCCC Droplet Countercurrent Chromatography

DCCD Dicyclohexylcarbodiimide

DCCE Department of Chemistry and Chemical Engineering

DCCEAS District of Columbia Council of Engineering and Architectural Societies [Washington, US]

DCCH 7-(Diethylamino)coumarin-3-Carbohydrazide

DC-Chol Cholesteryl 3β-(N-[Dimethylaminoethyl]carbamate) [Biochemistry]

DCCI Dipartimento di Chimica e Chimica Industriale [Department of Chemistry and Industrial Chemistry, of University of Pisa, Italy]

DCCI Dichlorohexylcarbodiimide

DCCI Dicyclohexylcarbodiimide

DCCI Dissociation-Consistent Configuration Interaction

DCCI Dorset Chamber of Commerce and Industry [UK]

DCCI-NHS Dicyclohexylcarbodiimide and N-Hydroxysuccinimide

DCC-MSF Direct Contact Condensation–Multistage Flash

DCCPS Division of Cancer Control and Population Science [of National Cancer Institute, Bethesda, Maryland, US]

DCCR Division of Cancer Control and Rehabilitation [of National Cancer Institute, Bethesda, Maryland] US]

DCCS Defense Communications Control System [US Air Force]

DCCS Digital Command Communications System

DCCS Distributed Capacity Computing System

DCCS Distributed Computer Control System

DCCT Diabetes Control and Complications Trial [US Study from 1983-93]

DCCU Data Communications Control Unit

DCCU Digital Command and Control Unit

DCCU Digital Communications and Control Unit

DCCU Display Computer Control Unit

DCD Data Carrier Detect

DCD Detector-Cooling Dewar

DCD Digital Coherent Detector

DCD Direct-Current Demagnetization (Remanence Curve) [also dcd]

DCD Double Channel Duplex

DCD Double Crystal (X-Ray) Diffractometer; Double Crystal (X-Ray) Diffractometry

DCD Dynamic Computer Display

DCDA Data Communication Dealers Association [US]

DCDA Dicyanodiamine

DCDAS Dicyanodiamine Sulfate

DCDB Digital Cadastral Data Base

DCDC Data Communication to Disk Control

DCDC Division for Communicable Disease Control [of Massachusetts Department of Public Health, US]

DCDC Double Cleavage Drilled Compression (Test Specimen) [Mechanics]

DC-DC Direct-Current to Direct-Current (Converter) [also DC/DC, D-C/D-C, DCDC, or d-c/d-c]

DCDD Dichloro-p-Bibenzodioxin

DCDES Dichlorodiethyl Sulfide

DCDF Dichlorodibenzofuran

DCDFM Dichlorodifluoromethane

DCDMA Diamond Core Drill Manufacturers Association [US]

dCDP 2'-Deoxycytidine Diphosphate [also d-CDP] [Biochemistry]

DCDR Data Collection and Data Relay

Dcdr Decoder

DCDRD Director of Chemical Defense Research and Development [UK]

DCDS Digital Control Design System

DCDS Double-Cotton, Double-Silk Covered (Magnet Wire) [also dcds]

DCDSTF Digital Cartographic Data Standards Task Force

DCDT Double Crystal (X-Ray) Diffraction Topography

dC/dt time rate of change (usually an increase) of concentration [also dc/dt]

–dC/dt time rate of decrease of concentration [also –dc/dt]

dC/dx concentration gradient (i.e., change in concentration with distance) [also dc/dx]

ΔC/Δx Concentration Gradient (i.e., Change in Concentration with Distance) [also dc/dx]

DCE Dallas Cotton Exchange [Texas, US]

DCE Data Circuit-Terminating Equipment

DCE Data Communications Equipment

DCE Data Conversion Equipment

DCE Department of Ceramic Engineering

DCE Department of Chemical Engineering

DCE Department of Civil Engineering

DCE Department of Conservation and Environment [Victoria State, Australia]

DCE 1,1-Dichloroethane; 1,2-Dichloroethene

DCE Dichloroethylene

DCE Die Casting Engineer [Metallurgy]

DCE Differential Coefficient of Expansion

DCE Digital Control Element

DCE Directorate of Communications Electronics

DCE Directorate of Conservation and Environment [of Department of National Defense, Canada]

DCE Dissociative Charge Exchange

DCE Distributed Computing Environment [of Open Software Foundation]

DCE Distributed Computing Equipment

DCE Division of Cancer Etiology [of National Cancer Institute, Bethesda, Maryland, US]

DCE Domestic Credit Expansion

δ-Ce Delta Cerium [Symbol]

DC-EAF Direct-Current Electric-Arc Furnace [also DC EAF, or DCEAF]

DCED Distributed Computing Environment Daemon

DCEE Department of Civil and Environmental Engineering

DCEE Dichloroethyl Ether

DCEL Direct Current Electroluminescent

DCEMC Dartmouth College Electron Microscopy Center [Hanover, New Hampshire, US]

DCEMS Department of Chemical Engineering and Materials Science

DCEN Direct Current Electrode Negative

DCEO Defense Communications Engineering Office [US]

DCEP Direct Current Electrode Positive

DCES Delaware Council of Engineering Societies [US]

DCF Data Communication Facility

DCF Data Compression Facility

DCF Data Count Field

DCF DeCarb-Free Product

DCF Dipole (Moment) Autocorrelation Function

DCF Direct Correlation Function

DCF Discounted Cash Flow (Analysis) [also dcf]

DCF Disk Control Field

DCF Disk Controller Formatter

DCF Distributing Center Facility

DCF Document Composition Facility

DCF (Lotus) Driver Configuration File

DCF Droplet Combustion Facility

Dcf Continental Taiga Climate, Severe Winter [Classification Symbol]

DCFEM Dynamic Crossed-Field Electron Multiplication

DCFL Direct-Coupled FET (Field-Effect Transistor) Logic

DCFN Developing Countries Farm Radio Network

DCFROR Discounted Cash Flow Rate of Return

D-CFZ 3-Chloro-2-Hydroxy-N,N,N-Trimethyl-1-Propanaminium Chloride

DCG Dependent Charge Group

DCG Derived Concentration Guide

DCG Design Coordination Group

DCG Dichromated Gelatin

DCG Diffusion-Controlled Growth [Materials Science]

DCG Diode Capacitor Gate

DCG Direct-Current Generator

DCG Doppler-Controlled Gain

DCGD Direct-Current Glow Discharge [also DC-GD]

DCGE Dicresyl Glycerin Ether

DCGEM Directorate of Clothing, General Engineering and Maintenance

DCH Data Channel

DCH Department of Canadian Heritage

DCH Dicyclohexane; Dicyclohexano-

DCH Dichlorohydrin

DCH Direct Contact Hydrogenation (Process) [Chemical Engineering]

DCH-6 Dicyclohexano-18-Crown-6

DCH-8 Dicyclohexano-18-Crown-8

δ-ch delta-chain specific (immunoglobulin)

DCHA Dicyclohexylamine

D-Channel Data Channel [also D-channel]

DCHBH Dicyclohexylborane

DCHD 1,6-Di(N-Carbazolyl)-2,4-Hexadiyne (Polydiacetylene)

DCHE Dicyclohexyl Ether

DChE Doctor of Chemical Engineering

DChem Doctor of Chemistry

DCHP Dicyclohexyl Phthalate

DCHTA 1,2-Diaminocyclohexane-N,N,N',N'-Tetraacetic Acid

DCI Data Communication Interrogate

DCI Defense Computer Institute [US]

DCI Desorption Chemical Ionization

DCI Dichloroisoproterenol

DCI 1-(3',4'-Dichlorophenyl)-2-Isopropylaminoethanol

DCI Direct Channel Interface

DCI Directorate of Chemical Inspection [UK]

DCI Display Control Interface

DCI Ductile Cast Iron

DCIA 7-(Diethylamino)-3,4-(Iodoacetyl)amino)phenyl-4-Methylcoumarin

DCIB Data Communication Input Buffer

DCID Director Central Intelligence Directive [US]

DCIEM Defense and Civil Institute of Environmental Medicine [of Department of National Defense, Canada]

DCIEM/ATG Defense and Civil Institute of Environmental Medicine/Air Transport Group [of Department of National Defense, Canada]

DCIM Display-System Computer Input Multiplexer

DCIO Direct Channel Interface Option

DCIP Dichlorophenol-Indilphenol

DCIS Department of Computing and Information Science

DCIU Digital Control and Interface Unit

Dckng Docking

DCKP Direct Current Key Pulsing

DCL Data Control Language

DCL DEC (Digital Equipment Corporation) Command Language

DCL Delayed Call Limit

DCL Demountable Cathode Lamp

DCL Depth of Cut Line

DCL Device Clear [Computers/Communications]

DCL Diamond Cut Lug (Refractory)

DCL Digital Command Language

DCL Digital Computer Laboratory

DCL Digital Control Laboratory

DCL Digital Control Logic

DCL Direct Coupled Logic

DCL Discontinuous Coarsened Lamellae [Metallurgy]

DCL Document Change List

DCL Dual Current Layer

DCl Deuterium Chloride

Dcl Declaration [also dcl]

DClAc Dichloro Acetic Acid

DCLC Drift Cyclotron Loss Cone (Instability)

DClChem Doctor of Clinical Chemistry

DCLT Differential Cathode Luminescence Topography

DCLU Digital Carrier Line Unit

DC-LVDT Direct-Current Linear Variable-Differential Transformer

DCM Data Communications Multiplexer

DCM DECOM (Decommutator) Control Memory

DCM Department of Coins and Medals [of British Museum]

DCM Diagnostic Control Module

DCM d,d-Dicampholylmethanate [also dcm]

DCM Dichloromethane

DCM 4-Dicyanomethylene-2-Methyl-6-p-Dimethylaminostyryl-4H-Pyran

DCM Difference-Curve Method

DCM Digital Capacitance Meter

DCM Digital Cartographic Model [Geographic Information System]

DCM Digital Command Language

DCM Dimensional Constraint Manager

DCM Direction Cosine Matrix

DCM Display (and) Control Module

DCM Direct-Current Magnetron (Sputtering)

DCM Direct Current Noise Margin [also dcm]

DCM Direct Current Motor

DCM Distinguished Conduct Medal

DCM Divisão de Ciências Meteorólogicas [Meteorological Sciences Division; of Brazilian National Institute for Space Research]

D&CM Dressed (One, or Two Sides) and Center Matched [Lumber]

D$^+$/cm² Positive Donors per square centimeter (in solid-state physics) [Symbol]

D$^-$/cm² Negative Donors per square centimeter (in solid-state physics) [Symbol]

DCMA Dessert and Cake Mixes Association [UK]

DCMA Dry Color Manufacturers Association [US]

DCMB Development Configuration Management Board

DCMC Discontinuously-Reinforced Ceramic-Matrix Composite

DCME Department of Chemical and Materials Engineering

DCME Department of Chemical and Metallurgical Engineering

DCME Department of Civil and Mineral Engineering

DCME Dichloromethyl Ether

DCMFM Dichloromonofluoromethane

DCMM Departmento de Ciencias de Materiais e Metallurgia [Department of Materials and Metallurgical Engineering; of Pontifica Universidade Catolica do Rio de Janeiro, Brazil]

DCMO 2,3-Dihydro-5-Carboxanilido-6-Methyl-1,4-Oxathiin

DCMO Dysprosium Calcium Manganese Oxide (Superconductor) [also DyCMO]

DCMP 2,3-Dichloro-3-Methylpentane

DCMP Division of Condensed Matter Physics [of American Physical Society, US]

dCMP 2'-Deoxycytidine-5'-Monophosphate [also d-CMP] [Biochemistry]

DCMPLX Double Complex [also Dcmplx]

DCMS Dedicated Computer Message Switching

DCMS Digital Chart Management System

Dcmt Decrement [also dcmt]

DCMX Dichloro-m-Xylenol

DCN Data Communications Network

DCN Design Change Notice

DCN Deuterated Hydrocyanic Acid

DCN Dichloronaphthalene

DCN Distributed Computer Network

DCN Document Change Notice

DCN Drawing Change Notice

DCNA Data Communication Network Architecture

DCNA 2,4-Dichloro-6-Nitroaniline

DCna Dicyanamide

DCNE Department of Chemistry and Nuclear Engineering

DCNO Deputy Chief of Naval Operations [of US Department of the Navy]

DCNP Document Change Notice Proposal

DCNQI N,N'-Dicyanoquinonediimine

DCNR Department of Conservation and Natural Resources [Victoria State, Australia]

DCO Data-Controlled Oscillator

DCO Dehydrated Castor Oil

DCO Detailed Checkout

DCO Digital Central Office

DCOC 2,4-Dichlorobenzoyl Chloride

ΔCOD Change in Crack Opening Displacement [Mechanics]

DCOM Distributed Component Object Model

DCOMP Division of Computational Physics [of American Physical Society, US]

DCOP Detailed Checkout Procedures

DCOP Displays, Controls and Operations Procedures

DCORE Deployer Computer-Operated Readiness Equipment [NASA]

DCOS Data Collection Operating System

DCOS Data Communication Output Selector

DCOS Direct-Coupled Operating System

DCP Data-Collecting Program

DCP Data Collection Platform

DCP Data Communications Processor

DCP Depot Condemnation Percent

DCP Design Change Package

DCP Design Criteria Plan

DCP DEU (Display Electronics Unit) Control Program

DCP Development Cost Plan

DCP Dibasic Calcium Phosphate; Dicalcium Phosphate

DCP Dicapryl Phthalate

DCP Dicetyl Phosphate

DCP Dichloropentane

DCP Dichlorophenol

DCP Dichloropropene

DCP 3,5-Dichloropyridine

DCP Dichromatic Pattern

DCP Dicumyl Peroxide

DCP Dicyclopentadiene; Dicyclopendadienyl [also dcp]

DCP Diffusion Controlled Pellistor

DCP Digital Computer Processor

DCP Digital Computer Programming

DCP Digital Control(ler) Programmer

DCP Direct-Current Argon Plasma

DCP Direct-Current Plasma

DCP Direct-Current Polarogram; Direct-Current Polarography

DCP Directly-Coupled Plasma

DCP Display Control Panel

DCP Distributed Communications Processor

DCP Division of Cancer Prevention [of National Cancer Institute, Bethesda, Maryland, US]

DCP Division of Chemical Physics [of American Physical Society, US]

DCP Division of Clinical Psychology [of American Psychological Association, US]

DCPA Defense Civil Preparedness Agency [formerly Office of Civil Defense; now Federal Emergency Management Agency, US]

DCPA Dichloropropionaniline

DCPA Dimethyl Ester 2,3,5,6-Tetrachloroterephthalic Acid

DCPA Dimethyl Tetrachloroterephthalate

DCPA Distribution Common Point

DCPAC Desertification Control Program Activity Center

DCPAES Direct-Current Plasma Atomic Emission Spectrometry

DC-PBH Double-Channel Planar-Buried Heterostructure [Solid-State Physics]

DC-PBH-LD Double-Channel Planar-Buried Heterostructure Laser Diode

DCPC Dichlorophenyl Methyl Carbinol; 4,4'-Dichlorodiphenyl Methyl Carbinol

DCPC Division of Cancer Prevention and Control [of National Cancer Institute, Bethesda, Maryland, US]

DCPC Dual Channel Port Controller

DCPCM Differentially Coherent Pulse Code Modulation

DC-PCVD Direct-Current Plasma-Enhanced Chemical-Vapor Deposition

DCPD Dicalcium Phosphate Dihydrate

DCPD Dicyclopentadiene

DCPD Direct Current Potential Drop [also dcpd]

dCpdA 2'-Deoxycytidylyl 2'-Deoxyadenosine [Biochemistry]

dCpdG 2'-Deoxycytidylyl 2'-Deoxyguanosine [Biochemistry]

DCPEEA Dicyclopentenyloxyethyl Acrylate

DCPEEMA Dicyclopentenyloxyethyl Methacrylate

DCPEI DEU (Display Electronics Unit) Control Program End Item

DCPES Direct-Current Plasma Emission Spectroscopy

DCPG Defense Communications Planning Group [US]

DCPI Dicyclopentadienyl Iron

DCPIC 2,6-Dichlorophenol-Indo-o-Cresol

DCPIP Dichlorophenolindophenol

DCP-LA Direct-Current Argon Plasma Laser Ablation

DCPM Di-(p-Chlorophenoxy)methane

DCPMS Dicyclopentyldimethoxysilane

DCP-OES Direct Current Plasma Optical Emission Spectrometry

DCPP Data Communications Preprocessor

DCPR Detailed Continuing Property Record

DCPS Digitally-Controlled Power Source

DCPS Direct-Current Plasma Spectrometry

DCPSK Differentially-Coherent Phase Shift Keying

DCPZ 2,5-Dicarboxypyrazine

DCQ Double Cascade Quench-Patenting [Metallurgy]

DCR Data Communication Read

DCR Data Conversion Receiver

DCR Data Coordinator and Retriever

DCR Defect-Controlled Relaxation (Model)

DCR Design Certification Review

DCR Design Change Recommendation

DCR Design Change Request

DCR Design Concern Report

DCR Detail Condition Register

DCR Dichlororesorcine

DCR Digital Cassette Recorder

DCR Digital Condition Register

DCR Digital Conversion Receiver

DCR Direct Conversion Reactor

DCR Direct Conversion Reactor Study [US]

DCR Direct-Current Restorer

DCR Division of Computing Research [of Commonwealth Scientific and Industrial Research Organization, Australia]

DCR Document Change Record

DCR Document Change Review

DCR Double Cold-Reduced; Double Cold Reducing [Metallurgy]

DCR Double-Stage Cold-Rolled; Double-Stage Cold Rolling [Metallurgy]

DCR Drawing Change Request

DCR Dual Channel Microwave Radiometer

DCR Dual-Cycle (Nuclear) Reactor

DCR Dynamic Charge Restoration

DCR Dynamic Routing Technology

Dcr Decrease [also dcr]

DCRA Dyers and Cleaners Research Association [UK]

DCRABS Disk Copy Restore and Backup System

DCRF Die Casting Research Foundation [now North American Die Casting Association, US]

DC/RF Direct Current/Radio Frequency (Measurement) [also DC/rf, or dc/rf]

DCRN Dashpot Cup Retention Nut

DCRP Direct-Current Reverse-Polarity

DCRP Double-Stage Cold-Rolled and Polished; Double-Stage Cold-Rolling and Polishing [Metallurgy]

DCRT Data Collection Receive Terminal

DCRT Division of Computer Research and Technology [of National Institutes of Health, Bethesda, Maryland, US]

DCS Dark Current Spectroscopy

DCS Data Collection System

DCS Data Communication Subsystem

DCS Data Communication System

DCS Data Conditioning System

DCS Data Control Service

DCS Data Control System

DCS Defense Communications System [of US Department of Defense]

DCS Deputy Chief of Staff [of US Department of the Army]

DCS Design Case Study

DCS Design Communication System

DCS Design Control Specifications

DCS Design Criteria Specification

DCS Desktop Color Separation

DCS Detail Checkout Specification(s)

DCS Diagnostic Control Store

DCS Dichlorosilane

DCS Die Casting Society [UK]

DCS Differential Cross-Section [Physics]

DCS Digital Camera System

DCS Digital Command Subsystem

DCS Digital Command System

DCS Digital Communication(s) System

DCS Digital Computer System

DCS Digital Cross-Connect System

DCS Direct-Coupled System

DCS Direct-Current Sputtering

DCS Dislocation Cell Size [Materials Science]

DCS Display (and) Control System

DCS Distributed Computing Services

DCS Distributed Computing System

DCS Distributed Control System

DCS Division of Chemical Sciences [of US Department of Energy]

DCS Doctor of Commercial Science

DCS Document Control System

DCS Dodecamethylcyclohexasilane

DCS Double Channel Simplex

DCS Double-Cotton, Single-Silk Covered (Magnet Wire)

DCS Double-Crystal Spectrometer; Double-Crystal Spectrometry

DCS Double-Current System

DCS Dual Checkout Station

D&CS Display and Controls Subsystem

D/CS Distributed Client/Server [Internet]

DCs Developed (or Developing) Countries

ΔC(s) Unwanted Deviation of Controlled Variable *C(s)* Caused by a Disturbance Input [Symbol]

DCSA 1-Docosanamine [also Dcsa]

DCSC Defense Construction Supply Center [of US Department of Defense]

DCSc Doctor of Commercial Science

DCSCS Data Code and Speech Conversion Subsystem

DCSD Département de Commande des Systèmes et Dynamique du Vol [Department of Systems Mechanics and Flight Dynamics, ONERA-CERT, Toulouse, France]

DCSEE Differential Cross-Sections of Emitted Electrons

DCS&H Department of Community Services and Health [Australia] [also DCSH]

DCSI Data and Control Signal Interface

dc-Si Diamond-Cubic Silicon (Phase)

DCSP Defense Communications Satellite Program [US]

DCSP Digital Control Signal Processor

DCSP Direct-Current Straight Polarity

DC SQUID Direct-Current Superconducting Quantum Interference Device [also DC-SQUID, dc SQUID, or dc-SQUID]

DCSS Defense Communication Terminal System

DCST Department of Chemical Science and Technology

DCSU Digital Computer Switching Unit

DCS-WSi$_x$ Dichlorosilane Based Tungsten Silicide

DCT Data Communications Terminal

DCT Depth Charge Thrower

DCT Destination Control Table

DCT Developmental Consulting Program

DCT Dichlorotoluene

DCT N,N-Diethyl Dithiocarbamate

DCT Direct-Current Test

DCT Direct-Current Transmission

DCT Discrete Cosine Transform

DCT Disk-Shaped Compact Tension (Test Specimen) [also DC-T] [Mechanics]

DCT Disk-Shaped Compact-Type (Test Specimen) [also DC-T] [Mechanics]

DCT Division of Chemical Toxicology [of American Chemical Society, US]

DCT Document [also Dct, or dct]

DCT Dust-Cloud Theory [Astronomy]

DC-T Disk-Shaped Compact Tension (Test Specimen) [also DCT] [Mechanics]

DC-T Disk-Shaped Compact-Type (Test Specimen) [also DCT] [Mechanics]

DC(T) Disk-Shaped, Cyclic Tension-Loaded (Test Specimen) [Mechanics]

.DCT Dictionary [File Name Extension] p[also .dct]

ΔC(t) Relaxation Rate in Concentration Difference

DCTA 1,2-Diaminocyclohexane-N,N,N',N'-Tetraacetic Acid

DCTB Detroit Convention and Tourist Bureau [Michigan, US]

DCTE Data Collection and Terrain Evaluation

ΔCTE Difference in Coefficient(s) of Thermal Expansion

DCTFE Dichlorotetrafluoroethane

Dctg Decorating [also dctg]

DCTL Direct-Coupled Transistor Logic

DCTN Defense Commercial Television Network

DCTP Division of Cancer Treatment and Diagnosis [of National Cancer Institute, Bethesda, Maryland, US]

dCTP 2'-Deoxycytidine 5'-Triphosphate [also d-CTP] [Biochemistry]

DCTS Data Communication Terminal System

DCTU Directly Corrected Test Unit

DCU Data-Cache Unit

DCU Data Capture Unit

DCU Data Collection Unit
DCU Data Command Unit
DCU Data Communications Unit
DCU Data Control Unit
DCU Decade Counting Unit
DCU Decimal Counting Unit
DCU Device Control Unit
DCU N,N-Dichlorourethane
DCU Digital Computer Unit
DCU Digital Control Unit
DCU Digital Counting Unit
DCU Disk Control Unit
DCU Display and Control Unit [of NASA Apollo Program]
DCU Display Control Unit
DCU Distributed Control Unit
DCU 1,3-Di-(2,2,2-Trichloro-1-Hydroxyethyl)urea
DCU Drum Control Unit
DCU Dublin City University [Ireland]
DCUP Dicumyl Peroxide
DCUTL Direct-Coupled Unipolar Transistor Logic
DCV Direct Current, Volts
DCV Directional Control Valve
DCV Double Concave (Lens)
DCV Double Cotton Varnished (Magnet Wire)
DCVD Division for Chemical Vapor Deposition
DCW Data Communication Write
DCW Data Control Word
DCW Differentially Compound-Wound
DCW Digital Chart of the World
DCW Dynamic Channel Exchange
Dcw Subarctic Climate, Coldest Month above–36.4°F (–38°C), Less than Four Months over 50°F (10°C), Dry Season in Winter [Classification Symbol]
DCWM Differentially Compound-Wound Motor
DCWT Division of Chemical and Wood Technology [of Commonwealth Scientific and Industrial Research Organization, Australia]
DCWV Direct-Current Working Voltage
DCX Direct Current Experiments [of Oak Ridge National Laboratory, Tennessee, US]
DCX Double Convex (Lens)
DCXRD Double Crystal X-Ray Diffraction
DCYPP 1,3-Bis(dicyclohexylphosphine)propane [also dcypp]
DCyTE trans-1,2-Diaminocyclohexane-N,N,N'N'-Tetraacetic Acid
DCZAA Direct-Current Zeeman Atomic Absorption [also DC-ZAA]
DD Dahlgren Division [of Naval Surface Warfare Center, Silver Springs, Maryland, US]
DD Dangerous Deck [Commerce]
DD Data Definition (Statement)
DD Data Demand
DD Data Depository
DD Data Description; Data Descriptor
DD Data Dictionary

DD Data Display
DD Data Division
DD Day [also dd] [Date]
DD Days after Date [also D/D]
DD Decimal Display
DD Decimal Divide
DD Decision Data
DD Decoder Driver
DD Dedicated Display
DD Deep-Donor (State) [Solid-State Physics]
DD Deep-Drawn; Deep Drawing [Metallurgy]
DD Deferred Delivery
DD Definitive Design
DD Degree Day
DD Degree of Dissociation
DD Delay Driver
DD Delayed Dormant [Pesticides]
DD Delivered [also Dd, or dd]
DD Delivered at Docks
DD Demand Draft [also D/D]
DD Density-Dependent
DD Dentural Doctor; Doctor of Denturism
DD Depth Dose
DD Descending Difference
DD Design Department [UK]
DD Destroyer [US Navy Symbol]
DD Destructive Distillation
DD Detail Drawing
DD Developmental Disability; Developmentally Disabled
DD Development Division
DD Device Driver
DD Dewey Decimal
DD β,β'-N-Diaminodiethylether-N,N'-Tetraacetic Acid
DD Dichloropropane-Dichloropropane
DD Differential Diagnosis [Medicine]
DD Digital Data
DD Digital Display
DD Digital-to-Digital
DD Dimensionality Domination
DD Dipolar Decoupling
DD Dipole-Dipole (Interaction)
DD Direct Dialing
DD Direct Drive
DD Directional Drilling
DD Directives Documentation
DD Disconnecting Device
DD Discontinuous Decomposition
DD Discontinuous Dissolution [Metallurgy]
DD Discriminating Digit
DD Dishonorable Discharge
DD Disk Drive
DD Display Double-Words (Command) [Pascal Programming]
DD Document Delivery

DD Double Deck (Screen)

DD Double Density

DD Double-Diode

DD Double-Distilled

DD Dowlish Developments Limited [UK [Now Part of High-Voltage Engineering Europa BV, Netherlands]

DD Downdraft

DD Drawing Deviation

DD Druyvesteyn Distribution

DD Dual Damping

DD Dual Density

DD Ductile Dimple [Metallurgy]

DD Due Date

DD Duplex Drive

D&D Decontamination and Decommissioning [Nuclear Engineering]

D&D Design and Development

D/D Days after Date [also DD]

D/D Demand Draft [also DD]

D/D Direct Debit

D/D Documentary Draft

D$_H$/D$_D$ E$_{cc}$ (Closed Circuit Potential) Critical Hydrogen Cracking Potential [Symbol]

D-D Dark-Dark (Fringe)

D-D Demand Draft

D-D Deuterium-Deuterium (Reaction)

D-D Dichloropropene—Dichloropropane (Mixture)

D-D Digital-to-Digital

Dd Subarctic Climates [Classification Symbol] [also Dc]

D/d Diameter Ratio [Symbol]

dD Distorted-Coherent Dislocation (in Materials Science) [Symbol]

dd dangerous deck [Commerce]

dd day [also DD] [Date]

dd *(de dato)* – from date of issue

dd delivered

d/d days after date

d-d deuterium-deuteron (process) [Symbol]

ΔD Density Difference (for X-Ray Film) [Symbol]

DDA Data Differential Analyzer

DDA Depreciation, Depletion and Amortization

DDA Depth-Duration-Area (Value) [Meteorology]

DDA Design Data Administration

DDA Deutscher Dampfkesselausschuß [German Boiler Committee]

DDA 4,4'-Dichlorodiphenylacetic Acid

DDA Didecyl Adipate

DDA Didecyl Amine

DDA Didecyldimethylammonium

DDA 2',3'-Dideoxyadenosine [Biochemistry]

DDA Die Makers and Die Cutters Association [now National Association of Die Makers and Die Cutters, US]

DDA Digital Dealers Association [US]

DDA Digital Differential Analyzer

DDA Digital Display Alarm

DDA Dimethyldioctadecylammonium Bromide

DDA Direct Data Attachment

DDA Direct Device Attachment

DDA Direct Differential Analyzer

DDA Direct Disk Attachment

DDA Discrete Dipole Approximation

DDA Distributed Data Access

DDA Division on Dynamical Astronomy [of American Astronomical Society, US]

DDA Document Departure Authorization

DDA Dodecenyl Acetate

DDA Domain-Defined Attribute

DDA Dynamic Dielectric Analysis

DDA Dynamics Differential Analyzer

DD&A Depreciation, Depletion and Amortization

E-8-DDA trans-8-Dodecenyl Acetate

Z-7-DDA cis-7-Dodecenyl Acetate

Dda 1-Dodecanamide

ddA 2',3'-Dideoxyadenosine [Biochemistry]

DDAB Didodecyl Dimethylammonium Bromide; Dodecyldimethylammonium Bromide [also DDABr]

D-Day Beginning of the Action Day [also D day]

DDAFP Diesel Driven Auxiliary Feedwater Pump

DDAO N,N-Dimethyldodecylamine-N-Oxide

DDAS Didecylaminesulfate

DDAS Digital Data Acquisition System [also D-DAS]

ddATP 2',3'-Dideoxyadenosine 5'-Triphosphate [Biochemistry]

DDAVP 1-Desaminocysteine-8-D-Arginine Vasopressin [Biochemistry]

DDB Deutscher Dolmetscherbund [German Confederation of Interpreters]

DDB Device Dependent Bitmap

DDB Device Descriptor Block

DDB 2,3-Dimethoxy-1,4-Bis(dimethylamino)butane

DDB Distributed Database

DDB Dodecylbenzene

DDB Double Declining Balance

DDBA Dodecylbenzenesulfonate

DDBG DIMDI Database Generator [Deutsches Institut für Medizinische Dokumentation und Information, Germany]

DDBMS Distributed Database Management System

DDBS Descriptor Database System

DDBS Digital Data Broadcast System

DDBSA Dodecylbenzenesulfonic Acid

δ(DBTT) Variation (or Change) in Ductile-to-Brittle Transition Temperature [Materials Science]

DDC Data Distribution (or Distributor) Center

DDC Deck Decompression Chamber

DDC Defense Documentation Center [of US Department of Defense at Cameron Station, Alexandria, Virginia]

DDC Defensive Driving Course

DDC Degree-Day in Celsius

DDC Device Development Center [of Hitachi Ltd., Japan]

DDC Dewey Decimal Classification

DDC Die Casting Development Council

DDC Dideoxycytidine; 2',3'-Dideoxycytidine [also ddC] [Biochemistry]

DDC N,N-Diethyl Dithiocarbamate

DDC Digital Data Cell

DDC Digital Data Channel [of Video Electronics Standards Association]

DDC Digital Data Communication

DDC Digital Data Conversion; Digital Data Converter

DDC Digital Display Conversion; Digital Display Converter

DDC Digital Dynamic Convergence

DDC Direct Digital Control(ler)

DDC Director Digital Control

DDC Display Data Channel

DDC District Development Communities

DDC DOPA (Dihydroxyphenylalanine) Decarboxylase [Biochemistry]

DDC Dual Diversity Comparator

DDC Dynamic Deformation and Characterization (Section) [of Los Alamos National Laboratory, New Mexico, US]

DDC Dynamic Differential Calorimetry

DDC1 Display Data Channel One

DD&C Data Dictionary and Catalogue (Database) [of Environmental Resources Information Network, Australia]

DD/C Dual Down-Converter

ddC Dideoxycytidine; 2',3'-Dideoxycytidine [also DDC] [Biochemistry]

DDC-BSA Dideoxycytidine-Bovine Serum Albumin (Immunogen) [also ddC-BSA]

DDCE Digital Data Conversion Equipment

DDCIP Dipolar-Decoupled Composite Inversion Pulse

DDCl Dimethyl(dichlorophenylhydroxy)ethyl Isopropylammonium Chloride

DDCM Digital Data-Communication Message [of Digital Equipment Corporation]

DDCMP Digital Data-Communication Message Protocol [of Digital Equipment Corporation]

DDCP Dibutyl-N,N-Diethylcarbamoylphosphonate

DDCS Data Definition Control System

DDCS Digital Data Calibration System

DDCS Distributed Database Connection Services

DDCS Double Differential Cross-Section [Physics]

DD&CS Dedicated Display and Control Subsystem

ddCTP 2',3'-Dideoxycytidine 5'-Triphosphate [Biochemistry]

ddCyd 2',3'-Dideoxycytidine [Biochemistry]

DDD Dangerous, Difficult and Dirty (Tundish Maintenance) [also 3-D] [Metallurgy]

DDD Defense, Description and Designation

DDD Defense Diamond Development [US]

DDD Design Definition Document

DDD Dichlorodiphenyl-1-(Dichlorophenyl)-2,2-Dichloroethane; Dichlorodiphenyl-1,1-Dichloroethane

DDD Digital Decoder Driver

DDD 6,6'-Dihydroxy-2,2'-Dinaphthylsulfide

DDD Direct Distance Dialing

DDD Display Decoder Drive(r)

DDD 6,6'-Dithiodi-2-Naphthol Disulfide

DDD Double-Density Disk [also 2DD]

DDD Double-Density Disk Drive

o,p'-DDD 1-(o-Chlorophenyl)-1-(p-Chlorophenyl)-2,2-Dichloroethane

DD/D Data Dictionary/Directory

2DD Double-Density Disk [also DDD]

.$$$ Temporary File [File Name Extension]

Δd/d change in diameter per unit diameter [Symbol]

DDDA 1,9-Diamino-3,7-Diazanonane-3,7-Diacetate [also ddda]

DDDA Dodecadienyl Acetate

DDDA Dodecanedioic Acid

(E,E)-8,10-DDDA trans-8, trans-10-Dodecadienyl Acetate

(E,Z)-7,9-DDDA trans-7, cis-9-Dodecadienyl Acetate

DDDAB Didodecyl Dimethylammonium Bromide

DDDL Digital Data Downlink

DDDM Dihydroxydichlorodiphenylmethane

DDDOL Dodecadienol [also Dddol]

(E,E)-8,10-DDDOL trans-8, trans-10-Dodecadienol

DDDP Detailed Design Data Package

DDDP Discrete Differential Dynamic Programming

dD/dt grain growth rate (in metallurgy) [Symbol]

DDDU Digital Decoder Driver Unit

DDE Decentralized Data Entry

DDE Dichlorodiphenyl Ethane

DDE 2,4'-Dichlorodiphenyl Dichloroethane; 4,4'-Dichlorodiphenyl Dichloroethane

DDE Dichlorodiphenyl-1,1-Dichloroethylene

DDE Difference-Differential Equation

DDE Direct Data Entry

DDE Direct Design Engineering

DDE Director (of) Design Engineering

DDE Distributed Data Entry

DDE (Microsoft) Dynamic Data Exchange

DDE Escort Destroyer [US Navy Symbol]

o,p'-DDE o,p'-Dichlorodiphenyl Dichloroethane

DDEF Department of Defense [Australia]

D-D EFC Dark-Dark Extreme Fringe Contrast

DDEL Display Device Engineering Laboratory [of Toshiba Corporation, Yokohama, Japan]

DDEML (Microsoft) Dynamic Data Exchange Manager Library

DDES Direct Data Entry Station

DDETA Ethyleneglycol-bis(2-Aminoethyl)ether-N,N,N',N'-Tetraacetate

DDF Database Definition File

DDF Degree Day in Fahrenheit

DDF Design Disclosure Format

DDF Digital Distribution Frame

DDF Dose Distribution Factor

DDF Dynamic Data Formatting

DDG Digital Data Generator

DDG Digital Data Group

DDG Digital Delay Generator

DDG Digital Display Generator

DDG Dixnier, Doi and Guinier (Method) [Metallurgy]

DDGE Digital Display Generator Element

DDGSR Division of the Director-General of Scientific Research [UK]

ddGTP 2',3'-Dideoxyguanosine 5'-Triphosphate [Biochemistry]

DDH 1,3-Dibromo-5,5-Dimethylhydantoin

DDH Dichlorodimethylhydantoin

DDH Digital Data Handling

DDH&DS Digital Data Handling and Display System

DD/HH:MM:SS Day/Hour:Minute:Second [also dd/hh:mm:ss]

DDH$_2$O Deionized Distilled Water; Double-Distilled Water

DDI Dedicated Display Indicator

DDI Depth Deviation Indicator

DDI Device Driver Interface

DDI Dibasic Diisocyanate

DDI Dideoxyinosine; 2',3'-Dideoxyinosine [also ddI] [Biochemistry]

DDI Digital Document Interchange

DDI Dipole-Dipole Interaction

DDI Direct Dial(ing)-In [Private Automatic Branch Exchange]

DDI Direct Digital Interface

DDI Discrete Data Input

DDI Discrete Digital Input

DDI Dislocation-Dislocation Interaction [Materials Science]

DDI Distributed Data Interface

ddI Dideoxyinosine; 2',3'-Dideoxyinosine [also DDI] [Biochemistry]

DDI-BSA Dideoxyinosine-Bovine Serum Albumin (Immunogen) [also ddI-BSA]

DDIE Direct Dialing Interface Equipment

ΔDiHA Hyaluronic Acid Disaccharide

DDIS Data Depository Index System

DDIS Development Drilling Incentive System [Mining]

DDIS Document Data Indexing Set

ddIno 2',3'-Dideoxyinosine [Biochemistry]

ddITP 2',3'-Dideoxyinosine 5'-Triphosphate [Biochemistry]

DDK (Microsoft Windows) Device Driver Kit

DDL Data Definition Language

DDL Data Description Language

DDL Data Downlink

DDL Digital Data Link

DDL Digital Design Language

DDL Diode-Diode Logic

DDL Dispersive Delay Line

DDL Document-Description Language

DDLC Data Description Language Committee [of CODASYL–Conference on Data System Languages]

DDLEED Double-Diffraction Low-Energy Electron Diffraction

DDLG Data Definition Language Group

DDLG Driven Diffuse Lattice Gas

DDM Data Demand Module

DDM Data Display Module

DDM Data Display Monitor(ing)

DDM Department of Data Management [US]

DDM Derived Delta Modulation

DDM Design, Drafting and Manufacturing

DDM Diaminodiphenylmethane

DDM β,β'-Dichlorodiethylmethylamine

DDM 2,4'-Dichlorodiphenylmethane; 4,4'-Dichlorodiphenylmethane

DDM Difference in (or of) Depth Modulation

DDM Digital Display Makeup

DDM Dihydroxydichlorodiphenylmethane

DDM Diphenyldiazomethane

DDM Direct Drive Motor

DDM Discrete Data Management

DDM Distributed Data Management

DDM Dodecyl Mercaptan

DDM GDR Mark [Currency of the Former German Democratic Republic]

DDMA Disk Direct-Memory Access

DDMC Design and Drafting Management Council

DDMI Division of Design, Manufacture and Industrial Innovation [of National Science Foundation, US]

DD/MM/YY Day/Month/Year [also dd/mm/yy]

DDMS 1,1'-(2-Chloroethylidene)bis[(4-Chloro)benzene]

DDMS Department of Defense Manned Spaceflight [US]

DDMS Department of Defense Manager for SSS (Space Shuttle Support) [US]

DDMS Deputy Daily Mission Scientist [Aerospace]

DDMU 1,1'-(Chloroethenylidene)bis[(4-Chloro)benzene]

DDN Defense Data Network [of US Department of Defense]

DDN Digital Data Network

DDNAME Data Definition Name

DDN NIC Defense Data Network Network Information Center [of US Department of Defense]

DDNP 2-Diazo-4,6-Dinitrophenol

DDN PMO Defense Data Network Program Management Office [of US Department of Defense]

DDNS Dynamic Domain Naming System

DDNTP Di(dicyclohexylammonium)-2-Naphthylthiophosphate

DDNU 1,1'-Ethenylidene-bis[(4-Chloro)benzene]

DDO Direct Dialing Overseas

DDO Discrete Digital Output

DDOCE Digital Data Output Conversion Element

DDOCE Digital Data Output Conversion Equipment

DDOL Dodecenol [also Ddol]

7-DDOL 7-Dodecen-1-ol

Z-7-DDOL cis-7-Dodecen-1-ol

D-DOPA D-3,4-Dihydroxyphenylalanine [also D-Dopa, or D-dopa] [Biochemistry]

DDP Data Development Program

DDP Data Distribution and Processing [also ddp]

DDP Department of Defense Production [now Supplies and Services Canada]

DDP Design Development Plan

DDP cis/trans-Diammine Dichloroplatinum

DDP Dichlorodiammineplatinum

DDP Didecylphosphate

DDP Didecylphthalate

DDP Digital Data Processing; Digital Data Processor

DDP Digital Diffraction Pattern

DDP Distributed Data Processing

DDP Dodecyl Phthalate

DDPA Dodecylphosphoric Acid

DDPC Digital Data Processing Center

DDPE Digital Data Processing Equipment

DDPF Dedicated Display Processing Function

DDPL Demand Deposit Program Library

DDPS Digital Data-Processing System

DDPU Digital Data-Processing Unit

DDQ Deep Drawing Quality (Steel) [Metallurgy]

DDQ 2,3-Dichloro-5,6-Dicyano-1,4-Benzoquinone

DDR Damage-to-Dose Ratio [Physics]

DDR Data Description Record

DDR Data Discrepancy Report

DDR Design Development Record

DDR Detail Design Review

DDR Device Dependent Routine

DDR Dialed Digit Receiver

DDR Digital Data Receiver; Digital Data Reception

DDR Digital Data Recorder

DDR Discontinuous Dynamic Recrystallization [Metallurgy]

DDR Double Data Rate

DDR Double Drift Region

DDR Dynamic Device Reconfiguration

DDR Radar Picket Destroyer [US Navy Symbol]

DD&R Decommissioning, Decontamination and Reutilization (Division) [of American Nuclear Society, US]

DDR&E Directorate of Defense Research and Engineering [of US Department of Defense]

DDRP Deutsche Demokratische Republik Patent [Patent of the German Democratic Republic]

DDRR Directional Discontinuity Ring Radiator [Antennas]

DDRS Digital Data Recording System

DDR-SDRAM Double Data Rate Synchronous Dynamic Random-Access Memory

DDRX Discontinuous Dynamic Recrystallization [Metallurgy]

DDS Data Dependent System

DDS Data Dictionary System

DDS Data Display System

DDS Data Dissemination DDS System

DDS Dataphone Digital Services

DDS Decision Support System

DDS Deep-Diving System

DDS Denys-Drash Syndrome [Medicine]

DDS Deployable Defense System

DDS Design Data Sheet

DDS Detailed Design Specification

DDS Det Danske Stoalvalsevaerk A/S [Danish Steel Rolling Mill located at Frederiksvaerk]

DDS Dewey Decimal System

DDS Diaminodiphenyl Sulfone; 4,4'-Diaminodiphenyl Sulfone [also DAPSONE]

DDS Differential and Derivative Spectrophotometry

DDS Digital Dataphone Service [of AT&T Company, US]

DDS Digital Data Service

DDS Digital Data Storage

DDS Digital Data System

DDS Digital Display Scope

DDS Digital Dynamics Simulator

DDS Dihydroxydiphenyl Sulfone

DDS Dimethyldichlorosilane

DDS Distributed Database Services

DDS Doctor of Dental Science

DDS Doctor of Dental Surgery

DDS Documentation Distribution System

DDS Dodecenylsuccinic Anhydride

DDS Doppler Detection System

DDS Dynamic Data System

DDSA Defense Development Sharing Agreement [US/Canada]

DDSA Digital Data Service Adapter

DDSA (2-Dodecen-1-yl)succinic Anhydride; Dodecenylsuccinic Anhydride

DDSA/CC Defense Development Sharing Agreement Coordinating Committee [US/Canada]

DDSc Doctor of Dental Science

DDSCS Double-Differential Scattering Cross Section

DDSD Deep Defect States Distribution

DD&Shpg Dock Dues and Shipping [Commerce]

DDSS Diffusion Dependence Shell Structure

DDT Data Description Table

DDT Defect Detection Trial

DDT Deflagration to Detonation Transition

DDT Delayed Dailing Tone

DDT Design Data Transmittal

DDT DIBOL (Digital's Interactive Business-Oriented Language) Debugging Technique [of Digital Equipment Corporation]

DDT 4,4-Dichlorodiphenyl-1,1,1-Trichloroethane

DDT Digital Data Terminal

DDT Digital Data Transmission; Digital Data Transmitter

DDT Digital Debugging Tape

DDT Dithiotreitol

DDT Doppler Data Translator

DDT Double-Diode-Triode

DDT Double Dual Tandem

DDT Dynamic Debugging Technique

DDT 1,1,1-Trichloro-2,2-bis[4-Chlorophenyl]ethane

o,p'-DDT 2,4'-Dichloro-α-(Trichloromethyl)diphenylmethane

ddt deduct

DDTA Derivative Differential Thermal Analysis

DDTC N,N-Diethyldithiocarbamate; N,N-Diethyldithiocarbamic Acid

DDTE Digital Data Terminal Equipment

DDT&E Design, Development, Test and Engineering

DDT&E Design, Development, Test and Evaluation

DDTESM Digital Data Terminal Equipment Service Module

DDTF Dynamic Docking Test Facility [Aerospace]

DDTS Digital Data Transmission System

DDTS Dynamic Docking Test System [Aerospace]

ddTTP 3'-Deoxythymidine 5'-Triphosphate [Biochemistry]

DDTU Digital Data Transfer Unit

DDTV Dry Diver Transport Vehicle

DDU Data Display Unit

DDU Decommutator Distribution Unit

DDU Digital Distribution Unit

DDU Display Driver Unit

DDV Deck Drain Valve

DDVP Dimethyl-(2,2-Dichlorovinyl)phosphate (or Dichlorvos)

DDW Dense Dislocation Wall [Solid-State Physics]

DDW Distilled (and) Deionized Water

DDW Doubly Deionized Water

DDW Drug Discovery Workbench (Computer Code)

Ddw Subarctic Climate, Coldest Month below –36.4°F (–38°C), Warmest Month below 50°F (10°C) [Classification Symbol]

DDX Digital Data Exchange

DDX Double Crystal X-Ray Diffractometry

DDXF DISOSS (Distributed Office Support System) Document Exchange Facility

DDX-P Digital Data Exchange Packet [of Nippon Telegraph and Telephone Corporation, Japan]

δ-Dy Delta Dysprosium [Symbol]

DDZ α,α-Dimethyl-3,5-Dimethoxybenzyloxycarbonyl

DE Damon-Eshbach (Magnetostatic Surface Mode)

DE Data Element

DE Data Encryption

DE Data Entry

DE Daughter Element [Physics]

DE Debye Effect [Physics]

DE Debye Equation [Solid-State Physics]

DE Decelerating Electrode

DE Decision Element

DE De-Emphasis

DE Defect Engineer(ing)

DE Defense Engineer(ing)

DE Defensive Expenditures

DE Delaware [US]

DE Delta Echo [Radio Engineering]

DE Dember Effect [Electronics]

DE Department of Ethnography [of British Museum]

DE Deposition Efficiency

DE Design Engineer(ing)

DE Design Evaluation

DE Destroyer Escort [US Navy Symbol]

DE Device End

DE Dextrose Equivalent [Biochemistry]

DE Diatomaceous Earth

DE Didier Engineering GmbH [Essen, Germany]

DE Dielectric(s)

DE Diesel-Electric (Drive)

DE Diesel Engine

DE Differential Equation

DE Digital Electronics

DE Digital Element

DE Digital Encoder

DE Digital Enhancement

DE Dimensional Engineering [Metrology]

DE Disintegration Energy [Nucleart Physics]

DE Displacement Energy

DE Display Electronics

DE Display Element

DE Display Equipment

DE Dissociation Energy [Physical Chemistry]

DE District Engineer

DE Division Entry

DE Division of Engineering

DE Doctor of Engineering

DE Domain Energy [Solid-State Physics]

DE Domestic Engineer(ing)

DE Donnan Effect [Physical Chemistry]

DE Doppler Effect [Physics]

DE Doppler Extractor

DE Dose Equivalent

DE Double Entry

DE Double Exchange (Interaction) [also D-E]

DE Drawing Error

DE Drive End

DE Drop Electrode

DE Droplet Epitaxy

DE Drude Equation [Physics]

DE Dry Etching

DE Dynamic Equation

DE Dynamics Explorer

DE N,N'-[1,2-Ethanediylbis(oxy-2,1-Ethanediyl)-bis[(N-Carboxymethyl)glycine](s)

DE Germany [ISO Code]

2DE Two-Dimensional Electrophoresis

6F-DE Dimethyl Ester of 6F-Dianhydride [also 6FDE]

D&E Dila(ta)tion and Evacuation [Induced Abortion]

D-E Double Exchange (Interaction) [also DE]

D-E Electric Displacement versus Electric Field (Diagram) [Symbol]

De Deborah Number (in Rheology) [Symbol]

De Deryagin Number (in Physics) [Symbol]

dE Energy Change (or Difference) [Symbol]

dE Potential Difference (or Change) [Symbol]

.de Germany [Country Code/Domain Name]

ΔE Energy Change (or Difference) [Symbol]

ΔE Energy Gap (in Solid-State Physics) [Symbol]

ΔE Resonance Halfwidth [Symbol]

$\Delta E°$ Standard Electromotive Force (or Standard Potential) [Symbol]

ΔE_A Change in Acceptor Level Energy (of n-type Semiconductor) [also ΔE_a] [Symbol]

ΔE_D Change in Donor Level Energy (of n-type Semiconductor) [also ΔE_d] [Symbol]

ΔE_c Conduction-Band Discontinuity (in Solid-State Physics) [Symbol]

ΔE_D Magnetostatic Energy Change (or Difference) [Symbol]

ΔE_g Band-Gap Discontinuity (in Solid-State Physics) [Symbol]

ΔE_Q Quadrupole Splitting [Symbol]

Δ_ε dielectric anisotropy [also $\Delta\varepsilon$] [Symbol]

$\Delta\varepsilon$ strain range (in fatigue) [also Δe] [Symbol]

DEA Daily Equivalent Amplitude

DEA Dark Etching Area

DEA Data Encryption Algorithm

DEA Department of Economic Affairs [UK]

DEA Department of Egyptian Antiquities [of British Museum]

DEA Department of External Affairs [US]

DEA Deployed Electronics Assembly

DEA Dielectric Absorption

DEA Dielectric Analysis

DEA Diethanolamine

DEA Diethylamine [also Dea]

DEA N,N-Diethylaniline

DEA Digital Electronic Automation

DEA Diplôme d'Etude Approfondi [French Diploma of Advanced Studies]

DEA Display Electronics Assemblies

DEA Dissociative Electron Attachment

DEA Divisão de Eletrônica Aerospacial [Aerospace Electronics Division; of Brazilian National Institute for Space Research]

DEA Division of Extramural Activities [of National Cancer Institute, Bethesda, Maryland, US]

DEA Drilling Engineering Association [US]

DEA Drug Enforcement Administration [of US Department of Justice]

DEAA Diethylacetic Acid

DEAA N,N-Diethylacetoacetamide

DEAB Diethylamine Borane

DEABN 4-Diethylaminobenzonitrile

DEAC Data Exchange Auxiliary Console

DEAC Diethylaluminum Chloride

DEACON Direct English Access and Control

Deact Deactivate [also deact]

DEAD Diethyl Azodicarboxylate

DEAE Diethylaminoethyl

DEAE-C Diethylaminoethyl Cellulose [also DEAE Cellulose]

DEAE-D Diethylaminoethyl Ether of Dextran [also DEAE Dextran]

DEAE-RNA Ribonucleic Acid, Diaminoethyl Salt [Biochemistry]

DEAGa Diethyl-μ-Amido Gallium

DEAH Diethylaluminum Hydride

DEAI Diethylaluminum Iodide

DEAL Data Entry Application Language

DEAL Decision Evaluation and Logic

DEAlCl Diethylaluminum Chloride

DEAM Diethylaminomethyl

Deamino DPN Deaminodiphosphopyridine Nucleotide [Biochemistry]

Deamino DPNH Deaminodiphosphopyridine Nucleotide, Reduced Form [Biochemistry]

Deamino NAD Nicotinamide Hypoxanthine Dinucleotide [Biochemistry]

Deamino NAD Nicotinic Acid Adenine Dinucleotide [Biochemistry]

Deamino NADH Nicotinamide Hypoxanthine Dinucleotide, Reduced Form [Biochemistry]

Deamino NADP Nicotinamide Hypoxanthine Dinucleotide Phosphate [Biochemistry]

DEANOL Dimethylaminoethanol; Dimethylethanolamine [also Deanol, or deanol]

DEAP 2,2-Diethoxyacetophenone

DEAP Diffused Eutectic Aluminum Process [Metallurgy]

DEAP Dryden Early Access Platform [of NASA at Edwards, California, US]

DEAS Deutsche Auslegeschrift [Intermediate Stage toward a German Patent]

DEAS Division of Engineering and Applied Science

DEASA N,N-Diethylaniline-3-Sulfonic Acid

DEB Data Event Block

DEB Data Extension Block

DEB Diethyl Benzene

DEB Digital European Backbone

DEB Discharge Electron Bremsstrahlung

DEB Division of Energy Biosciences [of US Department of Energy]

Deb Debenture [also deb]

DEBe Diethyl Beryllium

DEBEC Deutsche Betriebsgesellschaft für Drahtlose Telegraphie [German Society for Wireless Telegraphy]

DEBITS Deposition of Biogeochemically Important Trace Species

DEBNA Digital's Ethernet BI (Bus Interconnect) Network Adapter [of Digital Equipment Corporation]

Debs Debentures [also debs]

DEBU DREO (Defense Research Establishment Ottawa) Elint Browsing Utility [Canada]

DEBW Direct Electron Beam Writing

DEC Decrement (Function) [also Dec]

DEC Denmark Environment Center [Western Australia]

DEC Department of Environmental Conservation

DEC Development Education Center

DEC Device Clear

DEC Dielectric Constant

DEC Dielectric Current

DEC 2-Diethylaminoethyl Chloride (Hydrochloride)

DEC Diethylcarbamazine Citrate

DEC Diethylcarbinol

DEC Diffused Emitter-Collector

DEC Digital Electronic Circuit

DEC Digital Equipment Corporation [Maynard, Massachusetts, US]

DEC Digital Equipment of Canada

DEC Direct Energy Conversion

DEC Displacement Error Criterion

DEC Division for Engineering Cybernetics

DEC Donnan Exclusion Chromatography

DEC Double Electron Capture

DEC Double End Crosslinkable

DEC Duluth Engineers Club [Minnesota, US]

DEC Dynamic Environmental Conditioning

Dec December

Dec Decimal [also dec]

Dec Declaration [also dec]

Dec Declination [also dec]

Dec Decoder [also dec]

Dec Decompose; Decomposition [also dec]

Dec Decoration; Decorative [also dec]

Dec Decrease [also dec]

Dec Decrement [also dec]

dec deceased

dec decimeter [Unit]

dec decorated

%/dec percent(age) of moment decay per decade (for magnetic viscosity) [Symbol]

deca decahydronaphthalene

DECADE Digital Equipment Corporation Automatic Design

decaf decaffeinated

DECAL Desk Calculator

DECAL Digital Equipment Corporation's Adaptation of Algorithmic Language

Decal Decalcomania [also decal]

DECB Data Event Control Block

DEC-BCD Decimal-to-Binary-Coded Decimal (Encoder)

DEC-BIN Decimal-to-Binary (Conversion)

DEC-CEN Decadal-to-Centennial (Time-Scale Variability)

DEC/CMS Digital Equipment of Canada/Code Management System

DEC Comput DEC Computing [Published by Digital Equipmenty Corporation in the UK]

DECD Department of Economic and Community Development [Maryland, US]

DECd Diethylcadmium

Decdr Decoder [also decdr]

DECE Department of Electrical and Computer Engineering

DECEO Defense Communication Engineering Office [US]

DECENT Distribution of Exact Classical Energy Transfer

DECHEMA Deutsche Gesellschaft für Chemisches Apparatewesen [German Chemical Equipment Manufacturer] [also Dechema]

DEC-HEX Decimal-to-Hexadecimal (Conversion)

Dechema Monogr Dechema Monographien [Dechema Monographs; published by Deutsche Gesellschaft für Chemisches Apparatewesen]

deci- SI prefix representing 10^{-1}

DECIEM Defense and Civil Institute of Environmental Medicine

decim decimeter [Unit]

Decis Decision [also decis]

Decis Sci Decision Sciences [Publication of the American Institute for Decision Sciences]

Decis Support Syst Decision Support Systems [Dutch Publication]

Decit Decimal Digit [also decit]

DECL Direct Energy Conversion Laboratory [of NASA Johnson Space Center, Houston, Texas, US]

Decl Declination [also decl]

DECLAB Digital Equipment Corporation Laboratory [US]

DECM Deceptive Electronic Countermeasures

DECM Defense Electronic Countermeasures System [US Army]

Decml Decimal [also decml]

DEC/MMS Digital Equipment of Canada/Module Management System

DECnet Digital Equipment Corporation Network [also, DECNET, US]

DEC-OCT Decimal-to-Octal (Conversion)

DECOM Digital's Ethernet Communication Transceiver [of Digital Equipment Corporation]

Decom Decommutate; Decommutator [also decom]

Decomp Decompose; Decomposition [also decomp]

decompd decomposed [also decomp'd]

Decompg Decomposing [also decompg]

Decompn Decomposition [also decompn]

Decontn Decontamination [also decontn]

DECOR Digital Electronic Continuous Ranging

DEC Prof DEC Professional [Published by Digital Equipment Corporation in the US]

Decr Decrease [also decr]

Decr Decrement [also decr]

DECS Department of Electrical Engineering and Computer Science

DECS Civil Engineering Office at Vandenberg [of NASA]

DECS Data Entry Control System

DECT Digital European Cordless Telephone

DECTAB Decimal Align Tab [also DecTab]

DECU Data Exchange Control Unit

DECUS Digital Equipment Computer Users Society [US]

DECUS Proc DECUS Proceedings [of Digital Equipment Computer Users Society, US]

DED Data Element Dictionary

DED Data Element Directory

DED Design Engineering Directorate

DED Diesel-Electric Drive

DED Diesel Engine Drive; Diesel Engine Driven

DED Distant End Disconnect

DED Doctor of Environmental Design

DED Double Error Detection

DED Dutch Elm Disease

DED Dynamic Electron Diffraction

DEd Doctor of Education

Ded Dedendum (of gears) [also ded]

Ded Dedicate(d); Dedication [also ded]

Ded Deduct(ion) [also ded]

DEDA Data Entry and Display Assembly [of NASA Apollo Program]

DEDB Data Entry Database

DEDD Deformation Electron Density Difference

DEDD Deformation Electron Density Distribution

DED/D Data Element Dictionary/Directory

DEDDM Diethyl Diaminodiphenylmethane

dE/dH change in activation energy with (applied) magnetic field

dE/di polarization resistance (i.e. change in electric potential with current density)

dE/dl change in energy per unit length

$\Delta E/\Delta I$ electrochemical impedance

DEDMS Dietoxydimethylsilane

DEDPPE 1-(Diethylphosphino)-2-(Diphenylphosphino)ethane

DEDPU Diethyldiphenylurea

$(d\varepsilon/d\rho)_e$ electronic stopping power [also $d\varepsilon/d\rho$]

$(d\varepsilon/d\rho)_n$ nuclear stopping power [also $d\varepsilon/d\rho$]

DEDS Data Entry and Display System

DEDS Dual Exchangeable Disk Storage

DEDT Department of Economic Development and Tourism [Canada]

dE/dT change in energy with temperature

dE/dT change in Young's modulus with temperature

$d\varepsilon/dT$ time-dependent change in strain [also $d\varepsilon/dT$]

$d\varepsilon/dt$ creep rate [also $d\epsilon/dt$]

$\Delta\varepsilon/dt$ minimum creep rate [also $\Delta\epsilon/dt$]

$d\varepsilon/dt$ (nominal) strain rate [also $d\epsilon/dt$]

DEDUCOM Deductive Communicator

dE/dV energy density of electric field

dE/dx change in energy with distance

DEE Data Encryption Equipment

DEE Department of Electrical Engineering

DEE Department of Electronics Engineering

DEE Diethylether [also Dee]

DEE Digital Evaluation Equipment

DEE Digital Events Evaluator

DEE 2,4-Dimethylglutaric Acid Diethyl Ester

DEE Division of Electrical Engineering

DEE Drive End Entry

DEE Major Diameter of External Spline [Symbol]

$\Delta E/E$ change in elastic modulus (or Young's modulus defect)

DEECS Department of Electrical Engineering and Computer Science

DEEDDA N,N'-Diethylethylenediamine-N,N'-Diacetate [also deedda]

DEEDAA N,N'-Diethylethylenediamine-N,N'-Diacetic Acid [also deedda]

DEEE Department of Electrical and Electronic Engineering

DEEG Diplôme d'Etudes Economiques Générales [French Diploma of General Economic Studies]

DEEN N,N'-Diethylethylenediamine [also deen]

DEEP Data Exception Error Protection

DEEP Diethylethylenephosphonate

DEEP Di-2-Ethylhexyl-2-Ethylhexylphopsphonate

Deep-Sea Res A Deep-Sea Research, Part A [Journal published in the UK]

Deep-Sea Res B Deep-Sea Research, Part B [Journal published in the UK]

Deep-Sea Res I: Oceanog Res Pap Deep Sea Research, Part I: Oceanographic Research Papers

Deep-Sea Res II: Top Stud Oceanog Deep Sea Research, Part II: Topical Studies in Oceanography

DEER Directional Explosive Echo Ranging

DEET Department of Employment, Education and Training [Australia]

DEET N,N-Diethyl-m-Toluamide [also deet]

DEF Data Entry Facility

DEF Data Extension Frame

DEF Defense Specifications [of Ministry of Defense, UK]

DEF Define (Statement)

DEF Detector Efficiency Function

DEF 1,1'-Diethanoyl Ferrocene

DEF Diethyl Formamide

.DEF Defaults (or Default Parameters) [File Name Extension] [also .def]

.DEF Definitions [File Name Extension] [also .def]

Def Defect(ive) [also def]

Def Defendant [also def]

Def Defense [also def]

Def Defer(red) [also def]

Def Definition [also def]

Def Defogger; Defogging [also def]

Def Defoliant [also def]

Def Defroster; Defrosting [also def]

def definite(ly)

DEFA Direction et Etudes des Fabrications d'Armement [Management and Study of Ordnance, France]

DEFCOM Defense Command

DEFCON Defense Condition

DEFDBL Define Double-Precision Number (Statement) [Programming]

Def Diff Forum Defect and Diffusion Forum

DefE Defense Engineer [also Def Eng, or DefEng]

Def Electron Defense Electronics [Journal published in the US]

Def Eng Defense Engineer(ing)

DEF FN Define (User) Function (Statement) [Programming]

Defgr Deflagrate(s) [also defgr]

Defgrg Deflagrating [also defgrg]

Defgrn Deflagration [also defgrn]

Defib Defibrillation; Defibrillator [also defib]

DEFINE Di-Olefin Saturation (Process) [Chemical Engineering]

DEFINT Define Integer (Statement) [Programming]

Defl Deflect(ion); Deflector [also defl]

DEFM Demographic and Economic Forecasting Model [Geographic Information System]

Def Metall Res Lab Rep Defense Metallurgical Research Laboratory Reports [India]

DEFMI Doubly Enhanced Four-Wave Mixing

Deform Met Deformación Metálica [Spanish Journal on the Deformation of Metals]

Defrag Defragment(ation); Defragmentor [also defrag]

Def Res Defense Research(er) [also def res]

Def Res Abs Contractors Ed Defense Research Abstracts–Contractors Edition [US]

Def Sci Defense Science; Defense Scientist

Def Sci J Defense Science Journal [India]

DEF SEG Define)or Definition) Segment Address (Statement) [Programming]

DEFSNG Define Single-Precision Number (Statement) [Programming]

Def Std Defense Standard [also Def Stan]

DEFSTR Define String (Statement) [Programming]

DEFT Definite Time

DEFT Direct Epifluorescent Filter Techniques

DEFT Driven-Equilibrium Fourier Transform

DEFT Dynamic Error-Free Transmission

Def Tech Defense Technologist; Defense Technology [also Def Technol]

DEFTR Digital's Ethernet Frequency Translator [of Digital Equipment Corporation]

DEF USR Define User Program (Statement) [Programming]

DEG Diethanolglycine

DEG Diethylene Glycol

DEG Directorate of Environmental Geology

DEG Division of Environmental Geosciences [of American Association of Petroleum Geologists, US]

Deg (Academic) Degree [also deg]

deg degree [also degr, or °] [Symbol]

DEGA Diethylene Glycol Adipate

DEGaCl Diethylgallium Chloride

DEGB Diethylene Glycol Benzoate

deg/cm degree(s) per centimeter [also deg cm^{-1}]

deg/(cm·Oe) degree(s) per centimeter [also deg cm^{-1} Oe^{-1}]

DEGDN Diethylene Glycol Dinitrate

DEGEBO Deutsche Gesellschaft für Bodenforschung [German Research Society for Soil Mechanics]

DEGeCl$_2$ Diethylgermanium Dichloride

D2EGFK Phosphoric Acid Bis(2-Ethylhexyl)ester

deg/m degree(s) per meter [also deg m^{-1}]

DEGMA Diethylene Glycol Methacrylate

DEGMBE Diethylene Glycol Monobutyl Ether

DEGMEE Diethylene Glycol Monoethyl Ether

DEGMN Diethylene Glycol Mononitrate

DEGN Diethylene Glycol Dinitrate

Degrad Degradation [also degrad]

DEGS Diethylene Glycol Succinate

DEGS Division of Extension and General Studies [of University of Virginia, Charlottesville, US]

DEGS-PS Diethylene Glycol Succinate/Polystyrene

DEH Department of Environment and Heritage [Queensland, Australia]

ΔE(h) Magnetic Energy Barrier [Symbol]

DEHA N,N-Diethyl Hexanamine

DEHA Diethylhydroxylamine

D2EHA Di(2-Ethylhexyl)amine

DEHg Diethylmercury

D2EHHP Di-(2-Ethylhexyl)hydrogen Phosphate

DEHI Double Exposure Holographic Interferometry

DEHP Di(ethylhexyl)phosphate

DEHP Di(2-Ethylhexyl)phthalate

D2EHP Di-(2-Ethylhexyl)phosphate

DEHPA Di(2-Ethylhexyl)phosphoric Acid [also D2EHPA]

D2EHPP Di(2-Ethylhexyl)phenylphosphonate

D2EHPPA Di(2-Ethylhexyl)phenylphosphonic Acid

Dehyd Dehydrate(d) [also dehyd]

Dehydn Dehydration [also dehydn]

Dehydro-PAF(C$_{16}$) L-α-Phosphatidylcholine, β-Acetyl-γ-O-(Octadec-9-cis-Enyl) (Platelet-Activating Factor) [Biochemistry]

DEI Dastur Engineering International GmbH [Germany]

DEI Design Engineering Identification

DEI Development Engineering Inspection

DEI Dielectrics and Electrical Insulation (Society) [of Institute of Electrical and Electronics Engineers, US]

DEI Double Electrically Isolated

DEI Dutch East Indies

DEI Major Diameter of Internal Spline [Symbol]

DEIS Design Engineering Inspection Simulation

DEIS Design Evaluation Inspection Simulator

DEIS Dielectrics and Electrical Insulation Society [of Institute of Electrical and Electronics Engineers, US]

DEIS Draft Environmental Impact Statement

DEIS Dual Electron Injection Structure

DEIT Division of Electronics and Information Technology [of Korean Institute of Science and Technology, Seoul]

DEJ Dentin-Enamel Junction [Dentistry]

DEJF Double End Jig Feet

DEK Data Encryption Key [Internet]

deka- SI prefix representing 10

DEL Data Entry Language

DEL Delete (Character, or Command) [also del] [Data Communications]

DEL Delft Hydraulics Laboratory [Netherlands]

DEL Deorbit, Entry and Landing

DEL Diesel-Electric Locomotive

DEL Direct Exchange Line

Del Delaware [US]

Del Delay [also del]

Del Delegate; Delegation [also del]

Del Delete; Deletion [also del]

Del Delineation [also del]

Del Deliver(y) [also del]

del delete (character, or command) [also DEL] [Data Communications]

Del Ac Delayed Action

DELCL Delete Clause

DELCO Sierra Leone Development Corporation

deld delayed [also del'd]

deld delivered [also del'd]

DELDT Delivery Date

DELE Delete [also dele] [Data Communications]

DELEX Destroyer Life Extension Program [of Canadian Forces]

DELFIA Dissociation-Enhanced Lanthanide Fluoroimmunoassay [also Delfia]

Delft Prog Rep Delft Progress Report [Published by Delft University Press, Netherlands]

Delhi Alum Patrika Delhi Aluminium Patrika [Aluminum Journal published in India]

Deliq Deliquescence; Deliquescent [also deliq]

DELM Department of the Environment and Land Management (of South Australia)

Delmarva Delaware–Maryland–Virginia (Peninsula) [US]

DELOS Division for Experimentation and Laboratory-Oriented Studies

DELNI Digital Ethernet Local Network Interconnect [of Digital Equipment Corporation]

DELP Department of Environment, Land and Planning [Canberra, Australian Capital Territory]

DELPHI DEtector with Lepton, Photon and Hadron Identification (Experiment) [of CERN–European Laboratory for Particle Physics, Geneva, Switzerland]

Delq Deliquescent [also delq]

DELQA Digital's Ethernet LAN (Local-Area Network) Q-Bus Adapter [of Digital Equipment Corporation]

DELSTR Delete String [Restructured Extended Executor (Language)]

DELTA Descriptive Language for Taxonomy

DELTA Distributed Electronic Test and Analysis

Delta M Delta Modulation

DELTIC Delay-Line Time Compression

delvd delivered [also delv'd]

Dely Delivery [also dely]

DEM Delta Modulation

DEM Department Engineering Materials

DEM Department of Extractive Metallurgy

DEM Deutsche Mark [Currency of Germany]

DEM Differt, Essmann and Mughrabi (Fatigue Cracking Computer Simulation) [Materials Science]

DEM Digital Elevation Model

DEM Discrete Exchange Model

.DEM Demonstration [File Name Extension] [also .dem]

Dem Demand [also dem]

Dem Demodulation; Demodulator [also dem]

Dem Demolition [also dem]

Dem Demonstration [also dem]

Dem Demurrage [also dem]

DEMA Data Entry Management Association [US]

DEMA Diesel Engine Manufacturers Association [US]

DEMA Diving Equipment Manufacturers Association [US]

DEMa Departmento de Engenharia Materiais [Department of Materials Engineering; of Universidade Federal de São Carlos, Brazil]

Demecolcine N-Deacetyl-N-Methylcolchicine

Dement Geriat Cogn Disord Dementia and Geriatric Cognitive Disorders [International Journal]

DEMEX Demetallization by Extraction (Process)

DEMIST Design Methodology Incorporating Self-Test

DEMIZ Distance Early Warning Military Identification Zone

Deml Demolition [also deml]

DEMO Demonstration Fusion Power Reactor [US]

Demo Demolition [also demo]

Demo Demonstration [also demo]

Demod Demodulate; Demodulator [also demod]

Demogr Demographic; Demographer; Demography

Demogr Yrbk Demographic Yearbook [of United Nations Statistical Office] [also Demogr Ybk]

DEMON Decision Mapping via Optimum Network

DEMPR Digital Ethernet Multiport Repeater

DEMPS Directorate of Engineering and Maintenance Planning Standardization

DEMR Département d'Electromagnétique et Radar [Department of Electromagnetics and Radar, ONERA-CERT, Toulouse, France]

DEMR Department of Energy, Mines and Resources [now Natural Resources Canada]

DEMS Defensively Equipped Merchant Ship

DEMS Development Engineering Management System

DEMS Differential Electrochemical Mass Spectroscopy

DEMS Digital Electronic Message Service

demst demonstrate

DEMUX Demultiplexer; Demultiplexing [also Demux, or demux]

DEN Department of Energy [UK]

DEN N,N-Diethyl Nitrosamine

DEN Document Enabled Networking

DEN Double-Edge Notched (Test Specimen)

Den Denmark

Den Density [also den]

DENA N,N-Diethyl Nitrosamine

DENALT Density Altitude [Meteorology]

Denb Denbigshire [Wales]

Denchuken Rev Denchuken Review [Published by the Central Research Institute of Electric Power Industry, Tokyo, Japan]

Denelux *(Deutschland-Niederlande-Luxemburg)* – German for "Germany-Netherlands-Luxembourg"

DEng Doctor of Engineering [also DEngr]

DEngRMC Doctor of Engineering of the Royal Military College [Canada]

DEngS Doctor of Engineering Science [also DEngSc]

DENI Department of Education of Northern Ireland

DENISE Dense Negative Ion-Beam Surface Experiment

DENR Department of Energy and Natural Resources [Canada]

DENR Department of Environmental and Natural Resources [South Dakota, US]

DENR Department of Environment and Natural Resources [State of South Australia]

DENS Density [also Dens, or dens]

DENS Diffuse Elastic Neutron Scattering

DENSAL Densification of Aluminum Alloy Castings (Process) [Metallurgy]

DENT Double Edge Notch(ed) Tension (Test Specimen) [Mechanics]

Dent Dental; Dentist(ry) [also dent]

Dent Clin N Am Dental Clinics of North America [Journal]

Dent Hyg Dental Hygiene; Dental Hygienist

Dent Hyg Dental Hygiene [Journal]

Dent Mater Dental Materials [International Journal] [also Dent Mat]

Dent Mater J Dental Materials Journal [also Dent Mat J]

DEnv Doctor of Environment

DEnvDes Doctor of Environmental Design

DEO Data-Entry Operator

DEO Digital End Office

DEOA Diethanolamine

DEOF Double End of File

DEORB Deorbit [also Deorb]

DEOS Deutsche Offenlegungsschrift [First Stage towards a German Patent]

DEOT Disconnect End-of-Transmission

DEOXY-BIGCHAP N,N'-Bis(3-D-Gluconamidopropyl)-deoxycholamide [also Deoxy-BIGCHAP]

DEP Data Entry Panel

DEP Dedicated Experiment Processor

DEP Department of Employment and Productivity [UK]

DEP Department of Environment and Planning

DEP Department of Environmental Protection

DEP Department of Experimental Physics

DEP Design External Pressure

DEP Deutsches Patent [Patent of the Federal Republic of Germany]

DEP Diagnostic Executive Program

DEP Dielectrophoresis; Dielectrophoretic

DEP Diethylphosphate

DEP Diethyl Phthalate

DEP Diethyl Propanediol

DEP Diethyl Pyrocarbonate

DEP Direct Exposure Probe

Dep Depart(ure) [also dep]

Dep Department [also dep]

Dep Dependence; Dependent

Dep Dependency

Dep Deposit(ion) [also dep]

Dep Depot [also dep]

Dep Deputy [also dep]

2,4-DEP 2,4-Dichlorophenoxyethylphosphite; Tris(2,4-Dichlorophenoxy)ethylphosphite

DEPA Defense Electric Power Administration [US]

DEPA Di(2-Ethylhexyl)phosphoric Acid

DEPA-TOPO Di(2-Ethylhexyl)phosphoric Acid and Trioctylphosphine Oxide (Process)

DEPC Diethylaminopropyl Chloride Hydrochloride

DEPC Diethyl Phosphorocyanidate

DEPC Diethylphosphoryl Cyanide

DEPC Diethylpyrocarbonate

DEPCOS Departure Coordination System [Germany]

DEPE 1,2-Bis(diethylphosphino)ethane [also depe]

DEPE Double Escape Peak Efficiency [Physics]

Depend Dependency [also depend]

DEPI Differential Equations Pseudocode Interpreter

DEPIC Dual Expanded Plastic-Insulated Conductor

Depl Deploy(ment) [also depl]

Depr Depreciation [also Deprec, depr, or deprec]

Depr Depression [also depr]

Depress Depressurize; Depressurization [also depress]

DEPS Department of Earth and Planetary Sciences

DEPSK Differential-Encoded Phase Shift Keying

DEPT Distortionless Enhancement by Polarization Transfer

Dept Department [also dept]

DEPT-GL Distortionless Enhancement by Polarization Transfer–Grand Luxe

deptml departmental

DEPU Diethylphenylurea

Depy Deputy [also depy]

DEQ Department of Environmental Quality

DEQ Dequeue

DEQ Dose Equivalent

DE/Q Design Evaluation/Qualification

DEQUE Double-Ended Queue [also deque]

DER Declining Error Rate

DER Department of Engineering Research

DER Departure End of Runway

DER Destroyer Escort Radar (Vessel) [US Navy Symbol]

DER Diesel Engine Reduction-Drive

DER Director for Energy Research

DER Directly Executable Representation

DER Distinguished Encoding Rules

DER Drawing Error Report

Der Derivation; Derivative; Derive(d) [also der]

DERA Directory of Education and Research in Australia

Derb Derbyshire [UK] [also Derbs]

DERCOS Decay via Random Coordinate Selection (Photoreaction)

DERD Display of Extracted Radar Data

DERD-MC Display of Extracted Radar Data–Minicomputer

DERE Dounreay Experimental Reactor Establishment [Scotland]

DEREP Digital's Ethernet Repeater [of Digital Equipment Corporation]

Derevoobrabat Prom Derevoobrabatyvayushchaya Promyshlennost [Russian Journal]

Derigid Derigidization; Derigidize [also derigid]

Deriv Derivation; Derivative [also deriv]

Derivn Derivation [also derivn]

DERL Defense Electronics Research Laboratory [India]

Derm Dermatological; Dermatologist; Dermatology [also Dermatol, dermatol, or derm]

Dermatol Dermatology [International Journal]

Dermatol Online J Dermatology Online Journal [Published by the University of California at Davis, US]

DERMO Département d'Etudes et de Recherches en Micro-Ondes [Department of Microwaves Research and Studies, ONERA-CERT, Toulouse, France]

DERO Département d'Etudes et de Recherches en Optics [Department of Optics Research and Studies, ONERA-CERT, Toulouse, France]

DERS Data Entry Reporting System

DERT Direction des Etudes, Recherches et Techniques [Directorate for Research, Studies and Techniques; of Délégation Générale pour l'Armement, France]

DERT Division of Extramural Research and Training [of National Institute of Environmental Health Sciences, US]

DERTS Département d'Etudes et de Recherches en Technologie Spatiale [Department of Space Research and Technology, ONERA-CERT, Toulouse, France]

DERV Diesel Engine(d) Road Vehicle [also Derv, or derv]

DES Data Encryption Standard [Internet]

DES Data Entry Sheet

DES Data Entry System

DES Data Exchange System

DES Debt-Equity Swap

DES Deep Echo Spectrum

DES Department of Earth Sciences

DES Department of Education and Science [UK]

DES Department of Emergency Services [US]

DES Department of Employment Security

DES Design and Evaluation System

DES Desoxycholate

DES Diethylstilbestrol [Pharmacology]

DES Diethyl Sulfate

DES Differential Equation Solver

DES Digital Expansion System

DES Diplôme d'Etudes Supérieures [French Diploma of Higher Studies]

DES Directorate of Engineering Standardization

DES Discrete Event System

DES Division of Engineering and Science

DES Division of Environmental Studies

DES Doctor of Engineering Science

DES Doctor of Environmental Studies

DES Dynamic Environment Simulator

.DES Description [File Name Extension] [also .des]

Des Descend [also des]

Des Descent [also des]

Des Description; Descriptor [also des]

Des Desert [also des]

Des Design(er) [also des]

2,4-DES Sodium 2-(2,4-Dichlorophenoxy)ethyl Sulfate

Desal Desalination [also desal]

Desat Desaturation [also desat]

desat desaturated

DESC Data Entry System Controller

DESC Defense Electronics Supply Center [of Defense Logistics Agency, US]

DESC N,N-Diethylmonoselenocarbamate [also desc]

DESC Digital Equation Solving Computer

Desc Descent [also desc]

desc descent(ing)

DESCIM Defense Environmental Security Corporate Information Management [US Department of Defense]

DESCON Consultative Group for Desertification Control

DESCONAP (Research and Training Centers on) Desertification Control in Asia and the Pacific

Descr Description; Descriptive [also descr]

Descr Anat Descriptive Anatomy

Descr Astron Descriptive Astronomy

Descr Bot Descriptive Botany

Descr Climatol Descriptive Climatology

Descr Geom Descriptive Geometry

Descr Met Descriptive Meteorology

Descr Mineral Descriptive Mineralogy

Descrpn Description [also descrpn]

DESDMM Diethylsulfone Dimethyl Methane

DESe Diethylselenium

DesE Design Engineer [also Des Eng, or DesEng]

Des Eng Design Engineer(ing)

Des Eng (NY) Design Engineering (New York) [Journal published in the US]

Des Eng (Tor) Design Engineering (Toronto) [Journal published in Canada]

Des Eng (UK) Design Engineering (UK) [Journal published in the UK]

Desg Design [also Desgn, desgn, or desg]

Desgn Designation [also desgn]

Desgr Designer [also desgr]

DESI Duke Engineering and Services Hanford Inc. [Washington State, US]

Desic Desiccator [also desic]

Desig Designate; Designation [also desig]

Desktop Publ J Desktop Publishers Journal [of National Association of Desktop Publishers, US]

Desktop Publ Today Desktop Publishing Today [Journal published in the UK]

DESMI Doubly Enhanced Sum-Frequency Mixing

Des News Design News [Published in the US]

DESNW Duke Engineering Services Northwest Inc. [US]

DESO Diethyl Sulfoxide

DESO Double End Shutoff (Connector)

DESON Consultative Group for Desertification Control

DESONOX Degussa SOx-NOx (Process) [*Note:* SOx stands for Sulfuric Oxide and NOx for Nitric Oxide]

DESP Département Environnement Spatial [Department of Space Research and Technology, ONERA-CERT, Toulouse, France]

DESPOT DESign Performance OpTimization

DESPOT Driven-Equilbrium Single-Pulse Observation of T1 [Physics]

ΔESR Electron Spin Resonance Linewidth

DESRT Development and Demonstration of Site Remediation Technology (Program) [of Environment Canada] [also DeSRT]

DESS Di-(2-Ethylhexyl)sulfosuccinate Sodium

DESS Diplôme d'Etudes Supérieures Specialisées [French Diploma of Specialized Higher Studies]

DESS Digital Electronic Switching System

DEST Department of the Environment, Sport and Territories [now Department of Environment, Australia]

Dest Destination [also dest]

DESTA Digital's Ethernet Station Adapter [of Digital Equipment Corporation]

D-Ester 2,4-Dichlorophenoxy Acetic Acid Ester

Destn Destination [also destn]

DESU Druck-Elektro-Schlacke-Umschmelzverfahren [Pressure-Electroslag-Refining Technique] [Metallurgy]

DESULF Didier Engineering Sulfide (Process)

DESVA Digital's Ethernet Small VAX (Virtual Address Extension) Adapter [of Digital Equipment Corporation]

DESY Deutsches Elektronen-Synchrotron [German Electron Synchrotron, Hamburg, Germany]

DESY Comput Newsl DESY Computing Newsletter [Germany]

DESY USG DESY User Support Group [Germany]

DET Deep Electron Trap

DET Device Error Tabulation

DET Device Execute Trigger

DET Dielectric Testing

DET Diethyl Tartrate

DET Diethyl Telluride [also DETe]

DET Diethyltoluamide

DET N,N-Diethyltryptamine [Biochemistry]

DET 3,6-Dioxaoctane-1,8-Diamine

DET Displacement Energy Threshold

DET Divorced Eutectic Transformation [Metallurgy]

DET Double Exposure Technique

Det Detachment [also det]

Det Detail [also det]

Det Detection; Detector [also det]

Det Detent [also det]

Det Determinant; Determination; Determine(d) [also det]

Det Detonation; Detonator [also det]

det detached

det determinant (of a matrix) [Mathematics]

DETA Diaminoethylethertetraacetic Acid

DETA Dielectric Thermal Analysis; Dielectric Thermal Analyzer

DETA Diethylenetriamine [also Deta]

DETA N,N-Diethyl-3-Toluamide

DETAB Decision Table

DETAB Design Table

DETAB-X Design Table, Experimental

DETAL Detergent Alkylation (Process)

DETAP Decision Table Processor

DETAPAC Diethylenetriaminepentaacetic Acid

DETC Diethylthiacarbocyanine

detd determined [also det'd]

DETF Data Exchange Test Facility

DETF Double-End Tuning Fork

Detg Determining [also detg]

Detn Detection [also detn]

Detn Determination [also detn]

DETOC Decision Table to COBOL

DETOL De-Alkylation of Toluene (Process)

Deton Detonate; Dentonation [also deton]

detond detonated [also deton'd]

Deton Vel Detonation Velocity

Detox Detoxification; Detoxify [also detox]

DETP Data Entry and Teleprocessing

DETPA Diethylenetriamine Pentaacetic Acid

Detr Detector [also detr]

DETRAN Decision Translator

DETRESFA Distress Phase [Aeronautics]

Dets Detachments [also dets]

DETU N,N'-Diethyl Thiourea

DETWAD Divorced Eutectic Transformation With Associated Deformation [Metallurgy]

DEU Data Encryption Unit

DEU Data Entry Unit

DEU Data Exchange Unit

DEU Display Electronics Unit

DEUA Diesel Engine Users Association [UK]

DEUA Digitronics Equipment Users Association

DEUCE Digital Electronic Universal Calculating Engine

DEUG Diplôme d'Etudes Universitaires Générales [French Diploma of General University Studies]

DEUNA Digital Equipment (Corporation) Unibus Network Adapter

DEUS Data Entry, University of Saskatchewan [Canada]

Deut Med Wochenschr Deutsche Medizinische Wochenschrift [German Medical Weekly]

Deut Papierwirt Deutsche Papierwirtschaft [Publication on the German Paper Industry]

DEV Device [Computers/Communications]

Dev Develop(ment) [also dev]

Dev Deviation [also dev]

Dev Device [also dev]

Dev Devonshire [UK]

DEVCO Cape Breton (Mining) Development Corporation [Nova Scotia, Canada]

DEVCO Development Committee [of International Standards Organization]

Devel Develop(ment) [also devel]

Devel Development [Journal published by Company of Biologists Limited, UK]

Devel Biol Developmental Biologist; Developmental Biology

Devel Biol Developmental Biology [Journal published in the US]

Devel Biol Developments in Biology [European Journal]

Devel J Development Journal

Devel Neurosci Developmental Neuroscience; Developmental Neuroscientist

Devel Neurosci Developmental Neuroscience [International Journal]

Devel Pharmacol Therapeut Developmental Pharmacology and Therapeutics [International Journal]

Devel Psychobiol Developmental Psychobiology (Journal)

Devel Psychol Developmental Psychologist; Developmental Psychology

Develt Development [also develt]

DEVIL Direct Evaluation of Index Languages

DEVMIS Development Management Information System

Devn Deviation [also devn]

Devon Devonshire [UK]

DEVR Distortion-Eliminating Voltage Regulator

DEVSIS Development Sciences Information System [of International Development Research Center, Canada]

DEW Digital Electronic Watch

DEW Directed Energy Weapon

DEW Distant Early Warning

DEWIZ Distant Early Warning Identification Zone

DEW Tech Ber DEW Technische Berichte [Technical Reports published by DEW AG, Krefeld, Germany]

DEX Data Exchange

DEX Deferred Execution

DEX Dislocation Exciton [Solid-State Physics]

DEXA Database and Expert Systems Applications

DEXAN Digital Experimental Airborne Navigator

DEXPO DEC (Digital Equipment Corporation) Exposition [US]

DEXT Distant End Crosstalk

dext dextrinated

dextro dextrorotatory

DEZ Diethyl Zinc (Process)

DEZn Diethylzinc

DF Damage Free

DF Damped Frequency

DF Damping Factor

DF Darkfield (Micrograph) [Microscopy]

DF Data Field

DF Data Flow

DF Data Fusion

DF Dayton-Faber (Process)

DF Dean of the Faculty

DF Decimal Fraction

DF Decontamination Factor [Nuclear Engineering]

DF Deep Freeze; Deep Freezing

DF Deflection Factor [Electronics]

DF Defluorination

DF Degree of Freedom

DF Dellinger Fade(out) [Telecommunications]

DF Dengue Fever

DF Dense Flint (Glass)

DF Densification Factor

DF Density Formalism

DF Density Function [also df]

DF Density Functional (Calculation) [Physics]

DF Depth of Field

DF Depth of Flash

DF Derived Fuel

DF Describing Function

DF Desferrioxamine

DF Destination Field

DF Deuterium Fluoride

DF Deutschlandfunk [German Public Broadcasting Station]

DF Development Flight

DF Development-Forward

DF Device Flag

DF Die Forming

DF Dielectric Field

DF Diesel Fuel

DF Differential Flotation [Mining]

DF Digital Filter

DF Digital Fluoroscopy

DF Dilution Factor

DF Dirac-Fock (Theory) [Quantum Mechanics]

DF Direct-Filling (Resin)

DF Direct Fire [also df]

DF Direct Flow

DF Direction Finder; Direction Finding

DF Disappearing Filament

DF Disassembly Facility

DF Discharge Flow

DF Discontinuous Fiber

DF Disk File

DF Disk Filter

DF Disordered Flat

DF Dispersed Fluorescence (Spectroscopy)

DF Display Function

DF Dissipation Factor [Electronics]

DF Distilled Fraction (Process)

DF Distortion Factor [also df]

DF Distribution Feeder

DF Distribution Function

DF Diversity Factor (for Electricity Demand)

DF Division of Forests [of Department of Agriculture and Conservation, US]

DF Doctor of Forestry

DF Dorsal Fin [Marine Biology]

DF Dose Factor

DF Double Feeder

DF Double Flag [Computers]

DF Double Focusing

DF Double Frequency

DF Douglas Fir

DF Drift Flux [Physics]

DF Drinking Fountain

DF Drive Fit

DF Driving Force

DF Drop Forging

DF (Steam) Dryness Fraction

DF Dual Facility

DF Dual Firing

DF Dynamic Focus

DF Dynamic Fracture

DF Dynamic Field Braking

DF Field Forcing (Decreasing on Variable Voltage) [Controllers]

DF* Critical Densification Factor [Powder Metallurgy]

$F Fijian Dollar [Currency of Fiji]

D&F Determination and Findings

D/F Deadfreight [also D/f, or d/f]

Df Humid Cold (or Microthermal) Climate, No Dry Season [Symbol]

Df Dilution Factor (of Sample) [Symbol]

df density function [also DF]

df frequency change [also ff] [Symbol]

df distortion factor [Symbol]

df frequency change [Symbol]

ΔF Force Change [Symbol]

ΔF Free Energy Change [Symbol]

ΔF Frequency Swing (of Modulated Wave) [Symbol]

ΔF Incremental Tolerance [Symbol]

ΔF° Standard-State Change of Free Energy [Symbol]

ΔF$_{max}$ Frequency Deviation (of Modulated Wave) [Symbol]

ΔF$_{max}$/f_{max} Deviation Ratio (i.e., Ratio of Frequency Deviation to Maximum Modulation Frequency) [Symbol]

Δf radiation linewidth [Symbol]

DFA Data Flow Analyst; Data Flow Analysis

DFA Data Flow Architecture

DFA Density Functional Algorithm [Physics]

DFA Design for Assembly

DFA Deutsche Flug-Ambulanz [German Flight Ambulance Service]

DFA Deutsche Forschungsanstalt für Lebensmittelchemie [German Research Institute for Food Chemistry, Garching]

DFA Distributed Function Architecture

DFA Drop Forging Association [now Forging Industry Association, US]

DFACTT Data Fusion and Correlation Techniques Testbed

DFAG Double-Frequency Amplitude Grating

DFAIT Department of Foreign Affairs and International Trade [Canada]

DFAS Detailed Functional Application Subsystem [of Canadian Forces Supply System, Canada]

DFAST Dynamic File Allocation System

DFAT Department of Foreign Affairs and Trade [Australia]

DFAT Direct Fluorescent Antibody Technique [Immunology]

DFAW Direct Fire Antitank Weapon

DFB Data Flag Branch

DFB Deutsche Forschungsgesellschaft für Blechverarbeitung e.V. [German Sheetmetal Industry Association]

DFB Diffusion Brazing

DFB Difluorobenzene

DFB Distance from Quenched End (of Carburized Jominy Hardenability Bar) to First Appearance of Bainite [Metallurgy]

DFB Distributed Feedback

DFB Distribution Fuse Board

DFB/AIF Deutsche Forschungsgesellschaft für Blechverarbeitung e.V./Arbeitsgemeinigungen Industrieller Forschungsvereinigungen e.V. [German Sheetmetal Industry Association/Federation of Industrial Research Associations, Germany]

DFB Ber Deutsche Forschungsgesellschaft für Blechverarbeitung Berichte [Reports on the Sheetmetal Industry published by Deutsche Forschungsgesellschaft für Blechverarbeitung e.V., Germany]

DFBC Double-Layer Flexible Boundary Condition

DFB LD Distributed Feedback Laser Diode

DFBM Data Flag Branch Manager

DFBR Data Flag Branch Register

DFBT Dynamic Functional Board Tester

DFC Dairy Farmers of Canada

DFC Dark-Field Condenser

DFC Data Flow Computer

DFC Data Flow Control

DFC Design Field Change

DFC Diagnostic Flowchart

DFC Digital (Automatic) Flight Control

DFC Digital Fuel Controller

DFC Direct Field Costs

DFC Direct Fuel Cell

DFC Disk File Check

DFC Disk File Controller

DFC Distinguished Flying Cross

DFC Double Frequency Change (or Changing)

DFC Duty Free Confederation

ΔF(c) Concentration Dependent Free Energy Change

DFC/ADM Digital Fuel Controller/Advanced Development Model

DFCC Distorted Face-Centered Cubic (Model) [also dfcc] [Crystallography]

DFCE Direct-Fired Combustion Equipment

DFCM Dysprosium Iron Cobalt Molybdenum (Compound)

DFCNV Disk Data File Conversion Program

DFCS Digital-Automatic Flight Control System

DFCS Digital Flight Control Software (System)

DFCS Digital Flight Control System

DFCS Direction-Finding Control Station [also DF/CS]

DFCU Disk File Control Unit

DFD Data Flow Diagram

DFD (International) Data for Development (Association) [France]

DFD Deutsches Fernerkundigungs-Datenzentrum [German Remote-Sensing Data Center]

DFD Digital Flight Display

DFD Division of Fluid Dynamics [of American Physical Society, US]

DFDCM Difluorodichloromethane

DFDD 1,1'-(2,2-Dichloroethylidene)bis[4-Fluoro)benzene]

DFDD Difluorodiphenyldichloroethane

DFDR Digital Flight Data Recorder

dF/dr change of interatomic force with interatomic separation

DFDSG Direct-Fired Downhole Steam Generator

DFDT Difluorodiphenyltrichloroethane

DFDT 1,1'-(2,2,2-Trichloroethylidene)bis-[(4-Fluoro)benzene]

dF/ds change in force with distance

df/dt time rate of frquency change

DFE Data Flow Engineer

DFE Decision Feedback Equalizer

DFE Design for (the) Environment (Program) [of US Environmental Protection Agency]

DFE Desktop Functional Equivalent

DFE Direction Finding Equipment

DFE 1,2-Ethanediylbis[(diphenyl)phosphine]

DFE Form Diameter of External Spline [Symbol]

δ-Fe Delta Iron [Symbol]

DFET Depletion Field-Effect Transistor

δ-FET Delta-Doped Field-Effect Transistor

DFEU Disk File Electronics Unit

DFF Display Format Facility

ΔF/F Modulation Factor (i.e. Ratio of Frequency Swing to Frequency of Unmodulated Carrier Wave) [Symbol]

Δf/f change in eigenfrequency [Physics]

DFFR Dynamic Forcing Function Report

DFFT Double Forward Fourier Transformation

DFG Deutsche Forschungsgemeinschaft [German (Scientific) Research Foundation]

DFG Diode Function Generator

DFG Display Format Generator

DFG Division of Fish and Game [of Department of Agriculture and Conservation, US]

DFG Dry Film Gauge

DFGA Distributed Floating-Gate Amplifier

DFGS Digital Flight Guidance System

DFHC Dark Field Hollow Core

DFHP Dislocation-Free High-Purity [Materials Science]

DFI Darkfield Illumination; Darkfield Illuminator [Microscopy]

DFI Darkfield Image [Microscopy]

DFI Data File Interrogate

DFI Deep Foundations Institute [US]

DFI Development(al) Flight Instrumentation

DFI Diabetes-Forschungsinstitut (an der Heinrich-Heine-Universität Düsseldorf) [Diabetes Research Institute (of Heinrich Heine University at Dusseldorf), Germany]

DFI Diesel Fuel Injection

DFI Digital Facility Interface

DFI Disk File Interrogate

DFI Form Diameter of Internal Spline [Symbol]

DFKI Deutsches Forschungszentrum für Künstliche Intelligenz [German Artificial Intelligence Research Center, Saarbrücken, Germany]

DFL David Florida Laboratory [US]

DFL Deutsche Forschungsanstalt für Luft– und Raumfahrt [German Aeronautics and Astronautics Research Institute]

DFL Display Formatting Language

DFl Dutch Florin/Guilder [Currency of the Netherlands] [also Dfl]

$/fl Dollars per flasche (of mercury)

DFLD Data Field

DF-LDA Density-Functional Local Density Approximation (Method) [Physics]

DFLNH Difucosyllacto-N-Hexaose [Biochemistry]

DFLNH(a) Difucosyllacto-N-Hexaose a [Biochemistry]

DFM Data Flow Manager

DFM Department of Ferrous Metallurgy

DFM Design for Manufacturability; Design for Manufacturing

DFM Distinguished Flying Medal

DFM Distortion Factor Meter

DFM Division of Financial Management [of Office of Management, National Institutes of Health, Bethesda, Maryland, US]

DFM Doppler Flowmeter

DFM Dual-Phase Ferrite-Martensite [Metallurgy]

DFMA Design for Manufacture and Assembly

DF-MQW Diffused Multiple Quantum Well [Electronics]

DFMS Double-Focusing Mass Spectrometer

DFMSR Directorate of Flight and Missile Safety Research [of US Air Force]

DFN Data File Number

DFN Deutsches Forschungsnetz [German Research Network]

DFNB 1,5-Difluoro-2,4-Dinitrobenzene

DFNS Debt-for-Nature Swap

DFN-Verein Verein zur Förderung eines Deutschen Forschungsnetzes [Association for the Advancement of a German Research Network]

DFO Decade Frequency Oscillator

DFO Deferoxamine

DFO Department of Fisheries and Oceans [Canada]

DFO Deutsche Forschungsgesellschaft für Oberflächenbehandlung e.V. [German Surface Treatment Research Association]

DFO Directed Format Option

DFO Director of Flight Operations

DFO Disk File Optimizer

DF-ODMR Delayed Fluorescence Optically Detected Magnetic Resonance

DFOLS Depth of Flash Optical Landing System

DFOM Deferoxamine [Biochemistry]

DFO/NSERC Department of Fisheries and Oceans/Natural Sciences and Engineering Research Council (Program) [Canada]

DFP Deferiprone [Biochemistry]

DFP Diesel Fire Pump

DFP Diisopropyl Fluorophosphate

DFP Directed Fiber Preform

DFP Division of Forest Products [of Commonwealth Scientific and Industrial Research Organization, Australia]

DFPG Double-Frequency Phase Grating

DFPL Data Flow Programming Language

DFpLNnH Difucosyl-para-Lacto-N-neo-Hexaose [Biochemistry]

DFPM Disappearing Filament Pyrometer

DFPT Disk File Protection Table

DFQR Diffused Quantum Well (Structure) [Electronics]

DFR Decreasing Failure Rate

DFR Design for Recycling

DFR Director of Fuel Research [UK]

DFR Disk File Read

DFR Dounreay Fast Reactor [Scotland]

DFR Dropped from Rolls

Dfr Defrost(er); Defroster [also dfr]

Dfr Djibouti Franc [also DFR; Currency of Djibouti]

DFRA Drop Forging Research Association [UK]

DFRC (Hugh L.) Dryden Flight Research Center [of NASA at Edwards, California, US]

DFRF (Hugh L.) Dryden Flight Research Facility [of NASA at Edwards, California, US]

DFRL Differential Relay

Dfrn Differential [also dfrn]

Dfrn Press Differential Pressure [also dfrn press]

DFS Dansk Fysisk Selskab [Danish Physical Society]

DFS Darkfield Stop [Microscopy]

DFS Department of Food Science

DFS Depth-First Search

DFS Detailed Functional Specification

DFS Deutsche Forschungsanstalt für Segelflug [German Gliding Research Institute]

DFS Deutsches Fernmeldesystem [German Telecommunications System]

DFS Direction-Finding System

DFS Disk Filing System

DFS Distributed File System

DFS Dividends from Space [US]

DFS Doctor of Forest Science

DFS Doppler Frequency Shift

DFS Dynamic Flight Simulator

DFSG Direct Formed Supergroup [Telecommunications]

DFSK Double Frequency Shift Keying

DFSMS Data Facility Storage Management Subsystem

DF-SPE-A Shuttle Project Engineering Office [of NASA Kennedy Space Center, Florida, US]

DFSR Directorate of Flight Safety Research [US]

DF STEM Darkfield Scanning Transmission Electron Microscope; Darkfield Scanning Transmission Electron Microscopy [also DFSTEM, or DF-STEM]

DFSU Disk File Storage Unit

DFT Deaerating Feed Tank

DFT Degree of Fiber Treatment

DFT Density Functional Technique [Physics]

DFT Density Functional Theory [Physics]

DFT Department of Fire Technology [of Southwestern Research Institute, San Antonio, Texas, US]

DFT Department of Foundry Technology

DFT Design for Test(ability)

DFT Diagnostic Function Test

DFT Digital Facility Terminal

DFT Digital Fourier Transform(ation)

DFT Discrete Fourier Transform

DFT Distributed Function Terminal

DFT Double-Flow Turbine

DFT Dry Film Thickness

Dft Draft [also dft]

d4T 2',3'-Didehydro-3'-Deoxythymidine [Biochemistry]

Δf(T) temperature-dependent radiation linewidth [Symbol]

DFTBPN 3,3-Difluoro-2,3,3-Tribromopropanenitrile

DFTCE 1,2-Difluorotetrachloroethane

DF TEM Darkfield Transmission Electron Microscope; Darkfield Transmission Electron Microscopy [also DFTEM, or DF-TEM]

Dftg Drafting [also dftg]

Dftg Drifting [also dftg]

DFTI Distance from Touchdown Indicator

DFT-LDA Density Functional Theory/Local Density Approximation [Physics]

DFT-LSDA Density Functional Theory/Local Spin-Density Approximation [Physics]

DFTPP Decafluorotriphenylphosphine

DFTS Dispersive Fourier Transform Spectrometry

Dftsmn Draftsman [also dftsmn]

Dftswmn Draftswoman [also dftswmn]

DFU Data File Utility

DFU Disposable Filter Unit

5'dFUrd 5-Fluoro-5'-Deoxyuridine [Biochemistry]

DFV Dengue Fever Virus

DFVLR Deutsche Forschungs– und Versuchsanstalt für Luft– und Raumfahrt e.V. [German Aeronautics and Astronautics Establishment]

DFVLR-GSOC DFVLR/German Space Operations Center [Germany]

DFW Dallas-Forth Worth (Metroplex) [Texas, US]

DFW Diffusion Welding

DFW Disk File Write

DFWM Degenerate Four-Wave Mixing [Optics]

DFX Design for Manufacturability, Installability, Reliability, Safety and Other Considerations beyond Performance and Functionality (Program) [of AT&T Technologies, US]

DFZ Dislocation-Free Zone (Model) [Materials Science]

DFZ Dust-Free Zone

DG Automotive Crankcase Oil, Diesel Engine, General Service Conditions [American Petroleum Institute Classification]

DG Dangerous Goods

DG Dark Green

DG Datagram

DG Davisson-Germer (Experiment) [Quantum Mechanics]

DG Deal-Grove (Oxidation Model) [Physics]

DG *(Dei gratia)* – by the grace of God

DG Delay Generator

DG Department of Geology

DG Detonation Gun

DG Diacylglycerol

DG Dial Gauge

DG Diesel Generator

DG Differential Gain [Electronics]

DG Differential Gear

DG Differential Generator

DG Differential Geometry

DG Diffraction Grating

DG Diglyme

DG Diode Gate

DG Diphase Gel

DG Directed Grain

DG Directional Gyro

DG Directorate-General

DG Director General

DG Dispersible Granular

DG Display Generator

DG Dissolved Gas(es)

DG Double Girder (Crane)

DG Double-Glass

DG Double-Glass-Covered (Electric Wire)

DG Double Groove (Insulator)

DG Drill Gauge

DG Driver Gas

DG Dry Gas

DG Dry Goods

DG Dutch Guiana [now Suriname]

Dg Grain Density [Crude Oil]

dG deoxyguanosine [Biochemistry]

dG change in (Gibbs) free energy [Symbol]

dG geometric extent of a beam of radiation [Symbol]

dg decigram [Unit]

ΔG Change in Conductance [Symbol]

ΔG Gibbs Free Energy Difference [Symbol]

ΔG_{298} Free Energy at 298 Kelvin (or 25°C) [Symbol]

$\Delta G^{\dagger}$ Free Energy of Activation [Symbol]

$\Delta G°$ Standard Free Energy [Symbol]

$\Delta G°^{\dagger}$ Standard Free Energy of Activation (in Transition State Theory) [Symbol]

ΔG_c Coherency Strain Energy [Symbol]

ΔG_f Gibbs Free Energy of Formation [Symbol]

$\Delta G_f°$ Standard (Gibbs) Free Energy of Formation [Symbol]

ΔG^M Gibbs Free Energy Change due to Mixing [Symbol]

$\Delta \bar{G}_i^M$ Partial Molar Free Energy of Mixing of Component i [Symbol]

ΔG_m (Gibbs) Free Energy of Mixing [Symbol]

ΔG_V Volume Free Energy Change [Symbol]

δG Conductance Change (or Fluctuation) [Symbol]

Δg Normalized Free Energy Change [Symbol]

Δg^l Gibbs Energy of Mixing of Liquid [Symbol]

Δg^s Gibbs Energy of Mixing of Solid [Symbol]

Δg_c Coherency Strain Energy per Unit Volume Symbol]

$\Delta g_f°$ Gibbs Energy of Fusion of Pure Component [Symbol]

Δg_m Free Energy of Mixing per Unit Volume [Symbol]

DGA Dangerous Goods Advisor

DGA Délégation Générale pour l'Armement [French Armaments Directorate]

DGA Dense-Graded (Mineral) Aggregate

DGA Differential Geometry and Applications (Conference)

DGA Differential Gravimetric Analysis; Differential Gravimetric Analyzer

DGAC Dirección General Aviación Civil [Directorate-General for Civil Aviation, Spain]

DGAC Direction Générale d'Aviation Civile [Directorate-General for Civil Aviation, France]

DGB Deutsche Gesellschaft für Betriebswirtschaft [German Society for Industrial Engineering and Management]

DGB Deutscher Gewerkschaftsbund [German Federation of Trade Unions]

DGB Disk Gap Bond

DGB Distributed Gaussian Basis

DGBAS Directorate General of Budget, Accounting and Statistics [Taiwan]

DGBC Digital Geoballistic Computer

DGCI&S Directorate General of Commercial Intelligence and Statistics [India]

DG/CS Data General (Corporation)/Communication System

DGD Deutsche Gesellschaft für Dokumentation [German Society for Documentation]

Dgd Diguanide

DG/DBMS Data General (Corporation)/Database Management System

DGDME Diethyleneglycol Dimethylether

DGDP Double-Groove, Double-Petticoat (Insulators)

dGDP 2'-Deoxyguanosine-5'-Diphosphate [also d-GDP] [Biochemistry]

dγ/dT temperature coefficient of surface tension (or change of surface tension with temperature)

DGE Diglycidyl Ether

DGE Divisão de Geofísica Espacial [Space Geophysics Division; of Brazilian National Institute of Space Research]

Dge N,N-Dihydroxyglycine

DGEBA Diglycidyl Ether of Bisphenol A [also DGEBPA]

DGEBA/DDS Diglycidyl Ether of Bisphenol A/ 4,4'-Diaminodiphenyl Sulfone (Epoxy)

DGEBF Diglycidyl Ether of Bisphenol F [also DGEBPF]

DGEBPA Diglycidyl Ether of Bisphenol A [also DGEBA]

DGEBPF Diglycidyl Ether of Bisphenol F [also DGEBF]

DGEM Deutsche Gesellschaft für Elektronenmikroskopie [German Electron Microscopy Society]

DGES Division of Graduate Education in Science [of National Science Foundation, US]

DGF Demountable Growth Flange

DGF Deutsche Gesellschaft für Flugwissenschaften [German Society for Aeronautical Sciences]

DGF Distributed Gaussian Function

DGF Dynamic Gradient Freeze (Technique)

DGFH Deutsche Gesellschaft für Holzforschung [German Wood Research Society]

DGFK Diheptylphosphinic Acid

DGfL Deutsche Gesellschaft für Logistik [German Logistics Society]

DGFP Department of Game, Fish and Parks [South Dakota, US]

DGG Deutsche Geophysikalische Gesellschaft [German Geophysical Society]

DGG Deutsche Gesellschaft für Galvanotechnik [German Electroplating Society]

DGG Deutsche Glastechnische Gesellschaft [German Glass Society]

DGG Dynamic Grain Growth [Materials Science]

DGG Dysprosium Gallium Garnet [$Dy_3Ga_5O_{12}$ Garnet]

ΔG/G Shear Modulus Defect [Mechanics]

DGGM Director General of Guided Missiles [UK]

DGGS Division of Geological and Geophysical Surveys [Alaska, US]

DGI Dental Gold Institute [US]

DGI Department of Geographic Information [Queensland, Australia]

DGI Divisão de Geração de Imagens [Image Generation Division; of Brazilian National Institute for Space Research]

DGICYT Departmento Governmental de Investigaciónes Cientificas y Técnológicas [Department for Scientific and Technological Research of the Spanish Government]

DGIR Department of Scientific and Industrial Research [UK]

DGIS Direct Graphics Interface Standard

DGIWG Digital Geographic Information Working Group [of NATO]

DGK Deutsche Gesellschaft für Kybernetik [German Cybernetics Society]

DGKC Deutsche Gesellschaft für Klinische Chemie [German Society for Clinical Chemistry]

DGL Deglycyrrhizinated Licorice

DGL Deutsche Gesellschaft für Logistik [German Logistics Society]

DGL Digital Graphics Language

DG/L Data General (Corporation)/Programming Language

DGLA D-Gamma-Linoleic Acid [Biochemistry]

D-Glass Fiberglass with Low Dielectric Constant [also D-glass]

DGLR Deutsche Gesellschaft für Luft– und Raumfahrt [German Aeronautics and Astronautics Society]

DGLS Division of Geology and Land Survey [Missouri, US]

DGM Data-Grade Media

DGM Deformation of Glacial Materials (Conference)

DGM Deutsche Gesellschaft für Materialkunde [German Society for Materials] [formerly Deutsche Gesellschaft für Metallkunde]

DGM Deutsche Gesellschaft für Metallkunde [German Society for Metallurgy] [Name Changed to Deutsche Gesellschaft für Materialkunde]

DGM Differential Galvanometer

DGM Dry Gas Meter

DGMA Deutsche Gesellschaft für Meßtechnik und Automatisierung [German Society for Measurement Technology and Automation]

DGMK Deutsche Gesellschaft für Mineralölwissenschaft und Kohlechemie [Germany Society for Petroleum Science and Coal Chemistry]

DGMMP Department of Geology, Mining and Mineral Processing

dGMP 2'-Deoxyguanosine-5'-Monophosphate [also d-GMP] [Biochemistry]

DGN Dirección General de Normas [National Standards Directorate, Argentina]

DGN Dirección General de Normas [National Standards Directorate, Mexico]

Dgnl Diagonal [also dgnl]

DGO Deutsche Gesellschaft für Oberflächentechnik [German Society for Surface Technology]

DGO Director General's Office [of International Union for the Conservation of Nature (and Natural Resources)]

DGOF Director General of Ordnance Factories [UK]

DGON Deutsche Gesellschaft für Ortung und Navigation [German Society for Position-Finding and Navigation]

DGP Dangerous Goods Panel [of International Civil Aviation Organization]

DGPM Direction de la Géologie et de la Prospection Minière [Directorate for Geology and Prospecting, Ivory Coast]

DGPS Differential Global Positioning Satellite

DGPS Differential Global Positioning System

DGQ Deutsche Gesellschaft für Qualität e.V. [German Society for Quality]

DGQ Direcção-General de Qualidade [Directorate-General for Quality Standards, Portugal]

DGR Dark Gray

DG Rev DG Review [US]

DGROUP Data Group

DGRST Délégation Générale à la Recherche Scientifique et Technique [General Delegation on Scientific and Technological Research, France]

DGS Dallas Geological Society [Texas, US]

DGS Data Gathering System

DGS Data Generation System

DGS Data Ground Station

DGS Degaussing System

DGS Delaware Geological Survey [US]

DGS Denver Geophysical Society [Colorado, US]

DGS Department of Geological Sciences

DGS Display Generating System

DGS Distance-Gain-Size (System) [Ultrasonic Testing]

DGS Doctor of Geological Sciences

DGSC Defense General Supply Center [Richmond, Virginia, US]

DGSO Dihexyl Sulfoxide

DGSO Detonation-Gaseous Spray Deposition (Process)

DGSS Distributed Graphics Support Subroutine

DGT Directorate Général de Télécommunication [Directorate General of Telecommunication, France]

Dgt Digit [also dgt]

$\Delta G(T)$ Temperature-Dependent Gibbs Free Energy

DGTL Digital [also Dgtl, or dgtl]

DGTL GND Digital Ground [Data Communications]

dGTP 2-Deoxyguanosine 5'-Triphosphate [also d-GTP] [Biochemistry]

DG/TPMS Data General (Corporation)/Transaction Processing Management System

DGU Diethylene Glycol Urethane

D-Gun (Gas) Detonation Gun [also D-gun]

DGV Deutscher Giessereiverband [German Foundry Association]

DGWRD Director of Guided Weapons Research and Development [UK]

DGZ Desired Ground Zero [Atomic Physics]

DGZfP Deutsche Gesellschaft für Zerstörungsfreie Prüfung [German Society for Nondestructive Testing]

DH Data Handling

DH Dead Head

DH Debye-Hückel (Equation) [Physical Chemistry]

DH Decay Heat [Nuclear Engineering]

DH Decimal-to-Hexadecimal [also D-H]

DH Decision Height [Aeronautics]

DH De Havilland (Aircraft)

DH Dehydrogenation

DH Delayed Hypersensitivity [Immunology]

DH Denavit-Hartenberg Process

DH Dental Hygiene; Dental Hygienist

DH Device Handler

DH Diamond-Hexagonal (Crystal Structure) [also dh]

DH Dielectric Heating; Dielectric Heater

DH Differential Heating [Metallurgy]

DH Dirham [Currency of Morocco]

DH District Heating

DH Dividing Head [Machine Tools]

DH Document Handling

DH Dortmund-Hörder Hüttenunion (Vacuum Degassing Process) [also D-H] [Metallurgy]

DH Double Helix

DH Double Heterojunction [Solid-State Physics]

DH Double Heterostructure [Solid-State Physics]

DH Duane-Hunt (Law) [also D-H] [Physics]

DH Dynamic Head

D/H Direct Hit

D-H Decimal-to-Hexadecimal [also DH]

D-H Dortmund-Hörder Hüttenunion (Vacuum Degassing Process) [also DH] [Metallurgy]

D-H Duane-Hunt (Law) [also DH] [Physics]

Dh Dirham [Currency of Morocco]

Dh United Arab Emirates Dirham [Currency]

D9/h9 Free Running Fit; Basic Shaft System [ISO Symbol]

dH hardness change

dH Deutsches Härtegrad [German Degree of Hardness]

dh diamond-hexagonal (crystal structure) [also DH]

ΔH Enthalpy Change [Symbol]

ΔH Enthalpy of Formation (or Formation Enthalpy) [Symbol]

ΔH Hardness Change [Symbol]

ΔH Heat of Reaction [Symbol]

ΔH Linewidth [Symbol]

ΔH Magnetic Excursion Range [Symbol]

ΔH Magnetic Field Change [Symbol]

ΔH Peak-to-Peak Linewidth (in Spectroscopy) [Symbol]

ΔH^* Activation Energy (or Enthalpy) [Symbol]

$\Delta H°$ Standard Enthalpy [Symbol]

$\Delta H°^{\dagger}$ Standard Enthalpy of Activation (in Transition State Theory) [Symbol]

$\Delta H^{\dagger}$ Enthalpy (or Heat) of Activation [Symbol]

$(\Delta H)_{1/2}$ Linewidth at Half Maximum [Symbol]

$(\Delta H)_{1/10}$ Linewidth at Tenth Maximum [Symbol]

ΔH_{298} Heat of Formation at 298 Kelvin (or 25°C) [Symbol]

ΔH_a Enthalpy (or Heat) of Activiation [Symbol]

ΔH_c Change in Magnetic Coercive Force [Symbol]

ΔH_c Heat Exchange on Cooling [Symbol]

ΔH_c Heat of Combustion [Symbol]

$\Delta H°_c(t°C)$ Net Enthalpy Change of Combustion (at Standard Temperature and Pressure) [Symbol]

ΔH_{diss} Dissociation Enthalpy [Symbol]

ΔH_{ex} Heat of Explosion [Symbol]

ΔH^f Activation Enthalpy of Formation [also Δh_f] [Symbol]

ΔH_F Defect Formation Enthalpy (in Materials Science) [Symbol]

ΔH_f Enthalpy (or Heat) of Fusion [Symbol]

$\Delta H_f°$ Standard Enthalpy (or Heat) of Formation [Symbol]

ΔH_f Heat of Formation [Symbol]

ΔH_f Latent Heat of Fusion [Symbol]

$\Delta H_f°$ Standard Enthalpy (or Heat) of Formation [Symbol]

ΔH_{fus} Enthalpy of Fusion [Symbol]

ΔH_{FV} Vacancy Formation Enthalpy (in Materials Science) [Symbol]

ΔH_h Heat Exchange on Heating [Symbol]

ΔH_{ion} Ionization Enthalpy [Symbol]

ΔH^M Enthalpy Change due to Mixing [Symbol]

ΔH_M Defect Migration Enthalpy (in Materials Science) [Symbol]

$\Delta \overline{H}_i^M$ Partial Molar Enthalpy of Mixing of Component i [Symbol]

ΔH_{pp} Peak-to-Peak Linewidth (in Spectroscopy) [Symbol]

Δh_r Relaxation Enthalpy [Symbol]

ΔH_{SD} Activation Enthalpy of Volume Self-Diffusion (in Solid-State Physics) [Symbol]

ΔH_{sub} Sublimation Enthalpy [Symbol]

ΔH_v Enthalpy Change of Vacancy (in Materials Science) [Symbol]

ΔH_v Heat (or Enthalpy) of Vaporization [Symbol]

ΔH_v Heat of Transformation [Symbol]

Δh potential drop (in fluid mechanics) [Symbol]

$\Delta h°$ standard molar enthalpy [Symbol]

Δh_1 differential heat of dilution [Symbol]

Δh_2 differential heat of solution [Symbol]

$\Delta h_f°$ Standard Molar Enthalpy of Fusion [Symbol]

Δh_v^m activation enthalpy of vacancy migration (in materials science) [Symbol]

DHA Dehydrated Humulinic Acid

DHA Dehydroacetic Acid

DHA Dehydro-L(+)-Ascorbic Acid [Biochemistry]

DHA Dehydro Epiandrosterone [Biochemistry]

DHA Design Hazards Analysis

DHA 9,10-Dihydroanthracene

DHA Dihydroxyacetone

DHA District Health Authority [UK]

DHA Docosahexaenoic Acid

bis-DHA Dehydro-L-(+)-Ascorbic Acid [Biochemistry]

D&HAA Dock and Harbour Authorities Association [UK]

DHAE Department of Home Affairs and Environment [now Department of Environment, Australia]

DHAEMAE Disposable Hypodermic and Allied Equipment Manufacturers Association of Europe [UK]

DHAM Directional and High-Frequency-Sensing Acoustic-Research-Array Module

DHAP Dihydroxyacetone Phosphate

DHAQ Dihydroxy-bis(hydroxyethylaminoethylamino)-anthraquinone

DHAS Dairy Herd Analysis Service (Program)

DHB Dihydroxybenzene

DHB Dihydroxybenzoic Acid

DHBA Dihydroxybenzylamine

DHBA 3,4-Dihydroxybenzylamine Hydrobromide

DHBP 4,4'-Dihydroxybenzophenone

DHBSA 3,4-Dihydroxybenzenesulfonic Acid

DHBT Double Heterojunction Bipolar Transistor [also D-HBT]

DHBV Duck Hepatitis B Virus

DHC Data Handling Center

DHC Decimal-to-Hexadecimal Conversion

DHC Delayed Hydride Cracking

DHC Diamond High Council [Brussels, Belgium]

DHC Drive Half Cycle

DHC Druck-Hydrogenium-Cracken (Process) [Chemical Engineering]

DHCC Decay Heat Closed Cooling [Nuclear Engineering]

DHCC Direct Hot Charge (Steelmaking) Complex [of LTV Steel Company, Cleveland, Ohio, US]

DHCE Dynamic Helium Charging Experiment

DHCF Distributed Host Command Facility

DHCP Double Hexagonal Close-Packed (Crystal) [also dhcp]

DHCP Dynamic Host Configuration Protocol [Internet]

DHD Double Heat-Sink Diode

2HD Double High Density (Disk) [also DHD]

DHDAB Dihexadecyl Dimethylammonium Bromide

DHDBCMP Dihexyl-N,N'-Dibutylcarbamylmethylene-phosphate

dH/dc change in hardness with concentration [Mechanics]

dH/dE change in hardness with Young's modulus [Mechanics]

DHDECP Dihexyl-N,N'-Diethylcarbamylphosphonate

DHDEE Dihydroxydiethyl Ether

DHDMB 2,3-Dihydroxy-2,3-Dimethylbutanoic Acid

DHDP Dihexadecyl Phosphate

$\Delta H_c/dp$ pressure coefficient of magnetic coercivity (or change in magnetic coercivity with pressure)

DHDP-PC Dihexadecyl Phosphate–Phthalocyanine

dH/dσ stress-induced hardness change [Mechanics]

dH/dT change in energy with temperature

$\Delta H_c/dT$ temperature coefficient of magnetic coercivity (or change of coercivity with temperature)

dH/dt differential scanning calorimetry signal

dH/dt heat flow rate (or heat flux)

dH/dt time rate of change of magnetic field

$\Delta H/\Delta t$ Change in Height per Unit Time

DHE Data-Handling Equipment

DHE Dehydroepiandrosterone [Biochemistry]

DHE Down-Hole Emulsification

DHE Dump Heat Exchanger

DHEA Dehydroepiandrosterone [Biochemistry]

DHEAMP [Di(2-Hydroxyethyl)amino]methyl-2-Pyridine

DHEBA N,N'-Dihydroxyethylene-bis-Acrylamide

D-HEMT Depletion-Mode High-Electron Mobility Transistor

DHEW Department of Health, Education and Welfare [now Department of Health and Human Services, US]

DHF Department of Horticulture and Forestry

DHF Dirac-Hartree-Fock (Theory) [Quantum Mechanics]

DHF Downstream Heat Flow

DHF Dynamic Hartree-Fock [Quantum Mechanics]

DHFR Dihydrofolate Reductase [Biochemistry]

DHG Dihydroxyethylglycinate

DHG Double Henyey-Greenstein (Function) [Physics]

DHGLA Dihomo-Gamma-Linoleic Acid [Biochemistry]

DHH Dehydrohalogenation

$\Delta H/H_c$ Switching Field Distribution [Magnetic Recording Media]

DHHCS Department of Health, Housing and Community Services [Australia]

DHHS Department of Health and Human Services [US]

DHHS-PHS Department of Health and Human Services–Public Health Service [US]

DHHU Dortmund-Hoerder Huettenunion [Division of Hoesch AG, Germany]

DHI Deutsches Historisches Institut [German Historical Institute; Offices in Washington, London, Paris and Rome]

DHI Deutsches Hydrographisches Institut [German Hydrographic Institute]

DHI Door and Hardware Institute [US]

DHIA Dairy Herd Improvement Association [US]

Dhifi Deutsches High-Fidelity-Institute [German High Fidelity] Institute] [also dhifi]

DHL Dynamic Head Loading

DH LED Double Heterostructure Light-Emitting Diode

DHLLP Direct High-Level Language Processor

DHLW Defense High-Level (Nuclear) Waste

dHMCP 2'-Deoxy-5-Hydroxymethylcytidine Monophosphate [Biochemistry]

DHMS 4,4-Dihydroxy-α-Methylstilbene

DHMS Diploma in Homeopathic Medicine and Science

Δ**HMV** Change in Microhardness

DHN Decahydronaphthalene

DHN Dihydronaphthalene

DHN Dihydroxynaphthalene

DHN 5,12-Dihydronaphthacene

1,2-DHN 1,2-Dihydroxynaphthalene

1,3-DHN 1,3-Dihydroxynaphthalene

1,4-DHN 1,4-Dihydroxynaphthalene

1,5-DHN 1,5-Dihydronaphthalene

1,6-DHN 1,6-Dihydroxynaphthalene

2,3-DHN 2,3-Dihydroxynaphthalene

2,6-DHN 2,6-Dihydroxynaphthalene

2,7-DHN 2,7-Dihydroxynaphthalene

DHNSA 6,7-Dihydroxy-2-Naphthalenesulfonic Acid

DHO Debye-Hückel-Onsager (Equation) [Physical Chemistry]

DHO Displaced Harmonic Oscillator

DH-OB Dortmund-Hoerder Oxygen-Bottom Blown (Steelmaking Process) [Metallurgy]

D Horizon D-Horizon (Composed of Weathered Soil Parent Material) [also D horizon]

DHP Deoxidized, High-Residual Phosphorus (Copper)

DHP Diheptly Phthalate

DHP Dihexadecylphosphate

DHP Dihexyl Phthalate

DHP Dihydric Phenol

DHP Dihydropyridine

DHP 3,4-Dihydropyran; 2,3-Dihydro-4H-Pyran

DHP Dihydropyridine

DHP Dihydroxypropyl

DHP Document Handler Processor

.DHP Dr. Halo PIC [File Name Extension]

DHP Cu Deoxidized, High-Residual Phosphorus Copper

DHPG Ganciclovir [Antiviral Drug]

DHP-MP Dihydroxypropyl Methylpiperazine; 1,4-Bis(2-Hydroxpropyl)-2-Methylpiperazine

DHPP Division of High Polymer Physics [of American Physical Society, US]

DHPZ 3,6-Dihydroxypyridazine [also DHPz]

(DHQ)$_2$PHAL Hydroquinine 1,4-Phthalazinediyl Diether

(DHQD)$_2$PHAL Hydroquinidine 1,4-Phthalazinediyl Diether

DHR Department of Human Resources

DHR Druck-Hydrogenium-Raffination (Process) [Chemical Engineering]

$/hr Dollars per Hour

DHRC Douglas Hospital Research Center [Montreal, Canada]

DHRD Department of Housing and Regional Development [Australia]

DHRRT Decreased Hot Rolling Reduction Treatment

DHRS Decay Heat Removal System [Nuclear Engineering]

DHRS Direct Heat Removal System

DHRS Division of Human Resource Systems [of Office of Human Resources Management, US]

DHS Data Handling System

DHS Department of Health Sciences

DHS Department of Health Services

DHS Detroit Historical Society [Michigan, US]

DHS Dihexyl Sebacate

DHS Dihydrostreptomycin [Biochemistry]

DHS Discrete Horizon Sensor

DHS Domestic Heating Society [UK]

DHSD Diesel Hybrid System Design

DHSH Department of Human Services and Health [Australia]

dh-Si Diamond-Hexagonal Silicon (Phase)

DHSS Department of Health and Social Security [UK]

DHT Deep Hole Trap

DHT Dehydrothermal Treatment

DHT Dihydrotestosterone [Biochemistry]

DHT 5,6-Dihydroxytryptamine [Biochemistry]

DHT Discrete Hartley Transform

DHT Discrete Hilbert Transform

5,7-DHT 5,7-Dihydroxytryptamine [Biochemistry]

DHTML Dynamic HyperText Markup Language

DHTP 2,5-Dihydroxythiophenol

DHTR Dragon High-Temperature Reactor [UK]

DHU Document Handler Unit

DHUD Department of Housing and Urban Development [US]

DHV Design Hourly Volume [Highway Engineering]

DHV Deutscher Handels– und Industrieangestellten-Verband [German Union of Employees in Commerce and Industry]

DHV Duck Hepatitis Virus

dHvA De Haas-van Alphen (Effect) [also DHVA, or DhvA] [Solid-State Physics]

DHW Domestic Hot Water

DHW Double-Hung Windows

DHX Dump Heat Exchanger

DHXCS Dump Heat Exchanger Control System

DI Dark Ignition

DI Data In

DI Data Input

DI Data Integration; Data Integrator

DI Data Interface

DI Definite Integral

DI Degradation Increase

DI Deicer; Deicing

DI Deionization; Deionized

DI Deliverable Item

DI Demand Indicator

DI Densely Inhabited

DI Dental Implant

DI Department of Industry [now Department of Trade and Industry , UK]

DI Deposit Interest

DI Design International [US]

DI Destination Index

DI Destructive Interference [Optics]

DI Deuterium Iodide

DI Development Integration

DI Device Independent

DI Diabetes Insipidus

DI Dial Indicator

DI Diesel Index

DI Differential Interferometer

DI Diffracted Intensity

DI Digital Image

DI Digital Instruments [US]

DI Digital Input

DI Dipolar Ion [Chemistry]

DI Dipole-Ion (Reaction)

DI Direct Image; Direct Imaging

DI Direct Injection

DI Disability Insurance [US]

DI Discomfort Index [Meteorology]

DI Discrete Input

DI Display (Unsigned) Integer (Command) [Pascal Programming]

DI Display Interface

DI Dissolved Iron

DI Double Irradiation

DI Dynamic ISDN (Integrated Services Digital Network)

DI Interrupt Disabled (Flag) [Computers]

D(I) Ideal Diameter Control (in Steelmaking) [Metallurgy]

D&I Development and Improvement

D/I Distinct(ive)ness of Image

D-I Direct-Image (Offset System) [also d-i]

Di Dalasi [Currency of Gambia]

Di Diaspore [Mineral]

Di Didymium [Essentially a Mixture of Praseodymium and Neodymium]

Di Dioxane

dI intensity change

dI deoxyinosine [Biochemistry]

di Digital Instruments [US]

DIA Defense Intelligence Agency [of US Department of Defense]

DIA Deformation-Induced Amorphization

DIA Densely Inhabited Area

DIA Design Industries Association [UK]

DIA Differentiation-Inhibition Activity [Immunochemistry]

DIA Digital Image Analysis

DIA Digital Input Adapter

DIA Direct Interface Adapter

DIA Document Interchange Architecture [of IBM Corporation]

DIA Dual Interface Adapter

DiA 4-(p-Dihexadecylaminostyryl)-N-Methylpyridinium Iodide

Dia Diameter [also dia]

Diabetol Croat Diabetologia Croatica [Croatian Journal on Diabetology]

DIAC Defense Industry Advisory Control

DIAC Defense Industry Advisory Council [US]

DIAC Diode Alternating-Current Switch [also Diac, or diac]

Diactor Direct-Acting Regulator [also diactor]

DIAD Digital Image Analysis and Display

DIAD Diisopropyl Azodicarboxylate

DIAD Drum Information Assembler and Dispatcher

DIADEM Dynamic International Access to Databases and Economic Models [UK]

Diag Diagonal [also diag]

Diag Diagram [also diag]

Diagn Diagnosis [also diagn]

Diagn Diagnostic(s); Diagonistician [also diagn]

Diagn Imaging Clin Med Diagnostic Imaging in Clinical Medicine [Journal published in Switzerland]

Diagnostika Progn Razrusheniya Svarnykh Konstr Diagnostika i Prognozirovanie Razrusheniya Svarnykh Konstruktsii [Russian Journal on Diagnosis and Prognosis in Welding and Construction]

Diags Diagonals [also diags]

Diags Diagrams [also diags]

DIAL Data for Interchange at the Application Level

DIAL Differential-Absorption LIDAR (Light Detection and Ranging)

DIAL Display Interactive Assembly Language

DIAL Display Interface Assembly Language

DIAL Draper Industrial Assembly Language [Robotics]

DIAL Drum Interrogation, Alteration and Loading (System)

DIALGOL Dialect of ALGOL

Dialogoye Sist Dialogoye Sistemy [Latvian Journal on Dialog Systems]

DIAM Data-Independent Architecture Model

Diam Diameter [also diam]

Diamond Relat Mater Diamond and Related Materials [Journal]

DIAN Digital–Analog

DIAN 4,4-Dihydroxydiphenyl Dimethylmethane

DIAND Department of Indian Affairs and Northern Development [Canada]

DIANE Digital Integrated Attack (and) Navigation Equipment

DIANE Direct Information Access Network for Europe [of Commission of the European Communities, Belgium]

DIANS Digital Integrated Attack Navigation System

DIAP Digitally Implemented Analog Processing

DIAP Diisoamylphosphate

DIAPA Diisoamylphosphoric Acid

Diaph Diaphragm [also diaph]

DIARS 1,2-Phenylenebis(dimethylarsine)

DIAS DIMDI Administration System [of Deutsches Institut für Medizinische Dokumentation und Information, Germany]

DIAS Dublin Institute for Advanced Studies [Ireland]

DIAS Dynamic Inventory Analysis System

DIAT Desorption Ion Angular Distribution

DIAZEPAM 5-Chloro-1,3-Dihydro-1-Methyl-5-Phenyl-2H-1,4-Benzodiazepin-2-One [also Diazepam]

DIB Daily Intelligence Bulletin [UK]

DIB Data Input Bus

DIB Data Inspection Board

DIB Deutsches Institut für Betriebswirtschaft [German Institute for Business Administration, Berlin]

DIB Device Independent Bitmap

DIB Diiodobenzene

DIB Diiodobutane

DIB Diisobutylene

DIB 1,4-Diisocyanobenzene [also DIB]

DIB 1,3-Diphenylisobenzofuran

DIB Directory Information Base

DIB Dual Independent Bus

dib 1,4-diisocyanobenzene [also DIB]

DIBA Diisobutyl Adipate

DIBA Doctor of International Business Administration

DIBA Dot Immunobinding Assay [Biochemistry]

DIBAC Diisobutylaluminum Chloride

DIBAD Dual Ion-Beam-Assisted Deposition

DIBAH Diisobutylaluminum Hydride [also DIBA-H]

DIBAL Diisobutylaluminum

DIBAL-H Diisobutylaluminum Hydride

dibas dibasic

Dibit Di-Binary Digit (i.e., A Group of Two Bits) [also dibit]

DIBK Diisobutylketone

DIBM Diisobutylmethyl

DIBOL Digital's Interactive Business-Oriented Language [of Digital Equipment Corporation]

DIBP Disobutyl Phosphate

DIBP Disobutyl Phthalate

DIBS Digital's Integrated Business System [of Digital Equipment Corporation]

DIBS Dual Ion-Beam Sputtering

DIC Data Insertion Converter

DIC Detailed Interrogation Center

DIC Differential Interface Contrast

DIC Differential Interference Contrast

DIC Differential Isothermal Calorimeter; Differential Isothermal Calorimetry

DIC Diffusion Induced Creep [Metallurgy]

DIC Digital Concentrator

DIC Digital Input Control

DIC Digital Interchange Code

DIC Diisocyanate

DIC Diisopropylaminoethyl Chloride Hydrochloride

DIC N,N'-Diisopropylcarbodiimide

DIC 2-Dimethylaminoisopropyl Chloride Hydrochloride

DIC 5-(3,3-Dimethyl-1-Triazenyl) 1H-Imidazole-4-Carboxamide

DIC Diploma of the Imperial College (of Science and Technology) [UK]

DIC Disseminating Intravascular Coagulation [Medicine]

DIC Dissolved Inorganic Carbon

DIC Doctor of the Imperial College (of Science and Technology) [UK]

DIC Dual-In-Line Ceramic [Electronics]

.DIC Dictionary [File Name Extension] [also .dic]

DICA Daily Interest Checking Account

DICAM Data-System Interactive Communications Access Method

DICAMBA 3,6-Dichloro-o-Anisic Acid (Methoxydichlorobenzoic Acid)

Dicarb Dicarboximide

DICASS Directional Command-Activated Sonobuoy System

DICBM Defense Intercontinental Ballistic Missile

DICD Dispersion-Induced Circular Dichroism

DICE Dairy and Ice Cream Equipment Association [UK]

DICE Digital Intercontinental Conversion Equipment

DICE Dynamic Inner Correlation Efficiency [Physics]

DICEF Digital Communications Experimental Facility

DICHAN Dicyclohexylammonium Nitrite

DICHILL Dilute-Chill (Process)

DICIS Duane Information Center Indexing Service [US]

DICLOFENAC 2-[(2,6-Dichlorophenyl)amino]benzeneacetic Acid [also Diclofenac]

DICM Differential Interference Contrast Microscope; Differential Interference Contrast Microscopy

DI/CMOS Dielectrically Isolated Complementary Metal-Oxide Semiconductor [also DI-CMOS, or DI CMOS]

DICO Dissemination of Information through Cooperative Organization

Dicofol 4,4'-Dichloro-α-Trichloromethylbenzhydrol [also dicofol]

DICON Digital Communication through Orbiting Needles

DICORAP Directional-Controlled Rocket-Assisted Projectile

DICORS Diver Communication Research System

DICP Dalian Institute of Chemical Physics [Dalian, PR China]

DICR Deformation-Induced Continuous Recrystallization [Metallurgy]

Dicryl 3'4'-Dichloro-2-Methylacrylanilide [also dicryl

Dict Dictation [also dict]

Dict Dictionary [also dict]

DICTA District Council of Technical Associations [UK]

Dicta Dictaphone [also dicta]

Dictn Dictation [also dictn]

Dicty Dictionary [also dicty]

DID Data Item Description

DID Datamation Industry Directory

DID Densely Inhabited District

DID Detergent Ingredients Database

DID Device Identifier

DID Device-Independent Display

DID Digital Image Document

DID Digital Information Detection

DID Digital Information Display

DID Dipole-Induced Dipole

DID Direct In(ward) Dialing

DID Display Interface Device

DID Division of Isotopes Development [of US Atomic Energy Commission]

DID Double INDOR (Internuclear Double Resonance) Difference (Spectroscopy)

DID Double Isotope Derivative

DID Drainage and Irrigation Department [Malaysia]

DID Drum Information Display

DIDA Diisodecyl Adipate

DIDAA Dodecyliminodiacetic Acid

DIDACS Digital Data Communication System

DIDAD Digital Data Display

DIDAP Digital Data Processor

di/dE polarization admittance (i.e. change in current density with electric potential)

$\Delta I/\Delta E$ Electrochemical Admittance (i.e., Change in Current with Potential)

DIDM Document Identification and Description Macro

DIDO Data Input/Data Output [also DI/DO]

DIDO Digital Input/Digital Output [also DI/DO]

DIDOC Desired Image Distribution using Orthogonal Constraints

DIDOCS Device-Independent Display Operator Console Support

DIDOL Division of Information, Department of Lands [State of Queensland, Australia]

DIDP Diisodecyl Phthalate

DIDS Decision Information and Display System

DIDS Defense Integrated Data System [of Department of National Defense, Canada]

DIDS Digital Information Display System

DIDS Diisothiocyanatostilbene Disulfonic Acid; 4,4'-Diisocyanatostilbene-2,2'-Disulfonic Acid; 4,4'-Diisothiocyanatostilbene-2,2'-Disulfonic Acid, Disodium Salt

DIDS Distributed Intrusion Detection System

DIDS Domestic Information Display System [US]

DIDT 5,6-Dihydro-3H-Imidazo[2,1-c]-1,2,4Dithiazole-3-Thione

dI/dt time rate of change of (dc) electric current

di/dt time rate of change of (instantaneous) electric current

di_1/dt time rate of change of current in primary circuit

$\Delta i/\Delta t$ change of instantaneous current during time interval Δt

dI/dV (electrical) conductance (or change in current with voltage)

$\Delta I_a/\Delta V_g$ Mutual Conductance (i.e., Change in Anode (or Plate) Current with Grid Voltage) (of Electron Tubes) [Semiconductor Symbol]

DIE Danish Institute for the International Exchange of Scientific and Literary Publications

DIE Deutsches Institut für Entwicklungspolitik GmbH [German Development Institute]

DIE Deutsches Institut für Erwachsenenbildung [German Adult Education Institute, Frankfurt/Main, Germany]

DIE Direct-Injection Enthalpimetry

DIE Doctor of Industrial Engineering

DIE Minor Diameter of External Spline [Symbol]

DIEA N,N-Diisopropylethylamine

Die Cast Eng Die Casting Engineer [Journal of the North American Die Casting Association]

Die Cast Manage Die Casting Management [Publication the of North American Die Casting Associa-tion]

Diecast Met Mould Diecasting and Metal Moulding [Journal published in the UK]

DIEGME Diethylene Glycol Monomethyl Ether

DIEN Data Input Ensemble

DIEN Diethylenetriamine [also dien]

DIERS Design Institute for Emergency Relief Systems (Users Group) [US]

DIESA Department of International Economic and Social Affairs [of United Nations]

Diesel Gas Turbine Prog Diesel and Gas Turbine Progress [Journal published in the US]

Diesel Gas Turbine Prog Worldw Diesel and Gas Turbine Progress Worldwide [Journal published in the US]

Diesel Gas Turbine Worldw Diesel and Gas Turbine Worldwide [Journal published in the US]

Diesel Prog Engines Drives Diesel Progress Engines and Drives [Journal published in the US]

Diesel Prog North Am Diesel Progress North America [Journal published in the US]

DIET Desorption Induced by Electronic Transition; Desorption Induced by Electrostatic Transition

DIET N,N-Diethyl-p-Phenylenediamine

Diet Dietician; Dietics [also diet]

DIF Data Interchange Format

DIF Deutsches Institut zur Förderung des Industriellen Führungsnachwuchses [German Institute for the Advancement of Junior Executives]

DIF Device Input Format

DIF Directory Interchange Format

DIF Discrete Increment Filter

DIF Division of Interior Fusion [of US Department of Energy]

Dif Difference [also dif]

dif different

DIFACS Digital Facsimile Workstation

DIFAD Deutsches Institut für angewandte Datenverarbeitung [German Institute for Applied Data Processing]

DIFAR Direction-Finding and Ranging (Sonobuoy)

DIfE Deutsches Institut für Ernährungsforschung [German Nutritional Research Institute, Potsdam-Rehbruecke]

DIFET Double-Injection Field-Effect Transistor

DIFEX Dimethyl Formamide Extraction (Process) [Chemical Engineering]

DIFF Deutsches Institut für Fernstudienforschung (an der Universität Tübingen) [German Institute for Correspondence-Course Research (of the University of Tubingen)]

DIFF Development Import Finance Facility

Diff Difference; Differential [also diff]

diff different

diffc difficult [also diffc]

DIFFSENS Differential Sense

Diffus Diffusion [also diffus]

Diffus Defect Data Diffusion and Defect Data [Published in Vaduz, Liechtenstein; subdivided into Parts A and B]

Diffus Defect Data, Solid State Data A, Defect Diffus Forum Diffusion and Defect Data–Solid State Data, Part A, Defect and Diffusion Forum [Vaduz, Liechtenstein]

Diffus Defect Data, Solid State Data B, Solid State Phenom Diffusion and Defect Data–Solid State Data, Part A, Solid State Phenomena [Published in Vaduz, Liechtenstein]

Difr Difference [also ifr]

DIFFTR Differential Time Relay

DIFKIN Diffusion Kinetics

DIFLUNISAL 5-(2,4-Difluorophenyl)salicylic Acid [also Diflunisal]

DIFP Diisopropyl Fluorophosphate

DIFPAT (Electron) Diffraction Pattern

DIFPEC Differentially Pumped Environmental Chamber

DIFU Deutsches Institut für Urbanistik [German Institute for Urban Planning]

DIG Design Implementation Guide

DIG Digital Image Generated; Digital Image Generation

DIG Digital Input Gate

DIG Diplom-Interessengruppe der Funkamateure [Interest Group of Graduate Radio Amateurs, Germany]

Dig Digest [also dig]

Dig Digest(ion); Digestive [also dig]

Dig Digit(al) [also dig]

Dig Digitization [also dig]

DIGACC Digital Guidance and Control Computer

DIGAS Distributed Gas

DIGBM Diffusion-Induced Grain Boundary Migration [Materials Science]

DIGCOM Digital Computer

DIGENOR Dirección General de Normas y Sistemas de Calidad [General Directorate for Standards and Calibrated System]

DIGEST Digital Geographic Exchange Standard

Digest Digestion; Digestive [also digest]

Digest Digestion [International Journal]

Digest Diseases Digestive Diseases [International Journal]

Digest Diseases Sci Digestive Diseases and Science [Journal]

Digest Surgery Digestive Surgery [Journal published in Switzerland]

DIGGER Discrete Isolation from Gradient-Governed Elimination of Resonances

Digi Digital [also digi]

DIGICOM Digital Communications; Digital Communications System

DIGILIN Digital and Linear [also digilin]

Dig Int Conf Sens Actuators Digest of the International Conference on Sensors and Actuators [of International Coordinating Committee on Solid-State Sensors and Actuators Research, US]

DIGISET Digital Typesetting (System)

Digit Digital [also digit]

DIGITAC Digital Airborne Computer

DIGITAC Digital Tactical Automatic Control

Digit Process Digital Processes [Journal published in Switzerland]

Digit Rev Digital Review [US]

Digit Tech J Digital Technical Journal [Published by Digital Equipment Corporation, Maynard, Massachusetts, US]

Dig Jpn Ind Technol Digest of Japanese Industry and Technology [Japan]

DIGLYME Diethyleneglycol Dimethylether [also Diglyme]

DIGM Diffusion-Induced Grain-Boundary Migration; Diffusion-Induced Grain-Boundary Motion [Materials Science]

DIGOL Diethylene Glycol [also Digol]

DIGS Data and Information Gathering System

DIGS Defect-Induced Gap State [Solid-State Physics]

DIGS Diffusion-Induced Grain-Boundary Sliding [Materials Science]

DIGS Disorder-Induced Gap State [Solid-State Physics]

Dig Sig Prog Digital Signal Processing

DIH Diploma in Industrial Health

DIH Discrete Input High

di-H Di-Hydrogen [also Di-H]

ΔDiHA Hyaluronic Acid Disaccharide

Di-HETE Dihydroxyeicosatetraenoic Acid [also DiHETE]

(8R,15S)-Di-HETE (8R,15S)-Dihydroxy(5Z,9E,11Z,13E) Eicosatetraenoic Acid

(5S,6R)-Di-HETE (5S,6R)-Dihydroxy(7E,9E,11Z,14Z) Eicosatetraenoic Acid

(5S,6S)-Di-HETE (5S,6S)-Dihydroxy(7E,9E,11Z,14Z) Eicosatetraenoic Acid

(5S,15S)-Di-HETE (5S,15S)-Dihydroxy(6E,8Z,11Z,13E) Eicosatetraenoic Acid

DI-H₂O Deionized Water

DIHT Deutscher Industrie– und Handelstag [Association of German Chambers of Industry and Commerce]

DII Diesel Ignition Improver

DII Dynamic Input Indicator

DII Minor Diameter of Internal Spline [Symbol]

ΔI/I Change in Electric Current per Unit Current

DIIC Dielectrically Isolated Integrated Circuit

DIIP Direct Interrupt Identification Port

DIIS Diatomics in Ionic Systems (Formalism)

DIIS Direct Inversion in the Iterative Subspace

DIK 1,3-Diketone

DiK Dipotassium (Salt)

DIKEGULAC 2,3,4,6-Di-O-Isopropylidene-2-Keto-L-Gulonic Acid [Biochemistry]

DIKF Double-Iterated Kalman Filter

DIL Deliverable Items List

DIL Discrete Input Low

DIL Dual-in-Line (Package) [Electronics]

Dil Dilution [also dil]

dil dilute

dild diluted [also dil'd]

DILEP Digital Line Engineering Program

dilg diluting

DILGEA Department of Immigration, Local Government and Ethnic Affairs [Australia]

DiLi Dilithium (Salt)

DILIC Dual-in-Line Integrated Circuit

Diln Dilution [also diln]

DILS Depolarized Induced Light Scattering

DILS Doppler Inertial LORAN (Long-Range Navigation) System

DIM Data and Instruction Management

DIM Data Interpretation Module

DIM Design Interface Meeting

DIM Device Interface Module

DIM Diatomics-in-Molecules (Method)

DIM Differential Interference (Contrast) Microscope ; Differential Interference (Contrast) Microscopy

DIM Differential Isotopic Method

DIM Dimension (Statement) [Programming]

DIM 2,3-Dimethyl-1,4,8,11-Tetraazatetradecane-1,3-Diene

Dim Dimension [also dim]

dim dimension(ality) [Mathematics]

dim diminished

DIMA International Conference on Diffusion in Materials

DIMA Digital Image Analysis; Digital Image Analyzer

DIMA Direct Imaging Mass Analyzer

DIMACS Center of Discrete Mathematics and Theoretical Computer Science [of Rutgers University, Piscataway, New Jersey, US]

DIMAT International Conference on Diffusion in Materials

DIMATE Depot-Installed Maintenance Automatic Test Equipment

DIMDI Deutsches Institut für Medizinische Dokumentation und Information [German Institute for Medical Documentation and Information]

DIME Dual Independant Map Encoding [Geographic Information System]

DIMECO Dual Independent Map Encoding File of Countries [of Harvard University, Cambridge, Massachusetts, US]

DIMEDONE 5,5-Dimethyl-1,3-Cyclohexanedione [also Dimedone, or dimedone]

DIMEDONE 1,1-Dimethyl-3,5-Diketocyclohexane [also Dimedone, or dimedone]

Dimefox bis-(Dimethylamino)fluorophosphate [also dimefox]

DIMEN Dimension [also Dimen, or dimen]

Dimensions (Sandvik Tubul Prod) Dimensions (Sandvik Tubular Products) [Publication of Sandvik Steel Co., Scranton, Pennsylvania, US]

DIMES Defense-Integrated Management Engineering System

DIMES Delft Institute of Microelectronics and Submicron Technology [of Delft University of Technology, Netherlands]

DIMES Departmental Information Management Exchange System

DIMETA (Conference on) Diffusion in Metals and Alloys

Dimethyl-POPOP Dimethyl Phenyloxazolyl Phenyloxazolylphenyl [also dimethyl-POPOP]

Dimethyl-POPOP 1,4-bis[2-(4-Methyl-5-Phenyloxazolyl)]-benzene [also dimethyl-POPOP]

DIMETN N,N'-Dimethyl Trimethylenediamine

Dimn Diminution [also dimin]

Diminco National Diamond Mining Company [Sierra Leone]

DIMM Data and Instruction Management Machine

DIMM Dual In-Line Memory Module

Dimn Dimension [also dimn]

dimorph dimorphous

DIMOX Directed Melt Oxidation (Process)

DIMPLE Deuterium-Moderated Pile Low-Energy Reactor [of Atomic Energy Research Establishment, UK] [also Dimple]

DIMS Distributed Intelligence Microcomputer System

DIM-3C Diatomics-in-Molecules plus Three-Center Terms

DIMUS Digital Multibeam Steering

DIN Deutsche Industrienormen [German Industrial Standards]

DIN Deutsches Institut für Normung [German Standards Institute]

DIN Drug Identification Number

Din Dinar [Currency of All Former Yugoslavian States]

DINA Database Industry Association [of Japan Information Processing Development Center]

DINA Diisononyl Adipate [Biochemistry]

DINA Direct Noise Amplification

DINA Distributed Information Processing Network Architecture

DiNa Disodium (Salt)

Dinam Dislok Plast Mir Mosc Dinamika Dislokacii i Plastitchnost, Mir, Moskva [Journal on Dislocation and Plasticity Dynamics published in Moscow, USSR]

DINAP Digital Network Analysis Program

DIN EN Deutsches Institut für Normung–Europa-Norm [German Standards Institute –European Standard]

DIN-EN-ISO Deutsches Institut für Normung–Europa-Norm–International Standards Organization [German Standards Institute-European Standard–International Standards Organization]

DIN IEC Deutsches Institut für Normung–International Electrotechnical Commission [German Standards Institute–International Electrotechnical Commission [also DIN-IEC, or DIN/IEC]

DIN/ISO Deutsches Institut für Normung [German Standards Institute]–International Standards Organization [German Standards Institute–International Standards Organization] [also DIN-ISO, or DIN ISO]

Dinitro Dinitrophenol

Dinol Diazodinitrophenol

DINOSar 1,8-Dinitrosarcophagine [also dinosar]

DINOSEB 2-sec-Butyl-4,6-Dinitrophenol [also Dinoseb]

DINP Diisononyl Phthalate

Din Prochn Mash Dinamika i Prochnost Mashin [Russian Journal Mechanical Engineering]

DINS Diffuse Inelastic Neutron Scattering Spectroscopy

DInsp Detective Inspector

DINUPS DIMDI Input and Update System [of Deutsches Institut für Medizinische Dokumentation und Information, Germany]

DIO Data In/Out (Line)

DIO Data Inpt/Output [General-Purpose Interface Bus Line]

DIO Diode [also Dio, or dio]

DIO Direct Input/Output

DiO 3,3'-Dioctadecyloxacarbocyanine Perchlorate

DIOA Diisooctyl Adipate [Biochemistry]

DIOB Digital Input/Output Buffer

DIOC Digital Input/Output Control

DiOC7 (3) 3,3'-Diheptyloxacarbocyanine Iodide

DIOD Digital Input/Output Display (System)

DIOL 2-Amino-2-Methyl 1,3-Propanediol

Diol Dihydric Alcohol [also diol]

DIOM Diisooctyl Maleate

DIOMP Diisooctylmethylphosphonate

DIOP DDD (Double-Density Disk Drive) Input/Output Processor

DIOP Diisooctyl Phthalate

DIOP Diisopropylidene Dihydroxy-bis(diphenylphosphino)butane

DIOP Dimethyldioxolane

DIOS Diisooctyl Sebacate

DIOS Direct Iron-Ore Smelting (Reduction Process) [Metallurgy]

DIOS Distribution, Information and Optimizing System

DIOS DMA (Direct Memory Access) Input/Output System

DIOSS Distributed Office Support System

DIOX 1,4-Dioxane

DIOZ Diisooctyl Azelate

DIP Defense Industry Productivity (Program) [Canada]

DIP Depth Image Processing

DIP Designated Inspection Point

DIP Design Internal Pressure

DIP Dialup Internet Protocol [Internet]

DIP Digital Image Processing

DIP Diisopropyl

DIP Dimethylaminoisopropyl

DIP Direct Insert Probe

DIP Display Information Processor

DIP Display Input Processor

DIP Display Interface Processing; Display Interface Processor

DIP Distributed Information Processing

DIP Division of International Programs [of National Science Foundation, US]

DIP Document Image Processing

DIP Double Ionization Potential

DIP Dual-in-Line Package [Electronics]

DIP Dual-in-Line Pin [Electronics]

DIP Dual-in-Line Programmable [Electronics]

DiP Diethyl Phosphonate

Dip Diphtheria

Dip Diploma [also dip]

Dip Dipole [also dip]

DIPA 2,6-Dichloro Indophenyl Acetate

DIPA Diisopropanolamine

DIPA Diisopropylamine

DIPA Diamond Industrial Products Association [UK]

DipA Diploma in Aquaculture

DIPA-DCA Diisopropylammonium Dichloroacetate

DIPAMP 1,2-Ethanediylbis[(2-Methoxyphenyl)-phenylphosphine]

DIPAN Directoria da Produção Animal [Directorate for Animal Production, Brazil]

DIPB Diisopropylbenzene

DipBact Diploma in Bacteriology

DipBus Admin Diploma in Business Administration

DIPC 2-Dimethylaminoisopropyl Chloride

DIPCD Diisopropyl Carbodiimide

DIPCDI Diisopropylcarbodiimide

DIPD Double Inverse Pinch Device

DIPE Diisopropyl Ether

DIPEC Defense Industrial Plant Equipment Center [of US Department of Defense]

DipEd Diploma in Education

DipEng Diploma in Engineering

DipEngTech Diploma in Engineering Technology

DIPF Deutsches Institut für Internationale Pädagogische Forschung [German Institute for International Educational Research, Frankfurt/Main]

DIPF Digital Imagery Processing Facility

DIPF-S Deutsches Institut für Internationale Pädagogische Forschung–Serviceeinrichtung [German Institute for International Educational Research–Service Institution , Frankfurt/Main]

DIPHB Diisopropylbenzene Hydroperoxide

DIPHOS 1,2-Bis(diphenylphosphino)ethane; Diphenylphosphinoethane [also Diphos]

Di-phos Diphosphoryl

DIP-I Dialkyliodoboroane

DIPIC Pyridine-2,6-Dicarboxylic Acid [also Dipic, or dipic]

DIPL Display Initial Program Load

Dipl Diploma [also dipl]

Dipl Diplomate [also dipl]

dipl diplococcus [Microbiology]

DiplAcu Diploma in Acupuncture

Dipl-Berging *(Diplom-Bergingenieur)* – Title Equivalent to Master of Science in Mining

Dipl-Chem *(Diplom-Chemiker)* – Title Equivalent to Master of Science in Chemistry

DipLibTech Diploma in Library Technology

Dipl-Ing *(Diplom-Ingenieur)* – Title Equivalent to Master of Science in Engineering

Dipl-Ing HTL *(Diplom-Ingenieur der Höheren Technischen Lehranstalt)* – Title Equivalent to Graduate Engineer, Institute of Technology [Switzerland]

Dipl-Ing (TH) *(Diplom-Ingenieur der Technischen Hochschule)* – Title Equivalent to Graduate Engineer, Institute of Technology

Dipl-Ing (TU) *(Diplom-Ingenieur der Technischen Universität)* – Title Equivalent to Graduate Engineer, Technical University

Dipl-Kaufm *(Diplom-Kaufmann)* – Title Equivalent to Master of Science in Business [also Dipl-Kfm]

Dipl-Ldw *(Diplom-Landwirt)* – Title Equivalent to Master of Science in Agriculture

Dipl-Phys *(Diplom-Physiker)* – Title Equivalent to Master of Science in Physics

Dipl-Volksw *(Diplom-Volkswirt)* – Title Equivalent to Master of Science in Economics

Dipl-Wirtsch-Ing *(Diplom-Wirtschaftsingenieur)* – Title Equivalent to Master of Science in Engineering Economy

DipMA Diploma in Marine Affairs

DIPMB Diisopentylmethylphosphonate

DIPN Diisopropylnaphthalene

DIPOPE 1,2-Bis(diisopropoxyphosphino)ethane [also Dipope, or dipope]

DIPP Defense Industry Productivity Program [US/Canada]

DIPP Diisopentyl Phthalate

DIPP 2,6-Diisopropylphenol

DipPA Diploma in Public Affairs

DIPPE 1,2-Bis(diisopropylphosphino)ethane [also Dippe, or dippe]

DIPPP 1,3-Bis(diisopropylphosphino)propane [also Dippp, or dippp]

DIPPR Design Institute for Physical Property Data [of American Institute of Chemical Engineers, US]

DIPR Direct Interaction with Product Repulsion (Model)

DIPRA Ductile Iron Pipe Research Association [US]

DIP RAM Dual-in-Line Package Random-Access Memory

Dipr Eth Diisopropyl Ether

DIPS Development Information Processing System

DIPS N,N-Diisopropyl-2-Benzothiazole Sulfenamide

DIPS 3,5-Diisopropyl Salicylic Acid

DIPSO 3-[N,N-bis(2-Hydroxyethyl)amino]-2-Hydroxypropanesulfonic Acid [also Dipso, or dipso]

DIPT Diisopropyltartrate

DIPT 2,6-Diisopropylthiophenol

(+)-DIPT Diisopropyl-L-Tartrate

(–)-DIPT Diisopropyl-D-Tartrate

DiPTe Diisopropyltelluride

Dip/Tert/Pert Diphtheria, Tetanus, Pertussis (Vaccine) [also DTP]

DiQ 4-(p-Dihexadecylaminostyryl)-N-Methylquinolinium Iodide

DIQD Disk-Insulated Quad

DIR Data Input Register

DIR Defense Industrial Research (Program) [of Department of National Defense, Canada]

DIR Department of Industrial Relations [Australia]

DIR Development Inhibitor Releasing

DIR Diffusion-Induced Recrystallization [Metallurgy]

DIR Direct Internal Reforming

DIR Directory (Command) [also dir]

DIR Discipline Oriented Information Retrieval

DIR Division of Intramural Research [of National Institute of Environmental Health Sciences, US]

DIR Document Information Record

DIR Document Information Retrieval

DIR Downwelling Infrared

Dir Direction; Director [also dir]

Dir Directory [also dir]

dir direct

DIRA Danish Industrial Robot Association

DIRAC DImeson Relativistic Atom Complex (Experiment) [of CERN–European Laboratory for Particle Physics, Geneva, Switzerland]

DIRB Dissimilar Iron-Reducing Bacteria

DIRCOL Direction Cosine Linkage

DIR-CONN Direct-Connected [Communications]

dird directed [also dir'd]

DIRECT Document Information Retrieval and Evaluation for Computer Terminals

Direct Directory [also direct]

Direct Midrex Direct from Midrex [Publication of Midrex Inc., Charlotte, North Carolina, US]

DIRLD Dynamic Infrared Linear Dichroism

Dirn Direction [also dirn]

DIRP Defense Industrial Research Program [of Department of National Defense, Canada]

DIRS Defense Industrial Research Section [of Department of National Defense, Canada]

DIRS DIMDI Information Retrieval System [of Deutsches Institut für Medizinische Dokumentation und Information, Germany]

DIRS Dispersive Infrared Spectroscopy

Dirs Directors [also dirs]

DIS Data and Information System

DIS Deep Inelastic Scattering [Physics]

DIS Defense Investigative Service [US]

DIS Department of Information Science

DIS Digital Integration System

DIS Digital Interface Subsystem

DIS Dissertation Inquiry Service

DIS Distributed Information System

DIS Distributed Interactive Simulation [of Institute of Electrical and Electronics Engineers, US]

DIS Documentation Index System

DIS Documentation Inventory System

DIS Data and Information System

DIS Draft International Standard [of International Standards Organization]

DIS Ductile Iron Society [US]

DIS (CompuCom) Dynamic Impedance Stabilization

DIS International Workshop on Deep Inelastic Scattering (and Quantum Chromodynamics)

Dis Disagree(ment) [also dis]

Dis Discount [also dis]

Dis Disease [also dis]

Dis Disintegration [also dis]

Dis Dissolution; Dissolve [also dis]

dis disintegration(s) [Unit]

DISA Danish Information Science Association

DISA Data Interchange Standards Association

DISA Defense Information Systems Agency [of US Department of Defense]

DISA Direct Inward System Access

DISAC Digital Simulator and Computer

DISA Inf DISA Information [Published by the Danish Information Science Association]

Di-Salt Dimethyl Ammonium Nitrate

DISAM Direct and Index Sequential Access System

disap disapproved

Disassy Disassembly [also disassy]

disb disburse

Disbmt Disbursement [also disbmt]

disbn disband

DISC Defense Industrial Supply Center [US]

DISC Differential Isochronous Self-Collimating Counter

DISC Differential Scatter

DISC Digital International Switching Center

DISC Disability Information Services of Canada

DISC Domestic International Sales Corporation [US]

Disc Discone [also disc] [Antennas]

Disc Disconnect(ion); Disconnection; Disconnector [also disc]

Disc Discount [also disc]

Disc Discovered; Discovery [also disc]

Disc Discussion [also disc]

disc discrete

Disc Faraday Soc Discussions of the Faraday Society [of Royal Society of Chemistry, UK]

Disch Discharge [also disch]

Dis Chest Diseases of the Chest [Journal]

DIS-Cl 2,p-Chlorosulfonyl-3-Phenylindone; 2,3-Diphenylindenone Sulfonyl Chloride

DISCO Differences and Sums in COSY (Correlation Spectroscopy) Spectra

DISCOM Digital Selective Communications

DISCON Defense Integrated Secure Communications Network [Australia]

Discon Disconnect [also discon]

discond disconnected [also discon'd]

Discont Discontinuation; Discontinue [also discont]

discontd discontinued [also discont'd]

Disp Displacement [also disp]

Disp Display [also disp]

Disp Disposal [also disp]

Disp Imaging Technol Display and Imaging Technology [Journal published in the UK]

Disp Technol Appl Displays, Technology and Applications [Journal published in the UK]

Displ Displacement [also displ]

DISPLAY Digital Service Planning Analysis

DISR Descent Imager Spectrometer Radar

DISRX Disable Receive [also DIS RX]

DISS Digital Interface Switching System

DISS Distributed Information Processing Service System

Diss Dissertation [also diss]

Diss Dissociation [also diss]

dis/s disintegrations per second [also dis/sec, DPS, or dps]

Diss Abstr Int Dissertation Abstracts International [US]

dissd dissolved [also diss'd]

dis/sec disintegration(s) per second [also dis/s, DPS, or dps]

dissoc dissociate

dissocd dissociated [also dissoc'd]

Dissocn Dissociation [also dissocn]

DISSPLA Display Integrated Software System and Plotting Language

DIST Department of Industry, Science and Technology

DiST Dissolved Solids Tester

Dist Distance [also dist]

Dist Distillation; Distiller; Distillery [also dist]

Dist Distinguish(ed) [also dist]

Dist Distribution [also dist]

Dist District [also dist]

distd distilled [also dist'd]

Distg Distilling [also distg]

Distn Distillation [also distn]

distng distinguish

Distr Distribution; Distributor [also distr]

Distr District [also distr]

DISTRAM Digital Space Trajectory Measurement System

Distrg Distributing [also distrg]

distrib distributed

Distrib Comput Distributed Computing [Journal published in Germany]

Distrib Syst Eng Distributed Systems Engineering [Journal of the Institute of Physics, UK]

DISTRO Distribution Rotation

DISTX Disable Transmit [also DIS TX]

DISU Digital International Switching Unit

DIT Daido Institute of Technology [Nagoya, Japan]

DIT Data Identification Table

DIT Deep Inelastic Transfer [Nuclear Physics]

DIT Devry Instutite of Technology [Pheonix, Arizona, US]

DIT Division of Information Technology [of Commonwealth Scientific and Industrial Research Organization, Australia]

DIT Detroit Institute of Technology [Michigan, US]

DIT Diffuse Interface Theory [Materials Science]

DIT Diiodotyrosine [Biochemistry]

DIT Directory Information Tree

DIT Divison of Information Technology [of Commonwealth Scientific and Industrial Research Organiztion, Australia]

DIT Doctor of Industrial Technology

DIT Drexel Institute of Technology [Philadelphia, Pennsylvania, US]

DITAC Department of Industry, Technology and Commerce [Australia]

DITB Distributive Industry Training Board [UK]

DITC Department of Industry, Trade and Commerce [Canada]

DITC Diisothiocyanate

DITC Division of Information Technology and Dissemination [of US Office of Education]

DITC 1,4-Diisothiocyanatobenzene

DITE Divertor in Torus Experiment [Nuclear Engineering]

Dithio Dinitrophenol

DITP Diisotridecyl Phthalate

DITEC Digital Television Communication System (for Satellite Links) [also Ditec, US]

dITP 2'-Deoxyinosine 5'-Triphosphate [also d-ITP] [Biochemistry]

DITR Department of Industry, Technology and Resources [Victoria State, Australia]

DITRAN Diagnostic FORTRAN

DITSL Deutsches Institute für Tropische und Subtropische Landwirtschaft [German Institute for Tropical and Subtropical Agriculture]

DIU Data Interchange Utility

DIU Digital Input Unit

DIU Digital Interface Unit

DIU Documents Index Unit

DIU Dynamic Integrated Test

DIURON 3-(3,4-Dichlorophenyl)-1,1-Dimethylurea [also Diuron]

DIV Data in Voice

DIV Digital Input Voltage

DIV Divide [Computers]

Div Divergence; Diverter [also div]

Div Dividend [also div]

Div Divide; Divider; Division [also div]

div divergence (of a vector) [Symbol]

div divided

div integer division [Arithmetic Operator]

DIVA Data Input Voice Answerback

DIVA Data Inquiry Voice Answer

DIVA Dicyanovinylanisole

DIVA Digital Input Voice Answerback

DIVA Digital Inquiry Voice Answerback

DIVAD Division Air Defense

DIVAH Diagonally Corrected Vibrationally Adiabatic Hyperspherical (Model) [Physics]

DivCHED Division of Chemical Education [of American Chemical Society, US]

Divd Dividend [also divd]

Divds Dividends [also divds]

DIVE Direct Interface Video Extension

div E divergence of electric field (vector) E [also ∇E]

DIVERSITAS International Program on Biological Diversity [Joint Program of the International Union of Biological Sciences , the Scientific Committee on Problems of the Environment and the United Nations Educational, Scientific and Cultural Organization]

div grad Laplacian (or Laplace Operator) [also ∇^2, or Δ]

DIVIMP Diverter Impurity Study [of University of Toronto Institute of Aerospace Studies, Canada]

Divn Division [also divn]

DIVOT Digital-to-Voice Translator

DIVOTS Direct-Input Voice-Output Telephone System

Div Rep Division Report

Div Rep CANMET/MRL Division Report of Canada Center for Mineral and Energy Technology /Mining Research Laboratories [of Natural Resources Canada]

Divs Divisions [also divs]

DIW Deionized Water

DIW Deutsches Institut für Wirtschaftsforschung [German Economic Research Institute, Berlin]

DIW D-Inside Wire

DIX Digital, Intel and Xerox (Ethernet)

DIY Do It Yourself [also diy]

DIZ Deutsches Informationszentrum [German Information Center]

DIZ Deutsches Ingenieurzentrum [German Engineering Center–An Internet Service of FIZ Karlsruhe]

DIZ Development Impact Zone

.DIZ Description In Zip [File Name Extension] [also .diz]

DJ Diamond Jet

DJ Diffusion Junction

DJ Digital Junction

DJ District Judge

DJ Djibouti [ISO Code]

DJ *(Doctor Juris)* – Doctor of Laws

D&J December and June [Business Finance]

D/J Ratio of Local Anisotropy Field to (Ferromagnetic) Interaction [Solid-State Physics]

.dj Djibouti [Country Code/Domain Name]

ΔJ Change in Rotational Quantum Number [Symbol]

Δj change in rotational quantum number of electron [Symbol]

dJ/da unnormalized mixed-mode I/III tearing modulus [Mechanics]

dJ/dt time rate of change of current density

DJD Degenerative Joint Disease

DJF Djibouti Franc [also DjFr] [Currency of Djibouti]

DJIA Dow Jones Industrial Average [US]

DJNR Dow Jones News Retrieval [US]

DJourn Doctor of Journalism

DJSU Digital Junction Switching Unit

DJT Diffused Junction Transistor

DJTE Dynamical Jahn-Teller Effect [Physical Chemistry]

DJTL Discrete Josephson Transmission Line

DJV Deutscher Journalistenverband [German Journalist Association]

DK Denmark [ISO Code]

DK Dillamore and Karoth (Theory) [Metallurgy]

DK Direct Kinematics

DK Display/Keyboard

Dk Deck [also dk]

Dk Dock(ing) [also dk]

dk dark

dk deka- [SI Prefix]

dk donkey [Immunochemistry]

.dk Denmark [Country Code/Domain Name]

$d\kappa$ change (usually an increment) in curvature

ΔK Stress Intensity Factor Range (in Fatigue) [Symbol]

ΔK_I Stress Intensity Factor Range (in Fatigue) [Symbol]

ΔK_i Resolved Stress Intensity Factor Range in Mode I [Symbol]

ΔK_{ii} Resolved Stress Intensity Factor Range in Mode II [Symbol]

ΔK_{app} Applied Stress Intensity Factor Range [Symbol]

ΔK_e Effective Stress Intensity Factor Range [Symbol]

ΔK_{eff} Effective Stress Intensity Factor [Symbol]

ΔK_{th} Theoretical Stress-Intensity Range (in Fatigue) [Symbol]

ΔK_{th} Threshold Stress-Intensity Range (in Fatigue) [Symbol]

ΔK_{total} Stress Intensity Factor Range Calculated by Using Tension-plus-Compression Stress Range (in Fatigue) [Symbol]

DKA Diabetic Ketoacidosis [Medicine]

DKD Deutscher Kalibrierdienst [German Calibration Service]

DKE Deutsche Elektrotechnische Kommission [German Electrotechnical Commission; of Verein Deutscher Elektrotechniker and Deutsches Institut für Normung]

DKE Deutscher Kleinempfänger [German Midget Receiver]

DKE/IEC Deutsche Elektrotechnische Kommission/International Electrotechnical Commission [German Electrotechnical Commission/International Electrotechnical Commission]

DKFZ Deutsches Krebsforschungszentrum [German Cancer Research Center, Heidelberg]

DKG Deutsche Keramische Gesellschaft [German Ceramic Society]

dkg dekagram [Unit]

$/kg (US) Dollar(s) per kilogram [Unit]

DKI Data Key Idle

DKI Deutsches Kunststoffinstitut [German Plastics Institute]

DKI Deutsches Kupferinstitut e.V. [German Copper Institute]

DKK Danish Krone [Currency of Denmark, Greenland and the Faroe Islands]

DKK Denki Kagaku Kogyo K.K. [Japan]

dkL dekaliter [also dkl]

dkm dekameter [Unit]

dkm² square dekameter [Unit]

dkm³ cubic dekameter [Unit]

DKMM Deutsches Komitee für Meeresforschung und Meerestechnik [German Committee for Marine Research and Technology]

DKP Dipotassium Phosphate

DKP Dissolving Kraft Pulp

DKr Danish Krone [Currency of Denmark, Greenland and the Faroe Islands]

DKRZ Deutsches Klimarechenzentrum GmbH [German Climatic Data Center, Hamburg, Germany]

dks dekastere [Unit]

DKT Deutsche Kautschuk-Tagung [German Rubber Conference]

DKUHD Division of Kidney, Urologic and Hematologic Diseases [of National Institute of Diabetes and Digestive and Kidney Diseases, US]

DKV Deutscher Kältetechnischer Verein [German Refrigeration Association]

DKW Deutsches Kommission für Wirtschaft [German Economic Commission]

$/kW Dollar(s) per Kilowatt [Unit]

DL aresbury Laboratory [of Engineering and Physical Sciences Research Council, UK]

DL Data Language

DL Data Link [also D/L]

DL Data List

DL Data Log(ger); Data Logging

DL Dead Load

DL Delay Line

DL Delete [EBCDIC Code]

DL Delta Air Lines [US]

DL Depletion Layer [Solid-State Physics]

DL Deputy Lieutenant

DL Design Load

DL Destroyer Leader (or Frigate) [US Navy Symbol]

DL Detection Limit

DL Deuterium Labeling

DL Developed Length

DL Development-Left

DL Diagnostic Laboratory

DL Diesel Locomotive

DL Difference Lumen

DL Differential Locking [of ABS Brake System]

DL Diffusion Layer

DL Dimension Line

DL Diode Laser

DL Diode Logic

DL Dipole Loop [Solid-State Physics]

DL Discharge Lamp

DL Discrete Layer

DL Disk Loading [Aerospace]

DL Document List

DL Donor Level [Solid-State Physics]

DL Dotted Line

DL Doppler Lidar

DL *(Dosis letalis)* – Lethal Dose [Medicine]

DL Double Layer

DL Double Loop

DL Downlink [also D/L]

DL Download [also D/L]

DL Drawing List

DL Drude-Lorentz (Theory) [Physics]

DL Drude-Linhard (Dielectric Function)

DL Dry Location

DL Dual Language

DL Dwight-Lloyd (Roasting Process) [Mining]

DL Dye Laser

DL Dynamic Load(er); Dynamic Loading

DL- Prefix Denoting a Compound Containing D and L Stereoisomers in Equal Parts [also DL-]

DL/1 Data Language One [also DL/I]

DL/1 Data Manipulation Language 1 [of IBM Corporation, US]

DL50 *(Dosis Letalis 50%)* – Lethal Dose 50% [Medicine]

D/L Data Link [also DL]

D/L Demand Loan

D/L Deorbit/Landing

D/L Downlist

D/L Downlink [also DL]

D/L Download [also DL]

dL deciliter [also dl]

dl change in length [Symbol]

dl deciliter [also dL]

dl prefix indicating an optically inactive crystal, or racemic mixture [Chemistry]

dλ change in wavelength [Symbol]

ΔL Change in Azimuthal (or Angular Momentum) Quantum Number [Symbol]

ΔL Change in Length [Symbol]

ΔL$_c$ Total Elongation of Fiber-Reinforced Composite in Direction of Stress Application (Under Isostress Condition) [Symbol]

ΔL$_f$ Elongation of Fiber Components of Fiber-Reinforced Composite in Direction of Stress Application (Under Isostress Condition) [Symbol]

ΔL$_m$ Elongation of Matrix of Fiber-Reinforced Composite in Direction of Stress Application (under Isostress Condition) [Symbol]

Δl change in azimuthal (or angular momentum) quantum number of electron [Symbol]

Δl displacement change [Symbol]

Δl length change [Symbol]

$\Delta\lambda$ change in Compton wavelength (in quantum mechanics) [Symbol]

$\Delta\lambda$ limiting difference of wavelength (of prisms) [Symbol]

$\Delta\lambda$ spectral bandwidth (in wavelength units) [Symbol]

$\Delta\lambda$ (general) wavelength change [Symbol]

$\Delta\bar{\lambda}$ spectral bandwidth (in wavenumber units) [Symbol]

DLA Data Link Adapter

DLA Data Link Address

DLA Defense Logistics Agency [US]

DLA Diffusion-Limited Aggregate; Diffusion-Limited Aggregation [Metallurgy]

DLA Dilaurylamine

DLA Direct Lift Control

DLA Direct-Line Attachment

DLA Distributed-Lumped Active

DLA Division of Library Automation [of University of California, US]

DLA Duplex Line Control

DLASC Diamond-Like Atomic-Scale Composite

DLAT Destructive Lot Acceptance Testing

D Lat Difference of Latitude [also d lat]

D Layer Ionosphere Layer between about 31 and 62 Miles (50 and 100 km) above the Surface of the Earth which Reflects Frequencies below 50 Kilohertz [also D layer]

D Layer Lower Portion of the Earth's Mantle between a Depth of about 620 and 1,800 Miles (1,000 and 2,900 km) [also D layer]

DLB Data Lock Box

$/lb Dollars per pound [Unit]

DLC Data Link Control

DLC Decision Level Concentration

DLC Diamond-Like Carbon [also dlc]

DLC Digital Logic Circuit

DLC Digital Loop Carrier [Telecommunications]

DLC Direct Lift Control

DLC Distributed Loop Carrier

DLC Double Layer Capacitor

DLC Duplex Line Control

DLC Dynamic Load Characteristics

DLCA Diffusion-Limited Cluster Aggregation [Metallurgy]

DLCC Desert Locust Control Committee [of Food and Agricultural Organization]

DLCC Digital Load Cell Comparison

DLCF Data Link Control Field

DLCI Data Link Connection Identifier

DLCO Decade (Inductance-Capacitance) Oscillator

DLCO Desert Locust Control Organization [of United Nations]

DLCO-EA Desert Locust Control Organization for Eastern Africa [of United Nations in Ethiopia]

DLCN Distributed Look Computer Network

DLCS Data Line Concentration System

DLC-Si Diamond-Like Carbon with Silicon

DLD Dark-Line Defect (of Lasers)

DLD Delay-Lock Discriminator

DLD Display List Driver

dld delivered [also dl'd]

DLDD Deep-Level Delta Doping

dλ/dH magnetostrictive susceptibility (i.e., change of magnetostriction with (applied) magnetic field)
 linear dispersion (i.e., linear separation of two lines in a spectrum per unit difference of wavelength)

dL/dT change in latent heat with (absolute) temperature

dL/dt loading rate (i.e., time rate of change of loading)

dL/dt time rate of change of angular momentum

dl/dT change in dilatation with temperature

dλ/dx change in wavelength over the distance travelled

DLE Data Link Escape

DLE Direct Line Equipment

DLE Disseminated Lupus Erythematosus [Medicine]

DLEA Double Leg Elbow Amplifier

DLEC Data Link Escape Character

DLEED Dynamical Low-Energy Electron Diffraction

DLE ETB Data Link Escape End of (Text) Block

DLE ETX Data Link Escape End of Text

DLES Department of Labour and Employment Services [Canada]

DLE STX Data Link Escape Start of Text

DLF Development Loan Fund [US]

DLF Directed Light Fabrication

DLFW Department of Lands, Forests and Waters [Canada]

DLG Deutsche Landwirtschaftsgesellschaft [German Agricultural Society]

DLG Digital Line Graph

Dlg Dialog [also Dlg]

dL/g deciliter(s) per gram [also dL g^{-1}, or dl/g]

DLG-E Digital Line Graph–Enhanced

DLG-FOODTEC International Exhibition for Dairy Technology and Food Processing (of the Deutsche Landwirtschaftsgesellschaft) [Frankfurt/ Main, Germany]

DLGS Doppler Landing Guidance System

DLH Deutsche Lufthansa [German Lufthansa Corporation]

DLI Data Link Interface

DLI Defense Language Institute [US]

DLI Delay Indefinite

DLI Direct Liquid Inlet

DLI Dual Link Interface

DLIEC Defense Language Institute, East Coast [US]

D Line D Line [Fraunhofer Line for Sodium having a Wavelength of 589.3 Nanometers] [also D line] [Spectroscopy]

D$_1$ Line D$_1$ Line [Fraunhofer Line for Sodium having a Wavelength of 589.3 Nanometers] [also D$_1$ line] [Spectroscopy]

D$_2$ Line D$_2$ Line [Fraunhofer Line for Sodium having a Wavelength of 588.9 Nanometers] [also D$_2$ line] [Spectroscopy]

D$_3$ Line D$_3$ Line [Fraunhofer Line for Sodium having a Wavelength of 587.6 Nanometers] [also D$_3$ line] [Spectroscopy]

DLIR Depot Level Inspection and Repair

DLIR Downward Looking Infrared

DLIS Desert Locust Information Service [of Food and Agricultural Organization]

DLIS Doctor of Library and Information Science

DLit *(Doctor Litterarum)* – Doctor of Letters; Doctor of Literature [also Dlitt]

DLJ Down-Link Jamming [Electronic Warfare]

DLK Data Link

DLL Data Link Layer

DLL Days of Life Lived

DLL Delay-Locked-Loop

DLL Design Limit Load

DLL Dial Long Lines

DLL Dynamic Lattice Liquid (Algorithm) [Physics]

DLL Dynamic Link Library

.DLL Dynamic Link Library (File) [File Name Extension] [also .dll]

ΔL/L Relative Change in Length (or Length Change per Unit Length)

ΔL/L Relative (or Overall) Thermal Expansion

ΔL/L Magnetostriction

–ΔL/L Relative (or Overall) Thermal Contraction (or Shrinkage)

ΔL/L$_0$ Linear Change (or Fractional Change in Length)

Δl/l relative change in length (or change in length per unit length)

DLM Daily List of Mail

DLM Data Line Monitor

DLM Data Link Management

DLM Data-Link Mapping

DLM Digital Landscape Model [Geographic Information System]

DLM Digital Light Meter

DLM Distributed LAN (Local Area Network) Monitor

DLM Distributed Lock Manager

DLM Dominant Lethal Mutation [Genetics]

DLM Double Layer Metal

DLM Dwight-Lloyd-McWane (Process) [Metallurgy]

DLM Dynamic Link Module

DLMCP Distributed Look Message Communication Protocol

DLMF Drug Literature Microfilm File [of Iowa Drug Information Service , US]

DLN Diamond-Like Nanocomposite

DLN Digital Ladder Network

DLNS Deep-Level Noise Spectroscopy

DLO Dead-Letter Office

DLO Dispatch Loading Only [also dlo]

DLO Double Local Oscillator

DLOC Developed Lines of Code

D Long Difference of Longitude [also d long]

DLOS Deep-Level Optical Spectroscopy

DLOS Distributed Loop Operating System

DLOS Diver Lock-Out Submersible [also DL-OS]

DLP Data Link Processor

DLP Data Listing Program

DLP Delta Lattice Parameter (Model) [Physics]

DLP Deoxidized, Low-Phosphorus Copper

DLP Digital Light Processing

DLP Display List Processing

DLP Dynamic Low-Pass

DLPA Decorative Laminate Products Association [US]

DLPA DL-Phenylalanine [Biochemistry]

DLPE Department of Lands, Planning and Environment [Northern Territory, Australia]

DLPG DIMDI List Program Generator [of Deutsches Institut für Medizinische Dokumentation und Information , Germany]

DLPU Data Link Processor Unit

DLR Decision Level Count Rate

DLR Deutsche Forschungsanstalt für Luft– und Raumfahrt [German Aerospace Research Establishment, Cologne]

DLR Deutsches Zentrum für Luft– und Raumfahrt e.V. [German Aerospace Research Center, Cologne]

DLR DOS (Disk-Operating System) LAN (Local-Area Network) Requester

Dlr Dealer [also dlr]

DLRG Design Layout Report Date

DLS Dark-Line Spectrum

DLS Data Link Set

DLS Data Link Splitter

DLS Data Link Switching

DLS Department of Life Sciences

DLS Diffused Light Storage

DLS Digital Line System

DLS Di-Lepton Spectrometer

DLS Direct Least Squares

DLS Distance Least-Squares (Analysis Method)

DLS Division of Laser Science [of American Physical Society, US]

DLS Doctor of Library Science

DLS Dominion Land Surveyor [now Canada Land Surveyor]

DLS Dynamic Light Scattering

DLS Dynamic Limb Sounder

DLS Dynamic Load Simulator

DLSA Defense Legal Service Agency [US]

DLSC Defense Logistics Service Center [US]

DLSM Data Link Summary Message

DLSO Dial Line Service Observing

DLT Data Line Terminal

DLT Data Line Translator

DLT Data Link Terminal

DLT Data Link Translator

DLT Data-Loop Transceiver

DLT Decision Logic Table

DLT Decision Logic Translator

DLT Depletion-Layer Transistor

DLT Digital Linear Tape

DLT Digital Line Termination

DLTDP Dilaurythiodipropionate

DLTM Data Link Test Message

DLTMA Dynamic Load Thermomechanical Analysis

DLTR Data Link Terminal Repeater

DLTR Data Link Transmission Repeater

DLTS Deep-Level Transient Spectroscopy

DLTS Deep-Level Transient Spectrum

DLTU Digital Line and Trunk Unit

DLU Data Line Unit

DLU Digital Line Unit

DLU Dynamic Load/Unload [Disk Drives]

DLV Damped Linear Vibration

dlvd delivered [also dlv'd]

DLVL Diverted into Low Velocity Layers [Geophysics]

DLVO Deryaguin-Landau-Vervey-Overbeck (Theory) [Physical Chemistry]

Dlvr Delivery [also Dlvry, Dlvy, dlvry, dlvy, or dlvr]

Dly Delay [also dly]

DLZR Directional(ly) Levitation Zone Remelted; Directional(ly) Levitation Zone Remelting [Metallurgy]

DM Adamsite [10-Chloro-5,10-Dihydrophenarsazine]

DM (Conference on Sources and Detection of) Dark Matter (in the Universe)

DM Dasymeter

DM Data Management; Data Manager

DM Data Manipulation

DM Data Mark

DM Data Memory

DM Data Model(ling)

DM Decametric

DM Decimal Multiply

DM Decimetric

DM Deformation Microstructure [Materials Science]

DM Degree of Mixing

DM Delay Modulation

DM Delta Modulation

DM Demand Meter

DM Demagnification Field

DM Dense Medium

DM Dental Material

DM Density Matrix [Quantum Mechanics]

DM Department of Materials

DM Department of Management

DM Department of Manuscripts [of British Library]

DM Department of Mathematics

DM Department of Metallurgy

DM Department of Mining

DM Depleted Material [Nuclear Engineering]

DM Depressor Mandibularis [Amphibian Anatomy]

DM Design Manual

DM Design Memorandum

DM Des Moines [Iowa, US]

DM Destroyer Minelayer [US Navy Symbol]

DM Detonation Meter

DM Deutsche Mark; Deutschmark [Currency of Germany]

DM Deutsches Museum [German (National) Museum, Munich]

DM Developmental Instrumentation MDM (Manipulator Deployment Mechanism)–Mid [NASA Space Shuttle Program]

DM Development and Maintenance

DM Development Motor

DM Diabetes Mellitus

DM Diagonal Matrix

DM Dichroic Mirror

DM Differential Mode

DM Diffraction Microscope; Diffraction Microscopy

DM Digital Mechanics

DM Digital Modulation

DM Digital Modelling

DM Digital Module

DM Digital Monolithic

DM Diphenylamine Chloroarsine

DM Dipole Moment

DM Direct Memory (Access)

DM Directory Maintenance

DM Disassembly Manual

DM Disconnected Mode

DM Discrete Mathematics

DM Discussion Memorandum

DM Disk Memory

DM Dislocation Mechanics [Materials Science]

DM Dispersive Medium

DM Distance Measuring

DM Distributed Memory

DM Distributing Main

DM District Municipality

DM Diurnal Motion [Astronomy]

DM Document(ation)

DM Docking Mechanism

DM Dominica [ISO Code]

DM Dot Matrix

DM Double Monochromator

DM Drafting Machine

DM Drift Mining

DM Drift Mobility [Solid-State Physics]

DM Drive Motor

DM Drum Memory

DM Dry Matter

DM Dumas Method [Chemistry]

DM Dura Mater (of Brain) [Anatomy]

DM Dynamic-Ion-Beam Mixing (Process)

DM Dynamic Memory

DM Dynamic Meteorology

DM Dynamic Modulus

DM Dynamic Moment

DM Dzyaloshinsky-Moriya (Anisotropy) [Solid-State Physics]

DM- East Germany [Civil Aircraft Marking] [obsolete]

DM14NQ Dimethyl-1,4-Naphthoquinone

D&M Dressed and Matched [Lumber]

D/M Demodulation/Modulation

D/M Deuterium-to-Metal (Ratio)

dM magnetization change (or variation) [Symbol]

dM (magnetic) moment change [Symbol]

dm change in mass [Symbol]

dm decimeter [Unit]

.dm Dominica [Country Code/Domain Name]

dm^2 square decimeter [also sq dm]

dm^3 cubic decimeter [also cdm, or cu dm]

ΔM Change in Magnetic Quantum Number [Symbol]

ΔM Full-Width at Half-Maximum Peak Height of Ion Peak of Mass "M" [Symbol]

ΔM (Magnetic) Hysteresis Loop Width [Symbol]

ΔM Magnetization Change (or Difference) [Symbol]

Δm change in magnetic quantum number of electron [Symbol]

Δm mass change [Symbol]

Δm mass defect (in nuclear physics) [Symbol]

$\Delta\mu$ chemical potential difference [Symbol]

$\Delta\mu$ permeability change [Symbol]

$\Delta\mu$ thermodynamic supersaturation [Symbol]

DMA Defense Manufacturers Association [UK]

DMA Defense Mapping Agency [US]

DMA Dental Manufacturers of America [US]

DMA Department of Municipal Affairs [Canada]

DMA Differential Mobility (Particle) Analysis; Differential Mobility (Particle) Analyzer

DMA 3,4-Dihydroxymandelic Acid

DMA Dimethyl Acetal

DMA N,N-Dimethylacetamide

DMA 1,3-Dimethyl Adamantane

DMA Dimethyl Adipimidate [Biochemistry]

DMA Dimethylaluminum [also DMAl]

DMA 5-(N,N-Dimethyl)-Amiloride

DMA Dimethylamine [also Dma]

DMA N,N-Dimethylaniline; Dimethylaniline

DMA 9,10-Dimethyl Anthracene

DMA Dimethylarginine [Biochemistry]

DMA Direct Memory Access

DMA Direct Memory Address(ing)

DMA Direct Microassembly

DMA Disodium (Mono-)Methylarsonate

DMA Distributed Multipoles

DMA Document Management Alliance

DMA Double Mechanical Alloying (Technique) [also dMA] [Metallurgy]

DMA Double Metal Alkoxide

DMA Drive Motor Assembly

DMA Dynamic Mechanical Analysis

DMA Dynamic Memory Array

2,6-DMA 2,6-Dimethylaniline [also 2,6-Dma]

dMA Double Mechanical Alloying (Technique) [also DMA] [Metallurgy]

Δm/A mass change per unit area

DMAA N,N-Dimethylacetoacetamide

DMAA Dimethylallylamine

DMAAC Defense Mapping Agency Aerospace Center [US]

DMAAP 4-Dimethylaminoantipyrine [also dmaap]

DMAB Dimethylamineborane

DMAB 4-Dimethylaminobenzenecarboxaldehyde

DMAB Dimethylaminobenzoic Acid

DMABA p-Dimethylamino-Benzaldehyde

DMABN 4-(Dimethylamino Benzonitrile

DMAC Data Management and Computation

DMAC Defense Metals Information Center

DMAC Design and Manufacturing Automation Corporation

DMAC 7-(N,N-Diethylamino)-4-Methylcoumarin

DMAC N,N-Dimethyl Acetamide [also DMAc]

DMAC Direct-Memory Access Channel

DMAC Direct Memory Access Controller

DMAc N,N-Dimethyl Acetamide [also DMAC]

DMAc Dimethyl Acetic Acid

DMACP Direct Memory Access Communications Processor

DMACS Distributed Manufacturing Automation and Control Software

DMAD Dimethyl Acetylenedicarboxylate

DMA-DEA N,N-Dimethylacetamide Diethylacetal

DMAE Département Modèles pour l'Aérodynamique et l'Energétique [Department of Aerodynamic and Energetic Models, ONERA-CERT, Toulouse, France]

DMAE Department of Mechanical and Aerospace Engineering

DMAE Dimethylaminoethanol ; 2-(Dimethylamino)ethanol

DMAE Dimethylaminoethoxide

DMAEM Dimethyl Amino Ethyl Methacrylate

DMAEMA Dimethylaminoethyl Methacrylate; 2-(Dimethylamino)ethyl Methacrylate

DMAH Dimethylaluminum Hydride

DMAHC Defense Mapping Agency Hydrographic Center [US]

DMAI Direct Memory Access Interface

DMAL Dimethylacetal

DMAl Dimethylaluminum [also DMA]

DManSc Doctor of Management Sciences

DMAP 4-Dimethylaminophenol

DMAP 3-Dimethylaminopropylamine

DMAP Dimethylaminopyridine; 4-Dimethylaminopyridine

DMAPMA Dimethylaminopropyl Methacrylamide; 3-N,N-Dimethylaminopropyl Methacrylamide

DMAPN Dimethylaminopropionitrile; 3-Dimethylaminopropanenitrile; 3-(Dimethylamino)propionitrile

DMAPS Digital Marine Acquisition and Processing System

D-Mark Deutsche Mark; Deutschmark [also D-mark] [Currency of Germany]

DMAS Distribution Management Accounting System

DMAU Dnebropetrovsk Metallurgical Academy of Ukraine [formerly Dnebropetrovsk Metallurgical Institute]

DMB Data Management Block

DMB 4,4'-Dichloromethylbenzhydrol

DMB 2,3-Dihydroxy-N,N-Dimethylbenzamide

DMB Dimethoxybenzene

DMB 2,3-Dimethylbutane

DMB (3,3-Dimethylbutyl)dimethylsilyloxy-

DMB Disconnect and Make Busy

1,2-DMB 1,2-Dimethoxybenzene

1,3-DMB 1,3-Dimethoxybenzene

DMBA Dimethylbenzanthracene; 9,10-Dimethyl-1,2-Benzanthracene

DMBA Dimethylbenzylamine

DMBA α,β-Dimethylbutyric Acid

DMBAS 2,5-Dimethoxy-4'-Aminostilbene

DMBE Double Many-Body Expansion (Method) [Physics]

DMBM 2,5-Dihydroxy-4-Methylbenzyl Mercaptan

DMBPPD N-(1,3-Dimethylbutyl)-N'-Phenyl-p-Phenylene Diamine

25DMBQ 2,5-Dimethyl-1,4-Benzoquinone

DMBU 2,3-Dimethyl 1,3-Butadiene

DMC Data Management Computer

DMC Data Management Controller

DMC Data Management Coordinator

DMC Department of Medical Cybernetics

DMC Dichlorodiphenylmethyl Carbinol

DMC 4,4'-Dichloromethylbenzhydrol

DMC Diffusion Monte Carlo (Method) [Materials Science]

DMC Digital Microcircuit

DMC Dimethylaminoethyl Chloride Hydrochloride

DMC Direct Manufacturing Cost

DMC Direct Maintenance Cost

DMC Direct Memory Channel

DMC Direct Multiplexed Control

DMC Divisão de Mecânica Espacial e Controle [Space Mechanics and Control Division; of Brazilian National Institute of Space Research]

DMC Dough Molding Compound [Polymer Processing]

DMC Downstate Medical Center [of State University of New York, Brooklyn, US]

DMC Duff Moisture Code

DMC Dynamic Magnetic Compaction

DMCD Dimethylcyclohexanedicarboxilate

DMCd Dimethylcadmium

DMCH 2-Chloro-1-Methylcyclohexane

DMCH 1,3-Dimethylcyclohexane

DMCH 5,5-Dimethyl-1,3-Cyclohexanedione

DMCHA Dimethylcyclohexyladipate

DMchf N,N-Dimethylchloroformiminium Chloride

DMCL Device Media Control Language

DMCM Division of Metallurgy and Chemical Metallurgy

DMCM Dynamic Monte Carlo Method

dMCMP 2'-Deoxy-5-Methylcytidine Monophosphate [also d-MCMP] [Biochemistry]

DMCP Department of Mathematics and Computer Processing

DMCP Dimethylcyclopentane

DMCP 1,2-Dimethylcyclopropane

DMCP DOE (Department of Energy) Methods Compendium Program [US]

DMCS Dimethylchlorosilane; Dimethyldichlorosilane

DMCS Distributed Manufacturing Control System

DMCTMS Dimethyl Carbamic Acid Trimethylsilyl Ester

DMD (*Dentariae Medicinae Doctor*) – Doctor of Dental Medicine

DMD Digital Micromirror Device

DMD Dimethyloxozolidinedione

DMD Divisão de Modelagem e Desenvolvimento [Modelling and Development Division; of Brazilian National Institute for Space Research]

DMD Doctor of Dental Medicine

DMD Drug Metabolism and Disposition [International Journal]

DMDB18C6 Dimethyldibenzo-18-Crown-6

DMDB24C8 Dimethyldibenzo-24-Crown-8

DMDB30C10 Dimethyldibenzo-30-Crown-10

DMDCS Dimethyldichlorosilane

dμ/dε change in permeability with strain [also dμ/dϵ]

DMDF 2,5-Dimethoxy-2,5-Dihydrofuran

dM/dH magnetic susceptibility (i.e., change of magnetization with magnetic field strength)

dM$_r$/dH irreversible subsceptibility (i.e., change of magnetic remanence magnetic field strength)

DMDHEU Dimethyloldihydroxyethyleneurea

DMDNT Dimethyldinitrosoterephthalate

dM/dp change in magnetization with pressure

DMDPU Dimethyldiphenylurea

DMDS Dimethyl Disulfide

DMDT Dimethoxydiphenyltrichloroethane

DMDT 1,1,1-Trichloro-2,2-bis-[p-Methoxyphenyl]ethane (or Methoxychlor)

dM/dT change of magnetization with temperature

dM/dt time rate of change of magnetization

dM/dt time rate of change of (total) mass

Δm/Δt change in mass per unit time

dμ/dt time rate of change of magnetic moment

dμ/dx chemical potential gradient (i.e., rate of change of chemical potential with distance)

DME Data Measuring Equipment

DME Department of Materials Engineering

DME Department of Mechanical Engineering

DME Department of Metallurgical Engineering

DME Department of Mines and Energy [Northern Territory, Australia]

DME Department of Mining Engineering

DME Developing Market Economy

DME Digital Measuring Equipment

DME Digital Multiplex Equipment

DME 1,2-Dimethoxyethane

DME Dimethyl Ether

DME Direct Machine Environment

DME Direct Measurements Explorer (Satellite)

DME Direct Memory Execution

DME Directorate of Materials and Explosives [UK]

DME Directorate of Mechanical Engineering [UK]

DME Distance-Measuring Equipment

DME Distributed Management Environment

DME Division of Metallurgical Engineering

DME Division of Mechanical Engineering

DME Dropping Mercury Electrode [Physical Chemistry]

DME Dulbecco's Modified Eagle (Medium) [Biochemistry]

DMEA Dimethylethylamine

DMEAA Dimethylethylamine Alane

DME/COTAR Distance-Measuring Equipment/Correlation Tracking and Ranging

DMED Digital Message Entrance (or Entry) Device

DMEDA Dimethylethylenediamine

DMEDBA N,N'-Dimethylethylenediamine-N,N'-Di(α-Butyric Acid) [also dmedba]

DMEDDA N,N'-Dimethylethylenediamine-N,N'-Diacetate; N,N'-Dimethylethylenediamine-N,N'-Diacetic Acid [also dmedda]

DMEDDS N,N'-Dimethylethylenediamine-N,N'-Disuccinate [also dmedds]

DME/DP Dropping Mercury Electrode/Differential Pulse [Physical Chemistry]

DMedSc Doctor of Medical Science

DMEE Department of Mechanical and Environmental Engineering

DMEE Di(2-Methoxyethyl)ether [also dmee]

DMeFc Decamethylferrocenium

DMeFc TCNE Decamethylferrocenium Tetracyanoethanide [also DMeFc(TCNE)]

DMeFc TCNQ Decamethylferrocenium Tetracyanoquinodimethanide [also DMeFc(TCNQ)]

DMEK Dimethylethylketone

DMEM Department of Metallurgy and Engineering Materials

DMEM Dulbecco's Modified Eagle Medium [Biochemistry]

DMEMS Department of Mechanical Engineering and Materials Science

DMEMS Department of Metallurgical Engineering and Materials Science

DMEN N,N'-Dimethylethylenediamine

DMEP Data Network Modified Emulator Program

DMEP Dimethoxyethylphthalate

DME/RD Directorate of Materials and Explosives, Research and Development [UK]

DMERT Duplex Multiple-Environment Real-Time (Operating System)

DMES Dimethylethoxysilane

DMET Dimethyl(ethylenedithio)diselenadithiafulvalene

DMET Distance-Measuring Equipment TACAN (Tactical Air Navigation)

DMet Doctor of Metallurgy

DMEU N,N'-Dimethyl-N,N'-Ethyleneurea

DMEU 1,3-Dimethyl-2-Imidazolidinone

DMF Data Management Facility

DMF Decayed, Missing and Filled [Dentistry]

DMF Dihydrochloride Methyl Farnesote

DMF Dimethylformamide

DMF Discontinuous Metal Film

DMF Dissolution Microvoid Formation

DMF Distribution Media Format

DMFA Differential Matched Filter

DMFA Digital Matched Filter

DMFA N,N-Dimethylformamide

DMFA Dimethylfuran

DMFA Disk Management Facility

DMFC Direct Methanol Fuel Cell

DMF-DMA Dimethylformamide Dimethyl Acetal

DMFL Dingot Magnesium Fluoride Liner [Metallurgy]

DMG Data Management Group

DMG Deutsche Mineralogische Gesellschaft [German Mineralogical Society]

DMG Dimethyl Glycine (or Pangamic Acid)

DMG Dimethylglyoxide

DMG Dimethylglyoxime

DMG Division of Mines and Geology [California, US]

DMGeCl$_2$ Dimethylgermanium Dichloride

DMH Device Message Handler

DMH 2,2-Dimethylheptane

DMH Dimethylhydantoin

DMH Direct Manhours

DMH Drop Manhole

ΔM(H) Field-Dependent Change in Magnetization [also δM(H)]

DmHas 2,4-Dimethyl 8-Quinazolinol

DMHF Dimethylhydantoin Formaldehyde

DMHg Dimethylmercury

DMHPPD N,N'-Di(3-Methylheptyl)-p-Phenylene Diamine

DMHy Dimethylhydrazine

DMI Danish Meteorological Institute

DMI Department of Manpower and Immigration

DMI Department of Microbiology and Immunology

DMI Design Management Institute [US]

DMI Desktop Mangement Interface

DMI Desmethylimipramine

DMI Digital Multiplexed Interface

DMI 1,3-Dimethyl-2-Imidazolidinone

DMI Direct Memory Interface

DMI Director of Military Intelligence [UK]

DMI Dnepropetrovsk Metallurgical Institute [Ukraine] [now DMAU]

DMI Dynamic Memory Interface

DMIA Dual Multiplexer Interface Adapter

DMIC Defense Metals Information Center [at Battelle Memorial Institute, Columbus, Ohio, US]

MID 1,3-Dithiole-2-One-4,5-Dithiolate [also dmid]

DMIE Department of Mechanical and Industrial Engineering

DMIPSCl Dimethylisopropylsilylchloride

DMIRR Demand Module Integral Rocket Ramjet

DMIRS Double-Modulation Infrared Spectrophotometer

DMIS Data Management Information System

DMIS Dimensional Measuring Interface Specification

DMIS Directory Management Information System

DMIT 4,5-Dimercapto-1,3-Dithiole-2-Thionate [also dmit]

DMJTC Differential Multijunction Thermal Converter

DM/kg Deutsche Mark per kilogram [Unit]

DMK Deutsche Mark [Currency of Germany]

DMK Dimethylketone

DML Data Management Language

DML Data Manipulation Language

DML Developmental-Instrumentation MDM (Manipulator Deployment Mechanism)–Left [NASA Space Shuttle Program]

DML Diagnostic Microbiology Laboratory

DML Dimyristoyl Lecithin [Biochemistry]

Dml Demolition [also dml]

DMLA Department of Medieval and Latin Antiquities [of British Museum]

DMLS Doppler Microwave Landing System

DMM Data Manipulation Mode

DMM Defense Market Measures (Database) [of US Department of Defense]

DMM Deformation Mechanism Map [Metallurgy]

DMM Department of Materials and Metallurgy

DMM Department of Metallic Materials

DMM Department of Metallurgy and Materials

DMM Department of Metallurgy and Mining; Department of Mining and Metallurgy

DMM Digital Multimeter

DMM Dimethylmercury

DMM Direct Metal Mastering

DMM Discrete Micromechanical Model

DMM Division of Materials and Metrology

DMM Dynamical Material Modeling [Metallurgy]

Δμ/μ relative permeability (i.e., change in permeability per unit permeability)

DMMB Dissimilar Mixed Mode Bending

DMMC Discontinuously-Reinforced Metal-Matrix Composite

DMME Department of Materials and Metallurgical Engineering

DMME Department of Materials and Mining Engineering

DMME Department of Metallurgical and Mineral Engineering

DMME Department of Metallurgy and Materials Engineering

DMME Department of Metallurgy and Metallurgical Engineering

DMME Department of Metallurgy and Mineral Engineering

DMME Department of Metallurgy and Mining Engineering

DMME Department of Mining and Metallurgical Engineering

(Δμ/μ)/ε change in permeability per unit strain

DMMF Dry, Mineral-Matter Free [also dmmf]

DMMal Dimethylmalonate

DMMP Dimethyl Methylphosphonate

DMMS Department of Metallurgy and Materials Science

DMMS Dynamic Memory Management System

D.MMSS Degrees–Minutes–Seconds [Angle Notation]

DMMT Department of Metallurgy and Materials Technology

DMN De Machinefabriek Noordwykerhout BV [Dutch Machinery Manufacturer]

DMN Dimethylnaphthalene

DMN Dimethylnitrosamine

DMN Direction de la Météorologie Nationale [National Meteorology Directorate, France]

Dmn Dimension [also dmn]

δ-**Mn** Delta Manganese [Symbol]

DMNA N,N-Dimethyl 1-Naphthalenamine

DMNA Dimethylnitrosamine

2,7-DMNAPY 2,7-Dimethyl-1,8-Naphthayridine [also 2,7-DMNapy]

DMNE Department of Materials and Nuclear Engineering

DMNFHSCl Dimethyl-3,3,4,4,5,5,6,6-Nonafluorohexyl-chlorosilane

DMNP Dimethyl (1-Naphthalenyl)phosphine

DM14NQ Dimethyl-1,4-Naphthoquinone

DMNSC Digital Main Network Switching System

DMO Data Management Office

DMO Delocalized Molecular Orbital [Physical Chemistry]

DMO Diode Microwave Oscillator

DMO Director of Military Operations

DMOA N,N-Dimethyl Octanamine

DMOD Delta Modulation

DMON Discrete Monitoring

δ-**MoN** Delta Molybdenum Nitride

DMOS Data Management Operating System

DMOS Diffused Metal-Oxide Semiconductor

DMOS Diffusive Mixing of Organic Solutions

DMOS Discrete Metal-Oxide Semiconductor

DMOS Double-Diffused Metal-Oxide Semiconductor [also D-MOS] ,

DMOS DSA (Diffusion Self-Aligned) Metal-Oxide Semiconductor [also D/MOS]

DMOSFET DMOS (DSA Metal-Oxide Semiconductor) Field-Effect Transistor

DMP Department of Mathematics and Physics

DMP Department of Metal Physics

DMP Deployable Maintenance Platform

DMP DEU (Display Electronics Unit) Message Processor

DMP Dimethoxypropane; 2,2-Dimethoxypropane

DMP Dimethylaminomethylphenol; 2,4,6-tris[(Dimethylamino)methyl]phenol

DMP 2-(Dimethylaminomethyl)phenyl [also dmp]

DMP Dimethylphenol; 2,6-Dimethylphenol

DMP Dimethyl Phthalate

DMP Dimethyl Pimelimidate

DMP Direct Memory Processor

DMP Division of Management Policy [of Office of Management, National Institutes of Health, Bethesda, Maryland, US]

DMP Division of Materials Physics [of American Physical Society, US]

DMP Dot Matrix Printer

DMP Dump [also Dmp, or dmp] [Computers]

DMP-10 3-Dimethylaminomethylphenol

DMP-30 2,4,6-Tri(dimethylaminomethyl)phenol

DMPA 2,2-Bis(hydroxymethyl)propionic Acid

DMPA Defense Material Procurement Administration [US]

DMPA Dichlorophenylmethyl Isopropylphosphoramidothioate; O-2,4-Dichlorophenyl-O-Methylthiophosphoric Acid Isopropylamide

DMPA Digitally-Modulated Power Amplifier

DMPA Dimethoxyphenylethylamine

DMPA Dimethylolpropionic Acid

DMPA N,N-Dimethyl-1,3-Propanediamine

DMPA Direct Mail Producers Association [UK]

DMP3A 1,2-Diamino-2-Methylpropanetriacetate [also dmp3a]

DMPC 3-Dimethylaminopropyl Chloride Hydrochloride

DMPC Dimethyl Pyrocarbonate

DMPC Dimyristoyl Phosphatidyl Choline [Biochemistry]

DMPC Distributed Memory Parallel Computer

DMPD 2,2-Dimethyl 1,3-Propanediol

DMPD 4,5-Dimethyl-1,2-Phenylenediamine

DMPD2HCl N,N-Dimethyl-1,4-Phenylenediamine Dihydrochloride

DMPE 1,2-Bis(dimethylphosphino)ethane [also dmpe]

DMPE Dimethoxphenylethylamine [also DMPEA]

DMPhen 2,9-Dimethyl-1,10-Phenanthroline [also dmphen]

DMPL Depolarization Micropulse Lidar

DMPM Bis(dimethylphosphino)methane [also dmpm]

DMPN 3-Dimethylamino Propanenitrile

DMPN Dimethylamino Propionitrile

DMPO Dimethylphosphate

DMPO Dimethylphosphine Oxide

DMPO 2,2-Dimethyl-2H-Pyrrole N-Oxide; 5,5-Dimethyl-1-Pyrroline-N-Oxide

DMPP 1,1-Dimethyl-4-Phenylpiperazinium Iodide

DMPP Distributed Memory Parallel Processor

DMPPL 3,4-Dimethyl-1-Phenylphosphole

2MPPO Dimethyl-P-Phenylene Oxide

Dmpr Damper [also dmpr]

DMPRT Dual-Mode Personal Rapid Transport

DMPS Differential Mobility Particle Sizer

DMPS 2,3-Dimercapto-1-Propanesulfonic Acid

DMPS Dimethylpolysiloxane

DMPSCl Dimethyl Phenyl Silylchloride

DMPU N,N'-Dimethyl-N,N'-Propyleneurea

DMQ Dimethylaminoquinoline

DMQ Direct Memory Queue

DMQS Display Mode Query and Set

DMR Data Management Routine

DMR Demultiplexing/Mixing/Remultiplexing

DMR Developmental-Instrumentation MDM (Manipulator Deployment Mechanism)–Right [NASA Space Shuttle Program]

DMR Differential Microwave Radiometer

DMR Digital Microwave Radio

DMR Discharge Monitoring Report

DMR Distributed Message Router

DMR Division for Materials Research [of National Science Foundation, US]

DMR Dynamic Mechanical Rheometry

DMR Dynamic Modular Replacement

DMRE Diploma in Medical Radiology and Electrology

DMRG Density Matrix Renormalization Group [Physics]

DMRGE Design of Magnetic Resonance (Experiments) by Genetic Evolution

DMRL Defense Metallurgical Research Laboratory [Hyderabad, India]

DMR-MRL Division for Materials Research/Materials Research Laboratory [of National Science Foundation, US]

DMS Daily Mission Scientist [Aerospace]

DMS Dansk Metallurgisk Selskab [Danish Metallurgical Society]

DMS Database Management System

DMS Data Management Service

DMS Data Management Software

DMS Data Management System

DMS Data Multiplex Switching

DMS Defense Missile Systems

DMS Degrees/Minutes/Seconds

DMS *(Dehnungsmeßstreifen)* – German for "Strain Gauge"

DMS Dense-Media Separation; Dense-Media Separator

DMS Department of Macromolecular Science

DMS Department of Materials Science

DMS Department of Mathematics and Statistics

DMS Deputy Mission Scientist [Aerospace]

DMS Differential Maneuvering Simulator [Aerospace]

DMS Digital Management System

DMS Digital Memory Sizing

DMS Digital Multimeter System

DMS Digital Multiplexing Synchronizer

DMS Digital Multiplex Switching (System)

DMS Digital Multiscan Technology

DMS Diluted Magnetic Semiconductor

DMS meso-2,3-Dimercaptosuccinic Acid

DMS Dimethylsilyl

DMS Dimethyl Suberimidate

DMS Dimethylsulfate

DMS Dimethylsulfide

DMS Disk Monitor System

DMS Display Management System

DMS Distributed Maintenance Services

DMS Divisão de Meteorologia por Satélite [Satellite Meteorology Division; of Brazilian National Institute for Space Research]

DMS Division of Materials Science [of US Department of Energy]

DMS Docking Mechanism Subsystem [of NASA]

DMS Docking Mechanism System [of NASA]

DMS Docking Module Subsystem [of NASA]

DMS Doctor of Medical Science

DMS Documentation of Molecular Spectroscopy

DMS Document Management System

DMS Drawing Management System

DMS Duplex Microstructure [Metallurgy]

DMS Dynamic Mapping System

DMS Dynamic Mechanical Spectroscopy

DMS Dynamic Motion Simulator

DMSA meso-2,3-Dimercaptosuccinic Acid

DMSA α,β-Dimethylsuccinic Acid

DMSAT Director of Mobile Satellites [of Communications Canada]

DMSC N,N-Dimethylmonoselenocarbamate [also dmsc]

DMSc Doctor of Medical Science

DMSCC Direct Microscopic Somatic Cell Count [Biochemistry]

DMSD Digital Multi-Standard Decoding

DMSE Department of Materials Science and Engineering

DMSE Department of Mechanical Systems Engineering

DMSe Dimethylselenide

DMSM Department of Materials Science and Metallurgy

DMSM Department of Metallurgy and Science of Materials

DMSME Department of Materials Science and Mineral Engineering

DMSn Dimethyltin [also DMT]

DMSnCl$_2$ Dimethyltin Dichloride

DMSnO Dimethyltin Oxide

DMSnS Dimethyltin Sulfide

DMSO Dimethyl Sulfoxide

DMSO$_2$ Dimethyl Sulfate

d$_6$-DMSO$_2$ Dimethyl-d$_6$ Sulfate

DMSP Defense Meteorological Satellite Program [of US Department of Defense]

DMSP Dimethylsulfonate

DMSR Director of Missile Safety Research

DMSS Data Management System Simulator

DMSS Digital Multiplexing Subsystem

DMSS Digital Multiplex Switching System

DMSS Distributed Mass Storage System

DMST Department of Materials Science and Technology

DMST Division of Materials Science and Technology [of Commonwealth Scientific and Industrial Research Organization, Clayton, Australia]

DMST Division of Materials Science and Technology [of Council for Scientific and Industrial Research, Pretoria, South Africa]

DMT Data Movement Time

DMT Department of Materials Technology

DMT Department of Metals and Technology

DMT Decreed Moscow Time

DMT Derjaguin-Muller-Toporov (Theory) [Materials Science]

DMT Diffused-Mesa Transistor

DMT Digital Message Terminal

DMT Digital Multi-Tone

DMT 4,4'-Dimethoxytriphenylmethylchloride

DMT 4,4'-Dimethoxytrityl (Chloride)

DMT Dimethyl Terephthalate

DMT N,N-Dimethyltryptamine [Biochemistry]

DMT Dimethoxytrityl

DMT Dimethoxytrityl Chloride

DMT Dimethyl Terephthalate

DMT Dimethyltin [also DMSn]

DMT N,N-Dimethyltryptamine [Biochemistry]

DMT Direct Memory Transfer

DMT Discrete Multi-Tone

DMT Dispersive Mechanism Test

DMT Drop Modulation Technology

DMT Dynamic Mechanical Testing

DMTA Dynamic Mechanical Thermal Analysis;Dynamic Mechanical Thermal Analyzer

DMTC Digital Message Terminal Computer

DMTC Dimethoxytrityl Chloride

DMTC N,N-Dimethyl Dithiocarbamate [also dmtc]

DMT-Cl 4,4'-Dimethoxytrityl Chloride

DMTCNQ Dimethyl Tetracyanquinonedimethane

DMTCNQ Dimethylenetetracyanoquinoline

DMTD 2,5-Dimercapto-1,3,4-Thiadiazole

DMT-dI 5'-O-(4,4'-Dimethoxytrityl)-2'-Deoxyinosine [Biochemistry]

DMT-dU 5'-O-(4,4'-Dimethoxytrityl)-2'-Deoxyuridine [Biochemistry]

DMTF Desktop Management Task Force

DMTF N,N-Dimethylthioformamide

DMTI Digital Moving-Target Indicator

DMTI Double Moving-Target Indicator

DMTO Dimethyltin Oxide

DMTR Dounreay Materials Testing Reactor [Scotland]

DMTS Department of Mines and Technical Surveys [Canada]

DMTS Dimethyltin Sulfide

DMT-T 5'-O-(4,4'-Dimethoxytrityl)thymidine [Biochemistry]

$/MTU Dollar(s) per Metric Ton Unit [also $/mtu]

DMU Data Management Unit

DMU N'-(3,4-Dichlorophenyl)-N,N-Dimethyl Urea

DMU Dictionary Management Utility

DMU Diesel-Multiple Unit

DMU Digital Message Unit

DMU Dimethylolurea

DMU N,N-Dimethylurea

DMU Disk Memory Unit

DMU Distance Measurement Unit

DMU Distributed Microprocessor Unit

DMU Dual Maneuvering Unit

DMUS Data Management Utility System

DMUX Demultiplex(er); Demultiplexing [also Demux, or demux]

DMUXER Demultiplexer [also Demuxer, or demuxer]

DMV Department of Motor Vehicles [US]

DMV Deutsche Mathematikervereinigung [German Mathematical Society]

DMV Discrete Multivibrator

DMV Division of Motor Vehicles

DMVPA Dimethyl Vinylphosphonic Acid

DMW Decametric Wave

DMW Decimetric Wave [also dmW]

DMW Demineralized Water

DMW Department of Mechanical Working

DMW Digital Milliwatt [also DmW]

DMW Dissimilar-Metal Weld

dmW Decimetric Wave [also DMV]

DMWD Department of Miscellaneous Weapons Development [UK]

DMWG Data Management Working Group

DMX Data Multiplexer

DMX Direct Memory Exchange

DMXRD Director of Materials and Explosives Research and Development [UK]

DMY Day-Month-Year [also D-M-Y, or D/M/Y]

DMZ Demilitarized Zone

DMZn Dimethylzinc

DN Danmarks Naturfredningsforening [Danish Association of Nature Conservation]

DN Data Name

DN Decimal Number(ing)

DN Decrement/Flag Direction Down [Computers]

DN Delayed Neutron [Nuclear Physics]

DN Department of Nursing

DN Department of the Navy [Arlington, Virginia, US]

DN Digital Number

DN Dinar [Currency of All Former Yugoslavian States]

DN Dinitro-

DN Directory Number

DN Discipline Node

DN Discrepancy Notice

DN Display Number (Command) [Pascal Programming]

DN Document Number

DN Domain Name [Computers]

DN Down

DN Drawing Note

DN Drawing Number [also DWG NO, or Dwg No]

DN Gutmann Donor Number [Physics]

DN111 Dicyclohexylamine 4,6-Dinitro-2-Cyclohexyl-phenolate

D/N Debit Note [also d/n]

dn decineper [Unit]

dn delta amplitude [Elliptic Function]

dn down

ΔN Change in the Number of …(e.g., Elements, Objects, Particles, Parts, etc.) [Symbol]

Δn change in principal quantum number [Symbol]

Δn change (or difference) in refractive index [Symbol]

Δn_c change in critical concentration [Symbol]

$\Delta \nu$ (light) frequency difference (or change) [Symbol]

DNA Defense Nuclear Agency [US]

DNA Deutscher Normenausschuß [German Standards Committee]

DNA Deoxyribonucleic Acid [Biochemisty]

DNA Department of National Archives [Colombo, Sri Lanka]

DNA Digital Network Architecture

DNA Dinitroaniline

DNA Dinonyl Adipate

DNA Dinolylaniline

DNA Does Not Apply

DNAcet Dinitroacetone

DNAF 10-(2',4'-Dinitrophenylazo)-9-Phenanthrol

DNAG Decade of North American Geology

DNAM Data Network Access Method

DNAns Dinitroanisole

DNAP Dinitroaminophenol

DNAP 4-(2',4'-Dinitrophenylazo)phenol

DNAPL Dense Nonaqueous Phase Liquid

DNA/RNA Deoxyribonucleic Acid/Ribonucleic Acid [Biochemistry]

DNase Deoxyribonuclease [Biochemistry]

DNase I Deoxyribonuclease I [Biochemistry]

DNase II Deoxyribonuclease II [Biochemistry]

DNAT Deutsch-Niederländische Amateurfunker-Tage [German-Dutch Convention of Amateur Radio Operators]

DNB Departure from Nucleate Boiling

DNB Deutsches Nachrichtenbüro [German News Agency]

DNB Dictionary of National Bibliography

DNB Dinitrobenzene

DNB 3,5-Dinitrobenzoic Acid

DNB Dinitrobenzoyl

DNBA Dinitrobenzaldehyde

DNBAc Dinitrobenzoic Acid

DNBC Dinadic Benzoyl Chloride

DNBC 3,5-Dinitrobenzoyl Chloride

DNBP Dinitrobutylphenol

DNBPA 3,5-Dinitrobenzyl-N-(n-Propyl)amine

DNBPP Di-n-Butylphenylphosphonate

DNBR Departure from Nucleate Boiling Ratio

DNBS 2,4-Dinitrobenzenesulfonic Acid, Sodium Salt

DNBSC 2,4-Dinitrobenzenesulfenyl Chloride

DNC Delayed Neutron Counter; Delayed Neutron Counting

DNC Delayed Neutron Coupling

DNC Department of Naval Construction

DNC Dinitrocellulose

DNC 4,6-Dinitro-o-Cresol

DNC Direct Numerical Control

DNC Distributed Numerical Control

DNCB 2,4-Dinitrochlorobenzene

DNCC Data Network Control Center

DNCCC Defense National Communications Control Center [US]

DNCH Dinitrochlorohydrin

DNCIAWPRC Danish National Committee of the International Association on Water Pollution Research and Control

DNCP Diet, Nutrition and Cancer Programs [of National Cancer Institute, Bethesda, Maryland, US]

DNCP Dipolar-Narrowed Carr-Purcell (Sequence)

DNCPB Dinitrochlorobenzene

DNCPH Dinitrochlorohydrin

DNCrs Dinitrocresol

DNCS Distributed Network Control System

DND Department of National Defense [Canada]

DNDAP Dinitrodiazophenol [also DNDAPh]

dn/dC refractive index increment (i.e., change in refractive index with concentration [also dn/dc]

DND-CAT Dow (Company)-Northwestern (University)-Dupont (Company) Collaborative Access Team [US]

DND/CRAD Department of National Defense/Chief, Research and Development [Canada]

DND/DSS Department of National Defense/Department of Supply and Services (Contract) [Canada]

dN/dE Auger Electron Spectrum (or Auger Signal)

dn/dλ rate of change of refractive index with wavelength (i.e., dispersion)

DNDMOxm Dinitrodimethyloxamide [also DNDMeOxm]

DNDMSA Dinitrodimethylsulfamide [also DNDMeSA]

dNDP Desoxyribonucleoside Diphosphate [Biochemistry]

DNDPhA Dinitrodiphenylamine

DNDS Dinitrodiphenyl Disulfide

dN/dt rate of nucleation [Metallurgy]

dN/dt time rate of change of particle number

dn/dT change in refractive index with temperature

$\Delta N/\Delta t$ Time Rate of Change in the Number of …(e.g., Elements, Objects, Particles, Parts, etc.) [Symbol]

DNE Delayed Neutron Emission

DNE Department of Nursing Education

DNE Department of Nuclear Engineering

DNE Dinitroethane

dN(E)/dE change of electron energy distribution with energy (e.g., in Auger spectroscopy)

DNES Department of Nuclear Energy Sciences

DNEteU Dinitroethyleneurea [also DNEU]

DNF Dinitrofuran

DNFA 2,4-Dinitro-5-Fluoroaniline

DNFB 2,4-Dinitrofluorobenzene

DNFM Department of Nonferrous Metallurgy

DNFSB Defense Nuclear Facility Safety Board [US]

DNG Diglycerindinitrate

DNGcU Dinitroglycoluril

DNH&AS Dorset Natural History and Archeological Society [UK]

DNHR Dynamic Non-Hierarchical Routing

DNHW Department of National Health and Welfare [Canada]

DNI Data Network Interface

DNI Digital Noninterpolation

DNI Director of Naval Intelligence [of US Navy]

DNIBR Danish National Institute of Building Research

DNIC Data Net(work) Identification Code

DNIC Digital Network Interface Circuit

δ-NiIn Delta Nickel Indium (Compound)

DNIS Dialed Number Identification Service

DNL Differential Non-Linearity

DNL Dynamic Noise Limiter

DNLK Downlink [also Dnlk]

DNLT Downlist [also Dnlt]

DNM Delayed Neutron Monitor

DNM Dinitromethane [also DNMe]

DNMA Dinitromethylaniline [also DNMeA]

DNMES Dynamic Non-Member Economies [of Organization for Economic Cooperation and Development]

dNMP Desoxyribonucleoside Monophosphate [also d-NMP] [Biochemistry]

DNMR Double Nuclear Magnetic Resonance

DNMR Dynamic Nuclear Magnetic Resonance

DNMR Two-Dimensional Nuclear Magnetic Resonance Spectroscopy

D-NMR Deuterium Nuclear Magnetic Resonance

DNN 1,5-Dinitronaphthalene

DNN Dinitronaphthol

ΔN/N Fraction of Total Number of Gas Molecules whose Velocity Lies within a Given Range (in Maxwell Distribution Law) [Symbol]

Δν/ν relative uncertainty (of spectral lines) (i.e., change in the frequency of light per unit frequency) [Symbol]

DNNA 2,4-Dinitro-1-Naphthylamine

DNNSA Dinonyl Naphthalene Sulfonic Acid

DNO Descending Node Orbit

DNO Directorate of Naval Ordnance

DNOA Di-n-Octylamine

DNOC 4,6-Dinitro-o-Cresol

DNOCHP Dinitro-o-Cyclohexylphenol

DNODA Di(n-Octyl-n-Decyl)adipate

DNODP Di(n-Octyl-n-Decyl)phthalate

DN-ONR Department of the Navy–Office of Naval Research [Arlington, Virginia, US]

DNOSBP Dinitro-O-Sec-Butylphenol

DNP Deoxyribonucleoprotein [Biochemistry]

DNP 1,2-Diaminopropane

DNP Dinaphthylphosphate

DNP 2,4-Dinitrophenol

DNP N-Dinitrophenyl [also Dnp]

DNP 2,4-Dinitrophenylhydrazine

DNP Dinitropyrene

DNP Dinonyl Phosphate

DNP Dinonyl Phthalate

DNP Division of Nuclear Physics [of American Physical Society, US]

DNP Dynamic Nuclear Polarization

Dnp N-Dinitrophenyl

DNP-AA N-(2,4-Dinitrophenyl)amino Acid

DNPC 2,6-Dinitro-p-Cresol

DNP-BSA Dinitrophenyl–Bovine Serum Albumin [Biochemistry]

2,6-DNPC 2,6-Dinitro-p-Cresol

DNP-CP-MAS Dynamic Nuclear Polarization Cross-Polarization Magic-Angle Spinning (Spectroscopy) [also DNP-CP/MAS]

DNPD Dinaphthylphenylenediamine; N,N'-Di-β-Naphthyl-P-Phenylene Diamine

DNP-DP-MAS Dynamic Nuclear Polarization Direct-Polarization Magic-Angle Spinning (Spectroscopy) [also DNP-DP/MAS]

DNPF Bis(dinitropropyl)fumarate

DNPF 2,4-Dinitro Phenylfluoride

DNPH 2,4-Dinitrophenylhydrazine

DNPh Dinitrophenol

DNPM Departmento Nacional de Produção Mineral [National Department of Mineral Production, Brazil]

DNPMT N,N'-Dinitroso Pentamethylene Tetramine

DNPN Bis(dinitropropyl)nitramine

DNPO Bis(2,4-Dinitrophenyl)oxalate

2,4-DNPO 2,4-Dinitrophenyl Ester

DNPS Dinitropropylsuccinate

DNPT 2,6-Dinitro(bicyclo)pentamethylene-2,4,6,8-Tetramine

DNPT N,N'-Dinitroso Pentamethylene Tetramine

DNPT 3,7-Dinitro-1,3,5,7-Tetrazabicyclo[3,3,1]nonane

DNP-Taurine 2,4-Dinitrophenyltaurine

DNPTB Dinitropropyl Trinitrobutyrate

DNPV Differential Normal Pulse Voltammetry

DNPyr Dinitropyridyl

DNQX 6,7-Dinitroquinoxaline-2,3(1H,4H)-Dione

DNR Department of Natural Resources

DNR Deutscher Naturschutzring e.V.–Bundesverband für Umweltschutz [German Nature Conservancy Group–Federal Environmental Protection Association, Germany]

DNR Dinitroresorcinol

DNR Direct Neutron Radiography

DNR Do Not Resuscitate (Order) [Medicine]

DNR Double Non-Return (Valve)

DNR Dynamic Noise Reduction

DNRC Democritus Nuclear Research Center [Greece]

DNRE Department of Natural Resources and Energy [Canada]

DNREC Department of Natural Resources and Environmental Control [Delaware, US]

DNR LED Dynamic Noise Reduction Light-Emitting Diode

DNRZ Delayed Non-Return-to-Zero

DNS 5-Dimethylamino-1-Naphthalenesulfonic Acid

DNS 5-(Dimethylamino)-1-Naphthalene-1-Sulfonyl Chloride (or Dansyl Chloride)

DNS Distributed Network System

DNS Domain Name Server [Internet]

DNS Domain Name System (Server) [Internet]

DNS Doppler Navigation System

DNSA 5-Dimethylamino Naphthalene-1-Sulfonamide

DNSAPITC Dimethylamino Naphthalenesulfonylamino Phenylisothiocyanate

DNSC Data Network Service Center

DNSC Digital Network Service Center

DNSCl 5-(Dimethylamino)-1-Naphthalenesulfonyl Chloride

DNSE Dirección Nacional del Servicio Estadístico [National Directorate of Statistical Services, Argentina]

DNS-F 5-(Dimethylamino)-1-Naphthalenesulfonyl Fluoride

DNSR Director of Nuclear Safety Research

DNT Desmethylnortriplyline

DNT Digital Network Terminator

DNT Dinitrotoluene

2,4-DNT 2,4-Dinitrotoluene

2,6-DNT 2,6-Dinitrotoluene

DNTB Diethyl-p-Nitrophenyl Monothiophosphate

DNTB Dithio-bis-Nitrobenzoic Acid

DNTC 4-Dimethylamino-1-Naphthyl Isothiocyanate

DNT-Cl 4-Chloro-3,5-Dinitro-Trifluorotoluene

dNTP Deoxyribonucleoside Triphosphate [also d-NTP] [Biochemistry]

DNTS Digital Data Exchange Network Testing System

DNX Dinitroxylene

DNZ Darned Near Zero

DO Damped Oscillation

DO Dark Operated

DO Dashpot Relay

DO Data Out

DO Data Output

DO Decimal-to-Octal

DO Defense Order

DO Delivery Order

DO Department of Oncology

DO Design Objective

DO Design Optimization

DO Diesel Oil

DO Digital Output

DO Diode Outline

DO Discrete Output

DO Disordered (Phase) [Solid-State Physics]

DO Disorder-Order (Transformation) [Solid-State Physics]

DO Dissolved Oxygen

DO Distributed Objects

DO Division of Oceanography [of Commonwealth Scientific and Industrial Research Organization, Australia]

DO Doctor of Ophthalmology

DO Doctor of Osteopathy

DO Dominican Republic [ISO Code]

DO Draw Out

DO Drying Oil

DO Dynamic Oscillation

D$_2$O Deuterium Oxide [also DOD]

D&O Description and Operations

D/O Delivery Order

D-O Decimal-to-Octal

Do Dolomite [Mineral]

do *(ditto)* – the same

.do Dominican Reublic [Country Code/Domain Name]

/do dosseg (DOS segment order) option [MS-DOS Linker]

$\Delta\Omega$ Solid Angle of Acceptance (in Ion Spectroscopy Scattering) [Symbol]

$\Delta\Omega$* Free Energy Barrier [Symbol]

$\Delta\omega$ full width at half maximum (in spectroscopy) [Symbol]

DOA Date of Availability

DOA Dead on Arrival

DOA Department of Agriculture [US]

DOA Department of Oriental Antiquities [of British Museum]

DOA Differential Operational Amplifier

DOA Digital Output Adapter

DOA 2-(3',7'-Dimethyl-2',6'-Octadienylamino)ethanol

DOA Dioctadecyladipate (Crystal)

DOA Dioctyl Adipate

DOA (Deliver) Documents on Acceptance (of Draft)

DOA Dominant Obstacle Allowance

D1A Diapason [French Satellite]

DOAE Defense Operational Analysis Establishment [UK]

DOAZ Dioctyl Azelate

DOB Data Output Bus

DOB Date of Birth [also dob]

DOB Deluxe Option Board [Computers]

DOB Dimethoxy-4-Bromoamphetamine

DOB (±)-Dimethoxy-4-Bromoamphetamine

DOBAMBC P-Decyloxybenzylidene P'-Amino-2-Methyl Butyl Cinnamate (Ferroelectric Liquid Crystal)

DOBIS Dortmunder On-Line Bibliothekssystem [Dortmund On-Line Library System; of University of Dortmund, Germany]

DOBS Disk-On-Bearing System

DObst Doctor of Obstetrics

DObstRCOG Doctor of Obstetrics, Royal College of Obstetricians and Gynaecologists

DOC Data Optimizing Computer

DOC Decimal-to-Octal Conversion

DOC Degree of Cure

DOC Delayed-Opening Chaff

DOC Deoxycorticosterone [Biochemistry]

DOC Department of Commerce [US]

DOC Department of Communications [Canada]

DOC Depth of Cut

DOC Designated Operational Coverage

DOC Design Office Consortium [now Construction Industry Computing Association, UK]

DOC Desoxycortisone [Biochemistry]

DOC Deterministic Optimal Control

DOC Dichromate Oxygen Consumed

DOC 3,3'-Diethyloxacarbocyanine (Iodide)

DOC Digital Operations Control(ler)

DOC Digital Output Control

DOC 4,6-Dinitro-o-Cresol

DOC Direct-on-Column Injector

DOC Direct Operating Cost

DOC Direct Oxychlorination (Process) [Chemical Engineering]

DOC Disk-Oriented Computer System

DOC Dissolved Organic Carbon

DOC Distributed Operator Console

DOC Dynamic Overload Control

D1C Diademe One [French Geodetic Satellite]

.DOC Document(ation) [File Name Extension] [also .doc]

DoC Department of Commerce [US]

Doc Doctor [also doc]

Doc Document(ation); Documentor [also doc]

DOCA Documentation Automatique [Automatic Documentation Section; of Centre Européen de Traitement de l'Information Scientifique , France]

DOCA Deoxycorticosterone Acetate [Biochemistry]

DOCC DCA (Defense Communications Agency) Operations Center Complex [US]

Doc Chem Yugosl Documenta Chemica Yugoslavica [Yugoslav Chemical Journal; published by the Serbian Chemical Society]

DOCDEL Document Delivery [of Direct Information Access Network for Europe]

DOCFAX Document Facsimile Transmission

DOCLINE Document Ordering On-Line

DOCOCEAN Documentation Oceanique [Documentation on Oceanography; of Bureau National des Données Oceanique , France]

Doc Ophthalmol Documenta Ophthalmologica [Netherlands]

DOCP Deterministic Optimal Control Problem

Doc Proc Document Processing [also doc proc]

DOCS Disk-Oriented Computer System

DOCS Document Organization and Control System

Docs Documents [also docs]

Doc-Sci Inf Domentaliste–Sciences de l'Information [French Publication on Documentation and Information Sciences; published by Association des Documentalistes et des Bibliothécaires Spécialisés]

DOCSV Data Over Circuit-Switched Voice

DOCSYS Display of Chromosome Statistics System

DOCUS Display-Oriented Computer Usage System

DOD Degree of Dissociation

DOD Department of Defense [also DoD]

DOD Depth of Discharge

DOD Detroit Ordnance District [Michigan, US]

DOD Deuterium Oxide [also D_2O]

DOD Development Operations Division

DOD Digital Optical Disk

DOD Direct Outward Dialing

DOD Drop on Demand

D1D Diademe Two [French Geodetic Satellite]

ΔOD Change in Optical Density

DODA Dioctadecylamine

DODA Door and Operator Dealers Association [US]

DODC 3,3'-Diethyloxadicarbocyanine

DODCI Department of Defense Computer Institute [US]

DODCSC Department of Defense Computer Security Center [US]

DODD Department of Defense Directive [US]

DOD/DARPA Department of Defense/Defense Advanced Research Projects Agency [US]

DODECP Dioctyl-N,N'-Diethylcarbamylphosphonate

DODGE Department of Defense Gravity Experiment [US]

DODGE-M DODGE, Multipurpose [US]

Dodine N-Dodecylguanidine Acetate [also dodine] [Biochemistry]

DODIS Distribution of Oceanographic Data on Isotropic Levels

DODISS Department of Defense Index of Specifications and Standards [also DoDISS, US]

DODMAB Dioctadecyldimethylammonium Bromide

DODMS Department of Defense Materials Specifications

DOD/NASA Department of Defense/National Aeronautics and Space Administration (Joint Project) [US]

DOD/NASA/FAA Department of Defense/National Aeronautics and Space Administration/Federal Aviation Administration (Conference) [US]

DOD/ONR Department of Defense/Office of Naval Research [US]

DODPA 4,4'-Dioctyl Diphenyl Amine

DODS Different Orbitals for Different Spins [Physics]

DOD-STD Department of Defense Standard [US] [also DoD-Std]

DODT Display Octal Debugging Technique

$\Delta\omega/\Delta t$ rate of change of angular velocity

DOE Department of Education [Canada]

DOE Department of Electronics [India]

DOE Department of Energy [US and UK] [also DoE]

DOE Department of Optical Engineering

DOE Department of the Environment [Canada and UK]

DOE Design of Experiment(s)

DOE Diffractive Optical Element

DOE Distributed Objects Everywhere

DoE Department of Energy [US and UK] [also DoE]

DoE Department of the Environment [Australia]

DOE/AES Department of Environment/Atmospheric Environment Service [Canada]

DOE-AIC Department of Energy –Advanced Industrial Concepts [US]

DOE-BES Department of Energy/(Office of) Basic Energy Science [US]

DOE/FE Department of Energy–Fossil Energy [US]

DOE/HEP Department of Energy/High-Energy Physics [US]

DOE-HQ Department of Energy, Headquarters [Washington, DC, US]

DOEM Designated Officials for Environmental Matters

DOE-OBES Department of Energy–Office of Basic Energy Science [US]

DOE-OER Department of Energy–Office of Energy Research [US]

DOE-OIP Department of Energy–Office of Industrial Programs [US]

DOE-OTM Department of Energy–Office of Transportation Materials [US]

DOE Pulse Department of Energy Pulse [Publication of the US Department of Energy]

DOE-RL Department of Energy, Richland Operations Office [Washington State, US]

DOES Density of Electron States [Solid-State Physics]

DOE-SF Department of Energy, San Francisco Operations Office [California, US]

DOES N(E_F) Density of Electron States at Fermi Level [Symbol]

DOETS Dual-Object Electronic Tracking System

DOE(UK) Department of Energy (UK)

DOE(US) Department of Energy (US)

DOF Deep Ocean Floor

DOF Degree of Freedom

DOF Department of Finance [Australia]

DOF Department of Fisheries [US]

DOF Depth of Field

DOF Depth of Focus

DOF Device Operating Failure

DOF Device Output Format

DOF Di(2-Ethylhexyl)fumarate

DOF Dioctyl Fumarate

DOF Direction of Flight

DOF Disordered Flat

Dofasco Dominion Foundries and Steel Limited [Hamilton, Ontario, Canada]

Dofasco Illus News Dofasco Illustrated News [Published by Dofasco Inc., Hamilton, Ontario, Canada]

DOFIC Domain-Originated Functional Integrated Circuit

DOFL Diamond Ordnance Fuse Laboratory [Washington, DC, US]

DOFP Direct-on Finish Process

DOG Digital Operating Guide

DOG 1,2-Dioctanoyl Glycerol

DOG Division of Oil and Gas [California, US]

DoGaMI Department of Geology and Mineral Industries [Oregon, US]

Doga Turk Fiz Astrofiz Derg Doga Turk Fizik Astrofizik Dergisi [Turkish Journal on Physics and Astrophysics; published by the Scientific and Technical Research Council of Turkey]

Doga Turk Muehendis Cevre Bilimleri Derg Doga Turk Muehendislik ve Cevre Bilimleri Dergisi [Publication of the Scientific and Technical Research Council of Turkey]

DOGIT Deed of Grant in Trust

DOH Department of Health

DOH Department of Health [Washington State, US]

DOH Department of Health [Taiwan]

DOH Discrete Output High

DOHC Double Overhead Camshaft; Dual Overhead Camshaft

(DOH)2en Bis(diacetylmonoximeimino)ethane-1,2

(DOH)2pn Bis(diacetylmonoximeimino)propane-1,3

DOHS Diploma in Occupational Health and Safety

DOI Department of Industry [now Department of Trade and Industry, UK]

DOI Department of the Interior [also DoI, US]

DOI Descent Orbit Insertion

DOI Differential Orbit Improvement

DOI Distinct(ive)ness of Image

DOINET Department of the Interior Wide Area Network [US]

DOIO Directly Operable Input/Output

DOI-OB Department of the Interio r–Office of Budget [US]

DOIP Dial Other Internet Providers

DOIP Diisoocytl Isophthalate

DOIT Database Oriented Interrogation Technique

DO/IT Digital Output/Input Translator [also DOIT]

DO-IT Disabilities, Opportunities, Internetworking and Technology

DOJ Department of Justice [also DoJ, US]

DOJ Dipole Orientation Junction

Dokl Akad Nauk Azerb SSR Doklady Akademii Nauk Azerbaidzhanskoi SSR [Doklady, Proceedings of the Academy of Sciences of the Azerbaydjanian SSR]

Dokl Akad Nauk Belorusskoi SSR Doklady Akademii Nauk Belorusskoi SSR [Doklady, Prcccedings of the Academy of Sciences of the Byelorussian SSR]

Dokl Akad Nauk BSSR Doklady Akademii Nauk BSSR [Doklady, Proceedings of the Academy of Sciences of the Byelorussian SSR]

Dokl Akad Nauk SSSR Doklady Akademii Nauk SSSR [Proceedings of the Academy of Sciences of the USSR, Physics, Chemistry and Physical Chemistry]

Dokl Akad Nauk Tadzh SSR Doklady Akademii Nauk Tadzhikshoi SSR [Doklady, Proceedings of the Academy of Sciences of the Tadzik SSR]

Dokl Akad Nauk USSR Doklady Akademii Nauk USSR [Proceedings of the Academy of Sciences (of the USSR)]

Dokl Akad Nauk Uzb SSR Doklady Akademii Nauk Uzbekskoi SSR [Doklady, Proceedings of the Academy of Sciences of the Uzbek SSR]

Dokl AN Azerb SSR Doklady Akademii Nauk Azerbaidzhanskoi SSR [Doklady, Proceedings of the Academy of Sciences of the Azerbaydjanian SSR]

Dokl AN Belorusskoi SSR Doklady Akademii Nauk Belorusskoi SSR [Doklady, Proccedings of the Academy of Sciences of the Byelorussian SSR]

Dokl AN BSSR Doklady Akademii Nauk BSSR [Doklady, Proceedings of the Academy of Sciences of the Byelorussian SSR]

Dokl AN SSSR Doklady Akademii Nauk SSSR [Proceedings of the Academy of Sciences of the USSR, Physics, Chemistry and Physical Chemistry]

Dokl AN Tadzh SSR Doklady Akademii Nauk Tadzhikshoi SSR [Doklady, Proceedings of the Academy of Sciences of the Tadzik SSR]

Dokl AN USSR Doklady Akademii Nauk USSR [Proceedings of the Academy of Sciences (of the USSR)]

Dokl AN Uzb SSR Doklady Akademii Nauk Uzbekskoi SSR [Doklady, Proceedings of the Academy of

Dokl Chem Doklady Chemistry [English Translation of Selected Articles from: *Doklady Akademii Nauk SSSR*; published in US]

Dokl Chem Technol Doklady Chemical Technology [Russian Proceedings on Chemical Technology; published in the US]

Dokl Earth Sci Sect Doklady Earth Sciences Section [Proceedings of the Earth Sciences Section of the Academy of Sciences of the USSR]

Dokl Phys Chem Doklady Physical Chemistry [English Translation of Selected Articles from: *Doklady Akademii Nauk SSSR*; published in US]

dokt vĕt doktor vĕt [Doctor of Science; Czech Degree Equivalent to US Doctor's Degree]

DOL Department of Labor [also DoL, or USDOL]

DOL Director of Laboratories

DOL Discrete Output Low

DOL Display-Oriented Language

DOL Dynamic Octal Load

Dol Dollar [also dol]

DOLA Department of Land Information [Western Australia]

DOLARS Digital Off-Line Automatic Recording System

DOLARS Doppler Location and Ranging System

DOLE Data On-Line Editing System; Data On-Line Editor

DOLE Digital On-Line Editing System; Digital On-Line Editor

DOLOG Do Logic (Software System)

DOLPHIN Deep-Ocean Logging Profiler Hydrographic Instrumentation and Navigation

DOM Day of Month

DOM Department of Mines [US]

DOM Design of Maintenance

DOM Di(2-Ethylhexyl)maleate

DOM Digital Ohmmeter

DOM 2,5-Dimethoxy-4-Methylamphetamine

DOM Diocytl Maleate

DOM Direction of Magnetization

DOM Disk Operating Monitor

DOM Dissolved Organic Matter

DOM Document Object Model

DOM Drawn Over Mandrel

(±)-DOM (±)-Dimethoxy-4-Methylamphetamine [Pharmacology]

DoM Department of Mines [US]

Dom Dominion [also dom]

dom domestic

dom dominant

DOMA Dirección General de Medio Ambiente [Directorate General for the Environment, Spain]

DOMA Dokumentation Maschinenbau (Database) [Documentation on Mechanical Engineering; of Fachinformationszentrum Technik, Germany]

Dom App Domestic Appliance [also dom app]

DOMF Distributed Object Management Facility

DOMF Methylphosphonic Acid Bis(6-Methylheptyl)ester

DOMP Diisooctylmethylphosphonate

DOMPB Department of Oriental Manuscripts and Printed Books [of British Library]

Dom Rep Dominican Republic

DOMS Delayed Onset Muscle Soreness

DOMSAT Domestic Satellite [Australia] [also Domsat]

DOMSAT Domestic Satellite System

DOMSATS Domestic Satellite Systems

DON Delayed Order Notice

DON Deoxynivaleno-

DON Department of the Navy [Arlington, Virginia, US]

DON Deuterium-Moderated Organic-Cooled Nuclear Reactor [Spain]

DON 6-Diazo-5-Oxo L-Norleucine [Biochemistry]

DON Dioxynaphthalene

DONA Decentralized Open Network Architecture

DONG Dansk Olje Og Naturgas (Line) [Danish Oil and Natural Gas Pipeline]

DONS Dioctyl Sodium Sulfosuccinate

DONUT Direct Observation of the Nu Tau (Experiment) [of Fermilab, Batavia, Illinois, US]

DONUTS Dust Offers New Uses To Steelmaking (Team) [of Michigan Technological University, Houghton, US]

$\Delta\Omega/\Omega$ Molar Volume Change [Symbol]

DOP Deliver Documents on Payment of Draft

DOP Designated Overhaul Point

DOP Detailed Operating Procedure

DOP Developing-Out Paper

DOP Diisooctyl Phthalate

DOP Dilution of Position

DOP Dioctyl Phosphate

DOP Dioctyl Phthalate

DOP Diphenyl-Octyl-Phthalate

DOP Diver Operated Plug

DOP Dominican Peso [Currency of the Dominican Republic]

Dop Doppler

DOPA 3-(3,4-Dihydroxyphenyl)alanine [also Dopa] [Biochemistry]

DOPA Dioctylphenylphosphoric Acid

DOPA Di-n-Octylphosphoric Acid [also Dopa] [Biochemistry]

D-DOPA D-3,4-Dihydroxyphenylalanine [also D-Dopa] [Biochemistry]

DL-DOPA DL-3,4-Dihydroxyphenylalanine; 3-(3,4-Dihydroxyphenyl)-DL-Alanine [also DL-Dopa, or DL-dopa] [Biochemistry]

L-DOPA L-3,4-Dihydroxyphenylalanine [also L-Dopa] [Biochemistry]

DOPAC Dihydroxyphenylacetic Acid

Dop Akad Nauk Ukr SSR Dopovidi Akademii Nauk Ukrains'koi SSR [Publication of the Ukrainian Academy of Sciences]

Dop Akad Nauk Ukr RSR A, Fiz-Mat Tekh Dopovidi Akademii Nauk Ukrains'koi RSR, Seriya A, Fiziko-Matematichnita Tekhnichni Nauki [Publication of the Ukrainian Academy of Sciences on Physical and Mathematical Sciences and Technology]

Dop Akad Nauk Ukr RSR B, Geol, Khim, Biol Dopovidi Akademii Nauk Ukrains'koi RSR, Seriya B, Geologiya, Khimiya, Biologiya [Publication of the Ukrainian Academy of Sciences on Geology, Chemistry and Biology]

DOPAMINE Dihydroxyphenethylamine [also Dopamine] [Biochemistry]

DOPD N,N'-Bis-(1-Ethyl-3-Methylpentyl)-P-Phenylene Diamine

DOPE Dioleoyl Phosphatidylethanolamine

DOPE Discrete Orthogonal Polynomial Expansion

DOPIC Documentation of Program in Core

DOPIE Department of Primary Industries and Energy [Australia]

DOPLID Doppler Lidar

DOPLOC Doppler Phase Lock

DOPP Dioctyl Phenylphosphonate

DOPS Dansk Optisk Selskab [Danish Optical Society]

DOPS Digital Optical Protection System

DOPS (3,4-Dihydroxyphenyl)serine

DL-DOPS DL-3,4-Dihydroxyphenylserine [also DL-Dops, or DL-dops] [Biochemistry]

DOPSK Differential Offset Phase Shift Keying

DOQ Dopamine Quinone [Biochemistry]

DOQQ Digital-Color Orthophoto Quarter Quadrangle

DOR Data Output Register

DOR Deuterium-Moderated and Organic-Cooled Reactor

DOR Digital Optical Recorder; Digital Optical Recording

DOR Digital Output Relay

DOR Division of Operating Reactors

DOR Double Rotation [Nuclear Magnetic Resonance Technique]

DORA Directorate of Operational Research and Analysis [UK]

DORA Double Roll-Out Array [also Dora]

DORAN Doppler Range and Navigation [also Doran, or doran]

DORCMA Door Operator and Remote Controls Manufacturers Association [US]

DORE Defense Operational Research Establishment [of Defense Research Board, Canada]

DORF Diamond Ordinance Radiation Facility [US]

DORIS Détermination d'Orbite et Radiopositionnement Intégre par Satellite [Orbit Determination and Integral Radio-Positioning by Satellite]

DORIS Direct Order Recording and Invoicing System

DORIS Doppel-Ring-Speicher [Double Ring Storage, Germany]

DORK Diagnostically Optimizable Recursive Keyword

DORK Direct Order Recording Keyboard

DORK Direct On-Line Retrievable Knowledge

dorm dormant

DORM HRT Dormant Hard Return [also Dorm Hrt]

DORS Developers Open Resource Software

Dors Dorsetshire [UK]

DORV Deep Ocean Research Vehicle

DOS Days on Stream

DOS Decision Outstanding

DOS Degree of Sensitization

DOS Density of States [also dos] [Solid-State Physics]

DOS Department of Orthopedic Surgery

DOS Department of State [US] [also DoS]

DOS Digital Operating System

DOS Dioctyl Sebacate

DOS Director of Ordnance Services [UK]

DOS Disk Operating System

DOS Doctor of Optical Sciences

dos *(dosis)* – dose [Medical Prescriptions]

DOSAR Dosimetry Applications Research Facility

DOSD Density of States Distribution [Solid-State Physics]

DOSD Dipole Oscillator Strength Distribution

DOSEM Disk-Operating System Emulation

DOSENCO Decay On a Specific Nuclear Coordinate (Photoreaction)

Dosim Dosimeter; Dosimetry

DOSLI Department of Survey and Land Information [New Zealand]

DOSP Dalhousie Ocean Studies Program [of Dalhousie University, Halifax, Canada]

DOSP Dimensionality of SiO_4 Polymerization

DOSSEG (Microsoft) Disk Operating System Segment (Ordering) [also dosseg]

DOSV Deep Ocean Survey Vehicle

DOS/VS Disk Operating System/Virtual Storage

DOS/VS-AF Disk Operating System/Virtual Storage–Advanced Functions

DOS/VSE Disk Operating System/Virtual Storage Extended

DOT Deep Ocean Technology

DOT Deep Ocean Transponder

DOT Deep Ocean Trough

DOT Department of Trade [UK]

DOT Department of Transport [Canada]

DOT Department of Transportation [also DoT, US]

DOT Deployment Operations Team

DOT Designating Optical Tracker

DOT Detailed Operations Timeline

DOT Digital Optical Technology

DOT Digital Overlay Technique

DOT Direct Oxidation Test

DOT Domain-Tip Propagation (Memory)

DoT Department of Tourism [Australia]

DoT Department of Transportation [also DOT, US]

DOTA Département Optique Théoretique et Appliquée [Department of Theoretical and Applied Optics, ONERA-CERT, Toulouse, France]

DOTAC Doppler TACAN (Tactical Air Navigation System) [also Dotac, or dotac]

DOTAC Department of Transport and Communications [Australia]

DOTAP N-[1-(2,3-Dioleoyloxy)propyl]-N,N,N-Trimethylammonium Methyl Sulfate

DOTC Department of Transportation Classification [US]

DOTC 3,3'-Diethyloxatricarbocyanide Iodide

DOTC Dioctadecyldithiocarbamate

DOT/CIAP Department of Transportation/Climatic Impact Assessment Program [US]

DOT/FAA Department of Transportation/Federal Aviation Administration (Joint Program) [US]

DOTG Diorthotolylguanidine; 1,3-Di-o-Tolylguanidine [Biochemistry]

DOTP Diisooctylterephthalate

DOTP Dioctyl Terephthalate

DOTS Dynamic Ocean Track System [US]

DOTSYS Dot System

DOTT Di-O-Tolylthiourea

DOTT Documentation for Translation and Terminology [of Secretary of State, Canada]

dott *(dottore)* – Italian Degree Equivalent to US Doctor Degree

DOT/UMTA Department of Transportation Urban Mass Transportation Administration [at Washington, DC, US]

DOT/UN Department of Transportation/United Nations

Double Liaison-Chim Peint Double Liaison-Chimie des Peintures [French Publication on Paint Chemistry]

doubt doubtful

DOUBTFUL Double (Quantum) Transitions for Finding Unresolved Lines

Doug Fir Douglas Fir

Doug Fir-L Douglas Fir–Lumber

DOUSER Doppler Unbeamed Search Radar [also Douser]

DOUT Data-Out Line

DOV Data Over Voice

DOV Degree of Variance [Statistics]

DOV Distilled Oil of Vitriol

DOV (Bottle-Nose) Dolphinpox Virus

DOVAP Doppler Velocity and Position [also Dovap]

DOVETT Double Velocity Transit Time

DOVS Density of Valence States

DOW Day of Week

DOWB Deep Operating Work Board

DOWP Document-Oriented Word Processor

DOXYL Dimethyloxazolidine-N-Oxyl

DOXYL Dimethyloxazolinyloxy

DOY Day of Year

DOZ Diisooctyl Azelate

DOZ Dioctyl Azelate

doz dozen [Unit]

DP Damp-Proof(ing)

DP Darkfield Pattern [Microscopy]

DP Dash-Pot

DP Data Point(er)

DP Data Port

DP Data Probe

DP Data Processing; Data Processor [also dp]

DP Data Protection

DP Date of Publication

DP Daughter Product [Physics]

DP Dead Plate

DP Decentralized Peripherals

DP Decimal Point

DP Deck-Piercing

DP Decomposition Product

DP Deep Penetration

DP Deflection Plate [Electronics]

DP Deformation Process(ing)

DP (Office of) Defense Programs [of US Department of Energy]

DP Deflector Plate

DP Deformation Potential [Solid-State Physics]

DP Degree of Polymerization

DP Delayed Procurement

DP Delayed Proton

DP Delta Pile

DP Densely Populated

DP Department of Philosophy

DP Department of Physics

DP Deployable Payload

DP Deposition Potential

DP Depth of Penetration [Hardness Testing]

DP Depth Perception

DP Depth Profile

DP Derivative Polarography

DP Design Parameter

DP Design Pressure

DP Design Proof

DP Desolder Pump

DP Development Phase

DP Dew Point [also dp]

DP Diagnostic Program

DP Dial Pulsing

DP Diametral Pitch (of Gears) [also dp]

DP Diamond Pin

DP Diamond Powder

DP Diaphragm Pump

DP Diastolic (Blood) Pressure

DP Dichlorophenoxypropionate

DP Differential Phase [Electronics]
DP Differential Polarography
DP Differential Pressure [also dp]
DP Differential Pulse
DP Diffraction Pattern
DP Diffractive Physics
DP Diffusion Pump
DP Digital Plotter
DP Digital Pulse [Telecommunications]
DP Digit Present
DP 3,4-Dihydroxyphenylalanine [Biochemistry]
DP Diphosgene
DP Dipping Pyrolysis (Process) [also D/P]
DP Direct-Inversion of Pole-Figures [Crystallography]
DP Direct Polarization
DP Direct Potentiometry
DP Direct Probability
DP Disconnection Pending
DP Discontinuous Precipitation [Metallurgy]
DP Disk Pack
DP Dispersed Phase
DP Dispersible Powder
DP Displaced Person
DP Displacement Pump
DP Display Package
DP Distinguished Professor
DP Distributed Processing
DP Distribution Point
DP Doctor of Podiatry
DP Documentary Payment
DP Document Processing
DP Dominican Peso [Currency of the Dominican Republic]
DP Donnan Potential [Physical Chemistry]
DP Double Paper (Covered Magnet Wire)
DP Double-Pass
DP Double Play (Tape)
DP Double-Pole (Switch) [also dp]
DP Double Precision
DP Drafting Panel
DP Draft Proposal
DP Drillpipe [Mining]
DP Drip-Proof
DP Driving Power
DP Drum Plotter
DP Drum Printer
DP Drum Processor
DP Drying Plant
DP Dual Phase
DP Dual-Phase (Steel)
DP Dual-Port (Bus)
DP Dual Property
DP Dulong-Petit (Law) [Thermodynamics]
DP Duplex (Microstructure) [Metallurgy]

DP Durable Press (Fabrics)
DP Dustable Powder
DP Dyadic Product [Mathematics]
DP Dynamic Plasticity
DP Dynamic Positioning [Offshore Drilling Rigs]
DP Dynamic Pressure
DP Dynamic Programming [Mathematics]
D&P Developing and Printing
D&P Drain and Purge
D/P Documents against Payment [also d/p]
D-P Dipping-Pyrolysis (Process) [also DP]
Dp Depth [also dp]
Dp Dystrophin [Immunochemistry]
dP distorted-coherent perfect (crystal) lattice [Symbol]
dP pressure change [Symbol]
dp data processing; data processor [also DP]
dp depth [also Dp]
dp diametral pitch (of gears) [also DP]
dp differential pressure [also DP]
dp 3,4-dihydroxyphenylalanine [Biochemistry]
dp double point [Woven Fabrics]
dp double-pole (switch) [also DP]
d/p documents against payment [also D/P]
dΦ magnetic flux change [Symbol]
dφ entropy change [Symbol]
ΔP Load Range (in Fatigue) [Symbol]
ΔP Optical Path Difference
ΔP Polarization Difference [Symbol]
ΔP Pressure Difference (or Pressure Differential) [Symbol]
Δp change in momentum [Symbol]
Δp pressure difference [Symbol]
ΔΦ* difference in chemical potential [Physical Chemistry]
Δφ misorientation (in crystallography) [Symbol]
2,4-DP 2-(2,4-Dichlorophenoxy)propanoic Acid; 2,4-Dichlorophenoxypropionic Acid
2,4-DP 2,4-Dichlorprop
13D2P 1,3-Diamino-2-Propanol
DPA Data Processing Activities
DPA Data Processing Assembly
DPA Data Protection Act [UK]
DPA Data Protection Agency
DPA Defense Production Act [of US Department of Defense]
DPA Demand Protocol Architecture
DPA Densely Populated Area
DPA Deoxidized Phosphorus Copper, Arsenical
DPA Department of Physics and Astronomy
DPA Destructive Physical Analysis
DPA Deutsches Patentamt [German Patent Office]
DPA Dew-Point Analysis; Dew-Point Analyzer
DPA Dial Pulse Access
DPA Different Premises Address
DPA Diphenolic Acid
DPA Diphenyl Acetylene

DPA Diphenylamine

DPA 9,10-Diphenyl Anthracene [also dpa]

DPA Di(2-Picolinyl)amine

DPA Dipicrylamine

DPA Diploma in Pathological Anatomy

DPA Diploma in Public Administration

DPA Di-n-Propylamine

DPA Directly Productive Activities

DPA Directory Publishers Association [UK]

DPA Displacement per Atom [also dpa]

DPA Distributed Polarizability Analysis

DPA Division of Professional Affairs [of American Association of Petroleum Geologists, US]

DPA Doctor of Public Administration

DPA Document Printing Architecture

dpa Deutsche Presse-Agentur [German Press Agency]

dpa displacement per atom [also DPA]

DPAA Draught Proofing Advisory Association [UK]

DPAC Doubly Perturbed Angular Correlation

DPAF 7,10-Bis(2-Pyridyl) 8,9-Diazafluoranthene [also dpaf]

DPAG Dangerous Pathogens Advisory Group

DPAGE Device Page

dpa/hr displacement per atom per hour [also dpa hr^{-1}]

DPAM Demand Priority Access Method

DPAP Department of Physics and Applied Physics

DPAREN Data Parity Enable

dpa/s displacement per atom per second [also dpa s^{-1}, or dpa/sec]

DPASV Differential Pulse Anodic Stripping Voltammetry

DPAT (Di-n-Propylamino)tetralin

D-PAT Drum-Programmed Automatic Tester

DPB Data Processing Branch

DPB Deep Penetration Bomb

DPB Defense Policy Board [US]

DPB Department of Printed Books [of British Library]

DPB Detonation Physics Branch [of Naval Surface Warfare Center, US Navy at Silver Springs, Maryland]

DPB 1,4-Bis(diphenylphosphino)butane

DPB 1,4-Diphenyl-1,3-Butadiene

DPB Discounted Payback (Period)

DPB Division of Physics of Beams [of American Physical Society, US]

DPB Division of Physics of Beams [of US Department of Energy]

DPB Dodecyl Pyridinium Bromide

DPB Double Position(ing) (Grain) Boundaries [Materials Science]

DPB Drive Parameter Block

DPBC Double-Pole, Back-Connected

DPBCO Dysprosium Praseodymium Barium Copper Oxide (Superconductor)

DPB Newsl Division of Physics of Beams Newsletter [of US Department of Energy]

DPBP Diphenylbutylphosphate

DPBS Dulbecco's Phosphate Buffered Saline [also D-PBS] [Biochemistry]

DPC Data Processing Center

DPC Data Processing Control

DPC Defense Planning Committee [of NATO]

DPC Deformation Potential Coupling [Solid-State Physics]

DPC Demonstration Poloidal Coil

DPC Department of Physical Chemistry

DPC Department of Physics and Chemistry

DPC Departure Control

DPC Desert Protective Council [US]

DPC Design Professionals Coalition [US]

DPC Destination Point Code

DPC Diamond Phase Carbon

DPC Differential Phase Contrast

DPC Differential Photo-Calorimetry

DPC Diphenyl Carbinol

DPC Diphenylcresylphosphate

DPC Direct Program Control

DPC Display Processor Code

DPC Distributed Process Control (System)

DPC Division of Physical Chemistry [of American Chemical Society, US]

DPC Division of Polymer Chemistry [of American Chemical Society, US]

DPC Double Paper Covered (Electric Wire)

DPC Double-Paper, Single-Cotton (Covered Magnet Wire)

$/pc (US) Dollar per Piece

DPcA Dipicrylamine

DPCF Phosphoric Acid Methylphenyl Diphenyl Ester

DPClBHH Di-(2-Pyridyl)-p-Chlorobenzoylketohydrazone

DPCM Delta Pulse-Code Modulation

DPCM Differential Pulse-Code Modulation

DPCM Distributed Processing Communications Module

DPCM Double-Phase Conjugate Mirror (Device)

DPCMA Double-Pass Cylindrical Mirror Analyzer

DPCP trans-1,2-Bis(diphenylphosphino)cyclopentane [also dpcp]

DPCPX 8-Cyclopentyl-1,3-Dipropylxanthine

DPCS Data Processing Consultative Service [of International Social Security Association, Belgium]

DPCS Distributed Process Control System

DPCSD Department for Policy Coordination and Sustainable Development [of United Nations]

DPCSV Differential-Pulse Cathodic Stripping Voltage

DPCSV Differential-Pulse Cathodic Stripping Voltammetry

DPCT Differential Protection Current Transformer

DPCTG Database Program Conversion Task Group [of CODASYL—Conference on Data System Languages]

DPCX Distributed Processing Control Executive

DPD Data Processing Department

DPD Data Processing Director

DPD Data Processing Division

DPD Department of Prints and Drawings [of British Museum]

DPD N,N-Diethyl-4-Phenylene Diamine; N,N-Diethyl-p-Phenylene Diamine

DPD Diffusion Pressure Deficit

DPD Digital Phase Difference

DPD Digital Plane Driver

DPDA Deterministic Pushdown Automaton

DPDA Diperoxydodecanedioic Acid

dΦ/dA radiant flux leaving a surface element *dA*

DPDC Double-Paper, Double-Cotton (Covered Magnet Wire)

DP-DDV Poly(2,3-Diphenyl Phenylene Vinylene)

dP/dh change in (applied) load with displacement

dP/dh slope of unloading segment (i.e., change in load with depth of penetration) (in hardness testing)

DPDL Distributed Program Design Language

DPDM Digital Pulse Duration Modulation

DPDM Diphenyldiazomethane

DPDSe Diphenyldiselenide

DPDT Double-Pole, Double-Throw (Switch) [also dpdt]

dP/dT change in pressure with (absolute) temperature

dP/dt time rate of change of pressure

dp/dT change in (vapor) pressure with (absolute) temperature

dp/dt time rate of change of momentum

dp/dt time rate of change of (vapor) pressure

dΦ/dt time rate of change of magnetic flux [also dφ/dt]

DPDT DB Double-Pole, Double-Throw, Double-Break [also dpdt db]

DPDT SW Double-Pole, Double-Throw Switch [also dpdt sw]

dp/dx pressure gradient (or change in pressure with distance)

DPE Data Processing Equipment

DPE Delayed Proton Emission

DPE Department of Plastics Engineering

DPE Dipentaerythritol

DPE Diphenylethane

DPE 1,2-Bis(diphenylphosphino)ethane

DPE Dipiperidinoethane

DPE Direct Plate Exposure

DPE Distributed Processing Environment

DPEA 2-[2',4'-Dichloro-6'-Phenylphenoxy]ethanamide

DPed Doctor of Pedagogy

DPEHN Dipentaerythritolhexanitrate

DPEK Differential Phase-Exchange Keying

DPEN N,N'-Diphenyl 1,2-Ethanediamine

d-PEP Deuterated Polyethylenepropylene [also dPEP, or DPEP]

DPEX Distributed Processing Executive Program

DPF Data Processing Facility

DPF Dense Plasma Focus

DPF Diesel Particulate Matter

DPF Diisopropyl Fluorophosphate

DPF Division of Particles and Fields [of American Physical Society, US]

DPF Dual Polarized Frequency

DPF Dual Program Feature

DPFC Double-Pole, Front-Connected

DPF/DPB Division of Particles and Fields/Division of Physics of Beams (Conference) [of American Physical Society, US]

DPFGR Demersal and Pelagic Fisheries Research Group

DPFZ Destor-Porcupine Fault Zone [Canada]

DPG Deutsche Physikalische Gesellschaft e.V. [German Physical Society]

DPG Digital Pattern Generator

DPG Diphenylguanidine [Biochemistry]

DPG Diphosphoglycerate; Diphosphoglyceric Acid

DPG Dipropylene Glycol

DPG Dugway Proving Ground [Utah, US]

1,3-DPG 1,3-Diphosphoglyceric Acid

2,3-DPG 2,3-Diphosphoglyceric Acid

DPH Department of Public Health

DPH Diamond Point Hardness

DPH Diamond Pyramid Hardness [also dph]

DPH 1,6-Diphenyl-1,3,5-Hexatriene

DPH Diphenylhydantoin

DPH Diploma in Public Health

DPH Doctor of Public Health

DPH$_{300gm}$ Diamond Pyramid Hardness with 300-Gram Indentation Load [Metallurgy]

DPh Doctor of Philosophy [also DPhil]

DPhA Diphenylamine [also DphA]

D-Phase Decagonal Phase [also d-phase] [Materials Science]

DPhBP Diphenylbutylphosphonate

DPHE Department of Public Health and Environment [Colorado, US]

DPHg Diphenylmercury

DPhil Doctor of Philosophy [also DPh]

DPHN Diamond Pyramid Hardness Number

DPI Data Pathing, Inc. [Now Part of NCR Corporation]

DPI Data Processing Installation

DPI Department of Primary Industries [Australia]

DPI Department of Public Information [of United Nations]

DPI Detail Program Interrelationships

DPI Device/Programmer Interface

DPI Differential Pressure Indicator

DPI Different Premises Information

DPI Digital Pressure Indicator

DPI Digital Process Instrument

DPI Digital Pseudorandom Inspection

DPI Diphenyleneiodonium Chloride

DPI Diphenylimide

DPI Discretized Path Integral

DPI Distributed Protocol Interface

DPI Divisão de Processamento de Imagens [Image Processing Division; of Brazilian National Institute of Space Research]

DPI Dots per Inch [also dpi]

DPIA Diethylenetriamine Producers and Importers Alliance [US]

DPIBF Diphenylisobenzofuran

DPIE Department of Primary Industries and Energy [Australia]

DPIF Department of Primary Industries and Fisheries [Northern Territories, Australia]

DPIH Dual-Pulse Induction Heating

DP Inverse Differential Pulse Inverse Stripping

DPI/O Data Processing Input/Output

DPIP 2,6-Dichlorophenol Indophenol

DPIP Diphenyl Isophthalate

DPIS Differential Pressure Isolation Switch

DP/IS Data Processing/Information System

DPIV Differential Pulse Inverse Voltammetry

DPIV Digital Particle Imaging Velocimetry

DPK Diphenyl Ketimine

DPK Diphenylketone

DPK Di-2-Pyridylketone [also dpk]

DPL Data Processing Language

DPL Descriptor Privilege Level

DPL Detroit Public Library [Michigan, US]

DPL Dual-Polarization Lidar

DPL Dynamic Low Pass

Dpl Diploma [also dpl]

DPLCS Digital Propellant Level Control System

DPLL Digital Phase-Locked Loop

DPLM Domestic Public Land Mobile

DPLMRS Domestic Public Land Mobile Radio Service

Dplr Doppler

DPM Data Processing Machine

DPM Data Processing Management; Data Processing Manager

DPM Decays per Minute

DPM Department of Powder Metallurgy

DPM α,α'-Diaminopimelic Acid [also Dpm]

DPM Diesel Particulate Matter

DPM Differential Pressure Meter

DPM Digital Panel Meter

DPM Diphenylmethane

DPM Bis(diphenylphosphino)methane

DPM (2S,4S)-4-Diphenylphosphino-2-Diphenylphosphinomethylpyrrolidone

DPM Dipivaloylmethane

DPM Disintegrations per Minute [also dpm, or dis/min]

DPM Distributed Presentation Management

DPM Distributive Principle of Multiplication

DPM Division of Personnel Management [of Office of Management, National Institutes of Health, Bethesda, Maryland, US]

DPM Doctor of Podiatric Medicine

DPM Documents per Minute [also dpm]

DPM Dual-Port Memory

DPM Dynamic Preisach (Hysteresis) Model [Solid-State Physics]

dpm dipivaloylmethanoate

dpm disintegrations per minute [also DPM, or dis/min]

DPMA Data Processing Management Association [US]

DPMAS Direct-Polarization Magic-Angle Spinning (Spectroscopy) [also DP/MAS]

DPMC Discontinuously-Reinforced Polymer-Matrix Composite

DPMC Dual Port Memory Control

DPMI DOS (Disk-Operating System) Protected Mode Interface

DPMOAP (Society of) Data Processing Machine Operators and Programmers [US]

DPMM Dots per Millimeter [also dpmm]

d-PMMA Deuterated Polymethyl Methacrylate [also dPMMA, DPMMA, or D-PMMA]

DPMP Bis(diphenylphosphinomethyl)phenyl Phosphine

DPMP Dipentylmethylphosphonate

DPMS Data Project Management System

DPMS Display Power Management Support

DPMSCl Diphenylmethylsilylchloride

D5MTD Dipentamethylene Thiuram Disulfide

DPN Diamond Pyramid Hardness Number

DPN Diphosphopyridine Nucleotide [also βDPN] [Biochemistry]

Dpn Diplococcus pneumoniae [Microbiology]

d(PN) Deoxypolynucleotide [Biochemistry]

α-DPN α-Diphosphopyridine Nucleotide [Biochemistry]

β-DPN β-Diphosphopyridine Nucleotide [also DPN] [Biochemistry]

DPNase Diphosphonucleosidase [Biochemistry]

DPNDA Diphenylnaphthalene Diamine

DPNH Dihydrodiphosphopyridine Nucleotide, Reduced Form [Biochemistry]

DPNH β-Diphosphopyridinium Dinucleotide, Reduced Form [also β-DPNH] [Biochemistry]

α-DPNH α-Dihydrodiphosphopyridine Nucleotide [Biochemistry]

β-DPNH β-Diphosphopyridinium Dinucleotide, Reduced Form [also DPNH] [Biochemistry]

DPNL Distribution Panel [also Dpnl]

DP-NR Deproteinated Natural Rubber

DPNSS Digital Private Network Signaling System [UK]

DP-NTCI N,N'-Diphenylnaphthyltetracarboxylic Imide

DPO Data Phase Optimization

DPO Data Processing Operations

DPO Deployable Payloads Office [of NASA Kennedy Space Center, Florida, US]

DPO Dial Pulse Originating

DPO Diphenylene Oxide

DPO Diphenyloxazole

DPO Diphenyloxide

DPO Divisão de Produtos e Operação [Products and Operations Division; of Brazilian National Institute for Space Research]

DPO Double Pulse Operation

DPO Dynamic Process Optimization

DP ODF Direct-Inversion of Pole-Figures Orientation Distribution Function [Crystallography]

DPodM Doctor of Podiatric Medicine

DPOF Phosphoric Acid 2-Ethylhexyl Diphenyl Ester

DPOS District Planning Officers Society [UK]

DPP 1,3-Bis(Diphenylphosphino)propane

DPP 2,3-Bis(2-Pyridyl)pyrazine [also dpp]

DPP Decimal-Point Programming

DPP Definition Phase Review

DPP Deployment Pointing Panels

DPP Dextran Phosphate Precipitate

DPP Differential Pulse Polarography

DPP Digital Parallel Processor

DPP Diphenylolpropane

DPP 4,7-Diphenyl 1,10-Phenanthroline

DPP Diphenylphosphate

DPP Diphenylphosphinyl

DPP Diphenyl Phthalate

DPP 1,2-Diphenylpropene

DPP Disposable Plotter Pen

DPP Division of Plasma Physics [of American Physical Society, US]

DPP Division of Polar Programs [of National Science Foundation, US]

DPP Drip Pan Pot

DPPA Dipalmitoyl Phosphatidyl Chloride

DPPA 1,2-Bis(diphenylphosphino)acetylene [also dppa]

DPPA 2-(Diphenylphosphino)anisole

DPPA Diphenylphosphoryl Azide

DPPB 1,2-Bis(diphenylphosphino)benzene

DPPB 1,4-Bis(diphenylphosphino)butane [also dppb]

DPPBA 2-(Diphenylphosphino)benzaldehyde

DPPC Dipalmitoyl Phosphatidyl Choline

DPP-Cl Diphenylphosphinyl Chloride

DPPC MLV Dipalmitoyl Phosphatidyl Choline Multilamellar Vesicles

DPPD N,N'-Diphenyl-p-Phenylenediamine

DPPE Bis(diphenylphosphino)ethane; 1,2-Bis(diphenylphosphino)ethene [also dppe]

DPPEN 1,2-Bis(diphenylphosphino)ethene [also DPPEE, dppee, or dppen]

DPPENE 1,2-Bis(diphenylphosphino)ethylene

DPPEO 1,2-Bis(diphenylphosphino)ethane Monoxide

DPPF 1,1'-Bis(diphenylphosphino)ferrocene

DPPH Derivative Peak-to-Peak Height

DPPH 1,1-Diphenyl-2-Picryl Hydrazyl; α,α'-Diphenyl-β-Picryl Hydrazyl

DPPH 2,2-Di(4-tert-Octylphenyl)-1-Picrylhydrazyl; 2,2-Diphenyl-1-Picryl Hydrazyl

DPPHE 1,6-Bis(diphenylphosphino)hex-3-Ene

DPPM Bis(diphenylphosphino)methane [also dppm]

DPPN 3,6-Di(2'-Pyridyl)pyradazine [also dppn]

DPPP 1,3-Bis(diphenylphosphino)propane [also dppp]

dPPP Derivatized Poly(p-Phenylene)

DPPQ 8-(Diphenylphosphino)quinoline

DPPS Dipalmitoylphosphatidylcholine

DPPX Distributed Processing Programming Executive

DPQ 2,2',6,6'-Tetra-tert-Butyldiphenoquinone

DPR Democratic People's Republic (of South Korea) [also DPR Korea]

DPR α,β-Diaminopropionic Acid [also Dpr]

DPR Digital Process Reporter

DPR Division of Petroleum Resources [of Commonwealth Scientific and Industrial Research Organization, Australia]

DPR Double-Pure-Rubber-Lapped (Cable)

(D)PR Dual Precipitation Radar

DPRA Department of Prehistoric and Romano-British Antiquities [of British Museum]

DPrGcDN Dipropyleneglycol Dinitrate

DPRI Disaster Prevention Research Institute

DPR Korea Democratic People's Republic of Korea

DPRR Department of Parks and Renewable Resources [Canada]

DPRS Département Prospective et Synthèse [Department of Prospecting and Synthesis, ONERA-CERT, Toulouse, France]

DPRS/SAE Département Prospective et Synthèse/Système Aéronautiques [Department of Prospecting and Synthesis/Aeronautical Systems, ONERA-CERT, Toulouse, France]

DPRTF Drought Policy Review Task Force

DPS Data Processing Standards

DPS Data Processing Station

DPS Data Processing Subsystem

DPS Data Processing System

DPS Department of Photogrammetry and Surveying

DPS Department of Physical Sciences

DPS Department of Planetary Sciences

DPS Department of Polymer Science

DPS Department of Poultry Science

DPS Department of Probability and Statistics

DPS Descent Power System

DPS Deuterated Polystyrene [also d-PS]

DPS Different Premises Subscriber

DPS (1,3-Dihydro-5,6-Dimethoxy-1-Oxo-2H-Isoindol-2-yl)-4-Benzenesulfonyl

DPS Diphenylstilbene; 4,4'-Diphenylstilbene; trans-p,p'-Diphenylstilbene

DPS Disintegrations per Second [also dps, dis/s, or dis/sec]

DPS Disk Programming System

DPS Distributed Processing System

DPS Dividend per Share

DPS Division for Planetary Sciences [of American Astronomical Society, US]

DPS Doctor of Professional Studies

DPS Doctor of Public Service

DPS Document Processing System

DPS Double-Page Spread

DPS Double-Pole Snap (Switch)

DPS Double Pulse Shaping

DPS Dynamical Peierls Stress [Physics]

DPS Dynamic Plane Source (Technique)

DPS Dynamic Processing System

DP&S Data Processing and Software

D-PS Deuterium-Terminated Porous Silicon

d-PS Deuterated Polystyrene [also dPS, or DPS]

dps disintegrations per second [also DPS, dis/s, or dis/sec]

DPSA Data Processing Suppliers Association [US]

DPSA Deep Penetration Strike Aircraft

DPSA Defense Production Sharing Agreement [US/Canada]

DPSA Dynamic Programming by Successive Approximations

DPSC Defense Personnel Support Center [US]

DPSC Department of Polymer Science and Chemistry

DPSCA Department of Political and Security Council Affairs [of United Nations]

DPS-Cl (1,3-Dihydro-5,6-Dimethoxy-1-Oxo-2H-Isoindol-2-yl)-4-Benzenesulfonyl Chloride

dPS-COOH Carboxylic Acid Terminated Deuterated Polystyrene

DPSE Department of Polymer Science and Engineering

DPSE Piperidinemonoselenocarbamate [also dpse]

DPSe Diphenylselenide

DPSeCl$_2$ Diphenylselenium Dichloride

DPSK Differential Phase-Shift Keying

DPSn Diphenyltin [also DPT]

DPSnCl$_2$ Diphenyltin Dichloride

dPS-NH$_2$ Amine Terminated Deuterated Polystyrene

DPSnO Diphenyltin Oxide

DPSnS Diphenyltin Sulfide

DPSP Defense Production Sharing Program [US/Canada]

DPSP Deferred Profit Sharing Plan

dPS-PVP Diblock Copolymer of Deuterated Polystyrene and Polyvinyl Propylene

DPSS Data Processing Software System

DPSS Data Processing Subsystem

DPSS Data Processing System Simulator

DPSS Direct Program Search System

DPSS Double-Pole Snap Switch

DP&SS Data Processing and Software Subsystem

DPSSC Drugs and Poisons Schedule Standing Committee

DPSSL Diode-Pumped, Solid-State Laser

DPST Double-Pole, Single-Throw [also dpst]

DP Steel Dual-Phase Steel [also DP steel]

DPST SW Double-Pole, Single-Throw Switch

DPSV Differential Pulse Stripping Voltage

DP SW Double-Pole Switch [also DP Sw]

DPT Department of Pharmacology and Toxicology

DPT Design Proof Test

DPT Dew-Point Temperature [Meteorology]

DPT Diagnostic Pressure Tap

DPT Dial Pulse Terminating

DPT Differential Polarization Telegraphy

DPT Different Premises Telephone

DPT Diffuse Phase Transition [Materials Science]

DPT 2,6-Dinitro(bicyclo)pentamethylene-2,4,6,8-Tetramine

DPT 3,7-Dinitro-1,3,5,7-Tetrazabicyclo[3,3,1]nonane

DPT N,N'-Dinitroso Pentamethylene Tetramine

DPT Diphenyltin [also DPSn]

DPT Diphtheria, Pertussis, Tetanus (Vaccine)

DPT Dipropylenetriamine [also dpt]

DPT Ductile-Phase Toughened; Ductile-Phase Toughening [Metallurgy]

Dpt Department [also dpt]

DPTA 1,3-Diamino-2-Propanol-N,N,N',N'-Tetraacetic Acid

DPTA Di-O,O'-Pivaloyl-L-Tartaric Acid

DPTE Deoxidized Phosphorus Copper, Tellurium Bearing

DPTG Dipyrrolidinyl Thiuram Disulfide

DPTH Diphenylthiohydantoin

DPTMDS 1,3-Diphenyl-1,1,3,3-Tetramethyldisilazane

DPTO Diphenyltin Oxide

DPtoTP Display Coordinates to Tablet Coordinates (Conversion)

DPTS Diphenyltin Sulfide

DPTT Dipentamethlene Thiuram-Tetra(hexa)sulfide

DPTT Double-Pole, Triple-Throw

DPTT SW Double-Pole, Triple-Throw Switch [DPTT Sw]

DPTU N,N'-Diphenylthiourea [also dptu]

DPTX Distributed Processing Terminal Exchange

Dpty Deputy [also dpty]

DPU Data Processing Unit

DPU Department of Public Utilities [US]

DPU De Paul University [Chicago, Illinois, US]

DPU Digital Patch Unit

DPU Diphenylurea

DPU Dip Pick-Up

DPU Disk Pack Unit

DPU Dust Preparation Unit

DPU Dynamic Pulse Unit

δ-Pu Delta Plutonium [Symbol]

δ'-Pu Delta Prime Plutonium [Symbol]

DPUD Department of Planning and Urban Design [Victoria State, Australia]

DPV Differential Pulse Voltammetry

DPV Dry Pipe Valve

DPW Department of Public Works

DPWB Diensten Programmatie van het Wetenshapsbeleid van Belgie [Belgian Scientific Grants Service Program]

DPWM Double Pulsewidth Modulation

DPWM Double-Sided Pulsewidth Modulation

Dpy Deploy [also dpy]

DPZ Deutsches Primatenzentrum GmbH [German Primate Center, Goettingen]

DPZn Diphenylzinc

DQ Data Quality

DQ Directory Enquiry

DQ Direct Quenched; Direct Quenching [Metallurgy]

DQ Drawing-Quality (Steel) [Metallurgy]

DQ Dry Quenching

dQ change in heat quantity [Symbol]

dQ change in electron, or ion charge [Symbol]

dQ heat absorbed by a (thermodynamic) system [Symbol]

dQ_{irrev} differential of heat for an irreversible path between initial and final states [Symbol]

dQ_{rev} differential of heat for a reversible path between initial and final states [Symbol]

dq change in electron, or ion charge [Symbol]

ΔQ change in heat quantity [Symbol]

ΔQ heat absorbed by a (thermodynamic) system [Symbol]

Δq change in heat quantity [Symbol]

Δq (electron) charge transfer [Symbol]

δQ heat absorbed by a thermodynamic system [Symbol]

DQA Data Quality Assessment

DQA Design Quality Assurance

DQAB Defense Quality Assurance Board [UK]

DQAK Drawing Quality, Aluminum-Killed (Steel) [Metallurgy]

DQC Data Quality Control

DQC Decagonal Quasicrystal [Materials Science]

DQC Double-Quantum Correlation [Nuclear Magnetic Resonance]

DQCC Deuterium Quadrupole Coupling Constant

DQDB Distributed Queue Dual Bus

$\Delta Q/\Delta m$ Change of Charge with Mass

dQ/dT amount of energy required to produce a unit change in temperature (or time rate of change of amount of heat)

dQ/dt time rate of change of heat quantity (or time rate of heat transfer)

dQ/dt time rate of change of electric charge

dQ_e/dt time rate of flow of radiant energy

DQE Department of Quantum Engineering

DQE Detective (or Detector) Quantum Efficiency

DQENMR Deuterium Quadrupole Echo Nuclear Magnetic Resonance

DQF-COSY Double Quantum Filtered Correlated (or Correlation) Spectroscopy

DQL Data Query Language

DQL Dielectric Quilted Liner

DQM Data Quality Monitor

DQM Digital Quadrature Modulation

DQMC Diffusion Quantum Monte Carlo (Method)

DQN Diazonaphthoquinone Novolak

DQO Data Quality Objective

DQOPP Data Quality Objectives Planning Process

DQPSK Differential Quadrature Phase Shift Keying

DQR Digital Quality Reception

DQRS Drawing-Quality Rimmed Steel

DQSK Drawing-Quality, Special-Killed (Steel)

DQT Double Quantum Transitions

DQTST Detailed Quantum Transition State Theory

D-Quark Down Quark [also d-quark] [Particle Physics]

DQW Double Quantum Well [Solid-State Physics]

DQW-RIT Double Quantum Well Resonant Interband Tunneling [Electronics]

DR Danish Reactor [Denmark]

DR Darkroom

DR Data Rate

DR Data Received; Data Receiver

DR Data Recorder; Data Recording

DR Data Reduction

DR Data Register

DR Dead Reckoning

DR Deadroom

DR Debit [also Dr, or dr]

DR Decay Rate [Nuclear Physics]

DR Defense Research

DR Defined Readout

DR Dehydration Reaction

DR Dehydrogenation Recombination (Process)

DR Delta Ray

DR Depletion Region [Electronics]

DR Deposition Rate

DR Depth Resolution

DR Destructive Read

DR Dextrorotatory

DR Diagnostic Routine

DR Differential Rate

DR Differential Resistivity

DR Diffused Resistor

DR Diffuse Reflectance

DR Diffuse Reflection

DR Digital Radiography

DR Digital Radioscopy

DR Digital Recorder; Digital Recording

DR Digital Resolver

DR Digit Receiver [Telecommunications]

DR Digit Row

DR Dilution Refrigerator

DR Dining Room

DR Diploma in Radiology

DR Dipole Radiation

DR Direction Ranging

DR Directory Routing

DR Direct (Iron Ore) Reduction (Process) [Metallurgy]

DR Disaster Recovery

DR Discontinuous(ly) Reinforced; Discontinuous Reinforcement

DR Discrepancy Report

DR Discrete Regulator

DR Discrimination Radar

DR Disperse Red

DR Distant Range

DR Division of Research

DR Doctor of Radiology

DR Document Reader

DR Document Release

DR Dominican Republic

DR Doryl Resin

DR Double Reduction [Metalworking]

DR Double Resonance

DR Drafting Request

DR Drag Reduction

DR Drift Rate [Electronics]

DR Drilling Regulation

DR Drill Rig

DR Drill Rod

DR Driver's Report

DR Drug-Resistant (Virus)

DR Dynamic Range [Electronics]

DR Dynamically Recrystallized; Dynamic Recrystallization [Metallurgy]

DR-1 First Danish Reactor [Roskilde, Denmark]

DR$ Dominican (Republic) Peso [Currency]

D&R Development and Research

D&R Distiller and Rectifier

D/R Deposit Receipt

D/R Direct or Reverse

D/R DOS (Disk Operating System) Resident (Program)

D/R Downrange

D-R Direct-Reduced (Iron)

Dr Debit [also DR]

Dr Debtor [also dr]

Dr Divisor [also dr]

Dr Doctor

Dr Drachma [Currency of Greece]

Dr Drain [also dr]

Dr Drawer [also dr]

Dr Drill [also dr]

Dr Drive [also dr]

Dr Drum [also dr]

D(r) Radial Distribution Function [Symbol]

dr *(drachma)* – a drachm; a dram [Apothecary]

ΔR Change in Electrical Resistance [Symbol]

ΔR Magnetoresistance

Δρ change in electrical resistivity [Symbol]

Δρ density difference [Symbol]

Δρ electron density contrast (in solid-state physics) [Symbol]

DRA Data Release Authorization

DRA Data Requirements and Analysis

DRA Data Resource Administrator

DRA Dead Reckoning Analyzer

DRA Defense Research Agency [of Royal Aeronautical Establishment, Malvern, UK]

DRA Department of Radio Astronomy

DRA Design Rule Analysis

DRA Discontinuously Reinforced Aluminum (Alloy)

DRA Drachma [Currency of Greece]

DR&A Data Reduction and Analysis

Dra Deinococcus radiophilus [Microbiology]

Dra Draco [Astronomy]

Dra I Deinococcus radiophilus I [Microbiology]

Dra II Deinococcus radiophilus II [Microbiology]

Dra III Deinococcus radiophilus III [Microbiology]

Drac Draco [Astronomy]

DRADP Deuterated Rubidium Ammonium Dihydrogen Phosphate

DRADS Degradation of Radar Defense System

DRAE Defense Research Analysis Establishment

DRAFT Document Read and Format Translator

Drager Rev Drager Review [Published in US]

Dr agr *(Doctor agronomiae)* – Doctor of Agriculture

DRAI Dead Reckoning Analog Indicator; Dead Reckoning Analyzer Indicator

DRAM Digital Recorded Announcement Module

DRAM Dynamic Random-Access Memory [also D-RAM, dRAM, d-RAM, or D-ram]

DRAMA Digital Radio and Multiplex Acquisition

DRAO Dominion Radio Astrophysical Observatory [Canada]

dr ap dram, apothecaries' [Unit]

dr av dram, avoirdupois [Unit]

dr avdp dram, avoirdupois [Unit]

DRAW Direct Read After Write (Memory); Direct React and Write (Memory)

DRB Defense Research Board [Canada]

DRB Design Requirements Baseline

DRB Design Review Board

DRB 5,6-Dichlorobenzimidazole Riboside [Biochemistry]

DRB 5,6-Dichloro-1-β-D-Ribofuranosyl 1H-Benzimidazole

DRBC Defense Research Board of Canada

DR.BOND Dial-up Router Bandwidth On Demand [of Nippon Electric Company, Japan]

DRC Damage Risk Criterion

DRC Data Recording Control

DRC Data Reduction Center

DRC Data Reduction Compiler

DRC Defense Research Committee

DRC Defense Review Committee [of NATO]

DRC Denver Research Center [Colorado, US]

DRC Design Rules Check(ing)

DRC Diamagnetic Ring Current

DRC Discontinuous Reinforced Composite

DRC Discrete Rate Command

DRC Doctor of the Royal College [UK]

DRC Dolphin Research Center [in the Florida Keys, US]

DRC Domestic Resource Cost

DRC Dynamic Recoverable Creep [Mechanics]

DRCL Defense Research Chemical Laboratories [Canada]

DRCM Double Resonance with Coupled Multiplets

DRCMC Discontinuous(ly) Reinforced Ceramic-Matrix Composite

DrComSc Doctor of Commercial Science

DRCR Dynamic Recrystallization Controlled Rolling [**Metallurgy]**

DRCS Distress Radio Call System

DRCS Dynamically Redefinable Character Set

DRD Data Recording Device

DRD Data Requirement Description

DRD Data Requirements Document

DRD Department of Restorative Dentistry

DRD Design Research Division

DRD Detailed Requirements Document

DRD Differential Range De-Ramp

DRD Directorate of Research and Development [of US Air Force]

DRD Director of Research and Development

DRD Direct-Reading Dosimeter

DRD Document Requirement Description

DRD Draw-Redraw

DRD Drill Rig Duty (Motor)

DRDA Distributed Relational Database Architecture

DRDC Detector Research and Development Committee [of CERN–European Laboratory for Particle Physics, Geneva, Switzerland]

dR/dε change in resistance with strain [also dR/dϵ]

dρ/dε change in density with strain [also dρ/dϵ]

dρ/dε change in dislocation density with strain [also dρ/dϵ] [Materials Science]

DrDes Doctor of Design

D-RDF Densified Refuse-Derived Fuel [also -RDF]

DRD-FD Differential Range De-Ramp–Frequency Domain (Processor)

dρ/dγ evolution of dislocation density with strain [Materials Science]

dR/dH change in resistance with (applied) magnetic field

Dr disc pol (*Doctor disciplinarum politicarum*) – Doctor of Social Sciences

DRDO Defense Research and Development Organization [New Delhi, India]

Dr Dobb's J Softw Tools Dr. Dobb's Journal of Software Tools [US]

DR-DOS Digital Research-Disk Operating System [of Digital Research Inc., US]

dR/dT change of resistance with temperature

dR/dT signal sensitivity

dR/dt reducibility of burden [Metallurgy]

dR/dt time rate of change of (electrical) resistance

dr/dt cluster growth rate [Metallurgy]

dρ/dT change of resistivity with temperature

dρ/dt densification rate (i.c., time rate of change of density) [Symbol]

DRDT Division of Reactor Development and Testing [of Argonne National Laboratory, US]

DRDTO Detection Radar Data Takeoff

DRDW Direct Read During Write

DRE Data Recording Equipment

DRE Data Reduction Equipment

DRE Dead Reckoning Equipment

DRE Defense Research Engineering

DRE Defense Research Establishment [of Department of National Defense, Canada]

DRE Department of Resources and Energy

DRE Destruction and Removal Efficiency

DRE Digital Rebalance Electronics

DRE Digital Rectal Exam(ination)

DRE Directional Reservation Equipment

DRE Director of Research and Engineering

DRE Dokumentationsring Elektrotechnik (Database) [Documentation on Electrical Engineering] [of Zentralstelle für Dokumentation Elektrotechnik, Germany]

DRE Pin Diameter of External Spline [Symbol]

D/RE Disassembly/Reassembly Equipment

DREA Defense Research Establishment, Atlantic [of Department of National Defense, Halifax, Nova Scotia, Canada]

DREAC Drum Experimental Automatic Computer

DR/EAF Direct Reduction/Electric-Arc Furnace (Steelmaking Practice) [also DR-EAF] [Metallurgy]

DREAM Data Requirements, Evaluation and Management

DREAM Data Retrieval Entry and Management

DRECT Demonstration of Resource and Energy Conservation Technologies (Program) [of Environment Canada]

DRECT Development (and Demonstration) of Resource and Energy Conservation Technologies (Program) [of Environment Canada] [also D-RECT]

DRED Data Routing and Error Detecting

DRED Double Rocking Electron Diffraction

DREE Department of Regional Economic Expansion [later Department of Regional and Industrial Expansion, Canada]

DREF Data Reference System [of Environment Canada]

DRELL Sid Drell Symposium [at Stanford Linear Accelerator Center, Stanford, California, US]

DRENet Defense Research Establishment Network [Canada]

Dr Eng Doctor of Engineering

DREO Defense Research Establishment, Ottawa [of Department of National Defense, Canada]

DREP Data Replanner

DREP Defense Research Establishment, Pacific [of Department of National Defense, Victoria, British Columbia, Canada]

DRES Defense Research Establishment, Suffield [of Department of National Defense, Suffield, Alberta, Canada]

DRES Direct Reading Emission Spectrograph

DRESS Depth-Resolved Surface-Coil Spectroscopy

DRET Defense Research Establishment, Toronto [of Department of National Defense, Canada]

DRET Directorate des Recherches, Etudes et Techniques de Armement [French Ministry of Defense]

DRET Direct Reentry Telemetry

DR-Et Davy-Reagent Ethyl

DRETS Direct Reentry Telemetry System

DREV Defense Research Establishment, Valcartier [of Department of National Defense, Valcartier, Quebec, Canada]

DREWS Direct Readout Equatorial Weather Satellite

DRF Data Request Form

DRF Deafness Research Foundation [US]

DRF Deutsche Rezeptformeln [German Pharmacopoeia]

DRF Dielectric Response Function

DRF Digital, Radio Frequency

DRF Documentation Requisition Form

dr fl dram, fluid [Unit]

DRFM Digital Radio-Frequency Monitor

Dr forest *(Doctor scientiae forestalium)* – Doctor of Forestry Science

DRFR Division of Research Facilities and Resources [US]

DR/FTIR Diffuse Reflectance/Fourier Transform Infrared Spectroscopy

DRFV Deutscher Radio– und Fernsehfachverband [German Radio and Television Association]

Dr Fx Drill Fixture

DRG Diagnosis Related Group [Medicine]

DRG Digital Ranging Generator

DRG Division of Research Grants [of National Research Council, US] [now Center for Scientific Review]

DRG Dorsal Root Ganglia

DRGS Direct Readout Ground Station

DRH Diffusion-Reaction Hardening

Dr habil *(Doctor habilitatus)* – Habilitated Doctor [European Academic Title]

Dr hc *(Doctor honoris causa)* – Honorary Doctor

Dr Hd Drill Head

DRI Dairy Research Inc. [US]

DRI Data Rate Indicator

DRI Data Reduction Interpreter

DRI Dead Reckoning Indicator

DRI Defense Research Institute [US]

DRI Denver Research Institute [Colorado, US]

DRI Descent Rate Indicator

DRI Desert Research Institute [of University of Nevada, Reno, US]

DRI Differential Refractive Index Detector

DRI Digital Research Inc. [US]

DRI Direct-Reduced Iron [Metallurgy]

DRI Document Retrieval Index

DRI Dynamic Rotation Isomeric (Model)

DRI Pin Diameter of Internal Spline [Symbol]

DRIC Defense Research Information Center [UK]

DRID Direct Readout Image Dissector

DRIDAC Drum Input to Digital Automatic Computer

DRIE Department of Regional and Industrial Expansion [now Industry, Science and Technology Canada]

DRI-EAF Direct-Reduced Iron–Electric-Arc Furnace [Metallurgy]

DRIFT Diffuse Reflectance Infrared Fourier Transform (Spectroscopy)

DRIFT Diversity Receiving Instrumentation for Telemetry

DRIFT Diversity Reliability Instantaneous Forecasting Technique

DRIFTS Diffuse Reflectance Infrared Fourier Transform Spectroscopy

Drilling Compl Fluids Mag Drilling and Completion Fluids Magazine [US]

D-RIM Debinding–Reaction Injection Molding (Process)

DRINC Dairy Research, Inc. [US]

Dr-Ing. *(Doktor-Ingenieur)* – Doctor of Engineering (Sciences)

Dr-Ing-Diss *(Doktor-Ingenieur-Dissertation)* – Doctor of Engineering–Dissertation

DRIP (International Conference on) Defect Recognition and Image Processing

DRIR Direct Readout Infrared

DRIRU Dry Rotor Inertial Reference Unit

DRIS Dynamic Rotational Isomeric State (Model)

DRIVE Document Read Information Verify and Edit

DRIVE Document Review into Video Entry

DRIVPARM (Device) Driver Parameters (Command) [also drivparm]

DRK Deutsches Rotes Kreuz [German Red Cross]

DRK Display Request Keyboard

drk dark

DRKL Defense Research Kingston Laboratory [Ontario, Canada]

DRL Data Requirements Language

DRL Data Requirements List

DRL Data-Retrieval Language

DRL Daytime Running Lights (System)

DRL Defense Research Laboratory [US]

DRL Device Research Laboratory [of University of California at Los Angeles]

DRL Direct Retrieval Language

DRL Document Requirements List

DRL Drilling Research Laboratory

DRLC Double Resonance via Level Crossing

DRLF Double Resonance between the Rotating and Laboratory Frame

Drlg Drilling [also drlg]

DRLM Depolarized Reflected Light Microscope; Depolarized Reflected Light Microscopy

DRM Data Records Management

DRM Depositional Remanent Magnetization

DRM Design Reference Mission

DRM Detrital Remanent Magnetization

DRM Digital Radiometer

DRM Direction of Relative Movement [Navigation]

DRM Discontinuously Reinforced Magnesium (Alloy)

DRM Drafting Room Manual

DRM Drawing Requirements Manual

DRM Drought Relief Measure(s)

DRM Dynamic Recoil Mixing

DR-Me Davy-Reagent Methyl

Dr med *(Doctor medicinae)* – Doctor of Medicine

Dr med univ *(Doctor medicinae universae)* – Doctor of Universal Medicine

Dr med vet *(Doctor medicinae veterinariae)* – Doctor of Veterinary Medicine

DRML Defense Research Medical Laboratories [of Defense Research Board, Canada]

DRMMC Discontinuous(ly) Reinforced Metal-Matrix Composite

DRMO District Records Management Officer

Dr mont *(Doctor montanarium)* – Doctor of Mining

DRMS Data Resource Management System

DRMs Drought Relief Measures

Drn Drain [also drn]

drn drawn

Dr nat techn *(Doctor rerum naturalium technicarum)* – Doctor of Natural Science and Technology

DRNL Defense Research Northern Laboratories [Canada]

DRO Data Request Output

DRO Department of Radiation Oncology

DRO Destructive Readout

DRO Dielectric Resonator Oscillator

DRO Digital Readout

DRO Diesel Reference Standard Organics [Fuel]

DRO Document Release Order

DRO Doubly Resonant Oscillator

DROD Delayed Read-Out Detector

Dr oec *(Doctor oeconomiae)* – Doctor of Economics

Drog Drogue

DROM Decoder Read-Only Memory

DROS Direct Readout Satellite

DROS Disk Resident Operating System

DROP Distribution Register of Pollutants [of US Environmental Protection Agency]

DRP Data Reception Process

DRP Data Reduction Program

DRP Dead Reckoning Plotter

DRP Dense Random-Packed; Dense Random Packing [Crystallography]

DRP Deutsches Reichspatent [German Patent before 1945]

DRP Digital Recording Process

DRP Directional Radiated Power [Antennas]

DRP Distribution Requirements Planning

DRP Distribution Resource Planning

DRP Dystrophin-Related Protein [Biochemistry]

Dr paed *(Doctor paedagogiae)* – Doctor of Pedagogy

DRP angem Deutsches Reichspatent angemeldet [German Patent Pending before 1945]

DRPC Defense Research Policy Committee [UK]

DrPH Doctor of Public Health

Dr pharm *(Doctor pharmaciae)* – Doctor of Pharmacy

Dr phil *(Doctor philosopiae)* – Doctor of Philosophy

Dr phil hc *(Doctor philosopiae honoris causa)* – Honorary Doctor of Philosophy

Dr phil nat *(Doctor of philosophiae naturalis)* – Doctor of Natural Sciences

DRPHS Dense Random Packing of Hard Spheres [Crystallography]

DRPI Digital Rod Position Indicator

DRPMC Discontinuous(ly) Reinforced Polymer-Matrix Composite

DRPS Disk Real-Time and Programming System

DRPS Dynamic Memory Relocation and Protection System

DRQ Data Ready Queue

DRQ Data Request

DRR Data Recorder/Reproducer

DRR Data Review Room

DRR Department of Renewable Resources [Canada]

DRR Design Requirements Review

DRR Digital Radar Relay

DRR Document Release Record

$\Delta R/R$ Change in Reflectance per Reflection

$\Delta R/R$ Magnetoresistance Ratio (or Magnetoimpedance)

$\Delta R/R(0)$ Piezoresistance

$\Delta \rho/\rho$ magnetoresistance (or magnetoresistive effect)

$\Delta \rho/\rho$ relative change in electrical resistivity

$\Delta \rho/\rho_T$ transverse magnetoresistance

$\Delta \rho/\rho_{TL}$ magnetoresistance in transverse to long direction

Dr rer comm *(Doctor rerum commercialium)* – Doctor of Commercial Sciences

Dr rer hort *(Doctor rerum hortensium)* – Doctor of Horticultural Sciences

Dr rer mont *(Doctor rerum montanarium)* – Doctor of Mining Sciences

Dr rer nat *(Doctor rerum naturalium)* – Doctor of Natural Sciences

Dr rer techn *(Doctor rerum technicarum)* – Doctor of Technical Sciences

DRRF Double Resonance in the Rotating Frame

DRRL Digital Radar Relay Link

DRRS Doubly Resonant Raman Scattering

DRS Data Rate Selector

DRS Data Recording Set

DRS Data Relay Satellite

DRS Data Relay Station

DRS Data Retrieval System

DRS Depolarized Rayleigh Scattering

DRS Detection Ranging System

DRS Device Resource (File)

DRS Diffuse Reflectance Spectrometry; Diffuse Reflectance Spectroscopy

DRS Digital Radar Simulator

DRS Digital Range Safety

DRS Direct Reception System

DRS Direct Recoil Spectroscopy

DRS Direct Relay Satellite

DRS Disassembly/Reassembly Station

DRS Discrete Radio Source

DRS Distributed Resource System

DRS Document Registration System

DRS Document Retrieval System

DRS Dominion Research Station [of Health Canada]

DRS Double Radio Source

DRS Driver Resource (File)

DRS Dynamic Reflectance Spectroscopy

.DRS Driver Resource [WordPerfect File Name Extension]

Drs Doctors

Drs Dress(ing) [also drs]

DRSC Direct Radar Scope Camera

Dr sc *(Doctor scientiarum)* – Doctor of Science

Dr sc agr *(Doctor scientiarum agrariarum)* – Doctor of Agricultural Sciences

Dr sc math *(Doctor scientiarum mathematicarum)* – Doctor of Mathematical Sciences

Dr sc nat *(Doctor scientiarum naturalium)* – Doctor of Natural Sciences

Dr sc techn *(Doctor scientiarum technicarum)* – Doctor of Technical Sciences

DRSN Drifting Snow

DRSS Data Relay Satellite System

DRSS Discrepancy Report Squawk Sheet

DRS-VIS Diffuse Reflectance Visible Spectroscopy

DRT Data Reckoning Tracer

DRT Data Recording Terminal

DRT Dead Reckoning Tracer

DRT Del Rio, Texas [Meteorological Station Designator]

DRT Design Reference Timeline

DRT Device Reference Table

DRT Device Rise Time

DRT Digital Real-Time

DRT Diode Recovery Tester

DRT Ductile Recovery Temperature

DR-T Davy-Reagent p-Tolyl

dr t dram, troy [Unit]

$\Delta r^2(t)$ mean square (particle) displacement [Symbol]

DRTC Documentation Research and Training Center [India]

DRTE Defense Research Telecommunications Establishment [now Communications Research Center, Canada]

DRTL Diode-Resistor-Transistor Logic

DRTM Disk Real-Time Monitor

DRTR Dead Reckoning Trainer

DRTS Data Relay Test Satellite [of National Space Development Agency of Japan]

dr tr dram troy [Unit]

DRU Data Recording Unit

DRU Data Reference Unit

DRUC Disposition Record Unsatisfactory Condition

Drug Devel Ind Pharm Drug Development and Industrial Pharmacy [Journal published in the US]

Drug Metabol Dispos Drug Metabolism and Disposition [International Journal]

Drug-Nutr Interact Drug–Nutrient Interactions [International Journal]

Drug Res Drug Research [US Journal]

DRUMS Dynamically Responding Underwater Matrix Sonar

DRUPA Internationale Messe Druck und Papier [International Trade Fair for Printing and Papermaking Machinery and Equipment, Dusseldorf, Germany]

DRV Data Recovery Vehicle

DRV Deep-Diving Research Vehicle

DRV Dynamic Recovery [Metallurgy]

.DRV Device Driver [File Name Extension]

Drv Drive(r) [also drv]

.DRW Draw [File Name Extension]

.DRW Drawing [File Name Extension]

DRX Dynamic Recrystallization [Metallurgy]

Drying Technol Drying Technology [Journal published in the US]

dry qt dry quart [Unit]

DRZ Deutsches Rechenzentrum [German Supercomputer Center located at Darmstadt, Germany]

DS Automotive Crankcase Oil, Diesel Engine, Severe Service Conditions [American Petroleum Institute Classification]

DS (Hydroelectric) Dam Stability

DS Dansk Standardiseringsrad [Danish Standards Association]

DS Dark Space [Electronics]

DS Data Scan(ning)

DS Data Security

DS Data Segment

DS Data Select(ion)

DS Data Send

DS Data Server

DS Data Set

DS Data Sheet

DS Data Size

DS Data Storage; Data Store

DS Data Stream

DS Data Structure

DS Data Switch

DS Data System

DS Date Scale [Astronomy]

DS Days after Sight [also ds, or d/s]

DS Dead Sea [Israel/Jordan]

DS Dead Storage

DS Debugging System

DS Debye-Smoluchowski (Theory) [Physics]

DS Decimal Subtract(er)

DS Deep Screw

DS Deep Space

DS Default Segment [Computers]

DS Defect Scattering [Physics]

DS Define Storage

DS Deflection Spectrometer

DS Degree of Substitution [Cellulose Derivatives]

DS Dektra System

DS Delivery System

DS Denotational Semantics

DS Descent Stage [of NASA Apollo Program]

DS Design Selection

DS Desk Stand

DS Design System

DS Destination Store

DS Deterministic System

DS Deuterated Solvent

DS Deutschlandsender [State-Controlled Broadcasting Station of the German Democratic Republic] [Obsolete]

DS Device Selector

DS Dial System

DS Dibasic Salt

DS Diesel Specialist

DS Difference Spectrophotometry ; Difference Spectrophotometry ; Differential Spectrophotometry; Differential Spectrophotometry

DS Diffraction Scattering

DS Diffraction Spectrum

DS Diffuse Scattering

DS Diffuse Spectrum

DS Digestive System

DS Digital Signal

DS Digital Signature

DS Digital Sum

DS Digit Select [EBCDIC Code]

DS Diode Switch

DS Dip Soldering

DS Directing Station

DS Directionally Solidified; Directional Solidification [Metallurgy]

DS Directional Sensitivity

DS Direct Sequence

DS Disaccharide

DS Discarding Sabot

DS Disconnect Switch

DS Discontinuity Surface [Physics]

DS Discrete System

DS Disintegration Series

DS Disk Storage

DS Disk System

DS Dislocation Source [Materials Science]

DS Dislocation Substructure [Materials Science]

DS Dispersion-Strengthened; Dispersion Strengthening [Metallurgy]

DS Displacement Spike

DS Dissolved Solids

DS Distributed System

DS Distributed Synchronization

DS Distyryl

DS Division of Soils [Commonwealth Scientific and Industrial Research Organization, Australia]

DS Doctor of Science

DS Documents Signed [also ds]

DS Document Supply

DS Dodecyl Sulfate

DS Donor-Solvent (Process)

DS Door Switch

DS Doppler Shift

DS Doppler Sonar

DS Double Salt

DS Double Sided

DS Double Silk (Covered Magnet Wire)

DS Double Space

DS Double Stranded (Electric Wire)

DS Double-Strength (Glass)

DS Downspout

DS Down's Syndrome

DS Downstream

DS Downwelling Solar

DS Drawing Salt

DS Drillship

DS Drill Steel

DS Drill Stem [Oil Drilling]

DS Drive Surface [Numerical Control]

DS Drive System

DS Drop Siding

DS Drum Storage

DS Dual Sludge (Process)

DS Duplex Steel

DS Dynamic Stability

DS Dynamic Storage

DS Dynamic System

DS1 Digital Signal, Level 1

DS2 Digital Signal, Level 2

DS3 Digital Signal, Level 3

D&S Display and Storage

D/S Days after Sight [also d/s, DS, or ds]

D/S Dynamic/Static Analysis

Ds Humid Cold Climate, Dry Summer [Symbol]

dS Entropy Change [also ΔS]

dS Surface Element [Symbol]

ds basal spacing (of polymers) [Symbol]

ds decistere [Unit]

ds document signed [also DS]

d/s days after sight [also D/S, DS, or ds]

ΔS Long-Range Order Parameter Change (in Materials Science) [Symbol]

ΔS Thermoelectric Power (or Thermopower) [Symbol]

$\Delta S^{\dagger}$ Entropy of Activation (or Activation Entropy) [Symbol]

ΔS° Standard Entropy [Symbol]

$\Delta S^{\circ\dagger}$ Standard Entropy of Activation (in Transition State Theory) [Symbol]

ΔS_a Entropy of Activation per Mole of Reactant [Symbol]

ΔS^f Activation Entropy of Formation [Symbol]

ΔS_F Defect Formation Entropy (in Solid-State Physics) [Symbol]

$\Delta \bar{S}_i^M$ Partial Molar Entropy of Mixing of Component i [Symbol]

ΔS^M Entropy Change due to Mixing [Symbol]

ΔS_M Defect Migration Entropy (in Solid-State Physics) [Symbol]

ΔS_t Thermal Entropy of Mixing [Symbol]

ΔS_v Entropy of Vaporization [Symbol]

Δs displacement (or change in distance) [Symbol]

Δs path difference (of signals, waves, etc.) [Symbol]

Δs   phase difference (or phase displacement) [Symbol]

Δs_f° standard molar entropy of fusion [Symbol]

$\Delta \sigma$ change in stress [Symbol]

$\Delta \sigma$ (applied nominal) stress range (for fatigue-crack initiation) [Symbol]

$\Delta \sigma_B$ Bauschinger stress factor (in metallurgy) [Symbol]

$\Delta \sigma_{max}$ maximum stress range (for fatigue-crack initiation) [Symbol]

DSA   Dataroute Serving Area

DSA   Data Set Adapter

DSA   Defense Supply Agency [US]

DSA   Degree of Self-Accommodation [Metallurgy]

DSA   Deoxystreptamine [Biochemistry]

DSA   (Bureau Européen d'Information pour le) Développement de la Santé Animale [European Information Bureau for the Development of Animal Health Products, Belgium]

DSA   Device Specific Adapter

DSA   Dial Service Assistance

DSA   Diffusion Self-Aligned (Metal-Oxide Semiconductor)

DSA   Digital Serving Area

DSA   Digital Signal Analyzer

DSA   Digital Signature Algorithm

DSA   Digital Storage Architecture

DSA   Digital Subtraction Angiography

DSA   Digital Surface Analyzer

DSA   Dimensionally Stable Anode

DSA   Directory System Agent [Internet]

DSA   Direct Selling Association [UK]

DSA   Direct Spectrum Analysis

DSA   Direct Spiral Analysis

DSA   Direct Storage Access

DSA   Dispersion-Strengthened Alloy

DSA   Distributed Systems Architecture

DSA   Divisão de Operações de Satélites Ambientais [Environmental Satellite Operation Division; of Brazilian National Institute for Space Research]

DSA   Division of Scientific Affairs [of NATO]

DSA   Doppler Spectrum Analyzer

DSA   Dozenal Society of America [US]

DSA   Drilling and Sawing Association [UK]

DSA   Drillsite Supervisor Association [US]

DSA   Dynamic Screening Anomaly

DSA   Dynamic Spring Analysis

DSA   Dynamic Storage Allocation

DSA   Dynamic Strain Aging [Metallurgy]

D&SA   Dissolved and Special Alpha (Grade Pulp)

DSAA   Defense Security Assistance Agency [US]

DSAA   Driving School Association of America

DSAD   Data Systems and Analysis Directorate [of NASA Johnson Space Center, Houston, Texas, US]

DSAM   Doppler Shift Attenuation Method

DSAN   Debug Syntax Analysis

DSAP   Data Systems Automation Program

DSAP   Defense Systems Application Program [US]

DSAP   Destination Service Access Point

DSAP   Directory Scope Analysis Program

DSAR   Data Sampling Automatic Receiver

DSARC   Defense Systems Acquisitions Review Council [US]

DSAU   DSI (Digital Speech Interpolation) Signal Access Unit

DSB   Data Set Block

DSB   (Car) Dealer Service Bulletin

DSB   Defense Science Board [US]

DSB   Dial System B-Position; Dial System B-Switchboard

DSB   Direct Sounding Broadcast

DSB   Dispersion-Strengthened Brass

DSB   Dispersion-Strengthened Bronze

DSB   Document Status Bulletin

DSB   Double Schottky Barrier (Model) [Solid-State Physics]

DSB   Double Sideband (Modulation)

DSB   Drug Supervisory Body [Switzerland]

DSBAM   Double Sideband, Amplitude Modulation [also DSB-AM]

DSBAMRC   Double Sideband, Amplitude Modulation, Reduced Carrier [also DSB-AM-RC]

DSB Defect   Double Schottky Barrier with a Defect (Model) [Solid-State Physics] [also DSB defect]

DSBEC   Double Sideband, Emitted Carrier [also DSB-EC]

Dsbg   Disbursing [also dsbg]

DSBM   Double Sideband Modulation

DSBPP   Di-sec-Butylphenylphosphonate

DSBRC   Double Sideband, Reduced Carrier [also DSB-RC]

DSBSC   Double Sideband, Suppressed Carrier [also DSB-SC]

DSBSCAM   Double Sideband, Suppressed Carrier Amplitude Modulation [also DSB-SC-AM]

DSBSCAM w/QM   DSBSCAM with Quadrature Multiplexing [also DSB-SC-AM w/QM]

DSBSCASK   Double Sideband, Suppressed Carrier Amplitude Shift Keyed [also DSB-SC-ASK]

DSBTC   Double Sideband, Transmitted Carrier [also DSB-TC]

DSC   Danbury State College [Connecticut, US]

DSC   Data Separator Card

DSC   Data Service Center

DSC   Data Set Controller

DSC   Data Stream Compatibility

DSC   Data Synchronizer Channel

DSC   Dedicated Signal Conditioner

DSC   Deep Sound Channel

DSC   Delaware Safety Council [US]

DSC   Delaware State College [Dover, US]

DSC   Delta State College [Ckleveland, Mississippi, US]

DSC   Design Safety Criteria

DSC   Differential Scanning Calorimeter; Differential Scanning Calorimetry

DSC   Directionally Solidified Czochralski (Method) [Materials Science]

DSC   Direct Satellite Communications

DSC   Direct Semiconductor

DSC   Direct Strip Casting [Metallurgy]

DSC   Dispersion-Strengthened Copper

DSC Displacement Shift Complete (Lattice) [Materials Science]

DSC Distinguished Service Cross

DSC N,N'-Disuccinimidyl Carbonate

DSC District Switching Center

DSC Disuccinimidyl Carbonate

DSC Doctor of Commercial Science

DSC Doctor of Surgical Chiropody

DSC Document Supply Center

DSC Double-Silk-Covered (Electric Wire) [also dsc]

DSC Dry Storage Container

DSc Doctor of Science

Dsc Discount [also dsc]

DSCA Default System Control Area

DScA Doctor of Science in Agriculture

DScAdmin Doctor in Administrative Science

DSCB Data Set Control Block

DSCC Deep Space Communications Complex

DSCC Design Set Communications Controller

DScC Doctor of Commercial Science

DSCCD Discount Code

DScCom Doctor of Commercial Science

DScEng Doctor of Science and Engineering

DSCF Digital Switched Capacitor Filter

DSCG Disodium Chromoglycate

DSCH Dual Service Channel

DSCIM Display-System Computer Input Multiplexer

DSCL Displacement Shift Complete Lattice [Materials Science]

DScMil Doctor in Military Sciences

DSCN Dispersion-Strengthened Cupro-Nickel

DScNat Doctor of Natural Sciences

DSCNO Dysprosium Strontium Copper Niobium Oxide (Superconductor) [also DySCNO]

DSCO Dysprosium Strontium Cobalt Oxide (Superconductor) [also DySCO]

Dscrm Discriminator [also dscrm]

DSCS Defense Satellite Communications System [US]

DSCS Defense System Communications Satellite [US]

DSCS Deskside Computer System

DSCSII Defense Satellite Communications System, Phase II [US]

DSCT Double Secondary Current Transformer

DScTech Doctor of Science and Technology

DSC/TGA Differential Scanning Calorimeter/Thermogravimetric Analyzer; Differential Scanning Calorimetry/Thermogravimetric Analysis

DSD Dark-Spot Defect (of Lasers)

DSD Data-Set Definition

DSD Data-Systems Division [of International Telephone and Telegraph, US]

DSD Defense Systems Division

DSD Digital System Design

DSD Digital System Diagram

DSD Diamond-Square-Diamond (Mechanism)

DSD Direct Service Dialing

DSD Direct Stream Digital

DSD Downstream Detector

DSDD Double-Sided, Double-Density (Disk) [also DS/DD]

dσ/dε work hardening rate (or strain hardening rate) [also dσ/dϵ]

DSDL Data Storage Description Language

DSDMA Distearyl Dimethyl Ammonium

DSDMA OH Distearyl Dimethyl Ammonium Hydroxide

dsDNA Double-Stranded Deoxyribonucleic Acid (Molecule) [Biochemistry]

dΣ/dΩ Macroscopic Differential (Neutron) Scattering Cross-Section

DSDP Deep Sea Drilling Project [of Joint Oceanographic Institutions for Deep Earth Sampling; from 1964 to 1983]

DSDS Dataphone Switched Digital Service [US]

DSDSC Dual Sample Differential Scanning Calorimetry

DSDT Data-Set Definition Table

dS/dt time rate of change of entropy

ds/dt velocity (i.e., time rate of change with distance)

dσ/dT change in stress with temperature

Δs/Δt relative velocity (i.e., change in distance with time)

DSE Dartmouth Society of Engineers [US]

DSE Data Set Extension

DSE Data Storage Equipment

DSE Data-Switch(ing) Exchange

DSE Data Systems Engineering

DSE Debye-Sears Effect [Physics]

DSE Debye-Stokes-Einstein (Law) [Physics]

DSE Deutsche Stiftung für internationale Entwicklung [German Foundation for International Development]

DSE Directionally Solidified Eutectic [Metallurgy]

DSE Direct Switching Equipment

DSE Direct Switching Exchange

DSE Distributed Systems Engineering [Journal of Institute of Physics, UK]

DSE Distributed Systems Environment

DSE Domestic Sewage Exclusion

DSEA Data Storage Electronic Assembly

DSEA Display Station Emulation Adapter

DSECT Data Section [also Dsect, or dsect]

DSECT Dummy (Control) Section

DSEE Directorate for Science and Engineering Education [of National Science Foundation, US]

DSEG Current Value of Data Segment Register [also DSeg]

DSEG Data Systems Engineering Group

DSEM Danish Society for Engineering Metrology

DSEM Dynamic Scanning Electron Microscopy

DSEP Distribution System Expansion Program

DSF Data Secured File

DSF Directional Solidification Furnace [Metallurgy]

DSF Disk Storage Facility

DSF Double Stacking Fault [Materials Science]

DSF Dynamic Structure Factor

DSFS Discarding Sabot Fin Stabilized (Projectile)

DSG Daido Super Gear (Steel)

DSG Data Standards Group [of International Organization for Plant Information]

DSG Deutsche Schlafwagen– und Speisewagen-Gesellschaft [German Sleeping and Dining Car Company]

DSG Development Studies Group [Peru]

DSG Digital Signal Generation; Digital Signal Generator

DSG Double Strength Glass

Dsgn Design [dsgn]

DSgt Detective Sergeant

DSH Deactivated Shutdown Hours

DSHD Double-Sided, High-Density (Disk)

DSHO Derivative Spectrometry, Higher Order

$/SHTU Dollars per Short Ton Unit [also $/shtu]

DSI Dairy Society International [US]

DSI Data Storage Institute [of National University of Singapore]

DSI Depth-Sensing Indentation (Method) [Hardness Testing]

DSI Digital Speech Interpolation

DSI Digital Speech Interpretation

DSI Direct Sample Insertion

DSI Dislocation-Solute Interaction [Solid-State Physics]

DSI Don't Say It, Write It

DSI Duration of Sustained Injection

DSI Dynamic System Interchange

DSID Data Set Identification

DSID Direct Sample Insertion Device

DSIF Deep Space Instrument(ation) Facility [of NASA]

DSig Digital Signatures [Internet]

DSIMS Dynamic Secondary Ion Mass Spectrometry [also D-SIMS]

DSIMS Dynamic Secondary Ion Mass Spectroscopy [also D-SIMS]

DSIR Department of Scientific and Industrial Research [UK]

DSIR/TIDU DSIR Technical Information and Documentation Unit [UK]

DSIS Defense Scientific Information Service [Canada]

DSIS Department of Scientific Information Services [of Department of National Defense, Canada]

DSIS Digital Software Integration Station

DSIS Directorate of Scientific Information Services

DSIS Distributed Support Information Standard

DSIS 2-Selenoxo 1,3-Diselenole-4,5-Diselenoate [also dsis]

DSITMS Direct Sampling Ion Trap Mass Spectrometry

Dsk Disk [also dsk]

DSKFC Double Skeleton Fuel Cell

DSL Data Set Label

DSL Data Simulation Language

DSL Data Structures Language

DSL Data System Language

DSL Datura Stramonium Lectin [Biochemistry]

DSL Deep Scattering Layer

DSL Deep Submergency Laboratory [of Woods Hole Oceanographic Institution, US]

DSL Detroit Signal Laboratory [of US Army in Michigan]

DSL Digital Simulation Language

DSL Digital Subscriber Line

DSL Domestic Substances List [of American Chemical Society, US]

DSL Drawing and Specification Listing

DSL Dynamic Simulation Language

DSL Dynamic Super Loudness

Dsl Diesel [also dsl]

DSLAM Digital Subscriber Line Access Multiplexer

DSLC Data Subscriber Loop Carrier

DSLC Defense Logistics Services Center [of US Department of Defense]

DSLC Domestic Substances List of Canada

D-sleep Desynchronized (or Dreaming) Sleep [also known as REM sleep]

DSLET [D-Serine, Threonine] Leucine-Enkephalin [Biochemistry]

DSLO Distributed Systems Licensing Option

DSM Deep Submicron (Deposit) [Physics]

DSM Defense Suppression Missile

DSM Deutsches Schiffartsmuseum [German Maritime Museum, Bremerhaven]

DSM Development of Substitute Materials

DSM Diagnostic and Statistical Manual (of Mental Disorders) [of American Psychiatric Association] [*Note:* The Acronym "DSM" is Often Followed by a Roman Numeral Indicating the Edition, e.g., DSM-IV Denotes the 4th Edition of the Diagnostic and Statistical Manual (of Mental Disorders)]

DSM Digital Scanning (Electron) Microscopy

DSM Digital Simulation Model

DSM Directorate of Stockpile Management [of US Department of Defense]

DSM Direct Signal Monitoring

DSM Direct Stiffness Method

DSM Disk Space Management

DSM Distinguished Service Medal

DSM Divisione Scienza dei Materiali [Materials Science Division, Italy]

DSM Dutonium 236 Space Modulator

DSM Dutch State Mines

DSM Dynamic (Light) Scattering Mode

DSM-II Diagnostic and Statistical Manual (of Mental Disorders), Second Edition [of American Psychiatric Association]

DSM-III Diagnostic and Statistical Manual (of Mental Disorders), Third Edition [of American Psychiatric Association]

DSM-III-R Diagnostic and Statistical Manual (of Mental Disorders), Third Revised Edition [of American Psychiatric Association]

DSM-IV Diagnostic and Statistical Manual (of Mental Disorders), Fourth Revised Edition [of American Psychiatric Association]

DSM-V Diagnostic and Statistical Manual (of Mental Disorders), Fifth Revised Edition [of American Psychiatric Association]

DSM1 Dynamic (Light) Scattering Mode 1

DSM2 Dynamic (Light) Scattering Mode 2

D&SM Dressed (One, or Two Sides) and Standard Matched [Lumber]

DSMA Digital Sense Multiple Access [Telecommunications]

DSMA Disodium Methanearsenate

DSMA Door and Shutter Manufacturers Association [UK]

DSMA/CD Digital Sense Multiple Access with Collision Detect [Telecommunications]

DSMC Digital Servo Motor Control

DSMC Direct Simulation Monte Carlo (Method)

D/SMC Dough/Sheet Molding Compound

DSM/GSA Directorate of Stockpile Management and General Services Administration [US]

DSMIT Distributed System Management Interface Tool

DSMG Designated Systems Management Group [of US Air Force]

DSMO Dysprosium Strontium Manganese Oxide (Superconductor) [also DySMO]

DSMR Dual-Stripe Magnetoresistance (Head)

DSMS Depth-Selective Moessbauer Spectroscopy

DSMS Document Service Management System

DSMSB Die Set Manufacturers Service Bureau [US]

DSMT Directorate of Space and Missile Technology [of US Air Force at Edwards Air Force Base, California]

DSMZ Deutsche Sammlung von Mikroorganismen und Zellkulturen GmbH [German Collection of Microorganism and Cell Cultures, Braunschweig]

DSN Data Set Name

DSN Deep Space Network [of NASA]

DSN Delivery Service Notification

DSN Direct Strong Nitric (Process)

DSN Dispersion-Strengthened Nickel

DSN Distributed System Network

DSNAME Data Set Name

DS-NBT Distyryl Nitroblue Tetrazolium Chloride

DSO Data Set Optimizer

DSO Data Systems Office

DSO Detailed Supplementary Objective

DSO Digital Storage Oscilloscope

DSO Distinguished Service Order

DSO Distributed System Network

DSO Di(N-Succinimidyl)oxalate

DSO Dynamic Scattering Object

DSOBIPS Dynamic, Spatially-Oriented, Biophysical Inventory Projection System

DSOP Domain of Structurally Ordered Population (on Compounds) [Materials Science]

DSOR Domain of Structurally Ordered Reactions [Materials Science]

DSORG Data Set Organization

DSOM Distributed System Object Model

DSOP Draft Statement of Principles [of Australian Heritage Commission]

DSOS Data Switch Operating System

DSP Dedicated Sensor Process

DSP Defense Services Program

DSP Defense Support Program [of US Air Force]

DSP Densified with Small Particles [Materials Science]

DSP Desilication Product

DSP Diarrheic Shellfish Poisoning

DSP Dibasic Sodium Phosphate

DSP Digital Signal Processing; Digital Signal Processor

DSP Directory Synchronization Protocol

DSP Disodium Phosphate

DSP Display Systems Protocol

DSP Dissolving Sulfite Pulp

DSP Distinguished Service Professor

DSP Distributed System Program

DSP Division Simplification Plan

DSP Downstream Processing

DSP Dryland Salinity Program [South Australia State]

DSP Dual Substituent Parameter

DSP Dynamic Support Program

DSP Phenylphosphonodithious Acid Bis(2-Methylphenyl)ester

DSP-4 2-Bromo-N-(2-Chloroethyl)-N-Ethyl Benzenemethanamine

dsp (decessit sine prole) – died without issue

d-SPAN Deuterated Sulfonated Polyaniline

DSPB Digital Signal Processing Board

DSPC Direct Shell Production Casting [Metallurgy]

DSPC Dynamical Spin Pair Correlation

DSPES Depth-Selective PhotoelectronSpectroscopy

DSPI Digital Speckle-Pattern Interferometry

Dspl Display [also dspl]

DSPLAB Digital Signal Processing Laboratory

DSPLC Display Controller

DSPM Digital Signal Processing Multiprocessor

Dspn Dispensary [also dspn]

DSP-NLR Dual Substituent Parameter–Nonlinear Resonance

DSPP Disodium Phenylphosphate

DSPT Dressed State Perturbation Theory [Physics]

DSQD Double-Sided, Quad-Density (Disk)

DSR Data Scanning and Routing

DSR Data Set Ready

DSR Data Storage and Retrieval [also DS&R]

DSR Debt Service Ratio

DSR Degree of Shape Recovery

DSR Device Status Register

DSR Device Status Report

DSR Digit Storage Relay

DSR Digital Stepping Recorder

DSR Director of Scientific Research and Experiments Department [US Navy]

DSR Discriminating Selector Repeater

DSR Divisão de Sensoreamento Remoto [Remote Sensing Division; of Brazilian National Institute of Space Research]

DSR Dynamic Shear Rheometer

DS&R Data Storage and Retrieval [also DSR]

DSRC David Sarnoff Research Center [Subsidiary of SRI International, Princeton, New Jersey, US]

DSRI Danish Space Research Institute [Copenhagen]

DSRI Digital Standard Relational Interface

DS RMIC Directionally Solidified Refractory Metal Intermetallic Composite

dsRNA Double-Stranded Ribonucleic Acid (Molecule) [Biochemistry]

DSRO Directional Short-Range Order [Solid-State Physics]

DSRT Deletable Soft Return [also DSRt]

DSRV Deep-Sea Research Vessel

DSRV Deep-Sea Search and Rescue Vessel

DSRV Deep-Submergence Rescue Vehicle [US Navy]

DSRW Dry-Sand, Rubber-Wheel

DSS Data Support Section [of National Science Foundation, US]

DSS Data Switching System

DSS Decision Support Software

DSS Decision Support System

DSS Deep-Space Station

DSS Deep Submergence System

DSS Department of Social Security [Australia]

DSS Department of Supplies and Services [now Public Works and Government Services of Canada]

DSS Diastereoselectivity [Chemistry]

DSS Digital Satellite System

DSS Digital Signature Standard

DSS Digital Subsystem

DSS Digital Switching Subsystem

DSS 2,2-Dimethyl-2-Silapentane-5-Sulfonate; 2,2-Dimethyl-2-Silapentane-5-Sulfonic Acid

DSS Diploma of Specialized Studies

DSS Director of Statistical Services

DSS Direct Station Selection

DSS Discrete Sync System

DSS Disk Support System

DSS Divisão de Sistemas de Solo [Ground Systems Division; of Brazilian National Institute for Space Research]

DSS Division of Safety Studies

DSS Doctor of Social Science

DSS Document Storage System

DSS Double-Shell Slurry [Nuclear Engineering]

DSS Double-Sided Scrubber

DSS Dry Saturated Steam

DSS Duplex Stainless Steel

DSS Dynamic Support System

DSS 3-(Trimethylsilyl)-1-Propanesulfonic Acid

ΔS/S(0) Piezothermopower

DSSA Defense Security Assistance Agency

DSSA Dental System Suppliers Association [UK]

DSSB Double Single-Sideband

DSSC Data Storage System Center

DSSC Department of the Secretary of State of Canada

DSSC Double-Sideband Suppressed Carrier

DSSc Diploma in Sanitary Sciences

DSSCS Defense Special Secure Communications System

DSSD Double-Sided, Single-Density (Disk) [also DS/SD]

DSSE Department of Solid-State Electronics

DSSF Double-Shell Slurry Feed [Nuclear Engineering]

DSSI Digital Standard Systems Interconnect [of Digital Equipment Corporation]

DSSM Dynamic Sequencing and Segregation Model

DSSP Deep Submergence Systems Project [US Navy]

DSSP Defense Standardization and Specification Program [US]

DSSPTO Deep Submergence Systems Project Technical Office [US Navy]

DSSRG Deep Submergence Systems Research Group [US Navy]

DSSS Direct-Sequencing Spread Spectrum

DS-SSDA Direct Sequence-Spread Spectrum Multiple Access

DSSSL Document Style Semantics and Specification Language [Internet]

DSSV Deep Submergence Search Vehicle [US Navy]

DST Data Summary Tape

DST Data Support Tools [also dst]

DST Data System Test

DST Daylight Saving Time [also dst]

DST Department of Science and Technology [of Indian Institute of Science]

DST Diamond Science and Technology

DST Disk Storage Terminal

DST Division of Steel Technology

DST Double Set Trigger

DST Double-Shelled Tank

DST Double Summer Time

DST Drill Stem Test(ing)

Dst District [also dst]

DSTC Differential Screw Thread Compression

DSTCD District Code

DSTD Division of Sexually Transmitted Diseases [of Centers for Disease Control and Prevention, Atlanta, Georgia, US]

DSTE Data Subcarrier Terminal Equipment

DSTE Data Subscriber Terminal Equipment

D-STEM Double-Rocking Scanning Transmission Electron Microscope; Double-Rocking Scanning Transmission Electron Microscopy

Dstln Distillation [also dstln]

DSTN Double Supertwisted Nematic

Dstn Destination [also dstn]

DSTP Draft Site Treatment Plan

DSTO Defense Science and Technology Organization [Australia]

DSTOA Defense Science and Technology Organization of Australia

D-stoff Phosgene

Dstr Distributor [also dstr]

DStV Deutscher Stahlbau-Verband [German Association for Steel Construction]

DSU Data Service Unit

DSU Data Storage Unit

DSU Data Switching Unit

DSU Data Synchronization Unit; Data Synchronizer Unit

DSU Delta State University [US]

DSU Device Switching Unit

DSU Digital Service Unit

DSU Directorate Services Unit [of CERN–European Laboratory for Particle Physics, Geneva, Switzerland]

DSU Disk Storage Unit

DSU Drum Storage Unit

DSV Deep-Sea Vessel; Deep-Submergence Vehicle

DSV Densovirus

DSV Digital Sum Variation

DSV Diving Support Vessel

DSV Double Silk Varnished (Magnet Wire)

DSVD Digital Simultaneous Voice and Data

DSVT Digital Secure Voice Telephone

DSW Data Status Word

DSW Device Status Word

DSW Direct Step On Wafer

DSWA Dry Stone Walling Association (of Great Britain)

DSWL Downstream Water Level

DSWV Directorate of Special Weapons and Vehicles [UK]

DSX Digital Signal Cross-Connection (Equipment)

DSX Distributed System Executive

DT Data Tag

DT Data Terminal

DT Data Transfer

DT Data Translator

DT Data Transmission

DT Day Tracer

DT Dead Time

DT Debye Theory [Physics]

DT Decay Time

DT Decision Table

DT Decision Technology

DT Deflection Temperature

DT Deformation Twin(ning) [Metallurgy]

DT Delay Time

DT Delirium Tremens [also dt] [Medicine]

DT Depressed Temperature

DT Design Temperature

DT Design Test

DT Desk-Top

DT Destructive Test(ing)

DT Detection Theory

DT Deuterium-Tritium (Fuel)

DT Developed Template [Metallurgy]

DT Dial(ling) Tone

DT Differential Thermometer

DT Differential Time

DT Differential Transducer

DT Differentiating Transformer

DT Diffraction Theory

DT Diffused Transistor

DT Diffuse Transmittance

DT Digitizing Tablet

DT Digital Technique

DT Digital Translator

DT Digital Transmission

DT Digit Tube

DT Digroup Terminal

DT Dihydrotachysterol [Biochemistry]

DT Dipole Theory

DT Dip Test

DT Direct Tension [Mechanics]

DT Discharge Tube

DT Discrete Time

DT Disk-Tape

DT Displacement Transducer

DT Display Terminal

DT Distillation Test

DT Distillation Tower

DT Distributed Transaction

DT Document Title

DT Document Type

DT Domain Theory [Solid-State Physics]

DT Double-Throw

DT Double Torsion (Test Specimen) [Mechanics]

DT Double Triacontahedron [Crystallography]

DT Doubling Time [Nuclear Engineering]

DT Downtime

DT Drain Tube

DT Drawn Tube

DT Drift Tube [Nuclear Engineering]

DT Drilling Technician

DT Drilling Technologist; Drilling Technology

DT Drive Train

DT Drop Top

DT Drop Tower

DT Drop Tube

DT Drude-Tronstad (Polarimeter) [Physics]

DT Dry Ton [Unit]

DT Dual Tandem

DT Ductility Transition (Temperature) [Metallurgy]

DT Dummy Target

DT Dwell Time

DT Dynamic Tear (Test)

DT$_{50}$ Time for 50% Loss [Symbol]

D²T² Dye Diffusion Thermal Transfer (Printing) [also D2T2]

D(T) Temperature-Dependent Diffusion [Symbol]

D&T Design and Test(ing)

D/T Deuterium/Tritium (Ratio) [also D-T]

D/T Disk/Tape

D-T Deuterium-Tritium (Ratio) [also D/T]

Dt Date [also dt]

D(t) Time-Dependent Diffusion [Symbol]

D$_s$(t) Time-Dependent Self-Diffusion [Symbol]

D/t Tube (or Pipe) Ratio (i.e., Outside Diameter to Wall Thickness) [Symbol]

$/t Dollars per ton [Unit]

dT Deoxythymidine [Biochemistry]

dT 2'-Deoxyribsylthymine [Biochemistry]

dT$_c$ Change in Critical Temperature [Symbol]

d4T 2',3'-Didehydro-3'-Deoxythymidine [Biochemistry]

dt change in temperature [Symbol]

dt delirium tremens [also DT] [Medicine]

dt time interval (or change in time) [Symbol]

ΔT Difference (or Change) in (Absolute) Temperature [Symbol]

ΔT$_b$ Change in Boiling Point [Symbol]

ΔT$_f$ Change in Freezing Point [Symbol]

Δt time interval (or change in time) [Symbol]

Δt difference (or change) in (Celsius) temperature [Symbol]

Δt$_{ij}$ (acoustic-emission) arrival time interval [Symbol]

Δθ linewidth [Symbol]

Δθ temperature difference (or change) [also Δϑ]

DTA Detailed Traffic Analysis

DTA Detroit Tooling Association [US]

DTA Development Test Article

DTA Differential Thermoanalysis; Differential Thermal Analysis; Differential Thermal Analyzer

DTA Disk Transfer Area

DTA Dispersion-Toughened Alumina

DTA Dominion Traffic Association [Canada]

.DTA Data [File Name Extension] [also .dta]

1,2-DTA 1,2-Dithiane

DTABr Dodecyl Trimethyl Ammonium Bromide [also DTAB]

DTACK Data Transfer Acknowledgement

DTA/DTG Differential Thermal Analysis/Differential Thermal Gravimetry (Analysis); Differential Thermoanalysis/Differential Thermogravimetry (Analysis)

DTAF 5-([4,6-Dichlorotriazin-2-yl]amino)fluorescein (Dihydrochloride)

6-DTAF 6-([4,6-Dichlorotriazin-2-yl]amino)fluorescein (Dihydrochloride)

DTAF-Dextran 5-([4,6-Dichlorotriazin-2-yl]amino)fluorescein Dextran (Dihydrochloride)

DTAM Descend to and Maintain

DTAN 2,2'-Dithio-bis(1-Aminonaphthalene)

DTARS Digital Transmission and Routing System

DTAS Data Transmission and Switching System

DTA-TG Differential Thermoanalysis–Thermogravimetry; Differential Thermal Analysis–Thermal Gravimetry

DTAU Digital Test Access Unit

DTB Danmarks Tekniske Bibliotek [Technical Library of Denmark]

DTB Decimal-to-Binary

DTB Desaminotrosyl-Trosine Alkyl Butyl

DTB Dynamic Translation Buffer

δ-Tb Delta Terbium [Symbol]

DTBB18C6 Di-tert-Butylbenzo-18-Crown-6

DTBC 3,5-Di(t-Butyl)catechol

DTBP Dedicated Total Buried Plant

DTBP Dimethyldithiobis(propionimidate)

DTBP Di-tert-Butyl Peroxide

DTBP 2,6-Di-tert-Butylphenol

DTBPP 1,3-Bis(Di-t-Butylphosphino)propane [also dtbpp]

DTBS Di-tert-Butylsilane

DTBSCl2 Di-tert-Butyldichlorosilane

DTC DACS (Data Acquisition and Control System) Termination Cabinet

DTC Danish Trade Commission

DTC Data Transfer Control(ler)

DTC Data Transmission Center

DTC Data Transmission Control

DTC Dead-Time Compensation

DTC Dead-Time Correction

DTC Decision Threshold Computer

DTC Denver Technological Center [Colorado, US]

DTC Department of Technical Cooperation [UK]

DTC Department of Transport and Communications [Australia] [also DT&C]

DTC Depository Trust Company [of New York Stock Exchange, US]

DTC Desert Tortoise Council [US]

DTC Design-to-Cost (Process)

DTC Desk-Top Computer

DTC Desk-Top Conferencing

DTC Detection Threshold Computer

DTC Diagnostic Trouble Code

DTC Differential Thermal Coating

DTC Dimethyltrimethylene Carbonate

DTC Diode Transistor Compound

DTC Discrete-Time Control

DTC Dithiocarbamate

DTC Doppler Translation Channel

DT&C Department of Transport and Communications [Australia] [also DTC]

D2C Decimal To Character [Restructured Extended Executor (Language)]

DTCD Department of Technical Cooperation for Development [of United Nations]

1,5-DTCO 1,5-Dithiacyclooctane

DT Co Drug Trading Company Limited [Toronto, Canada]

DTCP Diode Transistor Compound Pair

DTCP Division of Tropical Crops and Pastures [of Commonwealth Scientific and Industrial Research Organization, Australia]

DTCS Digital Test Command System

DTCS Discrete-Time Control System

DTCS DynCorp Tri-Cities Services Inc. [Richland, Washington, US]

DTCU Data Transmission Control Unit

DTD Data Transfer Done

DTD Dimethyl Tin Difluoride

DTD Document Type Definition

dtd let such doses be sent [Medical Prescriptions]

DIII-D Doublet Three-D (Tokamak) [San Diego, California, US]

DTDA Di(tridecylamine)

DTDD Diheptyl Tetramethyl Dioxaundecanediamine

DTDP Di-Tridecyl Phthalate

DTDM Dithio Dimorpholine

dθ/dλ angular dispersion (or dispersive power) (i.e., rate of change of angle of diffraction, or refraction with wavelength)

DTDMA Distrubuted Time Division Multiple Access

DTDMAC Ditallowdimethylammonium Chloride

DTDP Ditridecyl Phthalate

dTDP Thymidine-5'-Diphosphate [Biochemistry]

5'-dTDP Thymidine-5'-Diphosphate [Biochemistry]

DTDPA Dithiodipropionic Acid

DTDS Digital Television Display System

DT/DT Drop Tube/Drop Tower (Facility)

dT/dx temperature gradient or profile (i.e., change of temperature with distance)

dT/dt time rate of change of (absolute) temperature

dθ/dt time rate of change of coverage

$\Delta\theta$/Δt time rate of change of angular displacement

$\Delta\theta$/Δt time rate of change of coverage

dτ/dT change in surface tension with temperature

DTE Data Terminal Emulator

DTE Data Terminal Equipment

DTE Data Transmission Equipment

DTE Deaminotyrosyl-Tyrosine Ethyl Ester [Biochemistry]

DTE Destructive Testing Equipment

DTE Development Test and Evaluation

DTE Differential Thermal Expansion

DTE Digital Test Executive

DTE Digital Tune Enable

DTE Dithioerythritol

DTE Dumb Terminal Emulator

DT&E Design, Testing and Evaluation

DTech Doctor of Technology

DTED Department of Trade and Economic Development [Washington State, US]

DTED Digital Terrain Elevation Data [Geographic Information System]

DTE/DCE Data Terminal Equipment/Data Communications Equipment (Interface)

DTEM Deep Transient Electromagnetic System [also DTEMS]

DTENT Date of Entry

DTF Dairy Trade Federation [UK]

DTF Date to Follow

DTF Default-the-File

DTF Define the File

DTF Definite Tape File

DTF Dial Tone First

DTF Dielectric Thin Film

DTF Distributed Test Facility

DTF Diving Test Facility

DTF Dynamic Track Follower

DTFDW Deciduous Tree Fruit Disease Workers [US]

DTFT Discrete Time Fourier Transform

DTG Date-Time Group

DTG Derivative Thermogravimetry

DTG Differential Thermal Gravimetry; Differential Thermogravimetry

DTG Display Transmission Generator

DTG Dynamically Tuned Gyroscope

DTGA Differential Thermogravimetric Analysis

DTGS Deuterated Triglycerine Sulfate

DTGS Deuterated Triglycine Sulfate

DTH Deaminotyrosyl-Tyrosine Hexyl Ether [Biochemistry]

DTHD 2'-Deoxyribosylthymine [Biochemistry]

DTI Data Transfer Interface

DTI Department of Trade and Industry [UK]

DTI Digital Temperature Indicator

DTI Display Terminal Interchange

DTI Distortion Transmission Impairment

DTIC Defense Technical Information Center [of US Department of Defense at Cameron Station, Alexandria, Virginia]

DTIC 5-(3,3-Dimethyl-1-Triazenyl)imidazole-4Carboxamide

DTICA Departmento Tecnico Interamericano de Cooperación Agricole [Interamerican Technical Department for Agricultural Cooperation, Chile]

DTIE Division of Technical Information Extension [of United States Atomic Energy Commission]

DTIM Département Traitement de l'Information et Modélisation [Department of Information Processing and Modelling, ONERA-CERT, Toulouse, France]

δ-TiN Delta Titanium Nitride

DTIP Digital Tune In Progress

DTL Detroit Testing Laboratory [Michigan, US]

DTL Development Test Laboratory

DTL Dialog Tag Language

DTL Diode(-Coupled) Transistor Logic

DTL Dynamitron-Tandem-Laboratory [University of Bochum, Germany]

DTLC Two-Dimensional Thin Layer Chromatography

DTLS Digital Television Lightwave System

DTL/TTL Diode-Transistor Logic/Transistor-Transistor Logic

DTLZ Diode Transistor Logic with Z Diodes

DTM Delay Timer Multiplier

DTM Device Test Module

DTM Digital Terrain Map [Geographic Information System]

DTM Digital Terrain Model(ling) [Geographic Information System]

DTM Digital Terrain Module

DTM Directorate of Torpedoes and Mines [UK]

DTM Double Torsion Method [Mechanics]

D2M Diethylene Glycol Dimethyl Ether

DTMA Diethylenetriamine-1-Acetate [also dtma]

DTMB David Taylor Model Basin [US Navy]

DTMF Data Tone Multiple Frequency

DTMF Dual-Tone, Multifrequency (Signaling) [Telecommunications]

DTμL Diode-Transistor Micrologic [also DTML]

DTMM Desktop Molecular Modeller

dTMP Thymidine-5'-Monophosphate [also 5'-dTMP] [Biochemistry]

DTMS Database and Transaction Management System

DTMS Digital Test Monitoring System

DTN Data Transporting Network

DTN Defense Telephone Network

DTNB 3,3'-Dithio-bis(6-Nitrobenzoic Acid); 5,5'-Dithio-bis(2-Nitrobenzoic Acid)

DTNE 1,2-Di(1,4,7-Triaza-1-Cyclononyl)ethane [also dtne]

DTNP 2,2'-Dithiobis(5-Nitropyridine)

DTNSRDC David Taylor Naval Ship Research and Development Center [US]

DTO Decentralized Toll Office

DTO Desaminotrosyl-Trosine Alkyl Octyl [Biochemistry]

DTO Deuterium-Tritium Oxide

DTO Development Test Objective

DTO Digital Testing Oscilloscope

D to A Digital to Analog

DTOS N-Dimethyl Dithiocarbamyl-N'-Oxidiethylene Sulfenamide

$/t oz Dollars per troy ounce

DTP Daily Transaction Reporting

DTP Data Transfer Protocol

DTP Department of Theoretical Physics

DTP Desk-Top Plotter

DTP Desk-Top Publishing

DTP Detail Test Plan

DTP Diagnostic-Therapeutic Pair

DTP Diphtheria, Tetanus, Pertussis (Vaccine) [also Dip/Tet/Pert]

DTP Directory Tape Processor

DTP Distributed Transaction Process

DTP Dithiophosphate

DTP Di-(p-Tolyl)phosphate

DTPA Diethylenetriaminepentaacetate; Diethylenetriaminepentaacetic Acid

DTPB Dimethyl 3,3'-Dithiobispropionimidate

DTPC Desert Tortoise Preserve Committee [US]

DTPD N,N'-Ditolyl-P-Phenylene Diamine

DTPE Division of Teacher Preparation and Enhancement [of National Science Foundation, US]

DTPL Domain Tip Propagation Logic

DTPMT Date of Payment (Code)

DTPP Demonstration Tokamak Power Plant

DTPR Detailed Test Procedures

DTPR Dithiaperylene

1,12-DTPR 1,12-Dithiaperylene

3,9-DTPR 3,9-Dithiaperylene

3,10-DTPR 3,10-Dithiaperylene

3,10-DTPR(Ph)2 2,11-Diphenyl 3,10-Dithiaperylene

1,6-DTPY 1,6-Dithiapyrene [also 1,6-DTPy]

1,8-DTPY 1,8-Dithiapyrene [also 1,8-DTPy]

DTQ Département de Tourisme du Québec [Quebec Department of Tourism, Canada]

DTR Daily Transaction Reporting

DTR Data Telemetering Register

DTR Data Terminal Reader

DTR Data Terminal Ready

DTR Data Transfer Rate

DTR Data Transfer Register

DTR Definite Time Relay

DTR Delivery Truck Ramp

DTR Demand-Totalizing Relay

DTR Departmental Telecommunications Representative

DTR Diffusion Transfer

DTR Digital Target Report

DTR Digital Telemetering Register

DTR Discharge-Tube Recifier

DTR Disposable Tape Reel

DTR Distribution Tape Reel

DTR Document Transmittal Record

DTR Downtime Radio

DTRC David Taylor Research Center [US Navy at Annapolis, Maryland]

DTRD Development Test Requirements Document

DTR/DSR Data Terminal Reader/Data Storage and Retrieval

DTRE Defense Telecommunications Research Establishment

DTRF Darlington Tritium Removal Facility [Ontario, Canada]

DTRF Data Transmittal and Routing Form

DTRI Dairy Training and Research Institute [Philippines]

DTRS Development Test Requirement Specification

DT/RSS Data Transmission/Recording Subsystem

DTS Dallas Transit System [Texas, US]

DTS Data Tape Service

DTS Data Transfer System

DTS Data Transmission System

DTS Defense Telephone Service [US]

DTS Defense Telephone System

DTS Dense Tar Surfacing

DTS Desk-Top System

DTS Diffusion Total System

DTS Digital Tandem Switch

DTS Digital Termination System

DTS Discrete-Time Series

DTS Discrete-Time Signal

DTS Discrete-Time System

DTS Double-Throw Switch

DTs Delirium Tremens [Medicine]

DTSA Discrete Time Series Analysis

Dtsch Ges Materialkd Fachber Deutsche Gesellschaft für Materialkunde Fachberichte [German Reports on Metallurgy, Metallography and and Materials Science; published by Deutsche Gesellschaft für Materialkunde]

Dtsch Ges Metallkd Fachber Deutsche Gesellschaft für Metallkunde Fachberichte [German Reports on Metallurgy, Metallography and and Materials Science; published by Deutsche Gesellschaft für Metallkunde]

Dtsch Papierwirt Deutsche Papierwirtschaft [Publication on the German Paper Industry]

Dtsch Roheisen Deutsches Roheisen [Journal on German Iron and Steel]

D2S&M Dressed Two Sides and (Center, or Standard) Matched [Lumber]

D2S&SM Dressed Two Sides and Standard Matched [Lumber]

DTSP 3,3'-Dithiobis(succinimidylpropionate)

DTSP 3,3'-Dithiobis(propionic Acid N-Hydroxysuccinimide Ester

DTSS Dartmouth Time-Sharing System

DTSSP 3,3'-Dithiobis(sulfosuccinimidyl Propionate)

DTS-W Defense Telephone Service, Washington [US]

DTSY Department of the Treasury [Australia]

DTT Data Transition Tracking

DTT Data Transmission Terminal

DTT Design Thermal Transient

DTT Dithiothreitol; 1,4-Dithiothreitol; DL-Dithiothreitol

DTT Drive-Through Teaching [Robotics]

DL-DTT DL-Dithiothreitol

L-DTT L-Dithiothreitol

ΔT/T Relative Change in Transmissivity

DTTC 3,3'-Diethylthiatricarbocyanine Iodide

DTTF Deviation(s) in Time to Failure

DTTF Digital Tape and Tape Facility

dTTP Deoxyribonucleoside 5'-Triphosphate; 3'-Deoxythymidine 5'-Triphosphate; Thymidine-5'-Triphosphate [Biochemistry]

5'-dTTP Thymidine-5'-Triphosphate [Biochemistry]

DTTT Dynamic Time-Temperature Transformation [Metallurgy]

D2T2 Dye Diffusion Thermal Transfer (Printing) [also D^2T^2]

DTTU Data Transmission Terminal Unit

DTU Data Terminating Unit

DTU Data Transfer Unit

DTU Data Transmission Unit

DTU Digital Tape Unit

DTU Digital Telemetry Unit

DTU Digital Transmission Unit

DTU Dual Toplogical Unitarization

DTUC David Thompson University Center [Canada]

DTUL Deflection Temperature under Load [Mechanics]

DTUTL Digital Tape Unit Tape Facility

DTV Desk-Top Video

DTV Deutscher Verband technisch-wissenschaftlicher Vereine [German Federation of Scientific and Technical Associations]

DTV Digital Television

DTV Digital-to-Television [also D-TV, or D/TV]

DTVC Desk-Top Video Conferencing

DTVC Digital Transmission and Verification Converter

DTVM Differential Thermocouple Voltmeter

DTW Dual Tandem Wheel (Landing Gear)

DTWX Dial Teletypewriter Exchange

D2X Decimal To Hexadecimal [Restructured Extended Executor (Language)]

DU Dalhousie University [Halifax, Nova Scotia, Canada]

DU Daegu University [South Korea]

DU Delay Unit

DU Delft University [Netherlands]

DU Denison University [Granville, Ohio, US]

DU Depleted Uranium

DU Derived Unit (of Measurement)

DU Dillard University [New Orleans, Louisiana, US]

DU Dimensioning Unit

DU Disk Unit

DU Disk Usage

DU Display Unit

DU Doshisha University [Kyoto, Japan]

DU Drake University [Des Moines, Iowa, US]

DU Drexel University [Philadelphia, Pennsylvania, US]

DU Duke University [Durham, North Carolina, US]

DU Dundee University [UK]

D/U Up/Down Input [Digital Counter]

Du D-Load Ultimate (for Concrete Pipes) [Symbol]

Du Duke

Du Dutch

dU Deoxyuridine [Biochemistry]

dU Internal Energy Change [also ΔU]

ΔU_0 Lattice Energy (of Ionic Crystals) [Symbol]

DUA Digital Uplink Assembly

DUA Directory User Agent [Internet]

ΛUA 4-Deoxy-L-Threo-Hex-4-Enopyranosyluronic Acid [Biochemistry]

DUAL Dynamic Universal Assembly Language

DUALABS Data Use Access Laboratories

DUAT Direct User Access Terminal

DUC Dense Upper Cloud

DUC Digital Uplink Command

DUC Dual-Access Utility Circuit

DUCE Denied Usage Channel Evaluator

Duct Ductile; Ductility [also duct]

Duct Ducting [also duct]

d-UDP 2'-Deoxyuridine 5'-Diphosphate [also dUDP] [Biochemistry]

dU/dI change in non-ohmic resistance with current

dU/dT change in internal energy per unit temperature

dv/dx velocity gradient (i.e., change in velocity with distance (in x-direction))

dv/dy change in velocity with (perpendicular) distance (in y-direction)

DUE Detection of Unauthorized Equipment

DUES Diplôme Universitaire d'Etudes Scientifiques [French University Diploma of Scientific Studies]

DUF Diffusion under Film

DUG Datapac User Group

DUH Dynamic Ultramicrohardness

DUI Driving Under the Influence (of Alcohol)

DÜI Deutsches Übersee-Institute [German Overseas Institute, Hamburg]

DUIP Division of University and Industry Programs [of US Department of Energy–Office of Energy Services]

Duke Math J Duke Mathematical Journal [Published by Duke University Press, Durham, North Carolina, US]

DUKES Digest of the United Kingdom Energy Statistics [of Department of Energy, UK]

Duke Univ Res Stud Educ Duke University Research Studies in Education [Published by Duke University Press, Durham, North Carolina, US]

DUKW Duck (Amphibian Vehicle)

DUL Deformation under Load

D/(uL) Dispersion Number [Symbol]

DUM Disk User Multi-Access (Unit)

DUM Dummy [also Dum]

DUMAND Deep Underwater Muon and Neutrino Detector

Dumb Dumbarton [Scotland]

DUMC Duke University Medical Center [Durham, North Carolina, US]

DUMD Deep Underwater Measuring Device

Dumf Dumfried [Scotland]

d-UMP 2'-Deoxyuridine 5'-Monophosphate [also dUMP] [Biochemistry]

DUMS Deep Unmanned Submersible

DUMU Disk User Multi-Access Unit

DUN Dial-Up Networking [Microsoft Windows]

DUNC Deep Underwater Nuclear Counter; Deep Underwater Nuclear Counting

DUNCE Dial-Up Network Connection Enhancement

DUNDEE Down Under Doppler and Electricity Equipment

DUniv Doctor of the University

DUNMIRE Dundee University Numerical Method Information Retrieval System [UK]

DUNS Data Universal Numbering System

DUO Datatron Users Organization

DUO DOS (Disk-Operating System) under OS (Operating System)

Duod Duodenum [also duod]

DUP Data User Port

DUP Data User Program

DUP Delft University Press [Netherlands]

DUP Disk Utility Program

DUP Distinguished University Professor

DUP Diundecyl Phthalate

DUP Duke University Press [Durham, North Carolina, US]

Dup Duplicate [also dup]

Dup Duplex [also dup]

Dupl Duplicate; Duplicating [also dupl]

DUPLX Duplex [also Duplx, or duplx]

DUPLXR Duplexer [also Duplxr, or duplxr]

Du Pont Inf Serv Du Pont Information Service [Geneva, Switzerland]

Du Pont Mag Du Pont Magazine [Published by Du Pont Co., Wilmington, Delaware, US]

Du Pont Mag Eur Ed Du Pont Magazine, European Edition [Published by Du Pont, Geneva, Switzerland]

Dur Duration [also dur]

Durability Build Mater Durability of Building Materials [Journal published in the Netherlands]

Dural Duraluminum

DURD Department of Urban and Regional Development [Australia]

DURP Distinguished University Research Professor

DUS Data User Station

DUS Diagnostic Utility System

$US/bu United States Dollar per Bushel [Unit]

DUSCOFS Depleted Uranium Silicate Container Fill System

DUSD (Office of the) Deputy Under-Secretary of Defense [US]

DUSD-ES (Office of the) Deputy Under-Secretary of Defense for Environmental Security [US]

$US/gal United States Dollar per Gallon [Unit]

$US/g United States Dollar per Gram [Unit]

$US/kW United States Dollar per Kilowatt [Unit]

$US/kWh United States Dollar per Kilowatthour [Unit]

$US/L United States Dollar per Liter [Unit]

$US/lb United States Dollar per Pound [Unit]

DUSM Duke University School of Medicine [Durham, North Carolina, US]

$US/ton United States Dollar per Ton [Unit]

$US/yr United States Dollar per year [Unit]

DUT Dalian University of Technology [Dalian, PR China]

DUT Delft University of Technology [Netherlands]

DUT Device under Test

DUT Diplôme Universitaire de Technologie [French University Diploma in Technology]

dUTP 2'-Deoxyuridine-5'-Triphosphate [Biochemistry]

DUV Damaging Ultraviolet

DUV Data under Voice

DUV Deep Ultraviolet

DUV-C Deep Ultraviolet C

DV Daily Value

DV Damped Vibration

DV Decomposition Voltage

DV De Hosson and Vitek (Equation) [Materials Science]

DV *(Deo volente)* – God willing

DV Designated Verification

DV Designee for Verification

DV Desired Value

DV Devitrification

DV Differential Velocity

DV Differential Viscometer

DV Differential Voltage

DV Dilution Ventilation

DV Dimer Vacancy

DV Direct Vision

DV Discrete Variable [Mathematics]

DV (Spin-Polarized) Discrete Variational (Method) [Physics]

DV Double Vibration

DV Drift Velocity [Electronics]

DV Dump Valve

.DV DESQview Script [File Name Extension]

D&V Diarrhea and Vomiting [also D and V]

dV distorted-coherent vacancy (in materials science) [Symbol]

dV voltage change (or difference)

dV volume element [Symbol]

ΔV Activation Volume [Symbol]

ΔV (Acoustic-Emission) Amplitude Difference (or Change) [Symbol]

ΔV Volume Difference (or Change) [Symbol]

ΔV Volumetric Strain [Symbol]

ΔV^o Overall Cell Potential [Symbol]

ΔV_F Defect Formation Volume (in Materials Science) [Symbol]

ΔV_M Defect Migration Volume (in Materials Science) [Symbol]

ΔV_i^m Activation Volume of Interstitial Migration (in Materials Science) [Symbol]

ΔV_{iv}^f Activation Volume of Vacancy Formation (in Materials Science) [Symbol]

ΔV_{iv}^m Activation Volume of Vacancy Migration (in Materials Science) [Symbol]

ΔV^{SD} Activation Volume of Self-Diffusion (in Materials Science) [Symbol]

Δv velocity change [Symbol]

Δv change in drift velocity (of Electrons) [Symbol]

Δv volume difference (or change) [Symbol]

DVA Department of Veterans Affairs

DVA Designed, Verified and Assigned Date

DVA Digital Voice Announcer

DVA Dynamic Visual Acuity

DVARS Doppler Velocity Altimeter Radar Set

DVB Department of Veterinary Biology

DVB Digital Video Broadcast(ing)

DVB Divinylbenzene

DVBST Direct View Bistable Storage Tube

DVB-T Digital Video Broadcast(ing)–Terrestrial

DVC Delivery Verification Certificate [International Trade]

DVC Desktop Video Conferencing

DVC Differential Voltage Contrast

DVC Digital Video Cassette

Dvc Device [also dvc]

DVCCS Differential Voltage-Controlled Current Source

DVCSA Delaware Valley College of Science and Agriculture [Doylestown, Pennsylvania, US]

DVD Delta Velocity Display

DVD Detail Velocity Display

DVD Digital Versatile Disk

DVD Digital Video Disk

DVD Directed Vapor Deposition

DVD Division of Viral Diseases [of Centers for Disease Control and Prevention, Atlanta, Georgia, US]

D/VD Data/Voice Data

DVDA Dollar Volume Discount Agreement

dV/dI electrical resistance (i.e., change in voltage with current)

$\Delta V_a / \Delta I_a$ Internal Resistance (i.e., Change in Anode (or Plate) Voltage with Anode (or Plate) Current) (of Electron Tubes)

dV/dP Change in Volume with Pressure

dV/dp change in volume with vapor pressure

DVD-R Digital Video Disk–Recordable

DVD-RAM Digital Versatile Disk–Random-Access Memory

DVD-ROM Digital Versatile Disk–Read-Only Memory

dV/dT change of voltage with temperature

dV/dt time rate of change of volume

dv/dt time rate of change of velocity

$\Delta V_g / \Delta V_a$ Change in Grid Voltage with Anode (or Plate) Voltage (of Electron Tubes)

dv/dx velocity gradient (i.e., change in velocity with distance (in x-direction))

dv/dy change in velocity with (perpendicular) distance (in y-direction))

DVE Digital Video Effect

DVES DOD (Department of Defense) Value Engineering Services [US]

DVESO DOD (Department of Defense) Value Engineering Services Office [US]

DVF Digital Variable Frequency

DVFO Digital Variable-Frequency Oscillator

DVFR Defense Visual Flight Rules [US]

DVGW Deutscher Verein der Gas– und Wasserfachleute [German Association of Gas and Water Specialists]

DVI Denton Vacuum Inc. [Moorestown, New Jersey, US]

DVI Digital Video Interactive [also DV-I]

DVI Digital Video Interface

DVI Direct Voice Input

.DVI Device Independent [File Name Extension]

DV-I Digital Video Interactive [also DVI]

DVL Deutsche Versuchsanstalt für Luftfahrt [German Aeronautics Research Institute]

Dvlpt Development [also dvlpt]

DVM Deutscher Verband für Materialprüfung [German Materials Testing Association] [Name Changed to Deutscher Verband für Materialforschung und –prüfung]

DVM Deutscher Verband für Materialforschung und –prüfung [German Materials Research and Testing Association]

DVM Digital Voltmeter

DVM Discrete Variational Method [Solid-State Physics]

DVM Displaced Virtual Machine

DVM Doctor of Veterinary Medicine

DVMRP Distance Vector Multicast Routing Protocol [Internet]

DVOM Digital Volt-Ohmmeter

δ-VOPO₄ Delta Vanadium Phosphate

DVOR Doppler VHF (Very High Frequency) Omnidirectional Radio Range

DVP Diagonal Vanishing Point

DVR Design and Verification Routine

DVR Discrete Variable Representation

.DVR Device Driver [File Name Extension]

DVR-DGB Discrete Variable Representation–Distributed Gaussian Basis

DVS Design Verification Specification

DVS Deutscher Verband für Schweißtechnik [German Welding Association]

DVS Digital Voice System

DVS Digital Video System

DVS Direct-Vision Spectroscope

DVS Divinyltetramethyldisiloxane

DVS Ber DVS Berichte [Welding Reports published by Deutscher Verband für Schweißtechnik]

DVSc Doctor of Veterinary Science [also DVS]

DVST Direct-View Storage Tube

DVT Deep Vein Thrombosis [also dvt]

DVT Design Verification Testing

DVT Deutscher Verband technisch-wissenschaftlicher Vereine [German Federation of Scientific and Technical Associations]

Dvtl Dovetail [also dvtl]

DVTMDS 1,3-Divinyl-1,1,3,3-Tetramethyl Disilazane

DVV Deutscher Verzinkerei-Verband [German Galvanizing Association]

DVV Deutscher Volkshochschul-Verband e.V. [German Adult Education Association]

ΔV/V Relative Volume Change (or Volume Change per Unit Volume)

DVWK Deutscher Verband für Wasserwirtschaft und Kulturbau [German Association for Water Resources and Agricultural Engineering]

DVWSR Deutscher Verband für Wohnungswesen, Städtebau und Raumplanung [German Association for Housing, Town Building and Urban Planning]

DVX Digital Voice Exchange

DV-Xα Discrete Variational X-Alpha (Method) [also DV-X$_\alpha$] [Physics]

DW Daisy Wheel

DW Data Warehousing

DW Data Word

DW Deadweight [also dw]

DW Debug Word (Values) [Operation Code]

DW Debye-Waller (Factor) [Solid-State Physics]

DW Deionized Water

DW Delta Wing (Aircraft)

DW Department of Welding

DW Destructive Write

DW Detonation Wave

DW Deutsche Welle [A Cologne-Based German Radio Station]

DW Developed Width

DW Die Welding

DW Disaster Warning

DW Display Word (Command) [Pascal Programming]

DW Distilled Water

DW Dock Warrant

DW Domain Wall [Solid-State Physics]

DW Double Word

DW Dried Weight

DW Drinking Water

DW Drop Wire

DW Dry Weight

DW Dual Wavelength

DW Dual Wheel (Landing Gear)

DW Dumbwaiter

DW Dynamic Wave

Dw Humid Cold Climate, Dry Winter [Symbol]

dW Change in Amount of Work [Symbol]

d/w days per week [Unit]

ΔW Work Done on a (Thermodynamic) System [also δW]

DWA Daily Weighted Average

DWA Designated Waiting Area

DWA Double-Wire Armor

DWAA Department of Western and Asiatic Antiquities [of British Museum]

D-Wave Detonation Wave [also D-wave]

DWB Designer's Workbench

DWB Development Workbook

DWBA Direct-Wire Burglar Alarm

DWBA Distorted-Wave Born Approximation [Physics]

DWC Deadweight Capacity

D/WC Diamond-Coated Tungsten Carbide

D/WC(Co) Diamond-Coated Cobalt-Tungsten Carbide

DWCM Dried Weight of Cell Mass

dW/dt power (i.e., time Rate of change with work)

DWD Deutscher Wetterdienst [German Weather Service]

DWE Department of Welding Engineering

DWE Director of Weapons and Equipment [UK]

DWE Division of Wildlife and Ecology [of Commonwealth Scientific and Industrial Research Organization, Australia]

DWED Department of Western Economic Diversification [Canada]

DWED Dry Well Equipment Drain

DWF Debye-Waller Factor [Solid-State Physics]

DWFD Dry Well Floor Drain

Dwg Drawing [also dwg]

Dwg No Drawing Number [also DN]

DWI Data Word In

DWI Driving While Intoxicated

DWI Drop Weight Index

DWI Dutch West Indies

DWIA Distorted Wave Impulse Approximation

DWICA Deep Water Isotopic Current Analyzer

DWIPS Digital Weather Imaging Processing System

DWK Deutsche Gesellschaft für Wiederaufarbeitung von Kernbrennstoffen [German Society for Nuclear Fuel Reprocessing]

DWL Design Water Line

DWL Dominant Wavelength

DWL Downwind Localizer

Dwl Dowel [also dwl]

DWM Destination Warning Marker

DWM Director of Women Marines [US]

DWMI Diamond Wheel Manufacturers Institute [US]

DW-MR Domain-Wall Magnetoresistance [also DWMR] [Solid-State Physics]

DWMT Discrete Wavelet Multitone

DWN Druckwechsel Nitrogen (Process) [Chemical Engineering]

dwn down

DWNP Department of Wildlife and National Parks [Botswana]

DWP Daisy Wheel Printer

DWP Digital Waveform Pattern

DWP Domain Wall Pinning [Solid-State Physics]

$/Wp Dollar(s) per Peak Watt [Unit]

DWPF Defense Waste Processing Facility [at Savannah River Site, South Carolina, US]

DWR Department of Water Resources [US]

DWR Director of Weapons Research [UK]

DWR Division of Water Resources [of Commonwealth Scientific and Industrial Research Organization]

DWR Dry Weight Rank Method

DWR(D) Director of Weapons Research (Defense) [UK]

DWS Deep Water Submersible

DWS Diffusing Wave Spectroscopy

DWS Disaster Warning Satellite

DWS Disaster Warning System

DWS Doppler Wind Sensor

DWS Drinking Water Standards

$WS Western Samoan Dollar [Currency]

DWSBA Distorted Wave Second Born Approximation [Physics]

DWSF Dry Well Storage Facility

DWSMC Defense Weapons Systems Management Center [of US Department of Defense]

DWST Demineralized Water Storage Tank

DWSV Deadweight Safety Valve

DWT Deadweight Ton(nage) [also dwt]

DWT Department of Welding Technology

DWT Drop-Weight Test

dwt deadweight ton(nage) [Unit]

dwt *(denarius)* – pennyweight [Unit]

DWTF Decontamination and Waste Treatment Facility [Nuclear Engineering]

dwt/L pennyweight per liter [also dwt/l]

DWTT Drop-Weight Tear Test

DWV Data With Voice

DWV Dielectric Withstanding Voltage

DWV Drain, Waste and Ventilation System

DWV Drain, Waste and Vent (Pipe)

ΔW/W Energy Loss

DX Destroyer Experimental [Navy Symbol]

DX Deuterated (or Labeled) Halide

DX Direct Current Switching

DX Direct Expansion Coil

DX Distance

DX Distance Reception

DX Document Transfer

DX Document Transmission

DX Drosophila X [Microbiology]

DX Duplex (Repeater)

DX Long Distance (Communication)

Dx Diagnosis

dX Change in (Radiographic) Exposure [Symbol]

dXo East-West Rate of Change [Astronomy]

dx differential of x [Mathematics]

D2X Decimal To Hexadecimal [Restructured Extended Executor (Language)]

d²x second differential of x [Mathematics]

ΔX Change in Magnitude of Quantity X [Symbol]

Δx class interval (in statistics) [Symbol]

Δx fineness of (object) resolution [Symbol]

$\overline{\Delta x^2}$ mean square displacement (of a molecule) [Symbol]

δ(x) (Dirac) delta function (in mathematics) [Symbol]

DXA Differential X-Ray Absorption (Method) [Electron Microscopy]

.DXB Drawing Interchange Binary [AutoCAD File Name Extension] [also .dxb]

DXC Data Exchange Control

dx/dλ linear dispersion (in spectroscopy) [Function]

dX/dt time rate of change of volume fraction

dx/dt first derivative of x with respect to time [also $\dot{x}$]

d²x/dt² second derivative of x with respect to time [also $\ddot{x}$]

DXE Dixylylethane

DXF Data Exchange File (Format) [AutoCAD Software]

DXF Data Transfer Facility

DXF Digital Exchange Format [AutoCAD]

DXF Document Exchange Facility

DXF Drawing Exchange Format

DXF Drawing Interchange File

DXFCNV Drawing Interchange File Convert

DXI Data Exchange Interface

DXRD Dynamic X-Ray Diffraction

DX/RSTS Document Transmission/Resource Time Sharing

DXS Data Exchange System

DXS/OS Data Exchange System/Operating System

DXS/ST Data Exchange System/Statement Translator

DXS/TL Data Exchange System/Transaction Language

DXV Drosophila X Virus

DY Deputy Director [NASA Kennedy Space Center Directorate, Florida, US]

DY Drell-Yan (Process) [Physics]

Dy Day [also dy]

Dy Deputy [also dy]

Dy Dynode [also dy]

Dy Dysprosium [Symbol]

Dy^{3+} Dysprosium Ion [also Dy^{+++}] [Symbol]

Dy-123 DyBa$_2$Cu$_3$O$_{7-x}$ [A Dysprosium Barium Copper Oxide Superconductor]

Dy-124 DyBa$_2$Cu$_4$O$_{7-x}$ [A Dysprosium Barium Copper Oxide Superconductor]

Dy-156 Dysprosium-156 [also ^{156}Dy, or Dy^{156}]

Dy-158 Dysprosium-158 [also ^{158}Dy, or Dy^{158}]

Dy-159 Dysprosium-159 [also ^{159}Dy, or Dy^{159}]

Dy-160 Dysprosium-160 [also ^{160}Dy, or Dy^{160}]

Dy-161 Dysprosium-161 [also ^{161}Dy, or Dy^{161}]

Dy-162 Dysprosium-162 [also ^{162}Dy, or Dy^{162}]

Dy-163 Dysprosium-163 [also ^{163}Dy, or Dy^{163}]

Dy-164 Dysprosium-164 [also ^{164}Dy, or Dy^{164}]

Dy-165 Dysprosium-165 [also ^{165}Dy, or Dy^{165}]

Dy-166 Dysprosium-166 [also ^{166}Dy, or Dy^{166}]

dy differential of y [Mathematics]

d/y days per year [also d/yr]

DYANA Dynamic Analyzer [Computer Program]

DyAs Dysprosium Arsenide (Semiconductor)

DyBCO Dysprosium Barium Copper Oxide (Superconductor) [also DBCO]

DyCMO Dysprosium Calcium Manganese Oxide (Superconductor) [also DCMO]

DYCMOS Dynamic Complementary Metal-Oxide Semiconductor [also DY/CMOS, or DY-CMOS]

Dy-Co Dysprosium-Cobalt (Alloy System)

Dy-Co CMF Dysprosium-Cobalt Composition(al) Modulated Film

DYCOMS Dynamics and Chemistry of Marine Stratocumulus

dY/dt time rate of change of mole fraction

dy/dx slope of a curve (i.e., the change in y with x) [$\Delta y/\Delta x$]

Dy-Fe Dysprosium-Iron (Alloy System)

Dy/Fe CMF Dysprosium-Iron Composition(al) Modulated Film

DyFeCo Dysprosium Iron Cobalt (Alloy)

Dy(FOD)$_3$ Tris(heptafluorodimethyloctanedionato) dysprosium

Dyg Dyeing [also dyg]

Dy-Gd Dysprosium-Gadolinium (System)

DyIG Dysprosium Iron Garnet

Dy:LiYF$_4$ Dysprosium-Doped Lithium Yttrium Fluoride (Crystal) [also Dy^{3+}:LiYF]

Dyn Dynamic(s) [also dyn]

Dyn Dynamite [also dyn]

Dyn Dynamo [also dyn]

dyn dyne [Unit]

Dynamic SIMS Dynamic Secondary Ion Mass Spectroscopy

Dynas Dynasty [also dynas]

DYNASAR Dynamic Systems Analyzer

DYNA2D Two-Dimensional Hydrodynamic Fine-Element Code with Interactive Rezoning [of Lawrence Livermore National Laboratory, Califormia, US]

Dyn Atmos Oceans Dynamics of Atmospheres and Oceans [Published in the Netherlands]

DYNAVIS Dynamic Video Display System

dyn-cm dyne centimeter [also dyn cm, or dyn·cm]

dyn/cm dyne(s) per centimeter [also dyn cm^{-1}]

dyn/cm² dyne(s) per square centimeter [also dyn cm^{-2}, or dyn/sq cm]

dyn/cm °C dyne(s) per centimeter degrees centigrade [also dyn cm^{-1} °C^{-1}]

dyn·cm/cm^3 dyne-centimeter(s) per cubic centimeter [also dyn·cm cm^{-3}]

dyn-cm/s dyne-centimeter per second [also dyn cm/s]

DYNDUMP Dynamic Dump [also dyndump]

Dy-Ni Dysprosium-Nickel (Alloy System)

DYNM Dynamotor [also Dynm, or dynm]

dyn-s dyne-second [also dyn s]

dyn-s/cm² dyne-second(s) per square centimeter [also dyn·s cm^{-2}, dyn-sec/cm², or dyn-sec/sq cm]

dyn/sq cm dyne(s) per square centimeter [also dyn/cm², or dyn cm^{-2}]

DYNSRC Dynamic Source

Dyn Sys Dynamic System [also dyn sys]

Dy(O-i-Pr) Dysprosium Isopropoxide [also DyOPr^i]

DYP Directory Yellow Pages

DyP Dysprosium Phosphide (Semiconductor)

Dyp Dipyridyl

$/yr Dollars per year [Unit]

d/yr day(s) per year [also d/y]

DYSAC Digitally Simulated Analog Computer

DYSAC Dynamic Storage Analog Computer [also DYSTAC]

DySb Dysprosium Antimonide (Semiconductor)

DySCNO Dysprosium Strontium Copper Niobium Oxide (Superconductor) [also DSCNO]

DySCO Dysprosium Strontium Cobalt Oxide (Superconductor) [also DySCO]

DySMO Dysprosium Strontium Manganese Oxide (Superconductor) [also DSMO]

DYSTAL Dynamic Storage Allocation Language [Computer Language]

Dy-Ta Dysprosium-Tantalum (Alloy System)

Dy-Ta CMF Dysprosium-Tantalum Composition(al) Modulated Film

Dy(TMHD)$_3$ Dysprosium Tris(2,2,6,6-Tetramethyl-3,5-Heptanedionate)

DYTRAN Dynamic Transient Analysis (Software)

DYU Dah-Yeh University [Chang-Hua, Taiwan]

Dy-Y Dysprosium-Yttrium (System)

DyZn Dysprosium Zinc (Alloy)

DZ Algeria [ISO Code]

DZ Dead Zone

DZ Deformation Zone

DZ Deformed Zone
DZ Depleted Zone
DZ Distillation Zone
DZ Doctor of Zoology
DZ Double Zeta Basis
DZ Dropping Zone
Dz Diazole
Dz Dizygotic [also dz] [Biology]
Dz Drizzle
dz dozen
ΔZ Impedance Change [Symbol]
Dz Pr Dozen Pairs [also dz pr]
DZA Doppler Zeeman Analyzer

DZC Dispersed Zirconia Ceramics
DZD Dinar [Currency of Algeria]
DZF Dokumentationszentrale Feinwerktechnik [Documentation Center for Precision Mechanics, Germany]
DZM Deutsche Zentralbibliothek für Medizin [German Library Center for Medicine, University of Cologne, Germany]
DZP Double Zeta Basis and Polarization Function(s)
DZ+P Double Zeta Plus Polarization (Basis Set)
DZQ 2,5-Diaziridinyl-1,4-Benzoquinone
DZT Deutsche Zentrale für Tourismus e.V. [German Tourism Center]

E   Activation Energy [Symbol]

E   Additional Designation of Emission (or Transmission) Type–Width, or Duration Modulated [Symbol] [also e]

E   Applied Voltage [Symbol]

E   Bonding Energy (of Ions) [Symbol]

E   Cold Polar Climate [Symbol]

E   Duane Unit (in X Ray Dosimetry) [Symbol]

E   Ear

E   Ear(drum)

E   Earth

E   Earthquake

E   East(ern)

E   Eccentric(ity)

E   Echinacea [Botany]

E   Echinococcus [Genus of Tapeworms]

E   Echo

E   Echo [Phonetic Alphabet]

E   Eclipse; Ecliptic

E   Ecological; Ecologist; Ecology

E   Ecuador(ian)

E   Edge

E   Edison (Screw)

E   Edit(ing); Edit(ion); Editor

E   Edmonton [Alberta, Canada]

E   Effect(ive); Effectivity; Effector

E   Effect (in Statistics) [Symbol]

E   Effective Throat [Symbol]

E   Efficiency Factor (of Machine Tools) [Symbol]

E   Efficient; Efficiency

E   Effort

E   Effusion

E   Egg

E   Egypt(ian)

E   Eigenvalue

E   Einsteinium [also Es]

E   Eject(ion); Ejector

E   Ekuele [Currency of Equatorial Guinea]

E   Elapse(d)

E   Elastic(ity)

E   Elastic Modulus [Symbol]

E   Elastic (or Shellac) Type (Grinding Wheel) Bond

E   Elastomer(ic)

E   Elbow

E   Electric(ity); Electrical

E   Electrical/Computer Engineering (Technology) [Discipline Category Abbreviation]

E   (Applied) Electric Field [Symbol]

E   Electric Field Intensity; Electric Field Strength [Symbol]

E   Electric Potential [Symbol]

E   Electrochemical Equivalent [Symbol]

E   Electrode

E   Electrode Potential [Symbol]

E   Electrolysis; Electrolytic

E   Electromotive Force [Symbol]

E   Electron

E   Electronic(s)

E   Electronic Energy [Symbol]

E   Element(al)

E   Elevation; Elevator

E   Eliminate; Elimination

E   Ellipse; Elliptical

E   Ellipsoid(al)

E   Ellipsometric; Ellipsometry

E   Ellipticity

E   Elongation

E   Eluviate; Eluviation

E   Emanation

E   Embrittle(ment)

E   Embryo(nic)

E   Embryological; Embryologist; Embryology

E   Emerald [Mineral]

E   Emergency

E   Emission

E   Emissive Power [Symbol]

E   Emissivity [Symbol]

E   Emit(ter)

E   Emitter [Semiconductor Symbol]

E   Empennage (or Tail Assembly) [Aeronautics]

E   Emulsifiable; Emulsification; Emulsifier; Emulsion

E   Enamel(ed)

E   Enamel [Dentistry]

E   Enamel(led) Wire

E   Enclosure

E   Encode(r); Encoding

E   End

E   Endangered; Endangerment

E   Endometriosis [Medicine]

E   Endoscope; Endoscopy

E   Endospore [Biology]

E Endamoeba [Biology]
E Endoplasma
E Endotoxin [Microbiology]
E Energy
E (Acoustic-Emission) Energy Counts [Symbol]
E Engage(d); Engagement
E Engine
E Engineer(ing)
E England; English
E Enlarge(ment)
E Entamoeba [Genus of Parasitic Amoeba]
E Enter
E Enterobius [Genus of Ringworms]
E Entrainment
E Entrance
E Environment(al); Environmentalist
E Enzymatic; Enzyme
E Eötvös [Unit]
E Epidemic
E Epidemiological; Epidemiologist; Epidemiology [Medicine]
E Epidermal; Epidermis
E Epidermophyton [Genus of Ringworms]
E Epiglottis [Anatomy]
E Epilepsy; Epileptic(al)
E Episode
E Epitaxial; Epitaxis; Epitaxy
E (Upper case) Epsilon [Greek Alphabet]
E Equalization; Equalizer
E Equation
E Equator(ial)
E Equilibrium
E Equip(ment)
E Equipotential
E Equivalence; Equivalent
E Equus [Genus Comprising Horses, Donkeys, and Zebras]
E Erase(d); Eraser; Erasing; Erasure
E Erepsin [Biochemistry]
E Ericksen Number (in Physics) [Symbol]
E Erlang [Unit]
E Erosion; Erosive
E Error
E Erupt(ion)
E Erythrocyte [Anatomy]
E (−) Erythrose
E Escherichia [Genus of Aerobic Bacteria]
E Eshelby's Tensor (for Composites) [Symbol]
E E Size Sheet (of Paper) [34" × 44"]
E Esophageal; Esophagus
E Essen [Germany]
E Ester(ification)
E Estimate; Estimation; Estimator
E Estimated Commercial Readability [Symbol]
E Estonia(n)

E Estrogen [Biochemistry]
E Ethanol
E Ethene
E Ethiopia(n)
E Ethylene
E Eucalyptus
E Euglena [Genus of Microorganisms]
E Euphrates (River) [Southwest Asia]
E Europe(an)
E Eutectic
E Evacuate; Evacuation
E Evaluate; Evaluation
E Evaporate; Evaporation; Evaporator
E Evaporativity [Meteorology]
E Event
E Eventual(ity)
E Everglades [Florida, US]
E Evidence
E Exa- [SI Prefix]
E Examine(r); Examination
E Excavate; Excavation; Excavator
E Excellence; Excellent
E Except(ion)
E Excess(ive)
E Exchange(r)
E Excitation
E Exciton [Solid-State Physics]
E Execution
E Exercise
E Exert(ion)
E Exhale; Exhalation
E Exhaust(ion)
E Exhibit(ion)
E Exotoxin [Microbiology]
E Expand(er); Expansion
E Expect(ation)
E Expectation [Statistics]
E Expected Value [Statistics]
E Experience
E Experiment(al); Experimentation
E Expert
E Explanation
E Exploit(ation)
E Explore(r); Exploration
E Explosion; Explosive
E Exponent
E Expose(d); Exposition; Exposure
E Express(ion)
E Extension
E Extend(er)
E Extensor (Muscle) [Anatomy]
E External
E Extinct(ion)

E   Extract(ion)

E   Extraordinary

E   Extrapolate; Extrapolation

E   Extrinsic

E   Extract(ion)

E   Extrude(d); Extrusion

E   Extruder

E   Eye

E   Eyepiece

E   Fraunhofer "E" Line [Green Fraunhofer Line for Iron Having a Wavelength of 526.9 Nanometers] [Spectroscopy]

E   (+) Glutamic Acid; Glutamyl [Biochemistry]

E   Green Strain (or Lagrangian Strain) [Symbol]

E   Illuminance; Illumination [Symbol]

E   Internal Energy [Symbol]

E   Irradiance [Symbol]

E   Lilangeni [Currency of Swaziland]

E   Mean Value of Function (e.g., E(A) is the Mean Value of Area) [Symbol]

E   Modulus of Elasticity [Symbol]

E   Oxidation Reduction Potential [Symbol]

E   Pitch Diameter (of Buttress Threads) [Symbol]

E   Polar Climates [Classification Symbol]

E   Potential Difference [Symbol]

E   Redox Potential [Symbol]

E   Replacement of "x" by "x + 1" (e.g., $E(x)^2 = (x + 1)^2$) (in Mathematics) [Symbol]

E   Sound Energy [Symbol]

E   Special Electronic Installation (Aircraft) [USDOD Symbol]

E   Special Electronic Missile, or Rocket [USDOD Symbol]

E   Spectral Solar Constant [Symbol]

E   Spring Modulus [Symbol]

E   Stiffness (of Pressure Vessel) [Symbol]

E   Strain Rate [Symbol]

E   Third Tone in the Scale of C Major [Acoustics]

E   trans- [Chemistry]

E   Voltage (or Potential) [Symbol]

E   Water Vapor Flux [Symbol]

E   Young's Modulus (or Elastic Modulus) [Symbol]

E-   Railroad Bridge Loading [Letter Symbol Usually Followed by a Numeral Denoting the Maximum Weight in 1,000 lbs on Each Driving Axle of Engine, e.g., E 60 Denotes a Maximum Weight of 60,000 lbs]

-E   Special Electronic Gear (Aircraft) [US Navy Suffix]

/E   Executable Directories Option [Turbo Pascal]

(E)   E Isomer (on Opposite Sides of Molecule)

$\bar{E}$ Mean Total Energy [Symbol]

$\bar{E}$ Molar Internal Energy [Symbol]

E°   Standard (Oxidation Reduction) Potential [Symbol]

E°′   Formal Potential [Symbol]

°E   Degree(s) Engler (of Water Hardness) [Unit]

$E_{\parallel}$ Parallel Energy (of Beam Electromagnetic Radiation) [Symbol]

$E_{\parallel}$ Young's Modulus in Parallel (or In-Plane) Direction [Symbol]

$E_{\perp}$ Illuminance at a Retroreflector on a Plane Normal to the Direction of the Incident Light [Symbol]

$E_{\perp}$ Normal Illuminance [Symbol]

$E_{\perp}$ Perpendicular Energy (of Incident Beam of Electromagnetic Radiation) [Symbol]

$E_{\perp}$ Young's Modulus in Perpendicular (or Normal) Direction [Symbol]

E*   Complex Modulus (in Tension, or Flexure) [Symbol]

E*   Composite Modulus [Symbol]

E*   Effective Young's Modulus [Symbol]

E′   Storage Modulus (in Tension, or Flexure) [Symbol]

E′   Plane-Strain Young's Modulus [Symbol]

E′   Plane-Stress Orthotropic Modulus [Symbol]

E′   Real Component of Young's Modulus [Symbol]

E′   Storage Modulus [Symbol]

E″   Imaginary Component of Young's Modulus [Symbol]

E″   Loss Modulus (in Tension, or Flexure) [Symbol]

E_0 Accelerating Voltage (or Potential) [Symbol]

E_0 Almost Circular Elliptical Galaxy [Hubble Classification]

E_0 Critical Electric Field [Symbol]

E_0 Electromotive Force of Applied Field [Symbol]

E_0 Electron Beam Energy [Symbol]

E_0 Energy of Incident Probe Ion Prior to Collision (in Ion Spectroscopy Scattering) [Symbol]

E_0 Incident Energy [Symbol]

E_0 Isolated Output [Symbol]

E_0 Mean Translational Energy of Ideal Gas Molecule at 0°C [Symbol]

E_0 Specific Energy [Symbol]

E_0 Starting Potential [Symbol]

E_0-E_7 Elliptical Type Galaxies [Hubble Classification of Galaxies; Number Subscripts Range from 0 to 7 with E_0 Denoting Almost Circular Galaxies, and E_7 Referring to Pointed, or Elongated Elliptical Galaxies] [Astronomy]

E^1 Lhotse [Himalayan Mountain Peak in Nepal; 8,501 m (27,890 ft)]

E_1 Primary Voltage [Symbol]

$E_{1, 2,...}$ Energy Level 1, 2, ... [Symbol]

E_2 Induced Secondary Voltage [Symbol]

E_7 Pointed Elliptical Galaxy [Hubble Classification] [Astronomy]

E_A Acceptor Level Energy (of p-type Semiconductor) [Symbol]

E_A Activation Energy [Symbol]

E_A Anneal(ing) Energy [Symbol]

E_A Anode Voltage (of Vacuum Tube) [Symbol]

E_A Attractive Energy (between Two Atoms) [Symbol]

E_A Kinetic Energy of Auger Electrons [Symbol]

E_a Acceptor Level Energy (of p-type Semiconductor) [Symbol]

E_a Activation Energy [Symbol]

E_a Anisotropy Energy [Symbol]

E_a Anode (or Plate) Voltage [Semiconductor Symbol]

E_a Average Energy [Symbol]

E_α Energy of Electronic State α [Symbol]

E_{AC} Alternating-Current Electromotive Force [Symbol]

E_{avg} Average Voltage [Symbol]

E_B Binding Energy [also E_b] [Symbol]

E_B (Television) B Signal Voltage [Symbol]

E_b Emissivity of a Blackbody [Symbol]

E_b^F Fermi-Level Binding Energy [Symbol]

E_{bb} Plate (or Anode) Supply Voltage [Semiconductor Symbol]

E_{BD} Dielectric Breakdown Field [also E_{bd}] [Semiconductor Symbol]

E_{bs} Band Structure Energy (in Solid-State Physics) [Symbol]

E_{BX} Bound Exciton Energy (in Solid-State Physics) [Symbol]

E_C Charging Energy [Semiconductor Symbol]

E_C Collector Voltage [Semiconductor Symbol]

E_C Conduction (Band) Energy [also E_c] [Symbol]

E_C Voltage Across Capacitor at Start of Discharge for RC Transient Circuit [Semiconductor Symbol]

E_c Charging Energy [Semiconductor Symbol]

E_c Coercive (Electric) Field Energy [Symbol]

E_c Cohesive Energy (in Solid-State Physics) [Symbol]

E_c Composite Modulus (i.e., Elastic Modulus of Composite) [Symbol]

E_c Core-Level Binding Energy [Symbol]

E_c Correlation Energy [Symbol]

E_c Creep Modulus [Symbol]

E_c Critical Excitation Energy [Symbol]

E_c Effective Cadmium Cutoff Energy [Symbol]

E_c Grid Supply Voltage [Semiconductor Symbol]

E_c In-Phase Quadrature Amplitude [Symbol]

E_{CB} Voltage between Collector and Base [Semiconductor Symbol]

E_{cc} Closed Circuit Potential (in Electrochemistry) [Symbol]

E_{cc} Grid Supply Voltage [Semiconductor Symbol]

E_{cc} Direct-Current Cutoff Grid Voltage [Semiconductor Symbol]

E_{CE} Voltage between Collector and Emitter [Semiconductor Symbol]

E_{cell} Measured Cell Potential [Symbol]

E_{const} Constant Cell Potential [Symbol]

E_{cl} Elastic Modulus of a Continuous Fiber Composite in Longitudinal (or Alignment) Direction [Symbol]

E_{corr} Corrosion (or Open-Circuit) Potential [Symbol]

E_{crit} Critical Potential [Symbol]

E_{cl} Elastic Modulus of a Continuous Fiber Composite in Trabsverse Direction [Symbol]

E_D Deuterium Energy [Symbol]

E_D Diffusion Energy [Symbol]

E_D Dipolar Energy [Symbol]

E_D Discharging Energy [also E_d] [Semiconductor Symbol]

E_D Donor Level Energy (of n type Semiconductor) [Symbol]

E_D Stored Energy [Symbol]

E_d Demagnetizing Energy [Symbol]

E_d Desorption Energy [Symbol]

E_d Dissociation Energy [Symbol]

E_d Donor Level Energy (of n type Semiconductor) [Symbol]

E_d Electron Delocalization Energy [Symbol]

E_{DC} Direct-Current Electromotive Force [Symbol]

E_{D-D} Mean Dipole-Dipole Energy [Symbol]

E_{des} Desorption Energy [Symbol]

E_{dyn} Effective Dynamic Modulus (of Elasticity) [Symbol]

E_e Irradiance (of a Point on a Surface Element) [Symbol]

E_e Maximum Kinetic Energy of Photoelectrons [Symbol]

E_ε Tangent Component of Electric Vector [also E_ϵ] [Symbol]

E_{eff} Effective Emissivity [Symbol]

E_{eff} Effective Energy [Symbol]

E_{ext} External Electric Field [Symbol]

E_F Fermi Energy (Level) [also E_f, E_f, $F(E)$, or $f(E)$] [Symbol]

E_f Fiber Modulus (i.e., Elastic Modulus of Composite Fibers) [also E_f] [Symbol]

E_f Filament (or Heater) Voltage [Semiconductor Symbol]

E_f Final Energy [Symbol]

E_f Flux Volts (of Magnetic Component Winding) [Symbol]

E_{F1} Density of (Filled) States just below Fermi (Energy) Level [Symbol]

E_{F2} Density of (Empty) States just above Fermi (Energy) Level [Symbol]

E_{fk} Voltage between Heater and Cathode [Semiconductor Symbol]

E_{flex} Flexural Modulus [Symbol]

E_G Band Gap Energy (in Solid-State Physics) [also E_g] [Symbol]

E_G Grid Voltage (of Vacuum Tube) [Symbol]

E_G (Energy) Ground State [Symbol]

E_G (Television) G Signal Voltage [Symbol]

E_g Grid Bias [Semiconductor Symbol]

E_γ Energy of Gamma Irradiation [Symbol]

E_{g1} Grid Bias [Semiconductor Symbol]

E_{g2} Screen Grid Bias [Semiconductor Symbol]

E_{gg} Grid Voltage (or Potential) [Semiconductor Symbol]

E_{gx} Peak Inverse Grid Voltage [Semiconductor Symbol]

E_H Hugoniot Energy [Symbol]

E_H Hydrogen Energy [Symbol]

E_H Magnetic Field Energy [Symbol]

E_H Reducing Intensities referred to Standard Hydrogen Electrode [Symbol]

E_h Oxidation Potential [Symbol]

E_h^n Elastic Modulus of High Modulus Phase of a Uniformly Dispersed Aggregate Composite (n = 1 for Isostrain Condition and - 1 for Isostress Condition) [Symbol]

E_k Elastic Strain Energy [Symbol]

E_I Interaction Energy [Symbol]

E_I (Television) I Signal Voltage [Symbol]

E_I Irreversible Energy [Symbol]

E_i Atomic Binding Energy [Symbol]

E_i Collimated Irradiance [Symbol],

E_i Initial Energy (of Beam of Radiation) [Symbol]

E_i Intrinsic Energy Level [Symbol]

E_i Ion Beam Energy [Symbol]

E_i Ionic Anisotropy Energy [Symbol]

E_{id} Ignitor Voltage Drop [Semiconductor Symbol]

E_{ind} Induced Energy [Symbol]

E_{int} Internal Electric Field [Symbol]

E_J Josephson Junction Energy [Semiconductor Symbol]

E_j Josephson Coupling Energy [Semiconductor Symbol]

E_j Liquid Function Potential [Symbol]

E_K Energy of the K Electron Level [Symbol]

E_k Band Structure (in Solid-State Physics) [Symbol]

E_k Eigenvalue [Symbol]

E_k Energy Spectrum (in Bardeen-Cooper-Schrieffer Superconductivity Theory] [Symbol]

E_k Kinetic Energy [also E_K, or E_{kin}] [Symbol]

E_k Magnetocrystalline Anisotropy [Symbol]

E_{ke} Kinetic Energy of Auger Electron [Symbol]

E_{kr} Kinetic Energy due to Rotation [also E_{KR}] [Symbol]

E_{kt} Kinetic Energy due to Translation [also E_{KT}] [Symbol]

E_L Energy of the L Electron Level [Symbol]

E_L London Dispersion Energy [also E_{London}] [Symbol]

E_L Longitudinal Young's Modulus (of a Composite) [Symbol]

E_l^n Elastic Modulus of Low Modulus Phase of a Uniformly Dispersed Aggregate Composite (n = 1 for Isostrain Condition and −1 for Isostress Condition) [Symbol]

E_λ Monochromatic Emissive Power (for Wavelength λ) [Symbol]

E_{loc} Local Magnetic Field [Symbol]

E_M Energy of the M Electron Level [Symbol]

E_m Amplitude of Carrier Wave [Symbol]

E_m Magnetostatic Energy [Symbol]

E_m Matrix Modulus (i.e., Elastic Modulus of Composite Matrix) [Symbol]

E_m Most Probable Value of Potential [Symbol]

E_m Vacancy Migration Energy (in Solid-State Physics) [Symbol]

E_m^{Cu} Vacancy Migration Energy for Copper [Symbol]

E_{MCA} Magnetocrystalline Anisotropy Energy [also E_{mca}] [Symbol]

E_{ME} Magnetoelastic Energy [also E_{me}] [Symbol]

E_N Energy of the N Electron Level [Symbol]

E_N Net Energy (between Atoms) (i.e., Sum of Attractive and Repulsive Force Components) [Symbol]

E^n Euclidean n Space (in Mathematics) [Symbol]

E_n Energy of the n-th State (of an Atom) [Symbol]

E_n Normal Energy [Symbol]

E_ν Electric Field Associated with Single Photon [Symbol]

$E_{n,l,m}$ Eigenvalue of wave function $\Psi_{n,l,m}$ [Symbol]

E_{oc} Open-Circuit (or Corrosion) Potential [Symbol]

E_{oo} (Electron) Tunneling Probability Parameter [Symbol]

E_{op} Operating Voltage [Symbol]

E_{out} Output Voltage [Symbol]

E°_{ox} Standard Potential Corresponding to Oxidation Half-Cell Reaction [Symbol]

E_P Primary Energy [Symbol]

E_p Effective (RMS) Voltage in Primary (Transformer) Coil [Symbol]

E_p Particulate Modulus (i.e. Elastic Modulus of Particles in Composite) [Symbol]

E_p Peak Voltage [Symbol]

E_p Plasma Energy [Symbol]

E_p Plate (or Anode) Voltage [Semiconductor Symbol]

E_p Polarization Field Strength [Symbol]

E_p Potential Energy [also E_P, or E_{pot}] [Symbol]

E_p Primary Electron Energy [Symbol]

E_p Proton Energy [Symbol]

E_{PE} Elastic Potential Energy (e.g., of Springs, etc.) [Symbol]

E_{PF} Potential Energy due to Position in Force Field [Symbol]

E_{pit} Pitting Potential (in Corrosion Science) [Symbol]

E_{pl} (Cumulative) Plastic Strain [Symbol]

E_{pp} Plate (or Anode) Potential [Semiconductor Symbol]

E_Q (Television) Q Signal Voltage [Symbol]

E_Q Quadrupole Splitting [Symbol]

E_q Electric Field Associated with Charge Wave [Symbol]

E_R Reconstruction Energy (in Solid-State Physics) [Symbol]

E_R Repulsive Energy (between Two Atoms) [Symbol]

E_R (Television) R Signal Voltage [Symbol]

E_R Reversible Energy [Symbol]

E_r Atom Superposition Energy [Symbol]

E_r Reduced (Elastic) Modulus [Symbol]

E_r (Viscoelastic) Relaxation Modulus [Symbol]

E_r Residual Energy [Symbol]

E_ρ Perpendicular Component of Electric Vector [Symbol]

E_{ref} Reference Electrode (in Physical Chemistry) [Symbol]

E°_{red} Standard Potential Corresponding to Reduction Half-Cell Reaction [Symbol]

E_S Secondary Energy [Symbol]

E_S Surface Energy [Symbol]

E_s Effective (RMS) Voltage in Secondary (Transformer) Coil [Symbol]

E_s Kinetic Energy for Scattered Probe Ion (in Ion Spectroscopy Scattering) [Symbol]

E_s Modulus of Elasticity in Shear (or Shear Modulus) [Symbol]

E_s Out-of-Phase Quadrature Amplitude [Symbol]

E_s Secant Modulus [Symbol]

E_s Solar Irradiance [Symbol]

E_s Surface Electric Field [Symbol]

E_s Uniaxial Magnetostatic Shape Anisotropy [Symbol]

E_σ Uniaxial Magnetoelastic Anisotropy [Symbol]

E_{sg} Screen Grid Voltage [Semiconductor Symbol]

E_{sig} Applied Signal Voltage [Symbol]

E_{SR} Activation Energy for Strain Relaxation [Symbol]

E_T Threshold (Electric) Field [Symbol]

E_T Total Energy [also E_t] [Symbol]

E_T Translational Energy [Symbol]

E_T Transverse Young's Modulus (of Composites) [Symbol]

E_t Theoretical Electromotive Force [Symbol]

E_{ta} Target Voltage [Symbol]

E_{Th} Thevenin's Equivalent Voltage [Symbol]

E_u Uniform Elongation [Symbol]

E_V Valence (Band) Energy [also E_v] [Symbol]

E_w Domain-Wall Energy (in Solid-State Physics) [Symbol]

E_x Emitted Electron [Symbol]

E_x Excited Photon Energy [Symbol]

E_x X-Ray Energy [Symbol]

E_z Ionization Breakdown (or Striking) Voltage [Semiconductor Symbol]

E1 European Data Transmission Standard [Bandwidth 2.048Mbps] [Symbol]

E1 Unimolecular Elimination [Symbol]

E2 Bimolecular Elimination [Symbol]

E3 European Data Transmission Standard [Bandwidth 57.344 Mbps] [Symbol]

E605 Phosphorothioic Acid O,O-Bis(ethyl)-O-(4 Nitrophenyl) ester

E-1 Private [US Marine Corps Grade]

E-1 Recruit [US Army and Air Force Grade]

E-1 Seaman Recruit [US Navy and Coast Guard Grade]

E-2 Airman Third Class [US Air Force Grade]

E-2 Private [US Army Grade]

E-2 Private First Class [US Marine Corps Grade]

E-2 Seaman Apprentice [US Navy and Coast Guard Grade]

E-3 Airman Second Class [US Air Force Grade]

E-3 Lance Corporal [US Marine Corps Grade]

E-3 Private First Class [US Army Grade]

E-3 Seaman [US Navy and Coast Guard Grade]

E-4 Airman/Airwoman First Class [US Air Force Grade]

E-4 Corporal [US Marine Corps Grade]

E-4 Corporal, Specialist 4 [US Army Grade]

E-4 Petty Officer Third Class [US Navy and Coast Guard Grade]

E-5 Petty Officer Second Class [US Navy and Coast Guard Grade]

E-5 Sergeant [US Marine Corps Grade]

E-5 Sergeant, Specialist 5 [US Army Grade]

E-5 Staff Sergeant [US Air Force Grade]

E-6 Petty Officer First Class [US Navy and Coast Guard Grade]

E-6 Staff Sergeant [US Marine Corps Grade]

E-6 Staff Sergeant, Specialist 6 [US Army Grade]

E-6 Technical Sergeant [US Air Force Grade]

E-7 Chief Petty Officer [US Navy and Coast Guard Grade]

E-7 Gunnery Sergeant [US Marine Corps Grade]

E-7 Master Sergeant [US Air Force Grade]

E-7 Sergeant First Class, Platoon Sergeant, Specialist 7 [US Army Grade]

E-8 Master Sergeant, First Sergeant [US Marine Corps Grade]

E-8 Master Sergeant, First Sergeant, Specialist 8 [US Army Grade]

E-8 Senior Chief Petty Officer [US Navy and Coast Guard Grade]

E-8 Senior Master Sergeant [US Air Force Grade]

E-9 Chief Master Sergeant [US Air Force Grade]

E-9 Master Chief Petty Officer [US Navy and Coast Guard Grade]

E-9 Master Gunnery Sergeant [US Marine Corps Grade]

E-9 Sergeant Major, Specialist 9 [US Army Grade]

E-64 trans Epoxysuccinyl-L-Leucylamido(4-Guanidino) butane

E-64c (2S,3S) trans Epoxysuccinyl-L-Leucylamido 3 Methylbutane

E-64d (2S,3S) trans Epoxysuccinyl-L-Leucylamido 3 Methylbutane Ethyl Ester

e additional designation of emission (or transmission) type–width, or duration modulated [Symbol] [also E]

e base of natural logarithm (i.e. 2.71828) [Symbol]

e battery voltage [Symbol]

e charged lepton (in particle physics) [Symbol]

e coefficient of restitution (in mechanics) [Symbol]

e diffusion potential (of electrolytic cells) [Symbol]

e east(ern)

e eccentric

e eccentricity of conic sections [Symbol]

e eccentricity of load application [Symbol]

e effective

e efficiency [Symbol]

e efficient

e elastic

e electric(al)

e electron(ic)

e elementary charge (or electron charge) (1.6028×10^{-19} C) [Symbol]

e elongation [Symbol]

e emitter

e emitter [Semiconductor Symbol]

e end and save (command) [Edlin MS-DOS Line Editor]

e energy

e engineering (or conventional) strain [Symbol]

e epsilon [English Equilvalent]

e erg [Unit]

e (random) error [Symbol]

e eutectic

e evaporation

e excellent

e exponent(ial)

e exposure

e Friedrich and Kroenig (Unit((in X-Ray Radiology) [Symbol]

e instantaneous (induced) voltage [Symbol]

e quantum of electricity [Symbol]

e saturation vapor pressure [Symbol]

e (nominal) strain [Symbol]

e void ratio (in a soil mass) [Symbol]

e volume dilatation [Symbol]

e- electronic (e.g., e mail)

/e exepack option [MS DOS Linker]

$\bar{e}$ eta [English Equilvalent]

$\bar{e}$ mean of all differences between observed (or measured) value and predicted (or accepted) value [Symbol]

e^+ positron (i.e., antielectron, or positively charged electron) [Symbol]

e^- electron [Symbol]

$\bar{e}$ electric vector [Symbol]

e_0 zero gauge length elongation (in tension testing) [Symbol]

e_2 instantaneous induced voltage in secondary circuit [Symbol]

e_2 vapor pressure 2 meters above the surface [Symbol]

e_b plate (or anode) voltage [Semiconductor Symbol]

e_c critical void ratio (in a soil mass) [Symbol]

e_{el} elastic strain [Symbol]

e_f engineering strain at fracture [Symbol]

e_g grid voltage [Semiconductor Symbol]

e_i difference between observed (or measured) value and predicted (or accepted) value [Symbol]

e_i input voltage [Symbol]

e_i saturated (water) vapor pressure with respect to ice [Symbol]

e_o output voltage [Symbol]

e_p voltage induced in primary (transformer) coil [Symbol]

e_{pl} plastic strain [Symbol]

e_s saturation vapor pressure (in hydrology) [Symbol]

e_s voltage induced in secondary (transformer) coil [Symbol]

e_t total creep at time "t" [Symbol]

e_w saturation (water) vapor pressure with respect to water [Symbol]

e_x strain in x direction [Symbol]

e_z strain in z direction [Symbol]

$\mathscr{E}$ Effective Value of Alternating-Current Voltage [Symbol]

$\mathscr{E}$ Electric Field (Strength) (or Electric Field Intensity) [Symbol]

$\mathscr{E}$ Electromotive Force [Symbol]

$\mathscr{E}^0$ Standard Cell Potential [Symbol]

$\mathscr{E}_c$ Electric Coercive Field [Symbol]

$\mathscr{E}_H$ Hall Voltage (in Solid-State Physics) [Symbol]

$\mathscr{E}_{max}$ Maximum Instantaneous Alternating-Current Voltage, or Electromotive Force [Symbol]

e difference between observed (or measured) value and predicted (or ccepted) value [Symbol]

$\bar{e}$ mean of all differences between observed (or measured) value and predicted (or accepted) value [Symbol]

e_i difference between observed (or measured) value and predicted (or accepted) value [Symbol]

e (random) error [Symbol]

e instantaneous (induced) voltage [Symbol]

ϵ anisotropy [Symbol]

ϵ attitude (of bearings) [Symbol]

ϵ base of natural logarithm (i.e. 2.71828) [Symbol]

ϵ bond energy [Symbol]

ϵ coefficient of kinematic viscosity [Symbol]

ϵ Darken coefficient (in metallurgy) [Symbol]

ϵ dielectric constant [Symbol]

ϵ eccentricity of a conic section [Symbol]

ϵ eccentricity ratio (of bearings) [Symbol]

ϵ electrode potential [Symbol]

ϵ emissivity (of a thermal radiator) [Symbol]

ϵ emittance [Symbol]

ϵ energy density [Symbol]

ϵ engineering (or conventional) strain [Symbol]

ϵ (lower-case) epsilon [Greek Alphabet]

ϵ epsilon phase (of a material) [Symbol]

ϵ epsilon vibration direction [Symbol]

ϵ exposure [Symbol]

ϵ excitation energy (in quantum mechanics) [Symbol]

ϵ extraordinary index of refraction (of an uniaxial crystal) [Symbol]

ϵ molar absorptivity [Symbol]

ϵ molar extinction coefficient [Symbol]

ϵ normal strain [Symbol]

ϵ (electric) permittivity [Symbol]

ϵ initial porosity (or void fraction) [Symbol]

ϵ positive number in limit process arguments [Symbol]

ϵ rotation angle [Symbol]

ϵ small value [Symbol]

ϵ (automatic control) system error [Symbol]

$\epsilon_{\parallel}$ dielectric permittivity in parallel, or in-plane direction [Symbol]

$\epsilon_{\parallel}$ strain in longitudinal, parallel, or in-plane direction [Symbol]

$\epsilon_{\perp}$ dielectric permittivity in perpendicular direction [Symbol]

$\epsilon_{\perp}$ strain in transverse, or perpendicular direction [Symbol]

ϵ_{∞} dynamic dielectric constant [Symbol]

ϵ_{∞} high-frequency dielectric constant [Symbol]

$\bar{\epsilon}$ mean energy (imparted to matter) (in radiography) [Symbol]

$\bar{\epsilon}$ mean kinetic energy (or translational energy of a molecule) [Symbol]

$\dot{\epsilon}$ steady state creep rate [Symbol]

$\dot{\epsilon}$ strain (rate) [Symbol]

$\dot{\epsilon}$ true strain [Symbol]

ϵ' nominal strain [Symbol]

ϵ_0 elementary creep strain [Symbol]

ϵ_0 low-frequency dielectric constant [Symbol]

ϵ_0 (electric) permittivity of vacuum (8.8542×10^{-12} F/m) [Symbol]

ϵ_0 (constant) strain level (in viscoelastic polymers) [Symbol]

ϵ_{45} normal strain in a direction 45° from horizontal axis [Symbol]

ϵ_A stopping cross section of constituent element A in a compound specimen composed of elements "A" and "B" [Symbol]

ϵ_a anisotropy energy density [Symbol]

ϵ_a dielectric anisotropy [Symbol]

ϵ_a loading amplitude (in fatigue) [Symbol]

$\dot{\epsilon}_a$ alternating load [Symbol]

$\dot{\epsilon}_a$ applied strain rate [Symbol]

ϵ_B Bauschinger stress factor (in metallurgy) [Symbol]

ϵ_B bulk strain [Symbol]

ϵ_B stopping cross section of constituent element B in a compound specimen composed of elements "A" and "B" [Symbol]

ϵ_c critical strain [Symbol]

ϵ_c electrode potential above standard calomel electrode [Symbol]

ϵ_c (total) strain on a fiber reinforced composite [Symbol]

ϵ_e exchange energy density [Symbol]

ϵ_{eff} effective dielectric function [Symbol]

ϵ_{eff}'' effective loss factor [Symbol]

ϵ_F Fermi energy of conduction electrons [Symbol]

ϵ_f strain on fibers of a fiber reinforced composite [Symbol]

ϵ_h electrode potential above standard hydrogen electrode [Symbol]

$\dot{\epsilon}_H$ hemispherical emittance [Symbol]

ϵ_{ij} energy of the i-j bond (in physics) [Symbol]

ϵ_{ij} strain tensor [Symbol]

ϵ_K Kerr ellipticity (in physics) [Symbol]

ϵ_{JT} Jahn-Teller strain [Symbol]

ϵ_L Labusch combined interaction parameter [Symbol]

ϵ_L longitudinal optic phonon frquency (in solid-state physics) [Symbol]

ϵ_L longitudinal strain (of a composite) [or ϵ_l] [Symbol]

ϵ_λ spectral emittance (or spectral emissivity) [Symbol]

ϵ_m mean load (in fatigue loading) [Symbol]

ϵ_m strain on matrix of a fiber-reinforced composite [Symbol]

ϵ_{max} maximum load (in fatigue loading) [Symbol]

ϵ_{min} minimum load (in fatigue loading) [Symbol]

ϵ_{mw} microwave dielectric constant [Symbol]

ϵ_N nucleation strain (in metallurgy) [Symbol]

ϵ_p peak strain [Symbol]

ϵ_p plastic shear stress around a particle [Symbol]

ϵ_p plastic true strain [Symbol]

ϵ_r relative permittivity (or relative dielectric constant) [Symbol]

ϵ_r replacement energy [Symbol]

ϵ_r' relative dielectric constant [Symbol]

ϵ_r'' (dielectric) loss index [Symbol]

ϵ_r^* relative complex permittivity (or relative complex dielectric constant) [Symbol]

ϵ_{rms} root-mean-square strain [Symbol]

$\dot{\epsilon}_s$ steady state creep rate [also $\dot{\epsilon}_{SS}$] [Symbol]

ϵ_T strain at (final) temperature T [Symbol]

ϵ_T transverse strain (of a composite) [Symbol]

ϵ_T true strain [Symbol]

$\dot{\epsilon}_T$ total emittance [Symbol]

$\dot{\epsilon}_T$ true strain [Symbol]

ϵ_t thickness (or transverse) strain [Symbol]

ϵ_θ tensile hoop strain (in upset testing) [Symbol]

$\dot{\epsilon}_{tr}$ true strain [Symbol]

ϵ_u monotonic strain-to-failure [Symbol]

ϵ_{uc} cyclic strain-to-failure [Symbol]

ϵ_w lateral (or width) strain [Symbol]

ϵ_x normal strain in horizontal or x-direction [Symbol]

$\bar{\epsilon}_x$ x-component of mean kinetic energy (or translational energy of a molecule) [Symbol]

ϵ_y normal strain in vertical or y-direction [Symbol]

$\bar{\epsilon}_y$ y-component of mean kinetic energy (or translational energy of a molecule) [Symbol]

ϵ_z compressive axial strain (in upset testing) [Symbol]

ϵ_z normal strain in z direction [Symbol]

$\bar{\epsilon}_z$ z-component of mean kinetic energy (or translational energy of a molecule) [Symbol]

ε angular distance (of telescope) [Symbol]

ε base of natural logarithm [Symbol]

ε dielectric constant [Symbol]

ε divergence (of differential interference contrast microscope) [Symbol]

ε eccentricity (of conic sections) [Symbol]

ε electric field strength [Symbol]

ε electronic charge [Symbol]

ε engineering (or conventional) strain [Symbol]

ε (lower case) epsilon (variant) [Greek Alphabet]

ε epsilon phase (of a material) [Symbol]

ε error (in statistics) [Symbol]

ε external electric field [Symbol]

ε perturbation [Symbol]

ε probability of a given atomic jump direction [Symbol]

ε reaction rate ratio [Symbol]

ε relative dielectric constant [Symbol]

ε sensitivity (of a mass spectrometer) [Symbol]

ε specific surface (free) energy [Symbol]

ε (true) strain [Symbol]

ε tilt (in transmission electron microscopy) [Symbol]

$[\varepsilon]$ error sum (in statistics) [Symbol]

ε^* complex magnetic permittivity [Symbol]

ε' real part of magnetic permittivity [Symbol]

ε'' imaginary part of magnetic permittivity [Symbol]

$\varepsilon_\parallel$ strain in longitudinal (parallel) direction [Symbol]

$\varepsilon_\perp$ strain in transverse (perpendicular) direction [Symbol]

ε_∞ dynamic dielectric constant [Symbol]

$\bar{\varepsilon}$ average molecular energy [Symbol]

$\bar{\varepsilon}$ equivalent strain [Symbol]

$\dot{\varepsilon}$ strain (rate) [Symbol]

$\dot{\varepsilon}$ true strain [Symbol]

ε_0 electric permittivity of vacuum (8.8542×10^{-12} F/m) [Symbol]

ε_0 energy of the ground level [Symbol]

ε_a strain amplitude (in fatigue) [Symbol]

$\dot{\varepsilon}_a$ alternating load [Symbol]

ε_{ca} cavitation strain [Symbol]

ε_{cr} creep strain [Symbol]

ε_b size misfit parameter (in materials science) [Symbol]

ε_e elastic deformation [Symbol]

ε_e elastic strain amplitude (in fatigue) [Symbol]

$\dot{\varepsilon}_e$ elastic true strain [Symbol]

ε_f engineering strain at fracture [Symbol]

ε_f' fatigue ductility coefficient [Symbol]

$\dot{\varepsilon}_f$ fracture ductility [Symbol]

ε_G (shear) modulus misfit parameter (in materials science) [Symbol]

$\dot{\varepsilon}_H$ hemispherical emittance [Symbol]

ε_i energy of the i th level [Symbol]

ε_λ spectral emittance (or spectral emissivity) [Symbol]

ε_{max} maximum permittivity [Symbol]

ε_N nucleation strain (in metallurgy) [Symbol]

ε_p plastic deformation [Symbol]

ε_p plastic strain amplitude (in fatigue) [Symbol]

ε_r relative permittivity (or relative dielectric constant) [also $\dot{\varepsilon}_r$] [Symbol]

$\bar{\varepsilon}_{rot}$ rotational energy (of molecules) [Symbol]

$\dot{\varepsilon}_s$ strain rate for steady state creep [Symbol]

ε_{SS} steady-state creep [also ε_{ss}] [Symbol]

ε_T strain at (final) temperature T [Symbol]

ε_T transverse strain [Symbol]

$\dot{\varepsilon}_T$ total emittance [Symbol]

$\dot{\varepsilon}_T$ true strain [Symbol]

ε_t total emissivity [Symbol]

ε_t true strain [Symbol]

$\varepsilon\theta$ tensile hoop strain (in upset testing) [Symbol]

ε_{th} thermal strain [Symbol]

$\dot{\varepsilon}_{tr}$ true strain [Symbol]

$\dot{\varepsilon}_{tr}$ strain rate for transient state creep [Symbol]

$\bar{\varepsilon}_{trans}$ translational energy (of molecules) [Symbol]

$\bar{\varepsilon}_{vib}$ vibrational energy (of molecules) [Symbol]

ε_u uniform strain [Symbol]

ε_x lateral elongation in x direction ($-\varepsilon_x$ indicates a "lateral contraction") [Symbol]

ε_{xx} strain in x-x plane [Symbol]

ε_y lateral elongation in y direction ($-\varepsilon_y$ indicates a "lateral contraction") [Symbol]

ε_{yp} yield point strain [Symbol]

ε_{yy} strain in y-y plane [Symbol]

ε_z axial elongation in z direction [Symbol]

ε_{zz} strain in z-z plane [Symbol]

η (coefficient of) absolute viscosity [Symbol]

η asymmetry parameter [Symbol]

η coherency strain [Symbol]

η coordinate [Symbol]

η efficiency [Symbol]

η electrochemical potential [Symbol]

η electrolytic polarization [Symbol]

η equilibrium roughness exponent [Symbol]

η (lower-case) eta [Greek Alphabet]

η eta meson (in particle physics) [Symbol]

η eta phase (of a material) [Symbol]

η hysteresis [Symbol]

η intrinsic dissipation [Symbol]

η intrinsic standoff ratio [Semiconductor Symbol]

η long-range order parameter (in solid-state physics) [Symbol]

η loss factor (or loss coefficient) (in electronics) [Symbol]

η misfit parameter (in materials science) [Symbol]

η optical transfer efficiency [Symbol]

η overpotential (or overvoltage) [Symbol]

η packing fraction (in nuclear physics) [Symbol]

η perturbation [Symbol]

η photovoltaic conversion efficiency [Symbol]

η quantum efficiency (in electronics) [Symbol]

η quantum yield (in physical chemistry) [Symbol]

η radiant efficiency [Symbol]

η Raman cross section ratio [Symbol]

η reduced viscosity (of plastics) [Symbol]

η refractive index [Symbol]

η relative density (or specific gravity) [Symbol]

η spin (in quantum mechanics) [Symbol]

η surface charge density [Symbol]

η thermal efficiency [Symbol]

η (shear) viscosity [Symbol]

$[\eta]$ reduced specific viscosity (of a solution) [Symbol]

$\%\eta$ percent of efficiency (of a system) [Symbol]

η^* complex viscosity coefficient [Symbol]

η' eta prime meson (in particle physics) [Symbol]

η^+ eta plus (i.e., positively charged) meson (in particle physics) [Symbol]

η^- eta minus (i.e., negatively charged) meson (in particle physics) [Symbol]

η_0 viscosity of (pure) solvent [Symbol]

η_0 zero shear viscosity (of polymer) [Symbol]

η_a activation polarization overvoltage (in electrochemistry) [Symbol]

η_c concentration polarization overvoltage (in electrochemistry) [Symbol]

η_e radiant efficiency [Symbol]

η_{gb} grain boundary viscosity (in materials science) [Symbol]

η_i electrochemical potential of component i [Symbol]

η_K Kerr ellipticity [Symbol]

η_n efficiency of n th component of a system [Symbol]

$\%\eta_{ov}$ overall percent of efficiency (of a system) [Symbol]

η_P population efficiency [Symbol]

η_{pl} plastic viscosity [Symbol]

η_T total luminescence efficiency [Symbol]

η_t thermodynamic efficiency [Symbol]

η_t (electrical) transfer efficiency (i.e., ratio of load power to source power in percent) [Symbol]

η_v amount of point defects (in solid-state physics) [Symbol]

EA East Asia(n)
EA Eastern Air Lines [US]
EA Easy Axis (of Magnetization) [also ea]
EA Eccentric Axis
EA Edgetone Amplifier
EA Edgewood Arsenal [Maryland, US]
EA Educational Age [Psychology]
EA Edwards-Anderson (Model of Spin Glasses) [Solid-State Physics]
EA Effective Address
EA Effective Absorption
EA Effective Aperture
EA Effective Area
EA Elastic Anisotropy
EA Elastic Axis
EA Electric Actuator
EA Electrically Alterable
EA Electric Arc
EA Electrics Association
EA Electroabsorption
EA Electroacoustic(s) [also E-A]
EA Electroanalytical; Electroanalysis
EA Electron Absorption
EA Electron Accelerator
EA Electron Acceptor
EA Electron Affinity
EA Electron Association [Mass Spectroscopy]
EA Electronic Assembly
EA Electrophilic Addition
EA Electrostatic Actuator
EA Elemental Analysis
EA Elevation Angle
EA Elliptical Aperture
EA Enamelled Asbestos Covered (Electric Wire)
EA Endometriosis Association [US]
EA Energy Absorption; Energy Absorbing
EA Energy Act [UK]
EA Energy Additive
EA Energy Analysis; Energy Analyzer
EA Energy Association [US]
EA Energy Audit
EA Energy Authority [New South Wales, Australia]
EA Engine Analysis; Engine Analyzer
EA Engineering Acoustics
EA Engineering Adhesive
EA Engineering Alloy
EA Engineering Analysis
EA Enterprise Analysis
EA Enthalpimetric Analysis
EA Enumeration Area
EA Environment Abstracts
EA Environment Agency [Japan]
EA Environmental Action
EA Environmental Assessment

EA Environment Australia
EA Equilibrium Altitude
EA Error Analysis [Computers]
EA Ethanolamine
EA Ethoxyacetylene
EA Ethylamine
EA Euclidian Aigorithm [Mathematics]
EA Europium Aluminate
EA Event Analysis
EA Excess Air
EA Excited Atom
EA Exhaust Air
EA Explicit Atom (Model) [Physics]
EA Extended Abstracts [of Materials Research Society, US]
EA Extended Area
EA Extended Attribute
EA (Department of) External Affairs [Canada]
EA Extinguishing Agent
EA7S Antibody Sensitized Sheep Erythrocytes [Immunology]
E[A] Mean Areal Value (or Mean Value of Area) [Symbol]
$E(A_B)$ Bond Energy of Atom A to Atom B [Symbol]
$E_a(A_B)$ Activation Energy of Reaction Resulting in Bonding of Atom A to a Surface $B(s)$ [Symbol]
E-A Electroacoustic(s) [also EA]
E-A Elimination-Addition (Mechanism)
Ea Ethanamine
E/a Energy per atom [Unit]
ea each
ea easy axis (of magnetization) [also EA]
e/a electron-to-atom (ratio)
EAA East Africa Association [UK]
EAA Edinburgh Architectural Association [Scotland]
EAA Electric Auto Association [US]
EAA Electricity Arbitration Association [UK]
EAA Engineer in Aeronautics and Astronautics
EAA Environmental Assessment Act
EAA Essential Amino Acid [Biochemistry]
EAA Ethyl Acetoacetate
EAA Ethylene Acrylic Acid (Copolymer)
EAA European Accounting Association [Belgium]
EAA European Acoustics Association
EAA European Aluminum Association [Belgium]
EAA Excitatory Amino Acid [Biochemistry]
EAA Experimental Aircraft Association [US]
EAAA European Association of Advertising Agencies
EAAAF Experimental Aircraft Association Aviation Foundation [US]
EAAC East African Airways Corporation
EAAC European Agricultural Aviation Center
EAAC Experimental Aircraft Association of Canada
EAAE European Association of Agricultural Economists [Belgium]
EAAFRO European African Agricultural and Forestry Research Organization [Kenya]

EAAP European Association for Animal Production [Italy]

EAAPS Electron Excited Auger Electron Appearance Potential Spectroscopy

EAASH European Academy of Arts, Sciences and Humanities [France]

EAB Economic Analysis Bureau

EAB Educational Activities Board [of Institute of Electrical and Electronics Engineers, US]

EAB Engineering Activity Board [of Society of Automotive Engineers, US]

EAB European American Bank

EAB Exclusion Area Boundary [Nuclear Engineering]

EAB Extramural Advisory Board [of National Institutes of Health, Bethesda, Maryland, US]

$\varepsilon(A_xB_y)$ stopping cross section of a compound specimen [Symbol]

EAB/RAB Educational Activities Board/Regional Activities Board [of Institute of Electrical and Electronics Engineers, US]

EABRD Electrically Activated Bank Release Device

EAC East African Community

EAC East African Countries

EAC East Australian Current [Pacific Ocean]

EAC Eastern Arizona College [Thatcher, US]

EAC Education Affairs Committee [of ASM International, US]

EAC EEC (European Economic Community) Advisory Council

EAC Electroanalytical Chemistry

EAC Electronic Analog Computer

EAC Energy Absorbing Capacity

EAC Energy Accomodation Coefficient

EAC Engineering Accreditation Commission [US]

EAC Engineering Affairs Council [of American Association of Engineering Societies, US]

EAC Engineering Applications Center [UK]

EAC Environmental Action Coalition [US]

EAC Environmentally Assisted Cracking

EAC Equivalent Annual Cost

EAC Error Alert Control

EAC Estimate at Completion

EAC Expect Approach Clearance

EAC Expedition Advisory Center [UK]

EAC Experiment Apparatus Container

EAC External Affairs Canada

Eac Ethylaminocarbonyl

EACA ε Amino n Caproic Acid

EAcAc Ethyl Acetoacetate

EAC AIA EEC (European Economic Community) Advisory Council of the Asbestos International Association [Belgium]

EACC Error Adaptive Control Computer

EACE East African Certificate of Education [Kenya]

EACG European Association of Exploration Geophysicists [Netherlands]

EACI Ecology Action Educational Institute

ε-**Acid** 8-Hydroxy 1,6-Naphthalenedisulfonic Acid [also ε-acid]

ε-**Acid** 1-Naphthylamine 3,8-Disulfonic Acid [also ε-acid]

EACM European Association for Composite Materials [France]

EACN European Air Chemical Network

EACO Engineers and Architects Council of Oregon [US]

EACPI European Association of County Planning Institutions [Belgium]

EACRO European Association of Contract Research Organization

EACRP European-American Committee on Reactor Physics [of European Nuclear Energy Agency, France]

EACS Electronic Automatic Chart System

EACSO East Africa Common Services Organization [Defunct]

EACVD Electron-Assisted Chemical Vapor Deposition

EAD Easy Axis Dispersion [Solid-State Physics]

EAD East Australian Daylight (Time) [Greenwich Mean Time −09:00]

EAD Electrically Alterable Device

EAD Electroacoustic Dewatering

EAD Electron Affinity Difference (Method)

EAD Encoded Archival Description Project

EAD Estimated Availability Date

EAD Excitation Absorbed Dose

EADAS Engineering and Administrative (or Administration) Data Acquisition System

EADB East African Development Bank [Uganda]

EADC Ethyl Acetylenedicarboxylate

EADC Ethylaluminum Dichloride

EADF Elliptical Aperture with Dynamic Focus [also EA-DF]

EADI Electronic Attitude Directional (or Director) Indicator

EADI European Association of Development, Research and Training Institutes [Switzerland]

EADP European Association of Directory Publishers [Belgium]

EAE Eastern Association of Electroencephalographers [US]

EAE Elastic After Effect

EAE Environmentally Assisted Embrittlement

EAE Extended Arithmetic Element

Eae Enterobacter aerogenes [Microbiology]

EAEC East African Economic Community

EAEC Essential Airborne Equipment Characteristic

EAEC European Airlines Electronic Committee

EAEC European Atomic Energy Commission

EAEG European Association of Exploration Geophysicists [The Hague, Netherlands]

EAEP Energy Action Educational Project

EAERE European Association of Environmental and Resource Economists

EAES Electron-Excited Auger Electron Spectroscopy [also EEAES]

EAES Electron-Induced Auger Electron Spectroscopy [also EIAES]

EAES European Atomic Energy Society

EAET Efficiency and Alternative Energy Technologies

EAETB Efficiency and Alternative Energy Technology Branch

EAETLFEM European Association for the Exchange of Technical Literature in the Field of Ferrous Metallurgy [Luxembourg]

EAF East African Federation

EAF Electric Arc Furnace [Metallurgy]

EAF Environmental Action Foundation [Takoma Park, Maryland, US]

EAF Equivalent Availability Factor [Telecommunications]

E Af East Africa(n)

EAFB Edwards Air Force Base [of US Air Force in California]

EAFB Eglin Air Force Base [of US Air Force near Fort Walton, Florida]

EAFB Ellington Air Force Base [of US Air Force near Houston, Texas]

EAFFRO East African Freshwater Fishery Research Organization

EAFR Extended Arc Flash Reactor

E Afr East Africa(n)

EAFRO East African Fishery Research Organization

EAF-VAR Electric Arc Furnace/Vacuum Arc Remelted (Products) [Metallurgy]

E-AG Europe Atlantic Group [UK]

Eag Enterobacter agglomerans [Microbiology]

EAGE Electrical Aerospace Ground Equipment

EAGE European Association of Geoscientists and Engineers

EAGGF European Agricultural Guidance and Guarantee Fund

EAGLE Elevation Angle Guidance Landing Equipment

EAGLE European Association for Grey Literature Exploitation [UK]

Eagle Bull Eagle Bulletin [South African Environmental Bulletin]

EAH Electric Arc Heating [Metallurgy]

EAH Engineering Association of Hawaii [US]

EAHC Essex Archeological and Historical Congress [UK]

EAHP Ecole d'Application des Hauts Polymères [School of High-Polymer Applications; of Université Louis Pasteur, Strasbourg, France]

EAHRA East African Railways and Harbours Administration

EAI Economic Abstracts International [Netherlands]

EAI Engineering Advance Information

EAI Engineers and Architects Institute [US]

EAI Expanded Academic Index

EAIMB East African Industries Management Board [Kenya]

EAIR Extended Area Instrumentation Radar

EAIRB East African Industrial Research Board [Kenya]

EAIRO East African Industrial Research Organization

EAK Ethyl Amyl Ketone

EAL Eastern Airlines

EAL Electromagnetic Amplifying Lens

EAL Equipment Air Lock

EAL Extended Average Level

EALM European Association of Livestock Markets [Belgium]

EAM Electric(al) Accounting Machine

EAM Electric Arc Melting

EAM Electron-Acoustic Microscopy

EAM Electronic Accounting Machine

EAM Electronic Automatic Machine

EAM Elementary Access Method

EAM Embedded Atom Method [Materials Science]

EAM Emergency Action Message

EAM Ethylene Acrylate Copolymer

EAMCBP European Association of Makers of Corrugated-Base Papers [Germany]

EAMD East African Meteorological Department

EAMFRO East African Marine Fisheries Research Organization

E-AMP-AA Amino Acyl Adenosine Monophosphate Enzyme (Complex) [Biochemistry]

EAMS Electron Attachment (Association) Mass Spectroscopy [also EA-MS]

EAMTC European Association of Management Training Centers

EAN Effective Atomic Number (Rule)

EAN Engineering Association of Nashville [Tennessee, US]

EAN European Article Number

EAN European Article Numbering Association [Belgium]

EANA European Alliance of News Agencies

EANC Elastically Active Network Chain

EANDC European American Nuclear Data Committee [of European Nuclear Energy Agency, France]

EANDRO Electrically Alterable Nondestructive Readout

EANI European Article Numbering Institute

EANHS East African Natural History Society [Kenya]

EANPG European Air Navigation Planning Group [of International Civil Aviation Organization]

EAO Egyptian Agricultural Organization

EAO Electrolysis Association of Ontario [Canada]

EAO Environmental Assessment Office [of Office of Nuclear Waste Isolation, US Department of Energy]

EAON Except As Otherwise Noted [also eaon]

EAOS Easy Access Ordering System

EAOS Environment Australia On-line Service

EAP Ecological Agriculture Project [Canada]

EAP Emergency Action Plan

EAP Employee Assistance Program

EAP Environment Actions Plan [Commonwealth]

EAP Environmental Accounting Project [of US Environmental Protection Agency]

EAP (Office of) Environmental Assurance, Permits and Policy [of US Department of Energy]

EAP Environment Assistance Program

EAP Equivalent Air Pressure

EAP Evoked Action Potential

EAP Expenditure Analysis Plan

EAP Experimental Activity Proposal

EAP Experimental Aircraft Program [of British Aerospace]

EAP Explosively Activated Press

EAP N (2 Hydroxyethyl)phosphoramidic Acid

EAPA European Asphalt Pavement Association [Sweden]

EAPD Electro Absorption Photodiode

EAPFBO European Association of Professional Fire Brigade Officers [Belgium]

EAPFS Electron Appearance Potential Fine Structure

EAPFS Extended Appearance Potential Fine Structure

EAPFSS Electron Appearance Potential Fine Structure Spectroscopy

EAPFSS Extended Appearance Potential Fine Structure Spectroscopy

EAPH East Africa Publishing House [Kenya]

EAPM European Association of Personnel Management [Germany]

EAPPM European Association for Product and Process Modelling in the Building Industry

EAPR European Association for Potato Research [Netherlands]

EAPROM Electrically Alterable Programmable Read-Only Memory

EAPS European Association for Population Studies [Netherlands]

EAPV Eastern Arctic Patrol Vessels [Canada]

EAR Electron Affinity Rule [Physics]

EAR Electronically Agile Radar

EAR Electronic Aural Responder

EAR Employee Attitude Research

EAR Engineering Analysis Report

EAR Erection All Risks (Insurance)

EAR Eroded Area Rate

EAR European Addiction Research [Journal]

EAR External Access Register

EARB European Airlines Research Bureau [now Association of European Airlines, Belgium]

EARC Elimination of Ambiguity in Radiotelephony Call Signs (Study Group)

EARC Extraordinary Administrative Radio Conference

EARCCUS East African Regional Committee for the Conservation and Utilization of the Soil

EARCOM East Asia Regional Council of Overseas Schools [US]

EAREC Essential Airborne Radio Equipment Characteristic

EARI Engineering Agency for Resources Inventories [US]

EARIC East African Research Information Center

EARISS Energy and Angle-Resolved-Ion Scattering Spectrometer

EARL Easy Access Report Language

EARMP East African Regional Mathematics Programme [Kenya]

EARN European Academic (and) Research Network [Now Part of TERENA]

EAROM Electrically Alterable Read Only Memory

EAROPH East Asia Regional Organization for Planning and Housing

EARP Environmental Assessment and Review Process

EARP Environmental Assessment and Review Program

EARS Electronic Access to Reference Services

EARS Electronic Authoring and Routing System

EARS Explicit Archive and Retrieval System

EARS Experimental Array Radar System

EARSeL European Association of Remote Sensing Laboratories [France]

Earth Isl J Earth Island Journal

Earth Miner Sci Earth and Mineral Science (Journal)

Earth Moon Planets Earth, Moon and Planets [Journal published in the Netherlands]

Earth Negot Bull Earth Negotiations Bulletin

Earth Obs Mag Earth Observation Magazine

Earth Plan Sci Lett Earth and Planetary Science Letters [Journal published in the Netherlands] [also Earth Planet Sci Lett]

Earth Sci Rev Earth Science Reviews [Journal published in the Netherlands]

Earth Surf Process Landf Earth Surface Processes and Landforms [Published in the UK]

Earthq Eng Struct Dyn Earthquake Engineering and Structural Dynamics [Published in the UK]

EARTS En Route Automated Radar Tracking System

EAS East Australian Standard Time [Greenwich Mean Time -10:00]

EAS Eastern Analytical Symposium and Exposition [US/Canada]

EAS Education Administation Specialist

EAS Egyptian Academy of Sciences [Cairo]

EAS Electroacoustic Sensor

EAS Electron Accelerator System

EAS Electron Attachment Spectroscopy

EAS Electronic Air-Suspension (System) [Automobile]

EAS Electronic Article System

EAS Electronic Automatic Switch

EAS Electrophilic Aromatic Substitution

EAS Equivalent Air Speed

EAS Estonian Academy of Sciences [Tallin]

EAS European Aquaculture Society

EAS European Astronomical Society

EAS Experimental Army Satellite

EAS Experiment Analysis System

EAS Experiment Assurance System

EAS Extended Area Service [Telecommunications]

EAS Extensive Air Shower

EASA Electrical Apparatus Service Association [US]

EASA Engineering Association of South Africa

EASB East African Settlement Board

EASC Ethylaluminum Sesquichloride

EASCOMINT Extended Air Surveillance Communications Intercept [US Air Force]

EASCON Electronics and Aerospace Systems Convention [of Institute of Electrical and Electronic Engineers, US]

EASE Electrical Automatic Support Equipment

EASE Engineering Automatic System for Solving Equations

EASE European Association of Science Editors [UK]

EASI Earth and Space Index (Database) [of American Geophysical Union,]

EASI Electrical Accounting for the Security Industry

EASI Estimate of Adversary Sequence Interruption

EASI European Association of Shipping Informatics [UK]

EASIAP Engineering and Applied Sciences Industrial Affiliates Program [University of Manitoba, Winnipeg, Canada]

EASINET European Area Sales and Information Network

EASL Engineering Analysis and Simulation Language

EASL Experimental Assembly and Sterilization Laboratory [of NASA]

EASOE European Arctic Stratospheric Ozone Experiment

EASP Electric Arc Spraying [Metallurgy]

EAST European Academy of Science and Technology

EAST European Academy of Surface Technology [US]

EAST Experimental Army Satellite, Tactical

East Afr Med J East African Medical Journal

EASTEC Eastern States Exposition Center [US]

EASTT Experimental Army Satellite Tactical Terminal

EASY Early Acquisition System

EASY Efficient Assembly System

EASY Engine Analyzer System

EAT Electroacoustic Test(ing)

EAT Electroacoustic Transducer

EAT Elsevier Advanced Technology [Oxford, UK]

EAT Ensemble de l'Aéronautique de Toulouse [Toulouse Aeronautical Group, France]

EAT Environmental Acceptance Test

EAT Estimated Arrival Time

EAT Expected Approach Time

EAT Experiments in Art and Technology [US]

EATA Enhanced Advanced Technology Bus Attachment

EATAO East Asia Technical Advisory Office [of United States Agency for International Development]

EATCS European Association for Theoretical Computer Science [Austria]

EATITU East African Tractor and Implement Testing Unit

EATJP European Association for the Trade in Jute Products [Netherlands]

EATMS Electro-Acoustic Transmission Measuring System

EATP European Association for Textile Polyolefins [France]

EATU Eastern African Telecommunications Union

EAU Engineer Aviation Unit

EAUG European Atex (Computer) Users Group [Netherlands]

EAUTC Engineer Aviation Unit Training Center

EAV Enamelled Asbestos-Covered Varnish-Treated (Electric Wire)

EAV Equine Arthritis Virus

EAVE Experimental Autonomous Vehicle

EAVE EAST Experimental Autonomous Vehicle–East

EAVRO East African Veterinary Research Organization [Kenya]

EAW Equivalent Average Words

EAW Exchange and Anisotropy Wall [Solid-State Physics]

EAWAG Eidgenössische Anstalt für Wasserversorgung, Abwasserreinigung und Gewässerschutz [Federal Institute for Water Supply, Wastewater Treatment and Water Pollution Control, Switzerland]

EAWS East African Wildlife Society [Kenya]

EAX Electronic Automated (or Automatic) Exchange

E(AX$_B$) Bond Energy of Molecule AX (Consisting of Metal Atom A and Halogen X) to (Non metal) Atom B [Symbol]

E$_a$(AX$_B$) Activation Energy of a Reaction Resulting in Bonding of Molecule AX (Consisting of Metal Atom A and Halogen X) to Surface $B(s)$ [Symbol]

EB Eastbound

EB Edge Brightness

EB Edison Battery

EB Electric Battery

EB Electric Brake; Electric Braking

EB Electrobiological; Electrobiologist; Electrobiology

EB Electrode per Bit [also E/B]

EB Electron Band

EB Electron Beam

EB Electron Bombardment

EB Electronics Board

EB Elementary Body

EB Emeraldine Base

EB Emitter Barrier [Electronics]

EB Encyclopedia Brittannica

EB Energy Balance

EB Energy Band (Theory of Solids) [Solid-State Physics]

EB Energy Barrier

EB Energy Bond

EB End of Block (Character)

EB Engel-Brewer (Theory) [Physics]

EB Enterobacteria [Microbiology]

EB Epstein-Barr (Virus)

EB Equal Brake

EB Equatorial Bond

EB Erase Band [also Eb]

EB Erythroblast [Anatomy]

EB Ethylbenzoate

EB Ethylene Butylene (Copolymer)

EB Evans Blue

EB Evolutionary Biology

EB Exabyte [Unit]

EB Executive Board

EB Exposure Back

EB Extended Born (Approximation) [Solid-State Physics]

EB Eye Bolt

E/B Electrode per Bit [also EB]

E$ Ethiopian Birr [Currency]

25E14B 2,5 Diethoxy 1,4 Benzoquinone

E(B$_A$) Bond Energy of an Atom B on Surface $A(s)$ [Symbol]

E$_a$(B$_A$) Activation Energy of Chemisorption of B from a Prereaction State on Surface $A(s)$ [Symbol]

E$_d$(B$_n$) Dissociation Energy of Molecule B_n [Symbol]

$E'(B_A)$ Bond Energy of the Physisorption of an Atom B in a Prereaction State on a Surface A(s) [Symbol]

Eb Ebony [also eb]

Eb Erase Band [also EB]

Eb Extrudable [also eb]

EBA Eisenbahn Bundesamt [Federal Department of Railways, Germany]

EBA Electric Boat Association [UK]

EBA Electron Beam Acceleration

EBA Electron Beam Analysis

EBA Environmental Ballistics Associates [US]

EBA Ethyl Benzylmalonic Acid

EBA Ethyl Bromoacetate

EBA 2 Ethylbutanoic Acid

EBA Europäische Baustoffindustrie Adreßbuch (Fachadreßbuch der baustofferzeugenden Industrie) [Directory of the European Construction Materials Industry]

EBA European Broadcasting Area

EBA Extended Born Approximation [Solid-State Physics]

EBAA European Business Aviation Association [Belgium]

EBAA Eye Bank Association of America

EBAC European Bank Advisory Committee

EBAM Electron Beam Accessible Memory

EBAM Electron Beam Addressable (or Adressed) Memory

EBAMEP 1,2 Ethanediylbis[1-Amino(1-Methyl)ethylphosphonic Acid]

EBARA Electron-Beam Ammonia Reaction (Process)

Eb Asb Ebony Asbestos

EBB Electronic Bulletin Board

EBB Extra Best Best (Wire)

EBBA Eastern Bird Banding Association [US]

EBBA Estuarine and Brackish Water Biological Association [now Estuarine and Brackish Water Sciences Association, UK]

EBBA 4 Ethoxybenzylidene 4 Butylaniline

EBBMT Electron-Beam Button Melt Test [Metallurgy]

EBC EISA (Extended Industry Standard Architecture) Bus Controller

EBC Electron Beam Channeling

EBC Electron Beam Curing

EBC Electron Beam Cutting [Metallurgy]

EBC Emulated Buffer Computer

EBC Enamelled-Bonded Single-Cotton-Covered (Electric Wire)

EBC Energy Band Calculations [Solid-State Physics]

EBC Environmental Business Committee

EBC Equivalent Boron Content [Metallurgy]

EBCA External Branch Condition Address

EBCD Extended Binary Coded Decimal

EBCDI Extended Binary Coded Decimal Interchange (Code)

EBCDIC Extended Binary Coded Decimal Interchange Code

EBCE Electron Beam Control Electronics

EBCHM Electron Beam Cold-Hearth Melting (Process) [Metallurgy]

EBCHR Electron Beam Cold-Hearth Refined; Electron Beam Cold Hearth Refining [Metallurgy]

EBCI External Branch Condition Input

EBCO Erbium Barium Copper Oxide (Superconductor) [also ErBCO]

EBCO-123 $ErBa_2Cu_3O_{7-\delta}$ [An Erbium Barium Copper Oxide Superconductor]

EBCO-211 Er_2BaCuO_5 [An Erbium Barium Copper Oxide Superconductor]

EBCO-57_12 $Er_5Ba_7Cu_{12}O_y$ [An Erbium Barium Copper Oxide Superconductor]

EBCO-459 $Er_4Ba_5Cu_9O_y$ [An Erbium Barium Copper Oxide Superconductor]

EBCSM East Bay Council on Surveying and Mapping

EBCT Electron Beam Computed Tomography

EBCTO Europium Barium Copper Titanium Oxide (Superconductor)

EBCVD Electron-Beam(Assisted) Chemical Vapor Deposition [also EB CVD]

EBD Effective Billing Date

EBD Electron-Beam Deflection

EBD Extrinsic Boundary Dislocation [Materials Science]

EBD Eyeballs Down (+GZ) [Aerospace]

EBDA Equilibrium Binding and Data Analysis

EBDC Enamelled Bonded Double-Cotton-Covered (Electric Wire)

EBDC Ethylenebisdithiocarbamate

EBDI Electronic Business Data Interchange [US]

EBDP Enamelled Double-Paper-Bonded (Electric Wire)

EBDP cis-Ethylene-1,2-Bis(diphenylphosphine)

EBDPM Methylenebis[bis(2-Ethylbutyl)phoshine]dioxide

EBDS Enamelled Bonded Double-Silk-Covered (Electric Wire)

EBE Ethyl tert-Butyl Ether

E-Beam Electron Beam [also E-beam, or e-beam]

EBELACTONE A 3,11-Dihydroxy-2,4,6,8,10,12-Hexamethyl-9-Oxo-6-Tetradecenoic Acid-1,3-Lactone [also Ebelactone A] [Biochemistry]

EBELACTONE B 2-Ethyl-3,11-Dihydroxy-4,6,8,10,12-Pentamethyl-9-Oxo-6-Tetradecenoic Acid-1,3-Lactone [also Ebelactone B] [Biochemistry]

EBEP Electron-Beam-Excited Plasma

EBER Equivalent Binary Error Rate

EBES Electron-Beam Exposure System

EBES Electronic Banking Economics Society [US]

EBEX Ethyl Benzene Extraction (Process)

EBF Electron-Beam Focusing

EBF Electron Bombardment Furnace

EBF Externally Blown Flap

EB&F Equipment Blockages and Failures

EBFS Enclosure Building Filtration System

EBG Elektroblechgesellschaft mbH [German Manufacturer of Electrical Sheet Steel Products]

EBG Electron-Beam Gun

EBH Electron-Beam Heating

EBHC Equated Busy Hour Call [Telecommunications]

EBHSS Electron-Beam High-Speed Scan(ning)

EBI Electron-Beam Interference

EBI Equivalent Background Input

EBI Ergosterol Biosynthesis Inhibitor [Biochemistry]

EBI Extended Background Investigation

EBI Eyeballs In (+GX) [Aerospace]

EBIA Electron Intramolecular Vibrational

EBIB Energy Bibliography and Index [of Center for Energy and Mineral Resources, Texas A&M University, US]

EBIC Electron-Beam Induced Conduction; Electron-Beam Induced Conductivity

EBIC Electron-Beam Induced Current

EBIC European Banks International Company

EBICON Electron-Bombardment Induced Conductivity

EBID Electron-Bombardment Induced Desorption

EBIDMP Ethylenebis[iminodi(methylphosphonic Acid)]

e billing electronic billing

EBIS EIS (Economic Information Systems) Business Information System [US]

EBIS Electron-Beam Ion Source

EBIS Ethylenebisisothiocyanate Sulfide

EBL Effective Beam Length

EBL Electron-Beam Laser

EBL Electron-Beam Lithography

EBL Electronic Bearing Line

EBL Eyeballs Left (+GY) [Aerospace]

EBLI Electronics Business Leading Indicator

EBM Die Eisen-, Blech- und Metallverarbeitende Industrie [Catalog, Directory and Buyer's Guide of the German Iron, Sheet and Metal Processing Industries]

EBM Electromagnetic Billetmaker [Metallurgy]

EBM Electron-Beam Machining

EBM Electron-Beam (Cold-Hearth) Melting [Metallurgy]

EBM Electron-Beam Microprobe

EBM Electronic Bearing Marker [Marine Radar]

EBM Energy Balance Model

EBM Extended Branch Mode

EBMA Electron-Beam Microprobe Analysis

EBMA Engine Booster Maintenance Area

EBMD Electron-Beam Mode Discharge

EBMF Electron-Beam Microfabricating System

EBMF Electron-Beam Microfabricator

EBMS Energy Balance Models [also EBMs]

EBMW Energy from Biomass and Wastes [also EBW]

EBNA Epstein-Barr Nuclear Antigen [Biochemistry]

EBNF Extended Backus Naur Form [Computers]

EBO Eyeballs Out (-GX) [Aerospace]

EBOA Eisenbahnbau -und Betriebsordnung für Anschlussbahnen [German Railway Construction and Operating Regulations]

E-Boat Enemy Boat (Torpedo) [also E-boat]

Ebone All-European Backbone Service [Internet]

EBOR Experimental Beryllium-Oxide Reactor [US]

EBOR-CX Experimental Beryllium Oxide Reactor–Critical Assembly

EBP Enamelled Single Paper-Bonded (Electric Wire)

EBP Energetic Branched Polyethers

EBP Environmental Biophysics

EBPA Electron-Beam Parametric Amplifier

EBPA 2-(2-Pyridyl) N,N-Bis(2-Pyridylmethyl)ethanamine [also ebpa]

EBPG Electron-Beam Pattern Generator

EBPVD Electron-Beam(-Assisted) Physical Vapor Deposition [also EB PVD]

EBR Electron-Beam Recorder; Electron-Beam Recording

EBR Electron-Beam Remelting [Metallurgy]

EBR Electronic Beam Recorder; Electronic Beam Recording

EBR Epoxy Bridge Rectifier

EBR Experimental Breeder Reactor

EBR Eyeballs Right (-GY) [Aerospace]

EBR-1 Experimental Breeder Reactor 1 [of US Atomic Energy Commission at Idaho Falls; Started up in 1951, Now Shut-Down and a National Historic Landmark] [also EBR-I, or CP-4]

EBR-2 Experimental Breeder Reactor 2 [of US Argonne National Laboratory West at Idaho Falls; Started-up in 1964] [also EBR-II, or CP-8]

E-BR Emulsion Butadiene Rubber

EB&RA Engineer Buyers and Representatives Association [UK]

EBROS Endless Billet Rolling System [Metallurgy]

EBRD Electron-Beam Rotating Disk (Process)

EBRD European Bank for Reconstruction and Development

EBS Educational Broadcasting Service [Jamaica]

EBS Electric Breakdown Strength

EBS Electron-Beam Semiconductor

EBS Electron-Bombarded Semiconductor

EBS Electronic Beam Squint Tracking System [of Satellite Communication Agency, US]

EBS Emergency Borating System

EBS Emergency Breathing Subsystem [of NASA]

EBS Emergency Broadcasting System

EBS Enamelled Bonded Single-Silk-Covered (Electric Wire)

EBS Extended Basis Set

EBSA Estuarine and Brackish Water Sciences Association [UK]

EBSA 4 Ethylbenzenesulfonic Acid

EBSD Electron Backscatter(ing) Diffraction

EBSD/OIM Electron Backscatter(ing) Diffraction/Orientation Imaging Microscopy

EBSELEN 2 Phenyl 1,2 Benzisoselenazol 3[2H]one [also Ebselen]

EBSP Electron Backscattered Spectroscopy

EBSP Electron Backscatter(ing) Pattern

EBSS Earle's Balanced Salt Solution

EBSV Exchange-Biased Spin Valve

EBT Eccentric Bottom Tapping [Metallurgy]

EBT Electroless Bath Treatment [Metallurgy]

EBT Electron-Beam Tomograph(y)

EBT Electronic Benefits Transfer

EBT Elmo Bumpy Torus (Fusion Device) [of US Department of Energy at Oak Ridge National Laboratory, Tennessee, US]

EBT ERIN (Environmental Resources Information Network) Biodiversity Team [of Environment Australia]

EBT Eriochrome Black T

EBT p-Ethylsulfonyl BenzaldehydeThiosemicarbazone

EBT Excentric Bottom Tapping [Metallurgy]

EBT-EAF Eccentric Bottom Tapping–Electric-Arc Furnace [Metallurgy]

EBTS ECC Bypass Test Facility

EBU European Broadcasting Union [Brussels, Belgium]

EBU Eyeballs Up (–GZ) [Aerospace]

EBU Rev Tech EBU Review, Technical [Publication of the European Broadcasting Union, Brussels, Belgium]

e-business electronic business

EBV Electron-Beam Vaporization

EBV Epstein-Barr Virus

EBV Eschweiler Berwerks-Verein AG [A German Mining Company]

EBV-1 Epstein-Barr Virus, Type 1

EBV-2 Epstein-Barr Virus, Type 2

EB-VCA Ebstein-Barr Viral Capsid Antigen

EBVD Electron-Beam Vapor Deposition (Process)

EBW Effective Bandwidth

EBW Electron-Beam Welder; Electron-Beam Welding

EBW Energy from Biomass and Wastes [also EBMW]

EBW Exploding Bridgewire

EBW-HV Electron-Beam Welding–High Vacuum

EBW-MV Electron-Beam Welding–Medium Vacuum

EBW-NV Electron-Beam Welding–Nonvacuum

EBWR Experimental Boiling-Water Reactor [of US Atomic Energy Commission at Argonne National Laboratory, Illinois; Startup in 1956; Now Decommissioned] [also CP-7]

E by N East by North

E by S East by South

EBZM Electron-Beam Zone Melting

EC Earth Council [Costa Rica]

EC Earth Current

EC East Coast

EC Eastern Cedar

EC Echo Controller

EC Eclipsed Conformation [Physical Chemistry]

EC Ecology Center [US]

EC Economy Cartridge

EC Economy Class

EQ Equador(ian) [also Ec]

EC Ecuador [ISO Code]

EC Eddy (Heat) Conduction

EC Eddy Current

EC Edge Clamp

EC Edge Connector

EC Edgeworth Cramer (Chromatography)

EC Effective Concentration

EC Elastic Center [Mechanics]

EC Elastic Collision [Mechanics]

EC Elastic Constant

EC Electrical Conductivity; Electrical Conduction; Electrical Conductor; Electrically Conductive

EC Electric Car

EC Electric Circuit

EC Electric Connection; Electric Connector

EC Electric(al) Contact

EC Electric Controller

EC Electric Current

EC Electricity Council [Now Part of the UK Department of Energy]

EC Electrification Council [US]

EC Electrocapillarity; Electrocapillary

EC Electrocatalyst; Electrocatalysis

EC Electroceramic(s)

EC Electrochemical; Electrochemist(ry)

EC Electrochromatograph(y)

EC Electrochromic(ity)

EC Electrocondensation

EC Electrocrystallization

EC Electrode Current

EC Electrolytic Capacitor

EC Electrolytic Cell

EC Electromagnetic Coupling

EC Electron(ic) Capture

EC Electron Channeling (Pattern)

EC Electron Cloud [Physics]

EC Electron Compound [Metallurgy]

EC Electron Counter; Electron Counting

EC Electron Coupler; Electron Coupling

EC Electron Current

EC Electron Cyclotron

EC Electronic Calculator

EC Electronic Ceramic(s)

EC Electronic Chart

EC Electronic Circuit

EC Electronic Commerce

EC Electronic Conductivity; Electronic Conduction

EC Electronic Coupled; Electronic Coupler

EC Elemental Carbon

EC Element Contractor

EC Element Count

EC Elmira College [Elmira, New York, US]

EC Elizabeth City [North Carolina, US]

EC Elon College [North Carolina, US]

EC Emerson College [Boston, Massachusetts, US]

EC Emission Control

EC Emulsifiable Concentrate

EC Enamel Covering [Electric Wires]

EC Enamelled Single-Cotton-Covered (Electric Wire)

EC End Cell [Storage Batteries]

EC Endothelial Cell [Anatomy]

EC Energy Commission [UK]

EC Energy Conservation; Energy Conserver; Energy Conserving

EC Energy Consumption
EC Energy Conversion; Energy Converter
EC Energy Cost
EC Energy Council [Southern Australia]
EC Engine Capacity
EC Engine Control
EC Engine Cutoff
EC Engineering Ceramics
EC Engineering Change
EC Engineering Coating
EC Engineering Construction
EC Engineering Controls
EC Engineering Council [UK]
EC Engineering Cybernetics
EC Engineers Club
EC English Channel
EC Enterocylopathic (Virus)
EC Entry Corridor
EC Environment Canada
EC Environmental Capacity
EC Environmental Chemist(ry)
EC Environmental Control
EC Environmental Corrosion
EC Environmental Cracking
EC Environment Center
EC Enzyme Code (Number)
EC Enzyme Commission [US]
EC EPD (Extraction and Processing Division) Congress [of AIME Minerals, Metals and Materials Society, US]
EC Epitaxial Coating
EC Epitaxial Crystallization
EC Equilibrium Condition
EC Equilibrium Constant
EC Equipment Cost
EC Erosion-Corrosion [Metallurgy]
EC Error Control [Telecommunications]
EC Error Correction
EC Error Counter
EC Erythrocyte [Anatomy]
EC Escherichia Coli [Microbiology]
EC Essentiality Code
EC Esterified Cholesterol
EC Estimated Concentration
EC Ethyl Cellulose
EC Ethyl Centralite
EC Ethyl Cinnamate
EC Euler Cauchy (Method)
EC Euro-City Train
EC European Chapter
EC European Code of Practice
EC European Commission
EC European Community [formerly European Economic Community; now European Union]
EC Evansville College [Indiana, US]

EC Event Code
EC Event Count(er)
EC Events Controller
EC Events Coupler
EC Exchange Capacity [Geology]
EC Executive Council
EC Exhaust Coefficient
EC Expansion Coefficient
EC Experimental Chemist(ry)
EC Experiment Computer
EC Explorers Club [US]
EC Explosives Company
EC Extended Control (Mode)
EC External Combustion [also ec]
EC Extinction Coefficient
EC Extracorporeal Circulation
EC Extrinsic Conductance; Extrinsic Conduction; Extrinsic Conductivity [Solid-State Physics]
EC- Spain [Civil Aircraft Marking]
E/C Encoder/Coupler
EC_{50} Median Effective Concentration (or Effective Concentration 50) [also EC50]
$E\perp c$ Electric Vector Perpendicular (or Normal) to c-Axis [Symbol]
.ec Ecuador [Country Code/Domain Name]
e.c. *(exempli causa)* – as an example
η/c reduced viscosity-to-concentration (ratio)
ECA Earth Crossing Asteroid(s)
ECA Economic Commission for Africa [of United Nations at Addis Abeba, Ethiopia]
ECA Economic Cooperation Administration [US]
ECA Economic Cooperation Agreement
ECA Educational Centers Association [UK]
ECA Electrical Contractors Association [UK]
ECA Electrically Conductive Adhesive
ECA Electrochemical Affinity
ECA Electrochemical Analysis; Electrochemical Analyzer
ECA Electrochemical Anode
ECA Electronic Confusion Area
ECA Electronics Control Assembly [of NASA Apollo Program]
ECA (Semi)Elliptical Crack Approximation
ECA Emergency Controlling Authority
ECA Engineering and Computer Science Association
ECA Engineering Change Analysis
ECA Engineering Contractors Association [US]
ECA Engineering Critical Assessment
ECA Environmental Contaminants Authority
ECA Environment Conservation Authority
ECA Environment Council of Alberta [Canada]
ECA Enzyme Chemical Analyzer
ECA Epoxy Curing Agent
ECA Equal Channel Angular (Pressing Technique)
ECA Erythrina Cristagalli Agglutinin [Biochemistry]
ECA Ethyl Chloroacetate

ECA Ethyl Cyanoacetate

ECA European Confederation of Agriculture [Switzerland]

ECA Exchange Carrier Association [now National Exchange Carrier Association, US]

ECAART European Conference on Accelerators in Applied Research and Technology

ECAB Economic Abstracts [of European Economic Commission, Brussels, Belgium]

ECAC Electromagnetic Compatibility Analysis Center [of US Department of Defense]

ECAC European Civil Aviation Conference [France]

ECAC/US-CRS European Civil Aviation Conference/United States Working Group on Computer Reservation Systems

ECAD Electrical Computer Aided Design;

ECAD Electronic Computer Aided Design

ECAD Existing Chemical Assessment Division [of National Institute of Environmental Health Sciences, US]

ECAF Endothelial Cell Attachment Factor [Biochemistry]

ECAFE Economic Commission for Asia and the Far East [of United Nations]

ECAFM Electrochemical Atomic Force Microscope; Electrochemical Atomic Force Microscopy [also EC-AFM]

ECAHTI European Committee for Agricultural and Horticultural Tools and Implements [France]

ECAL Enjoy Computing And Learn

ECALE Electrochemical Atomic Layer Epitaxy

ECAM Ecole Catholique d'Arts et Métiers [Catholic School of Arts and Crafts, Lyons, France]

ECAM Electronic Centralized Aircraft Monitor

ECAM Environmental Cost Analysis Methodology

ECAMA European Citric Acid Manufacturers Association [of Conseil Européen des Fédérations de l'Industrie Chimique, Belgium]

ECAO Environmental Criteria and Assessment Office

ECAP Electronic Circuit Analysis Program

ECAP Energy-Compensated Atom Probe

ECAP Environmental Cooperation with Asia Program

ECAP European Common Agriculture Policy [of European Community]

ECAP Equal Channel Angular Pressing

ECAPD European Conference on Applications of Polar Dielectrics

ECAR East Central Area Reliability Coordination Agreement

ECar$ East Caribbean Dollar [Currency of Antigua Barbuda, Dominica, Grenada, Montserrat, St. Christopher Nevis, St. Kitts, St. Lucia, St. Vincent, and the Grenadines]

ECARL Expandable Cluster Aircraft Rocket Launcher

ECARS Electronic Coordinator and Readout System

ECARS Enhanced Airline Communications and Reporting System

ECAS Electrical Contractors Association of Scotland

ECAS Experiment Computer Application Software

ECASIA European Conference on Applications of Surface and Interface Analysis

ECASS Electronically Controlled Automatic Switching System

ECAT Electronic Card Assembly and Test

ECAT European Center for Automatic Translation [Luxembourg]

ECATRA European Car and Truck Rental Association [Germany]

ECB Electrically Controlled Birefringence

ECB Electronic Codebook

ECB Engineering Change Board

ECB Environment Coordination Board [of United Nations]

ECB European Conference on Biomaterials

ECB Event Control Block

ECB Events Control Buffer

ECBA European Communities Biologists Association [Germany]

ECBO European Cell Biology Organization [UK]

ECBTE European Committee for Building Technical Equipment [France]

ECC Earth Continuity Conductor

ECC East Carolina College [Greenville, North Carolina, US]

ECC Economic Council of Canada

ECC Effective Conjugation Coordinate

ECC Electrocardiocorder

ECC Electrochemical Chromatography

ECC Electrochemical Corrosion

ECC Electrochemical Cell

ECC Electron Channeling Contrast

ECC Electronically Controllable Coupler

ECC Electronic Calibration Center [of National Institute for Standards and Technology, Gaithersburg, Maryland, US]

ECC Electronic Commerce Committee [US]

ECC Electronic Common Control

ECC Electronic Components Conference [US]

ECC Elliptic Curve Crypto

ECC Emergency Control Center

ECC Emergency Core Coolant; Emergency Core Cooling [Nuclear Engineering]

ECC Employees Compensation Commission [US]

ECC Energy Conservation Coalition

ECC Engineering Change Control

ECC Engineering Critical Component

ECC Environmental Control Council

ECC Equatorial Counter Current [Oceanography]

ECC Equivalent Carbon Content [Metallurgy]

ECC Error Check Code

ECC Error Checking and Correction

ECC Error Correction Code

ECC Error Correction Control

ECC Ethyl Chlorocarbonate

ECC European Communities Commission

ECC European Conservation Conference

ECC European Crystallographic Council

ECC Expanded Community Calling

Ecc Eccentric(ity) [also ecc]

ECCA European Coil Coaters Association [Belgium] [also ecca]

ECCAI European Coordinating Committee for Artificial Intelligence [France]

ECCAI Newsl ECCAI Newsletter [of European Coordinating Committee for Artificial Intelligence]

ECCANE East Coast Conference on Aerospace and Navigational Electronics [US]

ECCB Electronic Components Certification Board [US]

ECCB Engineering Change Control Board

ECCC European Communities Chemistry Committee [UK]

ECCC European Creep Collaborative Committee

ECCCO European Culture Collections Curators Organization [now European Culture Collections Organization, UK]

ECCE European Council of Civil Engineers

ECCF Equilibrium Charge–Charge Flux

ECCFPP European Conference on Controlled Fusion and Plasma Physics

ECCI Electron Channeling Contrast Imaging [Microscopy]

ECCLS European Committee for Clinical Laboratory Standards [UK]

ECCM Eastern Caribbean Common Market

ECCM Electronic Counter Countermeasures

ECCM European Conference on Composite Materials

ECCM-CTS European Conference on Composite Materials Testing and Standardization

ECCMF European Council of Chemical Manufacturers Federations

ECCNNR European Committee for the Conservation of Nature and Natural Resources

ECCO Engineering Command Control and Operation

ECCO Engineers Coordinating Council of Oregon [US]

ECCO European Conference of Conscripts Organization [Netherlands]

ECCO European Culture Collections Organization [UK]

ECCOMAS European Community on Computational Methods in Applied Science

ECCP Engineering Concepts Curriculum Project [US]

ECCREDI European Council for Construction, Research, Development and Innovation

ECCS Electrolytic Chromium Coated Steel

ECCS Electronic C (Hundred) Call Seconds

ECCS Emergency Core Cooling System [Nuclear Engineering]

ECCS Engine and Component Control System [Aeronautics]

ECCS European Commission for Coal and Steel

ECCS European Convention for Construction Steelwork

ECCSL Emitter-Coupled Current Steered Logic

ECCS-TC European Convention for Construction with Steel–Technical Conference

ECCTO European Committee for Cocoa Trade Organizations

ECD East Caribbean Dollar [Currency of Antigua Barbuda, Grenada, St. Christopher Nevis, St. Kitts, St. Lucia, St. Vincent, and the Grenadines]

ECD Electrochemical Deposition

ECD Electrochemical Detection; Electrochemical Detector

ECD Electrochromic Display

ECD Electroconductivity Detector

ECD Electron Capture Detection; Electron Capture Detector

ECD Energy Conversion Device

ECD Engineering Ceramics Division [of American Ceramic Society, US]

ECD Engineering Control Drawing

ECD Engineers Club of Dayton [Ohio, US]

ECD Enhanced Color Display

ECD Enhanced Compact Disk

ECD Equipment Configuration Data

ECD Error Control Device

ECD Estimated Completion Date

ECD Exchange Coupling Direction [Physics]

EC$ East Caribbean Dollar [Currency of Antigua Barbuda, Dominica, Grenada, Montserrat, St. Christopher Nevis, St. Kitts, St. Lucia, St. Vincent, and the Grenadines]

ECDC Electrochemical Diffused Collector (Transistor)

ECDC European Composites Development Center [Bad Homburg, Germany]

ECDIN Environmental Chemicals Data and Information Network [of EURATOM]

ECDIS Electronic Chart Display and Information System

ECDIS/CRT Electronic Chart Display and Information System/Cathode-Ray Tube

ECDL Exchange-Coupled (Magnetic) Double Layer

ECDM Electrochemical Discharge Machining

ECDMMRL European Committee for the Development of the Meuse and Meuse/Rhine Links [Belgium]

ECDMS Environmental Contaminants Data Management System [US]

ECD MS Electron Capture Detection Mass Spectroscopy

EC-DNA Escherichia Coli–Deoxyribonucleic Acid [Biochemistry]

ECDO Electronic Community Dial Office

ECE Echo Control Equipment

ECE Economic Commission for Europe [of United Nations at Geneva, Switzerland]

ECE Electrochemical-Chemical-Electrochemical (Reaction Sequence Theory)

ECE Electrochemical Electrode

ECE Electrochemical Engineer(ing)

ECE Electrochemical Equivalent

ECE Electron Correlation Energy

ECE Electron Cyclotron Emission

ECE Electronic Commerce Europe

ECE Electronic Communications Engineer(ing)

ECE Energy Conversion Efficiency

ECE Engineering Capacity Exchange

ECE European Commodities Exchange [France]

ECE Export Council for Europe [UK]

ECE External Combustion Engine [also ece]

ECEA Economic Community of Eastern Africa

ECEA Ethoxycarbonylethylamine [also ecea]

ECEC East Carolina Engineers Club [US]

EC/EDI Electronic Commerce/Electronic Data Interchange

ECELR Epithermal Critical Experiment Laboratory Reactor

ECEP Experiment Checkout Equipment Processor

ECERM Environment Code of Ethics for Rangeland Managers [Australia]

ECerS European Ceramic Society

ECETOC European Chemical Industry Ecology and Toxicology Center [Belgium]

EC-EURAM European Community('s) European Research on Advanced Materials (Program)

ECF Earth Crust Formation

ECF Echo Control Factor

ECF Electric Crystal Field

ECF Electronic Commerce Finland

ECF Elementally Chlorine Free

ECF Element Charge Factor

ECF Emergency Cooling Functionality [Nuclear Reactors]

ECF Emission Contribution Fraction

ECF Enhanced Connectivity Facilities

ECF Equivalency Capability File

ECF Ethyl Chloroformate

ECF European Caravan Federation [UK]

ECF European Commission on Forestry and Forest Products [Italy]

ECF European Community Form

ECF European Composites Forum

ECF European Conference on Fracture

ECF European Coordinating Facility

ECF Extented Care Facility

ECF Extracellular Fluid

ECFA European Committee for Future Accelerators [of CERN–European Laboratory for Particle Physics, Geneva, Switzerland]

ECFCI European Center of Federations of the Chemical Industry

ECFFP European Commission on Forestry and Forest Products [Italy]

ECFI East Caribbean Farm Institute

ECFM Eddy-Current Flow Meter

ECFP Effective Crystal Field Parameter

ECG Echocardiogram; Echocardiography

ECG Ecosystem Conservation Group

ECG Electrocardiogram; Electrocardiography

ECG Electrochemical Grinding

ECG Electro-Epitaxial Crystal Growth

ECG Engineering Consulting Group [of Ministry of Economic Affairs, Taiwan]

ECGD Export Credits Guarantee Department [of Board of Trade, UK]

ECGF Endothelial Cell Growth Factor [Biochemistry]

ECGF European Container Glass Federation [Belgium]

β-ECGF β-Endothelial Cell Growth Factor [Biochemistry]

ECGS Endothelial Cell Growth Supplement [Biochemistry]

ECH Earth-Coverage Horizon Measurement

ECH Echo Cancellation Hybrid

ECH Eddy-Current Heating

ECH Ethylene Chlorohydrin

Ech Echelon [also ech]

ε-ch epsilon-chain specific (immunoglobulin)

ECHO Electronic Case Handling in Offices

ECHO Electronic Computing, Hospital-Oriented

ECHO Enterocytopathic–Human–Orphan (Virus)

ECHO European Clearinghouses Organization

ECHO European Commission Host Organization [Internet]

ECHO European Community Humanitarian Office [of European Union]

ECHO Expanded Characteristics Option

ECHO Experimental Contract Highlight Operation

Echo Rech L'Echo des Recherches [Publication on Telecommunications Research; of Centre National d'Etudes des Télécommunications, France]

ECHO Virus Enterocytopathic–Human–Orphan Virus [also ECHO virus]

ECI Cast-Iron Electrode

ECI Earth-Centered Inertia(ls)

ECI Editions Commerciales and Industrielles SA [Brussels, Belgium]

ECI Eddy-Current Inspection

ECI Effective Cluster Interaction [Metallurgy]

ECI Electrochemical Interface

ECI Electron Channeling-Contrast Image (or Imaging)

ECI Emergency Cooling Injection [Nuclear Reactors]

ECI Energetic Cluster Impact

ECI Engine Component Improvement

ECI Equity Capital Investment

ECI European Construction Institute

ECI European Currency Institute

ECI Experimental and Clinical Immunogenetics [International Journal]

ECI Extension Course Institute [of US Air Force]

ECI External Call Interface

ECI External Charge Injector

ECIC European Chemistry at Interfaces Conference

ECIF Electronic Components Industry Federation [UK]

ECIM European Commission for Industrial Marketing [UK]

ECIN Expert Center for Intracenter Networking

ECIO European Conference on Integrated Optics

ECIO Experiment Computer Input/Output

EC-IOA European Committee of the International Ozone Association [France]

ECIS Electric Cell Substrate Impedance Sensing

ECIS European Community Information Service

ECIS European Council of International Schools [UK]

ECISS European Committee for Iron and Steel Standardization [Belgium]

ECITER Electron Cyclotron International Thermonuclear Experimental Reactor [also EC-ITER]

ECITO European Central Inland Transport Organization

ECIWA European Committee of Importers and Wholesalers Associations

ECJRC European Commission Joint Research Center

ECKC Engineers Club of Kansas City [US]

ECL East Coast Laboratory [of Environmental Science Services Administration, US]

ECL Ecole Centrale de Lyon [Lyons Central School, Ecully, France]

ECL Eddy-Current Loss

ECL Electrical Communication Laboratories [of Nippon Telegraph and Telephone Corporation, Tokyo, Japan]

ECL Electrochemistry Laboratory [of University of Pennsylvania, Pittsburgh, US]

ECL Electrogenerated Chemiluminescence

ECL Electroluminescence; Electroluminescent

ECL Electronic Components Laboratory

ECL Emitter-Coupled Logic

ECL End-Cap Calorimeter

ECL Entry Closed Loop

ECL Environmental Chemistry Laboratory

ECL Equipment Component List

ECL Equivalent Chain Length (Chromatography)

ECL Execution Control Language

ECL Executive Control Language

E²CL Emitter (to)Emitter Coupled Logic [also EECL]

ECLA Economic Commission for Latin America [of United Nations at Santiago, Chile]

ECLAC Economic Commission for Latin America and the Caribbean [of United Nations]

ECLAT European Computer Leasing and Trading Association

ECLAT European Conference on Laser Treatment (of Materials)

ECLG European Consumer Law Group [Belgium]

ECLIPS Experimental Cloud Lidar Pilot Study

ECLM Economic Community for Livestock and Meat [Burkina Faso]

ECLO Emergency Center for Locust Operations [of Food and Agricultural Organization]

ECLO Emitter-Coupled Logic Operator

ECLS Environmental Control and Life Support

ECLS Epitaxial Continued Layer Structure

ECLSS Environmental Control and Life Support System

ECL-TTL Emitter-Coupled Logic/Transistor-Transistor Logic

ECM Effective Calls Meter

ECM Electric Coding Machine

ECM Electrochemical Machining

ECM Electrochemical Metallizing

ECM Electrochemical Milling

ECM Electro-Chemo-Mechanical (Treatment)

ECM Electron Channeling Micrograph

ECM Electronically Commutated Motor; Electronic Commutation Motor

ECM Electronic Control Module

ECM Electronic Countermeasures

ECM Embedded Cluster Method

ECM Energy Conservation Measure

ECM Engine Control Module

ECM Engineers Club of Minnesota [US]

ECM Environmentally Conscious Manufacturing

ECM European Common Market

ECM European Crystallographic Meeting

ECM Europium Calcium Manganate

ECM Exciton Chirality Method [Solid-State Physics]

ECM Extended Core Memory

ECM Extracellular Matrix

ECMA Electronic Computer Manufacturers Association

ECMA Embalming Chemical Manufacturers Association [US]

ECMA Engineering College Magazines Associated [US]

ECMA European Carton Makers Association [The Hague, Netherlands]

ECMA European Catalysts Manufacturers Association [of Conseil Européen des Fédérations de l'Industrie Chimique, Belgium]

ECMA European Computer Manufacturers Association [Switzerland]

ECMA Bull ECMA Bulletin [of European Carton Makers Assiociation, The Hague, Netherlands]

ECMC Electric Cable Makers Confederation [UK]

ECMC European Container Manufacturers Committee [Italy]

ECME Economic Commission for the Middle East [of United Nations]

ECME Electronic Countermeasures Equipment

ECME European Conference on Molecular Electronics

ECML Exchange-Coupled (Magnetic) Multilayer

ECMO Europium Calcium Manganese Oxide

ECMP Electronic Countermeasures Program

ECMRA European Chemical Marketing Research Association [UK]

ECMS Electron-Capture Mass Sepectrometry

ECMSA Electronics Command Meteorological Support Agency [US Army]

ECMT Electro-Chemo-Mechanical Treatment

ECMT European Conference of Ministers of Transport [France]

ECMU Electronic Countermeasures Upgrade

ECMWF European Center for Medium-Range Weather Forecasts [UK]

ECN École Centrale de Nantes [Nantes Central School, France]

ECN Emergency Communication Network

ECN Energieonderzoek Centrum Nederlands [Netherlands Energy Research Foundation, Petten, Netherlands]

ECN Energy Center Netherlands

ECN Engineering Change Notice

ECN Equivalent Carbon Number [Metallurgy]

ECN Equivalent Chain Number

ECN European Cytokine Network [Journal published in France]

ECNAP Eastern Caribbean Natural Area Management Program

ECNASAP East Coast of North America Strategic Assessment Project [of Department of Fisheries and Oceans, Canada]

ECNC European Center for Nature Conservation

ECNDT European Council for Nondestructive Testing

ECNE Electric Council of New England [US]

ECNE Enterprise-Certified NetWare Engineer

ECNG East Central Nuclear Group [US]

ECNM Engineers Club of Northern Minnesota [US]

ECNR European Council for Nuclear Research

ECNSW Electricity Commission of New South Wales [Australia]

ECNT Environment Center of the Northern Territory [Australia] [also EC(NT)]

ECNU East China Normal University [Shanghai, PR China]

ECO Earth Communications Office

ECO Electron Coupled Oscillator

ECO Electronic Central Office [US]

ECO Electronic Checkout

ECO Electronic Contact Operator

ECO Emergency Control Officer

ECO Engine Combustion

ECO Engine Cutoff

ECO Engineering Change Order

ECO Engineering Control Office

ECO Engineers Club of Omaha [Nebraska, US]

ECO (Ontario) Environmental Commissioners Office [Canada]

ECO Environmental Communicators Organization [UK]

ECO Epichlorohydrin Copolymer

ECO European Coal Organization

E/CO Ethylene/Carbon Monoxide (Copolymer) [also E-CO]

Eco57 Escherichia coli, RFL57 [Microbiology]

ε-Co Epsilon Cobalt

Ecoat Electrocoat [also E-Coat, or E-coat]

ECOC Engineering Club of Oklahoma City [US]

ECOC European Conference on Optical Communication

ECOCO Ecological Consortium

E-CODE Canadian Electrical Code (on CD-ROM) [of Canadian Standards Association]

ECODU European Control Data Users Organization

ECODUG European Control Data Users Group

ECoG Electrocortiogram; Electrocortiography

Ecol Ecological; Ecologist; Ecology [also ecol]

Ecol Econ Ecological Economics [Journal of the International Society for Environmental Education]

EcolE Ecological Engineer [also Ecol Eng, or EcolEng]

Ecol Eng Ecological Engineer(ing)

Ecol Eng Ecological Engineering (The Journal of Ecotechnology)

Ecol Ent Ecological Entomology [Journal of the Royal Entomological Society of London, UK]

Ecol Food Nutr Ecology of Food and Nutrition [International Journal]

Ecol Geol Helv Ecologae Geologicae Helvetiae [Swiss Journal of Ecological Geology]

E coli Escherichi coli (Bacteria) [Microbiology]

Ecol Ind Reg Ecology of Industrial Regions (Journal)

Ecol Monogr Ecological Monographs [Publication of the Ecological Society of America]

ECOM Electronic Computer-Originated Mail (Service) [of United States Post Office]

ECOM Electronics Command [US Army]

E-COM Electronic Communication [also E-Com, or e-com]

ECOMA European Computer Measurement Association

ECOMAP The National Hierarchical Framework of Ecological Units Mapping Project [of US Department of Agriculture Forest Service]

e-commerce electronic commerce

ECOMS Early Capability Orbital Manned Station

ECON Electromagnetic Emission Control

EcoN Escherichia coli N [Microbiology]

Econ Economic(s); Economist; Economizer; Economy [also econ]

Econ Bull Economics Bulletin [of Chemical Industries Association, UK]

Econ Comput Econ Cybern Stud Res Economic Computation and Economic Cybernetics Studies and Research [Publication of the Centre of Economic Computation and Economic Cybernetics, Bucharest, Romania]

ECONET Ecological Network [of Institute for Global Communications, US] [also EcoNet]

Econ Geogr Economic Geographer; Economic Geography

Econ Geol Economic Geologist; Economic Geology

Econ Geol Economic Geology [Bulletin of the Society of Economic Geologists, US]

Econistor Economic Resistor [also econistor]

Econ Intell Unit Spec Rep The Economist Intelligence Unit Special Report [US]

EconLit International Economic Literature Database

ECOO Educational Computing Organization of Ontario [Canada]

ECOP Extension Committee on Organization and Policy [US]

ECOPETROL Empresa Columbiana de Petrólea [Columbian State Petroleum Company] [also Ecopetrol]

ECOPS European Committee on Ocean and Polar Sciences

ECorA Erythrina Corallodendron Agglutinin [Biochemistry]

EcoR Escherichia coli R [Microbiology]

EcoR I Escherichia coli 1 RY-13 I [Microbiology]

EcoR II Escherichia coli II R-245 [Microbiology]

EcoR V Escherichia coli V [Microbiology]

ECORQ Economic Order Quantity

ECOS Energy Conservation and Substitution

ECOS Experiment Computer Operating System

ECoSLO Environmental Center of San Luis Obispo [California, US] [also EcoSlo]

ECOSOC Economic and Social Committee

ECOSOC Economic and Social Council [of Organization of American States]

ECOSOC Economic and Social Council [of United Nations at Bangkok, Thailand]

Eco Social Rev Eco Socialist Review [Journal of the Environmental Commission of the Democratic Socialists of America]

Ecosph Ecosphere

ECOSS European Conference on Surface Science

ECOSY Extended Correlated Spectroscopy [also E-COSY]

Ecotoxicol Environ Saf Ecotoxicology and Environmental Safety [Published in the US]

ECOVEST Environmentally Sound and Sustainable Development Investment Program

ECOWAS Economic Commission of West African States

ECOWAS Economic Community of West African States

ECP Ecole Centrale de Paris [Paris Central Higher School, France]

ECP Effective Cable Pair

ECP Effective Core Potential

ECP Electric Current Perturbation

ECP Electrochemical Polishing

ECP Electrochemical Potential

ECP Electrochemical Processing

ECP Electromagnetic Capability Program [US Air Force]

ECP Electromagnetic Containerless Processing

ECP Electron Channeling Pattern

ECP Electronic Circuit Protector

ECP Empresa Colombiana de Petróleos [Colombian State Oil Company]

ECP Energy Conversion Program [of Canada Center for Mineral and Energy Technology, Natural Resources Canada, Canada]

ECP Engineering Change Program

ECP Engineering Change Proposal

ECP Engineering Control Proposal

ECP Engineers Club of Philadelphia [Pennsylvania, US]

ECP Enhanced Capabilities Port [Computers]

ECP EniChem Polimeri [Porto Marghera, Italy]

ECP Equipment Conversion Package

ECP Ethylcyclopentane

ECP Evaporative Casting Process

ECP Explicitly Coded Program

ECP Extended Capabilities Port [Computers]

ECP External Casing Packer

ECPA Electric Consumers Protection Act [US]

ECPA Electronic Communication Privacy Act [US]

ECPA Energy Conservation and Production Act [US]

ECPA Energy Consumers and Producers Association [US]

ECPD Engineers Council for Professional Development [now Accreditation Board for Engineering and Technology, US]

ECPE European Center for Public Enterprises

ECPE Extended Chain Polyethylene

ECPGB Entrance Cable Protector Ground Bar

ECPI Electronic Computer Programming Institute

EC-PINS Electronic Chart Precision Integrated Navigation System

ECPRD European Center for Parliamentary Research and Documentation [Luxembourg]

ECPS Electronic Compendium of Pharmaceuticals and Specialties [of Canadian Pharmaceutical Association]

ECPS Expanded Control Program Store

ECPS Extended Control Program Support

EC-PSEC European Community Photovoltaic Solar Energy Conference

ECQAC Electronic Components Quality Assurance Committee

ECR Educational Computing Research

ECR Effective Cleaning Radius

ECR Electric Current Relaxation

ECR Electrocyclic Reaction

ECR Electrochemical Reaction

ECR Electron Channeling Pattern

ECR Electron Cyclotron Resonance

ECR Electronic Cash Register

ECR Electronic Control Receiver

ECR Electronic Control Relay

ECR Emergency Cooling Recirculation [Nuclear Reactors]

ECR Engineering Change Request

ECR Engineering Change Requirement

ECR Error Control Receiver

ECR Estimate Change Request

ECR Executive Communication Region

ECR External Control Register

ECRA Engineering Change Request/Authorization

ECRC Electricity Council Research Center [UK]

ECRC Electronic Components Reliability Center

ECRC Electronic Components Research Center [US]

ECRC Engineering College Research Council [US]

ECR-CVD Electron Cyclotron Resonance–Chemical Vapor Deposition [also ECRCVD]

ECRD Eddy-Current Resonance Digitizing

E-CR Glass Electrical-Grade Fiberglass with Improved Corrosion Resistance [also E CR glass]

ECRH Electron Cyclotron Resonance Heating

ECRICE European Conference on Research in Chemical Education

ECRIE European Center for Research and Information Exchange [Belgium]

ECRL Eastern Caribbean Regional Library [West Indies]

ECRL Exxon Corporate Research Laboratory [US]

ECR-MBE Electron Cyclotron Resonance (Assisted) Molecular Beam Epitaxy [also ECR MBE, ECRMBE, or ECR/MBE]

ECR-MOMBE Electron Cyclotron Resonance–Metallorganic Molecular Beam Epitaxy [also ECR MOMBE, or ECR/MOMBE]

ECRO European Chemoreception Research Organization [Switzerland]

ECR-PACVD Electron Cyclotron Resonance Plasma-Assisted Chemical Vapor Deposition (Process)

ECR-PECVD Electron Cyclotron Resonance Plasma-Enhanced Chemical Vapor Deposition (Process)

ECRS European Conference on Residual Stresses

ECRYS International Workshop on Electronic Crystals

ECS East China Sea

ECS Echo Control Subsystem

ECS Ecuadorean Sucre [Currency]

ECS Editorial Coordination Services

ECS (Department of) Electrical, Computer and Systems (Engineering)

ECS Electrical and Communication System

ECS Electrical Control System

ECS Electrochemical Science

ECS Electrochemical Series

ECS Electrochemical Society [US]

ECS Electroconvulsive Shock [Medicine]

ECS Electromagnetic Compatibility Society [of Institute of Electrical and Electronics Engineers, US]

ECS Electron Captive Spectroscopy

ECS Electronically Controlled Suspension [Automobiles]

ECS Electronic Control Switch

ECS Embedded Computer System

ECS Emergency Coolant System [Nuclear Reactors]

ECS Emission Control System

ECS Energy Communication Services

ECS Energy-Corrected Sudden (Law)

ECS Engine Control System

ECS Enhanced Chip Set

ECS Environmental Conditioning System

ECS Environmental Conservation Service

ECS Environmental Control Shroud

ECS Environmental Control Subsystem

ECS Environmental Control System

ECS EOSDIS (Earth Observing System Data and Information System) Core System

ECS ESR (Electroslag Remelting) + CIG (Cold-Walled Induction Guide) + Spray Forming (Process)

ECS Etched Circuits Society [US]

ECS European Ceramic Society

ECS European Committee for Standardization

ECS European Communications Satellite

ECS Evaporation Control System

ECS Expanded Control Store

ECS Experimental Communication Satellite [Japan]

ECS Extended Core Storage

ECSA Estuarine and Coastal Sciences Association [UK]

ECSA European Chlorinated Solvent Association

ECSA European Communications Security Agency

ECSA European Computing Services Association

ECSA Exchange Carriers Standards Association [US]

ECSA Expanded Clay and Shale Association [now Lightweight Aggregate Producers Association, US]

ECSAMR Emergency Committee to Save America's Marine Resources [US]

ECSC East Central State College [Ada, Oklahoma, US]

ECSC Eddy-Current Signal Conditioning (Unit)

ECSC Electronic Communications Steering Committee [of Institute of Electrical and Electronics Engineers, US]

ECSC Energy Conservation and Solar Center [UK]

ECSC European Ceramic Society Conference

ECSC European Coal and Steel Community; European Community for Steel and Coal

ECSE Electrical Computer and Systems Engineering

ECSEL Engineering Schools for Excellence in Education and Leadership [of National Science Foundation, US]

ECSF Engineers Club of San Francisco [California, US]

ECSG Electronic Connector Study Group [US]

ECSL Engineers Club of St. Louis [Missouri, US]

ECSRM European Conference on SiC (Silicon Carbide) and Related Materials

ECSS European Committee for the Study of Salt [France]

ECSS European Conference on Surface Science

ECSSC European Conference on Solid-State Chemistry

ECSTASI The Electrochemical Society Tool for Abstract Submission [Internet]

ECSTASY Electrical Control for Switching and Telemetering Automobile Systems

ECSTM Electrochemical Scanning Tunneling Microscope; Electrochemical Scanning Tunneling Microscopy

ECSU Elizabeth City State University [North Carolina, US]

ECSW Extended Channel Status Word

EC-SW End Cell Switch [Batteries]

ECT Eddy Current Test(ing)

ECT Elastic Continuum Theory [Physics]

ECT Electroconvulsion (or Electroconvulsive) Therapy [Medicine]

ECT Electronic Control Technology

ECT Emission Controlled Tomography

ECT Engineer's Club of Toronto [Canada]

ECT Enteric Coated Tablet(s) [Pharmacology]

ECT Equicohesive Temperature

ECT Equivalent Crystal Theory

ECT Evans Clear Tunnel

ECTA Error Correcting Tree Automation

ECTA European Cutting Tools Association [UK]

ECTC Eastern Coal Transportation Conference [US]

ECTD Electrochemical Thermodynamics

ECTDMS Electrochemical Thermal Desorption Mass Spectroscopy

ECTEL European Telecommunications and Professional Electronics Industry [UK]

ECTEOLA Epichlorohydrin Triethanolamine

ECTEOLA C Epichlorohydrin Triethanolamine Cellulose

ECTFE Ethylene Chlorotrifluoroethylene (Copolymer) [also E-CTFE]

ECTI Electro-Technical Council of Ireland

ECTL Electronic Communal Temporal Lobe

ECTL Emitter-Coupled Transistor Logic

ECTL Exchange-Coupled (Magnetic) Trilayer

ECTMAC East Coast Trawl Management Advisory Committee

EC-TMR European Community Transverse Magnetoresistance (Research Network)

ECTN Eastern Canada Telemetered Network

ECTR Extended Connection Table Representation

ECTUNAMAC East Coast Tuna Management Advisory Committee [Australia]

ECU East Carolina University [Greenville, North Carolina, US]

ECU EISA (Extended Industry Standard Architecture) Configuration Utility

ECU Electrical Control Unit

ECU Electronic Control Unit

ECU Electronic Conversion Unit

ECU Electronic Coupling Unit

ECU Emission Control Unit

ECU Energy Conservation and Utilization

ECU Environmental Control Unit

ECU European Currency Unit [also known as Eurodollar] [also Ecu, or ecu]

ECU European Customs Union [European Community]

ECU Exabyte Control Unit

ECU Exposure Control Unit

Ecua Ecuador(ian)

E-CUBE Energy Conservation Utilizing Better Engineering [also E-Cube]

ECUFIN Council of Economic Ministers (of the European Community)

ECUCT East China University of Chemical Technology [PR China]

ECUT Energy Conservation and Utilization Technology Program [of US Department of Energy]

ECV Enamelled Single-Cotton-Covered Varnished (Electric Wire)

ECV Exchange Capacity per Unit Volume [Geology]

ECV Externally Controlled Thermodynamic Variable

ECWA Economic Commission for Western Africa [of United Nations]

ECWIM European Committee of Weighing Instrument Manufacturers [France]

ECWR Electron Cyclotron Wave Resonance

ECWU Energy and Chemical Workers Union

ECX Electronically Controlled Telephone Exchange

ECY European Conservation Year

E-CYCLE Execution Cycle [also E-Cycle, or E-cycle]

ED Earth Detector

ED Eddy Diffusivity

ED Edge Dislocation [Materials Science]

ED Edge Distance

ED Eddy Diffusion

ED Education Division [of NASA]

ED Effective Demand

ED Effective Depth

ED Effective Dose

ED Elastic Deformation

ED Elastic Design

ED Elastodynamic(s)

ED Election District

ED Electrical Differential

ED Electric Dipole

ED Electric Discharge

ED Electric Disintegration

ED Electrochargeable Liquid

ED Electrodecantation

ED Electrodeposit(ed); Electrodeposition

ED Electrodialysis; Electrodialyzer

ED Electrodynamic(s)

ED Electron Decay

ED Electron Density

ED Electron Detection; Electron Detector

ED Electron Device

ED Electron Devices (Society) [of Institute of Electrical and Electronics Engineers, US]

ED Electron Diffraction

ED Electron Donor

ED Electron Dose

ED Electron Drift

ED Electronic Data

ED Electronic Defense

ED Electronic Design

ED Electronic Device

ED Electronic Dummy

ED Electronics Division

ED Emergency Distance

ED Energy Density

ED Energy Development

ED Energy Dispersion; Energy Dispersive

ED Energy Distribution

ED Engine Drive

ED Engineering Department

ED Engineering Design(er)

ED Engineering Directive

ED Engineering Division

ED Engineering Draftsman; Engineering Drawing

ED Entry Date

ED Environmental Damage

ED Environmental Degradation

ED Epitaxial Deposit

ED Equilibrium Diagram

ED Equivalent Dose

ED Erase Display

ED ERIC (Educational Resources Information Center) Document [of United States Office of Education]

ED Error Detection

ED Erythrodextrin

ED Esaki Diode

ED (Electron) Escape Depth

ED Estimated Dose

ED Ethyldichloroarsine

ED Ewing-Donn (Hypothesis) [Geophysics]

ED Exact Differential

ED Executive Decision

ED Existence Doubtful

ED Expanded Display

ED Expansion Drum

ED Explosive Device

ED Explosives Division [of Canada Center for Mineral and Energy Technology, Natural Resources Canada, Ottawa]

ED External Deflector

ED External Device

ED Extractive Distillation [Chemical Engineering]

ED Extra-Low Dispersion (Glass)

ED Extrusion Direction

ED_{01} Effective Dose 1 (i.e., that affects 1% of study group) [also ED01]

ED_{50} Effective Dose 50 (i.e., that affects 50% of study group) [also ED50]

E&D Engineering and Development

E&D Engineering and Development Directorate [of NASA Johnson Space Center, Houston, Texas, US]

E-D Ewing-Donn (Hypothesis) [Geophysics]

E/D Enhancement/Depletion (Mode)

E/D Environmental Devices, Inc. [Sacramento, California, US]

Ed Edit(ion); Editor [also ed]

Ed Education; Educator [also ed]

E/d Young's Modulus per Relative Density [Symbol]

ed edit(ed)

EDA Economic Development Administration [of US Department of Commerce]

EDA Economic Development Agreement

EDA Educational Development Association [UK]

EDA Electrical Development Association [UK]

EDA Electron Donor-Acceptor

EDA Electronic Defense Association [US]

EDA Electronic Design Automation

EDA Electronic Differential Analyzer

EDA Electronic Digital Analyzer

EDA Electronic Display Assembly

EDA Elevation Drive Assembly

EDA Embedded Direct Analysis

EDA Embedded Document Architecture

EDA Emergency Declaration Area

EDA Energy-Dispersive Analysis; Energy-Dispersive Analyzer

EDA Engineering Design Activities (Agreement) [for International Thermonuclear Experimental Reactor Project]

EDA Engineering Design Automation

EDA Environmental Dental Association [US]

EDA Ethylenediamine [also Eda]

EDA Ethylenediammonium

EDA European Demolition Association [Netherlands]

EDA European Desalination Association [Scotland]

EDA Explorative (or Exploratory) Data Analysis

12EDA 1,2-Ethanediamine [also 12Eda]

ED3A Ethylenediaminetriacetate; Ethylenediaminetriacetic Acid

EDAC Error Detection and Correction

EDAC 1-Ethyl-3 (3-Dimethylaminopropyl)carbodiimide

EDAM Educational Distributors of Manitoba [Canada]

EDANA European Disposables and Nonwovens Association [Belgium]

EDANS 5-Ethylenediamine Naphthalene-1-Sulfonic Acid

1,5-EDANS 1,5-Ethylenediaminenaphthalenesulfonic Acid

EDAS Electrophoresis Documentation and Analysis System

EDAS El Dschamhurija el Arabija es Surija [Arabian Republic of Syria]

EDATS Environmental Data Acquisition/Transmission System

EDAX Energy-Dispersion Analyzer, X-Ray

EDAX Energy-Dispersive X-Ray (Fluorescence) Analysis

Edax Ed Edax Editor [Published in the US]

EDB 1,2-Dibromoethane

EDB Earned Depletion Base

EDB Economic Defense Board [US]

EDB Economic Development Board [US and Singapore]

EDB Educational Data Bank

EDB Energy Database [of Oak Ridge National Laboratory, Tennessee, US]

EDB Environmental Data Book

EDB Ethylene Dibromide

EdB Bachelor of Education

EDBA 2,2'-(Ethylenediimino)dibutyric Acid

EDBD Environmental Database Directory [of National Oceanographic Data Center, US]

EDB/DBCP Ethylene Dibromide/Dibromochloropropane (Calibration Standard)

edbl edible

EDBS Educational Database System

EDC 1,2-Dichloroethane

EDC Economic Development Commission

EDC Economic Development Committee [of National Economic Development Council, UK]

EDC Economic Development Corporation [US]

EDC Economic Dispatching Control

EDC Education Development Center

EDC Effective Dielectric Constant

EDC Electrochemical-Depolarized CO (Carbon Monoxide)

EDC Electrode Dark Current

EDC Electrodischarge Consolidation

EDC Electronic Data Collection

EDC Electronic Desk Calculator

EDC Electronic Digital Computer

EDC Electrooptic Directional Coupler

EDC Emergency Decontamination Center

EDC Enamelled Double-Cotton-Covered (Electric Wire)

EDC Energy-Dispersive Curve

EDC Energy-Distribution Curve

EDC Engine-Driven Compressor [also edc]

EDC Engineering Design Change

EDC Engineering Design Consultant

EDC Engineering Documentation Center

EDC Enhanced Data Correction

EDC EROS (Earth Resources Observation System) Data Center [of US Geological Survey at Sioux Falls, South Dakota]

EDC Error Detection and Correction

EDC Error Detection Code

EDC Estimated Date of Completion [also edc]

EDC Ethylcarbodiimide Chloride

EDC 1-Ethyl-3-(3-Dimethylaminopropyl)carbodiimide (Hydrochloride)

EDC Ethylene Dichloride

EDC European Danube Commission

EDC European Defense Community

EDC European Development Center [Leuven, Belgium]

EDC European Documentation Center [UK]

EDC Export Development Corporation [Canada]

EDC Extensive Data Communication

EDC External-Device Control

EDC External Disk Channel

EDC External Drum Channel

EDCC Electronic Data Council of Canada

EDCL Electric-Discharge Convection Laser

EDCl Ethylcarbodiimide Chloride

EDCl 1-Ethyl-3-(3-Dimethylaminopropyl)carbodiimide Chloride

EDCM Electrochemical-Depolarized CO (Carbon Monoxide) Module

EDCOM Editor and Compiler [also Edcom]

EDCP Engineering Design Change Proposal

EDCPF Environmental Data Collection and Processing Facility

EDCS Energy Distribution Curve Spectroscopy

EDCS Engineering Data Control System

EDCT Expect Departure Clearance Time

ED-Cu Electrodeposited Copper

EDCV Enamelled Double-Cotton-Covered Varnished (Electric Wire)

EDCW External Device Control Word

EDD Electronic Data Display

EDD Electronic Document Delivery

EDD Energy-Dispersive Diffractometry

EDD Energy Distribution Difference

EDD Engineering Development Directorate

EDD Engineering Development Division [of National Oceanic and Atmospheric Administration–Pacific Marine Environmental Laboratory, US]

EDD Envelope Delay Distortion

EDD Environmental Data Directory (Database) [of Environment Australia] [also known as Green Pages]

EDD Equipment Development Division [of Alcoa Corporation, US]

EDD Ethylenediaminedinitrate

EDD Expected Date of Delivery

EdD Doctor of Education

edd edited [also ed'd]

EDDA N,N'-Ethylenediaminediacetate; N,N'-Ethylenediaminediacetic Acid [also edda]

EDDAMS Ethylenediamine N,N-Diacetate-N'-Monosuccinate [also eddams]

EDDC Extended Distance Data Cable

EDDDA Ethylenediamine-N,N'-Di-3-Propionate-N,N'-Diacetate [also edddda]

EDDF Error Detection and Decision Feedback

EDDHA Ethylenediamine Di(o-Hydroxyphenylacetic Acid)

EDDIP Ethylenediamine Diisopropylphosphonic Acid [also EDDIF]

EDDP O-Ethyl-S,S-Diphenyl Dithiophosphate

EDDPI Ethylenediamine-N,N'-Di(methylenephosphinic Acid)

EDDPO Ethylenediamine-N,N'-Bis(methylenephosphonic Acid)

EDDQ Extra-Deep-Drawing Quality (Steel)

EDDR Electron Dipole-Dipole Reservoir

EDDS Ethylenediamine-N,N'-Disuccinate [also edds]

EDDUS Electronic Data Display and Update System

EDE Economic Development Foundation [Philippines]

EDE Effective Dose Equivalent

EDE Emergency Decelerating

EDE Emitter Dip Effect

EDE External Document Exchange

EDES Electric Discharge Emission Spectroscopy

EDETATE Ethylene Diamine Tetraacetic Acid

EDEW Enhanced Distant Early Warning [Radar]

ED-EX Excavation Damage Extensometer

EDEXAFS Energy-Dispersive EXAFS (Extended X-Ray Absorption Fine Structure) [also ED-EXAFS]

EDF Effective Damage Function

EDF Electric Discharge Forming

EDF Electricité de France

EDF Electrophoresis Duplicating Film

EDF Elongatable Dow Fiber

EDF Energy Density Fluctuation

EDF Energy Distribution Function

EDF Engineering Data File

EDF Engineering Data Form

EDF Environmental Defense Fund [US]

EDF Eosinophil Differentiating Factor [Biochemistry]

EDF European Defense Force

EDF European Development Fund

EDFA Erbium-Doped Fiber Amplifier

EDF/EDP Electrophoresis Duplicating Film/Electrophoresis Duplicating Paper (Photochemicals)

EDF Letter Environmental Defense Fund Letter [US]

EDFP Engine-Driven Fire Pump

EDFR Effective Date of Federal Recognition

EDFT Early Detection and Forecast of Tsunamis (Project) [of National Oceanic and Atmospheric Administration–Pacific Marine Environmental Laboratory, US]

EDG Edge [also Edg, or edg]

EDG Electrical Discharge Grinder; Electrical Discharge Grinding

EDG Electrochemical Discharge Grinding

EDG Electrodermatogram; Electrodermatography

EDG Electrodischarge Grinding

EDG Electrodynamic Gradient

EDG Emergency Diesel Generator

EDG Enamelled Double-Glass-Covered (Electric Wire)

EDG Ethyl Diglycol

EDG Exploratory Development Goals

EDGAR Electronic Data Gathering, Analysis and Retrieval

EDGE Electronic Data Gathering Equipment

EDGES Expert DMIS (Directory Management Information System) Graphical Editor and Simulator

EDGF Electrodynamic Gradient Freeze

EDH Extended Debye-Hückel (Equation) [Physical Chemistry]

EDHE Experimental Data Handling Equipment

EDHE External Data Handling Equipment

EDI Electromagnetic Discharge Imaging

EDI Electron Diffraction Instrument

EDI Electronic Data Interchange

EDI Electronic Document Interchange

EDI Electrodeionization

EDIAC Electronic Display of Indexing Association and Content

EDIB Ethyl Diiodobrassidate

EDIBUILD Pan-European User Group for the Construction Industry

EDIC Exploration Drilling Incentive Program [Canada]

EDICON Electronic Data Interchange Construction Industry Association [UK]

EDICT Engineering Departmental Interface Control Technique

EDICT Engineering Document Information Collection Technique

EDIF Electronic Design Interchange Format

EDIFACT Electronic Data Interchange for Administration Commerce and Transport

EDIM Embedded Diatomics in Molecules (Method)

EDIN Electronic Data Interchange Network

Edin Edinburgh [Scotland]

EDIP European Defense Improvement Program [of NATO]

EDIS Electronic Distributorless Ignition System

EDIS Elektronisches Dokumentations– und Informationssystem [Electronic Documentation and Information Retrieval System] [Germany and Switzerland]

EDIS Engineering Data Information System

EDIS Environmental Data and Information Service [now National Environmental Satellite, Data and Information Service] [of National Oceanic and Atmospheric Administration, US]

EDIS Exploratory Drill(ing) Incentives System

EDIT Engineering Development Integration Test

EDIT Energy Distrubution of Ionizing Transitions

EDIT Error Deletion by Iterative Transmission

Edit Edited; Edition; Editor [also edit]

EDITAR Electronic Digital Tracking and Ranging

EDL Edit Decision List

EDL Electrical Double Layer

EDL Electrodeless Discharge Lamp

EDL Electrodynamic Levitation

EDL Electron Devices Letters [of Institute of Electrical and Electronics Engineers, US]

EDL Electronic Differential Locking [Part of Automotive Anti-Lock Brake System]

EDL Engineering Development Laboratory [of Pacific Northwest National Laboratory, Richland, Washington, US]

EDL Entry, Descent and Landing (System) [of NASA]

EDLC Ethernet Data Link Control

EDLCC Electronic Data Local Communications Central

EDLIN Line Editor (Program) [also Edlin, or edlin]

EDLN Engineering Development Logic Network

EDM Effective Dynamic Modulus

EDM Electrical Discharge Machine(d); Electrical Discharge Machining

EDM Electric Dipole Moment

EDM Electronic Distance Measurement

EDM Electronic Defense Mechanism

EDM Engineering Data Management

EDM Entity Data Model

EdM Master of Education

EDMA Ethylenediaminemonoacetate [also edma]

EDMA Ethylene Glycol Dimethacrylate

EDMA Extended Direct Memory Access

EDME Electronic Distance Measurement Equipment

EDMF Electric Dipole Moment Function

EDMF Euclid-IS Data Management Facilities

EDM'ed Electrical Discharge Machined

EDMI Ethydimethylindium [also EDMIn]

EDMR Electrically Detected Magnetic Resonance

EDMS Electronic Device and Materials Symposium

EDMS Electronic Document Management System

EDMS Engineering Data Management System [of NASA Jet Propulsion Laboratory, Pasadena, California, US]

EDMS Engineering Data Microreproduction System

EDMS Extended Data Management System

Edn Edition [also edn]

EDNA Ethylenedinitramine

EdNA Educational Network Australia

EDNATOL Ethylenedinitramine and Trinitrotoluene [also Ednatol]

EDO Economic Development Office

EDO Effective Diameter of Objective

EDO Engineering Duties Only

EDO Ethyl-n,n-Dimethyloxamate

EDO Executive Director of Operations

EDO Extended Data Out

EDOC Effective Date of Change

EDOP ER-2 Doppler Radar [*Note:* ER-2 is a NASA Ames High Altitude Aircraft]

EDO-RAM Extended Data Out Random-Access Memory

EDOS Enhanced Disk-Operating System (for Windows)

EDOS EOS (Earth-Observing System) Data and Operations System

EDOS Extended Disk-Operating System

EDOS-MSO Extended Disk-Operating System–Multistage Operations

EDOT Effective Date of Training

EDP Educational Data Processing

EDP Effective Depth of Penetration [Mechanical Testing]

EDP Electronic Data Processing

EDP Electrophoresis Duplicating Paper

EDP Engineering Development Plan

EDP Enterprise Development Program

EDP Environmental Definition Program

EDP Ethylene Diamine Pyrocatechol

EDP Experimental Development Program

EDPA 2-Ethyl-3,3-Diphenyl-2-Propenamine

EDPA Ethylenediamine-N,N'-Diacetate-N,N'-Di-α-Propionate [also edpa]

EDPA Exhibition Designers and Producers Association [US]

EDPAA Electronic Data Processing Auditors Association [US]

EDPAF Electronic Data Processing Auditors Foundation [Carol Stream, Illinois, US]

EDP Audit J EDP Auditors Journal [of EDP Auditors Foundation, US]

EDPC Electronic Data Processing Center

EDPE Electronic Data Processing Equipment

ED&PG Energy Development and Power Generation

ED&PG Energy Development and Power Generation [Institute of Electrical and Electronics Engineers–Power Engineering Committee, US]

EDPI Electronic Data Processing Institute

EDP In Depth Rep EDP (Electronic Data Processing) In Depth Reports [Canada]

EDPM Electronic Data Processing Machine

EDPO 1,2-Bis(diphenylphosphine)ethane Dioxide

EDPS Electronic Data Processing System

EDPSA Energy-Dependent Phase-Shift Analysis

EDPTA Ethylenediphosphine Tetraacetic Acid

EDR Edge Distance Ratio

EDR Electric Dispersion Reactor

EDR Electrodermal Reaction; Electrodermal Response

EDR Electrodialysis Reversal

EDR Electron Dispersion Reactor

EDR Electronic Decoy Rocket

EDR Electrothermal Direct Reduction (Steelmaking Process) [Metallurgy]

EDR Engineering Design Review

EDR Engineering Document Release

EDR Environmental Data Registry [of US Environmental Protection Agency]

EDR Equivalent Direct Radiation

EDR Extended Dynamic Range

EDRA Environmental Design Research Association [US]

EDRAM Extended Dynamic Random Access Memory

EDRAW Erasable Direct Read After Write (Memory)

EDRCC Electronic Data Remote Communications Complex

EDRI Electronic Distributors Research Institute [US]

EDRI Endocrine Disruptors Research Initiative [of National Science and Technology Council, US]

EDRL Effective Damage Risk Level

EDRL Energy Diversification Research Laboratory [of Canada Center for Mineral and Energy Technology, Natural Resources Canada, Ottawa]

EDRP European Demonstration Reprocessing Plant [Nuclear Engineering]

EDRS Engineering Data Retrieval System

EDRS ERIC (Educational Resources Information Center) Document Reproduction Service [of US Office of Education]

EDS Early Deployment System

EDS Enamelled Double-Silk-Covered (Electric Wire)

EDS Electrical Distribution System

EDS Electric Discharge Sintering [Metallurgy]

EDS Electric Drive System

EDS Electrodialysis System

EDS Electron Devices Society [of Institute of Electrical and Electronics Engineers, US]

EDS Electron Dispersive X-Ray(s)

EDS Electronic Data(Switching) System

EDS Electronic Data Systems Corporation [US]

EDS Electronic Document Storage

EDS Emergency Detection System

EDS Energy Data System [of US Environmental Protection Agency]

EDS Energy-Dispersive (X Ray) Spectrometer; Energy-Dispersive (X Ray) Spectrometry; Energy-Dispersive (X Ray) Spectroscopy

EDS Energy Dispersive System

EDS Engineering Data Sheet

EDS Engineering Data System [of US Department of Defense]

EDS Environmental Data Service [now National Environmental Satellite, Data and Information Service, US]

EDS Exchangeable Disk Storage

EDS Expert Debugging System

EDS Explosive Device System

EDS Explosives Development Section [Picatinny Arsenal, Dover, New Jersey, US]

EdS Education Specialist (or Specialist in Education)

Eds Editors [also eds]

EDSA Expert Dataflow and Static Analysis Tool

EDSA Expert Debugging System for Ada

EDSAC Electronic Delay Storage Automatic Calculator

EDSAC Electronic Delay Storage Automatic Computer

EDSAC Electronic Discrete Sequential Automatic Computer

EDSC Engineering Data Service Center [US Air Force]

EDS-EELS Energy Dispersive Spectroscopy–Electron Energy Loss Spectroscopy [also EDS/EELS]

EDSG Economic Development Studies Group [of Hudson Institute, US]

EDSG Electrooptical Data Systems Group [US]

EDSL Electron Device System Laboratory [of Kanazawa Institute of Technology, Ishikawa, Japan]

EDSP Energy Distribution by Sputtered Particles

EDSS Environment Decisionmaking Support System [of Commonwealth Scientific and Industrial Research Organization–Divison of Information Technology]

EDSS Explosives Detection Security System

EDS-STEM Energy DispersiveSpectroscopy–Scanning Transmission Electron Microscopy [also EDS/STEM]

EDST Eastern Daylight Saving Time

EDST Elastic Diaphragm Switch Technology

EDSV Enamelled Double-Silk-Covered Varnished (Electric Wire)

EDS-WDX Energy Dispersive Spectroscopy–Wavelength Dispersive X Ray Analysis [also EDS/WDX]

EDT Early Decay Time

EDT Eastern Daylight Time [Greenwich Mean Time +6:00] [also edt]

EDT Edge Delamination Test

EDT Electric Discharge Testing

EDT Electric Discharge Texturing

EDT Electric Discharge Tube

EDT Electronic Data Transmission

EDT Electronic Design Transfer

EDT Energy Dissipation Test

EDT Engineer Design Test

EDT Ethylenediamine Tartrate

EDTA Ethylenediaminetetraacetate; Ethylenediaminetetraacetic Acid [also Edta, or edta]

EDTA^{4-} Quadridentate Ethylenediaminetetraacetic Acid [also Edta^{4-}, or edta^{4-}]

ED3A Ethylenediaminetriacetate; Ethylenediaminetriacetic Acid [also ed3a]

EDTAN Ethylenediaminetetraacetonitrile

EDTA Na$_4$ Ethylenediaminetetraacetic Acid, Tetrasodium Salt

EDTC Ethyldipropylthiocarbamate

EDTCC Electronic Data Transmission Communications Central

EDTN 1-Ethoxy-4-(Dichloro-1,3,5-Triazinyl)naphthalene

EDTP Ethylenediamine Tetramethylphosphonic Acid

EDTP Ethylenediamine Tetrapropionate [also edtp]

EDTPI Ethylenediamine-N,N,N',N'-Tetra-(methylenephosphinic Acid)

EDTRI Ethylenediamine-N,N',N'-Triacetic Acid

EDTV Extended Definition Television

EDU Educational Institutions [Internet Domain Name]

EDU Electronic Display Unit

EDU Engineering Development Unit

EDU Experimental Diving Unit

Educ Education(al); Educator [also educ]

Educ Comput Educational Computing [Journal published in the UK]

Educ Comput Education and Computing [Journal published in the Netherlands]

EDUCOM Educational Communication

EDUCOM Bull EDUCOM Bulletin [Published by the Association for Educational Communications and Technology, US]

EDUCOM Rev EDUCOM Review [Published by the Association for Educational Communications and Technology, US]

Educ Psychol Meas Educational and Psychological Measurements (Journal)

Educ Res Bull Educational Research Bulletin [US]

Educ Rev Educational Reviews [International Journal]

Educ Technol Educational Technology [Journal published in the US]

Educ Technol Res Dev Educational Technology, Research and Development [Publication of the Association for Educational Communications and Technology, US]

eductl educational

Educ + Train Education + Training [Journal published in the UK]

Educ Train Technol Int Educational and Training Technology International [Journal published in the UK]

EDUG European Data Manager Users Group

EDUK Educational Kinesiology

EDUNET Education Network [also Edunet]

EDV Electronic Depressurizing Valve

EDV Exponentially Damped Van der Waals (Model)

EDVAC Electron Discrete Variable Automatic Compiler

EDVAC Electronic Digital Vernier Analog Computer

EDVAC Electronic Discrete Variable Automatic Calculator

EDVAC Electronic Discrete Variable Automatic Computer

EDW Edwards [Tactical Air Navigation Station, Edwards, California, US]

EDW Edwards Air Force Base [NASA Deorbiting Station]

EDW Electron Density Wave

EDWC Electrical Discharge Wire Cutting

EDWG Endocrine Disruptor Working Group [of National Science and Technology Council, US]

EDWM Electrodynamic Wattmeter

EDX Energy-Dispersive X-Ray Analysis

EDX Energy-Dispersive X-Ray Fluorescence

EDX Energy-Dispersive X-Ray Spectroscopy

EDX Energy-Dispersive X-Ray System

EDX Event-Driven Executive [Computers]

EDXA Energy-Dispersive X-Ray Analysis [also EDXRA]

EDXD Energy-Dispersive X-Ray Diffraction [also EDXRD]

EDXRA Energy-Dispersive X-Ray Analysis [also EDXA]

EDXRD Energy-Dispersive X-Ray Diffraction [also EDXD]

EDXRF Energy-Dispersive X-Ray Fluorescence

EDXRS Energy-Dispersive X-Ray Spectrometer; Energy Dispersive X-Ray Spectrometry

EDXS Energy-Dispersive X-Ray Spectroscope; Energy Dispersive X-Ray Spectroscopy

EDX/WDX Energy-Dispersive X-Ray Analysis/Wavelength Dispersive X-Ray Analysis

EE Earthquake Engineer(ing)

EE Edge Effect

EE Edison Effect

EE   Effective Energy

EE   Einstein Equation [Statistical Mechanics]

EE   Electrical Engineer(ing)

EE   Electric Engine

EE   Electrically Erasable

EE   Electric Energy

EE   Electro-Etching

EE   Electron Energy

EE   Electron Emission; Electron Emitter

EE   Electronic-Electronic (Operation)

EE   Electronic Energy

EE   Electronics Engineer

EE   Emitter Electrode

EE   Enantiomeric Excess [also ee] [Chemistry]

EE   End-Effector

EE   Energy Efficiency

EE   (Office of) Energy Efficiency and Renewable Energy [of US Department of Energy]

EE   Energy Engineer(ing)

EE   Energy Equivalent

EE   Energy Extractive

EE   Engineering Education

EE   Enolether

EE   Environmental Embrittlement [Metallurgy]

EE   Environmental Engineer(ing)

EE   Equatorial Electrojet [Geophysics]

EE   Equine Encephalitis

EE   Equipment Engineer

EE   Error Ellipsoid

EE   Error Equation

EE   Errors Excepted [also ee]

EE   Estado Español [State of Spain]

EE   Estonia [ISO Code]

EE   Ettinghausen Effect [Physics]

EE   Evolutionary Ecology

EE   Exact Equation

EE   Exchange Energy

EE   Explosives Engineer(ing)

EE   Extended Edition

EE   External Environment

E_f/E_m Elastic Modulus Ratio of Composite Fibers to Composite Matrix [Symbol]

E_s/E_0 Energy Ratio (in Ion-Spectroscopy Scattering) [Symbol]

E-E   End-to-End

E2E   Electron-Electron Coincidence Spectroscopy [also E-2E, e2e, or e-2e]

ee   enantiomeric excess [also EE] [Chemistry]

ee   even even (nuclei) (i.e., atomic nuclei having even numbers of protons and neutrons)

.ee   Estonia [Country Code/Domain Name]

e-e   electron-electron (Coulomb interaction) [Symbol]

$\mathcal{E}[\varepsilon]$ expected value of error sum (in statistics) [Symbol]

$\varepsilon(E)$ permittivity as a function of electric field

$\eta^*(E)$ effective quantum efficiency [Symbol]

η/η_0 relative viscosity (i.e., ratio of viscosity of solution to viscosity of solvent) [Symbol]

EEA   Electron Energy Analysis

EEA   Electrical Engineering Abstracts [of INSPEC–Information Services Physics, Electrical and Electronics, and Computers and Control, UK]

EEA   Electronic Engineering Association [UK]

EEA   Emergency Employment Act [US]

EEA   Environmental Education Act [US]

EEA   Ethylene Ethyl Acrylate (Copolymer)

EEA   Euonymus Europaeus Acetone Powder [Biochemistry]

EEA   Euonymus Europaeus Agglutinin [Biochemistry]

EEA   European Economic Area (Treaty)

EEA   European Environmental Agency [of European Union]

EEAA   Environmental Education Advisers Association [UK]

EEAES   Electron-Excited Auger Electron Spectroscopy

EEB   European Environmental Bureau [Belgium]

EEB   European Export Bank

EEBP   Estação Experimental de Biologia e Piscicultura [Experimental Station for Biology and Fish Culture, Brazil]

EEC   Eastern European Country

EEC   Electronic (Automotive) Engine Control

EEC   End of Equilibrium Cycle

EEC   Engineering Education Center

EEC   Eurocontrol Experimental Center

EEC   European Economic Commission

EEC   European Economic Community [later European Community; now European Union]

EEC   Evaporative Emission Control

EEC   Extended Error Correction

EECA   European Electronic Component Manufacturers Association [Belgium]

EECE   Electrical Engineering and Computer Engineering

EECE   Emergency Ecomonic Commission for Europe

EEC/EURAM   European Economic Community/European Research on Advanced Materials

EECL   Emitter (to)Emitter Coupled Logic [also E²CL]

EECOM   Electrical, Environmental and Communications

EECOM   Electrical, Environmental, Consumables and Mechanical

EECOMP   Electrical, Environmental and Consumables Systems Engineer

EECS   Electrical Engineering and Computer Science [also EE&CS]

EECW   Emergency Exchanger Cooling Water [Nuclear Reactor]

EED   Economy and the Environment Division [of US Environmental Protection Agency]

EED   Electrical and Electronics Division

EED   Electrical Engineering Department

EED   Electrical Engineering Division

EED   Electrical Engineering Division [of Institute of Engineers, Calcutta, India]

EED   Electrical Explosive Device; Electroexplosive Device

EED Electron Energy Distribution

EED End-to-End Distance [Polymer Science]

EED Energy and Environment Division [of Lawrence Berkeley Laboratory, California, US]

EED Energy Efficiency Division [of Canada Center for Mineral and Energy Technology, Natural Resources Canada, Ottawa]

EED Engineering Education Division [of American Society for Engineering Education, US]

EED European Enterprises Development (Company)

EED Exterior Electron Densities

EEDA Edmonton Economic Development Authority [Alberta, Canada]

EEDB ERDA (Energy Research and Development Adminstration) Energy Database

EEDC Electron Energy Distribution Curve

EEDC Electronics Economic Development Committee

EEDF Electron Energy Distribution Function

EEDMP N-Ethyl Ethylenediamine-N'-Mono 3-Propionate

EEDR Electron-Electron Double Resonance

EEDQ 2-Ethoxy-N-Ethoxycarbonyl-1,2-Dihydroquinoline; N-Ethoxycarbonyl-2-Ethoxy-1,2-Dihydroquinoline; Ethyl-1,2-Dihydro-2-Ethoxy-1-Quinolinecarboxylate

EEDTA Ethyletherdiamine-N,N,N',N'-Tetraacetate; Ethyletherdiamine N,N,N',N' Tetraacetic Acid [also eedta]

EEE Eastern Equine Encephalitis (Virus)

EEE Electrical/Electronic/Electromechanical (Components)

EEE Electrical Engineering and Electronics

EEE Electrical Engineering Exposition

EEE Electronic, Electrical, Electromechanical

EEE Ethoxyethyl Ether

EEE European Engineering and Electronics

EEE Exoelectron Emission [Physics]

2-EEE Bis(2-Ethoxyethyl) Ether

EEEF Energy Economics Education Foundation [of International Association for Energy Economics, US]

EEEI Energy, Economics and Environment Institute

EEES End-Effector Exchange System

EEET Electronic Excitation Energy Transfer

EEEU End-Effector Electronics Unit

EEF Earth Ecology Foundation [US]

EEF Engineering Employers Federation [UK]

EEF European Environment Foundation

EEF Export Expansion Fund [Canada]

EEG Electroencephalogram; Electroencephalograph(y)

EEG Energy and Environment Group [India]

EEG Environmental Engineers Group [of University of British Columbia, Canada]

EEGS Environmental and Engineering Geophysical Society [US]

EEG Soc Electroencephalographic Society [UK]

EEH Electron Emission Holography

EEH EMU (Extravehicular Mobility Unit) Electrical Harness [of NASA]

EEI Edison Electric Institute [US]

EEI Electron-Electron Interaction [Physics]

EEI Environmental Equipment Institute

EEI Essential Elements of Information

EEIA Electrical and Electronic Insulation Association [UK]

EEIB Environmental Engineering Intersociety Board [US]

EEIC Electrical/Electronics Insulation Conference [US]

EEIC Environmental Education and Information Committee

EEIG European Economic Interest Grouping

EEIS End-to-End Information System

EEJ Equatorial Electrojet [Geophysics]

EEL Electrical Equipment List

EEL Electron Energy Level

EEL Electron Energy Loss

EEL Emitter-to-Emitter Coupled Logic [also E^2L]

EEL Epsilon Extension Language

EEL Exclusive Exchange Line [Telecommunications]

EELFS Extended Energy-Loss Fine Structure (Spectroscopy)

EELS Electron Energy-Loss Spectroscopy

EELS Electron Energy-Loss Spectrum

EELS Engineering Electronic Library, Sweden (Project) [of Royal Institute of Technology Library, Stockholm, Sweden]

EELS-EDX Electron Energy-Loss Spectroscopy/Energy Dispersive X-Ray Analysis

EELV Evolved Expendable Launch Vehicle

EEM Earth Entry Module [of NASA]

EEM Effective Equations of Motion (Model)

EEM Electron Emission Microscope; Electron Emission Microscopy

EEM Electronics Engineer's Master (Catalog)

EEM Electronic Engine Management

EEM Electronic Equipment Monitoring

EEM Emission Electron Microscope; Emission Electron Microscopy

EEM Emission Excitation Matrix

EEM Equine Encephalomyelitis

EEM Excitation Emission Matrix

EEM Excitonic Enhancement Model [Solid-State Physics]

EEM Extended Memory Management

EEMA Electrical and Electronic Manufacturers Association [US]

EEMAC Electrical and Electronic Manufacturers Association of Canada

EEMDA Electrical/Electronics Materials Distributors Association [US]

EEMJEB Electrical and Electronic Manufacturers Joint Education Board

EEMS Enhanced Expanded Memory Specification

EEMS European Environmental Mutagen Society [Netherlands]

EEMTIC Electrical and Electronic Measurement and Test Instruments Conference

EEMUA Engineering Equipment and Materials Users Association [UK]

EEN N-Ethylethylenediamine [also een]

EENR Economic Evaluation of Natural Resources

EENT Eye, Ear, Nose and Throat

EEO Electroendosmosis; Electroendosmotic

EEO Equal Employment Opportunity

EEO Executive Engineering Order

EEO/AA Equal EmploymentOpportunity/Affirmative Action (Employer)

EEO/AA/ADA Equal EmploymentOpportunity/Affirmative Action/Americans with Disabilities Act (Employer)

EEOC Equal Employment Opportunity Commission [US]

EEP East European Program [of International Union for the Conservation of Nature (and Natural Resources)]

EEP Electroencephalophone; Electroenchephalophony

EEP Electron Escape Probability

EEP Energy and Environment Program [of Royal Institute of International Affairs, UK]

EEPA Electromagnetic Energy Policy Alliance [US]

EEPA Environmental Expenditure on Protection and Abatement

EEPC (India) Engineering Export Promotion Council [US]

EEPD Electronic Equipment Production and Distribution

EEPLD Electrically Erasable Programmable Read-Only Memory [also E^{2}PLD]

EEPNL Estimated Effective Perceived Noise Level

EEPROM Electrically Erasable Programmable Read-Only Memory [also E^{2}PROM]

EEPROM Electronically Erasable Programmable Read-Only Memory [also E^{2}PROM]

EEQUANT Epistemological and Experimental Perspectives on Quantum Mechanics (Conference)

EER Early Emissions Reduction

EER Electrolyte Electroreflectance

EER Electron-Electron Repulsion

EER Energy Efficiency Ratio

EER Explosive Echo Ranging

EERA Electrical Equipment Representatives Association [US]

EERC Earthquake Engineering Research Center [of University of California, US]

EERC Ejector Expansion Refrigeration Cycle

EERC Energy and Environmental Research Center [of University of North Dakota, Grand Forks, US]

EERI Earthquake Engineering Research Institute [US]

EERJ External Expansion Ramjet

EERL Earthquake Engineering Research Laboratory [US]

EERL Electrical Engineering Research Laboratory

EERM Etablissement d'Etudes et de Recherches Météorologiques [Meteorological Research and Study Establishment, France]

EE&RM Elementary Electrical and Radio Material Training School

EEROM Electrically Erasable Read Only Memory [also E^{2}ROM]

EES Ejection Escape Suit

EES Electrical Engineering Science

EES Emergency Ejection Suit

EES Engineering Equation Solver

EES Engineering Experiment Station [Annapolis, Maryland, US]

EES Egypt Exploration Society

EES Environment Effects Statement

EES Escape Ejection Suit

EES Escrow Encryption Standard

EES Excited Electronic State

EESB Earth and Environmental Science Building [of Pacific Northwest National Laboratory, Richland, Washington, US]

EESC Erie Engineering Societies Council [US]

EESI Earth Environment Space Initiative [of Canadian Space Agency]

EESMB Electrical and Electronics Standards Management Board [of American National Standards Institute, US]

Eesti NSV Tead Akad Fuus Inst Uurim Eesti NSV Teaduste Akadeemia Fuusiki Instituudi Uurimused [Estonian Journal on Physics of the Academy of Sciences]]

Eesti NSV Tead Akad, Toim, Fuus Ma Eesti NSV Teaduste Akadeemia, Toimetised, Fuusiki, Matemaatika [Estonian Journal on Pysics and Mathematics of the Academy of Sciences]

EEUA Engineering Equipment Users Association [UK]

EET Eastern European Time [Greenwich Mean Time −01:00]

EET Electrical Engineering Technologist; Electrical Engineering Technology

EET Electronic Excitation Transfer

EET Electronic Exhaust Transducer

EET Electronics Engineering Technician

EET Electronics Engineering Technologist; Electronics Engineering Technology

EET Elevated Electron Temperature

EET Entry Elapsed Time

EET Environmental Engineering Technologist; Environmental Engineering Technology

EET Epoxyeicosatrienoic Acid

EET Equipment Engaged Tone

EET Estimated Elapsed Time

EET Excitation Energy Transfer

8,9-EET 8,9 Epoxy (5Z,11Z,14Z) Eicosatrienoic Acid

11,12-EET 11,12 Epoxy (5Z,8Z,14Z) Eicosatrienoic Acid

14,15-EET 14,15 Epoxy (5Z,8Z,11Z) Eicosatrienoic Acid

14(R),15(S)-EET 14(R),15(S) Epoxy (5Z,8Z,11Z) Eicosatrienoic Acid

14(S),15(R)-EET 14(S),15(R) Epoxy (5Z,8Z,11Z) Eicosatrienoic Acid

EETB Electronic-Electrical Termination Building

EETD Environmental Emergencies Technology Division [of Environmental Technology Center, Canada]

EETF Electronic Environmental Test Facility

EETLA Expanded Extended Three-Letter Acronym

EETPU Electrical, Electronic, Telecommunication and Plumbing Union [UK]

EEUA Engineering Equipment Users Association

EEV Effective Exhaust Velocity [Aerospace]

e⁻/eV/cm² electrons per electron volt per square centimeter [also e⁻ eV⁻¹ cm⁻², or e⁻/(eV·cm⁻²)]

EEVL Edinburgh Engineering Virtual Library [UK]

EEVT Electrophoresis Equipment Verification Test

EEW Epoxy per Equivalent Weight

EEWD Enhanced Exchange Wide Dial [Telecommunications]

EEXAPS Electron Excited-X Ray Appearance Potential Spectroscopy

EEZ Exclusive Economic Zone

EF Cold Polar Climate, Perpetual Frost [Symbol]

EF Each Face

EF Earth Fault (Current) [also E/F]

EF Earth First! [US Environmental Journal]

EF Eddy Flow

EF Eddy Flux

EF Edge Filter

EF Effective Fire

EF Effective Force

EF Eigenfunction

EF Elastic Force

EF Electric Field

EF Electric Flux

EF Electric Furnace

EF Electrofining

EF Electrofinishing

EF Electroforming

EF Electromagnetic Force [also EMF, or emf]

EF Electron Focusing

EF Electron Fractography

EF Electronic Filing

EF Electronic Force

EF Electroviscous Fluid

EF Elevation Finder

EF Elliptic Filter

EF Elliptic Function

EF Emitter Follower

EF Empirical Formula [Chemistry]

EF Energy Factor

EF Energy Filter(ed)

EF Energy Flux

EF Engineering Foundation [New York City, US]

EF Enhancement Factor

EF Enriched (Nuclear) Fuel

EF (Nuclear) Enrichment Factor

EF Environmental Forecasting

EF Epitaxial Film

EF Erlenmeyer Flask

EF Error Factor

EF Error-Free

EF Error Function

EF Erythrobastosis Fetalis [Medicine]

EF Eurodata Foundation [UK]

EF Evaluation Finder

EF Everett Formula

EF Exact Function

EF Exchange Force [Quantum Mechanics]

EF Expeditionary Force

EF Exponential Function

EF Extended Facility

EF Extra Fine (Thread)

EF Extremely Fine

EF Perpetual Frost (Ice Cap), All Months below 32°F (0°C) [Classification Symbol]

E/F Earth Fault (Current) [also EF]

E(f) Fiber-Reinforced E-Glass [Symbol]

EFA Effective Field Approximation [Materials Science]

EFA Epilepsy Foundation of America [Landover, Maryland, US]

EFA Erbium-Doped Fiber Amplifier

EFA Essential Fatty Acid

EFA Esterified Fatty Acid

EFA European Fighter Aircraft

EFA Experiment Flight Applications

EFA Extended File Attribute

EFAB Low-Cost, Automated Electrochemical Batch Fabrication of Arbitrary 3-D Microstructures

EFAPI Euro Market Federation of Animal Protein Importers

EFAPP Enrico Fermi Atomic Power Plant

EFAR Economic Feeder Administration and Relief

EFAS Electronic Flash Approach System

EFAS En-Route Flight Advisory Service

EFATCA European Federation of Air Traffic Controllers Associations

EFB Electric Flash Butt Welding

EFB Engineering Foundation Board [US]

EFB Error-Free Block

EFB European Federation of Biotechnology [Germany]

EFB European Foulbrood [Bee Disease]

EFBD Emergency Feed Baron Detector

EFB Newsl EFB Newsletter [of European Federation of Biotechnology, Germany]

EFBPBI European Federation of the Brush and Paint Brush Industries [Belgium]

EFBWW European Federation of Building and Woodworkers [Belgium]

EFC Electric Furnace Conference [of Iron and Steel Society, US]

EFC Electronic Frequency Control

EFC Equivalent Full Charges

EFC European Federation of Corrosion [Germany]

EFC European Forestry Commission

EFC Expect Further Clearance

EFC Extreme Fringe Contrast

EFCA European Federation of Engineering Consultancy Associations

EFCC European Federation of Conference Cities

EFCE European Federation of Chemical Engineering [Germany]

EFCE Newsl EFCE Newsletter [of European Federation of Chemical Engineering, Germany]

EFCGU European Federation of Chemical and General Workers Union [Belgium]

EFCM Erbium Iron Cobalt Molybdenum (Compound)

EFC Newsl EFC Newsletter [of European Federation of Corrosion, Germany]

EFCO 2-Methyl-trans-1,3-Pentadiene Iron Tricarbonyl

EFCS Earth-Fixed Coordinate System

EFCS European Federation of Cytology Societies [Italy]

EFCSM European Federation of Ceramic Sanitaryware Manufacturers [Italy]

EFCT European Federation of Conference Towns

EFCV Excess Flow Check Valve

EFD Electrofluid Dynamic (Generator)

EFD Electron Flux Detector

EFD Engineering Flow Diagram

EFD Energy Flux Density

EFD Epitaxial Film Deposition

EFD Equivalent Full Discharge

EFDA European Federation of Data Processing Associations

EFDAS Epsilon Flight Data Acquisition System

EFE Early Fuel Evaporation System

EFE Electric Field Effect

EFE Electron Field Emission

EFE Ethylene-Forming Enzyme

EFED Energy-Filtered Electron Diffraction

EFEM Energy-Filtering Electron Microscopy

EFEO Ecole Française d'Extrême Orient [French Far East School]

EFEO European Flight Engineers Organization

EFET Enhancement Mode Field-Effect Transistor

η-FeZn Eta Iron Zinc

EFF Electronic Frontier Foundation [Internet]

EFF European Furniture Federation

EFF Exoplasmic Fracture Face

EFF Extended Fringing Field

Eff Effect [also eff]

Eff Effective(ness) [also eff]

Eff Efficiency [also eff]

EFFBR Enrico Fermi Fast Breeder Reactor [US]

EFFBCG Entrained Flow/Fixed Bed Composite Gasification

Effcy Efficiency [also effcy]

EFFD Elektronik Fabricant Foreningen Danmark [Danish Association of Electronic Products Manufacturers]

EFFE European Federation of Flight Engineers

EFFGRO Efficient Growth (Computer Program)

Effl Efflorescence [also Efflor, efflor, or effl]

Effl Effluent [also effl]

Efflor Efflorescence; Efflorescent [also efflor]

Effl Water Treat J Effluent and Water Treatment Journal [UK]

EFFOST European Federation of Food Science and Technology [UK]

Effy Efficiency [also effy]

EFG Edge-Defined Film-Fed Growth (Technique)

EFG Electric Field Gradient [also efg]

EFG Elemental and Functional Group (Analysis)

EFG Engineering Flow Diagram

EFH Extended Fenske-Hall (Linear Combination of Atomic Orbitals Method) [Physical Chemistry]

EFI Educational Futures, Inc. [US]

EFI Electromechanical Frequency Interference

EFI Electronic Flight Instrument

EFI Electronic Forum for Industry [UK]

EFI Electronic Fuel Injection

EFI Electronics For Imaging

EFI Engineered, Furnished and Installed

EFI Enrico Fermi Institute (of Nuclear Studies) [of University of Chicago, Illinois, US]

EFI Environic Foundation International [US]

EFI Error-Free Interval

EFIA European Fertilizer Importers Association [Belgium]

EFID Electric Field In Process Dressing

EFIGS English/French/Italian/German/Spanish

EFIR European Flight Information Region

EFIS Electronic Flight Instrument System

EFISH Electric-Field-Induced Second Harmonic Generation

EFISP Enrico Fermi International School of Physics [Varenna, Italy]

EFJ Evaporated Film Junction

EFL Effective Focal Length

EFL Egptian Federation of Labour

EFL Emitter Follower Logic

EFL Emitter Function Logic

EFL English as a First Language

EFL English as a Foreign Language

EFL Equivalent Focal Length

EFL Error Frequency Limit [Computers]

EFLA Education Film Library Association [US]

EFLT Effective Frequency Low Temperature (Propagator) [Physics]

EFM Eight-to-Fourteen Modulation

EFM Elastic Fracture Mechanics

EFM Electric Field Monitor

EFM Electrostatic Force Microscopy

EFM Engineering Field Manual

EFM Extended Flygare Method

EFMA European Fertilizer Manufacturers Association [Belgium]

EFMA European Fittings Manufacturers Association [UK]

EFMC Elastic Fabric Manufacturers Council [of Northern Textile Association, US]

EFMC European Federation of Medical Chemistry [UK]

EFMCNTA Elastic Fabric Manufacturers Council of the Northern Textile Association [US]

EFMD European Foundation for Management Development [Belgium]

EFMS Experimental Flight Management System

EFNEA European Federation of National Engineering Associations [France]

EFNMR Electric Field on Nuclear Magnetic Resonance (Spectra) [also EF-NMR]

EFNS Educational Foundation for Nuclear Science [US]

EFO Engineers Foundation of Ohio [US]

EFOA European Fuel Oxygenates Association [of Conseil Européen des Fédérations de l'Industrie Chimique, Belgium]

EFOMP European Federation of Organizations in Medical Physics

EFOR Equivalent Forced Outage Rate

EFORD Experiment Functional Objectives Requirements Document

E-Form Electronic Form [also e-form]

EFP Electric Field Potential

EFP Electrofluid Dynamic Process

EFP Electronic Field Production

EFP Emission Flame Photometry

EFP Enhanced Flux Pinning

EFP Enrico Fermi Power Plant [Italy]

EFP Environmental Finance Program [of US Environmental Protection Agency]

EFP ESA (European Space Agency) Furnished Property

EFP European Federation of Parasitologists [Sweden]

EFP European Federation of Purchasing

EFP Eviscerated Fish Protein

EFP Expanded Function (Operator) Panel

EFP Explosively Formed Penetrator [Military]

EFP Explosively Formed Projectile

EFPD Equivalent Full Power Days

EFPG Ecole Française de Papeterie de Grenoble [French National School of Paper Processing of Grenoble; of Institut National Polytechnique de Grenoble]

EFPG/INPG Ecole Française de Papeterie de Grenoble/ Institut National Polytechnique de Grenoble [French National School of Paper Processing of Grenoble/ National Polytechnic Institute of Grenoble]

EFPH Equivalent Full Power Hours

EFPI European Federation of the Plywood Industry [France]

EFPIA European Federation of Pharmaceutical Industries Associations [Belgium]

EFPM Effective Full Power Month

EFPS End-Functionalized Polystyrene

EFPS European Federation for Productivity Services [Sweden]

EFPW European Federation for the Protection of Waters

EFR Environmental Flow Requirements (of Australia's Waterways)

EFR Emerging Flux Regions

EFR Eppley Foundation for Research [New York, US]

EFR Error-Free Region

EFR Experimental Free Radical

EFRA Electronic Forms, Routing and Authorization

EFRAP Exchange Feeder Route Analysis Program

EFS Electric-Furnace Steel

EFS Electrofinishing Standard

EFS Electronic Frequency Selection

EFS Error-Free Seconds

EFSAED Energy-Filtered Selected-Area Electron Diffraction

EFSIM Explosive Foam Shock Initiation Model

EFS-SPEC Electrofinishing Standard Specification [also EFS-Spec]

EFT Earliest Finish Time

EFT Electronic Funds Transfer

EFT Eno Foundation for Transportation [US]

EFTA Electronic Funds Transfer Act [US]

EFTA Electronic Funds Transfer Association [US]

EFTA European Free Trade Association

EFTAPA European Free Trade Association–Plastics Association

EFTC European Freight Timetable Conference

EFTEC European Fluorocarbon Technical Committee [Belgium]

EFTEM Energy-Filtered (or Energy Filtering) Transmission Electron Microscope; Energy Filtered (or Filtering) Transmission Electron Microscopy

EFTI Engineering Flight Test Inspector

EFTO Encrypted for Transmission Only

EFTP Error File Teaching Package

EFTPOS Electronic Funds Transfer at Point of Sale [also EFTPoS, or eftpos]

EFTR Ethernet Frequency Translator

EFTS Electronic Funds Transfer System

EFTS Elementary Flying Training School

EFTSU Eqivalent Full Time Student Unit

EFuRD Europäischer Funkrufdienst [European Radio-Paging Service]

EFVA Education Foundation for Visual Aids [UK]

EFW Energy from Waste Program [Canada]

EFW/TS Energy-from-Waste Transfer Station

EG Eccentric Gear

EG Echelette Grating [Spectroscopy]

EG Ecological Genetics

EG Economic Geographer; Economic Geography

EG Edge Grain [Lumber]

EG Egypt [ISO Code]

EG Electrical Grade (Glass Fiber)

EG Electric Generator

EG Electrodynamic Generator

EG Electrogalvanized; Electrogalvanizing

EG Electrogas

EG Electrographic; Electrography

EG Electron Gas

EG Electron Gun

EG Enamelled Single-Glass-Covered (Wire)

EG Encephalogram; Encephalography

EG End Group

EG Endocrine Gland(s)

EG Energy Gap [Solid-State Physics]

EG Engineering Geologist; Engineering Geology

EG Entry Guidance [Aerospace]

EG Environmental Geologist; Environmental Geology

EG Epitaxial Growth [Solid-State Physics]

EG Equatorial Guinea

EG Equipment Grant

EG Equipment Ground

EG Ergodic Theory

EG Ethylene Glycol

EG Euclidean Geometry

EG Exhaust Gas

EG Exploration Geochemist(ry)

EG Exploration Geophysicist; Exploration Geophysics

E&G Elliott and Glenister (Solidification Theory) [Metallurgy]

E-G Electrolyte-Glass (Interface)

Eg Egypt(ian)

e.g. *(exempli gratia)* – for example

.eg Egypt [Country Code/Domain Name]

EGA Effluent Gas Analysis

EGA Electrographic Analysis

EGA Electrogravimetric Analysis

EGA Emergency Gas Act [US]

EGA Energy Gap Anisotropy [Solid-State Physics]

EGA Enhanced Graphics Adapter

EGA Ethylglycolacetate

EGA Evolved Gas Analysis; Evolved Gas Analyzer

EGA Extended Graphics Array

EGA Extragalactic Astronomy

EGA/CGA Enhanced Graphics Adapter/Color Graphics Adapter

EGADS Electronic Guidance and Documentation System (Project) [of Alcoa Corporation, US]

EGAHi Enhanced Graphics Adapter, High (Mode)

EGALo Enhanced Graphics Adapter, Low (Mode)

EGA/MDA Enhanced GraphicsAdapter/Monochrome Display Adapter

EGA MonoHi Enhanced Graphics Adapter, Monochrome, High Mode

EGAMS Evolved Gas Analysis using Mass Spectrometry

EGAS European Group for Atomic Spectroscopy [Netherlands]

EGASF European General Aviation Safety Foundation

EGATS EUROCONTROL Guild of Air Traffic Controllers

EGA/VGA Enhanced Graphics Adapter/Visual Graphics Adapter

EGB Equilibrium Grain Boundary [Materials Science]

EGBD Extrinsic Grain Boundary Dislocation [Materials Science]

EGBE Ethylene Glycol Butyl Ether

EGC East Gippsland Coalition [Australia]

EGC Environmental Geochemist(ry)

EGC Experiments Ground Computer

EGC Exploration Geochemistry

EGCI Export Group for the Constructional Industries [UK]

EGCMA European Gas Control Manufacturers Association

EGCR Experimental Gas-Cooled Reactor [of Oak Ridge National Laboratory, Tennessee, US]

E/GCR Extended Group Coded Recording

EGD Electrogasdynamic(s)

EGD Environmental Graphic Designer

EGD Evolved Gas Detection

EGDN Ethylene Glycol Dinitrate

EGDS Extragalactic Distance Scale

EGE Enhanced Greenhouse Effect

E Ger East Germany

EGF Electrical, Grapple Fixture

EGF Electrodynamic Gradient Freeze

EGF Epidermal Growth Factor [Biochemistry]

EGF European Grassland Federation [Netherlands]

EGFR Epidermal Growth Factor Receptor [also EGF-R, or EGF-r] [Biochemistry]

EGG Electrogastrogram; Electrogastrography

EGG Epitaxial Grain Growth [Materials Science]

EGG Equiaxed Grain Growth [Materials Science]

EG&G Edgerton, Germeshausen and Grier [Physics]

EG&G EG&G Idaho, Inc. [Idaho Falls, US]

EGGA European General Galvanizers Association [UK]

EG/H₂O Ethylene Glycol/Water (Mixture)

EGI Electronic Gas(oline) Injection

EGI End-of-Group Indicator

EGI Eucalyptol–Guaiacol–Iodine

EGIF Equipment Group Interface

EGIL Electrical, General Instrumentation and Lighting

EGIPT Electrically-Generated Intramolecular Proton Transfer

EGIS European Geographical Information Systems (Symposia)

EGiS Esrange Geophysical information Services [Europe]

EGK Geological Survey of Estonia

EGL Electrogalvanizing Line

EGL Exponential Gap Law

E-Glass Electrical-Grade Fiberglass [also E-glass]

EGM Electrogel Machining

EGM Electronic Governor Module

EGM European Glass Container Manufacturers [UK]

EGM Extraordinary General Meeting

EGMA Emergency Gold Mining Assistance Act

EGMA Ethylene Glycidyl Methacrylate (Copolymer)

EGMBE Ethylene Glycol Monobutylether

EGME Ethylene Glycol Monomethylether [also EGMME]

EGMEE Ethylene Glycol Monoethylether

EGMME Ethylene Glycol Monomethylether [also EGME]

EGO Eccentric Geophysical Observatory

EGOCO European Group of Oil Companies

EGOLF European Group of Official Laboratories for Fire Testing

EGOS European Group for Organizational Studies [Netherlands]

EGP Egyptian Pound [Currency]

EGP Electrogasdynamic Generation of Power

EGP Environmental Genome Project [of National Institute of Environmental Health Sciences, US]

EGP Exploration Geophysicist; Exploration Geophysics

EGP Exterior Gateway Protocol [Internet]

EG£ Egyptian Pound [Currency]

EGPC Egypt General Petroleum Corporation

EGPS Extended General-Purpose Simulator

EGR Electronic Governor Regulator

EGR Exhaust Gas Recirculation

EGREP Extended Global Regular Expression Print

EGRESS Emergency Global Rescue, Escape and Survival System [of NASA]

EGRESS Evaluation of Glide Reentry Structural Systems

EGRET Energetic Gamma-Ray Explorer Telescope

Egrs Egress [also egrs] [Astronomy]

EGRS Extragalactic Radio Source

EGRSA Edible Gelatin Research Society of America [US]

EGS Environmental and Geographical Science

EGS Echelette Grating Spectrometer

EGS Electronic-Grade Silicon

EGS Ethylene Glycol-bis(Succinic Acid)

EGS European Geological Surveys

EGS European Geophysical Society [Germany]

EGSA Electrical Generating Systems Association [US]

EGSC Eastern Group Supply Council [UK]

EGSCCP European Graduate Summer Course on Computational Physics

EGSE Electrical Ground Support Equipment

EG-SEA-AI European Group for Structural Engineering Applications of Artificial Intelligence

EGSMA Electrical Generating Systems Marketing Association [now Electrical Generating Systems Association, US]

EGS-NHS Ethylene Glycol-bis(Succinic Acid-N-Hydroxysuccinimide Ester)

EGT Elapsed Ground Time

EGT Estimated Ground Time

EGT Ethylene Glycol-bis(trichloroacetate)

EGT Exhaust Gas Temperature

EGTA Ethylene Glycol Tetraacetic Acid

EGTA Ethylene-bis(oxyethylenenitrilo)tetraacetic Acid

EGTA Ethylene Glycol-bis(β-Aminoethylether)-N,N,N',N'-Tetraacetic Acid

EGTA European Group of Television Advertising [UK]

EGTS Emergency Gas Treatment System

E-Gun Electron Gun [also E-gun, or e-gun]

EGW Electrogas Weld(ing)

Egypt Egypt(ian) [also Egy]

Egypt Comput J Egyptian Computer Journal [Published by Cairo University, Institute of Statistical Studies and Research]

Egypt J Biomed Eng Egyptian Journal of Biomedical Engineering [of National Information and Documentation Center, Cairo]

Egypt J Solids Egyptian Journal of Solids [of Egyptian Society of Solid-State Science and Applications, Cairo]

Egyptol Egyptological; Egyptologist; Egyptology

EH Effective Height [Antennas]

EH Elastic Hysteresis

EH Electric Heat(ing)

EH Electrohydraulic(s)

EH Electron Hole

EH Engineering Hydrology

EH Entrance Hatch

EH Entrance Head

EH Environmental Health

EH Ethylene Hexylene (Copolymer)

EH Extended Hueckel (Linear Combination of Atomic Orbitals Method) [Physical Chemistry]

EH Extra-Hard (Temper) [Metallurgy]

EH Extra-High

EH Western Sahara [ISO Code]

$E_{\perp}H$ Electric Field Perpendicular (or Normal) to Magnetic Field [Symbol]

E-H E-Plane/H-Plane (T-Junction) [Waveguides]

E/H Young's Modulus-to-Hardness (Ratio) (or Elastic Modulus-to-Hardness (Ratio))

Eh Oxidation Reduction Potential (or Redox Potential) [Symbol]

.eh Western Sahara [Country Code/Domain Name]

e-h electron-(electron) hole (interaction) [Symbol]

e^2/h quantized Hall conductance (3.874046×10^{-5} s) [Symbol]

8-H Eight-Harness (Weave) [Woven Fabrics]

EHA Economic History Association [US]

EHA Electric Heating Association [US]

EHA Ethyl Hexanol

EHA Ethyl Hexyl Acetate

EHA Ethyl Hexyl Alcohol

EHA European Helicopter Association

EHAA Eagle's Ham's Amino Acids

EHAP Employee Health Assistance Program

EHAP Extremely Hazardous Air Pollutants

EHC Electrical Heating Control

EHC Electrochemical Hydrogen Cracking [Chemical Engineering]

EHC Electrohydraulic Crushing

EHC Electrohydraulic Control

EHC Electrohydrodynamic Convection

EHC Environmental Hazard Communication

EHC Environmental Health Center [of National Safety Council, US]

EHC Ethylenic Hydrocarbon

EHCAC Egyptian High Commission of Automatic Control

EHCO Extended Hueckel Crystal Orbital

EHD Elastohydrodynamic(s)

EHD Electrohydrodimerization (Process) [Chemical Engineering]

EHD Electrohydrodynamic(s)

EHD Electron Hole Drop

EHDA Electrohydrodynamic Atomization

EHDIMS Electrohydrodynamic Ionization Mass Spectroscopy

EHDPM Bis(di-2-Ethylhexylphosphinyl)methane

EHDTV Enhanced High-Definition Television

EHE Ecole des Hautes Etudes [School of Higher Studies, University of Paris, France]

EHE Effective Height of Emission

EHE External Heat Exchanger

EHE Extraordinary Hall Effect

EHEC Ethylhydroxyethyl Cellulose

E-HEMT Enhancement Mode High-Electron-Mobility Transistor

EHEP Experimental High-Energy Physics

EHAS East Hertfordshire Archeological Society [UK]

EHF Electrohydraulic Forming

EHF Engineers Hall of Fame

EHF Extended Hartree Fock [Quantum Mechanics]

EHF Extra-High Frequency

EHF Extremely High Frequency [Frequency: 30 to 300 gigahertz; wavelength: 1 mm to 1 centimeter] [also ehf]

EHFA Electric Home and Farm Authority [US]

EHFA Emergency Home Financing Act [US]

EHFB Electrical Historical Foundation Board

EHF-SATCOM Extremely-High-Frequency Satellite Communication [also EHF-Satcom]

EHHPP 2-Ethylhexyl Hydrogen Phenylphosphonate [also 2EHHPP]

EHI Electron Hole Interaction

EHIS Emission History Information System [of US Environmental Protection Agency]

EHL Electron Hole Liquid

EHLLAPI Emulator High-Level Language Application Programming Interface

EHM Engine Health Monitoring

EHM Extended Hueckel Method [Physical Chemistry]

EH/M Extension Hose/Mouthpiece

EHMO Extended Hueckel Molecular Orbital

EHMS Electrohydrodynamic Mass Spectrometry

EHO (Verein) Eisenhütte Oesterreich [Austrian Iron and Steel Society]

EHO Environmental Health Officer

EHÖ Verein Eisenhütte Österreich [Austrian Iron and Steel Society]

EHOE Effective Height of Emission [Antennas]

EHOG European Host Operators Group

(EHO)2(Φ)PO Bis(2-Ethylhexyl)phenylphosphonate

EHO(Φ)PO(OH) 2-Ethylhexyl Hydrogen Phenylphosphonate

EHOT External Hydrogen/Oxygen Tank

EHP Effective Horsepower [also ehp]

EHP Electric(al) Horsepower

EHP Electron Hole Pair

EHP Electron Hole Plasma

EHP Electron Hole Potential (Method)

EHP Environmental Health Perspectives [Journal of the National Institute of Environmental Health Sciences, US]

EHP Equivalent Habit Plane [Crystallography]

EHP Equivalent Horsepower

EHP (2 Ethylhexyl)phosphate

EHP Evaporated Hall Plate

EHPA Di(2 Ethylhexyl)phosphoric Acid

2-EH(Φ)PA 2-Ethylhexylphenyl Phosphoric Acid

EHPG Ethylenediamine-N,N'-bis(2 Hydroxyphenylacetic acid)dimethyl Ester

EHΦP (2 Ethylhexyl)phenylphosphate

EHPPA 2-Ethylhexyl Phenylphosphonate; 2 Ethylhexyl Phenylphosphonic Acid [also 2EHPPA]

EHPRG European High-Pressure Research Group [France]

EHR Earned Hour Ratio

EHR (Directorate for) Education and Human Resources [of National Science Foundation, US]

EHR Electron Hole Recombination

EHS Englebreth-Holm-Swarm (Mouse Sarcoma)

EHS Enhanced HOSC (Huntsville Operations Support Center) System [of NASA]

EHS Environmental Health and/or Safety [also EH&S]

EHS Environmental, Health and/or Safety

EHS Environmental Health Section

EHS Environmental Health Service [now US Environmental Protection Agency]

EHS Environmental Health Standard(s)

EHS European Hybrid Spectrometer (Experiment) [of CERN– European Laboratory for Particle Physics, Geneva, Switzerland]

EHS Extra High Strength

EHS Extremely Hazardous Substances

EH&S Environmental Health and Safety [also EHS]

EHSI Electronic Horizontal Situation Indicator [Aeronautics]

EHSP Environmental Health Sciences Program [School of Public Health, University of Massachusetts, Amherst, US]

EHT Extended Hueckel Theory [Physical Chemistry]

EHT Extra High Tension [also eht]

EHT Extremely-High-Tension (Rectifier)

E/H/T E-Plane/H-Plane T-Junction [Waveguides]

EHTPS Extra-High-Tension Power Supply

EHV Electric Hybrid Vehicle

EHV Equine Herpesvirus

EHV Erythema–Hypersensitivity–Vasculitis [Medicine]

EHV Extra High Voltage; Extremely High Voltage [also ehv]

EHV 1 Equine Herpesvirus 1

EHW Extremely High Water

EHX Experiment-Dedicated Heat Exchanger

EI Earth Inductor

EI Easter Island [Chile]

EI East Indies

EI Elastically Isotropic (Material)

EI Elastic Interaction (Model)

EI Electric(al) Instrument

EI Electrical Insulation

EI Electrically Induced; Electric Induction

EI Electrogist International [now National Electrical Contractors Association, US]

EI Electromagnetic Interference

EI Electronic Interface

EI Electronic Imaging

EI Electronic Instrument(ation)

EI Electronic Interference

EI Electronics Inspector

EI Electron Image

EI Electron Impact

EI Electron-Impact Ions

EI Electron Interferometer; Electron Interferometry

EI Electron Ionization

EI Ellis Island [New York City, US]

EI Employee Involvement

EI Employment Insurance

EI End Injection

EI End Item

EI End of Injection [Aerospace]

EI Engineering Index

EI Engineering Information (Database) [US]

EI Engineering Instruction

EI Engine Instrument

EI Entrainment Instability (Criterion)

EI Entry Interface

EI Environmental Impact

EI Environment Institute [at Ispra, Italy]

EI Enzyme Inhibitor (Complex)

EI Error Indicator

EI Escape Initiator [Aeronautics]

EI Ethyleneimine

EI Exchange Interaction

EI Experiments Incorporated [US]

EI Exposure Index

EI Interrupt Enabled (Flag) [Computers]

EI Lower Deviation (of Internal Thread) [ISO Symbol]

EI Ireland [Civil Aircraft Marking]

E/I Voltage to Current (Ratio)

ei lower deviation (of external thread) [ISO Symbol]

EIA Electrical Industries of America

EIA Electronic Industries Association [US]

EIA End Item Assembly

EIA Energetic Ion Analysis

EIA Energy Information Administration [of US Department of Energy]

EIA Engineering Industries Association [UK]

EIA Environmental Impact Analysis

EIA Environmental Impact Appraisal

EIA Environmental Impact Assessment

EIA Environmental Information Association [US]

EIA Environment Institute of Australia

EIA Enzyme Immunoassay

EIA Equine Infectious Anemia

EIA Equipment Interchange Association [US]

EIA Exercise-Induced Asthma

EIAA Electronic Industry Association of Alberta [Canada]

EIAC Electronic Industries Association of Canada

EIAC Ergonomics Information Analysis Center [UK]

EIAES Electron-Induced Auger-Electron Spectroscopy

EIAG Exeter Industrial Archeology Group [UK]

EIAJ Electronic Industries Association of Japan

EIAM Electronics Industry Association of Manitoba [Canada]

EIAP Environmental Impact Assessment Project

EIASM European Institute for Advanced Studies in Management

EIASN End-Item Assembly Sequence Number

EIA/TIA Electronic Industries Association/Telecommunications Industry Association

EIAV Equine Infectious Anemia Virus

EIB Electronics Installation Bulletin

EIB European Investment Bank [Luxembourg]

EIB Export-Import Bank

EIBA Ethylene Isobutyl Acrylate

EIBA European International Business Association [Belgium]

EIC Electret Ionization Chamber

EIC Electrical Insulation Conference [now Electrical/Electronics Insulation Conference, US]

EIC Electrostatic Interaction Chromatography

EIC Employment and Immigration Canada

EIC Energy Industries Council

EIC Engineer in Charge

EIC Engineering Institute of Canada

EIC Environmental Improvement Commission [Maine, US]

EIC Environmental Industries Council [US]

EIC Environment Information Center

EIC Equal Interface Composition (Contour) [Metallurgy]

EIC Equipment Identification Code

EIC Experimental Intercom

EICA East India Cotton Association

EICAS Engine Indication and Crew Alerting System [Aerospace]

EI/CI Electron Ionization/Chemical Ionization

EICP Extracted-Ion Current Profile [Mass Spectrometry]

EICR Eastern Interior Coal Region

EICV Engine Idling Control Valve

EID Electrode Immersion Depth

EID Electroimmunodiffusion [Biocemistry]

EID Electronic Instrument Digest [US]

EID Electron Impact Desorption

EID Electron-Induced Desorption

EID Electron-Stimulated Ion Desorption

EID End-Item Documentation

EID Entropy of Interacting Dimers [Chemistry]

EIDAP Emitter-Isolated Difference Amplifier Paralleling [also Eidap]

EIDE Enhanced Integrated Drive Electronics

EIDI Environmental Industries Development Initiative [of Manitoba Department of Industry, Trade and Tourism, Canada]

EIDLT Emergency Identification Light

EIDP Electronic Information Distribution Project

EIDP End-Item Data Package

EIDS Electronic Information Delivery System

EIED Electrically Initiated Explosive Device

EIEMA Electrical Installation Equipment Manufacurers Association [UK]

EIES Electronic Information Exchange System [US]

EIES Electron Impact Emission Spectroscopy

EIF Electric Induction Furnace

EIF Exhibition Industry Federation [UK]

EIFA Element Interface Functional Analysis

EIFAC European Inland Fisheries Advisory Commission [of Food and Agricultural Organization]

EIFG Environment Industries Focus Group

EIFI Electrical Industries Federation of Ireland

EIGA Engineering Industry Group Apprenticeship [UK]

EIH Error Interrupt Handler [Computers]

EIHSW European Institute of Hunting and Sporting Weapons [Belgium]

EII Electron-Ion Interaction

EIII (Association of the) European Independent Information Industry

EIIP Emission Inventory Improvement Program [US]

EIIS Energy Industry Information System

EIJC Engineering Institutions Joint Council [UK]

EIJHE East Indian Jute and Hessian Exchange

EIL Electron Injection Laser

EIL Environmental Impairment Liability

EILR Educational Institute for Learning and Research [US]

EIM European Institute for the Media [UK]

EIM Excitability Inducing Material

EIMA Electrical Insulating Materials Association

EIMA Exterior Insulation Manufacturers Association [US]

EIM&M Environmental Instrumentation Measurement and Monitoring

EIMO Electronic Interface Management Office [US Navy]

EI-Model Elastic Interaction Model [also EI-model]

EIMS Electron Impact Mass Spectrometry

EIMU Environmental Information Management Unit

EIN Education Information Network

EIN Employer Identification Number [US]

EIN European Informatics Network

E Ind East Indies

EINECS European Inventory of Existing Commercial Chemical Substances

E in EE Engineer in Electrical Engineering

EINIS European Integrated Network of Image and Services [France]

E in ME Engineer in Mechanical Engineering

EINS European Institute for Nuclear Research

EIO Extended Interaction Oscillator

EIP Electronic Implementation Procedure

EIP Electronic Information Processing

EIP Employment Improvement Program [Canada]

EIP Emulator Interface Program

EIP Environmental Innovation Program [of Environment Canada]

EIP Equipment Installation Procedure

EIP Export Industrial Park

EIPA 5-(N-Ethyl-N-Isopropyl) Amiloride

EIPC European Institute of Printed Circuits [Switzerland]

EIPM Electronic Information Processing Machine

EIR Eidgenössisches Institut für Reaktorforschung [Federal Institute for (Nuclear) Reactor Research, Switzerland] [Now Part of Paul Scherrer Institute]

EIR Effective Ionic Radius [Physical Chemistry]

EIR Engineering Information Report

EIR Engineering Investigation Request

EIR Environmental Impact Report

EIR Equipment Identification Register [Telecommunications]

EIR Extreme Infrared

EIRB European Investment Research Bureau

EIRD Engineering Information Report Date

EIRFT Extended Inversion Recovery Fourier Transform

EIRMA European Industrial Research Management Association [France]

EIRP Effective Isotropic(ally) Radiated Power

EIRP Electroluminescent Image Retaining Panel

EIRP Equivalent Isotropically Radiated Power

EIS Earth Information System [US]

EIS Economic Information System

EIS Educational Institute of Scotland

EIS Effluent Inventory System

EIS Electrical Induction Steel

EIS Electrical Integration System

EIS Electrochemical Impedance Spectroscopy

EIS Electrode Impedance Spectrometry

EIS Electromagnetic Intelligence System [US Air Force]

EIS Electronic Information Service

EIS Electron Impact Spectroscopy

EIS End Interruption Sequence

EIS End-Item Specification

EIS Engineering Index Service

EIS Environmental Impact Statement

EIS Environmental Information System

EIS Electrochemical Impedance Spectroscopy

EIS Energy and Industry Subgroup [of Intergovernmental Panel on Climate Change–Response Strategies Working Group]

EIS Environmental Impact Statement

EIS Environmental Information System

EIS Epidemiologic Intelligence Service [of Centers of Disease Control and Prevention, Atlanta, Georgia, US]

EIS Executive Information System

EIS Expanded Inband Signalling

EIS Extended Instruction Set [Computers]

EISA European Independent Steelworks Association [Belgium]

EISA Extended Industry Standard Architecture

EISA/ISA Extended Industry Standard Architecture/Industry Standard Architecture

EISA/MCA Extended Industry Standard Architecture/Microchannel Architecture

Eisenbahntech Rdsch Eisenbahntechnische Rundschau [German Magazine on Railroad Engineering]

EISI Elemental and Interplanar Spacing Index [Crystallography]

EISN Environmental Information and Support Network [of Australian National University]

EISO Environmental Information Society of Ontario [Canada]

E-ISO ISO (International Organization for Standardization) Name, English Spelling

EISU Eastern Illinois State University [Charleston, US]

EIT Economic Innovation and Technology (Fund) [of Economic Innovation and Technology Council, Manitoba, Canada]

EIT Effective Interfacial Torque [Materials Science]

EIT Electronics Installation Technician

EIT Energy and Information Technology

EIT Engineer-in-Training

EIT Entry Interface Time

EIT European Institute of Technology

EIT Extended Irreversible Thermodynamics

EITB Engineering Industry Training Board [UK]

EITC Economic Innovation and Technology Council [Manitoba, Canada]

EITC Eosin-5-Isothiocyanate

EITC Dextran Eosin-5-Isothiocyanate Dextran

EITE European Institute of Transuranium Elements [Germany]

EIU Eastern Illinois University [Charleston, US]

EIU Economic Intelligence Unit [UK]

EIU Economist Intelligence Unit [US]

EIU Engine Interface Unit

EIU Equipment Inventory Update

EIVT Electrical and Instrumentation Verification Test

EIVT Electrical Interface Verification Test

EIVT Electronic Installation Verification Test

EIVT European Institute for Vocational Training

EIW European Institute for Water [France]

EIWG EM (Environmental Restoration and Waste Management) International Working Group

EIX Electrochemical Ion Exchange

EJ Electronic Jamming

EJ Electronic Journalism

EJ Electronic Junction

EJ Emitter Junction

EJ- Ireland [Civil Aircraft Marking]

E-J Voltage–Current Density (Characteristics)

EJB European Journal of Biochemistry [of Federation of European Biochemical Societies]

EJC Engineers Joint Council [US]

EJC European Joint Committee

EJCC Eastern Joint Computer Conference [US]

EJCNC Engineers Joint Council Nuclear Congress [US]

EJCT Engineering Joint Council Thesaurus [US]

EJD European Journal of Dermatology [France]

Eject Ejector

EJF Estimated Junction Frequency

E-JFET Enhancement Mode Junction Field-Effect Transistor [also E-JFET]

EJM Electrolytic Jet Machining

EJMA Expansion Joint Manufacturers Association [US]

Ejn Ejection [also ejn]

e-journal electronic journal

EJP European Journal of Physics [of European Physical Society]

EJR Emitted Josephson Radiation [Solid-State Physics]

EJTF Electronic Journals Task Force [of University of California, US]

EK Electrokinetic(s)

EK Esaka and Kurz (Microsegregation Model) [Metallurgy]

EKAB Europäisches Keram Adreßbuch (Fachadreßbuch der Europäischen Keramikindustrie) [Directory of the European Ceramic Industry]

EKE Electrokinetic Effect

EKE Equatorial Kerr Effect

EKF Educational Kinesiology Foundation [Ventura, California, US]

EKF Electrokinetic Force

EKG Effective Kilogram [also ekg]

EKG Electrocardiogram; Electrocardiograph(y)

EKG Electrokymogram; Electrokymograph(y)

EKIAP East Kimberley Impact Assessment Program [Australia]

EKN Ecology of Knowledge Network [US]

Ekon Mat Obz Ekonomicko Matematicky Obzor [Publication on Mathematical Economics; published by the Czech Academy of Sciences]

EKP Electrokinetic Potential (or Zeta Potential)

E(k,ϕ) Elliptic Integral of the Second Kind [Symbol]

EKS Electrocardiogram Simulator

EKSC Eastern Kentucky State College [now Eastern Kentucky University]

Eksploat Masz Eksploataeja Maszyn [Polish Journal on the Utilization of Machinery]

EKT Electrokinetic Transducer

EKU Eastern Kentucky University [Richmond, US]

EKV (Gesellschaft) Entwicklung-Konstruktion-Vertrieb [Development, Design and Distribution Society of Verein Deutscher Ingenieure, Germany]

EKW Electrical Kilowatts

EKY Electrokymogram; Electrokymography

EL Eastern Laboratory [Gibbstown, New Jersey, US]

EL Eccentric Load

EL Eddy Loss

EL Education Level
EL Elastic Limit
EL Elastic Liquid
EL Electric Lamp
EL Electric Line
EL Electric Locomotive
EL Electroluminescence; Electroluminescent
EL Electromagnetic Levitation
EL Electron-Beam Lithograph(y)
EL Electronics Laboratory
EL Electronics Letters [Publication of the Institution of Electrical Engineers, UK]
EL Electron Lens
EL Electron Level
EL Electrostatic Lens
EL Electrotechnical Laboratory
EL Empirical Law
EL Emulsifiable Liquid
EL Endurance Limit
EL Energy Level
EL Energy Loss
EL Engineering Letter
EL Entry Level
EL Environmental Law
EL Epitaxial Layer
EL Eppley Laboratories [US]
EL Erase Line
EL Eucken's Law
EL Exchange Line [Telecommunications]
EL Excimer Laser
EL Excitation Limiter
EL Executive Level
EL Explosion Limit; Explosive Limit
EL Exposure Limit
EL Extended Lateral Cracks [also known as Extended Laterals]
EL Extension Line
EL Extra-Low
EL- Liberia [Civil Aircraft Marking]
%EL Percent Elongation [also %El]
E/L Entry/Landing
El Elasticity [also el]
El Elevate; Elevation; Elevator [also el]
El Elongation [also el]
El_{max} Maximum Elongation [Symbol]
El_t Total Elongation [Symbol]
El_u Uniform Elongation (or Uniform Strain) [Symbol]
E(l) Mean Value of Length "l" [Symbol]
$E_c(l)$ Lower Bound Elastic Modulus of Composite [Symbol]
$E(\lambda)$ Spectral Irradiance [Symbol]
$E_{ph}(\lambda)$ Photon Energy [Symbol]
$\varepsilon(\lambda)$ total emittance (function) [also $\epsilon(\lambda)$]
ELA Elastomer Lubricating Agent
ELA Electroacoustic(s)

ELA European Laser Association [Netherlands]
ELA Experimental Lakes Area (Project) [Near Kenora, Ontario, Canada; Administered by Department of Fisheries and Oceans]
ELAC All-Electric Aircraft Flight Control Actuation
ELACS Extended-Life Attitude Control System
ELADA Electronic Atlas of Agenda Twenty One [Canada]
ELADA-21 Electronic Atlas of Agenda Twenty One [of Canada Center for Remote Sensing and the (Canadian) International Development Research Center]
ELAF Electron Linear Acceleration Facility
ELAF Escuela Latinoamericana de Fisica [Latin American School of Physics, of Universi dad Nacional Autonómous de Mexico, Mexico City]
ELAH Earle's Lactalbumin Hydrolysate [Biochemistry]
Elam Ethanolamine [also elam]
ELAN Elementary Language
ELAN Emulated Local-Area Network
ELAN Environment in Latin-America Network
ELAS Electroacoustic Spectrometry
Elast Elastic(ity) [also elast]
Elast Elastomer [also elast]
Elast Noteb Elastomers Notebook [Switzerland]
ELAT (Projeto) Eletricidade Atmosférica/Relâmpagos [Atmospheric Electricity/Lightning (Project); of Brazilian National Institute for Space Research]
ELATS Expanded Litton Automated Test Set
E Layer Ionosphere Layer between about 50 and 70 Miles (80 and 113 km) above the Sur face of the Earth, and which Reflects Low Frequency Radio Waves [also known as Heaviside layer, or Kennelly Heaviside layer] [also E layer]
ELB Emergency Locator Beacon
ELB Exciton Lifetime Broadening [Solid-State Physics]
Elb Elbow [also elb]
ELBA Emergency Location Beacon Aircraft
ELBC European Lead Battery Conference
ELC Element Count
ELC Embedded Linking and Control
ELC Environmental Law Centre [of International Union for the Conservation of Nature and Natural Resources]
ELC Environmental Liaison Center [Kenya]
ELC Exchange Line Capacity [Telecommunications]
ELC Excimer Laser Crystallization
ELC Extra-Low-Carbon (Steel)
Elc Electrochemical; Electrochemistry [also elc]
Elc Electroconductivity [also elc]
ELCA European Landscape Contractors Association [Germany]
ELCAP End-Use Load and Conservation Assessment Program
ELCB Earth-Leakage Circuit Breaker; Earth-Leakage Contact Breaker
ELCCI East Lancashire Chamber of Commerce and Industry [UK]
ELCD Electrochemical Detection

ELCD Electroconductivity Detector

ELCD Electrolytic Conductivity Detector

ELCD Evaporative Loss Control Device

ELCI Environment Liaison Center International

EL-CID Electric-Core Imperfection Device

ELCl Extra-Low-Chloride; Extra-Low Chlorine

ELCO Electrostatic Coaxial

ELCO Eliminate and Count Coding

ELCOM Electronics and Computers

ELCON Electricity Consumers Resource Council [US]

ELD Economic Load Dispatching

ELD Edge-Lighted Display

ELD Energy Loss Distribution

ELD Electroless Deposition

ELD Electroluminescent Device

ELD Electroluminescent Display

ELD Encapsulated Light Diffusion

ELD Energy-Level Density

ELD Energy-Level Diagram

ELD Engineering Libraries Division [of American Society for Engineering Education, US]

ELD Extra-Long Distance

ELDC European Lead Development Committee [UK]

ELDF Effective Liquid Density Functional

ELDISC Electrical Disconnect

ELDMR Electroluminescence Detected Magnetic Resonance

ELDO European Launcher Development Organization [Now Part of European Space Agency] [also Eldo]

ELDOR Electron-Electron Double Resonance

ELDORA Electra Doppler Radar

ELDV Electrically Operated Depressurization Valve

ELE Equivalent Logic Element

ELEC European League for Economic Cooperation [UK]

Elec Electric(al); Electrician; Electricity [also elec]

Elec Electronic(s) [also elec]

Elec Gr Electronic Grade (Chemical)

ELECOM Electronic Computer [also Elecom]

ELECOMPS Electronic Components [also Elecomps]

Elect Electric(al); Electrician; Electricity [also elect]

Elect Electrolyte; Electrolytic [also elect]

Electn Electrician [also electn]

Electr Electric(al); Electrician; Electricity [also electr]

Electr Electronic(s) [also electr]

Electr China Electricity for China [Published in the UK]

Electr Commun Electrical Communication [Journal published in the UK]

Electr Commun Lab Tech J Electrical Communication Laboratories Technical Journal [Tokyo, Japan]

Electr Constr Maint Electrical Construction and Maintenance [US]

Electr Contract Electrical Contractor [Journal published in the UK]

Electr Des Electrical Design [Publication of the Chartered Institution of Building Services Engineers, UK]

ELECTRECON Electronics Research Conference

Electr Eng (Australia) Electrical Engineer (Australia) [Journal published in Australia]

Electr Eng (London) Electrical Engineer (London) [Journal published in the UK]

Electr Eng Jpn Electrical Engineering in Japan [US]

Electr Eng Rev Electrical Engineering Review [of West Pakistan University of Engineering and Technology, Lahore, Pakistan]

Electr Equip Electrical Equipment [Published in the UK]

Electr Furn Conf Proc Electric Furnace Conference Proceedings [of AIME Iron and Steel Society, US]

Electr Furn Steel Electric Furnace Steel [Journal published in Japan]

Electr India Electrical India [Publication of the Indian Electrical Manufacturers Association]

Electr Light Facil Railw Electric Lighting and Facilities in Railways [of Railway Electrification Association, Japan]

Electr Mach Power Syst Electric Machines and Power Systems [Published in the US]

Electr Manuf Electrical Manufacturing [Journal published in the US]

Electrochem Electrochemical; Electrochemist [also electrochem]

Electrochem Ind Process Biol Electrochemistry in Industrial Processing and Biology [Translation of: *Elektronnaya Obrabotka Materialov (USSR)*; published in the UK]

Electrochem Soc Ext Abstr The Electrochemical Society, Extended Abstracts [US]

Electrochem Soc Proc The Electrochemical Society Proceedings [US]

Electrochem Technol Electrochemical Technology [Journal published in the US]

Electrochim Acta Electrochimica Acta [of International Society of Electrochemistry]

Electrocomp Sci Technol Electrocomponent Science and Technology [Journal published in the US]

Electrodep Electrodeposition [also electrodep]

Electromag Electromagnet(ic); Electromagnetism; Electromagnetics [also electromag]

Electrometall Electrometallurgical; Electrometallurgist; Electrometallurgy [also electrometall]

Electron Electronic(s) [also electron]

Electron Aust Electronics Australia (Journal)

Electron Bank Finance Electronic Banking and Finance [Journal published in the Netherlands]

Electron Bus Electronic Business [Journal published in the US]

Electron Commun Eng J Electronics and Communication Engineering Journal [of Institution of Electrical Engineers, UK]

Electron Commun Jpn 1, Commun Electronics and Communications in Japan, Part 1, Communica tion [English Translation; published in US]

Electron Commun Jpn 2, Electron Electronics and Communications in Japan, Part 2, Electronics [English Translation; published in US]

Electron Commun Jpn 3, Fundam Electron Sci Electronics and Communications in Japan, Part 3, Fundamental Electronic Science [English Translation; published in US]

Electron Compon Appl Electronic Components and Applications [Publication of Philips, Eindhoven, Netherlands]

Electron Des Electronic Design [Journal published in the Netherlands]

Electron Des Autom Electronic Design Automation [Journal published in the UK]

Electron Educ Electronic Education [Journal published in the US]

Electron Eng Electronic Engineering [Journal published in the UK]

ELECTRONICA International Trade Fair for Components and Assemblies in Electronics [Germany] [also electronica]

Electron Imaging Electronic Imaging [Journal published in the US]

Electron Ind Electronics Industry [Journal published in the UK]

Electron Ind Electronique Industrielle [French Journal on Industrial Electronics]

Electron Inf Plan Electronics Information and Planning [Publication of the Information Planning and Analysis Group, New Delhi, India]

Electron Learn Electronic Learning [Published in the US]

Electron Lett (Australia)

Electronics Letters (Australia) [Published in Melbourne]

Electron Lett (UK) Electronics Letters (UK) [of Institution of Electrical Engineers]

Electron Libr Electronic Library [UK]

Electron Manuf Electronics Manufacturing [Publication of the Society of Manufacturing Engineers, US]

Electron Manuf Test Electronics Manufacture and Test [Journal published in the UK]

Electron Microsc Electron Microscopy [Publication of the Faculty of Medicine, Department of Anatomy, Kyoto University, Japan]

Electron Microsc Rev Electron Microscopy Review [US]

Electron Model Electronic Modeling [Translation of: *Elektronnoe Modelirovanie (Ukrainian SSR)*; published in the UK]

Electron Opt Bull Electron Optics Bulletin [Published by Philips-Norelco, Eindhoven, Netherlands]

Electron Opt Publ Rev Electronic and Optical Publishing Review [UK]

Electron Opt Report Electron Optics Reporter [Published by Philips Electronic Instruments Inc., Mahwah, New Jersey, US]

Electron Packag Prod Electronic Packaging and Production [Journal published in the US]

Electron Prod Electronic Production [Journal published in the UK]

Electron Prod Electronic Products [Journal published in the US]

Electron Prod Des Electronic Product Design [Journal published in the UK]

Electron Prog Electronic Progress [Published by Raytheon Co., Lexington, Massachusetts, US]

Electron Prop Electronic Properties [also electron prop]

Electron Publ Bus Electronic Publishing Business [Journal published in the US]

Electron Publ, Orig Dissem Des Electronic Publishing: Origination, Dissemination and Design [Published in the UK]

Electron Puissance Electronique de Puissance [French Journal on Power Electronics]

Electron Purch Electronics Purchasing [Journal published in the US]

Electron Syst Des Mag The Electronic System Design Magazine [US]

Electron Syst News Electronic Systems News [of Institution of Electrical Engineers, UK]

Electron Tech Ind Electronique Techniques et Industries [French Publication on Electronics and Industry]

Electron Technol Electron Technology [Publication of the Electronic Technology Institute, Warsaw, Poland]

Electron Technol Electronic Technology [Journal of the Society of Electronic and Radio Technicians, UK]

Electron Telecommun Lett Electronics and Telecommunications Letter [Published by the Telecommunications Research Institute, Warsaw, Poland]

Electron Test Electronics Test [Journal published in the US]

Electron Times Electronic Times [UK]

Electron Today Electronics Today [Published in India]

Electron Today Int Electronics Today International [Published in the UK]

Electron Wirel World Electronics and Wireless World [Journal published in the UK]

Electron Wkly Electronics Weekly [Journal published in the UK]

Electron World Wirel World Electronics World + Wireless World [Journal published in the UK]

Electro-Opt Electro-Optics [Journal published in the UK]

Electrostat Electrostatic(s) [also electrostat]

ElectroTech Electrotechnical Conference and Exposition [of International Society for Measurement and Control, US]

Electrotech Electrotechnical [also electrotech]

Electrotech News Electrotechnical News [Publication of the Association of Supervisory and Executive Engineers, UK]

Electro Technol Electro Technology [Publication of the Society of Electronic Engineers, India]

Electroteh Electron Autom, Autom Electron Electrotehnica, Electronica si Automatica, Automatica si Electrotehnica [Romanian Journal on Electrical Engineering and Automation]

Electroteh Electron Autom, Electroteh Electrotehnica, Electronica si Automatica, Electrotehnica [Romanian Journal on Electrical Engineering]

Electr Prop Electrical Properties [also electr prop]

Electr Power Syst Res Electric Power Systems Research [Journal published in Switzerland]

Electr Railw Electric Railways [Publication of the Railway Electrification Association, Japan]

Electr Rep Electrical Reports [Published by the Edison Electric Institute, US]

Electr Rev Electrical Review [UK]

Electr Rev (London) Electrical Review (London) [UK]

Electr Technol USSR Electric Technology USSR [Published in the UK]

Electr Times Electrical Times [UK]

Electr Veh Dev Electric Vehicle Developments [Journal published in the UK]

Electr Word Electric Word [Journal published in the Netherlands]

Electr World Electrical World [Journal published in the US]

Elecy Electricity [also elecy]

ELED Edge-Emitting Light-Emitting Diode

ELED Entry-Level Employee Development

ELEED Elastic Low-Energy Electron Diffraction

Elektr Bahnen Elektrische Bahnen [German Journal on Electric Railways]

Elektr Energy Tech Elektrische Energie Technik [German Journal on Electric Energy Engineering and Technology]

Elektr Masch Elektrische Maschinen [German Journal on Electric Machinery]

Elektro Anz Elektro-Anzeiger [German Gazette for the Electrical Industry]

Elektrokhim Elektrokhimiya [Russian Journal on Electrochemistry]

Elektromeist Dtsch Elektrohandw Elektromeister und Deutsches Elektrohandwerk [German Journal for Master Electricians and the German Electric Trades]

Elektron Entwickl Elektronik Entwicklung [German Journal on Developments in Electronics]

Elektroniker Der Elektroniker [Swiss Publication on Electronics]

Elektron Ind Elektronik Industrie [German Publication on the Electronics Industry]

Elektron Int Das Elektron International [Austrian International Publication on Electrons and Electronics]

Elektron J Elektronik Journal [German Electronics Journal]

Elektron Model Elektronnoe Modelirovanie [Ukrainian Journal on Electronic Modelling]

Elektron Obrab Mater Elektronnaya Obrabotka Materialov [Russian Journal on Surface Engineering and Applied Electrochemistry]

Elektron Prax Elektronik Praxis [German Journal on Practical Electronics]

Elektron Tekh Elektronnaya Tekhnika [Russian Journal on Electronics and Electronic Engineering]

Elektrotech Elektrotekhnika [Soviet Journal on Electrical Engineering]

Elektrotech Cas Elektrotechnický Casopis [Slovak Publication on Electrical Engineering]

Elektrotech Inf tech Elektrotechnik und Informationstechnik [Austrian Journal on Electrical Engineering and Information Technology]

Elektrotech Masch Bau Elektrotechnik und Maschinenbau [German Journal on Electrical and Mechanical Engineering]

Elektrotech Z Elektrotechnische Zeitung [German Journal of Electrical Engineering]

Elektrotech Obz Elektrotechnický Obzor [Czech Journal on Electrical Engineering]

Elektroteh Vestn Elektrotehniski Vestnik [Yugoslavian Publication on Electrical Engineering]

Elektrowärme Int A Elektrowärme International, Edition A [German International Journal on Electroheat]

Elektrowärme Int B Elektrowärme International, Edition B [German International Journal on Electroheat]

Elektrowärme Tech Elektrowärme im Technischen [German Journal on Electroheat in Technology; later split into *Elektrowärme International A & B*]

Elektr Stn Elektricheskie Stantsii [Russian Journal on Power Engineering]

Elem Element(al) [also elem]

Elems Elements [also elems]

ELEP Electronic Converter Electric Power

Elepltg Electroplating [also elepltg]

ELEPS Extended Low-Energy Photon Spectroscopy

Elettron Oggi Elettronica Oggi [Italian Electronics Journal]

Elettron Telecomun Elettronica e Telecomunicazioni [Italian Journal on Electronics and Telecommunications]

Elev Elevation [also Elevn, elevn, or elev]

Elevn Elevator [also elev]

Elex Electronics [also elex]

ELF Electroluminescent Ferroelectricity

ELF Electronic Location Finder [also elf]

ELF Electronic Logging Facility

ELF Electron Localization Function

ELF Ellipsometry, Low Field

ELF Executable and Linking Format

ELF Extensible Language Facility

ELF Extremely Low Frequency [Frequency: below 300 hertz] [also elf]

ELFA Electric Light Fittings Association [UK]

ELFC Electroluminescent Ferroelectric Cell

ELFE Extremely-Low-Frequency Effect

ELFEM Effective Liquid Free-Energy Model

ELF-MF Extremely-Low-Frequency Magnetic Field

ELG Electrolytic Grinding

ELGIH Eötvös Lóránd Geophysical Institute of Hungary [Budapest, Hungary]

ELGMT Ejector Launcher Guided Missile Transporter

ELGRA European Low-Gravity Research Association

ELH Enol Lactone Hydrolase [Biochemistry]

ELI Energy Loss Image

ELI English Language Interpreter

ELI Environmental Law Institute [US]

ELI Environment Liaison International [Kenya]

ELI Equitable Life Interpreter

ELI Extended Lubrication Interval

ELI Extensible Language I

ELI Extra Low Interstitial [Materials Science]

ELIAS Environmental Libraries Automated System [of Environment Canada]

E-Lidar Experimental Lidar [also E-lidar]

E Line E Line [Fraunhofer Line for Iron Having a Wavelength of 526.9 Nanometers] [also E line] [Spectroscopy]

ELINT Electromagnetic Intelligence [also Elint]

ELINT Electronic Intelligence [also Elint]

Elin Z Elin Zeitschrift [Elin Journal; published in Austria]

ELIP Electrostatic Latent Image Photography

ELISA Enzyme-Linked Immunosorbent Assay [Biochemistry]

ELISA/ACT Enzyme-Linked Immunosorbent Assay/American College Test

ELIT Extra-Low Interstitial Transfer (Pack Rolling) [Metallurgy]

ELK External Link

ELKP Elastic-Linear-Kinematic Hardening Phase [Metallurgy]

ELL Eccentric Leveling Lugs

Ell Ellipsometry [also ell]

ELLA Eastern Lamp and Lighting Association [US]

ELLA European Long-Lines Agency [of NATO]

Ellipt Elliptical; Ellipticity [also ellipt]

ELM Electrical Length Measurement

ELM (Department of) Environment and Land Management [South Australia]

ELM Extended Length Measure

ELM Extra-Low Mass

Elm Electronic Mail for UNIX

ELMA Electric Lamp Manufacturers Association [UK]

ELMAP Exchange-Line Multiplexing Analysis Program

ELME Emitter Location Method

ELMECH Electromechanical; Electromechanics [also Elmech]

ELMIG Electronic Library Membership Initiative Group [of American Library Association, US]

ELMINT Electromagnetic Intelligence

ELMO European Laundry and Dry Cleaning Machinery Manufacturers Organization [Germany]

ELMS Elastic Loop Mobility System

ELMS Engineering Lunar Model Surface

ELMS Experimental Library Management System

Elms Elements [also elms]

ELMU Environment Law and Machinery Unit [of United Nations Environmental Program]

ELN Electrochemical Noise Technique

ELNES Electron Loss Near Edge Structure

ELNES Energy Loss Near Edge Structure

ELNI Ethernet Local Network Interconnect

ELO Electro-Osmosis; Electro-Osmotic

ELO Epitaxial Lateral Overgrowth

ELOA Educational Leave of Absence

ELOC Environmental Locality

ELOISE European Large Orbiting Instrumentation for Solar Experiments

E Lon East Longitude [also E Long]

Elong Elongation [also elong]

ELOP Extended Logic Plan

ELOXAL Electrolytic Oxidation of Aluminum (Process) [also Eloxal]

ELP Electrolytic Polishing

ELP Electropolishing

ELP Element Processor

ELP El Paso, Texas [Meteorological Station Designator]

ELP English Language Program

ELP Ericksen-Leslie-Parodi (Theory) [Physics]

ELP Environmental Leadership Program [of US Environmental Protection Agency]

ElP Electric Primer

Elp Elliptical [also elp]

ELPC Electroluminescent Photoconductive; Electroluminescent Photoconductor [also EL PC]

ELPG Electric Light and Power Group

ELPH Elliptical Head [also ELP H, or Elp H]

ELPHR Experimental Low-Temperature Process Heat Reactor

ELPO Electrodeposition

ELR Energy Loss Rate

ELR Engineering Laboratory Report

ELR Environment(al) Lapse Rate

ELR Equal Listener Response Scale

ELR Existing Lapse Rate

ELR Extra-Long Range

ELRAC Electronic Reconnaissance Accessory Set

ELRAD Earth Limb Radiance

ELRO Electronics Logistics Research Office [US Army]

ELS Earth Landing System (Subsystem) [of NASA Apollo Program]

ELS Eastern Launch Site [of NASA]

ELS Economic Lot Size

ELS Electroless Solution

ELS Electroluminescence Screen

ELS Electron Energy Loss Spectroscopy

ELS Electrophoretic Light Scattering

ELS Element Symbol

ELS Elevon Load System [Aeronautics]

ELS Emitter Location System

ELS Endurance Limit Stress

ELS Energy Level Scheme

ELS Energy Loss Spectroscopy

ELS Enhanced Layout System

ELS Entry Level System

ELS Environmental Labelling Schemes

ELS Environmental Law Society [of Harvard University, Cambridge, Massachusetts, US]

ELS Equal Load Sharing

ELS Error Likely Situation

ELS Expert Laboratory Support (Study)

ELSA (Bonn) Electron Stretcher Apparatus [of Physikalisches Institut Bonn, Germany]

ELSA Ellipse Saddle

ELSA Energy Loss Spectral Analysis

ELSB Edge-Lighted Status Board

ELSBM Exposed Location Single-Buoy Mooring

ELSC Earth Landing Sequence Controller [of NASA]

ELSEC Electronic Security

ELSI Ethical, Legal and Social Implications [of the International Human Genome Project]

ELSI Ethical, Legal and Social Issues [of Environmental Genome Project]

ELSI Extra-Large-Scale Integration

ELSIE Electronic Letter Sorting and Indicating Equipment

ELSIE Electronic Signalling and Indicating Equipment

ELSIE Electronic Speech Information System

ELSL Electroluminescent Semiconductor Laser

ELSS Emergency Life Support System

ELSS Extravehicular Life Support System

EL-SSC Electronic Switching System Control

ELSPECS Electronic Specifications

ELT Electronic Level Transducer

ELT Electronic Typewriter

ELT Electrometer [also Elt, or elt]

ELT Emergency Location Transmitter; Emergency Locator Transmitter

ELT English Language Teaching

Elt Element [also elt]

ELTAD Emergency Location Transmitter, Automatic Deployable

ELTAF Emergency Location Transmitter, Automatic Fixed

ELTAP Emergency Location Transmitter, Automatic Portable

ELTEC Electrical Technology Trade Fair [Germany]

Eltek Med Aktuell Elektron Elteknik med Aktuell Elektronik [Swedish Journal on Electrical Engineering and Electronics]

$\varepsilon(\lambda,\theta,\phi)$ spectral emittance (function) [also $\epsilon(\lambda,\theta,\phi)$]

ELTR Emergency Location Transmitter/Receiver

ELU Eötvös-Lóránd University [Budapest, Hungary]

ELU Existing Carrier Line Up

Eluv Eluviate; Eluviation

ELV Earth Launch Vehicle

ELV Electrically Operated Valve

ELV Enclosed Frame Low-Voltage

ELV Expendable Launch Vehicle

ELV Extra-Low Voltage

ELVAC Electric Furnace Melting, Ladle Refining, Vacuum Degassing and Continuous Casting [Metallurgy]

elvn eleven

ELW Extended Linked Well

ELW Extreme Low Water

ELWAR Electronic Warfare

ELZ Extensive Land Use Zone (Australia)

EM Easy Magnetization

EM Effective Mass [Solid-State Physics]

EM Effective Medium

EM Effective Modulus

EM Efficiency Modulation

EM Eigenmatrix [Mathematics]

EM Ekman Motion [Meteorology]

EM Elastic Modulus; E Modulus; Modulus of Elasticity [Mechanics]

EM Elastomechanic(al); Elastomechanics

EM Electrical Material

EM Electric Machinery

EM Electric Machinery [Institute of Electrical and Electronics Engineers–Power Engineering Committee, US]

EM Electric Moment

EM Electrode Material

EM Electrolytic Machining

EM Electrolytic Meter

EM Electromagnet(ic); Electromagnetism

EM Electromechanical; Electromechanic(s)

EM Electrometallurgical; Electrometallurgist; Electrometallurgy

EM Electrometer; Electrometric

EM Electromigration

EM Electromotance; Electromotive

EM Electromotor

EM Electronic Mail

EM Electronic Material(s)

EM Electronics Manufacturing Group [of Society of Manufacturing Engineers, US]

EM Electron Metallography

EM Electron Micrograph

EM Electronic Microphone

EM Electron Microprobe

EM Electron Microscope; Electron Microscopy

EM Electron Mobility [Solid-State Physics]

EM Electron Multiplier

EM Electrophoretic Mobility [Biochemistry]

EM Elemental Metal

EM Emergency Medicine

EM Emission Microscope

EM Emission Pump

EM Empty Magnification

EM End-of-Medium (Character) [also EOM] [Data Communications]

EM Endothelial Medium [Anatomy]

EM Energy Management [Aerospace]

EM Energy Meter

EM Energy Method

EM Engineering Management

EM Engineering Management (Society) [of Institute of Electrical and Electronics Engineers, US]

EM Engineering Manual

EM Engineering Manufacture

EM Engineering Material

EM Engineering Mathematics

EM Engineering Mechanics

EM Engineering Memorandum

EM Engineering Metrology

EM Engineering Model [also E/M]

EM Engineer of Mines (or Mining Engineer)

EM Engineering of Mines

EM Engine Mount

EM Enhanced Monitoring

EM Enlisted Man

EM Environmental Management

EM Enzyme Mechanism

EM Epitaxial Mesa [Solid-State Physics]

EM Equimomental

EM Ethoxylated Monoglyceride

EM Evaluation Model

EM Evanescent Mode (of Acoustic Waveguide)

EM Exception Monitor

EM Excerpta Medica [Medical Journal]

EM Exhaustive Methylation (of Amines)

EM Expanded Memory

EM Expectation Maximization (Algorithm)

EM Extensible Machine

EM Extinguishing Medium

EM Extractive Metallurgist; Extractive Metallurgy

EM Eyepiece Micrometer

EM Mining Engineer (or Engineer of Mines)

E&M Ear and Mouth [Telecommunications]

E/M Electromechanical; Electromechanics

E/M Engineering/Management (Program)

E/M Engineering Model [also EM]

Em (Radioactive) Emanation [also Eman]

Em Emergency [also em]

Em Emulsion [also em]

Em Exameter [Unit]

em (emeritus) – retired

em emphasize(d)

e/m charge-to-mass ratio (of electron) [Symbol]

e/m specific ion charge [Symbol] [also $e \cdot m^{-1}$]

e/m² electrons per square meter [also $e\ m^{-2}$]

e/m_e specific electron charge (1.7595×10^{11} C/kg) [Symbol] [also $e \cdot m$]

EMA Easy Magnetization Axis

EMA Effective Mass Approximation [Solid-State Physics]

EMA (Bruggeman) Effective Medium Approximation (Theory) [Materials Science]

EMA Electromagnetic-Acoustic Transducer

EMA Electromagnetic Analysis

EMA Electronic Mail Association [US]

EMA Electronic Missile Acquisition

EMA Electronics Manufacturers Association [US]

EMA Electronics Materiel Agency [US Army]

EMA Electron Microprobe Analysis; Electron Microprobe Analyzer

EMA Ellipsoidal Mirror Analyzer

EMA Emergency Minerals Administration [US]

EMA Employment Management Association [US]

EMA Energy-Modified Adiabatic (Approximation)

EMA Engineered Materials Abstracts (Database) [Joint Index of ASM International in the US and the Institute of Materials in the UK]

EMA Engineers and Managers Association [UK]

EMA Engine Manufacturing Association [US]

EMA Enhanced Monitoring Array

EMA Enterprise Management Architecture

EMA Environmental Management Association [US]

EMA Environmental Mediation Association (International) [US]

EMA Equipment Market Abstracts [US]

EMA Ethyl α-Methylacrylate

EMA Ethylene-Maleic Anhydride (Copolymer)

EMA Ethylene Methyl Acrylate

EMA European Maritime Area

EMA European Marketing Association [UK]

EMA European Monetary Agreement

EMA Excavator Makers Association

EMA Extended Mercury Autocoder

EMAA Engineering Materials Achievement Award [of ASM International, US]

EMAA Envelope Manufacturers Association of America [US]

EMAA Ethylene-Co-Methacrylic Acid; Ethylene Methacrylic Acid

EMAA European Mastic Asphalt Association [France]

EMABC Electronics Manufacturers Association of British Columbia [Canada]

EMAC Educational Media Association of Canada

EMAC Electronics Manufacturers Association of Canada

EMAC Ethylene-Methyl Acrylate Copolymer

EMAC European Marketing Academy [Belgium]

EMaCC Energy Materials Coordinating Committee [of US Department of Energy]

EMACS Editing Macros [Unix Operating System]

EMAD Engine Maintenance Assembly and Disassembly Facility [of Nevada Test Site, NASA and US Department of Energy]

E-MAD Engine Maintenance Assembly Disassembly

EMAG Electron Microscopy Advisory Group

EMAG-MICRO Electron Microscopy Advisory Group Conference

EMAIA Electrical Meter and Allied Industries Association [Australia]

E-Mail Electronic Mail [also E-mail, or e-mail]

Email Mét Email Métal [French Journal on Enamelled Metallic Surfaces]

EMAL Electron Microscopy Analysis Laboratory [of University of Michigan, US]

EMal Ethyl Malonate

Eman (Radioactive) Emanation [also Em]

EMAP Electromagnetic Array Profiling

EMAP Environmental Monitoring and Assessment Program [of US Environmental Protection Agency]

EMAP-LC Environmental Monitoring and Assessment Program–Landscape Characterization [of US Environmental Protection Agency]

EMAR Experimental Memory-Address Register

EMARC Energy and Minerals Applied Research Center [of University of Colorado, US]

EMAS Electromagnetic Acoustic System

EMAS Electromagnetic Analysis System

EMAS Electron Microprobe Auger Spectroscopy

EMASAR Ecological Management of Arid and Semi-Arid Rangelands [Australia]

EMAT Center for Electron Microscopy of Materials Science [of University of Antwerp, Belgium]

EMAT Electromagnetic Acoustic (Wave) Transducer

e-MATH Web Site of the American Mathematical Society [*Note:* The "e" stands for "electronic"]

EMATS Emergency Mission Automatic Transmission Service

EMATS Emergency Mission Automatic Transmission System

EMaTT Energy Materials and Transportation Technology Program [of Lawrence Livermore National Laboratory, University of California at Livermore, US]

EMAV Electromagnetic Relief Valve

EMAW Evanescent Mode of Acoustic Waveguide

EMB Electrical Modernization Bureau

EMB Electromagnetic Brake; Electromagnetic Braking

EMB Energy Mobilization Board [US]

EMB Engineering in Medicine and Biology (Society) [of Institute of Electrical and Electronics Engineers, US]

EMB Eosin-Methylene-Blue (Agar) [Biochemistry]

EMB Experimental Model Basin [US Navy]

EMB Extended Memory Block

EMBA Electron Microbeam Analysis

EMBA Executive Master of Business Administration

Emballage Dig Emballage Digest [Publication on Packaging; published by Société Européene de Presse et d'Edition]

Emballages Mag Emballages Magazine [French Publication on Packaging Materials and Techniques]

EMBARC Electronic Mail Broadcast to a Roaming Computer

EMBC European Molecular Biology Conference

EMBERS Emergency Bed Request System

EMBET Error Model Best Estimate of Trajectory

EMBH Etched Mesa Buried Heterostructure [Solid-State Physics]

Embkn Embarkation [also embkn]

EMBL European Molecular Biology Laboratory [Heidelberg, Germany]

EMBnet European Molecular Biology Data Network

EMBO European Microbiology Organization [Heidelberg, Germany]

EMBO European Molecular Biology Organization [Heidelberg, Germany]

EMBO Eta-Maleimidocaproyloxysuccinimide

EMBO J European Microbiology Organization Journal [Published in Germany]

EMBR Electromagnetic Brake

EMBRAPA Empresa Brasileira de Pesquisa Agropecuária [Brazilian Cattle Raising Research Enterprise] [also Embrapa]

Embryol Embryological; Embryologist; Embryology [also embryol]

EMBS Engineering in Medicine and Biology Society [of Institute of Electrical and Electronics Engineers, US]

EMC Eastern Montana College [Billings, US]

EMC Elastomeric-Molding Compound

EMC Electromagnetic Capability

EMC Electromagnetic Compatibility (Group) [of Institute of Electrical and Electronics Engineers, US]

EMC Electromagnetically Cast; Electromagnetic Casting

EMC Electromagnetic Compatibility

EMC Electromagnetic Composite

EMC Electromagnetic Constant

EMC Electromagnetic Control

EMC Electromechanochemical

EMC Electronic Materials Committee [of The Minerals, Metals and Materials Society/American Institute of Mining, Metallurgical and Petroleum Engineers, US]

EMC Electron Microscopy Center

EMC Electron Microscopy Center [of Argonne National Laboratory, Illinois, US]

EMC E-Mail Connection

EMC Engineered (or Engineering) Military Circuit

EMC Engineering Manpower Commission [of American Association of Engineering Societies, US]

EMC Engineering Mockup Critical Experiment [Nuclear Engineering]

EMC Enhanced Memory Chip

EMC Ensemble Monte Carlo (Simulation) [Physics]

EMC Environmental Management Committee

EMC Equilibrium Moisture Content [Physical Chemistry]

EMC European Mathematical Council [UK]

EMC European Mechanics Colloquium

EMC European Mechanics Committee [also EuroMech, Germany]

EMC European Metals Conference

EMC European Microwave Conference

EMC European Military Communication

EMC European Muon Collaboration [Particle Physics]

EMC Excess Minority Carrier [Solid-State Physics]

EMC Export Management Company [US]

EMC Extended Math Coprocessor

EMC Extended Multiplier Channel

EMCA Electronic Motion Control Association [US]

EMCA Environment–Mechanism of Injury–Number of Casualties–Additional Resources [Accident Scene Management]

EMCAB Electromagnetic Compatibility Advisory Board

EMCC Electronically Modulated Converter Clutch

EMCC Emergency Mission Control Center [of NASA]

EMCCC European Military Communications Coordinating Committee [of NATO]

EMCD Electro-Mechanical Control Diagram ◆

EMCF European Monetary Cooperation Fund

EMC-FOM Electromagnetic Compatibility Figure of Merit

EMCI Engineering Model Configuration Inspection

EMCON Emission Control [also Emcon]

EMCP Electromagnetic Compatibility Program

EMCR (International Symposium on) Electrochemical Methods in Corrosion Research

EMCRF European and Mediterranean Cereal Rusts Foundation [UK]

EMCS Electromagnetic Compatibility Standard(ization)

EMCS Energy Management and Controls Society

EMCS Energy Monitoring and Control System

EMCS N-(Eta-Maleimidocaproyloxy)succinimide

EMCS Eta-Maleimidocaproic Acid N-Hydroxy)succinimide

EMCTP Electromagnetic Compatibility Test Plan

EMCV Encephalomyocarditis Virus

EMCWP European Mediterranean Commission on Water Planning [Italy]

EMD Easy Magnetization Direction

EMD Electric-Motor Driven

EMD Electromagnetic Damping

EMD Electron Moment Distribution

EMD Electronic Map Display

EMD Electron Momentum Density

EMD Energy Minerals Division [of American Association of Petroleum Geologists]

EMD Engineering Materials Department

EMD Entry Monitor Display

EMD European Machinery Directive

EMD Extractive Metallurgy Division [of American Institute of Mining, Metallurgical and Petroleum Engineers, US]

EMD Eyepiece Micrometer Divisions

EMDI Energy Management Display Indicator

EMDP Energy Management Development Program

EMDR Eye Movement Desensitization and Reprocessing

EME Earth-Moon-Earth (Technique)

EME Electromagnetic Energy

EME Electromechanical Energy

EME Environmental Mine Engineer(ing)

EME European Mercantile Exchange

EMEC Electromagnetic Effects Capability

EMEC Electromagnetic Effects Compatibility

EMEC Electronic Maintenance Engineering Center [of US Army]

EMECS Environmental Management of Enclosed Coastal Seas

EMEFS Eulerian Model Evaluation Field Study

EMEP European Monitoring and Evaluation Program (for Long-Range Transmission of Air Pollution)

EMEPS Electronic Message Privacy System

Emer Emerald [also emer]

Emer Emergency [also emer]

Emer Emeritus [also emer]

Emer Infect Diseases Emerging Infectious Diseases [Journal of the National Center for Infectious Diseases, Centers for Disease Control and Prevention, Atlanta, Georgia. US]

Emer Med Emergency Medicine [also emer med]

Emer Med Rep Emergency Medicine Reports

EMES Earth Monitoring Educational System

EMES Electrical, Mechanical and Environmental Systems

EMet Engineer of Metallurgy

EMETF Electromagnetic Environmental Test Facility [US]

EMEX Equatorial Mesoscale Experiment

EMF Effective Mass Filter

EMF Electromagnetic Field

EMF Electromagnetic Force [also emf]

EMF Electromagnetic Forming

EMF Electromotive Force [also emf]

EMF European Management Forum

EMF European Metalworkers' Federation

EMF European Monetary Fund

EMF Evolving Magnetic Features

EMF External Magnetic Field

EMFC Electromotive Force Compression

EMFD Electromagnetic Fields Division [of National Institute of Standards and Technology, Boulder, Colorado, US]

EMFF Electromagnetic Form Factor

EMFM Electromagnetic Flowmeter [also emfm]

EMFWK Eidgenössische Munitionsfabrik und Waffenkontrolle [Federal Munition Plant and Arms Inspection, Switzerland]

EMG Electromagnetic Generator

EMG Electromyogram; Electromyograph(y)

Emgcy Emergency [also emgcy]

EMH Expedited Message Handling

EMHS Electronic Message Handling System

EMI Early Manufacturing Involvement

EMI Ecole Mohammadia d'Ingénieurs [Mohammedia School of Engineering, Morocco]

EMI Electromagnetic Induction

EMI Electromagnetic Interference [also emi]

EMI Electromagnetic Iron

EMI End-of-Message Indicator

EMI Environment Management Industries

EMI Extractive Metallurgy Institute [US]

EMIAA Environment Management Industry Association of Australia

EMIC Electronic Materials Information Center [US]

EMIC Environmental Mutagen Information Center [of National Institute of Environmental Health Sciences, US]

EMICE Electromagnetic Interference Control Engineer

EMID Employee Identification

EMIG EM (Environmental Restoration and Waste Management) International Working Group

EMINT Electromagnetic Intelligence

EMINWA Environmentally Sound Management of Inland Waters [of United Nations Environmental Program]

EMIR Electromagnetic Interference Resolution

EMI/RFI Electromagnetic Interference/Radio Frequency Interference

EMIRS Electrochemically-Modulated Infrared Reflectance Spectroscopy

EMIRTEL Emirates Telecommunications Corporation [of United Arab Emirates]

EMIS Educational Management Information System

EMIS Electromagnetic Intelligence System

EMIS Electronic Markets and Information System

EMIS Electronic Materials Information Service [of the US Institute of Electrical and Electronics Engineers and the UK Institution of Electrical Engineers]

EMIS Datareview Ser Electronic Materials Information Service Datareview Series [of Information Services Physics, Electrical and Electronics, and Computers and Control, Institution of Electrical Engineers, UK]

Emis Emission [also emis]

EMIT Engineering Management Information Technique

EMIT Enzyme Multiplied Immunoassay Technique

E&MJ Engineering and Mining Journal [US]

EMK Emergency Medical Kit

EMK Endothelial Medium Kit [Biochemistry]

EMKO Ethyl Michler's Ketone Oxime [Biochemistry]

EML Electromagnetic Laboratory

EML Electromagnetic Levitation; Electromagnetic Levitator

EML Electron Microscopy Laboratory

EML Engineering Mechanics Laboratory [of National Institute of Standards and Technology, Gaithersburg, Maryland, US]

EML Environmental Measuring Laboratory

EML Equipment and Material Laboratories [of US Navy]

EML Equipment Modification List

EML Expected Measured Loss

EML Exterior Metal Loss

.EML Electronic Mail [File Name Extension] [also .eml]

EMLF Eastern Mineral Law Foundation [US]

EMLI Electromagnetic Level Indication (System) [Metallurgy]

EMLI Electromagnetic Level Indication–Mold [Metallurgy]

EMLI Electromagnetic Level Indication–Tundish [Metallurgy]

EMLI Electromagnetic Level Indication–Ladle Control [Metallurgy]

EM/LM Electron Microscopy/Light Microscopy (Sectioning)

EMM Ebers-Moll Model [Physics]

EMM Electromagnetic Measurement

EMM Electromagnetic Moment

EMM Electron Mirror Microscopy

EMM Enhanced Monitoring Methodology

EMM Ethyl Methylmalonate

EMM Expanded Memory Manager

EMMA Electron Manual Metal-Arc

EMMA Electron Microscope Microanalyzer; Electron Microscopy and Microanalysis

EMMA Equitorial Mount with Mirrors for Acceleration

EMMA European Magnetic Materials and Applications Conference

EMMA Extra MARC (Machine-Readable Catalog) Material

EMMA Eye Movement Measuring Apparatus

EMMAQUA Equitorial Mount with Mirrors for Acceleration with Water Spray

EMMET Ethyltrimethylolmethane Trinitrate

EMMG Electric Metal Makers Guild, Inc. [US]

EMMI Environmental Method Monitoring Index

EMML Electromechanical Machine Laboratory

EMMP Environmental Monitoring and Mitigation Plan [US]

EMMPDL Electron Microscopy and Microanalysis Public Domain Library [Joint Software Library of Argonne National Laboratory's Electron Microscopy Center and the Electron Microscopy Society of America]

EMMS Electronic Mail and Message System

EMMSA Envelope Makers and Manufacturing Stationers Association [UK]

EMN Electromagnetic Loudspeaker with Moving Coil and Neutralizing Winding

EMN Engineering Management Network

EMN 1-Ethyl-4-Methylnaphthalene

EMO Electromagnetic Oscillograph

EMO Electromechanical Oscillator

EMO Emergency Measures Organization

EMO Erdmann-Jesnitzer-Mrowka-Ouvier (Quench-Aging of Steel) [Metallurgy]

EMO European Machine Tool Exhibition [also Emo]

EMO World Exhibition of Metalworking [Hanover, Germany]

EMOC EOSDIS (Earth Observing System Data and Information System) Mission Operations Center

E mode Transverse Magnetic Mode [British Usage]

EMON Exception Monitoring

EMOP Escola de Minas de Ouro Preto [Ouro Preto School of Mines, Brazil]

em o Prof (*Emeritus ordentlicher Professor*) – Central European Academic Title Equivalent to a Professor Emeritus

EMOR Effective Modulus of Rupture

EMOR Emergency Mission Operations Room

EMOS Earth's Mean Orbital Speed

EMP École des Mines de Paris [Paris School of Mines; of Conseil National de la Recherche Scientifique, France]

EMP Elastomer-Modified Plastomer

EMP Electromagnetic Potential

EMP Electromagnetic Pulse

EMP Electromagnetic Pump

EMP Electromechanical Power

EMP Electromolecular Propulsion

EMP Electron Microprobe (Analysis)

EMP Emission Pattern

EMP End of Month Payment

EMP Engineering Materials Program

EMP Enhanced Monitoring Period

EMP Environmental Management Plan

EMP Environment Management Program

EMP Equipment Mounting Plate

EMP Escuela de Medicina de Ponce [Ponce School of Medicine, Puerto Rico]

EMP Eutectic Melting Porosity [Metallurgy]

EM&P Electronic Materials and Processes (Symposium)

Emp Emperor; Empire; Empress [also emp]

EMPA Eidgenössische Materialprüfungsanstalt [Federal Laboratory for Materials Research and Testing, Switzerland]

EMPA Electron Microprobe Analysis

EMPA European Maritime Pilots Association

EMPA Executive Master of Public Affairs

EMPACT Environmental Monitoring for Public Access and Community Tracking [of US Environmental Protection Agency]

EMPB Emergency Mobilization Preparedness Board [US]

EMPF Electronic Manufacturing Productivity Facility

EMPGS Electrical/Mechanical Power Generation Subsystem

EMPIRE Early Manned Planetary-Interplanetary Round-Trip Expedition

Empl Employee [also empl]

Emplt Employment [also emplt]

EMPMD Electronic, Magnetic and Photonic Materials Division [of Minerals, Metals and Materials Society of AIME, US]

EMPMD Monogr Ser Electronic, Magnetic and Photonic Materials Division Monograph Series [of Minerals, Metals and Materials Society of AIME, US]

EMPMD/SMD (Joint Committee of the) Electronic, Magnetic and Photonic Materials Division and the Structural Materials Division [of Minerals, Metals and Materials Society/American Institute of Mining, Metallurgical and Petroleum Engineers]

EMPP Elastomer Modified Polypropylene

EMPR Electromagnetic Pulse Radiation

EMPR Electromagnetic Pulse Response

EMPR Ethernet Multiport Repeater

EMPRA Emergency Multiple Person Rescue Apparatus

EMPS Electronic Message Privacy System

EMPS Ethernet Multiport Station

EMQ Economic Manufacturing Quantity

EMQ Electromagnetic Quiet

EMR Eddy-Making Resistance

EMR Electrolytic Metal Recovery

EMR Electromagnetic Radiation

EMR Electromagnetic Relay

EMR Electromagnetic Response

EMR Electromechanical Relay

EMR Electromechanical Research

EMR Electronics Manpower Register

EMR (Department of) Energy, Mines and Resources [now Natural Resources Canada]

EMR Engine Mixture Ratio

EMR Enhanced Metafile Record

EMR Executive Management Responsibility

EMRA Electrical Manufacturers Representative Association

EMRC European Medical Research Council [France]

EMRG Electronic Materials Research Group [of Los Alamos National Laboratory, Albuquerque, New Mexico, US]

EMRIC Educational Media Research Information Center [US]

EMRL Engineering Mechanics Research Laboratory [US]

EMRL Equipment Maintenance Requirements List

EMRLD Excimer Moderated-Power Raman-Shifted Laser Device

EMRP Effective Monopole Radiated Power [Antennas]

E-MRS European Materials Research Society [also EMRS]

EMRTC Energetic Materials Research and Testing Center [US]

EMS Earnings per Manshift

EMS Earth and Mineral Sciences

EMS Earthquake Monitoring System

EMS Econometric Society [US]

EMS Edinburgh Mathematical Society [UK]

EMS Efficient Microcanonical Sampling (Procedure)

EMS Electrochemical Mass Spectroscopy

EMS Electromagnetic Sensing; Electromagnetic Sensor

EMS Electromagnetic Separation; Electromagnetic Separator

EMS Electromagnetic Spectrum

EMS Electromagnetic Stirrer; Electromagnetic Stirring

EMS Electromagnetic Submarine

EMS Electromagnetic Surveillance

EMS Electromagnetic Susceptibility

EMS Electromechanical System

EMS Electromotive Series

EMS Electronic Mail System

EMS Electronic Management System

EMS Electronic Medical System

EMS Electronic Message Service

EMS Electronic Message System

EMS Electronic Micro System

EMS Electronic Monitoring System

EMS Electron Momentum Spectroscopy

EMS Electron Microscopy Simulation (Program)

EMS Electron Microscopy Society

EMS Emergency Management System

EMS Emergency Medical Services

EMS Energy Management System

EMS Engineering Management Society [of Institute of Electrical and Electronics Engineers, US]

EMS Engineering Master Schedule

EMS Engineering Modeling System

EMS Engine Management System [US Army]

EMS Enhanced Memory Specification

EMS Entry Monitor(ing) System [of NASA]

EMS Entry Monitor(ing) Subsystem [of NASA]

EMS Environmental Management Service [Canada]

EMS Environmental Magnetic Shield

EMS Environmental Management System

EMS Environmental Monitoring System

EMS Environmental Mutagen Society [at Oak Ridge National Laboratory, Tennessee, US]

EMS Eosinophiliamyaglia Syndrome [Medicine]

EMS Equilibrium Mode Simulator

EMS Ethylmethane Sulfonate

EMS European Monetary System

EMS Expanded Memory Specification

EMS Expanded Memory Support

EMS Export Marketing Service

EMS Extended Memory Specification

EMSA Electron Microscopy Society of America

EMSA European Marine Step Association

EMSA Electronic Materiel Support Agency [US Army]

EMSA Bull EMSA Bulletin [of Electron Microscopy Society of America]

EMSAPI Extended Messaging Services Application Programming Interface

EMSC Electrical Manufacturers Standards Council [of National Electrical Manufacturers Association, US]

EMSE Ecole des Mines de Saint-Etienne [Saint-Etienne School of Mines, France]

EMSEC Emanations Security

EMSI Electromagnetic Sensing and Interpretation

EMSI Electron Microscopy Society of India

EMSL Environmental Molecular Sciences Laboratory [US]

EMSL Environmental Monitoring Systems Laboratory [of US Environmental Protection Agency]

EM/SME Electronics Manufacturing Group of the Society of Manufacturing Engineers [US]

EMSS Electromagnetic Servoactuator System

EMSS Electron Microscopy Society of Switzerland

EMSS Emergency Manual Switching System

EMSS E. Mitchell Scientific Society [US]

EMSSA Electron Microscopy Society of South Africa

EMT Eastern Mediterranean Time [Greenwich Mean Time −2:00]

EMT Effective-Mass Theory [Solid-State Physics]

EMT Effective Medium Theory

EMT Electrical Mechanical Tubing

EMT Electrical Metallic Tubing

EMT Electromagnetic Technology

EMT Electromagnetic Test(ing)

EMT Electromagnetic Theory

EMT Electromagnetic Tubing

EMT Electromechanical Test

EMT Electromechanical Transducer

EMT Electronics Maintenance Technician

EMT Emergency Medical Team

EMT Emergency Medical Technologist; Emergency Medical Technology

EMT Emergency Medical Technician

EMT Emergency Medical Treatment

EMT Equivalent Megaton

EMT European Mediterranean Tropo

EMTA Electromedical Trade Association [UK]

EMTC Edison Materials Technology Center [Cincinnati, Ohio, US]

EMTC Environmental Management Technical Center [of US Geological Survey/Biological Resources Division at Onalaska, Wisconsin]

EMTD Electromagnetic Technology Division

EMTEC Edison Materials Technology Center [Kettering, Ohio, US]

EMTEC European Marine Trade Exhibition

EMTECH Electromagnetic Technology

EMTL Engineered Materials Technology Laboratory [of General Electric Aircraft Engines, Cincinnati, Ohio, US]

EMTP Electromagnetic Transients Program

EMTS Ethylmercury-P-Toluenesulfonanilide

EMTTF Equivalent Mean Time To Failure

EMTUG European Manufacturing Technology Users Group

EMU Eastern Michigan University [Ypsilanti, US]

EMU Economic and Monetary Union [European Community] [also emu]

EMU Electrical-Multiple Unit

EMU Electromagnetic Unit [also emu]

EMU Engineering Model Unit

EMU European Mineralogical Union

EMU European Monetary Union

EMU European Monetary Unit

EMU Expanded Memory Unit

EMU Export Marketing Unit [of Ontario Ministry of Agriculture and Food and Rural Affairs, Canada]

EMU Extended Memory Unit

EMU Extravehicular Maneuvering Unit

EMU Extravehicular Mobility Unit [of NASA]

Emu Emulator [also emu]

emu/cc electromagnetic unit(s) per cubic centimeter [also emu/cm^3, or emu cm^{-3}]

EMUF Einplatinencomputer für Universelle Festprogrammanwendung [Single-Board Computer for Universal Fixed-Program Applications]

emu/cm^3 electromagnetic unit(s) per cubic centimeter [also emu cm^{-3}, or emu/cc]

emu/g electromagnetic units per gram [also emu g^{-1}]

emu K/mol electromagnetic units kelvin per mole [also emu K mol^{-1}]

Emul Emulation [also emul]

emu/mol electromagnetic units per mole [also emu mol^{-1}]

emu/Oe electromagnetic units per oersted [also emu Oe^{-1}]

emu/sec electromagnetic units per second [also emu sec^{-1}]

EMU-TV Extravehicular Mobility Unit–Television [of NASA]

E-MUX Electrical Multiplex [also E-Mux, or e-mux]

EMV Effective (Particle) Migration Velocity [also emv]

EMV Electron Multiplier Voltage

EMV Internationale Fachmesse und Kongreß für Elektromagnetische Verträglichkeit [International Exhibition and Conference on Electromagnetic Compatibility, Germany]

EMW Electromagnetic Warfare

EMW Electromagnetic Wave

EMWAC European Microsoft Windows NT (New Technology) Academic Center

EMX Enterprise Messaging Exchange

EN Earthcare Network [US]

EN Electrical Noise

EN Electrochemical (Potential) Noise

EN Electroless Nickel

EN Electronegative; Electronegativity

EN Electroneutral(ity) [Physical Chemistry]

EN Electronic Neutrino

EN Electronic Noise

EN El Niño [Oceanography]

EN Emergency Nurse; Emergency Nursing

EN Enable (Input) [Computers]

EN Enforcement Notification

EN Enrolled Nurse

EN Epithermal Neutron

EN Equinoctial; Equinox

EN Equipment Number

EN Erythema Nodosum [Medicine]

EN Esbjerg-Nørskov (Parameter) [Physics]

EN Ethylene Diamine [also En, or en]

EN 2-Ethylnaphthalene

EN European Normal (Document)

EN Europa-Norm [European Standard]

EN Even Number; Even-Numbered

EN Exhaust Nozzle

En Enamel(ling) [also en]

En Recording Media Noise

en ethylenediamine (or 1,2-diaminoethane) [Abbreviation Used in Coordination Compound Formulas, e.g., Co(en)$_3$]

ENA Ecole Nationale d'Administation [National School of Administration, France]

ENA Electronic Networking Association [US]

ENA Enable [Computers]

ENA Epithelial Neutrophil Activating (Peptide) [Biochemistry]

ENA European Neuroscience Association [Netherlands]

ENAA Epithermal Neutron Activation Analysis

ENADS Enhanced Network Administration System

Enam Enamel [also enam]

Enantio-PAF(C$_{16}$) D-α-Phosphatidylcholine, β-Acetyl-γ-O-Hexadecyl [Platelet-Activating Factor used in Biochemistry]

Enantio-PAF(C$_{18}$) D-α-Phosphatidylcholine, β-Acetyl-γ-O-Octadecyl [Platelet-Activating Factor; used in Biochemistry]

ENAR Eastern North American Region

ENB Electronuclear Breeder

ENB English National Board [UK]

ENB Ethylidene Norbornene

EnBA Ethylene n-Butyl Acetate

ENBC Erbium Nickel Boron Carbide

Enbo Electroless Nickel-Boron (Plating Bath)

ENBRI European Network of Building Research Institutes

EnBW Energie Baden-Württemberg [Energy Company of Baden-Wurttemberg, Germany]

ENC East North Central [US Geographic Region]

ENC Electronic Navigational Chart

ENC Entanglement Network Coalition [US]

ENC Equivalent Noise Charge

.ENC Encoded [File Name Extension] [also .enc]

Enc Enclosure [also enc]

enc enclosed

enc encode(d)

ENCA European Naval Communications Agency [of NATO]

ENCC Ente Nazionale per la Cellulosa e per la Carta [National Agency for Cellulose and Paper, Italy]

ENCC Environment(al) Noise Control Committee

Encd Encode

Encdr Encoder

ENCIAWPRC Egyptian National Committee of the International Association on Water Pollution Research and Control

Encl Enclosure [also encl]

encl enclose(d)

Encls Enclosures [also encls]

ENCORD European Network of Construction Companies for Research and Development

ENCOSTEEL World Conference on Environmental Control in the Steel Industry [of International Iron and Steel Institute (IISI) and Association of German Iron and Steel Engineers (VDEh)]

ENCRyM Escuela Nacional de Conservación, Restauración y Museografia [National School of Conservation, Restoration and Museography, San Diego, Mexico]

Ency Encyclopedia [also Encyc, Encyc, ency, encyc, or encycl]

END Endnote [also End]

END Endrin [Organic Chemistry]

END Equivalent Neutral Density

ENDA-TM Environnement et Développement du Tiers Monde [Environment and Development of the Third World, Senegal]

ENDC Eastern Nigeria Development Corporation

ENDC Eighteen-Nation Disarmament Conference [of United Nations]

ENDEC Encoder/Decoder [also Endec]

End Eff End Effector

ENDEWAX National Chemical Dewaxing (Process)

ENDF Evaluated Neutron Data File

ENDF Evaluated Nuclear Data File

Endic Endo-cis-Bicyclo[2.21]-5-Heptane-2,3-Dicarboxylic (Anhydride)

END NUM End Number [also End Num]

Endocrin Endocrinological; Endocrinologist; Endocrinology [also endocrin]

Endocrin Endocrinology [International Journal]

ENDOR Electron-Nuclear Double Resonance

ENDORIESR Electron-Nuclear Double Resonance Induced Electron Spin Resonance [also ENDOR-IESR]

ENDS End Segment [also EndS]

ENDS Environmental Data Services Limited

ENDS European Nuclear Documentation System [of EURATOM]

ENE East-Northeast

ENE Enterprise Networking Event [US]

ENE Estimated Net Energy

ENEA Ente per le Nuove Tecnologie, l'Energia e l'Ambiente [Agency for New Technologies, Energy and the Environment, Rome, Italy]

ENEA European Nuclear Energy Agency [of Organization for Economic Cooperation and Development, France]

ENEA European Nuclear Energy Agreement

ENEA European Nuclear Energy Association

ENEL Ente Nazionale per l'Energia Elettrica [National Electric Energy Agency, Italy]

ENEL Entreprise National des Industries Electrotechniques [National Electrotechnical Industries Enterprises, France]

ENERDEMO Conservation and Renewable Energy Demonstration Program [Canada]

Energ Atomtech Energia es Atomtechnika [Hungarian Journal on Energy and Nuclear Engineering and Technology]

Energ Elettr Energia Elettrica [Italian Journal on Electrical Energy; published by Associazione Elettrotecnica ed Elettronica]

Energie Fluide Air Ind Energie Fluide l'Air Industriel [French Journal on Fluid Energy and Industrial Air]

Energiewirtsch Tagesfr Energiewirtschaftliche Tagesfragen [German Publication on the Power Supply Industry]

Energ Nucl Energia Nuclear [Spanish Journal on Nuclear Energy; published by Junta de Energia Nuclear]

Energ Nucl Energia Nucleare [Italian Journal on Nuclear Energy]

Energomashinostr Energomashinostroenie [Russian Journal on Energy Engineering and Technology]

Energy Autom Energy and Automation [Publication of Siemens AG, Erlangen, Germany]

Energy Build Energy and Buildings [Journal published in Switzerland]

Energy Convers Manage Energy Conversion and Management [Journal published in the UK]

Energy Des Update Energy Design Update [US Publication]

Energy Dev Energy Developments [Journal published in Germany]

Energy Eng J Energy Engineering Journal [of Association of Energy Engineers, US]

Energy Fuels Energy and Fuels [Journal of the American Chemical Society–Division of Fuel Chemistry]

Energy Inf Abstr Energy Information Abstracts

Energy J The Energy Journal [Published by the Energy Economics Education Foundation of the International Association for Energy Economics, US]

Energy J Energy Journal [New Zealand]

Energy Mag Energy Magazine [Publication of Organisación Latinoamericana de Energia, Ecuador]

Energy Manage Energy Management [Published in the UK]

Energy Prog Energy Progress [Journal published in the US]

Energy Res Abstr Energy Research Abstracts

Energy Res Rep Energy Research Report [US]

Energy Syst Policy Energy Systems and Policy [Journal published in the US]

ENET Engineering Network

ENF End-Notch(ed) Flexure (Test Specimen) [Mechanical Testing]

ENF End-Notch(ed) Fracture (Test Specimen) [Mechanical Testing]

ENFET Enzyme-Based Field-Effect Transistor; Enzyme-Sensitive Field-Effect Transistor

ENFIA Exchange Network Facilities for Interstate Access [US]

ENFOR Energy from the Forest (Program) [of Forestry Canada]

ENG Electronic News Gathering

ENG Electroneurography

ENG Electronystagmography

ENG European Nursing Group

EN(G) Enrolled Nurse (General) [also ENG]

Eng Engine [also eng]

Eng Engineer(ing) [also eng]

Eng England; English

Eng Engraving [also eng]

Eng Anal Engineering Analysis

Eng Anal Engineering Analysis [Journal published in the US]

Eng Appl Artif Intell Engineering Applications of Artificial Intelligence [Journal published in the UK]

Eng Aust Engineers Australia [Publication of the Institution of Engineers of Australia]

Eng Autom Engineering and Automation [Publication of the Informatics Center of the Hungarian Industry, Budapest, Hungary]

Eng Bull Engineering Bulletin [also eng bull]

Eng Cer Engineering Ceramic(s)

Eng Chem Engineering Chemist(ry)

Eng Comput Engineering Computers [Journal published in the UK]

Eng Comput Engineering with Computers [Journal published in the US]

Eng Cornell Q Engineering: Cornell Quarterly [Publication of Cornell University, Ithaca, New York, US]

Eng Costs Prod Econ Engineering Costs and Production Economics [Journal published in the Netherlands]

EngD Doctor of Engineering

Eng Des Engineering Design(er) [also eng des]

Eng Des Engineering Design [Published in Switzerland]

Eng Des Engineering Designer [Publication of the Institution of Engineering Designers, UK]

Eng Des Graphics J Engineering Design Graphics Journal [of American Society for Engineering Education, US]

Eng Dig Engineering Digest [Canada]

Eng Dig Engineer's Digest [UK]

Eng Econ The Engineering Economist [Publication of the American Society for Engineering Education, US]

Eng Educ Engineering Education [also eng educ]

Eng Educ Engineering Education [Publication of the American Society for Engineering Education, US]

Eng Fail Anal Engineering Failure Analysis [Published in the UK]

Eng Fract Mech Engineering Fracture Mechanics [Journal published in the UK]

Eng Gaz Engineering Gazette [Published by the Engineering Industries Association, UK]

Eng Geol Engineering Geology [Journal published in the Netherlands]

Eng Ind Engineering Industries [Journal]

Eng Inf Engineering Information (Database) [US]

Engin Engineer(ing) [also engin]

ENGINE Australian Engineering Database [US]

Eng J Engineering Journal [Canada]

Eng J Engineering Journal [of American Institute of Steel Construction, US]

Engl England; English

Eng Lang Ed English Language Edition

Eng Lasers Engineering Lasers [Journal published in the UK]

Eng Manage Int Engineering Management International [Publication of the American Society for Engineering Management, US]

Eng Mat Engineering Material [also Eng Mater]

Eng Mater Engineering Materials [Journal published in Japan]

Eng Mater Des Engineering Materials and Design [Journal published in the UK]

Eng Math Engineering Mathematics

Eng Mech Engineering Mechanics

Eng Med Engineering in Medicine [Journal published in the UK]

Eng Med Biol Mag Engineering in Medicine and Biology Magazine [Publication of the Institute of Electrical and Electronics Engineers, US]

Eng Met Engineering Metal

Eng Min J Engineering and Mining Journal [US]

Eng News Engineering News [Published in the UK]

Eng News-Rec Engineering News-Record [Published in the UK]

ENGO Environment Non-Governmental Organization

Eng Outlook Engineering Outlook [Published in the US]

Eng Phys Engineering Physics

Eng Plast Engineering Plastics

Eng Plast Engineering Plastics [Journal of the Association of Directors of European Centers for Plastics, UK]

Engr Engineer [also engr]

Engr Engraver; Engraving [also engr]

engr engraved

Eng Rep Engineering Report [also eng rep]

Eng Res Engineering Research [also eng res]

Engrg Engineering [also Engr'g, Engrg, or engr'g]

Engrs Engineers [also engrs]

Engs Engines [also engs]

EngScD Doctor of Engineering Science

Eng Sci Engineering Science [also eng sci]

Eng Sci Engineering and Science [US Journal]

Eng Sci Rep Kyushu Univ Engineering Sciences Reports, Kyushu University [Published by the Graduate School of Engineering Sciences, Kyushu University, Kasuga, Japan]

Eng Soc Engineering Society [also EngSoc]

Eng Struct Engineering Structures [Journal published in the UK]

Eng Tech Engineering Technician [Journal published in Singapore]

Eng Technol Engineering Technologist [Journal published in Singapore]

Eng Technol Engineering and Technology [Published by Kansai University, Osaka, Japan]

Eng Technol, Kansai Univ Engineering and Technology, Kansai University [Published by Kansai University, Osaka, Japan]

Eng Times Engineering Times [Published by the Society of Professional Engineers, US]

ENI Ecole Nationale d'Ingénieurs [National School of Engineering, France]

ENI Electron-Neutron Interaction

ENI Ente Nazionale Idrocarburi [National Hydrocarbon Agency, Italy]

ENI Equivalent Noise Input

ENI Ethiopian Nutrition Institute

ENIAC Electronic Numerical Integrator and Calculator

ENIC Voltage Negative Impedance Converter [Note: "E" is Used as a Symbol for Voltage]

ENIQ European Network for Inspection Qualification [of Institute for Advanced Materials, Netherlands]

ENIT Ente Nazionale Italiana di Turiosmo [Italian Government Travel Office]

ENL Equivalent Noise Level

ENL Erythema Nodosum Leprosum [Medicine]

Enl Enamel(ling) [also enl]

enl enlarge(d)

enl enlisted

EN(M) Enrolled Nurse (Mental) [also ENM]

e/nm² electron(s) per square nanometer [Unit]

EN(MH) Enrolled Nurse for the Mentally Handicapped [also ENMH]

ENMOD Environment Modification Convention

ENMR Electrophoretic Nuclear Magnetic Resonance

ENMU Eastern New Mexico University [Portales, US]

ENN Environmental News Network

ENNA Ecole Normale Nationale d'Apprentissage [National Normal School for Apprenticeship, France]

ENNG 1-Ethyl-3-Nitro-1-Nitrosoguanidine [Biochemistry]

ENNPA Extended Nearest Neighbor Pair Approximation

ENO Erbium Nickel Oxide

eNOS Endothelial-Cell-Derived Nitric Oxide Synthase [also e-NOS] [Biochemistry]

ENP Enable Input "P" [Digital Counter]

ENP Escuela Nacional Politecnico [National Polytechnic School, Quito, Ecuador]

ENQ Enqueue

ENQ Enquiry (Character) [Data Communications]

ENR Engineering News-Record [US Publication]

ENR Epoxidized Natural Rubber

ENR Equivalent Noise Resistance

ENR Excess Noise Ratio

ENRC European Nuclear Research Committee

ENRM Environment and Natural Resource Management [of Group of Seven (Nations)]

ENRT En-Route [also Enrt, or enrt]

ENRX Enable Receive [also EN RX]

ENRZ Enhanced Non-Return to Zero

ENS Ecole Normale Supérieure [Higher Normal School; of Université de Paris, France]

ENS Electrostatic Neutron Scattering

ENS Electrostatic Nonmetallic Separator

ENS European Nuclear Society [Switzerland]

Ens Ensign [Naval Rank] [also ENS]

ENSA Ecole Normale Supérieure Agronomique [Higher Normal School of Agriculture, France]

ENSAM Ecole Nationale Supérieure d'Arts et Métiers [National Higher School of Arts and Crafts, France]

ENSC Ecole Nationale Supérieure de Chimie [National Higher School of Chemistry, Montpellier, France]

ENSC Ecole Normale Supérieure de Cachan [Cachan Normal Higher School, France]

ENSCI Ecole Nationale Supérieure de Céramique Industrielle [National Higher School of Industrial Ceramics, Limoges, France]

ENSCM Ecole Nationale Supérieure de Chimie de Montpellier [Montpellier National Higher School of Chemistry, France]

ENSCM Ecole Nationale Supérieure de Chimie de Mulhouse [Mulhouse National Higher School of Chemistry, France]

ENSCP Ecole Nationale Supérieure de Chimie de Paris [Paris National Higher School of Chemistry, Université Pierre et Marie Curie, France]

ENSCPB Ecole Nationale Supérieure de Chimie et Physique de Bordeaux [Bordeaux National Higher School of Chemistry and Physics, France]

ENSCS Ecole Nationale Supérieure de Chimie de Strasbourg [Strasbourg National Higher School of Chemistry; of Conseil National de la Recherche Scientifique, France]

ENSCT Ecole Nationale Supérieure de Chimie de Toulouse [Toulouse National Higher School of Chemistry; of Conseil National de la Recherche Scientifique, France]

ENSDF Evaluated Nuclear Structure Data File [of Information System Karlsruhe, Germany]

ENSEARCH Environmental Management Association of Malaysia

ENSEC European Nuclear Steelmaking Club

ENSEEG Ecole Nationale Supérieure d'Electrochimie et d'Electrométallurgie de Grenoble [National Higher School of Electrochemistry and Electrometallurgy of Grenoble; of Institut National Polytechnique de Grenoble, France]

ENSEEG/INPG Ecole Nationale Supérieure d'Electrochimie et d'Electrométallurgie de Grenoble/Institut National Polytechnique de Grenoble [National Higher School of Electrochemistry and Electrometallurgy of Grenoble/National Polytechnic Institute of Grenoble, France]

ENSEEG/LIESG Ecole Nationale Supérieure d'Electrochimie et d'Electrométallurgie de Grenoble/Laboratoire d'Ionique et d'Electrochimie des Solides de Grenoble [National Higher School of Electrochemistry and Electrometallurgy of Grenoble/Grenoble Ionics and Solid Electrochemistry Laboratory, France]

ENSEEG/LTPCM Ecole Nationale Supérieure d'Electrochimie et d'Electrométallurgie de Grenoble/Laboratoire de Thermodynamique et de Physico-Chimie Métallurgiques/Ecole Nationale Supérieure d'Electrochimie et d'Electrométallurgie de Grenoble [National Higher School of Electrochemistry and Electrometallurgy of Grenoble/Laboratory for Thermodynamics and Physicochemical Metallurgy, of INPG France]

ENSG Ecole Nationale Supérieure de Géologie [National Higher School of Geology; of Université de Nancy, France]

ENSHMG Ecole Nationale Supérieure d'Hydraulique et Mécanique de Grenoble [National Higher School of Hydraulics and Mechanics of Grenoble; of Institut National Polytechnique de Grenoble, France]

ENSHMG/INPG Ecole Nationale Supérieure d'Hydraulique et d'Mécanique de Grenoble/Institut National Polytechnique de Grenoble [National Higher School of Hydraulics and Mechanics of Grenoble/National Polytechnic Institute of Grenoble, France]

Enshurin Ken Hok Enshurin Kenkya Hokoku [Research Bulletin of the College for Experimental Forests, Hokkaido University, Sapporo, Japan]

ENSI Ecole Nationale Supérieure d'Ingénieurs [National Higher School of Engineering, France]

ENSI Energia Nucleare Sud Italia [Nuclear Energy for Southern Italy (Project); of Società Elettronucleare Nazionale, Italy]

ENSI Environment and School Initiatives Project

ENSI Equivalent Noise Sideband Input

ENSIA Ecole Nationale Supérieure des Industries Agricoles et Alimentaires [National Higher School for the Agricultural and Food Industries, Massy, France]

ENSIP Engine Structural Integrity Program [of US Air Force]

ENSL Ecole Normale Supérieure de Lyon [Higher Normal School of Lyons, France]

ENS-LPS Ecole Nationale Supérieure/Laboratoire de Physique des Solides [Higher National School/Solid-State Physics Laboratory, France]

ENSM Ecole Nationale Supérieure de Mécanique [National Higher School of Mechanics, Nantes, France]

ENSM Ecole Nationale Supérieure des Mines [National Higher School of Mines, St. Etienne, France]

ENSMA Ecole Nationale Supérieure de Mécanique et d'Aérotechnique [National Higher School of Mechanics and Aerospace Engineering, Poitiers, France]

ENSMIM Ecole Nationale Supérieure de la Métallurgie et de l'Industrie des Mines [National Higher School of Metallurgy and Mining Engineering, of Institut National Polytechnique de Lorraine, Nancy, France]

ENSMP Ecole Nationale Supérieure des Mines de Paris [Paris National School of Mines, France]

ENSMSE Ecole Nationale Supérieure des Mines de Saint-Etienne [Saint-Etienne National School of Mines, France]

ENSO El Niño Southern Oscillation [Oceanography]

ENSORB Exxon Adsorption (Process) [Chemical Engineering]

ENSP Engineering Specification

ENSPG Ecole Nationale Supérieure de Physique de Grenoble [National School of Physics of Grenoble; of Institut National Polytechnique de Grenoble, France]

ENSPG/INPG Ecole Nationale Supérieure de Physique de Grenoble/Institut National Polytechnique de Grenoble [National School of Physics of Grenoble/National Polytechnic Institute of Grenoble, France]

ENSS Exterior Nodal Switching Subsystem [Internet]

ENSSA Empresa Nacional Siderúrgica SA [Spanish National Iron and Steel Company]

ENST Ecole Nationale de Science et Technologie [National School of Science and Technology, Paris, France]

ENSTA Ecole Nationale Supérieure de Techniques Avancées [National Higher School of Advanced Technologies, Paris, France]

ENT Ear, Nose and Throat

ENT Electric Network Theory

ENT Emergency Negative Thrust

ENT Enable Input "T" [Digital Counter]

ENT Enter [also Ent, or ent]

ENT Equivalent Noise Temperature

Ent Enter; Entrance; Entry [also ent]

Ent Entropy [also ent]

ENTA Ethylene Diamine Tetraacetic Acid

ENTAC Engine-Teleguide Anti-Char

ENTC Engine Negative Torque Control

entd entered [also ent'd]

ENTEA Ente per le Nuove Tecnologie l'Energia e l'Ambiente [New Technology Energy and Environmental Agency, Casaccia, Rome, Italy]

ENTELEC Energy Telecommunications and Electrical Association [US]

Entrep Rhone-Alpes (Metall) Entreprises Rhônes-Alpes (Metallurgia) [Metallurgical Publication of Enterprises Rhônes Alpes, Lyons, France]

ENTG European NATO Training Group

Entr Entrance [also entr]

Entsorga Internationale Fachmesse für Recycling und Entsorgung [International Trade Fair for Recycling and Disposal, Germany]

ENTO Division of Entomology [of Commonwealth Scientific and Industrial Research Organization]

Entom Entomological; Entomologist; Entomology [also Entomol, entomol, or entom]

ENTX Enable Transmit [also EN TX]

ENU N-Ethyl-N-Nitrosourea

Env Envelope [also env]

Env Environment(al); Environmentalist [also env]

env envelope [Virology]

EnvE Environmental Engineer [also Env Eng, or EnvEng]

ENVEC Environmental Economics

Env Eng Environmental Engineer(ing)

Envel Envelope [also envel]

Envir Environment(al) [also envir]

EnvirofACS Newsl Newsletter of the Division of Environmental Chemistry of the American Chemical Society [US]

ENVIRON Environment String Table (Statement) [BASIC Programming]

Environ Environment(al); Environmentalist [also environ]

Environ The Environmentalist

Environ Abstr Environment Abstracts

Environ Build News Environmental Building News

Environ Bull Environment Bulletin [Newsletter of the World Bank Environment Community]

Environ Bus Mag Environment Business Magazine

Environ Can Environ Update Environment Canada, Environment Update [Canada]

Environ Can Notice Publ Environment Canada, Notice of Publications [Canada]

Environ Devel Econ Environment and Development Economics

Environ Dig The Environmental Digest [Canada]

Environ Eng Environmental Engineering [Journal published in the UK]

Environ Ent Environmental Entomology [of US Environmental Protection Agency]

EnviroNET Australia Environmental Information Network of Australia

Environ Exp Bot Environmental and Experimental Botany [Journal published in the UK]

Environ Geol Environmental Geology [Journal published in Germany]

Environ Health Perspect Environmental Health Perspectives [Journal of the National Institute of Environmental Health Sciences, US]

Environ Hist Environment and History

Environ Inf Sci Environmental Information Science [Publication of the Center for Environmental Information Science, Japan]

Environ Int Environment International [Published in the US]

Environ Manage Environmental Management [Published in Germany]

Environ Matters Environment Matters [Magazine of the World Bank]

Environ Monit Assess Environmental Monitoring and Assessment [Journal published in the Netherlands]

Environ Mutagenesis Environmental Mutagenesis [Journal published in the US]

Environ News Environmental News

Environ Pollut Environmental Pollution [Published in the UK]

Environ Prof Environmental Professional [Publication of the National Association of Environmental Professionals, US]

Environ Prog Environmental Progress [Journal of the American Institute of Chemical Engineers]

Environ Prot Environmental Protection

Environ Prot Eng Environmental Protection Engineering [Published in the US]

Environ Res Environmental Research [Journal published in the US]

Environ Res Newsl Environmental Research Newsletter [Published by the Environment Institute of the Joint Research Centre of the European Commission]

Environ Rev Environmental Review [US]

ENVIRON\$ Environment String (Function) [BASIC Programming]

Environ Sci Technol Environmental Science and Technology [Journal of the American Chemical Society, US]

Environ Sci Technol A Environmental Science and Technology A [Journal of the American Chemical Society, US]

Environ Sci Technol B Environmental Science and Technology B [Journal of the American Chemical Society, US]

Environ Softw Environmental Software [Published in the UK]

Environ Sol Mag Environmental Solutions Magazine

Environ Technol Environment and Technology

Environ Technol Lett Environmental Technology Letters [UK]

Environ Today Environment Today [Daily Environment News on the Web]

Environ Toxicol Chem Environmental Toxicology and Chemistry [Journal]

Environ Values Environmental Values

ENVISAT Environmental Satellite Program

ENVITEC Technology for Environmental Protection–International Trade Fair and Congress [Duesseldorf, Germany]

ENVR Division of Environmental Chemistry [of American Chemical Society, US]

EO Earth Observation

EO Earth Orbit

EO Editorial Operations

EO Elasto-Optic(s) (also E-O)

EO Electrolytic Oxidation

EO Electron-Optical; Electron Optics

EO Electron Orbit(al)

EO Electro-Optic(s) [also E-O]

EO Electroosmosis

EO Elementary Operation

EO El Oro [Ecuador]

EO Emulsion, Water-in-Oil

EO End Office [Telecommunications]

EO End-On

EO Energy Origin

EO Engineering Order

EO Equal Opportunity

EO Equatorial Orbit

EO Equivalent Orifice [Mechanical Engineering]

EO Errors and Omissions [also E&O, or e&o]

EO Ethylene Oxide

EO Executive Office(r)

EO Executive Order

EO Executive Organ

EO Explosion Only

E&O Eastern and Oriental Express

E&O Errors and Omissions [also e&o, or EO]

E/O Engineering/Operations

E-O Elasto-Optic(s) (also EO)

E-O Electro-Optic(s) [also EO]

eo even-odd (nuclei) (i.e., atomic nuclei whose spin is equal to an odd multiple of $\frac{1}{2} \times \hbar$, $\hbar$ = Planck's constant $\div 2\pi$)

$\varepsilon(\omega)$ dielectric function [Symbol]

$\varepsilon_{\parallel}(\omega)$ dielectric response function parallel to (superconductor) c-axis [Symbol]

$\varepsilon_{\perp}(\omega)$ dielectric response function perpendicular to (superconductor) c-axis [Symbol]

EOA Economic Opportunity Act [US]

EOA Electrical, Optical and Acoustic (Properties)

EOA End of Address (Character)

EOA Essential Oil Association [now Fragrance Materials Association, US]

EOA Extensively Overaged; Extensive Overaging [Metallurgy]

Eoa Ethanolamine

EO/AA Equal Opportunity/Affirmative Action (Employer)

EOAP Earth Observation Aircraft Program [US]

EOAR European Office of Aerospace Research

EOARDC European Office of the Air Research and Development Command [US]

EOARD European Office of Aerospace Research and Development

EOB Electronic Order of Battle

EOB End of Block

EOBT Estimated Off-Block Time

EOC Earth Observation Center [of National Space Development Agency of Japan in Saitama]

EOC EOS (Earth Observing System) Operations Center

EOC Eastern Oregon College [La Grande, US]

EOC Electro-Optic Coefficient

EOC Electro-Optic Crystal

EOC Electro-Optical Camera [ER-2 NASA Ames High-Altitude Aircraft] [also EO-Cam]

EOC Emergency Operating Center

EOC Emergency Operations Center

EOC End of Contract

EOC End of Conversion

EOC End of Cycle

EOC Engine Order Capability

EOC Engine-Out Capability

EOC Epithelial Ovarian Cancer

EO-Cam Electro-Optical Camera [ER-2 NASA Ames High-Altitude Aircraft] [also EOC]

EOCCM Electrooptic Counter-Countermeasures

EOCI Electric Overhead Crane Institute [now Crane Manufacturers Association of America, US]

EOCR Experimental Organic-Cooled Reactor [of Idaho National Engineering Laboratory, Idaho Falls, US]

EO-CSF Eosinophil Colony Stimulating Factor [Biochemistry]

EOCT Electro-Optical Camera, Translinear Scanning [ER-2 NASA Ames High-Altitude Aircraft]

EOD End of Data

EOD End of Dialing

EOD End of Document

EOD Environmental Operations Division [of US Department of Energy, Richland Operations Office, Washington State, US]

EOD Erie Ordnance Depot [Port Clinton, Ohio, US]

EOD Estimated on Dock

EOD Every Other Day

EOD Explosive Ordnance Disposal

EODA Eastern Ontario Development Association [Canada]

EODB End of Data Block

EODC Earth Observation Data Center [UK]

EODC Eastern Ontario Development Corporation [Canada]

EODD Electro-Optical Digital Deflector

EODS Explosive Ordnance Disposal Service

EODS-NPP Explosive Ordnance Disposal Service–Naval Propellant Plant

EODT Explosives Ordnance Disposal Team [formerly Bomb Disposal Squad]

EOE Electro-Optic Effect

EOE End of Extent

EOE Equal Opportunity Employer

EOE Error and Omission Excepted [also eoe]

EOE European Options Exchange [Business Finance]

E&OE Errors and Omissions Excepted [also e&oe]

EOEA Executive Office of Environmental Affairs [Massachusetts, US]

EOEC End of Equilibrium Cycle

EOEC European Organization for Economic Cooperation

EOELD Electron Orbital Energy Level Diagram

EOEML Electrooptic and Electromagnetic Laboratory [of Georgia Institute of Technology, Atlanta, US]

EOF Electroosmotic Flow

EOF Empirical Orthogonal Function (Analysis)

EOF End of File

EOF Energy Optimizing Furnace (Process) [Metallurgy]

EOF Enterprise Objects Framework

Eof End-of-File (Status) [Pascal Programming]

EOG Electrooculogram; Electrooculograph(y)

EOG Electroolfactogram; Electroolfactograph(y)

EOG Enron Oil and Gas Company [US]

EOGR Enhanced Oil and Gas Recovery (Research Program) [of University of Texas, US]

EOHI Electro-Optic Holographic Interferometer; Electro-Optic Holographic Interferometry

EOHP Except Otherwise Herein Provided

EOHT External Oxygen and Hydrogen Tanks [Aerospace]

EOI Earth Orbit Insertion

EOI Electron Optics Instrument(ation)

EOI End of Information

EOI End of Inquiry

EOI End of Interrupt

EOI End of Irradiation

EOI End, or Identify (Signal)

EOI Expression of Interest

EOIC Ethylene Oxide Industry Council [US]

EO-ICWG Earth Observations International Coordination Working Group

EOIM Evaluation of Oxygen Interactions with Materials (Program)

EOIS Electro-Optical Imaging System

EOJ End of Job

EOL Earth Observatory Laboratory

EOL End-of-Life (Fuel Condition) [Nuclear Engineering]

EOL End of Line

EOL End of List

EOL Expression-Oriented Language

EOLI Electrooptics and Laser International (Trade Show) [UK]

EOLM Electrooptical Light Modulator

EOLN End of Line

Eoln End-of-Line (Status) [Pascal Programming]

EOLR Electrical Objective Loudness Rating

EOLT End of Logical Tape

EOM Electro(n)-Optic Method

EOM Electrooptic Material

EOM Electrooptic Modulator

EOM End of Medium

EOM End of Message

EOM End of Mission

EOM End of Month [also eom]

EOM Engineering Operations Manual

EOM Equation of Motion

EOMET Electro-Optics Meteorology (Program) [of Naval Research Laboratory/Naval Ocean Systems Center, US]

EOMF End of Minor Frame

EON End of Number

EONR European Organization for Nuclear Research [of CERN–European Laboratory for Nuclear Research, Geneva, Switzerland]

EOP Earth and Ocean Physics

EOP Emergency Operating Procedure

EOP Emergency Oxygen Pack

EOP End of Page

EOP End of Program

EOP End Output

EOP Equatorial Orthographic Projection

EOP Even-Odd Predominance [Nuclear Physics]

EOP Exchange Offering Prospectus

EOP Executive Office of the President (of the United States of America)

EOP Experiments of Opportunity

EOPAP Earth and Ocean Physics Application Program

EOPC Electro-Optical Phase Change

EOPM Electro-Optic Phase Modulation

EOPO Equal Opportunity Program Office

EOPP Earth Observation Preparatory Program [of Canadian Space Agency]

EOQ Economic Order Quantity

EOQC European Organization for Quality Control [Switzerland]

EOR Earth Orbit Rendezvous

EOR Electro-Optical Research

EOR End-of-Range

EOR End of Record

EOR End of Reel

EOR End of Run

EOR Enhanced Oil Recovery (Process) [Chemical Engineering]

EOR Exclusive, or [preferably XOR]

EOR Explosives Ordnance Reconnaissance

EORA Explosives Ordnance Reconnaissance Agent

EORC Earth Observation Research Center [of National Space Development Agency of Japan in Tokyo]

EOR/F Enhanced Orbiter Refrigerator/Freezer [of NASA Space Shuttle] [also EORF]

EORS Emergency Oil Spill Response System

EORTC European Organization for Research on the Treatment of Cancer

EOS Earth Observation Satellite

EOS Earth Observatory Satellite [of National Space Development Agency of Japan]

EOS Earth Observing Satellite

EOS Earth Observing System [of NASA]

EOS Earth Orbit Shuttle

EOS Earth-Oriented Satellite

EOS Egyptian Organization for Standardization

EOS Electrical Overstress

EOS Electronic Office Service

EOS Electronic Optical System

EOS Electro-Optical Sensor

EOS Electro-Optical System

EOS Electro-Organic Synthesis

EOS Electrophoresis Operations in Space

EOS Emergency Oxygen Supply

EOS Emergency Oxygen System

EOS End of Selection

EOS End of String

EOS Equation of State

EOS ERIN (Environmental Resources Information Network) On-line Service [now Environment Australia On-line Service]

EOS European Optical Society

EOS Extended Operating System

EOS Exxon Oil Spill [near Valdez, Alaska]

EOS-AM Earth Observing System–Ante Meridiem [NASA Satellite Series; First Launch Mid-1998, Morning Equator Crossings (Descending)]

EOSAT Earth Observation Satellite (Company) [Commercial Landsat Satellite System Operator]

EOSDIS Earth Observing System Data and Information System [of NASA]

EOS/ESD Electrical Overstress/Electrostatic Discharge (Protection)

EOSP Earth Observing Scanning Polarimeter

EOS-PM Earth Observing System–Post Meridiem [NASA Satellite Series; Second Launch Late 2000; Afternoon Equator Crossings (Ascending)]

EOS-PM1 Earth Observatory Satellite–Post Meridiem One [of National Space Development Agency of Japan]

EOS Trans Am Geophys Union EOS Transactions of the American Geophysical Union [US]

EOT Electrooptical Technology

EOT End Of Table

EOT End of Tape

EOT End of Task

EOT End of Text

EOT End of Track

EOT End of Transaction

EOT End of Transmission

EOT Engine Order Telegraph

EOTA European Organization for Technical Approvals

EOTC European Organization for Testing and Certification

EOTS Electronic Optical Tracking System

EOTS Epitaxial On-Top Structure [Solid-State Physics]

EOV End of Volume

EOW Energy over Weight

EOW Engineering Order Wire

EOW Engine over the Wing

EOW Equal Opportunities for Women

EOX Extractable Organic Halide; Extractable Organic Halogen

EOY End of Year

EP Echo Prospecting

EP (Ecole Polytechnique) – French for "Polytechnic Institute," or "Institute of Technology"

EP Ecole Polytechnique [Polytechnic Institute, Montreal, Quebec, Canada]

EP Ectoparasite [Biology]

EP Effective Permeability [Physical Chemistry]

EP Eigenvalue Problem [Mathematics]

EP Einstein-Planck (Law) [Physics]

EP Elasticoplastic (Solid)

EP Elastic Peak

EP Elastic-Plastic (Behavior)

EP Elastoplastic(ity)

EP Elbow Pitch

EP Electrically Poled
EP Electrical Porcelain
EP Electrical Power
EP Electrical Properties
EP Electric Polarization
EP Electric Potential
EP Electric Power
EP Electrode Potential
EP Electrolytic Polishing
EP Electromagnetic Pump
EP Electronic Packaging
EP Electronic Potentiostat
EP Electron Pair [Physical Chemistry]
EP Electron-Phonon (Interaction) [Solid-State Physics]
EP Electron Probe
EP Electrophony
EP Electrophorescence; Electrophorescent
EP Electrophoresis; Electrophoretic
EP Electrophotographic; Electrophotograph(y)
EP Electrophysical; Electrophysics
EP Electroplastic(ity)
EP Electroplate(d); Electroplater; Electroplating
EP Electropneumatic(s)
EP Electropolish(ed); Electropolishing
EP Electropolymer(ization)
EP Electropositive; Electropositivity
EP Electropulse
EP Elongated Punch
EP Elliptically Polarized (Light); Elliptical Polarization
EP El Paso [Texas, US]
EP Emergency Preparedness
EP Emission Power
EP Empirical Pseudopotential [Physics]
EP Emulation Program
EP Emulsion Polymer(ization)
EP End of Program
EP Endogenous Pyrogen [Biochemistry]
EP Endoparasite [Microbiology]
EP (Office of Domestic and International) Energy Policy [of US Department of Energy]
EP Energy Product(ion)
EP Engineering Personnel
EP Engineering Physics
EP Engineering Plastics
EP Engineering Polymer
EP Engine Performance
EP Engine Pod
EP Engine Pressure
EP English Patent
EP Enter and Proceed [also ep]
EP Environmental Parameter
EP Environmental Park
EP Environmental Physiology
EP Environmental Protection

EP Environmental-Protective Plan
EP Epitaxial Planar [Solid-State Physics]
EP Epoxy [also Ep]
EP Equipment Practice
EP Equipotential
EP Error Propagation [Data Communications]
EP Escape Peak
EP Escape Pinion
EP Escape Probability
EP Etched Plate
EP Etch Pit(ting) [Metallurgy]
EP Ethylene Propylene
EP European Parliament
EP European Patent
EP Even Parity [Computers]
EP Excitation Power
EP Exclusion Principle [Quantum Mechanics]
EP Executive Pilot
EP Exhaust Piston
EP Exhaust Port
EP Expanded Polystyrene
EP Expendable (Casting) Pattern [Metallurgy]
EP Experimental Physicist; Experimental Physics
EP Experimental Procedure
EP Explosion-Proof
EP Extended Performance
EP Extended Play
EP Extended Programmability
EP External Pressure
EP Extraction Procedure (Toxicity)
EP Extra-Protection
EP Extra Pulse
EP Extraterrestrial Physics
EP Extreme Pressure
EP1200 Epizootiological Database [US]
E4P Erythrose-4-Phosphate
EP- Iran [Civil Aircraft Marking]
E/P Ethylene/Propylene (Ratio)
E-P Electron/Proton [also e-p]
Ep Epoxy [also EP]
E/p Voltage-to-Pressure (Ratio)
E£ Egyptian Pound [Currency]
eφ work function [Symbol]
EPA Eastern Provincial Airways [Canada]
EPA Education Privacy Act [US]
EPA Effective Program Approval
EPA Effective Projected Area
EPA (5,8,11,14,17-)Eicosapentaenoic Acid
EPA Electronic Publishing Abstracts [US]
EPA Emergency Planning Association [UK]
EPA Enhanced Performance Architecture
EPA Environmental Protection Act [Canada]
EPA Environmental Protection Agency [Washington, DC, US]

EPA Environment Planning Authority

EPA Environment Protection Agency (of Australia) [Succeeded by Environment Protection Group, Environment Australia]

EPA Environment Protection Authority [In Several Australian States, such as New South Wales, Victoria and Western Australia]

EPA Escuela Panamericana de Agricultura [Pan-American School of Agriculture, Qamorano Valley, Honduras]

EPA Ethyl γ-Phenylbutyrate

EPA Europäisches Patentamt [European Patent Office, Munich, Germany]

EPA European Photochemistry Association [UK]

EPA European Productivity Agency [now Organization for Economic Cooperation and Development, France]

EPA Extended Pair Approximation [Physics]

EPA Expanded Polystyrene Association

EPA Extended Performance Architecture

EPAA Emergency Petroleum Allocation Act [US]

EPAA Epithermal Neutron Activation Analysis

EPAA European Primary Aluminum Association [Germany]

EPA/AWMA Environmental Protection Agency/Air and Waste Management Association (Project) [US]

EPABX Electronic Private Automatic Branch Exchange

EPAC Economic Planning Advisory Council

EPAC European Petroleum Application Center [Norway]

EPAC Expanded Polystyrene Association of Canada

EPAD Error Protection (Data) Packet Assembly/Disassembly

EPA/ETI Environmental Protection Agency's Environmental Technology Initiative [US]

EPA/GC Environmental Protection Agency/Gas Chromatography (Trace Analysis)

EPA J EPA Journal [of US Environmental Protection Agency]

EPALL Emergency Preparedness at Local Level

EPAM Elementary Perceiver and Memorizer

EPA NCC Environmental Protection Agency National Computer Center [US]

EPA Newsl EPA Newsletter [of European Photochemistry Association, UK]

E-PAPS Electronic Physics Auxiliary Publication Service [of American Physical Society, US]

EPA/RCRA Environmental Protection Agency–Resource Conservation and Recovery Act [also EPA-RCRA, or EPA RCRA, US]

EP(ARR) Environment Protection (Alligator Rivers Region) (Act) [Australia]

EPASYS European Patents Administration System [of European Patent Office, Munich, Germany]

EPA-TCLP Environmental Protection Agency–Toxicity Characteristic Leaching Procedure [also EPA TCLP]

EPA/TSCA Environmental Protection Agency–Toxic Substances Control Act [also EPA TSCA, US]

EPB Earth Physics Branch [of Department of Energy, Mines and Resources, Ottawa, Canada]

EPB Electro-Physics Branch [of NASA Lewis Research Center, Cleveland, Ohio, US]

EPB Environmental Periodicals Bibliography [of Environmental Studies Institute, US]

EP&BC Environment Protection and Biodiversity Conservation (Bill) [Australia]

EPBCO Europium Praseodymium Barium Copper Oxide (Superconductor)

EPBHM Ethyl-n-Propyl-n-Butyl-n-Hexylmethane

EP-BL Ethylene Propylene Block Copolymer

EPBM Enhanced Probability-Based Matching

EPBS Earth-Pressure Balanced Shield

EPBX Electronic Private Branch Exchange

EPC Earth Potential Compensation

EPC Easy Processing Channel (Furnace Black)

EPC Economic Policy Committee [of Organization for Economic Cooperation and Development, France]

EPC Editorial Processing Center

EPC Electric Power Club [US]

EPC Electric Propulsion Conference

EPC Electrolytic Photocell

EPC Electronic Power Conditioner

EPC Electronic Pressure Control

EPC Electronic Program Control

EPC Electroplated Copper

EPC Electropowder Coating

EPC Elementary Processing Center

EPC Emergency Planning Canada

EPC Emergency Preparedness Canada [of Department of National Defense]

EPC Engineering, Procurement and Construction

EPC Environmental Protection Control

EPC Environment Protection Council [Kuwait]

EPC Error-Position Code

EPC Error Protection Code

EPC European Patent Convention

EPC European Plastics Converters Federation [Belgium] [formerly Eutraplast]

EPC Evaporative Pattern Casting [Metallurgy]

EPC Executive Policy Committee [Western Australia]

EPC Expendable Pattern Casting (Process) [Metallurgy]

EPC Export Publicity Council [UK]

EPC External Power Contractor

EPCA Economic Planning and Coordination Authority [Hawaii, US]

EPCA Energy Policy and Conservation Act [US]

EPC Black Easy-Processing Channel Black [also EPC black]

EPCC Environment Policy Coordinating Committee [of Environment Australia]

EPCCS Emergency Positive Control Communications System

EPCDC Electrical Power Conditioning, Distribution and Control

EPCG Environment Priorities and Coordination Group [of Environment Australia]

EPCIA Expanded Polystyrene Cavity Insulation Association [UK]

EPCM Electropulse Chemical Machining

EPCM Engineering, Procurement and Construction Management

EPCMS El Paso County Medical Association [Colorado, US]

EPCO Engine Parts Coordinating Office

EPCR Emergency Planning and Community Right-to-Know (Act) [US]

EPCRA Emergency Planning and Community Right-to-Know Act [US]

EPD Earliest Possible Date

EPD Earliest Practicable Date

EPD Eidgenössisches Politisches Departement [Federal Department of External Affairs, Switzerland]

EPD Electric Potential Drop

EPD Electric(al) Power Database

EPD Electric(al) Power Distribution

EPD Electronic Proximity Detector

EPD Electron Pair Donor

EPD Electrophoretic Deposition

EPD Emergency Procedures Document

EPD Engineering Physics Department

EPD Environmental Protection Division

EPD Epitaxial Planar Device [Solid-State Physics]

EPD Etch Pit Density [Metallurgy]

EPD Ethyl Phosphorous Dichloride

EPD Excess Profits Duty

EPD Exchange Parameter Definition [Physics]

EPD Experimental Physics Department

EPD Export Programs Division [of Department of Foreign Affairs and International Trade, Canada]

EPD Extraction and Processing Division [of Minerals, Metals and Materials Society; of American Institute of Mining, Metallurgical and Petroleum Engineers, US]

EPDA European Plastics Distributors Association

EPDAN Elastic-Plastic Deformation Analysis

EPDB Electrical Power Distribution Box

EPDB Experiment Power Distribution Box

EPDC Electrical Power Distribution and Control

EPDC Electric Power Development Company [Japan]

EPDCC European Pressure Die Casting Committee [UK]

EPDCS Electrical Power Distribution and Control System

EPDIC European Powder Diffraction Conference

EPDM Electronic Product Design and Manufacture [of Engineering and Physical Sciences Research Council, UK]

EPDM Ethylene-Propylene-Dimonomer; Ethylene-Propylene Diene Monomer (Rubber)

EPD/MDMD Extraction and Processing Division/Materials Design and Manufacturing Division (Joint Committee) [of Minerals, Metals and Materials Society; of American Institute of Mining, Metallurgical and Petroleum Engineers, US]

EPDS Electrical Power Distribution Subsystem

EPDS Electrical Power Distribution System

EPDS Environmental Planning Database System

EPDT Estimated Project Duration Time

EPDU Electrical Power Distribution Unit

EPE Electroplastic Effect

EPE Energetic Particles Explorer Satellite [of NASA]

EPE Engineering Progress Exposition

EPE Ethyl Phenyl Ether

EPE European Conference on Power Electronics and Applications

EPE (Semiconductor) External Power Efficiency [Solid-State Physics]

EPEA Electrical Power Engineers Association [UK]

EPEC Emerson Programmer, Evaluator and Controller

EPEC Engineers' Precollege Education Council [of American Association of Engineering Societies, US]

EPEC Environmental Protection and Education Club [Kenya]

EPEI Enterprise Prince Edward Island [Canada]

EPES Elastic Peak Electron Spectroscopy

EPETMA European Polyester Terephthalate Manufacturers Association

EPEX Epoxy Resin Extender

EPF Easy-Processing Furnace (Black)

EPF *(Ecole Polytechnique Fédérale)* – French for "Federal Polytechnic Institute"

EPF Electrical Percussion Fuse

EPF Electron Phase Focusing

EPF Emery Paper Figure

EPF Empresa Petrolera Fiscal [State Petroleum Company of Peru]

EPF European Packaging Federation [Netherlands]

EPF European Polymer Federation [Switzerland]

EPFA Established Program Financing Act

EPFB Easy Processing Furnace Black

EPF Black Easy-Processing Furnace Black [also EPF black]

EPFM Elastic-Plastic Fracture Mechanics

EPFL Ecole Polytechnique Fédérale de Lausanne [Swiss Federal Institute of Technology at Lausanne]

E&P Forum (Oil Industry International) Exploration and Production Forum [UK]

EPF Q EPF Quarterly [Publication of the European Packaging Federation, Netherlands]

EPFRS Effective Potential (Approximately in the) Free-Particle Reference System

EPFS Extended Photoemission Fine Structure

EPFT Elastic-Plastic Fracture Toughness

EPFW Eidgenössische Pulverfabrik in Wimms [Federal Powder Plant at Wimms, Switzerland]

EPG Electrical Power Generator

EPG Electrostatic Particle Guide

EPG Emergency Procedure Guidelines

EPG Environment Protection Group [Australia]

EPG European Press Group

EPG Experimental Proving Ground

EPGA Emergency Petroleum and Gas Administration [US]

EPH Electric Process Heating [Institue of Electrical and Electronics Engineers–Industry Applications Society Committee]

E$_h$-pH Oxidation Potential versus pH-Value (i.e., Pourbaix Diagram)

e/ph electron(s) per photon [Symbol]

e-ph electron-phonon (interaction) [Symbol]

EPHA Ecole Practique des Hautes Etudes [School of Advanced Applied Studies, France]

EPhMRA European Pharmaceutical Marketing Research Association [UK]

EPI Earth Path Indicator

EPI Echo-Planar Imaging

EPI Effective Pair Interaction [Physics]

EPI Electric Power Institute [of Central Research Institute, Japan]

EPI Electric Power Industry

EPI Electronic Position Indicator

EPI Electron Probe Instrument

EPI Electro-Plasma Inc. [US]

EPI Electro-Pyrolysis, Inc. [US]

EPI Elevation Position Indicator

EPI Emergency Position Indicator

EPI Emulsion Polymers Institute [US]

EPI Energy Principle of Indentation [Metallurgy]

EPI Engineering Physics Institute

EPI Environmental Policy Institute [US]

EPI Epichlorohydrin [also epi]

EPI Epitaxial Products International Limited

EPI Expanded Program on Immunization [of World Health Organization]

epi epichlorohydrin [also EPI]

EPIA Electric Power Industry Abstracts [of Edison Electrical Institute, US]

EPIB Emergency Position-Indicating Beacon

EPIC Earth-Pointing Instrument Carrier

EPIC Edison Polymer Innovation Corporation [of Case Western Reserve University, Cleveland, Ohio, US]

EPIC Educational Program Innovations Center [US]

EPIC Elastic-Plastic Impact Calculation [Finite Element Code]

EPIC Electrical Properties Information Center [US]

EPIC Electronic Photochromic Integrating CRT (Cathode Ray Tube)

EPIC Electronic Printer Image Construction

EPIC Electronic Products Information Center

EPIC Electronic Properties Information Center [of Center for Information and Numerical Data Analysis (and Synthesis), US]

EPIC Environmental and Plastics Industry Council [Canada]

EPIC Environmentally Protected Integrated Circuit

EPIC Epitaxial Integrated Circuit

EPIC European Photon Imaging Camera

EPIC European Proliferation Information Center [UK]

EPIC Evaluator Programmer Integrated Circuit

EPIC Evidence Photographers International Council [US]

EPIC Exchange Price Information Computer [UK]

EPIC Explicitly Parallel Instruction Computing

EPIC Extended Performance and Increased Capability

EPIC European Product Information Cooperation

EPIC-II Elastic-Plastic Impact Calculation for Two-Dimensional Problems [Finite Element Code]

EPIDC East Pakistan Industrial Development Council

Epidemiol Epidemiological; Epidemiologist; Epidemiology

EPIE Educational Products Information Exchange (Institute) [US]

epi-GaAs Epitaxial Gallium Arsenide [also Epi-GaAs, or EPI-GaAs]

epi-GaN Epitaxial Gallium Nitride [also Epi-GaN, or EPI-GaN]

epi-GTX-8 epi-Gonyautoxin VIII [Biochemistry]

EPIJ Electric Power Industry of Japan

EPI-LPPS Electro-Plasma Inc.–Low-Pressure Plasma System

EPIMS EEE (Electronic, Electrical, Electromechanical) Parts Information Management System [of NASA Goddard Space Flight Center, Greenbelt, Maryland, US]

EPINS Electronic Parts Information Network System [of NASA Jet Propulsion Laboratory, Pasadena, California, US]

EP(IP) Environment Protection (Impact of Proposals) (Act) [Australia]

EPI Q Rep EPI Quarterly Report [of Environmental Policy Institute, US]

EPIRB Emergency Position-Indicating Radio Beacon

EPIRBS Emergency Position-Indicating Radio Beacon System

epi-Si Epitaxial Silicon [also Epi-Si, or EPI-Si]

EPISTLE European Process Industries

epi-WS$_2$ Epitaxial Tungsten Disulfide [also EpiWS$_2$, or EPI-WS$_2$]

EPJ A The European Physical Journal A

EPJ B The European Physical Journal B

EPJ C The European Physical Journal C

EPJ D The European Physical Journal D

EPJAP European Physical Journal–Applied Physics

EPK Ethyl Propyl Ketone

EPKDC European Polycystic Kidney Disease Consortium

EPL Effective Privilege Level

EPL Electrical Parts List

EPL Electrical Power Level

EPL Electromechanical Parts List

EPL Electronic Parts List

EPL Electrophotoluminescence; Electrophotoluminescent

EPL ELINT (Electronic Intelligence) Parameter Limits List

EPL Emergency Power Level

EPL Encoder Programming Language

EPL Environmental Protection Limit

EPL Exciter Power Logic

EPL Exclusive Prospecting License

EPL Extreme Pressure Lubricant

EP-L Electrophotographic (Cartridge) [Laser Printers]

EPLA Electronic Precedence List Agency

EPLAF European Planning Federation [UK]

EPLAN Elastic-Plastic Analysis

EPLANS Engineering, Planning and Analysis System

EPLD EISA (Extended Industry Standard Architecture) Programmable Logic Device

EPLD Electrically Programmable Logic Device

EPLD Enhanced Programmable Logic Device

EPLD Erasable Programmable Logic Device

EP/LP Extended-Play/Long-Play (Mode)

EPLSI Ecole Polytechnique Laboratoire des Solides Irradies [Polytechnic Institute Irradiated Solids Laboratory, Palaiseau, France]

EPM Earth-Probe-Mars [of NASA]

EPM Ecole Polytechnique de Montreal [Polytechnic Institute of Montreal, Quebec, Canada]

EPM Economic Performance Monitoring

EPM Electromagnetic Processing of Materials

EPM Electron Probe Microanalysis; Electron Probe Microanalyzer [EPMA Preferred]

EPM Electron Probe Microscopy

EPM Electrophoretic Mobility [Biochemistry]

EPM Elemental Powder Metallurgy

EPM Empirical Pseudopotential Method [Physics]

EPM Engineering Procedure Memorandum

EPM Enhanced Editor for Presentation Manager

EPM Enterprise Process Management

EPM Environmental Planning and Management

EPM Environmental Project Manager

EPM Equivalent per Million [also epm]

EPM Ethylene-Propylene Monomer

EPM Explosions per Minute [also epm]

EPM Explosive Powder Metallurgy

EPM External Polarization Modulation

EPMA Electronic Parts Manufacturers Association

EPMA Electron Probe Microanalysis; Electron Probe Microanalyzer [also EPM]

EPMA European Powder Metallurgy Association

EPMAU Expected Present Multi-Attribute Utility

EPMS Engineering Performance Management System

EPN Epoxidized Phenol Novolac; Epoxy Phenol Novolac

EPN Ethyl-P-Nitrophenyl Phenylphosphorothioate

EPN o-Ethyl-o-p-Nitrophenyl Phenylthionophosphate

EPN o-Ethyl-o,p-Nitrophenyl Phenylphosphonothioate

EPN External Priority Number

EPNdB Effective Perceived Noise Level in Decibels

EPNdB Equivalent Perceived Noise Level in Decibels

EPNdBA Effective Perceived Noise Level in dBA (decibels, adjusted)

EPNL Effective Perceived Noise Level

EPNL Equivalent Perceived Noise Level

EPNS Electroplated Nickel Silver

EPO EOS (Earth Observing System) Project Office

EPO Element Project Office

EPO Emergency Power Outage

EPO Energy Policy Office [US]

EPO Epoxidized Polyolefin

EPO Erythropoietin [also Epo] [Biochemistry]

EPO European Patent Office [Munich, Germany]

EPO European Patent Organization

EPO Examination Procedure Outline

Epo Erythropoietin [also EPO] [Biochemistry]

EPOA East Coast Petroleum Operators Association [US]

EPOC Environment Protection Policy [of Organization for Economic Cooperation and Development]

EPOC ESCAP (Economic and Social Commission for Asia and the Pacific) Pacific Operations Center [of United Nations]

EPOC External Payload Operations Center

EPOCS Eastern Pacific Ocean Climate Study

EPOP European Polar-Orbiting Platform

EPOS Electronic Point of Sale [also epos]

EPOS Engineered Plastics on Screen

EPOS European Physical Optical Society

EPOS European Polarstern Expedition [*Note:* The "Polarstern" is a German Polar Research Vessel]

EPP Ecole Polytechnique de Paris [Polytechnic Institute of Paris, France]

EPP Electromagnetic Wave Propagation Panel [of NATO Advisory Group for Aerospace Research and Development]

EPP Electron-Positron Pair [Physics]

EPP Electrophoretic Potential

EPP Elemental Pitting Probability [Corrosion]

EPP Empirical Pairwise Potential [Physics]

EPP End Plate Potential

EPP Enhanced Parallel Port [of Institute of Electrical and Electronics Engineers, US]

EPP Environmentally Preferable Purchasing (Program) [US]

EPP Environment Protection Program

EPP European Producer Price

EPP Expanded Precessive Plasma

EPP Extended Parallel Port

EPPG European Program Providers Group [UK]

EPPO Emergency Planning Office of Ontario [Canada]

EPPO European and Mediterranean Plant Protection Organization [France]

EPPP Employee Profit Participation Plan

EPPS Electrical Power/Pyrosequential System

EPPS Ethylpiperazinepropanesulfonic Acid

EPPS N-(2-Hydroxyethyl)-Piperazine-N'-3-Propanesulfonic Acid; 4-(2-Hydroxyethyl)-1-Piperazinepropanesulfonic Acid

EPR East Pacific Rise

EPR Effective Production Rate

EPR Einstein-Podolsky-Rosen (Experiment) [Quantum Mechanics]

EPR Electrical Pressure Regulator

EPR Electrochemical Potentiochemical Reactivation

EPR Electrochemical Potentiokinetic Reactivation (Test)

EPR Electronic Planning and Research

EPR Electron Paramagnetic Resonance

EPR Electron Proton Resonance

EPR Enantiometrically Pure Chemicals

EPR Endoplasmic Reticulum [Cytology]

EPR Engine-Pressure Ratio

EPR Equivalent Parallel Resistance

EPR Error Pattern Register

EPR Essential Performance Requirements

EPR Ethylene-Propylene Rubber

EPR Evaporator Pressure Regulator

EPR-DOS Electrochemical Potentiokinetic Reactivation–Degree of Sensitization

EPRI Electric Power Research Institute [Palo Alto, California, US]

EPRI Electric Power Research Institute [Korea]

EPRI-CMP Electric Power Research Institute–Center for Materials Production [also EPRI CMP, US]

EPRI J Electric Power Research Institute Journal [also EPRIJ, US]

EPRI-RDS Electric Power Research Institute–Research and Development Information System [also EPRI RDS, US]

EPRN Emergency Program Release Notice

EPROM Electrically Programmable Read-Only Memory

EPROM Erasable Programmable Read-Only Memory

EPRS Electron Paramagnetic Resonance Spectroscopy

EPRTCS Emergency Power Ride Through Capability System

EPS Early Production System

EPS Earnings per Share

EPS Electric(al) Power Storage

EPS Electric(al) Power Subsystem

EPS Electric(al) Power Supply

EPS Electric(al) Power System

EPS Electric(al) Propulsion System

EPS Electrochemical Photocapacitance Spectroscopy

EPS Electromagnetic Position Sensor

EPS Electron-Phonon Scattering [Solid-State Physics]

EPS Electrophoresis Power Supply

EPS Emergency Power Supply

EPS Encapsulated PostScript (File)

EPS Engineering Performance Standards

EPS Enterprise Print Service

EPS Environmental Protection Service [Canada]

EPS Equilibrium Problem Solver

EPS Equipotential Surface

EPS Equivalent Prior Sample

EPS European Physical Society [Switzerland]

EPS Even Parity Select [Computers]

EPS Expandable (or Expanded) Polystyrene

EPS Experimental Power Supply

.EPS Encapsulated PostScript [File Name Extension]

EPSC Elastic-Plastic Self-Consistent

EPSCG European Parliamentary and Scientific Contact Group [France]

EPSCoR Experimental Program to Stimulate Competitive Research [of National Science Foundation, US]

EPSCoR-NSF Experimental Program to Stimulate Competitive Research–National Science Foundation (Grant) [US]

EPSCS Enhanced Private Switched Communications Service

EPSD Electronics and Power Sources Directorate [of US Army Research Laboratory, Fort Monmouth, New Jersey]

EPSEC European Photovoltaic and Solar Energy Conference

EPSF Encapsulated PostScript Files

EPSG Epiphytic Plant Study Group [UK]

EPS-HEP (International) Europhysics Conference on High-Energy Physics [of European Physical Society] [also EPSHEP]

EPSI Electron Point Source Instrument

EPSIG Standard for Electronic Manuscript Preparation and Mark-up of the Electronic Publishing Special Interest Group

EPSL Emergency Power Switching Logic

EPSP Enolpyruvylshikinate Phosphate [Biochemistry]

EPSP Excitatory Postsynaptic Potential [Medicine]

EPSP Experiment Power Switching Panel

EPSRC Engineering and Physical Sciences Research Council [formerly Science and Engineering Research Council, UK]

EPSS Experimental Packet-Switched (or Switching) Service [UK]

EPT Electromagnetic Propagation Tool

EPT Electron Propagator Theory

EPT Electrostatic Printing Tube

EPT Emergency Procedure Trainer

EPT Erbium Plutonium Titanate

EPT Ethylene-Poly(propylene) Terpolymer

EPT Ethylene-Propylene Terpolymer

EPT Exclusive Polarization Transfer

EPT Extraction Procedure Toxicity

EPTA Electrophysiological Technologists Association [UK]

EPTA European Power Tool Association [Germany]

EPTA Expanded Program of Technical Assistance [of United Nations' Economic and Social Council]

Eptam Ethyl-N-,N-Di-n-Propylthiocarbamate

EPTC Ethylpropylthiocarbamate; S-Ethyl Di-N,N-Propylthiocarbamate; S-Ethyl-N,N-Dipropyl Thiocarbamate

EPTC European Petroleum Technical Corporation

EPTD Ethylphosphorous Thiodichloride

EPTE Existed Prior to Entry

EPTU Events per Time Unit

EPU Earth Protection Unit [Electrical Engineering]

EPU Economic Planning Unit [Malaysia]

EPU Electrical Power Unit

EPU Emergency Power Unit

EPU European Payment Union [France]

EPU European Picture Union [Sweden]

EPU Executive Processing Unit

EPU Expandable Processor Unit

ε-Pu Epsilon Plutonium [Symbol]

EPUB Electronic Publisher [WordPerfect]

e-publication electronic publication

e-publishing electronic publishing

EPUL Ecole Polytechnique de l'Université de Lausanne [Polytechnic Institute of the University of Lausanne, Switzerland]

EPUSP Escola Politécnica da Universidade de São Paulo [Polytechnic Institute of the University of São Paulo, Brazil]

EPUT Events per Unit Time

EPVA Entomopoxvirus A

EPVB Entomopoxvirus B

EPVC Entomopoxvirus C

EPWG Environmental Projects Working Group

EPWRL Equivalent Plane Wave Reverberation Level

EPX Echoplexing [also Epx, or epx]

EPXMA Electron-Probe X-Ray Microanalysis

EPY N-Ethylpyridinium [also EPy, or epy]

EPZ Export Processing Zone

EPZA Export Processing Zone Administration [Taiwan]

EQ Electric Quadrupole

EQ End-Quench [Metallurgy]

EQ Enquiry

EQ Environmental Quality

Eq Equal(ization); Equalizer [also eq]

Eq Equation [also eq]

Eq Equator(ial) [also eq]

Eq Equivalency; Equivalent [also eq]

eq equal

eq (gram) equivalent (weight) [Unit]

$\varepsilon(q)$ static dielectric function [Symbol]

EQC Environmental Quality Council [US]

EQCC Entry Query Control Console

EQCM Electrochemical Quartz Crystal Microbalance

EQCN Electrochemical Quartz Crystal Nanobalance

EQCO European Quality Control Organization

EQD Electrical Quality (Assurance) Directorate [UK]

EQD Embedded Quantum Dot

EQE Event Queue Element

EQE External Quantum Efficiency [Electronics]

EQEC European Quantum Electronics Conference

Eq Guin Equatorial Guinea

EQI Electric Quadrupole Interaction

EQID Exchange-Quadrupole-Induced Dipole

Equip Equipment [also equip]

equiv equivalent

EQL Estimated Quantitation Limit

Eql Equal(ity) [also eql]

eq/L (gram) equivalents per liter [also eq/l] [Unit]

EQM Electric Quadrupole Moment

Eqmt Equipment [also eqmt]

Eqn Equation [also eqn]

Eqns Equations [also eqns]

EQO Environmental Quality Objective

Eqpt Equipment [also eqpt]

EQQ Electric Quadrupole-Quadrupole Correlation (Functions)

EQR Equipotential Rings

EQRC Engineering Quality Reliability Center [at Sandia National Laboratory, Albuquerque, New Mexico, US]

EQS Environmental Quality Standard

EQS Environmental Quality Statement

EQS Equivalent-to-Sheated Explosive

EqS Equivalent to Sheathed (Explosive)

eq sp equally spaced

equ equate

EQUALANT International Cooperative Investigations of the Equatorial Atlantic [also Equalant]

EQUALANT I First International Cooperative Investigation of the Equatorial Atlantic from February to April 1963 [of Intergovernmental Oceanographic Commission] [also Equalant I]

EQUALANT II Second International Cooperative Investigation of the Equatorial Atlantic from August to September 1963 [of Intergovernmental Oceanographic Commission] [also Equalant II]

Equat Equator(ial) [also equat]

EQUATE Electronic Quality Assurance Test Equipment

Equil Equilibrium [also equil]

equim equimolecular

EQUIP Equipment Usage Information Program

Equip Equipment [also equip]

equiv equivalent

Equl Equuleus [Astronomy]

ER Easy to Reach

ER Echo Ranging

ER Economic Region

ER Edge Retention (Technique)

ER Effectiveness Report

ER Effective Resistance

ER Efficiency Ratio

ER Einstein-Rosen (Waves) [Relativity]

ER Elastic Recoil [Physics]

ER Elastic Recovery

ER Elastoresistance

ER Electric(al) Resistance; Electric Resistivity

ER Electric Rocket

ER Electrolytic Rectifier

ER Electrolytic Reduction

ER Electrolytic Refining (or Electrorefining)

ER Electronic Radiography

ER Electronic Ram

ER Electronic Reconnaissance

ER Electronic Relay

ER Electron Radiography

ER Electron Reflectance

ER Electron Reflectometer

ER Electrophilic Reaction; Electrophilic Reagent

ER Electrorefining (or Electrolytic Refining)

ER Electroreflectance

ER Electrorheological; Electrorheology

ER Emergency Repair

ER Emergency Response

ER Emegency Room

ER Emission Rate

ER Emitter Region [Electronics]

ER Energy Ratio

ER Endoplasmic Reticulum [Cytology]

ER Energy Reradiation

ER (Office of) Energy Research [of US Department of Energy]

ER Energy Research

ER Energy Resolution

ER Energy Resource(s)

ER Engineering Request

ER Engineering Requirements

ER Engineering Research

ER Engineering Resin

ER Engineering Route

ER Engine Room

ER Enhanced Radiation

ER Entre Ríos [Argentina]

ER Environmental Report

ER Environmental Requirement

ER Environmental Resource [now Environmental Action Foundation, US]

ER Environmental Resource(s)

ER Environmental Restoration

ER Epitaxial Relationship [Solid-State Physics]

ER Epithermal (Nuclear) Reactor

ER Epoxidized Rubber

ER Epoxy Resin

ER Equivalent Roentgen

ER Eritrea [ISO Code]

ER Error [also Er, or er]

ER Error Rate

ER Error Recovery [Computers]

ER Etch Rate

ER Evaporation Rate

ER Exchange Rate

ER Expanded Rubber

ER Expansion Ratio

ER Experimental Reactor

ER Explanation Report

ER Exploratory Research

ER Extended Range

ER Extended Resources

ER Extensional Rheometry

ER External Ratio

ER External Reflection

ER External Resistance

ER Extraction Replica

ER-2 Extended Resources-2 [NASA Ames High-Altitude Aircraft]

E&R Education and Research

E&R Engineering and Repair

E(R) Total Energy of Crystal at Lattice Spacing R [Symbol]

Er Erbium [Symbol]

Er Error [also ER, or er]

Er^{3+} Erbium Ion [also Er^{+++}] [Symbol]

Er-123 ErBa$_2$Cu$_3$O$_{7-\delta}$ [An Erbium Barium Copper Oxide Superconductor]

Er-162 Erbium-162 [also ^{162}Er, or Er^{162}]

Er-163 Erbium-163 [also ^{163}Er, or Er^{163}]

Er-164 Erbium-164 [also ^{164}Er, or Er^{164}]

Er-165 Erbium-165 [also ^{165}Er, or Er^{165}]

Er-166 Erbium-166 [also ^{166}Er, or Er^{166}]

Er-167 Erbium-167 [also ^{167}Er, or Er^{167}]

Er-168 Erbium-168 [also ^{168}Er, or Er^{168}]

Er-169 Erbium-169 [also ^{169}Er, or Er^{169}]

Er-170 Erbium-170 [also ^{170}Er, or Er^{170}]

Er-171 Erbium-171 [also ^{171}Er, or Er^{171}]

Er-211 Er$_2$BaCuO$_5$ [An Erbium Barium Copper Oxide Superconductor]

Er-459 Er$_4$Ba$_5$Cu$_9$O$_y$ [An Erbium Barium Copper Oxide Superconductor]

Er-5712 Er$_5$Ba$_7$Cu$_{12}$O$_y$ [An Erbium Barium Copper Oxide Superconductor]

E(r) Backscattered Power [Symbol]

(r) Mean Rotational Energy [Symbol]

E/ρ specific stiffness (i.e., ratio of Young's modulus to density)

E-ρ Young's modulus-to-density (chart)

ε_x(r) exchange potential (in physics) [Symbol]

$\varepsilon_{xc}(\rho)$ exchange and correlation energy per electron of a homogeneous electron gas of density ρ (in physics) [Symbol]

ERA Early Retirement Adjustment

ERA Ecological Risk Assessment

ERA Economic Regulatory Administration [of US Department of Energy]

ERA Effective Range Approximation

ERA Electrical Replaceable Assembly

ERA Electric Railroaders Association [US]

ERA Electric Research Association [UK]

ERA Electronic Reading Automation

ERA Electronic Research Announcements [Publication of the American Mathematical Society, US]

ERA Electronic Research Association

ERA Electronic Revision Approval

ERA Electronics Representatives Association [US]

ERA Electron Ring Accelerator [Particle Physics]

ERA Emergency Relief Administration [US]

ERA Energy Reduction Analysis

ERA Energy Reorganization Act [US]

ERA Energy Research Abstracts

ERA Energy Resources of Australia

ERA Engineering Research Associate

ERA Engineering Research Association

ERA Environmental Radioactivity

ERA Environmental Resources of Australia

ERA Equal Rights Amendment [US]

ERA European Regional Airlines [UK]

ERA Expedited Response Action

ERA Extended Registry Attributes

ERAB Energy Research Advisory Board [of US Department of Energy]

Er(Ac)$_3$ Erbium Acetate [also Er(ac)$_3$]

Er(ACAC)$_3$ Erbium Acetylacetonate [also Er(AcAc)$_3$, or Er(acac)$_3$]

ERACE Electrical Remediation at Contaminated Environments

ERAM Extended-Range Antiarmor Munition

ERAMA Enfield Royal Arms Manufacturing Arsenal [UK]

ERAP Earth Resources Aircraft Program [of NASA]

ERAP Error Recording and Analysis Procedure

ERAPS Expendable Reliable Acoustic Path Sonobuoy

ERAS Electronic Reconnaissance Access Set

ERAS Electronic Routing and Approval System [of Hughes Aircraft Company, US]

ERAS Extended Real Associated Solution (Model)

ErAs Erbium Arsenide (Semiconductor)

ERASER Elevated Radiation-Seeking Rocket

ERATO Exploratory Research for Advanced Technology [of Japan Research Development Corporation]

ERATO JRDC Exploratory Research for Advanced Technology–Japan Research Development Corporation [Yokohama, Japan]

ERAU Embry-Riddle Aeronautical University [Daytona Beach, Florida, US]

ERB Earth Radiation Budget (Instrument) [NASA NIMBUS Meteorological Research Satellite Series]

ERB Easily Retrievable Basket

ERB Educational Records Bureau [US]

ERB Engineering Review Board

ERB Equipment Review Board

ERB Experiment Review Board

ErBCO Erbium Barium Copper Oxide (Superconductor) [also EBCO]

ERBE Earth Radiation Budget Experiment [NASA Langley Research Center Earth Radiation Budget Satellite]

ERBM Extended-Range Ballistic Missile

ERBS Earth Radiation Budget Satellite [of NASA Langley Research Center, Hampton, Virginia, US]

E-RBS Enhanced Rutherford Backscattering (Spectroscopy)

ERBSS Earth Radiation Budget Satellite System [of NASA]

ERC Economic Research Council [UK]

ERC Education Relations Commission [Canada]

ERC Electronics Research Center [of NASA]

ERC Emergency Relocation Center

ERC Emergency Response Coordinator

ERC Energy Research Center

ERC Energy Research Center [of Australian National University, Canberra]

ERC Energy Resources Council

ERC Engineering Research Center

ERC Engineering Research Center [of US Bureau of Reclamation]

ERC Engineering Research Centers (Program) [of National Science Foundation, US]

ERC Engineering Research Council

ERC Engine Research Center [of University of Wisconsin at Madison, US]

ERC Environmentally Responsive Composite

ERC Environment and Resources Council [US]

ERC Environmental Resource Center [Cary, North Carolina, US]

ERC Environmental Resources Commission [Maine, US]

ERC Environmental Science Research Consortia

ERC Equatorial Ring Current

ERC Equivalent Radium Content

ERC Ercolessi (Potential)

ERC Error Retry Count

ERC Evaluation Review Committee

ERC Event Recorder

ERC Excessive Requirements Cost

ERCA Electrochemically Regenerable CO_2 (Carbon Dioxide) Absorber

ERCB Energy Resources Conservation Board [Canada]

ERCC Engine Requirement Coordinating Committee

ERCC En-Route Control Center

ERCC Error Checking and Correction [Computers]

ERCEM European Regional Conference on Electron Microscopy

ERCH Energy Research Clearing House [US]

ERCIM European Consortium on Informatics and Mathematics

ERCNSM Engineering Research Center for Net Shape Manufacturing [of Ohio State University, Columbus, US]

ERCO Electric Reduction Company (of Canada)

ERCOT Electric Reliability Council of Texas [US]

ERCP Endoscopic Retrograde Cholangiopancreatography [Medicine]

Er(Cp)$_3$ Tris(cyclopentadienyl)erbium

ERCR Electronic Retina Computing Reader

ERCS Emergency Rocket Command System

ERCS Emergency Rocket Communication System

ERCSS European Regional Communications Satellite System

ERCW Emergency Raw Cooling Water

ERD Elastic Recoil Detection [Physics]

ERD Electronic Research Directorate [US Air Force]

ERD Emergency Recovery Display

ERD Emergency Repair Disk

ERD Emergency Return Device

ERD Energy Research and Development Inventory (Database) [of Oak Ridge National Laboratory, Tennesse, US]

ERD Entity Relationship Diagram

ERD Entomology Research Division [of US Department of Agriculture]

ERD Environmental Regulatory Document

ERD Environmental Restoration Division [of US Department of Energy, Richland Operations Office]

ERD Equivalent Residual Dose

ERDA Economic and Regional Development Agreements [Canada]

ERDA Elastic Recoil Detection Analysis

ERDA Electrical and Radio Development Association [Australia]

ERDA Electronics Research and Development Activity [of US Army]

ERDA Energy Research and Development Administration [formerly US Atomic Energy Commission. Now Part of US Department of Energy]

ERDAA Energy Research and Development Administration Act [US]

ERDA Conf Proc Energy Research and Development Administration Conference Proceedings [US]

ERDAF Energy Research and Development in Agriculture and Food

ERDAM Energy Research and Development Adminstration Manual

ERDAS Earth Resources Digital Analysis System (Software)

ERDC Eastern Region Development Corporation [Nigeria]

ERDC Exploratory Research and Development Center [of Los Alamos National Laboratory, Albuquerque, New Mexico, US]

ER&DC Engineering Research and Development Center [of US Army, Vicksburg, Mississippi]

ERDE Electronics and Radar Development Establishment [India]

ERDE Explosives Research and Development Establishment [of Ministry of Supply, UK]

ERDEC Edgewood Research and Development and Engineering Center [of US Army at Aberdeen Proving Ground, Maryland]

Erde Int Erde International [of Aloe Technology Association, US]

ERDEV Device Error Value (Variable) [Programming]

ERDF Environmental Remediation (or Restoration) Disposal Facility

ERDF European Regional Development Fund [Belgium]

ERDL Electronics Research and Development Laboratory [US Army]

ERDL Engineering Research and Development Laboratory [Fort Belvoir, Virginia, US]

Erdoel Kohle Erdoel und Kohle [German Journal on Petroleum and Coal]

Erdoel Kohle Erdgas Erdoel Kohle Erdgas und Petrochemie [German Journal on Petroleum, Coal, Natural Gas and Petrochemicals]

ERDS Earth Resource Data System

ERDS Environmental Recording Data Set

ERD-TOF Elastic Recoil Detection–Time of Flight (Technique)

ERE Edison Responsive Environment

ERE Effective Range Expansion

ERE Electric Rocket Engine

ERE Energy Requirement for Energy

EREC Energy Efficiency and Renewable Energy Clearinghouse [US]

EREC Esso Research and Engineering Company

EREP Earth Resources Experiment(al) Package

EREP Environmental Recording, Editing and Printing

ERES Environmental Record Editing and Statistics

ERETS Experimental Rocket Engine Test Station

ERF Electric Resistance Furnace

ERF Electron Resonance Frequency

ERF Electrorheological Fluid

ERF Epoxy Resins Formulators Division [of Society of Plastics Engineers, US]

ERF Estuarine Research Foundation [US]

erf (Gaussian) error function [Symbol]

ERFA European Radio Frequency Administration [of NATO]

erfc complementary error function [Symbol]

ERFMI European Resilient Flooring Manufacturers Institute

Er(FOD)$_3$ Tris(6,6,7,7,8,8,8-Heptafluoro-2,2-Dimethyl-3,5-Octanedionato)erbium [also Er(fod)$_3$]

ERFPI Extended Range Floating Point Interpretative System

ERFR Eastern Rockies Forest Reserve [Canada]

ERG Electron Radiography

ERG Electroretinogram; Electroretinography

ERG Emergency Response Guidebook

ERG Energy Research Group [Canada]

ERG Engineering Research Grant [of Engineering Foundation, US]

ERG Ergometer; Ergometry

ERG Existence Needs–Relatedness Needs–Growth Needs (Theory) [Business Administration]

ERG Explosives Research Group [of Utah University, US]

erg/°C erg(s) per degree Celsius [also erg °C^{-1}]

erg/cc erg(s) per cubic centimeter [also erg/cm^3, erg cm^{-3}]

erg/cm erg(s) per centimeter [also erg cm^{-1}]

erg/cm² erg(s) per square centimeter [also erg cm^{-2}]

erg/cm³ erg(s) per cubic centimeter [also erg cm^{-3}, or erg/cc]

erg/cm²·s erg(s) per square centimeter second [also erg/cm²/s, erg cm^{-2}s^{-1}, or erg cm^{-2}sec^{-1}]

erg/cm²·s·°C⁴ erg(s) per square centimeter second degree Celsius to the fourth [also erg/cm²/s/°C⁴, erg cm^{-2} s^{-1} °C^{-4}, or erg cm^{-2} sec^{-1} °C^{-4}]

erg/cm²·s·K⁴ erg(s) per square centimeter second kelvin to the fourth [also erg/cm²/s/K⁴, erg cm^{-2} s^{-1} K^{-4}, or erg cm^{-2} sec^{-1} K^{-4}]

erg/G erg(s) per Gauss [also erg G^{-1}]

erg/G·mol erg(s) per Gauss mole [also erg G^{-1} mol^{-1}]

erg/K erg(s) per Kelvin [also erg K^{-1}]

Er:Glass Erbium-Doped Glass (Laser)

erg/mol·K erg(s) per mole kelvin [also erg/mol/K, or erg mol^{-1} K^{-1}]

ERGON Ergonomic(s) [also Ergon, or ergon]

erg/Oe erg(s) per oersted [also erg Oe^{-1}]

ERGS Electronic Route-Guidance System

ERGS Enroute Guidance System

erg/s erg(s) per second [also erg/sec, or erg s^{-1}]

erg-s erg-second [also erg·s, erg·sec, or erg-sec]

erg/s·cm² erg(s) per second square centimeter [also erg/s/cm², erg s^{-1} cm^{-2}]

erg/sec erg(s) per second [also erg/s, or erg s^{-1}]

erg-sec erg-second [also erg-s, erg·s, or erg·sec]

erg-s/emu erg-second(s) per electromagnetic unit [also erg-s emu^{-1}, or erg·s·emu^{-1}]

erg-s/esu erg-second(s) per electrostatic unit [also erg-s esu^{-1}, or erg·s·esu^{-1}]

erg-s/g erg-second(s) per gram [also erg-s g^{-1}, or erg·s·g^{-1}]

Er(HFC)$_3$ Erbium Tris[3-(Heptafluoropropylhydroxymethylene) camphorate] [also Er(hfc)$_3$]

ERI Earthquake Research Institute [of University of Tokyo, Japan]

ERI Elm Research Institute [Waldwick, New Jersey, US]

ERI Energy Research Institute [US]

ERI Engineering Research Institute

ERI Engineering Research Institute [of Tokyo University, Japan]

ERI Engineering Research Institute [of Tulane University, New Orleans, US]

ERI Engineering Research Institute [of University of Michigan, US]

Eri Eridanus [Astronomy]

ERIC Educational Research Information Center

ERIC Educational Resources Information Center [of United States Office of Education]

ERIC Electronic Remote and Independent Control

ERIC Energy Rate Input Controller

ERIC Environmental Research and Information Center

ERICA Experiment on Rapidly Intensifying Cyclones over the Atlantic [Canada]

Ericsson Rev Ericsson Review [Published by LM Ericsson, Stockholm, Sweden]

ERICZC Educational Reptiles in Captivity Zoological Compound [US]

Erid Eridanus [Astronomy]

ERIE Environmental Resistance Inherent in Equipment

ERIG Engineering Research Initiation Grant [of Engineering Foundation, US]

ErIG Erbium Iron Garnet

ErIG:Sc Scandium-Substituted Erbium Iron Garnet

ERIM Environmental Research Institute of Michigan [Ann Arbor, US]

ERIN Environmental Resources Information Network [Australia]

ERIP Energy-Related Inventions Program [of US Department of Energy]

ERIS Electrostatic Reflex Ion Source

ERIS Emergency Resources Identification Equipment

ERIS Engineering Resins Information System

ERIS Environmental Resources Information System [of Environmental Resources Information Network]

ERIS Exoatmospheric Reentry Vehicle Interceptor System

ERISA Employee Retirement Income Security Act [US]

ERISS Environmental Research Institute of the Supervising Scientist [Australia]

ERISTAR Earth Resources Information Storage, Transformation Analysis and Retrieval [of NASA]

Erit Eritrea(n)

ERJE Extended Remote Job Entry

ERL Earth Resources Laboratory [of Massachusetts Institute of Technology, Cambridge, US]

ERL Echo Return Loss

ERL Electronic Research Laboratory [of Nippon Steel Corporation, Kanagawa, Japan]

ERL Energy Research Laboratories [of Canada Center for Mineral and Energy Technology; Natural Resources Canada, Ottawa]

ERL Energy Research Laboratory

ERL Energy Research Laboratory [of Hitachi Limited, Ibaraki, Japan]

ERL Environmental Research Laboratory [of National Oceanic and Atmospheric Administration at Corvallis, US]

ERL Equipment Revision Level

ERL Error Line Number (Variable) [Programming]

ERL Error Location

ERL ESSA (Environmental Science Services Administration) Research Laboratories [US]

ERL Explosives Research Laboratory [US]

Erl Erlang [Unit]

Erlen Fl Erlenmeyer Flask [also Erlen fl]

Er:LiYF$_4$ Erbium-Doped Lithium Yttrium Fluoride [also Er^{3+}:LiYF$_4$]

ERLL Enhanced Run Length Limited

ERM Earth Reentry Module

ERM Earth Return Module

ERM Eigenchannel R-Matrix Method

ERM Elastic-Reservoir Molding

ERM Electronic Recording Machine

ERM Environmental Relaxation Model

ERM Exchange Rate Mechanism [European Monetary System]

ERMA Electrical Reproduction Method of Accounting

ERMA Electronic Recording Machine Accounting

ERMA Electronic Recording Method, Accounting

ERMA Engineering Reprographic Management Association [now Engineering Reprographic Society, US]

ERMA Environment (and) Resources Management Association [Canada]

ERMA Expansion Rate Measuring Apparatus

ERMC Environmental Restoration Management Contractor

ERMCO European Ready Mixed Concrete Organization [UK]

ERMP Environment Management and Review Program

ERMS Energy-Resolved Mass Spectra

ERMS Environmental Resources Mapping System (Software)

ERN Engineering Release Notice

ER News Energy Research News [of US Department of Energy, Office of Energy Research]

ERNIE Electronic Random Number Indicator Equipment

ERNO Entwicklungsring Nord [Subsidiary of Vereinte Flugtechnische Werke-Fokker, Germany; European Consortium in the Aerospace and Satellite Systems Industries; also a Spacelab Contractor]

ERO Emergency Repair Overseer

ERO Energy Research Office

ERO Engineering Release Operations

ERO Engineering Release Order

ERO European Regional Organization [of International Confederation of Free Trade Unions, Belgium]

EROC (Study Group on) En-Route Obstacle Clearance Criteria

Er(O-i-Pr) Erbium Isopropoxide [also ErOPr^i]

EROM Erasable Read-Only Memory [also E-ROM]

EROP Extensions and Restrictions of Operators

EROPA Eastern Regional Organization for Public Administration

ErOPr^i Erbium Isopropoxide [also Er(O-i-Pr)]

EROPS Extended Range Operations (Aircraft)

EROS Earth Resources Observation Satellite

EROS Earth Resources Observation System [of US Geological Survey]

EROS Earth Resources Orbiting Satellite

EROS Eliminate Range 0 (Zero) System

EROS Experimental Reflector Orbit Shot

EROSDC Earth Resources Observation Systems Data Center [of US Geological Survey at Sioux Falls, South Dakota]

EROW Executive Right of Way

ERP Economic Recovery Program

ERP Effective Radiated (or Radiative) Power (of Antennas) [also erp]

ERP Effective Rate of Protection

ERP Electronics Research Paper

ERP Elevated Release Point

ERP Emergency Recorder Plot

ERP Emergency Response Plan [US]

ERP Emergency Response Program [US]

ERP Endocochlear Resting Potential [Medicine]

ERP Endoscopic Retrograde Pancreatography

ERP End Response [Data Processing]

ERP Enterprise Resource Planning (System)

ERP Environmental Research Paper

ERP Environmental Research Project

ERP Equivalent Reduction Potential [also erp]

ERP Error Recovery Procedure [Computers]

ERP European Recovery Program [US Post-World War II Program; also known as the "Marshall Plan"]

ERP Eye Reference Point

ERP External Research Program

ERPD Elementary Rupture Probability Density

ERPDB Eastern Regional Production Development Board [Nigeria]

ERPG Environmental Response Planning Guidelines

ERPLD Extended-Range Phase-Locked Demodulator

ERPM Emission Reduction Planning Model (Software)

ERPM Engineering Requirements and Procedures Manual

ERPTUAC European Recovery Program Trade Unions Advisory Program

ERR Eastern Range Regulation

ERR Elk River Reactor [US]

ERR Energy Release Rate

ERR Error Code (Variable) [Programming]

Err Error [also err]

ERRC Eastern Regional Research Center [of US Department of Agriculture at Philadelphia, Pennsylvania]

ERRC Error Character

ERRC Expendability/Recoverability/Repair Capability

ER/RC Extended Result/Response Code

Err Cntr Error Counter

ERRL Eastern Regional Research Laboratory [US]

erron erroneous(ly)

ERS Earth Recovery Subsystem [of NASA]

ERS Earth Regeneration Society [US]

ERS Earth Remote-Sensing Satellite

ERS Earth Resources Satellite [of European Space Agency]

ERS East Radiography Station [of Argonne National Laboratory–West, Idaho Falls, US]

ERS Economic Research Service [of US Department of Agriculture]

ERS Educational Research Service [US]

ERS Elastic Recoil Spectrometer; Elastic Recoil Spectrometry

ERS Electric Railway Society [UK]

ERS Electric Resistant Strain

ERS Electronic Raman Scattering

ERS Emergency Reporting System

ERS Employee Relocation Service

ERS Engineering Release System

ERS Engineering Reprographic Society [US]

ERS Engineers Register Study

ERS Entry and Recovery Simulation

ERS Environmental Research Satellite [of NASA]

ERS Environmental Resource Studies

ERS Ergonomics Research Society [UK]

ERS ESA (European Space Agency) Remote-Sensing Satellite (System)

ERS European Radio Satellite

ERS European Remote-Sensing Satellite

ERS Experimental Research Society [US]

ERS External Reflection Spectrometry

ERS External Regulation System

ER&S Exploratory Research and Study

ERS-1 First European Remote-Sensing Satellite

ERS-1 First Earth Resources Satellite 1 [of European Space Agency]

ERS-1 First European Radio Satellite

Ers Eriochrome Blue Black R

ERSA Electronic Research Supply Agency

ERSDAC Earth Resources Satellite Data Analysis Center [Japan]

ERSI Elastomeric Reusable Surface Insulation

ERSI Environmental Research Systems Institute

ERSIR Earth Resources Shuttle Imaging Radar

ERSO Electronics Research and Service Organization [Taiwan]

ERSMK Element Rotation Stand Modification Kit

ERSOS Earth Resources Survey Operational Satellite

ERSP Earth Resources Survey Program

ERSP Encapsulated Ring-Shell Projector

ERSP European Remote Sensing Program

ERSR Equipment Reliability Status Report

ERSS Earth Resources Satellite System

ERSS Earth Resources Survey Satellite

ERSS Effective Resolved Shear Stress

ERSU Energy Research Support Unit [of Science and Engineering Research Council, UK]

ERT Effective Range Theory

ERT Electrical Resistance Thermometer

ERT Electrical Resistance Tomography

ERT Emergency Response Team

ERT Emergency Response Training

ERT Enhanced Readiness Test

ERT Energy Resources Technology

ERT Engineering Research and Technology

ERT Environment Round Table

ERT Estimated Repair Time

ERT European Roundtable [Belgium]

ERT Expected Run Time [Computers]

ERTA Economic Recovery Tax Act [US]

ErTeN Erythritol Tetranitrate

Er(TFC)$_3$ Erbium(III)Tris[(Trifluoromethylhydroxymethylene)-camphorate] (also Er(tfc)$_3$]

Er(TMHD)$_3$ Erbium(III) Tris(2,2,6,6-Tetramethyl-3,5Heptanedionate) (also Er(tmhd)3]

ERTOR Effective Radiational Temperature of the Ozone-Layer Region

ERTS Earth Resources Technology Satellite [Series of US Satellites]

ERTS-1 First Earth Resources Technology Satellite [US]

ERU Earth Rate Unit [Astronomy]

ERU Emergency Recovery Utility [Computers]

ERV Electronic Repair Vehicle

ERV Embu Virus

ERV Entry Research Vehicle

ERV Expiratory Reserve Volume [Medicine]

ERW Electric Resistance Welded; Electric Resistance Welding

ERW HSR Electric Resistance-Welded Hot-Stretch Reduction (Process)

ERWP European Railway Waggon Pool [Belgium]

ERX Electronic Remote Switching

Er-Y Erbium-Yttrium (System)

Er:YAG Erbium-Doped Yttrium Aluminum Garnet (Laser) [also Er^{3+}:YAG]

Er:YLF Erbium-Doped Yttrium Lithium Fluoride (Laser) [also Er^{3+}:YLF]

Er:YSGG Erbium-Doped Yttrium Scandium Gadolinium Garnet (Laser) [also Er^{3+}:YSGG]

Er:ZBLAN Erbium-Doped ZBLAN (Zirconium Tetrafluoride–Barium Difluoride–Lanthanum Trifluoride–Aluminum Trifluoride–Sodium Fluoride (Laser) [also Er^{3+}:ZBLAN]

ES And [Amateur Radio]

ES Earth Science(s); Earth Scientist

ES Earth Station

ES East Siberia(n)

ES Eastern Section

ES Eccentric Shaft

ES Echo Sounder; Echo Sounding

ES Echo Suppression; Echo Suppressor

ES Econometric Society [US]

ES Education Society [of Institute of Electrical and Electronics Engineers, US]

ES Education Specialist

ES Effective Surface

ES Efros-Shklovskii (Hopping) [Physics]

ES Eigensolution

ES Ejection Seat

ES Elastic Scattering

ES Electrical Schematic

ES Electrical Shutter

ES Electrical Stickout [Welding Electrodes]

ES Electrical Stimulation

ES Electric Shock

ES Electric Steel

ES Electrochemical Society [US]

ES Electroluminescent Screen

ES Electromagnetic Storage

ES Electromagnetic Switch(ing)

ES Electronic Science

ES Electronic Spectroscopy

ES Electronic Spectrum

ES Electronic Standard

ES Electronic Structure

ES Electronic Switch(ing)

ES Electron Scattering

ES Electron Shell

ES Electron Spectroscopy

ES Electron Spectrum

ES Electron Spin [Physics]

ES Electron Synchrotron

ES Electrophoresis Society [US]

ES Electroslag

ES Electrospray

ES Electrostatic(s)

ES Electrostatic Storage

ES Electrostenolysis; Electrostenolytic

ES Electrostriction; Electrostrictive
ES Element Signal
ES El Salvador(ean)
ES Embryonic Stem (Cells)
ES Emission Spectrometer; Emission Spectrometry
ES Emission Spectroscopy
ES Emission Spectrum
ES Emulsion for Seed Treatment [Agriculture]
ES Enamelled Single-Silk-Covered (Electric Wire)
ES Enantioselective; Enantioselectivity
ES Endocrine System
ES End Seal
ES Energy Sampling
ES Energy Shift
ES Energy Source
ES Energy Spectrum
ES Energy Stocks
ES Energy Storage
ES Energy System
ES Engineering Specification
ES Engineering Support
ES Environmentally Sound (Technology)
ES Environmental Safety [US]
ES Environmental Science; Environmental Scientist
ES Environmental Security
ES Environmental Services
ES Enviroene (Site) [of US Environmental Protection Agency]
ES Ephemeris Second [Astronomy]
ES Epigraphic Society [US]
ES Equalization System
ES Equilibrium State
ES Ergonomics Society [UK]
ES Error Satisfaction
ES Escape System
ES Ether Sulfone
ES Ethylene Sulfite
ES Euclidean Space [Mathematics]
ES Eutectic Solidification [Metallurgy]
ES Ewald Sphere [Crystallography]
ES Excited State [Physics]
ES Executive Secretariat [of National Institutes of Health, Bethesda, Maryland, US]
ES Exhaust Stroke
ES Experimental Station
ES Experiment Segment
ES Expert System
ES Extension Service [of US Department of Agriculture]
ES External Storage; External Store [Computers]
ES Extra Segment [Computers]
ES Extra Spring (Temper) [Metallurgy]
ES Extremely Susceptible; Extreme Susceptibility
ES Extrinsic Semiconductor
ES Extruded Shape

ES Spain [ISO Code]
ES Upper Deviation (of Internal Thread) [ISO Symbol]
Es Einsteinium [Symbol]
Es-246 Einsteinium-246 [also ^{246}Es, or Es^{246}]
Es-247 Einsteinium-247 [also ^{247}Es, or Es^{247}]
Es-248 Einsteinium-248 [also ^{248}Es, or Es^{248}]
Es-249 Einsteinium-249 [also ^{249}Es, or Es^{249}]
Es-250 Einsteinium-250 [also ^{250}Es, or Es^{250}]
Es-251 Einsteinium-251 [also ^{251}Es, or Es^{251}]
Es-252 Einsteinium-252 [also ^{252}Es, or Es^{252}]
Es-253 Einsteinium-253 [also ^{253}Es, or Es^{253}]
E(s) Actuating Signal [Automatic Control Symbol]
es upper deviation (of external thread) [ISO Symbol]
.es Spain [Country Code/Domain Name]
e^{-}/s electrons per second [also e^{-}/sec]
ESA Ecological Society of America [US]
ESA Ecological Society of Australia
ESA Economic Stabilization Act [US]
ESA Effective Surface Area
ESA Electric Sector Analyzer
ESA Electric Spark Alloying
ESA Electric Supply Authority
ESA Electronically Scanned Array (Antenna)
ESA Electronic Surge Arrester
ESA Electrostatic Accelerator
ESA Electrostatic Analyzer
ESA Electrostatic Atomization; Electrostatic Atomizer
ESA Electrostatic System Analyzer
ESA Employment Standards Act
ESA Employment Standards Administration [of US Department of Labor]
ESA Endangered Species Act [US]
ESA Energy Sudden Approximation
ESA Engineering Supply Area
ESA Engineering Support Assembly
ESA Engineers and Scientists of America [US]
ESA Engine Service Association [US]
ESA Enterprise System Architecture
ESA Entomological Society of America [US]
ESA Environmentally Sensitive Area
ESA Epithelial Specific Antigen [Immunology]
ESA Equipment Service Association [US]
ESA Ethernet Station Adapter
ESA EURATOM Supply Agency [of European Economic Community, Belgium]
ESA European Space Agency [Paris, France]
ESA Excited State Absorption [Physics]
ESA Experimental Stress Analysis
ESA Explosive Safe Area [at NASA Kennedy Space Center, Florida, US]
ESA Externally Specified Address
ESAA Electricity Supply Association of Australia
ESAAB Energy Systems Acquisition Advisory Board [of US Department of Energy]

ESAAPG Eastern Section of the American Association of Petroleum Geologists [US]

ESA Bull ESA Bulletin [of European Space Agency, Paris, France]

ESAB Escuela Superior d'Agricultura de Barcelona [Barcelona Higher School of Agriculture, Spain]

ESAC Electronic Systems Assistance Center

ESAC Endangered Species Advisory Committee

ESAC Environmental Studies Association of Canada

ESA-CSA Energy Sudden Approximation–Centrifugal Sudden Approximation

ESACT European Society for Animal Cell Technology [UK]

ESAD Earth Science and Applications Division [of NASA Headquarters, Washington, DC]

ESAEI Electric Supply Authority Engineers' Institute [Wellington, New Zealand]

ESAEINZ Electric Supply Authority Engineers' Institute of New Zealand [Wellington, New Zealand]

ESA/ESTEC European Space Agency/European Space Technical Center [Noordwijk, Netherlands]

ESA Feat ESA Features [of European Space Agency, Paris, France]

ESAH Essex Society for Archeology and History [UK]

ESAIRA Electronically-Scanned Airborne Intercept Radar

ESA-IRS European Space Agency–Information Retrieval System

ESA J ESA Journal [of European Space Agency Technical Center, Noordwijk, Netherlands]

ESANET European Space Agency Network

ESA Newsl Entomological Society of America Newsletter [US]

ESANZ Economic Society of Australia and New Zealand

ESAO European Society for Artificial Organs [Switzerland]

ESAO Excited State Atomic Oxygen

ESAP Economics and Systems Analysis Program

ESAP Environmental Self-Assessment Program

ESAPS Experimental Strain Analysis Processing System

ESAR Electronically-Scanned Array Radar

ESAR Electronically-Steerable Array Radar

ESARCC Endangered Species Act Reauthorization Coordinating Committee [US]

ESARS Earth Surveillance and Rendezvous Simulator

ESARS Employment Service Automated Reporting System

ESAS Engineered Safeguards Actuation System

ESA SP European Space Agency Special Publication [also ESA Spec Publ]

ESATC European Space Agency Technical Center [Noordwijk, Netherlands]

ESB Earth Sciences Branch [of Department of Energy, Mines and Resources, Ottawa, Canada]

ESB Economic Stabilization Board [PR China]

ESB Electrical Standards Board [of United States of America Standards Institute]

ESB Electrical Stimulation of the Brain

ESB Electricity Supply Board

ESB Engineering Services (or Support) Building [of Pacific Northwest National Laboratory, Richland, Washington, US]

ESB Engineering Societies Building

ESB Engineering Society of Baltimore [Maryland, US]

ESB Engineering Society of Buffalo [New York State, US]

ESB Essential Switching Box

ESB European Society for Biomaterials

ESB European Society for Biomechanics [Netherlands]

ESB Export Services Branch [of Board of Trade, UK]

ESB Spanish Peseta [Currency of Spain]

Esb Eriochrome Blue Black R

ESBCO Erbium Samararium Barium Copper Oxide (Superconductor)

ESBFCOA Eastern States Blast Furnace and Coke Oven Association [US]

ESBP European Society for Biochemical Pharmacology

E-SBR Emulsion Styrene Butadiene Rubber

ESC East South Central [US Geographic Region]

ESC Echo Suppressor Control

ESC Edinboro State College [Pennsylvania, US]

ESC Edison Screw Cap

ESC EISA (Extended Industry Standard Architecture) System Component

ESC Electrical Skin Conductivity

ESC Electronic Spark Control

ESC Electronic Structure Calculations

ESC Electronic Systems Command [US Navy]

ESC Electroslag Casting [Metallurgy]

ESC Electroslurry Coating [Metallurgy]

ESC Electrostatic Coating

ESC Electrostatic Compatibility

ESC Empire State College [of State University of New York at Saratoga Spings, US]

ESC Energy Security Corporation

ESC Engineering and Scientific Computing

ESC Engineering Service Circuit

ESC Engineering Society of Cincinnati [Ohio, US]

ESC Engineering Support Center

ESC Engineers and Scientists of Cincinnati [Ohio, US]

ESC Entomological Society of Canada

ESC Environmental Stress Cracking

ESC Environmental Stress Crazing

ESC Environment Sensitive Cracking

ESC Equipment Serviceability Criteria

ESC Escape (Character) [Data Communications]

ESC (Portuguese) Escudo [Currency of Portugal and Madeira]

ESC European Seismological Commission [Switzerland]

ESC European Shippers Council [Netherlands]

ESC European Space Conference

ESC Expedited Site Characterization (Project) [of US Department of Energy at Ames Laboratory, Iowa State University]

ESC Export Supply Center [Canada]

ESȼ El Salvador Colón [Currency]

Esc Escape [also esc]

Esc Escutcheon [also esc]

Esc (Portuguese) Escudo [Currency of Portugal and Madeira]

ESCA Electron Spectroscopy for Chemical Analysis

ESCA Endangered Species Conservation Act [US]

ESCA Exhibition Service Contractors Association [US]

ESCA Extended Source Calibration Area

ESCA X-Ray Photoelectron Spectroscopy for Chemical Analysis

E-SCAN Electronic Selected Current Aerospace Notices [of NASA Scientific and Technical Information (Program Office)]

ESCAP Economic and Social Commission for Asia and the Pacific [of United Nations at Bangkok, Thailand]

ESCAPE Expansion Symbolic Compiling Assembly Program for Engineering

ESCAR Experimental Superconducting Accelerating Ring

ESCB European System of Central Banks [European Community]

ESCC Engineering Standards Coordinating Committee [UK]

ESCC External Stress Corrosion Cracking

ESCD Extended System Contents Directory

ESCES Experimental Satellite Communication Earth Station [India]

ESCI European Society for Clinical Investigation [UK]

ESCLA Environmental Services for Canada and Latin America

ESCM Electroslag Crucible Melting [Metallurgy]

ESCM Extended Services Communications Manager

ESCMT Engineering Societies Committee for Manpower Training [US]

ESCNO Europium Strontium Copper Niobium Oxide (Superconductor)

ESCO Educational, Scientific and Cultural Organization [of United Nations]

ESCOE Engineering Societies Commission on Energy

ESCOM Electricity Supply Commission [South Africa]

ESCON Enterprise System Connection (Architecture)

ESCORT Error Self-Compensation Researched by Tau-Scrambling

ESC/P Epson Standard Code for Printers

ESCPB European Society of Comparative Physiology and Biochemistry [Belgium]

ESCR Environmental Stress-Cracking Resistance

ESCS Emergency Satellite Communications System

ESCSI Expanded Shale Clay and Slate Institute [US]

ESCU Extended Service and Cooling Umbilical

ESCWA Economic and Social Commission for Western Asia [of United Nations]

ESCWS Essential Services Cooling Water System

ESD Earth Science Division [of Lawrence Livermore National Laboratory, University of California at Livermore, US]

ESD Echo Sounding Device

ESD Ecologically Sustainable Development

ESD Effective Standard Deviation

ESD Electronic Semiconductor Device

ESD Electronic Software Distribution

ESD Electronic System Design (Magazine) [US]

ESD Electronic Systems Division [US Air Force]

ESD Electron Spectroscopic Diffraction

ESD Electron-Stimulated Desorption

ESD Electron-Stimulated Disorder

ESD Electrospark Deposition

ESD Electrostatic Discharge; Electrostatic Discharging

ESD Electrostatic Dissipation

ESD Electrostatic Storage Deflection

ESD Elongated Single Domain (Magnet) [Solid-State Physics]

ESD Emergency Shutdown

ESD Ending Sequence, Done

ESD Energy Storage Device

ESD Engineering Services Division

ESD Engineering Society of Detroit [Michigan, US]

ESD Entry Systems Division [of International Business Machines Inc., US]

ESD Environmental Science Division

ESD Environmental Sciences Division [of Oak Ridge National Laboratory, Tennessee, US]

ESD Environmental Secondary Detector [of Environmental Scanning Electron Microscopy]

ESD Environmental Services Division

ESD Environment Strategies Division [of Environment Australia]

ESD Estimated Standard Deviation

ESD Experiment Systems Division

ESD Export Supply Directorate

ESD Extension Shaft Disconnect

ESD External Symbol Dictionary

ESDAC European Space Data Center [Darmstadt, Germany]

ESDD Earth Science Data Directory [of US Geological Survey]

ESDED Electron-Stimulated Desorbed Ions

ESD Electron Syst Des Mag ESD: The Electronic System Design Magazine [US]

ESDI Electron-Stimulated Desorption of Ions

ESDI Enhanced Small Device Interface

ESDI Enhanced System Device Interface

ESDIAD Electron-Stimulated Desorption Ion Angle (or Angular) Distribution

ESDIED Electron-Stimulated Desorbed Ion Energy Distribution

ESDIIR Ecologically Sustainable Development Intersectoral Issues Report [Australia]

ESDIM Earth System Data and Information Management [US]

ESDIM Environmental Services Data and Information Management (Program) [of National Oceanic and Atmospheric Administration, US]

ESDMS Electron Stimulated Desorption Mass Spectroscopy

ESDN Electron-Stimulated Desorption of Neutrons

ESDRAM Enhanced Synchronous Dynamic Random-Access Memory

ESDS Entry-Sequenced Data Set

ESDS Environmental Satellite Data System

ESDSC Ecologically Sustainable Development Steering Committee

ESDU Event Storage and Distribution Unit

ESDX Environment and Safety Data Exchange [US]

ESE East-Southeast

ESE Ecole Supérieure d'Electricité [Higher School of Electricity, of University of Paris, Gif-sur-Yvette, France]

ESE Elastic Scattering of Electrons

ESE Electrical Support Equipment

ESE Electronic Support Equipment

ESE Electronic System Evaluator

ESE Electron Spin Echo [Physics]

ESE Electrostatic Energy

ESE Electrostatic Engine

ESE Emitted Source Energy

ESE Environment, Safety and Economics

ESE EVA (Extravehicular Activity) Support Equipment [of NASA]

ESEA Elementary and Secondary Education Act [US]

ESEAFE Environmental Safety and Economic Aspects of Fusion Energy [of International Energy Agency, France]

ESEC European Symposium on Engineering Ceramics

e⁻/sec electrons per second [also e⁻/s]

ESEEM Electron Spin Echo Envelope Modulation Spectroscopy

ESeH Selenol

ESELCO Edison Sault Electric Company [US]

ESEM Electron Spin Echo Modulation [Physics]

ESEM Electron Spin Envelope Modulation [Physics]

ESEM Electroscattering Electron Microscopy

ESEM Environmental Scanning Electron Microscope; Environmental Scanning Electron Microscopy

ESeOH Selenenic Acid

ESeSR Selenenyl Sulfide

ESET Poly(ethoxysilane)ethyltitanate (Copolymer)

ESF Ecole Supérieure Fonderie [Higher School for Foundry Engineering, Bagneux, France]

ESF Electrospark Forming

ESF Electrostatic Field

ESF Electrostatic Filter

ESF Electrostatic Focus(ing)

ESF Electrostatic Force

ESF Electrostrictive Force

ESF Engineered Safety Feature; Engineering Safety Feature

ESF Ethynylphenoxysulfone

ESF EURECA (European Research Coordination Agency) Software Factory

ESF European Science Foundation [France]

ESF Exploratory Studies Facility

ESF Explosive Safe Facility

ESF Export Success Fund

ESF Extended Super Frame

ESF Extrinsic Stacking Fault [Materials Science]

ESFA Emergency Solid Fuel Administration [US]

ESFAS Engineered Safety Features Actuation System

ESF Commun ESF Communications [of European Science Foundation]

ESFD Electron-Stimulated Field Desorption

ESFI Epitaxial Silicon-on-Insulator

ESFM Ecologically Sustainable Forest Management

ESFM Escuela Superior de Fisica y Matemáticas [Higher School of Mathematics and Physics

ESFM-IPN Escuela Superior de Fisica y Matematicas de Instituto Politecnico Nacional [Higher School of Physics and Mathematics of the National Polytechnic Institute, Mexico City, Mexico]

ESFN Elastic Scattering of Fast Neutrons

ESF Synchrotron Radiat News ESF Synchrotron Radiation News [of European Science Foundation]

ESFQO European Science Foundation Conference on Quantum Optics

ESG Ecole Supérieure de Guerre [Military Academy, Paris, France]

ESG Electrically-Suspended Gyroscope

ESG Electronic Sweep Generator

ESG Electrostatic Generator

ESG Electrostatic Gyroscope

ESG Empirical Study Group [of National Oceanic and Atmospheric Administration, US]

ESG Engineers and Scientists Guild [US]

ESG (Old) English Standard Gauge

ESG Ethnobotany Specialists Group [US]

ESG Exchange Software Generator

ESH Electric Strip Heater

ESH Environment, Safety and Health (Section) [of Fermilab, Batavia, Illinois, US]

ESH Environment, Safety and Health [also ES&H]

ESH Equivalent Standard Hours

ESHAC Electric Space Heating and Air Conditioning Committee [of Institute of Electrical and Electronics Engineers–Industry Applications Society, US]

ESHG Electric Field-Induced Second Harmonic Generation

ESHP Equivalent Shaft Horsepower

ESHPH European Society for the History of Photography [Belgium]

ESH/QA Environment, Safety, Health and Quality Assurance (Section) [of Argonne National Laboratory, US]

ESHT Electroslag Hot Topping [Metallurgy]

ESHU Emergency Ship-Handling Unit

ESI Ecole des Sciences de l'Information [School of Information Sciences]

ESI Ecological Studies Institute [UK]

ESI Electrical System Integration

ESI Electricity Supply Industry

ESI Electron Spectroscopic (or Spectroscopy) Imaging

ESI Electrospray Ionization

ESI Electrostatic Induction

ESI End of Segment Indicator

ESI Engineering and Scientific Interpreter

ESI Enhanced Serial Interface (Specification)

ESI Environmental Science Index [of Environment Information Center, US]

ESI Environmental Studies Institute [of International Academy at Santa Barbara, US]

ESI Equivalent-Step Index

ESI Erich Schmid Institut (für Festkörperphysik) [Erich Schmid Institute for Solid-State Physics, Loeben, Poland]

ESI Externally-Specified Index

e-Si Epitaxial Silicon

E Sib East Siberia(n)

ESIC Earth Science Information Center [of US Geological Survey]

ESIC Environmental Science Information Center [of National Oceanic and Atmospheric Administration, US]

ESIC European Steel Institutes Club

ESID Electron-Stimulated Ion Desorption

ESID External Symbolic Identidication (Number)

ESIE Electron-Stimulated Ion Emission

ESIG Environmental and Societal Impacts Group [of National Center for Atmospheric Research, US]

ESIG Eugenics Special Interest Group [US]

ESIP Employee Savings Investment Plan

ESIPT Excited-State Intramolecular Proton Transfer

ESIQIE Escuela Superior de Ingenería Química e Industrias Extractivas [Higher School of Chemical Engineering and Extractive Industries] [of IPN, Mexico City]

ESIQIE-IPN Escuela Superior de Ingeniería Química e Industrias Extractivas de Instituto Politecnico Nacional [Higher School of Chemical Engineering and Extractive Industries of the National Polytechnic Institute, Mexico City]

ESIS Environmentally Sensitive Investment System

ESIS European Shielding Information Service [of EURATOM]

ESIS European Space Information System [of European Space Agency]

ESIS European Structural Integrity Society

ESISS Electronic Submarine Integrated Sonar System

ESISS Experimental Submarine Integrated Sonar System

ESIT Egyptian Society for Information Technology

ESJCP Engineers and Scientists Joint Committee on Pensions

Esk Eskimo

ESL Electrical Services League

ESL Electron-Beam Switched Latch

ESL Engineering Societies Library

ESL English as a Second Language

ESL Essential Service Line

ESL Expected Significance Level

ESL Evans Signal Laboratory [US Army]

ESLAB European Space Research Laboratory [of European Space Research Organization]

ESLO European Satellite Launching Organization

ESM Ecole Spéciale Militaire [Special Military School, Paris, France]

ESM Elastomeric Shield Material

ESM Electronic Support Measures

ESM Electroslag Melting [Metallurgy]

ESM Electrospark Machining

ESM Engineering Scab Melter

ESM Engineers and Scientists of Milwaukee [Wisconsin, US]

ESM Equivalent Standard Minute

ESM Erbium Strontium Manganate

ESM Error Satisfaction Method

ESMA Electronic Sales and Marketing Association

ESMA Engraved Stationery Manufacturers Association [US]

ESMC Eastern Space and Missile Center [formerly US Air Force Eastern Test Range]

ESMC Empire State Materials Council [US]

ESMC Engineering Societies Monograph Committee

E-SMC Epoxy-Matrix Sheet Molding Compound

ESME Excited State Mass Energy [Physics]

ESM/ECM Electronic Support Measures/Electronic Countermeasures

ESMH Element-Specific Magnetic Hysteresis

ESMO Erbium Strontium Manganese Oxide

ESMR Electrically Scanning Microwave Radiometer

ESMR Electronically Scanned Microwave Radiometer

ESMR Enhanced Specialized Mobile Radio

ESMS East Sullivan Monzonitic Stock [Geology]

ESMST European Society of Membrane Science and Technology [Italy]

ESN Electronic Security Number

ESN Electronic Serial Number

ESN English Speaking Nations

ESN European Society for Nematologists [Scotland]

ESNE Engineering Societies of New England [US]

ESnet Energy Sciences Network

ESNSW Entomological Society of New South Wales [Australia]

ESNZ Entomological Society of New Zealand

ESO Ecole Supérieure d'Optique [Higher School of Optics, Orsay, France]

ESO Economic Stabilization Office

ESO Electrical Stickout [Welding Electrodes]

ESO Emergency Support Organization

ESO European Southern Observatory [on Cerro la Silla near La Serena, Chile]

ESO Event Sequence Override

ESOC Early-Stage Ovarian Cancer

ESOC ESRO (European Space Research Organization) Space Operations Center [at Darmstadt, Germany]

ESOC European Space Operations Center [at Darmstadt, Germany]

ESoCE European Society of Concurrent Engineering

ESOCS Electronic Structure of Closed-Packed Structures (Computer Program)

ESOMAR European Society for Opinion and Marketing Research [Denmark]

ES-OMCVD Electrospray Organometallic Chemical Vapor Deposition

ESONE European Standards of Nuclear Electronics

ESONEC European Standards of Nuclear Electronics Committee [Switzerland]

ESOP Employee Stock Ownership Plan

ESOR European Symposium on Organic Reactivity

ESO/ST-ECF European Southern Observatory/Space Telescope–European Coordinating Facility, Garching/Munich, Germany]

ESOW Engineering Statement of Work

ESP Edge-Supported Pulling

ESP Edit, Save and Plot Technology

ESP Electrical Submersible Pump

ESP Electronic Short Pathfinder

ESP Electron Spin Polarization

ESP Electrosensitive Programming

ESP Electroshock Protection

ESP Electrosonic Profiler

ESP Electrostatic Potential

ESP Electrostatic Precipitation; Electrostatic Precipitator

ESP Electrostatic Printer

ESP Elsevier Science Publishers [Netherlands]

ESP Emulation Sensing Processor

ESP Encapsulating Security Payload (Header)

ESP Endangered Species Program [of Australian Nature Conservation Agency]

ESP End of Sustained Pressure [Aerospace]

ESP Engine Service Platform

ESP Enhanced Serial Port

ESP Enhanced Service Provider

ESP Epoch Start Pulse

ESP Escudo [Currency of Portugal]

ESP Estimated Selling Price

ESP Evoked Synaptic Potential [Medicine]

ESP Exchangeable Sodium Percentage

ESP Experimental Solids Proposal

ESP Experiment Sensing Platform

ESP Expert System Prototype

ESP Expert Systems Program [of University of Southern California, Center for Accounting Research, Los Angeles, US]

ESP Extended Service Plan

ESP Externally Supported Processor

ESP Extrasensory Perception [Psychology]

ESP Extravehicular Support Pack [of NASA]

E&SP Equipment and Spare Parts

esp especially

ESPA Exhaust Systems Professional Association [US]

ESP Act Endangered Species Protection Act [Australia]

ESPAR Electronically Steerable Phased Array Radar

ESPB European Student Press Bureau [Netherlands]

ESPCI Ecole Supérieure de Physique et de Chimie Industrielles (de la Ville de Paris) [Higher School of Industrial Physics and Chemistry (of the City of Paris), France]

ESPERT Expanded Program Evaluation and Review Technique

ESPI Electronic Speckle-Pattern Interferometer; Electronic Speckle Pattern Interferometry

ESPL Electronic Switching Programming Language

ESPIP Efficient Separations/Processes Integrated Program

ESPOD Electronic Systems Precision Orbit Determination

ESPOL Executive System Problem-Oriented Language

ESPP Endangered Species Protection Program [of US Environmental Protection Agency]

ESPRIT Estimation of Signal Parameters by Rotational Invariance Techniques

ESPRIT European Strategic Program for Research and Development in Information Technology [also Esprit]

ESPRIT/WIRE European Strategic Program for Research (and Development) in Information Technology/Web Information Repository for the Enterprise

ESPS Engineering Societies Personnel Service

ESPS Experiment Segment Pallet Simulator

ESPT Excited-State Proton Transfer

ESQ Entomological Society of Queensland [Australia]

Esq Esquire [also Esqr]

ESR Early Site Review

ESR Early Storage Reserve

ESR Effective (or Equivalent) Series Resistance

ESR Effective Signal Radiated

ESR Effective Sunrise

ESR Electronic Scanning Radar

ESR Electronic Send Receive

ESR Electron Spin Resonance

ESR Electroslag Refined; Electroslag Refining [Metallurgy]

ESR Electroslag Remelted; Electroslag Remelting [Metallurgy]

ESR Engineering Support Request

ESR En-Route Surveillance Radar

ESR Equipment Supervisory Rack

ESR Erythrocyte Sedimentation Rate [Medicine]

ESR Event Service Routine

ESR Exchangeable Sodium Ratio

ESR Excision Resynthesis Repair

ESR Experimental Superheat Reactor

ESR Expert Systems Research

ESR External Standard Ratio

E-SR Emulsion Synthetic Rubber

ESRANGE European Sounding Rocket Range [Kiruna, Sweden] [also Esrange]

ESRANGE European Space Range [of European Space Research Organization] [also Esrange]

ESRB European Society for Radiation Biology [Belgium]

ESRC Economic and Social Research Council [UK]

ESRD End-Stage Renal Disease

ESRF Environmental Studies Revolving Fund

ESRF European Synchrotron Radiation Facility [of European Science Foundation at Grenoble, France]

ESRI Earth Sciences and Resources Institute [of University of Utah, Salt Lake City, US]

ESRI Economic and Social Research Institute [Ireland]

ESRI Electron Spin Resonance Imaging

ESRI Engineering and Statistical Research Institute

ESRI Environmental Systems Research Institute [Canada]

ESRI Environmental Systems Research Institute (Proprietary Limited) [Australia]

ESRIN European Space Research Institute [of European Space Research Organization, France]

ESRO European Space Research Organization [Now Part of European Space Agency] [also Esro]

ESRP Environmental Standard Review Plan

ESRR Early Site Review Report

ESRS Early Sites Research Society [US]

ESRT Electroslag Refining Technology [Metallurgy]

ESS Earle's Salt Solution [Biochemistry]

ESS Earth-Synchronous Satellite

ESS Earth System Sciences

ESS Echo Suppression System

ESS Ecologically Sustainable Society

ESS Effective Sunset

ESS Electricity Supply System

ESS Electronic Sequence Switching

ESS Electronic Switching System

ESS Electroslag Surfacing [Metallurgy]

ESS Electrostatic Separation; Electrostatic Separator

ESS Electrostatic Spray(ing)

ESS Emplaced Scientific Station [of NASA]

ESS Energy Survey Scheme [UK]

ESS Engineered Safety Systems

ESS Environmental Science Services

ESS Environmental Stress Screening

ESS Environmental Support System

ESS Environmental Survey Satellite

ESS Equilibrium Solvation State

ESS Equipment Support Section

ESS European Symposium for Stereology

ESS Event Scheduling System

ESS Experiment Subsystem Simulator

ESS Extended Serial Subsystem

ES&S Engineering Services and Safety

Ess Essex [UK]

ess essential(ly)

ESSA Ecole Supérieure de Soudure Autogène [Higher School of Autogenous Welding, Paris, France]

ESSA Economists, Sociologists and Statisticians Association

ESSA Electronic Scanning and Stabilizing Antenna

ESSA Emergency Safeguards System Activation

ESSA Endangered Species Scientific Authority

ESSA Entomological Society of Southern Africa [South Africa]

ESSA Environmental and Social Systems Analyst [Canada]

ESSA Environmental Science Services Administration [US]

ESSA Environmental Survey Satellite

ESSA European Single Service Association [Switzerland]

ESSC Earth System Science Center [of Penn State University, University Park, Pennsylvania, US]

ESSC Earth Systems Science Committee

ESSC East Stroudsburg State College [Pennsylvania, US]

ESSCO Engineering Student Societies Council of Ontario [Canada]

ESSD Environmentally Sound and Sustainable Development

ESSDERC European Solid-State Device Research Conference

ESSEC Ecole Supérieure des Sciences Economiques et Commerciales [Higher School of Economic and Commercial Sciences, France]

ESSF Edmonton Space Sciences Foundation [Alberta, Canada]

ESSG Engineer Strategic Studies Group [US Army]

ESSI European System and Software Initiative

ESSL Engineering and Scientific Subroutine Library

Esso Mag Esso Magazine [Publication of Esso Ltd., UK]

Esso Oilways Int Esso Oilways International [Publication of Esso Ltd., UK]

ESSP Earliest Scram Set Point [Aerospace]

ESSP Engineers Society of St. Paul [Minnesota, US]

ESSR Estonian Soviet Socialist Republic [USSR]

ESSRA Economic and Social Science Research Association [UK]

ESSS Endangered Species Scientific Subcommittee

ESSSSA Egyptian Society of Solid-State Science and Applications [also ES4A]

ess sup essential supremum [Symbol]

ESSU Electronic Selective Switching Unit

EST Earliest Start Time

EST Eastern Standard Time [Greenwich Mean Time +05:00] [also est]

EST Elastic-Solid Theory

EST Electrolytic Sewage Treatment

EST Electronic Science and Technology

EST Electronic Spark Timing

EST Electroshock Therapy; Electroshock Treatment [Medicine]

EST Electrostatic Storage Tube

EST Electrostatic Transducer

EST Elementary Scattering Theory

EST Emerging Sciences and Technologies

EST Engineering Support and Technologies (Division) [of CERN–European Laboratory for Particle Physics, Geneva, Switzerland]

EST Engineers Society of Tulsa [Oklahoma, US]

EST Enlistment Screening Test

EST Environmentally Sound Technology

EST Environment, Sport and Territories (Portfolio) [Australia]

EST (2S,3S)-trans-Epoxysuccinyl-L-Leucylamido-3-Methylbutane Ethyl Ester

EST Expressed Sequence Tag [Genetics]

Est Eriochrome Black T

Est Establish(ed); Establishment [also est]

Est Estate [also est]

Est Estimate; Estimation; Estimator [also est]

Est Estonia(n)

est establish(ed)

est estimate(d)

ESTA Earth Science Teachers Association [UK]

ESTA Egyptian State Tourist Administration

ESTA Electronic System Test Equipment

ESTA Escape System Test Article [Aerospace]

ESTA Ethernet Station Adapter

ESTA European Security Transport Association [Belgium]

estab established

ESTAC Environmental Science and Technology Alliance Canada [formerly Institute for Chemical Science and Technology, Canada]

ESTAR Electrically Scanning Thinned Array Radiometer

estb establish

estbd established [also estb'd]

Estbt Establishment [also estbt]

ESTC Electrochemical Science and Technology Center [of University of Ottawa, Canada]

ESTC European Space Tribology Center [UK]

ESTCO Electrochemical Science and Technology Center [of University of Ottawa, Canada]

ESTCP Environmental Security Technology Certification Program [US]

ESTD Engineering Science and Technology Division

estd estimated [also est'd]

ESTEC ESA (European Space Agency) Technical Center [Noordwijk, Netherlands]

ESTEC European Space Research and Technology Center [of European Space Research Organization]

Estg Estimating [also estg]

ESTI European Space Technology Institute

ESTL Electronic Systems Test Laboratory

Estn Estimation [also estn]

ESTO Electronic Systems Technology Office [of US Department of Defense]

ESTO European Science and Technology Observatory

Estonian SSR Estonian Soviet Socialist Republic [also Est SSR]

ESTRAC European Satellite Tracking, Telemetry and Telecommand Network [also ESTRACK, Estrac, or Estrac] [of European Space Research Organization, France]

ESTRACK ESRO (European Space Research Organization) Tracking Network [also Estrac]

ESTS Echo Suppressor Testing System

ESTSC Energy Science and Technology Software Center [of US Department of Energy]

Est SSR Estonian Soviet Socialist Republic [also Estonian SSR]

ESTU Electronic System Test Unit

Estuarine Coastal Shelf Sci Estuarine Coastal and Shelf Science [Journal published in the UK]

ESTV Error Statistics by Tape Volume

Est Wt Estimated Weight [also est wt]

ESU Electricity Supply Union

ESU Electrostatic Unit [also esu]

ESU Emergency Service Unit

ESU Emporia State University [Kansas, US]

ESU Empty Signalling Unit

ESU Endangered Species Unit [of Environment Australia–Threatened Species and Ecological Communities (Section)]

ESU English Speaking Union

esu/g electrostatic units per gram [also esu g^{-1}]

esu/mol electrostatic units per mole [also esu mol^{-1}]

ESV Emergency Support Vessel

ESV Enamelled Single-Silk-Covered Varnished (Electric Wire)

ESV Essential Service Value

ESV Emergency Shutoff Valve

ESV Experimental Safety Vehicle

ESV Extended State Vector [Quantum Mechanics]

ESVM Electrostatic Voltmeter

ESVS Escape Suit Ventilation System [Aerospace]

ESVS Escape System Ventilation System [Aerospace]

ESW Electroslag Welding

ESW Electrostatic Wave [Plasma Physics]

ESW Engine Status Word

ESW Enhanced Sludge Washing

ESW Error Status Word

ESWL Equivalent Single Wheel Load

ESWL Extracorporal Shock Wave Lithotripsy

ESWM Engineering Society of Western Massachusetts [US]

ESWP Engineers Society of Western Pennsylvania [US]

ESWS Earth Satellite Weapon System

ESWS Emergency Service Water System

ESYCAT Expert System for Catalyst Design

ET Cold Polar Climate, Short Cool Summer and Long Cold Winter [Symbol]

ET Early Transition Metal (Elements)

ET Earth Tide [Geophysics]

ET Eastern Time [also et]

ET Eddy-Current Test(ing)

ET Edge Thickness (of Lenses)

ET Edge Tone

ET Edge-Triggered (Flip-Flop)

ET Edge Twin [Metallurgy]

ET Educational Technician

ET Educational Technologist; Educational Technology

ET Effective Temperature

ET Effective Thickness

ET Elapsed Time

ET Electrical Technician

ET Electrical Time

ET Electrodynamic Transducer
ET Electrolytic Tank
ET Electronics Technician
ET Electronic Technology
ET Electronic Transformer
ET Electronic Transition
ET Electron Telescope
ET Electron Theory
ET Electron Transfer
ET Electron Transport [Biochemistry]
ET Electron Tube
ET Electron Tunneling [Quantum Mechanics]
ET Electrostrictive Transducer
ET Electrotherapeutic(s); Electrotherapist; Electrotherapy [Medicine]
ET Electrothermal; Electrothermic(s)
ET Electrotype; Electrotyping
ET Elevated Temperature
ET Embryo Transfer
ET EMF (Electromotive Force)–Temperature
ET Emission Tube
ET End of Tape
ET End of Text
ET Endothelial; Endothelium [Anatomy]
ET Endothelin [Biochemistry]
ET Endothermic
ET Energy Technologist; Energy Technology
ET Energy Transfer
ET Engaged Tone
ET Engage Test
ET Engineering Technician
ET Engineering Technologist; Engineering Technology
ET Engineering Test
ET Engineering Thermoplastics
ET Enhancement Technology
ET Environmental Technician
ET Environmental Technologist; Environmental Technology
ET Environmental Test(ing)
ET Ephemeris Time [Astronomy]
ET Epitaxial Transistor
ET Epithelial Tissue
ET Equal Taper
ET Equilibrium Thermodynamics
ET Equipment Transfer
ET Equivalent Temperature
ET Erector Transporter
ET Escape Tower [Aerospace]
ET Estimation Theory [Statistics]
ET Ethiopia [ISO Code]
ET Ethylenedithiotetrathiafulvalene (Salt)
ET Eustachian Tube [Anatomy]
ET Event Timer
ET Everhart-Thronley (Detector) [also E/T]
ET Exchange Terminal

ET Excitation Transformer
ET Exothermic
ET Exotoxin [Microbiology]
ET Explosives Technology
ET Extended Time
ET External Tank
ET External Tank [NASA Space Shuttle]
ET Extraterrestrial
ET Extruded Tube
ET Extrusion Technology
ET International Aluminum Extrusion Technology Seminar and Exposition
ET Tundra Climate, Warmest Month below 50°F (10°C), but above 32°F (0°C) [Classification Symbol]
ET-1 Endothelin-1 (Human, Porcine Derived) [Biochemistry]
ET-2 Endothelin-2 (Human Derived) [Biochemistry]
ET-3 Endothelin-3 (Human, Rat Derived) [Biochemistry]
ET- Ethiopia [Civil Aircraft Marking]
ET_A Endothelin Receptor [Biochemistry]
E(T) Temperature-Dependent Elastic Modulus
E/T Everhart-Thornley (Detector) [also ET]
Et Ethane [also et]
Et Ethidium
Et Ethyl [also et]
Et Ethylene [also et, Ete, or ete]
E(t) Time-Dependent Electric Field
E(t) Tensile Creep Modulus [Symbol]
$\bar{E}(t)$ Mean Translational Energy [Symbol]
$E_c(t)$ (Time-Dependent) Creep Modulus (of Viscoelastic Polymers) [Symbol]
$E_r(t)$ Viscoelastic Relaxation Modulus (i.e., Time-Dependent Elastic Modulus of Viscoelastic Polymers) [Symbol]
e-T electron concentration–temperature (curve) [Symbol]
e(t) input (or event) [Symbol]
.et Ethiopia [Country Code/Domain Name]
$\varepsilon(t)$ time-dependent strain (or strain as a function of time) [Symbol]
ε-T Strain-Temperature (Curve [also ε-T] [Symbol]
$\varepsilon(T)$ temperature-dependent permittivity [Symbol]
$\eta(T)$ temperature-dependent viscosity [Symbol]
ETA Educational Television Association [UK]
ETA Elektrowärme im Technischen Ausbau [Series of German Journals on Electroheat and Installation]
ETA Electrothermal Analysis; Electrothermal Analyzer
ETA Electrothermal Atomization; Electrothermal Atomizer
ETA Elevated Temperature Annealing [Metallurgy]
ETA Emanation Thermal Analysis
ETA Employment and Training Administration [of US Department of Labor]
ETA Environmental Test Article
ETA Environmental Transport Association [UK]
ETA Environment Teachers' Association
ETA Estimated Time of Arrival [also eta]
ETA Estimated Time to Acquisition

ETA European Tube Association

ETA European Tugowners Association [UK]

ETA Event Tree Analysis [Computers]

ETA Explosive Transfer Assembly

ETA External Tank Attachment [Aerospace]

ETAA Eastern Townships Agricultural Association

ETAA Electrothermal Atomic Absorption

ETA-AAS Electrothermal Atomization Atomic Absorption Spectrometry

ETACS Extended Total Access Communication System

ETAC Environment Technical Advisory Committee

ETAC Environmental Technical Applications Center [of US Air Force]

ETAC Equilibrium Transfer Alkylating Crosslink

ETAC Ethyl Acetoacetate [also etac]

ETAC European Trade Association for Composite Materials [Switzerland]

EtAc Ethyl Acetate

etac ethyl acetoacetate [also ETAC]

Et Acet Ethyl Acetate [also et acet]

ETAC-1 Equilibrium Transfer Alkylating Cross-link

ETAD Ecological and Toxicological Association of the Dyestuffs Manufacturing Industry [Switzerland]

ETADM Embedded Trainer Advanced Development Model

ETA Elektrowärme Tech Ausbau A ETA Elektrowärme im Technischen Ausbau, Edition A [German Journal on Electroheat and Installation]

ETA Elektrowärme Tech Ausbau B ETA Elektrowärme im Technischen Ausbau, Edition B [German Journal on Electroheat and Installation]

ETA-I Electronics Technicians Association, International [US]

et al (et alii; et aliae) – and others

Et_3Al Triethylaluminum

Et Alc Ethyl Alcohol [also et alc]

$EtAlCl_2$ Ethylaluminum Dichloride

Et_2AlCl Diethylaluminum Chloride

Et_2AlCN Diethylaluminum Cyanide

Et_2AlH Diethylaluminum Hydride

Et_2AlI Diethylaluminum Iodide

Et_2AlOEt Diethylaluminum Ethoxide

ETANN Electrically Trainable Analog Neural Network; Electrically Trainable Artificial Neural Network

ETAP Extended Technical Assistance Program [US]

ETAPC European Technical Association for Protective Coatings [Belgium]

ETB Emerging Technology Branch [of Philips Laboratory, Edwards Air Force Base, California, US]

ETB End of Text Block

ETB End of Transmission Block

ETB Energy Technology Branch [of Natural Resources Canada]

ETB Environmental Technology Building

ETB Estimated Time of Berthing

ETB Ethiopian Birr [Currency]

ETB Exchange-Transferred Bound

ETB Equipment Transfer Bag

Et_3B Triethyl Borane (or Boron Triethyl)

ETBE Ethyl Tertiary Butyl Ether

Et_2Be Diethylberyllium

ETB-NOE Exchange-Transferred Bound Nuclear Overhauser Effect [Physics]

EtBr Ethidium Bromide

EtBr Ethyl Bromide

Et_3BrGe Triethylbromogermane

EtBz Ethyl Benzene

Et_2Bz Diethylbenzene

Et_3Bz Triethylbenzene

ETC Earth Terrain Camera

ETC Electrical Trade Council [US]

ETC Electronic Thermal Conductivity

ETC Electronic Time Card

ETC Electronic Toll Collection

ETC Electronic Tuning Control

ETC Electrothermal (Integrated) Circuit

ETC English Translucent China

ETC Energy Technology Center [of Canada Center for Mineral and Energy Technology; Natural Resources Canada, Ottawa]

ETC Enhanced Throughput Cellular (Modem)

ETC Environmental Technology Center [Canada]

ETC Environmental Transport Code [Canada]

ETC Estimated Time of Completion

ETC Estimate to Completion

ETC Ethylene Carbonate

ETC Ethyl Technical Center [Baton Rouge, Louisiana, US]

ETC European Tool Committee [Germany]

ETC European Topic Center [of European Environment Agency]

ETC European Translations Center [Netherlands]

ETC Experimental Techniques Center [UK]

ETC Extended Text Compositor

etc (et cetera) – and so on; and so forth; and the rest

Et_2Cd Diethylcadmium

ETCE Energy-Sources Technology Conference and Exhibition [of American Society of Mechanical Engineers, US]

ETCG Elapsed-Time Code Generator

EtCl Ethyl Chloride

ETC/LC European Topic Center on Land Cover [of European Environment Agency]

Et_3ClGe Triethylchlorogermane

ETCO Equipment Transfer/Change Order

ETCS Electronic Traction Control System

Et_2CuLi Lithium Diethylcopper

ETD Economic Transformer Design

ETD Electrical Terminal Distributor

ETD Electromagnetic Technology Division [of National Institute of Standards and Technology, Gaithersburg, Maryland, US]

ETD Embedded Temperature Detector

ETD Energy Technology Division [of Argonne National Laboratory, Illinois, US]

ETD Engineering Technology Division [of Oak Ridge National Laboratory, Tennessee, US]

ETD Equivalent Transmission Density

ETD Estimated Time of Departure [also etd]

ETD Estimated Turnover Date

ETD Exploration Technology Division [of Mining Industry Technology Council of Canada]

ETD Exploratory Battery Technology Development and Testing

ETD External Tank Door [Aerospace]

ETDB Energy Technology Database [of Canada Center for Mineral and Energy Technology; Natural Resources Canada, Ottawa]

EtdBr Ethidium Bromide

ETDCFRL European Training and Development Center for Farming and Rural Life [Belgium]

ETDL Electronics Technical and Developmental Laboratory [of US Army at Fort Monmouth, New Jersey]

ET4DIEN N,N,N',N'-Tetraethyl Diethylene Triamine

E-TDMA Enhanced Time Division Multiple Access [Telecommunications]

ETDTPY Bis(ethylenedithio)-1,6-Dithiapyrene

ETE Electronic Transition Energy

ETE Electrothermal Engine

ETE Electrothermal Excitation

ETE End-to-End

ETE Environmental Technology Conference and Exposition

ETE Estimated Time Enroute

ETE Estimated Time of Ejection

Ete Ethylene [also ete, Et, or et]

ETEC Energy Technology Engineering Center [US]

ETEC Environment Technology Export Council

ETECG Electronics Test Equipment Coordination Group [US]

ETED Electronics and Telecommunications Engineering Division [of Institute of Engineers, Calcutta, India]

ETEGM Ethoxytriethyleneglycol Methacrylate

ETER Epichlorohydrin-Ethylene Oxide-Allylglycidyl Ether Terpolymer

E-TE-S European Television Service [also ETES]

ETF Electron Translation Factor

ETF Engine Test Facility

ETF Enriched Text Format

EtF Ethyl Fluoride

ETFA Ethyl Trifluoroacetate

ETFE Ethylene Tetrafluoroethylene (Copolymer); Poly(Ethylene-Co-Tetrafluoroethylene)

ETFRN European Tropical Forest Research Network

ETG Electronic Target Generator

ETG Electrotopography

ETG Energietechnische Gesellschaft [Energy Technology Society; of Verein Deutscher Elektrotechniker, Germany]

ETG-Fachber Energietechnische Gesellschaft Fachberichte [Reports of the Energy Technology Society; of Verein Deutscher Elektrotechniker, Germany]

ETG European Training Group [of NATO]

ET-G$_3$ 4,6-Ethylidene (G$_3$)

ET-G$_4$ 4,6-Ethylidene (G$_4$)

ET-G$_5$ 4,6-Ethylidene (G$_5$)

Et$_2$GaCl Diethylgallium Chloride

Et$_3$Ge Triethylgermane [also (Et)$_3$Ga]

Et$_4$Ge Tetraethylgermane [also (Et)$_4$Ge]

EtGeCl$_3$ Ethylgermanium Trichloride

Et$_2$GeCl$_2$ Diethylgermanium Dichloride

Et$_3$GeCl Triethylgermanium Chloride

EtGe(EtO)$_3$ Ethyltriethoxygermane

ETG-Fachber ETG (Energietechnische Gesellschaft) Fachberichte [Reports of the Energy Technology Society of the Verein Deutscher Elektrotechniker, Germany]

EtGly N-Ethyl Glycine

ET-G$_7$PNP 4,4-Ethylidene (G$_7$)-p-Nitrophenol (G$_1$)-α,D-Maltoheptaside

ETGS East Texas Geological Society [US]

ETGTS Electronic Text and Graphics Transfer System

ETH Eidgenössische Technische Hochschule [Swiss Federal Institute of Technology]

Eth Ether [also eth]

Eth Ethyl (Ether) [also eth]

Eth Ethiopia(n)

8th eighth

Eth Acet Ethyl Acetate [also eth acet]

ETHEL European Tritium Handling Experimental Laboratory [Italy]

Et$_2$Hg Diethylmercury

ETH Hönggerberg Eidgenössische Technische Hochschule Hönggerberg [Hoenggerberg Federal Institute of Technology, Switzerland]

ETHIC Electric Trace Heating Industry Council [UK]

Ethnohist Ethnohistory [International Journal]

Ethnol Ethnological; Ethnologist; Ethnology [also ethnol]

Ethnol Ethnology [International Journal]

ETHS East Tennessee Historical Society [US]

ETHTP Energy Technology for High-Temperature Processes (Program) [of Energy Technology Branch, Natural Resources Canada]

Ethyl PCT O,O-Diethylphpsphorochloridothioate [also ethyl PCT]

ETHZ Eidgenössische Technische Hochschule Zürich [Zurich Federal Institute of Technology, Switzerland] [also ETH Zürich]

ETHZ-PSI Eidgenössische Technische Hochschule Zürich–Paul Scherrer Institut [Zurich Federal Institute of Technology/Paul Scherrer Institute, Switzerland]]

ETH Zürich Eidgenössische Technische Hochschule Zürich [Zurich Federal Institute of Technology] [also ETHZ]

ETI Education Turnkey Institute [US]

ETI Elapsed Time Indicator

ETI Electrical Tool Institute [now Power Tool Institute, US]

ETI Electron Tomographic Imaging

ETI Engineering Technology Institute [US]

ETI Environmental Technology Initiative [of US Environmental Protection Agency]

ETI Equipment and Tool Institute [US]

ETI Expert Center for Taxonomic Identification [Netherlands]

ETI Extraterrestrial Intelligence

EtI Ethyl Iodide

$(ET)_2I_3$ Bis(ethylenedithio)tetrathiafulvalene Triiodine (Organic Superconductor)

ETIC Environmental Technology Information Center

ETIC European Technical Information Center [Birmingham, UK]

ETIM Elapsed Time

ETIP Environmental Technology Investment Program [of Western Economic Diversification Canada]

ETIPS Electrothermal Ice Protection System

ETIS European Technical Information Service

ETIS-MARFO ETIS (European Technical Information Service) in Machine-Readable Form

ETL Educational Technology Language [of University of Western Ontario, Canada]

ETL Effective Testing Loss

ETL Electrical Testing Laboratories [US]

ETL Electrotechnical Laboratory

ETL Electrotechnical Laboratory [of Agency of Industrial Science and Technology, Ibaraki, Japan]

ETL Emitter-Follower Transistor Logic

ETL Ending Tape Label

ETL Environmental Technology Laboratory [of National Oceanic and Atmospheric Administration, US]

ETL Etching by Transmitted Light

Etl Ethanol [also etl]

ETLA Extended Three Letter Acronym

ETLA Extended Time Limited Area

EtLi Ethyl Lithium

ETLOW External Tank Lift-Off Weight [Aerospace]

ETLP Environmental Technology Loan Program [of Western Economic Diversification Canada]

ETLS Extruded Tunnel Lining System

ETM Early Transition Metal

ETM Elapsed Time Meter

ETM Electronic Test and Measurement

ETM Enhanced Thematic Mapper [NASA Landsat Program]

ETM Enhanced Timing Module

ETM Engineering Test Model

ETM Engineering Test Module [of Superconductive Magnetic Energy Storage Initiative, US Pentafon]

ETM+ Enhanced Thematic Mapper Plus [NASA Landsat Program]

ETMA Ethyl Trimethylacetate

Et_3MeOGe Triethylmethoxygermane

ETMF Elapsed Time Multiprogramming Factor

EtMgBr Ethylmagnesium Bromide

EtMgCl Ethylmagnesium Chloride

EtMgI Ethylmagnesium Iodide

ETM/LTM Early Transition Metal/Late Transition Metal

ETMOD Environmental Tritium Model

ETMQ 6-Ethoxy-2,2,4-Trimethyl-1,2-Dihydroquinoline

ETMWG Electronic Trajectory Measurements Working Group

ETN Electronic Tandem Network

ETN Electronic Trading Network

ETN Equipment Table Nomenclature

Et_3N Triethylamine

Et_3N Triethyl Nitrogen

Et_4NAc Tetraethylammonium Acetate

Et_4NBF_4 Tetraethylammonium Tetrafluoroborate

Et_3N-BH_3 Borane-Triethylamine Complex

Et_4NBr Tetraethylammonium Bromide

Et_4NCl Tetraethylammonium Chloride

Et_4NCN Tetraethylammonium Cyanide

Et_4NF Tetraethylammonium Fluoride

Et_2NH Diethylamine

$Et_2NH-HBr$ Diethylamine Hydrobromide

$Et_2NH-HCl$ Diethylamine Hydrochloride

Et_4NI Tetraethylammonium Iodide

Et_4NOH Tetraethylammonium Hydroxide

ETO Energy Technology Office

ETO Estimated Time Off [Telecommunications]

ETO Estimated Time Over Significant Point

ETO European Transportation Organization

ETO Exponential-Type Orbital [Physics]

ETO Extended Time Observation

EtO Ethoxide (or Ethylate); Ethoxy-

EtO Ethylene Oxide

EtO Ethyl Oxide (or Ethyl Ether)

$(EtO)_2$ Diethoxy-

$(EtO)_3$ Triethoxy-

EtO_4 Ethyl Tetroxide

$(EtO)_4$ Tetraethoxy-

Et_2O Diethyl Oxide (or Ethyl Ether)

EtO-Alc 2-Ethoxyethanol [also Eto-alc]

$(EtO)_3B$ Triethyl Borate (or Triethoxyborane)

$(EtO)_2CO$ Diethyl Carbonate

EtOD Deuterated (or Labeled) Ethanol

E-to-E Electronic-to-Electronic

E-to-E End-to-End

ETOG European Technical Operations Group

EtOH Ethanol (or Ethyl Alcohol)

EtOK Potassium Ethoxide (or Potassium Ethylate)

EtOLi Lithium Ethoxide (or Lithium Ethylate)

ETOM Electron Trapping (Erasable) Optical Memory

EtONa Sodium Ethoxide (or Sodium Ethylate)

$(EtO)_3P$ Triethyl Phosphite

ETOPS Extended Range Twin Operations

ETOS Extended Tape Operating System

$(EtO)_2SO$ Diethyl Sulfite

(EtO)$_2$SO$_2$ Ethyl Sulfate (or Diethyl Sulfate)

EtOSO$_3$H Ethyl Hydrogen Sulfate

EtO-THP Tetrahydropyranyl Ether

ETOUSA European Theather of Operations, US Army

ETOX EPROM (Erasable Programmable Read-Only Memory) Tunnel Oxide (Device)

EtOx Ethylene Oxide

ETP Effluent Treatment Plant

ETP Electrolytic Tough Pitch (Copper)

ETP Electron Transfer Process

ETP Electrothermal Printer

ETP Emission Thermophotometry

ETP Engineering Thermoplastic(s)

ETP Equipment Test Plan

ETP Equivalent Top Product

ETP Estimated Time of Penetration [Aerospace]

Et$_3$P Triethylphosphine

ε(2π) hemispherical emittance (function) [also $\epsilon(2\pi)$] [Symbol]

ε(θ,φ) directional emittance (function) (or directional emissivity (function)) [also $\epsilon(\theta,\phi)$]

ETPAE Ethyl-Terminated Polyarylene Ether

ETPB 4-Ethyl-2,6,7-Trioxa-1-Phosphabicyclo[2.2.2]octane [also etpb]

Et$_4$Pb Tetraethyllead

Et$_4$PBr Tetraethylphosphonium Bromide

ETPC Electrolytic Tough Pitch Copper

Et$_4$PCl Tetraethylphosphonium Chloride

ETP Cu Electrolytic Tough Pitch Copper

ETPL Endorsed Tempest Products List

Et$_4$PI Tetraethylphosphonium Iodide

ETPO European Trade Promotion Organization

Etps Enterprise [also etps]

Etpyz Ethyl Pyrazine

etpyz ethylpyrazine [abbreviation used in coordination compound formulas]

ETQAP Education and Training in Quality Assurance Practices [of American Society for Quality Control, US]

ETR Eastern Test Range [of US Air Force]

ETR Electron Transfer Reaction

ETR Engineering Test Reactor [of US Department of Energy]

ETR Equal Time Rule

ETR Estimated Time of Restore

ETR Expected Time of Response

ETR Experimental Test Reactor

ETR Extended Temperature Range

ETRAC Educational Television and Radio Association of Canada

ETR-CX Engineering Test Reactor Critical Assembly [of US Department of Energy]

(ET)$_2$ReO$_4$ Bis(ethylenedithio)tetrathiafulvalene Rhenium Oxide (Organic Superconductor)

ETRI Electronics and Telecommunication Research Institute [Taejon, South Korea]

ETRI Electrotechnical Research Institute [South Korea]

ETRIPHOS Ethylidynetris[methyl(diethyl)phoshine]

ETR/ITER Engineering Test Reactor/International Thermonuclear Experimental Reactor

ETROD Eastern Test Range Operations Directive [of US Air Force]

ETRS European Textile Research Symposium

ETRTO European Tire and Rim Technical Organization [Belgium]

ETS East Tennessee Section [of Materials Research Society, US]

ETS Econometric Time Series

ETS Educational Technology System

ETS Educational Television Station

ETS Educational Testing Service [Princeton, New Jersey, US]

ETS Electrical Test Set

ETS Electrodepositors Technical Society [UK]

ETS Electronic Tandem Switching

ETS Electronic Telegraph System

ETS Electronic Translator System

ETS Electronic Test Set

ETS Electron Transmission (or Tunneling) Spectroscopy

ETS Electron Transport System [Biochemistry]

ETS Elevated Temperature Sintering

ETS Emergency Temporary Standard

ETS Energy Transfer System

ETS Engineering Test Satellite [of National Space Development Agency of Japan]

ETS Engineering Thermoset(s)

ETS Enquiry Terminal System

ETS Environmental Technical Specification

ETS Environmental Technology Seminar [US]

ETS Environmental Tobacco Smoke

ETS *(Escuela Tecnica Superior)* – Spanish for "Higher Technical School"

ETS External Tank Subsystem [Aerospace]

ETS External Tank System [Aerospace]

Et$_2$S Ethyl Sulfide

ETSA Electricity Trust of South Australia

ETSA Ethyltrimethylsilylacetate

ETSAL Electronic Terms for Space Age Language

ETSAP Energy Technology Systems Analysis Project

ETSC East Tennessee State College [Johnson City, US]

ETSC East Texas State College [Commerce, US]

ETSCO Engineering and Technical Societies Council [US]

ETSD Engineering and Technical Services Division [of Canada Center for Mineral and Energy Technology; Natural Resources Canada, Ottawa]

ETSD Enhanced Thermionically Supported Discharge

ET-SEP External Tank Separation [Aerospace]

et seq *(et sequentes; et sequentia)* – and the following

ETSH Extended Tensor Surface Harmonic (Theory) [Physics]

Et$_4$Si Tetraethylsilane

Et$_3$SiCF$_3$ Tetraethyl(trifluoromethyl)silane

Et₃SiCl Chlorotriethylsilane

Et₃SiH Triethylsilane

EtSiHMe₂ Dimethylethylsilane

ETSIM Escuela Tecnica Superior de Ingenieros de Minas [Higher Technical School for Mining Engineers, Oviedo, Spain]

Et₃SiOH Triethylsilanol

Et₃Sn Triethyltin

Et₄Sn Tetraethyltin

Et₃SnBr Triethyltin Bromide (or Bromotriethylstannate)

Et₂SO₂ Ethyl Sulfone (or Diethyl Sulfone)

Et₂SO₄ Ethyl Sulfate (or Diethyl Sulfate)

ETSQ Electrical Time, Superquick

ETSS Entry Time Sharing System

ETSS External Tank Separation Subsystem [Aerospace]

ETSU East Tennessee State University [Johnson City, US]

ETSU East Texas State University [Commerce, US]

ETSU Energy Technology Support Unit [of Department of Energy, UK]

ETT Electrothermal Thrusters

ETT Explosion Temperature Test

ETTA Eastern Townships Textile Association

ETTA 1,2-Ethanediylidenetetrakisacetic Acid

ETTA 2,2′,2″,2‴-(1,2-Ethanediylidenetetrakis[thio])-tetrakis-acetic Acid

Et₂Te Diethyltelluride

4EtTSC 4-Ethyl Thiosemicarbazide

ETU Electrical Trades Union

ETU Electronic Text Unit

ETU Enhanced Telephone Unit

ETU N,N'-Ethylene Thiourea

ETUC European Trade Union Confederation

Etudes Epistémol Génét Etudes d'Epistémologie Génétique [French Journal on Genetic Epistemology Studies]

ETUI European Trade Union Institute

ETV Educational Television [US]

ETV Electric Transfer Vehicle

ETV Electrothermal Vaporization

ETV Environmental Technology Verification (Program) [of US Environmental Protection Agency]

ETV Elevating Transfer Vehicle

ETVA External Tank Vent Arm [Aerospace]

ETVM Electrostatic Transistorized Voltmeter

ETW European Transonic Windtunnel [Joint Organization of France, Germany, Netherlands and UK] [also ETWT]

ETW European Trans-Sound Wind Tunnel

ETWT European Transonic Windtunnel [Joint Organization of France, Germany, Netherlands and UK] [also ETW]

ETWTC European Tire and Wheel Technical Conference [now European Tire and Rim Technical Organization, Belgium]

ETX End of Text

EtX Ethyl Halide

ETYA 5,8,11,14-Eicosatetraynoic Acid

Etym Etymologica; Etymologist; Etymology [also etym]

ETZ Elektrotechnische Zeitschrift [German Journal of Electrical Engineering; published by Verein Deutscher Elektrotechniker]

Etz 1-Ethyl 1H-Tetrazole [also ETZ]

ETZ Arch ETZ Archiv [German Journal of Electrical Engineering; published by Verein Deutscher Elektrotechniker]

Et₂Zn Diethylzinc

EU Edinburgh University [UK]

EU Ehime University [Japan]

EU Electronic Unit

EU Electron Unit [also eu]

EU Emory University [Atlanta, Georgia, US]

EU Endotoxin Unit [Microbiology]

EU End User

EU Engineering Unit

EU Enriched Uranium

EU Entropy Unit [also eu]

EU Enzyme Unit [Biochemistry]

EU Eötvös University [Budapest, Hungary]

EU European Union

EU Execution Unit

EU Expanding Universe

EU Expected Utility

EU Experimentally Undetermined [also eu]

EU Experimental Unit

Eu Euler Number (of Fluid Flow) [Symbol]

Eu Euronorm [European Standard]

Eu Europium [Symbol]

Eu Eutectic [also eu]

Eu²⁺ Divalent Europium Ion [also Eu⁺⁺] [Symbol]

Eu³⁺ Trivalent Europium Ion [Symbol]

Eu-123 Europium Barium Copper Oxide [EuBa₂Cu₃O₇₋ₓ] [also Eu123]

Eu-144 Europium-144 [also ¹⁴⁴Eu, or Eu¹⁴⁴]

Eu-145 Europium-145 [also ¹⁴⁵Eu, or Eu¹⁴⁵]

Eu-146 Europium-146 [also ¹⁴⁶Eu, or Eu¹⁴⁶]

Eu-147 Europium-147 [also ¹⁴⁷Eu, or Eu¹⁴⁷]

Eu-148 Europium-148 [also ¹⁴⁸Eu, or Eu¹⁴⁸]

Eu-150 Europium-150 [also ¹⁵⁰Eu, or Eu¹⁵⁰]

Eu-151 Europium-151 [also ¹⁵¹Eu, or Eu¹⁵¹]

Eu-152 Europium-152 [also ¹⁵²Eu, or Eu¹⁵²]

Eu-153 Europium-153 [also ¹⁵³Eu, or Eu¹⁵³]

Eu-154 Europium-154 [also ¹⁵⁴Eu, or Eu¹⁵⁴]

Eu-155 Europium-155 [also ¹⁵⁵Eu, or Eu¹⁵⁵]

Eu-156 Europium-156 [also ¹⁵⁶Eu, or Eu¹⁵⁶]

Eu-157 Europium-157 [also ¹⁵⁷Eu, or Eu¹⁵⁷]

Eu-158 Europium-158 [also ¹⁵⁸Eu, or Eu¹⁵⁸]

Eᶜ(u) Upper-Bound Elastic Modulus of Composite [Symbol]

eu electron unit [also EU]

eu entropy unit [also EU]

eu experimentally undetermined [also EU]

EUA Electrical Utilities Application

EUA European Unit of Account

EuBCO Europium Barium Copper Oxide (Superconductor)

EUC End-User Computing

EUC End-User Control

EUC Equivalent Uranium Content

EUC European Union of Coachbuilders [Belgium]

EUC Extended Unit Cell

EUC Extended Unix Code

EUCAPA European Capsules Association

EUCARPIA European Association for Research on Plant Breeding [Netherlands]

EUCATEL European Committee of Associations of Telecommunication Industries

EUCEPA European Liaison Committee for Pulp and Paper

EuCeSCNO Europium Cerium Strontium Copper Niobium Oxide (Superconductor)

EUCHEMAP European Committee of Chemical Plant Manufacturers

EUCLID Experimental Use Computer London Integrated Display

EUCOMED European Confederation of Medical Suppliers Associations [UK]

EUCREX European Cloud and Radiation Experiment

EUDISED European Documentation and Information System on Education [of Council of Europe]

EUE Expected Unserved Energy [Hydropower]

EUF Electro-Ultrafiltration

EUF End User Facility

EUF Equivalent Unavailability Factor

EUFMC Electric Utilities Fleet Managers Conference

Eu(FOD)₃ Europium Tris(1,1,2,2,3,3,3-Heptafluoro-7,7-Dimethyl-4,6-Octanedionate); Europium Tris(6,6,7,7,8,8,8-Heptafluoro-2,2-Dimethyl-3,5-Octanedionate) [also Eu(fod)₃]

EUFODA European Foodstuffs Distributors Association

EUG Electricity Utilization Group [of University of Cambridge, UK]

EUG European Union of Geosciences [Strasbourg, France]

EuGH Europäischer Gerichtshof [European Court of Justice, The Hague, Netherlands]

Eu(HFC)₃ Europium Tris[3-(heptafluoropropylhydroxy-methylene)camphorate] [also Eu(hfc)₃]

EUI End-User Interface

EUI European University Institute [Florence, Italy]

EUJS European Union of Jewish Students [Belgium]

EULA End-User License Agreement

EuLα Europium L-Alpha (Radiation) [also EuL_α]

EULER European Libraries and Electronic Resources in Mathematical Sciences

EUM Estados Unidos Mexicanos [United Mexican States]

EUM European Mediterranean Region [of International Civil Aviation Organization]

EUMABOIS Comité Européen des Constructeurs de Machines à Bois [European Committee of Woodworking Machinery Manufacturers, France]

EUM-AFTN European Mediterranean Aeronautical Fixed Telecommunications Network

EUMAPRINT European Committee of Associations of Printing and Paper Converting Machinery [France]

EUMC Enameled Utensil Manufacturers Council

Eu(MeOBB)₃ Europium(III) Methoxybenzoylbenzoate

EUMETSAT European Organization for the Exploitation of Meteorological Satellites [also Eumetsat]

EUMETSAT European Meteorological Satellite

EuO Europium(II) Oxide

EUP European Patent [also EuP]

EUPA European Patent Application

EUPC Electric Utility Planning Council

EUPEPTIC Evaluation of Unitary Programs for Effecting Plural Tasks in Index Construction

EUPOCO European Postgraduate Education in Polymer and Composites Engineering (Program)

EUPS Extreme Ultraviolet Photoemission Spectroscopy

EUR Erasmus Universiteit van Rotterdam [Erasmus University of Rotterdam, Netherlands]

EUR European Region [of International Civil Aviation Organization]

Eur Europe(an)

EURABANK European-American Bank [also Eurabank]

EURACA European Air Carrier Assembly

EurACS European Association of Classification Societies [Belgium]

Eur Addict Res European Addiction Research [Journal]

EURADH European Adhesion Congress and Exhibition

Eur Adhes Seal European Adhesives and Sealants [UK Journal]

EURALARM Association of European Manufacturers of Fire and Intruder Alarm Systems [Germany]

EURAM European Research on Advanced Materials (Program) [of European Economic Community] [also EuRam]

EURANP European Air Navigation Plan [of International Civil Aviation Organization] [also EUR ANP]

Eur Appl Res Rep European Applied Research Report [UK]

Eur Appl Res Rep Nucl Technol Sect European Applied Research Reports, Nuclear Science Technology Section [of Commission of the European Communities; published in US]

EURAS European Academy for Standardization

EURAS European Anodizers Association [Switzerland]

EURASAP European Association for the Science of Air Pollution Control [UK]

EURASIA European Asia [also Eurasia]

EurasiaRC Eurasia Research Center [Internet]

EURASIP European Association for Signal Processing [Switzerland]

EURATOM European Atomic Energy Community [also Euratom]

Eur Biophys J European Biophysics Journal [Germany]

EurChem European Chemist (Qualification)

Eur-Chem Europa-Chemie [German Publication on Chemistry in Europe; published by Verband der Chemischen Industrie]

Eur Cytokine Netw European Cytokine Network [Journal published in France]

EUREAU Union des Associations des Distributeurs d'Eau de Pays Membres des Communautés Européennes [Union of the Water Supply Associations from Countries of the European Community, Belgium]

EURECA European Research Coordination Agency [also Eureca]

EURECA European Retrievable Carrier

EUREL Association Européenne des Réserves Naturelles Libres [European Association for Free Nature Reserves, Belgium]

EUREL Convention of National Societies of Electrical Engineers of Western Europe [Switzerland]

EUREM European Conference on Electron Microscopy

EUREMAIL Conference Permanente de l'Industrie Europeenne de Produits Emailles [Permanent Conference of the European Enamelled Products Industry]

Eu-Resolve Tris(2,2,6,6-Tetramethyl-3,5-Heptanedionato-O,O'-) europium

EUREX Enriched Uranium Extraction

EURFCB European Frequency Coordinating Body [of International Civil Aviation Organization]

EURIFI European Association of Research Institutes for Furniture

EURIM European Conference on Research into Information Management

EURING European Union for Bird Ringing

EurIng European Engineering Qualification

EURIPA European Information Providers Association [UK]

EURIS European Information Service

EURISOTOPE EURATOM Radioisotope Information Bureau

Eur J Biochem European Journal of Biochemistry [of Federation of European Biochemical Societies]

Eur J Clin Pharmacol European Journal of Clinical Pharmacology

Eur J Dermatol European Journal of Dermatology [France]

Eur J Eng Educ European Journal of Engineering Education [UK]

Eur J Immunol European Journal of Immunology

Eur J Mech A Solids European Journal of Mechanics, Series A, Solids [France]

Eur J Mech B Fluids European Journal of Mechanics, Series B, Fluids [France]

Eur J Mech Eng European Journal of Mechanical Engineering [Belgium]

Eur J Mineral European Journal of Mineralogy [Joint Publication of Deutsche Mineralogische Gesellschaft, Société Française de Mineralogie et de Cristallographie, Società Italiana di Mineralogia e Petrologia and European Mineralogical Union]

Eur J Nucl Med European Journal of Nuclear Medicine [Germany]

Eur J Oper Res European Journal of Operational Research [Netherlands]

Eur J Orthodont European Journal of Orthodontics [Published by Oxford University Press, UK]

Eur J Paed European Journal of Paediatics

Eur J Pharmacol European Journal of Pharmacology

Eur J Phys European Journal of Physics [of Institute of Physics, UK]

Eur J Solid State Inorg Chem European Journal of Solid State and Inorganic Chemistry [France]

Eur Manage J European Management Journal

Eur Market Newsl European Marketing Newsletter [of European Commission for Industrial Marketing, UK]

Eur Neurol European Neurology [Journal]

EuroASIC European Conference on Application-Specific Integrated Circuits

EUROAVIA Association of European Aeronautical and Astronautical Students [Netherlands]

Eurobase European Database [also EUROBASE]

EUROBAT Association of European Battery Manufacturers [Switzerland]

Eurobeam European Beam Antenna [also EUROBEAM]

EUROBIT European Association of Manufacturers of Business Machines and Data Processing Equipment [Germany]

EUROBITUME European Bitumen Association [Belgium]

EURO-Blech International Sheet Metal Working Technology Exhibition [Hanover, Germany]

Eurobuild European Organization for the Production of New Techniques and Methods in Building [also EUROBUILD]

Eurobuild Inf Bull Eurobuild Information Bulletin [France]

EUROCAE European Organization for Civil Aviation Electronics [also Eurocae]

EUROCAM European Organization for Civil Aviation Electronics Manufacturers

EUROCEAN European Ocean Association [Monaco]

EUROCHIMIE Societé Européenne pour le Traitement Chimique des Combustibles Irradiés [European Society for the Chemical Treatment of Radiant Fuels]

EUROCOM European Coal Merchants Union [also Eurocom]

EUROCOMP European Computing Congress

EUROCOMSATE European Consortium Communications Satellite

EUROCON European Convention [of Institute of Electrical and Electronics Engineers, US]

Eurocontrol European Organization for the Safety of Air Navigation [also EUROCONTROL]

EUROCOPI European Computer Program Information Center [of European Atomic Energy Community Joint Research Center]

EUROCORD Fédération des Industries de Ficellerie et Corderie de l'Europe Occidentale [Federation of Western European Rope and Twine Industries, France]

EUROCORR (International) Corrosion Congress/Convention [also Eurocorr] [of European Federation of Corrosion]

EUROCVD European Conference on Chemical-Vapor Deposition [also EURO CVD and EuroCVD]

EuroDAC European Conference on Design Automation

EURODICAUTOM European Automatic Dictionary [also Eurodicautom] [of European Commission]

EURODOC European Joint Documentation Service [of European Space Research Organization and European Industrial Space Research Group]

Eurodollar European Currency Unit [also ecu]

EUROFAR European Future Advanced Rotorcraft

EUROFER European Steel Industry Federation [also Eurofer]

EUROFEU European Committee of the Manufacturers of Fire Protection and Safety Equipment and Fire Fighting Vehicles [Germany]

EUROFLAG European Future Large Aircraft Group

Euro Flexo Mag Euro Flexo Magazine [Netherlands]

EURO FOOD CHEM European Conference on Authenticity and Adulteration of Food–The Analytical Chemistry Approach

EUROFORGE European Committee of Forging and Stamping Industries [France]

Eurogel European Conference on Sol-Gel Processing

EUROGRAPHICS European Computer Graphics Association

Eurogress European Congress and Conference Center [Aachen, Germany]

EUROGYPSUM Working Community of the European Gypsum Industry [France]

EURO-HKG Europäische Gesellschaft zur Auswertung von Erfahrungen bei Planung, Bau und Betrieb von Hochtemperaturreaktoren [European High-Temperature Nuclear Power Station Society, Germany]

Eurolab European Laboratory

EUROLAIT Union Européenne du Commerce des Produits Laitièrs et Dérivés [European Union of Importers, Exporters and Dealers in Dairy Products, UK]

EUROLAT European Network on Lateritic Weathering and Global Environment [at TU Berlin, Germany]

EUROMAP European Committee of Machinery Manufacturers for the Plastics and Rubber Industries

EUROMAR European Marine Research and Technology Project [of European Research Coordination Agency]

Euromarket European Economic Community

EUROMAT European Conference on Advanced Materials and Processes [also EuroMat]

Euromat Euromaterials [Publication and Conference of the Federation of European Materials Societies]

EuroMBE European Conference on Molecular Beam Epitaxy and Related Growth Methods

EUROMECH European Mechanics Colloquium [Germany]

EuroMech European Mechanics Committee [also EMC, Germany]

EURO-MET European Metallographic Conference and Exhibition

EUROMICRO European Association for Microprocessing and Microprogramming [Netherlands]

EUROMIL European Organization of Military Associations [Germany]

Euromines European Association of the Mining Industries [Belgium]

EUROMOT European Committee of Associations of Manufacturers of Internal Combustion Engines [Netherlands]

EURONEM European Association of Netting Manufacturers [France]

EURONET European Public Data Network [also Euronet]

EuropaCat European Congress on Catalysis

EuroPACE European Network of Universities and their Partners in Education and Training

Europages European Business Directory [also EUROPAGES]

Europatent European Patent [also EUROPATENT]

EUROPEC European Offshore Petroleum Conference and Exhibition

EUROPECHE Association des Organisations Nationales d'Entreprises de Pêche de la Communauté Economique Européenne [Association of National Organizations of Fishing Enterprises in the European Economic Community, Belgium]

EUROPEX European Information Center for Explosion Protection [also EuropEx]

EUROPHOT Association Européenne des Photographes Professionels [European Association of Professional Photographers]

Europhysics European Physical Society [also EUROPHYSICS, Switzerland]

Europhys Lett Europhysics Letters [of European Physical Society, Switzerland]

Europhys News Europhysics News [of European Physical Society, Switzerland]

Europhys News Extra Europhysics News Extra [of European Physical Society, Switzerland]

EUROPILOTE European Organization of Airline Pilots Associations

EUROPLANT European Plantmakers Committee [Belgium]

Euro PM European Conference on Advanced Powder-Metallurgy Materials

EUROPOL Intra-European Air Transport Policy [of European Civil Aviation Conference, France]

EUROPTO Series Series of Topical Meetings of European Optical Society and SPIE (International Society for Optical Engineering)

EUROPUMP European Committee of Pump Manufacturers [UK]

EUROPUR European Association of Flexible Foam Block Manufacturers [Belgium]

EUROQSYM Euroconference on New Symmetries in Statistical Mechanics and Condensed Matter Physics

EURORAD European Association of Manufacturers of Radiators [Switzerland]

EUROSAC Fédération Européenne des Fabricants de Sacs en Papier à Grande Contenance [European Federation of Multiwall Paper Sack Manufacturers, France]

Eurospace European Industrial Space Research Group [also EUROSPACE, France]

EUROSTAR European Communications Satellite [UK and France]

Eurostat Statistical Office of the European Communities [also EUROSTAT, Luxembourg]

Eurostat Rev Eurostat Review [of Statistical Office of the European Community, Luxembourg]

EUROTALC Scientific Association of European Talc Industry [Belgium]

Eurotest European Committee on Testing and Materials [also EUROTEST, Belgium]

Eurotest Tech Bull Eurotest Technical Bulletin [of Eurotest, Belgium]

EUROTOX Comité Européen Permanent de Recherches sur la Protection des Populations contre le Risques de Toxicité [Permanent European Research Committee for the Protection of the Population against Toxic Substances]

EUROTRAC European Project on the Transport of Atmospheric Contaminants [of European Research Coordination Agency]

EUROTRIB (International) European Congress on Tribology

Eurotron European Proton Synchrotron [of CERN–European Laboratory for Particle Physics, Geneva, Switzerland] [also EUROTRON]

EuroVHDL European Conference on VHSIC (Very-High-Scale Integrated Circuit) Hardware Description Language

EUROVISION Union Européenne de Radiodiffusion [European Broadcasting Union, France] [also Eurovision]

Eur Packag Mag European Packaging Magazine [UK]

EurPhys European Physicist (Qualification)

Eur Phys J A The European Physical Journal A

Eur Phys J App Phys European Physical Journal–Applied Physics

Eur Phys J B The European Physical Journal B

Eur Phys J C The European Physical Journal C

Eur Phys J D The European Physical Journal D

Eur Phys J Direct European Physical Journal Direct

Eur Phys Lett European Physical Letters

EURPISO European Union of Public Relations International Service Organization [Italy]

Eur Pkg Mag European Packaging Magazine [UK]

Eur Plast News European Plastics News [UK]

Eur Polym J European Polymer Journal [UK]

EUR/RAN European Regional Air Navigation Meeting [of International Civil Aviation Organization]

Eur Rubb J European Rubber Journal [UK]

EUR/TFG European Traffic Forecasting Group [of Eurocontrol–European Organization for the Safety of Air Navigation]

Eur Semicond European Semiconductor [Journal published in the UK]

Eur Semicond Des Prod European Semiconductor Design and Production [Journal published in the UK]

Eur Semicond Prod European Semiconductor Production [Journal published in the UK]

Eur Surg Res European Surgical Research [Journal]

Eur Symp European Symposium

Eur Urol European Urology [International Journal]

EURYDICE Education Information Network in the European Community [Belgium]

EuS Europium Sulfide (Semiconductor)

EUSA Electrical Utilities Safety Organization [Canada]

EuSe Europium Selenide (Semiconductor)

EUSEC Conference of Representatives from European and United States (of America) Engineering Societies

EUSIDIC European Association of Scientific Information Dissemination Centers [UK]

EUSIREF European Association of Science Information Referral Centers

EUSIREF European Scientific Information Retrieval Working Group [of European Association of Scientific Information Dissemination Centers]

EUSJA European Union of Science Journalists Associations [Italy]

EUSM Emory University School of Medicine [Atlanta, Georgia, US]

EUSSG European Union for the Scientific Study of Glass [Belgium]

EUT Eindhoven University of Technology [Netherlands]

EUT Equipment under Test

EuTe Europium Telluride (Semiconductor)

EUTELSAT European Telecommunications and Satellite Organization [France]

Eu(TFC)$_3$ Europium Tris[3-(Trifluoromethylhydroxy-methylene) camphorate] [also Eu(tfc)$_3$]

Eu(THD)$_3$ Europium Tris(2,2,6,6-Tetramethyl-3,5-Heptane-dionate) [also Eu(thd)3]

Eu(TMHD)$_3$ Europium Tris(2,2,6,6-Tetramethyl-3,5-Heptane-dionate) [also Eu(tmhd)$_3$]

EUTP Enhanced, Unshielded Twisted Pair (Cable)

Eutraplast Committee of Plastic Converter Associations of Western Europe [also EUTRAPLAST] [now European Plastics Converters Federation, Belgium]

EUUG European Unix Users Group [UK]

EUV Extreme Ultraviolet

EUVE Extreme Ultraviolet Explorer [of NASA]

EUVEPRO European Vegetable Protein Federation [Belgium]

EUVITA Extreme Ultraviolet Telescope Array

Eu:Y$_2$O$_3$ Europium-Doped Yttria [also Eu^{3+}:Y$_2$O$_3$]

EV Eclipsing Variable [Astronomy]

EV Eddy Velocity

EV Eddy Viscosity

EV Effective Circular Space Width (of Splines) [Symbol]

EV Efficient Vulcanization; Efficient Vulcanizing (System)

EV Effluent Velocity

EV Eigenvalue

EV Elasticoviscosity; Elasticoviscous (Solid)

EV Elastic Vibration

EV Electric Vehicle

EV Electrovalence; Electrovalency; Electrovalent

EV Electroviscosity; Electroviscous

EV Energy Value

EV Engler Viscometer

EV Equalizer Valve

EV Error Voltage

EV Exhaust Valve

EV Exhaust Velocity

EV Expected Value

EV Expendable Vehicle [Aerospace]

EV Exposure Value

EV Extravehicular

E&V Endangered and Vulnerable

Ev Event [also ev]

(v̄) Mean Vibrational Energy [Symbol]

eV Electron-Volt [Unit] [also ev]

EVA Earned Value Analysis

EVA Electric Vehicle Association [UK]

EVA Electronic Velocity Analyzer

EVA Electrothermal Vaporization Analysis

EVA Employee Volunteer Action

EVA Error Volume Analysis

EVA Ethylene Vinyl Acetate (Copolymer)

EVA Extravehicular Activity

eV/Å Electronvolt per Angstrom [also eV Å^{-1}]

eV/Å^2 Electronvolt per square Angstrom [also eV Å^{-2}]

eV/Å^3 Electronvolt per cubic Angstrom [also eV Å^{-3}]

EVAC Evacuate [also Evac, or evac]

EVAC Electric Vehicle Association of Canada

EVAC Emergency Voice and Communication

E/VAC Ethylene/Vinyl Acetate Copolymer

EVAc Poly(ethylene-*co*-Vinyl Acetate)

evacd evacuated [also evac'd]

EVAD Electrohydraulic (Left) Ventricular Assist Device [Medicine]

EVAL Earth Viewing Applications Laboratory

Eval Evaluate; Evaluation; Evaluator [also eval]

Eval Eng Evaluation Engineering

Eval Eng Evaluation Engineering [Journal published in the US]

Evap Evaporate; Evaporation; Evaporator [also evap]

evapd evaporated [also evap'd]

Evapg Evaporating [also evapg]

Evapn Evaporation [also evapn]

eV/at electron volt(s) per atom [also eV/atom] [Unit]

EVATA Extravehicular Activity Translational Aid

EVATMI European Vinyl Asbestos Tile Manufacturers Institute

eV/atom electron volt(s) per atom [also eV/at] [Unit]

eV/au electron volt(s) per atomic unit [Unit]

EVB Empirical Valence Bond (Method)

EVC Ecological Vegetation Class

EVC Electric Vehicle Council

EVC Emergency Voice–Communication

EVC Extravehicular Communications

EVCA European Venture Capital Association [Belgium]

eV·cm electron-volt centimeter [Unit]

eV/cm^2 electron-volt per square centimeter [also eV cm^{-2}]

eV/cm^3 electron-volt per cubic centimeter [also eV cm^{-3}]

eV·cm^2/at electron-volt per square centimeter per atom [also eV cm^2/atom] [Unit]

EVCON Events Control (Subsystem)

EVCP Event Control Block

EVCS Extravehicular Communication(s) System [of NASA]

EVCU Extravehicular Communications Umbilical

EVD Electrochemical Vapor Deposition

EVD Explosive Vapor Detector

EVD External Visual Display

EVDE External Visual Display Equipment [of NASA]

EVDG Electric Vehicle Development Group [UK]

EVDS Electronic Visual Display Subsystem

EVE Electroviscous Effect

EVE (Data) Entry and Validation Equipment

EVE Ethyl Vinyl Ether

EVE European Videoconferencing Experimentation

EVE Exemplary Voluntary Effort

EVE Extensible VAX (Virtual Address Extension) Editor

EVE Extreme Value Engineering

EVES Environment for the Verification and Evaluation of Systems

EVES Environment for Verifying and Evaluating Software

EVEST Experimental Valleditos Superheat Reactor [US]

EVF Equipment Visibility File

EVFM Ex-Vessel Flux Monitor

eV-fu electron-volt functional unit

Evg Evening [also evg]

EVGA Extended Video Graphics Adapter

EVGA Extended Video Graphics Array

eV/GPa electron-volt(s) per gigapascal [Unit]

EVHM Ex-Vessel Handling Machine

EVIL Elevation Versus Integrated Log

EVIL Extensible Video Interactive Language

EVIS Ergometer Vibration Isolation System

EVIST Ethic and Values in Science and Technology Program

EVK Ethyl Vinyl Ketone

EVK Evaluation Kit

eV/K electron-volt per kelvin [also eV K^{-1}]

eV/kT electron-volt per kilotesla [also eV kT^{-1}]

Evln Evolution [also evln]

EVLSS Extravehicular Life Support System [of NASA]

EVM Earth Viewing Module

EVM Electronic Voltmeter

EVM Ethylene Vinylacetate Copolymer

EVM Extended Virtual Machine

eV·m electron-volt meter [Unit]

eV·m^2/kg electron-volt meter squared per kilogram [also eV m^2 kg^{-1}]

EVMS Eastern Virginia Medical School [Norfolk, US]

EVMU Extravehicular Mobility Unit

EVN European VLBI (Very-Large Baseline Interferometry) Network

EVN/JIVE European VLBI (Very-Large Baseline Interferometry) Network/Joint Institute for VLBI (Very-Long Baseline Interferometry) in Europe (Symposium)

eV/nm Electronvolt(s) per nanometer [also eV nm^{-1}]

eV/nm³ Electronvolt(s) per cubic nanometer [also eV nm^{-3}]

eV/nm³s Electronvolt(s) per cubic nanometer second [also eV nm^{-3} s^{-1}]

eV/nm/ion Electronvolt(s) per nanometer per ion [also eV nm^{-1} ion^{-1}]

EVO Engineering Verification Order

EVO Europium Vanadate (Crystal) [$EuVO_4$]

EVOH Ethylene-Vinyl Alcohol (Copolymer)

Evol Evolution [also evol]

Evol Comput Evolutionary Computation [Journal published by MIT Press, Cambridge, US]

Evoln Evolution [also evoln]

EVOM Electronic Voltohmmeter

EVOP Evolutionary Operation

EVPI Expected Value of Perfect Information

EVR Electronic Valve Rectifier

EVR Electronic Video Recorder; Electronic Video Recording

EVS Electro-Optical Viewing System

EVS Electroreflectance Vibrational Spectroscopy

EVS Energie-Versorgung Schwaben AG [German Energy Supply Company]

EVS Equipment Visibility System

EVS Ethics and Values Studies

EVS Extravehicular Suit

eV/s electron-volt(s) per second [also eV s^{-1}]

EVSC Extravehicular Suit Communications

eV/sec electron volt(s) per second [also eV sec^{-1}]

EVSS Extravehicular Space Suit

EVSU Extravehicular Space Unit

EVT Engineering Verification Test

EVT Equiviscous Temperature

EVT Extravehicular Transfer

eV/T electron-volt(s) per Tesla [also eV T^{-1}]

EVTCM Expected Value Terminal Capacity Matrix

EVTM Ex-Vessel Transfer Machine

EVVA Extravehicular Visor Assembly

EVX Electronic Voice Exchange

EW Early Warning (Radar)

EW Eightfold Way

EW Elastic Wave

EW Electric Wave

EW Electric Welded; Electric Welding

EW Electronic Warfare

EW Electron Wave [Quantum Mechanics]

EW Electroslag Welding

EW Electroweak (Matter) [Particle Physics]

EW Electrowinning [also E/W] [Metallurgy]

EW Emulsion, Oil-in-Water

EW Energy Well

EW Enlisted Woman

EW Entry Week

EW Epoxide Weight

EW Equivalent Weight

EW Erftwerk (Coating Process)

EW Escape Wheel

EW Expansion Wave

E/W Electrowinning [also EW] [Metallurgy]

E/W Energy-to-Weight (Ratio)

E/W Energy over Weight

EWA Estimated Warehouse Arrival

EWA European Welding Association [Netherlands]

EWADAT European Waste Data Bank

EWAN Emulator Without A Good Name [Internet]

EWASER Electromagnetic Wave Amplification by Simulated Emission of Radiation [also Ewaser]

EWC Electric Water Cooler

EWC European Waste Catalogue [of European Community]

EWCAS Early Warning and Control Aircraft System

EWCE Eastern Washington College of Education [Cheney, Washington State, US]

EWCS European Wideband Communications System

EWD Economic Warfare Division

EWE Emergency Window Escape

EWEA European Wind Energy Association [UK]

EWES Engineering Waterways Experiment Station [of US Army]

EWF Effective Work Function

EWF Electrical Wholesalers Federation [UK]

EWF Electron Wave Function

EWF Equivalent-Weight Factor

EWF Essential Work to Fracture

EWG Earth Works Group, Inc.

EWG Ethics Work Group [of International Union for the Conservation of Nature and Natural Resources]

EWG Executive Working Group

EWH Energieerzeugungswerke Helgoland [Power Generating Station of (the Island of) Heligoland, Germany]

EWHS Eastern Washington Historical Society [US]

EWI Edison Welding Institute [Columbus, Ohio, US]

EWI Executive Women International [US]

EWIA External Wall Insulation Association [UK]

EWICST European Workshop of Industrial Computer Systems–Technical Committee

EWIF Evanescent Wave-Induced Fluorescence

EWIFS Evanescent Wave-Induced Fluorescence Spectroscopy

EWL Exchange Work List

EWM Electroweak Matter [Particle Physics]

EWMA Exponentially Weighted Moving Average

EWNN Eisen-Weinsäure-Natriumsalz (Iron-Sodium Tartrate)

EWO Engineering Work Order

EWP Electrowinning Process [Metallurgy]

EWP Exploding Wire Phenomena

EWPCA European Water Pollution Control Association [Germany]

EWR Early Warning Radar

EWR Electromagnetic Wave Resistivity

EWR Engineering Work Request

EWRC European Weed Research Council [now EWRS, Germany]

EWRS European Weed Research Society [Germany]

EWRT Electrical Women's Round Table [US]

EWS Early Warning System

EWS Electronic Workstation

EWS Elektronisches Wählsystem [Electronic Dialing System of the German Post Office]

EWS Electrowinning System [Metallurgy]

EWS Emergency Weather Station

EWS Employee Written Software

EWS Engineering Workstation

EWSA European Wheat Starch Manufacturers Association [Germany]

EWSC Eastern Washington State College [Cheney, Washington, US]

EWSM Electronic Warfare Support Measures

EWST Elevated Water Storage Tank

EWT Eastern Winter Time

EWTAT Early Warning Threat Analysis Display

EWTMI European Wideband Transmission Media Improvement

EWTMIP European Wideband Transmission Media Improvement Program

EWTR Electronic Warfare Test Range [of US Air Force]

EWWS ESSA (Environmental Science Services Administration) Weather Wire Service [US]

EX Excitation Coil

EX Executive Management Office [NASA Kennedy Space Center Directorate]

EX Experimental (Steel) [AISI-SAE Designation]

E(X) Expectation of X [Statistics]

E(X) Mean Value of X [Statistics]

Ex Examination; Examine(d) [also ex]

Ex Example [also ex]

Ex Exception [also ex]

Ex Excess [also ex]

Ex Exchange(r) [also ex]

Ex Execute; Executive; Execution [also ex]

Ex Exercise [also ex]

Ex Exponent [also ex]

Ex Express [also ex]

Ex Exterior [also ex]

E(x) Expected (or Mean) Value of x [Statistics]

ex except(ed)

ex external

ex extra

exa- SI prefix representing 10^{18}

EXACT (International) Exchange of Authenticated Component Performance Test Data

EXAFS Extended X-Ray Absorption Fine Structure (Spectroscopy)

EXAFSS Extended X-Ray Absorption Fine Structure Spectroscopy

EXAM Elemental X-Ray Analysis of Materials

Exam Examiner; Examination [also exam]

exam examine(d)

examd Examinated [also exam'd]

Examg Examinating [also examg]

Examn Examination [also examn]

Examr Examiner [also examr]

EXAPS Electron-Excited X-Ray Appearance Potential Spectroscopy

ex aq *(ex aqua)* – in water [Medical Prescriptions]

EXAT Exotic Atoms, Molecules and Muon Catalyzed Fusion

EXC Experiment Computer

Exc Excavation [also exc]

Exc Excellency

Exc Exception [also exc]

Exc Excitation; Exciter [also exc]

Exc Excursion [also exc]

exc excellent

exc except

ExCA Exchangeable Card Architecture

ExCAc Ethoxycarbonylacetic Acid

EXCELS Expanded Communications Electronics System [of US Department of Defense]

EXCESS Extensible Expert System Shell (Technology)

Exch Exchange [also exch]

Exch Exchequer [also exch]

exch exchanged

Excimer Excited Dimer

EXCIPLEX Excited State Complex [Physics]

Excl Exclamation [also excl]

Excl Exclude; Excluding; Exclusion [also excl]

excl exclusive

EXCLASS Expert Job Classification Assistant; Expert Job Evaluation Assistant

excld excluded [also excl'd]

EXCO Executive Committee

EXCO Executive Council

EXCO Exfoliation Corrosion

EXCOM Executive Committee [of United Nations High Commissioner for Refugees]

ExCom Executive Committee [of Institute of Electrical and Electronics Engineers, US]

EXCP Execute Channel Program

Exctr Exciter [also exctr]

Excvtg Excavating [also excvtg]

EXD Exchange Degeneracy [Physics]

ex-d *(ex dividendum)* – without dividend

EXDAMS Extended Debugging and Monitoring System

EXDC External Data Controller

ex div *(ex dividendum)* – without dividend

EXDS Extra Diffuse Scattering

.EXE Executable (File) [File Name Extension] [also .exe]

EXEC Execute (Statement) [Computers]

EXEC Executive Statement

EXEC Executive System

Exec Execute; Execution; Executive; Executor [also exec]

ExecMBA Executive Master of Business Administration

ExecMPA Executive Master of Public Administration

ExecMS Executive Master of Science

ExecMSE Executive Master of Science in Engineering

Exec VP Executive Vice-President

EXEELFS Extended X-Ray Electron Energy Loss Fine Structure

EXELFS Extended Electron (Energy) Loss Fine Structure; Extended Energy Loss Fine Structure

EXEPACK Executable File Packing [also exepack]

EXER Executive Committee of Energy Research

Exer Exercise [also exer]

Exer Immunol Rev Exercise Immunology Review [Journal]

EXERSUG Executive Committee of Energy Research Supercomputer Users Group [US]

EXES Electron-Induced X-Ray Emission Spectroscopy

EXE2BIN Convert Executable (.EXE) Files to Binary Format (Command) [also exe2bin]

EXF Ex Factory [also exf]

EXF External Function

EXH Exhaust [also Exh, or exh]

Exhib Bull Exhibition Bulletin [UK]

EX-HY Extra-Heavy [also Ex-Hy]

EX-IM Export-Import Bank [also EXIM, or Ex-Im]

Eximbank Export-Import Bank of China

ex int ex interest [Commerce]

Exist Existence [also exist]

EXLST Exit List

EX Mag EX Magazine [Germany]

EXMETNET Experimental Meteorological Sounding Rocket Research Network

ex n ex new [Commerce]

EXNOR Exclusive nor [also EX-NOR]

EXO Experiment Operator

EX-O Executive Officer [also Ex O]

EXOR Exclusive Or [also EX-OR]

Exor Executor [also exor]

Exosph Exosphere [also exosph]

EXP Expontential (of Argument) [also Exp] [Programming]

Exp Expand; Expansion [also exp]

Exp Expense(s) [also exp]

Exp Experiment(al) [also exp]

Exp Explosion [also exp]

Exp Exponent(ial) [also exp]

Exp Export(ation); Exporter [also exp]

Exp Expose; Exposure [also exp]

Exp Express(ion) [also exp]

Exp Expulsion [also exp]

exp expired

exp exponential function [Mathematics]

EXPAC Explosion Prediction and Analysis Code

Exp Astron Experimental Astronomer; Experimental Astronomy

Exp Astron Experimental Astronomy [Journal published in the Netherlands]

Exp Cell Biol Experimental Cell Biology [Journal]

Exp Cell Res Experimental Cell Research [Journal published in the US]

Exp Chem Experimental Chemist(ry)

Exp Clin Endocrin Experimental Clinical Endocrinology [US Journal]

Exp Clin Immunogen Experimental and Clinical Immunogenetics [International Journal]

expen expendable

Expend Expenditures [also expend]

Expert Syst Expert Systems [Journal published in the UK]

Expert Syst Appl Expert System Applications [Journal published in the UK]

Expert Syst Rev Expert Systems Review [Published by the Center for Accounting Research, University of Southern California, Los Angeles, US]

Expert Syst User Expert Systems User [Published in the UK]

Exp Eye Res Experimental Eye Research [Journal published in the US]

Exp Gerontol Experimental Gerontology [Journal]

Exp Fluids Experiments in Fluids [Published in Germany]

Exp Heat J Experimental Heat Journal [US]

Exp Heat Transfer Experimental Heat Transfer [Journal published in the US]

EXPIO Expander Input/Output

Expl Explode; Explosive(s) [also expl]

Expl Explore(r) [also expl]

expld exploded [also expl'd]

Explg Exploding [also explg]

Expln Explosion [also expln]

Explor Exploration [also explor]

Explor Geophys Exploration Geophysics [Australian Publication]

Explor Min Geol Exploration and Mining Geology [Journal published in the Netherlands]

Expls Explosive(s) [also expls]

Exp Mech Experimental Mechanics

Exp Mech Experimental Mechanics [Journal of the Society for Experimental Stress Analysis, US]

Exp Mol Pathol Experimental and Molecular Pathology [Journal published in the US]

Exp Nephrol Experimental Nephrology [International Journal]

Exp Neurol Experimental Neurology [Journal published in the US]

Expo Exposition [also expo]

Expo-Optica International Exposition of Ophthalmic Optics, Seeing and Hearing Aids

Export Dig Export Digest [UK]

Expos Exposure [also expos]

Exp Parasitol Experimental Parasitology [Journal published in the US]

Exp Phys Experimental Physicist; Experimental Physics

Exp Physiol Experimental Physiology [Published by Cambridge University Press for the Physiological Society, UK]

Expr Exploder [also expr]

Exprf Explosionproof [exprf]

Exps Expenses [also exps]

Expt Experiment(al) [also expt]

Exp Tech Experimental Techniques [of Society for Experimental Mechanics, US]

Exp Tech Phys Experimentelle Technik der Physik [German Journal on Experimental Techniques in Physics]

Exp Therm Fluid Sci Experimental Thermal and Fluid Science [Journal published in the US]

exptl experimental

Exp Val Experimental Values [also exp val]

Expwy Expressway [also expwy]

EXR Execute and Repeat

exs ex ship [Commerce]

EXSA Exhibition Association of South Africa

EXSC Executive Standards Council [also ExSC]

exsec exterior secant

EXSTA Experimental Station

EXSY Exchange Spectroscopy

EXSYS Expert System (Software)

EXT Experiment Terminal

Ext Extend(ed); Extension [also ext]

Ext Exterior [also ext]

Ext External [also ext]

Ext Extinction [also ext]

Ext Extinguisher [also ext]

Ext Extract(ion) [also ext]

Ext Extract [Medical Prescriptions]

ext exterior

ext external(ly)

ext extinct

ext extra

extd extended [also ext'd]

exter external

EXTERR Extended Error Information (Function) [Programming]

EXTERRA Extraterrestrial Research Agency [US Army]

EXT.FM External Frequency Modulation

Extg Exterminating [also extg]

Extgh Extinguish [also extgh]

Extn Extraction [also extn]

EXTOXNET The Extension Toxicology Network

EXTR Extraction Code

Extr Extract [also extr]

Extr Extrusion [also extr]

EXTRADOP Extended Range DOVAP (Doppler Velocity and Position)

Extrapol Extrapolation [Journal of the Science Fiction Research Association, US]

extrd extracted [also extr'd]

Extrg Extracting [also extg]

Extr Met Extractive Metallurgist; Extractive Metallurgy

EXTRN External Reference

Extrn Extraction [also extrn]

Extrus Extrusion [also extrus]

Extrus Showc Extrusion Showcase [Published in the US]

EXW Explosion Welding

exw ex works [Commerce]

EXWIF Evanescent X-Ray Wave Induced Fluorescence

EXXPOL Exxon Polymerization (Process)

EY Energy Yield

EY Entry Year

EYE European Year of the Environment

EYH Expanding Your Horizons (Program) [US]

EZ Electrical Zero

E(z) Electrical Field Function [Symbol]

EZI European Zinc Institute [Netherlands]

E-ZINE Electronic Magazine [also E-Zine, or e-zine]

F Additional Designation of Emission (or Transmission) Type–Phase or Position-Modulated [Symbol] [also f]

F Amplitude [Symbol]

F As Fabricated [Basic Temper Designation for Aluminum and Magnesium Alloys]

F (Ionic) Bonding Force [Symbol]

F Calibration Factor (of Analytical Samples)

F Carbon-Tungsten Type Special-Purpose Tool Steels [AISI-SAE Symbol]

F Deflection (of Springs) [Symbol]

F Deformation Gradient [Symbol]

F (Number of) Degrees of Freedom [Symbol]

F Diamond (Crystal System) [Symbol]

F Driving Force [Symbol]

F Eddy Current Inspection Test Frequency [Symbol]

F (Applied) Electric Field [Symbol]

F Energy per Unit Weight [Symbol]

F Fabricate; Fabrication

F Face

F Face-Centered Space Lattice in Orthorhombic and Cubic Crystal Systems [Hermann-Mauguin Symbol]

F Faceted (Eutectics) [also f] [Metallurgy]

F Face Width (of Gears) [Symbol]

F Facilitate; Facilitator

F Facility

F Facing

F Facsimile

F Fact

F Factor(ial)

F Factory

F Fagus [Genus of Deciduous Trees Including the Beeches]

F Fahrenheit [Unit]

F Fail(ure)

F Fair [also f]

F Fairing

F Fall

F Family

F Farad [Unit]

F Faradaic; Faraday

F Faraday Constant (9.648531×10^4 C/mol) [Symbol]

F Fasciculus [Anatomy]

F Fasten(er); Fastening

F Fat(ty)

F Father

F Fatigue

F Fault

F Fauna

F F(isher) Distribution [Symbol]

F Feather(ing)

F February

F Fecal; Feces

F Feed; Feeder; Feeding

F Feed (Rate) [Symbol]

F Feline

F Felis [Genus Comprising Wild and Domestic Cats]

F Fellow

F Female [also f]

F Femur [Anatomy]

F Ferrite; Ferritic [Metallurgy]

F Ferrite [Solid-State Physics]

F Ferroelectric(s); Ferroelectricity

F Ferromagnet(ic); Ferromagnetism

F Fertile; Fertility

F Fetal; Fetus

F Fever(ish)

F (Fiat) – Let there be made; Make [Medical Prescriptions]

F Fiber [also f]

F Fiberglass

F Fibril

F Fibrin [Biochemistry]

F Fibrinogen [Biochemistry]

F Fibroid [Medicine]

F Fibrous

F Fibula [Anatomy]

F Ficus [Genus of Plants Including the Mulberries and Figs]

F Field

F Fight(er)

F Fighter (Aircraft) [US Air Force and US Navy Symbol]

F Figure of Merit [Electronic Symbol]

F Fiji

F Filament(ary) [also f]

F File

F Filial Generation [Genetics]

F Fill(ing); Filler

F Film(ing)

F Filter(ing); Filtration

F Fin

F Final [also f]

F Fine Line [Spectroscopy]

F   Finger
F   Finish(ing)
F   Finish of Weld [Symbol]
F   Finland; Finnish
F   Fine
F   Fir
F   Fire [also f]
F   Firing
F   Fissile; Fission
F   Fit(ting)
F   Fix
F   Fixture
F   Flag
F   Flagella [Biology]
F   Flake
F   Flame; Flammability; Flammable
F   Flame Spray(ing)
F   Flap
F   Flash
F   Flashing
F   Flat (Product)
F   Fleet
F   Flexible; Flexibility
F   Flexor (Muscle) [Anatomy]
F   Flexure; Flexural
F   F Line [Medium-Blue-Green Fraunhofer Line for Hydrogen Having a Wavelength of 486.1 Nanometers] [also H_β line] [Spectroscopy]
F   Flint (Glass)
F   Flint [Mineral]
F   Flint [Michigan, US]
F   Float(ing)
F   Flora
F   Florence [Italy]
F   Florida [US]
F   Floppy (Disk)
F   Flotation
F   Flow(ing) [also f]
F   Flowability; Flowable
F   Flowmeter
F   Fluctuate; Fluctuation
F   Fluid(al) [also f]
F   Fluorescein
F   Fluorescence; Fluorescent
F   Fluorescence Current [Symbol]
F   Fluoridation
F   Fluorine [Symbol]
F   Fluorite [Mineral]
F   Fluorochrome
F   Fluorometric; Fluorometry
F   Fluoroscope; Fluoroscopy
F   Flute Length (of Drills) [Symbol]
F   Flux
F   Flux (of Atoms, Molecules, etc.) [Symbol]

F   Fluxing [Metallurgy]
F   Foam(ing)
F   Focal Length; Focal Ratio [also f]
F   Focus
F   Focusing
F   Fog(ging)
F   Foil
F   Fold
F   Follicle
F   Food(stuff)
F   Foot
F   (Applied) Force [Symbol]
F   Force(d); Forcing [also F]
F   Foreign(er)
F   Forensic(s)
F   Forge; Forging
F   Form
F   Formable; Formability
F   Formaldehyde
F   Formal (Solution) [Note: A 1F Solution (e.g., 1F HCl) of a Strong Electrolyte Contains 1 Gram-Formula Weight of Solute per Liter of Solution]
F   Formality (of Solutions) [Chemistry]
F   Formamide
F   Format(ting)
F   Formation
F   Formation Factor [Symbol]
F   Former; Forming
F   Formica
F   Formicidae [Family Comprising the Ants]
F   Formosa [Taiwan]
F   Formula [also f]
F   Formulary [Pharmacology]
F   Formulation
F   Forward [also f]
F   Fossil
F   Foundry
F   Fourth Tone in the Scale of C Major [Acoustics]
F   Foxtrott [Phonetic Alphabet]
F   Fractal
F   Fractile [Statistical Distribution]
F   Fraction(al) [also f]
F   Fractionation
F   Fractograph(y)
F   Fracture [also f]
F   Fragile [also f]
F   Fragment(ation) [also f]
F   Frame(work)
F   France
F   Francisella [Genus of Bacteria]
F   Frankfurt [Germany]
F   Fraunhofer [Symbol]
F   Fraxinus [Genus of Deciduous Trees Including the Ashes]

F	Fredericton [New Brunswick, Canada]
F	Free(dom)
F	Free Energy [Symbol]
F	Freetown [Sierra Leone]
F	Freeze(r); Freezing
F	Freezing Index [Symbol]
F	French
F	French Franc [Currency of France, French Guinea, Guadeloupe, Martinique, Monaco, and St. Pierre and Miquelon]
F	Frequency
F	Fresno [California, US]
F	Friability; Friable
F	Friction(al) [also f]
F	Friction(al) Force [Symbol]
F	Friction Loss [Symbol]
F	Friday
F	Front(al)
F	Frost
F	Frosting
F	(−)-Fructose [Biochemistry]
F	F-Size Sheet (of Paper) [28" × 40"]
F	Fuel [also f]
F	Fuerstenau [X-Ray Dosimetry Unit]
F	Fulcrum
F	Full(y) [also f]
F	Fullerton [California, US]
F	Fume; Fuming
F	Function(al) [also f]
F	Function [Symbol]
F	Fund(ing)
F	Fundament(al)
F	Fundamental Line [Spectroscopy]
F	Fungal; Fungus
F	Fungicidal; Fungicide [also f]
F	Funnel(ing),
F	Furan
F	Furnace
F	Furrow
F	Fuse
F	Fuselage
F	Fusing; Fusion [also f]
F	Gibbs Free Energy [Symbol]
F	Grade for Failing Work
F	Individual Launched (Vehicle) [USDOD Symbol]
F	Intensity (of Beam) [Symbol]
F	Lithography Feature Size [Symbol]
F	(Concentrated) Load [Symbol]
F	Luminous Flux [Symbol]
F	Magnetic Potential (or Magnetomotive Force) [Symbol]
F	Noise Factor (or Figure) [Electronic Symbol]
F	(−)-Phenylalanine; Phenylalanyl [Biochemistry]
F	Principal Power of Lenses, etc. [Symbol]
F	Puncture Resistance [Symbol]
F	Scattering Factor [Symbol]
F	Schmid Factor (in Materials Science) [Symbol]
F	Sensitivity Factor (for Recording Media) [Symbol]
F	Single Vinyl-Acetal-Covered (Electric Wire)
F	Size of Focal Spot (in Radiography) [Symbol]
F	Star Group of Spectral Type F [Surface Temperature 6,650°C or 12,000°F] [Letter Designation]
F	Structure Factor (in Solid-State Physics) [Symbol]
F	Tearing Strength (of Fabric, Paper, etc.) [Symbol]
F	Thrust [Symbol]
F-	Fighter (Aircraft) [Followed by a Numeral, e.g., F-14, F-16, etc.]
F-	France [Civil Aircraft Marking],
$\bar{F}$	Effective Force (Transmitted by Soil or Rock Mass) [Symbol]
$\bar{F}$	Average Thrust [Symbol]
F*	Fano Number [Symbol]
F*	Elastic Deformation Gradient [Symbol]
F⁻	Fluorine Ion [Symbol]
F°	Rotary Deflection (of Springs) [Symbol]
F′	Second Derivative of Free Energy [Symbol]
F″	Second Derivative of Free Energy [Symbol]
°F	Degrees Fahrenheit
/F	Find Error Option [Turbo Pacal]
(°F)⁻¹	One per Degree Fahrenheit [Unit]
F_1	First Filial Generation [Genetics]
F_1, F_2, F_3, etc.	First, Second ,Third, etc., Filial Generation [Genetics]
F_1-F_2	Direct-Current Shunt Field [Controllers]
F_2	Molecular Fluorine [Symbol]
F_2	Second Filial Generation [Genetics]
F_A	Air Resistance [Symbol]
F_A	Attractive Force (Between Two Atoms) [Symbol]
F_A	Fraction of "A" Atoms in a Solid Solution [Symbol]
F_a	Anisotropy Energy per Unit Length [Symbol]
F_a	Applied Force [Symbol]
F_{AB}^X	Matrix Factor [Symbol]
F_B	Bloch Frequency [Symbol]
F_B	Stress Biaxiality Factor [Symbol]
F_C	Coriolis Force [Symbol]
F_c	Centrifugal Force [Symbol]
F_c	Composite Load (i.e., Load on Composite, not on Fibers or Matrix) [Symbol]
F_c	Corresponding Force [Symbol]
F_c	Coulomb(ic) Force [Symbol]
F_c	Depinning Threshold (in Solid-State Physics) [Symbol]
F_c	Force Constant [Symbol]
F_{cr}	Critical Load [Symbol]
F_D	Total Drag [Symbol]
F_d	Damping Force [Symbol]
F_E	Effort Force (for a Simple Machine) [Symbol]
$\bar{F}_E$	Electric Force on Electron in an Electric Field [Symbol]
F_F	Frictional Force [also F_f, or F_{fl}] [Symbol]

F_F Fundamental Reflection–X-Ray Structure Factor [Symbol]

$F_{\dot{f}}$ Charged Fluorine Molecule [Symbol]

F_f Fiber Load (of Composites) [also F_f] [Symbol]

F_f Load at Fracture [Symbol]

F_g Gravitational Force [Symbol]

F_H Structure Factor [Symbol]

F_{hkl} Structure Factor of Reflection (hkl) [Symbol]

F_i (Net) Force on i-th Atom of a System [Symbol]

F_K Johnson's Figure of Merit (in Semiconductor Physics) [Symbol]

F_K Keyes's Figure of Merit (in Semiconductor Physics) [Symbol]

F_L London Dispersion Force (in Physical Chemistry) [Symbol]

F_L Lorentz Force [Symbol],

F_m Magnetic Force [Symbol]

F_m Magnetostatic Energy per Unit Length [Symbol]

F_m Matrix Load (of Composites) [Symbol]

F_{max} Maximum Thrust [Symbol]

F_μ Figure of Merit for Magnetostrictive Strain Gauges [Symbol]

F_N Net Force (between Atoms) (i.e., Sum of Attractive and Repulsive Force Components) [Symbol]

F_N Normal Force [Symbol]

F^P Plastic Deformation Gradient [Symbol]

F_p Pinning Force (in Solid-State Physics) [Symbol]

F_p Reflected Flux [Symbol]

F_{ps} Positronium Yield [Symbol]

F_R Figure of Merit for Resistive Strain Gauges [Symbol]

F_R Radial Force (in Machining) [Symbol]

F_R Radial Component of Centrifugal Force [Symbol]

F_R Repulsive Force (Between Two Atoms) [Symbol]

F_R Resistance Force [Symbol]

F_R Resultant Force (of Two or More Forces) [Symbol]

F_R Sliding (or Coulomb) Friction [Symbol]

F_r Pulse Repetition Frequency [Symbol]

F_{Rmax} Static Friction [Symbol]

F_S Single-Electron Tunneling Frequency [Symbol]

F_S Superlattice Reflection–X-Ray Structure Factor [Symbol]

F_s Shearing Force [Symbol]

F_s Spring Force [Symbol]

F_s Structure Amplitude [Symbol]

F_σ Union of Countably Many Closed Sets (in Mathematics) [Symbol]

F_{sol} Downward Solar Radiation [Symbol]

F_T (Bridgman) Stress Triaxiality Factor [Symbol]

F_T Tangential Force [Symbol]

F_T Tangential Component of Centrifugal Force [also F_t] [Symbol]

F_T Transmitted Force [Symbol]

F_t Tangential Force [Symbol]

F_t Total Force [Symbol]

F^θ Thermal Deformation Gradient [Symbol]

F_{th} Thermophoretic Force [Symbol]

F_{ty} Tensile Yield Strength [Symbol]

F_W Van der Waals Force (in Physical Chemistry) [Symbol]

F_w Flow Index [Symbol]

F_w Weight Force [Symbol]

F_x Component of Force F along X-Axis [Symbol]

F_y Component of Force F along Y-Axis [Symbol]

F_z Component of Force F along Z-Axis [Symbol]

F1 Frequency-Shift Keying [Symbol]

F2 Frequency Modulation Telegraphy by Modulating Audio Frequency Keying [Symbol]

F3 Frequency Modulation Telephony [Symbol]

F4 Frequency Modulation Facsimile [Symbol]

F5 Frequency Modulation Television [Symbol]

F6 Frequency-Shift Keying–Twinplex, Duoplex [Symbol]

F9 Composite Emissions (or Transmissions) with Frequency Modulation [Symbol]

F-17 Fluorine-17 [also ^{17}F, or F^{17}]

F-18 Fluorine-18 [also ^{18}F, or F^{18}]

F-19 Fluorine-19 [also ^{19}F, F^{19} or F]

F-20 Fluorine-20 [also ^{20}F, or F^{20}]

$\mathscr{F}$ Faraday (9.648531×10^4 C/mol) [Symbol]

$\mathscr{F}$ Fourier Transform [Symbol]

$\mathscr{F}$ Magnetomotive Force [Symbol]

f additional designation of emission (or transmission) type–phase or position-modulated [Symbol] [also F]

f atomic scattering factor [also f] [Symbol]

f constantly moist, rainfall all through the year [Subtype of Climate Region, e.g., in Cd, Ddw, etc.]

f Coriolis parameter [Symbol]

f correction factor [Symbol]

f Fanning Friction Factor [Symbol]

f failure [also F]

f fair [also F]

f family [also F]

f farad [Unit]

f farthing

f fathom

f female [also F]

f feminine

f femto- [SI Prefix]

f ferromagnetic (phase)

f fetch

f f-factor [Symbol]

f (fiat) – let there be made; make [Medical Prescriptions]

f fiber [also F]

f filament [also F]

f final [also F]

f fine

f fire [also F]

f fixed

f flat

f florin (or guilder) [Currency of the Netherlands]

f flow(ing) [also F]

f fluid(al) [also F]

f   focal length; focal ratio [also F]

f   fog(ging) [also F]

f   foil [also F]

f   folio

f   following (page, etc.)

f   foot (or feet) [Unit]

f   f-orbital [Symbol]

f   force [also F]

f   formed

f   formula [also F]

f   forward [also F]

f   fraction(al) [also F]

f   fracture [also F]

f   fragile [also F]

f   fragment(ation) [also F]

f   freezing [also F]

f   frequency [Symbol]

f   friction(al) [also F]

f   from

f   fuel [also F]

f   fuel-to-air ratio [Symbol]

f   fugacity [Symbol]

f   full(y)

f   function(al) [also F]

f   fungicidal; fungicide [also F]

f   furlong [Unit]

f   fusion

f   recoil-free factor (in Mössbauer spectroscopy) [Symbol]

f   relative block volume fraction (of block copolymers) [Symbol]

f   sampling fraction [Symbol]

f   viewing azimuth angle [Symbol]

f^o folio [Symbol]

f/   f-number (or aperture ratio) [Symbol]

f_0 grain shape factor (in materials science) [Symbol]

f_0 natural (or resonant) frequency [Symbol]

f_{bcc} fraction of bcc (body-centered cubic) phases (in a sample) [Symbol]

f_c coercive field [Symbol]

f_c critical frequency [Symbol]

f_c cutoff frequency [Symbol]

f_d demagnetization force [Symbol]

f_{fcc} fraction of fcc (face-centered cubic) phases (in a sample) [Symbol]

f_H force of external magnetic field [Symbol]

f_{Hc} force on coercive magnetic field [Symbol]

f_{hfb} common-base small-signal short-circuit forward current transfer ratio cutoff frequency [Semiconductor Symbol]

f_{hfc} common-collector small-signal short-circuit forward current transfer ratio cutoff frequency [Semiconductor Symbol]

f_{hfe} common-emitter small-signal short-circuit forward current transfer ratio cutoff frequency [Semiconductor Symbol]

f^i image force (on dislocation) [Symbol]

f_L index of dynamic stressing (of roller bearings) [Symbol]

f_M Madelung constant (= 1.7476...) [Symbol]

f_{max} maximum frequency of oscillation [Semiconductor Symbol]

f_N Nyquist frequency [Symbol]

f_n speed factor (of roller bearings) [Symbol]

f_p average flux quantum per plaquette [Symbol]

f_p parallel-resonance (or antiresonance) frequency [Symbol]

f_p pinning force (in solid-state physics) [Symbol]

f_r feed per revolution (in machining) [Symbol]

f_s index of static stressing (of roller bearings) [Symbol]

f_s sampling rate [Symbol]

f_s superfluid fraction [Symbol]

f_T cutoff frequency [Symbol]

f_T transition frequency (i.e., frequency at which common-emitter forward current transfer ratio extrapolates to unity) [also f_t] [Semiconductor Symbol]

f_t feed per tooth (in milling) [Symbol]

f_w Bloch wall force (in solid state physics) [Symbol]

ƒ   acceleration [Symbol]

ƒ   activity coefficient (for molar concentration) [Symbol]

ƒ   apparent (or observed) frequency (e.g., Doppler equation) [Symbol]

ƒ   atomic scattering factor [also f] [Symbol]

ƒ   average composition (of polymer) [Symbol]

ƒ   coefficient of friction [Symbol]

ƒ   deflection (of a spring for one active coil) [Symbol]

ƒ   feed rate (of cutting tools) [Symbol]

ƒ   fiber

ƒ   final

ƒ   flattening (of the earth) [Symbol]

ƒ   florin (or guilder) [Currency of the Netherlands]

ƒ   fluidity [Symbol] ,

ƒ   (first) focal length [Symbol]

ƒ   f-orbital [Symbol]

ƒ   force [Symbol]

ƒ   fraction [Symbol]

ƒ   fraction of the system in an assemblage (in canonical distribution) [Symbol]

ƒ   fracture

ƒ   frequency [Symbol]

ƒ   frustration (in spin glasses) [Symbol]

ƒ   function (e.g., $y = f(x)$) [Symbol]

ƒ   maximum deflection [Symbol]

ƒ   misfit parameter (in crystallography) [Symbol]

ƒ   modulation frequency [Symbol]

ƒ   molecular partition function [Symbol]

ƒ   mole fraction [Symbol]

ƒ   parallax factor (in electron stereomicroscopy) [Symbol]

ƒ   partition function [Symbol]

ƒ   preferred structural orientation factor [Symbol]

ƒ   recoil-free fraction (in Mössbauer spectroscopy) [Symbol]

ƒ   resonance frequency [Symbol]

f   scattering factor [Symbol]

f   separation factor (in gas separation) [Symbol]

f   skin friction (between structural elements and soil) [Symbol]

f   slack (of a rope) [Symbol]

f   (effective) stress [Symbol]

f   transfer function [Symbol]

f   volume fraction [Symbol]

f   width of uncut chip (in boring and turning) [Symbol]

f′   apparent frequency (e.g., Doppler equation) [Symbol]

f′   second focal length [Symbol]

f′   wall friction [Symbol]

$f_{\parallel}$ magnetic in-plane anisotropy [Symbol]

$f_{\perp}$ magnetic out-of-plane anisotropy [Symbol]

f_0 cutoff frequency [Symbol]

f_0 fundamental frequency [Symbol]

f_0 resonant frequency [Symbol]

f_0 true frequency (e.g., Doppler equation) [Symbol]

f_1 focal length of objective (of optical instrument) [Symbol]

f_1 mole fraction of Substance 1 [Symbol]

f_2 focal length of eyepiece (of optical instrument) [Symbol]

f_a atomic scattering function [Symbol]

f_a' real part of dispersion correction [Symbol]

f_a'' imaginary part of dispersion correction [Symbol]

f_α volume fraction of alpha phase [Symbol]

f_{ar} antiresonance (or parallel-resonance) frequency [Symbol]

f_b beat frequency [Symbol

f_β volume fraction of beta phase [Symbol]

f_c carrier-wave frequency [Symbol]

f_c central frequency [Symbol]

f_c critical frequency [Symbol]

f_c cutoff frequency [Symbol],

f_c threshold value for composition [Symbol]

f_{cr} critical stress [Symbol]

f_d* electrical force on dislocation (in solid-state physics) [Symbol]

f_e emitted frequency [Symbol]

f_e focal length of eyepiece (of optical instrument) [Symbol]

f_{hkl} atomic form factor [Symbol]

f_{IM} intermodulation distortion frequency [Symbol]

f_{in} input frequency [Symbol]

f_{long} frequency of longitudinal vibrations [Symbol]

f_M Madelung constant (= 1.7476...) [Symbol]

f_m modulating frequency [Symbol]

f_{max} maximum clock frequency (for Schottky TTL logic) [Symbol]

f_{max} maximum frequency [Symbol]

f_{MC} magnetocrystalline energy density [Symbol]

f_{ME} magnetoelastic energy density [Symbol]

$f^{(n)}$ n-th derivative (in Maclaurin series) [Symbol]

f_n atomic scattering factor of the n-th atom in a unit cell [Symbol]

f_n natural frequency [Symbol]

f_n number fraction of drops of diameter less than D (in liquid particle statistics) [Symbol]

f_n resonant frequency [Symbol]

f_o focal length of objective (of optical instrument) [Symbol]

f_r received frequency [Symbol]

f_r resonant frequency [Symbol]

f_{rot} rotational partition function [Symbol]

f_s factor of safety [Symbol]

f_s frequency shift [Symbol]

f_s volume fraction of superconductivity [Symbol]

f_T cutoff frequency [Symbol]

f_T relative free volume [Symbol]

f_{tors} frequency of torsional vibrations [Symbol]

f_{trans} translational partition function [Symbol]

f_{vib} vibrational partition function [Symbol]

f_{yp} yield-point stress [Symbol]

FA   Factor Analysis

FA   Factory Automation

FA   Failure Analysis [also F/A]

FA   Fairchild Aircraft [US]

FA   False Acceptance [Statistics]

FA   False Alarm

FA   Fast Algorithm

FA   Fast Anneal(ing) [Metallurgy]

FA   Fast Axis [Optics]

FA   Fatigue Analysis

FA   Fatty Acid

FA   Febrile Agglutinin [Immunology]

FA   Fellgett's Advantage

FA   Ferro-Alloy

FA   Ferrous Alloy

FA   Fiber Axis

FA   Fibonacci Association [US]

FA   Field-Accelerating (Relay)

FA   Field Acceleration

FA   Field Address

FA   Field Adsorption

FA   Field Anisotropy

FA   Field Artillery

FA   Field Availability

FA   Fighter Aircraft [also F/A]

FA   File Access

FA   Final Approach [Aeronautics]

FA   Fine Adjustment

FA   Final Aperture

FA   Final Assembly

FA   Fine Aggregate

FA   Finish Allowance

FA   Fire Administration

FA   Fire Alarm

FA   First Aid

FA   Fixed Ammunition

FA   Flame Atomization; Flame Atomizer

FA   Flammable Aerosol

FA Flash Adsorption
FA Fleet Admiral [also FADM or F Adm]
FA Flight Activities
FA Flight Aft
FA Flight Assignment
FA Flora of Australia [Australian Biological Resources Study]
FA Flow Analysis
FA Flowing Afterglow
FA Fluorescent Antibody (Technique) [Immunology]
FA Fluoroapatite [also FAP, or FAp]
FA Folic Acid
FA Forced-Air (Cooling)
FA Forensic Analysis
FA Formic Acid
FA Foundry Alloy
FA Fourier Analysis; Fourier Analyzer
FA Fractal Analysis
FA Fracture Analysis
FA Frankford Arsenal [at Phila, Pennsylvania, US]
FA Franklin-Adams (Star Chart) [Astronomy]
FA Free Air
FA French Antilles
FA Frequency Agility
FA Friction Angle
FA Front Axle
FA Fuel Assembly [also F/A] [Nuclear Engineering]
FA Full Abstraction
FA Full Add(er)
FA Full Anneal(ing) [Metallurgy]
FA Fully-Automatic
FA Fumaric Acid
FA Functional Analysis
FA Furfuryl Alcohol
FA Furnace Annealed; Furnace Annealing [Metallurgy]
FA Furnace Atomizer
FA Fuse Alarm
FA Fusible Alloy
FA Iron Aluminide
FA Iron-Aluminum Alloy
F&A Facilities and Administration; Facilities and Administrative
F+A Ferrite and Austenite (Region) [Metallurgy]
F/A Failure Analysis [also FA]
F/A Ferrite/Austenite (Phase Boundary) [Metallurgy]
F/A Fighter Aircraft [also FA]
F/A Fuel/Air (Ratio)
F/A Fuel Assembly [also FA] [Nuclear Engineering]
Fa Fayalite [Mineral]
fA femtoampere [Unit]
fa fayalite [Mineral]
$f_s(A_B)$ Coverage (or Filling Factor) of Surface B with A Atoms [Symbol]
5A- Libya [Civil Aircraft Marking]

FAA Failure Analysis Activity
FAA False Alarm Avoidance
FAA Federación Agraria Argentina [Agrarian Federation of Argentina]
FAA Federal Aviation Act [US]
FAA Federal Aviation Agency [now Federal Aviation Administration]
FAA Federal Aviation Administration [US]
FAA Fluid Applied Asphalt
FAA Free of All Average [also faa]
FAA Furnace Atomic Absorption
FAAAS Fellow of the American Academy of Arts and Sciences
FAAAS Fellow of the American Association for the Advancement of Science
FAAB Frequency Allocation Advisory Board
FAACIA Fellow of the American Association of Clinical Immunology and Allergy
FAAD Forward Area Air Defense
FAAD-LOS Forward Area Air Defense–Line-of-Sight
FAAD-LOS(H) Forward Area Air Defense–Line-of-Sight Heavy
FAAEM Fellow of the American Academy of Environmental Medicine
FAAFPRS Fellow of the American Academy of Facial Plastic and Reconstructive Surgery
FAAFS Fellow of the American Academy Foot Surgeons
FAAOM Fellow of the American Academy of Medicine
FAAO Fellow of the American Academy of Optometrists
FAAP Federal Aid Airport Program [US]
FAAP Fundação Armando Alvares Penteado [Armando Alvares Penteado Foundation, Sao Paulo, Brazil]
FAAR Forward Area Alerting Radar
FAAS Fellow of the American Academy of Arts and Sciences
FAAS Flame Atomic Absorption Spectrometry
FAAS Flame Atomic Absorption Spectroscopy
FAAS/GFAAS Flame Atomic Absorption Spectroscopy/Graphite Furnace Atomic-Absorption Spectroscopy
FAATC Federal Aviation Administration Technical Center [Atlantic City, New Jersey, US]
FAB Fachvereinigung Auslandsbergbau e.V. [German Association for International Mining]
FAB Fast Atom Bombardment
FAB Federal Advisory Board [US]
Fab Fabricate; Fabrication [also fab]
F(ab) Fragment of Antibody [Immunochemistry]
Fabg Fabricating [also fabg]
FABI Fédération Royale des Associations Belges d'Ingénieurs [Royal Federation of Belgian Associations of Engineers]
FAB ISO Fabrication Isometric (Drawing)
FAB-MAS Fast Atom Bombardment Mass-Spectrometry; Fast Atom Bombardment Mass Spectroscopy [also FABMS]
FABMDS Field Army Ballistic Missile Defense System [US Army]

FABMS Fast Atom Bombardment Mass Spectrometry; Fast Atom Bombardment Mass Spectroscopy [also FAB-MAS]

Fabn Fabrication [also fabn]

FABS Formulated Abstracting Service

FAbT Fluorescent Antibody Technique [Immunology]

FABU Fuel Additive Blender Unit

FAC Fast Affinity Chromatography

FAC Fast As Can [also fac]

FAC Federal Airports Corporation [Australia]

FAC Federal Atomic Commission

FAC Federation of Agricultural Cooperatives [UK]

FAC (International) Federation of Automatic Control

FAC Field Accelerator

FAC File Access Channel

FAC File Access Code

FAC Floating-Point Accumulator

FAC Florida Administrative Code [US]

FAC Forward Air Control(ler) [US Air Force]

FAC Free Available Chlorine

FAC Frequency Allocation Committee

FAc Fluoroacetic Acid

Fac Facsimile [also fac]

Fac Facility [also fac]

Fac Factor [also fac]

Fac Factory [also fac]

Fac Faculty [also fac]

FACA Fédération Algerienne de la Cooperation Agricole [Algerian Federation for Agricultural Cooperation]

FACA Fellow of the Association of Certified Accountants

FACAC Fast As Can As Customary [also facac]

Facam 3-Trifluoroacetyl-D-Camphorate

FACC Federation of African Chambers of Commerce [Ethiopia]

FACC Fellow of the American College of Cardiology

FACCA Fellow of the Association of Certified and Corporate Accountants

FACCP Fellow of the American College of Clinical Pharmacology and Chemotherapy

FACD Fellow of the American College of Dentists

FACD Foreign Area Customer Dialling

FATE Fatal Accident Circumstances and Epidemiology

FACE (International) Federation of Associations of Computer Users in Engineering, Architecture and Related Fields [Australia]

FACE Federation of Associations on the Canadian Environment

FACE Field-Alterable Control Element

FACE Field Artillery Computer Equipment

FACerS Fellow of the American Ceramic Society

FACES FORTRAN Automated Code Evaluation System

FACET Fluid Amplifier Control Engine Test

Fachh/Bull Tech Fachhefte (Chemigraphie, Lithographie, Tiefdruck)/Bulletin Technique (Photogravure, Lithographie, Heliogravure) [Switzerland]

Fachber Hüttenprax Metallweiterverarb Fachberichte Hüttenpraxis Metallweiterverarbeitung [German Publication on Practical Metallurgy and Metal Processing]

Fachz Lab Fachzeitschrift für das Labor [German Laboratory Journal]

FACI Fellow of the American Concrete Institute

FACI First-Article Configuration Inspection

F-Acid 2-Amino-2-Naphthalenesulfonic Acid; 7-Hydroxy-2-Naphthalenesulfonic Acid; 2-Naphthol-7-Sulfonic Acid [also F-acid]

Facil Facility [also facil]

Facil Manager Facilities Manager [UK Publication; issued with *Office Equipment Index*]

FACISCOM Finance and Controller Information System Command

FACM Foundation for Advances in Chemistry and Medicine [now Foundation for Advances in Medicine and Science, US]

FACM Free Atom Comparison Method

FACO Fabrication and Acceptance Checkout

FACO Factory Acceptance Checkout

FACO Factory Assembly and Checkout

FACO Final Assembly Checkout

FACOG Fellow of the American College of Obstetricians and Gynecologists

FACOGAZ Union of European Manufacturers of Gas Meters [Germany]

FA/COSI Final Assembly and Closeout System Installation

FACP Fellow of the American College of Physicians

FACP Food and Agriculture Council of Pakistan

FACR Fellow of the American College of Radiology

FACS Facility Assignment Control System

FACS Federation of American Controlled Shipping [US]

FACS Fellow of the Academy of Cosmetic Surgery

FACS Fellow of the American Ceramic Society

FACS Fellow of the American College of Surgeons

FACS Field-Activated Combustion Synthesis

FACS Finance and Control System

FACS Fine Attitude Control System

FACS Floating-Decimal Abstract Coding System

FACS Fluorescence Activated Cell Sorter

FACS Frequency Allocation Coordinating Subcommittee

FACS Fully Automatic Compiling System

Facs Facsimile [also facs]

FACSC Frequency Allocation Coordinating Subcommittee, Canada

FACSFAC Fleet Area Control Surveillance Facility

FACSI Fast Access Coded Small Images

Facsim Facsimile [also facsim]

FACSS Federation of Analytical Chemistry and Spectroscopy Societies

FACT Facility for the Analysis of Chemical Thermodynamics [also F*A*C*T] [McGill Uinversity On-Line Computer System, Canada]

FACT Fast Access Current Text

FACT Federation Against Copyright Theft [UK]

FACT Federation of Automated Coding Technologies [US]

FACT Flexible Automatic Circuit Tester

FACT Flight Acceptance Composite Test

FACT Ford Anodized-Aluminum Corrosion Test

FACT Foundation for Advanced Computer Technology [US]

FACT Foundation for Advancement in Cancer Therapy [New York City, US]

FACT Fully Automatic Cataloguing Technique

FACT Fully Automatic Compiler-Translator

FACT Fully Automatic Compiling Technique

FACTA Food, Agriculture, Conservation and Trade Act [US]

F-Actin Fibrous Actin [also F-actin] [Biochemistry]

FACTS Facsimile Transmission System

FACTS Federally Assisted and Conducted Programs Tracking System [US]

FACTS Financing Alternative Computer Terminal System

FACTS Fuel Abstracts and Current Titles [of Institute of Energy, UK]

Facty Factory [also facty]

FAD Failure Assessment Diagram

FAD Field Ammunition Depot

FAD Filtered Arc Deposition (Source)

FAD Fish Aggregation Device

FAD Flavin Adenine Dinucleotide [Biochemistry]

FAD Final Approach Display

FAD Floating Add

FAD Floating And

FAD Fuel Advisory Departure

FADA Federation of Automobile Dealers Associations

FADAC Field Artillery Digital Automatic Computer [US Army]

FADC First Aid and Decontamination Center

FADEC Full Authority Digital Electronic Control

FADES Fuselage Analysis and Design Synthesis

FADH Flavine-Adenine-Dinucleotide, Reduced [Biochemistry]

FADIC Field Artillery Digital Computer

FADM Fleet Admiral [also F Adm or FA]

FADP Finnish Association for Data Processing

FADS Filtered Attitude Determination System

FADS Force Administration Data System

FADS FORTRAN Automatic Debugging System

FADSA Fellow of the American Dental Society of Anesthesia

FADTW Fourth Annual Diamond Technology Workshop

FADU File Access Data Unit

FAE Field Application Engineer

FAE Final Approach Equipment

FAE Fuel/Air Explosives

FAES Federated American Engineering Societies

FAES Flame Atomic Emission Spectrometry; Flame Atomic Emission Spectroscopy

FAETUA Fleet Airborne Electronics Unit, Atlantic

FAETUP Fleet Airborne Electronics Unit, Pacific

FAF Final Approach Fix

FAF Financial Accounting Federation [US]

FAF Financial (Student) Aid Form [US]

FAF First Aerodynamic Flight

FAF Fly Away Field

F-AF Ferromagnetic-Antiferromagnetic (Transition) [Solid-State Physics]

FA-FAME Fatty Acid/Fatty Acid Methyl Ester

FAFI Federation of the Austrian Food Industry

FAFM Field Artillery Field Manual

FAFPIC Forestry and Forest Products Industry Council

FAFPS Five-Axis Fiber Placement System

FAFS Flame Atomic Fluorescence Spectrometry; Flame Atomic Fluorescence Spectroscopy

FAFT Farm and Food Society [UK]

FAGC Fast Automatic Gain Control

FA-Gly-Leu-NH$_2$ N-(3-[2-Furyl]acryloyl)-Gly-Leu Amide (or N-(3-[2-Furyl]acryloyl)-Glycine-Leucine Amide)

FAGS Federation of Astronomical and Geophysical Data Analysis Services [France]

FAGS Federation of Astronomical and Geophysical Services [of International Association for the Physical Sciences of the Ocean, US]

FAGS Fellow of the American Geographical Society

FAGT Federation of Agricultural Group Traders [UK]

FAGU Fellow of the American Geophysical Union

FAH Facilitation Awards for Handicapped Scientists and Engineers [of National Science Foundation, US]

FAH$_2$ Dihydrofolic Acid

FAHO Gesellschaft von Freunden der Aachener Hochschule e.V. [Society of Friends of the Aachen Institute of Technology, Germany] [also Faho]

FAHQMT Fully-Automatic High-Quality Machine Translation

Fahr Fahrenheit

FAI Fail As Is

FAI Fédération Aéronautique Internationale [International Aeronautic Federation, France]

FAI Fertilizer Association of India

FAI First Aid Institute [of National Safety Council, US]

FAI Fresh Air Intake

FAIA Fellow of the American Institute of Architects

FAIC Fellow of the Agricultural Institute of Canada

FAIC Fellow of the Architectural Institute of Canada

FAICE Fellow of the Institute of Civil Engineers

FAID Fellow of the American Institute of Interior Designers

FAID Field-Assisted, Ion-Induced Desorption

FAIEE Fellow of the American Institute of Electrical Engineers [now Fellow of the Institute of Electrical and Electronic Engineers]

FAIM (International) Flexible Automation and Information Management (Conference)

FAIMME Fellow of the American Institute of Mining and Metallurgical Engineers

FAIR Fabrication, Assembly and Inspection Record

FAIR Fair Access to Insured Responsibility

FAIR Fast Access Information Retrieval

Fair Fairing

FAIRS Federal Aviation Information Retrieval System [US]

FAIRS Food and Agricultural Information Storage and Retrieval System [of Food and Agricultural Organization]

FAIRS Fully Automatic Information Retrieval System

FAITC Foreign Affairs and International Trade Canada

FAK File Access Key

FAK Freight All Kinds

FAL Bundesforschungsanstalt für Landwirtschaft [Federal Agricultural Research Institute, Germany]

FAL File Access Listener

FAL First Approach and Landing (Test)

FAL Frequency Allocation List

1-FAL 1-Fluoroallene

FALC (On-Board) Fatigue Life Counter

FALCON Fuel-Air Line Charge Ordnance Neutralizer

FALGPA Furylacryloyl-Leu-Gly-Pro-Ala (or Furylacryloyl-Leucine-Glycine-Proline-Alanine)

Falk Is Falkland Islands

FALP Flowing Afterglow Langmuir Probe (Technique) [Physics]

FALT FADAC (Field Artillery Digital Automatic Computer) Automatic Logic Tester

FALTRAN FORTRAN-to-ALGOL Translator

FAM Facilities Alcohol Management (Program)

FAM Family of Frequencies

FAM Fast Access Memory

FAM Fast Air Mine

FAM Fast Auxiliary Memory

FAM Fermentation Analysis Module

FAM Fibrous Aerosol Monitor

FAM File Access Manager

FAM Final Address Message

FAM Fire Apparatus Manufacturers Association [US]

FAM N-(3-Fluoroanthyl) Maleimide

FAM Frequency and Amplitude Modulation

Fam Familiar(ize); Familiarization; Family [also fam]

FAMA α-Amanitin Fluorescein Isothiocyanate

FAMA Fellow of the American Medical Association

FAMCC Florida Advanced Materials Chemistry Conference [US]

FAME Fatty Acid Methyl Ester

FAME Financial Assistance for Mineral Exploration (Program)

FAME Florida Association of Marine Explorers [US]

FAMECE Family of Military Engineer Construction Equipment

FAMEM Federation of Associations of Mining Equipment Manufacturers

Fameta International Trade Fair for Metalworking [Nuremburg, Germany]

FAMHM Federation of Associations of Materials Handling Manufacturers

FAMI Federation of Advanced Materials Industries [US]

FAMIS Factory Management Information System [UK]

Famophos O,O-Dimethyl-O-[p-(Dimethylsulfamoyl)phenyl]-Phosphorothioate [also famophos]

FAMOS Fast Multitask Operating System

FAMOS Fleet Application of Meteorological Observations for Satellites

FAMOS Flexibel Automatisiertes Montagesystem [Flexible Automated Assembly System]

FAMOS Flight Acceleration Monitor Only System

FAMOS Floating-Avalanche-Injection Metal-Oxide Semiconductor; Floating-Gate Avalanche-Injection Metal-Oxide Semiconductor

FAMOSS Fiberscopic Apparatus for Measurement of Surface Strain

FAMOUS French-American Mid-Ocean Underwater Study (Project)

FAMP Fire Alarm Monitoring Panel

FAMPA Ferro-Alloys and Metals Producers Association [UK]

Famphur O,O-Dimethyl-O-[p-(Dimethylsulfamoyl)phenyl]-Phosphorothioate [also famphur]

Fam Pract Family Practice [Journal published by Oxford University Press, UK]

FAMS Fellow of the American Microscopical Society

FAMS Forecasting and Modelling System

FAMS Foundation for Advances in Medicine and Science [formerly Foundation for Advances Chemistry and Medicine, US]

Fam Safety Health Family Safety and Health [Publication of National Safety Council, US]

FAMSO Formaldehyde Dimethyl Mercaptal-S-Oxide [also Famso]

FAMSO Methyl Methylsulfinylmethyl Sulfide

FAMT Fully Automatic Machine Translation

FAMU Florida A&M (Agricultural and Mechanical) University [Tallahassee, US]

FAMU Fuel Additive Mixture Unit

FAN Federation of Alberta Naturalists [Canada]

FAN First Aid Nurse

FANES Flameless Nonthermal Excitation Spectrometry

FANGIO Feedback Analysis for GCM (Global Circulation (or Climate) Model) Intercomparison and Observation

fANP Frog Atrial Natriuretic Peptide [also FANP]

FANS Future Air Navigation Systems (Committee) [of International Civil Aviation Organization]

FANTAC Fighter Analysis Tactical Air Combat

FAO Finish All Over

FAO Flight Activities Officer

FAO Food and Agricultural Organization [of United Nations in Rome, Italy]

FAO/ECE Food and Agricultural Organization/Economic Commission for Europe [of United Nations]

FAO/IAEO Food and Agricultural Organization/International Atomic Energy Organization (Symposium) [of United Nations]

FAOIP French Association of On-Line Information Providers

FAOMA Fellow of the American Occupational Medicine Association

FAO/WHO Food and Agricultural Organization/World Health Organization (Joint Program) [of United Nations]

FAP Facility Analysis Plan

FAP Failure Analysis Program

FAP Fault Analysis Process

FAP Federal Airway Plan [US]

FAP Field Application Panel

FAP File Access Protocol

FAP Filmmakers Assistance Program [of National Film Board of Canada]

FAP Final Approach Plane

FAP Final Approach Point

FAP Financial Advantage Program [of Institute of Electrical and Electronics Engineers, US]

FAP First Aid Post

FAP Floating-Point Arithmetic Package

FAP Fluoroapatite [also FAp]

FAP FORTRAN Assembly Program

FAP Forward Ammunition Point

FAP Frequency Allocation Panel

FAP Furylacryloylphenylalanine [Biochemistry]

FAPA Fellow of the American Psychological Association

FAPA Future Aviation Professionals of America [US]

FAPC Fatty Acids Producer's Council [US]

FAPCF Ferric Aquapentacyanoferrate

FAPERRGS Fundação de Amparo à Pesquisa do Estado do Rio Grande dol Suld [Research Foundation of the State of Rio Grande dol Sul, Brazil]

FAPERJ Fundação de Amparo à Pesquisa do Estado do Rio de Janeiro [Research Foundation of the State of Rio de Janeiro, Brazil]

FAPESP Fundação de Amparo à Pesquisa do Estado de São Paulo [Research Foundation of the State of Sao Paulo, Brazil]

FAPGG Furylacryloylphenylalanylglycylglycine; N-(3-[2-Furyl]acryloyl)-Phenylalanylglycylglycine [Biochemistry]

FAPI Family Application Program Interface

FAPIG First Atomic Power Industry Group [Japan]

FAPP Fractured-Area-Projection Plot

FAPRA Federation of African Public Relations Associations [Ghana]

FAPS Fellow of the American Physical Society

FAPUS Frequency Allocation Panel, United States

FAPUS-MCEB Frequency Allocation Panel, United States– Military Communications Electronics Board

FAQ Fair Average Quality

FAQ Fédération de l'Automation au Québec [Quebec Automation Federation, Canada]

FAQ Free Alongside Quay [also faq]

FAQ Frequently Asked Question(s) [Internet]

FAQ Functional Activities Questionnaire [Alzheimer's Disease]

FAQS Fast Queuing System

FAR Faculty Awards for Research

FAR Failure Analysis Report

FAR Federal Acquisition Regulation [US]

FAR Federal Airworthiness Regulations [US]

FAR Federal Aviation Regulations [US]

FAR Federation of Arab Republics

FAR File Address Register

FAR Final Acceptance Review

FAR First Assessment Report

FAR First Assesment Report [of Intergovernmental Panel on Climate Change]

FAR Flight Acceptance Review

FAR Flight Aptitude Rating

FAR Foundation of Applied Research [US]

FAR Fuel-Air Ratio [also F/AR]

F/Ar Fluorine/Argon (Mixture)

FARADA Failure Rate Data

Faraday Disc Faraday Discussions [Journal of the Royal Society of Chemistry, UK]

Faraday Disc Chem Soc Faraday Discussions of the Chemical Society [UK]

Faraday Symp Chem Soc Faraday Symposia of the Chemical Society [UK]

Faraday Trans Faraday Transactions [Journal of the Royal Society of Chemistry, UK]

Faraday Trans I Faraday Transactions I [of the Royal Society of Chemistry, UK]

Faraday Trans II Faraday Transactions II [Journal of the Royal Society of Chemistry, UK]

Farbe Lack Farbe und Lack [German Publication on Paints, Lacquers and Varnishes]

FARC Field Artillery Reserve Corporation

FARE Florida Aquanaut Research Experiment [of National Oceanic and Atmospheric Administration, US]

FARE Federal Acquisition Regulation [US]

Far East Econ Rev Far Eastern Economic Review [Hong Kong]

FAREGAZ Union of European Manufacturers of Gas Pressure Controllers [Germany]

FARET Fast Reactor Experiment Test [of Argonne National Laboratory, US] [also Faret]

FARET Fast Reactor Test Assembly (or Facility) [of Argonne National Laboratory, US] [also Faret]

Far-IR Far Infrared [also far-IR]

Farmacol Toksikol Farmacologiia i Toksikologiia [Journal on Pharmacology and Toxicology]

FARNET Federation of American Research Networks [Internet]

FARR Forward Area Refuelling and Rearming

FARS Facility for Atmospheric Remote Sensing [of University of Utah, Salt Lake City, US]

FARS Fatal Accident Reporting System

FAS Faculty of Administrative Studies

FAS Faculty of Applied Science

FAS Federal Airways System [of Federal Aviation Administration, US]

FAS Fédération des Architectes Suisses [Swiss Federation of Architects]

FAS Federation of American Scientists

FAS Federation of Astronomical Societies [UK]

FAS Ferrous Aluminum Sulfate

FAS Fetal Alcohol Syndrome

FAS Field Advisory Service

FAS File Access Subsystem

FAS Filtered Air Supply

FAS Final Assembly Schedule

FAS Financial Accounting Standard [US]

FAS Finnish Academy of Sciences [Helsinki]

FAS Firsts and Seconds [Hardwood Lumber]

FAS Flame Absorption Spectroscopy

FAS Floating-Point Arithmetic System

FAS Florida Academy of Sciences [US]

FAS Foreign Agricultural Service [of US Department of Agriculture]

FAS Forward Area Sight

FAS Frame Alignment Signal

FAS Free Alongside [also fas]

FAS Free Alongside Ship [also fas]

FAS Frequency Allocations Subcommittee

FAS Frequency Analysis System

FAS Frequency Assignment Subcommittee

FAS Fuel Availability System [of US Bureau of Mines]

Fas Fastener [also fas]

FASA Federation of ASEAN (Association of Southeast Asian Nations) Shipowners Associations [Malaysia]

FASAF Filipinas Americas Science and Art Foundation [US]

FASB Federal Accounting Standards Board [US]

FASB Financial Accounting Standards Board [US]

Fasc Fasciculus [also fasc] [Anatomy]

FASCE Fellow of the American Society of Civil Engineers

FASCOS Flight Acceleration Safety Cutoff System

FASE Federation of Acoustical Societies of Europe

FASE Force-at-Specified Elongation

FASE Fundamentally Analyzable Simplified English

FASEB Federation of American Societies for Experimental Biology [US]

FASEB J FASEB Journal [of Federation of American Societies for Experimental Biology]

FASi-4 1,2-Bis(methyldifluorosilyl)ethane

FASIC Function and Algorithm-Specific Integrated Circuit

FASII Federation of Associations of Small Industries in India

FASINEX Frontal Air-Sea Interaction Experiment

FASIS Fully Automatic Syntactically-Based Indexing System

FASLMS Fellow of the American Society of Liposuction Medicine and Surgery

FASM Fellow of the American Society for Materials International

FASME Fellow of the American Society of Mechanical Engineers

FASO Fellow of the American Society of Oculists

FASP Fault-Tolerant Array Signal Processor

FASP Final Average Sustained Pressure [Aerospace]

FASRA Foundation to Assist Scientific Research in Africa [Belgium]

FASS Federation of Associations of Specialists and Sub-Contractors [UK]

FASS Ford Aerospace Satellite Services,

FASS Forward Acquisition Sensor

FASSET Functional Advanced Satellite-Communication System for Evaluation and Test [of Department of Defense, Canada]

FASST Farming for Agriculturally Sustainable Systems in Tasmania

FAST Facility for Automatic Sorting and Testing

FAST Fairchild Advanced Schottky Technology

FAST Fast Access Storage Technology

FAST Fast Auroral Snapshot Explorer [of NASA]

FAST Fast Automatic Shuttle Transfer

FAST Federation Against Software Theft [UK]

FAST Federazione delle Associaziones Scientifiche e Tecniche [Federation of Scientific and Technical Associations, Italy]

FAST Field-Data Applications, Systems and Techniques

FAST Flexible Algebraic Scientific Translator

FAST Fluorescent Antibody Staining Technique

FAST Formal Auto-Indexing of Scientific Texts

FAST Formula and Statement Translator

FAST Forward/Aft Scanning Mode

FAST Forward Air Strike

FAST Fuel Assembly Stability Test

FAST Function Analysis System Technique

FASTAC Furnace Aerosol Sampling Technique with Autocalibration

FASTAR Frequency Angle Scanning, Tracking and Ranging

Fast Ferry Int Fast Ferry International [UK Publication]

Fasten Age Fastener Age [Publication of the Wire Association International, US]

FASTEX Fronts and Atlantic Storm-Track Experiment [Joint International Experiment]

FASTI Fast Access to Systems Technical Information

FASTOP Flutter and Strength Optimization Program

FASTS Federation of Australian Scientific and Technical Societies

FASTW First ASEAN (Association of Southeast Asian Nations) Science and Technology Week

FASWC Fleet Antisubmarine Warfare Command

FAT Factory Acceptance Test

FAT File Allocation Table [Computers]

FAT Final Approach Track

FAT Fixed Analyzer Transmission

FAT Flight Attitude Table

FAT Fluorescent Antibody Test

FAT Foreign Area Toll

FAT Foreign Area Translation

FAT Formula Assembler Translator

FATAL FADAC (Field Artillery Digital Automatic Computer) Automatic Test Analysis Language [of US Army]

FATAR Fast Analysis Tape and Recovery

FATCAT Film and Television Correlation Assessment Technique

FATDL Frequency and Time-Division Data Link

FATE Federation of Automatic Transmission Engineers [UK]

FATE Fel Accelerator Test Experiment [of Stanford Synchrotron Radiation Laboratory, Stanford, California, US]

FATE Fusing and Arming Test and Evaluation

FATE Fusing and Arming Test Experiment

fath fathom [Unit]

Fatigue Fract Eng Mater Struct Fatigue and Fracture of Engineering Materials and Structures [Journal published in the UK]

FATIMA Fatigue Indicating Meter Attachment

FATIPEC Fédération d'Associations de Techniciens des Industries des Peintures, Vernis, Emaux et Encres d'Imprimerie de l'Europe Continentale [Federation of the Associations of Technicians of the Paint, Varnish, Enamel and Printing Ink Industries of Continental Europe, France]

FATIS Food and Agriculture Technical Information Service [of Organization for Economic Cooperation and Development, France]

FATPA France-Asia Trade Promotion Association

FATR Fixed Autotransformer

FATT Fracture-Appearance Transition Temperature

FATT Fracture-Area Transition Temperature

FATUREC Federation of Air Transport User Representatives in the European Community

FAU Florida Atlantic University [Boca Raton, Florida, US]

FAU Frequency Allocation and Uses

FAU Friedrich-Alexander-Universität [Friedrich Alexander University, Erlangen-Nuremberg, Germany]

FAUL Five Associated University Libraries (Cooperative) [US]

FAUSTUS Frame-Activated Unified Story Understanding System

FAV Fast-Acting Valve

FAV Fowl Adenovirus

FAV Free Acid Value (Test)

FAV-1 Fowl Adenovirus, Type 1

FAW Fiber Areal Weight

FAW Fluid Acoustic Waveguide

FAW Frame Alignment Word

FAWG Flight Assignment Working Group

FAWPCA Florida Air and Water Pollution Control Act [US]

FAWS Flight Advisory Weather Service

FAX Facsimile (Transmission) [also Fax or fax]

.FAX Fax [File Name Extension] [also .fax]

FAXCOM Facsimile Communications [also Faxcom]

FAX/TEL Facsimile/Telephone

FAZ Frankfurter Allgemeine Zeitung [A Newspaper published in Frankfurt, Germany]

FB Belgian Franc [Currency]

FB (*Fachbereich*) – German for "University Department"

FB Faculty of Building [UK]

FB Feedback

FB Fermi Band [Solid-State Physics]

FB F-Format, Blocked Data Set

FB Fiber Bundle [Optics]

FB Fibroblast [Anatomy]

FB Fighter Bomber

FB File Block

FB Film Badge [Radiology]

FB Final Braking

FB Fine Business [Amateur Radio]

FB Fire Boat

FB Fire Box

FB Fire Brigade

FB Fish Bolt

FB Fixed Bed

FB Fixed Block [Computers]

FB Flame Black

FB Flash Back

FB Flash Bulb

FB Flatband [also fb]

FB Flat-Bed

FB Flat Bottom

FB Floating Bearing

FB Flowback

FB Fluid(ized) Bed

FB Fluorobenzene

FB Flying Boat

FB Focused Beam

FB Forest Biologist; Forest Biology

FB Forging Brass

FB Formal Bond

FB Forward Bias [Electronics]

FB Fragmentation Bomb

FB Frame Buffer

FB Freight Bill [also fb]

FB Fuel Battery

FB Furnace Black

FB Furnace Brazing

FB Fuse Block

FB Fuse Box

F&B Fire and Bilge

fb freight bill [also FB]

$f_s(B_A)$ Coverage (or Filling Factor) of Surface A with B Atoms [Symbol]

5B- Cyprus [Civil Aircraft Marking]

FBA Farm Building Association [UK]

FBA Federation of British Audio [UK]

FBA Fellow of the British Academy

FBA Fiber Box Association [US]

FBA First Born Approximation [Physics]

FBA Fixed-Block Architecture

FBA Fluoboric Acid

FBA Folded Band Aurorae

FBA Freshwater Biological Association [of Natural Environmental Research Council, UK]

FBA Furnace Bottom Ash

FBAS Federation of British Aquatic Societies [UK]

FBAS Fixed Base Aft Station

FBB Fast Blue BB

FBB Fluidized-Bed Boiler

FBB Free Balloon Barrage

FBB Fusion Breeder Blanket [Nuclear Engineering]

FBBM Federation of Building Block Manufacturers [UK]

FBC Federation of Brickwork Contractors [UK]

FBC Federation of Business Centers [UK]

FBC Feedback Carburetor

FBC Feedback Compensation; Feedback Compensator

FBC Feedback Control(ler)

FBC Final Boundary Conditions

FBC Fixed Boundary Conditions

FBC Fluidized-Bed Combustion

FBC Forward BGO (Bismuth-Germanium Oxide)-Crystal Calorimeter

FBC Fully-Buffered Channel

4FB15C5 4'-Formylbenzo-15-Crown-5

FBCAEI Federation of Builders, Contractors and Allied Employers of Ireland

FBCCI Franco-British Chamber of Commerce and Industry [UK]

FBCMA Fiber Bonded Carpet Manufacturers Association [UK]

FBCN Federation of British Columbia Naturalists [Canada]

FBCR Fixed-Bed Catalytic Reaction (Technology)

FBCR Fluidized-Bed Combustion Residue

FBCR Fluidized-Bed Control Rod

FBCS Feedback Control System

FBCS Fixed Base Crew Station [NASA Shuttle Mission Simulator]

FBCVD Fluidized-Bed Chemical-Vapor Deposition (Coating) [also FB-CVD or FB CVD]

FBD Feedback Device

FBD Full Business Day

FBD Functional Block Diagram

FBDB Federal Business and Development Bank [Canada]

FBDZF Farkas-Becker-Doering-Zelodovich-Frenkel (Equation) [Metallurgy]

FBE Faculty of Business Education [UK]

FBE Flat Bed Electrophoresis

FBE Fluidized Bed Electrowinning

FBE Fusion-Bonded Epoxy (Coating)

FBED Focused Beam Electron Diffraction

FBEM Fixed-Beam Electron Microscope

FBETM Federation of British Engineers' Tool Manufacturers [UK]

FBF Fast Brittle Failure

FBFM Feedback Frequency Modulation

FBFS Fuel Building Filter System

FBFS Furnace Belt, Flat Seat

FBFT Flow Bias Functional Test

FBG Fixed-Bed Gasifier

FBGT Fasting Blood Glucose Test [Medicine]

FBH Ferdinand-Braun-Institut (für Höchstfrequenztechnik) [Ferdinand Braun Institute (for Microwave Engineering), Berlin, Germany]

FBH Flat-Bottom Hole

FBHE Fetal Bovine Hearth Endothelial (Cells)

FBHTM Federation of British Hand Tool Manufacturers [UK]

FBHVC Federation of British Historical Vehicle Clubs [UK]

FBI Federal Building Initiative [Canada]

FBI Federal Bureau of Investigation [US]

FBI Federation of British Industries [now Confederation of British Industry, UK]

FBI Ferdinand-Braun-Institut [Ferdinand Braun Institute, Berlin, Germany]

FBIC Farm Building Information Center [UK]

FBIM Fellow of the British Institute of Management

F Bk Flat Back [also f bk] [Building Trade]

FBL Feedback Loop

FBL FIATA Bill of Lading

FBL Form Block Lines

FBLO Foreign Branch Liaison Office

FBM Feet Board Measure [also fbm, bd ft or ft bm]

FBM F-Format, Blocked Data Set with "M" (Machine) Control Character

FBM Fleet Ballistic Missile

FBM Foreground and Background Monitor

FBMA Food and Beverage Manufacturers Association [UK]

FBMP Fleet Ballistic Missile Program

FBMS Fleet Ballistic Missile Submarine

FBMWS Fleet Ballistic Missile Weapon System

FBN Food Business Network

FBN Forschungsinstitut für die Biologie landwirtschaftlicher Nutztiere [Research Institute for Biology and Agricultural Livestock, Dummerstorf, Germany]

FBN Fuel-Based Nitrogen (Generator)

FBNML Francis Bitter National Magnet Laboratory [of Massachusetts Institute of Technology, Cambridge, US]

FBO Fixed Base Operator

FBOA Fellow of the British Optical Association

FBP Final Boiling Point [also fbp]

FBP Flat-Bed Plotter

FBP Fluid-Bed Processing

FBP Fluidized-Bed Polymerization

FBP Fluidized-Bed Process

FBP Foreign Buyers Program

FBPS Forest and Bird Protection Society [New Zealand]

FBR Fast Blue RR

FBR Fast Breeder Reactor

FBR Fast Burst Reactor

FBR Feedback Resistance

FBR Finite Basis Representation

FBR First Bohr Radius [Atomic Physics]

FBR Fixed-Bed Reactor

FBR Fluid-Bed Roasting

FBR Fluidized-Bed Reactor

FBR Forskningsbibiloteksradet [Research Library Council, Sweden]

FBR Full Bibliographic Record

Fbr Fiber [also fbr]

FBRAM Federation of British Rubber and Allied Manufacturers [UK]

5-BRANES (Trieste Conference on) Super Five-Branes in 5+1 Dimensions [of International Center for Theoretical Physics, Trieste, Italy]

FBRL Final Bomb Release Line

FBS Fast Blue Salt

FBS Fasting Blood Sugar [Medicine]

FBS Feedback System

FBS Fetal Bovine Serum [Biochemistry]

FBS F-Format, Blocked Standard Data Set

FBS Fire Brigade Society [UK]

FBS Firefighters Breathing System

FBSA F-Format, Blocked Standard Data Set with "A" (ASCII) Control Character

FBSC Federation of Building Specialist Contractors [UK]

FBSC Fellow of the British Society of Commerce

FBSC Frederick Biomedical Supercomputing Center [of National Cancer Institute, Bethesda, Maryland, US]

FBSID Full Body Spinal Immobilization Device [Medicine]

FBSOA Forward Bias Safety Operating Area

FBT Face Bend Test

FBT Functional Board Tester

FBTR Fast Breeder Test Reactor

FBTR Federation of British Tape Recordists [UK]

FBU Federation of Broadcasting Unions [UK]

FBU Fire Brigades' Union [UK]

FBU Fluidized-Bed Catalytic Oxidation Unit

FBUA Franco-British Union of Architects [UK]

FBV Field Base Visit

FBV Fuel Bleed Valve

FBW Fly-by-Wire (System) [Aeronautics]

FBW Forschungsgemeinschaft Bauen und Wohnen [Research Association for Building and Housing, Germany]

FC Faceted Concentrator

FC Facilities Control

FC Faculty of Chemistry

FC Faculty of Commerce

FC Fail Closed

FC Fan Compressor

FC Farad(a)ic Current

FC Faraday Cell

FC Faraday Coil

FC Faraday Constant

FC Faraday Cup

FC Fast-Cooled; Fast Cooling

FC Fast Cycling

FC Fatigue Crack(ing)

FC Fault Current

FC Feature Control

FC Fenn College [Cleveland, Ohio, US]

FC Fermi Chopper [Physics]

FC Ferrite Core

FC Ferroconcrete

FC Fiala and Cadek (Creep) [Metallurgy]

FC Fiber Channel

FC Fiber Composite

FC Fiber Concrete

FC Fiber Glass Cover

FC Fibrocyte [Anatomy]

FC Fibrous Composite

FC Field Capacity [Hydrology]

FC Field Coil

FC Field-Cooled; Field Cooling [Solid-State Physics]

FC File Code

FC File Comparison

FC File Conversion

FC Filter Circuit

FC Filter Cloth

FC Final Condition

FC Finance Committee

FC Find Calling Party [Telecommunications]

FC Fine Control

FC Fire Code

FC Fire Control (Radar)

FC Fire Cracking

FC First Class

FC Fit Check [also F/C]

FC Fixed Carbon

FC Flame Cutter; Flame Cutting

FC Flash Chromatography

FC Flexible Connection

FC Flexible Coupling

FC Flight Computer

FC Flight Control(s)

FC Flight Controller [also F/C]

FC Flight Crew

FC Flight Critical

FC Flood Control

FC Florida Current [Oceanography]

FC Flotation Chemistry

FC Flow Calorimetry

FC Flowchart

FC Flow Control

FC Fluid-Coke (Process)

FC Fluid(ic) Control

FC Fluor Crown (Glass)

FC Fluorocarbon

FC Flux Change(s) [also fc]

FC Flux Concentrator

FC Foam Cell

FC Foam Core
FC Focusing Coil
FC Font Cartridge
FC Font Change (Character)
FC Food Chain
FC Footcandle [also fc, or ftc]
FC Foot Care
FC Foot Control
FC Force Control
FC Forced Convection
FC Forced Cooling
FC Ford (Viscosity) Cup
FC Forest Center
FC Forestry Canada
FC Forestry Commission [UK]
FC Forestry Commission [of New South Wales, Australia]
FC Fort Custer [Michigan, US]
FC Foundation Center [US]
FC Fourier Components
FC Fractionating Column
FC Fraction Collector
FC Franck-Condon (Principle) [Physical Chemistry]
FC Franklin College [Indiana, US]
FC Free Carbon [Metallurgy]
FC Free Cholesterol
FC Free Convection
FC Free Cursor
FC Freeze-Coating
FC French Community (of Africa) [Defunct]
FC Frequency Change(r)
FC Frequency Control(ler)
FC Frequency Conversion; Frequency Converter
FC Frequency Counter
FC Fretting Corrosion [Metallurgy]
FC Friction Clutch
FC Friedel-Crafts (Reaction)
FC Front Connected; Front Connection
FC Fuel Calorimeter; Fuel Calorimetry
FC Fuel Cell [also FC]
FC Fuel Channel
FC Fuel Cost
FC Fuel Cycle [Nuclear Reactors]
FC Full Constraint (Taylor Model) [Metallurgy]
FC Full Coordination
FC Function Code
FC Fundamental Constant [Physics]
FC Funnel Cloud [Meteorology]
FC Furnace Cooled; Furnace Cooling
FC Fuse Chamber
F2C Furan-2-Carboxylic Acid
F3C Furan-3-Carboxylic Acid
F&C Flammable and Combustible (Material)
F&C Front and Center (Memorandum)

F/C Fit Check [also FC]
F/C Flight Controller [also FC],
F/C Fluorine-to-Carbon (Ratio) [Fluoropolymers]
F/C Fuel Cell [also F/C]
F-C Free Energy–Composition (Curve) [also F/C]
Fc Fc Fragment Specific [[Immunochemistry]
fc file compare (command) [MS-DOS]
fc flux change(s) [also FC]
fc footcandle [Unit]
$f(c)$ directional order anisotropy [Symbol]
4/C Four Conductor
FCA Faraday Cup Array
FCA Farm Credit Administration [US]
FCA Fédération des Coopératives Agricoles [Federation of Agricultural Cooperatives, France]
FCA Federation of Commodity Associations [UK]
FCA Fellow of the Institute of Chartered Accountants [UK]
FCA Fencing Contractors Association [UK]
FCA FIRE (First ISCCP (International Satellite Cloud Climatology Project) Regional Experiment) Central Archive
FCA Fire Control Area
FCA Fluorescence Concentration Analyzer
FCA Fluids Control Assembly
FCA Food Casings Association [UK]
FCA Forward Controller Assembly
FCA Frequency Control Analysis; Frequency Control and Analysis
FCA Functional Compatibility Analysis
FCA Functional Configuration Audit
FCA(CAN) Fellow of the Canadian Institute of Chartered Accountants
FCAD Fellow of Canadian Academy of Denturism
FCAF Flight Crew Accommodations Facility
FCAFPRS Fellow of the Canadian Academy of Facial Plastic and Reconstructive Surgery
FCAH Fellow of the Canadian Association of Homeopathy
FCAI Federal Chamber of Automotive Industries
FC/AL Fiber Channel/Arbitrated Loop
FCAMRT Fellow of the Canadian Association of Medical Radiation Technologists
FCAP Flight Control Applications Program
FCAPO Fellow of the College of American Pathologists
FCAR Fonds pour la Formation de Chercheurs et l'Aide à la Recherche [Fund for Researcher Education and Research Aid, Quebec, Canada]
FCASI Fellow of the Canadian Aeronautics and Space Institute
FCAW Flux-Cored Arc Welding
FCAW-EG Flux-Cored Arc Welding–Electrogas
FCB Fibronectin Cell Binding (Peptide) [Biochemistry]
FCB File Control Block [also fcb]
FCB Fives-Cail Babcock [US]
FCB Flip-Chip Bonding
FCB Forms Control Buffer
FCB Frequency Coordinating Board

FCB Frequency Coordinating Body [of International Civil Aviation Organization]

FCB Function Control Block

a-FCB *α*-Fibronectin Cell Binding (Peptide) [Biochemistry]

FCBM Federation of Clinker Block Manufacturers [UK]

FCBS File Control Blocks [also fcbs]

FCC Face-Centered Cubic (Crystal) [also fcc]

FCC Farm Credit Corporation [Canada]

FCC Federal Communications Commission [US]

FCC Federal Construction Council [of National Academy of Science, US]

FCC Federal Coordinating Council [US]

FCC Fédération des Coopératives de Consommation [Federation of Consumer Cooperatives, France]

FCC Federation of Crafts and Commerce [UK]

FCC Feedforward/Cascade Control

FCC Field-Cooled Data on Cooling [Solid-State Physics]

FCC File Carbon Copy [also fcc]

FCC Flat Conductor Cable

FCC Flexible Current Carrier

FCC Flight Control Center

FCC Flight Control Computer

FCC Flight Crew Compartment

FCC Florida Citrus Commission [US]

FCC Fluid Catalytic Cracking (Process)

FCC Fluid-Cracking Catalyst

FCC Flux Current Control

FCC Food Chemical Codex

FCC Food Contaminants Commission [US]

FCC Frame Check Character

FCC Fuel-Cell Catalyst

Fcc File Carbon Copy [Internet]

f/cc fibers per cubic centimeter (of air) [also f cm^{-3} or f/cm^3]

FCCA Forestry, Conservation Communications Association [US]

FCCC Federation of Commonwealth Chambers of Commerce [UK]

FCCC Framework Convention on Climate Change

FCCI Fuel-Cladding Chemical Interaction [Nuclear Engineering]

FCCMG Fellow of the Canadian College of Medical Genetics

FCCP Firm Contract Cost Proposal

FCCP p-Trifluoromethoxy Carbonylcyanidephenylhydrazone

FCCPO Federal Contract Compliance Programs Office

FCCSET Federal Coordinating Council for Science, Engineering and Technology [Now Part of National Science and Technology Council, US]

FCCTS Federal COBOL Compiler Testing Service [US]

FCCU Fluid Catalytic Cracking Unit

FCD Failure-Correction Decoding

FCD Fine Control Damper

FCD Flight Control Division [of NASA Johnson Space Center, Houston, Texas, US]

FCD Foundation for Child Development [Washington, DC, US]

FCD Franck-Condon Density [Physical Chemistry]

FCD Frequency Compression Demodulator

FCDA Federal Civil Defense Administration [US]

FCDB Flight Control Data Bus

FCDE Federation of Clothing Designers and Executives [UK]

FCDM Flow Control Decision Message

FCDPA Field Committee for Development Planning in Alaska [US]

FCDR Failure Cause Data Report

FCE Flexible Critical Experiment

FCE Flight Control Equipment

FCE Flight Crew Equipment

FCE Fourier Conduction Equation

FCE Fuel-Cell Electrolyte

FCEA Federal Capital Equipment Authority

FCEA Fellow of Computerized Electrolysis Association

FCEC Federation of Civil Engineering Contractors [UK]

FCEF Flight Crew Equipment Facility

FCEI Facility Contractor End Item

FC/EL Fiber Channel/Enhanced Loop

FCEM Flow Control Execution Message

FCEV Fuel Cell Electric Vehicle

FCF First Captive Flight

FCF Flight Critical Forward

FCF Franck-Condon Factor [Physical Chemistry]

FCF Fuel Conditioning Facility [of Argonne National Laboratory-West, Idaho Falls, US]

FCF Fuel Cycle Facility [of US Atomic Energy Commission]

FCF Functional Check Flight

4CF Four Corners Section Fall Meeting [of American Physical Society, US]

F-C-F Flat-Corrugated-Flat (Fiberboard)

FCFA Focal Canadian Foundation for the Americas

FCFA Franc Communanuté Financière Africaine [African Financial Community Franc] [Currency]

FCFC Flat Conductor Flat Cable

F-C-F-C-F Flat-Corrugated-Flat-Corrugated-Flat (Fiberboard)

F-C-F-C-F-C-F Flat-Corrugated-Flat-Corrugated-Flat-Corrugated-Flat (Fiberboard)

FCFM Flight Combustion Facility Monitor

FCFO Full Cycling File Organization

FCFP Franc Communanuté Financière Pacifique [Pacific Financial Community Franc] [Currency]

FCFS First-Come, First-Served

FCFT Fixed Cost, Fixed Time Estimate

FCHF Frozen Core Hartree-Fock [Physics]

FCG False Cross or Ground

FCG Fatigue Crack Growth [Mechanics]

FCG Federal Coordination Group [US]

Fcg Facing [also fcg]

FCGA Fellow of the Canadian General Accountants Association

FCGR Fatigue Crack Growth Rate [Mechanics]

FCGS Four Corners Geological Society [US]

FCH Fachinformationszentrum Chemie GmbH [Chemical Information Center, Berlin, Germany]

FCH Flight Controllers Handbook

FCHC Fluorinated and Chlorinated Hydrocarbons

FCHL Flight Control Hydraulics Laboratory

FCI Face-Centered Icosahedral (Structure) [also fci]

FCI Fatigue Crack Initiation [Mechanics]

FCI Ferritic Cast Iron

FCI Flight Control Indicator

FCI Fluid Controls Institute [US]

FCI Flux Change(s) per Inch [also fci]

FCI Food Corporation of India

FCI Fuel-Cladding Interaction [Nuclear Engineering]

FCI Fuel Coolant Interaction

FCI Full Configuration Interaction

FCI Functional Configuration Identification

FCIA Foreign Credit Insurance Association [US]

FCIA Fur Conservation Institute of America

FCIB Foreign Credit Interchange Bureau [now FCIB/NACM, US]

FCIB/NACM Foreign Credit Interchange Bureau/National Association of Credit Management [US]

FCIC Federal Crop Insurance Corporation [US]

FCIC Fellow of the Chemical Institute of Canada

FCICA Floor Covering Installation Contractors Association [US]

FCIM Farm, Construction and Industrial Machinery

FCIM Flight Control Interface Module

FCIP Flight Cargo Implementation Plan

FCIR Fatigue Crack Initiation Resistance [Mechanics]

FCIT First Computer Interface Tester

FCL Feedback Control Loop

FCL Ferric Chloride Leach

FCL Ferromagnetic Correlation Length [Solid-State Physics]

FCL Fiber Composite Laminate

FCL Flight Crew Licensing

FCL Format Control Language

FCL Forward Calorimeter

FCL Freon Coolant Line

FCL Freon Coolant Loop

FCL Full Container Load

FCLA Fisheries Council of Latin America

Fclty Facility [also fclty]

FCM Fault Control Module

FCM Federation of Canadian Municipalities

FCM Firmware Control Memory

FCM First Class Mail

FCM Floating-Carrier Modulation

FCM Foot-Candle Meter [Unit]

FCM Fuel-Containing Mass

FCM Futures Commission Merchant [US]

f/cm^3 fibers per cubic centimeter (of air) [also f cm^{-3} or f/cc]

FCMA Fellow of the Institute of Cost and Management

FCMA Fiber Cement Manufacturers Association [UK]

FCMD Fire Command [also FCmd]

FCME First Class Marine Engineer

FCMI Federation of Coated Macadam Industries [UK]

FCMI Fuel-Cladding Mechanical Interaction [Nuclear Engineering]

FCMJ Federation of Canadian Manufacturers in Japan

fc/mm flux changes per millimeter [Unit]

FCMS Factory Control Management System

FCMU Foot Controlled Maneuvering Unit

FCMV Fuel Consuming Motor Vehicle

FCN Facilities Change Notice

FCN Field Change Notice

FCN Function [also Fcn]

FCNP Fire Control Navigation Panel

FCNSW Forestry Commission of New South Wales [Australia]

FCO Face-Centered Orthorhombic (Structure) [also fco]

FCO Field Change Order

FCO Final Checkout

FCO Firing Control Order

FCO Flight Clearance Office

FCO Foreign and Commonwealth Office [UK]

FCO Frequency-Change Oscillator

FCO Functional Checkout

fco *(franco)* – free [Commerce]

F Co Fair Copy

FCOH Flight Controllers Operational (or Operations) Handbook

4-C1N 4-Chloro-1-Naphthol

FCOS Flight Computer Operating System [of NASA Orbiter]

FCOS Flight Control Operational Software

FCOS Flight Control Operating System

FCP Failure Correction Panel

FCP Fatigue Crack Propagation

FCP Fellow of the College of Preceptors

FCP File Control Package

FCP File Control Procedure

FCP File Control Processor

FCP File Control Program

FCP Firm Cost Proposal

FCP Flat Concurrent Prolog

FCP Flight Correction Proposal

FCP Four Color Problem

FCP Franck-Condon Pumping [Physical Chemistry]

FCP Fuel Cell Power

1-FCP 1-Fluorocyclopropene

3-FCP 3-Fluorocyclopropene

fcp foolscap

FCPC Fleet Computer Programming Center

FCPP Fuel-Cell Power Plant

FCPR Fatigue Crack Propagation Rate

FCPR Fatigue Crack Propagation Resistance

FCPS Fuel-Cell Power System

FCPS Fuel-Cell Power Subsystem

FCPU Facility Central Processing Unit

FCR Facility Change Request

FCR Fast Ceramic Reactor

FCR Fast Cycle (Synthetic) Resin; Fast Cycling (Synthetic) Resin

FCR FIFO (First-In, First-Out) Control Register

FCR Final Configuration Review

FCR Fire Control Radar

FCR Flight Configuration Review

FCR Flight Control Room

FCR Floor Cavity Ratio

FCR Forwarding (Agents) Certificate of Receipt

FCR France Cables and Radio

FCR Fuse Current Rating

FCRA Fabric Care Research Association [UK]

FCRA Fecal Collection Receptacle Assembly [of NASA]

FCRDC Frederick Cancer Research and Development Center [of National Cancer Institute, Bethesda, Maryland, US]

FCRE Foundation for Cotton Research and Education [US]

FCREPA Florida Committee on Rare and Endangered Plants and Animals [US]

FCRP Fast Ceramic Reactor Program

FCRS Flight Crew Record System

FCRT Flight-Display Cathode-Ray Tube

FCS Facsimile Communications System

FCS Faculty of Chemical Sciences

FCS Farm Cooperative Service [US]

FCS Fecal Containment System

FCS Federal Communications System [US]

FCS Federation of Communication Services [UK]

FCS Fellow of the Chemical Society [UK]

FCS Fellow of the College of Sciences

FCS Fermi Chopper Spectrometer [Physics]

FCS Fiber Channel Standard

FCS Field Control Strain

FCS File Control Services

FCS Fire Control System

FCS First Critical Speed

FCS Fish Culture Station

FCS Fixed Control Storage

FCS Fixed Coordinate System

FCS Flexible Clamping System

FCS Flight Control Subsystem

FCS Flight Control System

FCS Flight Crew System

FCS Focused Collision Sequence

FCS Four-Crystal X-Ray Spectrometer

FCS Frame Check(ing) Sequence

FCS Free Crystalline Silica

FCS French Chemical Society

4CS Four Corners Section [of American Physical Society, US]

FCSC Florida and Caribbean Science Center [of US Geological Survey/Biological Resources Division at Gainesville, Florida]

FCSC FTPS Florida and Caribbean Science Center File Transfer Protocol Server [Internet]

FCSC HP Florida and Caribbean Science Center Home Page [Internet]

FCSC Foreign Claims Settlement Commission

FCSCC Fellow of the Canadian Society of Clinical Chemists

FCSE Fellow of the Canadian Society of Electroencephalographers

Fcsle Forecastle

FCSLT Fellow of the Canadian Society of Laboratory Technologists

FCSM Flight Combustion Stability Monitor

FCSPP French Committee for the Study of Paranormal Phenomena [France]

FCSR&CC Free of Capture, Seizure, Riots and Civil Commotion [also fcsr&cc]

FCSRT Fellow of the Canadian Society of Radiological Technicians

FCSS Fire Control Sight System

FCSS Fuel Cell Servicing System

FCST Federal Council for Science and Technology [US]

Fcst Forecast [also fcst]

FCT Face-Centered Tetragonal (Crystal) [also fct]

FCT Field-Controlled Thyristor

FCT Filament Center Tap [also fct]

FCT File Control Table

FCT Flight Control Team

FCT Flight Crew Trainer; Flight Crew Training

FCT Flip-Chip Transistor

FCT Flow-Controller Tester

FCT Focal-Conic Texture

FCT Fuel Cell Test

FCT Function [also Fct or fct]

FCTB Fibro-Caseous Tuberculosis [Medicine]

FCTB Flight Crew Training Building

FCTC Fuel Centerline Thermocouple

FCTF Fuel Cell Test Facility

Fctn Function [also fctn]

Fctry Factory [also fctry]

FCTS Flight Crew Trainer Simulator

FCTT Fuel Cladding Transient Tester

Fcty Factory [also fcty]

FCU File Control Unit

FCU Flight Control Unit

FCU Fluid Checkout Unit

FCU Fuel Conditioning Unit

FCU Fu-jen Catholic University [Taipei, Taiwan]

FCUG Fuel Cell User Group [US]

FCV Feline Calicivirus

FCV Free-Column Volume

FCVI Forced Chemical Vapor Infiltration (Technique)

FCVS FORTRAN Compiler Validation System

FCW Field Cooling and Warming [Vibrating Sample Magnetometer]

FCW Format Control Word

FCWA Fellow of the Institute of Cost and Works Accountants

FCWG Frequency Coordination Working Group [US]

fcy fancy
FD Family Doctor
FD Fault Detection; Fault Detector
FD Fault Diagnosis
FD F(isher's) Distribution [Statistics]
FD Federal District
FD Fermi-Dirac (Statistics) [Statistical Mechanics]
FD Fermi Distribution [Solid-State Physics]
FD Feynman Diagram [Quantum Mechanics]
FD Fiber Detector
FD Fiber Duct
FD Field-Decelerating; Field Deceleration
FD Field Depot
FD Field Desorption [Solid-State Physics]
FD Field Diaphragm
FD Field Discharge
FD Field Dose
FD Field-Time Waveform Distortion [Television]
FD Fiji Disease [Medicine]
FD File Definition
FD File Description
FD File Directory
FD Film Deposition
FD Finished Dialing
FD Finite-Deformation (Approach of Linear Elasticity)
FD Finite Difference(s)
FD Fire Damper
FD Fire Department
FD Firing Data
FD Fixed Distance
FD Fixture Drive
FD Flange Local Distance
FD Flash Desorption
FD Flash Distillation
FD Flavo(u)r Dilution
FD Flexible Disk
FD Flight Data
FD Flight Day
FD Flight Deck
FD Flight Director
FD Flight Dispatcher
FD Flight Dynamics
FD Floor Drain
FD Floppy Disk [also fd]
FD Floppy Drive
FD Flotation Device
FD Flow Diagram
FD Fluctuation Dissipation (Theory)
FD Fluid Drive
FD Fluid Dynamics
FD Flux Delta
FD Flux Density
FD Focal Distance

FD Forced Diffusion
FD Forced Draft
FD Forging Direction
FD Forward
FD Fractional Distillation
FD Free Discharge [Commerce]
FD Freeze Dried; Freeze Dryer; Freeze Drying
FD Frenkel Defect [Solid-State Physics]
FD Frenkel Disorder [Solid-State Physics]
FD Frequency Demodulator
FD Frequency Detection
FD Frequency Discriminator
FD Frequency Distance
FD Frequency Distribution
FD Frequency Diversity
FD Frequency Divider; Frequency Division
FD Frequency Domain
FD Frequency Doubler
FD Friction Drive
FD Fuel Demand
FD Full Duplex [also FDX, fd or fdx]
FD Fully Dense (Material)
FD Functional Derivative [Chemistry]
FD Functional Description
FD Functional Design
FD Functional Diagram
FD Function Designator
FD Furniture Design(er)
FD Fuse Delay
F$ Fijian Dollar [Currency]
4D Four-Dimensional [also 4-D]
5D Five-Dimensional [also 5-D]
F&D Facilities and Design
F&D Findings and Determination
F&D Freight and Demurrage
F/D Fill/Drain
F/D Focal Length to Diameter Ratio
Fd Feed [also fd]
Fd Fjord [also fd]
Fd Food [also fd]
Fd Forward [also fd]
Fd Fund [also fd]
Fd Ferredoxin [Biochemistry]
F-d Force-Penetration Depth (Curve) [Materials Testing]
fd farad [Unit]
fd free discharge [Commerce]
fd size factor [Component of the Continuity Index of Intermetallics]
$f_n(D)$ normal (drop size) distribution (i.e. number fraction of liquid drops of diameter less than D) [Symbol]
$f_v(D)$ Rosin-Rammler (drop size) distribution (i.e. volume fraction of liquid in drops of diameter less than maximum diameter) [Symbol]
FDA Fault Detection and Annunciation

FDA Feynman-Dyson Amplitude [Physics]

FDA Final Design Approval

FDA Finite Difference Analysis

FDA Finite Difference Approximation

FDA Flight Deck Assembly

FDA Fluorescein Diacetate

FDA Food and Drug Act [Canada]

FDA Food and Drug Administration [of US Department of Health and Human Services]

FDA Frequency-Domain Analysis

FDAA Federal Disaster Assistance Administration [Now Part of Federal Emergency Management Agency, US]

FDAA Fluorenyldiacetamide

FDAC Florida Department of Agriculture (and Consumer Services) [US]

FDA Drug Bull Food and Drug Administration's Drug Bulletin [US]

FDAI Flight Director Attitude Indicator

FDAR Federal Department of Agricultural Research [Nigeria]

FDAS Frequency Distribution Analysis Sheet

FDAU Flight Data Acquisition Unit

FDA/USDA Food and Drug Administration/United States Department of Agriculture [US]

FDB Fahrenheit Dry Bulb

FDB Field Descriptor Block

FDB Field Dynamic Braking

FDB File Data Block

FDB Forced Draft Blower

FDB Forestry Data Bank [of Forestry Canada]

FDB Functional Description Block

FDBR Fachverband Dampfkessel-, Behälter- und Rohrleitungsbau e.V. [Boiler, Pressure Vessel and Pipeline Construction Association, Germany]

FDC Facsimile Data Converter

FDC Film-Development Chromatography

FDC Film Development Corporation [Canada]

FDC Fire Department Connection

FDC Fire Direction Center

FDC Firing Data Computer

FDC Flight Director Computer

FDC Floppy-Disk Controller [also fdc]

FDC Florida Department of Citrus [US]

FDC Flow Duration Curve

FDC Fluid Die Compaction

FDC Fluid Drive Coupling

FDC Follicular Dendritic Cell

FDC Formation Density Content (Log)

FDC Frazer-Duncan-Collar [Matrix Algebra]

FDC Full-Duplex Circuit

FDC Functional Design Criteria

FDC Furniture Development Council [now Furniture Industry Research Association, UK]

FD&C Foods, Drugs, and Cosmetics Color [of US Food and Drug Administration]

FD&CA Food, Drug and Cosmetic Act [US] [also FDCA]

FDCD Fluorescence Detected Circular Dicroism

FDCM Fluorodichloromethane

FDCS Functionally Distributed Computing System

FDD Flexible Disk Drive

FDD Flight Definition Document

FDD Floppy Disk Drive

FDD Formatted Data Disk

FDD Frequency Division Duplex [Telecommunications]

FDDA Four Dimensional Data Assimilation

FDDB Function Designator Data Base

FDDI Fiber Digital Device Interface; Fiber Digital Data Interface; Fiber Distributed Data Interface

FDD/IDE Floppy Disk Drive/Interactive Data Entry (Interface)

FDDL Field Data Description Language

FDDL Frequency Division Data Link

FDDS Flight Data Distribution System

FDE Field Deceleration; Field Decelerator

FDEM Fuel Demand Evaluation Model

FDEMS Frequency-Dependent Electromagnetic Sensor

FDEP Flight Data Entry Panel

FDEP Florida Department of Environmental Regulation [US]

FDEP Formatted Data Entry Program

FDES Framework for Development of Environmental Statistics [of United Nations Statistical Office]

FDF Fermi Distribution Function [Solid-State Physics]

FDF Fiber Distribution Frame

FDF Flight Data File

FDF Forced-Draft Fan

F2F Frequency Double Frequency

FD/FF Flux Delta/Flux Flow

FDFM Flight Data and Flow Management Group [of International Civil Aviation Organization]

FDFM Frequency Division/Frequency Modulation

FDFR Federal Department of Forestry Research [Nigeria]

FDG Four-Dimensional Geodesy

FDG Fractional Doppler Gate

Fdg Fading [also fdg]

Fdg Funding [also fdg]

FDGB Freier Deutscher Gewerkschaftsbund [Free German Trade Union, East Germany]

FDH Fluor Daniel Hanford, Inc. [US]

FDHD Floppy Drive High Density

FDI Failure Detector Indicator

FDI Fault Detection and Identification

FDI Fault Detection and Isolation

FDI Fédération Dentaire Internationale [International Dental Federation]

FDI Field Discharge

FDI Flight Detector Indicator

FDI Foreign Direct Investment

FDI Frequency Domain Instrument

FDIC Federal Deposit Insurance Corporation [US]

FDIIR Fault Detection, Isolation, Identification and Recompensation

FDIIR Fault Detection, Isolation, Identification and Recovery/Recognition

FDIM Fluorescence Digital Imaging Microscope; Fluorescence Digital Imaging Microscopy

FDIR Fault Detection, Identification/Isolation and Recovery/Recognition

FDISK Fixed Disk [also Fdisk or fdisk]

FDISK Format (Hard) Disk [also fdisk]

FDL Fast Deployment Logistics

FDL Fault Detection and Localization

FDL Flight Director Loop

FDL Flight Dynamics Laboratory [of US Air Force]

FDLC Fluorinated Diamond Like Carbon

FDLI Food and Drug Law Institute [US]

FDLN Feedline [also Fdln]

FDLS Fast Deployment Logistics Ship [US Navy]

FDM Field-Desorption Microscope; Field-Desorption Microscopy

FDM Finite-Difference Method

FDM Flight Data Manager,

FDM Form Description Macro

FDM Frequency Data Multiplexer

FDM Frequency-Division Multiplex(ing); Frequency-Division Multiplexer [also fdm]

FDM Fused Deposition Method

FDM Fused Deposition Modeling (Process)

FDM Fused Deposition Molding

FDMA Frequency-Division Multiple Access [Telecommunications]

FDMA Frequency-Domain Multiple Access

FDMA Perfluoro-N,N-Dimethyl Cyclohexyl Methylamine

FDM/FM Frequency-Division-Multiplexed/Frequency-Modulated

FDMR Fluorescence Detected Magnetic Resonance

FDMS Factory Data-Management System

FDMS Federation of Deer Management Societies [UK]

FDMS Field Desorption Mass Spectrometry; Field-Desorption Mass Spectroscopy [also FD/MS]

FDMS Flash-Desorption Mass Spectrometry

FDMS Floppy Disk Management System

Fdn Foundation [also fdn]

FDNB 1-Fluoro-2,4-Dinitrobenzene

FDNC Frequency-Dependent Negative Conductance; Frequency-Dependent Negative Conductor

FDNR Florida Department of Natural Resources [US]

FDNR Frequency-Dependent Negative Resistance; Frequency-Dependent Neghative Resistor

FDNW Fluor Daniel Northwest [Hanford, Washington, US]

FDO Flight Dynamics Officer

FDO Fee Determination Official

FDO 6,6,7,7,8,8,8-Heptafluoro-2,2-Dimethyl-3,5-Octanedione [also fdo]

4DOF Four Degrees Of Freedom

5DOF Five Degrees Of Freedom

FDOR Flight Design Operations Review

FDOS Floppy Disk Operating System

FDOT Florida Department of Transportation [US]

FDP Fast Delivery Processor

FDP Fast Digital Processor

FDP Fibrin(ogen) Degradation Product

FDP Field-Developed Program

FDP Filter Drainage Protection

FDP Flight Data Processing

FDP Fluid Dynamics Panel [of NATO Advisory Group for Aerospace Research and Development]

FDP Flying Duty Period

FDP Form Description Program

FDP Forward Direction Post

FDP Frequency-Dependent Polarizability

FDP Fructose Diphosphate [Biochemistry]

FDP Full-Duplex Protocol

FDP Future Data Processor

FDPC Federal Data Processing Center(s) [US]

FDPS Flight Data Processing System

FDR Facility Development Research

FDR Federation of Drum Reconditioners [UK]

FDR File Data Register

FDR Final Design Review

FDR Flight Data Recorder

FDR Flood Damage Reduction (Program)

FDR Frequency Dependent Rejection

FDR Frequency Diversity Radar

FDR Frequency-Domain Reflectometer; Frequency-Domain Reflectometry

FDR Functional Design Requirements

FDR Functional Design Review

Fdr Feeder [also fdr]

F Dr Fire Door

f dr fluid drachm [Unit]

FDRAKE First Dynamic Response and Kinematics Experiment

FDRI Flight Director Rate Indicator

FDRI Fluid Dynamics Research Institute [of University of Windsor, Ontario, Canada]

F-Drive Front Drive [also F-drive]

FDRS Food Distribution Research Society [US]

Fdry Foundry [also fdry]

FDS Fallout Decay Simulation

FDS Fermi-Dirac-Sommerfeld (Law) [Physics]

FDS Fermi-Dirac Statistics [Physics]

FDS Field Desorption Spectroscopy

FDS Field-Discharge Switch

FDS Filter Difference Spectrometer

FDS Fixed-Disk Storage

FDS Flash Desorption Spectrum

FDS Flexible Disk System

FDS Flight Design and Scheduling

FDS Flight Design System
FDS Flight Dynamics Simulator
FDS Flight Dynamics Software
FDS Flight Dynamics System
FDS Floppy Disk System
FDS Fluid Density Sensor
FDS Fluid Distribution System
FDS Foreign Agriculture Service [of US Department of Agriculture]
FDS Frame Difference Signal
FDS Frequency Division Separator
FDS Functional Design Specifications
FDSC Flight Dynamics Simulation Complex
FDSC Flight Dynamics Situation Complex
FDSE Fractional Debye-Stokes-Einstein (Law)
FDSR Floppy Disk Send Receive
FDSS Fine Digital Sun Sensor
FDSSA Fine Digital Sun Sensor Assembly
FDSU Flight Data Storage Unit
FDT Fetal Diagnosis and Therapy [International Journal]
FDT Flowing-Gas Detonation Tube
FDT Fluctuation-Dissipation Theorem
FDT Formatted Data Tape
FDT Full Duplex Teletype
FDT Functional Description Table
FDT Fundación para el Desarollo Tecnológico [Technological Development Foundation, Argentina]
FDTA Fisheries Development Trust Account [Australia]
FDTC Fiber Drum Technical Council [US]
FDTD Finite-Difference Time-Domain (Method) [also FD-TD]
FDU Fairleigh Dickinson University [Rutherford, New Jersey, US]
FDU Flexible Disk Unit
FDU Floppy Disk Unit
FDU Fluid Distribution Unit
FDU Formatter and Drive Unit
FDU Form Description Utility
FDU Frequency Divider Unit [also fdu]
F-dU 5-Fluorodeoxyuridine [also F-dU] [Biochemistry]
FDV Fault Detect Verification
FDV Fiji Disease Virus
FDVR Federal Department of Veterinary Research [Nigeria]
Fd Wmr Food Warmer
FDX Full Duplex [also fdx, FDX, FD or fd]
Fdy Foundry [also fdy]
FE Faculty of Education
FE Faculty of Engineering
FE Faraday Effect
FE Far East
FE Feature Extraction
FE Feed Enrichment
FE Fermi Edge [Physics]
FE Fermi Energy [Physics]
FE Ferrielectric(ity)

FE Ferroelastic(ity)
FE Ferroelectret
FE Ferroelectric(ity)
FE Field Effect
FE Field Electron
FE Field Emission
FE Field Engineer
FE Field Equation
FE Finite Element
FE Fire Engineer(ing)
FE First Edition
FE Fixed End (Moment)
FE Flash Evaporation
FE Fleet Engineer
FE Fleet Equipment [Automobiles]
FE Flight Engineer
FE Fluid Energy
FE Fluidization Engineer(ing)
FE Fluid Engineer(ing)
FE Fluoroelastomer
FE Fluoroethylene
FE Focusing Electrode
FE Font Error
FE Forest Engineer(ing)
FE Format Effector (Character) [Data Communications]
FE Forum on Education [of American Physical Society, US]
FE Fossil Energy
FE Foundation Engineer(ing)
FE Foundry Engineer(ing)
FE Framing Error [Computers]
FE Free Electron
FE Free End
FE Free Energy
FE Free Exciton [Solid-State Physics]
FE Frenkel Exciton [Solid-State Physics]
FE Freshwater Ecology
FE Friends of the Everglades [US]
FE Front End
FE Fuel Element [Nuclear Engineering]
FE Fugative Emission
FE Fuller's Earth
FE Functional Entity
FE Functional Equation
FE Functional Expansion
FE Fundamental Equation
FE Fundamentals-of-Engineering Exam(ination)
FE Further Education
FE Fuse Element
FE Fusion Energy
FE Office of Fossil Energy [of US Department of Energy]
F&E Facility and Environment
F/E Full/Empty
Fe Ferrite

Fe *(ferrum)* – Iron [Symbol]

Fe^{2+} Divalent Iron (or Ferrous Iron) Ion [also Fe^{++}] [Symbol]

Fe^{3+} Trivalent Iron (or Ferric Iron) Ion [also Fe^{+++}] [Symbol]

Fe$_8$ Cluster of 8 Iron Ions

Fe-52 Iron-52 [also ^{52}Fe, or Fe^{52}]

Fe-53 Iron-53 [also ^{53}Fe, or Fe^{53}]

Fe-54 Iron-54 [also ^{54}Fe, or Fe^{54}]

Fe-55 Iron-55 [also ^{55}Fe, or Fe^{55}]

Fe-56 Iron-56 [also ^{56}Fe, Fe^{56}, or Fe]

Fe-57 Iron-57 [also ^{57}Fe, or Fe^{57}]

Fe-58 Iron-58 [also ^{58}Fe, or Fe^{58}]

Fe-59 Iron-59 [also ^{59}Fe, or Fe^{59}]

Fe II Iron II [Symbol]

F(ε) Normalized Lorentzian Function [Symbol]

f(E) Fermi Function (or Fermi Distribution) [also F(E) or f(E)]

FEA Fast Ethernet Alliance

FEA Failure Effect(s) Analysis

FEA Farmstead Equipment Association [of Farm and Industrial Equipment Institute, US]

FEA Federal Energy Administration [Now Part of US Department of Energy]

FEA Fédération Européenne des Association Aérosols [European Federation of Aerosol Associations, Belgium]

FEA Finite Element Analysis

FEA Florida Education Association [US]

FEA Fluctuation Exchange Approximation

FEA Fluids Experiment Apparatus

FEA Foreign Economic Administration

FEA French Equatorial Africa

FEA Front-End Alignment

FeAA Ferric Acetylacetonate

FEA Aerosol Bull FEA Aerosol Bulletin [Published by Fédération Européenne des Association Aérosols, Belgium]

FEAC Fusion Energy Advisory Committee [of US Department of Energy]

Fe(Ac)$_2$ Iron(II) Acetate [also Fe(ac)$_2$]

Fe(Ac)$_3$ Iron(III) Acetate [also Fe(ac)$_3$]

Fe(ACAC)$_3$ Iron(III) Acetylacetonate [also Fe(AcAc)$_3$, or Fe(acac)$_3$]

FEA/CAD Finite-Element Analysis/Computer-Aided Design

FEACO Fédération Européenne des Associations de Conseils en Organisation [European Federation of Management Consultant Associations, France]

FEAD Front End Accessory Belt-Drive [Automobiles]

FEA/FDA Finite-Element Analysis/Finite Difference Approximation

FEAES Field-Emission Auger Electron Spectroscopy [also FE-AES or FE AES]

FEAG Filter Expansion Aerosol Method

Fe/Ag Iron on Silver (Substrate)

Fe-Ag Iron-Aluminum (Alloy System)

Fe-Al Iron-Aluminum (Alloy System)

FeAl Iron Aluminide

FeAl(B) Boron-Doped Iron Aluminide

FEALC Federación Espeleológica de America Latina y el Caribe [Speleological Federation of Latin America and the Caribbean, Venezuela]

FeAl(C) Carbon-Doped Iron Aluminide

FeAl(Mo) Molybdenum-Doped Iron Aluminide

FeAlSi Iron Aluminum Silicon (Alloy)

FeAl(Zr) Zirconium-Doped Iron Aluminide

FEANI Fédération Européenne de l'Associations Nationales des Ingénieurs [European Federation of National Associations of Engineers]

FEAO Federation of European-American Organizations

FEARO Federal Environmental Assessment Review Office [Canada]

FeAs Iron Arsenide

FEAST Further Exploitation of Advanced Shell Technology (Process)

FEAT Final Engineering Acceptance Test

FEAT Frequency of Every Allowable Term

feath feathery

FEA TOOL Fault Effects Analysis Tool (Development)

Fe/Au Iron on Gold (Substrate)

FE Auger Field-Emission Auger (Technique)

FEB Forbidden Energy Bands

FEB Forward Equipment Bay

FEB Functional Electronic Block

FeB Iron Boride

FeB Iron-Boron (Alloy) (or Ferroboron)

Feb February

FEBA Federal Energy Bar Association [US]

FEBA Forward Edge of Battle Area

FEBAB Federação Brasiliera de Associacoes de Bibliotecarios [Brazilian Federation of Library Association]

FEBMA Federation of European Bearing Manufacturer Associations [Germany]

FeBN Iron Boronitride

FEBS Federation of European Biochemical Societies [France]

FEBSi Iron Boron Silicon (Alloy)

FEBS Lett FEBS Letters [of Federation of European Biochemical Societies, France]

FEC Faculty Exchange Center [US]

FEC Farm Electrification Council [US]

FEC Federal Election Commission

FEC Fédération Européenne de la Corrosion [European Federation of Corrosion]

FEC Field-Effect Capacitor

FEC Field Engineering Change

FEC Floating Error Code

FEC Food and Energy Council [now National Food and Energy Council, US]

FEC Forward Error Control

FEC Forward Error Correction

FEC Forward Events Controller

FEC (The) Foundation for Environmental Conservation

FEC Front-End Computer

FEC Fullerene Electrochromism

fec *(fecit)* – he (or she) made it

Fe-C Iron-Carbon (Structure)

Fe₃C Iron Carbide (or Cementite)

FECC Fédération Européene du Commerce Chimique [European Federation of Chemical Trade, Belgium]

FECC Inf Bull FECC Information Bulletin [Fédération Européene du Commerce Chimique, Belgium]

FECD Free Energy versus Composition Diagram

FECES Forward Error-Control Electronics System

FECHEM Federation of European Chemical Societies

FECL Feedback Emitter-Coupled Logic

FeCl₃GBC Iron(III) Chloride Graphite Bi-Intercalation Compound

FeCl₃GIC Iron(III) Chloride Graphite Intercalation Compound

FECN Forward Explicit Congestion Notification

FECO Fringes of Equal Chromatic Order

FeCo Iron Cobalt (Alloy)

Fe-Co Iron-Cobalt (Alloy System)

Fe-Co-Ln-B Iron-Cobalt-Lanthanide-Boron (Alloy)

FeCoPt Iron Cobalt Platinum (Alloy)

FeCoSiB Iron Cobalt Silicon Boron (Alloy)

FeCoZr Iron Cobalt Zirconium (Alloy)

FECP Front End Communications Processor

Fe(Cp)₂ Dicyclopentadienyliron (or Ferrocene)

Fe-Cr Iron-Chromium (Alloy System)

FeCrAl Iron-Chromium-Aluminum (Coating)

FeCrAlY Iron-Chromium-Aluminum-Yttrium (Coating)

FECS Fédération Européenne des Fabricants de Ceramiques Sanitaires [European Federation of Ceramic Sanitary Equipment Manufacturers]

FECS Federation of European Chemical Societies [UK]

FECS Foreign Exchange Counselling System [of European American Bank]

Fe-Cu Iron-Copper (Alloy System)

FeCuNbSiB Iron-Copper-Niobium-Silicon-Boron (Alloy)

FECT Federation of European Chemical Trade [Belgium]

Fe-C-X Iron-Carbon Ternary Alloy (e.g., Fe-C-Ni, Fe-C-V, etc.) [Metallurgy]

FED Faculty of Engineering and Design

FED Ferroelectric Domain [Solid-State Physics]

FED Field-Effect Diode

FED Field-Emission Deposition

FED Field-Emission Display; Field Emitter Display

FED Flight Events Demonstration

FED Freeze Etch(ing) Device

FED Fuel Examination Facility

FED Fusion Engineering Device

Fed Federal; Federated; Federation [also fed]

FEDA Farm Equipment Dealers Association

FEDA Food Service Equipment Distributors Association [US]

FEDAL Failed Element Detection and Location Instrument

FEDC Federal Economic Development Coordinator

FEDC Federation of Engineering Design Consultants [UK]

FEDC Fusion Engineering Design Center [of Oak Ridge National Laboratory, Tennessee, US]

FEDD For Early Domestic Dissemination [NASA Term]

FEDEMOA Federacion Mexicana de Organisaciones Agricoles [Mexican Federation of Agricultural Organizations]

FEDEREL Fédération des Relamineurs du Fer et de l'Acier [Federation of Iron and Steel Relaminators] [of European Economic Community, Belgium]

FEDES Fédération Européenne de l'Emballage Souple [European Federation for the Flexible Packaging Manufacturers, France]

FEDEX Federal Index [US]

FedEx Federal Express [also FEDEX, US]

FEDIOL Fédération de l'Industrie de l'Huilerie [Federation of Seed Crushers and Oil Processors] [of European Economic Community, Belgium]

FEDIT Font Editor (Program)

fedl federal [also fed'l]

FEDLINK Federal Library and Information Network [US]

FEDLTS Field-Effect Deep-Level Transient Spectroscopy

Fe-DMS Iron-Diluted Magnetic Semiconductor

Fedn Federation [also fedn]

FEDNET Federal Network [US]

FEDP Facility and Equipment Design Plan

FEDP Federal Executive Development Program [US]

FEDREG Federal Register [US]

FEDS Federal Energy Decision System [US]

FEDS Fixed and Exchangeable Disk Storage

Fed Spec Federal Specification

Fed Std Federal Standard

FEDSTRIP Federal Standard Requisitioning and Issue Procedure

Fed Wire Federal Reserve Board Wire [Electronic Clearing and Communication System of the US Federal Reserve Board]

FEE Field Electron Energy Spectroscopy

FEE Frog Embryology Experiment [of NASA]

FEED Field Electron Energy Distribution

FEED Field Emission Energy Distribution

FEED Floating Electrode Effect Development

FEEDS Field Emission Energy Distribution Spectroscopy

Fe EDTA Iron (III) Ethylenediaminetetraacetate

FEEF Federal Energy Efficiency Fund [US]

FEEM Field Electron Emission Microscopy

FEEM Fondazione ENI (Ente Nazionale Idrocarburi) Enrico Mattei [Enrico Mattei Foundation, Italy]

FEEP Field Emission Electric Propulsion System

FEES Field Electron Energy Spectroscopy

FEESP Field-Emitted Electron Spin Polarization

Fe(EtO)₃ Iron III Ethoxide

FEF Fast Extruding Furnace (Black)

FEF Field Emission Fluctuation

FEF Foundry Educational Foundation [US]

FEF Friends of the Earth Foundation [US]

FEF Fusion Energy Foundation [New York City, US]

FEFANA Fédération Européene des Fabricants d'Adjuvants pour la Nutrition Animale [European Federation of Manufacturers of Feed Additives, Germany]

FEF Black Fast-Extruding Furnace Black [also FEF black]

FEFC Far Eastern Freight Conference

FEFCEB Fédération Européenne des Fabricants de Caisses et Emballages en Bois [European Federation of Wooden Boxes and Packaging Materials]

FEFCO Fédération Européenne des Fabricants de Carton Ondule [European Federation of Corrugated Board Manufacturers, France]

Fe-FeO Iron-Ferrous Oxide (Cell)

FEFF Full Multiple Scattering [Extended X-Ray Absorption Fine Structure]

FEFF-EXAFS Full Multiple Scattering Extended X-Ray Absorption Fine Structure (Simulation)

FEFG International Conference on Fracture of Engineering Materials and Structures

FEFG/ICF International Joint Conference on Fracture of Engineering Materials and Structures/International Congress on Fracture

FEFN Field-Emission Flicker Noise

FEFO First-Ended, First-Out [also fefo]

FEFP Fuel Element Failure Propagation [Nuclear Engineering]

FEFPEP Fédération Européenne des Fabricants de Palettes et Emballages en Bois [European Association for the Flexible Packaging Industry, France]

FEFPL Fuel Element Failure Propagation Loop [Nuclear Engineering]

FEG Ferroelectric Generator

FEG Field Emission Gun

FEG AEM Field Emission Gun Analytical Electron Microscope; Field Emission Gun Analytical Electron Microscopy [also FEG-AEM or FEG/AEM]

FeGe Iron Germanide (Semiconductor)

FEG SEM Field Emission Gun Scanning-Electron Microscope; Field Emission Gun Scanning Electron Microscopy [also FEG-SEM or FEM/SEM]

FEG STEM Field Emission Gun Scanning Transmission Electron Microscope; Field Emission Gun Scanning Transmission Electron Microscopy [also FEG-STEM, or FEM/STEM]

FEG TEM Field Emission Gun Transmission Electron Microscope ; Field Emission Gun Transmission Electron Microscopy [also FEG-TEM or FEM/TEM]

Fe-Hf Iron-Hafnium (Alloy System)

Fe-His Iron Histidine [Biochemistry]

FEHM Finite Element Heat and Mass (Transfer)

FEHVA Federation of European Heating and Ventilating Associations [Netherlands]

FEI Field Electrons and Ions

FEI Field Engineering Instructions

FEI Financial Executive Institute [US]

FEI France-Europe International [France]

FEI Fundação de Ciencias Aplicadas/Escola de Engenharia Industrial [Foundation of Applied Sciences/School of Industrial Engineering, Sao Bernardo do Campo, Brazil]

FEIA Flight Engineers International Association

FEIC Fellow of the Engineering Institute of Canada

FEICA Fédération Européenne des Industries de Colles Adhesifs [Association of European Adhesives Manufacturers, Germany]

FEICRO Federation of European Industrial Cooperative Research Organizations [UK]

FEID Flight Equipment Interface Device

FEID Functional Engineering Interface Device

FEIEA Federation of European Industrial Editors Associations

FEIM Fédération Européenne des Importateurs de Machines et d'Equipements de Bureau [European Federation of Importers of Business Equipment, Belgium]

Feinwerktech Messtech Feinwerktechnik und Messtechnik [German Journal on Precision Mechanics and Measurement Technology]

FEIS Final Environmental Impact Statement

FEJB Forum of Environmental Journalists of Bangladesh

FEK Frequency Exchange Keying

FeKα Iron K-Alpha (Radiation) [also FeK$_\alpha$]

Fe/Kr Iron/Krypton (Interface)

FEL Fly's Eye Lens

FEL Free-Electron Laser

FeLα Iron L-Alpha (Radiation) [also FeL$_\alpha$]

Fe/La-214 Iron on Lanthanum Barium Copper Oxide Superconductor ($La_2BaCu_4O_5$)

FELACUTI Federación Latinoamericana de Usuarios del Transporte [Latin American Federation of Shippers Council, Colombia]

FeLaSoFi Federación Latinoamericana de Sociedades de Fisica [Latin American Federation of Physical Societies]

FELIPE Free-Electron Laser Internal Photoemission [also FEL-IPE]

FELIX Forward ELastic and Inelastic EXperiment [at the Large-Hadron Collider of CERN–European Laboratory for Particle Physics, Geneva, Switzerland]

Fe-Ln-B Iron-Lanthanide-Boron (Alloy)

Fe-Ln-Zr-B Iron-Lanthanide-Zirconium-Boron (Alloy)

FELP Free-Electron Laser Physics

FELP Free-Energy Laser Program [of University of Utah, Salt Lake City, US]

FeLV Feline Leukemia Virus

FEM Fédération Européenne de la Manutention [European Federation of Handling Industries, Switzerland]

FEM Ferroelectric Material

FEM Field-Effect Modified (Transistor)

FEM Field-Emission Microscope; Field-Emission Microscopy

FEM Final Effluent Monitor

FEM Financial Evaluation Program

FEM Finite Element Mesh

FEM Finite-Element Method

FEM Finite-Element Model(ing)

FEM Firmware Expansion Module

FEM Fluid-Energy Mill

FEM FORTRAN Enhancement Package

FEM Free Energy Model

FEM Fundação de Estudos do Mar [Ocean Studies Foundation, Brazil]

fem female

fem feminine

fem femoral [Medicine]

FEMA Farm Equipment Manufacturers Association [US]

FEMA Federal Emergency Management Agency [Washington, DC, US]

FEMA Fire Equipment Manufacturers Association [US]

FEMA Flavor and Extract Manufacturers Association [US]

FEMA Food Equipment Manufacturers Association [US]

FEMA Foundry Equipment and Materials Association

FEMA Foundry Equipment Manufacturers Association [now Casting Industry Suppliers Association, US]

FEMAC Fédération Européene des Fabricants d'Aliments Composés [European Federation of Compound Animal Feeding Stuff Manufacturers, Belgium]

FEMAP Finite-Element Mold-Filling Analysis (Computer Software Program)

FEMAR Fundação de Estudios do Mar [Ocean Studies Foundation, Brazil]

FEMCPL Facilities and Environment Measurement Compatibility Parts List [of NASA]

FEMD Fundación para el Estudio de los Materiales Dentales [Foundation for the Study of Dental Materials, Argentina]

Fe/Me Iron/Metal (Tape)

FEMF Floating Electronic Maintenance Facility [of Los Alamos Scientific Laboratory, New Mexico, US]

FEMF Foreign Electromotive Force

FEM/FEA Finite Element Method/Finite Element Analysis

FEM/FEA Finite-Element Modelling/Finite-Element Analysis

FEMFET Field-Effect Modified Field-Effect Transistor

FEMFM Federation of European Manufacturers of Friction Materials [France]

FEM GEN Finite-Element Method Generator

FEMIB Fédération Européenne de Synicats des Fabricants de Menuiseries de Bâtiment [European Federation of Building Joinery Manufacturers, Germany]

FEMIPI Fédération Européenne des Mandataires de l'Industrie en Propriété Industrielle [European Federation of Agents of Industrial Property, Germany]

FEMIS Federal Emergency Management Information System [US]

FEMM Free Energy Miniaturization Method

FeMn Iron Manganese (Alloy)

FeMnN$_2$ Iron Manganese Nitride

FEMO Free-Electron Molecular Orbital

Fe-Mo Iron-Molybdenum (Alloy)

FEMP Facility Effluent Monitoring Plan

FEMP Federal Ecological Monitoring Program [Canada]

FEMP Federal Energy Management Program [US]

FEMP Fernald Environmental Management Project [US]

FEMP Fusion Engineering Materials Program [of Ontario Hydro, Canada]

FEMS Federation of European Materials Societies [also fems]

FEMS Federation of European Microbiological Societies [UK]

FEMS/ASM Federation of European Materials Societies/ ASM (American Society of Materials) International (Joint Conference)

FEMS Circ FEMS Circular [of Federation of the European Microbiological Societies, UK]

FEMS Ecol FEMS Ecology [of Federation of the European Microbiological Societies, UK]

FEMS Microbiol FEMS Microbiology [of Federation of the European Microbiological Societies, UK]

FEMS Microbiol Lett FEMS Microbiology Letters [of Federation of the European Microbiological Societies, UK]

FEMS News FEMS News [Newsletter of the Federation of European Materials Societies; issued with *Euromaterials*]

femto- SI prefix representing 10^{-15}

FEMUR Fiber Extension Matrix Undercut Removal (Technique) [Materials Science]

FEMW Field Engineering and Mine Warfare [UK]

FEN Fédération de l'Education Nationale [National Education Federation, France]

FEN Frequency-Emphasizing Network

FeN Iron Nitride

Fe-N Iron-Nitrogen (Structure)

Fenac 2,3,6-Trichlorophenylacetic Acid [also fenac]

Fe-Nb Iron-Niobium (Alloy System)

Fenchol Fenchyl Alcohol [also fenchol]

Fe-Ni Iron-Nickel (Alloy System)

FENIX Fusion Engineering International Experimental Magnet Facility

Fenuron 3-Phenyl-1,1-Dimethylurea [also fenuron]

FENCO Foundation of Engineering Corporations [Canada]

FENDT Far East Conference on Nondestructive Testing

FEng Fellow of Engineering

Fe-57 NMR Iron-57 Nuclear Magnetic Resonance [also ^{57}Fe NMR]

FeNO Ferronitrate

Fenxi Huaxue (Anal Chem) Fenxi Huaxue (Analytical Chemistry) [Published in PR China]

Fenxi Shiyanshi (Anal Lab) Fenxi Shiyanshi (Analytical Laboratory) [Published in PR China]

FEO Facility Emergency Organization

FEO Federal Energy Office [US]

FEO Field Engineering Order

FeO Iron(II) Oxide (or Ferrous Oxide)

FeO$_x$ Iron Oxides (e.g., FeO, Fe$_2$O$_3$ or Fe$_3$O$_4$) [also FeOx]

FEOE Full Equalization of Orbital Electronegativity

FeOPr^i Iron(III) Isopropoxide

FEOTC Federal Exporters Overseas Transport Committee [Australia]

FEOV Force End of Volume

FeOx Iron Oxides (e.g., FeO, Fe$_2$O$_3$ or Fe$_3$O$_4$) [also FeO$_x$]

FEP Features, Events and Processes

FEP Field Emission Process

FEP Financial Evaluation Program

FEP Fluid Entrainment Pump

FEP Fluorinated Ethylene Propylene (Resin); Fluorinated Perfluoroethylene-Propylene

FEP Front-End Processor

FEP Foundation for Education with Production [Botswana]

FEP Fuse Enclosure Package

FEPA Fédération Européenne des Fabricants de Produits Abrasifs [European Federation of Abrasives Manufacturers, France] [also Fepa]

FEPACE Fédération Européenne des Producteurs Autonomes et des Consommateurs Industriels d'Energie [European Federation of Autoproducers and Industrial Consumers of Energy, Belgium]

FEPC Fair Employment Practices Committee [US]

FePc Iron Phthalocyanine

FEPCA Federal Environmental Pesticide Control Act [US]

FEP-CORDE Foundation for Education with Production–Cooperative Research, Development and Education [Botswana]

FEPD Fédération Européenne des Parfumeurs Detaillants [European Federation of Perfumery Retailers, France]

FePd Iron Palladium (Magnet)

FEPE Fédération Européenne de la Publicité Extérieure [European Federation of Outdoor Advertising, Belgium]

FEPE Fédération Européenne des Producteurs d'Enveloppes [European Association of Envelope Manufacturers, Switzerland]

FEPE Fédération Européenne pour la Protection des Eaux [European Federation for Water Pollution Control]

FEPE Full Energy Peak Efficiency

FEPEM Federation of European Petroleum Equipment Manufacturers [Netherlands]

FEPHM Field Emission Probe-Hole Microscopy

FEPI Front End Programming Interface

FePIH Ferric Pyridoxal Isonicotinoyl Hydrazone

FEPMA Federation of European Pencil Manufacturer Associations [Germany]

FEPROM Flash Erasable Programmable Read-Only Memory

FEPT Fluorinated Ethylene Polypropylene Lined Tygon [*Note:* "Tygon" is a Polyvinyl Polymer]

FEQL Food and Environmental Quality Laboratory [of Washington State University–Tri-Cities, Richland, US]

FER Federation of Engine Re-Manufacturers [UK]

FER Ferroelectric Retina

FER For [Amateur Radio]

FER Forward Engine Room

FER Foundation for Educational Research [of ASM International; also ASMFER]

FER Friends of Ecological Reserves

FER Fusion Engineering Reactor

FER Fusion Experimental Reactor [Japan]

Fer Ferritin

FERA Federal Emergency Relief Administration [US]

FERA Further Education Research Association [UK]

FERAM Ferroelectric Random-Access Memory [also FeRAM]

Ferbam Iron Tris(N,N-Dimethyldithiocarbamate

FERC Federal Energy Regulatory Commission [of US Department of Energy]

FERD Facility and Equipment Requirements Document

FERD Fuel Element Rupture Detection [Nuclear Engineering]

Fe-RE Iron Rare-Earth (Alloy)

FERETS Front-End Resolution Enhancement (Using) Tailored Sweeps

FeRh Iron Rhodium (Alloy)

FeRhIr Iron Rhodium Iridium (Alloy)

FERIC Forest Engineering Research Institute of Canada

FERIS Forest Environment and Resources Information System [of the State of Queensland, Australia]

FERMI Enrico Fermi Breeder Reactor Plant

Fermilab Fermi National Accelerator Laboratory [of US Department of Energy at Batavia, Illinois, US] [also FNAL]

Fermilab Comput News Fermilab Computing News [of Fermilab Computing Division, Batavia, Illinois, US]

Fermilab Inf Res Newsl Fermilab Information Resources Newsletter [of Fermilab, Batavia, Illinois, US]

FERN Forest Ecosystem Research Network

Fernmelde-Ing Fernmelde-Ingenieur [German Journal on Telecommunication Engineering]

Fernseh- & Kino-Tech Fernseh- und Kino-Technik [German Journal on Television Engineering and Cinematography]

Fernwärme Int Fernwärme International [German International Journal on District Heating]

FERP Field Emission Retarded Potential

FERPA Family Educational Rights and Privacy Act [US]

FERPIC Ferroelectric Ceramic Picture Device

FERPIC Ferroelectric Photoconductor Image Camera

FERPIC Ferroelectric Picture [also Ferpic]

FERROD Ferrite-Rod Antenna [also ferrod]

Ferroelectr Ferroelectric(al); Ferroelectricity; Ferroelectrics [also Ferroelec, ferroelec, or ferroelectr]

Ferroelectr Ferroelectrics [Journal published in the UK]

Ferroelectr Lett Ferroelectrics Letters [UK]

Ferroelectr Lett Sect Ferroelectrics Letters Section [UK]

Ferroin 1,10-Phenanthroline Iron(II) Sulfate Complex [also ferroin]

Ferr phos Iron Phosphate (Tissue Salt)

FERSI Flat Earth Research Society International [US]

FERSIM Fertigungssimulationssystem [A German Manufacturing Simulation System]

FERT Division of Fertilizer and Soil Chemistry [of American Chemical Society, US]

Fert Fertilizer [also fert]

Fertigungstech Betr Fertigungstechnik und Betrieb [German Publication on Manufacturing Technology]

Fe/Ru Iron/Ruthenium (Superlattice)

FeRuGaSi Iron Ruthenium Gallium Silicon (Alloy)

FES Fachnormenausschuss Eisen und Stahl [Standards Committee of the Iron and Steel Industry, of Deutsches Institut für Normung, Germany]

FES Faculty of Engineering Sciences

FES Far-End Suppressor

FES Federation of the European Trout and Salmon Industry [Scotland]

FES Fellow of the Entomological Society

FES Field Emission Spectrometry; Field Emission Spectroscopy

FES Final Environmental Statement

FES Finite Element Solver

FES Flame-Emission Spectroscopy

FES Flash Evaporator System

FES Flight Element Set

FES Florida Engineering Society [US]

FES Florida Entomological Society [US]

FES Flywheel Energy Storage

FES Foerster Energy Transfer

FES Fondation Européenne de la Science [European Science Foundation, France]

FES Foor Education Society [UK]

FES Forest Extension Services [Canada]

FES Forms Entry System

FES Fundamental Electrical Standards

FeS Iron(II) Sulfide (or Ferrous Sulfide)

FESA Federation of Engineering and Scientific Associations

FESA Finite-Element Stress Analysis

FESA Foundry Equipment and Supplies Association [UK]

FESDK Far East Software Development Kit

FESE Field-Enhanced Secondary Emission

Fe-Sc Iron-Scandium (Alloy System)

FeSe Ferrous Selenide (Semiconductor)

FE-SEM Field Emission Scanning-Electron Microscope; Field Emission Scanning Electron Microscopy [also FESEM or FE SEM]

FESI Fédération Européenne des Syndicats d'Entreprises d'Isolation [European Federation of Associations of Insulation Contractors, France]

FeSi Iron Monosilicide

Fe-Si Iron-Silicon (Alloy System)

FeSiAlN Iron Silicon Aluminum Nitrogen (Alloy)

FeSiB Iron Silicon Boron (Alloy)

FeSiBC Iron Silicon Boron Carbon (Alloy)

FESIC Far East Seed Improvement Conference

FESL Failure Effects Summary List

FeSn Iron Stannide

FeSO Ferrosulfate

FeSoCiMe Federación de Sociedades de Cientificas de México [Federation of Mexican Scientific Societies]

FESPA Federation of European Screen Printers Associations

FESS Facilities Engineering and Safety Services [of Canadian Fusion Fuels Technology Project]

FESS Facilities Engineering Services Section [of Fermilab, Batavia, Illinois, US]

FESS Flywheel Energy Storage System

FEST Federation for Education, Science and Technology [South Africa]

FEST Fuzzy Expert System Tool

Fe-S-X Iron-Sulfur-Metal (Alloy System)

FESYP Fédération Européenne des Syndicats de Fabricants de Panneaux de Particules [European Federation of Associations of Particleboard Manufacturers, Germany]

FET Federal Excise Tax [US]

FET Federation of Environmental Technologists [US]

FET Field-Effect Transistor

FET Field-Effect Triode

FET Field Emission Tube

FET Finite-Element Technique

FET Flight Elapsed Time

FET Fluid Engineering Technology

FET Free Electron Theory

F(E,T) Strain and Temperature Dependent Helmholtz Free Energy [Symbol]

FETA Federation of Environmental Trade Associations [UK]

FETA Fire Extinguishing Trades Association [UK]

FeTaC Iron Tantalum Carbide

Fetal Diagn Therapy Fetal Diagnosis and Therapy [International Journal]

FeTaN Iron Tantalum Nitride

FeTaNC Iron Tantalum Nitrocarbide

FETC Federal Energy Technology Center [of US Department of Energy]

FETDIP Fetlington Dual-in-Line Package

FE-TEM Field-Emission Transmission Electron Microscope; Field-Emission Transmission Electron Microscopy [also FETEM or FE TEM]

FETF Ferroelectric Thin Film

FETFE Fluorelastomer with Tetrafluoroethylene Additives

$Fe(TFAC)_3$ Iron(III) Trifluoroacetylacetonate [also $Fe(tfac)_3$]

FeTi Iron Titanium (Compound)

Fe-Ti Iron-Titanium (Alloy)

FeTNa Iron-Sodium Tartrate

FETO Factory Equipment Transfer Order

FETO Free Estimated Time of Overflight,

FETS Far-East Trade Service, Inc.

FETS Field-Effect Transistors [also FETs]

FETT Field-Effect Tetrode Transistor

FEUE Federación des Estudiantes Universitarios del Ecuador [Federation of University Students of Ecuador]

FEUS French Engineers in the United States

FEV Forced Expired Volume

Fe-V Iron-Vanadium (Alloy System)

FEVE Fédération Europénne du Verre d'Emballage [European Container Glass Federation, Belgium]

Fe-W Iron-Tungsten (Alloy System)

FEWA Farm Equipment Wholesalers Association [US]

Few-Body Syst Few-Body Systems [Published in Austria]

Few-Body Syst Suppl Few-Body Systems Supplementum [Issued with Few-Body Systems]

FEWG Flight Evaluation Working Group

FEWIA Federation of European Writing Instruments Associations [Germany]

FEWITA Federation of European Wholesale and International Trade Associations [Belgium]

FEWMA Federation of European Window Manufacturers Associations [Germany]

FeWN$_2$ Iron Tungsten Nitride

FEX Foreign Exchange

FeX Amorphous Binary Iron Alloy [X Represents One or More Metalloids]

FEXT Far-End Crosstalk

Fe/Y-123 Iron on Yttrium Barium Copper Oxide Superconductor ($Yba_2Cu_3O_x$)

FeZn Iron Zinc (Compound)

Fe-Zr Iron-Zirconium (Alloy System)

FF Far Field

FF Farm Foundation [US]

FF Fast-Fast Wave [Physics]

FF Fast Fission

FF Fast Forward

FF Fatigue Failure

FF Fatigue Fracture [Mechanics]

FF Federal Funding; Federally Funded

FF Fermi Function [Physics]

FF Ferrofluid

FF Fiber Forming

FF Field Function

FF Fill(ing) Factor

FF Film Flotation

FF Film Former; Film Forming

FF Filtration Fraction,

FF Fine Focus

FF Fine Furnace (Black)

FF Firefighter; Firefighting

FF Fission Fuel

FF Fixed Focus

FF Fleischer and Field (Analysis of Ductile-to-Brittle Transition of Ceramics)

FF Flight Forward

FF Flip-Flop [also F-F]

FF Fluid Flow

FF Fluid Friction

FF Flux Flow

FF Flying Fortress

FF Force Feed

FF Force Field

FF Ford Foundation [New York City, US]

FF Form Factor [also ff]

FF Form Feed (Character) [Data Communications]

FF Fossil Fuel

FF Fourier Frequencies

FF Fowler Flap [Aeronautics]

FF Fracture Face

FF Fracture Fatigue

FF Fragrances Foundation [US]

FF Free Fall

FF Free Flight

FF Freeze Fracture (Technique)

FF Freight Forwarder

FF French Franc [Currency of France, French Guinea, Guadeloupe, Martinique, Monaco, and St. Pierre and Miquelon]

FF Frequency Filter

FF Frictional Force

FF Friction Factor [Fluid Dynamics]

FF Friction Feed

FF Froth Flotation,

FF Fuel Flow(meter)

FF Full-Face (Respirator)

FF Full Field

FF Fundamental Frequency

FF Fusion Fuel

F&F Fire and Flushing

F&F Fieser and Fieser (Reagents for Organic Synthesis)

F/F Fill/Full; Full/Fill

F$_f$/F$_m$ Load Ratio of Composite Fibers to Composite Matrix [Symbol]

F-F Ferromagnetic-Ferromagnetic (Transition)

F-F Flip-Flop [also FF]

fF femtofarad [Unit]

ff folios

ff following (pages, sections, etc.)

ff shape factor [Component of the Continuity Index of Intermetallics]

ff form factor [also FF]

4×4 Four by Four (Truck)

FFA Federal Fisheries Act [Canada]

FFA Fiberglass Fabrication Association [US]

FFA Flammable Fabrics Act [US]

FFA Flygtekniska Forsoksanstalten [Aeronautical Research Institute, Bromma, Sweden]

FFA Forest Farmers Association [US]

FFA (South Pacific) Forum Fisheries Agency

FFA Free Fatty Acid

FFA Free from Alongside [also ffa]

FFA Free from Average

FFAG Fixed-Field Alternating Gradient (Accelerator) [Nuclear Engineering]

FFAR Folding Fin Air(craft) Rocket

FFAW Fishermen, Food and Allied Workers (Union) [Canada]

FFB French Forces Broadcasting

FFBD Functional Flow Block Diagram

FF Black Fine Furnace Black [also FF black]

FFC Fast Field-Cycling [Physics]

FFC Feed Forward Control

FFC Final Flight Certification

FFC Flagstaff Field Center [of US Geological Survey/ Geologic Division at Flagstaff, Arizona]

FFC Flexible Flatness Control

FFC Flip-Flop Complimentary

FFC Foster Findlay Corporation [UK]

FF III$_1$-C Fibronectin Fragment III$_1$-C [Microbiology]

F-F-C Flat-Flat-Corrugated (Fiberboard)

FFCA Federal Facilities Compliance Act [US] [also FFC Act]

FFC-NMR Fast Field-Cycling Nuclear Magnetic Resonance

FFCO 1,3-Cyclohexadiene Iron Tricarbonyl

FFCP Farm Financial Counselling Program [of the State of Queensland, Australia]

FFD Failure Flux Density

FFD Fixed Format Display

FFD Fluid Flow Dynamics

FFD FMS (Flight Management System) Flight Data

FFD Functional Flow Diagram

FFDC First Failure Data Capture

FFDCA Federal Food, Drug and Cosmetics Act [US]

FFDNB 1,5-Difluoro-2,4-Dinitrobenzene

FFE Ferric-Ferrous Electrode

FF&E Furniture, Furnishings and Equipment

FFEC Field-Free Emission Current

FFEM Freeze Fracture Electron Microscopy

FFF Farm Field Foundation [US]

FFF Fast-Fission Factor

FFF Field Flow Fractionation (Sedimentation)

FFF Film-Forming Foam

FFF Fire-Fighting Foam

FFF Flicker Fusion Frequency

FFF Flight Freedoms Foundation [US]

FFF Foundry Educational Foundation [US]

FFF Free Flux Flow

F/F/F Foil/Fiber/Foil (Technique) (for Metal-Matrix Composites)

FFFF Fusion-Fission Fuel Factory [PR China]

FFFP Film-Forming Fluoroprotein

FFG Fine-Fine Grain (or Superfine Grain)

FFG Flora and Fauna Guarantee (Act) [of the State of Victoria, Australia]

FFG Form and Finish Grinding

FFG Fission Fragment Generator

FFH For Further Headings

FFHR Fusion-Fission Hybrid Reactor

FFI Free Fluid Index

FFI Fuel Flow Indicator

FFI Full Field Investigation

FFIM Fully Frustrated Ising Model [Solid-State Physics]

FFIS Forests and Forest Industries Strategy

FFL Finished Floor

FFL First Financial Language

FFL Front Focal Length

FFLA Federal Farm Loan Association [US]

FFM Faculty of Ferrous Metallurgy

FFM Fast File Manager

FFM Flash-Filament Method

FFM Foundation for Microbiology [US]

FFM Free-Flying (Experiment) Module [of NASA]

FFM Frictional Force Microscope; Frictional Force Microscopy

FFM Fuel Failure Mockup (Facility)

FFM Fuel Flow Meter

FFM Full-Face Modular (Wheel)

FFMC Fine Fuel Moisture Code

FFMC Freshwater Fish Marketing Corporation

fF/μm^2 femtofarad(s) per square micrometer [also fF μm^{-2}]

FFMS Fast Fourier Mass Spectrometry

FFMS Free Flight Melt Spinning [Metallurgy]

FFOS Factory Floor Operator Station

FFP Farm Forestry Program

FFP Firm Fixed Price

FFP First Focal Point

FFP Fixed Fee Procurement

FFP Fresh Frozen (Human Blood) Plasma

FFPA Fauna and Flora Preservation Society [UK]

FFPA Free from Prussic Acid

FFPAF Forest and Forest Products Policy Advisory Forum

FFPB Flora and Fauna Protection Board [Australia]

FFPRI Forest and Forest Product Research Institute [Ghana]

FFPRI Forestry and Forest Product Research Institute [Japan]

FFPS Fauna and Flora Preservation Society [UK]

FFR Fauna and Flora Reserve [Australia]

FFR Fellow of the Faculty of Radiology

FFR Field-Free Region

FFR Flat Face Roller

FFR Flat Face Rolling (System)

FFR Fluid-Fuel Reactor

FFR Folded Flow Reactor

FFr French Franc [Currency]

FFRDC Federally Funded Research and Development Center [US]

FFRI Fruit and Food Research Institute [South Africa]

FFRL Fracture and Fatigue Research Laboratory [Georgia Institute of Technology, Atlanta, US]

FFRM Foundation for Fundamental Research on Matter [Netherlands]

FFRS Federal Forest Research Station [of International Union of Forestry Research Organizations, Austria]

FFS Family Financial Statement [US Student Aid]

FFS Fast File System

FFS Federal Financial System

FFS Fisheries Forms System

FFS Flame Fluorescence Spectroscopy

FFS Flash Filament Spectroscopy

FFS Flash File System [of Microsoft Corporation, US]

FFSA Field Functional System Assembly

FFSF Fossil Fired Steam Plant

FFSF Full Fat Soy Flour

FFSPN Fédération Française des Sociétés de Protection de la Nature [French Federation of Nature Conservation Societies] [also France Nature Environment]

446SS AISI-SAE 446 Ferritic Stainless Steel

FFST First Failure Support Technology

FF-Sulfone Bis(4-Fluoro-3-Nitrophenyl)-Sulfone

FFT Fast Fourier Transform

FFT Final Form Text

FFT Flash Fusion Technology

FFT Flicker Fusion Threshold

FFT Flux Flow Transistor

FFT Fossil Fuels Technology

FFT Fusion Fuels Technology

FFTA Fast Fourier Transform Analyzer

FFTA Film Flow Transfer Apparatus

FFTA Finnish Foreign Trade Association

FFTA Foundation of the Flexographic Technical Association [US]

FFTB Final Focus Test Beam (Collaboration) [Stanford Linear Accelerator and Fermilab, US]

FFTF Fast Flux Test Facility [of US Department of Energy at Hanford, Washington]

Fftg Firefighting

FFTFPO Fast Flux Test Facility Project Office [of US Atomic Energy Commission]

°F·ft²·h/Btu Degrees Fahrenheit square foot hour per British thermal unit [also °F/(Btu/h/ft²), F/(Btu/h/ft²), or F·ft²·h/Btu]

FFTO Free-Flying Teleoperator

FFTR Fast Flux Test Reactor

FFTR Federal Fuel Tax Rebate [Canada]

FFTS Fast Fourier Transform Spectrometry

FFTSB Federation of Trade Fairs and Trade Shows of Benelux [France]

442SS AISI-SAE 442 Ferritic Stainless Steel

FFTV Free Flight Test Vehicle

FFV Field Failure Voltage

FFV Fire Fighting Vehicle

FFV Flexible Fuel Vehicle(s)

FFVMA Fire Fighting Vehicles Manufacturers Association [UK]

FG *(Fachgemeinschaft)* – German for "Special Working Group"

FG Failed in Gage Section

FG Feature Group [Telecommunications]

FG Fermi Gas [Statistical Mechanics]

FG Fiberglass

FG Field Gun

FG Filament Ground

FG Fine Grain

FG Finite Geometry

FG First Generation (Computer) [also 1G]

FG Fifth Generation (Computer) [also 5G]

FG Fission Gas

FG Flake Graphite [Metallurgy]

FG Flame Gun

FG Flammable Gas

FG Flat Glass

FG Flat Grain [Wood]

FG Flint Glass

FG Floating Gate [Electronics]

FG Flow Gain

FG Flue Gas

FG Fluoride Glass

FG Fluxgate Gradiometer

FG Fog Gun

FG Foot Guard

FG Force Gauge

FG Foreground

FG *(Forschungsgruppe)* – German for "Research Group"

FG Forward Gate

FG Fourth Generation (Computer) [also 4G]

FG Frame Ground

FG Fraunhofer-Gesellschaft [Fraunhofer Society, Germany]

FG Free Gyro [Aeronautics]

FG French Guiana

FG Friction Glaze

FG Fuel Gas

FG Fuel Gauge

FG Fully Gamma (Microstructure) [Metallurgy]

FG Functional Group [Chemistry]

FG Function Generator

1G First Generation (Computer)

4G Fourth Generation (Computer)

5G Fifth Generation (Computer)

Fg Forging [also fg]

ƒ(G) function of geometry [Symbol]

ƒ(g) orientation distribution function (in materials science) [Symbol]

FGA Feature Group A [of Exchange Network Facilities for Interstate Access, US]

FGA Fellow of the Gemological Association

FGA Flue-Gas Analysis; Flue-Gas Analyzer

FGA Free of General Average

FGA Frozen Gas Approximation

FGAA Federal Government Accountants Association

FGAN Fertilizer Grade Ammonium Nitrate

FGB Feature Group B [of Exchange Network Facilities for Interstate Access, US]

FGC Feature Group C [of Exchange Network Facilities for Interstate Access, US],

FGC Fifth Generation Computer [also 5GC]

FGC Fish and Game Commission [California, US]

FGC Flat Glass Council [UK]

1GC First Generation Computer

4GC Fourth Generation Computer

5GC Fifth Generation Computer

FGCS Fifth Generation Computer System [also 5GCS]

1GCS First Generation Computer System

4GCS Fourth Generation Computer System

5GCS Fifth Generation Computer System [also FGCS]

FGCS Flight Guidance and Control System

FGD Fine-Grain Data

FGD Flue Gas Desulfurization (Process)

FGDC Federal Geographic Data Committee [US]

FG-EEPROM Floating Gate Type Electrically Erasable and Programmable Read-Only Memory

FGEPR Field-Gradient Electron Paramagnetic Resonance

FGETR Federal Gasoline Excise Tax Refund [Canada]

FGF Fibroblast Growth Factor [Biochemistry]

FGF-4 Fibroblast Growth Factor-4 [Biochemistry]

FGF-5 Fibroblast Growth Factor-5 [Biochemistry]

FGF-6 Fibroblast Growth Factor-6 [Biochemistry]

FGF-9 Fibroblast Growth Factor-9 [Biochemistry]

FGFAF Fraunhofer-Gesellschaft zur Förderung der Angewandten Forschung [Fraunhofer Society for the Advancement of Applied Research, Germany]

FGFWFC Florida Game and Fresh Water Fish Commission [US]

FGGE First GARP (Global Atmospheric Research Program) Global Experiment

FGGM Federation of Gelatine and Glue Manufacture

FGHAZ Fine-Grain(ed) Heat-Affected Zone [Metallurgy]

FGI Federation of German Industries

FGI Federation of Greek Industries

F-GIC Fluorine Graphite Intercalation Compound

FGIN-1-27 N,N-Dihexyl-2-(4-Fluorophenyl)indole-3-Acetamide

FGIPC Federation of Government Information Processing Councils [US]

FGIS Federal Grain and Inspection Service [of US Department of Agriculture]

FGJA Flat Glass Jobbers Association [now Flat Glass Marketing Association, US]

1GL First Generation Language

4GL Fourth Generation Language

5GL Fifth Generation Language

FGM Fission Gas Monitor

FGM Functional Gradient Material

FGM International Symposium on Functional Gradient Materials

FGM International Symposium on Structural and Functionally Gradient Materials

FGMA Flat Glass Manufacturing Association [UK]

FGMA Flat Glass Marketing Association [US]

FGMDSS Future Global Maritime Distress and Safety System

FGMM Fluxgate Magnetometer

Fgn Foreign [also fgn]

FGR Fast Gaseous Reaction

FGR Fertility and Genetics Research

FGR Fondazione Giorgio Ronchi [Giorgio Ronchi Foundation, Italy]

Fgr Finger [also fgr]

FGREP Fixed Global Regular Expression Print [Unix Operating System]

FGRP Fiberglass-Reinforced Plastic(s)

FGS Fellow of the Geographical Society

FGS Fellow of the Geological Society [UK]

FGSA Fellow of the Geographical Society of America

FGSA Fellow of the Geological Society of America

FGSE Field-Gradient Spin-Echo

FGSE Fluid Ground Support Equipment

FGT Fermat's Great Theorem

FGT Field-Gradient Tensor

FGTB Fédération Générale des Travailleurs de Belgique [General Federation of Belgian Workers]

FGTS Flammable Gas Tank Safety (Project) [US]

FGW Floor Ground Window

FH (Fachhochschule) – German for "Technical College", or "College of Technology"

FH Fermi Hole [Solid-State Physics]

FH Fiber Handle

FH File Handler; File Handling

FH Fiber Hub [Telecommunications]

FH Filter Heater

FH Fire Hazard

FH Fire Hose

FH First Harmonic (Generation) [also 1H]

FH Firth Hardometer

FH Flame-Hardened; Flame Hardening

FH Flame Holder

FH Flame Hydrolysis

FH Flat-Head (Screw)

FH Flex(ible) Hose

FH Flory-Huggins (Model) [Physics]

FH Foghorn

FH Fracture Hydrology

FH Franck-Hertz (Experiment) [Electrical Engineering]

FH Fraunhofer Hologram

FH French Hybrid

FH Frequency Hopping

FH Fresnel Hologram

FH Friction Head

FH Full Hard (Temper) [Metallurgy]

FH Full Hardening; Fully Hardened [Metallurgy]

FH Full Height

FH Full-Hole Mining

FH Fumurate Hydratase [Biochemistry]

FH Functional Hypoglycemia [Medicine]

1H First Harmonic (Generation) [also FH]

4H Four-Wheel Drive Position, High-Range

5H- Tanazania [Civil Aircraft Marking]

4-H Head, Hands, Heart and Health Club [Washington, DC, US]

5-H Five Harness (Weave) [Woven Fabrics]

F7/h8 Close Running Fit; Basic Shaft System [ISO Symbol]

FHA Farmers Home Administration [of US Department of Agriculture]

FHA Federal Highway Administration [US]

FHA Federal Housing Administration [US] [Defunct]

FHANG Federation of Heathrow Anti-Noise Groups [UK]

FHAZ Fusion Heat-Affected Zone [Metallurgy]

FHB Flat Head Brass (Screw)

FHB Fuel Handling Building

FHC Federal Housing Commissioner [of Office of the Assistant Secretary for Housing, US]

FHC Flight Half Coupling

FHC Fluid Hydrostatic Cell

FHC Fluorohydrocarbon

FHC Fire Hose Cabinet

FHCRC Fred Hutchinson Cancer Research Center [US]

FHD Ferrohydrodynamic(s)

FHD Fixed-Head Disk

F Hd Flat Head (Screw)

FHDA Fir and Hemlock Door Association [US]

FHDP Fluorinated High-Density Plasma (Oxide)

FHDS Fixed-Head Disk Storage

FHDU Fixed-Head Disk Unit

FhE *(Fraunhofer-Einrichtung)* – German for "Fraunhofer Establishment"

FHEO Fair Housing and Equal Opportunity [of Office of the Assistant Secretary, US]

FHER Foundation for Homeopathic Education and Research [US]

FHF First Horizontal Flight

°F·h·ft²/Btu Degrees Fahrenheit hour square foot per British thermal unit [also °F/h/ft²/Btu]

FHFW Federation of High Frequency Welders [UK]

FHG First Harmonic Generation [also 1HG]

1HG First Harmonic Generation [also FHG]

FhG Fraunhofer Gesellschaft (für Angewandte Forschung) [Fraunhofer Society (for Applied Research), Germany]

FhG Ber FhG Bericht [German Technical Report of Fraunhofer Gesellschaft]

FhG-IAF Fraunhofer-Institut für Angewandte Festkörperphysik [Fraunhofer Institute for Applied Solid-State Physics, Freiburg, Germany] [also FhIAF]

FhG-IAO Fraunhofer-Institut für Arbeitswissenschaft und Organisation [Fraunhofer Institute for Industrial Science and Organization, Stuttgart, Germany] [also FhIAO]

FhG-IBF Fraunhofer-Institut für Betriebsfestigkeit [Fraunhofer Institute for Operational Strength of Materials, Germany] [also FhIBF]

FhG-IBMT Fraunhofer-Institut für Biomedizinische Technik [Fraunhofer Institute for Biomedical Technology, St. Ingbert, Germany] [also FhIBMT]

FhG-ICT Fraunhofer-Institut für Chemische Technology [Fraunhofer Institute for Chemical Technology, Karlsruhe, Germany] [also FhICT]

FhG-IFAM Fraunhofer-Institut für Angewandte Materialforschung [Fraunhofer Institute for Applied Materials Research at Bremen and Dresden, Germany] [also FhIFAM]

FhG-IfS Fraunhofer-Institut für Silicatforschung [Fraunhofer Institute for Silicate Research, Wuerzburg, Germany] [also FhIfS]

FhG-IFW Fraunhofer-Institut für Festkörper- und Werkstofforschung [Fraunhofer Institute for Solid-State and Materials Research, Dresden, Germany] [also FhIFW]

FhG-IfzP Fraunhofer-Institut für zerstörungsfreie Prüfverfahren [Fraunhofer Institute for Nondestructive Testing, Saarbrücken, Germany] [also FhIfzP]

FhG-IKTS Fraunhofer-Institut für Keramische Technologien und Sinterwerkstoffe [Fraunhofer Institute for Ceramic Technologies and Sintered Materials, Dresden, Germany] [also FhIKTS]

FhG-IMT Fraunhofer-Institut für Mikrostrukturtechnik [Fraunhofer Institute for Microstructural Technology, Germany] [also FhIMT]

FhG-IPA Fraunhofer-Institut für Produktionstechnik- und Automatisierung [Fraunhofer Institute for Production Technology and Automation, Germany] [also FhIPA]

FhG-IPM Fraunhofer-Institut für Physikalische Messtechnik [Fraunhofer Institute of Physical Measurement Techniques, Germany]

FhG-IRB Fraunhofer-Informationszentrum Raum und Bau [Fraunhofer Information Center for Development Planning and Construction, Stuttgart, Germany] [also FhIRB]

FhG-ISE Fraunhofer-Institut für solare Energiesysteme [Fraunhofer Institute for Solar Energy Systems, Freiburg, Germany] [also FhISE]

FhG-ISI Fraunhofer-Institut für Systemtechnik und Innovationsforschung [Fraunhofer Institute for Systems Engineering and Innovation Research, Karlsruhe, Germany] [also FhISI]

FhG-ISiT Fraunhofer-Institut für Siliziumtechnologie [Fraunhofer Institute for Silicon Technology, Berlin, Germany] [also FhISiT]

FhG-ITW Fraunhofer-Institut für Transporttechnik und Warendistribution [Fraunhofer Institute for Transportation Engineering and Goods Distribution, Dortmund, Germany] [also FhITW]

FHH Frenkel-Halsey-Hill (Isotherm Equation) [Physics]

FHI Fritz-Haber-Institut (der Max-Planck-Gesellschaft) [Fritz Haber Institute (of the Max Planck Society), Berlin, Germany]

FhIAF Fraunhofer-Institut für Angewandte Festkörperphysik [Fraunhofer Institute for Applied Solid-State Physics, Freiburg, Germany] [also FhG-IAF]

FhIAO Fraunhofer-Institut für Arbeitswissenschaft und Organisation [Fraunhofer Institute for Industrial Science and Organization, Stuttgart, Germany] [also FhG-IAO]

FhIBF Fraunhofer-Institute für Betriebsfestigkeit [Fraunhofer Institute for Operational Strength, Germany] [also FhG-IBF]

FhIBMT Fraunhofer-Institut für Biomedizinische Technik [Fraunhofer Institute for Biomedical Technology, St. Ingbert, Germany] [also FhG-IBMT]

FhICT Fraunhofer-Institut für Chemische Technology [Fraunhofer Institute for Chemical Technology, Karlsruhe, Germany] [also FhG-ICT]

FhIFAM Fraunhofer-Institut für Angewandte Materialforschung [Fraunhofer Institute for Applied Materials Research at Bremen and Dresden, Germany] [also FhG-IFAM]

FhIfS Fraunhofer-Institut für Silicatforschung [Fraunhofer Institute for Silicate Research, Wuerzburg, Germany] [also FhG-IfS]

FhIFW Fraunhofer-Institut für Festkörper- und Werkstofforschung [Fraunhofer Institute for Solid-State and Materials Research, Dresden, Germany] [also FhG-IFW]

FhIfzP Fraunhofer-Institut für zerstörungsfreie Prüfverfahren [Fraunhofer Institute for Nondestructive Testing, Saarbrücken, Germany] [also FhG-IfzP]

FhIKTS Fraunhofer-Institut für Keramische Technologien und Sinterwerkstoffe [Fraunhofer Institute for Ceramic Technologies and Sintered Materials, Dresden, Germany] [also FhG-IKTS]

FhIMT Fraunhofer-Institut für Mikrostrukturtechnik [Fraunhofer Institute for Microstructural Technology, Germany] [also FhG-IMT]

FhIPA Fraunhofer-Institut für Produktionstechnik- und Automatisierung [Fraunhofer Institute for Production Technology and Automation, Germany] [also FhG-IPA]

FhIPM Fraunhofer-Institut für Physikalische Messtechnik [Fraunhofer Institute of Physical Measurement Techniques, Germany]

FhIRB Fraunhofer-Informationszentrum Raum und Bau [Fraunhofer Information Center for Development Planning and Construction, Stuttgart, Germany] [also FhG-IRB]

FhISE Fraunhofer-Institut für solare Energiesysteme [Fraunhofer Institute for Solar Energy Systems, Freiburg, Germany] [also FhG-ISE]

FhISI Fraunhofer-Institut für Systemtechnik und Innovationsforschung [Fraunhofer Institute for Systems Engineering and Innovation Research, Karlsruhe, Germany] [also FhG-ISI]

FhISiT Fraunhofer-Institut für Siliziumtechnologie [Fraunhofer Institute for Silicon Technology, Berlin, Germany] [also FhG-ISiT]

FhG-ITW Fraunhofer-Institut für Transporttechnik und Warendistribution [Fraunhofer Institute for Transportation Engineering and Goods Distribution, Dortmund, Germany] [also FhITW]

FHK Fourier-Hermite Kernel

FHKSU Fort Hays Kansas State University [US]

F(hkl) Structure Factor for an *hkl* (Crystal) Lattice Plane [Symbol]

F$_s$(hkl) Structure Amplitude of (Crystal) Lattice Plane Set *(hkl)* [Symbol]

FHL Ferrohydrodynamic Levitation

FHLB Federal Home Loan Bank [US]

FHLBB Federal Home Loan Bank Board [US]

FHLMC Federal Home Loan Mortgage Corporation [US]

FHM Forest Health Monitoring [US]

FH Münster Fachhochschule Münster [Munster University (of Applied Science), Germany]

FHNC Fermi Hypernetted Chain [Physics]

4HNE (E)-4-Hydroxy-2-Nonenal

FHP Flash Hydrogen Pyrolysis

FHP Fluid Hydrostatic Pressure

FHP Forum on History of Physics [of American Physical Society, US]

FHP Frisch-Hasslacher-Pomeau (Mode) [Fluid Flow Simulation]

FHPP Friction Hydro Pillar Processing

FHD Fractional Horsepower [also fph]

FHD Friction Horsepower [also fhp]

FHO Failed Handover

FHP Fuel High Pressure

FHR Fire Hose Rack

FHR First Harmonic Rejector

FHRF Finney-Howell Research Foundation [US]

°F·hr·ft²/Btu Degree Fahrenheit hour square foot per British thermal unit [also °F/hr/ft²/Btu]

FH&RM Fuel Handling and Radioactive Maintenance

FHS *(Fachhochschule)* – German for "University of Applied Science"

FHS Fan Heat-Sink

FHS Flat Head Steel (Screw)

FHS Florida Historical Society [US]

FHS Forest History Society [US]

FHS Forward Head Shield

FHS Furniture History Society [UK]

FHSA Federal Hazardous Substances Act [US]

FHSF Fixed-Head Storage Facility

FHSR Final Hazards Summary Report

FHSS Frequency-Hopping Spread Spectrum

FHT Full Heat Treatment; Fully Heat-Treated [Metallurgy]

FHTA Federated Home Timber Associations [UK]

FHTIC Florida High-Technology and Industry Council [US]

FHTS Fast Hadamard Transfer Spectroscopy

FHVA Fine Hardwood Veneer Association [US]

FHVA/AWMA Fine Hardwood Veneer Association/American Walnut Manufacturers Association [US]

FHWA Federal Highway Administration [US]

FHY Fire Hydrant

FI Faeroe Islands [Denmark]

FI Fail-in-Place

FI Falkland Islands

FI Fan-In [Electronics]

FI Farmland Industries [US]

FI Fault Identification

FI Feasibility Indicator

FI Ferris Institute [Big Rapids, Michigan, US]

FI Feynman Integral [Quantum Mechanics]

FI Field Intensity

FI Field Ion(s)

FI Field Ionization

FI Fighter-Interceptor

FI Figure of Insensitiveness

FI Final Inspection

FI Finland [ISO Code]

FI Fixed Interval

FI Fizeau Interferometer

FI Flight Instructor

FI Flight Instrument(ation)

FI Flow Indicator

FI Flow Injection; Flow Injector

FI   Fluorescent Indicator,

FI   Focused Infrared (Melt Fusion)

FI   Foresight Institute [US]

FI   Formaldehyde Institute [US]

FI   Formal Inspection

FI   Fort Irwin [California, US]

FI   Fourier Integral

FI   Franklin Institute [Philadelphia, Pennsylvania, US]

FI   Fraunhofer Institute [Germany]

FI   Free In [also fi] [Commerce]

FI   Fuel Igniter

FI   Fuel Injection; Fuel Injector

FI   Fuse, Instantaneous

FI   Fysisk Institutt [Institute of Physics, University of Oslo, Norway]

F&I   Focus and Intensity [also f&i]

F&I   Furnish and Install

Fi   Fighter [also fi]

fi   for instance

fi   free in [also FI] [Commerce]

.fi   Finland [Country Code/Domain Name]

FIA   Facilities Inventory Assessment

FIA   Factory Insurance Association

FIA   Federal Insurance Administration [US]

FIA   Federated Ironworkers Association of Australia

FIA   Fédération des Industries Agricoles et Alimentaires [Federation of the Agricultural and Food Industries, Belgium]

FIA   Fédération Internationale de l'Automobile [International Automobile Federation, France]

FIA   Fellow of the Institute of Actuaries

FIA   Field-Induced Anisotropy

FIA   Field Information Agency [US]

FIA   Financial Inventory Accounting

FIA   Fixed-Ion Approximation

FIA   Flame Ionization Analysis

FIA   Flow Injection Analysis

FIA   Fluorescence Immunoassay

FIA   Fluorescent Indicator Absorption

FIA   Fluoroimmunoassay

FIA   Forest Inventory and Assessment [US]

FIA   Forging Industry Association [US]

FIA   Free Interstitial Atom

FIA   Fruit Importers Association [UK]

FIA   Freund's Incomplete Adjuvant [Immunochemistry]

FIA   Full Interest Admitted

FIAA Fellow of the Incorporated Association of Architects and Surveyors

FIACC Five International Associations Coordinating Committee [Hungary]

FIAeS Fellow of the Institute of Aeronautical Sciences

FIAMS Flinders Institute for Atmospheric and Marine Studies [US]

FIANATM Fédération Internationale des Associations de Négociants en Aciers [International Federation of Associations of Steel, Tube and Metal Merchants, Switzerland]

FIAP Fédération Internationale des Architectes Paysagistes [International Federation of Landscape Architects, France]

FIAPM Fellow of the International Association of Podiatric Medicine

FIAR Failure Investigation Action Report

FIAS Fellow of the Institute of Aeronautical Sciences

FIAS Flow Injection Analysis System

FIAT Field Information Agency, Technical [US]

FIAT Forest Industries Association of Tasmania

FIATA Fédération Internationale des Associations de Transitaires et Assimilies [International Federation of Freight Forwarders Associations, Switzerland]

FIB   Fast Ion Bombardment (Mass Spectrometry)

FIB   Fédération des Industries Belges [Federation of Belgian Industries]

FIB   Fédération Internationale de Béton [International Concrete Federation]

FIB   Fellow of the Institute of Biology

FIB   File Information Block

FIB   Focused Ion Beam (Technology)

FIB   FORTRAN Information Bulletin [US]

FIB   Forward Indicator Bit

FIB   Free into Barge [also fib]

FIB   Free into Bunkers [also fib]

Fib   Fibrin [Biochemistry]

Fib   Fibrinogen [Biochemistry]

Fib   Fibula [Anatomy]

FIBCA Flexible Intermediate Bulk Container Association [UK]

FIB Cu Focused Ion Beam Copper (Micromachining)

FIBEC Federal Industrial Boiler Emission Control [Canada]

Fiber Integr Opt Fiber and Integrated Optics [US Publication]

Fiber Opt Mag Fiber Optics Magazine [US]

FIBiol Fellow of the Institute of Biology

FIB JJ Focused-Ion Beam Josephson Junction

FIB-MBE (Combined) Focused Ion Beam and Molecular Beam Epitaxy (System)

FIBOR Fiber Building Board Organization [UK]

FIBOR Frankfurt Interbank Offered Rate [also Fibor, Germany]

FIBP Federal Industrial Boiler Program [of Energy Technology Branch, Natural Resources Canada]

Fibr Fiber [also fibr]

FIBRC Federal Integrated Biotreatment Research Consortium [US]

Fibre Sci Technol Fibre Science and Technology [Journal published in the UK]

FIBS Flight Information Billing System

FIB/SEM Focused Ion-Beam Scanning Electron Microscope; Focused Ion-Beam Scanning Electron Microscopy

FIC   Fast Ion Chromatography

FIC Fast Ion Conductor

FIC Federal Information Centers [US]

FIC Fellow of the Institute of Chemists

FIC Fellow of the Institute of Commerce

FIC Field Ionization Cinetics

FIC Fire Industry Council [UK]

FIC First-in-Chain

FIC Flight Information Center

FIC Flow Indication and Control

FIC Fluoride-Ion Cleaning

FIC Flying Instructor Course

FIC (John E.) Fogarty International Center [of National Institutes of Health, Bethesda, Maryland, US]

FIC Forest Industries Council [US]

FIC Frequency Interference Control

FIC Fur Institute of Canada

FICA Federal Insurance Contribution Act [US]

FICA Forest Industries Campaign Association

FICB Federal Intermediate Credit Bank [US]

FICC Frequency Interference Control Center [of US Air Force]

FICCI Federation of the Indian Chambers of Commerce and Industry

FICE Fellow of the Institution of Civil Engineers

FICI Federation of the Irish Chemical Industries [UK]

FICON File Conversion

FICPI Fédération Internationale des Conseils en Propriété Industrielle [International Federation of Industrial Property Attorneys, Switzerland]

FICS Fellow of the International College of Surgeons

FICS Forecasting and Inventory Control System

FID Failure Identification

FID Fédération Internationale de Documentation [International Federation of Documentation]

FID Field Intelligence Division [US Military]

FID Field Ionization Detector

FID Flame Ionization Detection; Flame Ionization Detector

FID Flight Implementation Directive

FID Food Industry Division [of Ontario Ministry of Agriculture and Food and Rural Affairs, Canada]

FID Forecasts-In-Depth

FID Free Induction Decay

FID Fuse, Instantaneous Detonating

Fid Fidelity [also fid]

FIDA Federation of Industrial Development Organizations [UK]

FIDA Fonds International pour le Développement Agricole [International Fund for Agricultural Development; of United Nations]

FIDAC Film Input to Digital Automatic Computer

FIDACSYS Film Input to Digital Automatic Computer System

FIDAS Formular-Orientiertes Interaktives Datenbanksystem [A German Form-Oriented Interactive Database System]

FIDASE Falkland Islands Dependencies Aerial Survey Expedition

FIDER Foundation for Interior Design Education Research [US]

FIDIC Fédération Internationale des Ingénieurs Conseils [International Federation of Consulting Engineers]

FIDO Film Industry Defense Organization [UK]

FIDO Fog Investigation and Dispersal Operation

FIDO Frazer Island Defenders Organisation [Queensland, Australia]

FIDOH Flame Ionization Detector Oxygen-Hydrogen

FID/OM FID Committee on Operational Machine Techniques and Systems [of Fédération Internationale de Documentation, Netherlands]

FIDOR Fiber Building Board Development Organization [UK]

FIDS Flight Information and Display System

FID/TM FID Committee on Theory and Methods of Systems, Cybernetics and Information Networks [of Fédération Internationale de Documentation, Netherlands]

FID/TMO FID Committee on Theory, Methods and Operation of Information Systems and Networks [Fédération Internationale de Documentation, Netherlands]

FIE Faculty of Industrial Engineering

FIE Federation of Irish Employers

FIE Flight Instrumentation Engineer

FIE Fuel Injection Equipment

FIEA Fédération Internationale des Experts en Automobile [International Federation of Automobile Experts, Belgium]

FIEC Fédération des Industries Européene de Construction [European Construction Industry Federation]

FIED Field Ion(ization) Energy Distribution

FIEE Fellow of the Institute of Electrical Engineers

FIEEE Fellow of the Institute of Electrical and Electronics Engineers

FIEI Farm and Industrial Equipment Institute [US]

FIELS Foreign Information Exchange for Life Scientists [US]

FIEN Forum Italiano dell'Energia Nucléare [Italian Nuclear Energy Forum]

FIEP Forest Industry Energy Program [US]

FIER Foundation for Instrumentation Education and Research [of Instrument Society of America, US]

FIERA Ferrous Industry Energy Research Association [Canada]

FIERF Forging Industry Educational and Research Foundation [US]

FIES Fellow of the Illuminating Engineering Society

FIF Fractal Image Format

FIFC Fur Information and Fashion Council [US]

FIFDA Fellow of the International Furnishings and Design Association

FIFE Fédération Internationale des Associations de Fabricants de Produits d'Entretien [International Federation of Associations of Manufacturers of Household Products, Belgium]

FIFE First ISLSCP (International Satellite Land Surface Climatology Project) Field Experiment

FIFO First-In, First-Out [also fifo]

FIFO Floating Input/Floating Output

FIFRA Federal Insecticide, Fungicide and Rodenticide Act [US]

FIG Fast Ionization Gauge

FIG Federation Internationale des Geometres [International Federation of Surveyors, Finland]

FIG FORTH (Programming Language) Interest Group [San Jose, California, US]

Fig Figure [also fig]

fig figurative(ly)

FIGAS Falkland Islands Government Air Service

FIGE Field Inversion Gel Electrophoresis

FIGED Fédération Internationale des Grandes et Moyenne Entreprises de Distribution [International Federation of Retail Distributors, Belgium]

FIGIEFA Fédération Internationale des Grossistes, Importateurs et Exportateurs en Fournitures Automobiles [International Federation of Automotive Aftermarket Distributors, Belgium]

FIGLU Formimino-L-Glutamic Acid

Fig Num Figure Number

FIGS Figures Shift [Telecommunications]

FIGS FOCUS (Forum on Computing: Users and Services) Interactive Graphics Software [of Fermilab, Batavia, Illinois, US]

Figs Figures [also figs]

FIHCS Fisheries Information and Habitat Classification System

FII Federation of Irish Industries

FIIA Fellow of the Institute of Industrial Administration

FIIG Federal Item Identification Guide [US]

FIILS Full Integrity Instrument Landing System

FIIR Federal Institute of Industrial Research [Nigeria]

FIJ Fédération Internationale des Journalistes [International Federation of Journalists]

.FIL Hidden File [DOS File Name Extension]

Fil Filament [also fil]

Fil Fillet [also fil]

Fil Fillister (Screw)

FILA Fellow of the Institute of Landscape Architects

Fila Filament [also fila]

FILE Future Identification and Location Experiment

FilePos Current File Position [Pascal Programming Function]

FILEX File Exchange

Fil H Fillister Head (Screw) [also FilH]

Fil HB Fillister Head Brass (Screw) [also FilHB]

Fil HS Fillister Head Steel (Screw) [also FilHS]

Fill Filling [also fill]

FILO First-In, Last-Out [also filo]

FILS Flare-Scan Instrument Landing System

Fils, Tubes, Bandes Profilés Fils, Tubes, Bandes et Profilés [French Publication on Wires, Pipes/Tubes, Strips and Sections]

Filt Filter [also filt]

Filte Filtrate [also filte]

Filtn Filtration [also Filtr, filtr, or filtn]

Filtr Sep Filtration and Separation [Published in the UK]

FILU Four-Bit Interface Logic Unit

FIM Fault Isolation Meter

FIM Fellow of the Institute of Metals [now Fellow of the Institute of Materials]

FIM Fellow of the Institute of Materials

FIM Fellow of the Institute of Mining and Metallurgy [also FIMM]

FIM Field Inspection Manual

FIM Field Intensity Meter

FIM Field-Ion Micrograph; Field-Ion Microscope; Field-Ion Microscopy

FIM Finnish Markka [Currency]

FIM Full Indicator Movement

FIMA Fellow of the Industrial Medical Association

FIM-AP Field-Ion Microscopy with Atom Probe [also FIM/AP]

FIM-APS Field-Ion Microscopy Atom Probe Spectrometry

FIMATE Factory-Installed Maintenance Automatic Test Equipment

FIMC Fellow of the Institute of Management Consultants

FIMCEE Fédération Internationale des Marbriers de la Communauté Economique Européenne [International Federation of the Marble Industries in the European Economic Community, Belgium]

FIMCLA Farm Improvement and Marketing Cooperatives Loans Acts [Canada]

FIME Faculdad de Ingeniería Mecaníca y Electríca [Faculty of Mechanical and Electrical Engineering; of Universidad Autónoma de Nuevo León, Mexico]

FIMechE Fellow of the Institute of Mechanical Engineers [also FIME]

FIMF Fellow of the Institute of Metal Finishing

FIM/FEM Field-Ion Microscopy/Field-Emission Microscopy

FIML Full-Information Maximum Likelihood

FIMM Fellow of the Institute of Mining and Metallurgy [also FIM]

FIMOS Floating-Gate Ionization Injection Metal-Oxide Semiconductor

FIMS Facilities Information Management System

FIMS Field Ionization Mass Spectrometry

FIMS Financial Information Management System

FIMS Functionally-Identifiable Maintenance System

FIN Focused Information Network

FIN Futures Information Network [US]

Fin Finance; Financial [also fin]

Fin Finish [also fin]

Fin Finland; Finnish

FINAC Fast Interline Non-Active Automatic Control

FINAL Financial Analysis Language

FINANC Financial [also Financ or financ]

Financ Times Financial Times [UK]

Financ World Financial World [Published in the US]

FINAT Fédération Internationale des Fabricants et Transformateurs d'Adhésifs et Thermocollants sur Papier et Autres Supports [International Federation of Manufacturers and Converters of Pressure-Sensitive and Heat-Seals on Paper and Other Base Materials, Netherlands]

FINC Fast Integration of Nonlinear Circuits

FINCO Finance Committee [of International Standards Organization]

FINCOM Finance Committee [of Institute of Electrical and Electronics Engineers, US]

FIND File Interrogation of Nineteen Hundred Data

FIND File of Industrial Data [UK]

FINDAR Facility for Interrogating the National Directory of Australian Resources [of National Resource Information Center, Australia]

Fin Dev Finance and Development [Quarterly of the International Monetary Fund and the World Bank]

FINE Fighter Inertial Navigation System

Fine Ceram Fine Ceramics [Journal published in Japan]

FINEFTA Finland–European Free Trade Association

FINESSE Fusion Integrated Nuclear Experiment Strategy Study Effort (Program) [US]

FINEX Financial Instrument Exchange [New York, US]

FINFO First-In, Not Used, First-Out [also finfo]

FINGAL Fixation in Glass of Active Liquors (Process)

FINGLA Fission Products in Glass (Process)

Finish Finishing [also finish]

Finish Ind Finishing Industries [Journal published in the UK]

Finish Line Finishing Line [Publication of the Society of Manufacturing Engineers Association for Finishing Processes, US]

Finish Manage Finishers' Management [Journal published in the US]

FINISTRAT Finishing Strategies

Finite Elem Anal Des Finite Elements in Analysis and Design [Journal published in the Netherlands]

Finite Elem News Finite Elements News [UK]

Finn Finnish

Finn Chem Lett Finnish Chemical Letters [Published in Helsinki, Finland]

FINNIDA Finnish International Development Agency

Fin Off Financial Officer [also Fin Off'r]

Finommech–Mikrotech Finommechanika–Microtechnika [Hungarian Journal on Precision Mechanics and Microengineering]

Fin Plan Financial Plan

FINS Fishing Industry News Service [Australia]

Finshg Finishing [also finshg]

Fin State Financial Statement [also fin state]

FInstP Fellow of the Institute of Physics

FInstPet Fellow of the Institute of Petroleum

FINUDA Fisica Nucleare a DAFNE [Nuclear Physics at DAFNE, Italy]

FIO Federacion Internacional de Oleicultura [International Olive Oil Federation, Italy]

FIO Florida Institute of Oceanography [US]

FIO Food Investigation Organization [UK]

FIO Free In and Out [also fio]

FIO Furnished and Installed by Others

FIOA File Input/Output Area

FIOCES Fédération Internationale des Organisations de Correspondances et d'Echanges Scolaires [International Federation of Correspondence Organizations and Student Exchange]

FIOP FORTRAN Input/Output Package

FIOR Fluid Iron-Ore Reduction (Process)

FIOS Free In and Out Stowed [also fios]

FIP Factory Instrumentation Protocol

FIP Fairly Important Person

FIP Federal Implementation Plans [US]

FIP Federal Information Processing (Standards) [US]

FIP Fédération Internationale de la Précontrainte [International Prestressed Concrete Federation]

FIP Feline Infectious Peritonitis [Veterinary Medicine]

FIP Field-Induced Polarization

FIP Field Inspection Procedure

FIP File Processor Buffering

FIP Finance Image Processor

FIP Fluorescent Indicator Panel

FIP Foam(ed)-in-Place (Resin)

FIP Forestry Inceptive Program [US]

FIP For Information Purposes

FIP Forum on International Physics [of American Physical Society, US]

FIP Fractional Integer Programming

FIP Fully-Ionized Plasma

FIPACE Fédération Internationale des Producteurs Auto-Consommateurs Industriels d'Electricité [International Federation of Industrial Producers of Electricity for Own Consumption]

FIPAGO Fédération Internationale des Fabricants de Papiers Gommes [International Federation of Manufacturers of Gummed Paper, Netherlands]

FIPM Feed in Inches per Minute [Numerical Control]

FIPMT Fraunhofer Institute of Physical Measurement Techniques [Germany]

FIPO For Information Purposes Only

FIPP Fédération Internationale de la Presse Périodique [International Federation of the Periodicals Press]

FIPR Feed in Inches per Revolution [Numerical Control]

FIPS Federal Information Processing Standard [US]

FIPS Federal Item Procurement Specification [US]

FIPS Fission Product Solidification (Process)

FIPSCAC Federal Information Processing Standard Coordinating and Advisory Committee

FIQS Fellow of the Institute of Quantity Surveyors

1-FIQTSC 1-Formylisoquinoline Thiosemicarbazone

FIR Far Infrared

FIR Far Infrared (Plasma) Resonance

FIR File Indirect Register

FIR Finite-Duration (or Length) Impulse Response

FIR Finite-Impulse Response (Filter) [also fir]

FIR Flight Information Region

FIR Food Irradiation Reactor [US]

FIR Forschungsinstitut zur Rationalisierung [Research Institute for Rationalization; of RWTH Aachen, Germany]

FIR Fuel Indicator Reading

fir firkin [Unit]

FIRA Foreign Investment Review Act [Canada]

FIRA Foreign Investment Review Agency [Canada]

FIRA Furniture Industry Research Association [UK]

FIRA Bull FIRA Bulletin [of Furniture Industry Research Association, UK]

FIRC Fishing Industry Research Council

FIRC Forest Industries Radio Communications [now Forest Industries Telecommunications, US]

FIRCA Fogarty International Research Collaborations Award [Medicine]

FIRD Far Infrared Detector

FIRD Fast-Induced Radioactivity Decay

FIRDC Fishing Industry Research and Development Council

FIRE Fellow of the Institute of Radio Engineers

FIRE First ISCCP (International Satellite Cloud Climatology Project) Regional Experiment

FIRE Flight Investigation Reentry Environment

FIRE Focused Infrared Energy (Method)

FIRE Forest Industry Renewable Energy Program [of Natural Resources Canada]

FIRE I First ISCCP (International Satellite Cloud Climatology Project) Regional Experiment, Phase I (1984-1989)

FIRE II First ISCCP (International Satellite Cloud Climatology Project) Regional Experiment, Phase II (1990-1994)

FIRE III First ISCCP (International Satellite Cloud Climatology Project) Regional Experiment, Phase III (1995-)

FIRE 86 First FIRE Cirrus Field Experiment

FIRE 87 FIRE I Marine Stratocumulus IFO (Intensive Field Observations)

FIRE 91 Second FIRE Cirrus Field Experiment

FIRE 93 FIRE Pilot Tropical Cirrus Experiment [Tropical Ocean and Global Atmosphere/Coupled OceanAtmosphere Response Experiment]

FireE Fire Engineer [also Fire Eng or FireEng]

Fire Eng Fire Engineer(ing)

Fire Flammabl Bull Fire and Flammability Bulletin [UK]

FIRE IFO First ISCCP (International Satellite Cloud Climatology Project) Regional Experiment–Intensive Field Observations

FIRE IFO II FIRE IFO Second Cirrus Field Experiment

Fire Mater Fire and Materials [Published in the US]

FIRE MS First ISCCP (International Satellite Cloud Climatology Project) Regional Experiment Marine Stratus

Fire Saf J Fire Safety Journal [US]

FIRE Stratos First ISCCP (International Satellite Cloud Climatology Project) Regional Experiment Marine Stratocumulus IFO (Intensive Field Observations) (California 1987)

FIRETRAC Firing Error Trajectory Recorder and Computer

FIRFD Finite-Impulse Response Filter Design

FIRFT Fast Inversion-Recovery Fourier Transform

FIRI Fishing Industry Research Institute [South Africa]

FIR-IR Far-Infrared–Infrared

FIRL Faceted Information Retrieval System for Linguistics

FIRL Franklin Institute Research Laboratories [Philadelphia, Pennsylvania, US]

FIRM Far Infrared Maser

FIRM Financial Information for Resource Management

.FIRM Business or Firm [Internet Domain Name] [also .firm]

FIRMR Federal Information Resources Management Regulations [US]

FIRMS Far Infrared Molecular Spectroscopy

FIRMS Forecasting Information Retrieval of Management System

FIRO Finite Impulse Response Operator

FiRP Glulam Glulam Reinforced with Fiber-Reinforced Plastics [Note: "Glulam" stands for Glue-Laminated Wood] [also FiRP glulam]

FIRPRECAN Fire Prevention Association of Canada

FIRR Failure and Incidents Report Review Committee [of American National Standards Institute, US]

FIRR Far-Infrared Radiation

FIRR Federal Institute for Reactor Research [Switzerland]

FIRS Far Infrared Source

FIRS Far Infrared Spectroscopy

FIRST Fabrication of Inflatable Reentry Structures for Testing

FIRST Far-Infrared and Submillimeter Telescope

FIRST Fast Interactive Retrieval System Technology

FIRST Federal Information Research Science and Technology Network [of Committee on Scientific and Technical Information, US]

FIRST Financial Information Reporting System

FIRST Florida Integrated Research in Silicon Technologies (Center) [of University of Florida, US]

FIRST Forschungszentrum für Innovative Rechnersysteme und -technologie [Research Center for Innovative Computer Systems and Technology] [of Gesellschaft für Mathematik und Datenverarbeitung, Berlin, Germany]

FIRST Forum of Incident Response and Security Teams

FIRST Fragment Information Retrieval of Structures

FIRT Fertilizer Industry Round Table [US]

FIRTA Fishing Industry Research Trust Account

FIS Fachinformationssystem [Technical Information System, Germany]

FIS Forschungsinstitut und Naturmuseum Senckenberg [Senckenberg Research Institute and Nature Museum, Frankfurt/Main, Germany]

FIS Facility Interface Sheet

FIS Field Information System

FIS Field Injection Titrimetry

FIS Financial Information System

FIS Flight Information Service

FIS Floating-Point Instruction Set

FIS Forest Industry Strategy

FIS Functional Interference Specification

FISA Food Industries Suppliers Association [US]

FISA Forest Industries Safety Association [Canada]

FISC Foundation for International Scientific Coordination [France]

Fisc Fiscal [also fisc]

Fisc Yr Fiscal Year [also fisc yr]

FISDW Field-Induced Spin Density Wave [Solid-State Physics]

FISH First In, Stays Here [also fish]

FISH Fluorescent in-situ Hybridization [Genetics]

FISH Fully-Instrumented Submersible Housing

FISHROD Fiche Information Selectively Held and Retrieved on Demand

FISICA Conferência Portuguesa de Fisica [Portuguese Physics Conference]

F-ISO ISO (International Organization for Standardization) Name, French Spelling

FISS Fédération Internationale des Sociétes Scientifiques [International Federation of Scientific Societies]

FISSS Frequency Independent Strong Signal Suppression

FISST Fissile Solution Storage Tank

FIST Fault Isolation by Semi-Automatic Techniques

Fis Tecnol Fisica e Tecnologia [Italian Publication on Physics and Technology; published by Società Italiana di Fisica, Italy]

FISTM Field-Ion Scanning Tunneling Microscope; Field-Ion Scanning Tunneling Microscopy [also FI-STM]

FIT Failure in Time

FIT Failure(s) in 10^9 Device-Hours [Unit; e.g., 1 FIT Represents 1 Failure in 10^9 Device-Hours, 2 FIT(s) represent 2 Failures in 10^9 Device-Hours, etc.]

FIT Fashion Institute of Technology [US]

FIT Fast Installation Technique

FIT Fault Isolation Test

FIT Federal Income Tax

FIT Féderation Internationale des Traducteurs [International Federation of Translators, Belgium]

FIT Fiber Impregnated with Thermoplastics (Process)

FIT Field Investigation Team

FIT File Information Table

FIT File Inquiry Technique

FIT Flexible Interface Tool

FIT Florida Institute of Technology [Melbourne, US]

FIT Fluctuation-Induced Tunneling (Model of Photoconductivity)

FIT Forest Industries Telecommunications [US]

FIT Forschungsgesellschaft für Informationstechnik mbH [A German Information Technology Research Company located at Hildesheim]

FIT Fourier's Inversion Theorem

FIT Free and Independent Traveller

FIT Fukuoka Institute of Technology [Japan]

FIT Functional Integration Technology

FIT Institut für Angewandte Informationstechnik [Institute for Applied Information Technology; of Gesellschaft für Mathematik und Datenverarbeitung, Berlin, Germany]

FITA Federation of International Trade Associations [US]

FITB Fluorspar International Technical Bureau [Belgium]

FITC Fluorescein Isothiocyanate

FITC-BSA Fluorescein Isothiocyanate–Bovine Serum Albumin (Immunogen)

FITC-Dextran Fluorescein Isothiocyanate-Dextran

FITCE Fédération des Ingénieurs des Télécommunications de la Communauté Européenne [Federation of Telecommunications Engineers of the European Community, Brussels, Belgium]

FITC/PE Fluorescein Isothiocyanate/R-Phycoerythrin (Standard)

FITE Forward Interworking Telephony Event

FITH Fire-in-the-Hole [Mining]

FITI Fabric Inspection Testing Institute [South Korea]

FITS Flexible Image Transport System [of NASA]

FITS Trifluoromethanesulfonate

FITT Forum for International Trade Training [Canada]

FIU Facility Interface Unit

FIU Federal Information Users [US]

FIU Federation of Information Users [US]

FIU Fighter Interception Unit

FIU Fingerprint Identification Unit

FIU Florida International University [Miami, US]

FIV Feline Immunodeficiency Virus [Veterinary Medicine]

FIV Fuel Isolation Valve

FIW Flight Input Workstation

FIW Free in Waggon [also fiw]

FIX Fault Isolator and Exercizer

FIX Federal Information Exchange [US Government]

FIX Federal Internet Exchange [US]

Fix Fixture [also fix]

Fix BC Fixed Boundary Condition

FIXIT Flexible Information Exploitation Interpretative Transfer

Fixs Fixtures [also fixs]

FIZ (*Fachinformationszentrum*) – German for "Center for Scientific and Technical Information"

Fiz Atomn Yad Fizika Atomnoya Yadra [Russian Journal on Atomic Nuclei]

Fiz Elem Chast Atomn Yad Fizika Elementarnykh Chastits i Atomnoya Yadra [Russian Journal on (Elementary) Particles and (Atomic) Nuclei]

FIZ EPM Fachinformationszentrum Energie–Physik–Mathematik [Information Center for Energy, Physics and Mathematics, Karlsruhe, Germany]

Fiz Goren Vzryva Fizika Goreniya i Vzryva [Russian Journal on Combustion, Explosion and Shock Waves]

FIZ KA Fachinformationszentrum Karlsruhe (Gesellschaft für wissenschaftlich-technische Information mbH) [Karlsruhe Center for Scientific and Technical Information, Germany]

Fiz-Khim Mekh Mater Fiziko-Khimicheskaya Mekhanika Materialov [Ukrainian Journal on the Chemistry and Mechanics of Materials]

Fiz Khim Obrab Mater Fizika i Khimiya Obrabotki Materialov [Russian Journal on the Physics and Chemistry of Materials Treatment]

Fiz Khim Stekla Fizika i Khimiya Stekla [Russian Journal of Glass Physics and Chemistry]

Fiz Kristall Fizika Kristallizatsii [Russian Journal on the Physics of Crystals]

Fiz Met Metalloved Fizika Metallov i Metallovedenie [Russian Journal on the Physics of Metals and Metallography]

Fiz Nizk Temp Fizika Nizkikh Temperatur [Ukrainian Journal of Low-Temperature Physics]

Fiz Plazmy Fizika Plazmy [Russian Journal of Plasma Physics]

FIZ-Technik Fachinformationszentrum Technik [Information Center for Engineering and Technology, Germany]

Fiz Tekh Poluprovodn Fizika i Tekhnika Poluprovodnikov [Russian Journal on Semiconductors]

Fiz-Tekh Probl Razrab Polezn Iskop Fiziko-Tekhnicheskie Problemy Razrabotki Poleznykh Iskopaemykh [Russian Technical Physics Journal]

Fiz Tekh Vys Davlenii Fizika i Tekhnika Vysokikh Davlenii [Russian Journal on Technical Physics]

Fiz Tverd Tela Fizika Tverdogo Tela [Russian Journal on Solid-State Physics]

FIZ-W Fachinformationszentrum Werkstoffe e.V. [Materials Information Center, Germany]

FJ Fiji [ISO Code]

FJ Fluid Jet

FJ Fuel-Jet

FJCC Fall Joint Computer Conference [US]

FJD Fiji Dollar [Currency]

FJI Federal Job Information

FJMS Free-Jet Melt Spinning [Metallurgy]

FJSRL Frank J. Seiler Research Laboratory [US Air Force]

FJU Fu-Jen University [Taipeh, Taiwan]

FK Falicov-Kimball (Model) [Physics]

FK Falkland Islands (Malvibas) [ISO Code]

FK Florida Keys [US]

FK Frank-Kasper (Phase) [also F-K] [Metallurgy]

FK Frenkel-Kontorova (Hamiltonian) [Metallurgy]

FK Frenkel-Kuczynski (Model) [Metallurgy]

FK Fixture Key

F-K Frank-Kasper (Phase) [also FK] [Metallurgy]

F-K Fuller-Kinyon (Dry Pump)

F(κ) Magnetic Structure Factor [Symbol]

F$_s$(κ) Nuclear Unit Cell Structure Factor [Symbol]

f$_a$(κ) atomic scattering amplitude [Symbol]

FKB Flight-Display Keyboard

FKM Fluoroelastomer

FKBG Fourdrinier Kraft Board Group [of American Paper Institute, US]

FKBG-API Fourdrinier Kraft Board Group of American Paper Institute [US]

FKBI Fourdrinier Kraft Board Institute [US]

FKBP FK-Binding Protein [Biochemistry]

FKE Forschungsinstitut für Kinderernährung [Research Institute for Child Nutrition, Dortmund, Germany]

FKI Fachverband Klebstoffindustrie [Association of (European) Adhesives Manufacturers, Germany]

FKI Federation of Korean Industries

FKM Forschungskuratorium Maschinenbau [Research Board for Mechanical Engineering, Germany]

FKO Franz-Keldysh Oscillations

FKP Falkland Islands Pound [Currency]

F(k,φ) Elliptic Integral of the First Kind [Symbol]

FL Fail

FL Faraday's Law

FL Fatigue Life [Mechanics]

FL Fault Location; Fault Locator

FL Feed Line

FL Fermi Level [Statistical Mechanics]

FL Fermi Liquid

FL Fick's Law [Physical Chemistry]

FL Field Length

FL Field Lens

FL Field Loss

FL Finite Loading

FL Flame Laser

FL Flammability Limits; Flammable Limits

FL Flammable Liquid

FL Flat Root Buttress Thread [Symbol]

FL Flight Leg

FL Flight Level

FL Flight Lieutenant [also F/L]

FL Flood Light

FL Floor Line

FL Florida [US]

FL Flow Line

FL Fluorescent Lamp

FL Fluorescent Light(ing)

FL Fluoroleucine [Biochemistry]

FL Flux Leakage

FL Flux Line

FL Focal Length

FL Focal Line

FL Foot-Lambert [Unit]

FL Foreign Language

FL Foreign Listing

FL Formal Language

FL Formal Logic

FL Forming Limit [Metallurgy]

FL Fort Lauderdale [Florida, US]

FL Forward Link

FL Fraunhofer Line [Spectroscopy]

FL Free Layer

FL   Free Length (of Springs) [Symbol]

FL   Frenkel and Ladd Method [Metallurgy]

FL   Fresnel Lens

FL   Friedel's Law [Crystallography]

FL   Frontal Lobe (of Brain) [Anatomy]

FL   Full Liner

FL   Full Load

FL   Fully Lamellar (Microstructure) [Metallurgy]

FL   Function Language

FL   Fusible Link

FL/1   Function Language 1 [also FL/I]

F/L   Flight Lieutenant [also FL]

4L   Four-Wheel Drive Position, Low-Range

Fl   Fail [also fl]

Fl   Flanders [Belgium]

Fl   Flemish

Fl   Flash(ing) [also fl]

Fl   Floor [also fl]

Fl   Florin (or Guilder) [Currency of the Netherlands]

Fl   Fluid [also fl]

Fl   Fluorene [also fl]

Fl   Fluorescence; Fluorescent [also fl]

Fl   Fluorine [Abbreviation]

Fl   Fluorspar [Mineral]

Fl   *(Fluvius)* – River

fL   foot-lambert [also fl] [Unit]

fl   flasche [Unit of Measurement for Mercury; 1 flasche equals 76 lb, or 34.47 kg]

fl   flake(s)

fl   *(floruit)* – flourished

fl   *(fluidus)* – Fluid [Medical Prescriptions]

FLA   Fabric Laminators Association [US]

FLA   Fellow of the Library Association [UK]

FLA   First Lord of the Admiralty [UK]

FLA   Flash Lamp Annealing

FLA   Fluorescent Lighting Association

FLA   Four Letter Acronym

FLA   Frame-Lid Assembly

FLA   Fuel Lane Attendant [Autombiles]

Fla   Florida [US]

Fl-AAS   Flame Atomic Absorption Spectroscopy

FLAC   Florida Automatic Computer

FLAD   Fluorescence-Activated Display

FLAE   Faculty of Liberal Arts and Education

FLAG   Fleet Locating and Graphics

FLAG   FORTRAN Load and Go

FLAGS   Far North Liquids and Associated Gas System

FLAIR   FORTRAN Language in Core Rapid Translator

FLAJA   Federación Latinoamericana de Jóvenes Ambientialistas [Latin American Federation of Young Environmentalists, Panama]

Flak   *(Flug(zeug)abwehrkanone)* – German for "Anti-Aircraft Cannon"

FLAM   Forward-Launched Aerodynamic Missile

Flam   Flammable [also flam]

FLAMR   Flores Assembly Program

FLAMR   Forward-Looking Advanced Multilobe Radar

FLAP   Flight Application Software

FLAPHO   Flame Photometer [also Flapho]

FLAPW   Full-Potential Linearized Augmented Plane Wave (Method)

FLAR   Forward-Looking Airborne Radar

FLARE   Fluorinated Poly(arylene Ether)

FLAS   Faculty of Liberal Arts and Sciences [of University of Osaka Prefecture, Osaka, Japan]

FLASH   Fast Low-Angle Shot (Gradient-Echo Sequence) [Physics]

Flash P   Flash Point [also flash p]

FLASTO   Full-Potential Linearized Augmented Slater-Type Orbital (Method) [Physics]

Flav   Flavopiridol [Cancer Treatment Drug]

flav   *(flavus)* – yellow [Medical Prescriptions]

F Layer   Fermentation Layer [also F layer] [Soil Ecology]

F Layer   Ionosphere Layer about 150 Miles (240 km) Above the Surface of the Earth which Reflects Higher-Frequency Radio Waves [also F layer] [also known as Appleton layer]

F_1 Layer   Ionosphere Layer which Begins at About 112 miles (180 km) Above the Surface of the Earth [also F_1 layer]

F_2 Layer   Ionosphere Layer which Begins at About 186 Miles (300 km) from the Surface of the Earth [also F_2 layer]

FLB   Federal Land Bank [US]

FLB   Flow Brazing

FLB   Fondation Louis de Broglie [Louis de Broglie Foundation, France]

FLBE   Filter, Band Elimination [also FL-BE]

FLBIN   Floating-Point Binary

FLBP   Filter Bandpass

FLBR   Fusible Link Bottom Register

FLC   Federal Laboratory Consortium [US]

FLC   Federal Library Committee [now Federal Library and Information Center Committee, US]

FLC   Ferroelectric Liquid Crystal

FLC   Flame Chamber (Process)

FLC   Forming Limit Curve [Metallurgy]

FLC   Fort Lewis College [Durango, Colorado, US]

FLC   Forward Load Control(ler)

FLC   Frame Level Control

FLCA   Forward Load Control(ler) Assembly

FLCS   Fiberoptics Low-Cost System

FLD   Field [also Fld or fld]

FLD   Final Limit–Down [Controllers]

FLD   Fluidic Logic Device

FLD   Formability Limit Diagram; Forming Limit Diagram [Metallurgy]

FLD   Friends of the Lake District

FLDA   Federal Land Development Authority

FLDEC   Floating Point Decimal

FLDL Field Length

fl dr fluid dram [Unit]

FLE Firefly Lantern Extract [Biochemistry]

FLEA Flux Logic Element Array

FLEA Four Letter Extended Acronym

FLEC 1-(9-Fluorenyl)ethyl Chloroformate

(+)-FLEC (+)-1-(9-Fluorenyl)ethyl Chloroformate

(–)-FLEC (–)-1-(9-Fluorenyl)ethyl Chloroformate

FLECHT Full-Length Emergency Cooling Heat Transfer (Test) [Nuclear Reactors]

FLED Flexible Light-Emitting Diode

Fleep Feeble Beep [also fleep]

Flem Flemish

FLEX Flaw Examination

Flex Flexibility; Flexible [also flex]

flex flexible cord

Flex Hd Flexible Head (of Combines) [also flex hd]

FLF Final Limit–Forward [Controllers]

FLF Fixed Length Field

FLF Follow-the-Leader Feedback

Flg Flag [also flg]

Flg Flagship [also flg]

Flg Flange [also flg]

Flg Flooring [also flg]

flgd flanged

Flge Flange [also flge]

FLH Final Limit–Hoist [Controllers]

Fl H Flat Head (Screw) [also FlH]

Flhls Flashless [also flhls]

FLHP Filter, High Pass

FLI Field Length Indicator

FLI Flick [Computers]

FLI Fluorescence Lifetime Instrument

FLI Fluorescence Line Imager

FLI Free Language Indexing

FL/I Function Language One [also FL/1]

FLICC Federal Library and Information Center Committee [US]

FLIH First-Level Interrupt Handler

FLIM Fast Library Maintenance

FLIMBAL Floated Inertial Measurement Ball

F Line F Line [Medium-Blue-Green Fraunhofer Line for Hydrogen Having a Wavelength of 486.1 Nanometers] [also F line] [Spectroscopy]

FLINK Flash/Wink Signal

FLINT Floating Interpretative Language

FLIP Film Library Instantaneous Presentation

FLIP Flight-Launched Infrared Probe

FLIP Floated Lightweight Inertial Platform

FLIP Floating Index Point

FLIP Floating Instrument Platform [of Scripps Institution of Oceanography, US]

FLIP Floating Laboratory Instrument Platform

FLIP Floating Point Interpretative Program

FLIR Forward-Looking Infrared

FLIRT Federal Librarians Round Table [of American Library Association, US]

FLIRT Free Language Information Retrieval Tool

FLIS Functional-Level Information Systems

FLIT Fault Location by Interpretative Testing

FLITE Federal Legal Information through Electronics [US]

FLL Final Limit–Lower [Controllers]

FLL Flux Line Lattice [Metallurgy]

FLL Flux Locked Loop

FLL FoxPro Link Library

FLLK Frustum Lifting Lug Kit

FLM Flight Line Mechanic

FLM Flux Lattice Melting [Metallurgy]

Fl Mech Fluid Mechanical; Fluid Mechanics

FLMEM Floppy Disk Memory

FLMPTS Future Land Mobile Personal Telephone Service

FLMTO Full-Potential (Version of) Linear(ized) Muffin-Tin Orbital (Method) [Materials Science]

FLN Fluorescence Line-Narrowing

FLN Fuel Line

FLNH Fucosyllacto-N-Hexaose [Biochemistry]

FLNH III Fucosyllacto-N-Hexaose III [Biochemistry]

FLNPP Fedeal Library Network Prototype Project [US]

FLO Frederick Law Ohmsted [US Urban Planner]

FLO Functional Line Organization

FLOA Frederick Law Ohmsted Association [US]

FLOAD Font Load (Program)

floc flocculent

Flo Con Floating Container

FLODAC Fluid-Operated Digital Automatic Computer

FLOLS Fresnel Lens Optical Landing System

FLOOD Fleet Observation of Oceanographic Data [US Navy]

FLOP Floating Octal Point

FLOP Floating Point [also flop]

FLOP Floating-Point Operation [also flop]

FLOPS Floating-Point Operations per Second [also flops]

Flor Florida [US]

FLORICO Florida-Puerto Rico Submarine Cable

Flot Flotation [also flot]

FloTech Florida Institute of Technology [Melbourne, US]

FLOTOX Floating-Gate Tunnel-Injection Nonvolatile Memory

FLOTOX Floating Gate Tunnel Oxide

FLOTRAN Flowcharting FORTRAN

flour flourished [also flor]

FLOW Flow Welding

FLOWS FAA (Federal Aviation Administration) Lincoln Laboratory Operational Weather Studies [US]

FLOWS Forward Line Osmosis Water Sets

FLOX Liquid Fluorine/Liquid Oxygen (Mixture) [also flox]

fl oz fluid ounce [Unit]

FLP Filter, Low Pass

FLP Flux Line Pinning [Physics]

FLP Foreign Language Program

Fl P Flash Point [also fl p]

Flp Flameproof [also flp]

Flp Flap [also flp]

FLPAU Floating Point Arithmetic Unit

FLPDC Floppy Disk Controller

FLPE Fluorinated Polyethylene

FLPL FORTRAN List Processing Language

FLPP Fluorinated Polypropylene

Flpr Flameproof [also flpr]

FLPS Flight Load Preparation System

FLR Fermi Level Referencing

FLR Final Limit–Reverse [Controllers]

FLR Flora Reserve [Australia]

FLR Fluoroleucine Resistance [Biochemistry]

FLR Forward-Looking Radar

FLR Fukuyama-Lee-Rice (Model)

.FLR Folder [File Name Extension] [also .fdr]

Flr Failure [also flr]

Flr Flare [also flr]

FLRT Federal Librarian Round Table [US]

FLRT Flush Right [also FlRt]

Flry Flurry [also flry]

FLS Fair Labor Standards

FLS Flight Surgeon

FLS Flow Switch

FLS Free Line Signal

FLS Front Lamp Support

FLSA Fair Labor Standards Act [US]

FLSC Flexible Linear-Shaped Charge

Flshls Flashless [also flshls]

FLSH Flush [also Flsh]

FLSH RT Flush Right (Alignment) [also Flsh Rt]

FLSM Floquet-Liouville Supermatrix [Mathematics]

FLSP Flame Spraying [Metallurgy]

FLSP Fluorescein Labeled Serum Protein [Biochemistry]

FLT Fault [also flt]

FLT Fault Location Technology

FLT Fermat's Last Theorem [Mathematics]

FLT Fermi Level Tuning

FLT Fermi Liquid Theory

FLT First Law of Thermodynamics

FLT Fork Lift Truck

Flt Fault [also flt]

Flt Filter [also flt]

Flt Fleet [also flt]

Flt Flight [also flt]

Flt Float(ing) [also flt]

$f(\lambda, T)$ function of wavelength and (absolute) temperature [Symbol]

Flt Cad Flight Cadet [also FltCad]

FLTCAL Flight Calibration Procedures

FLTCK Flight Check

FLTF Field Lysimeter Testing Facility

Fltg Floating [also fltg]

Flt Lt Flight Lieutenant [also FltLt]

FLTS First Line Troubleshooting System

Flt Sgt Flight Sergeant [also FltSgt]

FLTR Fusible Link Top Register

Fltr Filter [also fltr]

FLTSATCOM Fleet Satellite Communications System [US Navy]

FLU Final Limit–Up [Controllers]

FLU Full Line-Up

flu influenza

FLUC Fluctuation [also Fluc or fluc]

Fluid Abstr Fluid Abstracts

Fluid Dyn Fluid Dynamics [Translation of: *Izvestiya Akademii Nauk SSSR, Mekhanika Zhidkosti i Gaza (USSR)*; published in the US]

Fluid Dyn Res Fluid Dynamics Research [Journal published in the Netherlands]

Fluidics Fluid Dynamics [also fluidics]

Fluidics Fluid Logic Technology [also fluidics]

Fluid Mech Sov Res Fluid Mechanics–Soviet Research [Journal published in the US]

Fluid/Particle Sep J Fluid/Particle Separation Journal [US]

Fluid Phase Equilib Fluid Phase Equilibria [Journal published in the Netherlands]

FLULC Forest Land Use Liaison Committee

FLUO Division of Fluorine Chemistry [of American Chemical Society, US]

Fluor Fluorescence; Fluorescent [also Fluores, fluores, or fluor]

Fluorexon Fluorescein Complexon

FLUTD Feline Lower Urinary Tract Disease [Veterinary Medicine]

Flux Luxembourg Franc [Currency]

FLV Freund Leukemia Virus

Flvr Flavo(u)r [also flvr]

FLW Feetlot Waste

FLW Fellows [Tactical Air Navigation Station]

FLWF Feetlot Waste Fiber

FLWF Frank Lloyd Wright Foundation [US]

flwg following

FM Face Measurement

FM Fachnormenausschuß Maschinenbau [Standards Committee on Mechanical Engineering, Germany]

FM Face Mill(ing)

FM Facilities Management

FM Factory Mutual (Testing Laboratory) [US]

FM Faculty of Materials

FM Faculty of Metallurgy

FM Failure Mechanism

FM Failure Mode

FM Fair Merchantable [also fm]

FM Fan Marker

FM Fast Memory

FM Fault Modelling

FM Federation of Malaysia

FM Feedback Mechanism
FM Feroba Magnet
FM Ferrimagnet(ic); Ferrimagnetism
FM Ferromagnet(ic); Ferromagnetism
FM Ferromagnetic Metal
FM Ferrous Material
FM Ferrous Metal
FM Ferrous Metallurgy
FM Fibroblast Medium [Biochemistry]
FM Field Magnet
FM Field Manual
FM Field Marshall
FM Field Matrix
FM Field Modulation
FM File Maintenance
FM File Management; File Manager
FM Filler Material
FM Filler Metal
FM Filter Medium
FM Financial Management
FM Fine Measurement
FM Finemet-Type (Ferromagnetic Alloy)
FM (Abrams) Fineness Modulus
FM Fire Main
FM Flexibility Matrix
FM Flexible Manufacturing
FM Flexural Modulus
FM Flexural Moment
FM Flight Manifest
FM Flight Model
FM Floating Magnet
FM Flotation Machine
FM Flour Mill(ing)
FM Fluorescence Microscope; Fluorescence Microscopy
FM Flowmeter
FM Fluid Mechanics
FM Focusing Magnet
FM Fonte Mince Process [A Casting Process for Making Thin-Walled Castings]
FM Foramen Magnum (of Brain) [Anatomy]
FM Force Majeure
FM Foreign Matter
FM Forensic Medicine
FM Forest Management; Forest Manager
FM Fracture Mechanics
FM Free Machining
FM Free Moisture [Mining]
FM Freezing Mixture
FM Frequency Management
FM Frequency Meter
FM Frequency Modulated; Frequency Modulation
FM Frequency Multiplex(ing); Frequency Multiplier
FM From [Amateur Radio]
FM Fuel Manifold

FM Full Moon
FM Fulminate of Mercury
FM Fumi Method
FM Iron Molybdate
FM Micronesia [ISO Code]
F&M Feinwerktechnik und Meßtechnik [German Journal on Precision Engineering and Measurement Technology published by Verein Deutscher Ingenieure/Verein Deutscher Elektrotechniker]
F&M Fokker F-27 and Merlin IV [Atlantic Stratocumulus Transition Experiment Airplanes]
F&M Force and Momentum (Computer) [Aeronautics]
F/M Female/Male (Connector)
Fm Fermium [Symbol]
Fm Formic Acid
Fm-253 Fermium-253 [also ^{253}Fm, or Fm^{253}]
Fm-254 Fermium-254 [also ^{254}Fm, or Fm^{254}]
Fm-255 Fermium-255 [also ^{255}Fm, or Fm^{255}]
Fm-256 Fermium-256 [also ^{256}Fm, or Fm^{256}]
Fm-257 Fermium-257 [also ^{257}Fm, or Fm^{257}]
F/m Farad per meter [also F m^{-1}]
fm fair merchantable [also FM]
fm fathom [Unit]
fm femtometer [Symbol]
FMA Fabricators and Manufacturers Association [now Fabricators and Manufacturers Association International, US]
FMA Failure Mode Analysis
FMA Fan Manufacturers Association [UK]
FMA Fertilizer Manufacturers Association [UK]
FMA Flexicore Manufacturers Association [US]
FMA Flight Mode Annunciation
FMA Food Machinery Association [UK]
FMA Forest Management Area
FMA Fragment Mass Analyzer
FMA Fragrance Materials Association [US]
FMA Free Mineral Acidity
FMAA Furniture Manufacturers' Association of Australia
FMAC Frequency Management Advisory Council
FM/AF Ferromagnet/Antiferromagnet [Solid-State Physics]
FM-AFM Ferromagnet(ic)–Antiferromagnet(ic) [Solid-State Physics]
FMAHTS Flight Manifest and Hardware Tracking System
FMAI Fabricators and Manufacturers Association International [US]
FM/AM Frequency Modulation/Amplitude Modulation
FMAR Ferromagnetic Antiresonance
FMAS Florida Marine Aquarium Society [US]
F-19 MAS NMR Fluorine-19 Magic Angle Spinning Nuclear Magnetic Resonance [also ^{19}F MAS NMR, F-19 MAS-NMR, or ^{19}F MAS-NMR]
FMB Federal Maritime Board
FMB Federation of Master Builders [UK]
FMB Ferromagnetic Bilayer [Solid-State Physics]
FMB Foundation for Microbiology

FMBRA Flour Milling and Baking Research Association [UK]

FMC Federal Maritime Commission [US]

FMC Felt Manufacturers Council [US]

FMC Ferromagnetic Ceramics

FMC Ferrous Metallurgy Committee [of The Minerals, Metals and Materials Society, US]

FMC Filler Metal Comparison (Chart) [of American Welding Society, US]

FMC First Muon Collider (Workshop) [at Fermilab, Batavia, Illinois, US]

FMC Fixed Message Code

FMC Flatness Measuring and Control (System) [Metallurgy]

FMC Flexible Machining Center

FMC Flexible Manufacturing Cell

FMC Flexible Manufacturing Center

FMC Florida Memorial College [St. Augustine, US]

FMC Flow Microcalorimeter

FMC Flutter Mode Control

FMC Food Machinery Corporation [US]

FMC Forces Mobile Command [of Canadian Forces]

FMC Ford Motor Company [Detroit, Michigan, US]

FMC Fort McCoy, Wisconsin [Meteorological Station Identifier]

FMC Forward Motion Compensation

FMC Franklin and Marshall College [Lancaster, Pennsylvania, US]

FMC Frequency-Modulated Cyclotron [of US Atomic Energy Commission]

FMC Friele-MacAdam-Chickering (Color Difference)

FMC-2 Friele-MacAdam-Chickering, Version 2 (Color Difference)

FMCE Federation of Manufacturers of Construction Equipment [UK]

FMCEC Federation of Manufacturers of Construction Equipment and Cranes [UK]

FMCF First Manned Captive Flight

FMCG Fast-Moving Consumer Goods [also fmcg]

FMCS Federal Mediation and Conciliation Service [US]

FMCS Flight Management Computer System

FMCW Frequency-Modulated Continuous Wave [also FM-CW]

FMD Ferromagnetic Domain [Solid-State Physics]

FMD Foot-and-Mouth Disease [Veterinary Medicine]

FMD Freely Migrating Defect [Metallurgy]

FMD Frequency Multiplexing Division

FMD Function Management Data

FMDC Flange-Mounted Disconnect

FMDM Flex Multiplexer/Demultiplexer

FMDS Fleet Management Demonstration System

FMDU Fast Multiply/Divide Unit

FMDV Foot-and-Mouth Disease Virus

FME Faculty of Mechanical Engineering

FME Field Moisture Equivalent (of Soil)

FME Flow Modulation Epitaxy

FME Frequency Measuring Equipment

FMEA Failure Modes and Effects Analysis

FMEC Forward Master Events Controller

FMECA Failure Modes, Effects and Criticality Analysis

FMEF Fuels and Materials Examination Facility [Nuclear Engineering]

FMEI Farm Management Extension Initiative [Western Australia]

FMEP Friction Mean Effective Pressure

FMES Full Mission Engineering Simulator

F-MET-PHE Formyl-Methionyl-Phenylalanine

FMEVA Floating Point Means and Variance

FMF Ferromagnetic Film

FMF Fleet Marine Forces [of US Marine Corps]

FMF Food Manufacturers Federation [UK]

FMF Freiburger Materialforschungszentrum [Freiburg Materials Research Center at the Albert-Ludwigs-Universität, Germany]

FMF Fuel Manufacturing Facility [of Argonne National Laboratory-West, Idaho Falls, US]

FMF Fysisch-Mathematische Faculteitsvereniging [Physical-Mathematical Faculty Association, Netherlands]

FMFB Frequency Modulation with Feedback

FM/FM Frequency Modulation/Frequency Multiplexing

FM-FM Frequency Modulation–Frequency Modulation (System)

FMG Ferrimagnetic Garnet

FMG Ferrous Metallurgy Grant [US]

FMG Food Machinery Group [UK]

FMG Malagasy Franc [Currency of Malagasy Republic]

FMGP Ferrous Metallurgy Grant Program [of Iron and Steel Society Foundation, US]

FMGS Flight Management and Guidance System

FMHA Farmers Home Administration [also FmHA]

FMHS Flexible Materials Handling System

FMI Ferritic Malleable Iron [Metallurgy]

FMI Finnish Meteorological Institute

FMI For Materials Ingenuity

FMI Frequency Modulation Intercity (Broadcasting)

FMIC Frequency Monitoring and Interference Control

FMICW Frequency-Modulated Intermittent Continuous Wave

FMICW Frequency-Modulated Interrupted Continuous Wave

FM/I/FM Ferromagneti/Insulator/Ferromagnet (Trilayer) [Solid-State Physics]

FMIG Food Manufacturers Industrial Group [UK]

°F/min Degree Fahrenheit per Minute [also °F min^{-1}]

FMIP Financial Management Improvement Program

FMIS Fiscal Management Information System

FMIT Fusion Materials Irradiation Test (Facility) [Nuclear Engineering]

FMK Fibroblast Medium Kit [Biochemistry]

FMK Finnish Markka [also Fmk] [Currency of Finland]

FML Fault Message Line

FML Feedback, Multiple Loop

FML File Manipulation Language

FML Final Materials List

FML Frequency-Modulated Laser

FML Functional Materials Research Laboratory [of NEC Corporation, Kawasaki, Kanagawa, Japan]

FMLF File Management Loading Facility

FMLS Full-Matrix Least Squares

FMM Faculty of Mining and Metallurgy

FMM Fast Multiple Method

FMM Ferromagnetic Material

FMM Field Mapping Method

FMM Financial Management Manual

FMM Fringed Micelle Model

FMM Fizika Metallov i Metallovedenie [Russian Journal on Metal Physics and Metallography]

FMMA Field-Modulated Microwave Absorption

FMN Flavin Mononucleotide [Biochemistry]

FMN 1-Fluoro-2-Methylnaphthalene

FMN Iron Manganese Nitride

Fmn Formation [also fmn]

FMNH Field Museum of Natural History [Chicago, Illinois, US]

FMNH$_2$ Flavin Mononucleotide, Reduced Form [Biochemistry]

FMNR Ferromagnetic Nuclear Resonance

FMO Flatland Meteorological Observatory

FMO Fleet Mail Office [US Military]

FMO Flight Medical Officer

FMO Fragment Molecular Orbital (Extended Hueckel Approximation) [Physical Chemistry]

FMO Free-Electron Molecular Orbital [Physical Chemistry]

FMOA First Moment of Area

FMOC 9-Fluorenylmethoxycarbonyl

N-FMOC N-Fluorenylmethoxycarbonyl

FMOC-Cl 9-Fluorenylmethoxycarbonyl Chloride

FMOC-Cl 9-Fluorenylmethyl Chloroformate

FMOC-Gly 9-Fluorenylmethoxycarbonyl Glycine

FMOC-OBT 9-Fluorenylmethylcarbonate-O-Benzotriazolyl

FMOC-ONSu 9-Fluorenylmethoxycarbonyl Succinimide

FMOF First Manned Orbital Flight

FMOFEVA First Manned Orbital Flight with EVA (Extravehicular Activity)

FMOFPL First Manned Orbital Flight with Payload

FMOI First Moment of Inertia

FMP Faculty of Mathematics and Physics

FMP Fellgett's Multiplex Principle

FMP Flight Mechanics Panel [of NATO Advisory Group for Aerospace Research and Development]

FMP 2-Fluoro-2-Methylpropane

FMP Formation Pressure

FMP Forschungsinstitut für Molekulare Pharmakologie [Research Institute for Molecular Pharmacology, Berlin, Germany]

FMP Fossil-Energy Materials Proceedings [of Oak Ridge National Laboratory, US Department of Energy]

FMP Fructose Monophosphate [Biochemistry]

FMP Full Metal Patched (Bullet)

FMPC Faculty of Metal Physics and Chemistry

FMPCW Frequency-Modulated Pulse Continuous Wave

FMPE Federation of Master Process Engravers [UK]

FM/PM Frequency Modulation/Phase Modulation

FMPP Flexible Multi-Pipeline Processor

FMPS FORTRAN Mathematical Programming System

FMPS Functional Mathematical Programming System

FMPT First Materials Processing Test [of NASA]

FMPTS Fluoromethylpyridinium-p-Toluenesulfonate

FMQ Frequency-Modulated Quartz-Circuit

FMR Ferromagnetic Resonance

FMR Ferromagnetoresistance

FMR Field Modification Request

FMR Frequency-Modulated Radar

FMR Frequency-Modulated Receiver

FMR Frequency-Modulated Refractometer

FMR Function Maximum Rate

fmr former

FMRC Factory Mutual Research Corporation [US]

FMRL Ford Motor Research Laboratories [US]

fmrly formerly

FMRR Financial Management Rate-of-Return

FMS Facilities Management System

FMS Facility Mapping System

FMS Facility Monitoring System

FMS Facsimile Mail System

FMS Farm Management System

FMS Federation of Materials Societies [US]

FMS Festkörper-Massenspektrometrie [Solid Mass Spectrometry]

FMS File Management Supervisor

FMS File Management System

FMS Financial Management System

FMS Fire Management System

FMS First Melt Sample [Metallurgy]

FMS Flexible Machining System

FMS Flexible Manufacturing System

FMS Flight Management System

FMS Flow Measuring System

FMS Flux Monitoring System

FMS Food Management Subsystem [of NASA]

FMS Food Management System

FMS Foreign Military Sales

FMS Forms Management System

FMS FORTRAN Monitor System

FMSA Fellow of the Mineralogical Society of America

FMSC Flexible Manufacturing System Complex

FM/SC Ferromagnet/Semiconductor (Interface)

FMSI Food Machinery Service Institute [US]

FMSI Friction Material Standards Institute [US]

FMSL Fort Monmouth Signal Laboratory [of US Army in New Jersey]

FMSO Fleet Material Supply Office [of US Military]

FMSP Frequency Modulation Signal Processor

FMSR Fast Mixed Spectrum Reactor

FMSR Finite Mass Sum Rule

FMSWR Flexible Mild Steel Wire Rope

FMT Flexible Manufacturing Technology

FMT Flight Management Team

FMT Flour Milling Technology

FMT Flush Metal Threshold

FMT Format [also Fmt, or fmt]

FMT Frequency-Modulated Transmitter

FMTA Farm Machinery and Tractor Trade Association [Australia]

FMTP File Management Transaction Processor

FMTS Field Maintenance Test Station

FMU Flow Management Unit

FMU Forward Muon Chamber [Physics]

FMV Fair Market Value

FMV Formation Volume Factor

FMV Full Motion Video

FMVE Perfluoro Methyl Vinyl Ether

FMVSS Federal Motor Vehicle Safety Standard [US]

FMW Formulated Molecular Weight

FMWA Fixed Momentum Wheel Assembly

FMX Frequency-Modulated Transmitter; Frequency Modulation Transmitter

FN False Negative [Medicine]

FN Family Name

FN Fast Neutron

FN Fecal Nitrogen

FN Ferrite Number [Metallurgy]

FN Fiber Node [Telecommunications]

FN Fibronectin [Biochemistry]

FN Filter Network

FN Fixed Needle

FN Fixed Nuclei (Approximation) [Physics]

FN Flange Nut

FN Flat-Nosed

FN Flat-Nose Projectile

FN Flight Nurse

FN 1-Fluoronaphthalene

FN Force Fit (or Shrink Fit) [ANSI Symbol]

FN Foil Normal (of Specimen)

FN Fowler-Nordheim (Equation) [Electron Tunneling]

FN Free Neutron

FN Fresnel Number [Physics]

FN Function [also Fn]

FN Functional Network

FN Futures Network [UK]

FN1 Light Drive Fit (H6/...) [ANSI Symbol]

FN2 Medium Drive Fit (H7/s6) [ANSI Symbol]

FN3 Heavy Drive Fit (H7/t6) [ANSI Symbol]

FN4 Force Fits (H7/u6) [ANSI Symbol]

FN5 Force Fits (H8/x7) [ANSI Symbol]

F/N Ferromagnetic/Nonferromagnetic (Boundary)

5N- Nigeria [Civil Aircraft Marking]

Fn Footnote [also fn]

Fn Human Fibronectin [Biochemistry]

$f(\nu)$ frequency distribution (of oscillators) [also $f(\nu)$]

FNA (Fachnormenausschuß) –German for "Standards Committee"

FNA Final Approach

FNA Fixed Nuclei Approximation

FNA Flora North America (Project) [of American Society of Plant Taxonomists]

FNA 4-Fluoro-3-Nitrophenyl Azide

FNA Fuming Nitric Acid

FNA Fujitsu Network Architecture

FNAA Fast Neutron Activation Analysis

FNAB Fédération Nationale des Artisans du Bâtiment [National Federation of Construction Workers, France]

FNACE Fellow of NACE (National Association of Corrosion Engineers) International

FNAI Florida Natural Areas Inventory [US]

FNAL Fermi National Accelerator Laboratory [of US Department of Energy at Batavia, Illinois, US] [also Fermilab]

FNB Fédération Nationale du Bâtiment [National Federation of Building Trades, France]

FNB Food and Nutrition Board [of National Research Council, US]

FNBIC Florida/NASA Business Incubation Center [at NASA Kennedy Space Center, Florida, US]

FNC Federal Networking Council [US Coordinating Group of Advanced Research Projects Agency, Department of Defense, Department of Energy, National Aeronautics and Space Administration, National Science Foundation, and (Department of) Health and Human Services]

FNC Ferritic Nitrocarburizing [Metallurgy]

FNCC Foreign Claims Commission

FNCI Financial News Composite Index

FNCIAWPRC Finnish National Committee of the International Association on Water Pollution Research and Control

FNCT Function [also Fnct or fnct]

FND Facility Need Date

fnd found

Fndn Foundation [also fndn]

Fndry Foundry [also fndry]

FNE Fachnormenausschuß Elektrotechnik [Standards Committee on Electrical Engineering, Germany]

FNERP Federal Nuclear Emergency Response Plan [Canada]

FNF First Normal Form

FNFM Faculty of Nonferrous Metallurgy

Fng Furnishing [also fng]

FNH Flashless Nonhygroscopic (Powder)

FNHP Federation of Nurses and Health Professionals [US]

FNIMC Florida Normal and Industrial Memorial College [St. Augustine, US]

fnl final

FNM Fachnormenausschuß Materialprüfung [Standards Committee on Materials Testing, of Deutsches Institut für Normung, Germany]

F-NMR Fluorinated Nuclear Magnetic Resonance

F-19 NMR Fluorine-19 Nuclear Magnetic Resonance [also ^{19}F NMR]

FNMRI Flow Nuclear Magnetic Resonance Imaging

FNN First Nearest Neighbor (Approximation) [Crystallography]

FNNE Fachnormenausschuß Nichteisenmetalle [Standards Committee for Nonferrous Metals] [of DIN–Deutsches Institut für Normung, Germany]

FNNPE Federation of Nature and National Parks of Europe

FNO Fixed Nuclear Orientation (Approximation) [Physics]

FNOC Fleet Numerical Oceanography Center

FNP Floating Nuclear Plant

FNP Front-End Network Processor

Fn P Fusion Point [also fn p]

FNPA Foreign Numbering Plan Area

FNPS 4-Fluoro-3-Nitrophenyl Sulfone

FNPS bis(4-Fluoro-3-Nitrophenyl) Sulfone

FNPT National Pipe Thread, Female [also (F)NPT]

Fn Pt Fusion Point [also fn pt]

FNR Fast-Neutron (Nuclear) Reactor

FNR Ferromagnetic Nuclear Resonance

FNR File Next Register

FNR Ford Nuclear Reactor [of University of Michigan, Detroit, US]

FNRC Food and Nutrition Research Center [Philippines]

FNRL Fixed Nitrogen Research Laboratory

FNRS Fonds National de la Recherche Scientifique [National Fund for Scientific Research, Belgium]

FNS Fast-Neutron Spectrometry

FNS File Nesting Store

FNS Food and Nutrition Service [of US Department of Agriculture]

FNS Fusion Neutron Source

F-N-S Ferromagnet–Normal Metal–Superconductor (Multilayer)

FNSCC Federation of Nuclear Shelter Consultants and Contractors [UK]

Fnsh Finish [also fnsh]

FNT Failure Notification Telex

FNT Finger-Nose Test [Medicine]

FNT Flint, Michigan [Meteorological Station Identifier]

FNT Font [also Fnt]

FNT Fowler-Nordheim Tunneling [Electronics]

FNT Fumaronitriledithiolate

FNT Fusion Nuclear Technology

.FNT Font [File Name Extension]

FNTU Federation of Norwegian Transport Users

FNUG Federation of NCR (Computer) Users Groups [US]

F-Number Feigenbaum Number (or Universal Number) [also F-number]

FNWF Fleet Numerical Weather Facility [US Navy]

FO Factory Order

FO Fail Open

FO Fail Operational

FO Fan-Out [Electronics]

FO Faroe Islands [ISO Code]

FO Fast Operate

FO Fatty Oil

FO *(Fernsprechordnung)* – German for "Telephone Regulations"

FO Fiber-Optic(s) [also F-O, or F/O]

FO Field Office(r)

FO Field Operations

FO Field Order

FO Field Ordnance

FO File Organization

FO Firing Order

FO Flare-Out [Aerospace]

FO Flashover

FO Flash Override

FO Flight Officer [also F/O]

FO Flight Operations

FO Flow Orientation

FO Flying Officer

FO Forced Oscillation

FO Forced-Oil (Cooling)

FO Foreign Object

FO Foreign Office [UK]

FO Forward Observer

FO Free Overside

FO Fuel Oil

FO Functional Objective

FO Furnace Oil

F&O Facilities and Operations [of Pacific Northwest National Laboratory, Richland, Washington, US]

F&O (Department of) Fisheries and Oceans [Canada]

F/O Flare-Out [Aerospace]

F/O Flight Officer [also FO]

F/O Fluorine-to-Oxygen (Ratio)

F/O Fuel/Oxidizer (Ratio)

F-O Fiber-Optic(s) [also FO or F/O]

Fo Forsterite [Mineral]

Fo Fourier Number (of Diffusion) [Symbol]

Fo$_f$ Fourier Number (of Fluid Mechanics) [Symbol]

fo folio

fo forsterite [Mineral]

fo Faroe Islands [Country Code/Domain Name]

fΩ femtoohm [also fOhm]

FOA Forced-Oil and Forced-Air Cooling

FOA Foreign Operations Administration [US]

FOA Forsvarets Forskningsanstalt [National Defense Research Institute, Stockholm, Sweden]

FOA Free of Average

FOA Free on Aircraft [also foa]

FOA Free on Board Aircraft [also foa]

FOA Friends of Animals [also FoA, US]

FOAF Friend of a Friend [Internet]

FOA Rep FOA Reports [Reports on Defense Research published by of Forsvarets Forskningsanstalt, Stockholm, Sweden]

FOB Federal Office Building

FOB Flight Operations Building

FOB Forward Operating Base

FOB Free off Board [also fob]

FOB Free on Board [also fob]

FOB Freight on Board

FOB Fuel-Oil Blend

FOB Fuel on Board

FOBFO Federation of British Fire Organizations [UK]

FOB/FOB Free on Board/Free off Board [also fob/fob]

FOBS Fractional Orbit(al) Bombardment System

FOC Chemical Flux Cutting [Metallurgy]

FOC Fiber Optic Cable

FOC Fiber Optic Communications

FOC Fiber Optic Converter

FOC Fluorinated Organic Compound

FOC Free of Charge [also foc]

FOC Free on Car [also foc]

FOC Full Operational Capability

Foc Focal; Focus [also foc]

FOCAL Formula Calculator

FO Can Fisheries and Oceans Canada [formerly Department of Fisheries and Oceans]

FOCH Forward Channel

FOCI Fisheries-Oceanography Coordinated Investigations (Program) [of National Oceanic and Atmospheric Administration–Pacific Marine Environmental Laboratory, US]

FOCI First Order Configuration Interaction

FOC/LAN Fiber Optic Communications/Local Area Networks (Conference and Exposition)

FOCOL Field of Fire, Observation Cover and Concealment, Obstacles, Lines of Communication

Fo'c'sle Forecastle [also fo'c'sle]

FOCSY Foldover-Corrected Spectroscopy

FOCUS Forschungszentrum für Offene Kommunikationssysteme [Research Center for Open Communication Systems; of Gesellschaft für Mathematik und Datenverarbeitung, Berlin, Germany]

FOCUS Forum on Computing: Users and Services [of CERN–European Laboratory for Particle Physics, Geneva, Switzerland]

FOCUS Forum of Control Data Users

FOD Fax On Demand

FOD Field Operations Division [of International Union for the Conservation of Nature and Natural Resources]

FOD Field Ordnance Depot

FOD Flight Operations Department [of Civil Aviation Authority, UK]

FOD Flight Operations Directorate [of NASA Johnson Space Center, Houston, Texas, US]

FOD Foreign-Object Damage

FOD Foreign-Object Debrish

FOD Fourth-Order Differencing (Method)

FOD Free of Damage [also fod]

FOD Functional Operational Design

FOD 1,1,1,2,2,3,3-Heptafluoro-7,7-Dimethyl-4,6-Octanedionate; 1,1,1,2,2,3,3-Heptafluoro-7,7-Dimethyl-4,6-Octanedionato- [also fod]

FODA Fiber-Optic Demonstration Array

FODIC Photoelectric Document Input to Computer Tape

FODWBA First-Order Distorted Wave Born Approximation [Physics]

FOE Figure-of-Eight (Reception)

FOE First-Order Effect

FOE Flight Operations Engineer

FOE Friends of the Earth [also FoE, US]

FOEB Fuel-Oil Equivalent Barrel

FOEC First-Order Elastic Constant

FOEC Fourth-Order Elastic Constant

FOEF Friends of the Earth–France [also FOE(F)]

FOEHK Friends of the Earth–Hong Kong [also FOE(HK)]

FOEI Friends of the Earth International

FoEng Fellowship of Engineering [UK]

FOEP Friends of the Earth–Portugal

Foerdmittel J Foerdermittel Journal [German Journal on Conveying Equipment]

FO-ESR Flow Orientation–Electron Spin Resonance

FOEUK Friends of the Earth–United Kingdom [also FOE(UK)]

FOF Factory of the Future

FOF First Operational Flight

FOF First Orbital Flight

FOF Flatness of Field

FOF Flight Operations Facilities

FO/FS Fail-Operational/Fail-Safe

FOG Fiber-Optic Gyro(scope)

FOG First Osborne Group [US]

FOG-M Fiber-Optic Guided Missile

FOGRA (Deutsche) Forschungsgesellschaft für Druck- und Reproduktionstechnik [German Research Society for Printing and Reproduction Technology]

FOGRA Lit FOGRA Literaturdienst [German Publication on Printing and Reproduction Technology; provided by Deutsche Forschungsgesellschaft für Druck- und Reproduktionstechnik]

FOGRA Mitt FOGRA Mitteilungen [German Communications on Printing and Reproduction Technology; published by Deutsche Forschungsgesellschaft für Druck- und Reproduktionstechnik]

FOH Forced Outage Hours

fOhm femtoohm [also fΩ]

FOHMD Fiber-Optic Helmet-Mounted Display

FOI Fiber-Optic Illumination (System)

FOI First-Order Interpolator

FOI Follow-On Interceptor

FOI Freedom of Information

FOIA Freedom of Information Act [US]

FOIF Free Oceanographic Instrument Float

FOIH Flight Operations Integration Handbook

FOIL File-Oriented Interpretative Language

FOIRL Fiber-Optic Inter-Repeater Link [of Institute of Electrical and Electronics Engineers, US]

FOK First-Order Kinetics

FOK Free of Knots [Lumber]

Fok Flavobacterium okeanokoites [Molecular Biology]

FOL Facility Operating License

FOL Field of Lens

FOL First-Order Language

FOL First-Order Logic

fol folio

fol following

FOLACL Federation of Local Authority Chief Librarians [UK]

Fold Folding [also fold]

Folding Carton Ind Folding Carton Industry [Published in the UK]

Folding Des Folding and Design [International Biology Journal]

FOLDOC Free On-Line Dictionary of Computing

Folia Fac Sci Nat Univ Purkyn Brun, Phys Folia Facultatis Scientarum Naturalium Universitatis Prkkynianae Brunensis, Physica [Czech Publications of the Faculty of Natural Sciences, University J.E. Purkyne, Brno, Physics Series]

Folia Pharmacol Japon Folia Pharmacologica Japonica [Pharmacological Journal published in Japan]

Folia Phoniat Logoped Folia Phoniatrica et Logopedia [Journal on Phoniatrics and Logopedics published in Switzerland]

Folia Primatol Folia Primatologica [International Journal on Primatology]

foll following

FOLR Foreign Ownership Land Register [of the State of Queensland, Australia]

fols folios

folw follow

folwg following

FOLZ First-Order Laue Zone [Crystallography]

FOM Factor of Merit

FOM Faculty of Medicine

FOM Fiber-Optic Material

FOM Fiber-Optic Meter

FOM Fiber-Optic Modem

FOM Fiber Optics Microscope; Fiber Optics Microscopy

FOM Figure of Merit [Electronics]

FOM Flight Operations Management

FOM (Stichting voor) Fundamenteel Onderzoeg der Materie [Foundation for Fundamental Research on Matter, Utrecht, Netherlands]

fΩm² femtoohm square meter [also fΩ·m² or fΩ-m²]

FOMBT First-Order Many-Body Theory [Physics]

FOMC Federal Open Market Commission [US]

FOMC Federal Open Market Committee [US]

FOMCAT Foreign Material Catalogue

FOMD First-Order Magnetic Phase Transition Temperature Calibration **Device**

FOMP First-Order Magnetization Process

FOMR First-Order Moment Reorientation

FOMR Flight Operations Management Room

FOMRC Fiber Optic Materials Research Center [of Rutgers-The State University of New Jersey, Piscataway, US]

FOM-Utrecht (Stichting voor) Fundamenteel Onderzoeg der Materie–Utrecht [Foundation for Fundamental Research on Matter, Netherlands]

FON Federation of Ontario Naturalists [Canada]

.FON Font [File Name Extension]

.FON Phone [File Name Extension]

.FON Phone Directory [File Name Extension]

FONA Friends of the National Arboretum [US]

FONASBA Federation of National Associations of Shipbrokers and Agents [UK]

FONDA First-Order Nondegenerate Adiabatic (Approximation) [Physics]

Fonderia Ital Fonderia Italiana [Italian Foundry Journal]

FOND-EX International Foundry Exhibition and Fair [Czech Republic]

FONSI Finding of No Significant Impact

FOO Frequency of Optimum Operation

501SS AISI-SAE 501 Heat-Resisting Stainless Steel

Food Drug Packag Food and Drug Packaging [US Publication]

FoodE Food Engineer [also Food Eng, or FoodEng]

Food Eng Food Engineer(ing)

Food Eng Food Engineering [International Journal]

Food Manuf Food Manufacture [Published in the UK]

Food Nutr Bull Food Nutrition Bulletin [International Journal]

Food Process Food Processing [Published in the US]

Food Technol Food Technology [Journal of the Institute of Food Technologists, US]

FOOS Force Out of Service

FOP First-Order Predictor

FOP Flight Operations Panel

FOP Flight Operations Plan

FOP Follow on Production

FOPDAC Federation of Overseas Property Developers and Consultants [UK]

FOPG Flight Operations Planning Group

FOPI Four Pi (i.e. π) (Experiment) [of Gesellschaft für Schwerionenforschung mbH, Darmstadt, Germany]

FOPL First-Order Predicate Logic

FOPP Follow on Parts Production

FOPPA First-Order Polarization Propagator (Theory)

FOPS Falling Object Protective Structure [Mining]

FOPS Flight Operations Planning Schedule

FOPT Fiber-Optic Photon Transfer

FOQ Free on Quay [also foq]

FOQLS Fiber-Optic Quasi Light Scattering

FOR First-Order Reaction [Physical Chemistry]

FOR First-Order Red

FOR Flight Operations Review

FOR Forced Outage Rate

FOR Free on Rail [also for]

FOR Friends of the River [US]

.FOR Form File [File Name Extension]

.FOR FORTRAN Source Code [File Name Extension]

For Force [also for]

For Foreign [also for]

For Forest; Forester; Forestry [also for]

For Formosa [now Taiwan]

For Formyl

FORAC Fisheries and Oceans Research and Advisory Council [Canada]

FORACS Fleet Operational Readiness Accuracy Check Site [US Navy]

FORAST Formula Assembler Translator

FORATOM Forum Atomique Européen [Association of European Atomic Forums, UK]

FORBLOC FORTRAN Compiled Block-Oriented Simulation Language

FORC Federation of Resource Centers [UK]

Forc Forcing [also forc]

FORC Formula Coder

FORCE FORTRAN Conversational Environment

FORD Federal Office of Regional Development [Canada]

FORD Floating Ocean Research and Development Station

FORDAC Fortran Data Acquisition Compiler [also Fordac]

FORDACS Fuel-Oil Route Delivery and Control System

FORDAP FORTRAN Debugging Aid Program

Fördern Heben Fördern und Heben [German Journal on Conveying and Lifting]

Fördmittel J Fördermittel Journal [German Journal on Conveying Equipment]

FORD-Q Federal Office of Regional Development–Quebec [Canada]

FORDS Floating Ocean Research and Development Station

FOREM File Organization Evaluation Model

Forensic Eng Forensic Engineering [Journal published in the UK]

FORESDAT Formerly Restricted Data

Forest Forestry [also forest]

Forest AIDS Forest Products Abstract Information Digest Service [of Forest Products Research Society, Madison, Wisconsin, US]

Forest Ecol Manage Forest Ecology Management [Journal published in the US]

Forest Genet Resour Inf Forest Genetic Resources Information [Published by the Food and Agricultural Organization of the United Nations]

Forest Ind Forest Industries [Published in the US]

Forest Prod Abstr Forest Products Abstracts [of Commonwealth Agricultural Bureaus, UK]

Forest Prod J Forest Products Journal [of Forest Products Research Society, US]

Forestry Abstr Forestry Abstracts [of Commonwealth Agricultural Bureaus, UK]

Forestry Chron Forestry Chronicle [Published by the Canadian Institute of Forestry]

Forestry Res Newsl Forestry Research Newsletter [of Great Lakes Forest Research Center, Sault Ste. Marie, Ontario, Canada]

Forest Sci Forest Science [Publication of the Society of American Foresters]

Forest Sci Monogr Forest Science Monographs [Publication of the Society of American Foresters]

Forest Trees People Newsl Forest, Trees and People Newsletter [US]

FOREX Foreign Exchange [also Forex or forex]

FOR/FOT Free on Rail/Free on Truck [also for/fot]

Forg Forging [also forg]

FORGE File Organization Generator

FORGO FORTRAN Load and Go

Forgs Forgings [also forgs]

Forg Top Forging Topics [US Publication]

FORIMS FORTRAN Oriental Information Management Program

FORJ Fiber-Optic Rotary Joint

FORKAT Förderungskatalog [On-Line Database of the Bundesministerium für Bildung, Wissenschaft, Forschung und Technologie Federal Ministry of Education, Science, Research and Technology, Germany]

form formic [Chemistry]

FORMAC Formula Manipulation Compiler

FORMAT Form of Material

FORMAT FORTRAN Matrix Abstraction Technique

FORML Forth (Programming Language) Modification Laboratory [Forth (Programming Language) Interest Group Committee]

Formn Formation [also formn]

FORMOX Formaldehyde by Oxidation (Process)

FORS Forschungsprojekte aus Raumordnung, Städtebau und Wohnungswesen [Research Projects in Urban Planning and Housing (Database), Germany]

FORS Fully Optimized Reaction Space

Forschber Landes Nordrh-Westfal Forschungsberichte des Landes Nordrhein-Westfalen [Research Reports of the German State of North Rhine Westphalia]

Forsch Ingenieurwes Forschung im Ingenieurwesen [German Publication on Engineering Research; published by Verein Deutscher Ingenieure]

Forschungsber Ind Logistik Forschungsberichte zur Industriellen Logistik [German Research Reports on Industrial Logistics; published by Deutsche Gesellschaft für Logistik]

Forst Forestry [also forst]

Fort Fortification [also fort]

Fort Fortress [also fort]

fort (fortis) – strong [Medical Prescriptions]

FORTH FOundation for Research and Technology–Hellas [Heraklion, Crete, Greece]

Forthcoming Int Sci Tech Conf Forthcoming International Scientific and Technical Conferences [Published in the UK]

Forth Dimens Forth Dimension [Publication of the Forth (Programming Language) Interest Group, San Jose, California, US]

FORTH-IESL Foundation for Research and Technology–Hellas International Electron Synchrotron Laboratory [Greece]

FORTRAN Formula Translator; Formula Translation [also Fortran] [Computer Language]

FORTRAN-S FORTRAN for Scientific Applications

FORTRANSIT FORTRAN and Internal Translator System

FORTRUNCIBLE FORTRAN Style Runcible

Fortschr-Ber VDI-Z Fortschritt-Berichte der VDI Zeitschrift [Progress Reports of the VDI Journal; published by Verein Deutscher Ingenieure, Germany]

Fortschr Chem Org Naturst Fortschritte der Chemie Organischer Naturstoffe [German Journal on the Progress in the Chemistry of Natural Organic Substances]

Fortschr Med Fortschritte der Medizin [German Journal on Progress in Medicine]

Fortschr Phys Fortschritte der Physik [German Publication on Progress in Physics]

Fortschr Verf-Tech Fortschritte der Verfahrenstechnik [German Publication on Progress in Process Engineering]

Fortschrber Dtsch Keram Ges Fortschrittsberichte der Deutschen Keramischen Gesellschaft [Progress Reports of the German Ceramic Society]

FOS Fiber-Optic Sensor

FOS Fiber-Optic Source(s)

FOS Fiber-Optic System

FOS Field of Science

FOS Final Offer Selection

FOS First-Order Spectrum

FOS Fisheries Organization Society [UK]

FOS Flight Operations Support

FOS Fluorescent Optrode Sensor

FOS Free on Ship [also fos]

FOS Free on Steamer [also fos]

FOS Fructooligosaccharide [Biochemistry]

FOS Fuel-Oxygen Scrap

FOSA Fellow of the Optical Society of America

FOSC Field Operations Support Center

FOSDIC Film Optical Scanning Device for Input (in)to Computers

FOSDIC Film Optical Sensing Device for Input (in)to Computers

FOSE Federal Office Systems Exposition [US]

FOSE First-Order Stark Effect

FOSFA Federation of Oils, Seeds and Fats Associations [UK]

FOS/GHRS Fiberoptic Sensor/Goddard High-Resolution Spectrograph [of NASA at Goddard Space Flight Center, Greenbelt, Maryland, US]

FOSI Format Option Specification Instance

FOSKOR South African Phosphate Deposit Development Program in Transvaal

FOSO Flight Operations Scheduling Officer

FOSP Flight Operations Support Personnel

FOSS Field Operations Support System

FOSS Follow-On Search and Rescue

FOSSIL Fido/Opus/Seadog Standard Interface Layer

FOST Flight Operations Support Team

FOT Fiber Optics Technology

FOT Fiber-Optic Thermometer

FOT First-Order Theory [Physics]

FOT First-Order Transition [Thermodynamics]

FOT First-Order Twin [Metallurgy]

FOT Flight Operations Team

FOT Forward Transfer

FOT Free on Truck [also fot]

FOT Frequency, Optimum Traffic [also F-OT]

FOT Fuel Oil Tank

FOTA Fuels Open Test Assembly

FOTO Forced Oscillation of a Tightening Oscillator

FOTP Fiber-Optic Test Procedure

FOTS Fiber-Optic Transmission System

403SS AISI-SAE 403 Martensitic Stainless Steel

502SS AISI-SAE 502 Heat-Resisting Stainless Steel

FOU Forward Observation Unit

Found Foundation [also found]

Found Control Eng Foundations of Control Engineering [Publication of the Institute of Control Engineering, Technical University of Poznan, Poland]

Found Phys Foundations of Physics [Journal published in the US]

Found Phys Lett Foundations of Physics Letters [US]

Foundry Manage Technol Foundry Management and Technology [Journal published in the US]

Foundry Pract Foundry Practice [Journal published in the UK]

Foundry Technol Foundry Technology [Journal published in PR China]

Foundry Trade J Foundry Trade Journal [Journal published in the UK]

Foundry Trade J Int Foundry Trade Journal International [UK Journal]

FOURA Forward Observation Unit, Royal Artillery [UK]

FOV Field of View

FOV Field of Vision

FOV First Orbital Vehicle

FÖV Forschungsinstitut für öffentliche Verwaltung bei der Deutschen Hochschule für Verwaltungswissenschaften [Research Institute for Public Administration (of the German University for Administrative Sciences), Speyer, Germany]

FOW First Open Water (Chartering) [also fow]

FOW Forced-Oil and Forced-Water (Cooling)

FOW Forge Welding

FOW Free on Wagon [also fow]

FOWCIS Forest and Wildlands Conservation Information System [of Food and Agricultural Organization]

FOWP Fertilizers from Organic Wastes Program

FOX Futures and Options Exchange [London, UK]

FOx Flowable Oxide
FP Fabry-Perot (Interferometer)
FP Faceplate
FP False Positive [Medicine]
FP Family Physician
FP Family Practitioner
FP Far Point
FP Fast Passage
FP Feedback, Positive
FP Feedback Potentiometer
FP Feeding Point
FP Feed Pump
FP Fermat's Principle [Optics]
FP Fermentation Product
FP Ferritic-Pearlitic (Cast Iron) [Metallurgy]
FP Fetoprotein [Biochemistry]
FP Fibrous Protein [Biochemistry]
FP Field Plate
FP Field Protective
FP Fighter Pilot
FP Filament Power Supply
FP File Processing
FP File Protect(ion)
FP Filter Paper
FP Filter Pump
FP Final Point
FP Fine Paper
FP Fine Particle
FP Fine Pitch
FP Fine Pointing
FP Finger Pin
FP Finite Perturbation
FP Fire Place
FP Fire Point
FP Fire Prevention
FP Fireproof
FP Fireproofing
FP Firing Point [Electronics]
FP Fish Plate
FP Fission Product
FP Fixed Point
FP Fixed Position
FP Fixed Price
FP Fixed Pulley
FP Flame Photometer; Flame Photometry
FP Flameproof
FP Flameproofing
FP Flashless Propellant
FP Flash Photolysis
FP Flash Point [also Fp]
FP Flat Pack (Integrated Circuit)
FP Flexible Pavements [US]
FP Flight Plan

FP Flight Procedures
FP Floating Point
FP Floating Potential [Electronics]
FP Flow Pattern
FP Fluid Power
FP Fluorinated Polymer
FP Fluorochrome/Protein [Biochemistry]
FP Fluorophosphate
FP Fluoroplastic(s)
FP Fluoropolymer
FP Flux Pinning
FP Focal Plane
FP Focal Point
FP Fokker-Planck (Equation) [Statistical Mechanics]
FP Food Processing; Food Processor
FP Foolproof
FP Foot Pedal
FP Forepump
FP Forest Park [Australia]
FP Forward Perpendicular
FP Four-Fold Points (for Grain Boundaries) [Materials Science]
FP Four-Pole [also F/P]
FP FoxPro [Computers/Communications]
FP Fracture Propagation
FP Free Path [also fp]
FP Free Piston
FP Freeze Protection
FP Freezing Point [also fp]
FP French Patent
FP French Polynesia(n)
FP Friction Power [also fp]
FP Fructose Phosphate [Biochemistry]
FP Fuel Pressure
FP Fuel Pump
FP Full Period
FP Full Potential
FP Fully Paid [also fp]
FP Functionalized Polymer
FP Function Path
FP Function Processor
FP Fundamental Parameter
FP Fusible Plug
FP Fusion Point
FP Fusion Power
FP Polycrystalline Alumina Fiber [Designation]
F-2,6-P Fructose-2,6-Biphosphate
F/P Flat Pattern
F/P Fluorine-to-Phosphorus (Ratio)
F/P Fluorochrome-to-Protein (Molar Ratio)
F/P Free-Energy to Process-Energy (Ratio)
Fp Flash Point [also FP]
F£ Falkland Islands Pound [Currency]
4P Four-Pole [also FP]

fp foot-pound [Unit]

fp fully paid [also FP]

fp position factor [Component of the Continuity Index of Intermetallics]

$f(\phi)$ function of phi [Symbol]

$f(\phi)$ normalized distribution function [Symbol]

FPA Federal Power Act [US]

FPA Federal Preparedness Agency [of General Services Administration] [Now Part of Federal Emergency Management Agency]

FPA Final Power Amplifier

FPA Fire Protection Association [UK]

FPA Flexible Packaging Association [US]

FPA Floating-Point Accelerator

FPA Floating-Point Arithmetic

FPA Focal Plane Array

FPA Focal Plane Assembly

FPA Foreign Press Association [UK]

FPA Forest Practices Act [Tasmania]

FPA Forest Products Association [of New South Wales, Australia]

FPA Free of Particular Average [also fpa]

FPA Freight Prepaid and Allowed [also FP&A]

FPA Fresh Produce Association [UK]

FPA Fusion Power Associates [US]

FP&A Freight Prepaid and Allowed [also FPA]

F&PA Flowers and Plants Association [UK]

FPAA Federación Panamericana de Asociaciónes de Arquitectos [Pan-American Federation of Architects Associations, Uruguay]

FPAC Flight Path Analysis and Command

FPAC Food Services Purchasing Association of Canada

FPAH Foundation for Preservation of the Archeological Heritage [US]

FPAN Fluorinated Polyaniline

FPAP Floating-Point Arithmetic Processor

FPAP Floating-Point Array Processor

FPAPA Forest Products Accident Prevention Association

FPASA Fire Protection Association of Southern Africa [South Africa]

FPB Forum of Private Business [UK]

FPB Four-Point Bending [also 4PB]

FPB Fuel Preburner

4PB Four-Point Bending [also FPB]

FPBOV Fuel Preburner and Oxidizer Valve

FPC Family Practitioner Committee [UK]

FPC Fatigue Pre-Crack(ing) [Mechanics]

FPC Federal Petroleum Commission [US]

FPC Federal Petroleum Council [US]

FPC Federal Power Commission [now Federal Energy Regulatory Commission, US]

FPC Federal Publisher's Committee [US]

FPC Filter-Paper Chromatography

FPC Final Processing Center

FPC Financial Planning Center

FPC Fire Prevention Code

FPC Fire Pump Control

FPC Fish Protein Concentrate [also fpc]

FPC Fixed Partial Charge (Model)

FPC Flat-Plate Collector

FPC Flexible Program Control

FPC Floating Point Calculation

FPC Flow Proportional Counter

FPC Fluid Power Center [UK]

FPC Fluid Pressure Control

FPC For Private Circulation

FPC Forward Power Controller

FPC Free-Programmable Controller

FPC Fresno Pacific College [Fresno, California, US]

FPC Full Protective Clothing

FPC Functional Progression Chart

FPC Fusion Power Core

FPCA Forward Power Control(ler) Assembly

FPCE Fission Products Conversion and Encapsulation

FPCE Floating-Point C Extension (Specification)

FP CG Iron Ferritic-Pearlitic Compacted-Graphite (Cast Iron) [also FP CG iron]

FPCH Foreign Policy Clearinghouse

FPCI Fluid Power Consultants International [US]

FPCIL Fukui Prefectural Ceramic Industry Laboratory [Japan]

FPCS Farm Planning Computer Service [UK]

FPCS First-Order Pyramidal Cross Slip [Material Science]

FPCS Free Polar Corticosteroid [Biochemistry]

FPCS Fuel Pool Cooling System

FPCSTL Fission Product Control Screening Test Loop

FPD Fixed Partial Denture

FPD Flame Photometric Detector; Flame Photometric Detection

FPD Flash Point Determination

FPD Flat Panel Display

FPD Focal Plane Deviation

FPD FoxPro for Disk-Operating System

FPD Freezing-Point Depression

FPD Full Power Days

FPDA Fluid Power Distributors Association [US]

FPDC Federation of Plastering and Drywall Contractors [UK]

FPDD Final Project Design Description

FPDI Food Processing Development Irradiator

4PDT Four-Pole, Double-Throw [also FPDT]

4PDT SW Four-Pole, Double-Throw Switch [also FPDT SW]

FPDU FTAM (File Transfer, Access, and Management) Protocol Data Unit

FPE Fire Protection Engineer(ing)

FPE Force-Producing Element

FPE Free-Piston Engine

FPE Functional Program Element

FPEEPM Floor Proximity Emergency Escape Path Marking

FPEG Fast-Pulse Electron Gun

4πemu/Oe·cm³ four pi electromagnetic unit(s) per Oersted cubic centimeter [also 4πemu Oe^{-1}·cm^{-3}]

FPF Fabry-Perot Filter

FPH Fondation pour le Progres de l'Homme [Foundation for the Progress of Man, France]

FPHB Flight Procedures Handbook

F-PI Fluorinated Polyimide [also FPI]

FPIAA Fire Protection Industry Association of Australia

FPP Federal Pests Program [Australia]

FPP Four-Point-Probe (Sheet Resistance)

FPT Ferranti-Packed Transformers [at St. Catharines, Ontario and Trois-Rivières, Quebec, Canada]

FPT Fine Particle Flotation

FPT First Principles Technique

FPF Four Point Flexure (Specimen) [Mechanics]

FPF Fundação Percival Farquhar [Percival Farquhar Foundation, Governador Valadares, Brazil]

FPFA Flexible Polyurethane Foam Association [now Polyurethane Foam Association, US]

FPG Flat-Pulse Generator

FPG Fuel-Pressure Gauge

FPGA Field-Programmable Gate Array [also fpga]

FPH Feet per Hour [also fph]

FPH Floating Point Hardware

FPhysS Fellow of the Physical Society

FPI Fabry-Perot Interferometer; Fabry-Perot Interferometry

FPI Fast Probability Integrator

FPI Fluorescent Penetrant Inspection

FPI Food Service Packaging Institute [US]

FPI Fuel Pump Injection

FPIA Fluorescence Polarization Immunoassay

FPIB Food Production Inspection Branch [of Agriculture Canada]

FPIF Fixed Price Incentive Fee

FP-IMS Fixed-Point Ion Mobility Spectrometer

FPIS Forward Propagation Ionospheric Scatter

FPL Feline Panleucopenia [Veterinary Medicine]

FPL Field Processing Language

FPL Filed Flight Plan

FPL Final Protective Line

FPL Fine-Pitch Leaded

FPL Florida Power and Light Company [US]

FPL Folia Phoniatrica et Logopedia [Journal on Phoniatrics and Logopedics published in Switzerland]

FPL Forest Products Laboratory [Madison, Wisconsin, US]

FPL Frequency Phase Lock

FPL Full Power Level

FPL Full Power Load

FPL Functional Problem Logging

FPLA Fair Packaging and Labeling Act [US]

FPLA Field-Programmable Logic Array [also fpla]

FPLA Fur Products Labeling Act [US]

FPLC Federal-Provincial Liaison Committee [Canada]

FPLF Field Programmable Logic Family

FPLMTO Full-Potential Linearized Muffin Tin Orbital [Physics]

FPM Faculté Polytechnique de Mons [Polytechnic Faculty of Mons, Belgium]

FPM Fast Page Mode

FPM Federal Personnel Manual

FPM Feet per Minute [also fpm]

FPM File Protect Mode

FPM Flashes per Minute [also fpm]

FPM Fluorocarbon Polymer

FPM Floppy-Disk Processor Mode

FPM Feet per Minute [also fpm]

FPM Folding Platform Mechanism

FPM Four-Photon Mixing

FPM FoxPro for Macintosh (Computers)

FPM Frames per Minute [also fpm]

FPM Frequency Position Modulation

FPM Functional Planning Matrices

fpm feet per minute [also FPM]

4πM Volume Magnetization [Symbol]

4πM$_s$ Saturation Magnetization [Symbol]

FP-MCSCF Finite Perturbation Multiconfiguration Self-Consistent Field [Physics]

FPMD First-Principles Molecular Dynamics

FPMH Failures per Million Hours [also fpmd]

FPMI Forest Pest Management Institute [of Canadian Forest Service]

FPMMC Federal-Provincial Mines Ministers' Conference [Canada]

FPM&SA Food Processing Machinery and Suppliers Association [US]

FPN Fixed Pattern Noise

FPO Federation of Professional Organizations [UK]

FPO Field Post Office

FPO Fleet Post Office

FPOV Fuel Preburner Oxidizer Valve

FPP Fabry-Perot Plates

FPP Fine Particle Processing

FPP First Principal Plane

FPP Fixed Path Protocol

FPP Fixed-Pitch Propeller

FPP Fixed-Point Protocol

FPP Floating-Point Package

FPP Floating-Point Processor

FPP Fluorphlogopite [Mineral]

FPP Freon Pump Package

1-FPP 1-Fluoropropyne

3-FPP 3-Fluoropropyne

FPPR Fast Passage in Paramagnetic Resonance

FPPR Fixed Price with Price Revision

FPPS Federal Payroll and Personnel System

FPPS Flight Plan Processing System

2FPy 2-Fluoropyridine

FPQA Fixed Portion Queue Area

FPQINA Federal Plant Quarantine Inspectors National Association [now National Association of Agricultural Employees, US]

FPR Farm Publications Report [US]

FPR Federal Procurement Regulations [US]

FPR Flat Plate Radiometer

FPR Flight Performance Reserve

FPR Flight Planned Route

FPR Floating Point Register

FPR Fluorescence Photobleaching Recovery

FPRAC Federal Prevailing Rate Advisory Committee [US]

FPRC Flying Personnel Research Committee [UK]

FPRF Fats and Proteins Research Foundation (Inc.) [Fort Myers Beach, Florida, US]

Fprf Fireproof [also fprf]

FPRF Forest Products Research Foundation [US]

FPRI Fellow of the Plastics and Rubber Institute

FPRI Forest Products Research Institute [Japan and Philippines]

FPRL Fish-Pesticide Research Laboratory [of US Department of the Interior]

FPRL Forest Products Research Laboratory [UK]

FPRO Federal-Provincial Relations Office [Canada]

FPROM Field-Programmable Read-Only Memory [also FROM]

FPRS Federal Property Resources Service [of General Services Administration, US]

FPRS Forest Products Research Society [US]

FPS Faceplate Starter

FPS Fauna Preservation Society [now Fauna and Flora Preservation Society, UK]

FPS Favorite Picture Selection

FPS Federation of Piling Specialists [UK]

FPS Feet per Second [also fps]

FPS Fellow of the Pathological Society

FPS Fellow of the Philosophical Society

FPS Fiber Placement System

FPS Film Performance Score

FPS Fine Particle Society [US]

FPS Finite Population Sampling [Statistics]

FPS Fire Protection Specialist

FPS Flight per Second

FPS Floating-Point System

FPS Flocculated Polystyrene

FPS Fluid Power Society [US]

FPS Focus Projection and Scanning

FPS Foot-Pound-Second (System) [also fps]

FPS Forest Products Society [US]

FPS Forum on Physics and Society [of American Physical Scoiety, US]

FPS Forward Power Supply

FPS Frames per Second [also fps]

fps feet per second [also FPS]

fps² feet per second squared [also FPS²]

FPSB Fisheries Prices Support Board

FPSE Foot-Pound-Second Electrostatic System [also fpse]

FPSE Free Piston Stirling Engine

FPSEP Flash-Photolysis Stimulated Emission Pumping

FP-SEP Flow Pattern–Secco Etch Pits [Metallurgy]

FPSK Frequency and Phase-Shift Keying

4PSK Quadra-Phase-Shift Keying

FPSL Fission Product Screening Loop

FPSM Foot-Pound-Second Electromagnetic System [also fpsm]

FPSP Federal Power Support Program

FPST Flash Photolysis–Shock Tube [also FPST]

4PST Four-Pole, Single-Throw [also FPST]

FPSTR Fund to Promote Scientific and Technical Research [Burkina Faso]

4PST SW Four-Pole, Single-Throw Switch [also FPST SW]

FPSU Foot-Pound-Second Unit [also fpsu]

4P SW Four-Pole Switch [also FP SW]

FPT Female Pipe Thread

FPT Fine Particle Technology

FPT Fine-Pitch Technology

FPT Finite Perturbation Theory

FPT Fire Protection Technician

FPT Focal-Plane Tomography (Technique)

FPT Forced Perfect Termination

FPT Frame Paperfeed Transport

FPT Fuel Processing Technology [Journal published in the US]

FPT Full Power Trial

FPT Functional Perturbation Theory

FPTA Forest Products Traffic Association [US]

FPTC Forest Products Trucking Council [US]

FPTF Fuel Performance Test Facility

FPTS Forward Propagation Tropospheric Scatter

FPU Feed Preparation Unit

FPU Fixed Pitch Unit

FPU Floating-Point Processor Unit

FPV Flow Proportioning Valve

FPV Fowlpox Virus

4PV 1,4-Bis[4-(3,5-Di-tert-Butylstyryl)styryl] Benzene

FPVPC Federation of Paint and Varnish Production Clubs [US]

FPW Flexible Printed Wiring,

FPW Flexural Plate-Wave (Sensor)

FPW FoxPro for Windows [Computers]

FPWT Fire Protection Water Tank

FPY Failures per Year [also fpy]

FPY First-Pass Yield

FPY First Polar Year [1882-83] [Geophysics]

2FPy 2-Fluoropyridine

FPYTSC Formylpyridine Thiosemicarbazone

2-FPYTSC 2-Formylpyridine Thiosemicarbazone

4-FPYTSC 4-Formylpyridine Thiosemicarbazone

FPZ Fracture Process Zone

FPZ Frontal Process Zone

FPZG Forest Park Zoological Gardens [St. Louis, Missouri, US]

FQ Field Quenching
FQ Flight Qualification
FQ Fluxoid Quantum
FQ Flux Quantification
FQ Flux Quantization
FQ Flux Quantum
FQ Free Quark [Particle Physics]
FQ Fundamental Quantity
FQ Fuse, Quick
F(q) Scattering Amplitude [Symbol]
f(q) form factor [Symbol]
FQC Forest Quality Class
Fqcy Frequency [also fqcy]
FQDN Fully Qualified Domain Name [Internet]
FQEDX Fully Quantitative Energy Dispersive X-Ray Analysis
FQFP Fine-Pitch Quad Flat Pack [Electronics]
FQH Fractional Quantum Hall [Physics]
FQHE Fractional Quantum Hall Effect [Physics]
FQHL Fractional Quantum Hall Liquid [Physics]
FQI Flight Qualification Instrumentation
FQL Formal Query Language
FQLM Fung's Quasi-Linear Model
FQLS Fiber-Optic Quasi Light Scattering
FQP Fundamental Questions Program [of Center for Resource and Environmental Studies, Australian National University, Canberra]
FQPA Food Quality Protection Act [of US Environmental Protection Agency]
FQPR Frequency Programmer
FQR Flight Qualification Recorder
FQR Flight Qualification Review
FQS Friendly Query System
FQT Formal Qualification Test
fqt frequent
FQW Fractional Quantum Wall [Solid-State Physics]
FR Fabric Reinforced; Fabric Reinforcement
FR Facsimile Record(ing)
FR Faculty of Radiology
FR Failure Rate
FR False Rejection [Statistics]
FR Faraday Rotation [Optics]
FR Fast Reactor
FR Fast Release
FR Fast-Release Relay
FR Fatigue Ratio
FR Federal Register [US]
FR Federal Reserve [US]
FR Feed Rate
FR Feed per Revolution (in Milling)
FR Fehrenbacher and Rice (Model of Superconductivity)
FR Felicity Ratio [Acoustic Emission Testing]
FR Fermi Resonance [Physical Chemistry]
FR Ferroresonance; Ferroresonant
FR Fiber(glass) Reinforced; Fiber Reinforcement

FR Field Regulator
FR Field Reversing
FR Field Rheostat
FR File Register
FR Film Radiography
FR Final Report
FR Fineness Ratio [Aerospace]
FR Fire-Refined; Fire Refining [Metallurgy]
FR Fire Resistance; Fire Resistant
FR Fire Retardancy; Fire Retardant
FR Fire Risk
FR Firing Room
FR Fission Reactor
FR Flame Resistance; Flame Resistant
FR Flame Retardancy; Flame Retardant
FR Flammability Rating
FR Flash Radiation
FR Flash Radiography
FR Flash Ranging
FR Flexural Rigidity
FR Flight Readiness
FR Flight Refuelling
FR Flight Rule
FR Floating Regulator
FR Flocculation Reaction [Medicine]
FR Flow Regulator
FR Flow Resistance
FR Fluence Rate
FR Fluorescent Radiation
FR Forbidden Reflection
FR Forced Release [Telecommunications]
FR Forest Ranger
FR Forest Reserve
FR (Forschungsreaktor) – German for "Research Reactor"
FR Forward Reaction [Chemistry]
FR Fossil Resin
FR Fractional Recovery
FR Frame Recognition
FR Frame Reset
FR France [ISO Code]
FR Frank-Read (Source) [Metallurgy]
FR Free Radical
FR Freezing Rain
FR Frege-Russell (Reduction) [Mathematics]
FR Frequency Ratio [Statistics]
FR Frequency Recorder; Frequency Recording
FR Frequency Response
FR Frictional Resistance
FR Fries Rearrangement [Chemistry]
FR Fuel Rating
FR (Reactor) Fuel Rod
FR Full Range
FR Full Rate

FR Fundamental Research

FR Fusion Reactor

FR Fusion Research

FR-1 Forschungsreaktor 1 [Research Reactor located at Karlsruhe, Germany]

FR-2 Forschungsreaktor 2 [Research Reactor located at Karlsruhe, Germany]

F&R Functions and Requirements

F/R Flared Rudder

F(R) Range Function [Symbol]

4R- Sri Lanka [Civil Aircraft Marking]

5R- Madagascar [Civil Aircraft Marking]

Fr Faraday Rotation [Symbol]

Fr Fragment [also fr]

Fr Frame [also fr]

Fr France; French

Fr Francium [Symbol]

Fr Franklin [Unit]

Fr Friday

Fr Frontier (or Boundary) [also fr] [Mathematics]

Fr Froude Number (of Fluid Flow) [Symbol]

Fr French Franc [Currency of France, French Guinea, Guadeloupe, Martinique, Monaco, and St. Pierre and Miquelon]

Fr Frequency [Statistics]

Fr Frontier (or Boundary) [also fr] [Mathematics]

Fr-212 Francium-212 [also ^{212}Fr, or Fr^{212}]

Fr-218 Francium-218 [also ^{218}Fr, or Fr^{218}]

Fr-219 Francium-219 [also ^{219}Fr, or Fr^{219}]

Fr-220 Francium-220 [also ^{220}Fr, or Fr^{220}]

Fr-221 Francium-221 [also ^{221}Fr, or Fr^{221}]

Fr-222 Francium-222 [also ^{222}Fr, or Fr^{222}]

Fr-223 Francium-223 [also ^{223}Fr, Fr^{223} , or AcK]

$F(\rho)$ Embedding Function [Symbol] [Note: ρ represents the electron density]

fr from

.fr France [Country Code/Domain Name]

$f(\rho)$ (surface) roughness function

$f(\rho)$ function of density

FRA Fatigue Risk Analysis

FRA Federal Radio Act [US]

FRA Federal Railroad Administration [US]

FRA Forward Rate Agreement

FRA Free-Radical Addition

FRA Frequency Response Analysis; Frequency Response Analyzer

FRA Full-Range Analyzer

FRA Future Rate Agreement

fra fractal [Cumulus Cloud Structure]

Frac Fraction(al) [also frac]

Frac Fractional Part (of Argument) [Pascal Function]

Frac Fractionated Antiserum [[Immunochemistry]

Frac Fractionated; Fractionation [also frac]

FRACR Fellow of the Royal Australasian College of Radiologists

Fract Mech Ceram Fracture Mechanics of Ceramics (Journal)

FRAD Frame Relay Access Device

FRAD Frame Relay Assembler/Disassembler

Fr Adv Sci Technol French Advances in Science and Technology [US]

FRAeS Fellow of the Royal Aeronautical Society

Fra Fys Verden Fra Fysikkens Verden [Publication of the Institute of Physics, University of Oslo, Norway]

FRAG Fragment(ation) [Computers]

Frag Fragment(ation) [also frag]

frag fragile

Fragm Fragment [also fragm]

Fragn Fragmentation [also fragn]

FRAGNET Fragmented Network

FRAIC Fellow of the Royal Architectural Institute of Canada

FRAM Fellow of the Royal Academy of Medicine

FRAM Ferroelectric Random-Access Memory

FRAMATOME Societé Franco-Americaine de Constructions Atomiques [French-American Nuclear Construction Company, France] [also Framatome]

FraMCoS Fracture Mechanics of Concrete Structures

FRAME Fund for the Replacement of Animals in Medical Experiments [UK]

FRANK Frequency Regulation and Networking Keying

FRAP FKBP12 (FK-Binding Protein 12) Rapamycin-Associated Protein [Biochemistry]

FRAP Fluorescence Recovery After Photobleaching

FRAP (Nuclear) Fuel Rod Analysis Program

FRAP-S Fuel Rod Analysis Program–Steady-State

FRAP-T Fuel Rod Analysis Program–Transient

FRAS Fellow of the Royal Astronomical Society

FRAS Fire Risk Assessment Standard [US]

FRASTA Fracture-Surface Topography Analysis

FRAT Fiber-Reinforced Advanced Titanium

FRAM Ferroelectric Random-Access Memory

FRB Burundi Franc [Currency of Burundi]

FRB Failure Review Board

FRB Federal Reserve Bank [US]

FRB Fire Research Board [UK]

FRB Fisheries Research Board [Canada]

FRB FKBP12 (FK-Binding Protein 12) Rapamycin Binding (Protein) [Biochemistry]

FRB Forschungsreaktor Berlin [Nuclear Research Reactor at Berlin; of Hahn-Meitner-Institut, Germany]

FRB Fuel Research Board [South Africa]

FrB Burundi Franc [Currency of Burundi]

fr bel from below

FrBr Burundi Franc [Currency of Burundi]

FRBS Fellow of the Royal Botanical Society

FRC Fabric-Reinforced Ceramics

FRC Fairchild Research Center [US]

FRC Fast Reaction Concept

FRC Fast Rescue Craft

FRC Federal Radiation Council [Now Part of US Environmental Protection Agency]

FRC Federal Radio Commission [US]

FRC Fiber-Reinforced Cement

FRC Fiber-Reinforced Ceramics

FRC Fiber-Reinforced Composite

FRC Fiber-Reinforced Concrete

FRC File Research Council [UK]

FRC Final Routing Center

FRC Flight Research Center [of NASA at Edwards Air Force Base, California, US]

FRC Food Research Center

FRC Foreign Relations Committee [of US Senate]

FRC Free Carrier

FRC Free Radical Chemistry

FRC Free-Radical Chlorination

FRC Full Route Clearance

FRC Funtional Redundancy Checking

FRC Functional Residual Capacity [Medicine]

FRC Future Requirements Committee [US]

frc free carrier

FRCA Fire Retardant Chemicals Association [US]

FRCAB Flat Roofing Contractors Advisory Board [UK]

Fr CFA CFA (Communauté Financière Africaine) Franc [Currency of Benin, Congo, Gabon, Ivory Coast, Niger, Senegal, Togo and Upper Volta]

Frchs Franchise [also frchs]

FRCI Fibrous Refractory Composite Insulation [of NASA Space Shuttle]

FRCM Fellow of the Royal College of Medicine

FRCMC Fiber-Reinforced Ceramic-Matrix Composite

FRCOphth Fellow of the Royal College of Ophthalmologists

FRCP Fellow of the Royal College of Physicians

FRCPath Fellow of the Royal College of Pathology

FRCPC Fellow of the Royal College of Physicians of Canada [also FRCP(C)]

FRCS Fellow of the Royal College of Surgeons

FRCS Forged Radius Clamp Straps

FRCS Forward Reaction Control Subsystem

FRCS Forward Reaction Control System

FRCSC Fellow of the Royal College of Surgeons of Canada

FRCS(Ed) Fellow of the Royal College of Surgeons (Education)

FRCTF Fast Reactor Core Test Facility [of United States Atomic Energy Commission]

FRCVS Fellow of the Royal College of Veterinary Surgeons

FRD Fast Reaction Dynamics

FRD Federal Reserve District [US]

FRD Fiber Resin Development

FRD Flight Requirements Document

FRD Foundation for Research (and) Development [South Africa]

FRD Functional Reference Device

FRDA Forest Resource Development Agreement [Canada]

FRDC Fisheries Research and Development Corporation [Australia]

FRDC Fusion Research and Development Center [of Japan Atomic Energy Research Institute]

FR-DLP Frame Recognition–Data Link-Processor

FRE Flight Related Element

FRE Format Request Element

FRE Free/Available Bytes (Function) [Programming]

FREC Forestry Research and Education Center [Sudan]

FREColl Forest Resources and Environment Collective

FREconS Fellow of the Royal Economical Society

FRED Fast Reactor Experiment Dounreay [Scotland]

FRED Fast Recovery Epitaxial Diode

FRED Fast Retrieval Enterprise Data

FRED Field Recovery Epitaxial Diode

FRED Figure-Reading Electronic Device

FRED Fractionally Rapid Electronic Device

FRED Frame Editor [Computers]

FRED Frequency Doubler; Frequency Doubling

FRED Front End for Databases [of General Telephone and Electronics Laboratories, US]

FRED Front-End to Dish

FRED Fund for Rural Economic Development [Canada]

FREDA Fuzzy Relational Environmental Design Assistant

FRED-FET Field Recovery Epitaxial Diode Field-Effect Transistor

FREDI Flight Range and Endurance Data Indicator

FREDOM Frequency Domain Propagation Model (Software) [of NASA]

FREE Fund for Renewable Energy and the Environment [now Renew America, US]

Freebd Freeboard [also freebd]

Freib Forschungsh Freiberger Forschungshefte [German Publication on Engineering Research; later split into Part A and Part B; published by the University of Freiburg]

Freib Forschungsh A Freiberger Forschungshefte Reihe A [German Publication on Engineering Research; published by the University of Freiburg]

Freib Forschungsh B Freiberger Forschungshefte Reihe B [German Publication on Engineering Research; published by the University of Freiburg]

Freib Forschungsh B, Metall Werkstofftech Freiberger Forschungshefte B, Metallurgie Werkstofftechnik [German Publication on Research in Metallurgy and Materials Engineering published by the University of Freiburg]

FREL Feltman Research and Engineering Laboratory [of US Army at Picatinny Arsenal, Dover, New Jersey, US]

Freq Frequency [also freq]

freq frequent(ly)

Fr Eq Afr French Equatorial Africa

Freq Conv Frequency Converter

Freq Div Frequency Divider

Freq Mult Frequency Multiplier

Freqy Frequency [also freqy]

FRES Federation of Recruitment and Employment Services [UK]

FRES Fellow of the Royal Economic Society

FRES Fellow of the Royal Entomological Society

FRES File Retrieval and Editing System

FRES Forward Recoil Energy Spectrometry; Forward Recoil Energy Spectroscopy

FRES Forward Recoil Emission Spectrometry; Forward Recoil Emission Spectroscopy

FRESC Forest and Rangeland Ecosystem Science Center [of US Geological Survey/ Biological Resources Division at Corvallis, Oregon]

FRESCAN Frequency Scanning

FRESCANAR Frequency Scanning Radar [also FRESCANNAR]

Fresenius J Anal Chem Fresenius Journal of Analytical Chemistry [English Edition]

Fresenius Z Anal Chem Fresenius Zeitschrift für Analytische Chemie [German Edition]

FRESH Foam Removal for Environmentally Safe Housing [US]

Freshwater Biol Freshwater Biology [Published in the UK]

FRET Freezing Rain Endurance Test

FRF Field Record Form

FRF Flight Readiness Firing

FRF Follicle-Releasing Factor [Biochemistry]

FRF Fragrance Research Fund [US]

FRF French Franc [Currency of France, French Guinea, Guadeloupe, Martinique, Monaco, and St. Pierre and Miquelon]

FRFA Federal Regulatory Flexibility Act [US]

FRFT Flight Readiness Firing Test

FRFWM Fully Resonant Four-Wave Mixing

FRG Federal Republic of Germany

FRG Forschungsreaktor Geesthacht [Geesthacht Research Reactor, Germany]

FRG Functional Renormalization Group [Physics]

FRG-1 Federal Republic of Germany Reactor 1 [at GKSS–Research Center, Geesthacht]

FRG-2 Federal Republic of Germany Reactor 2 [at GKSS–Research Center, Geesthacht]

$f(\rho,G)$ function of density and geometry [Symbol]

FRGS Fellow of the Royal Geographical Society

FRGSC Fellow of the Royal Geographical Society of Canada

Frgt Freight [also frgt]

Fr Gui French Guiana

FRH Fire Resistant Hydraulics (Program)

FRH Free-Radical Halogenization

Fr(H) Magnetic Field Dependent Faraday Rotation [Symbol]

FRHB Federation of Registered Housebuilders [UK]

FRHCS Field-Restored Highly Conducting State

FRHistS Fellow of the Royal Historical Society

FRHortS Fellow of the Royal Horticultural Society [also FRHS]

FRI Federal Relighting Initiative

FRI Flux Reversals per Inch [also FR/I, or FRPI]

FRI Food Research Institute [US]

FRI Forest Research Institute [of Commonwealth Scientific and Industrial Research Organization]

FRI Fuel Research Institute [South Africa]

FRI Fulmer Research Institute [UK]

Fri Friday

FRIBA Fellow of the Royal Institution of British Architects

FRIC Fellow of the Royal Institute of Chemistry

FRIC Forest Research Institute of Canada

Fricn Friction [also fricn]

FRICS Fellow of the Royal Institute of Chartered Surveyors

FRIL Fuzzy Relational Inference Language

FRIM Fermi Resonance Interface Mode

FRIM Forest Research Institute of Malawi

FRIMP Flexible Reconfigurable Interconnected Multiprocessor

Fr Indoch French Indochina

FRINGE File and Report Information Processing Generator

FRIP Fushun Research Institute of Petrochemistry [PR China]

FRIPHH Fellow of the Royal Institute of Public Health and Hygiene

FR IS Fast Reactions in Solution (International Conference)

FRISB Flathead River International Study Board

FRISA Fuel Research Institute of South Africa

Frisco San Francisco [California, US]

FRISNO French-Israeli Conference on Quantum and Nonlinear Optics [Eilat, Israel]

FRITALUX France, Italy, Benelux Economic Union [also Fritalux]

FRJ Forschungsreaktor Jülich [Nuclear Research Reactor at Julich; of Kernforschungsanlage Jülich,Germany]

FRJD Forward Reaction Jet Driver

FRL Activated Carbon Fiber, Long Phenolic Type [Designation]

FRL Feltman Research Laboratory [US Army]

FRL Filter, Regulator and Lubricator Unit

FRL Fisheries Research Laboratory [New Zealand]

FRL Frame Reference Line

FRL Fukui Research Laboratory [Japan]

FRL Fuselage Reference Line

FRM Fiber-Reinforced Metal

FRM Fiber-Reinforced MMC (Metal-Matrix Composite)

FRM Flow Rate Meter

FRM Forschungsreaktor München [Munich (Nuclear) Research Reactor, of University of Munich, Germany]

FRM Frequency Response Method

FRM II Forschungsreaktor München [Second Munich (Nuclear) Research Reactor, Garching, Germany]

Frm Farm [also frm]

Frm Framing [also frm]

FRMetS Fellow of the Royal Meteorological Society

fr/min frames per minute [Unit]

FRMM Flux Reversals per Millimeter [also FR/MM, FRPMM, or fr/mm]

Frmn Foreman [also frmn]

Frmr Farmer [also frmr]

FRMR Field Reversed Mirror Reactor

FRMS Fellow of the Royal Microscopical Society

Frmt Format [also frmt]

FRMZ Forschungsreaktor Mainz [Mayence Research Reactor, Germany]

FRN Federal Republic of Nigeria

FRN Fixed-Radix Numeration

FRN Floating Rate Note [Business Finance]

Frn Furniture [also frn]

FRNM Foundation for Research on the Nature of Man [US]

FRNP Federal Radio Navigation Plan [US]

FRNS Fellow of the Royal Numismatic Society

FRO Fire Research Organization [now Joint Fire Research Organization, UK]

FROG Friends of Research and Odd Gadgets [Canada]

FROM Factory Programmable Read-Only Memory

FROM Field Programmable Read-Only Memory [also FPROM]

FROM Fusible Read-Only Memory

Front Frontier [also front]

Front Frontispiece [also front]

Front Biosci Frontiers in Bioscience [International Journal]

Front Neuroendocrinol Frontiers in Neuroendocrinology [Journal published in the US]

FROPA Frontal Passage

FROSFC Frontal Surface

froz frozen

FRP Fares and Rates Panel [of International Civil Aviation Organization]

FRP Federal Radio Navigation Plan [US]

FRP Fiberglass-Reinforced Plastic(s)

FRP Fiberglass-Reinforced Plywood

FRP Fiberglass-Reinforced Polyester

FRP Fiber-Reinforced Pipe

FRP Fiber-Reinforced Plastic(s) [also F-RP]

FRP Fiber-Reinforced Polyester

FRP Fiber-Reinforced Polymer

FRP Forest Response Program [US]

FRP Free-Radical Polymerization

FRP French Patent [also FrP, or Fr Pat]

FRP Fuel Reprocessing Plant [Nuclear Engineering]

FRP Fuselage Reference Plane

F-RP Fiber-Reinforced Plastic(s) [also FRP]

FrP French Patent [also Fr Pat, or FRP]

Fr P Freezing Point [also fr p]

Fr Pat French Patent [also FrP, or FRP]

FRPC Fiber-Reinforced Polymer Composite

FRPI Flux Reversals per Inch [also FRI, or FR/I]

FRPMM Flux Reversals/Millimeter [also FRMM or FR/MM]

FRPS Flux Reversals per Second [also FRS or FR/S]

Fr Pt Freezing Point [also fr pt]

FRR Fast-Recovery Rectifier

FRR Fixed Retard Ratio

FRR Flight Readiness Review

FRR Flow Rate Ratio

FRR Functional Recovery Routine

FRR Rwandan Franc [Currency of Rwanda]

Fr Railw Rev French Railway Review [Translation of: *Revue Generale des Chemins de Fer (France)*; published in the UK]

FRRID Flight Readiness Review Item Description

FRRID Flight Readiness Review Item Disposition

FRRP Front Range Resources Project [of US Geological Survey at the Front Range Urban Corridor in Colorado and Wyoming]

FRRS Full Remaining Radiation Service [US]

FRS Activated Carbon Fiber, Short Phenolic Type [Designation]

FRS Federal Reserve System [US]

FRS Fellow of the Royal Society [UK]

FRS Ferrodoxin Reducing Substance [Biochemistry]

FRS Fiber-Reinforced Superalloy

FRS Financial Reporting System

FRS Fire Research Station

FRS Fisheries Research Society

FRS Flux Reversals per Second [also FR/S, frs, fr/s, or FRPS]

FRS Font Resource (File)

FRS Forced Rayleigh Scattering

FRS Forest Resource Survey [of Resource Assessment Commission, Australia]

FRS Forward Radar Station

FRS Forward Ready Signal

FRS Forward Recoil Spectroscopy

FRS Fragility Response Spectrum

FRS Fragment Separator [of Gesellschaft für Schwerionenforschung mbH, Darmstadt, Germany]

FRS Frank-Read Source [Materials Science]

FRS Free Radical (Plasma) Source

FRS Free-Radical Substitution

FRS Free Rotor States

FRS Free-Running Speed

FRS Fruit Research Station [New Zealand]

FRS Fuel Research Station

.FRS WordPerfect Graphics Driver [File Name Extension]

FRSA Fellow of the Royal Society of Arts

FRSB Frequency-Referenced Scanning Beam

FRSC Fellow of the Royal Society of Canada

FRSE Fellow of the Royal Society of Edinburgh [UK]

FRSH Fellow of the Royal Society of Health

FRSI Felt Reusable Surface Insulation [of NASA Space Shuttle]

FRSI Flexible Reusable Surface Insulation

FRSKGD Fauna Research Section of the Kenya Game Department

FRSL Fellow of Royal Society of London

FRSM Fellow of the Royal School of Mines

FRSM Fellow of the Royal Society of Medicine

FRSNA Fellow of the Royal School of Naval Architecture

Fr Speak Am French Speaking America

FRSS Fellow of the Royal Statistical Society

FRSTP Fire-Refined Tough Pitch Copper with Silver

FRT Finish Rolling Temperature [Metallurgy]

FRT Flight Readiness Test

FRT Flight Readiness Training

FRT Frequency Response Test

FrT Fragmentation Test

Frt Freight [also frt]

Fr Tech French Techniques [Published in Paris, France]

FRTEF Fast Reactor Thermal Engineering Facility

Frt Fwd Freight Forward [also frt fwd]

FRTISO Floating Point Root Isolation

FRTP Fiber-Reinforced Thermoplastic(s); Fiber-Reinforced Thermosetting Plastic(s)

FRTP Fire-Refined Tough Pitch (Copper)

FRTP Fraction of Rated Power

FRTP Cu Fire-Refined Tough Pitch Copper

FRTPE Fiber-Reinforced Thermoplastic Polyethylene

Frt Ppd Freight Prepaid [also frt ppd]

FRTO Flight Radio Telephony Operator

FRTV Forward Repair Test Vehicle

FRU Field Replaceable Unit

Fru D-Fructose [Biochemistry]

FRUGAL FORTRAN Rules Used as a General Applications Language

Fru (1,6)-P$_2$ D-Fructose 1,6-Diphosphate [Biochemistry]

Fru (2,6)-P$_2$ D-Fructose 2,6-Diphosphate [Biochemistry]

Fru (6)-P D-Fructose 6-Phosphate [Biochemistry]

FRUSA Flexible Rolled-Up Solar Array

FRUSTUM Frame-Based Unified Story-Understanding Model

FRV Fast Red Violet

FRV Feedwater Regulation Valve

FrV Fragment Velocity

FRW Friction Welding

Fr W Afr French West Africa

FRWI Framingham Relative Weight Index

Frwk Framework [also frwk]

Frwmn Forewoman [also frwmn]

Frwy Freeway [also frwy]

FRXD Fully-Automatic Reperforator-Transmitter Distributor

Frz Freeze [also frz]

Frz Frozen [also frz]

Frzr Freezer [also frzr]

FS Face Shield

FS Factor of Safety

FS Factor Sensitive; Factor Sensitivity

FS Factory System

FS Faculty of Science

FS Fail Safe

FS Fail to Sync

FS Fall Semester

FS Faraday Society [of Royal Society of Chemistry, UK]

FS Far Side [Aerospace]

FS Fast-Slow Wave

FS Fast Store

FS Fatigue Strength

FS Fault Summary

FS Feasibility Study

FS Federal Specification [US]

FS Federal Standard [US]

FS Feedback, Stabilized

FS Fehling's Solution [Chemistry]

FS Female, Soldered

FS Fermi Sea [Solid-State Physics]

FS Fermi Surface [Solid-State Physics]

FS Ferritic Steel

FS Fiber Society [US]

FS Fibonacci Sequence; Fibonacci Series [Mathematics]

FS Field Separator (Character) [Data Communications]

FS Field Service [UK]

FS Field Stop

FS Field Strength

FS File Segment

FS File Separator (Character) [Data Communications]

FS File Server

FS Field Service

FS File System

FS Filing System

FS Filled Stoping [Mining]

FS Filter Sand

FS Filter Spectrophotometer

FS Filter System

FS Filtration Society [UK]

FS Final Sector

FS Final Selector [Telecommunications]

FS Final Splice

FS Fine-Slipped (Ware)

FS Fine Structure [Physics]

FS Finite Size

FS Finnis-Sinclair (Model) [also F-S] [Metallurgy]

FS Fin-Stabilized; Fin Stabilization

FS Fire Sensor

FS Fire Suppression; Fire Suppressor

FS First Series

FS First Shell

FS First Surface

FS Fiscal Service

FS Fission Spectrum

FS Fixed Star

FS Flame Spectrometer; Flame Spectrometry

FS Flame Spectrophotometer; Flame Spectrophotometry

FS Flame Spectrum

FS Flame Spray(ing)

FS Flammable Solid

FS Flash Smelting [Metallurgy]

FS Flash Source

FS Flash Spectroscope; Flash Spectroscopy
FS Flash Spotting
FS Flat Screen
FS Flat Slip
FS Flexible Shaft(ing)
FS Flexural Strength
FS Flight Simulation; Flight Simulator
FS Flight System
FS Flotation Stability
FS Floating Sign
FS Fluorosilicone
FS Flow Stress
FS Flowable Concentrate for Seed Treatment [Agriculture]
FS Fluorescence Spectra
FS Fluorescent Screen
FS Fluid Seal
FS Fluid Statics
FS Flying Saucer
FS Fog Siren
FS Foreign Service
FS Forensic Scientist; Forensic Science
FS Foresight
FS Forest Service [of US Department of Agriculture]
FS Forged Steel
FS Formant Synthesizer
FS Forskningsstiftelsen Skogsarbeten [Forest Operations Institute, Sweden]
FS Fourier Series [Mathematics]
FS Fourier Spectroscopy
FS Fourier Spectrum
FS Fourier Synthesis
FS Fractional Softening [Metallurgy]
FS Fracture Stage
FS Fracture Strength
FS Frame Storage
FS Fraunhofer Society [Germany]
FS Fraunhofer Spectrum
FS Freeform Surface
FS Free Space
FS Free Sterols [Biochemistry]
FS Free Surface
FS Freeze Substitution
FS Freon Servicer
FS Frequency Selection; Frequency Selector
FS Frequency Shift
FS Frequency Spectrum
FS Frequency Standard
FS Frequency Synthesizer
FS Front Surface
FS Fuchs-Sondheimer (Conduction Model)
FS Fuel Society [UK]
FS Fuel Spray
FS Full Scale (also fs)
FS Full Section (View)

FS Full Size
FS Full Stabilization; Fully Stabilized
FS Full Subtract(or)
FS Functional Schematic
FS Functional Specification
FS Functional System
FS Function Select(or)
FS Fundamental Series [Spectroscopy]
FS Fundamental System
FS Furnace Soldering
FS Furnace Synthesis
FS Fused Salt
FS Fused Silica
FS Fuselage Station
FS Iron Sulfate
FS Swiss Franc [Currency of Liechtenstein and Switzerland]
F/S Fetch and Send
F-S Finnis-Sinclair (Model) [also FS] [Metallurgy]
Fs Fractostratus (Cloud)
Fs Fuse [also fs]
°F/s Degree Fahrenheit per second [also °F/sec]
F(s) General Function, Transformed [Symbol]
F(s) Kinematical Diffraction Amplitude [Symbol],
4S 4-Sulfate
4S Society for Social Studies in Science [US]
fs femtosecond [also fsec]
fs foot-second [also ft-s or ft-sec]
(s) Laplace transform of function $f(t)$ [Symbol]
FSA Farm Safety Association [Canada]
FSA Farm Security Administration [US]
FSA Fastress Stress Analyzer
FSA Federal Student Aid Application [US]
FSA Fédération Suisse des Architectes Indépendants [Swiss Federation of Independent Architects]
FSA Fellow of the Society of Actuaries
FSA Fellow of the Society of Arts
FSA Field Search Argument
FSA Financial Stationers Association [US]
FSA Fine Structure Analysis
FSA Finite Size Analysis
FSA Finnish Society of Automation
FSA Fluid Sealing Association [US]
FSA Fluosilicic Acid
FSA Food Security Act [US]
FSA Forward Sortation Area [Canada Post]
FSA Fracture Surface Analysis
FSA French Society of Acoustics
FSA Fuming Sulfuric Acid
fsa (*fiat secundum artum*) – make according to the art [Medical Prescriptions]
FSAA Fellow of the Society of Accountants and Actuaries
FSAA Flat Slips All Around
FSAA Flight Simulator for Advanced Aircraft

FSAAR Fraunhofer Society for the Advancement of Applied Research [Germany]

FSAC Finnish Society for Automatic Control

FSAE Fellow of the Society of Automotive Engineers

FSAI Farm Safety Association, Inc. [Ontario, Canada]

FSAR Final Safety Analysis Report

FSATU Federation of South African Trade Unions

FSAWDSP Flemish-South African Workshop (on the Physics of Waves in) Dusty, Solar and Space Plasmas [Leuven, Belgium]

FSB Federal Specifications Board [US]

FSB Federation of Small Business

FSB Field Selection Board

FSB Flat Slip on Bottom

FSB Functional Specification Block

FSB Fuel Storage Battery

fsbl fusible

FS-BW First Shell–Black/White (Model)

FSBZ Fermi Surface Brillouin Zone (Interaction) [also FsBz]

FSC Fairmont State College [West Virginia, US]

FSC Farm Safety Council [Canada]

FSC Fayetteville State College [North Carolina, US]

FSC Federal Stock Classification [US]

FSC Federal Supply Class [US]

FSC Federal Supply Classification [US]

FSC Ferris State College [Big Rapids, Michigan, US]

FSC Field Studies Council [UK]

FSC Fine Structure Constant

FSC Fisheries Science Centers [of National Oceanic and Atmospheric Administration, US]

FSC Fisher Scientific Company [US]

FSC Flame-Sprayed Coating

FSC Flight Safety Committee [UK],

FSC Florence State College [Alabama, US]

FSC Foreign Sales Corporation [US]

FSC Foundation for the Study of Cycles [US]

FSC Free Secretory Component

FSC Fresno State College [California, US]

FSC Frostburg State College [Maryland, US]

FSC Fulham-Simon-Carves (Process)

FSCB File System Control Block

FSCC Federal Surplus Commodities Corporation [US]

FSCC Ferrous Scrap Consumers Coalition [US]

FSCGC Fused Silica Capillary Gas Chromatography

FSCI Frequency Space Characteristic Impedance

FSCM Federal Supply Code for Manufacturers [US]

FSCP Fire Sensor Control Panel

FSCR Field Select Command Register

FSCS Forged Straight Clamp Strap

FSCT Federation of Societies for Coating Technology [US]

FSCT Floyd Satellite Communications Terminal

FSD Fiber Science Division

FSD Field Supply Depot

FSD File System Driver [IBM Operating System 2]

FSD Flight Simulation Division [of NASA Johnson Space Center, Houston, Texas, US]

FSD Flying Spot Digitizer

FSD Fourier Self-Deconvolution

FSD Full-Scale Deflection [also fsd]

FSD Full-Scale Development

FSD Full-Size Detail

FSDC Federal Statistical Data Center

FSDE Fission Suppressed Direct Enrichment [Nuclear Engineering]

FSDO Flight Standards District Office [US]

FS/FD Flux System/Flux Delta

FSE Faculty of Science and Engineering

FSE Fellow of the Society of Engineers

FSE Flight Simulation Engineer

FSE Four-Stroke Engine

FSE Frankfurt Stock Exchange [Germany]

FSE Full Screen Editor

FSE Fused Salt Electrolysis

FSEC Federal Software Exchange Center [US]

FSEC Florida Solar Energy Center [US]

°F/sec Degrees Fahrenheit per second [also °F/s]

fsec femtosecond [also fs]

FSEI Food Service Equipment Industry [now Food Service Equipment Distributors Association, US]

FSET FIRE (First ISCCP (International Satellite Cloud Climatology Project) Regional Experiment) Science Experiment Team

FSF Fading Safety Factor

FSF Fibrin-Stabilizing Factor [Biochemistry]

FSF First Static Firing

FSF Fixed Sequential Format

FSF Flash Smelting Furnace [Metallurgy]

FSF Flight Safety Foundation [US]

FSF Forensic Sciences Foundation [US]

FSF Free Software Foundation [Internet]

FSFE Flash Smelting Furnace with Electrodes [Metallurgy]

FS/FW Flow of Steam/Flow of Water

FSG Factoring Services Group [UK]

FSG Ferro(magnetic) Spin Glass [Solid-State Physics]

FSG First-Stage Graphitization [Metallurgy]

FSG Flight Study Group [of International Civil Aviation Organization]

FSG Fluorinated Silica(te) Glass

FS-G First Shell–Gray (Model) [also FS-Gray]

FSGB Foreign Service Grievance Board

FSGD Feasibility Study Guidance Document

FSGO Floating Spherical Gaussian Orbital [Physics]

FS-Gray First Shell–Gray (Model) [also FS-G]

FSGS Flare/Shallow Glide Slope

FSH Follicle-Stimulating Hormone [Biochemistry]

FSHDB (World) Fishing Catch Database [of Food and Agricultural Organization]

FSHRH Follicle-Stimulating Hormone-Releasing Hormone [Biochemistry]

FSI Federation of Sussex Industries [UK]

FSI Fellow of the Surveyors Institute

FSI Final Systems Installation

FSI Flight Simulator Instructor

FSIH Fellow of the Society of Industrial Hygiene

FSIM Functional Simulator

FSIS Fast Sample Insertion System

FSIS Food Safety and Inspection Service [of US Department of Agriculture]

FSIWG Flight System Interface Working Group

FSJ Fuel Society of Japan

FSK Frequency-Shift Keyed; Frequency-Shift Keying

FSL Fail-Safe Logic

FSL First Sea Lord [UK]

FSL Flexible Satellite Link

FSL Flexible System Link

FSL Flight Systems Laboratory

FSL Flight Simulation Laboratory

FSL Florida State Library [Tallahassee, US]

FSL Forecast Systems Laboratory [of National Oceanic and Atmospheric Administration, US]

FSL Formal Semantic Language

FSL French as a Second Language

FSL Full Stop Landing

FSLP First Spacelab Program

FSLIC Federal Savings and Loan Insurance Corporation [US]

FSLP First Spacelab Payload

FSLT First Sea Level Test

FSM Federated State of Micronesia

FSM Federation of Small Mines [UK]

FSM Field Strength Meter

FSM Finite Strip Method

FSM Finite State Machine

FSM Fixed Spin Moment (Procedure) [Physics]

FSM Fuel Supply Module

FSM Fused-Salt Chemistry

FSMA Frame Screen Manufacturers Association [now Screen Manufacturers Association, US]

FSMA Iron-Based Shape Memory Alloy

FSMAC Fellow of the Society of Management Accountants of Canada

FSMGB Federation of Small Mines of Great Britain

FSMWO Field Service Modification Work Order

FSN Federal Stock Number [US]

FSN Forward Sequence Number

FSN Forward Specification Number

FSN Full Service Network

FSNB Florida State News Bureau [US]

FSO Full-Scale Output

FSO Functional Supplementary Objective

FSOC Foreign Service Officer Corps [of United States Foreign Service]

FSOH Flight Support Operations Handbook

FSOS Free-Standing Operating System

FSP Fault Summary Page

FSP Fiber Saturation Point

FSP File Service Protocol

FSP Flight Safety Foundation [US]

FSP Ford Satellite Plan [US]

FSP Foundation for (the Peoples of) the South Pacific

FSP Fragment Simulated Projectile

FSP Frequency-Shift Pulsing

FSP Frequency Standard, Primary

FSP Full-Screen Processing

Fsp Fischerella species [Microbiology]

4 SPD Four-Speed Manual Transmission [also 4 spd]

5 SPD Four-Speed Manual Transmission [also 5 spd]

FSPE Federation of Societies of Professional Engineers

FSPE Fellow of the Society of Plastics Engineers

FSPPR Fast Supercritical Pressure Power Reactor [US]

FSPT Federation of Societies for Paint Technology [now Federation of Societies for Coating Technology, US]

FSQS Food Safety and Quality Service

FSR Farming Systems Research

FSR Fast Sodium-Cooled Reactor

FSR Fast Source (Nuclear) Reactor

FSR Feedback Shift Register

FSR Field Service Regulations

FSR Field Service Representative

FSR Field Strength Ratio

FSR File Storage Region

FSR Final System Release

FSR Financial Status Report (System)

FSR First Soviet Reactor [USSR]

FSR Flight Specific Requirements

FSR Force Sensing Resistor

FSR Forward Space Record

FSR Foundation for Scientific Relaxation

FSR Free Spectral Range

FSR Free System Resources

FSR Frequency Stability Rate

FSR Fundamental Scientific Research

FS&R Filling, Storage and Remelt System

FSRA Federal Sewerage Research Association [US]

FSRA Fine Spot Random Arc (Evaporation)

FS/RH Four Steps/Right-Hand (Convention for Burgers Vector Sign) [Crystallography]

FSRR Flight Software Readiness Review

FSRR Flight System Readiness Review

FSRS Flight Safety Research Specialist

FSRS Flight System Recording System

FSS Fatigue Striation Spacing [Mechanics]

FSS Federal Supply Service [of General Services Administration, US]

FSS Feet-Slug-Second (System) [also fss]

FSS Fellow of the (Royal) Statistical Society [UK]

FSS Ferritic Stainless Steel

FSS Field Spectrometer System

FSS Fine Structure Superplasticity
FSS Finite Size Scaling (Theory)
FSS Finnish Scientific Society
FSS Fire Sprinkler System
FSS Fire Suppression System
FSS Fixed Satellite Service
FSS Fixed Service Structure
FSS Flight Service Station
FSS Flight Standards Service
FSS Flight Support Station [for NASA Manned Mobility Unit]
FSS Flight Support Structure
FSS Flight Support System
FSS Flight Systems Simulator
FSS Floor Service Station
FSS Fluid Surface Stability
FSS Flying Spot Scanner
FSS Forensic Science Society [UK]
FSS Formatted System Services
FSS Frame Storage System [Computers]
FSS Frequency Selective Surfaces
FSSA Fire Suppression Systems Association [US]
FSSC Foreign Student Service Council [US]
FSSP Facilities Security and Safety Plan
FSSP Forward Scattering Spectrometer Probe [Cloud Droplet Measurements]
FSSR Functional Subsystem Software Requirements
FSSR Flight Systems Software Requirements
FST Faculty of Science and Technology
FST Federal Sales Tax [US]
FST Field Support Terminal
FST File Status Table
FST Final Steam Temperature
FST FIRE (First ISCCP (International Satellite Cloud Climatology Project) Regional Experiment) Science Team
FST Flat Slip on Top
FST Flat Square Tube (Monitor)
FST Flavonol 3-Sulfotransferase [Biochemistry]
FST Forged Steel
FST Frequency Shift Transmission
FST Frit Slurry Transport [Nuclear Engineering]
FST Fuel Science and Technology [Journal published in the US]
FST Full Scale Test
FST Full Screen Television
FST Functional Subassembly Tester
FST Functional Systems Test
FST Fuzzy Set Theory [Mathematics]
FS&T Federal Science and Technology (Budget) [US]
1st first
FSTA Food Science and Technology Abstracts (Database) [of International Food Information Service, UK]
FSTC Field Sound Transmission Class
FSTC Foreign Science and Technology Center [of US Army]
FSTE Factory Special Test Equipment

1st Lt First Lieutenant
FSTMA Firearm and Security Trainers Management Association [US]
Fstnr Fastener [also fstnr]
FSTO Federation of Scientific and Technical Organizations [Bulgaria]
FSTU Fluid Sealing Technology Unit
FSTV Fast-Scan Television
FSTV Flat-Square Television
FSU Facsimile Switching Unit
FSU Fiber (Tensile) Strength Utilization
FSU Field Select Unit
FSU Field Support Unit
FSU Final Signal Unit
FSU Florida State University [Tallahassee, US]
FSU Former Soviet Union
FSU Freon Servicing Unit
FSU Friedrich Schiller University [Jena, Germany]
FSUC Federal Statistics User Conference [US]
FSV Falling-Sphere Viscometer
FSV Fire Service Valve
FSV Fishery Survey Vessel
FSV Floating-Point Status Vector
FSVM Frequency Selective Voltmeter
FSW Feet of Seawater [also fsw]
FSW Flexible Steel Wire
FSW Flight Software
FSW Forward-Swept Wing
FSW Fourier Series Window
FSW Free Space Wave
FSWA Federation of Sewage Works Associations [US]
FSWR Flexible Steel Wire Rope
FSZ Fully Stabilized Zirconia
FT Faculty of Technology
FT Fallopian Tubes [Anatomy]
FT Fan Turbine
FT Fatigue Test(ing)
FT Fatigue Threshold
FT Fault Tolerance; Fault-Tolerant
FT Feedthrough
FT Fermentation Tank
FT Fermentation Tube [Microbiology]
FT Fermi Temperature [Statistical Mechanics]
FT Fermi Transition [Nuclear Physics]
FT Ferritic Transformation [Metallurgy]
FT Fibrous Tissue [Anatomy]
FT Field Test(ing)
FT Field Theory
FT Film Thickness
FT Financial Times
FT Fine Thermal (Carbon Black) [also FT Black or FT black]
FT Fine Tuned; Fine Tuning
FT Finishing Temperature [Metallurgy]
FT Fire Technologist; Fire Technology

FT   Fischer-Tropsch (Synthesis of Hydrocarbon Fuels)

FT   Firing Tables

FT   Fixed Target

FT   Flat Top

FT   Flame Temperature

FT   Flame Thrower

FT   Flame-Tight

FT   Flexural Test

FT   Flight Team

FT   Flight Test

FT   Float Tracing [Mining]

FT   Flotation Tank

FT   Fluorotoluene

FT   Flush Threshold

FT   Food Technologist; Food Technology

FT   Foil Thickness

FT   Forbidden Transition [Quantum Mechanics]

FT   Force Transducer

FT   Formal Training

FT   Form Tolerance

FT   Forward Transfer

FT   Foundry Technologist; Foundry Technology

FT   Fourier Theorem [Mathematics]

FT   Fourier Transform(ed); Fourier Transformation

FT   Fractionating Tower

FT   Fracture Toughness

FT   Frame Transfer

FT   Francis Turbine

FT   Free Troposphere

FT   Freezing Temperature [also ft]

FT   Frequency and Time Committee [of Institute of Electrical and Electronic Engineers–Instrumentation and Measurement Society, US]

FT   Frequency Tracker

FT   Fuel Tank

FT   Full-Term

FT   Full Terms [also ft]

FT   Full Text

FT   Full Time [also F/T]

FT   Fume-Tight

FT   Functional Test

FT   Function Table

FT   Fusion Technology

F&T   Fuel and Transportation

F/T   Full Time [also FT]

5T-   Mauritania [Civil Aircraft Marking]

Ft   Fallopian Tubes [Anatomy]

Ft   Forint [Currency of Hungary]

Ft   Fort

Ft   Freight [also ft]

Ft   Fruit [also ft]

F(t) Force as a Function of Time [Symbol]

F(t) Log-Normal Cumulative Distribution Function [Symbol]

F(t) N-Dimensional Generalized External Force [Symbol]

fT   femtotesla [Unit]

ft   *(fiat)* – let there be made; make [Medical Prescriptions]

ft   foot (or feet) [Unit]

ft   full terms [also FT]

ft^2 square foot [also sq ft]

ft^3 cubic foot [also cu ft]

ft^4 foot to the fourth power [Unit]

f(t) frequency as a function of time [Symbol]

f(θ) distribution function of the misorientation angle (in crystallography) [Symbol]

f(T) function of temperature [Symbol]

f(T) misfit parameter at temperature T (in materials Science) [Symbol]

f(t) failure rate at time t [Symbol]

f(t) frequency as a function of time [Symbol]

f(t) (general) function of time [Symbol]

f(t) log-normal probability density function [Symbol]

f(t) transducer output [Symbol]

f*(t) sampled-data form of a time function [Symbol]

f(θ) function of angle θ [Symbol]

$f_N(\theta)$ number distribution function of the misorientation angle (in crystallography) [Symbol]

$f_S(\theta)$ surface-weighted distribution function of the misorientation angle (in crystallography) [Symbol]

4×2   Four by Two (Truck)

FTA   Fatigue Test Article

FTA   Fault Tree Analysis

FTA   Federal Telecommunications Authority

FTA   Federation of Trade Associations [Ireland]

FTA   Field Test Algorithm

FTA   Field to Advice

FTA   Flexographic Technical Association [US]

FTA   Flight Test Article

FTA   Flotation Tank Association [US]

FTA   Fluorescent Treponemal Antibody (Venereal Disease Test)

FTA   Foreign Trade Association [Germany]

FTA   Free Trade Agreement [Canada/US]

FTA   Freight Transport Association [UK]

FTA   Frozen-Temperature Approximation [Metallurgy]

FTAA Free Trade Area of the Americas

FTAB Field Tab(ulator)

FTACCC Florida Technical Advisory Committee on Citrus Canker [US]

FTAM File Transfer, Access, and Management; File Transfer Access Management

FTAM File Transfer and Access Method

FTAS Fast Time Analysis System

FTase Farnesyltransferase [Biochemistry]

FTAT Furniture Timber and Allied Trades Union [UK]

FTB   Fibroid Tuberculosis [Medicine]

FTB   Fleet Torpedo Bomber

FT Black Fine Thermal (Carbon) Black [also FT black]

ft bm board foot meter [also bd ft or fbm]

FTC   Fast Time Constant

FTC Fast Time Control (Circuit)

FTC Fault Tolerant Compiler

FTC Federal Trade Commission [US]

FTC Firing Top-Center

FTC Fischer-Tropsch Catalyst

FTC Flight Test Conductor

FTC Florida Test Center [of McDonnell Douglas Astronautics Company, US]

FTC Fluid-Bed Thermal Cracking (Process)

FTC French Trade Commission

FTC Frequency Time Control

FTC Fuels Technology Center [of Argonne National Laboratory, Illinois, US]

FTC Fusion Technology Commission [of European Economic Community]

FT&C Formal Training and Certification

FT&C Functional Test and Calibration

ftc footcandle [also ft-c, or ft cd]

ft caps *(fiat capsula)* – let a capsule be made [Medical Prescriptions]

ft cd footcandle [also ftc, or ft-c]

FT-CIDEC Fourier Transform–Chemically Induced Dynamic Electron Polarization

ft/cm femtotesla per centimeter [also fT cm^{-1}]

FTCS Forward Time Center Space (Scheme)

FTD Fan Turbine Discharge

FTD Fission-Track Dating [Geology]

FTD Flame Thermionic Detector

FTD Flight Training Device

FTD Flying Training Division [of Air Training Command, US Air Force]

FTD Folded Triangular Dipole

FTD Foreign Technology Division [of US Air Force Systems Command]

FTD Frequency Translation Distortion

FTD Fuse Time Difference

FtD Fort Dix [New Jersey, US]

ft/day feet per day [Unit]

FTDB Full Text Data Base

FTDV Function Table Development and Verification

FTE Factory Test Equipment

FTE FFTF (Fast Flux Test Facility) Test Engineering [Nuclear Engineering]

FTE Flight Technical Error

FTE Flight Test Engineer

FTE Forced Test End

FTE Fracture Transition Elastic (Temperature) [Metallurgy]

FTE Fritted Trace Elements [Crops]

FTE Full-Time Employee

FTE Full-Time Equivalent

FTEMIRS Fourier Transform Electromodulated Infrared Spectroscopy

FTEP Full-Time Equivalent of Part-Time Teaching Faculty

FTES Fluorotriethoxysilane

FTESA Foundry Trades Equipment and Supplies Association [UK]

FTESE Fourier Transform Electron Spin Echo (Spectroscopy) [also FT-ESE]

FTESR Fourier Transform Electron Spin Resonance

FTET Full-Time Equivalent Terminals

FTF Fault Transfer Facility

FTF Flared Tube Fitting

FTFET Four-Terminal Field-Effect Transistor

434SS AISI-SAE 434 Ferritic Stainless Steel

Ftg Fitting [also ftg]

Ftg Footing [also ftg]

FTH Fourier Transform Hologram

FTH Full Tree Harvesting

fth fathom [also fthm]

ft/h feet per hour [also ft h^{-1}, ft/hr, FPH, or fph]

ft²/h square feet per hour [also ft^2 h^{-1}, ft²/hr, sq ft/h, or sq ft/hr]

ft³/h cubic feet per hour [also ft^3 h^{-1}, ft³/hr, cu ft/h, or cu ft/hr]

5th fifth

4th fourth

FTHA Fork Truck Hire Association [UK]

ft H$_2$O feet of water [also ftH$_2$O]

FTI Facing Tile Institute [US]

FTI Fan Turbine Inlet

FTI Federal Tax Included

FTI Fellow of the Textile Institute

FTI Foreign Traders Index [of US Department of Commerce]

FTI Fourier Transform Infrared

FTI Fourier Transform Interferometer

FTI Flux Transitions per Inch [also FT/I, or FTPI]

FTI Fullerene Technologies International [Tucson, Arizona, US]

FTI Fundação de Tecnologia Industrial [Industrial Technology Foundation, Lorena, Brazil]

FTIAP Footwear and Tanning Industry Adjustment Program

FTICR Fourier Transform Ion Cyclotron Resonance [also FT-ICR]

FTICRMS Fourier Transform Ion Cyclotron Resonance Mass Spectrometry

FTIR Fourier Transform Infrared [also FT-IR]

FTIR Fourier Transform Infrared Resonance

FTIR Fourier Transform Infrared Spectroscopy [also FT-IR]

FTIR-AA Fourier Transform Infrared Atomic Absorption

FTIRAS Federal Technical Institute of the Russian Academy of Sciences [St. Petersburg]

FTIRAS Fourier Transform Infrared Reflection-Absorption Spectroscopy [also FTIRRAS or FTIR-RAS]

FTIR-ATR Fourier Transform Infrared Attenuated Total Reflection [also FT-IR-ATR]

FTIR-GC Fourier Transform Infrared Gas Chromatography

FTIRRAS Fourier Transform Infrared Reflection Absorption Spectroscopy [also FTIR-RAS, FTIRAS or FT-IRRAS]

FTIRRS Fourier Transform Infrared Reflection Spectroscopy [also FTIR-RS or FTIRS]

FTIRS Fourier Transform Infrared Spectroscopy

FTIRS Fourier Transform Infrared Reflection Spectroscopy [also FTIRRS or FTIR-RS]

FTIS Flight Test Instrumentation System

FTIU Fault Transient Interface Unit

FTK Fertigungstechnisches Kolloquium [Coloquium on Production Engineering, Germany]

FtKn Fort Knox [Kentucky, US]

FTL Fast Transient Loader,

FTL Federal Telecommunications Laboratory [US]

FTL Field Transmission Loss

FTL Flash Transition Layer

FTL Flight Time Limitations

FTL Fourier Transform Lens

FTL Full Term License

FTL Full-Time Teaching Load

ft-L foot-Lambert [also ft L]

FTLB Flight Time Limitations Board

ft-lb foot-pound [also ft lb]

ft²/lb square foot per pound [also sq ft/lb]

ft³/lb cubic foot per pound [also cu ft/lb]

ft-lbf foot-pound force [also ft lbf]

ft-lbf/ft foot-pound force per foot [also ft lbf/ft]

ft-lbf/ft² foot-pound force per square foot [also ft, or lbf/ft²]

ft-lbf/hr foot-pound force per hour [also ft lbf/hr, ft-lbf/h, or ft lbf/hr]

ft-lbf/in foot-pound force per inch [also ft lbf/in²]

ft-lbf/in² foot-pound force per square inch [also ft lbf/in²]

ft-lbf/min foot-pound force per minute [also ft lbf/min]

ft-lbf/sec foot-pound force per second [also ft lbf/sec, ft-lbf/s, or ft lbf/s]

ft-lb/in foot-pound per inch [also ft lb/in]

ft-lb/lb·°F foot-pound per pound degree Fahrenheit [also ft lb/lb·°F, or ft-lb/lb/°F]

ft-lb/lb-mol·°R foot-pound per pound-mole degree Rankine [also lb ft/lb mol·°R, or lb ft/lb-mol/°R]

ft-lb/lb·°R foot-pound per pound degree Rankine [also ft-lb/lb/°R, or ft-lb/(lb·°R)]

ft-lb/hr foot-pound per hour [also ft lb/hr, ft-lb/h, or ft lb/h]

ft-lb/min foot-pound per minute [also ft lb/min]

ft-lb/sec foot-pound per second [also ft lb/sec, ftlb/s, or ft lb/s]

ft-lbs-sec² foot-pounds-second squared [Unit]

ft-lb wt foot-pound weight [also ft lb wt]

ft-lb wt/sec foot pound-weight per second [also ft lb wt/sec, ft-lb wt/s, or ft lb wt/s]

FTM Fischer Thickness Management (Software) [Thickness Measurement]

FTM Fixed Target Mode

FTM Flat Tension Mask (Monitor)

FTM Flight Test Missile

FTM Folded Triangular Monopole

FTM Frequency Time Modulation

FTM Functional Test Manager

FTMC Frequency and Time Measurement Counter

ft/min feet per minute [also FPM, or fpm]

ft³/min cubic feet per minute [also cu ft/min, CFM, cfm, or cf/min]

ft mist *(fiat mistura)* – mix; let a mixture be made [Medical Prescriptions]

FTMP Fault Tolerant Multiprocessor System

FTMS Federal Test Method Standards

FTMS Fourier Transform Mass Spectrometer; Fourier Transform Mass Spectrometry ,

FTMS Fourier Transform Mass Spectroscope; Fourier Transform Mass Spectroscopy

FTMT Final Thermomechanical Treatment [Metallurgy]

FTMW Fourier Transform Microwave

FTN Forschungsberichte aus Technik und Naturwissenschaft [Database of Scientific and Technical Research Reports; of Bundesministerium für Forschung und Technologie, Germany]

Ftn Footnote [also ftn]

Ftn Fountain [also ftn]

FTNIR Fourier Transform Near Infrared

FTNMR Fourier Transform Nuclear Magnetic Resonance [also FT-NMR or FT NMR]

FTN NUM Footnote Number [also Ftn Num]

FTNQR Fourier Transform Nuclear Quadrupole Resonance

FTNSC Francis T. Nicholls State College [Thibodaux, Louisiana, US]

FTO Failed to Open

FTO Ferric Trioxide

FTO Flight Test Objective

FTO Flying Training Organization

FTO Fruit Traffic Organization [UK]

FTO Functional Test Objective

FTOH Flight Team Operations Handbook

FTOH Flight Test Operations Handbook

FTOI Financial Times Ordinary (Stock) Index [UK]

F to F Face to Face

FTORAFUR 5-Fluoro-1-(Tetrahydro-2-Furfuryl)uracil

FTP Federal Test Procedure

FTP FFTF (Fast Flux Test Facility) Test Procedure [of US Atomic Energy Commission]

FTP Field Task Proposal

FTP Field Terminal Platform

FTP File Transfer Protocol [also ftp] [Internet]

FTP Fischer-Tropsch Process [Chemistry]

FTP Fixed Term Plan

FTP Flight Test Procedure

FTP Fourier Transform Polarography

FTP Fructose Triphosphate [Biochemistry]

FTP Fuel Transfer Port

FTP Fuel Transfer Pump

FTP Full Throttle Position

FTP Functional Test Program

FTP Function Test Progress

FTPD File Transfer Protocol Domain [Internet]

ft pdl foot-poundal [also ft-pdl]

FTPGSE Fourier Transform Pulsed-Gradient Spin Echo

FTPI Flux Transitions per Inch [also FTI, or FT/I]

FTPS File Transfer Protocol Server

FTR Fast Test Reactor

FTR Feather [also Ftr or ftr]

FTR Federal Telephone and Radio (Corporation) [US]

FTR Federal Travel Regulations

FTR Feed per Tooth per Revolution [Mechanical Engineering]

FTR Fischer-Tropsch Reaction

FTR Fixed Transom

FTR Flat Tile Roof

FTR Flight Test Requirement

FTR Forward Torpedo Room

FTR Frustrated Total Reflection

FTR Full Text Retrieval

FTR Functional Test Requirements

F(t)·r Feynman's Path Integral for Interaction (in Physics) [Symbol]

FT RA-IR Fourier Transform Reflectance-Absorption Infrared (Technique)

FTRC Federal Telecommunications Records Center

FTRD Flight Test Requirements Document

FTRD Functional Test Requirements Document

FTRIA Flow and Temperature Removable Instrument Assembly

FTRS Fourier Transform Raman Spectroscopy [also FT-RS]

FTS Facing Target Splitting (Apparatus)

FTS Facsimile Text Society [US]

FTS Federal Telecommunications System [US]

FTS Fiber Tensile Strength

FTS Field Test Set

FTS Fischer-Tropsch Synthesis

FTS Flexible Track System

FTS Flexible Transportation System

FTS Flight Telerobotic Servicer [of NASA]

FTS Flight Test Station

FTS Flight Test System

FTS Fluorotrimethylsilane

FTS Foot Switch

FTS Fourier Transform Spectrometer; Fourier Transform Spectrometry

FTS Fourier Transform Spectroscopy

FTS Free-Time System

FTS Frequency and Timing Subsystem

FTS Full Turbulence Simulation [Computerized Fluid Dynamics]

FTS Functional Test Specifications

ft/s feet per second [also ft/sec, FPS, or fps]

ft/s² feet per second squared [also ft/sec²]

ft²/s square feet per second [also ft²/sec]

ft³/s cubic feet per second [also ft³/sec]

ft-s foot-second [also fs, or ft-sec]

FtSam Fort Sam [Houston, Texas, US]

FTSC Fault Tolerant Spaceborne Computer

FTSC Federal Telecommunications Standards Committee [US]

FTSE Financial Times Stock Exchange (Index) [also FT-SE, UK]

ft/sec feet per second [also ft/s, FPS, or fps]

ft/sec² feet per second squared [also ft/s²]

ft²/sec square feet per second [also ft²/s]

f³/sec cubic feet per second [also ft³/s]

ft-sec foot-second [also ft-s, or fs]

ft/sec/sec feet per second per second [also ft/sec²]

410SS AISI-SAE 410 Martensitic Stainless Steel

430SS AISI-SAE 430 Ferritic Stainless Steel

436SS AISI-SAE 436 Ferritic Stainless Steel

FTT Flight Technical Tolerance

FTT Fachstelle Technologie-Transfer [Technology Transfer Center; of Paul Scherrer Institute, Switzerland]

FTTA Federal Technology Transfer Act [US]

FTTC Fiber To The Curb

FtTh Fort Thomas [Kentucky, US]

FTTS Flow Through Tube Sampler

2-FTTSC 2-Formylthiophene Thiosemicarbazone

FTTU Field Technical Training Unit

FTU First-Time Use(r)

FTU Fixed Treatment Unit

FTU Flight Test Unit

FTU Formazin Turbidity Unit

FTU Fortbildungszentrum für Technik und Umwelt [Educational Center for Technology and the Environment, Germany]

FTU Frascati Tokamak Upgrade [An Italian Nuclear Reactor]

ft ung *(fiat unguentum)* – let an ointment be made [Medical Prescriptions]

FTX Fault-Tolerant Unix [Computers]

FTXS FDDI (Fiber Distributed Data Interface) Transmission System

FTZ Fernmeldetechnisches Zentralamt [Central Telecommunications Office, Darmstadt, Germany]

FTZ Foreign Trade Zone

FTZ Free Trade Zone

FTZB Foreign-Trade Zones Board

FU Factor of Utilization

FU Finsen Unit [Physics]

FU First Unit

FU Flight Unit

FU Fluorouracil

FU Follow-Up

FU Fordham University [Bronx, New York, US]

FU Franklin University [US]

FU Free University

FU Fudan University [Shanghai, PR China]

FU Fuel Utilization

FU Fukui University [Fukui, Japan]

FU Fukuoka University [Fukuoka, Japan]

FU Fume Concentration

FU Functional Unit [also fu]

5-FU 5-Fluorouracil

5U- Niger [Civil Aircraft Marking]

Fu Fuse [also fu]

fu flux unit [Astrophysics]

fu functional unit [also FU]

FUB Forward Utility Bridge

FUB Free University of Berlin [Germany]

FUBAR Fouled Up Beyond All Recognition

FU Berlin Freie Universität Berlin [Free University of Berlin, Germany]

Fuc Fucose; Fucosyl [also fuc] [Biochemistry]

α-L-Fuc α-L-Fucose [also α-L-fuc] [Biochemistry]

FUD Fear, Uncertainty and Doubt

FUD First Use Date

FUDENA Fundación para la Defensa de la Naturaleza [Environmental Defense Foundation]

FUDR Failure and Usage Data Report

FUDR 5-Fluorodeoxyuridine [Biochemistry]

FUE Federated Union of Employers [UK]

FUE Fukuoka University of Education [Japan]

FuE (Forschung und Entwicklung) – German for "Research and Development"

FUEL Division of Fuel Chemistry [of American Chemical Society, US]

Fuel Chem News Fuel Chemistry News [Publication of American Chemical Society–Division of Fuel Chemistry]

Fuel Process Technol Fuel Processing Technology [Journal published in the US]

Fuel Sci Technol Fuel Science and Technology [Journal published in the US]

Fuel Sci Technol Int Fuel Science and Technology International [Journal published in the US]

Fuel Soc J Fuel Society Journal [UK]

FUES Follow-Up and Evaluation Section

FUF Fuel Utilization Factor

FUFO Fuel Fusing Option

FUFOR Fund for UFO (Unidentified Flying Object) Research [US]

FUI File Update Information

FUIF Fire Unit Integration Facility

Fuji Electr Rev Fuji Electric Review [Published by Fuji Electric Co. Ltd., Tokyo, Japan]

Fujikura Tech Rev Fujikura Technical Review [Published by Fujikura Cable Works Ltd., Tokyo, Japan]

Fujitsu Sci Tech J Fujitsu Scientific and Technical Journal [Published by Fujitsu Ltd., Kanagawa, Japan]

Fukuoka Univ Rev Technol Sci Fukuoka University Review of Technological Sciences [Published by the Central Research Institute of Fukuoka University, Japan]

Fulmer Newsl Fulmer Newsletter [UK]

Fum Fuming [also fum]

FumA Fumaric Acid

Fumg Fuming [also fumg]

FUMP Far Ultraviolet Molecular Photoproduct

FUNC Function [also Func, or func] [Computers]

Funct Function(al) [also funct]

FUNCTLINE Functional Line Diagram

Funct Mater Function and Materials [Japanese Publication]

Fund Funding [also fund]

Fundam Fundament(al) [also fundam]

Fundam Appl Toxicol Fundamental and Applied Toxicology [Journal of the Society of Toxicology, US]

Fundam Clin Pharmacol Fundamentals of Clinical Pharmacology

Fundam Cosm Phys Fundamentals of Cosmic Physics [Journal published in the UK]

Fundam Inform Fundamenta Informaticae [Computer Science Journal published in the Netherlands]

Fundam Math Fundamenta Mathematicae [Mathematical Journal published in Warsaw, Poland]

Fundição Matér Primas Fundição e Matérias Primas [Brazilian Journal on Primary Materials]

FUNET Finnish University and Research Network

Fungal Genet Biol Fungal Genetics and Biology [Journal published in the US]

FUNLIS Fundamentals of Library and Information Science

FUNOP Full Normal Plot

FUNY Free University of New York [US]

FUO Fever of Undetermined Origin [Medicine]

FUO Follow-up Output

Fur Furnish [also fur]

Fur Furring [also fur]

fur furlong [Unit]

FUrd 5-Flourouridine [Biochemistry]

Furn Furnish [also furn]

Furol Fuel and Road Oil (Viscosity) [also Furol]

FURZ Fairly Unreacted Zone

FUS FORTRAN Utility System

Fus Fuselage [also fus]

FUSAG First United States Army Group

FUSE Far Ultraviolet Spectroscopic Explorer

FUSE Federation for Unified Science Education [US]

Fusion Eng Des Fusion Engineering and Design [Journal published in Switzerland]

Fusion Technol Fusion Technology [Journal of the American Nuclear Society, US]

FUS/LF Fuselage, Lower Forward

Fuslg Fuselage [also fuslg]

Fusn Fusion [also fusn]

FUSRAP Formerly Utilized (Waste) Sites Remediation Action Program [US]

Fussboden Ztg Fussboden Zeitung [German Publication on Flooring Materials and Design]

FUS/UF Fuselage, Upper Forward

Fut Future [also fut]

FUTA Federal Unemployment Tax Act [US]

Fut Comput Syst Future Computing Systems [UK]

Fut Gener Comput Syst Future Generation Computer Systems [Published in the Netherlands]

Fut Life Future Life [US Publication]

Futures Res Q Rep, Aluminum Futures Research Quarterly Report, Aluminum [UK]

Futures Res Q Rep, Copper Futures Research Quarterly Report, Copper [UK]

Futures Res Q Rep, Lead Zinc Futures Research Quarterly Report, Lead and Zinc [UK]

Futures Res Q Rep, Precious Met Futures Research Quarterly Report, Precious Metals [UK]

Futures Res Q Rep, Tin Futures Research Quarterly Report, Tin [UK]

FUV Far Ultraviolet

FUVMP Far-Ultraviolet Molecular Process

FUVMS Far-Ultraviolet Molecular Spectroscopy

FUVR Far-Ultraviolet Radiation

FUW Farmers Union of Wales [UK]

Fuzzy Ball Hydrogenated (60-Carbon-Atom-) Buckminsterfullerene Molecule

Fuzzy Sets Syst Fuzzy Sets and Systems [Journal published in the Netherlands]

FV *(Fachverband)* – German for "Technical Association" [also Fv]

FV Fiber Volume

FV Final Value

FV Finite Volume

FV Flash Vaporization

FV Flexural Vibration [Crystallography]

FV Flight Version

FV Floor Valve

FV Flounder Virus

FV Fluid Volume

FV Flux Value

FV Forced Vibration

FV Food Vacuole [Cytology]

FV Foreline Valve

FV Fraser Valley [British Columbia, Canada]

FV Free Valence

FV Free Vibration

FV Free Volume

FV Front View

FV Full Voltage

FV Future Value

F(V) Cumulative (Acoustic-Emission) Amplitude Distribution [Symbol]

$F_t(V)$ Cumulative (Acoustic-Emission) Threshold Crossing Distribution [Symbol]

F/V Frequency-to-Voltage (Converter)

Fv *(Fachverband)* – German for "Technical Association" [also FV]

fV femtovolt [Unit]

f(V) differential (acoustic-emission) amplitude distribution [Symbol]

$f_t(V)$ differential (acoustic-emission) threshold crossing distribution [also $f_t(V)$] [Symbol]

ƒ(V) differential (acoustic-emission) amplitude distribution [Symbol]

ƒ(v) Maxwell distribution (function) [Symbol]

FV-3 Frog Virus, Type 3

FVA Floor Valve Adapter

FVA Forschungsvereinigung Antriebstechnik e.V. [Power Transmission Engineering Research Association, Germany]

FVB Fast Violet B

FVC Forced Vital Capacity [Medicine]

FVC Frequency-to-Voltage Converter

FVD Front Vortex Distance

FVF First Vertical Flight

FVLS Full Volume Least-Squares (Method)

FVM Fluorovinylmethylene

1-FVM 1-Fluorovinylmethylene

3-FVM 3-Fluorovinylmethylene

FVMS Fluid Volume Measurement System

FVN Failed Vector Numbers

FVPRA Fruit and Vegetable Preservation Research Association [US]

FvPM Fachverband Pulvermetallurgie [Powder Metallurgy Association, Germany]

FVR Feline Viral Rhinotracheitis [Veterinary Medicine]

FVR Fiber Volume Ratio

FVR Flexible Vocabulary Recognition

FVR Fuse Voltage Rating

FVRDE Fighting Vehicles Research and Development Establishment [of Ministry of Defense, UK]

FVS Fernmelde-Versuchssatellit [Experimental Communications Satellite, Germany]

FVS Finska Vetenska-Societeten [Finnish Scientific Society]

FVS Flight Vehicle Systems Committee [of Institute of Electrical and Electronic Engineers–Aerospace and Electronic Systems Society]

FVSC Fort Valley State College [Georgia, US]

FVT Flash-Vacuum Thermolysis

FVT Full Video Translation

FVU File Verification Utility

FVV Facility Verification Vehicle

FVW Forward Volume Wave

FW Face-Width [Gears]

FW Fachnormenausschuß Wärmebehandlungstechnik metallischer Werkstoffe [Standards Committee on Heat Treatment of Metallic Materials, Germany]

FW Falling Weight

FW Feed Water

FW Felix-Weil (Blood Test) [Medicine]

FW Field Weakening

FW Filament Winding; Filament-Wound

FW Firewall

FW Firmware

FW First Word

FW Fixed Width

FW Flag Word

FW Flash Welding

FW Flat Washer

FW Flexural Wave

FW Fog Whistle

FW Foot Wall [Geology]

FW Formula Weight

FW Fort Worth [Texas, US]

FW Fresh Water

FW Frost Weathering

FW Fuel Wasting

FW Full Wave

F&W Fall and Winter

5W- Western Samoa [Civil Aircraft Marking]

FWA Federal Works Agency [US]

FWA Feedwater Accumulator

FWA Ferromagnet with Wandering Axes [Solid-State Physics]

FWA File Work Area

FWA First Word Address

FWA Fluorescent Whitening Agent

FWA Forward Wave Amplifier

FWA French Water Study Association

FWA French West Africa

FWAD Fort Worth Army Depot [Texas, US]

FWB Fahrenheit Wet Bulb

FWB Four Wheel Brake [also 4WB]

FWBF Foster Wheeler Bergbau-Forschung (Process) [Mining]

FWAG Farming and Wildlife Advisory Group [UK]

FWAT Forest Workers Association of Tasmania

FWC Fault Warning Computer

FWC Fine-Cut WC (Tungsten Carbide)

FWC Fourdrinier Wire Council [US]

FWC Full-Wave Circuit

FWC Fully Loaded Weight and Capacity

FWCA Fish and Wildlife Coordination Act [US]

FWCC First World Climate Conference

FWCI Foundation of the Wall and Ceiling Institute [US]

FWD Federation of Wholesalers and Distributors

FWD Free-Wheel Diode

FWD Free Working Distance [Microscopy]

FWD Freight Forward [also fwd]

FWD Fresh Water Damage

FWD Front-Wheel Drive

fwd forward

4WD Four-Wheel Drive

FWDC Full-Wave Direct Current

Fwdg Forwarding [also fwdg]

FWD HT SHLD Forward Heat Shield [also Fwd Ht Shld]

Fwdr Forwarder [also fwdr]

FWE Finished with Engines

FWE Freshwater Ecology

FWERAT Fourth World Educational and Research Association Trust [UK]

FWF Fokker-Wheeler-Feynman Electrodynamics [Physics]

FWF Fond für Wissenschaft und Forschung [Austrian National Science Foundation]

FWG Facility Working Group

FWG Force Working Group

FWGE Fort Worth Grain Exchange [Texas, US]

FWH Feedwater Heater

FWHF Federation of World Health Foundations [Switzerland]

FWHH Full Width at Half-Height [also fwhh]

FWHH Full Width of Height of Spectral Peak

FWHM Full Width at Half-Maximum [also fwhm]

FWHP Flywheel Horsepower [also fwhp]

FWHFT Fast Walsh/Hadamard/Fourier Transform

FWI Fernwärme International [German International Journal on District Heating]

FWI Fire Weather Index

FWI French West Indies

FWIC Federated Women's Institute of Canada

FWISL Freshwater Institute Science Laboratory [Winnipeg, Manitoba, Canada]

FWIW For What It's Worth [Internet Jargon]

FWL Fixed Word Length

FWM Fachnormenausschuß Werkzeugmaschinen [Standards Committee on Machine Tools, Germany]

FWM Feedwater Meter

FWM Fourier Wave Mixing

FWM Four Wave Mixing

FWN Fermi Wave Number [Physics]

FWN Iron Tungsten Nitride

FWO Fond for Wetenschapellijk Onderzoeg [Scientific Research Fund, Belgium]

FWONA Free World Outside North America

F-Word Feed-Rate Word [Numerical Control]

FWP Fair-Witness Project [US]

FWP Federal Water Policy [Canada]

FWP Field-Work Proposal

FWPCA Federal Water Pollution Control Act [US]

FWPCA Federal Water Pollution Control Administration [US]

FWPRDC Forest and Wood Products Research and Development Corporation [US]

FWQA Federal Water Quality Administration [US]

FWR Feedwater Regulator

FWR Full-Wave Rectification; Full-Wave Rectifier

FWR Fusiform Wood Ray

FWRAP Federal Water Resources Assistance Program [US]

FWRC Federal Water Resources Council [US]

FWRM Federation of Wire Rope Manufacturers [UK]

FWRS Fish and Wildlife Reference Service [US]

FWS Fachnormenausschuß Werkzeuge and Spannzeuge [Standards Committee on Tools and Fixtures, Germany]

FWS Filter Wedge Spectrometer

FWS Fish and Wildlife Service [of US Department of the Interior, Washington, DC]

FWS Fourier-Walsh Series

4WS Four-Wheel Steering [also 4ws]

4W SW Four-Way Switch [also 4W Sw, FW SW, or FW Sw]

FWT Gesellschaft Feinwerktechnik [Society for Precision Engineering; of Verein Deutscher Ingenieure/Verein Deutscher Elektrotechniker, Germany]

FWTAO Federation of Women Teachers Associations of Ontario [Canada]

FWTM Full Width at Tenth Maximum [also fwtm]

4WTS Four Wire Terminating Set [also FWTS]

FWTT Fixed Wing Tactical Transport

FWW Food, Water and Waste

FWWM Food, Water and Waste Management

FWWMS Food, Water and Waste Management Subsystem

FWWMS Food, Water, and Waste Management System

FX Fixed Aera [Computers]

FX Foreign Exchange

FX France (European Territory) [ISO Code]

F/X Special Effects

4X- Israel [Civil Aircraft Marking]

5X- Uganda [Civil Aircraft Marking]

F(x) Function of x [Mathematics]

F(x) Distribution Probability Function

F$_i$(x) Newton Polynomial [Symbol]

f(x) function of x [Mathematics]

f'(x) derivative of $y = f(x)$ with respect to x [Mathematics]

ƒ(X) distribution function [Symbol]

ƒ(x) function of x [Mathematics]

ƒ(x) Gaussian distribution (or normal distribution) [Symbol]

ƒ$^{(n)}$(x) n-th derivative of function $f(x)$ (in Taylor series) [Symbol]

.fx France (European Territory) [Country Code/Domain Name]

FXC Ferroxcube [Soft Magnetic Material]

FX-CCSA Foreign Exchange–Common Circuit Switching Arrangement

FXD Flash X-Ray Diffraction

fxd fixed

FXR Flash X-Ray Radiography

FXRE Flash X-Ray Equipment

FXS Fine-Focus X-Ray Series

FXT Fixed Time Call

ƒ(x,y) function of x and y [Mathematics]

ƒ(x,y,z) function of x, y and z [Mathematics]

FY Feng Yun [Series of Chinese Geostationary Satellites]

FY Financial Year

FY Fiscal Year [also FYr]

FY Francon-Yamamoto (Interference Contrast)

FY-1 Feng Yun 1 [Chinese Geostationary Satellite]

FY-2 Feng Yun 2 [Chinese Geostationary Satellite]

5Y- Kenya [Civil Aircraft Marking]

ƒ(y) probability density function [Symbol]

FYDP Five Year Defense Program [US]

FYDP Future Years Defense Plan [of US Department of Defense]

FYE Full Year Equivalent

FYI For Your Information

FYM Farmyard Manure

FYP Five Year Plan

FYr Fiscal Year [also FY]

FYSIKDB Geofysisk Databas [Geophysical Database, of Swedish Geological Survey]

Fys Tidsskr Fysisk Tidsskrift [Danish Journal of Physics]

FYTD Fiscal Year to Date

FYWP Fiscal Year Work Plan

FZ Fixed Zero

FZ Flame Zone

FZ Floating Zero

FZ Float(ing) Zone

FZ Fluorozirconate

FZ Fracture Zone [Geology]

FZ Freezing [also Fz]

FZ Fresnel Zone

FZ Fusion Zone [Welding]

Fz Fuze [also fz]

FZB Forschungszentrum Borstel, Zentrum für Medizin und Biowissenschaften [Borstel Research Center for Medicine and Biosciences, Germany]

FZC Float(ing) Zone Crystal

FZDZ Freezing Drizzle [also FzDz]

FZES Float(ing) Zone Experiment System

FZF Float(ing) Zone Furnace

FZFG Freezing Fog [also FzFg]

FZG Fluorozirconate Glass

FZI Forschungszentrum Informatik [Research Center for Information Science, Karlsruhe, Germany]

FZK Forschungszentrum Karlsruhe [Karlsruhe (Nuclear) Research Center, Germany]

FZM Float(ing) Zone Method

FZP Fresnel Zone Plate

FZR Forschungszentrum Rossendorf e.V. [Rossendorf Research Center, Rossendorf, Germany]

FZRA Freezing Rain [FzRa]

FZS Fellow of the Zoological Society

FZS Forschungszentrum Strangpressen [Extrusion Research Center, Berlin, Germany]

FZ-UDS Floating-Zone Unidirectional Solidification (Technique)

G Agricultural Engineering (Technology) [Discipline Category Abbreviation]
G Amplification Factor [Symbol]
G AND Dependency [Digital Logic]
G Antenna Gain [Symbol]
G ASTM Grain-Size Number [Symbol]
G Crack-Extension Force [Symbol]
G Critical (Corrosion) Pit Size [Symbol]
G Defect Creation Rate [Symbol]
G de Genne's Factor [Symbol]
G Dose Rate [Symbol]
G Dutch Gulden [Currency]
G Edge Gradient (of Photographic Image) [Symbol]
G Effective Temperature Gradient [Symbol]
G Electric Conductance [Symbol]
G (Mechanical) Energy Release Rate [Symbol]
G Field Gradient [Symbol]
G Fifth Tone in the Scale of C Major [Acoustics]
G Force [Symbol]
G Fraunhofer "G" Line [Fraunhofer Line for Iron Having a Wavelength of 430.8 Nanometers] [Spectroscopy]
G Gabon(ese)
G Gaborone [Botswana]
G Ga(u)ge
G Gain
G (+)-Galactose [Biochemistry]
G Galactic; Galaxy
G Galvanic
G Galvanize(d); Galvanizing
G Galvanometer
G Gambia(n)
G Gamete [Biology]
G Gamma
G Ganges (River) [India/Pakistan]
G Ganglion(ic) [Anatomy]
G Gao (Model of Crack Bridging) [Metallurgy]
G Gap

G Gas(eous)
G Gasoline
G Gastric
G Gastrocnemius [Anatomy]
G Gastroenteritis [Medicine]
G Gastrula(tion) [Embryology]
G Gate
G Gate [Semiconductor Symbol]
G Gauge
G Gauss(ian)
G Gauss [Unit]
G Gdansk [Poland]
G Gear(ing)
G Geiger
G Gel
G Gelatin(ous)
G Gemmule [Biology]
G Gene
G General
G General Audiences [Motion-Picture Rating]
G Generation; Generator
G Generic
G Genetic(s)
G Geneva [Italy]
G Genital(ia) [Anatomy]
G Genoa [Italy]
G Genus
G Geodesic; Geodesist; Geodesy
G Geometric(al); Geometrician; Geometry
G Georgetown [Guyana]
G Georgia [US]
G Geriatric(ian); Geriatrics
G Germ
G German(y)
G Gerontological; Gerontologist; Gerontology
G Gestation(al)
G Get
G Geyser
G Ghana
G Giant
G Gibberellin [Biochemistry]
G Gibbs Free Energy (or Gibbs Function) [Symbol]
G Gibraltar
G Giga- [SI Prefix]
G Gilbert [Unit]
G Gill [Biology]
G Gimbal
G Girder
G Giza [Egypt]
G Gizzard [Ornithology]
G Gland(ular)
G Glass(y)
G Glide

G   Global; Globe
G   Globular; Globule
G   Globulin
G   Gloecapsa [Genus of Algae]
G   Glomerulus [Anatomy)
G   Glottis [Anatomy]
G   Glow
G   D-Glucose; (+)-Glucose) [Biochemistry]
G   Glycerin; Glycerol
G   Glycine
G   Glycolic Acid
G   Glycyl
G   Go [Computers]
G   Goal
G   Goiter [Medicine]
G   Gold [Abbreviation; Symbol: Au]
G   Golgi [Symbol]
G   Gomphonema [Genus of Algae]
G   Gonad [Anatomy]
G   Gonadotropin [Biochemistry]
G   Gonococcal; Gonococcus [Microbiology]
G   Gonorrhea
G   Good [also g]
G   Gorilla
G   Gorki [Russia]
G   Goss (Texture) [Metallurgy]
G   Göteborg [Sweden]
G   Gourde [Currency of Haiti]
G   Grade(d)
G   Gradient
G   Graduate(d)
G   Grain
G   Grain Size [Symbol]
G   Granada [Spain]
G   Granular; Granule
G   Granulocyte
G   Graph(ical)
G   Graphite
G   Grating
G   Gravel
G   Gravitation(al)
G   Gravitational Constant (6.67259 × 10^{-11} m^3/(s^2·kg)) [Symbol]
G   Gravity
G   Greece; Greek
G   Green
G   Greenland(ic)
G   Green's Function [Symbol]
G   Greenwich Time
G   Grenada [West Indies]
G   Grenade
G   Grid
G   Griffith (Toughening Model) [Metallurgy]
G   Grind(ing)

G   Grit
G   Groove; Grooving
G   Ground
G   Ground Thread [Symbol]
G   Grounding
G   Group(ing)
G   Grout(ing)
G   Grow(th)
G   Growth Function [Symbol]
G   Growth Rate (in Nucleation of Crystalline Solid) [Symbol]
G   Guadalajara [Mexico]
G   Guadeloupe [Lower Antilles]
G   Guam [US]
G   Guanine [Biochemistry]
G   Guanosine [Biochemistry]
G   Guarani [Currency of Paraguay]
G   Guard(ing)
G   Guatemala(n)
G   Guide(line)
G   Guinea
G   Gulf
G   Gum(my)
G   Gun(ner); Gunning
G   Guyana
G   Gypsum
G   Gyration; Gyratory; Gyro
G   Gyroid (Phase) [Symbol]
G   (Crystal) Lattice Factor [Symbol]
G   Length of Gear-Worm Thread Section [Symbol]
G   Magnetoresistance [Symbol]
G   Mass Velocity [Symbol]
G   Permanently Grounded (Aircraft) [USDOD Symbol]
G   Quantity of Gas [Symbol]
G   Shear Modulus (or Rigidity Modulus) [Symbol]
G   Single-Glass-Covered (Electric Wire)
G   Specific Gravity [Symbol]
G   Star Group of Spectral Type G [Surface Temperature 5,540°C, or 10,000°F] [Letter Designation]
G   Surface Attack (Vehicle) [USDOD Symbol]
G   Temperature Gradient [Symbol]
G   Torsional Modulus of Elasticity [Symbol]
G   Torsional Spring Constant [Symbol]
Ġ   Weight [Symbol]
G- Great Britain [Civil Aircraft Marking]
-G Search and Rescue (Aircraft) [US Navy Suffix]
.G Source Language Source File [File Name Extension] [also .g]
⟨G⟩ Cahn-Hagel Migration Rate (in Crystallization) [Symbol]
Ḡ   Generalized Strain Release Rate [Symbol]
Ḡ   Molar Free Energy [Symbol]
G* Complex Modulus (in Shear) [Symbol]
G° Standard Gibbs Free Energy [Symbol]

G′ Fraunhofer line for hydrogen having wavelength of 434.1 nanometers [Spectroscopy]

G′ Storage Modulus (in Shear) [Symbol]

G″ Loss Modulus (in Shear) [Symbol]

G$_□$ Conductance per Square [Symbol]

G$_↓$ Conductance for the Downspin Electrons [Symbol]

G$_↑$ Conductance for the Upspin Electrons [Symbol]

G$_0$ (Mechanical) Energy Release Rate at Zero Velocity [Symbol]

G$_0$ Intrinsic Fracture Energy (of Plastics) [Symbol]

G$_I$ Interlaminar Fracture Toughness, Mode I–Crack Propagation [Symbol]

G$_{Ic}$ Mode I–Fracture Energy [Symbol]

G$_{Ic}$ Interlaminar Fracture Toughness (Mode I– Peel; Mode II – Shear; Mode III –Scissor Shear) [Symbol]

G$_{II}$ Interlaminar Fracture Toughness, Mode II–Crack Propagation [Symbol]

G$_{IIc}$ Mode II–Fracture Energy [Symbol]

G$_{IIIc}$ Mode III–Fracture Energy [Symbol]

G$_{1c}$ Fracture Energy [Symbol]

G$_a$ Air-Space Ratio (of Rock, or Soil) [Symbol]

G$_a$ Apparent Specific Gravity [Symbol]

G$_b$ Gilbert [Symbol]

G$_c$ Critical Strain-Energy Release Rate [Symbol]

G$_c$ Energy per Unit Crack Area (of Plastics) [Symbol]

G$_c$ (Interface) Fracture Toughness (or Fracture Energy) [Symbol]

G$_c$ Critical Generalized Strain Release Rate [Symbol]

G$_D$ Film Gradient (in Radiography) [Symbol]

G$_δ$ Intersection of Countably Many Open Sets [Mathematical Symbol]

G$_{eq}$ Equivalent (Electrical) Conductance [Symbol]

G$_f$ Flake Graphite (in Cast Iron) [Symbol]

G$_G$ Gibbs Free Energy of Gas [Symbol] [also G$_g$]

G Shear Modulus in Crystal Direction [Symbol]

G$_i$ Gibbs Free Energy of Species i [Symbol]

G$_i$° Molar Free Energy [Symbol]

$\overline{G}_i$ Partial Molar Free Energy of Component i [Symbol]

G$_i^{xs}$ Excess Molar Free Energy of Mixing of Species i [Symbol]

G$_{if}$ Interfacial Fracture Energy [Symbol]

G$_L$ Gibbs Free Energy of Liquid [Symbol] [also G$_l$]

G$_ℓ$ Local Energy per Area (of Plastics) [Symbol]

G$_{LT}$ Longitudinal Shear Modulus (of a Composite) [Symbol]

G$_m$ Bulk Specific Gravity (or Specific Mass Gravity) [Symbol]

G$_m$ Mutual Conductance (or Transconductance) (of Electron Tubes) [Symbol]

G$_m$ Shear Modulus of Matrix [Symbol]

G$_n$ Nodular Graphite (or Graphite Nodules) (in Cast Iron) [Symbol]

G$_p$ Conductance Peak-Height [Symbol]

G$_p$ Shear Modulus of Particles [Symbol]

G$_{PB}$ Common-Base Large-Signal Insertion Power Gain [Semiconductor Symbol]

G$_{pb}$ Common-Base Small-Signal Insertion Power Gain [Semiconductor Symbol]

G$_{PC}$ Common-Collector Large-Signal Insertion Power Gain [Semiconductor Symbol]

G$_{pc}$ Common-collector Small-Signal Insertion Power Gain [Semiconductor Symbol]

G$_{PE}$ Common-Emitter Large-Signal Insertion Power Gain [Semiconductor Symbol]

G$_{pe}$ Common-Emitter Small-Signal Insertion Power Gain [Semiconductor Symbol]

G$_{pg}$ Common-Gate Small-Signal Insertion Power Gain [Semiconductor Symbol]

G$_{ps}$ Common-Source Small-Signal Insertion Power Gain [Semiconductor Symbol]

G$_{p,T}$ Gibbs Energy (of a Chemical System) [Symbol]

G$_R$ Crack-Extension Resistance [Symbol]

G$_r$ Graphite Rosettes (in Cast Iron) [Symbol]

G$_{RS}$ Enhancement Factor of Raman Scattering [Symbol]

G$_S$ Gibbs Free Energy of Solid [also G$_s$][Symbol]

G$_s$ Specific Gravity (of Soilds) [Symbol]

G$_s$ Surface Free Energy [Symbol]

G$_s$ Temperature Gradient in Solids [Symbol]

G$_T$ Total Conductance [Symbol]

G$_{TB}$ Common-Base Large-Signal Transducer Power Gain [Semiconductor Symbol]

G$_{tb}$ Common-Base Small-Signal Transducer Power Gain [Semiconductor Symbol]

G$_{TC}$ Common-Collector Large-Signal Transducer Power Gain [Semiconductor Symbol]

G$_{tc}$ Common-Collector Small-Signal Transducer Power Gain [Semiconductor Symbol]

G$_{TE}$ Common-Emitter Large-Signal Transducer Power Gain [Semiconductor Symbol]

G$_{te}$ Common-Emitter Small-Signal Transducer Power Gain [Semiconductor Symbol]

G$_{tg}$ Common-Gate Small-Signal Transducer Power Gain [Semiconductor Symbol]

G$_{ts}$ Common-Source Small-Signal Transducer Power Gain [Semiconductor Symbol]

G$_{TT}$ Transverse Shear Modulus (of a Composite) [Symbol]

G$_V$ Gibbs Free Energy of Vapor [Symbol]

G$_v$ Air-Void Ratio (of Rock, or Soil) [Symbol]

G$_v$ Skid Number-Speed Gradient (of Pavement Surface) [Symbol]

G$_x$ Torque Component about X-Axis [Symbol]

G^{xs} Excess Molar Free Energy of Mixing [Symbol]

G$_{xy}$ In-Plane Shear Modulus of Laminate [Symbol]

G$_y$ Torque Component about Y-Axis [Symbol]

G$_z$ Torque Component about Z-Axis [Symbol]

G-5 Group of Five [i.e., Five Most Highly Industrialized Nations – Britain, France, Germany, Japan and the United States]

G-7 Group of Seven [i.e., Seven Most Highly Industrialized Nations – Britain, Canada, France, Germany, Italy, Japan and the United States] [also G7]

G-8 Group of Seven plus Russian Federation [also G8]

G-10 Group of Ten (Nations) [also known as the Paris Club]

G-20 Group of Group of Twenty [Interim Committee of the International Monetary Fund]

G-24 Group of Twenty Four [Intergovernmental Group of Twenty Four Nations from Africa, Asia and Latin America on International Monetary Affairs]

G-77 Group of Seventy Seven (Developing Countries)

g   absolute frequency density (in statistics) [Symbol]

g   asymmetry factor [Symbol]

g   diffraction vector (in electron microscopy) [Symbol]

g   effective phonon coupling constant [Symbol]

g   electron-phonon coupling strength [Symbol]

g   elemental pitting probability (in corrosion science) [Symbol]

g   Fraunhofer line for calcium having wavelength of 422.7 nanometers [Spectroscopy]

g   function (especially solved functions, $x = g(y)$) [Symbol]

g   ga(u)ge

g   gallon [Unit]

g   gamma [English Equilvalent]

g   gamma (unit) [1 gamma = 10-5 gauss]

g   gap

g   gas(eous)

g   gate

g   gate [Semiconductor Symbol]

g   gauche [Chemistry]

g   gauss

g   genus

g   (Landé) g-factor [Symbol]

g   gilbert [Unit]

g   gilt

g   gluon (in particle physics) [Symbol]

g   good [also G]

g   grain [Unit]

g   gram [Unit]

g   gravitational acceleration (9.8067 m/s²) [Symbol]

g   gravity

g   green

g   grid

g   gross

g   group

g   gulf

g   gyromagnetic tensor [Symbol]

g   Kubo-Greenwood conductance [Symbol]

g   length (of an involute spline) [Symbol]

g   number of atoms in a cluster [Symbol]

g   optical interval (or optical tube length) [Symbol]

g   osmotic coefficient [Symbol]

g   reciprocal (crystal) lattice vector [Symbol]

g   reflection vector (in crystallography) [Symbol]

g   specific conductance [Symbol]

g   transfer constant [Symbol]

±g   greenness-redness (opponent-color scale) [OSA-UCS]

g°   gain (of amplification system) [Symbol] [Electron Microscopy]

g_0 degeneracy of the (energy) ground level [Symbol]

g_0 gravitational acceleration (of earth) at sea-level (9.81 m/s²) [Symbol]

g_{45} gravitational acceleration at lattitude 45°[Symbol]

g^E molar excess free energy [Symbol]

$g_e/2$ electron magnetic moment (or free electron g factor in Bohr magnetons) (1.0011597) [Symbol]

g_{fs} common-source small-signal forward transfer conductance [Semiconductor Symbol]

g_{hkl} diffraction vector in *hkl* crystallographic plane [Symbol]

g_i degeneracy of the *i*-th energy level [Symbol]

g_{is} common-source small-signal input conductance [Semiconductor Symbol]

g_m mutual conductance (or transconductance) (of electron tubes) [Symbol]

$g_\mu/2$ muon magnetic moment (or free muon g factor in Bohr magnetons) (1.0011592) [Symbol]

g_{max} maximum (value of) gravitational acceleration [Symbol]

g_{MB} common-base static transconductance [Semiconductor Symbol]

g_{MC} common-collector static transconductance [Semiconductor Symbol]

g_{ME} common-emitter static transconductance [Semiconductor Symbol]

g_n normal gravitational acceleration [Symbol]

g_{os} common-source small-signal output conductance [Semiconductor Symbol]

g_{rs} common-source small-signal reverse transfer conductance [Semiconductor Symbol]

g_s standard gravitational acceleration (9.8067 m/s²) [Symbol]

g_{TB} common-base small-signal reverse transfer conductance [Semiconductor Symbol]

0g zero gravity (or zero weight) [Symbol]

1g one gravity (or normal weight) [Symbol]

2g two times gravity (or twice normal weight) [Symbol]

Γ   Activity Coefficient (of Electrolytes) [Symbol]

Γ   Adatom Hopping Rate [Symbol]

Γ   Atomic Jump Frequency [Symbol]

Γ   Brillouin Zone [Symbol]

Γ   (Spectroscopic) Broadening Parameter [Symbol]

Γ   Capillarity Constant [Symbol]

Γ   Circulation (for Fluid Motion) [Symbol]

Γ   Coulomb Coupling Parameter [Symbol]

Γ   Dislocation Line Density (in Materials Science) [Symbol]

Γ   Flory-Huggins Parameter [Symbol]

Γ   Fracture Energy [Symbol]

Γ   (Upper-case) Gamma [Greek Alphabet]

Γ   Gamma Function [Symbol]

Γ   Gibbs-Thomson Coefficient [Symbol]

Γ   Interfacial Energy [Symbol]

Γ   (Magnetic Resonance) Linewidth [Symbol]

Γ   Magnetic Torque [Symbol]

Γ Normalized Energy Release Rate [Symbol]

Γ Order Fault Energy [Symbol]

Γ Phonon-Phonon Interaction Strength [Symbol]

Γ Photorefraction Gain [Symbol]

Γ Pitch Angle (of Spiral Bevel Gears) [Symbol]

Γ Plasma Parameter [Symbol]

Γ Propagation Constant [Symbol]

Γ Segregation Amount [Symbol]

Γ Surface Concentration Excess [Symbol]

Γ Thermal Transmittance (or Thermal Conductance) [Symbol]

Γ Torque [Symbol]

Γ Toughening Ratio [Symbol]

Γ (Electron) Tunneling Rate [Symbol]

Γ_0 Linewidth [Symbol]

Γ_{APB} Antiphase Boundary Energy (in Materials Science) [Symbol]

Γ_B Magnetic Grüneisen Parameter [Symbol]

$\Gamma_{E(C)}$ Tunneling Rate through Emitter-Collector Barrier [Semiconductor Symbol]

Γ_G Griffith Energy [Symbol]

Γ_i Gibbsian Interfacial Excess of Element i [Symbol]

Γ_i Surface Excess Concentration of Component i [Symbol]

Γ_{ij} Velocity of Interface Between Regions i and j [Symbol]

Γ_l Surface Concentration of Component l [Symbol]

Γ_m Magnetic Constant (or Permeability of Empty Space) [Symbol]

Γ_m Maximum Concentration of Molecules at the Surface [Symbol]

Γ_s Gibbs Excess Surface Energy [Symbol]

Γ_T Thermal Grüneisen Parameter [Symbol]

γ activity coefficient [Symbol]

γ (plane) angle [Symbol]

γ anisotropy parameter [Symbol]

γ austenite (phase) [Symbol]

γ center angle (of helical gearings) [Symbol]

γ circulating power ratio (of gear trains) [Symbol]

γ coefficient of cubical expansion [Symbol]

γ creep strain (in mechanics) [Symbol]

γ damping coefficient [Symbol]

γ degree of dissociation (of a solution) [Symbol]

γ degree of thermal expansion anisotropy [Symbol]

γ domain-wall energy (in solid-state physics) [Symbol]

γ duty cycle [Symbol]

γ effective gamma content [Symbol]

γ electrical conductivity [Symbol]

γ Euler's constant (0.5772...) [Symbol]

γ face angle (of milled bevel gears) [Symbol]

γ (lower-case) gamma [Greek Alphabet]

γ gamma magnetic unit [Symbol]

γ gamma phase (of a material) [Symbol]

γ gamma quantum [Symbol]

γ g-factor (or spectroscopic splitting factor) [Symbol]

γ Grüneisen constant (in solid-state physics) [Symbol]

γ gyromagnetic ratio (in physics) [Symbol]

γ interatomic distance [Symbol]

γ interfacial tension (in materials science) [Symbol]

γ kinetic energy transfer efficiency [Symbol]

γ largest principal refractive index of a biaxial crystal [Symbol]

γ lattice parameter–crystallographic angle between a and b axes of unit cell [Symbol]

γ magnetic loss angle [Symbol]

γ mass fraction [Symbol]

γ phase angle [Symbol]

γ photon [Symbol]

γ pitch angle (of spiral bevel gear pinions) [Symbol]

γ prefix used to indicate the position of a substituent atom, or radical [Chemistry]

γ propagation constant [Symbol]

γ quantum field [Symbol]

γ rake angle (in metal cutting) [Symbol]

γ ratio of specific heats (C_p/C_v) [Symbol]

γ reduced temperature (i.e. ratio of temperature to critical temperature, t/t_c) [Symbol]

γ shear rate; shear strain [Symbol]

γ slope of the straightline part of sensitometric curve (in photography) [Symbol]

γ specific conductance [Symbol]

γ specific gravity [Symbol]

γ stacking-fault energy (in solid-state physics) [Symbol]

γ (specific) surface energy [Symbol]

γ surface tension [Symbol]

γ third-direction angle of a line, angle with the z-axis [Symbol]

γ unit weight (or weight per unit volume) [Symbol]

γ work-hardening coefficient [Symbol]

γ' gamma prime phase (of a material) [Symbol]

γ' submerged (or buoyant) unit weight (of soil, or rock) [Symbol]

γ'' gamma double prime phase (of a material) [Symbol]

$\gamma_\pm$ mean ionic activity coefficient [Symbol]

γ_0 surface free energy of the minimum free energy configuration [Symbol]

γ_a interfacial free energy per atom [Symbol]

γ_B magnetic sensitivity [Symbol]

γ_{cl} catalytic-liquid interfacial free energy [Symbol]

γ_{cs} catalytic-solid (crystal) interfacial free energy [Symbol]

γ_d dry unit weight (of soil, or rock solids) [Symbol]

γ_e effective unit of weight (of soil, or rock solids) [Symbol]

γ_{eff} effective surface energy [Symbol]

γ_{exp} electronic specific heat coefficient [Symbol]

γ_f failure strain [Symbol]

γ_G saturated unit weight (of soil) [Symbol]

γ_g molar interfacial free energy [Symbol]

γ_i activity coefficient of component i [Symbol]

γ_L Lüders strain (in metallurgy) [Symbol]

γ_{L-N} energy of liquid/nucleus interphase [Symbol]

γ_{L-S} energy of liquid/substrate interphase [Symbol]

γ_{LT} shear strain in longitudinal plane (of composites) [Symbol]

γ_{LV} liquid/vapor interfacial tension [Symbol]

γ_m magnetic constant [Symbol]

γ_m submerged (or buoyant) unit weight (of soil, or rock) [Symbol]

γ_m surface energy of metal [Symbol]

γ_m wet (or mass) unit weight (of soil, or rock) [Symbol]

γ_{max} maximum unit weight (of soil, or rock) [Symbol]

γ_{max} shear deformation [Symbol]

γ_{mol} molecular polarizability [Symbol]

γ_p gyromagnetic ratio of protons in water (corrected for diamagnetism of water) ($2.6752213 \times 10^8 \ s^{-1} \ T^{-1}$) [Symbol]

γ_p plastic deformation energy [Symbol]

γ_p plastic flow [Symbol]

γ_{P-L} particle/liquid interfacial tension [Symbol]

γ_{P-S} particle/solid interfacial tension [Symbol]

γ_s (specific) surface energy (or interfacial energy) [Symbol]

γ_s surface energy of substrate [Symbol]

γ_s surface tension [Symbol]

γ_s zero air voids unit weight (of soil, or rock mass) [Symbol]

γ_{sat} saturated unit weight (of soil mass) [Symbol]

γ_{SL} solid/liquid interfacial tension [Symbol]

γ_{S-N} energy of substrate/nucleus interface [Symbol]

γ_{sub} submerged (or buoyant) unit weight (of soil, or rock solids) [Symbol]

γ_{SV} solid/vapor interfacial tension [Symbol]

γ_{TT} shear strain in transverse plane (of a composite) [Symbol]

γ_w unit weight of water [Symbol]

γ_{wet} wet (or mass) unit weight (of soil, or rock solids) [Symbol]

γ_{xy} shear strain in XY plane [Symbol]

γ_{xz} shear strain in XZ plane [Symbol]

γ_y yielding strain [Symbol]

γ_{yz} shear strain in YZ plane [Symbol]

γ_z zero air voids unit weight (of soil, or rock mass) [Symbol]

∇ Gradient [Symbol]

GA Gabon [ISO Code]

GA Gadolinium Aluminide

GA Gain of Antenna

GA Galactic Astronomy

GA Gallic Acid

GA Galvanic Activity

GA Galvanizers Association [UK]

GA Galvanneal(ed); Galvannealing

GA Gamma Absorption

GA Gap Analysis [Biodiversity]

GA Gas Absorption

GA Gas Act [UK]

GA Gas Adsorption

GA Gas Amplification

GA Gas Analysis; Gas Analyzer

GA Gas Atomization; Gas Atomizer [Powder Metallurgy]

GA Gastric Acid

GA Gate Array

GA Gaussian Algorithm

GA Geological Age

GA General Agent

GA General Analysis

GA General Assembly

GA General Assembly [of United Nations]

GA General Assistance

GA General Atomics [San Diego, California, US]

GA General Availability

GA General Average [also G/A]

GA General Aviation

GA Geographical Association [UK]

GA Geologists Association [UK]

GA Geomagnetic Anomaly

GA Georgia [US]

GA Geometric Analysis

GA Gibberellic Acid

GA Glancing Angle

GA Glassy Alloy

GA Glide Angle

GA Global Address

GA Glutamine Acid

GA Glyoxylic Acid

GA Go Ahead (Signal)

GA Good Afternoon [Amateur Radio]

GA Graphic Ammeter

GA Graphic Approximation

GA Graphics Adapter

GA Grapple Adapter

GA Gravimetric Analysis; Gravimetric Analyzer

GA Gravitational Acceleration

GA Grazing Angle

GA Greater Antilles

GA Green Alliance [UK]

GA Greening Australia

GA Gross Anatomy

GA Gypsum Association [US]

GA Gyro Assembly

GA Gyrotropic Anisotropy

GA3 Gibberellic Acid [Biochemistry]

G&A General and Administrative

G/A General Average [also GA]

G/A Ground-to-Air [also g/a]

G-A Graphite-Adhesive (Joint)

G-A Glass-Air (Interface)

G-A Ground-to-Air [also g-a]

Ga Ga(u)ge [also ga]

Ga Galileo Number (of Fluid Mechanics) [Symbol]

Ga Gallium [Symbol]

Ga Georgia [US]

Ga^{2+} Divalent Gallium Ion [also Ga^{++}] [Symbol]

Ga^{3+} Trivalent Gallium Ion [also Ga^{+++}] [Symbol]

Ga I Gallium I [Symbol]

Ga II Gallium II [Symbol]

Ga-65 Gallium-65 [also ^{65}Ga, or Ga^{65}]

Ga-66 Gallium-66 [also ^{66}Ga, or Ga^{66}]

Ga-67 Gallium-67 [also ^{67}Ga, or Ga^{67}]

Ga-68 Gallium-68 [also ^{68}Ga, or Ga^{68}]

Ga-69 Gallium-69 [also ^{69}Ga, or Ga^{69}]

Ga-70 Gallium-70 [also ^{70}Ga, or Ga^{70}]

Ga-71 Gallium-71 [also ^{71}Ga, or Ga^{71}]

Ga-72 Gallium-72 [also ^{72}Ga, or Ga^{72}]

Ga-73 Gallium-73 [also ^{73}Ga, or Ga^{73}]

G(a) Height-Difference Correlation Function [Symbol]

GAA Graduate Administrative Assistant

GAA Gravure Association of America [US]

GAA Greenhouse Action Australia

GAA Ground Antiaircraft Control

GAAB Graphic Arts Advisory Board [US]

Ga(ACAC)$_3$ Gallium(III) Acetylacetonate [also Ga(AcAc)$_3$, or Ga(acac)$_3$]

GAAEC Graphic Arts Advertisers and Exhibitors Council [US]

GAAG Gross Actual Generation

Ga-Al Gallium-Aluminum (Alloy System)

GaAlAs Gallium Aluminum Arsenide (Semiconductor)

GaAlAsSb Gallium Aluminum Arsenic Antimonide (Semiconductor)

GaAlP Gallium Aluminum Phosphide (Semiconductor)

GaAlSb Gallium Aluminum Antimonide (Semiconductor)

GAAP General Adjustment Assistance Program

GAAP Generally Accepted Accounting Principles

GAAS Generally Accepted Auditing Standards

GAAS Graphite-Furnace Atomic Absorption Spectroscopy

GaAs Gallium Arsenide (Semiconductor)

GaAsAl Gallium Arsenide Aluminide (Semiconductor)

GaAs/AlAs Gallium Arsenide/Aluminum Arsenide (Superlattice)

GaAs/AlAs SL Gallium Arsenide/Aluminum Arsenide Superlattice

GaAs ALE Gallium Arsenide Atomic Layer Epitaxy (Process)

GaAs/AlGaAs Gallium Arsenide/Aluminum Gallium Arsenide (Superlattice)

GaAs/AlGaAs SL Gallium Arsenide/Aluminum Gallium Arsenide Superlattice

GaAs:Be Beryllium-Doped Gallium Arsenide

GaAs:C Carbon-Doped Gallium Arsenide [also GaAs(C)]

GaAs:Cr Chromium-Doped Gallium Arsenide (Semiconductor)

GaAs:Cu Copper-Doped Gallium Arsenide (Semiconductor)

GaAs:Er Erbium-Doped Gallium Arsenide (Semiconductor)

GaAsFET Gallium Arsenide Field-Effect Transistor [also GaAs-FET]

GaAs/GaAs Gallium Arsenide on Gallium Arsenide

GaAs/GaP Gallium Arsenide/Gallium Phosphide (Heterojunction)

GaAs/Ge Gallium Arsenide/Germanium (Heterojunction)

GaAs HBT Gallium Arsenide Heterojunction Bipolar Transistor

GaAs:I Intrinsic Gallium Arsenide

GaAs IC Gallium Arsenide Integrated Circuit

GaAs IC Symp Gallium Arsenide Integrated Circuit Symposium [of Institute of Electrical and Electronics Engineers, US]

GaAsIMPATT Gallium Arsenide Impact Avalanche Transit Time [also GaAs-IMPATT]

GaAs/InAs Gallium Arsenide/Indium Arsenide (Superlattice)

GaAs/InAs SL Gallium Arsenide/Indium Arsenide Superlattice

GaAs/InGaAs Gallium Arsenide on Indium Gallium Arsenide (Substrate)

GaAs:Li Lithium-Doped Gallium Arsenide (Semiconductor)

GaAs MBE Gallium Arsenide Molecular Beam Epitaxy [also GaAs-MBE]

GaAsMESFET Gallium Arsenide Metal-Semiconductor Field-Effect Transistor [also GaAs-MESFET]

GaAsMITATT Gallium Arsenide Mixed Tunneling Avalanche Transit Time [also GaAs-MITATT]

GaAsMMIC Gallium Arsenide Monolithic Microwave Integrated Circuit [also GaAs-MMIC]

GaAsN Gallium Arsenide Nitride

GaAsP Gallium Arsenide Phosphide

GaAs:Sb Antimony-Doped Gallium Arsenide

GaAs:Se Selenium-Doped Gallium Arsenide

GaAs:Si Silicon-Doped Gallium Arsenide (Semiconductor)

GaAs/Si Gallium Arsenide on Silicon (Substrate)

GaAs:Sn Tin-Doped Gallium Arsenide

GaAs:Te Tellurium-Doped Gallium Arsenide (Semiconductor)

GaAs:X Extrinsic Gallium Arsenide

GaAs:Zn Zinc-Doped Gallium Arsenide

GaAs(Zn):H Hydrogenated Zinc-Doped Gallium Arsenide

GAATS Gander Automatic Air Traffic System [Canada]

GAB Graphic Adapter Board

GAB Grants Administration Branch [of National Institutes of Health, Bethesda, Maryland, US]

GAB Great Australian Bight

GAB Gusseted Angle Bracket

GAB World Glass Address-Book [Published in Germany]

GABCC Great Australian Bight Consultative Committee

GABCOM Gmelin Abbreviations of (Chemical) Compounds [of Gmelin Institute, Germany]

GABA Gamma-Amino-n-Butyric Acid

GABase Gamma-Amino-n-Butyrase [Biochemistry]

GABIA Great Australian Bight Industry Association

GABMET Gmelin Abbreviations of (Chemical and Physical) Methods [of Gmelin Institute, Germany]

GABOB Gamma-Amino-Beta-Hydroxybutyric Acid

GABTF Great Australian Bight Trawl Fishery

GAC Gas Adsorption Chromatography

GAC Geological Association of Canada

GAC Global Area Coverage [Advanced Very-High-Resolution Radiometer]

GAC Government Advisory Committee

GAC Granular Activated Carbon

GAC Gross Available Capacity

GAC Ground Attitude Control

GAC Grumman Aerospace Corporation [US]

G-Acetin Globular Actin [also G-acetin] [Biochemistry]

G-Acid 7-Hydroxy-1,3-Naphthalenesulfonic Acid [also G-acid]

G-Acid 2-Naphthol-6,8-Disulfonic Acid [also G-acid]

γ-Acid 6-Amino-4-Hydroxy-2-Naphthalenesulfonic Acid [also γ-acid]

γ-Acid 2-Amino-8-Naphthol-6-Sulfonic Acid [also γ-acid]

GaCl Gallium Chloride

GAC-MAC Geological Association of Canada–Mining Association of Canada

G/A CON General Average Contribution

GA&CS Ground Acquisition and Command Station

GACS Gun Alignment Control System

GACT Generally Achievable Control Technology

GACTAI General Arbitration Council of the Textile and Apparel Industry [US]

GACU Ground Air Conditioning Unit

GACU Ground Avionics Cooling Unit

GAD Glutamic Acid Decarboxylase [Biochemistry]

GADDS General Area Detector (X-Ray) Diffraction System

GaDECl Gallium Diethylchloride

G/A DEP General Average Deposit

GADEST (International Meeting) Gettering and Defect Engineering in Semiconductor Technology

GADL Ground-to-Air Data Link

GADO General Aviation District Office [US]

GADS Generating Availability Data System [of National Electric Reliability Council, US]

GAEC Greek Atomic Energy Commission

GAeF Gesellschaft für Aerosolforschung [Association for Aerosol Research, Germany]

GAELIC Grumman Aerospace Engineering Language for Instructional Checkout

GA&ES Georgia Architectural and Engineering Society

GaEt₃ Triethyl Gallium (or Gallium Triethyl)

Ga(EtO)₃ Gallium(III) Ethoxide

GAESDA Graphic Arts Equipment and Supply Dealers Association [US]

GAF Gas Amplification Factor

GAF General Aniline and Film (Corporation) [US]

GAFB Griffis Air Force Base [of US Air Force in New York State]

GAFTA Grain and Feed Trade Association [UK]

GAG Georg-Agricola-Gesellschaft (zur Förderung der Geschichte der Naturwissenschaften und der Technik) [Georg Agricola Society (for the Advancement of the History of Science and Technology), Dusseldorf, Germany]

GAG Glycosaminoglycan [Biochemistry]

GAG Gross Available Generation

GAG Group-Specific Antigen [Immunology]

GAGB Gemmological Association of Great Britain

γ-AgAl Gamma Silver Aluminide

γ-AgCd Gamma Silver Cadmium

GAGE Global Atmosphere Gases Experiment

Ga-Ge Gallium-Germanium (Alloy System)

GAGP Generalized Antisymmetrized Geminal Power

GAGR Group Automatic Gain Regulator

γ-AgZn Gamma Silver Zincide

Ga-H Gallium-Hydrogen (System)

g/Ah gram(s) per ampere-hour [also g/A·h]

GAHF Grapple Adapter Handling Fixture

GAHu Goat Anti-Human [Biochemistry]

GAI Generalization Area of Intersection

GAI Guaranteed Annual Income

GAI Guild of Architectural Ironmongers [UK]

GAIA Graphic Arts Industries Association

GAIBA γ-Aminoisobutyric Acid

GAIM Global Analysis, Interpretation and Modelling [Climatology]

GAIN Graphic Aids for investigating Networks

GaInAs Gallium Indium Arsenide (Semiconductor)

GaInAs/InP Gallium Indium Arsenide/Indium Phosphide (Superlattice)

GaInAsP Gallium Indium Arsenide Phosphide (Laser)

GaInN Gallium Indium Nitride (Semiconductor)

GaInP Gallium Indium Phosphide (Semiconductor)

GAINS Gimballess Analytic Inertial Navigation System

GaInSb Gallium Indium Antimonide (Semiconductor)

GaInSn Gallium Indium Stannide (Semiconductor)

GAIU Graphic Arts International Union

GAIT General Aviation Infrastructure Tariff [Australia]

GAJ Guild of Agricultural Journalists [UK]

GaK Galactokinase [Biochemistry]

GaK Gallium K-Absorption Edge [X-Ray Absorption Fine Structure]

GAL Generic Array Logic

GAL Global Address List

GAL Greening Australia Limited

GAL Grün-Alternative Liste [Electoral Pact of Green and Alternative Parties, Germany]

GAL Guggenheim Aeronautical Laboratory [of California Institute of Technology, Pasadena, US]

Gal D-Galactose; Galactosyl [also GAL] [Biochemistry]

Gal β-D-Galactopyranoside [Biochemistry]

Gal Galileo [Unit]

(Gal)₂ bis-Galactosyl (or Di-Galactosyl) [Biochemistry]

α-Gal α-Galactose [also α-gal]

β-Gal β-Galactose [also β-gal]

gal gallon [Unit]

GalA Galacturonic Acid [Biochemistry]

gal/A gallon(s of spray liquid) per acre [Unit]

gal/bag gallon(s of water) per bag (of cement)

GalC Galactocerebroside [Biochemistry]

GALCIT Guggenheim Aeronautical Laboratory of California Institute of Technology [US]

γ-AlCu Gamma Aluminum Copper

gal/d gallon(s) per day [also gal d^{-1}, GPD, gpd, or gal/day]

gal/ft²·min gallon(s) per square foot minute [Unit]

(Gal-GlcNAc)$_3$ tris(Galactosyl-N-Acetyl-D-Glucosaminyl) [Biochemistry]

(Gal-GlcNAc)$_4$ tetrakis(Galactosyl-N-Acetyl-D-Glucosaminyl) [Biochemistry]

gal/hr gallon(s) per hour [also gal/h, GPH, gph, or gal h^{-1}]

gal/hp-h gallon(s) per horsepower-hour [Unit]

Galilean Electrodyn Galilean Electrodynamics [US Publication]

Gall Gallery [also gall]

gall gallon [Unit]

gal/min gallon(s) per minute [also gal min^{-1}, GPM, or gpm]

GalN D-Galactosamine

GalNAc Galactosyl-N-Acetyl [also galNAc]

γ-Al$_2$O$_3$ Gamma Alumina [also Gamal]

γ-Al$_2$O$_3$:Er Erbium-Doped Gamma Alumina

Gal-1P Galactose-1-Phosphate

GALP Good Automated Laboratory Practices

GalP Galactose Phosphate [Biochemistry]

GALPAT Galloping Pattern Memory

Gal-PUT Galactose-1-Phosphate Uridyl Transferase [Biochemistry]

gal/sec gallon(s) per second [also gal/s, gal sec^{-1}, gal s^{-1}, GPS, or gps]

gal/tree gallon(s) of spray liquid per tree [Symbol]

Galv Galvanizing [also galv]

Galv Galvanometer [also galv]

GALVATECH International Conference on Zinc and Zinc Alloy Coated Sheet Steels [also Galvatech]

Galvano-Organo-Trait Surf Galvano-Organo-Traitements de Surface [French Journal on Galvanoorganic Surface Treatment]

Galvanotech Galvanotechnik [German Electroplating Journal]

Galvano-Tek Tidsskr Galvano-Teknisk Tidsskrift [Norwegian Electroplating Journal]

GAM Goat Anti-Mouse [Biochemistry]

GAM Graduate in Aerospace Mechanical Engineering

GAM Gram Atomic Mass [also gam]

GAM Graphic Access Method

GAM Ground-to-Air Missile

GAM Growth Associated Molecule

GAM Guided Aircraft Missile

Gam Gambia(n)

GAMA Gas Appliance Manufacturers Association [US]

GAMA General Aviation Manufacturers Association [US]

GAMA Graphics-Assisted Management Application

Gamal Gamma Alumina [also γ-Al$_2$O$_3$]

GAMBIT Gate Modulated Bipolar Transistor

GAME Graduate in Aerospace Mechanical Engineering

GAMESS General Atomic and Molecular Electronic Structure System

GAMI Gorham Advanced Materials Institute [US]

GAMIS General Analytical Methods Information Service [of Laboratory of the Government Chemists, UK]

GAMIS Graphic Arts Marketing and Information Service [of Printing Industries Association, US]

GAMLOGS Gamma-Ray Logs

GAMM German Association for Applied Mathematics and Mechanics

GAMM Gesellschaft für angewandte Mathematik und Mechanik [Association for Applied Mathematics and Mechanics, Germany]

(Ga,Mn)As Gallium Manganese Arsenide

GAMP Global Atmospheric Measurements Program [US National Center for Atmospheric Research Atmosphere Technology Division]

GAMPS Gander Automated Message Processing System [of International Civil Aviation Organization]

GAMS γ-D-Glutamylaminomethylsulfonic Acid [Biochemistry]

GAMS Groupement pour l'Avancement des Méthodes Spectrographiques [Working Group for the Advancement of Spectrographic Methods, France]

GAMT German Association of Medical Technologists

GAMTA General Aviation Manufacturers and Traders Association [UK]

GAN Generalized Activity Network

GAN Generating and Analyzing Network

GAN Generating and Assembly Network

GaN Gallium Nitride (Semiconductor)

GaN/AlN Gallium Nitride/Aluminum Nitride (Superlattice)

GaN/Al$_2$O$_3$ Gallium Nitride/Aluminum Oxide (Thin Film)

GaNAs Gallium Nitrogen Arsenide (Semiconductor)

GaN:Cd Cadmium-Doped Gallium Nitride

GaN-GaAs Gallium Nitride-Gallium Arsenide (System)

GaNi Gallium Nickel (Compound)

GANIL Grand Accelerator National A Ions Lourds

GaN:Mg Magnesium-Doped Gallium Nitride (Semiconductor)

GaNO Gallonitrate

GaNP Gallium Nitrogen Phosphide (Semiconductor)

GaN/SiC Gallium Nitride/Silicon Carbide (Superlattice)

GaN:Zn Zinc-Doped Gallium Nitride

GAO Gadolinium Aluminum Oxide

GAO General Accounting Office [of US Congress]

GAP Gap Analysis (Biodiversity) Program [US]

GAP Gas and Particle (Sampler)

GAP General Assembly Program

GAP Giant Array Project [of Fermilab, Batavia, Illinois, US]

GAP Glyceraldehyde Phosphate; D-Glyceraldehyde-3-Phosphate [Biochemistry]

GAP Glycidyl Azide Polymer

GAP GOAL (Ground Operations Aerospace Language) Automatic Procedure

GAP Graphics Application Program

GAP Growth Associated Protein [Biochemistry]

GAP Gun Aiming Point

GAP Gun Aiming Post

GaP Gallium Phosphide (Semiconductor)

GAPA Ground-to-Air Pilotless Aircraft

GAPAN Guild of Airline Pilots and Navigators [UK]

G&A PAN Gyroscope and Accelerometer Panel

GAPD Glyceraldehyde-3-Phosphate Dehydrogenase [also GAPDH] [Biochemistry]

GAPDH Glyceraldehyde-3-Phosphate Dehydrogenase [also GAPD] [Biochemistry]

GAPE Guyana Association of Professional Engineers

GAPEX Ground-Based Atmospheric Profiling Experiment

GaP-Ge Gallium Phosphide-Germanium (Heterojunction)

GAPI Gateway Application Programming Interface

GaP:I Intrinsic Gallium Phosphide

GAPL Group Assembly Parts List

GAPMB Ghana Agricultural Produce Marketing Board

GaP:N Nitrided Gallium Phosphide (Semiconductor)

GaPO Gallophosphate

GAPP Geometric-Arithmetic Parallel Processor

GAPS Giga (or Billions) of Actions per Second

γ-APS Gamma Aminopropyltriethoxysilane

GAPSFAS Graduate and Professional School Financial Aid Service

GaP/Si Gallium Phosphide/Silicon (Heterojunction)

GAPT Graphical Automatically Programmed Tools

GaP:Te Tellurium-Doped Gallium Phosphide

GAR Gas Atomization Reaction

GAR Goat Anti-Rabbit [Immunochemistry]

GAR Grain Aspect Radio [Metallurgy]

GAR Growth Analysis and Review

GAR Guided Air(craft) Rocket

GaR Garand Rifle

Gar Garage [also gar]

GARa Goat Anti-Rat [Innumochemistry]

GARBD Garboard [also Garbd, or garbd]

GARC Graphic Arts Research Center [US]

GARD General Address Reading Device

GARDAE Gather, Alarm, Report, Display and Evaluate

GARDENEX Federation of Garden and Leisure Equipment Exporters [UK]

GAREX Ground Aviation Radio Exchange System

GARF Graphic Arts Research Foundation [US]

GARMI General Aviation Radio Magnetic Indicator

GARP Global Atmospheric Research Program [of National Academy of Sciences, US]

GARP Globally Optimized Alternating Phase Rectangular Pulse [Nuclear Magnetic Resonance]

Garr Garrison [also garr]

GARS Gas Atomization Reaction Synthesis

GART Graphics Address Relocation Table

GARTEUR Group for Aeronautical Research and Technology in Europe

GAS General Accounting System

GAS General Aviation Services

GAS Georgia Academy of Science [Atlanta, US]

GAS Georgian Academy of Science [Tblisi, Georgian Republic]

GAS Getaway Special [of NASA]

GAS Glasgow Agricultural Society [Scotland]

GAS Glasgow Archeological Society [Scotland]

GAS Goods Acquisition System

GAS Graphics Attachment Support

GAS Greek Academy of Sciences [Athens]

GaS Gallium Sulfide (Semiconductor)

Gas Gasoline [also gas]

GaSb Gallium Antimonide (Semiconductor)

GaSb:Cr Chromium-Doped Gallium Antimonide (Semiconductor)

GASC Guggenheim Aviation Safety Center [at Cornell University, Ithaca, New York, US]

GASCan Getaway Special Canister [of NASA] [also GASCAN]

GASCO General Aviation Safety Committee [UK]

GASDA Gasoline and Automotive Service Dealers Association [US]

GaSe Gallium Selenide (Semiconductor)

Gas Energy Rev Gas Energy Review [of American Gas Association]

Gas Eng Manage Gas Engineering and Management [UK Publication]

GASFET Gas-Sensitive Field-Effect Transistor

GASGASGAS Guild of Ancient Suppliers of Gas Appliance, Skills, Gins, Accessories and Substances [US]

GASH Guanidinium Aluminum Sulfate Hexahydrate [Biochemistry]

GASL General Applied Science Laboratory

GASLAB Global Atmospheric Sampling Laboratory

GASM Guided Air-to-Surface Missile

GASMACO Gas Machinery Company (Process)

GASMPA Group for the Advancement of Spectroscopic Methods and Physicochemical Analysis [France]

Gaso Gasoline [also gaso]

GaSO Gallosulfate

GASP General Activity Simulation Program

GASP Generalized Academic Simulation Program

GASP Global Assimilation and Prognosis System

GASP Global Atmospheric Sampling Program

GASP Grand Accelerated Space Platform

GASP Graphic Applications Subroutine Package

GASPE Gated Spin Echo

GASPI Guidance Attitude Space Position Indicator

Gas Res Inst Dig Gas Research Institute Digest [US]

Gas Res Inst (Res Dev Results) Gas Research Institute (Research and Development Results) [US]

GASS Generalized Assembly System

GASS Grest-Anderson-Sahni-Srolovitz (Model) [Materials Science]

GASSAR General Atomic Standard Safety Analysis Report

Gas Sep Purif Gas Separation and Purification [Journal published in the US]

GASSP General Application Spectrum Synthesis Program [of York University, Toronto, Canada]

GAST Gas-Cooled Solar Tower (Power Plant)

GAST Globally Averaged Surface Temperature

Gastroenterol Gastroenterology [US Journal]

Gastrointest Endosc Gastrointestinal Endoscopy [International Journal]

Gas Turb World Gas Turbine World [Journal published in the US]

Gas Turb World Cogen Gas Turbine World and Cogeneration [Journal published in the US]

Gas Wärme Int Gas und Wärme International [German International Journal on Gas and Heating]

GAT Gas Aggregation Technique

GAT General Air Traffic

GAT General Aviation Transponder

GAT Generalized Algebraic Translator

GAT Georgetown Automatic Translator [of Georgetown University, Washington, DC, US]

GAT Glow-to-Arc Transition

GAT Graphic Arts Terminal

GAT Greenwich Apparent Time

GATAC General Assessment Three-Dimensional Analog Computer

GATB General Aptitude Test Battery

GATCO Guild of Air Traffic Control Officers [UK]

GATD Graphic Analysis of Three-Dimensional Data

GATE GARP (Global Atmospheric Research Program) Atlantic Tropical Experiment [of National Academy of Science, US]

GATE Generalized Algebraic Translator Extended

GATE Group to Advance Total Energy

GATF Graphic Arts Technical Foundation [US]

GATF World GATF World [Journal of the Graphic Arts Technical Foundation, US]

Ga(TMHD)₃ Gallium Tris(2,2,6,6-Tetramethyl-3,5-Heptanedionate) [(also Ga(tmhd)₃]

GATNIP Graphic Approach to Numerical Information Processing

GATP Ground Acceptance/Article Test Procedure

GATR Ground-to-Air Transmitter-Receiver

GATS General Acceptance Test Software

GATT Gate-Assisted Turnoff Thyristor

GATT General Agreement on Tariffs and Trade [of United Nations]

GATT Graphics Address Translation Table

GATT Ground-to-Air Transmitter Terminal

GATTIS Georgia Institute of Technology Technical Information Service [US]

GATT/TBT General Agreement on Tariffs and Trade/Technical Barriers to Trade

GATU Geophysical Automatic Tracker Unit

GATV Gemini-Agena Target Vehicle [of NASA]

GAU Gas Atomization Unit

GAU Georg-August University [Goettingen, Germany]

γ-AuCd Gamma Gold Cadmium

γ-AuZn Gamma Gold Zinc

GAV Gemeinschaftsausschuss Verzinken [Joint Galvanizing Committee, Germany]

GAW Global Atmospheric Watch [of World Meteorological Organization]

GAW Gram-Atomic Weight [also gaw]

GAW Gross Axle Weight

GAW Guaranteed Annual Wage

GAWG General Aviation Working Group [UK]

GAWR Gross Axle Weight Rating

Ga:YIG Gallium-Doped Yttrium Iron Garnet [also Ga-YIG, or GaYIG]

Gaz Gazette [also gaz]

Gaz Mat Gazeta de Matematica [Portuguese Mathematical Gazette]

Gazz Chim Ital Gazzetta Chimica Italiana [Italian Chemical Gazette; published by Società Chimica Italiana]

GB Gain × Bandwidth (Product)

GB Gall Bladder

GB Gamow Barrier [Nuclear Physics]

GB Gas Black

GB Gas Burner

GB Gear Bronze

GB General Background

GB Germanium Bolometer

GB Gibbs-Bogoliubov Principle [Chemical Physics]

GB Gigabyte [Unit]

GB Glacier Bay [Alaska, US]

GB Glide(r) Bomb

GB Grain Boundary [Materials Science]

GB Granular Bainite [Metallurgy]

GB Granular Bait

GB Gray Body

GB Great Britain

GB Great Britain [ISO Code]

GB Greenish Blue

GB Green Star, Blinker, Parachute [US]

GB Grid Bias [Electronics]

GB Guard Band [Electronics]

GB Guide Block

GB Guillain-Barré (Syndrome) [Medicine]

GB Guinea-Bissau [West Africa]

GB Gun Boat

GB Gun Branch

G+B Gordon and Breach Science Publishers [US]

G(B) Magnetic Induction Dependent Conductance [Symbol]

Gb Gigabit [also Gbit]

Gb gigabyte [GB preferred]

Gb Gilbert [also gb] [Unit]

gb Great Britain [Country Code/Domain Name]

γ-B Gamma Boron [Symbol]

GBA Global Biodiversity Assessment (Book)

GBA Grain Boundary Allotriomorph [Materials Science]

GBAA German Business Aviation Association

GBARC Great Britain Aeronautical Research Committee

GBB Groupement Belge du Béton [Belgian Concrete Society]

GBBA Glass Bottle Blowers Association [US and Canada]

GBC George Brown College [Canada]

GBC Grain-Boundary Character [Materials Science]

GBC Grain-Boundary Component [Materials Science]

GBC Grain-Boundary Corrosion [Materials Science]

GBC Grain-Boundary Cracking [Materials Science]

GBC Graphite Bi-Intercalation Compound

GBC Gray-to-Binary Conversion

GBC Green Bag Charge

GBCD Grain Boundary Character Distribution [Materials Science]

GBCD Grain Boundary Characteristics Design [Materials Science]

Gb/cm Gilbert per centimeter [Unit]

GBCNO Gadolinium Barium Copper Niobium Oxide (Superconductor)

GBCO Gadolinium Barium Copper Oxide (Superconductor)

GBCT Guild of British Camera Technicians [UK]

GBCTO Gadolinium Barium Copper Titanium Oxide (Superconductor)

GBD Grain Boundary Diffusion [Materials Science]

GBD Grain Boundary Dislocation [Materials Science]

GBD Grain Bridging Degradation [Materials Science]

GBDL Gesellschaft für Bibliothekswesen und Dokumentation des Landbaues [Association for Librarianship and Documentation in Agriculture, Germany]

GB-DOS Grain Boundary Density of States

GBE Butyl Glycolate

GBE Grain Boundary Engineered; Grain Boundary Engineering [Materials Science]

GBF Geographic Base File

GBF Gesellschaft für Biotechnologische Forschung [Society for Biotechnology Research, Darmstadt, Germany]

GBF Grain Boundary Flow [Materials Science]

GBF Grain Boundary Fracture [Materials Science]

GBF Great Bear Foundation [US]

gbf grain boundary orientation factor (in materials science) [Symbol]

GBFR Grain Boundary Fracture Ratio [Materials Science]

GBFS Glacier Bay Field Station [of US Geological Survey/Biological Resources Division]

GBH Gamma Benzene Hexachloride

GBH Grievous Bodily Harm

GBH Group Busy Hour

GBHA Glyoxal-bis(2-Hydroxyanil)

GBI Gesellschaft für Betriebswirtschaftliche Information [Society for Business Information, Germany]

GBI Grand Bahama Island

GBIC Graphite Bi-Intercalation Compound

GBIF Global Biodiversity Information Facility [of Organization for Economic Cooperation and Development]

Gb/in² Gigabits per square inch [also Gbit/in²]

Gbit Gigabit(s) [also Gb]

Gbit/s Gigabits per second [also Gb/s]

GBL Gibraltar Pound [Currency]

GBL Government Bill of Lading

GBL Great Bear Lake [Northwest Territories, Canada]

GBLC Grain Boundary Limited Conduction [Materials Science]

GBLIC Gaussian Band Limited Channel

GBM Glomerular Basement Membrane

GBM Grain Boundary Melting [Materials Science]

GBM Grain Boundary Migration [Materials Science]

GBM Grain Boundary Mobility [Materials Science]

GBM Grape Berry Moth

GBM Green Belt Movement [Kenya]

GBMD Grain Boundary Misorientation Distribution [Materials Science]

GBMS Grain-Boundary Misorientation Spectrum [Materials Science]

GBN Graphite-Like Boron Nitride [also gBN, or g-BN]

GBNE Guild of British Newspaper Editor [UK]

GBO Goods in Bad Order [also gbo]

GBO Government Business Opportunities [of Science and Professional Services Directorate, Canada]

GBO Grain Boundary Oxidation [Materials Science]

GBP Gain Bandwidth Product

GBP Grain Boundary Plane [Materials Science]

GBP Grain Boundary Precipitate [Materials Science]

GBP British Patent [also GB Pat]

GBP Guinea-Bissau Peso [Currency of Guinea-Bissau]

GBP Pound Sterling [Currency of the United Kingdom and Northern Ireland]

GB P Guinea-Bissau Peso [Currency of Guinea-Bissau]

GB Pat British Patent [also GBP]

GBPS Gigabits per Second [also gbps]

GBQ Generating Bearing Quality

Gbq Gigabecquerel [Unit]

GBq/mmol Gigabecquerel(s) per millimole [Unit]

GBR Glass Bead Rating

GBR Great Barrier Reef [Queensland, Australia]

GBRA Guadalupe-Blanco River Authority [US]

GB RAM Gigabytes of Random-Access Memory [also GB-RAM]

GBRCC Great Barrier Reef Consultative Committee [Queensland, Australia]

GBRMP Great Barrier Reef Marine Park [Queensland, Australia]

GBRMPA Great Barrier Reef Marine Park Authority [Queensland, Australia]

GBRN Global Basins Research Network [of Columbia University, New York, US]

GBRS Groupe Belge de Recherche Sousmarine [Belgian Submarine Research Group]

GBS Grain Boundary Sliding [Materials Science]

GBS Grain Boundary Slip [Materials Science]

GBS Ground Based Software

GBS Gütegemeinschaft Bauelement aus Stahlblech [Quality Assurance Group for Sheet Steel Components, Germany]

Gb/s Gigabits per second [also Gbit/s]

GBSL Grain Boundary Superlattice [Materials Science]

GBSR Graphite-Moderated Boiling and Superheating Reactor

GBSS Gey's Balanced Salt Solution

GBT Generalized Bernoulli Theorem [Physics]

GBT Generalized Burst Trapping

GBT Grain Boundary Thickness [Materials Science]

GBT Ground Based Test

GBT Ground Bounce Technique

GBX Glovebox

GByte(s) Gigabyte(s)

GBZ Glass Bonded Zeolite

GC Gadolinium Chromite

GC Gain Control

GC Galactic Center [Astronomy]

GC Galvanic Cell

GC Galvanic Corrosion

GC Gametocyte [Biology]

GC Gamma Counter

GC Gantry Crane

GC Garden City [New York State, US]

GC Gas Calorimeter; Gas Calorimetry

GC Gas Capacitor

GC Gas Cell

GC Gas Centrifuge (Process) [Nuclear Engineering]

GC Gas Chemisorption

GC Gas Chromatograph(y)

GC Gas Classification; Gas Classifier

GC Gas Compressor

GC Gas Constant

GC Gas Cooled; Gas Cooler; Gas Cooling

GC Gas Counter

GC Gate Circuit

GC Geiger Counter

GC General Circular

GC General Counsel

GC Genetic Code

GC Geneva Convention

GC Geochemical; Geochemist(ry)

GC Geochronologist; Geochronology

GC Germinal Center [Biochemistry]

GC Georgopoulos-Cohen (Method) [also G-C] [Metallurgy]

GC Gigacycle [Unit]

GC Glass Ceramics

GC Glass Chemist(ry)

GC Glassy Carbon

GC Global Change(s)

GC Glycerin; Glycerol

GC Golay Cell (Radiometer)

GC Golgi Cell [Anatomy]

GC Gonococcal; Gonococcus [Microbiology]

GC Gonocyte [Cytology]

GC Governing Council

GC Government of Canada

GC (Nonlocal) Gradient Corrected (Approximation) [Physics]

GC Graduated Cylinder

GC Grain Cube

GC Grain-Coarsened; Grain Coarsening [Metallurgy]

GC Grambling College [Grambling, Louisiana, US]

GC Grand Canyon [Arizona, US]

GC Grand Cayman [Cayman Islands]

GC Grant Conditions [National Science Foundation, US]

GC Granulocyte [Medicine]

GC Graphic Character

GC Graphitized Coke

GC Gravity Concentration

GC Great Circle

GC Green Community [Environment Concept]

GC Green Compact [Powder Metallurgy]

GC Grid Current [Electronics]

GC Griffith Crack [Mechanics]

GC Grinnell College [Grinnell, Iowa, US]

GC Ground Control(ler)

GC Grounded Collector

GC Group Code

GC Group Connector

GC Growth Cell

GC Grüneisen Constant [Solid-State Physics]

GC Guanine-Cytosine (Nucleotide) [Biochemistry]

GC Guatemala City

GC Guidance Computer

GC Guidance Control

GC Gulf Coast

GC Gun Control

GC Gun Cotton

GC Gyro-Compass(ing)

GC Gyroscopic Couple

GC² Glass Capillary Gas Chromatography

G&C Guidance and Control

G-C Georgopoulos-Cohen (Method) [also GC] [Metallurgy]

Gc Gigacycle [Unit]

Gc Glycol

γ-C Gamma Carbon [also C$^\gamma$]

GCA Gain Control Driver

GCA (Computer) Game Control Adapter

GCA General Control Approach

GCA Generator Coordinate Approximation

GCA Geneva Convention Act

GCA Glazing Contractors Association

GCA Government Contract Awards [US]

GCA Grains Council of Australia

GCA Graphic Communications Association [US]

GCA Ground-Controlled Approach (Radar)

GC-AED Gas Chromatography with Atomic Emission Detection [also GC/AED]

GCAGS Gulf Coast Association of Geological Societies [US]

GCAK Green Communities Assistance Kit [US Environmental Initiative]

g-cal gram-calorie [Unit]

g-cal/g/°C gram-calorie per gram per degree centigrade [also g-cal/g·°C]

g-cal/s/cm/°C gram-calorie per second per centimeter per degree centigrade [also g-cal/sec/cm/°C, g-cal/(s·cm·°C), or g-cal/(sec·cm·°C)]

GCAP Generalized Circuit Analysis Program

GCAW Gas Carbon-Arc Welding

GCB General Circuit Breaker

GCB German Congress-Bureau

GCB Ground Contamination Bomb

GCBR Gas-Cooled Breeder Reactor

GCC Gas Chromatograph Column

GCC Gasification Combined Cycle

GCC General Cerium Company [US]

GCC General Channel Coordinator

GCC Global Change Category

GCC Global Community Center

GCC Global Competitiveness Council

GCC GNU (Gnu's Not Unix) C-Compiler [Unix Operating System]

GCC Government Communications Center [UK]

GCC Grid Cooperating Center

GCC Ground Communications Coordinator

GCC Ground Control Center

GCC Group Classification Code

GCC Grove City College [Pennsylvania, US]

GCC Gulf Cooperation Council [of Arab Gulf States]

GCC Gulf Cooperative Countries (Region)

g/cc gram(s) per cubic centimeter [also g/cu cm, g/cm^3, $g\ cm^{-3}$, or gm/cc]

GCCA Graphic Communications Computer Association [now Graphic Communications Association, US]

GCCC Ground Control Computer Center

GCCD Glass-Passivated Ceramic Chip Diode

g/ccm gram(s) per cubic meter [also g/cu m, g/m^3, $g\ m^{-3}$, or g/ccm]

GCCMD General Conference of the Condensed Matter Division [of European Physical Society]

GCCNO Gadolinium Calcium Copper Niobium Oxide (Superconductor)

GCCO Ground Control Checkout

GCCS Gas Collision Cross-Section

GCD Greatest Common Denominator

GCD Gyro-Compass, Desired Cluster Orientation

Gcd Glutaconic Dialdehyde

gcd greatest common divisor [also GCD]

GCDA Geocentric Datum of Australia

GCDC Global Change Data Center [US]

GCDC Ground Checkout Display and Control

GCDCS Ground Checkout Display and Control System

GCDIS Global Change Data and Information System [US]

GCdN Glycol Dinitrate

GCDP Global Change Database Project

GCDRA Green Crop Driers Research Association [UK]

GCE Galveston Cotton Exchange and Board of Trade [Texas, US]

GCE General Certificate of Education

GCE General Consumers Electronics [US]

GCE Generator Conversion Efficiency

GCE Glassy Carbon Electrode

GCE Global Change Encyclopedia

GCE Grand Canonical Ensemble [Statistical Mechanics]

GCE Ground Checkout Equipment

GCE Ground Communication Equipment

GCE Ground Control Equipment

γ-Ce Gamma Cerium [Symbol]

GCE 2D Goddard Cumulus Ensemble–Two-Dimensional [NASA Goddard Space Flight Center Convective Scale Model]

GCEC Gold Coast Environment Center [Ghana]

GC-ECD Gas Chromatograph/Electron Capture Detector; Gas Chromatography with Electron Capture Detection [also GC/ECD]

GC-ELCD Gas Chromatograph/Electrolytic Conductivity Detector; Gas Chromatography with Electrolytic Conductivity Detection [also GC/ELCD]

GCEMC Grand-Canonical Emsemble Monte Carlo

GCEM Goddard Cumulus Ensemble Model [of NASA Goddard Space Flight Center, Greenbelt, Maryland, US]

GC ENV Government of Canada, Environment Canada

GCEP Governing Council for Environmental Programs

GCF Gas Correction Factor

GCF General Control Function

GCF Greatest Common Factor

GCF Gross Capacity Factor

GCF Ground Communications Facility

GCFAP Guidance and Control Flight Analysis Program

GCFBR Gas-Cooled Fast Breeder Reactor [also GCFR]

GCFI Gulf and Caribbean Fisheries Institute [US]

GC-FID Gas Chromatograph/Flame Ionization Detector; Gas Chromatography with Flame Ionization Detection [also GC/FID]

GC-FPD Gas Chromatograph/Flame Photometric Detector; Gas Chromatography coupled with Flame Photometric Detection [also GC/FPD]

GCFR Gas-Cooled Fast (Breeder) Reactor

GCFRE Gas-Cooled Fast (Breeder) Reactor Experiment

GC-FTIR Gas Chromatography Fourier Transform Infrared (Spectroscopy) [also GC/FTIR, or GC/FT-IR]

GCG Gamow-Condon-Gurney (Theory) [Nuclear Physics]

GCG Gas Chromatography

GCG Genetics Computer Group

GCGC Glass-Capillary Gas Chromatography [also GC^2]

GCGR Gas-Cooled Graphite-Moderated Reactor

GCH Gigacharacter [10^9 Characters]

γ-ch gamma-chain specific (immunoglobulin)

GCHAZ Grain-Coarsened Heat-Affected Zone

GC-HPLC Gas Chromatography/High-Performance Liquid Chromatography [also GC/HPLC]

GChM Global Chemistry Model

GCHQ Government Communications Headquarters [UK]

GCHWR Gas-Cooled Heavy Water Moderated Reactor

GCHX Ground Cooling Heat Exchanger

GCI Generalized Communications Interface

GCI Génie Climatique International [International Union of Heat and Air Conditioning and Ventilating Contractors, France]

GCI Getty Conservation Institute [Marina de Rey, California, US]

GCI Graphics Command Interpreter

GCI Gray Cast Iron

GCI Ground-Controlled Interception

GCI Ground-Controlled Intercept Radar

GCI Gyratory Compactibility Index

GCIL Ground Command Interface Logic

GCIL Ground Control Interface Logic

GCILC Ground Command Interface Logic Controller

GCILU Ground Control Interface Logic Unit

GC-IR Gas Chromatography Infrared (Spectroscopy)

GCIRC Groupe Consultatif International de Recherche sur le Colza [International Consultative Research Group on Rape Seed, France]

GCIS Ground Control Intercept Squadron

GC-ISS Gas Chromatography Isotope Separation System

GCIU Graphic Communications International Union

GCL Gas-Cooled Loop

GCL (International) Gas Flow and Chemical Lasers (Symposium)

GCL Generic Control Language

GCL Geosynthetic Clay Liner

GCL Gibbs Chemical Laboratory [of Harvard University, Cambridge, Massachusetts, US]

GCL Ground Control(led) Landing

GCL Ground Coolant Loop

GC-LC-MS Gas Chromatography/Liquid Chromatography/ Mass Spectrometry [also GC/LC/MS]

GCLISP Golden Common LISP (List Processor)

GCM Gadolinium Calcium Manganate

GCM General Circulation Model [Climatology]

GCM General Council Meeting

GCM General Court Martial

GCM Generator Coordinate Method

GCM Global Change Model(s)

GCM Global Circulation Model [Climatology]

GCM Global Climate Model

GCM Greatest Common Measure [also gcm]

GCM Ground Check Monitor

GCM Groupe des Couches Minces [Thin Film (Research) Group; of Ecole Polytechnique, Montreal, Quebec, Canada]

G·cm^2 Gauss centimeter squared [also G-cm^2]

G·cm^3 Gauss centimeter cubed [G-cm^3]

g-cm gram-centimeter [Unit]

g-cm^2 gram-centimeter squared [Unit]

g/cm^2 gram(s) per square centimeter [also g/sq cm, or g cm^{-2}]

g/cm^3 gram(s) per cubic centimeter [also g/cc, g/cu cm, gm/cc, or g cm^{-3}]

GCMA Government Contract Management Association of America

GCMC Glass Ceramic Matrix Composite

GCMC Grand-Canonical Monte Carlo (Method)

GCMD Global Change Master Directory [of NASA Goddard Space Flight Center, Greenbelt, Maryland, US]

g/cm^2-Hz gram(s) per square centimeter Hertz [also g cm^{-2} Hz^{-1}]

GCMI Glass Container Manufacturers Institute [now Glass Packaging Institute, US]

GCMR Graduate Center for Materials Research [of University of Missouri-Rolla, US]

GCMS Galactic Cosmic Mass Spectrometer

GCMS Gas Chromatography Mass Spectroscopy; Gas Chromatography Mass Spectrometry [also GC/MS, or GC-MS, gcms, gc-ms, or gc/ms]

g-cm/s gram-centimeter per second [Unit]

g-cm/s^2 gram-centimeter per second squared [also g-cm/sec^2, or g-cm s^{-2}]

g/cm s^2 gram per centimeter-second squared [also g/(cm s^2), g/(cm sec^2), or g cm^{-1} s^{-2}]

g cm^2/s gram centimeter squared per second [also g cm^2/sec, or g cm^2 s^{-1}]

g-cm^2/s^2 gram-centimeter squared per second squared [also g-cm^2/sec^2, or g-cm^2 s^{-2}]

g-cm^2/s^3 gram-centimeter squared per second cubed [also g-cm^2/sec^3, or g-cm^2 s^{-3}]

GCMSC George C. Marshall Spaceflight Center [of NASA at Huntsville, Alabama, US]

GC/MSD Gas Chromatography/Mass Selective Detector [also GC-MSD]

GCN Gauge Code Number

GCN Ground Communications Network [Air-to-Ground Worldwide Net]

GCN Ground Control Network [of NASA]

GCO Ground Checkout

gCO$_2$ grams of carbon dioxide equivalent

g-Co Granular Cobalt

GCOS General Comprehensive Operating Supervisor

GCOS General Comprehensive Operating System

GCOS General Computer Operational System

GCOS Global Climate Observation and Supervision (Program) [of World Health Organization]

GCOS Global Climate Observing System

GCOS Great Canadian Oil Sands (Process)

GCOS Ground Computer Operating System

GCOSP Great Canadian Oil Sands Plant [Alberta, Canada]

GCP Geometrically Close-Packed [also gcp] [Crystallography]

GCP Geometric Coordinate Planes

GCP Guidance and Control Panel [of NATO Advisory Group for Aerospace Research and Development]

GCPAS GeoComp Product Archive System

GC-PID Gas Chromatograph/Photoionization Detector; Gas Chromatography with Photoionization Detection [also GC-PID]

GCPM Green Compact Preparation Method [Powder Metallurgy]

GCPS Gigacycles per Second [also GCS, GC/s, or Gc/s]

GCR Galactic Cosmic Rays

GCR Galvanocutaneous Reaction

GCR Gas-Cooled Reactor

GCR General Component Reference

GCR General Control Relay

GCR Grid Control Ratio [Electronics]

GCR Ground Controlled Radar

GCR Group-Code(d) Recording

GCR Grumman Corporate Research [of Grumman Aerospace Corporation, US]

GCR Gyrocompass Repeater

GCRC General Clinical Research Center

GCRC Goodrich Research Center [Brecksville, Ohio, US]

GCRC Grumman Corporate Research Center [Bethpage, New York, US]

GCRDE Glassy Carbon Rotating Disk Electrode

GCRE Gas-Cooled Reactor Experiment [of Idaho National Engineering Laboratory, Idaho Falls, US]

GCRI German Carpet Research Institute

GCRI Glasshouse Crops Research Institute [UK]

GCRIO Global Change Research Infomation Office [US]

GCRL General Circulation Research Laboratory

GCRL Gulf Coast Research Laboratory [US]

GCRP Global Change Research Program [of US Geological Survey/Biological Resources Division]

GCS Game Conservation Society [US]

GCS Gate-Controlled Switch [Electronics]

GCS General Communications System

GCS General Computer Systems [US]

GCS Geocentric Coordinate System

GCS Gigacycles per Second [also GCPS, GC/s, or Gc/s]

GCS Graphic Compatibility System

GCS Ground Communications System

GCS Ground Control System

GCS Guidance Cutoff Signal

GCS Gulf Coast Section [of American Association of Petroleum Geologists, US]

Gc/s Gigacycles per Second [also GCS, GCPS, or GC/s]

GCSE General Certificate of Secondary Education [UK]

G-CSF Granulocyte Colony-Stimulating Factor [Biochemistry]

GC-SFC Gas Chromatography/Supercritical Fluid Chromatography [also GC/SFC]

GC-SFE Gas Chromatography/Supercritical Fluid Extraction [also GC/SFE]

GCSG Graphic Communications Societies Group [UK]

GCSQ Grand-Canonical Simulated Quenching [Metallurgy]

GCSS GEWEX (Global Energy and Water Cycle Experiment) Cloud System Study [of World Meteorological Organization–World Climate Research Program]

GCT Gadolinium Calcium Titanate

GCT General Classification Test

GCT Germ Cell Tumor [Medicine]

GCT Giant Cell Tumor [Medicine]

GCT Graphics Communications Terminal

GCT Greenwich Civil Time

GCT Greenwich Conservatory Time

GCT Guard Control System

GCTE Global Change and Terrestrial Ecosystems (Program) [of International Geosphere/Biosphere Program]

GcTNB Glycol Trinitrobutyrate

GCTS Gas Component Test Stand

GCTS Ground Communication(s) Tracking System

GCU General Control Unit

GCU Generator Control Unit

GCU Ground Cooling Unit

GCU Gyro Coupling Unit

g/cu cm gram(s) per cubic centimeter [also g/cm^3, $g\ cm^{-3}$, g/cc, or gm/cc]

g/cu m gram(s) per cubic meter [also g/m^3, $g\ m^{-3}$, g/ccm]

G-Curve Strain Energy Release Rate Curve [also G-curve]

γ-CuZn Gamma Copper Zinc

GCV Ganciclovirus

GCV GO_2 (Gaseous Oxygen) Control Valve

GCV Gross Calorific Value

GCVS General Catalogue of Variable Stars

GCW Global Chart of the World

GCW Gram-Combining Weight [also gcm]

GCW Gross Combined Weight (of Vehicle plus Trailer)

GCWM General Conference of Weights and Measures

GCWR Gross Combined Weight Rating (of Vehicle plus Trailer)

GD Galena Dating [Earth Sciences]

GD Gas Detection; Gas Detector

GD Gas Diffusion

GD Gas Discharge

GD Gas Diverter

GD Gas Dynamic(s)

GD Gated Decoupling [Electronics]

GD Gate Driver

GD Gaussian Distribution

GD Gelatin Dynamite

GD General Depot

GD Geodynamic(s)

GD Geologic Division [of US Geological Survey]

GD Geology Division [of Canadian Institute of Mining and Metallurgy]

GD Geometric Definition

GD Geometric Distribution

GD Geriatric Depression

GD Germanium Diode

GD Gibbs-Duhem (Equation) [Physical Chemistry]

GD Glissile Dislocation [Materials Science]

GD Glow Discharge

GD Goal-Driven

GD Golay Detector

GD Graphic Diagram

GD Graphic Display

GD Graphics Data

GD Graphics Device

GD Grenada [ISO Code]

GD Grenade Discharger

GD Grinding Direction

GD Ground Defense

GD Ground Detector

GD Group Delay

GD Grown Diffused

GD Guinier Diffractometer

GD Gunn Diode

GD Gyrodynamic(s)

G$ Guyana Dollar [Currency of Guyana]

Gd Gadolinium [Symbol]

Gd Good [also gd]

Gd Grade [also gd]

Gd Guard [also gd]

Gd^{3+} Gadolinium Ion [also Gd^{+++}] [Symbol]

Gd II Gadolinium II [Symbol]

Gd-123 $GdBa_2Cu_3O_x$ [A Gadolinium Barium Copper Oxide Superconductor]

Gd-148 Gadolinium-148 [also ^{148}Gd, or Gd^{148}]

Gd-149 Gadolinium-149 [also ^{149}Gd, or Gd^{149}]

Gd-150 Gadolinium-150 [also ^{150}Gd, or Gd^{150}]

Gd-151 Gadolinium-151 [also ^{151}Gd, or Gd^{151}]

Gd-152 Gadolinium-152 [also ^{152}Gd, or Gd^{152}]

Gd-153 Gadolinium-153 [also ^{153}Gd, or Gd^{153}]

Gd-154 Gadolinium-154 [also ^{154}Gd, or Gd^{154}]

Gd-155 Gadolinium-155 [also ^{155}Gd, or Gd^{155}]

Gd-156 Gadolinium-156 [also ^{156}Gd, or Gd^{156}]

Gd-157 Gadolinium-157 [also ^{157}Gd, or Gd^{157}]

Gd-158 Gadolinium-158 [also ^{158}Gd, or Gd^{158}]

Gd-159 Gadolinium-159 [also ^{159}Gd, or Gd^{159}]

Gd-160 Gadolinium-160 [also ^{160}Gd, or Gd^{160}]

Gd-161 Gadolinium-161 [also ^{161}Gd, or Gd^{161}]

Gd-211 Gd_2BaCuO_x [A Gadolinium Barium Copper Oxide Superconductor]

GDA General Development Agreement [now Economic and Regional Development Agreements, Canada]

GDA Gimbal Drive Actuator

GDA Gimbal Drive Assembly

GDA Global Data Area

GDA Glycol Diacetate

GDB Gene Database [of University College of London, UK]

GDB Genome Data Base [of the US Department of Energy and the US Department of Health and Human Services]

GDB Global Database

$Gd(Ac)_3$ Gadolinium Acetate

$Gd(ACAC)_3$ Gadolinium Acetylacetonate [also $Gd(AcAc)_3$, or $Gd(acac)_3$]

GdBCO Gadolinium Barium Copper Oxide (Superconductor)

GDBMS Generalized Database Management System

GDBS Generalized Data Base System

GDBS Grenada Bureau of Standards

GDC Gamma-Ray, Density, and Caliper Survey

GDC Garage Door Council [US]

GDC General Dental Council [UK]

GDC General Design Criteria

GDC General Dynamics/Convair [San Diego, California, US]

GDC Generalized Dynamic Charge

GDC Glow Discharge Condition

GDC Graphic Designers of Canada

GDC Graphics Display Controller

GDC Gross Dependable Capacity

GDC Gyro Display Coupler

GDCH Glycerol Dichlorohydrin

GDCh Gesellschaft Deutscher Chemiker [German Chemical Society]

GDCI Gypsum Drywall Contractors Association

Gd-Co Gadolinium-Cobalt (Alloy System)

GdCo Gadolium Cobalt (Alloy)

GGDD Gas Discharge Display

GDD Graphic Display Device

GDDA General Desert Development Authority [Egypt]

GDDE Gradually Doped Source-Drain Extension [Electronics]

GDDL Graphical Data Definition Language

GDDM Graphics Data Display Manager

GDE Gage Deviation (of Bearing Balls) [Symbol]

GDE Generalized Data Entry

GDE Generalized Diffusion Equation

GDE Genetic Data Environment

GDE Gibbs-Duhem Equation [Physical Chemistry]

GDE Ground Data Equipment

Gde Gourde [Currency of Haiti]

GDES Government Department for Electrical Specification [UK]

GDF Gas Dynamic Facility [US Air Force]

GDF Generalized Density Functional (Theory) [Physics]

GDF Geographic Data File [Geographic Information System]

GDF Group Distribution Frame

GDF Growth and Differentiation Factor

GdF Gaz de France [Gas Corporation of France]

GDFCF Gross Domestic Fixed Capital Formation

GdFeCo Gadolinium Iron Cobalt (Alloy)

$Gd(FOD)_3$ Tris(heptafluorodimethyloctandionato)-gadolinium

GDG Generation Data Group

GdGG Gadolinium-Gallium Garnet

GDH Glutamate Dehydrogenase [Biochemistry]

GDH Glycerophosphate Dehydrogenase [Biochemistry]

GDH Grammar, Dempsey and Hudson Company [US]

α-GDH Glycerophosphate Dehydrogenase [Biochemistry]

α-GDH-TPI Glycerophosphate Dehydrogenase Triosephosphate Isomerase [also GDH-TPI] [Biochemistry]

GDI Graphical (or Graphics) Device Interface

GDI Gross Domestic Investment

GdIG Gadolinium Iron Garnet

GDIFS Gray and Ductile Ironfounders Society [US]

GDIT Georgia Department of Industry and Trade [US]

GDL Gas Dynamic Laser

GDL Glow Discharge Lamp

GDL Graphic Display Library

GDL Grand-Duché de Luxembourg [Grand Duchy of Luxembourg]

g/dL gram(s) per deciliter [also g dL^{-1}]

GDLA Generalized Diffusion-Limited Aggregation (Model)

GDLC Generic Data Link Control

GDM Gestational Diabetes Mellitus

GDM Global Data Manager

GDMB Gesellschaft Deutscher Metallhütten– und Bergleute [German Society for Metallurgy and Mining]

GDME Gibbs-Duhem-Margules Equation [Physical Chemistry]

GDME Glycol Dimethyl Ether

GDMS Generalized Data Management System

GDMS Glow Discharge Mass Spectrometry; Glow Discharge Mass Spectroscopy

GDN Glycol Dinitrate

GDN Government Data Network

GdN Gadolinium Nitride

Gdn Garden [also gdn]

GDNÄ Gesellschaft Deutscher Naturforscher und Ärzte [German Society of Scientists and Physicians]

Gdnce Guidance [also gdnce]

GDNF Glial Cell-Line-Derived Nerve-Growth Factor [Biochemistry]

Gd-Ni Gadolinium-Nickel (Alloy System)

GDNP Glial-Cell-Derived Neurotrophic Factor [Biochemistry]

Gdnr Gardener [also gdnr]

Gdns Gardens

GDO Gesellschaft Deutscher Organisatoren [German Society of Organizers]

GDO Grid-Dip Oscillator

GDO Gross Domestic Output

GDO Guidance Officer

GDO Gunn Diode Oscillator

GDOES Glow Discharge Optical Emission Spectroscopy [also GD-OES]

Gd$_2$O$_3$(Eu) Europium-Doped Gadolinium Oxide [also Gd$_2$O$_3$:Eu]

GDOP Geometric Dilution of Precision

GDOS Glow Discharge Optical Spectroscopy

GdOS Gadolinium Oxysulfide

GDP Galvanostatic Double-Pulse (Method)

GDP Generalized Documentation Processor

GDP Generalized Drawing Primitive

GDP Glow Discharge Polymer

GDP Goal Directed Programming

GDP Graphic Data Processing

GDP Graphic Draw Primitive

GDP Gross Domestic Product

GDP Guanosine 5'-Diphosphate [Biochemistry]

5'-GDP Guanosine-5'-Diphosphate [Biochemistry]

GDP-β-S Guanosine-5'-[β-Thio]diphosphate] [Biochemistry]

GDPC Groupe de Dynamique des Phases Condensées [Condensed Phases Dynamics Group; of Université de Montpellier II, France]

GDPG Guanosine 5'-Diphosphoglucose

GDP-NHP Guanosine-5'-[β,γ-Imino]triphosphate (Sodium Salt)

GDPS Global Data Processing System

GDR General Design Requirement

GDR Geometric Design Rules

GDR Geometric Dynamic Recrystallization [Metallurgy]

GDR German Democratic Republic

GDR Giant Dipole Resonance

GDR Ground Delay Response

GD&R Grinning, Ducking and Running [Computer Jargon]

GDRC Gaines Dog Research Center [US]

GDRS Geoscience Data Referral System

GDRX Geometric Dynamic Recrystallization [Metallurgy]

GDS General Declassification Schedule

GDS Gao, Dregia and Shewmon (Calculation) [Metallurgy]

GDS Gas Discharge System

GDS Geriatric Depression Scale

GDS Global Digital Services

GDS Glow Discharge Spectrometer; Glow Discharge Spectrometry

GDS Glow Discharge Spectroscopy

GDS GNC (Guidance and Navigation Computer) Dynamic Simulator

GDS Goldstone, California [NASA Space Tracking and Data Network Site]

GDS Goods [also Gds, or gds]

GDS Graphic Data System

GDS Graphic Design System

GDS Graphic Display Station

GDS Graphic Display System

GDS Gross Debt Service

GdS Gadolinium Sulfide

Gds Guards [also gds]

GdSCNO Gadolinium-Strontium-Copper-Niobium Oxide (Superconductor)

GDSD Ground Data Systems Division [of NASA]

GdSe Gadolinium Selenide

GDSF Generalized Data Structure Definition Facility

GDSM Ground Data Systems Manager

GDSO Ground Data Systems Officer

GDSSR GDSD (Ground Data Systems Division) Staff Support Room

GDSU Global Digital Service Unit

GDT Gas Decay Tank

GDT Generator Development Tool

GDT Geometric Dimensioning and Tolerancing [also GD&T]

GDT Global Descriptor Table

GDT Glow-Discharge Tube

GDT Graphic Display Terminal

GDT Graphics Development Toolkit

GDT Grown-Diffused Transistor

GD&T Geometric Dimensioning and Tolerancing [also GDT]

Gd-Tb Gadolinium-Terbium (Alloy System)

GdTbCoFe Gadolinium Terbium Cobalt Iron (Alloy)

GdTbFe Gadolinium Terbium Iron (Alloy)

GdTe Gadolinium Telluride

Gd(THD)$_3$ Gadolinium Tris(2,2,6,6-Tetramethylheptane-dionate) [also Gd(thd)$_3$]

Gd(TMHD)$_3$ Gadolinium Tris(2,2,6,6-Tetramethylheptane-dionate) [also Gd(tmhd)$_3$]

GDU Graphic Display Unit

GDX Gated Diode Crosspoint

GDX Goldstone, California [NASA Space Tracking and Data Network Site, Second Antenna]

Gd-Y Gadolinium-Yttrium (System)

γ-Dy Gamma Dysprosium [Symbol]

GE Galvanic Element

GE Gamma Emission; Gamma Emitter

GE Gas Ejection

GE Gas Engine

GE Gas Engineer(ing)

GE Gas Enthalpimetry

GE Gas Generating Product

GE Gasoline Engine

GE Gastroenteritis [Medicine]

GE Gastroenterological; Gastroentrologist; Gastroenterology

GE Gastroesophageal [Medicine]

GE Gateway Exchange

GE Gaussian Elimination [Mathematics]

GE Gel Electrophoresis

GE General Electric Company [US]

GE Generic Element

GE Genetically Engineered; Genetic Engineer(ing)

GE Geoelectric(ity)

GE Geological Engineer(ing)

GE Georgia [ISO Code]

GE Geothermal Energy

GE Geothermal Engineer(ing)

GE Glass Electrode

GE Global Embrittlement

GE Good Evening [Amateur Radio]

GE Gorkov and Eliashberg (Theory) [Physics]

GE Graduate Engineer

GE Graphoepitaxial; Graphoepitaxy

GE Greater than, or Equal to

GE Greenhouse Effect

GE Gross Energy

GE Guanidoethyl

GE Gunn Effect

G/E Graphite Epoxy [also G/Ep]

G-E Glass-Electrolyte (Interface)

G(E) Green Function [Symbol]

G(E) Normalized Energy-Barrier Density Function [Symbol]

Ge Gauge [also ge]

Ge Germanium [Symbol]

Ge^{4+} Germanium Ion [Symbol]

Ge II Germanium II [Symbol]

Ge III Germanium III [Symbol]

Ge-66 Germanium-66 [also ^{66}Gd, or Gd^{66}]

Ge-67 Germanium-67 [also ^{67}Gd, or Gd^{67}]

Ge-68 Germanium-68 [also ^{68}Gd, or Gd^{68}]

Ge-69 Germanium-69 [also ^{69}Gd, or Gd^{69}]

Ge-70 Germanium-70 [also ^{70}Gd, or Gd^{70}]

Ge-71 Germanium-71 [also ^{71}Gd, or Gd^{71}]

Ge-72 Germanium-72 [also ^{72}Gd, or Gd^{72}]

Ge-73 Germanium-73 [also ^{73}Gd, or Gd^{73}]

Ge-74 Germanium-74 [also ^{74}Gd, or Gd^{74}]

Ge-75 Germanium-75 [also ^{75}Gd, or Gd^{75}]

Ge-76 Germanium-76 [also ^{76}Gd, or Gd^{76}]

Ge-77 Germanium-77 [also ^{77}Gd, or Gd^{77}]

Ge-78 Germanium-78 [also ^{78}Gd, or Gd^{78}]

.ge Georgia [Country Code/Domain Name]

GEA Garage Equipment Association [UK]

GEAE General Electric Aircraft Engines (Division) [Cincinnati, Ohio, US]

GEAE Rep General Electric Aircraft Engines (Division) Report [US]

Ge-Al Germanium-Aluminum (Alloy System)

Ge:Al Aluminum-Doped Germanium (Semiconductor)

GEANP General Electric Aircraft Nuclear Propulsion

GEAP General Electric Atomic Power

GEAP General Electric Atomic Products

Ge:As Arsenic-Doped Germanium (Semiconductor)

GeAsSe Germanium Arsenic Selenide (Semiconductor)

GEATP Group of Experts on Air Transport Policies [of International Civil Aviation Organization]

GEB General Education Board [US]

GEBCO General Bathymetric Chart of the Oceans

Ge(BuO)$_4$ Germanium Butoxide (or Tetrabutoxygermane)

GEC Gaseous Electronics Conference [of American Physical Society, US]

GEC General Electric Company [US]

GEC Generalized Equivalent Cylinder

GEC Global Environment Centre Foundation [of United Nations Environment Programme–International Environmental Technology Centre at Osaka City, Japan]

GEC Global Environmental Change

GeC Germanium Carbide

Ge-C Germanium-Carbon (System)

GECAP General Electric Computer Analysis Program

GEC Eng GEC Engineering [of General Electric Company Research Laboratories, UK]

GECENT General Electric Central Processor

GEC J Res GEC Journal of Research [of General Electric Company Research Laboratories, UK]

GEC J Res Inc Marconi Rev GEC Journal of Research Incorporating the Marconi Review [of General Electric Company Research Laboratories, UK]

GEC J Sci Technol GEC Journal of Science and Technology [of General Electric Company Research Laboratories, UK]

GEC Rev GEC Review [of General Electric Company Research Laboratories, UK]

GECRL General Electric Company Research Laboratories [Chelmsford, Essex, UK]

GECOM General(ized) Compiler

Gécomine Société Générale Congolaise des Minérais [Congo Mining and Minerals Company]

GECOS General Comprehensive Operating Supervisor

GECOS General Comprehensive Operating System

GECOS General Electric Comprehensive Operating System

GECOT Group of Experts on Costs and Tariffs

GECR Global Environment Change Report

GECRC General Electric Corporate Research Center [Schenectady, New York, US] [also GE CRC]

GECRD General Electric Corporate Research and Development (Center) [Schenectady, New York, US] [also GE CRD]

GEC Telecommun GEC Telecommunications [Publication of GEC Telecommunications Ltd., Coventry, UK]

GECUS Groupe d'Etudes de l'Urbanisme Souterrain [Study Group for Underground Town Planning, France]

GED Gas Electron Diffraction

GED General Educational Development

GED General Equivalency Diploma

GED Generalized Extreme-Value Distribution

GEDAC General Electric Detection and Automatic Correction

GEDEX Greenhouse Effect Detection Experiment

GEDIS Gesellschaft für Datenverarbeitung und Informationssysteme mbH [German Manufacturer of Data Processing and Information Systems]

GEDIS Geological, Exploration and Development Information System [Victoria State, Australia]

GEE Group of Economic Experts [of Organization for Economic Cooperation and Development]

GEEI General Electric Electronic Installation

GEEI General Electronics Engineering Installation

GEEIA Ground Electronics Engineering Installation Agency

GEEP General Electric Electronic Processor

GEEP Group of Experts on Environmental Pollutants

GEERS Groupe d'Etudes Européen des Recherches Spatiales [European Study Group for Space Research, France]

GEESE General Electric Electronic Systems Evaluator

Ge(EtO)$_4$ Germanium(IV) Ethoxide (or Tetraethoxygermane)

GEF Gaussian Error Function

GEF Global Environment Facility [of World Bank]

GEF Global Environment Fund [of World Bank]

GEF Gravure Education Foundation [US]

GEF Growth-Enhancing Factor

GEFA Gulf-European Freight Association [US]

GEFACS Groupement des Fabricants d'Appareils Sanitaires en Céramiques [Group of Ceramic Sanitary Equipment Manufacturers; of European Economic Community, Belgium]

GEFAP Groupement Européen des Associations Nationales de Fabricants de Pesticides [European Federation of National Associations of Pesticide Manufacturers]

GEFOAMS General Electric Foams Optimized through Analysis and Materials Selection

GEFRC General File/Record Control

GEG General Euclidian Geometry

GeGa Gesellschaft für Gastechnik Lotz mbH [German Gas Engineering Company, located at Hofheim-Wallau]

Ge:Ga Gallium-Doped Germanium (Semiconductor)

Ge/GaAs Germanium /Gallium Arsenide (Heterojunction)

GEGAS General Electric Gas (Process)

GEGB General Electricity Generating Board

Ge:H Hydrogenated Germanium

GeH$_4$/He Germanium Hydride/Helium (Mixture)

GEI Graphics Engine Interface

Ge:I Intrinsic Germanium

γ-EIC Equal-Interface-Composition Contour in the Gamma Single Phase Region [Metallurgy]

Ge:In Indium-Doped Germanium (Semiconductor)

GEIP Greenhouse Eduction and Information Program

GEIS General Electric Information Service

GEIS Generic Environmental Impact Statement

GEISA Gestion d'Etudes des Informations Spectrographiques Atmospheriques [Database on Spectroscopy of Planetary Atmospheres, France]

GEISHA Geodetic Inertial Survey and Horizontal Alignment

GEIXS Geological Electronic Information Exchange System (Project) [Joint Project of European Geological Surveys]

GEJ Group of Experts on Jurisdiction [of International Civil Aviation Organization]

GEK Geomagnetic Electrokinetograph

GeKα Germanium K-Alpha (Radiation) [also GeK$_\alpha$]

GE KAPL General Electric Knoll's Atomic Power Laboratory [Schenectady, New York, US] [also GE-KAPL]

GEL General Emulation Language

Gel Gelatine; Gelatinous [also gel]

gelat gelatinous

GeLi Germanium-Lithium (Diode) [also Ge-Li, or GELI]

Ge(Li) Lithium-Drifted Germanium (Detector)

GELIC Grainger Engineering Library Information Center [of University of Illinois at Urbana-Champaign, US]

GELOAD General Loader

GEM General Electronics Module

GEM General Epitaxial Monolith

GEM Generic Experiment Module

GEM Genetically Engineered Machine

GEM Geostatistical Evaluation of Mines

GEM Governmental Energy and Minerals Committee [of The Minerals, Metals and Materials Society of the American Institute of Mining, Metallurgical and Petroleum Engineers, US]

GEM Graduate Engineering for Minorities [of National Consortium for Graduate Degrees for Minorities in Engineering, US]

GEM Gram Equivalent Mass

GEM Graphic Engine Monitor

GEM Graphics Environment Manager (Program) [of Digital Research Inc., US]

GEM Ground-Effect Machine

GEM Ground Electromagnetic (Survey)

GEM Graphics Environment Manager

GEM Guidance Evaluation Missile

GEMAC General Electric Measurement and Control [also GE/MAC]

GEMC Grain Equipment Manufacturers Council [US]

GEMCQ Giddings-Eyring Extended by McQuarrin (Theory) [Physics]

GEMCOS Generalized Message Control System

Ge(MeO)$_4$ Germanium(IV) Methoxide (or Tetramethoxygermane)

GEMFET Gain-Modulated Field-Effect Transistor

GEMI Global Environmental Management Initiative

GEMM Generalized Electronics Maintenance Model

GEMOV General Electric Metal-Oxide Varistor [also GE-MOV]

GEMPPM Groupe d'Etudes de Métallurgie Physique et de Physique des Matériaux [Physical Metallurgy and Materials Physics Group; of Institut National des Sciences Appliquées de Lyon, Lyons, France]

GEMS General Education Management System

GEMS General Electrical and Mechanical Systems

GEMS General Electric Manufacturing Simulator

GEMS Global Environmental Monitoring Service [of World Health Organization]

GEMS Global Environmental Monitoring System [of United Nations Environmental Program]

GEMS Ground Electromagnetic Survey

GEMSA Guanidinoethyl Mercaptosuccinic Acid [Biochemistry]

GEMSD General Electric Medical Systems Division [Milwaukee, Wisconsin, US]

GEMSIP Gemini Stability Improvement Program [of NASA]

GEMT Group of European Metallurgical Thermodynamicists [Joint Group of German, British, French and Belgian Researchers]

Gen Gender [also gen]

Gen General(ity) [also gen]

Gen General [Military Rank]

Gen Generate; Generation; Generator [also gen]

Gen Geneva [Switzerland]

GenBank Genetic Codes Databank [of National Center for Biological Information, US]

Gen Compar Endocrinol General and Comparative Endocrinology [Journal published in the US]

GENDA General Data Analysis

GENDARME Generalized Data Reduction, Manipulation and Evaluation

GENDAS General Data Analysis and Simulation

Gen Dent General Dentistry [US Journal]

Gen Eng General Engineer [UK Publication for Mechanical Engineers]

GENERIC Generation of Integrated Circuits

Genes Devel Genes and Development [Journal published by Cold Spring Harbor Laboratory Press, US]

GENESIS GenRad's Environment for Strategy Independent Software

GENESSIS Generic Scene Simulation Software [Geographic Information System]

GENESYS General Engineering System

GENESYS Generalized System

GENESYS Graduate Engineering Education System [US]

Genet Genetics [Journal of the Genetics Society of America, US]

GENIE General Electric Network for Information Exchange

GENIE General Information Extractor

Génie Logiciel Syst Experts Génie Logiciel and Systèmes Experts [French Journal on Computer Software and Expert Systems]

GENIS Grundlegende Entwicklungen für die Netzwerkfähigkeit einer intelligenten Sensor-Aktorfamilie in Mikrosystemtechnik [Basic Development for the Networkability of an Intelligent Sensor-Actor Family in Micro System Technology (Program); of Bundesministerium für Forschung und Technology, Germany]

GENIUS Genetic Interactive Unix System

genl general [also gen'l]

Gen Mgr General Manager

Genom Genomics [Journal published in the US]

Genome Res Genome Research [Journal published by Cold Spring Harbor Laboratory Press, US]

Gen Pharmacol General Pharmacology [US Journal]

Gen Relativ Gravit General Relativity and Gravitation [Published in the US]

Genset Generator Set

GENSLA Generic Structure Language

Gen Supt General Superintendant

GENTEC Genetic Technology Databank [of Gesellschaft für Biotechnologische Forschung, Germany]

Gen Tech Rep General Technical Report [also Gen Tech Rept]

GEO Geoscience Electronics [Institute of Electrical and Electronics Engineers Group]

GEO Geosynchronous

GEO Geostationary Earth Orbit

GEO Global Environment Organization

GEO Global Environment Outlook (Database) [of United Nations Environment Programme]

GeO Germanium Monoxide

GEOARCHIVE Geology Archive (Database) [UK]

GeoBase Geography Database for the United States

Geobot Geobotanical; Geobotanist; Geobotany [also geobot]

Geocarto Int Geocarto International [Published in Hong Kong]

Geochem Geochemical; Geochemist(ry) [also geochem]

Geochem Int Geochemical International [Journal published in Russia]

Geochem News Geochemical News [of Geochemical Society, US]

Geochim Cosmochim Acta Geochimica et Cosmochimica Acta [Published in the UK]

GEOCOMP Geocoding and Compositing

Geod Geodesist; Geodesy; Geodetic [also geod]

GEODAS Geophysical Data Management System [of National Oceanic and Atmospheric Administration National Geophysical Data Center, US]

GEODE General Electric Organic Destruction (Process)

Geodin Acta Geodinamica Acta [International Journal on Geodynamics]

GEODIS Geographic Design and Implementation System [Australian Capital Territory]

GEODSS Ground-Based Electro-Optical Deep-Space Surveillance System

Geodyn Geodynamic(s) [also geodyn]

Geodyn Ser Geodynamics Series [Publication of the American Geophysical Union]

Geoelectr Geoelectric(ity) [also geoelectr]

Ge(OEt)$_4$ Germanium(IV) Ethoxide (or Tetraethoxygermane)

GEOF Geological Editor of Field Notes [Computer Software]

GEOFA Deutsche Fachmesse für alle Geobereiche [German Trade Fair and Exhibition for Earth Sciences and Related Fields]

Geofiz Appar Geofizicheskaya Apparatura [Russian Journal on Geophysical Equipment]

Geofiz Zh Geofizicheskii Zhurnal [Russian Geophysical Journal]

Geogr Geographer; Geographic(al); Geography [also Geog, geog, or geogr]

Geogr J Geographical Journal [of Royal Geographical Society, UK]

Geogr Rev Geographical Review [US]

GEO-IRS Geostationary Orbit–Infrared Sensor

GEOIS Geographic Information System

Geol Geologic(al); Geologist; Geology [also geol]

GeolE Geological Engineer [also Geol Eng, or GeolEng]

Geol Eng Geological Engineer(ing)

Geol Geofiz Geologiya i Geofizika [Russian Journal on Geology and Geophysics]

GEOLIS Geological Literature Search System [of Geological Survey of Japan]

Geol J Geological Journal [UK]

Geol Mag Geology Magazine

Geol Mijnb Geologie en Mijnbouw [Dutch Journal on Geology and Mining]

Geol Ore Dep Geology of Ore Deposits [Journal published in Russia]

Geol Pac Ocean Geology of the Pacific Ocean [International Journal]

Geol Runds Geologische Rundschau [German Geological Review]

Geol Sci Geological Science(s); Geological Scientist

Geol Soc Am Bull Geological Society of America Bulletin

Geol Soc Am Ann Mtg Geological Society of America Annual Meeting [US]

Geol Surv Bull Geological Survey Bulletin [US Government Printing Office, Washington, DC]

Geol Today Geology Today [International Journal]

Geom Geometric(al); Geometrician; Geometry [also geom]

Geomagn Geomagnetic(s) [also Geomag, geomag, or geomagn]

Geomagn Aeron Geomagnetism and Aeronomy [Translation of: *Geomagnetizm i Aeronomiya (USSR)*; published in the US]

Geomagn Ser Earth Phys Branch Geomagnetic Series, Earth Physics Branch [of Natural Resources Canada]

GEOMAR Forschungszentrum für Marine Geowissenschaften [Research Center for Marine Geosciences, University of Kiel, Germany]

Geo-Mar Lett Geo-Marine Letters [Journal published in Germany]

GeoMATE Geographic Map Attribute Enhancement [also GEOMATE]

Ge(OMe)$_4$ Germanium(IV) Methoxide (or Tetramethoxygermane)

Geomech Geomechanical; Geomechanics

Geominco Geological and Mining Company [Hungary]

Geomorph Geomorphologist; Geomorphology [also geomorph]

GEON Gyro-Erected Optical Navigation

Geophys Geophysical; Geophysicist; Geophysics [also geophys]

Geophys Geophysics [Journal of Society of Exploration Geophysicists, US]

Geophys Astrophys Fluid Dyn Geophysical and Astrophysical Fluid Dynamics [Published in the UK]

Geophys Explor Geophysical Exploration [Publication of the Society of Exploration Geophysicists of Japan]

Geophys J Geophysical Journal [Translation of: *Geofizicheskii Zhurnal (Ukrainian SSR)*; published in the UK]

Geophys J Int Geophysical Journal International [UK]

Geophys Mag Geophysical Magazine [of the Japan Meteorological Agency]

Geophys Monogr Ser Geophysical Monograph Series [Publication of the American Geophysical Union]

Geophys Nor Geophysica Norvegica [Geophysical Journal; published in Norway]

Geophys Prosp Geophysical Prospecting [Publication of the European Association of Exploration Geophysicists, Netherlands]

Geophys Res Lett Geophysical Research Letters [of the American Geophysical Union]

Geophys Space Phys Geophysics and Space Physics (Journal)

Geophys Trans Geophysical Transactions [Published by the Eötvös Lóránd Geophysical Institute, Budapest, Hungary]

GeoPOP Geosciences Project into Over-Pressure [UK]

GeOPr^i Germanium Isopropoxide

GEOPS Geodetic Estimates from Orbital Perturbations of Satellites

GEOREF (World) Geographic Reference System

GEOREF Geological Reference File [of American Geological Institute, US] [also GeoRef]

GEOREGISTER GeoRegisterdokumentdatabas [Geological Register Documentation Data Base, of Swedish Geological Survey] [also GeoRegister]

Georg Georgia [Former Republic of the USSR]

Georg SSR Georgian Soviet Socialist Republic [USSR] [also Georgian SSR]

GEORGE General Organizational Environment

GEOS Geodetic Earth Orbiting Satellite [of NASA]

GEOS Geosynchronous Earth Observation System

GEOS GSFC EOS (Earth Observing System) [NASA Goddard Space Flight Center Data Assimilation System]

GEOS (Geoworks) Graphic Environment Operating System

GEOSAR Geosynchronous Synthetic Aperture Radar

Geosat Geodesy Satellite

GEOSCAN Geological Survey of Canada

GEOSCAN Ground-Based Electronic Omnidirectional Satellite Communications Antenna

Geosci Geoscience; Geoscientist [also geosci]

Geosci Can Geoscience Canada [Canadian Journal]

GEOSECS Geochemical Ocean Sections Study

GEOSEPS Geosynchronous Solar Electric Propulsion Stage

Geotech Geotechnic(s) [also geotech]

Geotech News Geotechnical News [Published by the Canadian Geotechnical Society]

geotechnica Internationale Fachmesse und Kongreß für Geowissenschaften und Geotechnik [International Fair and Congress for Geosciences and Geotechnology, Germany]

Geotech Test J Geotechnical Testing Journal [Published by the American Society for Nondestructive Testing]

Geotect Geotectonics [Journal published in Russia]

Geotherm Geothermics [Journal published in the Netherlands]

GEOTp Geographic Township

GEP General Equivalence Point (Titration)

GEP Goddard Experiment Package [of NASA]

GEP Gradient Extremal Path

GEP Grasslands Ecology Program

G/Ep Graphite/Epoxy [also G/E]

GEPAC General Electric Process Automation Computer [also GE/PAC]

GEPAC General Electric Programmable Automatic Comparator

Ge/Pd Germanium/Palladium (Heterojunction)

GEPI Gyratory Elasto-Plastic Index

Ge-PN Germanium Positive-Negative (Diode) [also Ge-pn]

GEPVP Groupement Européen des Producteurs de Verre Plat [European Group of Flat Glass Manufacturers, Belgium]

GER Gastroesophageal Reflux [Medicine]

GER (Division of) Geology and Earth Resources [Washington State, US]

GER Gross Energy Requirement

Ger German(y)

GERD Gastroesophageal Reflux Disease [Medicine]

GERD Gross Expenditure on Research and Development

GE R&D Center General Electric Research and Development Center [also GERDC] [Niscayna, New York, US]

GEREP Generalized Equipment Reliability Evaluation Procedure

GE Rep General Electric Report [US]

GERG Groupe Européene de Recherche Gazières [European Gas Research Group, Netherlands]

GERL General Electric Research Laboratory [US]

Gerlands Beitr Geophys Gerlands Beiträge zur Geophysik [Gerlands' Contributions to Geophysics; German Publication]

German Res German Research [Publication of Deutsche Forschungsgemeinschaft, Germany]

Gerontol Geronological; Gerontologist; Gerontology

Gerontol Gerontology [International Journal]

GERSIS General Electric Range Safety Instrumentation System

GERT Graphical Evaluation and Review Technique

GERTS General Electric Range Tracking System

GERTS General Electric Remote Terminal Supervisor

GERTS General Electric Remote Terminal System

GERTS General Remote Terminal System

GES Goddard Experiment Support System [of NASA Goddard Space Flight Center, Greenbelt, Maryland, US]

GES Ground Equipment Storage

GeS Germanium Sulfide (Semiconductor)

Ge(S) Sulfur-Doped Germanium (Semiconductor)

GESA Gemeinschaft Experimentelle Spannungsanalyse [Study Group for Experimental Stress Analysis; of Gesellschaft für Meß– und Automatisierungstechnik, Germany]

GESAC General Electric Self-Adaptive Control System

GESAMP Group of Experts on the Scientific Aspects of Marine Pollution [of International Maritime Organization]

Ge:Sb Animony-Doped Germanium (Semiconductor)

GESC Government EDP (Electronic Data Processing) Standards Committee [Canada]

GESC Government Electronic Services Card [Canada]

GeSe Germanium Selenide (Semiconductor)

Ge$_x$Si$_{1-x}$ Germanium Silicide

Ge/Si Germanium/Silicon (Heterojunction)

GESIS Gesellschaft Sozialwissenschaftlicher Infrastruktureinrichtungen e.V. [Society for Infrastructure in the Social Sciences, Bonn, Germany]

Ge$_x$Si/Si Germanium Silicide/Silicon (Superlattice)

Ge$_x$Si$_{1-x}$/Si Germanium Silicide/Silicon (Superlattice)

GESMAR Geodetic Survey Marks Register [of Western Australia]

GESMO Generic Environmental Statement on Mixed Oxides

GESOC General Electric Satellite Orbit Control

GESS Grinding Energy Saving System

GESSAR General Electric Standard Safety Analysis Report

Gesundheits-Nachr Gesundheits-Nachrichten [Health Newsletter; published in Switzerland]

GET Gas Evaporation Technique

GET General Equivalence-Point Titration

GET Genetic Engineering Technologist; Genetic Engineering Technology

GET Geological Engineering Technologist; Geological Engineering Technology

GET Gesellschaft Energietechnik [Energy Technology Society; of Verein Deutscher Ingenieure, Germany]

GET Get Execute Trigger

GET Ground Elapsed Time

GETAC General Electric Telemetering and Control [also GE/TAC]

GeTe Germanium Telluride (Semiconductor)

GETEL General Electric Test Engineering Language

GETF Global Environmental and Technology Foundation

GETF Global Environmental Trust Fund [of Global Environment Fund–Core Fund, World Bank]

GETI Ground Elapsed Time of Ignition

GETIL Ground Elapsed Time of Landing

GETIS Ground Environment Team of the International Staff [of NATO]

GETOL General Electric Training Operational Logic

GETOL Ground Effect Takeoff and Landing

GETR General Electric Test(ing) Reactor

GETS Ground Equipment Test Set

GETURN General Electric Lathe Parts Programming Software

GET/VDI Gesellschaft Energietechnik im Verein Deutscher Ingenieure [Energy Technology Society of the Association of German Engineers]

GEUS Geological Survey of Denmark and Greenland

GEV Gross Energy Value

GEV Ground Effect Vehicle

GeV Gigaelectronvolt [Unit]

GEVIC General Electric Variable Increment Computer

GEW Gewerkschaft Erziehung und Wissenschaft [Union of Teachers and Scientists, Germany]

GEW Gram-Equivalent Weight [also gew]

GEWEX Global Energy and Water Cycle Experiment [of World Meteorological Organization–World Climate Research Program]

GEWS Group on Engineering Writing and Speech [of Institute of Electrical and Electronics Engineers, US]

Ge:X Extrinsic Germanium

GF Gadolinium Ferrite

GF (Strain) Gage Factor

GF Gap-Frame (Press)

GF Gaseous Fuels

GF Gas-Fired (Furnace)

GF Gas Flow

GF Gas Focusing [Electronics]

GF Gauge Field

GF Gaussian Function

GF Gel Filtration

GF Generalized Force

GF Generalized Function

GF Generator Field

GF Georg Fischer AG [Switzerland]

GF Geological Fault

GF Geomagnetic Field

GF Geostrophic Force

GF Germanium Filter

GF Giauque Function [Physics]

GF Glass Fiber

GF Glass Formation; Glass Former; Glass Forming

GF Gold-Filled; Gold Filling

GF Goodness-of-Fit (Test)

GF Gorilla Foundation [US]

GF Government-Furnished

GF Gradient Field

GF Gradient Freezing

GF Gradient Furnace

GF Grand Forks [North Dakota, US]

GF Graphite Fiber

GF Gravimetric Factor (in Chromatography)

GF Gravitational Field

GF Gravitational Force

GF Gravity Feed

GF Green's Function

GF Ground Fault [also G/F]

GF Group Frequency

GF Group Function (Theory) [Physics]

GF Growth Factor

GF Gun Fire

GF Guyana (French) [ISO Code]

GF Guyane Française [French Guiana]

GF Gypsum Binder and Fluorescent Material

G/F Ground Fault

G-f Conductance-Frequency (Characteristic) [also G-f]

gf generative force [Unit] [Actuators]

gf gram force [Unit]

GFA Gasket Fabricators Association [US]

GFA Genetic Functional Algorithm

GFA Glass Forming (or Formation) Ability

GFA Grain Futures Administration [US]

GFA Graphite Furnace Atomizer

GFAA Game Fishing Association of Australia

GFAA Graphite-Furnace Atomic Absorption

GFAAS Graphite-Furnace Atomic Absorption Spectroscopy [also GF-AAS]

GFAE Government-Furnished Aeronautical Equipment; Government-Furnished Aerospace Equipment; Government-Furnished Aircraft Equipment; Government-Furnished Avionics Equipment

GFAP Glial Fibrillary Acidic Protein [Biochemistry]

GFAPA Glycerine and Fatty Acid Producers Association [now Glycerine and Oleochemicals Association, US]

GFC Gas-Fan Cooled; Gas-Fan Cooling

GFC Gas-Filled Cable

GFC Gas-Flow (Radiation) Counter

GFC Gel Filtration Chromatography

GFC Geologiska Forskningscentralen [Geological Research Center, Finland]

GFCB Glassy-Film-Coated Boundary

GFCI Ground Fault Circuit Interrupter; Ground Fault Current Interrupter

GFCM General Fisheries Council for the Mediterranean [of Food and Agricultural Organization]

GFCM Green Function Cellular Method

GFCO Bicyclo[2.2.1]heptadiene-Iron-Tricarbonyl

GFCR Gas Filter Correlation Radiometry

GFCS Gun Fire Control System

GFD Gas-Filled Detector

GFD Gas-Filled Diode

GFD Gaussian Frequency Distribution

GFD Generalized Focusing Diffractometer

GFD Geophysical Fluid Dynamics

GFD Government-Furnished Data

GFDE Global Force-Displacement Equation

GFDI Geophysical Fluid Dynamics Institute [of Florida State University, Tallahassee, US]

GFDL Geophysical Fluid Dynamics Laboratory [of National Oceanic and Atmospheric Administration in Princeton, New Jersey, US]

GFDNA Grain and Feed Dealers National Association [now National Grain and Feed Association, US]

GFE Gibbs Free Energy [Thermodynamics]

GFE Government-Furnished Equipment

GFE *(Großforschungseinrichtung)* – German for "National Research Center"

GfE Gesellschaft für Elektrometallurgie mbH [German Electrometallurgical Company located in Duesseldorf]

γ-Fe Gamma Iron [Symbol]

GFE&M Government-Furnished Equipment and Material

γ-Fe$_2$O$_3$ Gamma Ferric Oxide (or Maghemite)

GFF Glass Fiber Filter

GFFAPA Grain, Feed and Fertilizer Accident Prevention Association

GFFAR Guided Folding Fin Aircraft Rocket

GFG Gyro Flux Gate (Compass)

GfG Gesellschaft für Gerätebau mbH [German Equipment Manufacturer]

GFI Gaseous Fuel Injection (System)

GFI Ground Fault Interrupter

GFI Guided Fault Isolation

GFID German Foundation for International Development

G Fis Giornale di Fisica [Italian Physics Journal; published by Società Italiana di Fisica]

GfKK Gesellschaft für Kältetechnik und Klimatechnik mbH [German Manufacturer and Supplier of Refrigeration and Air-Conditioning Equipment and Technology]

GFL Gas-Filled Lamp

GfL Gesellschaft für Labortechnik mbH [German Laboratory Equipment and Supplies Company]

GFLI General Federation of Labour in Israel

GFLOPS Giga (Billion) Floating Point Operations per Second [also Gflops]

GFM Gas Flowmeter

GFM Gesellschaft für Fertigungstechnik und Maschinenbau mbH [Austrian Production Engineering and General Mechanical Engineering Company, Steyr]

GFM Gesellschaft für Marktforschung [Market Research Society, Germany]

GFM Government-Furnished Material

GFM Graphics Function Monitor

GFMC Green Function Monte Carlo (Method)

GFMD Gold Film Mercury Detector

GFN Global Futures Network [India]

GFN Grain Fineness Number [Metallurgy]

GFO Gap-Filler Output

GFO Global Fallout

GFP Generalized File Processor

GFP Glass Fiber Pulling

GFP Government-Furnished Property

GF&P Gases, Fluids and Propellants

GFPBBD Groupement Française des Producteurs de Bases et Banques de Données [French Association of Database and Databank Producers]

GFPdNi Gold-Flash Palladium Nickel

GFQ Gradient Furnace with Quenching Device

GFR Gas-Filled Rectifier

GFR Gas-Filled Relay

GFR Gas Flow Rate

GFR Glomerular Filtration Rate [Medicine]

GFR Gardon Foil Radiometer

GFR Glass-Fiber Reinforced; Glass Fiber Reinforcement

GFR Glass-Forming Range

GFR Graphite-Fiber Reinforced; Graphite Fiber Reinforcement

GFR Ground Fault Monitor

GFR Groupe Français de Rhéologie [French Rheology (Research) Group, Paris, France]

GFRC Glass-Fiber-Reinforced Cement

GFR Epoxy Glass-Fiber-Reinforced Epoxy

GFR Nylon Glass-Fiber-Reinforced Nylon

GFRP Glass-Fiber-Reinforced Plastic(s) [also GFRP],

GFRP Glass-Fiber-Reinforced Polymer [also G-FRP]

GFRP Graphite-Fiber Reinforced Plastic(s)

GFRP Graphite-Fiber Reinforced Polymer

GFR PC Glass-Fiber-Reinforced Polycarbonate

GFR PEEK Glass-Fiber-Reinforced Polyether Ether Ketone

GFR PET Glass-Fiber-Reinforced Polyethylene Terephthalate

GFR PP Glass-Fiber-Reinforced Polypropylene

GFR PTFCE Glass-Fiber-Reinforced Polytrifluorochloroethylene

GFRS Great Falls Raceway System [US]

GFRTP Glass-Fiber-Reinforced Thermoplastic(s)

GFS Government Finance Statistics [of International Monetary Fund]

GFS Government-Furnished Software

GFS Gravity Feed System

GFSS Giant Fractional Shapiro Step [Physics]

GFT Gas-Filled Triode

GFT Gas-Filled Tube

GFT Glass-Forming Tendency

GfT Gesellschaft für Tribologie e.V. [Society for Tribology, Germany]

GFTU General Federation of Trade Unions [UK]

GFW Glass Filament Winding; Glass Filament Wound

GFW Gram-Formula Weight [also gfw]

GFW Ground-Fault Warning

GfW Gesellschaft für Weltraumforschung [Space Research Society, Germany]

GFY Government Fiscal Year

GFZ Geo-Forschungs-Zentrum Potsdam [Potsdam Geosciences Research Center, Germany]

GG Gamma Globulin [Biochemistry]

GG Gardner-Gatling (Pistol)

GG Gas Generator

GG Giga Giga [Symbol]

GG Glacial Geologist; Glacial Geology

GG Glycylglycine

GG Golden Gate [San Franscisco, California, US]

GG Grain Growth [Metallurgy]

GG Gravitational Geons

GG Gravity Gradient

GG Green-Green [Double Star Rocket]

GG Grey Group

GG Gross Generation

G/G Ground-to-Ground [also G-G, g/g, or g-g]

Gg Gigagram [Unit]

g-g green-green [Double Star Rocket]

g-g group-to-group

g$^\pm$g$^\mp$ gauche plus/minus–gauche minus/plus [Chemistry]

GGA Generalized Gradient Approximation (Method)

GGA Geowissenschaftliche Gemeinschaftsaufgaben [Joint Federally and State-Funded Research Institute for Applied Geosciences, Hannover, Germany]

γ-Ga Gamma Gallium [Symbol]

GGC Generalized Gradient Correction(s)

GGC Golden Gate College [San Francisco, California, US]

GGDPAC Government Geoscience Database Policy Advisory Committee [Commonwealth]

GGF Glass and Glazing Federation [UK]

GGG Gadolinium Gallium Garnet [also G^3]

GGG Midland, Texas [Meteorological Station Designator]

GGG:YIG Gadolinium Gallium Garnet Doped Yttrium Iron Garnet

g/g H$_2$O gram(s) per gram of water [Unit]

GGI GPS (Global Positioning System) Geoscience Instrument

GGIG Gadolinium Gallium Iron Garnet

GGM Glacial Geomorphologist; Glacial Geomorphology

GGMC Guyana Geology and Mines Commission

g/g-mol gram per gram-molecule [also g g-mol^{-1}]

GGMR Granular Giant Magnetoresistance [also G^{2}MR]

GGNA L-Gamma-Glutamyl-p-Nitroanilide

GGOR Gross Gas-Oil Ratio

GGP Gateway-Gateway Protocol [Internet]

GGP Gross Global Product

GGPD Gross Gas Produced

GGPNA Gamma-Glutamyl-P-Nitroanilide

GGR Gas Graphite Reactor

GGR Grain Growth Rate [Metallurgy]

G Gr Great Gross [also g gr] [Unit]

GGS Global Geospace Science

GGS GSE (Ground Support Equipment) Gateway Services [of NASA]

GGS Gyro Gunsight

GGSE Gravity-Gradient Stabilization Experiment [of NASA]

GGT γ-Glutamyltransferase [also G-GT] [Biochemistry]

GGT Global Geoscience Transects (Project) [US]

γ-GT γ-Glutamyltranspeptidase [also GGTP] [Biochemistry]

ggt drop(s) [Medical Prescriptions]

GGTP γ-Glutamyltranspeptidase [also γ-GT] [Biochemistry]

GGTS Gravity-Gradient Test Satellite [of NASA]

GGU Golden Gate University [San Francisco, California, US]

GGW Guaiacol-Glycerol-Water (Mixture)

GH Gamma Heating

GH *(Gesamthochschule)* – German for "Polytechnic Institute"

GH Ghana [ISO Code]

GH Ghosh and Hamilton (Model) [Metal Forming]

GH Gibbs-Helmholtz (Equation) [Physical Chemistry]

GH Greenhouse

GH Gross Head

GH Growth Hormone [Biochemistry]

GH Gun Howitzer

GH Gyro Horizon

GH$_2$ Gaseous Hydrogen

g/h gram(s) per hour [also g h^{-1}, g/hr, or g hr^{-1}]

GH-3 Gerovital H-3

G7/h6 Sliding Fit; Basic Shaft System [ISO Symbol]

GHA Grassland Heritage Association [US]

GHA Greenwich Hour Angle [Astronomy]

GHB Gamma-Hydroxybutyrate

GHb Glycated Hemoglobin

GHBC Ground-State Hydrogen-Bonded Complex

GHA General Housekeeping Area

g/ha gram(s) per hectare [Unit]

GHC Ghanian Cedi [Currency of Ghana]

GHC Ground Half Coupling

GHCN Global Historical Climate Network

GHCP Georgia Hospital Computer Group [US]

GH Duisburg Gesamthochschule Essen [Duisburg Polytechnic Institute, Germany]

g/100 cc gram(s) per (one) hundred cubic centimeters [also g/100 cm³]

g/100 cm³ gram(s) per (one) hundred cubic centimeters [also g/100 cc]

GHe Gaseous Helium

γ-He Gamma Helium [Symbol]

GH Essen Gesamthochschule Essen [Essen Polytechnic Institute, Germany]

GHF Generalized Hartree-Fock [Physics]

GHF Gradient Heating Facility

GHFF General Harmonic Force Field

GHF-NO-CI Generalized Hartree-Fock/Natural Orbital/ Configuration Interactions [Physics]

GHG Greenhouse Gases

GHIS Geostationary High-Resolution Interferometer Sounder

GHIUD Global Human Information Use per Decade

GH Kassel Gesamthochschule Kassel [Cassel Polytechnic Institute, Germany]

GHL Growth-Hormone Loaded

GH/LCD Guest-Host/Liquid Crystal Display

g/100 mL gram(s) per (one) hundred milliliter [also g/100 ml]

GHM Gesellschaft für Handwerksausstellungen und – messen [Trade Fairs and Exhibitions Company, Germany]

GHOST Global Horizontal Sounding Technique (Balloon)

GHPD Gated High Power Decoupling [Electronics]

GHQ General Headquarters [also Ghq]

GHRH Growth Hormone-Releasing Hormone [Biochemistry]

GHRI Guidance System, Hybrid Radio-Inertial

GHRIH Growth Hormone-Releasing Inhibiting Hormone [Biochemistry]

GHRM Generalized Hard Rod Model

GHRS Goddard High-Resolution Spectrograph [of NASA at Goddard Space Flight Center, Greenbelt, Maryland, US]

GHS *(Gesamthochschule)* – German for "Polytechnic Institute"

GHS Gill Hematoxylin Solution [Chemistry]

GH Siegen Gesamthochschule Siegen [Siegen Polytechnic Institute, Germany]

GHSR Governor's Highway Safety Representative [US]

GHT Gesellschaft für Hochdrucktechnik mbH [German High-Pressure Engineering Company]

GHT Glycosylated Hemoglobin Test [Medicine]

GHX Ground Heat Exchanger

GHz Gigahertz [Unit]

GHz/kOe Gigahertz per kilooersted [also Ghz kOe⁻¹]

GHz/T Gigahertz per Tesla [also GHz T⁻¹]

GI Galvanized Iron

GI Gas Industry

GI Gas Ionization; Gas Ionizer

GI Gastrointestinal (Tract) [also gi]

GI Gaussian Integration

GI Gegenions

GI General Ionex (Analytical Products Group) [Now Part of High Voltage Engineering Europa BV, Netherlands]

GI General Issue [US Military]

GI Geodesic Isotensoid

GI Geometric Isomer(ism)

GI Geophysical Institute [of University of Alaska at Fairbanks, US]

GI Gesellschaft für Informatik [Society for Informatics, Germany]

GI Gibralta [ISO Code]

GI Glycemic Index

GI Glissile Interface [Materials Science]

GI Gold Institute [US]

GI Governmental and Industrial

GI Government-Initiated; Government Initiative

GI Government Issue [US Military]

GI Graded Index

GI Graphical Integration

GI Grating Interferometer

GI Grazing Incidence

GI Greenpeace International [UK]

GI Gross Investment

GI Ground Intercept(ion)

GI Guest Investigator

GI Gyro Integrator

gi gill [Unit]

GIA Gemmological Institute of America [US]

GIA General Industrial Application

GIA General Industry Applications (Committee) [of Institute of Electrical and Electronics Engineers – Industry Applications Society, US]

GIA Geographical Information Analysis [Geographical Information System]

GIA GPC (General-Purpose Computer) Interface Adapter

GIA Gummed Industries Association [US]

GIAB Grazing Incidence Asymmetric Bragg (Scattering)

GIAM (Conference on) Global Impacts of Applied Microbiology

GIANT Genealogical Information and Name Tabulating System

GIAO Gauge-Independent Atomic Orbitals; Gauge-Invariant Atomic Orbitals

GIAR Grazing-Incidence Antireflection

GIAS Global Integration and Systhesis [of International Geosphere/Biosphere Program]

GIASTA Groupement International pour l'Avancement des Sciences et Techniques Alimentaires [International Working Group for the Advancement of Food Science and Technology, France]

Gib Gibraltar

GIBMS Guided-Ion-Beam Mass Spectroscopy

GIC Garbuny's Image Converter

GIC General Impedance Converter

GIC General Input Channel

GIC General Input/Output Channel

GIC Generalized Impedance Converter

GIC Graphite Intercalation Compound

GIC Guaranteed Investment Certificate

GICCW Government-Industry Conference against Chemical Weapons

GICQ General Investment Corporation of Quebec [Canada]

GICS Generic Intelligence Control System

GID Gesellschaft für Information und Dokumentation mbH [German Information and Documentation Company, Frankfurt/Main]

GID Grazing Incidence (X-Ray) Diffraction

.GID Topics [Microsoft File Name Extension] [also .gid]

GIDEON Global Infectious Disease and Epidemiology Network [also Gideon]

GIDEP Government-Industry Data Exchange Program [US]

Gidroliz Lesokhim Prom Gidroliznaya i Lesokhimicheskaya Promyshlennost [Russian Publication on Hydrolysis and Chemistry of Wood]

GIE Generator Isentropic Efficiency

GIE Giant Isotope Effect [Solid-State Physics]

GIE Ground Instrumentation Equipment

Giesserei-Erfahr Giesserei-Erfahrungsaustausch [German Publication on Foundry Know-How]

Giesserei-Prax Giesserei-Praxis [German Publication on Foundry Practices]

Giess-Tech Giesserei-Technik [German Publication on Foundry Technology]

GIEWS Global Information and Early Warning System [of Food and Agricultural Organization]

GIF Gatan (Inc.) Image (or Imaging) Filter

GIF German-Israeli Foundation (for Scientific Research and Development)

GIF Graphics Interchange Format [also gif]

GIF Gravito-Inertial Force

.GIF Graphics Interchange Format [File Name Extension] [also .gif]

GIFA Internationale Giessereifachmesse [International Foundry Trade Fair, Germany]

GIFAP Groupement International des Associations Nationales de Fabricants de Produits Agrochimiques [International Group of National Association of Manufacturers of Agrochemical Products, Belgium]

GIFAS Groupement des Industries Françaises Aéronautiques et Spatiales [Working Group of the French Aeronautics and Space Industries]

GIFS Generalized Interrelated Flow Simulation

GIFT Gas Insulated Flow Tube

GIFT General Internal FORTRAN Translator

GIFY Grazing-Incidence Fluorescence Yield

GIG Genetics Interest Group [UK]

GIG Geoscience Information Group [of University of Bristol, UK]

giga- SI prefix representing 10^9

Gigaflops Giga Floating-Point Operations per Second [Unit]

GIGI General Imaging Generator and Interpreter

GIGO Garbage-In, Garbage-Out [Computers]

GIGO Garbage-In, Gospel-Out [Computers]

GIGS Grazing Incidence Grating Spectrometer

GII Global Information Infrastructure

GII Grazing Incidence Interferometer

GIIR Grazing Incidence Infrared

GIIP Groupement International de l'Industrie Pharmaceutique [International Working Group of the Pharmaceutical Industry; of European Economic Community, Belgium]

GILS Government Information Locator Service [of NASA]

GILSP Good Industrial Large-Scale Practice

GIM General(ized) Information Management

GIM Generic Interface Module

GIM Ground Interface Module

GIMA Garden Industry Manufacturers Association [UK]

GIMC Grinding of Industrial Minerals Conference

GIMIC Guard-Ring Isolated Monolithic Integrated Circuit

GIML Generalized Information Management Language

GIMMS Geographic Information Mapping and Management System

GIMMS Global Inventory Modeling and Monitoring Study

GIMRADA Geodesy Intelligence and Mapping Research and Development Agency [US Army]

GIMS Global Integrated Monitoring System

GIMU Gimballess Inertial Measuring Unit

GIN Global Information Network [US]

GINA Fachnormenausschuß Giessereiwesen [Standards Committee for Foundry Technology, Germany]

GINETEX Groupement International d'Etiquetage pour l'Entretien des Textiles [International Association for Textile Care Labelling, Spain]

GINO Graphical Input/Output

GIO Gallium Indium Oxide

GIO Government Information Office

GIO Griffith, Irwin and Orowan (Model) [Metallurgy]

GIOC Generalized Input/Output Controller

GIOP General-Purpose Input/Output Processor

Giornale Mat Battaglini Giornale Matematica di Battiglini [Mathematical Journal of Battaglini, Italy]

GIP Gastric Inhibitory Peptide (or Polypeptide) [Biochemistry]

GIP Gibraltar Pound [Currency]

GIP General Information Program

GIP Genetic Improvement Programs [Queensland, Australia]

GIP Ground Instructor Pilot

GIP Ground-State Inversion Potential (Method) [Physics]

GIP Group Interface Processor

GIPGS Greenhouse Information Program Grants Scheme [Australia]

GIPME Global Investigation of Pollution in the Marine Environment [of UNESCO/Intergovernmental Oceanographic Commission]

GIPS Geographical Information Processing System

GIPS Giga-Instructions per Second [also gips]

GIPS Ground Information Processing System

GIPSE Gravity-Independent Photosynthetic Exchange

GIPSY General Information Processing System

GIPSY Generalized Information Processing System [of United States Geological Survey]

GIPSY GPS (Global Positioning System) Infrared Positioning System

GIR Glancing Incidence Reflection

GIR Grazing Incidence Reflection

GIR Grazing Internal Reflection

GIRD General Incentive for Research and Development [Canada]

GIRD Grants for Industrial Research and Development

GIRD Ground Integration Requirements Document

GIRI Government Industrial Research Institute [at Ikeda, Osaka and Nagoya, Japan]

GIRI Gray Iron Research Institute [now Iron Casting Research Institute, US]

GIRIN Government Industrial Research Institute at Nagoya [Japan]

GIRL Generalized Information Retrieval Language [of United States Defense Nuclear Agency]

GIRL German Infrared Laboratory

GIRL Graphic Information Retrieval Language

GIRLS General(ized) Information Retrieval and Listing System

GIS Gas Industry Standard [UK]

GIS Gas Insulated Substation

GIS Gas Insulated Switchgear

GIS Gas Insulated System

GIS Gated Isolated Structure

GIS Generalized Information System

GIS Geographic Information System

GIS (National) Geographic Information Systems (Program) [of US Environmental Protection Agency]

GIS Geological Information System

GIS Geophysical Incentive System

GIS Geoscience Information Society [of American Geological Institute, US]

GIS Global Information Solutions [of AT&T Company, US]

GIS Grant Information System [US]

GIS Grazing Incidence Spectrograph

GISC Grupo Interdisciplinar de Sistemas Complicados [Interdisciplinary Group for Complicated Systems; of Universidad Carlos III de Madrid, Spain]

GISD Geographic Information Systems Division [of Natural Resources Canada]

GISL Groupement des Industries Sidérurgiques Luxembourgeoises [Working Group of the Luxembourg Iron and Steel Industry]

GIS Newsl GIS Newsletter [of Geoscience Information Society, US]

GISP Greenland Ice Sheet Program

GISPRI Global Industrial and Social Progress Research Institute

GISS Goddard Institute for Space Studies [of NASA at New York City, US]

GIT Gastrointestinal Tract

GIT General Industrial Training

GIT Georgia Institute of Technology [Atlanta, US]

GIT Gesellschaft für Informationsvermittlung und Technologieberatung [Association for Information Brokerage and Technological Consultancy, Germany]

GIT Granton Institute of Technology [Toronto, Canada]

GIT Graph Isomorphism Tester

GITC Glucopyranosyl Isothiocyanate

GITC 2,3,4,6-Tetra-O-Acetyl-β-D-Glucopyranosyl Isothiocyanate

GITG Ground Interface Technical Group

GITL Gas Insulated Transmission Line

GITL Government/Industry Technical Liaison Committee

GITO Government of India Tourist Office

GITSIS Georgia Institute of Technology School of Information Science [US]

GITT Galvanostatic Intermittent Titration Technique

GIWEP Gesellschaft für industrielle Wärme–, Energie– und Prozeßtechnik mbH [Industrial Heat, Energy and Process Engineering Company, Germany]

GIWG Ground Interface Working Group

GIX Global Internet Exchange [Internet]

GIXD Grazing Incidence X-Ray Diffraction

GIXS Grazing Incidence X-Ray Scattering

GJ Gas Jet

GJ Gastric Juice

GJ Gauss-Jordan (Method) [also G-J] [Mathematics]

GJ Gigajoule [Unit]

GJ Grown Junction [Electronics]

GJC Gas Jet Cooled; Gas Jet Cooling

GJE Gauss-Jordan Elimination [Mathematics]

GJ/kg Gigajoule per kilogram (of fuel) [also GJ kg^{-1}]

GJ/m³ Gigajoule per cubic meter [also GJ m^{-3}]

GJP Graphic Job Processor

GJPC Grown-Junction Photocell

GJPO Grand Junction Project Office [US]

GJT Grown-Junction Transistor

GJ/t Gigajoule(s) per ton [Unit]

GK Gas Kinematic(s)

GK Gordon-Kim (Electron Gas Model) [Physics]

G/K Shear Modulus-to-Bulk Modulus (Ratio)

Gk Greek

GKA Gutzow, Kashchiev and Avramov (Nucleation Equation) [Metallurgy]

GKE Gesellschaft Konstruktion und Entwicklung [Society for Design and Development; of Verein Deutscher Ingenieure, Germany]

GKE-VDI Gesellschaft Konstruktion und Entwicklung im Verein Deutscher Ingenieure [Society for Design and Development of the Association of German Engineers]

GKH Gesellschaft für Krankenhaushygiene mbH [German Supplier of Hospital Hygiene Products, Hilden]

GKS Graphical (or Graphics) Kernel System [Computers]

GKSS Forschungzentrum Geesthacht [Geesthacht Research Center, Germany]

GKT Gas Kinetic Theory

GKV Gesamtverband Kunststoffverarbeitende Industrie e.V. [Federation of the Plastic Processing Industries, Germany]

GL Galvanoluminescence; Galvanoluminescent

GL Gas Lamp

GL Gas Laser

GL Gas Law(s)

GL Gas Leak(age)

GL Gas Lens

GL Gate Leads [Electronics]

GL General Labeling (of Cells) [Biochemistry]

GL General Ledger [also G/L]

GL Geodesic Line [Mathematics]

GL Geographic Location

GL Ginzburg-Landau (Theory of Superconductivity)

GL Glass-Type Loctal Tube

GL Glow Lamp

GL Glycolipid [Biochemistry]

GL Gothic Letter

GL Gouy Layer

GL Grade Line [Civil Engineering]

GL Granato-Lücke (Dislocation Theory) [also G-L] [Materials Science]

GL Graphic(s) Library

GL Graphics Language

GL Gray Level

GL Great Lakes [US/Canada]

GL Greenland [ISO Code]

GL Green Library (Organization) [France]

GL Green Light

GL Grenade Launcher

GL Grid Leak [Electronics]

GL Gross Leak

GL Ground Level

GL Ground Line

GL Ground Loop

GL Group Leader

GL Growth Ledge

GL Gun Laying (Radar)

GL Gun Limber

GL Full Linear Group [Symbol]

G/L Gaussian-Lorentzian (Sum Function) [Symbol]

G/L General Ledger [also Gl]

G-L Granato-Lücke (Dislocation Theory) [also GL] [Materials Science]

Gl Glacier [also gl]

Gl Glass [also gl]

Gl Glaze [also gl]

Gl Glucinium [now Beryllium]

Gl Glycine

G(λ) Photodiode Responsivity [Semiconductor Symbol]

g/L gram(s) per liter [also g/l, GPL, gpl, g L^{-1}, or g l^{-1}]

gl gill [Unit]

.gl Greenland [Country Code/Domain Name]

GLA Gamma Linolenic Acid

GLA Geologisches Landesamt [Geological Surveys of the German States of Bavaria, Hamburg, Mecklenburg-Vorpommern, North-Rhine Westphalia, Rhineland Palatinate and Saxon-Anhalt]

GLA Goddard Laboratory for Atmospheres [of NASA Goddard Space Flight Center, Greenbelt, Maryland, US]

γ-La Gamma Lanthanum [Symbol]

GLAADS Gun Low Altitude Air Defense System

Glac Glycol Alcohol

glac glacial

GLACS Ground Loop Attitude Control System

GLAG Ginzburg-Landau-Abrikosov-Gorkov (Theory) [Solid-State Physics]

Glam Glamorganshire [UK]

GLANCE Global Lightweight Airborne Navigation Computer Equipment

GLAS Geoscience Laser Altimeter System

γ-La$_2$S$_3$ Gamma Lanthanum Sesquisulfide

GLASOD Global Assessment on Soil Degradation

Glasphalt Glass Asphalt [also glasphalt]

Glas Mat I Glasnik Matematicki, Serija I [Yugoslavian Publication on Mathematics]

Glas Mat II Glasnik Matematematicki, Serija II [Yugoslavian Publication on Mathematics]

Glas Mat III Glasnik Matematematicki, Serija III [Yugoslavian Publication on Mathematics]

GLASS Germanium-Lithium Argon Scanning System

Glass Ceram Glass and Ceramics [Translation of: *Steklo i Keramika (USSR)*; published in the US]

Glass(Er) Erbium-Doped Glass (Laser)

Glass(Nd) Neodymium-Doped Glass (Laser)

Glas Srp Akad Nauka Umet Od Tekh Nauka Glas Srpske Akademije Nauka i Umetnosti, Odeljenje Tekhnichkikh Nauka [Yugoslavian Scientific Journal]

Glass Technol Glass Technology [Publication of the Society of Glass Technology, UK]

GLAST Gamma-Ray Large Area Space Telescope

GLASTEC Glastechnisches Kolloquium [Colloquium on Glass Technology, Germany]

GLASTEC International Trade Fair–Machinery–Equipment–Applications (for the Glass Industry) [Germany]

Glastech Ber Glastechnische Berichte [Reports on Glass Technology; published by Deutsche Glastechnische Gesellschaft]

glb greatest lower bound [also GLB]

GLBC Great Lakes Basin Commission

GLC Gambia Labour Congress

GLC Gas-Liquid Chromatography [also glc]

GLC Generator Line Contractor

GLC German Library Conference

GLC Greater London Council [UK]

GLC Great Lakes Commission [US]

GLC Great Lakes Corporation [US]

GLC Ground-Level Concentration [also glc]

Glc D-Glucose [Biochemistry]

Glc β-D-Glucopyranoside [Biochemistry]

GLCA Great Lakes Colleges Association [US]

GlcA D-Gluconic Acid [Biochemistry]

GLCC Great Lakes Composites Consortium, Inc. [US]

GLCM Ground-Launched Cruise Missile

GlcN D-Glucosamine [Biochemistry]

Glcn Gallocyanine

GlcNAc N-Acetyl-D-Glucosamine [Biochemistry]

(GlcNAc)$_2$ Di-(N-Acetyl-D-Glucosamine) [Biochemistry]

(GlcNAc)$_3$ Tri-(N-Acetyl-D-Glucosamine) [Biochemistry]

Glc (1)-P α-D-Glucose 1-Phosphate [Biochemistry]

Glc (6)-P D-Glucose 6-Phosphate [Biochemistry]

Glc (1,6)-P$_2$ α-D-Glucose 1,6-Diphosphate [Biochemistry]

Glc-(1,6)-P$_2$-4CHA α-D-Glucose-1,6-Diphosphate Tetracyclohexylammonium [Biochemistry]

GLCTS Global Land Cover Test Sites

GlcUA D-Glucuronic Acid [Biochemistry]

GLD Generalized Langevin Dynamics (Method) [Physics]

Gld Glide [also gld]

Gld Guilder (or Florin) [Currency of the Netherlands]

GLDC Glutamic Decarboxylase [Biochemistry]

GLDH L-Glutamate Dehydrogenase [Biochemistry]

GLDH Glutamic Dehydrogenase [Biochemistry]

L-GLDH L-Glutamic Dehydrogenase [Biochemistry]

GLDP Glutamic Dephosphatase [Biochemistry]

GLDS Ground Laser Designator Station

GLDV Gray Level Difference Vector

GLE Generalized Langevin Equation [Physics]

GLEAM Graphic Layout and Engineering Aid Method

GLEEP Graphite Low-Energy Experimental Pile [UK]

GLEER Graphite Low-Energy Experimental Reactor [UK]

GLERL Great Lakes Environmental Research Laboratory [of National Oceanic and Atmospheric Administration, US]

GLF Gaussian Lobe Function

GLFRC Great Lakes Forest Research Center [Sault Ste. Marie, Ontario, Canada]

GLGA Ginzburg-London-Garkow-Abrikosow (Superconductivity)

Glgst Geologist

GLHS Great Lakes Historical Society [US]

GLHK Granato-Lücke Theory for High Kelvin Temperatures

GLI Global Imager

GLI Great Lakes Initiative [US]

GLI Great Lakes Institute [Windsor, Ontario, Canada]

γ-Li Gamma Lithium [Symbol]

GLIAS Greater London Industrial Archeology Society [UK]

GLIM Generalized Linear Interactive Modelling (System) [Software]

GLIN Great Lakes Regional Information System [of US Environmental Protection Agency]

G Line G Line [Fraunhofer Line for Iron Having a Wavelength of 430.8 Nanometers] [also G line] [Spectroscopy]

GLINT Global Intelligence

GLIPAR Guide Line Identification Program for Antimissile Research

GLIS Global Land Information System [of US Geological Survey]

GLISP Great Lakes International Surveillance Plan

Glit Glitter(ing) [also glit]

GLL Galileo (Mission) [of NASA]

GLLD Ground Laser Locator Designator

GLLFAS Great Lakes Laboratory for Fisheries and Aquatic Sciences [Burlington, Ontario, Canada]

GLM General(ized) Linear Models [Computers]

GLMA Great Lakes Mink Association [US]

GLMI Great Lakes Maritime Institute [US]

Gln (+)-Glutamine; Glutaminyl [Biochemistry]

GLNPO Great Lakes National Program Office [of US Environmental Protection Agency]

GLO Goddard Launch Operations [of NASA]

Glob Globule; Globular [also glob]

Global Biodiv Global Biodiversity [Publication of the Canadian Museum of Nature]

Global Environ Fac Newsl Global Environment Facility Newsletter [Publication of the World Bank]

Global Environ Change Rep Global Environmental Change Report

Global Planet Change Global and Planetary Change [Journal published in the Netherlands]

GLOBE Global Learning by Observations to Benefit the Environment [Internet]

GLOBECOM Global Communications System [of US Air Force]

GLOBECOM Global Communications Conference [of Institute of Electrical and Electronic Engineers, US]

GLOC Gravity-Induced Loss of Consciousness

GLOCHANT Global Change and the Antarctica

GLOCOM Global Communications System

GlOH Glycol

GLOK Granato-Luecke Theory for 0 (Zero) Kelvin [Materials Science]

GLOMEX Global Meteorological Experiment

GLONASS Global Orbiting Navigation Satellite System [Russian Federation]

GLOPAC Gyroscopic Lower-Power Attitude Controller

GLORIA Geological Long-Range Inclined ASDIC (Antisubmarine Detection Investigation Committee)

Glos Gloucestershire [UK]

GLOSS Global Ocean Surveillance System

GLOSS Global Sea Level Observing System

Gloss Glossary [also gloss]

GLOTRAC Global Tracking Network

GLOTRAC Global Tracking System [of NASA, Apollo Project]

Glouc Gloucestershire [UK]

GLOW Gross Lift-Off Weight [also G LOW]

GLOW Ground Lift-Off Weight [also G LOW]

GLP Glucagon-Like Peptide [Biochemistry]

GLP Goal Language Processor

GLP Good Laboratory Practices (Standard)

GLPC Gas-Liquid-Partition Chromatography

GLP/GALP Good Laboratory Practices/Good Automated Laboratory Practices

GLPO Good Laboratory Practice Office [US]

GLPT Great Lakes Persistent Toxic (Substances) [US/Canada]

GLR Gas/Liquid Ratio

GLR General Laboratory Report

GLR Government Land Register [States of Queensland and Western Australia, Aus-tralia]

GLR Great Lakes Research

Glr Glare [also glr]

GLRC Grain Legume Research Council

GLRIS Great Lakes Regional Information System [US]

GLRS Geoscience Laser Ranging System

GLRS Geodynamics Laser Ranging System

GLRS-A Geoscience Laser Ranging System–Altimeter [now Geoscience Laser Altimeter System]

GLRS-R Geoscience Laser Ranging System–Ranger

Glr Shld Glare Shield

GLS Global Load Sharing

GLS Global Load Shearing

GLS Graduate Library School

GLS Ground Launch Sequencer

Gls Glass [also gls]

gls glass(y) (state) [Chemistry]

GLSA Great Lake Seaplane Association [US]

GLSC Great Lakes Science Center [of US Geological Survey/Biological Resources Division at Ann Arbor, Michigan]

Glt Glutarimide [Biochemistry]

Glta Glutaric Acid [Biochemistry]

GLTN Glycerinlactatetrinitrate [Biochemistry]

GLTS Gas Leakage Test Stand

GLU Great Lakes United [US]

Glu Gluconate [Biochemistry]

Glu Glucose [Biochemistry]

Glu (+)-Glutamic Acid; Glutamate [also glu]

Glu Glutamyl [Biochemistry]

D-(+)-Glu D-(+)-Glucose [Biochemistry]

α-D-(+)-Glu α-D-(+)-Glucose [Biochemistry]

β-D-(+)-Glu β-D-(+)-Glucose [Biochemistry]

Glu-Cys Glutamylcysteine [Biochemistry]

Glu-Cys-Gly Glutamylcysteinyglycine Glutathione [Biochemistry]

Glueckauf Forschungsh Glueckauf Forschungshefte [German Publication on Mining Research]

Glulam Glue-Laminated Wood [also glulam]

Glu(NH$_2$) Glutamine [Biochemistry]

Glut Max Gluteus Maximus [also glut max] [Anatomy]

Glut Med Gluteus Medius [also glut med] [Anatomy]

GLV Gemini Launch Vehicle [of NASA]

GLV Glove Valve

GLV Graphische Versuchs– und Lehranstalt [Institute for Graphical Research and Education, Austria]

GLV Gross Leukemia Virus

Glv Glove(s) [also glv]

GLWQA Great Lakes Water Quality Agreement [US/Canada]

GLWQI Great Lakes Water Quality Initiative [US/Canada]

Glwr Glassware [also glwr]

GLX Gel Liquid Extraction

Glx Glutamic Acid [Biochemistry]

Glx Glutamine [Biochemistry]

GLY Glycol

GLY Water-Glycol Mixture

.GLY Glossary [Microsoft Word Extension]

Gly Glycinate [also gly] [Biochemistry]

Gly Glycine; Glycyl [Biochemistry]

Gly-Ala Glycylalanine [Biochemistry]

Gly-Ala-Phe Glycylalanylphenylalanine [Biochemistry]

GLYME 1,2-Dimethoxyethane [also Glyme]

Glyc Glycerin; Glycerol

Glycobiol Glycobiology [Journal published by Oxford University Press, UK]

GM Gadolinium Molybdate

GM Galvanomagnetic(s); Galvanomagnetism

GM Galvanometer; Galvanometry

GM Gambia [ISO Code]

GM Gamow Maximum [Physics]

GM Gang Mill(ing)

GM Gaseous Mixture

GM Gas Maser

GM Gas Mechanic(s)

GM Gas Meter

GM Geiger-Mueller (Tube) [also G-M]

GM Gell-Mann (Relation) [Particle Physics]

GM General Manager

GM General Motors (Corporation) [US]

GM Genetically Modified (Food)

GM Genetic Modification

GM Geological Material

GM Geomagnetic(s); Geomagnetism

GM Geometric Mean

GM Geometric Model(ing)

GM Geomorphological; Geomorphologist; Geomorphology

GM Geostrophic Motion

GM Gilding Metal

GM Gifford-McMahon (Refrigerator) [Superconductor Science]

GM Gill-Morell (Oscillation) [Physics]

GM Gjønnes-Moodie (Electron Diffraction Lines) [also G-M] [Physics]

GM Glass Mat

GM Glycol Monomethylether

GM Go Monitor (Command) [Pascal Programming]

GM Good Morning [Amateur Radio]

GM Gradient Motion

GM Grain Matrix [Materials Science]

GM Grand Master

GM Granular Material

GM Granular Medium

GM Granulocyte-Macrophage [Anatomy]

GM Gravimetric; Gravimetry

GM Gravitational Mass

GM Green Mountains [Vermont, US]

GM Greenwich Meridan

GM Grid Modulation

GM Grinding Machine

GM Grinding Media

GM Group Mark

GM Group Multiplication

GM Guided Missile

GM Gulf of Mexico

GM Gun Metal

GM Gutzeit Mechanism [Chemistry]

GM Gyromagnet(ism); Gyromagnetic(s)

GM Gyrotropic Medium

G-M Geiger-Muller (Tube) [also GM]

G-M Graphite-Metal

G-M Gjønnes-Moodie (Electron Diffraction Lines) [also GM] [Physics]

Gm Gigameter [Unit]

gm gram [also g]

g/m² gram(s) per square meter [also g m^{-2}]

g/m³ gram(s) per cubic meter [also g/cu m, or g m^{-3}]

g(μ) density of (localized) states at the Fermi level [Symbol]

$\nabla\mu$ chemical potential gradient [Symbol]

GMA Gas Metal-Arc (Welding)

GMA Geomechanics Abstracts [of Royal School of Mines, UK]

GMA Gesellschaft für Meß– und Automatisierungstechnik [Society for Measurement and Automation Technology; of Verein Deutscher Ingenieure/Verein Deutscher Elektrotechniker, Germany]

GMA Glasgow Mathematical Society [UK]

GMA Glycidyl Methacrylate

GMA Glycol Methacrylate

GMA Gravimetric Analysis

GMA Grocery Manufacturers Association [US]

GMA Grocery Manufacturers of Australia

GMAC Gas Metal-Arc Cutting

GMAC General Motors Acceptance Corporation (of Canada Limited)

GMAC Genetic Manipulation Advisory Committee

GMAG Genetic Manipulation Advisory Group [UK]

GM-AGT General Motors Advanced Gas Turbine (Program) [US]

GMAL General Electric Macro Assembly Language

G-Man Government Man [FBI Agent]

GM and S General Medicine and Surgery [also GM&S]

GMAP Generalized Macroprocessor

GMAP General Macro Assembly Program

GMAP Geometric Modeling Application Program

GMAT Graduate Management Admission Test

GMAT Greenwich Mean Astronomical Time

GMAW Gas Metal-Arc Welding

GMAW-EG Gas Metal-Arc Welding, Electrogas

GMAW-MIG Gas Metal-Arc Welding/Metal Inert Gas (Welding)

GMAW-P Pulsed Gas Metal-Arc Welding

GMAW-S Short Circuiting Gas Metal-Arc Welding

GMB Good Merchantable Brand [of London Metal Exchange, UK]

GmbH Gesellschaft mit beschränkter Haftung [Limited (Liability) Company; Corporate Form in Central Europe]

Gmbl Gimbal

GMBS γ-Maleimidobutyric Acid N-Hydroxysuccinimide

GMBS N-(γ-Maleimidobutyrloxy)succinimide

GMC Geiger-Mueller Counter

GMC General Medical Council [UK]

GMC General Motors Corporation [US]

GMC Glass-Matrix Composite

GMC Graphite Matrix Composite

GMC Graphite Multi-Intercalation Compound

GMC Gross Maximum Capacity

GMC Ground Movement Controller

GMC Groundwater Management Caucus [US]

GMCC General Motors Ceramic Committee [US]

GMCC Geophysical Monitoring for Climate Change [of National Oceanic and Atmospheric Administration, US]

gm/cc gram(s) per cubic centimeter [also gm/cm³, g/cc, g/cu cm, g/cm³, or g cm^{-3}]

GMCL Ground Measurements Command List

GMCM Generalized Molecular Crystal Mode

gm/cm gram(s) per cubic centimeter [also gm/cc, g/cc, g/cu cm, g/cm³, or g cm^{-3}]

GMCM Guided Missile Countermeasures

GMCR Global Minimum Catchment Region

GM-CSF Granulocyte-Macrophage Colony-Stimulating Factor [Biochemistry]

GM-CSFR Granulocyte-Macrophage Colony-Stimulating Factor Receptor [Biochemistry]

GMD Gambian Dalasi [Currency of Gambia]

GMD Gesellschaft für Mathematik und Datenverarbeitung [Society for Mathematics and Data Processing, Darmstadt, Germany]

GMD Guided Missiles Division [US Air Force]

GMDA Groundwater Management Districts Association [now Groundwater Management Caucus, US]

GMDEP Guided Missile Data Exchange Program [US Navy]

GMDH Group Method of Data Handling

GMDSS Global Maritime Distress and Safety System

GME Galvanomagnetic Effect

GME Gelatin Manufacturers of Europe [of Conseil Européen des Fédérations de l'Industrie Chimique, Belgium]

GME Generalized Machine Equation

GME Generalized Magneto-Optical Ellipsometry

GME Generalized Master Equation

GME Generic Macro Expander

GME Gesellschaft für Mikroelektronik [Society for Microelectronics; of Verein Deutscher Ingenieure/ Verein Deutscher Elektrotechniker, Germany]

GME Global Metals Environment (Conference)

GME Glycol Methyl Ether

GME Gyromagnetic Effect

G-MEM General Memory

G-MEM GPC (General-Purpose Computer) Memory

G-MEM Glascow Modified Eagle's Medium [Biochemistry]

G Metal Guillaume Metal [also G metal]

GMF Galactic Magnetic Field [Astronomy]

GMFB Gang Mill Fixture Base

GMFC Gem and Mineral Federation of Canada

GMFCS Guided Missile Fire Control System

GMFP Geometrical Mean Free Path

GMG Gross Maximum Generation

g/m²·h gram(s) per square meter hour [also g/m²/h, or $\mathrm{g\ m^{-2}\ h^{-1}}$]

GMI Gelatin Manufacturers Institute [US]

GMI General Motors (Engineering and Management) Institute [Flint, Michigan, US]

GMI Geomagnetic Index

GMI Giant Magnetoimpedance

GMIH Giant Magnetoimpedance Hysteresis

g/min gram(s) per minute [also $\mathrm{g\ min^{-1}}$]

GMIS Generalized Management Information System

GMIS Government Management Information Sciences [US]

GMIX Gas Membrane–Ion Exchange (Process)

GMJ Gilding Metal Jacket [Bullet]

GML Generalized Mark-Up Language

GML General Measurement Loop

GML Generic Mark-Up Language

GML Graphic Machine Language

g/mL gram(s) per milliliter [also g/ml, or gm/ml]

GMLTeN Glycerin Monolactate Tetranitrate

GMM Galvanometric Method

GMM General Matrix Manipulator

GMM Gram Molecular Mass [also gmm]

GMM Guided Missile Mechanic

gm/m² gram per square meter [also $\mathrm{gm\ m^{-2}}$]

g/μm gram(s) per micrometer [also $\mathrm{g\ \mu m^{-1}}$]

g/μm² gram(s) per square micrometer [also $\mathrm{g\ \mu m^{-2}}$]

GMMC Ground Master Measurements List

gm/ml gram(s) per milliliter [also g/mL, or g/ml]

GMN Gorman [Tactical Air Navigation Station)

GMN Gell-Mann-Nishijima (Scheme) [Particle Physics]

γ-Mn Gamma Manganese [Symbol]

GMO Gell-Mann-Okubo (Mass Formula) [Particle Physics]

GMO General Meteorological Office [UK]

GMO Genetically Manipulated Organism; Genetically Modified Organism

GMO Glycol Monooleate

GMO Grants Management Officer

g/mol gram(s) per mole [also $\mathrm{g\ mol^{-1}}$]

g-mol gram-molecule

GMODS Glass Material Oxidation and Dissolution System

γ-Mo₂N Gamma Dimolybdenum Nitride

GMP Geometric Modelling Project [UK]

GMP Gesellschaft für Medizinische Praxis [Society for Medical Practice, Germany]

GMP Global Mobile Professional

GMP Good Manufacturing Practices

GMP Ground Movement Planner

GMP Guanosine Monophosphate; Guanosine 5'-Monophosphate [Biochemistry]

GMP Guild of Metal Perforators [UK]

GMP Gulf of Mexico Program [of US Environmental Protection Agency]

2'-GMP Guanosine-2'-Monophosphate [Biochemistry]

3'-GMP Guanosine-3'-Monophosphate [Biochemistry]

5'-GMP Guanosine-5'-Monophosphate [Biochemistry]

GMP-BSA Guanosine Monophosphate–Bovine Serum Albumin [Biochemistry]

GMP-PCP Guanosine-5'-[β,γ-Methylene]triphosphate [Biochemistry]

GMP-PNP Guanosine-5'-[β,γ-Imido]triphosphate [Biochemistry]

GMR Gas-to-Metal Ratio

GMR General Modular Redundancy

GMR General Motors Research (Division) [US]

GMR Gesellschaft Meß– und Regelungstechnik [Society for Measurement and Control Engineering; of Verein Deutscher Ingenieure/Verein Deutscher Elektrotechniker, Germany]

GMR Giant Magnetoresistance

GMR Giant Monopole Resonance

GMR Graphite-Moderated Reactor

GMR Greatest Meridional Radius

GMR Ground Mapping Radar

GMR Ground Movement Radar

GMR Gyromagnetic Ratio

GMRAM Giant Magnetoresistance Random-Access Memory

GMRC General Motors Research Center [US]

GMR&D General Motors Research and Development [US]

GMRDC General Motors Research and Development Center [Warren, Michigan, US]

GMRES Generalized Mininun (or Minimal) Residual Method

GMR Fachber Gesellschaft Meß– und Regelungstechnik Fachberichte [Reports of the Society for Measurement and Control Engineering; of Verein Deutscher Ingenieure/Verein Deutscher Elektrotechniker, Germany]

GMRL General Motors Research Laboratories [Warren, Michigan, US]

GMRL Group of Mathematicians of Romance Languages [Portugal]

GMRWG Guided Missile Relay Working Group [of US Navy]

GMS Gas Mass Spectrometer; Gas Mass Spectrometry

GMS Gemini Mission Simulator [of NASA]

GMS Generalized Main Scheduling

GMS General Maintenance System

GMS Geometric Modeling System

GMS Geostationary Meteorological Satellite [Japan]

GMS Gesellschaft für Medizinische Software mbH [German Supplier of Medical Software Products, Mannheim]

GMS Global Management System

GMS Global Messaging Service [of Novell Inc., US]

GMS Glycerol Monostearate

GMS Grants Management Specialist

GMS Ground Maintainance Support

GMS Ground Measurement System

GM&S General Medicine and Surgery [also GM and S]

gms grams [Unit]

GMSA Generalized Mean Spherical Approximation

GMSF Gulf Marine Support Facility [of National Oceanic and Atmospheric Administration at Pascagoula, Mississippi, US]

GMSFC George C. Marshall Space Flight Center [of NASA at Huntsville, Alabama, US]

GMSK Gaussian Filtered Minimum Shift Keying

GMSK Gaussian Mean Shift Keying [Telecommunications]

gms/m² grams per square meter [also gms m^{-2}]

GMSMA Generalized Mean Spherical Model Approximation

GMSS Graphical Modelling and Simulation System

GMT Geiger-Mueller Tube

GMT Generalized Machine Theory

GMT Generalized Multitasking

GMT Glass Mat Thermoplastics

GMT Global Money Transfer

GMT Grain Misorientation Texture [Materials Science]

GMT Graphics Mouse Technology

GMT Greenwich Mean Time

GMTC General Motors Technical Center [US]

GMU Gadjah Mada University [Jogjakarta, Indonesia]

GMU George Mason University [Fairfax, Virginia, US]

GMUG GeoQuest Metroplex User Group [US]

GMV Gram-Molecular Volume [also gmv]

GMV Guaranteed Minimum Value

GMW Gram-Molecular Weight [also gmw]

GM/WM Group Mark/Word Mark

GMWU General Municipal Workers Union [UK]

GN Gadolinium Niobate

GN Gain [also Gn]

GN Generic Name

GN Geonavigation

GN Global Name

GN Glomerulonephritis [Medicine]

GN Good Night [Amateur Radio]

GN Graduate Nurse

GN Gram-Negative (Bacteria) [Microbiology]

GN Gregory-Newton (Formula) [Mathematics]

GN Ground Network

GN Group Number [Oceanography]

GN Guinea [ISO Code]

G&N Guidance and Navigation

G/N Glucose-to-Nitrogen (Ratio)

GN$_2$ Gaseous Nitrogen

Gn Gain [also GN]

Gn Green [also gn]

γ-N Gamma Nitrogen [Symbol]

∇n concentration gradient [Symbol]

GNAT Grade-Nine Achievement Test [Jamaica]

GNAV Graphic Area Navigation

GNBS Guyana National Bureau of Standards

GNC Global Navigation Chart

GNC Graphic Numerical Control

GNC Guidance, Navigation and Control

GNC Guidance and Navigation Computer

GN&C Guidance, Navigation and Control

GNCFTS Guidance, Navigation and Control Flight Test Station

GNCIS Guidance, Navigation and Control Integration Simulator

GNCS Guidance and Navigation Control System

GNCTS Guidance, Navigation and Control Test Station

GND Geometrically Necessary Dislocation [Materials Science]

GND Ground [also Gnd, or gnd]

GND C/O Ground Checkout

Gnde Grenade [also gnde]

GNE Gross National Expenditure

GNF German Nuclear Forum

GNI Generation of New Ideas

GNI Gross National Investment

γ-Ni Gamma Nickel [Symbol]

GNIS (Online) Geographic Names Information Service [of US Geological Survey]

γ-NiZn Gamma Nickel Zinc

GNL Galanthus Nivalis Lectin (Agglutinate) [Biochemistry]

GNM Germanisches Nationalmuseum [Germanic National Museum, Nuremberg]

GNM Glycerophosphate, Nucleinate and Methylarsenate of Sodium

GN/m² Giganewton(s) per square meter [also GN m^{-2}]

GNMA Government National Mortgage Association [US]

GNMS Ground Network Management System

GNN Global Network Navigator [Internet]

GNO Gadolinium Niobate (or Gadolinium Niobium Oxide)

GNO Green Nickel Oxide

GNOME GNU (Gnu's Not Unix) Network Object Model Environment [Operating System]

GNP Geographical North Pole

GNP Glacier National Park [Montana, US]

GNP Grain Neutral Spirits

GNP Gross National Product

γ-Np Gamma Neptunium [Symbol]

GNR Geographical Names Register [of New South Wales, Australia]

GNR Giant Nuclear Resonance

GNR Glycidyl Novolac Resins

Gnr Gunner [also gnr]

GnRH Gonadotropin-Releasing Hormone [Biochemistry]

GNS Guinea Syli [Currency]

G&NS Guidance and Navigation Subsystem

GNSI Guild of Natural Science Illustrators [US]

GNSS Global Navigation Satellite System

GNU Gnu's Not Unix (Operating System)

GNVQ General National Vocational Qualification [UK]

GNWT Government of the Northwest Territories [Canada]

GO Gas Oil

GO Gaussian Orbitals [Physics]

GO General Office

GO General Order

GO Generated Output

GO Geodesic Ovaloid

GO Geometrical Optics

GO Glycerin Oleate

GO Glycol Oxalate (Process)

GO Gold Colored

GO Goniometer; Goniometry

GO Government of Ontario [Canada]

GO Grain-Oriented; Grain Orientation [Metallurgy]

GO Graphitic (or Graphite) Oxide

GO Greenwich Observatory [at Herstmonceux Castle, East Sussex, UK]

GO Group Orbitals

GO Gunn Oscillator

GO_2 Gaseous Oxygen

G/O Gas/Oil Ratio

$G\Omega$ Gigohm [Unit]

γ-O Gamma Oxygen [Symbol]

G&OA Glycerine and Oleochemicals Association [US]

GOAL Ground Operations Aerospace Language

GOALS Global Ocean-Atmosphere-Land-Surface Interactions

GOALS Global Ocean-Atmosphere-Land System

GOAM Government Owned and Maintained

GOB Good Ordinary Brand [also gob]

GOBR Group of Officials on Biotechnology Regulation

GOC Gas-Oil Concentration

GOC Gas/Oil Contact

GOC General Officer Commanding

GOC General Optical Council [UK]

GOC Graphic Option Controller

GOC Greatest Overall Coefficient

GOC Ground Operations Coordinator

GOCA Ground Operations Control Area

GOCI General Operator-Computer Interaction

GOCINC General Officer Commanding in Chief [also GOC in C]

GOCO Government-Owned, Contractor-Operated

GOCR Gated-Off Controlled Rectifier

GOd Glucose Oxidase [or G Od] [Biochemistry]

God Vissh Khimikotekhnol Inst Sofiya Godishnik na Visshiya Khimikotekhnologiya Institut Sofiya [Bulgarian Journal on Chemical Engineering published by the Sofia Institute for Chemical Technology]

God Vissh Minno-Geol Inst Sofiya Godishnik na Visshiya Minno-Geolozhki Institut Sofiya [Bulgarian Journal on Mining and Geology published by the Sofia Institute of Mining and Geology]

GOE Gaussian Orthogonal Ensemble

GOE General Overhauser Effect [Physics]

GOE Guinea Ekwele [Currency of Equatorial Guinea]

GOE Ground Operating Equipment

GOe Gauss-Oersted [Unit]

G/Oe Gauss(es) per Oersted [Unit]

GOES Geostationary Operational Environmental Satellite [Series of Weather Satellites of the US National Oceanic and Atmospheric Administration]

GOES Global Omnibus Environmental Survey [of Consortium for International Earth Science Information Networks]

GOES-A Geostationary Operational Environmental Satellite A [of National Oceanic and Atmospheric Administration, US]

GOES-NEXT Geostationary Operational Environmental Satellite, Next Generation [of National Oceanic and Atmospheric Administration, US]

GOEZS Global Ocean Euphotic Zone Study

GOF Goodness-of-Fit (Test)

GOFFEX German Options and Financial Futures Exchange

GOFO Geometry-Optimized Floating Orbitals [Physics]

GOFS Global Ocean Flux Study

GOGO Government-Owned, Government-Operated

GOI Gate Oxide Integrity

GOI Government of Indonesia

GOI Gynecology and Obstetric Investigation [International Journal]

GOIC (Persian) Gulf Organization for Industrial Consulting [Qatar]

GOJ Government of Japan

GOL Gasoline and Oil Testing Laboratory [of Alberta Research Council, Canada]

GOL General Operating Language

GOL Goal-Oriented Language

GOL Graphic On-Line Language

GOLD Graduates Of the Last Decade

Gold Bull Gold Bulletin [Switzerland]

GOLDOX Gold Oxidation (Process)

Goldschmidt Inf Goldschmidt Informiert [Newsletter published by Th. Goldschmidt AG Chemische Fabrik, Essen, Germany]

GOM Grain and Oil Seeds Marketing Incentives Program [Canada]

GOM Ground Operations Manager [NASA Kennedy Space Center]

GOM Group Occupancy Meter [Telecommunications]

GOMAC Government Microcircuit Applications Conference

GOMMS Ground Operations and Material Management System

GOMOS Global Ozone Monitoring by Occultation of Stars

GOMR Global Ozone Monitoring Radiometer

GOMS Ground Operations Management System

GONII Gas and Oil National Information Infrastructure [US]

GOOS Global Ozone Observing System

GOP General Operational Plot

GOP Giant Oxygen Precipitate

GOP Ground Operations Panel

GOP Gulf Ordnance Plant [Aberdeen, Mississippi, US]

GOPG Ground Operations Planning Group

GOR Gained Output Ratio

GOR Gas/Oil Ratio

GOR General Operational Requirement

GOR Gross Overriding Royalty

GOR Ground Operations Review

GORI Gross Overriding Royalty Interest

GORID Ground Optical Recorder for Intercept Determination

GORP Ground Operations Requirements Plan

GORS Ground Observer RF (Radio Frequency) System,

GORX Graphite Oxidation from Reactor Excursion

GOS Gate Operating System

GOS Generalized Oscillator Strength [Physics]

GOS Geomagnetic Observing System

GOS Global Observational System; Global Observing System [of World Meteorological Organization]

GOS Grade of Service

GOS Graphics Operating System

GOSEAC Group of Specialists on Environmental Affairs and Conservation

GOSIP Government OSI (Open Systems Interconnection) Profile [US]

GOSIPS Government OSI (Open Systems Interconnection) Procurement Specification [US]

GOSP Gas-Oil Separation Plant

GOSS Ground Operations (or Operational) Support System

GO Steel Grain-Oriented Steel [or GO steel] [Metallurgy]

GOSUB Go to Subroutine

GOT Glutamic-Oxalacetic Transaminase [Biochemistry]

Goth Goth(ic)

GOTH Gothic (Font) [also Goth]

GO Train Government of Ontario (Commuter) Train [Canada]

GO Transit Government of Ontario Transit System [Canada]

GOTS Graphically Oriented Time-Sharing System

GOTS Gravity-Oriented Test Satellite [of NASA]

GOU Gourde [Currency of Haiti]

GOV Gas Overvoltage

.GOV Governmental [Internet Domain Name] [also .gov]

Gov Government(al) [also gov]

Gov Governor [Title]

Gov Governor [Device] [also gov]

Gov Data Syst Government Data Systems [US Publication]

Gov-Gen Governor-General

Gov Inf Q Government Information Quarterly [UK Publication]

Gov Publ Government Publications [UK]

Gov Publ Rev Government Publication Review [US]

Gov Res Announc Government Research Announcements [US]

Govt Government [also govt]

Govtel Government Telegram

GovtPrtgOff Government Printing Office

GOW Gopher Ordnance Works [Rosemont, Minnesota, US]

GOW Guided Optical Wave

GOWG Ground Operations Working Group

GOX Gaseous Oxygen [also GOx, or gox]

GOx Glucose Oxidase [Biochemistry]

GP Galvanically Protected; Galvanic Protection

GP Gang Punch

GP Gas Phase

GP Gas Plasma

GP Gas Port

GP Gas Producer

GP Gauge Port

GP Gauge Pressure

GP Generalized Programming

GP General Practitioner

GP General Preferential Tariff

GP General Processor

GP General Protection (Coverall)

GP General Publication

GP General Purpose

GP Geodetic Parameter

GP Geological Prospecting; Geological Prospector

GP Geometric(al) Pitch

GP Geometric Processing

GP Geometric Progression

GP Geophysical; Geophysicist; Geophysics

GP Geopotential

GP Germ Plasm [Biology]

GP Getter Pump

GP Glans Penis [Anatomy]

GP Glass Physicist; Glass Physics

GP Glide Path

GP Glide Plane [Crystallography]

GP Globular Protein [Biochemistry]

GP Glow Plug [Diesel Engines]

GP Glycerophosphate

GP Glycoprotein [Biochemistry]

GP Gold Point

GP Gold Probe

GP Goniophotometer

GP Gopher-Protected (Electric Cable)

GP Goto Pair [Electronics]

GP Government Purchaser; Government Purchasing

GP Graduate in Pharmacy

GP Gram-Positive (Bacteria) [Microbiology]

GP Grand Potential

GP Granville-Phillips (Corporation) [Boulder, Colorado, US]

GP Graphic Plotter

GP Graphics Package

GP Graphics Processor

GP Graphite (or Graphitic) Powder

GP Great Plains

GP Green Party

GP Ground Pressure

GP Ground Protection

GP Group [also Gp, or gp]

GP Group Processing

GP Green Party

GP Grüneisen Parameter [Solid-State Physics]

GP Guadeloupe (French) [ISO Code]

GP Guanosine Phosphate [Biochemistry]

GP Guaranteed Performance (Grade)

GP Guard Plate

GP Guatemala Protocol (on Aviation Liability)

GP Guinea Pig

GP Guinier-Preston (Zone) [also G-P] [Metallurgy]

GP Gun Powder

GP Gyroplane

GP Gyroscopic Pendulum

GP I Guinier-Preston Zone I [Metallurgy]

GP II Guinier-Preston Zone II [Metallurgy]

GP2 (Laboratoire du) Génie des Procedes Papetiers [Paper Process Engineering of Institut National Polytechnique de Grenoble, France]

G-P Guinier-Preston (Zone) [also GP] [Metallurgy]

G-1-P Glucose-1-Phosphate; α-D-Glucose 1-Phosphate [Biochemistry]

G-6-P D-Glucose 6-Phosphate [Biochemistry]

G-5'-P Guanosine 5'-Monophosphate [Biochemistry]

G-1,6-P2 Glucose-1,6-Diphosphate [Biochemistry]

Gp Gas Pressure Relay

gp guinea pig [Immunochemistry]

∇P Pressure Gadient [Symbol]

GPA Gas Processors Association [US]

GPA General-Purpose Amplifier

GPA General-Purpose Analysis

GPA General-Purpose Array

GPA Glycerin Producers Association [US]

GPA Glycerophosphoric Acid

GPA Grade Point Average

GPA Graphical PERT (Production Evaluation and Review Technique) Analog

GPA Grit Pattern Analysis; Grit Pattern Analyzer

GPA Groupement de Productivité Agricole [Group for Agricultural Productivity, France]

GPa Gigapascal [Unit]

GPa^{-1} one per gigapascal [also 1/GPa]

GpA Guanylyladenosine

GPAC General-Purpose Analog Computer

GPAC Government and Public Affairs Committee [of ASM International, US]

GPAC Great Plains Agricultural Council [US]

GPAD Gallons per Acre per Day [also gpad]

GPALS Global Protection Against Limited Strikes

GPa·m^3/Mg Gigapascal cubic meter per Megagram [Gpa m^3 Mg^{-1}]

GPa$\sqrt{nm}$ Gigapascal(s) square root of nanometer [also GPa nm$^{\frac{1}{2}}$]

GPAS General-Purpose Airborne Simulator

γ-P-32 ATP γ-Phosphorus-32 Adenosine-5'-Triphosphate [also γ-^{32}P-ATP] [Biochemistry]

GPATS General-Purpose Automatic Test System

GPAX General-Purpose Automation Executive

GPB General Purpose Bomb

GPB Guinier-Preston-Bagaryatskii (Dissolution) [Metallurgy]

GPB Ground Power Breeder

γ-PBA γ-Phenylbutyric Acid

GPBIM General Purpose Buffer Interface Module

GPC Gas-Phase Condensation

GPC Gas-Pressure Cable

GPC Gas-Pressure Casting

GPC Gel-Permeation Chromatograph(y)

GPC Generalized Predictive Control

GPC General Peripheral Controller

GPC General Precision Connector

GPC General-Purpose Computer

GPC Georgia Peanut Commission [US]

GPC Global Petroleum Center

GPC Global Processing Center

GPC Granville-Phillips Corporation [Boulder, Colorado, US]

GPC Graphics Performance Committee [US]

GPC Graphics Performance Characterization

GPC Gross Profit Contribution

g/pc gram(s) per piece [Unit]

GpC Guanylylcytidine

GPCA General-Purpose Communications Adapter

GPCB GOAL (Ground Operations Aerospace Language) Program Control Block

GPCC Global Precipitation Climatology Center

GPCD Gallons per Capita per Day [also gpcd]

GPC-DV Gel Permeation Chromatography with Differential Viscometer

GPCL General-Purpose Closed Loop

GPCP Generalized Process Control Programming

GPCP Global Precipitation Climatology Project

GpCpC Guanylylcytidylylcytidine [Biochemistry]

GPD Gallons per Day [also gpd, gal d^{-1}, or gal/d]

GPD General Protocol Driver`

GPD General-Purpose Discipline

GPD Gimbal Position Display

gpd gallons per day [also GPD, gal d^{-1}, or gal/d]

gpd grams per denier [Unit]

GPDA Gypsum Products Development Association [UK]

GPDC General-Purpose Digital Computer [also GP/DC]

GPD/ft Gallons per day per foot [Unit]

GPDH Glucose-Phosphate Dehydrogenase [Biochemistry]

GPDH Glyceraldehyde Phosphate Dehydrogenase [Biochemistry]

G-6-PDH Glucose-6-Phosphate Dehydrogenase [also G6PD] [Biochemistry]

gpDm Geopotential Decameter

GPDS General-Purpose Display System

GPE Gas Phase Embrittlement

GPE General-Purpose Equipment

GPE Graphic Picture Enhancement

GpE Geophysical Engineer [also GpEng]

GPED Gas Phase Electron Diffraction

GpEng Geophysical Engineer [also GpE]

GPES Ground Proximity Extraction System

GPF Gas-Phase Fluorescence

GPF Generalized Production Function

GPF General Planning Forecast

GPF General Protection Fault

GPF General-Purpose Facilities

GPF General-Purpose Furnace (Black)

GPF Grand Partition Function

GPF Ground Processing Facility

GPF Black General-Purpose Furnace Black [also GPF black]

GPG Grains per Gallon [also gpg]

G[5']P2[5']G P^1, P^2-Di(guanosine-5'-) Diphosphate [Biochemistry]

G[5']P3[5']G P^1, P^3-Di(guanosine-5'-) Triphosphate [Biochemistry]

G[5']P4[5']G P^1, P^4-Di(guanosine-5'-) Tetraphosphate [Biochemistry]

G[5']P5[5']G P^1, P^5-Di(guanosine-5'-) Pentaphosphate [Biochemistry]

GpG Guanylylguanosine [Biochemistry]

gpg guinea pig [Immunochemistry]

GPGAP Great Plains Gasification Associates Project [US]

GPGL General-Purpose Graphic Language

GPH Gallons per Hour [also gph]

Gph Geophysics [also gph]

gph grains (of hardness) per gallon (of water) [Unit]

Gph Graphite [also gph]

GPI General Paralysis of the Insane [Psychology]

GPI Geologisch-Paleontologisches Institut [Institute of Geology and Paleontology, University of Munster, Germany]

GPI Gimbal Position Indicator

GPI Glass Packaging Institute [US]

GPI Glucose Phosphate Isomerase [Biochemistry]

GPI GOES (Geostationary Operational Environmental Satellite) Precipitation Index [of National Oceanic and Atmospheric Administration, US]

GPI Graphics Programming Interface

GPI Ground Point of Intercept

GPI Ground Position Indicator

GPIA General-Purpose Interface Adapter

GPIA Generic Pharmaceutical Industry Association [US]

GPIB General-Purpose Instrument(ation) Bus

GPIB General-Purpose Interface Bus

GPIB/IEEE General-Purpose Interface Bus/Insitute of Electrical and Electronics Engineers

GPIB-PC General-Purpose Interface Bus–Personal Computer

GPIDN Great Plains International Data Network [of US Environmental Protection Agency]

GPIO General-Purpose Input-Output

GPIS Gemini Problem Investigation Status [of NASA]

GPIS Groundwater Pumping Incentives Scheme [Victoria State, Australia]

GPL Generalized Programming Language

GPL General-Purpose Laboratory

GPL General-Purpose Language

GPL General-Purpose Loader

GPL Giant Pulse Laser

GPL Goal Processing Language

GPL Gram(s) per Liter [also gpl, g/L, g/l, g L^{-1}, or g l^{-1}]

GPL Group Processing Logic

GPL Immunoglobulin G Phospholipid Unit [Biochemistry]

Gpl Glyoxal-Bis(isopropylimine)

GPLAN Generalized Database Planning System

GPLP General-Purpose Linear Programming

GPM Gallons per Minute [also gpm, gal min^{-1}, or gal/min]

GPM Generalized Perturbation Method [Physics]

GPM General-Purpose Macrogenerator

GPM Geopotential Meter

GPM Ground Potential Model

GPM Groups per Message

GPM Groups per Minute

GPM2 (Laboratoire du) Génie Physique et Mécanique des Materiaux [Physical and Mechanical Materials Engineering Laboratory; of Institut National Polytechnique de Grenoble, France]

gpm gallons per minute [also GPM, gal min^{-1}, or gal/min]

GPMA German Productivity and Management Association [Germany]

GPMAS Gas-Phase Molecular Absorption Spectroscopy

GPME General-Purpose Mission Equipment

GPM2-ENSPG (Laboratoire du) Génie Physique et Mécanique des Materiaux–Ecole Nationale Supérieure de Physique de Grenoble [Physical and Mechanical Materials Engineering Laboratory–National School of Physics of Grenoble; of Institut National Polytechnique de Grenoble, France]

gpm/ft² gallon(s) per minute per square foot [also gpm/sq ft]

GPMS General-Purpose Microprogram Simulator

gpm/sq ft gallon(s) per minute per square foot [also gpm/ft²]

GPMU Graphical, Paper and Media Union [UK]

GPNA Glutamyl-P-Nitroanilide [Biochemistry]

GPO General Post Office

GPO Government Printing Office [Washington, DC, US] [also USGPO, or US GPO]

GPO-PIA Government Printing Office and Printing Industry of America [US]

GPOS General-Purpose Operating System

GPP Gas-Phase Polymerization

GPP Gas-Phase Polypropylene

GPP General Plant Project

GPP Geophysical Prospecting

GPP General Print and Punch

GPP Gross Provincial Product [Canada]

GPPD General-Purpose Powder Diffractometer

Gpp[NH]p 5'-Guanylylimidodiphosphate [Biochemistry]

GPPS General-Purpose Simulation System

Gppt Gel Precipitate [also GPPT]

GPR Gas Production Rate

GPR General-Purpose Radar

GPR General-Purpose Register [also gpr]

GPR General-Purpose Rubber

GPR Geophysical Research

GPR Government Property Register [of New South Wales, Australia]

GPR Ground-Penetrating Radar

GPR Ground Power Rise

GPR Ground-Probing Radar

GPRA Government Performance and Results Act [US]

GPRC Glass Passivated Rectifier Chip

GPRF General-Purpose Rocket Furnace

GPRF-G General-Purpose Rocket Furnace–Gradient

GPRF-I General Purpose Rocket Furnace–Isothermal

GPRMC Groupe des Plastiques Renforces et Materiaux Composites [Organization of Reinforced Plastics and Composite Materials, Belgium]

GPRN GOAL (Ground Operations Aerospace Language) Test Procedure Release Notice [of NASA]

GPS Gallons per Second [also gps, gal/s, gal/sec, or gal sec⁻¹]

GPS Gas-Pressure Sintered; Gas Pressure Sintering

GPS General Problem Solver

GPS General Processing Subsystem

GPS General Processing System

GPS Geological and Planetary Sciences

GPS Global Positioning Satellite

GPS Global Positioning System

GPS Good Practices Standard

GPS Gram(s) per Second [also gps, g/sec, g/s, or g s⁻¹]

GPS Graphic Programming Services

GPS Greater Pittsburgh Section [of Materials Research Society, US]

GPS Ground Power Supply

GPS Ground Processing Simulation

GPS Groupe de Physique des Solides [Solid-State Physics Group; of Ecole Normale Supérieure, Université de Paris, France]

GPS Grüne Partei der Schweiz [Green Party of Switzerland]

gps gallons per second [also GPS, gal/s, gal/sec, or gal sec⁻¹]

GPSA Gas Processors Suppliers Association [US]

GPS/ARGOS Global Positioning System/Atmospheric Research Geostationary Orbit Satellite

GPSCO Global Position System Consortium [Australian Consortium]

GPSCS General-Purpose Satellite Communications System

GPSDR Global Positioning System Demonstration Receiver

GPSDW General-Purpose Scientific Document Writer

GPS/GIS Global Positioning System/Geographical Information System

GPSIM Global Position System Integrity Monitoring (System)

GPSP General-Purpose Simulation Program

GPS PTT Global Positioning System Platform Transmitter Terminal [of US Army]

GPSS General Problem Statement Simulator

GPSS General Process Simulation Studies

GPSS General-Purpose Simulation System

GPSS General-Purpose System(s) Simulation; General-Purpose System(s) Simulator

GPSSN Gas Pressure Sintered Silicon Nitride

GPT Gas Phase Titration

GPT Gas Power Transfer

GPT Gemini Pad Test [of NASA]

GPT General Preferential Tariff

GPT General-Purpose Terminal

GPT Glutamic-Pyruvic Transaminase [Biochemistry]

GPT-C Glutamic-Pyruvic Transaminase-C [Biochemistry]

GPTE General Purpose Test Equipment

GPTM Gross Profit this Month

GPTY Gross Profit this Year

GPU Geopotential Unit [also gpu]

GPU Ground Power Unit

GpU Guanylyluridine [Biochemistry]

γ-Pu Gamma Plutonium [Symbol]

GpUpG Guanylyluridylylguanosine [Biochemistry]

GPUR GOAL (Ground Operations Aerospace Language) Test Procedure Update Request

GPWS Ground Proximity Warning System

GPX Generalized Programming Extended

GPZ Guinier-Preston Zone [Metallurgy]

GQ Equatorial Guinea [ISO Code]

GQ Gravitational Quantum
GQE Generalized Queue Entry
GQM Goal/Question/Metric
GQR Giant Quadrupole Resonance
GR Game Reserve
GR Gamma Radiation
GR Gamma Radiograph(y)
GR Gamma Ray
GR Gas-Cooled Reactor
GR Gas Ratio
GR Gas Regulator
GR Gate Region [Electronics]
GR Gattermann Reaction [Organic Chemistry]
GR General-Purpose Register
GR General Reconnaissance
GR General Register
GR General Relativity
GR General Reserve
GR Generation-Recombination
GR Genetic Resources
GR Geometrical Resonance
GR Germanium Rectifier
GR Glass Region
GR Glass-Reinforced; Glass Reinforcement
GR Glide Return
GR Glutathione Reductase [Biochemistry]
GR Government Reserve
GR Grain Refinement; Grain Refiner [Metallurgy]
GR Grand Rapids [Michigan, US]
GR Granules [Chemistry]
GR Grating Reflector
GR Gravitational Radiation
GR Greece [ISO Code]
GR Grenz Ray
GR Grid Ratio [Electronics]
GR Grid Resistor [Electronics]
GR Grid Return [Electronics]
GR Grignard Reaction; Grignard Reagent
GR Ground Role
GR Growth Rate
GR Growth Ring [Botany]
GR Guanidine Rhodanate
GR Guard Ring
GR Gunnery Range
GR Gyromagnetic Ratio
Gr Gear [also gr]
Gr Grade [also gr]
Gr Grain [also gr]
Gr Grain-force (or grain-weight) [Unit]
Gr Granular; Granules [also gr]
Gr Graphite [also gr]
Gr Grashof Number [Fluid Flow] [Symbol]
Gr Grecian

Gr Greece; Greek
Gr Group [also gr]
G(r) Orientational Pair Distribution Function [Symbol]
gr grain [Unit]
gr (granum) – a grain [Medical Prescriptions]
gr gram [Unit]
gr gray
gr gross
.gr Greece [Country Code/Domain Name]
g(r) radial distribution function [Symbol]
$g_1(r)$ one-dimensional interface correlation function (in physics) [Symbol]
$\gamma_1(r)$ one-dimensional correlation function (for polymers) [Symbol]
GRA Gamma-Ray Absorption
GRA Gamma-Ray Altimeter
GRA Gamma-Ray Astronomy
GRA Government Report Announcements (Database) [of National Technical Information Service, US]
GRACE Graphic Arts Composing Equipment
GRAD Generalized Remote Access Database
GRAD General Recursive Algebra and Differentiation
GRAD Graduate Resume Accumulation and Distribution
Grad Gradient [also Grad, or grad]
Grad Graduate(d); Graduation [also grad]
grad gradient [Mathematics]
GRADB Generalized Remote Access Database
GRADD Graphics Adapter Device Driver
GradDip Graduation Diploma
GRADS Generalized Remote Access Database System
Graec Graecia(n)
GRAF Graphic Addition to FORTRAN
GRAFTABL Display Extended Character Set in Graphics Mode (Command) [also graftabl]
Graf Tech Grafische Technik [German Journal on Graphic Arts; published by the Institut für Grafische Technik]
GRAIIM Groupe de Recherche sur les Applications de l'Informatique à l'Industrie Minérale [Research Group for Information Science Applications in the Mining Industry; of Laval University, Montreal, Canada]
GRAIN Graphics-Oriented Relational Algebraic Interpreter
Gr-Al Graphite-Aluminum (Composite) [also Gr/Al]
GRAM Global Reference Atmospheric Model (Software) [of NASA Marshall Space Flight Center, Huntsville, Alabama, US]
GRAMPA General Analytical Model for Process Analysis
GRAN Global Rescue Alarm Network
Gran Granular; Granule [also gran]
GRANADA Grammatical Non-Algorithmic Data
GRANAS Global Radio Navigation System [Germany],
GRAND Gamma-Ray Astrophysics at Notre Dame (Project) [of University of Notre Dame, Indiana, US]
GrA-Ni Graphite Aluminum-Nickel (Composite)
GRANIS Graphical Natural Inference System
Grants Newsl Grants Newsletter [UK]

GRAO Gamma-Ray Astronomy Observatory

GRAPDEN Graphic Data Entry Unit

GRAPE Gamma-Ray Attenuation Porosity Evaluator

Graph Graphic(s) [also graph]

GRAPHDEN Graphic Data Entry

Graphic Arts Lit Abstr Graphic Arts Literature Abstracts [US]

graPHIGS Advanced Graphics Programming Interface for PHIGHS (Programmers' Hierarchical Interactive Graphics System

Graphs Comb Graphs and Combinatorics [Published in Germany]

GRAPL Graphic Application Programming Language; Graphic Interactive Programming Language

GRAPWG Gamma-Ray Astronomy Program Working Group [of NASA]

GRARR Goddard Range and Range Rate [of NASA]

GRAS Generally Recognized as Safe (List) [of US Food and Drug Administration]

GRASER Gamma-Ray Amplification by Stimulated Emission of Radiation [also Graser, or graser]

GRASP Generalized Read and Simulate Program

GRASP Generalized Remote Acquisition and Sensor Processing

GRASP Generalized Retrieval and Storage Program

GRASP Graphical Robot Applications Stimulation Package

GRASP Graphic Service Program

GRASS Geographic(al) Resources Analysis Support System

GRATIS Generation, Reduction, and Training Input System

grav (acceleration of) gravity

GRB Gamma-Ray Burst [Astronomy]

GRB Gas Research Board [UK]

GRB Geophysical Research Board [of National Research Council, US]

GRB Government Reservation Bureau

GRB Green Bay, Wisconsin [Meteorological Station Identifier]

Gr Br Great Britain [also Gr Brit]

GRC Gamma-Ray Constant

GRC Gamma-Ray Counter

GRC Gas Recycling Center

GRC Geotechnical Research Center [Canada]

GRC Geothermal Resources Council [US]

GRC Germanium Rectifier Crystal

GRC (John) Glenn Research Center [of NASA]

GRC Gordon Research Conference [Series of Topical US Scientific Conferences]

GRC Graduate Research Center

GRC Graphite-Reinforced Composite

GRCA Glassfiber Reinforced Cement Association [UK]

GRCDA Governmental Refuse Collection and Disposal Association [US]

GR/CIDS Genetic Resources/Communication, Information and Documentation System [of Food and Agricultural Organization]

GRCSW Graduate Research Center of the Southwest [US]

Gr-Cu Graphite-Copper (Composite) [also Gr/Cu]

gr/cu ft grain(s) per cubic foot [also gr/ft³]

GRD Gamma-Ray Detector

GRD Greek Drachma [Currency of Greece]

GRD Ground Detector

GRD Ground-Resolved Distance

Grd Grind [also grd]

Grd Ground [also grd]

GRDA Grand River Dam Authority [Oklahoma, US]

GRDB Geoscientific Resource Data Base [Queensland, Australia]

GRDC Grains Research and Development Corporation

GRDF Gypsum Roof Deck Foundation [now National Roof Deck Contractors Association, US]

Grdn Garden [also grdn]

GRE Gamma-Ray Emission

GRE Gamma-Ray Explorer

GRE Gas Release Event

GRE Gaussian Resolution Enhancement

GRE General Research Equipment

GRE Graduate Record Examination (Board)

GRE Graduate Reliability Engineering

GRE Graphics Engine [Computers]

GRE Ground Radar Equipment

GRE Ground Run-Up Enclosure

GREA Groupe de Recherche en Environnement Agroalimentaire [Agro-Alimentary Environment Research Group; of Laval University, Montreal, Canada]

GREAT Graduate Research in Engineering and Technology

GREB Galactic Radiation Experiment Background (Satellite)

Grec Grecia(n)

GRED Generalized Random Extract Device

Green Greenland(ic)

GREMEX Goddard Research Engineering Management Exercise [of NASA]

GREP Global Regular Expression Print

GREP Graphite/Epoxy Composite [also Gr/Ep]

GREP Ground Control and Emergency Preparedness (in Ontario Mines) [Canada]

GREPCO Greenland Petroleum Consortium

GRESE Grants for Research and Education in Science and Engineering [of National Science Foundation, US]

GRETA Ground Radar Emissions Training Aviator

GRF Gesneriad Research Foundation [US]

GRF Grassland Research Foundation [US]

GRF Gravity Research Foundation

GRF Ground Repetition Frequency

GRF Growth-Hormone Releasing Factor [Biochemistry]

.GRF Graph [Extension] [also .grf]

GrF Graphite Fiber

GrF Grazing Fire

grf grain-force (or grain-weight) [Unit]

gr/ft³ grain(s) per cubic foot [also gr/cu ft]

GRG Gamma-Ray Ga(u)ge

gr/gal grain(s) per gallon [Unit]

gr/h·ft²·in Hg grain(s) per hour square foot inch of mercury [also gr/ft²/h/in Hg]

gr/100 ft³ grain(s) per (one) hundred cubic feet [Unit]

GRG Glass-Fiber Reinforced Gypsum

GRG (International Society for) General Relativity and Gravitation [Switzerland]

GRG Geodesy Research Group

GRG Graphical Rewriting Grammar

GRG Gross Reserve Generation

GRH Gas Recycle Hydrogenator (Process)

GRI Gage Research Institute [Toronto, Ontario, Canada]

GRI Gas Research Institute [Chicago, Illinois, US]

GRI Gear Research Institute [US]

GRI Geoscience Research Institute [Manchester, UK]

GRI Graduate Research Institute [of Baylor University, Dallas, Texas, US]

GRI Grassland Research Institute [US]

GRI Group Repetition Interval

gr.i. 1 grain [Apothecary]

GRID Gas Research Institute Digest [US]

GRID Global Resource Information Database [of United Nations Environmental Program]

GRID Graphic Interactive Display

GRID Graphic Retrieval and Information Display

gr.ii. 2 grains [Apothecary]

gr.ii.ss. 2½ grains [Apothecary]

gr.i.ss 1½ grains [Apothecary]

GRIN Graded (Refractive) Index (Lens); Gradient (Refractive) Index (Lens)

GRIN Graphic Input

gr/in² grain(s) per square inch [also gr/sq in]

GRINDER Graphic Interactive Network Designer

GRI Net Gas Research Institute Network [US]

GRIN Lens Graded (Refractive) Index Lens; Gradient (Refractive) Index Lens

GRINS General Retrieval Inquiry Negotiation Structure

GRIN-SCH Graded-Index Separate-Confinement Heterostructure

GRIP General Retrieval of Information Program

GRIP Graphics Interactive Program

GRIP Greenland Ice-Core Project

GRIPHOS General Retrieval and Information Processing for Humanities-Oriented Studies

GRIPS General Relation-Based Information Processing System

GRIS Geographic Resource Information System

GRIS Global Resources Information System

GRISA Groupe de Recherche de l'Information Scientifique et Automatique [Research Group on Scientific and Automatic Information; of European Atomic Energy Community]

GRIST Grazing-Incidence Solar Telescope

GRIT Graduated Reduction in Tension

gr.j. 1 grain [Apothecary]

GRJP Glandless Reciprocating Jet Pump

GRL Gamma-Ray Laboratory [Germany]

GRL Gamma-Ray Laser (or Graser)

GRL Greenland [ISO Code]

.grl Greenland [Country Code/Domain Name]

GRLS Glide Return to Landing Site; Glide Return to Launch Site

GRM Generalized Reed-Muller Code

GRM Generalized Rouse Model

GRM Global Range Missile

grm gram [Unit]

Gr-Mg Graphite-Magnesium (Composite) [also Gr/Mg]

Gr-MMC Graphite Metal-Matrix Composite [also Gr MMC]

grn green

Grnd Ground [also grnd]

grnsh greenish

GRO Gamma-Ray Observatory [of NASA]

GRO Gasoline Range Organics [Fuel]

GRO Geotechnical Research Office

GRO Ground Radio Operator

GRO Growth Related Cytokine [Biochemistry]

Gro Gross [also gro]

GRO-α Growth Related Cytokine α [Biochemistry]

GRO-β Growth Related Cytokine β [Biochemistry]

GRO-γ Growth Related Cytokine γ [Biochemistry]

GROATS Graphical Output Package for Atlas

GroEL Chaperonin 60 [Biochemistry]

GroES Chaperonin 10 [Biochemistry]

Grom Grommet [also grom]

GRO-α/MGSA Growth Related Cytokine α/Melanoma Growth Stimulating Activity [Biochemistry]

GROUT Graphical Output

GROWIAN Großwindanlage [A German Super Wind Turbine] [also Growian]

GRP Gastrin-Releasing Peptide [Biochemistry]

GRP Glass-Reinforced Plastic(s)

GRP Glucose-Regulating Protein [Biochemistry]

GRP Group Reference Pilot

GRP Group Reference Point

.GRP Group [File Name Extension] [also .grp]

Grp Group [also grp]

GRPA Generalized Random-Phase Approximation

GRR Gamma-Ray Radiography

GRR Ga(u)ge Repeatability and Reproducibility [also GR&R]

GRR Ga(u)ge Reproducibility and Reliability [also GR&R]

GRR Geneva Radio Regulations

GRR Greek Research Reactor [Greece]

GRR Guidance Reference Release

GR&R Ga(u)ge Repeatablilty and Reproducibility [also GRR]

GR&R Ga(u)ge Reproducibility and Reliability [also GRR]

G(r,R) Green Function

GRS Gamma-Ray Scattering

GRS Gamma-Ray Source

GRS Gamma-Ray Spectrometer; Gamma-Ray Spectrometry; Gamma-Ray Spectroscopy

GRS Gamma-Ray Spectrum

GRS Generalized Retrieval System

GRS General Radio Service [Canada]

GRS General Register Stack

GRS General Reporting System

GRS Geoscience and Remote Sensing (Society) [of Institute of Electrical and Electronics Engineers, US]

GRS German Research Satellite

GRS German Research Society [also Deutsche Forschungsgemeinschaft, Germany]

GRS Gesellschaft für Reaktorsicherheit mbH [German Nuclear Reactor Safety Company]

GRS Gesellschaft für Röntgen– und Strahlentechnik mbH [German X-Ray and Radiation Technology Company, Dortmund]

GRS Global Reference System

GR-S General-Purpose Synthetic Rubber

GR-S Government Rubber-Styrene [now SBR–Styrene Butadiene Rubber]

Grs Gears [also grs]

gr/SCF grain(s) per standard cubic foot [Unit]

gr/SCF$_{dry}$ grain(s) per standard cubic foot, dry [Unit]

gr/SCF$_{wet}$ grain(s) per standard cubic foot, wet [Unit]

gr/sq in grain(s) per square inch [also gr/in^2]

GRSS Geoscience and Remote Sensing Society [of Institute of Electrical and Electronics Engineers, US]

gr.ss ½ grain [Apothecary]

GRT Gamma-Ray Telescope [of NASA]

GRT Gamma-Ray Test(ing)

GRT General Reactor Technology

GRT General Relativity Theory

GRT Gross Register(ed) Ton(nage)

G$_s$(r,t) Van Howe Autocorrelation Function (in Physics) [Symbol]

GRTD Gate Resonant Tunneling Diode

Grtg Grating [also grtg]

GRTP Gamma-Ray Transition Probability

GRTS General Electric Remote Terminal Supervisor

GRTS General Electric Remote Terminal System

GRTS General Remote Terminal System

GRU Gyro Reference Unit

Grundl Landtech Grundlagen der Landtechnik [German Publication on Agricultural Engineering Fundamentals; published by Verein Deutscher Ingenieure]

Grv Groove [also grv]

GRVE Geometrically Representative Volume Element

GRW Generalized Random Walk

Grw Grower [also grw]

Gr Wt Gross Weight [also gr wt]

gr wt grain-weight (or grain-force) [Unit]

GS Gabriel Synthesis [Organic Chemistry]

GS Galvanic Series

GS Galvanized Steel

GS Galvanostatic(s)

GS Gaseous State

GS Gas Saturation (Method)

GS Gas Separation (Unit)

GS Gas Servicer

GS Gas Source

GS Gas Spectrum

GS Gauss-Seidel (Method) [Mathematics]

GS General Secretariat; General Secretary

GS General Service [UK]

GS General Semantics

GS General Solution

GS General Staff

GS General Studies

GS Generating Station

GS Geochemical Society [US]

GS Geodetic Satellite

GS Geodetic Survey(ing)

GS Geographical Society

GS Geological Science(s)

GS Geological Society [UK]

GS Geological Survey

GS Geometric Series

GS Geophysical Signal

GS Geostatistic(al); Geostatistician; Geostatistics

GS Geosyncline [Geology]

GS Geosynthesis; Geosynthetic(s)

GS Geotechnic Sounder

GS German Silver

GS Giemsa Stain [Biochemistry]

GS Ginzburg Superconductivity

GS Girdler Sulfide (Process)

GS Glaciological Society [now International Glaciological Society, UK]

GS Glass Science; Glass Scientist

GS Glide Slope

GS Global Section [Initial Graphics Exchange Specification Files]

GS Goddard Spectrograph [of NASA]

GS Gold Standard

GS Graduate School

GS Grain Size

GS Gram's Stain [Microbiology]

GS Graphic Scale [Cartography]

GS Graphic System

GS Graphite Support

GS Grating Spectrograph; Grating Spectrometer; Grating Spectroscope

GS Gray Scale

GS Green Strength [Powder Metallurgy]

GS Grid Spectrometer

GS Griffonia Simplicifolia [Immunology]

GS Grignard Synthesis [Organic Chemistry]

GS Ground Speed

GS Ground State
GS Ground Station
GS Ground Stopper
GS Ground Support
GS Ground System
GS Group Selector [Telecommunications]
GS Group Separator (Character) [Data Communications]
GS Guild of Surveyors [UK]
GS Gulf Stream
GS Gunnarson-Schonhammer (Theory) [Physics]
GS Gun Sight
GS Gyrostabilizer
GS Gyrostatic(s)
GS South Georgia & South Sandwich Islands [ISO Code]
G&S Grind and Sieve [Powder Preparation]
G&S Gunton and Saunders (Shear Modulus) [Metallurgy]
G/S Guided Steering
Gs Gauss [also gs] [Unit]
G(s) Forward Transfer Function (in Automatic Control) [Symbol]
G$_c$(s) Compensating Elements (in Automatic Control) [Symbol]
G$_N$(s) Normalized Transfer Function (in Automatic Control) [Symbol]
G^{-1}s^{-1} One per Gauss per second [also 1/G·s]
g/s gram(s) per second [also g/sec, GPS, gps, g s^{-1}]
γ-S Gamma Sulfur [Symbol]
GSA General Services Administration [US]
GSA General Services Agencies
GSA General Syntax Analyzer
GSA Genetics Society of America [US]
GSA Geographical Society of America
GSA Geological Society of America [US]
GSA Geological Society of Australia
GSA Geological Survey of Alabama [US]
GSA Greenhouse Supplier Association [US]
GSA Griffonia Simplicifolia Agglutinin [Immunology]
GSA Geophysical Signal Analysis
GSA Government Services Administration
GSA Graduate Student Awards [of Materials Research Society, US]
GSA Ground State Absorption
GSA Guaranteed Savings Account
GSA Bull Geological Society of America Bulletin
G-Salt 7-Hydroxy-1,3-Naphthalenedisulfonic Acid Salt [also G-salt]
GSAM Generalized Sequential Access Method
GSAM Generalized Standard Addition Method
GSAO Ground State Atomic Oxygen
GSAS General(ized) Structure (or Structural) Analysis System (Computer Code) [of Los Alamos Neutron Scattering Center, New Mexico, US]
GSAS Graduate School of Arts and Science
GSAT General Satellite [NASA Space Transportation System]

GSB Geographical Society of Baltimore [US]
GSB Glide Slope Beam
GSB Graphic Standards Board
GSB Gütegemeinschaft für stückbeschichtete Bauteile [Quality Assurance Group for Batch-Processed Coated Components, Germany]
GSC Gas-Solid Chromatography
GSC General Staff College [US Army Command, Fort Leavenworth, Kansas]
GSC General Staff Corps
GSC Genetics Society of Canada
GSC Geological Society of Chicago [Illinois, US]
GSC Geological Survey of Canada
GSC Georgia State College [Atlanta, US]
GSC Georgia Southern College [Statesboro, US]
GSC German Smaller Companies
GSC Glassboro State College [New Jersey, US]
GSC Glenville State College [West Virginia, US]
GSC Government Services Canada
GSC Grant Selection Committee
GSC Green Star, Cluster
GSC Group Switching Center [UK]
GSC Guaranteed Savings Certificate
GS/C Gray Scale/Color (Interface)
GSCC German Society for Clinical Chemistry
GSCCO Gallium Strontium Calcium Copper Oxide (Superconductor)
GSCNO Gadolinium Strontium Copper Niobium Oxide (Superconductor)
GSCD Ground State Combination Difference
GSCG Ground Systems Coordination Group
GSCL Gas-Solid Chemiluminescence
g/s·cm gram(s) per second centimeter [also g/sec·cm]
GSC-NCR Government Services Canada/National Capital Region [Canada]
GSCNO Gadolinium Strontium Copper Niobium Oxide (Superconductor)
GSCS Generalized Self-Consistent Scheme [Materials Science]
GSCU Ground Support Cooling Unit
GSD General Supply Depot
GSD General Systems Division
GSD Generator Starter Drive
GSD Generic Structure Diagram
GSD Genetically Significant Dose
GSD Geological Survey Division [Michigan, US]
GSD German Society for Documentation
GSDB Geophysics and Space Data Bulletin [US]
GSDE Ground Support Design Engineering
GSDL Ground Software Development Laboratory
GSDS General Status Display System
GSDS Goldstone Duplicate Standard [of NASA]
GSE Graphics Screen Editor
GSE Ground Servicing Equipment
GSE Ground-State Energy

GSE Ground-Support Equipment

gSE Spike Energy Unit

γ-Se Gamma Selenium [Symbol]

g/sec gram(s) per second [also g/s, GPS, gps, or g s⁻¹]

GSEL Ground Support Equipment List

GSEP General Summary Edit Program

GSES Graduate School of Engineering Sciences [of Kyushu University, Fukuoka, Japan]

GSEST Graduate School of Electronic Science and Technology [of Shizuoka University, Hamamatsu, Japan]

GSF Geological Survey of Finland

GSF Geometrical Stacking Fault [Materials Science]

GSF Gesellschaft für Strahlen– und Umweltforschung [Society for Radiological and Environmental Research, Germany] [Name changed to Forschungszentrum für Umwelt und Gesundheit (Center for Environmental and Health Research), Germany]

GSF Growth-Suppressing Factor [Biochemistry]

GSF Gyratory Strength Factor

GSFC Goddard Space Flight Center [of NASA at Greenbelt, Maryland, US]

GSFET Gas-Sensitive Field-Effect Transistor

GSG Galvanized Sheet Gauge

GSG Genuine Spin Glass [Solid-State Physics]

GSG Glass-Silicon-Glass

GSGG Gadolinium-Scandium-Gallium Garnet

GSGG Gallium-Scandium-Gadolinium Garnet

GSH Glutathione Peroxidase [Biochemistry]

GSH Glutathione-SH (Reduced Form) [Biochemistry]

GSH-OEt Glutathione-SH (Reduced Form) Ethyl Ester [Biochemistry]

GSHP Ground-Source Heat Pump

GSI General Server Interface

GSI Geographical Survey Institute [Tsukuba, Japan]

GSI Geological Survey of Ireland

GSI Gesellschaft für Schwerionenforschung mbH [German Heavy Ion Research Company, Darmstadt]

GSI Gigascale Integration

GSI Glasstech Solar, Inc. [US]

GSI Glide Slope Indicator

GSI Gold and Silver Institute

GSI Government Source Inspection

GSI Grand Scale Integration

GSI Graphic Structure Input

GSI Guinea Syli [Currency of Guinea]

GSI Gunite/Shotcrete Association [US]

GSI Gyratory Stability Index

GSIA Gloucestershire Society of Industrial Archeology [UK]

GSIA Graduate School of Industrial Administration [of Carnegie-Mellon University, Pittsburgh, Pennsylvania, US]

GSIE Graduate School of Information Engineering [Tamkang University, Taiwan]

GSIME Groupement Syndicat des Industries de Materials d'Equipement Electrique [Association of Electric Equipment Materials Industries, Paris, France]

GSIN Goods and Services Identification Number [Canada]

GSIS Geographic Snow Information System [Canada]

GSIN Group for the Standardization of Information Services [US]

GSIU Ground Standard Interface Unit

GSJ Geological Survey of Japan [of Agency of Industrial Science and Technology–Ministry of International Trade and Industry, Japan]

GSJV Green Street Joint Venture

Gskt Gasket [also gskt]

GSL Generalized Simulation Language

GSL Generation Strategy Language

GSL Geological Society of London [UK]

GSL Geotechnical Science Laboratories [Canada]

GSL German Society for Logistics

GSL Great Salt Lake [Utah, US]

GSLIS Graduate School of Library and Information Science [of University of Illinois, Champaign, US]

GSLQ Generalized Singular Linear Quadratic (Control)

GSLS Graduate School of Library Science

GSM Generalized Sequential Machine

GSM General Sales Manager

GSM General Syntactic Processor

GSM General Systems Model

GSM Geological Society of Malaysia

GSM Global Shared Memory

GSM Global System for Mobile-Communications (Network)

GSM Graduate School of Medicine [of University of Tokyo, Japan]

GSM Graphics System Module

GSM Greenough Stereomicroscope

GSM Ground-State Maser

GSM Groupe Speciale Mobile [Special Mobile Telephone Group, France]

gsm gram(s) per square meter [also GSM, or g/m²]

GSMB Graphic Standards Management Board [of American National Standards Institute, US]

GSMBE Gas-Source Molecular-Beam Epitaxy [also GS-MBE]

GSMT General Society of Mechanics and Tradesmen [US]

GSN Geological Society of Nevada [US]

GSNO S-Nitrosoglutathione [Biochemistry]

GS No Grain Size Number [also GS #] [Materials Science]

GSNU Gyeong Sang National University [South Korea]

GSNW Gateway Service for NetWare

GSO Gadolinium Orthosilicate

GSO General Staff Officer

GSO Geostationary Orbit

GSO Geosynchronous Orbit

GSO Graduate School of Oceanography [of University of Rhode Island, Providence, US]

GSO Graduate Service Overseas [UK]

GSO Ground Safety Office(r)

GSO Ground Speed Oscillator

GSO Ground Support Office(r)

GSO Ground Support Operations

GSO Ground Systems Operations

GSOC German Space Operations Center [of Deutsche Forschungs– und Versuchsanstalt für Luft– und Raumfahrt e.V.]

GSOP Guidance Systems Operations Plan

GsOWDir Directory of Geoscience Organizations of the World [of Geological Survey of Japan]

g/sq cm grams per square centimeter [also g/cm^2]

GSP Generalized System of Preferences [US]

GSP General Simulation Program

GSP General Syntactic Processor

GSP Generic Server Passer

GSP Geographical Society of Philadelphia [US]

GSP Geographical South Pole

GSP Global Studies Program [of Pacific Northwest National Laboratory, Richland, Washington, US]

GSP Goodwin-Skinner-Pettifor (Model)

GSP Graphic Subroutine Package

GSP Green Star, Parachute

GSP Guidance Signal Processor

GSPA Grain Sorghum Producers Association [US]

GSPC Gas Scintillation Proportional Chamber

GSPC Gas Scintillation Proportional Counter

GSPC Graphic Standards Planning Committee

GSPN Ground Systems Pneumatics

GSPO Gemini Spacecraft Project Office [of NASA]

GSPR Guidance Signal Processor Repeater

GSR Galvanic Skin Response

GSR Gas-Solid Reaction

GSR Glide Slope Receiver

GSR Global Shared Resources

GSR Gray Scale Recording

GSR Great Swamp Research [now Great Swamp Research Institute, US]

GSR Ground Surveillance Radar

GSRI Great Swamp Research Institute [US]

GSRI Gulf South Research Institute [US]

GSRP Graduate Student Research Program

GSRS General Support Rocket System

GSRT Hellenic Ministry for Research and Technology [Greece]

GSRTST German Society for Rocket Technology and Space Travel

GSS Galvanized Steel Strand

GSS Gamma Scintillation System

GSS Geostationary Satellite

GSS Geosynchronous Satellite

GSS Geosynchronous Space Station

GSS Gichner Shelter System

GSS Global Surveillance System

GSS Government Statistics Service [UK]

GSS Graduate Student Society

GSS Graphic Service System

GSS Graphic Support Software

GSS Ground Support Software

GSS Ground Support System

GSS Gyratory Static Shear

γ-SS Austenitic Stainless Steel

GSSA Grassland Society of Southern Africa [South Africa]

GSSA Ground Support Systems Activation

GSSC General Support Services Contractor

GSSC Ground Support Simulation Computer

GSSC Ground Support Systems Contractor

GSSDAF Gatineau Satellite Station Data Acquisition Facility [Quebec, Canada]

GSSG Glutathione-S-S-Glutathione (Oxidized Form) [Biochemistry]

GSSI Government and Social Science Information

GSSI Ground Support System Integration

GSSPS Gravitationally Stabilized Solar Power System

GSSR Geologicka Sluzba Slovenskej Republiky [Geological Survey of the Slovak Republic]

GSSR Georgian Soviet Socialist Republic [USSR]

GSSW Gas-Shielded Stud Welding

GST General Sales Tax

GST General System Theory

GST Gesellschaft Schweizerischer Tierärzte [Swiss Veterinary Society]

GST Glass Science and Technology

GST Glide Slope Transmitter

GST Glutathione-S-Transferase [Biochemistry]

GST Goods and Services Tax [Canada]

GST Greenwich Sidereal Time [Astronomy]

GST Ground System Test

GSTA Ground Surveillance and Target Acquisition

GSTC Gene Strull Technology Center [Baltimore, Maryland, US]

GSTDN Ground Spacecraft Tracking and Data Network; Ground Spaceflight Tracking and Data Network

GSTDN Ground Station Tracking Data Network

GSTF Ground Systems Test Flow

GSTN General Switched Telephone Network

GSTP General Support Technology Program [Canada]

GSTP General System of Tariff Preferences

GSU Gas Separation Unit

GSU Gas Servicer Unit

GSU General Service Unit

GSU Georgia State University [Atlanta, US]

GSU Graduate Student Union

GSU Grain Services Union

GSU Group Switching Unit

GSU Guaranteed Supply Unit

GSUG Gross Seasonal Unavailable Generation

G-Suit Gravity Suit [also G-suit]

GSV Golden Shiner Virus

GSV Governor Steam Valve

GSV Ground-to-Surface Vessel (Radar)

GSV Guided Space Vehicle

GSVP Ground Support Verification Plan

GSW Galvanized Steel Wire
GSW Ground Saucer Watch [US]
GSW Gunshot Wound
GSWA Geological Survey of Western Australia
GSWR Galvanized Steel Wire Rope
GSZ Geological Society of Zimbabwe
GT Gabay-Toulouse (Line) [Spectroscopy]
GT Gadolinium Tantalate
GT Gadolinium Titanate
GT Galilean Telescope
GT Galvanothermometer
GT Game Theory
GT Gamow-Teller (Interaction) [Nuclear Physics]
GT Garnet [also Gt]
GT Gas Technologist; Gas Technology
GT Gas Thermometer; Gas Thermometry
GT Gas Torque
GT Gas Turbine
GT Gaussian Transformation
GT Geissler Tube [Electronics]
GT Gear Train
GT Gemini-Titan [of NASA]
GT Gene Therapy [Medicine]
GT Genital Tract [Anatomy]
GT Geologic Time
GT Geometric Tolerance
GT Geotectonic(s)
GT Geothermal
GT Germanium Transistor
GT Geotextile
GT Gibbs-Thompson (Equation) [also G-T]
GT Gires Tournois (Compensated Laser)
GT Glan-Thompson (Prism)
GT Glass Technology
GT Glass Transition
GT Glass Tube
GT Glover Tower
GT Glutamylpeptidase [Biochemistry]
GT Glutamyltransferase [also G-GT, or GGT] [Biochemistry]
GT Goody-Thomas (Electronegativity)
GT Gopher Tape
GT Go Trace (Command) [Pascal Programming]
GT Graphic Tablet
GT Graphics Terminal
GT Graph Theory
GT Gravitational Theory
GT Grease Trap
GT Greater Than
GT Greninger-Troiano (Orientation Relationship) [Materials Science]
GT Grid Tube
GT Griffith's Theory [Mechanics]
GT Gross Ton(s) [Unit]
GT Gross Thrust

GT Ground Team
GT Ground Test
GT Ground Transmit
GT Group Technology [Industrial Engineering]
GT Group Theory
GT Growth Transformation [Metallurgy]
GT Guatemala [ISO Code]
GT Gun Turret
GT Gyro Torque
γ-GT γ-Glutamylpeptidase [also GGTP] [Biochemistry]
G/T Gain-to-Noise Temperature (Ratio)
G/T Grams of Plutonium per (Irradiated) Ton of Uranium
G-T Gibbs-Thompson (Equation) [also GT]
G-T Conductance-Temperature (Characteristic)
G(T) Shear Modulus at Temperature "T" [Symbol]
G^A(T) Gibbs Energy of Austenite at Temperature "T" [Symbol]
G^F(T) Gibbs Energy of Ferrite at Temperature "T" [Symbol]
G^M(T) Gibbs Energy of Martensite at Temperature "T" [Symbol]
G_n(T) Volumetric Gibbs Free Energy of the Normal State (in Solid-State Physics) [Symbol]
G_s(T) Volumetric Gibbs Free Energy of the Superconducting State [Symbol]
Gt Garnet [also GT]
Gt Gigatonne [Unit]
Gt Great [also gt]
G(t) Growth Rate [Symbol]
G(θ) Conductance at Angle θ [Symbol]
gt goat [Immunochemistry]
gt Guatemala [Country Code/Domain Name]
g/t gram(s) per ton(ne) [also g t^{-1}] [Unit]
g-t gross ton [Unit]
g(t) impulse response (of control systems) [Symbol]
g_e/2 electron magnetic moment (or free electron g factor in Bohr magnetons) (1.0011597) [Symbol]
g_μ/2 muon magnetic moment (or free muon g factor in Bohr magnetons) (1.0011592) [Symbol]
$\gamma(\theta)$ angle dependent surface tension [Symbol]
$\gamma(\theta)$ misorientation dependent grain-boundary free energy [Symbol]
∇T Temperature Gradient [Symbol]
GTA Gallotannic Acid
GTA Gamma Titanium Aluminide
GTA Gas Tungsten-Arc
GTA Gemini-Titan-Agena [of NASA]
GTA Genetic Toxicology Association [US]
GTA Glass Tempering Association [US]
GTA Glutaraldehyde
GTA Glycerol Triacetate
GTA Government Telecommunications Agency [of US Department of Commerce]
GTA Grading Terminal Assembly
GTA Graphite Tube Atomizer
GTA Graph Theoretic Algorithm

GTA Ground Test Access

GTA Ground Test Article

GTA Ground Torquing Assembly,

GTA Guam Telephone Authority

GT&A Ground Test and Acceptance

GTAA Greater Toronto Airport Authority [Canada]

GTAB Ground-Plane Tape-Automated Bonding

GTAC Gas Tungsten-Arc Cutting

GTAW Gas Tungsten-Arc Welding

GTAW-P Pulsed Gas Tungsten-Arc Welding

GTAW-TIG Gas Tungsten-Arc Welding/Tungsten Inert Gas (Welding)

γ-Tb Gamma Terbium [Symbol]

GTBA Gasoline Grade Tertiary Butylacetate

GTBOS Glad To Be Of Service

Gt Br Great Britain [also Gt Brit]

GTC Gain Time Constant

GTC Gain Time Control

GTC Gas Turbine Compressor

GTC General Trading Companies

GTC Genetic Thermal Cycler

GTC Geological Testing Consultant

GTC Gifu Technical College [Japan]

GTC Glycol Trim Console

GTC Golden Technologies Company, Inc. [US]

GTC Good Till Cancelled (Order) [US]

GTC Ground Test Conductor

GTCC Group Technology Characterization Code

GTCE Global Tropospheric Chemistry Experiment

GTCU Ground Thermal Conditioning Unit

GTD Gas Turbine Division [now International Gas Turbine Center of American Society of Mechanical Engineers, US]

GTD Gaussian-Type Distribution

GTD General Technology Division [of IBM Corporation at Hopewell Junction, New York, US]

GTD Geometrical Theory of Diffraction

GTD Graphic Tablet Display

gtd guaranteed [also gt'd]

GTDI Guidelines for (International) Trade Data Interchange

GTDPL Generalized Top Down Parsing Language

GTE Gas Turbine Engine

GTE Gas Turbine Engineer(ing)

GTE General Telephone and Electronics Corporation [US]

GTE Geotechnical Engineer(ing)

GTE Geothermal Energy

GTE Global Tropospheric Experiment

GTE Gépipari Tudományos Egyesület [Hungarian Mechanical Engineering Society]

GTE Ground Transport Equipment

GTE Group Translating Equipment

GTE Autom Electr World-Wide Commun J GTE Automatic Electric World-Wide Communications Journal [Published by GTE Communication Systems, Phoenix, Arizona, US]

GTEC Geothermal Energy Conversion (System)

GTE J Sci Technol GTE Journal of Science and Technology [Published by GTE International, Waltham, Massachusetts, US]

GTEL General Telephone and Electronics Laboratories [US]

GTEP General Telephone and Electronics Practice

GTEPS General Telephone and Electronics Data Services [US]

GTF Gaussian-Type Function

GTF Generalized Trace Facility

GTF Generalized Transformation Function

GTF Glucose Tolerance Factor [Medicine]

GTF Gravity Tube Feeder

GTG Gas-to-Gasoline (Process)

GTG Gas Technology Group [of Canada Center for Mineral and Energy Technology/Energy Diversification Research Laboratory, Canada]

GTG Gas Turbine Generator

GTG Gaussian-Type Geminals

GTG Genetic Technology Grade

GTH Gas-Tight High-Pressure Method

GTH Geothermal Heating

g/24 h·m²·mm Hg gram(s) per 24 hours per square meter per millimeter of mercury [Unit]

GTHTGR Gas Turbine High-Temperature Gas-Cooled Reactor [also GT-HTGR]

GTI Glass Technical Institute [US]

GTI Grand Turk Island [Bahamas]

GTI Ground Test Instrumentation

γ-TiAl Gamma Titanium Aluminide

GTIS Gloucestershire Technical Information Service [UK]

GTIS Government Telecommunications and Information Services [Canada]

GTIS Ground-Based Traffic Information System

GTK Geologian Tutkimuskeskus [Geological Survey of Finland]

GTK Gross Tonne Kilometer [Unit]

GTL Gas-Turbine Locomotive

GTL Geis-Twichell-Lyszczarz (Device)

GTL Global Title Translation [Telecommuncations]

GTL Gunning Transceiver Logic

γ-Tl Gamma Thallium [Symbol]

GTM Gadolinium Titanate Molybdate

GTM Ground Team Manager

GTM Ground Test Missile

GTM Ground Test Motor

GTM Gyratory Testing Machine

GTM Gyratory Test Machine [of US Corps of Engineers]

GTMA Gauge and Tool Makers Association [UK]

GTN German Television News [Berlin, Germany]

GTN Global Trends Network

GTN Government Telecommunications Network [UK]

GTO Gate Turn-Off (Switch)

GTO Gaussian-Type Orbital [Physics]

GTO Government Telecommunications Organization

GTO Guide to Operations

GTOS Global Terrestrial Observing System [International Organizational Joint Venture]

GTOW Gross Takeoff Weight

GTP General Test Plan

GTP Geometry Theorem Prover

GTP Geotechnical Project

GTP Graphics Transform Package

GTP Group Transfer Polymerization

GTP Guanosine Triphosphate; Guanosine 5'-Triphosphate [Biochemistry]

G-T-P Gibbs Free Energy–Temperature–Pressure (Diagram)

5'-GTP Guanosine-5'-Triphosphate [Biochemistry]

GTPC Gene Therapy Policy Conference [Medicine]

GTPD Geotechnical Project Design

GTP-β-S Guanosine-5'-[β-Thio]triphosphate [Biochemistry]

GTP-γ-S Guanosine-5'-[γ-Thio]triphosphate [Biochemistry]

GTP-g-S Guanosine-5'-O-(3-Thiotriphosphate) [Biochemistry]

GTQ Guatamala Quetzal [Currency of Guatemala]

GTR Gas Transmission Rate

GTR General Theory of Relativity

GTR Ground Test Reactor [of US Air Force]

GTRE Gas Turbine Research Establishment [Bangalore, India]

GTRI Georgia Tech Research Institute [of Georgia Institute of Technology, Atlanta, US]

GTRI Geotechnology Research Institute [of Houston Area Research Center, Texas, US]

GTRO Geotechnical Research Office

GTRR Georgia Institute of Technology Research Reactor [Atlanta, US]

GTS Gadolinium Titanate Stannate

GTS General Technical Service

GTS General Test Support

GTS Geologic Time Scale

GTS Geostationary Technology Satellite

GTS Geotechnical Sciences

GTS Global Telecommunication System [of World Meteorological Organization]

GTS GN&C (Guidance, Navigation and Control) Test Station

GTS Goldstone Tracking Station [of NASA Jet Propulsion Laboratory, Pasadena, California, US]

GTS Ground Telemetry Subsystem

GTS Guided Transportation System

GTSL Geotechnical Science Laboratories [Canada]

GTSM Group Theory Statistical Mechanics

GTSPP Global Temperature and Salinity Pilot Project

GTST Generalized Transition-State Theory [Physics]

GTT Glass Transition Temperature

GTT Glucose Tolerance Test [Medicine]

gtt *(guttae)* – drops [Apothecary]

GTTF Gas-Turbine Test Facility [US]

GTU Group Terminal Unit

GTU Glycol Trim Unit

GTUC Ghana Trades Union Congress

GTV Gas Turbine Vessel

GTV Gate Trigger Valve

GTV Gate Valve

GTV Ground Test Vehicle

GTW Gesellschaft für Technik und Wirtschaft [Society for Technology, Industry and Commerce, Germany]

GTW Gross Takeoff Weight

GTX Geographutoxin [Biochemistry]

GTX Gonyautoxin [Biochemistry]

GTX II Geographutoxin II [Biochemistry]

GTX-1 Gonyautoxin I [Biochemistry]

GTX-2 Gonyautoxin II [Biochemistry]

GTX-3 Gonyautoxin III [Biochemistry]

GTX-4 Gonyautoxin IV [Biochemistry]

GTX-5 Gonyautoxin V [Biochemistry]

GTX-6 Gonyautoxin VI [Biochemistry]

GTX-8 Gonyautoxin VIII [Biochemistry]

epi-GTX-8 epi-Gonyautoxin VIII [Biochemistry]

GTZ Deutsche Gesellschaft für Technische Zusammenarbeit [German Society for Technical Cooperation]

GU Gastric Ulcer

GU *(Gemeinschaftskommittee)* – German for "Joint Committee"

GU Generic Unit

GU Genitourinary [Medicine]

GU Georgetown University [Washington, DC, US]

GU Gifu University [Gifuken, Japan]

GU Glycosuria [Medicine]

GU Guam (US) [ISO Code]

GU Guarantee [also Gu, or gu]

Gu Guam [US]

Gu Guanidine [Biochemistry]

.gu Guam (US) [Country Code/Domain Name]

γ-U Gamma Uranium [Symbol]

GUA Guarani [Currency of Paraguay]

Guar Guarantee(d) [also guar]

GUARDSMAN Guidelines and Rules for Data Systems Management

Guat Guatemala(n) [also Guate]

GUCCIAAC General Union of Chambers of Commerce, Industry and Agriculture for Arab Countries [Lebanon]

GUCP Ground Umbilical Carrier Plate [Aerospace]

Guer Guernsey [UK]

GUD Good [Amateur Radio]

GuD Gas und Dampf (Process) [Gas and Vapor Process] [also GUD]

GUF General University Funds

GUG Geoshare User's Group [US]

GUGA Graphical Unitary Group Approach

GUHA General Unary Hypotheses Automation

GUH Georgetown University Hospital [Washington, DC, US]

GUI Graphical User Interface

Gui Guiana; Guianese

GUID Globally Unique Identifier; Global Universal Identifier

Guid Guidance [also guid]

GUIDE Guidance for Users of Integrated Data Processing Equipment

GuidO Guidance Officer [also Guid O]

GUILD Government-University-Industry-Laboratory Development (Concept)

Guin Guinea

GUK *(Gemeinschaftsunterkommittee)* – German for "Joint Subcommittee"

GuK Guanylate Kinase [Biochemistry]

GUL GSE (Ground-Support Equipment) Utilization List

GULP General Utility Library Program

GUM Guam [Meteorological Station Designator]

GUMC Georgetown University Medical Center [Washington, DC, US]

Gummi Fas Kunstst Gummi, Fasern, Kunststoffe [German Journal on Rubber, Plastics and Fibers]

GUMP Geophysical Unit of Menlo Park [of US Geological Survey in California]

G-UMPS Gyroscopic Unmanned Profiling System

Gun Gunnery [also gun]

GUNFO Gulf United Nuclear Fuels Corporation

Guo Guanosine [Biochemistry]

GURC Gulf Universities Research Consortium [US]

GUS β-Glucuronidase [Biochemistry]

GUSB Guided Unified S-Band

GUSM Georgetown University School of Medicine [Washington, DC, US]

GUSTO Guidance Using Stable Tuning Oscillations

GUT Genitourinary Tract

GUT Grand Unification Theory [Theoretical Physics]

GUT Grand Unified Theory [Astronomy]

GUTS Gothenburg University Terminal System [Sweden]

GUTS Guaranteed Up-Time Support

GUTs Grand Unification Theories [Theoretical Physics]

gutt *(guttia)* – a drop [Apothecary]

GUUG Gross Unit Unavailable Generation

GUY (French) Guyana Space Center [of European Space Center]

Guy$ Guyana Dollar [Currency of Guyana]

GV Gas Valve

GV Gate Valve

GV General Ventilation

GV Gentian Violet [Microbiology]

GV Giammarco-Vetrocoke (Process)

GV Glicksman-Vorhees (Model) [Metallurgy]

GV Gray Value

GV Grid Voltage

GV Group Velocity

GV Guard Valve

GV Guard Vessel

GV Guide Vane

G/V Gradient/Velocity (Ratio)

G/V Conductance-Voltage (Characteristics)

G-V Gravity-Velocity

G(V) Tunneling Conductance [Symbol]

g(V) logarithmic amplitude distribution (in acoustic emission) [Symbol]

GVA Gigavolt-Ampere [Unit]

GVA GOX (Gaseous Oxygen) Vent Arm

GVA Graphic Kilovolt-Ampere Meter

GVB Gelatin Veronal Buffer

GVB Generalized Valence Bond [Physical Chemistry]

GVB^{2+} Gelatin Veronal Buffer [Biochemistry]

GVB-PP Generalized Valence Bond (with) Perfect Pairing [Physical Chemistry]

GVC Gesellschaft Verfahrenstechnik und Chemieingenieurwesen [Society for Process and Chemical Engineering; of Verein Deutscher Ingenieure, Germany]

GVC Glazed Vitrified Clay

GVCO Grants to Voluntary Conservation Organizations [of Environment Australia]

GVC-VDI Gesellschaft Verfahrenstechnik und Chemieingenieurwesen im Verein Deutscher Ingenieure [Society for Process and Chemical Engineering of the Association of German Engineers]

GVD Gyroscopic Vibration Damper

GVDW Generalized Van der Waals (Theory) [also GvdW] [Physical Chemistry]

GVFF Generalized Valence Force Field [Physics]

GVH Graft-Versus-Host [Medicine]

GVHRR Geosynchronous Very-High-Resolution Radiometer

Gvl Gravel [also gvl]

GVMR Gross Vehicle Mass Rating

GVO Gaviota [Tactical Air Navigation Station]

γ-VOPO$_4$ Gamma Vanadium Phosphate

GVP Gross Value of Production

g·V^{-1}·s^{-1} gram(s) per Volt per second [also g/V/s]

GVSC GEOVann Sand Control [Note: *GEOVann* is a Tradename]

GVSU Grand Valley State University [Allendale, Michigan, US]

GVT Global Virtual Time

GVT Gravity Vacuum Tube System

GVT Ground Vibration Test

GVTA Ground Vibration Test Article

GVW Gross Vehicle Weight

GVWR Gross Vehicle Weight Rating

GW Gas Welding

GW General Warning

GW Geostrophic Wind [Meteorology]

GW Gigawatt [Unit]

GW Glass Wool

GW Global Warming (Conference)

GW Gravitational Water [Hydrology]

GW Gravitational Wave [Relativity]

GW Gravity Wave [Fluid Mechanics]

GW Grinding Wheel

GW Gross Weight [also G Wt]

GW Ground Water

GW Ground Wave

GW Guard Wire

GW Guided Weapon

GW Guinea Bissau [ISO Code]

GW Gull Wing (Airplane)

GW Gull Wing (Leads) [Electronics]

GWA General Work Area

GWA Gewässerschutz, Wasser und Abwasser [German Publication on Water, Sewage and Water Pollution Prevention]

GWA Gesamtverband Werbeagenturen [Federation of Advertising Agencies, Germany]

GW-Basic Gee Whiz Basic (Beginner's All-Purpose Symbolic Instruction Code) [of Microsoft Corporation, US]

GWC Gippsland Waters Coalition [Australia]

GWC Global Weather Center [US Air Force]

GWC Golden West Conventions [Australia]

GWC Ground Water Council

GW/cm² Gigawatt(s) per square centimeter [also GW cm^{-2}]

GWD Gaussian Wave-Packet Dynamics

GWDB Groundwater Development Bureau [Taiwan]

GWDG Gesellschaft für wissenschaftliche Datenverarbeitung mbH Göttingen [German Scientific Data Processing Company, Goettingen]

GWE Gravity Wave Energy [Fluid Mechanics]

GWe Gigawatt, electrical [Unit]

GWe Gigawatt of Electricity

GWEN Ground Wave Energy Network

GWF Generalized Wannier Function [Solid-State Physics]

GWF Groundwater Flow

GWF Gutzwiller Wave Function [Physics]

GWGF Generalized Wiener G-Functional(s)[Physics]

GWh Gigawatthour [also GWhr]

GWh/mo Gigawatthour per month [Unit]

GWI Grinding Wheel Institute [US]

GWI Groundwater Institute [US]

GWIC Global Warming International Center

GWIC-USA Global Warming International Center–United States of America

GWL Gesellschaft für Wartung und Lüftungsanlagen mbH [German Manufacturer and Supplier of Ventilation Equipment and Maintenance Services)

GWM Guam [NASA Space Tracking and Data Network]

GWO Guided Wave Optics

G-Word Preparatory Word (Code) [Numerical Control]

GWP Gaussian Wave Packet

GWP Global Warming Potential

GWP Greenhouse Warming Potential

GWP Guinea-Bissau Peso [Currency]

GWp Gigawatt peak [Unit]

GWPC Ground Water Protection Council [US]

GWp/yr Gigawatt(s) per year [Unit]

GWR Geostrophic Wind Relation [Meteorology]

GWR Ground Wave Radar

GWRDC Grape and Wine Research and Development Corporation

GWRI Ground Water Research Institute

GWS Groundwater Surveillance

GWT Ground Winds Tower

G Wt Gross Weight [also GW]

GWTD Ground Water Technology Division [now Association of Groundwater Scientists and Engineers, US]

GWth Gigawatt, thermal [Unit]

GWU Gambia Workers Union

GWU General Workers Union

GWU George Washington University [Washington, DC, US]

GWUMC George Washington University Medical Center [Washington, DC, US]

GWUP Gesellschaft für die Wissenschaftliche Untersuchung der Parawissenschaften [Society for the Scientific Investigation of Parascience, Germany]

GWVA Great War Veterans Association (of Canada)

GWVSS Ground Wind Vortex Sensing System

γ_a(WXY) WXY Auger Transition Probability Factor (in Physics) [Symbol]

G(x) Radiation Chemical Yield [Symbol]

γX Activity Coefficient of Component "X" (in Physical Chemistry) [Symbol]

Γ(x) Gamma Function (in Mathematics) [Symbol]

GXRE Goddard X-Ray Experiment [of NASA]

GY Gaussian Year (1 GY = 365.2569 Days) [Astronomy]

GY General Yielding (of Specimen) [Mechanics]

GY Green(ish)–Yellow

GY Green-Yellow (Double Star) [Astronomy]

GY Guyana [ISO Code]

Gy Gray [Unit]

Gy Gunnery [also gy]

.gy Guyana [Country Code/Domain Name]

γ-Yb Gamma Ytterbium [Symbol]

GYD Guyana Dollar [Currency]

GYFM General Yielding Fracture Mechanics

GYM Guaymas, Mexico [Remote Site]

Gyn Gynecological; Gynecologist; Gynecology [also gyn]

Gynecol Gynecological; Gynecologist; Gynecology [also gynecol]

Gynecol Obster Invest Gynecology and Obstetric Investigation [International Journal]

Gynecol Oncol Gynecologic Oncology [Journal published in the US]

Gyp Gypsum [also gyp]

Gyp Lime Gypsum and Lime [Published in the US]

Gyr Gyration [also gyr]

Gyro Gyroscope [also gyro]

GYRO A Gyro A [also Gyro A]

Gyrocomp Gyrocompass(ing) [also gyrocomp]

GZ GNU (Gzip) Compression [also gz]

GZ Ground Zero [Atomic Physics]

.GZ GNU (Gzip) Compression [File Name Extension]

G(z) "z" Transfer Function (of Control Systems) [Symbol]

.gz Compressed File [UNIX Extension in GNU gzip Compression/Decompression Program]

GZIP GNU (Gnu's Not Unix) Zip [Computers]

GZS Gesellschaft für Zahlungssysteme [Society for Funds Transfer Systems, Germany]

GZT Greenwich Zone Time [Astronomy]

H

H Altitude (of Aircraft) [Symbol]
H Bandpass Width [Symbol]
H Chemical Engineering (Technology) [Discipline Category Abbreviation]
H Cloud Depth [Symbol]
H Differential Heating in Zonal Direction [Symbol]
H Enthalpy (or Heat Content) [Symbol]
H (Upper-case) Eta [Greek Alphabet]
H (Relative) Exposure [Symbol]
H Gap Space (Between Two Electrodes) [Symbol]
H (Magnetic) Field Strength [Symbol]
H Fraunhofer "H" Line [Extreme-Violet Fraunhofer Line for Calcium Having a Wavelength of 396.8 Nanometers] [Spectroscopy]
H Habit
H Habitat(ion)
H Hadron [Particle Physics]
H Haemonchus [Genus of Nematodes]
H Haemophilus [Genus of Bacteria]
H Haifa [Israel]
H Haiti(an)
H Halide
H Halifax [Nova Scotia, Canada]
H Halo
H Halobacterium [Microbiology]
H Halogen(ation)
H Halt
H Hamburg [Germany]
H Hamilton [Bermuda]
H Hamiltonian (Function) [Symbol]
H Hammer(ing)
H Hampton [Virginia, US]
H Hand
H Hanoi [North Vietnam]
H Hanover [Germany]
H Harbin [PR China]
H Hard

H Hard (Temper) [Metallurgy]
H Hardened; Hardening
H Hardenability; Hardenable [Metallurgy]
H Hardener
H Hardening Rate [Symbol]
H Hardness
H Hardware
H Harm(ful)
H Harmonic(s)
H Harness (Weave) [Woven Fabrics]
H Harrisburg [Pennsylvania, US]
H Hartford [Connecticut, US]
H Hash(ing)
H Hatch
H Havana [Cuba]
H Hawaii(an)
H Hazard(ous)
H Haze
H Head
H Header
H Heading
H Head of Fluid [Symbol]
H Headquarters
H Headset
H Health
H Heart
H Hearth
H Heat [also h]
H Heater; Heating
H Heat Content (or Enthalpy) [Symbol]
H Heating Effect (of Electric Current) [Symbol]
H Heating Rate [Symbol]
H Heat(ing) Value [Symbol]
H Heat of Reaction (at Constant Pressure) [Symbol]
H Heavy(weight)
H Hebrides [Scotland]
H Height [Symbol]
H Helena [Montana, US]
H Helicopter
H Helicopter [US Air Force and US Navy Symbol]
H Heliobacter [Genus of Bacteria]
H Heliport
H Helix
H Helmet
H Helsinki [Finland]
H Helminth(ic) [Biology]
H Hemisphere; Hemispherical
H Hemlock
H Hemophilia(c)
H Hemophilus [Genus of Hemophilic Bacteria]
H Hemorrhage; Hemhorrhagic
H Hemorrhoid(al)
H Henry [Unit]

H   Hepatic

H   Hepatitis [Medicine]

H   Heptagon(al)

H   Herb

H   Herbicidal; Herbicide [also h]

H   Herbivore; Herbivorous

H   Hereditary; Heredity

H   Hermetic(s)

H   Hernia

H   Heroin

H   Heterakis [Gemus of Nematodes]

H   Heterogeneity; Heterogeneous

H   Heterolysis; Heterolytic

H   Heterotroph(ic) [Biology]

H   Heuristic(s)

H   Hexa-

H   Hexadecimal

H   Hexagon(al)

H   Hexagonal (Crystal System) [Symbol]

H   Hexapole

H   Hexode

H   Hexose [Biochemistry]

H   Hickory

H   Hierarchical; Hierarchy

H   High [also h]

H   High Address Byte [also HI]

H   Himalayas [Nepal]

H   Hiroshima [Japan]

H   Histidine; Histidyl [Biochemistry]

H   Histological; Histologist; Histology

H   Histomonas [Genus of Flagellate Protozoans]

H   Histoplasma [Genus of Fungi]

H   Hit

H   Hit(ting) Rate [Symbol]

H H_2O (Water) [Ceramics]

H   Hobart [Tasmania]

H   Hoist(ing)

H   Hokkaida [Japan]

H   Hole

H   Holland

H   Hollandite [Mineral]

H   Hollywood [California, US]

H   Hologram; Holographic; Holography

H   Holozoic [Zoology]

H   Holzknecht Unit [X-Ray Technology]

H   Home

H   Homing

H   Homogeneity; Homogeneous

H   Homolog [Chemistry]

H   Homologous; Homology

H   Homology [Mathematics]

H   Homolysis; Homolytic

H   Honduran; Honduras

H   Honolulu [Hawaii, US]

H   Honshu [Japan]

H   Hoop

H   Horizon(tal) [also h]

H   Hormonal; Hormone

H   Horn

H   Horsepower Rating [Symbol]

H   Horticultural; Hortoculture

H   Hospital

H   Host

H   Hot [also h]

H   Hotness

H   Hour(s)

H   House

H   Housing

H   Houston [Texas, US]

H   Howitzer

H   Human(ity) [also h]

H   Humerus [Anatomy]

H   Humid(ity)

H   Humidification; Humdifier

H   Humor [Anatomy]

H   Hungarian; Hungary

H   Huntsville [Alabama, US]

H   Husband(ry)

H   Hyderabad [India and Pakistan]

H   Hybrid(ization)

H   Hybridoma [Immunology]

H   Hydra [Zoology]

H   Hydrant

H   Hydraulic(s)

H   Hydro- [Prefix]

H   Hydrodynamic(s)

H   Hydrofoil

H   Hydrogen [also ^1H, or H-1] [Symbol]

H   Hydrogenated; Hydrogenation

H   Hydrophilic(s)

H   Hydrophobic(s)

H   Hymen [Anatomy]

H   Hymenolepis [Genus of Tapeworms]

H   Hyostrongylus [Genus of Nematode]

H   Hypha [Mycology]

H   Hyperfine Field (in Solid-State Physics) [Symbol]

H   Hypothesis; Hypothetical

H   Hypothalamus [Anatomy]

H   Hysteresis; Hysteretic

H   Hysteric(s); Hysteria

H   Irradiance (in Spectroscopy) [Symbol]

H   Linewidth [Symbol]

H   Magnetic Field (Intensity) [Symbol]

H   Magnetic Field Strength (or Magnetizing Force) [Symbol]

H   Magnetic Field Vector [Symbol]

H   Momentum [Symbol]

H   Product of Inertia [Symbol]

H   Radiant Exposure [Symbol]

H   Sensible Heat Flux [Symbol]

H   Shear Layer Depth [Symbol]

H   Soil Stratum Thickness When Drained on One Side (Soil Consolidation Theory) [Symbol]

H   Strain Hardened (Wrought Products Only) [Basic Temper Designation for Aluminum Alloys]

H   Maximum Tire Speed of 130 mph (or 210 km/h) [Rating Symbol]

H   Total Heat (or Enthalpy) [Symbol]

H-   Search and Rescue (Aircraft) [USDOD Symbol]

H-   Silo-Stored (Vehicle) [USDOD Missile Symbol]

-H   Ambulance (Aircraft) [US Navy Suffix]

.H   Header, or Include File [C File Name Extension] [also .h]

$\hat{H}$ Frontal-Plane Projection of Longitudinal Axis of Heart (in Electrocardiography) [Symbol]

$\overline{H}$ Molar Enthalpy [Symbol]

$\overline{H}$ Enthalpy of Mixing [Symbol]

$\hat{H}$ Quantum Harmonic Oscillator [Symbol]

H*   Pentration Field (of Superconductor) [Symbol]

H′   Apparent Magnetizing Force [Symbol]

H′   Fraunhofer Line for Calcium Having a Wavelength of 393.4 Nanometers [Spectroscopy]

H$^+$ Hydrogen-Ion (Concentration) [Symbol]

H$^-$ Hydride [Symbol]

H$_\parallel$ Magnetic Field Applied in Parallel Direction Symbol]

H$_\perp$ Magnetic Field Applied in Perpendicular Direction [Symbol]

H$_0$ Hammett Acidity Function [Symbol]

H$_0$ Harmonic Hamiltonian (in Physics) [Symbol],

H$_0$ Instrinsic Stress Hardening [Symbol]

H$_0$ Magnetic Field at Absolute Zero [Symbol]

H$_0$ Modulation Amplitude [Symbol]

H$_0$ Static Magnetic Field [Symbol]

H$_0$ Stress Induced Anisotropy [Symbol]

H$_0$ Unperturbed Hamiltonian (in Physics) [Symbol]

H$_0$ Zero-Order Hamiltonian (in Physics) [Symbol]

H^1 Hydrogen-1 (or Protium) [also ^{1}H, or H-1]

H^{1+} Proton [also P]

H^2 Hydrogen-2 (or Deuterium) [also ^{2}H, H-2, or D]

H$_2$ Histamine (Neurotransmitter) [Symbol]

H$_2$ (Molecular) Hydrogen [Symbol]

H^3 Hydrogen-3 (or Tritium) [also ^{3}H, H-3, or T]

H$_A$ Antiferromagnetic Anisotropy Constant [Symbol]

H$_a$ Anisotropy Field [Symbol]

H$_a$ Applied Magnetic Field [Symbol]

H$_a$ Applied Magnetizing Force [Symbol]

H$_a$ Magnetic Field Strength between Pole Shoes (of an Electric Machine), or in Air Gap (of a Magnetic Circuit) [Symbol]

H$^\alpha$ Alpha Hydrogen [also α-H] [Symbol]

H$_\alpha$ Fraunhofer "H$_\alpha$" Line [Strong-Red Fraunhofer Line for Hydrogen Having a Wavelength of 766.1 Nanometers] [Spectroscopy]

$\hat{H}_\alpha$ Non-Integratable Two-Spin Hamiltonian [Symbol]

H$_{ac}$ Alternating-Current Magnetic Field [Symbol]

H$_{app}$ Apparent Magnetic Field [Symbol],

H$_{app}$ Applied Magnetic Field [Symbol]

H$_b$ Biasing Magnetizing Force [Symbol]

H$_b$ Hard-Axis Bias Magnetic Field [Symbol]

H$^\beta$ Beta Hydrogen [also β-H] [Symbol]

H$_\beta$ Fraunhofer "H$_\beta$" Line [Medium-Blue-Green Fraunhofer Line for Hydrogen Having a Wavelength of 486.1 Nanometers [Spectroscopy]

H$_{BN}$ Barkhausen Noise Amplitude [Symbol]

H$_c$ (Normal Magnetic) Coercive Field [Symbol]

H$_c$ Coercive Force (or Coercivity) [Symbol]

H$_c$ Critical Height (of Soil Embankments) [Symbol]

H$_c$ Critical Magnetic Field (for Superconductivity) [Symbol]

H$_c$ Hardness of Composite Materials [Symbol]

H$_c$ Thermodynamic Critical Field [Symbol]

H$_c$′   Apparent Coercive Force [Symbol]

H$_{c\parallel}$ In-Plane (or Parallel) Coercivity (or Coercive Force) [Symbol]

H$_{c\perp}$ Perpendicular Coercivity (or Coercive Force) [Symbol]

H$_{c1}$ Lower Critical (Magnetic) Field (of a Type II Superconductor) [Symbol]

H$_{c2}$ Upper Critical (Magnetic) Field (of a Type II Superconductor) [Symbol]

H$_{ce}$ Easy-Axis (Magnetic) Coercivity [also H$_{c,e}$] [Symbol]

H$_{ch}$ Hard-Axis (Magnetic) Coercivity [also H$_{c,h}$] [Symbol]

H$_{ci}$ Intrinsic (Magnetic) Coercive Force (or Intrinsic Coercivity) [Symbol]

H$_{ci}$ Intrinsic Coercivity (or Coercive Force) [Symbol]

H$_{co}$ Residual Coercivity (or Coercive Force) [Symbol]

H$_{cr}$ Crossover Field (of Superconductor)

H$_{cs}$ (Magnetic) Coercivity [Symbol]

H^D Self-Magnetostatic Field [Symbol]

H$_d$ Demagnetizing Force (or Field) [Symbol]

H$_d$ Permeance of Space Occupied by Magnet [Symbol]

H$_\Delta$ Incremental Magnetizing Force [Symbol]

H$_\Delta$′   Incremental Apparent Magnetizing Force [Symbol]

H$_\delta$ Fraunhofer "H$_\delta$" Line [Weak-Extreme Violet Fraunhofer Line for Hydrogen Having a Wavelength of 410.2 Nanometers] [Spectroscopy]

H$_{dc}$ Direct-Current Magnetic Field [Symbol]

H$_{demag}$ Demagnetizing Field [Symbol]

H$_E$ Antiferromagnetic Exchange Field [Symbol]

H$_e$ Hamiltonian of Electrons [Symbol]

H$_e$ Hamiltonian of Excited State [Symbol]

H$_e$ Radiant Exposure (at a Point on a Surface) [Symbol]

H$_{easy}$ Easy-Axis Bias Magnetic Field [Symbol]

H$_{eL}$ Electron-Lattice Hamiltonian [Symbol]

H$_{eff}$ Effective Heat Transfer Coefficient [Symbol]

H$_{eff}$ Effective Magnetic Field [Symbol]

H$_{ext}$ External Magnetic Field [Symbol]

H$_f$ Formation Enthalphy (or Enthalpy of Formation) [Symbol]

H$_f$ Heat of Fusion [Symbol]

H_f Ferromagnetic Coupling Field [Symbol]

H_g Air-Gap Magnetizing Force [Symbol]

H_g Enthalpy of Gas [Symbol]

H_g Hamiltonian of Ground State [Symbol]

H_γ Fraunhofer "H_γ" Line [Weak-Violet Fraunhofer Line for Hydrogen Having a Wavelength of 434.1 Nanometers [Spectroscopy]

$\hat{H}_\gamma$ Quantum Harmonic Oscillator [Symbol]

H_{gs} Grain Size Hardening (in Metallurgy) [Symbol]

H_{hf} Hyperfine (Magnetic) Field [Symbol]

H_i Indicated Altitude (of Aircraft) [Symbol]

H_i Magnetic Field Strength in Iron Core (of a Magnetic Circuit) [Symbol]

H_{ij} Heisenberg Hamiltonian (in Physics) [Symbol]

H_{int} Internal Magnetic (or Magnetizing) Field [Symbol]

H_j Energy of Jog Formation [Symbol]

H_K Enthalpy of Mixing [Symbol]

H_K Uniaxial Anisotropy Field [Symbol]

H_k Effective Magnetic Anisotropy Field [Symbol]

H_{KB} Bulk (Uniaxial) Anisotropy Field [Symbol]

H_{KS} Surface (Magnetic) Anisotropy Field [Symbol]

H_L Alternating-Current Magnetizing Force in Terms of Peak Magnetizing Current [Symbol]

H_L Lattice Hamiltonian [Symbol]

H_l Enthalpy of Liquid [Symbol]

H_{loc} Local Magnetic Field [Symbol]

H^M Enthalpy of Mixing [also H^m][Symbol]

H_m Magnetizing Force [Symbol]

H_m Migration Enthalpy [Symbol]

H_m^i Migration Enthalpy of Interstitial [Symbol]

H_m^v Migration Enthalpy of Vacancy [Symbol]

H_{max} Maximum Magnetic Field (or Maximum Magnetizing Force) [Symbol]

H'_{max} Maximum Apparent Magnetizing Force [Symbol]

H^N N-Particle Hamiltonian (in Physics) [Symbol]

H_N Nucleation Field [also H_n] [Symbol]

H_o Ferromagnetic Offset Field [Symbol]

H_o Octahedrallly-Coordinated Hydrogen [Symbol]

H_p Alternating-Current Magnetizing Force in Terms of Measured Peak Exciting Current [Symbol]

H_p Pentration Field (of Superconductor) [Symbol]

H_p Propagation Field [Symbol]

H_{pp} Peak-to-Peak Linewidth (in Spectroscopy) [Symbol]

H_R Hruby Factor (of Glass Stability) [Symbol]

H_R Magnetic Resonance Field [Symbol]

H_R Pro-R Enantiotropic Hydrogen [Symbol]

H_r Heat of Reaction [Symbol]

H_{rf} Radio-Frequency Magnetic Field [Symbol]

H_{rms} (Root-Mean-Square) Microroughness [Symbol]

H_S Magnetic Shift Field [Symbol]

H_S Pro-S Enantiotropic Hydrogen [Symbol]

H_s Generalized Enthapy [Symbol]

H_s Hardness of Substrate [Symbol]

H_s Heat of Sublimation [Symbol]

H_s (Magnetic) Saturation Field [Symbol]

$H_{s\parallel}$ (Magnetic) Saturation for Parallel Field [Symbol]

$H_{s\perp}$ (Magnetic) Saturation for Perpendicular Field [Symbol]

H_{sh} Strain (or Work) Hardening (in Metallurgy) [Symbol]

H_{SK} Enthalpy of Segregation (in Metallurgy) [Symbol]

H_{SO} Spin-Orbit Coupling (in Quantum Mechanics) [Symbol]

H_{ss} Solid-Solution Hardening (in Metallurgy) [Symbol]

H_t Instantaneous Magnetizing Force [Symbol]

H_t True Airspeed [Symbol]

H_{th} Threshold Magnetic Field [Symbol]

H_V Vickers Hardness [also H_v] [Symbol]

H_v Enthalpy of Vapor [Symbol]

H_W Weiss (Magnetic) Field [Symbol]

H_Z Zeeman Magnetostatic Energy [Symbol]

H_z Alternating-Current Magnetizing Force in Terms of Root-Mean-Square Exciting Current [Symbol]

H1 Strain-Hardened Only (Wrought Products) [Temper Designation for Aluminum Alloys]

H1-H19 Chromium-Type Hot-Work Tool Steels [AISI-SAE Symbol]

H2 Strain-Hardened and Partially Annealed (Wrought Products) [Temper Designation for Aluminum Alloys]

H3 Strain-Hardened and Stabilized by Low Temperature Thermal Treatment (Wrought Products) [Temper Designation for Aluminum Alloys]

H10 Strain-Hardened [Temper Designation for Magnesium Alloys]

H11 Strain-Hardened [Temper Designation for Magnesium Alloys]

H20-H39 Tungsten-Type Hot-Work Tool Steels [AISI-SAE Symbol]

H23 Strain-Hardened and Annealed [Temper Designation for Magnesium Alloys]

H24 Strain-Hardened and Annealed [Temper Designation for Magnesium Alloys]

H26 Strain-Hardened and Annealed (Wrought Products) [Temper Designation for Magnesium Alloys]

H40-H59 Molybdenum-Type Hot-Work Tool Steels [AISI-SAE Symbol]

H-1 Hydrogen-1 (or Protium) [also ^1H, H^1, or H]

H-2 Hydrogen-2 (or Deuterium) [also ^2H, H^2, or D]

H-2 Deuterated (or Deuterium-Treated) [also ^2H]

H-3 Hydrogen-3 (or Tritium) [also ^3H, H^3, or T]

H-3 Tritiated (or Tritium-Treated) [also ^3H]

2H Two Hours

2H Two-Wheel Drive Position

2H- 2H Polytype [Ramsdell Notation: 2 Refers to Number of Layers Necessary to Produce a Unit Cell, and H Refers to the Hexagonal Symmetry, e.g., 2H-SiC]

4H Four-Wheel Drive Position, High-Range (Power Delivered for Increased Traction)

4H- 4H Polytype [Ramsdell Notation: 4 Refers to Number of Layers Necessary to Produce a Unit Cell, and H Refers to the Hexagonal Symmetry, e.g., 4H-SiC]

6H- 6H Polytype [Ramsdell Notation: 6 Refers to Number of Layers Necessary to Produce a Unit Cell, and H Refers to the Hexagonal Symmetry, e.g., 6H-SiC]

H   Product of Inertia [Symbol]

H   Radiant Exposure [Symbol]

H   Sensible Heat Flux [Symbol]

H   Shear Layer Depth [Symbol]

H   Soil Stratum Thickness When Drained on One Side (Soil Consolidation Theory) [Symbol]

H   Strain Hardened (Wrought Products Only) [Basic Temper Designation for Aluminum Alloys]

H   Maximum Tire Speed of 130 mph (or 210 km/h) [Rating Symbol]

H   Total Heat (or Enthalpy) [Symbol]

H-   Search and Rescue (Aircraft) [USDOD Symbol]

H-   Silo-Stored (Vehicle) [USDOD Missile Symbol]

-H   Ambulance (Aircraft) [US Navy Suffix]

.H   Header, or Include File [C File Name Extension] [also .h]

$\hat{H}$ Frontal-Plane Projection of Longitudinal Axis of Heart (in Electrocardiography) [Symbol]

$\overline{H}$ Molar Enthalpy [Symbol]

$\overline{H}$ Enthalpy of Mixing [Symbol]

$\hat{H}$ Quantum Harmonic Oscillator [Symbol]

H*   Pentration Field (of Superconductor) [Symbol]

H′   Apparent Magnetizing Force [Symbol]

H′   Fraunhofer Line for Calcium Having a Wavelength of 393.4 Nanometers [Spectroscopy]

H^+ Hydrogen-Ion (Concentration) [Symbol]

H^- Hydride [Symbol]

$H_{\parallel}$ Magnetic Field Applied in Parallel Direction Symbol]

$H_{\perp}$ Magnetic Field Applied in Perpendicular Direction [Symbol]

H_0 Hammett Acidity Function [Symbol]

H_0 Harmonic Hamiltonian (in Physics) [Symbol],

H_0 Instrinsic Stress Hardening [Symbol]

H_0 Magnetic Field at Absolute Zero [Symbol]

H_0 Modulation Amplitude [Symbol]

H_0 Static Magnetic Field [Symbol]

H_0 Stress Induced Anisotropy [Symbol]

H_0 Unperturbed Hamiltonian (in Physics) [Symbol]

H_0 Zero-Order Hamiltonian (in Physics) [Symbol]

H^1 Hydrogen-1 (or Protium) [also ^1H, or H-1]

H^{1+} Proton [also P]

H^2 Hydrogen-2 (or Deuterium) [also ^2H, H-2, or D]

H_2 Histamine (Neurotransmitter) [Symbol]

H_2 (Molecular) Hydrogen [Symbol]

H^3 Hydrogen-3 (or Tritium) [also ^3H, H-3, or T]

H_A Antiferromagnetic Anisotropy Constant [Symbol]

H_a Anisotropy Field [Symbol]

H_a Applied Magnetic Field [Symbol]

H_a Applied Magnetizing Force [Symbol]

H_a Magnetic Field Strength between Pole Shoes (of an Electric Machine), or in Air Gap (of a Magnetic Circuit) [Symbol]

H^a Alpha Hydrogen [also α-H] [Symbol]

H_α Fraunhofer "H_α" Line [Strong-Red Fraunhofer Line for Hydrogen Having a Wavelength of 766.1 Nanometers] [Spectroscopy]

$\hat{H}_\alpha$ Non-Integratable Two-Spin Hamiltonian [Symbol]

H_{ac} Alternating-Current Magnetic Field [Symbol]

H_{app} Apparent Magnetic Field [Symbol]

H_{app} Applied Magnetic Field [Symbol]

H_b Biasing Magnetizing Force [Symbol]

H_b Hard-Axis Bias Magnetic Field [Symbol]

H^β Beta Hydrogen [also β-H] [Symbol]

H_β Fraunhofer "H_β" Line [Medium-Blue-Green Fraunhofer Line for Hydrogen Having a Wavelength of 486.1 Nanometers [Spectroscopy]

H_{BN} Barkhausen Noise Amplitude [Symbol]

H_c (Normal Magnetic) Coercive Field [Symbol]

H_c Coercive Force (or Coercivity) [Symbol]

H_c Critical Height (of Soil Embankments) [Symbol]

H_c Critical Magnetic Field (for Superconductivity) [Symbol]

H_c Hardness of Composite Materials [Symbol]

H_c Thermodynamic Critical Field [Symbol]

$H_c{}'$ Apparent Coercive Force [Symbol]

$H_{c\parallel}$ In-Plane (or Parallel) Coercivity (or Coercive Force) [Symbol]

$H_{c\perp}$ Perpendicular Coercivity (or Coercive Force) [Symbol]

H_{c1} Lower Critical (Magnetic) Field (of a Type II Superconductor) [Symbol]

H_{c2} Upper Critical (Magnetic) Field (of a Type II Superconductor) [Symbol]

H_{ce} Easy-Axis (Magnetic) Coercivity [also $H_{c,e}$] [Symbol]

H_{ch} Hard-Axis (Magnetic) Coercivity [also $H_{c,h}$] [Symbol]

H_{ci} Intrinsic (Magnetic) Coercive Force (or Intrinsic Coercivity) [Symbol]

H_{ci} Intrinsic Coercivity (or Coercive Force) [Symbol]

H_{co} Residual Coercivity (or Coercive Force) [Symbol]

H_{cr} Crossover Field (of Superconductor)

H_{cs} (Magnetic) Coercivity [Symbol]

H^D Self-Magnetostatic Field [Symbol]

H_d Demagnetizing Force (or Field) [Symbol]

H_d Permeance of Space Occupied by Magnet [Symbol]

H_Δ Incremental Magnetizing Force [Symbol]

$H_\Delta{}'$ Incremental Apparent Magnetizing Force [Symbol]

H_δ Fraunhofer "H_δ" Line [Weak-Extreme Violet Fraunhofer Line for Hydrogen Having a Wavelength of 410.2 Nanometers] [Spectroscopy]

H_{dc} Direct-Current Magnetic Field [Symbol]

H_{demag} Demagnetizing Field [Symbol]

H_E Antiferromagnetic Exchange Field [Symbol]

H_e Hamiltonian of Electrons [Symbol]

H_e Hamiltonian of Excited State [Symbol]

H_e Radiant Exposure (at a Point on a Surface) [Symbol]

H_{easy} Easy-Axis Bias Magnetic Field [Symbol]

H_{eL} Electron-Lattice Hamiltonian [Symbol]

H_{eff} Effective Heat Transfer Coefficient [Symbol]

H_{eff} Effective Magnetic Field [Symbol]

H_{ext} External Magnetic Field [Symbol]

H_f Formation Enthalphy (or Enthalpy of Formation) [Symbol]

H_f Heat of Fusion [Symbol]

h_{ic} common-collector small-signal shortcircuit input impedance [Semiconductor Symbol]

h_{IE} common-emitter static input resistance [Semiconductor Symbol]

h_{ie} common-emitter small-signal short-circuit input impedance [Semiconductor Symbol]

$h_{ie(imag)}$ imaginary part of common-emitter smallsignal short-circuit input impedance [Semiconductor Symbol]

$h_{ie(real)}$ real part of common-emitter small-signal short-circuit input impedance [Semiconductor Symbol]

h_{JT} Jahn-Teller coupling term (in physical chemistry) [Symbol]

h_o object height (in optics) [Symbol]

h_{ob} common-base small-signal open-circuit output admittance [Semiconductor Symbol]

h_{oc} common-collector small-signal opencircuit output admittance [Semiconductor Symbol]

h_{oe} common-emitter small-signal open-circuit output admittance [Semiconductor Symbol]

$h_{oe(imag)}$ imaginary part of common-emitter smallsignal open-circuit output admittance [Semiconductor Symbol]

$h_{oe(real)}$ real part of common-emitter small-signal open-circuit output admittance [Semiconductor Symbol]

h_{rb} common-base small-signal open-circuit reverse voltage transfer ratio [Semiconductor Symbol]

h_{rc} common-collector small-signal opencircuit reverse voltage transfer ratio [Semiconductor Symbol]

h_{re} common-emitter small-signal open-circuit reverse voltage transfer ratio [Semiconductor Symbol]

$\mathcal{H}$ Direct-Current Electric Conductivity [Symbol]

$\mathcal{H}$ Hamiltonian (Operator) [Symbol]

$\mathcal{H}_\infty$ Direct-Current Electric Conductivity of Bulk Material [Symbol]

$\mathcal{H}_l$ Direct-Current Electric Conductivity of Fiber, or Thin Layer [Symbol]

$\mathcal{H}$ Tight-Binding Hartree Hamiltonian (in Physics) [Symbol]

$\mathcal{H}_{1e}$ One-Electron Hartree Hamiltonian (in Physics) [Symbol]

$\mathcal{H}_{e-e}$ Electron-Electron Hartree Hamiltonian (in Physics) [Symbol]

$\hbar$ Planck's Constant Divided by 2π ($1.0547266 \times 10^{-34}$ J·s) [Symbol]

HA Apogee Altitude

HA Hadronic Atom

HA Half Add(er)

HA Hall Accelerator [Plasma Physics]

HA Hall Angle [Electronics]

HA Haloalkane

HA Halstead Arsenal [UK]

HA Hard Anodizing

HA Hard Automation

HA Hard Axis (of Magnetization) [also ha]

HA Harmonic Analysis; Harmonic Analyzer

HA Harmonic Approximation

HA Hartree Approximation (Method) [Physics]

HA Hazard Analysis

HA Head Amplifier

HA Heat Absorber; Heat Absorbing; Heat Absorption

HA Heat-Activated; Heat Activation

HA Heat-Actuated; Heat Actuation

HA Heavy Artillery

HA Heavy-Atom (Method) [Physics]

HA Health Advisory

HA Health Authority [of National Health Service, UK]

HA Heine-Abarenkov (Pseudopotential) [Physics]

HA Helix Aspersa [Biochemistry]

HA Hemadsorption (Virus)

HA Hemagglutinin [Immunology]

HA Hepatic Artery [Anatomy]

HA Hepatitis A [Medicine]

HA Heptoic Aldehyde

HA Heterophile Antibody [Immunology]

HA Hierachical Approach

HA High Accuracy

HA High Africa(n)

HA High Altitude

HA High-Ambient

HA High Angle (Grain Boundary) [Materials Science]

HA High Asia(n)

HA High Availability

HA Hindered Amine(s)

HA Hiperco Alloy [*Note:* Hiperco is a Trade Name for High Magnetic Permeability Iron-Cobalt Alloys]

HA Historical Association [UK]

HA Hoe Automation

HA Hollow Anode

HA Home Address

HA Horizontal Alignment

HA Horse Artillery

HA Hot Air

HA Hour Angle [Astronomy]

HA Humic Acid

HA Hyaluronic Acid [Biochemistry]

HA Hydraulic Actuator

HA Hydraulic Amplifier

HA Hydraulic Association [UK]

HA Hydroaromatic(s)

HA Hydrogen Absorption

HA Hydroxamic Acid Ester

HA Hydroxyalkanoate

HA Hydroxyapatite [also HAP], or HAp]

HA- Hungary [Civil Aircraft Marking]

(±)-HA-966 (±)-3-Amino-1-Hydroxy-2-Pyrrolidone

(–)-HA-966 (–)-3-Amino-1-Hydroxy-2-Pyrrolidone

2HA Polyelectrolytic Acid

3HA 3-Hydroxyalkanoate

Ha Hahnium [Element 105]

Ha Hartmann Number (of Fluid Flow) [Symbol]

ha hard axis (of magnetization) [also HA]

ha hectare [Unit]

ha *(hoc anno)* – this year

HAA Acetylacetone

HAA Heavy Anti-Aircraft Artillery

HAA Height Above Airport

HAA Helicopter Association of America [US]

HAA Helix Aspersa Agglutinin [Immunochemistry]

HAA High Altitude Abort

HAA Historic Aircraft Association [UK]

HAA Hydroaromatic Compound

HAADF High-Angle Annular Darkfield (Microscopy)

HAART Highly Active Antiretroviral Therapy

HAATC High-Altitude Air Traffic Control

HAB Hanford Advisory Board [of US Department of Energy in Washington State]

HAB High Altitude Bombing

HAB High-Angle (Grain) Boundary [Materials Science]

HAB Home Address Block

HAb Heterophile Antibody [Immunology]

Hab Habitat(ion) [also hab]

H∥ab Magnetic Field Parallel to a-b Plane [Symbol]

H⊥ab Magnetic Field Perpendicular to a-b Plane [Symbol]

HABA 4'-Hydroxyazobenzene-2-Carboxylic Acid

HABA 2-(4-Hydroxyphenylazo)benzoic Acid

Habil Habilitation

habil *(habilitatus doctor)* – doctor with postdoctoral lecturing qualification

HABITAT Habitat and Human Settlements Foundation [of United Nations]

Habitat Int J Habitat International Journal [Published by the World Environment and Resources Council, Belgium]

HAC Hague Arbitration Convention

HAC Heading Alignment Circle

HAC Heading Alignment Cone

HAC Heading Alignment Cylinder

HAC Heat-Absorption Coefficient

HAC Herbicide Assessment Commission [US]

HAC High-Alumina Cement

HAC High-Alumina Ceramics

HAC Historical Artillery Corps [UK]

HAC Honorary Advisory Council

HAC Horizontal Aperture Correction

HAC Horticultural Advisory Council [UK]

HAC Hughes Aircraft Company [Los Angeles, California, US]

HAC Hydraulic Asphalt Concrete

HAC Hydrogen-Assisted Cracking

HAC Hydroxyapatite Crystal

HAc Acetic Acid [also Hac]

H-ACAC Acetylacetone [also H-acac, H-Acac, or H-AcAc]

HACCP Hazard Analysis and Critical Control Point

H-Acid 1-Amino-8-Hydroxy-2,7-Naphthalenedisulfonic Acid [also H-acid]

H-Acid 1-Amino-8-Naphthol-3,6-Disulfonic Acid [also H-acid]

HACL Harward Air Cleaning Laboratory [of Harvard University, Cambridge, Massachusetts, US]

HACMP High-Availability Cluster(ing) Multiprocessing

HACSIR Honorary Advisory Council for Scientific and Industrial Research [Canada]

HAD Half-Amplitude (Pulse) Duration

HAD Heat-Actuated Device

HAD High-Altitude Density (Rocket)

HAD High-Aluminum Defect [Metallurgy]

HAD Historical Astronomy Division [of American Astronomical Society, US]

HAD Hollow Anode Discharge

HAd Adrenaline [also Had]

HADC Holloman Air Development Center [at Holloman Air Force Base, near Alamogordo, New Mexico, US]

HADES High Acceptance Di-Electron Spectrometer (Experiment) [of Gesellschaft für Schwerionenforschung mbH, Darmstadt, Germany]

HADES Hypersonic Air Data Entry System

HADN Hexamine Dinitrate

HADR Hughes Air Defense Radar

HADRON Workshop on Hadron Physics in the 21st Century [Washington, DC, US]

Hadronic J Hadronic Journal [US]

Hadronic J Suppl Hadronic Journal Supplement [US]

HADS Hypersonic Air Data Sensor

HADS Hypersonic Air Data System

HAE Hot-Air Engine

Hae Haemophilus aegyptius [Microbiology]

Hae II Haemophilus aegyptius II [Microbiology]

Haemostas Haemostasis [International Medical Journal]

HAES Hawaii Agricultural Experiment Station [US]

HAES Helium(+)-Excited Auger-Electron Spectroscopy

HAES High Altitude Effects Simulation

HAF Heissenberg Antiferromagnet(ism) [Solid-State Physics]

HAF High-Abrasion Furnace (Black)

HAF High Altitude Fluorescence

HAF High-Alumina Fiber

HAFB Holloman Air Force Base [near Alamogordo, New Mexico, US]

HAF Black High-Abrasion Furnace Black [also HAF black]

HAFID Hydrogen Atmosphere Flame Ionization Detector

HAFM Helium Accumulation Fluence Monitor [Nuclear Reactors]

HAG Hold for Arrival of Goods

HAg Human Antigen [Immunology]

HAGB High-Angle Grain Boundary [Materials Science]

HAHA Hartmann-Hahn (Spectroscopy)

HAI Helicopter Association International [US]

HAI Hemagglutination Inhibition (Assay) [Immunology]

Hai Haiti(an)

HAIC Hearing Aid Industry Conference

HAIC Hetero-Atom in Context

HAINS High-Accuracy Inertial Navigation System

HAIRS High-Altitude Infrared Source

HAIT Hash Algorithm Information Table

HAK Hatch Access Kit [Aerospace]

HAK Horizontal Access Kit

Hakim Beyisr Hakimekai Beyisrael [Publication of the Israel Chemical Society]

HAL Hard Array Logic

HAL Hardware Abstraction Layer

HAL Harwell Automated Library [of United Kingdom Atomic Energy Authority]

HAL Heuristically Programmed Algorithmic (Computer)

HAL High-Accuracy Linear (Positioning System)

HAL High-Activity Locations

HAL Highly-Automated Logic

HAL High-Order Articulated Language

HAL High-Order Assembly Language

HAL Hindustan Aeronautics Limited [Bangalore, India]

HAL Hot Acid Leaching (Process) [Chemical Engineering]

HAL House-Programmed Array Logic

HAL Houston Aerospace Language [of NASA]

Hal Halide [also hal]

Hal Halogen [also hal]

HALDIS Halifax and District Information Service [UK]

HALE High-Altitude Long-Endurance (Aircraft)

HALL ED Hall Effect Device [also Hall ED]

HALO High-Altitude, Low-Opening Parachute

HALO High-Attitude Large Optics

HALOE Halogen Occultation Experiment

HALOMET Halogen Metal (Process)

HALS Hindered Amines Liquid Stabilizer

HAL/S High-Order Assembly Language/Shuttle Computer; High-Order Assembly Language for Shuttle (Flight) [NASA Space Shuttle Program]

HALSIM Hardware Logic Simulator

HALT Hydrate Addition at Low Temperature (Process)

HAM (Radio) Amateur [also Ham]

HAM Hardware Associative Memory

HAM Height Adjustment Maneuver

HAM Hierarchical Access Method [Computers]

HAM Hold-and-Modify Image

HAM Hydrogenic Atoms in Molecules

ham hamster [Immunochemistry]

Hamb Ärztebl Hamburger Ärzteblatt [Medical Journal published in Hamburg, Germany]

HAMCHAM Haitian-American Chamber of Commerce and Industry

HAMCHAM Honduran-American Chamber of Commerce

HAM-D Hamilton Depression Rating Scale [Psychology]

HAMMER Hanford's Hazardous Materials Management of Emergency Resources [of US Department of Energy in Washington State]

HAMP Heine-Abarenkov Model Pseudopotential [Physics]

HAMS High-Altitude Mapping System

HAMS Hour Angle of the Mean Sun [Astronomy]

HAMT Human-Aided Machine Translation

HAMTC Hanford Atomic Metal Trades Council [of US Department of Energy in Washington State]

HANDS High-Altitude Nuclear Detection Studies [of National Institute for Standards and Technology, US]

HANE High-Altitude Nuclear Effects

HANES Health and Nutrition Examination Survey [US]

HANFO Heavy Ammonium Nitrate and Fuel Oil

HAN/LCD Hybrid Assigned Nematic/Liquid Crystal Display

HANP Homeopathic Academy of Naturopathic Physicians [US]

H Antigen Flagellar Antigen [also H antigen] [Microbiology]

HAO High-Altitude Observatory [of National Center for Atmospheric Research, US]

HAOSS High-Altitude Orbital Space Station

HAP Hardware Allocation Panel

HAP Hazardous Air Pollutants

HAP High-Altitude Photography

HAP High-Altitude Platform [Aerospace]

HAP Host Access Protocol

HAp Hydroxyapatite [also HAP, or HA]

HAPAAP 4-N-(2'-Hydroxyacetophenylidene) aminoantipyrine

HAPDAR Hard-Point Demonstration Array Radar

HA-PE Hydroylapatite–Polyethylene

HAPEX Hydrological/Atmospheric Pilot Experiment

HAPI Helicopter Approach Plate Indicator System

HA-PLA Hydroxylapatite–Polylactic Acid

HAPO Hanford Atomic Products Operation [Washington State, US]

HAPP House Assessment Prescription Program [Canada]

HAPPE Honeywell Associative Parallel Processing Ensemble

HAPUB High-Speed Arithmetic Processing Unit Board

HAPUG Harbich-Pungs-Gerth Modulation Circuit

HAQO Hydroxyaminoquinoline Oxide

HAR Harbor Advisory Radar

HAR Harmonized Wire and Cable [CENELEC Standard]

HAR High-Amplitude Resonance

HAR High Aspect Ratio

HAR Home Address Register

HAR Hydrogen Absorption Reaction

H/Ar Hydrogen/Argon (Mixture)

HARA High-Altitude Radar Altimeter

HARA High-Altitude Resonance Absorption

HARAC High-Altitude Resonance Absorption Calculation

Harb Harbor [also harb]

HARC Houston Area Research Center [The Woodlands, Texas, US]

HARC Human Affairs Research Centers [of Battelle Seattle Research Center, US]

HARCO Hyperbolic Area Coverage

Hard PZT Lead Zirconate Titanate Doped for Coercive Field Increase

Hard Soft Hard and Soft [Journal on Computer Hardware and Software; published by Verein Deutscher Ingenieure, Germany]

HARDTS High-Accuracy Radar Transmission System

Hardwr Hardware [also hardwr]

HARLID High Angular Resolution Laser-Irradiation Detector

HARM High-Aspect-Ratio Micromachining

HARM High-Speed Anti-Radar Missile

Harmonic FC Harmonic Full Constraints (Model) [Materials Science]

HARMST High-Aspect-Ratio Microstructure Technology (Workshop)

HARP Halpern Antiradar Point [UK]

HARP High-Altitude Relay Point

HARP High-Altitude Research Probe

HARP High-Altitude Research Project

HARP High-Altitude Rocket Probe

HARP Hitachi Arithmetic Processor

HARP Hybrid Argon Recovery Process

Harper's Mag Harper's Magazine [Published in the US]

HARPS Heathrow Airport Radar Processing System [UK]

HART Height Area Rain Threshold

HART Highway Addressable Remote Transmitter (Interface)

HARTRAN Hardwell FORTRAN

Härt-Tech Mitt Härterei-Technische Mitteilung [German Technical Communication on Hardening and Heat Treatment]

Harv Bus Rev Harvard Business Review [Published by Harvard School of Business, Boston, US]

HARVEST Highly Active Residue Vitrification Experimental Studies (Process) [Nuclear Engineering]

HAS Hatch Access Structure [Aerospace]

HAS Hawick Archeological Society [UK]

HAS Heading Altitude System; Heading and Attitude System

HAS Helicopter Avionics System [US Air Force]

HAS Helium Atom Scattering

HAS High-Alloy Steel

HAS High-Angle Scattering

HAS Holddown Alignment Support

HAS Hungarian Academy of Sciences [Budapest, Hungary]

HAS Hydraulic Actuation System

HAS Hydrogen Actuation System

HAS Hydroxylamine Acid Sulfate

HASC Hanford Analytical Service Council [of US Department of Energy in Washington State]

HASCC Hydrogen-Assisted Stress-Corrosion Cracking [also HSCC]

HASCI Human Applications Standard Computer Interface

HASG Helicopter Airworthiness Study Group [UK]

HASL Health and Safety Laboratory [of US Department of Energy]

HASL Hot Air Solder Leveling

Hasler Rev Hasler Review [Technical Publication of Hasler Ltd., Berne, Switzerland]

HASM Hanford Analytical Services Program [of US Department of Energy in Washington State]

HASM Hanford Analytic Sample Management (Office) [of US Department of Energy in Washington State]

HAsO⁻ Monohydrogen Arsenate (Ion) [Symbol]

HASP Harmonically Analyzed Sensitivity Profile

HASP High-Altitude Sampling Plane

HASP High-Altitude Sampling Program

HASP High-Altitude Sounding Project

HASP High-Level Automatic Scheduling Program

HASP Houston Automatic Spooling Priority (System); Houston Automatic Spooling Processor; Houston Automatic Spooling Program

HASQ Hardware-Assisted Software Queue

HASQAP Hanford Analytic Services Quality Assurance Plan [of US Department of Energy]

HAST High-Altitude Supersonic Target

HAST Highly-Accelerated Stress Test

HASY Handling System

HASYLAB Hamburger Synchrotronstrahlungslabor [Hamburg Synchrotron Radiation Laboratory; Germany] [also Hasylab]

HAT Height Above Touchdown

HAT High-Altitude Testing (Rocket)

HAT Home Area Toll [Telecommunications]

HAT Hypoxanthine/Aminopterin/Thymidine Medium [Biochemistry]

HATA Hellenic Atlantic Treaty Association [Greece]

HATAPH Hexaalkyltriamidophosphazohydride

Hatb 2-[(2-Aminoethyl)thiomethyl]benzimidazole

Hati 4-/5-[(2-Aminoethyl)thiomethyl]imidazole

Hatmi 4-[(2-Aminoethyl)thiomethyl]-5-Methylimidazole

HATREMS Hazardous and Trace Emission System [of US Environmenral Protection Agency, US]

HATS Hardened Tactical Shelter

HATS Helicopter Advanced Tactical System

HATS High-Accuracy Targeting Subsystem

HATS High-Altitude Terrain Sensor

HATS Hour Angle of the True Sun [Astronomy]

HATT High-Speed Aeronautical Technologies Testbed

HATT-X High-Speed Aeronautical Technologies Testbed–Experimental

HAU Horizontal Arithmetic Unit

HAU Hybrid Arithmetic Unit

haust *(haustus)* – a portion [Medical Prescriptions]

HAV Heavy Armored Vehicle

HAV Hepatitis A Virus

HAV-1 Human Astrovirus, Type 1

HAV-1 Human Adenovirus, Type 1

HAV-2 Human Adenovirus, Type

hav haversine [Mathematics]

HAW Hawaii–Continental United States Submarine Cable

HAW Hawaii [NASA Space Tracking and Data Network]

HAW Heavy Anti-Armor Weapon

HAW High-Activity (Nuclear) Waste

Haw Hawaii(an)

HAWK Homing-All-the-Way Killer [A Rocket]

HAWS Heavy Anti-Armor Weapon System

HAXRD High-Angle X-Ray Diffraction

Haynes Alloy Dig Haynes Alloys Digest [Published by Haynes International, Inc., Kokomo, US]

Haynes Dig Haynes Digest [Published by Haynes International, Inc., Kokomo, US]

HAYSTAQ Have You Stored Answers to Questions? [Computers]

HAZ Heat-Affected Zone [Metallurgy]

Haz Hazard(ous) [also haz]

HAZAN Hazard Analysis [also Hazan]

Hazard Mater Control Mag Hazardous Materials Control Magazine [Published by the Hazardous Materials Control Research Institute, US]

HAZCHEM Hazardous Chemical(s) [also Haz Chem, HazChem, or Hazchem]

HAZCHEM Hazardous Material Identification System [also HazChem]

HAZ-COM Hazard Compliance (Standard) [US] [also Haz-Com]

HAZEL Homogeneous Assembly Zero Energy Laboratory [of Atomic Energy Research Establishment, UK]

HAZFILE Hazards File [of National Chemical Emergency Center, UK]

HAZMAT Hazardous Material(s) [also HAZ MAT, Haz Mat, HazMat, or Hazmat]

HAZMAT Hazardous Materials Management Conference [also HAZ-MAT, or HazMat]

Haz Mat Manage Hazardous Materials Management [Canadian Publication]

HAZ/MZ Heat-Affected Zone/Melted Zone [Metallurgy]

HAZOP Hazard and Operability (Study) [also Hazop]

HAZ PROD Hazardous Product(s) [also Haz Prod, HazProd]

H-3 AZT Tritated Azidothymidine [also ^{3}H-AZT] [Biochemistry]

Haz Subst Rev Hazardous Substances Review

Haztech Canada Environmental Control/Hazardous Waste Management Conference [Canada]

Haz W Hazardous Waste [also HAZW]

HAZWOPER Hazardous Waste Operations and Emergency Response (Regulations) [US]

HAZWRAP Hazardous Waste Remedial Action Program [of US Department of Energy]

HB Brinell Hardness

HB Handbook

HB Hard Black [On Pencils]

HB Hatchback

HB Heat Balance

HB Heat Barrier [Aerospace]

HB Heine-Borel (Theorem) [Mathematics]

HB Hepatitis B [Medicine]

HB Hering-Breuer (Reflex) [Medicine]

HB Hexachlorobutadiene

HB Hexadecimal-to-Binary [also H/B]

HB High Bay

HB Hollow Base (Bullet)

HB Hopf Bifurcation [Linear Analysis]

HB Horizontal Baffle

HB Horizontal Beam

HB Horizontal Bridgman (Crystal Growth) [Solid-State Physics],

HB Hose Bib

HB Hudson Bay [Canada]

HB Hybrid Bond

HB Hydraulic Brake

HB Hydrogenation of Benzene (Process)

HB Hydrogen Bomb

HB Hydrogen Bond(ing)

HB Hydrogen Bridge [Chemistry]

HB Hydropower Benefit

HB Hydroxybenzene

HB Hydroxybutyrate

HB- Switzerland and Liechtenstein [Civil Aircraft Marking]

3HB 3-Hydroxybutyrate

4HB 4-Hydroxybutyrate

25H14B 2,5-Dihydroxy-1,4-Benzoquinone

H/B Hexadecimal-to-Binary [also HB]

Hb Hemangioblastoma [Medicine]

Hb Hemoglobin

Hb A Hemoglobin A [also HbA]

Hb A$_1$ Glycerated Hemoglobin, Fraction A$_1$ [also Hb A$_1$]

Hb A1C Hemoglobin A1C

Hb C Hemoglobin C [also HbC]

Hb E Hemoglobin E [also HbE]

Hb F Fetal Hemoglobin [also HbF]

Hb H Hemoglobin H [also HbH]

Hb M Hemoglobin M [also HbM]

Hb S Hemoglobin S (or Sickle-Cell Hemoglobin) [also HbS]

H‖b Magnetic Field Parallel to b-Axis [Symbol]

HBA Home Builders Association [US]

HBA Honours Bachelor of Arts

HBA Host Bus Adapter

HBA Hydrogen Bond Acceptor

HBA Hydroxybenzoic Acid

HBA Hydroxybutyrate

HBA Hydroxybutyric Acid

α-HBA α-Hydroxybutyrate

β-HBA D-β-Hydroxybutyrate

D-β-HBA D-β-Hydroxybutyrate

Hb A Hemoglobin A [also HbA]

Hb A$_1$ Glycerated Hemoglobin, Fraction A$_1$ [also Hb A$_1$]

HBAAP 4-N-(2'-Hydroxybenzylidene)aminoantipyrine

HBABA 4-Hydroxyazobenzene-2-Carboxylic Acid

Hb A1C Hemoglobin A1C

hbar hectobar [Unit]

HBB Hardened Ball Bearing (Steel)

HBBA N-Benzoyl Benzamide

HBB Steel Hardened Ball Bearing Steel [also HBB steel]

HBC Hereditary Breast Cancer

HBC Hexadecimal-to-Binary Conversion

HBC Higher Binding-Energy Component

HBC Hudson Bay Company [Canada]

HBC Human Biology Council [US]

HbC$_2$ Carbohemoglobin

HBCCO Mercury Barium Calcium Copper Oxide (Superconductor)

HBCO Holmium Barium Copper Oxide (Superconductor)

HBCO Mercury Barium Copper Oxide (Superconductor)

HbCO Carboxhemoglobin

HBComm Honours Bachelor of Commerce

HBCU Historically Black Colleges and Universities [US]

HBD Hexabutyl Distannoxane

HBDH Hydroxybutyrate Dehydrogenase [Biochemistry]

α-HBDH α-Hydroxybutyrate Dehydrogenase [Biochemistry]

β-HBDH β-Hydroxybutyrate Dehydrogenase [Biochemistry]

HBE High Binding Energy

HBE High Bombardment Energy [also hbe] [Plasma-Assisted Vapor Deposition]

Hb E Hemoglobin E [also HbE]

HBEC Hanford Business Exchange Conference [Washington State, US]

HBEC High-Built Electrocoat (Process)

HBED N,N'-Di(2-Hydroxybenzyl)ethylenediamine

HBED N,N'-Bis(2-Hydroxybenzyl)ethylenediamine-N,N'-Diacetic Acid

H2B2EDP Di-n-Butylethane-1,2-Diphosphonic Acid

h-BeGa$_2$O$_4$ Hexagonal Beryllium Gallate

HBEDPO 1,2-Ethanediylbis[(2-Hydroxyphenylmethyl)aminomethyl]phosphornic Acid

HB-EGF Heparin-Binding Epidermal Growth Factor [Immunology]

HBEN High Byte Enable

HBF House Builders Federation [UK]

Hb F Fetal Hemoglobin [also HbF]

HBFC Hydrobromofluorocarbon

HBFP Hematoxylin Basic Fuchsin Pecric

HB-GAM Heparin-Binding Growth Associated Molecule [Immunology]

HBGF Heparin-Binding Growth Factor [Immunology]

HBGM Hypersonic Boost-Glide Missile

p-HBGP p-Hydroxybenzoylglycyl-L-Phenylalanine [Biochemistry]

Hb H Hemoglobin H [also HbH]

HBI Hot-Briquetted Iron [Metallurgy]

H2BITS 4-Benzamino-1-Satin-3-Thiosemicarbazone

HBL Horizontal Bloch Line [Solid-State Physics]

HBLV Human B-Cell Lymphotropic Virus

HBM Her/His Britannic Majesty

Hb M Hemoglobin M [also HbM]

HBMePz 1,1',1"-Borylidynetris[(4-Methyl)-1H-Pyrazole] [also Hbmepz]

HBMe3Pz 1,1',1"-Borylidynetris[(3,4,5-Trimethyl)1H-Pyrazole] [also Hbme3pz]

HBMS Hudson Bay Mining and Smelting Company Limited [Toronto, Canada]

HBN Barkhausen Noise Amplitude

HBN Brinell Hardness

HBN Hazard Beacon

HBN Health-Based Numbers

HBN Hexagonal Boron Nitride [also h-BN, or hBN]

HBO Heavy Batch Oven

HBO Home Box Office [US]

HBO Mercury Burner [An Arc Lamp]

HbO$_2$ Oxyhemoglobin [also Hb O$_2$]

HBOC Hereditary Breast and/or Ovarian Cancer [Medicine]

H-Bomb Hydrogen Bomb [also H-bomb]

HBP High Blood Pressure

HBPMP 2,6-Bis[bis(2-Pyridylmethyl)aminomethyl]-4-Methyl Phenol

HBPT Hexabutylphosphoric Acid Triamide

HBpz 1,1',1"-Borylidynetris(1H-Pyrazole) [also Hbpz]

HBR Harward Business Review [Published by Harward School of Business, Boston, US]

HBR High Bit Rate

HBr Brinell Hardness

HBr Hydrogen Bromide (or Hydrobromic Acid)

Hbr Harbour [also hbr]

HBRA Howitzer Battery Royal Artillery [UK]

HBS Hawaii Botanical Society [US]

Hb S Hemoglobin S (or Sickle-Cell Hemoglobin) [also HbS]

HBsAg Hepatitis B Surface Antigen [Immunology]

HBScF Honours Bachelor of Science in Forestry

HBSMAA Hack and Band Saw Manufacturers Association of America [US]

HBSS Hank's Balanced Salt Solution

HBSS Hank's Basic Salt Solution

HBT Heflex Bioengineering Test

HBT Heterojunction Bipolar Transistor [also HJBT]

HBT N-(P-Hexyloxybenzylidene)-P-Toluidene

HBTM Hybrid Bispectral Threshold Method

HBTU O-Benzotriazoyl-1-yl-N,N,N',N'-Tetramethyluronium Hexafluorophosphate

HBU Hollandische Bank-Unie NV [Netherlands]

HBV Hepatitis B Virus

HBW Hot Bridge Wire

HBWR Halden Boiling Water Reactor [Norway]

HBX High Blast Explosive

HBZ *(Hochschulbibliothekszentrum)* – German for "University Library Center"

Hbzac Benzoylacetone

Hbztfac Benzoyltrifluoroacetone

HC Hadron Collision [Particle Physics]

HC Half Cell [Chemistry]

HC Half-Cycle (Signal),

HC Hall Coefficient

HC Hall Conductivity

HC Hall Constant

HC Halocarbon
HC Halochromism; Halochromy
HC Hamilton College [Clinton, New York, US]
HC Hamming Code [Data Transmission]
HC Hand Control(ler)
HC Handling Capacity
HC Hanging Ceiling
HC Hard Copy
HC Hard-Cover (Format)
HC Hard Cradle (Balancer)
HC Hardware Capability
HC Harmonic Control
HC Harvard College [of Harvard University, Cambridge, Massachusetts, US]
HC Haversian Canal [Medicine]
HC Hazardous Chemical(s)
HC Head Changer
HC Head Count
HC Health Canada [formerly Health and Welfare Canada]
HC Health Care
HC Health Center
HC Heat Capacity
HC Heat Compensation
HC Heat Conduction; Heat Conductivity; Heat Conductor
HC Heat(ing) Coil
HC Heating Chamber
HC Helminthosporium Carbonum (Toxin)
HC Hemicellulose
HC Hemicrystal(line) [Petrology]
HC Hepatitis C [Medicine]
HC Hepatocyte
HC Heptachlor
HC Heritage Canada
HC Herpolhode Cone [Mechanics]
HC Hersch Cell
HC Heterogeneous Catalysis
HC Heterogeneous Chemistry
HC Heterolytic Chemistry
HC Heuristic Concepts
HC Hexachloroethane
HC Hexadecimal Code
HC Hierarchical Control
HC High Capacity
HC High Carbon
HC High Coercivity
HC High Compression
HC High Concentration
HC Hillig and Charles (Model) [Mechanics]
HC Holding Coil [Electrical Engineering]
HC (Electron) Hole Conduction [Solid-State Physics]
HC Holland College [Canada]
HC Hollerith Code [Computers]
HC Hollow Cathode
HC Homogeneous Catalysis

HC Homologous Chromosome [Genetics]
HC Homolytic Chemistry
HC Hopkins-Cole (Test) [Biochemistry]
HC Horizontal Cell
HC Horizontal Chromatography
HC Horizontal Control
HC Hose Connector
HC Host Computer
HC Hot Cathode
HC Hot-Cell [Nuclear Engineering]
HC Hot Corrosion
HC Hour Circle [Astronomy]
HC House Cable
HC Humidity Control
HC Hunter College [of City University of New York, US]
HC Hybrid Ceramic(s)
HC Hybrid Circuit
HC Hybrid Coil [Electrical Engineering]
HC Hybrid Computer
HC Hydraulic Circuit
HC Hydraulic Conductivity
HC Hydraulic Controller
HC Hydraulic Coupling
HC Hydrocarbon [also H/C]
HC Hydrocarbon Cleaner
HC Hydroconversion
HC Hydrocooling [Food Sciences]
HC Hydrocortisone [Biochemistry]
HC Hydrocracker; Hydrocracking [Chemical Engineering]
HC Hydrogen-Charged; Hydrogen Charging
HC Hydrogen Chemisorption
HC Hydrologic Cycle [Ecology]
HC Hyperconjugation [Physical Chemistry]
HC Hypercooling
HC Hysteresis Coefficient [Physics]
HC- Ecuador [Civil Aircraft Marking]
H&C Hot and Cold (Water) [also h&c]
H/C Hand Carry
H/C Hydrocarbon [also HC]
H/C Hydrogen-to-Carbon (Ratio)
Hc Hermitian conjugate [Mathematics]
H(c) Hardness as a Function of Concentration [Symbol]
H∥c Magnetic Field Parallel to c-Axis [Symbol]
H⊥c Magnetic Field Perpendicular to c-Axis [Symbol]
H11/c11 Loose Running Fit; Basic Hole System [ISO Symbol]
h/°C hours per degree Celsius (or centigrade) [also h °C^{-1}]
hc highest common factor [Symbol]
hc (honoris causa) – honorary
HCA Health Care Aide
HCA Hemicylindrical Auger Analyzer [Materials Science]
HCA Heptine Carbonic Acid
HCA Hexachloroacetone
HCA Hydrochloric Acid

HCA Hydroxylcarbonate Apatite

HCA 17α-Hydroxycorticosterone-21-Acetate

HCAA Hellenic Civil Aviation Authority [Greece]

HCAL Hadron Calorimeter [of CERN–European Laboratory for Particle Physics, Geneva, Switzerland]

Hcapt N-Carboamido-2-Pyrrolethioamide

HCAR Higher Committee for Agrarian Reform [Egypt]

HCB Hard Convex Body

HCB 5-Chloro-2-Hydroxybenzophenone

HCB Hemocytoblast

HCB Hexachlorobenzene [also HCBz]

HCB Hexachlorobutadiene

HCB Hollow-Cone Beam

HCB Hydrocarbon [also hcb]

HCB-CBED Hollow-Cone Beam/Convergent-Beam Electron Diffraction

HCBD 1,1,2,3,4,4-Hexachloro-1,3-Butadiene

H2Cbdmpz Bis(3,5-Dimethylpyrazolyl)methane

H2Cbpz Bis(1-Pyrazolyl)methane

HCBz Hexachlorobenzene [also HCB]

HCC Hardware Capability Code

HCC Hepatocellular Carcinoma [Medicine]

HCC Hermetic Chip Carrier

HCC Heterocyclic Compound

HCC γ-Hexachlorocyclohexane

HCC Homeopathic College of Canada [Toronto]

HCC Horizontal Continuous Casting [Metallurgy]

HCC Hydroconversion Catalyst

HCCH Hexachlorocyclohexane

HCD High-Capacity Disk

HCD High Crack Density [Mechanics]

HCD High Current Density

HCD Hollow-Cathode Discharge

HCD Hot-Carrier Diode

HCDA Hypothetical Core-Disruptive Accident [Nuclear Engineering]

HCDF Hollow Cathode Dark Field (Microscopy)

HCDF TEM Hollow Cathode Dark Field Transmission Electron Microscopy

HCDT Hollow-Cathode Discharge Tube

H4CDTA Cyclohexanediaminetetraacetic Acid

HCE Hamburg Consulting and Steel Engineering GmbH [of Neue Hamburger Stahlwerke, GmbH, Germany]

HCE Heater Control Electronics

HCE Heptachlor Epoxide

HCE Hexachloroethane

HCE Hollow-Cathode Effect

HCE Human Caused Error

HCED Hand Controller Engage Driver

Hcept N-Carboethoxy-2-Pyrrolethioamide

Hcett N-Carboethoxy-2-Thiophenethioamide

HCEX Hypercharge Exchange [Particle Physics]

HCF Hercules (Polyacrylonitrile) Carbon Fiber

HCF High Carbohydrate and (High) Fiber (Diet) [Medicine]

HCF High-Cycle Fatigue [Mechanics]

HCF Highest Common Factor [also hcf]

HCF HIM (Hardware Interface Module) Configuration File

HCF Host Command Facility

HCF Hydrocarbon Fuel

HCFA Health Care Financing Administration [of US Department of Health and Human Services]

HCFC Hydrochlorofluorocarbon

HCFS High-Cycle Fatigue Strength [Mechanics]

HCFTA Home and Contract Furnishing Textiles Association [UK]

HCG Half-Cycle Generator

HCG Hardware Character Generator

HCG Hex(agonal) Coupling

HCG Highly Crystallized Graphite (Film)

HCG Horizontal Location of Center of Gravity

HCG Hot Cathode Gun

hCG Human Chorionic Gonadotropin [also HCG] [Biochemistry]

α-HCG Human Chorionic Gonadotropin, α Subunit [Biochemistry]

β-HCG Human Chorionic Gonadotropin, β Subunit [Biochemistry]

HCH Hexachlorocyclohexane; 1,2,3,4,5,6-Hexachlorocyclohexane

HCH Hollow Cathode Heating

H-ch heavy-chain specific (immunoglobulin)

H_2/CH_4 Hydrogen/Methane (Gas Mixture)

HCHO Formaldehyde

$H_2/CH_4/O_2$ Hydrogen/Methane/Oxygen (Gas Mixture)

HCI Host Computer Interface

HCI Hylleraas Configuration Interaction

HCI Human-Computer Interaction

HCI Human-Computer Interface

HCIB Health Computer Information Bureau

H-CITE Horizontal-Cargo Integration Test Equipment

HCL Hardware Compatibility List

HCL Helium-Cadmium Laser

HCL High, Common, Low (Relay)

HCL High Cost of Living

HCL Hollow Cathode Lamp

HCL Horizontal Centerline

HCl Hydrogen Chloride (or Hydrochloric Acid)

HCLF Horizontal Cask Lifting Fixture [Nuclear Engineering]

HCM Hard Copy Module

HCM High Carbon Monoxide (Process)

HCM Hydraulic Core Mockup

HCM Hyundai Color Monitor

HCMM Heat Capacity Mapping Mission [NASA Earth Radiation Budget Satellite]

HCMM Highly Conductive Mold Media

HCMM/AEM-1 Heat Capacity Mapping Mission/ Applications Explorer Mission [NASA Earth Radiation Budget Satellite]

HCMOS High-Density Complementary Metal-Oxide Semiconductor

HCMOS High-Speed Complementary Metal-Oxide Semiconductor

HCMTS High-Capacity Mobile Telecommunications System

HCMV Human Cytomegaolvirus

HCN Hydrogen Cyanide (or Hydrocyanic Acid)

HCNB Hexacyanobenzene

HCO Harvard College Observatory [of Harvard University, Cambridge, Massachusetts, US]

HCO Formyl (Radical) [Symbol]

HCO Bicarbonate (Ion) [Symbol]

HC$_2$O Bioxalate (Ion) [Symbol]

HCOHSA Health Care Occupational Health and Safety Association [Canada]

HCOO⁻ Formate (Ion) [Symbol]

HCP Hard Copy Printer

HCP Health Care Provider

HCP Hepatocatalase Peroxidase [Biochemistry]

HCP Hexachlorocyclopentadiene

HCP Hexagonal Close-Packed (Crystal) [also hcp]

HCP Host Communications Processor

HCPA Highly-Conducting Polyacetylene

Hcppt N-Carbophenylamido-2-Pyrrolethioamide

HCPRU Hot Climate Physiological Research Unit [Nigeria]

Hcpt N-Carboethoxy-1-Pyrrolethioamide

HCR Half-Cell Reaction [Chemistry]

HCR Hardware Check Routine

HCR Heat-Curable Rubber

HCR Heterogeneous Chemistry Reviews

HCR High Cross Range

HCR Hot-Charge Rolling [Metallurgy]

HCRA Hazardous Chemicals Removal Act [US]

H-1 CRAMPS Hydrogen-1 Combined Rotation and Multiple-Pulse Spectroscopy [also ¹H CRAMPS]

HCRC Holland College Royalty Center [Canada]

HCRD Hoechst Celanese Research Division [Summit, New Jersey, US]

HCRL Hanford Cultural Resources Laboratory [Washington State, US]

HCRP Human Chronobiology Research Program [US]

HCRS Heritage Conservation and Recreation Service

HCS Hazard Communication Standard [of Occupational Safety and Health Administration, US]

HCS Helium Circulator Seal

HCS High-Carbon Steel

HCS Host Composition System

HCS Human Chorionic Somatotropin [also hCS] [Biochemistry]

HCS Hundred Call Seconds

HCS Hungarian Chemical Society

HCSC Chromone-3-Carboxaldehyde Semicarbazone

HCSDS High-Capacity Satellite Digital Service

HCSI Hughes Communications Services, Inc. [US]

HCSL Hybrid Computation and Simulation Laboratory

HCSR Hulburt Center for Space Research [of Naval Research Laboratory, Washington, DC, US]

HCSS Head Compartment Support Structure

HCSS High-Capacity Storage System

HCSS Hospital Computer Sharing System

HCST H. C. Starck Inc. [Thermal Spray Technology Company located in Newton, Massachusetts, US]

HCT Heater Center Tap

HCT Hematocrit [Medicine]

HCT Hot Cathode Tube

HCT Mercury Cadmium Telluride [Hg$_{1-x}$Cd$_x$Te]

HCTDS High Capacity Terrestrial Digital Service

HCTLS High-Speed Complementary Transistor Low-Power Schottky

HCtpz Tris(1-Pyrazolyl)methane

Hctt N-Carboethoxy-4-Toluenethioamide

HCU Home Computer User

HCU Homing Comparator Unit

HCU Hydraulic Charging Unit

HCU Hydraulic Control Unit

HCU Hydrocracking Unit [Chemical Engineering]

HC/UC Horizontal Control/Universal Crown (Rolling Mill) [Metallurgy]

HCV Hepatitis C Virus

HCV Hydrogen Check Valve

HCVS Historic Commercial Vehicle Society [UK]

HCW Home Computing Weekly [UK Publication]

HCZ Hydrogen Convection Zone

HD Half Duplex [also hd, HDX, or hdx]

HD Halide Detector

HD Halogen Detector

HD Harbor Defense

HD Hard Diatomics

HD Hard Disk

HD Hard Drawing; Hard-Drawn [Metallurgy]

HD Hardness Determination

HD Harley Davidson (Motorcycle)

HD Harmonic Distortion

HD Harmonization Documents [of CENELEC for Electronic Equipment]

HD Harper-Dorn (Creep) [also H-D] [Metallurgy]

HD Hazardous Decomposition

HD Head Diameter

HD Heat Drop

HD Heavy Distillate

HD Heavy Duty

HD Hemispherical Deflector

HD Hemodynamic(s) [Medicine]

HD Heterodyne Detection; Heterodyne Detector [Electronics]

HD Heuristic Development

HD Hexadecimal-to-Decimal [also H-D]

HD Hierarchical Direct

HD High Damping

HD High Definition

HD High Density

HD High Dislocation-Density (Substrate) [Materials Science]

HD High Duty

HD High Desirability; Highly Desirable

HD Hofmann Degradation [Organic Chemistry]

HD Holddown

HD (Electron) Hole Diffusion [Solid-State Physics]

HD Home Defense [UK]

HD Homeopathic Doctor

HD Hoover Dam [Arizona, US]

HD Horizontal Diaphragm

HD Horizontal Distance

HD Horizontal Drain

HD Horse-Drawn

HD Hot-Drawing; Hot Drawn [Metallurgy]

HD Hub Diameter

HD Huntington's Disease [Medicine]

HD Hydrodynamic(s)

HD Hydrogen Drain

HD Hydrogen Disproportion (Process)

HD Hysteretic Damping [Mechanics]

%HD Percentage of Harmonic Distortion [Symbol]

H-D Hexadecimal-to-Decimal [also HD]

H-D Harper-Dorn (Creep) [also HD] [Metallurgy]

H-D Hunter-Driffield (Curve) [Physics]

Hd Hand [also hd]

Hd Head [also hd]

Hd Hogshead [Unit]

H9/d9 Free Running Fit; Basic Hole System [ISO Symbol]

HDA Hard Disk Assembly

HDA Hardwood Distributors Association [US]

HDA Harris Daishowa Australia (Limited)

HDA Head/Disk Assembly [Computers]

HDA Hemispherical Deflection Analyzer

HDA Hexadecenyl Acetate

HDA Holistic Dental Association [US]

HDA Homochiral Derivatizing Agents

HDA Horizontal Danger Angle

HDA Housekeeping Data Acquisition

HDA Hydrodealkylation (Process)

Z-11-HDA cis-11-Hexadecenyl Acetate

Hda 1-Hexadecanamine

HDAL Hexadecenal

Z-9-HDAL cis-9-Hexadecenal

Z-11-HDAL cis-11-Hexadecenal

HDAM Hierarchical Direct Access Method

HDAOS N-(2-Hydroxy-3-Sulfopropyl)-3,5-Dimethooxyaniline

HDAP Dialkylphosphoric Acid

HDAP Di-n-Amylphosphoric Acid

H2dapd 2,6-Diacetylpyridinedioxime

HDAS Hybrid Data Acquisition System

H-3 dATP Tritiated 2'-Deoxyadenosine 5'-Triphosphate [also ^{3}H-dATP] [Biochemistry]

HDB High-Density Binary

HDB High-Density Bipolar (Code)

HDB-3 High-Density Bipolar (Code), Three Zeroes

HDBEP Di(butoxyethyl)phosphoric Acid

HDBF Heavy-Duty Business Forum [US]

HDBH High Day Busy Hour

Hdbk Handbook [also hdbk]

Hdbm Dibenzoylmethane

HDBP Dibutylphosphoric Acid [also HDBPA]

H2dbzdto N,N'-Dibenzyldithiooxamide

H2dmdto N,N'-Dimethyldithiooxamide

H2dpm Dopamine [Biochemistry]

HDC Half-Duplex Circuit

HDC Health Data Card

HDC Heavy Duty Clamp

HDC Helium Direct-Current

HDC Hexadecimal Code

HDC Hexadecimal-to-Decimal Conversion

HDC Hybrid Device Controller

HDC Hydrodynamic Chromatography

HDC Hydroelectric Design Center [of Army Corps of Engineers, US]

HDCD High-Definition Compatible Digital

Hdcdto N,N'-Dicyclohexyldithiooxamide

HDCM High-Density Ceramic Module

H(dcm) d,d-Dicampholylmethane

H4DCTA 1,2-Diaminocyclohexane-N,N,N',N'-Tetraacetic Acid

HDCU Hard Disk Control Unit

HDD Head-Down Display [Aeronautics]

HDD Heating Degree Day [Meteorology]

HDD High-Density Disk

HDD High Dislocation Density [Materials Science]

HDDA Hexadecadienyl Acetate

(Z,E)-7,11-HDDA cis-7, trans-11-Hexadecadienyl Acetate

(Z,Z)-7,11-HDDA cis-7, cis-11-Hexadecadienyl Acetate

H-3 ddC Tritiated Dideoxycytidine [also 3H-ddC] [Biochemistry]

HDDD High-Density Disk Drive

H-3 ddI Tritiated Dideoxyinosine [also ^{3}H-ddI] [Biochemistry]

HDDP Didecylphosphoric Acid

HD(DP) Decyldecylphosphoric Acid

HDDR High-Density Digital (Magnetic) Recorder; High-Density Digital (Magnetic) Recording

HDDR Hydrogenation, Disproportionation, Desorption, Recombination (Process) [Powder Metallurgy]

HDDS High-Density Data System

HDDT High-Density Digital Tape

HDDTC Hexadecyl Dithiocarbamate

HDDV High Dislocation Density Volume [Materials Science]

HDE Hanging-Drop Electrode

HDE Hauptgemeinschaft des Deutschen Einzelhandels e.V. [German Association of Retailers]

HDE Hydrodynamic Engine

HDEHP Di(ethylhexyl)phosphoric Acid

HD2EHP Di(2-Ethylhexyl)phosphoric Acid

HDEP Diethylphosphoric Acid

HDEP High-Density Electronic Packaging

HDF Hierarchical Data Format [of National Center for Supercomputing Applications, Champaign, Illinois, US]

HDF High-Density Flexible

HDF High Depth of Field (Microscope)

HDF High-Frequency Direction Finding

HDF Horizontal Distributing Frame

H/DF Human/Dolphin Foundation [US]

HDFD High-Density Floppy Disk

HDF/NetCDF Hierarchical Data Format/Net Common Data Format [also HDF/netCDF]

Hd Frz Hard Freeze [Meteorology]

HDG Hot-Dip Galvanizing; Hot Dip(ped) Galvanized [Metallurgy]

Hdg Heading [also hdg]

HDGA Hot Dip Galvanizers Association [US]

HDGAF Hot-Dip Galvanizing After Fabrication

HDGCP Human Dimensions of Global Change Program

HDGEC Human Dimensions of Global Environmental Change

HDGL Hot-Dip Galvanizing Line [Metallurgy]

HDH Hemidihydrate (Process)

HDH Hydrogenation-Dehydrogenation (Process)

HDHoEP Di[2-(n-Hexyloxy)ethyl]phosphoric Acid

HDHP Di-n-Hexylphosphoric Acid

HDI Head-Disk Interference [Computers]

HDI Hexamethylene Diisocyanate

HDI High-Density Inclusion [Metallurgy]

HDI High-Density Interconnect

HDI High Dose Implantation

HDI Human Development Index

HDIAP Diisoamylphosphoric Acid

HDiBP Di-Isobutylphosphoric Acid

HDIC High-Density Interconnect Circuit

HDIC High Digital Integrated Circuit

HDIRS Head of Defense Industrial Research Section [of Department of National Defense, Canada]

HDL Hardware Description Language

HDL Harry Diamond Laboratories [US Army at Adelphi, Maryland]

HDL Helmholtz Double Layer [Physics]

HDL High-Density Lipoprotein [Biochemistry]

HDL High-Level Data Link

HDL Hydrodynamics Laboratory [of California Institute of Technology, Pasadena, US]

HDL Hydrogen-Discharge Lamp

Hdl Handle [also hdl]

HDLC Hierarchical Data-Link Control

HDLC High-Level Data-Link Control

Hdlg Handling [also hdlg]

HDLS Helicopter Deck Landing Simulator

Hdls Headless [also hdls]

HDM Hardware Device Module

HDM Highly Dispersed Media

HDM Hill Determinant Method

HDM Humic Degradation Matter

HDM Hydrodemetallization

HDM Hydrodensimeter; Hydrodensimetry

HDM Hydrodynamic Machining

HDMA Hardwood Dimension Manufacturers Association [now National Dimension Manufacturers Association, US]

HDME Hanging Drop Mercury Electrode

HDML Handheld Device Markup Language

HDMR High-Density Moderated (Nuclear) Reactor

HDMR High-Density Multitrack Recording

HDMS Honeywell Distributed Manufacturing System

HDMSO Hexamethyldisiloxane

HDMSW High-Density Mach Shock Wave

HDN Hydrodenitrogenation (Process)

hdn harden

HDNP Dinaphthylphosphoric Acid

HDNP Dinonylphosphoric Acid

HDNP High-Density Nickel Powder

H-1 DNP Hydrogen-1 Dynamic Nuclear Polarization [also ^{1}H DNP]

H-1 DNP-NMR Hydrogen-1 Dynamic Nuclear Polarization Nuclear Mzagnetic Resonance [also ^{1}H DNP-NMR]

HDNSW High-Density Nuclear Shock Wave

HDO High Density Overlay [Plywood]

HDOAA Di-(n-Octyl)arsonic Acid

HDOC Handy Dandy Orbital Computer

HDODA 1,6-Hexanediol Diacrylate

HDOL Hexadecenol

Z-11-HDOL cis-11-Hexadecenol

HDOP Di-n-Octylphosphoric Acid

HDOS Hard Disk Operating System

HDOS Heath Disk Operating System

HDP Diamond Pyramid Hardness

HDP Half-Duplex Protocol

HDP Hazardous Decomposition Product(s)

HDP High-Density Plasma

HDP High-Density Polyethylene

HDP High Detonation Pressure

HDP High Discharge Pressure

HDP Holddown Post

HDP Horizontal Data Processing

HDP Human Dimensions of Global Environmental Change Program

HDPA Heptadecylphosphoric Acid

HDPA Hydroxydiphenylamine

HDPB Bis(di-n-Hexylphosphinyl)butane

HD-pCl-PP Di-(p-Chlorophenyl)phosphoric Acid

HDPCVD High-Density Plasma Chemical-Vapor Deposition [also HDP-CVD]

Hdpdma N,N'-Diphenyl Dithiomalonamide

HDPE Bis(di-n-Hexylphosphinyl)ethane
HDPE High-Density Polyethylene
HDPE-PA High-Density Polyethylene/Polyamide (or Nylon) (Blend) [also HDPE/PA]
HDPM Bis(di-n-Hexylphosphinyl)methane
Hdpm Dipivaloylmethane
HDPP Bis(di-n-Hexylphosphinyl)propane
HDPP Diphenylphosphoric Acid
Hdq Headquarters [also Hdqrs]
HDR Harwell Dilution Refrigerator [of United Kingdom Atomic Energy Agency]
HDR Header [Computers]
HDR High Data Rate
HDR High Definition Radiography
HDR High-Density Recording
HDR Hot-Direct Rolling [Metallurgy]
HDR Hot Dross Recycling [Metallurgy]
HDR Hot-Dry-Rock (Process) [Geothermal Heating]
Hdr Header [also hdr]
HDRA Heavy-Duty Representatives Association [US]
HDRA Henry Doubleday Research Association [UK]
HDRA High Data Rate Assembly
HDRI Hannah Dairy Research Institute [Scotland]
HDRM High Data Rate Multiplexer
HDRR High Data Rate Recorder
HDRSS High Data Rate Storage System
HDS Hampden Data Services
HDS Hardware Development System
HDS Hardware Descrption Sheet
HDS Headset
HDS Hermes Data System
HDS Hierarchical Data Structure
HDS High Density Sludge
HDS High Dislocation-Density Substrate [Materials Science]
HDS Huang Diffuse Scattering
HDS Humungous Development Syndrome [Medicine]
HDS Hybrid Development System
HDS Hydrodesulfurization (Process)
Hds Hundreds [also hds]
HDSAS 1-Hexadecanesulfonic Acid
HDSC High-Density Signal Carrier [of Digital Equipment Corporation]
HD/SCSI Hard Disk/Small Computer System Interface
HDSL HDR (High Data Rate) Digital Subscriber Line; High-Rate Digital Subscriber Link
HDSS Holographic Data Storage System (Program)
HDSS Hospital Decision Support System
HDST High-Density Shock Tube
Hdst Headset
HDT Heat Deflection Temperature
HDT Heat Deflection Test
HDT Heat Distortion Temperature
HDT Host Digital Terminal
HDT Hydrodynamic Technology

H/D/T Hydrogen/Deuterium/Tritium (Ratio)
HDTA Hexamethylenediaminetetraacetate [also HDTA]
HDTP Di-(p-Tolyl)phosphoric Acid
H5DTPA N,N,N',N'N"-Diethylenetriamine Pentaacetic Acid
HDTUL Heat-Deflection Temperature under Load
HDTV High-Definition Television
HDTV High-Density Television
HDU Hard Disk Unit
HDU Heat-Dissipation Unit
HDV Hepatitis Delta Virus
HDVD High-Definition Volumetric Display
HDVS High-Definition Video System
HDW Hardware [also Hdw, or hdw]
HDW Hanford Defense Waste [US Department of Energy]
HDW High-Pressure Demineralized Water
HDW Howaldtswerke–Deutsche Werft AG [German Shipyard; located at Hamburg and Kiel]
HDWC Hawaii Deep Water Cable [US]
Hdwd Hardwood [also hdwd]
Hdwe Hardware [also hdwe]
HDW-EIS Hanford Defense Waste Environmental Impact Statement [of US Department of Energy]
Hd Whl Hand Wheel [also hd whl]
HD WND Head Wind [also Hd Wnd]
Hdwre Hardware [also hdwre]
HDX Half Duplex [also hdx, HD, or hd]
HDZ Heat and Deformation Zone
HE Hall Effect [Electronics]
HE Hammett Equation [Chemistry]
HE Handling Engineer(ing)
HE Harmful Environment
HE Hartree Equation [Electronics]
HE Head End [Telecommunications]
HE Health Education
HE Heat Energy
HE Heat Engine
HE Heater Element
HE Heat Exchange
HE Heat Exchanger [also H/E]
HE Heating Element
HE Heavy Enamel(led) (Electric Wire)
HE Helium Embrittlement
HE Heating Engineer(ing)
HE Heavy Enamel (Electric Wire)
HE Heptachlorine Epoxide
HE Hermite Equation [Mathematics]
HE Heteroepitaxial; Heteroepitaxy
HE High Efficiency
HE High Elongation
HE High Energy
HE Higher Education
HE High Explosive
HE Highway Engineer(ing)
HE Hildebrand Equation [Physics]

HE   Hofmann Elimination [Chemistry]

HE   Hofmann-Erlewein (Equation) [also H-E] [Metallurgy]

HE   Hole Expansion (of Steel Sheeting)

HE   Homoepitaxial; Homoepitaxy

HE   Homogeneous Equation [Mathematics]

HE   Horizontal Equivalent

HE   Hot Extruded; Hot Extrusion [Metallurgy]

HE   Housekeeping Element [Computers]

HE   Human Engineer(ing)

HE   Hydraulic Engineer(ing)

HE   Hydroelastic(ity)

HE   Hydroelectric(ity)

HE   Hydrogen Electrode [Physical Chemistry]

HE   Hydrogen Embrittlement

HE   Hydrogen Energy

HE   Hydrostatic Extrusion

HE   Hyperelastic(ity)

HE   Hysteresis Effect

H/E   Hardness-to-Young's Modulus (Ratio)

H/E   Heat Exchanger [also HE]

H-E   Hofmann-Erlewein (Equation) [also HE] [Metallurgy]

H(E)   Energy Histogram [Symbol]

He   Helium [Symbol]

He   Hexane

He I   Helium I [Liquid Helium between 2.2 and 4.2 Kelvin] [Symbol]

He II   Helium II [Liquid Helium below 2.2 Kelvin] [Symbol]

He-3   Helium-3 [also ^{3}He, or He^3]

He-4   Helium-4 [also ^{4}He, or He^4]

He-6   Helium-6 [also ^{6}He, or He^6]

/he   help option [MS-DOS Linker]

h.e.   *(hoc est)* – that is

HEA   Ethanolamine

HEA   Heating Engineering Association [UK]

HEA   High-Efficiency Antireflection Coating

HEA   High-Energy Accelerator [also HEAC] [Physics]

HEA   High-Energy Approximation

HEA   High-Energy Astrophysics

HEA   Higher Education Act [US]

HEA   Horticultural Education Association [UK]

HEA   Horticultural Exhibitors Association [UK]

HEA   2-Hydroxyethyl Acrylate

HEA   Hydroxyethylamine

HEAC   High-Energy Accelerator [also HEA] [Physics]

HEAC   Hydrogen Environmentally Assisted Cracking

HEACC   (International Conference on) High-Energy Accelerators [Physics]

HEAD   High-Energy Astrophysics Division [of American Astronomical Society, US]

HEAL   Human Exposure Assessment Location [of World Health Organization/United Nations Environmental Program]

HEALS   Honeywell Error Analysis and Logging System

Health Educ J   Health Education Journal

Health Healing   Health and Healing [Journal published in Australia]

Health J   Health Journal [Published in the US]

Health Phys   Health Physics [Published in the UK]

Health Safety Work   Health and Safety at Work [Published in the UK]

Health Serv J   Health Services Journal

HealthSTAR   Health Services Technology, Administration and Research (Database) [of National Library of Medicine, US]

Health Technol Assess   Health Technology Assessment [Journal of the National Coordinating Center for Health Technology Assessment, UK]

Health Technol Assess (Rockv)   Health Technology Assessment (Rockville) [Journal of the Agency for Health Care Policy and Research, National Institutes of Health; published in Rockville, Maryland, US]

Health Technol Assess Rep   Health Technology Assessment Reports [of the Agency for Health Care Policy and Research, National Institutes of Health, US]

HEAO   High-Energy Astronomy (or Astronomical) Observatory

HEAP   Helicopter Extended Area Platform

HEAP   High-Energy Aim Point

HEAP   High-Explosive, Armor Piercing [also HE-AP]

Hear Res   Hearing Research [Journal published in the Netherlands]

HEART   Human Engineering Analysis and Requirements Tool

Heart Lung   Heart and Lung [US Journal]

HEAT   High-Energy Antimatter Telescope [at CERN– European Laboratory for Particle Physics, Geneva, Switzerland]

HEAT   High-Explosive, Antitank [also HE-AT]

HEAT   Hydroxyphenyl Ethyl Aminoethyl Tetralone

HEAT   dl-2-[β-(3-Iodo-4-Hydroxyphenyl)ethylaminoethyl]-tetralone

Heat   Heating [also heat]

Heat Air Cond J   Heating and Air Conditioning Journal [UK]

HEAT CAP   Heat Treating Certificate of Educational Achievement Program [of ASM International, US]

Heat/Combust Equip News   Heating/Combustion Equipment News [Published in the US]

Heat Process Dig   Heat Processing Digest [Published by ASM International, US]

Heat Recovery Syst CHP   Heat Recovery Systems and Combined Heat and Power [Journal published in the UK]

Heat Technol   Heat and Technology [Journal published in Italy]

Heat Technol   Heat Technology [Journal published in the US]

Heat Transf Eng   Heat Transfer Engineering [Journal published in the US]

Heat Transf Jpn Res   Heat Transfer–Japanese Research [Published in the US]

Heat Transf Sov Res   Heat Transfer–Soviet Research [Published in the US]

Heat Treat   Heat Treating [Journal published in the US]

Heat Treat Met Heat Treatment of Metals [Publication of the Wolfson Heat Treatment Center, Birmingham, UK]

Heat Treat Met (China) Heat Treatment of Metals (China) [Journal published in PR China]

HEB High-Efficiency Binding

HEBA 2-Hydroxy-2-Ethyl Butanoic Acid

HEBC Heavy Enamelled Bonded Single-Cotton-Covered (Electric Wire)

HEBD High-Explosive, Base Detonating [also HE-BD]

HEBDC Heavy Enamelled Bonded Double-Cotton-Covered (Electric Wire)

HEBDP Heavy Enamelled Double-Paper-Bonded (Electric Wire)

HEBDS Heavy Enamelled Bonded Double-Silk-Covered (Electric Wire)

HEBIS Hessisches Bibliothekssystem [Hessian Automated Library System, Germany]

HEBM High-Energy Ball Mill(ing)

HEBP Heavy Enamelled Single-Paper-Bonded (Electric Wire)

HEBS Heavy Enamelled Bonded Single-Silk-Covered (Electric Wire)

HEC Halstead Exploiting Center [UK]

HEC Hastings Environment Council [UK]

HEC Heavy Enamelled Single-Cotton-Covered (Electric Wire)

HEC High-Efficiency Cyclone

HEC High-Energy Collider [Particle Physics]

HEC High-Energy Corona

HEC Hollerith Electronic Computer

HEC Horowitz-Eastman-Crane Method

HEC Hydro-Electricity Commission [Tasmania]

HEC Hydrogen Embrittlement Cracking

HEC Hydrologic Engineering Center [of US Army Corps of Engineers]

HEC Hydroxyethylcellulose

HECD Hall Electroconductivity Detector

HECD Hall Electrolytic Conductivity Detector

HeCd Helium-Cadmium (Laser)

HECI Human-Interface Equipment Catalog Item

n-HeCl n-Hexyl Chloride

HECS Hazardous Energy Control Standard [of Occupational Safety and Health Administration, US]

HeCs Helium Cesium (Compound)

HECSAGON Horowitz-Eastman-Crane Symbol Array Governed by Orthodox Notation

hecto- SI prefix representing 10^2

hectol hectoliter [Unit]

HECTOR Heated Experimental Carbon Thermal Oscillator Reactor

HECTOR Hot Enriched Carbon-Moderated Thermal Oscillator Reactor

HECV Heavy Enamelled Single-Cotton-Covered Varnished (Electric Wire)

HECV Helium Check Valve [also HeCV]

HED Hall Effect Device

HED Hauteinheitsdosis (or Hauterythemdosis) [Skin Erythema Dose]

HED Horizontal Electric Dipole [also hed]

HED Hydroelectric Dam

HED N-Hydroxyethylethylenediamine-N,N',N'-Triacetic Acid [also Hed]

HEDA High-Explosive, Delayed Action [also HE-DA]

HEDC Houston Economic Development Council [Texas, US]

HEDC Heavy Enamelled Double-Cotton-Covered (Electric Wire)

HEDCOM Headquarters Command [of US Air Force in Washington, DC, US]

HEDCV Heavy Enamelled Double-Cotton-Covered Varnished (Electric Wire)

HEDI High Endoatmospheric Defense Interceptor

HEDIEN N-(2-Hydroxyethyl)diethylenetriamine

HEDL Hanford Engineering Development Laboratory [Washington State, US]

HEDP High Explosive, Dual Purpose

HEDP 1-Hydroxyethylidene-1,1-Diphosphonic Acid

HEDR Hanford Environmental Dose Reconstruction (Project) [Washington State, US]

HEDS Heavy Enamelled Double-Silk-Covered (Electric Wire)

HEDS High Energy Dislocation Structure [Materials Science]

HEDS Human Exploration and Development of Space (Enterprise) [of NASA]

HEDSV Heavy Enamelled Double-Silk-Covered Varnished (Electric Wire)

HEDTA N-Hydroxyethylethylenediamine Triacetate; N-(2-Hydroxyethyl)ethylenediaminetriacetic Acid

HED3A N-Hydroxyethylethylenediamine-N,N',N'-Triacetate [also Hed3a, or hed3a]

HEDTA³⁻ Tridentate Hydroxyethylethylethylenediaminetriacetic Acid

H₂EDTA²⁻ Bidentate Hydroxyethylethylenediaminetriacetic Acid

H₄EDTA Ethylethylenediaminetetraacetic Acid, Tetra Acid

H4EDTA Ethylenediaminetetraacetic Acid [also H4edta, or h4edta]

HEE Human Energy Expenditure

HEE Hydroelectric Energy

HEE Hydrogen Environment Embrittlement

HEEB High-Energy Electron Beam

HEEC (2-Hydroxyethyl)ethylcellulose

HEED High-Energy Electron Diffraction

HEEDTA N-Hydroxyethylethylenediaminetriacetic Acid

H4EEDTA N,N'-[Oxybis(2,1-Ethanediyl)bis[N-(Carboxymethyl) glycine] [also H4eedta, or h4eedta]

HEELS High-Energy Electron (Energy) Loss Spectroscopy

HEELS High-Resolution Electron Energy-Loss Spectroscopy

HEEP Health Effects of Environmental Pollution [of National Library of Medicine/Biosciences Information Service, US]

HEEP Highway Engineering Exchange Program [US]

HEF Heated Effluents (Database) [of Cornell University, Ithaca, New York, US]

HEF High-Elongation Furnace (Black)

HEF High-Energy Forging

HEF High-Energy Forming

HEF High-Energy Fuel

HEF High-Expansion Foam

HEF High-Explosive, Fragmentation (Bomb) [also HE-F]

HEF Hispanic Energy Forum [US]

He/F$_2$ Helium-to-(Molecular) Fluorine (Ratio)

HEF Black High-Elongation Furnace Black [also HEF black]

HEG Heavy Enamel (Single) Glass (Electric Wire)

HEG Helium Gage [also HeG]

H4EGTA Ethyleneglycol Bis(2-Aminoethylether)tetraacetic Acid [also H4edta, or H4edta]

HEGV Helium Gage Valve [also HeGV]

HEH High-Explosive, Heavy (Projectile) [also HE-H]

HEH(ClMP) 2-Ethylhexylhydrogenchloromethylphosphonic Acid

HEH(EHP) 2-Ethylhexylhydrogen(2-Ethylhexylphosphonic Acid)

HEHF Hanford Environmental Health Foundation [Washington State, US]

HEHIXE High-Energy Heavy Ions X-Ray Emission

HEHΦP 2-Ethylhexylphenylphosphoric Acid

H2EHP Mono(2-Ethylhexyl)phosphoric Acid

H2[EHP] (2-Ethyl)hexylphosphonic Acid

HEI Health Effects Institute [Cambridge, Massachusetts, US]

HEI Heat Exchange Institute [US]

HEI High-Energy Ignition

HEI High Explosive, Incendiary [also HE-I]

HEI Human Exploration Initiative [Canada and US]

HEIA High-Explosive, Immediate Action

HEIS Hanford Environmental Information System

HEIS High-Energy Ion Scattering

HEIS High-Energy Ion Backscattering Spectroscopy

HEI-T High-Explosive Incendiary with Tracer

HEIX Home Economics Information Exchange [of Food and Agricultural Organization]

HEL High-Energy Laser

HEL High-Explosive, Light (Shell)

HEL Hugoniot Elastic Limit [Mechanics]

HEL Human Engineering Laboratory [Aberdeen Proving Ground, Maryland, US]

Hel Helicopter [also hel]

HeLa Helen Lane (Cancer Cells) [Medicine]

HeLa Henrietta Lacks (Cancer Cells) [Medicine]

HELC High-Explosive, Long Case [also HE-LC]

HELCOM Baltic Marine Environment Protection Commission–Helsinki Convention [Finland]

HELEX Hydrogenous Exponential Liquid Experiment

HELINOISE Helicopter and Tilt-Rotor Aircraft Exterior Noise Research

HELIOPS Helicopter Operations Panel [of International Civil Aviation Organization]

HELMEPA Hellenic Marine Environmental Protection Association

HELMS Helicopter Multifunction System

HELP Hazardous Emergency Leaks Procedure [of Union Carbide Corporation, US]

HELP Helicopter Electronic Landing Path

HELP Helicopter Emergency Life-Saving Program

HELP Highly-Extendable Language Processor

HELP Highway Emergency Locating Plan [US]

HELPIS Higher Education Learning Programs Information Service [of Council for Educational Technology, UK]

HELS High-Energy Loss Spectrum

HELV Helvetica (Font) [also Helv]

Helv Helvetica [Switzerland, or Swiss]

Helv Chim Acta Helvetica Chimica Acta [Chemical Journal published in Switzerland]

Helv Phys Acta Helvetica Physica Acta [Published by Schweizerische Physikalische Gesellschaft, Switzerland]

HEM Harmonization of Environmental Measurement

HEM Hazardous Environment Machine

HEM Heat Exchange(r) Method

HEM Helicopter-Borne Electromagnetics

HEM Helicopter Electromagnetic Survey

HEM High-Energy (Ball) Mill(ing)

HEM Hitchhiker Experiment Module [of NASA]

HEM Hostile-Environment Machine

HEM Hybrid Electromagnetic (Wave)

HEM Hydrogen Embrittlement

Hem β-Hydroxyethylmercaptan

HEMA Hydroxyethyl Methacrylate

HEMA 2-Hydroxyethyl Methacrylate

HEMAC Hybrid Electromagnetic Antenna Coupler

HEMATOL Hematological; Hematologist; Hematology [also hematol]

Heme Hemoglobin–Myoglobin [also heme] [Biochemistry]

HEMEL Hexamethylmelamine [also HMM, Hemel, or hemel]

HEMF High-Efficiency Metal Fiber (Filter)

Hem Fir Hemlock Fir

HEMI Hemispherical (S-Band Antenna)

Hem Ind Hemijska Industriji [Yugoslavian Scientific Journal]

HEMIP High-Efficiency Microwave-Induced Plasma

HEMP High-Altitude Electromagnetic Pulse

HEMPA Hexamethylphosphoramide [also HMPA, Hempa, or hempa]

HEMS Helicopter Electromagnetic Survey

HEMS Helicopter Emergency Medical Services [UK]

HEMT High-Electron Mobility Transistor

HEMT High-Electron Movement Transistor

HEMV Helium Manual Valve [also HeM]

HeNe Helium-Neon (Laser) [also He-Ne]

He/NH$_3$ Helium/Ammonia (Mixture)

HENR Higher Energy Nuclear Reaction

HENRE High-Energy Neutron Reactions Experiment [of US Atomic Energy Commission]

He/N₂O Helium/Nitrous Oxide (Mixture)

HEO Hanford Environmental Oversight [US]

HEO High Earth Orbit (Satellite)

HEO High Energy Orbit

He/O₂ Helium/Oxygen (Ratio) [Symbol]

HEOD Dieldrin

HEOD Hexachloro Epoxy Octahydroendohexa Dimethanonaphthalene

HEOS Highly Eccentric Orbit Satellite [Series of Satellites of the European Space Research Organization]

HEOS Hildebrand Equation of State

HEOS-1 First Highly Eccentric Orbit Satellite [Series of Satellites of the European Space Research Organization]

HEP Habitat Evaluation Procedure

HEP Heterogeneous Element Processor

HEP High-Energy Particle

HEP High-Energy Physics

HEP High-Explosive, Penetrating

HEP High-Explosive Plastic

HEP Hochschulerneuerungsprogramm (für die neuen Länder) [University Renewal Program (for the New States), Germany]

HEP International Europhysics Conference on High-Energy Physics

HEP Homogeneous Element Processor

HEP Human Error Probability

HEP Hydroelectric Plant

HEP Hydroelectric Power

HEp Human Epithelial (Cells) [Medicine]

Hep Heptatitis [Medicine]

Hep Heptane

Hep Heptatoma [Medicine]

HEPA High-Efficiency Particle Accumulator

HEPA High-Efficiency Particle Air (Filter); High-Efficiency Particulate Absolute (Filter); High-Efficiency Particulate Air (Filter)

Hep A Hepatitis A [Medicine]

HEPAT High-Explosive Plastic Antitank (Charge)

Hepatol Hepatological; Hepatologist; Hepatology

Hepatol Hepatology [International Journal]

Hep B Hepatitis B [Medicine]

HEPC Hydro-Electric Power Commission [Canada]

Hep C Hepatitis C [Medicine]

HEPCAT Helicopter Pilot Control and Training

HEP-CCC High Energy Physics Computing Coordinating Committee [of CERN–European Laboratory for Particle Physics, Geneva, Switzerland]

HEPD Heat Engine Propulsion Division [of US Department of Energy]

HEPE Hydroxyeicosapentaenoic Acid

5-HEPE 5-Hydroxy-(6E,8Z,11Z,14Z,17Z)-Eicosapentaenoic Acid

(±)-5-HEPE (±)-5-Hydroxy-(6E,8Z,11Z,14Z,17Z)-Eicosapentaenoic Acid

15(S)-HEPE 15(S)-Hydroxy-(5Z,8Z,11Z,13E,17Z)-Eicosapentaenoic Acid

HEPES N-(2-Hydroxylethyl)piperazine-N'-(2-Ethanesulfonic Acid) [also Hepes]

HEPEX Heavy Element Partitioning by Extraction

HEPG High-Energy Physics Group

HEPI High-Energy Physics Index [of Deutsches Elektronen-Synchrotron, Hamburg, Germany]

HEPIC High-Energy Physics Information Center [Netherlands]

HEPIX (Meeting of) UNIX Users in High Energy Physics [of Fermilab, Batavia, Illinois, US] [also HEPiX]

HEPL High-Energy Physics Laboratory [of Stanford University, California, US]

HEPnet News High-Energy Physics Network News [of HEP Network Resource Center]

HEPP Hoffman Evaluation Program and Procedure

HEPP Hydroelectric Power Plant

HEPPS N-(2-Hydroxyethyl)piperazine-N'-(3-Propanesulfonic Acid)

HEPPSO N-(2-Hydroxyethyl)piperazine-N'-(2-Hydroxypropanesulfonic Acid)

HEPR Hard-Grade Ethylene Propylene Rubber

HEPS High-Energy Prespark

HEPS Hydroelectric Power Station

HEP-T High-Explosive, Plastic with Tracer

HEPTA N-Hydroxyethylethylenediaminetriacetic Acid

HEPVIS High Energy Physics Visualization and Data Analysis Workshop [at Stanford Linerar Accelerator Center, Stanford University, California, US] [also HEPVis]

HER HIM (Hardware Interface Module) Equipment Rack

HER Hot Electron Relaxation

HER Human Error Rate

HER Hydrogen Evolution Reaction

Her Hercules [Astronomy]

Her X-1 Hercules X-1 [Astronomy]

HERA (Hadron-Elektronen-Ring-Anlage) – German for "Hadron Electron Ring Accelerator"

HERA Hadron Electron Ring Accelerator [of Deutsches Elektronen-Synchrotron, Hamburg, Germany]

HERA Heritage Australia Information System [of Australian Heritage Commission]

HERA High-Explosive Rocket-Assisted

HERAC Health and Environmental Research Advisory Committee [of US Department of Energy]

HERALD Heterogenous Experimental Reactor, Aldermaston [of Atomic Weapons Research Establishment, UK]

HERALD Highly Enriched Reactor, Aldermaston [UK]

HERAS Hellenic Radar System [Greece]

Herbol Herbological; Herbologist; Herbology

HERBRECS Queensland Herbarium Plant Specimen Database [Australia]

He-RBS Helium Rutherford Backscattering (Spectroscopy) [or He RBS]

Herbs Health Herbs for Health [US Journal]

HERC Hercules (Graphics Driver) [Computers] [also Herc]

Herc Hercegovina; Hercegovinian

HercMonoHi Hercules (Graphics Driver), Monochrome, High Mode

HERD High-Energy X-Ray Diffraction

HERD Human Exposure Research Division [of National Exposure Research Laboratory, US]

Heref Hereford [UK]

HERF Hazards of Electromagnetic Radiation to Fuel

HERF High Energy Rate Forging

HERF High Energy Rate Forming

HERL Health Effects Research Laboratory [US]

Her Libr Sci Herald of Library Science [Published by Banaras Hindu University, Varanasi, India]

HERMAN Hierarchical Environmental Retrieval for Management Access and Networking [of Biological Information Service, US]

HERMES Heavy Element and Radioactive Material Electromagnetic Separator [UK]

HERMES Heuristic Mechanized Documentation Information Service [Romania]

HERMES High-Efficieny High-Resolution Measurements

Hermsdorfer Tech Mitt Hermsdorfer Technische Mitteilungen [Technical Communication on Ceramics; published in Hermsdorf, Germany]

HERO Hazards of Electromagnetic Radiation to Ordnance

HERO Heath Robot

HERO Hot Experimental Reactor of 0 (Zero) Power [UK]

HERP Hazards of Electromagnetic Radiation to Personnel

HERS Hardware Error Recovery System

HERSCP Hazardous Exposure Reduction and Safety Criteria Plan

Herts Hertfordshire [UK]

HERTIS Hertfordshire Technical Library and Information Service [UK]

Her X-1 Hercules X-1 [Astronomy]

HES Hall Effect Sensor

HES Hanford Engineering Service [Hanford, Washington State, US]

HES Hawaiian Entomological Society [US]

HES Heavy Enamelled Single-Silk-Covered (Electric Wire)

HES Hercules Experiment Station [Woodale, Delaware, US]

HES High Early Strength (Cement)

HES High-Energy Scattering

HES High-Explosive, Shell

HES High-Explosive, Smoke (Shell)

HES Homeopathic Education Services [US]

HES House Exchange System

HES Hydroxyethyl Starch

HESEC Hanford Environmental Science and Engineering Consortium [Washington State, US]

HESH High-Energy Scattering of Hadrons [Particle Physics]

HE/SH High-Explosive, Squashhead

HESS High Energy Squib Simulator

HESS History of Earth Sciences Society [US]

HESS Houston Engineering and Scientific Society [US]

HESV Heavy Enamelled Single-Silk-Covered Varnished (Electric Wire)

HET Heavy Equipment Transporter

HET Hexaethyltetraphosphate

HET Hole Expansion Test (for Steel Sheeting)

HET Hot Electron Transistor

HET Hydroelectric Turbine

HE-T High Explosive with Tracer

HETCOR Heteronuclear Correlation (Spectroscopy)

HETE High-Energy Transient Experiment [of Massachusetts Institute of Technology, Cambridge, Massachusetts, US]

HETE Hydroxyeicosatetraenoic Acid

(+)-5-HETE (+)-5(S)-Hydroxy-(6E,8Z,11Z,14Z)-Eicosatetraenoic Acid

(±)-5-HETE (±)-5-Hydroxy-(6E,8Z,11Z,14Z)-Eicosatetraenoic Acid

(−)-5-HETE (−)-5(R)-Hydroxy-(6E,8Z,11Z,14Z)-Eicosatetraenoic Acid

8(R)-HETE 8(R)-Hydroxy-(5Z,9E,11Z,14Z)-Eicosatetraenoic Acid

8(S)-HETE 8(S)-Hydroxy-(5Z,9E,11Z,14Z)-Eicosatetraenoic Acid

(±)-9—HETE (±)-Hydroxy-(5Z,7E,11Z,14Z)-Eicosatetraenoic Acid

9(S)-HETE 9(S)-Hydroxy-(5Z,7E,11Z,14Z)-Eicosatetraenoic Acid

11(R)-HETE 11(R)-Hydroxy-(5Z,8Z,12E,14Z)-Eicosatetraenoic Acid

11(S)-HETE 11(R)-Hydroxy-(5Z,8Z,12E,14Z)-Eicosatetraenoic Acid

12(R)-HETE
     11(R)-Hydroxy-(5Z,8Z,10E,14Z)-Eicosatetraenoic Acid

12(S)-HETE 12(S)-Hydroxy-(5Z,8Z,10E,14Z)-Eicosatetraenoic Acid

15(S)-HETE 15(S)-Hydroxy-(5Z,8Z,11Z,13Z)-Eicosatetraenoic Acid

20-HETE 20-Hydroxy-(5Z,8Z,11Z,14Z)-Eicosatetraenoic Acid

Heteroatomic Chem Heteroatomic Chemistry [International Journal]

heteroepi-Al Heteroepitaxial Aluminum [also Heteroepi-Al]

heteroepi-CdTe Heteroepitaxial Cadmium Telluride [also Heteroepi-CdTe]

heteroepi-Si Heteroepitaxial Silicon [also Heteroepi-Si]

heteroepi-SiGe Heteroepitaxial Silicon Germanium [also Heteroepi-SiGe]

HETLC High-Efficiency Thin-Layer Chromatography

HETM Hybrid Engineering Test Model

HETP Height Equivalent of a Theoretical Plate

HETP Hexaethyl Tetraphosphate

15(S)-HETRE 15(S)-Hydroxy-(8Z,11Z,13E)-Eicosatrienoic Acid

HETS Height Equivalent to a Theoretical Stage

He-II UPS Helium-II Ultraviolet Photoelectron Spectroscopy

HEU High(ly)-Enriched Uranium

HEU Hydroelectric Unit

HEU/Th Highly Enriched Uranium/Thorium Fuel

HEVAC Heating, Ventilating and Air Conditioning

HEVAC Heating, Ventilating and Air Conditioning Manufacturers Association [UK]

HEW Hamburgische Elektrizitätswerke [Electricity Company of Hamburg, Germany]

HEW (Department of) Health, Education and Welfare [now (Department of) Health and Human Services]

Hewlett-Packard J Hewlett-Packard Journal [US]

HEX Heat Exchanger [also HEx]

HEX Hexadecimal [also Hex, or hex]

HEX High-Energy Explosive

HEX High-Energy X-Rays

Hex Hexagon(al) [also hex]

hex Uranium Hexafluoride

HexA Hexanoic Acid

Hexa Hexanitrodiphenylamine

hexa hexamethylene tetramine [also Hexa, or HMTA]

HEX-BIN Hexadecimal-to-Binary (Conversion) palso Hex-Bin]

Hex Bolt Hexagonal Bolt [also hex bolt]

HEX-DEC Hexadecimal-to-Decimal (Conversion) [also Hex-Dec]

HEXFET Hexagonal Metal-Oxide Field-Effect Transistor

Hex Hd Hexagonal Head (Screw)

Hex Nut Hexagonal Nut [also hex nut]

HEX$ Hexadecimal-Value String [Programming]

Hex Scr Hexagonal Screw [also hex scr]

HEY High Electron Yield

HEZS Hyper-Environmental Test System [of US Air Force]

HF Hamiltonian Function [Mechanics]

HF Hand Feed

HF Harassing Fire

HF Hardenability Factor [Metallurgy]

HF Hard Failure [Computers]

HF Harmonic Frequency

HF Harmonic Function [Mathematics]

HF Hartree-Fock (Field) [Quantum Mechanics]

HF Haze Filter

HF Heat Flow

HF Heavy Fermion (System) [Particle Physics]

HF Heavy Fuel

HF Heavy Flint (Glass)

HF Heavy Vinyl-Acetal-Covered (Electric Wire)

HF Height Finder; Height Finding [Radar]

HF Heisenberg Force [Nuclear Physics]

HF Hellmann-Feynman (Theorem) [Quantum Mechanics]

HF Hemlock Fir

HF Heptaline Formate

HF Heredity Factor [Genetics]

HF Hermitian Formula [Mathematics]

HF High Fatigue

HF High Field

HF High Flow

HF High Frequency [also H-F, H/F, hf, h/f, or h-f]

HF High Friction

HF High Frontier [US]

HF History File

HF Hoar Frost

HF Holding Furnace [Metallurgy]

HF Homogeneous Flow

HF Horizontal Flight

HF Hot Filament

HF Hot Firing

HF Hot Forging

HF Human Factors

HF Hydraulic Fluid

HF Hydraulic Fracturing [Oil Exploration]

HF Hydraulic Friction

HF Hydrofining [Chemical Engineering]

HF Hydrofoil

HF Hydrogen Fill

HF Hydrogen Fluoride (or Hydrofluoric Acid)

HF Hyperfiltration

HF Hyperfine [also hf]

HF Hyperfine Field [also hf] [Physics]

H/F Hydrogen-to-Fluorine (Ratio)

Hf Hafnium [Symbol]

Hf-170 Hafnium-170 [also ^{170}Hf, or Hf^{170}]

Hf-171 Hafnium-171 [also ^{171}Hf, or Hf^{171}]

Hf-172 Hafnium-172 [also ^{172}Hf, or Hf^{172}]

Hf-173 Hafnium-173 [also ^{173}Hf, or Hf^{173}]

Hf-174 Hafnium-174 [also ^{174}Hf, or Hf^{174}]

Hf-175 Hafnium-175 [also ^{175}Hf, or Hf^{175}]

Hf-176 Hafnium-176 [also ^{176}Hf, or Hf^{176}]

Hf-177 Hafnium-177 [also ^{177}Hf, or Hf^{177}]

Hf-178 Hafnium-178 [also ^{178}Hf, or Hf^{178}]

Hf-179 Hafnium-179 [also ^{179}Hf, or Hf^{179}]

Hf-180 Hafnium-180 [also ^{180}Hf, or Hf^{180}]

Hf-181 Hafnium-181 [also ^{181}Hf, or Hf^{181}]

H(f) Frequency Response Function [Symbol]

hf half

hf hyperfine [also HF]

H8/f7 Close Running Fit; Basic Hole System [ISO Symbol]

h-f high frequency [also HF, H/F, H-F, hf, or h/f]

HFA Hard Fiber Association [US]

HFA Hartree-Fock Approximation [Quantum Mechanics]

HFA Hexafluoroacetone

HFA High Field Approximation [Physics]

HFA High Flow Alarm

HFA High Frequency Accelerometer

HFA High-Frequency Amplifier

HFA Homofolic Acid

HFA Hydrofluoric Acid

HFAA Heptafluorobutanoic Acid Anhydride

HFAC Hexafluoroacetylacetonate; Hexafluoroacetylacetonato-1,1,1,5,5,5-Hexafluoro-2,4-Pentanedione (or 1,1,1-Hexafluoroacetylacetone) [also hfac]

HFAC Hydrofluoric Acid Copper

HFAC Human Factors Association of Canada

(HFAC)Cu(BTMSA) 1,1,1,5,5,5-Hexafluoroacetylacetonate Copper (I) Bis(trimethylsilyl)acetylene [also (hfac)Cu(btmsa)]

(HFAC)Cu(1,5-COD) 1,1,1,5,5,5-Hexafluoroacetylacetone Copper (I) 1,5-Cyclooctadiene [also (hfac)Cu(1,5-COD)]

(HFAC)Cu(COD) 1,1,1,5,5,5-Hexafluoroacetylacetonate Copper (I) Cyclooctadiene [also (hfac)Cu(COD)]

(HFAC)Cu(PMe₃) 1,1,1,5,5,5-Hexafluoroacetylacetonate Copper (I) Trimethylphosphine Copper [also (hfac)Cu(PMe₃)]

(HFAC)Cu(TMVS) 1,1,1,5,5,5-Hexafluoroacetylacetonate Copper (I) Trimethylvinylsilane [also (hfac)Cu(tmvs)]

(HFAC)Cu(VTMS) 1,1,1,5,5,5-Hexafluoroacetylacetonate Copper (I) Vinyltrimethylsilane [also (hfac)Cu(vtms)]

HFACH 1,1,1,5,5,5-Hexafluoroacetylacetonate Acid [also hfacH]

HFAR Honduran Foundation for Agricultural Research

HF/Ar Hydrogen Fluoride/Argon (Gas)

HFAS Honeywell File Access System

HFB Hamburger Flugzeugbau [German Aircraft Manufacturer located at Hamburg]

HFB Hartree-Fock-Bogoliubov [Physics]

HFB Hexafluorobut-2-yne

HFB Hopper-Feeder-Bolter

HFBA Heptafluorobutanoic Acid

HFBA Heptafluorobutyric Acid

HFBA Heptafluorobutyric Anhydride

HFBC High-Frequency Broadcasting

Hf Bd Half Bound [also hf bd]

Hf-Be Hafnium-Beryllium (Alloy System)

HFBR High Flux Beam Reactor [of Brookhaven National Laboratory, Upton, New York, US]

HFBS High-Frequency Broadcasting Schedule [of International Telecommunications Union, Switzerland]

Hf(BuO)₄ Hafnium Butoxide

HFC Hard Fille Capsule(s) [Pharmacology]

HFC Heat Flow and Convection

HFC 3-(Heptafluoropropylhydroxymethylene)d-Camphorate [also hfc]

HFC High-Frequency Correction

HFC High-Frequency Current

HFC Hybrid Fiber-Coaxial (Cable)

HFC Hydraulic Flight Control

HFC Hydrofluorocarbon

HFC Hyperfine Coupling [Solid-State Physics]

HfC Hafnium Carbide

HFCE HFIR (High-Flux Isotope Reactor) Critical Experiment [of Oak Ridge National Laboratory, Tennessee, US]

HFCO Ethyl-trans,trans-2,4-Hexadienoate-Iron Tricarbonyl

Hf(Cp)₂Cl₂ Bis(cyclopentadienyl)hafnium Dichloride (or Hafnocene Dichloride)

HFCT Hydraulic Flight Control Test

HFCV Helium Flow Control Valve

HFCVD Hot-Filament Chemical Vapor Deposition

HFD Hartree-Fock Dispersion [Physics]

HFD High-Field Domain [Solid-State Physics]

HFD High-Frequency Deflection

HFD High-Frequency Discharge

HFD Hrvatsko Fizikalno Društvo [Croatian Physical Society]

HFDF High-Frequency Direction-Finder [also HF/DF]

HFDF High-Frequency Distribution Frame

HFDS Hydrogen Fluid Distribution System

HFE High-Frequency Executive

HFE Hot Field Emission

HFE Human Factors Engineering

HFE Human Factors Evaluation

HFE Human Factors in Engineering

HFEF Hot Fuel Examination Facility [of Argonne National Laboratory-West, Idaho Falls, US]

HFET Heterojunction Field-Effect Transistor

Hf(EtO)₄ Hafnium Ethoxide

HFETR High-Flux Engineering Testing Reactor [of US Department of Energy]

HFEV Higher Free Energy Value

HFF Hartree-Fock Field [Quantum Mechanics]

HFF High-Frequency Fluctuation

HFF High-Frequency Furnace

HFF Hochschulbibliothek für Film– und Fernsehen [University Library for Film and Television, University of Potsdam, Germany]

HFF Hyperfine Field [Solid-State Physics]

HFG Heavy Free Gas

HFG High-Frequency Generator

HFG Höhere Fachschule für Gestaltung [Higher School of Arts, Switzerland]

HFH High-Frequency Heating

Hf H Half Hard [also hf h] [Metallurgy]

HF/H₂O Hydrogen Fluoride-to Water (Ratio)

HFI Hair and Faces International

HFI Height Finding Instrument

HFI Helicopter Foundation International [US]

HFI High-Frequency Inductance

HFI High-Frequency Injection

HFI Hyperfine Interaction [also hfi] [Solid-State Physics]

HFIAW (International Association of) Heat and Frost Insulators and Asbestos Workers

HFIB Hexafluoroisobutylene

H-Field Magnetic Field [also H-field]

HFIF High-Frequency Induction Furnace

HFIH High-Frequency Induction Hardening [Metallurgy]

HFIM High-Frequency Induction Motor

HFIM High-Frequency Instruments and Measurements [Institute of Electrical and Electronics Engineers Instrumentation and Measurement Society Committee]

HFIP Hexafluoroisopropanol

HFIP 1,1,1,3,3,3-Hexafluoro-2-Propanol

HFIR High-Flux Isotope Reactor [of Oak Ridge National Laboratory, Tennessee, US]

HFIS HEPA (High-Efficiency Particle Accumulator) Filter Inspection System

HFITR High-Field Ignition Test Reactor

HFIW High-Frequency Induction Welding

HFL High Free Lift

HFL Holloman Field Laboratory [at Holloman Air Force Base, near Alamogordo, New Mexico, US]

HFl Florin (or Guilder) [Currency of the Netherlands]

HF LCAO Hartree-Fock Linear Combination of Atomic Orbitals (Method) [also HF-LCAO]

HFM Hazardous Fluids Module

HFM Heat-Flow Meter

HFM High Field Magnetism

HFM High-Frequency Microphone

HFM Hollow Fiber Membrane

HFMBPT Hartree-Fock Many-Body Perturbation Theory [Physics]

Hf-Mo Hafnium-Molybdenum (Alloy)

HFMR High-Field Magnetoresistance [Solid-State Physics]

HFMU High Fidelity Mockup

HfN Hafnium Nitride

Hf(NEt$_2$)$_4$ Tetrakis(diethylamino)hafnium

HFNRF High Flux Neutron Radiography Facility [of Sandia National Laboratories, Albuquerque, New Mexico, US]

HFO Hartree-Fock Orbital [Physics]

HFO Hartree-Fock-Overhauser (Theory) [Physics]

HFO High-Frequency Oscillator

Hf(OBu)$_4$ Hafnium Butoxide

HFOD 1,1,1,2,2,3,3-Heptafluoro-7,7-Dimethyl-4,6-Octanedionate [also hfod]; Heptafluorodimethyloctanedione [also hfod]

Hf(OEt)$_4$ Hafnium Ethoxide

HfOPr^i Hafnium Isopropoxide

HFORL Human Factors Operations Research Laboratory

HFORMAT (Disk) Head Format (Command) [also hformat]

HFOVR Hemispherical Field-of-View Radiometer

HFP Heavy Flavor Physics [Particle Physics]

HFP Hexafluoropropylene

HFP High-Field Phase [Solid-State Physics]

HFPD Hartree-Fock Proper Dissociation [Physics]

HFPE 1-Hydro-Penta-Fluoropropylene

HfPMCpCl$_3$ Pentamethylcyclopentadienylhafnium Trichloride

Hf(PMCp)$_2$Cl$_2$ Bis(pentamethylcyclopentadienyl)hafnium Dichloride

HFPO Hexafluoropropylene Epoxide

HFPRI Hokkaido Forest Products Research Institute [Japan]

HFPS High-Frequency Phase Shifter

HFPVF Hexafluoropropylene Vinylidene Fluoride

HFR High-Flux (Nuclear) Reactor

HFR High-Frequency Recombination [also Hfr] [Microbiology]

HFRI Humboldt Field Research Institute [Steuben, Maine, US]

HFRDF High-Frequency Repeater Distribution Frame

HFRSc Forged Roll Scleroscope Hardness Number, Model c

HFRSd Forged Roll Scleroscope Hardness Number, Model d

HFRT High-Frequency Resonance Technique

HFRW High-Frequency Resistance Welding

HFRX High-Frequency Receiver Controller

HFS Hartree-Fock-Slater [Physics]

HFS Hierarchical File System

HFS High-Field Superconductor

HFS High-Frequency Sparking

HFS Horizontal Flight Simulator

HFS Human Factors Society [US]

HFS Hyperfine Structure [also hfs] [Spectroscopy]

HFS Hypothetical Future Samples

HFSA Human Factors Society of America

HFSA Hydrofluosilicic Acid

HFSC Heavy-Fermion Superconductor

HF-SCF Hartree-Fock Self-Consistent Field [also HF/SCF] [Quantum Mechanics]

HFSF High-Flux Solar Furnace

hFSH Human Follicle Stimulating Hormone [also HFSH] [Biochemistry]

HfSi Hafnium Silicide

Hf/Si Hafnium on Silicon Substrate

HFSS High-Frequency Search System

HFSS High-Frequency Sounder System

HFSS Hyperfine Structure Spectrum [Spectroscopy]

HFST High-Flux Scram Trip [Nuclear Reactors]

HFT Haigh Fatigue Test

HFT Hartree-Fock Theory [Physics]

HFT Heat Flux Transducer

HFT High-Frequency Test(er)

HFT High-Frequency Triode

HFT High Function Terminal

HFT Horizontal Flight Test

HFT Hypercomplex Fourier Transformation

Hf-181/Ta PACS Hafnium-181/Tantalum Perturbed Angular Correlation Spectroscopy

HFTBA Heptacosafluoro Tri-N-Butylamine

HFTF Horizontal Flight Test Facility

Hf(TFAC)$_4$ Hafnium Trifluoroacetylacetonate [also Hf(tfac)$_4$]

h·ft·°F/Btu hour-foot-degree Fahrenheit per British thermal unit [also h·ft·F/Btu]

h·ft^2·°F/Btu-in hour-square foot-degree Fahrenheit per British thermal unit inch [also h·ft^2·°F/Btu-in]

H-1 FT-NMR Hydrogen-1 (or Proton) Fourier Transform Nuclear Magnetic Resonance (Spectra) [also ^{1}H FT-NMR, or H-FT-NMR]

H-FTPGSE Hydrogen-1 (or Proton) Fourier Transform Pulsed-Gradient Spin Echo

HFTRM High-Field Thermoremanent Magnetization

HFTS Horizontal Flight Test Simulator

HFW Header, Footer, Watermark (Occurrence) [Word Processing]

HFW High-Frequency Wave

HFW Hole Full of Water

HFW Horizontal Full Width

HFX High Frequency Transceiver

HG Hall Generator

HG Handelsgesellschaft [Commercial Company; Corporate Form in Central Europe]

HG Hand Generator

HG Haptoglobin [Biochemistry]

HG Harmonic Generation; Harmonic Generator

HG (Write) Head Gradient [Computers]

HG Height Gauge

HG Helical Gear

HG Henyey-Greenstein (Approximation) [Physics]

HG Heritage Group [of Environment Australia]

HG High Gain

HG High Grade (Material)

HG High Grade of Slab Zinc

HG Historical Geologist; Historical Geology

HG Homopolar Generator

HG Horizon Grow

HG Horse Guards

HG Hotchkiss Gun

HG Human Gastrin [Biochemistry]

HG Human Genetics

HG Human Genome,

HG Hydraulic Gate

HG Hydrogeologist; Hydrogeology

HG Hyperemesis Gravedorium [Medicine]

Hg *(Hydrargyrum)* – Mercury [Symbol]

Hg$^+$ Monovalent Mercury Ion [also Hg$^+$] [Symbol]

Hg^{2+} Divalent Mercury Ion [also Hg^{++}] [Symbol]

Hg II Mercury II [Symbol]

Hg-121 HgBa$_2$CuO$_4$ [A Mercury Barium Copper Oxide Superconductor]

Hg-189 Mercury-189 [also ^{189}Hg, or Hg^{189}]

Hg-190 Mercury-190 [also ^{190}Hg, or Hg^{190}]

Hg-191 Mercury-191 [also ^{191}Hg, or Hg^{191}]

Hg-192 Mercury-192 [also ^{192}Hg, or Hg^{192}]

Hg-193 Mercury-193 [also ^{193}Hg, or Hg^{193}]

Hg-195 Mercury-195 [also ^{195}Hg, or Hg^{195}]

Hg-196 Mercury-196 [also ^{196}Hg, or Hg^{196}]

Hg-197 Mercury-197 [also ^{197}Hg, or Hg^{197}]

Hg-198 Mercury-198 [also ^{198}Hg, or Hg^{198}]

Hg-199 Mercury-199 [also ^{199}Hg, or Hg^{199}]

Hg-200 Mercury-200 [also ^{200}Hg, or Hg^{200}]

Hg-201 Mercury-201 [also ^{201}Hg, or Hg^{201}]

Hg-202 Mercury-202 [also ^{202}Hg, or Hg^{202}]

Hg-203 Mercury-203 [also ^{203}Hg, or Hg^{203}]

Hg-204 Mercury-204 [also ^{204}Hg, or Hg^{204}]

Hg-205 Mercury-205 [also ^{205}Hg, or Hg^{205}]

Hg-1212 HgBa$_2$CaCu$_2$O$_6$ [A Mercury Barium Calcium Copper Oxide Superconductor]

Hg-1222 HgBa$_2$Ca$_2$Cu$_3$O$_8$ [A Mercury Barium Calcium Copper Oxide Superconductor]

Hg-1223 HgBa$_2$Ca$_2$Cu$_3$O$_8$ [A Mercury Barium Calcium Copper Oxide Superconductor]

H7/g6 Sliding Fit; Basic Hole System [ISO Symbol]

H$_2$(g) Hydrogen Gas [Symbol]

hg hectogram [Unit]

HGA (Read/Write) Head Gimbal Assembly [Computers]

HGA Hercules Graphics Adapter

HGA High Gain Antenna

HGA Hop Growers Association [US]

HGAA Hydride Generation followed by Atomic Absorption

HgAc Mercury(I) Acetate (or Mercurous Acetate)

Hg(Ac)$_2$ Mercury(II) Acetate (or Mercuric Acetate)

h-GaN Hexagonal Gallium Nitride

HGAS High Gain Antenna System

Hgb Hemoglobin

(Hg,Bi)-1212 (Hg,Bi)Sr$_2$(YCa)Cu$_2$O$_{7-\delta}$ [A Mercury Bismuth Strontium Yttrium Calcium Copper Oxide Superconductor]

HgBr Mercury(III) Bromide (or Mercurous Bromide)

HGC Hartford Graduate Center [Connecticut, US]

HGC Hercules Graphics Card

HGCC Human Genome Coordinating Committee [of US Department of Energy]

HgCdFeSe Mercury Cadmium Iron Selenide (Semiconductor)

HgCdTe Mercury Cadmium Telluride (Laser)

HgCdTe IR Mercury Cadmium Telluride Infrared (Detector)

HgCdTe MBE Mercury Cadmium Telluride Molecular Beam Epitaxy (Technology)

(Hg$_2$)$_3$Cit$_2$ Mercury(II) Citrate

Hg$_3$Cit$_2$ Mercury(I) Citrate

HGCP Hercules Graphics Card Plus

Hg-CTP 5-Mercuricytidine 5'-Triphosphate (Carbonate) [Biochemistry]

HGD Hob Generating Diameter (of Silent Chain) [Mechanical Engineering]

Hg-dCTP 5-Mercuri-2'-Deoxycytidine 5'-Triphosphate (Carbonate) [Biochemistry]

HGDS Hazardous Gas Detection System

Hg-dUTP 5-Mercuri-2'-Deoxyuridine 5'-Triphosphate (Carbonate) [Biochemistry]

HGE Hot Gas Efficiency

HGE Hydraulic Grade Elevations

HGF Hepatocyte Growth Factor [Biochemistry]

HGF Hyperglycemic-Glycogenolytic Factor [also HG-Factor, or HG-factor] [Biochemistry]

HgFeS Mercury Iron Sulfide (Semiconductor)

HgFeSe Mercury Iron Selenide (Semiconductor)

HgFeTe Mercury Iron Telluride (Semiconductor)

HGG Human Gamma Globulin [Biochemistry]

HGH High Grade Heat

HGH Human Growth Hormone [also hGH] [Biochemistry]

HGI Hardgrove Grindability Index (for Minerals and Ores)

HGL Hydraulic-Transient Grade Line

HgL$_{III}$ Mercury L-III (Absorption Edge) [X-Ray Scattering]

HGM Hot-Gas Manifold

HGM Human Gene Mapping

HGMD Human Gene Mutation Database [of University of Wales College of Medicine, Cardiff, UK]

HGMIS Human Genome Management Information System [of US Department of Energy at Oak Ridge National Laboratory]

HGMS High-Gradient Magnetic Separation

HgMnS Mercury Manganese Sulfide (Semiconductor)

HgMnSe Mercury Manganese Selenide (Semiconductor)

HgMnTe Mercury Manganese Telluride (Semiconductor)

HgNO Mercuronitrate

HgO Mercury(II) Oxide (or Mercuric Oxide)

HGP Hormonal Growth Promotant [Biochemistry]

HGP Human Genome Program [An Initiative of the US Department of Energy with International Participation]

HGP Human Genome Project [Project of the US Department of Energy, the National Institutes of Health and the National Human Genome Research Institute]

HGP Hydrogen Gas Porosity

(Hg,Pb)-1223 $(Hg,Pb)Ba_2Ca_2Cu_3O_x$ [A Mercury Lead Barium Calcium Copper Oxide Superconductor]

HGPRT Hypoxanthine-Guanine Phosphoribosyl Transferase [Biochemistry]

HGR Hangar [also Hgr, or hgr]

(Hg,Re)-1212 $(Hg,Re)Ba_2CaCu_2O_y$ [A Mercury Rhenium Barium Calcium Copper Oxide Superconductor]

(Hg,Re)-1223 $(Hg,Re)Ba_2Ca_2Cu_3O_y$ [A Mercury Rhenium Barium Calcium Copper Oxide Superconductor]

HGRF Hot Gas Radiating Facility

HGR&SPTFAC Hangar and Support Facility

HGS Houston Geological Society [Texas, US]

HGS Hydrogen Gas Saver

HgS Mercury(II) Sulfide (or Mercuric Sulfide)

HGSD Harvard Graduate School of Design [Cambridge, Massachusetts, US]

HgSe Mercury(II) Selenide (Semiconductor)

HgSeTe Mercury SeleniumTelluride (Semiconductor)

HgSO Mercurosulfate

HGT High Group Transmit

HGT Hypergeometric Group Testing

Hgt Height [also hgt]

HgTe Mercury(II) Telluride (Semiconductor)

HgTe/HgCdTe Mercury(II) Telluride/Mercury Cadmium Telluride (Heterojunction)

HgTe/MCT Mercury(II) Telluride/Mercury Cadmium Telluride (Heterojunction)

Hg(TFAC)₂ Mercuric Trifluoroacetate [also $Hg(tfac)_2$]

(Hg,Tl)-1212 $(Hg,Tl)Sr_2(Y_{1-x}Ca_x)Cu_2O_{7-\delta}$ [A Mercury Thallium Strontium Yttrium Calcium Copper Oxide Superconductor]

HGTP Hanford Grout Technology Program [of US Department of Energy]

Hg-UTP 5-Mercuriuridine 5'-Triphosphate [Biochemistry]

HGV Heavy Goods Vehicle

HGV Hydrogen Gas Valve

HGVT Horizontal Ground Vibration Test

HgZnSe Mercury-Zinc Selenide (Semiconductor)

HH Double Hard [On Pencils]

HH Half-Hard (Temper) [also ½H] [Metallurgy]

HH Half-Height

HH Hall-Héroult (Process) [Metallurgy]

HH Halohydrin

HH Hand-Held

HH Handhole

HH Hanging Handset

HH Health Hazard

HH Heavy Hydrogen

HH Her/His Highness

HH Hershey and Hosford (Criterion)

HH Hitchhiker [of NASA]

HH Houben-Hoesch (Synthesis)

HH Hour [also hh] [Time]

HH Hour Hand [Horology]

HH Hydraulic Hose

HH Hydrometer Heating

HH- Haiti [Civil Aircraft Marking]

½H Half-Hard (Temper) [also HH] [Metallurgy]

H7/h6 Locational Clearance Fit [ISO Symbol]

HHA Hexahydric Alcohol

Hha Haemophilus haemolyticus [Microbiology]

HHAS High-Resolution Helium Atom Scattering

HHB Hex(agonal) Head Brass (Screw)

HHB Hexahydroxybenzene

HHBT High-Temperature and High-Humidity-Storage Bias-Loading Test

HHC Hammer Head Crane

HHC Hand-Held Computer

HHC Higher Harmonic Control

HHC Halogenated Hydrocarbon

HHCP Home Health Care Provider

HHCP 1-Hydroperoxy-1'-Hydroxy Dicyclohexylperoxide

hhd hogshead [Unit]

HHDN Aldrin

HHDN Hexachlorohexahydro Dimethanonaphthalene

HHF Hartree-Hartree-Fock [Physics]

HHF Höhere Hauswirtschaftliche Fachschule [Higher School of Home Economics, Switzerland]

HHFAC Hexafluoroacetylacetone [also Hhfac, or hhfac]

HHFE Heavy-Hole Free-Exciton [Solid-State Physics]

HHFOD 1,1,1,2,2,3,3-Heptafluoro-7,7-Dimethyl-4,6-Octanedione [also Hhfod, or hhfod]

HH-G Hitchhiker, Goddard [of NASA]

H₁-H₂- H₃ Transformer Primary [Controllers]

HHHMU Hydrazine Hand-Held Maneuvering Unit

HHI Heinrich-Hertz-Institut (für Nachrichtentechnik) [Heinrich Hertz Institute (for Communication Engineering), Berlin, Germany]

HHI Hyundai Heavy Industries Company Limited [South Korea]

HHL High-Hazard Laboratory [of Environmental Technology Center, Canada]

HHLGCS (Department of) Health, Housing, Local Government and Community Services [Australia]

HHLR Horace Hardy Lestor (Nuclear) Reactor [US]

HHLT Hand-Held Logic Tool

HH-M Hitchhiker, MSFC (Marshall Space Flight Center) [of NASA]

HHMI Howard Hughes Medical Institute [US]

HH:MM:SS Hours:Minutes:Seconds [also hh:mm:ss]

HH:MM:SS.CC Hours:Minutes:Seconds:Centiseconds [also hh:mm:ss.cc]

HHMU Hand-Held Maneuvering Unit

HHP Hand Held Products [US Wand Manufacturer]

HHPA Hexahydrophthalic Anhydride

HHRR Hand-Held Rationing Radiometer

HHS Harris Hematoxylin Solution [Biochemistry]

HHS (Department of) Health and Human Services [US]

HHS Hex(agonal) Head Steel (Screw)

HHS Human Head Simulator

H$_2$/H$_2$S Hydrogen/Hydrogen Sulfide (Gas Mixture)

HHSI High Head Safety Injection

HHSNN 3-Hydroxy-4-[(2-Hydroxy-4-Sulfo-1-Naphthalenyl)-azo]2-Naphthalenecarboxylic Acid

HHT High Temperature and High Humidity (Storage Test)

HHT High-Temperature Reactor with Helium Turbine

HHT Huang, Hutchinson and Tvergaard (Examination) [Mechanical Testing]

HHT Hydrogen Heat Treatment [Metallurgy]

HHT Hydroxyheptadecatrienoic Acid,

12(S)-HHT 12(S)-Hydroxy-(5Z,8E,10E)-Heptadecatrienoic Acid

HHTK Hand-Held Test Kit

HHTU Hand-Held Teaching Unit

HHV High Heat Value; Higher Heating Value

HHV Human Herpes Virus

HHV-1 Human Herpes Virus, Type 1

HHV-2 Human Herpes Virus, Type 2

HHV-3 Human Herpes Virus, Type 3

HHV-4 Human Herpes Virus, Type 4

HHV-5 Human Herpes Virus, Type 5

HHV-6 Human Herpes Virus, Type 6

HHV-7 Human Herpes Virus, Type 7

HHV-8 Human Herpes Virus, Type 8

HHW Household Hazardous Waste

HHX Heavy Hole Exciton [Solid-State Physics]

HI Half-Interval (Method)

HI Hampton Institute [Virginia, US]

HI Hawaii [US]

HI Hawaiian Islands [US]

HI Health Insurance

HI Heard Island

HI Heat Input

HI Heat Insulation

HI Heavy Ion(s)

HI Height of Instrument

HI Heisenberg-Ising (Spin) [Solid-State Physics]

HI Hemagglutination Inhibition

HI Herder-Institut e.V. [Herder Institute, Marburg, Germany]

HI Hermitian Interpolation [Mathematics]

HI High [also Hi, or hi]

HI High Address Byte [also H]

HI High Intensity

HI (Electron) Hole Injection

HI Hologram Interferometer

HI Holographic Interferometer; Holographic Interferometry

HI Honeywell, Inc. [US]

HI Hospital Insurance

HI Human Interface

HI Hydraulic Institute [US]

HI Hydraulic Intensifier

HI Hydriodic Acid

HI Hydrogen Iodide (or Hydroiodic Acid)

HI Hydronics Institute [US]

HI Hyperfine Interaction [Physics]

HI Laughter [Amateur Radio]

HI- Dominican Republic [Civil Aircraft Marking]

H∥I Magnetic Field In-Plane (or Parallel) to Electric Current [Symbol]

H⊥I Magnetic Field Out-of-Plane (or Perpendicular) to Electric Current [Symbol]

Hi High [also hi]

Hi High-Order Byte (of Argument) [Pascal Programming Function]

HIA Hawaiian Irrigation Authority [US]

HIA Herzberg Institute of Astrophysics [of National Research Council of Canada]

HIA High-Intensity Atomizer

HIA Horological Institute of America [US]

HIA Housing Industry Association

HIA Human Interface Architecture

HIA Hydrogen-Induced Amorphization [Metallurgy]

HIAA Hydroxyindole Acetic Acid

5-HIAA 5-Hydroxyindoleacetic Acid

HIAC High Accuracy

HIALS High Intensity Approach Light System

HIAPP Hydrogen Ingress Analysis by Potentiostatic Pulsing

HIB High-Index (Grain) Boundary [Materials Science]

HIB High Iron Briquetting [Metallurgy]

Hi-B High-(Magnetic) Induction Grade (Silicon Steel)

Hib Hemophilus influenzae, type b [Microbiology]

HIBA Hydroxyisobutyric Acid

HIBCC Health Industry Bar Code Council

HIBEX High-Acceleration Boost Experiment

HIBEX High Impulse Booster Experiment

HIBS Heavy-Ion Backscattering

HIBS Heavy-Ion Backscatter(ing) Spectrometry

HIC Harding's Image Converter

HIC Hickam Air Force Base, Hawaii [NASA Deorbiting Site]

HIC High Integrity Container

HIC Human Interaction Component [Computer Programming]

HIC Hybrid Integrated Circuit

HIC Hydrogen-Induced Cracking

HIC Hydrophobic Interaction Chromatography

HI-C High Conversion Critical Experiment

HiC High Capacity

HICACOM High-Capacity Communications

HICAPCOM High-Capacity Communication System

HICAT High-Altitude Clear Air Turbulence

HICLASS Hierarchical Classification

HICOM High Carbon Monoxide (Process)

HICOM High Commission

HICON High Convection Annealing Technology [Metallurgy]

HICON/H$_2$ High Convection Annealing Technology with Hydrogen Atmosphere [Metallurgy]

HICS Hierarchical Information Control System

HICT Higher Institute of Chemical Technology [Sofia, Bulgaria]

HID Hardware Interface Device

HID High Density [also HiD]

HID High-Intensity Discharge (Lamp)

HID High-Interstitial Defect [Materials Science]

HID HIM (Hardware Interface Module) Interface Distributor

HID Housing Industry Dynamics (Database) [US]

HID Hydrogen-Induced Drift

HIDA N-(2,6-Dimethylphenylcarbamoylmethyl) iminodiacetic Acid

HIDA N-(2-Hydroxyethyl)iminodiacetic Acid

H2IDA Iminodiacetic Acid

HIDAM Hierarchical Indexed Direct Access Method

HIDB Highlands and Islands Development Board [UK]

HIDBF Hydrogen Induced Delayed Brittle Fracture

HIDC Hexamethylindodicarbocyanine

HIDC Hydrogen-Induced Delayed Cracking

HIDE Hydrogen-Induced Deformation Experiment

HIDEC Highly-Integrated Digital Electronic Control

HIDES High-Absorption Integrated Defense Electromagnetic System

HIDF Horizontal Side of an Intermediate Distribution Frame

HiD/LoD High Density/Low Density (Tariff) [Telecommunications]

HIDM High Information Delta Modulation

HIE Hibernation Information Exchange [now International Hibernation Society, US]

HIEF Hybrid Isoelectric Focusing

HIEME Higher Institute of Electrical and Mechanical Engineering [Bulgaria]

HIF Haus Industrieforum [Essen, Germany]

HIF High Integrity Flange

HIF Hokkaido International Foundation [of Hokkaido University, Japan]

HIF Hot Isostatically Forged; Hot Isostatic Forging

HIF Hyper-G Interchange Format

HIFAM High-Fidelity Amplitude Modulation

HIFAR High-Flux Australian Reactor [Australia]

HIFD High-Density Floppy Disk

HIFET Heterointerface Field-Effect Transistor

HiFi High Fidelity [also Hi-Fi, hifi, hi-fi, HI-FI, or HIFI]

HIFIT High-Frequency Input Transistor

HIFO Highest In, First Out (Method) [also hifo]

HIFP Hydrogen Internal Friction Peak

HIFT Hardware Implemented Fault Tolerance

HIFT Heard Island Feasibility Test [US/Canada]

HIG Hawaii Institute of Geophysics [of University of Hawaii at Honolulu, US]

HIG Hermetically-Sealed Integrated Gyro

HIgA Human Immmunoglobulin A [Biochemistry]

HIGB High-Index Grain Boundary [Materials Science]

HIgD Human Immmunoglobulin D [Biochemistry]

HIgE Human Immmunoglobulin E [Biochemistry]

HIGFET Heterostructure Insulated-Gate Field-Effect Transistor

HIgG Human Immmunoglobulin G [Biochemistry]

HIgG1 Human Immmunoglobulin G1 [Biochemistry]

HIgG2 Human Immmunoglobulin G2 [Biochemistry]

HIgG3 Human Immmunoglobulin G3 [Biochemistry]

HIgG4 Human Immmunoglobulin G4 [Biochemistry]

HIGGSS Higgs and Supersymmetry: Search and Discovery (Conference) [Particle Physics]

HIGGS-SUSY Higgs and Supersymmetry: Search and Discovery (Conference) [Particle Physics]

HIGH High Core Threshold

HIGHCOM High Fidelity Compander

High Energy Phys Nucl Phys High Energy Physics and Nuclear Physics [Journal published in the UK]

High Energy Phys Nucl Phys (China) High Energy Physics and Nuclear Physics (China) [Journal published in PR China]

HIGH GASSER High Geographic Aerospace Search Radar [also High Gasser]

High-Perform Ceram High-Performance Ceramics [Journal published in Spain]

High Perform Plast High Performance Plastics [Journal published in the UK]

High Perform Polym High Performance Polymers [Journal of the Institute of Physics, UK]

High Perform Syst High Performance Systems [Journal published in the US]

High Perform Text High Performance Textiles [Published in the UK]

High Purity Subst High Purity Substances [Published in the US]

HIGH RES High Resolution [also High Res, or high res]

High-Speed Surf Craft High-Speed Surface Craft [Published in the UK]

HIGH TECH High Technology [also High Tech, or high tech]

High Tech High Technology [Journal]

High Tech Ceram News High Tech Ceramics News

High Technol High Technology [Journal published in the US]

High Technol Bus High Technology Business [Journal published in the US]

High Temp High Temperature [Translation of: *Teplofizik Vysokikh Temperatur (USSR)*; published in the US]

High Temp–High Press High Temperatures–High Pressures [Journal published in the US]

High Temp Mater Process High Temperature Materials and Processes [Journal published in the UK]

High Temp Order Intermetall Alloys High-Temperature-Ordering in Intermetallic Alloys [Publication of Materials Research Society, US]

High Temp Sci High Temperature Science [Journal published in the US]

High Temp Technol High Temperature Technology [UK]

HIGHVISION (Japanese) High-Definition Telvision System

HIgM Human Immmunoglobulin M [Biochemistry]

HIHAT High-Resolution Hemispherical Reflector Antenna Technique

HIHCL High-Intensity Hollow-Cathode Lamp

HI-HICAT High High Altitude Clear Air Turbulence

HIHM Hollandsche Industrie und Handels Maatschappij [Dutch Chamber of Industry and Commerce]

HII Himalayan International Institute [Honesdale, Pennsylvania, US]

HIIDMS Heavy Ion Induced Mass Spectrometry

HIIS Honeywell Institute for Information Sciences [US]

HIJ Horological Institute of Japan [Tokyo, Japan]

Hi-K High Dielectric Constant

HIL Hazardous Industrial Liquid

HIL High-Intensity Lighting

HIL Human Interface Link [of Hewlett Packard]

hIL Human Interleukin [also HIL] [Biochemistry]

hIL-1α Human Interleukin-1α [Biochemistry]

hIL-1β Human Interleukin-1β [Biochemistry]

hIL-2 Human Interleukin-2 [Biochemistry]

hIL-3 Human Interleukin-3 [Biochemistry]

hIL-4 Human Interleukin-4 [Biochemistry]

hIL-4R Human Interleukin-4 Receptor [Biochemistry]

hIL-6R Human Interleukin-6 Receptor [Biochemistry]

hIL-5 Human Interleukin-5 [Biochemistry]

hIL-6 Human Interleukin-6 [Biochemistry]

hIL-7 Human Interleukin-7 [Biochemistry]

hIL-8 Human Interleukin-8 [Biochemistry]

hIL-9 Human Interleukin-9 [Biochemistry]

hIL-10 Human Interleukin-10 [Biochemistry]

hIL-11 Human Interleukin-11 [Biochemistry]

hIL-12 Human Interleukin-12 [Biochemistry]

hIL-13 Human Interleukin-13 [Biochemistry]

HILA Heavy-Ion Linear Accelerator

HILAC Heavy-Ion Linear Accelerator [also Hilac, or hilac]

HILAN High-Level (Computer) Language [also Hilan, or hilan]

HILDA Human Interleukin for DA-Type Cells [Biochemistry]

Hi-Lo High-Low (Check) [also Hi-Lo, or hi-lo]

HILM High-Induced (Crystal) Lattice Migration [Metallurgy]

HILSP High-Intensity Light Surface Preparation

HIM Hardware Interface Module

HIM Hazardous Industrial Material

HIM Hollow Injection Molding

HIM Hotel Institute Montreux [Switzerland]

HIMA Health Industry Manufacturers Association [US]

HIMAIL Hitachi Integrated Message and Information Library

HIMAT Highly Maneuverable Aircraft Technology [also HiMAT]

HIMDA N-(2-Hydroxyethyl)iminodiacetic Acid

Himda β-Hydroxyethyliminodiacetic Acid

2-HIMDA N-(2-Hydroxyethyl)iminodiacetic Acid

HIMEM High Memory

HIMES Highly Maneuverable Experimental Spacecraft [also HiMES]

HIMS Hanford Issues Management System

HIMSS High-Resolution Microwave Spectrometer Sounder [of NASA]

HIN Hybrid Integrated Network

HINAS Historic Naval Ships Association of North America [US]

Hinc Haemophilus influenzae Rc [Microbiology]

Hind Haemophilus influenzae Rd [Microbiology]

Hind Hindu(stan); Hindustani

Hinf Haemophilus influenzae serotype f [Microbiology]

HINIL High-Noise-Immunity Logic [also HiNIL]

HINT Hydrophobic Interaction

HIO High Input/Output

HIOMT Hydroxyindole-O-Methyltransferase [Biochemistry]

H-Ion Hydrogen Ion [also H-ion]

HIOS Heath/Zenith Instrument Operating Software

HIOS High Island Offshore System

HIP Hanford Isotope Plant [Washington State, US]

HIP Hardware Interface Program

HIP Heterojunction Internal Photoemission (Photodiode)

HIP High-Impact Polystyrene

HIP Hitachi Parametron Automatic Computer

HIP Host Information Processor

HIP Hot Isostatic(ally) Pressed [also HIPed, HIPped, or HIP'd]

HIP Hot Isostatic Pressing [also HIPing, or HIPping]

HIPAC Heavy-Ion Plasma Accelerator

HIPAR High-Power Acquisition Radar

HIPC High Pressure Chamber

HIP/CIP Hot Isostatic Pressing/Cold Isostatic Pressing [also HIP-CIP]

HIPD High-Intensity Powder Diffractometer

HIP'd Hot Isostatic(ally) Pressed [also HIPed, or HIPped]

HIPE n-Hexyl Isopropyl Ether

HIPE High Internal Phase Emulsion [Polymeric Foams]

HIPed Hot Isostatic(ally) Pressed [also HIP'd, or HIPped]

HIPERNAS High-Performance Navigation System

HIPIC Hot Isostatic(ally) Pressed Carbon Impregnation (Process)

HIPing Hot Isostatic Pressing [also HIPping, or HIP]

HIPIR High-Power Illuminating Radar

HIPO Hierarchical Input-Process-Output

HIPO Hierarchy plus Input-Process-Output (Chart)

HIPOT High-Potential [also Hipot, or hipot]

HIPOTT High-Potential Test [also Hipott, or hipott]

HIPped Hot Isostatic(ally) Pressed [also HIP'd, or HIPed]

HIPPI High-Performance Parallel Interface [also HiPPI, or Hippi]

HIPRBSN Hot-Isostatic Pressed Reaction-Bonded Silicon Nitride [also HIP-RBSN]

HIPRS High-Intensity Pulsed Radiation Source

HIPRSN Hot-Isostatic Pressed Reaction-Bonded Silicon Nitride [also HIP-RSN]

HIPS High-Impact Polystyrene

HIPSC Hot Isostatic(ally) Pressed Silicon Carbide [also HIP-SiC, HIPSiC, or HSC]

HIPSiC Hot Isostatic(ally) Pressed Silicon Carbide [also HIPSiC, HIPSC, or HSC]

HIPSN Hot Isostatic(ally) Pressed Silicon Nitride [or HSN]

HIPS-PE High-Impact Polystyrene–Polyethylene

HIPS-PPE High-Impact polystyrene/Polyphenylene Ether (Blend)

HIPS-PVC High-Impact Polystyrene and Polyvinyl Chloride

HIPS-PVDC High-Impact Polystyrene–Polyvinylidene Chloride

HIPS-PVDC-PE High-Impact Polystyrene–Polyvinylidene Chloride–Polyethylene

HIPS-PVDC-PP High-Impact Polystyrene–Polyvinylidene Chloride–Polypropylene

HIPSSN Hot Isostatic(ally) Pressed Sintered Silicon Nitride

HIQ Horizontal Internal Quenching (Furnace)

HIQSA Hydroxyiodoquinolinesulfonic Acid

HIR (Study Group on) Harmful Interference to Radio [of International Civil Aviation Organization]

HIR Honiara, Solomon Islands [Meteorological Station Designator]

HIR Horizontal Impulse Reaction

HIRAC High Random Access

HIRAM Highest-Position Random-Access Memory

HIRAN High-Precision Shoran [also Hiran, or hiran],

HiRAP High-Resolution Accelerometer Package

HIRD High-Intensity Radiation Device

HIRDLS High-Resolution Dynamics Limb Sounder

HIRDS High-Intensity Radiation Development Laboratory [US]

HI-REL High Reliability [also Hi-Rel, or hi-rel]

HIRES Hypersonic In-Flight Refueling System

HiRES High-Resolution EUV (Extreme Ultaviolet) Spectroheliometer

Hi-res high-resolution [also hi-res]

HIRF High-Intensity Reciprocity Failure

HIRIS High-Resolution Imaging Spectrometer

HIRIS High-Resolution Interferometer/Spectrometer

HIRL High-Intensity Runway Lights

HIRNS Helicopter Infrared Navigation System

Hiroshima J Med Sci Hiroshima Journal of Medical Science [Japan]

HIRS High-Impulse Retro-Rocket System

HIRS High-Resolution Infrared (Radiation) Sounder

HIS Heavy-Ion Source

HIS Helicopter Integrated System

HIS High-Resolution Interferometric Sounder [NASA Ames ER-2 High-Altitude Aircraft]

HIS Hochschul-Information-System GmbH [University Information System Company, e.g., Conducts Surveys of Students, Germany]

HIS Holographic Illumination System

HIS Home Information System

HIS Home Interactive System

HIS Homogeneous Information Set

HIS Honeywell Information System

HIS Hospital Information System

His Histidine; Histidyl [Biochemistry]

HISAM Hardware Initiated Stand-Alone Memory

HISAM Hierarchical Indexed Sequential Access Method

HISARS Hydrological Storage and Retrieval Information System [of North Carolina State University, Raleigh, US]

HISDAM Hierarchical Indexed Sequential Direct-Access Method

HISCC Hydrogen-Induced Stress-Corrosion Cracking [also HSCC]

HISPID Herbarium Information Standards and Protocols for the Interchange of Data (Standard)

HISR Hydrogen-Induced Structural Relaxation [Materials Science]

Hist Historian; Historical; History [also hist]

Hist Geol Historical Geologist; Historical Geology

Hist Metall Historical Metallurgy [Journal of the Historical Metallurgy Society, UK]

Histochem Histochemical; Histochemist(ry) [also histochem]

Histol Histological; Histologist; Histology [also histol]

Histopathol Histopathological; Histopathologist; Histopathology [also histopath]

Histophysiol Histophysiological; Histophysiologist; Histophysiology [also histophysiol]

HIT ERIN (Environmental Resources Information Network) Heritage Information Team [of Environment Australia]

HIT Harbin Institute of Technology [Harbin, PR China]

HIT Heavy-Ion Theory

HIT High Isolation Transformer

HIT Himeji Institute of Technology [Himeji, Japan]

HIT Hiroshima Institute of Technology [Japan]

HIT Hokkaido Institute of Technology [Sapporo, Japan]

HIT Houston International Teleport [Texas, US]

HIT Huazhong Institute of Technology [PR China]

HIT Hypersonic Interference Technique

HIT Hypervelocity Impact Test(ing)

HIT Workshop on Heavy Ion Theory [at CERN–European Laboratory for Particle Physics, Geneva, Switzerland]

HITC Hexamethylindotricarbocyanine

HITAC Hitachi Computer Services [of Hitachi Limited, Japan]

Hitachi Rev Hitachi Review [of Hitachi Limited, Japan]

Hitachi Zosen Tech Rev Hitachi Zosen Technical Review [Published by Hitachi Zosen Technical Research Institute, Osaka, Japan]

HITEC High-Temperature Emission Control System

HITEC Highway Innovative Technology Evaluation Center [US]

HI-TECH High Technology [also Hi-Tech, or hitech]

HITEMP High-Temperature Engine Materials Technology Program [of NASA]

HITEMP High-Temperature Turbine Engine Program [US]

HITEX High-Temperature Experiment [UK]

HITEX High-Temperature Isotope Exchange

HITL Human Interface Technology Laboratory [of University of Washington, Seattle, US]

HITRAC High-Technology Training Access

HITRESS High-Test Recorder and Simulator System

HITS HazMat (Hazardous Material) Inventory Tracking System

HITS Hierarchical Integrated Test Simulator

HITS Holloman Infrared Target Simulator [of Holloman Air Force Base, near Alamogordo, New Mexico, US]

HITS HRM (High Rate Multiplexer) Input/Output Test System

HITT Heterojunction Integrated Transit Time

HIU Headset Interface Unit

HIU Hydrostatic Interface Unit

HIV Helium Isolation Valve

HIV Human Immunodeficiency Virus

HIV-1 Human Immunodeficiency Virus, Type 1

HIV-2 Human Immunodeficiency Virus, Type 2

HIVAN Human Immunodeficiency Virus Associated Neuropathy

HIV/DNA Human Immunodeficiency Virus/Deoxyribonucleic Acid [also HIV DNA]

HIVIG Human Immunodeficiency Virus Immunoglobulin [also HIVIg]

HIVNET Human Immunodeficiency Virus Network for Prevention Trials

Hi Vol High Volume

Hi-Volt High Voltage [also hi-volt]

HIVOS High-Vacuum Orbital Simulation; High-Vacuum Orbital Simulator

HIW Hazardous Industrial Waste

HIW Homojunction Interfacial Workfunction

HIWT Hobart Institute of Welding Technology [Troy, Ohio, US]

HIXE Heavy Ion Induced X-Ray Emission

HIXE Helium-Induced X-Ray Emission

HIXSE Heavy-Ion-Induced X-Ray Satellite Emission

HJ Harris-Jones (Creep) [also H-J] [Metallurgy]

HJ Heterojunction [Solid-State Physics]

HJ Hose Jacket

HJ Hot Junction,

HJ Hybrid Junction

HJ Hydraulic Jack

H&J Hyphenation and Justification [Computers]

H-J Harris-Jones (Creep) [also HJ] [Metallurgy]

H∥j Magnetic Field In-Plane (or Parallel) to Current Density [Symbol]

HJBT Heterojunction Bipolar Transistor [also HBT]

HJD Heliocentric Julian Date [Astronomy]

HJFET Heterojunction Field-Effect Transistor

HJP Hydraulic Jet Propulsion

HK Hand Knob

HK Hefner-Kerze [Hefner Candle] [also Hk]

HK Hexokinase [Biochemistry]

HK Hong Kong

HK Hong Kong [ISO Code]

HK Horvath-Kawazoe (Theory) [Materials Science]

HK Housekeeping [Computers]

HK Human Kinetics

HK Hydrokinetic(s)

HK Knoop Hardness (Number)

HK- Colombia [Civil Aircraft Marking]

Hk Hecker [Unit of Atomic Bond Directionality] [Symbol]

H7/k6 Locational Transition Fit; Basic Hole System [ISO Symbol]

.hk Hong Kong [Country Code/Domain Name]

HKBU Hong Kong Baptist University [Kowloon]

HKCE Hong Kong Commodity Exchange

HKCS Hong Kong Computer Society

HK$ Hong Kong Dollar [Currency] [also HKD]

HKFE Hong Kong Futures Exchange

HKGCC Hong Kong General Chamber of Commerce

HKI Hans-Knöll-Institut (für Naturstofforschung) [Hans-Knoell Institute, Germany]

HKIE Hong Kong Institution of Engineers

{hkil} family of (crystallographic) planes with Miller-Bravais indices [Symbol]

(hkil) Miller-Bravais indices (for hexagonal crystal systems) [Symbol]

[HKL] Crystallographic Rotation Axis [Symbol]

h,k,l (crystal) lattice planes [Symbol]

[hkl] crystallographic direction [Symbol]

⟨hkl⟩ family of (crystallographic) directions [Symbol]

{hkl} family of (crystallographic) planes with Miller indices [Symbol]

(hkl) Miller indices (or crystallographic planes) [Symbol]

HKLA Hong Kong Library Association

HKM Henshall-Kassner-McQueen (Theory) [Materials Science]

HKN Knoop Hardness Number [Metallurgy]

HKPS Hong Kong Physical Society

HKPU Hong Kong Polytechnic University [Kowloon]

HKS Hohenberg-Kohn-Sham (Method) [Physics]

HKSM Henry Krumb School of Mines [of Columbia University, New York City, US]

HKTA Hong Kong Tourist Association

HKTC Honk Kong Trade Council

HKTDC Hong Kong Trade Development Council

H2KTS 2,2'-[1-(1-Methyl-2-Oxabutyl)-1,2-Ethanediylidene]bis (hydrazinecarbothiamide)

HKUST Hong Kong University of Science and Technology [Kowloon]

HL Cloud Base Height
HL Half Life
HL Hardline [also H/L]
HL Harwell Laboratory [UK]
HL Hazardous Liquid
HL Heap Leaching [Metallurgy]
HL Hearing Level
HL Heat Loss
HL Heaviside Layer [Geophysics]
HL Heavy Liquid
HL Heavy Load(ing)
HL Heel Line
HL Height-Length
HL Height Loss
HL Helium Laser
HL Hemorrhagic Lesion [Medicine]
HL Herpetologists League [US]
HL Hidden Line
HL High Leucite
HL High Level
HL Hind Lobe (of Brain) [Anatomy]
HL Hinge Line [Geology]
HL Hirth Lock
HL Holding Line
HL Hooke's Law [Mechanics]
HL Horizontal Line
HL Host Language
HL Hot Line
HL Hot Luminescence
HL Hwang and Liano (Model) [Metal Forming]
HL Hydraulic(s) Laboratory
HL Hydraulic Lift
HL Hydrogen Lamp
HL Hydrogen Laser
HL Hydrogen Line [Spectroscopy]
HL Hydrophile/Lipophile; Hydrophilic/Lipophilic [Chemistry]
HL Hydrostatic Loading
HL Hysteresis Loop [Physics]
HL Hysteresis Loss [Physics]
HL Lactic Acid
HL7 Health Industry Level 7
HL- South Korea [Civil Aircraft Marking]
H/L Hardline [also HL]
hL hectoliter [also hl]
hl (hoc loco) – here
hν_0 work function (in physics) [Symbol]
HLA Halifax Library Association [Nova Scotia, Canada]
HLA Helicopter Loggers Association [US]
HLA High-Level Architecture
HLA Human Leucocyte Antigen [also HL-A, or hLA] [Immunology]
HLAD Hearing-Lookout Assist Device
HLAF High-Level Arithmetic Function
HLAIS High-Level Analog Input System

HLAH Hank's Lactalbuminhydrolysate [Biochemistry]
HLAL High Level Assembler Language
HLAN Hanford Local Area Network [Washington State, US]
HLAO High-Lying Antibonding Orbital [Physics]
HLAS Hot Line Alert System
HLB Hydrophile/Lipophile Balance [Chemistry]
HLC Headwater Level Control
HLCO High Low Close Open
HLCV Hot Leg Check Valve
HLD Halide Leak Detector; Halogen Leak Detector
HLD Helium Leak Detector
HLD High-Lift Device [Aeronautics]
Hld Hold [also hld]
HLDA Hold Acknowledge
Hld Cyc Hold Cycle [also hld cyc]
Hldg Holding [also hldg]
HLDLC High-Level Data Link Control
HLDS Hydrogen Leak Detection System
HLDTL High-Level Data Transistor Logic
HLDW High-Level Defense Waste
hlf half
HlfB Hessisches Landesamt für Bodenforschung [Hessian State Office for Soil Research, Wiesbaden, Germany]
Hlg Hauling [also hlf]
HLH Heavy Lift Helicopter
HLH Heizung Lüftung/Klima Haustechnik [German Journal on Heating, Ventilation, Air-Conditioning and Domestic Engineering; published by Verein Deutscher Ingenieure]
HLH High-Level Heating
hLH Human Luteinizing Hormone [also HLH] [Biochemistry]
HLI High-Level Inversion [Meteorology]
HLI Host Language Interface
HLINE Horizontal Line [also HLine, or Hline]
H Line Fraunhofer Line for Calcium Having Wavelength of 396.8 Nanometers [Spectroscopy] [also H line]
H$_\alpha$ Line (Strong-Red) Fraunhofer Line for Hydrogen Having Wavelength of 766.1 Nanometers [Spectroscopy] [also H$_\alpha$ line, C Line, or C line]
H$_\beta$ Line (Medium-Blue-Green) Fraunhofer Line for Hydrogen Having Wavelength of 486.1 Nanometers [also H$_\beta$ line, F Line, or F line] [Spectroscopy]
H$_\gamma$ Line (Weak-violet) Fraunhofer Line for Hydrogen Having Wavelength of 434.1 Nanometers [also H$_\gamma$ line, G' Line, or G' line] [Spectroscopy]
H$_\delta$ Line (Weak-Extreme Violet) Fraunhofer Line for Hydrogen Having Wavelength of 410.2 Nanometers [also H$_\delta$ line, or h line] [Spectroscopy]
H' Line Fraunhofer Line for Calcium Having Wavelength of 393.4 Nanometers [Spectroscopy] [also H' line]
h line Fraunhofer line for hydrogen having wavelength of 410.2 nanometers [Spectroscopy] [also H$_\delta$ line]
HLIV Hot Leg Isolation Valve
HLK Spezialmesse für Heizungs–, Luft- und Klimatechnik [Trade Fair for Heating, Ventilation and Air Conditioning Engineering, Switzerland]

HLL High-Level (Programming) Language
HLL High-Level Logic
HLLAPI High Level Language Application Programming Interface
HLLV Heavy Lift Launch Vehicle
HLLW High-Level Liquid (Nuclear) Waste
HLLWT High-Level Liquid Waste Tank
HLM High Level Meeting
HLM Holding Line Marker
Hlm Hefner-Lumen [Unit]
HLMI High-Load Melt Index [Synthetic Resins]
HLML High-Level Microprogramming Language
HLMR Hunter-Liggett Military Reserve [of US Army in California, US]
Hlmt Helmet
HLN Hex(agonal) Long Nipple
HLOS Homodyne Local Oscillator System
HLP Help [also Hlp, or hlp]
HLP High-Level Programming
.HLP Help [File Name Extension] [also .hlp]
HLPI High-Level Programming Interface
HLQ High-Level Qualifier
HLQL High-Level Query Language
HLR High-Level Representation
HLR Home Location Register [Telecommunications]
Hlr Holder [also hlr]
HLRC High-Level Radiochemistry Facility
HLRW High-Level Radioactive Waste
HLRZ (Hochleistungsrechenzentrum)—German for "Supercomputer Center"
HLRZ Jülich Hochleistungsrechenzentrum Jülich [Julich Supercomputer Center, Kernforschungsanlage Julich, Germany]
HLS Heavy Logistic Support [US Air Force]
HLS High-Level Scheduler
HLS Hue, Lightness, Saturation; Hue, Luminance, Saturation
HLSC High-Level Service Circuit
HLSC Hindustan Lever Research Center [Bombay, India]
HLSE High-Level, Single-Ended
HLSM Homopolar Linear Synchronous Motor
HLSRS High Level Sisal Research Station [East Africa]
HLSTO Hailstone
HLSUA Honeywell Large System Users Association [US]
HLT Halt
HLT Headwater Level Transmitter
HLT Helium Leak Test(er); Helium Leak Testing
HLT Heterodyne Look-Through
HLT Highly Leveraged Transaction
Hlt Halt [also hlt]
HLTF High Level Task Force
HLTL High Level Test Language
HLTL High-Level Transistor Logic
HLTTL High-Level Transistor-Transistor Logic

HLU House Logic Unit
HLVW Heavy Logistics Vehicle Wheeled (Refueller)
HLW High-Level (Radioactive) Waste
HLWS High-Level Waste Solidification
Hlx Hefner-Lux [Unit]
HLYR Haze Layer Aloft
HM Half-Mask (Respirator)
HM Hall Mobility [Solid-State Physics]
HM Hall Moduluator [Electronics]
HM Hand-Made
HM Hard Magnetic(ic)
HM Hard Material
HM Harman's Method [Physical Chemistry]
HM Harmonic Mean
HM Harmonic Motion
HM Harvest Moon [Astronomy]
HM Hazardous Material
HM Heard and McDonald Islands [ISO Code]
HM Heater Middle
HM Heavy Medium
HM Heavy Metal
HM Heavy Mobile
HM Hectometer; Hectometric
HM Helimagnet(ism) [Solid-State Physics]
HM Her/His Majesty
HM Hermann-Mauguin (Symbols) [Crystallography]
HM Hermitian Matrix [Mathematics]
HM Heterogeneous Membrane
HM High Magnification
HM High-Melting
HM High Modulus
HM Hinge Mount
HM Histamine [Biochemistry]
HM Hogging Moment [Civil Engineering]
HM (Electron) Hole Mobility
HM Holomicrography
HM Home [also Hm, or hm]
HM Homogeneous Membrane
HM Hot Melt(ing)
HM Hot Mix
HM Hot Metal
HM Hubbard Model
HM Hybrid Magnet [Solid-State Physics]
HM Hydraulic Modelling
HM Hydraulic Motor
HM Hydrazine Mononitrate
HM Hydrogen Maser
HM Hydrogen Monitor
HM Hydromagnetic(s); Hydromagnetism
HM Hydromechanic(al); Hydromechanics
HM Hydrometallurgical; Hydrometallurgist; Hydrometallurgy
HM Hypermedia
H/M Hydrogen-to-Metal (Ratio)

H/m Henry per meter [also H m^{-1}]

H∥M Magnetic Field In-Plane (or Parallel) to Magnetization [Symbol]

H⊥M Magnetic Field Out-of-Plane (or Perpendicular) to Magnetization [Symbol]

hm hectometer [Unit]

hm² square hectometer [Unit]

hm³ cubic hectometer [Unit]

h/m$_e$ quantum of circulation (7.2738961 × 10–4 Js/kg) [Symbol]

HMA Hard Magnetic Alloy

HMA Hardwood Manufacturers Association [US]

HMA Headmasters Association

HMA Hemispherical Mirror Analyzer

HMA (Microsoft) High Memory Area

HMA Hoist Manufacturers Association [now Hoist Manufacturers Institute, US]

HMA Hot-Mix Asphalt

HMA Hub Management Architecture

HMA Hypergol(ic) Maintenance Area [Aerospace]

HMAC Hazardous Materials Advisory Council [US]

HMAC Hot-Mix Asphaltic Concrete

HMAC House Military Affairs Committee [US Whitehouse]

HMANA Hawk Migration Association of North America [US]

HMAT Hot-Mix Asphalt Technology [Publication of the National Asphalt Pavement Association, US]

HMB Hexamethylbenzene

HMB 2-Hydroxy-5-Methoxybenzaldehyde

HMB β-Hydroxy β-Methylbutyrate

HMBA Hydroxymercurbenzoate

HMBC H-Detected Multiple-Bond Correlation

HMBC Heteronuclear Multiple-Bond Correlation

H2MBP Monobutylphosphoric Acid

HMC Hamburg Messe und Congress [Hamburg Trade Fair and Congress, Germany]

HMC Health Ministers Council

HMC Hemimicelle Concentration

HMC Her/His Majesty's Customs and Excise [UK]

HMC High-Strength Molding Compound

HMC Horizontal Machining Center

HMC Horizontal Motion Carriage

HMC Hybrid Microcircuit

HMC Hypergolic Maintenance and Checkout [Aerospace]

HMC Hysteresis Measurement Control

HMCB 2-Hydroxy-4-Methoxy-4'-Chlorobenzophenone

HMCC Hypergolic Maintenance and Checkout Cell [Aerospace]

HMCF Hypergolic Maintenance and Checkout Facility [Aerospace]

HMCRI Hazardous Materials Control Research Institute [US]

HMCS Her/His Majesty's Canadian Ship

HMCSP Heavy Metals Contamination Soil Project [US]

HMD Head-Mounted Display

HMD Helmet-Mounted Display

HMD Hexamethylenediamine

HMD Horizontal Movement Detection

HMD Hot-Mix Design

HMD Humid [also Hmd, or hmd]

HMD Hyaline Membrane Disease [Medicine]

HMD Hybrid Metalloorganic Decomposition

HMD Hydraulic Mean Depth

HMDA Hexamethylenediamine

HMDAA Hydroxymethyl Diacetone Acrylamide

HMDB Hexamethyl "Dewar Benzene"

HMDBA Hollow Metal Door and Buck Association [US]

HMDE Hanging Mercury Drop Electrode

HMDE/ASV Hanging Mercury Drop Electrode/Anodic Stripping Voltage

HMDE/DPCSV Hanging Mercury Drop Electrode/ Differential-Pulse Continuous Stripping Voltage

HMDF Horizontal Side of Main Distribution Frame

HMDI Hydrogenated 4,4'-Diphenylmethane Diisocyanate

HMDIC Hexamethylenediisocyanate

HMD PZT Hybrid Metalloorganic Decomposition Lead Zirconate Titanate

HMDS Hexamethyldisilazane

HMDS Hexamethyldisiloxane [also HMDSO]

HME High-Vinyl Modified Epoxy

HMeH(Φ)P 1-Methylheptylphenylphosphonic Acid

H2MEHP Mono-2-Ethylhexylphosphoric Acid

HMEM Habitat Management Evaluation Method

HMF Half-Metallic Ferromagnetic(s)

HMF Heavy-Metal Fluoride

HMF High Magnetic Field

HMF High-Modulus Fiber

HMF High-Modulus Furnace (Black)

HMF His/Her Majesty's Force [UK]

HMF Horizontal Mating Facility

HMF 5-Hydroxymethyl-2-Furfural

HMF Hydroxymethylfuraldehyde

HMF Hypergol(ic) Maintenance Facility [Aerospace]

HMF Black High-Modulus Furnace Black [also HMF black]

HMFG Heavy Metal Fluoride Glass

HMG Hardware Message Generator

HMG Heavy Machine Gun

HMG Her/His Majesty's Government [UK]

HMG High-Magnesia Glass

HMG Historical Metallurgy Group [now Historical Metallurgy Society, UK]

HMG Human Menopausal Gonadotropin [Biochemistry]

HMG Human Molecular Genetics [Journal published by Oxford University Press, UK]

HMG 3-Hydroxy-3-Methyl Glutaric Acid

HMI Hahn-Meitner Institut Berlin GmbH [Hahn-Meitner Institute, Berlin, Germany]

HMI Halogen Metallide Iodide

HMI Hardware Monitor Interface

HMI Hardwood Manufacturers Institute [now Hardwood Manufacturers Associoation, US]

HMI Her/His Majesty's Inspector [UK]

HMI   Hexamethyleneimine

HMI   Hoisting Machinery Institute

HMI   Hoist Manufacturers Institute [US]

HMI   Horizontal Motion Index

HMI   Human-Machine Interface

H∥M∥I   Magnetic Field In-Plane (or Parallel) to Magnetization and (Electric) Current [Symbol]

H∥M⊥I   Magnetic Field In-Plane (or Parallel) to Magnetization and Out-of-Plane (or Perpendicular) to (Electric) Current [Symbol]

H⊥M∥I   Magnetic Field Out-of-Plane (or Perpendicular) to Magnetization and In-Plane (or Parallel) to (Electric) Current [Symbol]

H⊥M⊥I   Magnetic Field Out-of-Plane (or Perpendicular) to Magnetization and (Electric) Current [Symbol]

HMI Berlin GmbH   Hahn-Meitner Institut Berlin GmbH [Hahn-Meitner Institute, Berlin, Germany]

HMIG   Hazardous Materials Identification Guide

HMIPI   Her Majesty's Industrial Pollution Inspectorate [US]

HMIS   Hazardous Materials Identification System

HMIS   Hazardous Material Information System

H·m²/kg   Henry meter squared per kilogram [also H·m² kg^{-1}]

HML   Hard Mobile Launcher

HML   Hardware Modelling Library

HML   Hawaii Marine Laboratory [Coconut Island, Oahu, US]

HML   Hexagonally Modulated Lamellae

HMM   Hard Magnetic Material

HMM   Hardware Multiply Module

HMM   Heavy Meromyosin [Biochemistry]

HMM   Hexakis(methoxymethyl)melamine

HMM   Hexamethylmelamine [also HEMEL, Hemel, or hemel]

HMMC   Halogen-Bridged Mixed-Valence Metal Complex

H·m²/mol   Henry meter squared per mole [also H·m² mol^{-1}]

HMMP   Hyper-Media Management Protocol

HMMR   High-Resolution Multifrequency Microwave Radiometer

HMMS   Highway Maintenance Management System

HMMWV   High-Mobility Multipurpose Wheeled Vehicle

HMN   2,2,4,4,6,8,8-Heptamethylnonane

HMO   Hardware Microcode Optimizer

HMO   Health Maintenance Organization [US]

HMO   Hueckel Approximation for Molecular Orbitals [Physical Chemistry]

HMO   Hueckel Molecular Orbital (Method) [Physical Chemistry]

H2MOP   Mono-n-Octylphosphoric Acid

H2MOΦP   Mono[p-(Isooctyl)phenyl]phosphoric Acid

HMOS   High-Density Metal-Oxide Semiconductor

HMOS   High-Performance Metal-Oxide Semiconductor

HMOS   High-Speed Metal-Oxide Semiconductor

HMOS-E   High-Density Metal-Oxide Semiconductor–Erasable

HMOS-E   High-Speed Metal-Oxide Semiconductor–Erasable

HMP   Hanford Mission Plan [of US Department of Energy, Hanford Plant, Washington State]

HMP   (Advances in) Hard Materials Production

HMP   Hexametaphosphate

HMP   Hexose Monophosphate [Biochemistry]

HMP   High-Modulus Polymer

HMP   Host Monitoring Protocol [Internet]

HMP   Hydrazine Monopropellant [for NASA Reaction Control System]

HMPA   Hexamethylphosphoramide [also HEMPA, Hempa, or hempa]

HMPA   Hexamethylphosphotriamide

HMPA-THF   Hexamethylphosphoramide Tetrahydrofuran

HMPC   Her Majesty's Privy Council [UK]

HMPG   Hydroxymethoxyphenylglycol

HMPT   Hexamethylphosphoric Acid Triamide

HMPT   Hexamethylphosphorotriamide

HMQC   Heteronuclear Multiple-Quantum Coherence

HMR   Hazardous Materials Regulations [of Department of Transportation, US]

HMR   Homer [also Hmr]

HMR   Humidity-Mixing Ratio [Meteorology]

HMR   Hybrid Modular Redundancy

HMR   Hydrometallurgical Reduction

HMRB   Hazardous Materials Regulation Board [US]

HMRL   Harvard Materials Research Laboratory [of Harvard University, Cambridge, Massachusetts, US]

HMS   Hanford Meteorological Station [Washington State, US]

HMS   Hanford Meteorological System [Washington State, US]

HMS   Hanford Meteorology Survey [Washington State, US]

HMS   Harmonic Mode Scrambling

HMS   Harvard Medical School [Boston, Massachusetts, US]

HMS   Hazardous Materials System [of US Bureau of Explosives]

HMS   Her/His Majesty's Service [UK]

HMS   Her/His Majesty's Ship [UK]

HMS   Her/His Majesty's Steamer [UK]

HMS   Historical Metallurgy Society [UK]

HMS   History Memory System

HMSA   Hardware Manufacturers Statistical Association [now Builders Hardware Manufacturers Association, US]

HMSA   Hawk Mountain Sanctuary Association [US]

HMS(BOE)   Hazardous Materials System (Bureau of Explosives) [US]

HMSC   Hatfield Marine Science Center [of Oregon State University at Newport, US]

HMSC   Huntsman Marine Science Center [Canada]

HMSD   Helmet-Mounted Sight and/or Display

HMSF   Hexamethylenetetraselenafulvalene

HMSF   Honolulu Marine Support Facility [of National Oceanic and Atmospheric Administration in Hawaii, US]

HMSH   Hanford Museum of Science and History [Washington State, US]

HMSO   Her/His Majesty's Stationery Office [of Department of Trade and Industry, UK]

HMSS   Hospital Management Systems Society

HMSLD   Helium Mass Spectrometer Leak Detector

HMT Hand-Microtelephone

HMT Hazardous Materials Transportation

HMT Hexamethylenetetramine [also HMTA, HMTeA, or Hexa]

HMT High-Mobility T-Lymphocytes [Medicine]

HMT Hypoxanthine/Methotrexate/Thymidine Medium [Biochemistry]

HMTA Hazardous Materials Transportation Act [US]

HMTA Hexamethylenetetramine [also HMT, HMTeA, or Hexa]

HMTC Hexamethylenetetramine Camphorate

HMTD Hexamethylenetriperoxidediamine [also HMTPDA]

HMTeA Hexamethylenetetramine [also HMT, HMTA, or Hexa]

HMTP High-Modulus Thermoplastic Preform

HMTPDA Hexamethylenetriperoxidediamine [also HMTD]

HMTSeF Hexamethylenetetraselenafulvalene [also HMTSF]

HMTT HMT Technology Corporation [Fremont, California]

HMTTeF Hexamethylenetetratellurafulvalene

HMTTF Hexamethylenetetrathiofulvalene

HMTSF Hexamethylenetetraselenafulvalene [also HMTSeF]

HMTSF-TCNQ Hexamethylenetetraselenafulvalene Tetracyanoquinoline

HMTUSA Hazardous Materials Transportation and Uniform Safety Act [US]

HMU Hardware Mockup

HMV Hydrodynamic Modulation Voltammetry

HMV Hydrogen Manual Valve

HMV Vickers Microhardness

HMW Hectometric (Radio) Wave

HMW High Molecular Weight

HMWB High-Molecular-Weight Branched (Polymer)

HMW HDPE High Molecular Weight High-Density Polyethylene [also HMWHDPE]

HMWK High-Molecular-Weight Kininogen [Biochemistry]

HMWPE High-Molecular-Weight Polyethylene [also HMW PE]

HMWPS High-Molecular-Weight Polystyrene [also HMW PS]

HMX Cyclotetramethylene Tetranitramine

HMX Her/His Majesty's Explosive

HMX Homocyclonite

HMY Her/His Majesty's Yacht

HN Hard Nickel (Plating)

HN Havriliak-Negami (Function) [Physics]

HN Hemagglutinin Neuraminidase [Biochemistry]

HN Heterogeneous Nucleation

HN Hex(agonal) Nipple

HN Hex(agonal) Nut

HN Honduras [ISO Code]

HN Horizontal Needle

HN Host-to-Network

HN Hotter than No Flash [British Propellant]

HN$_2$ Mechlorethamine

H$_2$/N$_2$ Hydrogen/Nitrogen (Ratio)

Hn Horn [also hn]

H7/n6 (Accurate) Locational Transition Fit; Basic Hole System [ISO Symbol]

.hn Honduras [Country Code/Domain Name]

HNA Hierachical Network Architecture

HNA Hitachi Network Architecture

HNA Hydroxynaphthoic Acid

HNAAP 4-N-(2-Hydroxy-1'-Naphthylidene) Aminoantipyrine

HNAB Hexanitroazobenzene

H2nad Noradrenaline [Biochemistry]

H2napbu 2,2'-[1,4-Butanediylbis(nitrilomethylidyne)-bis(naphthalen-1-ol)

H2napdec 2,2'-[1,10-Decanediylbis(nitrilomethylidyne)-bis(naphthalen-1-ol)

H2napen 2,2'-[1,2-Ethanediylbis(nitrilomethylidyne)-bis(naphthalen-1-ol)

H2nappn 2,2'-[1,3-Butanediylbis(nitrilomethylidyne)-bis(naphthalen-1-ol)

H-NBR Hydrogenated Nitrile Butadiene Rubber

HNC Hand Numerical Control

HNC Higher National Certificate [UK]

HNC Hypernetted Chain (Approximation)

HNCbl Hexanitrocarbanilide

HNCC High Nuclearity Carbonyl Cluster

HNCIAWPRC Hungarian National Committee of the International Association on Water Pollution Research and Control

HNCIMU Hungarian National Committee for the International Mathematical Union

HNC/MS Hypernetted-Chain Spherical (Equation)

HND Higher National Degree [UK]

HND Higher National Diploma [UK]

Hnd Hundred [also hnd]

Hndbk Handbook [also hndbk]

Hndle Handle [also hndle]

Hndler Handler [also hndler]

HNDPh Hexanitrodiphenyl

HNDPhA Hexanitrodiphenylamine

HNDPhAEN Hexanitrodiphenylaminoethylnitrate

HNDPhBzl Hexanitrodiphenylbenzyl

HNDPhGu Hexanitrodiphenylguanidine

HNDPhSfi Hexanitrodiphenylsulfide

HNDPhSfo Hexanitrodiphenylsulfone

HNDPhU Hexanitrodiphenylurea

HNDT Holographic Nondestructive Test

4HNE (E)-4-Hydroxy-2-Nonenal

HNED Horizontal Null External Distance

HNEI Hawaii Natural Energy Institute [of University of Hawaii, US]

HNEt Hexanitroethane

H-NFE-TB Hybrid Nearly Free Electron Tight-Binding [Physics]

HNMnt Hexanitromannitol [Biochemistry]

H-NMR Hydrogen Nuclear Magnetic Resonance

H-1 NMR Hydrogen-1 (or Proton) Nuclear Magnetic Resonance [also ^{1}H NMR]

H-2 NMR Hydrogen-2 (or Proton) Nuclear Magnetic Resonance [also ^{2}H NMR]

HNG Hawaii National Guard

HNG Hydrin-Nitroglycerin

Hngr Hangar [also hngr]

HNH Hexanitroheptane

HNIL High-Noise-Immunity Logic

HNIRI Hokkaido National Industrial Research Institute [of Agency of Industrial Science and Technology, Japan]

HNK Human NK Cell [Medicine]

HNL Helium Neon Laser

HNL Honduran Lempira [Currency of Honduras]

HNM Helicopter Noise Model

HNM Hexanitromannite [Biochemistry]

HNMe$_2$ Dimethylamine

H-1 NMR Hydrogen-1 (or Proton) Nuclear Magnetic Resonance [also ^{1}H NMR]

H-2 NMR Hydrogen-2 (or Deuterium) Nuclear Magnetic Resonance [also ^{2}H NMR]

HNO Hexanitrooxaanilide

HNO Nitroxyl

HNO$_3$ Nitric Acid

HN-O(ClMP) n-Octylhydrogenchloromethylphosphonic Acid

HNO$_3$/H$_2$O Nitric Acid-to-Water (Ratio)

HNOE Heteronuclear Overhauser Effect [Physics]

HNOxn Hexanitrooxaanilide

HNP Human Neutrophil Peptide [Biochemistry]

HNPA Home Numbering Plan Area

HNPF Hallam Nuclear Power Facility [Nebraska, US] [Decommissioned]

HNR Hoffman-Nord-Ruedenberg (Gradient Extremals)

HNRC Human Nutrition Research Center [of US Department of Agriculture at Grand Forks, North Dakota, US]

HNRIMS Human Nutrition Research and Information Management System [of US Department of Agriculture, US Department of Health and Human Services and National Institutes of Health]

hnRNA Heterogeneous Nuclear Ribonucleic Acid [Biochemistry]

HNS Hexanitrostilbene

HNS High-Nitrogen Steel

HNS International Conference on High-Nitrogen Steels

HN(SiMe$_3$)$_2$ Bistrimethylsilylamine

H-1 NSLR Hydrogen-1 Nuclear Spin-Lattice Relaxation [also ^{1}H NSLR]

HNTD Highest Non-Toxic Dose

HNU Henan Normal University [Xinxiang, PR China]

HNVS Helicopter Night Vision System

HNVS Hughes Night Vision System

HO Hale Observatories [Now Mount Wilson Observatory and Palomar Observatory, US]

HO Hamiltonian Operator [Quantum Mechanics]

HO Hand Operation; Hand-Operated

HO Hard Overhung (Balancer)

HO Harmonic Oscillation; Harmonic Oscillator

HO Hartley Oscillator

HO Head Office

HO Heat Output

HO Heavy Oil

HO Heavy Oxygen

HO Helicopter, Observation [US Navy Symbol]

HO Hermitian Operator [Mathematics]

HO Herringbone Order

HO Hertzian Oscillator

HO High Orbit [also H/O]

HO High Order

HO Hofmann Orientation [Materials Science]

HO Home Office [UK]

HO Hybrid Orbital [Physics]

HO Hydrogen-Oxygen (Atmosphere)

HO Hydrographic Office [US]

HO Hydroxyl (Radical) [Symbol]

HO Rotary-Wing Observation (Aircraft) [US Army Symbol]

H$_2$O Water

H$_3$O$^+$ Hydronium (or Oxonium) Ion [Symbol]

H$_2$O$_2$ Hydrogen Peroxide

H/O Handover

H/O High Orbit [also HO]

Ho Holmium [Symbol]

Ho House [also ho]

Ho^{3+} Holmium Ion [also Ho^{+++}] [Symbol]

Ho II Holmium II [Symbol]

Ho-123 HoBa$_2$Cu$_3$O$_{7-x}$ [A Holmium Barium Copper Oxide Superconductor]

Ho-160 Holmium-160 [also ^{160}Ho, or Ho^{160}]

Ho-161 Holmium-161 [also ^{161}Ho, or Ho^{161}]

Ho-162 Holmium-162 [also ^{162}Ho, or Ho^{162}]

Ho-163 Holmium-163 [also ^{163}Ho, or Ho^{163}]

Ho-164 Holmium-164 [also ^{164}Ho, or Ho^{164}]

Ho-165 Holmium-165 [also ^{165}Ho, Ho^{165}, or Ho]

ho horse [Immunochemistry]

hΩ phonon energy (in solid-state physics) [Symbol]

h(ω) magnitude of displacement oscillation [Symbol]

HOA Highly Overaged; High Overaging [Metallurgy]

HOAB Heptyloxyazoxybenzene

HOAc Hydroxyacetic Acid (or Glycolic Acid)

HOAG Honiara Aircraft Operations Group [Solomon Islands]

H$_2$O/Alk Water-to-Alkoxide (Ratio)

H$_2$/O$_2$/Ar Hydrogen/Oxygen/Argon (Mixture)

HOARS Hands-On Annotated Recorded Search

HOB Homing on Offset Beacon

HOB Hot Ore Briquetting [Metallurgy]

Ho:BaY$_2$F$_8$ Holmium-Doped Lithium Yttrium Fluoride [also Ho^{3+}:BaY$_2$F$_8$]

HoBCO Holmium Barium Copper Oxide (Superconductor)

HOBF Higher-Order Bright Field (Microscopy)

HOBIS Hotel Billing Information System

HOBITS Haifa On-Line Bibilographic Text System [Israel]

HOBO Homing Official Bomb

HOBOS Homing Bomb System

HOBT 1-Hydroxy 1H-Benzotriazole; 1-Hydroxybenzotriazole [also HOBt]

HOC Halogenated Organic Compound

HoC Hollow Charge [Explosives]

Hoc Cyclohexyloxycarbonyl

H_2O/C_2H_5OH Water-Ethyl Alcohol (Mixture)

HOCl Hypochlorous Acid

HO(Cl2MP) n-Octylhydrogendichloromethylphosphonic Acid

HOCO Highest Occupied Crystal Orbital

HOD Higher-Order Differentiation

HODE Hydroxyoctadecadienoic Acid

13(S)-HODE 13(S)-Hydroxyoctadeca-(9Z,11E)Dienoic Acid

HODOS Hole Drilling Operating System

HODRAL Hokushin Data Reduction Algorithm Language

HODS Higher Order Derivation Spectroscopy

HOE Height-of-Eye

HOE Holographic Optical Element

HOE Homing Overlay Experiment

Hoechst High Chem Mag Hoechst High Chemistry Magazine [Published in the UK]

H2O2EDP Dioctyl-P,P'-Ethane-1,2-Diphosphonic Acid

HOESY Heteronuclear NOE (Nuclear Overhauser Enhancement) Sequence [Two-Dimensional Nuclear Magnetic Resonance]

HOF Head of Form

HOF Highly Oriented Film

HOF High-Octane Fuel

HOFNET High-Order Feedback Neural Network

$Ho(FOD)_3$ HolmiumTris(heptafluorodimethyloctanedionate) [also $Ho(fod)_3$]

HOFR Heat-Resistant, Oil-Resistant, Flame-Retardant

HOG Harley (Davidson) Owners Group

HOG Hall Oil Gasification (Process)

HOH Water

HOHAHA Homonuclear Hartmann-Hahn (Spectroscopy)

H_2O/HNO_3 Water-to-Nitric Acid (Ratio)

HOI Hypoiodous Acid

HOJ Home On Jamming

$Ho:KYF_4$ Holmium-Doped Potassium Yttrium Fluoride [also $H^{3+}:KYF_4$]

HOL High(er)-Order (Computer) Language

Hol Holiday [also hol]

Hol Holography [also hol]

hol hollow

HOLC High-Order Language Computer

Holdg Holding [also holdg]

$Ho:LiYF_4$ Holmium-Doped Lithium Yttrium Fluoride [also $Ho^{3+}:LiYF_4$]

Holl Holland

Hollow Sect Hollow Section [Published in the UK]

Hols Holidays [also hols]

HOLUG Houston On-Line Users Group [US]

HOLZ Higher Order Laue Zone [Crystallography]

Holzforsch Holzverwert Holzforschung und Holzverwertung [Austrian Journal on Wood Research and Wood Exploitation]

Holz Roh-Werkstoff Holz als Roh- und Werkstoff [German Publication on Wood as a Raw and Engineering Material]

Holztechnol Holztechnologie [German Journal on Wood Technology published by Institut für Holztechnologie und Faserbaustoffe]

HOM Higher-Order Mode

Hom Homogeneity [also hom]

Hom Homomorphism [Mathematics]

HOMA Heads of Marine Agencies

HOMC 7-Hydroxy-4-Methylcoumarin

HOMCOR Homonuclear Correlation

Homeop Homeopathic; Homeopathist; Homeopathy [also homeop]

Hommes Fonderie Hommes et Fonderie [French Journal on Man and Foundry]

HOMO Heads of Marine Organizations

HOMO Highest Occupied Molecular Orbital [Physical Chemistry]

Homo-DPT 3,7-Dinitro-1,5-Endoethylene-1,3,5,7-Tetrazacyclo-octane [also homo-DPT]

HOMO-LUMO Highest Occupied Molecular Orbital/Lowest Occupied Molecular Orbital [Physical Chemistry]

HON 2-Amino-5-Hydroxy-4-Oxo Pentanoic Acid

Hon Honorable

Hon Honorary [also hon]

Hon Honors [also hon]

HONB N-Hydroxy-5-Norbornene-2,3-Dicarbonic Acid Imide; N-Hydroxy-5-Norbornene-2,3-Dicarboximide

Hond Honduran; Honduras

Hong Kong Eng Hong Kong Engineer [Engineering Journal published in Hong Kong]

Hons Honours (Degree)

hon sec honorary secretary

Hoogovens Groep Bull Hoogovens Groep Bulletin [Dutch Technical Bulletin]

HO(OP) n-Octylhydrogen-n-Octylphosphonate

HOOPS Hierarchical Object-Oriented Picture System

HOP HEDL (Hanford Engineering Development Laboratory) Overpower

HOP Hoosier Ordnance Plant [Indiana Arsenal, Charleston, Indiana, US]

HOP House Operating Tape

HOP Hybrid Operating Program

HOP Hydrogen Overpotential [Metallurgy]

HoP Holmium Phosphide

HO(Φ)P n-Octylhydrogenphenylphosphonate

H2(OP) n-Octylphosphonic Acid

HOPE Hydrogen-Oxygen Primary Extraterrestrial [of NASA]

HOPE-X H-II Orbiting Plane–Experiment [of National Space Development Agency of Japan]

HOPG Highly-Oriented Pyrolytic Graphite

HOPL History of Programming Languages

HO(Φ)P n-Octylhydrogenphenylphosphonate

HOPS Heads of Public Services [of University of California and Stanford University, US]

HOQ Home Office Quote

HOR H_2O (Water) Reactor

HOR Heliocentric Orbit Rendezvous

HOR Hydrogen Oxidation Reaction

Hor Horizon [also hor]

hor horizontal

HORACE H_2O Reactor Aldermaston Critical Experiment [UK]

Horiz Horizon [also horiz]

horiz horizontal

Hormone Metabol Res Hormone and Metabolic Research [Journal]

Hormone Res Hormone Research [International Journal]

Hormones Behav Hormones and Behavior [Journal published in the US]

Horol Horological; Horologer; Horology [also horol]

Horol J Horological Journal [Joint Publication of the British Horological Institute and the British Clock and Watch Association]

HORSCERA House of Representatives Standing Committee on the Environment, Recreation and the Arts

HORSEC House of Representatives Standing Committee on Environment and Conservation

HORSES High-Order Raman Spectral Excitation Studies

Hort Horticultural; Horticulture; Horticulturist [also hort]

horz horizontal

HOS High-Order Software

HOS Horizontal Obstacle Sonar

HOSA Hydroxylamine-O-Sulfonic Acid

HOSC Huntsville Operations Support Center [of NASA Marshall Space Flight Center in Alabama, US]

HOSG Helicopter Operations Study Group [UK]

Hosp Hospital [also hosp]

Hosp Manage Hospital Management (Journal) [US]

Hosp Pract Hospital Practice [Journal published in the US]

Hosp Top Hospital Topics (Journal)

HOST Hot Section Technology

HOSu N-Hydroxysuccinimide

HOT Hand Over Transmitter

H(ωt) Sinusoidal (Magnetic) Field [Symbol]

HOTCE Hot Critical Experiment [US]

H_2O/TEOS Water-to-Tetraethylorthosilicate (Ratio)

Ho(THD)$_3$ Holmium Tris(tetramethylheptanedionate) [also Ho(thd)$_3$]

Ho(TMHD)$_3$ Holmium Tris(2,2,6,6-Tetramethyl-3,5-Heptanedione) [also Ho(tmhd)$_3$]

Ho:Tm:YLF Holmium and Thulium Doped Yttrium Lithium Fluoride (Laser)

HOTOL Horizontal Takeoff and Landing

HOTRAN Hover and Transition Simulator

HOTS Heads of Technical Services [of University of California and Stanford University, US]

HOTSHOT Hydrogen-Oxygen Turbine Superhigh Operating Temperature

HOTT Hot Off The Tree [Electronics Newsletter]

Hot Work Technol Hot Working Technology [Journal published in PR China]

Houst J Math Houston Journal of Mathematics [Published by the University of Houston, Texas, US]

HOV High Occupancy Vehicle (Lanes)

HOW Holston Ordnance Works [Kingsport, Tennessee, US]

H_2O/W Water-to-Tungsten (Ratio)

How Howitzer [also H]

HOWAQ Hot Water Quench (Process) [also Howaq] [Metallurgy]

HOX Hypohalous Acid (X = Halogen) [General Formula]

HOx Odd Hydrogen [e.g., (OH, HO_2, H_2O_2)]

Hox 5-Hydroxyquinazoline

Ho-Y Holmium-Yttrium (Alloy System)

Ho:YAG Holmium-Doped Yttrium Aluminum Garnet (Laser) [also Ho^{3+}:YAG]

HoYBCO Holmium Yttrium Barium Copper Oxide (Superconductor)

HP Habit Plane [Crystallography]

HP Hagen-Poiseuille (Constant) [Fluid Mechanics]

HP Half-Period

HP Hall-Petch (Equation) [also H-P] [Metallurgy]

HP Hall Plate

HP Hall Process [Metallurgy]

HP Hamilton's Principle [Mechanics]

HP Handling Procedure

HP Hand Pump

HP Hardcore Pinch

HP Hard Pipe

HP Harmonic Progression [Mathematics]

HP Hatfield Polytechnic [Hatfield, Hertfordshire, UK]

HP Hazardous Product

HP Healthcare Product

HP Health Physicist; Health Physics

HP Heatable Plastic

HP Heat Pipe

HP Heatproof(ing)

HP Heat Pump

HP Helicopter Pilot

HP Helix Pomatia [Biochemistry]

HP Helper Peak [Immunochemistry]

HP Hermite Polynomial

HP Heteroaromatic Polymer(ization)

HP Heterophase Polymer(ization)

HP Heusler Phase [Metallurgy]

HP Hewlett Packard Company [also H-P] [US]

HP Hiding Power

HP High-Pass (Filter)

HP High Performance

HP High Permeability

HP High Phosphorus

HP High Polymer
HP Highly Porous; High Porosity
HP High Positive
HP High Power
HP High Pressure [also H-P, h-p, or hp]
HP High Purity
HP Highway Patrol
HP Hire Purchase [also hp]
HP Hittorf's Phosphorus
HP Holding Pattern
HP Hollow Point (Bullet)
HP Home Page [Internet]
HP Homopolymer(ization)
HP Hope [Amateur Radio]
HP Horizontal Polarization
HP Horsepower [also hp]
HP Host Processor
HP Hot-Pressed; Hot Pressing [Powder Metallurgy]
HP HP-Type Structural Steel Shape [AISI/AISC Designation]
HP Huygens Principle [Optics]
HP Hunter (Reduction) Process
HP Hurwitz Polynomial
HP Hybridoma/Plasmacytoma [Medicine]
HP Hydraulic Power
HP Hydraulic Press(ing)
HP Hydraulic Pressure
HP Hydraulic Pump
HP Hydrogen Purifier
HP Hydroplastic(ity)
HP Hydropneumatic(s)
HP Hydropower
HP Hydroxyproline
HP Hydroxypropionate
HP Hyperpolarizability
HP Pyruvic Acid
HP- Panama [Civil Aircraft Marking]
HP-029 Hydroxytacrine (or (±)-9-Amino-1,2,3,4-Tetrahydroacridin-1-ol)
HP-1 Helper Peak-1 [Immunochemistry]
3HP 3-Hydroxypropionate
H-P Hall-Petch (Equation) [also HP] [Metallurgy]
H-P Hewlett-Packard Company [also HP] [US]
H-P High Pressure [also HP, hp, or h-p]
Hp Hematoporphyrin [Biochemistry]
H7/p6 Locational Interference Fit; Basic Hole System [ISO Symbol]
hP Pearson symbol for primitive (simple) space lattice in hexagonal crystal system (this symbol is followed by the number of atoms per unit cell, e.g. hP2, hP8, etc.)
HPA Hazardous Products Act [Canada]
HPA Helix Pomatia Agglutinin [Biochemistry]
HPA Heuristic Path Algorithm
HPA High-Performance Alloy
HPA High-Power Amplifier

HPA High-Pressure (Oxygen) Annealing [Metallurgy]
HPA Holding and Positioning Aid
HPA 2-Hydroxypropyl Acrylate
hPa hectopascal [Unit]
Hpa Haemophilus parainfluenzae [Microbiology]
Hpa I Haemophilus parainfluenzae I [Microbiology]
Hpa II Haemophilus parainfluenzae II [Microbiology]
HPAC Center for High-Performance Applied Computing [of Delft University of Technology, Netherlands] [also HpαC]
HPAE High-Pressure Air Extraction
HPB Health Protection Branch [of Health Canada]
HPB Health Protection Branch [of Canadian Pharmaceutical Association]
HPB Hydrogenated Polybutadiene
HPBC Homopolar Pulse Billet Heating
HPBC Hot-Pressed Boron Carbide
HPBN Hot-Pressed Boron Nitride
HPBW Half-Peak Beamwidth
HPBW (Antenna) Half-Power Beamwidth
HPC Handheld Personal Computer
HPC Hard-Processing Channel (Black)
HPC Hawaii Pacific College [Keneohe, Honolulu, US]
HPC High Pin Count
HPC High-Performance Ceramics
HPC High-Performance Chromatography
HPC High-Performance Composite
HPC High-Performance Computing
HPC High-Performance Concrete
HPC High-Pressure Chemistry
HPC High-Pressure Chromatography
HPC Horticultural Policy Council
HPC Hot Potassium Carbonate (Process)
HPC Hydro Power Control
HPC Hydroxypropylcellulose
HPCA High-Performance Communications Adapter
HPCA Housing Pressure Cold Advance (Regulator) [Automobiles]
HPC Black Hard-Processing Channel Black [also HPC black]
HPCC High-Performance Computing and Communications
HPCE High-Performance Capillary Electrophoresis
HPCI High-Pressure Coolant Injection
HPCL Hydro Petroleum Canada Limited
HPCN High-Performance Computer Network
HPCP High-Performance Computing Platform
HPCS High-Pressure Combustion Sintering
HPCS High-Pressure Core Spray; High-Pressure Core Spraying System [Nuclear Engineering]
HPCwire High-Performance Computing On-Line Publication
HPD Hearing Protective Device
HPD Highest Posterior Density
HPD High Power Density
HPD Horizontal Polar Diagram
HPD Hough-Powell Device
HpD Hematoporphyrin Derivative [Biochemistry]

HPDJ Hewlett-Packard Desk Jet (Printer)

HPDTA 2-Hydroxy-1,3-Propanediaminetetraacetic Acid

4-HPDTA Propylenediamine-N,N,N'N'-Tetraacetic Acid [also H4PDTA]

HPE Hexaphenylmethane

HPE High-Pressure Equipment

HPE Hydropower Engineer(ing)

HPEP High-Performance Electrophoresis

HPE-ROW High-Pressure and Roll Welding

HPES Hot-Pressing-Aided Exothermic Synthesis

HPETE Hydroperoxyeicosatetraenoic Acid

5(S)-HPETE 5(S)-Hydroperoxy-(6E,8Z,11Z,14Z)-Eicosatetraenoic Acid

15(S)-HPETE 15(S)-Hydroperoxy-(5Z,8Z,11Z,13E)-Eicosatetraenoic Acid

HPETP High-Performance Engineering Thermoplastics

HPF Heat Pipe Furnace

HPF Highest Probable Frequency

HPF Highest Possible Frequency

HPF High-Pass Filter

HPF High-Performance FORTRAN

HPF High-Power Factor [Electronics]

HPF Horizontal Processing Facility

HPF Hot-Pressed Ferrite [Metallurgy]

HPF Hydroplastic Forming

HPFD Hybrid Personal Floating Device

HPFL High-Performance Fuels Laboratory [US]

HPFS Hewlett Packard File System [also HP FS]

HPFS High-Performance File System

HPFT High-Pressure Fuel Turbopump [also HPFTP]

HPFZ High-Pressure Float Zone (Crystal Growing Method)

HPG Hard Page [also Hpg]

HPG Hewlett-Packard Graphics

HPG High-Performance Graphics

HPG High-Pressure Gas

HPG Homopolar Generator

HPG Hydroxypropyl Guar

HPGA High-Pressure Gas Atomization (Process) [Powder Metallurgy]

HPGe High-Purity Germanium [also HpGe]

HPGF High-Pressure Gas Facility

HPGL Hewlett-Packard Graphics Language [also HP-GL]

HPGS High Pressure Gas System

HPH High-Performance Hydrogen (Technology)

Hph Haemophilus parahaemolyticus [Microbiology]

hph horsepower hour [also HPH, Hphr, HP-hr, hp-hr, or hp hr]

HP/HIP Hot Pressing followed by Hot Isostatic Pressing

HPhr Horsepower-Hour [also HP-hr, hp-hr, hp hr, HPH, or hph]

HPHT High-Pressure, High-Temperature (Process)

HP/HVOF High-Pressure High-Velocity Oxyfuel (Technology)

HPHW High-Pressure Hot Water (Heating System)

HPI Hardwood Plywood Institute [now Hardwood Plywood Manufacturers Association, US]

HPI Height-Position Indicator [Radar]

HPI Heinrich-Pette-Institut (für Experimentelle Virologie und Immunologie an der Universität Hamburg) [Heinrich Pette Institute (for Experimental Virology and Immunology), University of Hamburg, Germany]

HPI High-Performance Imagery

HPI High-Performance Insulation

HPI High-Pressure Injection

HPI History of Present Illness [Medicine]

HPIA Hydroxyphenylisopropyladenosine [Biochemistry]

HPIA Iodo-4-Hydroxyphenylisopropyl)adenosine [Biochemistry]

HPIB Hewlett Packard Interface Bus [also HP-IB]

HPIC High-Performance Ion Chromatography

HPIEC High-Performance Ion Exchange Chromatography

HPIM High-Performance Image Analysis

hp/in³/min horsepower per cubic inch per minute [also hp/cu in/min]

HPIB Hanford Permanent Isolution Barrier

HPIR High-Probability-of-Intercept Receiver

HPIS High-Pressure Injection System

HPIT High-Performance Infiltrating Technology

HPIWF High-Pressure Industrial Water Facility

HPL Hexagonally Perforated Lamellae

HPL Hewlett-Packard Laboratories [Palo Alto, California, US]

HPL High-Power Laser

HPL High-Pressure Laminate(s)

HPL Hot Photoluminescence

HPL Human Performance Laboratory [of Ball State University, Muncie, Indiana, US]

HPL Human Placental Lactogen [also hPL] [Biochemistry]

HPLAC High-Performance Liquid Affinity Chromatography

hPLAP Human Placental Alkaline Phosphatase [Biochemistry]

HPLB High-Power Laser Beam

HPLC High-Performance Liquid Chromatography [also hplc]

HPLC High-Pressure Liquid Chromatography [also hplc]

HPLC International Symposium on High-Performance Liquid-Phase Separations and Related Techniques

HPLC-MS High-Pressure Liquid Chromatograpy with Mass Spectrometry; High-Performance Liquid Chromatography–Mass Spectrometry [also HPLC/MS]

HP-LCVD High-Pressure Laser-Assisted Chemical Vapor Deposition [also HPLCVD]

HP-LEC High-Pressure Liquid Encapsulated Czochralski

HPLJ Hewlett-Packard Laser Jet (Printer)

HP/LP High Pressure/Low Pressure,

HPLPLC High-Performance Low-Pressure Liquid Chromatography

HPLR Hinge Pillar [also H PLR, or H Plr]

HPLT High-Pressure Low-Temperature (Compaction) [Powder Metallurgy]

HPM Hard-Part Machining

HPM High-Performance Manufacturing

HPM High-Performance Motor

HPM High-Power Microwave

HPM High-Pressure Melting

HPM High-Pressure Molding

HPM High-Purity Metal

HPM Hybrid Phase Modulation

HPMA Hardwood Plywood Manufacturers Association [US]

HPMA High-Pressure Microwave-Plasma-Assisted (Chemical-Vapor Deposition System)

HPMA Hydroxypropyl Methacrylate

HPMACVD High-Pressure Microwave-Plasma-Assisted Chemical Vapor Deposition

HPMC Hydroxypropylmethylcellulose

HPML High-Pressure Mercury Lamp

HPMo 12-Molybdophosphoric Acid

HPMS High-Pressure Mass Spectrometry

HPMV High-Pressure Mercury Vapor

HPMVL High-Pressure Mercury-Vapor Lamp

HPMWA High-Power Microwave Amplifier

HPN High-Pass Notch [Electronics]

HPN Horsepower, Nominal

HP-NiAl High-Purity Nickel Aluminide

HPO Hatfield Polytechnic Observatory [UK]

HPO High-Pressure Oxidization

HPO Hydroxylamine Phosphate Oxime (Process)

HPO$^-$ Monohydrogen Phosphate (Ion) [Symbol]

H$_2$PO Dihydrogen Phosphate (Ion) [Symbol]

HPOF High-Pressure Oil-Filled (Cable)

13(S)-HPODE 13(S)-Hydroperoxy-(9Z,11E)-Octadecadienoic Acid

H Polynomial Hurwitz Polynomial [Mathematics]

HPOP High-Pressure Oxidizer Pump

HPOT Helipotentiometer [also Hpot, or hpot]

HPOT High Pressure Oxidizer Turbopump [HPOTP]

HPP Harward Project Physics [of Harvard University, Cambridge, Massachusetts, US]

HPP High-Performance Plastics

HPP High-Performance Polymer

HPP High-Performance Polymers [Publication of Institute of Physics, UK]

HPP High-Performance Process

HPP High-Polymer Physics

HPP High-Pressure Phenomenon

HPP High-Pressure Physics

HPP High-Pressure Process [Chemical; Engineering]

HPP High-Purity Polycrystalline (Ceramics)

HPP Holding Under Promise of Payment

HPP Hot Processing Plant

HPP 4-Hydroxypyrazolo[3,4-d]pyrimidine (or Allopurinol) [Pharmacology]

HPPA Hewlett-Packard Precision Architecture

HPPA High-Performance Pipe Association [UK]

HPPA 5-(4-Hydroxyphenyl)-5-Phenylhydantoin

HPPb$_2$ Hexaphenyldilead

HPPC High Performance Parallel Computing

HP-PCL Hewlett-Packard Printer Command Language

HPPD High Power Proton Decoupling [Physics]

HPPF Horizontal Payloads Processing Facility

HPPH 5-Hydroxylphenyl-5-Phenylhydantoin

HPPI High-Performance Parallel Interface

HPPLC High-Performance Planar Liquid Chromatography

HPPLC High-Performance Preparative Liquid Chromatography

HPPS High-Performance Paper Society [Japan]

HPR Hexazotized Pararosaniline

HPR High Performance Routing

HPRC Houston Petroleum Research Center [Texas, US]

HPRC Human Performance Research Center [of Utah State University, Logan, US]

H-Press High-Pressure [also h-press]

HPRI Highest Priority [Electronics]

HPRI/BCD Highest Priority Decimal to Binary-Coded Decimal (Encoder)

HPRL Hewlett-Packard Research Laboratory [US]

HPRL Hoechst Pharmaceutical Research Laboratories [US]

H2prot Isoproterenol

HPR Hydrogen Pressure Regulator

HP/RPM Horsepower per Revolution per Minute [also Hp/rpm, or hp/rpm]

HPRR Health Physics Research Reactor [of Oak Ridge National Laboratory, Tennessee, US]

HPRS High-Pressure Reaction Sintering (Process) [Metallurgy]

HPRS High-Purity Rotor Steel

HPRT Hypoxanthine Phosphoribosyl Transferase [Biochemistry]

HPRV Hydrogen Pressure Relief Valve

HPS Hanford Plant Standards [formerly Hanford Works Standard]

HPS Hardy Plant Society [UK]

HPS Hazardous Polluting Substance

HPS Health Physics Society

HPS Hernon Porosity Sealant

HPS Hepatic Portal System [Zoology]

HPS High-Performance Paper Society [Japan]

HPS High-Performance Steel

HPS High-Performance System

HPS High Pressure Sintering [Metallurgy]

HPS High-Pressure Sodium (Lamp)

HPS High Protein Supplement [Medicine]

HPS Horizontal Plane Slipmeter

HPS Horizontal Pull Slipmeter

HPS Hydraulic Power System

HPS Hydrogenated Polystyrene [also hPS, or h-PS]

HPS Hydropower Station

HPS Hydropower System

H-PS Hydrogen-Terminated Porous Silicon

HPSC High-Pressure Self-Combustion Sintering [Metallurgy]

HPSC Hot-Pressed Silicon Carbide [also HPSiC]

HPSe High-Purity Selenium [also HpSe]

HPSEC High-Performance Size-Exclusion Chromatography

HPSi High-Purity Silicon [also HpSi]

HPSiC Hot-Pressed Silicon Carbide [also HPSC]

HPSiC High-Pressure Silicon Carbide [also HPSC]

HPSIS High-Pressure Safety Injection System

HPSN Hot-Pressed Silicon Nitride

HPSn High-Purity Tin [also HpSn]

HPSTC Highest Point of Single (Gear) Tooth Contact

HPSW High-Pressure Service Water

HPT Head per Track [also HP/T]

HPT Health Physics Technician

HPT Heterojunction Phototransmitter

HPT Hexamethylphosphoramide

HPT Hexamethylphosphoric Acid Triamide

HPT High-Performance Textile

HPT High-Performance Titrimetry

HPT High-Performance Thermoset(s)

HPT High-Precision Thermostat

HPT High-Pressure Test

HPT High-Pressure Turbine

HPT Horizontal Plot Table

HPT Hydroxypyrenetrisulfonate; 8-Hydroxy-1,3,6-Pyrenetrisulfonic Acid

H PT High Point [also H Pt, or h pt]

HPTA High-Pressure Technology Association [UK]

HPTE High-Performance Turbine Engine

HPTET High-Performance Turbine Engine Technology

HPTLC High-Performance Thin-Layer Chromatography

HPTe High-Purity Tellurium [also HpTe]

HPTP High-Performance Thermoplastic(s)

HPTPC High-Performance Thermoplastic Composite

HPTS p-Toluenesulfonic Acid

HPU Hydraulic Power Unit

HPUX Hewlett Packard Unix (Operating System) [also HP-UX]

HPV Habit Plane Variant [Crystallography]

HPV Hepatic Portal Vein [Anatomy]

HPV High-Pressure Valve

HPV High Production Volume (Chemicals)

HPV Human Papilloma Virus

HPV Human Powered Vehicle

HPV-1 Human Parainfluenza Virus, Type 1

HPV Newsl HPV Newsletter [of International Human Powered Vehicle Association, US]

HPVT High-Pressure Physical Vapor Transport

HP-VUE Hewlett-Packard Visual User Environment [also HPVUE]

HPW High Performance Workstation

HPW High Purity Water

HPW Homopolar Pulse Welding

HPW Hot Pressure Welding

HPW 12-Tungstophosphoric Acid

HPZ Harris-Plishke-Zukermann (Model) [Materials Science]

HPZ Helicopter Protected Zone

HPZ Hydridopolysil(yl)azane

HQ Habitat Quality

HQ Headquarters [also Hq]

HQ Heavy Quantum [Physics]

HQ Heavy Quark [Particle Physics]

HQ (Workshop on) Heavy Quarks at Fixed Target [of Fermilab, Batavia, Illinois, US]

HQ High Quality

HQ Hydro-Quebec [Canada]

HQ Hydroquinone

HQ Hydroxyquinoline [also Hq]

8-HQ 8-Hydroxyquinoline [also Hq]

HQDA Headquarters Department of the Army [US]

HQC Headquarters Command [of US Air Force]

HQE Hydroquinone Electrode [Physical Chemistry]

HQEE Hydroquinone Di(β-Hydroxyethyl) Ether

HQG Hydrogen Quantum Generator

HQI Habitat Quality Index

HQI Hydro-Quebec International [Canada]

HQIR Hydro-Quebec Institute of Research [Canada]

HQO Hydroxyquinoline Oxide

HQP Highly Qualified Person(s)

HQP Highly Qualified Personnel

HQR Handling Qualities Rating

HQS High-Quality Sound

HQS 8-Hydroxyquinoline-5-Sulfonic Acid Monohydrate

HQT High-Q-Triplate (Line) [Electronics]

.HQX BinHex [Macintosh File Name Extension] [also .hxq]

HR Croatia [ISO Code]

HR Hagen-Rubens (Relation) [Optics]

HR Half-Reaction [Chemistry]

HR Half-Round [also ½R]

HR Hall Resistance [Electronics]

HR Handling Routine

HR Hand Reset

HR Hard to Research

HR Hard-Rolled [also hr]

HR Hartree-Roothaan [Physics]

HR Hazard Rating

HR Hear [Amateur Radio]

HR Heating Resistor

HR Heat(ing) Rate

HR Heat Reactor

HR Heat Recycle (Process)

HR Heat Regenerator

HR Heat Rejection

HR Heat-Resistant; Heat Resistance

HR Here [Amateur Radio]

HR Hertzsprung-Russell (Chart) [Astronomy]

HR Hessischer Rundfunk [Hessian State Broadcasting Station, Germany]

HR Heterogeneous Reaction; Heterogeneous Reactor

HR Heterolytic Reaction

HR Highly Resistant

HR High Range

HR High Reduction

HR High Resilience

HR High Resistance

HR High Resolution

HR High Resonance-Damping

HR High-Risk

HR Hindered Rotation [Chemistry]

HR Historical Record

HR Hit Ratio

HR Hofmann Rearrangement [Organic Chemistry]

HR Hoist Ring

HR Holding Register

HR Homogeneous Reaction; Homogeneous Reactor

HR Homolytic Reaction

HR Hot-Rolled; Hot Rolling [also hr]

HR Hueckel Rule [Physical Chemistry]

HR Human Relations

HR Human Resource(s)

HR Hume-Rothery (Rule) [Solid-State Physics]

HR Humidity, Relative

HR Hund's Rule [Physical Chemistry]

HR Rockwell Hardness

HR15N Rockwell '15N' Superficial Hardness (Diamond Cone Indenter; Applied Load 15 kg)

HR15T Rockwell '15T' Superficial Hardness (1/16 in. Steel Sphere Indenter; Applied Load 15 kg)

HR15W Rockwell '15W' Superficial Hardness (1/8 in. Steel Sphere Indenter; Applied Load 15 kg)

HR15X Rockwell '15X' Superficial Hardness (1/4 in. Steel Sphere Indenter; Applied Load 15 kg)

HR15Y Rockwell '15Y' Superficial Hardness (1/2 in. Steel Sphere Indenter; Applied Load 15 kg)

HR30N Rockwell '30N' Superficial Hardness (Diamond Cone Indenter; Applied Load 30 kg)

HR30T Rockwell '30T' Superficial Hardness (1/16 in. Steel Sphere Indenter; Applied Load 30 kg)

HR30W Rockwell '30W' Superficial Hardness (1/8 in. Steel Sphere Indenter; Applied Load 30 kg)

HR30X Rockwell '30X' Superficial Hardness (1/4 in. Steel Sphere Indenter; Applied Load 30 kg)

HR30Y Rockwell '30Y' Superficial Hardness (1/2 in. Steel Sphere Indenter; Applied Load 30 kg)

HR45N Rockwell '45N' Superficial Hardness (Diamond Cone Indenter; Applied Load 45 kg)

HR45T Rockwell '45T' Superficial Hardness (1/16 in. Steel Sphere Indenter; Applied Load 45 kg)

HR45W Rockwell '45W' Superficial Hardness (1/8 in. Steel Sphere Indenter; Applied Load 45 kg)

HR45X Rockwell '45X' Superficial Hardness (1/4 in. Steel Sphere Indenter; Applied Load 45 kg)

HR45Y Rockwell '45Y' Superficial Hardness (1/2 in. Steel Sphere Indenter; Applied Load 45 kg)

HR- Honduras [Civil Aircraft Marking]

½R Half-Round [also HR]

H&R Hughton and Richards [US]

H-R Herzsprung-Russell (Diagram) [Astronomy]

Hr Harbo(u)r [also hr]

hR Pearson symbol for primitive (simple) space lattice in rhombohedral crystal system (this symbol is followed by the number of atoms per unit cell, e.g. hR1, hR4, etc.)

hr (hora) – hour [Medical Prescriptions]

hr hour [also h]

.hr Croatia [Country Code/Domain Name]

HRA Hanford Remedial Action [US Department of Energy, Hanford Plant, Washington State]

HRA Health Resources Administration [Now Part of Health Resources and Services Administration, US]

HRA High Risk Area

HRA Hot-Rolled and Aged; Hot Rolling and Aging [Metallurgy]

HRA Hot-Rolled and Annealed; Hot Rolling and Annealing [Metallurgy]

HRA Human Reliability Analysis

HRA Human Resource Accounting

HRA Rockwell 'A' Hardness (Conical Diamond Indenter; Applied Load 60 kg)

HRAA High Rate Acquisition Assembly

HRAEM High-Resolution Analytical Electron Microscopy

HRAES High-Resolution Auger Electron Spectroscopy

HRAI Heating, Refrigerating and Air Conditioning Institute

HRB Highway Research Board [US]

HRB Rockwell 'B' Hardness (1/16 in. Dia. Steel Sphere Indenter; Applied Load 100 kg)

HRB Humboldt River Basin [Nevada, US]

H-RBS Hydrogen Rutherford Backscattering (Spectroscopy) [or H RBS]

HRC Hardwood Research Council [US]

HRC Harmonically Related Carrier Frequency

HRC Hasselblad Reflex Camera

HRC Health Resource(s) Center

HRC Herpes Research Center [of American Social Health Association at Research Triangle Park, North Carolina, US]

HRC High Resolution Control

HRC High Rupturing Capacity

HRC Horizontal Redundancy Check

HRC Howmet Research Center [of Howmet Corporation at Whitehall, Michigan, US]

HRC Human Resources Committee [of Canadian Institute of Mining and Metallurgy]

HRC Hypothetical Reference Circuit

HRC Rockwell 'C' Hardness (Conical Diamond Indenter; Applied Load 150 kg)

HRCG Hex(agonal) Reducing Coupling

HRD Heard [Amateur Radio]

HRD High Rate Dosimeter

HRD High-Resolution Digital

HRD Horizontal Resin Duct (in Wood)

HRD Human Resources Department

HRD Human Resources Development

HRD Hurricane Research Division [of National Oceanic and Atmospheric Administration, US]

HRD Hydraulic Research Department

HRD Hydrometeorology Research Division [Canada]

HRD Rockwell 'D' Hardness (Conical Diamond Indenter; Applied Load 100 kg)

H Rd Half Round [also 1/2 Rd]

hrd hard

HRDA High Rate Data Assembly

HRDC Human Resources Development Canada

HRDI High Resolution Doppler Imager

HRDM High Rate Demultiplexer

HRDP Hypothetical Reference Digital Path

HR/DQ/T Hot Rolling/Direct Quenching/Tempering (Process) [Metallurgy]

HRDR High Rate Digital Recorder

HRDS High Rate Data Station

Hrdwd Hardwood [also hrdwd]

Hrdwr Hardware [also hrdwr]

HRE Homogeneous Reactor Experiment [Nuclear Engiuneering]

HRE Hyper Raman Effect [Physics]

HRE Hypersonic Research Engine

HRE Rockwell 'E' Hardness ($\frac{1}{8}$ in. Dia. Steel Sphere Indenter; Applied Load 100 kg)

HREBL High-Resolution Electron-Beam Lithography

HRED High-Resolution Electron Diffraction

HREELS High-Resolution Electron Energy Loss Spectrometer; High-Resolution Electron Energy Loss Spectrometer

HREELS High-Resolution Electron Energy Loss Spectroscopy

HREELS High-Resolution Electron Energy Loss Spectrum

HRELS High-Resolution Energy-Loss Spectrometry

HREF Horticulture Research Experimental Farm [Canada]

HRELS High-Resolution Electron Loss Spectroscopy

HREM High-Resolution Electron Microscopy

HREM/EDS High-Resolution Electron Microscopy/Energy-Dispersive (X Ray) Spectroscopy

HREOC Human Rights and Equal Opportunity Commission

HRES High-Resolution Electron Spectroscopy

HRF Herb Research Foundation [US]

HRF Rockwell 'F' Hardness (1/16 in. Dia. Steel Sphere Indenter; Applied Load 60 kg)

HRFAX High-Resolution Facsimile

HR-FESEM High-Resolution Field Emission Scanning Electron Microscopy

HRFIMS High-Resolution Field-Ionization Mass Spectrometry

HRG Heregulin-α (Epidermal Growth Factor Domain) [Immunology]

HRG High-Resolution Graphics

HRG Horizontal Ribbon Growth [Metallurgy]

HRG Rockwell 'G' Hardness (1/16 in. Dia. Steel Sphere Indenter; Applied Load 150 kg)

HRGC High-Resolution Gas Chromatography

HRH His/Her Royal Highness

HRH Rockwell 'H' Hardness (1/8 in. Dia. Steel Sphere Indenter; Applied Load 60 kg)

HRHA Hydronic Radiant Heating Association [US]

HRI Height-Range Indicator

HRI Hierarchical (Biodiversity) Richness Index

HRI Horticultural Research Institute [US]

H-3 RIA Hydrogen-3 (or Tritium) Radioimmunoassay [also ^{3}H RIA]

HRIO Height-Range Indicator Operator

HRIO Horticultural Research Institute of Ontario [Canada]

HRIR High-Resolution Infrared

HRIR High-Resolution Infrared Radiometer; High-Resolution Infrared Radiometry

HRIS High Resolution Imaging Spectrometer

HRIS Highway Research Information Service [of Highway Research Board, US]

HRIS Human Resource Information System

HRIS Abstr Highway Research Information Service Abstracts [of Transportation Research Board, US]

HRK Hochschulrektorenkonferenz [Conference of University Presidents, Germany]

HRK Rockwell 'K' Hardness (1/8 in. Dia. Steel Sphere Indenter; Applied Load 150 kg)

HRL Hitachi Research Laboratory [of Hitachi Limited, Japan]

HRL Horizontal Reference Line

HRL Hughes Research Laboratories [Malibu, California, US]

HRL Human Resources Laboratory [US Air Force]

HRL Rockwell 'L' Hardness (1/4 in. Dia. Steel Sphere Indenter; Applied Load 60 kg)

HRLC High-Resolution Liquid Chromatography

HRLC Human Resources and Labour Canada

HR-LED High-Radiance Light-Emitting Diode [also HRLED]

HRLEED High-Resolution Low-Energy Electron Diffraction

HRLEELS High-Resolution Low-Energy Electron Loss Spectroscopy

HRLEL High-Radiation Level Examination Laboratory [of Oak Ridge National Laboratory, Tennessee, US]

HRM Hard-Rock Mine(r); Hard-Rock Mining

HRM Hardware Read-In Mode

HRM High Rate Multiplexer

HRM High (or Heavy) Reduction Machine [Metallurgy]

HRM Rockwell 'M' Hardness (1/4 in. Dia. Steel Sphere Indenter; Applied Load 100 kg)

HRMA High-Resolution Microanalysis

HRMA High Resolution Mirror Assembly [of NASA]

HRMAS High-Resolution Magic-Angle Spinning [also HR-MAS]

HRMAS NMR High-Resolution Magic-Angle Spinning Nuclear Magnetic Resonance [also HR MAS NMR]

HRMS High-Resolution Mass Spectrometer; High-Resolution Mass Spectrometry

HRMS High-Resolution Microwave Survey [of NASA]

HRMS Hot Roll Mill Steel [Metallurgy]

HRMS Human Resource Management System

HRMSI High-Resolution Multispectral Stereo Imager

HRN Hex(agonal) Reducing Nipple

HRNA Heterogeneous Ribonucleic Acid [Biochemistry]

HRNES Host Remote Node Entry System

HROI High-Resolution Optical Instrument

HROS Human Resources and Organizational Services [of US Environmental Protection Agency]

HRP Hand Retractable Plunger

HRP Heat-Resistant Phenolic

HRP Horseradish Peroxidase [Biochemistry]

HRP Hypergroup Reference Pilot

HRP Rockwell 'P' Hardness (¼ in. Dia. Steel Sphere Indenter; Applied Load 150 kg)

H&RP Holding and Reconsignment Point

HRPM Human Resources Program Manager

HRPO Horseradish Peroxidase [Biochemistry]

HRPS Hazard Reduction Precedence Sequence

HRPT High-Resolution Picture Transmission

HRR High-Range Resolution (Techniques)

HRR Hutchinson, Rice and Rosengren (Model) [Fracture Mechanics]

HRR Hutchinson-Riedel-Rice (Field) [Creep Theory]

HRR Rockwell 'R' Hardness (½ in. Steel Sphere Indenter; Applied Load 60 kg)

HRRC Human Resources Research Center

HRRL Human Resources Research Laboratory

HR/RQ/T Hot Rolling/Reheat Quenching/Tempering (Process) [Metallurgy]

HRRS High-Rate Reactive Sputtering

HRRTS High-Resolution Remote Tracking Sonar

HRS Hair Restoration Surgery

HRS Hazard Ranking System

HRS Heading Reference System

HRS High-Resolution Sensing

HRS High-Resolution Spectrography

HRS High-Resolution Spectrometer; High-Resolution Spectrometry

HRS Host Resident Software

HRS Hot-Rolled Steel

HRS Hovering Rocket System

HRS Hydraulics Research Station [UK]

HRS Hyper-Rayleigh Scattering

HRS Rockwell 'S' Hardness (½ in. Steel Sphere Indenter; Applied Load 100 kg)

hrs hours

HRSA Health Resources and Services Administration [of US Department of Health and Human Services]

hrs/°C hours per degree Celsius (or Centigrade) [Unit]

HRSEM High-Resolution Scanning Electron Microscopy [also HR-SEM]

hrs/°F hours per degree Fahrenheit [Unit]

HRSGC High-Resolution Subtraction Gas Chromatography

HRSI High-Temperature Reusable Surface Insulation [of NASA Space Shuttle]

HRSI High-Temperature Reusable Surface Insulator

HRSIM High-Resolution Selected Ion Monitoring

HRSM High-Resolution Scanning Magnetometer; High-Resolution Scanning Magnetometry

HRSP (Association of) Human Resource Systems Professionals [US]

HRSR High Resolution Scanning Radiometer [of National Oceanic and Atmospheric Administration, US]

HRSS Host Resident Software System

HRST High-Resolution Sensing Technology

HR Steel High-Resistance Steel [also HR steel]

HRSTEM High-Resolution Scanning Transmission Electron Microscopy

HRSV Human Respiratory Syncytial Virus

hrs/yr hours per year [Unit]

HRT Hard Return [also Hrt]

HRT Heat-Resistant Temperature

HRT High Rate Telemetry

HRT High-Resolution Tracker

HRT Homogeneous Reactor Test [Nuclear Engineering]

Hrt Heart [also hrt]

HRTEM High-Resolution Transmission Electron Microscope; High-Resolution Transmission Electron Microscopy

HRTF Head Related Transfer Function

Hrtwd Heartwood [also hrtwd]

HRU Heading Reference Unit

HRU Hydrological Research Unit

HRV Heat Recovery Ventilator

HRV High Resolution Video

HRV Hyperbaric Rescue Vehicle

HRV Hypersonic Research Vehicle

HRV Rockwell 'V' Hardness (½ in. Steel Sphere Indenter; Applied Load 150 kg)

HRV Human Rhinovirus

HRV-1A Human Rhinovirus, Type 1A

HRWR High-Range Water Reducing (Admixture) [also H-RWR] [Concrete]

HRX Hypothetical Reference Connection

HRXPS High-Resolution X-Ray Photoelectron Spectroscopy [also HR-XPS]

HRXRD High-Resolution X-Ray Diffraction; HighResolution X-Ray Diffractometer; High-Resolution X-Ray Diffractometry [also HR-XRD]

hr/yr hour(s) per year [Unit]

Hrzn Horizon

HS (Conference on) Hadron Structure [Particle Physics]

HS Halfshade

HS Half-Subtract(er)

HS Handle Screw

HS Handset

HS Hanke-Sham (Theory) [Physics]

HS Hard Sector(ed Diskette)

HS Hard Solder

HS Hard Sphere (Model)

HS Hashin and Shtrikman (Relation for Two-Phase Materials) [also H-S] [Materials Science]

HS Haworth Synthesis [Biochemistry]

HS Hazardous Substance

HS Headspace Sampler
HS Health Service
HS Health Surveillance
HS Heating Surface
HS Heat Seal(ing)
HS Heat-Sensitive; Heat Sensitivity
HS Heat Shield [also H/S]
HS Heat Shock
HS Heat Sink
HS Heat Source
HS Heat Stabilizer
HS Heat Storage [Oceanography]
HS Heat Stress
HS Heat Switch
HS Heavy Section
HS Heidenreich-Shockley (Partial Dislocation) [Materials Science]
HS Helicopter, Antisubmarine [US Navy Symbol]
HS Helium Spectrometer
HS Hemisphere; Hemispherical
HS Hemisuccinyl
HS Hermetic Seal(ing)
HS Herpes Simplex [Medicine]
HS Heterostructure [Solid-State Physics]
HS Heuristic Search
HS Hierarchically Structured
HS High School
HS High Sensitivity; Highly Sensitive
HS High Shear
HS High Silicon (Iron Alloy)
HS High Speed
HS High Spot
HS High Stability; Highly Stable
HS High-Stimulation
HS High Strain
HS High Strength
HS High-Strength Steel
HS High Structure
HS Hilbert Space [Mathematics]
HS Histochemical Society [US]
HS Hollow Sphere
HS Holographic Stereogram
HS Homologous Series [Chemistry]
HS Hoop Strain [Mechanics]
HS Hoop Stress [Mechanics]
HS Horizon Scanner
HS Horizon Sensor
HS Horizontal Spectrometer
HS Horizontal Stabilization; Horizontal Stabilizer
HS Horizontal Work-Roll Stabilization [Metallurgy]
HS Host to Satellite
HS Hot Spot
HS Hot Spring
HS Hot Stage

HS House Surgeon
HS Howe and Smith (Analysis) [Crystallography]
HS Huang Scattering [Physics]
HS Hubbard-Stratonowich (Decoupling)
HS Hughes Satellite [of Hughes Aircraft Company, US]
HS Human Serum (Protein) [Biochemistry]
HS Hydraulic System
HS Hydride Shift
HS Hydrofoil Ship
HS Hydrogen Spectrum
HS Hydrogen Storage
HS Hydrographic Society [UK]
HS Hydroseparation; Hydroseparator
HS Hydrostatic(s)
HS Hydroxylamine Sulfate
HS Hypersensitization; Hypersensitized
HS Hypersonic(s)
HS$^-$ Hydrosulfide (Ion) [Symbol]
HS- Thailand [Civil Aircraft Marking]
H{S} Ginzburg-Landau Free Energy Functional (in Solid-State Physics) [Symbol]
H&S Health and Society (Database)
H/S Hard/Soft (Ratio)
H/S Hardware/Software
H/S Hardwood/Softwood (Pulp Blend)
H/S Health and Safety
H/S Heat Shield [also HS]
H-S Hashin and Shtrikman (Relation for Two-Phase Materials) [also HS] [Materials Science]
H7/s6 Medium Drive Fit; Basic Hole System [ISO Symbol]
H(s) Feedback Elements [Symbol]
hs *(hora somnus)* – at bedtime [Medical Prescriptions]
HSA Hanford Strategic Analysis [of Hanford Site, Washington State, US]
HSA Hard-Sphere Approximation [Physical Chemistry]
HSA Hawker Siddeley Aviation [UK]
HSA Hazardous Substances Act [US]
HSA (Read/Write) Head Stack Assembly [Computers]
HSA Health Services Administration [Now Part of Health Resources and Services Administration, US]
HSA Hemispherical Analyzer
HSA Herb Society of America [US]
HSA Hidden Surface Algorithm
HSA Holly Society of America [US]
HSA Hot-Strip Annealed; Hot Strip Annealing [Metallurgy]
HSA Human Serum Albumin [Biochemistry]
HSA Hydroponic Society of America [US]
HSA Hydroxystearic Acid
12HSA 12-Hydroxystearic Acid [or 12-HSA]
HSAB 4-Azidobenzoic Acid N-Hydroxysuccinimide Ester
HSAB Hard-Soft Acid-Base (Theory) [Chemistry]
HSAC Health Safety and Analysis Center
HSAC Helicopter Safety Advisory Conference [US]
HSAC High-Speed Analog Computer

H3(sal)3tach 1,3,5-Tris(salicyclaldimino)cyclohexane

HSAM Hierarchical Sequential Access Method

HSAS Houldsworth School of Applied Science [of University of Leeds, UK]

HSB High-Speed Buffer

HSB Hue, Saturation, Brightness

Hsb Hefner-stilb [Unit]

HSBA High-Speed Bus Adapter

H-SB-BL Hydrogenated Styrene Butadiene Block Copolymer

HSBIP Hsinchu Science Based Industrial Park [Taiwan]

HSBR High-Speed Bombing Radar

HSBRAM Hanford Site Baseline Risk Assessment [of US Department of Energy]

HSC Hard Superconductor

HSC Hardware/Software Coordination

HSC Health and Safety Commission [UK]

HSC Health Service Commissioner [of National Health Service, UK]

HSC Health Sciences Center

HSC Hierarchical Storage Controller

HSC High-Conversion Soaker Cracking (Process) [Chemical Engineering]

HSC High Speed Channel

HSC High-Speed Collision

HSC High-Speed Concentrator

HSC High-Speed Cutter; High-Speed Cutting

HSC High-Spot Counting

HSC High-Temperature Salt Corrosion

HSC HIPed (Hot Isostatically Pressed) Silicon Carbide [also HIPSC, or HIPSiC]

HSC Hospital for Sick Children [Toronto, Canada]

HSC Hotter than Solventless Carbamite [British Propellant]

HSC Houston Science Center [of University of Houston, Texas, US]

HSC Hoyland Steel Company [US]

HSC Humboldt State College [Arcata, California, US]

HSC Hydrocarbon Subcommittee

HSC Hydrogen Stress Cracking

HS-C Hamilton Standard CO (Carbon Monoxide) Absorbent Material

HSc Scleroscope Hardness Number, Model c

HSCA Harvard-Smithsonian Center for Astrophysics [of Harvard University, Cambridge, Massachusetts, US]

HSCC Heavy Specialized Carrier Conference [US]

HSCC Hydrogen-Assisted Stress Corrosion Cracking [also HASCC]

HSCC Hydrogen-Induced Stress Corrosion Cracking [also HISCC]

HSCP High-Speed Card Punch

HSCT High-Speed Civil Transport (Program) [US]

HSCT High-Speed Commercial Transport

HSCT High-Speed Compound Terminal

HSCU Hydraulic Supply and Checkout Unit

HSD Doctor of Health and Safety

HSD Hamilton Standard Division [US]

HSD High-Speed Data

HSD High-Speed Displacement

HSD Hillock Space Distribution

HSD Horizontal Situation Display

HSD Hot Shutdown [Nuclear Reactors]

HSD Hydroxysteroid Dehydrogenase [Biochemistry]

HSd Scleroscope Hardness Number, Model d

3a-HSD 3α-Hydroxysteroid Dehydrogenase [Biochemistry]

HSDA High-Speed Data Acquisition [also HS-DA]

HSDB Hazardous Substances Data Bank

HSDB High-Speed Data Buffer

HSDC Hybrid Synchro-to-Digital Converter

HSDir Director of Health and Safety

HSDL High-Speed Data Line

HSDL High-Speed Data Link

HSDP Hawaii Scientific Drilling Project [US]

HSDS Hot Spot Detection System

HSDW Helical Spin Density Wave [also H-SDW, or H SDW] [Solid-State Physics]

HSE Health and Safety Executive [UK]

HSE Herpes Simplex Encephalitis [Medicine]

HSE High Sensitivity (Probe)

HSE Hydrostatic Equilibrium

HSE Hydrostatic Extrusion [Metallurgy]

Hse Homoserine; Homoserinyl [Biochemistry]

Hse House [also hse]

HSECT Hard Disk Sector (Command) [also hsect]

HSEL High-Speed Selector Channel

HSER Health, Safety and Environmental Research (Program) [of Canadian Fusion Fuels Technology Program]

HSES Hanford Scientific and Engineering System [of US Department of Energy, Hanford Site, Washington State, US]

HSES Health and Safety Executive Sheffield [of University of Sheffield, UK]

HSF Heart and Stroke Foundation

HSF Hepatocyte Stimulating Factor [Biochemistry]

HSF High-Speed Flow

HSF High-Strength Fiber

HSF Hot-Spot Factor

HSF Hydrostatic Forging

HSF Hypergol(ic) Servicing Facility [Aerospace]

HSF Hypersonic Flight

HSF Hypersonic Flow

HSFP Hanford Surplus Facilities Program [of US Department of Energy, Hanford Site, Washington State, US]

HSG Heat-Strengthened Glass

HSG High-Silica Glass

HSG High Sustained G's Acceleration [Aerospace]

HSG Horizontal Sweep Generator

Hsg Housing [also hsg]

HSGC Head-Space Gas Chromatography [also HS/GC]

HS/GC/MS Head-Space Gas Chromatography/Mass Spectrometry [also HSGS/MS]

HSGPC High-Speed Gel Permeation Chromatography

HSGT High-Speed Ground Transport(ation)

HSGTC High-Speed Ground Test Center

H₂S/H₂ Hydrogen Sulfide/Hydrogen (Mixture)

HS/HM High-Strength/High-Modulus

HSI Habitat Suitability Index (Modeling Program) [US]

HSI Hang Seng (Stock) Index [Hong Kong]

HSI Hardware-Software Integration (Test)

HSI Heat Stress Index

HSI Horizontal Situation Indicator

HSI Hue, Saturation, Intensity

HSI Hyperthermal Surface Ionization

HSIA Halogenated Solvents Industry Alliance [US]

HSIDI Hyperthermal Surface Induced Dissociative Ionization

HSi(OEt)₃ Triethoxysilane

HSIP Hsinchu Science-Based Industrial Park [Taiwan]

HS-IR-RAS High-Sensitivity Infrared Reflection Absorption Spectroscopy

HSIS Health Sciences Information Service [of University of California at Berkeley, US]

HSIS Human Settlements Information System

H/S IR Hardware/Software Integration Review

HSiW 12-Tungstosilicic Acid

HSK Handshake [Data Communications]

HSK Horizontal Sling Kit

HSK Housekeeping [also Hsk, or hsk]

HSKi Handshake Input [Data Communications]

HSKo Handshake Output [Data Communications]

HSL Hardware Simulation Laboratory

HSL Hazardous Substances List

HSL High-Speed Launch

HSL Highway Safety Literature [of National Highway Traffic Safety Administration, US]

HSL Hue-Saturation-Luminance (Monitor)

HSLA High-Speed Line Adapter

HSLA High-Strength Low-Alloy (Steel)

HSLA-i Isotropic High-Strength Low-Alloy (Steel)

HSLA Steel High-Strength Low-Alloy (Steel) [also HSLA steel]

HSLA Steels International Conference on High-Strength Low-Alloy Steels

HSLC Hierarchical Storage Manager

HSLC High-Speed Liquid Chromatography

HSLG Hard Sphere Lattice Gas

HSM Hard-Sphere Model

HSM Hard-Sphere Molecule

HSM Hierarchical Storage Management

HSM High-Speed Machining

HSM High-Speed Mechanics

HSM High-Speed Memory

HSM Hot-Stage Microscope; Hot-Stage Microscopy

HSM Hot Strip (Rolling) Mill [Metallurgy]

HSMB Hyperthermal Supersonic Molecular Beam

HSMHA Health Services and Mental Health Administration [US]

HSMRC Hazardous Substance Management Research Center [US]

HSMS High-Speed Mine Sweeper

HSN Highly Saturated Nitrile Rubber

HSN HIPed (Hot Isostatically Pressed) Silicon Nitride [also HIPSN]

HSN Hospital Satellite Network

Hsng Housing [also hsng]

HSNY Horticultural Society of New York [US]

HSO Hydroxylamine Sulfate Oxime (Process)

HSO Hydrosulfite (Ion) [Symbol]

HSO Hydrosulfate (Ion) [Symbol]

HSP Hard-Sphere Potential [Physical Chemistry]

HSP Health Stabilization Program

HSP Heat Shock Protein [Biochemistry]

HSP High-Speed Photography

HSP High-Speed Photometer

HSP High-Speed Plating

HSP High-Speed Printer

HSP High Speed Processor

HSP High-Strength Porcelain

HSP Hydrostatic Press(ing) [Powder Metallurgy]

HSP Hydrostatic Pressure

HSPA Hawaiian Sugar Planters Association [US]

HSPE High-Strength Polyethylene

HSPF Hydrologic Simulation Program in Forestry [British Columbia, Canada]

HSPG Heparan Sulfate Proteoglycan [Biochemistry]

HSPH Harvard School of Public Health [Boston, Massachusetts, US]

HSPN High-Speed Packet Network

HSPTR High-Speed Paper Tape Reader

HSQ Historical Society of Queensland [Australia]

HSQ Hydraulically Separated Quartz

HSQ Hydrogen Silsesquioxane

HSR Hardware Status Register

HSR Health Sciences Research [of Oak Ridge National Laboratory, Tennessee, US]

HSR High-Speed Radiography

HSR High-Speed Rail

HSR High-Speed Reader

HSR High-Strain Rate

HSR High-Strength (Synthetic) Resin

HSR Högskolan Senter Rogaland [Rogaland University Center, Norway]

HSR Hot Stretch Reduction (Process) [Metallurgy]

HSRA High Speed Rail Association [US]

HSRAM Hanford Site Risk Assessment Methodology [of US Department of Energy]

HSRC Health Sciences Resource Center [of Canada Institute for Scientific and Technical Information]

HSRC Highway Safety Research Center [US]

HSRCM Hanford Site Radiological Control Manual [of US Department of Energy]

HSRD Health Sciences Research Division [of Oak Ridge National Labotary, Tennessee, US]

HSRI High-Sensitivity Refractivity Index

HSRI Highway Safety Research Institute [US]

HSRIOP High-Speed RAD Input/Output Processor

HSRL High Spectral Resolution Lidar

HSRO High-Speed Repetitive Operation

HSRS High Strain Rate Superplastic(ity) [Metallurgy]

HSRTM High-Speed Resin-Transfer Molding

HSRTM High-Strength-Resin Transfer Molding

HSS Heavy-Section Steel

HSS Hierarchy Service System

HSS High-Speed Steel

HSS High-Speed Storage

HSS High Stress Strain

HSS History of Science Society [US]

HSS Hollow Structural Section

HSS Homogeneity Spoil Spectroscopy

HSS Hydraulic Subsystem Simulator

HSS Hypersonic Speed

hss *(hora somnus sumendus)* – to be taken at bedtime [Medical Prescriptions]

HSSA High-Speed Steel Association [UK]

HSSA Hydrographic Society of South Africa

HSSCC Hot-Salt-Induced Stress-Corrosion Cracking

HSSDS High-Speed Switched Digital Service

HSSI Heavy-Section Steel Irradiation

HSSI High-Speed Serial Interface

HSSR Hydrogeochemical and Stream Sediment Reconnaissance (Program) [US]

HSSSTP Hanford Site-Specific Science and Technology Plan [of US Department of Energy]

HSST Heavy Section Steel Technology (Program) [of National Research Council, US]

HST Handelskammer Schweiz-Tschechoslovakei [Chamber of Commerce Switzerland-Czechoslovakia]

HST Hawaiian Standard Time [Greenwich Mean Time +10:00]

HST Hawaiian Sugar Technologists [US]

HST Heat-Shrinkable Tubing

HST High-Speed Taxi-Way Turn-Off [Aeronautics]

HST High Speed Technology [of US Robotics]

HST High Stability (Probe)

HST History of Science and Technology (Database)

HST Horizontal Seismic Trigger

HST Hubble Space Technology

HST Hubble Space Telescope

HST Hydrostatic Test

HST Hypersonic Transport

HST Hypervelocity Shock Tunnel

.HST History [File Name Extension] [also .hst]

.HST Host [File Name Extension] [also .hst]

HSTA Hawaii State Teachers Association [US]

HSTAT Health Services and Technology Assessment Text

HSTCO High-Stability Temperature-Compensated Crystal Oscillator

HSTP Hard Stop [Computers]

HST-STIS Hubble Space Telescope–Space Telescope Imaging Spectrograph [of Space Telescope–European Coordinating Facility] [also HST STIS]

HSTU Hunan Science and Technology University [Beijing, PR China]

HSU Haile Selassie I University [Addis Ababa, Ethiopia]

HSU Helium Service Unit

HSU Humboldt State University [California, US]

HSV Herpes Simplex Virus

HSV Hue Saturation Value

HSV Hybrid Spin Valve [Physics]

HSV-1 Herpes Simplex Virus, Type 1

HSV-2 Herpes Simplex Virus, Type 2

HSVB High-Speed Video Bus

HSVD Hankel Singular Value Decomposition

HSVP High-Speed Vector Processor [Computers]

HSVPS High-Speed Vector Processing System

HSVPS High-Speed Vector Processor System

HSV-TK Herpes Simplex Virus Thymidine Kinase [Biochemistry]

HSW Hamburger Stahlwerke GmbH [German Steel Manufacturer locasted at Hamburg, Germany]

HSW Heat-Sensitive Wire

HSWA Hazardous and Solid Waste Amendments [to US Resource Conservation and Recovery Act]

HSWA Health and Safety at Work Act

HSY Hard-Sphere-Yukawa (Model) [Physics]

HSYNC Horizontal Synchronization

HSZD Hermetically-Sealed Zener Diode

HT Haiti [ISO Code]

HT Half Thickness [Physics]

HT Half Time

HT Half Tone

HT Half-Track(ed)

HT Handy Talkies [Small Portable Low-Range Transceivers]

HT Hardgrove Test [Mineral Technology]

HT Hardness Test(ing)

HT Head per Track [also HPT, or HP/T]

HT Hard Temper [Metallurgy]

HT Heat Transfer

HT Heat Transmission

HT Heat Treat(ing); Heat Treatment; Heat Treatability; Heat Treatable [Metallurgy]

HT Heavy Truck

HT Height of Target

HT Heliotropin [Organic Chemistry]

HT Hematoxylin [Organic Chemistry]

HT Hemotoxin [Biochemistry]

HT High Technology

HT High Temperature

HT High Tensile

HT High Tension

HT High Tide

HT Hinsberg Test [Analytical Chemistry]

HT Holding Time

HT (Electron) Hole Theory [Solid-State Physics]

HT Hold Time

HT Homing Transponder

HT Horizontal Tab(ulation Character) [Data Communications]

HT Hot Trap

HT Hubble Telescope

HT Hueckel Theory [Physical Chemistry]

HT Hybrid Transformer

HT Hydraulic Turbine

HT Hydrothermal (Process)

HT Hydro-Treating [Chemical Engineering]

HT Hydroxytryptamine [Biochemistry]

HT Hydroxytryptophan [Biochemistry]

HT Hypoxanthine/Thymidine Medium [Biochemistry]

HT Hysteresis Tester

HT Tritiated Hydrogen

5-HT 5-Hydroxy-L-Tryptamine [Biochemistry]

5-HT 5-Hydroxy-L-Tryptophan [Biochemistry]

HT + High Tension + (Line)

H/T Heat Treat

H/T High Temperature (Storage Test)

H(T) Heat Capacity [Symbol]

H(T*) Irreversibility Line (of Superconductors) [Symbol]

H(T) Temperature-Dependent (Magnetic) Field [Symbol]

$H_c(T)$ Temperature-Dependent Critical (Magnetic) Field [Symbol]

$H_{c1}(T)$ Temperature-Dependent Lower Critical (Magnetic) Field [Symbol]

$H_{c2}(T)$ Temperature-Dependent Upper Critical (Magnetic) Field [Symbol]

$H_{IRR}(T)$ Irreversibility Line (of Superconductors) [Symbol]

H-T Enthalpy-Temperature (Diagram)

H-T High-Temperature (Phase) [Materials Science]

H-T Magnetic Field–Temperature (Plane)

Ht Height [also ht]

Ht Heat [also ht]

h(t) height at time t [Symbol]

HTA *(Haupttelegraphenamt)* – German for "Central Telegraph Office"

HTA Health Technology Assessment

HTA Health Technology Assessment [Journal of the National Coordinating Center for Health Technology Assessment, UK]

HTA Heat-Transfer Agent

HTA Heavier-than-Air (Craft)

HTA High-Temperature Alloy(s)

HTA High-Temperature Amorphous (Resin)

HTA High-Temperature Anneal(ed); High-Temperature Annealing [Metallurgy]

HTA High-Temperature Approximation

HTA Horticultural Trades Association [UK]

HTA Hypophysiotropic Area

HTB Hexadecimal-to-Binary (Conversion)

HTB Hexagonal Tungsten Bronze

HTB Highway Tariff Bureau [now Associated Motor Carriers Tariff Bureau, US]

HTBM High-Temperature Birch-Murnaghan (Model) [Materials Science]

HTC Heat-Transfer Coefficient

HTC Heat-Transfer Control(ler)

HTC High T_c (Superconductor) [*Note:* T_c is the Critical Transition Temperature]

HTC High-Tech Ceramics, Inc. [Alfred, New York, US],

HTC High-Tech(nology) Ceramics

HTC High-Temperature Ceramics

HTC High-Temperature Chemistry

HTC High-Temperature Chlorination (Process)

HTC High-Temperature Combustion

HTC High-Temperature Corrosion

HTC High-Temperature Creep [Mechanics]

HTC High-Temperature Crystalline; High-Temperature Crystallizable; High-Temperature Crystallization

HTC Horizontal Toggle Clamp

HTC Houston-Tillotson College [Austin, Texas, US]

HTC Hybrid Technology Computer

HTC Hydrotalcite [Mineral]

HTC Hydro-Treating Catalyst [Chemical Engineering]

HTCE Heat Treating Conference and Exposition [of ASM International, US]

HTCE Historical Tank Content Estimate

HTCI High-Tensile Cast Iron

HTCL High-Temperature Corrosion Laboratory [of University of Michigan, US]

HT-CMC High-Temperature Ceramic Matrix Composite

HT-CMC International Conference on High-Temperature Ceramic Matrix Composites

HT-CVD High Temperature Chemical Vapor Deposition

HTD Hand Target Designator

HTD Helium Thermal Desorption (Technique)

HTD Heterojunction Tunneling Diode

HTD High-Temperature Defect

HTD High-Temperature Deformation

HTD High-Temperature Deposition

HTD High Torque Drive

HTD Hydrogen Thermal Desorption

HTDA High-Temperature Dilute Acid

HTDC High Tension, Direct Current [also HT-DC, or HT DC]

HTDO Hydrous Titanium Dioxide

HTDS Hanford Thyroid Disease Study [of Hanford Site, Washington State, US]

HTDS High-Temperature Drawing Salt

HTDTA 1,3-Diamino-2-Propanol-N,N,N',N'-Tetraacetic Acid

H3DTAP Three-Dimensional Finite Element Program for Thermal and Atomic Diffusion of Hydrogen

HTE Heat Transfer Efficiency

HTE Heat Treating Exposition

HTE High-Temperature Equilibrium

HTE Hypergroup Translating Equipment

H₂Te Hydrogen Telluride

H3TEA Triethanolamine

Ht Exch Heat Exchanger

HTF Heat Transfer Fluid

HTF High-Temperature Fatigue [Mechanics]

HTF High-Tenacity Fiber

HTF Horizontal Tube Feeder

HTFAC Trifluoroacetylacetone [also Htfac, or htfac]

HTFC High-Temperature Fuel Cell

HTFFR High-Temperature Fast-Flow Reactor

HTFMI Heat Transfer and Fluid Mechanics Institute

HTFMS Trifluoromethanesulfonic Acid

HTFS Heat Transfer and Fluid Flow Service [of United Kingdom Atomic Energy Agency]

Htftbd 4,4,4-Trifluoro-1-(2-Thienyl)butane-1,3-Dione

HTG Haitian Gourde [Currency]

HTG Hydrostatic Tank Gauging

Htg Heating [also htg]

HTGCR High-Temperature Gas-Cooled Reactor [Australia]

HTGE High-Temperature Gas Extrusion

HTGL High-Temperature Gas-Dynamics Laboratory [of Stanford University, California, US]

HTGPF High-Temperature General-Purpose Furnace

HTGR High-Temperature Gas(-Cooled) Reactor

HTGR-CX High-Temperature Gas Reactor Critical Experiment

HTGR SC/C HTGR Steam Cycle Cogeneration

HTH High-Temperature Hydrogenation

HTH High-Test Hypochlorite (Bleach)

HTH Hope This Helps [Internet Jargon]

100th (one) hundredth

H-THD Tetramethylheptadione [also H-thd]

HTHP High-Temperature/High-Pressure

HT/HTO Tritiated Hydrogen/Tritiated Hydrogen Oxide

HTI Hand Tools Institute [US]

HTI Headway Technologies Inc. [Milpitas, California, US]

HTI Height-Time

HTI Higher Technical Institute

HTI High-Technology Intensive

HTI High-Temperature Impact

HTI High-Temperature Insulation

HTIP Hanford Technology Integrated Program [of Pacific Northwest National Laboratory, Richland, Washington State, US]

HTIP Housing Technology Incentives Program [of Canada Mortgage and Housing Corporation, Canada]

HTIR Hadamard Transform Infrared Spectroscopy [also HT-IR]

HTIS Heat Transfer Instrument System

HTL Hanford Technical Library [Washington State, US]

HTL Heat Transfer Loop [Nuclear Reactors]

HTL Henderson Technical Laboratory [of Titanium Metals Corporation, Denver, Colorado, US]

HTL High-Temperature Limit

HTL High-Threshold Logic

HTL (Höhere Technische Lehranstalt) – German for "Institute of Technology" [Austria and Switzerland]

HTL Hydroxyl Terminated Liquid

HTLA Heat-Treatable Low-Alloy (Steel)

HTLA High-Temperature Low-Activity

HTLDC Hsinchu Tidal Land Development Planning Commission [Taiwan]

HTL-Dipl Ing (Diplomingenieur der Höheren Technischen Lehranstalt) – Austrian and Swiss Academic Degree from an Institute of Technology Equivalent to Graduate Engineer

HTLL High-Test-Level Language

HTLTR High-Temperature Lattice Test Reactor [US]

HTLV Human T-Cell Lymphotropic Virus; Human T-Cell Leukemia Virus

HTLV-1 Human T-Cell Lymphotropic (or Leukemia) Virus– Type 1 [also HTLV-I]

HTLV-2 Human T-Cell Lymphotropic (or Leukemia) Virus– Type 2 [also HTLV-II]

HTLV-3 Human T-Cell Lymphotropic (or Leukemia) Virus– Type 3 [also HTLV-III]

HTM Härterei-Technische Mitteilungen [Heat Treating Shop–Technical Communications; German Journal for Heat Treating and Materials Technology]

HTM Heat-Transfer Material; Heat-Transfer Medium

HTM High Tech(nology) Material(s)

HTM High-Temperature Material

HTM High-Temperature Melter

HTM High-Temperature Microscopy

HTM High-Temperature Molding

HTM Hybrid Transfer Mode

HTM Hypothesis Testing Model

.HTM HTML (Hypertext Markup Language) File [File Name Extension] [also .htm]

h/2m$_e$ quantum of circulation (3.6369481 × 10–4 Js/kg) [Symbol]

HTMA Hawaii Territorial Medical Association [US]

HTMA Hydraulic Tool Manufacturers Association [US]

HTML High-Temperature Materials Laboratory [of Oak Ridge National Laboratory, Tennessee, US]

HTML Hyper-Text Markup Language [also html]

HTML+ Hyper-Text Markup Language Plus

.HTML Hyper-Text Markup Language File [File Name Extension] [also .html]

HTMMC High-Temperature Metal-Matrix Composite

HTMOS High-Threshold Metal-Oxide Semiconductor [also HT-MOS]

HTMP Hydroxytetramethylpiperidine

H-TMS Honeywell Test Management System

HTN Heat Treatable Nodular [Metallurgy]

HTN (International) Heat Treating Network [Cleveland, Ohio, US]

HTO Heat Transfer Oil

HTO High-Temperature Oxidation

HTO Horizontal Takeoff

HTO Tritiated Hydrogen Oxide (or Tritiated Water)

HTO Hydrous Titanium Oxide

HTP High-Temperature Performance

HTP High-Temperature Photochemistry

HTP High-Temperature Physics

HTP High-Temperature Plastics

HTP High-Temperature Polymer

HTP High-Temperature Pretreatment

HTP High-Temperature Pyrolysis

HTP High Test Peroxide (Process)

HTP High Thermal Performance

HTP Hydrothermal Processing

HTP Hydroxytryptophane [Biochemistry]

5-HTP 5-Hydroxytryptophane [Biochemistry]

HTPB Hydroxy(l) Terminated Polybutadiene

H₂TPP 5,10,15,20-Tetra-Phenylporphyrin

HTPS High Tension Power Supply

H₂T(4-Py)P 5,10,15,20-Tetra(4-Pyridyl)porphyrin

HTR Hanford Test Reactor [of US Department of Energy in Washington State, US]

HTR Heat Treated; Heat Treatment

HTR High-Temperature Reaction

HTR High-Temperature (Nuclear) Reactor

HTR High-Temperature (Synthetic) Resin(s)

HTR Hitachi Training Reactor [Japan]

HTR Hydrothermal Reaction

Htr Heater [also htr]

HTRB High-Temperature Reverse Bias

HTRDA High-Temperature Reactor Development Associates [US]

HTRE Heat Transfer Reactor Experiment [US]

HTRE High-Temperature Reactor Experiment

HTRI Heat Transfer Research Institute [US]

HTRP Humane Trap Research Program [of Fur Institute of Canada]

HTS Hadamard Transfer Spectrometry

HTS Head Track and Selector

HTS Heat Transfer System

HTS Heat Transport Section

HTS Heat Treating Society [of ASM International, US]

HTS Heat-Treating System

HTS Height Telling Surveillance

HTS Higher Technical School

HTS High-Temperature Shift

HTS High-Temperature Silylation [Organic Chemistry]

HTS High-Temperature Sintering [Metallurgy]

HTS High-Temperature Stability

HTS High-Temperature Storage (Test)

HTS High-Temperature Superconducting; High-Temperature Superconductivity; High-Temperature Superconductor [also HTSC, or HTSc]

HTS High-Tensile Steel

HTS High Tensile Strength

HTS Host to Satellite

HTS Hubble Telescope Science

Hts Heights [Geography]

HTSC High-Temperature Superconductivity; High-Temperature Superconductor [also HTSc, or HTS]

HTSD High-Temperature Successive Deposition (Process)

HTSE High-Temperature Series Expansion

HTSEC High-Temperature Size-Exclusion Chromatography

HTSF High-Temperature Sodium Facility

HTSH Human Thyroid Stimulating Hormone [Biochemistry]

Ht Shld Heat Shield

HTSJ High-Temperature Society of Japan

HTSL Heat Transfer Simulation Loop [Nuclear Reactor]

HTS LSCO High-Temperature Superconducting Lanthanum Strontium Copper Oxide

HTSP High-Temperature Superplastic(ity)

HT/SPC Heat Treating/Statistical Process Control

HTS PSA High-Temperature Shift Pressure-Swing Adsorption [also HTS/PSA]

HTSS Hamilton Test Simulation System

HTSS Honeywell Time-Sharing System

HTSSE High-Temperature Superconducting Space Experiment

HTS SQUID High-Temperature Superconductor Superconducting Quantum Interference Device

HTST High-Temperature Short-Time (Method) [Pasteurization]

HTSUP Height Supervisor

HTS YBCO High-Temperature Superconducting Yttrium Barium Copper Oxide

HTT Heat Treatment Temperature [Metallurgy]

HTT High-Temperature Technology

HTT High-Temperature Test(ing)

HTT High-Temperature Tetragonal (Phase) [Materials Science]

HTT Hydrothermal Titration

HTTF High-Temperature Test Facility

HTTL High-Power Transistor-Transistor Logic [also HT²L]

HTTMT High-Temperature Thermomechanical Treatment

HTTP High-Temperature Thermoplastic(s)

HTTP HyperText Transfer Protocol [also http]

H-3 TTP Tritiated Thymidine 5'-Triphosphate [also ³H-TTP] [Biochemistry]

HTTP-NG HyperText Transfer Protocol–Next Generation

HTTP HyperText Transfer Protocol Secure

Ht Tr Heat Treat(ment)

HTTT High-Temperature Turbine Technology

HTTVMT High-Temperature Thermovibrational Mechanical Treatment

HTU Hamburger Transport Unternehmung GmbH [Hamburg Transport Company; of Hamburger Stahlwerke GmbH]

HTU Harbin Technical University [Harbin, PR China]

HTU Heat Transfer Unit

HTU Height of Transfer Unit [Gas Chromatography/Distillation]

HTV High Temperature and Velocity

HTV High-Temperature Vulcanizate; High-Temperature Vulcanization; High-Temperature Vulcanized

HTVB High-Temperature Vacuum Brazing

HTW High-Temperature Water

HTW High-Temperature Winkler (Process) [Chemistry]

HTX High-Temperature Crystalline; High-Temperature Crystallizable; High-Temperature Crystallization

HTXRD High-Temperature X-Ray Diffraction; High-Temperature X-Ray Diffractometry [also HT-XRD]

HU Hampton University [Virginia, US]

HU Hang-Up

HU Hanyang University [Korea]

HU Harvard University [Cambridge, Massachusetts, US]

HU Hebrew University [Jerusalem, Israel]

HU High Usage

HU Hiroshima University [Japan]

HU Hofstra University [Hempstead, New York, US]

HU Hokkaido University [Sapporo, Japan]

HU Hongshan University [PR China]

HU Hosei University [Tokyo, Japan]

HU Howard University [Washington, DC, US]

HU Humboldt-Universität [Humboldt University, Berlin, Germany]

HU Hungary [ISO Code]

HU Hydrostatic Unloading

Hu Human [Immunochemistry]

H7/u6 Force Fit; Basic Hole System [ISO Symbol]

.hu Hungary [Country Code/Domain Name]

HUB Humboldt-Universität zu Berlin [Humboldt University at Berlin, Germany]

HUC High Usage Circuit

HUD Head-Up Display

HUD (Department of) Housing and Urban Development [US]

HUDA Housing and Urban Development Act [US]

HUDAC Housing and Urban Development Association of Canada [now Canadian Home Builders Association]

HUDE Head Up Display Electronics

HUD-FDA (Department of) Housing and Urban Development–Federal Housing Administration [US]

HUDWAC Head-Up Display Weapons-Aiming Computer

HUF Hungarian Forint [Currency]

HUFF-DUFF High-Frequency Direction Finder

HUFSAM Highway Users Federation for Safety and Mobility [US]

HUG Hastech Users Group [US]

HUG Honeywell Users Group [US]

HUGHES-NEL Hughes Aircraft Company–Nuclear Electronics Laboratory [US]

HUGO Highly Unusual Geophysical Operation

HUGO (International) Human Genome Organization

HUH Hahnemann University Hospital [Philadelphia, Pennsylvania, US]

Hu IFN-α Human Interferon-α [Biochemistry]

Hu IFN-αLy Human Lymphoblastoid Interferon [Biochemistry]

Hu IFN-β Human Interferon-β [Biochemistry]

Hu IFN-γ Human Interferon-γ [Biochemistry]

HUL Hardware Utilization List

Hule Mex Plast Hule Mexicano y Plasticos [Mexican Publication on Cloth and Plastics]

HULTIS Hull Technical Interloan Scheme [UK]

HUM (Helicopter) Health and Usage Monitoring (Program)

Hum Human(ity) [also hum]

Hum Humidity [also hum]

Hum Ultra-Microhardness (Test) [also UMH]

Hum-Comput Interact Human-Computer Interactions [Published in the US]

Hum Devel Rep Human Development Report [of United Nations Development Program]

Hum Exp Toxicol Human and Experimental Toxicology (Journal)

Hum Factors Human Factors [Publication of the Human Factors Society, US]

Hum Geogr Human Geography

Hum Hered Human Heredity [International Journal]

Hum Mol Genet Human Molecular Genetics [Journal published by Oxford University Press, UK]

Hum Nutr Appl Nutr Human Nutrition: Applied Nutrition (Journal)

Hum Nutr Clin Nutr Human Nutrition: Clinical Nutrition (Journal)

Hum(P) Load-Dependent Ultra-Microhardness

Hum Reprod Human Reproduction [Journal published by Oxford University Press, UK]

HumRRO Human Resources Research Office [also HUMRRO]

Hum Syst Manage Human Systems Management [Published in the Netherlands]

Hung Hungarian; Hungary

Hung J Ind Chem Hungarian Journal of Industrial Chemistry

HUNO Harvard University News Office [Cambridge, Massachusetts, US]

Hunt Hunting

Hunts Huntingdonshire [UK]

HUP Harvard University Press [Cambridge, Massachusetts, US]

HUR Heat Up Rate

Hurcn Hurricane [also hurcn]

HURL Hawaii Undersea Research Center [at University of Hawaii-Manoa, Honolulu, US]

HUS Hemolytic Uremic Syndrome [Medicine]

HUS Hypergolic Umbilical System [Aerospace]

HUSAT Human Sciences and Advanced Technology Research Center [UK]

HUST Huazhong University of Science and Technology [Wuhan, PR China]

HUT Hard Upper Torso

HUT Helsinki University of Technology [Espoo, Finland]

HUT HEDL (Hanford Engineering Development Laboratory) Up Transient

HUT Hopkins Ultraviolet Telescope [of Johns Hopkins University, US]

Hutn Listy Hutnické Listy [Czech Metallurgical Journal]

HUVE Human Umbilical Vein Endothelial (Cells) [Medicine]

HUVEC Human Umbilical Vein Endothelial Cells [Medicine]

HV Half Value [Physics]

HV Hall Voltage [Solid-State Physics]

HV Hand-Operated Valve; Hand Valve

HV Hantaan Virus

HV Hardware Virtualizer

HV Have [Amateur Radio]

HV Health Visitor

HV Heating and Ventilation [also H&V]

HV Heat(ing) Value

HV Helvetica (Font)

HV Hepatic Vein [Anatomy]

HV Herpesvirus

HV Hidden Variables [Quantum Mechanics]

HV High Vacuum

HV High Velocity

HV High Viscosity

HV High Voltage [also hv]

HV High Volume

HV Hudson Valley [New York, US]

HV Hydraulic Valve

HV Hydraulic Vibrator

HV Hydrogen Vent

HV Hydroxyvalerate

HV Hypervelocity

HV Vickers Hardness (Number)

HV$_{100}$ Vickers Microhardness (Applied Indentation Load 100 kg) [Symbol]

3HV 3-Hydroxyvalerate

4HV 4-Hydroxyvalerate

HV(+) High Voltage, Positive

HV(–) High Voltage, Negative

HV(+/–) High Voltage, Positive, or Negative

H&V Heating and Ventilation [also HV]

H-V Horizontal-Vertical [also H/V]

hv heavy

HVA High Voltage Apparatus

HVA Homovanillic Acid

HVAB High-Volatile A Bituminous (Coal)

HVAC Heating, Ventilation (or Ventilating) and Air Conditioning

HVAC High Vacuum [also hvac]

HVAC High-Voltage Actuator

HVAC High-Voltage Alternating Current

HVACC High Voltage Apparatus Coordinating Committee [of American National Standards Instituite, US]

HVAF High-Velocity Air Fuel (Process)

HVAP Hypervelocity, Armor Piercing

HVAPDSFS Hypervelocity Armor-Piercing, Discarding Sabot, Fin Stabilized (Projectile)

HVAR High-Velocity Airborne Rocket

HVAR High-Velocity Aircraft Rocket

HVAST Hull Vibration and Strength Analysis

HVAT High-Velocity Antitank

HVAT Hypervelocity, Antitank

HVBB High-Volatile B Bituminous (Coal)

HVBO Heterojunction Valence-Band Offset [Solid-State Physics]

HVC High-Velocity Clouds

HVC High-Voltage Circuit

HVC Horizontal-Vertical Control (Rolling Technology) [Metallurgy]

HVC Hydrovac Process

HVCA Heating and Ventilating Contractors Association [UK]

HVCB High-Volatile C Bituminous (Coal)

HVC Horizontal-Vertical Control (Rolling Technology) [Metallurgy]

HVC/CVC Horizontal-Vertical Control/Continuously Variable Crown (Rolling Technology) [Metallurgy]

HVCMOS High-Voltage Complementary Metal-Oxide Semiconductor

HVCTEM High-Voltage Conventional Transmission Electron Microscopy

HVD Half-Value Depth [Radiography]

HVD High-Velocity Detonation

HVDC High-Voltage Direct-Current

HVDCT High-Voltage Direct-Current Transmission

HVDF High and Very-High Frequency Direction Finding

HVDS Hypergolic Vapor Detection System [Aerospace]

HVE High Voltage Engineering

HVE Horizontal Vertex Error

HVEE High Voltage Engineering Europa BV [Amersfoort, Netherlands]

HVEM High-Voltage Electron Microscope; High-Voltage Electron Microscopy

H&V Eng Heating and Ventilation Engineer [Published in the UK]

HVF High-Viscosity Fuel

HVFe Hauptverwaltung für Fernmeldewesen [Central Telecommunications Administration, Germany]

HVFu Hauptverwaltung für Funkwesen [Central Radio Communications Administration, Germany]

HVG High-Voltage Generator

HVG Hypervelocity Gun

HVHMD Holographic Visor Helmet-Mounted Display

HVI High-Velocity Impact

HVI Home Ventilating Institute [now Home Ventilating Institute Division of the Air Movement Control Association, US]

HVI Human Visual Inspection

HVIC High-Voltage Integrated Circuit

HVIDAMCA Home Ventilating Institute Division of the Air Movement Control Association [US]

HVIS Hypervelocity Impact System

HVL Half-Value Layer [Radiography]

HVLP High-Velocity Low Pressure

HVLP High-Volume, Low Pressure

HVM High-Voltage Microscopy

HVMF High Valency Metal Fluoride

HVMOS High-Voltage Metal-Oxide Semiconductor
HVMS High Voltage Mass Separator
HVN Vickers Hardness Number
HVOF High-Velocity Oxy-Fuel (Coating Process)
HVOF High-Velocity Oxygen Fuel (System)
HVOSM Highway Vehicle Object Simulation Model
HVP Hauptverwaltung für Postwesen [Central Postal Administration, Germany]
HVP High Video Pass [Electronic Warfare]
HVP High Volume Production
HVP High Voltage Pulser
HVP Horizontal and Vertical Position
HVPE Halide Vapor-Phase Epitaxy
HVPE Hydride Vapor-Phase Epitaxy
HVPF Hauptverwaltung für das Post– und Fernmeldewesen [Central Postal and Telecommunications Administration, Germany]
HVPM High-Vapor Pressure Metal
HVPS High-Voltage Power Supply
HVR Hard Vertical Rotating Balancer
HVR Hardware Vector to Raster
HVR High-Vacuum Rectifier
HVR High-Voltage Regulator
HVR High Volume Resistivity
HVR Home Video Recorder
HVR Hover [also Hvr, or hvr]
HVRA Hawaiian Volcano Research Association [US]
HVRA Heating and Ventilating Research Association [UK]
HVS Hard Vertical Static Balancer
HVS High-Temperature Vapor-Phase Synthesis
HVS High-Voltage Source
HVS High Vacuum Seal
HVSCR High-Voltage Selenium Cartridge Rectifier
HVSEM High-Voltage Scanning Electron Microscope; High-Voltage Scanning Electron Microscopy
HVSF Honeywell Verification Simulation Facility
HVSL Holidays, Vacation and Sick Leave
HVSTEM High-Voltage Scanning Transmission Electron Microscope; High-Voltage Scanning Transmission Electron Microscopy
HVT Half-Value Thickness [Physics]
HVT High-Vacuum Tube
HVT High-Voltage Test
HVT Hydraulic Variable Timing
HVTEM High-Voltage Transmission Electron Microscope; High-Voltage Transmission Electron Microscopy
HVTP High-Velocity Target Practice
HVTP Hypervelocity, Target Practice
HVTP-T Hypervelocity, Target Practice, Tracer
HVTR High-Voltage Track Rate
HVTS High-Volume Time Sharing
HVTS Hypervelocity Techniques Symposium
hvy heavy
HVZ Hell-Volhard-Zelinsky (Reaction) [Organic Chemistry]
HW Half Wave

HW Half-Width [Mathematics]
HW Half-Word [Computers]
HW Hammer Welding
HW Handset, Wall Model
HW Hanging Wall [Geology]
HW Hardware [also H/W]
HW Hard Water
HW Hard-Wired (System)
HW Hardwood
HW Harmonic Wave
HW Hazardous Waste
HW Head Wall
HW Head Water
HW Head Wave
HW Headwind
HW Heat Wave
HW Heavy Water
HW Heavy Weapon
HW Hertzian Wave
HW High Water
HW Highway [also Hwy]
HW High-Wing (Monoplane)
HW Hot Water
HW Hot Wire
HW Hot Work(ed); Hot Working [Metallurgy]
HW Hour Wheel [Horology]
HW How [Amateur Radio]
H/W Hardware [also HW]
H-W Heavy-Wall (Pipe)
Hw Highway [also Hwy]
hW Hectowatt [also hw]
hw hollow
HWA Hot-Wire Analysis; Hot-Wire Analyzer
HWC Half-Wave Circuit
HWC Half-Wave Current
HWC Hazardous Waste Containment
HWC Health and Welfare Canada [now Health Canada]
HWC Hot Water Circulation
H_2-WCl$_6$ Hydrogen-Tungsten Hexachloride (Gas Mixture)
HWCP Hardware Code Page [also hwcp]
HWCTR Heavy-Water Components Test Reactor [of US Atomic Energy Commission]
HWCVD Hot-Wire Chemical Vapor Deposition
HWD Half-Wave Dipole
HWD Hazardous Waste Disposal
HWD Height-Width-Depth [also hwd]
HWD Hot Wire Detector
HW&D Hynson, Westcott and Dunning, Inc. [Pharmaceutical Manufacturer located at Baltimore, Maryland, US]
HWE Hot Wall Epitaxy
HWERL Hazardous Waste Engineering Research Laboratory [of US Environmental Protection Agency]
HWF Hazardous Waste Federation [US]

HWF *(Höhere Wirtschaftsfachschule)* – German for "Higher School of Business"

HWF Houston Wafer Fabrication [of Texas Instruments, Inc., US]

HWGCR Heavy-Water(-Moderated) Gas-Cooled Reactor

HWGD Hot-Wall Glow Discharge Decomposition

HW-GTAW Hot-Wire Gas Tungsten-Arc Welding

HWHS Hot-Water Heating

HWHH Half-Width at Half-Height [also hwhh]

HWHM Half Width at Half Maximum [also hwhm]

HWHS Hot-Water Heating System

HWI Hardware Interpreter

HWI Hauptman-Woodward Medical Research Institute [Buffalo, New York, US]

HWIM Hear What I Mean

HWIR Hazardous Waste Identification Rule [of US Environmental Protection Agency]

HWL High Water Level

HWL High Water Line [Oceanography]

HWL Hot Water Limit (Sensor)

HWLC Head Water Level Control

HWLT Head Water Level Transmitter

HWLWR Heavy-Water-Moderated Light-Water-Cooled Reactor

HWM Half-Width Method

HWM Hazardous Waste Management

HWM High Water Mark [Computers]

HWM High Wet Modulus

HWMF Hazardous Waste Management Facility

h-WO₃ Hexagonal Tungstic Oxide

HWOCR Heavy Water Organic-Cooled Reactor [also HWOR]

HWOST High Water Ordinary Spring Tide

HWP Half-Wave Plate [Optics]

HWP Half-Wave Potential

HWP Heavy-Water Plant

HWR Half-Wave Rectification; Half-Wave Rectifier

HWR Heavy-Water (Nuclear) Reactor

HWRI Harbin Welding Research Institute [Harbin, PR China]

HWRV Half-Wave Retardation Voltage

HWS Hanford Works Specification [now Hanford Plant Standard]

HWS Hanford Works Standard [now Hanford Plant Standards]

HWS Hazardous Warning System

HWS Hot-Water System

HW/SW Hardware/Software (Package)

hwt hundredweight [Unit]

HWT Hazardous Waste Treatment

HWT Hydrid- und Wasserstofftechnik mbH [Hydride and Hydrogen Technology Company; of Mannesmann AG in Mülheim, Germany]

HWTC Hazardous Waste Treatment Council [US]

HWTF Hazardous Waste Treatment Facility

HWTU Hanford Waste Treatment Unit [of US Department of Energy, Washington State, US]

HWU Heriot-Watt University [Edinburgh, UK]

HWV *(Höhere Wirtschafts– und Verwaltungsschule)* – German for "Higher School of Business Economics and Administration" [Switzerland]

HWVP Hanford Waste Verification Plant [of US Department of Energy in Washington State, US]

HWW Hot and Warm Working [Metallurgy]

HWWA Hamburgisches Welt-Wirtschafts-Archiv [Hamburg Archives for World Economics and Commerce, Germany]

HWWS Hyperfiltration Wash Water Recovery System

Hwy Highway

HX Half Duplex [also hx]

HX Heat Exchange(r)

HX Hydrogen Halide,

HX Trans-1,4-Hexadiene

Hx Hydroxy-

H(x) Hardness as a Function of Indentation Depth [Symbol]

H(x) (Disk) Reader Microtrack Profile [Symbol]

h(x) microscopic magnetic field [Symbol]

HXIS Hard X-Ray Imaging Spectrometer

HXRS Hard X-Ray Spectrometer

HXSA Hexenylsuccinic Anhydride

HXSA n-Hexenyl Succinic Anhydride

HXTA N,N'-[2-Hydroxy-5-Methyl-1,3-Phenylenebis-(methylene)]bis[N-(Carboxymethyl)glycine

h(x,y) microscopic magnetic field [Symbol]

h(x,y) point spread function [Symbol]

HY High Yield

Hy Henry [also H, or h]

Hy Hydrazine

HYACS Hybrid Analog-Switching Attitude Control System

HYAN Hybrid Anaerobic (Process)

Hyb Hybrid (System) [also hyb] [Computers]

Hy Ball Hydraulic Ball

HYBLOC Hybrid Computer Block-Oriented Compiler

Hybrid Circuit Technol Hybrid Circuit Technology [Journal published in the US]

Hybrid-SV Hybrid Spin Valve

.HYC Hyphenation Dictionary (File) [Word Perfect File Name Extension] [also .hyc]

HYCATS Hydrofoil Collision Avoidance and Tracking System

HYCOL Hybrid Computer Link

HYCON Hydrogen Conversion (Process)

HYCOTRAN Hybrid Computer Translator

Hyd Hydraulic(s) [also hyd]

Hyd Hydrological; Hydrologist; Hydrology [also hyd]

HYDAC Hybrid Digital-Analog Computer

HYDAPT Hybrid Digital-Analog Pulse Timer

HYDAS Hydroacoustic Data Acquisition System

HYDEAL Hydrogen Dealkylation (Process)

HYDEC Hydride/Dehydride/Cast (Process)

HYDLAPS Hydrographic Data Logging and Plotting System

Hydr Hydraulic(s) [also hydr]

Hydr Hydrostatic(s) [also hydr]

Hydraul Hydraulic(s) [also hydraul]

Hydraul Pneum Hydraulics and Pneumatics [Journal published in the US]

Hydro Hydrodynamic(s) [also hydro]

Hydro Hydrostatic(s) [also Hydrostat, hydrostat, or hydro]

Hydrocarbon Process Hydrocarbon Processing [US Publication]

Hydrod Hydrodynamics [also Hydrodyn, hydrodyn, or hydrd]

Hydrogen Energy Prog Hydrogen Energy Progress [Journal of the International Association for Hydrogen Energy, US]

HYDRODB Hydrogeologiska Databas [Hydrogeology Database, of Swedish Geological Survey]

Hydrogr Hydrographer; Hydrographic; Hydrography [also hydrogr]

Hydrogr J Hydrographic Journal [of Hydrographic Society, UK]

Hydrol Hydrological; Hydrologist; Hydrology [also hydrol]

HYDROLANT Hydrographic Warning–Atlantic Ocean [of Naval Oceanographic Office, US]

Hydrol Process Hydrological Processes [UK Publication]

Hydrol Res Hydrological Research(er)

Hydrol Sci J Hydrological Sciences Journal [of International Association of Hydrological Sciences, Netherlands]

Hydrol Sci Technol Hydrological Sciences and Technology [Journal of the American Institute of Technology, US]

HYDROPAC Hydrographic Warning–Pacific Ocean [of Naval Oceanographic Office, US]

Hydrostat Hydrostatic(s) [also Hydro, hydro, or hydrostat]

Hydrotech Constr (USSR) Hydrotechnical Construction (USSR) [Published in the US]

D-β-Hydroxybutyryl-S-ACP D-β-Hydroxybutyryl-S-Acyl Carrier Protein [Biochemistry]

Hydrs Hydrostatics [also hydrs]

Hydrx Hydroxide [also Hydx, hydx, or hydrx]

HYFAC Hydrogenated Fatty Acid

HYFES Hypersonic Flight Environmental Simulator

Hyg Hygiene [also hyg]

hyg hygroscopic

HYGAS Hydrogasification (Process) [also Hygas] [Chemical Engineering]

hygl hypergolic [Aerospace]

hygr hygroscopic

Hygry Hygroscopicity [also hygry]

HYL Hojalata y Lamina (Direct-Reduction Ironmaking Process) [also HyL] [Metallurgy]

Hyl δ-Hydroxy-L-Lysine; (–)-Hydroxylysine [Biochemistry]

HYL Rep HYL (Hojalata Y Lamina) Report [Metallurgy]

HYLA Hybrid Language Assembler

HYLO Hybrid LORAN (Long-Range Navigation)

HYLSA Hojalata Y Lamina SA [Mexican Iron and Steel Works located at Monterrey]

HYMAT Hybrid Material [also Hymat]

HYMATS Hybrid Materials [also Hymats]

HYMETS Hypersonic Materials Environmental Test System [of NASA Langley Research Center, Hampton, Virginia, US]

HYP Harvard, Yale, and Princeton Universities [US]

HYP Hyphen (Character)

.HYP Hyphenation [File Name Extension] [also .hyp]

Hyp (–)-Hydroxyproline [Biochemistry]

Hyp Hyphenation [also hyp]

Hyp Hypotenuse [also hyp]

Hyp Hypothesis [also hyp]

3-Hyp 3-Hydroxyproline [Biochemistry]

4-Hyp 4-Hydroxyproline [Biochemistry]

HYPACE Hybrid Programmable Attitude Control Electronics

HyperCP Search for CP Violation in the Decays of Cascade Minus/Cascade Plus and Lambda/ Anti-Lambda Hyperons (Experiment) [of Fermilab, Batavia, Illinois, US] [Particle Physics]

Hyperf Interact Hyperfine Interactions [Publication of J.C. Baltzer AG, Basel, Switzerland]

HYPERONS International Conference on Hyperons, Charm and Beauty Hadrons [Particle Physics]

Hypertens Hypertension [Journal of the American Heart Association, US]

HYPER Hypertapes [Computers]

HYPERDOP Hyperbolic Doppler [also Hyperdop]

HYPERNET Hyper Network [also Hypernet]

Hyph Hyphenation [allso hyph]

hyp log hyperbolic logarithm [Mathematics]

HYPO High-Power Output (Boiling-Water) Reactor

hypo hypochondria [Medicine]

hypo hypodermic needle (or syringe); hypodermic injection [Medicine]

HYPSES Hydrographic Precision Scanning Echo Sounder

HYSOMER Hydroisomerization (Process) [Chemical Engineering]

HYSTAD Hydrofoil Stabilization Device

HYSTOR Hydrogen Storage

HYT High Yield Technology

HYTELNET Hypertext-Browser for Telnet Accessible Sites

HYTEX Hydrogen Texaco (Process) [Chemical Engineering]

HYV High-Yield Variety

HZ Haze [also Hz, or hz]

HZ Herpes Zoster [Medicine]

HZ Hot Zone

HZ- Saudi Arabia [Civil Aircraft Marking]

Hz Haze [also hz]

Hz Hertz [Unit]

Hz Hydrazo-

Hz^{-1} One per Hertz [also 1/Hz]

Hz/G Hertz per Gauss [also Hz G^{-1}]

HZIMP Horizontal Impulse

HZONE Hyphenation Zone [also H-Zone, or H-zone]

HZP Hot Zero Power

HZQC Homonuclear (Proton) Zero-Quantum Conference

Hz/T Hertz per Tesla [also Hz T^{-1}]

HZTRI Hitachi Zosen Technical Research Institute [Osaka, Japan]

HZV Herpes Zoster Virus

Hz/V Hertz per Volt [also Hz V^{-1}]

I

I AC (Alternating Current) Exciting Current [Symbol]

I Acoustic Intensity [Symbol]

I Additional Designation of Emission (or Transmission) Type–Single Sideband, Suppressed Carrier [Symbol] [also i]

I Bias Current [Symbol]

I Body-Centered Space Lattice in Orthorhombic, Tetragonal and Cubic Crystal Systems [Hermann-Mauguin Symbol]

I (Electric) Current

I Definite Integral [Symbol]

I Effective Value of AC Current [Symbol]

I Electron Current (in Thermionic Emission) [Symbol]

I Geometry Factor (of Gears) [Symbol]

I Grade Denoting Incomplete Work

I Hydraulic Gradient (of Rocks) [Symbol]

I Hypoxanthine [Biochemistry]

I Ice

I Iceland(ic)

I Icing

I Icosahedral [Crystallographic Symmetry Group Symbol]

I Idaho [US]

I Identity

I Identity Distance (of Crystals) [Symbol]

I Identity Matrix (in Mathematics) [Symbol]

I Identity Tensor [Symbol]

I Idle

I (–)-Idose [Biochemistry]

I Igneous [also i]

I Ignition

I Ignore

I Iguana [Zoology]

I Ilex [Genus of Trees Including the Hollies]

I Ileum [Anatomy]

I Illinois [US]

I Illite [Mineral]

I Illuminance

I Illumination; Illuminator

I Illuviation [Geology]

I Image(ry)

I Imaginary Number

I Imaging

I Imide

I Immediate

I Immersion

I Immiscibility; Immiscible

I Immun(ity)

I Immunization

I Immunological; Immunologist; Immunology

I Impact

I Impaction [Medicine]

I Impeller

I Imperial

I Imperfect(ion)

I Implant(ation)

I Impregnate(d); Impregnation

I Impress(ion)

I Improve(ment)

I Impulse

I Impure; Impurity

I Inaccuracy; Inaccurate

I Incandescence; Incandescent

I Incendiary

I Incenter (of a Triangle) [Symbol]

I Inchon [South Korea]

I Incidence; Incident

I Incident Beam [Symbol]

I Incisor [Dentistry]

I Inclination; Incline(d)

I Include(d); Inclusion

I Incoherence; Incoherent

I Incombustibility; Incombustible

I Increase

I Increment(al)

I Incubate; Incubation; Incubator

I Incus (of Ear) [Anatomy]

I Indent(ation)

I Indenter [Mechanical Testing]

I Independence; Independent

I Index

I India [Phonetic Alphabet]

I India(n)

I Indiana [US]

I Indianapolis [US]

I Indicate(d); Indication; Indicator

I Indonesia(n)

I Indoor(s)

I Induce(d)

I Inductance

I Induction

I Inductor

I Inductive(ly)
I Indus (River) [Tibet/Pakistan]
I Industrial; Industry
I Industrial/Manufacturing Engineering (Technology) [Discipline Category Abbreviation]
I Inelastic(ity)
I Inert(ance)
I Inertia(l)
I Infant(ile); Infantilism
I Infantry
I Infarct(ion) [Medicine]
I Infect(ed); Infection; Infectious
I Infer(ence)
I Infiltrate(d); Infiltration
I Infinite; Infinity
I Inflammability; Inflammable
I Inflame(d); Inflammation; Inflammatory
I Influence
I Influence Value [Symbol]
I Influenza
I Inform(ation)
I Infrared
I Ingest(ion)
I Ingot [Metallurgy]
I Inhalation; Inhale
I Inhibit(ion); Inhibitor
I Initial(ize); Initialization
I Initiate(d); Initiation; Initiator
I Inject(ion); Injector
I Injury
I Inlet
I Inoculation
I Inorganic
I Inosine [Biochemistry]
I Input
I Inquiry
I Insect
I Insecticidal; Insecticide
I Insectivora; Insectivorous [Biology]
I Insensitive; Insensitivity
I Insert(ion)
I Inside
I Insolubility; Insoluble
I Insomnia(c)
I Inspect(ion); Inspector
I Instability; Instable
I Install(ation)
I Instantaneous
I Instruct(ion); Instructor
I Instrument(al); Instrumentation
I Insulate(d); Insulation; Insulator
I Insulin [Biochemistry]
I Insurance
I Integer

I Integral; Integrand; Integrate(d); Integration; Integrator
I Integration Constant
I Integrity
I Intelligence; Intelligent
I Intensification; Intensify(ing)
I Intensity [Symbol]
I Intention
I Interaction; Interactive
I (Aerial) Intercept Vehicle [USDOD Symbol]
I Interchange(ability); Interchangeable
I Interconnect(ion)
I Interdiffusion
I Interest
I Interface; Interfacial
I Interfere(nce)
I Interferogram; Interferography
I Interferometer; Interferometry
I Intergalactic
I Intermetallic
I Intermittency; Intermittent
I Internal
I Interphase
I Interphone [Symbol]
I Interplanetary
I Interpolation; Interpolator
I Interpret(ation)
I Interrupt(ion)
I Interstellar
I Interstice; Interstitial(cy)
I Interval
I Intestinal; Intestine
I Intrinsic
I Introduction
I Intrude(r); Intrusion
I Intumescence; Intumescent
I Invariance; Invariant
I Invade; Invasion
I Invasive(ness)
I Invent(ion); Inventer
I Inventory
I Inverse; Inversion; Inverted
I Inverter
I Investigation
I Involute
I Iodine [Symbol]
I Iodopsin [Biochemistry]
I Ion(ic)
I Ionic Quantity; Ionic Strength [Symbol]
I Ionization; Ionize(d); Ionizer
I Ionomer
I (Upper-case) Iota [Greek Alphabet]
I Iowa [US]
I Iran(ian)

I   Iraq(i)
I   Ireland; Irish
I   Iridescence; Iridescent
I   Iris
I   Iron [Abbreviation; Symbol: Fe]
I   Irradiance
I   Irradiant; Irradiation
I   Irreversible; Irreversibility
I   Irritability; Irritable
I   Irrotational
I   Islamabad [Pakistan]
I   Island
I   Islands (in Homogeneous Connections of Atoms) [Symbol]
I   Isle
I   Iso- [Chemistry]
I   Isobutylene
I   Isoclinal; Isocline; Isoclinic
I   Isolability; Isolator
I   Isolate(d); Isolation
I   (+)-Isoleucine; Isoleucyl [Biochemistry]
I   Isomer(ism)
I   Isoprene
I   Isospin [Nuclear Physics]
I   Isostasy [Geophysics]
I   Isotherm(al)
I   Isotope; Isotopic
I   Isotropic; Isotropy
I   Isotropic Phase (of Liquid Crystal)
I   Israel(i)
I   Istanbul [Turkey]
I   Italian; Italy
I   Italics
I   Item(ize)
I   Iterate; Iteration; Iterative
I   Izmir [Turkey]
I   Luminous Intensity [Symbol]
I   Magnetization [CGS-EMU Symbol]
I   Moment of Inertia of Plane Area [Symbol]
I   Nucleation Rate (in Metallurgy) [Symbol]
I   Radiant Intensity (or Radiance) [Symbol]
I   Stoner Exchange Parameter (in Physics) [Symbol]
I   Threshold Intensity for Test Person [Symbol]
I   Transmitted Radiant Flux [Symbol]
I   Unit Vector (in Mathematics) [Symbol]
I-   Italy [Civil Aircraft Marking]
/I   Include Directories Option [Turbo Pascal]
/I   Instructions Option [MS-DOS Shell]
⟨I⟩   Average Intensity [Symbol]
I^{++} Plus-Plus Nonspin Flip Intensity (in Physics) [Symbol]
I^{+-} Plus-Minus Spin Flip Intensity (in Physics) [Symbol]
I^{-} Iodine Ion [Symbol]
I^{--} Minus-Minus Nonspin Flip Intensity (in Physics) [Symbol]
I^{-+} Minus-Plus Spin Flip Intensity (in Physics) [Symbol]

$I_{\parallel}$ Parallel Polarization [Symbol]
$I_{\perp}$ Cross(ed) Polarization [Symbol]
I_0 Brightfield Intensity (in Electron Microscopy) [Symbol]
I_0 Electron Current at Zero Potential (of Electron Tube) [Symbol]
I_0 Incident Radiant Flux [Symbol]
I_0 Initial Intensity [Symbol]
I_0 Intensity of Incident Electrons, Ions, etc. [Symbol]
I_0 Intensity of Primary Beam [Symbol]
I_0 Intensity of Sound at Origin [Symbol]
I_0 Maximum Current [Symbol]
I_0 Reverse Saturation Current [Semiconductor Symbol]
I_0 (Light) Scattering Intensity at Angle $\theta = 0°$ [Symbol]
I_0 Threshold Intensity for Normal Ear [Symbol]
$I_0{'}$ Intensity of Nonreflected Incident (Electromagnetic) Radiation [Symbol]
$I_{0.25t}$ Quarter-Second (Propellant) Impulse (in Aerospace Engineering) [Symbol]
I_1 Current in the Primary Circuit [Symbol]
I_2 Current in the Secondary Circuit [Symbol]
I_2 Molecular Iodine [Symbol]
I^3 Intelligent, Integrated and Interactive [also III]
I_A Anode Current [Semiconductor Symbol]
I_A Intensity of Absorbed Beam (of Electromagnetic Radiation) [Symbol]
$I_{A/2}$ Half Current [Symbol]
I_a Action Time (Propellant) Impulse (in Aerospace Engineering) [Symbol]
I_a Anode (or Plate) Current [Semiconductor Symbol]
I_a Armature Current [Symbol]
I_a Second Axial Moment of Area [Symbol]
I_α Auger Current (in Auger Spectroscopy) [Symbol]
I_{AC} AC Current [also I_{A-C}, or I_{a-c}] [Symbol]
I_{ac} Aerial (or Antenna) Current [also A_{ae}] [Symbol]
I_{ao} Zero Signal Current [Semiconductor Symbol]
I_B Base-Terminal Current, DC [Semiconductor Symbol]
I_B Bulk Current (i.e., the Current in the Bulk) [Symbol]
I_B Intensity of Bragg Peak (in Physics) [Symbol]
I_b Alternating Component (Root-Mean Square Value) of Base-Terminal Current [Semiconductor Symbol]
I_b Background Intensity [Symbol]
I_{B2} Interbase Current [Semiconductor Symbol]
$I_{B2(mod)}$ Interbase Modulated Current [Semiconductor Symbol]
I_{BEV} Base Cutoff Current, DC [Semiconductor Symbol]
I_{BF} Brightfield Intensity (in Electron Microscopy) [Symbol]
I_C Collector-Terminal Current, DC [Semiconductor Symbol]
I_C (Nominal) Control Current [Symbol]
I_c Capacitive Component of Leakage Current [Symbol]
I_c Capacitor Current [Semiconductor Symbol]
I_c AC Core-Loss Current [Symbol]
I_c Alternating Component (Root-Mean Square Value) of Collector-Terminal Current [Semiconductor Symbol]
I_c Coherent Atomic Scattering Intensity [Symbol]
I_c Collector Current [Semiconductor Symbol]

I_c Corrected Intensity [Symbol]

I_c Critical Current [Symbol]

I_c Opaque Cloud Radiance [Symbol]

I_c Relative Consistency (of Soil) [Symbol]

I_{CBO} Collector Cutoff Current (DC), Emitter Open [Semiconductor Symbol]

I_{CC} Average Supply Current [Semiconductor Symbol]

I_{CCH} Total Supply Current for HIGH Gate Output [Semiconductor Symbol]

I_{CCL} Total Supply Current for LOW Gate Output [Semiconductor Symbol]

I_{CEO} Collector Cutoff Current (DC), Base Open [Semiconductor Symbol]

I_{CER} Collector Cutoff Current (DC), Specified Resistance between Base and Emitter [Semiconductor Symbol]

I_{CES} Collector Cutoff Current (DC), Base Short-Circuited to Emitter [Semiconductor Symbol]

I_{CEV} Collector Cutoff Current (DC), Specified Voltage between Base and Emitter [Semiconductor Symbol]

I_{CEX} Collector Cutoff Current (DC), Specified Circuit between Base and Emitter [Semiconductor Symbol]

I_{ci} Critical Current of the i-th Contact [Symbol]

I_{cj} Critical Current of Junction [Semiconductor Symbol]

$I_{C(max)}$ Collector-Terminal DC Current [Semiconductor Symbol]

I_{CL} Cathodoluminescence Intensity [Symbol]

I_{CMOS} Current Flow in CMOS (Complementary Metal-Oxide Semiconductor) Inverter [Symbol]

I_{co} Cutoff Collector Current [Semiconductor Symbol]

I_{corr} Corrosion Current Density [Symbol]

I_D Diffuse Component of Intensity [Symbol]

I_D Diffusion Current [Symbol]

I_D Drain Current, DC [Semiconductor Symbol]

I_d Direct-Current [Symbol]

I_d Diode Anode Current [Semiconductor Symbol]

I_d Drive Current [Symbol]

I_d Relative Density (of Soil) [Symbol]

I_{DC} DC Current [Symbol]

I_{dc} Constant Current [Symbol]

I_{d-c} DC Current [Symbol]

I_{DF} Darkfield Intensity (in Electron Microscopy) [Symbol]

$I_{D(off)}$ Drain Cutoff Current [Semiconductor Symbol]

$I_{D(on)}$ On-State Drain Current [Semiconductor Symbol]

I_{DRM} Forward Breakover Current (for Thyristors and Trigger Devices) [Semiconductor Symbol]

I_{DS} Drain-Source Current [Semiconductor Symbol]

I_{DSS} Zero-Gate-Voltage Drain Current [Semiconductor Symbol]

I_E Electron Current at Potential E (of Electron Tube) [Symbol]

I_E Emitter-Terminal Current, DC [Semiconductor Symbol]

I_e Alternating Component (Root-Mean-Square Value) of Emitter-Terminal Current [Semiconductor Symbol]

I_e Electron Current for Ionization [Symbol]

I_e Radiant Intensity [Symbol]

I_e Incident Intensity of X-Ray Beam [Symbol]

I_{EB20} Emitter Reverse Current [Semiconductor Symbol]

I_{EBO} Emitter Cutoff Current (DC), Collector Open [Semiconductor Symbol]

$I_{EC(ofs)}$ Emitter-Collector Offset Current [Semiconductor Symbol]

I_{ECS} Emitter Cutoff Current (DC), Base Short-Circuited to Collector [Semiconductor Symbol]

I_{eff} Effective Current [Symbol]

$I_{E1E2(off)}$ Emitter Cutoff Current [Semiconductor Symbol]

I_F DC Forward Current (Without Alternating Component) (for Rectifier and Signal Diodes) [Semiconductor Symbol]

I_F DC Forward Current (for Voltage-Reference and Voltage Regulator Diodes) [Semiconductor Symbol]

I_f Alternating Component (Root-Mean-Square Value) of Forward Current [Semiconductor Symbol]

I_f Heater (or Filament) Current [Semiconductor Symbol]

I_f Flow Index (in Geology) [Symbol]

$I_{F(AV)}$ Forward Current, DC (with Alternating Component) [Semiconductor Symbol]

I_{FM} Maximum (Peak) Total Forward Current [Semiconductor Symbol]

$I_{F(OV)}$ Forward Current, Overload [Semiconductor Symbol]

I_{FRM} Maximum (Peak) Forward Current, Repetitive [Semiconductor Symbol]

$I_{F(RMS)}$ Total Root-Mean-Square Forward Current [Semiconductor Symbol]

I_{FSM} Maximum (Peak) Forward Current, Surge [Semiconductor Symbol]

I_G Gate Current, DC [Semiconductor Symbol]

I_g Darkfield Intensity (in Electron Microscopy) [Symbol]

I_g Grid Current (of Electron Tubes) [Symbol]

I_{g1} Control Grid Current [Semiconductor Symbol]

I_{g2} Screen-Grid Current [Semiconductor Symbol]

I_{GAO} Gate-Anode Current, Cathode Open [Semiconductor Symbol]

I_{GF} Forward Gate Current [Semiconductor Symbol]

I_{GR} Reverse Gate Current [Semiconductor Symbol]

I_{GSS} Reverse Gate Current, Drain Short-Circuited to Source [Semiconductor Symbol]

I_{GSSF} Forward Gate Current, Drain Short-Circuited to Source [Semiconductor Symbol]

I_{GSSR} Reverse Gate Current, Drain Short-Circuited to Source [Semiconductor Symbol]

I_H Holding (or Retaining) Current [Semiconductor Symbol]

I_h Helix Current [Symbol]

I_h Icosahedral (Crystal Structure) [Symbol]

I_{hk} Heater-Cathode Leakage Current [Symbol]

I_{hkl} Diffracted Intensity of (Crystal) Lattice Plane Set *(hkl)* [Symbol]

I_I Inflection-Point Current [Semiconductor Symbol]

I_I Input Current [Semiconductor Symbol]

I_i Incoherent Atomic Scattering Intensity [Symbol]

I_{IB} Input Bias Current (for Operational Amplifier) [Semiconductor Symbol]

I_{IH} HIGH Level Input Current [Semiconductor Symbol]

I_{IL} LOW Level Output Current [Semiconductor Symbol]

I_{in} Input Current [Semiconductor Symbol]

I_{IO} Input Offset Current (for Operational Amplifier) [Semiconductor Symbol]

I_j Intensity of Component "j" [Symbol]

I_k Cathode Current [Semiconductor Symbol]

I_L Inductor Current [Symbol]

I_L Latching Current [Semiconductor Symbol]

I_L Leakage Current [Symbol]

I_L Liquidity Index (or Water-Plasticity Ratio) [Symbol]

I_L Load Current (of Electric Circuit) [Symbol]

I_m AC Magnetizing Current [Symbol]

I_m Average Current [Symbol]

I_m Measured Intensity [Symbol]

I_{max} Maximum Illumination [Symbol]

I_{max} Maximum Instantaneous AC Current [Symbol]

I_{max} Maximum Regulated Output Current [Semiconductor Symbol]

I_{min} Minimum Illumination [Symbol]

I_N Normal Reflection (in Crystallography) [Symbol]

I_N Norton's Equivalent Current [Symbol]

I_O Average Forward Current, 60-Hz Half Sine Wave, 180° Conduction Angle [Semiconductor Symbol]

I_O Output Current [Semiconductor Symbol]

I_{OH} HIGH Level Output Current [Semiconductor Symbol]

I_{OL} LOW Level Output Current [Semiconductor Symbol]

I_{OS} Short-Circuit Output Current [Semiconductor Symbol]

I_{out} Output Current [Semiconductor Symbol]

I_{OZ} Off-State Output Current [Semiconductor Symbol]

I_p Peak-Point Current [Semiconductor Symbol]

I_p Effective Current in Primary Coil (of Transformer) [Symbol]

I_p Peak Current [Symbol]

I_p Peak Current Density [Symbol]

I_p Plasticity Index (for Rock, or Soil) [Symbol]

I_p Plate (or Anode) Current [Symbol]

I_p Primary Electron Current [Symbol]

I_p Pulsed Current [Symbol]

I_p Second Polar Moment of Area [Symbol]

I_{PMOS} Current Flow in PMOS (P-Type Metal-Oxide Semiconductor) Inverter [Symbol]

I_{pp} Phonon-Peak Current (in Solid-State Physics) [Symbol]

I_Q Current Through Transistor [Semiconductor Symbol]

I_{Q1} Current Through Transistor 1 [Semiconductor Symbol]

I_{Q2} Current Through Transistor 2 [Semiconductor Symbol]

I_R DC Reverse Current (without Alternating Component) (for Rectifier and Signal Diodes) [Semiconductor Symbol]

I_R DC Reverse Current (for Voltage-Reference and Voltage-Regulator Diodes) [Semiconductor Symbol]

I_R Intensity of Reflected Beam (of Electromagnetic Radiation) [Symbol]

I_R Remolding Index (in Geology) [Symbol]

I_R Resistive Component of Leakage Current [Symbol]

I_r Alternating Component (Root-MeanSquare Value) of Reverse Current [Semiconductor Symbol]

I_r Reverse Current [Semiconductor Symbol]

I_r Resistive Current [Symbol]

$I_{R(AV)}$ Reverse Current, Direct-Current (With Alternating Component) [Semiconductor Symbol]

I_{RB} Current Through Base Resistance [Semiconductor Symbol]

I_{ref} Reference Current [Symbol]

I_{rms} Root-Mean-Square Current [Symbol]

I_{RM} Peak (Maximum) Total Reverse Current [Semiconductor Symbol]

I_{Rp} Current through Pull-up Resistor [Semiconductor Symbol]

I_{RRM} Peak (Maximum) Reverse Current, Repetitive [Semiconductor Symbol]

$I_{R(RMS)}$ Total Root-Mean-Square Reverse Current [Semiconductor Symbol]

I_{rs} Resonator Current (of Reflex Klystron) [Semiconductor Symbol]

I_{RSM} Peak (Maximum) Reverse Current, Surge [Semiconductor Symbol]

I_S Source Current, DC [also I_s] [Semiconductor Symbol]

I_S Specular Component of Intensity [Symbol]

I_S Superlattice Reflection (in Crystallography) [Symbol]

I_s Effective Current in Secondary Coil (of Transformer) [Symbol]

I_s Intensity of Magnetization (or Magnetic Moment per Volume) [Symbol]

I_s Secondary Electron Current [Semiconductor Symbol]

I_s Specific Impulse (in Aerospace Engineering) [Symbol]

I_s Supercurrent Through Junction [Symbol]

I_{SAT} Reverse Saturation Current [Semiconductor Symbol]

I_{sc} Short-Circuit Current [Semiconductor Symbol]

I_{SD} Source-Drain Current [Semiconductor Symbol]

I_{SDS} Zero-Gate-Voltage Source Current [Semiconductor Symbol]

$I_{S(off)}$ Source Cutoff Current [Semiconductor Symbol]

I_{sp} Specific Impulse (of Propellant) [Symbol]

I_{sp}^0 Theoretical Specific Impulse (of Propellant) [Symbol]

I_{spd} Measured (or Delivered) Specific Impulse (of Propellant) [Symbol]

I_{spd}^0 Theoretical Delivered Specific Impulse (of Propellant) [Symbol]

I_{sps} Standard Deliverable Specific Impulse (of Propellant) [Symbol]

I_{sps}^0 Standard Theoretical Specific Impulse (of Propellant) [Symbol]

I_{st} Starter Current [Symbol]

I_T Intensity of Transmitted Beam (of Electromagnetic Radiation) [Symbol]

I_T Threshold Current [Semiconductor Symbol]

I_T Toughness Index (of Soil) [Symbol]

I_T' Intensity of Nonabsorbed (or Transmitted) (Electromagnetic) Radiation [Symbol]

I_t Average Discharge (Corona) Current [Symbol]

I_t Torsion(al) Resistance [Symbol]

I_t Total Diffracted Intensity (in X-Ray Analysis) [Symbol]

I_t Transmitted Intensity of X-Ray Beam [Symbol]

I_θ Scattering Intensity at Angle θ [Symbol]

I_{ta} DC Target Current [Semiconductor Symbol]

I$_{tot}$ Total (Propellant) Impulse (in Aerospace Engineering) [Symbol]

I$_V$ Valley-Point Current [Semiconductor Symbol]

I$_v$ Specific Impulse in Vacuum (in Aerospace Engineering) [Symbol]

I$_w$ Plasticity Index (for Rock, or Soil) [Symbol]

I$_x$ (Electric) Current in x Direction [Symbol]

I$_x$ (Auger) Intensity of Element x [Symbol]

I$_x$ Intensity of Sound at a Distance x Apart [Symbol]

I$_x$ Moment of Inertia for a Plane Area with Respect to x Axis [Symbol]

I$_x$ Transmitted Radiant Flux (after Traversing Distance x) [Symbol]

I$_{xy}$ Product (Area) Moment of Inertia [Symbol]

I$_y$ Moment of Inertia for a Plane Area with Respect to y Axis [Symbol]

I$_z$ Reference Current, Regulator Current [Semiconductor Symbol]

I$_z$ Zener Current [Semiconductor Symbol]

I$_z$ Intrinsic Isotopic Spin (in Physics) [Symbol]

I$_z$ Moment of Inertia of Plane Area for z Axis [Symbol]

I$_{ZK}$ Reference Current, Regulator Current (Direct-Current Near Breakdown Knee) [Semiconductor Symbol]

I$_{ZM}$ Maximum Permissible Zener Current [Semiconductor Symbol]

I$_{ZM}$ Reference Current, Regulator Current (DC Maximum Rated Current) [Semiconductor Symbol]

I$_{ZT}$ Zener Test Current [Semiconductor Symbol]

I(0) Intensity (of Substrates) for Clean Surfaces [Symbol]

I-5 Five-Cylinder In-Line (Automotive) Engine

I-6 Six-Cylinder In-Line (Automotive) Engine

I-8 Eight-Cylinder In-Line (Automotive) Engine

I-120 Iodine-120 [also ^{120}I, or I^{120}]

I-121 Iodine-121 [also ^{121}I, or I^{121}]

I-122 Iodine-122 [also ^{122}I, or I^{122}]

I-124 Iodine-124 [also ^{124}I, or I^{124}]

I-125 Iodine-125 [also ^{125}I, or I^{125}]

I-126 Iodine-126 [also ^{126}I, or I^{126}]

I-127 Iodine-127 [also ^{127}I, I^{127}, or I]

I-128 Iodine-128 [also ^{128}I, or I^{128}]

I-129 Iodine-129 [also ^{131}I, or I^{129}]

I-130 Iodine-130 [also ^{130}I, or I^{130}]

I-131 Iodine-131 [also ^{131}I, or I^{131}]

I-132 Iodine-132 [also ^{132}I, or I^{132}]

I-133 Iodine-133 [also ^{133}I, or I^{133}]

I-134 Iodine-134 [also ^{134}I, or I^{134}]

I-135 Iodine-135 [also ^{135}I, or I^{135}]

I-136 Iodine-136 [also ^{136}I, or I^{136}]

I-137 Iodine-137 [also ^{137}I, or I^{137}]

I-138 Iodine-138 [also ^{138}I, or I^{138}]

I-139 Iodine-139 [also ^{139}I, or I^{139}]

i additional designation of emission (or transmission) type–single sideband, suppressed carrier [Symbol] [also I]

i angle of incidence [Symbol]

i chemical constant [Symbol]

i (instantaneous) current [Symbol]

i current density [Symbol]

i heat content (or enthalpy) [Symbol]

i hydraulic gradient [Symbol]

i icosahedral (phase)

i identical

i igneous [also I]

i imaginary part (of complex number) [Symbol]

i inactive

i incendiary

i incisor [Dentistry]

i inclination

i incomplete

i independent

i index

i indicate(d)

i induce(d)

i industrial

i initial

i inner

i insecticide

i insert lines (command) [Edlin MS-DOS Line Editor]

i insoluble

i instantaneous

i instantaneous current [Symbol]

i institute; institution

i instrumental

i integration constant (in statistical mechanics for solid-vapor equilibria) [Symbol]

i intensity of rainfall (or rainfall rate) [Symbol]

i interest (rate)

i interface; interfacial

i interfer(ence) [Symbol]

i internal

i international

i interstitial

i intrinsic

i ionic

i iota [English Equivlalent]

i iron

i islands (in heterogeneous connections of atoms) [Symbol]

i iso- [Chemistry]

i isotactic [Polymers]

i isotopic

i number of rows of balls, or rollers (of bearings) [Symbol]

i order (in sample statistics) [Symbol]

i running index [e.g., $\sum_{i=1}^{n} a_i = a_1 + a_2 + \cdots + a_n$]

i sampling size interval (of liquid drops) [Symbol]

i summation index [Symbol]

i total heat [Symbol]

i unit vector along positive x-axis [Symbol]

i Van't Hoff coefficient [Symbol]

i- icosahedral (phase)

i_0 exchange current density [Symbol]

i_0 peak value of instantaneous current [Symbol]

i_a anodic pulse amplitude (in pulse reversal plating) [Symbol]

i_a intensity of absorbed electrons [Symbol]

i_B instantaneous total value of base-terminal current [Semiconductor Symbol]

i_b intensity of backscattered electrons [Symbol]

i_c corrosion current density [also i_{corr}] [Symbol]

i_c instantaneous total value of collector-terminal current [Semiconductor Symbol]

i_c cathodic pulse amplitude (in pulse reversal plating) [Symbol]

i_c critical hydraulic gradient [Symbol]

i_d diffusion current [Symbol]

i_{dep} deposition current density (in electroplating) [Symbol]

i_E instantaneous total value of emitter-terminal current [Semiconductor Symbol]

i_F instantaneous total forward current [Semiconductor Symbol]

i_f feedback current [Symbol]

i_g grid current [Symbol]

i_i input current [Symbol]

i_L limiting diffusion current density [Symbol]

i_m instantaneous magnetizing current [Symbol]

i_m measured current intensity [Symbol]

i_{max} maximum (value of) instantaneous current [Symbol]

i_p constant passive current density [Symbol]

i_p current in the primary (transformer) coil [Symbol]

i_p passive current density [Symbol]

i_R instantaneous total reverse current [Semiconductor Symbol]

i_s intensity of secondary electrons [Symbol]

i_t intensity of transmitted electrons [Symbol]

ι (lower-case) iota [Greek Alphabet]

ι unit vector [Symbol]

ι_d diffusion current [Symbol]

$\mathbb{I}$ Integer [Symbol]

$\Im$ Imaginary Part (of Complex Number) [also Im]

IA Ice Age [Geology]

IA Image Amplifier

IA Image Analysis; Image Analyzer

IA Image Array

IA Immediate Address(ing)

IA Immunoaffinity

IA Immunoassay

IA Incorporated Accountant

IA Incremental Analysis

IA Independent Assortment [Genetics]

IA Index(ed) Address(ing)

IA Indiana Arsenal [Charlestown, Indiana, US]

IA Indicated Altitude [Aeronautics]

IA Indirect Address

IA Indoor Air

IA Induced Anisotropy [Solid-State Physics]

IA Induction Accelerator

IA Induction Amplifier

IA Industrial Aerodynamics

IA Industrial Application

IA Industrial Artists; Industrial Arts

IA Industrial Automation

IA Industry Applications (Society) [of Institute of Electrical and Electronics Engineers, US]

IA Industry Association [US]

IA Inert Atmosphere

IA Information Analysis

IA Inherent Address

IA Initial Appearance

IA Initial Approach [Aeronautics]

IA Input Axis [Navigation]

IA Institute of Actuaries [UK]

IA Institute of Archeology [of University College of London, UK]

IA Institut d'Automatique [Automation Institute; of Ecole Polytechnique Fédérale de Lausanne, Switzerland]

IA Instituto de Astronomía [Institute of Astronomy, Mexico]

IA Instruction Address

IA Instrumental Analysis

IA Instrumentation Amplifier

IA Integrated Adapter

IA Inter-Agency

IA Interatomic

IA Interchange Address

IA Interciencia Association [Venezuela]

IA Interface Analysis [Materials Science]

IA Interfacial Amorphization [Materials Science]

IA Intermediate Amplifier

IA Intermediate Anneal(ing) [Metallurgy]

IA International Alphabet

IA International Angstrom

IA Interstitial Atom [Crystallography]

IA Invent America (Foundation) [US]

IA Inverter Assembly

IA Iodine Absorber

IA Ion Accelerator

IA Ionizing Avalanche

IA Iowa [US]

IA Iron Age [Archeology]

IA Irrigation Association [US]

IA Isotopic Analysis

IA Issuing Agency

IA Istituto di Automatica [Automation Institute; of University of Rome, Italy]

IA Itaconic Acid

I&A Indexing and Abstracting

I&A (Combined) Iron and Alumina [also I and A]

IAA Iberis Amara Agglutinin [Biochemistry]

IAA Indirect Absolute Addressing

IAA Indole-3-Acetic Acid; Indol-3-ylacetic Acid [Biochemistry]

IAA Institut Agricole d'Algérie [Agricultural Institute of Algeria]

IAA Institute for Alternative Agriculture [US]

IAA Institute of Administrative Accounting [UK]

IAA Instituto Antarctico Argentino [Argentinian Antarctic Institute, Buenos Aires]

IAA Inter-American Accountant Association [Dominican Republic]

IAA Interim Access Authorization

IAA International Academy of Architecture

IAA International Academy of Astronautics [France]

IAA International Academy of Astronautics [US]

IAA International Acetylene Association

IAA International Advertising Association [US]

IAA International Aerosol Association [Switzerland]

IAA International Aerospace Abstracts [of NASA]

IAA International Apple Association [US]

IAA International Asphalt Association [now International Waterproofing Association, Belgium]

IAA International Association of Agriculturalists

IAA International Association of Astacology [US]

IAA Internationale Automobil-Ausstellung [International Automobile Exhibition/Exposition, Frankfurt/Main, Germany]

IAA Inventors Association of America [US]

IAA Isoamyl Acetate

IAA Isoamyl Amine

IAAA International Airforwarders and Agents Association [US]

IAAB Inter-American Association of Broadcasters

IAAC Inter-American Accreditation Corporation

IAAC International Agricultural Aviation Center [UK]

IAAC International Air Cargo Consolidators

IAAC International Antarctic Analysis Center

IAAC International Association for Analog Computation [now International Association for Mathematics and Computers in Simulation]

IAAC Israel Association for Automatic Control

IAACC Ibero-American Association of Chambers of Commerce [Colombia]

IAACC Inter-Allied Aeronautical Commission of Control

IAAE Institution of Automotive and Aeronautical Engineers [now Society of Automotive Engineers of Australasia]

IAAE International Association of Agricultural Economists [US]

IAAEES International Association for the Advancement of Earth and Environmental Sciences [US]

IAAF International Agricultural Aviation Foundation [US]

IAAGR Institute of Arctic and Alpine Geochronological Research [of University of Colorado, Boulder, US]

IAAI International Airports Authority of India

IAAI International Association of Arson Investigators [US]

IAAIP Inter-American Association of Industrial Property [Argentina]

IAAJ International Association of Agricultural Journalists

IAALD International Association of Agricultural Librarians and Documentalists [Netherlands]

IAAM International Association of Auditorium Managers [US]

IAAM International Association of Automotive Modelers

IAAO International Association of Assessing Officers [US]

IAAPEA International Association Against Painful Experiments on Animals

IAAR Institute of Arctic and Alpine Research [of University of Colorado, Boulder, US]

IAARC International Administrative Aeronautical Radio Conference [UK]

IAARC International Association of Automation and Robotics in Construction

IAAS Incorporated Association of Architects and Surveyors [UK]

IAAS International Association of Agricultural Students [Sweden]

IAASE Inter-American Association of Sanitary Engineering [later IAASEE, US]

IAASEE Inter-American Association of Sanitary and Environmental Engineering [now IAASEES, US]

IAASEES Inter-American Association of Sanitary Engineering and Environmental Sciences [US]

IAASM International Academy of Aviation and Space Medicine [Portugal]

IAASP International Association of Airport and Seaport Police

IAAW&D Inspector of Anti-Aircraft Weapons and Devices [UK]

IAB Initiation Area Discriminator

IAB Inter-American Association of Broadcasters [Name Changed to International Association of Broadcasting, Uruguay]

IAB International Aquatic Board

IAB International Association of Broadcasting [Uruguay]

IAB Internet Architecture Board

IAB Interrupt Address to Bus [Computers]

IAB Intramural Advisory Board [of National Institutes of Health, Bethesda, Maryland, US]

IAB Isoamyl Benzoate

IAB Isoamyl Butyrate

IABA International Association of Aircraft Brokers and Agents [Norway]

IABC Insulation Applicators Association of British Columbia [Canada]

IABC International Association of Business Communicators [US]

IABD Ion-Assisted Beam Deposition

IABG Industrieanlagen Betriebsgesellschaft [Industrial Plant Operating Company, Ottobrunn, Germany]

IABG International Association of Botanic Gardens

IABLA Inter-American Bibliographical Library Association

IABM International Association of Broadcasting Manufacturers [UK]

IABO International Association of Biological Oceanography [France]

IABRM International Association for Bear Research and Management [Canada]

IABS International Association for Biological Standardization [Switzerland]

IABSE International Association for Bridge and Structural Engineering [Switzerland]

IABSOIW International Association of Bridge, Structural and Ornamental Iron Workers

IABTI International Association of Bomb Technicians and Investigators [US]

IAC Idle Air Control (Valve)

IAC Image Analysis Computer

IAC Indian Airlines Corporation

IAC Induced Auger Channeling

IAC Industries Assistance Commission [now Industry Commission, Australia]

IAC Industry Advisory Committee [US]

IAC Industry Advisory Committee (on Surveying and Mapping) [Queensland, Australia]

IAC Information Analysis Center [of US Department of Defense]

IAC Infrastructure Assurance Center [of Argonne National Laboratory, Illinois, US]

IAC Instituto de Astrofisica de Canarias [Institute of Astrophysics of the Canaries, Tenerife, Canary Islands, Spain]

IAC Instrument-on-a-Card

IAC Integrated Avionics Computer

IAC Intelligent Asynchronous Controller

IAC Interactive Array Computer

IAC Interface Analysis Center [of University of Bristol, UK]

IAC Integrated Avionics Computer (System)

IAC Integration, Assembly, Checkout

IAC Inter-American Conference

IAC Inter-American Council [of Organization of American States]

IAC (Macintosh) Inter-Application Communication

IAC Interference Absorption Circuit

IAC Interim Acceptance Criteria

IAC International Academy of Ceramics

IAC International Accounting Center

IAC International Activities Committee [of Minerals, Metals and Materials Society of American Institute of Mining, Metallurgical and Petroleum Engineers, US]

IAC International Advertising Consultants Limited [Toronto, Canada]

IAC International Advisory Committee

IAC International Aerological Commission

IAC International Agricultural Center

IAC International Agricultural Club [of US Department of Agriculture]

IAC International Algebraic Compiler

IAC International Alloy Conference

IAC International Analysis Code [Meteorology]

IAC International Apple Core

IAC International Association for Cybernetics [Namur, Belgium]

IAC Interrupted Accelerated Cooling [Metallurgy]

IAC Ion-Assisted Coating

IAC Isoamyl Caprate

IAc Iodoacetic Acid

IACA Inter-American College Association [US]

IACA International Air Carrier Association [Belgium]

IACA International Association for Classical Archeology [Italy]

I-ACAC Inter-American Commercial Arbitration Commission [US]

IACAHP Inter-African Advisory Committee for Animal Health and Production [Kenya]

IACB Inter-Agency Consultative Board

IACB International Advisory Committee on Bibliography [of UNESCO]

IACB International Association of Convention Bureaus [US]

IACBC International Advisory Committee on Biological Control

IACBDT International Advisory Committee on Bibliography, Documentation and Terminology [of UNESCO]

IACC IIM (Indian Institute of Metals)–ASM (American Society for Materials International) Cooperation Committee

IACC International Agricultural Coordination Commission

IACC International Association for Cell Cultures

IACC International Association of Cereal Chemistry

IACC International Association of Conference Centers [US]

IACChE Inter-American Confederation for Chemical Engineering

IACCI International Association of Computer Crime Investigators

IACCP Inter-American Council of Commerce and Production

IACD International Association of Clothing Designers

IACDT International Advisory Committee for Documentation and Technology [of UNESCO]

IACE International Association for Computing in Education [US]

IACED Interagency Committee on Environment and Development

IACH International Association for Colon Hydrotherapy

IACHE International Association of Cylindrical Hydraulic Engineers [US]

IACHR Inter-American Commission on Human Rights [of Organization of American States]

IACI Inter-American Children's Institute [of Organization of American States]

IA/CI Industry Applications/Cement Industry [Institute of Electrical and Electronics Engineers Committee, US]

I-Acid 7-Amino-4-Hydroxy-2-Naphthalenedisulfonic Acid; 6-Amino 1,3-Naphthalenedisulfonic Acid [also I-acid]

IACIT International Association of Conference Interpreters and Translators

IACITC International Advisory Committee of the International Teletraffic Congress [Denmark]

IACM International Association for Computational Mechanics

IACME Inter-American Committee for Mathematical Education

IACOMS International Advisory Committee on Marine Sciences

IACORDS International Association of Cold Region Development Studies

IACP Institute of Automation and Control Processes [of Academy of Sciences of the USSR]

IACP International Association of Chiefs of Police

IACP International Association of Computer Programmers

IACPS Inter-American Committee on Peaceful Settlement [of Organization of American States]

IACR Institute of Arable Crops Research [Rothamsted, UK]

IACR Inter-American College for Radiology

IACR International Association for Cryptologic Research [US]

IACRDVT Inter-American Center for Research and Documentation on Vocational Training [Uruguay]

IACRS Inter-Agency Committee on Remote Sensing

IACS Indian Association for the Cultivation of Science [India]

IACS Inertial Attitude Control System

IACS Integrated Armament Control System

IACS International Annealed Copper Standard

IACS International Association of Classification Societies [UK]

%IACS Percent International Annealed Copper Standard

IACSS International Association for Computer Systems Security [US]

IACST International Association for Commodity Science and Technology

IACT Institute for Applied Composites Technology [US]

IACT Inter-Association Commission on Tsunami [of International Union of Geodesy and Geophysics, Belgium]

IACUC Institutional Animal Care and Use Committee [US]

IACW Inter-American Commission of Women [of Organization of American States]

IAD Immediate Action Directive

IAD Industrial Aerodynamics

IAD Initial Address Designator

IAD Initiation Area Discriminator

IAD Integrated Automatic Documentation

IAD Interatomic Distance

IAD Interface Agreement Document

IAD Interface Analysis Document

IAD Internal Aerodynamics

IAD International Academy of Design

IAD Internationale Arbeitsgemeinschaft für Donauforschung [International Working Association for Danube Research, Austria]

IAD International Astrophysical Decade

IAD Inventory Available Date

IAD Ion Angular Distribution

IAD Ion-Assisted (Thin-Film) Deposition

IAD Ion-Beam Assisted Deposition [also IBAD]

IADA Independent Automotive Damage Appraisers Association [US]

IADB Inter-American Defense Board [of Organization of American States]

IADB Inter-American Development Bank [US]

IADC Inter-American Defense College

IADC Interdepartmental Advisory and Development Committee

IADC International Association of Dredging Companies [Netherlands]

IADC International Association of Drilling Contractors [US]

IADC Internationale Arbeitsgemeinschaft der Archiv-, Bibliotheks- und Graphikrestauratoren [International Association for Conservation of Books, Paper and Archival Material, Germany]

IADES Integrated Attitude Detection and Estimation System

IADIC Integration Analog-to-Digital Converter

IADIS Irish Association for Documentation and Information Services

IADIWU International Association for the Development of International World Universities [France]

IADL Instrumental Activity of Daily Living [Medicine]

IADLs Instrumental Activities of Daily Living [Medicine]

IADN Integrated Atmospheric Deposition Network [Canada/US]

IADP INTELSAT (International Telecommunications Satellite Organization) Assistance and Development Program [US]

IADP Intensive Agricultural District Program [India]

IADPC Inter-Agency Data Processing Committee

IADPPNW International Architects, Designers and Planners for the Prevention of Nuclear War

IADR International Association for Dental Research [US]

IADS International Agricultural Development Service [of US Department of Agriculture]

IAE Institut d'Administration des Entreprises [Institute of Business Administration, France]

IAE Institute for the Advancement of Engineering [US]

IAE Institute of Atomic Energy [PR China]

IAE Institute of Atomic Energy [of Kyoto University, Japan]

IAE Institute of Automobile Engineers [UK]

IAE Integral Absolute Error

IAE International Academy of Education

IAE International Atomic Exposition

IAE Ion-Beam Assisted Etching [also IBAE]

IAE Irregular Atomic Environment

IAEA Institute of Automotive Engineer Assessors

IAEA Inter-American Education Association

IAEA International Agricultural Exchange Association [Scotland]

IAEA International Atomic Energy Agency [of United Nations in Vienna, Austria]

IAEA International Atomic Energy Authority

IAEA Bull IAEA Bulletin [Quarterly Journal of the International Atomic Energy Agency]

IAEAC International Association of Environmental Analytical Chemistry [Switzerland]

IAEA-UNESCO International Atomic Energy Agency–United Nations Educational, Scientific and Cultural Organization

IA-ECOSOC Inter-American Economic and Social Council [of Organization of American States]

I-AEDANS Iodoacetamidoethyl Aminonaphthalenesulfonic Acid; N-Iodoacetyl-N'-(Sulfo-1-Naphthyl)ethylenediamine

1,5-I-AEDANS N-(Iodoacetylaminoethyl)-5-Naphthylamine-1-Sulfonic Acid; N-Iodoacetyl-5-Ethyldiamine 1-Naphthalenesulfonic Acid

1,8-I-AEDANS N-(Iodoacetylaminoethyl)-8-Naphthylamine-1-Sulfonic Acid; N-Iodoacetyl-8-Ethyldiamine 1-Naphthalenesulfonic Acid

IAEE International Association for Earthquake Engineering [Japan]

IAEE International Association for Energy Economics [US]

IAEE International Automotive Engineering Exposition

IAEE Israel Association of Environmental Engineers

IAEEA International Association for the Evaluation of Educational Achievement

IAEE Newsl International Association for Energy Economics Newsletter [US]

IAEG International Archive of Economic Geology [of University of Wyoming, US]

IAEG International Association for Engineering Geology [France]

IAEI International Association of Electrical Inspectors [US]

IAEKM International Association of Electronic Keyboard Manufacturers [US]

IAEL International Association of Electrical Leagues [now International League of Electrical Association, US]

IAEMS International Association of Environmental Mutagen Societies [Finland]

IAEO International Atomic Energy Organization [of United Nations]

IA/ES Industry Applications/Energy Systems [Institute of Electrical and Electronics Engineers Committee, US]

IAES Institute of Aeronautical Sciences [UK]

IAES International Academy for Environmental Safety

IAES Ion-Excited Auger Electron Spectroscopy; Ion-Induced Auger Electron Spectroscopy

IAESP Instituto Agronomico de Estado de São Paulo [Agricultural Institute of the State of Sao Paulo, Brazil]

IAESR Institute of Applied Economic and Social Research

IAESTE International Association for the Exchange of Students for Technical Experience

IAET Irregular Atomic Environment Type

IAETL International Association of Environmental Testing Laboratories [US]

IAEVG International Association for Educational and Vocational Guidance [Northern Ireland]

IAEVI International Association for Educational and Vocational Information

IAF Indirect-Arc Furnace [Metallurgy]

IAF Industrial Air Filtration

IAF Initial Approach Fix [Aeronautics]

IAF Institut Armand-Frappier [Laval, Quebec, Canada]

IAF (Fraunhofer-)Institut für Angewandte Festkörperphysik [(Fraunhofer) Institute for Applied Solid-State Physics, Freiburg, Germany]

IAF Institute for Alternative Futures [US]

IAF Interactive Facility

IAF Inter-American Foundation

IAF International Aeronautical Federation

IAF International Aquaculture Foundation [US]

IAF International Astronautical Federation [France]

IAF Intragranular Acicular Ferrite [Metallurgy]

IAF Iodoacetamidofluorescein

5-IAF 5-(Iodoacetamido)fluorescein

IAFC Inter-American Freight Conference [Argentina/Brazil/US]

IAFC International Association of Fire Chiefs [US]

IAFCI Inter-American Federation of the Construction Industry [Mexico]

IAFE International Association of Fairs and Exhibitions [US]

IAFE Instituto de Astronómia et Fisica Espaciales [Institute for Space Astronomy and Physics, Argentina]

IAFES International Association for the Economics of Self-Management

IAFF International Association of Fire Fighters

IAFG Ion-Beam-Assisted Film Growth

IAFMM International Association of Fish Meal Manufacturers [UK]

IAFP International Association for Financial Planning

IAFR Institute of Applied Forth Research [Rochester, New York State, US]

IAFraMCoS International Association of Fracture Mechanics of Concrete Structures

IAFS Integrated Air-Fuel Systems (Facility) [Windsor, Ontario, Canada]

IAFS International Association for Fire Safety

IAFS International Association of Forensic Sciences [Canada]

IAFSS International Association for Fire Safety Science

IAFWA International Association of Fish and Wildlife Agencies [US]

IAG IFIP (International Federation for Information Processing) Administrative Group [US]

IAG Instruction Address Generation

IAG Interagency Agreement Group

IAG Interagency Group

IAG International Applications Group [of International Federation for Information Processing]

IAG International Association of Geodesy [Denmark]

IAG Internationale Arbeitsgemeinschaft für den Unterrichtsfilm [International Association for Educational Films, Germany]

IAGA International Association of Geomagnetism and Aeronomy [Scotland]

IAGBN International Absolute Gravity Base Network

IAGC Instantaneous Automatic Gain Control

IAGC International Association of Geochemistry and Cosmochemistry [Canada]

IAGC International Association of Geophysical Contractors [US]

IAGLP International Association of Great Lakes Ports

IAGLR International Association for Great Lakes Research [US]

IAGOD International Association on the Genesis of Ore Deposits

IAGP International Antarctic Glaciological Project [US]

IAgr Institut Agronomique [Institute of Agriculture]

IAgrE Institution of Agricultural Engineers [UK]

IAGM Institute of Agricultural Management [also I Agr M] [Europe]

IAGM Instituto Avanzados de Geologia y Mineralogica [Institute of Advanced Geology and Mineralogy, Granada, Spain]

IAGM International Association of Garment Manufacturers

I Agr M Institute of Agricultural Management [also IAGM] [Europe]

IAGS Inter-American Geodetic Survey [[US]

IAgSA Institute of Agricultural Secretaries and Administrators [UK]

IAH Inter-American Highway

IAH International Association of Hydrogeologists

IAH Intersite Atomic Hopping [Physics]

IAHB International Association of Human Biologists [UK]

IAHE International Association for Hydrogen Energy [US]

IAHP International Association of Horticultural Producers

IAHR International Association for Hydraulic Research [Netherlands]

IAHR Ice Symp Proc International Association for Hydraulic Research–Ice Symposium Proceedings

IAHS International Academy of the History of Science [France]

IAHS International Association for Housing Science

IAHS International Association of Hydrological Sciences [formerly International Association of Scientific Hydrology, Netherlands]

IAHS Publ International Association of Hydrological Sciences Publication [Netherlands]

IAI Industrial Arts Index

IAI Informational Acquisition and Interpretation

IAI Initial Address Information

IAI International Apple Institute [US]

IAI International Alliance for Interoperability

IAI International Association for Identification [US]

IAI International Automotive Institute [Monaco]

IAIA International Association for Impact Assessment [US]

IAIABC International Association of Industrial Accident Boards and Commissions [US]

IAIALAR Ibero-American Institute of Agrarian Law and Agrarian Reform [Venezuela]

IAIAS Inter-American Institute of Agricultural Sciences [of Organization of American States]

IAIBS Ion-Assisted Ion-Beam Sputtering

IA/IC Industry Applications/Industrial Control [Institute of Electrical and Electronics Engineers Committee, US]

IAICU International Association of Independent Colleges and Universities

IAIDPA International Association of Information and Documentation in Public Administration [Belgium]

IAIE International Association for Intercultural Education

IAII Inter-American Indian Institute [of Organization of American States]

IAIIB International Association of Independent Information Brokers [now Association of Independent Information Professionals, US]

IAIN International Association of Institutes of Navigation [UK]

IAIS Industrial Aerodynamics Information Service [UK]

IAIS International Association of Independent Scholars [US]

IAK Internet Access Kit

IAL Icelandic Air Lines

IAL Immediate Action Letter

IAL Instrument Approach and Landing Chart

IAL International Algebraic Language [now Algorithmic (Oriented) Language]

IAL International Algorithmic Language

IAL Investment Analysis Language

IALA International Association of Lighthouse Authorities [France]

IALC International Aluminum-Lithium Conference

IALC International Arid Lands Consortium

IALD International Association of Lighting Designers [US]

IALE Instrumented Architectural Level Emulation

IALE International Association of Landscape Ecology

IALL International Association for Labour Legislation [US]

IALL International Association of Law Libraries

IALM International Association of Lighting Maintenance [now National Association of Lighting Management Companies, US]

i-AlCuFe Icosahedral Aluminum Copper Iron (Quasicrystal)

i-AlMn Icosahedral Aluminum Manganese (Quasicrystal)

i-AlPdMn Icosahedral Aluminum Palladium Managanese (Quasicrystal)

IALS International Agency Liaison Service [of Food and Agricultural Organization]

IALS International Association of Legal Science

IAM Indefinite Admittance Matrix

IAM Index Access Method

IAM Initial Address Message

IAM Institute for Advanced Materials [of Joint Research Center of the European Commission at Petten, Netherlands]

IAM Institute of Administrative Management [Orpington, Kent, UK]

IAM Institute of Agricultural Medicine [Lublin, Poland]

IAM Institute of Aviation Medicine [UK]

IAM Institut für Angewandte Mikroelektronik [Institute for Microelectronics, Braunschweig, Germany]

IAM Interactive Algebraic Manipulation

IAM Intermediate Access Memory

IAM International Academy of Management [US]

IAM International Association of Machinists (and Aerospace Workers)

IAM International Association of Meteorology

IAM Internationale Arbeitsgemeinschaft für Müllforschung [International Waste Research Association, Germany]

IAM Iodine–Arsenic–Manganese (Complex)

IAMA Incorporated Advertising Management Association [UK]

IAMACS International Association for Mathematics and Computers in Simulation [also IAMCS]

IAMAP International Association of Meteorology and Atmospheric Physics [of International Council of Scientific Unions]

IAMAS International Association of Meteorology and Atmospheric Sciences

IAMAW International Association of Machinists and Aerospace Workers

IAMB International Association of Microbiologists

IAMBE International Association of Medicine and Biology of Environment [France]

IAMC Institute for Advancement of Medical Communication [US]

IAMC Institute of Association Management Companies [US]

IAMCA International Association of Milk Control Agencies [US]

IAMCR International Association for Mass Communication Research [UK]

IAMCS International Association for Mathematics and Computers in Simulation [also IAMACS]

IAMDA International Alpha Micro Dealers Association

IAME International Association for Modular Exhibitry

IAMFE International Association on Mechanization of Field Experiments [Norway]

IAMFES International Association of Milk, Food and Environmental Sanitarians [Ames, Iowa, US]

IAMG International Association for Mathematical Geology [US]

IAMG Newsl IAMG Newsletter [of International Association for Mathematical Geology, US]

IAMHIST International Association for Audio-Visual Media in Historical Research and Education [Italy]

IAML International Association of Music Libraries

IAMLO International African Migratory Locust Organization [Mali]

IAMLT International Association of Medical Laboratory Technologists [UK]

IAMM Irish Association of Master Mariners

IAMO Institut für Agrarentwicklung in Mittel– und Osteuropa [Institute for Agricultural Development in Central and East Europe, Halle, Germany]

IAMP Institute for Advanced Materials Processing [of Tohoku University, Sendai, Japan]

IAMP Institute of Atomic and Molecular Physics [Netherlands]

IAMP International Association of Mathematical Physics [US]

IAMP News Bull IAMP News Bulletin [of International Association of Mathematical Physics, US]

IAMQS International Academy of Molecular and Quantum Sciences

IAMR Institute of Applied Manpower Research [India]

IAMR Institute of Arctic Mineral Resources [of University of Alaska, US]

IAMS Institute of Advanced Manufacturing Sciences [Cincinnati, Ohio, US]

IAMS Institute of Archeo-Metallurgical Studies [of Institute of Archeology, University College of London, UK]

IAMS Institute of Atomic and Molecular Science [Taipei, Taiwan]

IAMS Integrated Academic Information Management System

IAMS International Advanced Microlithography Society

IAMS International Association of Microbiological Societies [now International Union of Microbiological Societies, UK]

IAMSLIC International Association of Marine Science Libraries and Information Centers [US]

IAMTEC Institute of Advanced Machine Tool and Control Technology [also IAMTCT, US]

IAMWF Inter-American Mine Workers Federation

IAN Instituto Agrario Nacional [National Institute of Agriculture, Venezuela]

IAN Integrated Acoustic Network

IAN Integrated Analog Network

IAN Isoamyl Nitrate

IANA Internet Assigned Numbers Authority

IANAP Interagency Noise Abatement Program

IANC International Airline Navigators Council

IANC International Anatomical Nomenclature Committee

I and A (Combined) Iron and Alumina [also I&A]

I and C Installation and Checkout [also I&C]

I and C Instrumentation and Communication [also I&C]

I and D Information and Documentation [also I&D]

IANDS International Association for Near-Death Studies [US]

IANEC Inter-American Nuclear Energy Commission [of Organization of American States]

IANAL I Am Not A Lawyer (But...) [Computer Jargon]

IANI International Article Numbering Institute

IANI Israel Agency for Nuclear Information

IANRP International Association of Natural Resource Pilots [US]

IAO In and Out

IAO (Fraunhofer-)Institut für Arbeitswissenschaft und Organisation [(Fraunhofer) Institute for Industrial Science and Organization, Stuttgart, Germany]

IAO Institut Agricole d'Oka [Oka Agricultural Institute; of University of Montreal, Canada]

IAO Insurers Advisory Organization

IAO Internal Automation Operation

IAOD International Academy of Optimum Dentistry

IAOPA International Council of Aircraft Owner and Pilot Associations [US]

IAOS International Association for Official Statistics [Netherlands]

IAOS Irish Agricultural Organization Society

IAP Image Array Processor

IAP Imaging Atom Probe

IAP Information Associates Program

IAP Inhibitor of Actin Polymerization (Protein) [Biochemistry]

IAP Initial Approach Procedure [Aeronautics]

IAP *(Institut d'Astrophysique)* – French for "Institute of Astrophysics"

IAP Institute of Astrophysics

IAP Institute of Atmospheric Physics [of Russian Academy of Sciences]

IAP Institute of Australian Photographers

IAP Institut für Atmosphärenphysik (an der Universität Rostock) [Institute for Atmospheric Physics (of the University of Rostock, Germany]

IAP Institutional Assistance Program

IAP Institution of Analysts and Programmers [UK]

IAP Integrated Action Plan

IAP Internal Array Processor

IAP International Academy of Pathology

IAP International Association of Photoplatemakers [now International Photographers Association, US]

IAP International Association for Planetology [Belgium]

IAP International Association of Pteridologists

IAP Internet Access Provider

IAP Inventors Assistance Program

IAPA Industrial Accident Prevention Association [Canada]

IAPA Inter-American Press Association

IAPA International Accident Prevention Association

IAPA International Airline Passengers Association

IAPA International Association of Parametric Analysts [US]

IAPBPPV International Association of Plant Breeders for the Protection of Plant Varieties [Switzerland]

IAPC Inter-American Peace Committee [of Organization of American States]

IAPC International Association for Pollution Control

IAPC International Auditing Practices Committee

IAPCO International Association of Professional Congress Organizers [Belgium]

IAPD Institut für Angewandte Physik Dresden [Dresden Institute for Applied Physics, TU Dresden, Germany]

IAPDOI Italian Association for the Production and Distribution of On-Line Information

IAPH International Association of Ports and Harbors [Japan]

IAPHC International Association of Printing House Craftsmen [US]

IAPIP International Association for the Protection of Industrial Property

IAPM International Academy of Preventive Medicine

IAPM Irish Assocxiation of Paper Merchants [UK]

IAPMO International Association of Plumbing and Mechanical Officials [US]

IAPN International Association of Professional Numismatists

IAPO International Association of Physical Oceanography [now International Association for the Physical Sciences of the Ocean, US]

IAPP Indian Association for Plant Physiology

IAPP International Association for Plant Physiology [Australia]

IAPP Islet Amyloid Polypeptide [Biochemistry]

IAPR International Association for Pattern Recognition [UK]

IAPRI International Association of Packaging Research Institutes

IAPS International Association for the Properties of Steam [US]

IAPS Ion Appearance Potential Spectroscopy

IA/PSE Industry Applications/Power Systems Engineering [Institute of Electrical and Electronics Engineers–Power Engineering Society, US]

IAPSO International Association for the Physical Sciences of the Ocean [US]

IAPT International Association for Plant Taxonomists

IAPU Image Array Processing Unit

IAQ Indoor Air Quality

IAQ International Academy for Quality [Germany]

IAQ International Association for Quality

IAQC International Association for Quality Circles [US]

IAQDE Independent Association of Questioned Document Examiners [US]

IAQR International Association on Quaternary Research

IAR Institute for Aerospace Research [of National Research Council of Canada]

IAR Institute of Atomic Research [of US Department of Energy at Iowa State University, Ames, US]

IAR Instruction Address Register

IAR Interrupt Address Register

IAR Intersection of Air Routes

IAR Ion-to-Atom Arrival Ratio

IAR Isobaric Analog Resonance [Physics]

i(αr) modified spherical Bessel function of the first kind [Symbol]

IARA International Aerosol Research Assembly

IARC Independent Assessment and Research Center [UK]

IARC International Agency for Research on Cancer [of World Health Organization; located at Lyons, France]

IARC International Agricultural Research Center [of Consultative Group on International Agricultural Research, US]

IARD Information Analysis and Retrieval Division [of American Institute of Physics, US]

IARI Indian Agricultural Research Institute [Delhi]

IARIGAI International Association of Research Institutes for the Graphic Arts Industry [UK]

IARIW International Association for Research on Income and Wealth

IARM International Association of Ropeway Manufacturers

IARMCLRS International Agreement Regarding the Maintenance of Certain Lights in the Red Sea [UK]

IARR International Association for Radiation Research [Netherlands]

IARS Independent Air Revitalization System

IARU International Amateur Radio Union

IARUS International Association for Regional and Urban Statistics [Netherlands]

IARW International Association of Refrigerated Warehouses [US]

IAS Image Analysis System

IAS Immediate Access Storage

IAS Impact Assessment Sheet

IAS Impact Assessment Study

IAS Indiana Academy of Science [US]

IAS Indian Academy of Sciences [Bangalore, India]

IAS Indicated Air Speed

IAS Industry Analysis Division [of US Department of Commerce]

IAS Industry Applications Society [of Institute of Electrical and Electronics Engineers, US]

IAS Institute for Advanced Studies

IAS Institute for Advanced Study [Princeton, New Jersey, US]

IAS Institute for Aerospace Studies [of University of Toronto, Canada]

IAS Institute for Atmospheric Sciences [of Environmental Science Services Administration, US]

IAS Institute of Advanced Studies [US Army]

IAS Institute of Advanced Studies [of Australian National University, Canberra]

IAS Institute of Aeronautical Sciences [US]

IAS Institute of Aerospace Sciences [Now Part of American Institute of Aeronautics and Astronautics, US]

IAS Institute of Agricultural Sciences [of Washington State University, Pullman, US]

IAS Instituto Argentino de Siderurgia [Argentinian Institute of Iron and Steel Industries]

IAS Instrument Approach System [Aeronautics]

IAS Integrated Analytical System

IAS Integrated Aquatic Systems

IAS Integrated Avionics System

IAS Interactive Applications Supervisor

IAS Interactive Applications System

IAS International Accountants Society

IAS International Accounting Standards

IAS International Association of Sedimentologists [Denmark]

IAS International Association of Siderographists

IAS International Audio-Visual Society [US]

IAS International Auditing Standards

IAS Iowa Academy of Sciences [Cedar Falls, US]

IAS Irish Archeological Society

IAS Islamic Academy of Sciences

IAS Isobaric Analog State [Nuclear Physics]

IASA Insurance Accounting and Statistical Association

IASA International Air Safety Association

IASA International Alliance of Sustainable Agriculture [US]

IASB International Academy at Santa Barbara [California, US]

IASB International Aircraft Standards Bureau

IASC Indexing and Abstracting Society of Canada

IASC Inter-American Safety Council

IASC International Accounting Standards Committee [UK]

IASC International Association for Statistical Computing [Netherlands]

IASC International Association of Science Clubs

IASC International Association of Seed Crushers [UK]

IASCB Ibero-American Society for Cell Biology [Chile]

IASCC Irradiation-Assisted Stress Corrosion Cracking

IASD Industrial Automation Services Division

IASF Instrumentation in Aerospace Simulation Facilities [Institute of Electrical and Electronic Engineers–Aerospace and Electronic Systems Society Committee]

IASF International Atlantic Salmon Foundation

IASH International Association of Scientific Hydrology [now International Association of Hydrological Sciences, Netherlands]

IASI Improved Atmospheric Sounding Interferometer

IASI Inter-American Statistical Institute [of Organization of American States]

i-a-Si:H Intrinsic Hydrogenated Amorphous Silicon

IASLIC Indian Association for Special Libraries and Information Centers

IASM International Association of Structural Movers [US]

IASMART International Association of Structural Mechanics and Reactor Technology

IASMS Institute for Advanced Space Materials and Structures [US]

IASOS Institute of Antarctic and Southern Ocean Studies

IASP Initial Average Sustained Pressure [Aerospace]

IASP International Association for the Study of Pain

IASPEI International Association of Seismology and Physics of the Earth's Interior [UK]

IASPS International Association for Statistics in Physical Sciences

IASR Interruption Address Storage Register

IASS International Association for Shell and Spatial Structures [Spain]

IASS International Association of Security Service [US]

IASS International Association of Soil Science

IASS International Association of Survey Statisticians [France]

IASS Inverter/ATCS (Active Thermal Control System) Support Structure

IASSAR International Association of Structural Safety and Reliability

IASSIST International Association for Social Science Information Service and Technology [Canada]

IAST Ideal Adsorbed Solution Theory

IAST Institute for Advanced Science and Technology [Osaka Prefecture University, Osaka, Japan]

IASTED International Association of Science and Technology for Development [Canada]

IASU International Association of Satellite Users [now IASUS, US]

IASUS International Association of Satellite Users and Suppliers [US]

IASY International Active Sun Years [Astronomy]

IAT Import Address Table [Computers]

IAT Information Assessment Team

IAT Internal Average Temperature

IAT Inside Air Temperature

IAT Institute for Advanced Technology [of National Institute for Standards and Technology, Gaithersburg, Maryland, US]

IAT Institute for Applied Technology

IAT Institute of Advanced Technology

IAT Institute of Animal Technology [UK]

IAT Institute of Asphalt Technology [UK]

IAT Institute of Automatics and Telemechanics [Russia]

IAT Integrated Avionics Test

IAT International Air Transport

IAT International Association for Timekeeping

IAT International Atomic Time

IAT International Automatic Time

IAT Ion-Assisted Texturing [Materials Science]

IATA International Air Transport Agreement

IATA International Air Transport Association [Switzerland]

IATA International Appropriate Technology Association [US]

IATA International Association of Trade Associations

IATAE International Accounting and Traffic Analysis Equipment

IATAL International Association of Theoretical and Applied Limnology [US]

IATAP Intramural AIDS Targeted Antiretroviral Program [US]

IATC International Air Traffic Communications

IATC International Association of Tool Craftsmen

IATEL International Association of Testing and Environmental Laboratories

IATM International Association for Testing Materials

IATM International Association of Transport Museums [Switzerland]

IATME International Association of Terrestrial Magnetism and Electricity [US]

IATN International Association of Telecomputer Networks [US]

IATP International Agricultural Training Program

IATSS International Association of Traffic Safety Sciences

IATTC Inter-American Tropical Tuna Commission [US]

IATU Inter-American Telecommunications Union [US]

IATUL International Association of Technical University Libraries [Sweden]

IATUL Q IATUL Quarterly [Published by Oxford University Press, UK]

IAU Infrastructure Accounting Unit

IAU Institute for American Universities [France]

IAU Inter-American University [San Germán, Puerto Rico]

IAU International Academic Union [Belgium]

IAU International Association of Universities [France]

IAU International Astronomical Union

IAU International Astronomical Unit

IAUP International Association of University Presidents [Thailand]

IAUPD International Association for Urban Planning and Design

IAUPL International Association of University Professors and Lecturers [France]

IAUPR Inter-American University of Puerto Rico [San Germán]

IAUR Istituto di Automatica dell'Universita di Roma [Institute of Automation of the University of Rome, Italy]

IAUSD Inter-American Union for Scientific Development

IAV International Association for Video–VIDION [US]

IAV International Association of Volcanology

IAV Inventory Adjustment Voucher

IAVB International Association of Visceral Biomechanics

IAVC Instantaneous Automatic Video Control

IAVC Instantaneous Automatic Volume Control

IAVC International Audiovisual Center

IAVCEI International Association of Volcanology and Chemistry of the Earth's Interior [Germany]

IAVFH International Association of Veterinary Food Hygiene

IAVG International Association for Vocational Guidance

IAVGO Industrial Accident Victims Group of Ontario [Canada]

IAVS International Association for Vegetation Science [Germany]

IAVSD International Association for Vehicle Systems Dynamics [Netherlands]

IAVTC International Audio-Visual Technical Center

IAW In Accordance With [also iaw]

IAWA International Association of Wood Anatomists [Netherlands]

IAWA Bull IAWA Bulletin [of International Association of Wood Anatomists, Netherlands]

IAWCM International Association of Wiping Cloth Manufacturers [US]

IAWE International Association for Wind Engineering [Canada]

IAWGD Inter-Agency Working Group on Desertification

IAWL International Association of Water Law [Italy]

IAWPR International Association on Water Pollution Research [now IAWPRC, UK]

IAWPRC Indian Association on Water Pollution Research and Control

IAWPRC International Association on Water Pollution Research and Control [UK]

IAWR Internationale Arbeitsgemeinschaft der Wasserwerke im Rheineinzugsgebiet [International Association of Waterworks in the Rhine Basin Area, Netherlands]

IAWR Institute for Air Weapons Research

IAWRBA International Association of Waterworks in the Rhine Basin Area [Netherlands]

IAWS International Academy of Wood Science

IAWS Irish Agricultural Wholesale Society

IAYSEP Inter-American Young Scientist Exchange Program [of American Chemical Society, US]

IB Ice Box

IB Identifier Block [Computers]

IB Inboard

IB In Bond [Customs]

IB Inbound

IB Incendiary Bomb

IB Incentive Based

IB Incoherent Backscatter [Physics]

IB Induction Balance

IB Induction Brazing

IB Inert Building

IB Infinite Baffle [Acoustics]

IB Information Bank

IB Information Block

IB Information Bulletin

IB Information Bureau

IB Input Buffer

IB Input Bus

IB Institute of Bankers [London, UK]

IB Institute of Biology [UK]

IB Instruction Book

IB Instruction Bus

IB Instrumental Broadening [Physics]

IB Interface Board

IB Interface Bus

IB Intermediate Boson [Physics]

IB Internal Bus

IB International Baccalaureate

IB Inversion Boundary

IB Iodobenzene

IB Ion Beam

IB Ion Bombardment

IB Ionic Bond(ing) [Physical Chemistry]

I&B Ingram and Bell Limited [Canadian Pharmaceutical Manufacturer]

ib *(ibidem)* – in the same place

IBA Independent Broadcasting Authority [of Independent Television, London, UK]

IBA Indole-3-Butanoic Acid

IBA Indole-3-Butyric Acid

IBA Industrial Biotechnology Association [US]

IBA Institute for Briquetting and Agglomeration [US]

IBA Institute of British Architects

IBA Institute of Business Appraisers [US]

IBA Interacting Boson Approximation [Physics]

IBA International Bankers Association [US]

IBA International Bauxite Association [Jamaica]

IBA International Biographical Association

IBA International Biometric Association [US]

IBA International Broadcasting Authority

IBA International Bryozoology Association [France]

IBA International Conference on Ion Beam Analysis

IBA Investment Bankers Association

IBA Ion Beam Analysis (Technique)

IBA Isobornyl Acetate

IBA Isobutanol

IBA Isobutyl Acetate

IBA Isobutylamine [also Iba]

IBA Isobutyric Acid

IBAA International Bankers Association of America [US]

IBAA International Business Aircraft Association [US]

IBAC International Business Aviation Council [US]

IBAD Ion-Beam-Assisted Deposition [also IAD]

IBAE Institute of British Agricultural Engineers

IBAE Ion-Beam-Assisted Etching [also IAE]

IBAFG Ion-Beam-Assisted Film Growth

IBAH Inter-African Bureau for Animal Health

IBAHP Inter-African Bureau for Animal Health and Production [Kenya]

IBAM Institute of Business Administration and Management [Japan]

IBAM Instituto Brasileiro de Administração Municipal [Brazilian Institute for Municipal Administration]

IBAMS Ion-Beam-Assisted Magnetron Sputtering

IBAN Institut Belge pour l'Alimentation et la Nutrition [Belgian Institute for Food and Nutrition]

IBAP Ion-Beam-Assisted Polishing

IBA Q Rev IBA Quarterly Review [of International Bauxite Association, Jamaica]

IBA Rep IBA Reports [of Industrial Biotechnology Association, US]

IBA Rev IBA Review [of International Bauxite Association, Jamaica]

IBAS Ion-Beam Analysis System

IBA Tech Rev IBA Technical Review [of Independent Broadcasting Authority, UK]

IBB International Brotherhood of Boilermakers

IBB Invest in Britain Bureau [UK]

IBB Isobutyl Benzoate

IBBA Inland Bird Banding Association [US]

IBBD Instituto Brasileiro de Bibliografia e Documentação [Brazilian Institute of Bibliography and Documentation]

IBBH Internationaler Bund der Bau– und Holzarbeiter [International Federation of Building and Wood Workers, Switzerland]

IBBM Iron Body Brass Mounted

IBBM Iron Body Bronze Mounted

IBBS Integrated Broadband Service

IBC Industrialized Building Commission

IBC Initial Boundary Condition

IBC Insect Biotech Canada (Network)

IBC Institute for Building Control [UK]

IBC Institute of Business Counsellors [UK]

IBC Institutional Biosafety Committee [US]

IBC Instituto Bacteriológico de Chile [Chilean Institute of Bacteriology]

IBC Instituto Brasileiro de Café [Brazilian Institute of Coffee]

IBC Instrument Bus Computer

IBC Integrated Block Controller

IBC International Banking Center

IBC International Biographical Center [UK]

IBC International Biometric Conference [US]

IBC Institutional Biosafety Committee

IBC International Botanical Congress

IBC International Boundary Commission [US]

IBC International Broadcasting Convention

IBC International Business Center [US]

IBC Isobutyl Carbinol

IBCA Industry Bar Code Alliance [US]

IBCC International Building Classification Committee [Netherlands]

IBCC International Business Council of Canada

IBCCSE International Board of Computing in Civil and Structural Engineering

IBCD Ion-Beam Controlled Deposition

IBCG Internationaler Bund Christlicher Gewerkschaften [International Federation of Christian Trade Unions, Germany]

IBCS Intel Binary Compatibility Specification [also iBCS]

IBCS Intel Binary Compatibility Standard

IBCS International Bureau of Commercial Statistics

IBCS2 Intel Binary Compatibility Standard, Version 2

IBD Infectious Bursal Disease

IBD Institute of Business Designers [US]

IBD International Baccalaureate Diploma

IBD Intrinsic (Grain) Boundary Dislocation [Materials Science]

IBD Ion Beam Deposition

IBD Isobutyraladehyde

ibd *(ibidem)* – in the same place

IBDH Inboard Chromatography Data Handler

IBDI International Bureau of Documentation and Information

IBDU Isobutylidene Urea

IBDV Infectious Bursal Disease Virus

IBE Institut Belge de l'Emballage [Belgian Packaging Institute]

IBE Institute of British Engineers

IBE International Bureau of Education [of UNESCO]

IBE Ion Beam Epiplantation

IBE Ion-Beam Etching

IBEC International Bank for Economic Cooperation

IBEC International Body of Engineering Conferences

IBED Ion-Beam-Enhanced Deposition [also IED]

IBEDOC International Bureau of Education Documentation and Information Systems [of UNESCO]

IBEE International Builders Exchange Executives [US]

IBEF International Bio-Environmental Foundation [US]

IBELCO Institut Belge de Coopération Technique [Belgian Institute for Technical Cooperation]

IBEN Incendiary Bomb with Explosive Nose

IBETA Irish Business Equipment and Technology Association [UK]

IBEW International Brotherhood of Electrical Workers

IBEX International Biotechnology Exposition

IBF Incident Bright-Field [Microscopy]

IBF Institute of British Foundrymen

IBF (Fraunhofer-)Institute für Betriebsfestigkeit [(Fraunhofer) Institute for Operational Strength of Materials, Germany]

IBF Institut für Bildsame Formgebung [Institute for Plastic Deformation; of Rheinisch-Westfälische Technische Hochschule, Aachen, Germany]

IBF Internally Blown Flap

IBF International Banking Facility

IBF International Booksellers Federation [Austria]

IBF Ion Beam Facility [of University of Pennsylvania, Philadelphia, US]

IBFA Interacting Boson-Fermion Approximation [Physics]

IBFD International Bureau of Fiscal Documentation

IBFI International Business Forms Industries [US]

IBFMP International Bureau of the Federation of Master Printers

IBFO International Brotherhood of Firemen and Oilers Helpers

IBG Institute of British Geographers

IBG Interblock Gap [Computers]

IBG Internationales Büro für Gebirgsmechanik [International Bureau of Strata Mechanics, Poland]

IBG Iodobenzylguanidine [Biochemistry]

IBGE Instituto Brasileiro de Geografia e Estatistica [Brazilian Institute of Geography and Statistics, Rio de Janeiro]

IBHP Institut Belge des Hautes Pressions [Belgian High Pressure Institute]

IBI Institute of Biology in Ireland

IBI Intelligent Buildings Institute [US]

IBI Intergovernmental Bureau for Informatics [Italy]

IBI Intergovernmental Bureau for Information [of United Nations]

IBI International Broadcasting Institute [now International Institute of Communications, UK]

IBIB Isobutyl Isobutyrate

IBICNZ Instituto Brasileiro de Informação do Chumbo, Niquel e Zinco [Brazilian Institute for Information on Copper, Nickel and Zinc, Sao Paulo, Brazil]

IBICT Instituto Brasileiro de Informação em Ciencia e Tecnológia [Brazilian Institute for Information in Science and Technology, Brazil]

ibid *(ibidem)* – in the same place

IBIEC Ion-Beam-Induced Epitaxial (Re-)Crystallization

IBIIA Ion-Beam-Induced Interfacial Amorphization

IBI-ICC Intergovernmental Bureau for Informatics– International Computation Center

IBIM Instituto de Información y Documentación en Biomedicina [Institute for Biomedical Information and Documentation, Spain]

IBION Issue Based Indian Ocean Network

I-biq 2,2'-Isoquinoline [also i-biq]

IBIS ICAO (International Civil Aviation Organization) Bird-Strike Information System

IBIS Indonesian Biodiversity Information System

IBIS Integrated Blade Inspection System

IBIS Integrated Botanical Information System (Specimen Database of Australian National Botanic Gardens, Canberra]

IBIS Intelligent Business Information System

IBIS Intense-Bunched Ion Source

IBIS International Bank Information System

IBIS International Book Information Service [Netherlands]

IBIS International Bryological Information Service [of University of Duisburg, Germany]

IBIS I/O (Input/Output) Buffer Information Specification [of Electronic Industries Association, US]

IBJ Industrial Bank of Japan

IBL Ion Beam Laboratory

IBLA Inter-American Bibliographical and Library Association

IBLC Internal Boundary Layer Capacitor

IBM Indian Bureau of Mines

IBM Infinite Barrier Model

IBM Injection Blow Molding

IBM Institute of Builders Merchants [UK]

IBM Integrated Background Monitoring

IBM Interacting Boson Model [Physics]

IBM Intercontinental Ballistic Missile

IBM International Business Machines Corporation [US]

IBM Ion-Beam Microanalysis

IBM Ion Beam Mixing

IBM Ion-Beam Modification

IBMA Independent Battery Manufacturers Association [US]

IBM ADSTAR IBM (Corporation) Automatic Document Storage and Retrieval Facility [San José, California, US]

IBMARNR Instituto Brasileiro do Meio Ambiente e das Recursos Naturais Renováveis [Brazilian Institute of Environment and Renewable Natural Resources]

IBM CAD IBM (Corporation) Computer-Aided Design

IBM CGA IBM (Corporation) Color Graphics Adapter

IBM CIM IBM (Corporation) Computer-Integrated Manufacturing

IBM Corp International Business Machines Corporation [US]

IBM-DPD IBM (Corporation)–Data Processing Division [US]

IBMCUA IBM (Corporation) Computer Users Association [US]

IBM-DOS IBM (Corporation) Disk-Operating System

IBME Institute of Biomedical Engineering [of University of Toronto, Canada]

IBME Instituto de Biología y Medicina Experimental [Institute of Experimental Biology and Medicine, Argentina]

IBM-ESD IBM (Corporation)–Entry Systems Division [US]

IBM-GL IBM (Corporation) Graphics Language

IBM J R&D IBM Journal of Research and Development [Published by IBM Corporation, US] [also IBM J Res Devel]

IBML Ion-Beam Materials Laboratory [of Los Alamos National Laboratory, Albuquerque, New Mexico, US]

IBMM International Conference on Ion-Beam Modification of Materials

IBMM Ion-Beam Modification of Materials

IBM Nachr IBM Nachrichten [IBM News; published by IBM Deutschland GmbH, Stuttgart, Germany]

IBM PC IBM (Corporation) Personal Computer

IBM PC AT IBM (Corporation) Personal Computer Advanced Technology

IBM PC ATE IBM (Corporation) Personal Computer Advanced Technology Enhanced

IBM PC ATX IBM (Corporation) Personal Computer Advanced Technology Expanded

IBM PC jr IBM (Corporation) Personal Computer–Junior

IBM PC LAN IBM (Corporation) Personal Computer– Local-Area Network

IBM PC XT IBM (Corporation) Personal Computer Expanded Technology

IBM PS/2 IBM (Corporation) Personal System 2

IBMR International Bureau for Mechanical Reproduction

IBM RT PC IBM (Corporation) Real-Time Personal Computer

IBM Res Rep IBM Research Report [of IBM Corporation, US]

IBMS Ion Beam-Modified Surface

IBM SC IBM (Corporation) Scientific Center [Palo Alto, California, US]

IBM SSD IBM (Corporation) Solid-State Devices [San José, California, US]

IBM Syst J IBM Systems Journal [Published by IBM Corporation, US]

IBMT (Fraunhofer-)Institut für Biomedizinische Technik [(Fraunhofer) Institute for Biomedical Technology, St. Ingbert, Germany]

IBMT International Bureau of Mining and Thermophysics

IBM Tech Discl Bull IBM (Corporation) Technical Disclosure Bulletin

IBM TSS IBM (Corporation) Time-Sharing System

IBMX 3-Isobutyl-1-Methylxanthine

IBN Indigenous Biodiversity Network

IBN Institut Belge de Normalisation [Belgian Standards Institute]

IBN International Biosciences Network

IBN Inverse (Data) Block Number

IBN (Direct) Ion-Beam Nitridation

IBn Identification Beacon [also IBN]

IBNORCA Instituto Boliviano de Normalisación y Calidad [Bolivian Institute for Standards and Calibration]

IBO International Baccalaureate Organization [Switzerland]

IBO International Broadcasting Organization [now International Broadcasting and Television Organization]

IBO Ion-Beam-Induced Oxidation

IBOA Isobornyl Acrylate

IBOL Interactive Business-Oriented Language

IBOMA Isobornyl Methacrylate

IBOP International Brotherhood of Operative Pottery and Allied Workers

IBP Initial Boiling Point [also ibp]

IBP Institute for Business Planning [UK]

IBP International Biological Program [of National Research Council, US]

IBP Interphase Boundary Precipitates [Materials Science]

IBP Ion-Beam Processing

IBPAT International Brotherhood of Painters and Allied Trades

IBPA International Business Press Associates

IBPA Israel Book Publishers Association

IBPBAC Isobutyl-4-(4'-Phenylbenzylideneamino)cinnamate

IBPCS International Bureau for Physicochemical Standards

IBPF Ice-Breaking Permafrost

IBPGR International Board for Plant Genetic Resources [Rome, Italy]

IBPH 2,2'-Isobutylidene-Bis-(4-Methyl-6-tert-Butylphenol)

IBPSA International Building Performance and Simulation Association [France]

IBQ Iced Brine Quenched; Iced Brine Quenching [Metallurgy]

IBR Infectious Bovine Rhinotracheitis (Virus)

IBR Institute for Basic Research [of National Institute for Standards and Technology, Gaithersburg, Maryland, US]

IBR Integral Boiling Reactor [of United Kingdom Atomic Energy Authority]

IBR Integrated Bridge Rectifier

IBr Iodine Monobromide

IBRA Interim Biogeographical Regionalization for Australia (Program) [of Australian Nature Conservation Agency]

IBRA International Bee Research Association [UK]

IBRA Institut Belge pour la Robotique et l'Automation [Belgian Institute for Robotics and Automation]

IBr/Ar Iodine Monobromide/Argon (Mixture)

IBRD International Bank for Reconstruction and Development [of United Nations] [also known as World Bank]

IBRG International Biodeterioration Research Group [UK]

IBRL Initial Bomb Release Line

IBRO International Brain Research Organization

IBRRC International Bird Rescue Research Center [US]

IBS Institute for Basic Standards [of National Institute for Standards and Technology, Gaithersburg, Maryland, US]

IBS INTELSAT (International Telecommunications Satellite Organization) Business Services

IBS Intercollegiate Broadcasting System [US]

IBS Integrated Baseline System

IBS Integrated Broadband Service

IBS Interlibrary Borrowing Service

IBS International Biohydrometallurgy Symposium

IBS International Broadcasters Society

IBS International Business School

IBS Ion Backscattering Spectrometer; Ion Backscattering Spectrometry

IBS Ion-Beam Spectroscopy

IBS Ion-Beam Splitter

IBS Ion-Beam Sputter(ing) (Deposition)

IBS Ion-Beam Synthesis

IBS Irritable Bowel Syndrome

IBSA International Biotechnology Suppliers Association [US]

IBSAC Industrialized Building Systems and Components

IBS Asian Electron News IBS Asian Electronics News [Published by IBS Limited, Taipei, Taiwan]

IBSC Individual Benefits and Services Committee [of Institute of Electrical and Electronics Engineers, US]

IBSCA Ion-Beam Spectrochemical Analysis

IBSD Ion-Beam Sputter Deposited; Ion Beam Sputter Deposition

IBSE Ion Beam Sputter Etching

IBSFC International Baltic Sea Fishery Commission [Poland]

IBSHR Integral Boiling and Superheat Reactor [US]

IBSMW International Brotherhood of Sheet Metal Workers

IBSNAT International Benchmark Sites Network for Agrotechnology Transfers

IBSR Interactive Bibliographic Search and Retrieval

IBSS IPN (Interpenetrating Polymer Network) Build-Up Structure System

IBT Inclined Bottom Tank

IBT Industrial Bio-Test Laboratory [Canada]

IBT Integrated Bipolar Transistor

IBT Intrinsic-Barrier Transistor

IBT Ion-Beam Technology

IBTE Institution of British Telecommunication Engineers [London, UK]

IBTE International Bureau of Technical Education

IBTO International Broadcasting and Television Organization

IBTT International Bureau for Technical Training

IBTTA International Bridge, Tunnel and Turnpike Association [US]

IBTU International Bureau of Transport Users

IBU International Broadcasting Union [Switzerland]

IBu Isobutyl [also i-Bu]

iBu Isobutyric Acid

iBu Isobutyryl

i-Bu Isobutyrate

IBuBz Isobutylbenzene [also iBuBz]

iBu-DMT-dC N^4-Isobutyryl-5'-O-(4,4'-Dimethoxytrityl) 2'-Deoxycytidine [Biochemistry]

iBu-DMT-dG N^2-Isobutyryl-5'-O-(4,4'-Dimethoxytrityl) 2'-Deoxyguanosine [Biochemistry]

IBuMgBr Isobutyl Magnesium Bromide [also iBuMgBr]

IBuO Isobutoxide (or Isobutylate) [also iBuO, or I-BuO]

Ibuprofen α-Methyl-4-(Isobutyl)phenylacetic Acid [also ibuprofen]

Ibuprofen 2-(4-Isobutylphenyl)propionic Acid [also ibuprofen]

IBUPU International Bureau of the Universal Postal Union

IBV Internationale Buchhändler-Vereinigung [International Booksellers Federation, Austria]

IBVE Isobutylvinylether
IBW Impulse Bandwidth
IBW Ion Bernstein Wave [Physics]
IBWA International Bottled Water Association
IBWC International Boundary and Water Commission [US]
IBWM International Bureau of Weights and Measures [France]
IBWS International Bureau of Whaling Statistics
IBX Internal Telephone System [of Battelle, US]
IBY International Biological Year
IBZ Irreducible Brillouin Zone [Solid-State Physics]
IBz Iodobenzene
IC Ice Calorimeter; Ice Calorimetry
IC Ice Cap
IC Ice Crystal
IC Icing Condition [also I/C]
IC Identification Code
IC Ignition Coil
IC Ignition Current
IC Illinois College [Jacksonville, US]
IC Image Coding
IC Image Contrast
IC Image Conversion; Image Converter
IC Immediate Constituent
IC Immersion Cleaning [Metallurgy]
IC Immunochemical; Immunochemist(ry)
IC Imperial College [London, UK]
IC Impermeable Carbon
IC Impregnated Cable
IC Impregnated Carbon (Silicon Carbide)
IC Impulse Conductor
IC Inclusion Complex [Chemistry]
IC Incommensurate (Phase) [Materials Science]
IC Incomplete Combustion
IC Incremental Cost
IC (Dreyer's) Index Catalog (of Celestial Objects) [Astronomy]
IC Indirect Cost
IC Indirect Cycle [Nuclear Reactors]
IC Induced Current
IC Inductance-Capacitance
IC Induction Coil
IC Industrial Chemist(ry)
IC Industrial Control
IC Industry Commission [formerly Industries Assistance Commission, Australia]
IC Industry Canada
IC Inelastic Collision [Mechanics]
IC Inertial Component
IC Information Center
IC Information Circular
IC Information Code
IC Infrared Cell
IC Initial Condition

IC Initial Configuration
IC Injection Cooling
IC Inlet Chamber
IC Inner City
IC Inter-Core
IC Inorganic Carbon
IC Inorganic Chemical(s); Inorganic Chemistry
IC Inorganic Component
IC Input Circuit
IC Inscribed Circle [Mathematics]
IC Installed Capacity (of Hydroelectric Plants)
IC Institute/Center [National Institutes of Health, Bethesda, Maryland, US]
IC Institute of Ceramics [Stoke-on-Trent, UK]
IC Institute of Chemistry
IC Instruction Cell
IC Instruction Code
IC Instruction Counter
IC Instruction Cycle
IC Instrument Correction
IC Insulated Conductor
IC Insulated Conductors [Institute of Electrical and Electronics Engineers–Power Engineering Committee, US]
IC Integral Control
IC Integrated Circuit
IC Intelligent Color
IC Intelligent Control
IC Intelligent Copier
IC Intensive Care
IC Interaction Constant
IC Intercalation Compound
IC Interchange Center
IC Intercity (Train)
IC Intercom [also I/C]
IC Intercommunication
IC Intercomparison
IC Intercomputer
IC Intercomputer Communication
IC Interconversion; Interconvertibility
IC Intercrystalline [also ic]
IC Intercrystalline Corrosion
IC Interexchange Carrier
IC Interface Card
IC Interface Composition [Materials Science]
IC Interface Control
IC Interference-Causing (Equipment)
IC Interference Color
IC Interference Contrast
IC Intergranular Corrosion
IC Interim Change
IC Interior Communication
IC Intermediate Coupling
IC Intermetallic Compound
IC Internal Calorimeter

IC Internal Combustion [also ic]
IC Internal Computer
IC Internal Connection
IC Internal Conversion [Nuclear Physics]
IC International Center
IC International Classification
IC International Colloquium
IC International Conference
IC Interrupt Controller
IC Interrupted Current
IC Interspecies Communication [US]
IC Intracrystalline
IC Intrinsic Conductance; Intrinsic Conduction; Intrinsic Conductivity [Solid-State Physics]
IC Inventory Control
IC Inverse Conduction
IC Invested Capital
IC Investment Cast(ing) [Metallurgy]
IC Investment Commission
IC Investment Committee
IC Investment Cost
IC Ion Chamber
IC Ion Channeling
IC Ion Chromatogram
IC Ion Chromatograph(y)
IC Ion Collection; Ion Collector
IC Ion Conductor
IC Ion Counter
IC Ionic Charge
IC Ionic Compound
IC Ionic Conductance; Ionic Conduction; Ionic Conductor
IC Ionic Crystal
IC Ionic Current
IC Ionization Chamber
IC Iowa City [US]
IC Irrigation Consultant
IC Isochronous Cyclotron
IC Isocyanate
IC Ithaca College [New York State, US]
IC Ivory Coast
I^2C Inter-Integrated Circuit (Bus) [also IIC]
I&C Installation and Checkout [also I and C]
I&C Instrumentation and Communication [also I and C]
I&C Instrumentation and Control [also I+C]
I/C Icing Condition [also i/c]
I/C In Charge of [also I/c]
I/C Industrial/Commercial
I/C Incoming
I/C In Command
I/C Intercom [also IC]
I∥C Current Direction Parallel to the c-Axis (of Superconductor)
i-C Carbon Prepared by Ion-Assisted Techniques
i-C Inorganic Carbon

ic *(inter cibum)* – between meals [Medical Prescriptions]
ic intercrystalline [also IC]
ICA IBM (Corporation) Customer Agreement
ICA In-Circuit Analyzer
ICA Independent Crossing Approximation
ICA Industrial Communications Agency
ICA Industrial Communications Association
ICA Initial Cruise Altitude [Aeronautics]
ICA Institute of Canadian Advertising
ICA Institute of Chartered Accountants [UK]
ICA Institute of Company Accountants [UK]
ICA Integrated Circuit Amplifier
ICA Integrated Communications Adapter
ICA Intercompany Agreement
ICA Intercomputer Adapter
ICA Intermuseum Conservation Association [US]
ICA Intelligent Console Architecture
ICA Internal Cartoid Artery [Anatomy]
ICA International Cartographic Association [Australia]
ICA International Carwash Association [US]
ICA International Ceramic Association [US]
ICA International Chiropractors Association
ICA International Color Authority
ICA International Commission on Acoustics [US]
ICA International Commodities Agreement
ICA International Common Access
ICA International Communication Agency [now United States Information Agency]
ICA International Communication Association [US]
ICA International Conference in Asia [of International Union of Materials Research Societies]
ICA International Congress of Americanists
ICA International Control Agency
ICA International Cooperation Administration [now Agency for International Development, US]
ICA International Cooperative Alliance [Switzerland]
ICA International Copper Association [US]
ICA International Council on Archives
ICA International Council for ADP (Automatic Data Processing) [Madrid, Spain]
ICA Intra-application Communications Area
ICA Inventors Club of America [US]
ICA Inverse Cerenkov Acceleration [Particle Physics]
ICA Islamic Cement Association
ICA Islet Cell Autoantibodies [Immunology]
ICA Isocrotonic Acid
ICA Isocyanic Acid
ICA Item Change Analysis
ICA Item Control Area
ICAA Institute of Chartered Accountants in Australia
ICAA Institute of Chartered Accountants of Alberta [Canada]
ICAA Insulation Contractors Association of America [US]
ICAA International Civil Airports Association [France]

ICAA International Civil Aviation Authority [Canada]

ICAA International Conference on Aluminum Alloys

ICAC International Civil Aviation Committee

ICAC International Civil Aviation Convention

ICAC International Computer Science Convention

ICAC International Conference on Analytical Chemistry

ICAC International Conference on Automated Composites

ICAC International Cotton Advisory Committee [US]

ICAC Islamic Civil Aviation Council

ICACGP International Commission on Atmospheric Chemistry and Global Pollution

ICAC/IFAC International Computer Science Convention/ International Federation of Automatic Control (Symposium)

ICACM International Conference on Advances in Composite Materials

ICACSC International Conference on Amorphous and Crystalline Silicon Carbide

ICAD Integrated Computer-Aided Design

ICAD Integrated Control and Display

ICADD International Committee for Accessible Document Design

ICADEM Integrated Computer-Aided Design, Engineering and Manufacturing

ICADI Inter-American Center for Agricultural Documentation and Information [of Inter-American Institute of Agricultural Sciences, Organization of American States]

ICAE International Commission of Agricultural Engineering

ICAE International Commission on Atmospheric Electricity [UK]

ICAE International Conference of Agricultural Economists

ICAE Inverse Cerenkov (Laser) Accelerator (Experiment) [at the Accelerator Test Facility of Brookhaven National Laboratory, Upton, New York, US]

ICAEC International Confederation of Associations of Experts and Consultants [France]

ICAEN Iowa Computer-Aided Engineering Network [US]

ICAES International Congress of Anthropological and Ethnological Sciences

ICAF Industrial College of the Armed Forces [of National Defense University, Washington, DC, US]

ICAF International Committee on Aeronautical Fatigue [Netherlands]

ICAI Intelligent Computer-Assisted Instructional System

ICAI International Commission for Agricultural Industries

ICAI International Conference on Artificial Intelligence

ICA Inf ICA Information [of International Council for ADP (Automatic Data Processing) in Government Administration, Spain]

ICAIR International Centre for Antarctic Information and Research

ICAIT International Conference on Advanced Information Technology

ICAITI Instituto Centro-Americano de Investigaciones y Tecnología Industrial [Central American Institute for Industrial Research and Technology]

ICAK International College of Applied Kinesiology [US]

ICAL Intermediate-Scale Calorimeter

ICALEO International Congress on Applications of Lasers in Electrooptics

ICALEPS International Conference on Large Electron-Positron Colliders

ICALP International Colloquium on Automata, Languages and Programming

ICALPE International Center for Alpine Environments

ICALU International Confederation of Arab Labour Unions

ICAM Integrated Communications Access Method

ICAM Integrated Computer-Aided Manufacturing

ICAM International Center for the Advancement of Management Education [US]

ICAM International Confederation of Architectural Museums [France]

ICAM International Conference on Advanced Materials

ICAM International Congress on Applied Mineralogy

ICAME International Committee on Application of Mössbauer Effect

ICAMQ International Committee on Automation of Mines and Quarries [Hungary]

ICAMPS International Conference and Exhibition on Advances in Materials and Processes [India]

ICAMRS International Civil Aviation Message Routing System

ICAMS Industrial Central Atmosphere Monitoring System

ICAN Individual Circuit Analysis

ICAN Integrated Composite Analyzer (Computer Code)

ICAN International Committee on Air Navigation

ICAND Multiplicand [also icand]

ICAO Institute of Chartered Accountants of Ontario [Canada]

ICAO International Civil Aviation Organization [of United Nations in Canada]

ICAP Inductively-Coupled Argon Plasma

ICAP Industrial Conversion Assistance Program [of Natural Resources Canada]

ICAP International Conference on Atomic Physics

ICAP International Congress of Applied Psychology

ICAP International Computational Accelerator Physics (Conference)

ICAP Internet Calendar Access Protocol

ICAPE International Chemical and Petroleum Engineering Exhibition [UK]

ICAQUO Inventory of Contaminants in Aquatic Organisms (Database) [of Food and Agricultural Organization]

ICAR Indian Council of Agricultural Research [India]

ICAR Interface Control Action Request

ICAR International Conference on Advanced Robotics

ICAR Inventory of Canadian Agricultural Research

ICAR Investigation and Corrective Action Report

I-CAR Inter-Industry Conference on Auto Collision Repair [US]

ICARDA International Center for Agricultural Research in the Dry Areas [Aleppo, Syria]

ICARE International Center for the Advancement of Research Education

ICARVS Interplanetary Craft for Advanced Research in the Vicinity of the Sun

ICAS Independent Collision Avoidance System [Aeronautics]

ICAS Intermittent Commercial and Amateur Service

ICAS International Committee for Atmospheric Sciences [of Federal Council for Science and Technology, US]

ICAS International Conference on Amorphous Semiconductors Science and Technology

ICAS International Council of Aerospace Sciences

ICAS International Council of the Aeronautical Sciences [Germany]

ICAS Intel Communicating Applications Specifications

ICAS Institute of Chartered Accountants of Saskatchewan [Canada]

ICAS Intracavity Absorption Spectroscopy

ICASALS International Center for Arid and Semi-Arid Land Studies [also ICASLS]

ICASE Institute for Computer Applications in Science and Engineering [at NASA Langley Research Center, Hampton, Virginia, US]

ICASE International Council of Associations for Science Education [Hong Kong]

ICASLS International Center for Arid and Semiarid Land Studies [also ICASALS]

ICASSP International Conference on Acoustics, Speech and Signal Processing

ICAST International Center for Agricultural Science and Technology [Saskatoon, Saskatchewan, Canada]

ICAST International Conference on Amorphous Semiconductor Technology

ICAT International Center for Actuators and Transducer [of Pennsylvania State University, University Park, US]

ICATO Iran Civil Aviation Training Organization

ICATU International Confederation of Arab Trade Unions

ICAV Instantaneous Crankshaft Angular Velocity

ICB Incoming Calls Barred

ICB Institute of Canadian Bankers

ICB Interface Control Board

ICB Interference Calibration Blank

ICB Interim Change Bulletin

ICB Internal Common Bus

ICB International Commodity Body

ICB International Container Bureau

ICB Internet Citizen's Band

ICB Interrupt Control Bit

ICB Interrupt Control Block

ICB Ion(ized) Cluster Beam (Material Deposition Method)

ICBA Israel Cattle Breeders Association

ICBD International Center for Brewing and Distilling [UK]

ICBD Ion(ized) Cluster Beam Deposition

ICBE Ion(ized) Cluster Beam Epitaxy

ICBG International Cooperative Biodiversity Groups [of National Science Foundation, US]

ICBIC International Conference on Bioinorganic Chemistry

ICBM Intercontinental Ballistic Missile

ICBN International Code for Botanical Nomenclature

ICBO International Conference of Building Officials [US]

ICBP International Council for Bird Preservation [UK]

ICBRSD International Council for Building Research, Studies and Documentation [Netherlands]

ICBT Intercontinental Ballistic Transport

ICBWR Improved Cycle Boiling Water Reactor

ICC Ignition Control Compound

ICC Independent Channel Controller

ICC Indian Cryogenics Council [Calcutta, India]

ICC Industrial Communication Council [US]

ICC Information Commissioner of Canada

ICC Instrument Center Correction

ICC Instrument Control Center

ICC Intercomputer Channel

ICC Intercomputer Communication

ICC Intercrystalline Corrosion

ICC Interference Contrast Colloid

ICC Inuit Circumpolar Conference

ICC Interface Control Chart

ICC Intergovernmental Consultative Committee

ICC Interim Consultative Committee

ICC Intermarket Clearing Corporation

ICC International Catalysis Congress

ICC Internal Conversion Coefficient [Nuclear Physics]

ICC International Association for Cereal Chemistry [Austria]

ICC International Chamber of Commerce

ICC International Climatological Commission

ICC International Coffee Council

ICC International Committee for Conservation [of International Council of Museums, France]

ICC International Computation Center [Italy]

ICC International Computer Center

ICC International Computing Center [of United Nations]

ICC International Conference on Communications [of Institute of Electrical and Electronics Engineers, US]

ICC International Conference on Continuous Casting [Metallurgy]

ICC International Conference on Creep

ICC International Congress on Catalysis

ICC International Control Center

ICC International Coordinating Committee

ICC International Corrosion Congress

ICC International Corrosion Council [UK]

ICC Internationales Congress-Centrum [International Congress Center, Berlin, Germany]

ICC Interstate Commerce Commission [US]

ICC Intrinsic Carrier Concentration [Solid-State Physics]

ICC Invitational Computer Conference

ICC Iowa Conservation Commission [US]

ICC Island Coordinating Council [Torres Strait, Australia]

IC&C Invoice Cost and Charges [also ICC]

ICCA Independent Computer Consultants Association [US]

ICCA Instituto Centroamericano de Ciencias Agricolas [Central American Institute of Agricultural Sciences]

ICCA International Chemistry Conference in Africa [South Africa]

ICCA International Committee on Coordination for Agriculture

ICCA International Congress and Convention Association [Netherlands]

ICCA International Corrugated Case Association [France]

ICCA International Council for Commercial Arbitration [Austria]

ICCA International Council of Chemical Associations

ICCAD IEEE Conference on Computer-Aided Design [of Institute of Electrical and Electronics Engineers, US]

ICCAGRA Interagency Coordinating Committee for Airborne Geosciences Research and Applications [of NASA, National Oceanic and Atmospheric Administration, National Science Foundation and Navy Research Laboratory]

ICCAIA International Coordinating Council of Aerospace Industries Associations [US]

ICCAM Integrated Climate Change Analysis Model

ICCAT International Commission for the Conservation of Atlantic Tunas [Spain]

ICCATCI International Committee to Coordinate Activities of Technical Groups in the Coatings Industry [France]

ICCB Integrated Change Control Board

ICCBD Intergovernmental Committee on the Convention of Biological Diversity

ICCBE International Conference on Chemical Beam Epitaxy and Related Growth Techniques

ICCBM International Conference on the Crystallization of Biological Macromolecules

ICCC Inter-Council Coordination Committee [France]

ICCC International Color Computer Club

ICCC International Conference on Coordination in Chemistry

ICCC International Council for Computer Communication [Washington, DC, US]

ICCC Internally Cooled Cable in Conduit

ICC-CAPA International Chamber of Commerce–Commission on Asian and Pacific Affairs

ICCCS International Contamination Control Conference and Symposium [Switzerland]

ICCD Intensified Charge-Coupled Device (Camera)

ICCD International Conference on Computer Design

ICCE International Center for Conservation Education

ICCE International Commission on Continental Erosion

ICCE International Conference on Composites Engineering

ICCE International Council for Computers in Education [University of Oregon, Eugene, US]

ICCEC India Chemists and Chemical Engineers Club [now International Society of Indian Chemists and Chemical Engineers]

ICCEE International Classification Commission for Electrical Engineering

ICCET Imperial College, Center for Environmental Technology [UK]

ICCF International Computing and Control Facility

ICCF International Conference on Cold Fusion

ICCFS Imperial College, Center for Fusion Studies [UK]

ICCG International Conference on Coatings on Glass

ICCG International Conference on Crystal Growth

ICCH International Commodities Clearinghouse [UK]

ICCI International Conference on Composite Interfaces

ICCICA Interim Coordinating Coomittee on International Commodity Agreements [US]

ICCICE Islamic Chamber of Commerce, Industry and Commodity Exchange [Pakistan]

ICCL International Commission on Climate

ICCM International Committee for the Conservation of Mosaics [UK]

ICCM International Conference (and Exposition) on Composite Materials [of The Minerals, Metals and Materials Society, US]

ICCM Isoconcentration Contour Migration

ICCMB International Committee for the Conservation of Mud Brick [Italy]

ICCMS International Conference on Composite Materials and Structures

ICCO International Carpet Classification Organization [Belgium]

ICCOSS International Conference on the Chemistry of the Organic Solid State

ICCP Impressed Current Cathodic Protection

ICCP Information, Computers and Communications Policy

ICCP Institute for Certification of Computer Professionals [US]

ICCP Interface Coordination and Control Procedure

ICCP International Climate Change Partnership

ICCP International Cloud Climatology Program

ICCP International Commission for Coal Petrology [Belgium]

ICCP International Commission on Cloud Physics

ICCP International Conference on Cataloguing Principles

ICCPPS International Conference on Ceramic Powder Processing Science

ICCR International Committee for Coal Research [Belgium]

ICCR International Conference on Cosmic Rays

ICCROM International Centre for the Study, Preservation and Restoration of Cultural Property

ICCS Ice Center Communications System

ICCS Industrial Combustion Control System

ICCS Inter-Computer Communications System

ICCS International Center for Chemical Studies

ICCS International Commission of Control and Supervision

ICCS International Conference on Composite Structures [Scotland]

ICCS International Council on Civil Status [France]

ICCSSSAR International Coordinating Committee on Solid-State Sensors and Actuators Research [US]

ICCSTR International Coordinating Committee on Solid-State Transducers Research [US]

ICC-TM Interstate Commerce Commission–Transport Mobilization [US]

ICCU Interchannel Comparison Unit

ICCVD International Conference on Chemical Vapor Deposition

ICD In-Circuit Debugger

ICD Institute/Center/Division [National Institutes of Health, Bethesda, Maryland, US]

ICD Institute of Civil Defense [UK]

ICD Integrated Circuit Design

ICD Interface Control Document

ICD Interface Control Drawing

ICD International Center for Development

ICD International Center for the Disabled

ICD International Classification of Diseases

ICD International Code Designator

ICD International Congress for Data Processing [also ICDP]

ICD International Cooperation Department [of Ministry of Economic Affairs, Taiwan]

ICD Isocitrate Dehydrogenase [Biochemistry]

ICD-9-CM International Classification of Diseases, 9th Revision, Clinical Modification

ICDA International Chromium Development Association

ICDA International Compressor Distributors Association [US]

ICDA International Cooperative Development Association

ICDB Integrated Corporate Database

ICDD International Center for Diffraction Data [of Joint Committee on Powder Diffraction Standards at Swarthmore, Pennsylvania, US]

ICDDB International Control Description Database

ICDDR International Center for Diarrheal Disease Research [Dacca, Bangladesh]

ICDDS Institute of Civil Defense and Disaster Studies [UK]

ICDES Item Class Description

ICDF Inorganic Crystallographic Data File [of Information System Karlsruhe, Germany]

ICDF Intermediate Coupling Dirac-Fock [Quantum Mechanics]

ICDG Institute of Chemistry and Dynamics of the Geosphere [US]

ICDH Isocitrate Dehydrogenase [Biochemistry]

ICDL Integrated Circuit Description Language

ICDL Internal Control Description Language

ICDM Integrated-Circuit Diode Matrix

ICDM International Commission on Dynamic Meteorology

ICDO International Civil Defense Organization [Switzerland]

ICDP International Confederation for Disarmament and Peace

ICDP International Congress for Data Processing [also ICD]

ICDR Incremental Critical Design Review

ICDR Inward Call Detail Recording

ICDS Institutional Cooperation and Development Services [Canada]

ICDS International Conference on Defects in Semiconductors

ICDT Islamic Center for Development of Trade [Morocco]

ICDU Inertial Coupling Data Unit

ICDW Incommensurate Charge Density Wave [Solid-State Physics]

ICE Immediate Cable Equalizer

ICE In-Circuit Emulator

ICE Independent Cost Estimate

ICE Input-Checking Equipment

ICE Input Control Element

ICE Insertion Communications Equipment

ICE Institute for Chemical Education [US]

ICE Institute for Consumer Ergonomics [UK]

ICE Institute of Ceramic Engineers [US]

ICE Institute of Chartered Engineers [UK]

ICE Institute of Control Engineering [Technical University of Poznan, Poland]

ICE Institution of Chemical Engineers [Rugby, UK]

ICE Institution of Civil Engineers [UK]

ICE Instrument Checkout Equipment

ICE Insulated Cable Engineer

ICE Integrated Circuit Engineering

ICE Integrated Communications Architecture

ICE Integrated Computing (or Computer) Environment

ICE Integrated Cooling for Electronics

ICE Intelligent Component Expert

ICE Interactive Cost Estimating (System)

ICE Intercity Experimental (Train)

ICE Intercity Express [Germany]

ICE Interface Cancellation Equipment

ICE Intermediate Cable Equalizer

ICE Internal Combustion Engine [also ice]

ICE International Cirrus Experiment

ICE International Cometary Explorer

ICE International Congress of Ecology

ICE International Congress of Entomology [now Council for International Congresses of Entomology, UK]

ICE Internet Connections for Engineering [of Cornell University, Ithaca, US]

ICE Ion Chromatography Exclusion

Ice Island(ic)

Ice Icing [also ice]

ICEA Insulated Cable Engineers Association [US]

ICEA International Commission for Environmental Assessment

ICEA International Consulting Economists Association [UK]

ICEA International Consumer Electronics Association [UK]

ICEAM International Committee on Economic and Applied Microbiology [US]

ICEAS Intermittent Cycle Extended Aeration System

ICEC Ice Center Environment Canada

ICEC Institute of Chartered Engineers of Canada

ICEC International Conference on Electrical Contacts

ICEC International Cost Engineering Council [US]

ICEC International Cryogenic Engineering Committee [Germany]

ICEC International Cryogenic Engineering Conference

ICECAN Iceland-Canada Submarine Cable [also ICE-CAN]

ICEC/ICMC International Cryogenic Engineering Conference/ International Cryogenic Materials Conference [USSR]

Ice Col Ice Column (Model)

Ice Cream Rev Ice Cream Review [US]

ICED International Coalition for Energy Development

ICED International Conference on Engineering Design

ICED International Council for Educational Development

ICED International Council on Environmental Design [US]

ICEE International Control Engineering Exposition [US]

ICEF International Committee for Research and Study of Environmental Factors

ICEF International Conference(s) on Environmental Future

ICEF International Congress on Engineering and Food

ICEF International Council for Educational Films

ICEF International Federation of Chemical, Energy and General Workers Unions [Belgium]

ICEFs International Conferences on Environmental Future

ICEI Institution of Civil Engineers of Ireland

ICEI International Combustion Engine Institute [now Engine Manufacturing Association, US]

ICEL Insulating Ceramic Electroluminescent (Device)

ICEL International Council of Environmental Law [Germany]

Icel Island(ic)

ICEM Integrated Computer-Aided Engineering and Manufacturing

ICEM International Conference on Electronic Materials

ICEM International Conference on Electron Microscopy

ICEM International Congress on Electron Microscopy

ICEM International Congress on Experimental Mechanics

ICEM International Conference on Electronic Materials

ICEM International Council for Education Media

ICENAV Arctic Shipboard Radar Navigation System

ICEP Impedance Conversion and Error Processing

ICEPAK Intelligent Classifier Engineering Package

ICEPF International Commission for the Eriksson Prize Fund [Sweden]

ICER Information Center of the European Railways

ICER Infrared Cell, Electronically-Refrigerated

ICER Interdepartmental Committee on External Relations

ICES Institute for Complex Engineered Systems

ICES Integrated Civil Engineering System

ICES Integrated Civil Engineering System Users Group [US]

ICES Institution of Civil Engineering Surveyors [UK]

ICES Interference-Causing Equipment Standard

ICES International Conference of Engineering Societies

ICES International Conference of Engineering Studies

ICES International Conference on Environmental Systems

ICES International Council for the Exploration of the Sea [Denmark]

ICES Ion-Chromatography Eluation Suppression

IC/ES Intercommunication/Emergency Station

ICESC International Committee for European Security and Cooperation [Belgium]

ICESD Intergovernment Committee on Ecologically Sustainable Development

ICES J ICES Journal [of Integrated Civil Engineering System Users Group, US]

ICET Institute for the Certification of Engineering Technicians

ICET International Center for Earth Tides [Belgium]

ICET International Council on Education for Teaching [UK]

ICETK International Council of Electrochemical Thermodynamics and Kinetics

ICETT Industrial Council for Educational and Training Technology

Ice XAL Ice Crystal

ICF Incommunication Flip-Flop

ICF Incremental Cost per Foot

ICF Inertial Confinement Fusion [Nuclear Engineering]

ICF Infinite Continued Fraction

ICF Institute of Chartered Foresters [UK]

ICF Instrument Control Facility

ICF Integrated Control Facility

ICF Interacting Correlated Fragment (Method) [Biology]

ICF Interactive Communications Feature

ICF Intercommunication Flip-Flop

ICF Intercrystalline Failure

ICF Intercrystalline Fracture

ICF Interface Control Function

ICF International Casting Federation

ICF International Conference on Ferrites

ICF International Congress on Fracture [Japan]

ICF International Consultants Foundation [Sweden]

ICF International Cotton Federation

ICF International (Whooping) Crane Foundation [Bartaboo, Wisconsin, US]

ICFA Inland Commercial Fisheries Association [US]

ICFA Institute of Chartered Financial Analysts

ICFA International Committee on Future Accelerators [of CERN–European Laboratory for Particle Physics, Geneva, Switzerland]

ICFA Instrum Bull International Committee on Future Accelerators Instrumentation Bulletin [of Stanford Linear Accelerator Center, Stanford University, US]

ICFAP International Conference on Failure Analysis Principles

ICFBC International Conference on Fluidized Bed Combustion

ICFC International Center for Fairs and Congresses [Belgium]

ICFE International Conference on F-Elements

ICF Int Consult Newsl ICF International Consultants Newsletter [of International Consultants Foundation, Sweden]

ICFM Inlet Cubic Feet per Minute

ICFMH International Committee on Food Microbiology and Hygiene [Denmark]

ICFP International Conservation Financing Project [of World Resources Institute]

ICFR Interdepartmental Committee of Futures Research

ICFRM International Conference on Fusion Reactor Materials

ICFTA International Committee of Foundry Technical Associations

ICFTU International Confederation of Free Trade Unions [Belgium]

ICG Indocyanine Green

ICG In-Flight Coverall Garment

ICG Interactive Computer Graphics

ICG Intercharacter Gap

ICG Interface-Controlled Growth [Materials Science]

ICG International Commission on Glass

ICG International Congress on Glass

ICG International Congress of Genetics

Icg Icing [also icg]

ICGB International Cargo Gear Bureau [US]

ICGEB International Center for Genetic Engineering and Biotechnology [of United Nations]

ICGEBnet International Center for Genetic Engineering and Biotechnology Network

ICGG International Conference on Grain Growth (in Polycrystalline Materials)

ICG-GUD Integrated Coal Gasification–Gas und Dampf (Process)

ICGI International Conference on Geoscience Information

ICGIC Icing in Clouds

ICGICIP Icing in Clouds in Precipitation

ICGIP Icing in Precipitation

ICGM Intercontinental Guided Missile

ICGM International Colloquium about Gas Marketing [France]

ICGTI International Center for Gas Technology Information [UK]

ICGTMP International Colloquium on Group Theoretical Methods in Physics

ICGW International Commission on Groundwater

ICH Infectious Canine Hepatitis

ICH Interchanger [also Ich]

ICH International Center of Hydrology [of University of Padua, Italy]

ICH International Conference on Hydropower [US]

ICH International Course on Hydrology [of International Center of Hydrology, Italy]

ICH Ion Channeling

ICHAZ Intercritical Heat-Affected Zone [Metallurgy]

ICHC International Committee for Horticultural Congresses

IChC International Chemistry Celebration [of American Chemical Society–Office of International Activities]

ICHCA International Cargo Handling and Coordination Association [UK]

IChemE Institution of Chemical Engineers [UK]

ICHENP International Conference on High-Energy Nuclear Physics

ICHEP International Conference on High-Energy Physics

ICHIB Information Center of the Hungarian Industry in Budapest

ICHM International Committee on Historical Metrology

ICHMT International Center on Heat and Mass Transfer (in Manufacturing Processes)

ICHS Inter-African Committee for Hydraulic Studies [Burkina Faso]

ICHS International Congress of History of Science

ICHSP International Congress on High-Speed Photography [now ICHSPP, US]

ICHSPP International Congress on High-Speed Photography and Photonics [US]

Icht Ichthyological; Ichthyologist; Ichthyology [also Ichth, ichth, or icht]

ICHTM International Congress on the Heat Treatment of Materials

ICHTS International Conference on High-Temperature Superconductors

ICI Image Component Information

ICI Imperial Chemical Industries plc [UK]

ICI Industrial-Commercial-Institutional (Sector)

ICI Intelligent Communications Interface

ICI Inter-American Copyright Institute

ICI International Commission on Illumination

ICI Investment Casting Institute [Atlanta, Georgia, US]

ICIA Industrial, Commercial and Institutional Accountant

ICIA Information and Communications Industry Association [UK]

ICIA International Center of Information on Antibiotics [Belgium]

ICIA International Communications Industries Association [US]

ICIA International Credit Insurance Association

ICIA International Crop Improvement Association [US]

ICIANZ Imperial Chemical Industries of Australia and New Zealand

ICIAQC International Conference on Indoor Air Quality and Climate

ICIB Indian Commercial Information Bureau

ICIB International Cargo Inspection Bureau

ICIC International Cancer Information Center [of National Cancer Institute, Bethesda, Maryland, US]

ICIC International Copyright Information Center [of UNESCO]

ICID International Commission on Irrigation and Drainage [India]

ICIDI International Commission on International Development Issues (or Brandt Commission)

ICI Eng Plast ICI Engineering Plastics [Publication of Imperial Chemical Industries, UK]

ICIL Imperial Chemical Industries Limited [UK]

ICI Mag ICI Magazine [Published by Imperial Chemical Industries, UK]

ICIMOD International Center for Integrated Mountain Development

ICIO Interim Cargo Integration Operations

ICIP International Conference on Information Processing

ICIPE International Center of Insect Physiology and Ecology [Kenya]

ICI Polyurethane Newsl ICI Polyurethanes Newsletter [of Imperial Chemical Industries, UK]

ICIREPAT (Committee for) International Cooperation in Information Retrieval among Examining Patent Offices

ICIS Integrated Circuit Inspection System

ICIS International Conference on Ion Sources

ICISB Initial Crack Indentation Strength in Bending [Fracture Toughness]

ICISE International Conference on Interface Science and Engineering

ICISS Impact-Collision Ion-Scattering Spectrometry; Impact-Collision Ion-Scattering Spectroscopy

ICIT Instituto Cubano de Investigaciones Tecnologícas [Cuban Institute for Technological Research]

ICITA International Chain of Industrial and Technical Advertising Agencies [US]

ICITA International Cooperative Investigation of the Tropical Atlantic

ICITO International Committee of the International Trade Organization [of United Nations]

ICITP International Conference on Impact Treatment Processes

ICJ Incoming Junction

ICJ International Commission of Jurists [Switzerland]

ICJ International Court of Justice [of United Nations in The Hague, Netherlands]

ICL Idiopathic CD4 Lymphocytopenia [Medicine]

ICL Incoming Line

ICL Input Control Logic

ICL Inserted Connection Loss

ICL Interband Cascade Laser

ICL Intercommunication Logic

ICL Inter-Computer Communications Logic

ICL Interface Clear

ICL Interface Control Layer

ICL International Computers Limited

ICL International Conference on LOS (Luminous and Optical Spectroscopy)

ICL Interpretative Coding Language

ICL Irish Central Library

ICL Isentropic Condensation Level

ICl Iodine Monochloride

ICLA Indian College Library Association [India]

ICLA International Committee on Laboratory Animals

ICLAE International Council of Library Association Executives [US]

ICl/Ar Iodine Monochloride/Argon (Mixture)

ICLARM International Center for Living Aquatic Resources Management [Philippines]

ICLAS Intracavity Laser Absorption Spectroscopy

ICLASS International Conference on Liquid Atomization and Spray Systems

ICLCF International Conference on Low-Cycle Fatigue [Mechanics]

ICLCP International Conference on Large Chemical Plants [Belgium]

ICLD International Commission on Large Dams

ICLEI International Council for Local Environmental Initiatives [Toronto, Canada]

ICLID Incoming-Call Line Identification

ICLR International Committee for Lift Regulations [France]

ICL Tech J ICL Technical Journal [Publication of International Computers Ltd., UK]

ICM Image Color Matching

ICM Improved Capability Missile

ICM Incoming Message

ICM Injection Compression Molding

ICM Institute of Construction Management [UK]

ICM Instituto de Ciencia de Materiales [Materials Science Institute; of Consejo Superior de Investigaciónes Cientificas, Madrid and Sevilla, Spain]

ICM Instituto di Chimica dei Materiali [Materials Chemistry Institute; of Consiglio Nazionale delle Ricerche, Rome, Italy]

ICM Instruction Control Memory

ICM Instrumentation and Communications Monitor

ICM Integral Charge-Control Model

ICM Integrated Catchment Management

ICM International Conference on Magnetism

ICM International Conference on Mechanical Behaviour of Materials

ICM International Conference on Microlithography

ICM International Congress of Mathematicians

ICM International Congress on Mechanical Behavior of Materials

ICM Inverted Coaxial Magnetron [Electronics]

ICM Ion Chromatography Module

i/cm² ions per centimeter [also ions/cm²]

ICMA Institute of Certified Management Accountants [US]

ICMA Instituto de Ciencia de Materiales de Aragón [Aragon Materials Science Institute; of Universidad de Zaragoza, Spain]

ICMA International City Managers Association

ICMA International Congresses for Modern Architecture

ICMA International Congress on Metalworking and Automation

ICMAB Instituto de Ciencia de Materiales de Barcelona [Barcelona Materials Science Institute; of Consejo Superior de Investigaciónes Cientificas at Catalunya, Spain]

ICMAS International Conference on Modem Aspects of Superconductivity

ICMASA Intersociety Committee on Methods for Air Sampling and Analysis [US]

ICMB Institute of Clinical Molecular Biology [UK]

ICMB Instituto Ciencia de Materiales de Barcelona [Barcelona Institute Materials Science Institute; of Consejo Superior de Investigaciónes Cientificas at Catalunya, Spain]

ICMC Indian-Ocean Cable Management Committee

ICMC Institute of Chemical Machine Construction [Russia]

ICMC International Cokemaking Congress

ICMC International Conference on Materials Characterization

ICMC International Conference on Metallurgical Coatings

ICMC International Cryogenic Materials Conference

ICMC/ICTF International Conference on Metallurgical Coatings/International Conference on Thin Films

ICMCTF International Conference on Metallurgical Coatings and Thin Films

ICMD International Congress on Mine Design

ICME International Committee on Microbial Ecology [US]

ICME International Council on Metals and the Environment [Canada]

ICMEE Institution of Certified Mechanical and Electrical Engineers [Marshalltown, Transvaal, South Africa]

ICMEDC International Council of Masonry Engineers for Developing Countries [Scotland]

ICMF International Conference on (Advances in) Metal Forming Techniques

ICMG International Commission on Mibrobial Genetics

ICMI International Commission on Mathematical Instruction [UK]

ICMLT International Congress of Medical Laboratory Technicians

ICMM Instituto de Ciencia de Materiales de Madrid [Madrid Materials Science Institute, Spain]

ICMMA Industrial Cleaning Machine Manufacturers Association [UK]

ICMMB International Conference on Mechanics in Medicine and Biology [US]

ICMM-CSIC Instituto de Ciencia de Materiales de Madrid–Consejo Superior de Investigaciónes Cientificas [Madrid Materials Science Institute of the National (Scientific) Research Council, Spain]

ICMMP International Committee on Military Medicine and Pharmacy

ICMO Integrated Configuration Management Office

ICMOVPE International Conference on Metal-Organic Vapor Epitaxy

ICMP Interchannel Master Pulse

ICMP Internet Control Message Protocol

ICMP International Center for Materials Physics [of Chinese Academy of Sciences, Shenyang, PR China]

ICMPC International Conference on Materials and Process Characterization

ICMR Indian Council of Medical Research

ICMR International Center for Medical Research

ICMS Indirect Cost Management System

ICMS Integrated Crop Management Services [Canada]

ICMS International Commission on Mushroom Science

ICMS Inverted Cylindrical Magnetron Sputtering

i/cm²-s ions per square centimeter second [also ions/cm²-s]

ICMSF International Commission on Microbiological Specifications for Food [US]

ICMT Intercontract Material Transfer

ICMUA International Commission on Meteorology in Upper Atmosphere

ICMV Integrated Circuit Multivibrator

ICN Cyanogen Iodide

ICN Idle Channel Noise

ICN Instrument Communication Network

ICN Integrated Computer Network

ICN Interim Change Notice

ICN International Communications Network

ICN International Council of Nurses

ICNAA International Conference on Nucleation and Atmospheric Aerosols

ICNAF International Commission for the Northwest Atlantic Fisheries

IC/NATAS International Council–National Academy of Television Arts and Sciences [US]

ICNB International Committee on Nomenclature of Bacteria

ICNCP International Commission for the Nomenclature of Cultivated Plants [Netherlands]

ICNDST International Conference on the New Diamond Science and Technology

ICNDT International Committee on Nondestructive Testing

ICNDT International Conference on Nondestructive Testing

ICNI Integrated Communication, Navigation and Identification

ICNM International Committee on Nanostructured Materials

ICNND Interdepartmental Committee on Nutrition for National Defense [US]

ICNPPA International Commission on National Parks and Protected Areas [now Commission on National Parks and Protected Areas, Canada]

ICNV International Committee for the Nomenclature of Viruses

ICO Infinity Corrected Object

ICO Integrated Checkout

ICO Inter-Agency Committee on Oceanography [of Federal Council for Science and Technology, US]

ICO Intergovernmental Commission on Oceanography [UK]

ICO Interim Conservation Order

ICO International Carbohydrate Organization [Canada]

ICO International Chemistry Office

ICO International Cocoa Organization

ICO International Coffee Organization

ICO International Commission on Oceanography

ICO (Congress of the) International Commission on Optics [UK]

ICO International Computer Orphanage [US]

.ICO Icon [File Name Extension] [also .ico]

ICOA International Castor Oil Association [US]

ICOC Indian Central Oil Seeds Committee

ICOD International Center for Ocean Development

ICOE Instrument Checkout Equipment

ICOGRADA International Council of Graphic Design Associations [UK]

ICOH International Commission on Occupational Health [Switzerland]

ICOH International Congress on Occupational Health

ICOHTEC International Committee for the History of Technology [UK]

ICOIN Inland Waters, Coastal and Ocean Information Network

ICOIN/IMS Inland Waters, Coastal and Ocean Information Network/Information Management System

ICOL International Conference on Optics and Optoelectronics

ICOLD International Commission on Large Dams [France]

ICOLP International Cooperative for Ozone Layer Protection

ICOM International Council of Museums [France]

ICOMAT International Conference on Martensitic Transformation [US]

ICOMC International Conferences on Organometallic Chemistry

ICOME International Committee on Microbial Ecology [US]

ICOMIA International Council of Marine Industry Associations [UK]

ICOMOS International Council on Monuments and Sites [France]

ICOMP Intel Comparative Microprocessor Performance [also iCOMP]

ICON Integrated Chloride Oxidation (Process)

ICON Integrated Control

ICONA Instituto (Nacional) de la Conservación do Natura [National Institute for the Conservation of Nature, Spain]

ICONCLASS Iconography Classification

ICONS Information Center on Nuclear Standards [of American Nuclear Society, US]

ICONS Isotopes of Carbon, Oxygen, Nitrogen and Sulfur (Program) [of United States Atomic Energy Commission]

ICONTEC Instituto Colombiano de Normas Técnicas y Certificación [Colombian Institute for Technical Standards and Certification]

ICOP Imported Crude Oil Processing

ICOR Incremental Capital-Output Ratio

ICOR Intergovernmental Conference on Oceanic Research [of UNESCO]

ICorrST Institution of Corrosion Science and Technology [UK]

ICOS Impressa Construzioni Opere Specializzate [Italy]

ICOS Improved Crew Optical Sight

ICOS Interactive COBOL Operating System

ICOS International Committee on Onomastic Sciences

ICOS International Conference/Congress Organizing Service

IC OS Integrated Circuit One-Shot (Multivibrator)

ICOSN International Conference on Optical Engineering for Sensing and Nanotechnology [Joint Conference of SPIE and Optical Society of Japan]

ICOSO International Committee on Outer Space Onomastics

ICOSS Inertial-Command Offset System

ICOSSAR International Conference on Structural Safety and Reliability

ICOSSM International Conference On Shape-Memory Alloys

ICOT Institute for New Generation Computer Technology [Japan]

ICOTOM International Conference on Textures of Materials

ICOTT Industry Coalition on Technology Transfer [US]

ICOU Inertial Coupling Data Unit

ICP ICS (Intercommunication System) Control Program

ICP Incoherent Critical Point [Metallurgy]

ICP Incoming (Message) Process

ICP Indicator Control Panel

ICP Induction-Coupled Plasma; Inductively-Coupled Plasma

ICP Industrial Cooperation Program [of Canadian International Development Agency]

ICP Inherently Conducting Polymer

ICP Initial Connection Protocol

ICP Innovative Concepts Program

ICP Instrument Calibration Procedure

ICP Integrated Channel Processor

ICP Intelligent Communications Processor

ICP Interconnected Processing

ICP Interim Compliance Panel

ICP International Center of Photography [New York City, US]

ICP International Classification of Patents

ICP International Commission for Palynology [now International Federation of Palynological Societies, Netherlands]

ICP International Communication Planning

ICP International Computer Programs

ICP International Congress of Protozoology

ICP International Control Plan

ICP International Cooperative Program

ICP Intrinsically Conductive Plastics

ICP Inventory Control Point

ICP Investment Corporation of Pakistan

ICP Ion-Carburizing Calculation Program

ICPA Inductively Coupled Plasma Analysis

ICPA International Cooperative Petroleum Association [US]

ICPA International Cotton Producers Association

ICP-AAS Inductively-Coupled Plasma Atomic Absorption Spectroscopy [also ICPAAS]

ICP-AE Inductively-Coupled Plasma Atomic Emission

ICP-AES Inductively-Coupled Plasma Atomic Emission Spectroscopy [also ICPAES]

ICP-AES Inductively-Coupled Plasma Auger Electron Spectroscopy [also ICPAES]

ICPAM International Center for Pure and Applied Mathematics [France]

ICPB Insurance Crime Prevention Bureau [Canada]

ICPBR International Commission for Plant-Bee Relationships [Denmark]

ICPC International Cable Protection Committee [UK]

ICPC International Coal Preparation Congress

ICPC Interrange Communications Planning Committee

ICPCSH International Conference on Physics and Chemistry of Semiconductor Heterostructures

ICP-CVD Inductively-Coupled-Plasma-Enhanced Chemical Vapor Deposition

ICPD Induction (or Inductively) Coupled Plasma Deposition

ICPD Inductively-Coupled Plasma Discharge

ICPD International Conference on Population and Development [of United Nations Fund for Population Activities]

ICPDATA International Commodity Production Data [of United Nations Statistical Office]

ICP/DCP Inductively-Coupled Plasma/Directly Coupled Plasma (Analysis)

ICPE Institutul de Cercetare si Proiectare Pentru Industria Elektrotehnica [Scientific Institute for Control in the Electrotechnical Industry, Bucharest, Romania]

ICPE Intake-Compression-Power-Exhaust

ICPE International Commission on Physics Education [of International Union of Pure and Applied Biophysics]

ICPE International Conference in Production Engineering

ICPEMC International Commission for Protection against Environmental Mutagens and Carcinogens [Netherlands]

ICPEPA International Conference on Photo-Excited Processes and Applications

ICP-ES Inductively-Coupled Plasma Emission Spectroscopy [also ICPES]

ICPhS International Congress of Photonic Sciences

ICPI International Conference on Polyimides

ICPIC International Cleaner Production Information Clearinghouse

ICPIG International Conference on Phenomena in Ionized Gases

ICPKD Interdisciplinary Centers for Polycystic Kidney Disease Research

ICPLC International Commission for the Protection of Lake Constance [Switzerland]

ICPM International Commission for Plant Raw Materials

ICPM International Commission on Polar Meteorology

ICPM International Conference on Advanced Physical Metallurgy

ICPM International Conference on Polymers in Medicine

ICPMP International Commission for the Protection of the Moselle against Pollution [France]

ICP-MS Inductively-Coupled Plasma Mass Spectrometer; Inductively-Coupled Plasma Mass Spectrometry [also ICPMS]

ICP-MS Inductively-Coupled Plasma Mass Spectroscopy [also ICPMS]

ICPN International Council of Plant Nutrition [Netherlands]

ICPNS International Conference on the Physics of Noncrystalline Solids [UK]

ICPO International Criminal Police Organization [also INTERPOL, UK]

ICP-OES Inductively-Coupled Plasma Optical-Emission Spectrometer; Inductively-Coupled Plasma Optical-Emission Spectrometry [also ICPOES]

ICP-OES Inductively-Coupled Plasma Optical-Emission Spectroscopy [also ICPOES]

ICPP Idaho Chemical Processing Plant [of United States Atomic Energy Commission]

ICPP International Commission on Plasma Physics

ICPR International Conference on Production Research

ICPRAP International Commission for the Protection of the Rhine against Pollution [Germany]

ICPS Inductively-Coupled Plasma System

ICPS Industrial and Commercial Power Systems [Institute of Electrical and Electronics Engineers Industry Applications Society Committee]

ICPS Interconnected Power System [also IPS]

ICPS International Coalition for Procurement Standards

ICPS International Conference on the Properties of Steam

ICPS International Congress of Photographic Sciences

ICPS Institute of Chemical Process Fundamentals AS [Prague, Czech Republic]

ICPSI International Conference on Plasma Surface Interactions

ICPTUR International Conference for Promoting Technical Uniformity on Railways [Switzerland]

ICPVT International Council for Pressure Vessel Technology [US]

ICQSA International Conference on Quanititative Surface Analysis

ICR Ice Crystal Replicator

ICR Indirect Control Register

ICR Indirect-Cycle (Nuclear) Reactor (System)

ICR Inductance-Capacitance-Resistance

ICR Initial Concentrated Rubber

ICR Input Control Register

ICR Institute for Cancer Research [US]

ICR Institute for Cooperative Research

ICR Institute of Chemical Research [Kyoto University, Japan]

ICR Instruction Change Request

ICR Intelligent Character Recognition

ICR Interface Compatibility Record

ICR International Commission on Rheology

ICR International Congress of Radiology

ICR International Council for Reprography

ICR Interrupt Control Register

ICR Ion Cyclotron Resonance (Spectroscopy)

ICR Iron-Core Reactor

ICRA Industrial Chemical Research Association [US]

ICRA Industrial Copyright Reform Association [UK]

ICRA Interagency Committee on Radiological Assistance

ICRA International Center for Relativistic Astrophysics [Rome, Italy]

ICRA International Conference on Robotics and Automation

ICRAF International Council for Research in Agroforestry [Nairobi, Kenya]

ICRC Indian Cancer Research Institute

ICRC International Committee of the Red Cross [Name Changed to International Movement of the Red Cross and Red Crescent] [Geneva, Switzerland]

ICRC International Conference of the Red Cross

ICRC International Conference on Robotics in Construction [US]

ICRC International Cosmic Ray Conference

ICRCCM Intercomparison of Radiation Codes in Climate Models

ICRDA Independent Cash Register Dealers Association [US]

ICRF Imperial Cancer Research Fund [UK]

ICRF Ion Cyclotron Range of Frequencies

ICRH Institute for Computer Research in the Humanities [US]

ICRH Ion Cyclotron Resonance Heating

ICRI International Communications Research Institute [Tokyo, Japan]

ICRI International Crops Research Institute

ICRI Iron Casting Research Institute [US]

ICRICE International Center of Research and Information on Collective Economy [Switzerland]

ICRISAT International Crops Research Institute for the Semi-Arid Tropics [India]

ICRM Institute of Certified Records Managers [US]

ICRM International Committee on Radionuclide Metrology

ICRM Newsl ICRM Newsletter [of Institute of Certified Records Managers, US]

ICRMS Ion-Cyclotron-Resonance Mass Spectrometer; Ion-Cyclotron-Resonance Mass Spectrometry

ICRO International Cell Research Organization [France]

ICRP Institutional Collaborative Research Program [of University of California at Santa Barbara, US]

ICRP International Commission on Radiological Protection [UK]

ICRPMA International Committee for Recording the Productivity of Milk Animals [Italy]

ICRQM International Conference on Rapidly Quenched Metals [Metallurgy]

ICRS Index Chemicals Registry System [of Institute for Scientific Information, US]

ICRS International Commission on Radium Standards

ICRS International Conference on Residual Stresses

ICRSDT International Commission on Remote Sensing and Data Transmission

ICRU International Commission on Radiation Units and Measurements [US]

ICRU International Commission on Radiological Units

ICRU Rep ICRU Reports [of International Commission on Radiation Units and Measurements, US]

ICRW International Convention for the Regulation of Whaling

ICS Immunochemistry System

ICS Incident Command System (Manual)

ICS Indian Ceramic Society

ICS Indian Chemical Society

ICS Infinity Color-Corrected System; Infinitely Color-Corrected System [Microscopy]

ICS Information Control Station

ICS Instrumentation Control System

ICS Inland Computer Service [US]

ICS Input Control System

ICS Institut Charles Sadron [of Conseil National de la Recherche Scientifique, Strasbourg, France]

ICS Institute of Chartered Shipbrokers [UK]

ICS Institute of Chartered Surveyors

ICS Institute of Computer Science

ICS Institution of Corrosion Science [UK]

ICS Instrumentation and Control Society [Singapore]

ICS Integral Cross-Section

ICS Integrated Communication System

ICS Integrated Controller Software

ICS Integrated Control System

ICS Integration Control System

ICS Interactive Communications System

ICS Intercarrier Sound System

ICS Intercommunication System

ICS Inter-Computer Synchronizer

ICS Interconnecting Station

ICS Interface Control Specification

ICS Interference Check Standard

ICS Interlinked Computerized System

ICS Intermittent Control System

ICS Internal Countermeasures Set

ICS International Carbohydrate Symposium

ICS International Chamber of Shipping [UK]

ICS International Chemical Society

ICS International Chemical System

ICS International Chemometrics Society [Belgium]

ICS International Cogeneration Society [US]

ICS International College of Surgeons

ICS International Commission on Stratigraphy

ICS International Congress on Steelmaking

ICS International Coronelli Society [Austria]

ICS International Correspondence Schools [US]

ICS Interphone Control Station

ICS Interpretative Computer Simulation

ICS Intuitive Command Structure

ICS Iron Casting Society

ICS Isolation Containment Spray [Nuclear Reactors]

ICS Israel Chemical Society

ICSA In-Core Shim Assembly [Nuclear Reactors]

ICSA International Council for Scientific Agriculture

ICSAB International Civil Service Advisory Board

ICSAM International Conference on Superplasticity in Advanced Materials

ICSAPI Internet Connection Services Application Program Interface

ICSAR Interdepartment Committee on Search and Rescue

ICSB International Committee on Systematic Bacteriology [Canada]

ICSB International Council for Small Business [US]

ICSBA International Committee on the Study of Bauxite, Alumina and Aluminum

ICSC Indian Center for Superconductivity [of Materials Research Society of India]

ICSC Interim Communications Satellite Committee [now International Telecommunications Satellite Organization, US]

ICSC International Center for Safety Communication

ICSC International Civil Service Commission [US]

ICSC International Conference on Solution Chemistry

ICSC International Conference on Substrate Crystals (and High-Temperature Superconductor Thin Films)

ICSC International Computer Science Convention

ICSC Inter-Ocean Canal Study Commission [US]

ICSCA International Confederation of Societies of Composers and Authors

ICSCC Intercrystalline Stress Corrosion Cracking

ICSCI International Center for Soil Conservation Information

ICSCRM International Conference on Silicon Carbide and Related Materials

ICSCS International Conference on Surface and Colloid Science

ICSD Inorganic Crystal Structure Database [of Fachinformationszentrum Karlsruhe, Germany]

ICSD Inorganic Crystal Structure Database [of Canada Institute for Scientific and Technical Information] [also CRYSTIN]

ICSE Intermediate Current Stability Experiment [UK]

ICSEAF International Commission for the Southeast Atlantic Fisheries [Spain]

ICSEM International Commission for the Scientific Exploration of the Mediterranean Sea [Monaco]

ICSEM International Council for the Scientific Exploration of the Mediterranean

ICSEP International Center for the Solution of Environmental Problems [US]

ICSF International Conference on Spray Forming

ICSFS International Conference on Solid Films and Surfaces

ICSH Interstitial-Cell-Stimulating Hormone [Biochemistry]

ICSHB International Committee for Standardization in Human Biology

ICSHM International Conference on the Science of Hard Materials

ICSHT International Center for Science and High Technology [of International Center for Theoretical Physics, Italy]

ICSI International Commission on Snow and Ice [of International Association of Hydrological Sciences, Netherlands]

ICSI International Computer Science Institute [of University of California at Berkeley, US]

ICSI International Conference on Scientific Information

ICSI Intracystoplasmic Sperm Injection [Infertility Treatment]

ICSIC International Conference on Shallow Impurity Centers [Materials Science]

ICSICT International Conference on Solid-State and Integrated Circuit Technology of Materials

ICSID International Center for Settlement of Investment Disputes

ICSID International Council of Societies of Industrial Designers [Finland]

ICSISM Institute for Cooperative Study of International Seafood Markets [now International Institute of Fisheries Economics and Trade, US]

ICSM Intergovernmental Committee on Surveying and Mapping [formerly Intergovernmental Advisory Committee on Surveying and Mapping]

ICSM International Committee of Scientific Management

ICSM International Conference on Structural Mechanics

ICSM International Conference on Synthetic Metals

ICSMA International Conference on the Strength of Materials

ICSMA International Conference on the Strength of Metals and Alloys

ICSMM International Conference on Superlattices, Microstructures and Microdevices

ICSMP Interactive Continuous Systems Modelling Program

ICSMRT International Conference on Structural Mechanics in Reactor Technology

ICSOS International Conference on the Structure of Surfaces

ICSP Interagency Council on Standards Policy

ICSP International Committee on Science Photography

ICSP International Conference on Shot Peening

ICSPRO Inter-Secretariat Committee for Scientific Programs Relating to Oceanography

ICSPS International Council for Science Policy Studies

ICSPTF International Conference on Structure and Properties of Thin Films

ICSS Instrumentation and Control Society of Singapore

ICSS International Committee for Shell Structures [now International Association for Shell and Spatial Structures, Spain]

ICSS International Conference on Solid Surfaces

ICSS International Congress of Soil Science

ICSSDM International Conference on Solid-State Devices and Materials

ICSSUR International Conference on Squeezed States and Uncertainty Relations [Physics]

ICSSVM International Commission for Small Scale Vegetation Maps [India]

ICST Imperial College of Science and Technology [London, UK]

ICST Institute for Chemical Science and Technology [now Environmental Science and Technology Alliance Canada, Canada]

ICST Institute for Computer Sciences and Technologies [US]

ICST Institute of Corrosion Science and Technology [London, UK]

ICST International Center for Scientific and Technical Information

ICSTDCS International Conference on the Science and Technology of Defect Control in Semiconductors

ICSTI International Center for Scientific and Technical Information [Russian Federation]

ICSTI International Congress on the Science and Technology of Ironmaking [of Iron and Steel Society, US]

ICSTI International Council for Scientific and Technical Information [France]

ICSTI/ISS International Congress on the Science and Technology of Ironmaking of the Iron and Steel Society [US]

ICSTM Imperial College of Science, Technology and Medicine [London, UK]

ICSTM International Conference on Scanning Tunneling Microscopy

ICSU Independent Canadian Steelworkers Union

ICSU International Council of Scientific Unions

ICSUAB International Council of Scientific Unions Abstracting Board [now International Council for Scientific and Technical Information]

ICSU-CTS Committee on the Teaching of Science of the International Council of Scientific Unions

ICSW International Commission on Surface Water

ICSWOA International Center for Scientific Work Organization in Agriculture

ICT Image Creation Terminal

ICT In-Circuit Tester

ICT Incoherent Twin (Boundary) [Metallurgy]

ICT Incoming Trunk

ICT Indenter and Crack Tester [Mechanical Testing]

ICT Industrial Ceramics Technology

ICT Industrial Computed Tomography

ICT Influence Coefficient Tests

ICT Institute of Circuit Technology [UK]

ICT Institute of Clay Technology [UK]

ICT Institute of Computer Technology

ICT Institute of Concrete Technology [UK]

ICT (Fraunhofer-)Institut für Chemische Technology [(Fraunhofer) Institute for Chemical Technology, Karlsruhe, Germany]

ICT Instytutu Cybernetyki Technicznej [Institute of Technical Cybernetics; of Wroclaw Technical University, Poland]

ICT Insulating Core Transformer

ICT Insulin Coma Therapy

ICT Integrated-Circuit Technology

ICT Integrated Computer Telemetry

ICT Interaction Control Table

ICT Interactive Command Test

ICT International Coal Trade

ICT Internal Combustion Turbine [also ict]

ICT International Commission on Trichinellosis [Poland]

ICT International Computers and Tabulators

ICT International Conference on Thermoelectrics

ICT International Council of Tanners [UK]

ICT International Critical Tables

ICT Intramolecular Charge Transfer

IC/T Integrated Computer/Telemetry

ICTA International Center for Typographical Arts

ICTA International College of Tropical Agriculture [India]

ICTA International Confederation for Thermal Analysis [Israel]

ICTA International Confederation of Technical Agriculturalists

ICTAA Imperial College of Tropical Agriculture Association [UK]

ICTAM International Congress on Theoretical and Applied Mechanics

ICTB Incoherent Twin (Grain) Boundary [Materials Science]

ICTB International Conference on Tall Buildings [US]

ICTC Interdepartmental Committee on Toxic Chemicals

ICTC Interface Control Tooling Center

ICTC International Cooperative Training Center [US]

IC/TC Intelligent Color/Trash Coordinator

ICTDR International Centers for Tropical Disease Research

ICTE Inertial Component Test Equipment

ICTED International Cooperation in Transport Economics Documentation (Information Network) [of European Conference of Ministers of Transport, France]

ICTF International Cocoa Traders Federation

ICTF International Conference on Thin Films

ICtl Industrial Control [Institute of Electrical and Electronic Engineers–Industry Applications Society, US]

ICTP International Center for Theoretical Physics [Trieste, Italy]

ICTP International Center for Theoretical Physics [of International International Atomic Energy Agency-UNESCO]

ICTP International Conference on the Technology of Plasticity

ICTPA International Conference on Titanium Products and Applications [of Titanium Development Association, US]

ICTPDC Imperial College Thermophysical Properties Data Center [UK]

ICTP-IUPAP International Center for Theoretical Physics–International Union of Pure and Applied Physics

ICTR International Center for Technical Research

ICTS Intermediate Capacity Transport System

ICTS International Center for Transportation Studies [Italy]

ICTS Isothermal Capacitance Transient Spectroscopy

ICTSD International Center for Trade and Sustainable Development

ICTTM International Conference and Exhibition on Thermal Treatment of Materials (and Heat Treat Show)

ICTTT International Congress on Technology and Technology Transfer

ICTU Independent Canadian Transit Union

ICTU Irish Congress of Trade Unions

ICTV International Committee on Taxonomy of Viruses

ICTVdb International Committee on Taxonomy of Viruses Database

ICTVTR Islamic Center for Technical and Vocational Training and Research

ICU Indiana Central University [Indianapolis, US]

ICU Indicator Control Unit

ICU Inertial Components Unit

ICU Instruction-Cache Unit

ICU Instruction Control Unit

ICU Integrated Control Unit

ICU Intel Configuration Utility

ICU Intensive-Care Unit

ICU Interface Connection Unit

ICU Interface Control Unit

ICU International Chemistry Union

ICU International Communication Union

ICU International Communication Unit

ICU ISA (Industry Standard Architecture) Configuration Utility

ICUAE International Congress on University Adult Education

ICUMSA International Commission for Uniform Methods of Sugar Analysis [Australia]

ICV Initial Chaining Value

ICV Internal Calibration Verification

ICV Internal Correction Voltage

ICVA International Council of Voluntary Agencies

ICVAN International Committee on Veterinary Anatomical Nomenclature [Austria]

ICVGE International Conference on Vapor Growth Epitaxy

ICVM Integrated Circuit Valving Manifold

ICVM International Conference on Vacuum Metallurgy

ICVT Improved Canonical Variational (Transition State) Theory [Physics]

ICVT-LCG Improved Canonical Variational (Transition State) Theory (with) Large-Curvature Ground-State [Physics]

ICW In Connection With

ICW Initial Condition Word [Computers]

ICW Ion Cyclotron Whistler [Nuclear Engineering]

ICW Institute of Clayworkers [UK]

ICW Intake Cooling Water

ICW Interface Control Word [Computers]

ICW International Council for Women

ICW Interrupted Carrier Wave

ICW Interrupted Continuous Wave (Telegraphy) [also icw]

ICWA Institute of Cost and Works Accountants [UK]

ICWA International Coil Winding Association [US]

ICWE International Conference on Water and the Environment

ICWES International Conference of Women Engineers and Scientists

ICWG Interface Control Working Group

ICWG International Clubroot Working Group [Scotland]

ICWG International Coordination Working Group

ICWM International Committee of Weights and Measures

ICWP Interstate Conference on Water Policy [US]

ICWQ International Commission on Water Quality

ICWRS International Commission on Water Resources Systems

ICWRU Idaho Cooperative Wildlife Research Unit [US]

ICWS International Cooperative Wholesale Society

ICWU International Chemical Workers Union

ICX Integrated Circuit Extractor

ICXOM International Congress on X-Ray Optics and Microanalysis

ICY International Cooperation Year

I-Cycle Instruction Cycle [also I-cycle]

ICYYLM International Commission on Yeast and Yeast-Like Microorganisms [France]

ICZ International Congress of Zoology

ICZG Internal Centrifugal Zone Growth

ICZN International Code of Zoological Nomenclature

ICZN International Commission on Zoological Nomenclature [UK]

ID Idaho [US]

ID Idaho Operations Office [of US Department of Energy]

ID Ideal Dielectric(s)

ID Identification

ID Identification Data

ID Identification Division [of Federal Bureau of Investigations, US]

ID Identity

ID Image Digitization

ID Image Dipole

ID Image Distance

ID Image Disector

ID Immunodeficiency; Immunodeficient

ID Immunodiffusion

ID Improvement District

ID Incident Dose [X-Rays]

ID Indefinite Delivery

ID Indicating Device

ID Indicator Diagram

ID Indirect Detection

ID Individual Development

ID Indonesia [ISO Code]

ID Induced Dipole

ID Induced Draft

ID Induced Drag

ID Industrial Democracy

ID Industrial Design(er)

ID Industrial Development

ID Industrial Drive

ID Industrial Dryer; Industrial Drying

ID Inexact Differential

ID Infectious Disease

ID Infective Dose

ID Infinitesimal-Deformation (Linear Elasticity Approach) [Mechanics]

ID Information Distributor

ID Initial Deflection

ID Inner Diameter

ID Input Device

ID Inside Diameter [also id]

ID Inside Dimension

ID Instruction Decoder

ID Insufficient Data

ID Insulation Displacement

ID Integral Dose (of Radiation)

ID Integrated Demonstration

ID Intelligence Department

ID Intelligent Digitizer

ID Interactive Debugging

ID Interactive Device

ID Interdendritic [Metallurgy]

ID Interdigital

ID Interdisciplinary

ID Interface Document

ID Interferometer and Doppler

ID Interior Department

ID Interior Design(er)

ID Intermediate Description

ID Internal Damping

ID Internal Diameter [also id]

ID Interrupt-Driven

ID Interstitial Diffusion [Materials Science]

ID Intradermal

ID Intrinsic Dislocation [Materials Science]

ID Inversion Domain [Physics]

ID Ion Detection; Ion Detector

ID Ion-Dipole (Reaction)

ID Iraqi Dinar [Currency]

ID Iris Diaphragm [Optics]

ID Isodynamic(s)

ID Isothermal Desorption

ID Isotope Dating

ID Isotope Dilution

ID Item Description

ID Item Documentation

ID$_{50}$ Median Infective Dose (or Infective Dose 50) [also ID50, or MID]

I^{3}D Intelligent, Integrated, Interactive Design [also IIID]

I&D Information and Documentation

I&D Integrate and Dump (Detection)

I/D Incrementer/Decrementer

I/D Instruction/Data

I-D Internet-Draft

Id Idaho [US]

Id Idea [also id]

Id Identification;Identifier; Identify

id *(idem)* – the same

id incident dose [X-Rays]

id inside diameter [also ID]

id internal diameter [also ID]

.id Indonesia [Country Code/Domain Name]

IDA Iminodiacetate [also ida]

IDA Iminodiacetic Acid

IDA Independent Distributors Association [US]

IDA Industrial Developers Association

IDA Industrial Diamond Association [US]

IDA Input Data Assembler

IDA Institute for Defense Analysis [US]

IDA Integrated Digital Access

IDA Integrated Digital Avionics

IDA Integrated Disk Adapter

IDA Intelligent Data Access

IDA Intelligent Data Acquisition

IDA Intelligent Disk Array

IDA Intelligent Drive Array

IDA Interactive Debugging Aid

IDA Interconnect Device Arrangement

IDA Interface Display Assembly

IDA International Database Association

IDA International Desalination Association [US]

IDA International Development Assistance

IDA International Development Association [of United Nations]

IDA International Diamond Association (of America) [US]

IDA International Distribution Association [US]

IDA Intrusion Detection Alarm

IDA Ionospheric Dispersion Analysis

IDA Irish Dental Association [UK]

IDA Isotope-Dilution Analysis (with Mass Spectrometry)

Ida Idaho [US]

IDAC Import Duties Advisory Committee [UK]

IDAC Industrial Developers Association of Canada

IDAC Integrated Digital-Analog Converter

IDACA International District Heating and Cooling Association [US]

IDAF International Defense and Aid Fund

IDAFSA International Defense and Aid Fund for Southern Africa

IDAL Indirect Data Access List

IDAM Indexed Direct Access Method

IDAM Intelligent Data Management

IDAP Industrial Design Assistance Program

IDAP Interactive Data Access System

IDAPI Integrated Database Application Programming Interface

IDAPS Image Data-Processing System

IDAS Information Displays Automatic Drafting System

IDAS Integrated Data Acquisition System

IDAS Integrated Development Approval System

IDAS Intelligent Data Acquisition System

IDAST Interpolated Data and Speech Transmission [also Idast]

IDAW Indirect Data Address Word

IDB Industrial Development Board [of United Nations]

IDB Industrial Development Bureau [of Ministry of External Affairs, Taiwan]

IDB Input Data Buffer

IDB ln-Suit Drink Bag [Aerospace]

IDB Integrated Database

IDB Inter-American Development Bank

IDB Inversion Domain Boundary [Physics]

IDB Ion-Dipole Bond

IDB Islamic Development Bank [Saudi Arabia]

IDB Isolated Double Bond

IDBMS Integrated Database Management System

IDBMS International Data Base Management Association [US]

IDBP Industrial Development Bank of Pakistan

IDC Image Dissector Camera

IDC IMBLMS (Integrated Medical Behavioral Laboratory Measurement System) Digital Computer

IDC Imperial Defense College [UK]

IDC Impulse-Driven Clock

IDC In Due Course

IDC Industrial Development Corporation [South Africa]

IDC Instantaneous Deviation Control

IDC Instructor Development Course [of National Safety Council's First Aid Institute, US]

IDC Insulated Displacing Connector

IDC Insulation Displacement Connector

IDC Insulation Displacement Contact

IDC Intangible Drilling and Development Costs

IDC Integrated Database Connector

IDC Integrated Demonstration Coordinator

IDC Integrated Desktop Connector

IDC Integrated Displays and Controls

IDC Integrated Disk Control

IDC Interdendritic Corrosion [Metallurgy]

IDC Interdepartmental Committee

IDC Interface Document Control

IDC Interior Design Society [US]

IDC Intermediate Defect Configuration [Materials Science]

IDC Internal Data Channel

IDC International Dairy Committee

IDC International Dairy Congress

IDC International Data Corporation [Canada]

IDC International Design Conference

IDC International Diabetes Center [Minneapolis, Minnesota, US]

IDC International Diamond Council [Belgium]

IDC International Documentation Center [Sweden]

IDC Internationale Dokumentationsgesellschaft für Chemie [International Society for Chemical Documentation, Germany]

IDC Internet Database Connector

IDC Iowa Development Commission [US]

IDC Isothermal Diffusion Couple

IDCA International Development Cooperation Agency [US]

IDCA Inverter Distribution and Control Assembly [also ID&CA]

IDCAS Industrial Development Center for Arab States [Egypt]

IDCC Integrated Data Communications Controller

IDCC International Data Coordinating Center

IDCC International Data Communications Center

IDCCC International Dredging Conference Coordinating Committee [Netherlands]

IDCDA Independent Dealer Committee Dedicated to Action [US]

IDCHE Intergovernmental Documentation Center on Housing and Environment [France]

IDCJ International Development Center, Japan

IDCMA Independent Data Communications Manufacturers Association [US]

IDCNA Insulation Distributor Contractors National Association [now National Insulation Contractors Association, US]

IDCNS Interdivisional Committee on Nomenclature and Symbols [of International Union of Pure and Applied Chemistry]

IDCP International Data Collecting Platform

IDCR International Decade of Cetacean Research

IDCS Image Dissector Camera System

IDCS Instrumentation/Data Collection System

IDCS International Digital Channel Service

IDCSP Initial Defense Communication Satellite Program [US Air Force]

IDD Identification Data

IDD Industrial Development Division [of University of Michigan, Ann Arbor, US]

IDD Integrated Data Dictionary

IDD Interface Definition Document(s)

IDD International Direct Dialing

IDDA Interior Decorators and Designers Association [UK]

IDDD International Direct Distance Dialing [US]

IDDE Integrated Development and Debugging Environment

IDDF Intermediate Digital Distribution Frame

IDDM Insulin-Dependent Diabetes Mellitus [also known as Type I Diabetes Mellitus]

IDDN Integrated Digital Data Network

IDDP International Dairy Development Program [of Food and Agricultural Organization]

IDDRG International Deep-Drawing Research Group [UK]

IDDRG Congr International Deep-Drawing Research Group Congress

IDDS International Digital Data Service

IDE Imbedded Drive Electronics

IDE Initial Design Evaluation

IDE Institute for Developing Economics [Japan]

IDE Integrated Development Environment

IDE Integrated Drive Electronics

IDE Intelligent Drive Electronics

IDE Interactive Data Entry

IDE Interactive Design and Engineering

IDE Interface Design Enhancement

IDE Integrated Definition Electronics

IDEA Ideas Deserving Exploratory Analysis [of National Research Council, US]

IDEA Identification, Development, Exposure and Action; Identify, Develop, Expose and Act

IDEA Industrial Designers Excellence Award

IDEA Innovation Development for Employment Advancement

IDEA Innovative Document Engineering Application [Information Technology]

IDEA Instituto de Estrategias Agropecuarias [Institute for Agricultural Strategies, Spain]

IDEA Institut d'Epidémiologie Appliquée [Institute for Applied Epidemiology, France]

IDEA Institut pour le Développement et l'Education des Adultes [Adult Education and Training Institute, France]

IDEA Instrument Driver for Experimental Applications (Software)

IDEA Integrated Data for Enforcement Analysis (System)

IDEA Integrated Diesel European Action [of European Commission]

IDEA Integrated Digital Electric Aircraft

IDEA Interactive Data Entry Access

IDEA Interactive Digital Electronic Appliance

IDEA Inter-Institutional Directory of European Administrations

IDEA International Data Encryption Algorithm

IDEA International Data Exchange Association

IDEA International Decade of Energy Alternatives

IDEA Investment in Developing Export Agriculture [European Parliament]

IDEA Istituto per la Cura et la Prevenzione della Depressione e dell'Ansia [Institute for the Cure and Prevention of Depressions and Anxiety, Italy]

IDEA-FC Instrument Driver for Experimental Applications Field Control (Software)

IDEALS Ideal Design of Effective and Logical Systems

IDEA-MPS Instrument Driver for Experimental Applications Magnet Power Supply (Software)

IDEAS Incidence-Dependent Excitation for Auger Spectroscopy

IDEAS Integrated Design and Engineering Automated System

IDEAS Integrated Design/Engineering/Architectural System

IDEAS Interior Department Electronic Acquisition System [of US Department of the Interior]

IDEAS International Data Exchange for Aviation Safety [of International Civil Aviation Organization]

IDEA-VSM Instrument Driver for Experimental Applications Vibrating Sample Magnetometer (Software)

IDEC Interior Design Educators Council [US]

IDECO International Development Education Committee of Ontario [Canada]

IDEEA Information and Data Exchange Experimental Activities

IDEF ICAM (Integrated Computer-Aided Manufacturing) Definition Method

IDEF Integrated Data Exchange Facility [of Commonwealth Scientific and Industrial Research Organization, Australia]

IDEF Integrated System Definition Language

IDEF$_0$ ICAM (Integrated Computer-Aided Manufacturing) Definition, Version Zero

IDEF$_1$ ICAM (Integrated Computer-Aided Manufacturing) Definition, Version One

IDEF$_2$ ICAM (Integrated Computer-Aided Manufacturing) Definition, Version Two

IDEM Interdepartmental Electronic Mail [UK]

IDEMA International Disk Drive Equipment and Materials Association

IDEMA Insight IDEMA Insight [Publication of the International Disk Drive Equipment and Materials Association]

IDEMS Integrated Diagnostic Engine Monitoring System

IDEN Instituto de Engenharia Nuclear [Institute of Nuclear Engineering, Brazil]

IDEN Interactive Data Entry Network

Ident Identification; Identify [also ident]

IDENT VISION International Exhibition and Congress on Systems and Applications of Image Processing and Identification Technologies [Germany]

IDEP International Data Exchange Program [US]

IDEP Interservice Data Exchange Program [of US Department of Defense]

IDERA International Development Education Resources Association

IDES Integrated Defense System

IDES Integrated Design and Engineering System

IDEX Initial Defense Communication Satellite Program Experiment

IDEX Initial Defense Experiment

IDF Image Database File

IDF Image Description File

IDF Indicating Direction Finder

IDF Incident Dark-Field [Microscopy]

IDF Induced Draft Fan

IDF Industrial Development Fund

IDF Industrial Diesel Fuel

IDF Institut pour le Développement Forestier [Institute for the Development of Forestry, France]

IDF Integrated Data File

IDF Interior Design Institute [Canada]

IDF Intermediate Distribution Frame

IDF International Dairy Federation [Belgium]

IDF International Development Foundation [US]

IDF International Drilling Federation [US]

IDF Inverse Document Frequency

IdF Institut de France [French Institute, Paris]

IDFL Integral Diode FET (Field-Effect Transistor) Logic

IDF/MS Image Database File/Management System

IDFT Inverse Discrete Fourier Transform

IDFTA International Dwarf Fruit Trees Association [US]

IDG Individual Drop Glider

IDG Inspector of Degaussing

IDG Integrated Drive Generator

IDGE Isothermal Dendritic Growth Experiment [Metallurgy]

IDGTE Institution of Diesel and Gas Turbine Engineers [UK]

IDH Isocitrate Dehydrogenase [Biochemistry]

IDHA International District Heating Association [now IDHCA, US]

IDHCA International District Heating and Cooling Association [US]

IDHE Institute of Domestic Heating

IDHM Infinite-Dimensional Hubbard Model

IDI Improved Data Interchange

IDI Indian Development Institute

IDI Industrial Design Institute [US]

IDI Initial Domain Identifier

IDI Institut de Droit International [Institute of International Law, Paris, France]

IDI Instrumentation Data Items

IDI Intelligent Dual Interface

IDI International Development Institute

IDIA Irish Dairy Industries Association [UK]

I-Di-Acid 2-Amino-5-Hydroxy-1,7-Naphthalenedisulfonic Acid [I-Di-acid]

IDIAS Ice Data Integration Analysis System

IDIB Industrial Diamond Information Bureau [London, UK]

IDIC Industrial Development and Investment Center [of Ministry of External Affairs, Taiwan]

IDIC Islamic Documentation Information Center

IDICT Instituto de Documentación e Información Cientifica y Técnica [Institute for Scientific and Technical Information and Documentation, Cuba]

IDIDI Induced-Dipole-Induced-Dipole Interaction

IDIIOM Information Display Incorporated Input-Output Machine

IDIMS Interactive Data Integration and Management System

IDIN Identification Initial

IDIN International Development Information Network

IDIOT Instrumentation Digital On-Line Transcriber

IDIQ Indefinite Delivery/Indefinite Quantity

IDIS International Dairy Industries Society

IDIS Iowa Drug Information Service [of University of Iowa, US]

IDIV Integer Divide

IDK I Don't Know

IDK Internal Derangement of the Knee [Medicine]

IDL ICAD (Integrated Computer-Aided Design) Design Language

IDL Immuno-Diagnostic Lab(oratory) [San Leandro, California, US]

IDL Information Description Language

IDL Instruction Definition Language

IDL Instrument Design Laboratory [of University of Kansas, Lawrence, US]

IDL Instrument Detection Limit

IDL Interactive Data Language

IDL Interdisciplinary Research Laboratory

IDL Interface Definition Language

IDL International Date Line

IDL Isodynamic Lines [Geophysics]

IDLE International Date Line East [Greenwich Mean Time – 12:00]

IDLH Immediate Danger to Life and Health; Immediate(ly) Dangerous to Life and/or Health

IDLIS International Desert Locust Information Service

IDLRS International Data Library and Reference Service [also IDL&RS]

IDLS Intracavity Dye Laser Spectrometry

IDLW International Date Line West [Greenwich Mean Time +12:00]

IDM (International Workshop on the) Identification of Dark Matter

IDM Incrementally Deposited Material

IDM Industrial Data Management

IDM Information Distribution Manager

IDM Injector Drive Module [Diesel Engine]

IDM Integral and Differential Monitoring

IDM Intelligent Data Mapper

IDM Interactive Data Machine

IDM Interdepartmental Meeting

IDM Interdiction Mission [Millitary]

IDM (Potassium) Iodine–Dyphylline–Mepyramine Maleate

IDMA International Diamond Manufacturers Association

IDMAS Interactive Database Manipulator and Summarizer

IDMH Input Destination Message Handler

IDMM Intermediate and Depot Maintenance Manual

IDMRL Interdisciplinary Materials Research Laboratories [of Advanced Research Projects Agency, US Department of Defense]

IDMS Integrated Database Management System

IDMS Integrated Data Management System

IDMS Isotope Dilution (Analysis with) Mass Spectrometry [also ID-MS]

IDMS/R Integrated Data Management System/Relation

IDN Integrated Digital Network

IDN Intelligent Data Network

IDN Industrial Development Organization [of United Nations]

IDN International Directory Network

IDNR Indiana Department of Natural Resources [US]

IDNR Iowa Department of Natural Resources [US]

IDNX Integrated Digital Network Exchange

IDO Industry Development Office [of National Research Council of Canada]

IDO Interdivisional Operations

IDO International Development Office

IDO Iterative Discrete On-Axis (Encoding)

IDOC Inner Diameter of Outer Conductor

IDOC Internal Dynamic Overload Control

IDOC International Documentation and Communication Center

IDOCS Intrusion-Detection System

ID/OD Inside Diameter/Outside Diameter (Ratio) [also id/od]

IDOE International Decade of Ocean Exploration

IDOP Identification Operational

IDOS Integrated Density of States [Solid-State Physics]

IDOS Interactive Disk-Operating System

IDOS Interrupt Disk-Operating System

IDP Image Depth Profiling

IDP Industrial Data Processing

IDP Inosine Diphosphate; Inosine 5'-Diphosphate [Biochemistry]

IDP Input Data Processor

IDP Institute of Data Processing [London, UK]

IDP Integrated Data Processing; Integrated Data Processor

IDP Interdigit Pause

IDP Intermodulation Distortion Percentage

IDP International Development Program

IDP Isolated Dipole Pair

5'-IDP Inosine-5'-Diphosphate [Biochemistry]

IDPC Integrated Data Processing Center

IDPI International Data Processing Institute

IDPM Institute of Data Processing Management [UK]

IDPM Inf Manage IDPM Information Management [Publication of the Institute of Data Processing Management, UK]

IDPS Interface Digital Processor System

IDPS Interactive Direct Processing System

IDPS/LF Interactive Direct Processing System/Large File

IDQ Industrial Development Quotient

IDQA Individual Documental Quality Assurance

IDR Indonesian Rupiah [Currency]

IDR Industrial Data Reduction

IDR Information Dissemination and Retrieval

IDR Initial Design Review

IDR Inspection Discrepancy Report

IDR Institute of Delphinid Research [now Dolphin Research Center, US]

IDR Integrated Dry Route (Process)

IDR Intelligence Division Report

IDR Intelligent Document Recognition

IDR Interim Discrepancy Report

IDR Intermediate Design Review

IDR Isobaric Double Recycle (Process)

IDRC Industrial Development Research Council [US]

IDRC International Development Research Center [of Indiana University, Bloomington, US]

IDRC International Development Research Center [Ottawa, Canada]

IDRC International Display Research Conference

IDRD Information Definition Requirements Document

IDRD Internal Data Requirement Description

IDRI Information Display Research Institute [of Ajou University, Suwon, South Korea]

IDS Identification Section

IDS Idle Signal Unit

IDS Image-Dissector Scanner

IDS Inclined Driveshaft

IDS Indigodisulfonate

IDS Industrieverband Deutscher Schmieden e.V. [German Forging Industries Association]

IDS Infectious Disease Service [of Georgetown University Hospital, Washington, DC, US]

IDS Information Display System

IDS Inorganic Dielectric Substance

IDS Institut für Deutsche Sprache [German Language Institute, Mannheim]

IDS Instrument Development Section

IDS Integrated Data Store

IDS Intelligent Display System

IDS Intensity-Dependent (Spatial) Summation

IDS Interactive Design Software

IDS Interactive Display System

IDS Interdisciplinary Science; Interdisciplinary Scientist

IDS Interface Data Sheet

IDS Interim Decay Storage

IDS International Development Services

IDS Item Description Sheet

IDSA Iminodisuccinic Acid

IDSA Indian Dairy Sciences Association

IDSA Industrial Designers Society of America [US]

IDSA Industrial Development Subsidiary Agreement

IDSA Infectious Diseases Society of America

IDSA International Development Service of America [US]

IDSA International Diving Schools Association [US]

IDSC International Die Sinkers Conference

IDSCP Initial Defense Satellite Communications Project [US]

IDSCS Initial Defense Satellite Communications System

IDSD Institutional Data System Division [of NASA Johnson Space Center, Houston, Texas, US]

IDSO Interdivisional Sales Order

IDSS ICAM (Integrated Computer-Aided Manufacturing) Decision Support System

IDSS Integrated Documentation Support System

IDST Information and Documentation on Science and Technology

IDT Initial-Load Data Tape

IDT Instrumented Drop Tube

IDT Intelligent Data Terminal

IDT Interdigital Transducer

IDT Interface Design Tool

IDT Interface Display Terminal

IDT Interrupt Descriptor Table

IDT Isodensiotracer

IDTA Interdivisional Technical Agreement

IDTC International Driving Tests Committee [France]

ID-THY Immunodeficiency with Thymoma

IDTIQ Instituto de Desarrollo Tecnológico para la Industria Química [Technological Development Institute for the Chemical Industries, of Universidad Nacional del Litoral, Santa Fe, Argentina]

I-DTMA N,N-Bis(2-Aminoethyl)glycine, Ion(1–) [also I-dtma] [Biochemistry]

IDTS Instrumentation Data Transmission System

IDTS Integrated Demonstration Tracking System

IDTSC Instrumentation Data Transmission System Controller

IDU Idle Signal Unit

IDU Idoxuridine [Biochemistry]

IDU Industrial Development Unit

IDU Injecting Drug User

IDU Interface Data Unit

IDU Interface Demonstration Unit

IDU International Development Unit

IDU 5-Iodo-2'-Deoxyuridine; 5-Iododeoxyuridine [also I-dU, or IDUR] [Biochemistry]

IDW Institut für Dokumentationswesen [Institute for Documentation, Germany]

IDWA Interdivisional Work Authorization

IDWG Interdepartmental Working Group

IDWSSD International Drinking Water Supply and Sanitation Decade

IDX Intelligent Digital Exchange

.IDX Index [File Name Extension] [also .idx]

IE Ice Engineer(ing)

IE Illuminating Engineer(ing)

IE Image Enhancement

IE Immunoelectrophoresis; Immunoelectrophoretic

IE Impact Energy

IE Impulse Engine

IE Incoherent Emission

IE Index Error

IE Indo-European

IE Induced Emission

IE Inductive Effect

IE Industrial Electronics

IE Industrial Engineer(ing)

IE Industrial Ergonomics

IE Inference Engine

IE Information Engineer(ing)

IE Infrared Emission

IE Inhomogeneous Equation

IE Initial Enrichment

IE Initial Equipment

IE Initiating Explosive

IE Input Equipment

IE Inspection and Enforcement

IE Institute of Education

IE Institute of Energy [London, UK]

IE Institute of Export [UK]

IE Institution (or Institute) of Engineers [Barton, Australian Capital Territory]

IE Institution of Engineers [Calcutta, India]

IE Institution of Electronics [UK]

IE Instytutu Elektrotechniki [Institute of Electrical Engineering, Warsaw, Poland]

IE Instrument Engineer(ing)

IE Integral Equation

IE Integrated Electron (Theory)

IE Integrated Electronics

IE Interactive Environment

IE Interfering Element

IE Intermediate Electrode

IE Intermediate Electron

IE Internal Energy

IE Internet Explorer [Microsoft Browser]

IE Interrupt Enable

IE Ion Emission

IE Ion Engine [Aerospace]

IE Ion Etching

IE Ion Exchange; Ion-Exchanged; Ion Exchanger

IE Ion Excitation

IE Ionization Energy

IE Ireland [ISO Code]

IE Isentropic Efficiency [Thermodynamics]

IE Isoelectric

IE Isoelectronic(s)

IE Isothermal Expansion

IE Isotope Effect

IE Isotopic Exchange

IE4 Internet Explorer, Version 4.0 [Microsoft Browser]

IE5 Internet Explorer, Version 5.0 [Microsoft Browser]

I&E Information and Education

I&E Internal and External

I(E) Intensity versus Kinetic Energy

I(E) Signal Intensity (in Electron Energy Loss Spectrometry) [Symbol]

.ie Ireland [Country Code/Domain Name]

i.e. *(id est)* – that is

IEA Idaho Education Association [US]

IEA Industria Española del Aluminio [Spanish Aluminum Industries]

IEA Information Engineering Association

IEA Institute for Energy Analysis [UK]

IEA Institute of Economic Affairs [UK]

IEA Institute (or Institution) of Engineers of Australia [Barton, Australian Capital Territory]

IEA Institute of Environmental Action [Name Changed to Urban Initiatives, US]

IEA Institute of Environmental Assessment [UK]

IEA Instituto de Energia Atomica [Institute for Atomic Energy, Brazil]

IEA Integrated Electronic(s) Assembly

IEA International Economic Association

IEA International Electrical Association [UK]

IEA International Energy Agency [of Organization for Economic Cooperation and Development, France]

IEA International Entomological Association

IEA International Environment Assistance

IEA International Ergonomics Association [Finland]

IEA International Exchange Association [US]

IEA International Exhibitors Association [US]

IEA Irish Exporters Association [UK]

IEADN Institut d'Etudes Avancé de la Défense Nationale [Institute for Advanced National Defense Studies, France]

IEAES Ion-Excited Auger-Electron Spectroscopy

IEAR IEA (Instituto de Energia Atomica) Reactor [Brazil]

IEAust Institution (or Institute) of Engineers of Australia [also IE-AUST] [Barton, Australian Capital Territory]

IEB International Education Board

IEB International Energy Bank [UK]

IEB International Environmental Bureau

IEB International Executive Board

IE-B Institute of Engineers–Bangladesh

IEC Independent Electrical Contractors [US]

IEC Industrial Electrification Council [now The Electrification Council, US]

IEC Inelastic Collision [Mechanics]

IEC Information Exchange Center

IEC Institute of Employment Consultants [UK]

IEC Institute of Energy Conversion [of University of Delaware, Newark, US]

IEC Institute of Environmental Chemistry [of National Research Council of Canada]

IEC Integrated Electronic Component

IEC Integrated Equipment Components

IEC Intergraph Education Center

IEC Interexchange Carrier

IEC Interfacial Electrochemistry

IEC Interfering Element Correction

IEC International Education Center

IEC International Egg Commission [UK]

IEC International Electrochemical Commission

IEC International Electronics Commission

IEC International Electrotechnical Commission [Switzerland]

IEC Ion-Exchange Cell

IEC Ion-Exchange Chromatography

IEC Ion-Exchange Conference

IEC Ion-Exclusion Chromatography

IEC Isotropic Elastic Constant

I&EC Industrial and Engineering Chemistry (Research) [Publication of American Chemical Society, US]

IECA International Erosion Control Association [US]

IEC Bull IEC Bulletin [of International Electrotechnical Commission, Geneva, Switzerland]

IECC International Express Carriers Conference

IECE Institute of Electronic Communications Engineers

IECEC Intersociety Energy Conversion Engineering Conference [of Institute of Electrical and Electronics Engineers, US]

IECEE International Electrotechnical Commission System for Conformity Testing to Standards for Safety of Electrical Equipment

IECEJ Institute of Electronic Communications Engineers of Japan

IECEJ Tech Rep Institute of Electronic Communications Engineers of Japan Technical Reports

IECF International European Construction Federation [France]

IECI Industrial Electronics and Control Instrumentation [Institute of Electrical and Electronics Engineers Group, US]

IECM Induced Environment Contamination Monitor

IECM International Energy Conservation Month

IECM Ion Exchange Resin Method

IECN Instituto Ecuadoriano de Ciencias Naturales [Ecuador Institute of Natural Sciences, Quito, Ecuador]

IECO Inboard Engine Cutoff

IECO International Engineering Company [US]

IECP Instituto Espanol de Corrosión y Protección [Spanish Institute for Corrosion and Protection]

I&EC Process Des Dev IEC (Industrial and Engineering Chemistry), Process Design and Development [of American Chemical Society, US] [also IEC Process Des Dev]

I&EC Prod Res Dev IEC (Industrial and Engineering Chemistry), Product Research and Development [of American Chemical Society, US] [also IEC Prod Res Dev]

IECQA International Electrotechnical Commission Qualification Assessment

IECR Institute of Engineering Cybernetics and Robotics [Sofia, Bulgaria]

I&EC Res Industrial and Engineering Chemistry Research [Journal of the American Chemical Society, US]

IECS Igloo Environment Control Subsystem [of NASA]

IECT Impulsive Ergodic Collision Theory [Physics]

IEC/TC International Electrotechnical Commission/ Technical Committee

IED Individual Effective Dose

IED Industrial Engineering Design

IED Industrial Engineering Division

IED Institution of Engineering Designers [Westbury, Wiltshire, UK]

IED Intelligent Electronic Device(s)

IED Interacting Equipment Documents

IED Interactive Electronic Display

IED Ion-Beam-Enhanced Deposition [also IBED]

IED Ion Energy Distribution

IED Ionospheric Electron Density

IEDC International Energy Development Corporation

IEDD Institution of Engineering Draftsmen and Designers [UK]

IEDF Ion-Energy Distribution Function

IEDM International Electron Devices Meeting [of Institute of Electrical and Electronics Engineers, US]

IEDM Tech Dig International Electron Devices Meeting Technical Digest [of Institute of Electrical and Electronics Engineers, US]

IEDO Institution of Economic Development Officers [UK]

IEDP Independent Exploratory Development Program [of Naval Ocean Systems Center, San Diego, California, US]

IEDS International Environment and Development Service [of United States Agency for International Development/ World Environment Center]

IEE Induced Electron Emission

IEE Initial Environmental Evaluation

IEE Institute for Earth Education [US]

IEE Institute for Environmental Education [US]

IEE Institute of Electrical Engineers [Tokyo, Japan]

IEE Institution of Electrical Engineers [UK]

IEE International Electrology Education

IEE Iso-Electronic Element

IEEA Industrial Energy and Environmental Analysis

IEEC Industrial Electrical Equipment Council

IEEC Inter-African Electrical Engineering College

IEEE Institute of Electrical and Electronics Engineers [New York City, US]

IEEE 488 Institute of Electrical and Electronics Engineers Communications Interface Bus

IEEE 802.x IEEE Standards for Local-Area Network Protocol Definition

IEEE-1284/ECP IEEE Parallel Port Interface/Extended Capabilities Port

IEEE Aerosp Electron Syst Mag IEEE Aerospace and Electronic Systems Magazine [Publication of IEEE AESS, US]

IEEE AESS Institute of Electrical and Electronics Engineers–Aerospace and Electronic Systems Society [also IEEE-AESS, or IEEE/AESS]

IEEE ASSP Institute of Electrical and Electronics Engineers–Acoustics, Speech and Signal Processing Society [also IEEE-ASSP, or IEEE/ASSP]

IEEE ASSP Mag IEEE ASSP Magazine [Publication of IEEE ASSP Society, US]

IEEE CES Institute of Electrical and Electronics Engineers–Consumer Electronics Society [also IEEE-CES, or IEEE/CES]

IEEE CG&A IEEE Computer Graphics and Applications [Publication of IEEE CS, US]

IEEE CHMT Institute of Electrical and Electronic Engineers–Components, Hybrids and Manufacturing Technology (Transactions) [US]

IEEE CHMTS Institute of Electrical and Electronic Engineers–Components, Hybrids and Manufacturing Technology Society [US]

IEEE Circ Dev Mag IEEE Circuits and Devices Magazine [of Institute of Electrical and Electronics Engineers, US]

IEEE COM Institute of Electrical and Electronics Engineers–Communications Society [also IEEE-COM, or IEEE/COM]

IEEE Commun Mag IEEE Communications Magazine [of IEEE COM, US]

IEEE Compon Hybrids Manuf Technol IEEE Components, Hybrids and Manufacturing Technology (Transactions) [of Institute of Electrical and Electronics Society, US]

IEEE Comput Appl Eng Educ IEEE Computer Applications in Computer Engineering [Publication of Institute of Electrical and Electronics Engineers, US]

IEEE Comput Appl Power IEEE Computer Applications in Power [Publication of Institute of Electrical and Electronics Engineers, US]

IEEE Comput Graph Appl IEEE Computer Graphics and Applications [Publication of IEEE CS, US]

IEEE Control Syst Mag IEEE Control Systems Magazine [of Institute of Electrical and Electronics Engineers, US]

IEEE CPMT IEEE Components, Packaging and Manufacturing Technology (Society) [of Institute of Electrical and Electronics Engineers, US]

IEEE CS Institute of Electrical and Electronics Engineers–Computer Society [also IEEE-CS]

IEEE CS&E IEEE Computational Science and Engineering [Publication of IEEE CS, US]

IEEE DEIS Institute of Electrical and Electronics Engineers–Dielectrics and Electrical Insulation Society [also IEEE-DEIS, or IEEE/DEIS]

IEEE Des Test IEEE Design and Test [Publication of IEEE Computer Society, US]

IEEE Des Test Comput IEEE Design and Test of Computers [of IEEE Computer Society, US]

IEEE D&T IEEE Design and Test [Publication of IEEE Computer Society, US]

IEEE ECS Institute of Electrical and Electronics Engineers–Electromagnetic Compatibility Society [also IEEE-ECS, or IEEE/ECS]

IEEE EDL Institute of Electrical and Electronics Engineers–Electron Devices Letters [of IEEE EDS, US]

IEEE EDS Institute of Electrical and Electronics Engineers–Electron Devices Society [also IEEE-EDS, or IEEE/EDS]

IEEE Electr Insul Mag IEEE Electrical Insulation Magazine [of IEEE DEIS, US]

IEEE Electron Dev Lett IEEE Electron Devices Letters [of IEEE EDS, US]

IEEE ElectroTechnol Rev IEEE ElectroTechnology Review [of Institute of Electrical and Electronics Engineers, US]

IEEE EMS Institute of Electrical and Electronics Engineers–Engineering Management Society [also IEEE-EMS, or IEEE/EMS]

IEEE Eng Med Biol Mag IEEE Engineering in Medicine and Biology Magazine [of Institute of Electrical and Electronics Engineers, US]

IEEE GaAs IC Symp Tech Dig IEEE Gallium Arsenide Integrated Circuit Symposium Technical Digest [of Institute of Electrical and Electronic Engineers, US]

IEEE/GBIP Institute of Electrical and Electronics Engineers/General-Purpose Interface Bus

IEEE GRSS Institute of Electrical and Electronics Engineers–Geoscience and Remote Sensing Society [also IEEE-GRSS, or IEEE/GRSS]

IEEE IAS Institute of Electrical and Electronics Engineers–Industry Applications Society [also IEEE-IAS, or IEEE/IAS]

IEEE IAS Conf Rec IEEE IAS Conference Record [of IEEE IAS, US]

IEEE IECEC Institute of Electrical and Electronics Engineers–Intersociety Energy Conversion Engineering Conference [US]

IEEE IECI Institute of Electrical and Electronics Engineers–Industrial Electronics and Control Instrumentation (Group) [also IEEE-IECI, or IEEE/IECI]

IEEE IEDM Institute of Electrical and Electronics Engineers–International Electron Devices Meeting [IEEE-IEDM, or IEEE/IEDM]

IEEE IES Institute of Electrical and Electronics Engineers–Industrial Electronics Society [also IEEE-IES, or IEEE/IES]

IEEE IMS Institute of Electrical and Electronics Engineers–Instrumentation and Measurement Society [also IEEE-IMS, or IEEE/IMS]

IEEE Int Reliab Phys Symp Proc IEEE International Reliability Physics Symposium Proceedings [of Institute of Electrical and Electronics Engineers, US]

IEEE Int SOI Conf Proc IEEE International SOI (Silicon-on-Insulator) Conference Proceedings [of Institute of Electrical and Electronics Engineers, US]

IEEE Int Symp Appl Ferroelectr IEEE International Symposium on Application of Ferroelectrics [of Institute of Electrical and Electronics Engineers, US]

IEEE IRPS Institute of Electrical and Electronics Engineers–International Reliability Physics Symposium [also IEEE-IRPS, or IEEE/IRPS]

IEEE ITG Institute of Electrical and Electronics Engineers–Information Theory Group [also IEEE-ITG, or IEEE/ITG]

IEEE J Ocean Eng IEEE Journal of Oceanic Engineering [of IEEE OES, US]

IEEE J Quantum Electron IEEE Journal of Quantum Electronics [of Institute of Electrical and Electronics Engineers–Lasers and Electro-Optics Society, US]

IEEE J Robot Autom IEEE Journal of Robotics and Automation [of IEEE RAC, US]

IEEE J Sel Areas Commun IEEE Journal of Selected Areas in Communications [of Institute of Electrical and Electronics Engineers, US]

IEEE J Solid-State Circuits IEEE Journal of Solid-State Circuits [of Institute of Electrical and Electronics Engineers, US]

IEEE LEOS Institute of Electrical and Electronics Engineers–Lasers and Electro-Optics Society [also IEEE-LEOS, or IEEE/LEOS]

IEEE Manage Rev IEEE Management Review [of IEEE EMS, US]

IEEE MS Institute of Electrical and Electronics Engineers–Magnetics Society [or IEEE-MS, or IEEE/MS]

IEEE MTT-S IEEE Microwave Theory and Techniques Society [of Institute of Electrical and Electronics Engineers, US] [also IEEE MTTS]

IEEE MTT-S Symp Dig IEEE Microwave Theory and Techniques Society Symposium Digest [of Institute of Electrical and Electronics Engineers, US]

IEEE Multimed IEEE Multimed [of Institute of Electrical and Electronics Engineers, US]

IEEE Netw IEEE Network [of Institute of Electrical and Electronics Engineers, US]

IEEE NPSS Institute of Electrical and Electronics Engineers–Nuclear and Plasma Sciences Society [also IEEE-NPSS, or IEEE/NPSS]

IEEE OES Institute of Electrical and Electronics Engineers–Oceanic Engineering Society [also IEEE-OES, or IEEE/OES]

IEEE PCS Institute of Electrical and Electronics Engineers–Professional Communication Society [also IEEE-PCS, or IEEE/PCS]

IEEE P&DT IEEE Parallel and Distributed Technology [Publication of IEEE Computer Society, US]

IEEE PEDS IEEE Protective Equipment Decontamination Section [of Institute of Electrical and Electronics Engineers, US]

IEEE PES Institute of Electrical and Electronics Engineers–Power Engineering Society [also IEEE-PES, or IEEE/PES]

IEEE P&HEP Institute of Electrical and Electronics Engineers–Plasma and High-Energy Physics (Group) [also IEEE-P&HEP, or IEEE/P&HEP]

IEEE Photon Technol Lett IEEE Photonics Technology Letters [of IEEE LEOS, US]

IEEE Photov Spec Conf Institute of Electrical and Electronics Engineers–Photovoltaic Special Conference

IEEE PHP Institute of Electrical and Electronics Engineers–Parts, Hybrids, and Packaging (Group) [also IEEE-PHP, or IEEE/PHP]

IEEE Power Eng Rev IEEE Power Engineering Review [of IEEE PES, US]

IEEE Press Institute of Electrical and Electronics Engineers Press [US]

IEEE Proc IEDM Institute of Electrical and Electronics Engineers Proceedings of the International Electron Devices Meeting

IEEE PS&A Institute of Electrical and Electronics Engineers–Plasma Sciences and Applications (Society) [also IEEE-PS&A, or IEEE/PS&A]

IEEE PSIM Institute of Electrical and Electronics Engineers–Power System Instrumentation and Measurement (Committee) [also IEEE-PSIM, or IEEE/PSIM]

IEEE PSR Institute of Electrical and Electronics Engineers–Power System Relaying (Committee) [also IEEE-PSR, or IEEE/PSR]

IEEE PVSEC IEEE (International) Photovoltaics Science and Engineering Conference [of Institute of Electrical and Electronic Engineers, US]

IEEE RAC Institute of Electrical and Electronics Engineers–Robotics and Automation Council [also IEEE-RAC, or IEEE/RAC]

IEEE Reliab Soc Newsl IEEE Reliability Society Newsletter [of Institute of Electrical and Electronics Engineers, US]

IEEE RS Institute of Electrical and Electronics Engineers–Reliability Society [also IEEE-RS, or IEEE/RS]

IEEE-SA Institute of Electrical and Electronics Engineers–Standards Association [US]

IEEE/SEMI Institute of Electrical and Electronics Engineers/Semiconductor Equipment and Materials Institute (Symposium)

IEEE SEMI Int Semicond Manuf Symp Proc IEEE SEMI International Semiconductor Manufacturing Symposium Proceedings [of Institute of Electrical and Electronics Engineers and Semiconductor Equipment and Materials Institute, US]

IEEE SMCS Institute of Electrical and Electronic Engineers–Systems, Man and Cybernetics Society [also IEEE-SMCS, or IEEE/SMCS]

IEEE Softw IEEE Software [Publication of the Institute of Electrical and Electronics Engineers, US]

IEEE SOS/SOI Technol Conf Proc IEEE Silicon-on- Sapphire/Silicon-on-Insulator Technology Conference Proceedings

IEEE Spectr IEEE Spectrum [of Institute of Electrical and Electronics Engineers, US]

IEEE SSAP Institute of Electrical and Electronic Engineers–Statistical Signal and Array Processing Committee [also IEEE-SSAP, or IEEE/SSAP]

IEEE SSCC Institute of Electrical and Electronic Engineers–Solid-State Circuits Council [also IEEE-SSCC or IEEE/SSCC]

IEEE SMCS Institute of Electrical and Electronic Engineers–Systems, Man and Cybernetics Society [also IEEE-SMCS, or IEEE/SMCS]

IEEE Technol Soc Mag IEEE Technology and Society Magazine [of Institute of Electrical and Electronics Engineers, US]

IEEE TJMJ IEEE Translation Journal on Magnetics in Japan [of Institute of Electrical and Electronics Engineers, US]

IEEE Trans Acoust Speech Signal Process IEEE Transactions on Acoustics, Speech and Signal Processing [of Institute of Electrical and Electronics Engineers, US]

IEEE Trans Aerosp Electron Syst IEEE Transactions on Aerospace and Electronic Systems [of Institute of Electrical and Electronics Engineers, US]

IEEE Trans Antennas Propag IEEE Transactions on Antennas and Propagation [of Institute of Electrical and Electronics Engineers, UK]

IEEE Trans Appl Supercond IEEE Transactions on Applied Superconductivity [of Institute of Electrical and Electronics Engineers, US]

IEEE Trans Autom Control IEEE Transactions on Automatic Control [of Institute of Electrical and Electronics Engineers, US]

IEEE Trans Biomed Eng IEEE Transactions on Biomedical Engineering [of Institute of Electrical and Electronics Engineers, US]

IEEE Trans Broadcast IEEE Transactions on Broadcasting [of Institute of Electrical and Electronics Engineers, US]

IEEE Trans CHMT IEEE Transactions on Components, Hybrids and Manufacturing Technology [of Institute of Electrical and Electronics Engineers Society, US]

IEEE Trans CHMT A IEEE Transactions on Components, Hybrids and Manufacturing Technology, Part A [of Institute of Electrical and Electronics Society, US]

IEEE Trans CHMT B IEEE Transactions on Components, Hybrids and Manufacturing Technology, Part B [of Institute of Electrical and Electronics Society, US]

IEEE Trans Circuits Syst IEEE Transactions on Circuits and Systems [of Institute of Electrical and Electronics Engineers, US]

IEEE Trans Commun IEEE Transactions on Communications [of IEEE Communications Society, US]

IEEE Trans Compon Hybrids Manuf Technol IEEE Transactions on Components, Hybrids and Manufacturing Technology [of Institute of Electrical and Electronics Society, US]

IEEE Trans Compon Hybrids Manuf Technol A IEEE Transactions on Components, Hybrids and Manufacturing Technology, Part A [of Institute of Electrical and Electronics Society, US]

IEEE Trans Compon Hybrids Manuf Technol B IEEE Transactions on Components, Hybrids and Manufacturing Technology, Part B [of Institute of Electrical and Electronics Society, US]

IEEE Trans Comput IEEE Transactions on Computers [of Institute of Electrical and Electronics Engineers, US]

IEEE Trans Comput-Aided Des Instr Circuits Syst IEEE Transactions on Computer-Aided Design of Integrated Circuits and Systems [of Institute of Electrical and Electronics Engineers, US]

IEEE Trans Consum Electron IEEE Transactions on Consumer Electronics [of IEEE Consumer Electronics Society, US]

IEEE Trans Educ IEEE Transactions on Education [of Institute of Electrical and Electronics Engineers, US]

IEEE Trans Electr Insul IEEE Transactions on Electrical Insulation [of IEEE DEIS, US]

IEEE Trans Electromagn Compat IEEE Transactions on Electromagnetic Compatibility [of IEEE ECS, US]

IEEE Trans Electron Devices IEEE Transactions on Electron Devices [of IEEE EDS, US]

IEEE Trans Energy Convers IEEE Transactions on Energy Conversion [of IEEE PES, US]

IEEE Trans Eng Manage IEEE Transactions on Engineering Management [of IEEE EMS, US]

IEEE Trans Geosci Remote Sens IEEE Transactions on Geoscience and Remote Sensing [of IEEE GRSS, US]

IEEE Trans Ind Appl IEEE Transactions on Industry Applications [of IEEE IAS, US]

IEEE Trans Ind Electron IEEE Transactions on Industrial Electronics [of IEEE IES, US]

IEEE Trans Inf Theory IEEE Transactions on Information Theory [of IEEE ITG, US]

IEEE Trans Instrum Meas IEEE Transactions on Instrumentation and Measurement [of IEEE IMS, US]

IEEE Trans Knowl Data Eng IEEE Transactions on Knowledge and Data Engineering [of Institute of Electrical and Electronics Engineers, US]

IEEE Transl J Magn Jpn IEEE Translation Journal on Magnetics in Japan [of Institute of Electrical and Electronics Engineers, US]

IEEE Trans Magn IEEE Transactions on Magnetics [of Institute of Electrical and Electronics Engineers–Magnetics Society, US]

IEEE Trans Med Imaging IEEE Transactions on Medical Imaging [of Institute of Electrical and Electronics Engineers, US]

IEEE Trans Microw Theory Tech IEEE Transactions on Microwave Theory and Techniques [of Institute of Electrical and Electronics Engineers, US]

IEEE Trans Multimed IEEE Transactions on Multimedia [of Institute of Electrical and Electronics Engineers, US]

IEEE Trans Nucl Sci IEEE Transactions on Nuclear Science [of IEEE Nuclear and Plasma Sciences Society, US]

IEEE Trans Parts Hybrids Packag IEEE Transactions on Parts, Hybrids, and Packaging [of Institute of Electrical and Electronics Engineers, US]

IEEE Trans Pattern Anal Mach Intell IEEE Transactions on Pattern Analysis and Machine Intelligence [of Institute of Electrical and Electronics Engineers, US]

IEEE Trans PHP IEEE Transactions on Parts, Hybrids, and Packaging [of Institute of Electrical and Electronics Engineers, US]

IEEE Trans Plasma Sci IEEE Transactions on Plasma Science [of IEEE Nuclear and Plasma Sciences Society, US]

IEEE Trans Power Appar Syst IEEE Transactions on Power Apparatus and Systems [of Institute of Electrical and Electronics Engineers, US]

IEEE Trans Power Deliv IEEE Transactions on Power Delivery [of IEEE Power Engineering Society, US]

IEEE Trans Power Electron IEEE Transactions on Power Electronics [of Institute of Electrical and Electronics Engineers, US]

IEEE Trans Power Syst IEEE Transactions on Power Systems [of Institute of Electrical and Electronics Engineers, US]

IEEE Trans Prof Commun IEEE Transactions on Professional Communications [of IEEE Professional Communication Society, US]

IEEE Trans Reliab IEEE Transactions on Reliability [of IEEE Reliability Society, US]

IEEE Trans Robot Autom IEEE Transactions on Robotics and Automation [of Institute of Electrical and Electronics Engineers, US]

IEEE Trans Semicond Manuf IEEE Transactions on Semiconductor Manufacturing [of IEEE Electron Devices Society, US]

IEEE Trans Softw Eng IEEE Transactions on Software Engineering [of Institute of Electrical and Electronics Engineers, US]

IEEE Trans Sonics Ultrason IEEE Transactions on Sonics and Ultrasonics [of Institute of Electrical and Electronic Engineers, US]

IEEE Trans Syst Man Cybern IEEE Transactions on Systems, Man and Cybernetics [of IEEE Systems, Man and Cybernetics Society, US]

IEEE Trans Ultrason Ferroelectr Freq Control IEEE Transactions on Ultrasonics, Ferroelectrics and Frequency Control [of IEEE Ultrasonics, Ferroelectrics and Frequency Control Society, US]

IEEE Trans Veh Technol IEEE Transactions on Vehicular Technology [of IEEE Vehiclular Technology Society, US]

IEEE UFFC Institute of Electrical and Electronic Engineers–Ultrasonics, Ferroelectrics and Frequency Control (Society) [also IEEE-UFFC, or IEEE/UFFC]

IEEE UFFCS Institute of Electrical and Electronic Engineers–Ultrasonics, Ferroelectrics and Frequency Control Society [also IEEE-UFFCS, or IEEE/UFFCS]

IEEE-USA [of Institute of Electrical and Electronics Engineers–United States of America

IEEE VTS Institute of Electrical and Electronic Engineers–Vehicular Technology Society [also IEEE-VTS, or IEEE/VTS]

IEEIE Institution of Electrical and Electronic Incorporated Engineers [UK]

IEEJ Institute of Electrical Engineers of Japan [Tokyo, Japan]

IEEMA Indian Electrical and Electronics Manufacturers Association [India]

IEEMA J IEEMA Journal [of Indian Electrical and Electronics Manufacturers Association, India]

IEEP Institute for (or of) European Environmental Policy

IEEP International Environmental Education Program [of UNESCO]

IEE Proc A, Phys Sci Meas Instrum Manage Educ IEE Proceedings A, Physical Science, Measurement and Instrumentation, Management and Education [of Institution of Electrical Engineers, UK]

IEE Proc A, Phys Sci Meas Instrum Manage Educ, Rev IEE Proceedings A, Physical Science, Measurement and Instrumentation, Management and Education, Reviews [of Institution of Electrical Engineers, UK]

IEE Proc B, Electr Power Appl IEE Proceedings B, Electric Power Applications [of Institution of Electrical Engineers, UK]

IEE Proc C, Gener Transm Distrib IEE Proceedings C, Generation, Transmission and Distribution [of Institution of Electrical Engineers, UK]

IEE Proc, Circuits Devices Syst IEE Proceedings, Circuits, Devices and Systems [of Institution of Electrical Engineers, UK]

IEE Proc, Commun IEE Proceedings, Communications [of Institution of Electrical Engineers, UK]

IEE Proc, Comput Digit Tech IEE Proceedings, Computers and Digital Techniques [of Institution of Electrical Engineers, UK]

IEE Proc, Control Theory Appl IEE Proceedings, Control Theory and Applications [of Institution of Electrical Engineers, UK]

IEE Proc D, Control Theory Appl IEE Proceedings D, Control Theory and Applications [of Institution of Electrical Engineers, UK]

IEE Proc E, Comput Digit Tech IEE Proceedings E, Computers and Digital Techniques [of Institution of Electrical Engineers, UK]

IEE Proc, Electr Power Appl IEE Proceedings, Electric Power Applications [of Institution of Electrical Engineers, UK]

IEE Proc F, Commun Radar Signal Process IEE Proceedings F, Communications, Radar and Signal Processing [of Institution of Electrical Engineers, UK]

IEE Proc F, Radar Signal Process IEE Proceedings F, Radar and Signal Processing [of Institution of Electrical Engineers, UK]

IEE Proc G, Circuits Devices Syst IEE Proceedings G, Circuits, Devices and Systems [of Institution of Electrical Engineers, UK]

IEE Proc G, Electron Circuits Syst IEE Proceedings G, Electronic Circuits and Systems [of Institution of Electrical Engineers, UK]

IEE Proc, Gener Transm Distrib IEE Proceedings, Generation, Transmission and Distribution [of Institution of Electrical Engineers, UK]

IEE Proc H, Microw Antennas Propag IEE Proceedings H, Microwaves, Antennas and Propagation [of Institution of Electrical Engineers, UK]

IEE Proc I, Commun Speech Vis IEE Proceedings I, Communications, Speech and Vision [of Institution of Electrical Engineers, UK]

IEE Proc I, Solid-State Electron Devices IEE Proceedings I, Solid-State and Electron Devices [of Institution of Electrical Engineers, UK]

IEE Proc J, Optoelectron IEE Proceedings J, Optoelectronics [of Institution of Electrical Engineers, UK]

IEE Proc, Microw Antennas Propag IEE Proceedings, Microwaves, Antennas and Propagation [of Institution of Electrical Engineers, UK]

IEE Proc, Optoelectron IEE Proceedings, Optoelectronics [of Institution of Electrical Engineers, UK]

IEE Proc Radar Sonar Navig IEE Proceedings. Radar, Sonar and Navigation [of Institution of Electrical Engineers, UK]

IEE Proc Sci Meas Technol IEE Proceedings, Science, Measurement and Technology

IEE Proc Softw Eng IEE Proceedings, Software Engineering [of Institution of Electrical Engineers, UK]

IEER Institute of Energy and Earth Resources [Commonwealth Scientific and Industrial Research Organization at Port Melbourne, Australia]

IEE Rev IEE Review [of Institution of Electrical Engineers, UK]

IEES Imaging Electron Energy Spectrometer

IEETE Institution of Electrical and Electronics Technicians and Engineers [London, UK]

IEF Information Engineering Facility

IEF Institut d'Electronique Fondamentale [Institute of Fundamental Electronics; of Université Paris Sud, Orsay, France]

IEF Instruction Execution Function

IEF Interface Energy Function

IEF International Environmental Facility

IEF International Environment Forum [of World Environment Center, US]

IEF International Exchange of Fingerprints

IEF Inventory Exchange Format [Computers]

IEF Isoelectric Focusing [Physical Chemistry]

IEFC International Emergency Food Committee [of Food and Agricultural Organization]

IEFR International Emergency Food Reserve

IEFUA International Electronic Facsimile Users Association

IEG Information Exchange Group [of NATO]

IEH Industrial Electric Heating

IEHMO Iterative Extended Hueckel Molecular Orbital [Physical Chemistry]

IEHO Institution of Environmental Health Officers [UK]

IEHT Iterative Extended Hueckel (Molecular Orbital) Theory

IEI Implantation-Enhanced Interdiffusion [Materials Science]

IEI Industrial Education Institute [US]

IEI Institut d'Esthétique Industrielle [Institute for Industrial Design]

IEI Institute of Electrical Inspectors [Australia]

IEI Institution of Engineering Inspection [UK]

IEI Institution of Engineers of India [Calcutta, India]

IEI Institution of Engineers of Ireland [Dublin, Ireland]

IEI International Educator's Institute [US]

IEI International Enamellers Institute [UK]

IEI Ion-Electron Interaction

IEI Istituto di Elaborazione della Informazione [Information Specification Institute; of Centro Nazionale delle Ricerche, Italy]

IEIC Institution of Engineers-in-Charge [UK]

IEICE Institute of Electronics, Information and Communication Engineers [Tokyo, Japan]

IEICEJ Institute of Electronics, Information and Communication Engineers of Japan [Tokyo]

IEICE Trans Electron IEICE Transactions on Electronics [of Institute of Electronics, Information and Communication Engineers, Tokyo, Japan]

IEIJ Illuminating Engineering Institute of Japan [Tokyo, Japan]

IEIS Integrated Engine Instrument System

IEJE Institut Economique et Juridique de l'Energie [Institute for Economic and Legals Aspects of Energy, France]

IEK Institut für Experimentale Kernphysik [Institute for Experimental Nuclear Physics, University of Karlsruhe, Germany]

IEL IEEE/IEE (Institute of Electrical and Electronics Engineers/Institution of Electrical Engineers) Electronic Library [US/UK]

IEL Injection Electroluminescence

IELA International Exhibition Logistics Associates [Switzerland]

IEM Immersion Electron Microscope

IEM Institut Electrotechnique Montefiori [Montefiori Institute of Electrical Engineering, Liege, Belgium]

IEM Interim Examination and Maintenance

IEM Interstitial-Electron Model

IEM Ion Exchange Membrane

IEM Isocyanatoethylmethacrylate

IEM Itinerant Electron Metamagnetic (Transition)

IEMA Indian Electrical Manufacturers Association

IEMA International Explosive Metalworking Association

IEMB Indiana Environmental Management Board [US]

IEM CELL Interim Examination and Maintenance Cell

IEME Inspectorate of Electrical and Mechanical Equipment [UK]

IEMIS Integrated Emergency Management Information System

IEMM Incident Energy Modulation Method

IEMSI Interactive Electronic Mail Standard Identification

IEMT International Electronic Manufacturing Technology

IEMTF Interim Examination and Maintenance Training Facility

IEN Internet Engineering Notes

IEN Internet Experiment Note

IEN Istituto Elettrotecnico Nazionale (Galileo Ferraris) [Galileo Ferraris National Electrotechnical Institute, Torino, Italy

IENAA Instrumental Activation Analysis with Epithermal Neutrons

IENFG Istituto Elettrotecnico Nazionale Galileo Ferraris [Galileo Ferraris National Electrotechnical Institute, Torino, Italy]

IEnvSc Institution of Environmental Sciences [UK]

IEO Industry and Environment Office [of United Nations Environment(al) Program]

IEO Institute of Electro-Optics [of National Chiao Tung University, Hsinchu, Taiwan]

IEO Integrated Electronic Office

IEOS International Earth Observing (or Observation) System

IEOSM International Earth Observation Satellite Missions

IEP Image Edge Profile

IEP Immunoelectrophoresis; Immunoelectrophoretic

IEP Improved Effective Potential (Theory) [Physics]

IEP Industry and Environment Office [of International Union for the Conservation of Nature and Natural Resources]

IEP Instantaneous Effective Photocathode

IEP Institute for Ecological Policies

IEP Instrumentation for the Evaluation of Pictures

IEP International Energy Program

IEP Irish Pound [Currency]

IEP Isoelectric Point [also iep] [Physical Chemistry]

IEPA Illinois Environmental Protection Agency [US]

IEPA Independent Electron Pair Approximation

IE/PAC Industry and Environment Program Activity Center [of United Nations Environmental Program]

IEPG Independent European Program Group

IEPG Internet Engineering Planning Group

IEPRC International Electronic Publishing Research Center [UK]

IEPS International Electronics Packaging Society [US]

IEPS International Electronic Packaging Symposium

IEPS Isoelectric Point of the Surface [Physical Chemistry]

IEPS J IEPS Journal [of International Electronics Packaging Society, US]

IEQ Istituto di Elettronica Quantistica [Quantum Electronics Institute; of Consiglio Nazionale delle Ricerche at Firenze, Italy]

IEQ-CNR Istituto di Elettronica Quantistica–Consiglio Nazionale delle Ricerche [Quantum Electronics Institute of the National Research Council at Firenze, Italy]

IER Institute for Econometric Research [US]

IER Institute for Environmental Research [of Environmental Science Services Administration, US]

IER Institute of Employment Rights [UK]

IER Institute of Engineering Research [US]

IER Ion-Exchange Resin

IER Multiplier [also ier]

IERD Industry Energy Research and Development (Program) [of Energy Technology Branch, Natural Resources Canada]

IERE Institution of Electronic and Radio Engineers [Australia]

IERE Institution of Electronic and Radio Engineers [Now Part of Institution of Electrical Engineers, UK]

IERE International Electrical Research Exchange

IERE Conf Proc Institution of Electronic and Radio Engineers Conference Proceedings [UK]

IERF International Education and Research Foundation [US]

IERI Illuminating Engineering Research Institute [US]

IERN Institut Européen pour la Recherche Nucléaire [European Institute for Nuclear Research]

IERS International Earth Rotation Service

IES Illuminating Engineering Society [US, UK and Australia]

IES Incoming Echo Suppressor

IES Industrial Electronics Society [US]

IES Inelastic Scattering

IES Information Exchange System

IES Institute for Earth Sciences [of Environmental Science Services Administration, US]

IES Institute of Environmental Sciences [US]

IES Institute of Environmental Studies [Canada]

IES Institution of Engineers and Shipbuilders [UK]

IES Institution of Engineers of Singapore

IES Institution of Environmental Sciences [UK]

IES Instituto de Estudios Superiores [Institute for Higher Studies, Montevideo, Uruguay]

IES Integral Error Squared

IES Internal Environmental Simulator

IES International Ecology Society [US]

IES International Exchange Service [of Smithsonian Institution, Washington, DC, US]

IES Intrinsic Electric Strength

IES Inverted Echo Sounder

IES Iowa Engineering Society [US]

IES Isoelectronic Sequences [Spectroscopy]

IESC Information Exchange Steering Committee [of Department of Finance, Australia]

IESC International Executive Service Corps (Team)

IESG Internet Engineering Steering Group

IES J IES Journal [of Institution of Engineers, Singapore]

IESL Institution of Engineers of Sri Lanka

IESIS Institution of Engineers and Shipbuilders of Scotland

IES Monogr Ser IES Monograph Series [of Institution of Environmental Engineers, UK]

IESNA Illuminating Engineering Society of North America [US]

IESNEC Institution of Engineers and Shipbuilders of the Northeast Coast [UK]

IES Proc IES Proceedings [of Institution of Environmental Engineers, UK]

IESR Induced Electron Spin Resonance

IESS Inelastic Electron Scattering Spectroscopy

IESS Institution of Engineers and Shipbuilders in Scotland

IESS Instituto di Elettronica dello Stato Solido [Institute for Solid-State Electronics; of Consiglio Nazionale delle Ricerche, Rome, Italy]

IESS-CNR Instituto di Elettronica dello Stato Solido–Consiglio Nazionale delle Ricerche [Institute for Solid-State Electronics National Research Council, Rome, Italy]

IET Inelastic Electron Tunneling

IET Initial Engine Test

IET Institut de l'Engrenage et desTransmissions [Institute for Gears and Transmissions, France]

IET Institute of Engineers and Technicians [UK]

IET Instrumentation Engineering Technician

IET Instrumentation Engineering Technologist; Instrumentation Engineering Technology

IETA International Electrical Testing Association [US]

IETAAS Impact Electrothermal Atomic Absorption Spectroscopy

IETAM Instituto de las Estudios Tecnico et Avanzado de Monterrey [Monterrey Institute for Advanced Technical Studies, Mexico]

IETC Interagency Emergency Transportation Committee

IETC International Environmental Technology Centre [of United Nations Environment Program in Japan]

IETE Institution of Electronics and Telecommunications Engineers [New Delhi, India]

IETEJ Institute of Electronics and Telecommunications Engineers of Japan

IETE Tech Rev IETE Tech Rev [of Institution of Electronics and Telecommunication Engineers, India]

IETF Internet Engineering Task Force

IETG International Energy Technology Group [of Organization for Economic Cooperation and Development, France]

IETO Interagency Office of Environmental Technology [US]

IETS Inelastic Electron Tunneling Spectroscopy

IETS Inelastic Electron Tunneling Spectrum

IETT Institute of European Trade and Technology

IETTAB International Environment Technology Transfer Advisory Board

IEU International Ecosystems University–Forum International [US]

IEU International Electrical Unit

IEV Initial Entry Vehicle

IEV Institut für Energieverfahrenstechnik [Institute for Energy Process Engineering; of Kernforschunganlage Julich, Germany]

IEV International Electrotechnical Vocabulary [of International Electrotechnical Commission, Switzerland]

IE-WO₃ Ion-Exchanged Tungstic Oxide

IEX Ion Exchange

IEX Ion-Excited X-Ray

I/EX Instruction Execution

IEXC Ion Exchange Chromatography

IEXF Ion-Excited X-Ray Fluorescence

IExpE Institute of Explosives Engineers [UK]

IEXS Ion-Excited X-Ray Spectroscopy

IEXTRU International Conference on Extrusion

IF Ideal Fluid

IF Image Field

IF Image Furnace

IF Impact Force

IF Importance Factor

IF Incompressible Flow

IF Index Feed [Machining]

IF Index Fossil

IF Induced Fission

IF Induced Flow

IF Induction Field

IF Induction Furnace

IF Industrial Fluoroscopy

IF Industrial Furnace

IF Inertial Flow

IF Inertial Force

IF Information [also If, or if]

IF Information Feedback

IF Inlet Filter

IF Inorganic Fullerene-Like (Material)

IF Input Format

IF Instantaneous Frequency

IF Institut de France [French Institute, Paris]

IF *(Instituto de Fisica)* – Portuguese for "Institute of Physics"

IF Institute for Fermentation [Japan]

IF Instruction Field

IF Instrument Flight; Instrument Flying

IF Insufficient Funds

IF Integral Function

IF Integration Facility

IF Interchange Factor

IF Interface [also I/F]

IF Interfacial Film [Physics]

IF Interference Filter

IF Interference Fit

IF Interference Fringe

IF Interferon [Biochemistry]

IF Intermediate-Approach Fix [Aeronautics]

IF Intermediate Frequency [also I-F, if, or i-f]

IF Intermolecular Forces [Physical Chemistry]

IF Internal Friction

IF Interstitial-Free (Steel) [also I-F]

IF Inventrepreneur Forum [US]

IF Inverse Fission [Physics]

IF Inverted File

IF Inviscid Flow

IF Ion Focusing

IF Ionization Front [Astronomy]

IF Isoelectric Focusing [Physical Chemistry]

IF Isothermal Flow [Thermodynamics]

IF$_{int}$ Internal Friction, Intrinsic [Symbol]

IF$_{pt}$ Internal Friction, Phase Transition [Symbol]

IF$_{tr}$ Internal Friction, Transient [Symbol]

I/F Interface [also IF]

I-F Insecticide-Fertilizer (Mixtures)

I-F Intermediate Frequency [also IF, or if]

I-F Interstitial-Free (Steel) [also IF]

If Information [also If, or if]

if intermediate frequency [also IF, or I-F]

i-f intermediate-frequency [Adjective]

IFA Image-Based Failure Analysis

IFA Immunofluorescence Analysis

IFA Immunofluorescent Antibody Assay

IFA In-Flight Analyses

IFA Information Flow Analysis

IFA Institute for Atomic Energy [Norway]

IFA Institute of Field Archeologists [UK]

IFA Institute of Financial Accountants [UK]

IFA Institute of Foresters of Australia

IFA Institut Français d'Archéologie [French Archeological Institute]

IFA Integrated File Adapter

IFA Intensive Flux Array
IFA Interface Functional Analysis
IFA International Federation of Accountants [US]
IFA International Federation of Airworthiness [UK]
IFA International Fertilizer Industry Association [France]
IFA International Financial Accountants [UK]
IFA International Fiscal Association
IFA International Fructose Association
IFA Ionization Front Accelerator
IfA Institut für Arbeitsphysiologie an der Universität Dortmund [Institute for Occupational Physiology (of the University of Dortmund), Germany]
IFAA International Federation of Advertising Agencies [US]
IFAA International Flow Aids Association [US]
IFAA International Furniture Accessory Association
IFABC International Federation of Audit Bureaus of Circulation
IFAC Integrated Flexible Automation Center
IFAC International Federation of Accountants [US]
IFAC International Federation of Automatic Control [Austria]
IFAC International Food Additives Council [US]
IFACE Interface Element
IFAD International Fund for Agricultural Development [of United Nations Food and Agricultural Organization in Rome, Italy]
IFAF Institut für Atom- und Festkörperphysik [Institute for Atomic and Solid-State Physics; of FU Berlin, Germany]
IFAG Bundesamt für Kartographie und Geodäsie [Federal Office of Cartography and Geodesy, Germany]
IfAG Institut für Angewandte Geodäsie [Institute for Applied Geodesy, University of Potsdam, Germany]
IFAI Industrial Fabrics Association International [US]
IFALPA International Federation of Airline Pilot Associations [UK]
IFAM Initial–Final Address Message
IFAM (Fraunhofer-)Institut für Angewandte Materialforschung [Institute for Applied Materials Research] [of Fraunhofer Institute, Bremen and Dresden, Germany] [also IfaM]
IFAM Inverted File Access Method
IFAN Internationale Föderation der Ausschüsse Normenpraxis [International Federation for the Application of Standards, Switzerland]
IFAP International Federation of Agricultural Producers [France]
IfAP Institute für Angewandte Physik [Applied Physics Institute; of Zurich Federal Institute of Technology, Switzerland]
IFAPA International Foundation of Airline Passenger Associations [Switzerland]
IFAPWE Institute of Ferroalloy Producers in Western Europe [Switzerland]
IFARD International Federation of Agricultural Research Systems for Development [Netherlands]
IFAS Intelligent Fracture Analysis System

IFASC Integrated Functions Assessment Steering Committee [US]
IFAT Indirect Fluorescent Antibody Technique [Immunology]
IFAT Internationale Fachmesse für Entsorgung: Abwasser, Abfall, Städtereinigung, Straßenbetriebs– und Straßenwinterdienst [International Trade Fair for Waste Water and Waste Disposal, Municipal Cleaning and Snow Removal, Munich, Germany]
IFATCA International Federation of Air Traffic Controller Associations [Ireland]
IFATCC International Federation of Associations of Textile Chemists and Colorists
IFATE International Federation of Aerospace Technology and Engineering
IFATE International Federation of Airworthiness Technology and Engineering [now International Federation of Airworthiness, UK]
IFATSEA International Federation of Air Traffic Safety Electronic Associations [UK]
IFATU International Federation of Arab Trade Unions
IFAW International Fund for Animal Welfare
IFAWPCA International Federation of Asian and Western Pacific Contractors Associations [Philippines]
IFAX International Facsimile Service
IFAXA International Facsimile Association [also IFaxA, US]
IFB International Film Bureau
IFB Interrupt Feedback (Line)
IFB Inverse Fluidized Bed
IFB Invitation for Bid
IfB Institut für Bauforschung e.V. [Building Research Institute, Germany]
IFBA International Fire Buff Associates
IFBPW International Federation of Business and Professional Women
IFBS Industrieverband zur Förderung des Bauens mit Stahlblech e.V. [Association for the Promotion of the Sheet-Steel Construction Industry, Germany]
IFBSO International Federation of Boat Show Organizers [US]
IfBt Institut für Bautechnik [Structural Engineering Institute, Berlin, Germany]
IFBWW International Federation of Building and Wood Workers [Switzerland]
IFC Image Flow Computer
IFC Independent Fire Control
IFC Industrial Finance Corporation
IFC Information Collector
IFC Inner Field Compensation
IFC Instantaneous Frequency Correlation; Instantaneous Frequency Correlator
IFC Institute of Forest Conservation
IFC Institut Français du Caoutchouc [French Rubber Institute]
IFC Instructor's Free Copy
IFC Interface Clear [General-Purpose Interface Bus Line]
IFC International Facilitating Committee
IFC International Finance Corporation [of United Nations]

IFC International Fisheries Commission

IFC International Forging Congress

IFC International Formulation Committee

IFC Internet Foundation Classes

IFC Iodofluorocarbon

IFCAA International Fire Chiefs Association of Asia [Japan]

IFCATI International Federation of Cotton and Allied Textile Industries

IFCB International Federation of Cell Biology [Canada]

IFCC International Federation of Clinical Chemistry [Finland]

IFCC Newsl IFCC Newsletter [of International Federation of Clinical Chemistry]

IFCCTE International Federation of Commercial, Clerical and Technical Employees

IFCE Institut Français des Combustibles et de l'Energie [French Institute of Combustibles and Energy]

IFCE International Federation of Consulting Engineers

IFCF Integrated Fuel Cycle Facility [Nuclear Engineering]

IFCFAR Instantaneous Frequency Constant False Alarm Rate

IFCI Industrial Finance Corporation of India

IFCM International Federation of Christian Metalworkers

IFCMU International Federation of Christian Miner's Unions

IFCN Interfacility Communication Network

IFCP Institute for Financial Crime Prevention [US]

IFCPAR Indo-French Center for the Promotion of Advanced Research

IFCR International Foundation for Cancer Research

IFCS In-Flight Checkout System

IFCS International Federation of Computer Sciences

IFCS International Financial Services Center [Ireland]

IFCT Industrial Finance Corporation of Thailand

IFCT Institute of Fine Chemical Technology [Russian Federation]

IFCTU International Federation of Christian Trade Unions

IFD Image File Directory

IFD Immunofluorescence Detection

IFD Incipient Fire Detector

IFD Indentation Force Deflection [Materials Testing]

IFD Inland Fisheries Division [State of California, US]

IFD Instantaneous Frequency Discriminator

IFD Instrument Flow Diagram

IFD Internationale Föderation des Dachdeckerhandwerks [International Federation of Roofing Contractors, Germany]

IFD International Federation for Documentation [now International Federation for Information and Documentation, Netherlands]

IFDA International Furnishings and Design Association [Dallas, Texas, US]

IFDC International Fertilizer Development Center [US]

IFDEMS Imaging Field-Desorption Mass Spectrometry

IFDO International Federation of Data Organizations (for the Social Sciences) [Netherlands]

IFE Immunofixation Electrophoresis

IFE Institute for Fluitronics Education [US]

IFE Institute of Freshwater Ecology [of Natural Environmental Research Council, UK]

IFE Institution of Fire Engineers [UK]

IFE Intelligent Front End

IFE International Fastener Exposition and Conference [US]

IFE Ion Field Emission

IFEAT International Federation of Essential Oils and Aroma Trades [UK]

IFEES International Federation of Electroencephalographic Societies

IFEL Inverse Free-Electron Laser (Experiment) [at the Accelerator Test Facility of Brookhaven National Laboratory, Upton, New York, US]

IFEM Intergraph's Finite Element Modelling System [also I/FEM]

IFEMS International Federation of Electron Microscope Societies

IFenE Institute of Fence Engineers [UK]

IFEP Integrated Front-End Processor

IFER International Federation of Engine Reconditioners [France]

IFER International Federation of Railway Advertising Companies [UK]

IFERS International Flat Earth Research Society [now Flat Earth Research Society International, US]

IFES Image Feature Extraction System

IFES International Field Emission Society

IFES International Field Emission Symposium

IFF Identification Friend, or Foe [also IF/F]

IFF If And Only If [also iff]

IFF Institute for Fermentation [Japan]

IFF Institut für Fertigteiltechnik und Fertigbau [Institute for Finished Components Technology and Prefabrication, Weimar, Germany]

IFF Institut für Festkörperforschung [Institute for Solid-State Research; of Kernforschungsanlage Julich, Germany]

IFF Intensity Fluctuation Factor

IFF Interchange(able) File Format

IFF Intermediate Frequency Furnace

IFF International Flying Farmers [US]

IFF Ionized Flow Field

IFF Isoelectric Focusing Facility

IF/F Identification Friend/Foe [also IFF]

IFFA Indigenous Flora and Fauna Association

IFFA Institut für Forstliche Arbeitswissenschaft [Institute of Forestry Sciences, Germany]

IFFA Internationale Fleischwirtschaftliche Fachmesse [International Meat Trade Fair, Germany]

IFFA International Frozen Food Association [US]

IFF/ATCRBS Identification Friend, or Foe/Air-Traffic Control Radar Beacon System

IFFC Integrated Flight and Fire Control (System)

IFF/KFA Institut für Festkörperforschung/ Kernforschungsanlage Jülich [Institute for Solid-State Research of the Julich Nuclear Research Center, Germany]

IFF/SIF Identification Friend, or Foe/Selective Identification Feature

IFFT Inverse Fast Fourier Transform

IFG Incoming Fax Gateway

IfG Institut für Geophysik [Institute for Geophysics, TU Clausthal, Germany]

IFGI International Federation of Graphical Industries

IFGL Initial File Generation Language

IfGT Institut für Grafische Technik [Institute for Graphic Technology, Germany]

IFH International Foundation for Homeopathy [US]

IFH Human Interferon [Biochemistry]

IfH Institut für Halbleitertechnik [Institute for Semiconductor Technology; of RWTH Aachen, Germany]

IfH Institut für Halbleiterphysik und Optik [Institute for Semiconductor Physics and Optics, Germany]

IFHE International Federation of Hospital Engineering

a-IFH, Le Leucocyte Human Interferon [Biochemistry]

IFHP International Federation of Housing and Planning

IFHPM International Federation of Hydraulic Platform Manufacturers

IFHT International Federation for Heat Treatment (and Surface Engineering)

IFI Industrial Fasteners Institute [US]

IFI Information for Industry [US]

IFI International Fabricare Institute [US]

IFI International Fastener Institute

IFI International Federation of Interior Architects and Interior Designers [Netherlands]

IFIA Intermountain Forest Industry Association [US]

IFIA International Federation of Inventors Associations [Sweden]

IFIA International Federation of Ironmongers and Iron Merchants Associations

IFIA International Fence Industry Association [US]

IFIA International Fertilizer Industry Association [France]

IFIAS International Federation of Institutes of Advanced Studies

IFIC International Ferrocement Information Center

IFIC International Food Information Council [US]

IFIC Investment Funds Institute of Canada

IFID International Federation for Information and Documentation [Netherlands]

IFIEC International Federation of Industrial Energy Consumers [Switzerland]

IFIF International Federation of Interior Designers

IFIFR International Federation of International Furniture Removers [Belgium]

IFILE Interface File

IFIM Instream Flow Incremental Methodology

IFIMAT Instituto de Fisica de Materiales Tandil [Tandil Institute of Materials Physics, Argentina]

IFIMAT-CICPBA Instituto de Fisica de Materiales Tandil– Comisión de Investigaciones Cientificas de la Provincia de Buenos Aires [Tandil Institute of Materials Physics– Scientific Research Commission of the Province of Buenos Aires, Argentina]

IFIMUP Instituto de Fisica de Materiales do Universidad do Porto [Institute of Materials Physics of the University of Porto, Portugal]

IFIP International Federation for Information Processing [also Ifip]

IFIPC International Federation of Information Processing Congresses

IFIPS International Federation of Information Processing Societies [now International Federation for Information Processing]

IFireE Institution of Fire Engineers [UK]

IFIS Infrared Flight Inspection System

IFIS International Financial Intelligence Service [US]

IFIS International Food Information Service [of Commonwealth Bureau of Dairy Science and Technology, UK]

IFITU Indian Federation of Independent Trade Unions

IFJ International Federation of Journalists

IFK Internal Flow Kinematics

IFKF Institute für Festkörperforschung [Institute for Solid-State Research, of Kernforschungsanlage Julich, Germany]

IFKT International Federation of Knitting Technologists [Switzerland]

IFL Induction Field Locator

IFL Institute of Fluorescent Lighting [UK]

IFL Interfacility Link

IFL Internationale Fachmesse für Ledertechnik [International Leather Engineering Exhibition, Germany]

IFL International Frequency List

IfL Institut für Länderkunde e.V. [Institute for Regional Studies, Leipzig, Germany]

IfL Institut für Logistik [Institute for Logistics, Germany]

IfL-Mitt IfL-Mitteilungen [Communications of the Institute for Logistics, Germany]

IFLA International Federation of Landscape Architects [France]

IFLA International Federation of Library Associations (and Institutions)

IFLASC International Federation of Latin American Study Centers [Mexico]

IFLIPS Integrated Flight Prediction System

IFLOT Intermediate Focal Length Optical Tracer

IFM Induction Flowmeter

IFM In-Flight Maintenance

IFM Institutt for Fysikalsk Metallurgi [Institute for Physical Metallurgy, Norway]

IFM Institute of Fisheries Management [UK]

IFM Instytutu Fisyki Molekularij [Institute of Molecular Physics; of Polish Academy of Sciences]

IFM Integrating Frequency Meter

IFM Interactive File Manager

IFM Interfacial Force Microscope; Interfacial Force Microscopy

IFM Interference Film Microscope; Interference Film Microscopy

IfM Institute for Micromanufacturing [of Louisiana Technical University, Ruston, US]

IfM Institut für Meereskunde (an der Universität Kiel) [Institute for Oceanography (of the University of Kiel, Germany]

IfM Institut für Metallforschung (an der Westfälischen Wilhelms-Universität [Institute for Metals Research of the University of Muenster, Germany]

IFMA Independent Furniture Manufacturers Association [UK]

IFMA Industrial Furnace Manufacturers Association [now Industrial Heating Equipment Association, US]

IFMA International Facility Management Association [US]

IFMA International Farm Management Association [UK]

IFMA International Federation of Margarine Associations [Belgium]

IFMA International Food Service Manufacturers Association [US]

IFMBE International Federation for Medical and Biological Engineering [Canada]

IFME International Federation for Medical Electronics

IFME International Federation of Municipal Engineers [UK]

IFMIS Intelligent Fire Management Information System

IFMMS International Federation of Mining and Metallurgical Students

IFM-PAN Instytutu Fisyki Molekularij–Polskiej Akademii Nauk [Institute of Molecular Physics of the Polish Academy of Sciences] [also IFMPAN]

IFMR Instantaneous Frequency Measurement Receiver

IFMRF International Fusion Materials Radiation Facility

IfMS Institute für Metallische Sonderwerkstoffe [Institute for Special Metallic Materials, TU Dresden, Germany]

IFN Ice Forming Nuclei [Meteorology]

IFN Information [also Ifn, or ifn]

IFN Institut Française de Navigation [French Institute of Navigation, Paris, France]

IFN Interferon [Biochemistry]

IFN-α Interferon-α (or Leucocyte Interferon) [Biochemistry]

IFN-β Interferon-β [Biochemistry]

IFN-β_1 Interferon-β_1 [Biochemistry]

IFN-β_2 Interferon-β_2 [Biochemistry]

IFN-γ Interferon-γ [Biochemistry]

IfN Leibniz-Institut für Neurobiologie [Leibniz Institute for Neurobiology, Magdeburg, Germany]

IFNA International FidoNet (Electronic Mail System) Association [US]

IFNAES International Federation of the National Associations of Engineering Students

IFN-αLy Lymphoblastoid Interferon [Biochemistry]

IFNSA International Federation of the National Standardization Associations

IFO Identifiable Flying Object; Identified Flying Object

IFO Institut für Wirtschaftsforschung [Economic Research Institute, Germany]

IFO Intensive Field Observation(s)

IFO International FORTRAN Organization [US]

IFO Interplanetary Flying Object

ifo Institut für Wirtschaftsforschung e.V. [Economic Research Institute, Munich, Germany] [also IFO]

IFOAM International Federation of Organic Agriculture Movements [Germany]

IFOP Institut Français de l'Opinion Publique [French Institute of Public Opinion]

IFORS International Federation of Operational Research Societies [Denmark]

IFOS Ion Formation from Organic Solids (Conference)

IFOSA International Federation of Stationers Association

IFOT In-Flight Operations and Training

IFOV Instantaneous Field of View

IFOV Instrument Field of View

IFP Institut Français du Pétrole [French Petroleum Institute]

IFP *(Instituto de Formación Profesional)* – Spanish for "Institute (or College) of Vocational Education"

IFP Instruction Fetch Pipeline

IFP Integrated File Processor

IFP Intensive Field Program

IFP International Federation of Prestressing

IFP International Federation of Purchasing

IFPA IFPS (Integrated Flight Plan Processing System) Area

IFPA Industrial Fire Protection Association [UK]

IFPA Inter-American Federation of Personnel Administration

IFPA International Fire Photographers Association [US]

IFPAG Isoelectric Focusing in Polyacrylamide Gel

IFPAN Instytutu Fizyki–Polskiej Akademii Nauk [Institute of Physics–Polish Academy of Sciences, Warsaw]

IFPCW International Federation of Petroleum and Chemical Workers

IFPE Institute for Fluid Power Education [now Institute for Fluitronics Education, US]

IFPG International Frequency Planning Group

IFPI International Federation of the Phonographic Industry

IFPI International Federation of the Photographic Industry

IFPITB Inorganic Feed Phosphates International Technical Bureau [of Conseil Européen des Fédérations de l'Industrie Chimique, Belgium]

IFPL ICAO (International Civil Aviation Organization) Flight Plan

IFPM In-Flight Performance Monitor

IFPM International Federation of Physical Medicine

IFPMA International Federation of Pharmaceutical Manufacturers Associations [Switzerland]

IFPMM International Federation of Purchasing and Materials Management [Switzerland]

IFPO Institute of Fire Prevention Officers [UK]

IFPP International Federation of the Periodicals Press [UK]

IFQSC Instituto de Fisica e Química de São Carlos [Sao Carlos Institute of Physics and Chemistry; of Universidade de São Paulo, Brazil]

IFPR IFPS (Integrated Flight Plan Processing System) Region

IFPRA International Federation of Park and Recreation Administration [UK]

IFPRA International Federation of Public Relations Association

IFPRI International Food Policy Research Institute [Washington, DC, US]

IFPS Interactive Financial Planning System

IFPS International Federation of Palynological Societies [Netherlands]

IFPS Integrated Flight-Plan Processing System

IFPTA International Forest Products Transport Association

IFPTE International Federation of Professional Technical Engineers

IFPVS International Federation of Phonogram and Videogram Societies

IFPW International Federation of Petroleum Workers

IFPWA International Federation of Public Warehousing Associations [France]

IFPZ IFPS (Integrated Flight Plan Processing System) Zone

IFQSC Instituto de Fisica e Quimica de São Carlos [Sao Carlos Institute of Physics and Chemistry, Brazil]

IFR Image-to-Frame Ratio

IFR Increasing Failure Rate

IFR In-Flight Refueling

IFR Instantaneous Frequency Receiver

IFR Institute of Food Research [of American Food Research Council, US]

IFR Instrument Flight Rules

IFR Integral Fast Reactor [of Argonne National Laboratory, US]

IFR Interface Register

IFR Internal Function Register

IFR International Federation of Robotics

IFRA International Fragrance Association [Switzerland]

IFRA Spec Rep IFRA Special Report

IFRB International Frequency Registration Board [of International Telecommunication Union]

IFRC International Federation of Red Cross (and Red Crescent Societies)

IFRC International Fusion Research Council [of International Atomic Energy Agency]

IFREMER Institut Français de Recherche sur la Mer [French Ocean Research Institute, Brest]

IFRF International Flame Research Foundation [UK]

IF-RP Interstitial-Free Rephosphorized (Steel)

IFRU In-Flight Replacement Unit

IFRU Interference Frequency Rejection Unit

IFRU Interference Rejection Unit

IFRW Inertia Friction Welding

IFS Infrared Fourier Spectrometry

IFS Installable File System

IFS Institute of Fusion Studies [of University of Texas at Austin, US]

IFS Interactive File Sharing

IFS Interactive Flow Simulator

IFS Interchange File Separator

IFS Intermediate Frequency Strip

IFS International Federation of Surveyors

IFS International Financial Statistics [of International Monetary Fund]

IFS International Fluidics Services

IFS International Foundation for Science

IFS Investment Feasibility Studies

IFS Ionospheric Forward Scatter

IFSA International Fuzzy Systems Association [US]

IFSA Intumescent Fire Seals Association [UK]

IFSC Instituto de Fisica de São Carlos [Sao Carlos Institute of Physics, of University of Sao Paulo, Brazil]

IFSCC International Federation of Societies of Cosmetic Chemists [UK]

IFSEA International Federation of Scientific Editors Associations [US]

IFSEA International Food Service Executives Association [Las Vegas, Nevada, US]

IFSEM International Federation of Societies for Electron Microscopy [US]

IFSHLP Installable File System Helper

IFSL Interface Free Superlattice [Solid-State Physics]

IFSM Information Systems Management

IFSM Institute of Fracture and Solid Mechanics [Lehigh University, Bethlehem, Pennsylvania, US]

IFSM Inter-Fuel Substitution Model

IFSMA International Federation of Shipmasters Associations [UK]

IFSR International Federation for Systems Research [Austria]

IFSS Interfacial Shear Strength [Metallurgy]

IFSS International Flight Service Station

IFST Institute of Food Science and Technology [UK]

IFST International Federation of Shorthand and Typewriting

IFSTA International Fire Service Training Association [US]

IFSTAD Islamic Foundation for Science, Technology and Development [Saudi Arabia]

IF Steel Interstitial-Free Steel [also IF steel, I-F Steel, or I-F steel]

IFT Indiana Federation of Teachers [US]

IFT Indirect Fourier Transform(ation) (Technique)

IFT In-Flight Test

IFT Institute of Food Technologists [Chicago, Illinois, US]

IFT Interfacial Tension

IFT Intermediate Frequency Transformer

IFT International Federation of Translators [Belgium]

IFT International Foundation for Telemetering [US]

IFT International Foundation for Time-Sharing

IFT Inverse Fourier Transform(ation)

IfT Institut für Tieflagerung (radioaktiver Abfälle) [Institute for Underground Storage (of Radioactive Wastes), Braunschweig, Germany]

IfT Institut für Troposphärenforschung [Institute for Tropospheric Research, Leipzig, Germany]

IFTA In-Flight Thrust Augmentation

IFTA Institut Français des Transports Aériens [French Institute of Air Transport]

IFTA International Federation of Teachers Associations

IFTA International Federation of Television Archives [Spain]

IFTAC Inter-American Federation of Touring and Automobile Clubs [Argentina]

IFTAXION IFT Workshop on Axions [Particle Physics]

IFTC International Federation of Thermalism and Climatism [France]

IFTC International Film and Television Council [Italy]

IFTDO International Federation of Training and Development Organizations [UK]

IFTE Intermediate Forward Test Equipment

IFTF Immobilized Fuel Test Facility [of Whiteshell Nuclear Research Establishment, Manitoba, Canada]

IFTF Institute for the Future [US]

IFTF International Fur Trade Federation

IFTM In-Flight Test Maintenance

IFTR Institute of Fundamental Technological Research [Warsaw, Poland]

IFTS In-Flight Test System

IFToMM International Federation for the Theory of Machines and Mechanisms [Czech Republic]

IFTUTW International Federation of Trade Unions of Transport Workers [Belgium]

IFTWA International Federation of Textile Workers Associations

IFU Instruction Fetch Unit

IFU Interface Unit

IFUC Inter-Provincial Farm Union Council [Canada]

IFUNAM Instituto de Fisica del Universidad Nacional Autonóma de Mexico [Institute of Physics of the National Autonomous University of Mexico, Cuernavaca]

IF-UNICAMP Instituto de Fisica–Universidade Estadual Campinas [Institute of Physics of the Campinas State University, Brazil]

IFUNO Indian Federation of United Nations Association

IF-USP Instituto de Fisica–Universidade de São Paulo [Institute of Physics of the University of Sao Paulo, Brazil]

IFUW International Federation of University Women [Switzerland]

IFVME Inspectorate of Fighting Vehicles and Mechanical Equipment [of Ministry of Defense, UK]

IFVPA Independent Film, Video and Photographers Association [UK]

IFVTCC Internationale Föderation der Vereine der Textilchemiker und Coloristen [International Federation of Associations of Textile Chemists and Colorists, Switzerland]

IFW (Fraunhofer-)Institut für Festkörper– und Werkstofforschung [(Fraunhofer) Institute for Solid-State and Materials Research, Dresden, Germany]

IFW Institut für Fertigungstechnik und Spanende Werkzeugmaschinen [Institute for Production Engineering and Metalworking Machine Tools, University of Hanover, Germany]

IfW Institut für Weltwirtschaft (an der Universität Kiel) [Institute for World Economy (of the University of Kiel), Kiel, Germany]

IfW Institut für Werkstoffkunde [Institute of Materials Research, Nuclear Research Center at Karlsruhe, Germany]

IFW Dresden (Fraunhofer-)Institut für Festkörper– und Werkstofforschung Dresden [(Fraunhofer) Institute for Solid-State and Materials Research, Germany]

IFWEA International Federation of Workers Educational Associations

IFWRI Institute of the Furniture Warehousing and Removing Industry [UK]

IFWT Institut für Feinwerktechnik [Institute for Precision Mechanics, Technical University of Vienna, Austria]

IFYGL International Field Year for the Great Lakes

IfZ Institut für Zeitgeschichte [Institute for Contemporary History, Munich, Germany]

IfzP (Fraunhofer-)Institut für zerstörungsfreie Prüfverfahren [(Fraunhofer) Institute for Nondestructive Testing, Saarbrücken, Germany]

IG Ideal Gas

IG Igloo

IG Illuminating Gas

IG Image Gas

IG Immunogold [Immunochemistry]

IG Imperial Gallon [Unit]

IG Impulse Generator

IG Induction Generator

IG Inductor Generator

IG (*Industriegewerkschaft*) – German for "Industrial (Trade) Union"

IG Inert Gas

IG Inertial Guidance [Navigation]

IG Infrastructure Grant

IG Inner Glideslope

IG Inspection Ga(u)ge

IG Inspector General

IG Institute for Geophysics

IG Institution of Geologists [UK]

IG Instructor's Guide

IG Instrumentation Group

IG Instrument Ground

IG Interactive Graphics

IG (*Interessengemeinschaft*) – German for "Association," or "Trust"

IG Interest Group

IG Intergalactic

IG Intergovernmental

IG Intergranular

IG Internal Gear

IG Intragranular

IG Ion(ization) Gauge

IG Ion Gun

IG Island Grain [Metallurgy]

IG Isolated Gate (Transistor)

IG Isotope Geochronology

IG Isotope Geology

Ig Immunoglobulin [Biochemistry]

IGA Inert-Gas Atomization; Inert-Gas Atomized

IGA Inner Gimbal Angle

IGA Inner Gimbal Axis

IGA Institut de Génie Atomique [Atomic Engineering Institute, of Ecole Polytechnique Fédérale de Lausanne, Switzerland]

IGA Integrated Graphics Array

IGA Intergovernmental Agreement

IGA Intergranular Attack [Metallurgy]

IGA International General Aviation

IGA International Geographical Association

IGA International Geothermal Association [UK]

IgA Immunoglobulin A [Biochemistry]

IgA1 Immunoglobulin A1 [Biochemistry]

IgA2 Immunoglobulin A2 [Biochemistry]

IGAAS Integrated Ground/Airborne Avionics System

IGAC International Global Atmospheric Chemistry (Program) [of International Geosphere/Biosphere Program]

IGACSM Intergovernmental Advisory Committee on Surveying and Mapping [now Intergovernmental Committee on Surveying and Mapping]

IGADD Intergovernmental Authority on Drought and Desertification

IGAE Intergovernmental Agreement on the Environment [Australia]

IGAEA International Graphic Arts Education Association [US]

IgA-ED Immunoglobulin A Enhancing Factor [Biochemistry]

I Gal Imperial Gallon [also i gal]

IgA (λ) Immunoglobulin A Lambda Light Chains [Biochemistry]

IGAP International Global Aerosol Program

IGAP Inventors' Grant Assistance Program [of Alberta Department of Energy, Canada]

IGARSS International Geoscience and Remote Sensing Symposium

IGAS International Graphic Arts Society [US]

IGAS International Graphoanalysis Society

IGasE Institution of Gas Engineers [UK]

IGAX Inner Gimbal Axis

IGB Institut für Gewässerökologie und Binnenfischerei [Institute for Freshwater Water Ecology and Fisheries, Berlin, Germany]

IGB Intermediate Gearbox

IGB International Gravimetric Bureau [France]

IGBD Intrinsic Grain Boundary Dislocation [Materials Science]

IGBP International Geosphere/Biosphere Program

IGBP-DIS IGBP Data and Information System

IGBT Insulated Gate Bipolar Transistor

IGC Experimental Chamber Ion Gauge

IGC Incremental (Roll) Gap Control [Metallurgy]

IGC Inert Gas Catalysis

IGC Inert Gas Condensation (Process)

IGC Inspectorate General of Customs [of Ministry of Finance, Taiwan]

IGC Institute for Global Communications

IGC Institute for Graphic Communication [US]

IGC Institutional Grants Committee

IGC Integrated Graphics Controller

IGC Intergovernmental Committee

IGC Intergovernmental Conference

IGC Intergovernmental Copyright Committee [France]

IGC Intergranular Corrosion

IGC Intergranular Crack(ing)

IGC Intermagnetics General Corporation [US]

IGC Internal Gain Control [also igc]

IGC International Geochemical Congress

IGC International Geological Congress

IGC International Geophysical Committee

IGC International Geophysical Cooperation

IGC International Grassland Congress

IGCAR Indira Gandhi Center for Atomic Research [Kalpakkam, India]

IGCC Integrated Gasification Combined Cycle (Process)

IGCC Intergovernmental Copyright Committee

IGCC Insulating Glass Certification Council [US]

IGCC Integrated Gasification Combined Cycle

IGCG Inertial Guidance and Calibration Group [US Air Force]

IGCI Industrial Gas Cleaning Institute [US]

IGCP International Geological Correlation Program [of UNESCO]

IGD Institut de Géodynamique [Geodynamics Institute, Université Michel de Montaigne, Bordeaux, France]

IGD Inverse Gated Decoupling

IgD Immunoglobulin D [Biochemistry]

IGDC Interior Geographic Data Committee [US]

IGDL Institut für Geologie und Dynamik der Lithosphäre [Institute for Lithospheric Geology and Dynamics, University of Gottingen, Germany]

IGDN Inter-American Geospatial Data Network

IGDS Iodine Generating and Dispensing System

IGE Institution of Gas Engineers [UK]

IGE Interdisciplinary and General Engineering

IGE Isopropyl Glycidyl Ether

IgE Immunoglobulin E [Biochemistry]

IGES Initial Graphics Exchange Specification

IGES Institute of Geography and Earth Sciences [of University of Wales, Aberystwyth, UK]

IGES International Geochemical Congress

IGES International Graphics Exchange Standard

IGES/DXF Initial Graphics Exchange Specification/Data Exchange Format

IGES/PDES Initial Graphics Exchange Specification/Product Design Exchange Specification [also IGES-PDES]

IGESUCO International Ground Environment Subcommittee [of NATO]

IGEX International Germanium Experiment

IGF Inert Gas Fusion

IGF Innovative Growth Firm

IGF Insulin-Like Growth Factor [Biochemistry]

IGF Intergranular Fracture

IGF International Genetics Federation [US]

IGF International Graphical Federation

IGF I Insulin-Like Growth Factor I (One) [Biochemistry]

IGF II Insulin-Like Growth Factor II (Two) [Biochemistry]

IGFA International Game Fish Association

IGFBP Insulin-Like Growth Factor Binding Protein [Biochemistry]

IGFET Insulated-Gate Field-Effect Transistor [also IG-FET]

IgG Immunoglobulin G [Biochemistry]

IgG1 Immunoglobulin G1–Human, Mouse, or Rat [Biochemistry]

IgG2 Immunoglobulin G2–Human [Biochemistry]

IgG2a Immunoglobulin G2a–Mouse, or Rat [Biochemistry]

IgG2b Immunoglobulin G2b–Mouse, or Rat [Biochemistry]

IgG2c Rat Immunoglobulin G2c–Rat [Biochemistry]

IgG3 Immunoglobulin G3–Human, or Mouse [Biochemistry]

IgG4 Immunoglobulin G4–Human [Biochemistry]

IGGA International Grooving and Grinding Association [also IG&GA, US]

IgG Fc Immunoglobulin G Fc Fragment [Biochemistry]

IgG frac Immunoglobulin G Fraction of Antiserum [Biochemistry]

IgG (κ) Immunoglobulin G Kappa Light Chains [Biochemistry]

IgG (λ) Immunoglobulin G Lambda Light Chains [Biochemistry]

IGGG Institute for Geology, Geotechnics and Geophysics [Slovenia]

IGGT Institute for Guided Ground Transport

IGH Internationaler Gerichtshof [International Court of Justice, Netherlands]

IGHE Inspector General of Highway Engineering

IGHIA International Garden Horticultural Industry Association [US]

IGHP Innovative Guided Hypervelocity Projectile

IGI Industrial Guest Investigator

IGIA Interagency Group for International Aviation [of Federal Aviation Administration, US]

IGIC Isoconductive Gradient Ion Chromatography

IGIP Internationale Gesellschaft für Ingenieurpädagogik [International Society for Engineering Education]

IGIS Integrated Geographical Information System

IGKB Internationale Gewässerschutzkommission für den Bodensee [International Commission for the Protection of Lake Constance, Switzerland]

IGL Interactive Graphics Language

IGLC International Group for Lean Construction

IGLO Individual Gauge for Localized Orbitals [Physics]

IGLS Integrated Gradient and Least Squares (Algorithm)

IGM Gas Manifold Ion Gauge

IGM Instituto Geología y Metallurgía [Institute of Geology and Metallurgy; of Universidad de las Americas, Mexico]

IGM Interactive Guidance Mode

IGM Intergalactic Material

IGM Intergalactic Medium

IGM Internationale Gesellschaft für Moorforschung [International Society for Marsh Research, Germany]

IGM Iterative Guidance Mode

IgM Immunoglobulin M [Biochemistry]

IGMAA International Gas Model Airplane Association [US]

IgM (κ) Immunoglobulin G Kappa Light Chains [Biochemistry]

IGMP Internet Group Multicast Protocol

IGN Institut Géographique National [National Geographical Institute, France]

IGN International Geographic Institute

Ign Ignite(r); Ignition [also ign]

IGND Instrument Ground

IGNITOR Ignition Torus [also Ignitor]

IGNS Institute for Geological and Nuclear Sciences [Lower Hutt, New Zealand]

Ignt Ignite(s) [also ignt]

Igntg Igniting [also igntg]

IGO Impulse-Governed Oscillator

IGO Intergovernmental Organization

I-GOOS IOC (Intergovernmental Oceanographic Commission) Committee for Global Ocean Observing System

IGOR Interactive Generation of Organic Reactions

IGOR Intercept Ground Optical Recorder [also Igor]

IGORTT Intercept Ground Optical Recorder Tracking Telescope [also igortt]

IGOSS Integrated Global Ocean Services System

IGOSS Integrated Global Ocean Station System

IGP Interior Gateway Protocol [Internet]

IGPC Impregnated Gas-Pressure Cable

IGPCE International Great Plains Conference of Entomologists

IGPM Imperial Gallon per Minute [also igpm, or imp gal/min]

IGPM Institut für Geometrie und Praktische Mathematik [Institute for Geometry and Practical Mathematics; of Rheinisch-Westfälische Technische Hochschule, Aachen, Germany]

IGPP Institute of Geophysics and Planetary Physics [of Scripps Institute of Oceanography, University of California at San Diego, US]

IGPRAD Intergovernmental Panel on Radioactive Waste

IGR Insect Growth Regulator

IGR Institute for Groundwater Research [of University of Waterloo, Ontario, Canada]

IGR Insulated Gate Rectifier

IGRC International Gas Research Conference [of Gas Research Institute, US]

IGRC Proc International Gas Research Conference Proceedings [Gas Research Institute, US]

IGROUP Inspection Group

IGRP Interior Gateway Routing Protocol

IGRP International Genetic Resources Program [now Rural Advancement Fund International, US]

IGR&P Inert Gas Receiving and Processing

IGS Idaho Geological Survey [US]

IGS Immunogold Stain(ing)

IGS Indiana Geological Survey [US]

IGS Inert Gas-Shielded (Welding)

IGS Inert Gas System

IGS Inertial Guidance System

IGS Information Generator System

IGS Information Group Separator

IGS Inner Glideslope

IGS Instrument Guidance System

IGS Integrated Geophysical System

IGS Integrated Graphics System

IGS Interactive Graphics System

IGS Interchange Group Separator

IGS Intergalactic Space

IGS Internationale Gesellschaft für Stereologie [International Society for Stereology, Germany]

IGS International Glaciological Society [UK]

IGS Internet Go Server

IGS Irish Graphical Society

IGSB Iowa Geological Survey Bureau [of Iowa Department of Natural Resources, US]

IGSC Intergranular Stress-Corrosion

IGSC International Gold and Silver Conference

IGSCC Intergranular Stress-Corrosion Cracking

IGSE In-Space Ground Support Equipment

IgSF Immunoglobulin Superfamily [Biochemistry]

IGSHPA International Ground Source Heat Pump Association [US]

IGSPS International Gold and Silver Plate Society

IGSS Immunogold Silver Stain(ing) [Biochemistry]

IGT Impaired Glucose Tolerance [Medicine]

IGT Institute of Gas Technology [US]

IGT Institut für Geotechnik [Institute of Geotechnical Engineering, of Swiss Federal Institute of Technology at Zurich]

IGT Institut für Grafische Technik [Institute for Graphic Technology, Germany]

IGT Insulated Gate Transistor

IGT Intelligent Graphics Terminal

IGT International Gas Technology (Highlights) [Publication of Institute of Gas Technology, US]

IG/T Ionization Gauge/Thermocouple (Gauge)

IGTC International Gas Turbine Center [now International Gas Turbine Institute of the American Society of Mechanical Engineers, US]

IGTC International Gas Turbine Conference

IGT Highlights International Gas Technology Highlights [of Institute of Gas Technology, US]

IGTI International Gas Turbine Institute [of American Society of Mechanical Engineers, US]

IGU International Gas Union

IGU International Geographical Union

IGUPT Institute for Gas Utilization and Processing Technologies [US]

IGV Inlet Guide Valve

IGV Inlet Guide Vane

IGV Institut für Grenzflächenforschung und Vakuumphysik [Institute for Interfacial Research and Vacuum Physics; of Kernforschungsanlage Julich, Germany]

IGV/KFA Institut für Grenzflächenforschung und Vakuumphysik/Kernforschungsanlage Jülich [Institute for Interfacial Research and Vacuum Physics of Julich Nuclear Research Center, Germany]

IGWMC International Ground Water Modeling Center [US]

IGWR Institute for Groundwater Research [of University of Waterloo, Canada]

IGY International Geophysical Year [July 1, 1957 to December 31, 1958]

IGY-IGC International Geophysical Year–International Geophysical Cooperation

IGZ Institut für Gemüse– und Zierpflanzenbau [Institute for Vegetables and Ornamental Plants, Großbeeren/Erfurt, Germany]

IH Immersion Heater

IH Immunohistology

IH Indentation Hardness

IH Induction Heater; Induction Heating

IH Industrial Hygiene; Industrial Hygienist

IH Industrialized Housing

IH Infectious Hepatitis [Medicine]

IH Input Hopper

IH Inside Height

IH Integral Hologram

IH Interaction Handler

IH Inter-Height

IH Internet Hub

IH Interrupt Handler [Computers]

$I_d(H)$ Direct-Current Demagnetization [Symbol]

$I_r(H)$ Isothermal Remanent Magnetization [Symbol]

IHA Irish Hardware Association [UK]

IHAA International Hard Anodizing Association

IHACE International Heating and Air-Conditioning Exposition

IHAS Integrated Helicopter Avionics System

IHB Institute for Human Biology [of University of Utrecht, Netherlands]

IHB International Hydrographic Bureau [now International Hydrographic Organization, Monaco]

IHC Industrial Hygiene Conference [US]

IHC Institut d'Hydrologie et Climatologie [Institute of Hydrology and Climatology, France]

IHC Intergranular Hot Cracking

IHC Interstate Highway Capability [US]

IHC Inverse Heat Conduction (Method)

IHC Isovalent Hyperconjugation [Physical Chemistry]

IHCMCB International Hazard Control Manager Certification Board [US]

IHCP Induction Hardened Chrome Plating

IHCP International Handbook on Coal Petrology

IHCP Inverse Heat Conduction Problem

IHD Ischemic Heart Disease

IHD Integrated Help Desk [of IBM Corporation, US]

IHD International Hydrological Decade [1965-74]

IHDP International Human Dimensions Program (on Global Environmental Change)

IHDP Update International Human Dimensions Program Update

IHE Industry, Human Settlements and Environment (Division) [of United Nations Economic and Social Commission for Asia and the Pacific]

IHE Institute for the Human Environment [US]

IHE Institut für Höhere Elektrotechnik [Institute of Advanced Electrical Engineering, of ETH Zurich, Switzerland]

IHE Institution of Higher Education

IHE Institution of Highway Engineers [UK]

IHE (International) Institute for Hydrologic and Environmental Engineering

IHe Isohexyl [also iHe]

IHEA Industrial Heating Equipment Association [US]

IHEB International Heat Economy Bureau

IHeCl Isohexyl Chloride

IHEP Institute for High-Energy Physics [Protvino, Russia]

IHEP Institute for High-Energy Physics [Chengdu, PR China]

IHES Illinois Horticultural Experiment Stations [US]

IHET Industrial Heat Exchanger Technology

IHF Industrial Hygiene Foundation [US]

IHF Inhibit Halt Flip-Flop

IHF Institute of High Fidelity [US]

IHF Intermediate High Frequency

IHFA Industrial Hygiene Foundation of America [US]

IHFBC International High Frequency Broadcasting Conference

IHFC International Heat Flow Commission

IHFM Institute of High-Fidelity Manufacturers

IHGA International Hop Growers Association

IHHSF International Habitat and Human Settlement Foundation

IHI Ishikawajimi-Harima Heavy Industries Limited [Tokyo, Japan]

IHIA Institute of Highway Incorporated Engineers [UK]

IHI Bull IHI Bulletin [of Ishikawajimi-Harima Heavy Industries Limited, Tokyo, Japan]

IHIC Iso Hexa Imino Cryptand

IHI Eng Rev IHI Engineering Review [Published by Ishikawajimi-Harima Engineering, Tokyo, Japan]

IHIS Integrated Hospital Information System

IHK *(Industrie- und Handelskammer)* – German for "Chamber of Industry and Commerce"

IHK *(Internationale Handelskammer)* – German for "International Chamber of Commerce"

I (hkl) Measured Intensity from (hkl) (Crystallographic) Plane

I$_0$ (hkl) Theoretical Intensity from (hkl) (Crystallographic) Plane

ihl inhalation [Toxicology]

IHM Induction Heating Model

IHMA Industrial Housing Manufacturers Association [US]

IHM/CG Induction Heating Model with Complex Geometry

IHMM Institute of Hazardous Materials Management

IHN Interstitial Hypertrophic Neuritis

IHN 2-Isohexylnaphthalene

IHO Institute of Human Origins [US]

IHO International Health Organization

IHO International Hydrographic Organization [formerly Hydrographic Organization Bureau, Monaco]

IHospE Institute of Hospital Engineering

IHP Indicated Horsepower [also ihp]

IHP Inner Helmholtz Plane [Physics]

IHP Inositol Hexaphosphate [Biochemistry]

IHP Institut für Halbleiterphysik GmbH [Institute for Semiconductor Physics, Frankfurt/Oder, Germany]

IHP Institut Henri Poincaré [Henri Poincaré Institute, France]

IHP Internal Hydrogen Precipitation

IHP International Hydrological Program [of UNESCO]

ihp indicated horsepower [also IHP]

IHPA International Hardwood Products Association [US]

IHPH Indicated Horsepower-Hour [also IHP HR, IHPhr, or ihph]

IHPTEP Integrated High-Performance Turbine Engine Program [US]

IHPTET Integrated High Performance Turbine Engine Technology (Program) [of US Department of Defense]

IHPTET Intermediate High Performance Turbine Engine Technology

IHPVA International Human Powered Vehicle Association [US]

IHR Institute of Historical Research [UK]

IHR Institute of Human Relations [New York, US]

IHR Interstellar Hydroxyl Radicals

IHRD Institute for Human Resource Development

IHRH Instituto de Hidraulica e Recursos Hidricos [Institute of Hydraulics and Water Resources, Porto, Portugal]

IHS Indiana Horticultural Society [US]

IHS Indian Health Service [of US Department of Health and Human Services]

IHS Information Handling Service [US]

IHS Information Handling Services

IHS Institute for Home Safety [UK]

IHS Institute for Hydrogen Studies

IHS Integrated Hospital System

IHS Integration Hardware and Software [also IH/S]

IHS Intensity, Hue, Saturation

IHS International Hibernation Society [US]

IHS International Hydrofoil Society [UK]

IHS International Hydrogen Scale

IHS International Hydrology Symposium

IHS Interstate Highway System [US]

IH&S Industrial Hygiene and Safety

IH/S Integration Hardware and Software [also IHS]

IHSB Industrial Health and Safety Branch [of Ministry of Labour, Canada]

IHSBR Improved High-Speed Bombing Radar

IHSI Induction Heating Stress Improvement

IHSM Institute for High-Speed Mechanics [of Tohoku University, Sendai, Japan]

IH/SR Integration Hardware and Software Review

IHT Institution of Highways and Transportation [UK]

IHT Intercritical Heat Treatment [Metallurgy]

IHT Intermediate Heat Treatment [Metallurgy]

IHTS Intermediate Heat Transport System

IHTU Interservice Hovercraft Trials Unit [of Ministry of Defense, UK]

IHTV Interim Hypersonics Test Vehicle

IHV Independent Hardware Vendor

IHV Infectious Hepatitis Virus

IHVE Institution of Heating and Ventilating Engineers [UK]

IHVS Intelligent Highway Vehicle System

IHW Industrial and Hazardous Waste

IHW International Halley Watch [of NASA Jet Propulsion Laboratory, Pasadena, California, US]

IHX Interloop Heat Exchanger

IHX Intermediate Heat Exchanger

IHXGV Intermediate Heat Exchanger Guard Vessel

II Image Intensifier

II Impact Ionization

II Indefinite Integral

II Index of Inflammability

II Indirect Interaction

II Individual Immunity

II Industrial Imaging

II Information Interchange

II Innovators International [US]

II Input Indicator

II Institute of Inventors [UK]

II Interdisciplinary Investigator

II Intermediate Image [Microscopy]

II Intermolecular Interaction [Physical Chemistry]

II Interrupt Inhibit

II Intrinsic-Intrinsic-Type Layer (Interface) [also i/i] [Materials Science]

II Inverted Image

II Inversional Isomer

II Ion Implantation

II Isoionic

I/I$_0$ Intensity Ratio (i.e., Intensity of Transmitted Beam to Incident Beam (of Electromagnetic Radiation)) [Symbol]

i/i intrinsic-intrinsic-type layer (Interface) [also II] [Materials Science]

IIA Indirect Indexed Addressing [Computers]

IIA Information Interchange Architecture

IIA Information Industry Association [US]

IIA Institute of Internal Auditors [US]

IIA Instituto de Investigaciones Antropologicas [Anthropological Research Institute, Mexico]

IIA Instituto Internaciónal dos Americas [International Institute of the Americas, of World University, Hato Rey, Puerto Rico]

IIA Intelligence Industries Association [US]

IIA Intelligent Industrial Automation

IIA International ICSC (International Computer Science Convention) Symposium on Intelligent Industrial Automation

IIA International Institute of Agriculture [of Food and Agricultural Organization in Rome, Italy]

IIA International Institute of the Americas [of World University, Hato Rey, Puerto Rico]

IIA International Inventors Association

IIAA Institute of Interamerican Affairs

IIAES Ion-Induced Auger Electron Spectroscopy

IIAILS Interim Integrated Aircraft Instrumentation and Letdown System

IIAR International Institute of Ammonia Refrigeration [US]

IIAS Interactive Instructional Answering System

IIAS Inter-American Institute of Agricultural Sciences

IIAS International Institute of Administrative Sciences [Belgium]

IIASA International Institute of Applied Systems Analysis

IIASES International Institute for Aerial Survey and Earth Sciences [Enschede, Netherlands]

IIA-UK Institute of Internal Auditors–United Kingdom

IIA(WU) International Institute of the Americas (of the World University) [Hato Rey, Puerto Rico]

IIB Institut International des Brevets [International Patent Institute]

IIB Institute of International Bankers [US]

IIB International Investment Bank [Russian Federation]

IIB (International Conference on) Intergranular and Interphase Boundaries in Materials

IIBA International Institute for Bioenergetic Analysis

IIBA International Intelligent Buildings Association [US]

IIBBC Institute for Industrial Building and Building Construction

IIBBR International Institute for Biological and Botanical Research

IIBC International Institute of Biological Control [UK]

IIBCA Instituto de Investigaciones en Biomedicina y Ciencias Aplicadas [Institute Biomedical Research and Applied Science, Venezuela]

IIBEM Indian Institute of Biochemistry and Experimental Medicine

IIC Impact (Sound) Insulation Class

IIC India International Center

IIC Indirect Ion Chromatography

IIC Information Interchange

IIC Institute for Instrumentation and Control [Australia]

IIC Institute of Independent Consultants [UK]

IIC Institute of Inorganic Chemistry [of Russian Academy of Sciences]

IIC Instituto de Ingenieros de Chile [Chilean Institute of Engineers]

IIC Inter-Integrated Circuit (Bus) [also I²C]

IIC International Institute for Conservation of Historic and Artistic Work [UK]

IIC International Institute for Cotton [Belgium]

IIC International Institute of Communications [UK]

IIC International Institute of Conservation

IIC International Investment Corporation

IIC Integrated Information Center

IIC Irradiation Induced Creep

IIC Islamic Investment Company

IICA Inter-American Institute for Cooperation on Agriculture [Turrailba, Costa Rica]

IICAF Institute for International Collaboration in Agriculture and Forestry

IICB International Import Custom Brokers

IICC International Institute for Commercial Competition [Belgium]

IICCG International Institute of Conservation–Canadian Group

IICL Institute of International Container Lessors [US]

IICMFA Integrated Information Center of the Ministry of Foreign Affairs [Saudi Arabia]

IICS International Interactive Communications Society [US]

IICT Instituto de Investigaciones Cientificas y Tecnologicas [Institute for Scientific and Technical Investigations, Argentina]

IICY International Investment Corporation for Yugoslavia

IID Impurities Induced Disordering [Materials Science]

IID Independent Identically Distributed (Random Process) [also iid]

IID Institute for Infectious Diseases [of Tokyo University, Japan]

IID Intermittent-Integrated Doppler

IID Ion Impact Desorption

IID Ion-Induced Desorption

IIDA Individualized Instruction for Data Access

IIDA Instituto Interamericano de Direito de Autor [Inter-American Copyright Institute, Brazil]

IIDC Institute for International Development and Cooperation

IIDCT Instituto de Información y Documentación en Ciencia y Tecnologia [Institute for In-formation and Documentation in Science and Technology, Madrid, Spain]

IIDQ 2-Isobutoxy-1-Isobutoxycarbonyl-1,2-Dihydroquinoline

IIDQ Isobutyl 1,2-Dihydro-2-Isobutoxyl-1Quinoline Carboxylate

IIE Institute of Industrial Engineers [Ireland]

IIE Institute of Industrial Engineers [Norcross, Georgia, US]

IIE Institute of International Education [New York City, US]

IIE Instituto de Investigaciones Electricas [Electrical Research Institute, Mexico]

IIE International Institute of Energy

IIEA International Institute for Environmental Affairs

IIED International Institute for Environment and Development [now Center for International Development and Environment, US]

IIEE Ion-Induced Electron Emission

IIEHE Internation Institute for Energy and Human Ecology [Sweden]

IIEIC International Institute Examinations Inquiry Committee [UK]

IIEM Indian Institute of Experimental Medicine

IIES Industrial Innovation Extension Service [of New York State Science and Technology Foundation, US]

IIES International Institute for Environment and Society [of Wissenschaftszentrum Berlin, Germany]

IIES International Institute of the Environment Society

IIE Trans IIE Transactions [of Institute of Industrial Engineers, US]

IIExE Institution of Incorporated Executive Engineers [UK]

IIF Immediate Interface [also II/F]

IIF Institute of International Finance [US]

IIF Instituto de Investigaciónes Médicas [Medical Research Institute; of University of Granada, Spain]

IIF Interionic Force [Physical Chemistry]

IIF International Institute of Forecasters [US]

IIFET International Institute of Fisheries Economics and Trade [US]

IIFSO International Islamic Federation of Student Organizations [US]

IIFT Indian Institute of Foreign Trade

IIH N^1-Isonicotinyl-N^2-Isopropylhydrazine [Biochemistry]

IIHR Iowa University Institute of Hydraulic Research [US]

IIHS Insurance Institute for Highway Safety [US]

III Interstate Identification Index [of National Crime Information Center, US]

III Intelligent, Integrated and Interactive [also I³]

III International Institute of Interpreters [US]

IIIC Innovation-Technology-Transfer-Consulting

IIIC International Institute for Intellectual Cooperation

IIIC International Irrigation Information Center

IIICR International Institute of Interdisciplinary Cycle Research [US]

IIID Intelligent, Integrated, Interactive Design [also I³D]

IIIF Impurity-Induced Intergranular Fracture [Metallurgy]

IIIL Isoplanar Integrated Injection Logic [also I³L]

IIIS Internal Image Intensification System

IIL Institute of International Law [Paris, France]

IIL Integrated Injection Logic [also I²L]

IIL(s) Intermediate Inversion Layer(s) [Meteorology]

IILD Impurity-Induced Layer Disordering [Materials Science]

IILE Ion-Induced Light Emission

IILP International Institute for Lath and Plaster [US]

IILS International Institute for Labour Studies

IIM Indian Institute of Management

IIM Indian Institute of Materials [Calcutta, India]

IIM Indian Institute of Metals [now Indian Institute of Materials]

IIM Institute for Information Management [Campbell, California, US]

IIM Institution of Industrial Managers [UK]

IIM Instituto de Investigaciones en Materiales [Institute of Materials Research; of Universidad Nacional Autónoma de Mexico, Mexico City]

IIMA Industrial Instruments Manufacturing Association

IIMC International Information Management Congress [Fairport, New York State, US]

IIME Institut Interdépartemental de Microscopie Electronique [Interdepartmental Institute for Electron Microscopy, of Ecole Polytechnique Fédérale de Lausanne, Switzerland]

IIMI International Irrigation Management Institute [Kandy, Sri Lanka]

IIMT International Institute of Milling Technology [UK]

IIM-UNAM Instituto de Investigaciones en Materiales/ Universidad Nacional Autonóma de Mexico [Institute of Materials Research/Autonomous National University of Mexico, Mexico City]

IINCE International Institute for Noise Control Engineering

I Inf Sci Institute of Information Scientists [UK]

IINREN Interagency Interim National Research and Education Network

IINS Incoherent Inelastic Neutron Scattering

IINS Inter-University Institute of Nuclear Sciences [Belgium]

IINSE International Institute of Nuclear Science and Engineering

IIOC Intelligent Input/Output Channel

IIOE International Indian Ocean Expedition [from 1961-66]

IIOP Integrated Input/Output Processor

IIOP Internet Inter-Operability Protocol

IIP Immigrant Investor Program [of Employment and Immigration Canada, Canada]

IIP Implementation and Installation Plan

IIP Index of Industrial Production

IIP Indian Institute of Petroleum

IIP Instantaneous Impact Point

IIP International Ice Patrol [of US Coast Guard]

IIP Isoionic Point [Physical Chemistry]

IIPE Institution of Incorporated Plant Engineers [UK]

IIPER International Institution of Production Engineering Research

IIPP International Institute for Promotion and Prestige [Switzerland]

IIPS Interactive Instructional Presentation System

IIR Imaging Infrared

IIR Immediate Impulse Response

IIR Infinite(-Duration) Impulse Response

IIR Institute of Interdisciplinary Research [of University of Tokyo, Japan]

IIR Institute of Intergovernmental Relations

IIR Institute of Intermodal Repairers [US]

IIR Institute of International Research

IIR Integrated Instrumentation Radar

IIR International Institute of Refrigeration [France]

IIR International Inventors Registry [US]

IIR Ion-Induced (X-)Radiation

IIR Isobutylene-Isoprene Rubber

IIR Isotactic Isoprene Rubber

IIR Resistance Heating [also I^2R]

IIRA International Industrial Relations Association [Switzerland]

IIRB Institut International de Recherches Betteravières [International Institute for Sugar Beet Research, Belgium]

IIRC Interrogation and Information Reception Circuit

IIRFD Infinite-Impulse Response Filter Design

IIRI International Industrial Relations Institute

IIRS Institute for Industrial Research and Standards [US]

IIRS Ion Impact Radiation Spectroscopy

IIRS Ion-Induced Radiation Spectroscopy

IIS Index to International Statistics

IIS India Information Service

IIS Indian Institute of Science [Bangalore, India] [also IISc]

IIS Information and Intelligent Systems

IIS Inquiry into Science Program

IIS Inspection Item Sheet

IIS Institute of Industrial Science [of University of Tokyo, Japan]

IIS Institute of Industrial Supervisors

IIS Institute of Information Scientists [UK]

IIS Integrated Instrument System

IIS Interactive Instructional System

IIS International Isotope Society [US]

IIS Internet Information Server

IIS Intrinsic Instruction Set

IIS Item Information Service [of Public Works and Government Services of Canada]

IISBR International Institute of Sugar Beet Researchers [Belgium]

IISc Indian Institute of Science [Bangalore, India] [also IIS]

IISEE International Institute of Seismology and Earthquake Engineering [Japan]

IISI International Iron and Steel Institute [Brussels, Belgium]

IISL International Institute of Space Law [Netherlands]

IISLS Improved Interrogator Sidelobe Suppression

IISN Institut Interuniversitaire des Sciences Nucléaires [Inter-University Institute of Nuclear Sciences, Belgium]

IISO Institution of Industrial Safety Officers

IISP International Institute of Site Planning

IISR Indian Institute of Sugar Cane Research

IISRP International Institute of Synthetic Rubber Producers [US]

IISS Integrated Information Support System

IISS International Institute for Strategic Studies [UK]

IISS International Institute for the Science of Sintering [Yugoslavia]

IISSC International Industrial Symposium (and Exhibition) on the Supercollider

IISSM International Institute of Sports Science and Medicine [of Indiana University School of Medicine, Mooresville, US]

IIST Institute of Information Science and Technology [of Tokyo Denki University, Japan]

IIST Institution of Instrumentation Scientists and Technologists [India]

IIT Iligan Institute of Technology [of Mindanao State University, Philippines]

IIT Illinois Institute of Technology [Chicago, US]

IIT Index of Inflammability Test [UK]

IIT Indian Institute of Technology

IIT Indiana Institute of Technology [Fort Wayne, US]

IIT Industrial Information Transfer

IIT Information Technology Institute [Toronto, Canada]

IIT Institute of Industrial Technicians [UK]

IIT Institute of Instrumentation Technology [Shenyang, PR China]

IIT Instituto de Investigaciones Tecnologicas [Institute for Technological Investigations, Colombia]

IIT International Conference on Ion Implantation Technology

IIT Ion Implantation Technology

IIT (Technion) Israel Institute of Technology [Haifa]

IITA Inland International Trade Association [US]

IITA Institute of Information Theory and Automation [Czech Republic]

IITA International Institute of Tropical Agriculture [Ibadan, Nigeria]

IITAP International Institute of Theoretical and Applied Physics [US]

IITB Indian Institute of Technology, Bombay [India]

IITB Institut für Informationsverarbeitung in Technik und Biologie [Institute for Information Processing in Technology and Biology, Germany]

IITCS Igloo Internal Thermal Control Section

IITD Indian Institute of Technology, Delhi [India]

IITF Information Infrastructure Task Force

IITK Indian Institute of Technology, Kanpur [India]

IITM Indian Institute of Technology, Madras [India]

IITRAN Illinois Institute of Technology Translator

IITRC Illinois Institute of Technology Research Center [Chicago, US]

IITRI Illinois Institute of Technology Research Institute [Chicago, US]

IITS International Institute of Theoretical Sciences

IITV Image Intensified Television

IIU Instruction Input Unit

IIUG Internationales Institut für Umwelt und Gesellschaft [International Institute for Environment and and Society; of Wissenschaftszentrum Berlin, Germany]

IIW International Institute of Welding [UK]

IIWG International Industry Working Group [Canada]

IIWPA International Information/Word Processing Association [formerly International Word Processing Association; now Association of Information Systems Professionals]

IIX Ion-Induced X-Ray (Emission)

IIXE Ion-Induced X-Ray Emission

IIXS Ion-Induced X-Ray Spectroscopy

IJ Ink Jet

IJAF Instrumentcenter for Jordbaseret Astronomisk Forskning [Center for Terrestrial Astronomical Research, Denmark]

IJAJ International Jitter Antijam

IJC International Joint Commission [US/Canada]

IJC International Joint Conference

IJC Internet Journal of Chemistry

IJCAI International Joint Conference on Artificial Intelligence [US]

IJCNN International Joint Conference on Neural Networks

IJCRAB International Joint Commission Research Advisory Board

IJE International Journal of Epidemiology [Published by Oxford University Press, UK]

IJJU Intentional-Jitter Jamming Unit

IJGIS International Journal of Geographical Information Systems

IJMA Indian Jute Millers Association

IJMARI Indian Jute Millers Association Research Institute

IJNSR Internet Journal of Nitride Semiconductor Research [of Materials Research Society]

IJO International Jute Organization [Bangladesh]

IJP Ink-Jet Printer

IJP Internal Job Processing

IJPA International Jelly and Preserve Association [US]

IJS Institut J. Stefan [J. Stefan Institute, University of Ljubljana, Slovenia]

IJS Interactive Job Submission

IK Index Kewensis

IK Inverse Kinematics

IKAT Interactive Keyboard and Terminal

IKBI Inverse Kirkwood-Buff Integral (Method) [Physics]

IKBS Intelligent Knowledge-Based System

IKC International Kennel Club

IKCOC International Kyoto Conference on (New Aspects of) Organic Chemistry [Japan]

IKE Ion Kinetic Energy

IKEM Interactive Knowledge Based System

IKES Ion Kinetic Energy Spectroscopy

IKK Internationale Fachmesse Kälte- und Klimatechnik [International Trade Fair for Refrigeration and Air-Conditioning, Germany]

IKL Intersecting Kikuchi Lines [Crystallography]

IKO Institute van Kernph Ouder [Netherlands]

IKOFA Internationale Fachmesse der Ernährungswirtschaft [International Trade Fair for the Food Industry, Specialties, Shopfitting and Equipment, Germany]

IKP Institut für Kunststoffprüfung [Institute of Polymer Testing; of University of Stuttgart, Germany]

IKP Internet Keyed Payments

IKr Icelandic Króna [Currency of Iceland] [also Ikr]

IKRA International Kirlian Research Association [US]

IKS Instituut voor Kern- en Stralingsfysica [Institute for Nuclear and Radiation Physics, of Katholieke Universiteit Leuven, Belgium]

IKS-KUL Instituut voor Kern– en Stralingsfysica–Katholieke Universiteit Leuven [Institute for Nuclear and Radiation Physics–Catholic University of Leuven, Belgium]

IKSR Internationale Kommission zum Schutze des Rheins gegen Verunreinigung [International Commission for the Protection of the Rhine against Pollution, Germany]

IKT Internationale Kautschuk-Tagung [International Rubber Conference, Germany]

IKTS (Fraunhofer-)Institut für Keramische Technologien und Sinterwerkstoffe [(Fraunhofer) Institute for Ceramic Technologies and Sintered Materials, Dresden, Germany]

IKZ Institut für Kristallzüchtung [Institute for Crystal Growing, Berlin, Germany]

IL Idle [Data Communications]
IL Ignition Limit
IL Ignition Loss
IL Illinois [US]
IL Ilmenite [Mineral]
IL Imaging Library
IL Immiscible Liquid
IL Impact Level
IL Impact Load(ing)
IL Incandescent Lamp
IL Incident Light
IL Including Loading
IL Incoherent Light
IL Incorrect Length
IL Indication Lamp
IL Infinite Layer
IL Injection Laser
IL Injection Luminescence
IL In-Line
IL Inorganic Liquid
IL Insertion Loss [Electronics]
IL Instruction List
IL Instrument Luminosity
IL Insulating Layer
IL Intensity Level
IL Interface Loop
IL Interference Layer
IL Interior Length
IL Interlaminar
IL Interleave
IL Interleukin [Biochemistry]
IL Interline
IL Intermediate Language
IL Intermediate Layer
IL International Law
IL Interrupt Level
IL Intralaminar
IL Intra-Layer
IL Invariant-Line (Approach [Crystallography])
IL Inversion Layer [Meteorology]
IL Inverted Laboratory (Microscope)
IL Ioffe Lamp

IL Ion Laser
IL Ion(ic) (Crystal) Lattice
IL Ionoluminescence; Ionoluminescent
IL Iron Loss
IL Isotope Lamp
IL Israel [ISO Code]
I²L Integrated Injection Logic [also IIL]
IL-1 Interleukin-1 [Biochemistry]
IL-1α Interleukin-1α [Biochemistry]
IL-1β Interleukin-1β [Biochemistry]
IL-1R Interleukin-1 Receptor [Biochemistry]
IL-1sR Interleukin-1 Soluble Receptor [Biochemistry]
IL-2 Interleukin-2 [Biochemistry]
IL-3 Interleukin-3 [Biochemistry]
IL-4 Interleukin-4 [Biochemistry]
IL-4R Interleukin-4 Receptor [Biochemistry]
IL-4sR Interleukin-4 Soluble Receptor [Biochemistry]
IL-5 Interleukin-5 [Biochemistry]
IL-5sR Interleukin-5 Soluble Receptor [Biochemistry]
IL-6 Interleukin-6 [Biochemistry]
IL-6R Interleukin-6 Receptor [Biochemistry]
IL-6sR Interleukin-6 Soluble Receptor [Biochemistry]
IL-7 Interleukin-7 [Biochemistry]
IL-8 Interleukin-8 [Biochemistry]
IL-9 Interleukin-9 [Biochemistry]
IL-10 Interleukin-10 [Biochemistry]
IL-11 Interleukin-11 [Biochemistry]
IL-12 Interleukin-12 [Biochemistry]
IL-13 Interleukin-13 [Biochemistry]
IL-15 Interleukin-15 [Biochemistry]
I&L Installation and Logistics
I/L Import License
I-L Isotropic-to-Lamellar (Transition) [Metallurgy]
Il Illinium [Promethium]
.il Israel [Country Code/Domain Name]
i(λ) photodiode current [Symbol]
ILA Image Light Amplifier
ILA Institute of Landscape Architects [UK]
ILA Instrument Landing Approach
ILA Intelligent Line Adapter
ILA Inter-Laboratory Agreement
ILA Internationale Luft– und Raumfahrtausstellung [International Aeronautics and Astronautics Exhibition, Berlin, Germany]
ILA International Language for Aviation
ILA International Law Association
ILA International Longshoremen's Association
ILA Iterative Linear Algebra
ILAAS Integrated Light Aircraft Attack System
ILAAS Integrated Light Attack Avionics System
ILAAT Interlaboratory Air-to-Air Missile Technology
ILAB Bureau of International Labor Affairs [US]
ILAC International Laboratory Accreditation Conference

ILACS Integrated Library Administration and Cataloguing System [Netherlands]

ILAFA Instituto Latinoamericano del Fierra y el Acero [Latin American Institute of Iron and Steel, Santiago, Chile] [also Ilafa]

ILAI Italo-Latin American Institute [Italy]

ILAP Industry and Labour Adjustment Program

ILAP Integrated Local Area Planning

ILAR Institute of Laboratory Animal Resources [US]

ILAS Improved Limb Atmospheric Spectrometer

ILAS Instrument Landing Approach System

ILAS Interrelated Logic Accumulating Scanner

ILAS Isotrace Laboratory for Analytical Services [Canada]

ILASL Istituto Lombardo Accademia di Scienze e Lettere [Lombard Institute of the Academy of Science and Letters, Milan, Italy]

ILASS Institute for Liquid Atomization and Spray Systems

ILB Initial Load Block [Computers]

ILB Inner Layer Board

ILB Inner-Layer Bond [Tape-Automated Bonding]

ILB Inner Lead Bond

ILB International Labor Board [US]

ILB International Liaison Bureau [UK]

ILBM Interleave Bit Map

ILBT Interrupt Level Branch Table

ILC Initial Launch Capability

ILC Inrevocable Letter of Credit

ILC Institut Linguistique et Commercial [Institute for Linguistics and Commerce]

ILC Instruction Length Code

ILC Instruction Length Counter

ILC Instruction Location Counter

ILC Integrated Laminating Center

ILC Interference-Layer Contrast

ILC Interlayer Coupling

ILC International Labor Conference [of International Labor Organization]

ILC International Latex Corporation

ILC International Lifeboat Conference

ILC Ion/Liquid Chromatography

ILCA International Labor Communications Association [US]

ILCA International Launching Class Association

ILCA International Lightning Class Association

ILCA International Livestock Center for Africa [Addis Ababa, Ethiopia]

ILCCS Integrated Launch Control and Checkout System

ILCMP International Liaison Committee on Medical Physics

ILCOP International Liaison Committee of Organizations for Peace

ILD Indentation Load Deflection [Mechanical Testing]

ILD Injection Laser Diode

ILD Interlayer Dielectric(s)

ILD Interlevel Dielectric(s)

ILD International Labour Documentation [of International Labour Organization]

ILDA Industrial Lighting Distributors of America [US]

ILDA Independent Laboratory Distributors Association [Canada]

ILDIS International Legume Database and Information System

ILDM Institute of Logistics and Distribution Management [UK]

ILDM Intrinsic Low-Dimensional Manifold

ILE Institute of Locomotive Engineers [UK]

ILE Institution of Lighting Engineers [UK]

ILE Interfacing Latching Element

Ile (+)-Isoleucine; Isoleucyl [Biochemistry]

ILEA Inner London Education Authority [UK]

ILEA International League of Electrical Association [US]

ILEC International Lake Environment Committee [of United Nations Environment Programme–International Environmental Technology Centre at Shiga Prefecture, Japan]

ILEE Industrial Laser and Electronic Engineering AG [Switzerland]

ILEED Inelastic Low-Energy Electron Diffraction

ILEM Inter-Library Electronic Mail

ILENP Institute of Low-Energy Nuclear Physics [PR China]

Ileu Isoleucinate; Isoleucine [Biochemistry]

ILF Inductive Loss Factor

ILF Industrial Leathers Federation [UK]

ILF Infra-Low Frequency

ILF Input Loading Factor

ILF Integrated Lift Fan

ILF Interlaminar Failure

ILF Intralow Frequency [also ilf]

IL/F Imaging Library/FORTRAN

ILFI International Labour Film Institute

ILG Isotope Level Gauge

ILGB International Laboratory of Genetics and Biophysics

ILGWU International Ladies Garment Workers Union (of America)

ILHMFLT International Laboratory for High Magnetic Fields and Low Temperatures

IL-HP Interleukin Hybridoma/Plasmacytoma [Medicine]

IL-HP1 Interleukin Hybridoma/Plasmacytoma-1 [Medicine]

ILI Indiana Limestone Institute (of America) [US]

ILI Instant Lunar Ionosphere

ILI Institute for Land Information

ILI Inter-African Labour Institute

ILIC International Library Information Center [of University of Pittsburgh, Pennsylvania, US]

ILID Institut für Landwirtschaftliche Information und Dokumentation [Institute for Agricultural Information and Documentation, Germany]

ILIMA International Licensing Industry and Merchandisers Association [US]

ILIP In-Line Instrument Package

ILIR In-House Laboratories Independent Research

ILIS Intelligent Landscape Integrated System

ILL Institut Laue-Langevin [Laue-Langevin Institute, Grenoble, France]

ILL Inorganic Liquid Laser

ILL Inter-Library Loan

Ill Illinois [US]

Ill Illustration; Illustrator [also ill]

ill illustrate(d)

ILLIAC Illinois Institute of Advanced Computation [of University of Illinois, US]

ILLIAC (University of) Illinois Integrator and Automatic Computer [also Illiac]

ILLINET Illinois Library Network [US]

Illum Illumination [also illum]

illum illuminate(d)

Illumg Illuminating [also illumg]

Illus Illustration; Illustrator [also illus]

illus illustrate(d)

Illust Illustration; Illustrator [also illust]

illust illustrate(d)

ILM Incident Light Microscope; Incident Light Microscopy

ILM Independent Landing Monitor

ILM Information Logic Machine

ILM Institut für Lasertechnologien in der Medizin [Institute for Laser Technologies in Medicine; of University of Ulm, Germany]

ILM Intermediate Language Machine

ILM Intermediate Layer Model [Materials Science]

ILM Intrinsic Localized Mode

ILMA Independent Lubricant Manufacturers Association [US]

ILMAC International Congress and Fair for Laboratory, Measuring and Automation Techniques in Chemistry [also Ilmac]

ILMC Indiana Labor Management Council [US]

ILMC International Light Metal Congress

ILMS Integrated Library Management System

ILMS Intra-Layer Multiple Scattering

ILO Individual Load Operation

ILO Industry Liaison Office

ILO Injection-Locked Oscillator

ILO Interaction Localized Orbital

ILO International Labour Office [of United Nations in Switzerland]

ILO International Labour Organization [of United Nations]

I-Load Initial Load [also I-load]

ILP Improved Low-Pressure (Casting Process) [Metallurgy]

ILP Ingersoll (Numeric-Control) Lathe Program

ILP In-Line Processing

ILP Integrated Logistics Panel

ILP Intermediate Language Processor

ILP Intermediate Language Program

ILP International Lithosphere Program [Geology]

ILPA International Labor Press Association [now International Labor Communications Association, US]

ILPC International Linen Promotion Commission [US]

ILPD Intergranular Liquid-Phase Distribution [Metallurgy]

ILPF Ideal Low-Pass Filter

ILR Indentation Load Relaxation (Test) [Mechanical Testing]

ILR Industrial Laser Review [US]

ILR Institute of Library Research [of University of California, US]

ILR Instruction Location Register

ILRA Inbred Livestock Registry Association [US]

ILRA International Laboratory for Marine Radioactivity

ILRA International Lactic Acid Research Association

ILRAD International Laboratory for Research on Animal Diseases [Nairobi, Kenya]

ILRD Inter-Limb Resistance Device

ILRI Indian Lac Research Institute

ILRI International Institute for Land Reclamation and Improvement [Netherlands]

ILRT Integrated Leakage Rate Test

ILRT Intermediate Level Reactor Test [Nuclear Engineering]

ILS Ideal Liquidus Structures [Metallurgy]

ILS Imaging Line Scanner

ILS Incoherent-Light System

ILS Industrial Locomotive Society [UK]

ILS Institute of Life Sciences [UK]

ILS Instrument Landing System

ILS Instrument Line Shape (Function)

ILS Integrated Library System

ILS Integrated Logistics System

ILS Integrated Logistic(s) Support

ILS Interactive Laboratory System

ILS Interactive Layout System

ILS Interlamellar Spacing [Metallurgy]

ILS Interlaminar Shear [Mechanics]

ILS Interlevel Shorts

ILS International Language Support

ILS International Latitude Service

ILS International Line Selector

ILS International Lunar Society [US]

ILS Interrupt Level Subroutine

ILS Invariant Line Strain

ILS Ionization Loss Spectroscopy

ILS Israeli Shekel [Currency]

ILSAP Instrument Landing System Approach

ILSC International Language Schools of Canada

ILSDW Incommensurate Longitudinal Spin-Density Wave [Solid-State Physics]

ILSF Intermediate Level Sample Flow

ILSI International Life Sciences Institute

ILS/LAR Integrated Logistics System and Logistics Assessment Review

ILSMT Integrated Logistic Support Management Team

ILSO Incremental Life Support Operation

ILSP Integrated Logistics Support Plan

ILSRO Interstare Land Sales Registration Office [US]

ILSRP Instrument Landing System Reference Point

ILSS Interlaminar Shear Strength

ILSS Interlaminar Shear Stress

ILSSE Integrated Life Science Shuttle Experiments [of NASA]

ILSTAC Instrument Landing System and TACAN (Tactical Air Navigation)

ILSW Interrupt Level Status Word

ILT Inverse Laplace Transform(ation)

ILTA Independent Liquid Terminals Association [US]

Iltc Indolemonothiocarbanate

ILTMS International Leased Telegraph Message Switching (Service) [UK]

ILTRI Idaho Long-Term Research Initiative [US]

ILTS Institute of Low Temperature Science [of Hokkaido University, Sapporo, Japan]

ILU Illinois University [US]

ILU International Longshoremen's Union

ILUPA Institut für Lebensmitteluntersuchungen, Umwelthygiene und Pharmaka-Analytik [Institute for Food Testing, Environmental Hygiene and Pharmaceutical Analysis, Moenchengladbach, Germany]

ILW Industrial Liquid Waste

ILW Intermediate-Level (Radioactive) Waste

ILWIS Integrated Land and Watershed Management Information System

ILWU International Longshoremen's and Warehousemen's Union

ILZ Intensive Land-Use Zone [Australia]

ILZIC Indian Lead and Zinc Information Center [New Delhi]

ILZRO International Lead and Zinc Research Organization [Research Triangle Park, North Carolina, US]

ILZSG International Lead and Zinc Study Group [US]

IM Ideal Modulation

IM Identification Mark

IM Ignore Mouse (Parameter)

IM Industrial Material

IM Information Management; Information Manager

IM Image Matching

IM Imaginary Part

IM Impact Modulator

IM Imperial Measure

IM Impermeable Membrane

IM Induced Magnetism

IM Induction Melting [Metallurgy]

IM Induction Meter

IM Induction Motor

IM Industrial Management; Industrial Manager

IM Industrial Mathematician; Industrial Mathematics

IM Industry Motion Picture

IM Inert(ial) Mass

IM Inertial Motion

IM Inertia Matrix

IM Infectious Material

IM Infectious Mononucleosis [Medicine]

IM Information Management; Information Manager

IM Information Memory

IM Ingot Metallurgy [also I/M]

IM Inherent Moisture

IM Injection Mold(ing)

IM Inner Marker

IM Inorganic Material

IM Input Medium

IM Insoluble Matter

IM Installation and Maintenance

IM Institut de Mensurations [Institute of Surveying; of Ecole Polytechnique Fédérale de Lausanne, Switzerland]

IM (The) Institute of Materials [UK]

IM Institute of Mathematics

IM Institute of Metallurgy

IM Institute of Meteorology

IM Institute of Metrology [RussianFederation]

IM Institute of Micrographics [UK]

IM Institutet för Metallforskning [Institute for Materials Research, Stockholm, Sweden]

IM Instruction Memory

IM Instrumentation and Measurements (Group) [of Institute of Electrical and Electronics Engineers, US]

IM Insulating Material

IM Integrated Manufacturing

IM Integrated Modem

IM Integrating Meter

IM Integration Method

IM Intelligent Motion

IM Intensity Modulation

IM Interactive Mode

IM Interceptor Missile

IM Interchangeable Manufacture

IM Interface Management

IM Interface Module

IM Interference Microscope; Interference Microscopy

IM Intermediate Missile

IM Intermediate Modulus

IM Intermediate Moisture

IM Intermetal

IM Intermetallic(s) [also im]

IM Intermodulation [Electronics]

IM Internal Medicine

IM Internal Memory

IM International Mounting

IM Interrupt Mask

IM Intramuscular [also im]

IM Intrinsic Mobility [Solid-State Physics]

IM Inventory Management; Inventory Manager

IM Inverse Matrix [Mathematics]

IM Inverted Microscope

IM Ionic Mobility

IM Ion Microscope; Ion Microscopy

IM Ion Milled; Ion Milling

IM Ion Mixing

IM Ishibashi Mechanism

IM Ishikawa and McLellan (Method for Hydrogen Diffusion in Alloys) [Metallurgy]

IM Isle of Man [UK]

IM Isomagnetic; Isomagnetism [Geophysics]

IM Isometric; Isometry

IM Isospin Multiplet [Particle Physics]

IM Isotropic Mixing

IM Item Mark

I&M Inspection and Maintenance

I&M Inventory and Monitoring

I/M Ingot Metallurgy [also IM]

I/M Inspection/Maintenance (Test)

I/M Inspection and Maintenance (Program)

I-M Insulator-to-Metal (Transition)

Im Imaginary Part (of Complex Number) [also ℑ]

Im 1H-Imidazole

Im (Complete) Irregular Magellanic Galaxy (or Magellanic Irregular) [Astronomy]

im intermetallic(s) [also IM]

IMA Ideal Mechanical Advantage

IMA Indium Manganese Arsenide (Semiconductor)

IMA Industrial Marketing Association [UK]

IMA Industrial Medical Association [US]

IMA Input Message Acknowledgement

IMA Institute of Management Accountants [US]

IMA Institute of Mathematics and its Applications [UK]

IMA Instituto Magnetismo Aplicado [Institute for Applied Magnetism; of Consejo Superior de Investigaciónes Cientificas, Universidad Complutense, Madrid, Spain]

IMA Instituto Mexicano de Antropologia [Mexican Institute of Anthropology]

IMA Interdisciplinary Master of Arts

IMA International Amusement and Vending Machines Trade Fair

IMA International Magnesium Association [US]

IMA International Management Association [UK]

IMA International MIDI (Musical Instrument Digital Interface) Association [US]

IMA International Military Archives [US]

IMA International Milling Association

IMA International Mineralogical Association [Germany]

IMA International Mohair Association [UK]

IMA International Mycological Association [UK]

IMA Invalid Memory Address

IMA Inventory of Marine Activities (Database) [of Economic and Social Commission for Asia and the Pacific, United Nations, Thailand]

IMA Ion Microprobe Mass Analysis; Ion Microprobe Mass Analyzer

IMA Irish Medical Association

IMA Islamic Medical Association

IMAC Illinois Microfilm Automated Cataloging [of Illinois State Library, Springfield, US]

IMACA International Mobile Air Conditioning Association [US]

IMACS Image Management and Communication System

IMACS International Association for Mathematics and Computers in Simulation [Belgium]

IMADS Integrated Machinery Analysis and Diagnostic System

IMAFO Intelligent Manufacturing Foreman (Software)

IMAG Internationaler Messe– und Ausstellungsdienst GmbH [International Trade Fair and Exhibition Service Company, Munich, Germany]

imag imaginary

IMAGE Integrated Model for the Assessment of the Greenhouse Effect

IMAGE Interactive Menu-Assisted Graphics Environment

Image Process Image Processing [Journal published in the UK]

IMAGES Integrated Modular Avionics General Executive Software

Image Technol Image Technology [Journal of the British Kinematograph, Sound and Television Society]

Image Vis Comput Image and Vision Computing [Journal published in the UK]

IMAI Istituto di Metodologie Avanzate Inorganiche [Institute for Advanced Inorganic Methodology; of Consiglio Nazionale delle Ricerche at Monterotondo Scalo, Italy]

IM/AI Instrumentation and Measurement/Automated Instrumentation [Institute of Electrical and Electronics Engineers Committee, US]

IMA J Appl Math IMA Journal of Applied Mathematics [Publication of the Institute of Mathematics and Applications, UK]

IMA J Math Appl Med Biol IMA Journal of Mathematics Applied to Medicine and Biology [Publication of the Institute of Mathematics and Applications, UK]

IMA J Math Control Inf IMA Journal of Mathematical Control and Information [Publication of the Institute of Mathematics and Applications, UK]

IMA J Numer Anal IMA Journal of Numerical Analysis [Publication of the the the Institute of Mathematics and Applications, UK]

IMAM International Meeting on Advanced Materials [of Materials Research Society, US]

IManf Institute of Manufacturing [UK]

IMAP Institute for Materials and Advanced Processes [of University of Idaho, Moscow, US]

IMAP Institute for Materials and Advanced Processing [of University of Illinois, US]

IMAP Integrated Mechanical Analysis Project [US]

IMAP Internet Message Access Protocol

IMarE Institute of Marine Engineers [UK]

IMAS Image and Microanalysis System

IMAS Industrial Management Assistance Survey [of US Air Force]

IMAT International Masonry Apprenticeship Trust [now International Masonry Institute Apprenticeship and Training, US]

IMATS Issue Management and Tracking System

IMAW International Molders and Allied Workers Union

IMAX Maximum Image

IMB Institut für Molekulare Biotechnologie e.V. [Institute for Molecular Biotechnology, Jena, Germany]

IMB Irvine Michigan Brookhaven (Experiment) [of Brookhaven National Laboratory, Upton, New York, US]

IMB Intermodule Bus

IMB Internationale Messe für Bekleidungsmaschinen [International Clothing Machine Fair, Cologne, Germany]

IMB Institute of Molecular Biology

IMB Institut für Molekulare Biotechnologie [Institute for Molecular Biotechnology, Jena, Germany]

IMB Investigator of Micro-Biosphere

IMBA International Master of Business Administration

IMBDC International Marine Biodiversity Development Corporation [Canada]

IMBLMS Integrated Medical Behavioral Laboratory Measurement System

IMBM Institute of Maintenance and Building Management [UK]

IMC Ice Mass Content

IMC Image Motion Compensation

IMC Industrial Microcomputer

IMC Information Management Congress [US]

IMC Insertion-Mounted Component

IMC Institute of Macromolecular Chemistry [of Czech Academy of Sciences, Prague, Czech Republic]

IMC Institute of Management Consultants [London, UK]

IMC Institute of Measurement and Control [UK, Australia and New Zealand]

IMC Institute of Motorcycling [UK]

IMC Instructional Materials Center

IMC Instrument Meteorological Conditions

IMC Integrated Maintenance Concept

IMC Integrated Microcircuit

IMC Integrated Multiplexer Channel

IMC Intelligent Matrix Control

IMC Interactive Medical Communications

IMC Interactive Module Controller

IMC Interface Module Cabinet

IMC Intermediate Metal Conduit

IMC Intermetallic Composite

IMC Intermetallic Compound

IMC Intermetallic-Matrix Composite

IMC Internal Model Control

IMC International Information Management Congress [US]

IMC International Maintenance Center

IMC International Maritime Committee

IMC International Materials Conference

IMC International Microelectronics Conference

IMC International Micrographic Congress [Name Changed to International Information Management Congress, US]

IMC International Minerals and Chemical Corporation

IMC International Morse Code

IMC International Music Council

IMC Inverse Monte Carlo (Method)

IMC Israel Materials Conference

ImC Intensity Millicurie [Unit]

IMCA Internal Model Control Approach

IMCB Institute of Molecular and Cell Biology [of National University of Singapore]

IMCC Institute of Management Consultants of Canada

IMCC Integrated Mission Control Center

IMCC International Mineral and Chemical Corporation [Canada]

IMCC Interstate Mining Compact Commission [US]

IMCE Inter-Ministerial Committee for Environment

IMCEA International Military Club Executives Association [US]

IMC J IMC Journal [of International Information Management Congress, US]

ImCl 1-Methyl-3-Ethylimidazolium Chloride

IMC Newsl IMC Newsletter [of International Information Management Congress, US]

IMCO Institute of Management Consultants of Ontario [Canada]

IMCO Intergovernmental Maritime Consultative Organization [now International Maritime Organization]

IMCO International Maritime Consultative Organization [now International Maritime Organization]

IMCO International Maritime Countries Organization [now International Maritime Organization]

Im2CO 1,1'-Carbonylbis(1H-Imidazole)

IMCOS International Meteorological Consultant Service

IMCP Instituto de Madera, Celulosa y Papel [Institute of Wood, Cellulose and Paper, Mexico]

IMCP Integrated Monitor and Control Panel

IMCRA Interim Marine Regionalization of Australia

IMCS Image Motion Compensation System

IMCS Institute of Marine Coastal Science [at Rutgers University, New Brunswick, New Jersey, US]

IMCS International Metal Container Section [of Material Handling Institute, US]

IMCSMHI International Metal Container Section of the Material Handling Institute [US]

IMD Industrial Mineral Division [of Canadian Institute of Mining and Metallurgy]

IMD Industrial Minerals Division [of Society of Mining Engineers, US]

IMD Institute for Marine Dynamics [of National Research Council of Canada]

IMD Institute of Metals Division [of American Institute of Mining, Metallurgical and Petroleum Engineers, US]

IMD Institut für Maschinelle Dokumentation [Institute for Automated Documentation, Austria]

IMD Intercept Monitoring Display

IMD Intermetal Dielectric (Application)

IMD Intermodulation Distortion [Electronics]

IMD International MTM (Methods-Time Measurement) Directorate [Sweden]

imd intermediate

IMDA Iminodiacetic Acid [also Imda]

IMDA Indian Measurement and Design Association

IMDA International Map Dealers Association [US]

IMDB Integrated Maintenance Data Base

IMDC Interceptor Missile Direction Center

IMDE Institute of Medical and Dental Engineering [of Tokyo Medical and Dental University, Japan]

IMDG International Maritime Dangerous Goods Code

IMDM Iscove's Modified Dulbecco's Medium [Biochemistry]

IMDO Installation and Materiel District Office [of Federal Aviation Administration, US]

IMDP Iminodipropionic Acid [also Imdp]

2-IMDPT 1,11-Bis(imidazol-2-yl)-2,6,10-Triazaundecane

4-IMDPT 1,11-Bis(imidazol-4-yl)-2,6,10-Triazaundecane

IMDS Image Data Stream

IMDS International Meat Development Scheme [of Food and Agricultural Organization]

IMDS International Microform Distribution Service

imdt immediate

IME Input Method Editor

IME Institute of Makers of Explosives [US]

IME Institute of Marine Engineers [UK]

IME Institute of Mining Engineers [US]

IME Institute of Municipal Engineering [of American Public Works Association, US]

IME Institution of Mechanical Engineers [UK]

IME Instituto Militar de Engenharia [Military Institute of Engineering, Rio de Janeiro, Brazil]

IME Instytutu Metrologii Electrycznej [Electrical Metrology Institute, of Wroclaw Polytechnic, Wroclaw, Poland]

IME Interface Magnetoelastic Energy

IME International Magnetosphere Explorer

IME International Microcomputer Exhibition

Im(ε) Imaginary Part of Dielectric Function [Symbol]

IMEA Incorporated Municipal Electrical Associations [UK]

IME/APWA Institute of Municipal Engineering/American Public Works Association [US]

IMEBORON International Meeting on Boron Chemistry

IMEC Inter-University Microelectronics Center [Leuven, Belgium]

IMEC Israel Materials Engineering Conference

IMECA Independent Metallurgical Engineering Consultants of America

IMECC Independent Metallurgical Engineering Consultants of California [also IMECCA, US]

IMECO International Measurement Confederation [Hungary]

IMechE Institution of Mechanical Engineers [UK]

IMechIE Institution of Mechanical Incorporated Engineers [UK]

IMED Industrial Maintenance and Engineering Division [Picatinny Arsenal, Dover, New Jersey, US]

IMEG International Management Engineering Group

IMEKO Internationale Meßtechnische Konföderation [International Measurement Confederation, Hungary]

IMEKO Bull IMEKO Bulletin [of Internationale Meßtechnische Konföderation, Hungary]

IMEMME Institution of Mining Electrical and Mining Mechanical Engineers [Manchester, UK]

iMEMS Intergrated-Circuit Microelelectromechanical System (Technology)

IMEP Indicated Mean Effective Pressure [also imep]

IMET Improved Meteorological Instrumentation [of Woods Hole Oceanographic Institute, US]

IMET A. A. Baikov Institute of Metallurgy [of Russian Academy of Sciences]

IMet Institute of Metals [Name Changed to Institute of Materials, London, UK]

IMETS Imaging Technologies and Evolving Management Systems (Conference/Congress)

IME/USP Instituto Militar de Engenharia [Military Institute of Engineering, Rio de Janeiro, Brazil]

IMEX Image Format Conversion Software System

IMEX International Mail Exchange

Imexa Ispat Mexicana SA [Mexican Iron and Steel Works]

IMF Ice Mass Flux [also imf]

IMF Institute of Metal Finishing [UK]

IMF Intermodulated Fluorescence

IMF Intermolecular Force [Physical Chemistry]

IMF Internal Magnetic Field

IMF Internal Magnetic Focus Tube

IMF International Marketing Federation

IMF International Meeting on Ferroelectricity

IMF International Metalworkers Federation [Switzerland]

IMF International Monetary Fund [of United Nations]

IMF International Motorcycle Federation

IMF Interplanetary Magnetic Field

IMF Inventory Master File

IMF Ionic Magnetic Fluid

IMFET Internally Matched (Power) Field-Effect Transistor

IMFI Industrial Mineral Fiber Institute [US]

IMFP Inelastic Mean Free Path

IMFP International Meeting on Frontiers of Physics

IMG Institut de Mécanique de Grenoble [Grenoble Institute of Mechanics; of Institut National Polytechnique de Grenoble, France]

IMG Interferometric Monitor of Greenhouse Gases

Img Image [also img]

Img Immigration [also img]

IMGC Instituto di Metrologie "G. Coloneti" [G. Coloneti Institute of Metrology, Italy]

IMGCN Integrated Missile Ground Control Network

IMGE Ion-Selective Multisolvent Gradient Elution

IMGS Irrigation Management Grants Scheme [of Victoria State, Australia]

IMH Inlet Manhole

IMH Institute of Materials Handling [UK]

Im(h_{ie}) Imaginary Part of Common-Emitter Small-Signal Short-Circuit Input Impedance [Semiconductor Symbol]

Im(h_{oe}) Imaginary Part of Common-Emitter Small-Signal Open-Circuit Output Admittance [Semiconductor Symbol]

IMHEF Institut de Machines Hydrauliques et de Mécanique des Fluides [Institute of Hydraulic Machinery and Fluid Mechanics; of Ecole Polytechnique Fédérale de Lausanne, Switzerland]

IMHO In My Humble Opinion [Internet]

IMHTS International Meeting on High-Temperature Superconductivity [Rostov, Russia]

IMI Ignition Manufacturers Institute

IMI Imperial Metal Industries

IMI Improved Manned Interceptor

IMI Institute for Marine Information [US]

IMI Institute of the Motor Industry [UK]

IMI Intermediate Machine Instruction

IMI International Maintenance Institute [US]

IMI International Market Intelligence

IMI International Masonry Institute [US]

IMI International Metaphysical Institute [UK]

IMI Invention Marketing Institute [US]

IMI Irish Management Institute

IMI Israel Mining Industries (Process)

IMIA International Machinery Insurers Association [Germany]

IMIA International Medical Informatics Association [Canada]

IMIAT International Masonry Institute Apprenticeship and Training [US]

IMIC Industrial Minerals International Congress

IMIC Infrastructure Modernization Implementing Council [US]

IMIC International Medical Information Center [Japan]

IMIE Institution of Mechanical Incorporated Engineers [UK]

IMIF International Maritime Industries Forum [UK]

IMINCO Iran Marine International Oil Company [also Iminco]

IMinE Institution of Mining Engineers [UK]

IMIP Industrial Modernization Incentive Program

IMIR Interceptor Missile Interrogation Radar

IMIS Instituto Mexicano de Investigationes Siderúrgicas [Mexican Institute for Iron and Steel Research]

IMIS Integrated Management Information System

IMIT Institute of Musical Instrument Technology [UK]

IMIT Instituto Mexicano de Investigationes Tecnologicas [Mexican Institute for Technological Research]

IM/IT Information Management/Information Technology

IMITAC Image Input to Automatic Computer

IMJM Instytut Metalurgii Jnznierii Materialowej [Institute of Metallurgy and Materials Engineering, Krakow, Poland]

IMK Institut für Meteorologie und Klimaforschung [Institute for Meteorology and Climatology, Karlsruhe Research Research Center, Germany]

Imk Identification Mark

IML Incoming Matching Loss

IML Information Manipulation Language

IML Initial Machine Load

IML Initial Microcode Load

IML Inner Mold Line

IML Inside Mold Line

IML Interactive Maintenance Language

IML Intermediary Music Language

IML Intermediate Language

IML International Microgravity Laboratory [of NASA]

IML-1 First International Microgravity Laboratory [of NASA]

IMLS Institute of Medical Laboratory Sciences [UK]

IMM Injection Molding Machine

IMM Institute of Materials Management [UK]

IMM Institute of Mathematical Machines [Poland]

IMM Institute of Mechanics and Materials [of University of California at San Diego, La Jolla, US]

IMM Institute of Metals and Materials [Australasia] [also AusIMM]

IMM Institut für Metallkunde und Metallphysik [Institute of Metallurgy and Metal Physics [of Rheinish-Westfälische Technische Hochschule, Aachen, Germany]

IMM Institution of Mining and Metallurgy [UK]

IMM Integrated Maintenance Management

IMM Integrated Maintenance Manual

IMM Intelligent Memory Manager

IMM Interactive Multimedia

IMM International Monetary Market [Chicago, US]

IMMA International Motorcycle Manufacturers Association [France]

IMMA Ion Microprobe Mass Analysis; Ion Microprobe Mass Analyzer

IMM Abstr Institution of Mining and Metallurgy Abstracts [UK]

IMM Bull Institution of Mining and Metallurgy Bulletin [UK]

IMMC Intermetallic Metal-Matrix Composite

IMME Institute of Mining and Metallurgical Engineers

Imm Histo Immunohistology

immed immediate

immedy immediately

Immisc Immiscibility; Immiscible [also immisc]

IMMOA International Mercantile Marine Officers Associations [UK]

IMMP Integrated Maintenance Management Plan

IMMPS Institute for the Mechanics of Metal-Polymer Systems [Russian Federation]

IMMR Institut für Mineralogie und Mineralische Rohstoffe [Institute for Mineralogy and Mineralogical Resources, TU Clausthal, Germany]

IMMRC International Mass Media Research Center

IMMRN International Multi-Media Research Network

IMMS Institute of Metallurgy and Materials Science [of Shanghai University of Technology, PR China]

IMMS International Material Management Society [US]

IMMT Integrated Maintenance Management Team

Immun Immunity [also immun]

Immun Immunity [International Medical Journal]

Immunol Immunological; Immunologist; Immunology [also immunol]

Immunol Today Immunology Today [US Journal]

IMMW International Magnet Measurement Workshop [at Fermilab, Batavia, Illinois, US]

IMN Institut des Matériaux de Nantes [Nantes Institute of Materials, France]

IMNF Instituto do Metais Não Ferrosos [Institute for Nonferrous Metals, Sao Paulo, Brazil]

IMO In My Opinion

IMO Institut voor Materialonderzoeg [Institute for Materials Research, Limburg University Center, Deipenbeek, Belgium]

IMO International Maritime Organization [of United Nations]

IMO International Meteorological Organization [now World Meteorological Organization, Switzerland]

IMO International Miners Organization

imo image orthicon [Television]

IMOS Interactive Multiprogramming Operating System

IMOS Implant Metal-Oxide Semiconductor [also Imos]

IMOS Ion-Implanted Metal-Oxide Semiconductor

IMP Ice-Motion Package

IMP Imager for Mars Pathfinder [of NASA]

IMP Industrial Management Plan

IMP Industrial Mobilization Plan

IMP Information Management Program

IMP Information Message Processor

IMP Initial Maximum Pressure [Aerospace]

IMP Initial Memory Protection

IMP Injection into Microwave Products

IMP Inosine Monophosphate; Inosine 5'-Monophosphate [Biochemistry]

IMP Inosinic Acid [Biochemistry]

IMP Insoluble Metaphosphate

IMP Institute of Materials Processing [of Michigan Technological University, Houghton, US]

IMP Institute of Metal Physics [Russian Academy of Sciences–Ural Division, Ekaterinburg]

IMP Instituto Mexicano del Petróleo [Mexican Petroleum Institute, Mexico City]

IMP Integrated Message Processor

IMP Integrated Microprocessor

IMP Integrated Microwave Products

IMP Intelligent Machine Prognosticator

IMP Intelligent Message Processor

IMP Intensity-Modulated Photocurrent

IMP Interdiffused Multilayer Process

IMP Interface Message Processor

IMP Interindustry Management Program

IMP Interplanetary Monitoring Platform [of NASA]

IMP Interplanetary Monitoring Probe

IMP Intramembranous Particle

IMP Intrinsic Multiprocessing

IMP 1-Iodo-2-Methylpropane

IMP Ion Microprobe

IMP Ion-Moderated Partition

IMP Isothermal Melt Processing

5'-IMP Inosine-5'-Monophosphate [Biochemistry]

Imp Impact [also imp]

Imp Imperfect(ion) [also imp]

Imp Import(er) [also imp]

Imp Improvement [also imp]

Imp Impulse [also imp]

Imp Impurity [also imp]

imp impairment unit [Television]

imp imperial

imp import(ed)

imp important

IMPA Iminopropionicacetic Acid [also Impa]

IMPA Information Management and Processing Association

IMPA Intelligent Multi-Port Adapter

IMPA International Maritime Pilots Association [UK]

IMPA International Master Printers Association

IMPA International Motor Press Association [US]

IMPA Ion Microprobe Analysis; Ion Microprobe Analyzer

IMPAC Information for Management Planning Analysis and Coordination

IMPACT Immunization Monitoring Program, Active (Program) [of Health Canada]

IMPACT Improved Management of Personnel Administration through Computer Technology

IMPACT Integrated Managerial Programming Analysis Control Technique

IMPACT Inventory Management Program and Control Technique

IMPATT Impact Avalanche (and) Transit Time (Diode); Impact (Ionization) Avalanche Transit Time (Diode)

IMPC International Marine Pollution Conference

IMPC International Mineral Processing Congress

IMPCM Improved Capability Missile

IMPCM Instituto Mexicano del Petróleo, Catálisis y Materiales [Mexican Institute of Petroleum, Catalysis and Materials]

IMPCON Inventory Management and Production Control

Impd Impedance [also impd]

IMPE International Meeting on Petroleum Engineering

Imperf Imperfect(ion) [also imperf]

IMPEX Service des Importations et des Exportations [Importation and Exportation Services]

Impg Importing [also impg]

Impg Impregnation [also impg]

imp gal imperial gallon [Unit]

imp gal/min imperial gallon per minute [also IGPM or igpm]

IMPHOS Institut Mondial du Phosphate [World Phosphate Institute, Morocco]

IMPI International Microwave Power Institute [Clifton, Virginia, US]

IMPL Implementation Language

IMPL Initial Microprogram Load

Impl Implement(ation) [also impl]

Impls Impulse [also impls]

IMPMS Ion Microprobe Mass Spectrometer

Impn Importation [also impn]

IMPPT Intermolecular Moeller-Plesset Perturbation Theory [Physics]

imp pt imperial pint [Unit]

imp qt imperial quart [Unit]

Impr Improve(ment); Improved [also impr]

Impreg Impregnate(d); Impregnation [also Impreg or impreg]

Impreg Impregnated Wood [also impreg]

IMPRESS Interdisciplinary Machine Processing for Research and Education in the Social Sciences

IMPReSS IMage PeRimeters of Sky Surveys [of NASA Astrophysics Data Facility, Goddard Space Flight Center, US]

imprg impregnate

imprgd impregnated [also imprg'd]

IMPRINTA Internationaler Kongress und Ausstellung für Kommunikationstechniken [International Congress and Exhibition for Communication Technology]

Impr Nouv Imprimerie Nouvelle [French Publication on the Modern Printing Office]

imprt important

Imprv Improvement [also imprv]

IMPS Integrated Manufacturing Production System

IMPS Integrated Master Programming and Scheduling System

IMPS Integrated Modular Panel System

IMPS Intelligent Management Programming System

IMPS Intensity-Modulated Photocurrent Spectroscopy

IMPS Interface Message Processors [also IMPs]

IMPS International Microprogrammers Society

IMPS Interplanetary Measurement Probes

Imps Imports [also imps]

Imps Impurities [also imps]

Impt Implement [also impt]

impt important

Imptr Importer [also imptr]

IMPTS Improved Programmer Test Station

IMR Improved Military Rifle

IMR Innovations in Materials Research [International Journal published in Singapore] [also Innov Mat Res, or Innov Mater Res]

IMR Institute for Magnetics Research [of George Washington University, Washington, DC, US]

IMR Institute for Materials Research [of National Institute for Standards and Technology, Gaithersburg, Maryland, US]

IMR Institute for Materials Research [of McMaster University, Hamilton, Canada]

IMR Institute for Materials Research [of Tohoku University, Sendai, Japan]

IMR Institute of Materials Research [of Slovak Academy of Sciences, Slovak Republic]

IMR Institute of Medical Research [Malaysia]

IMR Institute of Metal Repair [US]

IMR Institute of Metals Research [of Chinese Academy of Sciences, Shenyang, PR China]

IMR Institute of Mineral Research

IMR Integrated Microscopy Resource [of University of Wisconsin, US]

IMR Intelligent Machine Research

IMR Internal Mold Release

IMR International Materials Review [UK]

IMR Internet Monthly Report

IMR Interruption Mask Register

IMR Inverse Magnetoresistance

IMR Irreducible Matrix Representation [Mathematics]

IMRA Incentive Manufacturers Representatives Association [US]

IMRA Independent Motocycle Retailers of America [US]

IMRA Industrial Marketing Research Association [UK]

IMRA Infrared Monochromatic Radiation

IMRA International Manufacturers Representatives Association [US]

IMRA International Marine Radio Association

IMRADS Information Management Retrieval and Dissemination System

IMRAMN International Meeting on Radio Aids to Marine Navigation

IMRAN International Marine Radio Aids to Navigation

IMRC International Marine Radio Committee

IMRC International Materials Research Committee

IMRE Institute for Materials Research [Singapore]

IM Rep Integrated Manufacturing Report [of Association for Integrated Manufacturing Technology, US]

IMRF Independent Manufacturers Representatives Forum [US]

IMRI Industrial Materials Research Institute [Boucherville, Quebec, Canada]

IMRI/NRC Industrial Materials Research Institute/National Research Council [Boucherville, Quebec, Canada]

IM/RIT Institute of Materials/Royal Institute of Technology [Stockholm, Sweden]

IMRL Integrated Materials Research Laboratory [of Sandia Nation Laboratories, Albuquerque, New Mexico, US]

IMRO Investment Management Regulatory Organization [UK]

IMRP International Meeting on Radiation Processing [Canada]

IMR SAS Institute of Materials Research of the Slovak Academy of Sciences [Košice, Slovak Republic]

IMS Incommensurately Modulated Structure [Materials Science]

IMS Independent Monomer State

IMS Indian Mathematical Society

IMS Industrial Management Services (Group) [of ORTECH–Ontario Research and Technology Foundation, Canada]

IMS Industrial Management Society [US]

IMS Industrial Management System

IMS Industrial Mathematics Society [US]

IMS Inertial Measuring Unit

IMS Information Management System

IMS Institut de Métallurgie Structurale [Structural Metallurgy Institute; of Université de Neuchâtel, Switzerland]

IMS Institute for Manpower Studies [UK]

IMS Institute for Marine Science [of University of Alaska at Fairbanks, US]

IMS Institute for Mathematical Statistics [of University of Michigan, Ann Arbor, US]

IMS Institute for Microstructural Sciences [of National Research Council of Canada]

IMS Institute for Molecular Science [Japan]

IMS Institute of Management Sciences [Providence, Rhode Island, US]

IMS Institute of Management Services [Enfield, Middlesex, UK]

IMS Institute of Management Specialists [UK]

IMS Institute of Materials Science

IMS Institute of Materials Science [of University of Connecticut, Stoors, US]

IMS Institute of Mathematical Sciences [of New York University, US]

IMS Institut für Mikroelektronische Schaltungen und Systeme [Institute for Microelectronic Circuits and Systems; of Fraunhofer Institute, Germany]

IMS Instructional Management System

IMS Instrument(ation) and Measurement Science

IMS Instrumentation and Measurement Society [of Institute of Electrical and Electronics Engineers, US]

IMS Integrated Mechanical System

IMS Integrated Meteorological System [of US Army]

IMS Intelligent Manufacturing System (Project)

IMS Interactive Market System [US]

IMS Intermagnetic Shield

IMS Intermediate Maintenance Standards

IMS International Magnetospheric Studies

IMS Information Management System

IMS International Metallographic Society [Materials Park, US]

IMS International Military Staff

IMS International Mountain Society [US]

IMS Intrinsic Monomer Stress

IMS Inventory Management and Simulator

IMS Inventory Management System

IMS Ion Mobility Spectrometer; Ion Mobility Spectrometry

IMS Ion Mobility Spectroscopy

IMS Isocratic Multisolvent

IMS Israeli Metallurgical Society

IMSA International Management Systems Association

IMSA International Medical Sciences Academy

IMSA International Motor Sports Association [US]

IMSA International Municipal Signal Association [US]

Imsa Industrias Monterrey SA [Monterrey Industries Company, Mexico]

IMS/ASM International Metallographic Society/American Society for Materials International (Program)

IMS Bull Institute of Mathematical Statistics Bulletin [US]

IMSC International Mass Spectrometry Conference

IMSC International Mobile Satellite Conference

IMSCO International Maritime Satellite Consortium

IMSE Institute for Materials Science and Engineering [of National Institute for Standards and Technology, Boulder, Colorado, US]

IMSE Institute for Materials Science and Engineering [of National Chiao Tung University, Hsinchu, Taiwan]

IMSE Interactive Materials Science and Engineering [CD-ROM Package published by John Wiley and Sons Inc., New York, US]

IMS EXPO International Manufacturing Software Exposition and Conference [of International Society for Measurement and Control, US]

IMSI Information Management System Interface

IMSL International Mathematical and Statistical Libraries

IMSM Interdigitated Metal-Semiconductor-Metal (Schottky-Barrier Photodiode)

IMSP Institute of Metals Superplasticity Problems [of Russian Academy of Sciences; located at Ufa]

IMSP Integrated Mass Storage Processor

IMSP Internet Message Support Protocol

IMSR Interplanetary Mission Support Requirements

IMSS Instituto Mexicana para el Securidad Sociale [Mexican Social Security Institute]

IMSSCE Interceptor Missile Squadron and Supervisory Control Equipment

IMSSS Interceptor Missile Squadron Supervisory Station

IMST International Mushroom Society for the Tropics [Hong Kong]

IMSTG Instrument and Measurement Science Topical Group [of American Physical Society, US]

IMSU Integrated Mass-Storage Unit

IMS/VS Information Management System/Virtual Storage

IMT Immersion Tank (Test) System

IMT Immersion Testing

IMT Industrial and Materials Technologies (Program) [Europe]

IMT Industrial Materials Technologist; Industrial Materials Technology

IMT Industrial Materials Technology, Inc.

IMT Injection Molding Technology

IMT Institute for Magnesium Technology [Sainte Foy, Quebec, Canada]

IMT Institute of Mechanical Technology [Pusan National University, South Korea]

IMT Institute of Metals and Technology [of Dalian Maritime University, PR China]

IMT Institute of Municipal Transport [UK]

IMT Institut für Mikrostrukturtechnik [Instiitute for Microstructural Technology; at Karlsruhe Research Center, Germany]

IMT Instituto Maua de Tecnologia [Maua Institute of Technology, São Caetano do Sul, Brazil]

IMT Integrated Micro-Image Terminal

IMT Inter-Machine Trunk

IMT Intermediate Tape

IMT Ion Microtomography

imt immediate

IMTA Intensive Military Training Area

IMTA International Mass Transit Association [US]

IMTA International Marine Transport Association [US]

IMTC Instrumentation/Measurement Technology Conference [of Institute of Electrical and Electronics Engineers, US]

IMTC International Multimedia Teleconferencing Consortium

IMTEG (European) ILS (Instrument Landing System)/MLS (Microwave Landing System) Transition Group [of International Civil Aviation Organization]

IMTF Ising Model (in a) Transverse Field [Solid-State Physics]

IMTM Istituto di Metodologie e Tecnologie per la Microelettronica [Institute for Microelectronics Methodology and Technology; of Consiglio Nazionale delle Ricerche at Università di Catania, Italy]

IMTNE International Meteorological Teleprinter Network, Europe

IMTS Improved Mobile Telephone Service; Improved Mobile Telephone System

IMTS International Machine Tool Show

IMTS International Manufacturing Technology Show

IMTV Interactive Multimedia Television

IMU Increment Memory Unit

IMU Inertial Measurement Unit

IMU Inertial Measuring Unit [of NASA Apollo Program]

IMU Instruction Memory Unit

IMU Internal Measurement Unit

IMU International Mathematical Union [Finland]

IMU International Metal Union [Switzerland]

IMU Iwaki-Meisei University [Fukushima, Japan]

IMU Bull IMU Bulletin [of International Mathematical Union]

IMUL Integer Multiply

IMun&CyE Institution of Municipal and County Engineers [UK]

IMunE Institution of Municipal Engineers [UK]

IMUX Inverse Multiplexer [also Imux]

IMVIC Indole, Methyl Red, Voges-Proskauer, and Citrate Test [Biotechnology]

IMVS Institute of Medical and Veterinary Science [Australia]

IMW International Map of the World

IMWA International Mine Water Association [Spain]

IMWoodT Institute of Machine Woodworking Technology [UK]

IMX Instruction Memory Exchange [US]

IMX International Metalworking Exhibition

Im(x) Imaginary Part of x [Symbol]

IMXA Ion Microprobe X-Ray Analysis

IN Ice Nucleus [Meteorology]

IN India [ISO Code]

IN Indiana [US]

IN Inertial Navigation; Inertial Navigator

IN Inflammatory Response

IN Information Network

IN Information Systems Directorate [of NASA Kennedy Space Center, Florida, US]

IN Input

IN Institute of Navigation [Washington, DC, US]

IN Insoluble Nitrogen

IN Interference-to-Noise (Ratio)

IN Internodal

IN Internuclear

IN Iodine Number

IN Iodonaphthalene

IN Ion Neutralization

IN Irrational Number

IN Isomer Number

IN Isonitrile

In Indium [Symbol]

In Indigo

In Inlet [also in]

In Input [also in]

In^{3+} Indium Ion [also In^{+++}] [Symbol]

In-107 Indium-107 [also ^{107}In, or In^{107}]

In-108 Indium-108 [also ^{108}In, or In^{108}]

In-109 Indium-109 [also ^{109}In, or In^{109}]

In-110 Indium-110 [also ^{110}In, or In^{110}]

In-111 Indium-111 [also ^{111}In, or In^{111}]

In-112 Indium-112 [also ^{112}In, or In^{112}]

In-113 Indium-113 [also ^{113}In, or In^{113}]

In-114 Indium-114 [also ^{114}In, or In^{114}]

In-115 Indium-115 [also ^{115}In, or In^{115}]

In-116 Indium-116 [also ^{116}In, or In^{116}]

In-117 Indium-117 [also ^{117}In, or In^{117}]

In-118 Indium-118 [also ^{118}In, or In^{118}]

In-119 Indium-119 [also ^{119}In, or In^{119}]

in interior

in internal

in inch(es) [also "]

in^2 square inch(es) [also sq in]

in^3 cubic inch(es) [also cu in]

in^4 inch(es) to the fourth power

.in India [Country Code/Domain Name]

INA Initial Approach [Aeronautics]

INA Institute of Nautical Archeology [US]

INA Instituto Nacional Agrarianos [National Agrarians Institute, Honduras]

INA Institution of Naval Architects [UK]

INA Integrated Network Architecture

INA International Normal Atmosphere

INA Isonicotinic Acid

INAA Integrated Neutron Activation Analysis

INAA Instrument(al) Neutron Activation Analysis

In(Ac)$_3$ Indium Acetate

In(ACAC)$_3$ Indium (III) Acetylacetonate [also In(AcAc)$_3$, or In(acac)$_3$]

inact inactive

INADEQUATE Incredible Natural Abundance Double Quantum Transfer Experiment [Nuclear Magnetic Resonance]

INAG Ionospheric Network Advisory Group

INAH Instituto Nacional de Antropologia Historia [National Institute for Historical Anthropology, Mexico City]

INAH Isonicotinic Acid Hydrazide; Isonicotinic Acid Hydrazine

INALCO International Conference on Aluminum Weldments

InAlAs Indium Aluminum Arsenide (Semiconductor)

InAlN Indium Aluminum Nitride (Semiconductor)

InAlP Indium Aluminum Phosphide (Semiconductor)

INAOE Instituto Nacional de Astrofísica, Optica y Electrónica [National Institute for Astrophysics, Optics and Electronics, Puebla, Mexico]

in aq *(in aqua)* – in water [Medical Prescriptions]

INAS Inertial Navigation and Attack System

InAs Indium Arsenide (Semiconductor)

InAs/GaAs Indium Arsenide/Gallium Arsenide (Superlattice)

InAs/GaSb Indium Arsenide/Gallium Antimonide (Superlattice)

InAs/InAsSb Indium Arsenide/Indium Arsenide Antimonide (Superlattice)

INASMET Centro Tecnologico de Materiales [Technological Center for Materials, San Sebastian, Spain]

InAsN Indium Arsenide Nitride (Semiconductor)

InAsSb Indium Arsenide Antimonide (Semiconductor)

InAsSbP Indium Arsenide Antimonide Phosphide (Semiconductor)

In-Au Indium-Gold (Alloy System)

INAV Instituto Nacional del Audiovisual [National Institute of Audiovisual Techniques, Spain]

INB Institut National du Bois [National Wood Institute, France]

INB Instituto Nacional de Bacteriologia [National Institute of Bacteriology, La Paz, Bolivia]

Inbd Inboard [also inbd]

Inbd Inbound [also inbd]

INBio Instituto Nacional de Biodiversidad [National Biodiversity Institute, Costa Rica] [also INBIO]

InBr Indium(I) Bromide (or Indium Monobromide)

INC In Cloud

INC Increment (Function) [also Inc]

INC Input Control System

INC Installation Notice Card

INC Installation Notification Certificate

INC Instituto Nacional de Consumo [National Consumers Institute, Spain]

INC Instrumentation and Communications

INC Intergovernmental Negotiating Committee

INC International Numismatic Commission

INC Intranuclear Cascade (Model) [Nuclear Physics]

Inc Incandescence; Incandescent [also inc]

Inc Incendiary [also inc]

Inc Inclination [also inc]

Inc Inclusion; Inclusive [also inc]

Inc Incoming [also inc]

Inc Incorporate(d); Incorporating; Incorporation [also inc]

Inc Increase [also inc]

Inc Increment [also inc]

INCA Independent National Computing Association [UK]

INCA International Newspaper and Color Association [Switzerland]

INCAL International Conference on Aluminum

INCAP Instituto Nutricióne Centroamericana y Pánama [Institute of Nutrition for Central America and Panama, Guatemala]

INCB International Narcotics Control Board [of United Nations]

INCC International Network Controlling Center

Incdt Incident [also incdt]

INCE Institute of Noise Control Engineering [Poughkeepsie, New York State, US]

INCE International Network in Chemical Education

Incend Incendiary [also incend]

Incep Interceptor [also incep]

INCERFA Uncertainty Phase [Aeronautics]

INC/FCCC Intergovernmental Negotiating Committee for a Framework Convention on Climate Change

INCH Indirectly-Bonded Carbon-Hydrogen (Shift Correlation)

INCH Integrated Chopper [also IN CH, or In Ch]

INCIAWPRC Indian National Committee of the International Association on Water Pollution and Control

Incin Incinerator [also incin]

INCINC International Copyright Information Center [US]

INCIRS International Communication Information Retrieval System [of University of Florida, US]

InCl Indium(I) Chloride (or Indium Monochloride)

Incl Inclosure [also incl]

Incl Inclusion [also incl]

incl include(d); including; inclusive

incld included [also incl'd]

inclg including

INCO Instrumentation and Communications Officer

INCO International Chamber of Commerce

Inco International Nickel Company (Process)

Inco Alloys Int Inco Alloys International [Publication of Inco Alloys International Ltd., Hereford, UK]

INCODA International Congress of Dealers Associations [US]

INCOFILT International Consortium of Filtration Research Group

incog incognito

INCOLSA Indiana Cooperative Library Services Authority [US]

INCOM International Symposium on Manufacturing Technology [Canada]

INCOMEX International Computer Exhibition

incompl incomplete

INCONCRYO International Congress on Cryogenics

INCOPAC International Consumer Policy Advisory Committee [also InCoPAC]

INCOR Indian National Committee on Oceanic Research [India]

INCOR Intergovernmental Conference on Oceanographic Research

INCOR Israel National Committee for Oceanographic Research

INCOSPAR Indian National Committee for Space Research [India]

INCO SPP Inco International Inc. Specialty Powder Products

INCOT In-Core Test Facility [Nuclear Reactors]

Incoterms International Commercial Terms

INPC International Nuclear Physics Conference

InCp Cyclopentadienylindium(I)

INCPEN Industrial Council for Packaging and Environment [UK]

Incpt Intercept [also incpt]

INCR Interrupt Control Register

Incr Increase [also incr]

Incr Increment [also incr]

incr increase(d); increasing

INCRA International Copper Research Association [US]

INCRA Res Rep INCRA Research Report [of International Copper Research Association, US]

Incre Increment [also incre]

Increm Increment(al) [also increm]

Increm Motion Control Syst Devices Newsl Incremental Motion Control Systems and Devices Newsletter [Published by the University of Illinois-Urbana, US]

INCREST Research Institute for Aircraft Materials [Romania]

INCSL Incoherent Superlattice [Solid-State Physics]

INCUM Indiana Computer Users Meeting [US]

IND International Number Dialing

IND Internuclear Distance [Physical Chemistry]

IND Investigative (or Investigational) New Drug

.IND Index [File Name Extension] [also .ind]

Ind Independence; Independent [also ind]

Ind Index [also ind]

Ind India(n)

Ind Indiana [US]

Ind Indication; Indicator [also ind]

Ind Indigo [also ind]

Ind Indonesia(n)

Ind Induction [also ind]

Ind Industrial(ist); Industry [also ind]

ind indirect

INDA International Nonwoven and Disposables Association [US]

Indag Math Indagationes Mathematicae [Mathematical Journal of the Koninklijke Nederlandse Akademie van Wetenschappen (Royal Dutch Academy of Sciences)]

INDAL Indian Aluminum Company Limited [also Indal]

Ind-Anz Industrie-Anzeiger [Publication on the German Industry]

indc indicate

Ind Carta Industria della Carta [Italian Publication on the Paper Industry]

Ind Céram Industrie Céramique [French Journal on the Ceramics Industry]

Ind Ceram Industrial Ceramics [Publication Issued with *Ceramics International;* published in Italy]

Ind Ceram (Paris) Industrie Ceramique (Paris) [French Journal on the Ceramics Industry]

Ind Commer Train Industrial and Commercial Training [Journal published in the UK]

Ind Comput Industrial Computing [Journal published in the UK]

Ind Corros Industrial Corrosion [Journal published in the UK]

Ind Datatek Industriell Datateknik [Swedish Publication on Industrial Data Systems Technology]

Ind Diamond Rev Industrial Diamond Review [UK]

INDE ITER (International Thermonuclear Experimental Reactor)/NET (Next European Torus) Design and Engineering

IndE Industrial Engineer [also Ind Eng, or IndEng]

INDECOPI Instituto Nacional de Defensa de la Competencia y de la Protección de la Propriedad Intelectual [National Defense Institute for the Competency and Protection of Intellectual Property, Argentina]

Ind Ed Industrial Education

indef indefinite

Ind Electr Inf Industrie–Electricité–Informations [French Newsletter on Industry and Electricity]

Ind Eng Industrial Engineer(ing) [also IndEng, or IndE]

Ind Eng Industrial Engineering [Journal of the American Institute of Industrial Engineers, US]

Ind Eng Chem Industrial and Engineering Chemistry [Publication of the American Chemical Society, US]

Ind Eng Chem, Fundam Industrial and Engineering Chemistry, Fundamentals [Publication of the American Chemical Society, US]

Ind Eng Chem, Process Des Dev Industrial and Engineering Chemistry, Process Design and Development [Publication of the American Chemical Society, US]

Ind Eng Chem, Prod Res Dev Industrial and Engineering Chemistry, Product Research and Development [Publication of the American Chemical Society, US]

Ind Eng Chem Res Industrial and Engineering Chemistry Research [Journal of the American Chemical Society, US]

Ind Engr Industrial Engineer(ing)

Ind Environ Industry and Environment [Publication of United Nations Environment Programme]

Indep Independence; Independent [also indep]

Indep Eye Independent Eye [Magazine of SGS United Kingdom Ltd.]

Indep Power Independent Power [Publication of Alternative Energy Sources Inc., US]

INDEPTH International Deep Profiling of Tibet and the Himalaya [Joint US-PR China Project]

INDERENA Instituto Nacional de los Recursos Naturales Renovables y del Ambient [National Institute for Renewable Natural Resources and the Environment, Colombia]

INDEX Inter-NASA Data Exchange

Ind Exp Tech Industrial and Experimental Techniques [Journal published in the US]

Ind Finish Industrial Finishing [Journal published in the US]

Ind Gomma Industria della Gomma [Italian Journal on the Rubber and Gum Industries]

Ind Health Industrial Health [Journal published in the US]

Ind Heat Industrial Heating [Journal published in the US]

INDIAN Interplanar Distances and Angles [Crystallography]

Indian Ceram Indian Ceramics [Journal published in India]

Indian Chem Eng Indian Chemical Engineer [Journal published in India]

Indian East Eng Indian and Eastern Engineer [Journal published in Bombay, India]

Indian Eng Chem Res Indian Engineering Chemistry Research (Journal)

Indian Foundry J Indian Foundry Journal

Indian J Chem Indian Journal of Chemistry [Published by the Council for Scientific and Industrial Research, India]

Indian J Chem A Indian Journal of Chemistry, Section A [Published by the Council for Scientific and Industrial Research, India]

Indian J Chem B Indian Journal of Chemistry, Section B [Published by the Council for Scientific and Industrial Research, India]

Indian J Cryog Indian Journal of Cryogenics [of Indian Cryogenics Council]

Indian J Mar Sci Indian Journal of Marine Sciences [of the Council of Scientific and Industrial Research, India]

Indian J Med Sci Indian Journal of Medical Science

Indian J Nutr Diet Indian Journal of Nutrition and Dietetics

Indian J Phys Indian Journal of Physics [of Indian Association for the Cultivation of Science]

Indian J Phys A Indian Journal of Physics, Section A [of Indian Association for the Cultivation of Science]

Indian J Phys B Indian Journal of Physics, Section B [of Indian Association for the Cultivation of Science]

Indian J Physiol Pharmacol Indian Journal of Physiology and Pharmacology

Indian J Power River Val Dev Indian Journal of Power and River Valley Development [India]

Indian J Pure Appl Math Indian Journal of Pure and Applied Mathematics [Published by the Indian National Science Academy]

Indian J Pure Appl Phys Indian Journal of Pure and Applied Physics [Published by the Council for Scientific and Industrial Research, India]

Indian J Radio Space Phys Indian Journal of Radio and Space Physics [Published by the Council for Scientific and Industrial Research, India]

Indian J Technol Indian Journal of Technology [Published by the Council for Scientific and Industrial Research, India]

Indian J Textile Res Indian Journal of Textile Research [Published by the Council of Scientific and Industrial Research, India]

Indian J Theor Phys Indian Journal of Theoretical Physics [Published by the Institute of Theoretical Physics, Calcutta, India]

Indian Min Eng J Indian Mining and Engineering Journal

Indian Sci Abstr Indian Science Abstracts [Published by the Indian National Scientific Documentation Center]

Indian Weld J Indian Welding Journal

indic indicated

INDIS Industrial Information System

INDITECNOR Instituto Nacional de Investigaciones Tecnologicas y Normalisación [National Institute for Technological Investigations and Standardization, Chile]

ind'l industrial

Ind Lab Industrial Laboratory [Translation of: *Zavodskaya Laboratoriya (USSR)*; published in the US]

Ind Lackier-Betr Industrie Lackier-Betrieb [German Journal on the Paint and Coating Industry]

Ind Lubr Tribol Industrial Lubrication and Tribology [Journal published in the UK]

Ind Manage Industrial Management [Journal of the Institute of Industrial Engineers, US]

Ind Manage + Data Syst Industrial Management + Data Systems [Journal published in the UK]

Ind Mgt Industrial Management

Ind Math Industrial Mathematics [Published by the Industrial Mathematics Society, US]

Ind Market Dig Industrial Marketing Digest [UK]

Ind Min (Brussels) Industrie Minérale (Brussels) [Belgian Publication on the Minerals Industry]

Ind Miner (London) Industrial Minerals (London) [UK Publication]

Ind Min, Mines Carr Industrie Minérale, Mines et Carrières [French Publication on the Minerals, Mining and Quarry Industries]

Ind Min, Mines Carr, Tech Industrie Minérale, Mines et Carrières, Les Techniques [French Publication on Methods and Techniques in the Minerals, Mining and Quarry Industries]

Ind Min (Paris) Industrie Minerale (Paris) [Publication on the Minerals Industry; published in Paris, France]

Ind Min (St Etienne) Industrie Minerale (St Etienne) [Publication on the Minerals Industry; published in St. Etienne, France]

Ind Min Suppl, Tech (St Etienne) Industrie Minerale Supplement, Les Techniques (St Etienne) [Supplement to *Industrie Minérale*; published in St. Etienne, France]

Indn Indication [also indn]

INDO Incomplete Neglect of Differential Overlap; Intermediate Neglect of Differential Overlap [Physics]

Ind O Indian Ocean

Indoch Indochina

INDOEX Indian Ocean Experiment

Indon Indonesia(n)

Indon J Nat Sci Indonesian Journal of Natural Science

INDOPROFEN α-Methyl-p-(1-Oxo-2-Isoindolinyl)-Benzeneacetic Acid [also Indoprofen, or indoprofen]

INDOR Internuclear Double Resonance

INDOTEC Instituto Dominicano de Tecnología Industrial [Dominican Institute of Industrial Technology]

Ind Phys The Industrial Physicist [Journal of the American Institute of Physics, US]

Ind Prod Industrial Production [Publication of the Statistical Office of the European Communities, Luxembourg]

Ind Recovery Industrial Recovery [Journal published in the UK]

INDREG Inductance Regulator

Ind Res Indian Reservation

Ind Res Industrial Research(er)

Ind Res Industrial Research (Magazine)

Ind Res Dev Industrial Research and Development [Journal published in the US]

Ind Robot Industrial Robot [Journal published in the UK]

INDS In-Core Nuclear Detection System [Nuclear Reactors]

Ind Sci Instrum Industrial and Scientific Instruments [Journal published in the UK]

Ind Sci Technol Industrial Science and Technology [Journal published in Japan]

Ind Soc Industrial Society [Publication of the Industrial Society, London, UK]

Ind Spec Industrial Specialties [also ind spec] [Pulp and Paper Industry]

Ind Tech Industries et Techniques [French Publication on Industries and Processes]

INDTR Indicator-Transmitter

Ind Trends Industrial Trends [Publication of the Statistical Office of the European Communities, Luxembourg]

Induced EMF Induced Electromotive Force [also induced emf]

industl industrial [also indust'l]

industr industrial

Industrial ACE Industrial Automation Conference and Exhibition [of Institute of Electrical and Electronics Engineers, US]

Indv Individual [also indv]

Indvdl Individual [also indvdl]

Ind Week Industry Week [Published in the US]

Indy Industry [also Ind'y, indy, or ind'y]

INE Inertial Navigation Equipment

INE Institut für Nukleare Entsorgungstechnik [Institute for Nuclear Waste Management, of Kernforschungszentrum Karlsruhe, GmbH, Germany]

INE Institution of Nuclear Engineers [London, UK]

INE Instituto Nacional de Estadística [National Statistical Institute, Spain]

IN&EA International Nuclear and Energy Association [also INEA, US]

INED Institut National d'Etudes Démographiques [National Institute for Demographic Studies, France]

INEEL Idaho National Engineering and Environmental Laboratory [of Lockheed Martin Idaho Technologies Company, Idaho Falls, US]

INEL Idaho National Engineering Laboratory [of US Department of Energy at Idaho Falls, US]

INEL International Exhibition of Industrial Electronics

Inel Inelastic(ity) [also inel]

INELEC Institut Nationale d'Electricité et d'Electronique [National Electrical and Electronic Institute, Algeria]

INEL CPS Idaho National Engineering Laboratory–Curved Position-Sensitive Detector [of US Department of Energy at Idaho Falls] [also INEL-CPS]

INEN Instituto Ecuatoriano de Normalisación [of Ecuadorean Standards Institute]

INENCO Center for International Environmental Cooperation [Russia]

INEOA International Narcotic Enforcement Officer Association [US]

INEPT Insensitive Nuclei Enhanced by Polarization Transfer

INEPTCR Insensitive Nuclei Enhanced by Polarization Transfer (under) Composite Refocusing [also INEPT-CR, or INEPT CR]

INER Institute of Nuclear Energy Research [Lungtan, Taiwan]

INER List of Pesticide Product Inert Ingredients

INESC Institute de Engenharia de Sistemas e Computadores [Institute for Systems and Computer Engineering, Portugal]

INET Institute of Nuclear Energy Technology [of Tsinghua University, Beijing, PR China]

INET Intelligent Network Simulator

INETI Instituto Nacional de Engenharia e Tecnologia Industrial [National Institute of Engineering and Industrial Technology, Portugal]

INEWS Integrated Electronic Warfare System

INF Instituto Nacional de Farmacologia [National Pharmacological Institute, Brazil]

INF Intermediate-Range Nuclear Forces (Treaty) [Between US and USSR]

INF Iwatani Naoji Foundation [Japan]

.INF Information [File Name Extension] [also .inf]

Inf Infantry [also inf]

Inf Information [also inf]

inf inferior

inf infimum [Mathematics]

inf *(infra)* – below

INFAC Instrumented Factory (for Gears Program) [of Illinois Institute of Technology Research Institute, US]

INFACON International Ferroalloys Congress

INFACT Irish National Federation Against Copyright Theft

InFACT Integrated Flexible Assembly Cell Technology

Inf Age Information Age [Journal published in the UK]

Infalum Bull Infalum Bulletin [Switzerland]

INFANT Iroquois Night Fighter and Night Tracker

Infar Informations-Funk-Auto-Radio [Radio Traffic Information Service, Germany]

Inf Bull Var Stars Information Bulletin on Variable Stars [Hungary]

Inf Bull Circ Informations Bulletinen Circulus [Bulletin on Life Quality]

INFCE International Nuclear Fuel Cycle Evaluation [US]

Inf Chim Informations Chimie [Chemistry Information published by Société d'Expansion Technique et Economique, France]

INFCO Information Committee [of International Standards Organization]

Inf Comput Information and Computation [US Publication]

Inf Decis Technol Information and Decision Technologies [Journal published in the Netherlands]

Inf Dev Information Development [Journal published in the UK]

Inf Disp Information Display [Journal published in the US]

Inf Econ Policy Information Economics and Policy [Published in the Netherlands]

Infect Immun Infection and Immunity [Journal of the American Society for Microbiology, US]

Inf Elektron Informacio Elektronika [Hungarian Publication on Electronic Information]

Inf Elettron Informazione Elettronica [Italian Publication on Electronic Information]

Inf Epigast Art Inferior Epigastric Artery [also inf epigast art] [Anatomy]

Inf Epigast Vein Inferior Epigastric Vein [also inf epigast vein] [Anatomy]

Inf Exec Information Executive [Publication of the Data Processing Management Association, US]

inf,glb infimum, greatest lower bound [Mathematics]

INFIC International Network of Feed Information Centers [Australia]

INFINA Informatik für die Industrielle Automation [Congress on Informatics for Industrial Automation, Germany]

Inf INT Informativo do INT [Publication of Instituto Nacional de Tecnologia, Brazil]

Inf Intell Online Newsl Information Intelligence Online Newsletter [Published by Information Intelligence Inc., Phoenix, US]

Inf IREQ Information IREQ [Publication of Institut de Recherche en Electricité du Québec, Varennes, Canada] [now Inf TAI]

INFIRS Inverted File Information Retrieval System [of Chemical Information Service, UK]

INFIS Institut für Internationale Sozialforschung e.V. [Institute for International Social Research, Berlin, Germany]

Infl Inflammable; Inflammability [also infl]

Infl Influence [also infl]

infl influenced

Inflam Inflammability; Inflammable [also inflam]

INFM Consorzio Interuniversitario di Fisica della Materia [Inter-University Consortium on Matter Physics; Laboratorio Tecnologie Avanzate Superfici e Catalisi, Trieste, Italy]

INFM Istituto Nazionale per la Fisica della Materia [National Institute for Matter Physics, Trieste, Italy]

Inf Manage Information and Management [Netherlands]

Inf Media Technol Information Media and Technology [Journal of CIMTECH, Hatfield, UK]

Inf Mkt Information Market [Publication of the Commission of the European Communities, Luxembourg]

INFN Instituto Nazionale de Fisico Nucleare [National Institute of Nuclear Physics, Italy]

Infn Information [also infn]

Inf Net Information Network [also inf net]

INFO Information Network and File Organization

.INFO Information Services [Internet Domain Name]

Info Information [also info]

INFOBASE International Data Bank Congress and Exhibition

INFOCEN Information Center

INFOCLIMA Climate Data Referral System [of World Meteorological Organization]

INFOCOM Information Communications [US]

INFODATA Dedicated Digital Data Information Service [of Canadian National/Canadian Pacific Telecommunications] [also Infodata]

INFOES In-Flight Operational Evaluation of Space Systems

INFOEXCHANGE Digital Circuit-Switching Information Exchange [of Canadian National/Canadian Pacific Telecommunications, Canada] [also Infoexchange]

INFOFILM International Information Film Service

INFOHOST Information on Hosts (Database) [Germany]

INFOL Information-Oriented Language

INFOLAC Information for Latin American Countries (Project) [of United Nations]

INFOMAT Information on Materials and Coatings

INFONET Information Network [UK]

INFOPAC Pacific Bell Information System [US]

INFOR Information-Oriented Language

Inform Informatique [French Journal on Information Science; published by Association Française pour le Cybernetique, Economique et Technique, France]

Informatologia Yugosl Informatologia Yugoslavica [Yugoslavian Journal on Information Science and Technology; published by the University of Zagreb]

Inform Diritto Informatica e Diritto [Italian Publication on Informatics]

Inform Forsch Entwickl Informatik Forschung und Entwicklung [German Publication on Information Science– Research and Development]

Inform Spectrum Informatik Spektrum [German Publication on Information Science]

Inform Theor Appl Informatique Theorique et Applications [French Journal on Theoretical Information/Computer Science and Its Applications]

INFOSWITCH Packet-Switching Data Information Network [of CNCP Telecommunications, Canada] [also Infoswitch]

INFOTERM International Information Center for Terminology [Austria]

INFOTERRA International Register for Sources of Environmental Information [of United Nations Environmental Program]

INFOTEX Information via Telex [US]

Inf Power Information Power [Journal published in the UK]

Inf Process Information Processing [Journal of the Information Processing Society of Japan]

Inf Process Lett Information Processing Letters [Netherlands]

Inf Process Manage Information Processing and Management [Journal published in the UK]

Inf Process Soc Jpn Information Processing Society of Japan [Japanese Journal]

INFRAL Information Retrieval Automatic Language

Infrared Phys Infrared Physics [UK]

Inf Recht Informatik und Recht [German Journal on Information Science and the Law]

Inf Referral, J Alliance Inf Referral Syst Information and Referral: Journal of the Alliance of Information Referral Systems [US]

Inf Resour Manage J Information Resources Management International Journal [US]

Inf Sci Information Science; Information Scientist

Inf Sci Information Sciences [Journal published in the US]

Inf Serv Use Information Service and Use [Published in the Netherlands]

Inf Soc Information Society [US]

Inf Softw Technol Information and Software Technology [Journal published in the UK]

Inf Stand Q Information Standards Quarterly [Published by the National Information Standards Organization, US]

Inf Strategy, Exec J Information Strategy: The Executive's Journal [US]

Inf Syst Informacne Systemy [Slovak Publication on Information Systems]

Inf Syst Information Systems [Journal published in the UK]

Inf Syst Res Information Systems Research [Publication of The Institute of Management Sciences, US]

Inf TAI Information TAI [Publication of Institut de Recherche en Electricité du Québec, Varennes, Canada] [formerly Inf IREQ]

Inf Tech–IT Informationstechnik–IT [German Journal on Information Technology]

Inf Technol Information Technologist; Information Technology

Inf Technol Dev Information Technology for Development [Published by Oxford University Press, UK]

Inf Technol Learn Information Technology and Learning [Journal published in the UK]

Inf Technol Libr Information Technology and Libraries [Published by the Library and Information Technology Association, US]

Inf Technol Public Policy Information Technology and Public Policy [Journal published in the UK]

Inf Times Information Times [Publication of the Information Industry Association, US]

Inf Today Information Today [Journal published in the US]

infus infusible

Inf Vena Cava Inferior Vena Cava [also inf vena cava] [Anatomy]

Infy Infantry [also infy]

ING Index Nominum Genericorum

ING Instituto Nacional de Geotécnica [National Geotechnical Institute, Spain]

ING Integrated Ground

ING Intense Neutron Generator

ING International Newspaper Group [US]

Ing *(Ingegnere)* – Italian for "Engineer"

Ing *(Ingeniero)* – Spanish for "Engineer"

Ing *(Ingénieur)* – French for "Engineer"

Ing *(Ingenieur)* – German for "Engineer"

INGA Interactive Graphics Analysis

INGAA Interstate Natural Gas Association of America [US]

InGaAl Indium Gallium Aluminide (Semiconductor)

InGaAlP Indium Gallium Aluminum Phosphide (Semiconductor)

InGaAs Indium Gallium Arsenide (Semiconductor)

InGaAs/GaAs Indium Gallium Arsenide/Gallium Arsenide (Heterojunction)

InGaAs/InAlAs Indium Gallium Arsenide/Indium Aluminum Arsenide (Quantum Well)

InGaAsP Indium Gallium Arsenic Phosphide (Semiconductor)

InGaAs QW Indium Gallium Arsenide Quantum Well

InGaAsSb Indium Gallium Arsenide Antimonide (Semiconductor)

InGaN Indium Gallium Nitride (Semiconductor)

InGaN/GaN Indium Gallium Nitride/Gallium Nitride (Superlattice)

InGaP Indium Gallium Phosphide (Semiconductor)

InGaP/GaAs Indium Gallium Phosphide/Gallium Arsenide (Superlattice)

InGaP/GaAs SL Indium Gallium Phosphide/Gallium Arsenide Superlattice

InGaP:Si Silicon-Doped Indium Gallium Phosphide (Semiconductor)

Ing-Arch Ingenieur-Archiv [German Engineering Journal]

Ing Automob Ingenieurs de l'Automobile [French Journal on Automotive Engineering]

Ing Electr Mec Ingenieria Electrica y Mecanica [Venezuelan Journal on Electrical and Mechanical Engineering; published by Asociación Venezolana de Ingenieros Electricos y Mecanicos]

Ing Estructural Ingenieria Estructural [Cuban Journal on Structural Engineering]

Ing Ind Ingenieria Industrial [Cuban Journal on Industrial Engineering]

Ing Mec Electr Inginieria Mecanica y Electrica [Mexican Journal on Mechanical and Electrical Engineering; published by Asociación Mexicana de Ingenieros Mecanicos y Electricistas]

INGO International Nongovernmental Organization

InGPS Indoleglycerolphosphate Synthetase [Biochemistry]

Ing Quim Ingenieria Quimica [Spanish Journal on Chemical Engineering]

INGRES Interactive Graphic and Retrieval System

Ing Sanit Ingenieria Sanitaria [Journal on Sanitary Engineering; published by Inter-American Association of Sanitary Engineering]

Ing Werkst Ingenieur Werkstoffe [German Journal on Engineering Materials]

INH Isonicotinic Acid Hydrazine (or Isoniazid)

INH Isonicotinic Hydrazide

In-H Indium-Hydrogen (System)

in Hg inch(es) of mercury [also inHg]

in H₂O inch(es) of water [also inH$_2$O]

INHIGEO International Commission on the History of the Geological Science [France]

INHS Illinois Natural History Survey [US]

INI Instituto Nacional de Industria [National Industrial Institute, Spain]

.INI Initialize [File Name Extension] [also .ini]

InI Indium(I) Iodide (or Indium Monoiodide)

INIBAP International Network for the Improvement of Banana and Plantain [Montpellier, France]

INIC Current Negative Impedance Converter [*Note:* Letter "I" is Used to Represent "Electric Current"]

INICHAR Institut National de l'Industrie Charbonnière [National Institute of the Coal Industry, Belgium]

INiDI Indian Nickel Development Institute [New Delhi]

INIFERTER Initiator–Transfer–Termination (Group) [Polymer Chemistry]

INIMS Isaac Newton Institute for Mathematical Sciences [Cambridge, UK]

ININ Instituto Nacional de Investigaciónes Nucleares [National Institute for Nuclear Research, Mexico]

in/in inch(es) per inch [Unit]

in²/in³ square inch(es) per cubic inch [Unit]

in/in/°F inch per inch per degree Fahrenheit [Unit]

INIS International Nuclear Information Service [of International Atomic Energy Agency, Austria]

INIS International Nuclear Information System

INIS International Nuclear-Induced Sensitization Symposium

INIS-GB International Nuclear-Induced Sensitization Symposium of Great Britain

Init Initial(ization); Initialize; Initiation [also init]

init initial

Inj Inject(ion) [also inj]

Injn Injection [also injn]

INKA Information System Karlsruhe [Germany]

INKA-CONF Information System Karlsruhe–Conference Announcements [Germany]

INKA-CORP Information System Karlsruhe–Corporations [Germany]

INKA-DATACOMP Information System Karlsruhe–Data Compilations [Germany]

INKA-MATH Information System Karlsruhe–Database on Mathematics [Germany]

INKA-NUCLEAR Information System Karlsruhe–Database on Nuclear Science and Technology [Germany]

INKA-PHYS Information System Karlsruhe–Database on Physics [Germany]

INKEY$ Input from Keyboard String (Statement) [Programming]

Ink Print Ink and Print [Journal published in the UK]

INL Internal Noise Level

INL International Falls, Minnesota [Meteorological Station Identifier]

INLA International Nuclear Law Association [Belgium]

in-lb inch-pound [also in lb]

in/lb inch per pound [Unit]

in³/lb cubic inch(es) per pound [also cu in/lb]

in-lb/Btu inch-pound(s) per British thermal unit [also in lb/Btu]

in-lb/in³ inch-pound(s) per cubic inch [also in lb/in³, or lb/cu in]

in-lbf/in² inch-pound force per square inch [also in lbf/in², or lbf/sq in]

in-lb/in² inch-pound per square inch [also in lb/in², or lb/sq in]

in⁵/lb-sec inch(es) to the fifth (power) per poundsecond [also in⁵/lb·sec, or in⁵/lb-s]

in-lb-sec² inch-pound-second squared [also in·lbsec²]

INLC Initial Launch Capability

in lim *(in limine)* – at the outset

in loc *(in locum)* – in the place of

INM Independent Normal Mode

INM Instantaneous Normal Mode

INM Institut für Neue Materialien [Institute for New Materials, Saarbrücken, Germany]

INM International Nautical Mile

i/nm² ion(s) per centimeter [also ion(s)/nm²]

INMAP Independent Microlelectronics Applications

INMARSAT International Maritime Satellite [also Inmarsat]

INMARSAT International Maritime Satellite Organization [also Inmarsat, UK]

INMC International Network Management Center

INMG Instituto Nacional de Meteorológica e Geofisica [National Institute of Meteorology and Geophysics, Portugal]

in/min inch(es) per minute [also in min⁻¹]

in²/min square inch(es) per minute [also in min⁻², or sq in/min]

in³/min cubic inch(es) per minute [also in min⁻³, or cu in/min]

INMM Institute of Nuclear Materials Management [Northbrook, Illinois, US]

InMnAs Indium Manganese Arsenide (Semiconductor)

INMR Inverse Nuclear Magnetic Resonance

INMS Ionized Neutral Mass Spectrometer; Ionized Neutral Mass Spectrometry

INN Instituto Nacional de la Nutrición [National Institute of Nutrition, Argentina]

INN Instituto Nacional de Normalisación [National Standards Institute, Chile]

INN Institut za Nuklearne Nauke [Institute for Nuclear Sciences, Belgrade, Serbia]

INM Integrated Network Management

INM Instituto Nacional Meteorología [National Meteorological Institute, Cordoba, Argentina]

i/nm² ions per square nanometer [also ions/nm⁻²]

INN International Negotiation Network

INN International Nonproprietary Name [by World Health Organization]

InN Indium Nitride (Semiconductor)

INND Internet News Daemon

InNi Indium Nickelide

In-115 NMR Indium-115 Nuclear Magnetic Resonance [also ¹¹⁵In NMR]

INNOMATA Innovation by Materials (Exhibition-Congress for Materials Technology and Materials Applications, Germany] [also Innomata]

Innov Innovation [also innov]

innov innovative

Innov Manage Innovation and Management [Publication of the Institute of Management Consultants, UK]

Innov Mat Res Innovations in Materials Research [International Journal published in Singapore] [also Innov Mater Res, or IMR]

INNS International Neural Network Society [US]

INNSE Innocenti Santeustachio [Italian Rolling Mill Equipment Manufacturer located at Milano]

INNSE Not/News Innocenti Santeustachio Notizie/News [INNSE Newsletter, Milano, Italy]

INO Indian Ocean Region [of International Civil Aviation Organization]

INO Institut National d'Optiques [Quebec, Canada]

INO Iterative Natural Orbital [Physics]

Ino Inosine [Biochemistry]

InO Indium(I) Oxide (or Indium Monoxide)

INOCI Iterative Natural Orbital Configuration Interaction [Physics]

Inop Inoperative [also inop]

inorg inorganic

Inorg Chem Inorganic Chemist(ry)

Inorg Chem Inorganic Chemistry [Journal of the American Chemical Society, US]

Inorg Chim Acta Inorganica Chimica Acta [Journal published in the US]

Inorg Mater Inorganic Materials [Translation of: *Izvestiya Akademii Nauk SSSR, Neorganicheskie Materialy (USSR)*; published in the US]

INOSHAC Indian Ocean and Southern Hemisphere Analysis Center

INP If Not Possible

INP Inert Nitrogen Protection

INP Input [also Inp, or inp]

INP (Read Byte from Machine) Input Port (Function) [Programming]

INP Institute of National Planning [Egypt]

INP Institute of Nuclear Physics

INP Institut für Niedertemperatur-Plasmaphysik e.V. [Institute for Low-Temperature Plasma Physics, Greifswald, Germany]

INP Instituto Nacional de Parasitologia [National Parasitology Institute; of University of Granada, Spain]

INP Instituto Nacional Politecnico [National Polytechnic Institute, Mexico City]

INP Integrated Network Processor

INP Intelligent Network Processor

INP International News Photo

InP Indium Phosphide (Semiconductor)

IN2P3 Institut Nationale des Physiques Nucléaires et Particules [National Institute of Nuclear and Particle Physics, France] [also IN_2P_3]

INPA Instituto Nacional de Pesquisas da Amazonia [National Institute for Amazon Research, Brazil]

INPADOC International Patent Documentation Center [Austria]

INPADOC INKA (Information System Karlsruhe)–Patent Documentation

INPC Isopropyl-N-Phenylcarbamate

InPd Indium Palladium (Compound)

In/Pd Indium/Palladium (Heterojunction)

INPE Instituto Nacional de Pesquisas Espaciais [National Space Research Institute, São José dos Campos, Brazil]

INPEA 2-Isopropylamino-1-(4-Nitrophenyl)ethanol

INPE-LAS Instituto Nacional de Pesquisas Espaciais–Laboratorio Associado de Sensores [National Space Research Institute–Associated Sensor Laboratory, São José dos Campos, Brazil]

INPFC International North Pacific Fisheries Commission [Canada]

InP:Fe Iron-Doped Indium Phosphide (Semiconductor)

INPG Institut National Polytechnique de Grenoble [National Polytechnic Institute of Grenoble, France]

InP/GaAs Indium Phosphide/Gallium Arsenide (Heterojunction)

InP/GaP Indium Phosphide/Gallium Phosphide (Superlattice)

InP/GaP SL Indium Phosphide/Gallium Phosphide Superlattice

INPG/ENIEG Institut National Polytechnique de Grenoble/Ecole Nationale d'Ingénieurs Electriciens de Grenoble [National Polytechnic Institute of Grenoble/National Higher School of Electrical Engineering of Grenoble, France]

INPG/ENSEEG Institut National Polytechnique de Grenoble/Ecole Nationale Supérieure d'Electrochimie et d'Electrométallurgie de Grenoble [National Polytechnic Institute of Grenoble/National Higher School of Electrochemistry and Electrometallurgy of Grenoble, France]

INPG/ENSHMG Institut National Polytechnique de Grenoble/Ecole Nationale Supérieure d'Hydraulique et d'Mécanique de Grenoble [National Polytechnic Institute of Grenoble/National Higher School of Hydraulics and Mechanics of Grenoble, France]

INPG/ENSPG Institut National Polytechnique de Grenoble/Ecole Nationale Supérieure de Physique de Grenoble [National Polytechnic Institute of Grenoble/National Higher School of Physics of Grenoble, France]

INPI Institut National de la Propriété Industrielle [National Institute of Industrial Property, France]

InPIMPATT Indium Phosphide Impact Avalanche Transit Time (Diode)

INPL Institut National Polytechnique de Lorraine [National Polytechnic Institute of Lorraine, Nancy, France]

INPL/ENSMIM Institut National Polytechnique de Lorraine/Ecole Nationale Supérieure de la Métallurgie et de l'Industrie des Mines [National Polytechnic Institute of Lorraine/National Higher School of Metallurgy and Mining Engineering, Nancy, France]

INPMA Instituto Nacional para la Preservación del Medio Ambiente [National Institute for the Preservation of the Environment, Uruguay]

InPN Indium Phosphide Nitride

INP NNC RK Institute of Nuclear Physics, National Nuclear Center, Republic of Kazakstan

INPO Institute of Nuclear Power Operations [US]

INPR In Progress

InP:S Sulfur-Doped Indium Phosphide (Semiconductor)

InP/Si Indium Phosphide/Silicon (Heterojunction)

InP:Sn Tin-Doped Indium Phosphide (Semiconductor)

INPT Institut National Polytechnique de Toulouse [National Polytechnic Institute of Toulouse, France]

INPT Instituto Nacional dos Patrones Tecnicos [National Technical Standards Institute, Paraguay]

INPT/ENSC Institut National Polytechnique de Toulouse/ Ecole Nationale Supérieure de Chimie [National Polytechnic Institute of Toulouse, National Higher School of Chemistry, Toulouse, France]

InP:Zn Zinc-Doped Indium Phosphide (Semiconductor)

Inq Inquire; Inquiry [also inq]

INQUA International Union for Quaternary Research [Switzerland]

INR Impact Noise Reduction

INR Indian Rupee [Currency of India]

INR Institute for Nuclear Research [Russia]

INR Institut National de Radiodiffusion [National Institute of Broadcasting, Belgium]

INR Interference-to-Noise Ratio

INRA Institute for Natural Resources in Africa

INRA Institut National de la Recherche Agraire [National Institute for Agrarian Research, Algeria]

INRA Institut National de la Recherche Agronomique [National Institute for Agronomic Research, France]

INRA Instituto Nacional de Reforma Agraria [National Institute of Agrarian Reform, Cuba]

INRA International Natural Rubber Agreement

in³/rad cubic inch(es) per radian [also cu in/rad]

INRAT Institute National de la Recherche Agronomique de Tunisie [Tunesian National Institute for Agricultural Research]

INRE Institute of Natural Resources and Environment [of Commonwealth Scientific and Industrial Research Organization, Australia]

INREQ Information on Request

in/rev inches per revolution [also IPR, or ipr]

INRIA Institut National de Recherche en Information et Automatique [National Institute for Research in Computer Science and Control, Le Chesnay, France]

INRNE Institute of Nuclear Research and Nuclear Energy [of Bulgarian Academy of Sciences, Sofia]

INRO International Natural Rubber Organization [Malaysia]

INRO International Naval Research Organization [US]

INROADS Information on Roads (Database) [of Australian Road Research Board, Australia]

INROWASP In Rotating Water Spinning Process [Metallurgy]

INRS Institut National de la Recherche Scientifique [National Institute for Scientific Research, Quebec, Canada]

INRSEM Institut National de la Recherche Scientifique Energie et Matériaux [National Institute for Scientific Research on Energy and Materials, Varennes, Quebec, Canada]

INRSO Institut National de la Recherche Scientifique en Océanologie [National Institute for Scientific Research in Oceanology, Rimouski, Quebec, Canada]

INRV Institut National des Recherche Vétérinaire [National Institute of Veterinary Research, Belgium]

INS Immigration and Naturalization Service [US]

INS Incoherent Neutron Scattering

INS Indian Navy Ship

INS Induced Neutron Scattering

INS Inelastic Neutron Scattering

INS Inertial Navigation Sensor

INS Inertial Navigation System

INS Information Network System [Japan]

INS Input String

INS Institute for Naval Studies

INS Institute of Noetic Sciences [US]

INS Institute for Nuclear Studies [of US Department of Energy at Oak Ridge National Laboratory, Tennessee]

INS Institute for Nuclear Studies [Japan]

INS Instituto Nacional de la Salud [National Institute of Health, Spain]

INS Integrated Networking System

INS Integrated Network Server

INS International Navigation System

INS International News Service [Now Part of United Press International]

INS International Notational System

INS Interstation Noise Suppression

INS Investors News Service

INS Iodine Number and Saponification Number

INS Ion Neutralization Spectroscopy

INS Iron Soldering

INS Izumi, Nakamura and Shiohara (Superconductor Model)

INS 6-[(4-Methylphenyl)amino] 2-Naphthalenesulfonate

Ins Insert(ion) [also ins]

Ins Insolubility; Insoluble [also ins]

Ins Inspection; Inspector [also ins]

Ins Insulation [also ins]

Ins Insurance [also ins]

ins inches [Unit]

in/s inches per second [also in/sec, IPS, ips, or in sec^{-1}]

in/s² inches per second squared [also in/sec²]

in²/s square inch(es) per second [also in²/sec, sq in/s, or sq in/sec]

in³/s cubic inches per second [also in³/sec]

INSA Indian National Science Academy [New Delhi, India]

INSA Institut National des Sciences Appliquées [National Institute of Applied Sciences, Toulouse, France]

INSA International Shipowners Association [Poland]

INSAL Institut National des Sciences Appliquées de Lyon [Lyons National Institute of Applied Sciences, France]

INSAR Institut National des Sciences Appliquées de Rennes [Rennes National Institute of Applied Sciences, France]

INSAR Instruction Address Register

INSA-SNCMP Institut National des Sciences Appliquées–Service National des Champs Magnetique Pulsés [National Institute of Applied Sciences–National Office for Pulsed Magnetic Fields, Toulouse, France]

INSAT India Satellite; Indian Geostationary Satellite

INSATECH International Conference on Intelligent Materials

INSATRAC Interception with Satellite Tracking

InSb Indium Antimonide (Semiconductor)

InSb:Si Silicon-Doped Indium Antimonide (Semiconductor)

INSC International Nuclear Safety Center [of Argonne National Laboratory, Illinois, US]

INSCIR Institut National Supérieur de Chimie Industrielle de Rouen [National Higher Institute for Industrial Chemistry of Rouen, France]

INSDOC Indian National Scientific Documentation Center [India]

InSe Indium Selenide (Semiconductor)

in/sec inches per second [also in/s, IPS, ips, or in sec^{-1}]

in/sec² inches per second squared [also in/s²]

in²/sec square inch(es) per second [also in²/s, sq in/s, or sq in/sec]

in³/sec cubic inches per second [also in³/s]

INSEE Institut National de la Statistique et des Etudes Economiques [National Institute for Statistics and Economical Studies, France]

INSERM Institut National de la Santé et de la Recherche Médicale [National Institute of Health and Medical Research, Strasbourg, France]

INSET Institut National Supérieur d'Enseignement Technique [National Higher Institute for Technical Education, Ivory Coast]

Insft insufficient

INSH o-Hydroxybenzal-Isonicotinoylhydrazine

INSIPID Inadequate Sensitivity Improvement by Proton Indirect Detection

INSIS Inter-Institutional Integrated Services Information System [of the Commission of the European Communities]

INSITE Integrated Sensor Interpretation Techniques

INSJ Institute for Nuclear Studies, Japan

InsLine Insert (Empty) Line [Pascal Programming]

InSn Indium Antimonide (Semiconductor)

insol insoluble

Insoly Insolubility [also insoly]

INSONA International Society of Naturalists

Insp Inspection; Inspector [also insp]

insp inspect(ed)

INSPEC Information Services Physics, Electrical and Electronics, and Computers and Control [of Institution of Electrical Engineers, UK]

INSPEC Data Information Services Physics, Electrical and Electronics, and Computers and Control Data [of Institution of Electrical Engineers, UK]

INSPEC/EMIS Information Services Physics, Electrical and Electronics, and Computers and Control/Electronic Materials Information Service [of Institution of Electrical Engineers, UK]

INSPEC/IEE Information Services Physics, Electrical and Electronics, and Computers and Control/Institution of Electrical Engineers [UK]

INSPEL International Journal of Special Libraries [Published by Technische Universität Berlin, Germany]

INSPEX Engineering Inspection and Quality Control Conference and Exhibition

Insp Gen Inspector General

Inspn Inspection [also inspn]

Inspr Inspector [also inspr]

Insp, Test, Anal, InSpeC Inspection, Testing, Analysis, InSpeC [Journal published in Japan]

Inst Installer

Inst Instant [also inst]

Inst Instructor [also inst]

Inst Instrument [also inst]

Inst Institute; Institution [also inst]

inst instant(aneous)

INSTAB Information Service on Toxicology and Biodegradability [of Water Pollution Research Laboratory, UK]

Instab Instability [also instab]

Instal Instalment [also instal]

INSTAR Inertialess Scanning, Tracking and Ranging

INSTARS Information Storage and Retrieval System

Instby Instability [also instby]

InstBE Institution of British Engineers

InstCE Institute of Civil Engineers

Inst Chem Eng Symp Ser Institution of Chemical Engineers Symposium Series [UK]

InstE Institute of Electronics

Inst Eng Aust Chem Eng Trans Institution of Engineers of Australia, Chemical Engineering Transactions

Inst Eng Aust Civ Eng Trans Institution of Engineers of Australia, Civil Engineering Transactions

Inst Eng Aust Electr Eng Trans Institution of Engineers of Australia, Electrical Engineering Transactions

Inst Eng Aust Mech Eng Trans Institution of Engineers of Australia, Mechanical Engineering Transactions

Inst Forum Institute Forum [Published in the UK]

InstFs Instantaneous Fuse [also InstFz]

INSTINET Institutional Networks Corporation Automatic Communications System [also Instinet]

Instl Installation [also instl]

Instl Instalment [also instl]

Instln Installation [also instln]

InstM The Institute of Materials [London, UK]

InstM The Institute of Metals [Name Changed to The Institute of Materials, London, UK]

InstMC Institute of Measurement and Control [UK]

InstME Institute of Mechanical Engineers [UK]

InstME Institute of Mining Engineers [UK]

InstMM Institute of Mining and Metallurgy [UK]

INSTN Institut National des Sciences et Techniques Nucléaires [National Institute of Nuclear Science and Technology, Saclay, France]

Instn Institution [also instn]

Instn Instruction [also instn]

Inst Nat Sci Tech Nucl Pres Univ France Institut National des Sciences et Techniques Nucléaires, Presses Universitaires de France National Institute of Nuclear Science and Technology, University Press of France, Saclay] [also INSTN Pres Univ France]

INSTN Pres Univ France Institut National des Sciences et Techniques Nucléaires, Presses Universitaires de France National Institute of Nuclear Science and Technology, University Press of France, Saclay] [also Inst Nat Sci Tech Nucl Pres Univ France]

INSTOC Institute for the Study of the Continents [of Cornell University, Ithaca, New York, US]

INSTOP Institut National Scientifique et Technique d'Océanographie et de Pêche [National Scientific and Technical Institute of Oceanography and Fishery, France]

InstP Institute of Physics [London, UK]

InstPet Institute of Petroleum [UK]

Inst Phys Institute of Physics [London, UK]

Inst Phys Conf Ser Institute of Physics Conference Series [UK]

INSTR Interpreter Search String (Function) [Programming]

Instr Instruct(ion); Instructor [also instr]

Instr Instrument(ation); Instrumental [also instr]

InstR Institute of Refrigeration [UK]

INSTRAW International Research and Training Institute for the Advancement of Women [of United Nations]

Inst R Meteorol Belg Bull Trimest Obs Ozone Institut Royal Météorologique de Belgique–Bulletin Trimestriel Observations d'Ozone [Royal Meteorological Institute of Belgium–Quarterly Bulletin of Ozone Observations]

Instrn Instruction [also Instrn, or instrn]

INSTRU Instrument [also instru]

Instrum Instrument(ation) [also instrum]

Instrum Control Eng Instrumentation and Control Engineering [Journal published in Japan]

Instrum Exp Tech Instruments and Experimental Techniques [Translation of: *Pribory i Tekhnika Eksperimenta (USSR)*; published in the US]

Instrum India Instruments India [Published in India]

Instrum Technol Instrumentation Technology [Journal published in the US]

InstSMM Institute of Sales and Marketing Management

InstWE Institute of Water Engineers [UK]

InstWP Institute of Word Processing and Systems Technology [UK]

INSU Institut National des Sciences de l'Univers [National Institute for Cosmic Sciences, France]

Insulation J Insulation Journal [Journal published in the UK]

Insuln Insulation [also insuln]

Insur Insurance [also insur]

INT Institute for Nuclear Theory [of University of Washington, Seattle, US]

INT Institut National des Télécommunications [National Telecommunications Institute, Evry, France]

INT Instituto Nacional de Tecnologia [National Institute of Technology, Brazil]

INT Instituto Nacional de Tecnologia [National Institute of Technology, Portugal]

INT Integer (Function) [Programming]

INT Iodonitrotetrazolium Violet

INT p-Iodonitrotetrazolium

INT 2-(4-Iodophenyl)-3-(4-Nitrophenyl)-5-Phenyl-2H-Tetrazolium Chloride; 2-(4-Iodophenyl)-3-(4-Nitrophenyl)- 5-Phenyltetrazolium Chloride

INT Isaac Newton Telescope

INT Iterative Numerical Technique

Int Integer [also Int, or int]

Int Integrate(d); Integration [also int]

Int Intensity; Intensive [also int]

Int Interest [also int]

Int Interference [also int]

Int Intermediate [also int]

Int Interphone [also int]

Int Interrogation [also int]

Int Interrupt(ion) [also int]

Int Intersection [also int]

Int Interval [also int]

int interior

int intermediate

int internal

int International

INTA Instituto Nacional de Tecnica Aerospacial [National Institute of Aerospace Technology, Spain] [also Inta]

INTA Interrupt Acknowledge

Inta/Cienc Tec Aerosp Inta/Ciencia y Tecnica Aerospacial [Spanish Journal on Aerospace Science and Technology; published by Instituto Nacional de Tecnica Aerospacial]

Inta/Conie Inf Aerosp Inta/Conie Información Aerospacial [Spanish Journal on Aerospace Science and Technology; published by Instituto Nacional de Tecnica Aerospacial]

Int Aerosp Abstr International Aerospace Abstracts [of NASA]

INTAG International Technology Management Advisory Group

INTAMEL International Association of Metropolitan City Libraries

INTAMIC International Microcircuit Card Association [France]

Int Angiol International Angiology [Journal]

Int Arch Allergy Immunol International Archives of Allergy and Immunology [Journal]

Int Arch Occup Environ Health International Archives of Occupational and Environmental Health [Journal published in the US]

INTAS International Association for the Promotion of Cooperation with Scientists (from the Independent States) of the Former Soviet Union

INTASAT Instituto Nacional de Tecnica Aeroespacial Satélite [National Institute for Aerospace Technology Satellite] [NASA-Launched Spanish Satellite]

Int Astron Union Circ International Astronomical Union Circular [US]

Int Biodeterior International Biodeterioration [Journal published in the UK]

Int Biodeterior Biodegrad International Biodeterioration and Biodegradation [Journal published in the UK]

Int Broadcast International Broadcasting [Journal published in the UK]

Int Broadcast Eng International Broadcast Engineer [Journal published in the UK]

Int Bus Equip International Business Equipment [Published in Belgium]

Int Bus Rev International Business Review

INTC Industrial Nuclear Technology Conference

Int Cart Art Internal Cartoid Artery [also int cart art] [Anatomy]

Int Cast Met J International Cast Metals Journal [of American Foundrymen's Society, US]

Int Cell Biol International Cell Biology [Journal of the International Federation of Cell Biology, Canada]

Int Chem Eng International Chemical Engineering [Journal of the American Institute of Chemical Engineers]

Intchg Interchange(ability) [also intchg]

Intchgr Interchanger [also intchgr]

Int Civil Eng Abstr International Civil Engineering Abstracts

Int Classif International Classification [Published in Germany]

INTCO International Code of Signals

INTC/O Integrated Checkout

Int Coal Rev International Coal Review [Published by the National Coal Association, US]

Int CODATA Conf International Committee on Data for Science and Technology Conference [of International Council of Scientific Unions]

INTCOM International Liaison Committee [of Institute of Electrical and Electronics Engineers, US]

Int Comet Q International Comet Quarterly [Published by the Appalachian State University, Boone, North Carolina, US]

Int Commun Heat Mass Transf International Communications in Heat and Mass Transfer [UK]

Int Comput Law Advis International Computer Law Adviser [Published in the US]

Int Conf Electrochem Soc International Conference of the Electrochemical Society [US]

Int Conf Peaceful Uses Atom Energ Proc International Conference on the Peaceful Uses of Atomic Energy Proceedings

Int Constr International Construction [Journal published in the UK]

Int Copper Inf Bull International Copper Information Bulletin [UK]

INTCP Radio Intercept Station [also intcp]

Intcp Intercept(ion) [also intcp]

Int Cryog Mater Conf Ser International Cryogenic Materials Conference Series [Published by the International Cryogenic Materials Conference, US]

INTD Institut National des Techniques de la Documentation [National Institute of Documentation Techniques, France]

Int Def Rev International Defense Review [Switzerland]

INTEBRID (Monolithic) Integrated and Hybrid Circuitry

INTEC Interface Technology

INTECO Instituto de Normas Técnicas de Costa Rica [Costa Rican Standards Institute]

INTECOL International Association for Ecology [US]

INTECOL Bull INTECOL Bulletin [Published by the International Association for Ecology, US]

INTECOL Newsl INTECOL Newsletter [Published by the International Association for Ecology, US]

INTECOM International Council for Technical Communication [UK]

INTECOM International Society of Technical Communication [Denmark]

Integ Integration [also integ]

Integr Integrate(d); Integration [also integr]

Integr Manuf Rep Integrated Manufacturing Report [Published by the Association for Integrated Manufacturing Technology, US]

Integr VLSI J Integration, The VLSI (Very-Large-Scale Integration) Journal [Netherlands]

INTEL Integrated Electronics

Intel Intelligence; Intelligent [also intel]

INTELECT International Electrical Conference and Exhibition

INTEL-ED International Tele-Education [US]

INTELEVENT International Televent [US]

Intell Intelligence; Intelligent [also intell]

Intell Instrum Comput Intelligent Instruments and Computers [Journal published in the US]

INTELNAV Electronic Navigation System [Canada]

INTELPOST International Telecommunications Post [of INTELSAT]

INTELSAT International Telecommunications Satellite [Series of International Global Communications Satellites]

INTELSAT International Telecommunications Satellite Organization [also Intelsat, US]

INTEMA Instituto de Investigaciones en Ciencia y Tecnológia de Materiales [Institute for Materials Science and Technology; Universidad de Mar del Plata, Argentina]

Inten Intensity [also inten]

Int Environ Bull International Environmental Bulletin [of the Stockholm Environment Institute, Sweden]

Inter Interrogation [also inter]

Inter Interruption [also inter]

inter intermediate

inter intermittent

Interact Interact(ive); Interaction [also interact]

Interact Learn Int Interactive Learning International [Published in the UK]

INTERALIS International Advanced Life Information System

Inter-Am Aff Inter-American Affairs [Journal]

INTERATOM Internationaler Atomreaktorbau [International Nuclear Reactor Construction Company, Germany]

Interavia Aerosp Rev Interavia Aerospace Review [UK]

Interceram Interceramics [International Ceramics Journal]

Interch Interchangeability; Interchangeable [also interch]

Interchimie International Conference on Industrial Chemistry [France]

INTERCOLOR International Commission for Fashion and Textile Colors [France]

Intercom Intercommunicating System; Intercommunication(s) [also intercom]

Intercomp Intercompany [also intercomp]

INTERCON International Convention [of Institute of Electrical and Electronics Engineers, US]

Inter Connex Inter Connexion [French Journal published by Groupement Syndicat des Industries de Materials d'Equipement Electrique]

INTERCOOP International Association of Consumer Cooperatives [Denmark]

Interdiscip Sci Rev Interdisciplinary Science Reviews [UK]

INTERDOC Interdisciplinary Document Processing (Project) [of Fachinformationszentrum Karlsruhe, Germany]

Interface, Comput Educ Q Interface: The Computer Education Quarterly [US]

INTERFINISH Surface Finishing Congress

Interfinish International Conference on Metal Finishing

INTERFRIGO International Railway-Owned Companies for Refrigerated Transport [Switzerland]

INTERGALVA International Galvanizing Conference [also InterGalva, or Intergalva]

INTERGU Internationale Gesellschaft für Urheberrechte [International Copyright Society, Germany]

Interhospital International Hospital Congress [Germany]

INTERKAMA Internationaler Kongress mit Ausstellung für Mess- und Automatisierungstechnik [Market of Innovations in Measurement and Automation, Dusseldorf, Germany]

INTERLAINE Comité des Industries Lainières [Committee of the Wool Textile Industry; of European Economic Community, Belgium]

INTERMAG International Magnetics Conference [of Institute of Electrical and Electronics Engineers, US]

INTERMAG International Magnetism and Magnetic Materials Conference [also Intermag]

INTERMAG Conf International Magnetism and Magnetic Materials Conference [also Intermag Conf]

INTERMARC International Machine-Readable Catalogue

INTERMARGEO International Organization for Marine Geology [Russian Federation]

Intermet Intermetallics [Journal published in the UK]

intern internal

Internal EMF Internal Electromotive Force [also internal emf]

internat international

INTERNET International Project Management Association [Switzerland]

InterNIC Internet Network Information Center

Interp Interpreter [also interp]

INTERPACK International Intersociety Electronic/Photonic Packing Conference [Pacific Rim and American Society of Mechanical Engineers]

INTERPACK Internationale Messe für Verpackungsmaschinen, Packmittel, Süßwarenmaschinen [International Fair for Packaging Machinery, Packing Materials and Candy Machinery, Dusseldorf, Germany] [also Interpack]

INTERPLAN International Group for Studies in National Planning

Interpol International (Criminal) Police Organization [UK]

INTERPLAS International Plastics Exhibition and Conference

INTERQ Interrupt Request

Interrog Interrogate; Interrogation [also interog]

INTERSPUTNIK International Organization of Space Communication [Russia]

INTERTANKO International Association of Independent Tanker Owners [Norway]

INTERTEL International Television Federation

Intertool Internationale Fachmesse für Metallbearbeitung, Automatisierung und Fertigungstechniken [International Trade Fair for Machining, Automation and Production Techniques, Austria]

Intervirol Intervirology [Publication of the International Union of Microbiological Societies, UK]

INTERVISION International Radio and Television Transmission [of Organisation Internationale de Radiodiffusion et Télévision] [also Intervision]

Interwire International Wire Conference [of Wire Association International, US]

INTEX International Futures Exchange

Intf Interface [also Intf]

Int F Internal Frosted (Light Bulb)

Intfc Interface [also intfc]

intfcl interfacial

In(TFA)$_3$ Indium Trifluoroacetate [also In(tfa)$_3$]

In(TFAC)$_3$ Indium Trifluoroacetylacetonate [also In(tfac)$_3$]

INT.FM Internal Frequency Modulation

Int Forum Inf Doc International Forum on Information and Documentation [Publication of the All-Union Institute of Scientific and Technical Information, Moscow, Russian Federation]

INTFU Interface Unit

Intg Integral; Integrated; Integration [also intg]

INTGEN Interpreter Generator

Int Geol Rev International Geology Review

Int Hydrogr Bull International Hydrographic Bulletin [of Bureau Hydrographique International, Monaco]

Int Hydrogr Rev International Hydrographic Review [of Bureau Hydrographique International, Monaco]

INTI Institut Nauchnoi i Tekhnicheskoi Informatsii [Institute for Scientific and Technical Information, Moscow, Russian Federation]

INTI Instituto Nacional de Tecnologia Industrial [National Institute of Industrial Technology, Argentina]

INTIM Interrupt and Timing

INTIME Interactive Textual Information Management Experiment

Int Immunol International Immunology [Journal published by Oxford University Press, UK]

Int Inf Commun Educ International Information, Communication and Education [Journal published in India]

INTIP Integrated Information Processing

INTIPS Integrated Information Processing System

Int Iron Steel Inst Bull International Iron and Steel Institute Bulletin [Belgium]

INTIST International Institute for Science and Technology

Int J Adapt Control Signal Process International Journal of Adaptive Control and Signal Process [UK]

Int J Adhes Adhes International Journal of Adhesion and Adhesives [UK]

Int J Adv Counsel International Journal for the Advancement of Counselling

Int J Adv Manuf Technol International Journal of Advanced Manufacturing Technology [UK]

Int J Ambient Energy International Journal of Ambient Energy [UK]

Int J Appl Electromag Mater International Journal of Applied Electromagnetics in Materials [Netherlands]

Int J Appl Eng Educ International Journal of Applied Engineering Education [UK]

Int J Approx Reason International Journal of Approximate Reasoning [Published by the North American Fuzzy Information Processing Society, US]

Int J Biochem International Journal of Biochemistry [Published in the US]

Int J Biol Macromols International Journal of Biological Macromolecules [UK]

Int J Bio-Med Comput International Journal of Bio-Medical Computing [Netherlands]

Int J Biomet International Journal of Biometeorology [Published by the International Society of Biometeorology, Switzerland]

Int J Bulk Solids International Journal of Bulk Solids [US]

Int J Bulk Solids Storage Silos International Journal of Bulk Solids, Storage in Silos [US]

Int J Cancer International Journal of Cancer

Int J Cem Compos Lightweight Concr International Journal of Cement Composites and Lightweight Concrete [UK]

Int J Circuit Theory Appl International Journal of Circuit Theory and Applications [UK]

Int J Climatol International Journal of Climatology [UK]

Int J Clin Monit Comput International Journal of Clinical Monitoring and Computing [Netherlands]

Int J Clin Pharmacol International Journal of Clinical Pharmacology

Int J Coal Geol International Journal of Coal Geology

Int J Comput Adult Educ Train International Journal of Computers in Adult Education and Training [Published by the Department of Adult and Continuing Education, University of Keele, UK]

Int J Comput Appl Technol International Journal of Computer Applications in Technology [Switzerland]

Int J Comput Integr Manuf International Journal of Computer Integrated Manufacturing [UK]

Int J Comput Math International Journal of Computer Mathematics [UK]

Int J Comput Math Electr Electron Eng International Journal for Computation and Mathematics in Electrical and Electronic Engineering [Ireland] [also COMPEL]

Int J Comput Vis International Journal of Computer Vision [Netherlands]

Int J Control International Journal of Control [UK]

Int J Dermatol International Journal of Dermatology

Int J Digit Analog Cabled Syst International Journal of Digital and Analog Cabled Systems [UK]

Int J Digit Libr International Journal on Digital Libraries [Published in Germany]

Int J Earthquake Eng Struct Dyn International Journal of Earthquake and Structural Dynamics [Japan]

Int J Eating Disorders International Journal of Eating Disorder

Int J Electr Eng Educ International Journal of Electrical Engineering Education [Published by Manchester University Press, UK]

Int J Electron International Journal of Electronics [UK]

Int J Electr Power Energy Syst International Journal of Electrical Power and Energy Systems [UK]

Int J Energy Res International Journal of Energy Research [UK]

Int J Energy Syst International Journal of Energy Systems [US]

Int J Eng Fluid Mech International Journal of Engineering Fluid Mechanics [US]

Int J Eng Sci International Journal of Engineering Science [UK]

Int J Environ Anal Chem International Journal of Environmental Analytical Chemistry [of International Association of Environmental Analytical Chemistry, Switzerland]

Int J Environ Stud International Journal of Environmental Studies [US]

Int J Epidemiol International Journal of Epidemiology [Published by Oxford University Press, UK]

Int J Expert Syst Res Appl International Journal of Expert Systems Research and Applications [US]

Int J Fatigue International Journal of Fatigue [UK]

Int J Food Microbiol International Journal of Food Microbiology [Published by the International Union of Microbiological Societies, UK]

Int J Forecast International Journal of Forecasting [Netherlands]

Int J Fract International Journal of Fracture

Int J Fusion Energy International Journal of Fusion Energy [Published by the Fusion Energy Foundation, US]

Int J Game Theory International Journal of Game Theory [Austria]

Int J Gen Syst International Journal of General Systems [UK]

Int J Geogr Inf Syst International Journal of Geographical Information Systems [UK]

Int J Glob Energy Issues International Journal of Global Energy Issues [Switzerland]

Int J Heat Fluid Flow International Journal of Heat and Fluid Flow [UK]

Int J Heat Mass Transf International Journal of Heat and Mass Transfer [UK]

Int J High Technol Ceram International Journal of High Technology Ceramics [UK]

Int J High Temp Ceram International Journal of High Temperature Ceramics [UK]

Int J Hybrid Microelectron International Journal of Hybrid Electronics [of International Society for Hybrid Microelectronics, US]

Int J Hydrog Energy International Journal of Hydrogen Energy [UK]

Int J Hydrop Dams The International Journal on Hydropower and Dams

Int J Impact Eng International Journal of Impact Engineering [UK]

Int J Infect Diseases International Journal of Infectious Diseases [US]

Int J Inf Manage International Journal of Information Management [UK]

Int J Infrared Millim Waves International Journal of Infrared and Millimeter Waves [US]

Int J Intell Syst International Journal of Intelligent Systems [US]

Int J Joining Mater International Journal for the Joining of Materials [Denmark]

Int J Low Energy Sust Build International Journal of Low Energy and Sustainable Buildings [Published by the Royal Institute of Technology, Stockholm, Sweden]

Int J Mach Tools Manuf International Journal of Machine Tools and Manufacture [UK]

Int J Man-Mach Stud International Journal of Man-Machine Studies [UK]

Int J Manuf Technol International Journal of Manufacturing Technology [US]

Int J Mass Spectrom Ion Phys International Journal of Mass Spectrometry and Ion Physics

Int J Mass Spectrom Ion Process International Journal of Mass Spectrometry and Ion Processes

Int J Mater Charact International Journal of Materials Characterization

Int J Mater Eng Appl International Journal of Materials in Engineering Applications [UK]

Int J Mater Prod Technol International Journal of Materials and Product Technology [Switzerland]

Int J Math Educ Sci Technol International Journal of Mathematical Education in Science and Technology [UK]

Int J Mech Sci International Journal of Mechanical Sciences [UK]

Int J Microcomput Appl International Journal of Microcomputer Applications [of International Society for Mini- and Microcomputers]

Int J Microgr Video Technol International Journal of Micrographics and Video Technology [UK]

Int J Min Eng International Journal of Mining Engineering [US]

Int J Miner Process International Journal of Mineral Processing [Netherlands]

Int J Mine Water International Journal of Mine Water [of International Mine Water Association, Spain]

Int J Mini Microcomput International Journal of Mini and Microcomputers [US]

Int J Model Simul International Journal of Modelling and Simulation [US]

Int J Mod Phys A International Journal of Modern Physics A [Singapore]

Int J Mod Phys B International Journal of Modern Physics B [Singapore]

Int J Multiph Flow International Journal of Multiphase Flow [UK]

Int J Museum Manage Curator International Journal of Museum Management and Curatorship

Int J Non-Linear Mech International Journal of Non-Linear Mechanics [US]

Int J Numer Anal Methods Geomech International Journal for Numerical and Analytical Methods in Geomechanics [UK]

Int J Numer Methods Eng International Journal for Numerical Methods in Engineering [UK]

Int J Numer Methods Fluids International Journal for Numerical Methods in Fluids [UK]

Int J Numer Model, Electron Netw Devices Fields International Journal of Numerical Modelling: Electronic Networks, Devices and Fields [UK]

Int J Obesity International Journal of Obesity

Int J Oper Prod Manage International Journal of Operations and Production Management [UK]

Int J Opt Comput International Journal of Optical Computing

Int J Optoelectron International Journal of Optoelectronics [UK]

Int J Parallel Program International Journal of Parallel Programming [US]

Int J Pattern Recognit Artif Intell International Journal of Pattern Recognition and Artificial Intelligence [Singapore]

Int J Peptide Protein Res International Journal of Peptide and Protein Research

Int J Pharm International Journal of Pharmaceutics

Int J Phys Distrib Mater Manage International Journal of Physical Distribution and Materials Management [UK]

Int J Plast International Journal of Plasticity [UK]

Int J Policy Inf International Journal on Policy and Information [Published by the Graduate School of Information Engineering, Tamking University, Taiwan]

Int J Polym Mater International Journal of Polymeric Materials [UK]

Int J Powder Metall International Journal of Powder Metallurgy [Metal Powder Industries Federation, US]

Int J Powder Metall Powder Technol International Journal of Powder Metallurgy and Powder Technology [of Metal Powder Industries Federation, US]

Int J Pressure Vessels Piping International Journal of Pressure Vessels and Piping [US]

Int J Prod Res International Journal of Production Research [UK]

Int J Proj Manage International Journal of Project Management [UK]

Int J Rock Mech Min Sci International Journal of Rock Mechanics and Mineral Science

Int J Rock Mech Min Sci Geomech Abstr International Journal of Rock Mechanics and Mineral Science–Geomechanics Abstracts

Int J Psych Med International Journal of Psychiatry in Medicine

Int J Quantum Chem International Journal of Quantum Chemistry [US]

Int J Quantum Chem, Quantum Biol Symp International Journal of Quantum Chemistry, Quantum Biology Symposium [US; issued with *International Journal of Quantum Chemistry*]

Int J Quantum Chem, Quantum Chem Symp International Journal of Quantum Chemistry, Quantum Chemistry Symposium [US; issued with *International Journal of Quantum Chemistry*]

Int J Radiat Biol International Journal of Radiation Biology [UK]

Int J Radiat Oncol Biol Physiol International Journal of Radiation, Oncology and Biological Physiology [UK]

Int J Rapid Solidif International Journal of Rapid Solidification [UK] [also Int J Rapid Solid]

Int J Refract Hard Mater International Journal of Refractory Metals and Hard Materials [UK]

Int J Refrig International Journal of Refrigeration [of International Institute of Refrigeration, France]

Int J Remote Sens International Journal of Remote Sensing [UK]

Int J Robot Autom International Journal of Robotics and Automation [US]

Int J Robot Res International Journal of Robotics Research [Published by MIT Press, Cambridge, US]

Int J Rock Mech Min Sci International Journal of Rock Mechanics and Mining Sciences [UK]

Int J Rock Mech Min Sci Geomech Abstr International Journal of Rock Mechanics and Mining Sciences and Geomechanics Abstracts [UK]

Int J Satell Commun International Journal of Satellite Communications [UK]

Int J Sci Educ International Journal of Science Education [UK]

Int J Self-Propag High-Temp Synth International Journal on Self-Propagating High-Temperature Synthesis

Int J Sol Energy International Journal of Solar Energy [Switzerland]

Int J Solids Struct International Journal of Solids and Structures [US]

Int J Spec Libr International Journal of Special Libraries [Published by Technische Universität Berlin, Germany]

Int J Sport Nutr International Journal of Sport Nutrition [Published by Human Kinetics USA]

Int J Sports Med International Journal of Sports Medicine

Int J Supercomput Appl International Journal of Supercomputer Applications [Published by MIT Press, Cambridge, US]

Int J Syst Bacteriol International Journal of Systematic Bacteriology [of International Union of Microbiological Societies, UK]

Int J Syst Sci International Journal of Systems Science [UK]

Int J Technol Manage International Journal of Technology Management [Switzerland]

Int J Theor Phys International Journal of Theoretical Physics [US]

Int J Thermophys International Journal of Thermophysics [US]

Int J Tissue React International Journal on Tissue Reactions

Int J Turbo Jet-Engines International Journal of Turbo and Jet-Engines [UK]

Int Jug V Internal Jugular Vein [also int jug v] [Anatomy]

Int J Veh Des International Journal of Vehicle Design [Switzerland]

Int J Vit Nutr Res International Journal of Vitamin and Nutrition Research

intl internal

intl international [also int'l]

Int Lab International Laboratory [Journal published in the US]

Intlk Interlock [also intlk]

Int Manage International Management [Switzerland]

Int Mater Rev International Materials Reviews [of ASM International, US, and Institute of Materials, UK] [also Inter Mat Rev]

Int Met Rev International Metals Reviews [of ASM International, US, and Institute of Materials, UK]

Int Min International Mining [Published in Singapore]

Int Mod Foundry International Modern Foundry [UK Publication]

intmt intermittent

INTN Instituto Nacional de Tecnología y Normalisación [National Institute for Standards and Technology]

IN/TN Insoluble Nitrogen-to-Total Nitrogen

intnl international

Int Nursing Bull International Nursing Bulletin [of International Council of Nurses]

INTO Interrupt if Overflow (Occurs)

INTO Irish National Teachers Organization

in-ton inch-ton [Unit]

Intoolex Fachmesse für industrielle Werkzeugtechnik [Trade Fair for Industrial Tool Engineering, Zurich, Switzerland]

INTOP International Operations Simulation

INTOR International Tokamak (Nuclear) Reactor

INTOR International Torus [Nuclear Engineering]

Intp Interpret [also intp]

Int Pbd Ind International Paperboard Industry [Published in the US]

Int Peat J International Peat Journal [of International Peat Society, Finland]

Intper Interpreter [also intper]

INTPHTR Interphase Transformer

Int Pkg Abstr International Packaging Abstracts [Journal published in the UK and the US]

Int Polym Process International Polymer Processing [Journal published in the US]

Int Polym Sci Technol International Polymer Science and Technology [of Rubber and Plastics Research Association, UK]

Int Power Gener International Power Generation [Journal published in the UK]

Int QC Forum International Quality Control Forum [Journal published in the US]

Intr Interrupt [also intr]

Intr Software Interrupt [Pascal Programming]

intr interior

INTRAFAX Closed Circuit Facsimile System

INTRAN Input Translator

INTREDIS International Tree Disease Register [of United States Forestry Service]

Int Reinf Plast Ind International Reinforced Plastics Industry [Journal published in the UK]

Int Rev Internal Revenue

Int Rev International Review

Int Rev Cytol International Reviews in Cytology [US]

Int Rev Phys Chem International Reviews in Physical Chemistry [UK]

INTREX Information Transfer Experiment (Project) [of Massachusetts Institute of Technology, Cambridge, US]

Intrg Interrogate; Interrogator [also intrg]

Intrlvr Interleaver

Intrn Introduction [also intrn]

Intro Introduction; Introductory [also intro]

Introd Introduction; Introductory [also introd]

Intrp Interrupt(ion) [also intrp]

Intrpt Interrupt [also intrpt]

Int Rubb Dig International Rubber Digest [Published by the International Rubber Study Group, UK]

ints intense

Int SAMPE Tech Conf Ser International SAMPE Technical Conference Series [of Society for the Advancement of Materials and Process Engineering, US]

Intsf Intensify [also Intsfy, intsfy, or intsf]

Int Spectr International Spectrum [US]

Int Stat Rev International Statistical Review [of International Statistical Institute, Netherlands]

Int Sugar J International Sugar Journal [US]

Int Symp International Symposium

Int Symp Forest Sci International Symposium on Forest Science [of National Academy of Sciences, US]

Int Symp SOFC International Symposium on Solid Oxide Fuel Cells

INTSYS International Workshop on Classical and Quantum Integrable Systems

Int Technol International Technology [of International Technology Institute, US]

Int Tech Transf Bus International Tech(nology) Transfer Business [UK]

Int Trans Oper Res International Transactions in Operational Research

INTUG International Telecommunications Users Group [UK]

INTV Interim Hypersonics Test Article

Intvl Interval [also intvl]

Intvlm Intervalometer

Int Water Power Dam Constr International Water Power and Dam Construction [UK]

int X interior of X [Mathematics]

INU Inertial Navigation Unit

INucE Institution of Nuclear Engineers [UK]

Inv Invention; Inventor; Inventory [also inv]

Inv Inversion; Inverter [also inv]

Inv Investment [also inv]

Inv Invoice [also inv]

inv invariable

inv a Involute function of pressure angle (of involute splines) [Symbol]

inval invalid

Invas Metastas Invasion and Metastasis [Journal published in Switzerland]

Inverse Probl Inverse Problems [Journal of the Institute of Physics, UK]

Invest Investigate [also invest]

Invest Investment [also invest]

investd investigated [also invest'd]

Investg Investigating [also investg]

Investn Investigation [also investn]

Invest Tec Papel Investigacion y Tecnica del Papel [Spanish Journal on Paper Research and Technology; published by Asociación de Investigación Tecnica de la Industria Papelera Espanola]

Inv Math Inventiones Mathematicae [Published in Germany]

Inv Mgt Inventory Management

INVOF In the Vicinity Of

inv ϕ involute function of pressure angle (of a gear) [Symbol]

Invrn Inversion [also invt]

Invt Inventory [also invt]

Invt Investment [also invt]

INWATS Inward Wide Area Telephone Service

INWG International Network Working Group [of International Federation of Information Processing Societies]

INWG Internet Network Working Group [of NASA]

in wg inch water gauge [Unit]

INX Index Character

in/yr inches per year [also in/y, or in y^{-1}]

Inz Apar Chem Inzyniera i Aparatura Chemiczna [Polish Journal on Chemical Engineering and Equipment]

Inz Chem Procesowa Inzyniera Chemiczna i Procesowa [Polish Journal on Chemical and Process Engineering]

Inzh-Fiz Zh Inzhenerno-Fizicheskii Zhurnal [Russian Journal of Engineering Physics]

Inz Mater Inzyniera Materialowa [Polish Journal on Materials Engineering]

IO British Indian Ocean Territories [ISO Code]

IO Inboard-Outboard (Engine)

IO Indium Oxide

IO Information Organization

IO In Order

IO Input/Output [also io, I/O, i/o, I-O, or i-o]

IO Institute for Oceanography [of Environmental Science Services Administration, US]

IO Institut Océanographique [Oceanographic Institute, France]

IO Integrated Optic(al); Integrated Optics

IO Intelligence Office

IO Interactive Operation

IO Intermediate Output

IO Internal Oxidation [Metallurgy]

IO Inter-Ocular

IO Interpretive Operation

IO Intra-Ocular

IO Inverted Optics

IO Isotopic Osmosis

I²O Intelligent Input/Output [also I2O]

I/O Input/Output [also i/o, IO, io, I-O, or i-o]

Io Ionium [also Th-230, ^{230}Th, or Th^{230}]

Io Iowa [US]

.io British Indian Ocean Territories [Country Code/Domain Name]

IOA Input/Output Adapter

IOA Input/Output Address

IOA Input/Output Analysis

IOA Input/Output, or Assembly

IOA Institut de Génie Atomique [Institute for Nuclear Engineering; of Ecole Polytechnique Fédérale de Lausanne, Switzerland]

IOA Institute of Acoustics [UK]

IOA Institute of Outdoor Advertising [US]

IOA International OMEGA (Radio Navigation System) Association [US]

IOA International Ozone Association [US]

IoA Institute of Astronomy [Cambridge, UK]

IOAP Internally-Oxidized Alloy Powder [Metallurgy]

IOAT Input/Output Allocation Table

IOAU Input/Output Access Unit

IOB Input/Output Bound

IOB InPut/Output Box

IOB Input/Output Buffer

IOB Input/Output Bus

IOB Institute of Bankers [London, UK]

IOB Institute of Biology [UK]

IOB Institute of Building [UK]

IOB Interorganization Board [of United Nations]

IOBB International Organization for Biotechnology and Bioengineering [Guatemala]

IOBC Indian Ocean Biological Center

IOBC International Organization for Biological Control of Noxious Animals and Plants [France]

IOBFR Input/Output Buffer

IOBPS Input/Output Box and Peripheral Simulator

IOBS Input/Output Buffering System

IOC Inclusion of the Overlap Charges [Physics]

IOC Indian Ocean Commission [Mauritius]

IOC Indirect Operating Costs

IOC Initial Operational Capability; Initial Operating Capability

IOC Initial Orbiting Capability [of NASA]

IOC Inorganic Chemical(s)

IOC Input/Output Channel

IOC Input/Output Connector

IOC Input/Output Control(ler) [also I/OC, or ioc]

IOC Input/Output Converter

IOC Institute of Carpenters [UK]

IOC Institute of Ceramics [Stoke-on-Trent, UK]

IOC Institute of Chemistry [UK]

IOC Institute of Commerce [UK]

IOC Institute of Oriental Culture [of Tokyo University, Japan]

IOC Integrated Optical Circuit

IOC INTELSAT (International Telecommunications Satellite Organization) Operations Center [US]

IOC Intergovernmental Oceanographic Commission [of UNESCO]

IOC International Organization Committee

IOC International Ornithological Congress [New Zealand]

IOC Inter-Office Channel

IOC Inter-Office Communication

IOC Iodized Organic Compound

IOC Iron Ore Company (of Canada)

I/OC Input/Output Controller [also IOC, or ioc]

IOCA Independent Oil Compounders Association [now Independent Lubricant Manufacturers Association, US]

IOCARIBE IOC (Intergovernmental Oceanographic Commission) Regional Subcommission for the Caribbean and Adjacent Regions

IOCC Input/Output Channel Converter

IOCC Input/Output Control Center

IOCC Input/Output Control Command

IOCC Input/Output Controller Chip

IOCC Interstate Oil Compact Commission [US]

IOCD International Organization for Chemical Sciences in Development [Mexico]

IOCE Input/Output Control Element

IOCG Industrial Oil Consumers Group [US]

IOCG International Organization on Crystal Growth

IOCM Input/Output Control Module

IOCM Interim Operational Contamination Monitor (Experiment) [NASA Space Shuttle]

IOCOM India-Malaysia Submarine Cable

IOC-ω Inclusion of the Overlap Charges in the Omega [Physics]

IOCP Input/Output Connection Panel

IOCS Input/Output Control Subroutine

IOCS Input/Output Control System

IOCTF Input/Output Control TDMA (Time-Division Multiple Access) Facility

IOCTL Input/Output Control

IOCTL Send Input/Output Control Data [Programming]

IOCTL$ Read Input/Output Control Data String [Programming]

IOCTR Input/Output Controller

IOCU Input/Output Control Unit

IOCU International Organization of Consumers Unions [Netherlands]

IOCWMC International Organizing Committee of the World Mining Congress

IOCV International Organization of Citrus Virologists [US]

IOD Identified Outward Dialing

IOD Information on Demand

IOD Input/Output Device

IOD Institute of Directors [UK]

IOD Inter-Ocular Distance

IODE Integrated Development and Debugging Environment

IODE International Oceanographic Data (and Information) Exchange

IODS International Ocean Disposal Symposium [US]

IOE Indian Ocean Expedition

IOE Industrial and Operations Engineer

IOE Input/Output Error (Log Table)

IOE Institute of Energy [London, UK]

IOE International Organization of Employers

IOEH Institute of Occupational and Environmental Health [Canada]

IOF Independent Order of Foresters

IOF Initial Operational Flight

IOF Input/Output Forcing

IOF Input/Output Front-End

IOF Institute of Fuel [now Institute of Energy UK]

IOF Interactive Operations Facility

IOF International Oceanographic Foundation [US]

IOFC Indian Ocean Fishery Commission [Italy]

IOFI International Organization of the Flavour Industry [Switzerland]

I of M Isle of Man [UK]

I of W Inspector of Works

I of W Isle of Wight [UK]

IOGA Industry-Organized, Government-Approved

IOGCC Interstate Oil and Gas Compact Commission [US]

IOGEN Input/Output Generation

IOH Input/Output Handler

IOH Institute of Horticulture [UK]

IOH Institute of Housing [UK]

IOI International Ocean Institute [Malta]

IOI International Olympiad in Information Science [Germany]

IOI International Ozone Institute [now International Ozone Association, US]

IOI Israel Office of Information

IOIE International Organization of Industrial Employers

IOIH Input/Output Interrupt Handler

IOITBAG International Oil Industry TBA (Tires, Batteries, Accessories) Group [Canada]

IOJ (Chartered) Institute of Journalists [UK]

IOL Instantaneous Overload

IOL Intra-Ocular Lens

IOLA Input/Output Line Adapter

IOLA Input/Output Link Adapter

IOLC Input/Output Link Controller

IOLIM International On-Line Information Meeting [UK]

IOLL Infinite-Order Local Linearization [Physics]

IOLM International Organization for Legal Metrology

IOLS Input/Output Label System

IOM Including Other Minerals

IOM Index and Option Market

IOM Individual Ordinary Member(s)

IOM Inorganic Material

IOM Input/Output Module [also iom]

IOM Input/Output Multiplexer

IOM Institute of Materials [formerly Institute of Metals, London, UK] [also IoM]

IOM Institute of Medicine [US]

IOM Institute of Metals [Name Changed to Institute of Materials, London] [also IoM]

IOM Institute of Occupational Medicine [UK]

IOM Institute of Office Management [UK]

IOM Institut für Oberflächenmodifizierung e.V. [Institute for Surface Modification, Leipzig, Germany]

IOM Instrument of the Month

IOM International Organization for Mycoplasmology [US]

IOM Isle of Man [UK]

I/OM Input/Output Multiplexer

IoM Institute of Materials [London, US]

IOMA International Oxygen Manufacturers Association [US]

IOMAC Indian Ocean Marine Affairs Cooperation

IOMH Institute of Occupational Medicine and Hygiene [of Yale University, New Haven, Connecticut, US]

IOMMMS International Organization of Minerals, Metals and Materials Society

IOMP Input/Output Microprocessor

IOMP International Organization for Medical Physics [UK]

IOMS Input/Output Management System

IOMTR International Organization for Motor Trades and Repairs [Netherlands]

IOMVM International Organization of Motor Vehicle Manufacturers [France]

ION Institute of Navigation [Washington, DC, US] [also IoN]

ION Ionosphere and Aural Phenomena Advisory Committee [of European Space Research Organization]

Ion Ionic(s) [also ion]

Ionosph Ionosphere [also ionosph]

ions/cm² ions per square centimeter [also i/cm^{-2}]

ions/cm²-s ions per square centimeter second [also i/cm²-s]

ions/nm² ions per square nanometer [also i/nm^{-2}]

IOO Input/Output Operation

IOOC International Olive Oil Council [Spain]

IOOP Input/Output Operation

IOP Ibero-American Organization of Pilots [Mexico]

IOP Industrial Opportunities Program

IOP Innovative Onderzoegprogramm [Innovative Research Program, Netherlands]

IOP In-Orbit Plane

IOP Input/Output Port

IOP Input/Output Processor

IOP Institute for Oncology Problems [of Ukrainian Academy of Sciences, Kiev]

IOP Institute of Packaging [UK]

IOP Institute of Petroleum [UK]

IOP Institute of Physics [London, UK]

IOP Institute of Plumbing [UK]

IOP Institute of Printing [UK]

IOP Integrated Operation Plan

IOP Integrated Optics Processor

IOP Intensive Observing Period [Tropical Ocean and Global Atmosphere/Coupled Ocean Atmosphere Response Experiment]

IOP International Organization of Paleobotany [UK]

IOP Intra-Ocular Pressure

IOP Iowa Ordnance Plant [Burlington, Iowa, US]

IoP Institute of Physics [UK]

IOPAB International Organization for Pure and Applied Biophysics

IOPB International Organization of Plant Biosystematists [Switzerland]

IOPC International Oil Pollution Compensation Fund

IOPG Input Output Processor Group

IOPH International Office of Public Health

IOPI International Organization for Plant Information

IOPIDSG International Organization for Plant Information Data Standards Group [also IOPI/DSG, or IOPI-DsGS]

IOPIISC International Organization for Plant Information Information Systems Committee [also IOPI/ISC, or IOPI-ISC]

IOPKG Input/Output Package

IOPL Input/Output Privilege Level

IOPL Integrated Open Problem List

IOPN International Office for the Protection of Nature

IOPP Institute of Physics Publishing [UK] [also IOP Publishing]

IOPS Input/Output Programming System

IOQ Input/Output Queue

IOQ Institute of Quarrying [UK]

IOQE Input/Output Queue Element

IOR Index of Refraction

IOR Indian Ocean Region

IOR Industrially Oriented Research (Grant)

IOR Input/Output Rail

IOR Input/Output Register

IOR Institute for Operational Research [UK]

IOR Interoperable Object Reference

IÖR Institut für Ökologische Raumentwicklung e.V. [Institute for Ecological Regional Development, Dresden, Germany]

IoR Institute of Roofing [UK]

IORB Input/Output Record Block [also IORCB]

IOResult Integer Value Status of Last input/Output Operation [Pascal Function]

IORQ Input/Output Request [also IOREQ]

IORT Incremental Oil Revenue Tax

IOS Indian Ocean Ship [Space Tracking and Data Network]

IOS Infinite-Order Sudden (Approximation) [Physics]

IOS Input/Output Scan

IOS Input/Output Selector

IOS Input/Output Subsystem

IOS Input/Output Supervision; Input/Output Supervisor

IOS Input/Output System

IOS Institute of Oceanographic Sciences [UK]

IOS Institute of Ocean Sciences [Sidney, British Columbia, Canada]

IOS Institute of Statisticians [UK]

IOS Instructor Operator Station

IOS Interactive Operating System

IOS International Organization for Standardization

IOS International Organization for Succulent Plant Study [UK]

IOS Investors Overseas Service

IOSA Infinite-Order Sudden Approximation [Physics]

IOSA Integrated Optic Spectrum Analyzer

IOSA International Oil Scouts Association [US]

IOSA Irish Offshore Services Association

IOSC Integrated Operations Support Center

IOSC International Oxygen Steelmaking Congress

IOSCD International Organization for Scientific Cooperation and Development

IOSEWR International Organization for the Study of the Endurance of Wire Ropes [France]

IOSGA Input/Output Support Gate Array

IOSH Institute for Occupational Safety and Health [of US Department of Health, Education and Welfare]

IOSH Institution of Occupational Safety and Health [UK]

IOSPE International Offshore and Polar Engineering Conference

IOSS Input/Output Subsystem

IOSTE International Organization of Science, Technology and Education

IOT Indian Ocean Territory

IOT In-Orbit Test Antenna

IOT Input/Output and Transfer

IOT Input/Output Terminal

IOT Input/Output Transfer

IOT Input/Output Trunk

IOT Institute of Transport [UK]

IOTA Information Overload Testing Apparatus

IOTA Institut d'Optique Théorique et Appliquées [Institute of Theoretical and Applied Optics, Orsay, France]

IOTA International Occultation Timing Association [US]

IoTA Institute of Transport Administration [UK]

IOTA/ES International Occultation Timing Association–
European Section [Germany]

IOTCG International Organization for Technical
Cooperation in Geology [Russian Federation]

IOT&E Initial Operation Test and Evaluation

IOTG Input/Output Task Group [of CODASYL– Conference
on Data System Languages]

IOTs Indian Ocean Territories

IOU Immediate Operation Use

IOU Input/Output Unit [also iou]

IOU Input/Output Utility

IOUBC Institute of Oceanography, University of British
Columbia [Vancouver, Canada]

IOUS Input/Output Utility Subsystem

IOVC In the Overcast

IOVST International Organization for Vacuum Science and
Technology

IOW In Other Words

IOW Institute of Welding

IOW Institut für Ostseeforschung Warnemünde (an der der
Universität Rostock) [Warnemunde Institute for Baltic
Research (of the University of Rostock,
Rostock-Warnemunde, Germany]

Iox Isoxazole

IOZ Inorganic Zinc Coating

IOZ Internal Oxidation Zone [Metallurgy]

IP Ice Point [also ip]

IP Identification Point

IP Identification of Position [also I/P]

IP Igloo Pallet [NASA]

IP Igneous Petrology

IP Ignition Point [also ip]

IP Image Plane

IP Image Processing; Image Processor

IP Imaginary Part (of a Complex Number)

IP Imaging Plate

IP Imipramine

IP Impact Parameter [Nuclear Physics]

IP Impact Physics

IP Impact Point [Aerospace]

IP Impact Pressure

IP Impact Printer

IP Implementation Plan

IP Implicit Parallelism

IP Incisoproximal [Medicine]

IP Inclined Plane

IP Incommensurate Phase [Physics]

IP Incubation Period

IP Index of Performance [Industrial Engineering]

IP Index Point

IP Induced Polarization

IP Induction Period [Physical Chemistry]

IP Inductive Probability

IP Industrial Park

IP Industrial Plant

IP Industrial Production

IP Inertial Processing

IP Information Policy

IP Information Processing; Information Processor

IP Information Professional

IP Information Provider [also ip]

IP Initial Phase

IP Initial Point

IP Injection Pump

IP Inner Product

IP Inosine Phosphate [Biochemistry]

IP In-Patient

IP In-Phase

IP In-Process

IP Input Processing; Input Processor

IP Institute of Physics [Taiwan]

IP Institute of Petroleum [UK]

IP Institut Pasteur [Pasteur Institute, Paris, France]

IP Instruction Pointer

IP Instruction Processor

IP Instructor Pilot

IP Instrumentation Payload

IP Instrumentation PCM (Pulse-Code Modulation) Data Bus

IP Instrument Panel [also I/P]

IP Insulating Phase

IP Intake Port

IP Integer Programming

IP Integrated Program

IP Intellectual Property

IP Intelligent Peripheral

IP Interactive Plotting

IP Interactive Programming

IP Interchange Point

IP Interdigital Pause

IP Interdisciplinary Panel

IP Interface Processor

IP Interfacial Potential

IP Interference Points

IP Intermediate Pallet

IP Intermediate Phase [Metallurgy]

IP Intermediate Pressure [also I-P, ip, or i-p]

IP Intermediate Product

IP Intermolecular Potential

IP Internal Pressure

IP Internet Protocol (Address)

IP Interphalangeal (Joint) [Anatomy]

IP Interplanetary

IP Interpolymer

IP Interrupted Projection

IP Interrupt Prodecure

IP Intraperitoneal(ly) [also ip] [Anatomy]

IP Invasion Percolation (of Fluids and Porous Media)

IP Inverted Population

IP Ionic Polymer(ization)

IP Ionization Potential

IP Ionizing Power

IP Ion(ic) Pair

IP Ion Plating

IP Ion Probe

IP Ion Pump

IP Isometric Projection

IP Isopentane

IP Isophorone

IP Isopropyl

IP Item Processing

IP Powder Injection

I(P) Ivantsov Function of the Peclet Number [Symbol]

I&P Indexed and Paged

I&P Inerting and Preheating

I/P Identification of Position [also IP]

I/P Input [also i/p]

I/P Instrument Panel [also IP]

I/P Irregular Input Process

I/P I (Current) to P (Pressure)

I-P Intermediate-Pressure [also IP, ip, or i-p]

Ip Isopropylidene

ip ice point [also IP]

ip ignition point [also IP]

ip information provider [also IP]

ip intermediate pressure [also IP, I-P, or i-p]

i/p intrinsic-p-type layer (interface) [Solid-State Physics]

I£ Irish Pound [Currency]

I-5'-P Inosine 5'-Monophosphate [Biochemistry]

2iP N^6-(2-Isopentenyl)adenine [Biochemistry]

IPA Illinois Pharmaceutical Association [US]

IPA Image Power Amplifier

IPA Independent Petroleum Association [US]

IPA Indiana Pharmaceutical Association [US]

IPA Indian Physics Association [India]

IPA Indole-3-Propionic Acid

IPA Industrial Perforators Association [US]

IPA Information Processing Architecture

IPA Information Processing Association [Israel]

IPA Information-Technology Promotion Agency [Japan]

IPA Institute for Polyacrylate Absorbents [US]

IPA Institute of Public Administration [US]

IPA Institute of Public Administration [Ireland]

IPA Institute of Public Affairs [Australia]

IPA (Fraunhofer-)Institut für Produktionstechnik und Automatisierung [(Fraunhofer) Institute for Production Technology and Automation, Stuttgart, Germany]

IPA Institutul de Cercetari si Proiectari Automatizar [Scientific Institute for Automatic Control, Romania]

IPA Integrated Peripheral Adapter

IPA Integrated Photodetection Assembly

IPA Integrated Printer Adapter

IPA Intermediate Power Amplifier

IPA International Paleontological Association [US]

IPA International Patent Agreement

IPA International Permafrost Association

IPA International Pharmaceutical Abstracts [of American Society of Hospital Pharmacists, US]

IPA International Phonetic Alphabet

IPA International Phonetic Association [UK]

IPA International Photographers Association [US]

IPA International Pipe Association [US]

IPA International Platform Association

IPA International Prepress Association [US]

IPA International Press Association

IPA International Publishers Association [Switzerland]

IPA Interpass Absorption (Process)

IPA Inverse Perturbation Analysis

IPA Iowa Pharmaceutical Association [US]

IPA Isepentenyl Adenosine [Biochemistry]

IPA Isophthalic Acid

IPA Isopropyl Acetate

IPA Isopropanol; Isopropyl Alcohol

IPA Isopropyl Amine [also Ipa]

IPAA Independent Petroleum Association of America [US]

IPAA 5-Aminomethyl-3,3,5-Trimethyl Cyclohexanol

IPAA International Pesticide Applicators Association [US]

IPAC Independent Petroleum Association of Canada

IPAC Infrared Processing and Analysis Center [of NASA at Jet Propulsion Laboratory, Pasadena, California, US]

IPAC Institute of Public Administration of Canada

IPAC Institut Polytechnique de l'Afrique Centrale [Polytechnic Institute of Central Africa]

IPAC International Peace Academy Committee [US]

IPACS Integrated Power and Attitude Control System

IPACS Integrated Probabilistic Assessment of Composite Structures

IPAD Integrated Program for Aerospace Vehicle Design

IPAD International Plastics Association Directors

IPADAE Integrated Passive Action Detection Acquisition Equipment

IPAE Isopropylaminoethanol

Ipae Diisopropylaminoethanol

IPAG Information, Planning and Analysis Group [of University of New Delhi, Department of Electronics, India]

IPAI International Primary Aluminum Institute [UK]

IPA/IAO (Fraunhofer-)Institut für Produktionstechnik und Automatisierung/(Fraunhofer-)Institut für Arbeitswissenschaft und Organisation [(Fraunhofer) Institute for Production Technology and Automation/ [(Fraunhofer) Institute for Industrial Science and Organization, Stuttgart, Germany]

IPAL Integrated Program on Arid Lands

IPAM Isopropylacrylamide

IPA-MS Independent Petroleum Association of Mountain States [US]

IPA-NM Independent Petroleum Association of New Mexico [US]

IPAR Intra-Pulse Analysis Receiver

IPARA International Publishers Advertising Representatives Association

IPARC International Pesticide Application Research Center [UK]

IPAS Institute of Physics Academia Sinica [Taipeh, Taiwan]

IPB Illuminated Push Button

IPB Illustrated Parts Breakdown

IPB Information Policy Board [of Government of Queensland, Australia]

IPB Institut für Pflanzenbiochemie [Institute for Plant Biochemistry, Halle, Germany]

IPB Institutului Politehnic Bucuresti [Bucharest Polytechnic Institute, Romania]

IPB Integrated Processor Board

IPB Interphase (Grain) Boundary [Materials Science]

IPB Interprocessor Buffer

IPB Isopropyl Benzene

IPBAM International Permanent Bureau of Automotive Manufacturers

IPBC International Power Beam Conference

IPBF Installed Peripheral Base Flexibility

IPC Idaho Potato Commission [US]

IPC Incipient Percolation Cluster

IPC Independent Control Point

IPC Indiana Port Commission [US]

IPC Indicative Planning Council

IPC Indirect Photometric Chromatography

IPC Industrial Personal Computer

IPC Industrial Process Control

IPC Industry Planning Council [US]

IPC Industry Policy Council [US]

IPC Information Processing Center

IPC Information Processing Code

IPC In-Line Process Control

IPC Institute for Interconnecting and Packaging Electronic Circuits [US]

IPC Institute for Personal Computing [Miami, Florida, US]

IPC Institute of Paper Chemistry [Appleton, Wisconsin, US]

IPC Institute of Paper Conservation [UK]

IPC Institute of Physical Chemistry [of Academy of Sciences, Moscow, Russian Federation]

IPC Institute of Printed Circuits [Name Changed to Institute for Interconnecting and Packaging Electric Circuits, US]

IPC Instituto de Plásticos y Caucho [Institute for Plastics and Rubber, Spain]

IPC Institutul Politehnic Cluj [Cluj Polytechnic Institute, Romania]

IPC Instructions per Clock

IPC Instrumented Precracked Charpy (Test) [Mechanics]

IPC Integrated Peripheral Channel

IPC Integrated Pollution Control

IPC Integrated Process Control

IPC Integrated Program Coordinator

IPC Integrated Program for Commodities [of United Nations]

IPC Intermediate Processing Center

IPC Intermittent Position Control

IPC Intermittent Positive Control

IPC International Parachuting Commission

IPC International Patent Classification

IPC International Petroleum Corporation [US]

IPC International Photobiology Committee

IPC International Photosynthesis Committee [Sweden]

IPC International Poplar Commission [Italy]

IPC International Pressure Conference

IPC International Programmable Control Conference (and Exposition)

IPC Interpenetrating Phase Composite

IPC Interphase Precipitation of Cementite [Metallurgy]

IPC Interprocess Communication

IPC Interprocess Controller

IPC Interprocess Coupler

IPC Interprocessor Communication

IPC Intrinsic Photoconductivity

IPC Iraq Petroleum Company

IPC Ion-Pair Chromatography

IPC Isoproyl Carbinol

IPC Isopropyl Chloride

IPC Isopropyl Cresol

IPC Isopropyl-N-Phenylcarbamate

IP-C IEE Proceedings, Communications [of Institution of Electrical Engineers, UK]

Ipc Isopinocampheyl

IPCA Industrial Pest Control Association [UK]

IPCA International Petroleum Cooperative Alliance

IPCAVD Integrated Preclinical/Clinical AIDS Vaccine Development

Ipc$_2$BCl Chlorodiisopinocampheylborane

(IpcBH$_2$)·TMEDA Monoisopinocampheylborane Tetramethylethylenediamine

IPCC Intergovernmental Panel on Climate Change

IPCCS Information Processing in Command and Control Systems

IP-CDS IEE Proceedings, Circuits, Devices and Systems [of Institution of Electrical Engineers, UK]

IP-CDT IEE Proceedings, Computers and Digital Techniques [of Institution of Electrical Engineers, UK]

IPCE Incident (Monochromatic) Photon-to-Current Conversion Efficiency

IPCE Integrated Protective Clothing and Equipment (Program) [of Canadian Forces]

IPCEA Insulated Power Cable Engineers Association [now Insulated Cable Engineers Association, US]

IPCE/TD Integrated Protective Clothing and Equipment Technology Demonstrator [of Canadian Forces]

IPCF Interprocess Communication Facility

IPCL Institut du Pétrole, des Carburants et Lubrifiants [Institute of Petroleum, (Motor) Fuels and Lubricants, France]

IPCL Instrumentation Program and Component List

IPCM Institut des Plasmas et des Couches Minces [Plasma and Thin-Film Institute; of Conseil National de la Recherche Scientifique, Université de Nantes, France]

IPCMS Institut de Physique et Chimie des Matériaux de Strasbourg [Strasbourg Institute for Materials Physics and Chemistry, France]

IPCN Institutului Politechnic Cluj-Napoca [Polytechnic Institute of Cluj-Napoca, Romania]

IP-COM IEE Proceedings, Communications [of Institution of Electrical Engineers, UK]

IPCP Internet Protocol Control Protocol

IPCR Institute of Physical and Chemical Research [Saitama, Japan]

IPCRI Israel-Palestine Center for Research Information

IP-CRSP IEE Proceedings F, Communications, Radar and Signal Processing [of Institution of Electrical Engineers, UK]

IPCS Institution of Professional Civil Servants [UK]

IPCS Interactive Problem Control System

IPCS International Program for Chemical Substances

IPCS International Program on Chemical Safety [of United Nations Environmental Program]

IPCS International Program on Chemical Study [of World Health Organization]

IP-CSV IEE Proceedings I, Communications, Speech and Vision [of Institution of Electrical Engineers, UK]

IP-CTA IEE Proceedings, Control Theory and Applications [of Institution of Electrical Engineers, UK]

IPD Information Processing Division

IPD Insertion Phase Delay

IPD Institute of Professional Designers [UK]

IPD Integrated Product Definition (Environment)

IPD Intellectual Property Division

IPD Intermittent Peritoneal Dialysis [Medicine]

IPD Interplanar Distance [Crystallography]

IPD Interplanetary Dust

IPD Isophoronediamine

IPDC International Program for Development of Communications [of UNESCO]

IPDD Initial Project Design Description

IPDE Identify, Predict, Decide, Execute

IPDG Isopropyl-(β-D-)Thiogalactopyranoside

IPDH In-Service Planned Derated Hours

IPDI Isophorone Diisocyanate

IPDMUG International Product Data Management Users Group

IPDP Industrial Programmed Data Processor

IPDP Intervals of Pulsations of Diminishing Period

IPDR Incremental Preliminary Design Review

IPDS IBM Personal Dictation System [of IBM Corporation, US]

IPDS Intelligent Printer Data Stream

IPDU Internet Protocol Data Unit

IPE Industrial and Production Engineering (Journal) [Germany]

IPE Industrial Plant Equipment

IPE Information Processing Equipment

IPE Institut de Physique Expérimentale [Institute of Experimental Physics, Université de Lausanne, Switzerland]

IPE Institute of Power Engineers

IPE Institution of Plant Engineers [London, UK]

IPE Institution of Production Engineering [UK]

IPE Institution of Professional Engineers [Wellington, New Zealand]

IPE Internal Photoemission

IPE International Petroleum Exchange [London, UK]

IPE International Political Economy

IPE Interpret Parity Error

IPE Intrinsic Photoemission

IPE Inverse Photoemission

IPE Inverse Photoemission Experiment

IPE Ion Photon Emission

IPE Isopropoxyethanol

IPe Isopentyl [also iPe, or i-Pe]

IPeAc Isopentyl Acetate [also iPeAc, or i-PeAc]

IPeCl Isopentyl Chloride [also iPeCl, or i-PeCl]

IP-ECS IEE Proceedings G, Electronic Circuits and Systems [of Institution of Electrical Engineers, UK]

IPEE Institut pour une Politique Européene de l'Environnement [Institute for European Environmental Policy]

IPEF Instituto de Pesquisas e Estudos Florestais [Institute of Flower Research, Brazil]

IPEI Ionospheric Plasma and Electrodynamics Instrument

IPE Int Ind Prod Eng IPE International Industrial and Production Engineering [Journal published in Germany]

IPEN Instituto Peruano de Energía Nuclear [Peruvian Institute of Nuclear Energy, Lima, Peru]

IPEN Instituto de Pesquisas Energéticas e Nucleares [Institute of Energy and Nuclear Research, Sao Paulo, Brazil]

IPEN-CN Instituto Peruano de Energía Nuclear–Centro Nuclear [Peruvian Institute of Nuclear Energy– Nuclear Center, Lima, Peru]

IPEN/CNEN Instituto de Pesquisas Energéticas e Nucleares/Centro Nacional de Energia Nuclear [Institute of Energy and Nuclear Research/National Nuclear Energy Center, Sao Paulo, Brazil]

IPENZ Institution of Professional Engineers of New Zealand [Wellington]

IPEP International Permanent Exhibition of Publications

IP-EPA IEE Proceedings, Electric Power Applications [of Institution of Electrical Engineers, UK]

IPES Inverse Photoemission Spectroscopy

IPES Inverse Photoemission Spectrum

IPETE International Petroleum Equipment and Technology Exhibition

IPETEX Institute of Petroleum Working Group on Petroleum Exploration Training [UK]

IPEX International Photographic Exposition

IPF Indicative Planning Figures

IPF Inherent Power Factor

IPF International Pain Foundation [US]

IPF International Pharmaceutical Federation

IPF International Powerlifting Federation

IPF Institut für Polymerforschung e.V. [Polymer Research Institute, Dresden, Germany]

IPF Ionic Packing Factor [Physics]

IPF Irish Printing Federation [UK]

IPF IUS (Inertial Upper Stage) Processing Facility [of NASA]

IPFA International Professional Security Association [UK]

IPFC Indo-Pacific Fisheries Commission [Thailand]

IPFC Information Presentation Facility Compiler

IPFM Integral Pulse Frequency Modulation

IPG Independent Publishers Guild [UK]

IPG Inferior Parathyroid Gland [Anatomy]

IPG Information Policy Group [of Organization for Economic Cooperation and Development, France]

IPG Information Planning Group

IPG In-Plane-Gated (Transistor)

IPG Institute of Professional Goldsmiths [UK]

IPG Institut für Petrographie und Geochemie [Institute for Petrography and Geochemistry, University of Karlsruhe, Germany]

IPG Interactive Presentation Graphics

IPG Isopropyl Glycol

IPGP Institut de Physique du Globe de Paris [Geophysical Institute of Paris, France]

IPGS Industrial Postgraduate Scholarship [of Natural Sciences and Engineering Research Council, Canada]

IP-GTD IEE Proceedings, Generation, Transmission and Distribution [of Institution of Electrical Engineers, UK]

IPH Inches per Hour [also iph]

IPH Institut de Paléontologie Humaine [Institute of Human Paleontology, France]

IPH International Association of Paper Historians [Germany]

IPH Inform IPH Information [Publication of the International Association of Paper Historians, Germany]

IPhA Isophthalic Acid

i-phase icosahedral phase [Materials Science]

IPHC International Pacific Halibut Commission [US/Canada]

IPHE Institution of Public Health Engineers [UK]

IPHS Isopropyl Hydrogen Sulfate

IPI Individually Presented Instruction

IPI Industrial Production Index

IPI In-Process Inspection

IPI Institute of Patentees and Inventors [UK]

IPI Institute of Polymer Industry [Japan]

IPI Institutului Politehnic din Iaşi [Iasi Polytechnic Institute, Romania]

IPI Instrument Principal Investigator

IPI Integrated Permits and Inspections

IPI Intelligent Peripheral Interface

IPI Intelligent Printer Interface

IPI Interchemical Printing Ink

IPI International Pesticide Institute [US]

IPI International Potash Institute [Switzerland]

IPI International Prognostic Index

IPI International Press Institute

IPI ITO (Indium-Tin Oxide)–PZT (Lead Zirconate Titanate)–ITO (Indium-Tin Oxide)

IpI Inosylylinosine [Biochemistry]

IPIB Indian Press Information Bureau

IPIE Institute of Profit Improvement Executives

IPIECA International Petroleum Industry Environmental Conservation Association [UK]

IPIN Instituto Panamericano de Ingeniera Naval [Pan-American Institute of Naval Engineers, Brazil]

IPIP International Personhood of Iliterate Programmers [US]

IPIRA Indian Plywood Industries Research Association

IPIRI Indian Plywood Industries Research Institute

IPIX Intelligent (Multi-)Parameter Imaging X-Band

IPK Institut für Pflanzengenetik und Kulturpflanzenforschung [Institute for Plant Genetics and Cultivated Plant Research, Gatersleben, Germany]

IPL Image Processing Laboratory [of NASA]

IPL Indentured Parts List

IPL Information Processing Language

IPL Initialize Program Load; Initial Program Load(er); Initial Program Loading

IPL Inner Plexiform Layer

IPL Inorganic Protective Layer

IPL Integrated Payload

IPL Interprocessor Link

IPL Interprovincial Pipe-Line Limited [Canada]

IPL Interrupt Priority Level

IPL Ion Projection Lithography

IPL Isotope Products Laboratory [of US Department of Energy]

IPLA Interstate Producers Livestock Association [US]

IPlantE Institution of Plant Engineers [UK]

IPLC International Private Leased Circuit [UK]

IPLCA International Pipeline Contractors Association [France]

IPLO Institute of Professional Librarians of Ontario [Canada]

IPL-V Information Processing Language V

IPM Ideal Paramagnetic

IPM Illuminations per Minute

IPM Images per Minute [also ipm]

IPM Impulses per Minute

IPM Inches per Minute [also ipm]

IPM Inches (of) Penetration per Month [Unit]

IPM Incidental Phase Modulation

IPM Incident Power Meter

IPM Independent Particle Method (Approximation)

IPM Independent Particle Model

IPM Industrial Productivity Monitoring

IPM Initial Pretreatment Module

IPM Input Position Map(per)

IPM Insect Pest Management

IPM Institute for Practical Mathematics [Germany]

IPM Institute of Personnel Management [UK]

IPM Institute of Physical Metallurgy [of Czech(oslovak) Academy of Sciences at Brno]

IPM Instruction per Minute

IPM Instytut Podstaw Metallurgii [Podstaw Institute of Metallurgy; of Polish Academy of Sciences at Krakow]

IPM Integrated Pest Management

IPM Integrated Program Manager

IPM Intelligent Power Manager

IPM Intelligent Processing of Materials

IPM Interference Prediction Model

IPM Internal Polarization Modulation

IPM International Pest Management

IPM Interpersonal Message

IPM Interruptions per Minute

ipm inches per minute [also IPM]

IPMA In-Plant Printing Management Association [US]

IPMA International Personnel Management Association [US]

IP-MAP IEE Proceedings, Microwaves, Antennas and Propagation [of Institution of Electrical Engineers, UK]

IPM CAS Institute of Physical Metallurgy of the Czechoslovak Academy of Sciences [Brno, Czech Republic]

IPME Institute of Problems for Mechanical Engineering [of Russian Academy of Sciences in St. Petersburg]

IPMI International Powder Metallurgy Institute [US]

IPMI International Precious Metals Institute [US]

IPMM (Australasia-Pacific Forum on) Intelligent Processing and Manufacturing of Materials

IPMP Integrated Pest Management Programs

IPM-PAN Instytut Podstaw Metallurgii–Polskiej Akademii Nauk [Podstaw Institute of Metallurgy of the Polish Academy of Sciences, Krakow]

IPMR Institute of Physical Medicine and Rehabilitation [of New York University, US]

IPMS Institute for Problems of Materials Science [Ukraine]

IPMS Institution of Professionals, Managers and Specialists [UK]

IPMS International Polar Motion Service

IPMSSF IPM School on Cosmology: Large Scale Structure Formation [Babolsar, Iran]

IPN Infectious Pancreatic Necrosis

IPN Initial Priority Number

IPN Initial Processing Number

IPN Inspection Progress Notification

IPN Instant Private Network

IPN Institut de Physique Nucléaire [Nuclear Physics Institute, Orsay, France]

IPN Institut für die Pädagogik der Naturwissenschaften (an der Universität Kiel) [Institute for Science Education (of the University of Kiel), Germany]

IPN Instituto Politecnico Nacional [National Polytechnic Institute, Mexico City]

IPN Internal Priority Number

IPN Inter-Penetrating Network; Interpenetrating Polymer Network

IPN Isophorone Nitrile

IPN 1-Isopropylnaphthalene

IPN-ESIQIE Instituto Politecnico Nacional–Escuela Superior de Ingeniería Química e Industrias Extractivas [National Polytechnic Institute–Higher School of Chemical Engineering and Extractive Industries, Mexico City, Mexico]

IPNG Internet Protocol Next Generation [also Ipng]

IPNJ Industrial Photographers of New Jersey [US]

IPNL Institut de Physique Nucléaire de Lyon [Lyons Institute of Nuclear Physics, Université Claude Bernard, Villeurbanne, France]

IPNL Integrated Perceived Noise Level

IPNS Intense Pulsed Neutron Source [Established at Argonne National Laboratory, US in 1981]

IPNS/LANSCE Intense Pulsed Neutron Source/Los Alamos Neutron Scattering Center (Joint Program Advisory Committee)

IPNV Infectious Pancreatic Necrosis Virus

IPO IGES-PDES (Initial Graphics Exchange Specification/ Product Design Exchange Specification) Organization [US]

IPO Indolephenoloxidase [Biochemistry]

IPO Interim Protection Order

IPOC Idaho Potato and Onion Commission [US]

IPOC International Partner Operations Center

IPOD International Phase of Ocean Drilling

IPOEE Institution of Post Office Electrical Engineers [UK]

IPOL Institute of Polarology [UK]

IP/OP Input/Output Interface

IP-OPT IEE Proceedings, Optoelectronics [of Institution of Electrical Engineers, UK]

IPOS Institute for Polymers and Organic Solids [of University of California at Santa Barbara, US]

IPOs Interim Protection Orders

IPOT Inductive Potentiometer [also I-POT, ipot, or i-pot]

IPOTMS Isopropenyloxytrimethyl Silane

IPP Imaging Photopolarimeter

IPP Impact Prediction Point

IPP Industrial Partners Program [of Natural Resources Canada]

IPP Injury Prevention Program

IPP In-Phase Particles (of Intermetallics)

IPP In-Plant Plus Program [of ASM International, US]

IPP Institut für Plasmaphysik [Institute for Plasma Physics; of Max-Planck Institute at Garching, Germany]

IPP Institute of Plasma Physics

IPP Integrated Plotting Package

IPP Integrated Program Plan

IPP Interface Package Process

IPP International Phototelegraph Position

IPP Inter-Processor Process

IPP Ion Pair Partition

IPP Isopropyl Percarbonate

IPP Isotactic Polypropylene [also i-PP]

IPPA Independent Pair Potential Approximation [Physics]

IPPA Independent Professional Painting Contractors Association (of America) [US]

IPPA Indo-Pacific Prehistory Association [Australia]

IPPA International Printing Pressmen and Assistants Union

IPPB Intermittent Positive Pressure Breathing

IPPC Industrial Promotion and Productivity Center

IPPC Integrated Pollution Prevention and Control

IPPC Isopropyl-N-Phenylcarbamate

IPPD Integrated Product and Process Development

IPPD N-Isopropyl-N'-Phenyl-P-Phenylene Diamine

IPPDSEU International Plate Printers, Die Stampers and Engravers Union [Canada and US]

IPPE Isophthalic Polyester

IPPF Instruction Preprocessing Function

IPPJ Institute of Plasma Physics, Japan [of Nagoya University] [also IPP(J)]

IPPNW International Physicians for the Prevention of Nuclear War [UK]

IPPP Inner Projections of the Polarization Propagator

IPPP Institute of Plant Production and Processing [of Commonwealth Scientific and Industrial Research Organization, Australia]

IPPP Isopentenyl Pyrophosphate

IPP-PA11 Isotactic Polypropylene–Polyamide 11 (Blend)

IPPPM Iterative Pariser-Parr-Pople (Method with) Matago-Nishimoto Approximation [Physics]

IPPPO Iterative Pariser-Parr-Pople (Method with) Ohno Approximation [Physics]

IPPS Indian Point Power Station [New York, US]

IPPS Institute of Physics and the Physical Society [UK]

IPPS International Plant Propagators Society

IPPTA Indian Pulp and Paper Technical Association [India]

IPQ Instituto Portugèsa da Qualidade [Portuguese Quality Institute]

IPQC In-Process Quality Control

IPR Inches per Rack [Textiles]

IPR Inches per Revolution [also ipr, or in/rev]

IPR Indirect Photoreactivation

IPR Inflow Performance Relationship

IPR In-Progress Review

IPR In-Pulse-to-Register

IPR Institute for Polymer Research [Waterloo, Ontario, Canada]

IPR Institute of Population Registration [UK]

IPR Institute of Public Relations [UK]

IPR Intellectual Property Rights

IPR Interim Problem Report

IPr Isopropyl [also iPr, or i-Pr]

(IPr)₂ Diisopropyl [also (iPr)₂, or (i-Pr)₂]

ipr intraperitoneal [Toxicology]

IPRA International Peace Research Association [Netherlands]

IPRA International Public Relations Association [Switzerland]

IPrAc Isopropyl Acetate [also i-PrAc]

IPrBr Isopropyl Bromide [also i-PrBr]

IPRC International Personal Robot Congress (and Exposition)

IPrCl Isopropyl Chloride [also i-PrCl]

IPRE International Professional Association for Environmental Affairs [Belgium]

IPrF Isopropyl Fluoride [also i-PrF]

IPRG International (Construction) Procurement Research Group

IPrMgCl Isopropylmagnesium Chloride [also iPrMgCl]

IPRO International Patent Research Office

IPrO Isopropoxide (or Isoproylate) [also i-PrO]

IProdE Institution of Production Engineers [UK]

IPrOH Isopropyl Alcohol [also i-PrOH]

IPRs Intellectual Property Rights

IP-RSN IEE Proceedings, Radar, Sonar and Navigation [of Institution of Electrical Engineers, UK]

IPRT Institute for Physical Research and Technology [of Iowa State University, Ames, US]

IPrX Isopropyl Halide [also iPrX, or i-PrX]

IPS Image Processing System

IPS Impact Polystyrene

IPS Impact Predictor System [Aerospace]

IPS Impulses per Second [also ips]

IPS Inches per Second [also ips]

IPS Incorporated Phonographic Society [UK]

IPS Induction Plasma Spraying

IPS Information Processing Society [Japan]

IPS In-Pulse-to-Sender

IPS Installation Performance Specification

IPS Institute of Polymer Science [of Auburn Science and Engineering Center, Akron, Ohio, US]

IPS Institute of Purchasing and Supply [Stamford, Lincolnshire, UK]

IPS Instructions per Second

IPS Instrumentation Power Supply

IPS Instrumentation Power Subsystem

IPS Instrumentation Power System

IPS Instrument Pointing Subsystem

IPS Instrument Pointing System

IPS Integrated Photosector

IPS Integrated Power System

IPS Integrated Process System

IPS Integrated Production System

IPS Intelligent Printing System

IPS Intelligent Programming System

IPS Interceptor Pilot Simulator

IPS Interconnected Power System

IPS Interconnected Power Systems [US]

IPS Interface Problem Sheets

IPS Interim Policy Statement

IPS Interior Pipe Size

IPS Intermolecular Potential-Energy Surface

IPS International Paleontological Society [US]

IPS International Palm Society [US]

IPS International Peat Society [Finland]

IPS International Phycological Society [US]

IPS International Physical Society [Italy]

IPS International Pipe Standard [also ips]

IPS International Planetarium Society [US]

IPS International Pressure Society [US]

IPS International Publishers Service, Inc. [US]

IPS Interplanetary Space

IPS Interpretative Programming System

IPS Interruptions per Second

IPS Invariant Plane Strain [Metallurgy]

IPS Inverse Photoemission Spectroscopy

IPS Inverter Power Supply

IPS Ionospheric Prediction Service

IPS Ion Photon Spectroscopy

IPS Iron Pipe Size

IPS Israel Physical Society [Hebrew University, Jerusalem]

IPS Isotropic Proton Shift

IPs Intrinsic Pseudoelasticity

i-PS Isotactic Polystyrene

ips inches per second [also IPS]

IPSA Institute of Public Service Administrators [UK]

IPSA International Passenger Ship Association

IPSB Interprocessor Signal Bus

iPSC Intel Parallel Supercomputer

IPSD Integrating Position-Sensitive Detector

IPSE Integrated Programming Support Environment

IPSE Integrated Project Support Environment

IP-SEN IEE Proceedings, Software Engineering [of Institution of Electrical Engineers, UK]

IPSEP International Project for Soft Energy Paths

IPSF International Pharmaceutical Students Federation [Israel]

IPSFC International Pacific Salmon Fisheries Commission

IPSJ Information Processing Society of Japan

IPSL Interface Problem Status Log

IPSM Institute of Physical Sciences in Medicine [York, UK]

IPSMCB International Product Safety Management Certification Board [US]

IP-SMT IEE Proceedings, Science, Measurement and Technology [of Institution of Electrical Engineers, UK]

IPSO Interface Peripheral Standard Olivetti

IPSO International Program Support Office

IPSOC Information Processing Society of Canada

IPSP Inhibitory Postsynaptic Potential [Medicine]

IPSS International Packet Switching Service [UK]

IPSS International Packet Switching System

IPSS International Packet Swichstream (Service) [of British Telecom]

IPSS International Power Sources Symposium [UK]

IPSS Inter-Processor Signaling System

IPSSB Infomation Processing Systems Standards Board [of American National Standards Institute, US]

IP-SSED IEE Proceedings, Solid-State and Electron Devices [of Institution of Electrical Engineers, UK]

IPST Institute of Paper Science and Technology [US]

IPST International Practical Scale of Temperature

IPST Israel Program for Scientific Translations

IPT Inches per Tooth [also ipt]

IPT Indexed, Paged and Titled

IPT In-Plant Transporter

IPT Integrated Product Team

IPT Internal Pipe Thread

IPT International Pipe Thread

IPT Ion-Phonon Transition

IPT Isopropyl Toluene

IPTA International Patent and Trademark Association [US]

IPTC International Press Telecommunications Council [UK]

IPTD Sulfa-5-Isopropyl-1,3,4-Thiadiazole

IPTEC Division of Interinstitutional Cooperation in Science and Technology [of Indonesian Institute of Science]

IPTG Isopropyl-β-D-Thiogalactopyranoside; Isopropyl-β-D-1-Thiogalactopyranoside [Biochemistry]

IPTI Interpulse Time Interval

IPTM Interval Pulse Time Modulation

IPTO Information Processing Technologies Office [of Advanced Research Projects Agency, US]

IPTS International Practical Temperature Scale [of National Institute for Standards and Technology, US]

IPTS-48 International Practical Temperature Scale of 1948

IPTS-68 International Practical Temperature Scale of 1968

IPTS-90 International Practical Temperature Scale of 1990

Iptz Isopropyl Tetrazole [also iptz]

IPU Indian Point Unit [US Nuclear Reactor]

IPU Institute of Public Utilities [US]

IPU Instruction Processing Unit

IPU International Paleontological Union [now International Paleontological Society, US]

IPU Inter-Parliamentary Union

IPU International Postal Union

IPU Interprocessor Unit

IPU Islamic Press Union

IPU-1 Indian Point Unit 1 [US Nuclear Reactor]

IpU Inosylyluridine [Biochemistry]

IPV Improve [also Ipv, or ipv]

IPv4 Internet Protocol Version 4

IPv5 Internet Protocol Version 5

IPv6 Internet Protocol Version 6

IPVC Irradiated Polyvinyl Chloride

IPW Induction-Pressure Welded; Induction-Pressure Welding

IPWF International Public Works Federation

IPX Internetwork Packet Exchange

IPY Inches (of) Penetration per Year [Unit]

IPY Inches per Year [also ipy]

IPY Internal Photoinjection Yield

IPY International Polar Year [1882 and 1932]

2IPy 2-Iodopyridine

IQ Image Quality

IQ Image Quantification

IQ Indefinite Quantity

IQ Institute of Quarrying [UK]

IQ Insured Quality

IQ Intelligence Quotient

IQ Iraq [ISO Code]

IQ Isoquinoline

I(Q) Neutron Scattering Intensity [Symbol]

$I_{hkl}(Q)$ Intensity Contribution for (hkl) Plane [Symbol]

I(q) Scattered Intensity Distribution [Symbol]

I(q) Integrated Intensity [Symbol]

iq *(idem quod)* – the same as

.iq Iraq [Country Code/Domain Name]

IQA Institute of Quality Assurance [London, UK]

IQA Instituto de Qualidade Alimentar [Institute of Food Quality, Lisboa, Portugal]

IQA Instituto de Quimico Agricola [Institute of Agricultural Chemistry, Brazil]

IQA Instituto de Quimica Ambiental [Institute of Environmental Chemistry, Spain]

IQA Instituto de Quimica de Araraquara [Araraquara Institute of Chemistry, of Universidade Nacional de Engenharia de São Paulo, Araraquara, Brazil]

IQA Irish Quality Association

IQB Instituto de Quimico Biológica [Institute of Biological Chemistry, Brazil]

IQC Icosahedral Quasicrystal

IQC International Quality Center [of European Organization for Quality Control, Switzerland]

IQD Iraqi Dinar [Currency]

IQE Internal Quantum Efficiency [Electronics]

IQEC International Quantum Electronics Conference [of Institute of Electrical and Electronics Engineers, US]

iqed *(id quod erat demonstrandum)* – that which was to be proved

IQF Image Quality Filter

IQF Interactive Query Facility

IQHE Integral Quantum Hall Effect [Electronics]

IQI Image Quality Indicator

IQL Incompressible Quantum Liquid

IQL Indifference Quality Level

IQL Information Query Language

IQL Interactive Query Language

IQN Inner Quantum Number

IQM Input Queue Manager

IQMF Ion Quadrupole Mass Filter

IQMH Input Queue Message Handler

IQPF International Quick Printing Foundation

IQPP Interactive Query Preprocessor

IQR Interquartile Range [Statistics]

IQS Institute of Quantity Surveyors [UK]

IQSC Instituto de Quimica de São Carlos [Sao Carlos Institute of Chemistry, of University of Sao Paulo, Brazil]

IQS International Quality Study [Canada]

IQSY International Quiet Sun Year [1964-65]

I(q,0) Wave-Vector Dependent Scatter(ing) Intensity at Time t = 0 [Symbol]

I(q,t) Wave-Vector and Time Dependent Scatter(ing) Intensity [Symbol]

IR Igneous Rock [Geology]

IR Image Reconstruction

IR Image Restoration

IR Image Rotation

IR Imaginary Root

IR Immersion Refractometer

IR Impact Resistance

IR Inception Rate

IR Inclination of the Ascending Return [Aeronautics]

IR Independent Research

IR Index Register

IR Induced Radioactivity

IR Inductive Resistor

IR Industrial Relations

IR Industrial Research(er)

IR Industrial Robot

IR Inertial Resistance

IR Informal Report

IR Information Research

IR Information Recording

IR Information Resources

IR Information Retrieval

IR Infrared [also ir]

IR Infrared Radiation

IR Initial Report

IR Injection Refining

IR Inland Revenue

IR Input Ready

IR Inside Radius

IR Insoluble Residue

IR Inspection Routine

IR Instantaneous Relay

IR Instruction Register

IR Instrument Rating [also I/R]

IR Instrument Reading

IR Insulation Resistance

IR Intelligence Ratio

IR Intelligent Robot

IR Interest Rate

IR Interference Refractometer; Interference Refractometry

IR Interim Report

IR Intermediate (Nuclear) Reactor

IR Intermediate Register

IR Internal Reflection; Internal Reflectance

IR Internal Representation

IR Internal Resistance

IR Internet Registry

IR Interrogator-Responder

IR Interrupt Register

IR Interrupt Request

IR Intrinsic Resistance

IR Inverse Reaction

IR Ionizing Radiation

IR Current–Resistance

IR Iran [ISO Code]

IR Irish Republic

IR Isolation Router

IR Isoprene Rubber

IR Isoprene Rule

IR Isothermal Reaction

IR Isotope Ratio

I&R Intelligence and Reconnaissance

I&R Interchangeability and Replacement

I/R Instrument Rating [also I/R]

Ir Iridium [Symbol]

Ir Ireland; Irish

Ir-187 Iridium-187 [also ^{187}Ir, or Ir^{187}]

Ir-188 Iridium-188 [also ^{188}Ir, or Ir^{188}]

Ir-190 Iridium-190 [also ^{190}Ir, or Ir^{190}]

Ir-191 Iridium-191 [also ^{191}Ir, or Ir^{191}]

Ir-192 Iridium-192 [also ^{192}Ir, or Ir^{192}]

Ir-193 Iridium-193 [also ^{193}Ir, or Ir^{193}]

Ir-194 Iridium-194 [also ^{194}Ir, or Ir^{194}]

Ir-195 Iridium-195 [also ^{195}Ir, or Ir^{195}]

Ir-196 Iridium-196 [also ^{196}Ir, or Ir^{196}]

Ir-197 Iridium-197 [also ^{197}Ir, or Ir^{197}]

.ir Iran [Country Code/Domain Name]

IRA Individual Retirement Account

IRA Induced Radioactivity

IRA Information Resource Administration

IRA Infrared Absorption

IRA Infrared Analysis; Infrared Analyzer

IRA Input Reference Axis

IRA *(Institut de Recherches Appliquées)* – French for "Institute of Applied Research"

IRA Institute of Registered Architects [UK]

IRA Institute Research Awards

IRA Internal Reflection Attachment

IRA International Reprographics Association

IRA International Rubber Association [Malaysia]

IR-A Infrared, Range A [from 780 to 1400 nm]

IRABOIS Institut de Recherches Appliquées au Bois [Institute for Applied Wood Research, France]

IRAC Information Resource and Analysis Center [US]

IRAC Infrared Array Camera [of NASA Space Infrared Telescope Facility]

IRAC Interdepartmental Radio Advisory Committee [US]

IRAC International Records Administration Conference [US]

Ir(ACAC)$_3$ Iridium(III) Acetylacetonate [also Ir(AcAc)$_3$, or Ir(acac)$_3$]

IRACQ Infrared Acquisition Radar

IRACQ Instrumentation Radar and Acquisition Panel

IRAD Independent Research and Development

IRAH Infrared Alternate Head

IrAl Iridium Aluminide

IRAM Indexed Random Access Memory

IRAM Institute of Research and Application of Development Methods [France]

IRAM Instituto Argentino de Normalisación [Argentinian Standards Institute]

IRAMS Infrared Automatic Mass Screening

IRAN Inspection and Repair(s) as Necessary

Iran Iranian

IRANDOC Iranian Documentation Center

Iran J Sci Technol Iranian Journal of Science and Technology [UK]

IRANSAT Iranian Government Communications Satellite

IRAP Industrial Research Aid Program

IRAP Industrial Research Assistance Program [of National Research Council of Canada]

IRAP Integrated Risk Assessment Project

IRAP Interagency Radiological Assistance Plan [of US Atomic Energy Commission]

IRAP Interagency Radiological Assistance Program

Iraqi J Sci Iraqi Journal of Science [Published by College of Science, Baghdad, Iraq]

IRAS Information Retrieval Advisory Services [UK]

IRAS Infrared Absorption Spectrophotometry

IRAS Infrared Absorption Spectroscopy

IRAS Infrared Astronomical Satellite [of NASA]

IRAS Infrared Reflection Absorption Spectroscopy

IRASA International Radio Air Safety Association

Iraser Infrared Amplification by Stimulated Emission of Radiation [also iraser]

Ir Astron J Irish Astronomical Journal [Published by Armagh Observatory, UK]

IRATA Industrial Rope Access Trade Association [UK]

IRATE Interim Remote Air Terminal Equipment

IR-ATR Infrared Absorption using Attenutated Total Reflection [also IR ATR]

Ir-Au Iridium-Gold (Alloy)

IRB Infinitely Rigid Bear

IRB (Fraunhofer) Informationszentrum Raum und Bau [(Fraunhofer) Information Center for Development Planning and Construction, Stuttgart, Germany]

IRB Institut Rugjer Boskovic [Rugjer Boskovic Institute, Zagreb, Croatia]

IRB Infrared Brazing

IRB Institute of Radiation Breeding [Japan]

IRB Institutional Review Board

IRB Interruption Request Block

IR-B Infrared, Range B [from 1.4 to 3 μm]

IRBEL Indexed References to Biomedical Engineering Literature [of National Institute for Medical Research, UK]

IRBIC Infrared Beam Induced Contrast

IRBIC Infrared Beam Induced Current

IRBM Intermediate-Range Ballistic Missile

IRBO Infrared Homing Bomb

IRC Incremental Related Carrier

IRC Industrial Relations Center

IRC Industrial Relations Councelors [US]

IRC Industrial Relations Council [Canada]

IRC Industrial Research Center

IRC Industrial Research Chairs (Program) [of Natural Sciences and Engineering Research Council, Canada]

IRC Industrial Reorganization Corporation

IRC Industry Relations Committee [of Institute of Electrical and Electronics Engineers, US]

IRC Information Resource Center

IRC Information Retrieval Center

IRC Infrared Camera

IRC Infrared Catastrophe [Quantum Mechanics]

IRC Infrared Communications

IRC Initial Rate of Climb

IRC Inland Revenue Commissioner [UK]

IRC Instant Response Chromatography

IRC Institut de Recherche sur la Catalyse [Catalysis Research Institute, France]

IRC Institute for Research in Construction [of National Research Council of Canada]

IRC Interdisciplinary Research Center

IRC Interdisciplinary Research Center (in Superconductivity) [of Imperial College of Science, Technology and Medicine, Cambridge, UK]

IRC Interdisciplinary Research Center (in Materials) [Birmingham/Swansea, UK]

IRC Interest Rate Cap

IRC Intermediate Routing Center

IRC Internal Revenue Code [US]

IRC International Radiation Commission [UK]

IRC International Record Carriers

IRC International Red Cross

IRC International Reference Center (for Community Water Supply and Sanitation) [Netherlands]

IRC International Rescue Commission

IRC International Research Council

IRC International Resource Company [US]

IRC International Rice Commission [Italy]

IRC International Rubber Conference

IRC Internet Relay Chat (Server)

IRC Intrinsic Reaction Coordinate

IRC Iron Ring Compressor

IR-C Infrared, Range C [from 3μ to 1 mm]

IRCA Immigration Reform and Control Act [US]

IRCA International Radio Club of America

IRCA International Railway Congress Association [Belgium]

IRCA International Remodeling Contractors Association [US]

IRCC Instrument Repair and Calibration Center [Thailand]

IRCC International Radio Consultative Committee [US]

IRCD Infrared Circular Dichroism [Optics]

IRCFE Infrared Communications Flight Experiment

IRCHMB International Research Center on Hydraulic Machinery, Beijing [PR China]

IRCIHE International Referral Center for Information Handling Equipment [of UNESCO]

IRCL Infrared Chemiluminescence

IRCL International Research Center on Lindane [Belgium]

IRCM Infrared Countermeasures

IRCN Institut de Recherche de la Construction Navale [Naval Construction Research Institute, France]

IRCO International Rubber Conference Organization [UK]

IRCOBI International Research Council on the Biokinetics of Impacts [France]

IRCOL Institute for Information Retrieval and Computational Linguistics [Israel]

Ir Comput Irish Computer [Published in Dublin, Ireland]

IRCS International Research Center in Superconductivity [UK]

IRCS International Research Communications System [UK]

IRCSS Interdisciplinary Research Chair in Surface Science [of University of Liverpool, UK]

IRCT Institut de Recherche du Coton et des Textiles Exotiques [Research Institute for Cotton and Exotic Textiles, France]

IRCWD International Reference Center for Water Disposal

IRD Information Requirements Document

IRD Infrared Detector

IRD Infrared Dryer; Infrared Drying

IRD Initiating Reference Document

IRD Integrated Receiver/Descrambler

IRD Internal Research Development

IRD International Resource Development [US]

IRD International Road Dynamics [Saskatoon, Saskatchewan, Canada]

IR&D Independent Research and Development

IR&D Industrial Research and Development

IRDA Industrial Research and Development Authority [UK]

IRDA Infrared Data Association

IRDAC Industrial Research and Development Advisory Committee

IRDB Information Retrieval Databank

IRDB Integrated Regional Data Base [of Australian Bureau of Statistics]

IRDC Infrared Data Committee [Japan]

IRDC International Research Development Center

IRDC International Rubber Development Committee [UK]

IRDF Indentation Residual Deflection Force [Mechanics]

IRDIA Industrial Research and Development Incentives Act [Canada]

IRDL Infrared Diode Laser

IRDO Intermediate Retention of Differential Overlap [Physics]

Irdome Infrared Dome [also irdome]

IRDP Industrial and Regional Development Program [Canada]

IRDP Industrial Research Development Program

IRDR Infrared Double Resonance (Spectroscopy)

IRDS Information Resource Dictionary System

IRDS International Road Documentation Scheme [of Organization for Economic Cooperation and Development, France]

IRDS Primary Irritation Dose [Toxicology]

IRE Infrared Emission

IRE Institute of Radio Engineers [now Institute of Electrical and Electronics Engineers, US]

IRE Institute of Radiophysics and Electronics [Calcutta, India]

IRE Institute of Refractories Engineers [UK]

IRE Institution of Radio Engineers [UK]

IRE Instrument Rating Examiner

IRE Internal Reflectance Ellipsometry

IRE Internal Reflection Element

IRE Internal Reflection Ellipsometry

Ire Ireland

IREB Intense Relativistic Electron Beam

IREC International Rare Earths Conference

IREC Irrigation Research and Extension Advisory Committee [Australia]

IRED Infrared-Emitting Diode

Ired Infrared [also ired]

IREDA International Radio Electrical Distributors Association

IREE Institute of Radio and Electric Engineers [Sydney, Australia]

IREEA Institute of Radio Engineering, Electronics and Automation [USSR]

IREP Integrated Reliability Evaluation Program

IREP Interim Reliability Evaluation Program

IREQ Institut de Recherche d'Hydro Québec [Quebec Hydro Research Institute, Canada]

IRES Infrared Emission Spectroscopy

IREE Institute of Radio Engineering and Electronics [Moscow, Russia]

IRET Interrupt Return

IRE Trans Ultrason Eng IRE Transactions on Ultrasonic Engineering [of Institution of Radio Engineers, UK]

IREX International Research and Exchanges Board [US]

IREX Iron-Ring Experiment

IRF Individual Request for Patent Family [of International Patent Documentation Center]

IRF Industrial Research Fellow

IRF Industrial Research Fellowship [of Natural Sciences and Engineering Research Council, Canada]

IRF Infrared Filter

IRF Infrared Furnace

IRF Input Register Full

IRF Instantaneous Radiative Flux

IRF Intermediate Routing Function

IRF International Road Federation [US]

IRF Interrogation Recurrence Frequency

IRF Inventory Record File

IRFA Institut de Recherches sur les Fruits et Agrumes [Research Institute for Fruits and Tropical and Subtropical Fruits, France]

IRFB International Radio Frequency Board

IRFC Intermediate Range Function Test

IRFITS Infrared Fault Isolation Test System

IRFNA Inhibited Red Fuming Nitric Acid

IRFPA Infrared Focal-Plane Array (Producibility Program) [US Army] [also IR FPA]

IRFPA Infrared Focal Plane Assembly [also IR FPA]

IRFPP Infrared Focal Plane Productivity (Program)

IRFT Inversion-Recovery Fourier Transform

IRG Industry Resources Group [of Auto/Steel Partnership Program, US]

IRG Inertial Rate Gyro

IRG Information Retrieval Group

IRG Infrared Glass

IRG Initial Review Group

IRG Inner Roll Gimbal

IRG Interest Rate Guarantee

IRG International Research Group

IRG Interrange Instrumentation Group

IRG Interrecord Gap [Computers]

IRGA Infrared Gas Analyzer

IRgA International Reprographics Association [US]

IRGCHAZ Intercritically Reheated Grain-Coarsened Heat-Affected Zone [Metallurgy]

IRGRD International Research Group on Refuse Disposal

IRGWP International Research Group on Wood Preservation [Sweden]

IRH Inductive Recording Head

IRH Infrared Heater; Infrared Heating

IRHA International Rural Housing Association

IRHD International Rubber Hardness Degrees

IRI IBEC (International Bank for Economic Cooperation) Research Institute [US]

IRI Industrial Research Institute [US]

IRI Industrial Research Institute [Hiroshima, Japan]

IRI Institution of the Rubber Industry [UK]

IRI Interfaculty Reactor Institute [of Delft University of Technology, Netherlands]

IRI International Robomation/Intelligence

IRI Iranian Rial [Currency]

IRI Islamic Research Institute

IRI Istituto per la Riconstruzione Industriale [Industrial Reconstruction Institute Company, Italy]

IRIA Infrared Information and Analysis Center [US]

I-125 RIA Iodine-125 Radioimmunoassay [also ^{125}I RIA]

IR-ICBM Infrared Intercontinental Ballistic Missile

IRICON Infrared Vidicon Tube

Irid Iridescence; Iridescent [also irid]

Iridesc Iridescence [also Iridesc, or iridesc]

IRIG Inertial Rate Integrating Gyro; Inertial Reference Integrating Gyro

IRIG Interrange Instrumentation Group [of US Department of Defense]

IRIG-B Interrange Instrumentation Group B [of US Department of Defense]

IRII Industrial Research Institute of Ishikawa [Kanazawashi, Japan]

IRIP Idaho Research Initiative Program [US]

IRIP Industrial Research Institutes Program

IRIRDR Infrared-Infrared Double Resonance
IRIS Imaging of Radicals Interacting with Surfaces
IRIS Industrial Relations Information Service
IRIS Information, Robotics and Intelligent Systems [US]
IRIS Infrared Information Symposia [US]
IRIS Infrared Interferometer Spectrometer
IRIS Inorganic Rings International Symposium
IRIS Instant Response Information System
IRIS Institute for Robotics and Intelligent Systems [Canada]
IRIS Instructional Resources Information System [of Ohio State University, Columbus, US]
IRIS Integrated Radar Imaging System
IRIS Integrated Reconnaissance Intelligence System
IRIS Interactive Radar Information System
IRIS Interactive Recorded Information Service [UK]
IRIS Interactive Remote Instructional System [of Wright State University, Dayton, Ohio, US]
IRIS International Research Information Service [US]
IRIS Interrogation and Location
IRIS Italian Research Interim Stage
IRIS-PRECARN Institute for Robotics and Intelligent Systems/Precarn Associates Inc. (Conference) [Canada]
IrisSOLE Iris Systems Operator Loading Evaluation
IRIT Infrared Imaging Tube
IRJE Interactive Remote Job Entry
Ir J Food Sci Technol Irish Journal of Food Science and Technology
IRL Industrial Research Laboratory
IRL Information Retrieval Language
IRL Infrared Lamp
IRL Infrared Laser
IRL Initiating Reference Letter
IRL Interactive Reader Language
IRL Interface Requirement List
IRL Inter-Repeater Link
IRL Irreversibility Line
IRLA Independent Research Libraries Association [US]
IRLAP Infrared Link Access Protocol
IRLCO International Red Locust Control Organization
IRLCO-CSA International Red Locust Control Organization for Central and Southern Africa [Zambia]
IRLDS Infrared Linear Dichroism Spectroscopy
IRLED Infrared Light-Emitting Diode [or IR-LED]
IR-LESR Infrared Light-Induced Electron Spin Resonance [Physics]
IRLS International Red Locust Control Service
IRLS Interrogation, Recording and Location Subsystem
IRLS Interrogation, Recording and Location System [of NASA]
IRM Image Reject(ion) Mixer
IRM Induction Reductance Motor
IRM Information Recording Material
IRM Information Recording Medium
IRM Information Resources Management [Westinghouse Hanford Company, Washington State, US]

IRM Infrared Maser
IRM Infrared Measurement
IRM Infrared Microscope; Infrared Microscopy
IRM Inherent Rights Mask
IRM Inorganic Reaction Mechanism
IRM Institute for Resource Management
IRM Institute of Risk Management [UK]
IRM Interim Research Memorandum
IRM Intermediate Range Monitor
IRM International Research Monitoring
IRM International Resource Management [US]
IRM Iodine Radiation Monitor
IRM Isothermal Remanent Magnetization
IRM Isotopic Reference Material
IRMA Immunoradiometrical Assay
IRMA Indian Refractory Makers Association [India]
IRMA Information Resources Management Association [US]
IRMA Information Revision and Manuscript Assembly
IRMA Infrared Miss-Distance Approximeter
IRMA International Rail Makers Association
IRMB Institute Royal Météorologique de Belgique [Royal Meteorological Institute of Belgium, Brussels]
IRMC Information Resources Management Council [of National Institutes of Health, Bethesda, Maryland, US]
IRMC International Radio-Maritime Committee [UK]
IRMCFC Internal Reforming Molten Carbonate Fuel Cell
IRME Initiator Resistance Measuring Equipment
Ir Med J Irish Medical Journal
IRMI Industrial Research Materials Institute
IRMI International Risk Management Institute
IRMIS Integrated Resource Management Information System [Canada]
Ir-Mn Iridium Manganese (Alloy)
IrMn SV Iridium Manganese Spin Valve
IRMP Infrared Measurement Program
IRMP Infrared Multiphoton
IRMP Interservice Radiation Measurement Program
IRMPA Infrared Multi-Photon Photon Absorption
IRMPD Infrared Multi-Photon Decomposition
IRMPD Infrared Multi-Photon Dissociation
IRMPE Infrared Multi-Photon Excitation
IRMS Information Retrieval and Management System
IRN Interface Revision Notice
IRN Internal Routing Network
IRN International Rivers Network [US]
Ir-Nb Iridium-Niobium (Alloy)
IRNP Identifying Research Needs Program [of American Society of Mechanical Engineers, US]
IRO Industrial Relations Office(r)
IRO Intermediate Range Order(ing) [Solid-State Physics]
IROD Instantaneous Readout Detector
IRODP International Registry of Organization Development Professionals [US]

IROE Istituto di Ricerca sulle Onde Elettromagnetiche [Research Institute for Electromagnetic Waves, Florence, Italy]

Iron Age, Met Prod Iron Age, Metals Producer [Journal published in the US]

Iron Steel (China) Iron and Steel (China) [Journal published in PR China]

Iron Steel Eng Iron and Steel Engineer [Journal of the Association of Iron and Steel Engineers, US]

Iron Steel Ind Iron and Steel Industry [Journal published in Japan]

Iron Steel Inst Iron and Steel Institute [UK]

Iron Steel Inst Spec Rep Iron and Steel Institute Special Report [UK]

Iron Steel Int Iron and Steel International [UK]

Ironmaking Conf Proc Ironmaking Conference Proceedings [of AIME Iron and Steel Society, US]

Ironmaking Steelmaking Ironmaking and Steelmaking [Journal of the Institute of Materials, UK]

Iron Steelmaker Iron and Steelmaker [Journal of the AIME Iron and Steel Society, US]

IROR Incremental Rate of Return

IROR Internal Rate of Return

IROS Increased Reliability Operational System

IROS Instant Response Order System [US]

IRP Image Reconstruction Procedure

IRP Infrared Photography

IRP Infrared Pyrometry

IRP Initial Receiving Point

IRP Innovative Research Project [of NASA]

IRP Integrated Resource Planning

IRP International Research Program

IRP International Routing Plan

IRP Intrinsic Reaction Path

IRP Islamic Republic of Pakistan

IR£ Irish Pound [Currency of Irish Republic]

IRPA Irrigation Pump Administration [Philippines]

IRPA International Radiation Protection Association [France]

IRPA Bull IRPA Bulletin [of International Radiation Protection Association, France]

IRPAS Infrared Photoacoustic Spectroscopy

IRPBDS Infrared Photothermal Beam Deflection Spectroscopy

IRPC Infrared Photocell

IRPC Infrared Photoconductor

IRPD Infrared Photodesorption

IRPG Interactive Research Projects Grants

IRPI Individual Rod Position Indicator

IRPL Infrared Photoluminescence [also IR PL]

IRPM Infrared Physical Measurement

IRPME Infrared Phase Modulated Ellipsometer; Infrared Phase Modulated Ellipsometry

IRPP Institute for Research on Public Policy

IRPP Integral Rupture Probability Density [Mechanics]

IRPS International Reliability Physics Symposium [of Institute of Electrical and Electronics Engineers, US]

Ir-Pt Iridium-Platinum (Alloy)

IRPTC International Register of Potentially Toxic Chemicals [of United Nations Environmental Program]

IRQ Interrupt Request

IRR Infrared Radiation

IRR Infrared Radiometer

IRR Infrared Rays

IRR Infrared Reflectance

IRR Infrared Retina

IRR Inspection Rejection Report

IRR Institute for Reactor Research [Switzerland]

IRR Integral Rocket Ramjet

IRR Internal Rate of Return

IRR Interrupt Return Register

IRR Iranian Rial [Currency]

IRR Irritant Effects (of Hazardous Materials)

IRR Israel Research Reactor

irr irradiated

IRRA Industrial Relations Research Association [US]

IRRAD Infrared Range and Direction (Equipment)

IRRAD Infrared Range and Direction Detection

Irrad Irradiation [also irrad]

IRRAMP Infrared Analysis, Measurements and Modeling Program [of US Navy]

IRRAS Infrared Reflection-Absorption Spectroscopy [also IR-RAS]

IRRB International Rubber Research Board

IRRC International Rice Research Center [Philippines]

IRRD International Road Research Documentation (Database) [of Organization for Economic Cooperation and Development, France]

IRRDB International Rubber Research and Development Bureau

IRRDB International Rubber Research Development Board [UK]

irreg irregular(ly)

IRREP Irreducible Representation [Mathematics]

IRRI International Rice Research Institute [Manila, Philippines]

IRRL Information Retrieval Research Laboratory [of University of Illinois, US]

IRRMP Infrared Radar Measurement Program

IRRO Information Resource for the Release of Organisms into the Environment [of United Nations Environmental Program]

IRRS Infrared Reflectance Spectroscopy; Infrared Reflection Spectroscopy

IR-RS Infrared Reflow–Solderable

IRRT Institution of Rail and Rapid Transit [UK]

Ir-Ru Iridium-Ruthenium (Alloy)

IRS Independent Research Service

IRS Industrial Robot System

IRS Inertial Reference System

IRS Infinitely Rigid System

IRS Information Receiving Station

IRS Information Retrieval Service
IRS Information Retrieval System
IRS Infrared Reflectance Spectroscopy
IRS Infrared Scan(ner)
IRS Infrared Soldering
IRS Infrared Source
IRS Infrared Spectro(photo)meter
IRS Infrared Spectrometry; Infrared Spectroscopy
IRS Infrared Spectrum
IRS Inquiry and Reporting System
IRS Institut für Regionalentwicklung und Strukturplanung e.V. [Institute for Regional Development and Infrastructure Planning, Erkner, Germany]
IRS Insulated Return System
IRS Interchange Record Separator
IRS Intermediate Reference System
IRS Internal Reflectance Spectroscopy; Internal Reflection Spectroscopy
IRS Internal Revenue Service [US]
IRS International Referral System (Database) [of United Nations Environmental Program]
IRS International Repeater Station
IRS Interrecord Separator [Computers]
IRS Inverse Raman Spectroscopy
IRS Investment Removal Salt (for Heat Treating) [Metallurgy]
IRS Irrigation Research Station [New Zealand]
IRS Isotope Removal Service
IRS Isotope Removal System
IrS Iridium Monosulfide
IRSA Infrared Science Archive [of Infrared Processing and Analysis Center, NASA Jet Propulsion Laboratory, Pasadena, California, US]
IRSA Institute for Remote Sensing Applications
IRSA International Radiator Standards Association [US]
IrSbSe Iridium Antimony Selenide (Semiconductor)
IRSC Inter-Regional Subject Coverage
IRSCS Inter-Regional Subject Coverage Scheme
IRSE Infrared Spectroscopic Ellipsometry [also IR-SE]
IRSE Institution of Railway Signal Engineers [London, UK]
IRSFC International Rayon and Synthetic Fibers Committee [France]
IRSG Information Retrieval Specialists Group [of British Computer Society]
IRSG International Rubber Study Group [UK]
IRSG Internet Research Steering Group
IrSi Iridium Silicide
IRSIA Institut pour l'Encouragement de la Recherche Scientifique dans l'Industrie et l'Agriculture [Institute for the Promotion of Scientific Research in Industry and Agriculture, Belgium]
IRSID Institut de Recherches de la Sidérurgie [Research Institute of the Iron and Steel Industry, Saint Germain, France]
IRSJ Infrared Society of Japan
IRSO Information Resources Security Officer

IRSO Institute of Road Safety Officers [UK]
IRSP Infrared Spectrophotometry
IR Spect Infrared Spectrometer
IRSR Infrared Spectroradiometry
IRSS Inertial Reference Stabilization System
IRSS Intelligent Remote Station Support
IRST Infrared Search and Track (System) [of US Navy]
IRST Istituto per la Ricerca Scientifica e Tecnologica [Institute for Scientific and Technical Research, Trent, Italy]
IRSTD Infrared Search and Target Designation System
IRSTS Infrared Search-Track System
IRT Index Return (Character)
IRT Infrared Telescope
IRT Infrared Television
IRT Infrared Thermometer
IRT Infrared Tracker
IRT Infrared Transducer
IRT In-Reactor Thimble [Nuclear Engineering]
IRT Institut de Recherche des Transports [Institute for Transportation Research, France]
IRT Institut für Rundfunktechnik [Radio Engineering Institute, Hamburg, Germany]
IRT Institute of Reprographic Technology [UK]
IRT Interrogator-Responder-Transponder
IRT Interrupted Ring Tone
IRTA Independent Research Training Award [US]
IRTA International Reciprocal Trade Association [US]
IRTA Irreducible Tensor Analysis [Mathematics]
IRTC Infrared Thermocouple
IRTC International Road Tar Conference
IRTC International Round Table Conference
IRTCES International Research Training Center for Erosion and Sedimentation
IRTCM Integrated Real-Time Contamination Monitor
IRTE Institute of Road Transport Engineers [UK]
IRTEMP Istituto di Ricerca e Tecnologia delle Materie Plastiche [Institute for the Research and Technology of Plastic Materials, Naples, Italy]
IRTF Internet Research Task Force
IRTO International Radio and Television Organization [UK]
IRTP Infrared Target Processor
IRTS Infrared Temperature Sounder
IRTS Interim Recovery Technical Specification
IRTS International Radio and Television Society [US]
IRTS Irish Radio Transmitters Society [UK]
IRTU Industrial Research and Technology Unit [Northern Ireland]
IRTU Intelligent Remote Terminal Unit
IRTU International Radio and Television University
IRTV Information Retrieval Television [US]
IRTWG Interrange Telemetry Working Group
IRU Inertial Reference Unit
IRU Information Research Unit
IRU Intel Real-Time Users Group [also iRUG]

IRU International Radium Unit

IRU International Raiffeisen Union [Germany]

IRU International Reference Unit

IRU International Road Transport Union [Switzerland]

IRUS Intrusion-Resistant Underground Structure [of Atomic Energy of Canada Limited at Chalk River, Ontario]

IR-UT Injection Refining with Temperature Raising (Process)

IR/UV Infrared/Ultraviolet

IR-UV-Vis Infrared-Ultraviolet-Visible [also IR-UV-vis]

IRV Infrared Vidicon

IRV Interrupt Request Vector

IRV Isotope Reentry Vehicle

IRVICON Infrared-into-Visible Converter [also Irvicon]

IR-Vis Infrared-to-Visible (Upconversion) [also IR-vis]

IRVR Instrumented Runway Visual Range

IRW Indirect Reference Word

IRW Institute of Rural Water

IRW Internationale Fachmesse für Reinigung und Wartung [International Trade Fair for Cleaning and Maintenance, Cologne, Germany]

IRW Inverted Rib Waveguide

IRWA International Right of Way Association [US]

IRX Information Retrieval Experiment

IRX Interactive Resource Executive

Ir-Zr Iridium-Zirconium (Alloy)

IS Iceland [ISO Code]

IS Ice Sheet

IS Ideal Solid

IS Ignition Switch

IS Image Space [Optics]

IS Image Storage

IS Imaging Science

IS Imaging System

IS Immersion Scan(ning)

IS Immunosuppression; Immunosuppressive

IS Impact Strength [Mechanics]

IS Incoherent Scattering

IS Incomplete Sequence (Relay)

IS Indentation Strength (Method) [Metallurgy]

IS Indexed Sequential

IS Indian Standard

IS Induction Soldering

IS Inductive Sensor

IS Industrial Society [London, UK]

IS Industrial Source

IS Industry Standard

IS Inelastic Scattering

IS Inertial System

IS Information Science; Information Scientist

IS Information Separation; Information Separator

IS Information Service

IS Information Storage

IS Information System

IS Infrasonic(s)

IS Infrasound

IS Inorganic Semiconductor

IS Input Signal

IS Input System

IS In Service

IS Installation Start

IS Installation Support

IS Institute of Science

IS Institute of Statisticians [UK]

IS Institut Solvay [Solvay Institute, France]

IS Institut Suisse (de la Promotion de la Sécurité) [Swiss Institute for the Promotion of Safety and Security]

IS Instruction Set

IS Instruction Sheet

IS Insulated System

IS Intake Stroke

IS Integrated System

IS Integrating Sphere [Optics]

IS (Office of) Intelligence and National Security [US]

IS Intelligent System

IS Intensifying Screen

IS Interactive System

IS Interference Spectroscope; Interference Spectroscopy

IS Interference Spectrum

IS Interference Suppressor

IS Interferometric Spectrometer; Interferometric Spectrometry

IS Interferometric Spectroscope; Interferometric Spectroscopy

IS Intragranular Slip [Metallurgy]

IS Interlamellar Spacing [Metallurgy]

IS Intermediate School

IS Intermediate State

IS Intermediate System [Internet]

IS Internal Shield [Electronics]

IS Internal Standard

IS International Standard

IS Internal Storage

IS Internal Stress [Mechanics]

IS Internal Structure

IS International Symposium

IS Interrupt Signal [Computers]

IS Interrupt Status

IS Interstellar Space

IS Interstitial Solution

IS Interval Signal

IS Intragranular-Dislocation Strain [Materials Science]

IS Intrazonal Soil [Geology]

IS Intrinsic Safety

IS Intrinsic Semiconductor

IS Involute Spline

IS Ion-Beam Synthesis

IS Ionization Spectrometer; Ionization Spectrometry

IS Ionization Spectroscope; Ionization Spectroscopy

IS Ion Scattering

IS Ion Source

IS Ion Spectroscopy

IS Ion Spectrum

IS Ion Storage

IS Inelastic Scattering

IS Inquiry Station

IS Isobaric Spin [Nuclear Physics]

IS Isolating Switch

IS Isomer(ic) Shift [Physical Chemistry]

IS Isostatic(s)

IS Isotopic Spin [Nuclear Physics]

IS Israeli Shekel [Currency]

I(S) Integrated Intensity [Symbol] [*Note:* "S" Represents Reflection]

I&S Inspection and Survey

Is Island [also is]

Is Isle [also is]

.is Iceland [Country Code/Domain Name]

ISA Image Sequence Analysis

ISA Imaging Sensor Autoprocessor

ISA Independent Scholar of Asia

ISA Independent Signcrafters of America [US]

ISA Industry Standard Architecture

ISA Information Science Abstracts [of American Chemical Society, US]

ISA Information System Access

ISA Information Systems Association [US]

ISA Inorganic Sampling and Analysis

ISA Input Spooling Area [Computers]

ISA Instruction Set Architecture

ISA Instrument Society of America [Name Changed to International Society for Measurement and Control, US]

ISA Instrument Subassembly

ISA Interim Stowage Assembly

ISA Intersecting Storage and Acceleration [Nuclear Physics]

ISA Internal Standard Addition

ISA International Seabed Authority [of United Nations]

ISA International Schools Association

ISA International Silk Association

ISA International Silo Association [US]

ISA International Society for Measurement and Control [Formerly Instrument Society of America]

ISA International Society of Appraisers [US]

ISA International Society of Arboriculture [US]

ISA International Standard Atmosphere

ISA International Studies Association [US]

ISA International Sugar Agreement

ISA International Symposium on Aerogels

ISA Interrupt Storage Area

ISA Ion Scattering Analysis

ISA Islamic Shipowners Association

ISAAS Indian Society for Afro-Asian Studies

ISAB Institute for the Study of Animal Behavior [UK]

ISAB International Scholastic Advisory Bureau [UK]

ISABGR International Society for Animal Blood Group Research [Australia]

ISAC Industrial Sector Advisory Committee [US]

ISAC International Scientific Agricultural Council

ISAC International Security Affairs Committee [US]

ISAD Information Science and Automation Division [of American Library Association, US]

ISADS Innovative Strategic Aircraft Design Studies

ISA EXPO International Society for Measurement and Control Exposition

ISAE Indian Society of Agricultural Economics

ISAF Institut für Schweißtechnik und Trennende Fertigungsverfahren [Institute for Welding Engineering and Cutting Processes, TU Clausthal, Germany]

ISAF Intermediate Super-Abrasion Furnace (Black)

ISAF International Symposium on Applications of Ferroelectrics [of Institute of Electrical and Electronics Engineers, US]

ISAF Black Intermediate Super-Abrasion Furnace Black [also ISAF black]

ISAGA International Simulation and Gaming Association [US]

ISAGUG International Software AG Users Group [US]

ISAH Integrated System for Automated Hydrography

ISA/IEC Instrument Society of America/International Electrotechnical Commission [now International Society for Measurement and Control/International Electrotechnical Commission]

ISAL Information System Access Line

ISAM Indexed Sequential-Access Method

ISAM Integrated Switching and Multiplexing

ISAM International Society for Aerosols in Medicine [Austria]

ISAMA International Scientific Association for Micronutrients in Agriculture

ISAMS Improved Stratospheric and Mesospheric Sounder

I/S Anal I/S Analyzer [Publication of United Communications Group, Bethesda, Maryland, US]

ISAO International Society for Artificial Organs

ISAP Information Sort and Predict

ISAP International Society for Asphalt Pavements

ISAPI Internet Server Application Program Interface

ISAR Information Storage and Retrieval

ISAR International Society for Astrological Research

ISAR Inverse Synthetic Aperture Radar

ISARD International School for Agriculture and Resource Development

ISAS Illinois State Academy of Science [Springfield, US]

ISAS Institut für Spektrochemie und angewandte Spektroskopie [Institute for Spectrochemistry and Applied Spectroscopy, Dortmund, Germany]

ISAS Institute of Space and Aeronautical Science [of University of Tokyo, Japan]

ISAS Institute of Space and Astronautical Science [Kanagawa, Japan]

ISAS Institute of Space and Atmospheric Science [Canada]

ISAS International School for Advanced Studies [Trieste, Italy]

ISAS International Schools Association [Switzerland]

ISAS International Screen Advertising Association

ISAS International Society of African Scientists [US]

ISAS Intelligent Systems and Semotics [of National Institute for Standards and Technology, US]

ISASD International Symposium on Aeroelastics and Structural Dynamics

ISASI International Society of Air Safety Investigators

ISASST International School on the Application of Surface Science Techniques

ISAST International Society for the Arts, Sciences and Technology [US]

ISAT Interrupt Storage Area Table

ISATA International Symposium on Advanced Transportation Applications

ISATA International Symposium on Automotive Technology and Automation

ISA Trans ISA Transactions [Published by the International Society for Measurement and Control (formerly known as the Instrument Society of America)]

ISAW International Society of Aviation Writers

ISB Indentation-Strength-in-Bending (Method) [Mechanics]

ISB Independent Sideband

ISB Information Systems Branch

ISB Information Systems Building

ISB Infrared Security Barrier

ISB Institute of Small Business [UK]

ISB International Society of Biomechanics

ISB International Society of Biometeorology [Switzerland]

ISB International Society of Biorheology [Germany]

ISB International Symposium on Bioceramics

ISBA Incorporated Society of British Advertisers [UK]

ISBA Indiana School Board Association [US]

ISBA International Sea-Bed Authority [of United Nations]

ISBA International Society of British Advertisers

IS:BA Information Storage : Basic and Applied [Section of the Journal of Magnetism and Magnetic Materials]

ISBB International Society of Bioclimatology and Biometeorology

ISBC International Small Business Congress

ISBD International Standard Bibliographic Description

ISBF Interactive Search of Bibliographic Files

ISBL Information System Base Language

ISBN International Standard Book Number

ISBO Islamic States Broadcasting Organization [Saudi Arabia]

ISBP International Society for Biochemical Pharmacology

ISBP International Symposium on Biological Physics

ISBS Integrated Small Business Software

ISC Idle Speed Control (Motor)

ISC Indiana State College [Terre Haute, US]

ISC Indirect Semiconductor

ISC Information Society of Canada

ISC Information Systems Committee [also International Organization for Plant Information]

ISC Initial Slope Circuit

ISC Initial Software Configuration (Map) [also isc]

ISC Instruction Set Computer

ISC Instruction Staticizing Control

ISC Integrated Storage Control

ISC Intercompany Services Coordination

ISC International Seismological Center [UK]

ISC International Sericultural Commisson [France]

ISC International Service Carrier

ISC International Society for Chronobiology [US]

ISC International Society of Chemotherapy [Germany]

ISC International Society of Citriculture [US]

ISC International Society of Cryptozoology [US]

ISC International Standards Council

ISC International Student Conference

ISC International Sugar Council

ISC International Switching Center

ISC Inter-State Commission [US]

ISC Intersystem Communication

ISC Iowa Safety Council [US]

ISC ISOLDE (Isotope Separator On Line Device) Experiments Committee [of CERN–European Laboratory for Particle Physics, Geneva, Switzerland]

ISCA Independent Safety Consultants Association [UK]

ISCA Industrial Specialty Chemical Association [US]

ISCA International Standards Coordination Association

ISCA International Standards Steering Committee for Consumer Affairs

ISCA Inter-Society Committee on Methods for Ambient Air Sampling and Analysis

ISCAC International Superconductor Applications Convention

ISCAN Inertialess Steerable Communication Antenna

ISCANI International Symposium on Cluster and Nanostructure Interfaces

ISCAS International Symposium on Circuits and Systems [of Institute of Electrical and Electronics Engineers, US]

ISCB International Society for Clinical Biostatistics [UK]

ISCB International Society of Cell Biology

ISCC Intercrystalline Stress Corrosion Cracking

ISCC Intergranular Stress Corrosion Cracking

ISCC International Semiconductor Conference

ISCC International Semiconductor Converter Conference [of Institute of Electrical and Electronics Engineers, US]

ISCC International Service Coordination Center

ISCC International Strata Control Conference

ISCC Inter-Society Color Council [US]

ISCC Inter-Society Committee on Corrosion

ISCC Interstate Solar Coordination Council [US]

ISCC-NBS Inter-Society Color Council–National Bureau of Standards (Color Name) [now ISCC-NIST]

ISCC-NIST Inter-Society Color Council–National Institute of Standards and Technology (Color Name) [formerly ISCC-NBS]

ISCCP International Satellite Cloud Climatology Project

ISCD International Symposium on Chiral Discrimination

ISCDS International Stop Continental Drift Society [US]

ISCE International Society of Chemical Ecology [US]

ISCED International Standard Classification of Education [of UNESCO]

ISCE Newsl ISCE Newsletter [of International Society of Chemical Ecology, US]

ISCES International Society of Complex Environmental Studies

ISCES International Symposium on Condensation and Evaporation of Solids

ISCET International Society of Certified Electronics Technicians [US]

ISCEV International Society for Clinical Electrophysiology and Vision

ISCFB International Society of Craniofacial Biology

ISCIE Institute of Systems, Control and Information Engineers [Kyoto, Japan]

ISCLC International Symposium of Column Liquid Chromatography

ISCLT International Society of Clinical Laboratory Technologists

ISCM International Society of Cybernetic Medicine

ISCMA International Superphosphate and Compound Manufacturers Association

ISCO International Soil Conservation Organization

ISCO International Standard Classification of Occupation

ISCOM Indian Satellite for Communication Technology [India]

ISCOR Iron and Steel Corporation [South Africa] [also Iscor]

ISCOTT Iron and Steel Company of Trinidad and Tobago [also Iscott]

ISCP Improved Self-Consistent Phonon (Theory) [Physics]

ISCP International Society for Clinical Pathology

ISCR Institute of Scientific Computing Research [of Lawrence Livermore National Laboratory, University of California at Livermore, US]

ISCRE International Symposium on Chemical Reaction Engineering

ISCRP International Society of City and Regional Planners

ISCS (Department of) Information Systems and Computer Science [of National University of Singapore]

ISCS International Scientific Cooperative Service

ISCS International Symposium on Compound Semiconductors

ISCT Inner Seal Collar Tool

ISCT Intersublattice Charge Transfer [Materials Science]

IScT Institute of Science and Technology

ISC/USO Intercompany Services Coordination/Universal Service Order

ISD Image Section Descriptor

ISD Induction System Deposit

ISD Information Structure Design

ISD Information System Development

ISD Information Systems Division

ISD Initial Selection Done

ISD Institute for Security Design [New York, US]

ISD Institute of Sustainable Development

ISD Instructional Systems Design (Process)

ISD Instructional Systems Development [of US Air Force]

ISD Intermediate Storage Device

ISD Internal Symbol Dictionary

ISD International Society of Differentiation [US]

ISD International Subscriber Dialing

ISD International Symbol Dictionary

ISD Ion-Stimulated Desorption

IS&D Integrate Sample and Dump

ISDA International Systems Dealers Association [US]

ISDB International Society of Developmental Biologists [France]

ISDD Information Systems Development Division

ISDD Integrated Systems Development Department

ISDF Intercrystalline Structure Distribution Function [Metallurgy]

ISDF Intermediate Sodium Disposal Facility

ISDG Information Science Discussion Group [UK]

ISDN Integrated Services Digital Network

ISDN International Society for Developmental Neuroscience [US]

ISDN International Standard Data Network

ISDOS Information System Design by Optimization System

ISDS Inadvertent Separation and Destruct System

ISDS Integrated Ship Design System [US Navy]

ISDS Integrated Software Development System

ISDS Integrated Switched Data Service

ISDS International Serials Data System [of United Nations Information System in Science and Technology]

ISDSI Insulated Steel Door Systems Institute [US]

ISDSN Integrated Services Digital Satellite Network

ISDT Institute of Shaft Drilling Technicians [US]

ISDT Integrated Services Digital Terminal

ISDW Incommensurate Spin-Density Wave [also i SDW, or I-SDW] [Solid-State Physics]

ISDW-CSDW Incommensurate Spin-Density Wave–Commensurate Spin-Density Wave [Solid-State Physics]

ISDW-P Incommensurate Spin-Density Wave–Paramagnetic (Transition) [Solid-State Physics]

ISDX Integrated Services Digital Exchange

ISE Indentation Size Effect [Materials Science]

ISE Induced Secondary Electron

ISE Inelastic Scattering of Electrons

ISE Initial-State Energy

ISE Institute for Software Engineering [now Institute for Information Management, US]

ISE Institute of Space Engineering [Canada]

ISE (Fraunhofer-)Institut für Solare Energiesysteme [(Fraunhofer) Institute for Solar Energy Systems, Freiburg, Germany]

ISE Institution of Sanitary Engineers [UK]

ISE Institution of Structural Engineers [UK]

ISE In-System Evaluator

ISE Integral Square Error [Control Engineering]

ISE Interface Science and Engineering (International Conference)

ISE IGM (Intergovernmental Meeting) of Scientific Experts on Biological Diversity

ISE International Society of Electrochemistry [Austria]

ISE International Society of Electrostimulation

ISE International Stock Exchange

ISE International Submarine Engineering

ISE Interrupt System Enable

ISE Intersystem Emulator

ISE Inverse Square Exchange [Physics]

ISE Inverse Stark Effect [Spectroscopy]

ISE Ion-Selective Electrode

ISE Ion Separation Exchange

ISE Ion-Specific Electrode

ISE Isolated Silicon Epitaxy (Process) [Solid-State Physics]

I&SE Installation and Service Engineering,

ISEA Industrial Safety Equipment Association [US]

ISEA International Society of Exposure Analysis

ISEARCH Information Services Electronic Archives

ISEC International Solvent Extraction Conference

ISEC International Statistics Educational Center

ISEC International Superconductor Electronics Conference

ISEC International Symposium on Engineering Ceramics

ISECSI International Society for Educational, Cultural and Scientific Interchanges

ISEE International Society for Environmental Education

ISEE International Society for Environmental Ethics

ESEE International Society for Eyesight Education

ISEE International Sun-Earth Explorer

ISEEL Inner Shell Electron Energy Loss [Physics]

ISEELS Inner Shell Electron Energy Loss Spectroscopy

ISEL Institute of Shipping Economics and Logistics [Germany]

ISEM International Society for Ecological Modeling [Denmark]

ISEM Irish Society for Electron Microscopy

ISEO Institute of Shortening and Edible Oils [US]

IS-EOS International Symposium on Electro-Organic Synthesis

ISEP International Scientific Exchange Program [of NATO]

ISEP International Society for Educational Planning

ISEP International Society for Evolutionary Protistology [UK]

ISEP International Standard Equipment Practice

ISEP International Symposium on Environmental Pollution

ISEP Ion Separation Exchange Process

ISEPP International Sun-Earth Physics Program

ISEPS International Sun-Earth Physics Satellites Program

ISER Institute of Social and Economic Research [of Memorial University of Newfoundland, St. John's, Canada]

ISER Integral Systems Experimental Requirements

ISES International Ship Electric Service Association [UK]

ISES International Society of Explosives Specialists

ISES International Solar Energy Society [Australia]

ISES International Solar Energy Society [US]

ISER International Symposium on Experimental Robotics

ISESCO Islamic Educational, Scientific and Cultural Organization

ISET Institut für Solare Energieversorgungstechnik [Institute for Solar Energy (Supply) Technology, Cassel, Germany]

IS&EU International Stereotypers and Electrotypers Union

ISF Imperial Smelting Furnace

ISF Indexed Sequential File

ISF Individual Store and Forward

ISF Industrial Space Facility

ISF Information Systems Factory

ISF Instituto i Scienze Fisiche [Physical Sciences Institute, Italy]

ISF Interaction Site Formalism [Physics]

ISF International Science Foundation [Washington, DC, US]

ISF International Shipping Federation [UK]

ISF International Society of Fat Research

ISF Intrinsic Stacking Fault [Materials Science]

ISFA Institute of Shipping and Forwarding Agents [UK]

ISFA International Scientific Film Association

ISFC International Scholarship Fund Committee

ISFD Integrated Software Functional Design

ISFE International Societies of Flying Engineers [US]

ISFET Ion-Selective Field-Effect Transistor; Ion-Sensitive Field-Effect Transistor

ISFL International Scientific Film Library

ISFM Indexed Sequential File Manager

ISFMC International Symposium on Fracture Mechanics of Materials

ISFMS Indexed-Sequential File Management System

ISFNT International Symposium on Fusion Nuclear Technology [Japan]

ISFPP International Symposium on Fine Particles Processing

ISFR International Society for Fluoride Research [US]

ISFSI Independent Spent Fuel Storage Installation [Nuclear Engineering]

ISFSI International Society of Fire Service Instructors

ISG Imperial Standard Gallon

ISG Inland Shipping Group [UK]

ISG Instrumentation Selection Guide

ISG Integrated Survey Grid

ISG Inter-Subblock Gap [Computers]

ISG Isolated Ground

ISGA International Study Group for Aerograms

ISGE International Society for Geothermal Engineering

ISGI International Service of Geomagnetic Indices

ISGMP Institut de Science et Génie des Matériaux et Procédés [Institute for Materials and Processes Science and Engineering , Font-Romeu, France]

ISGMP Institut Supérieur de Génie Mécanique et Productique [Higher Institute for Mechanical and Production Engineering, Université de Metz, France]

ISGN International Symposium on Gallium Nitride (and Related Materials)

ISGO International Society of Geographic Ophthalmology

ISGOTT International Safety Guide for Oil Tankers and Terminals

ISGS Illinois State Geological Survey [US]

ISGSH International Study Group on Steroid Hormones

ISGSR International Society for General Systems Research [now International Society for the Systems Sciences, US]

ISH Information Super Highway

ISH In-situ Heating

ISH In-situ Hybridization [DNA Research]

ISH International Society of Hematology

ISH Internationale Fachmesse Sanitär–Heizung–Klima [International Trade Fair: Sanitation, Heating and Air Conditioning, Germany]

ISHAM International Society of Human and Animal Mycology

ISHC International Society of Heterocyclic Chemistry

Ishikawajima-Harima Eng Rev Ishikawajima-Harima Engineering Review [Published by Ishikawajima-Harima, Tokyo, Japan]

ISHM International Society of Hybrid Microelectronics [Reston, Virginia, US]

ISHM International Symposium on Hybrid Microelectronics

ISHM JTC International Society of Hybrid Microelectronics Joint Technology Conference

ISHRA Iron and Steel Holding and Realization Agency [UK]

ISHS Idaho State Historical Society [US]

ISHS Institute for Self-Proagating High-Temperature Synthesis [of School of Ceramic Engineering and Sciences, New York State College of Ceramics, Alfred University, US]

ISHS International Society for Horticultural Science [Netherlands]

ISI Indian Standards Institution [India]

ISI Indian Statistical Institute [New Delhi, India]

ISI Initial Spread Index

ISI Initial Systems Installation

ISI In-Service Inspection

ISI Institute for Scientific Information [Philadelphia, Pennsylvania, US]

ISI Institut für Schicht- und Ionentechnik [Institute for Film and Ion Technology, Forschungzentrum Julich, Germany]

ISI (Fraunhofer-)Institut für Systemtechnik und Innovationsforschung [(Fraunhofer) Institute for Systems Engineering and Innovation Research, Karlsruhe, Germany]

ISI Instrumentation Support Instruction

ISI Internally Specified Index

ISI International Safety Institute

ISI International Standards Institute

ISI International Statistical Institute [Netherlands]

ISI Intersymbol Interference

ISI Ion Signal for Imaging

ISI Iron and Steel Institute [UK]

ISI Israel Standards Institute

ISIA International Snowmobile Industry Association [US]

ISIAQ International Society of Indoor Air Quality

ISIC International Solvay Institute of Chemistry [France]

ISIC International Standard Industrial Classification [of United Nations]

ISIC International Symposium on Intelligent Control

ISICCE International Society of Indian Chemists and Chemical Engineers

ISIC International Symposium on Intelligent Control/ International Symposium on Computational Intelligence in Robotics and Automation/Intelligent Systems and Semotics [of National Institute for Standards and Technology, US]

ISID International Society of Interior Designers [US]

ISIF Intermediate Standard Transfer Format

ISIF International Symposium on Integrated Ferroelectrics

ISIFM International Society of Industrial Fabric Manufacturers [US]

ISHMFA International Symposium on Hydrodynamics of Magnetic Fluids and its Applications

ISI/ISTP&B Institute for Scientific Information/Index to Scientific and Technical Proceedings and Books [US]

ISIJ Iron and Steel Institute of Japan [Tokyo, Japan]

ISIJ Int ISIJ International [Journal of the Iron and Steel Institute of Japan]

ISIL Interim Support Items List

ISILT Information Science Index Language Text

ISIM Inhibit Simultaneity [Data Processing]

ISIM Institut des Sciences de l'Ingénieur de Montpellier [Montpellier Institute of Engineering Sciences, France]

ISIMM International Society for the Interaction of Mechanics and Mathematics [Germany]

IS Ind Soc Mag IS Industrial Society Magazine [Published by the Industrial Society, UK]

ISINET Institute for Scientific Information Networks [US]

ISIP Indexed Security Investment Plan

ISIP Instantaneous Shut-In Pressure

ISIPBMP International Symposium on Interfacial Phenomena in Biotechnology and Materials Processing

ISI-PV Institut für Schicht– und Ionentechnik–Photovoltaik [Institute for Film and Ion Technology–Photovoltaics, Kernforschungsanlage Julich GmbH, Germany]

ISIR Institute of Scientific and Industrial Research [of Osaka University, Japan]

ISIR International Society for Interferon Research [Biochemistry]

ISIR International Society of Invertebrate Reproduction [UK]

ISIR International Symposium on Industrial Robots

ISIS Ice Services Integrated System

ISIS Image-Selected in vivo Spectroscopy

ISIS Institute of Surface and Interface Science [of University of California at Irvine, US]

ISIS Integrated Scientific Information System

ISIS Integrated Set of Information System [of International Labour Organization]

ISIS Integrated Systems and Information Services

ISIS Integrated Strike and Interceptor System

ISIS Internally Switched Interface System

ISIS International Satellite for Ionospheric Studies [US/ Canada]

ISIS International Science Information Service [US]

ISIS International Shipping Information Service

ISIS International Student Information System

ISIS Ion-Beam Synthesis in Semiconductors

IS-IS Intermediate System-Intermediate System [Internet]

ISISS International Summer Institute in Surface Science

ISI Spec Rep Iron and Steel Institute Special Report [UK]

ISiT (Fraunhofer-)Institut für Siliziumtechnologie [(Fraunhofer) Institute for Silicon Technology, Berlin, Germany]

ISITB Iron and Steel Industry Training Board [UK]

ISJ Infrared Society of Japan

ISJTA Intensive Student Jet Training Area [US]

ISK Insert Storage Key

ISK Instruction Space Key

ISK Internacia Scienca Kolegio [International College of Scientists, Italy]

ISK Icelandic Krona [Currency]

ISKA International Saw and Knife Association [US]

ISKM Internet Starter Kit for the Macintosh

ISL Illinois State Library [Springfield, US]

ISL Indiana State Library [Indianapolis, US]

ISL Inertial Systems Laboratory

ISL Information Search Language

ISL Information System Language

ISL Information Systems Laboratory [of Michigan State University, East Lansing, US]

ISL Institut für Seeverkehrswirtschaft und Logistik [Institute of Ocean Shipping Economics and Logistics, Germany]

ISL Instructional Systems Language

ISL Instrumentation and Sensing Laboratory [of US Department of Agriculture in Beltsville, Maryland]

ISL Integrated Schottky Logic

ISL Intelligent Systems Laboratory [of Palo Alto Research Center, Xerox Corporation, US]

ISL Interactive Simulation Language

ISL Interactive System Language

ISL Internal Standard Line [Spectroscopy]

ISL Intersatellite Link

ISL Intersystem Link

Isl Island [also isl]

ISLA International Survey Library Association [US]

Islamabad J Sci Islamabad Journal of Sciences [Published by Islamabad University, Pakistan]

Island Tel Island Telephone Company Limited [Prince Edward Island, Canada]

ISLEWTT International Symposium on Low-Cost and Energy-Saving Wastewater Treatment Technologies

ISLIC Israel Society for Special Libraries and Information Centers

ISLLSS International Society for Labour Law and Social Security [of International Labour Organization]

ISLM Integration Shop/Lab Manager

ISLS Improved Side-Lobe Suppression

ISLS Interrogation-Path Side-Lobe Suppression

Isls Islands

ISLSCP International Satellite Land Surface Climatology Project

ISLTC International Society of Leather Trades Chemists

ISM Induction Synchronous Motor

ISM Industrial, Scientific and Medical (Equipment)

ISM Information Services Management (Corporation) [of Open Bidding Services, Canada]

ISM Information System Management

ISM Information Systems for Management

ISM In-Situ Microfusion (Liquid-Phase Sintering Process) [Metallurgy]

ISM Institute of Sanitation Management [now Environmental Management Association, US]

ISM Institute of Supervisory Management [UK]

ISM Insulation System Module

ISM Interactive Surface Modeling

ISM International Safety Management

ISM International Software Marketing [US]

ISM Internet Service Manager

ISM Interpretive Structural Modeling

ISM Inverse Scattering Method [Physics]

ISM Inverse-Speed Motor

ISM Ion-Selective Membrane

ISM Istituto di Spettroscopi Moleculare [Institute of Molecular Spectroscopy; of Consiglio Nazionale delle Ricerche, Bologna, Italy]

ISM Istituto di Struttura della Materia [Institute for the Structure of Matter, of Centro Nazionale delle Ricerche, Frascati, Italy]

I&SM Iron and Steelmaker (Magazine) [Publication of the Iron and Steel Society/American Institute of Mining, Metallurgical and Petroleum Engineers, US]

ISMA International Satellite Monitoring Agency

ISMA International Securities Market Association [UK]

ISMA International Shipmasters Association

ISMA International Superphosphate Manufacturers Association

ISMA International Symposium on Mechanical Alloying

ISMA International Symposium on Microalloying

ISMA International Symposium on Mining in the Arctic

ISMAN Institute of Structural Macrokinetics [of Russian Academy of Sciences]

ISMANAM International Symposium on Metastable, Mechanically Alloyed and Nanocrystalline Materials

ISMB International Society of Mathematical Biology [France]

ISMB International Symposium on Mining with Backfill

ISMC International Switching Maintenance Center

ISMC International Seminar on Mathematical Cosmology

ISMCM Institut Supérieur des Matériaux et de la Construction Mécanique [Higher Institute of Materials and Mechanical Engineering, Saint Ouen, France]

ISMD International Symposium on Multiparticle Dynamics

ISMDA Independent Sewing Machine Dealers Association [US]

ISME Institute of Sheet Metal Engineering [UK]

ISMEC Information Service in Mechanical Engineering

ISMED International Symposium on Molecular Electronic Devices

ISMEX International Shoe Machinery Exhibition

ISMF Interactive Storage Management Facility

ISMH Input Source Message Handler

ISMH International Society of Medical Hydrology and Climatology [France]

ISML Intermediate System Mockup Loop

ISML Istituto Sperimentali Metalli Leggeri [Light Metal Research Institute, Milan, Italy]

ISMLS Interim Standard Microwave Landing System

ISMM International Society for Mini- and Microcomputers [Canada]

ISMMS Intrinsically Safe Mine Monitoring System

ISMR International Symposium for Metal Recycling

ISMRA Institut des Sciences de la Matière et du Rayonnement [Institute of Matter and Radiation Sciences; of Centre des Matériaux Supraconducteurs, University of Caen, France]

ISMS Image Store Management System

ISMS International Society for Mushroom Science [UK]

I/SMT Interconnect/Surface-Mount Technology

ISMX Integrated Subrate Data Multiplexer

ISN Information Systems Network

ISN Institut des Sciences Nucléaires de Grenoble [Grenoble Institute of Nuclear Science, France]

ISN Internal Statement Number [also isn]

ISN International Society of Neurochemistry

ISNAR International Service for National Agricultural Research [Singapore]

ISNET Inter-Islamic Network in Space Sciences and Technology

ISNTA International Staple, Nail and Tool Association [also ISN&TA, US]

ISO Individual System Operation

ISO Infrared Space Observatory [of European Space Agency]

ISO Insurance Services Office [New York, US]

ISO Intergalactic Sysop Alliance [US]

ISO International Science Organization

ISO International Standards Organization [Name Changed to International Organization for Standardization, Switzerland]

ISO International Sugar Organization [UK]

ISO Ion-Selective Optrode

Iso Isolation [or iso]c

iso isotropic (structure) [Crystallography]

Iso-APPA Iso-Amyl Phenylphosphonate [also isoAPPA]

ISO/ANSI International Standards Organization/American National Standards Institute

ISO/ASA International Standards Organization/American Standards Association

Iso-Bu Iso-Butyl [also iso-Bu]

ISOC Individual System/Organization Cost

ISOC Integrated Spectrum Observation Center [Canada]

ISOC The Internet Society

ISoCaRP International Society of City and Regional Planners [Netherlands] [also ISOCaRP]

ISO CASE International Standards Organization/Common Application Service Element (Kernel)

ISO-CMOS Isolated Fully Recessed Complementary Metal-Oxide Semiconductor

ISODATA Iterative Self-Organizing Data Analysis Technique

ISODE ISO (International Standards Organization) Development Environment [Internet]

ISOF International Society of Ocular Fluorophotometry

ISO FTAM International Standards Organization/File Transfer and Access Method

ISO/IEC International Standards Organization/International Electrotechnical Commission

ISO-KEL Isomerization-Kellogg (Process) [Chemical Engineering]

Isol Isolate; Isolation [also isol]

isold isolated [also isol'd]

ISOLDE Isotope Separator On-Line Device [of CERN–Laboratory for Particle Physics, Geneva, Switzerland]

ISOM International Society for Orthomolecular Medicine

Isom Isomer(ic) [also isom]

Isom Isometric; Isometry [also isom]

ISOMAR Isomerization of Aromatics (Process) [Chemistry]

ISONET International Standards Organization Network Committee

ISO/OSI International Standards Organization/Open System Interconnect(ion) (Standard) [also ISO-OSI]

ISOPA European Isocyanate Producers Association

iso-PAPS 2'-Phosphoadenosine-5'-Phosphosulfate [Biochemistry]

Iso-Pr Iso-Propyl [also iso-Pr]

ISORID International Information System on Research in Documentation [of UNESCO]

ISOS International Southern Ocean Studies

ISOSC International Society for Soilless Culture [Netherlands]

ISOSIV Isomer Separation by Molecular Sieves (Process)

ISO STEP International Standards Organization–Standard for the Exchange of Product Model Data [also ISO/STEP]

ISOT International School of Offshore Technology

ISOT International Scientific Oversight Team

Isot Isotope [also isot]

ISO/TC International Standards Organization/Technical Committee

ISO/TC/WG International Standards Organization/Technical Committee/Working Group

Isoth Isothermal [also isoth]

Isot Radiat Res Isotope and Radiation Research [Publication of the Middle Eastern Regional Isotope Center for the Arab Countries, Cairo, Egypt]

Iso Wd Isolation Ward

ISO/WG International Standards Organization/Working Group

ISP Imperial Smelting Process [Metallurgy]

ISP Independent Study Program

ISP Indexed Sequential Processor

ISP Industrial Security Plan

ISP Initial Specific Impulse [Aerospace]

ISP In-Line Strip Production (Process) [Metallurgy]

ISP Institute of Sewage Purification [UK]

ISP Instruction Set Processor

ISP Instrumentation Support Plan

ISP Integrated Support Plan

ISP Integrated System Peripheral

ISP Interferometer Software Package

ISP Internationally Standardized Profile

ISP International School of Physics

ISP International Society for Photogrammetry [now International Society of Photogrammetry and Remote Sensing, Japan]

ISP International Society of Postmasters [Canada]

ISP International Society of Psychophysics

ISP Internet Service Provider

ISP Interoperable Systems Project [US]

ISP Interrupt Stack Pointer

ISP Interrupt Status Port

ISP Ion-Sheath Phenomenon [Physics]

ISP Isotopic Spin Population [Nuclear Physics]

ISP Italian Society of Physics

ISPA International Small Printers Association

ISPA International Society of Parametric Analysts [US]

ISPABX Integrated Services Private Automatic Branch Exchange

ISPC International Sound-Program Center

ISPC International Symposium on Plasma Chemistry

ISPCA Irish Society for the Prevention of Cruelty to Animals

ISPE Illinois Society of Professional Engineers [US]

ISPE Industrial Safety (and Protective Equipment) Manufacturers Association [UK]

ISPE Institute of Swimming Pool Engineers [UK]

ISPE Institutului de Studii si Proiectari Energetice [Institute of Energetic Project Studies, Bucharest, Romania]

ISPE International Society of Pharmaceutical Engineers [US]

ISPE International Society of Planetarium Educators [now International Planetarium Society, US]

ISPEC Insulation Specification

ISPEMA Industrial Safety Personnel Equipment Manufacturers Association

ISPF Interactive System Productivity Facility

ISPF Interactive System Programming Facility

ISPF International Science Policy Foundation [UK]

ISPF/PDF Interactive System Productivity Facility/Program Development Facility

ISPG Institute of Sedimentary and Petroleum Geology [of Natural Resources Canada]

ISPG Institutional Support Planning Group

ISPhS International Society of Phonetic Sciences [US]

ISPIM International Symposium on Product Innovation Management

ISPM International Society of Plant Morphologists [India]

ISPM International Solar Polar Mission

ISPM International Symposium on the Physics of Materials

ISPMAS International Symposium on Plasticity of Metals and Alloys

ISPMB International Society for Plant Molecular Biology [US]

ISPO International Society for Prosthetics and Orthodontics [Denmark]

ISPO International Statistical Programs Office [of US Department of Commerce]

ISPP Information System for Policy Planning

ISPP In-Service Professional Program

ISPR International Standard Payload Rack

ISPRS International Society of Photogrammetry and Remote Sensing [Japan]

ISPRS J Photogramm Remote Sens ISPRS Journal of Photogrammetry and Remote Sensing [of International Society of Photogrammetry and Remote Sensing]

ISPS Instruction Set Processor Specifications

ISPS Intense Slow Positron Source

ISPT Improved Scaled Particle Theory

ISPT Istituto Superiore Poste e Telecomunicazioni [Higher Institute for Post and Telecommunications, Rome, Italy]

ISPWP International Society for the Prevention of Water Pollution [UK]

ISR Image Storage Retrieval

ISR Index to Scientific Reviews [of Institute for Scientific Information, US]

ISR Induction Skull Remelting [Metallurgy]

ISR Inductive Source Resistivity

ISR Information Storage and Retrieval [also IS&R]

ISR Infrared Scanning Radiometer

ISR Initial System Release

ISR Innovative Systems Research

ISR In-situ-Reinforced; In-situ-Reinforcement

ISR In-situ Remediation

ISR Institute for Standards Research [of American Society for Testing and Materials, US]

ISR Institute of Seaweed Research [UK]

ISR Interdisciplinary Systems Research

ISR Interrupt Service Routine

ISR Interrupt Status Register

ISR Intersecting Storage Rings [of CERN–European Laboratory for Particle Physics, Geneva, Switzerland]

ISR International Society of Radiology [US]

ISR Istituto di Struttura della Materia [Institute for Structural Matter Research, of Consiglio Nazionale delle Ricerche, Italy]

IS&R Information Storage and Retrieval [also ISR]

Isr Israel(i)

ISRC International Standard Recording Group

ISRC International Student Research Center

ISRC Inter-University Semiconductor Research Center [Seoul, South Korea]

ISRCSC Inter-Service Radio Components Standardization Committee [UK]

ISRD Information Storage, Retrieval and Dissemination

ISRF International Sugar Research Foundation

ISRG International Space Research Group [US]

ISRHAI International Secretariat for Research on the History of Agricultural Implements [Denmark]

ISRI Industrial Science Research Institute [of Iowa State University, Ames, US]

ISRI Institute of Scrap Recycling Industries [Washington, DC, US]

ISRIC International Soil Reference (and) Information Center

ISRI Commod ISRI Commodities [Publication of the Institute of Scrap Recycling Industries, US]

ISRIP In-situ Redox Integrated Program

ISRI Rep ISRI Report [Publication of the Institute of Scrap Recycling Industries, US]

Isr J Chem Israel Journal of Chemistry [Published by Weizmann Science Press of Israel, Jerusalem]

Isr J Earth-Sci Israel Journal of Earth-Sciences [Published by Weizmann Science Press of Israel, Jerusalem]

Isr J Med Sci Israel Journal of Medical Science

Isr J Technol Israel Journal of Technology [Published by Weizmann Science Press of Israel, Jerusalem]

ISRM Information System Resource Manager

ISRM International Society for Rock Mechanics [Portugal]

ISRO Indian Space Research Organization [India]

ISRP International Society of Respiratory Protection

ISRRS International Symposium on Research Reactor Safety

ISRRT International Society of Radiographers and Radiological Technicians [Canada]

ISRS International Safety Rating System

ISRS Impulsive Stimulated Raman Scattering [Physics]

ISRS International Symposium on the Reactivity of Solids

ISR SiC In-Situ-Reinforced Silicon Carbide

ISRSM International Symposium on Rockbursts and Seismicity in Mines

ISRT Invisible Soft Return [also ISRt]

ISRU International Scientific Radio Union

ISS Ideal Solidus Structures [Metallurgy]

ISS Image Sharpness Scale

ISS Imaging Science Subsystem [of NASA]

ISS Impulsive Stimulated (Light) Scattering

ISS Index of Specifications and Standards [of US Department of Defense]

ISS Inertial Subsystem

ISS Information Sending Station

ISS Information Services Seminar [US]

ISS Information Sharing System

ISS Information Storage System

ISS Information Systems Specialists (Office) [of Library of Congress, Washington, DC, US]

ISS Inhibit/Override Summary Snapshot (Display)

ISS Input Subsystem

ISS In-situ Sampling

ISS Installation Support Services

ISS Institute for Strategic Studies

ISS (Swiss) Institute (for the Promotion) of Safety and Security

ISS Institute of Systems Science [of National University of Singapore]

ISS Instruction Summary Sheet

ISS Instrument Society of Sweden

ISS Integrated Sounding System [of National Center for Atmospheric Research, US]

ISS Integrated Storage System

ISS Integrated System Schematic

ISS Intelligent Support System

ISS Interagency Sedimentation Subcommittee [of Federal Energy Regulatory Commission, US]

ISS Interfacial Shear Stress [Materials Science]

ISS Interlaminar Shear Strength

ISS International Schools Services [US]

ISS International Seaweed Symposium [Norway]

ISS International Seismological Summary

ISS International Society for Stereology [Germany]

ISS International Space Station [Cooperative Program of the USA, Canada, the UK, Germany, France, Italy, Spain, Sweden, Norway, Denmark, Switzerland, the Netherlands, Belgium, Japan, Russia and Brazil]

ISS International Steamboat Society

ISS International Student Service [US]

ISS International Summer School

ISS International Switching Symposium

ISS International Symposium on Superconductivity

ISS Interrupt Safety System

ISS Interrupt Service Subroutines

ISS Interstellar Space

ISS Ionosphere Sounding Satellite [Japan]

ISS Ion-Scattering Spectrometer; Ion-Scattering Spectrometry

ISS Ion-Scattering Spectroscopy

ISS Ion-Scattering Spectrum

ISS Ion Surface Scattering

ISS Iron and Steel Society [of American Iron and Steel Institute, US]

ISS Isotope Separation System [Physics]

ISS Istituto Sperimentale per la Selvicoltura [Silvicultural Research Institute, Italy]

Iss Issue [also iss]

ISSA Information Systems Security Association [US]

ISSA International Sanitary Supply Association [US]

ISSA International Ship Suppliers Association [UK]

ISSA International Slurry Seal Association [US]

ISSA International Social Security Association [Belgium]

ISSA International Society of Scientists and Artists [now International Society for the Arts, Sciences and Technology, US]

ISSA International Society of Stress Analysts [US]

ISS-AIME Iron and Steel Society of the American Institute of Mining, Metallurgical and Petroleum Engineers [US]

ISSC Institute of Solid-State Chemistry [Russia]

ISSC Interdisciplinary Surface Science Conference

ISSC International Ship Structures Conference

ISSCB International Society for Sandwich Construction and Bonding

ISSCC International Solid-State Circuits Conference [of Institute of Electrical and Electronics Engineers, US]

ISSCG International Summer School on Crystal Growth

ISSCO Integrated Software Systems Corporation [US]

ISSCT International Society of Sugar Cane Technologists [Brazil]

ISSDA Indian Stainless Steel Development Association

ISSE International Sun-Earth Explorer

ISS&E International SAMPE (Society for the Advancement of Materials and Process Engineering) Symposium and Exposition

ISSF Interfacial Shear Stress related to Friction [Materials Science]

ISSI International Symposium on Structural Intermetallics [of The Minerals, Metals and Materials Society of the American Institute of Mining, Metallurgical and Petroleum Engineers, US]

ISSL Initial Spares Support List

ISSLIC Israel Society of Special Libraries and Information Centers [Tel Aviv]

ISSLS International Symposium on Subscriber Loops and Services

ISSM Institute of Safety and Systems Management [of University of Southern California, Los Angeles, US]

ISSMB Information Systems Standards Management Board [of American National Standards Institute, US]

ISSMFE International Society for Soil Mechanics and Foundation Engineering [UK]

ISSN Integrated Special Services Network

ISSN International Standard Serial Number

ISSO Information System Security Officer

ISSO International Small Satellite Organization

ISSOL International Society for the Study of the Origin of Life [US]

ISSP Institute of Solid-State Physics [of University of Tokyo, Japan]

ISSP International Summer School of Physics

ISSPIC International Symposium on Small Particles and Inorganic Clusters [US]

ISSR Independent Secondary Surveillance Radar

ISSR Institute of Statistical Studies and Research [of Cairo University, Giza, Egypt]

ISSS International Society for the Study of Symbols

ISSS International Society for the Systems Sciences [US]

ISSS International Society of Soil Science [Netherlands]

ISSS International Symposium on Surface Science

ISSSB International Symposium on Separation Science and Biotechnology

ISSSC International Summer School on Solidification and Casting

ISSSC International Symposium on Solid-State Chemistry

ISSSE International Society of Statistical Science in Economics [US]

ISS-SIM Student/Industry Message Board of the Iron and Steel Society [of American Institute of Mining, Metallurgical and Petroleum Engineers, US]

ISS-SIMS Ion-Scattering Spectrometer–Secondary-Ion Mass Spectrometry

ISST International Society for the Study of Time [US]

ISSUE Information System Software Update Environment

Issues Sci Technol Issues in Science and Technology [Journal published in the US]

ISSX International Society for the Study of Xenobiotics

IST Impact Surface Treatment

IST Incredibly Small Transistor(s)

IST Incremental Step Test [Mechanical Testing]

IST Indian Standard Time

IST Industrial Science and Technology; Industry, Science and Technology

IST Information Science and Technology

IST Information Storage Technology

IST Information Systems and Technology

IST Innovative Science and Technology

IST (Fraunhofer) Institute for Thin-Film and Surface Technology [Hamburg, Germany]

IST Institute of Science and Technology

IST Institute of Science and Technology [UK]

IST Institute of Science and Technology [of University of Michigan, US]

IST Institute of Science and Technology [of University of Tokyo, Japan]

IST Instituto Superior Técnico [Higher Technical Institute, Lisbon, Portugal]

IST Instrument Support Terminal

IST Insulator-Superconductor Transition

IST Integrated Services Telephone

IST Integrated Switching and Transmission

IST Integrated Switching Technique

IST Integrated System Test

IST Integrated System Transformer

IST Interconnect Stress Test

IST International Sensor Technology, Inc. [Pullman, Washington, US]

IST International Skelton Tables

IST International Society on Toxicology [US]

IST International Standard Thread

IST International Strategy Team [of ASM International, US]

IS&T Imaging Science and Technology

IS&T Information Science and Technology

α-Ist α-Isopropyltropolone [Biochemistry]

β-Ist β-Isopropyltropolone [Biochemistry]

ISTA Independent Software Testing Association [now Association for Software Testing and Evaluation, US]

ISTA Indian Scientific Translators Association

ISTA Information, Science and Technology Agency [Victoria, British Columbia, Canada]

ISTA In-Situ Thermal Annealing [Metallurgy]

ISTA International Seed Testing Association [Switzerland]

ISTA International Society for Technology Assessment

ISTA International Special Tooling Association [Germany]

ISTA International Steel Trade Association [UK]

ISTA Intertank Structural Test Assembly

ISTAHC International Society for Technology Assessment in Health Care

ISTAR Image Storage, Translation and Reproduction

ISTAT International Society of Transport Aircraft Traders [US]

ISTB Integrated Subsystem Test Bed

ISTC Industry, Science and Technology Canada [formerly Department of Regional and Industrial Expansion, Canada]

ISTC Institute of Scientific and Technical Communicators [UK]

ISTC Integrated System Test Complex

ISTC International SAMPE (Society for the Advancement of Materials and Process Engineering) Technical Conference

ISTC International Science and Technology Center

ISTC International Steam Table Calorie

ISTC International Steam Tables Conference [now International Conference on the Properties of Steam]

ISTC International Student Travel Confederation [Switzerland]

ISTC International Switching and Testing Center [UK]

ISTC Iron and Steel Trades Confederation [UK]

ISTC/EC Industry, Science and Technology Canada/ Environment Canada (Program)

IST-CSS Incremental Step Test–Cyclic Stress-Strain (Curve) [Mechanical Testing]

ISTDA Institutional and Service Textile Distributors Association [US]

ISTE International Society for Tropical Ecology [India]

ISTEA Iron and Steel Trades Employers Association [UK]

ISTEA Intermodal Surface Transportation Efficiency Act [US]

ISTEC International Superconductivity Technology Center [Japan]

ISTEC J International Superconductivity Technology Center Journal [Japan]

ISTEM Internal Scanning Transmission Electron Microscopy

ISTERH International Society of Trace Element Research in Humans

ISTF Integrated Servicing and Test Facility

ISTF Integrated System Test Flow

ISTF International Society of Tropical Foresters [US]

ISTFA International Society for Testing and Failure Analysis

ISTFA International Symposium for Testing and Failure Analysis

Isth Isthmus [also isth]

ISTIC Institute of Scientific and Technical Information of China

ISTIM Interchange of Scientific and Technical Information in Machine Language

ISTIS International Science and Technology Information Service

ISTM International Society for Testing Materials

ISTN Integrated Switching and Transmission Network

ISTO Innovative Science and Technology Office [US]

IST/ONR Innovative Science and Technology (Program)/ Office of Naval Research [US]

ISTP Innovative Science and Technology Program [of Office of Naval Research, US]

ISTP International Solar Terrestrial Physics

ISTP&B Index to Scientific and Technical Proceedings and Books [of Institute for Scientific Information, US]

ISTR Indexed Sequential Table Retrieval

ISTRA Interplanetary Space Travel Research Association [UK]

ISTRC International Society of Tropical Root Crops

ISTRO International Soil Tillage Research Organization [Netherlands]

IStructE Institution of Structural Engineers [UK]

ISTS Impulsive Stimulated Thermal Scattering

ISTS Institute for Space and Terrestrial Science [North York, Ontario, Canada]

ISTS International Shock Tube Symposium

ISTS International Symposium on Space Technology and Science

ISTSC Isatin Thiosemicarbazone

IST/SDIO-ONR Innovative Science and Technology (Program)/ Strategic Defense Initiative Organization–Office of Naval Research [US]

ISTSIMM International Scientific-Technical Society of Instrument Makers and Metrologists [Russia]

ISTU Islamic States Telecommunications Union

ISTUM Industrial Energy End-Use Technology Model

ISTVS International Society for Terrain-Vehicle Systems [US]

ISU Idaho State University [Pocatello, US]

ISU Illinois State University [Normal, US]

ISU Independent Signal Unit

ISU Indiana State University [Terre Haute, US]

ISU Inertial Sensor Unit

ISU Initial Signal Unit

ISU Instruction Storage Unit

ISU Instrumentation (and) Services Unit [of Indian Institute of Science, Bangalore]

ISU Integral Serial Unit

ISU Interface Sharing Unit

ISU Interface Switching Unit

ISU International Scientific Union

ISU International Space University [Strasbourg, France]

ISU International System of Units

ISU Iowa State University (of Science and Technology) [Ames, US]

ISUE Indiana State University at Evansville [US]

ISUP Integrated Services User Part [Telecommunications]

ISUST Iowa State University of Science and Technology [Ames, US]

ISUTH Indiana State University at Terre Haute [US]

ISV Independent Software Vendor

ISV In situ Vitrification

ISV Instantaneous Speed Variation

ISV International Scientific Vocabulary

ISV International Society of Videographers [US]

ISVD Information System for Vocational Decisions

ISVR Institute of Sound and Vibration Research [UK]

ISVS In situ Vapor Sampling

ISVS In situ Vapor Stripping

ISVS International Society for Vegetation Science [now International Association for Vegetation Science, Germany]

ISVTNA International Symposium on Vacuum Technology and Nuclear Applications

ISW Industrial Solid Waste

ISW Interagency Sedimentation Workgroup [of Interagency Sedimentation Subcommittee, Federal Energy Regulatory Commission, US]

ISWA Insect Screening Weavers Institute [US]

ISWA International Science Writers Association

ISWA International Solid Wastes and Public Cleansing Association [France]

ISWG Imperial Standard Wire Gauge

ISWM International Society of Weighing and Measurement [US]

ISWL Isolated Single Wheel Load

ISWS Illinois State Water Survey [US]

ISWU International Society of Wang (Computer System) Users [US]

ISWWE Integrability: the Seiberg-Witten and Whitham Equations (Conference)

ISX Impurity Study Experiment

ISY International Space Year

IT Identification Transponder

IT Ignition Temperature

IT Image Tube

IT Imaging Technology

IT Immediate Transportation

IT Immunotherapeutical; Immunotherapist; Immunotherapy

IT Impact Test(er); Impact Testing [Mechanics]

IT Impulse Telegraphy

IT Impulse Turbine

IT Incoherent Tunneling [Electronics]

IT Incomplete Translation

IT Indent Tab(ulation Character)

IT Index Terms

IT Induction Theorem

IT Induction Tooling

IT Industrial Technician

IT Industrial Technologist; Industrial Technology

IT Industry Telephone

IT Infantry Tank

IT Information Technologist; Information Technology

IT Information Theory

IT Information Theory (Group) [of Institute of Electrical and Electronics Engineers, US]

IT Information Transmission

IT Infrared Transducer

IT Initial Tension (of Springs) [Symbol]

IT (Positive) Inner Tetrahedron [Crystallography]

IT Input Terminal

IT Input Translator

IT Installation Test

IT Institute of Technology

IT Institute of Tribology [UK]

IT Instruction Time

IT Insulating Transformer

IT Instrumentation Technician

IT Instrumentation Technologist; Instrumentation Technology

IT Instrument Test

IT Instrument Tree

IT Instrument Transformer

IT Integral Transform [Mathematics]

IT Intelligent Terminal

IT Interface Tool

IT Interfacial Tension

IT Interfacial Torque

IT Intergranular Tear [Metallurgy]

IT Internal Thread

IT Internal Translator

IT International Table

IT International Tolerance (Grade)

IT Interrogator-Transponder

IT Intersubband Transition

IT Intervalence Transition

IT Inter-Toll (Trunk)

IT Intestinal Tract [Anatomy]

IT In-Transit

IT Inverse Transformation [Materials Science]

IT Inversion Theorem

IT Inversion Temperature

IT Iodoform Test

IT Iodotoluene

IT Ionization Threshold

IT Ion Temperature

IT Ion Trap

IT Isometric Transition [Crystallography]

IT Isothermal

IT Isothermal Transformation

IT Isotropic Tracer

IT Italy [ISO Code]

IT Item Transfer

I&T Inspection and Test

I-T Isothermal Transformation (Diagram) [Metallurgy]

It Italian; Italy

It Italic [also It, or it]

I(t) Signal Intensity [Symbol]

I(t) Current as a Function of Time

$I_0(t)$ Time-Dependent Brightfield Intensity (in Electron Microscopy) [Symbol]

$I_g(t)$ Time-Dependent Darkfield Intensity (in Electron Microscopy) [Symbol]

I-t Current-Time (Diagram)

I-t Enthalpy-Temperature (Diagram)

$I(\theta)$ Intensity at Coverage (for Substrates) [Symbol]

$I_i(\theta)$ Scattered Ion Intensity from Species i at Scattering Angle θ (in Ion Spectroscopy Scattering) [Symbol]

.it Italy [Country Code/Domain Name]

ITA Independent Telephone Association [US]

ITA Independent Television Authority [UK]

ITA Industry and Trade Administration [now International Trade Administration, US]

ITA Indium Tantalum Antimonide

ITA Industrial Technical Adviser

ITA Industrial Technology Adviser

ITA Industrial Transport Association

ITA Industrial Truck Association [US]

ITA Industry and Trade Administration

ITA Information Technologies Association [now Information Technology Acquisition and Marketing Association, US]

ITA Institut de Transport Aérien [Air Transport Institute, France]

ITA Institute of Traffic Administration [UK]

ITA Institute for Telecommunication and Aeronomy [of Environmental Science Services Administration, US]

ITA Instrumentation Technology Associates

ITA Integrated Test Area

ITA Integrated Test Article

ITA Interface Test Adapter

ITA Intermodal Transportation Association [US]

ITA International Tape/Disk Association [US]

ITA International Tar Conference

ITA International Teleconferencing Association

ITA International Telegraph Alphabet

ITA International Textile-Trade Arrangement

ITA International Thermographers Association [US]

ITA International Tin Association

ITA International Tire Association [US]

ITA International Trade Administration [of US Department of Commerce]

ITA International Traders Association [US]

ITA International Tube Association [UK]

ITA International Tunneling Association [France]

ITA International Typographic Association

ITA Interstate Towing Association [US]

ITAA International Textile and Apparel Association

ITABC Independent Telephone Association of British Columbia [Canada]

ITAC Information Technology Association of Canada

ITAC Inter-Agency Textile Administrative Committee [US]

ITAC International Trade Affairs Committee

ITACS Integrated Tactical Air-Control System

ITAE Integral of Time Multiplied Absolute Error

ITAE Integrated Time and Absolute Error

ITAI Institute of Traffic Accident Investigators [UK]

ITAL Information Technology for Libraries [Publication of the Library and Information Technology Association, US]

ITAL Instituut voor Toepassing van Atoomenergie in de Landbouw [Institute for Nuclear Energy in Agriculture, Netherlands]

Ital Italian; Italy

ital italic(ized)

Ital P Italian Patent [also ItalP]

Ital Technol Italian Technology [Journal published in Italy]

ITAMA Information Technology Acquisition and Marketing Association [US]

ITAP Information Technology Advisory Panel [UK]

ITAP Institut für Theoretische und Angewandte Physik [Institute for Theoretical and Applied Physics; of University of Stuttgart, Germany]

ITAP Integrated Tactical Air Picture

ITAP Integrated Technical Assessment Panel

IT&AP Inspection/Test and Analysis Plan

ITAR International Traffic in Arms Regulations

ITARS Integrated Terrain Access and Retrieval System

ITAS Industrial Training Association of Scotland

ITAS Infrared Transmission-Absorption Spectroscopy

ITAS Institut für Technikfolgenabschätzung und Systemanalyse [Institute for Technological Impact Assessment and System Analysis, Karlsruhe, Germany]

ITAVS Integrated Testing Analysis and Verification System

ITB Iceland Tourist Board

ITB Incoherent Twin (Grain) Boundary [Materials Science]

ITB Incoming Trunk Busy

ITB Industry Training Board [UK]

ITB Infantry Training Bomb

ITB Information Technology Branch

ITB Information Technology Budget

ITB Intergrowth Tungsten Bronze

ITB Intermediate Test Block [Computers]

ITB Intermediate Text Block (Character)

ITB Intermediate Transmission Block

ITB Internal Transfer Bus

ITB International Time Bureau [Paris, France]

ITB Irish Tourist Bureau

ITBTP Institut Technique du Bâtiment et des Travaux Publics [Technical Institute of Buildings and Public Works, France]

ITC Igloo Thermal Control [NASA]

ITC Image Technology Center

ITC Immense Technology Commitment

ITC Inclusive Tour Charter

ITC Independent Telephone Company

ITC Independent Television Commission [UK]

ITC Industrial Technology Center

ITC Industrial Training Council [UK]

ITC (Ministry of) Industry, Trade and Commerce [Canada]

ITC Information Technology Committee

ITC Information Transfer Center [US]

ITC Inland Transport Committee [of United Nations]

ITC Institute of Technical Cybernetics [of Wroclaw Technical University, Poland]

ITC Instituto Tecnológico de Chihuahua [Chihuahua Institute of Technology, Mexico]

ITC Instructional Telecommunications Consortium [US]

ITC Integral Tube Components

ITC Integrated Temperature Control System

ITC Integrated Terminal Controller

ITC Intense Training Course

ITC Interagency Testing Committee

ITC Inter-American Telecommunications Network

ITC Interdata Transaction Controller

ITC Intermediate Toll Center

ITC International Tar Conference [France]

ITC International Tea Committee

ITC International Technology Council [of Building Research Board, US]

ITC International Telecommunications Convention

ITC International Telemetering Conference

ITC International Teletraffic Congress

ITC International Television Center

ITC International Test Conference [US]

ITC International Tin Council [UK]

ITC International Toki Conference (on Plasma Physics) [Japan]

ITC International Tomography Center [Novosibirsk, Russia]

ITC International Trade Center

ITC International Trade Center [Geneva, Switzerland]

ITC International Trade Commission [US]

ITC International Trade Council [of United Nations]

ITC International Training Center

ITC International Translators Center [Netherlands]

ITC International Tunneling Congress

ITC International Typeface Corporation

ITC Interval Time Control

ITC Investment Tax Credit

ITC Ionic Thermal Current

ITC Ionic Thermoconductivity

ITC Ionic Thermocurrent(s)

ITC Isothiocyanate

ITC Italian Tile Center

IT&C Industry, Trade and Commerce

ITCA International Technical Caramel Association [US]

ITCA International Typographical Composition Association [now Typographers International Association, US]

ITCA Isothiocyanic Acid

IT/CA International Tele/Conferencing Association [US]

IT cal International Steam Table Calorie

ITCAS International Training Center for Aerial Survey

ITCB International Textiles and Clothing Bureau

ITCC International Technical Communications Conference [of Society for Technical Communication, US]

ITCC Interstate Truckload Carriers Conference [US]

ITCG International Trade Communications Group [Canada]

ITC J ITC Journal [of International Institute for Aerial Survey and Earth Sciences, Netherlands]

ITCI International Tree Crops Institute [US]

ITCL Item Class

ITCLC Item Class Code

ITCPN International Technical Conference on the Protection of Nature [France]

ITC/REE (Ministry of) Industry, Trade and Commerce and Regional Economic Expansion [now Industry, Science and Technology Canada]

ITCS Industrial Trade and Consumer Show

ITCS Industrial Trade and Consumer Shows, Inc. [Toronto, Canada]

ITCS Integrated Temperature Control System

ITCS Integrated Thermionic Circuits

ITCSA Institute of Technical Communicators of Southern Africa

ITCWRM International Training Center for Water Resource Management

ITCZ Intertropical Convergence Zone(s) [Meteorology]

ITD Industrial Technology Development Center [of Argonne National Laboratory, Illinois, US]

ITD Institute of Training and Development [UK]

ITD Ion Trap Detector

ITD Isothermal Desorption

ITDA Independent Truck Drivers Association [US]

ITDA International Tape/Disk Association

ITDB International Trade Data Bank

ITDC International Trade Development Committee [US]

ITDE Interchannel Time Displacement Error

ITDG Intermediate Technology Development Group [UK]

ITDM Intelligent Time-Division Multiplexer

ITDN Integrated Telephone and Data Network

ITDP Innovative Technology Development Plan

ITDP Innovative Technology Development Program [of US Department of Energy]

ITDSC Item Description

ITDR Institute for Training and Demographic Research [Cameroon]

ITE Information Technology Equipment

ITE Institute of Telecommunications Engineers

ITE Institute of Television Engineers

ITE Institute of Terrestrial Ecology [of Natural Environmental Research Council, UK]

ITE Institute of Traffic Engineers [Name Changed to Institute of Transportation Engineers, US]

ITE Institute of Transportation Engineers [US]

ITE Instrumentation Test Equipment

ITE Instytutu Technologii Electronowej [Institute of Electron Technology, of Wroclaw Polytechnic, Poland]

ITE Integrated Test Equipment; Integration Test Equipment

ITE Intermediate Temper Embrittlement [Metallurgy]

ITE Inter-Site Transportation Equipment

ITE Intercity Transportation Efficiency

ITE Interface-Layer Thermionic Emission

ITE Intermediate Temperature Embrittlement [Metallurgy]

ITE International Telephone Exchange

ITEA International Technology Education Association [US]

ITEA International Test and Evaluation Association [US]

ITEA-CS International Technology Education Association–Council for Supervisors [US]

ITEA J Test Eval ITEA Journal of Test and Evaluation [of International Test and Evaluation Association, US]

ITEC Information Technology Center

ITEJ Institute of Television Engineers of Japan [Tokyo]

ITE J ITE Journal [Published Institute of Transportation Engineers, US]

ITEM Interference Technology Engineer's Master

ITEMS Imaging Technologies and Evolving Management Systems

ITEMS Incoterm Transaction Entry Management System

ITEO International Trade and Employment Organization

ITEP Institute of Theoretical and Experimental Physics [Moscow, Russia]

ITEPWS ITEP (Institute of Theoretical and Experimental Physics) Winter School of Physics [Moscow, Russia]

ITER International Thermonuclear Experimental Reactor

ITER EDA ITER Engineering Design Activities (Agreement)

ITES Instituto Tecnológico de Estudios Superiores [Technical Institute of Higher Studies, Mexico]

ITESM Instituto Tecnológico de Estudios Superiores de Monterrey [Monterrey Institute of Technology and Higher Learning, Mexico]

ITEWS Integrated Tactical Electronic Warfare System

ITF Input Tube Feeder

ITF Institut Textile de France [Textile Institute of France]

ITF Integrated Test Facility

ITF Interactive Test Facility

ITF Interactive Terminal Facility

ITF Interface File

ITF International Transfer Format

ITF International Transport Workers Federation [UK]

ITF Interstitial Transfer Facility

ITF Isothermal Forging; Isothermally Forged [Powder Metallurgy]

ITFS Instructional Television Fixed Service [US]

ITFTRIA Instrument Tree Flow and Temperature Removal Instrument Assembly

ITG Informationstechnische Gesellschaft [Information Technology Society; of Verein Deutscher Elektrotechniker, Germany]

ITG Information Theory Group [of Institute of Electrical and Electronics Engineers, US]

ITG International Teledata Group

ITG-Fachber ITG-Fachberichte [Reports on Information Technology; published by Verein Deutscher Elektrotechniker/Informationstechnische Gesellschaft, Germany]

ITGLWF International Textile, Garment and Leather Workers Federation

ITH Instituto Technológico de Hermosillo [Hermosillo Institute of Technology, Mexico]

ITH In-the-Hole Drilling

ITh Intrinsic Thermoelasticity

ITI Industrial Technology Institute [US]

ITI Industrial Training Institute

ITI Information Technology Institute [Singapore]

ITI Inspection and Test Instruction

ITI Inspection/Test Instruction

ITI Institute for Technical Interchange [Hawaii, US]

ITI Institute of Translation and Interpreting [UK]

ITI Interactive Terminal Interface

ITI Intermittent Trouble Indication

ITI International Technology Institute [US]

ITI International Training Institute

ITIA Industrial Tungsten Industry Association [UK]

ITIA International Tungsten Industry Association

ITIC International Tsunami Information Center [Hawaii, US]

ITIR Infrared Thermal Imaging Radiometer [now Advanced Spaceborne Thermal Emission and Reflection Radiometer]

ITIR Intermediate Thermal Infrared Radiometer

ITIRC IBM (Corporation) Technical Information Retrieval Center [US]

ITIS Insect Toxicologists Information Service [Netherlands]

ITIS Interagency Taxonomic Information System

ITKR Institut Tekhnicheskata Kibernetika i Robotikata [Institute of Engineering Cybernetics and Robotics, Sofia, Bulgaria]

ITL Industrial Test Laboratory [US Navy]

ITL Information Technology Laboratory [of National Institute of Standards and Technology, Gaithersburg, Maryland, US]

ITL Integrated Transfer Launch; Integrate, Transfer and Launch

ITL Integrity Test Laboratory [Downsview, Ontario, Canada]

ITL Intent-to-Launch

ITL Intermediate Text Language

ITL Intermediate Transfer Language

ITL Inverse Time Limit

ITL Italian Lira [Currency of Italy and Vatican City]

ITM Indirect Tag Memory

ITM Industry Trade Fair Malaysia

ITM Information Transfer Module

ITM Innertransition Metal

ITM Institut de Technologie Magnésium [Institute of Magnesium Technology, Ste. Foy, Quebec, Canada]

ITM Institute for Theoretical Metallurgy [of Rheinisch-Westfälische Technische Hochschule, Aachen, Germany]

ITM Institute of Travel Management [UK]

ITM Instituto Tecnológico de Morelia [Morelia Institute of Technology, Mexico]

ITMA Institute of Trade Mark Agents [UK]

ITMA Instituto Tecnológico de Materiales de Asturias [Asturias Institute of Materials Technology; of Consejo Superior de Investigaciónes Cientificas; located at Oviedo, Spain]

ITMA International Tanning Manufacturers Association

ITMA Irradiation Test Management Activity

ITMC International Transmission Maintenance Center

ITME Instytutu Technologii Materialov Electronowej [Institute of Electronic Materials Technology, Warsaw, Poland]

ITMF International Textile Manufacturers Federation [Switzerland]

ITMG Integrated Thermal/Micrometeoroid Garment [Aerospace]

ITMJ Incoming Trunk Message Junction

ITMS Ion Trap Mass Spectrometer

ITMT Intermediate Thermomechanical Treatment [Metallurgy]

ITN Identification Tasking and Networking

ITN Independent Television News (Service) [of Independent Television, London, UK]

ITN Integrated Teleprocessing Network

ITNAA Instrumental Activation Analysis with Thermal Neutrons

itnl internal

ITNS Integrated Tactical Navigation System

ITNSA Item Net Sales Amount

ITO India Tourist Office

ITO Indium-Tin Oxide

ITO Instituto Tecnológico de Oaxaca [Oaxaca Institute of Technology, Mexico]

ITO Integration and Test Order

ITO International Trade Organization [of United Nations]

ITOA Independent Terminal Operators Association [US]

Itogi Nauki Tekh, Issled Kosm Prostran Itogi Nauki i Tekhniki, Issledovanie Kosmicheskogo Prostranstva [Russian Journal on Cosmic Research and Science]

Itogi Nauki Tekh, Kompoz Mater Itogi Nauki i Tekhniki, Kompozitsionnye Materialy [Russian Journal on Composite Materials]

Itogi Nauki Tekh, Metalloved Term Obrab Itogi Nauki i Tekhniki, Metallovedenie i Termicheskaya Obrabotka [Russian Journal on Metal Science and Heat Treatment]

Itogi Nauki Tekh, Metall Teplotekh Itogi Nauki i Tekhniki, Metallurgicheskaya Teplotekhnika [Russian Metallurgical Journal]

Itogi Nauki Tekh, Metal Tsvetn Metall Itogi Nauki i Tekhniki, Metallurgiya Tsvetnykh Metallov [Russian Journal on Metallurgical and Metal Engineering]

Itogi Nauki Tekh, Poroshk Metall Itogi Nauki i Tekhniki, Poroshkovaga Metallurgiya [Russian Journal on Powder Metallurgy]

Itogi Nauki Tekh, Proizv Chugina Stali Itogi Nauki i Tekhniki, Proizvodstvo Chugina i Stali [Russian Journal on Iron and Steel]

Itogi Nauki Tekh, Prokatnoe Volchil'noe Proizv Itogi Nauki i Tekhniki, Prokatnoe i Volochil'noe Proizvodstvo [Russian Technical Journal]

Itogi Nauki Tekh, Svarka Itogi Nauki i Tekhniki, Svarka [Russian Journal on Welding]

I-TOOA Independent Truck Owner/Operator Association [US]

ITOPF International Tanker Owners Pollution Federation [UK]

ITOS Improved Television and Infrared Observation Satellite

ITOS Improved TIROS (Television and Infrared Observation Satellite) Operational Satellite [of National Oceanic and Atmospheric Administration, US]

ITOS Interactive Terminal Operating System

ITOS Iterative Time Optimal System

ITO/Si Indium-Tin Oxide/Silicon (Junction)

ITP Idiopathic Thrombocytopenic Purpura [Medicine]

ITP Information Technology Program [of Public Works and Government Services Canada]

ITP Inosine Triphosphate; Inosine 5'-Triphosphate [Biochemistry]

ITP Inspection Test Procedure

ITP Institute for Theoretical Physics [of University of California at Santa Barbara, US]

ITP Institute of Theoretical Physics [Calcutta, India]

ITP Institute of Theoretical Physics [Moscow, Russia]

ITP Integral Thermal Process

ITP Integrated Test Program

ITP Interactive Terminal Protocol

ITP Internal Target Physics (Experiment) [Netherlands]

ITR Internet Talk Radio

ITP Interstitial Thickening Process

ITP Isotachophoresis [Physical Chemistry]

5'-ITP Inosine-5'-Triphosphate [Biochemistry]

ItP Italian Patent [also It Pat, or ITP]

ITPA Independent Telephone Pioneers Association

ITPA International Truck Parts Association [US]

ITPA Irish Trade Protection Association

ITPA Italian Patent Application

It Pat Italian Patent [also ItP, or ITP]

ITPC International Television Program Center

ITPI International Transfer Printing Institute [US]

ITPO International TOGA (Tropical Ocean Global Atmosphere) Project Office [of World Meteorological Organization]

ITPP Institute of Technical Publicity and Publications

ITPR Infrared Temperature Profile Radiometer

ITPS Integrated Teleprocessing System

ITQ Individual Transferable Quota

ITQC Induction Tooling Quick Change

ITR Ignition Test Reactor

ITR In-Core Thermionic Reactor

ITR Institute for Telecommunications Research [Canada]

ITR Instrument Test Rig

ITR Instytutu Tele- i Radiotechnicznego [Institute of Television and Radio Engineering, Warsaw, Poland]

ITR Integrated Telephone Recorder

ITR INTELSAT (International Telecommunications Satellite Organization) Test Record

ITR Interactive Teleprocessing System

ITR Interactive Text Processing System

ITR Interkantonales Technikum Rapperswil [Intercantonal College of Technology at Rapperswil, Switzerland]

ITR Interrupt Control Unit

ITR Inverse Time Relay

ITR Ion Transfer Reaction

ITR Isolation Test Routine

ITRC International Tin Research Council [UK]

ITRDS Integrated Test Requirements Document(ation)

ITRI Industrial Technology Research Institute [Hsinchu, Taiwan]

ITRI International Tin Research Institute [UK]

ITRIA Instrument Tree Removable Instrument Assembly

ITRM Inverse Thermoremanent Magnetization

ITRPF International Tire, Rubber and Plastics Federation [UK]

ITS Design Engineering and Automation Components Trade Fair [Germany]

ITS Idaho Test Station [US]

ITS Indigotrisulfonate

ITS Industrial Trade Show

ITS Inelastic Tunneling Spectroscopy

ITS Information Theory Society [now Information Theory Group of the Institute of Electrical and Electronics Engineers, US]

ITS Information Transmission System

ITS Insertion Test Signal

ITS Institute of Telecommunication Sciences [US]

ITS Instituto Tecnologia de la Seguridad [Institute for Safety Technology, Spain]

ITS Instituto Tecnologico de Saltillo [Saltillo Institute of Technology, Mexico]

ITS Instrumentation Telemetry Station

ITS Insulation Test Specification

ITS (Bovine) Insulin–(Human) Transferrin–Sodium Selenite (Mixture) [Biochemistry]

ITS In-Tank Solidification

ITS Integrated Test System

ITS Integrated Tracking System

ITS Integrated Trajectory System

ITS Integrated Truss Structure [of International Space Station]

ITS Interim Teleprinter System

ITS Intelligent Terminal System

ITS Intelligent Test System

ITS Interactive Terminal Support

ITS Interactive Training System

ITS Interface Test Set

ITS Intermarket Trading System [US]

ITS International Telecommunications Society

ITS International Temperature Scale

ITS International Tesla Society [US]

ITS International Trade Secretariat

ITS International Trade Show

ITS International Turfgrass Society [US]

ITS Inter-Time Switch

ITS Invitation to Send

ITS Ion Trap System

ITS-90 International Temperature Scale of 1990 [also T_{90}]

I&TS Information and Technology Services

ITSA Independent Tank Storage Association [UK]

ITSA Institute for Telecommunication Sciences and Aeronomy [of Environmental Science Services Administration, US]

ITSA Institute of Trading Standards Administration [UK]

ITSA International Thermal Spray Association

ITSC International Telecommunications Satellite Consortium [US]

ITSC International Telecommunications Services Complex

ITSC International Telephone Services Center

ITSC International Thermal Spray Conference

ITSG International Tin Study Group

ITSL Institute of Transport Studies

IT/SP Instrument Tree/Spool Piece [Computers]

ITSS International Team for Studying Sintering [later International Institute for the Science of Sintering]

ITSU Information Technology Standards Unit [of Department of Trade and Industry, UK]

ITT Impact Transition Temperature

ITT Indian Institute of Technology [Bombay]

ITT Institute of Textile Technology [Charlottesville, Virginia, US]

ITT International Telephone and Telegraph [US]

ITT International Towing Tank Conference [Japan]

ITT Interrogator-Transponder Technique

ITT Intertoll Trunk

ITT Invitation to Tender

ITTA Insurance Technology Trade Association [UK]

ITTA International Tropical Timber Agreement

ITTC International Towing Tank Conference [Japan]

ITTC International Tropical Timber Council

ITTFL International Telephone and Telegraph Federal Laboratories [US]

ITTLP Information Technology and Telecommunication Investment Program [of Western Economic Diversification Canada]

ITTIP Information Technology and Telecommunication Loan Program [of Western Economic Diversification Canada]

ITTO International Tropical Timber Organization

ITU Ikutoku Technical University [Japan]

ITU Institute for Transuranium Elements [of Joint Research Center at Karlsruhe, Germany]

ITU Integrated Terrain Unit [Geographical Information System]

ITU International Technical University

ITU International Telecommunication Union [of United Nations]

ITU International Telegraphic Union

ITU International Turbidity Unit

ITU International Typographical Union

ITU Istanbul Technical University [Turkey]

ITUC Instituto de Tecnológico de Universidade Católica [Institute of Technology of the Catholic University, Rio de Janeiro, Brazil]

ITUE Instuitute for Transuranium Elements

ITUG International Tandem (Computer) Users Group

ITUG International Telecommunications User Group

ITUSA Information Technology Users Association [UK]

ITU-T International Telecommunication Union–Telecommunication Standards Section [of United Nations] [Formerly Comité Consultatif International Télégraphique et Téléphonique] [also ITU-TSS]

ITU-TIES International Telecommunication Union–Telecom Information Exchange Services [of United Nations]

ITU-TSS International Telecommunication Union–Telecommunication Standards Section [of United Nations] [Formerly Comité Consultatif International Télégraphique et Téléphonique] [also ITU-T]

ITV Improved TOW (Tube-Launched, Optically-Tracked, Wire-Guided) Vehicle

ITV Independently Targeted Vehicle

ITV Independent Television [London, UK]

ITV Industrial Television

ITV Instructional Television

ITV Interactive Television

ITVA International Television Association [US]

ITVS International Television Symposium

ITW Institut für Spanende Technologien und Werkzeugmaschinen [Institute for Machining Technologies and Machine Tools, Germany]

ITW (Fraunhofer-)Institut für Transporttechnik und Warendistribution [(Fraunhofer) Institute for Transportation Engineering and Goods Distribution, Dortmund, Germany]

ITWS International Tsunami Warning System

ITX Intermediate Text Block

ITZ Instituto Tecnologico de Zacatecas [Zacatecas Institute of Technology, Mexico]

ITZ Interfacial Transition Zone [Materials Science]

ITZ Intertidal Zone [Oceanography]

ITZN International Trust for Zoological Nomenclature

IU Ibaraki University [Japan]

IU Image Understanding

IU Indiana University [Bloomington, US]

IU Indian Union

IU Information Unit

IU Inhibitary Unit [Biochemistry]

IU Input Unit

IU Instruction Unit

IU Instrument(ation) Unit

IU Integer Unit

IU Interface Unit

IU Interference Unit

IU International Unit [also iu]

IU International University [New York City, US]

IU Intrauterine [also iu]

IU *(In Utero)* – Within the Uterus [also iu]

IU Iwate University [Morioka, Japan]

iu *(in utero)* – within the uterus [also IU]

IUA Inertial Unit Assembly

IUA Interface Unit Adapter

IUA International Union of Architects

IUAA International Union of Advertisers Associations [now World Federation of Advertisers, Belgium]

IUAI International Union of Aviation Insurers [UK]

IUAJ International Union of Agricultural Journalists

IUAO International Union for Applied Ornithology [Germany]

IUAP Internet User Account Provider

IUAP Inter-University Attraction Poles [Belgium]

IUAPPA International Union of Air Pollution Prevention Associations [UK]

IUAPPA Newsl IUAPPA Newsletter [of International Union of Air Pollution Associations, UK]

IUAS International Union in Agricultural Sciences

IUB International Union of Biochemistry [US]

IUB International Universities Bureau

IUBMB International Union for Biochemistry and Molecular Biology [UK]

IUBS International Union of Biological Sciences [France]

IUC International Union of Chemistry

IUC International Union of Crystallography [UK]

IUC International University Consortium [US]

IUC Inter-University Council

IUCADC Inter-Union Commission of Advice to Developing Countries [Canada]

IUCAF Inter-Union Commission on Allocation of Frequencies

IUCC Inter-University Committee on Computing

IUCD Intrauterine Contraceptive Device

IUCED Inter-Union Commission of European Dehydrators [France]

IUCF Indiana University Cyclotron Facility [Bloomington, US]

IUCFA Inter-Union Commission on Frequency Allocations (for Radio Astronomy and Space Science) [US]

IUCN International Union for Conservation of Nature and Natural Resources [now World Conservation Union]

IUCN Bull IUCN Bulletin [of International Union for Conservation of Nature and Natural Resources

IUCNGA IUCN General Assembly

IUCNPSG International Union for Nature and Natural Resources–Primate Specialists Group

IUCr International Union of Crystallography [UK]

IUCRM Inter-Union Commission on Radio Meteorology

IUCS Instrumentation Unit Update Command System

IUCS Instrumentation Update Command System

IUCS Inter-Union Commission on Spectroscopy

IUCST Inter-Union Commission on Science Teaching

IUCSTP Inter-Union Commission on Solar Terrestrial Physics [now Scientific Committee on Solar Terrestrial Physics, US]

IUCSTR Inter-Union Commission on Solar Terrestrial Relationships [US]

IUD Institute for Urban Design [US]

IUD Intrauterine Device

IUDH In-Service Unplanned Derated Hours

IUDH1 In-Service Unplanned Derated Hours, Class 1

IUDR Iododeoxyuridine [Biochemistry]

IUDZG International Union of Directors of Zoological Gardens [Canada]

IUE Interface Unit Error-Count Table

IUE International Ultraviolet Explorer (Mission) [of US, UK and European Space Agency in 1978]

IUE International Union of Electrical, Radio and Machine Workers

IUEC International Union of Elevator Constructors [UK]

IUEF International University Exchange Fund

IUF Institut Universitaire de France [of University Institute of France]

IUF International University Foundation [US]

IUFoST International Union of Food Science and Technology [Ireland]

IUFRO International Union of Forestry Research Organizations [Austria]

IUFRO News IUFRO News [of International Union of Forestry Research Organization, Austria]

IUG ICES (Integrated Civil Engineering System) Users Group [US]

IUG Intercom Users Group [US]

IUGB International Union of Game Biologists [Poland]

IUGG International Union of Geodesy and Geophysics [of American Geological Union, US]

IUGG International Union of Geology and Geophysics [of International Council of Scientific Unions]

IUGRI International Union of Graphic Reproduction Industries [France]

IUGS International Union of Geological Sciences

IUHPS International Union for History and Philosophy of Science

IUHS International Union of History of Science

IUI International Union of Interpreters

IUIN International Union for Inland Navigation [France]

IUIS International Union of Immunological Studies

IUIS/WHO International Union of Immunological Studies/ World Health Organization (Study)

IUJ International University of Japan

IUKT Internationale Universitätswochen für Kern– und Teilchenphysik [International University Week for Nuclear and Particle Physics, University of Graz, Austria]

IUL Information Utilization Laboratory [of University of Pittsburgh, Pennsylvania, US]

IULCS International Union of Leather Chemists Societies

IULD International Union of Lorry Drivers [Germany]

IULEC Inter-University Labor Education Committee [US]

IULTCS International Union of Leather Technologists and Chemists Societies

IU/mg Inhibitary Unit per Milligram [Biochemistry]

IUMI International Union of Marine Insurance [Scotland]

IUMP International Union of Master Painters [Germany]

IUMP International Upper Mantle Project [Geology]

IUMRS International Union of Materials Research Societies

IUMRS-ICA International Union of Materials Research Societies–International Conference in Asia

IUMRS-ICAM International Union of Materials Research Societies–International Conference on Advanced Materials

IUMRS-ICEM International Union of Materials Research Societies–International Conference on Electronic Materials

IUMS International Union of Microbiological Societies [UK]

IUMSBD International Union of Microbiological Societies– Bacteriological Division [UK]

IUNDH In-Service Unit Derated Hours

IUNS International Union of Nutritional Sciences

IUOE International Union of Operating Engineers [UK]

IUP Indiana University of Pennsylvania [US]

IUP Indiana University Press [Bloomington, US]

IUP Installed User Program

IUP International University Press [New York City, US]

IUP Irish University Press

IUP Israel University Press [Jerusalem]

IUPAB International Union of Pure and Applied Biophysics [Hungary]

IUPAB Biophys Ser IUPAP Biophysics Series [of International Union for Pure and Applied Biophysics, Hungary]

IUPAC International Union of Pure and Applied Chemistry [UK]

IUPAC-WAM International Union of Pure and Applied Chemistry–Workshop on Advanced Materials

IUPAP International Union of Pure and Applied Physics [Sweden]

IUPAP-ICTP Symp International Union of Pure and Applied Physics–International Center for Theoretical Physics Symposium

IUPESM International Union for Physical and Engineering Sciences in Medicine [Canada]

IUPHAR International Congress of International Union of Pharmacology

IUPHAR International Union of Pharmacology [UK]

IUPIP International Union for the Protection of Industrial Property

IUPIW International Union of Petroleum Industrial Workers

IUPN International Union for the Protection of Nature

IUPPS International Union of Prehistoric and Protohistoric Sciences [Belgium]

IUPS International Union of Physiological Sciences [France]

IUPsyS International Union for Psychological Sciences

IUPU Indiana University–Purdue University [Indianapolis, US] [also IU-PU]

IUQR International Union of Quarternary Research

IUR International Union of Radioecologists [Belgium]

IUR International Union of Railways [also Union Internationale des Chemins de Fer, France]

IUR Inventory Update Rule [of US Environmental Protection Agency]

IURC International Union for Research of Communication [Switzerland]

IUREP International Uranium Resources Evaluation Project

IURP Integrated Unit Record Processor

IURS International Union of Radio Sciences [US]

IUS Inertial Upper Stage [Aerospace]

IUS Information Unit Separator

IUS Interchange Unit Selector

IUS Interchange Unit Separator

IUS Interim Upper Stage [Aerospace]

IUS Interim Use Sheet

IUS Intermediate Upper Stage [Aerospace]

IUS International Ultraviolet Explorer

IUS International Union of Speleology [Austria]

IUS International Union of Students

IUS Inter-University Seminar

IUSF International Union for Surface Finishing [UK]

IUSF International Union of Societies of Foresters [Canada]

IUS/ITB Interchange Unit Separator/Intermediate Transmission Block

IUSM Indiana University School of Medicine [Indianapolis, US]

IUSM International Union of Surveys and Mapping

IUMSWA Industrial Union of Marine and Shipbuilding Workers of America [US]

IUSF International Union of Societies of Foresters

IUSO Institute of University Safety Officers [UK]

IUSSI International Union for the Study of Social Insects [Netherlands]

IUSSP International Union for the Scientific Study of Population [Belgium]

IUT Institut Universitaire de Technologie [University Institute of Technology, France]

IUTAM International Union of Theoretical and Applied Mechanics [Germany]

IUTAM/IAHR Symp International Union of Theoretical and Applied Mechanics/International Association for Hydraulic Research Symposium (on Ice-Structure Interaction)

IUTDMM International Union of Tool, Die and Mold Makers

IUTOX International Union of Toxicology

IUTS Inter-University Transit System [Canada]

IUVSTA International Union for Vacuum Science, Technique and Applications [US]

IUWA International Union of Women Architects [France]

IUWPM Independent University, Washington–Paris–Moscow

IV Independent Verification

IV Induced Voltage

IV Industrial Ventilation

IV *(Industrieverband)* – German for "Industrial Association"

IV Influenza Virus

IV Initial Value [Mathematics]

IV Initial Velocity

IV Intake Valve

IV Integrated Vehicle

IV Interactive Video

IV Interface Volume [Materials Science]

IV Intermediate Vacuum

IV Intermediate Valence

IV Intermediate Voltage [also iv]

IV Intervehicular [Aerospace]

IV Intravehicular [Aerospace]

IV Intravenous(ly) [also iv]

IV Intrinsic Viscosity

IV Inverse Voltage

IV Inverter [also Iv, or iv]

IV Invisibility; Invisible

IV Iodine Value

IV Isolation Valve

IV Isovalerianic Acid

IV Current/Voltage [also C/V, or C-V]

I/V Instrument/Visual

I-V Current-Voltage [also CV, or C/V]

IVA Industrieverband Agrar e.V. [Association of Agricultural Industries, Germany] [also iva]

IVA Intravehicular Activity [Aerospace]

IV-A Influenza Virus, A-Strain

IV-A$_2$ Influenza Virus, Asian Strain

IVAC International Vitamin A Consultative Group

IVAR Insertion Velocity Adjust Routine [of NASA]

IVAR Internal Variable
IVAS International Veterinary Acupuncture Society
IVB Intermediate Vector Boson [Particle Physics]
IV-B Influenza Virus, B-Strain
IVBC Integrated Vehicle Baseline Configuration
IVC Inferior Vena Cava [Anatomy]
IVC Interactive Videodisk Consortium [US]
IVC Intermediate Velocity Cloud
IVC International Vacuum Congress
IVC International Veterinary Congress
IVC Intervehicular Communications
IV-C Influenza Virus, C-Strain
IVCAL Interactive Video Computer-Assisted Learning
IVCE International Video and Communications Exhibition
IVC/ICSS International Vacuum Congress/International Conference on Solid Surfaces
Iv Cst Ivory Coast
IVST Intervalence Charge Transfer [Materials Science]
IVD Image Velocity Deceptor
IVD Innovative Vehicle Design
IVD Interactive Videodisc (Technology)
IVD Ion Vapor Deposition
IVDG Innovative Vehicle Design Group
IVDP Inter-Valley Deformation Potential
IVDS Independent Variable Depth Sonar
IVDS Induced Valence Defect Structure [Materials Science]
IVDT Integrated Voice-Data Terminal
IVDW Integrated Voice/Data Workstation
IVE Institute of Vitreous Enamellers [UK]
IVE Interface Verification Equipment
IVE Isobutyl Vinyl Ether
IVE Isocyanate Vinyl Ester
IVEM Institute of Virology and Environmental Microbiology [of Natural Environmental Research Council, UK]
IVEM Intermediate Voltage Electron Microscope; Intermediate Voltage Electron Microscopy
IVESC International Vacuum Electron Sources Conference
IVETA International Vocational Education and Training Association
IVF Institut foer Verkstadsteknisk Forskning [Institute for Laboratory Technique Research, Sweden]
IVF In-vitro Fertilization
IVFRC In VFR (Visual Flight Rules) Condition
IVG Interrupt Vector Generator
IVGA Israel Vegetable Growers Association
IVHM Integrated (Space) Vehicle Health Monitoring
IVHM In-Vessel Handling Machine [Nuclear Reactors]
IVHM-EM In-Vessel Handling Machine–Engineering Model [Nuclear Reactors]
IVHS Intelligent Vehicle Highway System
IVHU In-Vessel Handling Unit [Nuclear Reactors]
IVI Incremental Velocity Indicator
IVI Interactive Visual Interface

IVIA Interactive Video Industry Association [US]
IVIA International Videotext Industry Association
IVIC Instituto Venezolano de Investigaciones Cientificas [Venezuelan Institute for Scientific Research, Caracas]
IVICS Interactive Visual Image Classification System
IVIPA International Videotext Information Providers Association
IVIS Inertial Vibration Isolation System
IVIS Interactive Video Information System
IvKS Instituut voor Kern– en Stralingsfysica [Institute for Nuclear and Radiation Physics, of Katholieke Universiteit Leuven, Belgium] [also IKS]
IVL Independent Vendor League
IVL Intel Verification Laboratory [of Intel Corporation, US]
IVL Intervalometer [also Ivl]
IVM Initial Virtual Memory
IVM Interactive Volume Modeling
IVM Interface Virtual Machine
IVMC International Vacuum Microelectronics Conference
IVMS Integrated Voice Messaging System
IVMU Inertial Velocity Measurement Unit
ivn intravenous
IVNAA In Vivo Neutron Activation Analysis
IVO Improved-Virtual-Orbital (Method)
I-V-O (Self-)Interstitial-Vacancy-Oxygen (Ternary System) [Materials Science]
IVP Initial Value Problem [Mathematics]
IVP Initial Vapor Pressure
IVP Installation Verification Procedure
IVP Installation Verification Program
IVP Intravenous Pyelogram; Intravenous Pyelography
IVPA Independent Video Programmers Association [US]
IVPO Inside Vapor-Phase Oxidation
IVR Induction Voltage Regulator
IVR Instrumental Visual Range
IVR Integrated Voltage Regulator
IVR Interactive Voice Response
IVR Intramolecular Vibrational Relaxation
IVR Intramolecular Vibrational (Energy) Redistribution
IVRG International Verticillium Research Group [of University of Guelph, Ontario, Canada]
IVRI Indian Veterinary Research Institute
IVRRF In Vivo Radioassay and Research Facility
IVS Indian Vacuum Society
IVS (Informationsvermittlungsstelle) – German for "Information Brokerage Agency"
IVS Interactive Videodisk System
IVS International Vacuum Society
IVS In-Vessel Storage [Nuclear Reactors]
IVSA International Veterinary Students Association [Netherlands]
IVSI Inertial-Lead Vertical Speed Indicator
IVSI Instantaneous Vertical Speed Indicator

IVSM In-Vessel Storage Module [Nuclear Reactors]

IVSU International Veterinary Students Union

IVT Integrated Video Terminal

IVT Interrupt Vector Table

IVT Intervehicular Transfer

IVT Current–Voltage–Temperature (Method)

I-V/T Current–Voltage versus Temperature (Diagram)

IVV Independent Verification and Validation [also IV&V]

IVVF Independent Validation and Verification Facility [of NASA at Fairmont, West Virginia, US] [also IV&VF]

IVVS In-Vessel Vehicle System [Nuclear Reractors]

IVW Informationsstelle über die Verbreitung von Werbeträgern [Information Office for the Distribution of Advertising Materials and Media, Germany]

IW Ice Water

IW Idler Wave

IW Illuviation Weathering [Geology]

IW Index Word

IW Induction Welding

IW Industrial Waste

IW Inertia Wave

IW Inside Wire

IW Insolation Weathering [Geology]

IW Institute der deutschen Wirtschaft [German Economic Institute]

IW Instruction Word

IW International Workshop

IW Isle of Wight [UK]

IW Isotopic Weight

IWA Inland Waterways Association [UK]

IWA International Waterproofing Association [Belgium]

IWA International Wheat Agreement

IWA International Woodworkers of America

IWAA International Workshop on Accelerator Alignment [Physics]

IWAHMA Industrial Warm Air Heater Manufacturers Association

IWAP International Watershed Advocacy Project [now International Rivers Network, US]

Iwatsu Tech Rep Iwatsu Technical Report [Published by Iwatsu Electric Co. Ltd., Tokyo, Japan]

I-WAY Information Highway [also I-Way]

IWB Instruction Word Buffer

IWBS Indirect Work Breakdown Structure

IWC Ice Water Content

IWC Intermetallic-Bonded Tungsten Composite

IWC Internationale Ausstellung Wäscherei–Chemischreinigung [International Exhibition for Laundry and Dry Cleaning, Germany]

IWC International Welding Conference

IWC International Whaling Commission [UK]

IWC International Whaling Convention [US]

IWC International Wheat Council [UK]

IWC Interim Wilderness Committee

IWC International Wildlife Coalition [US]

IWCA Inside Wiring Cable

IWCA International Workshop on Cometary Astronomy

IWCC International Wrought Copper Council [UK]

IW CCL International Workshop on the Critical Current Limitations (in High-Temperature Superconductors)

IWCI Industrial Wire Cloth Institute [now American Wire Cloth Institute, US]

IWCS Integrated Wideband Communications System

IWCS/SEA Integrated Wideband Communications System/Southeast Asia

IWD Inland Waters Directorate [Canada]

IWDA Independent Wire Drawers Association

IWE Institut für Werkstoffe der Energietechnik [Institute for Energy Technology Materials; of Kernforschungsanlage Julich, Germany]

IWE Institution of Water Engineers [UK]

IWE Inverse Wiedemann Effect [Solid-State Physics]

IWEM Institution of Water and Environmental Management [UK]

IWES International Waste Energy System

IWF Industrial Water Facility

IWF Industry Workers Federation [San Marino]

IWF Institut für den Wissenschaftlichen Film GmbH [Scientific Film Institute, Gottingen, Germany]

IWF Institut für Werkstoff-Forschung [Institute of Materials Research; of Deutsche Forschungs– und Versuchsanstalt für Luft– und Raumfahrt e.V., Cologne, Germany]

IWFAC International Workshop on Fullerenes and Atomic Clusters

IWFNA Inhibited White Fuming Nitric Acid

IWFS Industrial Waste Filtering System

IWG Impacts Working Group [of Intergovernmental Panel on Climate Change]

IWG Interagency Working Group

IWG Interface Working Group

IWG International Working Group

IWG Investigator Working Group

IWG Iron Wire Gauge

IWGA International Wheat Gluten Association [US]

IWGDMGC Interagency Working Group on Data Management for Global Change

IWGDMGC International Working Group on Data Management for Global Change

IWGGCDM International Working Group on Global Change and Data Management

IWGGE Interdepartmental Working Group on the Greenhouse Effect [Northern Territory, Australia]

IWH Institut für Wirtschaftsforschung Halle [Halle Institute for Economic Research, Halle Germany]

IWHM Institution of Works and Highways Management [UK]

IWHS Institute of Works and Highways Superintendents [UK]

IWI Inventors Workshop International [now Inventors Workshop International Education Foundation, US]

IWIEF Inventors Workshop International Education Foundation [US]

IWIM Institut für Wissenschaftsinformation in der Medizin [Institute for Scientific Information in Medicine, Germany]

IWL Institut für gewerbliche Wasserwirtschaft und Luftreinhaltung e.V [Institute for Industrial Water Supply and Air Pollution Control, Cologne, Germany]

IWM Industrial Waste Management

IWM Institute of Waste Management [UK]

IWM Institute of Works Managers [UK]

IWM Institut für Werkstoffmechanik [Institute for the Mechanics of Mechanics; of Fraunhofer-Institut, Germany]

IWM Intermediate Wet Modulus

IWMA Institute of Weights and Measures Administration [UK]

IWMA International Wire and Machinery Association [UK]

IWMB Integrated Waste Management Board [of California Environmental Protection Agency, US]

IWO Institution of Water Officers [UK]

IWOF Instantaneous Work of Fracture

IWOMP International Workshop on Metastable Phases [Materials Science]

IWOP Integration Within an Ordered Product of Operators

IWP Ice Water Path

IWP Intelligent Work in Process

IWP Interim Working Party

IWP International Wildlife Park [Grand Prairie, Texas, US]

I-WP I-Graph Wrapped Package (for Triply Periodic Minimal Surfaces) [Materials Science]

IWPA International Word Processing Association [later International Information/Word Processing Association; now Association of Information Systems Professionals, US]

IWPA International Work Platform Association

IWPC Institute of Water Pollution Control [UK]

IWPO International Word Processing Organizations

IWPPA Independent Waste Water Processors Association [UK]

IWQ Ice Water Quench(ing) [Metallurgy]

IWQ Index of Wilderness Quality

IWR Institute for Water Resources [of Department of Army, US]

IWR Institute for Wildlife Research [US]

IWR Institute of Wood Research [of Michigan Technological University, Houghton, US]

IWR Isolated Word Recognition

IWRA International Water Resources Association [US]

IWRA International Wild Rice Association [US]

IWRB International Waterfowl Research Bureau [now International Waterfowl and Wetlands Research Bureau, UK]

IWRC IBM (Corporation) Watson Research Center [Yorktown Heights, New York, US]

IWRC Independent Wire Rope Core

IWRC International Wildlife Rehabilitation Council [US]

IWRI International Waterfowl Research Institute [UK]

IWRP Industrial Waste Reduction Program [US]

IWRPF International Waste Rubber and Plastics Federation

IWRS International Wood Research Society

IWS Industrial Water Society [UK]

IWS Industrial Workstation

IWS Institute of Water Study [UK]

IWS Institute of Wood Science [UK]

IWS Institute of Work Studies

IWS Integrated Work Sequence

IWS Intelligent Workstation

IWS International Wool Secretariat [UK]

IWSA International Water Supply Association [UK]

IWSc Institute of Wood Science [UK]

IWSG International Wool Study Group [UK]

IWS/IT Integrated Work Sequence/Inspection Traveller

IWSP Institute of Work Study Practitioners

IWSS International Weed Science Society [US]

IWTA Inland Water Transport Authority [Pakistan]

IWtC International Wheat Council

IWTF Intractable Wastes Task Force

IWTO International Wool and Textile Organization [Belgium]

IWTRC International Wool Textile Research Conference

IWU Illinois Wesleyan University [Bloomington, US]

IWU Isolation Working Unit

IWW Industrial Workers of the World [US]

IWWA International Water Works Association [Israel]

IWWA International Wild Waterfowl Association [US]

IWWK Institut für Weltwirtschaft an der Universität Kiel [Kiel University Institute for World Economy, Germany]

IWWRB International Waterfowl and Wetlands Research Bureau [formerly International Waterfowl Research Bureau, UK]

IX Index Register

IX Image Exchange

IX Inter-Exchange [Telecommunications]

IX Internet Exchange

IX Intersystem Crossing

IX Inverted Index

IX Ion Exchange(r)

I(x) Indefinite Integral [Symbol]

I(x) Intensity at Point x [Symbol]

IXA Ion-Excited X-Ray Analysis

IXAE International X-Ray Astrophysics Explorer

IXC Interchange Carrier

IXC Inter-Exchange Channel [Telecommunications]

IXEE International X-Ray and Extreme Ultraviolet Explorer

IXSD International Telex Subscriber Dialing

IXSS X-Ray Scattering Spectroscopy

IXT Interaction Crosstalk

I₀(x,y) Displacement-Dependent Brightfield Intensity (in Electron Microscopy) [Symbol]

$I_g(x,y)$ Displacement-Dependent Darkfield Intensity (in Electron Microscopy) [Symbol]

IYA Irish Yachting Association [UK]

IYBCO Indium Yttrium Barium Copper Oxide (Superconductor)

IYF International Youth Federation (for Environmental Studies and Conservation) [Denmark]

IYKWIM If You Know What I Mean

IYKWIMAITYD If You Know What I Mean And I Think You Do

IYNF International Young Nature Friends

IYSWIM If You See What I Mean

IZ Isolation Zone

IZA International Zeolite Association

IZAA Independent Zinc Alloyers Association [US]

IZE International Association of Zoo Educators [US]

IZE Inverse Zeeman Effect [Spectroscopy]

IZFP (Fraunhofer-)Institut für zerstörungsfreie Prüfverfahren [(Fraunhofer) Institute for Nondestructive Testing, Saarbrücken, Germany]

Izmer Tekh Izmeritel'naya Tekhnika [Russian Journal on Measurement Techniques]

i-ZnSe Intrinsic Zinc Selenide (Semiconductor)

IZO Indium-Doped Zinc Oxide

IZSBC Indium Fluoride–Zinc Fluoride–Strontium Difluoride–Barium Difluoride–Calcium Fluoride (Glass) [InF_3–ZnF_3–SrF_2–BaF_2–CaF_2]

IZSBCNGd Indium Fluoride–Zinc Fluoride–Strontium Difluoride–Barium Difluoride–Calcium Fluoride–Sodium Fluoride–Gadolinium Fluoride (Glass) [InF_3–ZnF_3–SrF_2–BaF_2–CaF_2–NaF–GdF_3]

IZSBGdCN Indium Fluoride–Zinc Fluoride–Strontium Difluoride–Barium Difluoride–Gadolinium Trifluoride–Calcium Fluoride–Sodium Fluoride (Glass) [InF_3–ZnF_3–SrF_2–BaF_2–GdF_3–CaF_2 –NaF]

IZSBGdGaN Indium Fluoride–Zinc Fluoride–Strontium Difluoride–Barium Difluoride–Gadolinium Trifluoride–Gallium Trifluoride–Sodium Fluoride (Glass) [InF_3–ZnF_3–SrF_2–BaF_2–GdF_3–GaF_3–NaF]

IZSBGdL Indium Fluoride–Zinc Fluoride–Strontium Difluoride–Barium Difluoride–Gadolinium Trifluoride–Lanthanum Trifluoride (Glass) [InF_3–ZnF_3–SrF_2–BaF_2–GdF_3–LaF_3]

IZSBGdN Indium Fluoride–Zinc Fluoride–Strontium Difluoride–Barium Difluoride–Gadolinium Trifluoride–Sodium Fluoride (Glass) [InF_3–ZnF_3–SrF_2–BaF_2–GdF_3–NaF]

IZSBGdY Indium Fluoride–Zinc Fluoride–Strontium Difluoride–Barium Difluoride–Gadolinium Trifluoride–Yttrium Trifluoride (Glass) [InF_3–ZnF_3–SrF_2–BaF_2–GdF_3–YF_3]

IZSBNGaGd Indium Fluoride–Zinc Fluoride–Strontium Difluoride–Barium Difluoride–Sodium Fluoride–Gallium Trifluoride–Gadolinium Trifluoride (Glass) [InF_3–ZnF_3–SrF_2–BaF_2–NaF–GaF_3–GdF_3]

IZSBPbCCd Indium Fluoride–Zinc Fluoride–Strontium Difluoride–Barium Difluoride–Lead Fluoride–Calcium Fluoride–Cadmium Difluoride (Glass) [InF_3–ZnF_3–SrF_2–BaF_2–PbF_2–CaF_2–CdF_2]

IZT Imidazoline-2-Thione

Izv Akad Nauk Arm SSR, Fiz Izvestiya Akademii Nauk Armyanskoi SSR, Fizika [Journal on Contemporary Physics; published by the Academy of Sciences of the Armenian SSR]

Izv Akad Nauk Arm SSR, Mekh Izvestiya Akademii Nauk Armyanskoi SSR, Mekhanika [Journal on Mechanics; published by the Academy of Sciences of the Armenian SSR]

Izv Akad Nauk Arm SSR, Tekh Nauk Izvestiya Akademii Nauk Armyanskoi SSR, Tekhnicheskaya Nauk [Journal on Technical Sciences; published by the Academy of Sciences of the Armenian SSR]

Izv Akad Nauk Azerb SSR, Fiz-Tekh Mat Nauk Izvestiya Akademii Nauk Azerbaidzhanskoi SSR, Seriya Fiziko-Tekhnicheskikh i Matematicheskh Nauk [Journal on Technical Physics and Mathematical Sciences; published by the Academy of Sciences of the Azerbaidzhanian SSR]

Izv Akad Nauk BSSR Izvestiya Akademii Nauk BSSR [Journal of the Academy of Sciences of the Byelorussian SSR]

Izv Akad Nauk Est SSR, Fiz Mat Izvestiya Akademii Nauk Estonskoi SSR, Fizichesko, Matematicheskaya [Journal on Physical and Mathematical Sciences; published by the Academy of Sciences of the Estonian SSR]

Izv Akad Nauk Est SSR, Khim Izvestiya Akademii Nauk Estonskoi SSR, Khimicheskaya [Journal on Chemical Sciences; published by the Academy of Sciences of the Estonian SSR]

Izv Akad Nauk Gruz SSR, Khim Izvestiya Akademii Nauk Gruzinskoi SSR, Seriya Khimicheskaya [Journal on Chemical Sciences; published by the Academy of Sciences of the Georgian SSR] [also Inv Akad Nauk GSSR]

Izv Akad Nauk Kazakh SSR Izvestiya Akademii Nauk Kazakhskoi SSR [Journal of the Academy of Sciences of the Kazakh SSR]

Izv Akad Nauk Kazakh SSR, Fiz-Mat Izvestiya Akademii Nauk Kazakhskoi SSR, Fiziko-Matematicheskaya [Journal on Physical and Mathematical Sciences; published by the Academy of Sciences of the Kazakh SSR]

Izv Akad Nauk Kazakh SSR, Khim Izvestiya Akademii Nauk Kazakhakoi SSR, Khimicheskaya [Journal on Chemical Sciences; published by the Academy of Sciences of the Kazakh SSR]

Izv Akad Nauk Kirgiz SSR Izvestiya Akademii Nauk Kirgizkoi SSR [Journal of the Academy of Sciences of the Kirghizian SSR]

Izv Akad Nauk Latv SSR, Fiz-Tekh Izvestiya Akademii Nauk Latviyskoi SSR, Fiziko-Tekhnicheskikh [Journal on Technical Physics; published by the Academy of Sciences of the Latvian SSR]

Izv Akad Nauk Latv SSR, Khim Izvestiya Akademii Nauk Latviyskoi SSR, Khimicheskikh [Journal on Chemistry; published by the Academy of Sciences of the Latvian SSR]

Izv Akad Nauk Mold SSR, Fiz-Tekh Mat Nauk Izvestiya Akademii Nauk Moldavskoi SSR, Fiziko-Tekhnicheskikh i Matematicheskikh Nauk [Journal on Technical Physics and Mathematical Sciences; published by the Academy of Sciences of the Moldavian SSR]

Izv Akad Nauk SSSR Izvestiya Akademii Nauk SSSR [Journal Series published by the Academy of Sciences of the USSR]

Izv Akad Nauk SSSR, Energ Transp Izvestiya Akademii Nauk SSSR, Energetika i Transport [Journal on Power Engineering; published by the Academy of Sciences of the USSR]

Izv Akad Nauk SSSR, Fiz Izvestiya Akademii Nauk SSSR, Fizicheskaya [Journal on Physics; published by the Academy of Sciences of the USSR]

Izv Akad Nauk SSSR, Fiz Atmos Okeana Izvestiya Akademii Nauk SSSR, Fizika Atmosfery i Okeana [Journal on Atmospheric and Ocean Physics; published by the Academy of Sciences of the USSR]

Izv Akad Nauk SSSR, Fiz Zemli Izvestiya Akademii Nauk SSSR, Fizika Zemli [Journal on the Physics of the Solid Earth; published by the Academy of Sciences of the USSR]

Izv Akad Nauk SSSR, Khim Izvestiya Akademii Nauk SSSR, Khimicheskaya [Journal on Chemistry; published by the Academy of Sciences of the USSR]

Izv Akad Nauk SSSR, Mekh Tverd Tela Izvestiya Akademii Nauk SSSR, Mekhanika Tverdogo Tela [Journal on the Mechanics of Solids; published by the Academy of Sciences of the USSR]

Izv Akad Nauk SSSR, Mekh Zhidh Gaza Izvestiya Akademii Nauk SSSR, Mekhanika Zhidkosti i Gaza [Journal on Fluid Dynamics; published by the Academy of Sciences of the USSR]

Izv Akad Nauk SSSR, Met Izvestiya Akademii SSSR, Metally [Journal on Russian Metallurgy; published by the Academy of Sciences of the USSR]

Izv Akad Nauk SSSR, Neorg Mater Izvestiya Akademii Nauk SSSR, Neorganicheskie Materialy [Journal on Inorganic Materials; published by the Academy of Sciences of the USSR]

Izv Akad Nauk SSSR Otd Tekh Nauk Izvestiya Akademii Nauk SSSR, Otdeleniya Tekhncheskikh Nauki [Journal on Applied Physics; published by the Academy of Sciences of the USSR]

Izv Akad Nauk SSSR Tekh Kibern Izvestiya Akademii Nauk SSSR, Tekhncheskikh Kibernetika [Journal on Technical Cybernetics; published by the Academy of Sciences of the USSR]

Izv Akad Nauk Turkm SSR Izvestiya Akademii Nauk Turkmenskoi SSR [Journal of the Academy of Science of the Turkomanian SSR]

Izv Akad Nauk Turkm SSR, Fiz-Tekh Khim Geol Nauk

Izvestiya Akademii Nauk Turkmenskoi SSR, Fiziko-Tekhnicheskikh Khimicheskikh i Geologicheskikh Nauk [Journal on Technical, Chemical and Geological Physics; published by the Academy of Sciences of the Turkomanian SSR]

Izv Akad Nauk USSR Ser Matemat Izvestiya Akademii Nauk USSR Seriya Matematiki [Journal on Mathematics; published by the Academy of Sciences of the USSR]

Izv Akad Nauk Uzb SSR, Fiz-Mat Nauk Izvestiya Akademii Nauk Uzbekskoi SSR, Fiziko-Matemati-cheskikh Nauk [Journal on Physical and Mathematical Sciences; published by the Academy of Sciences of the Uzbek SSR]

Izv Akad Nauk Uzb SSR, Tekh Nauk Izvestiya Akademii Nauk Uzbekskoi SSR, Tekhnicheskikh Nauk [Journal on Technical Sciences; published by the Academy of Sciences of the Uzbek SSR]

Izv Akad Nauk UzSSR, Fiz-Mat Nauk Izvestiya Akademii Nauk Uzbekskoi SSR, Fiziko-Matematicheskikh Nauk [Journal on Physical and Mathematical Sciences; published by the Academy of Sciences of the Uzbek SSR]

Izv Akad Nauk UzSSR, Tekh Nauk Izvestiya Akademii Nauk Uzbekskoi SSR, Tekhnicheskikh Nauk [Journal on Technical Sciences; published by the Academy of Sciences of the Uzbek SSR]

Izv Acad Sci USSR, Atmos Ocean Phys Izvestiya Academy of Sciences USSR, Atmospheric and Oceanic Physics [Translation of: *Izvestiya Akademii Nauk SSSR, Fizika At-mosfery i Okeana (USSR)*; published in the US]

Izv Acad Sci USSR, Phys Solid Earth Izvestiya Academy of Sciences USSR, Physics of the Solid Earth [Translation of: *Izvestiya Akademii Nauk SSSR, Fizika Zemli (USSR)*; published in the US]

Izv AN GSSR, Khim Izvestiya Akademii Nauk Gruzinskoi SSR, Seriya Khimicheskaya [Journal on Chemical Sciences; published by the Academy of Sciences of the Georgian SSR] [also Inzv Akad Nauk GSSR]

Izv AN SSSR, Energ Transp Izvestiya Akademii Nauk SSSR, Energetika i Transport [Journal on Power Engineering; published by the Academy of Sciences of the USSR]

Izv AN SSSR, Fiz Izvestiya Akademii Nauk SSSR, Fizicheskaya [Journal on Physics; published by the Academy of Sciences of the USSR]

Izv AN SSSR, Fiz Atmos Okeana Izvestiya Akademii Nauk SSSR, Fizika Atmosfery i Okeana [Journal on Atmospheric and Oceanic Physics; published by the Academy of Sciences of the USSR]

Izv AN SSSR, Fiz Zemli Izvestiya Akademii Nauk SSSR, Fizika Zemli [Journal on the Physics of the Solid Earth; published by the Academy of Sciences of the USSR]

Izv AN SSSR, Khim Izvestiya Akademii Nauk SSSR, Khimicheskaya [Journal on Chemistry; published by the Academy of Sciences of the USSR]

Izv AN SSSR, Mekh Tverd Tela Izvestiya Akademii Nauk SSSR, Mekhanika Tverdogo Tela [Journal on the Mechanics of Solids; published by the Academy of Sciences of the USSR]

Izv AN SSSR, Mekh Zhidh Gaza Izvestiya Akademii Nauk SSSR, Mekhanika Zhidkosti i Gaza [Journal on Fluid Dynamics; published by the Academy of Sciences of the USSR]

Izv AN SSSR, Met Izvestiya Akademii SSSR, Metally [Journal on Russian Metallurgy; published by the Academy of Sciences of the USSR]

Izv AN SSSR, Neorg Mater Izvestiya Akademii Nauk SSSR, Neorganicheskie Materialy [Journal on Inorganic Materials; published by the Academy of Sciences of the USSR]

Izv AN SSSR Otd Tekh Nauk Izvestiya Akademii Nauk SSSR, Otdeleniya Tekhncheskikh Nauki [Journal on Applied Physics; published by the Academy of Sciences of the USSR]

Izv AN SSSR Tekh Kibern Izvestiya Akademii Nauk SSSR, Tekhncheskikh Kibernetika [Journal on Cybernetics; published by the Academy of Sciences of the USSR]

Izv Gl Astron Obs Pulkove Izvestiya Glavnoi Astronomicheskoi Observatorii v Pulkove [Russian Bulletin of the Pulkove Astronomical Observatory]

Izv Khim Izvestiya po Khimiya [Journal on Chemistry; published by the Bulgarian Academy of Sciences]

Izv Khim, Bulg Akad Nauk Izvestiya po Khimiya, Bulgarska Akademiya na Naukite [Journal on Chemistry; published by the Bulgarian Academy of Sciences]

Izv Krym Astrofiz Obs Izvestiya Krymskoi Astroficheskoi Observatorii [Bulletin of the Crimean Astrophysical Observatory]

Izv RAN, Ser Fiz Izvestiya Russkoi Akademii Nauk, Seriya Fizitcheskaya [Physics Journal of the Russian Academy of Sciences]

Izv Ser Mat Izvestiya Seriya Matematicheskaya [Mathematical Journal of the Academy of Sciences of the USSR]

Izv Sibir Otd Akad Nauk SSSR, Khim Izvestiya Sibirskogo Otdeleniya Akademii Nauk SSSR, Khimicheskikh [Journal on Chemistry; published by the Academy of Sciences of the USSR]

Izv Sibir Otd Akad Nauk SSSR, Tekh Izvestiya Sibirskogo Otdeleniya Akademii Nauk SSSR, Tekhnicheskikh [Journal on Applied Physics; published by the Academy of Sciences of the USSR]

Izv VMEI 'Lenin' Izvestiya na VMEI 'Lenin' [Bulgarian Scientific Journal]

Izv VUZ Aviats Tekh Izvestiya Vysshikh Uchebnykh Zavedenii, Aviatsionnaya Tekhnika [Russian Journal on Aeronautics]

Izv VUZ Chernaya Metall Izvestiya Vysshikh Uchebnykh Zavedenii, Chernaya Metallurgiya [Russian Journal on Steel and Metallurgy]

Izv VUZ Elektromekh Izvestiya Vysshikh Uchebnykh Zavedenii, Elektromekhnika [Russian Journal on Electromechanics]

Izv VUZ Energ Izvestiya Vysshikh Uchebnykh Zavedenii, Energetika [Russian Journal on Energy Science and Engineering]

Izv VUZ Fiz Izvestiya Vysshikh Uchebnykh Zavedenii, Fizika [Russian Physics Journal]

Izv VUZ Khim Khim Tekhnol Izvestiya Vysshikh Uchebnykh Zavedenii, Khimiya i Khimicheskaya Tekhnologiya [Russian Journal on Chemistry and Chemical Technology]

Izv VUZ, Lesnoi Zh Izvestiya Vysshikh Uchebnykh Zavedenii, Lesnoi Zhurnal [Russian Journal on Forestry]

Izv VUZ Mashinostr Izvestiya Vysshikh Uchebnykh Zavedenii, Mashinostroenie [Russian Journal on Engineering Research]

Izv VUZ Mat Izvestiya Vysshikh Uchebnykh Zavedenii, Matematika [Russian Journal on Mathematics]

Izv VUZ Radioelektron Izvestiya Vysshikh Uchebnykh Zavedenii, Radioelektronika [Russian Journal on Radioelectronics and Communication Systems]

Izv VUZ Radiofiz Izvestiya Vysshikh Uchebnykh Zavedenii, Radiofizika [Russian Journal on Radio-Physics and Quantum Electronics]

Izv VUZ Tekhnol Legkoi Prom Izvestiya Vysshikh Uchebnykh Zavedenii, Tekhnologiya Legkoi Promyshlennosti [Ukrainian Journal]

Izv VUZ Tekhnol Tekstil Prom Izvestiya Vysshikh Uchebnykh Zavedenii, Tekhnologiya Tekstil'noi Promyshlennosti [Russian Journal on Textile Technology]

Izv VUZ Tsvetn Metall Izvestiya Vysshikh Uchebnykh Zavedenii, Tsvetnaya Metallurgiya [Russian Journal on Nonferrous Metals Research]

Izv Vyssh Uchebn Zaved Aviats Tekh Izvestiya Vysshikh Uchebnykh Zavedenii, Aviatsionnaya Tekhnika [Russian Journal on Aeronautics]

Izv Vyssh Uchebn Zaved Chernaya Metall Izvestiya Vysshikh Uchebnykh Zavedenii, Chernaya Metallurgiya [Russian Journal on Steel and Metallurgy]

Izv Vyssh Uchebn Zaved Elektromekh Izvestiya Vysshikh Uchebnykh Zavedenii, Elektromekhnika [Russian Journal on Electromechanics]

Izv Vyssh Uchebn Zaved Energ Izvestiya Vysshikh Uchebnykh Zavedenii, Energetika [Russian Journal on Energy Science and Engineering]

Izv Vyssh Uchebn Zaved Fiz Izvestiya Vysshikh Uchebnykh Zavedenii, Fizika [Russian Physics Journal]

Izv Vyssh Uchebn Zaved Khim Khim Tekhnol Izvestiya Vysshikh Uchebnykh Zavedenii, Khimiya i Khimicheskaya Tekhnologiya [Russian Journal on Chemistry and Chemical Technology]

Izv Vyssh Uchebn Zaved Lesnoi Zh Izvestiya Vysshikh Uchebnykh Zavedenii, Lesnoi Zhurnal [Russian Journal on Forestry]

Izv Vyssh Uchebn Zaved Mashinostr Izvestiya Vysshikh Uchebnykh Zavedenii, Mashinostroenie [Russian Journal on Engineering Research]

Izv Vyssh Uchebn Zaved Mat Izvestiya Vysshikh Uchebnykh Zavedenii, Matematika [Russian Journal on Mathematics]

Izv Vyssh Uchebn Zaved Radioelektron Izvestiya Vysshikh Uchebnykh Zavedenii, Radioelektronika [Russian Journal on Radioelectronics and Communication Systems]

Izv Vyssh Uchebn Zaved Radiofiz Izvestiya Vysshikh Uchebnykh Zavedenii, Radiofizika [Russian Journal on Radio-physics and Quantum Electronics]

Izv Vyssh Uchebn Zaved Tekhnol Legkoi Prom Izvestiya Vysshikh Uchebnykh Zavedenii, Tekhnologiya Legkoi Promyshlennosti [Ukrainian Journal]

Izv Vyssh Uchebn Zaved Tekhnol Tekstil Prom Izvestiya Vysshikh Uchebnykh Zavedenii, Tekhnologiya Tekstil'noi Promyshlennosti [Russian Journal on Textile Technology]

Izv Vyssh Uchebn Zaved Tsvetn Metall Izvestiya Vysshikh Uchebnykh Zavedenii, Tsvetnaya Metallurgia [Russian Journal on Nonferrous Metals Research]

IZW Informationszentrum Wärmepumpen und Kältetechnik [Information Center on Heat Pumps and Refrigeration Technology–An Information Service of Fachinformationszentrum Karlsruhe, Germany]

IZW Institut für Zoo– und Wildtierforschung [Institute for Zoo and Wild Animal Research, Berlin, Germany]

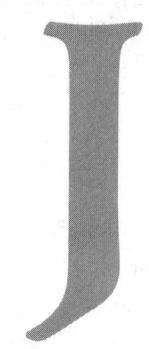

J   Angular Momentum [Symbol]

J   Coupling Constant (in Nuclear Magnetic Resonance Spectroscopy) [Symbol]

J   (Electric) Current Density [Symbol]

J   Diffusion Flux [Symbol]

J   Elastic-Plastic Crack Extension Force [Symbol]

J   Exchange Coupling [Symbol]

J   Ferromagnetic Interaction [Symbol]

J   Gram-Equivalent Weight [Symbol]

J   Ion Dose [Symbol]

J   Jack [also j]

J   Jacket

J   Jacobian [Mathematics]

J   Jade [Mineral]

J   Jam [Controllers]

J   Jamb

J   January

J   Japan(ese)

J   Jaundice [Medicine]

J   Jaw

J   Jejunum [Medicine]

J   Jelly

J   Jet [also j]

J   Jig

J   J-Integral [Symbol]

J   Job

J   Join(ing)

J   Joint

J   Joist

J   Joule [Unit]

J   Joule's Constant [Symbol]

J   Journal(ism); Journalist(ic)

J   Judge(ment)

J   Juglans [Genus of Deciduous Trees Including the Walnut and Butternut]

J   Juliet [Phonetic Alphabet]

J   Jump(er)

J   Jump (Command) [Computer Programming]

J   Junction

J   Juniper

J   Juniperus [Genus of Coniferous Trees Including the Junipers and Redcedars]

J   Jurassic [Geology]

J   Justice

J   Jute

J   J-Values [Spectroscopy]

J   Magnetic Polarization [Symbol]

J   Mechanical Equivalent of Heat [Symbol]

J   Nucleation Rate [Symbol]

J   Number of Equivalents [Symbol]

J   Polarization [Symbol]

J   Polar Moment of Inertia [Symbol]

J   Power Loss Factor (of a Thrust Bearing) [Symbol]

J   Radiant Intensity [Symbol]

J   Rotational Quantum Number [Symbol]

J   Second Polar Moment of Area [Symbol]

J   Seepage Force (for Soil or Rock) [Symbol]

J   Shear Compliance [Symbol]

J   Special Test, Temporary (Aircraft, Missile, or Rocket) [USDOD Symbol]

J   Spin-Spin Coupling Constant [Symbol]

J   Supercurrent (or Superfluid Current) [Symbol]

J   Superexchange Constant (in Solid-State Physics) [Symbol]

J   Tensile Modulus (for Geotextiles) [Symbol]

J   Thickness of Tang (of Drills [Symbol]

J   (Gear) Tooth Form Factor [Symbol]

J   Total Angular Momentum (of Orbital Electrons) [Symbol]

J   Tunneling Matrix Element (in Electronics) [Symbol]

J   Vertex Distance (of Bevel Gears) [Symbol]

-J   Target Tug (Aircraft) [US Navy Suffix]

J^* Complex Shear Compliance [Symbol]

J^* Nucleation Rate [Symbol]
     Diffusive Flux [Symbol]

J_0 Offset Tensile Modulus (for Geotextiles) [Symbol]

J_0 Saturation Current Density at Zero Applied Voltage [Symbol]

J_0 Zero-Order Bessel Function [Symbol]

J_{Ic} Plain-Strain Fracture Toughness (in Mechanics) [Symbol]

J_1 First-Order Bessel Function [Symbol]

J_a Atom Flux [Symbol]

J_c Collector Current Density [Semiconductor Symbol]

J_c Critical Current Density [Symbol]

J_c Equilibrium Volume Fraction of Chains in Bulk Polymer [Symbol]

J_{ct} Transport Critical Current Density (of Superconductors) [Symbol]

J_e Electron Flux [Symbol]

J_{ec} Jominy Equipment Cooling [Symbol]

J_H Hund's Coupling Parameter (in Solid-State Physics) [Symbol]

J_i Initial Tensile Modulus [Symbol]

J_i Ion Flux [Symbol]

J_{ij} Exchange Integral [Symbol]

J_{ij} Exchange Interaction Between Nearest Neighbor Particles ij [Symbol]

J_λ Intensity of Monochromatic Radiation of Wavelength λ [Symbol]

J_M Polar Moment of Inertia of Masses [Symbol]

J_N Molecule Flow Rate Density [Symbol]

J_q Heat Flux [Symbol]

J_R Crack-Extension Resistance (for J-Integral) [Symbol]

J_r Remanent (Magnetic) Polarization [Symbol]

J_s Spontaneous (Magnetic) Polarization [Symbol]

J_{sc} Short Circuit Current [Symbol]

J_{sec} Secant Tensile Modulus (for Geotextiles) [Symbol]

J_{TQ} Critical J-Integral [Symbol]

J_x Diffusion Flux in x Direction [Symbol]

J_x Polar Moment of Inertia with Respect to X-Axis [Symbol]

J_y Polar Moment of Inertia with Respect to Y-Axis [Symbol]

J_z Polar Moment of Inertia with Respect to Z-Axis [Symbol]

j   (electric) current density [Symbol]

j   imaginary quantity (e.g., $j = \sqrt{-1}$; $j^2 = -1$) [Symbol]

j   ion dose rate [Symbol]

j   jack [also J]

j   jet

j   join

j   journal

j   magnetic dipole moment [Symbol]

j   rotational quantum number of electron [Symbol]

j   running index [e.g., $\sum_{j=1}^{n} a_j = a_1 + a_2 + \cdots + a_j$][Symbol]

j   square root of –1 (for complex variable) [Symbol]

j   summation index [Symbol]

j   unit vector along positive y-axis [Symbol]

j   vertex distance at small end of bevel gear [Symbol]

j_∞ current density at time "t" approaching infinity [Symbol]

$\pm j$ yellowness-blueness (opponent-color scale) [Optical Society of America Uniform Color Scales]

j^* dimensionless current density [Symbol]

j_0 exchange current density [Symbol]

j_c capacitive current density [Symbol]

j_c ion current density [Symbol]

j_{eh} Jominy equivalent hardness [Symbol]

j_F faradaic current density [Symbol]

j_{gg} limiting current density [Symbol]

j_m average current density [Symbol]

j_p pulsed current density [Symbol]

j_t total current density [Symbol]

j_v vacancy (or point defect) generation rate [Symbol]

JA   Japan Academy [Tokyo]

JA   Jet Aircraft; Jet Airplane

JA   Jiles-Atherton (Hysteresis Model) [Solid-State Physics]

JA   Johnston Atoll [North Pacific Island]

JA   Joint Account [also J/A]

JA   Joliet Arsenal [Joliet, Illinois, US]

JA   Journal A [Engineering Journal of Koninklije Vlaamse Ingenieurs-Vereniging, Belgium]

JA   Journal Announcement

JA   Journal Article

JA   Judge Advocate

JA   Jump Address

JA   Jump if Above

JA-   Japan [Civil Aircraft Marking]

J/A   Joint Account [also JA]

Ja   January

JAA   Japan Art Academy

JAA   Joint Airworthiness Authority [UK]

JAAF   Joint Army–Air Force

JAALD   Japan Association of Agricultural Librarians and Documentalists

JAAS   Journal of Analytical Atomic Spectrometry [of Royal Society of Chemistry, UK]

JAAVSO   Journal of the American Association of Variable Star Observers

JAB   Japan Accreditation Board (for Quality System Registration)

J Abn Soc Psychol   Journal of Abnormal and Social Psychology

JABPPC   Joint Animal By-Products Committee[UK]

JAC   Joint Advisory Committee

JAC   Joint Astronomy Center [Hawaii, US]

JACADS   Johnston Atoll Chemical Agent Disposal System

JACC   Joint Automatic Control Conference [US]

J Account EDP   Journal of Accounting and Electronic Data Processing [US]

JACE   Joint Alternate Command Element

JACH   Journal of the American College of Health

JACI   Journal of Allergy and Clinical Immunology [US]

J-Acid   7-Amino-4-Hydroxy-2-Naphthalenesulfonic Acid

J-Acid   2-Amino-5-Naphthol-7-Sulfonic Acid

JACM   Journal of the Association for Computing Machinery [US]

JACOLA   Joint Airports Committee of Local Authorities

J Acoust   Journal d'Acoustique [French Journal of Acoustics]

J Acoust Emiss   Journal of Acoustic Emission [Published by the Acoustic Emission Group, US]

J Acoust Soc Am   Journal of the Acoustical Society of America

J Acoust Soc India   Journal of the Acoustical Society of India

J Acoust Soc Jpn   Journal of the Acoustical Society of Japan

JACS   Journal of the American Chemical Society [US]

JAD   Jamaican Dollar [Currency]

JAD   Joint Analysis and Design

JAD   Joint Application Design

JADA   Joint Agency Data Agreement

JADA Journal of the American Dental Association [US]

JADA Journal of the American Dietetic Association [US]

Jad Energ Jaderna Energie [Czech Publication on Energy Technology]

J Adhes Journal of Adhesion [UK]

J Adhes Sci Technol Journal of Adhesion Science and Technology [Netherlands]

J Adolesc Journal of Adolescence [Published in the US]

J Adv Transp Journal of Advanced Transportation [of Advanced Transit Association, US]

J Adv Mater Journal of Advanced Materials [of Society for the Advancement of Material and Process Engineering, US]

J Adv Med Journal of Advancement in Medicine [of American College for the Advancement of Medicine]

JAE Jump if Above or Equal

JAEC Japan Atomic Energy Commission

JAEIC Joint Atomic Energy Intelligence Committee

JAEIP Japan Atomic Energy Insurance Pool

JAERI Japan Atomic Energy Research Institute

J Aeronaut Mater Journal of Aeronautical Materials [PR China]

J Aeronaut Sci Journal of Aeronautical Sciences [US]

J Aeronaut Soc India Journal of the Aeronautical Society of India [Published by the Indian Institute of Science, India]

J Aerosol Med Journal of Aerosol Medicine [of International Society for Aerosols in Medicine, Austria]

J Aerosol Res Journal of Aerosol Research [Japan]

J Aerosol Res Jpn Journal of Aerosol Research Japan

J Aerosol Sci Journal of Aerosol Science [of Gesellschaft für Aerosolforschung, Germany]

J Aerosp Sci Journal of Aerospace Science [US]

JAF Job Accounting Facility

JAFA Japan Auto-Focus Association

JAFAE Japan Auto-Focus Association in Europe

J Affect Disorders Journal of Affective Disorders [US]

JAFNA Joint Air Force-NASA [US]

J Afr Earth Sci Journal of African Earth Sciences [Published in the UK]

JAG Judge Advocate General [US]

JAGC Judge Advocate General Corps [US]

JAGOS Joint Air-Ground Operations System

J Agric Eng Res Journal of Agricultural Engineering Research

J Agric Food Chem Journal of Agricultural and Food Chemistry [of American Chemical Society, US]

JAGS Journal of the American Geriatrics Society [US]

Jahangirnagar Rev A, Sci Jahangirnagar Review, Part A, Science [Bangladesh]

JAI Job Accounting Interface

JAIC Journal of the American Institute of Conservation [US]

JAICI Japan Association for International Chemical Information

JAIEG Joint Atomic Information Exchange Group [of US Department of Defense]

JAIF Japan Atomic Industrial Forum

JAIST Japan Advanced Institute of Science and Technology [Ishikawa]

J Aircr Journal of Aircraft [of American Institute of Aeronautics and Astronautics, US]

J Air Pollut Control Assoc Journal of the Air Pollution Control Association [US]

J Air Waste Manage Assoc Journal of the Air Waste Management Association [US]

JAJO January-April-July-October (Disbursements) [Plant Engineering]

JAL Japan Air Lines

J Algorithms Journal of Algorithms [US]

J Allergy Clin Immunol Journal of Allergy and Clinical Immunology [US]

J Alloy Phase Diagrams Journal of Alloy Phase Diagrams [India]

J Alloys Comp Journal of Alloys and Compounds [Published in the US] [formerly J Less-Common Met]

JALPAS Japan Airlines Passenger Autoprocessing System

JALPG Joint Automatic Language Processing Group

JAM Japanese Association for Microbiology

JAM Journal of Advancement in Medicine [US]

JAM Journal of Applied Meteorology [of American Meteorological Society]

Jam Jamaica(n)

JAMA Japan Automobile Manufacturers Association

JAMA Journal of the American Medical Association

J Am Acad Dermat Journal of American Academy of Dermatology [US]

J Am Acad Orthoped Surg Journal of the American Academy of Orthopedic Surgeons [US]

JAMASS Japanese Medical Abstracts Scanning System [of International Medical Information Center, Japan]

J Am Assoc Var Star Obs Journal of the American Association of Variable Star Observers [US]

JAMBA Japan-Australia Migratory Birds Agreement

J Am Ceram Soc Journal of American Ceramic Society [US]

J Am Chem Soc Journal of the American Chemical Society [US]

J Am Coll Cardiol Journal of the American College of Cardiology [US]

J Am Coll Health Journal of the American College of Health [US]

J Am Coll Nutr Journal of the American College of Nutrition [US]

J Am Coll Toxicol Journal of the American College of Toxicology [US]

Jam$ Jamaican Dollar [Currency]

J Am Dent Assoc Journal of the American Dental Association [US]

J Am Diet Assoc Journal of the American Dietetic Association [US]

J Am Geriatr Soc Journal of the American Geriatrics Society [US]

JAMGIS Jamaica Geographic Information System

J Am Helicopter Soc Journal of the American Helicopter Society [US]

J Am Leather Chem Assoc Journal of the American Leather Chemists Association [US]

J Am Math Soc Journal of the American Mathematical Society [US]

J Am Med Assoc Journal of the American Medical Association [US]

J Am Oil Chem Soc Journal of the American Oil Chemists Society [US]

J Am Podiat Med Assn Journal of the American Podiatric Medical Association [US]

JAMS Japan Association for Mathematical Sciences

JAMS Job Activities Management System

JAMS Journal of the American Mathematical Society [US]

JAMSAT Japanese Satellite for Amateur Radio

J Am Soc Brew Chem Journal of the American Society of Brewing Chemists [US]

J Am Soc Echocardiogr Journal of the American Society of Echocardiography [US]

J Am Soc Hort Sci Journal of the American Society for Horticultural Sciences [US]

J Am Soc Inf Sci Journal of the American Society for Information Science [US]

J Am Soc Mar Eng Journal of the American Society for Marine Engineering [US]

JAMTS Japan Association of Motor Trade and Service

J Am Vet Med Assoc Journal of the American Veterinary Medical Association [US]

J Am Water Works Assoc Journal of the American Water Works Association [US]

JAN Job Accomodation Network [US]

JAN Joint Army-Navy (Program) [US]

Jan January

JANAF Joint Army-Navy-Air Force [US]

JANAIR Joint Army-Navy Aircraft Instrument Research [of US Department of Defense]

J Anal Appl Pyrol Journal of Analytical and Applied Pyrolysis

J Anal At Spectrom Journal of Analytical Atomic Spectrometry [of Royal Society of Chemistry, UK]

J Anal Math Journal d'Analyse Mathématique [Journal of Mathematical Analysis; published in Jerusalem, Israel]

JANAP Joint Army-Navy-Air Force Publication [US]

Jane's Def Wkly Jane's Defence Weekly [UK]

JANET Joint Academic Network [UK]

JANIS Joint ANZECC/MCFFA (Australian and New Zealand Environment and Conservation Council/Ministerial Council on Forestry, Fisheries and Aquaculture) NFPS (National Forest Policy Statement) Implementation Subcommittee

JANMB Joint Army-Navy Munition Board [US]

JANNAF Joint Army-Navy-NASA-Air Force (Commitee) [also JAN-NAF]

JANOT Joint Army-Navy Ocean Terminal [US]

JANP Joint Army-Navy Manual (for Electronic Equipment)

JANS Joint Army-Navy Specification [of US Department of Defense]

JANTAP Joint Army and Navy Technical Aeronautical Board [US]

J Antibiot Journal of Antibiotics [US]

J Antimicrob Chemother Journal of Antimicrobial Chemotherapy

JAOAC Journal of the Association of Official Analytical Chemists [US]

JAP Journal of Applied Physics [of American Institute of Physics, US]

Jap Japan(ese)

Japan GCR Japan Gas-Cooled Reactor

Japan Pulp Paper Japan Pulp and Paper [Publication]

Japan TAPPI Japan Technical Association of the Pulp and Paper Industry [also Japan Tappi]

Japan TAPPI J Japan TAPPI Journal [of Japan Technical Association of the Pulp and Paper Industry]

JAPATIC Japan Patent Information Center

JAPCA Journal of the Air Pollution Control Association [now Journal of the Air Waste Management Association, US]

JAPEIC Japan Power Engineering and Inspection Corporation

JAPEX Japan Petroleum Exploration

JAPIA Japan Auto Parts Industries Association

J Apic Res Journal of Apicultural Research [US]

JAPIO Japan Patent Information Organization

JapP Japanese Patent [also Jap P]

J Appl Bacteriol Journal of Applied Bacteriology [of Society for Applied Bacteriology, UK]

J Appl Biochem Journal of Applied Biochemistry [of International Union of Biochemistry, US]

J Appl Biomater Journal of Applied Biomaterials [of Society for Biomaterials, US] [also J Appl Biomat]

J Appl Chem Journal of Applied Chemistry [of Society of Applied Chemistry, US]

J Appl Chem USSR Journal of Applied Chemistry of the USSR [Translation of: *Zhurnal Prikladnoi Khimii*; published in the US]

J Appl Cryst Journal of Applied Crystallography [Denmark] [also J Appl Crystallogr]

J Appl Electrochem Journal of Applied Electrochemistry [UK]

J Appl Environ Microbiol Journal of Applied Environmental Microbiology

J Appl Geophys Journal of Applied Geophysics

J Appl Math Phys Journal of Applied Mathematics and Physics [Switzerland]

J Appl Mech Journal of Applied Mechanics [of American Society of Mechanical Engineers, US]

J Appl Mech Tech Phys Journal of Applied Mechanics and Technical Physics [Translation of: *Zhurnal Prikladnoi Mekhaniki i Tekhnicheskoi Fiziki (USSR)*; published in the US]

J Appl Mech (Trans ASME) Journal of Applied Mechanics (Transactions of the American Society of Mechanical Engineers) [US]

J Appl Metalwork Journal of Applied Metalworking [US]

J Appl Meteorol Journal of Applied Meteorology [of American Meteorological Society, US]

J Appl Microbiol Biotech Journal of Applied Microbiology and Biotechnology [Sweden]

J Appl Nutrition Journal of Applied Nutrition [US]

J Appl Phys Journal of Applied Physics [of American Institute of Physics, US]

J Appl Physiol Journal of Applied Physiology [of American Physiological Society, US]

J Appl Polym Sci Journal of Applied Polymer Science [US]

J Appl Probab Journal of Applied Probability [UK]

J Appl Psychol Journal of Applied Psychology

J Appl Social Psychol Journal of Applied Social Psychology

J Appl Spectrosc Journal of Applied Spectroscopy [Translation of: *Zhurnal Prikladnoi Spektroskopii (USSR)*; published in the US]

J Appl Syst Anal Journal of Applied Systems Analysis [UK]

J Approx Theory Journal of Approximation Theory [US] [also JAPT]

JAR Japanese Association of Refrigeration

JAR Java Archive [Data File Format]

JAR Joint Airworthiness Requirements

JAR Jump Address Register [Computers]

JARC Joint Avionics Research Committee

J Archit Plan Res Journal of Architectural and Planning Research [US]

JARD Journal of the Academy of Rheumatoid Diseases [US]

JARI Japan Association of Railway Industries

J Arid Environ Journal of Arid Environments [US]

JARL Japan Amateur Radio League

JARRP Japan Association for Radiation Research on Polymers

J Artheroscler Res Journal of Atherosclerosis Research

JAS Japan Academic Society

JAS Japanese Amateur Radio Satellite

JAS Joint Airmiss Section [UK]

JAS Journal of Aerospace Science [US]

JAS Journal of Atmospheric Science [of American Meteorological Society, US]

JAS Junior Astronomical Society [UK]

JASA Joint Airworthiness Steering Committee [Europe]

JASA Journal of the American Statistical Society [US]

JASC Japan Academic Societies Center

JASC Japan–America Student Conference [US]

JASC Japan Sea Cable

JASDF Japan Air Self-Defense Force

J Asian Afr Stud Journal of Asian and African Studies

J Asian Earth Sci Journal of Asian Earth Sciences

JASIN Joint Air-Sea Interaction Experiment

JASIS Journal of the American Society for Information Science [US]

JASO Japan Standards Organization

J Assoc Am Med Coll Journal of the Association of American Medical Colleges

J Assoc Comput Mach Journal of the Association for Computing Machinery [US]

J Assoc Explor Geophys Journal of Association of Exploration Geophysicists [of Center of Exploration Geophysics, India]

J Assoc Lunar Planet Obs Strolling Astron Journal of the Association of Lunar and Planetary Observers, Strolling Astronomer [US]

JAST Japan Association of Sugar Technologists

JAST Joint Advanced Strike Technology

J Astronaut Sci Journal of the Astronautical Sciences [of American Astronautical Society, US]

J Astron Soc Egypt Journal of the Astronomical Society of Egypt

J Astrophys Astron Journal of Astrophysics and Astronomy [of Indian Academy of Sciences]

JAT Job Accounting Table Information

JATCRU Joint Air Traffic Control Radar Unit

J At Energy Soc Jpn Journal of the Atomic Energy Society of Japan

JATIS Japan Technical Information Service [Tokyo]

J Atmos Chem Journal of Atmospheric Chemistry [US]

J Atmos Oceanic Technol Journal of Atmospheric and Oceanic Technology [US]

J Atmos Sci Journal of Atmospheric Science [of American Meteorological Society, US]

J Atmos Terr Phys Journal of Atmospheric and Terrestrial Physics [UK]

JATO Jet-Assisted Takeoff (Engine)

J Audio Eng Soc Journal of the Audio Engineering Society [US]

J Aust Ceram Soc Journal of the Australian Ceramics Society [Australia]

J Aust Geol Geophys Journal of Australian Geology and Geophysics [of Bureau of Mineral Resources, Australia]

J Aust Inst Met Journal of the Australasian Institute of Metals

J Autoimmunol Journal of Autoimmunology [Published in the US]

J Autom Chem Journal of Automatic Chemistry [UK]

J Autom Reasoning Journal of Automated Reasoning [Netherlands]

Jav Javanese

.JAVA Java File [File Name Extension] [also .java]

JAWC Joint Animal Welfare Council

JAWG Joint Airmiss Working Group [UK]

JB Jervis Bay [New South Wales, Australia]

JB Jet Barrier

JB Jet-Propelled Bomb

JB Journal Bearing

JB Junction Box

JBA John Burroughs Association [US]

J Bacteriol Journal of Bacteriology [of American Society for Microbiology, US]

J Bangladesh Acad Sci Journal of the Bangladesh Academy of Sciences [Bangladesh]

J Basic Eng Trans AIME Journal of Basic Engineering, Transactions of the American Institute of Mining, Metallurgical and Petroleum Engineers

JBBL Jamming Beacons and Blind Landing

JBC Journal of Biological Chemistry [of American Society for Biochemistry and Molecular Biology]

JBD James Brake Decelerometer

JBE Japanese B Encephalitis (Virus)

JBE Jump if Below or Equal

J Beijing Univ Iron Steel Technol Journal of the Beijing University of Iron and Steel Technology [PR China]

JBI James Brake Index

J Biochem Journal of Biochemistry [US]

J Biochem Biophys Methods Journal of Biochemical and Biophysical Methods [Netherlands]

J Biochem (Japan) Journal of Biochemistry (Japan) [Published in Tokyo]

J Bioelectr Journal of Bioelectricity [US]

J Bioenerg Biomembr Journal of Bioenergetics and Biomembranes [US]

J Bioeng Journal of Bioengineering [UK]

J Bioeng (Trans ASME) Journal of Bioengineering (Transactions of ASME) [of American Society of Mechanical Engineers, US]

J Biol Chem Journal of Biological Chemistry [of American Society for Biochemistry and Molecular Biology]

J Biol Phys Journal of Biological Physics [US]

J Biol Stand Journal of Biological Standardization [of International Union of Microbiological Societies, UK]

J Biolumin Chemilumin Journal of Bioluminescence and Chemiluminescence [UK]

J Biomater Appl Journal of Biomaterials Applications [US]

J Biomater Sci Journal of Biomaterials Science [US]

J Biomech Journal of Biomechanics [UK]

J Biomech Eng (Trans ASME) Journal of Biomechanical Engineering (Transactions of the American Society of Mechanical Engineers) [US]

J Biomed Eng Journal of Biomedical Engineering [UK]

J Biomed Mater Res Journal of Biomedical Materials Research [of Society for Biomaterials, US] [also J Biomed Mat Res]

J Biomed Sci Journal of Biomedical Sciences [International Journal]

J Biotechnol Journal of Biotechnology [US]

J Biotechnol Appl Biochem Journal of Biotechnology and Applied Biochemistry [of International Union of Biochemistry, US]

J Birla Inst Technol Sci Journal of the Birla Institute of Technology and Science [India]

JBIS Journal of the British Interplanetary Society [UK]

JBMA Japan Brass Makers Association

JBMA John Burroughs Memorial Association [now John Burroughs Association, US]

JBMMA Japan Business Machine Makers Association

JBNQA James Bay and Northern Quebec Agreement [Canada]

J Bone Jt Surg Journal of Bone and Joint Surgery [US]

J Bone Miner Res Journal of Bone and Mineral Research [US]

JBP Journal of Biological Physics [US]

J Br Astron Assoc Journal of the British Astronomical Association

J Br Interplanet Soc Journal of the British Interplanetary Society

J Br IRE Journal of the British Institution of Radio Engineers

JBS Jamaica Bureau of Standards

JBS Japanese Biochemical Society

JBS Journal of Biomedical Sciences [International Journal]

J Bus Econ Stat Journal of Business and Economic Statistics [of American Statistical Association, US]

J Bus Strat Journal of Business Strategy [US]

JC Jack Connection

JC JEDEC (Joint Electron Device Engineering Council) Committee

JC Jefferson City [Missouri, US]

JC Jersey City [New Jersey, US]

JC Jet Coefficient

JC Jet Condenser

JC Jib Crane

JC Job Control [Computers]

JC Job Corps

JC Johnson Counter

JC Joint Commission

JC Joint Committee

JC Joint Conference

JC Joint Control

JC Josephson Contact [Electronics]

JC Josephson Current [Electronics]

JC Jump if Carry

JC Junction Capacitor

JCA Joint Commission on Accreditation

JCAB Japanese Civil Aviation Bureau

JCAE Joint Commission on Adult Education [Canada]

JCAE Joint Committee on Atomic Energy [US]

JCAH Joint Commission on Accreditation of Hospitals [US]

JCAM Joint Commission on Atomic Masses

JCAM Journal of Climate and Applied Meteorology [US]

J Can Ceram Soc Journal of the Canadian Ceramic Society

J Cancer Educ Journal of Cancer Education

J Can Dent Assoc Journal of the Canadian Dental Association

J Can Diet Assoc Journal of the Canadian Dietetic Association

J Can Pet Technol Journal of Canadian Petroleum Technology

J Capacity Manage Journal of Capacity Management [of Institute of Software Engineering, US]

JCAR Joint Commission on Applied Radioactivity [France]

J Carbohyd Chem Journal of Carbohydrate Chemistry [US]

JCASR Joint Committee on Avionic Systems Research

J Catal Journal of Catalysis [US]

JCAU Joint Commission on Accreditation of Universities

JCB Job Control Block [Computers]

JCB Joint Coal Board [Australia]

JCB Joint Communications Board

JCB Journal of Cell Biology [of American Society of Cell Biology, US]

JCC Job Control Card [Computers]

JCC Joint Communications Center

JCC Joint Computer Conference [US]

JCC Joint Conference Committee

JCC Joint Consultative Committee [UK]

JCC Joint Control Center

JCC Junior Chamber of Commerce

JCCCOMNET Joint Coordination Center Communications Network

JCDA Journal of Canadian Dental Association

JcDSV Junonia coenia Densovirus

JCDT Jamaica Conservation and Development Trust

JCE Journal of Chemical Education [of American Chemical Society, US]

JCEC Joint Communications Electronics Committee [US]

JCEG Joint Communications Electronics Group

J Cell Biol Journal of Cell Biology [of American Society of Cell Biology, US]

J Cell Plast Journal of Cellular Plastics [US]

J Cell Physiol Journal of Cellular Physiology [US]

J Cell Sci Journal of Cell Science [UK]

J Cem Concr Res Journal of Cement and Concrete Research

JCENS Joint Communications-Electronics Nomenclature System

J Cent China Norm Univ (Nat Sci) Journal of Central China Normal University (Natural Sciences) [PR China]

J Cent S Inst Min Metall Journal of Central-South Institute of Mining and Metallurgy [PR China]

J Ceram Soc Jpn Journal of the Ceramic Society of Japan

JCET Joint Committee on Educational Television [US]

JCEWG Joint Communications Electronics Working Group [of NATO]

JCF Joint Coordinating Forum

JCG Joint Coordinating Group

J/°C·g Joule(s) per Degree Celsius gram [also J/g/°C, or J °C^{-1} g^{-1}]

JCGP Journal of the College of General Practitioners [UK]

JCGS Joint Center for Graduate Study

JCHARS Joint Commission on High Altitude Research Stations [Switzerland]

J Chem Crystallogr Journal of Chemical Crystallography [US]

J Chem Ecol Journal of Chemical Ecology [of International Society of Chemical Ecology, US]

J Chem Educ Journal of Chemical Education [of American Chemical Society, US]

J Chem Eng Journal of Chemical Engineering

J Chem Eng Data Journal of Chemical Engineering Data [of American Chemical Society, US]

J Chem Eng Jpn Journal of Chemical Engineering Japan

J Chem Ind Eng Journal of Chemical Industry and Engineering [PR China]

J Chem Inf Comput Sci Journal of Chemical Information and Computer Sciences [of American Chemical Society, US]

J Chem Phys Journal of Chemical Physics [of American Institute of Physics, US]

J Chem Res Journal of Chemical Research [of Société Française de Chimie, France]

J Chem Res Journal of Chemical Research [of Royal Society of Chemistry, UK]

J Chem Soc Journal of the Chemical Society [UK]

J Chem Soc A Journal of the Chemical Society A [UK]

J Chem Soc, Chem Commun Journal of the Chemical Society, Chemical Communications [of Royal Society of Chemistry, UK]

J Chem Soc, Dalton Trans Journal of the Chemical Society, Dalton Transactions [of Royal Society of Chemistry, UK]

J Chem Soc, Faraday Trans Journal of the Chemical Society, Faraday Transactions [of Royal Society of Chemistry, UK]

J Chem Soc, Faraday Trans I Journal of the Chemical Society, Faraday Transactions I [of Royal Society of Chemistry, UK]

J Chem Soc, Faraday Trans II Journal of the Chemical Society, Faraday Transactions II [of Royal Society of Chemistry, UK]

J Chem Soc Jpn Journal of the Chemical Society of Japan

J Chem Soc, Perkin Trans Journal of the Chemical Society, Perkin Transactions [of Royal Society of Chemistry, UK]

J Chem Soc, Perkin Trans I Journal of the Chemical Society, Perkin Transactions I [of Royal Society of Chemistry, UK]

J Chem Soc, Perkin Trans II Journal of the Chemical Society, Perkin Transactions II [of Royal Society of Chemistry, UK]

J Chem Technol Biotechnol Journal of Chemical Technology and Biotechnology [of Society of Chemical Industry, UK]

J Chem Thermodyn Journal of Chemical Thermodynamics [UK]

J Chim Phys Journal de Chimie Physique [Journal of Physical Chemistry; of Société Française de Chimie, France]

J Chim Phys Phys-Chim Biol Journal de Chimie Physique et de Physico-Chimie Biologique [Journal of Physical Chemistry and Biological Physical Chemistry; of Société de Chimie Physique, France]

J China Inst Commun Journal of the China Institute of Communications [PR China]

J Chin Ceram Soc Journal of the Chinese Ceramic Society [PR China]

J Chin Chem Soc Journal of the Chinese Chemical Society [PR China]

J Chin Electron Microsc Soc Journal of Chinese Electron Microscopy Society [PR China]

J Chin Foundrymen's Assoc Journal of Chinese Foundrymen's Association [Taiwan]

J Chin Inst Chem Eng Journal of the Chinese Institute of Chemical Engineers [PR China]

J Chin Inst Eng Journal of Chinese Institute of Engineers [Taiwan]

J Chin Math Soc Journal of the Chinese Mathematical Society [PR China]

J Chin Silic Soc Journal of the Chinese Silicate Society [PR China]

J Chin Soc Mech Eng Journal of the Chinese Society of Mechanical Engineers [Taiwan]

J Chin Rare Earth Soc Journal of the Chinese Rare Earth Society [PR China]

JCHPME Joint Commission on Higher Professional Medical Education

J Chromat Journal of Chromatography [Netherlands]

J Chromat Sci Journal of Chromatographic Sciences [US]

J Chron Diseases Journal of Chronic Diseases

JCHSEEP Joint Council for Health, Safety and Environmental Education of Professionals [Des Plaines, Illinois, US]

JCI Joint Communications Instruction

JCI Journal of Clinical Investigation [Published by Rockefeller University Press, US]

JCIA Japan Chemical Industries Association

JCIE Joint Center for International Exchange

JCII Japan Camera Inspection Institute

JCIST Japan Information Center of Science and Technology

JCL Job-Command Language [Computers]

JCL Job-Control Language [Computers]

JCLA Journal of Current Laser Abstracts [US]

J Clim Journal of Climate [of American Meteorological Society, US]

J Clim Appl Meteorol Journal of Climate and Applied Meteorology [US]

J Climatol Journal of Climatology [UK]

J Clin Comput Journal of Clinical Computing [US] ,

J Clin Endocrinol Journal of Clinical Endocrinology [US]

J Clin Eng Journal of Clinical Engineering [US]

J Clin Invest Journal of Clinical Investigation [Published by Rockefeller University Press, US]

J Clin Lab Invest Journal of Clinical and Laboratory Investigations [US]

J Clin Microbiol Journal of Clinical Microbiology [of American Society for Microbiology, US]

J Clin Nutr Journal of Clinical Nutrition [US]

J Clin Pathol Journal of Clinical Pathology [US]

J Clin Periodont Journal of Clinical Periodontology [US]

J Clin Pharmacol Journal of Clinical Pharmacology [US]

J Clin Psychol Journal of Clinical Psychology [US]

J Cluster Sci Journal of Cluster Science [US]

JCM Journal of Chemical Research [UK]

JCM Journal of Clinical Microbiology [of American Society for Microbiology, US]

J/cm² Joule(s) per square centimeter [also J cm^{-2}]

J/cm³ Joule(s) per cubic centimeter [also J cm^{-3}]

JCMB Joint Committee on Medicine and Biology [of Institute of Electrical and Electronics Engineers and International Society for Measurement and Control]

J/cm²K Joule(s) per square centimeter kelvin [also J cm^{-2} K^{-1}, J/cm²·K or J/cm²-K]

JCMR Japan Congress on Materials Research

JCMT James Clerk Maxwell Telescope [of Joint Astronomy Center, Hawaii, US]

Jcn Junction [also jcn]

JCNMT Joint Committee of Nordic Marine Technology [Finland]

JCNNSRC Joint Committee of the of the Nordic Natural Science Research Council [Sweden]

JCNPS Joint Committee on Nuclear Power Standards

J Coast Res Journal of Coastal Research [of Coastal Education and Research Foundation, US]

J Coated Fabrics Journal of Coated Fabrics [US]

J Coat Technol Journal of Coatings Technology [of Federation of Societies for Coatings Technology, US]

J Cogn Neurosci Journal of Cognitive Neuroscience [Published by MIT Press, Cambridge, US]

J Coll Eng, Nihon Univ, A Journal of the College of Engineering, Nihon University, A [Japan]

J Coll Eng, Nihon Univ, B Journal of the College of Engineering, Nihon University, B [Japan]

J Colloid Interface Sci Journal of Colloid Interface Science [US]

J Coll Sci Teach Journal of College Science Teaching [US]

J Comb Theory Journal of Combinatorial Theory [US]

J Comb Theory A Journal of Combinatorial Theory A [US]

J Comb Theory B Journal of Combinatorial Theory B [US]

J Commun Res Lab Journal of the Communications Research Laboratory [Japan]

J Compar Physiol Psychol Journal of Comparative and Physiological Psychology [US]

J Complex Journal of Complexity [US]

J Compos Mater Journal of Composite Materials [US]

J Compos Technol Res Journal of Composites Technology and Research [of American Society for Testing and Materials, US]

J Comput Journal of Computing [of Operations Research Society of America]

J Comput Appl Math Journal of Computational and Applied Mathematics [Netherlands]

J Comput Assist Learn Journal of Computer Assisted Learning [UK]

J Comput Assist Microsc Journal of Computer Assisted Microscopy [US]

J Comput Assist Tomogr Journal of Computer Assisted Tomography [US]

J Comput-Based Instr Journal of Computer-Based Instruction [of Association for the Development of Computer-Based Instructional Systems, US]

J Comput Chem Journal of Computational Chemistry [US]

J Comput Inf Syst Journal of Computer Information Systems [of Society of Data Educators, US]

J Comput Math Journal of Computational Mathematics [PR China]

J Comput Math Sci Teach Journal of Computers in Mathematics and Science Teaching [of Association for Computers in Mathematics and Science Teaching, US]

J Comput Phys Journal of Computational Physics [US]

J Comput Sci Technol Journal of Computer Science and Technology [PR China]

J Comput Sci Technol (Engl Lang Ed) Journal of Computer Science and Technology (English Language Edition) [PR China]

J Comput Syst Sci Journal of Computer and System Sciences [US]

J Cond Monit Journal of Condition Monitoring [UK]

J Constr Steel Res Journal of Constructional Steel Research [UK]

J Consult Psychol Journal of Consulting Psychology [US]

J Contam Hydrol Journal of Contaminant Hydrology

J Contr Rel Journal of Controlled Release [of Controlled Release Society, US]

J Cost Manage Manuf Ind Journal of Cost Management for the Manufacturing Industry [US]

JCP Job-Control Program [Computers]

JCP Joint Committee on Printing [of US Congress]

JCP Journal of Chemical Physics [of American Institute of Physics, US]

JCP Junction Call Processing [Communications]

JCPDS Joint Committee on Powder Diffraction Standards [US]

JCPDS-ICDD Joint Committee on Powder Diffraction Standards–International Center for Diffraction Data [US]

JCPS Junction Call Processing Subsystem [Communications]

JCPT Journal of Canadian Petroleum Technology [of Canadian Institute of Mining and Metallurgy]

JCPWG Joint Certification Procedures Working Group [Europe]

J Craniof Surg Journal of Craniofacial Surgery

J Craniom Disorders Journal of Craniomandibular Disorders [US]

JCRR Joint Commission on Rural Reconstruction [PR China/US]

JCRT Joint Center for Radiation Therapy [of Harvard Medical School, US]

JCRT Joint Commission on Radiologic Technology [of American Medical Association, US]

J Cryptol Journal of Cryptology [US]

J Crystallogr Soc Jpn Journal of Crystallographic Society of Japan [US]

J Crystallogr Spectrosc Res Journal of Crystallographic and Spectroscopic Research [now Journal of Chemical Crystallography, US]

J Cryst Growth Journal of Crystal Growth [Published in the Netherlands]

JCS Japan(ese) Ceramic Society

JCS Jersey Cattle Society [UK]

JCS Job-Control Statement [Computers]

JCS Joint Chiefs of Staff [of US Department of Defense]

JCS Joint Coordinate System

JCS Journal of Cell Science [UK]

JCS Journal of the Chemical Society [of Royal Society of Chemistry, UK]

JCSAN Joint Chiefs of Staff Alerting Network [of US Department of Defense]

JCSC Jersey City State College [New Jersey, US]

JCS Chem Commun Journal of the Chemical Society, Chemical Communications [of Royal Society of Chemistry, UK]

JCS Dalton Journal of the Chemical Society, Dalton Transactions (Inorganic Chemistry) [of Royal Society of Chemistry, UK]

JCS Faraday I Journal of the Chemical Society, Faraday Transactions I (Physical Chemistry in Condensed Phases) [of Royal Society of Chemistry, UK]

JCS Faraday II Journal of the Chemical Society, Faraday Transactions II (Molecular and Chemical Physics) [of Royal Society of Chemistry, UK]

JCS Perkin I Journal of the Chemical Society, Perkin Transactions I (Organic and Bioorganic Chemistry) [of Royal Society of Chemistry, UK]

JCS Perkin II Journal of the Chemical Society, Perkin Transactions II (Physical Organic Chemistry) [of Royal Society of Chemistry, UK]

JCSS Journal of Computer and Systems Science [US]

JCSTR Joint Commission on Solar and Terrestrial Relationships

JCT Jerusalem College of Technology [Israel]

JCT Job Control Table [Computers]

JCT Journal Control Table

Jct Junction [also jct]

JCTFI Joint Committee for Training in the Foundry Industry

Jctn Junction [also jctn]

JCU James Cook University (of North Queensland) [Australia]

JCU John Carroll University [Cleveland, Ohio, US]

JCULS Joint Committee on the Union List of Serials [US]

JCUNQ James Cook University of North Queensland [Australia]

JCV JC Virus

J Cytom Journal of Cytometry [of Society for Analytical Cytology, US]

JD Jet Diode

JD Jet Dryer

JD Job Description

JD Joint Distribution [Statistics]

JD John Deere Company [US]

JD Jordanian Dinar [Currency]

JD Junction Diode

JD Julian Date [Astronomy]

JD Julian Day [Astronomy]

JD *(Juris Doctor)* – Doctor of Jurisprudence

JD *(Jurum Doctor)* – Doctor of Laws

JD Justice Department

J$ Jamaican Dollar [Currency]

Jd Joined [also jd]

JDA Japan(ese) Defense Agency

J Dalian Univ Technol Journal of Dalian University of Technology [PR China]

JDBC Java Database Connectivity

JDC Job Description Card

J Dent Disease Journal of Dental Disease [US]

J Dent Res Journal of Dental Research [US]

JDES Joint Density of Electronic States

J Dev Journal of Development [of Conference on Data System Languages, US]

JDFC Joint Danube Fisheries Commission [Czech Republic]

J Diff Eq Journal of Differential Equations [US] [also J Differ Equ]

J Disp Sci Technol Journal of Dispersion Science and Technology [US]

JDK Java Development Kit [Computers]

JDL Japan Digital Laboratory

JDL Job Description Library [Computers]

JDL Job Description Language; Job Descriptor Language [Computers]

JDM Journal of Data Management [US]

J Doc Journal of Documentation [of Association for Information Management, UK]

JDOS Joint Density of States [Solid-State Physics]

JDREMC Joint Departmental Radio and Electronic Measurements Committee [UK]

JDS Job Data Sheet

J Dyn Syst Meas Control (Trans ASME) Journal of Dynamic Systems, Measurement and Control (Transactions of the American Society of Mechanical Engineers) [US]

J Dyn Syst (Trans ASME) Journal of Dynamic Systems (Transactions of the American Society of Mechanical Engineers) [US]

JE Japan Encephalitis

JE Jet Engine

JE Job Entry

JE Joint Efficiency

JE Josephson Effect [Electronics]

JE Josephson Equation [Electronics]

JE Joule Effect [Thermodynamics]

JE Jump if Equal

JE Junction Efficiency

JE Junction Exchange

J-E Current Density–Electric Field (Curve)

Je June

JEA Japan Encephalitis, Type A

JEA Jet Engine Alloy

JEA Joint Endeavor Agreement

JEA Jordan Engineers Association [Amman]

JEAN JOSS (Johnniac Open-Shop System) Based Expression Analyzer for Nineteen-Hundred

JEB Japan Encephalitis, Type B

JEB Journal of Experimental Biology [UK]

JEBM Jet Engine Base Maintenance

JEBM-RR Jet Engine Base Maintenance Return Rate

JEC Japan Electric Standard Committee (Standard)

JEC Japan energy Company [Tokyo]

JEC Joint Economic Committee [US]

JECA Joint Engineers Council of Alabama [US]

JECC Japan Electronic Computer Center

JECC Joint Egyptian Cotton Committee

JECC Joint Electronic Components Conference

JECFA Joint Expert Committee on Food Additives [of Food and Agricultural Organization and World Health Organization]

J E China Inst Chem Technol Journal of the East China Institute of Chemical Technology [PR China]

JECL Job Entry Control Language

J Econ Dyn Control Journal of Economic Dynamics and Control [Netherlands]

J Econ Ent Journal of Economic Entomology [Published by the Entomological Society of America]

J Econ Hist Journal of Economic History [International Journal]

JECS Job Entry Central Service

JECS Joint Equalization Control System

JED Julian Ephemeris Date [Astronomy]

JEDEC Joint Electron(ic) Devices Engineering Council [US]

JEDTC Joint Electron Device Tube Council [now Tube Engineering Panel Advisory Council, US]

J Educ Comput Res Journal of Educational Computing Research [US]

J Educ Libr Inf Sci Journal of Education for Library and Information Science [of Association for Library and Information Science Education, US]

J Educ Psychol Journal of Educational Psychology [US]

J Educ Res Journal of Educational Research [US]

J Educ Sociol Journal of Educational Sociology

J Educ Technol Syst Journal of Educational Technology Systems [US]

JEE Journal of Electronic Engineering [Japan]

JEEP Joint Establishment Experimental Pile [Norway]

JEF Jacobian Elliptic Function [Mathematics]

JEIA Japan Electroplating Industry Association

JEIA Joint Export-Import Agency [US/UK]

JEIDA Japan(ese) Electronic(s) Industry Development Association

JEIPAC JICST Electronic Information Processing Automatic Computer [of Japan Information Center of Science and Technology, Tokyo]

JEL Johnson Elastic Limit [Mechanics]

J Elast Journal of Elasticity [Netherlands]

J Elastomers Plast Journal of Elastomers and Plastics [US]

J Electr Electron Eng Aust Journal of Electrical and Electronics Engineering, Australia [of Institution of Engineers]

J Electr Eng Journal of Electrical Engineering [of Chinese Society of Electric Engineering, PR China]

J Electroanal Chem Journal of Electroanalytical Chemistry [Switzerland]

J Electroanal Chem Interfacial Electrochem Journal of Electroanalytical Chemistry and Interfacial Electrochemistry [Switzerland]

J Electrochem Soc Journal of the Electrochemical Society [US]

J Electrochem Soc India Journal of Electrochemical Society of India [of Indian Institute of Sciences]

J Electromagn Waves Appl Journal of Electromagnetic Waves and Applications [Netherlands]

J Electron Journal of Electronics [PR China]

J Electron Comput Res Journal of Electronics and Computers Research [of Scientific Research Council, Iraq]

J Electron Def Journal of Electronic Defense [of Association of Old Crows, US]

J Electron Eng Journal of Electronic Engineering [Japan]

J Electron Mater Journal of Electronic Materials [US]

J Electron Microsc Journal of Electron Microscopy [of Japanese Society of Electron Microscopy]

J Electron Microsc Technique Journal of Electron Microscopy Technique [US]

J Electron Microsc Technol Journal of Electron Microscopy Technology [US]

J Electron Micry Journal of Electron Microscopy [of Japanese Society of Electron Microscopy]

J Electron Micry Technique Journal of Electron Microscopy Technique [US]

J Electron Packag (Trans ASME) Journal of Electronic Packaging (Transactions of the American Society of Mechanical Engineers) [US]

J Electron Spectrosc Journal of Electron Spectroscopy [Netherlands]

J Electron Spectrosc Relat Phenom Journal of Electron Spectroscopy and Related Phenomena [Netherlands]

J Electrost Journal of Electrostatics [Netherlands]

JEM Japan Electric Machinery (Industry Committee Standard)

JEM Japanese Experiment Module [of National Space Development Agency of Japan]

JEM Jet Engine Modulation

JEM Joint Endeavor Manager

JEM Journal of Electronic Materials [Joint Publication of the Institute of Electrical and Electronics Engineers and The Minerals, Metals and Materials Society, US]

JEM Journal of Experimental Medicine [Published by Rockefeller University Press, US]

JEMC Joint Engineering Management Conference

J Emer Nursing Journal of Emergency Nursing [US]

JEMIC Japan Electric Meters Inspection Corporation [Tokyo]

JEMIC Tech Rep JEMIC Technical Report [of Japan Electric Meters Inspection Corporation]

JEMIMA Japan Electric Measuring Instruments Manufacturers Association

Jernkontorets Ann Jernkontorets Annaler [Annals of Steel Production; published by the Swedish Steel Producers Association]

Jernkontorets Forsk Jernkontorets Forskning [Steel Research Journal; published by the Swedish Steel Producers Association]

JEMMSE Journal of Educational Modules for Materials Science and Engineering

Jemna Mech Opt Jemna Mechanika a Optika [Czech Journal on Mechanics and Optics]

JEMRB Joint European Medical Research Board

JEN Junta de Energía Nuclear [Nuclear Energy Authority, Madrid, Spain]

JENAM Joint European and National Astronomy Meeting

Jena Rev Jena Review [Published by Carl Zeiss, Jena, Germany]

J Endocrinol Journal of Endocrinology [of Society for Endocrinology, US]

JENER Joint Establishment for Nuclear Energy Research

J Energy Resour Technol (Trans ASME) Journal of Energy Resources Technology (Transactions of the American Society of Mechanical Engineers) [US]

J Eng Gas Turbines Power (Trans ASME) Journal of Engineering for Gas Turbines and Power (Transactions of the American Society of Mechanical Engineers, US]

J Eng Educ Journal of Engineering Education [of American Society for Engineering Education, US]

J Eng Ind (Trans ASME) Journal of Engineering for Industry (Transactions of the American Society of Mechanical Engineers, US]

J Eng Mater Technol (Trans ASME) Journal of Engineering Materials and Technology (Transactions of the American Society of Mechanical Engineers) [US]

J Eng Math Journal of Engineering Mathematics [Netherlands]

J Eng Mech Journal of Engineering Mechanics [of American Society of Civil Engineers, US]

J Eng Mech Div ASCE Journal of Engineering Mechanics, Division of the American Society of Civil Engineers [US]

J Eng Phys Journal of Engineering Physics [Translation of: *Inzhenerno-Fizicheskii Zhurnal (USSR)*; published in the US]

J Eng Power (Trans ASME) Journal of Engineering for Power (Transactions of the ASME) [of American Society of Mechanical Engineers, US]

J Eng Sci Journal of Engineering Sciences [Saudi Arabia]

J Eng Sci King Saud Univ Journal of Engineering King Saud University [Saudi Arabia]

J Eng Technol Journal of Engineering Technology [of American Society for Engineering Education, US]

J Eng Technol Manage Journal of Engineering and Technology Management [Netherlands]

J Environ Devel The Journal of Environment and Development [US]

J Environ Econ Manage Journal of Environmental Economics and Management [of Association of Environmental and Resource Economists, US]

J Environ Eng Journal of Environmental Engineering [of American Society of Civil Engineers, US]

J Environ Health Journal of Environmental Health

J Environ Polym Degrad Journal of Environmental Polymer Degradation [US]

J Environ Qual Journal of Environmental Quality [of Soil Science Society of America, US]

J Environ Radioact Journal of Environmental Radioactivity [UK]

J Environ Sci Journal of Environmental Sciences [of Institute of Environmental Sciences]

JEOCN Joint European Operations Communications Network

JEOL Japan Electro(n)-Optics Laboratories, Inc. [Tokyo]

JEOL News Anal Instrum JEOL News, Analytical Instrumentation [Published by JEOL, Tokyo, Japan]

JEOL News Electron Opt Instrum JEOL News, Electron Optics Instrumentation [Published by JEOL, Tokyo, Japan]

JEOS Japanese Earth Observing Satellite

JEP Joint Energy Program [now Energy and Environment Program; of Royal Institute of International Affairs, UK]

JEPI Joint Electronic Payment Initiative

JEPIA Japan Electronic Parts Industry Association

JEPOSS Javelin Experimental Protection Oil Sands System

JEPP Joint Emergency Planning Program

JEPS Job Entry Peripheral Service [Computers]

JEQ Jump Equal

J Equip Electr Electron Journal de l'Equipement Electrique et Electronique [French Journal on Electrical and Electronic Equipment]

JERC Japan Economic Research Center

JERC Japan-Europe Economic Research Center

JERC Joint Electronic Research Committee

JERS Japanese Earth Resources Satellite

JERS-1 First Japanese Earth Resource Satellite

JES Job Entry Subsystem; Job Entry System [Computers]

JES John Ericsson Society [US]

JES Joint Environmental Service

JES Joint Environmental Simulator

JES Joint Equalization System

JESAC Joint Engineering Student Activity Committee

JESSI Joint European Silicon Submicron Initiative [of European Research Coordination Agency]

JESSI Junior Engineers and Scientists Summer Institute [US]

JET Joint Enroute Terminal

JET Joint European Torus (Reactor)

JETDS Joint Electronics Type Designation System

JETEC Jet Engine Titanium Quality Committee

JETEC Joint Electron Tube Engineering Council [US]

J Ethnobiol Journal of Ethnobiology [of Society of Ethnobiology, US]

J Ethnopharmacol Journal of Ethnopharmacology [US]

JETP Journal of Experimental and Theoretical Physics [Translation of: *Zhurnal Eksperimental'noi i Teoreticheskoi Fiziki (USSR)*; published by the American Institute of Physics, US and MAIK Nauka, Russia]

JETP Lett JETP (Journal of Experimental and Theoretical Physics) Letters [Translation of: *Pis'ma v Zhurnal Eksperimental'noi i Teoreticheskoi Fiziki (USSR)*; published by the American Institute of Physics, US and MAIK Nauka, Russia]

JETR Japan Engineering Test Reactor [Japan]

JETRO Japan External Trade Organization

JETS Job Executive and Transport Satellite

JETS Joint Electronics Type (Designation) System

JETS Junior Engineering Technical Society [US]

Jett Jettison

J Eur Ceram Soc Journal of the European Ceramic Society [UK]

J Exp Anal Behav Journal of the Experimental Analysis of Behavior [International Journal]

J Exp Biol Journal of Experimental Biology [UK]

J Exp Bot Journal of Experimental Botany [of Society for Experimental Biology, UK]

J Exp Child Psychol Journal of Experimental Child Psychology [Published in the US]

J Exp Educ Journal of Experimental Education [US]

J Exp Med Journal of Experimental Medicine [Published by Rockefeller University Press, US]

J Exp Psychol Journal of Experimental Psychology [of American Psychological Society, US]

J Exp Theor Phys Journal of Experimental and Theoretical Physics [Translation of: *Zhurnal Eksperimental'noi i Teoreticheskoi Fiziki (USSR)*; published by the American Institute of Physics, US]

J Exp Theor Phys Lett Journal of Experimental and Theoretical Physics Letters [Translation of: *Pis'ma v Zhurnal Eksperimental'noi i Teoreticheskoi Fiziki (USSR)*; published by the American Institute of Physics in the US]

J Expl Eng Journal of Explosives Engineering [of Society of Explosives Engineers, US]

JF Jet Fighter

JF Jet Flap

JF Jet Flow

JF Jet Fuel

JF Joukowski Formula [Fluid Mechanics]

JF Jump Frequency

JF Junction Frequency

JF Junctor Frame [Electrical Engineering]

JFA Japan Fisheries Agency

J Fac Agr, Hokkaido Univ Journal of the Faculty of Agriculture, Hokkaido University [Japan]

J Fac Eng, Chiba Univ Journal of the Faculty of Engineering, Chiba University [Japan]

J Fac Eng, Ibaraki Univ Journal of the Faculty of Engineering, Ibaraki University [Japan]

J Fac Eng, Shinshu Univ Journal of the Faculty of Engineering, Shinshu University [Japan]

J Fac Eng, Univ Tokyo A Journal of the Faculty of Engineering, University of Tokyo, Series A [Japan]

J Fac Eng, Univ Tokyo B Journal of the Faculty of Engineering, University of Tokyo, Series B [Japan]

JFAP Joint Frequency Allocation Panel

JFCA Japan Fine Ceramics Association

JFCB Job File Control Block [Computers]

JFC Java Foundation Classes [Computers]

JFC Job Fair Center [US]

JFCC Japan Federation of Culture Collections

JFCC Japan Fine Ceramics Center [of Nagoya University]

JFEA Japan Federation of Employers Associations

J Ferrocement Journal of Ferrocement [Thailand]

JFET Junction Field-Effect Transistor [also J-FET]

JFF Jet Flow Function

JFI James Franck Institute [of University of Chicago, Illinois, US]

JFIF JPEG (Joint Photographic Experts Group) File Interchange Format

J Fire Sci Journal of Fire Sciences [US]

J Fiz Malays Jurnal Fizik Malaysia [Journal of Physics; of Malaysian Institute of Physics] ,

JFKSC John F. Kennedy Space Center [also Kennedy Space Center, Florida, US]

JFL Joint Frequency List

J Fluid Control Journal of Fluid Control [US]

J Fluid Mech Journal of Fluid Mechanics [UK]

J Fluids Eng (Trans ASME) Journal of Fluids Engineering (Transactions of the American Society of Mechanical Engineers) [US]

J Fluids Struct Journal of Fluids and Structures [UK]

JFMIP Joint Financial Management Improvement Program [US]

JFN Job File Number

J Food Eng Journal of Food Engineering [US]

J Food Process Eng Journal of Food Processing Engineering [US]

J Food Prot Journal of Food Protection

J Food Sci Journal of Food Science [of Institute of Food Technologists, US]

J Forensic Sci Journal of Forensic Sciences [of American Academy of Forensic Science, US]

J Forestry Journal of Forestry [of Society of American Foresters, US]

J Forth Appl Res Journal of Forth Application and Research [of Institute of Applied Forth Research, US]

J Four Electr Journal du Four Electrique et des Industries Electrochimiques [French Journal of Electric Furnaces and Electrochemical Industries]

JFP Joint Frequency Panel

JFR Jet and Fuel Resistance; Jet and Fuel Resisting

JFR Jet-Fuel Resistance; Jet-Fuel-Resistant

J Franklin Inst Journal of the Franklin Institute [Philadelphia, Pennsylvania, US]

J Fr Electrothermie Journal Français de l'Electrothermie [French Journal of Electrothermancy]

JFRO Joint Fire Research Organization [UK]

J Frottement Ind Journal du Frottement Industriel [French Journal of Industrial Friction]

JFS Japan Foundrymen's Society

JFS Journaled File System

JFS Jumbo (Group) Frequency Supply

JFSE Japan Federation of Small Enterprises

JFTOT Jet Fuel Thermal Oxidation Test

JFU Jersey Farmers Union [UK]

J Fuel Soc Jpn Journal of the Fuel Society of Japan

J Funct Anal Journal of Functional Analysis [US]

J Fusion Energy Journal of Fusion Energy [US]

JG Jena Glass

JG Jastrow-Gutzwillwe (Wavefunction) [Physics]

JG Joint Group

JG Jump if Greater

JG Junction Grammar

JG Junior Grade [also jg] [Military]

J/g Joule(s) per gram [Unit]

JGA Jojoba Growers Association [US]

JG-APP Joint Group on Acquisition Pollution Prevention

JGC Joint Generation Control

JGCC Japan Gas-Chemical Company (Process)

JGCR Japan Gas-Cooled Reactor

JGCS Joint Generation Control System

JGE Jump if Greater or Equal

J Gemmol Journal of Gemmology [of Gemmological Institute of America]

J Gen Chem Journal of General Chemistry [US]

J Gen Chem (USSR) Journal of General Chemistry of the USSR [Published in the US]

J Genet Psychol Journal of Genetic Psychology [US]

J Gen Microbiol Journal of General Microbiology [US]

J Gen Physiol Journal of General Physiology [Published by Rockefeller University Press, US]

J Gen Psychol Journal of General Psychology [US]

J Geochem Explor Journal of Geochemical Exploration [Netherlands]

J Geod Journal of Geodesy [Germany]

J Geodyn Journal of Geodynamics [UK]

J Geolectr Journal of Geoelectricity

J Geol Journal of Geology [of University of Chicago, Illinois, US]

J Geol Educ Journal of Geological Education [US]

J Geol Soc Journal of the Geological Society [London, UK]

J Geol Soc Jam Journal of the Geological Society of Jamaica

J Geom Phys Journal of Geometry and Physics [Italy]

J Geomagn Geoelectr Journal of Geomagnetism and Geoelectricity [Japan]

J Geophys Res Journal of Geophysical Research [of American Geophysical Union, US]

J Geophys Res A Journal of Geophysical Research A [of American Geophysical Union, US]

J Geophys Res B Journal of Geophysical Research B [of American Geophysical Union, US]

J Geophys Res C Journal of Geophysical Research C [of American Geophysical Union, US]

J Geotech Eng Journal of Geotechnical Engineering [US]

J Gerontol Journal of Gerontology

JGF Junctor Grouping Frame [Electrical Engineering]

JGI Japanese Geographical Institute

J/g·K Joule(s) per gram kelvin [also J/g-K,J/g/K or g^{-1} K^{-1}]

JGL Jackson, Gilmer and Leamy (Model) [Materials Science]

J Glac Journal of Glaciology [of International Glaciological Society] [also J Glaciol]

J Glacial Geol Geomorph Journal of Glacial Geology and Geomorphology [of Queen's University at Belfast, Northern Ireland]

J Glaciol Journal of Glaciology [of International Glaciological Society] [also J Glac]

JGN Junction Gate Number

JGOAL Java PCGOAL (Personal Computer Ground Operations Aerospace Language) [of NASA]

JGOFS Joint Global Ocean Flux Study (Program) [of International Geosphere/Biosphere Program]

JGP Journal of General Physiology [Published by Rockefeller University Press, US]

JGR Journal of Geophysical Research [of American Geophysical Union, US]

JGR A Journal of Geophysical Research A [of American Geophysical Union, US]

JGR B Journal of Geophysical Research B [of American Geophysical Union, US]

JGR C Journal of Geophysical Research C [of American Geophysical Union, US]

J Graph Theory Journal of Graph Theory [US]

J Grey Syst Journal of Grey Systems [UK]

JGRIP Japanese Government and Public Research in Progress

JGS Journal of the Geological Society [London, UK]

J Gt Lakes Res Journal of Great Lakes Research [of International Association for Great Lakes Research, US]

J Guid, Control Dyn Journal of Guidance, Control and Dynamics [of American Institute of Aeronautics and Astronautics, US]

JH Jackson and Hunt (Analysis) [Materials Science]

JH Joule Heat(ing) [Thermodynamics]

JH Juvenile Hormone

JH-I Juvenile Hormone I

JH-II Juvenile Hormone II

JH-III Juvenile Hormone III

$J_c(H)$ Magnetic Field Dependent Critical Current Density (of Superconductors) [Symbol]

$J_{ct}(H)$ Magnetic-Field-Dependent Transport Critical Current Density (of Superconductors) [Symbol]

J-H Current Density–Magnetic Field (Hysteresis Loop)

JHAPL Johns Hopkins Applied Physics Laboratory [of Johns Hopkins University at Silver Spring, Maryland, US]

J Harbin Inst Technol Journal of the Harbin Institute of Technology [PR China]

J Hard Mater Journal of Hard Materials [of Institute of Physics Publishing, UK]

J Hazard Mater Journal of Hazardous Materials [Netherlands]

J Hazard Waste Hazard Mater Journal of Hazardous Waste and Hazardous Materials [of Hazardous Materials Control Research Institute, US]

JHC Journal of Histochemistry and Cytochemistry [of Histochemical Society, US]

JHCAAD Johns Hopkins Center for Asthma and Allergic Disease [Baltimore, Maryland, US]

JHD Joint Hypocenter Determination

J Head Trauma Rehab Journal of Head Trauma Rehabilitation [US]

J Health Prom Journal of Health Promotion

J Heat Recovery Syst Journal of Heat Recovery Systems [UK]

J Heat Transf (Trans ASME) Journal of Heat Transfer (Transactions of the American Society of Mechanical Engineers) [US]

J Heat Treat Journal of Heat Treating [of ASM International, US]

J Heterocyc Chem Journal of Heterocyclic Chemistry

JHG Joule Heat Gradient [Thermodynamics]

Jh/g Joule-hour(s) per gram [also J-h/g, or Jh g^{-1}]

JHH Johns Hopkins Hospital [Baltimore, Maryland, US]

J High Energy Phys Journal of High-Energy Physics [US]

J Higher Educ Journal of Higher Education [US]

J High Temp Soc Jpn Journal of the High-Temperature Society of Japan

J Hist Astron Journal of the History of Astronomy [UK]

J Hist Met Soc Journal of the Historical Metallurgy Society [UK]

J Histochem Cytochem Journal of Histochemistry and Cytochemistry [of Histochemical Society, US]

JHN Japanese Helicopter Network

J Hokkaido Forest Prod Res Inst Journal of the Hokkaido Forest Products Research Institute [Japan]

J Horol Inst Jpn Journal of the Horological Institute of Japan

JHP Johns Hopkins Press [of Johns Hopkins University, Baltimore, Maryland, US]

JHPS Japan Hydraulics and Pneumatics Society

JHS Junior High School

JHT Journal of Heat Treating [of ASM International, US]

$J_c(H,T)$ Magnetic-Field and Temperature-Dependent Critical Current Density (of Superconductors) [Symbol]

$J_{ct}(H,T)$ Magnetic-Field and Temperature-Dependent Transport Critical Current Density (of Superconductors) [Symbol]

JHTR Journal of Head Trauma Rehabilitation [US]

JHU Johns Hopkins University [Baltimore, Maryland, US]

JHU/APL Johns Hopkins University, Applied Physics Laboratory [Silver Spring, Maryland, US]

J Huazhong Inst Technol Journal of Huazhong Institute of Technology [PR China]

J Huazhong Inst Technol (Engl Ed) Journal of Huazhong Institute of Technology (English Edition) [PR China]

J Hum Ecol Journal of Human Ecology [International Journal]

J Hum Evol Journal of Human Evolution [US]

J Hum Nutr Journal of Human Nutrition

J Hunan Sci Technol Univ Journal of Hunan Science and Technology University [PR China]

JHU/ORO Johns Hopkins University, Operations Research Office [US]

JHUSM Johns Hopkins University School of Medicine [Baltimore, Maryland, US]

J Hydraul Eng Journal of Hydraulic Engineering [US]

J Hydraul Res Journal of Hydraulic Research [of International Association for Hydraulic Research, Netherlands]

J Hydr Eng Journal of Hydraulic Engineering [US]

J Hydrol Journal of Hydrology [US]

J Hydrosci Hydraul Eng Journal of Hydroscience and Hydraulic Engineering [US]

J Hydr Res Journal of Hydraulic Research [of International Association for Hydraulic Research, Netherlands]

J Hyg Chem Journal of Hygiene Chemistry [US]

J/Hz Joule(s) per Hertz [also J Hz^{-1}]

JI Joint Implementation

JIA Jute Importers Association [UK]

J IAPM Journal of the International Academy of Preventive Medicine

JIAWG Joint Integrated Avionics Working Group [of US Air Force]

JIB Job Information Block

JIB Joint Intelligence Bureau [UK]

JIBEI Joint Industry Board of the Electrical Industry [US]

JIC Japan Industry Committee (Standard)

JIC Japan Information Center

JIC Jewelry Industry Council

JIC Joint Ice Center

JIC Joint Industrial Council [US]

JIC Joint Industry Committee

JIC Joint Industry Conference

JIC Joint Iron Council [UK]

JIC Just-in-Case

JICA Japan International Cooperation Agency

JICE Japan Institute of Control Engineering

JICNARS Joint Industry Committee for National Readership Surveys [UK]

JICST Japan Information Center of Science and Technology [Tokyo]

JICTAR Joint Industry Committee for Television Advertising Research [UK]

JIDA Japan Industrial Designers Association

JIDC Jamaica Industrial Development Corporation

JIE Junior Institution of Engineers [UK]

JIEE Journal of the Institution of Electrical Engineers [UK]

JIEI Japanese Industrial Engineering Institute

JIFA Japan Institute of Foreign Affairs

JIFDATS Joint In-Flight Data Transmission System [of US Department of Defense]

JIFTS Joint In-Flight Transmission System

Jiguang Zazhi (Laser J) Jiguang Zazhi (Laser Journal) [PR China]

JIII Japan Institute of Invention and Innovation

JIIM Journal of Information and Image Management [US]

JIISC Joing Industrial Investment Service Center [of Ministry of Economic Affairs, Taiwan]

JILA Joint Institute for Laboratory Astrophysics [Boulder, Colorado, US]

J Illum Eng Inst Jpn Journal of the Illuminating Engineering Institute of Japan

J Illum Eng Soc Journal of the Illuminating Engineering Society [US]

JILM Japan Institute of Light Metals [of Japan Light Metal Association, Tokyo]

JIM Japan Institute of Metals [Sendai]

JIMA Japan Industrial Management Association

J Imaging Sci Journal of Imaging Science [of Society of Photographic Scientists and Engineers, US]

J Imaging Technol Journal of Imaging Technology [of Society of Photographic Scientists and Engineers, US]

JIMAR Joint Institute for Marine and Atmospheric Research [of University of Hawaii, Honolulu, US]

JIMCS Japan Institute of Metals Conference Series

JIMIS Japan Institute of Metals International Symposium

J Immunol Journal of Immunology [of American Association of Immunologists, US]

J Immunol Meth Journal Immunological Methods [Japan]

JIMS Journal of Indian Mathematical Society [India]

JIMTOF Japan International Machine Tool Fair

JIN Japan Institute of Navigation [Tokyo]

J/in Joule(s) per inch [J in^{-1}]

J Ind (Trans ASME) Journal of Industry (Transaction of the American Society of Mechanical Engineers) [US]

J Ind Ecol Journal of Industrial Ecology [US]

J Ind Eng (Trans ASME) Journal of Industrial Engineering (Transaction of the American Society of Mechanical Engineers) [US]

J Ind Fabr Journal of Industrial Fabrics [US]

J Indian Inst Sci Journal of the Indian Institute of Science

J Indian Inst Sci A Journal of the Indian Institute of Science, Series A

J Indian Inst Sci B Journal of the Indian Institute of Science, Series B

J Indian Inst Sci C Journal of the Indian Institute of Science, Series C

J Indian Math Soc Journal of the Indian Mathematical Society

J Indian Refract Makers Assoc Journal of Indian Refractory Makers Association

J Indian Waterworks Assoc Journal of the Indian Waterworks Association

J Ind Eng Des Journal of Industrial Engineering Design [of American Society for Engineering Education, US]

J Ind Irr Technol Journal of Industrial Irradiation Technology [US]

J Ind Microbiol Journal of Industrial Microbiology [of Society for Industrial Microbiology, US]

J Ind (Trans ASME) Journal of Industry (Transactions of the American Society of Mechanical Engineers) [US]

J Infect Diseases Journal of Infectious Diseases [Published by University of Chicago Press, Illinois, US]

J Inf Manage Journal of Information Management [of Life Office Management Association, US]

J Inf Optim Sci Journal of Information and Optimization Sciences [India]

J Inf Process Journal of Information Processing [of Information Processing Society of Japan]

J Inf Process Cybern Journal of Information Processing and Cybernetics [Germany]

J Inf Rec Mater Journal of Information Recording Materials [Germany]

J Inf Sci Journal of Information Science [of Institute of Information Scientists, UK]

J Inf Sci Princ Pract Journal of Information Science, Principles and Practice [Netherlands]

J Inf Syst Manage Journal of Information Systems Management [US]

J Inf Technol Journal of Information Technology [UK]

J Inorg Nucl Chem Journal of Inorganic Nuclear Chemistry

J Inorg Organomet Polym Journal of Inorganic and Organometallic Polymers [US]

JINR Joint Institute for Nuclear Research [Dubna, Russia]

J Insect Physiol Journal of Insect Physiology

J Inst Electron Inf Commun Eng Journal of the Institute of Electronics, Information and Communication Engineers [Japan]

J Inst Electron Radio Eng Journal of the Institution of Electronic and Radio Engineers [UK]

J Inst Electron Telecommun Eng Journal of the Institution of Electronics and Telecommunication Engineers [India]

J Inst Energy Journal of the Institute of Energy [UK]

J Inst Eng (India), Chem Eng Div Journal of the Institution of Engineers (India), Chemical Engineering Division

J Inst Eng (India), Electr Eng Div Journal of the Institution of Engineers (India), Electrical Engineering Division

J Inst Eng (India), Electron Telecommun Eng Div Journal of the Institution of Engineers (India), Electronics and Telecommunication Engineering Division

J Inst Eng (India), Environ Eng Div Journal of the Institution of Engineers (India), Environmental Engineering Division

J Inst Eng (India), Interdiscip Gen Eng Journal of the Institution of Engineers (India), Interdisciplinary and General Engineering

J Inst Eng (India), Interdiscip Panels Journal of the Institution of Engineers (India), Interdisciplinary Panels

J Inst Eng (India), Mech Eng Div Journal of the Institution of Engineers (India), Mechanical Engineering Division

J Inst Eng (India), Metall Mater Sci Div Journal of the Institution of Engineers (India), Metallurgy and Materials Science Division

J Inst Eng (India), Min Metall Div Journal of the Institution of Engineers (India), Mining and Metallurgy Division

J Inst Ind Sci, Univ Tokyo Journal of Institute of Industrial Science, University of Tokyo [Japan]

J Inst Nav Journal of the Institute of Navigation [UK]

J Inst Refract Eng Journal of the Institute of Refractories Engineers [UK]

J Inst Telev Eng Jpn Journal of the Institute of Television Engineers of Japan

J Inst Water Eng Environ Manage Journal of the Institution of Water Engineers and Environmental Management [UK]

J Inst Wood Sci Journal of the Institute of Wood Science [UK]

J Int Acad Prev Med Journal of the International Academy of Preventive Medicine

J Int Appl Cobalt Journal Internationale des Applications du Cobalt [International Journal of Applications of Cobalt; published in Brussels, Belgium]

J Intell Mater Sys Struct Journal of Intelligent Material Systems and Structures

J Intell Robot Syst, Theory Appl Journal of Intelligent and Robotic Systems: Theory and Applications [Netherlands]

J Intell Syst Journal of Intelligent Systems [UK]

J Int Market Market Res Journal of International Marketing and Marketing Research [of European Commission for Industrial Marketing, UK]

J Int Med Res Journal of International Medical Research

J Invertebr Pathol Journal of Invertebrate Pathology [Published in the US]

JIO Joint Intelligence Organization

JIOA Joint Intelligence Objectives Agency

J Iowa Acad Sci Journal of the Iowa Academy of Sciences [US]

JIP Joint Input Processing

JIPDEC Japan Information Processing Development Center [Tokyo]

JIPDEC Rep JIPDEC Report [of Japan Information Processing Development Center, Tokyo]

JIPS JANET (Joint Academic Network) Internet Protocol Service

JIR Job Improvement Request

JIRA Japan Industrial Robot Association

JIRC Journal of Information Research Communications [UK]

J Iron Steel Inst Journal of the Iron and Steel Institute [UK]

J Iron Steel Inst Jpn Journal of the Iron and Steel Institute of Japan

JIS Jamaica Information Service

JIS Japanese Industrial Standard

JIS Job Information System

JIS Job Input System

JIS Joint Integrated Simulation

JIS Journal of Information Science [of Institute of Information Scientists, UK]

JISAO Joint Institute for the Study of the Atmosphere and Ocean [University of Washington, Seattle, US]

JISC Japanese Industrial (or Industry) Standards Committee

JISF Japan Iron and Steel Foundation

JISHA Japan Industrial Safety and Health Association

JISI Japan Iron and Steel Institute

JIT Job Instruction Training

JIT Just-In-Time

JITA Japan Industrial Technology Association

JITF Joint Informatics Task Force [of US Department of Energy and US Department of Health and Human Services]

JIT/TQC Just-In-Time/Total Quality Control

JIU Joint Inspection Unit

JIVE Joint Institute for VLBI (Very-Long Base-line Interferometry) in Europe [Manchester, UK]

JJ Josephson Junction [Electronics]

J&J Johnson and Johnson Limited [US Pharmaceutical Manufacturer]

J_i/J_a Ion-to-Atom Flux Ratio [Symbol]

J-J J-J (Coupling) [also j-j] [Atomic Physics]

JJA Josephson Junction Array [Electronics]

JJAP Japanese Journal of Applied Physics [of Physical Society of Japan]

JJP Japanese Journal of Physiology [Published by University of Tokyo Press]

J Jpn Air Clean Assoc Journal of the Japan Air Cleaning Association

J Jpn Ceram Soc Journal of the Japan Ceramic Society

J Jpn Compos Mater Journal of Japan Composite Materials

J Jpn Hydraul Pneum Soc Journal of the Japan Hydraulics and Pneumatics Society

J Jpn Inst Light Met Journal of the Japan Institute of Light Metals [of Japan Light Metal Association]

J Jpn Inst Met Journal of the Japan Institute of Metals

J Jpn Inst Navig Journal of the Japan Institute of Navigation

J Jpn Sewage Works Assoc Journal of the Japan Sewage Works Association

J Jpn Soc Aeronaut Space Sci Journal of the Japan Society for Aeronautical and Space Sciences

J Jpn Soc Air Pollut Journal of the Japan Society of Air Pollution

J Jpn Soc Artif Intell Journal of the Japanese Society for Artificial Intelligence

J Jpn Soc Civ Eng Journal of the Japan Society of Civil Engineers

J Jpn Soc Colour Mater Journal of the Japan Society of Color Materials

J Jpn Soc Compos Mater Journal of the Japan Society for Composite Materials

J Jpn Soc Heat Treat Journal of the Japan Society of Heat Treatment

J Jpn Soc Lubr Eng Journal of the Japan Society of Lubrication Engineers

J Jpn Soc Powder Powder Metall Journal of the Japan Society of Powder and Powder Metallurgy

J Jpn Soc Precis Eng Journal of the Japan Society of Precision Engineering

J Jpn Soc Simul Technol Journal of the Japan Society for Simulation Technology

J Jpn Soc Strength Fract Mater Journal of the Japanese Society for Strength and Fracture of Materials

J Jpn Soc Technol Plast Journal of the Japan Society for Technology of Plasticity

J Jpn Soc Tribol Journal of the Japanese Society of Tribologists

J Jpn Water Works Assoc Journal of the Japan Water Works Association

J Jpn Weld Soc Journal of the Japan Welding Society

J Jpn Wood Res Soc Journal of the Japan Wood Research Society,

JJSLE Journal of the Japan Society of Lubrication Engineers

J JSLE, Int Ed Journal of the Japan Society of Lubrication Engineers, International Edition [Japan]

JK J and K Input (Flip Flop) [Electronics]

J/K Joule(s) per kelvin [also J K^{-1}]

Jk Jack [also jk]

J/Kcm² Joule(s) per square centimeter kelvin [also J cm^{-2} K^{-1} or J/K·cm² or J/K-cm²]

J/kg Joule(s) per kilogram [also J kg^{-1}]

J/kg·°C Joule(s) per kilogram degree Celsius (orCentigrade) [also J/kg/°C or J kg^{-1} $°C^{-1}$]

J/kg·K Joule(s) per kilogram kelvin [also J/kg/Kor J kg^{-1} K^{-1}]

J/K·mol Joule(s) per kelvin mole [also J/K/mol or JK^{-1} mol^{-1}]

J Korea Inf Sci Soc Journal of the Korea Information Science Society [South Korea]

J Korea Inst Electron Eng Journal of the Korea Institute of Electronics Engineers [South Korea]

J Korean Ceram Soc Journal of the Korean Ceramic Society [South Korea]

J Korean Inst Chem Eng Journal of the Korean Institute of Chemical Engineers [South Korea]

J Korean Inst Met Journal of the Korean Institute of Metals [South Korea]

J Korean Inst Miner Min Eng Journal of the Korean Institute of Mineral and Mining Engineers [South Korea]

J Korean Inst Telemat Electron Journal of the Korean Institute of Telematics and Electronics [South Korea]

J Korean Nucl Soc Journal of the Korean Nuclear Society [South Korea]

J Korean Phys Soc Journal of the Korean Physical Society [South Korea]

JKR Johnson-Kendall-Roberts (Technique) [Solid-State Physics]

JKRS Johnson-Kendall-Roberts-Sperling (Thin-Film Theory) [Solid-State Physics]

JL Jefferson Laboratory [Newport News, Virginia, US] [also known as Thomas Jefferson National Accelerator Facility]

JL Jones and Laughlin (Steel Company) [US]

JL Jump if Less

JL Jump Line (Command) [Pascal Programming]

JL Junction Laser

Jl July

JLAB Jefferson Laboratory [Newport News, Virginia, US] [also known as Thomas Jefferson National Accelerator Facility] [also Jlab]

J Lab Clin Med Journal of Laboratory and Clinical Medicine [US]

JLAWF James F. Lincoln Arc Welding Foundation [Cleveland, Ohio, US]

JLC Joint Load Control

JLE Jump if Less than or Equal to

J Less-Common Met Journal of Less-Common Metals [now J Alloys Comp]

J Light Met Weld Constr Journal of Light Metal Welding and Construction [Japan]

J Light Vis Environ Journal of Light and Visual Environment [of Illuminating Engineering Institute of Japan]

J Lightwave Technol Journal of Lightwave Technology [of Institute of Electrical and Electronics Engineers, US]

JLIP Joint Level Interface Protocol

J Lipid Res Journal of Lipid Research [of Federation of American Societies for Experimental Biology]

J Liquid Chromat Journal of Liquid Chromatography [US]

JLMA Japan Light Metal Association

JLMA Lett JLMA Letters [of Japan Light Metal Association]

JLMS Journal of the London Mathematical Society [UK]

J Log Program Journal of Logic Programming [US]

J Lond Math Soc Journal of the London Mathematical Society [UK]

J Low Freq Noise Vib Journal of Low Frequency Noise and Vibration [UK]

J Low Temp Phys Journal of Low Temperature Physics [US]

J Lubr Technol (Trans ASME) Journal of Lubrication Technology (Transactions of the American Society of Mechanical Engineers) [US]

J Luminesc Journal of Luminescence [Published in the Netherlands]

JLP Jig Leg Plate

JLR Japan Laser Report

JLR Journal of Lipid Research [of Federation of American Societies for Experimental Biology]

JLS Jones and Laughlin Steel Corporation [US]

JLSC Joint Electronics Standardization Committee [UK]

J Lumin Journal of Luminescence [Published in the Netherlands]

JM Jacobian Matrix [Mathematics]

JM Jamaica [ISO Code]

JM Jar Mill(ing)

JM Jet Mixer; Jet Mixing

JM Jet Molding; Jet Molder

JM Johnson-Mehl (Grain Growth Model) [Metallurgy]

JM Journal of Micrographics [US]

J/M Jettison Motor

J/m Joule(s) per meter [also J m^{-1}]

J/m² Joule(s) per square meter [also J m^{-2}]

J/m³ Joule(s) per cubic meter [also J m^{-3}]

.jm Jamaica [Country Code/Domain Name]

JMA Japan Management Association

JMA Japan Meteorological Agency [Tokyo]

JMA Japan Microphotography Association

JMA Johnson-Mehl-Avrami (Theory) [Metallurgy]

J Macromol Sci Journal of Macromolecular Science [US]

J Macromol Sci A Journal of Macromolecular Science, Part A [US]

J Macromol Sci B Journal of Macromolecular Science, Part B [US]

J Macromol Sci C Journal of Macromolecular Science, Part C [US]

J Macromol Sci Chem Journal of Macromolecular Science–Chemistry [US]

J Macromol Sci Phys Journal of Macromolecular Science–Physics [US]

J Macromol Sci Rev Macromol Chem Phys Journal of Macromolecular Science–Reviews in Macromolecular Chemistry and Physics [US]

JMAF Japanese Ministry for Agriculture, Forestry and Fisheries

J Magn Magn Mater Journal of Magnetism and Magnetic Materials [Published in the Netherlands]

J Magn Reson Journal of Magnetic Resonance [US]

JMAJ Japan Management Association Journal

JMAK Johnson, Mehl-Avrami-Kolmogorov (Theory) [Metallurgy]

J Manage Inf Syst Journal of Management Information Systems [US]

J Manuf Oper Manage Journal of Manufacturing and Operations Management [US]

J Manuf Syst Journal of Manufacturing Systems [of Society of Manufacturing Engineers, US]

JMAPI Java Management Application Program Interface

J Mater Chem Journal of Materials Chemistry [of Royal Society of Chemistry, UK]

J Mater Energy Syst Journal of Materials for Energy Systems [US]

J Mater Eng Journal of Materials Engineering [US]

J Mater Process Technol Journal of Materials Processing Technology [Netherlands]

J Mater Res Journal of Materials Research [of Materials Research Society, US]

J Mater Sci Journal of Materials Science [UK]

J Mater Sci Lett Journal of Materials Science Letters [UK]

J Mater Sci Soc Jpn Journal of the Materials Science Society of Japan

J Mater Shaping Technol Journal of Materials Shaping Technology [of ASM International, US]

J Mater Synth Process Journal of Materials Synthesis and Processing [US]

J Math Anal Appl Journal of Mathematical Analysis and Application [US]

J Math Biol Journal of Mathematical Biology [Germany]

J Math Phys Journal of Mathematical Physics [of American Institute of Physics, US]

J Math Phys Journal of Mathematics and Physics [Published by the Massachusetts Institute of Technology, Cambridge, US]

J Math Phys Sci Journal of Mathematical and Physical Sciences [of Indian Institute of Technology]

J Math Purés Appl Journal de Mathématique Purés et Appliquées [Journal of Pure and Applied Mathematics; published by Centre National de la Recherche Scientifique, France]

J Math Soc Jpn Journal of the Mathematical Society of Japan

JMC Jefferson Medical College [Philadelphia, Pennsylvania, US]

JMC Joint Maritime Commission

JMC Joint Maritime Congress [US]

JMC Joint Meteorological Committee

JMC Journal of Materials Chemistry [of Royal Society of Chemistry, UK]

J/m³/°C Joule(s) cubic meter per degree Celsius [also J m^{-3} °C^{-1}]

JMD Jamaican Dollar [Currency]

JMD Journal of Mechanical Design (Transactions of the American Society Mechanical Engineers) [US]

JMD Jungle Message Decoder

JME Journal of Molecular Endocrinology [of Society for Endocrinology, US]

JME Jungle Message Encoder

J Meas Control (Trans ASME) Journal of Measurement and Control (Transactions of the American Society of Mechanical Engineers) [US]

J Mech Behav Mater Journal of the Mechanical Behavior of Materials [UK]

J Mech Des (Trans ASME) Journal of Mechanical Design (Transactions of the American Society Mechanical Engineers) [US]

J Mech Eng Lab Journal of Mechanical Engineering Laboratory [Japan]

J Mech Eng Sci Journal of Mechanical Engineering Science [UK]

J Mech Phys Solids Journal of the Mechanics and Physics of Solids [UK]

J Mech Transm Autom Des (Trans ASME) Journal of Mechanisms, Transmissions and Automation in Design (Transactions of the American Society of Mechanical Engineers) [US]

J Mech Work Technol Journal of Mechanical Working Technology [Netherlands]

J Méc Theor Appl Journal de Mécanique Théorique et Appliquée [French Journal of Theoretical and Applied Mechanics]

JMED Jungle Message Encoder-Decoder

J Med Journal of Medicine [International Publication]

J Med Chem Journal of Medical Chemistry [of American Chemical Society, US]

J Med Educ Journal of Medical Education [US]

J Med Eng Technol Journal of Medical Engineering and Technology [UK]

J Méd Nucl Biophys Journal de Médecine Nucléaire et Biophysique [French Journal of Nuclear Medicine and Biophysics]

JMEE Japan Manufacturing Engineer Exchange (Program) [Canada]

J Membr Biol Journal of Membrane Biology

J Membr Sci Journal of Membrane Science [Netherlands]

JMEP Journal of Materials Engineering and Performance [of ASM International, US]

J Met Journal of Metals [of AIME Minerals, Metals and Materials Society, US]

J Metamorph Geol Journal of Metamorphic Geology

J Meteorol Journal of Meteorology [US]

J Met Fin Soc Jpn Journal of the Metal Finishing Society of Japan

J Met Fin Soc Korea Journal of the Metal Finishing Society of Korea

JMF Java Media Framework [Computers]

JMI Japan Metals Institute

JMI John Muir Institute (for Environmental Studies) [US]

JMICB Joint Management Integrated Control Board

J Microbiol Methods Journal of Microbiological Methods [Netherlands]

J Microcomput Appl Journal of Microcomputer Application [UK]

J Microcomput Syst Manage Journal of Microcomputer Systems Management [US]

J Microencapsulation Journal of Microencapsulation [US]

J Micromech Microeng Journal of Micromechanics and Microengineering [of Institute of Physics, UK]

J Micropalaeont Journal of Micropalaeontology [of Geological Society, London, UK]

J Microsc Journal of Microscopy [UK]

J Microsc Spectrosc Electron Journal de Microscopie et de Spectroscopie Electroniques [French Journal of Electron Microscopy and Spectroscopy; published by Société Française de Microscopie Electronique]

J Microw Power Journal of Microwave Power [of International Microwave Power Institute, US]

J Microw Power Electromagn Energy Journal of Microwave Power and Electromagnetic Energy [of International Microwave Power Institute, US]

J Mine Vent Soc S Afr Journal of the Mine Ventilation Society of South Africa

J Min Mater Process Inst Jpn Journal of the Mining and Materials Processing Institute of Japan

J Min Metall Inst Jpn Journal of the Mining and Metallurgical Institute of Japan

J/m³K² Joule(s) per cubic meter kelvin squared [also J m^{-3} K^{-2}]

Jm²/kg Joule-meter squared per kilogram [also J·m²·kg^{-1}]

J/m²K·s Joule(s) per square meter kelvin second [also J m^{-2} K^{-1} s^{-1}]

JMKU Journal of Mathematics of Kyoto University [Japan]

JMMM Journal of Magnetism and Magnetic Materials

J Mod Opt Journal of Modern Optics [UK]

J/mol Joule per mole [also J mol^{-1}]

J/mol at Joule(s) per mole of atoms [also J/(mol at)$^{-1}$]

J Mol Biol Journal of Molecular Biology [US]

J Mol Cat Journal of Molecular Catalysis [US]

J Mol Cell Cardiol Journal of Molecular and Cellular Cardiology [US]

J Mol Electron Journal of Molecular Electronics [Now Part of Advanced Materials for Optics and Electronics]

J Mol Endocrinol Journal of Molecular Endocrinology [of Society for Endocrinology, US]

J Mol Evol Journal of Molecular Evolution [US]

J/mol·K Joule per mole kelvin [also J/mol/K or Jmol^{-1} K^{-1}]

J Mol Med Journal of Molecular Medicine

J Mol Sci Journal of Molecular Science [PR China]

J Mol Sci (Int Ed) Journal of Molecular Science (International Edition) [PR China]

J Mol Spectrosc Journal of Molecular Spectroscopy [US] [also J Mol Spectry]

J Mol Struct Journal of Molecular Structure [Netherlands]

JMOS Job Management Operations System [Computers]

JMP Journal of Mathematical Physics

Jmp Jump [also jmp]

JMPCF Joint Master Plan for Commercial Fisheries

JMPR Joint Meeting on Pesticide Residues and the Environment [of Food and Agricultural Organization's Panel of Experts and the World Health Organization's Expert Group]

JMR Journal of Materials Research [of Materials Research Society, US]

JMRA Journal of Materials Research Abstracts [of Materials Research Society, US]

JMRPS Joint Meteorological Radio Propagation Subcommittee [UK]

JMS Journal of Materials Science [UK]

JMS Lett Journal of Materials Science Letters [UK]

J/m²·s Joule(s) per square meter second [also J/m²/s or J m^{-2} s^{-1}]

JMSAC Joint Meteorological Satellite Advisory Committee

JMSJ Journal of Magnetic Society of Japan

JMSJ Journal of the Mathematical Society of Japan

J/m·s·K Joule(s) per meter second kelvin [or J/m/s/K or J m^{-1} s^{-1} K^{-1}]

JMSPO Joint Meteorological Satellite Program Office

JMS Rev Macromol Chem Phys Journal of Materials Science, Revue of Macromolecular Chemistry and Physics [UK]

JMST Journal of Materials Shaping Technology

JMSX Job Memory Switch Matrix

JMU James Madison University [US]

J Multivariate Anal Journal of Multivariate Analysis [US]

JMX Jumbo (Group) Multiplex

JN Jam Nut

JN Jet Nozzle

Jn Junction [also jn]

JNA Jump if Not Above

JNACC Joint Nuclear Accident Coordinating Center [US]

JNAE Jump if Not Above or Equal

J Nanjing Inst Technol Journal of Nanjing Institute of Technology [PR China]

J Natl Cancer Inst Journal of the National Cancer Institute [US]

J Natl Cancer Inst Monogr Journal of the National Cancer Institute Monographs [US]

J Natl Chem Lab Ind Journal of the National Chemical Laboratory for Industry [Japan]

J Natl Res Counc Thail Journal of the National Research Council of Thailand [of Thai National Documentation Center]

J Natl Tech Assoc Journal of the National Technical Association [US]

J Nat Prod Journal of Natural Products [of American Chemical Society, US]

J Nat Rubber Res Journal of Natural Rubber Research [of Rubber Research Institute of Malaysia]

J Navig Journal of Navigation [of Royal Institute of Navigation, UK]

JNB Jump if Not Below

JNBE Jump if Not Below or Equal

JNC Japanese National Commission

JNC Jet Navigation Chart

JNCASR Jawaharlal Nehru Center for Advanced Scientific Research [New Delhi, India]

JNCI Journal of the National Cancer Institute [US]

J NCI Monogr Journal of the National Cancer Institute Monographs [US]

JND Just Noticeable Difference [also jnd]

JNDI Journal of Nondestructive Inspection [Japan]

JNE Jump Not Equal

J Neurochem Journal of Neurochemistry [UK]

J Neurol Neurosurg Psych Journal of Neurology, Neurosurgery and Psychiatry

J Neurol Sci Journal of Neurological Science

J Neurophysiol Journal of Neurophysiology [of American Physiological Society, US]

J Neurosci Journal of Neuroscience [of Society for Neuroscience, US]

J Neurosci Res Journal of Neuroscience Research [US]

J New Gener Comput Syst Journal of New Generation Computer Systems [Germany]

JNG Jump if Not Greater

JNGE Jump if Not Greater or Equal

JNI Japan-Netherlands Institute

JNI Java Native Interface [Computers]

JNI Jet Nozzle Inlet

JNICT Junta Nacional de Investigação Cientifica e Tecnologica [National Scientific and Technological Research Authority, Portugal]

.JNK Junk [File Name Extension] [also .jnk]

Jnl Journal [also jnl]

JNLE Jump if Not Less or Equal

Jnlst Journalist [also jnlst]

JNM Journal of Nuclear Materials

JNMM Journal of Nuclear Materials Management [of Institute of Nuclear Materials Management, US]

JNMW Jump No Matter What

JNO Jet Nozzle Outlet

JNO Jump if No Overflow

JNOC Japan National Oil Corporation

J Non-Cryst Solids Journal of Non-Crystalline Solids [Published in the Netherlands]

J Nondestr Eval Journal of of Nondestructive Evaluation [US]

J Non-Equilib Thermodyn Journal of Non-Equilibrium Thermodynamics [Germany]

J Nonlinear Sci Journal of Nonlinear Science (includes Nonlinear Science Today) [US]

J Non-Newton Fluid Mech Journal of Non-Newtonian Fluid Mechanics [Netherlands],

J Northeast Univ Technol Journal of Northeast University of Technology [PR China]

JNP Jump if No Parity

JNR Japan(ese) National Railways

Jnr Junior [also jnr]

JNS Jump if No Sign

JNSC Japan Nuclear Safety Commission

JNSC Joint Navigation Satellite Committee

JNSDA Japan Nuclear Ship Development Agency

JNT Joint Network Team

Jnt Joint [also jnt]

JNTO Japan National Tourist Organization

Jnt Stk Joint Stock

JNU Jeonbuk National University [South Korea]

J Nucl Mater Journal of Nuclear Materials [Netherlands]

J Nucl Mater Manage Journal of Nuclear Materials Management [of Institute of Nuclear Materials Management, US]

J Nucl Med Journal of Nuclear Medicine [of Society of Nuclear Medicine, US]

J Nucl Med Technol Journal of Nuclear Medicine Technology [Society of Nuclear Medicine, US]

J Nucl Sci Technol Journal of Nuclear Science and Technology [Japan]

J Number Theory Journal of Number Theory [US]

J Nutr Journal of Nutrition [of American Institute of Nutrition, US]

J Nutr Educ Journal of Nutrition Education

JNZ Jump if Not Zero

J NZDA Journal of the New Zealand Dietetic Association

J NZIC Journal of the New Zealand Institute of Chemistry

JO Job Order

JO Johnson and Oh (Model) [Materials Science]

JO Joint Organization

JO Jordan [ISO Code]

JO Junction Office

.jo Jordan [Country Code/Domain Name]

JOAO JOSA A (Journal of the Optical Society of America A) Online

JOB NO Job Number

JOBO JOSA B (Journal of the Optical Society of America B) Online

JOB LIB Job Library

JOBS Job Opportunities in the Business Sector (Program) [of National Alliance of Businessmen, US]

JOB SEQ Job Sequence (Number)

JOC Joint Operation(s) Center

JOC Journal of Organic Chemistry

JOCCA Journal of the Oil and Colour Chemists Association [UK]

J Occup Med Journal of Occupational Medicine [US]

JOCV Japan Overseas Cooperation Volunteers

JOD Joint Occupancy Data

JOD Joint Occupancy Date

JOD Jordanian Dinar [Currency]

JODC Japan Ocean Data Center

JODC Japan Overseas Development Corporation

JODOS Joint Optical Density of States [Solid-State Physics]

JOE Java Objects Everywhere [Computers]

J Offshore Mech Arctic Eng Journal of Offshore Mechanics and Arctic Engineering

Jog Joggle [also jog]

JOG Jones Oil Gasification (Process)

JOHNNIAC John von Neumann Integrator and Automatic Computer [also Johnniac]

Johns Hopkins APL Tech Dig Johns Hopkins Applied Physics Laboratory Technical Digest [of Johns Hopkins University, Silver Spring, Maryland, US]

JOI Joint Oceanographic Institution

JOIDES Joint Oceanographic Institutions for Deep-Earth Sampling [of American Association of Petroleum Geologists, US]

J Oil Colour Chem Assoc Journal of the Oil and Colour Chemists' Association [UK]

Joining Mater Joining and Materials [UK]

JOIP Joint Operations Interface Procedure

JOIS Japan On-Line Information System [of Japan Information Center of Science and Technology]

JOL Job Organization Language [Computers]

JOLA Journal of Library Automation [of Library and Information Technology Association, US]

JOM (International Conference on the) Joining of Materials

JOM Journal of Metals [of AIME Minerals, Metals and Materials Society, US]

JOME Journal of Materials Engineering

JOM-e Electronic Supplement to the Journal of Metals [of AIME Minerals, Metals and Materials Society, US]

JONSWAP Joint North Sea Wave Project

JOP Joint Operating Procedure

JOP Journal of Physics–Condensed Matter [of Institute of Physics, UK]

JOP Jupiter Orbiter Probe-Galileo [of NASA]

JOPA Journal of Physics A–Mathematical and General [of Institute of Physics, UK]

JOPB Journal of Physics B–Atomic, Molecular and Optical Physics [of Institute of Physics, UK]

JOPC Journal of Physics C–Solid State Physics [of Institute of Physics, UK]

JOPD Journal of Physics D–Applied Physics [of Institute of Physics, UK]

JOPE Journal of Physics E–Scientific Instruments [of Institute of Physics, UK]

J Oper Manage Journal of Operations Management [of American Production and Inventory Control Society]

J Oper Res Soc Journal of the Operational Research Society [UK]

J Oper Res Soc Jpn Journal of the Operations Research Society of Japan

JOPF Journal of Physics F–Metal Physics [of Institute of Physics, UK]

JOPG Journal of Physics G–Nuclear and Particle Physics [of Institute of Physics, UK]

JOPL Journal of Paleo-Limnology [of University of Manitoba, Winnipeg, Canada]

JOPSE Journal of Petroleum Science and Engineering

J Opt Commun Journal of Optical Communications [Germany]

J Opt (France) Journal of Optics (France)

J Optim Theory Appl Journal of Optimization Theory and Applications [US]

J Opt (India) Journal of Optics (India) [of Optical Society of India]

J Opt Soc Am Journal of the Optical Society of America [US]

J Opt Soc Am A, Opt Image Sci Journal of the Optical Society of America A, Optics and Image Science [US]

J Opt Soc Am B, Opt Phys Journal of the Optical Society of America B, Optical Physics [US]

J Opt Technol Journal of Optical Technology

JOR Job Order Request

Jor Jordan(ian)

J Oral Med Journal of Oral Medicine [US]

JORDDB Jordartsgeologisk Databas [Quaternary Geology Database, of Swedish Geological Survey]

J Organomet Chem Journal of Organometallic Chemistry [US]

J Org Chem Journal of Organic Chemistry [of American Chemical Society, US]

J Orth Res Journal of Orthopedic Research [US]

J Orthopsych Journal of Orthopsychiatry [US]

JOSA Journal of the Optical Society of America [US]

JOSA A Journal of the Optical Society of America A [US]

JOSA B Journal of the Optical Society of America B [US]

JOSS Job Sharing System [Computers]

JOSS Johnniac Open-Shop System [Computer Language]

JOSS Joint Overseas Switchboard [US Military]

Jour Journal(ist) [also jour]

Jour Q Journalism Quarterly [US]

JOVIAL Jule's Own Version International Algebraic Language; Jule's Own Version of International Algorithmic Language

JP Japan [ISO Code]

JP Jaw Pin

JP Jet Propulsion; Jet Propellant; Jet Propelled

JP Jet Pump

JP Jig Pin

JP Job Processing; Job Processor [Computers]

JP Journal de Physique [Journal of Physics; published in France]

JP Junction Point [Electrical Engineering]

JP Junction Potential [Electronics]

JP Justice of the Peace

JP Jute-Protected; Jute Protection [Electric Cables]

J&P Joists and Planks [Building Trades]

J/Ψ J/psi Particle [Symbol]

J$_c$(P) Pressure-Dependent Critical Current Density [Symbol]

.jp Japan [Country Code/Domain Name]

JPA Job Pack Area

J Packag Technol Journal of Packaging Technology [US]

J Pain Symptom Manage Journal of Pain and Symptom Management

J Pak Med Assoc Journal of the Pakistan Medical Association

J Paleolimnol Journal of Paleolimnology [of University of Manitoba, Winnipeg, Canada]

J Paleont Journal of Paleontology

J Palliat Care Journal of Palliative Care

JPAM Journal of Pascal, Ada and Modula-2 [US]

J Parallel Distrib Comput Journal of Parallel and Distributed Computing [US]

J Parametr Journal of Parametrics [of International Association of Parametric Analysts, US]

J Parent Enter Nutr Journal of Parenteral and Enteral Nutrition

J Part Instrum Journal of Particle Instrumentation [US]

J Pascal Ada Modula-2 Journal of Pascal, Ada and Modula-2 [US]

JPB Joint Planning Board [US]

JPC Japan Productivity Center

JPC Jet Propulsion Center

JPC Joint Power Condition

JPCD Just Perceptible Color Difference [also jpcd]

JPCI Journal of the Prestressed Concrete Institute [US]

JPD Just Perceptible Difference [also jpd]

JPDC Japan Petroleum Development Corporation

JPDR Japan Power Demonstration Reactor

JPE Jump if Parity Even

J Ped Journal of Pediatrics [US] [also J Pediatr]

J Ped Psychol Journal of Pediatric Psychology

JPEG Joint Photographic Experts Group

J Perinatal Med Journal of Perinatal Medicine

J Personality Journal of Personality

JPET Journal of Pharmacology and Experimental Therapeutics [US]

J Pet Geol Journal of Petroleum Geology

J Petrol Journal of Petrology [of Oxford University Press, UK]

J Petrol Geol Journal of Petrology and Geology [US]

J Pet Sci Eng Journal of Petroleum Science and Engineering

J Pet Technol Journal of Petroleum Technology [of Society of Petroleum Engineers, US]

JPG Jefferson Proving Ground [Madison, Indiana, US]

.JPG JPEG (Joint Photographic Experts Group) Graphic File [File Name Extension] [also. jpg]

J Pharmacol Exp Ther Journal of Pharmacology and Experimental Therapeutics [US]

J Pharm Market Manage Journal of Pharmaceutical Marketing and Management [of American Association of Pharmaceutical Scientists, US]

J Pharm Pharmacol Journal of Pharmacy and Pharmacology [US]

J Pharm Sci Journal of Pharmaceutical Sciences [of American Pharmaceutical Association, US]

J Pharm Soc Jpn Journal of the Pharmaceutical Society of Japan

J Phase Equilib Journal of Phase Equilibria [of ASM International, US]

J Photogramm Remote Sens Journal of Photogrammetry and Remote Sensing [of International Society of Photogrammetry and Remote Sensing, Netherlands]

J Photogr Sci Journal of Photographic Science [of Royal Photographic Society, UK]

J Phycol Journal of Phycology [US]

J Phys Journal de Physique [Journal of Physics; published in France]

J Phys Journal of Physics [of Institute of Physics, UK]

J Phys I Journal de Physique I [Journal of Physics; published in France]

J Phys II Journal de Physique II [Journal of Physics; published in France]

J Phys III Journal de Physique III [Journal of Physics; published in France]

J Phys A, Math Gen Journal of Physics A–Mathematical and General [of Institute of Physics, UK]

J Phys B, At Mol Opt Phys Journal of Physics B–Atomic, Molecular and Optical Physics [of Institute of Physics, UK]

J Phys C, Solid State Phys Journal of Physics C–Solid State Physics [of Institute of Physics, UK]

J Phys Chem Journal of Physical Chemistry [of American Chemical Society, US]

J Phys Chem A Journal of Physical Chemistry A [of American Chemical Society, US]

J Phys Chem B Journal of Physical Chemistry B [of American Chemical Society, US]

J Phys Chem Ref Data Journal of Physical and Chemical Reference Data [of American Institute of Physics and American Chemical Society, US]

J Phys Chem Solids Journal of Physics and Chemistry of Solids [UK]

J Phys Colloq Journal de Physique Colloque [France]

J Phys, Condens Matter Journal of Physics, Condensed Matter [of Institute of Physics, UK]

J Phys D, Appl Phys Journal of Physics D–Applied Physics [of Institute of Physics, UK]

J Phys Earth Journal of Physics of the Earth [Japan]

J Phys E, Sci Instrum Journal of Physics E–Scientific Instruments [of Institute of Physics, UK]

J Phys F, Met Phys Journal of Physics F–Metal Physics [of Institute of Physics, UK]

J Phys G, Nucl Part Phys Journal of Physics G–Nuclear and Particle Physics [of Institute of Physics, UK]

J Physiol Journal of Physiology [Published by Cambridge University Press, UK]

J Phys Lett (Orsay) Journal de Physique Lettres (Orsay) [Journal of Physics Letters; published in Orsay, France]

J Phys Oceanogr Journal of Physical Oceanography [of American Meteorological Society]

J Phys Org Chem Journal of Physical Organic Chemistry [UK]

J Phys (Orsay) Journal of Physics (Orsay) [Published in Orsay, France]

J Phys Soc Journal of the Physical Society [Japan]

J Phys Soc Jpn Journal of Physical Society of Japan

JPIA Japan Plastics Industry Association

JPIC Joint Program Integration Committee

J Pipelines Journal of Pipelines [Netherlands]

J Pkg Technol Journal of Packaging Technology [US]

JPL Jet Propulsion Laboratory [of NASA at Pasadena, California, US]

J Plant Growth Regul Journal of Plant Growth Regulation [Germany]

J Plant Physiol Journal of Plant Physiology [Germany and US]

J Plasma Phys Journal of Plasma Physics [UK]

J Plast Film Sheet Journal of Plastic Film and Sheeting [US]

JPMA Japan Powder Metallurgy Association

Jpn Japan(ese)

Jpn Alum News Japan Aluminum News

Jpn Chem Week Japan Chemical Week

Jpn Circul J Japanese Circulation Journal [Medical Journal]

Jpn Comput Q Japan Computer Quarterly [of Japan Information Processing Development Center]

Jpn Environ Monitor Japan Environment Monitor [Periodical published in Japan]

Jpn Environ Q Japan Environment Quarterly [Publication of the Environment Agency]

Jpn Heart J Japanese Heart Journal

Jpn Ind Technol Bull Japan Industrial and Technological Bulletin

Jpn Iron Steel Inst Japan Iron and Steel Institute [also JISI]

Jpn J Appl Phys Japan Journal of Applied Physics

Jpn J Appl Phys 1, Regul Pap Short Notes Japan Journal of Applied Physics, Part 1, Regular Papers and Short Notes

Jpn J Appl Phys 2, Lett Japan Journal of Applied Physics, Part 2, Letters

Jpn J Appl Phys Suppl Japan Journal of Applied Physics, Supplement

Jpn J Cancer Res Japanese Journal of Cancer Research

Jpn J Ergon Japanese Journal of Ergonomics

Jpn J Freezing Drying Japanese Journal of Freezing and Drying

Jpn J Ind Health Japanese Journal of Industrial Health

Jpn J Malac Japanese Journal of Malacology

Jpn J Math Japanese Journal of Mathematics [of National Research Council of Japan]

Jpn J Med Electron Biol Eng Japanese Journal of Medical Electronics and Biological Engineering [of Japan Society of Medical Electronics and Biological Engineering]

Jpn J Physiol Japanese Journal of Physiology [Published by University of Tokyo Press]

Jpn Q Japan Quarterly [Journal]

Jpn J Water Pollut Res Japanese Journal of Water Pollution Research [Japan]

Jpn J Water Res Japan Journal of Water Research [Japan]

JPNL Judged Perceived Noise Level [also jpnl]

Jpn Met Bull Japan Metal Bulletin

Jpn Plast Age Japan Plastics Age [Journal]

Jpn Plast Ind Ann Japan Plastics Industry Annual

Jpn Pulp Paper Japan Pulp and Paper [Japan]

Jpn Railw Eng Journal of Railway Engineering [of Japan Railway Engineers Association]

Jpn Steel Bull Japan Steel Bulletin

Jpn Steel Works Japan Steel Works [Journal]

Jpn Steel Works Tech Rev Japan Steel Works Technical Review

Jpn TAPPI Japan TAPPI [now Jpn TAPPI Journal]

Jpn TAPPI J Japan TAPPI Journal [of Japan Technical Association of the Pulp and Paper Industry]

Jpn Telecommun Rev Japan Telecommunications Review

JPO Jump if Parity Odd

JPO Junior Professional Officer

JPOC JSC (Johnson Space Center) Payload Operations Center [of NASA at Houston, Texas, US]

J Polym Eng Journal of Polymer Engineering [UK]

J Polym Sci Journal of Polymer Science [US]

J Polym Sci A, Polym Chem Journal of Polymer Science, Part A, Polymer Chemistry [US]

J Polym Sci B, Polym Phys Journal of Polymer Science, Part B, Polymer Physics [US]

J Polym Sci C, Polym Lett Journal of Polymer Science, Part C, Polymer Letters [US]

J Polym Sci Lett Journal of Polymer Science Letters [US]

J Polym Sci Polym Symp Journal of Polymer Science, Polymer Symposia [US]

JPOP Japanese Polar Orbiting Platform

J Porous Mater Journal of Porous Materials [Published in the US]

JPOS Journal of the Patent Office Society [US]

J Powder Bulk Solids Technol Journal of Powder and Bulk Solids Technology

J Power (Trans ASME) Journal of Power (Transactions of the American Society of Mechanical Engineers) [US]

J Power Sources Journal of Power Sources [Switzerland]

JPP Japanese Patent

JPP Joint Program Plan

JPPA Japanese Patent Application

JPPS Joint Petroleum Products Subcommittee

JPRA Japan Phonographic Record Association

J Press Vessel Technol (Trans ASME) Journal of Pressure Vessel Technology (Transactions of the American Society of Mechanical Engineers) [US]

J Proc R Soc New South Wales Journal and Proceedings of the Royal Society of New South Wales [Australia]

J Prod Agric Journal of Production Agriculture [of Soil Science Society of America, US]

J Prod Innov Manage Journal of Product Innovation Management [US]

J Prop Power Journal of Propulsion and Power [of American Institute of Aeronautics and Astronautics, US]

J Prost Dent Journal of Prosthetic Dentistry [US]

J Prot Coatings Linings Journal of Protective Coatings and Linings [US]

J Protein Chem Journal of Protein Chemistry [US]

J Protozool Journal of Protozoology [US]

JPRRI Japan Public Relations Research Institute

JPRS Joint Publications Research Service [US]

JPS Japanese Physical Society

JPS JICST (Japan Information Center of Science and Technology) Photoduplication Service [Japan]

JPSA Japanese Plating Suppliers Association

JPSJ Journal of the Physical Society of Japan

JPT Journal of Petroleum Technology [of Society of Petroleum Engineers, US]

JPTF Joint Parachute Test Facility [of US Department of Defense]

JPU Job Processing Unit [Computers]

J Pulp Pap Sci Journal of Pulp and Paper Science [of Canadian Pulp and Paper Association]

J Purch Mater Manage Journal of Purchasing and Materials Management [of National Association of Purchasing Management, US]

J Pure Appl Ultrason Journal of Pure and Applied Ultrasonics [of Ultrasonic Society of India]

JPW Job Processing Word [Computers]

JPY Japanese Yen [Currency]

J-Q J-Integral versus Q-Stress [Mechanics]

JQE Journal of Quantum Electronics [of Institute of Electrical and Electronics Engineers Lasers and Electro-Optics Society, US]

$j_0(QR)$ Zero-Order Bessel Function [Symbol]

J Qual Particip Journal for Quality and Participation [of Association for Quality and Participation, US]

J Qual Technol Journal of Quality Technology [of American Society for Quality Control, US]

J Quant Spectrosc Radiat Transf Journal of Quantitative Spectroscopy and Radiative Transfer [UK]

JR Japan Rail [also JNR]

JR Jet Relay

JR Jet Route

JR Joint Research

JR Josephson Radiation [Electronics]

JR Junction Rectification; Junction Rectifier [Electronics]

JR Jute-Reinforced; Jute Reinforcement

J-R J-Integral to Crack Extension (Curve) [Mechanics]

J(R) Real Space Exchange Coupling (in Physics) [Symbol]

Jr Journal [also jr]

Jr Junior [also jr]

J(r) Interaction Potential [Symbol]

J(r) Total Current Density [Symbol]

$J_m(r)$ Macroscopic Current Density [Symbol]

j(r) current density [Symbol]

J Radiat Res Journal of Radiation Research [of Japan Radiation Research Society]

J Radioanal Nucl Chem Journal of Radioanalytical and Nuclear Chemistry [Switzerland]

J Radioanal Nucl Chem, Artic Journal of Radioanalytical and Nuclear Chemistry, Articles [Switzerland]

J Radioanal Nucl Chem, Lett Journal of Radioanalytical and Nuclear Chemistry, Letters [Switzerland]

J Radiol Prot Journal of Radiological Protection [of Institute of Physics, UK]

J Radio Res Lab Journal of the Radio Research Laboratories [Japan]

JRAI Journal of the Royal Anthropological Institute [UK]

J Raman Spectrosc Journal of Raman Spectroscopy [UK]

J R Anthrop Inst Journal of the Royal Anthropological Institute [UK]

J R Astron Soc Journal of the Royal Astronomical Society [UK]

J R Astron Soc Can Journal of the Royal Astronomical Society of Canada

JRATA Joint Research and Test Activities

J Rat Mech Anal Journal of Rational Mechanics and Analysis [of University of Indiana, US]

JRB Jig Rest Button

JRB Joint Radio Board [US]

JRB Joint Review Board

JRC Japan Radio Company Limited [Tokyo]

JRC Japan Research Center

JRC Japan Research Council

JRC Joint Research Center [of the European Communities; located in Belgium (Brussels and Geel), Germany (Karlsruhe), Italy (Ispra) and the Netherlands (Petten)]

JRC Joint Research Commission

JRC Junior Red Cross

JRCAT Joint Research Center for AtomTechnology [of National Institute for Advanced Interdisciplinary Research, Japan]

JRCAT-NAIR Joint Research Center for AtomTechnology of the National Institute for Advanced Interdisciplinary Research [Japan]

JRC-ISPRA Joint Research Center at Ispra [Italy]

JRCITL Joint Research Center Ispra Tritium Laboratory [Italy]

JRCM Japan R&D (Research and Development) Center for Metals

J R Coll GP Journal of the Royal College of General Practitioners [UK]

JRC Rev JRC Review [Published by the Japan Radio Co. Ltd., Tokyo]

JRD Joint Research and Development

JRDB Joint Research and Development Board [of US Army and Navy]

JRDC Japan Research Development Corporation

JRDC Joint Research and Development Center

JRDOD Joint Research and Development Objectives Document

JREA Japan Railway Engineering Association

J Rech Atmos Journal de Recherches Atmospheriques [French Journal of Atmospheric Research]

J Reinf Plast Compos Journal of Reinforced Plastics and Composites [US]

J R Electr Mech Eng Journal of the Royal Electrical and Mechanical Engineers [UK]

J Reprod Med Journal of Reproductive Medicine

J Res Journal of Research [of Steel Castings Research and Trade Association, UK]

J Res Comput Educ Journal of Research on Computing in Education [of International Association for Computing in Education, US]

J Res Dev Educ Journal of Research and Development in Education [US]

J Res Inst Catal, Hokkaido Univ Journal of the Research Institute for Catalysis, Hokkaido University [Japan]

J Res Inst Sci Technol, Nihon Univ Journal of the Research Institute of Science and Technology, Nihon University [Japan]

J Res Natl Bur Stand Journal of Research of the National Bureau of Standards [now J Res Natl Inst Stand Technol]

J Res Natl Inst Stand Technol Journal of Research of the National Institute of Standards and Technology [US]

J Res NBS Journal of Research of the National Bureau of Standards [now J Res NIST]

J Res NIST Journal of Research of the National Institute of Standards and Technology [US]

J Res Onoda Cem Co Journal of Research of the Onoda Cement Company [Japan]

JRF Janssen Research Foundation [Belgium]

J R Geogr Soc Journal of the Royal Geographical Society [UK]

J Rheol Journal of Rheology [of Society of Rheology, US]

JRIA Japan Radioisotope Association

JRMA Japan Rubber Manufacturers Association

JRNBS Journal of Research, National Bureau of Standards [now JRNIST]

JRNIST Journal of Research, National Institute of Standards and Technology [US]

J Rob Mech Journal of Robotic Mechanics

J Robot Syst Journal of Robotic Systems [US]

JRP Joint Research Program

JRP Journal of Radiological Protection [of Institute of Physics, UK]

JRP Jute-Reinforced Plastics

JRPS Japan Reinforced Plastics Society,

JRR Japan Research Reactor [of Japan Atomic Energy Research Institute at Tokai]

JRR-2 Japan Research Reactor 2 [of Japan Atomic Energy Research Institute at Tokai]

JRR-3 Japan Research Reactor 3 [of Japan Atomic Energy Research Institute at Tokai]

JRR-3M Japan Research Reactor 3M [of Japan Atomic Energy Research Institute at Tokai]

JRR-4 Japan Research Reactor 4 [of Japan Atomic Energy Research Institute at Tokai, Japan]

J(r,r′) Nearest-Neighbor Exchange Interaction [*Note:* r and r′ are Spin Sites] [Symbol]

JRRS Japan Radiation Research Society

JRS Junction Relay Set [Electrical Engineering]

J R Signals Inst Journal of the Royal Signals Institution [UK]

J R Soc Arts Journal of the Royal Society of Arts [UK]

J R Soc Health Journal of the Royal Society of Health [UK]

J R Soc Med Journal of the Royal Society of Medicine [UK]

J R Soc NSW Journal of the Royal Society of New South Wales [Australia]

j(r,t) concentration current density [Symbol]

JRTP Job Readiness Training Program [Canada]

JRV Joint Research Venture

JS Jam Strobe

JS Jaswal-Sharma (Theory) [Physics]

JS Jessop Steel Company [US]

JS Jet Spinning

JS Jet Stream

JS Job Shop [Management Engineering]

JS Joint Services

JS Jump if Sign

J/S Jamming-to-Signal (Ratio)

Js Joule-second [also J·s or J-sec]

J/s Joule(s) per second [also J/sec or J s^{-1}]

JSA Japanese Space Agency

JSA Japanese Standard Association

JSA Japan Silk Association

JSA Jesuit Seismological Association [US]

JSA Job Safety Analysis

J Safety Res Journal of Safety Research [of National Safety Council, US]

J S Afr Acoust Inst Journal of the South African Acoustics Institute

J S Afr Inst Min Metall Journal of the South African Institute of Mining and Metallurgy

JSAI Japan Society for Artificial Intelligence

JSAM JES (Job Entry Subsystem) Spool Access Method [Computers]

J S Am Earth Sci Journal of South American Earth Sciences

JSAP Japan Society of Applied Physics

JSASS Japan Society for Aeronautical and Space Sciences

JSC Jackson State College [Mississippi, US]

JSC Jacksonville State College [Alabama, US]

JSC Jamaica Schools Certificate

JSC Jamaica Schools Commission

JSC Japan Science Council

JSC (Lyndon B.) Johnson Space Center [of NASA at Houston, Texas, US]

JSC Joint Scientific Committee

JSC Joint Security Control

JSC Joint Steering Committee

Js/C Joule-second per Coulomb [also Js C^{-1}]

JSCA Japan Structural Consultants Association

JSCC Japan Society for Composite Materials

JSCE Japan Society of Civil Engineers

JSCE Japan(ese) Society of Corrosion Engineering

JSCE Japan Society of Corrosion Engineers

JSCFA Japan Steel Castings and Forgings Association

J Sci Comput Journal of Scientific Computing [US]

J Sci Educ Technol Journal of Science Education and Technology [US]

J Sci Food Agric Journal of the Science of Food and Agriculture [of Society of Chemical Industry, US]

J Sci Hiroshima Univ Journal of Hiroshima University [Japan]

J Sci Hiroshima Univ A Journal of Hiroshima University, Series A [Japan]

J Sci Hiroshima Univ B Journal of Hiroshima University, Series B [Japan]

J Sci Ind Res Journal of Scientific and Industrial Research [India]

J Sci Instr Journal of Scientific Instruments [US]

J Sci Technol Journal of Science and Technology [UK]

J Sci Res Banaras Hindu Univ Journal of Scientific Research of the Banaras Hindu University [Varanasi, India]

J Sci Soc Thail Journal of the Science Society of Thailand

JSCLC Joint Standing Committee on Library Cooperation [UK]

JSCM Japan Society for Composite Materials

JSCM Johnson Space Center Manual [of NASA]

JSC/WRCP Joint Scientific Committee for the World Climate Research Program [of World Meteorological Organization and International Council of Scientific Unions]

JSD Justification Service Digit

J-sec Joule-second [also Js or J·s]

J/sec Joule(s) per second [also J/s or J s^{-1}]

J Sediment Petrol Journal of Sedimentary Petrology [US]

J Sediment Res Journal of Sedimentary Research [of NOAA National Geophysical Data Center, US]

JSEDM Japan Society of Electrical Discharge Machining

JSEE Japan Society for Engineering Education

J Seism Explor Journal of Seismic Exploration [Japan]

J Seismol Soc Jpn Journal of the Seismological Society of Japan

JSEM Japan Society for Electron Microscopy

JSEM Japan Society of Electrical Discharge Machining

J Semant Journal of Semantics [Netherlands]

J Semicond Journal of Semiconductors [PR China]

J Semicust ICs Journal of Semicustom ICs (Integrated Circuits) [UK]

JSEP Joint Services Electronics Program [of Cornell University, Ithaca, New York, US]

J Sep Process Technol Journal of Separation Process Technology [US]

J Serb Chem Soc Journal of the Serbian Chemical Society

JSF Junctor Switch Frame [Electrical Engineering]

JSFM Japan Society for Strength and Fracture of Materials

JSHA Job Safety and Health Analysis

J Shanghai Jiaotong Univ Journal of the Shanghai Jiaotong University [PR China]

J Sheffield Univ Metall Soc Journal of the Sheffield University Metallurgical Society [UK]

J Ship Prod Journal of Ship Production [of Society of Naval Architects and Marine Engineers, US]

J Ship Res Journal of Ship Research [of Society of Naval Architects and Marine Engineers, US]

JSHS Junior Science and Humanities Symposium

JSHT Japan Society of Heat Treatment

JSIA Joint Service Induction Area

JSIC Joint Securities Industry Committee

JSIF Japan Shipbuilding Industry Foundation

J Singap Natl Acad Sci Journal of the Singapore National Academy of Science

JSIS Japan Society of Iron and Steel

Js/kg Joule-second per kilogram [also Js kg^{-1}]

JSL Jet Select Logic

JSL Job Specification Language [Computers]

JSLE Japan Society of Lubrication Engineers

JSLS Japan Society of Library Science

JSLWG Joint Spacelab Working Group

JSME Japan Society of Mechanical Engineers,

JSME Joint Soil Moisture Experiment

JSMEBE Japan Society of Medical Electronics and Biological Engineering

JSME Int J JSME International Journal [of Japan Society of Mechanical Engineers]

JSME Int J I, Solid Mech Strength Mater JSME International Journal, Series I, Solid Mechanics, Strength of Materials [of Japan Society of Mechanical Engineers]

JSME Int J II, Fluids Eng Heat Transf Power Combust Thermophys Prop JSME International Journal, Series II, Fluids Engineering, Heat Transfer, Power, Combustion, Thermophysical Properties [of Japan Society of Mechanical Engineers]

JSME Int J III, Vib Control Eng Eng Ind JSME International Journal, Series III, Vibration, Control Engineering, Engineering for Industry [of Japan Society of Mechanical Engineers]

J/s·m·K Joule per second meter kelvin [also J s^{-1}m^{-1} K^{-1} or J/s/m/K)]

JSMS Japan Society of Materials Science

JSNDI Japanese Society of Nondestructive Inspection

JSOAPC Japan Society of Oil/Air Pressure Control

J Soc Cosmet Chem Journal of the Society of Cosmetic Chemists [US]

J Soc Dyers Color Journal of the Society of Dyers and Colorists [UK]

J Soc Environ Eng Journal of the Society of Environmental Engineers [UK]

J Soc Fiber Sci Technol Journal of the Society of Fiber Science and Technology [Japan]

J Soc Instrum Control Eng Journal of the Society of Instrument and Control Engineers [Japan]

J Soc Mater Sci Jpn Journal of the Society of Materials Science of Japan

J Soc Nav Archit Jpn Journal of the Society of Naval Architects of Japan

J Soc Occup Med Journal of the Society of Occupational Medicine [US]

J Soc Photogr Sci Technol Jpn Journal of the Society of Photographic Science and Technology of Japan

J Soc Psychol Journal of Social Psychology

J Soc Res Adm Journal of the Society of Research Administrators [US]

J Soc Rheol Journal of the Society of Rheology [US]

J Soc Rheol Jpn Journal of the Society of Rheology of Japan

J Soc Rubber Ind Journal of the Society of Rubber Industry [Japan]

J Soil Sci Journal of Soil Science [US]

J Sol Energy Journal of Solar Energy [US]

J Sol Energy Eng (Trans ASME) Journal of Solar Energy Engineering (Transactions of the American Society of Mechanical Engineers) [US]

J Sol Energy Res Journal of Solar Energy Research [of Solar Energy Research Center, Iraq]

J Solid State Chem Journal of Solid State Chemistry [US]

J Sound Vib Journal of Sound and Vibration [UK]

J South Afr Acoust Inst Journal of the South African Acoustics Institute [South Africa]

J South Afr Inst Min Metall Journal of the South African Institute of Mining and Metallurgy [South Africa]

J Sov Laser Res Journal of Soviet Laser Research [US]

JSOW Joint Statement of Work

JSP Jackson Structured Programming

J Space Astron Res Journal of Space and Astronomy Research [Iraq]

J Spacecr Rockets Journal of Spacecraft and Rockets [of American Institute of Aeronautics and Astronautics]

JSPE Japan Society of Precision Engineering

J Spectrosc Soc Jpn Journal of the Spectroscopical Society of Japan

J Speech Disorders Journal of Speech Disorders

J Speech Hearing Res Journal of Speech and Hearing Research [US]

J Sports Med Journal of Sports Medicine

JSPP Japan Society of Plant Physiologists

JSPPM Japan(ese) Society of Powder and Powder Metallurgy

JSPS Japan Society for the Promotion of Science

JSPS Joint Strategic Planning System [of NATO]

JSPSR Japan Society for the Promotion of Scientific Research

JSRU Joint Speech Research Unit [UK]

JSS Java-Script Style Sheet [Computers]

JSS Joint Surveillance System [of Federal Aviation Administration and US Air Force]

JSSC Journal of Solid-State Circuits [of Institute of Electronic and Electrical Engineers Solid-State Circuits Council, US]

JSSFM Japan Society for Strength and Fracture of Materials

JSST Jamaican Society of Scientists and Technologists

JSST Japan Society for Simulation Technology

JST Japan Science and Technology Corporation [Tokyo]

JST Japan(ese) Standard Time [Greenwich Mean Time –09:00]

JST Japan Society of Tribologists

JST Joint Systems Test

JSTARS Joint Surveillance Target Attack Radar System

J Stat Phys Journal of Statistical Physics [US]

J Stat Plan Inference Journal of Statistical Planning and Inference [Netherlands]

JSTC Joint Scientific and Technical Committee

JSTC Joint Scientific and Technical Committee [of Global Climate Observing System]

JSTCC Joint Science and Technology Cooperation Committee [European Union]

J Steroid Chem Journal of Steroid Chemistry

JS&TIC Joint Scientific and Technical Intelligence Committee [UK]

J Stoch Process Appl Journal of Stochastic Processes and Their Applications [of Bernoulli Society for Mathematical Statistics and Probability, Netherlands]

JSTP Japan Society for Technology of Plasticity

J Strain Anal Eng Des Journal of Strain Analysis for Engineering Design [UK]

J Struct Biol Journal of Structural Biology [Published in the US]

J Struct Chem Journal of Structural Chemistry [Translation of: *Zhurnal Strukturnoi Khimii (USSR)*; published in the US]

J Struct Eng Journal of Structural Engineering [of American Society of Civil Engineers, US]

J Struct Geol Journal of Structural Geology [UK]

J Stud Alcohol Journal of Studies on Alcohol [US]

JSU Jackson State University [Mississippi, US]

J Supercomput Journal of Supercomputing [Netherlands]

J Supercond Journal of Superconductivity [US]

J Surf Sci Soc Jpn Journal of the Surface Science Society of Japan

J Surg Oncol Journal of Surgical Oncology [US]

J Surg Res Journal of Surgical Research [US]

JSWA Japan Sewage Works Association

JSWPR Japan Society on Water Pollution Research

J Symb Comput Journal of Symbolic Computation [UK]

J Symb Log The Journal of Symbolic Logic [of Association for Symbolic Logic, US]

J Syst Manage Journal of Systems Management [of Association for Systems Management, US]

J Syst Softw Journal of Systems and Software [US]

JSZT Japan Society of Zoological Science

JT Jahn-Teller (Effect) [also J-T] [Physical Chemistry]

JT Japanese Tokamak (Reactor)

JT Josephson Tunneling [Electronics]

JT Junction Temperature

JT Junction Transistor

J/T Joule(s) per Tesla [also J T^{-1}]

J-T Jahn-Teller (Effect) [also JT] [Physical Chemistry]

Jt Joint [also jt]

JTA Jamaica Teachers' Association

JTA Job Task Analysis

JTAC Joint Technical Advisory Committee [US]

JTAC Joint Telecommunications Advisory Committee

JTAG Joint Test Action Group

JTAM Job Transfer and Management

JTAPI Java Telephony Application Programming Interface [Computers]

JTAPPIK Journal of the Technical Association of the Pulp and Paper Industry of Korea [South Korea]

JTB Joint Transportation Board [US]

JTC Joint Technical Committee

JTC Joint Technology Conference

JTC Joint Telecommunications Committee

JTC Joint Trade Committee

JTCMF Joint Technical Committee on Marine Front

JTDS Joint Track Data Storage

JTE Jahn-Teller Effect [Physical Chemistry]

JTE Joule-Thomson Effect [Thermodynamics]

JTE Junction Tandem Exchange

JTEC Japanese Technology Evaluation Center

JTEC Japan Telecommunications Engineering and Consultancy

J Technol Journal of Technology [India]

J Tech Phys Journal of Technical Physics [of Polish Academy of Sciences]

J Telecommun Netw Journal of Telecommunication Networks [US]

J Terramech Journal of Terramechanics [of International Society for Terrain-Vehicle Systems, US]

J Test Eval Journal of Test and Evaluation [of International Test and Evaluation Association, US]

J Test Eval Journal of Testing and Evaluation [of American Society for Testing and Materials, US]

J Text Inst Journal of the Textile Institute [UK]

J Text Stud Journal of Texture Studies [US]

JTF Japan Trade Fair

JTF Joint Task Force

JTFC Japan Trade Fair Commission

JTG Joint Training Group

J$_c$(T,H) Temperature and Magnetic-Field-Dependent Critical Current Density [Symbol]

J Theor Biol Journal of Theoretical Biology [Published in the US]

J Therm Anal Journal of Thermal Analysis [UK]

J Therm Biol Journal of Thermal Biology [UK]

J Therm Eng Journal of Thermal Engineering [India]

J Therm Insul Journal of Thermal Insulation [US]

J Thermophys Heat Transf Journal of Thermophysics and Heat Transfer [of American Institute of Aeronautics and Astronautics, US]

J Thermosetting Plast, Jpn Journal of Thermosetting Plastics, Japan

J Therm Spray Technol Journal of Thermal Spray Technology [of ASM International, US]

J Therm Stresses Journal of Thermal Stresses [US]

J Thoracic Cardiovasc Surgery Journal of Thoracic and Cardiovascular Surgery [US]

JTIDS Joint Tactical Information Distribution System [of US Air Force]

J Time Ser Anal Journal of Time Series Analysis [UK]

JTIS Japan Technical Information Service

J Tissue Cult Meth Journal of Tissue Culture Methods [of Tissue Culture Association, US]

jtly jointly

JTM Job Transfer and Manipulation [Computers]

JTMP Job Transfer and Manipulation Protocol

J Tongji Univ Journal of Tongji University [PR China]

J Toxicol Journal of Toxicology

J Toxicol Environ Chem Journal of Toxicological and Environmental Chemistry [Switzerland]

J Toxicol Environ Health Journal of Toxicology and Environmental Health

JTPS Job and Tape Planning System

J Transp Res Forum Journal of the Transportation Research Forum [US]

J Tribol (Trans ASME) Journal of Tribology (Transactions of the American Society of Mechanical Engineers) [US]

J Trop Geogr Journal of Tropical Geography [of International Geographical Union]

JTRU Joint Services Tropical Research Unit

JTS JICST (Japan Information Center of Science and Technology) Translation Service [Japan]

J Tsinghua Univ Journal of Tsinghua University [PR China]

JTT Japan Telephone and Telegraph (Corporation)

JTTA Japan(ese) Technology Transfer Association

JTSS Japan Thermal Spraying Society

JTST Journal of Thermal Spray Technology

JTSTR Jet Stream [also JT STR or Jt Str]

JTTC Japanese Transport Technology Council

JTU Jackson Turbidity Unit

JTUAC Joint Trade Union Advisory Committee

J Turbomach (Trans ASME) Journal of Turbomachinery (Transactions of the American Society of Mechanical Engineers) [US]

Jt Vent Joint Venture

JU Jacksonville University [Florida, US]

JU Jack-Up (Oil Drilling Unit)

JU Jadavpur University [Calcutta, India]

JU Jahangirnagar University [Dacca, Bangladesh]

JU Joensuu University [Finland]

JU Joint User

JUDGE Judged Utility Decision Generator

Judge Adv Gen Judge Advocate General [US]

JUG Joint Users Group [US]

jug jugular [Anatomy]

JUGFET Junction-Gate Field-Effect Transistor

JUGHEAD Jonzy's Universal Gopher Hierarchy Excavation And Display [Internet]

JUL Joint University Libraries [US]

Jul July

Jül-Conf Jülich International Conference on Vacancies and Interstitials in Metals [Germany]

JULIE Joint Utility Locating Information for Excavators

J Ultrastruct Mol Struct Res Journal of the Ultrastructure and Molecular Structure Research [US]

J Ultrastr Res Journal of Ultrastructural Research [US]

Jun June

Jun Junior [also jun]

Junc Junction [also Juncn, juncn, or junct]

J Undergrad Res Phys Journal of Undergraduate Research in Physics [of Society of Physics Students, US]

JUNE Joint Utility Notification for Excavators

J Univ Kuwait Journal of the University of Kuwait

J Univ Sci Technol Beijing Journal of the University of Science and Technology Beijing [PR China]

Junr Junior [also junr]

JUPITER Juvenescent Pioneering Technology for Robots [Japan]

Jura Jurassic [Geology]

JURG Joint Users Requirements Group

J Urol Journal of Urology

JUSCANZ Japan, United States, Canada, Australia and New Zealand, Norway and Switzerland

JUSE Japanese Union of Scientists and Engineers

JUSE-AESOPP Japanese Union of Scientists and Engineers–An Estimator of Physical Properties (Database)

JUSMAPG Joint United States Military Advisory and Planning Group

JUSMG Joint United States Military Group

Just Justice [also just]

Just Justification; Justify [also just]

JUST LIM Justification Limit(s) [also Just Lim]

JUTEM Japan Ultrahigh Temperature Materials Research Center

JUTSS Joint Unattended Tactical Sensor System

JUUA Japan Univac Users Association

Juv Juvenile [also juv]

JV Jet Velocity

J-V Current Density–Voltage (Characteristic)

JVA Jordan Valley Authority [Israel]

J Vac Sci Technol Journal of Vacuum Science and Technology [of American Vacuum Society, US]

J Vac Sci Technol A, Vac Surf Films Journal of Vacuum Science and Technology A, Vacuum, Surfaces and Films [of American Vacuum Society, US]

J Vac Sci Technol B, Microelectron Process Phenom Journal of Vacuum Science and Technology B, Microelectronics Processing and Phenomena [of American Vacuum Society, US]

J Vac Soc Jpn Journal of the Vacuum Society of Japan

J Vasc Res Journal of Vascular Research [International Journal]

J Vasc Surgery Journal of Vascular Surgery [US]

JVC Japan Volunteer Center

JVD Jet Vapor Deposition

J Vib Acoust Stress Reliab Des (Trans ASME) Journal of Vibration, Acoustics, Stress, and Reliability in Design (Transactions of the American Society of Mechanical Engineers) [US]

J Vinyl Technol Journal of Vinyl Technology [of Society of Plastics Engineers, US]

J Virol Journal of Virology [of American Society for Microbiology, US]

J Vitaminol Journal of Vitaminology

J VLSI Comput Syst Journal of VLSI (Very-Large-Scale Integration) and Computer Systems [US]

JVM Java Virtual Machine [Computers]

J Volcanol Geotherm Res Journal of Volcanology and Geothermal Research

JVR Journal of Vascular Research [International Journal]

JVST Journal of Vacuum Science and Technology [of American Vacuum Society, US]

JVSTA Journal of Vacuum Science and Technology A (Vacuum, Surfaces and Films) [of American Vacuum Society, US]

JVSTB Journal of Vacuum Science and Technology B (Microelectronics Processing and Phenomena) [of American Vacuum Society, US]

JW Jacket Water

JW Johnson-Williams (Theory) [also J-W] [Physics]

J Water Pollut Control Fed Journal of the Water Pollution Control Federation [US]

J Water Waste Journal of Water and Waste [US]

J Weather Modif Journal of Weather Modification [of Weather Modification Association, US]

JWG Joint Working Group

J Wind Eng Ind Aerodyn Journal of Wind Engineering and Industrial Aerodynamics [of International Association for Wind Engineering, Canada]

JWL Jones-Wilkins-Lee (Equation of State) [Solid-State Physics]

J Wood Chem Technol Journal of Wood Chemistry and Technology [US]

JWPT Jersey Wildlife Preservation Trust [UK]

JWRI Japan Welding Research Institute

JWRS Japan Wood Research Society

JWS Japan Welding Society

JWS John Wiley and Sons, Inc. [New York City, US]

JWTC Joint Workplace Training Committee [of Ontario Ministry of Education and Training, Canada]

$J_n(x)$ Bessel Function of Order n [Symbol]

$J_v(x)$ Bessel Function [Symbol]

J Xiamen Univ (Nat Sci) Journal of Xiamen University (Natural Science) [PR China]

J Xian Inst Metall Constr Eng Journal of Xian Institute of Metallurgy and Construction Engineering [PR China]

J X-Ray Technol Journal of X-Ray Technology [of Society of X-Ray Technology, UK]

JY- Jordan [Civil Aircraft Marking]

J-Y Jobin Yvon (Optical Systems)

Jy July

$J(y)$ Longitudinal Momentum Distribution Function [Symbol]

JZ Jones Zone

JZ Jump if Zero

$J_n(z)$ Bessel Function [Symbol]

J Zhejiang Univ Journal of Zhejiang University [PR China]

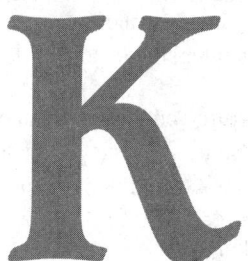

K Angular Addendum (of Bevel Gears) [Symbol]

K (Magnetic) Anisotropy Constant [Symbol]

K Bulk Modulus (of Elasticity) (or Compression Modulus) [Symbol]

K Cathode [Semiconductor Symbol]

K Cell Constant [Symbol]

K Cellophane-Covered (Electric Wire) [Symbol]

K Characteristic Impedance [Symbol]

K Click [Telegraphy]

K Coefficient of Thermal Conductivity [Symbol]

K Coefficient of Earth Pressure [Symbol]

K Compression Modulus (or Bulk Modulus of Elasticity) [Symbol]

K Constant [Symbol]

K Correction Factor [Symbol]

K Coupling Coefficient (between Electric Coils) [Symbol]

K Curvature Stress Correction Factor (of Springs) [Symbol]

K Dielectric Constant [Symbol]

K Distribution Coefficient (of Ion-Exchange Resin) [Symbol]

K Ellipticity (of Polarized Light) [Symbol]

K (Chemical) Equilibrium Constant [Symbol]

K Equilibrium Partition Ratio (in Welding) [Symbol]

K Film Thickness Factor (of Thrust Bearings) [Symbol]

K Fraunhofer Line for Calcium Having Wavelength of 393.4 Nanometers [Spectroscopy]

K Gain (of Control Elements, or Systems) [Symbol]

K Growth Coefficient (in Crystallography) [Symbol]

K Hall Coefficient (in Electronics) [Symbol]

K Hydrostatic Modulus (or Bulk Modulus of Elasticity) [Symbol]

K Ionization Constant (in Physical Chemistry) [Symbol]

K *(Kalium)* – Potassium [Symbol]

K Kalsilite [Mineral]

K Kangaroo

K Kaolin

K (Upper-case) Kappa [Greek Alphabet]

K Karat [Unit]

K *(Kathode)* – Cathode

K Keg

K Keel [Naval Architecture]

K K-Electron Captive [Symbol]

K Kelp

K Kelvin [formerly °K]

K Kenya(n)

K Keratin

K Kerf

K Kerma (in Nuclear Physics) [Symbol]

K Kernel [Symbol]

K Kerosine

K Ketone

K Ketose [Biochemistry]

K Kevlar

K Key

K Kidney

K Kiln

K Kilo-

K Kilo [Phonetic Alphabet]

K Kilobit

K Kilobyte

K Kilocycle [Unit]

K Kilohm [Unit]

K Kilometer [Unit]

K Kina [Currency of Papua New Guinea]

K Kinase [Biochemistry]

K Kinematic(s)

K Kinematical Hardening [Symbol]

K Kinetic(s)

K Kinetic Energy [Symbol]

K King [also k]

K Kininogen [Biochemistry]

K Kink

K Kip [Unit]

K Kip [Currency of Laos]

K Kitchen [also k]

K Klebsiella [Genus of Bacteria]

K K-Meson (in Particle Physics) [also kaon]

K Knight Shift (Value) [Symbol]

K Knock(ing)

K Knot

K Knowledge

K Koala

K *(Konstante)* – Constant

K Korea(n)

K K-Radiation

K K Shell (of Electron) [Symbol]

K Kubelka-Munk Absorption Coefficient (in Physics) [Symbol]

K Kuwait(i)

K Kwacha [Currency of Malawi and Zambia]

K Kymograph(y) [Medicine]

K　Length of (Drill) Tang [Symbol]

K　Load Factor (of Gears) [Symbol]

K　(+)-Lysine; Lysyl [Biochemistry]

K　Moment of Intertia [Symbol]

K　Minor Diameter (of Buttress Threads) [Symbol]

K　One Thousand

K　Permeability Coefficient [Symbol]

K　Pitch Factor (of Gears) [Symbol]

K　Pressure Constant (of Worm Gears) [Symbol]

K　Ratio of Heat Capacity at Constant Pressure to Heat Capacity at Constant Temperature [Symbol]

K　Reaction Constant [Symbol]

K　Scattering Vector [Symbol]

K　Segregation Coefficient (in Zone Refining) [Symbol]

K　Shape Factor (in Physics) [Symbol]

K　Spring Constant [Symbol]

K　Star Group of Spectral Type K [Surface Temperature 3,870°C, or 7,000°F] [Letter Designation]

K　Steady-State Creep Rate (in Mechanics) [Symbol]

K　Stiffness (of Constructional Elements) [Symbol]

K　Strength Coefficient (in True Stress versus True Strain Relationship) [Symbol]

K　Stress Concentration Factor (in Mechanics) [Symbol]

K　Stress Intensity Factor (or Stress Intensity Parameter) [Symbol]

K　Tanker (Aircraft) [US Air Force Symbol]

K　Thermal Conductivity [Symbol]

K　Thermoelastic Coefficient [Symbol]

K　Uniaxial Anisotropy [Symbol]

K　Wahl Factor (of Helical Springs) [Symbol]

K　Wavevector [Symbol]

K　X-Rays

-K　Target Drone (Aircraft) [US Navy Suffix]

$\bar{K}$　Average Thermal Conductivity [Symbol]

°K　Degree Kelvin [now K]

K%　Knight Shift (in Percent) [Symbol]

K*　Complex Modulus (in Compression) [Symbol]

K′　Cyclic Strength Coefficient [Symbol]

K′　Storage Modulus (in Compression) [Symbol]

K″　Loss Modulus (in Compression) [Symbol]

K^+　K-Plus Meson (i.e., Positively Charged K Meson) (in Particle Physics) [Symbol]

K^+　Potassium Ion [Symbol]

K^-　K-Minus Meson (i.e., Negatively Charged K Meson) (in Particle Physics) [Symbol]

$K^†$　Reaction Constant for Transition State [Symbol]

$K_{\parallel}$　In-Plane (or Parallel) Magnetic Anisotropy Constant [Symbol]

$K_{\perp}$　Out-of Plane (or Perpendicular) Magnetic Anisotropy [Symbol]

K_I　Stress Intensity Factor for Mode I [Symbol]

K_{II}　Stress Intensity Factor for Mode II [Symbol]

K_{III}　Stress Intensity Factor for Mode III [Symbol]

K_{Ia}　Indentation Critical Stress Intensity Factor [Symbol]

K_{Ia}　Plane-Strain Crack-Arrest Fracture Toughness [Symbol]

K_{Ic}　Critical Stress Intensity Factor (for Static Loading) [Symbol]

K_{Ic}　Mode I Critical Stress-Intensity Factor [Symbol]

K_{Ic}　Plane-Strain Fracture Toughness (Mode I) [Symbol]

K_{ICISB}　Fracture Toughness–Initial Crack Indentation Strength in Bending [Symbol]

K_{Id}　Critical Stress Intensity Factor (for Dynamic Loading) [Symbol]

K_{Id}　Dynamic Fracture Toughness [Symbol]

K_{IIc}　Fracture Toughness Parameter, Mode II [Symbol]

K_{Iv}　Short-Rod Plane-Strain Fracture Toughness [Symbol]

K^0　K-Zero Meson (i.e., Neutral K Meson) (in Particle Physics) [Symbol]

$\bar{K}^0$　K-Zero Meson, Antiparticle (i.e., Neutral K Meson, Antiparticle) (in Particle Physics) [Symbol]

K_0　At-Rest Coefficient of Earth Pressure [Symbol]

K_0　Displacement Kerma [Symbol]

K_0　Equilibrium Distribution Coefficient [Symbol]

K_0　Extinction Coefficient [Symbol]

K_0　Initial Crack (or Fracture) Resistance [Symbol]

K_0　Macroscopic Uniaxial Anisotropy Constant [Symbol]

K_0　Propagation Vector [Symbol]

K_0　Stress Intensity Factor at Crack Initiation (in Mechanics) [Symbol]

K_0　Threshold Stress Intensity Factor (in Mechanics) [Symbol]

K_1^0　Fast Decaying K-Zero Meson (in Particle Physics) [Symbol]

K_2^0　Slow-Decaying K-Zero Meson (in Particle Physics) [Symbol]

K_L^0　K-Zero-Long Meson (in Particle Physics) [Symbol]

K^{-1}　Reciprocal Kelvin (or One per Kelvin) [also 1/K]

K_1　Crystalline Anisotropy Constant [Symbol]

K_1　Magnetocrystalline Anisotropy (or First Order Cubic Anisotropy Constant) [Symbol]

K_1　Propagation Vector in Free Space [Symbol]

K_1　Stress Intensity Factor for Mode 1 [Symbol]

K_2　Second Anisotropy Constant (or Next Order Cubic Anisotropy Constant) [Symbol]

K_2　Stress Intensity Factor for Mode 2 [Symbol]

K_3　Stress Intensity Factor for Mode 3 [Symbol]

K_4　Magnetic In-Plane Anisotropy [Symbol]

K_A　Active Coefficient of Earth Pressure [Symbol]

K_A　Applied Stress Intensity Factor [Symbol]

K_a　Acid Dissociation Constant [Symbol]

K_a　Acidity Constant [Symbol]

K_a　Apparent Ionization Constant (in Physical Chemistry) [Symbol]

K_a　Bulk Modulus of Amorphous Phase [Symbol]

K_a　Crack-Arrest Fracture Toughness [Symbol]

K_a　K-Alpha Radiation [also Kα]

K_{abs}　Absorption Factor [Symbol]

K_b　Base Dissociation Constant [Symbol]

K_b　Bacicity Constant [Symbol]

K_b　Bending Energy Coefficient [Symbol]

K$_b$ Molal Boiling Point Elevation Constant [Symbol]

K$_b$ Thermodynamic Protolysis Constant [Symbol]

K$_{b(a)}$ Thermodynamic Protolysis Constant for Ammonia in H$_2$O at 25°C [Symbol]

K$_\beta$ K-Beta Radiation [also Kβ]

K$_c$ Equilibrium Constant in Terms of Concentration [Symbol]

K$_c$ Fracture Toughness at Onset of Brittle Fracture [Symbol]

K$_c$ Bulk Modulus of Crystalline Phase [Symbol]

K$_c$ Cubic Anisotropy Constant [Symbol]

K$_c$ Magnetocrystalline Anisotropy [Symbol]

K$_c$ Plane-Stress Fracture Toughness [Symbol]

K$_d$ Distribution Coefficient [Symbol]

K$_d$ Shape Anisotropy [Symbol]

K$_\delta$ K-Delta Radiation [also Kδ]

K$_e$ Back-EMF (Electromotive Force) Constant [Symbol]

K$_e$ Effective Distribution Coefficient [Symbol]

K$_e$ Crack-Tip Resistance (or Dislocation Generation Toughness) [Symbol]

K$_{eff}$ Effective Neutron Multiplication Factor (for a Nuclear Reactor) [Symbol]

K$_{eq}$ Equilibrium Constant [Symbol]

K$_f$ Fatigue Notch Factor [Symbol]

K$_f$ Fatigue-Strength Reduction Factor [Symbol]

K$_f$ Molal Freezing Point Depression Constant [Symbol]

K$_f$ Actual Stress Concentration Factor [Symbol]

K$_\gamma$ K-Gamma Radiation [also Kγ]

K$_h$ Henry's Law Constant [Symbol]

K$_{HIn}$ Dissociation Constant of Indicator Acid [Symbol]

K$_i$ Fracture Toughness of Interface [Symbol]

K$_i$ Ionization (or Dissociation) Equilibrium Constant [Symbol]

K$_i$ Thermal Conductivity of Interface [Symbol]

K$_{ICSR}$ Critical Stress Intensity Factor [also K$_{icsr}$] [Symbol]

K$_{IRLC}$ Ripple-Load Cracking Threshold Stress Intensity Factor [also K$_{irlc}$] [Symbol]

K$_{ISCC}$ Critical (or Threshold) Stress Intensity Factor for Intergranular Stress Corrosion Cracking [Symbol] [also K$_{iscc}$]

K$_{ISLC}$ Sustained-Load Cracking Threshold Stress Intensity Factor [also K$_{islc}$] [Symbol]

K$_J$ J-Integral Stress Intensity Factor [Symbol]

K$_{J0.2}$ Fracture Toughness after 0.2 mm of Blunting and Crack Extension [Symbol]

K$_{Jc}$ Fracture Toughness for Onset of Brittle Fracture After Elastic-Plastic Deformation [Symbol]

K$_{Ji}$ Fracture Toughness at Onset of Ductile Crack Initiation [Symbol]

K$_{Jm}$ Value of J-Integral Stress Intensity Factor at Maximum Load for Stable Fracture Behavior and Nonlinear Test Load [Symbol]

K$_{Ju}$ Fracture Toughness for Onset of Brittle Fracture and Some Amount of Ductile Tearing [Symbol]

K$_L$ Activity Product (or Solubility Product) of Electrolytes [Symbol]

K$_l$ Thermal Conductivity of Liquid [Symbol]

K$_\lambda$ Spectral Extinction Coefficient [Symbol]

K$_\lambda$ Magnetoelastic Anisotropy [Symbol]

K$_M$ Luminous Efficacy [Symbol]

K$_m$ Mass Transfer Coefficient [Symbol]

K$_m$ Michaelis Constant [Symbol]

K$_{MC}$ Magnetocrystalline Anisotropy [Symbol]

K$_{max}$ Maximum Stress-Intensity Factor [Symbol]

K$_{max}$ Maximum Kinetic Energy (of Photoemitted Electrons) [Symbol]

K$_{min}$ Minimum Stress-Intensity Factor [Symbol]

K$_n$ Stepwise Formation Constant [Symbol]

K$_{op}$ Crack Closure Stress Intensity Factor [Symbol]

K$_P$ Passive Coefficient of Earth Pressure [Symbol]

K$_P$ Preston (Polishing) Coefficient [Symbol]

K$_p$ Equilibrium Constant in Terms of Partial Pressure [Symbol]

K$_p$ Specific Permeability [Symbol]

K$_p$ Wavevector at First Peak in Diffuse X-Ray Scattering Curve [Symbol]

K$_Q$ Invalid Fracture Toughness Values [Symbol]

K$_R$ Crack (or Fracture) Extension (or Growth) Resistance [Symbol]

K$_S$ Shape Anisotropy Constant [Symbol]

K$_s$ Fracture Toughness of Substrate [Symbol]

K$_s$ Solubility Product (in Physical Chemistry) [Symbol]

K$_s$ Thermal Conductivity of Solid [Symbol]

K$_{s0}$ Solubility Product for Salts in Solution [Symbol]

K$_\sigma$ Magnetoelastic Stress Anisotropy Constant [Symbol]

K$_\sigma$ Stress Concentration Factor [Symbol]

K^{sc} Supercurrent Intensity Factor (in Solid-State Physics) [Symbol]

K$_{SP}$ Solubility Product (Constant) [Symbol] [also K$_{sp}$]

K$_{SS}$ Steady-State Stress Intensity Factor [also K$_{ss}$] [Symbol]

K$_{SSB}$ Intrinsic Steady-State Stress Bridging (Toughness) [also K$_{ss}$] [Symbol]

K$_T$ Isothermal Compressibility [Symbol]

K$_T$ Total Stress Intensity Factor [Symbol]

K$_t$ (Theoretical) Stress Concentration Factor (or Geometric Factor) [Symbol]

K$_t^\infty$ Stress Concentration Factor for Infinite Plate [Symbol]

K$_\tau$ Relative Reaction Time [Symbol]

K$_{th}$ Thermodynamic Equilibrium Constant [Symbol]

K$_{th}$ Threshold Crack Tip Stress-Intensity Factor [Symbol]

K$_{TT}$ Transverse Bulk Modulus (of Composites) [Symbol]

K$_u$ Uniaxial Magnetic Anisotropy Constant [Symbol]

K$_u$ Magnetocrystalline Anisotropy Energy [Symbol]

K$_v$ Precracked Charpy (in Mechanical Testing) [Symbol]

K$_v$ Valve Constant [Symbol]

K$_v$ Volume Anisotropy [Symbol]

K$_{vp}$ Molal Vapor Pressure Lowering Constant [Symbol]

K$_w$ Ion Product Constant for Water [Symbol]

K$_{wps}$ Crack Tip Stress Intensity with Warm Pressing [Symbol]

K$_x$ Equilibrium Constant in Terms of Mole Fractions [Symbol]

K2 Second Highest Mountain Peak in the World [8,611 m (28,250 ft); located in the Karakoram Mountain Range in Northern India and Pakistan]

K-6 Elementary School Level [US]

K-12 Kindergarten Through Grade 12 [US]

K-16 Kindergarten Through College/University [US]

K-37 Potassium-37 [also ^{37}K, or K^{37}]

K-38 Potassium-38 [also ^{38}K, or K^{38}]

K-39 Potassium-39 [also ^{39}K, or K^{39}]

K-40 Potassium-40 [also ^{40}K, or K^{40}]

K-41 Potassium-41 [also ^{41}K, or K^{41}]

K-42 Potassium-42 [also ^{42}K, or K^{42}]

K-43 Potassium-43 [also ^{43}K, or K^{43}]

K-44 Potassium-44 [also ^{44}K, or K^{44}]

k absorption coefficient (in optics) [Symbol]

k Archard (wear) constant [Symbol]

k Boltzmann's constant (1.380658×10^{-23} J/K) [Symbol]

k coagulation coefficient [Symbol]

k coefficient of conductivity [Symbol]

k coefficient of coupling [Symbol]

k coefficient of (soil) permeability [Symbol]

k coefficient (or modulus) of subgrade reaction [Symbol]

k coefficient of torsion (of wires, threads, etc.) [Symbol]

k cold and dry; at least one month below 32°F (0°C) [Subtype of Climate Region, e.g., in BSk, Bwk, etc.]

k complex propagation coefficient [Symbol]

k conductivity tensor [Symbol]

k die-opening factor (in bending) [Symbol]

k distortion factor [Symbol]

k effective distribution coefficient [Symbol]

k elastic constant [Symbol]

k (acoustic) elasticity [Symbol]

k extinction coefficient [Symbol]

k gain (of control elements, or systems) [Symbol]

k kappa [English Equivlalent]

k karat [Symbol]

k *(kathode)* – cathode [Symbol]

k kayser [Unit]

k key

k k-factor (i.e., relative x-ray intensity of instrument) [Symbol]

k kilo- [SI Prefix]

k kilobit [Unit]

k kilobyte [Unit]

k kilogram [Unit]

k kilohm [Unit]

k king [also k]

k kip [Unit]

k kitchen [also k]

k knit [Woven Fabrics]

k knot [Unit]

k k-space (in solid-state physics) [Symbol]

k local stress intensity (factor) [Symbol]

k magnetomechanical coupling factor [Symbol]

k (Hall-Petch) microstructural stress intensity (in metallurgy) [Symbol]

k modulus (of elliptic integrals) [Symbol]

k momentum [Symbol]

k multiplication factor (for nuclear reactors) [Symbol]

k notch sensitivity factor [Symbol]

k number of oscillations separating two amplitude measurements [Symbol]

k over [Teleprinter]

k photoelectron wave vector [Symbol]

k piezoelectric coupling coefficient [Symbol]

k propagation constant (of electromagnetic waves) [Symbol]

k proportionality factor in variation (e.g., $y = kx^2$) [Symbol]

k proportionality constant [Symbol]

k (principal) radius of gyration [Symbol]

k rate constant (in physical chemistry) [Symbol]

k ratio of specific heats (c_p/c_v) [Symbol]

k relative sensitivity [Symbol]

k running index [e.g., $\sum_{k=1}^{n} a_k = a_1 + a_2 + \cdots + a_k$] [Symbol]

k sample number (in statistics) [Symbol]

k spring constant [Symbol]

k (acoustic) stiffness [Symbol]

k surface diffusion coefficient [Symbol]

k thermal conductivity [Symbol]

k total inelastic collision efficiency [Symbol]

k unit vector along positive z-axis [Symbol]

k velocity constant (of chemical reaction) [Symbol]

k wave number [Symbol]

k wave vector [Symbol]

k wear coefficient [Symbol]

k' capacity factor (in chromatography) [Symbol]

k' coupling coefficient (in electronics) [Symbol]

$k^†$ reaction rate for transition state [Symbol]

$k_{||}$ thermal conductivity parallel to (composite) fiber direction [Symbol]

$k_{||}$ wavevector parallel to surface [Symbol]

$k_\perp$ thermal conductivity perpendicular to (composite) fiber direction [Symbol]

$k_\perp$ wave vector perpendicular to surface [Symbol]

k_0 drag coefficient [Symbol]

k_0 equilibrium partition coefficient (or partition ratio) [Symbol]

k_0 prefactor [Symbol]

k_0 radius of gyration with respect to axis of rotation [Symbol]

k_0 rate constant of neutral solution [Symbol]

k_1 magnetocrystalline anisotropy [Symbol]

k_a adsorption rate constant [Symbol]

k_a apparent thermal conductivity [Symbol]

k_B Boltzmann's constant (1.380658×10^{-23} J/K) [Symbol]

k_c curvature elastic constant [Symbol]

k^D rate constant of deuterium [Symbol]

k_d desorption rate constant [Symbol]

k_e electron thermal conductivity [Symbol]

k_e equivalent spring constant [Symbol]

k_F Fermi wave vector [Symbol]

k_f fatigue-strength reduction factor [Symbol]

k_f forward reaction rate [Symbol]

k_f fatigue notch factor [Symbol]

k_f Fermi momentum [Symbol]

k_i incident wavevector [Symbol]

k^H rate constant of hydrogen [Symbol]

k_l (crystal) lattice vibration conductivity [Symbol]

k_{LG} local Griffith stress intensity [Symbol]

k_s scattered wavevector [Symbol]

k^T rate constant of tritium [Symbol]

k_t theoretical stress concentration factor [Symbol]

k_{TF} Thomas-Fermi screening length [Symbol]

k_x wave vector in x-direction [Symbol]

k_y wave vector in y-direction [Symbol]

k_z wave vector in z-direction [Symbol]

κ absorption index (in optics) [Symbol]

κ attenuation index (in molecular spectroscopy) [Symbol]

κ Boltzmann's constant [Symbol]

κ compensation ratio [Symbol]

κ compressibility [Symbol]

κ (thermal) conductivity (of a solution) [Symbol]

κ coupling coefficient [Symbol]

κ curvature (of a curve) [Symbol]

κ curvature ratio (of roller bearings) [Symbol]

κ (relative) dielectric constant [Symbol]

κ diffraction vector [Symbol]

κ electrical (volume) conductivity [Symbol]

κ extinction coefficient [Symbol]

κ extrinsic bending rigidity [Symbol]

κ Ginzburg-Landau parameter (in solid-state physics) [Symbol]

κ ionic conductivity [Symbol]

κ (lower-case) kappa [Greek Alphabet]

κ kappa phase (of a material) [Symbol]

κ (crystal) lattice strain [Symbol]

κ magnetic (volume) susceptibility [Symbol]

κ (corrosion) pit generation rate [Symbol]

κ ratio of specific heats (i.e., C_p/C_v) [Symbol]

κ relative permittivity [Symbol]

κ specific conductance (of a solution) [Symbol]

κ specific inductive capacity [Symbol]

κ (longitudinal) stiffness ratio [Symbol]

κ surface stiffness coefficient [Symbol]

κ thermal conductivity [Symbol]

κ thermal diffusivity [Symbol]

κ wave number [Symbol]

κ^* relative complex permittivity (or relative complex dielectric constant) [Symbol]

κ' relative permittivity (or relative dielectric constant) [Symbol]

κ'' (relative) dielectric loss index (or factor) [Symbol]

κ_+ conductivity of cations [Symbol]

κ_- conductivity of anions [Symbol]

$\kappa_{\parallel}$ in-plane (or parallel) thermal conductivity [Symbol]

$\kappa_{\perp}$ out-of plane (or perpendicular) thermal conductivity [Symbol]

κ_0 initial susceptibility [Symbol]

κ_0 mean lattice strain [Symbol]

κ_0 permittivity of free space (vacuum) [Symbol]

κ_{300} room-temperature thermal conductivity [Symbol]

κ_a thermal conductivity along a-axis (of crystal) [Symbol]

κ_b thermal conductivity along b-axis (of crystal) [Symbol]

κ_l thermal conductivity of solid [Symbol]

κ_{mol} molar susceptibility [Symbol]

κ_p pinning potential restoring force constant [Symbol]

κ_ρ mass susceptibility [Symbol]

κ_s thermal conductivity of solid [Symbol]

$\varkappa$ isentropic exponent [Symbol]

$\varkappa$ (lower-case) kappa (variant) [Greek Alphabet]

KA K-Absorption

KA Kaman Aircraft (Corporation) [US]

KA Ketoacidosis [Medicine]

KA King Air (Airplane)

KA Kolmogorov-Avrami (Formula) [Metallurgy]

KA Kondo Alloy

KA Korean Air(lines)

KA Kynurenic Acid

K-A Potassium-40 (K-40) to Argon-40 (Ar-40) (Ratio) [also KA] [Geochronology]

K/Å Kelvin per Angstrom [Unit]

Ka Kina [Currency of Papua New Guinea]

Kα K-Alpha (Radiation) [also K_α]

kA kiloampere [Unit]

kÅ kiloangstrom [Unit]

ka *(kathode)* – cathode

KAAC Korea(n) Association of Automatic Control [South Korea]

KAAP Kellogg Advanced Ammonia Process

KAB Keep America Beautiful [US]

Ka band Microwave Frequency Band of 10.90 to 36.00 Gigahertz

KABLE Ka-Band Link Experiment

KAc Potassium Acetate

K(ACAC) Potassium Acetylacetonate [also K(AcAc), or K(acac)]

KACHAPAG Karlsruhe Charged Particle Group [of Information System Karlsruhe, Germany]

kA/cm² kiloampere(s) per square centimeter [also kA cm^{-2}]

K-Acid 4-Amino-5-Hydroxy-1,7-Naphthalenedisulfonic Acid; 1-Amino-8-Naphthol-4,6-Disulfonic Acid [also K-acid]

KACST King Abdul Aziz City for Science and Technology [Riyadh, Saudi Arabia]

KAD Kadena AB, Ryuku Islands [NASA Deorbiting Site]

KAEA Korea(n) Association of Electronics and Automation [South Korea]

KAEDS Keystone Association for Educational Data Systems

KAERI Korea Atomic Energy Research Institute [Taejon, South Korea]

KAERI/CD Korea Atomic Energy Research Institute/Contractor Report [Taejon, South Korea]

KAERI/TR Korea Atomic Energy Research Institute/Technical Report [Taejon, South Korea]

KAF Keratinocyte Autocrine Factor [Biochemistry]

KAFB Kirtland Air Force Base [New Mexico, US]

KAI Kasaskii Aviatsionnyu Institut [Kazakh Aeronautical Institute, Russian Federation]

KAICA Korea Auto Industries Cooperation Association [South Korea]

KAIST Korea Advanced Institute of Science and Technology [Taejon, South Korea]

KAIT Korean Academy of Industrial Technology [Seo-Ku Inchon, South Korea]

KAK Key-Auto-Key

Kal Kalamein [Mineral]

KALDAS Kidsgrove ALGOL Digital Analog Simulation

KALDO Kalling-Domnarvet (Steelmaking Process) [also Kaldo] [Metallurgy]

Kali phos Potassium Phosphate (Tissue Salt)

Kali sulph Potassium Sulphate (Tissue Salt)

KAM Keep Alive Memory

KAM Kolmogorov-Arnold-Moser (Theorem) [Physics]

kA/m kiloampere(s) per meter [also kA m^{-1}]

Kan Kansas [US]

kan kanamycin (antibiotic) [Microbiology]

kan/carb kanamycin to carbenicillin (ratio) [Microbiology]

kand vět *(kandidát vět)* – Candidate of Science; Czech Degree Equivalent to US Master's Degree

Kans Kansas [US]

KANUPP Karachi Nuclear Power Project [Pakistan]

KAO Krymskoi Astrofizicheskoi Observatorii [Crimean Astrophysical Observatory, Ukraine]

kaon K meson (i.e., K$^+$, K$^-$, or K^0 meson) [Particle Physics]

KaoS Kaon Spectrometer (Experiment) [of Gesellschaft für Schwerionenforschung mbH, Darmstadt, Germany]

KAP Potassium Acid Phthalate; Ortho-Phthalate Potassium Hydrogen (Crystal)

KAPL Knoll's Atomic Power Laboratory [of General Electric Company, Schenectady, New York, US]

KAPL KSC (Kennedy Space Center) Approved Parts List [of NASA]

K-APM KSC (Kennedy Space Center) Automated Payloads Plan/Requirement [of NASA]

K-APN KSC (Kennedy Space Center) Automated Payloads Notice [of NASA]

K-APPS KSC (Kennedy Space Center) Automated Payloads Project Specification [of NASA]

KAPWA Kite Aerial Photography Worldwide Association

KAR Kodak Automated Retrieval

K-Ar Potassium-Argon (Age Determination) [Geochronometry]

K$_l$(aR) Modified Spherical Bessel Function of the Second Kind [Symbol]

Karachi Univ J Sci Karachi University Journal of Science [Pakistan]

Karak ASSR Karakalpak Autonomous Soviet Socialist Republic [USSR]

KARC Kingston Area Recycling Corporation [Ontario, Canada]

KARI Korea Aerospace Research Institute [Daejeon, South Korea]

KARIN Karlsruhe Ring-Ion-Source Neutron (Generator) [Germany]

KARLDAP Karlsruhe Automatic Data Processing System [Germany]

KARP Korea Association for Radiation Protection [South Korea]

KAS Kent Archeological Society [UK]

KAS Kentucky Academy of Science [Louisville, US]

KAS Kenya Academy of Sciences [Nairobi, Kenya]

KAS Korean Astronomical Society [South Korea]

KAS Potassium Aluminosilicate

KAS Potassium Oxide–Alumina–Silica (Mineral)

KAS114 $K_2O \cdot Al_2O_3 \cdot 4SiO_2$ [A Potassium Aluminosilicate (Compound)]

KAS115 $K_2O \cdot Al_2O_3 \cdot 5SiO_2$ [A Potassium Aluminosilicate (Compound)]

KAS116 $K_2O \cdot Al_2O_3 \cdot 6SiO_2$ [A Potassium Aluminosilicate (Compound)]

KAS117 $K_2O \cdot Al_2O_3 \cdot 7SiO_2$ [A Potassium Aluminosilicate (Compound)]

K/at Kelvin(s) per atom [also K/atom]

K/at% Kelvin(s) per atomic percent [Unit]

KASC Knowledge Availability Systems Center [of University of Pittsburgh, Pennsylvania, US]

KASCADE Karlsruhe Shower Core and Array Detector [of Kernforschungszentrum Karlsruhe, Germany]

KASS Kent Automated Serials System [of Kent State University, Ohio, US]

KAST Kanagawa Academy of Science and Technology [Japan]

KASTO Korea Association of Standards and Testing Organizations [South Korea]

KATEE Kentra Anoteras Technikis ke Epangelmatikis Ekpedeuseos [Center for Higher Technical and Professional Studies, Greece]

KATI Kabeltext-Informationssystem [German Cable Text Information System]

KATS Kennedy (Space Center) Avionics Test Set [of NASA]

KATZ Kunststoff-Ausbildungs– und Technologiezentrum [Center for Plastics Education and Technology, Zurich, Switzerland]

KAU Key-Station Adapter Unit

KAU Kilo Accounting Unit

Kauch Rezina Kauchuk i Rezina [Russian Journal on Synthetic and Natural Rubber and Resins]

Kautsch Gummi Kunstst Kautschuk und Gummi, Kunststoffe [German Journal on Synthetic and Natural Rubber and Plastics]

KAVAA Kenya Audio-Visual Aid Association

KAW Koninklijke Akademie van Wetenschappen [Royal Academy of Sciences, Netherlands]

Kawasaki Rozai Tech Rep Kawasaki Rozai Technical Report [Japan]

Kawasaki Steel Bull Kawasaki Steel Bulletin [Japan]

Kawasaki Steel Newsl Kawasaki Steel Newsletter [Japan]

Kawasaki Steel Tech Bull Kawasaki Steel Technical Bulletin [Japan]

Kawasaki Steel Tech Rep Kawasaki Steel Technical Report [Japan]

Kawasaki Steel Tech Rep (Overseas) Kawasaki Steel Technical Report (Overseas) [Japan]

Kawasaki Tech Rep Kawasaki Technical Report [Japan]

KAWB Koninklijke Academie voor Wetenschappen van Belgie [Royal Belgian Academy of Sciences, Brussels, Belgium]

KAWLSKB Koninklijke Academie voor Wetenschappen, Letteren en Schone Kunsten van Belgie [Royal Belgian Academy of Schiences, Letters and Arts, Brussels, Belgium]

Kazakh Kazakhstan

Kazakh SSR Kazakh Soviet Socialist Republic

KB Kauri-Butanol (Value) [also kb]

KB Keyboard

KB Kilobyte [also kb, Kbyte, or kbyte]

KB Kirkpatrick-Baez (Type Spherical Mirror) [also K-B]

KB Knowledge Base(d)

KB Koninkrijk België [Kingdom of Belgium]

KB Kungliga Biblioteket [Royal Library/National Library, Stockholm, Sweden]

K-B Kirkpatrick-Baez (Type Spherical Mirror) [also KB]

Kb kilobit [also KB, Kbit, or kbit]

Kβ K-Beta (Radiation) [also K_β]

kb kilobar [also kbar]

kb kilobase [Unit of DNA fragment length equal to 10^3 nucleotides]

kb kilobyte [also KB, kbyte, or Kbyte]

KBA 3-Ketobutyraldehyde 1-Dimethyl Acetal

KBA Kraftfahrt-Bundesamt [Federal Motor Vehicle Office, Germany]

KBAC Kennedy (Space Center) Booster Assembly Contractor [NASA]

K band Microwave Frequency Band of 10.90 to 36.00 gigahertz

kbar kilobar [also kb]

KBBO Potassium Barium Bismuth Oxide (Superconductor)

KBD Keyboard [also Kbd]

KBD$ Keyboard [IBM Operating System/2]

KBDSC Keyboard and/or Display Controller

KBE Knowledge-Based Engineering

KBENC Keyboard Encoder

KBES Knowledge-Based Expert Systems

KBGIS Knowledge-Based Geographical Information System

KBI Kilobit(s) per inch [also Kbit/in, or Kbit in^{-1}, kbit/in, kbi, KB/i, kb/i, KBPI, or kbpi]

KBIM Keyboard Interface Module

Kbit(s) Kilobit(s) [also kbit(s), or Kb]

Kbit/in Kilobit(s) per inch [also Kbit/in, or Kbit in^{-1}, kbit/in, KBI, kbi, KB/i, kb/i, KBPI, or kbpi]

Kbit/sec Kilobit(s) per second [also Kbit/s, Kbit s^{-1}, kbit/s, KBS, kbs, KB/s, kb/s, KBPS, or kbps]

KBM Knowledge Base Management

KBMS Knowledge Base Management System

K-BOP Kawasaki Basic Oxygen Process (Converter) [Metallurgy]

KBPI Kilobits per Inch [also Kbit/in, kbit/in, KBI, kbi, KB/i, kb/i, or kbpi]

KBPRC Keyboard and Printer Controller

KBPS Kilobits per Second [also Kbit/s, kbit/s, KBS, kbs, KB/s, kb/s, or kbps]

KBPS Kilobytes per Second [also Kbps, KBps, or kbps]

kBq kilobecquerel [Unit]

KBr Potassium Bromide

KB RAM Kilobytes of Random-Access Memory

KBS Kilobit(s) per Second [also Kbit/s, kbit/s, kbs, KB/s, kb/s, KBPS, or kbps]

KBS Knowledge-Based System

KBS Knowledge-Based Systems [UK Publication]

KB/s Kilobits per second [also kb s^{-1}, Kbit/s, kbit/s, KBS, kbs, kb/s, KBPS, or kbps]

KBSC Knowledge-Based Systems Center [UK]

KBTG Keep Britain Tidy Group [UK]

KBU Keyboard Unit

KBuO Potassium Butoxide

KBVE Koninklijke Belgische Vereniging der Elektrotechniker [Royal Belgian Society of Electrical Engineers]

KBW Koppers, Babcock and Wilcox (Process)

KBWP Kernkraftwerk Baden-Württemberg Planungsgesellschaft [Nuclear Power Plant in Baden Württemberg, Germany]

Kbyte(s) Kilobyte(s) [also kbyte(s), KB, or kb]

KC Kajitani and Cook (Method) [also K-C] [Metallurgy]

KC Kansas City [Missouri, US]

KC Kármán Constant [Fluid Mechanics]

KC Kiln Car

KC Kilocharacter(s)

KC Kilocycle [also kc]

KC Kinetic Current

KC King's College [of University of London, UK]

KC King's Counsel

KC Knox College [Galesburg, Illinois, US]

KC Knudsen Cell

KC Kozeny-Carmen (Constant) [Fluid Mechanics]

KC Krebs Cycle [Biochemistry]

KC Tanker Transport [US Air Force Symbol]

KC$_x$ Binary Graphitide [x = 8, 16, 24, etc.]

K-C Kajitani and Cook (Method) [also KC] [Metallurgy]

kC kilocurie [also kc]

Kc Koruna [Currency of Czech Republic and Slovakia]

kc kilocurie [also kC]

kc kilocycle(s) [Unit]

kcal kilocalorie [also kg-cal, or Cal]

kcal/g/at kilocalorie(s0 per gram per atom [also kcal/g/atom]

kcal/kg kilocalorie(s) per kilogram [also kcal kg^{-1}]

kcal/kg·°C kilocalorie(s) per kilogram degree Celsius (or Centigrade) [also kcal kg^{-1} °C^{-1}]

kcal/m² kilocalorie(s) per square meter [also kcal m^{-2}]

kcal/m³ kilocalorie(s) per cubic meter [also kcal m^{-3}]

kcal/m³·°C kilocalorie(s) per cubic meter degree Celsius, or centigrade [also kcal m^{-3} °C^{-1}]

kcal/m³·°F kilocalorie(s) per cubic meter degree Fahrenheit [also kcal m^{-3} °C^{-1}]

kcal/m·h kilocalorie(s) per meter hour [also kcal m^{-1} h^{-1}]

kcal/m·h·°C kilocalorie(s) per meter hour degree celsius (or centigrade) [also kcal m^{-1} h^{-1} °C^{-1}]

kcal/m²·h·°C kilocalorie(s) per square meter hour degree Celsius (or Centigrade) [also kcal m^{-2} h^{-1} °C^{-1}]

kcal/mol kilokalorie(s) per mole [also kcal mol^{-1}]

kcal/s kilocalorie(s) per second [also kcal/sec]

KCARTCC Kansas City Air Regional Traffic Control Center [US]

KCAS Kent County Agricultural Society [UK]

KCAS Knots Calibrated Air Speed

KCB-S Krupp Combined Blowing for Stainless Steelmaking [Metallurgy]

KCBT Kansas City Board of Trade [US]

KCC Potassium Chlorochromate

KCC Keyboard Common Contact

KCC Knoxville Chamber of Commerce [Tennessee, US]

KCDQ Key Centre for Design Quality [University of Sydney, Australia]

KCE Kennel Club of England [UK]

KCE Kenya Certificate of Education

KCGL Koffolt Computer Graphics Laboratory [of Ohio State University, Columbus, US]

KCH Keramchemie Aktiengesellschaft [German Ceramics Company located at Siershahn]

kCi kilocurie [Unit]

K-CITEM KSC (Kennedy Space Center) CITE (Cargo Integration Test Equipment) Plan/Requirement [of NASA]

KCL Kirchhoff's Current Law [Electrical Engineering]

KCl Potassium Chloride (or Sylvite)

K/cm Kelvin per centimeter [also Kcm^{-1}]

K/cm kOe Kelvin(s) per centimeter per kilooersted [also K cm^{-1} kOe^{-1}]

KCN Potassium Calcium Niobate

KCN Potassium Cyanide

KCNA Korean Central News Agency [South Korea]

KCNN Potassium Calcium Sodium Niobate

KCNS Potassium Thiocyanate

KCO Keep Cost Order

KCP Koala Conservation Program [of Australian National Conservancy Agency]

KCP Keyboard-Controlled Phototypesetter

KCP Potassium Tetracyanoplatinate(II) [K$_2$Pt(CN)$_4$]

KCPE Kenya Certificate of Primary Education [formerly CPE]

KCPS Kilocycles per Second [also kcps, KC/s, Kc/s, kc/s, or kc/sec]

KCR Key Call Receiver

KCRT Keyboard Cathode Ray Tube

KCS Keyboard Configuration Studies

KCS Kilo (One Thousand) Characters per Second

KCS Kinki Chemical Society [Japan]

KCS Korean Ceramic Society [South Korea]

KCS Korean Chemical Society [South Korea]

Kčs Koruna (or Crown) [Currency of Czech Republic and Slovakia]

kc/s kilocycles per second [also KC/s, KCS, kc/sec, KCPS, or kcps]

KCTRA Kansas City Terminal Railroad Association [Missouri, US]

KCU Keyboard Control Unit

KD Key-to-Disk (System); Key-to-Diskette (System)

KD Khor-DasSarma (Potential) [Physics]

KD Kidney Disease [Medicine]

KD Kiln-Dried; Kiln Dryer; Kiln Drying

KD Kinematical Diffraction

KD Kink Diffusion (Process) [Materials Science]

KD Kjeldahl Determination [Chemistry]

KD Knocked Down [also kd]

KD Kongeriget Danmark [Kingdom of Denmark]

KD Kuwaiti Dollar [Currency]

Kδ K-Delta (Radiation) [also K$_\delta$]

kD kilodalton [Unit] [also kd]

**K$** Thousand Dollars (e.g., 75K$ stands for $75,000)

kDa kilodalton [Unit]

K Dan Vidensk Selsk Mat-Fys Medd Konglige Danske Videnskabernes Selskab Matematisk-Fysiske Meddelelser [Communications of the Royal Danish Academy of Sciences, Mathematics and Physics] [also Kgl Dan Vid Selsk Mat-Fys Medd]

KDB Kenya Dairy Board

KDB Korea Development Bank [South Korea]

KDD Kokusai Denshin Denwa Company Limited [Japan]

KDD Tech J KDD (Kokusai Denshin Denwa Company Limited) Technical Journal [Published by the International Communications Research Institute, Japan]

KDDP Potassium Dideuterophosphate

KDE Keyboard Data Entry

KDED Kansas Department of Economic Development [US]

KDEM Keyboard Data Entry Machine

KDEM Kurzweil Data Entry Machine

KDFC Korea Development Finance Corporation [South Korea]

KDLOC One Thousand Developed Lines of Code

KDMS Kennedy (Space Center) Data Management System [of NASA]

KDN 3-Deoxy-D-Glycero-D-Galacto-2-Nonulosonic Acid [Biochemistry]

KDN Kinetically Designed Nozzle

KDNBF Potassium Dinitrobenzofuroxan

KDO 2-Keto-3-Deoxyoctonate

K-DODM KSC (Kennedy Space Center) DOD (Department of Defense) Plan/Requirement

KDOS Key Display Operated System

KDOS Key-to-Disk Operating System

KDP Kalabagh Dam Project [Pakistan]

KDP Keyboard Display and Printer

KDP Key Development Plan

KDP Potassium Dihydrogen Phosphate

KΔΠ Kappa Delta Pi (Student Society)

KDPR Kentucky Department of Public Relations [US]

KDPI Kentucky Department of Public Information [US]

K-DPM KSC (Kennedy Space Center) DOD (Department of Defense) Payloads Plan/Requirement

K-DPN KSC (Kennedy Space Center) DOD (Department of Defense) Payloads Notice

K-DPPS KSC (Kennedy Space Center) DOD (Department of Defense) Payloads Project Specification

KDR Keyboard Data Recorder

KDS Keyboard Display Station

KDS Key Data Station

KDS Key Display System

KDS Key-to-Diskette System; Key-to-Disk System

KDS Kinetic-Dynamic Simulation

KDT Kentucky Department of Tourism [US]

KDT Keyboard and Display Test

KDT Key Data Terminal

KDT Key Definition Table

KDT Kinematical Diffraction Theory

KdT Kammer der Technik [Board of Engineering, Berlin, Germany]

KDU Keyboard and Display Unit

KDU Key-to-Disk Unit

KdV Korteweg-deVries (Soliton) [Physics]

KDVA Kongelige Danske Videnskabernes Akademie [Royal Danish Academy of Sciences, Copenhagen, Denmark]

KDVS Kongelige Danske Videnskabernes Selskab [Royal Danish Scientific Society]

KDVS Mat fys Medd Kongelige Danske Videnskabernes Selskab Matematisk-Fysiske Meddelelser [Communications on Mathematics and Physics of the Royal Danish Scientific Society]

KE Change Factor, External Spline [Symbol]

KE Kaiser Effect [Acoustics]

KE Kenya [ISO Code]

KE Kerr Effect [Optics]

KE Key Equipment

KE Kinetic Ellipsometry

KE Kinetic Energy

KE Kirchhoff Equations (of Thermodynamics)

KE Kirkendall Effect [Metallurgy]

KE Knife Edge

KE Knowledge Engineer(ing)

KE Kondo Effect [Metallurgy]

KE Korf Engineering GmbH [Duesseldorf, Germany] Average Kinetic Energy [Symbol]

K&E King's Evil [Medicine]

K/E Kevlar/Epoxy [also K/Ep]

.ke Kenya [Country Code/Domain Name]

$\kappa_0(\varepsilon_0)$ electric constant (or permittivity of free space) [Symbol]

KEA Kinetic Energy Ablation

KEAS Knots Equivalent Air Speed

KE(Auger) Kinetic Energy of Auger Electron

KEE Knowledge Engineering Environment (Software)

KEFIR Key Findings Reporter [of General Telephone and Electronics Corporation, US]

KEH Kaiser Engineers Hanford [of ICF Kaiser Hanford Company, Washington State, US]

KEI Kinetic Energy Implantation

Keio Sci Technol Rep Keio Science and Technology Reports [Published by Keio University, Faculty of Science and Technology, Kanagawa, Japan]

KEIS Kentucky Economic Information Systems [US]

KEK Koh-Enerugii Butsurigaku Kenkyuusho [National Laboratory for High-Energy Physics, Tokyo, Japan]

KEK Comput Newsl KEK (Koh-Enerugii Butsurigaku Kenkyuusho) Computing Newsletter [Japan]

KEKTANA KEK-Tanashi International Symposium on Physics of Hadrons and Nuclei [Tokyo, Japan]

KELL Kinetic, EDBA (2,2' (Ethylenediimino)dibutyric Acid), Ligand, Lowry (Radioligand Analysis)

KEM Kinetic Energy Metallization

Kema Sci Tech Rep Kema Scientific and Technical Reports [Published by NV KEMA, Arnheim, Netherlands]

KEMIDB Geokemisk Databas [Geochemical Database, of Swedish Geological Survey]

Kem Ind Kemija u Industriji [Yugoslavian Publication on the Chemical Industries]

KEMRI Kenya Medical Research Institute

Ken Kentucky [US]

Ken Kenya

KENGO Kenya Energy and Environment Association

Kent Rev Kent Review [Published by Brown Boveri Kent Ltd., Luton, Bedfordshire, UK]

Kent Tech Rev Kent Technical Review [Published by Brown Boveri Kent Ltd., Luton, Bedfordshire, UK]

KEO Kolloquium Eigenspannungen und Oberflächenverfestigung [Colloquium on Residual Stresses and Surface Strengthening, Germany]

KEP Key Entry Processing

KEP Kinetic Energy Polymerization

K/Ep Kevlar/Epoxy [also K/E]

KEPC Korea Electric Power Corporation

KE(PE) Kinetic Energy of Photoelectron

KEPZ Kaohsiung Export Processing Zone [Taiwan]

KER Kinetic Energy Release

KER Kolbe Electrosynthesis Reaction [Chemistry]

Keram Z Keramische Zeitschrift [German Ceramics Journal]

KERD Kinetic Energy Release Distribution

KERN Kerning [also Kern] [Typesetting]

KES Kansas Engineering Society [US]

KES Kenyan Shilling [Currency]

KES Korea Electronics Show [South Korea]

KESC Karachi Electric Supply Corporation [Pakistan]

KESS Kinetic Energy Storage System

KET Kernel Estimation Techniques

KET Krypton Exposure Technique

KEtO Potassium Ethoxide

KEV Kommunale Energieversorgung [Municipal Energy Supply (Project), of FIZ Karlsruhe, Germany]

keV kiloelectronvolt [also kev]

keV/at kiloelectronvolt(s) per atom [also keV/atom]

keV/cm³ kiloelectronvolt per cubic centimeter [also keV cm^{-3}]

keV/mg/cm² kiloelectronvolt(s) per milligram per square centimeter [also keV mg^{-1} cm^{-2}]

keV/μg kiloelectronvolt(s) per microgram [also keV μg^{-1}]

keV/nm kiloelectronvolt(s) per nanogram [also keV ng^{-1}]

KEW Kilogram-Equivalent Weight [also kew]

KEW Kinetic Energy Weapon

KEWB Kinetic Experiment on Water Boilers [US]

KEYB Load Keyboard Program (Command) [also keyb]

KEYBBE KEYBoard Program–Belgium

KEYBBR KEYBoard Program–Brazil

KEYBCF KEYBoard Program–Canadian-French

KEYBCZ KEYBoard Program–Czech Republic

KEYBD Keyboard [also Keybd, or keybd]

KEYBDK KEYBoard Program–Denmark

KEYBFR KEYBoard Program–France

KEYBGR KEYBoard Program–Germany

KEYBHU KEYBoard Program–Hungary

KEYBIT KEYBoard Program–Italy

KEYBLA KEYBoard Program–Latin America

KEYBNL KEYBoard Program–Netherlands

KEYBNO KEYBoard Program–Norway

KEYBPL KEYBoard Program–Poland

KEYBPO KEYBoard Program–Portugal

KEYBSF KEYBoard Program–Swiss-French

KEYBSG KEYBoard Program–Swiss-German

KEYBSL KEYBoard Program–Slovakia

KEYBSP KEYBoard Program–Spain

KEYBSU KEYBoard Program–Finland

KEYBSV KEYBoard Program–Sweden

KEYBUK KEYBoard Program–United Kingdom

KEYBUS KEYBoard Program–United States

KEYBYU KEYBoard Program–Yugoslavia

Key Eng Mater Key Engineering Materials [Journal published in Switzerland]

KEYTECT Keyword Detection

KF Kaiser-Fleischer (Ring) [Medicine]

KF Kaldo Furnace [Metallurgy]

KF Kalman Filter [Controllers]

KF Kaolin Fiber

KF Kern Front (Oilfield) [Southern California]

KF Key Field

KF Kiliani-Fischer (Synthesis) [Chemistry]

KF Kinematic Fluidity

KF Kinetic Friction

KF Kirchhoff Formula (of Thermodynamics)

KF Knudsen Flow [Physics]

KF Kouvel-Fisher (Method) [Magnetic Anisotropy]

KF Kruger Flap

KF Kurz-Fisher (Microsegregation Model) [also K-F] [Metallurgy]

KfA Kernforschungsanlage Jülich GmbH [Julich Nuclear Research Center, Germany] [also KFA]

KfA/ISI Kernforschungsanlage Jülich GmbH/Institut für Schicht– und Ionentechnik [Julich Nuclear Research Center/Institute for Film and Ion Technology, Germany] [also KFA-IS]

KfA Jülich Kernforschungsanlage Jülich GmbH [Julich Nuclear Research Center, Germany] [also KFA] Jülich]

KfA Rep Kernforschungsanlage Jülich GmbH Report [Report of the Julich Nuclear Research Center, Germany] [also KFA Rep]

KFAED Kuwait Fund for Arab Economic Development

KFAS Keyed File Access System

KFAS Kuwait Foundation for the Advancement of Sciences

KFCI Kilo Flux Change(s) per Inch [also KFCPI, kfci, or kfcpi]

KFCMM Kilo Flux Change(s) per Millimeter [also KFC/MM, kfcmm, or kfc/mm]

KFF Potassium Iron Fluoride

K-Field Stress Intensity Factor Field [also K-field] [Fracture Mechanics]

KfK Kernforschungszentrum Karlsruhe GmbH [Karlsruhe Nuclear Research Center, Germany]

KfK/INE Kernforschungszentrum Karlsruhe GmbH/Institut für Nukleare Entsorgungstechnik [Karlsruhe Nuclear Research Center/Institute for Nuclear Waste Management]

KFLOPS Kilo-Floating-Point Operations per Second [also Kflops]

KFR Karl Fischer Reagent

KFRI Kilo-Flux Reversals per Inch [also KFRPI, kfrpi, or kfri]

KFRP Kevlar Fiber-Reinforced Plastics

KFRMM Kilo-Flux Reversals per Millimeter [also KFR/MM, kfrmm, or kfr/mm]

KFS Konjunkturforschungsstelle [Office for Economic Research; of Eidgenossische Technische Hochschule Zurich, Switzerland]

KFT Karl Fischer Titration

KFTCIC Kuwait Foreign Trade Contracting and Investment Company

KFTI Khar'kovskii Fiziki-Tekhnicheskii Institut [Kharkov Institute of Technical Physics, Kharkov, Ukraine]

KFU King Fahd University [Dhahram, Saudi Arabia]

KFUPM King Fahd University of Petroleum and Minerals [Dhahram, Saudi Arabia]

KfW Kreditanstalt für Wiederaufbau [Bank for Reconstruction, Germany]

KG Ketoglutarate [Biochemistry]

KG Kish Graphite [Metallurgy]

KG Knudsen Ga(u)ge

KG *Kommanditgesellschaft*—Limited Partnership; Corporate Form in Central Europe

KG Kyrgyz Republic [ISO Code]

α-KG α-Ketoglutarate [Biochemistry]

K/G Bulk Modulus-to-Shear Modulus (Ratio)

Kγ K-Gamma (Radiation) [also K$_\gamma$]

kG kilo-gauss [Unit]

kg kilogram [Unit]

kg$_f$ kilogram-force [also kgf]

kg/ac kilogram(s) per acre [Unit]

kg ai/animal kilogram(s) of active ingredient per animal [Unit]

kg ai/ha kilogram(s) of active ingredient applied per hectare treated [Unit]

kg ai/plant kilogram(s) of active ingredient per plant [Unit]

kgal/min kilogallons per minute [Unit]

kg/animal kilogram(s) (of chemical) per animal [Unit]

kg/A-y kilogram(s) per ampere-year [also kg/A-yr]

KGB Komitet Gosudarstvennoy Bezopasnosti [Committee of State Security, USSR]

KGC Kawasaki Gas Control (Converter) [Metallurgy]

kg-cal kilocalorie [also kcal, or Cal]

kg/cm kilogram(s) per centimeter [also kg cm^{-1}]

kg/cm² kilogram(s) per square centimeter [also kg/sq cm, or kg cm^{-2}]

kg/cm³ kilogram(s) per cubic centimeter [also kg/cu cm, or kg cm^{-3}]

kg/cu m kilogram(s) per cubic meter [also kg/m³]

KGD Known Good Dice (Problem) [Intergrated Circuit Manufacture]

kg/d kilogram(s) per day [also kg d^{-1}]

kg/dm³ kilogram(s) per cubic decimeter [also kg dm^{-3}]

KGe Potassium Germanide

kgf kilogram-force [Unit]

kgf kilogenerative force [Unit]

kgf/cm² kilogram-force per square centimeter [also kgf cm^{-2}]

kgf/cm² kilogenerative force per square centimeter [also kgf cm^{-2}]

kgf/in² kilogram-force per square inch [also kgf/sq in]

kgf-m kilogram-force meter [also kgf·m]

kgf/m² kilogram-force per square meter [also kgf mm^{-2}]

kgf/mm² kilogram-force per square millimeter [also kgf mm^{-2}]

kgf·s²/m kilogram-force second squared per meter [also kgf·s² m^{-1}]

kgf/sq in kilogram-force per square inch [also kgf/in²]

KGF Keratinocyte Growth Factor [Biochemistry]

KGFST Korea(n) General Federation of Science and Technology [South Korea]

kG/K kilogauss(es) per kelvin [also kG K^{-1}]

kg/h kilogram(s) per hour [also kg/hr, or kg h^{-1}]

kg/ha kilogram(s) per hectare [also kg ha^{-1}]

kg/hL kilogram(s) per hectoliter [also kg hL^{-1}]

kg/h·kW kilogram(s) per hour per kilowatt [also kg h^{-1} kW^{-1}, or kg/h/kW]

kg/100m² kilograms per hundred square meters [Unit]

kg/hr kilogram(s) per hour [also kg/h, or kg h^{-1}]

kg/Hz kiligram(s) per Hertz [also kg Hz^{-1}]

K-GIC Potassium Graphite Intercalation Compound

kg/J kilogram per Joule [also kg J^{-1}]

kg·K kilogram-kelvin [Unit]

kg/kg kilogram(s) per kilogram [also kg kg^{-1}]

kg/kmol kilogram(s) per kilomole [also kg kmol^{-1}]

kg/kW kilogram(s) per kilowatt [also kg kW^{-1}]

kg/L kilogram(s) per liter [also kg L^{-1}]

Kgl Dan Vidensk Selsk Mat-Fys Medd Konglige Danske Videnskabernes Selskab Matematisk-Fysiske Meddelelser [Communications of the Royal Danish Academy of Sciences, Mathematics and Physics] [also K Dan Vid Selsk Mat-Fys Medd]

KGM Key Generator Module

kgm kilogram-meter [also kg-m]

kg/m kilogram(s) per meter [also kg m^{-1}]

kgm² kilogram(s) meter squared [also kg-m², or kg·m²]

kg/m² kilogram(s) per square meter [also kg/sq m, or kg m^{-2}]

kg/m³ kilogram(s) per cubic meter [also kg/cu m, or kg m^{-3}]

kg/(m²h) kilogram per square meter per hour [kg m^{-2} h^{-1}]

kg/min kilogram(s) per minute [also kg min^{-1}]

kg/mm² kilogram(s) per square millimeter [also kg/sq mm, or kg mm^{-2}]

kg²/mm³ kilogram(s) squared per cubic millimeter [also kg³ mm^{-3}]

kg/mm²/1/$\sqrt{nm}$ kilogram(s) per square millimeter per one over the square root of nanometer [also kg mm^{-2} nm$^{-1/2}$]

kg/mol kilogram(s) per mole [also kg mol^{-1}]

kgm/s kilogram-meter(s) per second [also kgm/s kg·m/s, kg-m/s, or kg-m/sec]

kgm/s² kilogram-meter(s) per second squared [also kg-m/s², kg·m/s², or kgm s^{-2}]

kgm²/s kilogram-meter squared per second [also kgm² s^{-1}, or kgm²/sec]

kgm²/s² kilogram-meter squared per second squared [also kgm² s^{-2}, or kgm²/sec²]

kgm²/s³ kilogram-meter squared per second cubed [also kgm² s^{-3}, or kgm²/sec³]

kg/ms² kilogram per meter-second squared [also kg/m-sec², or kg m^{-1} s^{-2}]

kg/m²s kilogram per meter squared second [also kg m^{-2} s^{-1}, or kg/m² sec]

kgm²/sec kilogram-meter squared per second [also kgm²/s, or kgm² s^{-1}]

kgm²/sec² kilogram-meter squared per second squared [also kgm² s⁻², or kgm²/s²]

kgm²/sec³ kilogram-meter squared per second cubed [also kgm² s⁻³, or kgm²/s³]

kg/mthm kilogram(s) per metric ton of hot metal [Unit]

kg/MW kilogram(s) per megawatt [Unit]

K/GPa Kelvin(s) per gigapascal [K GPa⁻¹]

kg/Pa·s·m kilogram(s) per pascal second meter [also kg Pa⁻¹ s⁻¹ m⁻¹]

kg/Pa·s·m² kilogram(s) per pascal second meter squared [also kg Pa⁻¹ s⁻¹ m⁻²]

kg/pc kilogram(s) per piece [Unit]

kgps kilogram(s) per second [Unit]

KGS Kansas Geological Survey [US]

KGS Kearny Generating Station [New Jersey, US]

kgs kilograms [Unit]

kg/s kilogram(s) per second [also kg s⁻¹, kg sec⁻¹, or kg/sec]

kg/sec kilogram(s) per second [also kg sec⁻¹, kg/s, or kg s⁻¹]

kg/sq cm kilogram(s) per square centimeter [also kg/cm², or kg cm⁻²]

kg/sq m kilogram(s) per square meter [also kg/m², or kg m⁻²]

kg/sq mm kilogram(s) per square millimeter [also kg/mm², or kg mm⁻²]

KGT Kurz-Giovanola-Trivedi (Model) [Metallurgy]

kg/1000L kilograms per thousand liters [Unit]

kg/ton kilogram(s) per ton also kg ton⁻¹]

kg-wt kilogram weight [also kgf]

kGy kilogray [Unit]

kGy/h kilogray per hour [also kGy/hr]

kg/yr kilogram(s) per day [also kg yr⁻¹]

KH Cambodia [ISO Code]

KH Kawasaki-Hitachi (Caster) [Metallurgy]

KH Kelvin-Helmholtz (Waves)

KH Ketohexose [Biochemistry]

KH Kinetic Head

KH Knoop Hardness

KH Koppers-Hasche (Process)

KH Potassium Hydride

Kh Horizontal Permeability [Mining]

K7/h6 Locational Transition Fit; Basic Shaft System [ISO Symbol]

K/h Kelvin(s) per hour [also K h⁻¹, or K/hr]

kH kilohenry [Unit]

k/h kilometer(s) per hour [Unit]

κ(H) magnetic field dependent thermal conductivity [Symbol]

KHB KSC (Kennedy Space Center) Handbook [of NASA]

KH-GIC Potassium Hydride Graphite Intercalation Compound

KHI Knoop Hardness Indenter

KHI Kunsthistorisches Institut [Institute of Art History, Germany]

KHIC Korea Heavy Industries and Construction Company Limited

Khim Drev Khimiya Drevesiny [Latvian Chemistry Journal]

Khim Fiz Khimicheskaya Fizika [Russian Journal of Physical Chemistry]

Khim Ind Khimiya i Industriya [Bulgarian Journal on Industrial Chemistry]

Khim Neft Mashinostr Khimicheskoei Neftyanoe Mashinostroenie [Russian Journal on Chemical Engineering and Technology]

Khim Prirod Soed Khimiya Prirodnykh Soedinenii [Uzbekistan Journal on Chemistry and Polymer Related Subjects]

Khim Prom Khimicheskaya Promyshlennost [Russian Chemistry Journal]

Khim Tekhnol Khimicheskaya Tekhnologiya [Ukrainian Journal on Chemical Technology]

Khim Tekhnol Masel Khimi'ia i Teknologiia i Masel [Journal on the Chemistry and Technology of Fuel Oils]

Khim Volokna Khimicheskie Volokna [Russian Chemistry Journal]

KHM Kunsthochschule für Medien [College of Media Arts, University of Cologne, Germany]

KHN Knoop Hardness Number

KHP Potassium Hydrogen Phthalate

KHR Kampuchean Riel [Currency of Kampuchea]

K/hr Kelvin(s) per hour [also K/h, or K/h⁻¹]

KHS Kentucky Historical Society [US]

KHS Knurled Head Screw

KHS Potassium Hydrosulfide

KHT Knoop Hardness Test

KHU Kyung Hee University [Seoul, South Korea]

kHz kilohertz [Unit]

KI Change Factor, Internal Spline [Symbol]

KI Karolinska Institutet [Carolinian Institute, Sweden]

KI Ketenimide

KI Keyboard Input

KI Kiribati [ISO Code]

KI Knowledge Index (Database)

KI Knowledge Integration

KI Korrosionsinstitutet [Corrosion Institute, Sweden]

KI Potassium Iodide

Ki Kirpichev Number [Symbol]

KI:Ag Silver-Doped Potassium Iodide

KIAS Knots Indicated Air Speed

Kibern Vychisl Tekh Kibernetika i Vychislitel'naya Tekhnika [Ukrainian Journal on Cybernetics and Computer Technology]

KI:Br Bromine-Doped Potassium Iodide

KIC Knowledge Integration Center

KICE Korean Institute of Chemical Engineers [South Korea]

KI:Cl Chlorine-Doped Potassium Iodide

KI:Cs Cesium-Doped Potassium Iodide

KICU Keyboard Interface Control Unit

KID Knowledge Intensive Designer (Project) [of Alcoa Corporation, US]

KIDC Kansas Industrial Development Commission [US]

Kidney Blood Press Res Kidney and Blood Pressure Research [International Journal]

KIE Kenya Institute of Education

KIE Kinetic Isotope Effect

KIE Kirklees Information Exchange

KIEE Korean Institute of Electrical Engineers [Seoul, South Korea]

KIEE Korean Institute of Electronics Engineers [Seoul, South Korea]

KIEEME Korean Institute of Electrical and Electronic Materials Engineers [South Korea]

KIEMP Kenya Industrial Energy Management Program

KIER Korea Institute of Energy and Resources [South Korea]

KIESEC Korea International Exchange Society for Education and Culture [South Korea]

KIET Korea Institute for Industrial Economics and Technology [South Korea]

KIF Knowledge Interchange Format

KIFIS Kollsman Integrated Flight Instrumentation System

KIGAM Korea Institute of Geology, Mining and Materials

KII Koninklijke Institut van Ingenieurs [Royal Institute of Engineers, Netherlands]

KIIET Korean Institute for Industrial Economics and Technology [Seoul, South Korea]

KIIS Korean Institute for International Studies [Seoul, South Korea]

KIL Keyed Input Language

KIL Krypton Ion Laser

KIGAM Korea Institute of Geology, Mining and Materials [Taejon, South Korea]

kilo kilogram [Unit]

kilo kilometer [Unit]

kilo- SI prefix representing 10^3

kiloflops kilo floating-point operations per second [Unit]

KIM Kenya Institute of Management

KIM Korean Institute of Metals [now Korean Institute of Metals and Materials]

KIM Korean Institute of Metals and Materials [Seoul, South Korea]

Kimcode Kimble Controlled-Devacuation (Method)

KIMM Korea Institute of Machinery and Metals [Changwon, South Korea]

KIMME Korean Institute of Mineral and Mining Engineers [South Korea]

KIMM-MRL Korea Institute of Machinery and Metals–Materials Research Laboratory [Kyungnam, South Korea]

Kin Kinescope; Kinescopy

KINAM Rev Fis KINAM Revista de Fisica [Mexican Review on Physics]

Kinematics Phys Celest Bodies Kinematics and Physics of Celestial Bodies [Translation of: *Kinematika i Fizika Nebesnykh Tel (Ukrainian SSR)*; published in the US]

Kinematika Fiz Nebesnykh Tel (Ukrainian SSR) Kinematika i Fizika Nebesnykh Tel (Ukrainian SSR)

KINET Kienzle Network [of Mannesmann AG, Germany] [also Kinet]

King Kingdom [also king]

KINGMAP King's Music Analysis Package [of King's College, UK]

KIO Kuwait Investment Office

KIOPI Kienzle Input/Output Processor Interface [of Mannesmann AG, Germany]

KIP Keyboard Input Processor

Kip Kip [Currency of Laos]

kip kilopound [Unit]

kip ft kilo foot-pounds [also kip-ft]

kip/in kilopound per inch [Unit]

kip/in² kilopound per square inch [Unit]

KIPH Korea Institute for Population and Health [South Korea]

KIPO Keyboard Input Printout

KIPS Kilowatt Isotope Power System

KIPS Kilo (10^3) Instructions per Second [also kips]

KIPS Kilopounds per Square Inch [also kips]

KIPS Knowledge Information Processing System

KIPT Kharkov Institute of Physical Technology [Ukraine]

KI:Rb Rubidium-Doped Potassium Iodide

KIRDI Kenya Industrial Research and Development Institute

Kirghiz Kirghizia(n)

Kirghiz SSR Kirghiz Soviet Socialist Republic [USSR]

KIS Keyboard Input Simulation

KIS Kiepenheuer-Institut für Sonnenphysik [Kiepenheuer Institute of Solar Physics, Freiburg, Germany]

KIS Knowbot Information Service [Internet]

KISA Korean International Steel Association [South Korea]

KI Shell Knowledge Integration Shell (Computer Software)

KISR Kuwait Institute of Scientific Research [Safait]

KIS Kitting Instruction Sheet

KISS Keep it Simple and Short; Keep it Simple Sam; Keep it Simple Sir; Keep it Simple Stupid [Computer Jargon]

KISS Keyed Indexed Sequential Search

KISS Korea Information Science Society [Seoul, South Korea]

KISS Kossel Internal Stress Method [Physics]

KIST Korean Institute for Science and Technology [Seoul, South Korea]

KIST Kwangji Institute of Science and Technology [PR China]

KIT Kanazawa Institute of Technology [Ishikawa, Japan]

KIT Kent Information Technology Conference

KIT Key Issue Tracking

KIT Kitami Institute of Technology [Kitami, Japan]

KIT Kunming Institute of Technology [Kunming, PRChina]

KIT KWIC (Keyword in Context) Interactive Tagger

KIT Kurume Institute of Technology [Kurume, Japan]

KIT Kyoto Institute of Technology [Japan]

KIT Kyushu Institute of Technology [Kitakyushu, Japan]

KITE Korean Institute of Telematics and Electronics [Seoul, South Korea]

KITE Kuiper Infrared Technology Experiment

KITI Korea Inspection and Testing Institute [South Korea]

K-IUSM KSC (Kennedy Space Center) IUS (Inertial Upper Stage) Plan/Requirement [of NASA]

K-IUSN KSC (Kennedy Space Center) IUS (Inertial Upper Stage) Notice [of NASA]

K-IUSPS KSC (Kennedy Space Center) IUS (Inertial Upper Stage) Project Specification [of NASA]

KIvI Koninklijke Institute van Ingenieurs [Royal Institute of Engineers, Netherlands]

KJ Knuckle Joint

KJ Kutta-Joukowski (Formula) [Fluid Mechanics]

kJ kilojoule [also kj]

kJ/°C·mol kilojoule(s) per degree Celsius mole [also kJ °C^{-1} mol^{-1}1, or kJ/°C/mol]

kJ/g kilojoule(s) per gram [also kJ g^{-1}]

kJ/g·mol kilojoule(s) per gram mole [also kJ g^{-1} mol^{-1}]

kJh/g kilojoule-hour(s) per gram [also kJ-h/g, or kJ h g^{-1}]

kJ/in kilojoule(s) per inch [Unit]

K-JIST Kwang-Ju Institute of Science and Technology [South Korea]

kJ/K kilojoule(s) per kelvin [also kJ K^{-1}]

kJ/kg kilojoule(s) per kilogram [also kJ kg^{-1}]

kJ/kg·°C kilojoule(s) per kilogram per degree celsius (or centigrade) [also kJ kg^{-1} °C^{-1}]

kJ/m kilojoule(s) per meter [also kJ m^{-1}]

kJ/m² kilojoule(s) per square meter [also kJ m^{-2}]

kJ/m³ kilojoule(s) per cubic meter [also kJ m^{-3}]

KJMA Kolmogorov-Johnson-Mehl-Avrami (Theory) [Metallurgy]

kJ/m³·°C kilojoule(s) per cubic meter degree Celsius (or centigrade) [also kJ m^{-3} °C^{-1}]

kJ/m·h kilojoule(s) per meter hour [also kJ m^{-1} h^{-1}]

kJ/m·h·°C kilojoule(s) per meter hour degree celsius (or centigrade) [also kJ m^{-1} h^{-1} °C^{-1}]

kJ/m²·h·°C kilojoule(s) per square meter hour degree Celsius (or centigrade) [also kJ m^{-2} h^{-1} °C^{-1}]

kJ/mm kilojoule(s) per millimeter [also kJ mm^{-1}]

kJ/mol kilojoule(s) per mole [also kJ mol^{-1}]

kJ/mol at kilojoule(s) per mole of atoms [also kJ/(mol at)$^{-1}$, or kJ/mol·atom]

kJ/s kilojoule(s) per second [also kJ/sec, or kJ s^{-1}]

kJ/sec kilojoule(s) per second [also kJ/s, or kJ s^{-1}]

KK Kink-Kink (interaction Mode) [Materials Science]

KK Kramers-Kronig (Transform) [also K-K] [Physics]

KK One Million [K (= 10^3) × K (= 10^3)]

k^H/k^D hydrogen isotope effect (i.e., rate constant of hydrogen to that of tritium)

k^H/k^T tritium isotope effect (i.e., rate constant of hydrogen to that of deuterium)

KKB Kernkraftwerk Brunsbüttel [Nuclear Power Plant at Brunsbüttel, Germany]

K/kbar Kelvin(s) per kilobar [Unit]

KKF Kunststoff-Kautschuk-Fasern (Datenbank) [German Database on Plastics, Rubber and Synthetic Fibers]

KKG Kernkraftwerk Grafenrheinfeld [Nuclear Power Plant at Grafenrheinfeld, Germany]

KKK Kernkraftwerk Krümmel [Nuclear Power Plant at Krümmel, Germany]

KK-Model Kink-Kink (Interaction) Model [Materials Science]

KKP Kernkraftwerk Philippsburg [Nuclear Power Plant at Philippsburg, Germany]

KKR Korringa-Kohn-Rostocker (or Kohn-Korringa-Rostoker) (Theory) [Atomic Bond-Calculation Method]

KKR-ASA-CPA Korringa-Kohn-Rostoker/Atomic Sphere Approximation/Coherent Potential Approximation

KKR-CPA Korringa-Kohn-Rostoker Coherent Potential Approximation

KKR-CPA-GPM Korringa-Kohn-Rostoker—Coherent Potential Approximation–Generalized Perturbation Method

K/ks Kelvin(s) per kilosecond [also K ks^{-1}]

KKT Kramers-Kronig Transformation [Optics]

KKW *(Kernkraftwerk)* – German for "Nuclear Power Plant"

KL *c-kit* Ligand [Chemistry]

KL Kalmeyer-Laughlin (Wave Function) [Physics]

KL Keck Labotarory [of California Institute of Technology, Pasadena, US]

KL Key Length

KL Khachaturyan and Laughlin (Theory) [also K-L] [Metallurgy]

KL Kikuchi Line(s) [Crystallography]

KL Klebs-Loeffler (Bacillus) [Microbiology]

KL Kleinmann-Low (Nebula) [Astronomy]

KL Koehlhoff-Luecke (Etching Technique) [Metallurgy]

KL Kohn-Luttinger (Type Hamiltonian) [Physics]

KL Kuala Lumpur [Malaysia]

KL Kuppers-Lorz (Instability)

K/L K-Shell/L-Shell (Emission Ratio)

K-L Khachaturyan and Laughlin (Theory) [also KL] [Metallurgy]

K-L Koehler-Lehoczky (Model) [Physics]

kL kiloliter [also kl]

k(λ) extinction coefficient as a function of wavelength [Symbol]

KLA Klystron Amplifier

kL/d Kiloliter per day [also kl/d]

KLEED Kinematic Low-Energy Electron Diffraction

KLEM Keck Laboratory of Engineering Materials [Pasadena, California, US]

K Line Fraunhofer Line for Calcium Having Wavelength of 393.4 Nanometers [also K line] [Spectroscopy]

KLG Kimble Laboratory Glassware [Toledo, Ohio, US]

KLH Keyhole Limpet Hemocyanin [Biochemistry]

KLIC Key Letter in Context

Klimaforum Bull Klimaforum Bulletin [German Bulletin on the Climate]

Klin Wochenschr Klinische Wochenschrift [German Clinical Weekly]

KLM K-Line, L-Line, M-Line (Markers) [Spectroscopy]

KLM Kommission Lärmminderung [Noise Abatement Commission; of Verein Deutscher Ingenieure, Germany]

KLM Koninklijke Luchtvaart Maatschappij [Royal Dutch Airlines]

KLN Potassium Lanthanum Niobate

KLN Potassium Lithium Niobate

KLNMS Key Largo National Marine Sanctuary [of US Department of Commerce in the Florida Keys]

KLO Klystron Oscillator

KLOC One Thousand Lines of Code

KLT Potassium Lanthanum Titanate

KLU Key and Lamp Unit

KM Comoros [ISO Code]

KM Kansallis Museo [Kansallis Museum, Helsinki, Finland]

KM Keratinocyte Medium [Biochemistry]

KM Kilomega [also kM]

KM Kinetic Mechanism

KM Kinetic Momentum

KM Kink Migration [Metallurgy]

KM Kirchoff Method [Physics]

KM Kjeldahl Method [Chemistry]

KM Kobayashi-Maskawa (Matrix) [Physics]

KM Krigar-Menzel (Law) [Metallurgy]

KM Kubelka-Munk (Theory) [Optics]

KM Kumar and McShane (Theory of Serrated Flow)

KM Potassium Mananate

K-M Kossel-Mollenstedt (Diffraction Pattern) [Physics]

$(Km)^{-1}$ One per Kelvin meter [also 1/Km, 1/K·m, or 1/(K·m)]

K/m Kelvin(s) per meter [also K m^{-1}]

kM kilomega [also KM]

kM kilomole(s) [also kmol]

km kilometer [Unit]

km^2 square kilometer [also sq km]

km^3 cubic kilometer [also cu km]

K-MAC Kerr-McGee Ash Concentrate [Metallurgy]

KMAL Kinetic Mass Action Law

KMAZ Krishtal-Mokrov-Akomov-Zakharov (Analysis) [Materials Science]

KMC Kinetic Monte Carlo (Simulation Method)

KMC King-Maries-Crossley (Formula) [Physics]

kMc kilomegacycle [also KMc/s, or kmc]

KM-CAL Kawasaki Multi-Purpose Continuous Annealing Line [Metallurgy]

kMc/s kilomegacycle(s) per second [also kMc s^{-1}]

KMDP Kossel-Moellenstedt Diffraction Pattern [Physics]

KME Kinetic Master Equation

KMeO Potassium Methoxide

KMER Kodak Metal Etch Resist

km/h kilometer(s) per hour [also km/hr, kmph, kph, or km h^{-1}]

KMI Kessler Marketing Intelligence [Newport, Rhode Island, US]

KMI KSC (Kennedy Space Center) Management Instruction [of NASA]

K/min Kelvin per minute [also K min^{-1}]

KMK Keranocyte Medium Kit [Biochemistry]

KMK Kuznetsky Metallurgical Kombinat [Siberian Soviet Socialist Republic]

km/L kilometers per liter [also km/l, or km L^{-1}]

K/min Kelvin(s) per minute [also K min^{-1}]

KMM Kilogram Molar Mass [also kmm]

K/mm Kelvin per millimeter [also K mm^{-1}]

K/μm Kelvin per micrometer [also K μm^{-1}]

K-MMSEM KSC (Kennedy Space Center) MMSE (Multiuse Mission Support Equipment) Plan/Requirement [of NASA]

K-MMSEN KSC (Kennedy Space Center) MMSE (Multiuse Mission Support Equipment) Notice [of NASA]

K-MMSEPS KSC (Kennedy Space Center) MMSE (Multiuse Mission Support Equipment) Project Specification [of NASA]

kmol kilomole(s) [also kM]

kmol/m^3 kilomole(s) per cubic centimeter [also kmol m^{-3}]

KMON Keyboard Monitor

KMOP Kings Mills Ordnance Plant [Ohio, US]

KMP Kangaroo Management Program [of Australian National Conservancy Agency]

kmph kilometer(s) per hour [also kph, km/h, or km/hr]

kmps kilometer(s) per second [also kps, km/s, or km/sec]

KMR Kwajalein Missile Range [Marshall Islands]

KMS Kato, Mori and Schwartz (Coherency Strain Calculation) [Metallurgy]

KMS Keysort Multiple Selector

KMS Kloeckner-Maxhütte-Stahl [Pig Iron Production Process]

KMS Kloeckner-Metacon Steel Converter [Metallurgy]

KMS Kloeckner Metallurgy Scrap (Process)

KMS K-Words X Millions of Seconds

km/s kilometer per second [also kmps, kps, km/sec, or km s^{-1}]

km/sec/mpc kilometer per second per one million parsecs increase of distance [Unit]

KMT Kinetic-Molecular Theory

KMU Karl Marx University [at Leipzig, Germany]

K·m^2/W Kelvin meter squared per watt [also K·m^2 W^{-1}, K W^{-1} m^2, or K/W/m^2]

km^3/yr cubic kilometer(s) per year [also km^3 yr^{-1}]

KMZ Kangaroo Management Zone [South Australia]

KN Kitting Notice

KN Knurled Nut

KN Kongeriket Norge [Kingdom of Norway]

KN Koninkrjk der Nederlanden [Kingdom of the Netherlands]

KN Potassium Niobate

KN Saint Kitts, Nevis, Anguilla [ISO Code]

K/N Potassium-to-Nitrogen (Ratio) [Fertilizer]

Kn Knudsen Number [Symbol]

kN kilonewton [Unit]

kn knot [Unit]

K/Na Potassium/Sodium (Ratio)

K-Na Potassium-Sodium (Alloy System)

KNVA Kongelige Norske Videnskaps Akademi [Royal Norwegian Academy of Sciences, Trondheim]

KNAW Koninklijke Nederlandse Akademie van Wetenschappen [Royal Netherlands Academy of Sciences, Amsterdam]

KNCCI Kenya National Chamber of Commerce and Industry

KNCIAWPRC Korean National Committee of the International Association on Water Pollution Research and Control [South Korea]

KNCIAWPRC Kuwaiti National Committee of the International Association on Water Pollution Research and Control

kN/cm² kilonewton per square centimeter [also kN cm^{-2}]

KNCT Kagoshima National College of Technology [Japan]

KNCV Koninklijke Nederlandse Chemische Vereniging [Royal Dutch Chemical Association]

KNEC Kenya National Examinations Council

KNFC Kenya National Federation of Cooperatives

KNH Knoop Hardness (Value)

KNH$_{50}$ Knoop Hardness (Value) at Test Load of 50 Grams

KNHS Koninklijke Nederlandsche Hoogovens en Staalfabrieken NV [Dutch Blast Furnace and Steel Manufacturing Company]

KNIQUEST Knowledge-Based, Integrated Quotation System for Electrical Equipment

KNIRI Kyushu National Industrial Research Institute [of Agency of Industrial Science and Technology, Japan]

KNLS Kenya National Library Services

KNM Kenya National Museum

KNM-ER Kenya National Museum–East Lake Rudolf [Archeological Catalog Number, Usually Followed by a Number, e.g., KNM-ER 1470]

kNm kilonewton-meter [Unit]

kN/m² kilonewton(s) per meter squared [also kN m^{-2}, or kN/sq m]

kN/sq m kilonewton(s) per square meter [also kN m^{-2}, or kN/m²]

KNO KSC (Kennedy Space Center) Notice [of NASA]

KNO Potassium Niobate (Ferroelectrics)

KNOC Korea National Oil Company [South Korea]

K Nor Vidensk Selsk Forh Kongelige Norske Videnskabers Selskabs Forhandlinger [Proceedings of the Royal Norwegian Scientific Society]

K Nor Vidensk Selsk Skr Kongelige Norske Videnskabers Selskabs Skrifter [Publication of the Royal Norwegian Scientific Society]

knots/sq cm knots per square centimeter [Unit in Carpetmaking] [also knots/cm²]

knots/sq in knots per square inch [Unit in Carpetmaking] [also knots/in²]

KNOW Knowledge Network of the West

Knowl Knowledge [also knowl]

Knowl-Based Syst Knowledge-Based Systems [UK Publication]

Knowl Eng Knowledge Engineer(ing)

Knowl Eng Rev Knowledge Engineering Review [Published by Cambridge University Press, UK]

Knowl Eng Knowledge Engineer(ing)

KNP Kakadu National Park [Northern Territory, Australia]

KNP Kosciusko National Park [New South Wales, Australia]

KNPC Kuwait National Petroleum Company

KNS Korean Nuclear Society [Seoul, South Korea]

KNT Potassium Niobium Titanate

KNT Short-Cycle Gas Nitriding [Metallurgy]

KNTO Potassium Niobium Tantalum Oxide (Superconductor)

Kn Ved Inf Kniznice a Vedecke Informacie [Czech Technical Journal]

KNU Kongju National University [South Korea]

KNU Kyungbuk National University [South Korea]

KNUST Kwame Nkrumah University of Science and Technology [Ghana]

KNUT Kenya National Union of Teachers

KNVS Kongelige Norske Videnskabers Selskap [Royal Norwegian Scientific Society]

KO Kanaya-Okayama (Range)

KO Kick-Off [Controllers]

KO Knock-On (Effect) [Solid-State Physics]

KO Knockout

KO K Orbital

KO Kulik-Omel'yanchuk (Superconductivity Model) [also K-O]

kΩ kilohm [also kohm]

KOAc Potassium Acetate

KOBAS Konstanzer Bibliotheksautomatisierungssystem [Constance Library Automation System, Germany]

Kobelco Tech Bull Kobelco Technical Bulletin [Published by Kobelco, Tokyo, Japan]

Kobelco Technol Rev Kobelco Technology Review [Published by Kobelco, Tokyo, Japan]

Kobe Res Dev Kobe Research and Development [Publication of Kobe Steel Ltd., Tokyo, Japan]

Kobe Steel Rep Kobe Steel Report [Published by Kobe Steel Ltd., Tokyo, Japan]

Kobe Steel Rep (Tech Kobesteel) Kobe Steel Reports (Technique Kobesteel) [Published by Kobe Steel Ltd., Tokyo, Japan]

KOBM Kombinierter OBM (Oxygen-Bodenblas-Maxhütte) Prozess [Combined OBM (Oxygen-Bottom-Blown Maxhuette) Steelmaking Process] [also K-OBM] [Metallurgy]

KOBM-S Kombinierter OBM (Oxygen-Bodenblas-Maxhütte) Prozess–Stainless [Combined OBM (Oxygen- Bottom-Blown Maxhuette) Steelmaking Process–Stainless] [also K-OBM-S] [Metallurgy]

KOC Kuwait Oil Company

K Ω^{-1} cm^{-1} Kelvin per ohm per centimeter [also K/Ω/cm, or K(Ω^{-1}cm^{-1})]

kΩcm kilohm centimeter [also kΩ-cm]

kΩ/cm² kilohm(s) per square centimeter [also kΩ cm^{-2}]

KOCN Potassium Cyanate

KOD Potassium Deuteroxide

Kōdai Math Sem Rep Kōdai Mathematical Seminar Reports [of Tokyo Institute of Technology, Japan]

kOe kilo-oersted [Unit]

KOEBES Kölner Bibliothekserschließungssystem [Cologne Automated Library Information Retrieval System, Germany]

kOe/s kilo-oersted per second [also kOe/sec, or kOe s^{-1}]

KOEt Potassium Ethoxide (or Potassium Ethylate)

KOEX Korea Exhibition Center [South Korea]

Koezl M Tud Akad Musz Fiz Kut Intez Koezlemenyei Magyar Tudomanyos Akademia Muszaki Fizikai Kutato Intezetenek [Publication on Technical Physics of the Hungarian Academy of Sciences]

Koezl M Tud Akad Szamtech Autom Kut Intez Koezlemenyei Magyar Tudomanyos Akademia Szamitastechnikai es Automatizalasi Kutato Intezete [Publication on Computer Engineering and Automation of the Hungarian Academy of Sciences]

KOH Potassium Hydroxide

kohm kilohm [also kΩ]

KOI Kano, Nigeria [Remote Site]

KOI KSC (Kennedy Space Center) Operation Instruction [of NASA]

Koks Khim Koks i Khimiya [Russian Journal on Coke and Chemistry]

Kolloidchem Kolloidchemie [German Journal of Colloid Chemistry]

Kolloidchem Beih Kolloidchemie Beihefte [Supplement to German Journal of Colloid Chemistry]

Kolloid Z Kolloid Zeitschrift [German Colloid Journal] [also Kolloid Zs, or Kolloid Zeitschr]

Kolloid Zh Kolloidnyi Zhurnal [Colloid Journal of the USSR]

KOM KSC (Kennedy Space Center) Organizational Manual [of NASA]

KOMAF Korean Machining Fair [South Korea]

Komi ASSR Komi Autonomous Soviet Socialist Republic [USSR]

kΩ/μm^2 kiloohm(s) per square micrometer [also k$\Omega\mu$m^{-2}]

KOMMTECH Europäische Kongressmesse für Technische Automation [European Congress/Fair for Technical Automation, Germany]

Kompleksni Sist Upr Kompleksni Sistemi za Upravlenie [Publication on Control Systems of Bulgarian Academy of Sciences]

Kompoz Polim Mater Kompozitsionnye Polimernye Materialy [Ukrainian Journal on Composite and Polymeric Materials]

Koncar Strucne Inf Koncar Strucne Informacije [Yugoslav Engineering Publication]

Konstr Giessen Konstruieren und Giessen [German Journal on Casting and Design; published by Zentrale für Gussverwendung, Germany]

KOP Kansas Ordnance Plant [Parsons, Kansas, US]

KOP Kingsbury Ordnance Plant [LaPorte, Indiana, US]

KOPS Kilo (i.e. One Thousand) Operations per Second [also kops]

Kor Korea(n)

Korea Inf Sci Soc Rev Korea Information Science Society Review [South Korea]

Korean Appl Phys Korean Applied Physics [Journal of the Korean Physical Society, South Korea]

Korean J Chem Eng Korean Journal of Chemical Engineering [South Korea]

Korean J Mater Res Korean Journal of Materials Research [of Materials Research Society of Korea]

KORF Korf Oxy-Refining Fuel (System) [Metallurgy]

Koroze Ochr Mater Koroze a Ochrana Materialu [Czech Publication on the Corrosion and Deterioration of Materials]

Korrosionsinst Rapp Korrosionsinstitutet Rapport [Report of the Corrosion Institute; published in Stockholm, Sweden]

Korróz Figy Korróziós Figyeló [Hungarian Journal on Corrosion]

Korroz Zashch Korroziya i Zashchita v Neftegazovoi Promyslennoist [Russian Journal on Korrosion and Deterioration of Materials]

KORSEF Korea Science and Engineering Foundation

KORSO Verfahren zur Sicherstellung des Korrekten Verhaltens von Software [Process to Guarantee the Correct Behaviour of (Computer) Software (Project); of Gesellschaft für Mathematik und Datenverarnbeitung, Germany]

KORSTIC Korea Scientific and Technological Information Center [South Korea]

Kos Koseavan, Kosevo [Yugoslavia]

KOSAA Korea Shipping Agencies Association [South Korea]

KOSAMI Korea Society for the Advancement of Machine Industry [also KSAMI, South Korea]

KOSEF Korea Science and Engineering Foundation [South Korea]

Kosm Isssled Kosmicheske Issledovaniya [Russian Journal on Cosmic Research]

KOSTE Kohlenstaub Einblas(verfahren) [Pulverized Coal Injection Process] [Metallurgy]

KOTFA Korea World Travel Fair [South Korea]

KOTRA Korean Organization for Trade Advancement [South Korea]

KOTRA Korea Trade Promotion Corporation [South Korea]

Kov Mater Kovové Materiály [Slovak Journal on Metallic Materials]

KOW Kankakee Ordnance Works [Joliet, Illinois, US]

KOW Keystone Ordnance Works [Meadville, Pennsylvania, US]

KOW Knock on Wood [Computer and Internet Jargon]

KP Kaldo (Steelmaking) Process [Metallurgy]

KP Kasper Phase [Materials Science]

KP Kennedy and Pancu (Circle)

KP Ketopentose [Biochemistry]

KP Key Pulsing

KP Key Punch

KP Kick Plate [Building Trades]

KP Kikuchi Pattern [Crystallography]

KP Kinematic Potential

KP Kinetic Potential [Mechanics]

KP Korea (North) [ISO Code]

KP Kossel (Diffraction) Pattern [Physics]

KP Kraft Paper

KP Kroll Process [Metallurgy]

KP Kronig-Penney (Potential) [Solid State Physics]

KP Potassium Phosphate

K(P) Initial Uploading Slope as a Function of Peak Loading [Symbol]

Kp Kip [Currency of Laos]

kp kilopond [Obsolete Unit]

KPA Key Pulse Adapter

KPA Klystron Power Amplifier

kPa kilopascal [Unit]

kPag kilopascal gauge (pressure) [also kPa gauge]

kPa·m/s kilopascal-meter(s) per second [also kPa·m s^{-1}]

KPB Keshishev, Parshin and Bakin (Crystallization Wave) [Physics]

K-Pb Potassium-Lead (Alloy System)

KPC Kakuda Propulsion Center [of National Space Development Agency of Japan]

KPC Keyboard Printer Control

KPC Knudsen Permeability Constant

KPD KSC (Kennedy Space Center) Program Directive [of NASA]

KP&D Kick Plate and Drip

KPE Kenya Preliminary Examination

KPEG Potassium Polyethylene Glycol (Process)

KPF Key Pulse on Front Cord

KPF Kink Pair Formation [Materials Science]

KPH Keystrokes per Hour [Unit]

kph kilometers per hour [also KPH, kmph, km/h, or km/hr]

KPI Kernel Programming Interface

KPIC Key Phrase in Context

K-Plastic Kautsch Ztg K-Plastic and Kautschuk Zeitung [German Publication on Plastics and Rubber]

KPM Key-Punch Machine

Kpn Klebsiella pneumoniae [Microbiology]

KPNO Kitt Peak National Observatory [Arizona, US]

KPO Key Punch Operator

KPO Kinematic Phase Object

KPO Potassiophosphate

KPPS Kilo (i.e. One Thousand) Pulses per Second

KPR Kodak Photoresist

Kpr Keeper [also kpr]

KPRC Keane Physics Research Center [of Catholic University of America, Washington, DC, US]

KPRD KSC (Kennedy Space Center) Program Requirements Document [of NASA]

KPS Korean Physical Society [Seoul, South Korea]

kps kilometer per second [also kmps, km/s, or km/sec]

KPSI Kilopounds per Square Inch [also kpsi]

KPSM Klystron Power Supply Modulator

KPV Khantha-Pope-Vitek (Model) [Metallurgy]

KPW North Korean Won [Currency]

KPZ Kardar-Parisi-Zhang (Equation)

KQML Knowledge Query and Manipulation Language

KR Kellogg-Rust (Process)

KR Kelvin Resistance

KR Kerr Rotation [Physics]

KR Key Register

KR Kiloroentgen [also kR, or kr]

KR Kings Regulations [UK]

KR Knowledge Representation

KR Kohlereduktionsverfahren [German Coal Reduction Process for Pig Iron Manufacture]

KR Kolbe Reaction [Chemistry]

KR Kondo Resonance [Solid-State Physics]

KR Korea (South) [ISO Code]

KR K Radiation

KR Krupp-Renn (Process) [Metallurgy]

Kr Krypton [Symbol]

Kr$^+$ Krypton Ion [Symbol]

Kr-77 Krypton-77 [also ^{77}Kr, or Kr^{85}]

Kr-78 Krypton-78 [also ^{78}Kr, or Kr^{78}]

Kr-79 Krypton-79 [also ^{79}Kr, or Kr^{79}]

Kr-80 Krypton-80 [also ^{80}Kr, or Kr^{80}]

Kr-81 Krypton-81 [also ^{81}Kr, or Kr^{81}]

Kr-82 Krypton-82 [also ^{82}Kr, or Kr^{82}]

Kr-83 Krypton-83 [also ^{83}Kr, or Kr^{83}]

Kr-84 Krypton-84 [also ^{84}Kr, or Kr^{84}]

Kr-85 Krypton-85 [also ^{85}Kr, or Kr^{85}]

Kr-86 Krypton-86 [also ^{86}Kr, or Kr^{86}]

Kr-87 Krypton-87 [also ^{87}Kr, or Kr^{87}]

Kr-88 Krypton-88 [also ^{88}Kr, or Kr^{88}]

Kr-89 Krypton-89 [also ^{89}Kr, or Kr^{89}]

Kr-90 Krypton-90 [also ^{90}Kr, or Kr^{90}]

Kr-91 Krypton-91 [also ^{91}Kr, or Kr^{91}]

Kr-92 Krypton-92 [also ^{92}Kr, or Kr^{92}]

Kr-93 Krypton-93 [also ^{93}Kr, or Kr^{93}]

Kr-94 Krypton-94 [also ^{94}Kr, or Kr^{94}]

Kr-95 Krypton-95 [also ^{95}Kr, or Kr^{95}]

Kr-97 Krypton-97 [also ^{97}Kr, or Kr^{97}]

kR Kiloroentgen [also KR, or kr]

.kr Korea (South) [Country Code/Domain Name]

krad kilorad(s) [Unit]

KRAM Kilobytes of Random-Access Memory

KRAS Keyworded References to Archeological Science [UK]

KRB Kernreaktor FWE (Fernwärmeenergie) Bayernwerk [Nuclear Power Reactor (District Heating Energy) Bayernwerk, Germany]

K/Rb Potassium/Rubidium (Ratio) [Geology]

KrBr Krypton Bromide (Nanocrystal)

KRC Kaufman-Radcliffe-Cohen (Model) [Metallurgy]

KRDC Kingston Research and Development Center [of Alcan International Ltd., Kingston, Ontario, Canada]

KREEP Potassium-, Rare-Earth-Element- and Phosphorus-Rich [Moon Rocks]

KREI Korea Rural Economy Institute [South Korea]

KRES (Arbeitskreis) Kunststoffrecycling und Entsorgung Schweiz [Swiss Association for Plastics Recycling and Disposal; of Arbeitskreis Schweizerische Kunststoff-Industrie, Switzerland]

KRF Korea Research Foundation

KrF Krypton Fluoride (Excimer Laser)

KrI Krypton Iodide

KRICT Korea Research Institute of Chemical Technology [Taejon, South Korea]

KRIPES K-Resolved Inverse Photoemission Spectroscopy [also KRIPS]

KRIS Korea Research Institute of Standards [South Korea]

KRISS Korea Research Institute of Standards and Science [Taejon, South Korea]

Kristall Kristallografiya [Russian Journal of Physical Crystallography]

Kristall Tech Kristall und Technik [German Journal on Crystal Research and Technology]

KRL Kjeller Research Laboratories [Norway]

KRL Knowledge Representation Language

KRL Kodak Research Laboratory [US]

KRL Kurt Rossmann Laboratory [US]

KRL Kyoto Research Laboratory [of Matsushita Electronics Corporation, Japan]

KRM Kurzweil Reading Machine

KRP Kevlar-Reinforced Plastics

KRR Kansai Research Reactor [Japan]

KRS Knowledge Retrieval System

KRW South Korean Won [Currency]

KS Kansas [US]

KS Kaposi's Sarcoma [Medicine]

KS Kekulé Structure [Organic Chemistry]

KS Ketosteroid [Biochemistry]

KS Killed Steel

KS Kinetic Stress

KS Kirkendall Shift [Metallurgy]

KS Klockner Steelmaking (Process) [Metallurgy]

KS Knee Switch

KS Knight Shift

KS Knife Switch

KS Knowledge Source

KS Kohn-Sham (Approximation) [Physics]

KS Kolmogorov-Smirnov (Test) [also K-S] [Materials Science]

KS Konungariket Sverige [Kingdom of Sweden]

KS Korean Standard

KS Kossel-Sommerfeld (Law) [Spectroscopy]

KS Kuramoto-Sivashinsky (Equation) [Physics]

KS Kurdjumov-Sachs (Orientation Relationship) [also K-S] [Materials Science]

17-KS 17-Ketosteroid [Biochemistry]

K/S Kick Stage

K-S Kolmogorov-Smirnov (Test) [also KS] [Materials Science]

K/s Kelvin(s) per second [also K/sec, or K s^{-1}]

ks kilosecond [also ksec]

κ/σ Wiedemann-Franz ratio (i.e., thermal conductivity to electrical conductivity ratio) [also k/σ] [Symbol]

KSA Kirkwood Superposition Approximation [Physics]

KSA Knob Shoe Assembly

KSA Ku-Band Single Access

KSAM Keyed Sequential Access Method

KSAM Key Field Sequential Access Method

KSAMI Korea Society for the Advancement of Machine Industry [also KOSAMI, South Korea]

KSB Klein, Schanzlin & Becker AG [Frankenthal, Germany]

KSB Tech Ber KSB Technische Berichte [Technical Reports published by Klein, Schanzlin & Becker AG, Frankenthal, Germany]

KSC Kagoshima Space Center [Kyushu Island, Japan]

KSC Kawasaki Steel Corporation [Japan]

KSC Kearney State College [Nebraska, US]

KSC Keene State College [New Hampshire, US]

KSC (John F.) Kennedy Space Center [of NASA in Florida, US] [also JFKSC]

KSC Kentucky State College [Frankfort, US]

KSCAP Kennedy Space Center Area Permit [of NASA]

KSCN Potassium Thiocyanate

KSCP Kansas State College of Pittsburg [US]

KSDN Kennedy Switched Data Network [of NASA Kennedy Space Center]

KSDS Key-Sequenced Data Set

KSE Keyboard Source Entry Program

KSE Kuwait Society of Engineers

KSEA Korean Scientists and Engineers Association

K/sec Kelvin(s) per second [also K/s, or K s^{-1}]

ksec kilosecond [also ks]

KSeCN Potassium Selenocyanate

KSF Kato Science Foundation [Japan]

KSF Kilopound(s) per Square Foot [also ksf]

KSH Key Stroke(s) per Hour

KSH Korn Shell (Program) [Unix Operating System]

.KSH K-Shell Script [File Name Extension] [also .ksh]

KSh Kenyan Shilling [Currency]

KSHS Kansas State Historical Society [US]

HSHV Kaposi's Sarcoma Herpes Virus

KSI Kilopound(s) per Square Inch [also ksi]

ksi$\sqrt{\text{in}}$ kilopound(s) per square inch square root of inch [also ksi·in$^{1/2}$]

K-SLM KSC (Kennedy Space Center) Spacelab Plan/Requirement [of NASA]

K-SLN KSC (Kennedy Space Center) Spacelab Notice [of NASA]

KSLOC One Thousand Source Lines of Code

K-SLPS KSC (Kennedy Space Center) Spacelab Project Specification [of NASA]

K-SM KSC (Kennedy Space Center) Shuttle Management Document [of NASA]

KSME Korean Society of Mechanical Engineers

KSN Potassium Strontium Niobate

KSO Potassiosulfate

KSP Kamagawa Science Park [Tokyo, Japan]

KSPH Keystrokes Per Hour [also ksph]

K-SPN KSC (Kennedy Space Center) Shuttle Project Notice [of NASA]

K-SPS KSC (Kennedy Space Center) Shuttle Project Specification [of NASA]

KSR Kendall Square Research [US Supercomputer Manufacturer]

KSR Keyboard Select Routing

KSR Keyboard, Send/Receive

KSR Keyboard Send Receiver

KSS Potassium Antimonyl Silicate

ks/s kilosymbols per second [Unit]

KSSS KSC (Kennedy Space Center) Station Set Specification [of NASA]

K-SSS KSC (Kennedy Space Center) SPS (Shuttle Project Station) Set Specification [of NASA]

KST Klöckner Stahltechnik GmbH [Steel Engineering Company located at Hamburg, Germany]

Kst Keyseat [also kst]

KSt Potassium Stearate

KSTC Kenya Science Teachers College

KSTR Kema Suspension Test Reactor [Netherlands]

K-STSM KSC (Kennedy Space Center) STS (Shuttle Test Station) Plan/Requirement [of NASA]

K-STSN KSC (Kennedy Space Center) STS (Shuttle Test Station) Notice [of NASA]

K-STSPS KSC (Kennedy Space Center) STS (Shuttle Test Station) Project Specification [of NASA]

KSU Kansas State University (of Agriculture and Applied Science) [Manhattan, US]

KSU Kent State University [Ohio, US]

KSU Key Service Unit

KSU Kharkov State University [Ukraine]

KSU Kim Il Sung University [Pyongyang, North Korea]

KSU King Saud I University [Riyadh, Saudi Arabia]

KSU Kyoto Sangyo University [Japan]

KSV Khantha-Cserti-Vitek (Deformation Model)

KSV Kungliga Svenska Vetenskapsakademien [Royal Swedish Academy of Science, Stockholm]

KSW Krupp Südwestfalen AG [German Manufacturer of Industrial Equipment]

KSWPRC Korean Society of Water Pollution Research and Control [South Korea]

KSZE Konferenz über Sicherheit und Zusammenarbeit in Europa [Conference on Security and Cooperation in Europe]

KT Kaplan Turbine

KT Katz and Thompson (Theory of Porous Media)

KT Key-to-Tape (System)

KT Keyword Term

KT Kinetic Temperature

KT Kinetic Theory [Physics]

KT Kinetic Treatment

KT Kirchhoff Theory [Optics]

KT Kosterlitz-Thouless (Theory) [also K-T] [Materials Science]

KT Potassium Tantalate

KT Potassium Titanate

K(T) Temperature-Dependent Anisotropy [Symbol]

K/T Kelvin per Tesla [also K T^{-1}]

K-T Kosterlitz-Thouless (Theory) [also KT] [Materials Science]

Kt Karat [also kt]

Kt Kit [also kt]

Kt Knight [also kt]

Kt Knot [also kt]

kT kilotesla [Unit]

kT noise power per cycle per second bandwidth [Symbol]

k(T) rate constant (i.e., temperature-dependent rate of reaction)

Kt Kyat [Currency of Burma]

kt kiloton [Unit]

κ(T) surface stiffness coefficient [Symbol]

κ(T) thermal conductivity (or temperature-dependent conductivity) [Symbol]

$\kappa(T_{max})$ maximum thermal conductivity [Symbol]

KTA Kerntechnischer Ausschuss [Nuclear Engineering Committee; Germany] [also KtA]

KTA Key Telephone Adapter

KTA Korea Tourist Association [South Korea]

KTA Potassium Titanyl Arsenate

KTAS Knots Air Speed

KTB Kontinentales Tiefbohrprogramm [Continental Deep Drilling Program, Germany]

KTB Kosterlitz-Thouless-Berezinski (Transition) [Materials Science]

KTC Temperature Coefficient of Sensitivity to Strain

KTCS Katsuura Tracking and Communication Station [of National Space Development Agency of Japan]

KTDS Key-to-Disk Software

kT/e kilotesla per electron [Unit]

KTEC Korea Technical Exchange Company [South Korea]

KTeV Kaons at the Tevatron (Experiment) [of Fermilab, Batavia, Illinois, US] [also KTEV]

KTF Kanagawa (High) Technology Foundation [Japan]

KTF Potassium Tantalum Fluoride

KTF Potassium Titanium Fluoride

KTFR Kodak Thin Film Resist

KTG Kerntechnische Gesellschaft [Nuclear Engineering Society] [also KtG, Germany]

KTG Kinetic Theory of Gases [Physical Chemistry]

KTH Kinetic Theory of Heat [Thermodynamics]

KTH Kungliga Tekniska Högskolan [Royal Institute of Technology, Stockholm, Sweden]

KTHB Kungliga Tekniska Högskolan Biblioteket [Royal Institute of Technology Library, Stockholm, Sweden]

KTH-Electrum Kungliga Tekniska Högskolan Electrum [of Royal Institute of Technology, at Kista, Sweden]

K(THF)xC_{24} Potassium Tetrahydrofuran Fullerene [x = 1.7, 2.5, etc.]

KTI Kommission für Technologie und Innovation [Technology and Innovation Commission, Switzerland]

KTIC Korea Toy Industry Cooperation [South Korea]

KTL Kinetic Theory of Liquids [Physical Chemistry]

KTM Key Transport Module

KTMS Knapp Time Metaphor Scale

KTN Potassium Tantalate Niobate

KTN Kuratorium für die Tagungen der Nobelpreisträger [Standing Committee for Nobel Prize Winners Congresses, Germany]

KTNO Potassium-Tantalum-Niobium Oxide

KTO Potassium Titanyl Oxalate

kton kiloton [Unit]

KTP Potassium Titanyl Phosphate

KTPC Korea Trade Promotion Corporation [South Korea]

KTpCPB Potassium Tetrakis(4-Chlorophenyl)borate

KTPI Kilotrack(s) per second [also ktpi]

KTPP Potassium Tripolyphosphate

KTR Keyboard Typing Reperforator

KTS Key Telephone System

KTS Key-to-Tape System

kts knots [Unit]

KTSA Kahn Test of Symbol Arrangement

KTSP Knots True Air Speed

KTT Kuhn-Tucker Theory [Physics]

KTTC Kenya Technical Teachers College

KTU Key Telephone Unit

KTU Key-to-Tape Unit

KTU Kyushu Tokai University [Kumamoto, Japan]

KU Kabul University [Afghanistan]

KU Kagoshima University [Japan]

KU Kanazawa University [Japan]

KU Kansai University [Osaka, Japan]

KU Karachi University [Pakistan]

KU Keio University [Kanagawa, Japan]

KU Ketonuria [Medicine]

KU Keyboard Unit

KU Kinski University [Osaka, Japan]

KU Kobe University [Japan]

KU Kochi University [Japan]

KU København Universitet [University of Copenhagen, Denmark]

KU Kogakuin University [Tokyo, Japan]

KU Kokushukan University [Japan]

KU Kumamoto University [Japan]

KU Korea University [Chochiwon, South Korea]

KU Kyoto University [Japan]

KU Kyushu University [Fukuoka, Japan]

Ku Kurchatovium

KUB Kidney, Urether and Bladder

Kubota Tech Rep Kubota Technical Reports [Published by Kubota, Osaka, Japan]

KUC Kenyatta University College [Mombasa, Kenya]

KUL Katholieke Universiteit Leuven [University of Leuven, Belgium] [also KU Leuven]

Ku LNB Ku-Band Low-Noise Block Downconverter

KULSAA Karachi University Library Science Alumni Association [Pakistan]

Kumamoto J Math Kumamoto Journal of Mathematics [Publication of Kumamoto University, Japan]

KUN Katholieke Universiteit Nijmegen [University of Nijmegen, Netherlands]

Kunstst Bau Kunststoffe im Bau [German Journal on Plastics for the Construction Industry]

Kunstst Ger Plast Kunststoffe, German Plastics [German Journal]

Kunstst J Kunststoff Journal [German Plastics Journal]

Kunstst Plast Kunststoffe–Plastics [Journal on Plastics; published in Switzerland]

KUR Kyoto University (Research) Reactor [Research Reactor Institute, Kumatori-Osaka, Japan]

KURASEP Kuraray Separation (Process)

KUSP Ku-Band Signal Processor

Kuznechno-Shtampov Kuznechno-Shtampovochnoe Proizvodstvo [Russian Publication]

KV Kinematic Viscosity

KV Kirkaldy-Venugopalan (Model) [Materials Science]

KV Potassium Vanadate

Kv Vertical Permeability [Mining]

kV kilovolt [also kv]

KVA Kungliga Vetenskapsakademin [Royal Academy of Sciences, Stockholm, Sweden]

kVA kilovolt-Ampere [also KVA, or kva]

KVAC Kilovolt, Alternating Current [also kVAC, or kVac]

kVAh Kilovolt-Ampere-hour [also KVAh, KVA-h, or kVA-h]

kVAhm Kilovolt-Ampere-hour Meter

Kvant Elektron Kvantovaya Elektronika [Russian Journal on Quantum Electronics]

Kvant Elektron Kiev Kvantovaya Elektronika, Kiev [Journal on Quantum Electronics published in Kiev, Ukraine]

Kvant Elektron Mosk Kvantovaya Elektronika, Moskva [Journal on Quantum Electronics published in Moscow, Russia]

kvar kilovar; reactive kilovolt-ampere [Unit]

kvarh kilovarhour [Unit]

KVAWLSB Koninklijke Vlaamse Academie voor Wetenschappen, Letteren en Schone Kunsten van België [Royal Flemish Academy of Sciences, Letters and Fine Arts, Belgium]

kV/cm kilovolt(s) per centimeter [also kV cm^{-1}]

KVB Koninklijke Vereniging van België [Royal Society of Belgium]

kV/cm kilovolt per centimeter [also kV cm^{-1}]

KVDC Kilovolt, Direct Current [also kVDC, or kVdc]

KVG Kavieng, Papua, New Guinea [Meteorological Station Designation]

KVIV Koninklijke Vlaamse Ingenieurs-Vereniging [Royal Flemish Association of Engineers, Belgium]

KVL Kirchhoff's Voltage Law

kV/m Kilovolt(s) per meter [also kV m^{-1}]

kV/mm kilovolt(s) per millimeter [also kV mm^{-1}]

KVP Kilovolt Power

kVp kilovolt(s) peak

KVS Kármán Vortex Street [Fluid Mechanics]

KVS Kunststoffverband Schweiz [Swiss Plastics Association]

KVVS Kungliga Vetenskaps- och Vitterhets Samhälleti [Royal Society of Science and Letters, Goteborg, Sweden]

KW Keyword

KW Kear-Wilsdorf (Dislocation Locking Mechanism) [also K-W] [Materials Science]

KW Kilo-Word [1,024 Computer Words]

KW Kimmelstiel-Wilson (Disease) [Medicine]

KW Kinematic Wave

KW Kuwait [ISO Code]

KW Kwanza [Currency of Angola]

kW kilowatt [also kw]

.kw Kuwait [Country Code/Domain Name]

KWAC Keyword and Context

KWADE Keyword As a Dictionary Entry

kW/cm² kilowatt(s) per square centimeter [also kW cm^{-2}]

KWD Kilowatt Peak Demand

KWD Kuwaiti Dollar [Currency]

kWe Kilowatt of Electric Power

KWF Kommission zur Förderung der Wissenschaftlichen Forschung [Commission for the Promotion of Scientific Research, Switzerland]

kW/ft kilowatt(s) per foot [Unit]

KWG Kernkraftwerk Grohnde [Nuclear Power Plant at Grohnde, Germany]

kWh kilowatthour [also KWH, kWhr, kW-hr, kwh, kwhr, or kw-hr]

kWh/kg kilowatthour(s) per kilogram [also kWh kg^{-1}]

kWhm kilowatthour meter [also kwhm]

kWhr kilowatthour [also kW-hr, kw-hr, kwhr, or kWh]

kWh/t kilowatthour(s) per (metric) ton [Unit]

kWh/t/d kilowatthour(s) per (metric) ton per day [also kWh/t/day]

kWh/yr kilowatthour(s) per year [also kWh yr^{-1}]

KWI Karst Waters Institute [US]

KWIC Keyword in Context

KWIP Keyword in Permutation

KWIT Keyword in Title

kW/m kilowatt per meter [Unit]

K/W·m² Kelvin(s) per Watt square meter [also K W^{-1} m²]

kW/m²·K kilowatt(s) per square meter kelvin [also kW m^{-2} K^{-1}, or kW/m²/K]

kW/m³·K⁴ kilowatt(s) per cubic meter kelvin to the fourth [also kW m^{-3} K^{-4}, or kW/m³/K⁴]

KWO Kernkraftwerk Obrigheim [Nuclear Power Plant at Obrigheim, Germany]

KWOC Keyword out of Context

KWOT Keyword out of Title

KWR Know-How Repeating

KWS Kenya Wildlife Service

K/wt% Kelvin per weight percent [Unit]

kWt Kilowatt of Thermal Energy

KWU Kraftwerk Union Aktiengesellschaft [German Manufacturer of Thermal and Nuclear Power Plants and Related Equipment (e.g., Turbines, Turbogenerators, etc.)]

KWU Kraftwerk Union Reaktor [A German Nuclear Reactor]

KWUC Keyword and Universal Decimal Classification

KWW Kernkraftwerk Würgassen [Nuclear Power Station at Würgassen, Germany]

KWW Kohlrausch-Williams-Watts (Model) [Materials Science]

Kwy Keyway [also kwy]

KX Kilo-X-Unit [also kX]

K-XANES K-Edge X-Ray Absorption Near-Edge Spectroscopy

KXFe Cliff-Lorimer Factor for Carbon According to Iron [Symbol]

KXU Kilo-X-Unit [also kXU]

KXV (Gray) Kangaroopox Virus

KY Cayman Islands [ISO Code]

KY Kaplan-Yorke (Conjecture)

KY Kentucky [US] [also Ky]

Ky Key [also ky]

Ky Kyat [Currency of Burma]

.ky Cayman Islands [Country Code/Domain Name]

KYAC Knowledge for Youth About Careers (Program) [of Employment and Immigration Canada]

KYBD Keyboard [also Kybd, or kybd]

KYCl Potassium Yttrium Chloride

KYD Cayman Islands Dollar [Currency]

KYF Potassium Yttrium Fluoride

KYI Potassium Yttrium Iodide

KYS Kloeckner-Youngstown Steelmaking Process [Metallurgy]

KZ Kazakhstan [ISO Code]

KZ Kwanza [also Kz; Currency of Angola]

.kz Kazakhstan [Country Code/Domain Name]

KZF Potassium Zinc Ferrite

KZF Potassium Zirconium Fluoride

K-Zn Potassium-Zinc (Alloy System)

KZV Koninklijke Zoologische Vereniging [Royal Zoological Society, Netherlands]

L

L   Active Length (of Springs) [Symbol]

L   Angular Momentum [Symbol]

L   Azimuthal Quantum Number [Symbol]

L   Camera Length [Transmission Electron Microscopy]

L   Cellulose Acetate-Covered (Electric Wire)

L   Cloud Height [Symbol]

L   Coefficient of Induction [Symbol]

L   Cold Weather Mission (Aircraft) [USDOD Symbol]

L   Conductance [Symbol]

L   Conductivity [Symbol]

L   Cone Distance (of Bevel Gears) [Symbol]

L   Crosshead Velocity (in Mechanical Testing) [Symbol]

L   Distance [Symbol]

L   Drawdown (for Soil, or Rock) [Symbol]

L   Distance between Entrance Aperture and Image Plane (in Neutron Radiography) [Symbol]

L   Ga(u)ge Length (in Mechanical Testing) [Symbol]

L   Grand-Canonical Heat Function (or Hill Energy) (in Physics) [Symbol]

L   Elastic Stiffness Tensor [Symbol]

L   Enthalpy of Fusion [Symbol]

L   Fifty [Roman Numeral]

L   Forging Direction [Symbol]

L   Heat Function [Symbol]

L   Heat of Fusion [Symbol]

L   Inductance [Symbol]

L   Label(ling)

L   Labia [Medicine]

L   Labile

L   Laboratory

L   Labo(u)r(er)

L   Labrusca [Botany]

L   Labyrinth

L   *(Lac)* – French for "Lake"

L   Lack

L   Lacquer

L   Lactobacillus [Microbiology]

L   Lactase [Biochemistry]

L   Lacteal(s) [Anatomy]

L   (+)-Lactose [Biochemistry]

L   Ladder

L   Ladle

L   Lag(ging)

L   *(Lago)* – Italian for "Lake"

L   Lagos [Nigeria]

L   Lagomorpha [Order of Mammals Including Rabbits and Hares]

L   Lagoon

L   Lagrangian (Function) (in Mechanics) [Symbol]

L   Lahore [Bangladesh]

L   Lake [also l]

L   Lambert [Unit]

L   Lamella(r)

L   Lamina(r)

L   Laminate; Lamination

L   Lamp

L   Lance

L   Land

L   Landing [also l]

L   Landing Time [Symbol]

L   Landscape (Font Orientation)

L   Langevin Function (in Solid-State Physics) [Symbol]

L   Langmuir (Unit) [Symbol]

L   Language

L   Lansing [Michigan, US]

L   Laos

L   Laparoscopy [Medicine]

L   Lap

L   Lapping

L   Larch

L   Lard

L   Large [also l]

L   Larix [Genus of Coniferous Trees Including the Larches]

L   Larva

L   Larynx [Anatomy]

L   Laser

L   Latency; Latent

L   Latent Heat [Symbol]

L   Lateral

L   Latex (Rubber)

L   Lathe

L   Latin

L   Latitude [also l]

L   Lattice

L   Latvia(n)

L   Launch(ing)

L   Lausanne [Switzerland]

L   Law

L   Lay

L   Lay (of Surface), Multidirectional [Symbol]
L   Layer(ing)
L   Leach(ing)
L   Lead [Abbreviation; Symbol: Pb]
L   Lead (of Spirals, Screw Threads, Helical Gears, Worm Threads, etc.) [Symbol]
L   Leaf; Leaves
L   Leak(age)
L   Leakage Coefficient (of Electric Machinery) [Symbol]
L   Learn(er); Learning
L   Leather
L   Leave
L   Lebanese; Lebanon
L   Lecture(r)
L   Ledeburite; Ledeburitic [Metallurgy]
L   Ledger
L   Leeds [UK]
L   Left [also l]
L   Leg
L   Legal
L   Legume [Botany]
L   Leiden [Netherlands]
L   Leipzig [Germany]
L   Leishmania [Genus of Protozoans]
L   Length
L   Length of Cut (in Machining) [Symbol]
L   Length of Stroke [Symbol]
L   Length of Weld [Symbol]
L   Lengthwise
L   Leningrad [USSR]
L   Lens
L   Lepidoptera [Order of Insects Including Butterflies, Moths and Skippers]
L   Leptospira [Genus of Bacteria]
L   Lesotho
L   Leu [Currency of Romania]
L   (–)-Leucine; Leucyl [Biochemistry]
L   Leucocyte [Anatomy]
L   Leukemia [Medicine]
L   Level(er); Leveling
L   Lever(age)
L   Levorotation; Levorotatory [Symbol]
L   Lewisite
L   Lhasa [Tibet]
L   Liaison Aircraft [US Army Symbol]
L   Liberia(n)
L   (Libra) – pound
L   Library
L   Libration [Astronomy]
L   Libreville [Gabon]
L   Libya(n)
L   License
L   Lichen [Botany]
L   Liechtenstein

L   Liège [Belgium]
L   Life
L   Lift(ing)
L   Lift Force [Symbol]
L   Ligament [Medicine]
L   Ligand
L   Ligation; Ligature [Medicine]
L   Light(ing) [also l]
L   Light Meson (in Particle Physics) [Symbol]
L   Lightning [also l]
L   Likelihood [Symbol]
L   Lille [France]
L   Lima [Peru]
L   Lima [Phonetic Alphabet]
L   Limit (in Mathematics) [Symbol]
L   Limit(er); Limiting
L   Line [also l]
L   Line [Occasionally Followed by a Subscripted Number, e.g., L_1 denotes Line 1] [Electrical Engineering]
L   Linear(ity)
L   Link(age) [also l]
L   Linnaeus [Botany]
L   Liouville Superoperator [Symbol]
L   Lipase [Biochemistry]
L   Lipid [Biochemistry]
L   Lipoprotein [Biochemistry]
L   Lipschitz (Condition) [Mathamatics]
L   Liquefier; Liquefy; Liquefaction
L   Liquid
L   Liquidus [Metallurgy]
L   Liquid Water Content [Symbol]
L   Liquor
L   Lira [Currency of Italy]
L   Lisbon [Portugal]
L   List
L   List (Command) [Edlin MS-DOS Line Editor]
L   Listen(er); Listening
L   Listening Post [Aeronautics]
L   Liter [also l, or ℓ]
L   Litmus
L   Lithuania(n)
L   Live
L   Liver [Anatomy]
L   Lizard
L   Loa [Genus of Filarial Worms]
L   Load(er); Loading
L   Lobe
L   Lobule [Medicine]
L   Local [also l]
L   Locator; Location
L   Loch
L   Locomotive
L   Locus
L   Lodz [Poland]

L   Logarithm(ic)
L   Logic(s)
L   Loire (River) [France]
L   Lomé [Togo]
L   London [UK]
L   Long [also l]
L   Long Wave
L   Longeron [Aeronautics]
L   Longitude [also l]
L   Longitudinal (Direction) [Metallurgy]
L   Loop(er); Looping
L   Lorentz Force (in Solid-State Physics]
L   Lorentzian [Symbol]
L   Lorentz Polarization Factor [Symbol]
L   Lorentz Number [Symbol]
L   Loss
L   Lot
L   Loti [Currency of Lesotho]
L   Loud(ness)
L   Louisiana [US]
L   Louisville [Kentucky, US]
L   Love Wave (or Surface Wave) [Seismology]
L   Low [also l]
L   Low-Alloy Type Special-Purpose Tool Steels [AISI-SAE Symbol]
L   Low-Address Byte
L   Lowering [Controllers]
L   Low-Power
L   L-Radiation [Physics]
L   L-Shell (of Electron) [Symbol]
L   Luanda [Angola]
L   Lubbock [Texas, US]
L   Lubricant; Lubricate; Lubrication; Lubricator
L   Lumbar [Anatomy]
L   Lumbricus [Genus of Earthworms]
L   Lumen
L   Luminance (or Brightness) [Symbol]
L   Luminescence; Luminescent
L   Lump
L   Lunar
L   Lunar Magnetic Variation [Symbol]
L   Lung
L   Lusaka [Zambia]
L   Lüta [PR China]
L   Luxembourg
L   Luzon [Philippines]
L   Lye
L   Lymph(atic); Lymphoid [Medicine]
L   Lymphocyte [Medicine]
L   Lynx [Zoology]
L   Lyons [France]
L   Lyophilic; Lyophilization; Lyophilize(d)
L   Lysin [Immunology]
L   Lysozome [Biochemistry]

L   (–)-Lyxose [Biochemistry]
L   Mean Linear Intercept Distance (in Quantitative Metallography) [Symbol]
L   Overall Lenth (of Drills) [Symbol]
L   Plain Lead-Covered (Cable) [Symbol]
L   Pound (Sterling) [UK]
L   Radiance [Symbol]
L   Rating Life (of Rolling Bearings) [Symbol]
L   Self-Inductance [Symbol]
L   Silo-Launched (Rocket) [USDOD Symbol]
L   Solubility Product (in Physical Chemistry) [Symbol]
L   Sound Pressure Level [Symbol]
L   Spectral Radiance [Symbol]
L   Steel Angle [AISI/AISC Designation]
L   (Length of) Stroke (of Piston) [Symbol]
L   Thickness [Symbol]
L   Twist Pitch Distance of the Composite Wire [Symbol]
L   Unit Length of Dislocation Line (in Materials Science) [Symbol]
L-   Lockheed (Aircraft) [Followed by a Numeral, e.g., L-1011]
L-   Prefix Denoting the Left-Handed Enantiomer of an Optical Isomer [also L-]
$\bar{L}$ Average Intercept Distance (in Millimeters) at 100 Magnification (in Metallography) [Symbol]
$\bar{L}$ Mean Areal Intercept (in Quantitative Metallography) [Symbol]
$\bar{L}$ Mean Chord Length (of Grain Boundary) (in Materials Science) [Symbol]
$\bar{L}$ Molecular Length [Symbol]
$\bar{L}$ Root-Mean-Square Length (of Polymeric Molecules) [Symbol]
±L Lightness (Opponent-Color Scale) [OSA-UCS]
L*   Characteristic Length [Symbol]
L_0 Initial Inductance [Symbol]
L_0 Latent Heat of Vaporization of Molecules at Absolute Zero [Symbol]
L_0 Length of Specimen at Lower Temperature [Symbol]
L_1 Core Inductance [Symbol]
L_1 Inductance of Electric Circuit 1 [Symbol]
L_1 Length of Panel in Span Direction for Flat Concrete Slabs [Symbol]
L_1 Length of Specimen at Higher Temperature [Symbol]
L_1 Liquid 1 [Symbol]
L_1 Liquid Phase 1 [Symbol]
L_2 Width of Panel at Right Angles to Span Direction for Flat Concrete Slabs [Symbol]
L_2 Mean Intercept Length of Planar Figures in Quantitative Metallography) [Symbol]
L_3 Mean Intercept Length of Three-Dimensional Bodies (in Quantitative Metallography) [Symbol]
$\bar{L}_3$ Mean Intercept Length (in Metallography) [Symbol]
L_{10} Rolling Bearing Rating Life (90% Reliability) [Symbol]
L_A Total Perimeter (i.e., Sum of Lengths of Linear Features Divided by Total Test Area) (in Quantitative Metallography) [Symbol]

L_a L-Alpha (Radiation) [also Lα]

L_a Lamellar Alpha Phase (of Liquid Crystal) [Symbol]

L_a Lipid Alpha Phase (in Biochemistry) [Symbol]

L_b Length Between (in Textile Unevenness Testing) [Symbol]

L_β L-Beta (Radiation) [also Lβ]

L_β Lipid Beta Phase (in Biochemistry) [Symbol]

L_c Conversion Loss [Semiconductor Symbol]

L_c Correlation Length [Symbol]

L_c Reinforcing-Fiber Transfer Length (i.e., Critical Length of Reinforcing Fiber) [Symbol]

L_c Unit Length of Instrument (in Textile Unevenness Testing) [Symbol]

L_D Debye Length (in Plasma Physics) [Symbol]

L_Δ Incremental Inductance [Symbol]

L_e Linear Expansion (of a Soil Mass) [Symbol]

L_e Radiance [Symbol]

$L_{e(\varepsilon=1)}$ Thermal Radiance (of a Full Radiator) [Symbol]

L_{eq} Equivalent Inductance [Symbol]

L_{ext} Extended Length (of Polymeric Molecules) [Symbol]

L_f Fluidity Length (in Casting) [Symbol]

L_f Latent Heat of Fusion [Symbol]

L_g Gate Length [Semiconductor Symbol]

L_γ L-Gamma (Radiation) [also Lγ]

L_i Intrinsic (Ferric) Inductance [Symbol]

L_i (X-Ray) Source-to-Image Distance (in Radiography) [Symbol]

L_k Kinetic Inductance [Symbol]

L_L Lineal Fraction (i.e., Sum of Linear Intercept Lengths Divided by Total Test Line Length) (in Quantitative Metallography) [Symbol]

L_m Mutual Inductance (of Two Electric Circuits) [Symbol]

L_o (X-Ray) Source-to-Object Distance (in Radiography) [Symbol]

L_p Characteristic Length [Symbol]

L_p Lorentz-Polarization Factor [Symbol]

L_p Mean Perimeter Length per Planar Figure (in Quantitative Metallurgy) [Symbol]

L_p Parallel Equivalent Inductance [Symbol]

L_p Phosphorus Distribution Ratio (of Slags) (in Metallurgy) [Symbol]

L_p Primary Coil Inductance [Symbol]

L_p Sound Pressure Level [Symbol]

$\overline{L}_p$ Mean-Free Path Between Particles [Symbol]

L_{ph} Photovoltaic Transport Length (of Electron) [Symbol]

L_s Linear Shrinkage (of a Soil Mass) [Symbol]

L_s Series Inductance [Symbol]

L_s Secondary Coil Inductance [Symbol]

L_T Total Inductance [Symbol]

L_T Total Length of Test Line (in Quantitative Metallography) [Symbol]

L_t Contact Transfer Length [Symbol]

L_{Th} Thouless Length (in Solid-State Physics) [Symbol]

L_V Length of Features per Test Volume (in Quantitative Metallography) [Symbol]

L_v Latent Heat of Vaporization [Symbol]

L_W Sound Power Level [Symbol]

L_w Length Within (in Textile Unevenness Testing) [Symbol]

L_w Liquid Limit (of Soil) [Symbol]

L_w Winding Inductance [Symbol]

L_z Orbital Angular Momentum Along z Axis [Symbol]

L1 1,2-Dimethyl-3-Hydroxypyridin-4-one

L1, L2, L3 Line 1, Line 2, Line 3, etc [Controllers]

4L Four-Wheel Drive Position, Low-Range (at Reduced Speeds)

l average rate of occurrence of defects per sample [Symbol]

l azimuthal (or angular momentum) quantum number of electron [Symbol]

l coefficient of induction [Symbol]

l constant length of random short test lines (in quantitative metallography) [Symbol]

l distance [Symbol]

l effective height of column (in civil engineering) [Symbol]

l effective span of beam, or slab (in civil engineering) [Symbol]

l equivalent ionic conductance [Symbol]

l first-direction cosine of a line (e.g., l = cos α) [Symbol]

l lake [also L]

l lambda [English Equilvalent]

l landing [also L]

l large [also L]

l last term of arithmetic, or geometric progression [Symbol]

l late

l latent heat per gram [Symbol]

l latitude [also L]

l lead (of gear-worms) [Symbol]

l league

l left [also L]

l length [Symbol]

l leu [Currency of Romania]

l levorotatory [Chemistry]

l *(liber)* – book

l *(libra)* – pound

l light [also L]

l lightning [also L]

l line [also L]

l linear dimension [Symbol]

l liner

l link(age) [also L]

l liquid

l liter [also L, or ℓ]

l local [also L]

l long [also L]

l low [also L]

l lumen [Unit]

l mean free path (of molecule) [Symbol]

l measured length of long diagonal of Knoop (hardness) indentation [Symbol]

l mixing length (in fluid flow) [Symbol]

l   object distance (in optics) [Symbol]

l   stroke length [Symbol]

/l   line numbers option [MS-DOS Linker]

$\bar{l}$ mean free path [Symbol]

l-   prefix denoting a substance as being levorotatory

l′   image distance (in optics) [Symbol]

l_0 characteristic dimension (of a pipe, etc.) [Symbol]

l_0 length at temperature $t = 0°C$ [Symbol]

l_0 optical absorption length [Symbol]

l_0 original (gauge) length (in mechanical testing) [Symbol]

l_c capillarity length [Symbol]

l_c critical (fiber) length (in composite engineering) [Symbol]

l_s diffusion length [Symbol]

l_t length at temperature t ($t \neq 0°C$) [Symbol]

l_t thermal diffusion length [Symbol]

l_x unit length of quantity x [Symbol]

l_y unit length of quantity y [Symbol]

l_z unit length of quantity z [Symbol]

Λ Compton Wavelength of Electron ($2.4263106 \times 10^{-12}$ m) [Symbol]

Λ Eigenvalue [Symbol]

Λ Equivalent Conductance (of an Electrolyte) [Symbol]

Λ (Diffraction) Grating Spacing [Symbol]

Λ (Upper-case) Lambda [Greek Alphabet]

Λ Lambda Particle (or Lambda Hyperon) (in Particle Physics) [Symbol]

Λ Logarithmic Decrement [Symbol]

Λ Loudness Level [Symbol]

Λ Modulation Wavelength [Symbol]

Λ Permeance [Symbol]

Λ Superlattice Wavelength (in Solid-State Physics) [Symbol]

$\Lambda°$ Limiting Equivalent Conductance of an Electrolyte [Symbol]

Λ^0 Lambda-Zero Particle (i.e., Neutral) (in Particle Physics) [Symbol]

$\overline{\Lambda}^0$ Lambda-Zero Antiparticle (i.e., Neutral) (in Particle Physics) [Symbol]

Λ_0 Equivalent Conductance at Infinite Dilution (of an Electrolyte) [Symbol]

Λ_0 Largest Eigenvalue (in Physics) [Symbol]

Λ_c Critical Wavelength of Superlattice (in Solid State Physics) [Symbol]

Λ_c Critical Modulation Wavelength (in Physics) [Symbol]

λ (lower-case) lambda [Greek Alphabet]

λ angle between the direction of the applied force and the slip direction (in crystallography) [Symbol]

λ attenuation constant (in physics) [Symbol]

λ azimuth of reflected light (e.g., in ellipsometry) [Symbol]

λ Burgers vector angle (crystallography) [Symbol]

λ celestial, or geographical longitude (in astronomy) [Symbol]

λ characteristic time [Symbol]

λ coefficient of absorption [Symbol]

λ (electron-phonon) coupling constant [Symbol]

λ Debye length (in plasma physics) [Symbol]

λ (radioactive) decay constant [Symbol]

λ diffusion distance [Symbol]

λ effective variation (of involute splines) [Symbol]

λ eigenvalue (e.g., λ_1 represents the first eigenvalue, λ_2 the second eigenvalue, etc.) [Symbol]

λ escape depth (of Auger electrons) [Symbol]

λ heat of vaporization [Symbol]]

λ intercept free (edge-to-edge) distance in a lamellar structure (in quantitative metallography) [Symbol]

λ interrod spacing (in metallography) [Symbol]

λ Lagrangian multiplier (in mathematics) [Symbol]

λ Lamé's constant (in mechanics) [Symbol]

λ lambda point [Symbol]

λ lead angle (of gear-worms) [Symbol]

λ Lyapunov exponent (in physics) [Symbol]

λ magnetostriction constant [Symbol]

λ mean free path [Symbol]

λ migration parameter (in thin-film technology) [Symbol]

λ modulation length [Symbol]

λ parameter (in mathematics) [Symbol]

λ (London) penetration depth (of superconductors) [Symbol]

λ (corrosion) pit generation rate [Symbol]

λ primary dendrite-arm spacing (in metallurgy) [Symbol]

λ push-rod ratio (in mechanical engineering) [Symbol]

λ relaxation time [Symbol]

λ superconducting coupling constant [Symbol]

λ thermal conductivity [Symbol]

λ toughness ratio [Symbol]

λ wavelength [Symbol]

λ Weiss constant (in solid-state physics) [Symbol]

λ^+ equivalent conductance of cations [Symbol]

λ^- equivalent conductance of anions [Symbol]

λ_+^0 limiting equivalent conductance of cations [Symbol]

λ_-^0 limiting equivalent conductance of anions [Symbol]

$\lambda_\parallel$ in-plane magnetostriction [Symbol]

$\lambda_\parallel$ parallel (longitudinal) penetration depth (of superconductors) [Symbol]

$\lambda_\perp$ perpendicular (or transverse) magnetostriction [Symbol]

$\lambda_\perp$ perpendicular (or transverse) penetration depth (of superconductors) [Symbol]

λ_0 resonant wavelength (of electromagnetic radiation) [Symbol]

λ_0 short-wavelength cutoff (in x-ray generation) [Symbol]

λ_0 wavelength (of light) in vacuum [Symbol]

λ_2 secondary dendrite-arm spacing (in metallurgy) [Symbol]

λ_{100} magnetostrictive constant in easy direction of magnetization [Symbol]

λ_a apparent thermal conductivity [Symbol]

λ_{ab} in-plane penetration depth (of superconductor) [Symbol]

λ_C Compton wavelength of electron ($2.4263106 \times 10^{-12}$ m) [Symbol]

λ_c critical wavelength [Symbol]

$\lambda_{C,n}$ Compton wavelength of neutron ($1.3195911 \times 10^{-15}$ m) [Symbol]

$\lambda_{C,p}$ Compton wavelength of proton ($1.3214100 \times 10^{-15}$ m) [Symbol]

λ_E interlamellar spacing (in metallurgy) [Symbol]

λ_e elastic mean free path [Symbol]

λ_i (general) eigenvalue (i = 1, 2, 3, etc.) [Symbol]

λ_i inelastic mean free path [Symbol]

λ_J Josephson length (or Josephson penetration depth) (of superconductor) [also λ_j] [Symbol]

λ_j magnetostriction coefficient [Symbol]

λ_K wavelength near K-absorption edge [Symbol]

λ_L London penetration depth (of superconductor) [Symbol]

λ_L lower confidence limit (in statistics) [Symbol]

λ_m wavelength of maximum monochromatic blackbody radiance at given temperature [Symbol]

λ_{max} wavelength of maximum spectral emissivity [Symbol]

λ_n wavelength medium with refractive index n [Symbol]

λ_R Rayleigh wave (or surface wave) [Symbol]

λ_R wavelength of reflected light [Symbol]

λ_s planar perturbation wavelength [Symbol]

λ_s magnetostrictive anisotropy [Symbol]

λ_s saturation magnetostriction (coefficient) [Symbol]

λ_{sf} spin-flip scattering length (in physics) [Symbol]

λ_U upper confidence limit (in statistics) [Symbol]

λ_x x-ray wavelength [Symbol]

$\mathscr{L}$ Absolute Luminosity (in Astronomy) [Symbol]

$\mathscr{L}$ (Magnetic) Flux Linkage [Symbol]

$\mathscr{L}$ Laplace Transform (in Mathematics) [Symbol]

$\mathscr{L}$ Total Number of Molecules per Unit Time [Symbol]

$\mathscr{L}_1$ Flux Linkage in the Primary Circuit [Symbol]

$\mathscr{L}_2$ Flux Linkage in the Secondary Circuit [Symbol]

$\mathscr{L}_m$ Mutual Flux Linkage [Symbol]

ℓ lattices (in heterogeneous connection of atoms) [Symbol]

ℓ length of conductor [Symbol]

ℓ liter [also L, or l]

ℓ (magnetic) flux path length [Symbol]

ℓ interfacial displacement field [Symbol]

ℓ perpendicular momentum transfer [Symbol]

ℓ_1 effective flux path length [Symbol]

ℓ_d Kolmogorov dissipation length (in physics) [Symbol]

ℓ_g magnetic gap length [Symbol]

LA Laboratoire d'Aérologie [Aerology Laboratory, France]

LA Laboratory Accreditation

LA Laboratory Automation

LA Lactic Acid [Biochemistry]

LA Landscape Architect; Landscape Architecture

LA Lanthanum Aluminate

LA Lanthanum Aluminide

LA Laos [ISO Code]

LA Large Angle

LA Large Aperture

LA Laser Ablation

LA Laser Amplifier

LA Laser Annealing [Metallurgy]

LA Laser-Gyro Axis

LA Lateral Aberration (in Optics)

LA Latex Agglutination

LA Latin America(n)

LA Lattice "Animals" [Materials Science]

LA Launch Abort

LA Launch Aft

LA Launch Area

LA Launch Azimuth

LA Lauric Acid

LA Lead-Alkali (Glass)

LA Lead Angle

LA Lead Azide

LA Left Auricle (of the Heart) [Medicine]

LA Lending Account

LA Length Analysis; Length Analyzer

LA Lesser Antilles

LA Letter of Authority

LA Lewis Acid

LA Library Administration

LA Library Association [UK]

LA Licensed Aircraft

LA Light Aircraft; Light Airplane

LA Light Alloy

LA Light Amplification; Light Amplifier

LA Light Armor

LA Light Artillery

LA Lighter than Air (Aircraft)

LA Lightning Arrester

LA Light-Wire Armored (Cable)

LA Limited Area

LA Line Adapter

LA Lineal Analysis

LA Linear Acceleration; Linear Accelerator

LA Linear Acoustics

LA Linear Algebra

LA Linear Amplifier

LA Linear (Antenna) Array

LA Link Allotter

LA Linoleic Acid (or Omega-6-Polyunsaturated Fatty Acid) [Biochemistry]

LA Liquid Air

LA Liquid Alloy

LA Liquid Atomization [Metallurgy]

LA Listed Address

LA Lithium Aluminate

LA Load Accumulator

LA Local Action

LA Local Address [Computers]

LA Local Agent

LA Local Area

LA Localizer Antenna [Aeronautics]

LA Logarithmic Amplifier

LA Logic Analysis; Logic Analyzer

LA Logical Address

LA Longitudinal Aberration (in Optics)

LA Longitudinal Acoustic (Wave)

LA Longitudinal Anneal(ing) [Mechanics]

LA Loop Antenna

LA Los Alamos [New Mexico, US]

LA Los Angeles [California, US]

LA Louisiana [US]

LA Low Africa

LA Low Altitude

LA Low Angle

LA Lower Atmosphere

LA_6 Lanthanum Hexaaluminate

L/A Letter of Advice

La Lanthanum [Symbol]

La Lempira [Currency of Honduras]

La Louisiana [US]

La^{3+} Lanthanum Ion [also La^{+++}] [Symbol]

La-131 Lanthanum-131 [also ^{131}La, or La^{131}]

La-132 Lanthanum-132 [also ^{132}La, or La^{132}]

La-133 Lanthanum-133 [also ^{133}La, or La^{133}]

La-134 Lanthanum-134 [also ^{134}La, or La^{134}]

La-135 Lanthanum-135 [also ^{135}La, or La^{135}]

La-136 Lanthanum-136 [also ^{136}La, or La^{136}]

La-137 Lanthanum-137 [also ^{137}La, or La^{137}]

La-138 Lanthanum-138 [also ^{138}La, or La^{138}]

La-139 Lanthanum-139 [also ^{139}La, or La^{139}]

La-140 Lanthanum-140 [also ^{140}La, or La^{140}]

La-141 Lanthanum-141 [also ^{141}La, or La^{141}]

La-142 Lanthanum-142 [also ^{142}La, or La^{142}]

La-143 Lanthanum-143 [also ^{143}La, or La^{143}]

La-144 Lanthanum-144 [also ^{144}La, or La^{144}]

La-211 La_2BaCuO_5 [A Lanthanum Barium Copper Oxide Superconductor]

La-214 $La_2BaCu_4O_5$ [A Lanthanum Barium Copper Oxide Superconductor]

La-221 $(LaBa)_2CuO_5$ [A Lanthanum Barium Copper Oxide Superconductor]

Lα L-Alpha (Radiation) [also L_α]

L-α Lyman-Alpha (Radiation) [Spectroscopy]

λ(Å) wavelength in angstroms

λ/α minimum thermal distortion (i.e., ratio of thermal conductivity to thermal expansion) [Symbol]

LAA Lab(oratory) Animal Allergy

LAA Laser Association of America [US]

LAA Library Association of Alberta [Canada]

LAA Library Association of Australia

LAA Light Anti-Aircraft Artillery

LAA Liters of Absolute Alcohol

$La(Ac)_3$ Lanthanum (III) Acetate

$La(ACAC)_3$ Lanthanum Acetylacetonate [also $La(AcAc)_3$, or $La(acac)_3$]

LAAD Latin American Agrobusiness Development Corporation

LAAD Los Angeles Aircraft Division [of Rockwell International Corporation, US]

LAAFTA Latin American Association of Freight and Transport Agents [Paraguay]

LAAPI Latin American Association of Pharmaceutical Industries [Argentina]

LAAPS Latin American Association of Physiological Sciences [Argentina]

LAAR Liquid Air Accumulator Rocket

LAAS Laboratoire d'Automatique et d'Analyse des Systèmes [Automation and Systems Analysis Laboratory, of Conseil National de Recherche Scientifique at Toulouse, France]

LaAs Lanthanum Arsenide (Semiconductor)

LAAV Light Airborne ASW (Antisubmarine Warfare) Vehicle

LAAW Light Assault Antitank Weapon

LAB Laboratory Animals Bureau [UK]

LAB Lead Acid Battery

LAB Linear Accelerator Breeder

LAB Low-Altitude Bombing

LAB Low-Angle (Grain) Boundary [Materials Science]

LaB_6 Lanthanum Hexaboride

Lab Laboratory [also lab]

Lab Labo(u)r [also lab]

Lab Labrador [Canada]

LABAT (International) Lead Acid Battery (Conference)

LaBCCO Lanthanum Barium Calcium Copper Oxide (Superconductor) [also LBCCO]

LABCI Latin American Bank of the Construction Industry

LaBCO Lanthanum Barium Copper Oxide (Superconductor) [also LBCO]

LABCOM Laboratory of Communications [of US Army, Electronics Technology and Devices Laboratory, Fort Monmouth, New Jersey]

LaBCSO Lanthanum Barium Copper Tin Oxide (Superconductor) [also LBCSO]

LaBCTO Lanthanum Barium Copper Titanium Oxide (Superconductor) [also LBCTO]

Lab Dev Eng Mater, Kanazawa Univ Laboratory for Development of Engineering Materials, Kanazawa University [Japanese Publication]

Lab Equip Dig Laboratory Equipment Digest [UK]

LABEX Laboratory Exposition [Canada]

Lab Forum Laboratory Forum [Published in the US]

Lab Hazards Bull Laboratory Hazards Bulletin [Published by the Royal Society of Chemistry, UK]

LABIL Light Aircraft Binary Information Link

Lab Invest Labotatory Investigations [US Journal]

Lab Microcomput Laboratory Microcomputer [UK Publication]

LABORDOC Labor Documentation (Database) [of International Labour Organization]

LABORELEC Laboratoire de l'Industrie Electrique [Electrical Industry Laboratories, Belgium]

Lab PHASE Laboratoire de Physique et Applications des Semiconducteurs [Semiconductor Physics and Applications Laboratory, Strasbourg, France]

Lab PICM Laboratoire de Physique des Interfaces et Couche Minces [Interface and Thin-Film Physics Laboratory; Ecole Polytechnique, Palaiseau, France] [also Lab PCM]

Lab Pract Laboratory Practice [Published in the UK]

Lab PS Laboratory Power Supply

LABS Linear Alkylbenzene Sulfonate

LABS Low-Altitude Bombing System

Labs Laboratories [also labs]

LA/BSA Linoleic Acid/Bovine Serum Albumin [Biochemistry]

LabVIEW Laboratory Virtual Instrument Engineering Workbench

Lab Wkbk Laboratory Workbook

LAC Laboratório Associado de Computação e Matemática Aplicada [Associate Laboratory of Computing and Applied Mathematics; of Brazilian National Institute of Space Research]

LAC LANSCE (Los Alamos Neutron Scattering Center) Advisory Committee [US]

LAC Latin America and the Caribbean

LAC Leading Aircraftman

LAC Lemon Administrative Committee [US]

LAC Lesotho Agricultural College [Maseru]

LAC Library Advisory Council [now Library and Information Service Council, UK]

LAC Library Assistant Certificate

LAC Lifshitz-Allen-Cahn (Law) [Metallurgy]

LAC Linear Absorption Coefficient

LAC Linear Array Camera

LAC Liquid Adsorption Chromatography

LAC Load Accumulator

LAC Local Arrangement Committee

LAC Local Area Coverage

LAC Long-Run Average Cost

LAC Loudness Adaption Control

LAC Lunar Aeronautical Chart

LAC Lux-Ampere Characteristics

LAc Lactic Acid [also Lac] [Biochemistry]

LaC Lance-Corporal

L/ac liters per acre [Unit]

LACAC Latin American Civil Aviation Commission [Peru]

LACAP Latin American Cooperative Acquisition Project

LACB Landing Aids Control Building [at NASA (Space) Shuttle Landing Facility at Kennedy Space Center]

LACBED Large-Angle Convergent-Beam Electron Diffraction

LACBPE Los Angeles Council of Black Professional Engineers [US]

LACBWR LaCrosse Boiling Water Reactor [Wisconsin, US]

LACC Los Angeles Convention Center [US]

LACCB Latin American Confederation of Clinical Biochemistry [Colombia]

LACE Library Advisory Council for England [now Library and Information Service Council, UK]

LACE Local Automatic Circuit Exchange

LACES London Airport Cargo EDP (Electronic Data Processing) System [UK]

LACES Los Angeles Council of Engineering Societies [US]

LACFFP Latin American Commission on Forestry and Forestry Products

Lachr Lachrymator [also lachr]

L-Acid 5-Hydroxy-1-Naphthalenesulfonic Acid; 1-Naphthol-5-Sulfonic Acid [also L-acid]

LACIE Large Area Crop Inventory Experiment [US]

LACIP Large Area Crop Inventory Program [US]

LACMA Latin American and Caribbean Movers Association [Panama]

LACP Latin American Center for Physics [Brazil]

La(Cp)$_3$ Tris(cyclopentadienyl)lanthanum

LACS Large-Area Chemical Sensor

LACSA Latin America and the Caribbean SA (Airlines)

LACT Lease Automatic Custody Transfer

Lact ON Lactose Octanitrate [Biochemistry]

LACTRAC Los Angeles County Transportation Commission [US]

La$_2$CuO$_4$:Sr Strontium-Doped Lanthanum Cuprate

LACV Logistics Air-Cushion Vehicle

LAD Lactic Dehydrogenase [Biochemistry]

LAD Laser Ablation Deposition

LAD Lexington Army Depot [of US Army at Creech Army Air Field, Lexington, Kentucky, US]

LAD Light-Assisted Desorption

LAD Lithium Aluminum Deuteride

LAD Local Anodic Dissolution

LAD Location Aid Device

LAD Logical Aptitude Device

LAD Lookout Assist Device

Lad Ladder [also lad]

LADAR Laser Detection and Ranging [also Ladar, or ladar]

LADB Laboratory Animal Database [of National Library of Medicine, Bethesda, Maryland, US]

LADD Lens Antenna Deployment Demonstration

LADDER Language Access to Distributed Data with Error Recovery

LADDR Layered Device Driver Architecture [of Microsoft Corporation, US]

LADEF Los Alamos Density Exponent Formula [Physics]

LADM Local Average Density Model

LADR Linear Accelerator-Driven Reactor

LADS Large Area Diamond System

LADS Local Area Data Set

LADSIRLAC Liverpool and District Scientific, Industrial and Research Library Advisory Council [UK]

LADT Local Area Data Transport

LADWP Los Angeles Department of Water and Power [US]

LAE Lead Angle Error

LAE Left Arithmetic Element

LAE   Licensed Aircraft Engineer

LAE   London Association of Engineers [UK]

LAED   Low-Angle Electron Diffraction [Physics]

LAEDP   Large-Area Electronic Display Panel

LAEM   Linear Anisotropic Elastic Model

LAER   Lowest Achievable Emission Rate

LAF   Landscape Architecture Foundation [US]

LAF   Laser Atomic Fluorescence (Spectroscopy)

LAF   Ligne Aéronautique de France [French Airlines]

LAF   Long Address Form

LAF   Lymphocyte Activating Factor [Biochemistry]

LaF   Lanthanum Fluoride (Crystal)

LAFC   Latin American Forestry Commission

LaF(Eu)   Europium-Doped Lanthanum Fluoride

LAFIS   Local Authority Financial Information System

LAFS   Laser Ablative Fluxless Soldering

LAFTA   Latin American Free Trade Association

LAG   Least-Action Ground State Approximation [Physics]

LAG   Library Automation Group [Australia]

LAG   Light Automatic Gun

LAG   Load and Go (Technique) [Computers]

Lag   Lagrangian (Function) [Mechanics]

LAGB   Large-Angle Grain Boundary [Materials Science]

LAGB   Library Association of Great Britain

LAGB   Low-Angle Grain Boundary [Materials Science]

LAGE   Los Angeles Grain Exchange [US]

LAGEOS   Laser Geodetic Earth Orbiting Satellite

LAGEOS   Laser Geodynamic Satellite [US]

LAGS   Laser-Activated Geodetic Satellite

LAGS   Launch Abort Guide Simulation

LAH   Lithium Aluminum Hydride

LAH   Logical Analyzer of Hypothesis

LAHCG   Lookahead Carry Generator [Computers]

LAHS   Leicestershire Archeological and Historical Society [UK]

LAHS   Low-Altitude, High-Speed

La-HTSC   Lanthanum High-Temperature Superconductor

LAI   Leaf Area Index [Ecology]

LAIA   Latin American Industrialists Association [Uruguay]

LAIEC   Latin American Institute of Educational Communication [Mexico]

LAIG   Library Association's Industrial Group [UK]

LAINS   Low-Altitude Inertial Navigation System

L Air   Liquid Air

LAIS   Laboratory of Atomic Imaging of Solids [PR China]

LAIS   Library Acquisitions Information System [of Library Automation Research and Consulting Association, US]

LAIT   Library Association Information Technology Group [UK]

LAIU   Launch Abort Interface Unit

LAIXR   Low-Angle of Incidence X-Ray Reflectivity

LAK   Laos Kip [Currency of Laos]

LAK   Lymphokine-Activated Killer (Cell) [Biochemistry]

LAL   Landcare Australia Limited

LAL   Limulus Amoebocyte Lysate [Biochemistry]

LaLα   Lanthanum L-Alpha (Radiation) [also LaL$_\alpha$]

L-Ala(P)   L(−)-1-Aminoethylphosphonic Acid [Biochemistry]

LALL   Longest Allowed Lobe Length

LALLS   Low-Angle Laser-Light Scattering

LALO   Low-Altitude Observation

LALR   Lookahead, Left-to-Right (Scan)

LALR(1)   Left-to-Right Scan, One Lookahead Token

LALS   LaGuardia Automated Library System [New York City, US]

LAM   Laminate Material

LAM   Laser-Assisted Machining

LAM   Liquid and Amorphous Metals Conference

LAM   Lloyd Aéreo Boliviano [Lloyd Bolivian Air Lines]

LAM   Load Accumulator with Magnetization

LAM   Load Accumulator with Magnitude

LAM   Logical Acknowlegdement Message

LAM   Longitudinal Acoustic Mode

LAM   Loop Adder and Modifier

Lam   Laminate; Lamination [also lam]

lam   laminate(d)

LAMA   Latin American Manufacturers Association

LAMA   Light Aircraft Manufacturers Association [US]

LAMA   Local Automatic Message Accounting

LAMA   Locomotive and Allied Manufacturers Association [UK]

LAMAS   London and Middlesex Archeological Society [UK]

LAMBD   Laser-Assisted Molecular Beam Deposition

LAMC   Language and Mode Converter

LAMCO   Liberian-American Swedish Minerals Company

LAMCS   Latin American Communications System

LAME   Licensed Aircraft Maintenance Engineer

LAMEF   Los Alamos Medium Energy Facility [New Mexico, US]

LAMES   Laser Micro-Emission Spectroscopy

LAMF   Los Alamos Meson Facility [New Mexico, US]

Lamin   Lamination [also lamin]

LAMIS   Local Authority Management Information System [UK]

LAMMA   Laser Microprobe Mass Analysis; Laser Microprobe Mass Analyzer

LAMMA   Laser Microprobe Microanalysis

LAMMS   Laser Micro-Mass Spectrometry

LaMnO$_3$ Lanthanum Manganate

LAMP   Laser Advanced Materials Processing (Conference)

LAMP   Laser and Maser Patents

LAMP   Low-Altitude Manned Penetrator

LAMPF   Los Alamos Meson Physics Facility [New Mexico, US]

LAMPRE   Los Alamos Molten Plutonium Reactor Experiment [US]

LAMPS   Laser Material Processing Program [Canada]

LAMPS   Light Airborne Multipurpose System

LAMPS   Limited-Area Mesoscale Prediction System

LAMS   Light Aircraft Maintenance Scheme [UK]

LAMS Load Alleviation and Mode Stabilization

LAMS London Association of Master Stonemasons [UK]

LA/MS Laser Ablation/Mass Spectrometry

LAMSAC Local Authority Management Services and Computer Committee [UK]

LAN Landing Aid

LAN Landscape (Font Orientation)

LAN Lateral Access Network

LAN Local-Area Network [also lan]

LaN Lanthanum Nitride

LANAC Laminar, Air Navigation and Collision [also Lanac, or lanac]

LANAC Laminar Navigation and Anti-Collision (System) [also Lanac, or lanac]

LANACS Local-Area Network Asynchronous Connection Server

Lanc Lancashire [UK]

LANCET Library Association, National Council for Educational Technology [Joint UK/US Project]

Lancet The Lancet [US Medical Journal]

Lancs Lancashire [UK]

LAND Lagrangian Ambient Noise Drifter

LAND Large Area Neutron Detector (Experiment) [of Gesellschaft für Schwerionenforschung mbH, Darmstadt, Germany]

LANDASSESS Land Degradation Assessment Program [of Commonwealth Scientific and Industrial Research Organization, Australia]

LANDATA Land Division Data Base [of Victoria State, Australia]

LANDCARE Land Degradation Management Improvement Program [of Commonwealth Scientific and Industrial Research Organization, Australia] [also Landcare]

LANDCARENET LANDCARE Telecommunication Network [of Commonwealth Scientific and Industrial Research Organization, Australia]

L and D Loss and Damage [also L&D]

LANDP LAN (Local-Area Network) Distributed Platform

LANDREST Cost-Effective Land Restoration Technologies Program [of Commonwealth Scientific and Industrial Research Organization, Australia]

LANDS Language-Oriented Development System

LANDSAT Earth Rersources Technology Satellite(s) [also Landsat] [Series of NASA Scientific Satellites]

LANDSAT-TM LANDSAT Thematic Mapper [also Landsat-TM]

Landscape Arch Q Landscape Architecture Quarterly [American Society of Landscape Architects, US]

LANE Local-Area Network Emulation

Lang Language [also lang]

Lang Soc Language in Society [International Journal]

LANICA Líneas Aéreas de Nicaragua [Nicaraguan Air Lines]

LANL Los Alamos National Laboratory [of US Department of Energy at Los Alamos, New Mexico, US]

LANL-MD Los Alamos National Laboratory–Materials Laboratory [of US Department of Energy at Los Alamos, New Mexico, US]

LANNET Large Artificial Nerve Net; Large Artificial Neuron Network

LANS Large-Angle Nuclear Scattering

LANS Load Alleviation and Stabilization

LANS LORAN (Long-Range Navigation) Airborne Navigation System

LANSCE Los Alamos Neutron Scattering Center [New Mexico, US]

Lanthanide Actinide Res Lanthanide and Actinide Research [Published in the Netherlands]

LANTIRN Low-Altitude Navigation and Targeting Infrared for Night [of US Air Force]

LAN-WAN Local-Area Network to Wide-Area Network Communications

LAO Localized Atomic Orbital [Physics]

La(OAc)$_3$ Lanthanum (III) Acetate

LAOCOON Least-Squares Adjustment of Calculation on Observed NMR (Nuclear Magnetic Resonance)(Spectra)

LAOD Los Angeles Ordnance District [Pasadena, California, US]

La(OH)$_3$ Lanthanum (III) Hydroxide

LAOI Laboratory of Atmosphere and Oceanology Institute [Russia]

LAOMC Laboratoire d'Acoustique et d'Optique de la Matière Condensée [Consensed Matter Acoustics and Optics Laboratory; of Université Pierre et Marie Curie, Paris, France]

La(OPr^i)$_3$ Lanthanum(III) Isopropoxide [also La(O-iPr)$_3$]

LAP Laboratório Associado de Plasmas [Associated Plasma Laboratory; of Brazilian National Institute of Space Research]

LAP Laboratory Accrediation Program

LAP Laboratory Analysis Package

LAP Latency Associated Peptide [Biochemistry]

LAP Lesson Assembly Program

LAP Leucine Aminopeptidase [Biochemistry]

LAP Life Analysis Package

LAP Limited-Access People

LAP Line Access Point

LAP Linea Aérea Paraguaya [Paraguayan Air Lines]

LAP Link Access Procedure [Computers]

LAP Link Access Protocol

LAP List Assembly Program

LaP Lanthanum Phosphide (Semiconductor)

LAPA Leucocyte Alkaline Phosphatase Activity [Biochemistry]

LAPA Lightweight Aggregate Producers Association [US]

LAPADS Lightweight Acoustic Processing and Display System

LAPB Link Access Procedure Balanced [also LAP-B]

LAPB Link Access Protocol Balanced [Computers]

LAPD Limited Axial Power Distribution

LAPD Link Access Procedure on the D-Channel [also LAP-D]

LAPD Los Angeles Police Department [US]

LAPDOG Low-Altitude Pursuit Dive On Ground

LAPES Low-Altitude Parachute Extraction System

LAPL Lafayette Association of Petroleum Landmen [Louisiana, US]

LAPM Link Access Procedure for Modems [also LAP-M]

LaPO₄ Lanthanophosphate

LaPO₄:Ce Cerium-Doped Lanthanophosphate

LApp Los Alamos Polycrystalline Plasticity (Code) [also LAPP] [of Los Alamos National Laboratory, Albuquerque, New Mexico, US]

LApp FC Los Alamos Polycrystalline Plasticity (Code), Full Constraint (Taylor) Model [of Los Alamos National Laboratory, Albuquerque, New Mexico, US]

LApp RC Los Alamos Polycrystalline Plasticity (Code), Relaxed Constraint (Taylor) Model [of Los Alamos National Laboratory, Albuquerque, New Mexico, US]

LAPPES Large Power Plant Effluent Study [of Tennessee Valley Authority, US]

LAPS Land Acquisition Priority System

LAPS Left Aft Propulsion Subsystem

LAPS Left Aft Propulsion System

LAPW Linearized Augmented Plane Wave [Physics]

Laq Lacquer [also laq]

LAR Limit Address Register

LAR Load Access Rights

LAR Low-Amplitude Resonance

LAR Low-Angle Reentry

LARA Latin American Railways Association [Argentina]

LARA Light-Armed Reconnaissance Aircraft

LARAM Line-Addressable Random-Access Memory

LARC Library Automation Research and Consulting Association [US]

LARC Light Amphibious Resupply Cargo Vessel

LARC Linear Aromatic Condensation [Polymer Science]

LARC Livermore Automatic Research Computer [of Lawrence Radiation Laboratory, California, US]

LARC Liverpool Automatic Research Calculator [UK]

LaRC Langley Research Center [of NASA at Hampton, Virginia, US]

LARCT Last Radio Contact

LARC-TPI Linear Aromatic Condensation–Thermoplastic Imide

LAREC Los Alamos Reactor Economics Code [of Los Alamos National Laboratory, New Mexico, US]

LARF Latin American Reserve Fund

Large Scale Syst Inf Decis Technol Large Scale Systems in Information and Decision Technologies [Published in the Netherlands]

L-Arg-MCA L-Arginine-4-Methylcoumaryl-7-Amide [Biochemistry]

LARIAT Laser Radar Intelligence Acquisition Technology

LARIS Laser Ablation and Resonance Ionization Spectrometry

LARM Low-Alloy Reference Material

LARN Laboratoire d'Analyses par Reactions Nucléaires [Nuclear Reaction Analysis Laboratory, Facultes Universitaires Notre Dame de la Paix, Namur, Belgium]

LARP Large Aspect Ratio Particle

LARP Local and Remote Printing

LARP Local Approvals Review Program

LARPS Local and Remote Printing Station

LARR Linear Accelerator-Regenerator Reactor

LARS Laminar Angular Rate Sensor

LARS Laser Angular Rate Scanner

LARS Laser Angular Rate Sensor

LARS Laser Articulated Robot System

LARS Library Accessions and Retrieval System

LARS Lower Airspace Radar Advisory Service [UK]

LA RSIS Library Association–Reference, Special and Information Section [also LA-RSIS, UK]

LARSSYAA Laboratory for Application of Remote Sensing System for Aircraft Analysis

LARU Latin American Research Unit

LAS Laboratoire d'Astronomie Spatiale [Space Astronomy Laboratory, Marseilles, France]

LAS Laboratório Associado de Sensores e Materiais [Associate Laboratory of Sensors and Materials Research; of Brazilian National Institute of Space Research]

LAS Laboratory Accreditation System

LAS Laboratory Automation Software

LAS Laboratory Automation System

LAS Laboratory of Atmospheric Sciences [of National Science Foundation, US]

LAS Land Agents Society [UK]

LAS Land Analysis System [Geographic Information System]

LAS Landing and Approach System

LAS LANSCE (Los Alamos Neutron Scattering Center) Advisory Committee [New Mexico, US]

LAS Large-Angle (Light) Scattering

LAS Large Astronomical Satellite [of European Space Research Organization]

LAS Large Autoclave System

LAS Laser Absorption Spectroscope; Laser Absorption Spectroscopy

LAS Latvian Academy of Sciences [Riga]

LAS Launch Auxiliary System

LAS League of Arab States

LAS Library Association of Singapore

LAS Light Absorption Sensitizer

LAS Light-Activated Switch

LAS Light-Active Sensor

LAS Linear Alkylate Sulfate

LAS Linear Alkylbenzene Sulfonate

LAS Linear Alkyl Sulfonate

LAS Lithia-Alumina-Silica [$Li_2O-Al_2O_3SiO_2$]

LAS Lithium Aluminosilicate (or Lithium Aluminum Silicate)

LAS Lithuanian Academy of Sciences [Vilna]

LAS Local Address Space [Computers]

LAS Logic Analysis System

LAS Louisiana Academy of Sciences [US]

LAS Low-Alloy Steel

LAS Low-Altitude (Observation) Satellite

LAS Low-Angle Scattering

LASA Laboratory Animals Science Association [UK]

LASA Large-Aperture Seismic Array [US]

LASA Laser-Aided Surface Alloying

LASA Latin American Shipowners Association

LASA Long-Acting Sulfonamide

LaSb Lanthanum Antimonide (Semiconductor)

LASC Louisiana Arts and Science Center [Baton Rouge, US]

LASCA Los Angeles State and County Arboretum [US]

LaSCNO Lanthanum Strontium Copper Niobium Oxide (Superconductor)

LASCO Latin American Science Cooperation Office

LASCO Lockheed Aeronautical Systems Company [Burbank, California, US]

LASCOT Large Screen Color Television System

LASCR Light-Activated Silicon-Controlled Rectifier

LASCS Light-Activated Silicon-Controlled Switch

LASD Los Angeles School of Dentistry [of University of California, US]

LASE (High-Power) Laser Applications in Science and Engineering

LASE Lidar Atmospheric Sensing Experiment [of NASA]

LASE Load-at-Specified Elongation [Mechanics]

LASE Logic Analysis and Slave Emulation

LASER Light Amplification by Stimulated Emission of Radiation [also Laser, or laser]

Laser Chem Laser Chemistry [Journal published in the UK]

LASERCOM Laser Computer Output Microfilm [also Lasercom]

Laserdisk Prof Laserdisk Professional [Published in the US]

Laser Focus/Electro-Opt Laser Focus/Electro-Optics [Journal published in the US]

Laser Optoelektron Laser und Optoelektronik [German Journal on Lasers and Optoelectronics]

Laser Part Beams Laser and Particle Beams [Published by Cambridge University Press, UK]

Laser Phys Laser Physics [International Journal]

Lasers Appl Lasers and Applications [Journal published in the US]

Lasers Life Sci Lasers in the Life Sciences [Journal published in the UK]

Lasers Optronics Lasers and Optronics [Journal published in the US]

LASH Laser Antitank Semi-Active Homing

LASH Lighter-Aboard Ship [also Lash]

LASIE Library Automated Systems Information Exchange [Australia]

LASIK Laser-Assisted In-Situ Keratomileusis [Eye Surgery]

LASIL Land and Sea Interaction Laboratory

LASL Los Alamos Scientific Laboratory [New Mexico, US]

LASL Los Angeles Shock Laboratory [of University of California, US]

LASO Large Amplitude Self-Oscillation

LASP Laboratory for Astronomy and Solar Physics [of NASA]

LASP Local Attached Support Processor

LASPAU Latin American Scholarship Program of American Universities [US]

LASR Laboratories for Astrophysics and Space Research [US]

LASR Litton Airborne Search Radar

LASR Low Altitude Surveillance Radar

(La,Sr)-212 $(La,Sr)_2CaCu_2O_6$ [A Lanthanum Strontium Calcium Copper Oxide (Superconductor)]

LASRA Leather and Shoe Research Association [New Zealand]

LASRM Low-Altitude Short-Range Missile

LASRM Low-Altitude Supersonic Research Missile

LASS Laser Processing of Superconductors in Space

LASS Local-Area Signaling Services

LASS Logistics Analysis Simulation System

LASS Low-Angle Silicon Sheet

LAS-SiC_f Silicon Carbide Fiber Reinforced Lithium Aluminosilicate

LASSO Laser Search and Secure Observer

LASSO Laser Synchronization from Stationary Orbit

LASSO Landing and Approach System, Spiral-Oriented

LASSOS Library Automation System and Services Options Study [UK]

LASSP Laboratory of Atomic and Solid-State Physics [of Cornell University, Ithaca, New York, US]

LASST Laboratory for Surface Science and Technology [of University of Maine, Orono, US]

LAST Large Aperture Scanning Telescope

LAST Light Applique System Technique (Armor)

LAST Lithia–Alumina–Silica–Titania (Gel) [$Li_2O–Al_2O_3–SiO_2–TiO_2$]

LAST Local-Area Storage Transport (Protocol)

LAST Lowest Anticipated Service Temperature

LASTO Linear-Augmented Slater-Type Orbital (Calculation) [Physics]

LASWMM London Association of Scale and Weighing Machine Manufacturers [UK]

LASV Low-Altitude Supersonic Vehicle

LAT Laboratoire d'Aérothermique (du Centre de la Recherche Scientifique) [Aerothermic Laboratory (of the Scientific Research Center), France]

LAT Laboratorium für Analysen und Trenntechnik GmbH [Laboratory for Analysis and Separation Technology Company, Germany]

LAT LADD (Lens Antenna Deployment Demonstration) Test

LAT Large Angle Torque

LAT Laser Acquisition and Tracking

LAT Licensed Aircraft Technologist

LAT Local Access Terminal

LAT Local Apparent Time [Astronomy]

LAT Local Area Transport

LAT Lot Acceptance Test

Lat Latch [also lat]

Lat Latin

Lat Latitude [also lat]

lat lateral

LATA Local Access and Transport Area [Telecommunications]

LATAF Logistics Activation Task Force

Lat Am Latin America(n) [also Lat Amer]

LATAR Laser Augmented Target Acquisition and Recognition

LATB Lithium Aluminum Tri-tert-Butoxyhydride

LATCA Los Angeles Traffic Control Area [US]

LATCC London Air-Traffic Control Center [UK]

LATEOR Laser Texturing Experiment on Rocket

La(THD)$_3$ Lanthanum Tris(2,2,6,6-Tetramethyl-3,5-Heptanedionate) [also La(thd)$_3$]

Lat Ht Latent Heat [also lat ht]

LATID Large Angle Tilt Implanted Drain (Process) [Electronics]

LATINCON Latin American Convention [of Institute of Electrical and Electronics Engineers, US]

LATIRN Low Altitude Navigation Targeting Infrared for Night Flying

LATIS Loop Activity Tracking Information System

LATM Los Angeles Testing Machine

La(TMHD)$_3$ Lanthanum Tris(2,2,6,6-Tetramethyl-3,5-Heptanedionate) [also La(tmhd)$_3$]

L atm/K·mol liter atmosphere(s) per kelvin mole [also L atm K^{-1} mol^{-1}]

LATP Lithium Aluminum Titanium Orthophosphate [Li$_{1-x}$Al$_x$Ti$_{2-x}$(PO$_4$)$_3$]

LATS LDEF (Long-Duration Exposure Facility) Assembly and Transportation System [of NASA]

LATS Litton Automated Test Set

LATS Long-Acting Thyroid Simulator [Biochemistry]

Lattiss Dorsi Lattissimus Dorsi [also lattiss dorsi] [Anatomy]

Latv Latvia(n)

Latv PSR Zinat Akad Vestis Latvijas PSR Zinatnu Akademijas Vestis [Publication of the Latvian Academy of Sciences]

Latv PSR Zinat Akad Vestis, Fiz Teh Zinat Ser Latvijas PSR Zinatnu Akademijas Vestis, Fizikas un Tehnisko Zinatnu Serija [Publication of the Latvian Academy of Sciences, Physical Sciences Series]

Latv SSR Latvian Soviet Socialist Republic [USSR]

LAU Laurate (Crystal)

LAU Line Access Unit

LAU Line Adapter Unit

LAU Lobe Access Unit

LAUA Lloyd's Aviation Underwriters Association [UK]

LAUNS Local-Area Underwater Navigation System

laur *(laurea)* – Italian Degree equivalent to US Master's Degree

LAV Landmaschinen– und Ackerschlepper-Vereinigung [Association for Agricultural Machines and Tractors; of Verband Deutscher Maschinen- und Anlagenbau, Germany]

LAV Light-Armor(ed) Vehicle

LAV Linea Aérea Venezolana [Venezuelan Air Lines]

LAV Lymphadenopathy-Associated Virus

LAVA Laser Vaporization

LAVA Linear Amplifier for Various Applications

LAVC Local-Area VAX (Virtual Address Extension) Cluster

LAVE L'Association Volcanologique Européenne [European Volcanological Association, France]

LAW Laser-Assisted Arc Welding

LAW Leading Aircraftwoman; Leading Air Woman

LAW Light Antiarmor Weapon

LAW Light Antitank Weapon

LAW Logic Analysis Workstation

LAW Low-Activity (Nuclear) Waste

Law Lawyer [also law]

LAWB Los Alamos Water Boiler [New Mexico, US]

LAWG Latin American Working Group

LAWN Local-Area Wireless Network

LAWRS Limited Air Weather Reporting Certificate

LAWS Laser Atmospheric Wind Sounder

Law/Technol Law/Technology [World Association of Lawyers, of the World Peace Through Law Center, Washington, DC, US]

LAXRR Low-Angle X-Ray Reflectometry

LAXS Large-Angle X-Ray Scattering

LAXS Large-Angle X-Ray Spectroscopy

LB Labrador [Canada]

LB Lambert-Beer (Law) [Analytical Chemistry]

LB Lamellar Body

LB Landauer-Büttiker (Formalism) [Physics]

LB Langmuir-Blodgett (Film) [also L-B] [Physical Chemistry]

LB Laser Beam

LB Laser Brazing

LB Lateral Bud [Botany]

LB Launch Boost

LB Launch Bus

LB Lebanon [ISO Code]

LB Lecture Bottle

LB Left Boundary

LB Left Button

LB Lewis Base

LB Liebermann-Burchard (Reaction) [Chemistry]

LB Light Beacon

LB Light Bomb

LB Light Bombardment

LB Light Button

LB Line Breadth

LB Line Broadening [Spectroscopy]

LB Line Buffer [Computers]

LB Lineweaver-Burk (Plot) [Physics]

LB Linoleum Base

LB Load Bank

LB Load-Bearing

LB Local Battery

LB Localizer Beam [Aeronautics]

LB Locating Button

LB Logical Block

LB Long Barrel

LB Long Beach [California, US]

LB Loosely Bound

LB Lorentz-Berthelot (Rules) [Physics]

LB Low Bay

LB Lower Bound [Mathematics]

LB Lower Brace

LB Lowry-Brønsted (Acids and Bases) [Chemistry]

LB Lueders Bands [Metallurgy]

L-B Langmuir-Blodgett (Film) [also LB] [Physical Chemistry]

Lb Label [also lb]

Lβ L-Beta (Radiation) [also L$_\beta$]

lb *(libra)* – pound [Unit]

lb$_f$ pound-force [Unit]

.lb Lebanon [Country Code/Domain Name]

LBA Lactobionic Acid [Biochemistry]

LBA Large-scale Land Processes in the Basin of the Amazon

LBA Laser Beam Analysis; Laser Beam Analyzer

LBA Lima Bean (*Paseolus limensis*) Agglutinin [Biochemistry]

LBA Linear Bounded Automaton

LBA Linear Buckling Analysis

LBA Limestone-Building Algae

LBA Logical Block Addressing [Computers]

LBA Local Bus Adapter

LBA Louis Berger and Associates [US]

LBA Ludwig Boltzmann Association [of Society for the Promotion of Scientific Research, Austria]

LBA Luftfahrt-Bundesamt [Federal Office for Aeronautics, Germany]

LBA$_4$ Lima Bean (*Paseolus limensis*) Agglutinin, Component III [Biochemistry]

LBA$_8$ Lima Bean (*Paseolus limensis*) Agglutinin, Component II [Biochemistry]

lb/A pound(s) per acre [Unit]

Lb/ac pounds per acre [Unit]

LBaF Lithium Barium Fluoride

lb/Ah pound(s) per ampere-hour [also lb/A·h]

lb AI pound(s) of active ingredient [also lb ai]

lb AI/A pound(s) of active ingedient per acre [also lb ai/A]

lb AI/animal pound(s) of active ingredient per animal [also lb ai/animal]

lb AI/plant pound(s) of active ingredient per plant [also lb ai/plant]

L-band Microwave Frequency Band of 0.39 to 1.55 gigahertz [also L-Band]

lb ap pound, apothecaries' [also lb apoth]

L-BAPA Nα-Benzoyl-L-Arginine-p-Nitroanilide [also L-Bapa, or L-bapa] [Biochemistry]

LBA-SPSR Ludwig Boltzmann Association–Society for the Promotion of Scientific Research [Austria]

lb av pound, avoirdupois [Unit]

LBB Laboratory for Bioanalysis and Biotechnology [of Washington State University, Pullman, US]

lb/1000 bbl pounds per thousand barrels (of oil)

lb bhp^{-1} hr^{-1} pounds per brake horsepower-hour [also lb/BHPhr]

LBBP Laboratory of Blood and Blood Products [US]

LBBS Logitech Bulletin Board Service [of Logitech Inc., US]

lb/bu pounds per bushel [Unit]

LBCO Lanthanum Barium Copper Oxide (Superconductor) [also LaBCO]

LBCCO Lanthanum Barium Calcium Copper Oxide (Superconductor) [also LaBCCO]

LBCSO Lanthanum Barium Copper Tin Oxide (Superconductor) [also LaBCSO]

LBCTO Lanthanum Barium Copper Titanium Oxide (Superconductor) [also LaBCTO]

LBC Laser-Beam Cutting

LBC Linear Block Code

LBC Local Bus Controller

LBC Lower Binding-Energy Component [Physics]

LBCC Low Brightness Calcined Clay

lbCHU Pound Centigrade Heat Unit [also lb CHU, or lb-CHU]

LBCM Locator at the Back Course Marker

LBCNO Lanthanum Barium Copper Niobium Oxide (Superconductor)

LBCO Lanthanum-Barium-Copper Oxide (Superconductor)

LBCTO Lanthanum Barium Copper Titanium Oxide (Superconductor)

lb/cu ft pounds per cubic foot [also lb/ft³, or lb per cu ft]

lb/cu in pounds per cubic inch [also lb/in³, or lb per cu in]

lb/cu yd pounds per cubic yard [also lb/yd³, or lb per cu yd]

LBD Libyan Dinar [Currency]

LBDT Low Bay Dolly Tug

LBE Lance-Bubbling-Equilibrium (Process) [Metallurgy]

LBE Lean Burn Engine

LBE Low Binding Energy [Physics]

LBE Low Bombardment Energy [also lbe] [Plasma-Assisted Vapor Deposition]

LBeF Lithium Beryllium Fluoride

LBEN Low Byte Enable

LBF Lactobacillus Bulgaricus Growth Factor [Biochemistry]

LBF Langmuir-Blodgett Film [Physical Chemistry]

LBF Lithium Barium Fluoride

lbf pound-force [also lb$_f$]

lbf-ft pound-force foot [also lbf ft, lbf·ft, lb-ft, or lb ft]

lbf/ft pound-force per foot [Unit]

lbf/ft² pound(-force) per square foot [also PSF, or psf]

lbf-ft/in pound-force foot per inch [also lbf ft/in, or lbf·ft/in]

lbf/in pound-force per inch [Unit]

lbf/in² pound(-force) per square inch [also psi or PSI]

lbf-in pound-force inch [also lbf in, lbf·in, lb-in, or lb in]

lbf/in² abs pound(-force) per square inch absolute [also PSIA, or psia]

lbf-in/in pound-force inch per inch [also lbf in/in or lbf·in/in]

lbf/lb pound-force per pound [Unit]

lbf-s pound-force second [also lbf·sec, or lbf·s]

lbf-sec pound-force second [also lbf·sec, or lbf-s, lbf·s]

lbf-sec/ft² pound-force second per square foot [also lbf·sec/ft², lbf·s/ft², lbf·s/sq ft, or lbf·sec/sq ft]

lbf-sec/in² pound-force second per square inch [also lbf·sec/in², lbf·s/in², lbf·s/sq in, or lbf·sec/sq in]

lbf-sec/sq in pound-force second per square inch [also lbf·sec/sq in, lbf·s/sq in, lbf·s/in², or lbf·sec/in²]

lbf-s/ft² pound-force second per square foot [also lbf·sec/ft², lbf·s/sq ft, or lbf·sec/sq ft]

lbf-s/in² pound-force second per square inch [also lbf·s/in², lbf·sec/in², lbf·s/sq in, or lbf·sec/sq in]

lbf-s/sq in pound-force second per square inch [also lbf·s/sq in, lbf·sec/sq in, lbf·s/in², or lbf·sec/in²]

lb/ft pound(s) per foot

lb$_f$/ft pound-force per foot

lb/ft² pound(s) per square foot [also lb/sq ft, or lb per sq ft]

lb/1000 ft² pound(s) per 1000 square feet [Unit]

lb$_f$/ft² pound(-force) per square foot [also PSF, or psf]

lb/ft³ pound(s) per cubic foot [also lb/cu ft, or lb per cu ft]

lb-ft pound-foot [also lb ft, or lb·ft]

lb$_f$-ft pound-force foot [also lb$_f$ ft, lb$_f$·ft, lb-ft, or lb ft]

lb-ft² pound-foot squared [also lb·ft²]

lb/ft-h pound(s) per foot hour [also lb/ft·h, or lb/ft/h]

lb/ft-hr pound(s) per foot hour [also lb/ft·hr, or lb/ft/hr]

lb$_f$-ft/in pound-force foot per inch [also lb$_f$ ft/in, or lb$_f$·ft/in]

lb-ft/s pound-foot per second [also lb ft/s, lb·ft/s, lbft/sec, or lb ft/sec]

lb/ft²·s² pound(s) per square foot second squared [also lb/ft²/s², lb/ft²·sec², or lb/ft²/sec²]

lb/ft·sec pound(s) per foot second [also lb/ft/sec]

lb/ft²·sec² pound(s) per square foot second squared [also lb/ft²/sec², lb/ft²·s², or lb/ft²/s²]

LBG Load Balancing Group

lb/gal pound(s) per gallon [Unit]

LBH Laser-Beam Hardening

L-BH Laxmanan-Corrected Burden and Hunt (Microsegregation Model) [Metallurgy]

lb/h pound(s) per hour [also lb/hr]

lb H$_2$O pound(s) of water [Unit]

lb/100 gal pound(s) per hundred gallons [Unit]

lb/h·kW pound(s) per hour kilowatt [also lb/h/kW, or lb h^{-1} kW^{-1}]

lb/HP pound(s) per horsepower [also lb/hp, PHP, or php]

lb/HP·h pound(s) per horsepower-hour [also lb/hp·h]

lb/HP·hr pound(s) per horsepower-hour [also lb/hp·hr]

lb/hr pound(s) per hour [also lb/h]

LBIC Light Beam Induced Current (Technique)

LBIDI Liberian Bank for Industrial Development and Investment

LBIM Logic Based Information Modelling

lb/in pounds per inch [also lb per in]

lb$_f$/in pound-force per inch [Unit]

lb/in² pounds per square inch [also lb per sq in, lb/sq in, PSI, or psi]

lb$_f$/in² pound(-force) per square inch [also psi or PSI]

lb/in³ pound(s) per cubic inch [also lb per cu in, or lb/cu in]

lb-in pound-inch [also lb in]

lb$_f$-in pound-force inch [also lb$_f$ in, lb$_f$·in, lb-in, or lb in]

lb-in² pound-inch squared [also lb·in²]

lb$_f$/in² abs pound(-force) per square inch absolute [also PSIA, or psia]

lb$_f$-in/in pound-force inch per inch [also lb$_f$ in/in, or lb$_f$·in/in]

lb-in/sec pound-inch per second [also lb in/sec, lb in/s, or lb in/s]

lb-in-sec/rad pound-inch-second per radian [also lb-in-s/rad]

LBIS Longitudinal Biased Initial Susceptibility

LBK Langmuir-Blodgett-Kuhn (Film) [Physical Chemistry]

LBK Left Bank [Geography]

lb/ksi pound(s) per kilopound per square inch [Unit]

LBL Label [Computers]

LBL Laminar Boundary Layer [Fluid Mechanics]

LBL Langmuir-Blodgett Layers [Physical Chemistry]

LBL Lawrence Berkeley Laboratory [of US Department of Energy at the University of California at Berkeley]

LBL Left Buttock Line

LBL Lima (*Paseolus limensis*) Bean Lectin [Biochemistry]

LbL Layer-by-Layer (Chemical Vapor Deposition Technique)

Lbl Label [also lbl]

lb$_f$/lb pound-force per pound [Unit]

LBLG Lawrence Berkeley Laboratory Group [of University of California at Berkeley, US]

LBLOCA Large Break Loss-of-Coolant Accident [Nuclear Reactors]

Lblty Liability [also lblty]

LBM Langmuir-Blodgett Multilayers [Physical Chemistry]

LBM Lanthanum Barium Manganate

LBM Langer, Bar-on and Miller (Coarse Microstructure Model) [Metallurgy]

LBM Laser-Beam Machined; Laser-Beam Machining

LBM Lattice Block Material(s) [Materials Science]

LBM Lever of Breech Mechanism

LBM Linear Bubble (Grain-Growth) Model [Metallurgy]

LBM Load Buffer Memory

LBM Locator Back Marker

lbm pound-mass [also lb$_m$]

LBMA Lumber and Building Materials Association [US]

LBMAC Lumber and Building Materials Association of Canada

LBMAO Lumber and Building Materials Association of Ontario [Canada]

LBMC Land Based Marine Pollution

lb/min pound(s) per minute [Unit]

lb/mol pound(s) per molecule [Unit]

lbm/sec·ft pound, mass per second per foot [also lbm/sec/ft, lbm/s·ft, or lbm/s/ft)

LBN Line Balancing Network

LBN Logic Bucket Number

lbn pounds, net [Unit]

LBNL Lawrence Berkeley National Laboratory [of University of California at Berkeley, US]

LBNO Long Baseline Neutrino Oscillation

LBNO Newsl Long Baseline Neutrino Oscillation Newsletter [of Argonne National Laboratory, US]

LBNP Lower Body Negative Pressure (Boots)

LBNPD Lower Body Negative Pressure Device

lb/NT pound(s) per net ton [Unit]

LBO Leveraged Buyout

LBO Light-Beam Oscillograph

LBO Line Buildout (Network)

LBO Lithium-Barium Oxide

LBO Lithium-Boron Oxide

LBP Laser-Beam Printer; Laser-Beam Printing

LBP Lebanese Pound [Currency]

LBP Length between Perpendiculars [Naval Architecture]

LBP Low Blood Pressure

LBPA 2,6-Lutidinyl-Bis(2-Picolinyl)amine [also lbpa]

lb per bhp hr pound(s) per brake horsepower-hour [also lb bhp^{-1} hr^{-1}, or lb/BHPhr]

lb per cu ft pound(s) per cubic foot [also lb/cu ft, or lb/ft^3]

lb per cu in pound(s) per cubic inch [also lb/in^3, or lb/cu in]

lb per cu yd pound(s) per cubic yard [also lb/cu yd, or lb/yd^3]

lb per ft pound(s) per foot [also lb/ft]

lb per in pound(s) per in [also lb/in]

lb per sq ft pound(s) per square foot [also lb/sq ft, or lb/ft^2]

lb per sq in pound(s) per square inch [also lb/sq in, lb/in^2, PSI, or psi]

lb per yd pound(s) per yard [also lb/yd]

LBR Librarian [Computers]

LBR Laser Beam Recorder; Laser Beam Recording

LBR Low Bit Rate

.LBR Library [File Name Extension] [also .lbr]

Lbr Librarian; Library [also Lbr, or lbr]

Lbr Lumber [also lbr]

LBRI Lake Biwa Research Institute

LBRM Large Basin Runoff Model [Hydrology]

LBRV Low Bit-Rate Voice

LBS Laser-Beam Sorting

LBS Load Balance System

LBS Lysine-Binding Site [Biochemistry]

LBS I Lysine-Binding Site I [Biochemistry]

LBS II Lysine-Binding Site II [Biochemistry]

lbs pounds

lb/s pound(s) per second [also lb/sec]

lb_f-s pound-force second [also $lb_f \cdot$ sec, or lbf·s]

LBSC Long Beach State College [California, US]

lb/SCF pound(s) per standard cubic foot [Unit]

lb/SCF$_{dry}$ pound(s) per standard cubic foot, dry [Unit]

lb/SCF$_{wet}$ pound(s) per standard cubic foot, wet [Unit]

lb/sec pound(s) per second [also lb/s]

lb_f-sec pound-force second [also $lb_f \cdot$ sec, or lb_f-s, lb_fs]

lb_m/sec·ft pound, mass per second per foot [also lb_m/sec/ft, lb_m/s·ft, or lb_m/s/ft]

lb-sec/ft² pound-second per square foot [also lbs/ft², or lb·s/ft²]

lb_f-sec/ft² pound-force second per square foot [also $lb_f \cdot$ sec/ft², $lb_f \cdot$ s/ft², $lb_f \cdot$ s/sq ft, or lb_f sec/sq ft]

lb-sec²/ft pound-second squared per foot [also lb-s²/ft, or lb·s²/ft]

lb-sec²/ft⁴ pound-second squared per foot to the fourth power [lb·sec²/ft⁴, or lb s²/ft⁴]

lb-sec/in pound-second(s) per inch [lb·sec/in, or lbs/in]

lb_f-sec/in² pound-force second per square inch [also $lb_f \cdot$ sec/in², $lb_f \cdot$ s/in², $lb_f \cdot$ s/sq in, or $lb_f \cdot$ sec/sq in]

lb-sec²/in pound-second squared per inch [lb·sec²/in, or lb s²/in]

lb-sec²/in⁴ pound-second squared per inch to the fourth power [lb·sec²/in⁴, or lb s²/in⁴]

lb_f-sec/sq in pound-force second per square inch [also $lb_f \cdot$ sec/sq in, lb_fs/sq in, lb_fs/in², or lb_f sec/in²]

lb_m/s·ft pound, mass per second foot [also lb_m/s/ft, lb_m/sec·ft, or lb_m/sec/ft]

lb-s/ft² pound-second per square foot [also lb·s/ft², or lb sec/ft²]

lb_f-s/ft² pound-force second per square foot [also $lb_f \cdot$ sec/ft², $lb_f \cdot$ s/sq ft, or $lb_f \cdot$ sec/sq ft]

lb-s²/ft pound-second squared per foot [also lb·s²/ft, or lb·sec²/ft]

lb-s²/ft⁴ pound-second squared per foot to the fourth power [also lb-sec²/ft⁴, or lb sec²/ft⁴]

lbs/100 gal pounds per hundred gallons [Unit]

lbs/hr pounds per hour [also lbs hr⁻¹]

lb-s/in pound-second(s) per inch [lb·sec/in, or lb-sec/in]

lb_f-s/in² pound-force second per square inch [also $lb_f \cdot$ s/in², $lb_f \cdot$ sec/in², $lb_f \cdot$ s/sq in, or $lb_f \cdot$ sec/sq in]

lb-s²/in pound-second squared per inch [lb·sec²/in, or lb-sec²/in]

lb-s²/in⁴ pound-second squared per inch to the fourth power [lb·sec²/in⁴, or lb-sec²/in⁴]

lb/sq ft pound(s) per square foot [also lb/ft², or lb per sq ft]

lb/sq in pound(s) per square inch [also lb/in², PSI, or psi]

lb_f-s/sq in pound-force second per square inch [also $lb_f \cdot$ s/sq in, $lb_f \cdot$ sec/sq in, $lb_f \cdot$ s/in², or $lb_f \cdot$ sec/in²]

LBT Langmuir-Blodgett Theory [Physical Chemistry]

LBT Listen Before Talk(ing)

LBT Local-Battery Talking

LBT Low Bit Test

lb t pound, troy [Unit]

lb/t pound(s) per ton [Unit]

LBT CBS Local-Battery Talking, Common-Battery Signaling

LBTM Layer Bispectral Threshold Method

lb/ton pound(s) per ton [Unit]

lb tr pound, troy [Unit]

LBTS Land-Based Test Site

lb/UK gal pound(s) per United Kingdom gallon [Unit]

lb/US gal pound(s) per United States gallon [Unit]

LBV Luminous Blue Variables (Meeting) [Astronomy]

LBW Laser-Beam Welding

LBWA Library Board of Western Australia

LBWR LaCrosse Boiling Water Reactor [Wisconsin, US]

lb-wt pound-weight [also lb wt]

lb-wt/ft² pound-weight per square foot [also lb wt/ft², or lb-wt/sq ft]

lb-wt/in² pound-weight per square inch [also lb wt/in², or lb-wt/sq in]

lb-wt/sec pound-weight per second [also lb wt/sec, lb-wt/s, or lb wt/s]

lb-wt/sq ft pound-weight per square foot [also lb wt/sq ft, or lb-wt/ft²]

lb-wt/sq in pound-weight per square inch [also lb wt/sq in, or lb-wt/in²]

LBX Local Bus Extension

lb/y pound(s) per year [also lb/yr]

lb/yd pounds per yard [also lb per yd]

lb/yd³ pounds per cubic yard [also lb per cu yd, or lb/cu yd]

lb/yr pound(s) per year [also lb/y]

LBZ Local Brittle Zone

LC Inductance–Capacitance; Inductor–Capacitor [also L-C]

LC Laboratoire Céramique [Ceramics Laboratory; of Ecole Polytechnique Fédérale de Lausanne, Switzerland]

LC Labour Canada

LC Ladle Car [Metallurgy]

LC Lafayette College [Easton, Pennsylvania, US]

LC Lake Champlain [Vermont/New York State, US]

LC Lamb Committee [of National Livestock and Meat Board, US]

LC Laminate Composite

LC Land Cover

LC Landing Circle

LC Landing Craft

LC Landscape Characterization [Part of the US Environmental Protection Agency's Environmental Monitoring and Assessment Program]

LC Langford-Cohen (Subgrain Strengthening) [Materials Science]

LC Lanthanide Contraction

LC Lanthanum Chromite

LC Lanthanum Cobaltate

LC Lanthanum Cuprate

LC Laser Cavity

LC Laser Chemistry

LC Laser Curvature (Technique)

LC Laser Cutting

LC Last (Punched) Card

LC Late Commitment

LC Lattice Constant [Crystallography]

LC Lattice Correspondence [Crystallography]

LC Launch Complex

LC Launch Control

LC Launch Countdown

LC Launch Critical

LC Lead Coating

LC Lead-Covered (Cable)

LC Leakage Coefficient (of Electric Machinery)

LC Learning Curve

LC Leather Chemist(ry)

LC Left Center

LC Lehmann College [of City University of New York, US]

LC Lens Cleaner

LC Lens Coating

LC Leptocyte [Medicine]

LC Lethal Concentration

LC Letter of Credit [also L/C, or l/c]

LC Level Control

LC Liaison Committee

LC Library of Congress [Washington, DC, US]

LC Life Cycle

LC Light Case (Chemical)

LC Lignocellulose; Lignocellulosic(s)

LC Limited Coordination

LC Lincoln College [Christchurch, New Zealand]

LC Lindlar Catalyst [Chemistry]

LC Linear Circuit

LC Linear Collider [Nuclear Engineering]

LC Linear Control

LC Linear Convergence

LC Line Carrying

LC Line Circuit

LC Line Concentrator [Telecommunications]

LC Line Conditioner

LC Line Connector

LC Line Construction (Tool)

LC Line Control(ler)

LC Line Current

LC Line of Communication

LC Line of Contact

LC Link Circuit

LC Lipschitz Condition [Mathematics]

LC Liquid Chromatography

LC Liquid Coated; Liquid Coating

LC Liquid Concentrate

LC Liquid-Condensed (Phase)

LC Liquid Controller

LC Liquid Crystal(line)

LC Lithium Cobaltate

LC Lithium Chlorate

LC Lithium Cuprate

LC Location Counter

LC Load Carrier

LC Load Cell

LC Load Center

LC Load Circuit

LC Load Compensation

LC Load Curve

LC Loading Coil

LC Local Controller

LC Localized Corrosion

LC Locational Clearance Fit [ANSI Symbol]

LC Locomotive Club [UK]

LC Logical Clock

LC Logic Circuit

LC Logic Clip

LC Logic Corporation

LC Logistics Command

LC Lomer-Cottrell (Dislocation) [Materials Science]

LC Long Case

LC Long-Chain (Molecule)

LC Loschmidt's Constant

LC Low-Melting Component

LC Low Carbon

LC Low Coercivity

LC Low Concentration

LC Low Cost

LC Lower Case [also lc]

LC Lower Cassette

LC Luggin Capillary [Chemistry]

LC Lummus-Crest (Process)

LC Lunar Calendar

LC Luxury Car

LC Lyman Continuum [Spectroscopy]

LC Lymphocyte

LC Saint Lucia [ISO Code]

LC_{50} Median Lethal Concentration (or Lethal Concentration 50) [also LC50]

LC_{Lo} Lethal Concentration Low [also LC_{Lo}]

LC^3 Low-Cost Ceramic Composite (Program)

%LC Percent Linear Change

LC1 Locational Clearance Fit (H6/h5) [ANSI Symbol]

LC2 Locational Clearance Fit (H7/h6) [ANSI Symbol]

LC3 Locational Clearance Fit (H8/h7) [ANSI Symbol]

LC4 Locational Clearance Fit (H10/h9) [ANSI Symbol]

LC5 Locational Clearance Fit (H7/g6) [ANSI Symbol]

LC6 Locational Clearance Fit (H9/f8) [ANSI Symbol]

LC7 Locational Clearance Fit (H10/e9) [ANSI Symbol]

LC8 Locational Clearance Fit (H10/d9) [ANSI Symbol]

LC9 Locational Clearance Fit (H11/c10) [ANSI Symbol]

LC10 Locational Clearance Fit (H12/...) [ANSI Symbol]

LC11 Locational Clearance Fit (H13/...) [ANSI Symbol]

LC-1 Liquid Chromatography with Reversed-Phase Trimethylsilyl Column

LC-8 Liquid Chromatography with Reversed-Phase Octyldimethylsilyl Column

LC-8-DB Liquid Chromatography with Deactivated Reversed-Phase Octyldimethylsilyl Column

LC-18 Liquid Chromatography with Reversed-Phase Octadecyldimethylsilyl Column

LC-18-DB Liquid Chromatography with Deactivated Reversed-Phase Octadecyldimethylsilyl Column

LC-18-T Liquid Chromatography with Octadecylsilyl Column for Nucleotides

LC-318 Liquid Chromatography with Reversed-Phase, 300Å Pore-Size Octadecyldimethylsilyl Column

L&C Laboratory and Checkout

L/C Letter of Credit [also l/c]

L-C Inductance–Capacitance; Inductor–Capacitor [also LC]

lc (loco citato) – in the place cited

lc lower case

.lc Saint Lucia [Country Code/Domain Name]

LCA Lake City Arsenal [Independence, Missouri, US]

LCA Landing Craft Assault (Ship)

LCA Launch Control Amplifier

LCA Lead Contractors Association [UK]

LCA Leading Concept Ammonia (Process)

LCA Left Coronary Artery [Anatomy]

LCA Leucocyte Common Antigen [Immunology]

LCA Life Cycle Analysis

LCA Life Cycle Assessment (Method)

LCA Lightweight Cycle Association [UK]

LCA Line Control Adapter

LCA Liverpool Cotton Association [UK]

LCA Load Control(ler) Assembly

LCA Local Communications Adapter

LCA Lotus Communications Architecture

LCA Lower Case Alphabet

LCAC Landing Craft Air Cushion

LCAHRS Low-Cost Attitude and Heading Reference System [also LCA HRS]

LCAO Linear Combination of Atomic Orbitals [Physical Chemistry]

LCAO-MO Linear Combination of Atomic Orbitals–Molecular Orbitals [Physical Chemistry]

LCAofGB Lightweight Cycle Association of Great Britain

LCAP Loop Carrier Analysis Program

LCAP Loosely Coupled Array of Processors

LCAPAS Linear Combination of Antisymmetrized Products of Atomic Substrates

LCAS Land Capability Assessment Strategy

LCAS Linear Combination of Atomic Substrates

LCAS Lithia-Calcia-Alumina-Silica [Li_2O-$CaOAl_2O_3$-SiO_2]

LCAST Lithia-Calcia-Alumina-Silica-Titania [Li_2OCaO-Al_2O_3-SiO_2-TiO_2]

LCB Launch Control Building

LCB Least Common Bit

LCB LHCC (Large Hadron Collider Committee) Computing Board [of CERN–European Laboratory for Particle Physics, Geneva, Switzerland]

LCB Light Case (Chemical) Bomb

LCB Line Control Block

LCB Logic Control Block

LCB London Center for Biotechnology [UK]

LCB Low-Cost Beta (Titanium Alloy)

LCBC Lake Chad Basin Commission [Africa]

LCBF Low-Carbon Bainite Ferrite [Metallurgy]

LCBO Linear Combinations of Bond Orbitals [Physics]

LCBOD Liquid Crystal Bistable Optical Device

LCBP Lake Champlain Basin Program [US Environmental Program]

LCC Land Care Committee

LCC Land Conservation Council [of Victoria State, Australia]

LCC Landing Craft, Control [US Navy Symbol]

LCC Late Choice Call

LCC Launch Commit(ment) Criteria

LCC Launch Complex Center

LCC Launch Control Center

LCC Launch Control Center [at NASA Kennedy Space Center, Florida, US]

LCC Launch Control Console

LCC Lead-Coated Copper

LCC Lead-Covered Cable

LCC Leadless Chip Carrier

LCC Least-Cost (Satisfactory Melting) Charge [Metallurgy]

LCC Lewis and Clark College [Portland, Oregon, US]

LCC Life-Cycle Cost(ing)

LCC Line Choking Coil

LCC Liquid Crystal Cell

LCC Load-Carry Capability [Aerospace]

LCC Loading Coil Case

LCC Local Communications Complex

LCC Local Communications Console

LCC Logistics Coordination Center

LCC London County Council [UK]

LCC Lost Calls Cleared

LCC Louisville Chamber of Commerce [Kentucky, US]

LCCA Lead Center for Communications Architecture [of NASA]

LCCC Leadless Ceramic Chip Carrier

LCCC Library of Congress Computer Catalogue [US]

LCCC Locular Countercurrent Chromatography

LCCCN Library of Congress Catalogue Card Number [US]

LCCD Launch Commit Criteria Document

LCCD Linearized Coupled Cluster Doubles

LCCD Low Complexity Color Display

LCCI London Chamber of Commerce and Industry [UK]

LCCM Late Choice Call Meter

LCCMARC Library of Congress Current Machine-Readable Catalog (File) [US]

LC-CN Liquid-Chromatography with Bonded Normal-Phase Cyanopropyldimethylsilyl Column

LCCNO Lanthanum Calcium Copper Niobium Oxide (Superconductor)

LCCO Lanthanum Calcium Copper Oxide (Superconductor)

LCCOGA Liaison Committee of Cooperating Oil and Gas Associations [US]

LCCP Laser-Controlled Chemical Processing

LCCs Land Care Committees

LCCT LaQue Center for Corrosion Technology

LCD Launch Countdown

LCD Letter Carrier Depot

LCD Life Cycle Design

LCD Liquid Crystal Device

LCD Liquid Crystal Diode (Display)

LCD Liquid Crystal Display

LCD Low Crack Density

lcd least common denominator [also LCD]

lcd lowest common denominator [also LCD]

LC-DABS Liquid Chromatography with [4-{(4-Dimethylaminophenyl)azo}benzene]sulfonyl (Column)

LCDC Laboratory Centre for Disease Control [of Health Canada]

LCDDS Leased Circuit Digital Data Service [UK]

LCDE Liquid Crystal Display Element

LC-Diol Liquid Chromatography with Bonded Normal-Phase Glycerylpropylsilyl Column

LC-DP Liquid Chromatography with Reversed-Phase Diphenylmethylsilyl Column

LCdr Lieutenant Commander [also LCDR, or Lcomdr]

LCDS Low-Cost Development System [US]

LCDTL Load-Compensated Diode Transistor Logic

LCDTL Low-Current Diode Transistor Logic

LCE Land-Covered Earth

LCE Launch Complex Equipment

LCE Linked-Cluster Expansion

LCE Lipiodol Chemoembolization [Cancer Treatment]

LCE Liquid Crystal Elastomer

LCE Load Circuit Efficiency

LCE Local Chemical Equilibrium

LCE London Commodity Exchange [UK]

LCE Lyapunov Characteristic Exponent [Physics]

LCEC Liquid Chromatography Electrochemistry

LCEC Liquid Chromatography with Electrochemical Detection

LCEECSTI Liaison Committee of the European Economic Community Steel Tube Industry [France]

LCEF Local Concentration Enrichment Factor

LCES Least Cost Estimating and Scheduling

LCF Language Central Facility

LCF Launch Control Facility

LCF Light Control Film

LCF Lithium Calcium Fluoride

LCF Loading–Cooling under Load–Fracture at Lower Temperature (Test)

LCF Logical Channel Fill

LCF Low-Cost Fiber

LCF Low-Cycle Fatigue [Mechanics]

LCFC Linear Combination of Fragment Configuration

LCFS Last Come, First Served

LCFS Low-Cycle Fatigue Strength [Mechanics]

LCFT Lower Critical Flocculation Temperature

LCFTU Lesotho Congress of Free Trade Unions

LCG Laboratory for Computer Graphics [of Harvard University, Cambridge, Massachusetts, US]

LCG Landing Craft, Gun

LCG Large-Curvature Ground-State (Approximation) [Physics]

LCG Linear Congruential Generator

LCG Liquid-Cooled Garment

LCG Liquid Crystal Graph

LCG Load Classification Group

LCGB Locomotive Club of Great Britain

LCG(M) Landing Craft Gun (Medium)

LCGO Linear Combination of Gaussian Orbitals [Physics]

LCGO-LSD Linear Combination of Gaussian Orbitals–Local Spin Density (Method) [Physics]

LCGTO Linear Combination of Gaussian Orbitals [Physics]

LCGTO-DF Linear Combination of Gaussian Orbitals Density Functional (Method) [Physics]

LCGTO-LSD Linear Combination of Gaussian Orbitals–Local Spin Density (Method) [Physics]

Lchr Laucher [also lchr]

LCHS Large Component Handling System

LCH Lens Culinaris Agglutinin [also LcH] [Immunology]

LCH Lost Calls Held

Lch Launch [also lch]

Lch Latch [also lch]

LCH-A Lens Culinaris Agglutinin A [also LcH-A] [Immunology]

L Chain Light Chain [also L chain] [Biochemistry]

LCH-B Lens Culinaris Agglutinin B [also LcH-B] [Immunology]

LC-HINT Liquid Chromatography/Hydrophobic Interaction (Column)

LCHOP Linear Combination of Harmonic Oscillator Products [Physics]

LCHTF Low-Cycle High-Temperature Fatigue [Mechanics]

LCI Landing Craft, Infantry [US Navy Symbol]

LCI Life Cycle Inventory

LCI Livestock Conservation Institute [US]

LCI Load-Commutated Inverter

LCI Logical Channel Identifier

LCIC Leverhulme Centre for Innovative Catalysis [of Surface Science Research Center, University of Liverpool, UK]

LC-IC Liquid-Chromatography/Ion Chromatography

LCIE Laboratoire Central des Industries Electriques [Central Laboratory for the Electrical Industries, France]

LCIGB Locomotive and Carriage Institution of Great Britain

LCI(L) Large Landing Craft, Infantry [US Navy Symbol]

LCJ Low-Contaminant Jarosite (Process)

LCK (Microsoft) Library Construction Kit

Lckr Locker [also lckr]

LCL Lanthanum Copper Lithiate

LCL Less than Carload [also lcl]

LCL Less than Carload Lot [also lcl]

LCL Less than Container Load

LCL Library Control Language

LCL Lifting Condensation Level [Meteorology]

LCL Limited Channel Log-Out

LCL Linkage Control Language

LCL Local [also Lcl, or lcl]

LCL Local Climb Locking (Mechanism) [Materials Science]

LCL Low Camera Length [Electron Microscopy]

LCL Low Capacity Link

LCL Lower Catastrophic Limit

LCL Lower Confidence Limit [Statistics]

LCL Lower Control Limit [Statistics]

LCL Lowest Car Load

LCL$_c$ Lower Control Limit for c Control Chart (in Statistics) [Symbol]

LCL$_p$ Lower Control Limit for p Control Chart (in Statistics) [Symbol]

LCL$_R$ Lower Control Limit for R Control Chart (in Statistics) [Symbol]

LCL$_{Rm}$ Lower Control Limit for Moving-Average (R_m) Control Chart (in Statistics) [Symbol]

LCL$_u$ Lower Control Limit for u Control Chart (in Statistics) [Symbol]

LCL Lower Control Limit for Control Chart (in Statistics) [Symbol]

LCl Low Chloride

lcl local

LCL-CBED Low Camera Length–Convergent Beam Electron Diffraction

LCLU Landing Control Logic Unit

LC/LU Land Cover/Land Use [of US Geological Survey]

LCLV Liquid-Crystal Light Valve

LCM Laboratoire de Chimie des Matériaux [Materials Chemistry Laboratory, Rennes, France]

LCM Laboratoire de Chimie Minérale [Minerals Chemistry Laboratory; of Université Claude Bernard, Villeurbanne, France]

LCM Lanthanum Calcium Manganate

LCM Large Core Memory

LCM Late Change Message

LCM Last Calls Meter

LCM Lead-Coated Metal

LCM Leading Concept for Methanol (Process)

LCM Line Concentrator Module

LCM Line Control Module

LCM Liquid Composite Molding

LCM Liquid Contamination Monitor

LCM Liquid Curing Medium

LCM Liquid Curing Method

LCM Live-Cell Microscopy

LCM Long-Chain Molecule

LCM Lost Circulation Material

lcm least common multiple [also LCM]

lcm lowest common multiple [also LCM]

LCMARC Library of Congress Machine-Readable Catalogue (Files) [US]

LCMA Lightweight Cycle Manufacturers Association [UK]

LCMBPT Linked-Cluster Many-Body Perturbation Theory [Physics]

LCMC Laboratoire de Chimie de la Matière Condensée [Condensed Matter Chemistry Laboratory; of Université Pierre et Marie Curie, Paris, France]

LCMCFC Liaison Committee for Mediterranean Citrus Fruit Culture [Spain]

LC-Me Liquid-Chromatography/Methyl

LCMFO Lanthanum Calcium Manganese Iron Oxide (Superconductor)

LMCI Laboratoire des Champs Magnétiques Intenses [Intense Magnetic Field Laboratory, Grenoble, France]

LCMI Laboratoire de Chimie des Matériaux Inorganiques [Inorganic Materials Chemistry Laboratory, Université Paul Sabatier, Toulouse, France]

LCMI Licentiate of the Cost and Management Institute

LCML Low-Level Current Mode Logic

LCMM Life Cycle Management Model

LCMO Lanthanum Cerium Manganate

LCMO Lanthanum Calcium Manganate

LCMO Linear Combination of Molecular Orbitals [Physical Chemistry]

LCMS Library Collection Management System [of University of Toronto Library Automation System, Canada]

LCMS Liquid Chromatography Mass Spectrometry [also LC/MS, or LC-MS]

LCMS Logistics Command Management System

LCMS Low-Cost Modular Spacecraft

LCMTO Linear Combination of Muffin Tin Orbitals [Physics]

LCN Lanthanum Copper Nickelate

LCN Load Classification Number

LCN Local Computer Network

LCN Logical Channel Number

LCn Rotate Left through Carry n Number of Bits (Function)

LC-NH$_2$ Liquid-Chromatography with Aminopropylsilyl Column

LCNMR Liquid-Crystal Nuclear Magnetic (Resonance) (Spectroscopy)

LCNO Lanthanum Calcium Nickel Oxide

LCNO Lanthanum Cerium Nickel Oxide

LCNR Liquid Core Nuclear Rocket

LCNT Link Celestial Navigation Trainer

LCNTR Location Counter

LCO Laboratoire de Compilation [Compiler Laboratory of Swiss Federal Institute of Technology]

LCO Laboratorio di Chimica Organica [Organic Chemistry Laboratory, University for Ferrara, Italy]

LCO Lanthanum Calcium Oxide (Superconductor)

LCO Lanthanum Copper Oxide (Superconductor)

LCO Liquid Chromatographic Oxygenates

LCO Limited Condition of Operability

LCO Light Cycle Oil [Chemistry]

LCO Lipochitooligosaccharide [Biochemistry]

LCO Lithium Cobalt(III) Oxide

LCO Limiting Conditions for Operations

LCO Load Control and Optimization

LCO-214 La$_2$CuO$_4$ [A Lanthanum Copper Oxide Superconductor]

LCol Lieutenant-Colonel

LCOM Liquid Crystalline Optical Material

LCOM Logic Control Output Module

LComdr Lieutenant-Commander [also Lcdr]

LCOS Lanthanum Chromium Oxysulfide

LCP Laboratoire de Chimie des Plasmas [Plasma Chemistry Laboratory; of University of Limoges, France]

LCP Laboratório Associado de Combustão e Propulsão [Associate Laboratory of Combustion and Propulsion; of Brazilian National Institute for Space Research]

LCP Landing Craft, Personnel

LCP Language Conversion Program

LCP Large Coil Program

LCP Laser-Induced Chemical Processing

LCP Launch Control Post

LCP Least Cost Planning

LCP Lechuguilla Cave Project [of US National Park Service at Carlsbad Caverns National Park]

LCP Licentiate of the College of Preceptors [UK]

LCP Link Control Procedure

LCP Link Control Protocol

LCP Liquid Crystal Polymer

LCP Liquid Cyclone Process

LCP Local Control Point

LCP Logic-Controlled Protocol(ling)

LCP Lower Collector Plate

LC-PAH Liquid-Chromatography Column for Polyaromatic Hydrocarbon

LCPC Laboratoire Central des Ponts et Chaussées [Central Laboratory for Roads and Bridges, at Paris and Nantes/Bouguenais, France]

LC-Phe Liquid-Chromatography/Phenyl (Phase)

L-Cpl Lance-Corporal [also L/Cpl]

LCPO Lanthanum Copper Phosphorus Oxide

LCPP Land Capability Planning Program [South Australia]

LCPU Liquid-Crystal Polyurethane

2,4-LCPU-6 2,4-Liquid-Crystal Polyurethane-6

LCR Landing Craft, Rocket

LCR Laser Crystallization

LCR Inductance–Capacitance–Resistance

LCR Least Cost Routing

LCR Letter Carrier Route

LCR Life Consumption Ratio [Fracture Mechanics]

LCR Light Cold Rolled; Light Cold Rolling [Metallurgy]

LCR Line Control Register

LCR Lithium-Cooled (Nuclear) Reactor

LCR Load-Compensating Relay

LCR Log Count Rate

LCR Low-Cross Range

L/Cr Letter of Credit

LCRA Lower Colorado River Authority [US]

LCRAMP Linear Combination of Radical and Angular Momentum Products [Physics]

LCRE Lithium-Cooled Reactor Experiment [of Idaho National Engineering Laboratory, Idaho Falls, US]

LCRO Linear Combination of Rydberg Orbitals [Physics]

LCRP Laboratory for Cosmic Ray Physics [of Naval Research Laboratory, Washington, DC, US]

LCRS Leachate Collection and Recovery System

LCRU Lunar Communications Relay Unit [of NASA]

LCS Laboratoire de Chimie des Solides [Solids Chemistry Laboratory; of Conseil National de Recherche Scientifique at Université Paris-Sud, Orsay, France]

LCS Laboratoire de Chimie des Solides [Solids Chemistry Laboratory; of Conseil National de Recherche Scientifique at Nantes, France]

LCS Laboratoire de Chimie des Solides [Solids Chemistry Laboratory; of Conseil National de Recherche Scientifique at Talence, France]

LCS Laboratoire de Chimie des Solides [Solids Chemistry Laboratory; of Université Blaise Pascal Clermont Ferrand III, Aubière, France]

LCS Laboratoire de Chimie du Solide [Solids Chemistry Laboratory, Bordeaux, France]

LCS Laboratory Control Sample

LCS Lanthanum Copper Stannate

LCS Large-Capacity Storage

LCS Large-Capacity Core Storage

LCS Large Core Storage

LCS Larsen Control System

LCS Launch Control System

LCS Library Computer System

LCS Limited-Coordination Specification

LCS Limiting Creep Stress [Mechanics]

LCS Lincoln Calibration Sphere

LCS Linear Collision Sequence [Physics]

LCS Liquid Control System

LCS List of Command Signals

LCS Loadable Control Storage

LCS Local Conservation Strategy

LCS Loop Control System

LCS Loudness-Contour Selector

LCS Low-Carbon Steel

LC-SAX Liquid-Chromatography/Strong Anion Exchange (Column)

LCSC Legislative Council Select Committee

LCS-CNRS Laboratoire de Chimie des Solides–Centre National de Recherche Scientifique [Solid-State Chemistry Laboratory–National Center for Scientific Research, Université Paris-Sud, Orsay, France]

LCSCO Lanthanum Calcium Strontium Copper Oxide (Superconductor)

LC-SCX Liquid-Chromatography/Strong Cation Exchange (Column)

LCSD Laminate Chip Signal Diode

LCSE Laboratoire de la Commission des Substances Explosives [Laboratory of the Commission on Explosive Substances, France]

LCSE Laser Communications Satellite Experiment [of NASA]

LCSH Library of Congress Subject Headings [US]

LC-Si Liquid-Chromatography/Silica (Phase)

LCSM Laboratoire de Chimie du Solide Minérale [Laboratory for Solid Mineral Chemistry; of Conseil National de la Recherche Scientifique at Vandœuvre-lès-Nancy, France]

LCSMO Lanthanum Calcium Strontium Manganese Oxide (Superconductor)

LCSO Laboratoire de Chimie Structurale Organique [of Structural Organic Chemistry Laboratory; Université Paris Sud, Orsay, France]

LCSO Local Communications Service Order

LCSPM Linear Combination of Symmetry-Adapted Product of Morse Function [Physical Chemistry]

LCSRM Loop Current Step Response Method

LCSs Local Conservation Strategies

LCSSAP Low-Cost Silicon Solar Array Project

LCST Lower Critical Solution Temperature [Physical Chemistry]

LCSU Local Concentrator Switching Unit

LCT Laboratoire Central de Télécommunications [Central Laboratory of Telecommunications, France]

LCT Laboratoire des Composites Thermostructuraux [Thermostructural Composites Laboratory; of Conseil National de Recherche Scientifique at Pessac, France]

LCT Landing Craft, Tank [US Navy Symbol]

LCT Large Coil Test

LCT Laser Cutting Tool

LCT Linear Combination Technique [Physics]

LCT Liquid Crystalline Thermoset

LCT Logical Channel Determination

lctd located [also lct'd]

LCTE Linear Coefficient of Thermal Expansion

LCTF Large Coil Test Facility

LCTI Large Components Test Installation

LCTL Large Components Test Loop

LCT(R) Landing Craft, Tank (Rocket)

LCTS Laboratoire des Composites Thermostructuraux [Thermostructural Composites Laboratory; of Société Européenne de Propulsion, Pessac, France]

LCTV Liquid Crystal Television [also LC-TV]

LCU Landing Craft [US Navy Symbol]

LCU Last Cluster Used

LCU Line Control Unit

LCU Line Coupling Unit

LCU Link Control Unit

LCUC Letter Carriers Union of Canada

LCUG Liquid Cooling under Garment

LCUP Least Cost Utility Planning

LCUV Liquid Chromatography with Ultraviolet Detection

LCV Landing Craft, Vehicle

LCV League of Conservation Voters [Washington, DC, US]

LCV Light Commercial Vehicle

LCV Lymphocytic Choriomeningitis Virus

LCVD Laser-Assisted Chemical Vapor Deposition; Laser Chemical Vapor Deposition

LCVD Least Voltage Coincidence Detection

LCVG Liquid Cooling and Ventilation Garment

LCVIP Licensee Contractor Vendor Inspection Report

LCW Line Control Word

LCW Lock-Crisp-West (Analysis of Radiation)

LC-WCX Liquid-Chromatography/Weak-Cation Exchange (Column)

lcz localize

Lczr Localizer [also lczr]

LD Laboratory Design

LD Lactate Dehydrogenase [Biochemistry]

LD Lactic Dehydrogenase [Biochemistry]

LD Ladder Diagram [Controllers]

LD Landing [Controllers]

LD Landau Damping [Plasma Physics]

LD Landau-Devonshire (Theory) [Physics]

LD Langevin Dynamics (Method) [Physics]

LD Large Diameter

LD Largest Diameter

LD Laser Deposition

LD Laser Desorption

LD Laser Diode

LD Lateral Diffusion

LD Lateral Direction

LD (Crystal) Lattice Defect

LD Lattice Dislocation [Materials Science]

LD Lawrence-Doniach (Model) [Physics]

LD Leak Detection; Leak Detector

LD Learning Disability; Learning Disabled

LD Lending Division [of British Library],

LD Lethal Dose

LD Letter Description

LD Level Detector

LD Level Discriminator

LD Libyan Dinar [Currency]

LD Lift-Drag Ratio

LD Light-Dark Cycle

LD Light Distillate

LD Light Drawing; Light-Drawn [Metallurgy]

LD Light Duty

LD Lighting Director

LD Linear Decision

LD Linear Demodulation

LD Linear Density

LD Linear Dependence

LD Linearly Dependent (Row)

LD Linear Dichroism (Technique)

LD Line Dislocation [Materials Science]

LD Line Drawing

LD Line of Departure

LD Line-Time Waveform Distortion

LD Linker Directive

LD Linz-Donawitz (Steelmaking Process) [Metallurgy]

LD Lipid Droplet [Biochemistry]

LD Liquid Delivery

LD List of Drawings

LD Loading Dock

LD Local Density

LD Localized Dryout (Concept)

LD Logarithmic Decrement

LD Logical Design

LD Logic Driver

LD Long Delay

LD Long Distance

LD Long Duration

LD Longitudinal Direction

LD Loop Disconnect

LD Low Density

LD Luminous Discharge

LD$_1$ Lethal Dose 1 [Lethal to 1% of Test Subjects] [also LD1]

LD$_{10}$ Lethal Dose 10 [Lethal to 10% of Test Subjects] [also LD10]

LD$_{30}$ Lethal Dose 30 [Lethal to 30% of Test Subjects] [also LD30]

LD$_{50}$ Median Lethal Dose (or Lethal Dose 50) [also LD50, or MLD]

LD$_{50/30}$ Median Lethal Dose in 30 Days [also LD50/30]

LD$_{50\,Time}$ Median Lethal Dose after Indicated Time [also LD50 Time]

LD$_{99}$ Lethal Dose 99 [Lethal to 99% of Test Subjects] [also LD99]

LD$_{HI}$ Lethal Dose, High

LD$_{LO}$ Lethal Dose, Low

L&D Landing and Deceleration

L&D Loss and Damage

L/D (Ratio of) Distance between Entrance Aperture and Image Plane to Diameter of Entrance Aperture (in Neutron Radiography) [Symbol]

L/D Length-to-Diameter (Ratio) [also l/d]

L/D Lift-to-Drag (Ratio)

L$ Liberian Dollar [Currency]

Ld Land

Ld Load [also ld]

L/d Liter(s) per day [also l/d, or ℓ/d]

L/d$_f$ (Reinforcing-)Fiber Aspect Ratio

ld levo- and dextrorotatory

ld limited

ld log dualis [also log$_2$]

l/d length-to-diameter (ratio)

LDA Landing Distance Available

LDA Laser Diode Array

LDA Laser-Doppler Analysis

LDA Laser-Doppler Anemometer; Laser-Doppler Anemometry

LDA Lead Development Association [UK]

LDA Linear Density Approximation [Physics]

LDA Linear Diode Array

LDA Line Data Area

LDA Line Driving Amplifier

LDA Lithium Diisopropylamide

LDA Load in Accumulator

LDA Local Data Administrator

LDA Local Density Approximation (Method) [Physics]

LDA Localizer-Type Directional Aid

LDA Locate Drum Address

LDA Logical Device Address

L-DAB L-2,4-Diamino Butanoic Acid [also L-Dab]

LD-AC Linz-Donawitz/ARBED-CRM [Steelmaking Process Developed Jointly by ARBED (Aciéries Réunies de Burbach-Eich-Dudelange) and CRM (Centre de Recherches Métallurgiques)]

LDA-KKR-CPA Local Density Approximation–Korringa-Kohn-Rostoker–Coherent Potential Approximation [Physics]

LDAM Linked Direct Access Method [Computers]

LDA-MTO Local Density Approximation Muffin-Tin Orbital [Physics]

LDAO Lauryldimethylamine Oxide

LDAP Lightweight Directory Access Protocol [Internet]

L/day Liters per day [also l/day]

LDB Launch Data Bus

LDB Light Distribution Box

LDB Logical Database

LDB Logistics Data Bank

LDBA Leakage Design Basis Accident [Nuclear Engineering]

LD-BOP Linz-Donawitz Basic Oxygen (Steelmaking) Process [Metallurgy]

LDBS Land Data Bank System

LDC Land Development Corporation [Prince Edward Island, Canada]

LDC LASA (Large Aperture Seismic Array) Data Center

LDC Latitude Data Computer

LDC Least Developed Country [also ldc]

LDC Less Developed Country [also ldc]

LDC Light Direction Center

LDC Linear Detonating Cord

LDC Line Drop Compensator

LDC Linguistics Documentation Center [of University of Ottawa, Canada]

LDC Liquid Dynamic Compaction

LDC Load Duration Curve [Mechanics]

LDC Local Display Controller

LDC Logical Device Coordinate

LDC London (Waste) Dumping (and Marine Pollution Prevention) Convention [UK]

LDC Long Distance Communications

LDC Lotus Development Corporation [US]

LDC Lower Dead Center

LDC Low-Speed Data Channel

LDCC Large Diameter Component Cask [Nuclear Engineering]

LDCS Long Distance Control System [US]

LDD Lifetime Drain Doping [Electronics]

LDD Lightly-Doped Drain [Electronics]

LDD Local Data Distribution

LDD Logical Data Description

LDD Logic Design Data

LDD Luminaire Dirt Depreciation (Factor) [Electrical Engineering]

LDDC London Docklands Development Corporation [UK]

LDDK Linear Driver Development Kit

LDDM Laser Doppler Displacement Meter

LDDS Limited Distance Data Set

LDDS Long-Distance Discount Services (Company) [US]

LDDS Low-Density Data System

LDE Laminar Defect Examination

LDE Language-Directed Editor

LDE Large Debond Energy (of Composite Materials)

LDE (Coated) Layered Dispersion Element

LDE Linear Differential Equation

LDE London Derivatives Exchange [UK]

LDE Long-Delay Echo

LDE Long Duration Exposure

LDEC Lunar Docking Events Controller [of NASA]

LDEF Long-Duration Exposure Facility [of NASA]

LDEM Laboratory for Development of Engineering Materials [of Kanazawa University, Japan]

LDEO Lamont-Doherty Earth Observatory [of Columbia University, New York, US]

LDEQ Louisiana Department of Environmental Quality [US]

LDF Land Disposal Facility

LDF Light Distillate Feedstock

LDF Linear Discriminate Function

LDF Load Factor

LDF Local Density Fluctuation [Physics]

LDF Local Density Formalism [Physics]

LDF Local Density Function [Physics]

LDF Local Density Functional (Approximation) [Physics]

LDF Long Discontinuous Fiber(s)

LD/FT/ICR Laser Desorption Fourier-Transform Ion Cyclotron Resonance (Mass Spectrometry)

LDG Linz-Donawitz (Converter) Gas [Metallurgy]

Ldg Landing [also ldg]

Ldg Leading [also ldg]

Ldg Ledger [also ldg]

Ldg Loading [also ldg]

LDGO Lamont-Doherty Geological Observatory [of Columbia University, New York, US]

LDH Lactic Dehydrogenase [Biochemistry]

LDH Layered Doube Hydroxide

LDH Limiting Dome Height

LDH Linearized Debye-Hueckel (Theory) [Physical Chemistry]

LDH L-Lactate Dehydrogenase; L-Lactic Dehydrogenase [Biochemistry]

LDH-A Lactic Dehydrogenase A [Biochemistry]

LDH-B Lactic Dehydrogenase B [Biochemistry]

LDH/PK Lactic Dehydrogenase/Pyruvate Kinase [Biochemistry]

LDI Landing Direction Indicator

LDI Lossless Discrete Integrator

LDI Low-Density Inclusion [Metallurgy]

LDIN Lead-In Light System

LDIU Launch Data Interface Unit

LD-KGC Linz-Donawitz-Kawasaki Gas Control Converter [Metallurgy]

LDL Language Description Language

LDL Low-Density Lipoprotein [Biochemistry]

$\lambda/\Delta\lambda$ energy resolution (in x-ray absorption fine structure) [Symbol]

$\lambda/\Delta\lambda$ resolving power (or resolution) (of an interferometer, or spectroscope) [Symbol]

LDLA Limited Distance Line Adapter

LDLC Low Dispersion Liquid Chromatography

LD Loop Long-Distance Loop [also LD loop] [Communications]

LDM Laser Diode Mount

LDM Limited-Distance Modem

LDM Linear Delta Modulation

LDM Local Data Manager

LDM Long Distance Modem

LDME Laser Doppler Microelectrophoresis

Ldmk Landmark [also ldmk]

LD-MOCVD Liquid-Delivery Metalorganic Chemical Vapor Deposition

LDMS Laser Desorption Mass Spectrometry; Laser Desorption Mass Spectroscopy

LDN Listed Directory Number

LDN Long Distance with Normal Focal Plane

LDNG Louis Dreyfus Natural Gas

LDNR Lead Dinitroresorcinolate

LDNS Laser Doppler Navigation System

LD-OB Linz-Donawitz Oxygen Bottom-Blown (Converter) [Metallurgy]

L-DOPA 3-(3,4-Dihydroxyphenyl)-L-Alanine [also L-Dopa, or L-dopa] [Biochemistry]

LDOS Local Density of States [also LDoS] [Solid-State Physics]

LDOT Low-Disorder-to-Order Transition [Materials Science]

LDP Language Data Processing

LDP Least Distance Programming (Algorithm)

LDPE Low-Density Polyethylene

LDPM Laser Diode Power Module

LDR Land Disposal Restriction(s) [US]

LDR Large Deployable Reflector

LDR Light-Dependent Resistor

LDR Limiting Draw(ing) Ratio [Metallurgy]

LDR Limiting Dome Ratio

LDR Linear Depolarization Ratio

LDR Linear Dynamic Range

LDR Line Driver-Receiver

LDR Low Data Rate

Ldr Leader [also ldr]

Ldr Loader [also ldr]

LDRD Laboratory-Directed Research and Development

LD-RH-OB (Two-Stage) Linz-Donawitz Basic-Oxygen Furnace Combined with Ruhrstahl Heraeus Oxygen Blowing (Process) [Metallurgy]

LDRI Low Data Rate Input

LDRPA Local Density-Based Random Phase Approximation [Physics]

LDRR Laboratory of Dignostic Radiology Research [of Office of the Director, National Institutes of Health, Bethesda, Maryland, US]

LDRS LEM (Logical End of Medium) Data Reduction System

LDRTA Long Distance Road Transport Association [Australia]

LDS Landing/Deceleration Subsystem

LDS Landing, Deservicing and Safing

LDS Large Dark Spot Effect

LDS Large Disk Storage

LDS Large-Format Document Scanner

LDS Large-Format Drawing Scanner

LDS Laser Deposition System

LDS Library Delivery System

LDS Light Distillate Spirit

LDS Linear Dichroism Spectrometry

LDS Liquid Delivery System

LDS Lithium Disilicate

LDS Lithium Dodecyl Sulfate

LDS Logic Digital Switch

LDS Local Distribution Service

LDS Local Distribution System

LDS Logic Design Translator

LDS Long-Distance Transmission

LDS Low-Dimensional Structures [Physics]

LDS Low Dislocation-Density Substrate [Materials Science]

Lds Loads [also lds]

Ldscpg Landscaping [also ldscpg]

LDSD Library and Documentation Services Division [of Canada Center for Mineral and Energy Technology, Natural Resources Canada, Ottawa]

LDSD (International Conference on) Low Dimensional Structures and Devices [Physics]

LDSU Local Digital Service Unit

LDT Language-Dependent Translator

LDT Laser Display Technology

LDT Linear Differential Transformer

LDT Linear Displacement Transducer

LDT Local Descriptor Table

LDT Logic Design Translator

LDT Long-Distance Telephone

LDT Long Distance Transmission

LDT Low-Density Telephony

LDTA Leak Detection Technology Association [US]

LDTP Long-Distance Thrift Pak

LDTS Linear-Discrete Time System

LDTS Low-Density Telephony Service

LDTV Low-Definition Television

LDU Local Delivery Unit

LDUA Light-Duty Utility Arm

LDV Laser-Doppler Velocimeter; Laser-Doppler Velocimetry

LDWT Low-Density Wind Tunnel

LDX Long-Distance Xerography

LE Labour Exchange [UK]

LE Ladder Element

LE Langevin Equation [Physics]

LE Langmuir Effect [Solid-State Physics]

LE Langmuir Equation [Physical Chemistry]

LE Laplace Equation

LE Large End

LE Laser Effect

LE Laser Energetic(s)

LE Laser Extensometer; Laser Extensometry

LE Launch(ing) Equipment

LE Launch Escape

LE Laundry Engineer(ing)

LE Lead Equivalent [Nuclear Engineering]

LE Leading Edge

LE Lee-Enfield (Rifle)

LE Length

LE Length of Engagement (of Screw Thread)

LE Less than, or Equal to

LE Light Emission

LE Light Energy

LE Light Equipment

LE Limit of (Measurement) Error

LE Linear Elasticity

LE Linear Equation

LE Line Equalization; Line Equalizer

LE Line Equipment

LE Linkage Editor (Program)

LE Liquid Equilibrium

LE Local Embrittlement

LE Local Exchange

LE Local(ly) Excited

LE Locomotive Engineer(ing)

LE Logic Element

LE Logic Evaluator

LE Logistics Engineer(ing)

LE Long Endurance

LE Loop Extender

LE Low Efficiency

LE Lower Envelope (Tray) [Laser Printers]

LE Low Explosive

LE Lubrication Engineer(ing)

LE Luminous Efficiency

LE Lunar Eclipse [Astronomy]

LE Lupus Erythematosus [also le] [Medicine]

LE Lyapunov Exponent [Physics]

L/E Displacement to (Magnetic) Field (Ratio) [Semiconductors]

Le Length

Le Length of Engagement (of a Buttress Thread)

Le Leone [Currency of Sierra Leone]

Le Lewis Number (in Physics) [Symbol]

Le$_{turb}$ Turbulent Lewis Number (in Physics) [Symbol]

l(E$_F$) mean free path of electrons [Symbol]

£E Egyptian Pound [Currency]

LEA Laboratory for Experimental Astrophysics [of Lawrence Livermore National Laboratory, University of California at Livermore, California, US]

LEA Light Element Analysis

LEA Linear Electron Accelerator

LEA Load Effective Address

LEA Local Education Agencies [US]

LEA Local Education Authority [UK]

LEA Logistics Engineering Analysis

LEA Long Endurance Aircraft

LEA Longitudinally Excited Atmosphere

LEA Lycopersicon Esculentum Agglutinin [Immunology]

Lea Leather [also lea]

LEAA Law Enforcement Assistance Administration

LEAC Linear Elution Adsorption Chromatography

LEAD Learn, Execute and Diagnose

LEAD Local Employment Assistance and Development

LEADER Lehigh Automatic Device for Efficient Retrieval [of Lehigh University, Bethlehem, Pennsylvania, US]

LEADERMART Lehigh Automatic Device for Efficient Retrieval–Mechanical Analysis and Retrieval of Text [of Lehigh University, Bethlehem, Pennsylvania, US]

LEADEX Lead Experiment

Lead Res Dig Lead Research Digest [US]

LEADS Law Enforcement Automated Data System

LEADS List of Educational Aids and Direct Sources [of ASM International, US]

Lead Zinc Lead and Zinc [Journal published in Japan]

Lead Zinc Q Rev Lead and Zinc Quarterly Review [US]

LEAF Law Enforcement Access Field

LEAF LISP (List Processing) Extended Algebraic Facility

Leaf Leaflet [also leaf]

LEAFS Laser-Excited Atomic Fluorescence Spectrometry

LEANS Lehigh Analog Simulator [of Lehigh University, Bethlehem, Pennsylvania, US]

LEAP Landcare and Environment Action Plan [Australia]

LEAP Language for the Expression of Associative Procedures

LEAP Laser and Electro-Optics Applications Program [of Conference (and Trade Show) on Lasers and Electrooptics]

LEAP LDC (Less Developed Country) Energy Alternatives Planning (System)

LEAP Lift-Off Elevation and Azimuth Programmer

LEAP Local Employment Assistance and Development Program

LEAP Londoners for the Safe Elimination of All PCB's (Polychlorinated Biphenyls) [London, Ontario, Canada]

LEAP (Biennial Conference on) Low Energy Antiproton Physics [also Leap]

LEAP Lunar Escape Ambulance Pack [of NASA]

LEAPP Locally Enclosed Area, Particulate Protected (Concept)

LEAR Logistics Evaluation and Review Technique

LEAR Low-Energy Anti-Photon Ring [of CERN–European Laboratory for Particle Physics, Geneva, Switzerland]

LEAR/LEIR Low-Energy Anti-Photon Ring/Low-Energy Ion Ring [of CERN–European Laboratory for Particle Physics, Geneva, Switzerland]

LEAS Lease Electronic Accounting System

LEAS Lower Echelon Automatic Switchboard

LEASAT Leased Satellite [US]

Least Sq St Dev Least Squares Standard Deviation

Leasg Leasing [also leasg]

LEB LHCC (Large Hadron Collider Committee) Electronics Board [of CERN–European Laboratory for Particle Physics, Geneva, Switzerland]

LEB Lower Equipment Bay

Leb Lebanese; Lebanon

LEBCO Low Energy Building Council

LEC LAN (Local-Area Network) Emulation Client

LEC Launceston Environment Center [Tasmania, Australia]

LEC Light Energy Converter

LEC Liquid-Encapsulated Czochralski (Method) [Crystal Growing]

LEC Liquid Exclusion Chromatography

LEC Local Exchange Carrier

LEC Lockheed Electronics Company [US]

LEC Low-Energy Channeling

LEC Low-Energy Cure

LECA Laboratoire d'Electrochimie de Chimie Analytique [Electrochemistry and Analytical Chemistry Laboratory, of Ecole Normale Supérieure–Université de Paris, France]

LECA Lightweight Expanded Clay Aggregate

LECBD Low-Energy Cluster Beam Deposition (Technique) [Physics]

LECC Lake Erie Cleanup Committee [US]

LECE Ligue Européenne de Coopération Economique [European League for Economic Cooperation, UK]

LECO Local Engineering Control Office

Lect Lecture(r) [also lect]

Lect Ser–Von Karman Inst Lecture Series–Von Karman Institute [Published in the US]

LECVD Low-Energy Chemical Vapor Deposition

LECVD Laser-Excited Chemical Vapor Deposition

LED Laser Energy Density

LED Light-Emitting Diode

LED Linear Electric Device

LED Low-Energy Dislocation [Materials Science]

LEDA Local Economic Development Assistance (Program) [Canada]

LEDA Low-Energy De-Asphalting (Process) [Civil Engineering]

LEDC Low-Energy Detonating Cord

LEDIS Local Economic Development Information Service [UK]

LEDS Low-Energy Dislocation Structure [Materials Science]

LEDT Limited-Entry Decision Table

LEEA Lifting Equipment Engineers Association [UK]

LEEBI Low Energy Electron Beam Irradiation

LEEC London Environmental Economics Center [UK]

LEED Laser-Energized Explosive Device

LEED Low-Energy Electron Diffraction

LEEDOA Low-Energy Electron Diffraction Optics Analyzer

LEED/STM Low-Energy Electron Diffraction/Scanning Tunneling Microscopy

Lee Is Leeward Islands

LEEIXS Low-Energy Electron-Induced X-Ray Spectroscopy

LEELS Low-Energy Electron Loss Spectrometry

LEELS Low-Energy Electron Loss Spectroscopy

LEEM Low-Energy Electron Microscopy

LEEM/PEEM Low-Energy Electron Microscopy/Photo-Electron Emission Microscopy

LEEP Library Education Experimental Project [of University of Syracuse, New York, US]

LEER Low-Energy Electron Reflection

LEES Low-Energy Electron Spectrometry

LEET Low-Energy Electron Transmission

LEF Laser-Excited Fluorescence (Spectroscopy)

LEF Light-Emitting Film

LEF Linear Electric Field

LEF Line Expansion Function

LEFD Large Experimental Fusion Device

LEFD Local Enhanced Field Desorption

LEFI Local Electric Field Instrument

LEFM Linear Elastic Fracture Mechanics

LE-FRES Low-Energy Forward Recoil Spectroscopy

Left Rect Abd Left Rectus Abdominus [also left rect abd] [Anatomy]

LEG Laboratoire d'Electrotechnique de Grenoble [Electrical Engineering Laboratory; of Institut National Polytechnique de Grenoble, France]

Leg Legal(ity) [also leg]

Leg Legislation; Legislative; Legislator; Legislature [also leg]

LEG Library Education Group [now Training and Education Group; of Library Association, UK]

LEGBAC Limited Exploratory Group on Broadcasting-to-Aeronautical Compatibility

Legis Legislation; Legislative; Legislator; Legislature [also legis]

LEGOL Legally Oriented Language

LEGR Low-Energy Gamma Radiation

LEGS Laser Electron Gamma-Ray Source (Experiment) [at Brookhaven National Laboratory, Upton, New York, US]

Leg Wt Legal Weight [also leg wt]

LEH Launch/Entry Helmet

LEHR Laboratory for Energy-Related Health Research

LEI Laser-Enhanced Ionization

LEIB Low-Energy Ion Beam

LEIBAD Low-Energy Ion Bombardment Angular Distributions

Leics Leicestershire [also Leic, UK]

LEID Low-Energy Ion Detector

LEIID Low-Energy Ion Implantation Deposition (System)

LEIM Laser Evaporation of Intact Molecules [Mass Spectrometry]

LEIR Low-Energy Ion Ring [of CERN–European Laboratory for Particle Physics, Geneva, Switzerland]

LEIS Laboratory for Electronic Intelligent Systems [of Research Institute of Electrical Communication [Tohoken University, Sendai, Japan]

LEIS Low-Energy Ion Scattering

LEISS Low-Energy Ion Scattering Spectroscopy

LEJ Longitudinal Expansion Joint

LEK Lek [Currency of Albania]

LEK Liquid Encapsulated Kyropoulos (Method) [Crystal Growing]

Lekcji Strukt Anal Lekcji po Strukturnomu Analizu [Lectures on Structural Analysis; published by Kharkov University, Ukraine]

LEL Large Engineering Loop [of NASA Lewis Research Center, Cleveland, Ohio, US]

LEL Lebanese Pound [Currency]

LEL Link, Embed and Launch-to-edit [Computers]

LEL Lower Explosion Level

LEL Lower Explosive (or Explosion) Limit

LELU Launch Enable Logic Unit

LEM Laboratoire d'Etude des Microstructures [Laboratory for Microstructural Studies, Office National d'Etudes et de Recherches Aérospatiales, Châtillon, France]

LEM Laboratory Environment Model

LEM Laboratory of Electromodeling

LEM Language Extension Module

LEM Laser Exhaust Measurement

LEM Launch Escape Monitor

LEM Lempira [Currency of Honduras]

LEM Linear Electric Machine

LEM Logical End of Medium [also LEOM]

LEM Logical Enhanced Memory

LEM Lorentz Electron Microscopy

LEM Low-Energy (Ball) Mill(ing)

LEM Lunar Excursion Module [of NASA, Apollo Program]

Lem Lemon [also lem]

LEMA Lifting Equipment Manufacturers Association

LEMBS Low-Energy Molecular Beam Scattering

LEMD Low-Energy Misfit Dislocation [Materials Science]

LEMIC LHC (Large Hadron Collider) Experimental Requirements Committee [of CERN–European Laboratory for Particle Physics, Geneva, Switzerland]

LEMIT Laboratorio de Entrenamiento Multidisciplinario para la Investigación Tecnologica [Multidisciplinary Laboratory for Technological Research, La Plata, Argentina]

LEMO Lowest Empty Molecular Orbital [Physical Chemistry]

LEM-ONERA Laboratoire d'Etude des Microstructures–Office National d'Etudes et de Recherches Aérospatiales [Laboratory for Microstructural Studies–National Office for Aerospace Studies and Research, Châtillon, France]

LEMP Lightning Electromagnetic Pulse

LEMS Low-Energy Molecular Beam Scattering

LEMUF Limit of Error on Material Unaccounted For

LEN Length (Function) [Programming]

LEN Low Entry Networking

Len Length [also len]

LENP Low-Energy Nuclear Physics

Lenzinger Ber Lenzinger Berichte [Lenzing Report; published by Lenzing AG, Austria]

LEO Lasers and Electro-Optics

LEO Librating Equidistant Observer

LEO Local Earth Observation (Project) [Canada]

LEO Low Earth Orbit (Satellite)

LEO Lyon's Electronic Office [Computer System]

LEOE Linear Electrooptic Effect

LEOM Logical End of Medium [also LEM]

LEOMA Laser and Electro-Optics Manufacturers' Association [US]

LEOS Lasers and Electro-Optics Society [of Institute of Electrical and Electronics Engineers, US]

LEP Large Electron-Positron (Collider) [of CERN–European Laboratory for Particle Physics, Geneva, Switzerland] [also Lep]

LEP Licensed Environmental Professional [Connecticut, US]

LEP Low-Energy Photon [Solid-State Physics]

LEP Low-Energy Physics

LEP Lower End Plug

LEPA Laboratoire d'Electronique et de Physique Appliquée [Electronics and Applied Physics Laboratory, France]

LEPC Large Electron-Positron (Collider) (Experiments) Committee [of CERN–European Laboratory for Particle Physics, Geneva, Switzerland]

LEPC Local Emergency Planning Committee

LEPD Low-Energy Photon Detector

LEPD Low-Energy Positron Diffraction

LEPES Laboratoire d'Etudes des Propriétés Electroniques des Solids [Laboratory for Solids Electronic Properties Studies, Grenoble, France]

LEPETH Lead Covering and (Sprayed-On) Polyethylene Terephthalate Coating (Cable) [also lepeth]

LEPS Launch Escape Propulsion System

LEPS London-Eyring-Polanyi-Sato (Potential Energy Surface)

LEPS Low-Energy Photon Spectroscopy

LEPSi Light-Emitting Porous Silicon [also LepSi]

LEPT Long Endurance Patrolling Torpedo

LEQ Line Equipped

LEQ Line of Equipment

LEQSF Louisiana Educational Quality Support Fund [US]

LER Licensee Event Report

LER Long Eye Relief (Optics)

LERA Limited Employee Retirement Account

LERC Laramie Energy Research Center [Wyoming, US]

LeRC Lewis Research Center [of NASA at Cleveland, Ohio, US]

LEREFLEOS Lead Recovery From Lead Oxide Secondaries (Project) [of European Union]

LERD Low-Energy X-Ray Diffraction

LERLS Lake Erie Regional Library System [US]

LERM Laboratoire de Electronique et de Résonance Magnétique [Electronics and Magnetic Resonance Laboratory; of Université Blaise Pascal, Aubiére, France]

LERM Laboratoire d'Etudes et de Recherche sur les Matériaux [Materials Research Laboratory, France]

LERM Localized Electron Spin Microscopy

LERMAT Laboratoire d'Etudes et de Recherches sur les Matériaux [Materials Research Laboratory; of Institut des Sciences de la Matière et du Rayonnement, Caen, France]

LERMAT-ISMRA Laboratoire d'Etudes et de Recherches sur les Matériaux–Institut des Sciences de la Matière et du Rayonnement [Materials Research Laboratory of the Institute of Matter and Radiation Sciences, University of Caen, France]

LERS Low-Energy Recoil Spectroscopy

LERT Linear-Extensional, Rotation and Twist Robot

LERTS Laboratoire d'Etudes et Recherches en Télédétéction Spatial [Research Laboratory for Spaceborne Remote Sensing, France]

LES LAN (Local- Area Network) Emulation Server

LES Large Eddy Simulation (Model) [Fluid Dynamics]

LES Launch Enabling System

LES Launch Equipment Shop [of NASA]

LES Launch Escape Subsystem; Launch Escape System

LES Law of Economies of Scale

LES Light Emission Sensitizer

LES Light-Emitting Switch

LES Light Exposure Speed

LES Limited Early Site

LES Lincoln Experimental (Communications) Satellite [Series of US Military Satellites]

LES Local Excitatory State

LES Loop Error Signal

LES Louisiana Engineering Society [US]

LES Lower Esophageal Sphincter [Anatomy]

LESA Lake Erie Steam Association [US]

LESA Lunar Exploration System for Apollo [of NASA]

LESC Light-Emitting Switch Control

LESHOUCHES Quantum Field Theory: Perspective and Prospective Meeting [of NATO Advanced Studies Institute Meeting held at Les Houches, France]

Lesnaya Prom Lesnaya Promyshlennost [Russian Publication on the Forestry Industry]

Lesnoe Khoz Lesnoe Khozuaistvo [Russian Publication on Forestry Operations]

LESR Light-Induced Electron Spin Resonance

LESR Limited Early Site Review

LESS Laser-Excited Shpol'skii Spectrometer

LESS Lateral Epitaxy by Speeded Solidification

LESS Leading Edge Structure Subsystem [Aerospace]

LESS Leading Edge Subsystem [Aerospace]

LESS Least Cost Estimating and Scheduling

LESSC Louisville Engineering and Scientific Societies Council [US]

L ès Sc *(Licencié ès Sciences)* – Licentiate in Science

LEST Low-Energy Speed Transmission

LET Launch Equipment Test

LET Launch Escape Tower

LET Leading Edge Trigger

LET Letters Shift

LET Light Emission Theory

LET Linear Energy Transfer

LET Lincoln Experimental Terminal [of NASA]

LET Logical Equipment Table

LET Low-Energy Temperature

.LET Letter [File Name Extension] [also .let]

Let Letter [also let]

LETAM Laboratoire d'Etude des Textures Appliquées aux Matériaux [Materials Textures Research Laboratory, Université de Metz, France]

LETB Local Exchange Test Bed

LETF Launch Equipment Test Facility [of NASA Kennedy Space Center, Florida, US]

LETIS Leicestershire Technical Information Service [UK]

LETOPS Lightweight Enhanced Trench Overhead Protection System

LeTRS Lewis Technical Reports Server [of NASA Lewis Research Center, Cleveland, Ohio, US]

LETS Law Enforcement Teletypewriter System

LETS Layout/(Cost) Estimation/Tool Design and Setup (Computer Graphics System)

Lett Letter [also lett]

Lett Appl Microbiol Letters in Applied Microbiology [Published by the Society for Applied Bacteriology, UK]

Lett Math Phys Letters in Mathematical Physics [Netherlands]

Lett Nuovo Cimento Lettere al Nuovo Cimento [Italian Physics Journal published by Societa Italiana de Fisica]

LEU Launch Enabling Unit

LEU Leu [Currency of Romania]

LEU Leucocyte [Biochemistry]

LEU Low-Enriched Uranium

Leu (–)-Leucine; Leucyl [Biochemistry]

LEU-CAM Leucocyte Adhesion Molecules [Biochemistry]

Leu-MCA L-Leucine-4-Methyl-7-Cumarinylamide [Biochemistry]

LEU/TH Low Enriched Uranium/Thorium Fuel

LEV Loader/Editor/Verifier

LEV Low-Emission Vehicle

LEV Lunar Excursion Vehicle [NASA Apollo Program]

Lev Level [also lev]

lev *(levis)* – light

Levacar Levitation Car

Levapad(s) Levitation Pad(s)

LEX Lexicon [Computers]

LEX Line Exchange

LEX Low-Energy X-Ray Spectral Analysis

.LEX Speller Lexicon (File) [WordPerfect File Name Extension]

Lex Lexicon [also lex]

Lexicog Lexicographer; Lexicographical; Lexicography

LEXIS Lexicography Information Service [Germany]

LEY Low Electron Yield

LEZM Liquid Encapsulated Zone Melting [Crystal Growing]

LF Laboratory Furnace

LF Lactoferrin [Biochemistry]

LF Ladle Furnace

LF Lagrangian Function [Mechanics]

LF Lamé Function [Mechanics]

LF Lamellar Field

LF Laminar Flow

LF Landé Factor [Physics]

LF Landfill

LF Land Forces

LF Landing Flap [Aeronautics]

LF Langevin Function [Physics]

LF Lang-Firsov (Ground-State Energy) [Physics]

LF Langmuir Frequency [Plasma Physics]

LF Lanthanum Ferrite

LF Large File

LF Larmor Frequency [Electrical Engineering]

LF Lattice Fringe [Materials Science]

LF Launch Facility

LF Launch Forward

LF Leadframe [also L/F]

LF Leapfrog Configuration

LF Left Forward

LF Legendre Function [Mathematics]

LF Leveling Foot

LF Liapunov's Function [Mathematics]

LF Lightface [also lf]

LF Light-Fast (Stain)

LF Light Filter

LF Light Flint (Glass)

LF Limiting Fragmentation

LF Linear File

LF Linear Foot

LF Linear Function

LF Line Feed (Character) [Data Communications]

LF Line Finder

LF Line Focus [Optics]

LF Line Frequency

LF Line(s) of Force

LF Linoleum Floor

LF Liquid Film

LF Liquid Fluorine

LF Lithium Ferrite

LF Load Factor

LF Logical File

LF Logic(al) Function

LF Longitudinal Fiber(s)

LF Loss Factor

LF Lower Fuselage

LF Low Fatigue [Mechanics]

LF Low Field [also lf]

LF Low Frequency [frequency: 30 to 300 kilohertz; wavelength: 1,000 to 10,000 meters] [also L-F, L/F, lf, l/f, or l-f]

LF Low Friction

L/F Leadframe [also LF]

Lf Leaflet [also lf]

Lf Limit of Flocculation [Pharmacology]

lf leaflet [also Lf]

lf lightface [also LF]

lf low field [also LF]

lf low frequency [frequency: 30 to 300 kilohertz; wavelength: 1,000 to 10,000 meters] [also LF, L-F, L/F, l/f, or l-f]

LFA Load Flow Analysis

LFA Lobster Fishing Area

LFA Local Flying Area

LFA Low Flow Alarm

LFAF Low-Frequency Accelerometer Flutter

LFAM Low-Frequency Accelerometer Modes

LFAP Lightweight Flow Admission Protocol

LFAP Low-Frequency Accelerometer Pogo [Aerospace]

LFB Linear Feedback

LFBB Liquid Fly-Back Booster

LFBC Livestock Feed Board of Canada

LFBR Liquid Fluidized Bed (Nuclear) Reactor

LFBR-CX Liquid Fluidized Bed Reactor Critical Experiment

LFC Laminar Flow Control [Aeronautics]

LFC Laminar Forced Convection

LFC Land Forces Command [of Department of National Defense, Canada]

LFC Large Format Camera

LFC Level of Free Convection

LFC Load Frequency Control

LFC Local Forms Control

LFC Lost-Foam Casting [Metallurgy]

LFC Low-Frequency Cycle

LFC Lowest-Feasible Concentration

LFCB Load to Forms Control Buffer

LFCC Laboratorio de Fisica Cosmico de Chacaltaya [Chacaltaya Laboratory of Cosmic Physics, Bolivia]

LFCC Laboratory Frame (Rotational) Close Coupling (Approximation) [Physics]

LFCF Laminar Free Convective Flow

LFCI Ligand Field Configuration Interaction

LFCM Lava-Like Fuel-Containing Mass [Nuclear Reactors]

LFCM Liquid-Fed Ceramic Melter

LFCM Low-Frequency Cross-Modulation

LFD Langmuir Film Deposition [Physical Chemistry]

LFD Light-Front Dynamics

LFD Line Fault Detector

LFD Liquefaction Fractional Distiller

LFD Local Frequency Disturbance

LFD Low-Frequency Disturbance

LFDF Low-Frequency Direction Finder [also LF/DF]

LFDS Labour Force Development Strategy

LFE Laboratory for Electronics
LFE Landau Free Energy [Physics]
LFE Life-Fatigue-Endurance (Test) [Mechanics]
LFE Local Field Effect [Physics]
LFE Local Field Emission (Current) [Physics]
LFEM Laboratoire Fédéral d'Essai des Matériaux [Federal Laboratory for Materials Testing, Switzerland]
LFER Linear Free Energy Relationship
LFEV Lower Free Energy Value
LFF Land– und Forstwirtschaftlicher Forschungsrat [Research Committee on Agriculture and Forestry, Germany]
LFF Low-Frequency Fluctuation
LFFET Low-Frequency Field-Effect Transistor
LFFP Laser Fusion Feasibility Project [of Laboratory for Laser Energetics, University of Rochester, New York, US]
LFG Lacher-Fowler-Guggenheim (Model) [Metallurgy]
LFG Landfill Gas
LFGS Low-Flow Gas Saver
LFI Laser-Fiber Illuminator
LFI Last File Indicator
LFI Laxforskningsinstitutet [Salmon Research Institute, Sweden]
LFIF Low-Frequency Induction Furnace
LFIM Low-Frequency Instruments and Measurements [Committee of the Institute of Electrical and Electronics Engineers–Instrumentation and Measurement Society]
LFIRS Low-Frequency Infrared Spectroscopy
LFIS Land Forces Information System
LFJ Local Feed Junctor [Electrical Engineering]
LFL Lower Flammability Limit; Lower Flammable Limit
LF LORAN Low-Frequency LORAN (Long-Range Navigation) [also LF Loran, or LF loran]
LFLU Liberia Federation of Labour Unions
LFM Large File Management
LFM Lateral Force Microscope; Lateral Force Microscopy
LFM Lateral Force Mode [Atomic Force Microscopy]
LFM Layer Fluctuation Model
LFM Limited-Area Fine-Mesh Model
LFM Linear Feet per Minute [also lfm]
LFM Linear Frequency Modulation
LFM Liquid Film Migration [Materials Science]
LFM Liquid Flowmeter
LFM Local File Manager
LFM Lower Felsic Metapyroclastics [Geology]
LFM Low-Frequency Microphone
LFM Low-Powered Fan Marker [Aeronautics]
LFM+ Large File Management Plus
LFMA Low-Field Microwave Absorption
LFMM Liquid-Fed Minimelter
LFMMA Low-Field Microwave Magnetoabsorption
LFMOP Linear Frequency Modulation on Pulse
LFMR Low-Field Magnetoresistance [Solid-State Physics]
LFN Line Format Number
LFN Logical File Name
LFN Long File Name

LFN Low-Frequency Noise
LFO Lithium Iron(III) Oxide
LFO Local Fallout
LFO Low-Frequency Oscillator
LFOP Landing and Ferry Operations Panel
LFP Laser Flash Photography
LFP Lithium Iron Phosphate
LFP Low-Field Phase [Materials Science]
LFPS Low-Frequency Phase Shifter
LFQ Light Foot Quantizer
LFR Lower Fuselage Radar
LFR Low-Flux (Nuclear) Reactor
LFR Low-Frequency Radio-Range [also LFRR]
LFr Luxembourg Franc [Currency]
LFRAP Long Feeder Route Analysis Program
LFRD Load-Factor Resistance Design
LFRD Lot Fraction Reliability Definition
LFRD Lot Fraction Reliability Deviation
LFRS Liquid Flat Reference Surface
LFRR Low Frequency Radio Range [also LFR]
LFS Laboratory of Forensic Science [of Scotland Yard, London, UK]
LFS Labour Force Survey [of Statistics Canada]
LFS Launch Facility Simulator
LFS LEED (Low-Energy Electron Diffraction) Fine Structure
LFS Local Format Storage
LFS Logical File Structure
LFS Low-Frequency Spectrum
LFS Lutetium Iron Silicide (Ceramics)
LFSE Ligand Field Stabilization Energy
LFSR Linear-Feedback Shift Register
LFSS Landing Force Support Ship [US Navy]
LFSW Landing Force Support Weapon [US Navy]
LFT Leaflet [also Lft, or lft]
LFT Ligand Field Theory [Chemistry]
LFT Live-Fire Test (Program) [of Wright Patterson Air Force Base, Dayton, Ohio, US]
LFT Low Function Terminal
Lft Leaflet [also lft]
lft left
LFTD Laser Flash Thermal Diffusivity
Lfts Leaflets [also lfts]
LFT TAB Left-Aligned Tab [also Lft Tab]
LFU Least Frequency Unit
LFU Least Frequently Used [Computers]
LFV Low-Frequency Vibration
LFV Lunar Flying Vehicle [of NASA Apollo Program]
LFVF Local Free Volume Fraction
LFW Linear Friction Welding
LFWA Listing of Fish and Wildlife Advisories (Database) [US]
LFZ Laser Floating Zone
LG β-Lactoglobulin [Biochemistry]
LG La Grange [Georgia, US]

LG Laminated Glass
LG Landau-Ginzburg (Theory of Superconductivity)
LG Landing Gear
LG Landing Ground
LG Lanthanum Gallate
LG Large Grain
LG Laser Glass
LG Lattice Gas (Model) [Physics]
LG (Crystal) Lattice Geometry
LG Leg [Apparatus]
LG Lethal Gene
LG Lewis Gun [UK]
LG Liaison Group
LG Light Green
LG Light Guide
LG Light Gun [Electronics]
LG Limit Ga(u)ge
LG Linearized Gradient (Approximation) [Physics]
LG Lindemer-Guy (Analysis) [Materials Science]
LG Linear Gate
LG Line Generator
LG Line Graph [Mathematics]
LG Liquefied Gas
LG Liquid Gas
LG Lithium Gallate
LG Local Government
LG Logic Gate
LG Low German
LG Low(er) Grade
LG Lubricating Grease
LG Lymph Gland [Anatomy]
L&G Lawn and Garden (Equipment)
Lg Length [also lg]
Lg Lodging [also lg]
Lg Logging [also lg]
Lg Looping [also lg]
Lγ L-Gamma (Radiation) [also L$_\gamma$]
lg common logarithm (i.e., to the base 10) [Symbol]
lg long
LGA Land Grid Array
LGA Lattice Gas (Cellular) Automata (Method) [Fluid Flow Simulation]
LGA Leadless Grid Array [Electronics]
LGA Light-Gun Amplifier
LGA Local Government Agency
LGA Local Government Area
LGA Local Government Authority
LGA Low Gain Antenna
LGA Low Gloss/Automotive
L-GAP Linear Glycidyl Azide Polymer
LGB Lamellar Grain Boundary [Materials Science]
LGC Laboratory of the Government Chemists [UK]
LGC LEM (Lunar Excursion Module) Guidance Computer [of NASA Apollo Program]

LGC Liquid Gas Carrier
LGCP Lexical-Graphical Composer Printer
LGD Local Government District
LGDP Local Government Development Program
LGDT Load Global Descriptor Table
LGE Lunar Geological Equipment [of NASA Apollo Program]
lge large
LGEIES Local Government Environmental Information Exchange Scheme [Australia]
LGen Lieutenant-General
LGEP Laboratoire de Génie Electrique de Paris [Paris Electrical Engineering Laboratory; of Université de Paris VI and XI, France]
LGF Lithium Gadolinium Fluoride (Laser)
LGG Light Gas Gun
LGG Light-Gun (Pulse) Generator
LGGE Laboratoire de Glaciologie et de Géophysique de l'Environnement [Glaciology and Environmental Geophysics Laboratory, St. Martin d'Hères, France]
LGHM Low Gravity Heavy Media
LGIO Local Government Information Office [UK]
LGIU Laser Gyro Interface Unit
LGK Lipton, Glicksman and Kurz (Model) [Metallurgy]
LGL Lamont Geological Laboratory [of Columbia University, New York, US]
LGM Laboratoire de Génie des Matériaux [Materials Engineering Laboratory; of Centre d'Etudes Nucléaires de Grenoble, France]
LGM Little Green Men
LGN Line Generator Number
LGN Logical Group Number
LGMA Lattice Girder Floor Association [UK]
LGO Lamont Geological Observatory [of Columbia University, New York, US]
LGO Lanthanum Gallium Oxide
LGov Lieutenant-Governor
LGP Low Ground Pressure (Tractor)
LGQ Laboratoire de Géologie du Quaternaire [Quaternary Geology Laboratory, of Conseil National de la Recherche Scientifique at Meudon-Belleville, France]
LGR Loop-Gap Resonator
LGR Light Gray
Lgr Ligroin [also lgr]
LGRB Landesamt für Geologie, Rohstoffe und Bergbau Baden-Württemberg [Baden Württemberg State Office for Geology, Natural Resources and Mining, Freiburg/Breisgau, Germany]
LGRB Landesamt für Geowissenschaften und Rohstoffe Brandenburg [Brandenburg State Office for Geosciences and Natural Resources, Kleinmachnow, Germany]
LGRP Liaison Group for International Educational Exchange [US]
LGS Lafayette Geological Society [Louisiana, US]
LGS Lamella Gravity Settler
LGS Landing Guidance System
LGS Langevin-Gioumousis-Stevenson (Model) [Physics]

LGS Lanthanum Germanium Sulfide

LGS Large-Grain Size [Metallurgy]

LGS LDB (Launch Data Bus) Gateway Services

LGS Liquid-Gas-Solidification (Processing)

LGS Louisiana Geological Survey [US]

LGST Lamella Gravity Settler/Thickener

LGT Low Group Transmit

Lgt Light [also lgt]

lg t long ton [Unit]

Lgth Length [also lgth]

LGU Liechtensteinische Gesellschaft für Umweltschutz [Liechtenstein Society for Environmental Protection]

LGV Large Granular Vesicle

LGV Lymphogranuloma Venereum [also lgv] [Medicine]

LGWS Laser-Guided Weapons System

LH Laser Head

LH Latent Heat

LH Left-Hand(ed) [also lh]

LH Leybold Heraeus GmbH (Hanau, Germany)

LH Light Helicopter

LH Linde-Hampson (Process) [Chemical Engineering]

LH Linear Hybrid

LH Lingkugan Hidup [Ministry of State for Environment, Indonesia]

LH Liquid Hydrogen

LH Litter Hook

LH Local Harmonic (Model) [Physics]

LH Local Horizontal

LH Locating Head

LH Lock Hopper

LH Loss-of-Head

LH Lower Half [also lh]

LH Lower Hybrid (Plasma) [Physics]

LH Low-Noise, High-Output [Electronics]

LH Lufthansa German Airlines

LH Luteinizing Hormone [Biochemistry]

LH$_2$ Liquid Hydrogen

L/H Low-to-High

L/h Liter(s) per hour [also L/hr, L h^{-1}, l/h, l/hr, lh^{-1}, ltr h^{-1} ℓ/hr, or ℓ/h]

LHA Lanthanum Hexaaluminate

LHA Linear Hypervertex Approximation

LHA Local Hour Angle [Astronomy]

LHA Louisiana Historical Association [US]

L/ha Liter(s) per hectare [also L ha^{-1}, l/ha, l ha^{-1}, or ℓ/ha]

LH-AMC Leybold Heraeus–Automatic Melt-Rate Control

LHANA Log House Association of North America [now Log House Builder's Association of North America, US]

LHASA Logic and Heuristics Applied to Synthetic Analysis

LHB Linear Hydrogen Bond [Physical Chemistry]

LHBANA Log House Builder's Association of North America [US]

LHBC Liquid Hydrogen Bubble Chamber [of Argonne National Laboratory, US]

LHC Large Hadron Collider [of CERN–European Laboratory for Particle Physics, Geneva, Switzerland]

LHC Left-Hand Circular (Polarization) [Physics]

LHC Light-Harvesting Complex [Biochemistry]

LHC Log Home Council [now North American Log Homes Council, US]

LHC-B Study of CP (Charge-Parity) Violation in B-Meson Decays at the Large Hadron Collider [of CERN–European Laboratory for Particle Physics, Geneva, Switzerland]

LHCC Large Hadron Collider (Experiments) Committee [of CERN–European Laboratory for Particle Physics, Geneva, Switzerland]

LHCD Lower Hybrid (Plasma) Current Drive [of Tokamak de Varennes, Quebec, Canada]

LHChl Light-Harvesting Chlorophyll (Protein Complex) [Biochemistry]

LHCP Left-Hand Circularly Polarized; Left-Hand Circular Polarization [Physics]

LHD Left-Hand Drive [also lhd]

LHD Left Hepatic Duct [Anatomy]

LHD *(Litterarum Humaniorum Doctor)* – Doctor of Humanities

LHD Load-Haul-Dump (Technique) [Mining]

LHDC Lateral Homing Depth Charge

LHE Light Harvesting Efficiency [Solar Cells]

LHE Low-Hydrogen Electrode [Arc Welding]

LHE Low-Hydrogen Embrittlement (Plating)

LHe Liquid Helium

LHEA Laboratory for High-Energy Astrophysics [of NASA at Goddard Space Flight Center, Greenbelt, Maryland, US]

LHED Lateral Hot Electron Device

LHEP Laboratory for High Energy Physics [of University of Berne, Switzerland]

LHF List Handling Facility

LHGR Linear Heat Generation Rate

LHI Laser Holographic Interferometer; Laser Holographic Interferometry

LHI Lord Howe Island [New South Wales, Australia]

L/100K Liters per 100 Kilometers [also L/100 km]

LHM Loop Handling Machine

LHMC London Hospital Medical College [UK]

LHMEL Laser-Hardened Materials Evaluation Laboratory [of Wright-Patterson Air Force Base, Dayton, Ohio, US]

LHNC Linearized Hypernetted Chain

LHO Linear Harmonic Oscillator

LHOTS Long-Haul Optical Transmission Set

LHP Left Hand Panel

LHP Longnet-Higgins and Pople (Hard-Sphere Model) [Metallurgy]

LHP Lower Half Plane

LHPG Laser-Heated Pedestal Growth

LHPR Laboratory for Human Performance Research [of Pennsylvania State University, University Park, US]

LHPS Lead Hydrogen Purge System

LHR Lower Hybrid Resonance

LHR Low Heat Rejection

l-hr lumen-hour [Unit]

L/hr Liter(s) per hour [also l/hr, L/h, L h^{-1}, l/h, lh^{-1}, ltr h^{-1}, ℓ/hr, or ℓ/h]

LHRH Lower Hybrid Resonance Heating

LHRH Luteinizing-Hormone Releasing Hormone [also LH-RH] [Biochemistry]

LHRH-BSA Luteinizing-Hormone Releasing Hormone–Bovine Serum Albumin (Immunogen) [Biochemistry]

LHS Laser Heterodyne Spectroscopy

LHS Lauryl Hydrogen Sulfate

LHS Lawrence Hall of Science [of University of California at Berkeley, US]

LHS Left-Hand Side [also lhs]

LHS Loop Handling System [Nuclear Reactors]

LHS Louisiana Historical Society [US]

LHS Low-Cost High-Strength

LHSC Lock Haven State College [Pennsylvania, US]

LHSI Low Head Safety Injection

LH-Stir Leybold-Heraeus Stirring (Process) [Metallurgy]

LHSV Liquid Hourly Space Velocity

LHTES Latent Heat Thermal Energy Storage System

LHTR Lighthouse Transmitter-Receiver

LHV Low Heat Value; Lower Heating Value

LH-VIM Leybold-Heraeus Vacuum Induction Melting [Metallurgy]

LHWA Linearized Hot-Wire Anemometer

LI Labeling Index

LI Landscape Institute [UK]

LI Langmuir Isotherm [Physical Chemistry]

LI Large Intestine [Anatomy]

LI Laser Inspection

LI Laser Interferometer; Laser Interferometry

LI Laser Ionization

LI Latent Image

LI Lawn Institute [US]

LI Leachability Index

LI Learned Information

LI Leeward Islands

LI Level Indicator

LI Level of Intensity [Acoustics]

LI Left In (Place)

LI Length Indicator

LI Liechtenstein [ISO Code]

LI Light Infantry

LI Line

LI Linear Independence

LI Linearly Independent (Row)

LI Linear Interpolation [Mathematics]

LI Link

LI Liquid Infiltration (Technology)

LI Liquid Injection

LI Load Indicator

LI London Institute [UK]

LI Long Island [Brooklyn, New York, US]

LI Low Inertia

LI Low Intensity

LI Lýdhveldidh Island [Republic of Iceland]

L-I Light-Current (Characteristics) [Electronics]

Li Lignite [Geology]

Li Lilangeni [Currency of Swaziland]

Li Lithium [Symbol]

Li$^+$ Lithium Ion [Symbol]

Li-6 Lithium-6 [also ^{6}Li, or Li^6]

Li-7 Lithium-7 [also ^{7}Li, or Li^7]

Li-8 Lithium-8 [also ^{8}Li, or Li^8]

Li-9 Lithium-9 [also ^{9}Li, or Li^9]

.li Liechtenstein [Country Code/Domain Name]

LIA Laser Industries Association [Name Changed to Laser Institute of America, US]

LIA Laser Institute of America [US]

LIA Lead Industries Association [US]

LIA Leather Industries of America [US]

LIA Libration-Induced Anisotropy [Physics]

LIA Linear Integrated Amplifier

LIA Lock-In Amplifier

LIA Loop Interface Address

LIA Luminescence Immunoassay [Immunology]

LIA Lysine Iron Agar [Biochemistry]

Liab Liability [also liab]

Liabs Liabilities [also liabs]

LIAC Legal Industry Advisory Council [US]

LiAc Lithium Acetate

Li(ACAC) Lithium Acetylacetonate [also Li(AcAc), or Li(acac)]

LIADA Liga Iberoamericana de Astronomía [Ibero-American Astronomy League, Venezuela]

LiAH Lithium Aluminum Hydride

LIAM Legal Institutional Assessment Model

LIAS Library Information Access System

LiAs Lithium Arsenide

Lias Liaison [also lias]

Lias Liassic [Geology]

LIB Laboratoire d'Immunologie et Biochimie [Immunology and Biochemistry Laboratory; Petit, France]

LIB Left Inboard

LIB Line Interface Base

LIB Low-Index (Grain) Boundary [Materials Science]

.LIB Library [File Name Extension] [also .lib]

Lib Liberia(n)

Lib Librarian; Library [also lib]

lib *(liber)* – book

LIBA Long Island Biological Association [US]

LIBCEPT Library Information System Intercept [Sweden]

LIBER Ligue des Bibliothèques Européennes de Recherche [League of European Research Libraries, Austria]

LIBID London Interbank Bid Rate [UK]

LiBO Lithoborate

Li₂B₄O₇:Mn Manganese-Activated Lithium Borate (Thermoluminescence Dosimeter)

LIBOR London Interbank Offered Rate [also Libor, UK]

LIBORS Laser-Ionization Based On Resonance Saturation

Li-BR Butadiene Rubber Based on Lithium Catalyst

LiBr Lithium Bromide

Libr Librarian; Library [also libr]

Libr Comput Syst Equip Rev Library Computer Systems and Equipment Review [US]

Lib Res Library Research

Libr Hi Tech Library Hi Tech [Published in the US]

Libr Inf News Library Information News [UK]

Libr Inf Sci Library and Information Science [Publication of the Mita Society of Library and Information Science, Japan]

LIBRIS Library Information System [Sweden]

Libr Micromation News Library Micromation News [UK]

Librn Librarian [also librn]

Libr Q Library Quarterly [US]

Libr Res Library Research(er)

Libr Resour Tech Serv Library Resources and Technical Services [Published by the American Library Association, US]

Libr Sci Library Science; Library Scientist

Libr Softw Rev Library Software Review [US]

Libr Sys Library System

Libr Trends Library Trends [Publication of the University of Illinois Graduate School of Library and Information Science, US]

LIBS Laser-Induced Breakdown Spectroscopy

Lib Sci Library Science; Library Scientist

Lib Sys Library System

LiBuO Lithium Butoxide

LIC Land Information Center [of (Department of) Conservation and Land Management, Bathurst, New South Wales, Australia]

LIC Leiden Institute of Chemistry [of Leiden University, Netherlands]

LIC Linear Integrated Circuit

LIC Line Interface Coupler

LIC Load Interface Circuit

LIC Loop Insertion Cell

⁷Li-¹³C Lithium-7-Carbon-13 (Double Resonance)

Lic License [also lic]

Lic Licentiate

LICA Land Improvement Contractors Association [US]

LiCAF Lithium-Calcium-Aluminum Fluoride (Laser Material) [LiCaAlF₆]

LiCAF(Cr) Chromium-Doped Lithium-Calcium-Aluminum Fluoride (Laser Material)

LicCien (Licenciado en Ciencias) – Spanish for "Licentiate in Science"

Li-Cd Lithium-Cadmium (Alloy System)

Lichtbogen Der Lichtbogen [Publication on Electric Arcs of Chemische Werke Hüls AG, Marl, Germany]

LiCl Lithium Chloride

LiCN Lithium Cyanide

LICOF Land Lines Communication Facilities

LICOR Lightning Correlation

LiCp Lithium Cyclopentadienyl

LICRY Liquid Crystal [also Li Cry]

LICRY-NMR Liquid Crystal Nuclear Magnetic Resonance (Spectroscopy)

LICS Lotus International Character Set

LICVD Laser-Induced Chemical Vapor Deposition

LID Land Information Division [of Melbourne Metropolitan Board of Works, Victoria State, Australia]

LID Laser-Induced Desorption

LID (Crystal) Lattice Invariant Deformation

LID Leadless Inverted Device [also lid]

LID Liberian Dollar [Currency]

LID Light-Induced Drift

LID Line Isolation Device

LID Liquid Development [also lid]

LID Local Improvement District

LID Local Issue Data

LID Locked-In Device

LID Logical Input Device

LID Logistics Identification Document

LiD Lithium Deuteride

LIDAR Laser Infrared Radar [also Lidar, or lidar]

LIDAR Laser Intensity Direction and Ranging [also Lidar, or lidar]

LIDAR Light Detection and Ranging [also Lidar, or lidar]

LIDC Lead Industries Development Council [UK]

LIDC London Interior Designers Center [UK]

LIDF Line Intermediate Distribution Facility

LiDEA Lithium Diethylamide

LiDMA Lithium Dimethylamide

LIDQA Landsat Image Data Quality Analysis [of NASA]

LIDS Laboratory for Information and Decision Systems [of Massachusetts Institute of Technology, Cambridge, US]

LIDS Landed Immigrant Data System [Canada]

LIDS Laser-Induced Decohesion Spectroscopy

LIDT Load Interrupt Descriptor Table

LIE Left Inboard Elevon [Aeronautics]

LIED Linkage Editor

LIEF Launch Information Exchange Facility [of NASA]

LIEP Large Internet Exchange Packet

LiErF₄ Lithium Erbium Fluoride (Laser)

LIESG Laboratoire d'Ionique et d'Electrochimie des Solides de Grenoble [Grenoble Ionics and Solid Electrochemistry Laboratory; of Institut National Polytechnique de Grenoble, France]

LIESG-ENSEEG /Laboratoire d'Ionique et d'Electrochimie des Solides de Grenoble–Ecole Nationale Supérieure d'Electrochimie et d'Electrométallurgie de Grenoble [Grenoble Ionics and Solid Electrochemistry Laboratory–National Higher School of Electrochemistry and Electrometallurgy of Grenoble, France]

LIESST Light-Induced Excited Spin-State Trapping [Physics]

LIET Lesotho In-Service Education for Teachers Certificate

Liet TSR Mokslu Akad Darb A Lietuvos TSR Mokslu Akademijas Darbai, Serija A [Russian Journal]

Liet TSR Mokslu Akad Darb B Lietuvos TSR Mokslu Akademijas Darbai, Serija B [Russian Journal]

Lieut Lieutenant

Lieut Cdr Lieutenant Commander [also Lieut Cmdr or Lieut Comdr]

Lieut Col Lieutenant-Colonel

Lieut Comdr Lieutenant-Commander [also Lieut Cdr, or Lieut Cmdr]

Lieut Gen Lieutenant General

Lieut Gov Lieutenant Governor

Lieut(N) Lieutenant (Navy)

LIF Laser-Induced Fluorescence

LIF Laser-Induced Fusion

LIF Leach Ion-Exchange Flotation

LIF Leukemia Inhibitory Factor [Immunology]

LIF Lighting Industry Federation [UK]

LIF Low Insertion Force

LiF Lithium Fluoride (Crystal)

LIFAC Limestone in Furnace and Added Calcium (Process)

LiF:Dy Dysprosium-Activated Lithium Fluoride (Thermoluminescence Dosimeter)

Life Sci Life Science

Life Sci Life Sciences [Journal]

LIFETECH Long-Term Funding for the Environment Through Technology

LIFFE London International Financial Futures Exchange [UK]

LIFLN Laser-Induced Fluorescence Line Narrowing

LiF:Mg Magnesium-Activated (or -Doped) Lithium Fluoride (Phosphor)

LiF:(Mg,Ti) Magnesium and Titanium Activated (or Doped) Lithium Fluoride (Phosphor)

LiF:Mn Manganese-Activated Lithium Fluoride (Thermoluminescence Dosimeter)

LIFMOP Linearly Frequency-Modulated Pulse

LIFO Last In, First Out [also lifo]

LIFR Low Instrument Flight Rules

LIFS Laser-Induced Fluorescence Spectroscopy

LIFT Link Intellectual Functions Tester

LIFT Logically Integrated FORTRAN Translator

LiF:Ti Titanium Activated (or Doped) Lithium Fluoride (Phosphor)

LiFZnS Lithium Fluoride Zinc Sulfide (Crystal)

LiFZnS(Ag) Silver-Doped Lithium Fluoride Zinc Sulfide (Crystal)

Lig Ligament [also lig]

Lig Lignite [also lig]

LIGA (*Lithographie, Galvanoformung und Abformung (Prozess)*) – German for "Lithography, Electroplating and Molding (Process)"

LIGAMUMPs LIGA Multi-User MEMS (Microelectromechanical System) (Process)

LIGB Low-Index Grain Boundary [Materials Science]

LIGDG Long Island Gastrointestinal Disease Group [Merrick, New York, US]

Li:Ge Lithium-Doped Germanium

Light Lighting [also light]

Light Aust Lighting in Australia [Australian Publication]

Light Met Age Light Metal Age [Published in the US]

Light Res Technol Lighting Research and Technology [Publication of the Chartered Institution of Building Services Engineers, UK]

LIGNA Internationale Fachmesse für Maschinen und Ausrüstung der Holz– und Forstwirtschaft [World Trade Fair for Machinery and Equipment for the Lumber and Forestry Industries, Hannover, Germany]

LIGNOL Lignin Phenol (Process)

LIGO Laser Interferometer Gravitational-Wave Observatory (Project) [Joint Project of CalTech and Massachusetts Institute of Technology, US]

Ligr Ligroin [also ligr]

LIH Lehrstuhl für Ingenieurgeologie und Hydrogeologie [Chair for Engineering Geology and Hydrogeology, RWTH Aachen, Germany]

LIH Light Intensity High

LIH Line Interface Handler

LiH Lithium Hydride

LIHS Long Island Historical Society [Brooklyn, US]

LIHS Long Island Horticultural Society [Brooklyn, US]

LIL LEP (Large Electron Positron) (Collider) Injector Linac [of CERN–European Laboratory for Particle Physics, Geneva, Switzerland]

LIL Large Ionic Lithophile

LIL Light Intensity Low

LIL Lunar International Laboratory

LiI Lithium Iodide

LILCO Long Island Lighting Company [New York State, US]

LILO Last In, Last Out [also lilo]

Li:LSCO Lithium-Doped Lanthanum Strontium Copper Oxide (Superconductor)

LIM Laboratoire d'Ingéniérie des Matériaux [Materials Engineering Laboratory, Villetaneuse, France]

LIM Lamellar Injection Molding

LIM Land Information Management

LIM Language Interpretation Module

LIM Light Intensity Medium

LIM Limit in the Mean [also lim]

LIM Linear Induction Motor

LIM Line Insulation Monitor

LIM Line Interface Module

LIM Liquid Injection Molding

LIM Locator Inner Marker [Aeronautics]

LIM Lotus/Intel/Microsoft (Memory Specification)

Lim Limit(ation); Limiter [also lim]

lim limited

lim limit (value) (in mathematics) [Symbol]

l̄im limit superior (in mathematics) [Symbol]

l̲im̲ limit inferior (in mathematics) [Symbol]

LIMA Laboratoire d'Instrumentation Météorologique Aéroportée [Laboratory for Airborne Meteorological Instrumentation, France]

LIMA Laser-Induced Ion Mass Analysis; Laser-Induced Ion Mass Analyzer

LIMA Laser-Induced Mass Analysis; Laser-Induced Mass Analyzer

LIMA Laser Ionization Mass Analyzer/Spectrometric Analysis

LIMA Logic in Memory Array

LIMA Lotus/Intel/Microsoft/AST [*Note:* "AST" Stands for AST Research, Inc.]

LIMAC Large Integrated Monolithic Array Computer

LIMB (Comprehensive) List of Molecular Biology Databases [of United Nations International Center for Genetic Engineering and Biotechnology]

LIMB Limestone Injection (into a) Multistage Burner (Process) [Metallurgy]

LIMB Liquid Metal Breeder (Reactor)

LiMCA Liquid Metal Cleanness Analyzer

LIMDO Light Intensity Modulation Direct Overwrite

LIME Laser-Induced Magnetically-Enhanced

LIM EMS Line Interface Module/Expanded Memory Specification

LIM EMS Lotus/Intel/Microsoft/Expanded Memory Specification

LiMeO Lithium Methoxide

LIMEX Labrador Ice Margin Experiment [Canada]

LIMFS Laser-Induced Molecular Fluorescence Spectrometry

lim inf limit inferior (in mathematics) [Symbol]

LIMIRIS Laser Induced Modulation of Infrared in Silicon

LIMIT Lot-Size Inventory Management Interpolation Technique

LIML Limited-Information Maximum Likelihood

LIMM Light Intensity Modulation Method

LIMMS Laboratory for Integrated MicroMecatronic Systems [of Conseil National de la Recherche Scientifique, France]

Limnol Limnological; Limnologist; Limnology

Limnol Oceanogr Limnology and Oceanography [Journal published in the US]

LIMP Language-Independent Macroprocessor

LIMP Lunar Interplanetary Monitoring Platform

LIMPOL Leningrad Institute for Mathematics and Political Sciences [USSR]

LIMR Liquid Injection Moldable Rubber

LIMRC LRU (Line Replaceable Unit) Identification and Maintenance Requirements Catalog

LIMS Laboratory Information Management Software

LIMS Laboratory Information Management System

LIMS Laser Ionization Mass Spectrometer; Laser Ionization Mass Spectrometry

LIMS Laser Ionization Mass Spectroscopy

LIMS Library Information Management System

LIMS Limb Infrared Monitor of the Stratosphere

LIMS Logistics Inventory Management System

LiMS Lithium-Metal Sulfide (Battery)

LIMS/SM Laser Ionization Mass Spectrometer Sample Management (Package)

lim sup limit superior (in mathematics) [Symbol]

Lim Sw Limit Switch

LIMTV Linear Induction Motor Test Vehicle

LIN Laser-Induced Nucleation

LIN Library Information Network

LIN Lindane

LIN Liquid Nitrogen [also LiN]

Lin Linearity [also lin]

lin lineal

lin linear

lin (*linimentum*) – liniment [Medical Prescriptions]

LINAC Linear Accelerator [also Linac, or linac]

LINAC2 Linear Accelerator 2 [Proton Accelerator at CERN– European Laboratory for Particle Physics, Geneva, Switzerland]

LINAC3 Linear Accelerator 3 [Lead Ion Accelerator at CERN– European Laboratory for Particle Physics, Geneva, Switzerland]

LINAS Laser Inertial Navigation Attack System

LiNb(OEt)$_5$ Lithium Niobium Double Ethoxide (or Lithium Niobium Double Ethylate)

LiNbO$_3$:Er^{3+} Titanium-Doped Erbium Metaniobate (Crystal)

LiNbO$_3$:Eu^{3+} Titanium-Doped Europium Metaniobate (Crystal)

LiNbO$_3$:Fe^{3+} Titanium-Doped Iron Metaniobate (Crystal)

LiNbO$_3$:Hf^{4+} Titanium-Doped Hafnium Metaniobate (Crystal)

LiNbO$_3$:In^{3+} Titanium-Doped Indium Metaniobate (Crystal)

LiNbO$_3$:Li^{1+} Titanium-Doped Lithium Metaniobate (Crystal)

LiNbO$_3$:Lu^{3+} Titanium-Doped Lutetium Metaniobate (Crystal)

LiNbO$_3$:Mn^{2+} Titanium-Doped Manganese Metaniobate (Crystal)

LiNbO$_3$:Nd^{3+} Titanium-Doped Neodymium Metaniobate (Crystal)

LiNbO$_3$:Ni^{2+} Titanium-Doped Nickel Metaniobate (Crystal)

LiNbO$_3$:Ta^{5+} Titanium-Doped Tantalum Metaniobate (Crystal)

LiNbO$_3$:Ti^{4+} Titanium-Doped Titanium Metaniobate (Crystal)

LINC Laboratory Instrument Computer

LINC Local Interactive Network Communications [of Vector Graphic, US]

LINCAGES Linkage Interactive Computer Analysis and Graphically Enhanced Synthesis (Package) [also Lincages]

LinCMOS Linear Complementary Metal-Oxide Semiconductor

Linc Lab J Lincoln Laboratory Journal [Publication of the Lincoln Laboratory at Massachusetts Institute of Technology, US]

Lincompex Linked Compander Expander [also lincompex]

Lincompex Linked Compressor and Expander

LINCS Language Information Network and Clearinghouse System [of Center for Applied Linguistics, US]

Lincs Lincolnshire [UK]

Linde Rep Sci Technol Linde Reports on Science and Technology [Published by Linde AG, Wiesbaden, Germany]

LINDOX Linde-Oxidation (Process)

Linear Algebr Appl Linear Algebra and Its Applications [Published in the US]

Lines Commun Lines Communication [Published in the UK]

LINET Legal Information Network

lin ft linear foot [Unit]

Ling Linguist; Linguistic(s) [also ling]

LINJET Liquid Injection Electric Thruster

LINK Library Information Network

LINLOG Linear-Logarithmic

lin m lineal meter [Unit]

LiNMe₂ Dimethylamidolithium

Linn Linnaeus [Botany]

LiNO Lithionitrate

Lino Linoleum [also lino]

LiNbO₃:Cu Copper-Doped Lithium Niobate

LiNbO₃:Fe Iron-Doped Lithium Niobate

LiNbO₃:Mg:Fe Magnesium and Iron-Doped Lithium Niobate

LINP Laser-Induced Nuclear Polarization

LINPOR Linde Porous Media (Process)

LINS Laser Inertial Navigation System

LINS LORAN (Long-Range Navigation) Inertial System

LINUP Laser-Induced Nuclear Polarization

LINUS Logical Inquiry and Update System

LIO Land Information Office [of Australian Capital Territory]

LIOD Laser In-Flight Obstacle Detector

LiOD Lithium Deuteroxide

LiOCl Lithium Hypochlorite

LIOCS Logical Input/Output Control System

Li(OEt) Lithium Ethoxide

LiOH Lithium Hydroxide

LiOMe Lithium Methoxide

LION Light Ion Fusion Facility [of Cornell University, Ithaca, New York, US]

LIONS Liberty, Intelligence, Our Nations' Safety (Organization) [now LIONS International, US]

LIOP Life In One Position

LIOP Logical Input/Output Processor

Li(OPrⁱ) Lithium Isopropoxide

Li(OR) Lithium (I) Alkoxide [General Formula; R represents an Alkyl Group]

Li₂O-SiO₂ Lithium Oxide-Silicon Dioxide (System)

LiOTa(OEt)₄ Lithium Tantalum (IV) Oxyethoxide

Li₂O-TeO₂ Lithium Oxide-Tellurium Oxide (System)

LIP Lance Injection Process [Metallurgy]

LIP Large Internet Packet

LIP Laser Integrated Prototype

LIP Lightning Instrument Package

LIP Limited Installation Program [US]

LIP Local Interested Party

Lip_α Lipschitz Class of Functions [also Lip α] [Mathematics]

LIPI Lembaga Ilmu Pengetahun Indonesia [Indonesian Institute of Science]

LIPN Latex Interpenetrating (Polymer) Network

LiPO Lithophosphate

LIPP Liquid Polymerization of Polypropylene

LIPP-SHAC Liquid Polymerization of Polypropylene with Superhigh Activity Catalyst (Process)

LIPL Linear Information Processing Language

LIPL Linear Information Programming Language

LIPS Lightweight Internet Person Schema

LIPS Logical Inferences per Second

Liq Liquid [also liq]

Liq Liquor [also liq]

liq (liquidus) – liquid; solution [Medical Prescriptions]

Liq Cryst Liquid Crystal

Liq Cryst Liquid Crystals [Published in the UK]

liq pt liquid pint [Unit]

liq qt liquid quart [Unit]

LIQT Liquid Transient (Procedure)

LIR Line Integral Refractometer

LIRA Linen Industry Research Association [Ireland]

LIRAD Lidar and Infrared Radiometer [also Lirad, or lirad]

LIRDP Luangwa Integrated Resource Development Project

LiREF₄ Lithium Rare-Earth Fluoride (Laser)

LIRES Literature Retrieval System

LIRES-MS Literature Retrieval System–Multiple Searching

LIRF Laser-Induced Resonance Fluorescence

LIRF Low-Intensity Reciprocity Failure [Photography]

LIRG Library and Information Research Group [UK]

LIRI Leather Industries Research Institute [South Africa]

LIRIMS Laser-Induced Resonance Ionization Mass Spectroscopy

LIRL Low-Intensity Runway Lights

LIRTS Large Infrared Telescope System

LIS Land Information System

LIS Lanthanide Induced Shifts

LIS Large Interactive Surface

LIS Laser-Induced Separation

LIS Laser-Induced Isotope Separation; Laser Isotope Separation

LIS (Crystal) Lattice Invariant Shear

LIS Library and Information Science

LIS Library and Information Service

LIS Lightning Imaging Sensor

LIS Line Information Store

LIS Line Isolation Switch

LIS List and Index Society [UK]

LIS Lithium 3,5-Diiodosalicylate

LIS Lockheed Information Systems [US]

LIS Logical Inferences per Second [also lis]

LIS Long Island Sound [Southeastern New York State, US]

LIS Long Island Sound (Program) [of US Environmental Protection Agency]

LIS Loop Input Signal

LIS Low-Inductance Stripline

LISA Leather Industry Suppliers Associates [UK]

LISA Library and Information Science Abstracts [UK]

LISA Library Systems Analysis

LISA Linked Indexed Sequential Access

LISA Locally Integrated Software Architecture

LiSAF Lithium-Strontium-Aluminum Fluoride (Laser Material)

LiSAF(Cr) Chromium-Doped Lithium-Strontium-Aluminum Fluoride (Laser Material)

LISARD Library and Information Service Automated Retrieval of Data

LISARD Library Information Search and Retrieval Data System [US Navy] [also LISARDS]

LISB Lenkungs– und Informationssystem, Berlin [Guidance and Information System, Berlin, Germany]

LISC Land Information System Committee [of State of Queensland, Australia]

LISC Library and Information Service Council [of Department of Education and Science, UK]

LiSCN Lithium Thiocyanate

LISD Library and Information Services Division [of National Oceanic and Atmospheric Administration, US]

LiSGAF Lithium-Strontium-Gallium-Aluminum Fluoride (Laser Material) [LiSrGaAlF$_6$]

LiSGAF(Cr) Chromium-Doped Lithium-Strontium Gallium-Aluminum-Fluoride (Laser Material)

Li-Si Lithium-Silicon (Alloy)

LISIC Library and Information Service to Industry and Commerce [UK]

LISN Line-Impedance Stabilization Network

Li-SO$_2$ Lithium-Sulfur Dioxide

LISP Laser Isotope Separation Program

LISP List Processing (Computer Language) [also Lisp]

LISP Littoral Investigation of Sediment Properties [Geology]

Li-SPS Sulfonated Polystyrene, Lithium Salt

LISP Symb Comput LISP and Symbolic Computation [Journal published in the Netherlands]

LISR Line Information Storage and Retrieval [of NASA]

LISSA Life Insurance Software Systems of America

LIST Library and Information Services for Teesside [UK]

LIST Linearly Increasing Stress Test [Mechanics]

LiSt Lithium Stearate

List Listening [also list]

LISTAR Lincoln Information Storage and Associative Retrieval System [of Lincoln Laboratory at Massachusetts Institute of Technology, US]

LISTSERV List Server [Internet]

Listy Cukrov Listy Cukrovarnicke [Czech Publication]

LIT Laboratório de Integração e Testes [Integration and Test Laboratory; of Brazilian National Institute of Space Research]

LIT Light Interchange Technology

LIT Light Interface Technology

LIT Linkoping Institute of Technology [Sweden]

LIT Liquid Injection Technique

LIT Little Rock, Arkansas [Meteorological Station Designator]

LIT Local Intelligent Terminal

LIT Logic Integrity Test

Lit Italian Lira [Currency]

Lit Literary [also lit]

Lit Literature [also lit]

lit literally

lit little

LITA Library and Information Technology Association [of American Library Association, US]

LiTa(OEt)$_6$ Lithium Tantalum Hexaethoxide

LITAS Low-Intensity Two-Color Approach Slope

LITASTOR Light Tapping Storage

LitB *(Literarum Baccalaureus)* – Bachelor of Letters; Bachelor of Literature

LITC Library and Information Technology Centre [of Polytechnic of Central London, UK]

LITCA Licensing, Innovation and Technology Consultants Association [Liechtenstein]

LiTCA Lithium Tetrachloroaluminate

LiTCG Lithium Tetrachlorogallate

LITCO Lockheed Idaho Technologies Company [Idaho Falls, US]

LITCO/INEL Lockheed Idaho Technologies Company/Idaho National Engineering Laboratory [Idaho Falls, US]

LITD Laser-Induced Thermal Desorption

LitD *(Literarum Doctor)* – Doctor of Letters; Doctor of Literature

LITE Laser-Induced Transmission Experiment

LITE Legal Information through Electronics

LITE Library and Information Technology Centre [of Polytechnic of Central London, UK]

LITE Lidar In-Space Technology Experiment [NASA Langley Research Center, Hampton, Virginia, US]

Liteinoe Proizvod Liteinoe Proizvodstvo [Russian Journal on Casting Technology]

Liter Literary [also Liter, or liter]

Liter Literature [also Liter, or liter]

Lith Lithographic; Lithograph; Lithography

Lith Lithological; Lithologist; Lithology

Lith Lithuania(n)

Li(THD) Lithium 2,2,6,6-Tetramethyl-3,5-Heptanedionate [also Li(thd)]

Lith J Phys Lithuanian Journal of Physics

Lith Min Resour Lithology and Mineral Resources [Journal published in the US]

Litho Lithographer; Lithographic; Lithography [also Lithog, lithog, or litho]

Lith SSR Lithuanian Soviet Socialist Republic [USSR]

Lit Linguist Comput Literary and Linguistic Computing [Publication of Oxford University Press, UK]

Litov Fiz Sb Litovskii Fizicheski Sbornik [Russian Collection of Physics Publications]

LITP Landau Institute for Theoretical Physics [Moscow, Russian Federation]

LITR Low-Intensity Test Reactor [of Oak Ridge National Laboratory, Tennessee, US]

Lit Rev Literature Review [also lit rev]

LITS Light Interface Technology System

LittB *(Litterarum Baccalaureus)* – Bachelor of Letters; Bachelor of Literature

LittD *(Litterarum Doctor)* – Doctor of Letters; Doctor of Literature

LITVC Liquid Injection Thrust Vector Control

LIU Labourer's International Union [US and Canada]

LIU LAN (Local-Area Network) Interface Unit

LIU Line Interface Unit

LIU Long Island University [Brooklyn, New York, US]

LIUNA Labourer's International Union of North America

LIV Low Investment Vehicle

L-I-V Light-Current-Voltage (Characteristics)

Living Rev Relat Living Reviews in Relativity [US Journal]

LIWB Livermore Water Boiler [of University of California, US]

LiWO Lithiotungstate

LiZn Lithium Zinc (Ferrite)

Li-Zn Lithium-Zinc (Alloy System)

LJ Lagrangian-Jaumann (Formulation) [Physics]

LJ Lap Joint

LJ Leaf Jig

LJ Left Justify; Left Justification

LJ Lennard-Jones (Potential) [also L-J] [Physical Chemistry]

LJ Liquid Junction [Physical Chemistry]

LJA London Jute Association [UK]

LJAO London Joint Area Organization [UK]

LJE Local Job Entry

LJP Lennard-Jones Potential [Physical Chemistry]

LJSU Local Junction Switching Unit

LK Laplace-Kelvin (Theory) [Physics]

LK Luttinger-Kohn (Hamiltonian) [Mechanics]

LK Sri Lanka [ISO Code]

Lk Lek [Currency of Albania]

Lk Leak [also lk]

Lk Link [also lk]

Lk Lock [also lk]

Lk Look [also lk]

.lk Sri Lanka [Country Code/Domain Name]

L/100K Liters per 100 Kilometers [also L/100 km]

LKE Language Knowledge Examination

LKE Longitudinal Kerr Effect [Optics]

LKKR Layer(ed) Korringa-Kohn-Rostoker (Method) [Solid-State Physics]

LKM Low Key Maintenance

L/100 km Liters per 100 Kilometers [also L/100K]

LKR Sri Lankan Rupee [Currency]

LKr Lithium Kryptonite

Lkr Locker [also lkr]

LK ROUTE Looking for Route [also Lk Route]

LKS Lithium Potassium Metasilicate

LKSV Laboratorium voor Kristallografie en Studie van de Vastestof [Crystallography and Solid-State Research Laboratory; Rijksuniversiteit Gent, Ghent, Belgium]

LKT Lipton, Kurz and Trivedi (Model) [Metallurgy]

Lk Wash Lock Washer

LKY Lifschitz-Kalman-Yakubovitch [Physics]

LL Lamp Lumen

LL Landau (Energy) Level [Solid-State Physics]

LL Landau-Lifshitz (Equation) [Solid-State Physics]

LL Landing Light

LL Land-Line

LL Laser Light

LL Lateral Line [Zoology]

LL Laterlog [Geology/Mining]

LL Launch and Landing

LL Launch Left

LL Leased Line

LL Lebanese Pound [Currency]

LL Left Lung [Anatomy]

LL Leftmost LALR (Lookahead, Left-to-Right Scan)

LL Lever Lock

LL Lifeline

LL Light Line

LL Light Load(ing)

LL Limit Load

LL Lincoln Laboratory [of Massachusetts Institute of Technology, US]

LL Line Leg

LL Line Link

LL Lines [also ll]

LL Liquid Laser

LL Liquid Limit (of Soil) [Geology]

LL Live Load

LL Load Line [Electronics]

LL Local Line

LL Local Loopback

LL Long-Lead (Material)

LL Long Life

LL Long Line

LL Looseleaf

LL Loudness Level

LL Lower Left

LL Lower Level

LL Lower Limit

LL Low Level

LL Low Leucite

LL Luer Lock

LL Luminescent Layer

L&L Launch and Landing

L/L Latitude/Longitude

L/L Liquid-in-Liquid (Solution)

L/L Liter per Liter [also L L^{-1}, l/l, l l^{-1}, ltr/ltr, or ltr ltr^{-1}, or ℓ/ℓ]

L-L Line-to-Line

L-L Liquid-Liquid (Interface)

ll lines [also LL]

ll *(loco laudato)* – in the place cited

λL Camera Constant [Transmission Electron Microscopy]

LL(1) Leftmost LALR(1) (Leftmost Left-to-Right Scan, One Lookahead)

LLA Leased Line Adapter

LLA Lend-Lease Administration [US]

LLA Lin-Lin-An (Theory) [Materials Science]

LLA Low Level Access

LLA Low-Level Attack

LLAD Low-Level Air Defense

LLAR Local Line Automatic Routing

LLB Laboratoire Léon Brillouin [Leon Brillouin Laboratory, Saclay, France]

LLB Lowest Lower Bound [Mathematics]

LL B *(Legum Baccalaureus)* – Bachelor of Laws

LLC Langmuir Liquid Crystal

LLC Link Layer Control

LLC Liquid Level Control

LLC Liquid-Liquid Chromatography

LLC Liquid Live Culture

LLC Logical Link Control

LLC Low-Level Circuit

LLC Lunar Launch Complex [of NASA Apollo Program]

LLCF Launch and Landing Computational Facilities

LLD Lamp Lumen Depreciation (Factor)

LLD Legislative and Liaison Division [US Army]

LLD Limit-Load Design

LLD Linear Low Density [Polymer Engineering]

LLD Load Line Displacement [Mechanics]

LLD Lower Limit of Detection

LLD Low-Level Discriminator

LL D *(Legum Doctor)* – Doctor of Laws

LLDC Least Developed Country [also lldc]

L2D2 LORAN (Long-Range Navigation) Digital Dropwindsonde [also L^{2}D^2]

LLDPE Linear Low-Density Polyethylene

LLDT Load Local Descriptor Table

LLE Laboratory for Laser Energetics [of University of Rochester, New York, US]

LLE Liquid-Liquid Equilibrium

LLE Liquid-Liquid Extraction [Chemical Engineering]

LLE Long Line Equipment

LLEG Lindhurst Laboratory of Experimental Geophysics [of California Institute of Technology, Pasadena, US]

LLF Line Link Frame

LLF Low Level Format

LLFF Low Level File Functions

LLFM Low-Level Flux Monitor

LLG Landau-Lifshitz-Gilbert (Magnetization Equation) [Solid-State Physics]

LLG Lanthanum Lithium Germanate

LLG Landcare Liaison Group [Australia]

LLG Logical Line Group

LLGA Leadless Land Grid Array

LLH Low-Level Heating

LLI Limited Life Item

LLI Low-Level Interface

LLI Lower Large Intestine

LLIL Long Leadtime Items List

LLIST Current Program in Memory to Line Printer (Command) [Programming]

LLL Landau-Lifshitz-Looyenga (Formula) [Physics]

LLL Lawrence Livermore Laboratory [at University of California at Livermore, US]

LLL Low-Level (Programming) Language

LLL Low-Level Logic [also L^3]

L1, L2, L3 Line 1, Line 2, Line 3, etc [Controllers]

LLLLLL Laboratories Low-Level Linked List Language [also L6, or L^6]

LLLTV Low Light-Level Television [also L^{3}TV]

LLLWT Low-Level Liquid Waste Tank

LLM Low-Level Multiplexer

LL M *(Legum Magister)* – Master of Laws

LLMC (Convention on) Limitation of Liability for Maritime Claims

LLMW Low-Level Mixed (Radioactive) Waste

LLN Laboratoire Louis Néel [of Centre National de la Recherche Scientifique, Grenoble, France]

LLN Line Link Network

LLN Local Line Network

LLN-CNRS Laboratoire Louis Néel–Centre National de Recherche Scientifique [Louis Neel Laboratory– National Center of Scientific Research, Grenoble, France]

LLNL Lawrence Livermore National Laboratory [of US Department of Energy at the University of California at Livermore]

LLO Light Lubricating Oil

LLOS Landmark Line-of-Sight [Navigation]

LLP Launch and Landing Project

LLP Lightning Location and Prediction

LLP Line Link Pulse

LLP Linux Lab(oratory) Project

LLP Liquid-Liquid Partition (Chromatography)

LLPL Low Low Pond Level

LLPMM Linear Lines per Millimeter [also llpmm]

LLPO Launch and Landing Project Office

LLQ Left Lower Quadrant (of Abdomen) [Medicine]

LLR Line of Least Resistance

LLR Lloyd's Register of Shipping [UK]

LLR Load-Limiting Resistor [also LLRES]

LLR Lower Luminaire Retainer

LLR Low-Level Radioactivity

LLRF Lunar Landing Research Facility [of NASA Apollo Program]

LLRMW Low Level Radioactive Mixed Waste

LLRT Low Level (Nuclear) Reactor Test

LLRV Lunar Landing Research Vehicle [of NASA Apollo Program]

LLRW Low Level Radioactive Waste

LLS Laser Light Scattering

LLS Laterolog, Shallow Investigation [Geology]

LLS Launch and Landing Site

LLS Liquid Level Switch

LLS Local Library System

LLS Local Load Sharing [Mechanics]

LLS Local Load Shearing

LLS Lunar Logistic System [of NASA Apollo Program]

LLS ATD Laser Light Scattering Advanced Technology Development (Program) [of NASA]

LLSBA Leicester Longwool Sheep Breeders Association [UK]

LLSE Linear Least Squares Estimation

LLSU Low Level Signaling Unit

LLT Lanthanum Lithium Titanate

LLT Long Leadtime

LLTHV Laboratorium voor Lage Temperaturen en Hoge-Veldenfysika [High-Temperature and High-Field Physics Laboratory; of KU Leuven, Belgium]

LLTR Large Leak Test Rig

LLTT Landline Teletype

LLTV Low Light(-Level) Television

LLTV Lunar Landing Training Vehicle [of NASA Apollo Program]

LLV Low-Level Vault [Nuclear Engineering]

LLV Lunar Logistics Vehicle [of NASA Apollo Program]

LLW Leaky Lamb Wave [Physics]

LLW Lower Low Water

LLW Low-Level (Radioactive) Waste

LLWI Lake Louise Winter Institute [at Lake Louise, Alberta, Canada]

LLWS Low-Level Wind Shear

LLWSAS Low-Level Wind Shear Alert System

LM Laboratoire de Métallurgie [Metallurgy Laboratory; of Conseil National de Recherche Scientifique at Marseilles, France]

LM Ladle Metallurgy

LM Lagrange Multiplier [Mathematics]

LM Laminated Metal

LM Landing Module [of NASA Apollo Program]

LM Land Mine

LM Landscape Management

LM Langmuir-McLean (Method) [Physical Chemistry]

LM Lanthanum Manganite

LM Laser Machining

LM Laser Material

LM Laser Medicine

LM Laser Memory

LM Laser Microprobe

LM Lateral Magnification

LM Lead Mechanic

LM Lean Mixture

LM Lee-Metford (Rifle)

LM Left Mid

LM Leg Multiple

LM Length Measurement

LM Life Member

LM Light Material

LM Light Metal

LM Light Micrograph; Light Microscope; Light Microscopy

LM Lime Mortar

LM Linear Modulation

LM Linear Molecule

LM Linear Monolithic

LM Linear Motor

LM Link Manager

LM Liquidity-Money (Curve)

LM Liquid Metal

LM Liquid Mix(ture)

LM Liquid Mix (Process) [Ceramic Powder Preparation]

LM List of Materials [also L/M]

LM Lithium Manganate

LM Lithium Molybdate

LM Littrow Mount(ing) [Spectroscopy]

LM Loading Machine [Mining]

LM Loadmeter [Aeronautics]

LM Load Module [Computers]

LM Local Memory

LM Logical Module

LM Logic Modelling

LM Logic Monitor

LM Longitudinal (Recording) Medium

LM Long Module

LM Loop Multiplexer

LM Lorentz Microscope; Lorentz Microscopy

LM Low Magnification

LM Low-Melting

LM Lunar Model

LM Lunar (Excursion) Module [of NASA Apollo Program]

LM Lunar Month [Astronomy]

LM Lurgi-Mitterberg (Process) [Metallurgy]

LM Lutetium Manganate

LM$_\Delta$ Laue Monotonic (X-Ray) Scattering [Symbol]

L/M Lines per Minute [also LPM, or lpm]

L/M List of Materials [also LM]

L/M L-Shell/M-Shell (Emission Ratio) [Nuclear Physics]

L/m Liter per meter [also l/m, L m^{-1}, l m^{-1}, or ℓ/m]

lm logarithmic mean area

lm lumen [Unit]

λM Sample Magnetization [Solid-State Physics]

LMA Laminating Materials Association [US]

LMA Laser Microspectral Analyzer

LMA Limited Motion Antenna

LMA Livestock Marketing Association [US]

LMA Logsplitters Manufacturers Association [US]

LMAGB Locomotive Manufacturers Association of Great Britain

LMAS Lithia-Magnesia-Alumina-Silica (Ceramics) [Li_2O-MgO-Al_2O_3-SiO_2]

LMAS Lithium-Magnesium Aluminosilicate (Glass)

LMB Laboratory of Molecular Biology [Cambridge, UK]

LMB Left Mouse Button

LMBC Liverpool Marine Biology Committee [UK]

LMBCS Lotus Multi-Byte Character Set

LMBI Local Memory Bus Interface

LMBW Lovett-Mou-Buff-Wertheim (Equation) [Materials Science]

LMC Laboratoire de Minéralogie–Cristallographie [Laboratory for Mineralogy and Crystallography, of Université de Paris VI et VII, France]

LMC Labor and Management Center [of Yale University, New Haven, Connecticut, US]

LMC Laminated Metal Composite

LMC Large Magellanic Cloud [Astronomy]

LMC Laser Machining Center

LMC Lattice Monte Carlo (Model) [Materials Science]

LMC Least Material Condition

LMC Life Members Committee [of Institute of Electrical and Electronics Engineers, US]

LMC Lime-Magnesium Carbonate

LMC Liquid-Metal Cooling [Metallurgy]

LMC Liquid-Metal Cracking

LMC London Medical Center [UK]

LMC Low-Pressure Molding Compound

LMCA Land Management Cooperative Agreement

LMCC Land Mobile Communications Council [US]

LLCC Leadless (Ceramic) Chip Carrier

LMCH Leadless Multiple-Chip Hybrid

LMCPE Lipid-Modified Carbon Paste Electrode

LMCT Ligand-to-Metal Charge Transfer

LMCTS Laboratoire de Matériaux Céramiques et Traitements de Surface [Ceramic Materials and Surface Treatment Laboratory; of Ecole Nationale Supérieure de Céramique Industrielle, Limoges, France]

LMCTS-CNRS Laboratoire de Matériaux Céramiques et Traitements de Surface–Conseil National de Recherche Scientifique [Ceramic Materials and Surface Treatment Laboratory–National Scientific Research Council at Ecole Nationale Supérieure de Céramique Industrielle, Limoges, France]

LMD Laboratorie de Météorologie Dynamique [Laboratory for Dynamic Meteorology, France]

LMD Laboratory Management Division

LMD Last Modification Date

LMD Light Metals Division [of Minerals, Metals and Materials Society of the American Institute of Mining, Metallurgical and Petroleum Engineers, US]

LMD Log Mean (Temperature) Difference

LMD Low-Molecular-Weight Dextran

LMDPE Linear Medium-Density Polyethylene

LMDR Laser Microwave Double Resonance

LMDS Local Multipoint Distribution Service

LME Large Marine Ecosystem

LME Launch Monitor Equipment

LME Liquid Membrane Electrode

LME Liquid Metal Embrittlement

LME London Metal Exchange [UK]

LMEC Liquid Metal Engineering Center [US]

LMEF Light Metal Educational Foundation, Inc. [Osaka, Japan]

LMEIC Life Member of the Engineering Institute of Canada

LMF Ladle Metallurgy Facility

LMF Lamb-Mössbauer Factor [Physics]

LMF Language Media Format

LMF Linear Magnetic Field

LMF Linear Matched Filter

LMF Liquid Metal Film

LMF Lithium Magnesium Fluoride

LMF Low and Medium Frequency [also L/MF]

LMF Lower Mid-Fuselage

LMFA Laboratoire de Mécanique des Fluides et d'Acoustique [Fluid Mechanics and Acoustics Laboratory, Lyon, France]

LMFA Light Metal Founders Association [UK]

LMFBR Liquid-Metal Fast Breeder Reactor

LMFC Linear Model-Following Control

LMFC Liquid-Metal Fuel Cell

LMFC Lower Mississippi Forecast Center [US]

LMFI Liquid Metal Field Ionization

LMFR Liquid-Metal Fuel Reactor

LMFRE Liquid-Metal Fuel Reactor Experiment

lm/ft² lumen(s) per square foot [also lm/sq ft]

LMG Left Main Gear [Aerospace]

LMG Light Machine Gun

LMGP Laboratoire de Matériaux et Génie Physique [Materials and Engineering Physics Laboratory; of Institut National Polytechnique de Grenoble, France]

LMGP-INSPG Laboratoire de Matériaux et Génie Physique–Institut National Polytechnique de Grenoble [Materials and Engineering Physics Laboratory–National Polytechnique Institute of Grenoble, France]

lm/h lumen per hour [also lm/hr, or lm h⁻¹, or lm hr⁻¹]

lm-h lumen-hour [also lmh, or lm-hr]

LMHC Lockheed Martin Hanford Company [Washington State, US]

LMHK Low-Magnesia High-Potassium (Glass)

lm/hr lumen per hour [also lm hr⁻¹, lm/h, lm/hr, or lm h⁻¹]

lm-hr lumen-hour [also lmh, or lm-hr]

LMHX Liquid Metal Heat Exchanger

LMI Laboratoire de Microscopie Ionique [Ion Microscopy Laboratory; Université of Rouen, Mont Saint Aignan, France]

LMI Lawn Mower Institute [now Outdoor Power Equipment Institute, US]

LMI Link Management Interface

LMI Liquid Metal Infiltration

LMI Liquid Metal Ion

LMI Liquid Metal (Field) Ionization

LMI Local Management Interface

LMI Local Memory Image

LMIC Liquid Metals Information Center [US]

LM-ICP-MS Laser-Microprobe Inductively-Coupled-Plasma Mass Spectrometry

LMIE Liquid Metal Induced Embrittlement

LMIM Large Motion (Vibration) Isolation Mount [of Canadian Space Agency]

Lmin Laxmanan (Microsegregation Model) Corrected for Minimum Undercooling [Metallurgy]

L/min Liter(s) per minute [also l/min, L min^{-1}, l min^{-1}, or ℓ/min]

L/min·mol Liter(s) per minute mole [also L min^{-1} mol^{-1}, l/min·mol, or l min^{-1} mol^{-1}]

LMIS Liquid Metal Ion Source

LM-K Langer and Müller-Krumbhaar (Theory) [Physics]

Lmk Landmark

LML Laboratoire de Mécanique de Lille [Lille Mechanics Laboratory; of Ecole de Lille, Villneuve d'Ascq, France]

LML Logical Memory Level

LMLC Low Molecular-Weight Liquid Crystal

LMLN Laboratoire de Magnétisme Louis Néel [Louis Néel Magnetism Laboratory for Magnetism, Grenoble, France]

LMLR Load Memory Lockout Register

LMM Laboratoire des Matériaux Minéraux [Mineral Materials Laboratory; of Ecole Nationale Supérieure de Chimie de Mulhouse, France]

LMM Laboratoire de Métallique Mécanique [Laboratory of Metal Mechanics; of Ecole Polytechnique Fédérale de Lausanne, Switzerland]

LMM Laboratoire de Microstructures et de Microélectronique [Laboratory for Microstructures and Microelectronics, of Conseil National de la Recherche Scientifique at Bagneux, France] [also L2M]

LMM Lactobacillus Maintenance Medium [Microbiology]

LMM Length Measuring Machine

LMM Light Meromyosin [Biochemistry]

LMM Locator at the Middle Marker [Aeronautics]

L2M Laboratoire de Microstructures et de Microélectronique [Laboratory for Microstructures and Microelectronics, of Conseil National de la Recherche Scientifique at Bagneux, France] [also LMM]

lm/m² lumen per square meter [also lm m^{-2}]

LMMB Lake Michigan Mass Balance (Study) [of Great Lakes National Program Office, US Environmental Protection Agency]

LMM/CNRS Laboratoire de Microstructures et de Microélectronique–Conseil National de la Recherche Scientifique [Laboratory for Microstructures and Microelectronics–National Scientific Research Council at Bagneux, France] [also L2M/CNRS]

LMMP Laboratoire des Matériaux-Mécanique Physique [Laboratory for Mechanical and Physical Materials Properties; of Ecole Centrale de Lyon, Ecully, France]

LMMP Laboratoire de Métallurgie des Matériaux Polycrystallin [Polycrystalline Materials Metallurgy Laboratory; University of Metz, France] [also LM2P]

LMMS Local Message Metering Service

LMN Lyman, Oklahoma [Meteorological Station Designator]

LMNA Long-Range Multipurpose Naval Aircraft

LMNR Lead Mononitroresorcinate

LMO Laboratoire de Magnétisme et d'Optique [Magnetism and Optics Laboratory; of University de Versailles, France]

LMO Lens-Modulated Oscillator

LMO Life Members Organization

LMO Linear Master Oscillator

LMO Lithium Manganese(IV) Oxide

LMO Local(ized) Molecular Orbital [Physical Chemistry]

LMOA Locomotive Maintenance Officers Association [US]

L/mol Liter per mole [also l/mol, L mol^{-1}, ltr/mol, ltr mol^{-1}, or ℓ/mol]

L/mol·min Liter(s) per mole minute [also L mol^{-1} min^{-1}, l/mol·min, or l mol^{-1} min^{-1}]

L/mol·s Liter(s) per mole second [also L mol^{-1} s^{-1}, l/mol·s, or l mol^{-1} s^{-1}]

LMOS Loop Maintenance Operations Service

LMP Laboratoire de Métallurgie Physique [Physical Metallurgy Laboratory; Poitiers, France]

LMP Laboratoire de Métallurgie Physique [Physical Metallurgy Laboratory; Université de Lille, France]

LMP Laboratoire de Métallurgie Physique [Physical Metallurgy Laboratory; Université Paris-Sud, Orsay, France]

LMP Laboratoire de Métallurgie Physique [Physical Metallurgy Laboratory, Ecole Polytechnique Fédérale de Lausanne, Switzerland]

LMP Laboratory of Manufacturing and Productivity [of Massachusetts Institute of Technology, Cambridge, US]

LMP Larson-Miller Parameter [Mechanics]

LMP Laser Microprobe

LMP Large-Volume Microwave Plasma Generator

LMP List of Measurement Points

LMP Lunar Module Pilot [of NASA Apollo Program]

Lmp Lamp [also lmp]

LMPA London Master Plasterers Association [UK]

LMPC Laser-Microwave Photoconductance (Technique)

LMPCM Laboratoire de Métallurgie et Physico-Chimie des Materiaux [Metallurgy and Materials Physical Chemistry; of Conseil National de Recherche Scientifique, France]

LMPE Low-Melting-Point Eutectic [Metallurgy]

LMPM Laboratoire de Mécanique et Physique des Matériaux [Laboratory for Materials Mechanics and Physics; Ecole Nationale Supérieure de Mécanique et d'Aérotechnique, Poitiers, France]

LMPM Laboratoire de Mécanique Productique et Matériaux [Laboratory for Production and Materials Mechanics; Université de Maine, Le Mans, France]

LMPM Laboratoire des Matériaux et des Procédés Membranaires [Membrane Materials and Processes Laboratory; of Ecole Nationale Supérieure de Chimie, National Higher School of Chemistry, Montpellier, France]

LMPRT Locally Most Powerful Rank Test

LMPS Low-Moisture Part Skim (Mozarella Cheese)

Lmps Lamps [also lmps]

LMR Land Mobile Radio

LMR Laser Magnetic Resonance

LMR Lateral Magnetoresistor [Solid-State Physics]

LMR Light Machine Rifle

LMR Liquid Metal (Nuclear) Reactor

LMR Liquid Molding Resin

LMR Longitudinal Magnetoresistance [Solid-State Physics]

LMR Lowest Maximum Range

LMRCT Localized Multireference Configuration Interaction (Calculation) [Physics[

LMRL Light Metals Research Laboratories [of University of Kentucky, Lexington, US]

LMRP Low-Medium Redox Potential [Physical Chemistry]

LMRU Library Management Research Unit [UK]

LMS Laboratoire de Métallurgie Structurale [Structural Metallurgy Laboratory, Université du Paris-Sud, Orsay, France]

LMS Laboratory for Microsensors Structures [Slovenia]

LMS Laboratory Management Standards

LMS Land Mobile Satellite (Service)

LMS Laser Mass Spectrometry

LMS Lead Mission Scientist

LMS Least Mean Square

LMS Level Measuring Set

LMS Library Management System

LMS Life and Microgravity Spacelab

LMS Linear Multistep

LMS List Management System

LMS Lithium Metasilicate

LMS Load Measurement System

LMS Local Measured Service

LMS Logistics Master Schedules

LMS London Mathematical Society [UK]

LMS Lotus Messaging Switch

lm-s lumen-second [also lm·s, lm-sec, or lm·sec]

LMSA Laser Microemission Spectroanalysis

LMSA Laser Microspectral Analysis

LMSA Le Gressus-Massignot-Sopiret-Auger [Physics]

LMSC Lockheed Missiles and Space Corporation [Palo Alto, California, US]

lm-sec lumen-second [also lm·sec, lm-s, or lm·s]

LMS/ENSCP Laboratoire de Métallurgie Structurelle, Ecole Ecole Nationale Supérieure de Chimie de Paris [Structural Metallurgy Laboratory–Paris National Higher School of Chemistry, France

LMSEO Labour Management Standards Enforcement Office [US]

LMSI Lockheed Martin Services Inc. [US]

LMSL Lateral Monolayer Superlattice [Materials Science]

lm/sq ft lumen(s) per square foot [also lm/ft²]

LMSS Lunar Mapping and Survey System

LMSV Loading Mooring Storage Vessel

LMSW Lockheed Martin Skunk Works [US]

LMSW Load Machine Status Word

LMT Land Mine Treaty

LMT Local Mean Time

LMT Logical Mapping Table

LMT Log Mean Temperature

LMT Low-Mobility T-Lymphocytes [Medicine]

Lmt Limit [also lmt]

LMTD Logarithmic Mean Temperature Difference

LMTEC Liquid Metal Thermoelectric Converter

LMTO Linear(ized) Muffin-Tin Orbital [Physics]

LMTO-ASA Linear Muffin-Tin Orbitals–Atomic Sphere Approximation [Physics]

LMTO-TB Linear Muffin-Tin Orbitals–Tight-Binding (Calculation) [Physics]

Lmtr Limiter [also lmtr]

LMU LAN (Local-Area Network) Management Utilities

LMU LAN (Local-Area Network) Manager for Unix

LMU Lincoln Memorial University [Harrogate, Tennessee, US]

LMU Line Monitor Unit

LMU Logical Mining Unit

LMU Ludwig-Maximilian Universität [Ludwig Maximilian University, Munich, Germany]

LMUA Lloyd's Motor Underwriters Association [UK]

LMW Laser Microwelding

LMW Low Molecular Weight

lm/W lumen per Watt [also LPW, lpw, or lpW]

LMWA Low-Molecular-Weight (Organic) Acid

LMWHC Low-Molecular-Weight Hydrocarbon

LMX L-Type Multiplex

LM/X LAN (Local-Area Network) Manager for Unix

LN Ladder Network

LN La Niña [Oceanography]

LN Lanthanum Nickelate

LN Laval Nozzle [Aerospace]

LN Line [also Ln, or ln]

LN Line Number

LN Linear Network

LN Liquid Nitrogen

LN Lithium Niobate

LN Load Number

LN Locational Interference Fit [ANSI Symbol]

LN Logarithm, Natural (Function)

LN Logic Network

LN Log-Normal (Distribution) [Statistics]

LN Long Nose

LN Lot Number

LN Lotus Notes [Computers]

LN Low Nitrogen

LN Low Noise

LN Lymph Node [Anatomy]

LN_2 Liquid Nitrogen

LN1 Interference Locational Fit (H6/n5) [ANSI Symbol]

LN2 Interference Locational Fit (H7/p6) [ANSI Symbol]

LN3 Interference Locational Fit (H7/r6) [ANSI Symbol]

LN- Norway [Civil Aircraft Marking]

L-N Longitudinal-Normal (Direction)

Ln Lanthanide (Metal)

Ln Line [also ln]

Ln Loan [also ln]

Ln Natural Logarithm [Programming]

Ln^{3+} Trivalent Lanthanide Ions

ln Naperian (or natural) logarithm [Symbol]

LNA Launch Numerical Aperture

LNA (Alpha) Linolenic Acid (or Omega-3-Superunsaturated Fatty Acid) [Biochemistry]

LNA Local Numbering Area [Telecommunications]

LNA Low-Noise Amplifier

α-LNA Alpha Linolenic Acid (or Omega-3-Superunsaturated Fatty Acid) [Biochemistry]

LnA Lanthanide Acetylacetonate

$Ln(ACAC)_3$ Lanthanide Acetylacetonate [also $Ln(AcAc)_3$, or $Ln(acac)_3$]

Ln-Al-TM Lanthanide-Aluminum-Transition Metal (Alloy)

LNAP Low Non-Essential Air Pressure

LNB Local Name Base

LnBCO Lanthanide Barium Copper Oxide (Superconductor)

LNC Low-Noise Converter

$Ln_{1-x}Ca_xMnO_3$ Lanthanide Calcium Manganate

Lnch Launch [also lnch]

LnCCO Lanthanum Cerium Copper Oxide (Superconductor)

LND Local Number Dialing

LND Log-Normal Distribution [Statistics]

Lndg Landing [also lndg]

LN:DI Lotus Notes:Document Imaging

LNDO Local Neglect of Differential Overlap [Physics]

LNE Laboratoire National d'Essais [National Test(ing) Laboratory, France]

LNE Local Network Emulator

LNF Laboratori Nazionali di Frascati [Frascati National Laboratory; of Instituto Nazionale de Fisico Nucleare, Italy]

LNF Liposoluble Neutral Fraction [Biochemistry]

LNFB Linear Negative Feedback

LNG Liquefied Natural Gas

lng long

LNGC Liquefied Natural Gas Carrier

LNGS Laboratori Nazionali del Gran Sasso [Gran Sasso National Laboratory, Italy]

LNI Local Network Interconnect

Lnk Link [also lnk]

LNL Lanthanum Nickel Lithiate

$Ln:LiYF_4$ Lanthanide-Doped Lithium Yttrium Fluoride [also $Ln^{3+}:LiYF_4$]

LNLS Laboratorio Nacional de Luz Sincrotron [National Synchrotron Laboratory, Unicamp, Campinas, Brazil]

λ(nm) Wavelength in nanometers

LNMRB Laboratory of Nuclear Medicine and Radiation Biology [of University of California at Los Angeles, US]

LNO Lanthanum Nickel Oxide (Superconductor)

LNO Lithium Niobate

LN-NR Low-Nitrogen Nitrile Rubber

LN NUM Line Number(ing) [also Ln Num]

LNO Lanthanum Nickel Oxide (Superconductor)

LNO Lithium Niobate

LNOC Libya National Oil Corporation

LNP Local Network Program

LNP Loss of Normal Power

LNPR Line Printer [also LN PR, or Ln Pr]

LNR Liquid Nitrate Rectifier

LNR Low-Noise Receiver

LNRD Land Natural Resources Division [of US Department of Justice]

LNS Laboratory of Nuclear Science [of Massachusetts Institute of Technology, Cambridge, US]

LNS Lesch-Nyhan Syndrome [Medicine]

LNS Linked Numbering Scheme

LNS Low Nitrogen and Sulfur Oxides

LNS Luxembourg Naturalists Society

$Ln_{1-x}Sr_xCoO_3$ Lanthanide Strontium Cobaltate

$Ln_{1-x}Sr_xMnO_3$ Lanthanide Stronium Manganate

LNSU Library Network *SIBIL* (Automation System) Users [Switzerland]

LNT Liquid Nitrogen Temperature

Ln_2TiO_7 Lanthanide Titanate [General Formula]

LNTSU Leo N. Tolstoi State University [Russian Federation]

LNTWTA Low-Noise Traveling-Wave Tube Amplifier

LNWT Low and Nonwaste Technology

LNYV Lettuce Necrotic Yellows Virus

LO Lake Ontario [US/Canada]

LO Landing Operations

LO Large Overshoot [Magnetic Recording]

LO Larkin-Ovchimikov (Theory) [Physics]

LO Launch Operations

LO Layout

LO Liaison Office(r)

LO Lick Observatory [Mount Hamilton, California, US]

LO Lift-Off [also L/O]

LO Light Operated

LO Linear Optics

LO Linear Oscillation

LO Line Occupancy

LO Linseed Oil

LO Local Order

LO Local Oscillator

LO Location Code

LO Lock-On

LO Lock-Out

LO Longitudinal Optic(al) (Mode) [Physics]

LO Long-Range Order(ing) [Solid-State Physics]

LO Low [also Lo, or lo]

LO Low-Address Byte

LO Lowell Observatory [Lowell, Arizona, US]

LO Low Order

LO Lubricating Oil

LO Lunar Orbiter [of NASA Apollo Program]

LO$_1$ Longitudinal Optic(al), Rocking Vibration (Mode) [Physics]

LO$_2$ Longitudinal Optic(al), Symmetric Stretching Vibration (Mode) [Physics]`

LO$_2$ Liquid Oxygen

LO$_3$ Longitudinal Optic(al), Asymmetric Stretching Vibration (Mode) [Physics]

LO2 Large Overshoot with Ringing [Magnetic Recording]

L/O Lift-Off [also LO]

Lo Lowest Reported [Chemistry]

Lo Low-Order Byte (of Argument) [Pascal Function]

LOA Landing Operations Area

LOA Lathyrus Odoratus Agglutinin [Immunology]

LOA Launch Operations Area

LOA Leave of Absense

LOA Length Overall (of a Ship, or Boat)

LOA Letter of Agreement [also LoA]

LOA Line of Assurance

LOA Log-Out Analysis

LOAC Low Accuracy

LOAD Laser Optoacoustic Detection

LOADS Low-Altitude Defense System

LOAMP Logarithmic Amplifier

LOAP List of Applicable Publications

LOAS Liftoff Acquisition System

LOB Launch Operations Branch [of NASA]

LOB Launch Operations Building [of NASA]

LOB Left Outboard

LOB Line of Balance [Industrial Engineering]

LOB Line of Bearing [Navigation]

LOB Line of Business

L-O^{t}Bu Leucine tert-Butoxy-

LOBAR Long-Baseline Radar

LOBC Locked On By Combination

LOBTA Lunar Orbiter Block Triangulation [of NASA]

LOBW Loss of Backwall (Gate)

LOC Laboratory Office Computer

LOC Lane Occupancy Control

LOC Large Optical Cavity (Diode)

LOC Launch Operations Center [of NASA]

LOC Launch Operations Complex

LOC Letter of Committment

LOC Level of Concern

LOC Liaison Officers Committee

LOC Library of Congress [US]

LOC Lines of Code

LOC Linked Object Code

LOC Load on Call

LOC Localizer [Aeronautics]

LOC Location Counter [Computers]

LOC Loop On-Line Control

LOC Loss of Coolant [Nuclear Engineering]

LOC Oxygen Lance Cutting

LoC Letter of Credit

Loc Local(ity) [also loc]

Loc Location; Locator [also loc]

LOCA Loss-of-Coolant Accident [Nuclear Reactors]

LOCAL Load on Call

LOCAP Low Capacitance [also Locap]

LOCAP Low Capacity [also Locap]

LOCAS Local Cataloguing Service [UK]

LOCATE Library of Congress Automation Techniques Exchange [US]

LOCATS Lockheed Optical Communications and Tracking System

LOCC Launch Operations Control Center

loc cit *(loco citato)* – in the place cited

LOCE Loss-of-Coolant Experiment [Nuclear Reactors]

LOCF Loss-of-Coolant Flow [Nuclear Reactors]

LOCH London Options Clearinghouse [UK]

LOCI Logarithmic Computing Instrument

LOCIS Library of Congress Information System

LOCMOS Local Oxidation Complementary Metal-Oxide Semiconductor [also LO-CMOS, or LO/CMOS]

Locom Locomotive [also locom]

LOCOS Local Oxidation of Silicon (Process) [Solid-State Physics]

LOCP Loss-of-Coolant Protection [Nuclear Reactors]

LOCS Librascope Operations Control System

LOCS Logic and Control Simulation

Loc Sci Location Science [Journal]

LOD Launch Operations Directorate [of NASA]

LOD Length of Day

LOD Letterkenny Ordnance Depot [Chambersburg, Pennsylvania, US]

LOD Level Of Detail

LOD Lima Ordnance Depot [Ohio, US]

LOD Limit of Detection

LOD Line of Direction

LOD Locally One-Dimensional

LOD Location Dependent

LOD Locked-On Device

LOD Low Density [also LoD]

LOD Low-Order Detonation

LODAR Long-Range Detection and Ranging [also Lodar, or lodar]

LODEM Loading Dock Equipment Manufacturers Association [US]

LODESMP Logistics Data Element Standardization and Management Process

LODESR Longitudinally Detected Electron Spin Resonance

LODESTAR Logically-Organized Data Entry, Storage and Recording

LODP Leveling-Off Degree of Polymerization

LODSB Load String Byte

LODTM Large Optical Diamond Turning Machine [of US Department of Energy at Lawrence Livermore National Laboratory, University of California at Livermore]

LOE Left Outboard Elevon [Aeronautics]

LOE Level of Effort

LOE Line of Effort

LOEC Lowest Observable Effect Level

LOERO Large Orbiting Earth Resources Observatory

LOES Laser-Optical Emission Spectroscopy

LOEX Library Orientation Instruction Exchange [US]

LOF Lack of Fusion

LOF Length of File

LOF Libbey-Owens-Ford (Company) [US]

LOF Line of Fire

LOF Line(s) of Force

LOF Local Oscillator Frequency

LOF Loss of Flow [Nuclear Reactors]

LOF Loss of Fluid

LOF Lowest Observable Frequency

LOF Lowest Operating Frequency

LOFA Loss-of-Flow Accident [Nuclear Reactors]

LOFAR Low-Frequency Acquisition and Ranging [also Lofar, or lofar]

LOFAR Low-Frequency Analysis (and) Recording [also Lofar, or lofar]

LOFF Larkin-Ovchinikov-Fulde-Ferrel (Physical State)

LOFI Last Out, First In [also Lofi]

LO-FIN Last Out, First In (Process)

LOFO Low Frequency Oscillation

LOFT Loss-of-Flow Test [Nuclear Reactors]

LOFT Loss-of-Fluid Test

LOFTI Low-Frequency Trans-Ionospheric Satellite [of US Department of Defense] [also Lofti]

LOFW Loss of Feedwater

LOG Laser Optogalvanic (Effect)

LOG Logarithm (Common) [Computers]

Log Logic(al); Logics [also log]

log common logarithm (base 10) [Symbol]

log$_e$ Napierian (or natural) logarithm (base e) [Symbol]

log$_2$ log dualis (base 2) [also ld] [Symbol]

log$_{10}$ common logarithm (base 10) [Symbol]

.LOG Activity Log [File Name Extension]

LOGACS Low-Gravity Accelerometer Calibration System

LOGALGOL Logical Algorithmic Language

LOGAMP Logarithmic Amplifier [also Logamp, or logamp]

LOGANAL (Digital Well) Log Analysis System

LOGANDS Logical Commands

LOGBALNET Logistics Ballistic Missile Network [of US Air Force]

LOGCOMNET Combat Logistics Network [of US Air Force]

Log Dec Logarithmic Decrement [also log dec]

LOGE Laser Optogalvanic Effect

LOGEL Logic Generating Language

LOGFED Log File Editor

LOGFTC Logarithmic Fast Time Constant [also LOG-FTC]

Logging Sawmill J Logging and Sawmilling Journal [US]

LOGIPAC Logical Processor and Computer

Logist Logistic(s) [also logist]

Logist Today Logistics Today [Publication of the Cranfield Institute of Technology, UK]

Logist Transp Rev Logistics and Transportation Review [Published by the University of British Columbia–Faculty of Commerce, Canada]

LOGIT Logical Interference Tester

LOGLAN Logical Language [also Loglan]

log MTD Logarithmic Mean Temperature Difference

LOGO Limit of Government Obligation

LOGRAM Logical Program

LOGREC (Current) Logical Record Number [BASIC Programming]

Logs Logistics [also logs]

Log Scale Logarithmic Scale [also log scale]

LOGTAB Logic Table [also LOG TAB, or Log Tab]

LOG-VIDEO Logarithmic Video [also LOG-Video, Log-Video, or log-video]

LOH Light Observation Helicopter

LOH Loss of Head

LOHAP Light Observation Helicopter Avionics Package

LOI Limited (or Limiting) Oxygen Index

LOL Limit of Liability

LOL Limited Operational Life

LOL Lot of Laughs [Internet Jargon]

LOLI Limited Operational-Life Items

LOI Land-Ocean Interaction

LOI Loss of Ignition

LOI Loss on Ignition

LOI Lunar Orbit Insertion

LoI Letter of Intent

LOICZ Land-Ocean Interactions in the Coastal Zone

LOL Laugh(ing) Out Loud [Internet Jargon]

LOL Length of Lead

LOL Loss of Load

LOLA Library On-Line Acquisition

LOLA London On-Line Local Authorities [UK]

LOLA Low-Level Oil Alarm

LOLA Lunar Orbit Landing Approach

LOLC Library Developed On-Line Catalogue

LOLEX Low-Level Extraction

LOLITA Language for the On-Line Investigation and Transformation of Abstractions

LOLITA Library On-Line Information and Text Access [of Oregon State University, Corvallis, US]

LOLO Lift-On/Lift-Off (Vessel) [also lo/lo]

LOLP Loss of Load Probability

LOM Laboratory of Oral Medicine [of National Institutes of Health, Bethesda, Maryland, US]

LOM Laminated Object Manufacturing

LOM Light Optical Micrograph; Light Optical Microscope; Light Optical Microscopy

LOM Locator at the Outer Marker [Aeronautics]

LOM Low-Frequency Outer Marker [Aeronautics]

LOMA Life Office Management Association [US]

LOMAC Low-Level Macroprocessor Language

LOMAR Local Manual Attempt Recording

LOMC Laboratoire d'Optique de la Matière Condensée [Condensed Mater Optics Laboratory; Université de Paris VI, France]

LOMC Lenape Ordnance Modification Center [Newark, Delaware, US]

L-OMe Leucine Methoxy-

LOMI Low Oxidation Metal Ions (Process)

LOMUSS Lockheed Multiprocessor Simulation System

LON Local Operating Network

Lon London [UK]

Lon Longitude [also lon]

LONA Low-Noise Amplifier

LONAL Local Off-Net Access Line

Lond London [UK]

Lond Math Soc Proc Ser 1 London London Mathematical Society Proceedings Series 1 [UK]

Lond Math Soc Proc Ser 2 London London Mathematical Society Proceedings Series 2 [UK]

L'Onde Elect L'Onde Electrique [French Publication on Electric Waves]

Long Longitude [also long]

Longn Longeron [Aeronautics]

LOOP Louisiana Offshore Oil Port [US]

LOOPE Loop while Equal

LOOPNE Loop while Not Equal

LOOPNZ Loop while Not Zero

LOOPZ Loop while Zero

LOOPS LISP (List Processing) Object-Oriented Programming System

LOOSP Loss of On-Site Power

LOP Lack of Penetration

LOP Line(s) of Position

LOP Line-Oriented Protocol

LOP Local Operational Plot

LOP Logical Operation

LOP Longitudinal Optic Phonon(s) [Solid-State Physics]

LOP Loss of Offsite Power

LOP Louisiana Ordnance Plant [Shreveport, US]

LOPAD Logarithmic Outline Processing System for Analog Data

LOPAR Low-Power Acquisition Radar [also Lopar, or lopar]

LOPI Loss of Piping Integrity

LOPO Low-Power Homogeneous (Nuclear) Reactor

LOPO Low-Power Water Reactor [US]

LOPOS Local Oxidation of Polycrystalline Silicon

LOPP Lunar Orbiter Photographic Project [of NASA]

LOPROX Low-Pressure Wet Oxidation (Process)

LOPS Line(s) of Position

LOPT Line Output Transformer

LOPT Lowest-Order Perturbation Theory [Physics]

LOQ Limit of Quantification

loq *(loquitur)* – he (or she) speaks

LOQ Lowest Obtainable Quantification

LO-QG Locked Oscillator-Quadrature Grid

LOR Laboratory for Oil Recovery

LOR Large Optical Reflector

LOR Large Outrigger

LOR Lining-over-Refractory

LOR Loss on Reduction

LOR Low-Frequency Omnirange [now MOR]

LOR Lubricating Oil Residue

LOR Lunar Orbit(al) Rendezvous

LORA Level of Repair Analysis

LORAC Long-Range Accuracy (Radar System) [also Lorac, or lorac]

LORAD Long-Range Active Detection [also Lorad, or lorad]

LORAD Long-Range Detection and Ranging [also Lorad, or lorad]

LORADAC Long-Range Active Detection and Communications System

LORAN Long-Range Navigation(al System) [also Loran, or loran]

LORAN-A Long-Range Navigation(al System) A [also Loran-A]

LORAN-C Long-Range Navigation(al System) C (Third Generation) [also Loran-C]

LORAN-D Long-Range Navigation(al System) D [also Loran-D]

LORAPH Long-Range Passive Homing [also Loraph, or loraph]

LORBI Locked-On Radar Bearing Indicator

LORDS Licensing On-Line Retrieval Data System

LORE Line Oriented Editor

LOREC Long-Range Earth Communications [also Lorec, or lorec]

LORENTZ International Workshop on Lorentz Group, CPT (Charge, Parity, Time) and Neutrinos

Lorentz TEM Lorentz Transmission Electron Microscope; Lorentz Transmission Electron Microscopy

LORG Localized Orbital Local Origin (Method)

LORL Large Orbital Research Laboratory [of NASA]

LORLS Lake Ontario Regional Library System [Canada]

LORO Lobe-On-Receive Only

LORPGAC Long-Range Proving Ground Automatic Computer

LORS Lunar Orbital Rendezvous System

LORTAN Long-Range and Tactical Navigation [also Lortan, or lortan]

LORV Low-Observable Reentry Vehicle

LOS Laser-Rangefinder Optical Sight

LOS Law of the Sea

LOS Layered Oxide Superconductor

LOS Length of Stay

LOS Lift-Off Simulator

LOS Line of Sight

LOS Line Out of Service

LOS Lock-Out Submersible

LOS Loss of Signal

LOS Loss of Sync

LOS Lunar Orbiter Spacecraft [of NASA]

LOSC Law of the Sea Conference

LOSP Loss of Offsite Power

LOSP Loss of System Pressure

LOSR Limit of Stack Register

LOSS Landing Observer Signal System

LOSS Large Object Salvage System

LOST Letdown Storage Tank

LOT Load-On Top (System)

LOT Large Orbital Telescope

LOT Lift-Off Time

LOT Light-Operated Typewriter

LOT Lock On Target

LOT Long Open Time

LOT Low Observable Technologies

LOT Polskie Linie Lotnicze [Polish Airlines]

lot *(lotionis)* – Lotion [Medical Prescriptions]

LOTA Low Observable Technology and Application

LOTEM Long Offset Transient Electromagnetic (Process)

LOTIS Logical Timing Sequencing

LOTIS Logic, Timing, Sequencing (Language)

LOTS Land Ownership and Tenure System [South Australia's Cadastral Land Information System]

LOTS Logistic Over The Shore (Vehicle)

LOTUS Logic of Temporal Ordering System

Lotus A Lotus Tetrasgonalobus Agglutinin [also LOTUS A] [Immunology]

LOV Limit of Visibility

LOV Loss of Visibility

LOW Lift-Off Weight

LOW Longhorn Ordnance Works [Marshall, Texas, US]

LOW Low Core Threshold

LOWG Landing Operations Working Group

LOWL Low-Level Language

LOWNOX Low-Emission Combustor Technology

LOWS Lake Ontario Winter Storms (Project) [Canada]

Low Temp Phys Low Temperature Physics [Russian Translation Journal Jointly Published by the American Institute of Physics (US) and MAIK Nauka (Russia)]

Low Temp Sci A, Phys Sci Low Temperature Science, Series A, Physical Sciences [Published by the Institute of Low Temperature Science, Hokkaido University, Sapporo, Japan]

LOWTRAN Low Resolution Atmospheric Radiance and Transmittance Model [of Air Force Geophysical Laboratory, US]

LOX Liquid Oxygen [also LOx, or lox]

LOX Liquid Oxygen Explosive

LOZ Liquid Ozone [also LOz, or loz]

LP Labial Palp [Zoology]

LP Labial Pipe

LP Laboratory Preparation

LP Laboratory Program

LP Ladder Polymer

LP Lagrangian Polynomial [Mathematics]

LP Landplane

LP Langmuir Probe [Plasma Physics]

LP Language Processor

LP La Paz [Bolivia]

LP La Plata [Argentina]

LP Larmor Precession [Physics]

LP Laser Printer

LP Laser Probe

LP Laser Processing

LP Laser Pump

LP Latency Period

LP (Crystal) Lattice Parameter

LP (Crystal) Lattice Pitch

LP (Crystal) Lattice Plane

LP (Crystal) Lattice Point

LP Launch Package

LP Launch Pad

LP Legendre Polynomial [Mathematics]

LP Laves Phase [Metallurgy]

LP International Symposium on Lepton-Photon Interactions [Particle Physics]

LP Lepton-Photon (Interaction) [Particle Physics]

LP Lever Principle

LP Library of Parliament [Ottawa, Canada]

LP Library Program

LP License Program

LP Life Prediction

LP Lifting Power

LP Light Pen

LP Linearly Polarized; Linear Polarization

LP Linear Polymer

LP Linear Positioning

LP Linear Prediction

LP Linear Processing

LP Linear Programming [Mathematics]

LP Line Printer

LP Link Plate (of Chains)

LP Lipoprotein [Biochemistry]

LP Liquefied Petroleum

LP Liquid Phase

LP Liquid Pressure

LP Liquid Prism

LP Liquid Propellant

LP Liquid Protection (Coverall)

LP List Price

LP Lithium Phosphate

LP Litmus Paper

LP Livens Projector

LP Load Point

LP Lodgepole Pine

LP Logic Probe

LP Logic Programming

LP Logic Pulser; Logic Pulsing

LP Log Periodic (Antenna)

LP Longitudinal Parity

LP Long Period

LP Long Player

LP Long Play(ing) (Record) [also lp]

LP Long Primer

LP Loop(ing)

LP Loop Control(ler)

LP Lorentz Polarization

LP Low-Pass (Filter)

LP Low Phosphorus

LP Low Point

LP Low Power

LP Low Pressure [also lp]

LP Low-Speed Printer

L-P Lorentz-Polarization Factor

L-P Low-Pressure [also LP]

Lp Lempira [Currency of Honduras]

L£ Lebanese Pound [Currency]

£P Palestine Pound [Currency]

LPA Laser-Diode Parametric Analyzer

LPA Leather Producers Association [UK]

LPA Limulus Polyphemus Agglutinin [Immunology]

LPA Linear Power Amplifier

LPA Linear Pulse Amplifier

LPA Line Profile Analysis

LPA Link Pack Area

LPA Liquid Petroleum Air

LPA Log Periodic Antenna

LPA Long Path Absorption

LPA Low-Pressure Alarm

LPA Low Profile Additive

LPAC Launching Program Advisory Committee [of European Space Research Organization]

LPAPB Long-Period Antiphase (Grain) Boundary [Materials Science]

LPAR Logic Programming and Automated Reasoning

LPAS Low-Pressure Arc Spraying [Metallurgy]

LPATS Lightning Position and Tracking System

LPB Long Packed Bed (Reactor)

LPB Low Pressure Boiler

LPBCO Lutetium Praseodymium Barium Copper Oxide (Superconductor)

LPC Laboratory Pasteurized Count

LPC Laboratory Precision Connector

LPC Language Products Center [Canada]

LPC Laser Particle Counter

LPC League of Professional Craftsmen [UK]

LPC Linear Power Controller

LPC Linear Predictive (or Prediction) Coding [Telecommunications]

LPC Linear Predictive Coefficients

LPC Local Procedure Call

LPC Local Processing Center [Cable Television]

LPC Log-Periodic Antenna, Coaxial-Fed

LPC Loop Control

LPC Loop Preparation Cask [Nuclear Engineering]

LPC Lower Pump Cubicle

LPC Low-Performance Chromatography

LPC Low-Pressure Chromatography

LPC Low-Pressure Counter

LPCMO Lanthanum Praseodymium Calcium Manganese Oxide (Superconductor)

LPCVD Low-Pressure Chemical Vapor Deposition

LPC-HS Laser Particle Counter–High Sensitivity

LPCI Low-Pressure Coolant Injection [Nuclear Reactors]

LPCIS Low-Pressure Coolant Injection System [Nuclear Reactors]

LPCM Laboratoire de Physico-Chimie des Matériaux [Materials Physics and Chemistry Laboratory; of Conseil National de Recherche Scientifique at Meudon, France]

LPCM Laboratoire de Physico-Chimie Moléculaire [Molecular Physical Chemistry Laboratory, Université de Pau et des Pays de l'Adour, Pau, France]

LPCM Laboratoire des Physique et des Couches Minces [Thin-Film and Physics Laboratory, of Ecole Polytechnique, Palaiseau, France]

LPCM Laboratoire des Plasmas et des Couches Minces [Plasma and Thin-Film Laboratory, of Institut des Plasmas et des Couches Minces, Université de Nantes, France]

LPCM Linear Phase Code Modulation

LPCM-CNRS Laboratoire de Physico-Chimie des Matériaux–Conseil National de Recherche Scientifique [Materials Physics and Chemistry Laboratory of the National Scientific Research Council at Meudon, France]

LPCML Laboratoire de Physico-Chimie des Matériaux Luminescents [Physical Chemistry Laboratory for Luminescent Materials, France]

LPCRS Low-Pressure Coolant Recirculation System [Nuclear Reactors]

LPCS Laboratoire de Physique des Composants à Semiconducteurs [Laboratory of Semiconductor Component Physics, of Institut National Polytechnique de Grenoble, France]

LPCS Laboratory of Polymer Condensed States [of Institute for Chemical Research, Kyoto University, Japan]

LPCS Low-Pressure Core Spray [Nuclear Reactors]

LPCVD Liquid-Phase Chemical Vapor Deposition

LPCVD Low-Pressure Chemical Vapor Deposition

LPD Amphibious Transport Dock [US Navy Symbol]

LPD Labeled Plan Display (Unit)

LPD Landing Point Designator

LPD Language Processing and Debugging

LPD Launch Procedure Document

LPD Light Point Defect(s) [Materials Science]

LPD Linear Phasing Device

LPD Linear Polarization Degree

LPD Linear Power Density

LPD Line Printer Daemon (Protocol) [of University of California at Berkeley, US]

LPD Load Point Displacement [Mechanics]

LPD Log Periodic Dipole (Antenna)

LPD Low Power Difference

LPD Low Pressure Difference

LPDA Log-Periodic Dipole Array

LPDM List of Physical Dimensions

LPDR Local Public Document Room

LPDT Low-Power Distress Transmitter

LPDTL Low-Power Diode Transistor Logic

LPDTμL Low-Power Diode Transistor Micrologic [also LPDTML]

LPDU Link-Layer Protocol Data Unit

LPE Lambda Pointing Experiment [of NASA]

LPE Linear Polyethylene

LPE Liquid-Phase Epitaxial; Liquid-Phase Epitaxy

LPE Loop Preparation Equipment

LPEC Lumped Parameter Equivalent Circuit

LPEE Liquid-Phase Electro-Epitaxial; Liquid-Phase Electro-Epitaxy

LPES Laboratory de Physique d'Etat Solide [Solid-State Physics Laboratory, University of Mons, Belgium]

LPF Leaching, Precipitation and Flotation; Leach Precipitation Float

LPF Liquid Pressure Forming

LPF Low-Pass Filter

LPF Low Pressure Forming

LPF Low Pressure Fuel

LPF Lymphocytosis-Promoting Factor [Biochemistry]

LPFD Log-Periodic Folded-Dipole (Antenna)

LPFL Low Pinchoff-Voltage FET (Field-Effect Transistor) Logic

LPFT Low Pressure Fuel Turboprop

LPFT Low Pressure Fuel Turbopump [also LPFTP]

LPG Laser Pedestal Growth (Method)

LPG Last Page Generator

LPG Liquefied Petroleum Gas

LPG Liquid Propellant Gun

LPG List Program Generator

LPG Low-Pressure Gas

LPGA Leadless Pad Grid Array

LPGA Liquefied Petroleum Gas Association

LPGC Liquefied Petroleum Gas Carrier

LPGITA Liquefied Petroleum Gas Industry Technical Association [UK]

LPGITC Liquefied Petroleum Gas Industry Technical Committee

LPGS Liquid Pathway Generic Study

LPGSASA Liquefied Petroleum Gas Safety Association of South Africa

LPGTC Liquefied Petroleum Gas Industry Technical Committee

LPH Amphibious Assault Ship [US Navy Symbol]

LPI LEP (Large Electron Positron)(Collider) Pre-Injector [of CERN–European Laboratory for Particle Physics, Geneva, Switzerland]

LPI Lepton-Photon Interaction [Particle Physics]

LPI Lightning Protection Institute [US]

LPI Lines per Inch [also lpi]

LPI Linus Pauling Institute [Menlo Park, California, US]

LPI Longitudinally-Applied Paper Insulation

LPI Louisiana Polytechnic Institute [Ruston, US]

LPI Low Power Injection

LPI Low Probability of Intercept

LPI Low Probability Intercept (Radar)

LPI Lunar and Planetary Institute [of NASA Johnson Space Center, Houston, Texas, US]

LPIA Label Printing Industries of America [US]

LPIA Liquid Propellants Information Agency [US]

LPID Logical Page Identifier

LPIS Land Protection Incentives Scheme [of Victoria State, Australia]

LPIS Launch Package Integration Stand

LPIS Low-Pressure Injection System

LPL Log-Periodic Ladder (Antenna)

LPL Laser-Pumped Laser

LPL Light Proof Louver

LPL Linear Programming Language

LPL Lipoprotein Lipid [Biochemistry]

LPL List Processing Language [also LISP]

LPL Local Processor Link

LPL Log-Periodic Ladder (Antenna)

LPL Long Particular Meter

LPL Long-Path Laser

LPL Lotus Programming Language [Lotus 1-2-3 Software]

LPL Lunar and Planetary Laboratory [of University of Arizona, Tucson, US]

LPLA Long-Path Laser Absorption

L-PLA L-Polylactic Acid

L/plant Liter(s) of Spray Liquid per Plant [Unit]

LPLB Low-Power Laser Beam

LPLCR Liquid Phase Laser Crystallization [also LP-LCR]

LP-LEC Low-Pressure Liquid Encapsulated Czochralski [Crystal Growing]

LPLF Low-Pressure Liquid-Filled (Electric Cable)

LPLS Low-Pressure Laser Spraying [Metallurgy]

LPLWS Launch Pad Lightning Warning System

LPM Laboratoire de Physique Moléculaire [Molecular Physics Laboratory; of University of Besançon, France]

LPM Laboratoire de Physique de la Matière [Matter Physics Laboratory; of Institut National des Sciences Appliquées, Villeurbanne, France]

LPM Laboratoire de Physique des Matériaux [Materials Physics Laboratory, Marseilles, France]

LPM Laboratoire de Physique des Matériaux [Materials Physics Laboratory, Meudon, France]

LPM Laboratoire de Physique des Matériaux [Materials Physics Laboratory, Ecole Normale Supérieure, Rabat, Morocco]

LPM Lanthanum Lead Manganate

LPM Laser Phase Macroscope

LPM Licensing Project Manager

LPM Linearly Polarized Mode

LPM Lines per Minute [also lpm]

LPM Low-Pressure Molding

LPM Lunar Portable Magnetometer [of NASA]

Lpm Liters per minute [also lpm]

LPMA Loose Parts Monitor Assembly

LPMC Laboratoire de Physique de la Matiére Condensée [Consensed Matter Physics Laboratory; of Institut National des Sciences Appliquées–Service National des Champs Magnetique Pulsés, Toulouse, France]

LPMC Low-Pressure Molding Compound

LPM-INSA Laboratoire de Physique des Matériaux–Institut National des Sciences Appliquées [Materials Physics Laboratory-National Institute of Applied Sciences, Toulouse, France]

LPMM Laboratoire de Physique et Mécanique des Matériaux [Laboratory of Materials Physics and Mechanics; of Université de Metz, France]

LPMM Line Pairs per Millimeter [also lpmm]

LPMM Lines per Millimeter [also lpmm]

LPMOCVD Low-Pressure Metal-Organic Chemical Vapor Deposition [also LP-MOCVD]

LPMS Laboratoire de Physicochimie de Matériaux Solide [Solid Materials Physics and Chemistry Laboratory, Montpellier, France]

LPMS Laser-Probe Mass Spectrometry

LPMTM Laboratoire de Propriétés Mécaniques et Thermodynamique des Matériaux [Mechanical Thermodynamic Properties Laboratory; of Université Paris Nord, Villetaneuse, France]

LPMTM-CNRS Laboratoire de Propriétés Mécaniques et Thermodynamique des Matériaux–Conseil National de la Recherche Scientifique [Mechanical and Thermodynamic Properties Laboratory–National Scientifique Research Council, of Université Paris Nord, Villetaneuse, France]

LPN Laboratory Product News [Canadian Publication]

LPN Licensed Practical Nurse

LPN Liquid Phase Nucleation [Metallurgy]

LPN Logical Page Number

LPN Low-Pass Notch

LPO Lanthanum Phosphate

LPO Likely Preferred Options

LPO Low-Power Output

LPOS Current Position of Line Printer Print Head (Function) [Programming]

LPOT Low-Pressure Oxidizer Turbopump [also LPOTP]

LPOX Low-Pressure Oxygen [also LPOx]

LPOXO Low-Pressure Oxidation (Process)

LPP Letter Processing Plant

LPP Liquid Plasma Process

LPP Liquid Precursor Processing [Metallurgy]

LPP Low-Power Physics

LPP Low-Purity Polycrystalline (Ceramics)

LPPD Low-Pressure Plasma Deposition

LPPM Laboratoire de Photophysique et Photochimie Moléculaire [Molecular Photophysics and Photochemistry Laboratory, France]

LPPP Ladder-Type Poly(paraphenylene) (Polymer)

LPPQFP Low-Profile Plastic Quad Flat Pack [Electronics]

LPPS Low-Pressure Plasma Spray(ing) [Metallurgy]

LPPS Low-Pressure Plasma System

LPPT Low Pressurization Pressure Test

Lppt Liquid Precipitate [alsp LPPT]

LPR Landau-Placzek Ratio [Physics]

LPR (Crystal) Lattice Parameter Ratio

LPR Linear Polarization Resistance

LPR Line Printer

LPR Line Printer Remote

LPR Liquid-Phase Reduction

LPR Lynchburg Pool Reactor [of Babcock and Wilcox, Inc., Virginia, US] [Shutdown]

LPRC Library Public Relations Council [US]

LPRE Liquid Propellant Rocket Engine

LPRINT Lookup Dictionary Print Program

LPRINT Print Data at Line Printer (Statement)

LPRM Low Power Range Monitor

LPRS Local Power Recirculation System

LPS Laboratoire de Physico-Chimie des Surfaces [Laboratory for Surface Physics and Chemistry, Paris, France]

LPS Laboratoire de Physique des Solides [Solid-State Physics Laboratory, Université Paris-Sud, Orsay, France]

LPS Laboratoire de Physique des Solides [Solid-State Physics Laboratory, Université Paul Sabatier, Toulouse, France]

LPS Laboratoire de Physique des Solides [Solid-State Physics Laboratory; of Consil National de Recherche Scientifique, Meudon, France]

LPS Laboratory for Physical Sciences [of University of Maryland, College Park, US]

LPS Laboratory of Plasma Studies

LPS Laser Printing System

LPS Launch Processor System; Launch Processing System

LPS Letter Processing Stream

LPS Light Proof Shades

LPS Linear Power Supply

LPS Linear Programming System

LPS Linear Pulse Sector

LPS Lines per Second [also lps]

LPS Lipopolysaccharide [Biochemistry]

LPS Liquid-Phase Sintered; Liquid-Phase Sintering

LPS Liters per Second [also Lps, or Lps]

LPS Log-Periodic Sheet (Antenna)

LPS Long-Period Superstructure; Long-Period Superlattice [Solid-State Physics]

LPS Low-Power Schottky [Electronics]

LPS Low-Pressure Scram [Nuclear Reactors]

LPS Low-Pressure Sodium [Nuclear Reac tors]

LPS Low-Pressure System

LPSB Laboratoire de Physique des Solides de Bellevue [Bellevue Laboratory for Solid-State Physics, Meudon, France]

LPS/CDS Launch Processing System/Central Data Subsystem

LPS-CNRS Laboratoire de Physique des Solides/Centre National de la Recherche Scientifique [Solid-State Physics Laboratory/National Center for Scientific Research, Université Paul Sabatier, Toulouse, France]

LPSD Logically Passive Self-Dual

LPSENS Laboratoire de Physique Statistique de l'Ecole Normale Supérieure [Statistical Physics Laboratory of the Higher Normal School, France]

LPSi Luminescent Porous Silicon

LP-Si$_3$N$_4$ Liquid Phase Silicon Nitride

LPSIRS Linear Potential Sweep Infrared (Reflectance) Spectroscopy

LPSO Long-Period Stacking Order [Solid-State Physics]

LPSR Low Protein Serum Replacement [Biochemistry]

LPSSC Liquid-Phase-Sintered Silicon Carbide

LPSTC Lowest Point of Single Tooth Contact [Gears]

LPS-UPS Laboratoire de Physique des Solides–Université Paris-Sud [Solid-State Physics Laboratory–Université Paris-Sud, Orsay, France]

LPSV Linear Potential Sweep Voltammetry

LPSVD Linear Prediction Singular Value Decomposition

LPSW Low-Pressure Service Water

LPT Laminated Plate Theory

LPT Linear Planarization Tool

LPT Line Printer

LPT Link Plate Thickness (of Chains)

LPT Low-Pressure Turbine

LPT1 First Parallel Printer Port

LPT2 Second Parallel Printer Port

LPT3 Third Parallel Printer Port

LPTB London Passenger Transport Board [UK]

LPTENS Laboratoire de Physique Théorique de l'Ecole Normale Supérieure [Theoretical Physics Laboratory of the Higher Normal School, Paris, France]

LPThM Laboratoire de Physique Théoretique et Modélisation [Theoretical Physics and Modelling Laboratory; of Université de Cergy-Pontoise, France]

LP TiAl$_3$ Long-Period Titanium Trialuminide

LPTF Low Power Test Facility

LPTR Livermore Pool-Type Reactor [of Lawrence Radiation Laboratory, Unikversity of California, US]

LPTTL Low-Power Transistor-Transistor Logic [also LP-TTL, LPT^{2}L, or LP-T^{2}L]

LPTV Large Payload Test Vehicle

LPTV Low-Power Television

LPT Log-Periodic Trapezoidal-Tooth (Antenna)

LPU Language Processor Unit

LPU Life Preserver Unit

LPU Line Processing Unit

LPU Local (Dislocation) Pinning-Unzipping (Model) [Materials Science]

LPU Local Processing Unit

LPU Log-Periodic Unipole (Antenna)

LPUG Lasers in Publishing Users Group [US]

L Pulm Vein Left Pulmonary Vein [also l pulm vein] [Anatomy]

LPV Laboratorio de Patologia Veterinaria [Laboratory for Veterinary Pathology, Portugal]

LPV Leporipoxvirus

LPV Light Proof Vent

LPV Log-Periodic Vee (Antenna)

LPVD Laser Physical Vapor Deposition

LPVT Large Print Video Terminal

LPW Lumen(s) per Watt [also lpW, lpw, or lm/W]

LPY Log-Periodic Yagi (Antenna)

LPZ Low Population Zone

LQ Lateral Quadrupole

LQ Letter-Quality (Printing)

LQ Limiting Quality

LQ Liquid Quench(ing)[Metallurgy]

LQ Low Quality

LQ- Argentina [Civil Aircraft Marking]

Lq Liquid [also lq]

Lq Liquor [also lq]

lq liquid (state) [Chemistry]

LQA Local Quadratic Approximation

LQD Low-Q Diffractometer

Lqd Liquid [also lqd]

LQDC Lowest Quantitatively-Determined Concentration

Lqds Liquids [also lqds]

LQG Linear Quadratic Gaussian (Regulator) [Controllers]

LQGC Linear Quadratic Gaussian Control

LQL Limiting Quality Level

LQM Link Quality Monitoring (Protocol)

LQP Letter-Quality Printer

lq pt liquid pint [Unit]

LR Label(l)ing Reagent

LR Landing Radar

LR Language Reference (Manual)

LR Lanthanide Research

LR Lapse Rate [Meteorology]

LR La Rioja [Argentina]

LR Laser Raman (Spectroscopy)

LR Laser Reflowed; Laser Reflowing

LR Laser Research

LR Laser Retroreflector

LR Laser Rod

LR Latex Rubber

LR Lattice Rotation [Crystallography]

LR Launch Right

LR Lawesson Reagent [Chemistry]

LR Leak(age) Rate

LR Left to Right
LR Left-Right (Matrix Transformation)
LR Left Rudder
LR Level Recorder
LR Level Reduction
LR Lever Rule
LR Liberia [ISO Code]
LR Life Ratio
LR Light Ray
LR Likelihood Ratio [Statistics]
LR Limited Recoverable
LR Limit Register
LR Linear Regression [Statistics]
LR Linear Response
LR Line Radiation
LR Line Receiver
LR Line Relay
LR Link Register
LR Little Rock [Arkansas, US]
LR Living Room
LR Lloyd's Register (of Shipping) [UK]
LR Local/Remote [also L/R]
LR Load Ratio
LR Load-Resistor Relay
LR Logical Record
LR Longitudinal Resonance
LR Long Range
LR Long Run [Building Trades]
LR Los Rios [Ecuador]
LR Lossen Rearrangement [Chemistry]
LR Lower Facet of Reverse-Position Orientation [Mechanical Testing]
LR Lower Right
LR Low Reduction
LR Low Resistor
LR Lucas Reagent
LR Lunar Rover [of NASA Apollo Program]
LR Lurgi-Ruhrgas (Process) [Metallurgy]
L&R Left and Right [also L/R]
L/R Inductance/Resistance
L/R Left and Right [also L&R]
L/R Local/Remote [also LR]
L/R Locus of Radius
L/R Luminence (Intensity) to Raman Line Intensity (Ratio) [Spectroscopy]
L/R Time Constant of an Inductor [Symbol]
Lr Labo(u)r [also lr]
Lr Lawrencium
Lr Lira [Currency of Italy]
Lr-256 Lawrencium-256 [also ^{256}Lr, or Lr^{256}]
Lr-257 Lawrencium-257 [also ^{257}Lr, or Lr^{257}]
lr lower
LRA Laser Gyro Reference Axis
LRA Laser Release Analysis

LRA Laser Research Association
LRA Laser Retroreflector Array
LRA Linear Regression Analysis [Statistics]
LRA Local Radio Association
LRA Logical Record Address
LRA Low Risk Area
LRAPW Linearized Relativistic Augmented Plane Wave (Method)
LRASV Long-Range Air-to-Surface Vessel (Radar)
LRB Labour Relations Board
LRB Local Reference Beam
LRB Long Range Bomb
LRC Langley Research Center [of NASA at Hampton, Virginia, US] [*Note:* LaRC preferred]
LRC Large Rock Cavern
LRC Lighting Research Center
LRC Light, Rapid, Comfortable (Train)
LRC Light Reflective Capacitor
LRC Linguistics Research Center [US]
LRC Lipid Research Center [of George Washington University School of Medicine, Washington, DC, US]
LRC Loading Rate Change [Mechanics]
LRC Load Ratio Control
LRC Local Register Cache [Computers]
LRC Longitudinal Redundancy Check
LRC Long-Range Cruise
LRC Lynchburg Research Center [of Babcock and Wilcox Inc., Virginia, US]
LRCC Library Resources Coordinating Committee [UK]
LRCC Longitudinal Redundancy Check Character
LRCM Long-Range Cruise Missile
LRCO Limited Remote Commications Outlet
LRCP Licentiate of the Royal College of Physicians [UK]
LRCS League of Red Cross Societies
LRCS Long-Range Cruise Speed
LRD Liberian Dollar [Currency]
LRD Longitudinal Research Database [of United States Census Bureau]
LRD Long-Range Data
LRDC Land Resources Development Center [UK]
LRDC London Research and Development Corporation [Ontario, Canada]
LRE Laser Radar Equation
LRE Linear Resource Extension
LRECL Logical Records of Fixed Length
LREE Light Rare-Earth Element
LREIS Laboratory for Resource and Environmental Information Systems
LREP Laser Rotating Electrode Process
LRF Laser Rangefinder
LRF Laser-Reduced Fluorescence
LRF Liquid Rocket Fuel
LRF Long-Range Facility
LRF Lumber Recovery Factor
LRF Luteal-Releasing Factor [Biochemistry]

LRFAX Low-Resolution Facsimile

LRFD Load and Resistance Factor Design

LRFS Long-Range Forecasting System

LRG Landscape Research Group [UK]

LRG Lightweight Recoilless Gun

LRG Long Range Gun

LRI Laser Reflection Interferometer; Laser Reflection Interferometry

LRI Lighting Research Institute [US]

LRI Long-Range Radar Input

LRI Lower Respiratory Infection

LRIA Level Removable Instrument Assembly

LRIM Long Range Input Monitor

LRIR Limb Radiance Inversion Radiometer

LRIR Low-Resolution Infrared Radiometer

LRIS Land Registration Information Service [Canada]

LRIS Land-Related Information System

LRIS Land Resources Information System

LRL Lawrence Radiation Laboratory [of University of California, US]

LRL Least Recently Loaded

LRL Linking Relocating Loader

LRL Livermore Research Laboratory [University of California, US]

LRL Logical Record Length

LRL Logical Record Location

LRL Lunar Receiving Laboratory [for NASA Apollo Program; was located at Houston, Texas]

LR/LD Line Receiver/Line Driver

LRLS Learning by Recursive Least Squares

LRLTRAN Lawrence Radiation Laboratory Translator [of University of California, US]

LRM Language Reference Manual

LRM Laser Raman Microscopy

LRM Least Recently-Used Master

LRM Limited Register Machine

LRM Linear Reluctance Motor

LRM Liquid Reaction Molding

LRM Long-Reach Manipulator

LRM Lunar Reconnaissance Module [of NASA Apollo Program]

LRMA Laser Raman Microanalysis

L/R MAR Left and Right Margin [also L/R Mar]

LR MCO Long Run, Mill Culls Out [Building Trades]

LRME Laboratoire de Reconnaissance des Matériaux dans leur Environnement [Laboratory for the Identification of Materials in their Environment, Université de Marne la Vallee, France]

LRMF Laboratoire de Recherche des Musées de France [Research Laboratory of the French Museum, Paris]

LRMS Low-Resolution Mass Spectroscopy

LRMTS Laser Rangefinder and Marked Target Seeker

LRO Large Radio Observatory

LRO Long-Range Order(ed); Long-Range Ordering [also lro] [Solid State Physics]

LROCSCM Long-Range Optimized (Heteronuclear) Chemical Shift Correlation Method [Physical Chemistry]

LROO Long-Range Orientational Order(ing) [Solid-State Physics]

LROP Long-Range Order Parameter [Solid-State Physics]

LRP Land Reclamation Program

LRP Linear Reaction Path

LRP Loan Repayment Program

LRP Long-Range Order Parameter [Solid-State Physics]

LRP Long-Range Path

LRP Low-Rank Perturbation

LRPA Long-Range Patrol Aircraft

LRPDS Long-Range Position-Determining System

LRPG Long-Range Proving Ground

LRPL Liquid Rocket Propulsion Laboratory [US Army]

LRPT Low-Resolution Picture Transmission

LRQA Lloyd's Register Quality Assurance

LRR Launch Readiness Review

LRR Loop Regenerative Repeater

LRRA Litter and Recycling Research Association

LRRP Lowest Required Radiating Power

LRRR Laser Ranging Retro-Reflector

LRRS Laboratoire de Recherche sur la Réactivité des Solides [Solids Reactivity Research Laboratory, University of Bourgogne, Dijon, France]

LRRS Long-Range Radar Site

LRS Language Resource (File)

LRS Laser Raman Scattering

LRS Laser Raman Spectrophotometer

LRS Linguistic Research System

LRS Liquid Radwaste System

LRS Log and Reporting System

LRS Long-Range Search

.LRS Language Resource [WordPerfect File Name Extension]

LRSI Low-Temperature Reusable Surface Insulation [of NASA Space Shuttle]

LRSM Laboratory for Research on the Structure of Matter [of University of Pennsylvania, Philadelphia, US]

LRSM Long-Range Seismographic Measurements

LRSOM Long-Range Stand-Off Missile

LRSS Laboratoire de Réactivité de Surface et Structure [Surface Reactivity and Structure Laboratory; of Université Pierre et Marie Curie, Paris]

LRSS Long-Range Survey System

LRT Leak Rate Test

LRT Light Rail Transport

LRT Light Rail Transit

LRT Light Railway Transport

LRT Light Rapid Transit

LRT Linear Response Theory

LRT Long-Range Transport

LRT Lower Respiratory Tract

LRTA Light Rail Transit Association [UK]

LRTAP Long-Range Transboundary Air Pollution (Convention)

LRTAP Long-Range Transport of Acidic Pollutants

LRTAP Long-Range Transport of Air(borne) Pollutants

LRTAPP Long-Range Transport of Acidic Pollutants Program [Canada]

LRTAPP Long-Range Transport of Air(borne) Pollutants Program [Canada]

LRTF Long-Range Technical Forecast

LRTL Light Railway Transport League

LRTM Long-Range Training Mission

LRTP Long-Range Training Program

LRTP Long-Running Thermal Precipitator

LRU Least Recently Used

LRU Least Replaceable Unit

LRU Line-Replaceable Unit

LRU Little Rock University [Arkansas, US]

LRV Launch Readiness Verification

LRV Layout Rule Violation (Program)

LRV Leishmania RNA (Ribonucleic Acid) Virus

LRV Light Rail Vehicle

LRV Light Rapid Vehicle

LRV Long Range Video

LRV *(Luftreinhalteverordnung)* – Swiss Air Pollution Control Regulations

LRV Lunar Roving Vehicle [of NASA, Apollo Program]

LRV 1-1 Leishmania RNA (Ribonucleic Acid) Virus 1-1

LRW Liquid Radioactive Waste

LRWE Long-Range Weapons Establishment [Australia]

LRX Large (Nuclear) Reactor Critical Facility [US]

LS Laboratoire de Solides [Solids Laboratory; of University of Burgundy, France]

LS Laboratory Standard [also L/S]

LS Laboratory System

LS Lamb Shift [Atomic Physics]

LS Lamellar Structure [Metallurgy]

LS Landing Ship [US Navy Symbol]

LS Landing Site [also L/S]

LS Land Subsidence

LS Land Service

LS Land Survey(ing); Land Surveyor

LS Language Specification

LS Language System

LS Lanthanum Antimonate

LS Laparoscopic Sterilization [Gynecology]

LS Large-Scale

LS Laser Science

LS Laser Scanner

LS Laser Sintering [Metallurgy]

LS Laser Source

LS Laser Spectrometer; Laser Spectrometry

LS Laser Spectroscopy

LS Laser Spectrum

LS Laser Storage

LS Laser Surgeon; Laser Surgery

LS Launch Sequence

LS Launch Services

LS Laser Stabilization; Laser Stabilizer

LS Laser System

LS Leading Seaman [Canadian Forces]

LS Least Significant

LS Least Squares (Method) [Statistics]

LS Left Side

LS Legal Status

LS Length of Seed [Agriculture]

LS Lens Stop

LS Lesotho [ISO Code]

LS Letter Signed

LS Level Switch

LS Levitation System

LS Library Science

LS Library Service

LS Library System

LS Life Science

LS Lifshitz and Slyozov (Theory) [Metallurgy]

LS Lift System

LS Ligand Substitution [Chemistry]

LS Light Scattering

LS Light-Sensitive; Light Sensitivity

LS Light Shield

LS Light Source

LS Light Stabilizer

LS Limit Switch

LS Linear System

LS Line Scan(ner)

LS Line Spectrum [Spectroscopy]

LS Line Switch

LS Link Switch

LS Liquid Scintillation

LS Liquid Starter

LS Lithium Silicate

LS Loading Splice

LS Load Sensing

LS Load(ing) Stress

LS Lobe Switching [Electrical Engineering]

LS Lockscrew

LS Local Store

LS Logic Symbol

LS Longitudinal Section

LS Long Shot [Television]

LS Lot Sample

LS Lot Size

LS Loudspeaker

LS Louisiana State [US]

LS Low-Power Schottky (Logic)

LS Low-Silicon (Iron Alloy)

LS Low Speed

LS Low Structure Black

LS Low Sulfur

LS Lumbo-Sacral [Anatomy]

LS Luminescent Screen

LS Lunar Science

LS Lung Surfactant

LS Luxury Sedan

LS Lyman Series [Spectroscopy]

LS Lymphatic System [Anatomy]

LS Syrian Pound [Currency]

L&S Logistics and Support

L/S Landing Site [also LS]

L/S Lines per Second [also LPS, or Lps]

L/S Load System

L-S Liquid-Solid (Interface) [also L/S]

L/s Liter(s) per second [also L/sec, l/s, l/sec, L s^{-1}, l s^{-1}, ltr/s, ltr/sec, ltr s^{-1}, or ℓ/sec, or ℓ/s]

LSA Labor Surplus Area

LSA LAN (Local-Area Network) and SCSI (Small Computer Systems Interface) Adapter

LSA Large Scale Area

LSA Large-Scale Soundings Array

LSA Large Surface Alloying

LSA Large Surface Area

LSA Laser-Supported Absorption

LSA Laser Surface Alloying

LSA Launch Services Agreement

LSA Lead Sheet Association [UK]

LSA Least Squares Analysis

LSA Least Squares Approximation

LSA Library Services Act [US]

LSA Licentiate in Agricultural Science

LSA Licentiate of the Society of Apothecaries

LSA Lignosulfonic Acid

LSA Limited Space-Charge Accumulation (Diode)

LSA Linear Stability Analysis [Mechanics]

LSA Linear System Analysis

LSA Line Sharing Adapter

LSA Lithium Strontium Aluminate

LSA Local Security Authority

LSA Locksmith Security Association [US]

LSA Logging Safety Association [Canada]

LSA Logistic(s) Support Analysis

LSA London Sisal Association [UK]

LSA Longitudinal Spherical Aberration

LSA Low-Cost Solar Array

LSA Low-Specific Activity [Radioactivity]

LSA Low-Sulfur and Aluminum

LSAA Library Services Authority Act [US]

LSAC Local Section Activities Committee [of American Chemical Society, US]

LSAH Launch Site Accommodations Handbook [of NASA]

LSALT Lowest Safe Altitude

LSAO Lanthanum Strontium Aluminum Oxide (Superconductor)

LSAP Launch Sequence Applications Program

LSAP Link-Layer Service Access Point

LSAPI License Services Application Program Interface

LSAS Longitudinal Stability Augmentation System

LSAT Law School Admissions Test

LSASV Linear Sweep Anodic Stripping Voltammetry

LSB Laboratory of Structural Biology [of Division of Computer Research and Technology, US]

LSB Large-Scale Bridging [Materials Science]

LSB Least Significant Bit

LSB Lower Sideband

LSB Low-Speed Breaker

LSBR Large Seed-Blanket Reactor

LSC Lanthanum Strontium Chromite (or Strontium-Doped Lanthanum Chromite)

LSC Large-Scale Computer

LSC Laser Semiconductor

LSC Laser-Supported Combustion

LSC Laser Surface Cladding

LSC Least Significant Character

LSC Leetown Science Center [of US Geological Survey/ Biological Resources Division at Kearneysville, West Virginia]

LSC Liberian Shipowners Council

LSC Life Sciences Collection

LSC Linear Shaped Charge

LSC Liquid Scintillation Counter; Liquid Scintillation Counting

LSC Liquid-Solid Chromatography

LSC Livingston State College [Alabama, US]

LSC Local Switching Center

LSC Loop Station Connector

LSC Low Shear Continuous

LSC Low-Speed Concentrator

LSC Luminescent Solar Concentrator

LSC Lunar Science Conference

LSc Licentiate in Science

LSCA Library Services and Construction Act [US]

LSCA Logistics Support Cost Analysis

LSCC Line-Sequential Color Composite

LSCC Local Servicing Control Center

LSCCO Lanthanum Strontium Calcium Copper Oxide (Superconductor)

LScComm Licentiate in Commercial Science

LSCE Launch Sequence and Control Equipment

LSCFO Lanthanum Strontium Copper Iron Oxide (Superconductor)

LSCFO Lanthanum Strontium Cobalt Iron Oxide (Superconductor)

LSCNO Lanthanum Strontium Copper Niobium Oxide (Superconductor)

LSCO Labour Safety Council of Ontario [Canada]

LSCO Lanthanum Strontium Cobalt Oxide (Superconductor)

LSCO Lanthanum Strontium Copper Oxide (Superconductor)

LSCP Low-Speed Card Punch

LSCPP Lead Sulfate Carrier Precipitation Process

LSCT Lamar State College of Technology [Beaumont, Texas, US]

LSCU Local Servicing Control Unit

LSCVO Lanthanum Strontium Copper Vanadium Oxide (Superconductor)

LSD Landing Ship, Dock [US Navy Symbol]

LSD Language for Systems Development

LSD Large-Scale Domain

LSD Large Screen Display

LSD Large Signal Diode

LSD Laser-Supported Detonation

LSD Last Significant Digit

LSD Launch Systems Data

LSD Least Significant Digit

LSD Life Sciences Division [of Los Alamos National Laboratory, New Mexico, US]

LSD Lightermen, Stevedores and Dockers

LSD Light-Sensing Device

LSD Line-Sharing Device

LSD Line Signal Detector

LSD Liquid Scintillation Detector

LSD Local Shear Deformation [Mechanics]

LSD Local Spin Density [Solid-State Physics]

LSD Logarithmic Series Distribution

LSD Logistics Support Division

LSD Loop Shutdown

LSD Low-Speed Data

LSD Lunar Surface Magnetometer

LSD Lysergic Acid Diethylamide

LSD *(librae, solidi denarii)* – pounds, shillings and pence [also lsd, or £sd]

LSDA Local Spin-Density Approximation [Solid-State Physics]

LSDA+U Local Spin-Density Approximation with the On-Site Coulomb Interaction

LSDC Least-Squares Differential Correction

L-SDDA L-Stilbenediamine-N,N'-Diacetate [also l-sdda]

LSDF Large Sodium Disposal Facility [Nuclear Engineering]

LSDF Local Spin-Density Function(al) [Solid-State Physics]

LSDR Local Storage Data Register

LSDSIC Local Spin Density Approximation with Self-Interaction Correction

LSDU Link-Layer Service Data Unit

LSDW Longitudinal Spin-Density Wave [Solid-State Physics]

LSE Language Sensitive Editor

LSE Launch Sequencer Equipment

LSE Launch Support Equipment

LSE Least Squares Estimation

LSE Levitt-Suter-Ernst (Pulse Sequence)

LSE Life Support Equipment

LSE Linear Stark Effect [Atomic Physics]

LSE Line Signaling Equipment

LSE Liquid-Solvent Extraction (Process)

LSE London School of Economics [UK]

LSE London Stock Exchange [UK]

LSE Longitudinal Section Electric

LSE Luxury Sedan Euro-Style

L/sec Liter(s) per second [also L/s, l/s, l/sec, L s^{-1}, l s^{-1}, ltr/s, ltr/sec, ltr s^{-1}, ℓ/sec, or ℓ/s]

LSECS Life Support and Environmental Control System

LSEM Lifshitz-Wagner Encounter Modified (Model) [Metallurgy]

LSEP Lunar Surface Experimental Package [of NASA]

LSF Laboratory Simulation Facility

LSF Least-Squares Fitting

LSF Linear Structural Formula

LSF Line Spread Function

LSF Line Switch Frame

LSF Lithium Strontium Fluoride

LSF Load Sheet Fuel

LSF Local Spin Fluctuation [Solid-State Physics]

LSF Lumped Selection Filter

L&SF Lemonthyme and Southern Forests [Tasmania, Australia]

LSFC Lewis Space Flight Center [of NASA at Cleveland, Ohio, US]

LSFE Life Sciences Flight Experiment

LSFES Lake States Forest Experiment Station [of US Department of Agriculture at St. Paul, Minnesota]

LSFFAR Low-Spin Folding Fin Aircraft Rocket

LSFO Lanthanum Strontium Iron Oxide (Superconductor)

LSFR Local Storage Function Register

LSFT Local Spin Fluctuation Theory [Solid-State Physics]

LSFT Low Steamline Flow Test

LSG Language Structure Group [of CODASYL–Conference on Data System Languages]

LSG Low-Stress Grinding

Lsg Leasing [also lsg]

LSGA Laminators Safety Glass Association [US]

LSGAECF Liquid Scintillation Gross Alpha Efficiency Calibration Form

LSG2M Laboratoire de Science et Génie des Matériaux Métalliques [Metallic Materials Science and Engineering Laboratory, Ecole de Mines, Nancy, France] [also LSGMM]

LSGO Lanthanum Strontium Gallium Oxide (Superconductor)

LSGS Laboratoire de Science et Génie des Surfaces [Surface Science and Engineering Laboratory, Ecole de Mines, Nancy, France]

LSH Lamb Shift in Hydrogen [Atomic Physics]

LSH Low Superheat (Casting) [Metallurgy]

LSHA Louisiana State Horticultural Association [US]

LSHI Large-Scale Hybrid Integration

LSHTM London School of Hygiene and Tropical Medicine [UK]

LSI Laboratoire des Solides Irradiés [Irradiated Solids Laboratory; of Ecole Polytechnique, Palaiseau, France]

LSI Langelier Saturation Index

LSI Large-Scale Integrated; Large-Scale Integration

LSI Large-Scale Investment

LSI Liquid-Surface Interferometer; Liquid-Surface Interferometry

LSIA Lamp and Shade Industries of America [US]

LSIC Large-Scale Integrated Circuit

LSID Launch Sequence and Interlock Document

LSID Local Session Identification

LSIG Line Scan Image Generator

LSI-MOS Large-Scale Integrated Metal-Oxide Semiconductor

LSIR Limb Scanning Infrared Radiometer

LSIS Laser Scanner Inspection System

LSI/VLSI Large-Scale Integration/Very-Large-Scale Integration

LSJ Laser Society of Japan

LSL Ladder Static Logic

LSL Lateral Superlattice [Solid-State Physics]

LSL Life Science Laboratory

LSL-II Life Science Laboratory II [of Pacific Northwest National Laboratory, Richland, Washington, US]

LSL Link Support Layer

LSL Linnean Society of London [UK]

LSL Link and Selector Language

LSL Load Segment Limit

LSL Louisiana State Library [US]

LSL Lower Specification Limit

LSL Lower Specified Limit

LSL Low-Speed Logic

LSL$_x$ Lower Specification Limit for Value "x" [Symbol]

LSLA Low-Speed Line Adapter

LSL-A Life Science Laboratory–Chemical Storage and Transfer Facility [of Pacific Northwest National Laboratory, Richland, Washington, US]

LSLD Local Store Loop Driver

LSLE Life Sciences Laboratory Equipment

LSLE R/F Life Sciences Laboratory Equipment Refrigerator/Freezer

LSM Landing Ship, Medium [US Navy Symbol]

LSM Lanthanum Strontium Manganite

LSM Laser Scan(ning) Microscope; Laser Scan(ning) Microscopy

LSM Laser Surface Melting

LSM Launcher Status Multiplexer

LSM Layered Synthetic Microstructure (Catalyst)

LSM Lesotho Maloti [Currency]

LSM Life Science Module

LSM Linearized Simulation Model,

LSM Linear Synchronous Motor

LSM Line Selection Module

LSM Line Switch Marker

LSM Liquid Surfactant Membranes

LSM Local Synchronous Modem

LSM Logic-State Map

LSM Logistics Support Management Information

LSM Longitudinal Section Magnetic

LSM Long Span Mezzanine [Architecture]

LSM Long Static Mixer

LSM Low-Speed Modem

LSM Lymphocyte Separation Medium [Biochemistry]

LSM Strontium-Doped Lanthanum Manganite

LSMA Low-Speed Multiplexer Arrangement

LSMI Lake Superior Mining Institute [US]

Lsmith Locksmith [also lsmith]

LSMM Laboratoire de Structure des Matériaux Métalliques [Metallic Materials Structure Laboratory, of Université du Paris-Sud, Orsay, France]

LSMNO Lanthanum Strontium Manganese Nickel Oxide (Superconductor)

LSMO Lanthanum Strontium Manganate

LSMO Lanthanum Strontium Molybdenum Oxide (Superconductor)

L/s·mol Liter(s) per second mole [also L s^{-1} mol^{-1}, l/s·mol, or l s^{-1} mol^{-1}]

LSMO/STO Lanthanum Strontium Manganese Oxide/Strontium Titanate

LSM(R) Landing Ship, Medium, with Rocket Launchers [also LSM-R]

LSMS Large System Multiple Scattering (Method)

LSMT Localized Soft Mode Theory

LSMU Lasercom Space Measurement Unit

LSN Linear Sequential Network

LSN Line Stabilization Network

LSN Low-Solids, Nondispersoid Mud

LSND Liquid Scintillator Neutrino Detector (Experiment) [Los Alamos National Laboratory, Los Alamos, New Mexico, US]

LSNFO Lanthanum Strontium Nickel Iron Oxide (Superconductor)

LBNGO Lanthanum Strontium Nickel Gallium Oxide (Superconductor)

LSNO Lanthanum Strontium Nickel Oxide (Superconductor)

LSNSW Linnean Society of New South Wales [Australia]

LSNY Linnean Society of New York [US]

LSO Landing Signal Officer

LSO Landing Support Officer

LSO Large Solar Observatory

LSO Line Signaling Oscillator

LSO Lutetium Orthosilicate

LSOP Lone Star Ordnance Plant [Texarkana, Texas, US]

LSP Laboratoire des Spectrométrie Physique [Physical Spectrometry Laboratory; of Université J. Fourier Grenoble I, France]

LSP Large-Scale Production

LSP Laser Shock Processing

LSP Laser Speckle Photography

LSP Laser Surface Processing

LSP Least Significant Portion

LSP Library Software Package

LSP Licensed Site Professional [Massachusetts, US]

LSP Licensed State Professional

LSP Ligand Substitution Process

LSP Linked Systems Project [US]

LSP Liquid-State Physics

LSP Local Store Pointer

LSP Loop Splice Plate

LSP Low-Speed Printer

LSPA Licensed Site Professional Association [also LSP Association] [Massachusetts, US]

LSPB Large Scale Prototype Breeder (Reactor)

LSPDF Life Science Payloads Development Facility

LSPE Lateral Solid Phase Epitaxial (Growith) [Metallurgy]

LSPES Laboratoire de Structure et Propriétés de l'Etat Solide [Solid-State Structure and Properties; of Conseil National de Recherche Scientifique at Université des Sciences et Technologie de Lille, Villeneuve d'Ascq, France]

LSPES-CNRS Laboratoire de Structure et Propriétés de l'Etat Solide–Conseil National de Recherche Scientifique [Solid-State Structure and Properties–National Scientific Research Council at Université des Sciences et Technologie de Lille, Villeneuve d'Ascq, France]

LSPET Lunar Samples Preliminary Examination Team [of NASA]

Lspk Loudspeaker

LSPMO Lanthanum Strontium Lead Manganese Oxide (Superconductor)

LSPN Ligue Suisse pour la Protection de la Nature [Swiss League for Nature Protection]

LSPS Local Service Planning System

LSPTR Low-Speed Paper Tape Reader

LSP-UJF Laboratoire des Spectrométrie Physique–Université J. Fourier Grenoble I [Physical Spectrometry Laboratory–J. Fourier University Grenoble I, France]

LSQ Least Squares (Fit)

LSQ Line Squall [Meteorology]

LSQA Local System Queue Area

LSR Laboratory for Space Research

LSR Land Sea Rescue

LSR Lanthanide Shift Reagent

LSR Launch Site Recovery

LSR Least-Squares Regression

LSR Liquid Silicon Rubber

LSR Load Shifting Resistor

LSR Local-Shared Resources

LSR Local Storage Register

LSR Low Spatial Resolution

LSR Low-Speed Reader

LSR Lynchburg Source Reactor [of Babcock and Wilcox Ltd., Virginia, US]

LSRA Lead Smelters and Refiners Association [UK]

LSRA Least-Squares Regression Analysis

LSR-AES Low Spatial Resolution–Auger Electron Spectroscopy

LSRH Laboratoire Suisse de Recherches Horlogères [Swiss Laboratory for Horological Research, Neuchâtel]

LSRM Life Science Research Module

LSRP Local Switching Replacement Planning

LSS Laboratory for Surface Studies [of University of Wisconsin-Milwaukee, US]

LSS Laboratory Services Section [of Fermilab, Batavia, Illinois, US]

LSS Laboratory Support System

LSS Ladder Support Software (Package)

LSS Lamb-Shift Source [Atomic Physics]

LSS Landing Separation Simulator

LSS Language for Symbolic Simulation

LSS Large-Scale System

LSS Large-Sized Specimen Stage

LSS Large Space Structure(s)

LSS Large Space System

LSS Laser-Storage System

LSS Launch Status Summarizer

LSS Life Support Subsystem

LSS Life Support System

LSS Livesaving Service [US]

LSS Limited-Specification Standard

LSS Lindhard-Scharff-Schioett (Theory)

LSS Liquid-Solid Separation

LSS Local Synchronization Subsystem

LSS Loop Surge Suppressor

LSS Loop Switching System

LSS Lunar Soil Simulator [of NASA]

LSSD Laboratory and Scientific Services Directorate

LSSD Late Stage Spinodal Decomposition (Model) [Materials Science]

LSSD Level-Sensitive Scan Design

LSSD Level-Sensitive Scan Device

LSSF Life Sciences Support Facility

LSSG Lateral Studies Sub-Group [of (Specialist Panel on) Navigation and Separation of Aircrafts]

LSSL Lateral Surface Superlattice [Solid-State Physics]

LSSL Life Sciences Space Laboratory [of NASA]

LSSM Launch Site Support Manager

LSSM Lunar Scientific Survey Module [of NASA Apollo Program]

LSSMO Lanthanum Strontium Strontium Manganese Oxide (Superconductor)

LSSP Latest Scram Set Point [Nuclear Reactors]

LSSP Launch Site Support Plan

LSSRC Life Sciences Shuttle Research Centrifuge [of NASA]

LSSS Limiting Safety System Setting

LSSST Laboratory for Solid-State Science and Technology [Syracuse, New York, US]

LSST Launch Site Support Team

LS/ST Light Shield/Star Tracker

LST Landing Ship, Tank [US Navy Symbol]

LST Large Space Telescope

LST Large Stellar Telescope

LST Laser Scanning Technique

LST Laser Surface Treatment

LST Last (Register)
LST Launch Support Team
LST Light Scattering Technique
LST Light Scattering Tomography
LST Liquid Storage Tank
LST List [also Lst] [Computers/Communications]
LST Local Sidereal Time
LST Local Solar Time
LST Local Standard Time [also lst]
LST Loud Speaking Telephone
LST Lyddane-Sachs-Teller (Relation) [Solid-State Physics]
LST N-Acetylneuramin-Lacto-N-neo-Tetraose [Biochemistry]
LST Standardized Light Source
.LST List [File Name Extension] [also .lst]
LSt Lead Styphnate
Lst List [also lst]
LSTA London Shellac Trade Association [UK]
Lstab Laxmanan (Microsegregation Model) Corrected for Marginal Stability [Metallurgy]
LSTB Low-Power Schottky TTL (Transistor-Transistor Logic) Bipolar
LST c N-Acetylneuramin-Lacto-N-neo-Tetraose c [Biochemistry]
LSTCO Lanthanum Strontium Terbium Copper Oxide (Semiconductor)
LSTD Laser Scattering Tomography Defect
LSTE Launch Site Transportation Equipment
LSTF Large Scale Test Facility
LSTG Laser Science Topical Group [of American Physical Society, US]
LSTPC List Price [also LST PC, or Lst Pc]
LSTTL Large-Scale Transistor-Transistor Logic
LSTTL Low-Power Schottky Transistor-Transistor Logic [also LS-TTL, or LS TTL]
LSTTL LSI Low-Power Schottky Transistor-Transistor Logic–Large-Scale Integration [also LS-TTL-LSI]
LSU Leading Signal Unit
LSU Library Storage Unit
LSU Life Support Umbilical [Aerospace]
LSU Line Selection Unit
LSU Line Sharing Unit
LSU Load Storage Unit
LSU Local Storage Unit
LSU Local Switching Unit
LSU Local Synchronization Utility
LSU Lone Signalling Unit
LSU Louisiana State University [Baton Rouge and New Orleans, US]
LSUBR Louisiana State University, Baton Rouge [US]
LSUNO Louisiana State University, New Orleans [US]
LSUSM Louisiana State University School of Medicine [New Orleans, US]
LSV Left Subclavian Vein [Anatomy]
LSV Levitated Sphere Viscometer
LSV Lift-Spray Vacuum (Pressing Technique)

LSV Linear Sweep Voltammetry
LSV Line Status Verifier
LSV Liquid Sampling Valve
LSV Low Signature Vehicle
LSV Lunar Surface Vehicle [of NASA Apollo Program]
LSW Lifshitz-Slyozov-Wagner (Theory) [Metallurgy]
LSY Large-Scale Yielding [Mechanics]
LSZ Lime-Stabilized Zirconia
LT Laboratory Technologist; Laboratory Technology
LT Laboratory Test
LT Lamellae Twin [Metallurgy]
LT Landau Theory (of Superconductivity)
LT Language Translation; Language Translator
LT Language Transmission
LT Lanthanum Tantalate
LT Lanthanum Titanate
LT Laplace Transform [Mathematics]
LT Laser Technologist; Laser Technology
LT Late Transition-Metal (Elements)
LT (Crystal) Lattice Theory
LT Laue Theory [Crystallography]
LT Leadthrough
LT Lead Time
LT Leak Test(ing)
LT Leather Technologist; Leather Technology
LT Leidenfrost Transport [Thermodynamics]
LT Less Than [also lt]
LT Letter Telegram
LT Leucotriene [Biochemistry]
LT Library Technician
LT Library Technologist; Library Technology
LT Life Test
LT Lifetime
LT Light-Tight
LT Light Truck
LT Limit Theorem
LT Linear Transformation [Mathematics]
LT Line Telecommunications
LT Line Telegraphy
LT Line-Tension (Model) [Materials Science]
LT Line Terminator
LT Link Terminal
LT Link Trainer
LT Lithium Tantalate
LT Lithium Tellurate
LT Lithium Titanate
LT Lithuania [ISO Code]
LT Load Tape
LT Load Test [also L/T]
LT Localizer Transmitter [Aeronautics]
LT Local Time
LT Logical Terminal
LT Logic Theorist; Logic Theory

LT Longitudinal Plane (of Composite) [Symbol]
LT Long-Tail Vehicle
LT Long Term
LT Long Ton [Unit]
LT Long-Transverse (Direction) [also L-T]
LT Loop Test
LT Low Temperature
LT Low Tide
LT Low Tension
LT Low Torque
LT Lubrication Technologist; Lubrication Technology
LT Lucas Test
LT Luer Tip
LT Lurgi/Thyssen (Process) [Metallurgy]
LT Transition Locational Fit [ANSI Symbol]
LT Turkish Lira [Currency]
LT1 Transition Locational Fit (H7/js6) [ANSI Symbol]
LT2 Transition Locational Fit (H8/js7) [ANSI Symbol]
LT3 Transition Locational Fit (H7/k6) [ANSI Symbol]
LT4 Transition Locational Fit (H8/k7) [ANSI Symbol]
LT5 Transition Locational Fit (H7/n6) [ANSI Symbol]
LT6 Transition Locational Fit (H7/n7) [ANSI Symbol]
L&T Laboratory and Test
L&T Letter and Telegram
L&T Link and Terminal
L/T Load Test [also LT]
L-T Long-Term
L-T Longitudinal-Transverse (Direction) [also LT]
£T Turkish Pound [Currency]
Lt Left [also lt]
Lt Lieutenant [also L, LT, or Lieut]
Lt Light [also lt]
L(t) Time-Dependent Gauge Length [Function]
l(T) mean free path [Symbol]
lt long ton [Unit]
.lt Lithuania [Country Code/Domain Name]
$\lambda_{max}T$ Wien's displacement law (constant) (i.e., 2.8978 $\times$ 10^{-3} mK) [Symbol]
$\lambda(T)$ temperature-dependent magnetic penetration depth [Symbol]
$\lambda(T)$ temperature-dependent wavelength [Symbol]
$\lambda_L(T)$ London penetration depth (in solid-state physics) [Symbol]
$\lambda(t)$ time-dependent magnetic penetration depth [Symbol]
LTA Lead Tetraacetate
LTA Library-Media Technical Assistant
LTA Lighter-than-Air (Craft)
LTA Linen Trade Association [US]
LTA Literary Translators Association [Canada]
LTA Logical Transient Area
LTA Long-Term Agreement
LTA Long-Term Arrangement
LTA Long-Time Annealing [Metallurgy]
LTA Lower Torso Assembly

LTA Low-Temperature Aging [Metallurgy]
LTA Low-Temperature Anneal(ing) [Metallurgy]
LTA Low-Temperature Ashing
LTA$_4$ Leucotriene A$_4$ [Biochemistry]
LtAA Light Anti-Aircraft Artillery
LTAB Line Test Access Bus
LTAC Low-Temperature Additive Conductor (Technology)
LTACS Land Tactical Area Communications System
LTAS Lighter-than-Air Ship
LTASC Laboratorio Tecnologie Avanzate Superfici e Catalisi [Laboratory for Advanced Surface and Catalysis Technology, Trieste, Italy]
LTASC-CIFM Laboratorio Tecnologie Avanzate Superfici e Catalisi–Consorzio Inter-universitario di Fisica della Materia [Laboratory for Advanced Surface and Catalysis Technology–Inter-University Consortium on Matter Physics, Trieste, Italy]
LTB Last Trunk Busy
LTB Lithium Tetraborate
LTBO Linear Time Base Oscillator
LTC Laboratory Training Coordination [Pacific Northwest National Laboratory, Richland, Washington, US]
LTC Lieutenant Colonel [also Lt Col]
LTC Lieutenant Commander [also Lt Cmdr, Lt Cdr, Lt Com, or Lt Comdr]
LTC Line Terminating Circuit
LTC Line Time Clock
LTC Line Traffic Coordinator
LTC Local Telephone Circuit
LTC Local Terminal Controller
LTC Local Test Cabinet
LTC Long Time Constant
LTC Low-Temperature Chlorination (Process)
LTC Low-Temperature Control
LTC Low-Temperature Creep [Mechanics]
LTCB Long-Term Credit Bank (of Japan)
LTCDA Low-Temperature Coal Distillers Association [UK]
Lt Cdr Lieutenant Commander [also Lt Cmdr, or Lt Comdr]
LT-CEMS Low-Temperature Conversion Electron Moessbauer Spectroscopy
LTCESFS Low-Temperature Constant Energy Synchronous Fluorescence Spectroscopy
LT(Cert) Certificate in Laboratory Technology
Lt Cmdr Lieutenant Commander [also LTC, Lt Cdr, Lt Com, or Lt Comdr]
Lt Col Lieutenant Colonel [also LTC]
Lt Col(N) Lieutenant Colonel (Navy)
Lt Comdr Lieutenant Commander [also LTC, Lt Com, Lt Cdr, or Lt Cmdr]
LTCP Laboratoire de Technologie des Composites et Polymères [Laboratory for Composite and Polymer Technology; of Ecole Polytechnique Fédérale de Lausanne, Switzerland]
LT Creep Low-Temperature Creep [Mechanics]
LTD Lamella Thickness Distribution
LTD Laser Target Designator

LTD Lift-to-Drag (Ratio)

LTD Line Transfer Device

LTD Local Test Desk

LTD Long Term Debt

LTD Low-Temperature Detector

LTD Low-Temperature Disintegration

Ltd Limited [also ltd]

lt/d long tons per day [Unit]

LTD$_4$ Leucotriene D$_4$ [Biochemistry]

LTDE Local Thermodynamic Equilibrium

LTDP Long-Term Defense Program

LTDR Low-Temperature Dynamic Recrystallization [Metallurgy]

LTDS Laser Target Designation System

LTDS Launch Trajectory Data System

LTDS Low-Temperature Drawing Salt [Metallurgy]

LTE Land Trust Exchange [US]

LTE Line Termination Equipment

LTE Local Telephone Exchange

LTE Local Thermal Equilibrium (Theory)

LTE Local Thermodynamic Equilibrium

LTEC Linear Thermal Expansion Coefficient

LTEC Low-Temperature Engineering and Cryogenics Conference

LTEL Long-Term Exposure Limit(s)

LTEM Lorentz Transmission Electron Microscope; Lorentz Transmission Electron Microscopy

LTEM Low-Temperature (Fluorescence Phosphorescence) Electron Microscopy

LTER Long-Term Ecological Research

LTF Laboratory Testing Facilities

LTF Lithium Terbium Fluoride

LTF Lithographic Technical Foundation [now Graphic Arts Technical Foundation, US]

LTF Low-Temperature Fluorescence

L2F Layer Two Forwarding [Computers]

LTFO Long-Term Fallout

LTFRD Lot Tolerance Fraction Reliability Deviation

LTFS Low-Temperature Fluorescence Spectroscopy

LTFT Low-Temperature Flow Test

LTFV Less than Fair Value

LTG Lieutenant General [also Lt Gen]

LTG Linear Targent Guidance

LTG Low-Temperature Growing; Low-Temperature-Grown [Semiconductor]

LTG Lufttechnische GmbH [German Pneumatic Engineering Company located in Stuttgart]

Ltg Lighting [also ltg]

Ltg Lightning [also ltg]

LT GaAs Low-Temperature Gallium Arsenide [also LT-GaAs]

LTGCC Lightning, Cloud to Cloud

LTGCCCG Lightning, Cloud to Cloud and Cloud to Ground

LTGCG Lightning, Cloud to Ground

LTGCW Lightning, Cloud to Water

Lt Gen Lieutenant General [also LTG]

LTG GaAs Low-Temperature-Grown Gallium Arsenide [also LTG-GaAs]

LTGH Lightening Hole [Civil Engineering]

LTGIC Lightning in Clouds

LTG InGaAs Low-Temperature-Grown Indium Gallium Arsenide [also LTG-InGaAs]

LTG InP Low-Temperature-Grown Indium Phosphide [also LTG-InP]

Lt Gov Lieutenant Governor

LTH Lactogenic Hormone [Biochemistry]

LTH Logical Track Header

LTH Low-Temperature Herschel [Physics]

LTH Lund Tekniska Högskola [Lund Institute of Technology, Sweden]

LTH Luteotropic Hormone [Biochemistry]

LTHA Low-Temperature High-Activity

Lthr Leather [also lthr]

LTHS Long-Term Hydrostatic Strength [Mechanics]

Lt Ho Lighthouse

LTI Laboratory Testing Inc. [Dublin, Pennsylvania, US]

LTI Light Transmission Index

LTI Lowell Technological Institute [Massachusetts, US]

LTI Low-Technology Intensive

LTI Low-Temperature Isotropic (Carbon)

LTIB Lead Technical Information Bureau [UK]

LTIC Laminated Timber Institute of Canada

Lt (jg) Lieutenant, junior grade [also LTJG]

LTk Light Tank

LTL Less than Truckload [also ltl]

LTL Lot Truck Load

LTL Lower Threshold Limit

LTL Lower Tolerance Limit

ltl little

Ltl Cg Little Change [also LTLCG, or LTL CG]

LTM Late Transition Metal

LTM Lead Time Matrix

LTM Live Traffic Model

LTM Long-Term Memory

LTM Long-Term Monitoring

LTM Low-Temperature Mixing

LTM Low-Temperature Molding

L2M Laboratoire de Microstructures et de Microélectronique [Laboratory for Microstructures and Microelectronics, of Conseil National de la Recherche Scientifique at Bagneux, France]

LTMA London Terminal Control Area [UK]

LTMBE Low-Temperature Molecular Beam Epitaxy [also LT-MBE]

LTMOCVD Low-Temperature Metal-Organic Chemical Vapor Deposition [also LT-MOCVD]

LTMP Low-Temperature Microprobe

LTMR Laser Target Marker Ranger

LTN London Telecom Network [Canada]

Lt(N) Lieutenant (Navy)

l tn long ton [Unit]

LTNO Lanthanum Thorium Nickel Oxide

LTNO Low-Temperature Nuclear Orientation [Physics]

l tn wt long ton-weight [Unit]

LTO Land Titles Office

LTO Lanthanum-Titanium Oxide (or Lanthanum Titanate)

LTO Lithium Tellurium Oxide (or Lithium Tellurate)

LTO Lithium-Tantalum Oxide (or Lithium Tantalate)

LTO Low-Temperature Orthorhombic (Phase) [Crystallography]

LTO Low-Temperature Oxidation

LTOC Laser-Trimmed On-Chip [Electronics]

LTOH Less Than One Hour [also ltoh]

LTO-HTT Low-Temperature Orthorhombic/High-Temperature Tetragonal [also LTO/HTT] [Crystallography]

LTOM Less Than One Minute [also ltom]

LTOM London Traded Options Market [UK]

LTOS Less Than One Second [also ltos]

LTP Laboratoire de Technologie des Poudres [Powder Technology Laboratory; of Ecole Polytechnique Fédérale de Lausanne, Switzerland]

LTP Laboratory for Theoretical Physics [of University of Puerto Rico, San Juan]

LTP Landline Telephone

LTP Language Translator Program

LTP Library Technology Project [of American Library Association, US]

LTP Liquid Transition Point

LTP Line and Trunk Peripheral (Cabinet)

LTP Lithium Titanium Orthophosphate

LTP Local Tracking Problem

LTP Long-Tail(ed) Pair [Electronics]

LTP Lower Threshold Point

LTP Low-Temperature Physics

LTP Low-Temperature Plastic(s)

LTP Low-Temperature Processing

$L_{e,th}(\theta,\phi)$ Thermal Radiance (of a Radiator in a Specified Direction) [Symbol]

LTPCM Laboratoire de Thermodynamique et de Physico-Chimie Métallurgiques [Laboratory for Metallurgical Thermodynamics and Physical Chemistry, of Ecole Nationale Supérieure d'Electrochimie et d'Electrométallurgie de Grenoble/Institut National Polytechnique de Grenoble, France]

LTPCM/ENSEEG Laboratoire de Thermodynamique et de Physico-Chimie Métallurgiques/Ecole Nationale Supérieure d'Electrochimie et d'Electrométallurgie de Grenoble [Laboratory for Metallurgical Thermodynamics and Physical Chemistry/National Higher School of Electrochemistry and Electrometallurgy of Grenoble, France]

LTPD Lot Tolerance Percent Defective

LTPD Low-Temperature Particle Detector

LTPEN 2,6-Lutidinyl-tris(2-Picolinyl)ethylenediamine [also ltpen]

LTPL Long-Term Procedural Language

LTPP Low-Temperature Physics Program [of National Science Foundation, US]

LTPWHT Low-Temperature Postweld Heat Treatment [Metallurgy]

LTQ Light-Triggered and (Light-)Quenched; Light-Triggered and (Light-)Quenching [Metallurgy]

LTR Lattice Test Reactor [US]

LTR Left-To-Right

LTR Letter [Computers/Communications]

LTR Load Task Register

LTR Lockheed Training Reactor [US]

LTR Long Terminal Repeat

LTR Long-Term Revitalization

LTR Loop Transfer Recovery

LTR Low-Temperature Recovery

L-TR Licensing Technical Review

Ltr Letter [also ltr]

ltr lighter

ltr liter [also L, l, or ℓ]

ltr/hr liter(s) per hour [also L/h, L/hr, L h^{-1}, l/h, l/hr, l h^{-1}, ltr/h, or ltr h^{-1}]

ltr/ltr liter(s) per liter [also L/L, L L^{-1}, l/l, l l–1 or ltr ltr^{-1}]

ltr/min liter(s) per minute [also L/min, L min^{-1} or l/min]

ltr/mol liter(s) per mole [also L/mol, l/mol, L mol^{-1}, or ltr mol^{-1}]

LTRMP Long-Term Resource Monitoring Program [US]

LTRS Letter Shift

LTRS Low-Temperature Research Station [UK]

Ltrs Letters

ltrs liters

ltr/s liter(s) per second [also ltr/sec, ltr s^{-1}, L/s, L s^{-1}, L/sec, l/s, l/sec, or l s^{-1}]

LTS Laboratoire de Technologie des Surfaces [Surface Technology Laboratory; of Ecole Central de Lyon, Ecully, France]

LTS Land Tenure System

LTS Launch Telemetry Station

LTS Launch Tracking System,

LTS Line Transient Suppression

LTS London Topographical Society [UK]

LTS Long-Term Stability

LTS Low-Temperature Science

LTS Low-Temperature Superconductivity; Low-Temperature Superconductor

LTS Low-Temperature Sensitization

LTS Low-Temperature Shift

LTS Low-Temperature State

Lts Lights [also lts]

LTSC Low-Temperature Superconductivity; Low-Temperature Superconductor

LTSC Light Thermal Structures Center [of University of Virginia, Charlottesville, US]

LTSP Long-Term Space Plan [Canada]

LTSP Low-Temperature Spin Polarization [Solid-State Physics]

LTSP Low-Temperature Superplastic (Material)

LT-STM Low-Temperature Scanning Tunneling Microscope; Low-Temperature Scanning Tunneling Microscopy

LTT Landline Teletypewriter

LTT Life-Time Test

LTT Low-Temperature Technology

LTT Low-Temperature-Temper(ed); Low-Temperature Tempering [Metallurgy]

LTT Low-Temperature Tetragonal (Phase) [Crystallography]

LTT Low-Temperature Thermometry

LTTAT Long-Tank Thrust-Augmented Thor [Aerospace]

LTTL Low-Power Transistor-Transistor Logic [also LT^{2}L]

LTTMT Low-Temperature Thermomechanical Treatment [Metallurgy]

L2TP Layer Two Tunneling Protocol

LTU LaTrobe University [Melbourne, Victoria, Australia]

LTU Line Termination Unit

LTU Louisiana Technical University [Ruston, US]

LTWA Log Tape Write Ahead

LTV Large Test Vessel

LTV Light Vessel

LTV Ling-Temco-Vought (Aircraft)

LTV Load Threshold Value

LTV Long Tube Vertical (Evaporator)

λ(T)/ξ(T) temperature-dependent penetration depth to temperature-dependent coherence length (ratio) [Solid-State Physics]

LU Lakehead University [Thunder Bay, Ontario, Canada]

LU Lancaster University [UK]

LU Land Use

LU Laurentian University [Sudbury, Ontario, Canada]

LU Laval University [Sainte Foy, Quebec, Canada]

LU Lehigh University [Bethlehem, Pennsylvania, US]

LU Leningradskogo Universiteta [Leningrad University, USSR] [now St. Petersburg University, Russian Federation]

LU Lincoln University [Jefferson City, Missouri, US]

LU Line Unit [Electronics]

LU Liverpool University [UK]

LU Logic(al) Unit

LU Look-Up [Computers]

LU Loughborough University [UK]

LU Lowell University [Massachusetts, US]

LU Lower Diagonal/Upper Diagonal (Decomposition)

LU Loyola University [Chicago, New Orleans, and Los Angeles, US]

LU Luleå University [Sweden]

LU Luxembourg [ISO Code]

Lu Lutetium [Symbol]

Lu^{3+} Lutetium Ion [also Lu^{+++}] [Symbol]

Lu-170 Lutetium-170 [also ^{170}Lu, or Lu^{170}]

Lu-171 Lutetium-171 [also ^{171}Lu, or Lu^{171}]

Lu-172 Lutetium-172 [also ^{172}Lu, or Lu^{172}]

Lu-173 Lutetium-173 [also ^{173}Lu, or Lu^{173}]

Lu-174 Lutetium-174 [also ^{174}Lu, or Lu^{174}]

Lu-175 Lutetium-175 [also ^{175}Lu, Lu^{175}, or Lu]

Lu-176 Lutetium-176 [also ^{176}Lu, or Lu^{176}]

Lu-177 Lutetium-177 [also ^{177}Lu, or Lu^{177}]

lu lumen [Unit]

.lu Luxembourg [Country Code/Domain Name]

LUA Lloyd's Underwriters Association [UK]

LUA Logical Unit Application (Interface)

Lu(Ac)$_3$ Lutetium Acetate

LUB Least Upper Bound [also lub] [Mathematics]

LUB Length of Unused Bed

LUB Logical Unit Block

LUB Lowest Upper Bound [Mathematics]

Lubr Lubricant; Lubricate; Lubrication; Lubricator [also Lub, lub, or lubr]

LubrE Lubrication Engineer [also Lubr Eng]

Lubr Eng Lubrication Engineer(ing)

Lubr Eng Lubrication Engineering [Journal of the Society of Tribologists and Lubrication Engineers, US]

LUC Land-Use Commission

LUC London University Center [UK]

LUCAS LUrgi-Claus-Abgas-Schwefelgewinnung (Process) [Lurgi-Claus Waste Gas Sulfur Extraction Process]

LUCCO Land Utilization Coordination Committee [Cyprus]

LUCF Loading–Unloading–Cooling–Fracture at Lower Temperature (Process) [Mechanics]

LUCID Language for Utility Checkout and Instrumentation Development

LUCK Logical Unit and Checker

LUCO Lowest Unoccupied Crystal Orbit

LUCOM Lunar Communication System [of NASA Apollo Program]

LUDA Land Use Data Analysis (Program) [of US Geological Survey]

LUE Linear Unbiased Estimator

LUE Link Utilization Efficiency

LUEMA Land Use and Environmental Management Authority

LUF Lowest Usable Frequency; Lowest Useful Frequency

LUF Luxembourg Franc [Currency]

LUFA Lipotropes-Unsaturated Fatty Acid [Biochemistry]

LUFORO London Unidentified Flying Objects Research Organization [UK]

LUFS Land Use and Forest Resource Survey [Taiwan]

Luft Kältetech Luft- und Kältetechnik [German Journal on Ventilation and Refrigeration]

LUG Light Utility Glider

LUG Local User Group [of Digital Equipment of Canada]

LUG LOCAS (Local Cataloguing Service) Users Group [UK]

Lug Luggage [also lug]

LUHF Lowest Usable High Frequency; Lowest Useful High Frequency

LUI Local User Input

LUIS Library User Information Service

LUIS Library User Information System [of Northwestern University, Evanston, Illinois, US]

LULA Loyola University of Los Angles [California, US]

LULS Lunar Logistics System [of NASA Apollo Program]

LULU Locally Unacceptable Land Use

LU-LU Logical Unit to Logical Unit (Session) [also LULU]

Lum Lumber [also lum]

lum luminous

LUMAS Lunar Mapping System

LUME Light Utilization More Efficient

LUMO Lowest Unoccupied Molecular Orbital [Physical Chemistry]

LUN Logical Unit Number

LUNA Laboratory for Underground Nuclear Astrophysics [of Laboratori Nazionali del Gran Sasso, Italy]

LUNAC Luminar Navigation and Anticollision (System) [also Lunac, or lunac]

Lunar Planet Inf Bull Lunar and Planetary Information [Publication of the Universities Space Research Association, US]

LUNR Land Use and Natural Resources

Lu$_2$O$_3$ Lutetium Oxide

LUP Liverpool University Press [UK]

LUPIS Land Use Planning Information System [of Commonwealth Scientific and Industrial Research Organization, Australia]

LuPO$_4$ Lutetium Phosphate (or Lutetium Orthophosphate)

LU Press Lehigh University Press [Bethlehem, Pennsylvania, US]

LUQ Left Upper Quadrant (of Abdomen) [Medicine]

LURE Laboratoire pour l'Utilisation du Rayonnement Electromagnetique [Electromagnetic Radiation Utilization Laboratory, Université de Paris-Sud, Orsay, France]

LURS London Underground Railway Society [UK]

LURSOT Laser Ultrasonic Remote Sensing of Oil Thickness

LUS Large Ultimate Size

LUS Liquid Upper Stage

LUSI Local User Interface

LUSI Lunar Surface Inspection

LUSS Linear Ultrasonic Scanning System

LUST Leaking Underground Storage Tank

lust lustrous

LUT Launcher Umbilical Tower [Aerospace]

LUT Local User Terminal

LUT Look-Up Table

LUT Loughborough University of Technology [Leicestershire, UK]

LUVO Lunar Ultraviolet Observatory [of NASA]

LUW Landbouwuniversiteit Wageningen [Wageningen Agricultural University, Netherlands]

LUW Logical-Unit-of-Work

Lux Luxembourg [also Luxem]

LUXIBOR Luxembourg Interbank Offered Rate [also Luxibor]

lu/W lumens per watt [also lu W^{-1}]

LUZ Land-Use Zone

LV Large-Volume

LV Laser Velocimeter

LV Laser Vision

LV Las Vegas [Nevada, US]

LV Latvia [ISO Code]

LV Launch Vehicle

LV Leak(age) Valve

LV Left Ventricle; Left Ventricular [Anatomy]

LV Lift Vector [Aerospace]

LV Light and Visible

LV Light Vehicle

LV Limit Value [Mathematics]

LV Linear Velocity

LV Line of Vision

LV Line Voltage

LV Liquid Volume

LV Lithium Vanadate

LV Loading Valve

LV Load Vertical

LV Local Ventilation

LV Local Vertical

LV Logical Volume

LV Longitudinal Vibration

LV Lotka-Volterra (Equation) [Mathematics]

LV Low Vacuum

LV Low Velocity

LV Low Viscosity

LV Low Voltage [also lv]

LV Low Volume

LV Lumbar Vertebrae [Anatomy]

LV Lymph Vessel [Anatomy]

LV- Argentina [Civil Aircraft Marking]

L/V Dislocation Line Length per Unit Volume (in Materials Science) [Symbol]

L-V Liquid-Vapor (Interface)

Lv Leave [also lv]

Lv Lev [Currency of Bulgaria]

.lv Latvia [Country Code/Domain Name]

LVA Large Vertical Aperture

LVA Launch Vehicle Availability

LVA Local Virtual Address

LVA Logarithmic Video Amplifier

LVA Low-Voltage Accelerator

LVAD Left-Ventricle Assist Device

LVAD Low-Velocity Air Delivery

LVB Low-Volatile Bituminous

LVC Laser Vaporization-Condensation (Technique)

LVCD Least Voltage Coincidence Detector

LVCT Linear Variable Capacitance Transformer

LVD Large Volume Detector (Experiment) [of Laboratori Nazionali del Gran Sasso, Italy]

LVD Liquid-Crystal Visual Display

LVD Low-Velocity Detonation

LVD Low-Velocity Dynamite

LVDA Launch Vehicle Data Adapter

LVDC Launch Vehicle Digital Computer

LVDT Linear Variable Differential Transformer
LVDT Linear Velocity Displacement Transformer
LVDT Linear Voltage Differential Transformer
LVE Linear Viscoelasticity
LVE Low-Voltage Engineering
LVECC Light Vehicles Energy Consumption Committee
LVF Left Ventricular Failure [Medicine]
LVFS Lake Victoria Fisheries Service [Kenya]
LVH Vickers Hardness under Load [Mechanical Testing]
LVHRSEM Low-Voltage High-Resolution Scanning Electron Microscopy
LVHV Low-Volume High-Velocity
LVL Low-Velocity Layer [Geophysics]
Lvl Level [also lvl]
LV/LH Local Vertical/Local Horizontal [also LVLH]
LVLSH Level Shifter
LVM Large Volume Mailer
LVM Local(ized) Vibrational Mode(s) [Physics]
LVM Logical Volume Management
LVM Long Vertical Mark
LVMH Louis Vuitton, Moet Hennessy
LVN Licensed Vocational Nurse
LVO Lithium-Vanadium Oxide
LVOD Launch Vehicles Operations Division [of NASA]
LVOF Low-Velocity Oxy-Fuel (Process)
LVP Launch Vehicle Programs
LVP Low-Voltage Protection
LVPS Low-Voltage Power Supply
LVR Ladle Vacuum Refinement [Metallurgy]
LVR Large Volume Receiver
LVR Linear Video Recorder
LVR Longitudinal Video Recorder; Longitudinal Video Recording
LVR Low-Voltage Relay
LVR Low-Voltage Release
LVR Low Volume Receiver
LVRE Low-Voltage Release Effect
LVRJ Low-Volume Ramjet
LVRT Linear Variable Reluctance Transducer
LVS Low-Accelerating Voltage Control System
LVS Low-Velocity Scanning
LVS Low-Vision System
Lvs Leaves [also lvs]
LVSEM Low-Voltage Scanning Electron Microscopy
LVSM Laboratorium voor Vaste Stoffysika en Magnetisme [Solid-State Physics and Magnetism Laboratory, of KU Leuven, Belgium]
Lvstk Livestock [also lvstk]
LVT Landing Vehicle, Tracked [US Navy Symbol]
LVTR Low VHF (Very-High Frequency) Transmitter-Receiver
LVZ Low-Velocity Zone
LVZ Low-Viscosity Zone
LW Lamb Wave [Physics]
LW Laser Welding

LW Launch Window
LW Least Work (Theory) [Metallurgy]
LW Leave Word
LW Lecher Wire [Electrical Engineering]
LW Lifshitz-Wagner (Equation) [Metallurgy]
LW Light Water
LW Light Wave
LW Lightweight
LW Linewidth
LW Liquid Waste
LW Liquid Water
LW Longwall [Mining]
LW Long Wave
LW Longitudinal Wave
LW Low Water
LW Low-Wing (Monoplane)
L/W Lumens per Watt [also l/W, LPW, or lpw]
Lw Lawrencium [Symbol]
l/w length to width (ratio) [Symbol]
LWA Last Word Address
LWA Light Wire Armored
LWA Limited Work Authorization
LWB Long Wheelbase
LWB Lower Bound [Mathematics]
LWBR Light-Water Breeder Reactor
LWC Lightwave Communications
LWC Lightweight-Coated (Paper)
LWC Lightweight Concrete
LWC Liquid Waste Containment
LWC Liquid Water Concentration [Hydrology]
LWC Liquid Water Content [Hydrology]
LWC Loop Wiring Concentrator
L&WCF Land and Water Conservation Fund [US]
LWCS Lightwave Communication System
LWD Larger Word
LWD Launch Window Display
LWD Left Wing Down
LWD Lightwave Device
LWD Long Working Distance [Microscopy]
LWDF Liquid Waste Disposal Facility
LWF Lightweight Fighterplane
LWF Luminous Wall Firing
LWG Logistics Working Group
LWHCR Light Water High Converter Reactor
LWHS Light-Weight Headset
LWIC Lightweight Insulating Concrete
LWIR Long-Wave Infrared
LWL Limited War Laboratory [of US Army]
LWL Load Water Line
LWL Low Water Level
LWM Liquid Waste Management
LWM Liquid Waste Monitor
LWM Low Water Mark

LWOP Leave Without Pay
LWOST Low Water of Spring Tide
LWP Liquid Water Path
LWP Long-Wave Phase [Earthquakes]
LWR Laser Warning Receiver
LWR Laser Write Read
LWR Light-Water (Nuclear) Reactor
LWR Liquid Waste Release
lwr lower
LWRHU Lightweight Radioisotope Heater Unit
LWRRDC Land and Water Resources Research and Development Corporation [Australia]
LWS Lightning Warning System
LWS Light Warning Set
LWS Liquid Wave Spectra
LWS Loire-Wendel-Sprunck (Process) [Metallurgy]
LWS Long-Wave Spectrometry
LWST Liquid Waste Storage Tank
LWT Lightwave Technology
LWT Light-Weight Tank
LWTC London World Trade Center [UK]
LWTS Low-Level (Nuclear) Waste Treatment System
LWTT Liquid Waste Test Tank
LWV Lightweight Vehicle
LX- Luxembiurg [Civil Aircraft Marking]
L(x) Langevin function (in solid-state physics) [Symbol]
L$_i$(x) Lagrangian Polynomial [Symbol]
lx lux [Unit]
λ/ξ (superconducting) penetration depth to coherence length (ratio) [Symbol]
LXe Lithium Xenonite
LXMAR Load External Memory Address Register
LY Leap Year
LY Lethal Yellowing [Biology]
LY Libya [ISO Code]
LY Light Year
Ly Lyophile; Lyophilize(d)

.ly Libya [Country Code/Domain Name]
LYbF Lithium Ytterbium Fluoride
LYCMO Lanthanum Yttrium Calcium Manganese Oxide (Superconductor)
LYD Libyan Dinar [Currency]
LYF Lithium Yttrium Fluoride
LYNX RP LYNX Replacement Project [Canada]
LYON Liquid Yield Option Note
LYP Lower Yield Point [Mechanics]
.LYR Lyrics (Music) [File Name Extension] [also .lyr]
Lyr Layer [also lyr]
LYRIC Language for Your Remote Instruction by Computer
LYS Lower Yield Stress [Mechanics]
Lys (+)-Lysine; Lysyl [Biochemistry]
Lys(OH) δ-Hydroxy-L-Lysine [Biochemistry]
LZ Landau-Zener (Problem) [Physics]
LZ Landing Zone
LZ Leading Zero(s)
LZ Levitation Zone [Metallurgy]
LZ Lithium Zirconate
LZ- Bulgaria [Civil Aircraft Marking]
LZCN Lithium Zinc Calcium Nitride
LZF Lithium Zinc Ferrite
LZFO Lithium Zinc Iron Oxide
LZG Lanthanum Zinc Germanate
LZH Laser-Zentrum Hannover [Hannover Laser Center, Germany]
LZO Lanthanium Zirconium Oxide
LZR Levitation Zone Refining [Metallurgy]
LZS Landau-Zener-Stückel-Berg (Method) [Physics]
LZT Laser-Zone-Textured; Laser-Zone Texturing [Computer Disks]
LZT Lead-Zinc-Tin
LZT Lead Zirconate Titanate
LZT Local Zone Time
LZTC Lead-Zinc-Tin Committee
LZW Lempel-Ziv-Welch (Algorithm) [Mathematics]

M

M Arithmetic Mean [Symbol]

M Atomic Weight [Symbol]

M Bending Moment [Symbol]

M Bierbaum Microcharacter Hardness Number [Symbol]

M Central Mass [also *M*] [Astronomy]

M Centroid (of a Triangle) [Symbol]

M Combined Machine-Specimen Compliance [Mechanical Testing]

M Degree of Enlargement (in Radiography) [Symbol]

M Figure of Merit [Semiconductor Symbol]

M Guided Missile [USDOD Symbol]

M Horsepower per Square Inch (for Thrust Bearings) [Symbol]

M In-Cloud Parcel [Symbol]

M Macao

M Mach [Symbol]

M Machinability; Machine(ry); Machining; Machinist

M Machine Gun [Followed by a Number, e.g., M60, etc.]

M Mach Number [Symbol]

M Mackarel

M Mackenzie (River) [North Western Canada]

M Macrophage [also m] [Biology]

M Madagascar

M Madelung (Energy) (in Solid-State Physics) [Symbol]

M Madison [Wisconsin, US]

M Madras [India]

M Madrid [Spain]

M Magenta

M Magma(tic)

M Magnaflux [Nondestructive Testing]

M Magnesia (or Magnesium Oxide) [Ceramics]

M Magnet(ic); Magnetics; Magnetism

M Magnetic Moment [Symbol]

M Magnetic Quantum Number [Symbol]

M Magnetic Vector [Symbol]

M Magnetite [Symbol]

M (Mass) Magnetization [SI Symbol]

M Magneto

M Magneton [Physics]

M Magnetron [Electronics]

M Magnification [Symbol]

M Magnification Factor [Symbol]

M Magnolia

M Mail

M Main

M Main Channel [Stereo Sound System]

M Maine [US]

M Maintainability; Maintainable; Maintenance

M Majesty

M Major(ity)

M Majorca

M Major Cycle [Computers]

M Make

M Malachi

M Málaga [Spain]

M Malaysia(n)

M Maldives

M Male [also m]

M Malfunction(ing)

M Mali

M Malleability; Malleable

M Malleus (of Middle Ear) [Anatomy]

M Malta; Maltese

M Maltase [Biochemistry]

M (+)-Maltose [Biochemistry]

M Mammal; Mammary

M Mammililae (of Avian Eggs) [Zoology]

M Man; Men

M Management; Manager

M Managua [Nicaragua]

M Manchester [UK]

M Mandrel

M Mandible; Mandibular [Anatomy]

M Manhattan [New York City, US]

M Manifold

M Manila [Philippines]

M Manipulation; Manipulator

M Manitoba [Canada]

M Mansonia [Genus of Mosquitoes]

M Mantissa

M Mantle

M Manual

M Manufacture(r); Manufacturing

M Manuscript

M Maputo [Mozambique]

M Maracaibo [Venezuela]

M Marble

M March

M Margarine
M Margin(al)
M Marijuana [Botany]
M Marine
M Maritime [also m]
M Mark; Marker; Marking [also m]
M Marrakesh [Morocco]
M Marrow
M Marseille(s) [France]
M Marsupial(ia) [Zoology]
M Martensite; Martensitic [Metallurgy]
M Maryland [US]
M Maser
M Maseru [Lesotho]
M Mask(ing)
M (Total) Mass [Symbol]
M Massachusetts [US]
M Massage; Masseur; Masseuse
M Master
M Mastic
M Mastigophora [Zoology]
M Material
M Matrix
M Matter
M Maui [Hawaii]
M Mauretania(n)
M Mauritius
M May
M Maximum [also m]
M Maxwell [Unit]
M McCartney (Model of Crack Bridging) [Metallurgy]
M Mean
M Measure(ment); Measuring
M Mecca [Saudi Arabia]
M Mechanic(s); Mechanical
M Mechanical Engineering (Technology) [Discipline Category Abbreviation]
M Mechanism
M Mecistocirrus [Genus of Nematodes]
M Median [Symbol]
M Medical; Medicine
M Medieval
M Mediterranean
M Medium [also M]
M Medulla [Anatomy]
M Mega- [SI Prefix]
M Megabyte
M Mekong (River) [Southeast Asia]
M Melanesia(n)
M Melanocyte
M Melanoma [Medicine]
M Melbourne [Australia]
M Meloidogyne [Genus of Nematodes]
M Melt(er); Melting

M Member
M Membrane
M Memory
M Memphis [Tennessee, US]
M Meninges (of Brain) [Anatomy]
M Meningitis [Medicine]
M Meniscus [Anatomy]
M Menopon [Genus of Lice]
M Menstruate; Menstruation
M Menu
M Mercury [Abbreviation; Symbol: Hg]
M Merge
M Meridian,
M (Meridies) – Noon
M Merismopedia [Genus of Algae]
M Merit
M Mesh
M Mesophyll [Botany]
M Mesozoic [Geological Era]
M Message; Messaging
M (Charles) Messier (Number) [Followed by a Numeral to Identify Star Clusters and Nebulae, e.g., M31 Denotes the Andromeda Galaxy] [Astronomy]
M Metabolic; Metabolism
M Metacarpal; Metacarpus [Anatomy]
M Metafile [Computers]
M Metal(lic)
M Metallization; Metallize(d)
M Metallograph(er); Metallographic; Metallography
M Metallurgical; Metallurgist; Metallurgy
M Metastasis [Physics]
M Metastasis; Metastatize [Medicine]
M Metazoa [Zoology]
M Metatarsal; Metatarsus [Anatomy]
M Meteorological; Meteorologist; Meteorology
M Meter [Instrument]
M Methanol
M (–)-Methionine; Methionyl [Biochemistry]
M Method(ology)
M Metical [Currency of Mozambique]
M Metopium [Genus of Trees Including the Poisonwood]
M Metric (System)
M Metric Screw Thread [Symbol]
M Meuse (River) [Western Europe]
M Mexican; Mexico
M Miami [Florida, US]
M Mica
M Micelle
M Michigan [US]
M Micrasterias [Genus of Algae]
M Microbe; Microbial
M Microcystis [Genus of Algae]
M Microemission
M Microfiche

M Microfilm(er)

M Micronesia(n)

M Microphone

M Microprobe

M Microscope; Microscopist; Microscopy

M Microspectral; Microspectrum

M Microsphere [Symbol]

M Microspore; Microsporum [Botany]

M Microstructural; Microstructure

M Microtome; Microtomy [Biology]

M Microtubule [Cytology]

M Middle [also m]

M (Stereo) Midplane Signal [Symbol]

M Migration [Symbol]

M Mike [Phonetic Alphabet]

M Milax [Genus of Slugs]

M Mildew

M Milan [Italy]

M Mile [Unit]

M Milk(er); Milking

M Mill [also m]

M Milling

M *(Mille)* – Thousand

M Million

M Milwaukee [Wisconsin, US]

M Mine(r); Mining

M Mineral(ogical); Mineralogist; Mineralogy

M Minimum

M Minneapolis [Minnesota, US]

M Minnesota [US]

M Minsk [Byelorussia]

M M-Integral

M Mirror(ing)

M *(Miscere)* – Mix; Let a Mixture be Made [Medical Prescriptions]

M Mischmetal

M Misorientation; Misorient(ed)

M Miss(ing) [also m]

M Missile

M Missile Carrier (Aircraft) [USDOD Symbol]

M Mississippi [US River and State]

M Mississippian [Geological Period]

M Missouri [US River and State]

M Mist [also m]

M Mitochondria [Cytology]

M Mix(er); Mixing; Mixture

M Mobile

M Mobile [Alabama, US]

M Mobile-Launched (Vehicle) [USDOD Symbol]

M Mobility

M Mockup

M Mode

M Mode Dependency [Digital Logic]

M Model

M Modem

M Moderation; Moderator

M Modifier; Modification

M Modulation; Modulator

M Module

M Modulo [Mathematics]

M Modulus [also m]

M Modulus (of a System of Logarithms) [Symbol]

M Modulus of Elasticity (of Soil, or Rock Masses) [Symbol]

M Mogadishu [Somalia]

M Moisture

M Molal; Molality [Symbol]

M Molar [Dentistry]

M Molarity, (of Solutions) [Symbol]

M Molar Mass [Symbol]

M Molar Solution [Symbol]

M Molar Volume [Symbol]

M Molar Weight [Symbol]

M Mold(ing)

M Mold [Biology]

M Mold Constant (in Metallurgy) [Symbol]

M Mole [Symbol]

M Mole [Zoology]

M Molecule; Molecular

M Molecular Weight [Symbol]

M Mollusk [Zoology]

M Molybdenum-Type High-Speed Tool Steels [AISI-SAE Symbol]

M Moment [Symbol]

M Moment of Force (or Torque) [Symbol]

M Momentum

M Monaco

M Monday

M Monel

M Mongolia(n)

M Mongolian; Mongoloid [Anthropology]

M Monitor(ing)

M Monday

M Monkey

M Mono-

M Monochromatic; Monochromator

M Monochrome

M Monoclinic (Crystal System) [Symbol]

M Monolith(ic)

M Monoplane

M Monrovia [Liberia]

M *(Monsieur)* – French for "Mister", "Sir", or "Gentleman"

M Monsoon

M Montana [US]

M Monterrey [Mexico]

M Montevideo [Uruguay]

M Montgomery [Alabama, US]

M Montpelier [Vermont, US]

M Montreal [Canada]

M   Montserrat [West Indies]

M   Month [also m]

M   Moon [also m]

M   Moraxella [Genus of Bacteria]

M   Moroccan; Morocco

M   Mortar

M   Moscow [Russia]

M   Mother

M   Motile; Motility [Biology]

M   Motion(al)

M   Motor

M   Mount(ing)

M   Mountain

M   Mountain [US Geographic Region]

M   Mouse

M   Mouth

M   Move(r); Movement; Moving

M   Mozambique

M   M-Shell (of Electron) [Symbol]

M   M-Type Structural Steel Shape [AISI/AISC Designation]

M   (Upper-case) Mu [Greek Alphabet]

M   Mud

M   Mullite [Mineral]

M   Multi-

M   Multiceps [Genus of Taeniid Tapeworms]

M   Multilayer

M   Multiplet [Physics]

M   Muliplex(er); Multiplexing

M   Multiplier; Multiplication; Multiplicity; Multiply

M   Multipole

M   Multitubular Propellant [Symbol]

M   Mumps [Medicine]

M   Munich [Germany]

M   Mus [Genus of Mammals Comprising Mice]

M   Muscle; Muscular

M   Mushroom

M   Music(ality)

M   Mutant; Mutation

M   Mutual Inductance [Symbol]

M   Mutualism [Ecology]

M   Mycoplasma [Microbiology]

M   Mycosis [Medicine]

M   Myelin [Biochemistry]

M   Myosin [Biochemistry]

M   Myriophyllum [Biology]

M   One Thousand [Roman Numeral]

M   Quantity of Adsorbent [Symbol]

M   Radiant Exitance [Symbol]

M   Star Group of Spectral Type M [Surface Temperature 1,760°C, or 3,200°F] [Letter Designation]

M   Time Constant [Symbol]

M   Torque (or Moment of Force) [Symbol]

M   Total Number of Pixels along X-Axis (in Image Analysis) [Symbol]

M   Turning Moment (in Mechanics) [Symbol]

/M   Make Option [Turbo Pascal]

-M   Missile Launcher (Aircraft) [US Navy Suffix]

M-   Magnetic Sextet Spectra [Letter Symbol Followed by a Numeral, e.g., M-1, M-2, etc.]

[M]   Molar Magnetic Rotation [Symbol]

$\overline{M}$ Average Molecular Weight (or Average Molar Mass) [Symbol]

$\overline{M}$ Average (or Mean) Size [Symbol]

$\overline{M}$ One Million [Roman Numeral]

$\overline{\overline{M}}$ One Billion [Roman Numeral]

$\overline{\overline{\overline{M}}}$ One Trillion [Roman Numeral]

M*   Magnetic Moment at Zero Field [Symbol]

M*   Magnification [Symbol]

M′   Storage Modulus (in Physics) [Symbol]

M″   Loss Modulus (in Physics) [Symbol]

M^{+} Magnetization at Increased Magnetic Field [Symbol]

M^{+} Molecular Ion [Symbol]

M$_{\infty}$ Relaxed Modulus (in Mechanics) [Symbol]

M$_{\infty}$ Saturation Magnetization [Symbol]

$\bar{M}$ Magnetization Vector [Symbol]

M$_{\parallel}$ Magnetization Parallel to Plane [Symbol]

M$_{\perp}$ Magnetization Perpendicular to Plane [Symbol]

M^{0} Neutral Metal (i.e., in Metal in Neutral State) [Symbol] [*Note:* The letter "M" May be Replaced by a Metallic Element, e.g., Fe^{0} is Neutral Iron, Zn^{0} is Neutral Zinc, etc.)

M$_{0}$ Free Stream Mach Number [Symbol]

M$_{0}$ Magnetization at Zero Kelvin [Symbol]

M$_{0}$ Mass of Probe Ion (in Ion Spectroscopy Scattering) [Symbol]

M$_{0}$ Spontaneous Magnetization [Symbol]

M$_{0}$ Unrelaxed Modulus (in Mechanics) [Symbol]

M^{1+} Monovalent Metal [also Me^{+}, or M^{I}] [Symbol]

M$_{1}$ Mass of Target Atom (in Ion Spectroscopy Scattering) [Symbol]

M^{2+} Divalent Metal [also Me^{++}, or M^{II}] [Symbol]

M^{3+} Trivalent Metal [also Me^{+++}, or M^{III}] [Symbol]

M^{4+} Tetravalent Metal [also Me^{IV}] [Symbol]

M^{5+} Pentavalent Metal [also M^{V}] [Symbol]

M^{6+} Hexavalent Metal [also M^{VI}] [Symbol]

M$_{A}$ Molar Mass of Component *A* [Symbol]

M$_{a}$ Molecular Weight of Air [Symbol]

M$_{\alpha}$ M-Alpha (Radiation) [also Mα]

M$_{b}$ Bending Moment [Symbol]

M$_{\beta}$ M-Beta (Radiation) [also Mβ]

M$_{c}$ Critical Entanglement Molecular Weight (of Polymers) [Symbol]

M$_{D}$ Diffusional Mobility [Symbol]

M$_{e}$ Elastic Moment [Symbol]

M$_{e}$ Radiant Exitance (at a Point on a Surface) [Symbol]

M$_{e(\varepsilon=1)}$ Thermal-Radiant Exitance (of a Full Radiator) [Symbol]

M$_{eff}$ Effective Magnetization [Symbol]

$M_{e,s}$ Self-Radiant Exitance [Symbol]

$M_{e,th}$ Thermal-Radiant Exitance [Symbol]

M_F Flow Moment [Symbol]

M_F Free Saturation Magnetization [Symbol]

M_f Martensite Finish Temperature (in Metallurgy) [Symbol]

M_{FC} Field-Cooled Magnetization [Symbol]

M_i Mass of the *i*-th Atom of a System [Symbol]

M_i Mean Molecular Weight within Polymer-Molecule Size Range *i* [Symbol]

M_i Mutual Inductance [Symbol]

M_m Mass-Average Molecular Weight (of Polymers) [also M_M, or $_m$] [Symbol]

M_m Maximum Magnitude Ratio [Symbol]

M^{n+} Positively Charged Metal Ion [Symbol] [*Note:* "n" Indicates the Valence, e.g., M^{2+} is a Divalent Metal, M^{3+} is a Trivalent Metal, etc.; "M" Can be Replaced by a Metallic Element, e.g., Pb^{2+} is Divalent Lead, Au^{3+} is Trivalent Gold, etc.]

M_n Information Bit (n = 1, 2, 3, etc. is the Position Number) [Symbol]

M_n Number-Average Molecular Weight (of Polymers) [also M_N, or $_n$] [Symbol]

M_o Monomer Molecular Weight (of Polymers) [Symbol]

M_p Pinned Saturation Magnetization [Symbol]

M_R Remanence (or Remanent Magnetization) [Symbol]

M_r Relative Molecular Mass [Symbol]

M_r Remanence (or Remanent Magnetization) [Symbol]

M_{RC} Major Loop Magnetization Value at Remanent Coercivity [Symbol]

$M_{R(H)}$ Isothermal Remanent Magnetization [Symbol]

M^s Surface Magnetization [Symbol]

M_s Martensite Start Temperature (in Metallurgy) [Symbol]

M_s Saturation Magnetization [also M_S] [Symbol]

M_s Spontaneous Magnetization [Symbol]

M_{si} Saturation Magnetization of the i-th Layer [Symbol]

M_T Magnetization at Measurement Temperature [Symbol]

M_t Molecular Weight of Tracer Gas [Symbol]

M_t Twisting (or Torsional) Moment [also M_T] [Symbol]

M_θ Magnetization at Anneal(ing) Temperature [Symbol]

M_v Viscosity Average Molecular Weight (of Polymers) [Symbol]

M_{ve} Viscoelastic Modulus [Symbol]

M_w Weight-Average Molecular Weight (of Polymers) [also M_W, or $_w$] [Symbol]

M_x Magnetization Component in Horizontal Direction [Symbol]

M_x Moment about X-Axis due to Force Component F_x [Symbol]

M_y Magnetization Component along Vertical In-Plane Direction [Symbol]

M_y Moment about X-Axis due to Force Component F_y [Symbol]

123-M_y $YBa_2Cu_{3-y}M_yO_{6+2x}$ [A Yttrium Barium Copper Metal Oxide Superconductor] [General Formula; M Represents Metallic Elements, e.g., Aluminum, Iron, Cobalt, etc.]

M_z Magnetization Component in Out-of-Plane Direction (or Magnetic Field Along z Axis) [Symbol]

M_z Moment about X-Axis due to Force Component F_z [Symbol]

M_{ZFC} Zero-Field-Cooled Magnetization [Symbol]

M1,M2,M3,M4 Alternating-Current Secondary [Controllers]

M3 Measurements, Modeling and Monitoring [also MMM]

M3 Third Medium-Sized Mission (of Horizon 2000) [Aerospace]

M50 Mean of 1950 (Coordinate System)

M-1 Mach 1 [Speed of 1,220 km/hr, or 760 mph at Sea Level and Standard Conditions of Humidity and Temperature]

M-2 Mach 2 [Speed of 2,440 km/hr, or 1,520 mph at Sea Level and Standard Conditions of Humidity and Temperature]

M-3 Mach 3 [Speed of 3,660 km/hr, or 2,280 mph at Sea Level and Standard Conditions of Humidity and Temperature]

3M Microsensors, Microactuators and Microsystems [also MMM]

3M Minnesota Mining and Manufacturing Company [St. Paul, US]

m clearance modulus (of journal bearings) [Symbol]

m (sound) dissipation constant [Symbol]

m electrical dipole moment [Symbol]

m equilibrium segregation coefficient [Symbol]

m germ (in metallurgy) [Symbol]

m hydraulic radius [Symbol]

m induced dipole moment [Symbol]

m integer [Symbol]

m macrophage [also M] [Biology]

m magnetic

m magnetic (dipole) moment [Symbol]

m magnetic quantum number (of electron) [Symbol]

m magnetic unit of pole strength [Symbol]

m magnification [Symbol]

m magnified; magnify

m male [also M]

m mandatory

m maritime [also M]

m mark(er) [also M]

m masculine

m mass [Symbol]

m matrix [Symbol]

m maximum [also M]

m mean

m measure(ment); measuring [also M]

m median (of a triangle) [Symbol]

m medium

m menstruate; menstruation [also M]

m meridian [also M]

m *(meridies)* – noon [also M]

m mesh [also M]

m meta- [Chemistry]

m   metastable

m   metal(lic) [also M]

m   meter [Unit]

m   middle

m   mil [Unit]

m   mile [Unit]

m   mill [also M]

m   *(mille)* – thousand

m   milli- [SI Prefix]

m   million

m   mine

m   minim [Unit]

m   minimum

m   minor(ity)

m   minute [Unit]

m   *(misce)* – mix [Medical Prescriptions]

m   miscellaneous

m   miss(ing)

m   mist [also M]

m   mix(ture)

m   mobile [also M]

m   mobility (of dislocations) (in materials science) [Symbol]

m   moderate

m   modular ratio (i.e. ratio of elastic moduli of steel and concrete) [Symbol]

m   modulation [also M]

m   modulation index [Symbol]

m   module (of gears, or splines) [Symbol]

m   modulus [also M]

m   molal

m   molality [Symbol]

m   molal solution [Symbol]

m   molar (tooth)

m   molar(ity)

m   mole

m   molecular mass [Symbol]

m   molecular weight [Symbol]

m   monoclinic (crystal system) [Symbol]

m   monsoon rain; short dry season, but total rainfall sufficient to support rainforest [Subtype of Climate Region, e.g., in Am, etc.]

m   month(ly)

m   moon [also M]

m   mother [also M]

m   move (command) [Edlin MS-DOS Line Editor]

m   mu [English Equilvalent]

m   multiplet [Physics]

m   multiplicity factor [Symbol]

m   number of bonds [Symbol]

m   number of final pulse during period of time over which discharges occur [Symbol]

m   number of revolutions required for rendezvous (in aerospace) [Symbol]

m   order of spectrum [Symbol]

m   pixel index (in image analysis) [Symbol]

m   radio-frequency component of magnetization [Symbol]

m   reduced magnetization [Symbol]

m   Schmid factor [Symbol]

m   second-direction cosine of a line (e.g., $m = \cos \beta$) [Symbol]

m   slope (of a line) [Symbol]

m   strain rate sensitivity factor [Symbol]

m   strain resolution factor [Symbol]

m   stranding factor (of a stranded conductor) [Symbol]

m   Taylor-type orientation factor (in physical metallurgy) [Symbol]

m   vapor flux [Symbol]

m   velocity factor (of worm gears) [Symbol]

m   Weibull modulus (in fracture theory) [Symbol]

/m   map option [MS-DOS Linker]

m-   meta- [Chemistry]

$\overline{m}$ mer molecular weight (of polymers) [Symbol]

$\dot{m}$ mass flux [Symbol]

m^* effective mass (of electron) [Symbol]

m^* Sachs-type orientation factor (in physical metallurgy) [Symbol]

m_+ molality of positive ions (i.e., cations) [Symbol]

m_- molality of negative ions (i.e., anions) [Symbol]

m_0 free electron mass [Symbol]

m_0 mass of electron at low velocity [Symbol]

m_0 rest mass (of a particle) [Symbol]

m^{-1} one per meter [also 1/m]

m_1 active mass (for magnetic core evaluation) [Symbol]

m_1 mass of body 1, particle 1, etc. [Symbol]

m^2 square meter [also sq m]

m^{-2} one per square meter [also $1/m^2$]

m^3 cubic meter [also cu m]

m^{-3} one per cubic meter [also $1/m^3$]

m^4 meters to the fourth power [Unit]

m_A mass of component A [Symbol]

m_a mass of (dry) air [Symbol]

m_a median from vertex A to side a of a triangle [Symbol]

m_α mass of alpha phase (in materials science) [Symbol]

m_β mass of beta phase (in materials science) [Symbol]

m_{bo} mass (of rocket motor) at burnout [Symbol]

m_e electron rest mass (9.10939×10^{-31} kg) [Symbol]

m_e equivalent mass [Symbol]

m_G gain margin (in automatic control) [Symbol]

m_g gravitational mass [Symbol]

m_γ mass of gamma phase (in materials science) [Symbol]

m_H mass of hydrogen atom [Symbol]

m_i inert mass [Symbol]

m_i molality of the i-th component (in a solution) [Symbol]

m_i thermally averaged spin at i-th site (in physics) [Symbol]

m_L mass of liquid phase [Symbol]

m_l magnetic quantum number [Symbol]

m_m (rocket) motor gross mass [Symbol]

m_μ muon rest mass (1.883533×10^{-23} kg) [Symbol]

m_n neutron rest mass (1.67493×10^{-27} kg) [Symbol]

m_p (rocket) propellant mass [Symbol]

m_p proton rest mass (1.67262×10^{-27} kg) [Symbol]

m^S Schmid Tensor (in materials science) [Symbol]

m_s spin quantum number [Symbol]

m_{ss} mass of solid-solution phase [Symbol]

m_t transverse mass (of a particle) [Symbol]

m_u atomic mass (unit) (1.66056×10^{27} kg) [Symbol]

m_v coefficient of (soil) volume compressibility (or modulus of soil volume change) [Symbol]

m_v mass of water vapor [Symbol]

m_x x-component of magnetic moment [Symbol]

m_y y-component of magnetic moment [Symbol]

m_z z-component of magnetic moment [Symbol]

℧ mho [Unit]

μ absolute viscosity [Symbol]

μ absorption coefficient (for radiation) [Symbol]

μ amplification factor (of electron tube) [Symbol]

μ chemical potential [Symbol]

μ coefficient of (sliding) friction [Symbol]

μ coefficient of sound absorption [Symbol]

μ damping ratio [Symbol]

μ dipole moment [Symbol]

μ direct-current permeability [Symbol]

μ Fermi energy (or Fermi level) [Symbol]

μ fugacity [Symbol]

μ Hall mobility (in solid-state physics) [Symbol]

μ ionic strength (of a solution) [Symbol]

μ Joule-Thomson coefficient (in thermodynamics) [Symbol]

μ Lamé's constant (in mechanics) Symbol]

μ linear absorption coefficient [Symbol]

μ linear attenuation coefficient [Symbol]

μ magnetic moment (of an electric coil) [Symbol]

μ magnetic momentum (of an orbital electron) [Symbol]

μ (population) mean (of a distribution) [Symbol]

μ micro- [SI Prefix]

μ micron [also mu]

μ micron phase (in materials science) [Symbol]

μ mobility (of plastic flow) [Symbol]

μ mobility (of charge carriers in semiconductors) [Symbol]

μ molecular (electrical) conductivity [Symbol]

μ molecular weight [Symbol]

μ moment about the mean (in statistics) [Symbol]

μ (lower-case) mu [Greek Alphabet]

μ mu meson (in particle physics) [also muon] [Symbol]

μ mu phase (of a material) [Symbol]

μ normal (direct-current) permeability [Symbol]

μ (magnetic) permeability constant [Symbol]

μ Poisson's ratio [Symbol]

μ population mean (in statistics) [Symbol]

μ refractive index [Symbol]

μ reduced mass (of a particle) [Symbol]

μ reluctance [Symbol]

μ repassivation rate (in metallurgy) [Symbol]

μ sensitivity (of microscopes)

μ shape factor (in shear modulus calculations) [Symbol]

μ shear modulus [Symbol]

μ specific energy [Symbol]

μ Thomson heat (in physics) [Symbol]

μ variance (of a distribution) [Symbol]

μ^* complex magnetic permeability [Symbol]

μ^* Coulomb interaction parameter [Symbol]

μ^* effective repulsive coupling interaction constant (for electrons) [Symbol]

μ^* effective shear modulus [Symbol]

μ' real part of magnetic permeability [Symbol]

μ' relative (magnetic) permeability [CGS-EMU Symbol]

μ'' imaginary part (or loss component) of magnetic permeability [Symbol]

μ^+ mu-plus meson (i.e., positively charged muon) (in particle physics) [Symbol]

μ^- mu-minus meson (i.e., negatively charged muon) (in particle physics) [Symbol]

μ° standard chemical potential [Symbol]

μ_0 coefficient of static friction [Symbol]

μ_0 initial absorption coefficient [Symbol]

μ_0 initial permeability [Symbol]

μ_0 magnetic permeability of vacuum ($12.566370614 \times 10^{-7}$ H/m) [Symbol]

μ_{0d} initial dynamic (alternating-current) permeability [Symbol]

μ_a atomic absorption coefficient [Symbol]

μ_a ideal (magnetic) permeability [Symbol]

μ_{abs} absolute direct-current permeability [Symbol]

μ_B Bohr magneton (9.274015×10^{-24} A·m²) [Symbol]

μ_c copolymer chemical potential [Symbol]

μ_c critical (or peak) mobility (of electrons) [Symbol]

μ_d differential permeability [Symbol]

μ_Δ incremental (direct-current) permeability [Symbol]

$\mu_{\Delta i}$ incremental intrinsic (direct-current) permeability [Symbol]

$\mu_{\Delta L}$ incremental alternating-current inductance permeability [Symbol]

$\mu_{\Delta z}$ incremental alternating-current impedance permeability [Symbol]

μ_e effective (magnetic) permeability [Symbol]

μ_e electron magnetic moment (9.284770×10^{-24} A·m²) [Symbol]

μ_e electron mobility [Symbol]

μ^ε (magnetic) permeability under constant strain conditions [Symbol]

μ_{eff} effective channel mobility (of electrons) [Symbol]

μ_{eff} effective (magnetic) circuit permeability [Symbol]

μ_f shear modulus of (thin) film [Symbol]

μ_{FE} field-effect mobility [Symbol]

μ_H chemical potential of hydrogen [Symbol]

μ_H hypothesized mean (in Student's T-statistic) [Symbol]

μ_h (electron) hole mobility [Symbol]

μ_i chemical potential of component i [Symbol]

μ_i^o standard chemical potential of component i [Symbol]

μ_i initial (relative magnetic) permeability [Symbol]

μ_i intrinsic (magnetic) permeability [Symbol]

μ_k coefficient of kinetic friction [Symbol]

μ_L alternating-current inductance permeability [Symbol]

μ_m maximum permeability [Symbol]

μ_m mean absorption coefficient [Symbol]

μ_μ muon magnetic moment ($4.4904514 \times 10^{-26}$ A·m²) (in particle Physics) [Symbol]

μ_N nuclear magneton (5.050787×10^{-27} A·m²) [Symbol]

μ_n mobility of negative charge carriers (of intrinsic semiconductors) [Symbol]

μ_n nuclear magneton [Symbol]

μ_p alternating-current peak permeability [Symbol]

μ_p mobility of positive charge carriers (of intrinsic semiconductors) [Symbol]

μ_p proton magnetic moment (1.410608×10^{-26} A·m²) [Symbol]

μ_{ps} instrinsic pseudoelasticity [Symbol]

μ_r relative (magnetic) permeability [SI Symbol]

μ_{rev} reversible direct-current permeability [Symbol]

μ_s coefficient of static friction [Symbol]

μ_s order moment (of superconductors) [Symbol]

μ_s saturation magnetization [Symbol]

μ_s shear modulus of substrate [Symbol]

μ^σ (magnetic) permeability at constant stress [Symbol]

μ^T (magnetic) permeability under constant stress conditions [Symbol]

μ_t instantaneous (magnetic) permeability [Symbol]

μ_{th} instrinsic thermoelasticity [Symbol]

μ_v (magnetic) permeability of unoccupied space (vacuum) [Symbol]

μ_z alternating-current root-mean-square impedance permeability [Symbol]

MA Mach Angle

MA Machine Address

MA Machining Allowance

MA Macroanalysis

MA Macro Assembler

MA *(Magister Artium)* – Master of Arts

MA Magnesium Aluminate

MA Magnesium Association [now International Magnesium Association, US]

MA Magnetic Aftereffect

MA Magnetic Alloy

MA Magnetic Amplifier

MA Magnetic Analysis; Magnetic Analyzer

MA Magnetic Anisotropy

MA Magnetic Aperture

MA Magnetic Axis

MA Magnetoabsorption

MA Magnetoacoustic(s)

MA Main Amplifier

MA Maintenance Ability

MA Maleic Acid

MA Maleic Anhydride

MA Manganese Aluminate

MA Manufactured Article

MA Maritime Administration [of US Department of Transportation]

MA Markovnikov Addition [Organic Chemistry]

MA Martensite/Retained Austenite [Metallurgy]

MA Massachusetts [US]

MA Mass Analysis; Mass Analyzer

MA Master

MA Master Alarm

MA Master Alloy [Metallurgy]

MA Master Antenna

MA Master of Arts

MA Masters Abstracts

MA Matched Angles

MA Material Analysis

MA Material Authorization

MA Mathematical Analysis

MA Mathematical Association [UK]

MA Matrix Algebra

MA Matrix Analysis

MA Matrix Array

MA Mechanical Advantage

MA Mechanical Alloying; Mechanically Alloyed [Metallurgy]

MA Mechanical Analysis

MA Mechanical Arm

MA Mechanical Arts

MA Mechanical Attrition

MA Medical Assistance; Medical Assistant

MA Medium Artillery

MA Megampere [Unit]

MA Memory Address

MA Memory Alloy

MA Memory Array

MA Mental Age

MA Mercury Arc

MA Message Assembler

MA Metabolic Analyzer

MA Metal Alkoxide

MA Meter Angle

MA Methods Analysis

MA Methyl Acetate

MA Methyl Acrylate

MA Methylamine

MA Metric Association [now United States Metric Association]

MA Michael Addition [Organic Chemistry]

MA Michaelis-Arbuzov (Reaction) [Organic Chemistry]

MA Microalloy(ing)

MA Microanalysis; Microanalyzer

MA Microbial Allergy [Medicine]

MA Microfilm Address

MA Microphone Amplifier

MA Microprobe Analysis

MA Microstructural Analysis

MA Mid-Atlantic (Ridge)

MA Middeck Act

MA Middle Ages

MA Middle Atlantic [US Geographic Region]

MA Migratory Aptitude [Chemistry]

MA Milan Arsenal [Tennessee, US]

MA Military Academy

MA Military Aircraft

MA Military Attaché

MA Military Automation

MA Military Aviation

MA Mill Anneal (Process) [Metallurgy]

MA Milliammeter

MA Mining Association [UK]

MA Ministry of Aviation [UK]

MA Misorientation Axis (Vector) [Crystallography]

MA Missed Approach

MA Mission Analysis

MA Mixed Acid

MA Mixed Anhydride

MA Modify Address

MA Modulation Analysis; Modulation Analyzer

MA Moisture Absorption

MA Molecular Absorption

MA Molecular Adsorption

MA Monoamine [Organic Chemistry]

MA Monoclonal Antibody [Immunology]

MA Montreal Agreement [of International Air Transport Association]

MA Morocco [ISO Code]

MA Motion Analysis [Industrial Engineering]

MA Motor Analyzer

MA Mountain Artillery

MA Moving Average

MA Multi-Access

MA Multi-Address

MA Multichannel Analyzer

MA Multiple Access

MA Muscular Atrophy [Medicine]

MA Music Analysis

MA Mycobacterium Avium [Microbiology]

MA Myristic Acid

M&A Mergers and Acquisitions

M/A Martensite/Austenite [Metallurgy]

M-A Martensite-to-Martensite (Transformation) [also M→A] [Metallurgy]

Ma Mach Number [Symbol]

Ma Machinery [also ma]

Ma Masurium [now Technetium]

Ma Methanamine

Mα M-Alpha (Radiation) [also M_α]

M[α] Molecular Rotatory Power [Symbol]

mA milliampere [also ma]

mÅ milli-angstrom [Unit]

ma major

.ma Morocco [Country Code/Domain Name]

μA microammeter

μA microampere [also μa, μamp, or muA]

μÅ microangstrom [Unit]

MAA Maackia Amurensis Agglutinin [Immunology]

MAA Macroaggregate

MAA Maleic Acid Anhydride

MAA Manitoba Architectural Association [Canada]

MAA Master of Administrative Arts

MAA Master of Aeronautics and Astronautics

MAA Master of Applied Arts

MAA Master of Arts in Architecture

MAA Material Access Area

MAA Mathematical Association of America [US]

MAA Maximum Authorized Altitude

MAA Mechanical Arm Assembly

MAA Medieval Academy of America [Cambridge, Massachusetts, US]

MAA Medium Anti-Aircraft (Artillery)

MAA Mercaptoacetic Acid

MAA Methacrylic Acid

MAA Methanearsonic Acid

MAA Mixed Aluminum Alloy

MAA Modeling Association of America

MAA Moped Association of America [US]

MA-A Medium Abrasive-Abrasive

mAaBA m-Acetamidobenzoic Acid

MAAB Materials Application Advisory Board

MAAC Mid-Atlantic Area Council

MAAC Model Aeronautics Association of Canada

MAACS Member of the American Academy of Cosmetic Surgery

MAACS Multi-Address Asynchronous Communication System

MAAD Material Availability Date

MAAE Master of Arts in Applied Economics

MAALOX Magnesium-Aluminum Hydroxide [also Maalox]

MAAOM Master of Arts in Applied Organizational Management

MAAR Monthly Associate Administrator's Review

MAARC Magnetic Annular Arc

MAArch Master of Arts in Architecture

MAAW Medium Antitank Assault Weapon

MAB Man and the Biosphere (Program) [of UNESCO]

MAB Master Acquisition Bus

MAB Materials Advisory Board [now National Materials Advisory Board, US]

MAB Materials Applications Board

MAB Mechanical Automation Breadboard

MAB Metallic Access Bus

MAB Metropolitan Area Business (Line)

MAB Micro-Address Bus

MAB Missile Assembly Building

MAB Modified Adiabatic Bend (Approximation)

MAB Monomethylaminoazobenzene

MAb Monoclonal Antibody [also mAb] [Immunology]

$\mu(\alpha,\beta)$ Preisach function (of Preisach-type magnetic hysteresis transducers) [Note: α is the switch-up value and β the switch-down value]

MABA Meta-Aminobenzoic Acid

MAB-BRIM Man and Biosphere–Biosphere Reserve Integrated Monitoring

MABC Mining Association of British Columbia [Canada]

MABL Marine Atmospheric Boundary Layer

MABR Member of the American Board of Radiologists

MABS Maritime Application Bridge System

MABS Methyl Methacrylate/Acrylonitrile-Butadiene-Styrene (Copolymer)

$\mu(\alpha,\beta,\theta)$ vector-valued Preisach function (of Preisach-type magnetic hysteresis transducers) [Symbol]

MAC Machine-Aided Cognition

MAC Macintosh (Computer) [of Apple-Macintosh Computers, Inc., US] [also Mac]

MAC Magnification Autocorrection; Magnification Autocorrector

MAC Main Display Console

MAC Maintenance Advisory Committee

MAC Maintenance Allocation Chart

MAC Management Advisory Committee

MAC Man and Computer

MAC Man against Computer

MAC Manufacturing Advisory Committee

MAC Mapping Applications Center [of US Geological Survey]

MAC Massay Agricultural College [New Zealand]

MAC Master of Art Conservation

MAC Master of Arts in Communications

MAC Material Accounting Center

MAC Material Acquisition and Control

MAC Matrix-Array Camera

MAC Maximum Acceptable Concentration

MAC Maximum Admissible Concentration

MAC Maximum Allowable Concentration

MAC Mean Aerodynamic Chord

MAC Measurement and Analysis Center

MAC Media Access Control

MAC Medical Advisory Council

MAC Medium Access Control

MAC Membrane Applications Center [UK]

MAC Memory Access Controller

MAC Message Act Concellation

MAC Message Authentication Code

MAC Metal Aerosphere Composite

MAC Metal-Arc Cutting

MAC Methyl Allyl Chloride

MAC Micro-Array Computer

MAC Microgravity Advisory Committee

MAC Middle Atlantic Conference [US]

MAC Military Aircraft

MAC Military Airlift Command [of US Air Force at Scott Air Force Base, Illinois, US]

MAC Mineralogical Society of Canada

MAC Mining Association of Canada

MAC Ministers Advisory Council [Canada]

MAC Mobile Attenuation Code [Telecommunications]

MAC Momentum Accomodation Coefficient

MAC MOS (Metal-Oxide Semiconductor) Associated Circuit

MAC Moves, Adds and Changes [Computers]

MAC Multi-Access Computer; Multi-Access Computing

MAC Multi-Application Computer

MAC Multiple Access Computer; Multiple Access Computer

MAC Multi-Attribute Ceramic

MAC Multiple Access Computer

MAC Multiplexed Analog Component

MAC Mycobacterium avium Complex [Microbiology]

M-A-C Martensite-(Residual) Austenite-Cementite (Structure) [Metallurgy]

.MAC MacPaint [Macintosh File Name Extension]

.MAC Macro [File Name Extension]

MAc Master of Accounting (or Accountancy)

Mac Macintosh (Computer) [of Apple-Macintosh Computers, Inc., US] [also MAC]

Mac Macrocycle [Organic Chemistry]

MACA Master of Arts in Computer Applications

mAcBA m-Acetylbenzoic Acid

MACC Manitoba Agricultural Credit Corporation [Canada]

MACC Modular Alter and Compose Console

MAcc Master of Accountancy

Macch Autom Compon Macchine Automazione & Componenti [Italian Journal on Automation]

MACCS Manufacturing and Cost Control System

MACCS Manufacturing Cost Collection System

MACCS Molecular Access System

MACCT Multiple Assembly Cooling Cask Test [of Oak Ridge National Laboratory, Tennessee, US]

MAcct Master of Accounting

MAcct Master of Accountancy [also MAccy]

MACDAC Machine Communication with Digital Automatic Computer

MACDD Marine and Coastal Data Directory of Australia [Also known as the "Blue Pages"]

MACDIF Map and Chart Data Interchange Format

MACDS Monitor and Control Display System

MACE Management Applications in Computer Environments

MACE Mechanical Antenna Control Electronics

MACE Melt Attack and Coolability Experiment [of Argonne National Laboratory, Illinois, US]

MACE Mexican Association of Corrosion Engineers

MACE MPLM (Mini-Pressurized Logistics Module) Access Certification Equipment

Maced Macedonia(n)

MACEd Master of Arts in Continuing Education

MACEF Mastic Asphalt Council and Employers Federation [UK]

MACH Multilayer Actuator Head

Mach Machine; Machinery; Machining; Machinist [also mach]

Mach Des Machine Design(er)

Mach Des Machine Design [Journal published in the US]

Mach Des Mag Machine Design Magazine [US]

Mach Desgr Machine Designer

Mach Korea Machinery Korea [South Korean Publication]

Mach Learn Machine Learning [Journal published in the Netherlands]

Mach Mod Machine Moderne [French Publication on Modern Machinery]

MACHO MAssive Compact Halo Object (Experiment) [of University of California and Lawrence Livermore National Laboratory, US]

Mach Outil Prod Machine Outil Produire [French Publication on Production Machine Tools]

Mach Prod Machines Production [French Publication on Production Machinery]

Mach Prod Eng Machinery and Production Engineering [Journal published in the UK]

Mach Steel, Austria Machinery and Steel, Austria [Journal published in Austria]

Mach Tool Blue Book Machine and Tool Blue Book [US]

Mach Transl Machine Translation [Journal published in the Netherlands]

Mach Vib Monit Anal Proc Machinery Vibration Monitoring and Analysis Proceedings [Published by the Vibration Institute, US]

Mach Vis Machine Vision

Mach Vis Appl Machine Vision and Applications [Journal published in the US]

Machy Machinery [also machy]

MACI Ministerio de Agricultura, Comercio e Industria [Ministry of Agriculture, Commerce and Industry, Panama]

MACI Monitor, Access and Control Interface

M-Acid 8-Amino-4-Hydroxy-2-Naphthalenesulfonic Acid; 1-Amino-5-Naphthol-7-Sulfonic Acid [also M-acid]

MACIE Matrix-Controlled Interference Engine

MACM (International Conference on) Marine Applications of Composite Materials

MA/cm² Megampere(s) per square centimeter [also MA cm^{-2}]

mA/cm milliampere(s) per centimeter [also mA cm^{-1}]

mA/cm² milliampere(s) per square centimeter [also mA/sq cm, or mA cm^{-2}]

µA/cm² microampere(s) per square centimeter [also µA/sq cm, or mA cm^{-2}]

MACMA Military and Aerospace Connector Manufacturers Association [US]

MACMIS Maintenance and Construction Management Information System

MACN Museo Argentino de Ciencias Naturales [Argentinian Museum of Natural Sciences, Buenos Aires]

MACO Major Assembly Checkout

MAComm Master of Arts in Communication

MACON Matrix Connector (Punched Card Programmer)

MAC OS Macintosh Operating System [of Apple-Macintosh Computers, Inc., US]

Macplas Int Macplas International [Italian Chemical Publication]

Macr N-Methylacridinium [also macr]

MACRF Macro Instruction Form

MACRO International Symposium on Macromolecules

MACRO Macrocode

MACRO Macroprogram

MACRO Merge and Correlate Recorded Output (Program)

MACRO Monopole, Astrophysics and Cosmic Ray Observatory [of Laboratori Nazionali del Gran Sasso, Italy]

MACROCAL Macro Version of Common Assembler Language

MACROL Macro-Based Display Oriented Language

MACROL Macrolanguage

Macromol Macromolecular; Macromolecule [also macromol]

Macromol Macromolecules [Publication of the American Chemical Society, US]

Macromol Chem Macromolecular Chemist(ry)

Macromol Chem Macromolecular Chemistry [Journal published in the US]

Macromol Chem Rapid Commun Macromolecular Chemistry Rapid Communications [Journal published in the US]

Macromol Chem Suppl Macromolecular Chemistry Supplement [Journal published in the US]

Macromol Phys Macromolecular Physics; Macromolecular Physicist

Macromol Rep Macromolecular Reports

Macromol Rev Macromolecular Reviews [US]

Macromol Sci Macromolecular Science; Macromolecular Scientist

MACS (Sample) Manipulation and Beam Collimation System [of Los Alamos National Laboratories, New Mexico, US]

MACS Master of Arts in Communication Studies

MACS Master of Arts in Computer Science

MACS Media Account Control System

MACS Medium Altitude Communications Satellite

MACS Member of the American Chemical Society

MACS Metering and Accounting System

MACS Military Aeronautical Communication System

MACS Mobile Air Conditioning Society [US]

MACS Mobile Automated Cleaning System

MACS Monitoring and Control Station

MACS Multiple-Technique Analytical Computer System

MACS Multiproject Automated Control System

MACS Multipurpose Accelerated Cooling System

MACS Multipurpose Acquisition Control System

MACSS Medium-Altitude Communication Satellite System

MACSMB Measurement and Automatic Control Standards Management Board [of American National Standards Institute, US]

MACSYMA Mac's Symbol Manipulator

MACT Master of Arts in College Teaching

MACT Maximum Achievable Control Technology

MACTM Master of Applied Communication Theory and Methodology

MACU Material Cost per Unit

MACU Michigan Association of Colleges and Universities [US]

MACV Multipurpose Airmobile Combat-Support Vehicle

MACVD Microwave-Assisted Chemical-Vapor Deposition

MAD Machine ANSI (American National Standards Institute) Data

MAD Madison, Wisconsin [Meteorological Station Designator]

MAD Madrid, Spain [NASA Space Tracking and Data Network]

MAD Magnetic Airborne Detector

MAD Magnetic Anomaly Detection; Magnetic Anomaly Detection

MAD Magnetic Azimuth Detector

MAD Magnetoaerodynamic(s)

MAD Maintenance Analysis Data

MAD Maintenance, Assembly and Disassembly

MAD Many Acronymed Device [the HEMT (High-Electron Mobility Transistor)]

MAD Maroccan Dirham [Currency]

MAD Mass Analyzer Detector

MAD Michigan Algorithmic Decoder

MAD Military Aircraft Division

MAD Military Application Division [of US Atomic Energy Commission]

MAD Mixed Analog and Digital

MAD Multi-Aperture Device

MAD Multiple Access Device

MAD Multiple Access Drive

MAD Multiple Audio Distribution

MAD Multi(ple)-Wavelength Anomalous Diffraction

MAD Multi(ple)-Wavelength Anomalous Dispersion

MAD Multiply and Add

MAD Mutually Assured Destruction

MAd Master of Administration

MA'd Mechanically Alloyed (Powder) [also MA'ed]

MaD Materials Design

Mad Madagascan; Madagascar

MADA Multiple-Access Discrete Address

MADAEC Military Application Division of the Atomic Energy Commission [US]

MADAM Moderately Advanced Data Management

MADAM Multipurpose Automatic Data Analysis Machine

MADAP Maastricht Automatic Data Processing System

MADAR Malfunction Analysis Detection and Recording

MADAR Malfunction Data Recorder

MADC Multiplexer Analog to Digital Converter

MADDAM Macromodule and Digital Differential Analyzer Machine

MADDDC Manufacturers of Aerial Devices and Digger-Derricks Council [US]

MADDIDA Magnetic Drum Digital Differential Analyzer

MADE Magnetic Device Evaluator

MADE Manufacturing and Automated Design Engineering

MADE Manufacturing Automation and Design Engineering (Program) [US]

MADE Microalloy Diffused Electrode

MADE Minimum Airborne Digital Equipment

MADE Multichannel Analog-to-Digital Data Encoder

MADEE Malonic Acid Diethylester

MA DEP Massachusetts Department of Environmental Protection [US] [also MADEP]

MADER Management of Atmospheric Data for Evaluation and Research [of National Oceanic and Atmospheric Administration, US]

MADESS (Progetto Finalizzato) Materiali e Dispositivi per l'Elettronica a Stato Solido [Materials and Equipment for Solid-State Electronics (Program); of Consiglio Nazionale delle Ricerche, Italy]

MADGE Microwave Aircraft Digital Guidance Equipment

MADGE Microwave Automatic Digital Guidance Equipment

MADIS Millivolt Analog-Digital Instrumentation System

Mad Is Madeira Islands

MADM Manchester Automatic Digital Machine [of Manchester University, UK]

MAdm Master of Administration [also MAdmin]

MAdmMgt Master of Administration Management

MADO Mulliken Approximation for Differential Overlap [Physics]

MADP Main Air Display Plot

MADR Microprogram Address Register

MADRE Magnetic Drum Receiving Equipment

MADRE Manufacturing Data Retrieval System

MADRE Martin Automatic Data-Reduction Equipment

MADREC Malfunction Detection and Recording

MADS Machine-Aided Drafting System

MADS Missile Attitude Determination System

MADS Modular Auxiliary Data System

MADT Microalloy Diffused-Base Transistor

MADT Microalloy Diffusion Technique Transistor

MADT Modified Anomalous Diffraction Theory [Physics]

MADW Military Air Defense Warning

MADYLAM Magnetodynamique des Liquides–Applications à la Métallurgie [Magnetodynamics of Liquids–Metallurgical Applications; of Ecole Nationale Supérieure d'Hydraulique et d'Mécanique de Grenoble/Institut National Polytechnique de Grenoble, France]

MADYMO Mathematical Dynamic Modeling

MAE Magnetic After-Effect

MAE Magneto-Acoustic Effect

MAE Magneto-Acoustic Emission

MAE Magnetcrystalline Anisotropy Energy

MAE Magnetomechanical Acoustic Emission

MAE Maine Association of Engineers [US]

MAE Manchester Association of Engineers [UK]

MAE Maritime Advisory Exchange

MAE Maryland Association of Engineers [US]

MAE Master of Aeronautical Engineering

MAE Master of Aerospace Engineering

MAE Master of Agricultural Economics

MAE Master of Agricultural Education

MAE Master of Agricultural Engineering

MAE Master of Arts in Education

MAE Mechanical After-Effect

MAE Mechanical and Aeronautical Engineering

MAE Memory Address Register

MAE Metropolitan Area Ethernet

MAE Microalloying Element

MAE Ministère des Affaires Etrangères [Ministry of Foreign Affairs, France]

MAE Ministry of Atomic Energy [Russia]

MAE Missions–Activities–End Products (Data) [Battelle Memorial Institute, Columbus, Ohio, US]

MAECO Multi-Unit Architects, Engineers and Construction Officers [US]

MAECON Mid-America Electronics Conference

MAED Micro Area Electron Diffraction

MAEd Master of Arts in Education

MAed Mechanically Alloyed (Powder) [also MA'd]

MAEd Master of Arts in Education

MAEDP Master of Arts in Economic Developmental Programming

MAEE Marine Aircraft Experimental Establishment [UK]

MAELV Mutual Atomic Energy Liability Underwriters

MAEP Minimum Auto-Land Entry Point

MAEP Monoazaetioporphyrin [Biochemistry]

MAEP II Monoazaetioporphyrin II [Biochemistry]

MAEP IV Monoazaetioporphyrin IV [Biochemistry]

MAERC Minority Access to Energy Related Research Centers

MAeroE Master of Aeronautical Engineering

MAES Master of Arts in Environmental Sciences

MAES Mexican-American Engineering Society [US]

MAES Minnesota Agricultural Experiment Station [US]

MAES Multielement Atomic Emission Spectrometry

MAESTRO Machine-Assisted Educational System for Teaching by Remote Operation

MAESTRO Method for Analysis, Evaluation, Optimization (Code)

MAF Magic Angle Flipping [Physics]

MAF Major Academic Field

MAF Management Accounting Factor

MAF Manpower Authorization File

MAF Mass Air Flow (Sensor)

MAF Mean Average Flow

MAF Michoud Assembly Facility [of NASA in New Orleans, Louisiana, US]

MAF Mineral-Ash Free

MAF Minimum Audible Field

MAF Ministry of Agriculture and Food [Canada]

MAF Mission Aviation Fellowship

MAF Mixed Amine Fuel

MAF Mode Amplitude Function

MAF Moisture and Ash-Free

MAF Multi-Apertured Fabrics

MAFD Minimum Acquisition Flux Density

MAFF Ministry of Agriculture, Fisheries and Food [UK]

MAFIS Master of Accountancy and Financial Information Systems

mA/ft² milliampere per square foot [also mA/sq ft]

MAFVA Miniature Armored Fighting Vehicles Association [UK]

MAG Macrogeneration; Macrogenerator

MAG Magnetics (Society) [of Institute of Electrical and Electronics Engineers, US]

MAG Management Accounting Guidelines

MAG Management Advisory Group

MAG Maritime Air Group [of Canadian Forces]

MAG Master of Applied Geography

MAG Materials Analysis Group

MAG Maximum Available Gain [Electronics]

MAG Medical Association of Georgia [US]

MAG Metal Active-Gas (Welding)

MAG Million Ampere Generator

MAG Mine Advisory Group [UK]

MAG Motorcycle Action Group [UK]

MAg Master of Agriculture

Mag Magazine [also mag]

Mag Magnesium [also mag]

Mag Magnet; Magnetic(s); Magnetism [also mag]

Mag Magneto [also mag]

Mag Magnetometer [also mag]

Mag Magnetron [also mag]

Mag Magnitude [also mag]

MAGAMP Magnetic Amplifier [also Magamp, or magamp]

MAGB Microfilm Association of Great Britain

Mag Concr Res Magazine of Concrete Research [UK]

Mag Des Prod Magazine of Design and Production

MAGE Marine and Aerosol Gas Exchange

MAGE Mechanical Aerospace Ground Equipment

MAgE Master of Agricultural Engineering [also MAgEng]

MAgEc Master of Agricultural Economics

MAgEd Master of Agricultural Education

MAgEng Master of Agricultural Engineering [also MAgE]

MAGFET Magnetic (Field-Sensitive) Field-Effect Transistor [also MagFET, or magFET]

MAGG Modular Alphanumeric Graphics Generation

MAGGS Modular Advanced Graphics Generation System

MAGHEL (Projeto) Presquisa Magnetosfera e Heliosfera [Magnetospheric and Heliospheric Research Project; of Brazilian National Institute of Space Research]

MAGI Multi-Array Gamma Indicator

MAGIC Machine-Aided Graphic Input to Computer

MAGIC Machine-Aided Graphics for Illustration and Composition

MAGIC Machine for Automatic Graphics Interface to a Computer

MAGIC Magnetic Integrated Circuit

MAGIC Matrix Algebra General Interpretative Coding

MAGIC Michigan Automatic General Integrated Computation

MAGIC MIDAC (Michigan Digital Automatic Computer) Automated General Integrated Computation

MAGICS Manufacturing, Accounting and General Information Control System

MAGIS Marine Air-Ground Intelligence System [US]

Maglev Magnetic Levitation (Train) [also maglev]

Mag-Lith Magnesium-Lithium (Alloy) [also mag-lith]

MAGLOC Magnetic Logic Computer

MAGME Methylacrylamido Glycolate Methyl Ether

MAGMOD Magnetic Modulator

Magn Magnet; Magnetic(s); Magnetism [also magn]

Magn Magnetron [also magn]

Magn Magnitude [also magn]

MAGNAS3 Multiphase Analysis of Groundwater, Nonaqueous Phase Liquid, Chemical and Soluble in Three Dimensions

Magnesium Mon Rev Magnesium Monthly Review [US]

Magn Gidrodin Magnitnaya Gidrodinamika [Latvian Journal on Magnetohydrodynamics]

Magnistor Magnetotransistor [also magnistor]

Magn Lett Magnetism Letters [US]

Magn Prop Magnetic Properties [also magn prop]

Magn Reson Imaging Magnetic Resonance Imaging [Journal published in the UK]

Magn Reson Med Magnetic Resonance in Medicine [Journal published in the US]

Magn Reson Rev Magnetic Resonance Review [UK]

Magn Sep News Magnetic Separation News [UK]

Mag phos Magnesium Phosphate (Tissue Salt)

MAGPIE Machine Automatically Generating Production Inventory Evaluation

MAgr Master of Agriculture

MAgrDevEc Master of Agricultural Development Economics

MAG REC Magnetic Recording [also Mag Rec]

MAgrEc Master of Agricultural Economics [also MAgrEcon]

MAgrSc Master of Agricultural Science [also MAgricSc, or MAgSc]

MAgrSt Master of Agricultural Studies

MAGS Measurement and Graphics System

Mags Magazines [also mags]

Mags Magnitudes [also mags]

MAGSAT Magnetometer Satellite

Magsat Magnetic Field Satellite

MAgSc Master of Agricultural Science [also MAgrSc, MAgricSc]

MAGTAPE Magnetic Tape [also MAG TAPE, or Mag Tape]

MAGTC Magnetic Tape Controller

MagTG Magnetic Texturing Growth

Magy Alum Magyar Aluminum [Hungarian Publication on Aluminum]

Magy Kem Foly Magyar Kemiai Folyoirat [Hungarian Chemical Journal]

Magy Kem Lapja Magyar Kemikusok Lapja [Hungarian Chemical Journal]

Magy Textiltech Magyar Textiltechnika [Hungarian Publication on Textile Technology]

MAH Maackia Amurensis Hemagglutinin [Immunology]

MAH Magic Angle Hopping [Physics]

MAH Maleic Anhydride

M_4AH_{13} Tetramagnesium Aluminate Hydrate [$4MgO \cdot Al_2O_3 \cdot 13H_2O$]

mAh milliampere-hour [Unit]

MAHA Maximum Acceptable Hole Angle

MAHLOVS Middle and High Latitudes Oceanic Variability Study

MAHPS Maleic Anhydride Terminated Polystyrene

MAHRM Master of Arts in Human Resources Management

MAHS Master of Arts in Human Services

MAHT Machine-Aided Human Translation

MAHT Master of Arts in History Teaching

MAI Machine-Aided Index(ing)

MAI Many-Atom Interaction [Physics]

MAI Master of Agricultural Industries

MAI Mean Annual Increment

MAI Mobile Arctic Island

MAI Multiple Access Interface

MAI Multiple Angle Incidence

MAI Multiple Applications Interface

.MAI Mail [File Name Extension]

MAIA Magnetic Antibody Immunoassay

MAIA Master of Arts in International Affairs

MAIA Master of Arts in Industrial Art

MAIBC Member of the Architectural Institute of British Columbia [Canada]

MAID Master of Arts in Interior Design

MAID Multiple Aircraft Identification Display

MAIDS Management Automated Information Display System

MAIDS Modular, Adaptive, Incremental Detection System

MAIDS Multi-Purpose Automatic Inspection and Diagnostic System

MAIEE Member of the American Institute of Electrical Engineers

MAIG Matsushita Atomic Industrial Group [Japan]

MAIKES Mass-Analyzed Ion Kinetic Energy Spectroscopy

MAIM Member of the American Institute of Management

MAIME Member of the American Institute of Mining and Metallurgical Engineers

MAIN Mid-America Interpool Network

Main Maintenance [also main]

mA/in milliampere per inch [also mA in^{-1}]

mA/in² milliampere per square inch [also mA/sq in, or mA in^{-2}]

MA in LS Master of Arts in Library Science

Maint Maintenance [also maint]

MAIntDes Master of Arts in Interior Design

Maint Eng Maintenance Engineering [Journal published in the UK]

Maint J Maintenance Journal [Published by the International Maintenance Institute, US]

Maint Manage Int Maintenance Management International [Journal published in the Netherlands]

MAinURP Master of Arts in Urban and Regional Planning

MAIP Matrix Algebra Interpretative Program

MAIPS Master of Arts in International Policy Studies

MAIR Manufacturing and Inspection Record

MAIR Master of Arts in Industrial Relations

MAIR Molecular Airborne Intercept Radar

MAIS Master of Arts in Interdisciplinary Studies

MAIS Master of Arts in International Studies

MAIS Member of the Association of Industrial Surgeons

MAJ Majuro, Federated State of Micronesia [Meteorological Station Designator]

MAJ Master of Arts in Journalism

Maj Major [also MAJ] [Military Rank]

Maj Major(ity) [also maj]

MAJAC Maintenance Anti-Jam Console [of US Air Force]

MAJAC Monitor, Anti-Jam and Control

MAJC Master of Arts in Journalism and Communication

Maj Gen Major General

MAK Methylated Albumin Kieselguhr

.MAK Make File [File Name Extension] [also .mak]

Makromol Chem Makromolekulare Chemie [Swiss Journal on Macromolecular Chemistry]

Makromol Chem Macromol Symp Makromolekulare Chemie, Macromolecular Symposia [Swiss Journal on Macromolecular Chemistry, Macromolecular Symposia]

Makromol Chem Rapid Commum Makromolekulare Chemie, Rapid Communications [Swiss Journal on Macromolecular Chemistry, Rapid Communications]

MAL Macro Assembly Language

MAL Malfunction [also Mal]

MAL Maltese Pound [Currency]

MAL Material Allowance List

MAL Memory Access Logic

MAL Mercury-Arc Lamp

MAL Meta-Access Language

MAL Mobile Airlock

MAL Molecular Absorption Spectroscopy with Line Source

MAl Metal Aluminide (e.g., Nickel Aluminide, or Iron Aluminide)

MAl Monoaluminide

Mal Malachi

Mal Malagasy (Republic)

Mal Malay(a); Malayan

Mal Malfunction

Mal Malonate [also mal]

Mal Malonic Acid

MALAGOC Mutual Assistance of the Latin American Government Oil Companies [Uruguay]

Malaysian Rubb Rev Malaysian Rubber Review [Published by the Malaysian Rubber Research and Development Board]

MALCAP Maryland Academic Library Center for Automated Processing [of University of Maryland, College Park, US]

MALD Master of Arts in Landscape Design

MALDI Matrix-Assisted Laser Desorption/Ionization (Mass Spectroscopy)

MALDI-TOF Matrix-Assisted Laser Desorption/Ionization Time-of-Flight

MALE Multi-Aperture Logic Element

Mal Fr Malagasy Franc [Currency]

MALIS Master of Arts in Library and Information Science

Mall Malleability [also mall]

MALLAR Manned Lunar Landing and Return

MALMARC Malaysian Machine-Readable Catalogue

Malonyl CoA Malonyl Coenzyme A [Biochemistry]

Malonyl-S-ACP Malonyl-S-Acyl Carrier Protein [Biochemistry]

MALR Moist Adiabatic Lapse Rate [Meteorology]

MALS Master of Arts in Library Science

MALS Maximum-Intensity Approach Light System

MALS Medium-Intensity Approach Light System

MALSF Medium-Intensity Approach Light System with Sequenced Flashing Lights

MALSR Medium-Intensity Approach Light System with Runway Alignment Indicator Lights

MALT Manitoba Association of Library Technicians [Canada]

MALT Mnemonic Assembly Language Translator

MALT Mobile Analytical Laboratory Truck

MAM Management and Administration Manual

MAM Master of Administrative Management

MAM Master of Agriculture and Management

MAM Master of Animal Medicine

MAM Master of Applied Mechanics

MAM Master of Aviation Management

MAM Memory Access Multiplexer

MAM Memory Allocation Manager

MAM Memory Allocation Map

MAM Message Access Method

MAM Methylazoxymethanol

MAM Modified Atoms-in-Molecules (Method)

MAM Monoaminomesitylene

MAM Multiple Access to Memory

MAM Multi-Application Monitor

MA/m Megampere(s) per meter [also MA m^{-1}]

mA/m^2 milliampere per square meter [also mA/sq m]

MAMA Methylazoxylmethanol Acetate [also MAMAc, or MAM Acetate]

MAMA Monoammonium Methanearsonate

MAMAc Methylazoxymethanol Acetate [also MAMA, or MAM Acetate]

MAMB Master of Applied Molecular Biology

MAMBA Martin Armoured Main Battle Aircraft

MAMBO Mediterranean Association for Marine Biology and Oceanography [Malta]

MAMC Master of Arts in Mass Communication

MAMF Masters' Association of Metals Finishers

MAMgt Master of Arts in Management

MAMI Machine-Aided Manufacturing Information

MAMI Modified Alternate Mark Inversion

MAMIE Magnetic Amplification of Microwave Integrated Emissions

MAMIS Mandatory Modification and Inspection Summary

MAMM Master of Agricultural Marketing Management

MAMMAX Machine-Made and Machine-Aided Index

MAMMOS Magnetic Amplifying Magnetooptical System

MAMOS Marine Automatic Meteorological Observing Station

μamp microampere [also μA, μa, or muA]

MAMRD Master of Agricultural Management and Resource Development

MAMS Master of Applied Mathematical Science

MAMS Multispectral Atmospheric Mapping Sensor

MAMSA Managing and Marketing Sales Association [UK]

m-AMSA 4-(9-Acridinylamino)-N-(Methanesulfonyl)-m-Anisidine

MAMU Milli-Atomic Mass Unit [also mamu]

MAMU Millimass Unit [also mamu]

mA/mW/cm² milliamperere per milliwatt per square centimeter [Unit]

μA/mW/cm² microamperere per milliwatt per square centimeter [Unit]

MAN Manufacturers Association of Nigeria

MAN Maschinenfabrik-Augsburg-Nürnberg [Major German Manufacturer of Industrial Plant, Machinery and Equipment]

MAN Methylacrylonitrile

MAN Metropolitan Area Network

MAN Microwave Aerospace Navigation

MAN Molecular Anatomy

.MAN Manual Page [File Name extension [also .man]

Man Manage; Managing; Manager [also man]

Man Manitoba [Canada]

Man D-Manno- [Biochemistry]

Man D-Mannose [Biochemistry]

Man Manual [also man]

Man Methylacrylonitrile

(Man)₃ Mannotriose [Biochemistry]

(Man)₅ Mannopentaose [Biochemistry]

(Man)₉ Mannononaose [Biochemistry]

α-Man α-Mannose [also α-man] [Biochemistry]

α-D-Man α-D-Mannose [also α-D-man] [Biochemistry]

man manual(ly)

MANA Manufacturers Agents National Association [US]

Manage Autom Managing Automation [Journal published in the US]

Manage Decis Management Decision [Journal published in the UK]

Manage Inf Management and Informatica [Journal on Management and Information; published in Italy]

Manage Inf Managing Information [Journal published in the UK]

Manage Inf Syst Q Management Information Systems Quarterly [Publication of the Society for Management Information Systems, US]

Manage Rev Management Review [UK]

Manage Sci Management Science [Publication of The Institute of Management Sciences, US]

Manage Serv Management Services [Publication of the Institute of Management Services, UK]

Manage World Management World [Publication of Administrative Management Society, US]

MANAV Maneuvering and Navigation System

MANAV Marine Integrated Navigation

MANC Mexican Association for the Nature Conservancy

Manch Manchuria(n)

MANDATE Multi-Line Automatic Network Diagnostic and Transmission Equipment

Man Dir Managing Director [also ManDir]

MANDOP Doppler Radar Analysis with Polynomial Decomposition of Three-Dimensional Wind Field

MANDRO Mechanically Alterable NDRO (Nondestructive Readout)

MANEB Manganese Ethylene-Bis(dithiocarbamate) [also Maneb, or maneb]

Manf Manifold [also manf]

MANFEP University of Manitoba Finite Element Program [Canada]

Mang Manufacturing [also mang]

MAN-GHH Maschinenfabrik-Augsburg-Nürnberg–Gutehoffnungshütte GmbH [Major German Manufacturer of Industrial Machinery and Equipment headquartered in Oberhausen]

(Man)₃(GlcNAc)₂ Mannotriose-Di-(N-Acetyl)-D-Glucosamine [Biochemistry]

(Man)₅(GlcNAc)₂ Mannopentaose-Di-(N-Acetyl)-D-Glucosamine [Biochemistry]

(Man)₉(GlcNAc)₂ Mannononaose-Di-(N-Acetyl)-D-Glucosamine [Biochemistry]

(Man)₅(GlcNAc)₂Asn Mannopentaose-Di-(N-Acetyl)-D-Glucosamine-Asparagine [Biochemistry]

Man Hr Man Hour [also man hr]

MANI Ministry of Agriculture of Northern Ireland

MANIAC Mathematical Analyzer Numerical Integrator and Computer

MANIAC Mechanical and Numerical Integrator and Computer

MANIFILE University of Manitoba File of the World's Nonferrous Metallic Deposits [Canada]

MANIP Manual Input

Manit Manitoba [Canada]

MANMAR Manual for Marine Weather Observation

MANMAX Machine-Made and Machine-Aided Index

Man Mo Man Month [also man mo]

Mannesmann Forschungsber Mannesmann Forschungsberichte [Research Reports of Mannesmann, Dusseldorf, Germany]

MANOP Manual of Operation

MANOVA Multivariate Analysis of Variance [Statistics]

MANPAGE Manual Page [Unix Operating System]

Manpr Manpower [also manpr]

MANR Ministry of Agriculture and Natural Resources [Nigeria]

MAN Res Eng Manuf MAN Research, Engineering and Manufacturing [Publication of Maschinenfabrik Augsburg-Nürnberg AG, Augsburg, Germany]

MAN-Salt Methylamine Nitrate [also Man-Salt]

MANSCETT Manitoba Society of Certified Engineering Technicians and Technologists [Canada]

ManTech Manufacturing Technology (Program) [of US Air Force at Wright Aeronautical Laboratory, Dayton, Ohio] [also MANTECH]

ManTech J Manufacturing Technology [US]

MANTIS Manchester Technical Information Service [UK]

MANTRAC Manual Angle Tracking Capability

Manuf Manufacture(r); Manufacturing [also manuf]

Manuf Chem Manufacturing Chemist(ry)

Manuf Chem Manufacturing Chemist [Journal published in the UK]

manufd manufactured [also manuf'd]

ManufE Manufacturing Engineer [also Manuf Eng]

Manuf Eng Manufacturing Engineer(ing)

Manuf Eng Manufacturing Engineering [Journal of the Society of Manufacturing Engineers, US]

Manuf Eng (London) Manufacturing Engineer (London) [Journal published in the UK]

Manuf Eng Mag Manufacturing Engineering Magazine [Publication of MAP-TOP Users Group, US]

Manufg Manufacturing [also Manuf'g, manufg, or manuf'g]

Manuf Process Autom Manufacturing and Process Automation [Canadian Journal]

Manuf Rev Manufacturing Review [US]

Manuf Syst Trans Manufacturing Systems Transactions [Publication of the Society of Manufacturing Engineers, US]

Manuf Technol Manufacturing Technology; Manufacturing Technologist

Manuf Technol Horiz Manufacturing Technology Horizons [Published in the US]

Manuscr Geod Manuscripta Geodaetica [Journal on Geodesy; published in Germany]

Manuscr Math Manuscripta Mathematica [Journal on Mathematics; published in Germany]

Manuv Maneuvering [also manuv]

Man Wk Man Week [also man wk]

MAO Methylaluminoxane

MAO Monoamine Oxidase [Biochemistry]

MAODS Mechanically Alloyed, Oxide Dispersion-Strengthened [also MA/ODS]

MAOI Monoamine Oxidase Inhibitor [Medicine]

MAOP Maximum Allowable Operating Pressure

MAOP Musée Astronomique de l'Observatoire de Paris [Astronomical Museum of the Paris Observatory, France]

MAOS Metal Alumina Oxide Semiconductor; Metal Aluminum-Oxide Semiconductor

MAOS Metal/Aluminum Oxide/Silicon

MAOT Maximum Allowable Operating Temperature

MAOT Maximum Allowable Operating Time

MAP Machinist Apprentice Program

MAP Macao Pataca [Currency of Macao]

MAP Macro Arithmetic Processor

MAP Macroassembly Program

MAP Magnetically-Confined Anode Plasma

MAP Magneto-Abrasive Powder

MAP Maintenance Analysis Procedure

MAP Maintenance Analysis Program

MAP Management Assistance Program

MAP Manifold Absolute Pressure

MAP Manifold Air Flow (Sensor)

MAP Manual Assistance Position

MAP Manufacturing Automation Protocol

MAP Map Analysis Package [Software]

MAP Marketing Assistance Program

MAP Mathematical Analysis without Programming

MAP Maximum a posteriori Probability [Statistics]

MAP Mediterranean Action Plan [of United Nations Environmental Program Regional Seas]

MAP Memory Allocation and Protection

MAP Memory Allocation Processor

MAP Message Acceptable Pulse; Message Acceptance Pulse

MAP Metal Analysis Probe

MAP Methoxyacetophenone

MAP 6α-Methyl-17α-Acetoxy-4-Pregnene-3,20-Dione

MAP Microprocessor Application Project [of Department of Trade and Industry, UK]

MAP Microprogrammed Array Processor

MAP Microtubule-Associated Protein [Biochemistry]

MAP Microwave-Assisted Process(ing)

MAP Middle Atmosphere Program

MAP Military Assistance Program

MAP Minimum Acceptable Performance

MAP Minimum Audible Pressure

MAP Ministry of Aircraft Production [UK]

MAP Missed Approach Point [also MAPt]

MAP Mission Application Program

MAP Mitogen-Activated Protein [Biochemistry]

MAP Model and Program

MAP Modular Analysis Processor

MAP Monoaluminumphosphate

MAP Monobasic Ammonium Phosphate; Monoammonium Phosphate

MAP Multiple Aim Point

MAP Multiple Allocation Procedure

MAP Multiple Antigen(ic) Peptide [Biochemistry]

MAP Murine Antibody Production [Biochemistry]

.MAP Linker Map [File Name Extension] [also .map]

MAP4 (S)-2-Amino-2-Methyl-4-Phosphonobutyric Acid

MA&P Maintenance Analysis and Planning

Map Mapping [also map]

mAP m-Acetytl Phenol [also maP, or map]

MAPA Master of Arts in Public Administration

MAPA Master of Arts in Public Affairs

MAPA 3-(Methylamino)propylamine

MAPAM Mines Accident Prevention Association of Manitoba [Canada]

MAPAO Mines Accident Prevention Association of Ontario [Canada]

MAPC Metropolitan Area Planning Committee

MAPCHE Mobile Automatic Programmed Checkout Equipment

MAPCON Microprocessor Applications Consultancy

MAPED Machine-Aided Program for Preparation of Electrical Designs

MAPEP Mixed Analyte Performance Evaluation Program

MAP-EPA Manufacturing Automation Protocol–Enhanced Performance Architecture

MAPHYA Trends in Mathematical Physics (Meeting)

MAPI Machinery and Allied Products Institute [US]

MAPI Mail/Messaging Applications Programming Interface

MAPICS Manufacturing, Accounting and Production Information Control System

MAPID Machine-Aided Program for Preparation of Instruction Data

MAPKK Mitogen-Activated Protein Kinase Kinase [Biochemistry]

MAPLE Madelung Part of Lattice Energy [Solid-State Physics]

MAPLHG Maximum Average Planar Linear Heat Generation; Maximum Average Planar Linear Heat Generator

MAPLHGR Maximum Average Planar Linear Heat Generation Rate

MApMa Master of Applied Mathematics

MAPO tris[1-(2-Methyl)aziridinyl]phosphine Oxide

MAPOLE Magnetic Dipole Spark Transmitter

MAPORD Methodology Approach to Planning and Programming Operational Requirements, Research and Development [US Air Force]

MAPP Mathematical Analysis of a Perception and Preference

MAPP Methylacetylene Propadiene [also Mapp]

MAPPER Maintaining, Preparing and Producing Executive Reports

MAPPS Management Association for Private Photogrammetric Surveyors [US]

MApplSc Master of Applied Science [also MapplS, or MAppSc]

MApplStat Master of Applied Statistics [also MAppStat, or MApStat]

MAPRAT Maximum Power Ratio

MAPS Management Accounting and Payroll System

MAPS Manufacturers Automation Peripheral System

MAPS Measurement of Air Pollution from Satellites; Measurement of Atmospheric Pollution from Satellites

MAPS Medical Application Processing Services [Gold River, California, US]

MAPS Mesoscale Analysis and Prediction System

MAPS tris[1-(2-Methyl)aziridinyl]phosphine Sulfide

MAPS Methylacetylene Propadiene, Stabilized

MAPS Middle Atlantic Planetarium Society [US]

MAPS Modular Acoustic Processor System

MAPS Modular Automatic Preparation System

MAPS Multicolor Automatic Projection System

MAPS Multivariate Analysis and Prediction of Schedules

MApStat Master of Applied Statistics [also MApplStat, or MAppStat]

MAPt Missed Approach Point [also MAPt]

MAPTAC Methacrylamidopropyltrimethylammonium Chloride

MAP/TOP Manufacturing AutomationProtocol/Technical Office Protocol [also MAP-TOP, or MAPTOP]

MAP/TOP Interf Tech J MAP/TOP Interface Technical Journal [Published by the MAP-TOP Users Group, US]

MAPW Master of Arts in Professional Writing

MAPW Modified Augmented Plane Wave [Physics]

MAQ Macquarie Island [Tasmania, Australia]

MAQ 2-Methyl-9,10-Anthraquinone

MAq Master of Aquaculture

MAQ-Br 2-Bromomethylanthraquinone

MAQL Mine Air Quality Laboratory

MAR Machine-Readable

MAR Macro Address Register

MAR Magic-Angle Rotation [Physics]

MAR Master of Arts in Research

MAR Mauritian Rupee [Currency]

MAR Maintenance and Refurbishment

MAR Malfunction Array Radar

MAR Margin [also Mar]

MAR Mercury-Arc Rectifier

MAR Memory Address Register

MAR Micro-Analytical Reagent

MAR Microautoradiograph(y)

MAR Microprogram Address Register

MAR Minimum Angle of Resolution

MAR Miscellaneous Apparatus Rack

MAR Mission Analysis Room

MAR Multifunction(al) Array Radar

MA(R) Master of Arts (Research)

Mar Margin [also mar]

Mar March

Mar Marine [also mar]

Mar Maritime [also mar]

Mar Market [also mar]

MARAD Marine Administration [US]

MARAD Maritime Administration [of US Department of Transportation]

MARAIRMED Maritime Air Forces Mediterranean Command [of NATO]

MARAS Middle Airspace Radar Advisory Service [US]

MARATS Mid-Atlantic Region Administrative Telephone System

MARB Methanotropic Airtight Rotating Bioreactor

Mar Biol Marine Biologist; Marine Biology

MARC Machine-Readable Catalog(ing)

MARC Material Accountability and Recoverability Code

MARC Melbourne Advanced Research Center [of University of Melbourne, Australia]

MARC Micro Analytical Research Center [of University of Melbourne, School of Physics, Australia]

MARC Modern Asian Research Center

MARC Monitoring and Assessment Research Center [UK]

MARC Multi-Axial Radial Circuit

MARCA Mid-Continent Area Reliability Coordination Agreement

MARCAS Maneuvering Reentry Control and Ablation Studies

MARCEP Maintainability and Reliability Cost Effectiveness Program

MArch Master of Architecture

MArchE Master of Architectural Engineering [also MArchEng]

Mar Chem Marine Chemistry [Journal published in the Netherlands]

MArchH Master of Architectural History

MArchUD Master of Architecture and Urban Design

MArchUD Master of Architecture in Urban Design

MARCIA Mathematical Analysis of Requirements for Career Information Appraisal

MARC IS Machine-Readable Catalog–Israel

MARC LC Machine-Readable Catalog–Library of Congress [US]

MARCO Machine Referenced and Coordinated Outline

MARCOGAZ Union de l'Industrie Gaz du Marché Commun [Union of the Gas Industry of the Common Market Countries, Belgium]

MARCOM Maritime Command Operations [Canada]

MARCOM Microwave Airborne Communications Relay

MARCOT Maritime Coordinated Training Exercises

MARCS MELCOM (Computer) Allround Adaptive Consolidated Software

MARC UK Machine-Readable Catalogue–United Kingdom

MARDAN Marine Differential Analyzer

MARDI Malaysian Agricultural Research and Development Institute

MarE Marine Engineer [also Mar Eng, or MarEng]

MAREA Member of the American Railway Engineering Association

Mar Ecol Prog Marine Ecology Progress [US Journal]

MARECS Marine Communications Satellite [of European Space Agency]

MARECS Maritime Communications Satellite System [also Marecs]

MarEng Marine Engineer [also Mar Eng, or MarE]

Mar Eng Marine Engineer(ing)

Mar Eng Bull Marine Engineers Bulletin [Published by the Institute of Marine Engineers, UK]

Mar Eng Rev Marine Engineers Review [Published by the Institute of Marine Engineers, UK]

MARFO Machine-Readable Form

Marg Margarine [also marg]

Marg Margin [also marg]

MARGEN Management Report Generator

Mar Geod Marine Geodesy; Marine Geodesist

Mar Geod Marine Geodesy [Journal published in the US]

Mar Geol Marine Geologist; Marine Geology

Mar Geol Marine Geology [International Journal]

Mar Geophys Res Marine Geophysical Research [Journal published in the Netherlands]

Mar Georesour Geotechnol Marine Georesources and Geotechnology [International Journal]

MARGIE Memory Analysis Response Generation and Interference in English

MARIN Marine Research Institute of the Netherlands

MARINDB Maringeologisk Databas [Marine Geology Database, of Swedish Geological Survey]

MARIOM Matrix Auto-Reinforced Organic Material

MARISAT Maritime Satellite (Service) [also Marisat, MARSAT, or Marsat]

Mark Market [also mark]

MARKAR Mapping and Reconnaissance Ku-Band Airborne Radar

MARKSIM (Computerized) Marketing Management Simulation

MARLIS Multi-Aspect Relevance Linkage Information System

MarMechE Marine Mechanical Engineer [also MarMechEng]

Mar Met Marine Meteoroloist; Marine Meteorology

Mar Micropaleont Marine Micropaleontology [International Journal]

MARN Manitoba Association of Registered Nurses [Canada]

MARNA Marine Navigation (Database) [Netherlands]

MARNAF Marquardt Navair Fuel

MAROTS Maritime Orbital Test Satellite

MARPAC Maritime Forces Pacific [of Department of National Defense, Canada]

MARPAL Marseilles-Palo Submarine Cable [France/Italy]

Mar Pet Geol Marine and Petroleum Geology [Journal published in the UK]

MARPIC Marine Pollution Information Center [UK]

MARPLOT Mapping Application for Response, Planning and Local Operational Task

MARPO Marine Pollution

MARPOL International Convention for the Prevention of Marine Pollution from Ships

MARPOS Marine Pollution System

Mar Pollut Bull Marine Pollution Bulletin [Published by the International Ocean Disposal Symposium, US]

Mar Pollut Bull Marine Pollution Bulletin [UK]

Mar Pollut News Marine Pollution News [Published by the International Ocean Disposal Symposium, US]

MARR Minimum Attractive Rate of Return

MAR REL Margin Release [also Mar Rel]

MARS Machine-Assisted Reference Section

MARS Machine Retrieval System

MARS Magnetic Airborne Recording System

MARS Management Analysis Reporting Service

MARS Manned Aerodynamic Reusable Spaceship

MARS Manned Astronomical Research Station

MARS Marconi Automatic Relay System

MARS Market Analysis Research System

MARS Martin Automatic Reporting System

MARS Materials Analysis Research Station

MARS Member of the Association of Railway Surgeons

MARS Memory-Address Register Storage

MARS Meteorological Airborne Radar Data System [of National Oceanic and Atmospheric Administration, US]

MARS Mid-Air Recovery System [also Mars] [for Recovery of Remotely Piloted Vehicles]

MARS Military Affiliated Radio System [US]

MARS Military Amateur Radio System

MARS Mirror Advanced Reactor Study

MARS Mitigation and Adaption Research Strategies

MARS Mobile Analytical Reconnaissance System

MARS Modular Airborne Recorder System

MARS Multi-Access Retrieval System

MARS Multi-Aperture Reluctance Switch

MARS Multiple Artillery Rocket System [US]

MARSAT Maritime Satellite (Service) [also Marsat, MARISAT, or Marisat]

MArSci Master of Arts and Sciences [also MArSc]

MARSE Measured Area under the Rectified Signal Envelope

MARSL Machine-Readable Shelf List

MART Maintenance Analysis Review Technique

MART Market Assessment of Research and Technology (Program) [of Science Council of British Columbia, Canada]

MART Mean Active Repair Time

MART Mechanical Analysis and Retrieval of Text

MART Mississippi Aerial River Transit System [US]

MART Mobile Automatic Radiation Tester

Mart Martian [also mart]

Mart Martinique

MARTD Multidisciplinary Association for Research and Teaching in Demography [Peru]

MARTEC Martin Thin-Film Electronic Circuit

MARTECH Center for Materials Research and Technology [of Florida State University, Tallahassee, US]

Mar Technol Marine Technology; Marine Technologist

Mar Technol Marine Technology [Published by the Society of Naval Architects and Marine Engineers]

Mar Technol Marine Technology [Journal of the Verein Deutscher Ingenieure, Germany]

Mar Technol Soc J Marine Technology Society Journal [US]

MARTEL Marseilles-Tel Aviv Submarine Cable [France/Israel]

MARTEL Missile Antiradar and Television [also Martel]

MARTI Machine-Readable Telegraphic Input

Martm Maritime [also martm]

MARTOS Multi-Access Real-Time Operating System

MARV Maneuverable (or Maneuvering) Antiradar Vehicle

MARV Maneuverable Reentry Vehicle

MARV Multi-Element Articulated Research Vehicle

MARVEL Machine-Assisted Realization of the Virtual Electronic Library [of Library of Congress, US]

MARVEL Managing Resources for University Libraries

Mar Watch Marine Watch [Quarterly Journal of the Marine Watch Institute, US]

MAS Macro-Assembler

MAS Macedonian Academy of Sciences

MAS Magic-Angle Spinning [Physics]

MAS Magnesia-Alumina-Silicate

MAS Magnesium Aluminosilicate

MAS Magnesium Aluminum Silicate (Ceramics)

MAS Magnetic Angle Spinning [Physics]

MAS Malaysian Airlines System

MAS Management Advisory Services

MAS Maraging Steel

MAS Maryland Academy of Sciences [US]

MAS Master of Accounting Science

MAS Master of Administrative Science

MAS Master of Aeronautical Science

MAS Master of Applied Science

MAS Master of Applied Statistics

MAS Master of Archival Studies

MAS Material Acquisition Specification

MAS May Aerial Survey [US Geological Survey]

MAS Mechanically Activated Synthesis (Process) [High-Energy Milling]

MAS Medium-Alloy Steel

MAS Merseyside Aviation Society [UK]

MAS Metal-Alumina Semiconductor

MAS Metal-Alumina-Silicon

MAS Microalloyed Steel

MAS Microbeam Analysis Society [US]

MAS Microfilm Advisory Service

MAS Military Agency for Standardization [of NATO]

MAS Military Alert System

MAS Millimeter-Wave Atmosphere

MAS Mississippi Academy of Science [Jackson, US]

MAS MODIS (Moderate Resolution Imaging Spectroradiometer) Airborne Simulator

MAS Modular Application System

MAS Moldavian Academy of Sciences [Kishinev, Russian Federation]

MAS Molecular Absorption Spectroscopy

MAS Momentary-Action Switch

MAS Moessbauer Absorption Spectroscopy

MAS Multi-Aspect Signaling

MA/s Megampere per second [also MA/sec, or MA s^{-1}]

mAs milliampere-second [Unit]

m³/As cubic meter(s) per ampere-second [Unit]

MASA Main Store Arrays

MASA Marine Accessories and Services Association [US]

MASA Master of Advanced Studies in Architecture

MASA Medical Acronyms, Symbols and Abbreviations (Dictionary)

MASA Military Accessories Service Association [US]

MASA Military Automotive Supply Agency [US]

MASB Motor Anti-Submarine Boat

MASC Main Store Controller

MASC Methylaluminum Sesquichloride

MASC Mountain Administrative Support Center [of National Oceanic and Atmospheric Administration, US]

MASc Master of Agricultural Sciences

MASc Master of Applied Science

masc masculine

Masch-Bau-Tech Maschinenbautechnik [German Journal on Mechanical Engineering Technology]

Masch Elektrotech Maschinenwelt Elektrotechnik [German Publication on Machinery and Equipment in Electrical Engineering]

Masch-Schad Maschinenschaden [Publication on Machine Failure; published by Allianz Versicherungs-AG, Munich, Germany]

MAScMechEng Master of Applied Science in Mechanical Engineering

MAScMetEng Master of Applied Science in Metallurgical Engineering

MASCON Mass Concentration [also mascon]

MASCOT Modular Approach to Software Construction, Operation and Test

MASCOT Motorola Automatic Sequential Computer Operator Test

MASCP&T Member of the American Society for Clinical Pharmacology and Therapeutics

MASDC Military Aircraft Storage and Deposition Center

MASE Materials, Applications and Services Exposition [of ASM International/The Minerals, Metals and Materials Society, US]

MASE/ACCE Materials, Applications and Services Exposition/Advanced Composites Conference and Exposition [of ASM International, US]

MASER Materials Science Experiment Rocket

MASER Microwave Amplification by Stimulated Emission of Radiation [also Maser, or maser]

MASER Molecular Amplification by Stimulated Emission of Radiation [also Maser, or maser]

MA/sec Megampere per second [also MA/s, or MA s^{-1}]

MASFET Metal-Alumina-Silicon Field-Effect Transistor

MASG Missile Auxiliary Signal Generator

MASH Manned Antisubmarine Helicopter

MASH Mobile Army Surgical Hospital [US]

MASHTEC International Conference on Materials Science for High Technologies

MASIF Mobile Atmospheric Sampling and Identification Facility

MASIS Management and Scientific Information System

MASK Maneuvering and Seakeeping

MASL Meters above Sea Level [also masl]

MASLS Member of the American Society for Liposuction Surgery

MASM Macro Assembler

MASM Member of the American Society for Metals

MASM Member of ASM International (American Society for Materials International)

MASM Meta-Assembler

MASM Microsoft's Macro Assembler

μA·s/m microampere second per meter [also μA·s m^{-1}]

MASME Member of the American Society of Mechanical Engineers

MASME Methyl Methacrylate

MAS-NMR Magic Angle Spinning Nuclear Magnetic Resonance [also MAS NMR, or MASNMR]

MASPAC Microfilm Advisory Service of the Public Archives of Canada

mA/sq cm milliampere per square centimeter [also mA/cm², or mA cm^{-2}]

mA/sq in milliampere per square inch [also mA/in², or mA in^{-2}]

mA/sq m milliampere per square meter [also mA/m²]

μA/sq cm milliampere per square centimeter [also μA/cm², or mA cm^{-2}]

MASRT Marine Air Support Radar Teams

MASRU Marine Air Support Radar Unit [of US Department of Defense]

MASS Magic-Angle Sample Spinning [Physics]

MASS Maintenance Activities Subsea Surface

MASS MARC (Machine Readable Catalog) Based Automated Serials System

MASS Master of Arts in Special Studies

MASS Maximum Availability and Support Subsystem

MASS Michigan Automatic Scanning System

MASS Mobile Army Sensor System

MASS Monitor and Assembly System

MASS Multi-Axial Span System

MASS Multiple Access Sequential Selection

MASS Multiple Access Switching System

Mass Massachusetts [US]

MASSDAR Modular Analysis, Speedup, Sampling, and Data Reduction

MASS-NMR Magic-Angle Sample Spinning Nuclear Magnetic Resonance

mass ppm mass parts per million

Mass Spectrom Bull Mass Spectrometry Bulletin [Published by the Royal Society of Chemistry, UK]

Mass Spectrosc Mass Spectroscopy [Published by the Mass Spectroscopy Society of Japan]

MASSTER Mobile Army Sensor System Test Evaluation and Review

MAST Magnetic Annular Shock Tube

MAST Marine Applications of Science and Technology [of European Community]

MAST Material Status

MAST Missile Automatic Supply Technique

MAST Monterey Area Ship Track Experiment [US]

MAST Multiple Aircraft Simulation Terminal

MASTER Matching Available Student Time to Educational Resources

MASTER MODIS/ASTER (Moderate Resolution Imaging Spectroradiometer/Advanced Spaceborne Thermal Emission and Reflection Radiometer) Airborne Simulator

MASTER Multiple Access Shared Time Executive Routine

MASTIR Microfilmed Abstracts System for Technical Information Retrieval [of Illinois Institute of Technology, Chicago, US]

MASTMAASIS Modified ASTM Acetic Acid Salt Intermittent Spray (Exfoliation Test) [of American Society for Testing and Materials, US]

MaSTS Marine Strategy for the Torres Strait [Australia]

MASU Mandalay Arts and Science University [Burma]

MASU Metal Alloy Separation Unit

MASUA Mid-America State Universities Association [US]

MASV Maximum Allowable Stress Value

MAT Machine-Aided Translation

MAT Maintenance Access Terminal

MAT Master of Arts in Teaching

MAT Maximized Average Torque

MAT Mechanical Aptitude Test

MAT Mechanical Assembly Technique

MAT Mechanically-Agitated Tank

MAT Memory Address Translator

MAT 2-(N-Methylamino)thiazole

MAT Meteorological Atmospheric Turbulence

MAT Metropolitan Area Trunks

MAT Microalloy Transistor

MAT Miller Analogies Test

MAT Mobile Automatic Telephony

MAT Monitor, Alarm and Trend Module

MAT Multiple Access Technique

MAT Multiple Actuator Test

MA&T Manufacturing Assembly and Test

Mat Material [also mat]

Mat Matrix [also mat]

MATA Michigan Aviation Trades Association [US]

Mat Apl Comput Matematica Aplicada e Computacional [Brazilian Journal on Applied and Computational Mathematics; published by the Sociedade Brasileira de Matematica Aplicada e Computacionall]

MatDB Materials Database (Management System) [of ASM International, US]

MATC Master of Arts in Textiles and Clothing

MATCALS Marine Air Traffic Control and Landing System

Mat Chem Materials Chemist(ry)

MATCO Materials Analysis, Tracking and Control

MATCON Microwave Aerospace Terminal Control [of US Air Force]

MATD Mine and Torpedo Detector

MatDC Materials Property Database [of ASM International, US]

Mat Des Materials and Design [Journal published in the UK]

MATE Manufacturing Advanced Technology Exchange

MATE Measuring and Test Equipment

MATE Memory-Assisted Terminal Equipment

MATE Modular Automatic Test Equipment

MATE Multiple-Access Time-Division Experiment

MATE Multisystem Automatic Test Equipment

MatE Materials Engineer [also Mat Eng, or MatEng]

MaTech Neue Materialien für die Schlüsseltechnologien des 21. Jahrhunderts [Novel Materials for the Key Technologies of the Twenty-First Century; of Bundesministerium für Forschung und Technologie, Germany; Replacement for MatFo]

MatEng Materials Engineer [also Mat Eng, or MatE]

Mat Eng Materials Engineer(ing)

Mat Eng Materials Engineering [UK/US Publication]

Mat Eng Des Materials Engineering Design

Mater Material [also mater]

Mater Australas Materials Australasia [Journal published in Austrialia]

Mater Charact Materials Characterization [Journal published in the US]

Mater Chem Materials Chemistry [Journal published in Italy]

Mater Chem Phys Materials Chemistry and Physics [International Journal of the Chinese Society for Materials Science]

Mater Constr Materiales de Construcción [Spanish Journal on Construction Materials]

Mater Des Materials and Design [Journal published in the UK]

Mater Edge Materials Edge [Journal published in the UK]

Mater Eng Materials Engineering [Journal published in the UK and the US]

Mater Eng (Cleveland) Materials Engineering (Cleveland) [Journal published in the US]

Mater Eng (Surrey) Materials Engineering (Surrey) [Journal published in the UK]

Mater Eval Materials Evaluation [Journal of the American Society for Nondestructive Testing, US]

Mater Flow Material Flow [Journal published in the Netherlands]

Mater Forum Materials Forum [Joint Publication of the Institute of Metals and Materials Australasia, the Australian Chemical Institute and the Australian Ceramic Society]

Mater Handl Eng Material Handling Engineering [Journal published in the US]

Mater Handl Outlook Material Handling Outlook [Publication of the International Material Management Society, US]

Mater Lett Materials Letters [Journal published in the Netherlands]

Mater Manuf Materials and Manufacture [Journal published in the UK]

Mater Manuf Process Materials and Manufacturing Processes [Journal published in the US]

Mater Mech Eng Materials for Mechanical Engineering [Journal published in PR China]

Mater News Int Materials News International [Journal published in Belgium]

Mater Ogniotrw Materialy Ogniotrwale [Polish Materials Publication]

Mater Perform Materials Performance [Publication of the National Association of Corrosion Engineers, US]

Mater Plast Materiale Plastice [Romanian Publication on Plastics]

Mater Plast Elast Materie Plastiche ed Elastomeri [Italian Publication on Plastics and Elastomers]

Mater Process Rep Materials and Processing Report [US]

Mater Prot Materials Protection [Journal published in PR China]

Mater-Prüf Materialprüfung [Journal on Materials Testing of Deutscher Verband für Materialprüfung, Germany]

Mater Reclam Wkly Materials Reclamation Weekly [UK]

Mater Res Bull Materials Research Bulletin [of Materials Research Society, US]

Mater Res Soc Int Mtg Adv Mater Materials Research Society International Meeting on Advanced Materials [US]

Mater Res Soc Symp Proc Materials Research Society Symposia Proceedings [US]

Mater Res Stand Materials Research and Standards (Journal)

Mater Res Symp Proc Materials Research Symposium Proceedings [US]

Mater Sci Materials Science [Published by Wroclaw Polytechnic, Poland]

Mater Sci Cit Ind Materials Science Citation Index [of Materials Research Society, US]

Mater Sci Commun Materials Science Communications [Part of the International Journal "Materials Chemistry and Physics" of the Chinese Society for Materials Science]

Mater Sci Eng Materials Science and Engineering [Journal published in Switzerland]

Mater Sci Eng A, Struct Mater, Prop Microstruct Process Materials Science and Engineering A, Structural Materials: Properties, Microstructure and Processing [Journal published in Switzerland]

Mater Sci Eng B, Solid-State Mater Adv Technol Materials Science and Engineering B, Solid-State Materials for Advanced Technology [Journal published in Switzerland]

Mater Sci Eng C Materials Science and Engineering C, Biomimetic Materials, Sensors and Systems [Journal published in Switzerland]

Mater Sci Eng R Materials Science and Engineering R [Journal published in Switzerland]

Mater Sci Forum Materials Science Forum [Published in Switzerland],

Mater Sci Monogr Materials Science Monographs [Published in the Netherlands]

Mater Sci Rep Materials Science Reports [Published in the Netherlands]

Mater Sci Res Materials Science Research [Journal published in the US]

Mater Sci Technol Materials Science and Technology [Publication of the Institute of Metals, UK]

Mater Soc Materials and Society [Journal published in the US]

Mater Struct Materials and Structures [Journal published in the UK]

Mater Syst Materials Systems [Journal published in Japan]

Mater Tech Material und Technik [Swiss Publication on Materials and Technology]

Matér Tech (Paris) Matériaux et Techniques (Paris) [French Publication on Materials and Technology]

Mater Technol Material Technology [Journal published in Japan]

Mater Tekhnol Materialoznanie i Tekhnologiya [Bulgarian Publication on Materials Engineering and Technology]

Mater Trans, JIM Materials Transactions, JIM [Published by the Japan Institute of Metals]

Mater wiss Werkst tech Materialwissenschaft und Werkstofftechnik [German Journal on Materials Science and Technology]

Mat Eval Materials Evaluation

MATFAP Metropolitan Area Transmission Facility Analysis Program

MATFO Materialforschung [Materials Research`Program of the Bundesministerium für Forschung und Technologie, Germany; Superceded by MaTech] [also MatFo]

Mat Fiz Nelineinaya Mekh Matematicheskaya Fizika i Nelineinaya Mekhanika [Ukrainian Journal on Mechanics and Mathematical Physics]

Mat-Fys Medd K Dan Vidensk Selsk Matematisk-Fysiske Meddelelser Konglige Danske Videnskabernes Selskab [Communications of the Royal Danish Academy of Sciences, Mathematics and Physics]

MATH Mathematics and Computer Laboratory [of US Department of Energy]

Math Mathematical; Mathematician; Mathematics [also math]

Math Anal Mathematical Analysis [also math anal]

Math Ann Mathematische Annalen [Annals of Mathematics; published in Germany]

Mat Handl Materials Handling [also mat handl]

Mat Handl Eng Materials Handling Engineer(ing)

Math Balk Mathematica Balkanica [Mathematics Journal; published in Yugoslavia]

Math Biol Mathematical Biologist; Mathematical Biology

Math Comput Mathematics of Computation [Publication of the American Mathematical Society, US]

Math Comput Educ Mathematics and Computer Education [Journal published in the US]

Math Comput Model Mathematical Computer Modelling [Journal published in the UK]

Math Comput Sim Mathematics and Computers in Simulation [Publication of the International Association for Mathematics and Computers in Simulation, Belgium]

Math Control Signals Syst Mathematics of Control, Signals and Systems [Journal published in the US]

MathD Doctor in Mathematics

MATHDI Didaktik der Mathematik und Informatik (Datenbank) [Didactics of Mathematics and Information Sciences (Database), Germany]

Math Geol Mathematical Geologist; Mathematical Geology

Math Geol Mathematical Geology [Publication of the International Association for Mathematical Geology, US]

Math Japon Mathematica Japonicae [Japanese Mathematical Journal]

MATHLAB Mathematical Laboratory

Math Ling Mathematical Linguistics [also math ling]

Math Mag Mathematics Magazine [Published by the Mathematical Association of America, US]

Math Methods Appl Sci Mathematical Methods in the Applied Sciences [Journal published in the UK]

Math Mod Mathematical Modelling [also math model]

Math Model Numer Anal Mathematical Modelling and Numerical Analysis [Journal published in France]

Math Nachr Mathematische Nachrichten [Mathematical Journal; published by the Deutsche Akademie der Wissenschaften, Berlin, Germany]

Math Numer Sin Mathematica Numerica Sinica [Journal published in PR China]

Math Oper Res Mathematics of Operations Research [Published by the Operations Research Society of America]

MATHPAC Mathematical and Statistical Program Package

Math Phys Mathematical Physicist; Mathematical Physics

Math Proc Camb Philos Soc Mathematical Proceedings of the Cambridge Philosophical Society [UK]

Math Prog Mathematical Programming; Mathematical Programmer [also Math Program]

Math Program Mathematical Programming [Journal published in the Netherlands]

Math Program A Mathematical Programming, Series A [Journal published in the Netherlands]

Math Program B Mathematical Programming, Series B [Journal published in the Netherlands]

Math Rep Toyama Univ Mathematics Reports of the Toyama University [Japan]

Math Rev Mathematical Reviews [of American Mathematical Society, US]

Maths Mathematics [also maths]

Math Scand Mathematica Scandinavica [Mathematical Journal; jointly published by the Mathematical Societies of Denmark, Finland, Island, Norway and Sweden]

Math Sci Mathematical Science; Mathematical Scientist

MathSciNet Mathematical Sciences Network [US]

Math Slovaca Mathematica Slovaca [Journal on Mathematics of the Slovak Academy of Sciences]

Math Soc Sci Mathematical Social Sciences [Journal published in the Netherlands]

Math Spectr Mathematical Spectrum [Published in the UK]

Math Stat Mathematical Statistician; Mathematical Statistics

Math Surv Mathematical Surveys [Publication of the American Mathematical Society, US]

Math Syst Theory Mathematical Systems Theory [Journal published in the US]

Math Z Mathematische Zeitschrift [German Mathematical Journal]

MATIC Multiple Area Technical Information Center

MATICO Machine Applications to Technical Information Center Operations

MATILDA Metering and Totalizing Instrument for Load Demand Assessment

MATized Mechanical Assembly Technique Type (Circuitry) [also matized]

Matl Material [also Mat'l, matl, or mat'l]

MATLAB Matrix Laboratory (Software)

MATLAN Matrix Language

Mat Lett Materials Letters [Published in the Netherlands]

Matls Materials [also matls]

Matl Sci Materials Science; Materials Scientist

MATM Master of Arts in Teaching of Mathematics

m-atm meter-atmosphere [Unit]

Mat Meas Materials Measurement [also mat meas]

Mat Med Pol Materia Medica Polona [Polish Medical Journal]

Mat News Int Materials News International [Journal published in Belgium]

MATO Military Air Traffic Operations

MA-TPM Maritime Administration–Transport Planning Mobilization [of US Department of Transportation]

Mat Phys Materials Physics; Materials Physicist

Mat Plast Materiale Plastice [Romanian Journal on Plastics]

Mat Plast Elast Materie Plastiche ed Elastomeri [Italian Journal on Plastics and Elastomers]

Mat Proc Materials Processing [also mat proc]

MATPS Machine-Aided Technical Processing System

MATR Management Access to Records

MATR Multiple Attenuated Total Reflection

Mat Res Materials Research(er)

Mat Res Symp Proc Materials Research Symposium Proceedings [of Materials Research Society, US]

Matric Matriculation [also matric]

MATRS Miniature Airborne Telemetry Receiving Station

MATS Master of Arts in Teaching of Science

MATS Materials Testing System

MATS Microprocessor-Assisted Testing System

MATS Military Air Transport(ation) Service [of US Air Force]

MATS Multi-Axial (Compression) Testing Machine

MATS Multiple Address Telex Service [Canada]

Mat Sbornik Matematicheskii Sbornik [Mathematical Journal published by the Academy of Sciences of the USSR]

Mat Sci Materials Science; Materials Scientist

Mat Sci Cit Ind Materials Science Citation Index [of Materials Research Society, US]

Mat Sci Div Materials Science Division

Mat Sci Eng Materials Science and Engineering

Mat Sci Eng Materials Science and Engineering [Journal published in Switzerland]

Mat Sci Eng A, Struct Mater, Prop Microstruct Process Materials Science and Engineering A, Structural Materials: Properties, Microstructure and Processing [Journal published in Switzerland]

Mat Sci Eng B, Solid-State Mater Adv Technol Materials Science and Engineering B, Solid-State Materials for Advanced Technology [Journal published in Switzerland]

Mat Sci Eng C Materials Science and Engineering C, Biomimetic Materials, Sensors and Systems [Journal published in Switzerland]

Mat Sci Eng R Materials Science and Engineering R [Journal published in Switzerland]

Mat Sci Forum Materials Science Forum [Published in Switzerland],

Mat Sci Rep Materials Science Reports [Published in the Netherlands]

Mat Sci Res Materials Science Research [Journal published in the US]

Mat Sci Technol Materials Science and Technology

Mat Struct Materials and Structures [Journal published in the UK]

Matsushita Electr Works Tech Rep Matsushita Electric Works Technical Report [Osaka, Japan]

Mat Syn Materials Synthesis [also mat syn]

MATT Missile ASW (Antisubmarine Warfare) Torpedo Target

MATTEK Materials Technology Division [of Council for Scientific and Industrial Research, Pretoria, South Africa]

Mat Tidsskr Matematisk Tidsskrift [Mathematical Journal; published by Mathematisk Forening i København, Copenhagen, Denmark]

MATURTEC Maturing Technological Environment

MATV Master Antenna Television

Mat-wiss Werkst tech Materialwissenschaften und Werkstofftechnik [German Publication on Materials Science and Technology]

MATZ Military Aerodrome Traffic Zone

MATZ Military Airfield Traffic Zone

MAU Marine Amphibious Unit

MAU Medium (or Media) Access Unit

MAU Medium (or Media) Adapter Unit

MAU Memory Access Unit

MAU Million Accounting Units

MAU Multi-Attribute Utility

MAU Multiple Access Unit

MAU Multistation Access Unit

Mau Mauritius

MAUA Master of Arts in Urban Affairs

MAUD Master of Arts in Urban Design

MAud Master of Audiology

MAUDE Morse Automatic Decoder

MAUDEP Metropolitan Association of Urban Designers and Environmental Planners [US]

MAuE Master of Automotive Engineering

MAUFS Municipal Arborists and Urban Foresters Society [US]

MAUK Mining Association of the United Kingdom

Maur Mauritania(n)

Maur Mauritius

Mau Rs Mauritian Rupee [also Mau R] [Currency]

MAURP Master of Arts in Urban and Regional Planning

MAUS Messtechnische Autonome Experimente Unter Schwerelosigkeit [Autonomous Measurement Experiments at Zero Gravity; of DLR Deutsche Forschungsanstalt für Luft– und Raumfahrt]

MAUS Mobile Automated Ultrasonic Scanner

MAUTEL Microminiaturized Autonetics Telemetry

MAV Mastadenovrius

MAV Medium-Armored Vehicle

MAV Municipal Association of Victoria [Australia]

mA/V milliampere per volt [also mA V^{-1}]

MAVAR Microwave Amplification by Variable Reactance [also Mavar, or mavar]

MAVAR Modulating Amplifier using Variable Reactance [also Mavar, or mavar]

MAVART (Mathematical) Model for the Analysis of Vibrations and Acoustic Radiation of Transducers

MAVDM Multiple Application Virtual DOS (Disk-Operating System) Machine

MAVEd Master of Administration in Vocational Education

MAVES Manned Mars and Venus Exploration Studies

MAVICA Magnetic Video Card

MAVIN Machine-Assisted Vendor Information Network

MAW Manual Arc Welding

MAW Marine Aircraft Wing

MAW Master of Arts in Writing

MAW Medium-Activity (Nuclear) Waste

MAW Medium Anti-Armor Weapon

MAW Medium Antitank Assault Weapon

MAW Metal-Arc Welding

MAW Mission-Adaptable Wing (Aircraft)

MAW Mission Adaptive Wing (Aircraft)

MAWP Maximum Allowable Working Pressure

MAWS Minimum Additive Waste Stabilization

MAWS Missile Approach Warning System [of Defense Research Establishment Ottawa, Canada]

MAWS Mobile Aircraft Weighing System

MAX Metal and Explosive (Mixture) [also max]

MAX Mobile Automatic Exchange

MAX Modular Applications Executive

max maximum

MaxAvail Maximum Available (Size of Largest Contiguous Free Block) [Pascal Function]

mAxBa m-Acetoxybenzoic Acid

Max Cap Maximum Capacity [also max cap]

MAXCOM Modular Applications Executive for Communications

MaxEnt Maximum Entropy (for Density of States) [Solid-State Physics]

MAXIT Maximum Interference Threshold

MAXNET Modular Applications Executive Network

Max Wt Maximum Weight [also max wt]

MAYDAY International Distress Signal of Aircraft and Ships [French for "m'aider" – help me] [also Mayday]

MB Bachelor of Medicine

MB Machine Bolt

MB Magnetic Brake; Magnetic Braking

MB Magnetic Bubble [Solid-State Physics]

MB Magnetron Branch

MB Mailer with Bottle

MB Main Battery

MB Main Bus

MB Maintenance Busy

MB Management Baseline

MB Manganese Borate

MB Manitoba [Canada]

MB Many-Body (Theory) [Physics]

MB Marine Biologist; Marine Biology

MB Marine Board [US]

MB Martynov-Batsanov (Electronegativity) [Physical Chemistry]

MB Master Builder

MB Material Balance

MB Mathematical Biologist; Mathematical Biology

MB Maxwell-Boltzmann (Distribution) [Statistical Mechanics]

MB Mechanical Behaviour

MB Mechanical Bond

MB *(Medicinae Baccalaureus)* – Bachelor of Medicine

MB Medium Besa [British Machine Gun]

MB Medium Bomber

MB Megabit [also Mbit, or Mb]

MB Megabyte [also Mbyte]

MB Megaloblast [Medicine]

MB Membrane

MB Memory Buffer

MB Mercedes-Benz

MB Mercury Barometer

MB Messmer and Briant (Model)

MB Metal Boride

MB Metallic Bond

MB Methyl Bromide

MB Methylene Blue

MB Metric Board

MB Miami Beach [Florida, US]

MB Microbalance

MB Microband [Metallurgy]

MB Microbeam

MB Mid-Brain [also mb]

MB Middle Button

MB Missile Battalion

MB Missile Bomber

MB Mixing Box

MB Mobile [also Mb, or mb]

MB Mobile Base (Equipment) [Early Warning]

MB Model Builder

MB Molecular Beam

MB Molecular Biologist; Molecular Biology

MB Monkey Board [Oil Drilling]

MB Monoblock

MB Motor Boat

MB Mountain Battery

MB Moving Bed [Chemical Engineering]

MB Moving Boundary (Reaction) [Chemistry]

MB Multibus

MB Music Background [Programming]

MB1 First Generation Microbands [Metallurgy]

MB2 Second Generation Microbands [Metallurgy]

MB15C5 Methylbenzo-15-Crown-5

MB18C6 Methylbenzo-18-Crown-6

23M14B 2,3-Dimethyl-1,4-Benzoquinone

25M14B 2,5-Dimethyl-1,4-Benzoquinone

26M14B 2,6-Dimethyl-1,4-Benzoquinone

M-B Marrhews and Blakeslee (Theory) [Metallurgy]

M-B Make-Break

Mb Earthquake Magnitude by Body Wave Method [Symbol]

Mb Megabase [Unit of DNA Fragment Length equal to 10^6 Nucleotides]

Mb Megabit [also MB, Mbit, or Mbit]

Mb Million bases (e.g., 2.5 Mb stands for 2.5 million bases) [Molecular Biology]

Mb Myoglobin [Biochemistry]

Mβ M-Beta (Radiation) [also M_β]

mb megabyte [also MB, Mbyte, or mbyte]

mb millibar [also mbar]

mb millibarn [Unit]

μb microbarn [Unit]

MBA Maltobionic Acid

MBA Manufacturered Buildings Association [US]

MBA Marine Biological Association [US and UK]

MBA Mass-Bus Adapter

MBA Master of Business Administration

MBA Material Balance Area

MBA Merion Bluegrass Association [US]

MBA Methoxybenzoic Acid

MBA Methoxybutyl Acetate

MBA 2-Methylbutanoic Acid

MBA β-Methylbutyric Acid

MBA Methyl-Bis-(β-Chloroethyl)amine

MBA Methylenebisacrylamide

MBA Methylene-Blue Absorptiometric (Method) [Biochemistry]

MBA Microbeam Analysis

MBA Modulated Bayard-Alpert (Ionization) Gauge

MBA Molecular-Beam Apparatus

MBA Multiple Beam Antenna

MBA Mung Bean Agglutinin [Immunology]

mBa m-Bromoaniline

MBAA Master Brewers Association of the Americas

MBAA Master of Business Administration in Aviation

MBAAgric Master of Agricultural Business and Administration

MBAC Marshall Booster Assembly Contractor [NASA Marshall Space Flight Center, Huntsville, Alabama, US]

MBAD Maximum Biologically Active Dose

MBAE Master of Biological and Agricultural Engineering

MBAIB Master of Business Administration in International Business

MBAIT Master of Business Administration in International Trade

mbar millibar [also mb]

μbar microbar [Unit]

MBARI Monterey Bay Aquarium Research Institute [California, US]

mbar L millibar liter [also mbar L, or mbar ltr]

mbar L/K millibar liter per kelvin [also mbar L K^{-1}, or mbar ltr K^{-1}]

mbar L/kg·K millibar liter per kilogram kelvin [also mbar L/kg/K, mbar L kg^{-1} K^{-1}, or mbar ltr kg^{-1} K^{-1}]

mbar L/mol·K millibar liter per mole kelvin [also mbar L/mol/K, mbar L mol^{-1} K^{-1}, or mbar ltr mol^{-1} K^{-1}]

mbar L/s millibar liter per second [also mbar L/sec, mbar L s^{-1}, or mbar ltr s^{-1}]

MBAS Methylene Blue Active Substance

MBASIC Microsoft BASIC (Beginner's All-Purpose Symbolic Instruction Code)

MBAUK Marine Biological Association of the United Kingdom

MBB Messerschmitt-Bölkow-Blohm GmbH [Major German Manufacturer of Airplanes, Helicopters, Military Equipment, Space Vehicles, etc.]

MBBA N-(4-Methoxybenzylidene)-4-Butylaniline; N-(p-Methoxybenzylidene)-p-Butylaniline

mBBA m-Bromo Benzoic Acid [also mBbA, or mbba]

MBC Main Beam Clutter [Radar]

MBC Malawi Broadcasting Corporation

MBC Marine Biology Center [of University of California at Santa Barbara, US]

MBC Mass Bias Correction

MBC Master of Building Construction

MBC Maximum Breathing Capacity [Medicine]

MBC Memory Bus Controller

MBC Migratory Bird Convention

MBC Model-Based Control

MBC Multiple Basic Channel

MBCD Modified Binary-Coded Decimal

MB/CD Thousand Barrels per Calender Day [Unit]

MbCO CO (Carbon Monoxide)-Ligated Myoglobin Complex [Biochemistry]

MBCCO Mercury Barium Calcium Copper Oxide (Superconductor)

MBCO Metal Barium Copper Oxide (Superconductor)

MBCS Motion-Base Crew Station [NASA Shuttle Mission Simulator]

MBCS Multi-Byte Character Set

MBD Magnetic Bubble Device

MBD Manual Board [also M BD, or M Bd]

MBD 4-(Methoxybenzylamino)-7-Nitro-2,1,3-Benzooxadiazole

MBD Milling and Baking Division [of American Association of Cereal Chemists, US]

MBD Million Barrels per Day [also mbd]

MBD Minimal Brain Dysfunction

MBD Molecular Beam Diffraction

MBD Molecular Biology Department

MB/D Thousand Barrels per Day [Unit]

M Bd Manual Board [also M BD]

mbd million barrels per day [also MBD]

MBDA Metal Building Dealers Association [now Systems Builders Association, US]

MBDA Minority Business Development Agency [of US Department of Commerce]

MBDAACC Milling and Baking Division of American Association of Cereal Chemists [US]

M bd ft Thousand Board Feet [also MBF, or MBM]

MBDS Modular Building Distribution System

MBE Many-Body Expansion [Physics]

MBE Master of Business Economics

MBE Master of Business Education

MBE Member of the British Empire

MBE 2-Methyl 3-Butene-2-ol

MBE Methyl tert-Butyl Ether

MBE Missile-Borne Equipment

MBE Molecular Beam Epitaxy

MBE Moving-Boundary Electrophoresis

MBED Micro-Beam Electron Diffraction

MBEd Master of Business Education

MBE-Ge Molecular Beam Epitaxial Grown Germanium

MBE/MOCVD Molecular Beam Epitaxy/Metallorganic Chermical Vapor Deposition

MBE/MOMBE Molecular Beam Epitaxy/Metallorganic Molecular Beam Epitaxy

MBEO Minority Business Enterprise Office

MBER Molecular Beam Electric Resonance

MBES Model-Based Expert System

MBE-Si Molecular-Beam Epitaxial-Grown Silicon

MBF Materials Business File [of Materials Information, ASM International, US]

MBF Metglas Brazing Foil

MBF Molecular Beam Facility

MBF Thousand Board Feet [also MBM, or M bd ft]

MBF 2,3,3a,4,5,6,7,7a-Octahydro-7,8,8-Trimethyl-4,7-Methanobenzofuran-2-yl [also Mbf]

MBFP Manufacturing Build and Flow Plan

MBFR Mutual Balanced Forces Reduction (Project) [of NATO]

MBG Metal Burner Gas

MBG Missouri Botanical Garden [St. Louis, US]

MBG Montreal Botanical Gardens [Canada]

MBGA Mini Ball Grid Array

MBH Thousand BTU (British Thermal Units) per Hour [also mbh]

MBHA 4-Methylbenzhydrylamine

MBTH 3-Methyl-2-Benzothiazolinone Hydrazone Hydrochloride Monohydrate

MBTH-S 3-Methyl-2-Benzo-(2'-Sulfo)thiazolino Hydrazone, Potassium Salt

MBI Max-Born-Institut (für Nichtlineare Optik und Kurzzeitspektroskopie) [Max Born Institute (for Nonlinear Optics and Short-Time Spectroscopy), Berlin, Germany]

MBI Memory Bank Interface

MBI 2-Mercaptobenzimidazole

MBI Meson-Baryon Interaction [Particle Physics]

MBI 2-Methyl-1,3-Butadiene Isoprene

MBI 2-Methyl 3-Butyne-2-ol

MBI Multibeam Image; Multibeam Imaging

MBI Multibeam Interferometer; Multibeam Interferometry

MBI Multi-Bus Interface

MBI Multiple-Beam Interference

MBI Multiple-Beam Interferometer; Multiple-Beam Interferometry

MBIAC Missouri Basin Inter-Agency Committee [US]

MBiChem Master of Biological Chemistry

MBiEng Master of Biological Engineering

MBIM Member of the British Institute of Management

MBIO Microprogrammable Block Input/Output

MBiomedE Master of Biomedical Engineering

MBiPhy Master of Biological Physics

MBiS Master of Biological Science

Mbit Megabit [also Mb]

MBITS Monitored Burn-In Test System

Mbit/s Megabit(s) per Second [also Mbit s^{-1}, Mbit/sec, MBS, MBPS, Mbps, or Mb/s]

μbit microbit [also mubit]

MBK Medications and Bandage Kit

MBK Multiple Beam Klystron [Electronics]

MBL Marine Biological Laboratory [Woods Hole, Massachusetts, US]

MBL Marine Biology Laboratory [of University of Delaware; located at Roosevelt Inlet, US]

MBL Marine Boundary Layer

MBL Mobile Unit [Tactical Air Navigation Station]

Mbl Marble [also mbl]

mbl mobile

MBLCC Marine Boundary-Layer Convective Complexes

MBldg Master of Building Construction

MBM Magnetic Bubble Memory

MBM Master of Business Management

MBM Metal-Barrier-Metal

MBM Molecular Beam Maser

MBM Molecular Beam Measurements

MBM Multipurpose Boring Machine

MBM Thousand Feet Board Measure [also Mbm, MBF, or M bd ft]

mb/meV/sr/fu millibarn(s) per millielectron-volt/steradian/formula unit [also mb meV^{-1} sr^{-1} fu^{-1}]

MBMA Master Boiler Makers Association [US]

MBMA Metal Building Manufacturers Association [US]

MBMC Monobutyl-M-Cresol

MBMG Montana Bureau of Mines and Geology [US]

MBMS Modulated Beam Mass Spectroscopy

MBMS Molecular Beam Mass Spectroscopy

MBN Magnetic Barkhausen Noise [Physics]

M-B-N Metal-Boron-Nitride (System)

MBO Magnetic Blow-Out

MBO Management Buy-Out

MBO Management by Objective(s)

MBO Manpower Branch Offices [Canada]

MBO Million Barrels of Oil

MBO Mulliken Bond Order [Physics]

Mbo Moraxella bovis [Microbiology]

Mbo I Moraxella bovis I [Microbiology]

Mbo II Moraxella bovis II [Microbiology]

MBOH Minimum Break-Off Height [Aerospace]

MBONE Multicast Backbone [also mbone] [Internet]

MBOS Multi-User Business Operating System

MBOTAD 4-(4-[6-Methoxy-2-Benzoxazolyl]phenyl)-1,2,4-Triazolidine-3,5-Dione

MBP Maltose Binding Protein [Biochemistry]

MBP Many-Body Problem [Mechanics]

MBP Mechanical Balance Package

MBP Mid-Boiling Point

MBP Monobutylphosphate

MBP Myelin Basic Protein [Biochemistry]

mBP m-Bromophenol

MBPA α-Methyl-N,N-Bis(2-Picolinyl) 2-Picolinamine [also mbpa]

MBPA Mono-n-Butylphosphoric Acid

MBPD Million Barrels per Day [also mbpd]

mBPIP m-Bromophenol Indophenol

MBPS Megabit(s) per Second [also Mbit s^{-1}, Mbit/s, MBS, Mb/s, or Mbps]

MBPS Megabyte(s) per Second [also MBps, Mbyte(s)/s, or MB/s]

MBPS Milk and Beef Production System

MBPS Million Bits per Second [also MBS, or Mbps]

MBPT Many-Body Perturbation Theory [Physics]

MBQ Modified Binary Code

235MBQ 2,3,5-Trimethyl-1,4-Benzoquinone

2M14BQ 2-Methyl-1,4-Benzoquinone

MBq Megabecquerel [Unit]

MBq/μm Megabecquerel(s) per micrometer [MBq μm^{-1}]

MBQT Many Body Quantum Theory [Physics]

MBR Marker Beacon Receiver

MBR Mars Balloon Relay [of NASA]

MBR Master Boot Record

MBR Material Balance Report

MBR Memory Base Register

MBR Memory Buffer Register

MBr$_x$ Metal Bromide [General Formula]

Mbr Member [also mbr]

MB RAM Megabyte(s) of Random-Access Memory [also MB-RAM]

MBRE Memory Buffer Register, Even [also MBR-E]

MBRL Million Barrels (of Oil) [also mbrl]

MBRO Memory Buffer Register, Odd [also MBR-O]

MBRS Molecular Beam Reactive Scattering

MBRS Molecular Beam Relaxation Spectrometry

MB-RSPT Many-Body Rayleigh-Schroedinger Perturbation Theory [Physics]

MBRV Maneuverable Ballistic Reentry Vehicle

MBS 1-Benzothiazole-N-Sulfene Morpholide

MBS Magnetic Bubble Storage

MBS Magnetron Beam Switching

MBS Maleimidobenzoyl-N-Hydroxysuccinimide

MBS m-Maleimidobenzoic Acid N-Hydroxysuccinimide Ester

MBS Maleimidobutyric Succinimide

MBS Manchester Business School [of University of Manchester, UK]

MBS Master of Basic Science

MBS Master of Building Science

MBS Master of Business Studies

MBS Megabit per Second [also MBPS, Mbps, Mb/s, Mbit/s, or Mbit s^{-1}]

MBS Methacrylate-Butadiene-Styrene; Methyl Methacrylate Butadiene Styrene (Terpolymer)

MBS Minimal Basis Set

MBS Mobile Base System

MBS Molecular Beam Scattering

MBS Monumental Brass Society [UK]

MBS Multi-Band Superconductivity

MBS Multi-Block Synchronization

MBS Multiple Batch Station

MBS Mutual Broadcasting System [US]

Mb/s Megabits per Second [also MBPS, Mbps, MBS, or Mbit s^{-1}, or Mbit/s]

MBSA Methylated Bovine Serum Albumin [Biochemistry]

MBSc Master of Business Sciences

MBSHC Mediterranean-Black Sea Hydrographic Commission

MBSI Master of Business Science and Information Management

MBSM Master of Building Science Management

MBS-PVC Methacrylate-Butadiene-Styrene/Polyvinyl Chloride (Blend)

MBSS Molecular Beam Surface Scattering

MBSS 2-Morpholinodithio Benzothiazole

MBT Main Battle Tank

MBT Meltback Transistor

MBT 2-Mercaptobenzothiazole

MBT Metal-Base Transistor

MBTFA N-Methyl-bis(trifluoroacetamide)

MBTH 3-Methyl-2-Benzothiazolinone Hydrazone

MBTS 2-Mercaptobenzothiazole Disulfide

MBTS Mercaptobenzothiazyl Disulfide

MBTS Moving Belt Transfer System

MBTU Million British Thermal Units [also MBtu, or Mbtu]

MBU Magnetic Bubble Unit

MBU Memory Buffer Unit

MBusEd Master of Business Education

MBV Main Base Visit

MBV Modified Bauer-Vogel (Oxide Conversion Coating Process)

MBV Mushroom Bacilliform Virus

MBVS Model-Based Vision System

MBWO Microwave Backward-Wave Oscillator

MBWP Memory-Based Word Processor

MBX Mailbox [also Mbx]

MBX (International Conference on) Many-Body Theories

Mbyte(s) Megabyte(s) [also MB]

Mbyte(s)/s Megabytes per second [also Mbyte(s)/sec]

MBZ Mandatory Broadcast Zone

MC Machinable Carbide

MC Machinable Ceramics

MC Machine Carbine

MC Machine Code

MC Machine Console

MC Machine Cycle

MC Machining Center

MC Macromolecular Chemistry

MC Macrocyte [Anatomy]

MC Madelung Constant [Solid-State Physics]

MC Madison College [Harrisonburg, Virginia, US]

MC Magellanic Cloud [Astronomy]

MC *(Magister Chirurgiae)* – Master of Surgery

MC Magnesium Chromite

MC Magnetic Card

MC Magnetic Circuit

MC Magnetic Clutch

MC Magnetic Coating

MC Magnetic Coil

MC Magnetic Compass

MC Magnetic Confinement [Plasma Physics]

MC Magnetic Cooling

MC Magnetic Core

MC Magnetic Course

MC Magnetic Cyclone

MC Magnetocaloric

MC Magnetochemical; Magnetochemist(ry)

MC Magnetoconductance

MC Main Circuit

MC Main Channel

MC Main Chamber

MC Main Constituent

MC Main Contact

MC Main Current
MC Major Component
MC Making-Current
MC Manganese Center [France]
MC Manhattan College [of City University of New York, US]
MC Manifold Control
MC Manned Control
MC Manual Code
MC Manual Control
MC Manufacturing Cell
MC Manufacturing Cluster
MC Mapping Camera
MC Marginal Checking [Electronics]
MC Marginal Cost
MC Marine Corps [US]
MC Marine Corrosion
MC Maritime College [of State University of New York in Bronx, US]
MC Maritime Commission [US]
MC Markov Chain [Mathematics]
MC Mast Cell [Medicine]
MC Master Catalog
MC Master Clock
MC Master Control
MC Master Cylinder
MC Master of Counseling
MC Master of Communication
MC Materials Characterization
MC Materials Chemist(ry)
MC Materials Conservation
MC Materials Control [Industrial Engineering]
MC Materials Cycle
MC Mathematical Crystallography
MC Matrix Calculus
MC Matrix Column
MC Mean Calorie
MC Mean Curvature
MC Mechanical Classification; Mechanical Classifier
MC Mechanochemical; Mechanochemist(ry)
MC Medical Center
MC Medical Chemist(ry)
MC Medical Corps
MC Medical Cybernetics
MC Medium Capacity
MC Medium Curing
MC Megalocyte
MC Melter Cell
MC Member of Council
MC Memory Cell
MC Memory Chip
MC Memory Circuit
MC Memory Configuration
MC Memory Control(ler)
MC Menlo College [Menlo Park, California, US]

MC Menigococcal; Menigococcus [Microbiology]
MC Mentor Current (of Pacific Ocean) [Oceanography]
MC Metacarpal; Metacarpus [Anatomy]
MC Metacenter [Fluid Mechanics]
MC Metal Carbide
MC Metal Ceramic(s)
MC Metal-Clad (Cable)
MC Metal(lic) Composite
MC Metal Cutting
MC Metal(lic) Coating
MC Metallic Conduction; Metallic Conductor
MC Metallized Capacitor
MC Metallized Ceramic(s)
MC (International) Metallography Conference [of ASM International/International Metallographic Society, US]
MC Metallurgical Coke
MC Meter-Candle [Unit]
MC Meter Change
MC Methoxychlor
MC Methylcellulose
MC Methylene Chloride
MC Methyl Chloride
MC Metric Carat [Unit]
MC Mexico City
MC Michael Condensation [Organic Chemistry]
MC Microbial Corrosion
MC Microcalorimeter; Microcalorimetry
MC Microcell(ular) [Telecommunications]
MC Microchemical; Microchemistry
MC Microchip
MC Microcircuit
MC Microcluster [Chemistry]
MC Microcode
MC Microcomputer [also μC]
MC Microcontroller
MC Microcrack(ing)
MC Microcyte [Anatomy]
MC Microscope Camera
MC Middlebury College [Vermont, US]
MC Military Committee [of NATO]
MC Military Computer
MC Military Cross
MC Milling Cutter
MC Mineral Chemist(ry)
MC Minicomputer
MC Mining College
MC Miscellaneous (Steel) Channel [AISI/AISC Designation]
MC Missile Control
MC Mission Capability
MC Mission Completion
MC Mission Continuation
MC Mission Control
MC Mixed Culture [Microbiology]
MC Mixing Chamber

MC Mobilization Cost

MC Modified Carbonate

MC Modified Cordite [British Explosive]

MC Modulation-Compensation (System) [also M-C]

MC Module Control

MC Moisture Content

MC Molded Component

MC Molding Compound

MC Molding Cycle

MC Molecular Conductivity

MC Molten Carbonate

MC Momentary Contact

MC Monaco [ISO Code]

MC Monitor Call

MC Monitoring Controller

MC Monmouth College [West Long Branch, New Jersey, US]

MC Monochlorination

MC Monolithic Circuit

MC Monte Carlo [Monaco]

MC Monte Carlo (Method) [Statistics]

MC Morningside College [Sioux City, Iowa, US]

MC Morse Code

MC Motion Control(ler)

MC Motor Car

MC Motor Carrier

MC Motor Control(ler)

MC Motor Converter

MC Motorcycle

MC Mott-Cabrera (Theory) [Physics]

MC Moving Coil

MC Multichannel (Plates)

MC Multichamber

MC Multichip

MC Multichromatic

MC Multicomponent

MC Multiconfiguration

MC Multiple Choice

MC Multiple Contact

MC Multiple Copy

MC Municipal Corporation

MC Munitions Command

MC Muon Collider [Particle Physics]

MC Muzzle Cap

M_xC_{60} Metallofullerene with 60 Carbon Atoms [General Formula]

M_xC_{70} Metallofullerene with 70 Carbon Atoms [General Formula]

M_xC_{80} Metallofullerene with 80 Carbon Atoms [General Formula]

M_2C_{60} Divalent Metal Carbide Buckminsterfullerene [Formula]

M_3C_{60} Trivalent Metal Carbide Buckminsterfullerene [Formula]

M_4C_{60} Tetravalent Metal Carbide Buckminsterfullerene [Formula]

M_6C_{60} Hexavalent Metal Carbide Buckminsterfullerene [Formula]

M_xC_y Metal Carbide [General Formula]

M&C Maintenance and Checkout

M&C Manufacturing and Control

M&C Metals and Ceramics Division [of Oak Ridge National Laboratory, Tennessee, US]

M&C Monitor and Control

M-C Medium-Curing (Asphalt)

M-C Modulation-Compensation (System) [also MC]

$M_{\parallel}c$ Magnetization In-Plane with (or Parallel to) the c-Axis

$M_{\perp}c$ Magnetization Out-of-Plane (or Perpendicular) the c-Axis

MC2 Manufacturing Cluster 2 [of Europractice]

MC-21 Metallic Composites for the 21st Century, Inc. [US]

Mc Megacycle [Uniy]

mC m-Cresol

mC milli-Celsius [Unit]

mC millicoulomb [Unit]

mC Pearson symbol for base-centered space lattice in monoclinic crystal system (this symbol is followed by the number of atoms per unit cell, e.g., mC12, mC28, etc.)

mc (mensis currentis) – this month

mc millicurie [also mC]

mc millicycle [Unit]

.mc Monaco [Country Code/Domain Name]

μC microcomputer [also MC]

μC microcoulomb [also muC]

μc microcrystalline

μc microcurie [Unit]

MCA Macrocell Array

MCA Magnetocrystalline Anisotropy [Solid-State Physics]

MCA Malaysian Commercial Association

MCA Manufacturing Chemists' Association [US]

MCA Marine Corps Association [US]

MCA Master Control Assembly

MCA Master of Communication Arts

MCA Material Control and Accountability

MCA Material Coordinating Agency

MCA Maternity Center Association [US]

MCA Maximum Credible Accident [Nuclear Reactors]

MCA Maximum Crossing Altitude

MCA Mechanical Contractors Association

MCA Metal Chelate Affinity

MCA Metal Construction Association [US]

MCA Methylcholanthrene

MCA Microcanonical Assembly [Statistical Mechanics]

MCA Micro-Channel Adapter

MCA Micro-Channel Architecture (System) [Computers]

MCA Middle Cerebral Artery [Anatomy]

MCA Military Coordinating Activity

MCA Minimum Crossing Altitude

MCA Mixed Carbonic Anhydride

MCA Model Cities Agency [US]

MCA Modified Clenshaw Algorithm

MCA Modified Conventional Alloy

MCA Momordica Charantia Agglutinin [Immunology]

MCA Monitoring and Control Assembly

MCA Monochloroacetic Acid

MCA Monoclonal Antibody [Immunology]

MCA Motor Control Assembly

MCA Multichannel Analyzer

MCA Multicomponent Alloy

MCA Multicomponent Analysis

MCA Multi-Criteria Analysis

MCA Multiple Crevice Assembly

MCA Multiprocessor Communications Adapter

MCA Music Corporation of America [US]

MC&A Material Control and Accountability

mCa m-Chloroaniline [also mCA]

MCAA Marine Corps Aviation Association [US]

MCAA Mason Contractors Association of America [US]

MCAA Mechanical Contractors Association of America [Chevy Chase, Maryland, US]

MCAA Monochloroacetic Acid

MCAC Metal Chelate Affinity Chromatography

MCAD Mechanical Computer-Aided Design

MCAD Mechanical Computer-Aided Design/Drafting

MCAD Mechanical Computer-Aided Drafting

MCAD Minneapolis College of Arts and Design [US]

MCAE Mechanical Computer-Aided Engineering

MCAF Marine Corps Air Facility [of US Marine Corps]

MCAF Mauritius Cooperative Agricultural Federation

MCAF Monocyte Chemotactic and Activitating Factor [Biochemistry]

MCAI Microcomputer-Assisted Instruction

Mcal Megacalorie(s) [Unit]

Mcal/mthm Megacalories per metric ton of hot metal [Unit]

μc-Al Microcrystalline Aluminum

μc-Al$_2$O$_3$ Microcrystalline Alumina

MCALS Minnesota Computer-Aided Library System [of University of Minnesota, Minneapolis, US]

MCAMC Monte Carlo with Absorbing Markov Chains (Method) [Mathematics]

MCANW Medical Campaign Against Nuclear Weapons [UK]

MCAP Material Control and Accountability Plan

MCAR Machine Check Analysis and Recording

MCAS Marine Corps Air Station [of US Marine Corps]

MCAS Micro-Controlled Airflow System

MCASI Member of the Canadian Aeronautics and Space Institute

MCAT Medical College Admissions Test

McAUTO McDonnell Douglas Automation Company [St. Louis, Missouri, US]

MCB Master Car Builder

MCB Medium-Capacity Bomb

MCB Memory Control Block

MCB Microcomputer Board

MCB Miniature Circuit Breaker

MCB Molecular and Cellular Biology [of American Society for Microbiology, US]

MCB Monochloroborazine

MCBA Magnesite and Chrome Brickmakers Association [UK]

mCBA m-Chlorobenzoic Acid

MCBF Mean Cycles between Failures

MCBIC Michigan/Canadian Bigfoot Information Center [US]

MCC Magnetic Crystal Class

MCC Main Combustion Chamber

MCC Main Communications Center

MCC Main Control Circuit

MCC Maintenance Control Center

MCC Maintenance Control Circuit

MCC Malawi Correspondence College

MCC Manchester Conference Center [UK]

MCC Management Computer Control

MCC Management Control Center

MCC Manned Control Car [of US Atomic Energy Commission]

MCC Master Control Center

MCC Master of Clinical Chemistry

MCC Materials Characterization Center [of Pacific Northwest National Laboratory, Richland, Washington, US]

MCC Materials Characterization Center (Test)

MCC Member Conduct Committee [of Institute of Electrical and Electronics Engineers, US]

MCC Mesoscale Cellular Convection [Meteorology]

MCC Mesoscale Convective Complex [Meteorology]

MCC Metal/Ceramic Composite

MCC Metal Chemistry Conference [Part of Canadian Materials Science Conference]

MCC Method of Cluster Components

MCC Metric Commission of Canada

MCC Microcomputer Control

MCC Microelectronics and Computer Technology Corporation [Austin, Texas, US]

MCC Microcrystalline Cellulose

MCC Miniature Chip Carrier

MCC Mini-Cylinder Core

MCC Miscellaneous Common Carrier

MCC Missile Control Center

MCC Mission Control Center [of NASA Johnson Space Center, Houston, Texas, US]

MCC Mission Control Center [of US Department of Defense]

MCC Missouri Conservation Commission [US]

MCC Mixing Cross-Bar Connectors

MCC Molecular Coupled Cluster

MCC Money Market Certificate

MCC Montana Chamber of Commerce [US]

MCC Motor Control Center

MCC Multichannel Carrier

MCC Multichannel Communications

MCC Multichannel Communications Control

MCC Multi-Chip Circuit

MCC Multicomponent Circuit

MCC Multicore Cable

MCC Multi-Crossover Cryotron

MCC Multiple Chip Carrier

MCC Multiple Computer Complex

MCC Munitions Carriers Conference [US]

MCCA Mercado Común Centroamericano [Common Central American Market]

MCCB Molded-Case Circuit Breaker

MCCC Mission Control and Computing Center

MCCD Multiplexed Charge-Coupled Device

MCC-DOD Mission Control Center–Department of Defense [US]

MCCGL Message Conveying Computers General License [UK]

MCC-H Mission Control Center–Houston [of NASA Johnson Space Center, Texas, US]

MCCHF Multiconfiguration-Coupled Hartree-Fock (Method) [Physics]

MCCI Manchester Chamber of Commerce and Industry [UK]

MCCI Molten Core–Concrete Interaction [Nuclear Reactors]

MCCIS Military Command, Control and Information System

MCCISWG Military Command, Control and Information Systems Working Group [of NATO]

MCC-K Mission Control Center-Kennedy [of NASA Kennedy Space Center, Florida, US]

μ**C/cm²** microcoulomb(s) per square centimeter [also μC cm^{-2}]

MCCN Marine and Coastal Community Network

MCC-NASA Mission Control Center-National Aeronautics and Space Administration [US]

μ**c-Co** Microcrystalline Cobalt

MCCP Mountain Cloud Chemistry Project [US]

MCCR Ministry of Consumer and Commercial Relations [Canada]

MCCS Military Command and Control System

MCCS Mine Countermeasures Control System

MCCS Mission Control Center Simulation

MCCS Monte Carlo Computer Simulation

MCCS Multichannel Carrier System

MCCU Multichannel Control Unit

MCCU Multiple Communications Control Unit

MCD Magnetic Circular Dichroism [Physics]

MCD Magnetic Crack Detection

MCD Manipulative Communications Deception

MCD Mast Cell Degranulating (Peptide) [Biochemistry]

MCD Master of Civic Design

MCD Materials and Components Division [of Argonne National Laboratory, Illinois, US] [Formerly part of the Materials Science and Technology Division]

MCD Measurement Control and Display

MCD Metals and Ceramics Division [of US Air Force Materials Directorate, Wright Laboratory, Dayton, Ohio] [also M&CD]

MCD Metals and Ceramics Division [of Oak Ridge National Laboratory, Tennessee, US]

MCD Microcoulometric Detector

MCD Minimum Charge Duration

MCD Monte Carlo Dynamics

MCD Months for Cyclical Dominance

MCD Multichannel Detector

MCD Multicomponent Distillation

MCD Multiple Concrete Duct

mcd millicurie-day [also mCd]

mcd millicurie-destroyed [also mCδ]

MCDA Magnetic Circular Dichroism of Absorption [Physics]

MCDBSU Master Control and Data Buffer Storage Unit

MCDD Marine and Coastal Data Directory (of Australia) [Also known as the "Blue Pages"]

MCDF Multiconfiguration Dirac-Fock [Physics]

MCDP Microprogrammed Communication Data Processor

MCDS Maintenance Control and Display System

MCDS Management Control Data System

MCDS Metaconta Data System

MCDS Microcomputer Development System

MCDS Multifunction CRT (Cathode-Ray Tube) Display System

MCDU Multifunction CRT (Cathode-Ray Tube) Display Unit

MCDW Matrix Continuum Distorted Wave (Model) [Physics]

MCDW Multichannel Distorted Wave (Model)

MCE Magnetocaloric Effect

MCE Malawi Certificate of Education

MCE Marshall, Cox and Evans (Crack Bridging Model) [Materials Science]

MCE Master of Civil Engineering

MCE Mapping and Charting Establishment [UK]

MCE Mechanocaloric Effect

MCE Memphis Cotton Exchange [US]

MCE Mercury Cathode Electrolysis

MCE Microcircuit Engineer(ing)

MCE Modular Clean Environment

MCE Montgomery Cotton Exchange [Alabama, US]

MCEB Military Communications Electronics Board [US]

MCEd Master of Continuing Education

MCEI Modulus of Complete Elliptic Integral

MCEL Machine Check Extended Log-Out

MCEng Master of Civil Engineering

MCerE Master of Ceramic Engineering

MCES Member of the Civil Engineering Society

MCESS Milwaukee Council of Engineering and Scientific Societies [Wisconsin, US]

MCET Microcomputer Engineering Technologist; Microcomputer Engineering Technology

MCEWG Military Communications Electronics Working Group [of NATO]

MCF Maintenance and Checkout Facility

MCF Malignant Catarrhal Fever [Veterinary Medicine]

MCF Mean Carrier Frequency

MCF Metal-Coated Fiber

MCF Micro-Complement Fixation [Immunology]

MCF Military Computer Family

MCF Million Cubic Feet [also mcf]

MCF Mission Control Facility

MCF Monolithic Crystal Filter

MCF Multicomponent Flow

MCF Muon Catalyzed Fusion [also μCF]

MCF Multiple Component Facilities

MCF Mutual Coherence Function

MCF Thousand Cubic Feet [also Mcf]

μCF Muon Catalyzed Fusion [also MCF]

MCFC Molten Carbonate Fuel Cell

MCFD Million Cubic Feet per Day [also MCF/D, or mcf/d]

MC/FE Monte Carlo/Finite Element (Approach) [Materials Science]

MCFET Monte Carlo Finite-Element Technique

MCFET Multichannel Field-Effect Transistor

MCFFA Ministerial Council on Forestry, Fisheries and Aquaculture [Australia]

MCFH Million Cubic Feet per Hour [also MCF/h, or mcf/h]

MCFR Magnetic Confinement Fusion Reactor

MCFS Mercury Cadmium Iron Selenide (Semiconductor)

MCG Magnetocardiogram; Magnetocardiographer; Magnetocardiography

MCG Man-Computer Graphics

MCG Master-Clock Generator

MCG Materials Coordinating Group

MCG Medical College of Georgia [Augusta, US]

MCG Metal-Coated Graphite

MCG Microwave Command Guidance

MCG Mobile Command Guidance

mcg microgram [also μg]

MCGF Mast-Cell Growth Factor [Biochemistry]

MCGA Multicolor Graphics Array

MCGAHi Multicolor Graphics Array, High Mode

MCGAMed Multicolor Graphics Array, Medium Mode

μC-Ge Microcrystalline Germanium

MCGS Microwave Command Guidance System

McGU McGill University [Montreal, Quebec, Canada]

McG-Q Press McGill-Queen's Press [of McGill University, Montreal and Queen's University, Kingston, Canada]

MCH Machine Check Handler

MCH Magnesium Chloride Hexahydrate

MCH Maternal and Child Health

MCH Mean Corpuscular Hemoglobin

MCH 5-Methyl-1,3-Cyclohexadiene

MCH Methylcyclohexane

MCH Methylenecyclohexane

MCH Microcomputer Hierarchy

MCH Monochlorhydrin

MCH Moveable Cultural Heritage

MCh (Magister Chirurgiae) – Master of Surgery

Mch Metal Chelon [Chemistry]

Mch Machine [also mch]

mch millicurie-hour [also mchr, mCh, mCi/h, mCi/hr, or mCi h^{-1}]

μ-ch mu-chain specific (immunoglobulin) [Immunochemistry]

MCHA Methylcyclohexyladipate

MCHA Moveable Cultural Heritage Act [Australia]

MChA Master of Applied Chemistry

MCHC Mean Cellular Hemoglobin Concentration

MChE Master of Chemical Engineering

MCHF Marine Corps Historical Foundation [US]

MCHF Multiconfiguration Hartree-Fock [Physics]

MCHFR Minimum Critical Heat Flux Rate

MCHMPCOSY Multibond Coupling Optimized Heteronuclear Multispin Coherence H-Detected COSY (Correlated Spectroscopy)

MCHR Minnesota Center for Health and Rehabilitation [Minneapolis, US]

mchr millicurie-hour [also mch, mCh, mCi/h, mCi/hr, or mCi h^{-1}]

M Chs Thousands of Characters [also M chs]

MCI Machine Check Interrupt

MCI Master of Clinical Immunology

MCI Malicious Call Identification

MCI Malleable Cast Iron [Metallurgy]

MCI Media Control Interface

MCI Microwave Communications, Inc. [now MCI Communications Corporation, US]

MCI Ministère du Commerce et de l'Industrie [Ministry of Industry and Commerce, France]

MCi Megacurie [also Mci]

mCi millicurie [also mci] [Unit]

μCi microcurie [also muCi] [Unit]

MCIA Methyl Chloride Industry Association [US]

MCIA Microcomputer Investors Association [US]

MCIA Mirror Class International Association

MCIA Motor Cycle Industry Association [UK]

MCIC Machine Check Interruption Code

MCIC Member of the Chemical Institute of Canada

MCIC Metals and Ceramics Information Center [of ASM International, US]

MCIC Multichip Integrated Circuit

MCIC Rep Metals and Ceramics Information Center Report [of ASM International, US]

MCID Multipurpose Concealed Intrusion Detector

McIDAS Man-computer Interactive Data Access System [also MCIDAS]

McIDAS-X Man-computer Interactive Data Access System for Unix (Operating System)

MCIF Member of the Canadian Institute of Forestry

mCi/h millicurie-hour [also mch, mchr, mCi/hr, or mCi h^{-1}]

mCi/hr millicurie-hour [also mch, mchr, mCi/h, or mCi hr^{-1}]

MCIM Master of Clinical Immunology and Microbiology

MCIM Member of the Canadian Institute of Mining and Metallurgy [also MCIMM]

mCi/m³ millicurie(s) per cubic meter [also mCi/cu m, or mCi m^{-3}]

μCi/m³ millicurie(s) per cubic meter [also μCi/cu m, or μCi m^{-3}]

mCi/mL millicurie(s) per milliliter [also mCi mL^{-1}]

mCi/mmol millicurie(s) per millimole [also mCi mmol^{-1}, mCi/mM, or mCi mM^{-1}]

Mcin 4-Methyl-8-Cinnolinol

MCIP Member of the Canadian Institute of Planners

MCIS Master of Computer and Information Science

MCIS Maintenance Control Information System

MCIS Materials Compatibility in Sodium

MCIU Manipulator Controller Interface Unit

MCIU Master Control and Interface Unit

MCIU Mission Control and Interface Unit

MCIVO Multiconfigurational Improved Virtual Orbital (Method) [Physics]

MCJ Master of Communication and Journalism

MCK Multicavity Klystron [Electronics]

MCL Magnetic Corrected Lamp

MCL Manufacturing Control Language

MCL Marine Corps League [US]

MCL Mass Change Log

MCL Master Configuration List

MCL Materials and Corrosion Laboratory [South Korea]

MCL Materials Characterization Laboratory [Scotia, New York State, US]

MCL Materials Characterization Laboratory [of Motorola Inc., Mesa, Arizona, US]

MCL Mathematics Computation Laboratory [US]

MCL Maximum Concentration Limit

MCL Maximum Contaminant Level; Maximum Contamination Level

MCL Maximum Crack Length

MCL Memory Control and Logging

MCL Microcathodoluminescence; Microcathodoluminescent

MCL Microprogram Control Logic

MCL Microsoft Compatibility Laboratories [US]

MCL Mid-Canada Line [Radar]

MCl$_x$ Metal Chloride [Formula]

MCLA Micro-Coded Communications Line Adapter

MCLA Monetary Centre for Latin America

MCLG Maximum Contaminant Level Goal(s); Maximum Contamination Level Goal(s)

MCLK Master Clock

MCLOS Manual Command Line-of-Sight

MCLR Multiconfigurational Linear Response

MCLT Master of Clinical Laboratory Technology

MCLT Maximum Cruise Level Thrust

MCM Machine for Coordinating Multiprocessing

MCM MADS (Modular Auxiliary Data System) Control Module

MCM Magnetic Core Memory

MCM Maintenance Control Module

MCM Markov Chain Method [Mathematics]

MCM Master of Construction Management

MCM Memory Control Module

MCM Menigococcal Meningitis [Medicine]

MCM Microwave Circuit Module

MCM Million Cubic Meters

MCM Mine Countermeasures

MCM Ministerial Council Meeting [of Organization for Economic Cooperation and Development]

MCM Monte Carlo Method

MCM Moving-Coil Motor

MCM Multicavity Magnetron

MCM Multichip Module

MCM Thousand Circular Mils [also Mcm]

MCM Thousand Cubic Meters

mC/m^2 millicoulomb per square meter [also mC m^{-2}]

m^2/cm^3 square meter(s) per cubic centimeter [also m^2 cm^{-3}]

MCMA Methoxycarbonylmethylamine

MCMC Mid-Continent Mapping Center [of US Geological Survey]

MCMC Multichip Microcircuit

MC/MD Monte Carlo/Molecular Dynamics (Technique)

MCMES Member of the Civil and Mechanical Engineering Society

MCMG Man-Carrying Motion Generator

MCMG Military Committee Meteorological Group [of NATO]

MCMM Management Control–Material Management

MCMST Montana College of Mineral Science and Technology [Butte, US]

MCMV Mine Countermeasure Vessel

MCMV Maize Chlorotic Mottle Virus

MCMV Mouse Cytomegalovirus

MCMV-1 Mouse Cytomegalovirus, Type 1

MCN Master Change Notice

MCN Museum Computer Network [US]

[M(CN)$_x$]$^{-x/2}$ Metal Cyanate [Note: x Indicates the Coordination Nnumber of the Metal Ion, e.g., [Ni(CN)$_4$]$^{2-}$ Denotes the Tetracyanonickelate(II) Ion]

McN A-343 4-(N-[3-Chlorophenyl]carbamoyloxy)-2-Butynyltrimethylammonium

mCNa m-Cyanoaniline

mCNBA m-Cyanobenzoic Acid

MCNC Microelectronics Center of North Carolina [of Center for Microelectronics, Research Triangle Park, US]

MCNE Master Certified Novell Engineer

μc-Ni Microcrystalline Nickel

MCNP Monte Carlo N-Particle Transport Code [Physics]

mCNP m-Cyanophenol

MCO Magnetic-Field Controlled Oscillator

MCO Maximum Concentration of Organics

MCO Medical College of Ohio [Toledo, US]

MCO Mill Culls Out [Building Trades]

MCO Mission Control Operations

MCO Molding and Cost Optimization

MCO Multiple Criteria Optimization

M(CO)$_x$ Metal Carbonyl [Formula] [*Note:* x Indicates the Coordination Number of the Metal Ion, e.g., Ni(CO)$_4$ is Nickel Tetracarbonyl]

M-COAT Multidither Coherent Adaptive Optical Technique

MCoEN Master of Computer Engineering [also MCOEN]

MCOGA Mid-Continent Oil and Gas Association [US]

MCOM Mathematics of Computation

MCOM Michigan College of Osteopathic Medicine [US]

MComm Master of Commerce [also MCom]

MComm Master of Communication

MCommAdmin Master of Commercial Administration

MCompSc Master of Computer Science [also MCompS, or MCompSci]

MComSc Master of Commercial Science

MCOP Mission Control Operations Panel

MCOQ Multiple Choice Objective Question

MCOS Member of Canadian Oculoplastic Society

MCOS Microprogrammable Computer Operating System

$\mu'/\cot\gamma$ magnetic figure of merit (in electronics) [Symbol]

MCP Magnetic Compton Profile [Physics]

MCP Main Call Process

MCP Main Control Program

MCP Major Crown Project [Canada]

MCP Massachusetts College of Pharmacy [Boston, US]

MCP Massachusetts Contingency Plan [US]

MCP Master Change Proposal

MCP Master Computer Program

MCP Master Control Program

MCP Master of (or in) City Planning

MCP Master of Community Planning

MCP Materials Control Plan

MCP M-Cresol Purple

MCP Measurements Control Procedure

MCP Medical College of Pennsylvania [Philadelphia, US]

MCP Memory-Centered Processor

MCP Message Control Program

MCP Metacarpophalangeal (Joint) [Anatomy]

MCP Metastable Crystalline Phase [Metallurgy]

MCP Meteorological Communications Package

MCP 2-Methyl-4-Chlorophenoxyacetic Acid

MCP Methylcyclopentane

MCP Microchannel Plate [Electronics]

MCP Microsoft Certified Professional

MCP Military Construction Program

MCP Mission Control Programmer

MCP Monitoring and Control Panel

MCP Monobasic Calcium Phosphate; Monocalcium Phosphate

MCP Monocyte Chemotactic Protein [Biochemistry]

MCP Monte Carlo Particle

MCP Multichannel Communications Program

MCP Multichannel Plate (Detector)

MCP Multichip Package

MCP Multicomponent Process

MCP Muon Collider Project [of Fermilab, Batavia, Illinois, US]

mCP m-Chlorophenol

μCP Microcontact Printing

MCPA Mechanically Cut Polyamide

MCPA 2-Methyl-4-Chlorophenoxyacetic Acid

MCPB 4-(2-Methyl-4-Chlorophenoxy)butanoic Acid

MCPB Methylchlorophenoxybutyric Acid

MCPBA m-Chloroperoxybenzoic Acid [also mCPBA, or m-CBPA]

MCPC Manipulator Controller Power Conditioner

MCPESCF Multiconfiguration Paired Excitation Self-Consistent Field [Quantum Mechanics]

MCPF Modified Coupled Pair Functional [Physics]

MCPG Medium Conversion Program Generator

MCPG α-Methyl-(4-Carboxyphenyl)glycine [Biochemistry]

(+)-MCPG (+)-α-Methyl-(4-Carboxyphenyl)glycine [Biochemistry]

($\pm$)-MCPG ($\pm$)-α-Methyl-(4-Carboxyphenyl)glycine [Biochemistry]

MC-PGA Metallized Ceramic–Pin-Grid Array

MCPHA Multichannel Pulse-Height Analyzer

MCPL Magnetic Circularly Polarized Luminescence

MCpl Master Corporal [Canadian Forces]

MCPM Monocalcium Phosphate Monohydrate

MCPM Multicomponent Protein Mixture [Biochemistry]

MCPO Master Chief Petty Officer [US Navy and Coast Guard]

MCPP 2(4-Chloro-2-Methylphenoxy)propanoic Acid

MCPP Methylchlorophenoxypropionic Acid

MCPR Maximum Critical Power Ratio

MCPS Maintenance Control and Statistics Process

MCPS Major Cost Proposal System

MCPS Mechanical-Copyright Protection Society [UK]

MCPS Microsoft Certified Product Specialist

MCPS Mini-Core Processing Subsystem

Mcps Megacycles per Second [also Mcs, or Mc/s]

MCPU Multiple Central Processing Unit

M-CPZ Modulated Carbon Precipitation Zone [Metallurgy]

MCQ Multiple Choice Questionnaire

MC-QFP Metallized Ceramic–Quad Flat Pack [Electronics]

MCR Magnetic Card Reader

MCR Magnetic Character Reader

MCR Magnetic Character Recognition

MCR Main Control Room

MCR Marine Corps Reserve [US]

MCR Master Change Record

MCR Master Control Register

MCR Master Control Relay

MCR Master Control Routine

MCR Maximum Continuous Rating

MCR Measurement Conversion and Regulation (Module)

MCR Memory Control Register

MCR Military Compact (Nuclear) Reactor [of US Atomic Energy Commission]

MCR Minimum Creep Rate [Mechanics]

MCR Modem Control Register

MCR Multispectral Cloud Radiometer

MCrAl Monoaluminide-Chromium-Aluminum (Overlay Coating)

MCrAlY Monoaluminide-Chromium-Aluminum-Yttrium (Overlay Coating)

MCRFP Monitoring of Chemical Residues in Food Products

MCROA Marine Corps Reserve Officers Association [US]

MCRP Master of City and Regional Planning

MCRPA Multiconfigurational Random Phase Approximation [Materials Science]

MCRPQFP Molded-Carrier Ring Plastic Quad Flat Pack [Electronics]

MCRPUD Master of City and Regional Planning and Urban Development

MCRR Machine Check Recording and Recovery

MCRS Micrographics Catalogue Retrieval System

MCRT Maximum Cruise Thrust

MCRT Multichannel Rotary Transformer

MCS Magnesium Calcium Silicate

MCS Magnetic-Core Storage

MCS Magnetic Crystal Symmetry

MCS Maintenance and Checkout Station

MCS Maintenance Control Subsystem

MCS Maleimidocaproic Succinimide

MCS Management Control System

MCS Maneuver Control System

MCS Manufacturing and Consulting Service

MCS Manufacturing Control System

MCS Marine Communication Subsystem [of INTELSAT–International Telecommunications Satellite Organization, US]

MCS Marine Corps School [Quantico, Virginia, US]

MCS Maritime Communications System

MCS Master Control System

MCS Master of Clinical Science

MCS Master of Commercial Science

MCS Master of Communications Studies

MCS Master of Computer Science

MCS Measurements Calibration System

MCS Mechanical Cultivation Service

MCS Medical Computer Services

MCS Medium-Carbon Steel

MCS Mellon College of Science [of Carnegie-Mellon University, Pittsburgh, Pennsylvania, US]

MCS Mesoscale Convective System [Meteorology]

MCS Message Control System

MCS Method of Constant Stimuli [Acoustics]

MCS Method of Corresponding Solutions

MCS Methylchlorosulfonate

MCS Microchannel Scaling

MCS Microcomputer Simulation

MCS Microcomputer System

MCS Microinstruction Control Store

MCS Microwave Carrier Supply

MCS Military College of Science [UK]

MCS Military Communications Station

MCS Missile Control System

MCS Mobile Calibration Station

MCS Mobile Communications System

MCS (Pacific Rim International Conference on) Modeling of Casting and Solidification Processes [Metallurgy]

MCS Modular Computer Systems

MCS Moisture Control System

MCS Monte Carlo Simulation

MCS Monte Carlo Step(s) [Materials Science]

MCS Motor Carrier Safety

MCS Motor Circuit Switch

MCS Motor Current Signature

MCS Multicathode Spot

MCS Multichannel Communications System

MCS Multichannel Scaler; Multichannel Scaling

MCS Multichannel Spectrometer

MCS Multichannel Switch

MCS Multiconsole System

MCS Multinational Character Set

MCS Multiple Cloning Site [Genetics]

MCS Multiple Console Support

MCS Multiprogrammed Computer System

MCS Multipurpose Communications and Signaling

Mc/s Megacycles per second [also Mcps, or Mc/sec]

mC/s milli-Celsius per second [also mC/sec, or mC s^{-1}]

MCSA Microcomputer Software Association [of Association of Data Processing Service Organizations, US]

MCSA Motor Coach Safety Association [Canada]

MCSA Motor Current Signature Analysis

MCS/at Monte Carlo Step(s) per Atom [Materials Science]

MCSC Medical College of South Carolina [Charleston, US]

MCSC Member/Customer Service Center [of ASM International, US]

MCSC Monell Chemical Senses Center [Philadelphia, Pennsylvania, US]

MCSc Master of Computer Science

MCSCF Multiconfiguration Self-Consistent Field [also MC-SCF] [Physics]

MCSCO Multiconfiguration Self-Consistent Orbital [also MC-SCO] [Physics]

MCSD Materials and Chemical Sciences Division [of Lawrence Berkeley Laboratory, University of California at Berkeley, US]

MCSD Microsoft Certified Solution Developer

MCSE Microsoft Certified Software Engineer

Mc/sec Megacycles per second [also Mcps, or Mc/s]

mC/sec milli-Celsius per second [also mC/s, or mC s^{-1}]

M-CSF Macrophage-Colony Stimulating Factor [Biochemistry]

μc-Si Microcrystalline Silicon

μc-SiC Microcrystalline Silicon Carbide

μc-SiC:H Microcrystalline Hydrogenated Silicon Carbide

μc-Si:H Hydrogenated Microcrystalline Silicon

μc-Si:H:B Boron-Doped Hydrogenated Microcrystalline Silicon

μc-Si:H:P Phosphorus-Doped Hydrogenated Microcrystalline Silicon

μc-SiO$_2$ Microcrystalline Silicon Dioxide (or Silica)

MCSL Microcomputer Simulation Laboratory

MCSM Microcomputer Systems Management

MCSP Manpower Consultative Service Program [Canada]

MCSP Microsoft Certified Solution Provider

MCSR Mississippi Center for Scientific Research [US]

MCSS McDonnell Center for the Space Sciences [of Washington University, St. Louis, Missouri, US]

MCSS Military Communications Satellite System

MCST Magnetic Card Selectric Typewriter [also MC/ST]

MCST Manchester College of Science and Technology [UK]

MCSTEP Multiconfiguration Spin Tensor Electron Propagator (Method) [Physics]

MCSW Mining Club of the Southwest [US]

MCT Magnetically Coupled Transformer

MCT Magnetic Card and Tape (Unit)

MCT Magnetic Carrier Technology

MCT Mass Culturing Technique [Microbiology]

MCT Maximum Climb Thrust

MCT Maximum Continuous Thrust

MCT Mechanical Comprehension Test

MCT Medium Chain Triglyceride [Biochemistry]

MCT Mercury Cadmium Telluride (Semiconductor)

MCT Microcomputer Technology

MCT Mission Control Table

MCT Mobile Communications Terminal

MCT Mode Coupling Theory

MCT Modified Compact-Tension (Specimen) [Mechanics]

MCT MOS (Metal-Oxide Semiconductor) Controlled Thyristor

MCTA Milling Cutter and Tool Bit Association [UK]

MCTAMIA Motor Cars, Tractor and Agricultural Machinery Importers Association [Cyprus]

MCTC Maritime Cargo Transportation Conference [of Maritime Transportation Research Board, US]

MCTC Metal Casting Technology Center [of University of Alabama, Tuscaloosa, US]

MCTD Materials and Components Technology Division [of Argonne National Laboratory, Illinois, US]

MCTDHF Multiconfiguration Time-Dependent Hartree-Fock (Method) [also MC-TDHF] [Physics]

MCTDSCF Multiconfiguration Time-Dependent Self-Consistent Field [also MC-TDSCF] [Physics]

MC-TdT Monoclonal Antibody to Terminal Deoxynucleotidyl Transferase [Biochemistry]

MCTEX Marine Continental Thunderstorm Experiment; Maritime Continent Thunderstorm Experiment

MCTI Metal Cutting Tool Institute

MCTI Monte Carlo Transfer Interval (Method) [Physics]

MCT IR Mercury Cadmium Telluride Infrared (Detector) [also MCT-IR]

MCTR Message Center

MCTRAP Mechanized Customer Trouble Report Analysis Plan

MCTRF Manitoba Cancer Treatment (and) Research Foundation [Canada]

MCTS Master Central Timing System

MCTST Monte Carlo Transition State Theory [Physics]

MCU Machine Control Unit

MCU Magnetic Card Unit

MCU Main Control Unit

MCU Maintenance Control Unit

MCU Master Control Unit

MCU Medium Close Up

MCU Memory Control Unit

MCU Message Construction Unit

MCU Microcellular Urethane

MCU Microcomputer Unit

MCU Microcontrol(ler) Unit

MCU Microprocessor Control Unit

MCU Microprogram Control Unit

MCU (Ontario) Ministry of Colleges and Universities [Canada]

MCU Mission Control Unit

MCU Modular Concept Unit

MCU Module Control Unit

MCU Multi-Chip Unit

MCU Multicoupler Unit

MCU Multiplexer Control Unit

MCU Multipoint Control Unit

MCU Multiprocessor Communications Unit

MCU Multiprogrammed Control Unit

MCU Multisystem Communications Unit

MCUG Military Computer Users Group

MCV Magnetic Cushion Vehicle

MCV Master Control Valve

MCV Mean Corpuscular Volume [Medicine]

MCV Medical College of Virginia [Richmond, US]

MCV Movable Closure Valve

MCVD Modified Chemical Vapor Deposition

MCVF Multichannel Voice Frequency

MCVFT Multichannel Voice Frequency Telegraphy

MCVI Vapor Infiltration with Microwave Heating

MCVP Materials Control and Verification Program

MCW Medical College of Wisconsin [Milwaukee, US]

MCW Memory Card Writer

MCW Modulated Carrier Wave

MCW Modulated Continuous Wave

MC&W Master Caution and Warning

MCWA Massey College Wood Association [New Zealand]

MCWCS Ministerial Conference of West and Central African States on Maritime Transportation [Ivory Coast]

MCX Minimum-Cost Estimate

MCXα Multiconfiguration $X\alpha$ [Physics]

MCXD Magnetic Circular X-Ray Dichroism

Mcy Methyl Cyclohexane

Mcy Motorcycle [also mcy]

MCZ Museum of Comparative Zoology [of Harvard University, Cambridge, Massachusetts, US]

MCZDO Multi-Center Zero Differential Overlap [Physics]

MD Biomedical Office [of NASA Kennedy Space Center Directorate, Florida, US]

MD Machine Direction

MD Madre de Dios [Peru]
MD Magnetic Detector
MD Magnetic Device
MD Magnetic Dipole [Physics]
MD Magnetic Disk
MD Magnetic Domain [Solid-State Physics]
MD Magnetic Drum
MD Magnetization Derivative
MD Magneto-Damping
MD Magnetodynamic(s)
MD Main Drum
MD Maintenance Depot
MD Make Directory (Command) [also md]
MD Malfunction Detection
MD Management Domain
MD Managing Director
MD Manhattan District [New York, US]
MD Manual Damper
MD Manual Data
MD Manual Direct
MD Manual Disconnect
MD Many-Body Density [Physics]
MD Marginal Distribution [Statistics]
MD Maryland [US]
MD (Recursive) Mask and Deposit (Process)
MD Master Dimension
MD Master Directory
MD Material Damping
MD Material Degradation
MD Materials Directorate
MD Matrix Division
MD Maximum (Electricity) Demand
MD McLellan-Dunn (Model) [Metallurgy]
MD Mean Deviation
MD Measured Discard
MD Mechanical Deformation
MD Mechanical Design(er)
MD Mechanical Drafting; Mechanical Draftsman;
        Mechanical Draftswoman; Mechanical Draftsperson
MD Medical Department
MD Medical Doctor
MD (Medicinae Doctor) – Doctor of Medicine
MD Medium Dense
MD Meeting Date
MD Megadalton [Unit]
MD Menu-Driven
MD Mesoscopic Disorder [Crystallography]
MD Message Data
MD Message Digest
MD Messages per Day
MD Metal Deformation
MD Metal Detector
MD Metals Datafile [of ASM International, US] [also MDF]
MD Meteorology Department

MD Methanol Dispersible
MD Methyldichloroarsine
MD Microdensitometer; Microdensitometry
MD Microdiffraction
MD Microdiffractometer
MD Microdot
MD Micro-Drill(ing)
MD Middle Distillate
MD Military District
MD Mine Depot
MD Mine Design
MD Mine Detector
MD Mine Disposal
MD Mineral Dressing
MD Minidisk
MD Minimum Dose
MD Misfit Dislocation [Materials Science]
MD Mission Director
MD Model-Driven
MD Mohave (or Mojave) Desert [Southern California, US]
MD Moldova [ISO Code]
MD Molecular Diffusion
MD Molecular Dynamic(s)
MD Moment Distribution (Method)
MD Monitor Display
MD Monochrome Display
MD Month after Date [also M/D, or m/d]
MD Moore Drive
MD Motion Detection; Motion Detector
MD Motor Drive
MD Movement Directive
MD Multidomain [Solid-State Physics]
MD Multidrop
MD Multinominal Distribution
MD Multiple Dissemination
MD Municipal District
MD Muscular Dystrophy [Medicine]
MD Music Data
MD- McDonnell Douglas (Aircraft) [Followed by a Numeral, e.g., MD-11]
M/D Month after Date [also MD, or m/d]
M-D Modulation-Demodulation; Modulator-Demodulator
M-D Multi-Domain [Solid-State Physics]
M³D Mechanics and Mechanisms of Materials Damping [American Society for Testing and Materials Special Technical Publication, US] [also MMMD]
Md Earthquake Magnitude by Duration Method [Symbol]
Md Maryland [US]
Md Median (in Statistics) [Symbol]
Md Mendelevium [also Mv] [Symbol]
Md-256 Mendelevium-256 [also ^{256}Md, Md^{256}, or Md]
Md-257 Mendelevium-257 [also ^{257}Md, or Md^{257}]
Md-258 Mendelevium-258 [also ^{258}Md, or Md^{258}]
M$ Malaysian Dollar [Currency]

**M$** Million Dollars (e.g., 75M$ stands for $75,000,000)

md millidarcy [also md]

md *(modus dictum)* – as directed [Medical Prescriptions]

.md Moldova [Country Code/Domain Name]

m/d month after date [also M/D, or MD]

m³/d cubic meter(s) per day [also cu m/d]

MDA Magnetic Deflection Analyzer

MDA Mail Delivery Agent [Internet]

MDA Main Distribution Assembly

MDA Maintainability Design Approach

MDA Malonic Dialdehyde

MDA Manitoba Department of Agriculture [Canada]

MDA Manufacturing Defects Analyzer

MDA Marketing Device Association [now Marketing Device Association International, US]

MDA Master Diversion Aerodrome

MDA Master of Development Administration

MDA Materials Dispersion Apparatus

MDA McDonnell Douglas Aerospace [US]

MDA Mechanized Directory Assistance

MDA Medical Devices Act [US]

MDA 1,8-p-Menthane Diamine

MDA Metal Deactivator

MDA Methylene Dianiline ; 4,4'-Methylene Dianiline

MDA Methylene Dioxyamphetamine

MDA Middeck Assembly

MDA Mineral Development Agreement [between Canada and Newfoundland]

MDA Minimum Daily Allowance

MDA Minimum Descent Altitude

MDA Minimum Detectable Activity

MDA Misuse of Drugs Act [UK]

MDA Monochrome Display Adapter

MDA Motor Drive Amplifier

MDA M-Phenylenediamine

MDA Multidimensional Access

MDA Multidimensional Analysis

MDA Multidimensional Array

MDA Multidocking Adapter

MDA Multiple Digit Absorbing

MDA Multiple Docking Adapter [of NASA]

MDa Megadalton [Unit]

mda *(modus dictum applicatus)* – to be applied as directed [Medical Prescriptions]

MDAA Muscular Dystrophy Association of America [US]

MDAA Mutual Defense Assistance Act

MDAB 4,4'-Bismaleimidodiphenylmethane [also MDA BMI]

MDAC McDonnell Douglas Astronautics Company [US]

MDAC Methyldiethylaminocoumarin

MDAC Multiplying Digital-to-Analog Converter

MDAC Muscular Dystrophy Association of Canada

MDAE Multidisciplinary Accident Engineering

MDA/EGA Monochrome Display Adapter/Enhanced Graphics Adapter

MDAFWP Motor Driven Auxiliary Feedwater Pump

MDAI Marketing Device Association International [US]

MDAK o-Methyldiarylketone

MDAL McDonnell Douglas Aerophysics Laboratory [St. Louis, Missouri, US]

MDAP Machine and Display Application Program

MDAP Mutual Defense Assistance Program

MDAR Malfunction Detection, Analysis and Recording

MDAS Meteorological Data Acquisition System

MDAS Mission Data Acquisition System

M-Day Mobilization Day [also M-day]

MDB Maritime Development Board

MDB 4,5-Methylenedioxy-1,2-Benzenediamine

MDB Mission Data Book

MDB Multilateral Development Bank

MDB Murray-Darling Basin [Australia]

MDBC Murray-Darling Basin Commission [Australia]

MDBF Measured Depth Below Formation

MDBK Madin-Darby Bovine Kidney (Cells)

MDBFTT Miniaturized Ring-on-Ring Disk-Bend Fracture Toughness Test

MDBMC Murray-Darling Basin Ministerial Council [Australia]

MDBT Miniaturized Disk-Bend Test [Mechanical Testing]

MDC Machinability Data Center [US]

MDC Main Display Console

MDC Main Distribution Center

MDC Maintenance Data Collection

MDC Maintenance Dependency Chart

MDC Manitoba Development Corporation [Canada]

MDC Market Development Center

MDC Materials Dissemination Center

MDC Materials Distribution Center

MDC Max-Delbrück-Centrum (für Molekulare Medizin) Max-Delbruck Center (for Molecular Medicine), Berlin, Germany]

MDC Maximum Dependable Capacity

MDC McDonnell Douglas Aerospace Corporation [St. Louis, Missouri, US]

MDC Membership Development Committee [of Institute of Electrical and Electronics Engineers– Regional Activities Board, US]

MDC Memory Disk Controller

MDC Metropolitan District Commission [US]

MDC Microprocessor Development Center [US]

MDC Miljödatacentrum [Environmental Satellite Data Center, Kiruna, Sweden]

MDC Minimum Detectable Concentration

MDC Missile Development Center [of US Air Force]

MDC Mission Director Center

MDC Mission Duty Cycle

MDC Multiduty Collector

MDC Multiple Device Control(ler)

MDC Multiple Drone Control

MDCA Main Distribution Control Assembly

MDCC Master Data Control Console

MDCCS Mourning Dove Call Count Survey [US]

MDCE Monitoring and Duplicate Control Equipment

MDCI Maine Department of Commerce and Industry [US]

MDCK Madin-Darby Canine Kidney (Cells)

MDC Rep McDonnell Douglas Aerospace Corporation Report [St. Louis, Missouri, US]

MDCS Malfunction Display and Control System

MDCS Maintenance Data Collection System

MDCS Master Digital Command System

MDCS Material Data Collection System

MDCT Mechanical Draft Cooling Tower [Nuclear Reactors]

MDCU Magnetic Disk Control Unit

MDD Machine Dependence Data

MDD Magnetic Disk Drive

MDD Mate/Demate Device [Aerospace]

MDD Materials Development Division [of Indira Gandhi Center for Atomic Research, Kalpakkan, India]

MDD Meteorological Data Distribution

MDD Milligram per Square Decimeter per Day [also mdd]

MDDBMS Multidimensional Data Base Management System

MDDC Microdiffractometer Data Collection

MDDDA 4-Methyl-1,9-Diamino-3,7-Diazanonane-3,7-Diacetate [also mddda]

MDDR Mean Depth of Deformation Rate

MDDS Material Directory Data Sheet

MDE Magnetic Decision Element

MDE Master of Developmental Economics

MDE Mercury-Dropping Electrode [Physical Chemistry]

MDE Missile Display Equipment

MDE Mission-Dependent Equipment

MDE Mission-Dependent Experiment

MDE (±)-3,4-Methylenedioxy-N-Ethylamphetamine

MDE Modular Design of Electronics

MDE Modular Display Electronics

MDEA Methyldiethanolamine

MDEB N-Dodecyl-N-Methylephedrinium Bromide

MDED Maine Department of Economic Development [US]

MDEFWP Motor-Driven Emergency Feedwater Pump

mdeg millidegree [Unit]

MDEIPO 4-Methyl-3,3-Diethyl-5-Isopropyloctane

MDEP Maine Department of Environmental Protection [US]

MDES Multiple Data Entry System

MDes Master of Design

MDesS Master of Design Studies [also MDesStud]

MDF Macro-Defect-Free (Cement)

MDF Main Distribution Frame; Main Distributing Frame

MDF Manipulator Development Facility

MDF Manual Direction Finder

MDF Mating/Demating Facilities [Aerospace]

MDF Mean Daily Flow

MDF Medium-Density Fiber

MDF Medium-Density Fiberboard

MDF Medium-Frequency Direction Finder; Medium-Frequency Direction Finding

MDF Menu Definition File

MDF Metals Data File [of ASM International, US] [also MD]

MDF Microcomputer Development Facilities

MDF Microdose Focusing

MDF Mild-Detonating Fuse

MDF Misorientation Distribution Function [Crystallography]

MDF Multi-Degree of Freedom

.MDF Menu Definition File [File Name Extension]

M/D-F Mixture and Derived From

MDFA Magnet Distributors and Fabricators Association [US]

MDFNA Maximum Density Fuming Nitric Acid

MDFRC Murray-Darling Freshwater Research Center [Australia]

Mdfy Modify [also mdfy]

MDG Major Donors Group

MDG Methyldiglycol

MDGC Multidimensional Gas Chromatography

MDH Malate Dehydrogenase [Biochemistry]

MDH Malic Dehydrogenase [Biochemistry]

MDH Mannesmann-Demag Hüttentechnik [Metallurgical Engineering Division of Mannesmann-Demag AG, Germany]

MDH Maximum Diameter Heat

MDH Mechanical Dividing Head

MDH Michigan Department of Health [US]

MDH Minimum Descent Height

Mdh Moroccan Dirham [Currency]

MDI Magnetic Direction Indicator

MDI Manual Data Input

MDI Maryland Department of Information [US]

MDI Maximum (Electricity) Demand Indicator

MDI Medium-Dependent Interface

MDI Memory Display Interface

MDI Menu-Driven Interface

MDI Methane Diisocyanate

MDI Methane Diphenyl Diisocyanate

MDI Methylene Diisocyanate

MDI 4,4'-Methylenedi(phenylisocyanate)

MDI Microdielectrometry

MDI Miss Distance Indicator

MDI Multiple Document Interface

MDIC Manchester Decoder and Interface Chip

MDIC Microwave Dielectric Integrated Circuit

MDIC Multilateral Disarmament Information Center [UK]

MDiEng Master of Diesel Engineering

MDIF Manual Data Input Function

MDIF Marine Department of Inland Fisheries

MDIS McDonnell Information System [of McDonnell Aerospace Corporation, St. Louis, Missouri, US]

MDI-PA 4,4'-Methylenedi(phenylisocyanate)–Polyamide

MDI-PPO 4,4'-Methylenedi(phenylisocyanate)–Polyphenylene Oxide

MDITT Manitoba Department of Industry, Trade and Tourism [Canada]

MDIU Manual Data Input Unit

MDK Multimedia Developers Kit

MDL Macro Description Language

MDL Maintenance and Diagnostic Logic Display

MDL Maritime Dynamics Laboratory

MDL Master Data Library

MDL Materials Design Laboratory [of Korean Institute for Science and Technology, Seoul, South Korea]

MDL Mechanical Development Laboratory [US]

MDL Medical Diagnostic Laboratory

MDL Method Detection Limit

MDL Mine Defense Laboratory [of US Department of Defense]

MDL Minimum Detectable Leakage

MDL Minimum Detectable Level

MDL Minimum Detectable Limit; Minimum Detection Limit

MDL Mitsubishi Denki Laboratory [Amagasaki, Japan]

Mdl Module [also mdl]

MDLA Methyldilaurylamine

MDLP Mobile Data Link Protocol

MDM Magnetic Dipole Moment

MDM Magnetic-Domain Memory

MDM Manipulator Deployment Mechanism

MDM Matrix Displacement Method

MDM Maximum Design Meter

MDM Medical Monitor

MDM Metal-Dielectric-Metal (Filter)

MDM Minimum Detectable Mass

MDM Multiplexer/Demultiplexer

.MDM Modem [File Name Extension]

MDMA Methylene Dioxymethamphetamine; (±)-3,4-Methylene Dioxymethamphetamine [Pharmacology]

MD/MC-CEM Many-Body Density Functional-Based Corrected Effective Medium (Theory) [Physics]

MDMD Materials Design and Manufacturing Division [of Minerals, Metals and Materials Society of American Institute of Mining, Metallurgical and Petroleum Engineers, US]

MDMD/EPD Materials Design and Manufacturing Division/ Extraction and Processing Division (Joint Committee) [of Minerals, Metals and Materials Society of American Institute of Mining, Metallurgical and Petroleum Engineers, US]

MDMG Microelectronics Dimensional Metrology Group [of National Institute for Standards and Technology, Gaithersburg, Maryland, US]

MDMOS Multiple-Drain Metal-Oxide Semiconductor [also MD-MOS]

MDMRS Multi-Departmental Mobile Radio System

MDMS Magnetic-Deflection Mass Spectrometer

MDMS Methylene Dimethanesulfonate

MDMS Multiple Database Management System

MDMV Maize Dwarf Mosaic Virus

MDN Malodinitrile

MDN Managed Data Network

MDN Ministère de la Défense Nationale [Ministry of National Defense, France]

MD/NC Mechanical Drafting/Numerical Control

MDNR Machinery Dealers National Association [US]

MDNR Maryland Department of Natural Resources [US]

MDNS Managed Data Network Services

Mdnt Midnight [also mdnt]

MDO MARC (Machine-Readable Catalog) Development Office [of Library of Congress, Washington, DC, US]

MDO Medium-Density Overlay

MDOA Methyldioctylamine

MDOF Multi-Degree of Freedom

MDOS Motorola Disk-Operating System

MDOT Minnesota Department of Transportation [US]

MDP Main Data Path

MDP Maintenance Diagnostic Processor

MDP Materials Development Program [of US Department of Energy]

MDP Mean Depth of Penetration [Hardness Testing]

MDP Memory Data Register

MDP Mercury Diffusion Pump

MDP Message Discrimination Process

MDP Methylenedioxyphenyl

MDP Methylenediphosphonic Acid

MDP Missile Data Processor

MDP Multidivisional Program

MDP Multi-Divisional Program [of Commonwealth Scientific and Industrial Research Organization, Australia]

MDP Multidomain Particle

MDP Muramyl Dipeptide [Biochemistry]

MDPA Methyldiphenylamine

MDPA Monochrome Display and Printer Adapter

MDPE Medium-Density Polyethylene

MDPF 2-Methoxy-2,4-Diphenyl-3(2H)-Furanone

MDPH Massachusetts Department of Public Health [US]

M(DPM)₂ Metal Dipyvaloylmethane

MDPR Mean Depth-of-Penetration Rate

MDPS Multiple-Workstation Direct-Processing System

MDQ Mininum Detectable Quantity

MDQW Modulation-Doped Quantum Well [Electronics]

MDR Magnetic Dipole Radiation

MDR Magnetic Disk Recorder

MDR Magnetic Field-Dependent Resistor

MDR Maintenance Demand Rate

MDR Major Design Review

MDR Manual Data Room

MDR Market Data Retrieval

MDR Memory Data Register

MDS Milestone Description Sheet

MDR Minimum Daily Requirement

MDR Minimum Design Requirement

MDR Minimum Detectable Count Rate

MDR Minor Discrepancy Repair

MDR Miscellaneous Data Recorder

MDR Missing Data Report

MDR Mission Data Reduction

MDR Monthly Director's Review

MDR Multichannel Data Recorder

MDR Multi-Disk Reader

MDR Multidrug-Resistant (Disease)

MDRD Mission Data Requirements Document

MDRI Multidetector Retention Index

MDRL McDonnell-Douglas Research Laboratory [St. Louis, Missouri, US]

MDRP Mean Depth Rate of Penetration

MDRS Manufacturing Data Retrieval System

MDRS Mission Data Retrieval System

MDR-TB Multidrug-Resistant Tuberculosis [Medicine]

MDS Magnetic Detector System

MDS Magnetic Drum Storage

MDS Main Device Scheduler

MDS Maintenance Data System

MDS Malfunction Detection System

MDS Management Data System

MDS Management Decision System

MDS Mannesmann-Demag-Sack AG [German Manufacturer of Metallurgical Plant and Equipment; located at Duesseldorf]

MDS Master Development Schedule

MDS Master Drum Sender

MDS Master of Decision Sciences

MDS Master of Dental Surgery

MDS Materials Data Sheet

MDS Matrix Diagonalization Sudden

MDS Medium Deep Survey [of International Hubble Space Telescope Project]

MDS Melt Directional Solidification [Metallurgy]

MDS Memory Disk System

MDS Metal-Dielectric-Semiconductor

MDS Metastable (Helium Atom) Deexitation Spectroscopy

MDS Micro Disk Storage

MDS Microprocessor Development System

MDS Minimum Detectable Signal

MDS Minimum Discernible Signal

MDS Minimum Discernible System

MDS Minimum Dose System

MDS *(Miscere, Dispendere, Signare)* – Mix, Dispense, Label [Medical Prescriptions]

MDS Mission Development Simulator

MDS Mission Definition Study [of NASA]

MDS Mission Demonstration-Test Satellite [of National Space Development Agency of Japan]

MDS Mobile Data Service

MDS Modern Data System

MDS Modular Disk Storage

MDS Mopar Diagnostic System [Automobiles]

MDS Multidimensional Scaling [Geographic Information System]

MDS Multidimensional Signal

MDS Multiple Data Set System

MDS Multipoint Distribution Service

MDS Multipoint Distribution System

MDS Multiprocessor Distributed System

mds millidarcies [Unit]

mds *(modus dictum sumendus)* – to be taken as directed [Medical Prescriptions]

MDSD Mate/Demate Stiff-Leg Derrick

Mdse Merchandise [also mdse]

MDSF Manipulator Development and Simulation Facility

Mdsg Merchandising [also Mdsing, mdsing, or mdsg]

MDSMB Medical Devices Standards Management Board

MDSS Meteorological Data Sounding System

MDSS Mission Data Staging System [of NASA]

MDSS Microprocessor Development Support System

MDT Maintenance Demand Time

MDT Maximum Diameter of the Thorax [Anatomy]

MDT Mean Detonating Time

MDT Mean Down Time

MDT Measurement Descriptor Table

MDT Mobile Data Terminal

MDT Modified Data Tag

MDT Mountain Daylight Time [Greenwich Mean Time +08:00]

MDT Movchan, Demchishin and Thornton (Microstructure) [Metallurgy]

MDT Multidimensional Tasking

mdt moderate

MDTA Manpower Development and Training Act [US] [Repealed]

MDTA Micro-Distillation Tube Assembler

MDTA/D Micro-Distillation Tube Assembler/Disassembler

MDTD Maryland Division of Tourist Development [of Department of Economic and Community Development, US]

MDTD Minimum Detectable Temperature Difference

MDTF Marine Dynamic Test Facility

MDTS Megabit Digital-to-Troposcatter Subsystem

MDTS Modular Data Transmission System

MDTSCO McDonnell Douglas Technical Services Company [US]

MD-TST-MC Molecular Dynamics, Transition State Theory, Monte Carlo (Method) [Physics]

MDU Maintenance Diagnostic Unit

MDU Message Decoder Unit

MDU Mine Disposal Unit

mdu *(modus dictum uti)* – to be used as directed [Medical Prescriptions]

MDUS Medium Data Utilization Station

MDV Doctor of Veterinary Medicine

MDV Midivariant [Biochemistry]

MDV Mucosal Disease Virus

MDV-1 Midivariant 1 [Biochemistry]

MDW Measured Daywork

MDW Military District, Washington [District of Columbia, US]

MDW Multiple Drop Wire

MDW-7-R 4-[(R)-2-Chloro-3-Methylbutyryloxy]phenyl 4-(Decyloxy)benzoate

MDW-7-S 4-[(S)-2-Chloro-3-Methylbutyryloxy]phenyl 4-(Decyloxy)benzoate

MD-WRA Maryland State Water Resources Administration [US]

MDX Modular Digital Exchange

mdx dystrophic mouse [Immunochemistry]

MDY Month/Day/Year [also m/d/y]

ME Machine Equation

ME Macromolecular Engineer(ing)

ME Magnetic Elasticity

ME Magnetic Estimation

ME Magnetoelastic(ity)

ME Magnetoelectric(ity)

ME Maine [US]

ME Main Engine [Aerospace]

ME Main Entry

ME Maintenance Engineer(ing)

ME Male Elbow (Pipe Section)

ME Malonic Ester

ME Management Engineer(ing)

ME Managing Editor

ME Manual-Feed Envelope

ME Manufacturing Engineer(ing)

ME Marine Engineer(ing)

ME Martini-Enfield (Rifle)

ME Masonry Engineer(ing)

ME Master Electrician

ME Master of Education

ME Master of Engineering

ME Materials Engineer(ing)

ME Materials Evaluation

ME Mathieu Equation [Mathematics]

ME Matteucci Effect [Physics]

ME Mature Equivalent [Dairy Cattle]

ME Maximum Effort

ME Maximum Entropy (Analysis Method)

ME Maxwell Equations [Physics]

ME Mechanical Efficiency

ME Mechanical Engineer(ing)

ME Mechanical Equivalent

ME Mechanoelectronic(s)

ME Medical Electronics

ME Medical Engineer(ing)

ME Medical Examiner

ME Meissner Effect [Solid-State Physics]

ME Melt Extraction [Metallurgy]

ME Melt Extractor [Chemical Engineering]

ME Memory Element

ME Mercaptoethanol

ME Message Element

ME Metabolizable Energy

ME Metal Evaporated; Metal Evaporation

ME Metalloenzyme

ME Metallurgical Engineer(ing)

ME Metals Extraction

ME Microcircuit Engineer(ing)

ME Microelectronic(s)

ME Microenvironment [Ecology]

ME Middle English

ME Migdal-Eliashberg (Theory) [Physics]

ME Military Engineer(ing)

ME Mineral Engineer(ing)

ME Mineral Exploration

ME Mining Engineer(ing)

ME Miscellaneous Equipment

ME Modulator Electrode

ME Moessbauer (or Mössbauer) Effect [Nuclear Physics]

ME Molecular Electronics

ME Molecular Engineering

ME Momentum Equation

ME Monitoring Equipment

ME Montreal Exchange [Canada]

ME Motion Estimation

ME Municipal Engineer(ing)

ME Muzzle Energy

ME Myalgic Encephalitis

MI Myocardial Infarction

.ME Opening Information [File Name Extension, e.g., READ.ME]

M&E Material and Equipment

M&E Music and Sound Effects

Me Maine [US]

Me Mercury [Abbreviation; Symbol: Hg]

Me Metal

Me Metalloid

Me Methyl [also me]

Me Microfiche

Me^{1+} Monovalent Metal [also Me^{++}, or M^I] [Symbol]

Me^{2+} Divalent Metal [also Me^{++}, or M^{II}] [Symbol]

Me^{3+} Trivalent Metal [also Me^{+++}, or M^{III}] [Symbol]

Me^{4+} Tetravalent Metal [also Me^{IV}] [Symbol]

Me^{5+} Pentavalent Metal [also M^V] [Symbol]

Me^{6+} Hexavalent Metal [also M^{VI}] [Symbol]

me milliequivalent [also meq]

m/e mass-to-charge ratio (of ions) [Symbol]

MEA Main Electronics Assembly

MEA Maintenance Engineering Analysis

MEA Malt Extract Agar [Biochemistry]

MEA Master of Engineering Administration

MEA Master of Engineering Architecture

MEA Materials Experiment Assembly [of NASA]

MEA Mechanical Engineering Abstracts [US]

MEA Mercaptoethanolamine

MEA β-Mercaptoethylamine [also Mea]

MEA Michigan Education Association [US]

MEA Middle East Airlines

MEA Minimum En-Route Altitude

MEA Moisture Evolution Analysis

MEA Monoethanolamine

MEA More Electric Aircraft (Technology)

2MEA 2-Mercaptoethylamine [also 2Mea]

MEAB Maintenance Engineering Analysis Board

MEAC Manufacturing Engineering Application Center

MeAc Methyl Acetate

3-MeACAC 3-Methylpentane-2,4-Acetylacetonate; 3-Methylpentane-2,4-Dione [also 3MeAcAc, or 3Meacac]

MEACON Masking Beacon [also Meacon, or meacon] [Radar]

MEAD Acid cis-5,8,11-Eicosatrienoic Acid [also MEAD acid]

MEAL Master Equipment Allowance List

MEAL Master Equipment Authorization List

MeAl Metal Aluminide

Me$_3$Al Trimethylaluminum

Me$_2$AlCl Dimethylaluminum Chloride

MEAM Modified Embedded Atom Method [Materials Science]

MEAN Ministry of Economic Affairs of the Netherlands

MEAN Multipole-Extracted Adiabatic Nuclei (Approximation) [Physics]

MeAN Methylamine Nitrate

Me2[15]aneN2O3 7,13-Dimethyl 1,4,10-Trioxa-7,13-Diazacyclopentadecane

Me3[12]aneN3 2,2,4-Trimethyl 1,5,9-Triazacyclodedecane

Me3[9]aneN3 1,4,7-Trimethyl 1,4,7-Triazacyclononane

2-Me[9]aneN3 2-Methyl 1,4,7-Triazacyclononane

2-Me[9]aneN2O 2-Methyl 1-Oxa-4,7-Diazacyclononane

MEAR Maintenance Engineering Analysis Report

MEAR Maintenance Engineering Analysis Request

MEARA Manitoba Environmental Assessment and Review Agency [Canada]

MEAS Materials Engineering and Applied Science (Department)

MeAs Metal Arsenide

Meas Measure(ment); Measuring [also meas]

Meas Control Measurement and Control [Journal of the Institute of Measurement and Control, UK]

Meas Insp Technol Measurement and Inspection Technology [Journal published in the UK]

Meas Sci Measurement Science; Measurement Scientist

Meas Sci Technol Measurement Science and Technology

Meas Sci Technol Measurement Science and Technology [Journal of the Institute of Physics, UK]

Meas Tech Measurement Techniques [Translation of: *Izmeritel'naya Tekhnika (USSR)*; published in the US]

Meas Technol Measurement Technologist; Measurement Technologist

MEASUCORA Measurement, Control Regulation and Automation

MEB Main Electronics Box

MEB Materials-Energy Balance

MEB Memory Expansion Board

MEB Modem Evaluation Board

MeB Metal Boride

MEBA Methylethyl-n-Butylamine

MEBB Microbial Ecology and Biotechnology Branch [of US Environmental Protection Agency at Gulf Breeze, Florida]

MeBB Methylenebenzoylbenzoate

Me$_2$BBr Dimethylboron Bromide (or Bromodimethylborane)

MeBe Metal Beryllide

MeBi Metal Bismuthide

MeBl Methylene Blue

MEBO Main Engine Burnout [Aerospace]

MeBr Metal Bromide

MeBr Methyl Bromide

Me$_3$BrGe Trimethylbromogermane

MEBS Marketing, Engineering and Business Services

Me$_3$Bz Trimethylbenzene

MeBuOH 2-Methyl-1-Butanol

MEC Main Engine Controller [Aerospace]

MEC Market Economy Country

MEC Marshalltown Engineering Club [Iowa, US]

MEC Maschinenbau Entwicklung Consulting GmbH [German Mechanical Engineering, Development and Consulting Company located at Eschweiler]

MEC Master Event Controller

MEC Materials Education Council [US]

MEC Materials Engineering Center

MEC Matsushita Electric Company [Osaka, Japan]

MEC Member of the Executive Council

MEC Metastable Epitaxial Compound

MEC Meteorology Engineering Center [of Naval Ocean Systems Center, US]

MEC Minimum Energy Conformer

MEC Ministério da Educação e Ciência [Ministry for Education and Science, Portugal]

MEC Ministerio de Educación y Ciencia [Ministry for Education and Science, Spain]

MEC Mission Events Controller

MEC Mobile Examination Center

MEC Molecular Electron Correlation

MEC Multiexciton Complex [Solid-State Physics]

MEc Master of Economics

MeC Metal Carbide

Me$_3$C Trimethyl Carbide

MECA Main Engine Controller Assembly [Aerospace]

MECA Maintainable Electronic Component Assembly

MECA Manufacturers of Emission Controls Association [US]

MECA Matsushita Electric Corporation of America

MECA Molecular Emission Cavity Analysis

MECAM 1,3,5-tris[[(2,3-Dihydroxybenzoyl)amino]methyl] benzene

MECB (Programa) Missão Espacial Completa Brasileira [Brazilian Complete Space Mission (Program); of Brazilian National Institute of Space Research]

MECC Micellar Electrokinetic Capillary Chromatography

MECC Minnesota Educational Computing Consortium [US]

MECCA Master Electrical Common Connector Assembly

MECCA Mechanized Catalog

MECCA Model Evaluation Consortium for Climate Assessment

Mecc Ital Meccanica Italiana [Italian Publication on Mechanics]

Mecc Mod Meccanica Moderna [Italian Publication on Modern Mechanics]

MECF Main Engine Computational Facilities [NASA Space Shuttle]

Me-C:H Metal-Bearing Amorphous Hydrogenated Carbon

Mech Mechanic(s); Mechanical; Mechanism; Mechanization [also mech]

Mech Ageing Devel Mechanisms of Ageing Development [UK Journal]

Mech Autom Adm Mechanizace Automatizace Administrativy [Czech Publication on Mechanization and Automation in Administration]

Mech Compos Mater Mechanics of Composite Materials [Translation of: *Mekhanika Kompozitnykh Materialov (USSR)*; published in the US]

Mech Corros Prop A, Key Eng Mater Mechanical and Corrosion Properties A, Key Engineering Materials [Journal published in Switzerland]

Mech Corros Prop B, Single Cryst Prop Mechanical and Corrosion Properties A, Single Crystal Properties [Journal published in Switzerland]

Mech Devel Mechanisms of Development [International Journal]

MechE Mechanical Engineer [also Mech Eng, or MechEng]

MEChemEng Master of Engineering in Chemical Engineering [also MEChemE, or MSChE]

MECHEN Mechanical Engineering (Database) [of Royal Institute of Technology Libraries, Sweden]

Mech Eng Mechanical Engineer(ing)

Mech Eng Mechanical Engineering [Journal of the American Society of Mechanical Engineers, US]

Mech Eng Bull Mechanical Engineering Bulletin [Published by the Central Mechanical Engineering Research Institute, India]

Mech Eng News Mechanical Engineering News [Published by the American Society for Engineering Education, US]

Mech Eng (NY) Mechanical Engineering (New York) [Journal of the American Society of Mechanical Engineers, US]

Mech Eng Sci Mechanical Engineering Science

Mech Eng Technol Mechanical Engineering Technologist; Mechanical Engineering Technology

Mech Eng Technol Mechanical Engineering Technology [Journal published in the UK]

Mech Homog Catal Mechanics of Homogeneous Catalysis [US Journal]

Mech Inc Eng Mechanical Incorporated Engineer [Journal published in the UK]

Mech Mater Mechanics of Materials [Journal published in the Netherlands] [also Mech Mat]

Mech Met Mechanical Metallurgist; Mechanical Metallurgy

Mech Prop Mechanical Properties [also mech prop]

Mech Res Commun Mechanical Research Communications [Journal published in the UK]

Mech Rigid Bodies Mechanics of Rigid Bodies [Translation of: *Mekhanika Tverdogo Tela (Ukrainian SSR)*; published in the US]

Mech Solids Mechanics of Solids [Translation of: *Izvestiya Akademii Nauk SSSR, Mekhanika Tverdogo Tela (USSR)*; published in the US]

Mech Struct Mach Mechanics of Structures and Machines [Journal published in the US]

Mech Syst Signal Process Mechanical Systems and Signal Processing [Journal published in the UK]

Mech Teor Stosow Mechanika Teoretyczna i Stosowana [Journal on Theoretical Mechanics and Related Subjects; published by the Polish Academy of Science]

MECI Manufacturing Engineering Certification Institute [US]

MECL Motorola Emitter-Coupled Logic

MECL Multi-Emitter Coupled Transistor Logic

MECL Musashino Electrical Communication Laboratory [Japan]

MeCl Dichloromethane

MeCl Metal Chloride

MeCl Methyl Chloride

Me₃ClGe Trimethylchlorogermane

MECLIZINE 1-[(4-Chlorophenyl)phenylmethyl]-4-[(3-Methylphenyl)methyl]piperazine [also Meclizine]

Mec Mater Electr Mécanique Matériaux Electricité [French Publication on the Mechanics of Materials used in Electrical Engineering]

MeCN Methyl Cyanide (or Acetonitrile)

MECNY Municipal Engineers of the City of New York [US]

MECO Main Engine Cutoff [Aerospace]

MECOMSAG Mobility Equipment Command Scientific Advisory Group

MEcon Master of Economics

MeCp(CO)₃ Methylcyclopentadienylmanganese Tricarbonyl

MECR Maintenance Engineering Change Request

MeCr Metal Chromide

MECS Molecular Electron Correlation Spectrometer; Molecular Electron Correlation Spectrometry

MECS Multiple Expansion Cluster Source

MECU Master Engine Control Unit

MECU Million European Currency Units

Me₂CuLi Lithium Dimethylcopper

1-MeCYT 1-Methylcytosine [Biochemistry]

MED Manual Entry Device

MED Marine Emergency Duties

MED Materials Engineering Department

MED Materials Engineering Design

MED Materials Explosives Division [UK]

MED Mechanical Engineering Department

MED Mechanical Engineering Division

MED Mechanical Engineering Division [of Institute of Engineers, India]

MED Metallurgical Engineering Department

MED Microelectronic Device

MED Microwave Plasma Emission Detector

MED Minimum Energy Dwelling

MED Minimum Essential Design

MED Mining and Exploration Division [of Society of Mining Engineers, US]

MED Mobile Energy Depot

MED Molecular Electronic Device

MEd Master of Education

Med Median [also med]

Med Medical; Medicine [also med]

Med Mediterranean

Med Medium [also med]

med medieval

MEDA Military Emergency Diversion Aerodrome

MEDA Mineral Economic Development Agreement [Canada]

MEDA Mineral Exploration Depreciation Allowance

MEDA Mining Exploration Depletion Allowance

MEDAB Methyl Dimethylaminoazobenzene

MEDAC Medical Electronic Data Acquisition and Control

MEDAC Medical Equipment Display and Conference

MEDAL Micromechanical Engineering Data for Automated Logistics; Micromechanized Engineering Data for Automated Logistics

MEDARATEL Mediterranean and Arabic Telecommunications Network

MedArty Medium Artillery

Med Biol Eng Comput Medical and Biological Engineering and Computing [Publication of the International Federation for Medical and Biological Engineering, Canada]

MEDC Microelectronics Educational Development Center [UK]

MEDC Moessbauer Effect Data Center [of University of North Carolina, US]

Med Chem Medical Chemist(ry)

MEDDA Mechanical (or Mechanized) Defense Decision Anticipation

MEDDS N-Methylethylenediamine-N,N'-Disuccinate [also medds]

MedE Medical Engineer [also Med Eng, or MedEng]

Med Ed Medical Education

Meded K Acad Wet Lett Schone Kunsten Belg Mededelingen van de Koninklijke Academie voor Wetenschappen, Letteren en Schone Kunsten van België [Communications of the Royal Belgian Academy of Sciences, Letters and Fine Arts]

Med Energy Res Rep Medical and Energy Research Reports [Published by the Oak Ridge Associated Universities, Tennessee, US]

Med Eng Medical Engineer(ing)

MEDes Master of Environmental Design

MEDGE Management Edge (Database) [Part of Business Australia Database]

MedGV *(Medizinische Geräte-Verordnung)* – German Regulations on Medical Devices

MEDH Microgravity Environment Description Handbook [of NASA Lewis Research Center, Cleveland, Ohio, US]

Med Hyp Medical Hypotheses (Journal)

MEDI Marine Environmental Data Information System [of Intergovernmental Oceanographic Commission]

MEDIA Magnavox Electronic Data Image Apparatus

Me₄DIB Tetramethyl 1,4-Diisocyanobenzene [also me₄dib]

Medicaid Medical Care Aid Program [US Federal-State Health Program for Low-Income Earners] [also medicaid]

Medicare Medical Care Insurance Program [US and Canadian Government Medical Care, or Health Insurance Program] [also medicare]

MEDICO Medical Institute Committee [US]

MEDICO Model Experiment in Drug Indexing by Computers [of Rutgers University, Piscataway, New Jersey, US]

MEDICS Medical Information and Communication System

MEDICS Medical Information Computer System

ME4DIEN N,N,N'N'-Tetramethyl Diethylene Triamine

MEDINA Methylenedinitramine

Med Inform Medical Informatics [Journal published in the UK]

Med Instrum Medical Instrumentation [Journal published in the US]

MEDIRS Marine Environmental Data Information Referral System [of Intergovernmental Oceanographic Commission]

MEDIS Modular Engineering Document Imaging System

Medit Mediterranean

Med J Aust Medical Journal of Australia

Med Lab Sci Medical Laboratory Sciences

MEDLARS Medical Literature Analysis and Retrieval System [of National Library of Medicine, Bethesda, Maryland, US]

Med Lett Medical Letters [US Journal]

MEDLINE MEDLARS On-Line System [of National Library of Medicine, Bethesda, Maryland, US] [also Medline]

MEDNET Mediterranean School on Science and Technology [Italy]

Med News Medical News

MEDO Multipole Expansion of Diatomic Overlap [Physics]

Med Obl Medulla Oblongata (of Brain) [also med obl] [Anatomy]

MeDOC Multimediale elektronische Dokumente [Multimedia Electronic Document Supply (Project), of Fachinformationszentrum, Karlsruhe, Germany]

MeDPA N-Methyl-bis(2-Pyridylmethyl)amine

Med Phys Medical Physicist; Medical Physics

Med Phys Medical Physics [Published for the American Association of Physicists in Medicine by the American Institute of Physics, US]

MEDPOL Mediterranean Action Plan Pollution Monitoring and Research Program [of United Nations Environmental Program]

Med Prog Through Technol Medical Progress Through Technology [Journal published in the Netherlands]

Med Radiol Meditsinskaya Radiologiya [Russian Journal on Medical Radiology]

Med Res Eng Medical Research Engineering [Journal published in the US]

Med Ref Medical Reference [also med ref]

Med Res Medical Research(er) [also med res]

Med Res Eng Medical Research Engineer(ing)

MEDRISK Medical Device Risk

MEDS Marine Ecological Data System (Program) [of National Marine Fisheries Service, US]

MEDS Marine Environmental Data Service

MEDS Medical Evaluation Data System

MEDS Medium Energy Dislocation Structure [Materials Science]

MedScD Doctor of Medical Science

Med Sci Sports Exerc Medicine and Science in Sports and Excercise [Journal]

MEDSERV Medical Service Corps [US]

MEDSMB Medical Standards Management Board [US]

MEDSPECC Medical Specialist Corps [US]

MEDTA N-Methylethylenediamine Triacetate [also Medta, or medta]

MED3A N-Methylethylenediamine-N,N',N'-Triacetate [also Med3a, or med3a]

Med Tek Medicinsk Teknik [Swedish Journal on Medical Technology]

Med Tekh Meditsinskaya Tekhnika [Russian Journal on Biomedical Engineering]

Med Times Medical Times [International Journal]

Med Trib Medical Tribune [US Journal]

MEDTT (Ontario) Ministry of Economic Development, Trade and Tourism [Canada]

MEduc Master of Education

Med Welt Medizinische Welt [German Medical Journal]

MEE Master of Electrical Engineering

MEE Microelectronic Engineer(ing)

MEE Migration-Enhanced Epitaxy [Materials Science]

MEE Mining Electrical Engineer(ing)

MEE Mission Essential Equipment

MEECI Mouvement des Etudiants et Elèves de la Côte d'Ivoire [Students and Pupils Movement of the Ivory Coast]

MEECN Minimum Essential Emergency Communications Network [US]

MEED Medium Energy Electron Diffraction

MEED/AES Medium Energy Electron Diffraction/Auger Electron Spectroscopy (Method)

MeEDNA N-Methylethylenedinitramine

MEEN N,N'-Dimethyl 1,2-Ethanediylbis(2-Aminoethanethiol)

MEEN N-Methyl 1,2-Ethanediamine

ME2EN N,N-Dimethyl 1,2-Ethanediamine

Me4[12]eneN3 2,4,4,9-Tetramethyl 1,5,9-Triazacyclododec-1-ene

MEEng Master of Electrical Engineering

MEEP Poly(bis-(2(2-Methoxyethoxy)ethoxy)phosphazene

Meerestech Meerestechnik [German Journal on Marine Technology of Verein Deutscher Ingenieure]

MEES Middle East Economic Survey

MEETAT Maximum Improvement in Electronics Effectiveness Through Advanced Techniques

MEF Maximum Expected Flight

MEF Multiple Effect Flash Evaporator

MeF Metal Fluoride

MeF Methyl Fluoride

MeFe$_{12}$O$_{19}$ Metal Hexaferrite [General Formula]

Me$_x$FeO$_4$ Cubic Spinel Type Ferrites [General Formula; *Me* Represents a Metal]

MEFIT Multielectron Fit

MEFTA Metalworking Industries of European Free Trade Association

MEFV Maintenance Equipment Floor Valve

MEG Magnetoelectric Generator

MEG Magnetoencephalogram; Magnetoencephalograph(y)

MEG Materials Engineering Group

MEG Megger

MEG Message Expediting Group

MEG Mineral Exploration Geophysics

MEG Modified Energy Gap [Solid-State Physics]

MEG Monoethyleneglycol

Meg Megabyte [also meg]

meg mega-

meg megohm [Unit]

MEGA Methylglucamide [Biochemistry]

MEGA Muon into an Electron plus a GAmma-Ray (Experiment) [of Los Alamos Meson Physics Facility, New Mexico, US]

MEGA-7 Heptanoyl-N-Methyl Glucamide [Biochemistry]

MEGA-8 N-Octanoyl-N-Methyl Glucamide [Biochemistry]

MEGA-9 N-Nonanoyl-N-Methyl Glucamide [Biochemistry]

MEGA-10 N-Decanoyl-N-Methyl Glucamide [Biochemistry]

MEGA-12 n-Dodecanoyl-N-Methyl Glucamide [Biochemistry]

(Me$_3$)Ga Trimethylgallium

Me-GaAs Metal-Gallium Arsenide (Couple)

(Me$_3$)Ge Trimethylgermane

Me$_3$GeBr Bromotrimethylgermane (or Trimethyl Germanium Bromide)

Me$_3$GeCl Chlorotrimethylgermane (or Trimethyl Germanium Chloride)

Megg Merging [also megg]

Me D-Glu Methyl D-Glucoside [Biochemistry]

Me α-D-Glu Methyl α-D-Glucoside [Biochemistry]

Me β-D-Glu Methyl β-D-Glucoside [Biochemistry]

MeGly N-Methyl Glycine [Biochemistry]

MEGW Megawatt [also MW, or mw]

MEH McCombie-Elcock-Huntington (Mechanism)

MeH Metal Hydride

MEHDPO Methylenebis[bis(2-Ethylhexyl)phosphine] Dioxide

Me$_2$Hg Dimethylmercury

MeHgCl Methylmercury(II) Chloride

MeHgI Methylmercury(II) Iodide

MeHgOH Methylmercury(II) Hydroxide

MEHP Mono(2-Ethylhexyl)phosphate

MEH(Φ)P (1-Methylheptyl)phenylphosphonate

MEHPA Monoethylhexylphosphoric Acid

MEH PPV Polymethoxy Ethyl Hexyloxy-p-Phenylenevinylene [or MEH-PPV]

MEH PPV Poly(2-Methoxy-5-(2-Ethyl Hexyloxy)-p-Phenylenevinylene [also MEH PPV]

7-MeHYP 7-Methylhypoxanthine

MEHQ Methyl Ether of Hydroquinone

Mehran Univ Res J Eng Technol Mehran University Research Journal of Engineering and Tech nology [Pakistan]

MEHT Minimum Pilot Eye-Height over Threshold

MEI Main Economic Indicators

MEI Manual of Engineering Instructions

MEI Marginal Efficiency of Investment

MEI Master Inspection Item

MEI Materials Engineering Institute [of ASM International, US]

MEI Matsushita Electric Industrial Company [Kyoto, Japan]

MEI Metals Engineering Institute [Name Changed to Materials Engineering Institute, US]

MEI Minimum Enroute Instrument (Altitude)

MEI Ministry of the Electronics Industry [PR China]

MEI 2-Morpholinoethylisocyanide

MeI Metal Iodide

MeI Methyl Iodide

MEIC Materials Engineering Institute Council [of ASM International, US]

MEIC Member of the Engineering Institute of Canada

MEIC 1-Methyl-3-Ethyl Imidazolium Chloride

Meiden Rev Meiden Review [Published by Meidensha Electric Manufacturing Co. Ltd., Tokyo, Japan]

Meiden Rev (Int Ed) Meiden Review (International Edition) [Published by Meidensha Electric Manufacturing Co. Ltd., Tokyo, Japan]

MEIG Main Engine Ignition [Aerospace]

MEIU Main Engine Interface Unit [Aerospace]

MeIle N-Methyl Isoleucine [Biochemistry]

1-MeIMID 1-Methyl Imidazole

(Me)₃In Trimethylindium

MEIP Mechanical, Electrical, Instrumentation and Control, Power Equipment

MEIP Mineral Exploration Incentives Program [Canada]

MEIR Medium-Energy Ion Reflection

MEIS Medium-Energy Ion Scattering (Spectroscopy)

MEIS Metal-Epitaxial Insulator-Semiconductor

MEIS Military Entomology Information Service [of US Department of Defense]

MEIS Multispectral Electrooptical Imaging Scanner

MEISFET Metal-Epitaxial Insulator-Semiconductor Field-Effect Transistor

MeISTSC 1-Methyl Isatin Thiosemicarbazone

MEIT Multi-Element Integrated Test

MEITC Manitoba Economic Innovation and Technology Council [Canada]

MEIU Mobile Explosives Investigation Unit

MEJ Ministry of Education of Japan

MEK MAP (Mitogen-Activated Protein) Activated Kinase [Biochemistry]

MEK Methyl Ethyl Ketone

Mekh Avtom Proizvod Mekhanizatsiya i Avtomatizatsiya Proizvodstva [Russian Journal on Mechanization and Automation]

Mekh Kompoz Mater Mekhanika Kompozitnykh Materialov [Latvian Journal on the Mechanics of Composite Materials]

Mekh Tverd Tela Mekhanika Tverdogo Tela [Ukrainian Journal on the Mechanics of Rigid Bodies]

MEKP Methyl Ethyl Ketone Peroxide

MEKTS Modular Electronic Key Telephone System

MEL Magnesium Elektron Limited [Manchester, UK]

MEL Many-Element Laser

MEL Marine Ecology Laboratory [Canada]

MEL Marine Engineering Laboratory [US Navy]

MEL Master Equipment List

MEL Materials Evaluation Laboratory [of Institute for Materials Research, National Institute for Standards and Technology, US]

MEL Maximum Excess Loss

MEL Mechanical Engineering Laboratory

MEL Mechanical Engineering Laboratory [of Agency of Industrial Science and Technology, Tokyo, Ibaraki, Japan]

MEL Minimum Equipment List

MEL Multi-Engine, Land [Aeronautics]

MEL Multiprogram Energy Laboratories [of US Department of Energy]

Mel Melamine

Mel Melbourne [Australia]

MELA Middle East Librarians Association

MELA Modified Effective Liquid Approximation

MelA Mellitic Acid

MELAB Michigan English Language Accreditation Board

MELBA Multipurpose Extended Lift Blanket Assembly

MELCU Multiple External Line Control Unit

MELEC Microelectronic(s)

MELEM Microelement [Electronics]

MElEng Master of Electrical Engineering

MELEO Materials Exposure in Low Earth Orbit [NASA Space Shuttle Experiment]

MELEO/IOCM Materials Exposure in Low Earth Orbit/Interim Operational Contamination Monitor [NASA Space Shuttle]

MELF Metal Electrode Face-Bonded; Metal Electrode Face-Bonding

MEL-FS Multiprogram Energy Laboratories–Facilities Support [of US Department of Energy]

MELI Master Equipment List Index

MELI Minimum Equipment List Index

MeLi Methyllithium

MELSIR Melbourne Specimen Information Register [of National Herbarium of Victoria (Database), Australia]

MELU Melbourne University [Australia]

MELVA Military Electronic Light Valve

MEM Mars Excursion Module [of NASA]

MEM Master Equation Method [Metallurgy]

MEM Master of Engineering Management

MEM Maximum Entropy (Analysis) Method

MEM Display (Used and Free) Memory (Command) [also mem]

MEM Meteoroid Exposure Module

MEM Method of Exponential Multipliers

MEM β-Methoxyethoxymethyl

MEM Methoxyethoxymethyl

MEM Methoxyethylmercury

MEM Microelectromechanical; Microelectromechanics

MEM Microelectronic Manufacturing

MEM Microelectronic Material

MEM Micro-Erosion Meter

MEM Mineral and Electrolytic Metabolism [International Journal]

MEM Minimum Essential Medium [Biochemistry]

MEM Mirror Electron Microscope; Mirror Electron Microscopy

MEM Modified Eagle's Medium [Biochemistry]

MEM Modified Effective Modulus

MEM Module Exchange Mechanism

Mem Member [also mem]

Mem Memoirs [also mem]

Mem Memorandum [also mem]

Mem Memorial [also mem]

Mem Memory [also mem]

MEMA Machinery and Equipment Manufacturers Association [US]

MEMA Marine Engine and Equipment Manufacturers Association [UK]

MEMA Microelectronic Modular Assembly

MEMA Motor and Equipment Manufacturers Association [US]

MEMAC Machinery and Equipment Manufacturers Association of Canada

MEMAC Marine Emergency Mutual Aid Center

Mem AAAS Memoirs of the American Academy of Arts and Sciences [US]

Mem Am Math Soc Memoirs of the American Mathematical Society [US]

Mem Am Phil Soc Memoirs of the American Philosophical Society [US]

Mem AMS Memoirs of the American Mathematical Society [US]

Mem APS Memoirs of the American Philosophical Society [US]

Mém Artillerie Fr Mémorial de l'Artillerie Française [Memorial of the French Artillery]

MemAvail Memory Available (Sum of Free Blocks) [Pascal Function]

Memb Membrane [also memb]

Mém BRGM Mémoire du Bureau de Recherches Géologiques et Minières [Memoirs of the Bureau for Geological and Mining Research, France]

MEMC Methoxyethylmercury Chloride

MEM Chloride β-Methoxyethoxymethyl Chloride

Mém Cl Sci Acad R Belg Mémoires de la Classe des Sciences de l'Academie Royale de Belgique [Memoirs of the Science Class of the Royal Academy of Belgium]

Mem Coll Agric, Kyoto Univ Memoirs of the College of Agriculture, Kyoto University [Japan]

Mem Coll Eng, Chuba Univ Memoirs of the College of Engineering, Chuba University [Japan]

MEM-D-Val Minimum Essential Medium (Eagle), D-Valine Modification [Biochemistry]

MEME Multiple Entry–Multiple Exit

Me-Me Metal-Metal (Reaction)

Mem Ent Soc Can Memoirs of the Entomological Society of Canada

Mém Etud Sci Rev Metall Mémoires et Etudes Scientifiques de la Revue de Metallurgie [French Memoirs and Scientific Studies in Metallurgy]

Mem Fac Eng Des, Kyoto Inst Technol Memoirs of the Faculty of Engineering and Design, Kyoto Institute of Technology [Japan]

Mem Fac Eng Des, Kyoto Inst Technol, Ser Sci Technol Memoirs of the Faculty of Engineering and Design, Kyoto Institute of Technology, Series of Science and Technology [Japan]

Mem Fac Eng, Fukui Univ Memoirs of the Faculty of Engineering, Fukui University [Japan]

Mem Fac Eng, Hiroshima Univ Memoirs of the Faculty of Engineering, Hiroshima University [Japan]

Mem Fac Eng, Hokkaido Univ Memoirs of the Faculty of Engineering, Hokkaido University [Japan]

Mem Fac Eng, Kagoshima Univ Memoirs of the Faculty of Engineering, Kagoshima University [Japan]

Mem Fac Eng, Kobe Univ Memoirs of the Faculty of Engineering, Kobe University [Japan]

Mem Fac Eng, Kumamoto Univ Memoirs of the Faculty of Engineering, Kumamoto University [Japan]

Mem Fac Eng, Kyoto Univ Memoirs of the Faculty of Engineering, Kyoto University [Japan]

Mem Fac Eng, Kyushu Univ Memoirs of the Faculty of Engineering, Kyushu University [Japan]

Mem Fac Eng, Miyazaki Univ Memoirs of the Faculty of Engineering, Miyazaki University [Japan]

Mem Fac Eng, Nagoya Univ Memoirs of the Faculty of Engineering, Nagoya University [Japan]

Mem Fac Eng, Osaka City Univ Memoirs of the Faculty of Engineering, Osaka City University [Japan]

Mem Fac Eng, Tamagawa Univ Memoirs of the Faculty of Engineering, Tamagawa University [Japan]

Mem Fac Eng, Yamaguchi Univ Memoirs of the Faculty of Engineering, Yamaguchi University [Japan]

Mem Fac Sci, Kochi Univ A Memoirs of the Faculty of Science, Kochi University, Series A [Japan]

Mem Fac Sci, Kochi Univ B Memoirs of the Faculty of Science, Kochi University, Series B [Japan]

Mem Fac Sci, Kyoto Univ Memoirs of the Faculty of Science, Kyoto [Japan]

Mem Fac Sci, Kyoto Univ, Ser Phys Astrophys Geophys Chem Memoirs of the Faculty of Science, Kyoto, Series of Physics, Astrophysics, Geophysics and Chemistry [Japan]

Mem Fac Sci, Kyushu Univ Memoirs of the Faculty of Science, Kyushu University [Japan]

Mem Fac Sci, Kyushu Univ A Memoirs of the Faculty of Science, Kyushu University, Series A [Japan]

Mem Fac Sci, Kyushu Univ B Memoirs of the Faculty of Science, Kyushu University, Series B [Japan]

Mem Fac Sci, Kyushu Univ C Memoirs of the Faculty of Science, Kyushu University, Series C [Japan]

Mem Fac Technol, Kanazawa Univ Memoirs of the Faculty of Technology, Kanazawa University [Japan]

Mem Fac Technol, Tokyo Metrop Univ Memoirs of the Faculty of Technology, Tokyo Metropolitan University [Japan]

MeMgBr Methylmagnesium Bromide

MeMgCl Methylmagnesium Chloride

MeMgI Methylmagnesium Iodide

Mem Hokkaido Inst Technol Memoirs of the Hokkaido Institute of Technology [Japan]

MEMIC Medical Microbiological Interdisciplinary Committee

Mem Inst Sci Ind Res, Osaka Univ Memoirs of the Institute of Scientific and Industrial Research, Osaka University [Japan]

Memistor Memory Register Storage Device [also memistor]

Memistor Resistor with Memory Effect [also memistor]

Mem Kitami Inst Technol Memoirs of the Kitami Institute of Technology [Japan]

Mem Kyushu Inst Technol Eng Memoirs of the Kyushu Institute of Technology, Engineering [Japan]

Meml Memorial [also meml]

Mem Natl Def Acad Memoirs of the National Defense Academy [Japan]

Memo Memorandum [also memo]

MeMo Metal Molybdide

MEMP Many-Electron, Many-Photon (Nonperturbative Theory) [Physics]

MEMP Maximization of Expected Maximum Profits

MEMP Morpholino-Ethyl-2 Methyl-4 Phenyl-6 Pyridazole-3 Chlorohydrate [also Ag246]

MEMRA Mechanical Equipment Manufacturers Representatives Association [US]

MEMRB Middle East Marketing Research Bureau

Mem Res Inst Sci Eng Ritsumeikan Univ Memoirs of the Research Institute of Science and Engineering, Ritsumeikan University [Japan]

MEMS Master of Engineering in Manufacturing Systems

MEMS Mechanism Modeling System

MEMS Microelectromechanical System

MEMS Micromachined Electromechanical Components and Systems

MEMS Multiple-Effect Multiple-Stage (Evaporation)

Mems Memoirs [also mems]

Mem Sch Eng Okayama Univ Memoirs of the School of Engineering, Okayama University [Japan]

Mem Sch Sci Eng, Waseda Univ Memoirs of the School of Science and Engineering, Waseda University [Japan]

Mém Sci Rev Metall Mémoires Scientifiques de la Revue de Metallurgie [French Scientific Memoirs and Review in Metallurgy]

Mem Sect Stiint Memoriile Sectiilor Stiintificeria [Memoirs of the Scientific Section; published by Academia Republicii Socialiste Romania]

Mem Soc Astron Ital Memorie della Societa Astronomica Italiana [Memoirs of the Italian Astronomical Society]

MEM(TCNQ)$_2$ Methoxyethoxymethyl Bis(Tetracyanoquinodimethane)

Mem Tohoku Inst Technol I, Sci Eng Memoirs of the Tohoku Institute of Technology, Series I, Science and Engineering [Japan]

memu milli-electromagnetic unit(s)

μemu micro-electromagnetic unit(s)

memu/cm² milli-electromagnetic unit(s) per square centimeter

MEN N-Methylethylenediamine [also men]

MEN Ministère de l'Education Nationale [Ministry of National Education, France]

MEN Multiple Endocrine Neoplasia [Medicine]

MEN Multiple Event Network

.MEN Menu [File Name Extension] [also .men]

MeN Metal Nitride

MeN Methyl Nitrate

Me$_3$N Trimethylamine

M(en)$_2$ Metal 1,2-Diaminoethane

MENA Methyl Ester of Naphthaleneacetic Acid

Me$_4$NAc Tetramethylammonium Acetate

MeNb Metal Niobide

Me$_3$N·BH$_3$ Borane-Trimethylamine Complex

Me$_4$NBF$_4$ Tetramehylammonium Tetrafluoroborate

Me$_4$NBH$_4$ Tetramehylammonium Borohydride

Me$_4$NBr Tetramehylammonium Bromide

MeNC Methyl Isonitrile

Me$_4$NCl Tetramehylammonium Chloride

Me$_4$NCN Tetramehylammonium Cyanide

MEND Mine Environment Neutral Drainage (Program) [formerly RATS]

MEND Mineral Environment Neutral Drainage (Program) [Canada]

Mendeleev Comm Mendeleev Communications [Journal of the Royal Society of Chemistry, UK]

MeNENA 1-Nitroxytrimethylene-3-Nitramine

MeNEt$_3$Br Triethylmethylammonium Bromide

MENEX Maintenance Engineering Exchange

Me$_4$NF Tetramehylammonium Fluoride

MEng Master of Engineering [also MEngr]

MEngMgt Master of Engineering Management

MEngSc Master of Engineering and Science [also MEngSci]

MEngStud Master of Engineering Studies

Me$_2$NH Dimethylamine

Me$_2$NH·BH$_3$ Borane-Dimethylamine Complex

Me$_3$N·HCl Trimethylammonium Chloride (or Trimethylamine Hydrochloride)

Me$_4$NHF$_2$ Tetramethylammonium Hydrogen Difluoride

Me$_4$NI Tetramehylammonium Iodide

Me$_4$NOH Tetramehylammonium Hydroxide

MENOsar 1-Methyl-8-Nitro-3,6,10,13,16,19-Hexaazabicyclo-[6.6.6]eicosane

MENU (International Symposium on) Meson Nucleon Physics and the Structure of the Nucleon

Me$_4$NPF$_6$ Tetramehylammonium Hexafluorophosphate

MEnv Master of Environmental Studies

MEnvDes Master of Environmental Design

MEnvSc Master of Environmental Science [also MenvS]

M(en)$_2$X$_2$ Metal (1,2-Diaminoethane) Divalent Halogen [e.g., Pt(en)$_2$Cl$_2$]

MEO Major Engine Overhaul

MEO Management Engineering Office

MeO Metal Oxide

MeO Methoxide (or Methylate)

MeO Methoxy-

Me$_2$O Methyl Oxide (or Dimethyl Ether)

Me$_p$O$_q$ Produced Oxide [Oxide/Metal Composites]

MeO Alc 2-Methoxyethanol

(MeO)$_3$B Methyl Borate (or Trimethyl Borate)

MeOBB Methoxybenzoylbenzoate

MeO·6Fe$_2$O$_3$ Metal Hexaferrite [General Formula]

MeOH Methanol (or Methyl Alcohol)

MeOK Potassium Methoxide (or Potassium Methylate)

MeOLi Lithium Methoxide (or Lithium Methylate)

MEOP Maximum Expected Operating Pressure

(MeO)$_3$P Methyl Phosphite (or Trimethyl Phosphite)

MeOr Methyl Orange

MEOS Magnetic Equation of State [Solid-State Physics]

MEOS Multiband Electrooptical Scanner

(MeO)$_2$SO Methyl Sulfite (or Dimethyl Sulfite)

(MeO)$_2$SO$_2$ Dimethyl Sulfate

MEP O,O-Dimethyl-O-(3-Methyl-4-Nitrophenyl)-phosphorothionate

MEP Main Engine Propellant [Aerospace]

MEP Main Entry Point

MEP Management Engineering Plan

MEP Management Engineering Program

MEP Manufacturing Extension Partnership [of National Institute for Standards and Technology, US]

MEP Master of Engineering Physics

MEP Master of Environmental Planning

MEP Mean Effective Potential

MEP Mean Effective Pressure [also mep]

MEP Medical Education Program

MEP Member of the European Parliament

MEP Methyl Ethyl Phenol

MEP 2-Methyl-5-Ethyl Pyridine

MEP Microelectronic Packaging

MEP Microelectronic Education Program [UK]

MEP Microfile Enlarger Printer

MEP Minimum Entry Point

MEP Minority Engineering Programs [at US Colleges and Universities]

MEP Molecular Electrostatic Potential

MEP Mission Effects Projector [of NASA]

MEP Motor-Evoked Potential

MeP Metal Phosphide

MeP Mexican Peso [Currency]

MEPA Marine and Estuarine Protected Area

MEPA Master of Engineering in Public Administration

MEPA Meteorological and Environmental Protection Administration

[Me$_3$P·AgI]$_4$ (Trimethylphosphine)silver Iodide Complex

MEPAS Multimedia Environmental Pollution Assessment System

Me$_3$PAuCl (Trimethylphosphine)gold(I) Chloride Complex

MEPC Marine Environment Protection Committee

MEPC Maritime Environmental Protection Committee [of International Maritime Organization]

MEPC Master of Environmental Pollution Control

MEPCB Mechanical Earth Pressure Counterbalance

MEPD Master of Education–Professional Development

MEPED Medium Energy Proton and Electron Detector

MEPF Multiple Experiment Processing Furnace

MEPHISTO Materials for the Study of Interesting Phenomena of Solidification on Earth and in Orbit

MePhs Methyl Phosphate

Me$_3$PhGe Trimethylphenylgermane

2-MePIPDTC 2-Methyl-Piperidine Dithiocarbamate

3-MePIPDTC 3-Methyl-Piperidine Dithiocarbamate

4-MePIPDTC 4-Methyl-Piperidine Dithiocarbamate

MEPN N,N'-Dimethyl-N,N'-Bis(2-Mercaptoethyl)-1,3-Propanediamine

MEPP Mineral Economics and Policy Program

MEPPV Methoxy Ethylhexyloxy Paraphenylenevinylene

MEPS Millions of Events per Second

MePVTSC Methyl Pyruvate Thiosemicarbazone

MePy N-Methyl Pyridinium [also Mepy, mepy or Me-py]

(Mepy)2Pytren N,N-Bis[2-{((6-Methyl-2-Pyridyl)methylene) amino}ethyl]-N'-(2-Pyridylmethylene) 1,2-Ethanediamine [also (mepy)2pytren]

(Mepy)(Py)2tren N'-[(6-Methyl-2-Pyridyl)methylene]-N,N-bis [2-{(2-Pyridylmethylene)amino}ethyl]-1,2-Ethanediamine [also (mepy)(py)2tren]

(Mepy)3tren N'-[(6-Methyl-2-Pyridyl)methylene]-N,N-bis [2-{((6-Methyl-2-Pyridyl)methylene)-amino}ethyl]-1,2-Ethanediamine [also (mepy)(py)2tren]

Mepyz Methylpyrazine [also mepyz]

Me$_2$pyz Dimethylpyrazine [also me$_2$pyz]

mEq milliequivalent(s) [also meq]

mEq/g milli-equivalent(s) per gram [also meq/g]

mEq/L milli-equivalent(s) per liter [also meq/L, mEq/l, or meq/l]

mEq/mL milli-equivalent(s) per milliliter [also meq/mL, mEqm/l, or meq/ml]

μEq/mL micro-equivalent(s) per milliliter [also μeq/mL, μEq/ml, or μeq/ml]

mEqs milliequivalents [also meqs]

MEQUIN 2-Methyl-8-Hydroxyquinoline

mEq wt milliequivalent weight [also meq wt]

MER Manned Earth Reconnaissance

MER Master of Energy Resources

MER Materials and Electrochemical Research Corporation [Tucson, Arizona, US]

MER Maximum Efficient Rate

MER Mediated Electrochemical Reaction

MER Message Error Rate

MER Minimum Energy Requirement

MER Mission Evaluation Room

MER Multiple Ejector Rack

MeR Methyl Red

Mer Mercantile [also mer]

Mer Merchandise; Merchandising; Merchant [also mer]

Mer Meridian [also mer]

MERA Molecular Electronics for Radar Applications

MERC Meat Export Research Center [of Iowa State University, Ames, US]

MERC Microelectronics Research Center [US]

Merc Mercantile [also merc]

Merch Merchandising [also merch]

MERCI Multimedia European Research Conferencing Integration

MER Corp Materials and Electrochemical Research Corporation [Tucson, Arizona, US]

MERDC Mobility Equipment Research and Development Center [US]

MERDL Medical Equipment Research and Development Laboratory [US Army]

MEREA Member of the Electrical Railway Engineering Association [US]

Meres Autom Meres es Automatika [Hungarian Publication on Automation]

MERGE Mechanized Retrieval for Greater Efficiency

MERGV Martian Exploratory Rocket Glide Vehicle

Meri Merionetsshire [Wales]

MERIC Middle Eastern Regional Isotope Center (for the Arab Countries) [Cairo, Egypt]

MERIE Magnetically-Enhanced Reactive Ion Etch (Tool)

MERIP Middle East Research Information Project

MERIS Medium Resolution Imaging Spectrometer

MERIT Michigan Educational Research Information Triad [US]

MERL Magnetic and Electronic Research Laboratory [of Hitachi Limited, Ibaraki, Japan]

MERL Materials Engineering Research Laboratory

MERL Materials Equipment Requirement List

MERL Mechanical Engineering Research Laboratory [UK]

MERL Municipal Environmental Research Laboratory [US]

MERLIN Machine-Readable Library Information

MERLIN Medium-Energy Research Light-Water-Moderated Industrial Nuclear Reactor [UK]

MERLIN-JULICH Medium-Energy Research Light-Water-Moderated Industrial Nuclear Reactor at Julich [Germany]

MERM Material Evaluation Rocket Motor

MERMUT Mobile Electronic Robot Manipulator and Underwater Television

MERP Minimum Energy Reaction Path

MERR Morse Exponential Repulsion Representation

MERRCAC Middle Eastern Regional Radioisotope Center for Arab Countries

Mer Regul Mereni a Regulace [Publication on Automation and Control; published by the Research Institute for Automation Systems, Prague, Czech Republic]

MERS Mobility Environmental Research Study

MERS Multiple Electron Resonance Spectroscopy

MERSAT Meteorology and Earth Observation Satellite

MERT Modified Effective Range Theory

MERT Multiple Environment Real Time

MERU Maharishi European Research University

MERU Mechanical Engineering Research Unit [South Africa]

MERU Milliearth Rate Unit

Meruzeler Dokl Meruzeler Doklady [Russian Journal]

MES Main Engine Start [Aerospace]

MES Manual Entry Subsystem

MES Manufacturing Execution System

MES Manufacturing Execution Systems Association [US]

MES Marine Engineering Society [Japan]

MES Master of Engineering Science

MES Master of Environmental Sciences

MES Master of Environmental Studies

MES Mated Elements Simulator

MES Mated Events Simulator

MES McMaster Engineering Society [of McMaster University, Hamilton, Canada]

MES Mechanical Engineering Science

MES Mechanical Engineering Section

MES Mechanical Engineering Society

MES Medium Energy Source

MES Metal-Gate Schottky

MES Metal-Semiconductor Device

MES Michigan Engineering Society [US]

MES Minerals Engineering Society [UK]

MES Ministry of Education of Singapore

MES Mission Events Sequence

MES Mobile Earth Station

MES Moessbauer Effect Spectra

MES Moessbauer Effect Spectroscopy

MES Molecule Emission Spectrometry

MES Morpholineethanesulfone

MES 2-(N-Morpholino)ethanesulfonic Acid

MES 4-Morpholinoethanesulfonic Acid

MeS Metal(lic) Sulfide

Me$_2$S Methyl Sulfide (or Dimethyl Sulfide)

Me$_2$S$_2$ Methyl Disulfide

Mes Mesitylene

Mes Mesocolon [also mes] [Anatomy]

MESA Manned Environmental System Assessment [of NASA]

MESA Marine Ecosystems Analysis

MESA Marine Education Society of Australia

MESA Mathematics, Engineering and Science Achievement

MESA 2-Mercaptoethanesulfonic Acid [also mesa]

MESA Miniature (or Miniaturized) Electrostatic Accelerometer

MESA Mining Enforcement and Safety Administration [US]

MESA Modular Equipment Stowage Assembly

MESA Modular Experiment Platform for Science and Applications

MESA MOPS (3-[N-Morpholino]propanesulfonic Acid) EDTA (Ethylenediaminetetraacetic Acid) Sodium Acetate (Buffer)

MESAEP Mediterranean Scientific Association for Environmental Protection

Me2-salen N,N'-Bis(α-Methylsalicylidene)ethylenediamine [also Me2-SALEN]

Mes asc Mesocolon ascendum [also mes asc] [Anatomy]

MeSb Metal Antimonide

Me$_2$S·BBr$_3$ Dimethyl Sulfide-Tribromoborane (or Boron Tribromide-Methyl Sulfide Complex)

Me$_2$S·BCl$_3$ Dimethyl Sulfide-Trichloroborane (or Boron Trichloride-Methyl Sulfide Complex)

Me$_2$S·BF$_3$ Dimethyl Sulfide-Trifluoroborane (or Boron Trifluoride-Methyl Sulfide Complex)

Me$_2$S·BH$_3$ Borane-Methyl Sulfide Complex

MESBP Management Excellence in Small Business Program [of Industry, Science and Technology Canada]

MESC Master Events Sequencer Controller

MESC Mid-Continent Ecological Science Center [of US Geological Survey/Biological Resources Division at Fort Collins, Colorado]

MESC Ministry of Education, Science and Culture [Japan]

MESc Master of Engineering Sciences

MESCJ Ministry of Education, Science and Culture of Japan

Mes desc Mesocolon descendum [also mes desc] [Anatomy]

MeSe Metal Selenide

Me$_2$Se$_2$ Methyl Diselenide

MESEC Multi-Cultural Environmental Science Education Centers

MES/ERP Manufacturing Execution Systems/ Enterprise Resource Planning

MESET Minority Enrichment Seminar in Engineering Training

MESF Molecules of Equivalent Soluble Fluorochrome

MESFET Metal-Gate Schottky Field-Effect Transistor

MESFET Metal-Semiconductor Field-Effect Transistor

MESG Maximum Experimental Safe Gap

MESG Micro-Electrostatically Suspended Gyro

MESG Minimum Electrical Spark Gap

Mesg Message [also mesg]

MESH Medical Subject Headings

MESHA Manager Environmental Safety and Health Awareness

MESI Modified, Exclusive, Shared and Invalid (Protocol)

Me$_3$SI Trimethylsulfonium Iodide

MeSi Metal Silicide

Me$_3$Si Trimethylsilyl

Me$_4$Si Tetramethylsilyl

Me$_x$Si$_y$ Metal Silicide

Me-Si Metal-Silicon (Contact)

Me$_3$SiBr Bromotrimethylsilane (or Trimethylsilyl Bromide)

MeSiCl$_3$ Trichloromethylsilane (or Methylsilyl Trichloride)

Me$_3$SiCl Chlorotrimethylsilane (or Trimethylsilyl Chloride)

Me$_2$SiCl$_2$ Dichlorodimethylsilane (or Dimethylsilyl Dichloride)

Me$_3$SiCl$_3$ Methyltrichlorosilane (or Trichlorotrimethylsilane)

Me$_3$SiCN Cyanotrimethylsilane (or Trimethylsilyl Cyanide)

Me$_3$SiF Fluorotrimethylsilane (or Trimethylsilyl Fluoride)

Me$_3$SiI Iodotrimethylsilane (or Trimethylsilyl Iodide)

MESJ Marine Engineering Society of Japan

MESL Merchants Exchange of St. Louis [Missouri, US]

Me$_2$Sn Dimethyltin

Me$_3$Sn Trimethyltin

Me$_4$Sn Tetramethyltin

MESNA 2-Mercaptoethanesulfonic Acid

Me$_2$SnBr$_2$ Dimethyltin Dibromide (or Dibromodimethyltin)

Me$_3$SnBr Trimethyltin Bromide (or Bromotrimethyltin)

MeSnCl$_3$ Methyltin Trichloride

Me$_2$SnCl$_2$ Dimethyltin Dichloride (or Dichlorodimethyltin)

Me$_3$SnCl Trimethyltin Chloride (or Chlorotrimethyltin)

Meso Mesoscale [also meso]

MeSO$_2$ Methyl Sulfoxide (or Dimethyl Sulfoxide)

Me$_2$SO$_2$ Methyl Sulfone (or Dimethyl Sulfone)

Mesop Mesopotamia(n)

Mesosp Mesosphere [also mesosp]

MESP Mineral Exploration Stimulation Program [Canada]

Mesrt Measurement [also mesrt]

MESS Monitor Event Simulation System

MESS Succinic Acid Maleimidoethyl N-Hydroxysuccinimide Ester

Messen Prüfen Autom Messen Prüfen Automatisieren [German Journal on Measurement, Testing and Automation]

MESSR Multispectral Electronic Self-Scanning Radiometer

Messrs *(Messieurs)* – French for "Gentlemen"

MEST (Ontario) Ministry of Energy, Science and Technology [Canada]

MESTIND Measurement Standards Instrumentation Division [of Instrument Society of America, US]

Mes trans Mesocolon transversus [also mes trans] [Anatomy]

MESUCORA Measurement, Control Regulation and Automation

MESYL Methanesulfonyl

MET Management Engineering Team

MET Marine Engineering Technician

MET Marine Engineering Technologist; Marine Engineering Technology

MET Master Events Timer

MET Mechanical Engineering Technologist; Mechanical Engineering Technology

MET Mechanoelectronic Transducer

MET Medical Engineering and Technology

MET Memory Enhancement Technology

MET Metal Evaporated Tape [Recording Media]

MET Meteorological Broadcast

MET Microelectronic Technologist; Microelectronic Technology

MET Microelectronic Testing

MET Middle European Time [Greenwich Mean Time –01:00]

MET (Ontario) Ministry of Education and Training [Canada]

MET Mission Elapsed Time

MET Mission Events Timer

MET Mobile Earth Terminal [Mobile Satellite Program, US Department of Commerce]

MET Modified Expansion Tube

MET Multielectrode Tube

MET Multi-Emitter Transistor

.MET Metafile [File Name Extension] [also .met]

Met Metal(lic) [also met]

Met Metallurgical; Metallurgist; Metallurgy [also met]

Met Metaphor [also met]

Met Metaphysics [also met]

Met Meteorological; Meteorologist; Meteorology [also met]

Met Meteorological Office [UK]

Met Methane [also met]

Met (–)-Methionine; Methionyl [Biochemistry]

Met Methyltropolone [Biochemistry]

Met Metronome [also met]

Met Metropole; Metropolitan [also met]

α-Met α-Methyltropolone [Biochemistry]

β-Met β-Methyltropolone [Biochemistry]

META Maintenance Engineering Training Agency [of US Army]

META Method for Extracting Text Automatically

META (The) Micro-Electrostatic Torsional Actuator

M&ETA Marine and Engineering Training Association [UK]

MeTa Metal Tantalide

Metabol Metabolism (Journal)

METADB Metadatabas [Meta Data Base, of Swedish Geological Survey]

METADEX Metals Abstracts Index [Joint Publication of ASM International in the US and the Institute of Metals in the UK] [also Metadex]

METAG Meteorological Advisory Group [of International Civil Aviation Organization]

METAL Metallurgical Fairs and Symposia [Ostrava, Czech Republic]

Metal Metallurgical; Metallurgist; Metallurgy [also metal]

Metal ABM Metalurgia–Associação Brasiliera de Metalurgia e Materiais [Journal of the Brazilian Association for Metallurgy and Materials]

Metalcaster Mag Metalcaster Magazine [Published by the American Cast Metals Association]

Metal Electr Metalurgia y Electricidad [Spanish Journal on Metallurgy and Electricity]

Metal Fabr News Metal Fabricating News [Publication of the Metal Fabricating Institute, US]

Metalforming Dig Metalforming Digest [Published by ASM International, US]

Metalforming Mach Metalforming Machinery [Chinese Publication]

Metalforming Mach Tools Metalforming Machine Tools [Journal published in PR China]

Metal Int Metalurgia International [International Metallurgical Journal published in Brazil]

Metall Metallurgical; Metallurgist; Metallurgy [also metall]

Metall Anal Metallurgical Analysis [Journal published in PR China]

Metalleiologika Metall Chron Metalleiologika Metallourgika Chronika [Greek Chronicles on Metallography and Metallurgy]

Metall Eng-IIT, Bombay Metallurgical Engineer-IIT, Bombay [Journal of the Indian Institute of Technology at Bombay]

Metall Equip Metallurgical Equipment [Journal published in PR China]

Metall Gornorudn Prom-st Metallurgicheskaya i Gornorudnaya Promyshlennost' [Russian Journal on Metallurgical Subjects]

Metall-Handw Tech Metall-Handwerk und Technik [German Journal on Metal Trades and Technology]

Metall Ital Metallurgia Italiana [Italian Journal on Metallurgy]

Metall Koksokhim Metallurgiya i Koksokhimiya [Russian Journal on Metallurgy and Coke Chemistry]

Metall Mater Technol The Metallurgist and Materials Technologist [UK Journal]

Metall Metalloved Metallurgiya i Metallovedenie [Russian Journal on Metallurgy and Metal Science]

Metallofiz, Akad Nauk Ukrainokoj SSR Metallofizika, Akademii Nauk Ukrainokoj SSR [Journal on Metal Physics published in the Ukrainian Soviet Socialist Republic]

Metalloved Legk Splav Metallovedenie Legkikh Splavov [Russian Journal on Light Alloy Metallurgy]

Metalloved Obrab Titan Zharop Spav Metallovedenie i Obrabotka Titanovih i Zharoprochnih Splavov [Russian Journal on Metallurgy of Titanium and Related Alloys]

Metalloved Term Obrab Met Metallovedenie i Termicheskaya Obrabotka Metallov [Russian Journal on Metal Science and Heat Treatment]

Metalloved Vys Splav Metallovedenie Vysokodempfiruyuschih Splavov [Russian Publication on Metal and Alloy Science]

Metall Plant Technol Metallurgical Plant and Technology [Published by Verein Deutscher Eisenhüttenleute, Dusseldorf, Germany]

Metall Plant Technol Int Metallurgical Plant and Technology International [Published by Verein Deutscher Eisenhüttenleute, Dusseldorf, Germany]

Metall Rev Metallurgical Reviews [Published in the US]

Metall Sci Technol Metallurgical Science and Technology [Journal published in Italy]

Metall Trans Metallurgical Transactions [Published by ASM International, US]

Metall Trans A, Phys Metall Mater Sci Metallurgical Transactions A, Physical Metallurgy and Materials Science [Published by ASM International and the American Institute of Mining, Metallurgical and Petroleum Engineers]

Metall Trans B, Process Metall Metallurgical Transactions, Process Metallurgy [Published by ASM International and the American Institute of Mining, Metallurgical and Petroleum Engineers]

Metal Odlew Metalurgia i Odlewnictwo [Polish Journal on Metallurgy and Related Subjects]

Metalozn Obrób Ciepl Metaloznawstwo i Obróbka Cieplna [Polish Metals Journal]

Metalozn Obrób Ciepl, Inz Powierzchni Metaloznawstwo, Obróbka Cieplna, Inzynieria Powierzchni [Polish Metals Journal]

Metal Proszków Metalurgia Proszków [Polish Publication on Metallurgy]

Metal Tsvetn Redk Met Metalurgiya Tsvednykh i Redkikh Metallov [Russian Journal on Metallurgy and Metals]

Metalwork Eng Mark Metalworking Engineering and Marketing [Journal published in Japan]

Metalwork Interfaces Metalworking Interfaces [US Publication]

Metalwork News Metalworking News [Published in the US]

Metalwork Prod Metalworking Production [Journal published in the UK]

Metaph Metaphysics [also metaph]

METAPLAN Methods of Extracting Text Automatically Programming Language

METAPLAN Methods of Extracting Text Autoprogramming Language

METAR Meteorological Aeronautical Code

Metastasis Rev Metastasis Review [International Journal]

METASYMBOL Metalanguage Symbol

Met Australas Metals Australasia [Journal]

Métaux-Corros-Ind Métaux-Corrosion-Industrie [French Journal on Metals, Corrosion and Industry]

Métaux Deform Métaux Deformation [French Journal on the Deformation of Metals]

METAV Markt für die Metallverarbeitung [Trade Fair and Exhibition for the Metalworking Industries, Germany]

METAV Messe für Fertigungstechnik und Automatisierung [Trade Fair for Production Engineering and Automation, Germany]

Met Build Rev Metal Building Review [US]

Met Bull Metal Bulletin [UK]

Met Bull Mon Metal Bulletin Monthly [UK]

METC Morgantown Energy Technology Center [of US Department of Energy in West Virginia]

Met Cast Metals and Castings [Australian Publication]

Met Cast (Australas) Metals and Castings (Australasia)

Met Chem Metal Chemist(ry)

Met Constr Metal Construction [Journal published in the UK]

METCOM Mechanical and Metal Trades Confederation [UK]

METCOM Metropolitan Consortium for Minorities in Engineering

METCUT Metal Cutting Exhibition

METD Dimethyldiethyl Thiuram Disulfide

Met Data Meteorological Data [also met data]

MeTe Metal Telluride

MetE Metallurgical Engineer [also Met Eng, or MetEng]

Mete Methylene [also mete]

METEC International Exhibition for Metallurgical Technology [Germany]

MetEng Metallurgical Engineer [also Met Eng, or MetE]

Met Eng Metallurgical Engineer(ing)

Met Eng Q Metals Engineering Quarterly

METEO Direction de la Météorologie Nationale [National Meteorology Office, France]

Meteorol Meteorological; Meteorologist; Meteorology [also Meteor, meteor, or meteorol]

Meteorol Atmos Phys Meteorology and Atmospheric Physics [Journal published in the Austria]

Meteorol Gidrol Meteorologiya i Gidrologiya [Russian Journal on Meteorology and Hydrology]

Meteorol Mag Meteorological Magazine [UK]

Meteorol Monogr Meteorological Monographs [Published by the US National Weather Association]

Meteorol Rundsch Meteorologische Rundschau [German Meteorological Review]

Meteor Planet Sci Meteoritics and Planetary Science [of University of Arkansas, Fayetteville, US]

METEOSAT (Geosynchronous) Meteorological Satellite [of European Space Agency]

METEOSAT Meteorology Satellite

Met Fabr News Metal Fabricating News [Journal published in the US]

Met Fin Metal Finisher; Metal Finishing

Met Finish Metal Finishing [Journal published in the US]

Met Finish Pract Metal Finishing Practice [Journal published in Japan]

Met Form Metal Forming [Journal published in the US]

Met Forum Metals Forum [Published in Australia]

Meth Method(ic) [also meth]

Meth Methamphetamine [also meth] [Biochemistry]

Meth Methyl [also meth]

Meth Al Methyl Alcohol [also meth al]

2-MeTHF 2-Methyl Tetrahydrofuran

Meth Cell Biol Methods of Cell Biology

Meth Immunol Immunochem Methods in Immunology and Immunochemistry [Published in the US]

Meth Inf Med Methods of Information in Medicine [Journal published in Germany]

Meth Oper Res Methods of Operations Research [Journal published in Germany]

Meth Org Synth Methods of Organic Synthesis [Publication of the Royal Society of Chemistry, UK]

Methyl-DAST (Dimethylamino)sulfurtrifluoride [Biochemistry]

Methyl-GAG Methylglyoxal bis-(Guanylhydrazone) [Biochemistry]

Methyl-L-DOPA 3-(3,4-Dihydroxyphenyl)-2-Methyl-L-Alanine [Biochemistry]

Met Ind Metal Industries [Journal published in Taiwan]

Met Ind (China) Metal Industries (China) [Journal published in Taiwan]

Met Ind News Metals Industry News [Published in the UK]

Me$_2$TiO$_4$ Metal (II) Titanate

Met Kunstst Metaal & Kunststof [Dutch Publication on Metals and Plastics]

METL Multi-Element, Two-Layer

METLO Meteorological Electronic Technical Liaison Office [US Navy]

metlrg metallurgical

Met Mark Wkly Rev Metal Markets Weekly Review [UK]

Met Mat Metallic Material

Met Mater Metals and Materials [Publication of the Institute of Metals, UK]

Met Mater (Czech) Metallic Materials (Czechoslovakia) [Translation of: *Kovove Materialy*; published in the UK]

Met Mater Process Metals, Materials and Processes [Journal published in India]

Met Miner Int Metals and Minerals International [UK]

Met News Metal News [Published in India]

Met oberfl: Angew Elektrochem Metalloberfläche: Angewandte Elektrochemie [German Journal on Metallic Surfaces and Applied Electrochemistry]

MetPack Bus MetPack Business [Journal published in the UK]

Met Powder Rep Metal Powder Report [UK]

Met Phys Metal Physics; Metal Physicist

Met Prog Metal Progress [Published by ASM International, US]

METR Materials Engineering Test Reactor [Belgium]

Metr Metriol

ME6TREN 2,2',2"-Tris(N,N-Dimethylamino)triethylamine

MetrTN Metriol Trinitrate

METRIA Metropolitan Tree Improvement Alliance [US]

METRIC Multi-Echelon Technique for Recoverable Item Control

Metrol Metrological; Metrologist; Metrology [also metrol]

Metrol Apl Metrologia Aplicata [Journal on Applied Metrology; published in Romania]

METROMEX Metropolitan Meteorological Experiment [US]

Metrop Metropole; Metropolitan [also metrop]

METS Modular Engine Test System

METS Modular Environmental Test System

METSAT Meteorological Satellite [also Metsat, or metsat]

Met Sci Metal Science; Metal Scientist

Met Sci Metal Science [Journal of the The Metals Society, UK]

Met Sci Heat Treat Metal Science and Heat Treatment [Translation of: *Metallovednie i Termicheskaya Obrabotka Metallov (USSR)*; published in the US]

Met Sci J Metal Science Journal

Met Sci Technol Metal Science and Technology [International Journal]

Met Soc World Metals Society World [Publication of the The Metals Society, UK]

Met Stamp Metal Stamping [Journal published in the US]

MET STN Meteorological Station [also Met Stn]

Met Technol Metals Technology; Metals Technologist

Met Technol Metals Technology [Publication of The Metals Society, UK]

Met Technol (Jpn) Metals and Technology (Japan) [Journal published in Japan]

Met Trans Metallurgical Transactions [Published by ASM International and the American Institute of Mining, Metallurgical and Petroleum Engineers]

Met Trans A, Phys Metall Mater Sci Metallurgical Transactions A, Physical Metallurgy and Materials Science [Published by ASM International and the American Institute of Mining, Metallurgical and Petroleum Engineers]

Met Trans B, Process Metall Metallurgical Transactions, Process Metallurgy [Published by ASM International and the American Institute of Mining, Metallurgical and Petroleum Engineers]

METTP Marine Engineering Technician Training Plan

METU Middle East Technical University [Turkey]

METVC Main Engine Thrust Vector [Aerospace]

Met Week Metals Week [Published in the US]

Metwk Metalwork(ing) [also metwk]

Met Wkly Rev Metals Weekly Review [UK]

Met World Metal World [Published in China]

Met World (China) Metal World (China) [Published in China]

4-MeTZ 4-Methyl Thiazole

MEU Medium Enriched Uranium

MEU Memory Expansion Unit

MEUF Micellar-Enhanced Ultra-Filtration

MEU/TH Medium Enriched Uranium/Thorium Fuel

MeV Megaelectronvolts [also mev]

MeV Metal Vanadide

MeV Million Electronvolts [also MEV, Mev, or mev]

μeV microelectronvolt [Unit]

meV/Å² millielectronvolt per square angstrom [also meV Å^{-2}]

MeVal N-Methyl Valine

MeV Ci Megaelectronvolt-Curie [Unit]

meV/GPa millielectronvolt per Gigapascal [also meV Gpa^{-1}]

meV/kbar millielectronvolt per kilobar [also meV kbar^{-1}]

MEVVA Metal(lic) Evaporation Vapor Vacuum Arc (Source)

MEW Magnetoelastic Wave [Solid-State Physics]

MEW Mean Equivalent Wind

MEW Microwave Early Warning

MeW Metal Tungstide

MEWG Maintenance Engineering Working Group

MEWS Missile Electronic Warfare System

MEWT Matrix Electrostatic Writing Technique

MEWU Mining and Energy Workers Union [Germany]

MEX Microcell Extender [Telecommunications]

MEX Military Exchange

MeX Methyl Halide

Mex Mexican; Mexico

mExa m-Ethoxyaniline

MEXE Military Engineering Experimental Establishment [of Ministry of Defense, UK]

Mexican MRS Mexican Materials Research Society

Mex$ Mexican Peso [Currency]

MExtEd Master of Extension Education

MEY Maximum Economic Yield

meV/au^3 millielectronvolt per atomic unit cubed [also meV au^{-3}]

Mezhvuz Sb Nauch Tr, Ekol Zashchita Lesa Mezhvuzovskii Sbornik Nauchnykh Trudov, Ekologiya i Zashchita Lesa [Russian Journal on Ecology and Forest Protection]

Mezhvuz Sb Nauch Tr, Ekon Prob Lesoobrabat Prom Mezhvuzovskii Sbornik Nauchnykh Trudov, Ekonomicheskie Problemy Leso-Obrabatyvayushchei Promyshlennosti [Russian Journal on Economic Problems of the Wood-Processing Industry]

Mezhvuz Sb Nauch Tr, Khim Pererab Drev Mezhvuzovskii Sbornik Nauchnykh Trudov, Khimicheskaya Pererabotka Drevesiny [Russian Journal on the Chemical Processing of Wood]

Mezhvuz Sb Nauch Tr, Khim Tekhnol Tsellyul Mezhvuzovskii Sbornik Nauchnykh Trudov, Khimiya i Tekhnologiya Tsellyulozy [Russian Journal on the Chemistry and Technology of Pulp]

Mezhvuz Sb Nauch Tr, Lesosech Lesosklad Raboty Transport Lesa Mezhvuzovskii Sbornik Nauchnykh Trudov, Lesosechnye, Lesoskladskie Raboty i Transport Lesa [Russian Journal on Felling, Log Storage Operations and Wood Transport]

Mezhvuz Sb Nauch Tr, Lesovod Lesn Kul'tury Pochvoved Mezhvuzovskii Sbornik Nauchnykh Trudov, Lesovodstvo, Lesnye Kul'tury i Pochvovedenie [Russian Journal on Forest Management, Silviculture and Soil Science]

Mezhvuz Sb Nauch Tr, Mashiny Orud Mekh Lesozagot Lesn Khoz Mezhvuzovskii Sbornik Nauchnykh Trudov, Mashiny i Orudiya dlya Mekhanizatsii Lesozagotovok i Lesnogo Khozyaistva [Russian Journal on Machines and Tools for Mechanization of Logging and Forestry Operations]

Mezhvuz Sb Nauch Tr, Stanki Instrum Derevoobrabat Proizv Mezhvuzovskii Sbornik Nauchnykh Trudov, Stanki i Instrumenty Derevoobrabatyvayushchikh Proizvodstvo [Russian Journal on the Machines and Tools of Wood-Processing Plants]

Mezhvuz Sb Nauch Tr, Technol Oborud Derevoobrabat Proizv Mezhvuzovskii Sbornik Nauchnykh Trudov, Tekhnologiya i Oborudovanie Derevoobrabatyvayushchikh Proizvodstvo [Russian Journal on the Technology and and Equipment of Wood-Processing Plants]

Me$_2$Zn Dimethylzinc

MeZnCl Methylzinc Chloride

Mezz Mezzanine

MF Magnesium Ferrite

MF Magnetic Field

MF Magnetic Fluid

MF Magnetic Focusing

MF Magnetic Foot [Unit]

MF Magnetomotive Force

MF Mainframe

MF Mains Frequency

MF Maintenance Factor [also M/F]

MF Maintenance Free

MF Mali Franc [Currency of Mali]

MF Major Function

MF Mandatory Frequency

MF Manganese Ferrite

MF Manual Feed

MF Mass Flow

MF Mass Fragmentography

MF Master File

MF Master of Forestry

MF Master Frame

MF Matched Filter(ing)

MF Mate and Ferry [Aerospace]

MF Mating Factor

MF Matrix Flexibility

MF Mean Field (Approximation) [Physics]

MF Measurement Frequency

MF Measuring Force

MF Mechanical Forming

MF Medium Frequency [Frequency: 300 to 3,000 kilohertz; wavelength: 100 to 1,000 meters] [also M-F, M/F, mf, m/f, or m-f]

MF Melamine Formaldehyde

MF Membrane Filter

MF Menninger Foundation [Topeka, Kansas, US]

MF Mercuric Fulminate

MF Metal Fabrication

MF Metal Finishing

MF Metal Forming

MF Methylformate

MF Microfabrication

MF Microfiche

MF Microfilm(er)

MF Microfiltration

MF Micro Floppy

MF Microform

MF Mid-Deck Forward

MF Middle French

MF Mid-Fuselage

MF Ministère des Finances [Ministry of Finance, France]

MF Mixed Flow

MF Mixed Frequency (Process)

MF Modulation Frequency

MF Molecular Field [Solid-State Physics]

MF Molecular Filter

MF Molecular Flow

MF Molecular Formula

MF Motor Field

MF Motor Fleet

MF Motor Fuel

MF Multifiber

MF Multifocal

MF Multifrequency

MF Multifunction

MF Muscle Fiber [Anatomy]

MF Music Foreground [Programming]

MF$_x$ Metal Fluoride [Formula]

M&F Male and Female [Fasteners, etc.]

M&F Materials and Facilities

M/F Maintenance Factor [also MF]

M/F Male/Female (Connector)

M-F Medium Frequency [Frequency: 300 to 3,000 kilohertz; wavelength: 100 to 1,000 meters] [also MF, M/F, mf, m/f, or m-f]

M(f) Fiber-Reinforced M-Glass [Symbol]

mF millifarad [Unit]

μF microfarad [also μf, μfd, or muF]

MFA Magnetic Field Annealing [Solid-State Physics]

MFA Manned Flight Awareness

MFA Manual of Field Artillery

MFA Mean Field Approach [Physics]

MFA Mean Field Approximation [Physics]

MFA Metal Finishing Association [UK]

MFA Minimum Flight Altitude

MFA Ministry of Foreign Affairs [Saudi Arabia]

MFA Moffett Federal Airfield [of NASA at Mountain View, California, US]

MFA Multifiber Agreement [United Nations Agreement for the Textile Industry]

MFA Multifont Adapter

MFA Multifurnace Assembly

MFA Trisammonium 12-Molybdophosphate

MFAI Metal Finishing Association of India

mFa m-Fluoroaniline

MFB Mill Fixture Base

MFB Mixed Functional Block

MFB Motional Feedback

mFBA m-Fluorobenzoic Acid

MFBM Thousand Foot Board Measure

MFBP Manufacturing Flow and Building Plan

MFBS Multifrequency Binary Sequence

MFC Mader, Feder and Chaudhari (Experiments) [Solid-State Physics]

MFC Magnetic-Tape Field Scan

MFC Manual Frequency Control

MFC Mass Flow Control(ler)

MFC Microfunctional Circuit

MFC Microsoft Foundation Class

MFC Molecular Field Coefficient [Soilid-State Physics]

MFC Monochromator Focusing Circle

MFC Multifrequency Code

MFC Multifrequency (Signalling), Compelled

MFC Multifunction Center

MFC Multifunction Controller

MFC Multiple Flight Computer

MFC Multiple Flight Controller

MFc Megaflux change [Unit]

MFCA Multifunction Communications Adapter

MFCES Montana Forest and Conservation Experiment Station [of Montana State University at Blackfoot Valley, US]

MFCI Molten Fuel Coolant Interaction [Nuclear Reactors]

MFC/LB Multifrequency Code/Local Battery

MFCM Multifunction Card Machine

mF/cm² millifarad(s) per square centimeter [also mF cm^{-2}]

μF/cm² microfarad(s) per square centimeter [also μF cm^{-2}]

MFc/s Megaflux change per second [Unit]

MFCU Multifunction Card Unit

MFD Magnetic Field Disturbance

MFD Magnetic Frequency Detector

MFD Magnetofluid Dynamics

MFD Main Feed

MFD Malfunction Detection

MFD Malfunctioning Display

MFD Master File Directory

MFD Microelectronic Functional Device

MFD Micro Floppy Disk

MFD Multifunction Detector

MFD Multifunction Display

MFD Multiple Faulted Defect [Materials Science]

MFD Multiple Function Detector

mfd manufactured

mfd microfarad [Unit]

μfd microfarad [also μF, μf, or muF]

MFDF Medium-Frequency Direction Finder [also MF/DF]

MFDP Mesoscale Frontal Dynamics Project

MFDS Maxey Flats Disposal Site [US]

MFDSU Multifunction Data Set Utility

MFE Magnetic Field Experiment

MFE Magnetic Fusion Energy

MFE Master of Forest Engineering

MFE Material Fuel Equivalent

MFE Mercury Film Electrode

MFE Mid-Frequency Executive

MFED Maximum Flat Envelope Delay

MFEng Master of Forestry Engineering

MFES Minnesota Federation of Engineering Societies [US]

MFET Magnetic Fusion Energy Technology

MFF MDM (Manipulator Deployment Mechanism) Flight Forward [NASA]

MFF Molecular Fixed Frame

MFFS Microsoft Flash File System

MFG Major Functional Group [Chemistry]

MFG Message Flow Graph

MFG Molded Fiber Glass

MFG Multifunction Generator

Mfg Manufacturing [also mfg]

Mfg T Manufacturing Technology [also Mfg Technol]

MFH Maximum Fork Height (of Forklift Trucks)

MFI Malta Federation of Industries

MFI Melt Flow Index(er)

MFI Metal Fabricating Institute [US]

MFI Metal Finishing Industry

MFI Mixed Fission Product

MFI Mobile Fuel Irradiator

MFI Modified Fast Inversion

MFI Multi-Point Fuel Injection

MFIE Multi-Factor Interaction Equation

MF-IFGR Michael Fund–International Foundation for Genetic Research

MFIRFT Modified Fast Inversion–Recovery Fourier Transform

MFJ Modified Final Judgment

MFJSA Mass Finishing Job Shops Association [US]

MFK Matematisk Forening i København [Mathematical Association of Copenhagen, Denmark]

MFK Mill Fixture Key

MFK 12-Molybdophosphoric Acid

MFKE Mean Field Kinetic Equation [Physics]

MFKP Multifrequency Key Pulsing

MFL Magnesium Fluoride Liner

MFL Magnetic Flux Leakage

MFL Mantitoba Federation of Labour [Canada]

MFL Marginal Fermi Liquid

MFL Master Fishermen's Licence

MFL Mobile Foundry Laboratory

MFLD Message Field

MFLOPS Mega-Floating Point Operations per Second; Million Floating-Point Operations per Second [also Mflops, or Mflop]

MFLP Multifile Linear Programming

MFLS Microtime Fault Locator System

MFM Magnetic-Field Meter

MFM Magnetic-Force Microscope; Magnetic-Force Microscopy

MFM Mass Flow Meter

MFM Mass Flow Monitor

MFM Master Frequency Meter

MFM Medium-Frequency Module

MFM Modified Frequency Modulation

MFM Multifrequency Module

MFM Multistage Frequency Multiplier

MFMA Maple Flooring Manufacturers Association [US]

MFMA Metal Framing Manufacturers Association [US]

MFmP m-Formyl Phenol

MFMR Multifrequency Microwave Radiometer

MFN Magnetic Flux Noise

MFN Most Favoured Nation (Tariff)

MFO Magnesium Ferrite [MgFe$_2$O$_4$]

MFO Major Function Overlay

MFO Mixed Function Oxidase [Biochemistry]

MFOD Manned Flight Operations Division

MFOM Member of the Faculty of Occupational Medicine

MFor Master of Forestry

MForSc Master of Forestry Science

MFOV Multi-Field-of-View (Radiometer)

MFOV Multiple-Fields-of-View

MFOWO Most Frequently Observed Whisker Orientation

MFP Mean Free Path [also mfp]

MFP Minimum Flight Path

MFP Multiform Printer

MFP Multi-Function Peripheral

MFP Multi-Function Product

mFP m-Fluorophenol

MFPA Massachusetts Forest and Park Association [US]

MFPB Mineral Fiber Products Bureau

MFPC Multifunction Protocol Converter

MF PECVD Mixed-Frequency Plasma-Enhanced Chemical Vapor Deposition

MFPG Mechanical Failure Prevention Group [US]

MFPG Mixed Fission Products Generator

MFPI Multifunction Peripheral Interface

MFPT Mean First Passage Time

MFR Manipulator Foot Restraint [of NASA]

MFR Mass Flow Rate

MFR Master of Forest Resources

MFR Maximum Flight Rate

MFR Mean Fuel Rating

MFR Melt Flow Rate

MFR Mirror Fusion Reactor

MFR Modified Frank-Read (Mechanism) [Metallurgy]

MFR Multi-Filter Radiometer

MFR Multifrequency Receiver

MFR Multifunctional Receiver

MFR Multifunctional Review

MFr Mali Franc [Currency of Mali] [also MFR]

Mfr Manufacturer [also mfr]

MFRC Maritime Forest Research Center [Fredericton, New Brunswick, Canada]

MFRC Master of Forest Resources and Conservation

MFRC Metal Finishing Research Center [Canada]

Mfrs Manufacturers [also mfrs]

MFRSR Multi-Filter Rotating Shadow-Band Radiometer

MFS Macintosh File System

MFS Magnetic Field Sensor

MFS Magnetic-Tape Field Search

MFS Magnesium Fluosilicate

MFS Magnetic Field Sensor

MFS Magnetic-Tape Field Search

MFS Manned Flying System

MFS Marine Finish Slate

MFS Master of Food Science

MFS Master of Forensic Studies

MFS Master of Forest Studies

MFS Memory File System

MFS Mercury Iron Selenide (Semiconductor)

MFS Metal-Ferroelectric-Semiconductor

MFS Metal Finishing Society

MFS Microwave Frequency Stabilizer

MFS Modified Filing System

MFS Molecular Fluorescence Spectrometry; Molecular Fluorescence Spectroscopy

MFS Multifrequency Signalling

MFS Multifunction Sensor

MFSA Metal Finishing Suppliers Association [US]

MFSJ Metal Finishing Society of Japan

MFSK Metal Finishing Society of Korea

MFSK Multiple-Frequency Shift Keying

MFSS Motor Fleet Safety Service

MFSS Multi-Frequency Signalling System

MFT Magnetic Flux Trapping

MFT Mainframe Termination

MFT Manual Fine Tuning

MFT Master File Table

MFT Master of Foreign Trade

MFT Mean Field Theory [Physics]

MFT Mean Flight Time

MFT Melter Feed Tank

MFT Mercury Iron Telluride (Semiconductor)

MFT Metallic Facility Terminal

MFT Miniature Fluorescent Tube

MFT Minimum Film (Formation) Temperature

MFT Mixed-Flow Turbine

MFT Monolayer Formation Time

MFT Muffin Tin (Energy) [Physics]

MFT Multiposition Frequency Telegraphy

MFT Multiprogramming with a Fixed Number of Tasks

MFT Thousand Feet

MFTF Mirror Fusion Test Facility

MFTG Manufacturing Technology Group [of Institute of Electrical and Electronics Engineers, US]

Mftl Milli-Foot-Lamberts [Unit]

MFTAD Master Flight Test Assignment Document [of NASA Johnson Space Center, Houston, Texas, US]

MFTF Mirror Fusion Test Facility

MFTM Mean Field Transfer Matrix (Method) [Physics]

MFTR Multifrequency Transmitter

MFTRS Magnetic Flight Test Recording System

MFU Maritime Fisherman's Union [Canada]

MFU McGill Faculty Union [of McGill University, Montreal, Canada]

MFU Mobile Floating Unit

μ_B/fu Bohr magneton(s) per functional unit [also μ_B fu^{-1}]

MFV Main Fuel Valve

MFV Maintenance Floor Valve

MFV Military Flight Vehicle

μF-V Capacitance-Volt

MG Machine-Glazed (Paper)

MG Machine Gun

MG Macroglobulin [Biochemistry]

MG Madagascar [ISO Code]

MG Magnesium Gallate

MG Magnesium Germanate

MG Magnetic Gyro

MG Maintenance Group

MG Major-General

MG Mammary Gland [Anatomy]

MG Management Guide

MG Manual Gain Control

MG Marginal Relay

MG Marine Geodesy

MG Marine Geologist; Marine Geology

MG Master Generator

MG Master-Group [Telecommunications]

MG Mathematical Geology

MG Maxwell-Garnett (Formula) [Physics]

MG May-Grunwald (Stain) [Biochemistry]

MG Mechanical Grinding; Mechanically Ground

MG Medical Geneticist; Medical Genetics

MG Megagauss [Unit]

MG Melt Growth

MG Message Generator

MG Metal-Glass (Type Tube)

MG Methylene Glycol

MG Methyl Glycol

MG Microgranule [Chemistry]

MG Middle Gimbal

MG Military Government

MG Minas Gerais [Brazil]

MG Mixed Grain [Materials Science]

MG Mobile Generator

MG Modulation Grid

MG Modulator Gauge

MG Molecular Geneticist; Molecular Genetics

MG Monkman-Grant (Constant) [Mechanics]

MG Morris Garages [US]

MG Motor-Generator

MG Mucous Gland [Anatomy]

MG Multiga(u)ge

MG Myasthenia Gravis [Medicine]

MG Mycoplasma Gallisepticum [Microbiology]

Mg Magnesium [Symbol]

Mg Megagram [Unit]

Mg Messenger [also mg]

Mg Mining [also mg]

Mg^{2+} Magnesium Ion [also Mg^{++}] [Symbol]

Mg-23 Magnesium-23 [also ^{23}Mg, or Mg^{23}]

Mg-24 Magnesium-24 [also ^{24}Mg, or Mg^{24}]

Mg-25 Magnesium-25 [also ^{25}Mg, or Mg^{25}]

Mg-26 Magnesium-26 [also ^{26}Mg, or Mg^{26}]

Mg-27 Magnesium-27 [also ^{27}Mg, or Mg^{27}]

Mg-28 Magnesium-28 [also ^{28}Mg, or Mg^{28}]

mG milligauss [Unit]

mg milligram [also mgm]

.mg Madagascar [Country Code/Domain Name]

m²/g square meter(s) per gram [also m^2 g^{-1}]

μg milligravity (1/1,000 of g$_0$)

μg microgram [also mug]

μg microgravity (1/1,000,000 of g$_0$)

MGA Master of Government Administration

MGA Meteorological and Geoastrophysical Abstracts (Database) [of American Meteorological Society, US]

MGA Methylene Glycol Anion

MGA Methyl Glycol Acetate

MGA Middle Gimbal Angle

MGA Monochrome Graphics Adapter

MGA Multiple Gas Analyzer

MGA Mushroom Growers Association [UK]

MGA Mycoplasma Gallisepticum Agglutinin [Immunology]

mg AB/mL milligrams of antibodies per milliliter [also mg Ab/mL, or mg AB/ml]

Mg(Ac)$_2$ Magnesium(II) Acetate

Mg(ACAC)$_2$ Magnesium (II) Acetylacetonate [also Mg(AcAc)$_2$, or Mg(acac)$_2$]

mg ai/kg milligram(s) of active ingredient per kilogram [also mg ai kg^{-1}]

mGal milligal [Unit]

mgal/d million gallons per day [also MGAL/D, or mgal d^{-1}]

MgAl$_2$O$_4$/Al Aluminum on Magnesium Aluminate (Substrate)

MgAl$_2$O$_4$/Al-Mg Aluminum-Magnesium on Magnesium Aluminate (Substrate)

Mg[(AlOPr)$_4$]$_2$ Magnesium (II) Aluminum Propoxide

MgAl$_2$(OPr^i)$_8$ Magnesium Dialuminum Isopropoxide

M-Gas Motor Gasoline [also M-gas]

MGB Machine Gun Belt

MGB Master Ground Bar

MGB Motor Gunboat

MGBNVCP Murray Geological Basin Native Vegetation Clearance Policy [Australia]

MGC Malachite Green Carbinol

MGC Manual Gain Control

MGC Microscopically-Shielded Griffith Criterion (Model) [Fracture Mechanics]

MGC Missile Guidance and Computer

MGC Missile Guidance and Control

MGC Modified Gouy-Chapman (Theory)

mg/C milligram per Coulomb [also mg C^{-1}]

Mg-Ca Magnesium-Calcium (Alloy System)

MGCC Missile Guidance and Control Computer

Mg-Cd Magnesium-Cadmium (Alloy System)

Mg$_3$Cit$_2$ Magnesium Citrate

mg/cm² milligram(s) per square centimeter [mg cm^{-2}]

μg/cm² microgram(s) per square centimeter [mg cm^{-2}]

mg/cm² h milligram(s) per square centimeter hour [also mg cm^{-2} h^{-1}]

mg/cm² min milligram(s) per square centimeter minute [also mg cm^{-2} min^{-1}]

μg/cm² min milligram(s) per square centimeter minute [also μg cm^{-2} min^{-1}]

Mg(Cp)$_2$ Bis(cyclopentadienyl)magnesium

MGCR Maritime Gas-Cooled Reactor

MGCR-CX Maritime Gas-Cooled Reactor Critical Experiment

MGCS METEOSAT (Meteorological Satellite) Ground Computer Equipment

MGD Magnetogasdynamic(s)

MGD Mean Gain Deviation

MGD Million Gallons per Day [also mgd]

mg/d milligram per day [also mg/day]

mg/dL milligram(s) per deciliter [also mg/dl]

μg/dL microgram(s) per deciliter [also μg/dl]

mg dm^{-2} day^{-1} milligram(s) per square decimeter per day [also mg/dm²/day]

MGE Maintenance Ground Equipment

MGE Master of Geological Engineering

MGE Minneapolis Grain Exchange [Minnesota, US]

MGE Modular GIS (Geographic Information System) Environment

MgEGTA Magnesium Ethylene Glycol bis(2-Amino-ethylether)tetraacetic Acid

MGen Major General [also M Gen]

Mgemt Management [also mgemt]

MGeoE Master of Geological Engineering [also MGeoEng]

M Geofiz Magyar Geofizika [Hungarian Journal on Geophysics]

MGeolE Master of Geological Engineering [also MGeolEng]

MGEOS Mie-Grüneisen Equation of State [Thermodynamics]

MGET Multiple Get [Unix Operating System]

Mg(EtO)$_2$ Magnesium Ethoxide (or Magnesium Ethylate)

MGF Malagasy Franc [Currency of Madagascar]

MGF Mast-Cell Growth Factor [Biochemistry]

MGG Memory Gate Generator

mg/g milligram(s) per gram [also mg g^{-1}]

μg/g microgram(s) per gram [also μg g^{-1}]

MGGA Midland General Galvanizers Association [UK]

MGGB Modular Guided Glide Bomb

Mg$_2$Ge Magnesium Germanide

MGH Massachusetts General Hospital [Boston, US]

MGH Montreal General Hospital [Canada]

mgh milligram-hour [also mg-h, or mg-hr]

mg/h milligram(s) per hour [also mg h^{-1}]

Mg(HFAC)$_2$ Magnesium Hexafluoracetylacetonate [also Mg(hfac)$_2$]

mg/hr milligram(s) per hour [also mg hr^{-1}]

mg-hr milligram-hour [also mg-h, or mgh]

Mg-In Magnesium-Indium (Alloy System)

MGIR Motor Glider Instructor Rating

MGIS Military Geographic Information System

MgKα Magnesium K-Alpha (Radiation) [also MgK$_\alpha$]

MgKβ Magnesium K-Beta (Radiation) [also MgK$_\beta$]

mg/kg milligram per kilogram [also mg kg^{-1}]

mg/kg/d milligram per kilogram per day [also mg kg^{-1} d^{-1}]

MGL Malachite Green Leucocyanite

MGL Matrix Generator Language

mg/L milligram per liter [also mg L^{-1}]

μg/L microgram(s) per liter [also μg L^{-1}, MPL, or mpl]

Mg-Li Magnesium-Lithium (Alloy System)

Mg-Ln-TM Magnesium-Lanthanide-Transition Metal (Alloy)

mg/L O$_2$ milligram per liter of oxygen [also mg L^{-1} O$_2$]

MGLS Minerals and Geotechnical Logging Society [US]

Mg-LT-ET Magnesium Late-Transition-Metal–Early-Transition-Metal (Alloy)

MGM Mechanics of Granular Materials

Mg/m^3 Megagram(s) per cubic meter [also Mg m^{-3}]

m-gM Multi-Gas Monitor

mgm milligram [also mg]

mg/m milligram(s) per meter [also mg m^{-1}]

mg/m^2 milligram(s) per square meter [also mg m^{-2}]

mg/m^3 milligram(s) per cubic meter [also mg m^{-3}]

μg/m^3 microgram(s) per cubic meter [also mug/m^3, or μg m^{-3}]

Mg(MeO)$_2$ Magnesium Methoxide (or Magnesium Methylate)

MGMI Mining, Geological and Metallurgical Institute [India]

mg/min milligram(s) per minute [also mg min^{-1}]

MGML Molecular Graphics and Modeling Laboratory [of University of Kansas, Lawrence, US]

mg/mL milligram(s) per milliliter [also mg mL^{-1}, or mg/ml]

μg/mL microgram(s) per milliliter [also μg mL^{-1}, or μg/ml]

mg/mm^2 milligram(s) per square milliliter [also mg mL^{-2}]

Mgmt Management [also mgmt]

MGN 2-Methylene Glutaric Dinitrile

MGN Multi-Grounded Neutral

Mg[Nb(OEt)$_6$]$_2$ Magnesium (II) Niobium Ethoxide

MgNO Magnesionitrate

MGO Magnesium Gallium Oxide

MGO Magnesium Germanium Oxide

MGO Malachite Green Oxalate

MGO Mega-Gauss Oersted [also MGOe]

MGO Mount Graham Observatory [US]

MgO Magnesium Oxide (Ceramics)

MgO/Al Aluminum on Magnesia (Substrate)

MgO-Al$_2$O$_3$ Magnesium Oxide–Aluminum Oxide (System)

MgO-Al$_2$O$_3$-SiO$_2$ Magnesium Oxide–Aluminum Oxide–Silicon Dioxide (System)

MgO-CaO-SiO$_2$ Magnesium Oxide–Calcium Oxide–Silicon Dioxide (System)

MgO(Cr) Chromium-Doped Magnesium Oxide

MGOe Megagauss-Oersted [also MG-Oe, or MG·Oe]

MgOEP 2,3,7,8,12,13,17,18-Octaethylporphine Magnesium(II)

Mg(OEt)$_2$ Magnesium Ethoxide (or Magnesium Ethylate)

MgO(Fe) Iron-Doped Magnesium Oxide

MgO-NiO Magnesium Oxide-Nickel Oxide (System)

Mg(OPr)$_2$ Magnesium Propoxide

MgO-PSZ Magnesium Oxide-Partially Stabilized Zirconia (Composite)

MgO-SiO$_2$ Magnesium Oxide-Silicon Oxide (System)

MgO(V) Vanadium-Doped Magnesium Oxide

MgO-ZrO$_2$ Magnesium Oxide-Zirconium Oxide (System)

MGP Magnetic and Graphic Products

MGP Magnetogenetic Province

MGP Manufactured Gas Plant

MGP Marine Geophysicist; Marine Geophysics

MGP Methyl Green-Pyronin

MGP Mountain Gorilla Project [of African Wildlife Foundation, US]

MGP Multiple Goal Programming

MGPAPD Million Gallons per Acre per Day [also mgpapd]

Mg$_2$Pb Magnesium Lead (Compound)

Mg-Pb Magnesium-Lead (Alloy System)

mg/pc milligram(s) per piece [Unit]

μg/pc microgram(s) per piece [Unit]

MgPO Magnesiophosphate

Mg(PrO)$_2$ Magnesium Propoxide

Mg(PMCp)$_2$ Bis(pentamethylcyclopentadienyl)magnesium

Mg-PSZ Magnesium Partially Stabilized Zirconia

MGPT Model-Generalized Perturbation Theory [Physics]

MGPT Model Generalized Pseudo-Potential Theory [Physics]

MGR Magnetogyric Ratio

Mgr Manager [also mgr]

Mgr *(Monseigneur)* – French for "My Lord"

Mgr *(Monsignor)* – Spanish for "Mister"

MGRA Master Geographical Reference Area

Mg-RE-Zr Magnesium-Rare Earth-Zirconium (Alloy System)

MGS Maine Geological Survey [US]

MGS Market-Grade Stainless Steel

MGS Maritime Geoscience Services [Canada]

MGS Mars Global Surveyor [of NASA]

MGS Maryland Geological Survey [US]

MGS Master of General Studies

MGS Master Gunnery Sergeant [US Marine Corps]

MGS Minnesota Geological Survey [US]

MGS Monolithic Gallium Arsenide/Silicon

MgS Magnesium Sulfide

MGSM Mayo Graduate School of Medicine [Rochester, Minnesota, US]

MGSA Melanoma Growth Stimulating Activity [Biochemistry]

MGSE Mechanical Ground Support Equipment

MG SET Motor Generator Set

MgSiAlON Magnesium Silicon Aluminum Oxynitride

MgSiN$_2$ Magnesium Silicon Nitride

MgSO Magnesiosulfate

MGSgt Master Gunnery Sergeant [US Marine Corps]

MGT Major Ground Test

MGT Master-Group Translator [Telecommunications]

MGT Mean Greenwich Time [Astrronomy]

Mgt Management [also mgt]

MgTe Magnesium Telluride (Semiconductor)

Mg(TFAC)$_2$ Magnesium Trifluoroacetylacetonate [also Mg(tfac)$_2$]

Mg-Th Magnesium-Thorium (Alloy System)

MgTl Magnesium Thallide

Mg(TMHD)$_2$ Magnesium Bis(2,2,6,6-Tetramethyl-3,5-Heptanedionate) [also Mg(tmhd)$_2$]

MGTP Mount Graham Telescope Project [US]

Mg-TZP Magnesia Tetragonal Zirconia Polycrystal

MGU Mid-Course Guidance Unit

MGU Mining and Geology University [Sofia, Bulgaria]

MGVT Mated Ground Vibration Test

MgWO Magnesiotungstate

μGy microgray [Unit]

MgY-TZP Magnesia-Yttria Tetragonal Zirconia Polycrystal

Mg-Zn Magnesium-Zinc (Alloy System)

Mg-Zn-RE Magnesium-Zinc-Rare-Earth (Alloy System)

Mg-Zr Magnesium-Zirconium (Alloy System)

MH Magnetic (Read/Write)Head

MH Main Hatch

MH Maleic Hydrazide

MH Manhole

MH Man Hour [Unit]

MH Manual Hold [Telecommunications]

MH Marshall Islands [ISO Code]

MH Martini-Henry (Rifle)

MH Master Herbalist

MH Material(s) Handling

MH Material Hopper

MH Medium Hardness

MH Mental Health

MH Message Handler

MH Metal Halide

MH Metal Hydride

MH Minute Hand [Horology]

MH Mobile High-Power Plant [US Army]

MH Modified Huffman [Computers]

MH Modified Hydrodynamic(s)

MH Mohs Hardness

MH Molar Heat

MH Mott-Hubbard (Theory) [Physics]

MH$_2$ Metal Hydride [General Formula]

M(H) Field-Dependent Magnetization [Symbol]

M$_\phi$/H$_\phi$ Circular Magnetic Hysteresis Loop [Symbol]

M$_\phi$/H$_Z$ Mattucci-Effect Magnetic Hysteresis Loop [Symbol]

M$_Z$/H$_Z$ Axial Magnetic Hysteresis Loop [Symbol]

M-H Magnetization versus Applied Magnetic Field (Loop) (or Hysteresis Loop)

M-H Microwave-Hydrothermal (Technique) [Materials Science]

M∥H Magnetization Parallel to (or In-Line with) Applied Magnetic Field [Symbol]

M⊥H Magnetization Perpendicular to (Out-of-Line with) Applied Magnetic Field [Symbol]

mH millihenry [also mh]

m/h meter(s) per hour [also m/hr]

m^3/h cubic meter(s) per hour [also m^3/hr, or m^3 h^{-1}]

μH microhenry [also μh, or muH]

μ(H) magnetic-field-dependent permeability [Symbol]

MHA Manila Hemp Association [UK]

MHA Marine Historical Association

MHA Master of Health Administration

MHA Master of (or in) Hospital Administration

MHA Maximum Hypothetical Accident [Nuclear Reactors]

MHA Mercaptohexadecanoic Acid

MHA Methionine Hydroxy Analog, Calcium Salt [Biochemistry]

MHA Minimum Holding Altitude

MHA Modified Handling Authorized

MHA Mueller Hinta Agar [Biochemistry]

MHA Multiple Headset Adapter

MHAMS Master of Historical Administration and Museum Studies

MHb Methemoglobin [Biochemistry]

MHBI Methyl-4-Hydroxybenzimidate

MHC Major Histocompatibility Complex [Immunology]

MHC Manipulator Hand Controller

MHC Martin Hard Coat (Process)

MHC Maui Historic Commission [Hawaii, US]

MHC Mean Horizontal Candlepower

MHC Medium-Heat Content

MHC Modified Huffman Coding [Computers]

MHC Morris Harvey College [Charleston, West Virginia, US]

MHC Mount Holyoke College [South Hadley, Massachusetts, US]

MHCP Mean Horizontal Candlepower [also mhcp]

MHD Magnetohydrodynamic(s)

MHD Master of Human Development

MHD Maximum Hub Diameter

MHD Movable Head Disk

MHD Moving Head Disk

MHD Multi-Head Disk; Multiple Head Disk

MHDE Magnetohydrodynamic Equation

MHDF Medium and High Frequency Direction Finding

MHDPO Methylene-Bis(di-n-Hexylphosphine Oxide)

MHDS Movable-Head Disk Storage

MHDU Movable-Head Disk Unit

MHDW Magnetohydrodynamic Wave

MHE Master of Home Economics

MHE Material(s) Handling Engineer(ing)

MHE Material(s) Handling Equipment

MHEA Material Handling Engineers Association [UK]

MHEA Mechanical Handling Engineers Association [UK]

MHEDA Material Handling Equipment Distributors Association [US]

MHEEd Master of Home Economics Education

MHEF Material Handling Education Foundation [Charlotte, North Carolina, US]

MHF Massive Hydraulic Fracturing [Oil Exploration]

MHF Medium High Frequency

MHF Message Handling Facility

MHF Mixed Hydrazine Fuel

MHFR Maximum Hypothetical Fission Product Release

MHG Middle High German

μHg micrometer of mercury [also μ Hg, muHg, or mu Hg]

MHH Medizinische Hochschule Hannover [Hanover University of Medicine, Germany]

MHH Microwave Hybrid Heating

MHHPA Methylhexahydrophthalic Anhydride

MHHW Mean Higher High Water

MHI Manufacturered Housing Institute [US]

MHI Material Handling Institute [US]

MHI (Fachgemeinschaft) Montage–Handhabung–Industrieroboter [Working Group on Assembly, Handling and Industrial Robots; of Verband Deutscher Maschinen- und Anlagenbau, Germany]

MHI Mitsubishi Heavy Industries [Tokyo, Japan]

MHiE Master of Highway Engineering [also MHiEng]

MHK Methylhexylketone

MHKW Midvale-Heppenstall-Klöckner-Werke (Steelmaking Process) [Metallurgy]

MHL Medium Heavy Load(ing)

MHL Messo Heated Ladle (Process) [of Messo GmbH, Germany] [Metallurgy]

MHL Magnetic Hysteresis Loop

MHLG Ministry of Housing and Local Government [UK]

MHLW Mean Higher Low Water

MHMA Mobile Home Manufacturers Association [now Manufacturered Housing Institute, US]

MHN Musée d'Histoire Naturelle [Museum of Natural History, Paris, France]

MHNC Modified Hypernetted Chain (Approximation) [Physics]

MHNI Michigan Headache and Neurological Institute [Ann Arbor, US]

MHortSc Master of Horticultural Science

MHOS Master of Human Organizational Science

MHP Master of Heritage Preservation

MHP Mental Health Professional

MHP Message Handling Processor

MHP Metric Horsepower

MHP Mini Hydropower

MHPG 3-Methoxy-4-Hydroxyphenylethyleneglycol

mHPPH 5-(3-Hydroxyphenyl)-5-Phenylhydantoin [also m-HPPH]

MHR Master of Human Resources

MHR Ministry of Human Resources [Canada]

MHRI Mental Health Research Institute [of University of Michigan, Ann Harbor, US]

MHRIP Mining Human Resource Investment Program [Canada]

MHRM Master of Human Resources Management

MHROD Master of Human Resources and Organization Development

MHRS Modified Hazard Ranking System

MHRST Medical and Health Related Sciences Thesaurus

MHS Magnetic Hyperfine Structure [Solid-State Physics]

MHS Malayan Historical Society

MHS Manitoba Historical Society [Canada]

MHS Maryland Historical Society [US]

MHS Massachusetts Historical Society [US]

MHS Massachusetts Horticultural Society [US]

MHS Master Handling System

MHS Master of Health Science

MHS Material Handling System

MHS Mayer's Hematoxylin Solution [Biochemistry]

MHS Message Handling Service

MHS Message Handling System

MHS Microwave Humidity Sounder

MHS Minnesota Historical Society [US]

MHS Military Historical Society [UK]

MHS Missouri Historical Society [US]

MHS Montana Historical Society [US]

MHS Multi-Aircraft Horizontally Separated

MHS Multiple Host Support

MHSA Mars Horizon Sensor Assembly [of NASA Mars Global Surveyor]

MHSA Ministry of Health and Social Affairs [South Korea]

MHSB Mining Health and Safety Branch [of Ministry of Labour, Canada]

MHSC Manipulator Handset Controller

MHSc Master of Health Science

MHSDC Multiple High-Speed Data Channel

MHSCP Mean Hemispherical Candlepower

MHT Magnetohydrodynamic Theory

MHT Malleablizing Heat Treatment [Metallurgy]

MHT Mean High Tide

MHT Mechanical Horizontal Tensile (Tester)

MHT Mechanical Horizontal Testing (Machine)

MHT Medium-High Temperature

MHT Microhardness Testing; Microhardness Tester

MHT Mild Heat Treatment

MHT Motorola High-Threshold Logic

MHT Museum of History and Technology [US]

M(H,T) Field and Temperature-Dependent Magnetization

MHTA Metal Heat Treating Association

MHTGR Modular High-Temperature Gas(-Cooled) Reactor

MHTS Main Heat Transport System

MHV Manned Hypersonic Vehicle

MHV Medium Heating Value

μHV Vickers Microhardness

MHVDF Medium, High and Very High Frequency Direction Finding

MHW Mean High Water

MHW Multi-Hundred Watt

MHWN Mean High-Water Neaps

MHWS Mean High Water Springs

mHy millihenry [also mH, or mh]

MHz Megahertz [Unit]

mHz millihertz [Unit]

μHz millihertz [Unit]

MHz/Oe Megahertz per Oersted [also MHz Oe^{-1}]

MI Machine-Independent

MI Machine Intelligence

MI MacKay Icosahedron [Metallurgy]

MI Magnetic Induction

MI Magnetoimpedance

MI Main Injector

MI Malleable Iron

MI Maloti [Currency of Lesotho]

MI Management Information

MI Management Interface

MI Manual Input

MI Marianas Islands [US Territory]

MI Marquesas Islands [French Polynesia]

MI Marshall Islands [US Territory]

MI Master of Instruction

MI Materials Information [of ASM International, US]

MI Mathematisches Institut [Mathematical Institute, University of Cologne, Germany]

MI Matrix Inertia

MI Medical Imaging

MI Medical Informatics

MI Medical Instrument(ation)

MI Medium Intensity

MI Mellon Institute [of Carnegie-Mellon University, Pittsburgh, Pennsylvania, US]

MI Melt Index [Polymer Processing]

MI Memory Information

MI Memory Interface

MI Metabolic Index

MI Metal Interface (Transistor)

MI Metal-Insulator (Transition)

MI Metal Ion

MI Metals Information [now Materials Information]

MI Michigan [US]

MI Micro-Injector

MI Micro-Instruction

MI Midway Islands [US Territory]

MI Mile [also mi]

MI Military Intelligence

MI Miller Indices [Crystallography]

MI Miller Integrator [Electronics]

MI Mineral-Insulated (Cable)

MI (The) Mineralogical Institute [of University of Mysore, India]

MI Ministère de l'Information [Ministry of Information, France]

MI Ministère de l'Intérieur [Ministry of the Interior, France]

MI Mitral Incompetence; Mitral Insufficiency [Medicine]

MI Mode Indicate

MI Molecular Ion

MI Moment of Inertia

MI Multiple Integral

MI Multiple Interaction

MI Myocardial Infarction [Medicine]

MI-5 Military Intelligence 5 (or Security Service) [UK]

MI-6 Military Intelligence 6 (or Secret Intelligence Service) [UK]

2M5I14 2-Methyl-5-Isopropyl-1,4-Benzoquinone

M/I Minimum Impulse

M-I Metal-Insulator (Transition)

Mi Mill [also mi]

mi mile [Unit]

mi million

mi² square mile [also sq mi]

MIA Malleable Ironfounders Association [UK]

MIA Manitoba Institute of Agrologists [Canada]

MIA Marble Institute of America [US]

MIA Master of International Administration

MIA Master of International Affairs

MIA Metal Interface Amplifier

MIA Methylisatoic Anhydride

MIA 5-(N-Methyl-N-Isobutyl)-Amiloride

MIA Minimum IFR (Instrument Flight Rules) Altitude

MIA Multiplex(er) Interface Adapter

MIA Murrumbidgee Irrigation Area [New South Wales, Australia]

mIa m-Iodoaniline

MIAB Magnetically Impelled Arc Butt (Welding)

MIAC Metals Information Analysis Center [of US Department of Defense]

MIAC Minimum Automatic Computer

MIACF Meander Inverted Autocorrelated Function

MIACS Manufacturing Information and Control System

MIAD Milwaukee Institute of Art and Design [Wisconsin, US]

MIADMB Murrumbidgee Irrigation Area and Districts Management Board [New South Wales, Australia]

MIAK Methyl Isoamyl Ketone

MIALS Medium-Intensity Approach Light System

MIAP Mineral Industry Assistance Program [Canada]

MIAP Molteno Institute of Animal Parasitology [of Cambridge University, UK]

MIArch Master of Interior Architecture

MIAS Marine Information and Advisory Service [of Natural Environmental Research Council, UK]

MIAS Member of the Institute of Aeronautical Sciences

MIAS Microprobe Image Analysis System

MIAS Monterrey Institute of Advanced Studies [Carmel, California, US]

MIAS Municipal Industrial Abatement Strategy [Canada]

M/IAS Mach Indicated Airspeed [also MIAS]

MIASI Moore Institute of Art, Science and Industry [Philadelphia, Pennsylvania, US]

MIB Management Information Base [Internet]

MIB Master Interconnect Board

MIB Master of International Business

MIB Manual Input Buffer

MIB Medical Information Bus [of Institute of Electrical and Electronics Engineers, US]

MIB Metal Information Bureau [UK]

MIB Micro-Instruction Bus

MIB Multilayer Interconnection Board

MIBA Master of International Business Administration

mIBA m-Iodobenzoic Acid

MIBC Methylisobutyl Carbinol

MIBERS Multi-Ion-Beam Reactive Sputtering (System)

MIBF Montreal International Book Fair [Canada]

MIBG m-Iodobenzylguanidine [Biochemistry]

MIBK Methyl Isobutyl Ketone

MIBL Masked Ion Beam Lithography

MIBL Michigan Ion-Beam Laboratory [US]

MIBS Master of International Business Studies

MIBS Multi-Ion-Beam Sputtering (System)

MIBSA Missouri-Illinois Bi-State Authority [US]

MIC Macro Instruction Compiler

MIC Macro Interpretive Command

MIC Management Information Center

MIC Maritime Information Center [Netherlands]

MIC Mass-Impregnated Cable

MIC Master Interrupt Control

MIC Maximum Immission Concentration

MIC Medical Information Center

MIC Message Identification Code

MIC Metal-Induced Crystallization

MIC Methyl Isocyanate

MIC Michigan Instructional Computer

MIC Microbial-Induced Corrosion; Microbiologically Induced (or Influenced) Corrosion

MIC Microelectronic Integrated Circuit

MIC Micrometer [Instrument]

MIC Microphone [also Mic, or mic]

MIC Microwave Integrated Circuit

MIC Minimal Inhibitory Concentration [also mic]

MIC Minimum Ignition Current

MIC Minimum Inhibiting (or Inhibitory) Concentration [also mic]

MIC Missing Interruption Character

MIC Missing Interruption Checker

MIC Mode Indicate Common

MIC Modular Industrial Computer

MIC Monitoring, Identification and Correlation

MIC Monolithic Integrated Circuit

MIC Motorcycle Industry Council [US]

MIC Multilayer Interconnection Computer

MIC Mutual Interference Chart

Mic Microphone [also mic]

Mic Microscope; Microscopic; Microscopist [also mic]

MICA Macro-Instruction Compiler Assembler

MICA Major Incidents Computer Application

MICAM Microammeter

MICAPP Microcomputer-Assisted Process Planning Program

MICC Meetings Industry Council of Canada

MICC Mineral-Insulated Copper-Covered (Cable)

MICE Megavolt Iron Coil Experiment

MICE Member of the Institute of Civil Engineers

MICELEM Microphone Element

MICG Mercury Iodide Crystal Growth

Mich Michigan [US]

MIChemE Member of the Institute of Chemical Engineers

MICIS Material Information Control and Information System

MICNS Modulator Integrated Communication and Navigation System

MICOM Missile Command [of US Army]

Miconex International Fair for Measurement, Instrumentation and Automation

MICOS Multifunctional, Infrared Coherent Optical Sensor

MICR Magnetic-Ink Character Reader; Magnetic-Ink Character Recognition

micr microscopic

MICRO International Symposium on Microscopy

MICRO Microcontamination (Conference and Exhibition)

MICRO Multiple Indexing and Console Retrieval Operations

Micro Microcomputer [also micro]

Micro Microwave [also micro]

micro- prefix denoting one millionth, or 10^{-6}

Microb Microbe; Microbial [also microb]

Microb Pathogen Microbial Pathogenesis [Journal published in the US]

Microbeam Anal J The Microbeam Analysis Journal [of Microbeam Analysis Society, US]

Microbiol Microbiological; Microbiologist; Microbiology [also microbiol]

Microbiol Mol Biol Rev Microbiological and Molecular Biology Reviews [of American Society for Microbiology, US]

Microbiol Rev Microbiological Reviews [Published by the American Society for Microbiology, US]

Microbiol Sci Microbiological Sciences [Published by the International Union of Microbiological Societies]

microCAD Computer-Aided Design for Microcomputers

microCADD Computer-Aided Design and Drafting for Microcomputers

MICROCAT Micro-Catalog(ue)

Microchem Microchemical; Microchemist(ry) [also microchem]

Microchem J Microchemical Journal [Published by the American Microchemical Society, US]

Microchim Acta Microchimica Acta [Journal published in Germany]

Microcomp Microcomputer; Microcomputing [also microcomp]

Microcomput Appl Microcomputer Applications [Publication of the International Society of Mini- and Microcomputers]

Microcomput Civ Eng Microcomputers in Civil Engineering [Journal published in the US]

Micro Comput Colleg Micro Computer Colleg [Journal on Microcomputers; published in Germany]

Microcomput Inf Manage Microcomputers for Information Management [Journal published in the US]

Microcomput Rev Microcomputer Review [Journal published in the US]

Micro Decis Micro Decision [Journal published in the UK]

Microelectron Microelectronic(s) [also microelectron]

Microelectron Eng Microelectronic Engineering [Journal published in the Netherlands]

Microelectron J Microelectronics Journal [UK]

Microelectron Manuf Test Microelectronic Manufacturing and Testing [Journal published in the US]

Microelectron Reliab Microelectronics and Reliability [Journal published in the UK]

MICROFABTECH Center for Microfabrication Technology [of University of Florida, Gainesville, US]

Microform Rev Microform Review [US]

Microgr Micrographic(s) [also microgr]

Microgravity Sci Technol Microgravity Science and Technology [Journal published in Germany]

Microgr Mark Place Micrographics Market Place [Journal published in the UK]

micro-in micro-inch [also μ-in, or mu-in]

MICROM Micro-Instruction Read-Only Memory

Micromin Microminiature [also micromin]

MICRONET Microcomputer Network [also Micronet]

Micron Microsc Acta Micron and Microscopia Acta [Journal published in the UK]

MICROPAC Micromodule Data Processor and Computer

Micropaleont Micropaleontology [International Journal]

Microprocess Microprogr Microprocessing and Microprogramming [Journal published in the Netherlands]

Microprocess Microsyst Microprocessors and Microsystems [Journal published in the UK]

MICROPSI Microcomputer Printed Subject Index

MICROSSATÉLITE Microsatélite Franco-Brasileiro [French-Brazilian Scientific Micro-Satellite; of Brazilian National Institute for Space Research]

Microsc Microscope; Microscopist; Microscopic [also microsc]

Microsc The Microscope [Journal published in the US]

Microsc Electron Biol Cel Microscopia Electronica y Biologia Celular [Journal on Electron Microscopy and Cellular Biology; published by the Ibero-American Society for Cell Biology, Chile]

Microsc Microanal Microstr Microscopical and Microanalytical Microstructures (Journal)

Microsc Res Tech Microscopical Research and Technology (Journal)

Microsoft Syst J Microsoft Systems Journal [Publication of Microsoft Corporation, US]

Micro Syst Micro System [Journal published in France]

Microvasc Res Microvascular Research [Journal published in the US]

Microw Microwave [also microw]

Microw J Microwave Journal [US]

Microw Opt Technol Lett Microwave and Optical Technology Letters [US]

Microw RF Microwave and RF (Radio Frequency) [Journal published in the US]

Microw RF Eng Microwave and RF (Radio Frequency) Engineering [Journal published in the UK]

MICS Macro Interpretive Commands

MICS Maintenance Inventory Control System

MICS Management Information and Control System

MICS Management Information Control System

MICS Microprocessor Intertie and Communication System

MICS Mineral-Insulated Copper-Sheathed (Cable)

MICSD Mathematical, Information and Computational Sciences Division [of US Department of Energy]

MICT Ministry for Industry, Commerce and Technology [Canada]

MICT Ministry for Industry, Commerce and Tourism [Canada]

MICT Multicomponent Ideal Chemical Theory (Metal Solution)

MICU Message Interface and Clock Unit

Micuma Société des Mines de Cuivre de la Mauritanie [Mauretanian Copper Mines Company]

MICV Mechanized Infantry Combat Vehicle [US Army]

MICYT Ministeria de Ciencia y Tecnologia [Ministry of Science and Technology, Spain]

MID Magnetically Insulated Diode

MID Maritime Identification Digits [SARSAT–Search and Rescue Satellite]

MID Master of Industrial Design

MID Master of Interior Design

MID Median Infectious Dose (or Infectious Dose 50) [also ID_{50}, or ID50]

MID Message Input Description

MID Microbially Influenced Degradation

MID Midbody [Aerospace]

MID Minimal Infective Dose

MID Mission Integration Document

MID Multiple Ion Detection

Mid Midland [UK]

mid middle

MIDA Maritime Industrial Development Area

MIDAC Management Information for Decision Making and Control

MIDAC Michigan Digital Automatic Computer

MIDAM Mid-America Commodity Exchange [US]

MIDAR Microwave Detection and Ranging [also Midar, or midar]

MIDAS Map Information Display and Analysis System

MIDAS Materials Information Database Automated System [of ASM International, US]

MIDAS Measurement Information Data Analytic System

MIDAS Memory Implemented Data Acquisition Systems

MIDAS Microdiagnosis for Analysis and Repair

MIDAS Micro-Imaged Data Addition System

MIDAS Microprogrammable Integrated Data Acquisition System

MIDAS Microprogramming Design-Aided System

MIDAS Missile Defense Alarm System

MIDAS Missile Intercept Data Acquisition System

MIDAS Modified Integration Digital-to-Analog Simulator

MIDAS Modulator Isolation Diagnostic Analysis System

MIDAS Modular International Dealing and Accounting System

MIDAS Multimode International Data Acquisition Service [Australia]

MIDAS Multiple Index Data Access System

MIDAS Munitions Items Disposition Action System (Database) [of Argonne National Laboratory, Illinois, US]

MIDC Metal Industries Development Center [Kaohsiung, Taiwan]

MIDDLE Microprogram Design and Description Language

Middle East Electr Middle East Electricity [Published in the UK]

Middx Middlesex [UK]

Mid East Middle East

MIDEF Microprocedure Definition

MIDES Missile Detection System

MIDI Musical Instrument Digital Interface (File)

MIDIP Military Industry Data Interchange Procedure

MIDIST Mission Interministérielle de l'Information Scientifique et Technique [Interministerial Mission on Scientific and Technical Information, France]

MIDLNET Midwest Library Network [US]

MIDMS Machine-Independent Data Management System

Midn Midnight [also midn]

MIDOP Missile Doppler [also Midop, or midop]

MIDOR Miss Distance Optical Recorder

MIDOT Multiple Interferometer Determination of Trajectories

MIDP Microwave Induced Delayed Phosphorescence

MIDR Musceloskeletal Imaging Diagnostic Radiology [of Warren Grant Magnuson Clinical Center, National Institutes of Health, Bethesda, Maryland, US]

MIDREX Midland-Ross Direct Reduction (Process) [Metallurgy]

MIDS Movement Information Distribution Station

MIDS Multifunction Information Distribution System

MIDS Multimode Information Distribution System

MIDWEEK Manager Integrated Dictionary Week (Users Group) [US]

Midx Middlesex [UK]

MIE Magnetically-Enhanced Ion Etch

MIE Major Item of Equipment

MIE Master of Industrial Engineering

MIE Master of Irrigation Engineering

MIE Mechanical Incorporated Engineer

MIE Metal-Induced Embrittlement

MIEC Mixed Ionic-Electronic Conducting

MIEE Member of the Institute of Electrical Engineers

MIEEE Member of the Institute of Electrical and Electronics Engineers

MIEHM Modified Iterative Extended Hueckel Method [Physical Chemistry]

MIER Malaysia Institute for Economic Research

MIES McMaster (University) Institute for Energy Studies [Hamilton, Canada]

MIES Metastable Impact Electron Spectroscopy

MIF Management Information File

MIF Management Information Format

MIF Micro-Indentation Fracture (Technique)

MIF Miners International Federation [Belgium]

.MIF Management Information File [File Name Extension] [also .mif]

MIFAS Mechanized Integrated Financial Accounting System

MIFASS Marine Integrated Fire and Air Support System

Miferma Société des Mines de Fer de la Mauritanie [Mauretanian Iron Mines Company]

MIFI Missile Flight Indicator

MIFL Master International Frequency List

MIFR Master International Frequency Register

MIFR Monitored International Frequency Register [of International Telecommunications Union]

MIFS Multiplex Interferometric Fourier Spectroscopy

MIG Magnetic Injection Gun

MIG Mercury-in-Glass (Thermometer)

MIG Metal Inert-Gas (Welding)

MIG Metal-In-Gap (Head) [Recording Media]

MIG Mikoyan and Gurevish [Russian Fighterplane] [also MiG]

MIGA Multilateral Investment Guarantee Agency [of World Bank]

MIGS Metal-Induced Gap State(s)

MIH Method of Intermediate Hamiltonians [Physics]

MIH Missing Interrupt(ion) Handler

MIH Multiplex Interface Handler

mi/h mile(s) per hour [also mi/hr, MPH, or mph]

MII Microsoft/IBM Corporation/Intel

MII Mineral Information Institute [US]

MIIA Mine Inspectors Institute of America [US]

MIIR Mellon Institute of Industrial Research [Pittsburgh, Pennsylvania, US]

MIIS Metal-Insulation-InsulationSemiconductor

MIIS Monterey Institute of International Studies [Monterey, California, US]

MIIT Materials Interface Interactions Test (Program) [US Nuclear Waste Management]

MIJ Master of International Journalism

MIJ Member of the Institute of Journalists

MIJI Meaconing, Intrusion, Jamming and Interference [Radio Navigation]

MIK Methyl Isobutyl Ketone

MIKAS Mikrofichekatalogsystem [Microfiche Cataloguing System, Switzerland]

MIKE Mass-Analyzed Ion Kinetic Energy

MIKE Measurement of Instantaneous Kinetic Energy

MIKE Microphone [also Mike, or mike]

MIKER Microbalance Inverted Knudsen Effusion Recoil

MIKES Mass-Analyzed Ion Kinetic Energy Spectrometry; Mass-Selected Ion Kinetic Energy Spectrometry

Mikro- & Kleincomput Mikro- und Kleincomputer [Swiss Publication on Micro- and Minicomputers]

Mikrocomput Z Mikrocomputer Zeitschrift [German Microcomputer Journal]

Mikrowellen Mag Mikrowellen Magazin [German Microwave Magazine]

MIL GSFC (Goddard Space Flight Center) Space (Flight) Tracking and Data Network Station [of NASA]

MIL Machine Interface Layer

MIL Micro-Implementation Language

MIL Module Interconnection Language

.MIL Military [Internet Domain Name] [also .mil]

mIL Mouse Interleukin [also MIL] [Biochemistry]

mIL-1α Mouse Interleukin-1α [Biochemistry]

mIL-1β Mouse Interleukin-1β [Biochemistry]

mIL-1R Mouse Interleukin-1 Receptor [Biochemistry]

mIL-2 Mouse Interleukin-2 [Biochemistry]

mIL-3 Mouse Interleukin-3 [Biochemistry]

mIL-4 Mouse Interleukin-4 [Biochemistry]

mIL-5 Mouse Interleukin-5 [Biochemistry]

mIL-6 Mouse Interleukin-6 [Biochemistry]

mIL-6R Mouse Interleukin-6 Receptor [Biochemistry]

mIL-7 Mouse Interleukin-7 [Biochemistry]

mIL-9 Mouse Interleukin-9 [Biochemistry]

mIL-10 Mouse Interleukin-10 [Biochemistry]

mIL-12 Mouse Interleukin-12 [Biochemistry]

mIL-13 Mouse Interleukin-13 [Biochemistry]

Mil Mileage [also mil]

Mil Military [also mil]

Mil Milling [also mil]

mil mil [Unit Equal to 0.001 inch]

mil² square mil(s) [also sq mil]

MILA Merritt Island Launch Area [of NASA in Florida, US]

MILADGRU Military Advisory Group [US]

Mil Att Military Attaché

Mil Av Military Aviation

MILC Metal-Induced Lateral Crystallization

MILC Midwest Inter-Library Center [Chicago, Illinois, US]

MILCOMSAT Military Communication Satellite

MILDIP Military Industry Logistics Data Interchange Procedure

MILE Military Electronics

MilE Military Engineer [also MilEng]

MILECON Military Electronics Conference [also MIL-E-CON]

MilEng Military Engineer [also Mil Eng, or MilE]

Mil Eng Military Engineer(ing)

Mil Eng Military Engineer [Publication of the Society of Military Engineers, US]

MILEPOST Middlesborough Initiative Local Electronic Provision of Service Technique

MILES Multiple Integrated Laser Engagement System

mil-ft mil-foot [also mil ft]

Mil Gov Military Government

MIL-HDBK Military Handbook [also Mil-Hdbk]

mill million

MIL-I Military Specification on Interfaces

MIL-I Military Specification on Interference

MILIC Microwave Insulated Line Integrated Circuit

mil-in mil-inch [also mil in]

MILIPAC Military Portable Computer

millihg millimeter of mercury [also mmHg]

millirem one-thousandth rem (roentgen equivalent man) [Unit]

Milnet Military (Data) Network [of US Department of Defense]

MIL P Military Post [also Mil P]

MILR Master of Industrial and Labor Relations

MILREP Military Representative [also Milrep, or milrep]

MILS Missile Impact Location System

MILSATCOM Military Satellite Communications [also Milsatcom]

MIL-SPEC Military Specification [also Mil-Spec]

MILSTAAD Military Standard Activity Address

MIL-STD Military Standard [also Mil-Std]

MILSTICCS Military Standard Item Characteristics Coding Structure

MILSTRIP Military Standard Requisition(ing) and Issue Procedure [US Army]

MILUS Mining Industry Land Use Strategy [Canada]

mil(s)/yr mil(s) per year [also mil(s)/y, MPY, or mpy]

MIM Magnetic Induction Method

MIM Maintenance Interface Machine

MIM Manitoba Institute of Management [Canada]

MIM Maryland Institute of Metals [US]

MIM Master of Industrial Management

MIM Master of International Management

MIM Medical Instrument Marketing

MIM Message Input Module

MIM Metal Injection Molding

MIM Metal-Insulator-Metal (Semiconductor)

MIM Microgravity Isolation Mount

MIM Molecules in Molecule (Calculation)

MIM Minimum [also Mim, or mim]

MIM Modem Interface Module

MIM Modified Index Method

MIM Mount Isa Mines Holdings Ltd. [Queensland, Australia]

Mim Minimum [also MIM, or mim]

MIMA Metal Injection Molding Association [US]

MIMA Mineral Insulation Manufacturers Association [US]

MIMarE Member of the Institute of Marine Engineers

MIMC Member of the Institute of Management Consultants

MIMD Multiple Input, Multiple Data

MIMD Multiple-Instruction (Stream)/Multiple-Data (Stream)

MIME Member of the Institution of Mining Engineers

MIME Missing Mode Effect (in Electronic Spectroscopy)

MIME Multi-Purpose Internet Mail Extension

MIMechE Member of the Institute of Mechanical Engineers

MIMI Michael-Michael (Ring) [Organic Chemistry]

MIMI-ARC Michael-Michael Aldol Ring Closure [Organic Chemistry]

MIMIC Microwave Monolithic Integrated Circuit

MIMIC Millimeter-Wave Monolithic Integrated Circuit

MIMIC Monolithic Microwave Integrated Circuit

MIMIC Multiparameter Input Mechanism for Interactive Control

MI/MIC Mode Indicate/Mode Indicate Common

MIMinE Member of the Institute of Mining Engineers

MIMI-MIRC Michael-Michael-Michael Ring Closure [Organic Chemistry]

MIMLC Microinfiltrated Macrolaminated Composite

MIMM Member of the Institute of Mining and Metallurgy

MIMO Man In–Machine Out

MIMO Miniature Image Orthicon [also Mimo, or mimo]

MIMO Multiple Input, Multiple Output

MIMOLA Machine-Independent Microprogramming Language

MIMOS Malaysian Institute of Microelectronic Systems

MIMOSA Mission Modes and Space Analysis

MIMR Madison Integrated Microscope Resource [of University of Wisconsin, US]

MIMR Magnetic Ink Mark Recognition

MIMR Multifrequency Imaging Microwave Radiometer

MIMS Medical Information Management System

MIMS Membrane Inlet Mass Spectrometry

MIMS Multi-Item Multi-Source

MIMUG Meetings Industry Microcomputer Users Group [US]

MIN Mobile Identification Number [Telecommunications]

MIn Metal Indicator [Chemistry]

Min Mineral(ogical); Mineralogist; Mineralogy [also min]

Min Mining [also min]

Min Minister; Ministry [also min]

Min Minor(ity) [also min]

min minim [Unit]

min minimum

min minor

min minute (60 seconds) [also ′] [Symbol]

μin micro-inch [also μ-in, mu-in, or micro-in]

MINA Member of the Institute of Naval Architects

MINAC Miniature Linear Accelerator

MINAC Miniature Navigation Airborne Computer

MINCYT Ministerio de Industria, Comercio y Turismo [Ministry of Industry, Commerce and Tourism, Spain]

min/cm minute(s) per centimeter [also min cm^{-1}]

MINDAC Marine Inertial Navigation Data Assimilation Computer

MIND Mass-Impregnated Non-Draining (Compound)

MIND Modular Interactive Network Designer

MIND Multiple Instrument Network Distribution

MInd Metal Indicator [Chemistry]

MINDD Minimum Due Date

MIndD Master of Industrial Design [also MIndDes]

MIndE Master of Industrial Engineering

MINDO Modified Incomplete Neglect of Differential Overlap [Physics]

MINDO Modified Intermediate Neglect of Differential Overlap [Physics]

MINE Microbial Information Network Europe

MinE Mining Engineer [also Min Eng, or MinEng]

MINEAC Miniature Electronic Autocollimator

MInEd Master of Industrial Education [also MIndEd]

MinEng Mining Engineer [also Min Eng, or MinE]

Min Eng Mining Engineer(ing)

Min Eng Mining Engineering [Publication of the Society of Mining Engineers, US]

Min Eng (Colorado) Mining Engineering (Colorado) [Publication of the Society of Mining Engineers, US]

Min Eng (London) Mining Engineer (London) [UK Publication]

Min Eng (NY) Mining Engineering (New York) [Publication of the Society of Mining Engineers, US]

Mine Quarry Mine and Quarry [Journal published in the US]

Miner Mineral(ological); Mineralogist; Mineralogy [also miner]

Mineral Mineralogical; Mineralogist; Mineralogy [also mineral]

MINERALEX Mining and Mineral Exploration Exposition [also Mineralex]

Miner Dep Mineralium Deposita [Journal on Mineral Deposits published in Germany]

Miner Electrol Metabol Mineral and Electrolytic Metabolism [International Journal]

Miner Eng Minerals Engineering [Journal published in the US]

Miner Ind Int Minerals Industry International [Published in the UK]

Miner Ind Surv, Alum Mineral Industry Surveys, Aluminum [Journal published in the US]

Miner Ind Surv Bauxite Mineral Industry Surveys, Bauxite [Journal published in the US]

Miner Ind Surv, Chromium Mineral Industry Surveys, Chromium [Journal published in the US]

Miner Ind Surv, Tungsten Mineral Industry Surveys, Tungsten [Journal published in the US]

Miner J Mineralogical Journal [US]

Miner Mag Mineralogical Magazine [Published by the Mineralogical Society, UK]

Miner Mat Minerals and Materials [Journal published in the US]

Miner Metal Mineração Metalurgia [Brazilian Publication on Minerals and Metallurgy]

Miner Metall Process Minerals and Metallurgical Processing [Publication of the Society of Mining Engineers, US]

Miner Met Rev Minerals and Metals Review [India]

Miner Petrol Mineral and Petrology [Journal published in Austria]

Miner Process Extract Metall Rev Mineral Processing and Extractive Metallurgy Review [Published in the US]

Miner Resour Eng Mineral Resources Engineering [Published in the UK]

Miner Sci Eng Minerals Science and Engineering [Published in South Africa]

Miner Soc Bull Mineralogical Society Bulletin [UK]

Miner Yrbk Minerals Yearbook [Publication of the US Department of the Interior]

Mines Metall Mines et Metallurgie [French Journal on Mining and Metallurgy]

MINET Medical Information Network

Minet Military (Data) Network, Europe [of US Department of Defense]

MINEYC Ministerio de Educación y Ciencia [Ministry of Education and Science, Spain]

MINGEL Martin Integrated Neutral Graphics and Engineering Language

Min Geol Mineria y Geologia [Cuban Publication on Mining and Geology]

Mini Minicomputer [also mini]

MINIAPS Minimum Accessory Power Supply

MINIC Minimal Input Cataloguing

MINICS Minimal Input Cataloguing System

MINICS/PDS Minimal Input Cataloguing System–Periodicals Data System

MINIDIP Miniature Dual-in-Line Package [also MiniDIP] [Electronics]

minim minimum

MINIMARS Mini-Mirror Advanced Reactor Study

MINIMAX Minimizing the Maximal Error

Mini Micro Mag Mini-Micro Magazin [German Magazine on Mini- and Microcomputers]

Mini-Micro Syst Mini-Micro Systems [Journal published in the US]

μ**in/in·°C** micro-inch per inch per degree Celsius [also μin/in/°C, or μin/in-°C]

μ**in/in·°F** micro-inch per inch per degree Fahrenheit [also μin/in/°F, or μin/in-°F]

MINIRAR Minimum Radiation Requirements

MINI-SUBLAB Miniature Submarine Laboratory

MINIT Minimum Interference Threshold

MINITEX Minnesota Interlibrary Telecommunications Exchange [US]

Min J Mining Journal [UK]

Minl Mineral [also Min'l, minl, or min'l]

MinlE Mineral Engineer [also MinlEng]

Minl Eng Mineral Engineer(ing)

Min Mag Mining Magazine [UK]

MIN MC Minimum Material Condition [also Min MC]

Min Metall Mining and Metallurgy [Journal published in the US]

Minn Minnesota [US]

MINNEMAST Minnesota Mathematics and Science Teaching Project [US]

Min News Lett Mining News Letter [Published by the United Mining Councils of America]

Minor Planet Circ Minor Planet Circulars [Publication of the Smithsonian Astrophysical Observatory, Cambridge, US]

MINOS Main Injector Neutrino Oscillations Search (Experiment) [of Fermilab, Batavia, Illinois, US]

MINPOLDB Mineralförsörjningsdatabasen [Minerals Research Databases of the Swedish Geological Survey]

Minprex International Congress on Extractive Metallurgy and Minerals Processing

MINPRT Miniature Processing Time

MINPROC Mineral Processing [also MIN PROC, Min Proc, or min proc]

MINPROC Mineral Processing Database [of Canada Center for Mineral and Energy Technology, Natural Resources Canada]

Min Proc Technol Mineral Processing Technology (Journal) [UK]

M in Psych Nurs Master in Psychiatric Nursing

Min Q Mining Quarterly [Australia]

μ**in-rms** microinches, root-mean-square [also μinrms, or mu-in-rms]

MINS Miniature Inertial Navigation System

mins minutes

Min Sci Technol Mining Science and Technology [US Publication]

MINSD Minimum Planned Start Date

MINSOP Minimum Slack Time per Operation

MInstCE Member of the Institution of Civil Engineers

MInstEE Member of the Institution of Electrical Engineers

MInstME Member of the Institute of Mining Engineers

MInstMM Member of the Institute of Mining and Metallurgy

MInstWE Member of the Institute of Water Engineers

MINT (Center for) Materials and Information Technology [of University of Alabama, Tuscaloosa, US]

MINT (Center for) Micromagnetics and Information Technology [of University of Minnesota, Minneapolis, US]

MINT Magnetic Information Technology

MINT Malaysian Institute for Nuclear Technology and Research [Kajang]

MINT Material(s) Identification and New Item Control Technique

MINTEC Mining Technology (Database) [Canada Center for Mineral and Energy Technology, Natural Resources Canada]

MINTECH Ministry of Technology [UK]

Min Tech Mining Technology; Mining Technologist

Min Technol Mining Technology [Publication of the Institution of Mining Electrical and Mining Mechanical Engineers, UK]

MINTEK Mining Technology [South African Publication]

MINTEK Rep MINTEK Report [South Africa]

MINTEK Res Dig MINTEK Research Digest [South Africa]

MINTEK Rev MINTEK Review [South Africa]

MINUET Minnesota Internet Users Essential Tools

MINW Master Interface Network

MINWR Merritt Island National Wildlife Refuge [Florida, US]

Min Wt Minimum Weight [also min wt]

MINX Multimedia Information Network Exchange

Miny Ministry [also miny]

MIO Magnesium Indium Oxide

MIO Management Integration Office

MIO Map Information Office [of United States Geological Survey]

MIO Multiple Input/Output

MIOC Multifunction Integrated Optical Circuit

MIOP Master Input/Output Processor

MIOP Multiplexing Input/Output Processor

MIOS Modular Input/Output System

MIOS Multi-IMU (Inertial Measurement Unit) Operation System

MIP Machine Construction Processor

MIP Macrophage Inflammatory Protein [Biochemistry]

MIP Malaysian Institute of Physics [of University of Malaysia, Kuala Lumpur]

MIP Mandatory Inspection Point

MIP Manual Input Processing

MIP Manual Input Program

MIP Manufacturers of Illuminating Products [US]

MIP Marine Insurance Policy [also mip]

MIP Master of Intellectual Property

MIP Matrix Inversion Program

MIP Mean Indicated Pressure

MIP Mercury Intrusion Porosimeter; Mercury Intrusion Porosimetry

MIP 1,4-Bis(1-Methyl-2-Imidazoyl)phthalazine

MIP Microscopic Interdendritic Porosity [Metallurgy]

MIP Microprocessor

MIP Microwave Induced Plasma

MIP Minimum Impulse Pulse

MIP Ministry of Industrial Production [India]

MIP Missile Impact Predictor

MIP Mixed-in-Place

MIP Mixed Integer Programming

MIP MMU (Manned Maneuvering Unit) Integration Plan [of NASA]

MIP Modification Instruction Package

MIP Most Important Person

MIP Multiband Imaging Photometer

MIP Museums Informatics Project [of University of California, US]

MIP-1α Macrophage Inflammatory Protein-1α [Biochemistry]

MIP-1β Macrophage Inflammatory Protein-1β [Biochemistry]

mIP m-Iodophenol

MIPA Monoisopropanolamine

MIPA Multiple Intensity Profile Analysis

MIPAC Microprocess Analysis Code

MIPAS Michelson Interferometric Passive Atmosphere Sounder

MIPC Methylisopropylcarbinol

MIPC Methylisopropylphenylcarbamate

MIPE Magnetic Induction Plasma Engine

MIPE Modular Information Processing Equipment

MIPE Moskow Institute of Power Engineers [Russian Federation]

MIPIR Missile Precision Instrumentation Radar

MIPK Methyl Isopropyl Ketone

MIPP Master of International Public Policy

MIPPT McMaster Institute for Polymer Production Technology [Hamilton, Canada]

MIPR Military Intergovernmental Procurement (or Purchase) Request [US]

MIPS Map and Image Processing System

MIPS Merritt Island Press Site [of NASA in Florida, US]

MIPS Microcomputer Image Processing System

MIPS Million Instructions per Second [also mips]

MIPS Mini Image Processing System [of US Geological Survey]

MIPS Minimum Inventory Production System

MIPS Missile Impact Predictor Set

MIPS Morton Institute of Polymer Science [Akron, Ohio, US]

MIPS Multiband Imaging Photometer System

MIPT Metal-Insulator Phase Transition

MIR Magnetoimpedance Ratio

MIR Malfunction Investigation Report

MIR Master of Industrial Relations

MIR Medium-Range Infrared Spectroscopy

MIR Memory-Information Register

MIR Memory Input Register

MIR Micro-Instruction Register

MIR Microwave Imaging Radiometer [ER-2 NASA Ames High-Altitude Aircraft]

MIR Mid-Infrared

MIR Model Incident Report

MIR Multiple Internal Reflectance; Multiple Internal Reflection

MIR Multiwavelength Isomorphous Replacement

MIR Music Information Retrieval

Mir Mirror [also mir]

MIRA Mechanical Industrial Relations Association

MIRA Motor Industry Research Association [UK]

MIRACL Management Information Report Access without Computer Languages

MIRACODE Microfilm Retrieval Access Code

MIRAGE Microelectronic Indicator for Radar Ground Equipment

MIRAN Missile Ranging [also Miran, or miran]

MIRC Michael Ring Closure [Chemistry]

MIRCEN Microbiological Resources Center

MIRD Medical Internal Radiation Dose

MIRE Member of the Institute of Radio Engineers

MIRED Micro-Reciprocal-Degree (Kelvin) [also mired]

MIRF Multiple Instantaneous Response File

MIRFAC Mathematics in Recognizable Form Automatically Compiled

MIRFTIRS Multiple Internal Reflection Fourier Transform Infrared Spectroscopy

MIRG Metabolic Imaging Research Group [of Queen's University, Kingston, Ontario, Canada]

MIR(H) (External) Magnetic Field Dependent Magnetoimpedance Ratio [Symbol]

MIRIAD Maritime Institute for Research and Industrial Development [US]

MIR-IR Multuiple Internal Reflectance Infrared Spectroscopy

MIRL Medium-Intensity Runway Lights

MIRL Micro Instrumentation Research Laboratory [of University of Utah, Salt Lake City, US]

MIRO Mineral Industry Research Organization [UK]

MIRO Mining Industry Research Organization

MIROC Mineral Industry Research Organization of Canada

MIROC Mining Industry Research Organization of Canada

MIROS Modulation Inducing Retrodirective Optical System

MIRPE Multiple Infrared Photon Excitation [Solid-State Physics]

MIRPS Multiple Information Retrieval by Parallel Selection

MIRR Material Inspection and Receiving Report

MIRROS Modulation Inducing Reactive Retrodirective Optical System

MIRS Manpower Information Retrieval System

MIRS Medical Information Retrieval Service [US]

MIRT Molecular Infrared Track

MIRU Move-In, Rig-Up

MIRU Myocardial Infarction Rescue Unit

MIRV Multiple Independent Reentry Vehicle

MIRV Multiple Independently Targeted (or Targetable) Reentry Vehicle

MIRV Multiple Individually-Targeted Reentry Vehicle

MIS Macroscopic Internal Stress

MIS Management Information Science

MIS Management Information Service

MIS Management Information System

MIS Management Integrated System

MIS Manipulated Identification of Spins [Physics]

MIS Manipulator Jettison Subsystem; Manipulator Jettison System

MIS Marine Information System [US]

MIS Market Intelligence Service [of Industry Canada]

MIS Master of Arts in Information Systems

MIS Master of Information Science

MIS Materials Information Service [UK]

MIS Membrane Inlet System [also mis]

MIS Merchandise Information System

MIS Mercury-in-Steel (Thermometer)

MIS Metal-Insulator-Semiconductor

MIS Metal-Interface Layer-Semiconductor

MIS Metering Information System

MIS Midlands Industrial Association [UK]

MIS Mining-Induced Seismic

MIS Mining Institute of Scotland

MIS Minority Investigator Supplement

MIS Mission Information Subsystem

MIS Modified Initial System

MIS Modified in situ Retorting

MIS Moody Institute of Science [US]

MIS Multimedia Information Sources [Internet]

Mis Missing [also mis]

mis membrane inlet system [also MIS]

MISA Municipal and Industrial Strategies for Abatement

MISAM Multiple Indexed Sequential Access Method

misc miscellaneous; miscellany

misci miscible

MISCON Midwest Superconductivity Consortium [at Purdue University, West Lafayette, Indiana, US]

Misc Pub Miscellaneous Publication

MISD Multiple-Instruction (Stream)/Single-Data (Stream)

MISDAS Mechanical Impact System Design for Advanced Spacecraft

MISDF Measured Internal Strain Distribution Function

mi/sec miles per second [also mi sec^{-1}]

MISER Minimum Size Executive Routines

MISFEED Misaligned Feed

MISFET Metal-Insulator-Semiconductor Field-Effect Transistor

MISFET Metal-Insulator-Silicon Field-Effect Transistor

MISHAP Missiles High-Speed Assembly Program

MISIM Metal–Insulator–Semiconductor–Insulator–Metal

MISM Metal-Insulator-Semiconductor-Metal

MISMDS Multiple Instruction Stream–Multiple Data Stream

MISO Multiple Input, Single Output

MISP Medical Information Systems Program

MISP Microelectronics Industry Support Program [UK]

MISP Mineral Investment Stimulation Program [Canada]

MISR Multi-Angle Imaging Spectroradiometer

MISR Multiple Input Signature Register

MISS Mechanical Interruption Statistical Summary

MISS Missile Intercept Simulation System

MISS Mobile Integrated Support System

MISS Multi-Item, Single-Source

Miss Mission [also miss]

Miss Mississippi [US]

MISSDS Multiple Instruction Stream–Single Data Stream

MISSIL Management Information System Symbolic Interpretative Language

MISSION Manufacturing Information System Support Integrated On-Line [of NCR Corporation, US]

Missouri Tech Missouri Institute of Technolgy [Kansas City, US]

MIST Metal-Insulator-Semiconductor Field-Effect Transistor

MIST Microbursts and Severe Thunderstorms Experiment

MIST Minor Isotopes Safeguards Techniques

MIST Modular Intermittent Sort and Test (Device)

MIST Music Information System for Theorists [of Indiana University, Bloomington, US]

mist *(mistura)* – a mixture [Apothecary]

MISTRAM Missile Trajectory Measurement System

MIStructE Member of the Institution of Structural Engineers

MISUM Monthly Intelligence Summary

MIT Master Instruction Tape

MIT Master of Industrial Technology

MIT Market-If-Touched (Order) [Securities Exchange]

MIT Materials Interaction Test

MIT Maupa Institute of Technology [Manila, Phillipines]

MIT Massachusetts Institute of Technology [Cambridge, US]

MIT Master Instruction Tape

MIT Metal-(to-)Insulator Transition

MIT Metastable Intermolecular Composite

MIT 2-Methyl-4-Isothiazolin-3-One

MIT Milled in Transit

MIT Ministry of Industry and Tourism [Canada]

MIT Ministry of Industry and Trade [now Ministry of Industry, Trade and Technology, Canada]

MIT Modular Industrial Terminal

MIT Modular Intelligent Terminal

MIT Monoiodotyrosine [Biochemistry]

MIT Muroran Institute of Technology [Sapporo City, Japan]

MITA Microcomputer Industry Trade Association

MITAC Micro-TACAN (Tactical Air Navigation System)

MITAC-M-D/A Micro-TACAN (Tactical Air Navigation System) Multiple Equipment with Digital/Analog Converter

MITATT Mixed Tunneling Avalanche Transit Time

MITC Methylisothiocyanate

MITE Microelectronic Integrated Test Equipment

MITE Miniaturized Integrated Telegraph Equipment

MITE Miniaturized Integrated Telephone Equipment

MITE Missile Integration Terminal Equipment

MITE Multiple Input Terminal Equipment

MITEA Modular Integrated Towed Electronics Array

MITEC Mining Industry Technology Council of Canada [also Mitec]

Mitel SCC Mitel Semiconductor Corporation [Kanata, Ontario, Canada]

MITER Modular Installation of Telecommunications Equipment Racks

MITF Musser International Turfgrass Foundation [US]

MITI Ministry of International Trade and Industry [Japan]

MITI-AIST Ministry of International Trade and Industry–Agency of Industrial Science and Technology [Ibaraki, Japan]

MITJP Massachusetts Institute of Technology–Japan Program Technical Publication Series

MITM Military Industry Technical Manual

MITO Minimum Interval Takeoff

MITOC Missile Instrumentation Technical Operations Communications

MITOL Machine-Independent Telemetry-Oriented Language

MitoMap Mitochondrial Genome Database [of Emory University, Atlanta, Georgia, US]

MIT-PFC Massachusetts Institute of Technology–Plasma Fusion Center [Cambridge, US]

MIT Press Massachusetts Institute of Technology Press [Cambridge, US]

MITR Massachusetts Institute of Technology Reactor [US]

MITS Management Information and Text System

MITS Materials Information Translation Service [of ASM International, US]

MITS Micro Instrumentation Telemetry System [US]

Mitsubishi Cable Ind Rev Mitsubishi Cable Industries Review [Osaka, Japan]

Mitsubishi Denki Lab Rep Mitsubishi Denki Laboratory Reports [Amagasaki, Japan]

Mitsubishi Electr Adv Mitsubishi Electric Advance [Tokyo, Japan]

Mitsubishi Heavy Ind Tech Rev Mitsubishi Heavy Industries Technical Review [Tokyo, Japan]

Mitsubishi Steel Manuf Tech Rev Mitsubishi Steel Manufacturing Technical Review [Tokyo, Japan]

Mitsubishi Tech Bull Mitsubishi Technical Bulletin [Tokyo, Japan]

Mitsui Zosen Tech Rev Mitsui Zosen Technical Review [Published by Mitsui Engineering and Shipbuilding Co. Ltd., Tokyo, Japan]

MITT Ministry of Industry, Trade and Technology [Canada]

Mitt AGEN Mitteilungen AGEN [Communications of AGEN; published by Geschäftsstelle des AGEN Instituts für Technische Physik, Eidgenössische Technische Hochschule-Aussenstation, Hönggerberg, Switzerland]

Mitt Astron Ges Mitteilungen der Astronomischen Gesellschaft [Communications of the German Astronomical Society]

Mitt Bundesforsch Forst-Holzwirtsch Mitteilungen der Bundesforschungsanstalt für Forst- und Holzwirtschaft [Communications of the German Federal Research Institute for Forestry and Timber Industry]

Mitt Math Ges Mitteilungen der Mathematischen Gesellschaft [Communications of the Mathematical Society, Germany]

MITTSE Mobile lgor Tracking Telescope System

Mitt Tech Univ Carolo-Wilhelmina Mitteilungen der Technischen Universität Carolo-Wilhelmina [Communications of the Carolo-Wilhelmina University of Technology, Braunschweig, Germany]

MIU Machine Interface Unit

MIU Malfunction Insertion Unit

MIU Manpower and Immigration Union

MIU Message Injection Unit

MIU MET (Mobile Earth Terminal) Interface Unit [Mobile Satellite Program, US Department of Commerce]

MIU Microalgae International Union [US]

MIU Model Interface Unit

MIU Multiplex Interface Unit

MIU Multistation Interface Unit

MIUH Meteorologisches Institut der Universität Hamburg [Meteorological Institute of the University of Hamburg, Germany]

MIUS Modular-Integrated Utility System

MIV Mosquito Iridescent Virus

MIVC Magnetically Induced Velocity Change

Mivometer Millivoltmeter [also mivometer]

MIVPO Modified Inside Vapor Phase Oxidation

MIW Micro-Instruction Word

MIWAC Marine and Inland Waters Advisory Committee [of Australian and New Zealand Environment and Conservation Council]

MIWS Magnetically Induced Wigner Solid [Solid-State Physics]

MIWT Member of the Institute of Wireless Technology

MIX Magnetic Ionization Experiment

MIX Master Index

MIX Member Information Exchange

MIX 1-Methyl-3-Isobutylxanthine [Biochemistry]

Mix Mixture [also Mixt, mixt, or mix]

MIZEX Marginal Ice Zone Experiment

MJ Mandibular Joint [Anatomy]

MJ Master of Journalism

MJ Mechanical Joint

MJ Megajoule [Unit]

MJ Metal Jacketed (Bullet)

MJ Mineral Jelly

M&J Menley and James Laboratories [Pharmaceutical Manufacturer]

M∥J Magnetization Parallel to (or In-Plane with) Current Density [Symbol]

M⊥J Magnetization Perpendicular to (or Out-of-Plane with) Current Density [Symbol]

mJ millijoule [Unit]

μJ microjoule [Unit]

m³/J cubic meter(s) per Joule [also m³ J^{-1}]

MJCA Midbody Jettison Control Assembly

mJ/cm² millijoule(s) per square centimeter [also mJ cm^{-2}]

μJ/cm² microjoule(s) per square centimeter [also μJ cm^{-2}]

MJD Modified Julian Date [Astronomy]

mJ/K microjoule(s) per Kelvin [also μJ K^{-1}]

mJ/K² microjoule(s) per Kelvin squared [also μJ K^{-2}]

μJ/K microjoule(s) per Kelvin [also μJ K^{-1}]

MJ/kg Megajoule(s) per kilogram [also MJ kg^{-1}]

MJ/m² Megajoule(s) per square meter [also MJ m^{-2}]

mJ/m millijoule(s) per meter [also mJ m^{-1}]

MJ/m³ Megajoule(s) per cubic meter [also MJ m^{-3}]

mJ/m² millijoule(s) per square meter [also mJ m^{-2}]

mJ/m²K millijoule(s) per square meter kelvin [also mJ m^{-2} K^{-1}]

mJ/mol·K millijoule per mole kelvin [also mJ/mol/K, mJ/mol-K, or mJ mol^{-1} K^{-1}]

mJ/mol·K² millijoule per mole kelvin squared [also mJ/mol/K², mJ/mol-K², or mJ mol^{-1} K^{-2}]

MJO Madden and Julian Oscillation

MJR McDonald Jelly Roll [Superconducting Cable Construction]

MJR Modified Jelly Roll

m⁴/Js meter to the fourth (power) per Joule-second [also m⁴ /J·s, or m⁴ J^{-1} s^{-1}]

mJ/s millijoule(s) per second [also mJ/sec, or mJ s^{-1}]

MJSA Manufacturing Jewelers and Silversmiths of America

mJ/sec millijoule(s) per second [also mJ/s, or mJ s^{-1}]

mJ/s/mg millijoule(s) per second per milligram [also mJ/sec/mg, or mJ s^{-1} mg^{-1}]

MJU Myong Ji University [Seoul, South Korea]

MK Macedonia [ISO Code]

MK Malawi Kwacha [Currency]

MK Manual Clock [Computers]

MK Mauna Kea [Hawaiian Mountain]

MK Mouse Keratinocyte [Biochemistry]

MK Potassium Metaphosphate Polymer

M-K Marciniak-Kuczynski (Model for Forming-Limit Curve) [Metallurgy]

Mk Mask [also mk]

Mk Mark [also mk]

Mk Mark [Currency]

M(κ) Local Magnetization Vector (in Solid-State Physics) [Symbol]

mK millikelvin [Unit]

m-K meter-Kelvin [also m-K]

mk Macedonia [Country Code/Domain Name]

mk mike (10^{-6} inch) [Unit]

μK microkelvin [Unit]

MKA Machine Knife Association [US]

MKD$ Convert Double-Precision Number to 8-Byte String (Function) [Programming]

MKDIR Make Directory (Command) [also MkDir, or mkdir]

MKG Meter-Kilogram (System) [also mkg, or m-kg]

Mkg Marking [also mkg]

m/kg meter(s) per kilogram [also m kg^{-1}]

m-kg meter-kilogram [Unit]

m²/kg square meter(s) per kilogram [also m² kg^{-1}]

m³/kg cubic meter(s) per kilogram [also m³ kg^{-1}]

m-kgf meter-kilogram-force [Unit]

MKH Multiple Key Hashing

MKI Mineralogisch-Kristallographisches Institut [Institute for Mineralogy and Crystallography, University of Goettingen, Germany]

MKI$ Convert Integer to 2-Byte String (Function) [Programming]

MKL Mask Left (Function)

mKLH Maleimido-Activated Keyhole Limpet Hemocyanin [Biochemistry]

MKMA Machine Knife Manufacturers Association [now Machine Knife Association, US]

MKP Monopotassium Phosphate

MKR Mask Right (Function)

MKR Radio Marker Beacon

Mkr Marker [also mkr]

MKS Meter-Kilogram-Second (System) [also mks]

MKSA Meter-Kilogram-Second-Ampere (System) [also mksa]

MKS$ Convert Single-Precision Number to 4-Byte String (Function) [Programming]

Mkt Market [also mkt]

Mktg Marketing [also mktg]

MKTI Mission Kit Technical Instruction

Mkt Val Market Value [also mkt val]

MKw Malawi Kwacha [Currency]

m²K/W square meter Kelvin per Watt [also m²K W⁻¹]

MkWh Thousand Kilowatthour(s) [Unit]

ML Automotive Crankcase Oil, Spark-Ignition Engine, Light Service Conditions [American Petroleum Institute Classification]

ML Machine Language

ML *(Magister Legum)* – Master of Laws

ML Magnetic Latching

ML Magnetic Lattice

ML Magnetic Leakage

ML Magnetic Levitation

ML Magnetic Link

ML Magnetic Multilayer

ML Magnetoluminescence; Magnetoluminescent

ML Main Lobe (of Antenna)

ML Maintenance Loop [also M/L]

ML Major Lobe (of Antenna)

ML Mali [ISO Code]

ML Manipulation Language

ML Marine Life

ML Material List

ML Materials Letters [Published in the Netherlands]

ML Materials Laboratory

ML Mathematical Linguistics

ML Mathematical Logics

ML Mauna Loa [Hawaiian Mountain]

ML Maximum Likelihood (Method) [Statistics]

ML Mean Level

ML Mean-Life

ML Measure Line

ML Memory Location

ML Message Length

ML Meta-Language

ML Metal-Lined

ML Metastable Liquid

ML Methods of Limit

ML Michelson Laboratory [US]

ML Microlithography

ML Microprogramming Language

ML Midcable Loop

ML Mid-Deck Left

ML Mineral Liberation

ML Mixed Lengths

ML Mobile Launcher

ML Mobile Low-Power Plant [US Army]

ML Mold Line

ML Molecular Lattice

ML Monodisperse Latex

ML Monolayer

ML Monotonically (Increasing) Load

ML Moseley's Law [Spectroscopy]

ML Motor Launch

ML Multilayer

ML Multipoint Logger

ML Music Legato [Programming]

ML Muzzle Loader; Muzzle Loading

M²L Laboratory of Micromachining and Microtransducers [also M2L, or MML]

M/L Maintenance Loop [also ML]

M/L Metal-to-Ligand (Ratio) [Organometallics]

M-L Metallic-Longitudinal

M-L Metal-Ligand (Complex) [Organometallics]

Ml Local Earthquake Magnitude (i.e. Original Richter Magnitude) [Symbol]

mL millilambert [Unit]

mL milliliter [also ml]

.ml Mali [Country Code/Domain Name]

µL microliter [also µl, or muL]

MLA Manitoba Library Association [Canada]

MLA Marine Librarians Association [UK]

MLA Master of Landscape Architecture

MLA Master Locksmiths Association

MLA Matching Logic and Adder

MLA MDM (Manipulator Deployment Mechanism) Launch Aft

MLA Medical Library Association [US]

MLA Mercantile Library Association [St. Louis, Missouri, US]

MLA Mineral Liberation Analysis

MLA Mixed Lead Alkyl

MLA Modern Language Association [US]

MLA Modulated Laser Absorption

MLA Moellenstedt Lens Analyzer

MLA Monochrome Lens Assembly

MLA Multilayer Ceramic Actuator

MLA Multi-Layered Alloy

MLA Multiple Line Adapter

MLAB Modeling Laboratory

MLandArch Master of Landscape Architecture

MLAP Metallic Line Access Port

MLAPI Multilingual Application Programming Interface

MLArch Master of Landscape Architecture

MLArchUD Master of Landscape Architecture and Urban Design

MLAS Master of Laboratory Animal Science

MLAUD Master of Landscape Architecture and Urban Design

MLB Magnetic Linear Birefringence

MLB Multilayer Board

MLC Magnesia-Doped Lanthanum Chromite

MLC Main Lobe Cancellation (of Antenna)

MLC Maneuver Load Control

MLC Michigan Library Consortium [US]

MLC Microprogram Location Counter

MLC Mississippi Library Commission [US]

MLC Mixed Leucocyte Culture [Microbiology]

MLC Mixed Lymphocyte Culture [Microbiology]

MLC Mobile Launcher Computer

MLC Molecular Luminescence Spectrometry with Continuum Source

MLC Multilayer Capacitor

MLC Multilayer Ceramic

MLC Multilayer Circuit

MLC Multilevel Cell (Program) [Internet]

MLC Multiline Control

MLCA Multiline Communications Adapter

MlcA Maleic Acid

MLCAEC Military Liaison Committee to the Atomic Energy Commission [US]

MLCB Multilayer Circuit Board

MLCC Monolithic Ceramic Capacitor

MLCD Microwave Liquid Crystal Display

MLCP Multiline Communications Processor

MLCPT Multicomponent Linear Chemical Physical Theory

MLCT Metal-to-Ligand Charge Transfer [Organometallics]

MLD Machine Language Debugger

MLD Magnetic Linear Dichroism

MLD Masking Level Difference

MLD Mean Lethal Dose

MLD Median Lethal Dose [also LD50, or LD_{50}]

MLD Microcrack Length Distribution

MLD Middle Landing [Controllers]

MLD Minimum Lethal Dose

MLD Minimum Line of Detection

MLD Mixed Layer Depth

MLD Multilayer Dielectric

MLD Multi-Loop Digital Controller

MLDAP Magnetic Linear Dichroism in Angle-Resolved Photoemission [Physics]

MLDes Master of Landscape Design

Mldg Molding [also mldg]

MLE Magazine Lee-Enfield (Rifle)

MLE Maximum Likelihood Estimate [Statistics]

MLE Medium Local Exchange

MLE Melt-Leach Evaporation

MLE Mesoscale Lightning Experiment [Meteorology]

MLE Molecular-Layer Epitaxy

MLE Multi-Line Editor

MLEK Magnetic Liquid Encapsulated Kyropoulos (Method) [Crystal Growing]

MLEV Malcolm-Levitt (Decoupling) [Nuclear Magnetic Resonance]

MLEV Manned Lifting Entry Vehicle

MLF Mali Franc [Currency]

MLF MDM (Manipulator Deployment Mechanism) Launch Forward [NASA]

MLF Medium Longitudinal Fascicule

MLF Multilateral Force

MLF Multilayer Film

MLG Main Landing Gear

MLG Methyl-L-Glutamate [Biochemistry]

MLG Middle Low German

MLGS Microwave Landing Guidance System

MLGW Maximum Landing Gross Weight

MLH Maximum Likelihood [Statistics]

ML/h Monolayer(s) per Hour [also ML/hr]

mL/ha milliliter(s) per hectare [also mL ha^{-1}, ml/ha, or ml ha^{-1}]

MLHCP Mean Lower Hemispherical Candlepower [also mlhcp]

MLHR Master of Labour and Human Resources

MLHW Mean Lower High Water

MLI Machine Language Instruction

MLI Magnetic Level Indicator

MLI Marker Light Indicator

MLI Message Level Interface

MLI Microwave Laboratories Inc. [Raleigh, New Jersey, US]

MLI Minimum Line of Interception

MLI Ministry of Local Industry [Russian Federation]

MLI Multi-Layer Insulation (Blanket) [of NASA]

MLIA Multiplex Loop Interface Adapter

MLib Master of Librarianship

MLID Multi-Link Interface Driver

MLIM Matrix Log-In Memory

MLIP Message Level Interface Port

MLIR Master of Labour and Industrial Relations

MLIS Master of Library and Information Science

MLIS Molecular Laser Isotope Separation

MLIS Multilingual Information Society (Program) [of European Commission]

mL/kg milliliter per kilogram [Unit]

MLL Manned Lunar Landing

MLL MDM (Manipulator Deployment Mechanism) Launch Left [NASA]

MLL Microlaterlog [Oil Exploration]

MLL Middle-Level (Programming) Language

MLL Molecular Luminescence Spectrometry with Line Source

mL/L milliliter per liter [also mL L^{-1}]

MLLE Medium Large Local Exchange

Mlle *(Mademoiselle)* – French for "Miss"

MLLP Manned Lunar Landing Program [of NASA Apollo Program]

MLLW Mean Lower Low Water

MLM Magazine Lee-Metford (Rifle)

MLM Mailing List Manager [Internet]

MLM Master of Library Media

MLM Microemulsion Liquid Membrane

MLM Molecular Layering Method

MLM Multilayer Metallization

MLM Multilevel Marketing

MLM Multilevel Metallization

MLMA Metal Ladder Manufacturers Association [US]

ML/min Monolayer(s) per minute [also ML min^{-1}]

mL/min milliliter(s) per minute [also mL min^{-1}]

mL/min/100g milliliter(s) per miniute per 100 grams [Unit]

MLMIS Minnesota Land Management Information System [US]

MLNSC Materials Laboratory and Nuclear Science Center

Mln Malononitrile

MLO Manned Lunar Orbiter [of NASA Apollo Program]

MLO Mauna Loa Observatory [Hawaii]

MLO Medium Lubricating Oil

MLP Machine Language Program(ming)

MLP Mean Life Period

MLP Mobile Launch(er) Platform

MLP Multiple Line Printing

MLPC Multiplicand [also Mlpc, or mlpc]

MLPCB Multilayer Printed Circuit Board

ML PED Mobile Launcher Pedestal [of NASA]

MLPH Ministry of Lands, Parks and Housing [Canada]

MLPP Multi-Level Precedence and Preemption

MLPPP Multilink Point-to-Point Protocol [Internet]

MLPR Multiplier [also Mlpr, or mlpr]

MLPWB Multilayer Printed Wiring Board

MLR Main Line of Resistance

MLR Maintenance Loop Recorder [also M/LR]

MLR MDM (Manipulator Deployment Mechanism) Launch Right

MLR Mechanized Line Record

MLR Memory Lockout Register

MLR Minimum Lending Rate

MLR Mixed Leucocyte Reaction [Biochemistry]

MLR Monodisperse Latex Reactor

MLR Multilayer Reflector

MLR Multiple Linear Regression [Statistics]

MLR Multiply and Round

M/LR Maintenance Loop Recorder [also MLR]

MLRG Marine Life Research Group [of Scripps Institute of Oceanography, University of California at San Diego, US]

MLRG Muzzle-Loading Rifled Gun

MLRHR Master of Labour Relations and Human Resources

MLRS Monodisperse Latex Reactor System

MLRS Multiple Launch Rocket System

MLS Machine Literature Search

MLS Manned Lunar Surface [NASA Apollo Program]

MLS Master Laboratory Station

MLS Master of Library Science

MLS Master of Library Services

MLS Master of Life Sciences

MLS Metal-Langmuir Semiconductor

MLS Microwave Landing System

MLS Microwave Limb Sounder [for NASA Upper Atmosphere Research Satellite]

MLS Missile Location System

MLS Mudline Suspension System

MLS Multilanguage System

MLS Multilayered Structure

MLS Multilevel Secure

MLS Multi-Loop System

MLS Multiple Listing Service

ML/s Monolayer(s) per second [also ML/sec]

Mls Multilayers

mL/s milliliter per second [also mL s^{-1}, or mL/sec]

MLSE Maximum Likelihood Sequence Estimation

mL/sec milliliter per second [also mL/s, or mL/s]

ML/SFA Metal Lath/Steel Framing Association Division [of National Association of Architectural Metal Manufacturers, US]

MLSFET Metal-Langmuir Semiconductor Field-Effect Transistor

MLSIT Master of Library Science and Instructional Technology

MLSNPG Microwave Landing System National Planning Group [US]

MLSO Mode-Locked Surface-Acoustic-Wave Oscillator

MLSS Manual Large-Sample Stage (Microscope)

MLSS Mixed Liquid Suspended Solids

MLSS Mixed Liquor Suspended Solids [Activated Sludge Treatment]

MLSW Modified Lifshitz, Slyozov and Wagner (Theory) [Metallurgy]

MLT Mass-Length-Time

MLT Mean Length of Turn

MLT Mean Length per Turn

MLT Mean Logistical Time

MLT Mean Low Tide

MLT Mechanized Line Testing

MLT Mechanized Loop Test

MLT Median Lethal Time

MLT Medical Laboratory Technician

MLT Medical Laboratory Technologist; Medical Laboratory Technology

MLT Micro-Layer Transistor

MLT Multi-Layered Cloud System

Mlt Maltose [Biochemistry]

MLTA Multiple Line Terminal Adapter

MLTF Multilayer Thick Film

MltON Maltose Octanitrate [Biochemistry]

MLTS Micro-Level Test Set

MLTY Multiply [also Mlty, or mlty]

MLU Memory Loading Unit

MLU Memory Logic Unit

MLU Multiple Logic Unit

Mlu Micrococcus luteus [Microbiology]

MLV Medium Launch Vehicle

MLV Medium Logistics Vehicle

MLV Multilamellar Vesicles

MLV Murine Leukemia Virus

MLVSS Mixed Liquor Volatile Suspended Solids [Activated-Sludge Treatment]

MLVW Medium Logistics Vehicle, Wheeled

MLW Maximum Landing Weight

MLW Mean Low Water

MLW Medium-Level (Nuclear) Waste

MLWA Maximum Landing Weight Authorized

MLWN Mean Low-Water Neaps

MLWS Mean Low Water Springs

Mlwk Millwork [also mlwk]

MLY Multiply [also Mly, or mly]

MM Automotive Crankcase Oil, Spark-Ignition Engine, Moderate Service Conditions [American Petroleum Institute Classification]

MM Machmeter

MM Magnemill (Magnetic Powder)

MM Magnetic Material

MM Magnetic Moment

MM Magnetomechanic(al); Magnetomechanics

MM Main Memory

MM Main Module

MM Maintenance Manual

MM Major Mode

MM Man-Made

MM Man Month [Unit]

MM Mass Market

MM Mass Memory

MM Master Monitor

MM Master of Management

MM Material Management

MM Materials Measurement

MM Mathematical Model(ling)

MM Matrix Mechanics [Quantum Mechanics]

MM Matrix Multiplication

MM Measuring Microscope

MM Mechanical Metallurgist; Mechanical Metallurgy

MM Mechanical Milling

MM Mechanochemical Mixing

MM Medical Material

MM Medical Microbiologist; Medical Microbiology

MM Megamega

MM Melanocyte Medium [Biochemistry]

MM Memory Management

MM Memory Module

MM Memory Multiplexer

MM Merchant Marine

MM Mesoscale Model [of National Center for Atmospheric Research, US]

MM *(Messieurs)* – French for "Gentlemen"

MM Metallurgical Microscope

MM Metals Museum [Sendai, Japan]

MM Methyl Mercaptan

MM Methyl Methacrylate

MM Michelson-Morley (Experiment) [Optics]

MM Micro-Manipulator [Microscopy]

MM Micromechanical; Micromechanics

MM Middle Marker [Aeronautics]

MM Miedema's Model

MM Million

MM Military Medal

MM Mine Manager; Mine Management

MM Mineral Matter

MM Mining Machine(ry)

MM Ministère de la Marine [French Ministry of the Navy]

MM Minutes [also mm]

MM Mirror Machine [Plasma Physics]

MM Mischmetal

MM Mission Manager

MM Mission Module [Aerospace]

MM Modified Mercalli (Scale) [Geophysics]

MM Molecular Mass

MM Molecular Mechanics

MM Molecular Modelling

MM Moment Method [Electromagnetics]

MM Monostable Multivibrator; Monovibrator

MM Month [also mm]

MM MOS (Metal-Oxide Semiconductor) Monolithic

MM Motor Meter

MM Moving Magnet

MM Mucous Membrane [Anatomy]

MM MultiMate (Font Software)

MM Multi-Media (System)

MM Multimeter

MM Myanmar [ISO Code]

MM1 Magnemill Resin Bonded Permanent Magnet

MM2 Magnemill Isotropic Compacted Permanent Magnet

MM3 Magnemill Magnetically Anisotropic Compacted Magnet

M&M Materials and Maintenance

M&M Mining and Metals

M&M Mowatt and Moore Limited [Canadian Pharmaceutical Manufacturer]

M-M Matrix-Matrix (Grain Boundary) [Materials Science]

M-M Metal-Metal (Composite)

M$_R$/M$_S$ Remanence Ratio (i.e., Ratio of Remanent Magnetization to Saturation Magnetization)

M_r/M_s Squareness of Magnetization Loop (i.e., Ratio of Remanent Magnetization to Saturation Magnetization)

$\overline{M}_w/\overline{M}_n$ Polydispersity (i.e., Ratio of the Weight-Average Molecular Weight and the Number-Average Molecular Weight of a Polymer) [also M_W/M_N, or M_w/M_n]

Mm Megameter [Unit]

mM millimole [Unit]

mm *(millia)* – thousands

mm millimeter [Unit]

mm minutes [also MM]

mm month [Date]

mm^{-1} one per millimeter [also 1/mm]

mm^2 square millimeter [Unit]

mm^{-2} one per square millimeter [also $1/mm^2$]

mm^3 cubic millimeter [Unit]

mm^{-3} one per cubic millimeter [also $1/mm^3$]

mm^4 millimeter to the fourth power [Unit]

m/m proportion by mass [Symbol]

m/m^2 meter(s) per square meter [also $m\ m^{-2}$]

m/m^3 meter(s) per cubic meter [also $m\ m^{-3}$]

m^2/m^2 square meter(s) per square meter [also $m^2\ m^{-2}$]

m^2/m^3 square meter(s) per cubic meter [also $m^2\ m^{-3}$]

m^3/m^3 cubic meter(s) per cubic meter [also $m^2\ m^{-3}$]

m_μ/m_e muon mass to electron mass ratio (206.7683) [Symbol]

m_p/m_e proton mass to electrom mass ratio (36.2) [Symbol]

$m\mu$ millimicron [Unit]

μM micromole [also μM]

μm micrometer [also mu m]

$\mu m^{-1/2}$ one per square root of micrometer [also $1/\mu m^{1/2}$]

μm^{-1} one per micrometer [also $1/\mu m$]

μm^2 square micrometer [Unit]

μm^{-2} one per micrometer [also $1/\mu m^2$]

μm^3 cubic micrometer [Unit]

μm^{-3} one per cubic micrometer [also $1/\mu m^3$]

$\mu\mu$ micromicron [also mu mu]

μ_e/μ_B electron magnetic moment (or free electron g factor in Bohr magnetons) (1.0011597) [Symbol]

μ_e/μ_p electron and proton magnetic moment ratio (658.2107) [Symbol]

μ_μ/μ_p muon and proton magnetic moment ratio (3.183346) [Symbol]

μ_p/μ_B proton magnetic moment in Bohr magnetons (1.5210322×10^{-3}) [Symbol]

μ_p/μ_N magnetic moment of protons in water corrected for diamagnetism of water (2.7928474) [Symbol]

μ/μ_N magnetic moment of protons in water in nuclear magnetons (2.7927756) [Symbol]

MMA Magnesium Methyl Carbonate

MMA Magnetically Modulated Microwave

MMA Magnetosensitive Microwave Absorption

MMA Manitoba Medical Association [Canada]

MMA Manual Metal-Arc (Welding)

MMA Master of Marine Affairs

MMA Master of Manpower Administration

MMA Master of Media Arts

MMA Materials Marketing Association [US]

MMA Merchant Marine Academy [at Kings Point, New York State, US]

MMA Metal Mining Association [Canada]

MMA Meter Manufacturers Association [UK]

MMA Methyl Methacrylate

MMA Microcomputer Managers Association [US]

MMA Microgravity Measurement Assembly

MMA Microtome Manufacturers Association [UK]

MMA Mining and Metallurgical Association [Japan]

MMA Monomethylamine

MMA Monorail Manufacturers Association [US]

MMA Multiple Module Access

mMa m-Methylaniline

MMAD Mass Median Aerodynamic Diameter

MMACS Maintenance Management and Control System

MMACS Maintenance, Mechanical Arm and Crew Systems (Engineer) [NASA]

MMAE Master of Mechanical and Aerospace Engineering

μ**MAG** Micromagnetic Modeling Activity Group [of National Institute of Standards and Technology, Gaithersburg, Maryland, US]

MMAJ Metal Mining Agency of Japan

MMAP Marine Mammal Action Plan

MMAP Microgravity Measurement and Analysis Project

MMAP Molecularly Modified Alkoxide Precursor

MMAR Main Memory Address Register

MMAS Material Management Accountability System; Material Management Accounting System

MMath Master of Mathematics

MMatSE Master of Material Science and Engineering

MMAU Master Multiattribute Utility

MMB Master of Medical Biochemistry

MMB Minimum Monthly Balance

MMB Modulated Molecular Beam

MMB Molecular Microbiology

$M(\mu B)$ Bohr Magneton Dependent Magnetization [Symbol]

MMBA Mutation Mink Breeders Association [US]

mMBA m-Methylbenzoic Acid

MMB/D Millions of Barrels per Day [also mmb/d]

MMBI Methyl-2-Mercapto Benzimidazole

MMBR Microbiological and Molecular Biology Reviews [of American Society for Microbiology, US]

MM BTU/hr Million British Thermal Units per hour [also MM Btu/hr]

MMC Magnesium Methyl Carbonate

MMC Magnetic Microscopy Center [of University of Minnesota, Minneapolis, US]

MMC Malaysia Mining Corporation

MMC Manganese Metal Company [South Africa]

MMC Man-Machine Communication

MMC Manufacturing Management Council [US]

MMC Martin Marietta Corporation [US]

MMC Matched Memory Cycle

MMC Master of Mass Communications

MMC Maximum Material Condition

MMC Maximum Metal Condition

MMC Meharry Medical College [Nashville, Tennessee, US]

MMC Memory Management Controller

MMC Metal-Matrix Carbide

MMC Metal-Matrix Composite

MMC Methoxymagnesium Methylcarbonate

MMC Metropolis Monte Carlo (Algorithm)

MMC Microcomputer Marketing Council [US]

MMC Micro-Machining Center

MMC Microsoft Management Console

MMC Mid Motor Controller

MMC Mini Magellanic Cloud

MMC Mining Monitoring and Control System

MMC Mission Management Center

MMC Mitsubishi Motor Corporation [Japan]

MMC Money Market Certificate

MMC Monolithic Multicomponent Ceramic

MMC Monopolies and Mergers Commission

MMC Multipart Memory Controller

MMCA Midbody Motor Control Assembly [Aerospace]

MMCA Mid Motor Controller Assembly

MMCA Multi-Media Compliance Assessment

MMCD Multi-Media Compact Disk

MMCF Million Cubic Feet [also mmcf]

MMCFD Million Cubic Feet per Day [also mmcfd]

MMCIAC Metal-Matrix Composites Information Analysis Center [of US Department of Defense]

MMCL Master Measurement and Control List

$\mu m/cm^2$ micrometer per square centimeter [also $\mu m\ cm^{-2}$]

MMCo$_5$ Mischmetal Cobalt (Permanent Magnet)

MMCP Micro-Master Control Processor

MMCR Millimeter Cloud Radar

MMCS Mass Memory Control Subsystem

MMCT Moment-Modified Compact Tension [Mechanical Testing]

MMCX Multimedia Communication Exchange

MMD Manual of the Medical Department

MMD Mass Median Diameter

MMD Master of Mechanical Engineering

MMD Microgravity Measuring Device

MMD Micromicrodiffraction [also $\mu\mu D$]

MMD Mines and Minerals Division [of Ministry of Northern Development, Canada]

MMD Mining and Metallurgy Department

MMD Moving Map Display [Aeronautics]

$\mu\mu D$ Micromicrodiffraction [also MMD]

MMDA Money Market Deposit Account

MMDB Mass Memory Data Base

MMDB Master Measurement Data Base

MMDB Molecular Modelling Database [of National Center for Biological Information, US]

MM-DD-YY Month–Day–Year [also mm-dd-yy]

MMDF Mission Model Data File

MMDS Martin Marietta Data Systems [of Martin Marietta Corporation, US]

MME Manufacturing Machines and Equipment

MME Master of Manufacturing Engineering

MME Master of Mechanical Engineering

MME Master of Mining Engineering

MME Materials and Metallurgical Engineering

MME Metallurgical and Materials Engineering

MME Metrolum Multimode Electrode

MME Mining Mechanical Engineer(ing)

MME Ministry of Minerals and Energy [Australia]

M&ME Materials and Metallurgical Engineering

M&ME Metallurgical and Materials Engineering

M-Me Metal-Metalloid (Compound)

Mme (Madame) – French for "Madam," or "Lady"

MMechEng Master of Mechanical Engineering [also MMechE]

MMEL Master Minimum Equipment List

MMEP Marine Mammal Events Program [US]

MMES Member of the Mechanical Engineering Society

MMES MSFC (Marshall Space Flight Center) Mated Element Systems [of NASA]

MMet Master of Metallurgy

MMetE Master of Metallurgical Engineering [also MMetEng]

MMetMatE Master of Metallurgical and Materials Engineering [also MMetMatEng]

MMetMatSc Master of Metallurgical and Materials Science [also MMetMatS]

MMF Magnetomotive Force [also mmf]

MMF Man-Made Fiber

MMF Materials Manufacturing Forum [US]

MMF Microwave Modulation Frequency

MMF Fleet Minelayer [US Navy Symbol]

MMF Minimum Mass Fraction

MMF Moving Magnetic Features

MMF Multiple Mainframe

mmf magnetomotive force [also MMF]

mmf micromicrofarad [also $\mu\mu F$, $\mu\mu f$, mmF, or mmfd]

mμf millimicrofarad [also mμF]

MMFC Major Materials Facilities Commission [of National Academy of Sciences, US]

MMFM Modified Modified Frequency Modulation [also M^{2}FM, or M2FM]

MMFITB Man-Made Fibers Industry Training Board

MMFPA Man-Made Fiber Producers Association [now American Fiber Manufacturers Association, US]

MMFS Manipulating Message Format Standard

MMFS Manufacturing Message Format Standard

MMG Magnetic Materials Group

MMG Metallurgy and Materials Group [of Indira Gandhi Centre for Atomic Research, Kalpakkam, India]

MMG Motor-Motor Generator

μ/mg micron(s) per milligram [Unit]

MMgt Master of Management [also MMgmt]

MMgtE Master of Management Engineering [also MMgtEng]

MMgtS Master of Management Science [also MmgtSc]

MMH Maintenance Man-Hour

MMH Methylmercury Hydroxide

MMH Monomethylhydrazine

mm/h millimeter(s) per hour [also mm h^{-1}, mm/hr, or mm hr^{-1}]

μm/h micrometer(s) per hour [also $μm\ h^{-1}$, μm/hr, or $μm\ hr^{-1}$]

MMHE Master of Materials Handing Engineering

mm Hg millimeter(s) of mercury [also mmHg]

mμ Hg millimicron(s) of mercury [also mμHg]

mm Hg L millimeter(s) of mercury liters [also mmHgL]

μmho micromho [also μ-mho, or mu-mho]

mm H_2O millimeter(s) of water [also mmH$_2$O]

μmho/cm micromho per centimeter [also $μmho\ cm^{-1}$]

μm Hg micrometer(s) of mercury [also μmHg, mu Hg, or muHg]

μm/hr micrometer per hour [also μm/h, or $μm\ h^{-1}$]

MMHS Master of Management in Human Services

μm·Hz micrometer hertz [Unit]

MMI Main Memory Interface

MMI Man-Machine Interaction

MMI Man-Machine Interface

MMI Materials Management International [UK]

MMI Michigan Molecular Institute [Midland, US]

MMI Modified Mercalli Intensity [Geophysics]

MMI Multimessage Interface

MMIC Monolithic Microwave Integrated Circuit; Microwave Monolithic Integrated Circuit

MMIC Motorcycle and Moped Industry Council [US]

MMIC MESFET Monolithic Microwave Integrated Circuit Metal-Semiconductor Field-Effect Transistor

MMIJ Mining and Materials Processing Institute of Japan

MMIJ Mining and Metallurgical Institute of Japan [now Mining and Materials Processing Institute of Japan]

m/min meters per minute [also mpm, or MPM]

m³/min cubic meters per minute [also cmpm, or CMPM]

MMinEng Master of Mining Engineering [also MMinE]

MMIS Man-Machine Interface Interface Subsystem

MMIS Master of Management Information Systems

MMIS Materials Management Information System

MMIU Multipart Memory Interface Unit

MMJ Modified Modular Jack

MMK Melanocyte Medium Kit [Biochemistry]

MMK Methyl Mesityl Ketone

MMK Multiprocessor Multitasking Kernel

m/m·K meter per meter kelvin [also m/(m·K), m/m/K, or $m\ m^{-1}\ K^{-1}$]

MML Manitoba Motor League [Canada]

MML Man-Machine Language

MML Martin Marietta Laboratories [Baltimore, Maryland, US]

MML Mathematics Markup Language

MML Master Measurement(s) List

MML Material Mechanics Laboratory

MML Mechanically Mixed Layer

MML Metallurgical Materials Laboratory [of University of Maryland, College Park, US]

MML Laboratory of Micromachining and Microtransducers [also M2L, or M²L]

MMLC Multimedia Learning Center

MMLS Microgravity Materials Science Laboratory

MMLS Model-Modes-Loads-Stresses [also M-M-L-S]

MML-TR Martin Marietta Laboratories–Technical Report [Baltimore, Maryland, US] [also MML TR]

MMM (Annual Conference on) Magnetism and Magnetic Materials

MMM Maintenance and Material Management

MMM Manned Mars Mission [of NASA]

MMM Mars Mission Module [of NASA]

MMM Mercury Motor Meter

MMM Mesoscale and Microscale Meteorology Division [of National Center for Atmospheric Research, US]

MMM (Automation in) Mining, Mineral and Metal Processing (Conference) [of Verein Deutscher Ingenieure/Verband Deutscher Elektrotechniker– Gesellschaft für Meß– und Automatisierungstechnik, Germany]

MMM Model Microfield Method [Spectroscopy]

MMM Multimission Module

mm³/m cubic millimeter(s) per meter [also $mm^3\ m^{-1}$]

MMMA Magnetically Modulated Microwave Absorption

MMMA Meat Machinery Manufacturers Association [US]

MMMA Metalforming Machinery Makers Association [UK]

MMMA Micromagnetic Multiparameter Microstructure and Residual Stress Analysis [also 3MA]

MMMA Milking Machine Manufacturers Association [UK]

MMMC Milking Machine Manufacturers Council [US]

MMMC Minnesota Mining and Manufacturing Company [St. Paul, US] [also 3M]

μm/m·°C micrometer per meter degree Celsius [also μm/m/°C, or $μm\ m^{-1}\ °C^{-1}$]

MMMD Mechanics and Mechanisms of Materials Damping [American Society for Testing and Materials Special Technical Publication, US]

MMMI Meat Machinery Manufacturers Institute [US]

mm/min millimeter(s) per minute [also $mm\ min^{-1}$]

mm³/min cubic millimeter(s) per minute [also $mm^3\ min^{-1}$]

μm/min micrometer(s) per minute [also $μm\ min^{-1}$]

MMMIS Maintenance and Material Management Information System

μm/m·K micrometer(s) meter kelvin [also μm/m/K, or $μm\ m^{-1}\ K^{-1}$]

M1,M2,M3,M4 Alternating-Current Secondary [Controllers]

mm/mm millimeters per millimeter [also $mm\ mm^{-1}$]

mm /mol cubic millimeter(s) per mole [also $mm^3\ mol^{-1}$]

MMMS Minerals, Metals and Materials Society [US]

MMMSocAm Member of the Mining and Metallurgical Society of America

mm/μs millimeter(s) per microsecond [also $mm\ μs^{-1}$]

mm^{-1} μs^{-2} one per millimeter per microsecond squared [also 1/mm μs^{-2}, mm^{-1} μsec^{-2}, or 1/mm μs^2]

mm/μs K millimeter(s) per microsecond Kelvin [also mm μs^{-1} K^{-1}]

mm/mW millimeter per milliwatt [also mm mW^{-1}]

MMN Manitoba Marketing Network [of Manitoba Department of Industry, Trade and Tourism, Canada]

μm/N micrometer(s) per Newton [also μm N^{-1}]

MMNC Mesoporous Magnetic Nanocomposite

mm^3/Nm cubic millimeter per Newton-meter [also mm^3 N^{-1} m^{-1}]

MMO Magnesium Manganese Oxide

MMO Materials and Manufacturing Ontario [of (Ontario) Ministry of Economic Development, Trade and Tourism, Canada]

MMO Metal-Matrix Oxide

M/MO Metal/Metal Oxide (Interface)

MMOA Mobile Modular Office Association [US]

MMOD Micromodule

mmol millimole [Unit]

m^3/mol cubic meter(s) per mole [also m^3 mol^{-1}]

μmol micromole [Unit]

mmol/dm^3 millimole(s) per cubic decimeter [also mmol dm^{-3}]

μmol/dm^3 micromole(s) per cubic decimeter [also mmol dm^{-3}]

μmol/J micromole(s) per joule [also μmol J^{-1}]

mmol/L millimole(s) per Liter [also mmol L^{-1}, or mmol/l]

μmol/L micromole(s) per Liter [also μmol/l or μmol L^{-1}]

μmol/m^2 micromole(s) per square meter [also μmol m^{-2}]

μmol/mg·h micromole(s) per milligram hour [also μmol mg^{-1} h^{-1}]

mmol/m^2·h micromole(s) per square meter hour [also μmol m^{-2} h^{-1}]

μmol S m^{-2} micromole(s) Siemens per square meter [Unit]

MMOS Message Multiplexer Operating System

MMOS Multimode Optical Sensor

MMP Maintenance Management Process

MMP Maintenance Management Program

MMP Mass Market Paper

MMP Master of Marine Policy

MMP Materials Management Plan

MMP Microphysical Measurement Package

MMP Minimum Miscibility Pressure

MMP Modified Modular Plug

MMP Module Message Processor

MMP Monitor Metering Panel

MMP Multiplex Message Processor

mMP m-Methoxy Phenol

MMPA Magnetic Material Producers Association [US]

MMPA Marine Mammal Protection Act [US]

MMPA Marine Mammal Protection Association [US]

M/MPC Minimum, or Maximum Permissible (or Permitted) Concentration [also m/mpc]

MMPD 4-Methoxy-1,3-Benzenediamine

MMPD Minerals and Metallurgical Processing Division [of Society of Mining Engineers of the American Institute of Mining, Metallurgical and Petroleum Engineers, US]

MMPF Microgravity and Materials Processing Facility

MMPI Mining and Materials Processing Institute

MMPI Minnesota Multiphasic Personality Inventory

MMPIJ Mining and Materials Processing Institute of Japan

MMPM Multi-Media Presentation Manager

MMPO Monomethyl Phosphate

MMPP Magnesium Monoperoxyphthalate

MMPP Mechanized Market Programming Procedures

MMPQFP Multilayer Molded-Plastic Quad Flat Pack [Electronics]

MMPR Missile Manufacturers Planning Reports

MMPS Multimachine Power System

MMPSE Multiuse Mission Payload Support Equipment [of NASA]

MMPT Man-Machine Partnership Translation

MMPU Memory Manager and Protect Unit

MMR Magnetic Memory Record

MMR Main Memory Register

MMR Massachusetts Military Reservation [US]

MMR Master of Marketing Research

MMR Maximum Metal Reflector

MMR Measles, Mumps and Rubella (Vaccine) [Medicine]

MMR Meteorological Measurement Radiometer

MMR Modified Modified Read

MMR Multiple Match Resolver

MMRA Maritime Marshland Rehabilitation Association [Canada]

MMRBM Mobile Medium-Range (or Mid-Range) Ballistic Missile

MMRC Mars Mission Research Center [Greensboro, North Carolina, US]

MMRIM Mat-Molding Reaction Injection Molding

MMRRI Mining and Mineral Resources Research Institute [US]

MMRS Mississippi-Missouri River System [US]

M-MRS Mexican Materials Research Society

MMS Malayan Mathematical Society

MMS Massachusetts Medical Society [US]

MMS Magnetic Median Surface

MMS Maimonides Medical Center [New York City, US]

MMS Maintenance Manipulator System

MMS Man-Machine System

MMS Manufacturing Message Service

MMS Manufacturing Message Specification

MMS Massachusetts Medical Society [US]

MMS Mass Memory Store

MMS Master of Management Science

MMS Master of Management Studies

MMS Master of Marketing Science

MMS Master of Materials Science

MMS Materials Management System

MMS Melbourne Medical School [Australia]

MMS Memory Management System

MMS Mercury Manganese Selenide (Semiconductor)

MMS Metallurgy and Materials Science

MMS Meteorological Measurement System

MMS Methyl Methanesulfonate

MMS DL-Methylmethioninsulfonium Chloride

MMS Microfiche Management System

MMS Minerals Management Service [of US Department of the Interior]

MMS Missile Monitoring System

MMS Mission Modular Spacecraft

MMS Modular Multiband Scanner

MMS Module Management System

MMS Moskow Mathematical Society [Russia]

MMS Multi Measuring System

MMS Multi-Media System

MMS Multimission Modular Spacecraft

MMS Multi-Mission Spacecraft

MMS Multipart Memory System

mm/s millimeter(s) per second [also mm/sec, or mm s^{-1}]

mm²/s square millimeter(s) per second [also mm²/sec, or mm² s^{-1}]

μm/s micrometer(s) per second [also μm s^{-1}· or μm/sec]

μm²/s square micrometer per second [also μm² s^{-1}, or μm²/sec]

MMSA Materials and Methods Standards Association [US]

MMSA Mining and Metallurgical Society of America [US]

mMSa m-Methylthioaniline

MMSC Manchester Materials Research Center [UK]

MMSC Materials and Molecular Simulation Center [of California Institute of Technology, Pasadena, US]

MMSC (Biennial) Microgravity Materials Science Conference

MMSc Master of Medical Science

MMSCF Million Standard Cubic Feet [Unit]

MMSCFD Million Standard Cubic Feet per Day [Unit]

MMSCFH Million Standard Cubic Feet per Hour [Unit]

MMSCFM Million Standard Cubic Feet per Minute [Unit]

MMSD Metallurgy and Material Science Division

MMSD Metallurgy and Material Science Division [of Institute of Engineers, India]

MMSE Master of Manufacturing Systems Engineering

MMSE Mini-Mental State Examination

MMSE Minimum Mean Squared Error

MMSE Multiuse Mission Support Equipment

mm/sec millimeter(s) per second [also mm s^{-1}]

mm²/sec square millimeter(s) per second [also mm²/s]

mμsec millimicrosecond [Unit]

μm/sec micrometer(s) per second [also μm/s, or μm s^{-1}]

μm²/sec square micrometer per second [also μm²/s, or μm² s^{-1}]

mMSfa m-Methylsulfonylaniline

MMSI Medium Metal-Support Interaction

MMSL Microgravity Materials Science Laboratory [of NASA]

MMSocAm Mining and Metallurgical Society of America

MMSP Modeling in Materials Science and Processing

mMSP m-Methylthiophenol

mMsP m-Methylsulfonylphenol

MMSt Master of Museum Studies [also MMStud]

MMST/YR Million Short Tons per Year [also MMst/yr]

MMT Mass Memory Test

MMT Mercury Manganese Telluride (Semiconductor)

MMT 4-Methoxytriphenylmethylchloride

MMT Methylcyclopentadienyl Manganese Tricarbonyl

MMT Methyl-m-Tyrosine [Biochemistry]

MMT Mont Megantic Telescope

MMT Multiple Mirror Telescope

α-MMT Methyl-m-Tyrosine [Biochemistry]

MMTC Mechanical Maintenance Training Center

MMTC Microelectronics and Materials Technology Center [of Royal Melbourne Institute of Technology, Victoria, Australia]

MMtlE Master of Metal Engineering [also MMtlEng]

MMTS S-Methylmethanethiosulfate

MMTS Methyl Methylmercaptomethyl Sulfoxide

MMTS Methyl Methylthiomethyl Sulfoxide

MMTT Multiple Mechanicothermal Treatment

MMTV Mouse Mammary Tumor Virus

MMU Main Memory Unit

MMU Manchester Metropolitan University [UK]

MMU Manned Maneuvering Unit

MMU Mass Memory Unit

MMU Memory Management Unit

MMU Memory Mapping Unit

MMU Metered Message Unit

MMU Milli-Mass Unit [also mmu]

MMU Minimum Mapping Unit

MMU Modular Maneuvering Unit

MMU Multi-Message Unit

mmu milli-mass unit [also MMU]

m mu millimicron [also mμ]

MMV Mice Minute Virus

MMV Muromegalovirus

MMVF Multimedia Video File

MMW Main Magnetization Winding

MMW Millimeter Wave

MMWR Morbidity and Mortality Weekly Report [of Centers for Disease Control and Prevention, Atlanta, Georgia, US]

MMX (Intel) Matrix Manipulation Extensions

MMX Multimedia Exchange

MMX Multimedia Extensions

M′M″X₃ Ceramic Compound of Metallic Element M′, Metallic Element M″ and Nonmetallic Element X in Which the Ratio of the Cations of M′ to the Cations of M″ to the Anions X is 1:1:3 (e.g., $BaTiO_3$, $CaTiO_3$, etc.)

M′MX₄ Ceramic Compound of Metallic Element M′, Metallic Element M″ and Nonmetallic Element X in Which the Ratio of the Cations of M′ to the Cations of M″ to the Anions X is 1:2:4 (e.g., $MgAl_2O_4$, $MgFe_2O_4$, $MnFe_2O_4$, etc.)

mMxA m-Methoxyaniline

mMxBa m-Methoxybenzoic Acid

mMxCa m-Methoxycarbonylaniline

mm/y millimeter(s) per year [also mm/yr]

μm/y micrometer(s) per year [also μm/yr, mu m/y, or mu m/yr]

MN Macronucleus (of Protozoans) [Zoology]

MN Magnesium Niobate

MN Magnetic North

MN Main Network

MN Master of Nursing

MN Matrix Notation

MN Meganewton [Unit]

MN Mendeleev Number [Chemistry]

MN Merchant Navy

MN Metal Nitride

MN Methylnaphthalene

MN Micronucleus (of Protozoans) [Zoology]

MN Minnesota [US]

MN Mnemonic Symbol

MN Mongolia [ISO Code]

MN Mononitro-

MN Motor Neuron [Anatomy]

MN Motor Nerve

MN Motor Number

MN Music Normal [Programming]

M/N M-Shell/N-Shell (Emission Ratio) [Nuclear Physics]

Mn Main [also mn]

Mn Manganese [Symbol]

Mn Manual [also mn]

Mn Mnemonic(s) [also mn]

Mn Month [also mn]

Mn Molecular Weight, Number-Average

Mn^{2+} Divalent Manganese Ion [also Mn^{++}] [Symbol]

Mn^{3+} Trivalent Manganese Ion [also Mn^{+++}] [Symbol]

Mn^{4+} Tetravalent Manganese Ion [Symbol]

Mn-51 Manganese-51 [also ^{51}Mn, or Mn^{51}]

Mn-52 Manganese-52 [also ^{52}Mn, or Mn^{52}]

Mn-53 Manganese-53 [also ^{53}Mn, or Mn^{53}]

Mn-54 Manganese-54 [also ^{54}Mn, or Mn^{54}]

Mn-55 Manganese-55 [also ^{55}Mn, or Mn^{55}, or Mn]

Mn-56 Manganese-56 [also ^{56}Mn, or Mn^{56}]

Mn-57 Manganese-57 [also ^{57}Mn, or Mn^{57}]

mN millinewton [Unit]

.mn Mongolia [Country Code/Domain Name]

m/n mass-to-charge ratio [Mass Spectrometry]

μN micronewton [Unit]

MNA Master of Nursing Administration

MNA Methoxyneuraminic Acid

MNA Methyl Nadic Anhydride

MNA Methyl Nitroaniline

MNA Methylnonylacetaldehyde

MNA Methyl-5-Norbornene-2,3-Dicarboxylic Anhydride

MNA 2-Methyl-4-Nitroaniline (Crystal)

MNA Ministry of Northern Affairs [Canada]

MNA Mononitroaniline

MNA Multi-Network Area

MNA Multi-Share Network Architecture

MNA Museo Nacional de Antropologia [National Anthropological Museum, Mexico]

MNAA Museo Nacional de Antropologia y Archeologia [National Anthropological and Archeological Museum, Lima, Peru]

Mn(Ac)$_2$ Manganese(II) Acetate

Mn(Ac)$_3$ Manganese(III) Acetate

Mn$_{12}$-Ac Manganese$_{12}$ Acetate

Mn(ACAC)$_2$ Manganese(II) Acetylacetonate [also Mn(acac)$_2$]

Mn(ACAC)$_3$ Manganese(III) Acetylacetonate [also Mn(acac)$_3$]

MnAlC Manganese Aluminum Carbon (Permanent Magnet)

MNAns Mononitroanisole

MnAs Manganese Arsenide

MNatSci Master of Natural Science [also MNatS, or MNatSc]

MNB 2-Methyl-2-Nitroso Butan-3-one

MNB Mononitrobenzene

Mn-B Manganese-Boron (Alloy) (or Manganese Boron)

MNBA Mononitrobenzaldehyde

mNBAAP 4-N-(3'-Nitrobenzylidene) Aminoantipyrine [also m-NBAAP]

MNBAc Mononitrobenzoic Acid

Mn-Bi Manganese-Bismuth (Alloy System)

MNBLE Modified Nearly Best Linear Estimator

Mn-Bronze Manganese Bronze

MNC Mediterranean Nonmember Countries

MNC Multinational Company

MNC Multinational Corporation

MNC Multiplicative Noise Compensator

MNC Museo Nacional de Chile [National Museum of Chile, Santiago]

MNCIAWPRC Malaysian National Committee of the International Association on Water Pollution Research and Control

Mn(Cp)$_2$ Bis(cyclopentadienyl)manganese (or Manganocene)

MnCp(CO)$_3$ Cyclopentadienylmanganese Tricarbonyl

MNCrs Mononitrocresol

Mn-Cu Manganese-Copper (Alloy System)

MnCu(pbaOH)(H$_2$O)$_3$]$_n$ Manganese Copper 2-Hydroxy-1,3-Propylenebis(oxamate) Trihydrate

MNCS Multipoint Network Control System

MND Ministry of National Defense

MND Ministry of Northern Development [Canada]

MND Multinomial Distribution

MNDDA Methyl-n-Didecylamine

MNDDO Modified Neglect of Diatomic Differential Overlap [Physics]

MNDEP Mineral Deposits Data Base (Database)

MNDM (Ontario) Ministry of Northern Development and Mines [Canada]

Mn-DMS Manganese Diluted Magnetic Semiconductor

MNDO Modified Neglect of Differential Overlap; Modified Neglect of Diatomic Overlap [Physics]

MNDO-C Modified Neglect of Differential Overlap Correlation; Modified Neglect of Diatomic Overlap Correlation [Physics]

MNDX Mobile Nondirector Exchange

MNE Master of Nuclear Engineering

M&NE (Department of) Materials and Nuclear Engineering

MNF Multisystem Networking Facility

Mng Managing [also mng]

Mng Dir Managing Director

Mngt Management [also mngt]

MNH Magnesium Nitrate Hexahydrate

MNH Museum of Natural History

MNHC Micro/Nano Hybrid Composite

Mn(HFAC)₂ Manganese (II) Hexafluoroacetylacetonate [also Mn(hfac)₂]

MNHN Museo Nacional de Historia Natural [National Museum of Natural History, Santiago, Chile]

M(Ni) Nickel (Salt) Molarity [Symbol]

M-NiAl Martensitic Nickel Aluminide Phase [Metallurgy]

Mn-In Manganese-Indium (Alloy System)

MNIND Mineral Industry Data Base (Database)

MNK Methyl Naphthyl Ketone

MNK Methylnonylketone

MnKα Manganese K-Alpha (Radiation) [also MnK$_\alpha$]

MNL Minnesota National Laboratory [US]

Mnl Manual [also mnl]

Mnl Mineral [also mnl]

Mnld Mainland [also mnld]

MnlE Mineral Engineer [also Mnl Eng, or MnlEng]

Mnl Eng Mineral Engineer(ing)

Mnl Proc Mineral Processing

MNM Miyazaki-Nakamura-Mori (Theory) [Metallurgy]

MNM Minimum [also mnm]

MNM Mononitromethane

MNM Museum of New Mexico [Santa Fe, US]

MN/m² Meganewton(s) per square meter [also MN m^{-2}]

MN/√m³ Meganewton(s) per square root of cubic meter [also MN/m$^{3/2}$]

mN/m millinewton(s) per meter [also mN m^{-1}]

MNM-C Miyazaki-Nakamura-Mori–Compression (Theory) [Metallurgy]

MNMeA Mononitromethylaniline

Mn(MeO)₂ Manganese(II) Methoxide (or Manganese (II) Methylate) [also Mn(OMe)₂]

Mn-Mg-Di Manganese-Magnesium-Didymium (Ferrite)

Mn-Mg-Zn Manganese-Magnesium-Zinc (Ferrite)

MN/mm² Megaton(s) per square millimeter [also MN mm^{-2}]

mN/μm millinewton(s) per micrometer [also mN μm^{-1}]

MNM-T Miyazaki-Nakamura-Mori–Tension (Theory) [Metallurgy]

MNN Mononitronaphthalene

Mnn Mannose [Biochemistry]

MNNG 1-Methyl-3-Nitro-1-Nitrosoguanidine; N-Methyl-N'-Nitro-N-Nitrosoguanidine [Biochemistry]

MnnHN Mannose Hexanitrate [Biochemistry]

Mn-Ni Manganese-Nickel (Alloy System)

mN/nm millinewton per nanometer [also mN nm^{-1}]

MnNO Manganonitrate

MNNR Marine National Nature Reserve

MNNS Mississippi Museum of Natural Science [Jackson, US]

MNO N,N'-Dinitrodimethyloxamide

MNO Magnesium Nickel Oxide

MnO Manganese(II) Oxide (or Manganous Oxide)

MnO Permanganate (Ion) [Symbol]

MnOₓ Manganese Oxide (e.g., MnO, MnO₂, Mn₂O₃, or Mn₃O₄) [also MnOx]

mNOa m-Nitroaniline

MnO-Al₂O₃ Manganese Oxide-Aluminum Oxide (System)

mNOBA m-Nitrobenzoic Acid

MnO-B₂O₃ Manganese Oxide–Boric Oxide (System)

MnO-CaO-SiO₂ Manganese Oxide–Calcium Oxide–Silicon Dioxide

mNOP m-Nitrophenol

MNOS Metal-Nitride-Oxide Semiconductor

MNOS Metal-Nitride-Oxide Silicon

mNOS Macrophage-Derived Nitric Oxide Synthetase [Biochemistry]

MNOS-AD Metal-Nitride-Oxide-Silicon Avalanche Diode

MNOSFET Metal-Nitride Oxide Semiconductor Field-Effect Transistor

MnO-SiO₂ Manganese Oxide-Silicon Dioxide (System)

MnO-TiO₂ Manganese Oxide-Thorium Dioxide (System)

MnO-ZrO₂ Manganese Oxide-Zirconium Oxide (System)

MNP 2-Methoxy-4-(2-Nitrovinyl)phenyl

MNP 2-Methyl-2-Nitrosopropane

MNP Methyl-N-Pyrrolidine

MNP Microcom Networking Protocol [Internet]

MnP Manganese Phosphide

MNP-Gal 2-Methoxy-4-(2-Nitrovinyl)phenyl β-D-Galactopyranoside [Biochemistry]

MNP-Glc 2-Methoxy-4-(2-Nitrovinyl)phenyl β-D-Glucopyranoside [Biochemistry]

MNPK Methyl-p-Nitrophenyl Ketone

Mn(PMCp)₂ Bis(pentamethylcyclopentadienyl)manganese

MNPS Minimum Navigation Performance Specifications

MNPT Medium National Pipe Thread

MNQ 2-Methyl-1,4-Naphthoquinone

MNR Maximum Normed Residual (Statistics)

MNR McMaster (University) Nuclear Reactor [Hamilton, Ontario, Canada]

MNR Minimum Noise Route

MNR (Ontario) Ministry of Natural Resources [Canada]

M(NR₂)ₓ Transition Metal Dialkylamido Complex [General Formula]

M(NR$_2$)$_4$ Tetrakis(dialkylamido)metal Complex [Formula]

MNRC McClellan Nuclear Radiation Center [of McClellan Air Force Base, Sacramento, California, US]

Mnrl Mineral [also mnrl]

MNRM Master of Natural Resources Management

MNRPRA Malaysia Natural Rubber Producers Research Association

M(NRR′)$_x$ Alkylamide [General Formula]

MNRU Modulated Noise Reference Unit

MNS Malayan Nature Society

MNS Manitoba Naturalists Society [Canada]

MNS Master of Natural Science

MNS Master of Nutritional Science

MNS Metal-Nitride Semiconductor

MNS Metal-Nitride-Silicon (Capacitor)

MnS Manganous Sulfide (or Manganese(II) Sulfide)

μN/s micronewton(s) per second [also μN s^{-1}]

MnSb Manganese Antimonide

MNSC Main Network Switching Center

MnSe Manganese Selenide (Semiconductor)

MNSFET Metal-Nitride Semiconductor Field-Effect Transistor

MnSi Manganese Silicide

MnSi$_2$ Manganese Disilicide

Mn-Sn Manganese-Tin (Alloy System)

MnSO Manganosulfate

MNT Mongolian Tugrik [Currency]

MNT Mononitrotoluene

MNT Mount (Request)

Mnt Maleodinitrile Dithiol [also mnt]

Mnt Maleonitrile Dithiolate [also mnt]

Mnt Mannitol [Biochemistry]

Mnt Monitor(ing) [also mnt]

Mnt Mount [also mnt]

MNT-Cl 4-Chloro-3-Nitrobenzotrifluoride

MnTe Manganese Telluride (Semiconductor)

MntHN Mannitol Hexanitrate [Biochemistry]

Mn(TMHD)$_3$ Manganese Tris(2,2,6,6-Tetramethyl-3,5-Heptanedionate) [also Mn(tmhd)$_3$]

mntn maintain

Mntr Monitor [also mntr]

MNU Methyl Nitrosourea

.MNU Menu [File Name Extension] [also .mnu]

MNucSci Master of Nuclear Science [also MNucS, or MNucSc]

MNurs Master of Nursing

Mnvr Maneuver [also mnvr]

MNX Mononitroxylene

Mn-Zn Manganese-Zinc (Alloy System)

MO Macau [ISO Code]

MO Magneto-Optic(s); Magneto-Optical [also M-O]

MO Mail Order

MO Maintenance and Operations [also M&O]

MO Major Objective

MO Manned Orbiter

MO Manual Operation

MO Manual Orientation

MO Manual Output

MO Manufacturing Order

MO Markovnikov Orientation

MO Masonry Opening

MO Mass Observation

MO Master Oscillator

MO Matrix Operation

MO Maximum Operation

MO Medical Officer

MO Medulla Oblongata (of Brain) [also mo] [Anatomy]

MO Membrane Osmometry

MO Memory Output

MO Mesityl Oxide

MO Metal(lo)-Organic(s) [also M-O]

MO Metal Oxide

MO Meteorological Office [UK]

MO Micro-Organism

MO Middeck Overhead

MO Mid-Ocean (Ridge) [Geology]

MO Military Observant

MO Mine Office

MO Mineral Oil

MO Mining Office(r)

MO Misorientation [Crystallography]

MO Missouri [US]

MO Mission Operations

MO Mixed Oxide

MO Mobile Object

MO *(Modus Operandi)* – Method (or Manner) of Working

MO Molecular Orbital [Physical Chemistry]

MO Molecular Optics

MO Money Order [also M/O, m/o, or mo]

MO Morse Oscillator

MO Move [also mo]

MO Multifunction Optical (Disk Drive)

M&O Maintenance and Operations [also MO]

M&O Management and Operations

M/O Money Order [also m/o, MO, or mo]

M-O Magneto-Optic(s); Magneto-Optical [also MO]

M-O Metal(lo)-Organic(s) [also MO]

M-O Metal-Oxide (Surface)

Mo Missouri [US]

Mo Mode (in Statistics) [Symbol]

Mo Molecule [also mo]

Mo Molybdenum [Symbol]

Mo Monday

Mo Month [also mo]

Mo Motor [also mo]

Mo Move [also mo]

Mo^{2+} Divalent Molybdenum Ion [also Mo^{++}] [Symbol]

Mo^{3+} Trivalent Molybdenum Ion [also Mo^{+++}] [Symbol]

Mo^{4+} Tetravalent Molybdenum Ion [Symbol]

Mo^{5+} Pentavalent Molybdenum Ion [Symbol]

Mo^{6+} Hexavalent Molybdenum Ion [Symbol]

Mo-91 Molybdenum-91 [also ^{91}Mo, or Mo^{91}]

Mo-92 Molybdenum-92 [also ^{92}Mo, or Mo^{92}]

Mo-93 Molybdenum-93 [also ^{93}Mo, or Mo^{93}]

Mo-94 Molybdenum-94 [also ^{94}Mo, or Mo^{94}]

Mo-95 Molybdenum-95 [also ^{95}Mo, or Mo^{95}]

Mo-96 Molybdenum-96 [also ^{96}Mo, or Mo^{96}]

Mo-97 Molybdenum-97 [also ^{97}Mo, or Mo^{97}]

Mo-98 Molybdenum-98 [also ^{98}Mo, or Mo^{98}]

Mo-99 Molybdenum-99 [also ^{99}Mo, or Mo^{99}]

Mo-100 Molybdenum-100 [also ^{100}Mo, or Mo^{100}]

Mo-101 Molybdenum-101 [also ^{101}Mo, or Mo^{101}]

Mo-102 Molybdenum-102 [also ^{102}Mo, or Mo^{102}]

Mo-103 Molybdenum-103 [also ^{103}Mo, or Mo^{103}]

Mo-104 Molybdenum-104 [also ^{104}Mo, or Mo^{104}]

Mo-105 Molybdenum-105 [also ^{105}Mo, or Mo^{105}]

Mö Mössbauer Effect (in Nuclear Physics) [Symbol]

MΩ Megohm [Unit]

M(ω) Magnitude Ratio (in Automatic Control) [Symbol]

M[ω] Molecular Magnetic Rotatory Power [Symbol]

M*(ω) Dielectric Modulus [Symbol]

mO mid oxygen (site) (in materials science) [Symbol]

mo mouse [Immunochemistry]

.mo Macau [Country Code/Domain Name]

mΩ millohm [Unit]

$\mu\Omega$ microhm [also μohm, or mu-ohm]

$\mu(\omega)$ (angular) frequency dependent (particle) mobility [Symbol]

MOA Make on Arrival

MOA Manufacturing Operations Analysis

MOA Matrix Output Amplifier

MOA Memorandum of Agreement

MOA Mesoxalic Acid

MOA Military Operating (or Operations) Area

MOA Ministry of Aviation [UK]

MOA Minute-of-Angle

MoA Museum of Australia

MOAA Member of the Office Administrators Association

MOABP Monooctyl-α-Anilinobenzylphosphonate

Mo(Ac)$_2$ Molybdenum Acetate

Mo-Ag Molybdenum-Silver (Alloy System)

Mo-Al Molybdenum-Aluminum (Alloy System)

Mo$_3$Al Molybdenum Aluminide

Mo/Al$_2$O$_3$ Molybdenum(-Fiber) Reinforced Alumina (Composite)

MOB Master of Organizational Behaviour

MOB Memory-Order Buffer

MoB Molybdenum Monoboride

mob mobile

MOBAT Modified Battalion Antitank

MOBIDAC Mobile Data Acquisition System

MOBIDIC Mobile Digital Computer

MOBL Macro-Oriented Business Language

Mobn Mobilization [also mobn]

MOBO Mother Board

MOBOL Mohawk Business-Oriented Language

MOBOT Marine Robot [also Mobot, or mobot]

MOBS Multiple Orbit Bombardment System

MOBULA Model Building Language

MOC Magnetic-Optic Converter

MOC Marine Operation Center

MOC Mars Orbiter Camera [of NASA]

MOC Master Operational Controller

MOC Master Operations Console

MOC Master Operations Control

MOC Memory Operating Characteristic

MOC Minimum Obstacle Clearance

MOC Minimum Operational Characteristics [Statistics]

MOC Mission Operation(s) Computer

MoC Molybdenum Carbide

MOCA Methylene-bis-o-Chloroaniline

MOCA Minimum Obstruction Clearance Altitude

MOCABP Monooctyl-α-(2-Carboxyanilino)benzylphosphonate

MOCAc 7-Methoxycoumarin-4-Acetyl [Biochemistry]

MOCAc-Pro-Leu 7-Methoxycoumarin-4-Acetyl-Proline-Leucine [Biochemistry]

MOCAc-Pro-Leu-Gly 7-Methoxycoumarin-4-Acetyl-Proline-Leucine-Glycine [Biochemistry]

MOCC METEOSAT (Meteorological Satellite) Operations Control Center

MOcE Master of Oceanographic Engineering [also MOcEng]

MOCF Mission Operations Computational Facilities

M(OCHMe$_2$)$_4$ Metal (IV) Isoproxide

MOCIC Molecular Orbital Constraint of Interaction Coordinates [Physics]

MΩcm Megohm-centimeter [also Mohm-cm]

MΩ/cm Megohm per centimeter [also MΩ cm^{-1}, or Mohm/cm]

mΩcm milliohm centimeter [also m$\Omega \cdot$cm, mΩ-cm, or mohm-cm]

$\mu\Omega$cm microhm centimeter [also $\mu\Omega \cdot$cm, $\mu\Omega$-cm, or μohm-cm]

$\mu\Omega$cm^2 microhm square centimeter [also $\mu\Omega \cdot$cm^2, or $\mu\Omega$-cm^2]

$\mu\Omega$/cm^3 microhm per cubic centimeter [also $\mu\Omega$ cm^{-3}]

$\mu\Omega$-cm/°C microhm square centimeter(s) per degree Celsius [also $\mu\Omega$-cm °C^{-1}, or $\mu\Omega$cm/°C]

Mo(Cp)$_2$Cl$_2$ Bis(cyclopentadienyl)molybdenum Dichloride

MOCR Mission Operations Control Room [of NASA]

Mo-Cr Molybdenum-Chromium (Alloy System)

MOCS Multichannel Ocean Color Sensor

Mo-Cu Molybdenum-Copper (Alloy System)

MOCV Manual O$_2$ (Oxygen) Control Valve

MOCVD Metal(lo)-Organic Chemical Vapor Deposition

MOCVD Metal-Oxide Chemical Vapor Deposition

MOCVD-GaAs Metalloorganic Chemical Vapor Deposited Gallium Arsenide

MOCVD/MOVPE Metal(lo)-Organic Chemical Vapor Deposition/Metal-Organic Vapor Phase Epitaxy

MOD Magnetooptical Data Recorder; Magnetooptical Data Recording

MOD Master of Organizational Development

MOD Message Output Description

MOD Metal(lo)-Organic Decomposition

MOD Metal(lo)-Organic Deposition

MOD Ministry of Defense [UK] [also MoD]

MOD Mission Objectives Document

MOD Mission Operations Director(ate)

MOD Modulation-Doped; Modulation Doping

MOD Moving Domain

Mod Model [also mod]

Mod Moderation; Moderator [also mod]

Mod Modification; Modifier; Modify [also mod]

Mod Modular; Modulation; Modulator [also mod]

Mod Module [also mod]

Mod Modulus [also mod]

mod moderate

mod modern

mod modified

mod modulo [Mathematical Symbol Usually Followed by a Numeral, e.g, mod 7]

mod modulus [Arithmetic Operator Symbol, e.g., mod $(3 - 4i) = 5$]

MODA Motion Detector and Alarm

MODAC Mountain System Digital Automatic Computer

MODACS Modular Data Acquisition and Control System

MODART Methods of Defeating Advanced Radar Threats

MODAS Modell-Datenbanksystem [Model Database System; of Gesellschaft für Mathematik und Datenverarbeitung, Germany]

Mod Cast Modern Casting [Journal published in the US]

MODCOMP Modular Computer Systems, Inc. [US]

MOD/DEMOD Modulating/Demodulating

mod dict *(modus dictum)* – as directed [Medical Prescriptions]

MODE Mid-Ocean Dynamics Experiment

MODEL Modelling [also Model, or model]

Model Identif Control Modeling, Identification and Control [Publication of the Royal Norwegian Council for Scientific and Industrial Research]

Model, Math Anal Numer Modelisation, Mathematique et Analyse Numerique [French Journal on Modelling, Mathematics and Numerical Analysis; published by Association Française pour le Cybernetique, Economique et Technique]

Model Simul Control A Modelling, Simulation and Control A [Journal published in France]

Model Simul Control B Modelling, Simulation and Control B [Journal published in France]

Model Simul Control C Modelling, Simulation and Control C [Journal published in France]

Model Simul Mater Sci Eng Modelling and Simulation in Materials Science and Engineering [Journal of Institute of Physics, UK]

Modem Modulator-Demodulator [also modem]

MODE S Mode Select (System)

MODEST Missile Optical Destruction Technique

MODF Misorientation Distribution Function [Crystallography]

MODFET Modulation-Doped Field-Effect Transistor

Mod Geol Modern Geology [UK Journal]

MODI Modified Distribution Method

MODI Modular Optical Digital Interface

MODICON Modular Digital Controller

MODICON Modular-Dispersed Control

Modif Modification

modif modified

modifd modified [also modif'd]

Modifn Modification [also modifn]

MODILS Modular Instrument Landing System

MODIS Moderate Resolution Imaging Spectroradiometer

MODIS-N Moderate Resolution Imaging Spectroradiometer–Nadir

MODIS-T Moderate Resolution Imaging Spectroradiometer–Tilt

Mod Lang J Modern Language Journal

Mod Mach Shop Modern Machine Shop [Journal published in the US]

Mod Mater Handl Modern Materials Handling [Journal published in the US]

Mod Med Modern Medicine [US Journal]

Mod Met Modern Metals [Journal published in the US]

Mod Meth Synth Modern Methods of Synthesis [Journal]

Mod Off Technol Modern Office Technology [Journal published in the US]

Mod Paint Coat Modern Paint and Coatings [Journal published in the US]

Mod Phys Lett Modern Physics Letters [Singapore]

Mod Phys Lett A Modern Physics Letters A [Singapore]

Mod Phys Lett B Modern Physics Letters B [Singapore]

Mod Plast Modern Plastics [Journal published in the US]

Mod Plast Int Modern Plastics International [Journal published in Switzerland]

MODPOT Modulator Potential [also Modpot, or modpot]

Mod Power Syst Modern Power Systems [Journal published in the UK]

MODR Microwave-Optical Double Resonance

MODS Major Operation Data System

MODS Manned Orbital Development System

MODS Manufacturing Operating Documentation System

Mods Modifications [also mods]

Mod Steel Constr Modern Steel Construction [Journal of the American Institute of Steel Construction]

Mod Synth React Modern Synthetic Reactions [US Journal]

Mod Tire Dealer Modern Tire Dealer [Journal published in the US]

MODTRAN Moderate Resolution Atmospheric Radiance and Transmittance Model [of Air Force Geophysical Laboratory, US]

MODU Mobile Offshore Drilling Unit

Modul Modular(ity) [also modul]

MODUS Mobile Offshore Drilling Units

Modus Modula-2 (Programming Language) Users Association

MODUSSE Manufacturers of Domestic Unvented Supply Systems Equipment [UK]

Mod Vet Pract Modern Veterinary Practice [US Journal]

MOE Master of Occupational Education

MOE Master of Ocean Engineering

MOE Measure of Effectiveness

MOE Meissner-Ochsenfeld Effect [Solid-State Physics]

MOE 2-Methoxyethanol

MOE Ministry of Education [UK]

MOE Ministry of Energy [Canada]

MOE Modulus of Elasticity

2-MOE 2-Methoxyethanol

mOe millioersted [Unit]

MOEA Ministry of Economic Affairs [Taiwan]

MOED meso-1,2-bis(4-Methoxyphenyl)ethylenediamine

MOEE (Ontario) Ministry of the Environment and Energy [Canada]

mOe/MPa millioersted(s) per Megapascal [mOe MPa^{-1}]

MOEMS Microoptoelectromechanical System

MOERO Medium Orbiting Earth Resources Observatory

MOESS Moessbauer Spectroscopy [also Moess]

MOF Manned Orbital Flight

MOF Maximum Observed Frequency

MOF Maximum Operating Frequency

MOF Ministry of Finance [Taiwan]

MOF Ministry of Forestry

MOF Momentum of Force (or Torque) [Symbol]

MOF Multiple Option Facility

MOFA Ministry of Foreign Affairs [Taiwan]

M of C Master of Commerce [also MofC]

Mo-Fe-Mn Molybdenum-Iron-Manganese (Alloy System)

M of F Method of Filling

μ of Hg micron of mercury [also μHg]

MOG Machinery of Government

MOG Mauritanian Ouguiya [Currency]

MOG Mesh of Grind

MOG Mills-Olney-Gaither (Procedure)

MOGA Michican Oil and Gas Association [US]

MOGA Microwave and Optical Generation and Amplification

MOGAS Motor Gasoline [also Mogas, or mogas]

MoGe Molybdenum Germanide (Semiconductor)

MOGR Moderate, or Greater

MOH Master of Occupational Health

MOH Maximum Operating Hours

MOH Medical Officer of Health

MOH Metal Hydroxide [General Formula]

MOH Ministry of Health [UK]

MOH (Ontario) Ministry of Health [Canada]

MO/2HA Metal Oxide to Polyelectrolytic Acid (Ratio)

mOHBA m-Hydroxybenzoic Acid

MOHLL Machine-Oriented High-Level Language

mohm mechanical ohm [Unit]

μohm microohm [also $\mu\Omega$, or mu-ohm]

μohm-cm microohm-centimeter [also $\mu\Omega$-cm]

Moho Mohorovičić Discontinuity [also MOHO] [Geology]

MOI Makapuu Oceanic Institute [Hawaii, US]

MOI Mars Orbit Insertion

MOI Ministry of Information [UK]

MOI Ministry of the Interior [Taiwan]

MOI Moment of Inertia

MOI Multiplicity of Infection

MOIG Master of Occupational Information and Guidance

Mo-Ir Molybdenum-Iridium (Alloy System)

Moist Moisture [also moist]

MOIT Ministry of Industry and Trade [Israel]

MOJ Material On Job Date

MOJ Metering Over Junction

MoKα Molybdenum K-Alpha (Radiation) [also MoK$_\alpha$]

MOKE Magneto-Optic Kerr Effect

MOKIM Dynamics: Models and Kinetic Methods for Nonequilibrium Many Body Systems (Conference) [at Lorentz Center, Leiden, Netherlands]

Mo/Kr Molybdenum/Krypton (Interface)

MOL Machine-Oriented Language

MOL Manned Orbiting Laboratory [of US Air Force]

MOL Maximum Output Level

MOL Ministry of Labour [Canada]

MOL Multiple On-Line Programming

Mol Molecular; Molecule [also mol]

mol molar

mol mole [Unit]

mol^{-1} one per mole [also 1/mol]

mol% mole percent [also mol %, or mol-%]

MOLA Mars Orbiter Laser Altimeter [of NASA]

MOLAB Mobile (or Moving) Lunar Laboratory [of NASA]

MOLAP Multidimensional On-Line Analytical Processing

MOLARS Meteorological Office Library Accessions and Retrieval System [UK]

Mol Biol Molecular Biologist; Molecular Biology

Mol Biol Cell Molecular Biology of the Cell [of American Society for Cell Biology, US]

Mol Cell Biol Molecular and Cellular Biology [of American Society for Microbiology, US]

Mol Cell Neurosci Molecular and Cellular Neurosciences [Journal published in the US]

Mol Cell Probes Molecular and Cellular Probes [Journal published in the US]

Mol Chem Molecular Chemist(ry)

mol/cm^2 mole per square centimeter [also mol cm^{-2}]

Mol Compd Molecular Compound [also mol compd]

Mol Cryst Molecular Crystal [also mol cryst]

Mol Cryst Liq Cryst Molecular Crystals and Liquid Crystals [UK Journal]

Mol Cryst Liq Cryst Lett Sect Molecular Crystals and Liquid Crystals, Letters Section [UK Journal]

Mol Cryst Liq Cryst Sci Technol A, Mol Cryst Liq Crys Molecular Crystals and Liquid Crystals, Section A: Molecular Crystals and Liquid Crystals [UK Journal]

Mol Cryst Liq Cryst Sci Technol B, Nonlin Opt Molecular Crystals and Liquid Crystals, Section B: Nonlinear Optics [UK Journal]

Mol Cryst Liq Cryst Sci Technol C, Mol Cryst Molecular Crystals and Liquid Crystals, Section C: Molecular Crystals [UK Journal]

Mol Cryst Liq Cryst Sci Technol D, Display Imaging Molecular Crystals and Liquid Crystals, Section D: Display and Imaging [UK Journal]

Mol Cryst Liq Cryst Suppl Ser Molecular Crystals and Liquid Crystals, Supplement Series [UK Journal]

mol/dm³ mole(s) per cubic decimeter [also mol dm^{-3}]

MOLDS Management On-Line Data System

MOLDS Multiple On-Line Debugging System

MoldSSR Moldavian Soviet Socialist Republic [USSR]

MOLE Molecular Optical Laser Excimer

MOLE Molecular Optical Laser Excitation

Molectronics Molecular Electronics [also molectronics]

MolE Molecular Engineer [also Mol Eng, or MolEng]

Mol Eng Molecular Engineer(ing)

Mol Gen Genet Molecular and General Genetics [Journal published in Japan]

Mol Ht Molecular Heat [also mol ht]

Mol Immun Molecular Immunology [Journal]

mol/J mole(s) per Joule [also mol J^{-1}]

mol/kg mole(s) per kilogram [also mol kg^{-1}]

moll *(mollis)* – soft [Medical Prescriptions]

mol/L mole(s) per liter [also mol L^{-1}, mol/l, mol/ltr, or mol ltr^{-1}]

Mol Liq Molecular Liquid [also mol liq]

mol/m² mole(s) per square meter [also mol m^{-2}]

mol/m³ mole(s) per cubic meter [also mol m^{-3}]

mol/m³h mole(s) per cubic meter per hour [also mol m^{-3} h]

Mol Microbiol Molecular Microbiologist; Molecular Microbiology

Mol Microbiol Molecular Microbiology [International Journal]

Mol Opt Molecular Optics [also mol opt]

Mol Pharmacol Molecular Pharmacology [Journal published in the US]

Mol Phylogenet Evol Molecular Phylogenetics and Evolution [Journal published in the US]

Mol Phys Molecular Physicist; Molecular Physics

Mol Phys Molecular Physics [UK Journal]

mol/s mole(s) per second [also mol s^{-1}, or mol/sec]

MOLSCAT Monolayer Scattering

Mol Sci Molecular Science; Molecular Scientist

mol/sec mole(s) per second [also mol/s, or mol s^{-1}]

MOLSINK Molecular Sink [also Molsink]

Mol Vis Molecular Vision [US Journal]

Mol Wt Molecular Weight [also mol wt]

mol/yr mole(s) per year [also mol/y]

Molysulfide Newsl Molysulfide Newsletter [US]

MOM Magnetooptic Method

MOM Manufacturing Operations Management

MOM Message-Oriented Middleware

MOM Message Output Module

MOM Metal(lo)-Oxide-Metal

MOM Methoxymethyl

MOM Microsoft Office Manager

MOM Middle of Month [also mom]

MOM Mission Operations Manager

MOM Moment of Momentum

Mom Moment [also mom]

μΩm microohm-meter [also $\mu\Omega \cdot m$, or $\mu\Omega$-m]

MOMBE Metal(lo)-Organic Molecular Beam Epitaxy

MOMM Molecular-Orbital-Based Molecular Mechanics

MΩ/μm² Megohm(s) per square micrometer [also M$\omega\mu$m^{-2}]

MOMO Maximum Overlap Molecular Orbital

Mo/MoSi₂ Molybdenum Reinforced Molybdenum Disilicide (Composite)

MOMS Metalorganic Magnetron Sputtering

MOMS Modular Optoelectronic Multi-Wavelength Scanner

MON Motor Octane Number

MoN Molybdenum Nitride

Mon Monday [also Mon]

Mon Monitor(ing) [also mon]

Mon Monmouthshire [UK]

Mon Montana [US]

Mon Month(ly) [also mon]

Mon Monument [also mon]

MONAB Mobile Operational Naval Air Base [UK]

Monatsh Chem Monatshefte für Chemie [Chemistry Monthly; published in Vienna, Austria]

Monatsh Math Monatshefte für Mathematik [Mathematics Montlhly; published by the Austrian Mathematical Society]

Mo-Nb Molybdenum-Niobium (Alloy System)

MONEG Monsoon Numerical Experimental Group [of World Meteorological Organization working on the Tropical Ocean and Global Atmosphere Experiment]

MONET Multiwavelength Optical Networking

Mo(NEt₂)₄ Tetrakis(diethylamido)molybdenum

MONEX Monsoon Experiment

Mong Mongolia(n)

Mon-H Mono-Hydrogen [also mon-H]

Mo-Ni Molybdenum-Nickel (Alloy System)

Monit Monitor(ing) [also monit]

Mon J Jpn Brass Makers Assoc Monthly Journal of Japan Brass Makers Association

MO/NM Magnetooptic/Nonmagnetic

Mon Not Astron Soc South Afr Monthly Notes of the Astronomical Society of South Africa

Mon Not R Astron Soc Monthly Notes of the Royal Astronomical Society [UK]

Mono (Infectious) Mononucleosis [also mono]

mono monaural

mono monostable

monobas monobasic

monocl monoclinic

Monogr Monograph [also monogr]

Monogr Soc Res Child Devel Monographs of the Society for Research in Child Development

Mono-phos Monophosphoryl

MONOS Metal-Oxide-Nitride-Oxide-Silicon

MONOS Monitor Out of Service

Mono-Si Monocrystalline Silicon [also mono-Si, or mono Si]

MonoXPS Monochromated X-Ray Photoelectron Spectroscopy

Mons Monmouthshire [UK]

Mon Stat Rev Monthly Statistical Review [UK]

Mont Montana [US]

Mon Tech Rev Monthly Technical Review [Germany]

Monten Montenegro

Montgom Montgomeryshire [Wales]

Mon Weather Rev Monthly Weather Review [Published by the American Meteorological Society] [also Mon Wea Rev]

MOO Multiple-Objective Optimization

MOO MUD (Multi-User Domain) Object Oriented [Internet]

Mo-O Molybdenum-Oxygen (Alloy System)

MoO$_x$ Molybdenum Oxides (e.g., MoO_2, MoO_3, or Mo_2O_3)

MoOPH Molybdenum Oxodiperoxypyridine Hexamethylphosphoramide

Mo-Os Molybdenum-Osmium (Alloy System)

MOOSE Manned Orbital Operations Safety Equipment

MOOSE Man Out of Space Easiest

MOOT Method of Optimal Truncation

MOP Macao Pataca [Currency of Macao]

MOP Magneto-Operational Amplifier

MOP Maintenance Operations Protocol

MOP Melt Optimization Program

MOP Memory Organization Packet

MOP Metastable Ordered Phase

MOP Mission Operations Plan

MOP Michoud Ordnance Plant [New Orleans, Louisiana, US]

MOP Modulation On Pulse

MOP Mono-n-Octylphosphate

MOP Multiple On-Line Processing

MOP Muriate of Potash

MOP Muskegon Ordnance Plant [Muskegon, Michigan, US]

8-MOP 8-Methoxypsoralene

MOΦP Mono[p-(iso-Octyl)phenyl]phosphate

MoP Molybdenum Phosphide

MOPA Master Oscillator Power Amplifier

MOPA Methoxyphenylacetic Acid

MOPAFD Master Oscillator Power Amplifier Frequency Doubler

MOPAP Phosphoric Acid Mono[4-(1,1,3,3-Tetramethylbutyl)phenyl]ester

MOPAR Master Oscillator Power Amplifier Radar

MOPB Manually-Operated Plotting Board

Mo-Pd Molybdenum-Palladium (Alloy System)

MOPEG 4-Hydroxy-3-Methoxyphenylglycol

MOPEG 3-Methoxy-4-Hydroxyphenylethyleneglycol

MOPI Maximum-Rate Output Initiator

MOPITT Measurement of Pollution in the Troposphere

MOPP Mechlorethamine–Vincristine–Procarbazine–Prednisone (Anticancer Drug)

MOΦP Mono[p-(iso-Octyl)phenyl]phosphate

MOPR Manner of Performance Rating

MOPR Mission Operations Planning Review

MOPR Mission Operations Planning Room

MOPS Military Operation(s) Phone System

MOPS Million Operations per Second [also mops]

MOPS Minimum Operational Performance Standards

MOPS Missile Operations

MOPS Mission Operations Planning System

MOPS 3-(N-Morpholino)propanesulfonic Acid

MOPS 3-(4-Morpholinyl) 1-Propanesulfonic Acid

MOp/s Mega-Operations per Second

MOPS-EDTA 3-(N-Morpholino)propanesulfonic Acid–Ethylenediaminetetraacetic Acid (Sodium Acetate Buffer)

MOPSO 3-(N-Morpholino)-2-Hydroxypropanesulfonic Acid; 3-(4-Morpholinyl)-2-Hydroxypropanesulfonic Acid

MOPT Mean One-Way Propagation Time

MOpt Master of Optometry

Mo-Pt Molybdenum-Platinum (Alloy System)

MOPTAR Multi-Object Phase Tracking and Ranging System

MOPTS Mobile Photographic Tracking Station

MOR Low-Frequency Omnirange [formerly LOR]

MOR Magneto-Optical Rotation

MOR Magneto-Optical Rotation (Spectroscopy)

MOR Malate Oxidoreductase [Biochemistry]

MOR Management Oversight and Risk Tree

MOR Manufacturing Operation Record

MOR Mars Orbital Rendezvous

MOR Memory Output Register

MOR Meteorological Optical Range

MOR Mid-Ocean Rift [Geology]

MOR Mission Operations Room

MOR Modulus of Rupture

MOR Molten Alkali Resistance

MOR Moment of Resistance

MOR Monthly Operating Report

.MOR Main Dictionary File for Spell Checker [WordPerfect File Name Extension]

M(OR)$_x$ Metal Alkoxide (R = Hydrocarbon Group, e.g., Ethyl, Methyl, Propyl, etc.; x = Oxidation State of Metal M, e.g., M = 1, 2, 3, 4) [General Formula]

Mor Moroccan; Morocco

Mor Mortar [also mor]

MORA Minimum Off-Route Altitude

Mo-Re Molybdenum-Rhenium (Alloy System)

MOREPS Monitor Station Reports

MORBID Morse Oscillator Rigid Bender Internal Dynamics (of Triatomic Molecules)

MORD Magnetic Optical Rotations Dispersion

MORD Medical Operations Requirements Document

MORD Mission Operations Requirements Document

MORE Maintenance, Repair, Overhaul Excellence

Mo-Rh Molybdenum-Rhodium (Alloy System)

MORI Market and Opinion Research Institute [UK]

MORIS Magneto-Optical Recording International Symposium

MORL Manned Orbital Research Laboratory

Mormtc Morpholinemonothiocarbamate

MORP Meteorite Observation and Recovery Project [of National Research Council of Canada]

Morph Morphactin

Morpho-CDI Cyclohexyl-2-(4-Methylmorpholino) ethylcarbodiimide Tosylate

MORS Mandatory Occurence Reporting Scheme [of Civil Aviation Authority, UK]

MORS Military Operations Research Society [US]

MORS Military Operations Research Symposium

MORST Melt Overflow Rapid Solidification Technology [Metallurgy]

MORT Management Oversight and Risk Tree

Mort Mortar [also mort]

Mor T Morse Taper [Tool Design]

Mo-Ru Molybdenum-Ruthenium (Alloy System)

MOS Machinery and Occupational Safety Act [US]

MOS Magneto-Optic Storage

MOS Management Operating System

MOS Manufacturing Operating System

MOS Margin of Safety

MOS Marine Observation Satellite [Japan]

MOS Master Operating System

MOS Mathematical Off-Print Service [of American Mathematical Society, US]

MOS Material Ordering System

MOS Mean Opinion Score

MOS Memory-Oriented System

MOS Metal-Oxide on Substrate

MOS Metal-Oxide Semiconductor

MOS Metal-Oxide Silicon

MOS Microprogram Operating System

MOS Military Occupational Specialty

MOS Minimum Operating System

MOS Ministry of Supply [UK]

MOS Mission Operating (or Operations) System [of NASA]

MOS Modular Operating System

MOS (International Conference on Physics and Chemistry of) Molecular Oxide Superconductor(s)

MOS Multiprogramming Operating System

Mos Months [also mos]

MOSA Medical Officers Schools Association [UK]

MOSAIC Macro Operation Symbolic Assembler and Information Compiler

MOSAIC Metal-Oxide Semiconductor Advanced Integrated Circuit

MOSAIC Ministry of Supply Automatic Integrator and Computer

MOSAIC Mobile System for Accurate ICBM (Intercontinental Ballistic Missile) Control

MOSAR Modulation Scan Array Receiver

Mosc Univ Comput Math Cybern Moscow University Computational Mathematics and Cybernetics [Translation of: *Vestnik Moskovskogo Universiteta, Vychislitel'naya Matematika i Kibernetika (USSR)*; published in the US]

Mosc Univ Phys Bull Moscow University Physics Bulletin [Translation of: *Vestnik Moskovskogo Universiteta, Fizika-Astronomiya (USSR)*; published in the US]

MOSD Ministry of Science and Development [Israel]

MOS DMOS Metal-Oxide Semiconductor Double-Diffused Metal-Oxide Semiconductor [also MOS-DMOS]

Mo-Se Molybdenum-Selenium (Alloy System)

MOSEL Monolithic Surface Emitting Diode Laser

MOS EPROM Metal-Oxide Semiconductor Erasable Programmable Read-Only Memory

MOSERD Ministry of State for Economic and Regional Development

MOSES Mission, Objectives and Strategies Execution Plan and Success Indicators

MOSFET Metal-Oxide Semiconductor Field-Effect Transistor

MOSFET Metal-Oxide Silicon Field-Effect Transistor

MOSI Microprocessor Operating Systems Interface

Mo/Si Molybdenum on Silicon (Substrate)

MoSi$_2$/Nb Molybdenum Disilicide (Fiber) Reinforced Niobium (Composite)

MoSi$_2$/Ta Molybdenum Disilicide (Fiber) Reinforced Tantalum (Composite)

MOSIC Metal-Oxide Semiconductor/Integrated Circuit [also MOS IC]

MOSIC Metal-Oxide Silicon/Integrated Circuit [also MOS IC]

Mo-Si-Co Molybdenum-Silicon-Cobalt (Alloy System)

Mo-Si-Cr Molybdenum-Silicon-Chromium (Alloy System)

MOSID Ministry of Supply Inspection Department [UK]

MOSIGT Metal-Oxide Semiconductor Insulated Gate Transistor

MoSi$_2$/Si Molybdenum Disilicide/Silicon (Multilayer)

MoSi$_2$/SiC Molybdenum Disilicide Reinforced Silicon Carbide (Composite)

MoSi$_2$/TiC Molybdenum Disilicide Reinforced Titanium Carbide (Composite)

Mo-Si-TM Molybdenum-Silicon-Transition Metal (Alloy System)

MOS/LSI Metal-Oxide Semiconductor for Large-Scale Integration

MOSM Metal-Oxide Semimetal

mOsm milli-osmole [Unit]

mOsm/kg milli-osmole per kilogram [Unit]

MOSPO Mobile Satellite Photometric Observatory

MOSPF Multicast Open Shortest-Path First [Internet]

MOSRAM Metal-Oxide Semiconductor Random Access Memory [also MOS RAM]

MOSROM Metal-Oxide Semiconductor Read-Only Memory [also MOS ROM]

MOSS Map Overlay and Statistics System [Geographic Information System]

MOSS Multibeam Optical Stress Sensor

MOSS Multilayer Optical Scanning Spectrometer

MÖSS Moessbauer Spectroscopy [also Möss]

MOSST Ministry of State for Science and Technology [Canada]

MOST Management Operation System Technique

MOST Metal-Oxide Semiconductor Technology

MOST Metal-Oxide Semiconductor Transistor

MOST Metal-Oxide Silicon Transistor

MOST Military Operational and Support Truck

MOST Ministry of Science and Technology [South Korea]

MOST Modular Operating System Tool

MOST Multipurpose Observation Sizing Technique

MOSY Modular Selective Analysis System

MOT Management of Technology [also MoT]

MOT Manned Orbital Telescope [of NASA]

MOT Master of Occupational Therapy

MOT Maximum Operating Temperature

MOT Ministry of Overseas Trade [UK]

MOT Ministry of Transport [also MoT, UK]

MOT Ministry of Transport [Commonwealth]

MOT (Ontario) Ministry of Transportation [Canada]

MOT Molecular Orbital Theory [Physical Chemistry]

MOT Mozambique Meticai [Currency]

MOT Multistate Orbital Treatment

Mot Motor(ing) [also mot]

MOTA Materials Open Test Assembly

MOTA Mail Order Trades Association [UK]

Mo-Ta Molybdenum-Tantalum (Alloy System)

MOTACC Manufacturers of Telescoping and Articulating Cranes Council [of Farm and Industrial Equipment Institute, US]

MOTARDES Moving Target Detection System

Mo-Tc Molybdenum-Technetium (Alloy System)

Mot Control Motor Control [International Journal]

MOTD Message Of The Day

MOTI Management of Technological Innovation

Mo-Ti Molybdenum-Titanium (Alloy System)

MO TiN Metalorganic Titanium Nitride (Film)

Motional EMF Motional Electromotive Force [also motional emf]

MOTIS Message-Oriented Text Interchange System

MOTMS Methoxytrimethylsilane

MOTMX Methoxy Trimethylxanthine

MOTNE Meteorological Operational Telecommunications Network of Europe

MOTOR Mobile-Oriented Triangulation of Reentry

Motor Motoring [also motor]

Motorola Tech Dev Motorola Technical Developments [Published by Motorola Inc., US]

Motorola Tech Discl Bull Motorola Technical Disclosure Bulletin [Published by Motorola Inc., US]

Motortech Z Motortechnische Zeitschrift [German Journal on Engine Design and Technology]

MOTS Mobile Operations Training Simulator

MOTS Module Test Set

Mot Ship Motor Ship [Journal published in the UK]

Mot Transp Newsl Motor Transportation Newsletter [of National Safety Council, US]

MOTU Mobile Optical Tracking Unit

MOU Memorandum of Understanding [also MoU]

MOUG Maryland On-Line User Group [US]

MOUSE Minimum Orbital Unmanned Satellite of the Earth

MOUTH Modular Ouput Unit for Talking to Humans

MOV Main Oxidizer Valve [Aerospace]

MOV Metal-Oxide Varistor

MOV Motor Operated Valve

MOV Move [also Mov]

MOV Quick-Time Movie

Mo/V Molybdenum on Vanadium (Epitaxial Film)

Mo-V Molybdenum-Vanadium (Alloy System)

MOVB Molecular Orbital Valence Bond (Theory) [Physical Chemistry]

MOVECAP Movement Capability

MOVEET Ministers of Vocational Education, Employment and Training

MOVPE Metallo-Organic Vapor Phase Epitaxy

MOVS Move String [Computers]

MOVSB Move Byte String

MOVSW Move Word String

Movt Movement [also movt]

MOW Maumelle Ordnance Works [Little Rock, Arkansas, US]

MOW Mission Operations Wing

Mo-W Molybdenum-Tungsten (Alloy System)

MOWASP Mechanization of Warehousing and Shipment Processing

MOX Metal, Oxidizer and Explosive [An Explosive Mixture]

MOX Mixed Oxide (Fuel) [also Mox]

MOX$_2$ Metal Oxyhalide [Formula]

MO$_2$X$_2$ Metal (II) Oxyhalide [Formula]

Moz Mozambique

MP Machine Pistol

MP Macromolecular Physics

MP Macroparticle

MP Macrophage [Anatomy]

MP Macroporosity; Macroporous

MP Macroprocessor

MP Magnetic Pole

MP Magnetic Properties

MP Magnetophotophoresis; Magnetophotophoretic

MP Magnetoplumbite

MP Magnifying Power [Symbol]

MP Main Phase
MP Maintenance Period
MP Maintenance Point
MP Maintenance Process
MP Maintenance Programmer; Maintenance Programming
MP Major Planet(s) [Astronomy]
MP Management Package
MP Management Process
MP Manhattan Project [US Atomic Research Project]
MP Manifold Pressure
MP Manufacturing Process
MP Marine Police
MP Marine Pollution
MP Martin und Pagenstecher GmbH [German Manufacturer of Fused Silica Products, located in Cologne] [also M&P]
MP Massively Parallel (Computer)
MP Mass Point
MP Mass Properties
MP Master of Planning
MP Master Painter
MP Master Patternmaker [Metallurgy]
MP Master Plumber
MP Master Printer
MP Master Pulse
MP Material Performance
MP Materials Performance (Journal)
MP Materials Physics
MP Materials Processing
MP Mathematical Physicist; Mathematical Physics
MP Mathematical Probability
MP Mathematical Programming
MP Matrix Product
MP Maugis and Pollock (Model) [Metallurgy]
MP Mayer Problem [Physics]
MP Mean (Effective) Pressure [also MEP, or mep]
MP Measurement Pragmatic
MP Measuring Point
MP Mechanical Part
MP Mechanical Processing
MP Mechanical Properties
MP Medium Pressure [also M-P, m-p, or mp]
MP Melamine Phenolic(s)
MP Melting Point [also mp]
MP Melt Powder
MP Member of Parliament
MP Membrane Potential [Electrobiology]
MP Membrane Process
MP Memory Protection
MP Mercaptopurine
MP Mercator Projection [Cartography]
MP Mercury Porosimeter
MP Mercury Pump
MP Metacarpophalangeal (Joint) [Anatomy]
MP Metallized Paper (Capacitor)

MP Metallurgical Process(ing)
MP Metal Particle
MP Metal Physicist; Metal Physics
MP Metal Powder
MP Metal Processing
MP Metamorphic Petrology
MP Metal Physics; Metal Physicist
MP Metaphysical; Metaphysics
MP Metastable Phase [Physical Chemistry]
MP Meteorology Panel
MP Methapyrilene
MP Methylpentane
MP Methylpurine
MP Metropolitan Police
MP Mexican Peso [Currency]
MP Microperoxidase [Biochemistry]
MP Microphotograph(y)
MP Microplastic(ity)
MP Microprobe
MP Microprocessing; Microprocessor [also μP]
MP Microprogram(ming)
MP Military Police(man)
MP Mine Planning
MP Mineral Processing
MP Minimal Polynomial [Mathmatics]
MP Minimum Phase
MP Ministry of Power
MP Mirror Plane
MP Miscellaneous Paper
MP Miscellaneous Publication
MP Missing Pulse
MP Mission Planner
MP Mitogenic Protein [Biochemistry]
MP Mobile Particle
MP Module Processor
MP Moeller-Plesset (N-Order Perturbation Approach Theory) [Physics]
MP Moiré Pattern [Optics]
MP Molecular Physics
MP Molecular Pump
MP Monitoring Program
MP Monoplane
MP Montreal Protocol (on the Environment)
MP Morse Potential [Physical Chemistry]
MP Motion Picture
MP Mounted Police
MP Mucoprotein [Biochemistry]
MP Multicritical Point
MP Multiperforated Propellant
MP Multiphase
MP Multiple Personality [Psychology]
MP Multiple Processors
MP Multiplier Phototube
MP Multiply

MP Multipoint

MP Multipole

MP Multiport

MP Multiprocessing; Multiprocessor

MP Multiprogramming

MP Multiprotocol

MP Multipulse

MP Multipurpose [also M-P]

MP Myeloperoxidase [Biochemistry]

MP Northern Mariana Islands [ISO Code]

M&P Martin und Pagenstecher GmbH [German Manufacturer of Fused Silica Products, located in Cologne] [also MP]

M&P Materials and Processes

M&P Materials and Processing

M/P Main Parachute

M-P (Reverse Transformation of) Martensite to Parent Phase [also M/P] [Metallurgy]

M-P Medium Pressure [also MP, m-p, or mp]

M-P Metal-Polymer (Interface) [also M/P]

M-P Multi-Purpose [also MP]

mP maritime polar air [Symbol]

mP millipoise [Unit]

mP Pearson symbol for primitive (simple) space lattice in monoclinic crystal system (this symbol is followed by the number of atoms per unit cell, e.g. mP12, mP16, etc.)

mp medium pressure [also MP, M-P, or m-p]

m/p month(s) after payment [also mp]

μP micropoise [also muP] [Unit]

μP microprocessor [also MP]

M£ Maltese Pound [Currency of Malta]

6-MP 6-Mercaptopurine

26M4P 2,6-Dimethyl 4(4H)-Pyranone

MPA Maclura Pomifera Agglutinin [Immunology]

MPA Magazine Publishers Association [US]

MPA Major Projects Association [UK]

MPA Management Professionals Association [India]

MPA Manufacturing and Process Automation [Canadian Journal]

MPA Marine Protected Area

MPA Maritime Patrol Aircraft

MPA Marketing and Promotion Association [UK]

MPA Master of Professional Accountancy; Master of Professional Accounting

MPA Master of Public Administration

MPA Master of Public Affairs

MPA Master Photographers Association [UK]

MPA (Staatliche) Materialprüfungsanstalt [(State) Materials Testing Institute, Stuttgart, Germany]

MPA Maximum Permissible Amount

MPA Mechanical Packing Association [now Fluid Sealing Association, US]

MPA Medium-Pressure Air

MPA Metal Powder Association [now Metal Powder Industries Federation, US]

MPA Mercaptopropanoic Acid

MPA Methoxyphenylethane

MPA Methoxypropylamine

MPA Methylphosphoric Acid

MPA Microprocessor Architecture

MPA Microwave-Plasma-Assisted (Vapor Deposition)

MPA 12-Molybdophosphoric Acid

MPA Mortar Producers Association [UK]

MPA Mouse Protective Antigen [Immunology]

MPA Multi-Period Average; Multiple-Period Average

MPA Multiphase Alloy

MPA Multiphoton Absorption [Solid-State Physics]

MPA Multiple-Photon Absorption [Atomic Physics]

MPA Multiple Peripheral Adapter

MPA Multiple-Use Planning Area

MPA Multipoint Asynchronous

MPA Multi-Purpose Additive

3-MPA 3-Methoxypropylamine

MPa Megapascal [Unit]

MPa Mercaptopropanoic Acid

MPa^{-1} Reciprocal Megapascal (or One per Megapascal) [also 1/MPa]

mPa millipascal [Unit]

μPa micropascal [Unit]

2MPa 2-Mercaptopropanoic Acid

MPAA Motion Picture Association of America

MPa/at% Megapascal(s) per atomic percent [Unit]

MPAcc Master of Professional Accountancy; Master of Professional Accounting [also MPAcct]

MPa/cm² Megapascal(s) per Square Centimeter [also Mpa cm^{-2}]

MPACS Management Planning and Control System

MPACVD Microwave-Plasma-Assisted Chemical Vapor Deposition

MPAD Maximum Permissible Accumulated Dose [Medicine]

MPAD Mission Planning and Analysis Division [of NASA Johnson Space Center, Houston, Texas, US]

MPADP Methyl(1-Phenylethyl)amino Diphenylphosphine

MPAff Master of Public Affairs

MPAI Maximum Permissible Annual Intake

MPa$\sqrt{m}$ Megapascal(s) square root of meter [also MPA m^{½}]

MPa/mm³ Megapascal(s) per cubic meter [also Mpa·mm^{-3}]

MPa μm Megapascal(s) micrometer [Unit]

MPa$\sqrt{\mu m}$ Megapascal(s) square root of micrometer [also MPA μm^{½}]

MPa$^{2/3}$m³/Mg Megapascal(s) cubic meter per megagram [Unit]

MPa·m³/s Megapascal(s) cubic meter per second [also MPa·m³ s^{-1}]

MPAR Microprogram Address Register

MPAS Metal Powder Association Standard

mPa s millipascal-second [also mPa-s]

m/Pa·s meter(s) per pascal-second [also m Pa^{-1} s^{-1}]

MPAURP Master of Public Affairs and Urban and Regional Planning

MPaz N-Methyl Phenazinium [also Mpaz]

MPB Maintenance Parts Breakdown

MPB Material Performance Branch [of US Air Force]

MPB Mean Point of Burst [also mpb]

MPB Methoxypropenylbenzene

MPB Methylpropylbenzene

MPB Momentary Pushbutton Switch

MPB Monobutyl Phosphate

MPB Morphotropic Phase Boundary [Materials Science]

MPBB Maximum Permissible Body Burden [Medicine]

MPBE Molten Plutonium Burn-Up Experiment [Nuclear Engineering]

MPBP (International Union of) Metal Polishers, Buffers and Platers

MPBW Ministry of Public Building and Works [UK]

MPC Manitoba Power Commission [Canada]

MPC Manitoba Press Council [Canada]

MPC Manufacturing Planning and Control

MPC Manual Pointing Controller

MPC Marginal Propensity to Consume

MPC Marker Pulse Conversion

MPC Master of Public Communication

MPC Materials Preparation Center [of US Department of Energy at Ames Laboratory, Iowa State University, US]

MPC Materials Processing Center [of Massachusetts Institute of Technology, Cambridge, US]

MPC Materials Properties Council [US]

MPC Maximum Permissible (or Permitted) Concentration [also mpc]

MPC Medium Processing Channel Black

MPC Memory Protection Check

MPC Metal Phthalocyanine [also MPc]

MPC Metal Physics Conference [Part of Canadian Materials Science Conference]

MPC Metal Process Control [Metallurgy]

MPC Metal Properties Council [now Materials Properties Council, US]

MPC Meteorological Prediction Center

MPC Methylpropylcarbinol

MPC 5-Methyl Pyrazole 3-Carbxylic Acid

MPC Microprocessor Control

MPC Microprogram Control

MPC Midbody Pyro-Controller

MPC Mid Power Controller

MPC Military Police Corps [of US Army]

MPC Miniature Protector Connector

MPC Minimum Permissible (or Permitted) Concentration [also mpc]

MPC Minor Planet Center [of Smithsonian Astrophysical Observatory, Cambridge, Massachusetts, US]

MPC Modular Peripheral-Interface Converter

MPC Modulated Photocurrent

MPC Montana Power Company [US]

MPC Multimedia Personal Computer

MPC Multipath Channel

MPC Multiplier Photocell

MPC Multipurpose Communications

MP&C Maintenance Planning and Control

MPc Metal Phthalocyanine [also MPC]

mpc (one) million parsecs [Unit]

MPCA Methylphenylcarbinol Acetate

MPCA Mid Power Controller Assembly

MPCAG Military Parts Control Advisory Group

MPCC Multiprotocol Communications Controller

MPCD Master in Planning and Community Development

MPCD Maximum Permissible Cumulative Dose [Medicine]

MPCI Multiport Programmable Communications Interface

MPCM Microprogram Control Memory

MP Corps Military Police Corps [of US Army]

MPCS Mission Planning and Control Station (Software)

MPCS Multi-Party Connection Subsystem

MPCU Marine Pollution Control Unit [of Department of Trade, UK]

MPCVD Microwave Plasma Chemical Vapor Deposition [also MP-CVD]

MPD Magnetic Permeability Disaccommodation

MPD Magnetoplasmadynamic(s)

MPD Main Power Distributor (Assembly)

MPD Map Pictorial Display

MPD Master of Planning Design

MPD Master of Product Design

MPD Materials Physics Division [of US Air Force]

MPD Materials Processing Division [of Vikram Sarabhai Space Center, Trivandrum, India]

MPD Materials Property Data

MPD Maximum Permissible Dose [Medicine]

MPD Maximum (Corrosion) Pit Depth [Metallurgy]

MPD Melting Point Depressant

MPD 2-Methyl Pentamethylene Diamine

MPD Methylphosphonous Dichloride

MPD Mineral Processing Division [of Society of Mining Engineers, US]

MPD m-Phenylenediamine

MPD Multiple Personality Disorder [Psychology]

MPD Multiphoton Decomposition [Atomic Physics]

MP&D Materials Processing and Design

MPDA m-Phenylenediamine

MPDC Mechanical Properties Data Center [US]

MPDE Maximum Permissible Dose Equivalent [Medicine]

MPDI Marine Products Development Irradiator

MPDI Poly(m-Phenylene Isophthalate)

MPDR Microwave Plasma Disk Reactor

MPDR Mobile Pulse Doppler Radar

MPDS Methylpolydisilylazane

MPDS Multi-Purpose Display System

MPDSM Micro Powder Diffraction Search/Match

1,3-MPDTA 1,3-Methylpropylenediamine-N,N,N',N'-Tetraacetate [also 1,3-mpdta]

1,3-MPD3A 1,3-(N-Methyl)propanediamine-N,N',N'-Triacetate [also 1,3-mpd3a]

MP-DV Multiply-Divide

MPE Mathematical and Physical Sciences and Engineering

MPE Mannesmann-Pfannen-Entschwefelung [Mannesmann Ladle Desulfurization Process] [Metallurgy]

MPE Marine Port Engineer

MPE Master Process Engraver

MPE Motion Picture Engineer(ing)

MPE Maximum Permissible Exposure

MPE Memory Parity Error

MPE Methyl Phenyl Ether

MPE Microwave Plasma Etching

MPE Mineral Processing Engineer(ing)

MPE Ministry of Planning and Environment

MPE Mission Peculiar Equipment

MPE Mission to Planet Earth [of NASA]

MPE Mouvement Populaire pour l'Environnement [Popular Movement for the Environment, Switzerland]

MPE Multiphase Extraction

MPE (Hewlett-Packard) Multiple Programming Executive

MPE Multiprogramming Executive

μPE Microprocessor Element [also μPe]

MPECVD Microwave Plasma Enhanced Chemical-Vapor Deposition

MPeEng Master of Petroleum Engineering [also MPetE, or MPetEng]

MPEG Methoxypolyethylene Glycol

MPEG Mixtures of Polyethyleneglycol Monomethylether

MPEG Motion Picture Experts Group

MPEMA 2-Ethyl-2-(P-Tolyl)malonamide

MPEMA M-Phenylethylmalonamide

MPEP Manual of Patent Examining Procedures

MP/EP Missing Pulse/Extra Pulse

MPER Master of Personnel and Employees Relations

MPESS Mission Peculiar Equipment Support Structure

MPetE Master of Petroleum Engineering [also MPetEng, or MPeEng]

MPF Mars Pathfinder (Mission) [of NASA]

MPF Maximum Probable Flood

MPF Metallized Paper Capacitor with Plastic Film Dielectric

MPF Multiphase Flow [Chemical Engineering]

MPFC Multi-Point Fuel Injection

MPFM Master of Public Financial Management

MPFP Melt Processible Fluoropolymer

MPFS Minimum Physical Fitness Standards

MPG Magnetic Point Group

MPG Manifold Pressure Gauge

MPG Max-Planck-Gesellschaft (zur Förderung der Wissenschaften) [Max Planck Society (for the Advancement of Science), Germany]

MPG Medical Products Group [of Hewlett Packard Company, US]

MPG 3-Mercaptoprpylguanidine [Biochemistry]

MPG Microwave Pulse Generator

MPG Miles per Gallon [also mpg]

MPG Miniature Precision Gyrocompass

MPG Motion Picture Experts Group [also MPEG]

MPG Multiple Point Ground; Multipoint Grounding

MPG Multiprogramming

MPG Museum of Practical Geology [London, UK]

MPGHM Mobile Payload Ground Handling Mechanism

MPGS Microprogramming Generating System

MPH Master of Public Health

MPH Methyl Hydropterine

MPH Miles per Hour [also mph]

MPH Moving Plasmoid Heater [Physics]

MPh Master of Philosopy [also MPhil]

6-MPH₄ DL-6-Methyl-5,6,7,8-Tetrahydropterine

MPhA m-Phenylaniline

MPharm Master of Pharmacy

mPhDA m-Phenylenediamine

MPHE Master of Public Health Education

MPHE Material and Personnel Handling Equipment

MPHEng Master of Public Health Engineering

MPhil Interdisciplinary Research Masters Degree

MPhil Master of Philosopy [also MPh]

Mphl Morpholine

MPho Master of Photography

MPhP m-Phenylphenol

MPHT Multiphasic Health Testing

MPhys Master of Physics [also MPhy]

MPI Magnesium Products Industries

MPI Magnetic Particle Inspection

MPI Mannose Phosphate Isomerase [Biochemistry]

MPI Manufacturing Processing Instruction

MPI Max-Planck-Institut (zur Förderung der Wissenschaften) [Max Planck Institute (for the Advancement of Science), Germany]

MPI Mean Point of Impact

MPI Meeting Planners International [US]

MPI Message Passing Interface

MPI Microprocessor Interface

MPI Mission Payload Integration

MPI Multi-Point Injection

MPI Multiprecision Integer

MPIA Master of Public and International Affairs

MPIA Multiple Peak Intensity Analysis

MPIAS Max Planck Institute for the Advancement of Science [Germany]

MPI-BPC Max-Planck-Institut für Biophysikalische Chemie [Max Planck Institute for Biophysical Chemistry, Goettingen, Germany]

MPIC Message Processing Interrupt Count

MPID Maximum Permissible Integrated Dose

MPI-EF Max-Planck-Institut für Eisenforschung [Max Planck Institute for Iron Research, Duesseldorf, Germany] [also MPI-Eisenforschung]

MPIF Metal Powder Industries Federation [US]

MPI-FKP Max-Planck-Institut für Festkörperphysik [Max Planck Institute for Solid-State Physics, Stuttgart, Germany] [also MPI-Festkörperphysik]

MPIIN Modification Procurement Instrument Identification Number

MPI-KF Max-Planck-Institut für Kohlenforschung [Max Planck Institute for Coal Research, Muelheim an der Ruhr, Germany] [also MPI-Kohlenforschung]

MPI-KP Max-Planck-Institut für Kernphysik [Max Planck Institute for Nuclear Physics, Heidelberg, Germany] [also MPI-Kernphysik]

MPI-MP Max-Planck-Institut für Metallphysik [Max Planck Institute for Metal Physics, Stuttgart, Germany] [also MPI-Metallphysik]

MPI-MSP Max-Planck-Institut für Mikrostrukturphysik [Max Planck Institute for Microstructural Physics, Halle/Saale, Germany] [also MPI MSP]

MPIO Mission and Payload Integration Office [of NASA]

MPIOR Multispectral Pushbroom Imaging Radiometer

MPIP Mineral Processing Investigations Program

MPIP Miniature Precision Inertial Platform

MPIP Multiple Picture-in-Picture

MPI-P Max-Planck-Institut für Physik [Max Planck Institute for Physics, Munich, Germany] [also MPI-Physik]

MPI-PF Max-Planck-Institut für Polymerforschung [Max Planck Institute for Polymer Research, Mayence, Germany] [also MPI-Polymerforschung]

MPI-Physik Max-Planck-Institut für Physik [Max Planck Institute for Physics, Munich, Germany] [also MPI-P]

MPI-PP Max-Planck-Institut für Plasmaphysik [Max Planck Institute for Plasma Physics, Garching, Germany] [also MPI-Plasmaphysik]

MPI-RA Max-Planck-Institut für Radioastronomie [Max Planck Institute for Radio Astronomy, Germany] [also MPI-Radioastronomie]

MPIS Multilateral Project Information System

MPIS Multi-Port Interconnect System

MPixel Mega-Pixels

MPixel Million Pixels

MPK Methyl Phenyl Ketone

MPK Methyl Propyl Ketone

MPL Immunoglobulin M Phospholipid Unit [Immunology]

MPL Magnetophotoluminescence; Magnetophotoluminescent

MPL Maintenance Parts List

MPL Manipulation Positioning Latches

MPL Master of Patent Law

MPL Maximum Permissible Level

MPL Maximum Permissible Limit

MPL Maximum Print Line

MPL Mechanical Parts List

MPL Message Passing Library

MPL Message Processing Language

MPL Micrograms per Liter [also mpl, μg/L or μg/l]

MPL Micropneumatic Logic

MPL Microprogramming Language

MPL Milwaukee Public Library [US]

MPL Minimum Power Level

MPL Monophosphoryl Lipid [Biochemistry]

MPL Multiple [also Mpl, or mpl]

MPL Multipurpose Processing Language

MPL Multischedule Private Line

MPL Multischedule Private Link

MPl Master of (Urban) Planning

MPLA Mask Programmable Logic Array

MPLC Medium-Performance Liquid Chromatography

MPLC Medium-Pressure Liquid Chromatography

μPLD Microcomputer Programmable Logic Device

MPLEA Multipurpose Long Endurance Aircraft

MPLM Mini-Pressurized Logistics Module

MPLN Maintenance Planning (Data Base)

MP-LPC Multipulse Linear Predictive Coding [Telecommunications]

MPLR Maximum Permissable Leakage Rate

MPLR Mininum Pressure Live Roller (Conveyor)

MPLX Multiplex(er) [also Mplx, or mplx]

MPLXR Multiplexer [also Mplxr, or mplxr]

MPM Magnetic Property Measurement

MPM Manipulator Positioning Mechanism [of NASA]

MPM Massobria, Pontikis and Martin (Potential)

MPM Master of Personnel Management

MPM Master of Pest Management

MPM Master of Professional Management

MPM Master of Public Management

MPM Materials Processing and Manufacturing

MPM Mechanical Properties Microprobe

MPM Microphotometer

MPM Microphysical Model

MPM Microprocessor Module

MPM Microprogram Memory

MPM Microscope Photometer; Microscope Photometry

MPM Microstructural Path Methodology [Metallurgy]

MPM Microstructural Path Modeling [Metallurgy]

MPM Monocycle Position Modulation

MPM Multiphase Material

MPM Multiport Modem

MPM Multiprogramming Monitor

MPM Multistand Pipe (Rolling) Mill [Metallurgy]

MP/M Multiprogramming Control Program for Microprocessors

mpm meters per minute [also MPM, or m/min]

MPMA Metal Packaging Manufacturers Association [UK]

MPMA Microprobe Microanalyzer

MPMA Montford Point Marine Association [US]

MPMC Magnetic Property Measurement System

MPMD Multiple Processor/Multiple Data

MPME Materials Processing and Manufacturing Institute [of Massachusetts Institute of Technology, Cambridge, US]

MPMC Microprogram Memory Control

MPMG Melt Powder-Melt Growth (Process)

MPMG Melt Powder-Melt Growth Group [of International Superconductivity Technology Center, Japan]

MPMP Mass Properties Management Plan

MPMS Magnetic Property Measurement System

MPMSE Multiuse Payload and Mission Support Equipment

MPMV Mason-Pfizer Monkey Virus

MPN Macroparticle Number

MPN Most Probable Number

MPN Multiprocessor Network

MPNBT Methylated Poly(nonylbithiazole)

MPO Manufacturing Production Order

MPO Maryland Point Observatory [of US Naval Research Laboratory]

MPO Maximum Power Output

MPO Memorandum Purchase Order

MPO Memory Protect Override

MPO Mono Power Amplifier

MPO Myeloperoxidase [Biochemistry]

MPOA Multi-Protocol Over ATM (Asynchronous Transfer Mode)

MPOD Mean Planned Outage Duration

MPOL Marine Pollutants List

MPP O,O-Dimethyl-O-(4-Methylthio-3-Methylphenyl) thiophosphate

MPP Massively Parallel Processing (Computer); Massively Parallel Processor

MPP Material Processing Procedure

MPP Medium-Purity Polycrystalline (Ceramics)

MPP Memory Parity and Protect

MPP Message Posting Protocol

MPP Message Processing Program

MPP Microwave Processing Project

MPP Mission Planning Program

MPP Morse Pair Potential [Physical Chemistry]

MPP Moly-Permalloy Powder

MPP Most Probable Position [Aeronautics]

MPP Multiphase Polymer

MPP Multi-Phonon Process [Solid-State Physics]

MPP Multi-Photon Process [Atomic Physics]

MP&P Materials Protection and Performance (Journal)

M&PP Materials and Plant Protection

MPPA Maritime Professional Photographers Association

MPPA Mechanically Processed Polyamide

MPPA Metal Powder Producers Association [US]

MPP a Pyropheophorbide a Methyl Ester [Biochemistry]

MPPCF Million(s) of Particles per Cubic Foot (of Air) [also mppcf]

MPPG (±)-α-Methyl-(4-Phosphonophenyl)glycine [Biochemistry]

MPPH 5-(4-Methylphenyl)-5-Phenylhydantoin; 5-(p-Methylphenyl)-5-Phenylhydantoin

MPPL Multipurpose Processing Language

MPPP Multilink Point-to-Point Protocol

MPPPM Master of Plant Protection and Pest Management

MPPS Most Penetrating Particle Size

MPPSE Multipurpose Payload Support Equipment

MPPT Maximum Power Point Tracker

MPQP Multi-Protocol Quad Port

MPR Magnetophonon Resonance

MPR Maintainability Problem Report

MPR Management Program Review

MPR Martin Package (Nuclear) Reactor

MPR Materials and Processing Report

MPR Mechanical Pressure Regulator

MPR Message Processing Region

MPR Metal Powder Report [Journal published in the US]

MPR Microseismic Processor Recorder

MPR Minimum Performance Recommendation

MPR Mission Planning Room

MPR Mockup Purchase Request

MPR Multiplanar Reconstruction

MPR Multiplier [also Mpr, or mpr]

MPR Multiport Repeater

MPR Multi Protocol Router

MPrA Master of Professional Accountancy [also MPrAcc]

MPRC McGill Pulp Research Centre [of McGill University, Montreal, Canada]

MPRE Medium Power Reactor Experiment

MPRE Multipurpose Research Reactor

MPRF Modulated Pulse Recurrence Frequency

MPRF Modulated Pulse Repetition Frequency

MPRF Multipole Radiation Field

MPRF Multi-Products Recycling Facility

MPrGph Master of Professional Geophysics

MPrMet Master of Professional Meteorology

MPROM Mask Programmable Read-Only Memory

MPRR Multipurpose Research Reactor

MPRTM Master of Parks, Recreation and Tourism Management

MPS Magnetic Power Supply

MPS Maine Public Service [US]

MPS Main Propulsion Subsystem; Main Propulsion System

MPS Manpower System

MPS Many-Particle System [Physics]

MPS Master of Personnel Service

MPS Master of Professional Studies

MPS Master of Public Service

MPS Master Production Schedule; Master Production Scheduling

MPS Master Productivity Specialist

MPS Master Program Schedule

MPS Material Processing Specification

MPS Material Processing System

MPS Materials Processing in Space

MPS Mathematical Programming Society [Netherlands]

MPS Mathematical Programming System

MPS Max Planck Society (for the Advancement of Science) [Germany]

MPS Median Period of Survival

MPS Medical Protection Society [UK]

MPS Megabit(s) per Second [also Mps, MBPS, or Mbps]

MPS 3-Mercapto 1-Propanesulfonic Acid

MPS Member of the Pharmaceutical Society [UK]

MPS Meters per Second [also mps, m/sec, or m/s]

MPS Methylacetylene Propadiene Stabilized

MPS Methyl Phenyl Sulfide

MPS Microanalysis Particle Sampler

MPS Microprobe Spectrometry

MPS Microprocessor System

MPS Microwave Power Supply

MPS Mineral Processing System

MPS Minimum Performance Specification

MPS Minimum Performance Standard

MPS Mission Planning and Scheduling

MPS Mission Preparation Sheet

MPS Modular Photomicrographic System

MPS Mucopolysaccaride [Biochemistry]

MPS Multiphase System

MPS Multiple Partition Support

MPS Multiple Protective Structure

MPS Multiprocessing System

MPS Multiprocessor Specification

MPS Multiprogramming Periodic Tasking System

MPS Multiprogramming System

MPS Multi-Purpose System

MPSC Materials and Process Sciences Center [of Sandia National Laboratories, Albuquerque, New Mexico, US]

MPSCC Multiprotocol Serial Communications Controller

MPSI Thousand Pounds per Square Inch [also mpsi]

MPSK Minimum Phase Shift Keying

MPSK Multiple Phase Shift Keying

MPSR Mission Profile Storage and Retrieval

MPSR Multi-Purpose Support Room

MPSS Main Parachute Support Structure

MPST Multi-Purpose Support Team

MPSX Mathematical Programming System, Extended

MPsych Nurs Master in Psychiatric Nursing

MPT Main Propulsion Test

MPT Male Pipe Thread

MPT Materials Processing Technology

MPT Memory Processing Time

MPT Metallurgical Plant and Technology (Journal)

MPT Microprogramming Technique

MPT Minimum Pressurization Temperature

MPT Ministry of Post and Telecommunications [Tokyo, Japan]

MPT Ministry of Post and Telegraph

MPT Mission Planning Team

MPT Mission Planning Tool [of Canadian Space Agency]

MPT Mouse-Protection Test

MPT Multichannel Pitting-Corrosion Test(ing)

MPT Multi-Phonon Transition [Solid-State Physics]

MPT Multiple Pure Tones

MPT Multi-Purpose Terminal

M Pt Melting Point [also m pt]

MPTA Main Propulsion Test Article

MPTA Mechanical Power Transmission Association

MPTA Metal Powder Technology Association [US]

MPTA α-Methoxy-α-Phenyl Trifluoropropanoic Acid

MPTD Dimethyl Diphenyl Thiuram Disulfide

MPTD Methylphosphonothioic Dichloride

MPTED Mechanical Power Transmission Equipment Distributors Association [now PTDA, US]

MPTF Main Propulsion Test Facility

MPTI Mechanical Power Transmission Institute [US]

MPTi Medium Purity Titanium

MPTN Multi-Protocol Transport Network(ing)

MPTP Main Propulsion Test Program

MPTP 1-Methyl-4-Phenyl-1,2,3,6-Tetrahydropyridine; 1,2,3,6-Tetrahydro-1-Methyl-4-Phenyl Pyridine

MPTR Mobile Position Tracking Radar

MPTS Mobile Photographic Tracking Station

MPTS Multi-Protocol Transport Services

MPTS Multipurpose Tool Set

MPTT Ministère des Postes, Télégraphes et Téléphones [Ministry of Post, Telegraph and Telephone, France]

MPU Main Processor Unit

MPU Memory Protection Unit

MPU Microprocessing Unit; Microprocessor Unit

MPU Multiple Partial Unloading [Mechanical Testing]

MPubAdm Master of Public Administration

MPUC Maine Public Utility Commission [US]

MPUrbDes Master of Planning and Urban Design

MPV Marmosetpox Virus [*Note:* Marmosets are South American Primates]

MPV Murine Polyomavirus

MPV Multi(ple) Passenger Vehicle

MPVA Main Propellant Valve Actuator

M-PVC Mass Polyvinyl Chloride

MPVM Master of Preventive Veterinary Medicine

MPW Macintosh Programmer's Workshop

MPW Male Pipe Weld

MPW Master of Public Works

MPW Modified Plane Wave [Physics]

MPW Multiple Project Wafer

MPWD Machine-Prepared Wiring Data

MPWP Maximum Permissible Working Pressure

M-PYROL 1-Methyl-2-Pyrrolidinone

MPX Mapped Programming Executive

MPX Multiplex(er); Multiplexing [also Mpx, or mpx]

MPX Multiprogramming Executive

mPxBA m-Phenoxybenzoic Acid

MPX LED Multiplex Light-Emitting Diode

MPX/R Relay Multiplexer

MPY Mil(s) per Year [also mpy, or mil/yr]

MPY Multiply [also Mpy, or mpy]

MPy Methylpyridinium [also mpy]

2MPy 2-Methyl Pyridine

3MPy 3-Methyl Pyridine

4MPy 4-Methyl Pyridine

mpy mil(s) per year [also MPY, or mil/yr]

mpy milliinch(es) per year [Unit]

MQ Magnequench [Metallurgy]

MQ Magnetic Quenching

MQ Martinique (French) [ISO Code]

MQ Melt Quench(ing) [Metallurgy]

MQ Multiple Quantum

MQ Multiplier/Quotient

MQ1 Magnequench Resin Bonded Magnet

MQ2 Magnequench Fully Dense (Hot-Pressed Isotropic) Magnet

MQ3 Magnequench Magnetically Aligned, Anisotropic (Hot-Worked) Magnet

M/q Ion Mass-to-Charge Ratio [Symbol]

MQAE 6-Methoxyquinolinium-1-Acetic Acid Ethyl Ester

MQB Multiquantum Barrier [Physics]

MQC Manufacturing Quality Control

MQC Macroscopic Quantum Coherence [Physics]

MQC Multi-Quantum Coherence [Physics]

MQDT Multichannel Quantum Defect Theory [Physics]

MQE Message Queue Element

MQE Molecular Quantum Electronics

MQES Metastable Quenched Electron Spectroscopy

MQF-COSY Multiple-Quantum-Filtered Correlated Spectroscopy

MQI Message Queue Interface

MQL Medical Query Language

MQM Magnetic Quadrupole Moment

MQN Magnetic Quantum Number

MQ-NMR Multiple-Quantum-Nuclear Magnetic Resonance

MQORC Modified Quadrature Overlapped Raised Cosine

MQPPMG Melt-Quenching-and-Pressurized-Partial-Melt-Growth (Processing)

MQR Multiplier-Quotient Register

MQRC Microelectronic Quality/Reliability Center (Affiliates Program) [of Sandia National Laboratories, Albuquerque, New Mexico]

MQS Mail Questionnaire Survey

MQS Metastable Quenching Spectroscopy

MQS Multiprogrammed Queued Tasking System

MQT Macroscopic Quantum Tunneling [Electronics]

MQT Magnetic Quantum Tunneling

MQT Melt-Quench Technique [Metallurgy]

MQT Multiple-Quantum Transition

MQW Multi(ple) Quantum Well [Electronics]

MQW APD Multi(ple) Quantum Well Avalanche Photodiode [also MQW-APD]

MQW HS Multi(ple) Quantum Well Heterostructure [also MQW-HS]

MQW IR Multi(ple) Quantum Well Infrared (Device) [also MQW-IR]

MQW IT Multi(ple) Quantum Well Intersubband Transition (Detector) [also MQW-IT]

MQW LD Multi(ple) Quantum Well Laser Diode [also MQW-LD]

MQWS Multiple Quantum Well Structures

MR Compensator–Running [Controllers]

MR Machine-Readable

MR Machine Records

MR Machine Rifle

MR Macrorheological; Macrorheology

MR Magazine Records

MR Magnetic Recording

MR Magnetic Refrigerator

MR Magnetic Resistance

MR Magnetic Resonance

MR Magnetic Rotation

MR Magnetogyric Ratio

MR Magnetoresistance; Magnetoresistor

MR Magnetoresistive; Magnetoresistivity

MR Magnetorheological; Magnetorheology

MR Magnetorotons [Physics]

MR Mail Room

MR Main Ring [Physics]

MR Malagasy Republic

MR Manganese Rhenate

MR Map Reference

MR Markovnikov's Rule [Chemistry]

MR Marqusee and Ross (Theory) [Metallurgy]

MR Mars Relay [of NASA]

MR Mask Register

MR Master Reset

MR Matching Record (Indicator)

MR Material Review

MR Materials Research(er)

MR Mate's Receipt [also mr] [Commerce]

MR Mathematical Reviews [of American Mathematical Society, US]

MR Matrix Receptance

MR Matrix Row

MR Matthiessen's Rule [Solid-State Physics]

MR Mauretania [ISO Code]

MR Mauritian Rupee [Currency]

MR Measured Range

MR Measurement Range

MR Mechanical Rectifier

MR Mechanical Relaxation

MR Medical Register [of General Medical Council, UK]

MR Medical Research(er)

MR Medium Reduction

MR Megaroentgen [Unit]

MR Membrane Radiometer

MR Memorandum Report

MR Memory Reclaimer

MR Memory Register

MR Mercury Rectifier

MR Mercury-Redstone [of NASA Mercury Program]

MR Message Rate

MR Message Repeat

MR Metallic Rectifier

MR Metamorphic Rock [Geology]

MR Methylresorcinol

MR Microradiograph(y)

MR Microradiometer

MR Mill Run [Mining]

MR Mineral Rubber

MR Mine Run [Mining]

MR Miscellaneous Report

MR Missile Room

MR Mission Report

MR Mississippi River [US]

MR Missouri River [US]

MR Mixture Ratio [also M/R]

MR Modem Ready

MR Moderately Resistant; Moderate Resistance

MR Moderating Ratio [Atomic Physics]

MR Modified Read

MR Modular Redundancy

MR Modulus of Rupture

MR Moisture Resistance; Moisture-Resistant

MR Molasses Residuum

MR Molecular Rays

MR Molecular Rearrangement

MR Molecular Recognition

MR Monitor Recorder

MR Morbidity Rate [Medicine]

MR Morse Code

MR Mortality Rate [Medicine]

MR Moving Range

MR Multiple Regression [Statistics]

MR Multiple Request(ing)

MR Multiplier Register

MR Multipoint Recorder

MR Multipole Radiation

MR Murray River [Australia]

MR Mutarotation [Chemistry]

MR$_n$ Metal Neodecanoates [Formula; n Represents the Oxidation State of the Metal]

M(R) Contribution from Multiple Scattering

M(R) Rotation Matrix [Crystallography]

M&R Maintenance and Refurbishment [also M+R]

M&R Maintenance and Repair [also M+R]

M/R Mixture Ratio [also MR]

M-R (Reverse Transformation of) Martensite to R-Phase [also M/R] [Metallurgy]

Mr Macroreticular [Chemistry]

Mr Marble [also mr]

Mr March

Mr Mercury [Abbreviation; Symbol: Hg]

Mr Mister

M(r) Magnetic Moment Density [Symbol]

M(r) Magnetic Vector [Symbol]

M(r) Mass of a Given Spherical Volume of Fractal Material with Radius r [Symbol]

M$ Malaysian Ringgit [Currency]

mR milliroentgen [also mr]

μR microroentgen [also μr, or muR]

μ_{en}/ρ mass energy absorption coefficient [Symbol]

MRA Machine Readable Archives Division [of Packaging Association of Canada]

MRA Manufacturers Representatives of America [US]

MRA Master of Recreational Administration

MRA Master of Resource Administration

MRA Materials Reliability Assurance

MRA Mechanical Readiness Assessment

MRA Medical Research Assistant

MRA Microgravity Research Association

MRA Minimum Reception Altitude

MRAA Marine Retailers Association of America [US]

MRAALS Marine Remote Approach and Landing System

MRAC Meter-Reading Access Circuit

MRAC Model Reference Adaptive Control

MRAD Mass Random Access Disk

MRAD Messen–Regeln–Automatisieren Datenbank, Deutsche Ausgabe [Measuring–Controlling–Automation Online Database, German-Language Edition]

MRad Master of Radiology

mrad millirad [Unit]

MRADS Mass Random Access Data Storage

mrad/s millirad(s) per second [also mrad s^{-1}, or mrad/sec]

MRAE Messen–Regeln–Automatisieren Datenbank, Englische Ausgabe [Measuring–Controlling–Automation Online Database, English-Language Edition]

MRaEng Master of Radio Engineering

MRAIC Member of the Royal Architectural Institute of Canada

MRAM Magnetoresistive Random Access Memory

MRAS Model Reference Adaptive System

MR-ATOMIC Multiple Rapid Automatic Test of Monolithic Integrated Circuit

MRB Magnetic Recording Borescope

MRB Maintenance Review Board [UK]

MRB Malaysian Rubber Bureau

MRB Marshall Richards Barco Limited [Crook, UK]

MRB Material Review Board

MRB Materials Research Bulletin [US]

MRB Microcircuit Reliability Bibliography [US Air Force, Rome Air Development Center, Griffiss Air Force Base, New York State]

MRBM Medium-Range Ballistic Missile

MRC Machine Readable Code

MRC Maintenance Requirement Card

MRC Manitoba Research Council [Canada]

MRC Manufacturing Research Corporation [Canada]

MRC Marconi Research Centre [of General Electric Company Research Laboratories, Chelmsford, Essex, UK]

MRC Marine Resources Council [US]

MRC Materials Research Center

MRC Materials Research Corporation [Orangeburg, New York, US]

MRC Mathematics Research Center [of US Army]

MRC Measurement Requirements Committee

MRC Medical Research Center

MRC Medical Research Council [Canada, UK and South Africa]

MRC Memory Request Controller

MRC Metals Reserve Company [US]

MRC Meteorological Research Committee [UK]

MRC Microelectronics Research Centre [London, UK]

MRC Microelectronics Research Center [of New Jersey Institute of Technology, Newark, US]

MRC Mississippi River Commission [US]

MRC Missouri Resources Commission [US]

MRC Mold Release Casting (Technology)

MRC Monsanto Research Corporation [Miamisburg, Ohio, US]

MRC Mount Royal College [Montreal, Quebec, Canada]

MRC Multiple Register Counter

MRCA Multirole Combat Aircraft

MRCC Maritime Rescue Coordination Center [UK]

MRCC Medical Research Council of Canada

MRCC Multireference Coupled-Cluster (Method) [Physics]

MRCD Movement for Responsible Coastal Development

MRCF Microsoft Real-Time Compression Format

MRCGP Member of the Royal College of General Practioners [UK]

MRCI Microsoft Realtime Compression Interface

MRCI Multireference Configuration Interaction [Physics]

MRCISD Multireference Configuration Interaction Single plus Double (Replacement) [Physics]

MRCM Master of Real Estate and Construction Management

MRCMC Murrumbidgee Regional Catchment Management Committee [Australia]

MRCO Manufacturing Research Corporation of Ontario [Canada]

M(RCO$_2$)$_x$ Metal Carboxylate [General Formula]

MRCP Master of Regional and City Planning

MRCP Master of Regional and Community Planning

MRCP Member of the Royal College of Physicians

MRCPath Member of the Royal College of Pathologists

MRcPk Master of Recreation and Parks

MRCR Measurement Requirement Change Request

MRCS Member of the Royal College of Surgeons

MRCS Multiple Report Creation System

MRCV Multirole Combat Vehicle

MRCVS Member of the Royal College of Veterinary Surgeons

MRD Machine-Readable Document

MRD Manual Ringdown [Communications]

MRD Maritime Research Department [of Webb Institute of Naval Architecture, Glen Cove, New York, US]

MRD Material Review Disposition

MRD Materials Reliability Division [of National Institute of Standards and Technology, Boulder, Colorado, US]

MRD Materials Research Diffractometer

MRD Materials Research Division

MRD Medical Records Department [of Warren Grant Magnuson Clinical Center, National Institutes of Health, Bethesda, Maryland, US]

MRD Medical Research Division

MRD Microbiological Research Department

MRD Mining Research Directorate [Sudbury, Ontario, Canada]

MRD Mission Requirements Document

MRD Multiple-Reference Double Excitation

MRD Multireference Density [Physics]

MRD Multiresponse Data

MRDA Media Research Directors Association [US]

MRDC Materials Research and Development Center

MRDC Materials Research and Development Center [of Chung Shan Institute of Science and Technology, Lung-Tan, Taiwan]

MRD-CI Multireference Density–Configuration Interaction [Physics]

MRD-CI Multireference with Double Excitations Configuration Interaction [Physics]

MRDE Mining Research and Development Establishment [UK]

MRDE Multiple-Reference Double Excitation [Physics]

MRDF Machine-Readable Data File

MRDF Moisture-Resistant Densified Fuel

MR&DF Malleable Research and Development Foundation [US]

MR-DOS Mapped Real-Time Disk-Operating System [also MRDOS, or MR DOS]

MRDR Material Review Disposition Record

MRDS Man-Carried Radiological Data System

MRE Magnetorefractive Effect

MRE Magnetoresistive Element

MRE Measured Over 2 Pins, External Spline [Mechanical Engineering]

MRE Medical Research Engineer(ing)

MRE Microbiological Research Establishment [of Ministry of Defense, UK]

MRE Mining Research Establishment [of National Coal Board, UK]

MRE Multiple-Response Enable

MRE Microbiological Research Establishment [UK]

MREC Malaviya Regional Engineering College [Jaipur, India]

MRED Master of Real Estate Development

MREDA Marine Resources and Engineering Development Act [US]

MREF Michigan Research Excellence Fund [US]

MRELB Malaysian Rubber Exchange and Licensing Board

MREM Minerals Resources Engineering and Management

MRERF Manufacturers Representatives Educational Research Foundation [US]

MRes Master of Research

MRF Maintenance Responsibility File

MRF Materials Recycling Facility

MRF Materials Recovery Facility

MRF Maximum Retarding Force

MRF Measurements/Stimuli Request Form

MRF Medical Research Foundation [France]

MRF Message Refusal

MRF Meterological Research Facility [UK]

MRF Multipath Reduction Factor

MRF Multiple Reaction Force

M&RF Maintenance and Refurbishing Facility

MRFAC Manufacturers Radio Frequency Advisory Committee [US]

MRFIT Multiple Risk Factors Intervention Trial [of US National Institutes of Health, Bethesda, Maryland, US]

MRFL Master Radio Frequency List

MRFT Material Removal and Forming Technology

MRFT Multiple Reaction Force Technique

MRFT Multiple Reaction Force Testing

MRFV Maize Rayado Fini Virus

MRG Macroscopic Renormalization Group

MRG Management Research Group [UK]

MRG Materials Research Grant [of National Science Foundation, US]

MRG Materials Research Group

MRG Medium Range

MRG Merge [also Mrg]

MRG Methane-Rich Gas

MRGA Manhattan Ryegrass Growers Association [US]

MRG CMND Merge Command [also Mrg Cmnd]

mrgl marginal

MR/GMR Magnetoresistance/Giant Magnetoresistance (Recording Head) [Electronics]

MRGS Mesabi Range Geological Society [Minnesota, US]

MRH Magnetoresistive Head

MRH Mechanical Recording Head

MRH Mobile Remote Handler

MRH Monitoring River Health

MRH MSH (Melanocyte-Stimulating Hormone) Releasing Hormone [Biochemistry]

MR(H) Magnetic Field Dependent Magnetoresistance [Symbol]

MR-H Magnetoresistance–Magnetic Field (Curve)

mR/h milliroentgen(s) per hour [also mR/hr, mR h^{-1}, or mR hr^{-1}; formerly also mr/hr]

μ-RHEED Micro Reflected High Energy Electron Diffraction

MRHI Monitoring River Health Initiative [of Commonwealth Environment Protection Agency, Australia]

MRI Machine Records Installation

MRI Magnetic Resonance Image; Magnetic Resonance Imaging

MRI Magnetic Rubber Inspection

MRI Maritime Research Institute [Netherlands]

MRI Mass Retailing Institute [US]

MRI Master of the Royal Institute

MRI Material Receiving Instruction

MRI Materials Research Institute [of University of Texas at El Paso, US]

MRI Materials Research Instrument(ation)

MRI Mean Recurrence Interval

MRI Measured Between 2 Pins, Internal Spline [Mechanical Engineering]

MRI Measurement Requirements and Interface

MRI Medical Research Institute [US Navy]

MRI Member of the Royal Institute

MRI Memory Reference Instruction

MRI Metal Recovery Industry

MRI Meteorological Research Institute [Japan]

MRI Midwest Research Institute [Kansas City, Missouri, US]

MRI Miscellaneous Radar Input

MRI Monopulse Resolution Improvement

MRi Materials Resources International [Blue Bell, Pennsylvania, US]

MRIE Magnetic Reactive Ion Etching

MRIE Magnetic Resonance Imaging (by the) Earth('s) (Magnetic Field)

MRIH MSH (Melanocyte-Stimulating Hormone) Release-Inhibiting Hormone [Biochemistry]

MRIIUS Metal Recovery Industries Inc. of the United States

MRINDO Modified Rydberg Intermediate Neglect of Differential Orbital [Physics]

MRIPS Multimodality Radiology Image Processing System

MRIR Market Research Information System

MRIR Medium-Resolution Infrared Radiometer

MRIS Maritime Research Information Service [of National Academy of Sciences, US]

MRJE Multiple Remote Job Entry

MRK Mark [Currency of Former East Germany]

Mrk Mark [also mek]

mrkd marked

MRL Machine Representational Language

MRL Maggi-Righi-Leduc (Effect) [Physics]

MRL Manipulator Retention Latch; Manipulator Retention Lock [NASA]

MRL Manufacturing Reference Line

MRL Material Requirements List

MRL Materials Reference Library

MRL Materials Research Laboratories [of National Science Foundation, US]

MRL Materials Research Laboratories [of Industrial Technology Research Institute, Taiwan]

MRL Materials Research Laboratories, Inc [Struthers, Ohio, US]

MRL Materials Research Laboratory

MRL Medical Research Laboratory

MRL Maximum Residue Limit

MRL Meteorological Research Laboratory

MRL Microwave Research Laboratory [of George Mason University, Fairfax, Virginia, US]

MRL Mining Research Laboratories [of Canada Center for Mineral and Energy Technology, Natural Resources Canada]

MRL Multiple Rocket Launcher

MRL Multipoint Recorder/Logger

MRL Bull Res Dev MRL (Materials Research Laboratories) Bulletin of Research and Development [Taiwan]

MRLC Multi-Resolution Land Characteristics (Consortium)

MRL/ITRI Materials Research Laboratory of Industrial Technology Research Institute [Japan]

MRM Magnetic Ring Modulator

MRM Master of Resource Management

MRM Metabolic Rate Monitor

MRM Militarized Reconfigurable Multiprocessor [of Department of National Defense, Canada]

MRM Most Recently-Used Master

MRMBPT Multireference Many-Body Perturbation Theory [Physics]

mR/min milliroentgen per minute [also mR min^{-1}; formerly also mr/min]

MRMU Mobile Radiological Measuring Unit

MRMU Mobile Remote Manipulating Unit [US Air Force]

MRN Meteorological Rocket Network [of NASA]

MRN Minimum Reject Number

MRN Modified Random Network (Model)

mRNA Messenger Ribonucleic Acid [also M-RNA, or m-RNA]

MRNC Model Reference Nonlinear Controller

Mrng Morning [also mrng]

MRO Maintenance, Repair and Operating

MRO Material Release Order

MRO Mauritanian Ouguiya [Currency]

MRO Medium-Range Order(ed); Medium-Range Ordering [also mro] [Solid-State Physics]

MRO Memory Resident Overlay

MRO Midrange Objectives

MRO Multi-Region Operation

MROA Magnetic Raman Optical Activity

MROD Mount Rainier Ordnance Depot [Tacoma, Washington, US]

MROM Macro Read-Only Memory

MRP Machine-Readable Passport

MRP Malfunction Reporting Program

MRP Management Resource Planning

MRP Manufacturer's Recommended Price

MRP Manufacturing Requirements Planning

MRP Manufacturing Resources Planning

MRP Master of (or in) Regional Planning

MRP Material Requirements Planning

MRP Materials Requirement Program

MRP Message Routing Process

MRP Metal Refining Process [of Mannesmann-Demag, Germany] [Metallurgy]

MRP Meteorological Reporting Point

MRP Mineral Resources Program [of US Geological Survey]

MRP Minimum-Energy Reaction Path

MRP Mouth Reference Point

MNP Manganese Nickel-Rich Precipitate

MRPC Mercury Rankine Power Conversion [of US Atomic Energy Commission]

MRPII Manufacturing Resources Planning II [also MRP II]

MRPL Main Ring Path Length

MRPPS Maryland Refutation Proof Procedure System [US]

MRPRA Malaysian Rubber Producers Research Association [UK]

MRPS Manufacturing Resources Planning System

MRPT Magnetoresistive Phase Transition

MRQC Microelectronics Reliability Quality Center [of Sandia National Laboratories, Albuquerque, New Mexico, US]

MRR Mechanical Reliability Report [of Federal Aviation Administration, US]

MRR Medical Research Reactor [of Battelle Laboratory, US]

MRR Material Removal Rate

MRR Mission Reconfiguration Request

MRR Multiple Response Resolver

MRR Multiple Restrictive Requirement Quality

MRRC Mineral Resources Research Center [of University of Minnesota, Minneapolis, US]

MRRDB Malaysian Rubber Research and Development Board

MRRV Maneuvering Reentry Research Vehicle

MRS Magnetic Resonance Sensor

MRS Magnetic Resonance Spectroscopy

MRS Magnetic Resonance Spectrum

MRS Magnetoresistive Sensor

MRS Magnification Reference Standard

MRS Management Reporting System

MRS Management Review System

MRS Manned Repeater Station

MRS Market Research Society [UK]

MRS Master Reference Source

MRS Materials Research Society [US]

MRS Mathematics Research Center

MRS Mean Random Spacing [also mrs] [Metallurgy]

MRS Mean Relative Strain

MRS Media Recognition System

MRS Media Resource Service [of Scientists Institute for Public Information, US]

MRS Medical Record System [UK]

MRS Medical Research Society [UK]

MRS Medium-Range Search

MRS Micro-Raman Spectroscopy [also μRS]

MRS Microwave Reflectance Spectroscopy

MRS Mobile Radio Station

MRS Mobile Radio System

MRS Mobile Roof Support

MRS Monitored Retrievable Storage [Radioactive Waste]

Mrs Mistress

μRS Micro-Raman Spectroscopy [also MRS]

MRSA Mandatory Radar Service Area

MRS Bull MRS Bulletin [Published by the Materials Research Society, US] [also MRSB]

MRSC Maritime Rescue Sub-Center [UK]

MRSC Member of the Royal Society of Canada

MRSDCI Multi-Reference Single plus Double Excitation Configuration Interaction [Physics]

MRSEC Materials Research Science and Engineering Center (Program) [of National Science Foundation Division for Materials Research, US]

MRSH Materials Referral System and Hotline [US]

MRS-HQ Materials Research Society–Headquarters [Pittsburgh, Pennsylvania, US]

MRS-I Materials Research Society of India

MRS-I/ICSC Materials Research Society of India–Indian Center for Superconductivity [also MRSI-ICSC]

MRS-I Newsl Materials Research Society of India Newsletter

MRS-J Materials Research Society of Japan [formerly AMSES]

MRS-J News Materials Research Society of Japan News

MRS-K Materials Research Society of Korea [also MRS-Korea]

MRS Newsl MRS Newsletter [now Materials Research Society Bulletin]

MRS Symp Proc MRS Proceedings [Published by the Materials Research Society, US]

MRS-Russia Materials Research Society of Russia

MRSS Maximum Resolved Shear Stress

MRS Int Mtg Adv Mater Materials Research Society International Meeting on Advanced Materials [US]

MRS Symp Proc MRS Symposia Proceedings [Published by the Materials Research Society, US]

MRST Materials Research and Strategic Technologies [of Motorola Inc., Mesa, Arizona, US]

MRS-T Materials Research Society of Taiwan [also MRS-Taiwan]

MRT Maximum Recommended Temperature

MRT Mean Radiant Temperature

MRT Mean Repair Time

MRT Medical Radiation Technologist; Medical Radiation Technology

MRT Medical Research Technology

MRT Ministère de la Recherche et de la Technologie [Ministry of Research and Technology, France]

MRT Mobile Radar Target

MRT Modified Rhyme Test

MRT Multiple Requesting Terminal Program

Mrt Media Remanent-Moment-Thickness Product [Magnetic Media]

MRTC Multiple Real-Time Command

MRTD Minimum Resolvable Temperature Difference

Mrtg Mortgage [also mrtg]

Mrtm Maritime [also mrtm]

MrTmag (Magnetic) Remanence Thickness Product

MRTP Master of Rural and Town Planning

MRTS Microwave Repeater Test Set

MRU Machine Records Unit

MRU Material Recovery Unit

MRU Maximum Receive Unit

MRU Message Retransmission Unit

MRU Meteorological Research Unit [UK]

MRU Ministère de la Reconstruction et de l'Urbanisme [Ministry of Reconstruction and Town Planning, France]

MRU Mobile Radio Unit

MRU Mountain Rescue Unit

MRV Maneuverable Reentry Vehicle

MRV Maximum Relative Variation

MRV Minimum Rated Value (of Capacitor)

MRV Multiple Reentry Vehicle

MRWC Multiple Read-Write Compute

mRy millirydberg [Unit]

mRy/at millirydberg per atom [also mRy at^{-1}]

mRy/fu millirydberg per formula unit [also mRy fu^{-1}]

MRZ Metal Reaction Zone [Metallurgy]

MS Automotive Crankcase Oil, Spark-Ignition Engine, Severe Service Conditions [American Petroleum Institute Classification]

MS Compensator–Starting [Controllers]

MS Machine Screw

MS Machine Selection

MS Machine Shop

MS Machine Steel

MS Macromodular System

MS Macromolecular Science

MS Macroscopic Segregation [Metallurgy]

MS Magnesium Silicate (or Olivine)

MS Magnetic Saturation

MS Magnetic Scattering

MS Magnetic Screen(ing)

MS Magnetic Semiconductor

MS Magnetic Separation; Magnetic Separator

MS Magnetic Spectrograph(y)

MS Magnetic Spectrometer; Magnetic Spectrometry

MS Magnetics Society [of Institute of Electrical and Electronics Engineers, US]

MS Magnetic Storage

MS Magnetic Survey [Geophysics]

MS Magnetic Susceptibility

MS Magnetic Switching

MS Magnetostatic(s)

MS Magnetostriction; Magnetostrictive

MS Magnet(ic) Steel

MS Mail Steamer

MS Main Storage

MS Main Switch

MS Maintenance Schedule

MS Maintenance Staff

MS Management Science

MS Manganese Silicate

MS Manganese Steel

MS Manned Satellite

MS Manned Spacecraft

MS Manufacturing System

MS Manuscript [also Ms, or ms]

MS Margin of Safety

MS Marine Stratocumulus [Meteorology]

MS Mark Sense; Mark Sensing [Computers]

MS Mash-Seam (Welding)

MS Mass Spectrograph(y)

MS Mass Spectrometer; Mass Spectrometry

MS Mass Spectroscope; Mass Spectroscopy

MS Mass Spectral; Mass Spectrum

MS Mass Storage

MS Master of Science

MS Master Seaman

MS Mastersizer [Metallurgy]

MS Master-Slave (Flip-Flop)

MS Master Station [Radio Navigation]

MS Master Switch

MS Material Specification

MS Materials Science; Materials Scientist

MS Materials Selection

MS Materials Synthesis

MS Mathematical Science(s)

MS Mathematical Series

MS Mathematical Statistician; Mathematical Statistics

MS Mating Sequence

MS Maximum Shielding [Nuclear Reactors]

MS Maximum Stress

MS Mean Square

MS Mean Sun [Astronomy]

MS Measured Service (Pricing)

MS Measurement Science; Measurement Scientist

MS Mechanical Separation; Mechanical Separator

MS Mechanized Scheduling

MS Media Society [UK]

MS Medical Survey

MS Mediterranean Sea

MS Medium Setting

MS Medium Steel

MS Megasample

MS Megasiemens [Unit]

MS Melt-Spun; Melt Spinning

MS Membrane Science

MS Memory Stack

MS Memory System

MS Mercury Spectrum

MS Mercury Switch

MS Mesa [also Ms, or ms]

MS Message Store

MS Message Switch(ing)

MS Messaging System

MS Metallosilicate

MS Metallurgical Society [of Canadian Institute of Mining and Metallurgy]

MS Metal-Semiconductor (Junction) [also M-S]

MS Metal Spinning

MS Metal Spray(er); Metal Spraying

MS Metastable; Metastability

MS Meteoritical Society [US]

MS Meteor Shower [Astronomy]

MS Methanesulfonate Salt

MS Methods Study [Industrial Engineering]

MS Metric System

MS Microgravity Services

MS Micro-Scale

MS Microseism(ic)

MS Microsoft Corporation [US]

MS Microspectrograph(y)

MS Microspectrometer; Microspectrometry

MS Microspectroscope; Microspectroscopy

MS Microsphere

MS Microspherical Silica Alumina

MS Microstructure

MS Microstrip

MS Microsurgery

MS Microsymposium

MS Microsystem

MS Mid-Toughness Structure

MS Midwest-Sunset (Oilfield) [Southern California, US]

MS Mild Steel

MS Milestone

MS Military Satellite

MS Military Science; Military Scientist

MS Military Secretary

MS Military Standard

MS Mineralogical Society [UK]

MS Mine Safety

MS Minesweeper [also M/S]

MS Mining System

MS Ministry of Supply [UK]

MS Minkowski Space [Relativity]

MS Misfit Strain [Materials Science]

MS Mission Scientist

MS Mission Specialist

MS Mission Station

MS Mission System

MS Mississippi [US]

MS Mitral Stenosis [Medicine]

MS Mobile Service

MS Mobile System

MS Moderately Susceptible; Moderate Susceptibility

MS Modulation Spectroscopy

MS Moessbauer Spectroscope; Moessbauer Spectroscopy

MS Moessbauer Spectrum

MS Molar Substitution

MS Molecular Sieve

MS Molecular Spectroscopy

MS Molecular Spectrum

MS Molecular Structure

MS Molecular Stuffing

MS Moles of Substituent Combined [Cellulose Drivatives]

MS Molten Salt

MS Monitoring System

MS Monosaccharide

MS Monostable

MS Montserrat [ISO Code]

MS Mosaic Spread

MS Motion Sensing; Motion Sensor

MS Motor Ship

MS Motor-Starter

MS Motor Switch

MS Mouse Serum (Protein) [Biochemistry]

MS Mullins-Sekerka (Instability) [Physics]

MS Mullite Substrate

MS Multichannel (Television) Sound

MS Multiple Scattering (Theory) [Physics]

MS Multiple Sclerosis [Medicine]

MS Multiplet Splitting [Physics]

MS Multispectral

MS Multi-Stage

MS Multi-String

MS Multisystem [Computers]

MS Murashige and Skoog (Basal Salt Mixture) [Biochemistry]

MS Music Staccato [Programming]

MS Music Synthesis

MS Myelin Sheath [Anatomy]

MS² Mass-Spectrometry-Coupled Mass Spectrometry [also MS2, or MSMS]

M₂Sₓ Metal Polysulfide [Formula]

MS-222 3-Aminobenzoic Acid Ethyl Ester, Methanesulfonate Salt

M(S) S-th Message [Data Communications]

%MS Percent of Maximum Shielding [Nuclear Reactors]

M&S Maintenance and Supply

M/S Mark-to-Space Ratio [Electronics]

M/S Measurement Stimuli

M/S Minesweeper [also MS]

M/S Mold/Sample (Interface)

M-S Metal-Semiconductor (Junction) [also MS]

Ms Earthquake Magnitude by Surface Wave Method [Symbol]

Ms Manuscript [also MS, or ms]

Ms Martensite [Metallurgy]

Ms Megasecond [also Msec]

Ms mesa [also ms]

Ms Mesyl [Chemistry]

Ms Miss

M(s) Manipulated Variable [Symbol]

mS millisiemens [Unit]

ms mesa [also Ms]

ms millisecond [also msec]

ms months after sight

m⁻²s⁻¹ one per square meter per second [also 1/m²/s, or 1/m²·s]

m/s meter(s) per second [also MPS, mps, m/sec, or m s⁻¹]

m/s² meter(s) per second squared [also m/sec², or m s⁻²]

m²/s square meter(s) per second [also m²/sec, or m² s⁻¹]

m³/s cubic meter(s) per second [also m³/sec, or m³ s⁻¹]

μS microsiemens [Unit]

μs microsecond [also μsec, or mu s]

MSA Magnetic Sector Analyzer

MSA Major Statistical Area(s)

MSA Malayan Scientific Association

MSA Malaysia-Singapore Airlines [Defunct]

MSA Maritime Safety Agency [Tokyo, Japan]

MSA Mass Storage Adapter

MSA Master of Science in Accounting

MSA Master of Science in Administration

MSA Master of Science in Agriculture

MSA Material Service Area

MSA Material Surveillance Assembly

MSA Mean-Sphere (or Spherical) Approximation [Physics]

MSA Mechanical Signature Analysis

MSA Message System Agent

MSA Methanesulfonic Acid

MSA Methanesulfuric Acid

MSA N-Methyl-N-(Trimethylsilyl) Acetamide

MSA Microgravity Science and Applications

MSA Metropolitan Serving Area [Telecommunications]

MSA Metropolitan Statistical Area [US] [formerly Standard Metropolitan Statistical Area]

MSA Military Service Act

MSA Mineralogical Society of America [US]

MSA Mine Safety Appliance

MSA Minimum Safe Altitude

MSA Minimum Sector Altitude

MSA Minimum Surface Area

MSA Modem Signal Analyzer

MSA Molecular Self-Assembly

MSA Most Seriously Affected

MSA Multibus Systems Architecture

MSA Multichannel Sound Adapter

MSA Multiplication Stimulating Activity

MSA Multi-Subsystem Adapter

MSA Mutual Security Agency [US]

MSA Mycological Society of America [US]

msa *(misce secundem artem)* – mix skillfully [Medical Prescriptions]

MSAA Master of Science in Astronautics and Aeronautics

MSAAE Master of Science in Aeronautical and Astronautical Engineering

MSAC Most Seriously Affected Country

MSAc Methylthio Acetic Acid

MSAcct Master of Science in Accounting

MSACSS Middle States Association of Colleges and Secondary School

MSACM Microsoft Audio Compression Manager

MSAD Materials Summary Acceptance Document

MSAD Microgravity Science and Applications Division [of NASA]

MSAdm Master of Science in Administration

MSAE Master of Science in Aerospace Engineering [also MSAeE, or MSAeEng]

MSAE Master of Science in Architectural Engineering

MSAeE Master of Science in Aerospace Engineering [also MSAeEng, or MSAE]

MSAg Master of Science in Agriculture

MSAgE Master of Science in Agricultural Engineering [also MSAgEng]

MSAI Moscow Steel and Alloys Institute [Russia]

MSAM Master of Science in Applied Mathematics

MSAM Master of Science in Applied Mechanics

MSAM Minus-S-Admittance Multiplicator

MSAM Multiple Sequential Access Method

MSAN Multiport Ship Agencies Network

MSANS Multiple Small-Angle Neutron Scattering

MSAOR Master of Science in Applied Operations Research

MSAP Master of Science in Applied Physics

MSAP Minislotted Alternating Priorities

MSAR Mine Safety Appliance Research

MSArch Master of Science in Architecture

MSArchSt Master of Architectural Studies [also MSArchStud]

MSArchTech Master of Science in Architectural Technology

MSAT Master of Science in Advanced Technology

MSAT Mobile Satellite [Canada]

MSAT Mobile Satellite Program [of US Department of Commerce]

MSAT Multipurpose Satellite

MSATA Motorcycle, Scooter and Allied Trades Association [US]

MSAU Multi-Station Access Unit

MSAUSC Muslim Students Association of the United States and Canada

MSAV Microsoft Anti-Virus

MSAW Magneto-Surface Acoustic-Wave (Element)

MSAW Minimum Safe Altitude Warning

MSB Macroscopic Shear Band [Metallurgy]

MSB Macroscopic Slip Band [Metallurgy]

MSB Master of Science in Business

MSB Maritime Subsidy Board [of US Maritime Administration]

MSB Mass Spectrometry Bulletin [Published by the Royal Society of Chemistry, UK]

MSB Mass Spectrometry Bulletin (Database) [of Mass Spectrometry Data Center, UK]

MSB Methylstyrylbenzene

MSB Micro Shear Band [Metallurgy]

MSB Minesweeping Boat [US Navy Symbol]

MSB Mining Standards Board

MSB Most Significant Bit [also msb]

MSB Mutual Savings Bank

bis-MSB 1,4-bis(2-Methylstyryl)benzene; p-bis(o-Methylstyryl)benzene

MSBA Master of Science in Business Administration

MSBA Medium and Small Business Administration [of Ministry of External Affairs, Taiwan]

m²/s/bar square meter per second per bar [also $m^2\ s^{-1}\ bar^{-1}$]

MSBC Master of Science in Building Construction

MSBE Master of Science in Biomedical Engineering

MSBE Molten Salt Breeder (Reactor) Experiment

MSBF Mean Swaps Between Failures

MSBLS Microwave Scan(ning) Beam Landing System

MSBLS Microwave Scanning Beam Land Station

MSBLS-GS MSBLS (Microwave Scan(ning) Beam Landing System) Ground Station

MSBM Master of Science in Business Management

MSBME Master of Science in Biomedical Engineering

MSBR Maximum Storage Bus Rate

MSBR Molten Salt Breeder Reactor

MSBTA Ministry of Small Business, Tourism and Culture [British Columbia, Canada]

MSBus Master of Science in Business

MSBusEd Master of Science in Business Education

MSBusMgt Master of Science in Business Management

MSBVW Magnetostatic Backward Volume Wave

MSBY Most Significant Byte

MSC Coastal Minesweeper [US Navy Symbol]

MSC MacNeal-Schwendler Corporation [US Manufacturer of Scientific Software]

MSC Macro Selection Compiler

MSC Maine Safety Council [US]

MSC Main Switching Center

MSC Mankato State College [Minnesota, US]

MSC Manitoba Safety Council [Canada]

MSC Manned Spacecraft Center [now NASA Johnson Space Center, Houston, Texas, US]

MSC Manpower Services Commission [UK]

MSC Mansfield State College [Pennsylvania, US]

MSC Marine Corps School [US]

MSC Maritime Safety Committee [of International Maritime Organization]

MSC Mass Storage Controller

MSC Master of Science in Commerce

MSC Master Sequence Controller

MSC Materials Science Center

MSC Materials Science Center [of Cornell University, Ithaca, New York, US]

MSC Materials Service Center

MSC Mayville State College [North Dakota, US]

MSC Mediterranean Society of Chemotherapy [Italy]

MSC Mediterranean Sub-Commission [of Food and Agricultural Organization]

MSC 2-Mesitylenesulfonyl Chloride

MSC Message Sequence Chart

MSC Message Switching Center

MSC Message Switching Computer

MSC Message Switching Concentration

MSC Metallizing Service Contractors [Name Changed to International Thermal Spray Association]

MSC Meteorological Service of Canada

MSC Meteorological Synthesizing Center [Cooperative Program for Monitoring and Evaluation of Long Range Transmission of Air Pollution in Europe]

MSC Methylamines Sector Group [of Conseil Européen des Fédérations de l'Industrie Chimique, Belgium]

MSC Metropolitan State College [Denver, Colorado, US]

MSC Microscopical Society of Canada

MSC Microsystems Center [UK]

MSC Midwest Science Center [US]

MSC Mile(s) of Standard Cable

MSC Military Sealift Command [US]

MSC Military Staff Committee [of United Nations Security Council]

MSC Millersville State College [Pennsylvania, US]

MSC Ministerio de Sanidad y Consumo [Ministry of Health and Consumption, Spain]

MSC Minnesota Safety Council [US]

MSC Minot State College [North Dakota, US]

MSC Minimum Spanning Circle

MSC Mississippi Southern College [Hattiesburg, US]

MSC Mobile Servicing Center [of Mobile Servicing System]

MSC Mobile Switching Center

MSC Moding Sequencing and Control

MSC Molecular Sieve Chromatography

MSC Monolithic Crystal Filter

MSC Montana State College [Bozeman, US]

MSC Montclair State College [Upper Montclair, New Jersey, US]

MSC Moorhead State College [Minnesota, US]

MSC Morehead State College [Kentucky, US]

MSC Morgan State College [Baltimore, Maryland, US]

MSC Most Significant Character

MSC Motor Speed Changer

MSC Motor Speed Control(ler)

MSC Multistage Compression; Multistage Compressor

MSC Multi-Strand Chain

MSC Multistrip Coupler

MSC Multisystem Coupling

MSC Murray State College [Kentucky, US]

MSc Marine Stratocumulus [Meteorology]

MSc Master of Science

Msc Micrococcus species [Microbiology]

MSCA Mixed Spectrum Critical Assembly [US]

MScA Master of Applied Science

MScAdm Master of Science in Administration

MScBusAdmin Master of Science in Business Administration

MSCC Manned Spaceflight Control Center [US Air Force]

MSCC Multistate Curve Crossing (Method)

MScC Master of Science in Commerce

MScCE Master of Science in Civil Engineering

MScComm Master in Commercial Science

MScCS Master of Science in Computer Sciences

MScD Master of Medical Science

MSCDEX Microsoft Compact Disk Extensions

MSCE Main Storage Control Element

MSCE Master of Science in Civil Engineering

MSCE Master of Science in Clinical Engineering

MSCE Master of Science in Computer Engineering

MScE Master of Science and Engineering [also MScEng]

MSCED Master of Science in Community Economic Development

MScEd Master of Science in Education

MSCEM Master of Science in Civil Engineering Management

MScEng Master of Science in Engineering [also MScE]

MSCer Master of Science in Ceramics

MSCerE Master of Science in Ceramic Engineering [also MSCerEng]

MSCerSc Master of Science in Ceramic Science [also MSCerS, or MSCerSci]

MSCF Materials Science Center Facility [of University of Minnesota, Minneapolis, US]

MSCF Thousand Standard Cubic Feet [also Mscf] [Unit]

MScF Master of Science in Forestry

MSCFD Thousand Standard Cubic Feet per Day [Unit]

MScFE Master of Science in Forest Engineering

MSCFH Thousand Standard Cubic Feet per Hour [Unit]

MSCFM Thousand Standard Cubic Feet per Minute [Unit]

MSChE Master of Science in Chemical Engineering [also MSChemE, or MSChemEng]

MSChemEng Master of Science in Chemical Engineering [also MSChemE, or MSChE]

MSc(Hons) Master of Science (Honours)

MSCI Materials Science Citation Index [of Materials Research Society, US]

MSCI Minnesota Supercomputer Institute [US]

MSCI Mission Scientist [also MSci]

MSCI Molten Steel Coolant Interaction [Metallurgy]

M/SCI Mission/Safety Critical Item

MSc in AgrEng Master of Science in Agricultural Engineering

MSc in CE Master of Science in Civil Engineering

MSc in EE Master of Science in Electrical Engineering

MSc in ME Master of Science in Mechanical Engineering

MSc in Pharm Master of Science in Pharmacy

MSCIS Master of Science in Computer Information Science

MSCIS Master of Science in Computer Information Systems

MsCl Mesyl Chloride

MSCLS Master of Science in Clinical Laboratory Studies

mS/cm millisiemens per centimeter [also mS cm^{-1}]

μS/cm microsiemens per centimeter [also μS cm^{-1}]

μ**s/cm** microsecond per centimeter [also μs cm^{-1}]

MScMed Master of Medical Science

MScMnlProcEng Master of Science in Mineral Processing and Engineering [also MScMnlProcE]

MScMetEng Master of Science in Metallurgical Engineering [also MScMetE]

MSCons Master of Science in Conservation [also MSConv]

MSContEd Master of Science in Continuing Education

MSCP Mass Storage Control Protocol

MSCP Master of Science in Community Planning

MSCP Mean Spherical Candlepower

MSCP Motor Short-Circuit Protector

MSCpE Master of Science in Computer Engineering

MScPharm Master of Science in Pharmacy [also MSPhm]

MScPl Master of Science in Planning

MSCR Measurement/Stimuli Change Request

MScr Machine Screw [also M Scr]

MSCRP Master of Science in City and Regional Planning

MSCRP Master of Science in Community and Regional Planning

MSCS Management Scheduling and Control System

MSCS Mass Storage Control System

MSCS Master of Science in Computer Science

MSCSE Master of Science in Computer Science and Engineering

MScStud Master of Science Studies [also MSciStud, or MScSt]

MSCT Mass Storage Control Table

MScT Master of Science in Teaching

MSCTC Mass Storage Control Table Create

MScTech Master of Science and Technology

MSCU Modular Store Control Unit

MSCV Mass Spectrometric Cyclic Voltamogram

MSCW Marked Stack Control Word

MSC-Xα Multiple Scattering X-α (Method) [Physics]

MSD Doctor of Medical Science

MSD Machining Systems Division

MSD Main Storage Database

MSD MARS Supplemental Data

MSD Mass-Selective Detector

MSD Mass Spectrometric (or Spectral) Detector

MSD Mass-Storage Device

MSD Master of Science in Dentistry

MSD Materials Science Department

MSD Materials Science Division

MSD Materials Science Division [of Argonne National Laboratory, US] [Formerly part of the Materials Science and Technology Division]

MSD Materials Science Division [of ASM International, US]

MSD Maximum Tolerated Dose

MSD Mean Solar Day [Astronomy]

MSD Mean-Square(d) Displacement

MSD Medical Systems Division

MSD Merck Sharp and Dohme of Canada Limited [Pharmaceutical Manufacturer]

MSD Metropolitan Sanitary District

MSD Microgravity Science Division [of NASA Lewis Research Center, Cleveland, Ohio, US]

MSD Microsoft System Diagnostics

MSD Ministry Skills Development

MSD Missile Systems Division [US Air Force]

MSD Modem Sharing Device

MSD Molecular Structure Distribution

MSD Most Significant Digit

MSD Multi-Sensor Display

MSD Multisource Deposition (Method)

MSDB Main Storage Database

MSDC Mass Spectrometry Data Center [of Royal Society of Chemistry, UK]

MSDD Master of Science in Design and Development

MSDent Master of Science in Dentistry

MSDL Microsoft Download Service

MSDN (International) Microbial Strain Data Network

MSDN Microsoft Developer Network

MS-DOS Microsoft–Disk Operating System [also MSDOS, or MS DOS]

MSDR Multiplexed Streaming Data Request

MSDS Marconi Space and Defense System

MSDS Master of Science in Decision Systems

MSDS Material Safety Data Sheet

MSDS Message Switching Data Service

MSDS Microsoft Developer Support

MSDS Multispectral Scanner and Data System

MSD/SMD Materials Science Division/Structural Materials Division (Joint Committee) [of Minerals, Metals and Materials Society of the American Institute of Mining, Metallurgical and Petroleum Engineers, US]

MSDT Maintenance Strategy Diagramming Technique

2,7-MSDTPY 2,7-Bis(methylseleno)-1,6-Dithiapyrene

MSE Magnetostatic Energy

MSE Manufacturing Systems Engineering

MSE Master of Sanitary Engineering

MSE Master of Science Education

MSE Master of Science in Education

MSE Master of Science in Engineering

MSE Master of Software Engineering

MSE Materials Science and Engineering

MSE Mean Square Error

MSE Mechanical Systems Engineering

MSE Microbeam Society of Europe

MSE Midwest Stock Exchange [Chicago, Illinois, US]

MSE Milwaukee School of Engineering [Wisconsin, US]

MSE Modern Shipping Equipment

MSE Macromolecular Science and Engineering

MSE Maintenance Support Equipment

MSE Measuring and Stimuli Equipment

MSE Mechanical Support Equipment

MSE Medical Support Equipment

MSE Mission Staff Engineer

MSE Montreal Stock Exchange [Canada]

MSE Multi-Stage Expansion

MS&E Materials Science and Engineering

MSEA Mass Specific Energy Absorption

MSE A Materials Science and Engineering A (Journal)

MSE B Materials Science and Engineering B (Journal)

MSEC Macromolecular Science and Engineering Center [of University of Michigan, Ann Arbor, US]

MSEC Master of Science in Economic Aspects of Chemistry

MSEC Master Separation Events Controller

MSEC Materials Science and Engineering Council [US]

Msec Megasecond [also Ms]

msec millisecond [also ms]

m/sec meter per second [also MPS, mps, m/s or m s^{-1}]

m²/sec square meter per second [also m²/s, or m³ s^{-1}]

m³/sec cubic meter per second [also m³/s, or m³ s^{-1}]

μsec microsecond [also μs, or mu s]

μ/sec micron(s) per second [also μ/s, or μ s^{-1}]

MSEcon Master of Science in Economics

MSED Materials Science and Engineering Division [of Argonne National Laboratory, Illinois, US]

MSEd Master of Science in Education

MSED Minimum Signal Element Duration

MSED Ministry of State for Economic Development

MSEE Master of Science in Electrical Engineering

MSEE Master of Science in Environmental Engineering

MSEE Mean Square Error Efficiency

MSEH Master of Science in Environmental Health

MSEI Mean Square Error Inefficiency

MSEL Materials Science and Engineering Laboratory [of National Institute for Standards and Technology, Gaithersburg, Maryland, US]

MSEM Master of Science in Engineering Management [also MSEMgt]

MSEM Master of Science in Engineering Mechanics [also MSEMech]

MSEM Master of Science in Engineering of Mines

MSEMech Master of Science in Engineering Mechanics [also MSEM]

MSEMgt Master of Science in Engineering Management [also MSEM]

MSEMPR Missile Support Equipment Manufacturers Planning Reports

MSEMRL Materials Science and Engineering and Materials Research Laboratory [of University of Illinois, US]

MSEng Master of Science in Engineering [also MSEngr, or MSE]

MSEngSc Master of Science in Engineering Sciences

MSEnt Master of Science in Entomology

MSEnvEng Master of Science in Environmental Engineering [also MSEnvE]

MSEnvSc Master of Science in Environmental Science

MSEP Macromolecular Science and Engineering Program [of University of Michigan, Ann Arbor, US]

MSEP Manufacturing Systems Engineering Program [US]

MSEP Materials Science and Engineering Program

MSER Master of Science in Energy Resources

MSERD Ministry of State for Economic and Regional Development

MS/ERDE Ministry of Supply, Explosives Research and Development Establishment [UK]

MSerg Master Sergeant [also M Serg, M Sgt, MSgt, or M/Sgt]

MSES Master of Science in Engineering Science

MSES Master of Science in Environmental Studies

MSES Minnesota Surveyors and Engineers Society [US]

MSESM Master of Science in Environmental Systems Management

MSET Master of Special Education Technology

MSF Fleet Minesweeper [US Navy Symbol]

MSF Magnetostrictive Force

MSF Maintenance Source File

MSF Manned Space Flight

MSF Mass Spectrometry Facility [of University of Texas at Austin, US]

MSF Mass Storage Facility

MSF Master of Science in Forestry

MSF Master of Science in Finance

MSF Master Source File

MSF Metascience Foundation [US]

MSF Michigan Society of Fellows [of University of Michigan, Ann Arbor, US]

MSF Motorcycle Safety Foundation [US]

MSF Multicomponent Spectral Fitting

MSF Multiscan Function

MSF Multistage Flash (Plant)

MSF Murata Science Foundation [Japan]

MSFB Magnetically Stabilized Fluidized Bed

MSFB Minnesota State Forestry Board [US]

MSFC (George C.) Marshall Space Flight Center [of NASA at Huntsville, Alabama, US]

MSFC-SSL Marshall Space Flight Center–Space Sciences Laboratory [of NASA at Huntsville, Alabama, US]

MSFD Mixed Spectral Finite Difference

MSFD Multistage Flash Distillation

MSFET Metal Semiconductor Field-Effect Transistor

MSFH Manned Spaceflight Headquarters [of NASA]

MSFM Master of Science in Financial Management

MSFM Master of Science in Forest Management

MSFN Manned Spaceflight Network [of NASA]

MSFor Master of Science in Forestry

MSFP Manned Space Flight Program

MSFR Minimum Security Function Requirements

MSFS Master of Science in Forensic Science

MSFS Main Steam and Feedwater System

MSFuelSc Master of Science in Fuel Science

MSFVW Magnetostatic Forward Volume Wave

MSG Manufacturers Standard Gauge

MSG Mapper Sweep Generator

MSG Maximum Stable Gain

MSG Mechanical Subsystem Group

MSG Message [Telecommunications]

MSG Miscellaneous Simulation Generator

MSG Mission Support Group

MSG Modular Steam Generator

MSG Monosodium Glutamate [Biochemistry]

.MSG Program Message [File Name Extension] [also .msg]

Msg Message [also msg]

Msg Missing [also msg]

MSGBI Mineralogical Society of Great Britain and Ireland

MSGC Montana Space Grant Consortium [of NASA]

MSGE Master of Science in Geological Engineering [also MSGeoE, MSGeolE, MSGeolEng, or MSGeoEng]

MSGeolE Master of Science in Geological Engineering [also MSGE, MSGeoE, MSGeolEng, or MSGeoEng]

MSGM Master of Science in Government Management

MSGMgt Master of Science in Game Management

MSGS Message Switch

MSgt Master Sergeant [also M Sgt, M/Sgt, MSerg, or M Serg]

MSG/WNG Message Warning

MSH Marginal Stability Hypthesis [Materials Science]

MSH Master of Science in Horticulture

MSH Master of Science in Hygiene

MSH Melanocyte-Stimulating Hormone [Biochemistry]

MSH 2-Mesitylenesulfonic Acid Hydrazide

MSH Mesitylenesulfonyl Hydrazide

MSH Mount Sinai Hospital [New York City, US]

MSHA Master of Science in Hospital Administration

MSHA Marine Safety and Health Administration [US]

MSHA Mine Safety and Health Administration [of US Department of Labor]

MSHA Mine Safety and Health Agency

MSHE Master of Science in Home Economics [also MSHEc]

MSHFBA N-Methyl-N-Trimethylsilyl Heptafluorobutyramide

MSHG Magnetization-Induced Second Harmonic Generation

MSHort Master of Science in Horticulture

MSHR Master of Science in Human Resources

MSHRM Master of Science in Human Resource Management

MSHRMD Master of Science in Human Resource Management and Development

MSHS Master of Science in Health Systems

MSHS Master of Science in Health and Safety

MSHS Montana State Historical Society [US]

M²S-HTSC (International Conference on) Materials and Mechanisms of Superconductivity–High- Temperature Superconductors

MSHyg Master of Science in Hygiene

MSI Maintenance Significant Items

MSI Manned Satellite Inspector

MSI Marine Science Institute [of University of Texas at Austin; located at Port Aransas, US]

MSI Materials Science International Services GmbH [Stuttgart, Germany]

MSI Medium-Scale Integrated; Medium-Scale Integration

MSI Megapound(s) per Square Inch [also msi]

MSI Metal-Semiconductor Interface

MSI Metal-Support Interaction

MSI 110-Foot Minesweeper [US Navy Symbol]

MSI Minnesota Supercomputer Institute [US]

MSI Multi-Spectral Imagery

MSI Museum of Science and Industry [Chicago, US]

M₂Si Metal Silicide

m-Si Monocrystalline Silicon

MSIA Master of Science in Industrial Administration

MSIA Master of Science in International Administration

MSIA Master of Science in International Affairs

MSIA Multispectral Image Analyzer

MSIB Master of Science in International Business

MSIBD Mass-Separated Ion-Beam Deposition

MSID Mass Spectrometric Isotope Dilution

MSID Mean Square Induced Dipole Moment (per Molecule)

MSID Measurement Stimulation Identification

MS-IDA Mass Spectrometric Isotope Dilution Analysis

MSIE Master of Science in Industrial Engineering

MSIE Microsoft Internet Explorer

MSIFO Marine Stratocumulus Intensive Field Observations [of First ISCCP (International Satellite Cloud Climatology Project) Regional Experiment, Phase I]

MSIH Master of Science in Industrial Hygiene

M-Si-N Metal-Silicon-Nitride (Alloy System)

MSILR Master of Science in Industrial and Labour Relations

MSIM Manufacturing System Implementation Methodology

MSIM Master of Science in Information Management

MSIM Minus-S-Impedance Multiplicator

MSIN Mail-Stop Identification Number

MSIN Multi-Stage Interconnection Network

M-Si-N Metal Silicon Nitride (Alloy System)

MS in AE Master of Science in Aerospace Engineering

MS in AgrEc Master of Science in Agricultural Economics

MS in AgrEd Master of Science in Agricultural Education

MS in CerE Master of Science in Ceramic Engineering

MS in Com Master of Science in Communications

MS in EE Master of Science in Electrical Engineering

MS in GE Master of Science in General Engineering

MS in GpEngr Master of Science in Geophysical Engineering

MS in Home Ec Master of Science in Home Economics

MS in HR Master of Science in Human Relations

MS in IM Master of Science in Industrial Management

MS in IndEd Master of Science in Industrial Education

MS in LS Master of Science in Library Science

MS in ME Master of Science in Mechanical Engineering

MS in MinE Master of Science in Mining Engineering

MS in NEd Master of Science in Nursing Education

MS in NT Master of Science in Nuclear Technology

MS in NuclE Master of Science in Nuclear Engineering

MS in Nurs Master of Science in Nursing

MS in Nutr Master of Science in Nutrition

MS in PA Master of Science in Public Administration

MS in PH Master of Science in Public Health

MS in Path Master of Science in Pathology

MS in PetE Master of Science in Petroleum Engineering

MS in SE Master of Science in Sanitary Engineering

MS in TextEng Master of Science in Textile Engineering

MS in Trans Master of Science in Transportation

MSIO Mass Storage Input/Output

MSIP Magnetron Sputtering Ion Plating

MSIPA Master of Science in International Public Administration

MSIR Master of Science in Industrial Relations

MSIRI Mauritius Sugar Industry Research Institute

MSIS Master of Science in Information Science

MSIS Master of Science in Information Systems

MSIS Master of Science in Interdisciplinary Studies

MSIS Manned Satellite Inspection System

MSIS Manufacturing Systems Integration Service

MSIST Master of Science in Information Systems Technology

MSIT Materials Science International Team [of Materials Science International Services GmbH, Stuttgart, Germany]

MSIUS Military Service Institution of the United States (of America)

MSIV Main Steam Isolation Valve

MSIVLCS Main Steam Isolation Valve Leakage Control System

MSJ Master of Science in Journalism

MSJ Mathematical Society of Japan

MSK Manual Select Keyboard

MSK Mask [also Msk, or msk]

MSK Minimum (Phase) Shift Keying

MSK Muscular-Sketetal Effects (of Hazardous Materials)

m/s/K² meter per second per Kelvin [also m s^{-1} K^{-2}]

m³/s²·kg cubic meter per second squared kilogram [also m³/(s²·kg), m³ s^{-2} kg^{-1}]

m² s K/J square meter-second-Kelvin per Joule [also m² s K J^{-1}]

MSKM Minimum Shift Keyed Modulation

MSL Machine Specification Language

MSL Map Specification Library

MSL Marine Sciences Laboratory [Seqium, Washington, US]

MSL Mass Spectrometry Laboratory [of University of Kansas, Lawrence, US]

MSL Master of Science in Linguistics

MSL Master of Science Librarianship

MSL Materials Science Laboratory

MSL Mean Sea Level [also msl]

MSL Mechanical System(s) Laboratory

MSL Mineral Sciences Laboratories [of Canada Center for Mineral and Energy Technology, Natural Resources Canada]

MSL Minesweeping Launch [US Navy Symbol]

MSL Mirrored Server Link

MSL Missile [also Msl, or msl]

MSL Missouri State Library [Jefferson City, US]

MSL Molecular Structure Laboratory [of University of Arizona, Tucson, US]

msl mean sea level [also MSL]

MSLD Mass Spectrometer Leak Detector

MSLDH Monomer Sequence Length Distribution of the Hard Segment [Polymer Engineering]

MSLIS Mita Society of Library Information Science [Japan]

MSLOC Million Source Lines of Code

MSLS Master of Science in Library Science

MSM Mackay School of Mines [of University of Nevada at Reno, US]

MSM Magnetostrictive Material

MSM Manned Support Module

MSM Master of Science in Management

MSM Master of Service Management

MSM Master/Slave Manipulator

MSM Mean Spherical Model

MSM Memory Storage Module

MSM Message Stream Modification

MSM Metal-Semiconductor-Metal

MSM Methylsulfonylmethane

MSM Minus-S-Multiplicator

MSM Modified Source Multiplication

MSM Montana School of Mines [Butte, US]

MSM Mystic Seaport Museum [US]

MSM Thousand Feet Surface Measure [also Msm]

MS/m Megasiemens per Meter [also MS m^{-1}] [Unit]

MSMA Margarine and Shortening Manufacturers Association [UK]

MSMA Moist Soil Management Advisor

MSMA Monosodium Acid Methanearsonate

MSMA Murashige and Skoog Shoot Multiplication, Medium A [Biochemistry]

MSMA Murrell-Shaw-Musher-Amos (Theory) [Physics]

MSMA 3-(Trimethoxysilyl)propyl Methacrylate

MSMatEng Master of Science in Materials Engineering [also MSMatE]

MSMatSc Master of Science in Materials Science [also MSMatSci, or MSMatS]

MSMatScEng Master of Science in Materials Science and Engineering [also MSMatSciEng or MSMatSE]

MSMB Mass-Spectrometric Molecular Beam

MSMB Mechanical Standards Management Board [of American National Standards Institute, US]

MSMB Murashige and Skoog Shoot Multiplication, Medium B [Biochemistry]

mSmBA m-Sulfamyl Benzoic Acid

MSMC Master of Science in Marketing Communications

MSMC Master of Science in Mass Communications

MSMC Mount Sinai Medical Center [New York City, US]

MSMC Murashige and Skoog Shoot Multiplication, Medium C [Biochemistry]

MSMCS Master of Science in Management and Computer Science

MSME Master of Science in Mechanical Engineering

MSME Materials Science and Materials Engineering [also MS&ME]

MS&ME Materials Science and Mineral Engineering [also MSME]

MSMEA Multiwall Sack Manufacturers Employers Association [UK]

MSMER Master of Science in Management/Employee Relations

MSMet Master of Science in Metallurgy

MSMetE Master of Science in Metallurgical Engineering [also MSMetEng]

MSMetMatS Master of Science in Metallurgy and Materials Science [also MSMetMatSc]

MSMetr Master of Science in Meteorology

MSMetS Master of Science in Metallurgical Sciences [also MSMetSc]

MSMFSE Master of Science in Manufacturing Systems Engineering

MSMgt Master of Science in Management

MSMgtE Master of Science in Management Engineering

MSMinE Master of Science in Mining Engineering [also MSMinEng]

MSMinrlE Master of Science Mineral Engineering [also MSMinrlEng]

MSMIS Master of Science in Management Information Systems

MSMLCS Mass Service Main Line Cable System

MSML Multiscalar Microlaminate

MSMLT Master of Science in Medical Laboratory Technology

MSMM (International Conference on) Modeling and Simulation in Metallurgical Engineering and Materials Science

MSMM Master of Science in Manufacturing Management

mS/mm millisiemens per millimeter [also mS mm^{-1}]

MSMnlProcEng Master of Science in Mineral Processing and Engineering [also MSMnlProcE]

MSMO Murashige and Skoog Basal Salts with Minimal Organics [Biochemistry]

MSMOB Master of Science in Management and Organizational Behavior

MSMPA Master of Science in Management/Public Administration

MSMPD Metal-Semiconductor-Metal Photodetector [also MSM-PD, or MSM PD]

MSMPD Metal-Semiconductor-Metal Photodiode [also MSM-PD, or MSM PD]

MSMR Multispectral–Multiresolution (Method)

MSMS Mass Spectrometry Coupled Mass Spectrometry

MSMS Master of Science in Management Science

MS/MS Multistage (or Tandem) Mass Spectrometry

MS/MS Material Science and Manufacturing in Space

MSMSA Master of Science in Management Systems Analysis

MSMSc Master of Science in Management Science

MSMSE Master of Science in Manufacturing Systems Engineering

MSMSE Master of Science in Materials Science and Engineering

MSMT Master of Science in Medical Technology

MSMtE Master of Science in Materials Engineering [also MSMtEng]

MSMV Monostable Multivibrator

MSN Magnesium Silicon Nitride

MSN Master of Science in Nursing

MSN Microsoft Network [of Microsoft Corporation, US]

MSN Microwave Systems News (and Communications Technology) [Journal published in the US]

MSN Mission [also Msn, or msn]

MSN Microw Syst News Commun Technol MSN Microwave Systems News and Communications Technology [Journal published in the US]

MSNE Master of Science in Nuclear Engineering

MSNF Multisystem Network(ing) Facility (Software)

MSNI 1-(Mesitylene-2-Sulfonyl)-4-Nitroimidazole

MSNJ Medical Society of New Jersey [US]

MSNS Master of Science in Natural Science

MSNS Mining Society of Nova Scotia [Canada]

MSNS Multiple Sclerosis National Society [US]

MSNT 1-(Mesitylene-2-Sulfonyl)-3-Nitro-1,2,4-Triazole

MSNucE Master of Science in Nuclear Engineering [also MSNucEng]

MSO Magnesium Silicate (Ceramics) [$MgSiO_3$]

MSO Magnetostriction Oscillator

MSO Mesityl Oxide

MSO Methionine Sulfoximine

MSO Model for Spare(s) Optimization

MSO Mount Stromlo Observatory [near Canberra, Australia]

MSO Multiple Systems Operator [Telecommunications]

MSO Multistage Optimization

MSO Ocean Minesweeper [US Navy Symbol]

MS&O Mission Systems/Operations (Utilization Directorate) [of Canadian Space Agency]

MSOAc Methylsulfonyl Acetic Acid

MSOB Manned Spacecraft Operations Building [now Operation(s) and Checkout (Building), NASA]

MSOB Master of Science in Organizational Behavior

MSOC (2-Methanesulfonyl)ethoxycarbonyl; 2-Methylsulfonylethyloxycarbonyl

MSOCC Multisatellite Operations Control Center

N-MSOC N-(2-Methanesulfonyl)ethoxycarbonyl

N-MSOC L-Phe N-(2-Methanesulfonyl)ethoxycarbonyl-L-Phenylalanine [Biochemistry]

N-MSOC L-Pro N-(2-Methanesulfonyl)ethoxycarbonyl-L-Proline [Biochemistry]

MSOD Master of Science in Organizational Development

MSOE Master of Science in Ocean Engineering

MSOE Milwaukee School of Engineering [Wisconsin, US]

MSOFC Monolithic Solid Oxide Fuel Cell

MSOIN Minor Subcontractor, or IDWA (Interdivisional Work Authorization) Notification

MSOM Master of Science in Organization and Management

MSOM Master Station Output Module

MSOR Master of Science in Operations Research [also MSOpR, or MSOpRes]

MsOR Mesylate

MSOrnHort Master of Science in Ornamental Horticulture

MSORS Mechanized Sales Office Record System

MSOS Mass Storage Operating System

MSOS Modified Solid-on-Solid (Model) [Solid-State Physics]

MSOSH Master of Science in Occupational Safety and Health

MSOT Master of Science in Occupational Therapy

MSP Macroscopic Shrinkage Porosity [Metallurgy]

MSP Maintenance Support Plan

MSP Maintenance Surveillance Procedure

MSP Mass Storage Processor

MSP Materials Synthesis and Processing

MSP Microgravity Sciences Program

MSP Medium Speed Printer

MSP Microelectronics Support Program

MSP Microscope System Processor

MSP Modular Switching Peripheral

MSP Modular System Program

MSP Molecular Science Program

MSP Monitor Signal Processor

MSP Monobasic Sodium Phosphate; Monosodium Phosphate

MSP Multipolar Surface Plasmon [Solid-State Physics]

MSP Multi-Scale Profiler

MSP Multispectral Projector

MSP Munitions Supply Program

.MSP Microsoft Paint [File Name Extension] [also .msp]

MS&P Materials Synthesis and Processing (Initiative) [of National Science Foundation, US]

Msp Moraxella species [Microbiology]

MSPE Matrix Solid-Phase Dispersion

MSPetEng Master of Science in Petroleum Engineering [also MSPetE]

MSPH Master of Science in Public Health

MSPharm Master of Science in Pharmacy [alsp MSPhar]

MSPHE Master of Science in Public Health Engineering

MSPHEd Master of Science in Public Health Education

MSPHR Master of Science in Pharmacy

MSPhys Master of Science in Physics

MSPhysOp Master of Science in Physiological Optics

MSPlmSc Master of Science in Polymer Science

MSPLT Master Source Program Library Tape

MSPN Multi-Stage Packet Network

MSPoly Master of Science in Polymers

MSPP Ministère de la Santé Publique et de la Population [Ministry of Public Health and Population, France]

MSPR Master Spares Positioning Resolver

MSPR Minimum Space Platform Rig [Oil Exploration]

MSPS Master of Science in Planning Studies

MSR Machine Status Register

MSR Machine Stress Rating

MSR Magnetic Slot Reader

MSR Magnetically-Induced Super-Resolution

MSR Magnetic Sound Recorder; Magnetic Sound Recording

MSR Main Storage Ring [Nuclear Physics]

MSR Malaysian Society of Radiographers

MSR Management Summary Report

MSR Mark Sense Reading

MSR Mark Sheet Reader

MSR Mass Storage Resident

MSR Material Status Report

MSR Memory Select Register

MSR Messen–Steuern–Regeln [Measuring–Controlling–Regulating; German Publication]

MSR Meta Stepwise Refinement (Method)

MSR Misroute [also Msr, or msr]

MSR Missile Site Radar

MSR Module Support Rack

MSR Molten-Salt (Nuclear) Reactor

MSR Multiple Specular Reflectance

MSR Muon Spin Resonance [also μSR]

MS(R) Master of Science (Research)

μSR Muon Spin Relaxation [also μ-SR]

μSR Muon Spin Resonance [also μ-SR, or MSR]

μSR Muon Spin-Rotation

μ^+SR (Positively Charged) Muon Spin Relaxation [also μ^+-SR]

MSRA Metal Roofing Systems Association [US]

MSRad Master of Science in Radiology

MSRadSci Master of Science in Radiation Science; Master of Science in Radiological Science [also MSRadSc]

MSRC Master of Science in Resource Conservation

MSRC Molecular Science Research Center [of Pacific Northwest Laboratory, Richland, Washington, US]

MSRD Mean-Square Relative Displacement

MSRE Molten-Salt Reactor Experiment [of Oak Ridge National Laboratory, Tennessee, US]

MSREF Metal Scrap Research and Education Foundation

MSR-FTIR Multiple Specular Reflectance Fourier Transform Infrared Spectroscopy

MSRI Mass Spectroscopy of Recoiled Ions

MSRL Marine Science Research Laboratory [of Memorial University of Newfoundland, St. John's, Canada]

MSRM Main Steam Radiation Monitor

MSR, Mess Steuern Regeln MSR Messen–Steuern–Regeln [Measuring–Controlling–Regulating; German Publication]

MSRMP Master of Science in Radiological Medical Physics

MSRP Manufacturer's Suggested Retail Price

MSRP Missile, Space and Range Pioneers [US]

MSRS Main Steam Radiation System

MSRS Multiple Stylus Recording System

MSRT Missile System Readiness Test

MSS Managed Storage System

MSS Manuscripts [also mss]

MSS Mass Spectroscopy Society

MSS Mass Storage System

MSS Master of Social Science

MSS Main Steam System

MSS Main Support Structure

MSS Maintenance Status System

MSS Management Science Systems

MSS Management Statistics Subsystem

MSS Manufacturers Standardization Society (of the Valve and Fittings Industry) [US]

MSS Manuscripts [also Mss, or mss]

MSS Martensitic Stainless Steel

MSS Mass Storage Subsystem

MSS Mass Storage System

MSS Master of Science in Safety

MSS Materials Science Section [of US Naval Postgraduate School, Monterey, California]

MSS Mechanical Support Systems

MSS Metal Science Society [Czech Republic]

MSS Military Supply Standard [of US Department of Defense]

MSS Mission Specialist Station

MSS Mission Status Summary

MSS Mixed Spectrum Superheater [Nuclear Engineering]

MSS Mobile Satellite Service

MSS Mobile Satellite System

MSS Mobile Service Structure

MSS Mobile Servicing Station

MSS Mobile Servicing System [of Canadian Space Agency]

MSS Mobile Service Structure

MSS Mobile Space Station

MSS Multiprotocol Switched Services

MSS Multi-Spectral Scanner (System) [of NASA Landsat Program]

MSS Multitask Single-Stream System

MSS Music Sensor System

MS/s Megasamples per second

mss manuscripts [also MSS]

MSSA Midland Steel Stockholders Association [UK]

MSSC Mass Storage System Communication

MSSC Missouri Southern State College [Joplin, US]

MSSC Multiple Sclerosis Society of Canada

MSSc Master of Sanitary Science

MSSCE Mixed Spectrum Superheater Critical Experiment [Nuclear Engineering]

MSSCS Manned Space Station Communications System

MSSE Master of Science in Sanitary Engineering [also MSSEng]

MSSE Master of Science in Systems Engineering

MSSEL Michigan Solid-State Electronics Laboratory [US]

MSS EMC Mobile Space Station Electromagnetic Capability

MSSG Message [also Mssg; or mssg]

MSSJ Mass Spectroscopy Society of Japan

MSSM Master of Science in Science Management

MSSM Master of Science in Systems Management

MSSM Mount Sinai School of Medicine [of City University of New York, US]

MSSR Mars Soil Sample Return

MSSR Mixed Spectrum Superheat (Nuclear) Reactor

MSSR Moldavian Soviet Socialist Republic [USSR]

MSSR Monopulse Secondary Surveillance Radar

MSSS Mass Spectral Search System; Mass Spectrometry Search System

MSSS Multisatellite Support System

MSST Ministry of State for Science and Technology [Canada]

MSStat Master of Science in Statistics

MSStEng Master of Science in Structural Engineering [also MSStE]

MSSW Magnetostatic Surface Wave

MSSySc Master of Science in Systems Science

MSSysE Master of Science in Systems Engineering [also MSSysEng]

MST Magnetization Step

MST Magnetostrictive Transducer

MST Manufacturers Sales Tax

MST Master of Science and Technology

MST Master of Science in Teaching

MST Master Station [Radio Navigation]

MST Materials Science and Technology

MST Materials Shaping Technology

MST Mathematical Systems Theory

MST Mathematics, Science and Technology

MST Mean Solar Time [Astronomy]

MST Measurement Science and Technology

MST Measurement Science and Technology [Journal of the Institute of Physics, UK]

MST Measurement Status Table

MST Measurement Systems Test-Bed

MST Medium-Scale Technology

MST 1-(Mesitylene-2-Sulfonyl)-1H-1,2,4-Triazole

MST Microphase Separation Transition

MST Minimum Spanning Tree (Method) [Metallurgy]

MST Ministry of Science and Technology [New Delhi, India]

MST Mission Science Team

MST Mission Selection Team

MST Mobile Service Tower

MST Module Service Tool

MST Monolithic Systems Technology

MST Mountain Standard Time [Greenwich Mean Time +07:00] [also mst]

MST Multiple Scattering Theory [Physics]

MST Multisubscriber Time Sharing

MST Museum of Science and Technology [Birmingham, UK]

MS&T Manufacturing Science and Technology (Program) [of US Department of Defense]

MS&T Methodical Structures and Textures

MSTA Maryland State Teachers Association [US]

MStat Master of Statistics

MSTD Materials Science and Technology Division

MSTD Materials Science and Technology Division [of American Nuclear Society, US]

MSTD Materials Science and Technology Division [of Argonne National Laboratory, Illinois, US] [Now Subdivided into the Materials Science Division and the Materials and Components Division]

MSTD Materials Science and Technology Division [of Los Alamos National Laboratory, New Mexico, US]

MSTD Materials Science and Technology Division [of Naval Research Laboratory, Washington, DC, US]

MSTD Mean-Stress-Triggered Dilatation

MSTDS Mass Spectrometry Thermal Desorption Spectra

MSTE Master of Science in Transportation Engineering

MSTE Mathematics, Science and Technology Education

MSTEd Master of Science in Technical Education

MSText Master of Science in Textiles

MSTextChem Master of Science in Textile Chemistry

MSTFA N-Methyl-N-(Trimethylsilyl)trifluoroacetamide

MSTG Mass Storage Task Group [of CODASYL (Conference on Data System Languages)]

MSTGS Marine Sciences and Technologies Grants Scheme [Australia]

MsTh Mesothorium [also Ms-Th]

MsTh$_1$ Mesothorium 1 [also Ra-228, ^{228}Ra or Ra^{228}]

MsTh$_2$ Mesothorium 2 [also Ac-228, ^{228}Ac, or Ac^{228}]

MSTJ Multilayer Superconducting Tunnel Junction

MSTK Ministry of Science and Technology of Korea

MSTM Master of Science in Technology Management

MSTM Master of Science in Teaching Mathematics

MSTP Manned Space Transport Program [of European Space Agency]

Mstr Master [also mstr]

Mstr Moisture [also mstr]

MSTransE Master of Science in Transportation Engineering [also MSTransEng]

MSTrPl Master of Science in Transportation Planning

MSTS Mean Standard Toxicity Score

MSTS Microsoft Terminal Server

MSTS Military Sea Transport(ation) Service [of US Navy]

MSTS Multiphase Subsurface Transport Simulator

MSTS Multisubscriber Time Sharing System

MSU Main Store Update

MSU Main Switching Unit

MSU Maintenance and Status Unit

MSU Maintenance Signal Unit

MSU Management Signal Unit

MSU Mass Storage Unit

MSU Memphis State University [Tennessee, US]

MSU Measuring Stimuli Units

MSU Message Switching Unit

MSU Metallic Service Unit

MSU Metropolitan State University [Saint Paul, Minnesota]

MSU Michigan State University [East Lansing, US]

MSU Microwave Sounding Unit

MSU Mindanao State University [Iligan City, Philippines]

MSU Mississippi State University [State College, US]

MSU Montana State University [Missoula, US]

MSU Morgan State University [Baltimore, Maryland, US]

MSU Moscow State University [Russia]

MSU Multiblock Synchronization Unit

MSU Multiple Signal Unit

MSU Murray State University [Kentucky, US]

MSU CFMR Michigan State University–Center for Fundamental Materials Research [East Lansing, US]

MSU CSM Michigan State University–Center for Sensor Materials [East Lansing, US]

MSUD Master of Science in Urban Design

MSUDC Michigan State University Discrete Computer

MSUESM Master of Science in Urban Environmental Systems Management

MSU-IIT Mindanao State University–Iligan Institute of Technology [Philippines]

MSUP Master of Science in Urban Planning

MSUP Michigan State University Press [East Lansing, US]

MSUS Master of Science in Urban Studies

MSV Maize Streak Virus

MSV Mass Storage Volume

MSV Mean Square Velocity

MSV Mean Square Voltage

MSV Monitored Sine Vibration (Test)

MSV Mouse Sacroma Virus

MSV Multifunction Support Vessel

mSv millisievert [Unit]

μSv microsievert [Unit]

mSv/y millisievert(s) per year [also μSv/yr]

μSv/y microsievert(s) per year [also μSv/yr]

MSVC Master of Science in Vocational Counseling

MSVC Mass Storage Volume Control

MSVI Mass Storage Volume Inventory

MSVP Master (Space) Shuttle Verification Plan

mSv/yr millisievert per year [Unit]

MSW Machine Status Word

MSW Magnetostatic Wave

MSW Marine Shear Wave (System)

MSW Master of Social Work

MSW Master Switch

MSW Microswitch

MSW Municipal Solid Waste

MSWL Municipal Solid Waste Landfill(s)

MSWLF Municipal Solid Waste Landfill(s)

MSWRE Master of Science in Water Resource Engineering

MSX Midcourse Space Experiment [of Infrared Processing and Analysis Center, NASA-Jet Propulsion Laboratory, Pasadena, California, US]

MSXα Multiple Scattering X Alpha [Physics]

MSY Maximum Sustainable (or Sustained) Yield

MSYNC Master Synchronization

MSYT Managed Stand Yield Tables [Forestry]

MSZ Magnesia-Stabilized Zirconia

MSZH Magyar Szabvarnyugyi Hivatal [Hungarian Office for Standardization]

MT Machine Technology

MT Machine Theory

MT Machine Tool

MT Machine Translation

MT Magnetic Core/Transistor (Module)

MT Magnetic-Particle Testing

MT Magnetic Tape

MT Magnetic Test(ing)

MT Magnetic Tube

MT Magnetization Transfer

MT Magnetotelluric(s)

MT Magnetotransport

MT Magpighian Tube [Entomology]

MT Mail Transfer

MT Maki-Thompson (Process) Physics]

MT Male Tee (Pipe Section)

MT Malta [ISO Code]

MT Management Technologist; Management Technology

MT Manganese Telluride (Semiconductor)

MT Manganese Titanate

MT Manual Transmission

MT Manufacturing Technologist; Manufacturing Technology

MT Marine Technologist; Marine Technology

MT Martensitic Transformation [Metallurgy]

MT Martensitic Transition [Metallurgy]

MT Masking Tape

MT Massage Therapist; Massage Therapy

MT Massive Transformation

MT Mass Transfer

MT Mass Transport

MT Master of Teaching

MT Master of Textiles

MT Master Terminal

MT Master Timer

MT Master Tool

MT Materials Technologist; Materials Technology

MT Materials Testing

MT Materials Transport

MT Matrix Transfer

MT Maximum Torque

MT Mean Tide

MT Mean Time [Astronomy]

MT Measured Time

MT Measurement Technologist; Measurement Technology

MT Measuring Device for Travel Time

MT Mechanical Technician

MT Mechanical Technologist; Mechanical Technology

MT Mechanical Test(ing)

MT Mechanical(ly) Texture(d) (Surface) [Disk Drives]

MT Mechanical Time

MT Mechanical Translation

MT Mechanical Transport

MT Medical Technologist; Medical Technology

MT Medical Textile

MT Medium-Frequency Telegraphy

MT Medium Temperature

MT Medium Thermal (Carbon Black)

MT Megaton [also Mt, or Mton]

MT Melting Temperature [also mt]

MT Melt-Through [Nuclear Engineering]

MT Mercury Telluride

MT Message Table

MT Message Transfer

MT Metallurgical Technologist; Metallurgical Technology

MT Metallurgical Thermodynamics

MT Metal Technologist; Metal Technology

MT Metatarsal; Metatarsus [Anatomy]

MT Metathesis [Chemistry]

MT Metering Time

MT Methyltyrosine [Biochemistry]

MT Metric Ton [Unit]

MT Microtubule Bundle [Cytology]

MT Micro-Twin [Metallurgy]

MT Mining Technologist; Mining Technology

MT Millon Test [Chemistry]

MT Mission Time

MT Mission Trajectory

MT Mobile Terminal

MT Moderately Tolerant; Moderate Tolerance

MT Moderate Temperature

MT Mode Transducer

MT Modified Tape

MT Moessbauer Transition [Spectroscopy]

MT Molecular Theory [Statistical Mechanics]

MT Montana [US]

MT Mother Twin [Metallurgy]

MT Motor Tanker

MT Mountain Time [also mt]

MT Mucous Tissue [Anatomy]

MT Muffin Tin (Sphere) [Materials Science]

MT Multiple Transfer

MT Multitask(ing)

MT Muscle Tissue [Anatomy]

MT Music Technologist; Music Technology

MT Structural Tee (Cut from M-Type Steel Shape) [AISI/AISC Designation]

α-MT α-Methyltyrosine [Biochemistry]

M(T) Temperature-Dependent Magnetization [Symbol]

M$_s$(T) Temperature-Dependent Saturation Magnetization

M/T Mail Transfer

M-T Magnetization–Temperature (Curve)

M-T Matrix-Twin (Grain Boundary) [Materials Science]

M-T Monoclinic-to-Tetragonal (Phase Transformation) [also M→T] [Materials Science]

Mt Megaton [also MT, or Mton]

Mt *(Mont)* – French for "Mount(ain)"

Mt Mount(ain) [also mt]

M(t) Time-Dependent Magnetization

(t) Weight-Average Cluster Mass [Symbol]

mT maritime tropical air [Meteorology]

mT millitesla [Unit]

m⁻¹ T⁻¹ one per meter per Tesla [also 1/(m·T)]

mt metric ton [Unit]

.mt Malta [Country Code/Domain Name]

m(t) remanant magnetic moment [Symbol]

μ(T) carrier mobility (in solid-state physics) [Symbol]

MTA Machining Technology Association [of Society of Manufacturing Engineers, US]

MTA Magnetic Tape Accessory

MTA Magnetothermal Analysis

MTA Magyar Tudományos Akadémia [Hungarian Academy of Sciences, Budapest]

MTA Maintenance Task Analysis

MTA Major Test Article

MTA Major Trading Areas [of Rand McNally, US]

MTA Manitoba Trucking Association [Canada]

MTA Manual Target Acquisition

MTA Marine Trades Association [UK]

MTA Maritime Transit Association [now International Marine Transport Association, US]

MTA Mass-Spectroscopic Thermal Analysis

MTA Mass Thermal Analysis

MTA Mass Transfer Accelerator [Physics]

MTA Material Transfer Agreement

MTA Medium Temperature Anneal(ing) [Metallurgy]

MTA *(Medizinisch-technische(r) Assistent(in))* – German for "Medical Technician"

MTA Message Transfer Agent

MTA Metallurgical Transaction A [of ASM International, US]

MTA 5'-S-Methyl-Thioadenosine [Biochemistry]

MTA Metriol Triacetate

MTA Metropolitan Transport Authority [US]

MTA Mica Trades Association [UK]

MTA Military Training Area

MTA Mobility Test Article

MTA Modified Tape-Armored (Cable)

MTA Motion/Time Analysis [Industrial Engineering]

MTA Moving Target Analyst System

MTA Moving Target Analyzer

MTA Multiple Terminal Access

MTA Multiterminal Adapter

mT/A millitesla(s) per ampere [also mT A⁻¹]

MTAA Motor Trades Association of Australia

MTAC Mathematical Tables and other Aids to Computation

MTAE Message Transfer Agent Entity

mT/A m⁻¹ millitesla per Ampere per meter [also mT/A/m]

MTAPA Metal Trades Accident Prevention Association

MTA/SME Machining Technology Association/Society of Manufacturing Engineers [US]

MTB Maintenance of True Bearing [Navigation]

MTB Materials Testing Branch

MTB Materials Transportation Bureau [US]

MTB Metallic Test Bus

MTB Metallurgical Transaction B [of ASM International, US]

MTB Methyl-tert-Butyl Ether

MTB Methyl Thymol Blue

MTB Mineral Technology Branch [of Canada Center for Mineral and Energy Technology, Natural Resources Canada]

MTB Motor Torpedo Boat

MTB Mueller-Takashige-Bednorz [Physics]

MTBB Mean Time Between Breakdowns

MTBC Mean Time Between Collisions [Physics]

MTBD Mean Time between Demand

MTBD 7-Methyl-1,5,7-Triazabicyclo[4.4.0]dec-5-ene

MTBCF Mean Time between Confirmed Failures

MTBD Mean Time between Defects

MTBD Mean Time between Degradations

MTBE Mean Time between Errors

MTBE Methyl-tert-Butyl Ether

MTBF Mean Time between Failures

MTBI Mean Time between Interrupts

MTBJ Mean Time Between Jams

MT Black Medium-Thermal Black [also MT black]

MTBM Mean Time between Maintenance

MTBM Mean Time between Malfunctions

MTBMA Mean Time between Maintenance Actions

MTBO Mean Time between Outages

MTBR Mean Time Between Repairs

MTBR Mean Time between Replacement

MTBSTFA Methyl-tert-Butyldimethylsilyl Trifluoroacetamide

MTBSTFA N-Methyl-N-(tert-Butyldimethylsilyl) trifluoroacetamide

MTC 2,3-Diphenyl-5-Methyl-2H-Tetrazolium Chloride

MTC Machine Tool Conference

MTC Magnetics Technology Center [of Carnegie Mellon University, Pittsburgh, Pennsylvania, US]

MTC Magnetics Technology Centre [Singapore]

MTC Magnetic Tape Cartridge

MTC Magnetic Tape Cassette

MTC Magnetic Tape Channel

MTC Magnetic Tape Control(ler)

MTC Maintenance Time Constraint

MTC Main Trunk Circuit

MTC Manual Traffic Control

MTC Manufacturing Technology Center [of National Research Council of Canada]

MTC Manufacturing-to-Cost (Process)

MTC Maritime Transport Committee [of Organization for Economic Cooperation and Development, France]

MTC Master of Textile Chemistry

MTC Master Tape Control

MTC Master Thrust Control(ler)

MTC Materials Technology Center
MTC Measurement Technology Conference
MTC Memory Test Computer
MTC Message Transmission Controller
MTC Metallurgical Thermochemistry
MTC Metalworking Trade Coalition [US]
MTC N-Methyl-O-Ethylthiocarbamate
MTC Michigan Tourist Council [US]
MTC Microelectronics Technology Center [Newbury Park, California, US]
MTC Midwestern Telecommunication Conference [US]
MTC Million Tons of Coal [Unit]
MTC Ministry of Transportation and Communications
MTC Missile Test Center
MTC Mission and Traffic Control
MTC Missouri Tourist Commission [US]
MTC Moderator Temperature Coefficient [Nuclear Engineering]
MTC Modulation Transfer Curve
MTC Momentum Transfer Control(ler)
MTC Montreal Transport Commission [Canada]
MTC Multimatic Technical Center
M-T-C Moore-Thompson-Clinger Limited [Canadian Pharmaceutical Manufacturer]
MTCA Minimum Terrain Clearance Altitude
MTCA Ministry of Transport and Civil Aviation [UK]
MTCA Monitor and Test Control Area
MTCA Multiple Terminal Communications Adapter
MTCC Master Timing and Control Circuit
MTCC Metropolitan Toronto Convention Center [Canada]
MTCC Million Tons of Clean Coal [Unit]
MTCD Microvolume Thermal Conductivity Detector
MTCE Million Tons of Coal Equivalent
MT&CE Materials Testing and Characterization Equipment
Mtce Maintenance [also mtce]
MTCF Mean Time to Catastrophic Failure
MTCF Multilateral Technical Cooperation Fund [of NATO Central European Treaty Organization]
MTCH Magnetic Tape Channel
MTCNQ Metal Tetracyanoquinodimethane
MTC/NRC Manufacturing Technology Center of the National Research Council [Canada]
MTCOECD Maritime Transport Committee of the Organization for Economic Cooperation and Development [France]
MTCR Missile Technology Control Regime
MTCS Masuda Tracking and Communication Station [of National Space Development Agency of Japan]
MTCS Maximum Teleprocessing Communications System
MTCU Magnetic Tape Control Unit
MTCV Main Turbine Control Valve
MT-CVD Medium Temperature–Chemical Vapor Deposition
MTCXO Mathematically Temperature-Compensated Crystal Oscillator
MTD Magnetic Tape Drive
MTD Manufacturing Technology Division [US Air Force]

MTD Marine Technology Directorate [UK]
MTD Martensitic Transformation Diagram [Metallurgy]
MTD Master of Transport Design
MTD Maximum Tolerated (or Tolerable) Dose
MTD Mean Temperature Difference
MTD Methyltetracyclododecane
MTD Microwave Technology Division [of Hewlett Packard at Santa Rosa, California]
MTD Minimal Toxic Dose
MTD Moscow Time Daylight [Greenwich Mean Time –02:00]
MTD Moving Target Detector
MTD M-Tolylenediamine
MTD Multiple Target Detection
MTD Multiple Tile Duct
mtd mounted
MTDA Methyl Trimethylsilyl Dimethylketene Acetal
MTDATA Metallurgical and Thermochemical Databank [of National Physical Laboratory, Teddington, Middlesex, UK]
MTDM Machinery, Tool, Die and Mold (Industry Sector)
MTDR Machine Tool Design and Research
MTDS Manufacturing Test Data System
MTDS Marine Corps Tactical Data System [of US Navy]
MTDSK Magnetic Tape Disk
MTDTPY Bis(methylthio)dithiapyrene
2,7-MTDTPY 2,7-Bis(methylthio)-1,6-Dithiapyrene
3,8-MTDTPY 3,8-Bis(methylthio)-1,6-Dithiapyrene
MTE Maintenance Test Equipment
MTE Materials Testing and Evaluation [also MT&E]
MTE Maximum Tracking Error
MTE Maximum Transferable Energy
MTE Michigan Test of English
MTE Ministry of Treasury and Economics [Canada]
MTE Multiple Terminator Emulator
MTE Multisystem Test Equipment
MT&E Materials Testing and Evaluation [also MTE]
M&TE Measuring and Test Equipment
MTEC Maintenance Test Equipment Catalog
Mtech Master of Technology
MTEE Maintenance Test Equipment–Electrical
MTEEC Maintenance Test Equipment–Electronic
MTEF Maintenance Test Equipment–Fluid(ic)
MTEM Maintenance Test Equipment Module
MTEM Maintenance Test Equipment–Mechanical
MTEO Maintenance Test Equipment–Optical
MTEOS Methyltriethoxysilane [also MTES]
MTF Magnetic Thin Film
MTF Main Time to Failure
MTF Manual Transmission Fluid
MTF Manufacturing Technology Fellowship (Program) [US]
MTF Mean Time to Failure
MTF Mechanical Time Fuse
MTF Medium Time to Failure
MTF Metastable Time-of-Flight Technique

MTF Microsoft Tape Format

MTF Mississippi Test Facility [of NASA; now National Space Technology Laboratories]

MTF Modulation Transfer Function

MTF Multilayer Thin Film

MTFA Modulation Transfer Function Analyzer

MTFE Mercury Thin-Film Electrode

MTFF Mean Time to First Failure

MTFF Mileage to First Failure [Automobiles]

MTFM Magnetic Thin-Film Memory

mTFMa m-Trifluoromethylaniline

MTFO Modular Training Field Option

MTG Main Traffic Group

MTG Melt-Textured Growth (Method); Melt-Textured Grown [Solid-State Physics]

MTG Methane-to-Gasoline

MTG Methanol-to-Gasoline (Process)

MTG Methyl β-D-Thiogalactoside [Biochemistry]

MTG Multiple-Trigger Generator

Mtg Meeting [also mtg]

Mtg Mortgage [also mtg]

Mtg Mounting [also mtg]

Mtge Mortgage [also mtge]

MTGP Monitor Table Generator Program

MTGW Maximum Total Gross Weight

MTG YBCO Melt-Textured Grown Yttrium Barium Copper Oxide (Superconductor)

MTH Magnetic Tape Handler

MTH Methylthiohydantoin

Mth Month [also mth]

5MTH 5-Methyltetrahydro-

1,000,000th (one) millionth

5MTHFA 5-Methyltetrahydrofolic Acid

MTHM Metric Tonne(s) of Hot Metal [also mthm]

M(T,H,t) Magnetization versus Temperature T, Time t and Magnetic Field H Parallel to c-Axis (of Superconductor) [Symbol]

MTHPA Methyltetrahydrophtalic Anhydride

MTHSR Mass Transportation/High Speed Rail

MTK Mechanical Time Keeping

MTK Medium Tank [also Mtk]

MTI Machine Tool Industry

MTI Management Technology International [US]

MTI Marine Training Institute [of Memorial University of Newfoundland, St. John's, Canada]

MTI Marked Temperature Inversion

MTI Materials Technology Institute [US]

MTI Medium-Technology Intensive

MTI Metal Treating Institute [US]

MTI Metalworking Technologies Inc. [of University of Pittsburgh, Pennsylvania, US]

MTI Ministry of Trade and Industry [Nigeria]

MTI Moving Target Indication; Moving Target Indicator

MTIA Metal Trades Industry Association

MTIAC Manufacturing Technology Information Analysis Center [US]

MTIC Moving Target Indicator Coherent

MTIL Maximum Tolerable Insecurity Level

MTIRA Machine Tool Industry Research Association [UK]

MTIS Multi-Port Interconnect System

MTJ Magnetic Tunnel Junction [Electronics]

MTJ MRAM Magnetic Tunnel Junction Magnetic Random-Access Memory [Electronics]

MTK Mongolian Tugrik [Currency]

MTL Main Transfer Line

MTL Manufacturing and Technology Laboratory

MTL Manufacturing Technologies Laboratory (Program) [of National Center for Manufacturing Sciences, US]

MTL Mass Transport Limited (Kinetics)

MTL Master Tape Loading

MTL Master Testing List [of US Environmental Protection Agency]

MTL Materials Technology Laboratory [US Army]

MTL Merged-Transistor Logic

MTL Message Transfer Layer

MTL Metals Technologies Laboratories [of Canada Center for Mineral and Energy Technology, Natural Resources Canada]

MTL Microsystems Technology Laboratory [of Massachusetts Institute of Technology, Cambridge, US]

MTL Microtrabecular Lattice [Cytology]

MTL Mineral Technology Laboratories

MTL Minimum Triggering Level

MTL Mobiltherm Light

Mtl Material [also mtl]

Mtl Montreal [Canada]

M2L Laboratory of Micromachining and Microtransducers [US]

MTL-CANMET Metals Technologies Laboratories–Canada Center for Mineral and Energy Technology [Natural Resources Canada]

MT-LKKR Muffin Tin Layer(ed) Korringa-Kohn-Rostoker (Method) [Physics]

MTLP Master Tape Loading Program

MTLP Monitor Table Listing Program

Mtls Materials [also mtls]

MTM Master in the Teaching of Mathematics

MTM Mechanical Testing Machine

MTM Methods-Time Measurement [Industrial Engineering]

MTM Metric Tonne(s) of Metal [also mtm]

MTM Multiple Terminal Manager

MTM Multiple Termination Module

MTM Multitasking Monitor

MTM Multiterminal Monitor

MTMA Methods-Time Measurement Association [US]

MTMA Military Terminal Control Area

MTMASR Methods-Time Measurement Association for Standards and Research [US]

MTMC 4-(Methylthio)-m-Cresol

MTMC M-Tolyl-N-Methylcarbamate

MTMF Multiple Task Management Feature

MT-MF Magnetic Tape to Microfilm

MTMH Master of Tropical Medicine and Hygiene

MTMP N-Bromomagnesium-2,2,6,6-Tetramethylpiperadine

MTMS Methyltrimethoxysilane

MTMSAP 6-Methyl-2-Trimethylsilylaminopyridine [also mtmsap]

MTMTS Military Traffic Management and Terminal Service [of US Department of Defense]

MTN Manitoba Television Network [Canada]

MTN Metriol Trinitrate

MTN M-Tolunitrile

MTN Multinational Trade Negotiations

Mtn Mountain [also mtn]

MTNS Metal Thick Nitride Semiconductor

Mtns Mountains [also mtns]

MTO Made to Order

MTO Magnetic Tape Operator

MTO Master Terminal Operator

MTO Master Timing Oscillator

MTO Maximum Time Out

MTO Metal Turnover

MTO Methyltrioxorhenium

MTO Mission, Task, Objective

MTO Modification Task Outline

MTO Muffin Tin Orbital [Physics]

MTO Multilateral Trade Organization [General Agreement on Tariffs and Trade]

MTOE Million Tons of Oil Equivalent

MTOGW Maximum Take-Off Gross Weight

MtOH Mannitol [Biochemistry]

Mton Megaton [also MT ot Mt]

MTOP Molecular Total Overlap Population

mTorr millitorr [Unit]

MTOS Magnetic Tape Operating System

MTOS Metal Thick Oxide Semiconductor

MTOS Metal Thick Oxide Silicon

MTOS Multitasking Operating System

MTOW Maximum Take-Off Weight

MTox Master of Toxicology

MTP Machine Tool Program

MTP Magnetic Tape Processor

MTP Magneto-Thermopower

MTP Maltese Pound [Currency]

MTP Manufacturing Technical Procedure

MTP Master Test Plan

MTP Mechanical Thermal Pulse

MTP Mercaptothiophosphate

MTP Message Transfer Part

MTP Message Transmission Part

MTP Metatarsophalangeal (Joint) [Anatomy]

MTP 4-(Methylthio)phenol

MTP Methyl Triphenoxyphosphonium

MTP Minimum Time Path

MTP Mission Test Plan

MTP Modular Terminal Processor

MTP Multiply Twinned Particle [Metallurgy]

MTPA α-Methoxy-α-(Trifluoromethyl)phenylacetic Acid

MTPA Mobile Transponder Performance Analyzer

(+)-MTPA (+)-α-Methoxy-α-(Trifluoromethyl)phenylacetic Acid

(–)-MTPA (–)-α-Methoxy-α-(Trifluoromethyl)phenylacetic Acid

MTPA-Cl Methoxy(trifluoromethyl)phenylacetic Acid Chloride

MTPCNA Metal Tube Packaging Council of North America [now Tube Council of North America, US]

MTPD Metric Tons per Day [also mtpd]

MTPE Mission To Planet Earth (Program) [of NASA]

MTPF Maximum Total Peaking Factor

MTPF Minimum Total Processing Time

MTPI Methyl Triphenoxyphosphonium Iodide

MTPP Methylenetriphrenylphosphorane

MTPS Magnetic Tape Programming System

MTPS Million Transitions per Second [also MTS]

MTPT Minimal Total Processing Time

MTPTT Ministère des Travaux Publics, Transports et Tourisme [Ministry of Public Works, Transport and Tourism, France]

MTPW Master of Technical and Professional Writing

MTPY Million (Short) Tons per Year [also Mtpy]

MTQ Exhibition on Materials Testing and Quality Assurance

MTQ International Exhibition on Measuring and Testing in Quality Safety [Germany]

MTR Magnetic Tape Reader

MTR Magnetic Tape Recorder

MTR Mass Transit Railway

MTR Materials Test(ing) Reactor [Commissioned by the US Atomic Energy Commission in 1950]

MTR Materials Testing Report

MTR Mean Time to Restore

MTR Membrane Technology and Research, Inc. [Menlo Park, California, US]

MTR Migration Traffic Rate

MTR Missile Tracking Radar

MTR Ministry of Tourism and Recreation [Canada]

MTR Mini-Temperature Recorder

MTR Model Tee Reader (Program)

MTR Moving Target Reactor

MTR Multiple-Track Radar (Range)

MTR Multiple Tracking Range

Mtr Monitor [also mtr]

Mtr Motor [also mtr]

MTRB Maritime Transportation Research Board [US]

MTRCA Metropolitan Toronto and Region Conservation Authority [Canada]

MTRE Magnetic Tape Recorder End

Mtrg Monitoring [also mtrg]

MTRI Marine Technology Research Institute [Norway]

MTrk Motor Truck

MTRP Maximum Transfer Rate Performance

MTRS Magnetic Tape Recorder Start

MTS Machine-Tractor Station

MTS Magnetic Tape Station

MTS Magnetic Tape Storage

MTS Magnetic Tape System

MTS Maintenance Test Station

MTS Main Trunk System

MTS Malta Test Station [Schenectady, New York, US]

MTS Manitoba Telephone System [Canada]

MTS Marine Technology Society [US]

MTS Mass Termination System

MTS Master Timing System

MTS Materials Testing System

MTS Maximum Tangential Stress

MTS Mean True Spacing [also mts] [Metallurgy]

MTS Mechanical Testing System

MTS Mechanical Threshold Stress

MTS Medium-Temperature Shift

MTS Member of Technical Staff

MTS Message Telecommunication Service

MTS Message Telephone Service

MTS Message Toll Service

MTS Message Transfer Service

MTS Message Transfer System

MTS Message Transmission Subsystem

MTS Message Transport Service

MTS Meteoroid Technology Satellite

MTS Meter-(Metric) Ton-Second (System) [also mts]

MTS Methods Time Sharing

MTS Metric Time System

MTS Methyltrichlorosilane

MTS Mica-Type Silicate

MTS Michigan Terminal System

MTS Microprocessor Training System

MTS Microsoft Transaction Server

MTS Microwave Temperature Sounder [ER-2 NASA Ames High-Altitude Aircraft]

MTS Million Transitions per Second [also MTPS]

MTS Missile Tracking System

MTS Mobile Telephone Service

MTS Mobile Tracking Station [of NASA]

MTS Modem Test Set

MTS Modem Test System

MTS Module Tracking System

MTS Molecular Transmission Spectrometry

MTS Moscow Time Standard [Greenwich Mean Time –03:00]

MTS Motor-Operated Transfer Switch

MTS Muffin-Tin Sphere [Physics]

MTS Multichannel Television Sound

MTS Multiple Terminal System

Mts Mountains

MTSC Magnetic Tape Selective Composer [also MT/SC]

MTSC Master of Technical and Scientific Communication

MTSC Middle Tennessee State College [now Middle Tennessee State University]

MTSD Military Transmission Systems Department [of North American Aerospace Defense Command]

MTSE Magnetic Trap Stability Experiment

MTSF Mileage to Subsequent Failure [Automobiles]

MTS-H$_2$ Methyltrichlorosilane–Hydrogen (Mixture)

MTSMB Material and Testing Standards Management Board

MTSO Mobile Telephone Switching Office

MT/SOT Mother Mother/Second-Order Twin (System) [Metallurgy]

MTSQ Mechanical Time, Superquick

MTSR Mean Time to Service Restoration

MTSS Military Test Space Station

MTSS Ministère du Travail et de la Sécurité Sociale [Ministry of Labour and Social Security, France]

MT/ST Magnetic Tape Selectric Typewriter [of IBM Corporation, US]

MTSU Middle Tennessee State University [Murfreesboro, US]

MTS/WATS Mobile Telephone Service/Wide Area Telephone Service

MTT Magnetic Tape Terminal

MTT Magnetic Tape Transport

MTT Maximum Touch Temperature

MTT Mechanical Thermal Treatment; Mechanicothermal Treatment

MTT Message Transfer Time

MTT 3-(4,5-Dimethylthiazol-2-yl)-2,5-Diphenyltetrazolium Bromide

MTT Microwave Theory and Techniques [Society of the Institute of Electrical and Electronics Engineers Society, US]

MTT Ministerial Trade Talks

MTT Multi-Transaction Timer

MTT Museum of Transport and Technology [Auckland, New Zealand]

MTT Thiozolyl Blue Tetrazolium Bromide

MT&T Maritime Telegraph and Telephone Company Limited [Canada]

M(t,T) Time and Temperature Dependent Magnetization

MTTA Machine Tool Technologies Association [UK]

MTTA Machine Tool Trades Association

MTTA Mean Time to Accomplish(ment)

MTTA Multi-Tenant Telecommunications Association [US]

MTT-Br Methylthiazolyldiphenyltetrazolium Bromide

MTTC Melinite-Tolite-Trinitrocresol [Mixture of Explosives]

MTTD Mean Time to Diagnosis

MTTE Magnetic Tape Terminal Equipment

MTTE Mean Time to Exchange

MTTF Mean Time to Failure

MTTFF Mean Time to First Failure

MTT FORMAZAN 1-(4,5-Dimethylthiazol-2-yl)-3,5-Diphenylformazan [also MTT Formazan]

MTTPO Mean Time to Planned Outage

MTTR Maximum Time to Repair

MTTR Mean Time to Repair

MTTR Mean Time to Restore

MTTS Microwave Theory and Techniques Society [of Institute of Electrical and Electronics Engineers, US] [also MTT-S]

MTTS Multitask Terminal System

MTTUO Mean Time to Unplanned Outage

MTU Magnetic Tape Unit

MTU Manchester Terminal Unit

MTU Master Terminal Unit

MTU Master Timing Unit

MTU Maximum Transmission Unit [Internet]

MTU Memory Transfer Unit

MTU Methylthiouracil

MTU Metric Ton(nes) of Uranium [Unit]

MTU Metric Ton Unit [also mtu]

MTU Michigan Technological University [Houghton, US]

MTU Mobile Training Unit

MTU Motoren– und Turbinenunion München GmbH [Munich-Based German Manufacturer of Vehicle and Aircraft Engines, Drive Mechanisms, Industrial Turbines, etc.]

MTU Multiplexer and Terminal Unit

MTU Multi-Terminal Unit

MTU Munich Technical University [Germany]

MTV Marginal Terrain Vehicle

MTV Master of Television

MTV Music Television

MTVAL Master Tape Validation

MTVC Manual Thrust Vector Control

MTW Male Tube Weld

MTW Maximum Taxi Weight

MTW Maximum Trailer Weight

MTW Mountain Wave [Meteorology]

M(t,w) Time and normalized excitation vector dependent response matrix (in transmission electron microscopy)

MTWA Maximum Total Weight Authorized

MTWR Maximum Trailer Weight Rating

MTWX Mechanized Teletypewriter Exchange

MTX Melinite-Tolite-Xylite [Mixture of Explosives]

MTX Methotrexate

MTX Methylpteroylglutamic Acid [Biochemistry]

MTX Mobile Telephone Exchange

MTXα Muffin-Tin X-α [Physics]

MTZ 1-Methyl 1H-Tetrazole [also mtz]

mtzd motorized

MU Mache Unit [Atomic Physics]

MU Machine Unit

MU Machine Utilization

MU Macquarie University [Sydney, Australia]

MU Make Up

MU Management Unit

MU Manchester University [UK]

MU Marquette University [Milwaukee, Wisconsin, US]

MU Marshall University [Huntington, West Virginia, US]

MU Massey University [Rotorua, New Zealand]

MU Master Unit

MU Mauritius [ISO Code]

MU McGill University [Montreal, Quebec, Canada]

MU McMaster University [Hamilton, Ontario, Canada]

MU Mehran University [Jamshoro, Pakistan]

MU Meiji University [Meijo, Nagoya, Japan]

MU Melbourne University [Australia]

MU Memory Unit

MU Methylene Unit

MU β-Methylumbelliferone; 4-Methylumbelliferyl

MU Miami University [Oxford, Ohio, US]

MU Microwave Ultrasonics

MU Middlesex University [London, UK]

MU Mie University [Mie-ken, Japan]

MU Missouri University [Columbia, US]

MU Mition Understanding

MU Miyazaki University [Japan]

MU Mobile Unit

MU Mock-Up [also M-U, or M/U]

MU Monash University [Melbourne/Clayton, Victoria, Australia]

MU Monitor Unit [also M/U]

MU Moskovskoga Universiteta [Moscow University, Russia]

MU Moskow University [Russia]

MU Muenster University [Germany]

MU Multiple Unit

MU Multi-User

MU Murdoch University [Perth, Australia]

4-MU 4-Methylumbelliferone

M/U Mock-Up [also MU, or M-U]

Mu Millimicron [Unit]

Mu Mullite [Mineral]

Mu Muonium [Symbol]

Mu* Anomalous Muonium [Particle Physics]

mu micron [also μ]

.mu Mauritius [Country Code/Domain Name]

MUA Machinery Users Association [UK]

MUA Mail User Agent

MUA Master of Urban Affairs

MUA Master of Urban Architecture

MUA Material(s) Usage Agreement

MUA Maximum Usable Altitude

MUA Ministry of State for Urban Affairs

muA microampere [also μA, μa, or μamp]

muÅ microangstrom [also μÅ]

MUART Multifunction Universal Asynchronous Receiver Transmitter

MUBIS Multiple Beam Interval Scanner

mubit microbit [also μbit]

MUC Makerere University College [Kampala, Uganda]

MUC Mesoscale Uncinus Complex

MUC Molecular Unit Cell

muC microcoulomb [also μC]

μCF Muon Catalyzed Fusion [Physics]

muCi microcurie [also μCi]

MUCG Management-Union Consultative Group [of Environment Australia]

MUCHA Multiple Channel Analysis

MUCO Diethylmuconate

MUCTC Montreal Urban Community Transport Commission [Canada]

MUD Master of Urban Design

MUD Multi-User Device

MUD Multi-User Dialog

MUD Multi-User Dimension

MUD Multi-User Domain [Internet]

MUD Multi-User Dungeon [Internet Game]

MUDISM Multidimensional Stochastic Method

MUDPAC Melbourne University Dual-Package Analog Computer [Australia]

MUDWNT Make Up Demineralizer Waste Neutralizer Tank

MUF Machine Utilization Factor

MUF Materials Unaccounted For

MUF Material Utilization Factor

MUF Maximum Usable Frequency; Maximum Useful Frequency

MUF Melamine-Urea-Formaldehyde

muF microfarad [also μF, μf, or μfd]

MUFON Mutual UFO (Unidentified Flying Object) Network [US]

MUFop Maximum Usable Frequency Operation

MUG MARC (Machine-Readable Catalog) User Group

MUG 4-Methylumbelliferyl β-D-Galactoside; 4-Methylumbelliferyl β-D-Glucoronide [Biochemistry]

MUG Minimum Useful Gradient

MUG Multiset Users Group [US]

MUG MUMPS (Massachusetts General Hospital Utility Multi-Programming System) Users Group [Boston, US]

mug microgram [also μg]

MUGB 4-Methylumbelliferyl p-Guanidinobenzoate [Biochemistry]

mug/m^3 microgram per cubic meter [also μg/m^3]

MUG Q MUG Quarterly [Publication of MUMPS (Massachusetts General Hospital Utility Multi-Programming System) Users Group, US]

muH microhenry [also μH, or μh]

muHg micrometer of mercury [also μHg, μ Hg, μm Hg]

MUI Module Interface Unit

mu-in micro-inch [also μin, μ-in, or micro-in]

mu-in-rms micro-inch root mean square [also μinrms or μ-in-rms]

MUL Melbourne University Library [Australia]

MUL Montanuniversität Leoben [Leoben University of Mining and Metallurgy, Austria]

MUL Multiply [also Mul, or mul]

muL microliter [also μL, or μl]

MULDEM Multiplexer/Demultiplexer [also Muldem, or muldem]

MULDEX Multiplexing/Demultiplexing [also Muldex, or muldex]

MULPIC Multipurpose Interrupted Cooling Process [also Mulpic] [Metallurgy]

MULSAM Multispectral Auger Microscope

MULSE Modified Universal Soil Loss Equation

MULSLI Multi-Slice (Calculation)

MULT Multiple; Multiplication; Multiply [also Mult, or mult]

MULTEWS Multiple Electronic Warfare Surveillance

MULTICS Multiplexed Information and Computing Service [US]

Multinatl Mon Multinational Monitor [Environmental Journal]

Multiphase Sci Technol Multiphase Science and Technology [Journal published in the US]

MULTP Multiplier [also Multp, or multp]

MuLV Murine Leukemia Virus

MUM Mass Memory Unit Manager

MUM Maximum Useful Magnification

MUM Methodology for Unmanned Manufacturing (Project) [Japan]

MUM Multiple-Unit Message

MUM Multi Unit Message

MUM Multi-User Message

MUM Multi-User Monitor

mu m micrometer [also μm]

mu-mho micromho [also μmho, or μ-mho]

mu-ohm microohm [also μohm, μ-ohm, or $\mu\Omega$]

MUMMS Marine Corps Unified Materiel Management System [US]

MUMPS Massachusetts General Hospital Utility Multi-Programming System [US]

MUMPs Multi-User MEMS (Microelectromechanical System) Processes [US Infrastructure Program Supported by Defense Advanced Research Projects Agency]

MUMS Mobile Utility Module System

MUMS Multiple Use MARC (Machine-Readable Catalog) System

mu mu micromicron [also $\mu\mu$]

mu m/y micrometer per year [also μm/yr, μm/y, or mu m/yr]

MUN Memorial University of Newfoundland [St. John's, Canada]

MUN Mineworkers Union of Namibia

Mun Municipal(ity) [also mun]

Mun Munition [also mun]

MunBd Munition Board [US]

Münch Med Wochenschr Münchener Medizinische Wochenschrift [Medical Weekly published in Munich, Germany]

MUNDAT Municipal Waterworks and Wastewater Systems Inventory [of Environmental Protection Service, Canada]

Mundo Electron Mundo Electronico [Spanish Publication on Worldwide Electronics]

Munic Municipal(ity) [also munic]

Muns Munitions [also muns]

MUO Miami University of Ohio [Oxford, US]

MUO Municipal University of Omaha [Nebraska, US]

MUOD Mean Unplanned Outage Duration

muon mu meson (i.e. μ^+, or μ^- meson) [also μ meson] [Particle Physics]

Muon Catal Fusion Muon Catalyzed Fusion [Swiss Publication]

MUP Manchester University Press [UK]

MUP Master of Urban Planning

MUP Melbourne University Press [Australia]

MUP Methylumbelliferyl Phosphate

MUP Multiple Utility Peripheral

4-MUP 4-Methylumbelliferyl Phosphate

muP micropoise [also μP]

MUPF Modified Ultraspherical Polynomial Filter

MUPI Multi-Photon Ionization (Mass Spectrometry)

MUPID Multi-Purpose Universal Programmable Intelligent Decoder

MUPP Master of Urban Planning and Policy

MUPPET Maryland University Project in Physics and Educational Technology (Utilities Software) [US]

MUPS Mechanized Unit Property System

MUPS Million Updates per Second [also mups]

MUPS Multiple Utility Peripheral System

MUR Management Update and Retrieval System

MUR Manpower Utilization Report

MUR Mauritian Rupee [Currency]

MUR Message Unit, Radio Relay

MUR Mock-Up Reactor [of NASA]

MUR Module Ultra-Rapid (Gate)

muR Microroentgen [also μR, or μr]

MURA Midwestern University Research Association [US]

MURATREC Multi-Radar Track Reconstruction

MURB Multiple Unit Residential Building

MURC Missouri University Research Center [Columbia, US]

MURC Murrumbidgee Users Rehabilitation Committee [New South Wales, Australia]

MURKS Multi-User Remote Keying System

MURP Master of Urban and Regional Planning

MURP Master of Urban and Rural Planning

MURR Missouri University Research Reactor [Columbia, US]

MURST Ministero dell'Università e delle Ricerca Scientifica e Tecnologica [Ministry for Universities and Scientific and Technological Research, Italy]

MURST/CNR Ministero dell'Università e delle Ricerca Scientifica e Tecnologica/Consiglio Nazionale delle Ricerche [Ministry for Universities and Scientific and Technological Research/National Research Council, Italy]

MUS Manual Update Service

MUS Master of Urban Studies

MUS Muskogee, Oklahoma [Meteorological Station Designator]

MUS Multiprogramming Utility System

Mus Museum [also mus]

Mus Music(al); Musician [also mus]

Mus Muslim

mus mouse [Immunochemistry]

mu s microsecond [also mu sec, μs, or μsec]

MUSA Multiple-Unit Steerable Antenna; Multiple-Unit Steerable Array [also musa]

MUSAP Multisatellite Augmentation Program

MUSAT Multi-Purpose UHF (Ultrahigh Frequency) Satellite [Canada]

MUSC Medical University of South Carolina [Charleston, US]

MUSC Microgravity User Support Center

MUSE Monitor of Ultraviolet Solar Energy

MUSE Multiple Stream Evaluator

MUSE Multiple Sub-Nyquist Sample Encoding

MUSE Multi-User Shared Environment

mu sec microsecond [also mu s, μs, or μsec]

Mushroom J Mushroom Journal [of Mushroom Growers Association, UK]

Mushroom Sci Mushroom Science (Journal) [US]

MUSIC Multiple Signal Characterization

MUSICOL Musical Instruction Composition Oriented Language

Music Technol Music Technology [UK Journal]

MUSIL Multiprogramming Utility System Interpretative Language

MUSS Manchester University Software System

MUSS Module Utility Support Structure

MUST Man Undersea Science and Technology (Office) [of National Oceanic and Atmospheric Administration, US]

MUST Melton Valley Storage and Evaporator Concentrate Storage Tank [of US Department of Energy at Oak Ridge National Laboratory, Tennessee]

MUSTARD Museum and University Storage and Retrieval of Data [of Smithsonian Institute, Washington, DC, US]

MUSTRAN Music Translation

MUSysE Master of Urban Systems Engineering [also MUSysEng]

MUT Mean Up Time

MUT Monitor Under Test

Mut Mutagen [Genetics]

Mut Mutant; Mutation [Genetics]

mut mutual

MUTA Military Upper Traffic Control Area

Mutagen Mutagenesis [Publication of Oxford University Press, UK]

Mutat Res Mutation Research [Journal published in the Netherlands]

MUTEX Multi-User Transaction Executive

mutl mutual

MUTMAC 4-Methylumbelliferyl p-Trimethylammonium Cinnamate Chloride

MUUG Manitoba Unix User's Group [Canada]

MUV Manned Underwater Vehicle

muV microvolt [also μV pr μv]

muV/m microvolts per meter [also μV/m, or μv/m]

muW microwatt [also μW, or μw]

MUX Multiplex(er); Multiplexing [also Mux, or mux]

MUXER Multiplexer [also Muxer, or muxer]

MV Machine Vision

MV Magnetic Viscosity

MV Main Valve

MV Maldives [ISO Code]

MV Manual Valve

MV Manufacturing Verification

MV Marburg Virus

MV Marine Vessel

MV Market Value

MV Matrix Vesicles

MV Mean Value

MV Mean Variation

MV Mean Velocity

MV Measured Value

MV Measured Vector

MV Measured Voltage

MV Mechanical Vehicle

MV Medium Vacuum

MV Medium Voltage [also mv]

MV Megavolt [also Mv]

MV Mesh Voltage

MV Methyl Violet

MV Million Volts

MV Minivan

MV Mississippi Valley [US]

MV Mitral Valve (of Heart) [Anatomy]

MV Mixed Valent (Behavior)

MV Molar Volume

MV Molecular Velocity

MV Mooney Viscosity [Chemical Engineering]

MV Motor Vehicle

MV Motor Vessel

MV Move [Computers]

MV Moving Vessel

MV Multivibrator

MV Multivitamin [Biochemistry]

MV Mumps Virus

MV Murray Valley [Australia]

MV Muzzle Velocity

Mv (The) Maldives

Mv Mendelevium [also Md]

mV millivolt [also mv]

mV$_{SCE}$ Saturated Calomel Electrode Potential in Millivolt [Symbol]

m/V magnetic moment per unit volume [Symbol]

m/V mass per volume [Symbol]

m/V mean-to-variance (ratio) [Statistics]

.mv Maldives [Country Code/Domain Name]

μV microvolt [also μv, or muV]

μV$_{pp}$ microvolt peak-to-peak [Symbol]

μV$_{rms}$ microvolt root-mean-square [Symbol]

MVA Machine Vision Association [of Society of Manufacturing Engineers, US]

MVA Main Valve Actuator

MVA Manufacturing Value Added [also mva]

MVA Master of Visual Arts

MVA Mean Vertical Acceleration [also mva]

MVA Megavolt-Ampere [Unit]

MVA Mevalonic Acid

MVA Million Volt-Ampere

MVA Missouri Valley Authority [US]

MVA Modified Vaccinia Virus, Ankara Strain

MVA Motor Vehicle Accident

MVA Multivariate Analysis [Statistics]

MVAS Multipurpose Ventricular Actuating System

MVA/SME Machine Vision Association of the Society of Manufacturing Engineers [US]

MVB Medium-Volatile Bituminous (Coal)

MVB Multimedia Viewer Book

MVB Multivesicular Body

MVB Multivibrator [also MVBR, Mvbr, mvbr, Mvb, or mvb]

MVC Manual Volume Control

MVC Master Volume Control

MVC Microvoid Coalescence

MVC Monomeric Vinylchloride

MVC Multimedia Viewer Compiler

MVC Multiple Variate Counter

MVC Multivariable Control(ler)

mV/°C millivolt per degree Celsius [also mV °C^{-1}]

μV/°C microvolt per degree Celsius [also μV °C^{-1}]

MVCD Magnetic Vibrational Circular Dichroism

mV/cd millivolt(s) per candela [also mV cd^{-1}]

MV/cm Megavolt(s) per centimeter [also MV cm^{-1}]

mV/cm millivolt(s) per centimeter [also mV cm^{-1}]

μV/cm microvolt per centimeter [also μV cm^{-1}]

MVCS Mind and Vision Computer Systems

MV-CTD Moving Vessel–Conductivity, Temperature, Depth

MVD Map and Visual Display

MVD Mechanical Vapor Deposition

MVD Medium Velocity Dynamite

MVD *(Ministerstvo Vnutrennikh Del)* – Russian for "Ministry of Internal Affairs"

MVD Multivariable Distribution

MVDA Memory Volume Discount Addendum

MVDA Motor Vehicle Dismantlers Association [UK]

MVDF Medium and Very-High Frequency Direction Finding

MVDL Mercury-Vapor Discharge Lamp

MVDM Multiple Virtual DOS (Disk-Operating System) Machines

MVDS Modular Video Data System

MVE Master of Vocational Education

MVE Methyl Vinyl Ether

MVE Motor Vehicle Event

MVE Multivariate Exponential Distribution

MVEd Master of Vocational Education

MVetSc Master of Veterinary Science

MVF Manned Vertical Flight

MVF Martensite Volume Fraction [Metallurgy]

MVF Miles Value Foundation [US]

MVFR Marginal Visual Flight Rules

MVG Manufacturing Vision Group [US]

Mvg Moving [also mvg]

MVGA Manitoba Vegetable Growers Association [Canada]

MVGA Monochrome Video Graphics Array

MVGVT Mated Vertical Ground Vibration Test

mV/h millivolt per hour [also mV h^{-1}]

MVHA Mississippi Valley Historical Association [US]

MVI Motor Vehicle Inspection

MVI Multivitamin Infusion [Medicine]

MVI Muzzle Velocity Indicator

MVIP Multi-Vendor Integration Protocol

MVISIP Murray Valley Irrigation and Salinity Investigation Program [Australia]

MVK Methyl Vinyl Ketone

mV/K millivolt(s) per kelvin [also mV K^{-1}]

μV/K microvolt(s) per Kelvin [also mV K^{-1}]

μV^2/kfci square microvolt(s) per kiloflux change per inch [also mV kfci^{-1}]

MVL Mercury Vapor Lamp

MVL Metal Vapor Laser

MVL Minimal Visible Lesion [Medicine]

MVL Molecular Virology Laboratory [of University of Madison, Wisconsin, US]

MVL Multiple-Valued Logic

MVLDC Murray Valley League for Development and Conservation [Australia]

MVLIFCT M. V. Lomonosov Institute of Fine Chemical Technology [St. Petersburg, Russia]

MVLUE Minimum Variance Linear Unbiased Estimate; Minimum Variance Linear Unbiased Estimator

MVM Mariner Venus/Mercury [Aerospace]

MVM Minimum Virtual Memory

MVM Mini-Vogel Mount (Cathode)

MVM Modified Volume Module; Modifying Volume Module

MVm Megavolt-meter [Unit]

MV/m Megavolt(s) per meter [also MV m^{-1}]

mV/m millivolt(s) per meter [also mV m^{-1}]

μV/m microvolt per meter [also μV m^{-1}, or muV/m]

MVMA Motor Vehicle Manufacturers Association [US]

mV/μbar millivolt(s) per microbar [also mV μbar^{-1}]

MVMC Motor Vehicle Maintenance Conference

MVm/C·K Megavolt-meter per Coulomb Kelvin [also MVm C^{-1} K^{-1}]

MV/m Megavolt(s) per meter [also MV m^{-1}]

mV/mG millivolt(s) per milligauss [also mV mG^{-1}]

mV/min millivolt(s) per minute [also mV min^{-1}]

μV/mm microvolt(s) per millimeter [also μV mm^{-1}]

mV/μm millivolt(s) per micrometer [also mV μm^{-1}]

μV$_{pp}$/μm microvolt(s) peak-to-peak per micrometer [also μV$_{pp}$ μm^{-1}]

Mvmt Movement [also mvmt]

MVOCC Mixed-Valent Oxide-Catalytic Carbonization (Process)

MVP Master Verification Plan

MVP Matrix-Vector Product

MVP Matrox Vision Processor

MVP Mechanical Vacuum Pump

MVP Megavolt of Power [also Mvp]

MVP 2-Methyl-5-Vinylpyridine

MVP Multiline Variety Package

MVP Multimedia Video Processor

MVP Multiple Virtual Processing

MVP Multivariable Process(es)

MVP Multi View Progressive (Lens)

MVp Megavolt peak

MVPCB Motor Vehicle Pollution Control Board [US]

MVR Maneuver [also Mvr, or mvr]

MVR Manual Voltage Regulator

MVR Mercury-Vapor Rectifier

MVR Monthly Variance Report

Mvr Maneuver [also mvr]

MvRe Maldivian Rupee [Currency]

MVRH Mott Variable Range Hopping (Model) [Physics]

mV RMS Millivolts, Root-Mean-Square [also mV rms, or mv rms]

MVS Machine Vision System

MVS Magnetic Voltage Stabilizer

MVS Master of Valuation Sciences

MVS Middle Value Select

MVS Minimum Visual Signal

MVS Motor Vehicle Safety

MVS Motor Vehicle Specification

MVS Multi-Aircraft Vertically-Stacked (Mission)

MVS Multiple Virtual Storage

MVS Multiprogramming with Virtual Storage

MVS Multivariable System

MVS Multivariate Statistics

mV/s millivolt(s) per second [also mV s^{-1}, or mV/sec]

m^2/V·s square meter per volt-second [also m^2/Vs, m^2 V^{-1}s^{-1}, or m^2/V·sec]

μV/s microvolt per second [also μV/sec, or μV s^{-1}]

MVSA Motor Vehicle Safety Act

MVSA Motor Vehicle Safety Association

MVSC Mississippi Valley State College [US]

MVSc Master of Veterinary Science

mV/sec millivolt(s) per second [also mV s^{-1}, or mV/s]

m^2/V·sec square meter per volt-second [also m^2/Vs, or m^2 V^{-1}s^{-1}]

μV/sec microvolt per second [also μV/s, or μV s^{-1}]

MVS/ESA Multiple Virtual Storage/Enterprise System Architecture

MVSMA Mechanical Vibrating Screen Manufacturers Association [US]

MVSS Motor Vehicle Safety Standard

MVSSA Mine Ventilation Society of South Africa

MVS/SE Multiple Virtual Storage/System Extension

MVS/SP Multiple Virtual Storage/System Product

MVT Mission Verification Test [of NASA]

MVT Moisture Vapor Transmission

MVT Multiprogramming with a Variable Number of Tasks

μVT Grand Canonical (Ensemble) [Statistical Mechanics]

MVTA Motor Vehicle Transport Act [Canada]

MVTC Manpower Vocation Training Commission [Canada]

MVTC Motor Vehicle Test Center

MVTEd Master of Vocational and Technical Education [also MVTE]

MVTL Modified Variable Threshold Logic

MVT/TSO Multiprogramming with a Variable Number of Tasks/Time-Sharing Option

MVU Minimum Variance Unbiased

MVUE Minimum Variance Unbiased Estimate

MVULE Minimum Variance Unbiased Linear Estimate

MW Mach Wave [Fluid Mechanics]

MW Magnetic Wave

MW Magnet Wire

MW Makeup Water [Chemical Engineering]

MW Malawi [ISO Code]

MW Manual Word

MW Man Week [Unit]

MW Matter Wave [Quantum Mechanics]

MW Measured Weight

MW Mechanical Working

MW Medium Wave

MW Medium-Wing (Monoplane)

MW Megawatt [also Mw]

MW Metallurgical Waste

MW Metalwork(er); Metalworking

MW Metric Wave

MW Microwave [also M/W]

MW Microweighing

MW Microwelding

MW Milky Way (System) [Astronomy]

MW Million Words

MW Mineral Water

MW Mineral Wool

MW Minute Wheel [Horology]

MW Miscellaneous Weapons

MW Mixed Waste

MW Mixed Width [also mw]

MW Modulated Wave

MW Molar Weight

MW Molecular Weight [also mw]

MW Multiwire

MW Municipal Waste

M/W Microwave [also MW]

Mw Molecular Weight, Weight-Average

Mw Earthquake Magnitude by Moment Method [Symbol]

mW milliwatt [also mw]

μW microwatt [also μw, or muW]

μw microwave [also μW]

MWA Microwave Antenna

MWAA Metropolitan Washington Airports Authority [US]

MWARA Major World Air Route Area

μ-wave microwave

MWB Master Work Book

MWB Metropolitan Water Board

MWB Minimum Wage Board

MWB Multiprogram Wire Broadcasting

μWb microweber [Unit]

MWC Mechanized Wire Centering

MWC Municipal Waste Combustor

MW/cm² Megawatt(s) per square centimeter [also MW cm^{-2}]

mW/cm² milliwatt(s) per square centimeter [also mW cm^{-2}]

μW/cm² microwatt(s) per square centimeter [also μW cm^{-2}]

MWCO Molecular Weight Cutoff

MWCS Microwave Communication System

MWCVD Microwave Plasma-Assisted Chemical Vapor Deposition

MWD Measurement While Drilling

MWD Megawatt Day [also MWd]

MWD Megaword

MWD Metropolitan Water District

MWD Microwave Device

MWD Molecular Weight Distribution

MWd Megawatt Day [also MWD]

MWDA Modified Weighted Density Approximation

MWDB Micro World Data Bank

MWDD Miscellaneous Weapons Development Department [UK]

MWDDEP Mutual Weapons Development Data Exchange Program

MWDP Mutual Weapons Development Program

MWD/t Megawatt Day(s) per Ton [also MWd/t]

MWE Manufacturer's Weight Empty

MWE Megawatt of Electricity [also MW(E)]

MWE Microwave Electromagnetics [also MWEM]

MWE Mott-Wannier Exciton [Solid-State Physics]

MWe Megawatt, electrical (power) [Unit]

MWF Marine Workers Federation

MWF Microwave Frequency

mW/F-F Milliwatt per Flip-Flop [Unit]

MWFT Microwave Fourier Transform

MWFTS Microwave Fourier Transform Spectroscopy

MWG Maintenance-Analyzer Working Group

MWG Music Wire Gauge

mW/g millivolt per gram [also mW g^{-1}]

MWH Megawatt of Heat [also MW(H)]

MWH Microwave Heating

MWH Microwave Holography

MWh Megawatt-hour [also MWhr]

MWHGL Multiple Wheel Heavy Gear Load

MWh/kg Megawatt-hour per kilogram [also WMhr/kg]

MWhr Megawatt-hour [also MWh]

MWI Marine Watch Institute [US]

MWI Message-Waiting Indicator

MWI Microwave Interferometry

MWIA Manitoba Wholesale Implement Association [Canada]

MWIC Microwave Integrated Circuit

MWIP Mixed Waste Integrated Program

MWIR Medium Wave Infrared

MWIR Mid-Wave Infrared

MWIR Mixed Waste Inventory Report

MWIR FPA Mid-Wave Infrared Focal Plane Array

MWIS (National Network of) Minority Women in Science [US]

MWK Malawi Kwacha [Currency]

MWL Master Warning Light

MWL Mean Water Level

MWL Milliwatt Logic

MWLID Mixed Waste Landfill Integrated Demonstration

MWM Materials World Modules (Program) [of National Science Foundation, US]

MWM Microwave Maser

MWM Motoren-Werke Mannheim AG [German Engine Manufacturer located at Mannheim]

MW/m² Megawatt(s) per square meter [also MW·m⁻²]

MW/m³ Megawatt(s) per cubic meter [also MW·m⁻³]

mW/m² milliwatt(s) per square meter [also mW·m⁻²]

μW/MHz microwatt(s) per megahertz [also μW Mhz⁻¹]

mW/mm² milliwatt(s) per square millimeter [also mW·mm⁻²]

μW/mm² microwatt(s) per square millimeter [also μW·mm⁻²]

mW/mV milliwatt(s) per millivolt [also mW·mm⁻¹]

MWO Maintenance Work Order

MWO Master Warrant Officer (or Chief Petty Officer, Second Class) [Canadian Forces]

MWO McGill Weather Observatory [of McGill University, Montreal, Canada]

MWO Medicine White Oil

MWO Meteorological Watch Office

MWO Mount Wilson Observatory [of Carnegie Institution of Washington; located near Los Angeles, California, US]

m-WO₃ Monoclinic Tungstic Oxide

M-Word Miscellaneous Function Word [Numerical Control]

MWP Maneuvering Work Platform [of NASA]

MWP Maximum Working Pressure

MWP Microwave Plasma [Physics]

MWP Microwave Power

MWP Mixed Waste Project [US]

MWp Megawatt peak [Unit]

Mwr Mower [also mwr]

MWPC Multiwire Proportional Counter

MWPR Monthly Work Package Report

MWPS Master of Wood and Paper Science

MWR Magnetic Tape Write (Memory)

MWR Mean Width Ratio

MWR Microwave Radiation

MWR Microwave Radiometer

MWR Multiwafer Reactor

MWRR Microwave Radio Relay

MWRT Microwave Radio Telephony

MWS Microwave Scatterometer

MWS Microwave Sintering

MWS Microwave Station

MWS Microwave Spectrometer; Microwave Spectrometry

MWS Microwave Spectroscope; Microwave Spectroscopy

MWS Microwave Spectrum

MWS Multi-Way Sorption (Filter)

MWS Multi-Workstation

MW-s Megawatt-second [also MWs]

MWSC Minimum Wage Study Commission

MWSC Missouri Western State College [Saint Joseph, US]

MW-SDS Molecular Weight–Sodium Dodecylsulfate

MWSP Mechanical Working and Steel Processing (Division) [of Iron and Steel Society, US]

MWSSD Microwave Solid-State Device

MWT Make-Up Water Treatment

MWT Master of Wood Technology

MWT Megawatt Thermal [also MW(T), MWt, or MW(Th)]

MWT Microwave Technologist; Microwave Technology

MWT Microwave Thermography

MWt Megawatt of Thermal Energy [also MWth, MWTor MW(Th)]

MW(Th) Megawatt Thermal [also MWth, MWT, MW(T), or Mwt]

MWTIP Mixed Waste Treatment Program [of US Department of Energy]

MW/tonU Megawatt per Ton of Uranium [also MW/tonneU]

MWU Midwestern University [Wichita Falls, Texas, US]

MWV Maximum Working Voltage [also mwv]

MWV Mineralölwirtschaftsverband [Mineral Oil Industries Federation, Germany]

MWW Man with Wrench (Factor)

MWYE Megawatt Year of Electricity

MW-yr Megawatt-year [also MWyr]

MW-yr/m² Megawatt-year(s) per square meter [MW-yr m⁻²]

MX Ceramic Compound of One Metallic Element and One Nonmetallic Element in which the Metal Ion to Nonmetal Ion Ratio is 1:1 (e.g., CsCl, NaCl, CaO, MgO, ZnO,ZnS, etc.)

MX 3-Chloro-4-(Dichloromethyl)-5-Hydroxy-2-(5H)-Furanone

MX Halogen-Bridged Metal

MX Mail Exchange (Record) [Internet]

MX Mail Exchanger [Internet]

MX Metal (I) Halide [Formula]

MX Mexico [ISO Code]

MX Missile Experimental [Symbol]

MX Multiplex [also Mx]

MX Metal-Nonmetal

MX₂ Ceramic Compound of One Metallic Element and One Nonmetallic Element in Which the Metal on to Nonmetal Ion Ratio is 1:2 (e.g., CaF_2, SiO_2, UO_2, etc.)

MX₂ Metal Dihalide [Formula]

MX₃ Metal Trihalide [Formula]

MX₄ Metal Tetrahalide [Formula]

MX₆ Metal Hexahalide [Formula]

M₂X₃ Ceramic Compound of One Metallic Element and One Nonmetallic Element in which the Metal Ion to Nonmetal Ion Ratio is 2:3 (e.g., Al_2O_3, Cr_2O_3, etc.)

M(X) Molarity of Metallic Element "X" [Symbol]

Mx Matrix [also mx]

Mx Maxwell [also mx] [Unit]

Mx Mix [also mx]

.mx Mexico [Country Code/Domain Name]

MxAc Methoxy Acetic Acid

MXC Multiplexer Channel

MXCD Magnetic X-Ray Circular Dichroism

mxd mixed

MXD M-Xylylene Diamine [also MXDA]

MXE Mobile Electronic Exchange

Mxea Methoxyethanamine

MXP Mexican Peso [Currency]

Mxp Methoxypyridine

2Mxp 2-Methoxypyridine

3Mxp 3-Methoxypyridine

4Mxp 4-Methoxypyridine

MXPS Monochromatic X-Ray Photoelectron Spectroscopy

MXR Mask Index Register

MXS Microsoft Exchange Server

Mx turn Maxwell turn [Symbol]

MXU Mobile Exhibition Unit

MY Malaysia [ISO Code]

MY Man Year

MY Medinat Yisrael [Republic of Israel]

MY Meeting Year

MY Molecular Yield [Solid-State Physics]

MY-5445 1-(3-Chlorophenylamino)-4-Phenylphthalazine

M/Y Motor Yacht

My Maintainability [also my]

My May

.my Malaysia [Country Code/Domain Name]

Mycol Mycological; Mycologist; Mycology [also mycol]

MYD Muller, Yushchenko and Derjaguin (Model) [Metallurgy]

m/yd meter(s) per yard (ratio)

MYF San Diego, California [Meteorological Station Designator]

myg myriagram [Unit]

myL myrialiter [also myl]

mym myriameter [Unit]

MYOB Mind(ing) Your Own Business

MY1DL M. Yoshida One-Dimensional Lagrangian (Hydrodynamic Code) [of New Mexico Institute of Mining and Technology, Socorro, US]

MYOP Multi-Year Operations Plan

MYPP Multiyear Program Plan

MYR Malaysian Ringgit [Currency]

MYTA Maintainability Task Analyses

MYVAL Maintainability Evaluation

MZ Mach-Zehnder (Interferometer) [Optics]

MZ Manganese Zirconate

MZ Melted Zone; Melting Zone

MZ Methylimidazole

MZ Mozambique [ISO Code]

MZ Mullite-Zirconia (Ceramics)

2-MZ Methylimidazole

Mz p-Methoxyphenylazobenzyloxycarbonyl

Mz Monozygotic [also mz]

.mz Mozambique [Country Code/Domain Name]

m/z mass per charge [Symbol]

MZA Mullite-Zirconia-Alumina (Ceramics)

MZF Manganese Zinc Ferrate

MZFR Mehrzweck-Forschungsreaktor [Multi-Purpose (Nuclear) Research Reactor at Kernforschunganlage Karlsruhe, Germany]

MZFW Maximum Zero Fuel Weight

MZI Mach-Zehnder Interferometer [Optics]

MZM Mozambique Meticai [Currency]

MZO Magnesium Zirconium Oxide

MZP Magnesium Zirconium Phosphate

MZP Monoclinic Zirconia Polycrystal [also $mZrO_2$]

MZR Mach-Zehnder Refractometer [Optics]

MZR Multiple Zone Recording

m-ZrO₂ Monoclinic Zirconia (Polycrystal) [also MZP]

MZS Master of Zoological Science [also MZSc]

MZS Mercury-Zinc Selenide (Semiconductor)

MZT Mechanical-Zone Texture(d) (Magnetic Disk)

MZT Mercury-Zinc Telluride (Semiconductor)

N

N American National 8, 12 and 16 (Screw) Thread Series [Symbol]

N (+)-Asparagine; Asparaginyl [Biochemistry]

N Atomic Fraction [Symbol]

N Atomic Number [Symbol]

N Audio-Frequency Power Output [Symbol]

N Avogadro's Number [Symbol]

N Chain Length (of Polymer) [Symbol]

N Complex Refractive Index [Symbol]

N Degree of Polymerization [Symbol]

N Demagnetization Factor [Symbol]

N Describing Function (in Automatic Control) [Symbol]

N Dislocation Density (in Materials Science) [Symbol]

N Droplet Concentration (in Meteorology) [Symbol]

N (Acoustic-Emission) Event Count Rate [Symbol]

N Factor of Safety [Symbol]

N Fatigue Life (in Mechanics) [Symbol]

N f-Number (in Photography) [Symbol]

N Fringe Order [Symbol]

N (ASTM) Grain-Size Number (in Metallurgy) [Symbol]

N Group Number (of Periodic System of Elements) [Symbol]

N Half Crack Length (in Crack Theory) [Symbol]

N Haploid Number of Chromosomes [Symbol]

N Landau (Energy) Level (of Electrons) [Symbol]

N Length of Near Field (in Ultrasonic Inspection) [Symbol]

N Lot Size (in Acceptance Sampling) [Symbol]

N Mole Fraction [Symbol]

N Nacelle

N Nagoya [Japan]

N Nagpur [India]

N Naira [Currency of Nigeria]

N Nairobi [Kenya]

N Name [also n]

N Namibia(n)

N Nanking (or Nanjing) [PR China]

N Nanocrystal(line); Nanocrystallinity

N Nantes [France]

N Naples [Italy]

N Narcotic(s)

N Narrow [Geography]

N Narrow (Design) [Symbol]

N Nashville [Tennessee, US]

N Nassau [Bahamas]

N Nation(al)

N Natural; Nature

N Nauru [Pacific Island]

N Nausea; Nauseous

N Nautical; Nautic(s)

N Naval

N Navigation(al); Navigator

N Navy

N Neat

N Nebraska [US]

N Nebula [Astronomy]

N Nebulize(r)

N Necator [Species of Hookworms]

N Neck

N Necking [Metallurgy]

N Neck Length (of Drills) [Symbol]

N Needle

N Negation; Negative

N Negentropy (or Information Content) [Communications]

N Negev (or Negeb) [Israeli Desert]

N Negroid [Anthropology]

N Neisseria [Genus of Bacteria]

N Nematic Phase (of Liquid Crystal)

N Nematicide [also n]

N Nematocyst [Zoology]

N Nematode [Zoology]

N Neolithic [Anthropology]

N Neon

N Nepal(ese)

N Neper [Unit]

N Nephelometer; Nephelometry [Optics]

N Nepheloscope; Nepheloscopy [Optics]

N Nephila [Genus of Spiders]

N Nephridium [Zoology]

N Nephron [Urology]

N Nerve; Nervous(ness) [also n]

N Nest

N Nesting

N Net (Weight) [also n]

N Net [also n]

N Network

N Netherlands

N Net

N Nets (in Homogeneous Connections of Atoms) [Symbol]

N Neuron [Anatomy]
N Neurotic(s); Neurosis [Psychology]
N Neutral(ize); Neutralizer; Neutralization
N Neutral Gear [Symbol]
N Neutrino [Physics]
N Neutron
N Neutronic(s)
N Neutron Number [Symbol]
N Nevada [US]
N New
N Newark [New Jersey, US]
N Newcastle [UK]
N Newfoundland [Canada]
N Newton [Unit]
N Newtonian Heat Transfer [Symbol]
N Ngultrum [Currency of Bhutan]
N Niamey [Niger]
N Nicaragua(n)
N Nice [France]
N Nickel [Abbreviation; Symbol: Ni]
N Nicosia [Cyprus]
N Nicotine
N Niger
N Nigeria(n)
N Night
N Nile (River) [East Africa]
N Niobium (V) Oxide
N Nipple
N Nitrate
N Nitridation; Nitriding
N Nitride
N Nitrile
N Nitro-
N Nitrogen [Symbol]
N Noble
N Nodal; Node
N No
N No Flash (Propellant)
N Noise
N Nominal
N Nomogram; Nomograph(y)
N (Magnetic) Noncoupling (State) [Symbol]
N Nonrotatable; Nonrotating; Nonrotational
N Noon [also n]
N Norfolk [Virginia, US]
N Norm(al)
N Normal Direction [Metallurgy]
N Normal Force [Symbol]
N Normalism
N Normality [also N] [Chemistry]
N Normalization; Normalize(d); Normalizing
N Normal Solution [also N] [Chemistry]
N North(ern)

N North Pole (of a Magnet)
N Norwegian; Norway
N Nose
N Nostril
N Notch
N Note
N Notice
N Nova [Astronomy]
N November
N November [Phonetic Alphabet]
N Noxious(ness)
N Nozzle
N N-Shell (of Electron) [Symbol]
N (Upper-case) Nu [Greek Alphabet]
N Nuclear
N Nuclear Engineering (Technology) [Discipline Category Abbreviation]
N Nucleate; Nucleation
N Nuclei Size (in Metallurgy) [Symbol]
N Nucleofugality [Organic Chemistry]
N Nucleofuge; Nucleofugic [Organic Chemistry]
N Nucleon; Nucleonic(s)
N Nucleophile [Physical Chemistry]
N Nucleus [also n]
N Number
N Number Density [Symbol]
N Number of Active Coils (of Springs) [Symbol]
N Number of Atoms (per Unit Volume) (e.g., N_{Fe} is the Number of Iron Atoms in an Alloy) [Symbol]
N Number of Charge Carriers per Unit Volume [Symbol]
N Number of Cycles (to Failure) [Symbol]
N Number of Electromagnetic Sources [Symbol]
N Number of Electrons [Symbol]
N Number of Features (in Quantitative Metallography) [Symbol]
N Number of Gear Teeth [Symbol]
N Number of Ions [Symbol]
N Number of Interfering Wave Trains [Symbol]
N Number of Lines per Unit Distance (of Diffraction Gratings) [Symbol]
N Number of Measurements [Symbol]
N Number of Molecules [Symbol]
N Number of Neutrons [Symbol]
N Number of (Fundamental) Particles [Symbol]
N Number of Revolutions (per Unit Time) [Symbol]
N Number of Sampling Units (in Statistics) [Symbol]
N Number of Species in Thermodynamic Systems [Symbol]
N Number of Spot, or Projection Welds [Symbol]
N Number of Strokes (of Piston) [Symbol]
N Number of Theoretical Plates (in Chromatography) [Symbol]
N Number of (Screw) Threads [Symbol]
N Number of Turns (of Magnetic Coils) [Symbol]
N (Total) Number of Unique Test, or Trial Results [Symbol]

N   Number of Workpieces [Symbol]

N   Numeric(al); Numerics

N   Nuphar [Ecology]

N   Nurse; Nursing

N   Nusselt Number (or Biot Number) [Symbol]

N   Nutate; Nutation

N   Nutrient; Nutrition(ist); Nutritious

N   Nylon

N   Nyssa [Genus of Deciduous Trees Including the Tupeloes]

N   Particle Number [Symbol]

N   Polymer Chain Length [Symbol]

N   Polymerization Index (of Homopolymer) [Symbol]

N   Probe [USDOD Symbol]

N   Satellite Field of View [Symbol]

N   Special Test, Permanent (Aircraft, Missile, or Rocket) [USDOD Symbol]

N   Spindle Speed [Symbol]

N   Summation Index [Symbol]

N   Total Cloud Cover Code

N   Total Frequency (in a Statistical Distribution) [Symbol]

N   Total Molar Amount of Matter [Symbol]

N   Total Molar Flux [Symbol]

N   Turns Ratio (of Transformer) [Symbol]

N   Yarn Number (in an Indirect System) [Symbol]

N- Nitrogen [Prefix Used with Organic Compounds]

N- United States of America [Civil Aircraft Marking]

-N Night-Operation (Aircraft) [US Navy Suffix]

$\overline{N}$ Average Noise in the Wavelength Interval λ_1 to λ_2 [Symbol]

$\dot{N}$ Nucleation Rate (of a Crystalline Solid) [Symbol]

N* Critical-Sized Nuclei (in Metallurgy) [Symbol]

N′ Number of Bohr Magnetons per Unit Volume [Symbol]

N^+ Number of Electrons with Positive (Polarization) Spin Components [Symbol]

N_+ Nucleation Site (in Metallurgy) [Symbol]

N^- Number of Electrons with Negative (Polarization) Spin Components [Symbol]

N_∞ Rydberg's Universal Series Constant [Symbol]

$N_\parallel$ Demagnetizing Factor Parallel to (or Along) c-Axis [Symbol]

$N_\perp$ Demagnetizing Factor Perpendicular to c-Axis [Symbol]

$N_\perp$ Number of Dislocations (in Materials Science) [Symbol]

N↑ Majority-Spin Component (in Physics) [Symbol]

N↑ Number of Valence Electrons with Spin Up [Symbol]

N↓ Minority-Spin Component (in Physics) [Symbol]

N↓ Number of Valence Electrons with Spin Down [Symbol]

N_0 Effective Cycles to Crack Initiation [Symbol]

N_0 Loschmidt's Number (6.022×10^{23} mol^{-1}) [Symbol]

N_0 Number of Molecules in (Energy) Ground Level [Symbol]

N_0 Number of Vertices (of Polyhedron) [Symbol]

N_0 Zero Crossings (in Fatigue Loading) [Symbol]

N_1 Number of Edges (of Polyhedron) [Symbol]

N_1 Primary Winding (of an Inductor) [Symbol]

N_2 Molecular Nitrogen [Symbol]

N_2 Number of Faces (of Polyhedron) [Symbol]

N   Charged Nitrogen Molecule [Symbol]

N^{3-} Nitride [Symbol]

N_3 Triazo (or Azido) (Radical) [Symbol]

N_3^- Azide Ion [Symbol]

N^{5+} Pentavalent Nitrogen Ion [Symbol]

N_{50} Number of Cycles for a 50 μm-Long Crack [Symbol]

N_{200} Number of Cycles for a 200 μm-Long Crack [Symbol]

N_A Acceptor-Atom Concentration (or Number of Acceptors) [Semiconductor Symbol]

N_A Areal Density (i.e., Number of Interceptions of Features Divided by Total Test Area) (in Quantitative Metallography) [Symbol]

N_A Avogadro Constant (6.022×10^{23} mol^{-1}) [Symbol]

N_A Number of Grains per Unit Area (in Metallography) [Symbol]

N_a Acceptor-Atom Density [Semiconductor Symbol]

N_α Atomic Density of Element α [Symbol]

N_b Number of Branches in Electric Circuit [Symbol]

N_{BO} Number of Bonding Orbitals per Atom [Symbol]

N_c Collector Dissipation [Semiconductor Symbol]

N_c Distance Between (Polymer) Crosslinking Points [Symbol]

N_c Effective Density of States in Conduction Band [Symbol]

N_c Nematic Phase along c-Axis [Symbol]

N_c Number of Effective Spring Coils [Symbol]

N_D Demagnetizing Factor [Symbol]

N_D Donor-Atom Concentration (or Number of Donors) [Semiconductor Symbol]

N_d Average Number of d-Electrons per Atoms [Symbol]

N_d Donor-Atom Density [Semiconductor Symbol]

N_d Linear Demagnetization Factor [Symbol]

N_d Number of Domains (in Solid-State Physics) [Symbol]

$N_d^\uparrow$ Average Number of Up-Spin d-Electrons per Atoms [Symbol]

$N_d^\downarrow$ Average Number of Down-Spin d-Electrons per Atoms [Symbol]

N_{dB} Power Gain, or Loss in decibels [Symbol]

N_{dBm} Power Gain, or Loss in decibel milliwatts [Symbol]

N_e Entanglement Threshold Chain Length (of Polymer) [Symbol]

N_e (Acoustic-Emission) Event Count [Symbol]

$\dot{N}_e$ (Acoustic-Emission) Event Count Rate [Symbol]

N_F Number of Particles per Square Millimeter [Symbol]

N_f Fatigue Life (i.e., Number of Cycles to Failure) [Symbol]

N_f Fatigue Life (i.e., Number of Cycles to Failure) [Symbol]

N_f Number of Cycles to Final Failure [Symbol]

N_{Foh} Fourier Number (of Heat Transfer) [Symbol]

N_{Fom} Fourier Number (of Mass Transfer) [Symbol]

N_{Fr1} Froude Number 1 (of Fluid Mechanics) [Symbol]

N_{Fr2} Froude Number 2 (of Fluid Mechanics) [Symbol]

N_G Griffith Crack Length (in Crack Theory) [Symbol]

N_{Gr} Grashof Number [Symbol]

N_h Helix Dissipation (in Electrical Engineering) [Symbol]

N_i Number of Cycles for Crack Initiation [Symbol]

N_i Number of Molecules in the *i*-th Energy Level [Symbol]

N_i Number of Scattering Centers of Species "i" per Unit (of Area, or Volume) (in Ion Spectroscopy Scattering) [Symbol]

N_j Number of Junctions (or Nodes) in Electric Circuit [Symbol]

N_K Knudsen Number [Symbol]

N_L Lineal Density (i.e., Number of Interceptions of Features Divided by Total Test Line Length) (in Quantitative Metallography) [Symbol]

N_L Loschmidt's Number (6.022×10^{23} mol^{-1}) [Symbol]

N_l Nematic Phase in Longitudinal Direction [Symbol]

N_l Number of Loops in Electric Circuit [Symbol]

N_{Le} Lewis Number [Symbol]

N_{Lo} Lorentz Number (in Physics) [Symbol]

N_m Avogadro's Number (6.022×10^{23} mol^{-1}) [Symbol]

N_m Media Noise (in Electronics) [Symbol]

N_{Ma} Mach Number [Symbol]

N_{NBO} Number of Nonbonding Orbitals per Atom [Symbol]

N_{Nu} Nusselt Number (of Forced Convection) [Symbol]

N_{Num} Nusselt Number (of Mass Transfer) [Symbol]

N_p Number of Cycles for Crack Propagation [Symbol]

N_p Number of Particles [Symbol]

N_p Number of Primary Turns (of Transformer) [Symbol]

N_ϕ Flow Value [Symbol]

N_ϕ Flux [Symbol]

N_{Pr} Prandtl (Convection) Number [Symbol]

N_{Ra1} Rayleigh Number 1 (of Fluid Mechanics) [Symbol]

N_{Re} Reynolds Number [Symbol]

N_s Number of Secondary Turns (of Transformers) [Symbol]

N_s Specific Speed (of Pumps and Turbines) [Symbol]

N_s Stability Factor (of Soil Embankment) [Symbol]

N_s Surface-Charge Density [Semiconductor Symbol]

N_{SH} Sherwood Number (of Mass Transfer) [Symbol]

N_{SH*} Modified Sherwood Number (of Mass Transfer) [Symbol]

N_{SS} Surface State Density [Symbol]

N_{Sc} Schmidt Number 1 [Symbol]

N_{Sh} Sherwood Number (of Mass Transfer) [Symbol]

N_{tf} Total Fixed Charge Density [Symbol]

N_v Number of Vacancies (per Unit Area) (in Physical Metallurgy) [Symbol]

N_V Number of Grains per Unit Volume (in Metallography) [Symbol]

N_V Total Number of Particles per Unit Volume (in Quantitative Metallography) [Symbol]

N_V Volumetric Density (i.e., Number of Features per Test Volume) (in Quantitative Metallography) [Symbol]

N_v Effective Density of States in Valence Band [Symbol]

N_v Electron-Vacancy Number [Symbol]

N_v Number of Vacancies (per Unit Area) (in Physical Metallurgy) [Symbol]

$(N_V)_j$ Number of Particles per Unit Volume in the j-th Class Interval (with j = 1 to 12) (in Quantitative Metallography) [Symbol]

N_{We1} Weber Number 1 (of Surface Tension) [Symbol]

N_{We2} Weber Number 2 (of Surface Tension) [Symbol]

N_{We3} Weber Number 3 (of Chemical Engineering) [Symbol]

N_x Linear Demagnetization Factor Along Short Axes of an Ellipsoid of Revolution [Symbol]

N_z Linear Demagnetization Factor Along Long Axes of an Ellipsoid of Revolution [Symbol]

N-13   Nitrogen-13 [also ^{13}N, or N^{13}]

N-14   Nitrogen-14 [also ^{14}N, N^{14}, or N]

N-15   Nitrogen-15 [also ^{15}N, or N^{15}]

N-16   Nitrogen-16 [also ^{16}N, or N^{16}]

N-17   Nitrogen-17 [also ^{17}N, or N^{17}]

2N   Diploid Number of Chromosomes [Symbol]

n   amount of substance [Symbol]

n   bond number [Symbol]

n   cavity number [Symbol]

n   charge (carrier) density [Symbol]

n   concentration [Symbol]

n   degree of polymerization [Symbol]

n   density of conduction electrons (in intrinsic semiconductors) [Symbol]

n   (Rosin-Rammler) dispersion factor [Symbol]

n   electrons (in extrinsic semiconductors) [Symbol]

n   exponent of pressure (in burning-rate equation) [Symbol]

n   factor of safety [Symbol]

n   foil normal (of specimen) [Symbol]

n   (natural) frequency [Symbol]

n   frequent fog [Subtype of Climate Region]

n   fringe number [Symbol]

n   gear ratio [Symbol]

n   (ASTM) grain-size number (in metallurgy) [Symbol]

n   ideality factor [Semiconductor Symbol]

n   index of refraction (or refractive index) [Symbol]

n   integer [Symbol]

n   molecular concentration [Symbol]

n   name [also N]

n   nano- [SI Prefix]

n   nanocrystal(line); nanophase; nanostructured

n   *(natus)* – born

n   neat

n   negative

n   nematicide [also N]

n   nerve; nervous(ness) [also N]

n   net [also N]

n   nets (in homogeneous connections of atoms) [Symbol]

n   neuter

n   neutron [Symbol]

n   neutron density [Symbol]

n   noncoherent (defect) [Metallurgy]

n   noon [also N]

n   normal [Organic Compounds]

n   normalize(d)

n   note [also N]

n   noun

n   nu [English Equilvalent]

n   nucleus [also N]

n   number [also N]

n   number density of (gas) molecules [Symbol]

n   number of active spring coils [Symbol]

n   number of atoms per unit cell [Symbol]

n   number of crystal defects (dislocations, vacancies, intersitials, etc.) [Symbol]

n   number of degrees of freedom [Symbol]

n   number of electron shells [Symbol]

n   number of electrons in an electrochemical reaction [Symbol]

n   number of conducting electrons per cubic meter [Symbol]

n   number of (stress) cycles (endured) [Symbol]

n   number of degrees of freedom (of molecules) [Symbol]

n   number of grains per square inch at magnification of $100\times$ (in metallurgy) [Symbol]

n   number of flux quanta per unit area [Symbol]

n   number of full length spring leaves [Symbol]

n   number of grains per linear inch (on abrasive wheels) [Symbol]

n   number of individual results (in statistics) [Symbol]

n   number of moles (in a system) [Symbol]

n   number of observations (in statistics) [Symbol]

n   number of particles (e.g., atoms, electrons, ions, molecules, photons, etc.) [Symbol]

n   number of revolutions [Symbol]

n   number of teeth (in machining) [Symbol]

n   number of teeth (of pinion) [Symbol]

n   number of terms in a finite series, or progression [Symbol]

n   number of tests [Symbol]

n   number of thread (of gear-worms) [Symbol]

n   number of turns (e.g. of shafts, etc.) [Symbol]

n   number of turns per unit length (of solenoid) [Symbol]

n   number of vibrations [Symbol]

n   number of years [Symbol]

n   number of x-ray photons [Symbol]

n   numeric(al)

n   order (of an interference peak) [Symbol]

n   order of reflection (in Bragg equation) [Symbol]

n   partial-discharge (corona) pulse rate [Symbol]

n   periodicity [Symbol]

n   polytropic exponent (in thermodynamics) [Symbol]

n   position number [Symbol]

n   principal quantum number [Symbol]

n   refractive index (or index of refraction) [Symbol]

n   resonance frequency [Symbol]

n   revolutions per unit of time [Symbol]

n   running index (e.g., the n-th term of a sequence, or series) [Symbol]

n   sample size (in statistics) [Symbol]

n   (liquid drop) size span (in Rosin-Rammler distribution) [Symbol]

n   (rotational) speed [Symbol]

n   spin (of particles) [Symbol]

n   (cyclic) strain-hardening exponent [Symbol]

n   third-direction cosine of a line (e.g., $n = \cos \gamma$) [Symbol]

n   total number of moles of a gas in given mixture [Symbol]

n   transport number (in physical chemistry) [Symbol]

n   (transformer) turns ratio (i.e., ratio of number of primary turns to number of secondary turns) [Symbol]

n   unit vector (normal to surface) [Symbol]

n   viscosity [Symbol]

n-   nanocrystalline (nanophase, or nanostructured) [Usually Followed by a Chemical Element, or Compound, e.g., n-Ni is Nanocrystalline Nickel, etc.]

n-   normal [Chemistry]

n-   n-type (semiconductor) [Solid-State Physics]

$\bar{n}$ antineutron [Symbol]

$\bar{n}$ average sample size (in statistics) [Symbol]

ñ   complex refractive index [Symbol]

n*   complex refractive index [Symbol]

n*   effective principal quantum number [Symbol]

n′   infrequent fog, but high humidity and low rainfall [Subtype of Climate Region]

n′   number of formula units per unit cell (for ceramics) [Symbol]

n^+ n-plus-type (semiconductor layer) [Symbol]

n^+ number of spin-up valence electrons [Symbol]

n_+ density of positive charge carriers (or electron holes) (of semiconductors) [Symbol]

n_+ positive ion density [Symbol]

n^- n-minus-type (semiconductor layer) [Symbol]

n^- number of spin-down valence electrons [Symbol]

n_- density of negative charge carriers (or electrons) (of semiconductors) [Symbol]

n_- negative ion density [Symbol]

n↑   density of states for up-spin electrons [Symbol]

n↑   up-spin (of particle) [Symbol]

n↓   density of states for down-spin electrons [Symbol]

n↓   down-spin (of particle) [Symbol]

n!   factorial of n [Symbol]

n_0 initial thrust-to-weight ratio (of space vehicle) [Symbol]

n_0 number of lattice atoms per unit volume (of semiconductor) [Symbol]

n_0 refractive index of (pure) solvent [Symbol]

n_0 stellar magnitude (of comets) [Symbol]

n_0 zero crossings (in fatigue loading) [Symbol]

n_1 molecular concentration of phase 1 (e.g. the solvent) [Symbol]

n_1 refractive index of medium 1 [Symbol]

$n_{1,2}$ relative refractive index from medium 1 to medium 2 [Symbol]

n^2 (transformer) impedance ratio (i.e., ratio of impedance in primary coil to impedance in secondary coil) [Symbol]

n^{20} refractive index at 20°C [Symbol]

n_{21} relative refractive index from medium 2 to medium 1 [Symbol]

n_{23} relative refractive index from medium 2 to medium 3 [Symbol]

n^{25} refractive index at 25°C [Symbol]

n_A number of moles of component A [Symbol]

n_A refractive index of Fraunhofer spectral line of terrestrial origin at 766.1 nanometers [Symbol]

n_a anion transport number (in physical chemistry) [Symbol]

n_a number density of atoms [Symbol]

n_a number of moles of (dry) air [Symbol]

n_{av} average refractive index [Symbol]

n_B (Bohr) magneton number (or saturation moment) [Symbol]

n_B refractive index of Fraunhofer spectral line of terrestrial origin at 686.7 nanometers [Symbol]

n_C refractive index of Fraunhofer spectral line for hydrogen at 656.3 nanometers [Symbol]

n_c critical concentration [Symbol]

n_{cal} calculated refractive index [Symbol]

n_D refractive index of Fraunhofer spectral line for sodium at 589.3 nanometers [Symbol]

n_D^{20} refractive index of Fraunhofer sodium D line (589.3 nm) at 20°C [Symbol]

n_{D1} refractive index of Fraunhofer spectral line (D_1 line) for sodium at 589.3 nanometers [Symbol]

n_{D2} refractive index of Fraunhofer spectral line (D_2 line) for sodium at 588.9 nanometers [Symbol]

n_{D3} refractive index of Fraunhofer spectral line (D_3 line) for sodium at 587.6 nanometers [Symbol]

n_E refractive index of Fraunhofer spectral line for iron at 526.9 nanometers [Symbol]

n_e effective (drainage) porosity (of a rock, or soil mass) [Symbol]

n_e electron density (in conductors) [Symbol]

n_e extraordinary refractive index [Symbol]

n_e number density of electrons (in semiconductor) [Symbols]

n_e plasma particle density [Symbol]

n_{eff} effective refractive index [Symbol]

n_F refractive index of Fraunhofer spectral line for hydrogen at 486.1 nanometers [Symbol]

n_f number of cycles to failure [Symbol]

n_G refractive index of Fraunhofer spectral line for iron at 430.8 nanometers [Symbol]

n_G' refractive index of Fraunhofer spectral line for hydrogen at 434.1 nanometers [Symbol]

n_H refractive index of Fraunhofer spectral line for calcium at 396.8 nanometers [Symbol]

n_H' refractive index of Fraunhofer spectral line for calcium at 393.4 nanometers [Symbol]

n_h number density of (electron) holes (in semiconductor) [Symbols]

n_i imaginary part of complex refractive index [Symbol]

n_i intrinsic (electron) carrier density (of semiconductor) [Symbol]

n_i number density of ions [Symbols]

n_i number of drops in sampling size interval [Symbol]

n_i number of moles of component i in a (gas) mixture [Symbol]

n_i valence of the i-th ion [Symbol]

n_i^s composition of the surface of component i [Symbol]

n_j recurrence rate of the j-th partial-discharge (corona) pulse [Symbol]

n_k cation transport number (in physical chemistry) [Symbol]

n_L number of grains intersected per unit length (in metallurgy) [Symbol]

n_m number of molecules [Symbol]

n_m number of moles (of a chemical element) [Symbol]

n_μ accepting phonon occupation number (in solid-state physics) [Symbol]

n_n density of negative charge carriers (or conduction electrons) (of intrinsic semiconductor) [Symbol]

n_n number-average degree of polymerization [Symbol]

n_n number of neutrons [Symbol]

n_o ordinary refractive index [Symbol]

n_p density of positive charge carriers (or electron holes) (of intrinsic semiconductor) [Symbol]

n_p number of protons [Symbol]

n_q number density of atoms in state q [Symbol]

n_R real part of complex refractive index [Symbol]

n_r real part of complex refractive index [Symbol]

n_s (charge) carrier density of sample [Symbol]

n_s density of superconducting electron pairs [Symbol]

n_s density of superfluid electrons [Symbol]

n_s refractive index of substrate [Symbol]

n_s synchronous speed (of synchronous electrical machines) [Symbol]

n_v number of moles of water vapor [Symbol]

n_v number of valence electrons [Symbol]

n_w weight-average degree of polymerization [Symbol]

n_{ws} Wigner-Seitz boundary (in solid-state physics) [Symbol]

ν Abbe number (or nu value) [Symbol]

ν accomodation coefficient (for gas transport) [Symbol]

ν amount of substance [Symbol]

ν (Landau electron level) filling factor [Symbol]

ν frequency (of electromagnetic radiation) [Symbol]

ν kinematic viscosity [Symbol]

ν molar volume [Symbol]

ν molecular density [Symbol]

ν moment about arbitrary origin (in statistics) [Symbol]

ν neutrino (in particle physics) [Symbol]

ν (lower-case) nu [Greek Alphabet]

ν number of molecules per unit area [Symbol]

ν Poisson's ratio [Symbol]

ν real part of complex refractive index [Symbol]

ν (magnetic) reluctivity [Symbol]

ν velocity [Symbol]

ν (atomic) vibration frequency [Symbol]

ν wave frequency [Symbol]

$\bar{\nu}$ wavelength [Symbol]

$\bar{\nu}$ wave number (of electromagnetic radiation) [Symbol]

ν_∞ Rydberg's fundamental frequency [Symbol]

$\nu^\dagger$ (vibrational) frequency for transition state [Symbol]

$\nu_\parallel$ Poisson's ratio for longitudinal (parallel) plane [Symbol]

$\nu_\perp$ Poisson's ratio for transverse (perpendicular) plane [Symbol]

$\nu_\perp$ roughening exponent [Symbol]

ν_0 basic nuclear magnetic resonance frequency [Symbol]

ν_0 characteristic frequency [Symbol]

ν_c critical Poisson's ratio [Symbol]

ν_c critical molar volume [Symbol]

ν_D Debye frequency (in solid-state physics) [Symbol]

ν_e (Landau) electron filling factor [Symbol]

ν_e nu neutrino (from β decay) (in particle physics) [Symbol]

$\bar{\nu}_e$ nu antineutrino (from β decay) (in particle physics) [Symbol]

ν_f Poisson ratio of film [Symbol]

ν_h (electron) hole filling factor [Symbol]

ν_{LT} Poisson's ratio for longitudinal plane (of a composite) [Symbol]

ν_m maximum frequency [Symbol]

ν_μ mu neutrino (from μ decay) (in particle physics) [Symbol]

$\bar{\nu}_\mu$ mu antineutrino (from μ decay) (in particle physics) [Symbol]

ν_{max} maximum frequency [Symbol]

ν_o oscillation frequency [Symbol]

ν_R resonant frequency of reference substance (in nuclear magnetic resonance spectroscopy) [Symbol]

ν_r reduced molar volume [Symbol]

ν_s Poisson ratio of substrate [Symbol]

ν_t propagation velocity [Symbol]

ν_t transverse wave [Symbol]

ν_τ tau neutrino (in particle physics) [Symbol]

$\bar{\nu}_\tau$ tau antineutrino (in particle physics) [Symbol]

$\mathbb{N}$ Natural Number [Symbol]

∇ Nabla (or Del Operator) [Symbol]

NA   Namibia [ISO Code]

NA   Naphthalene (Monomer)

NA   Naphthylamine

NA   (American) National Acme Thread

NA   National Arboretum [Washington, DC, US]

NA   National Archives (of Canada) [Ottawa]

NA   National Archives (of New Zealand) [Wellington]

NA   National Army

NA   National Association

NA   Natural Aging; Naturally Aged [Metallurgy]

NA   Naturally Aspirated (Engine)

NA   Natural Amber

NA   Nautical Almanac

NA   Naval Architect(ure)

NA   Naval Aviation; Naval Aviator

NA   Needle Roller Bearing, With Cage, Machined Ring, Lubrication Groove and Hole in Outer Diameter, Metric-Dimensioned [Symbol]

NA   Negative Acceleration

NA   Neodymium Aluminate

NA   Neolithic Age [Anthropology]

NA   Netherlands Antilles [West Indies]

NA   Network Analysis

NA   Network Application

NA   Network Architecture

NA   Neutral Atmosphere

NA   Neutral Area

NA   Neutral Axis

NA   Neutron Absorber; Neutron Absorption

NA   Neutron Activation

NA   Next Action

NA   Nickel Aluminide

NA   Nicotinamide

NA   Nicotinic Acid

NA   Nitric Acid

NA   No Access

NA   No Auxiliary Carry (Flag) [Computers]

NA   Node Access

NA   Nomina Anatomica

NA   Nonabrasive

NA   Nonacosadiynoic Acid

NA   Non-American

NA   Nonaqueous

NA   Non-Australian

NA   Noradrenaline [Biochemistry]

NA   Norddeutsche Affinerie AG [German Copper Producer located at Hamburg]

NA   Normal Acceleration

NA   Normal Atmosphere

NA   *(Normenausschuss)* – German for "Standards Committee"

NA   North Africa(n)

NA   North America(n)

NA   North Atlantic

NA   Not Adjustable; Not Adjusted

NA   Not Applicable [also N/A, or n/a]

NA   Not Assigned

NA   Not Authorized

NA   Not Available [also na]

NA   Now Available

NA   Nuclear Absorption

NA   Nuclear Alignment

NA   Nucleating Agent

NA   Nucleic Acid [Biochemistry]

NA   Nucleophilic Addition [Physical Chemistry]

NA   Number of Acquisition

NA Numerical Analysis

NA Numerical Aperture

NA Nurse's Aid

NA Nursing Administration

$N_s(A)$ Surface Density of A Atoms [Symbol]

N&A Nominations and Appointments Committee [of Institute of Electrical and Electronics Engineers, US]

N/A Next Assembly

N/A No Account [also n/a]

N/A Not Applicable [also n/a]

N/A Not Available [also NA]

Na *(Natrium)* – Sodium [Symbol]

Na^+ Sodium Ion [Symbol]

Na-20 Sodium-20 [also ^{20}Na, or Na^{20}]

Na-21 Sodium-21 [also ^{21}Na, or Na^{21}]

Na-22 Sodium-22 [also ^{22}Na, or Na^{22}]

Na-23 Sodium-23 [also ^{23}Na, Na^{23}, or Na]

Na-24 Sodium-24 [also ^{24}Na, or Na^{24}]

Na-25 Sodium-25 [also ^{25}Na, or Na^{25}]

nA nanoampere [Unit]

nA^2 square nanoampere [Unit]

na not applicable [also NA, N/A, or n/a]

na not available [also NA]

.na Namibia [Country Code/Domain Name]

NAA α-Naphthaleneacetic Acid; 1-Naphthaleneacetic Acid

NAA Naphthylacetic Acid

NAA National Academy of Arbitrators [US]

NAA National Aeronautic Association [US]

NAA National Aerosol Association [US]

NAA National Aggregates Association [US]

NAA National Airports Authority [UK]

NAA National Arborist Association [US]

NAA National Ash Association [now American Coal Ash Association, US]

NAA National Association of Accountants [US]

NAA Neutron Activation Analysis

NAA Nitriloacetic Acid

NAA North American Aviation

NAA North Atlantic Assembly [of NATO]

NAAA National Agricultural Aviation Association [US]

NAAA National Alarm Association of America [US]

NAAA National Auto Auction Association [US]

NAAASE National Association for Applied Arts, Science and Engineering [US]

NAAB National Architectural Accrediting Board [US]

NAAB National Association of Animal Breeders [US]

NAAC National Association of Agricultural Contractors [UK]

NaAc Sodium Acetate

Na(ACAC) Sodium Acetylacetonate [also Na(acac)]

NAACP National Association for the Advancement of Colored People

NAAD National Association of Aluminum Distributors [US]

NAAD Nicotinic Acid Adenine Dinucleotide [Biochemistry]

NAADC North American Air Defense Command

NAADP Nicotinic Acid Adenine Dinucleotide Phosphate [Biochemistry]

β-NAADP Nicotinic Acid Adenine Dinucleotide Phosphate [Biochemistry]

NAAE National Association of Aeronautical Examiners [US]

NAAE National Association of Agricultural Employees [US]

NAAEE North American Association for Environmental Education [US]

NAAFI Navy, Army, Air Force Institutions [also Naafi, UK]

NAAG NATO Army Armament Group

NAAG Nordic Association of Applied Geophysics [Sweden]

NAAFI Navy, Army, and Air Force Institute [US]

NAAI North American Aviation, Inc. [Downey, California, US]

NAAIDT National Association of Advisers for Design and Technology [UK]

NAAJ National Association of Agricultural Journalists [US]

NAAL North American Aerodynamic Laboratory [US]

n-α-Al_2O_3 Nanostructured Alpha Alumina

NAAM National Association of Architectural Metal Manufacturers [also NAAMM, US]

NAAMS North American Automotive Metric Standards [of A/SP (Auto/Steel Partnership) Consortium, US]

NAAMSA National Association of Automobile Manufacturers of South Africa

NAAN National Advertising Agency Network [US]

NAAO North Africa Area Office [of United Nations Children's Fund]

NAAQO National Ambient Air Quality Objectives [US]

NAAQS National Ambient Air Quality Standards [US]

NAAR National Association of Advertising Representatives [UK]

NAARS National Automated Accounting Research System [US]

NAAS National Agricultural Advisory Service [UK]

NAAS National Anorexic Aid Society [US]

NAAS National Association of Academies of Science [US]

NAAS Naval Auxiliary Air Station [of US Navy]

NAAS New Academic Appointments Scheme [UK]

NAASS North American Association of Summer Sessions [US]

NAATP New African Air Transport Policy

Na_2ATP Adenosine 5'-Triphosphate, Disodium Salt [Biochemistry]

NAATS National Association of Air Traffic Specialists [US]

NAATS National Association of Auto Trim Shops [US]

NAAWER National Association of Arc Welding Equipment Repairers [UK]

NAAWS NATO Anti-Air Warfare System

NAAWS NORAD (North American Aerospace Defense Command) Automatic Attack Warning System

NAB National Aircraft Beacon

NAB National Alliance of Businessmen [US]

NAB National Assistance Board [UK]

NAB National Association of Bank Servicers [US]

NAB National Association of Bookmakers [UK]

NAB National Association of Broadcasters [US]

NAB Naval Amphibious Base [of US Navy]

NAB Navigational Aid to Bombing

NAB Nuclear Assembly Bulding

NABADA National Barrel and Drum Association [US]

Nabam Sodium N,N-Dimethyldithiocarbamate [also nabam]

NABAS National Association of Balloon Suppliers [UK]

NA Bau Normenausschuss Bauwesen [Standards Committee for the Building and Construction Industry; of DIN Deutsches Institut für Normung, Germany]

NABBP North American Bird Banding Program [of US Geological Survey/Biological Resources Division and Canadian Wildlife Service]

NABC North American Blueberry Council [US]

NABCE National Association of Black Consulting Engineers [US]

NABCO National Association of the Building Co-ops Society [Ireland]

NABD National Association of Brick Distributors [US]

NABD Normenausschuß Bibliotheks- und Dokumentationswesen [Standards Committee on Library Science and Documentation, Germany]

NABDC National Association of Blueprint and Diazo Coaters [now Association of Reproduction Materials Manufacturers, US]

NABE National Association for Business Education [US]

NABER National Association of Business and Educational Radio [US]

NABET National Association of Broadcast Employees and Technicians [Canada]

NABGG National Association for Black Geologists and Geophysicists [US]

NaBi Sodium Bismuth (Compound)

NABIS National Association of Business and Industrial Saleswomen [US]

NABM National Association of Boat Manufacturers [US]

NABM National Association of British Manufacturers [now Confederation of British Industry, UK]

NABMA National Association of British Market Authorities [UK]

NaBO Sodioborate

NABP National Association of Black Professors [US]

NABP National Association of Boards of Pharmacy [US]

NABR National Association for Biomedical Research

NaBr Sodium Bromide (Compound)

NABS National Association of Black Students [US]

NABS North American Benthological Society [US]

NABS Nuclear-Armed Bombardment Satellite

NaBS Sodium Benzenesulfonate

NABSC National Association of Building Service Contractors [now Building Service Contractors Association International, US]

NABST National Advisory Board on Science and Technology [Canada]

NABT National Association of Biology Teachers [US]

NABTS North American Basic Teletext Specification

NABTS North American Broadcast Teletext Specification

NaBuO Sodium Butoxide (or Sodium Butylate)

NAC N-Acetylcysteine [Biochemistry]

NAC National Academy of Consiliators [US]

NAC National Accelerator Center [Faure, South Africa]

NAC National Action Committee

NAC National Advisory Committee

NAC National Agency Checks

NAC National Agricultural Center [UK]

NAC National Archives of Canada

NAC National Asbestos Council [US]

NAC National Asthma Center [US]

NAC National Aviation Club [US]

NAC Naval Avionics Center [of US Navy]

NAC Network Access Controller

NAC Network Adapter Card

NAC Network Advisory Committee [of Library of Congress, Washington, DC, US]

NAC Nitroacetylcellulose

NAC Noise Advisory Council [UK]

NAC Non-Aligned Countries

NAC Non-Aβ Component Amyloid Protein [Biochemistry]

NAC North American (Welding Research) Conference [US]

NAC North Atlantic Council [of NATO]

NAC North Atlantic Current [Oceanography]

Nac Nacelle

Nac Nitroacetic Acid

NACA National Academy of Code Administrators [US]

NACA National Advisory Committee for Aeronautics [Now Part of NASA]

NACA National Agricultural Chemicals Association [US]

NACA National Air Carrier Association [US]

NACA National Armored Car Association [US]

NACA National Association for Clean Air [South Africa]

NACA National Association of Campus Activities [US]

NACA National Association of Cellular Agents [US]

NACA National Autosound Challenge Association [US]

NACAA National Association of Computer-Assisted Analysis [US]

NACAA National Association of County Agricultural Agents [US]

NACAB National Agricultural Center Advisory Board [UK]

NACAC National Association of College Admission Counselors [US]

NACAM National Association of Corn and Agricultural Merchants [UK]

NACAR National Advisory Committee on Aeronautical Research [South Africa]

NACA Rep National Advisory Committee for Aeronautics Report [US]

NACA RM National Advisory Committee for Aeronautics Research Memorandum [US]

NACAS National Advisory Committee on Agricultural Services [Canada]

NaCaSiON Sodium Calcium Silicon Oxynitride

NACAT National Association of College Automotive Teachers [US]

NACATTS North American Clear Air Turbulence Tracking System

NACB National Accreditation of Certification Bodies

NACC National Advisory Cancer Council [US]

NACC National Anti-Counterfeiting Committee [Taiwan]

NACC National Automatic Controls Conference [US]

NACC North American Calibration Cooperation

NACC North American Coal Corporation [US]

NACC North American Control Committee

NACCG National Association of Crankshaft and Cylinder Grinders [UK]

NACCS National Association of Commodity Cargo Superintendents [UK]

NACD National Association of Chemical Distributors [US]

NACD National Association of Computer Dealers [US]

NACD National Association of Conservation Districts [US]

NACD National Association of Container Distributors [US]

NACDG North American Contact Dermatitis Group

NACDRAO National Association of College Deans, Registrars and Admission Officers [US]

NACE National Association of Corrosion Engineers [now NACE International, US]

NACE National Association of County Engineers [US]

NACE North American Commission on the Environment

NACE(E) National Association of Corrosion Engineers (Europe) [UK]

NACE Expo Annual Conference of the National Association of Corrosion Engineers International

NACEIC National Advisory Council for Education, Industry and Commerce [UK]

NACE Int'l NACE (National Association of Corrosion Engineers) International [formerly National Association of Corrosion Engineers, US]

NACEO National Advisory Council on Economic Opportunity

NAcGu N-Acetyl Guanidine [Biochemistry]

NACHA National Automated Clearinghouse Association [US]

NACHDC National Advisory Child Health and Human Development Council [US]

Nachr Dok Nachrichten für Dokumentation [Publication on Documentation of Deutsche Gesellschaft für Dokumentation, Germany]

Nachr Electron + Telemat Nachrichten Elektronik + Telematik [German Journal on Communication, Electronics and Telematics]

Nachr tech Nachrichtentechnik [German Journal on Communication Engineering]

Nachr tech Elektron Nachrichtentechnik Elektronik [German Journal on Communication Engineering and Electronics]

NaCHS Sodium Cyclohexanesulfonate

NACIS North American Cartographic Information Society [US]

NACISO NATO Communications and Information Systems Organization

NACK National Advisory Committee on Kangaroos [now National Consultative Committee on Kangaroos of the Australian Nature Conservancy Agency]

NACK Negative Acknowledge(ment) [also NAK]

NACL National Advisory Commission Libraries [US]

NaCl Sodium Chloride (or Halite) (Structure)

NACLE National Association of Chimney Lining Engineers [UK]

NACM National Association of Chain Manufacturers [US]

NACM National Association of Colliery Managers [UK]

NACM National Association of Cotton Manufacturers [UK]

nA/cm² nanoampere(s) per square centimeter [also nA cm^{-2}]

NACMA National Armored Cable Manufacturers Association [US]

NACME National Action Council for Minorities in Engineering [US]

NaCN Sodium Cyanide

NACO National Association of Colliery Overmen [UK]

NACO Navy Cool Propellants

NACOA National Advisory Committee on Oceans and Atmosphere [US]

NACOLADS National Council on Libraries, Archives and Documentation Services [Jamaica]

NACOM National Communication System [of US Department of Defense]

NACOSH National Advisory Committee on Occupational Safety and Health [US]

NACOSS National Approval Council for Security Systems [UK]

NACPD National Association of County Planning Directors [US]

NACR National Association of Chemical Recyclers [US]

NACS National Advisory Committee on Semiconductors [US]

NACS North American Catalysis Society

NACS Northern Area Communications System

NACSA North American Computer Service Association [US]

NACSC National Association of Cold Storage Contractors [US]

NACSIC National Association of Cold Storage Insulation Contractors [now NACSC, US]

NACSM North American Commission on Stratigraphic Nomenclature

NACT National Association of Cycle Traders [UK]

NACTA National Association of Colleges and Teachers of Agriculture [US]

NACUBO National Association of College and University Business Officers [US]

NACUFS National Association of College and University Food Services [US]

NAD National Academy of Design [US]

NAD Naval (or Navy) Ammunition Depot [US]

NAD β-Nicotinamide Adenine Dinucleotide [also β-NAD, NADIDE, or Nadide] [Biochemistry]

NAD No-Acid Descaling

NAD Node Administration

NAD Noise Amplitude Distribution

NAD North Atlantic Deep

NAD Nuclear Accident Dosimetry

NAD Nuclear Adiabatic Demagnetization

NaD Sodium Deuteride

α-NAD α-Nicotinamide Adenine Dinucleotide [Biochemistry]

β-NAD β-Nicotinamide Adenine Dinucleotide [also NAD, NADIDE, or Nadide] [Biochemistry]

Nad Noradrenaline [also nad] [Biochemistry]

NADA National Automobile Dealers Association [US]

NAD-ADH Nicotinamide Adenine Dinucleotide/Alcohol Dehydrogenase [Biochemistry]

NADAF National Association of Decorative Architectural Finishes [US]

NADase NAD (Nicotinamide Adeninen Dinucleotide) Glycohydrolase [Biochemistry]

NADase Nicotinamide Adenine Dinucleosidase [Biochemistry]

NADB National Air Data Branch [US]

NADC National Association of Demolition Contractors [US]

NADC National Association of Dredging Contractors [US]

NADC NATO Air Defense Committee

NADC Naval Air Development Center [of US Navy at Johnsville/Warminister, Pennsylvania, US]

NADC Nothern Alberta Development Council [Canada]

NADCA National Animal Damage Control Association [US]

NADCA National Air Duct Cleaners Association [Canada]

NADCA North American Die Casting Association [US]

NADCAP National Aerospace and Defense Contractors Accreditation Program [US]

NADCI North American Die Casting Institute [US]

NADD National Association of Die Makers and Die Cutters [US]

NADD Labeled Nicotinamide Adenine Dinucleotide [Biochemistry]

NADDRG North American Deep Drawing Research Group

NADE National Association for Design Education [UK]

NADEC National Association of Development Education Centers [UK]

NaDEC Sodium Diethyl Dithiocarbamate

NADEEC NATO Air Defense Electronic Environment Committee [Part of NATO Air Defense Committee]

NADEFCOL NATO Defense College

NADES National Association of Drop Forgers and Stampers [UK]

NADF National Arbor Day Foundation [US]

NADGE NATO Air Defense Ground Environment

NADGECO NATO Air Defense Ground Environment Corporation

NADH Hydrogenated Nicotinamide-Adenine Dinucleotide [also NADH₂] [Biochemistry]

NADH α-Nicotinamide Adenine Dinucleotide, Reduced Form [also α-NADH, or NADH₂] [Biochemistry]

α-NADH α-Nicotinamide Adenine Dinucleotide, Reduced Form [Biochemistry]

β-NADH β-Nicotinamide Adenine Dinucleotide, Reduced Form [also NADH] [Biochemistry]

NADH₂ β-Nicotinamide Adenine Dinucleotide, Reduced Form [also NADH, or β-NADH] [Biochemistry]

NADH-FMN β-Nicotinamide Adenine Dinucleotide, Reduced Form–Flavin Mononucleotide (Oxidoreductase) [Biochemistry]

NADI National Association of Display Industries [US]

NADIBO North American Defense Industrial Base Organization

NADIDE β-Nicotinamide Adenine Dinucleotide [also Nadide, NAD, or β-NAD] [Biochemistry]

NADL National Animal Disease Laboratory [US]

NaDMC Sodium Dimethyl Dithiocarbamate

NADMR National Association of Diversified Manufacturers Representatives [now National Association of General Merchandise Representatives, US]

NADMC Naval Air Development and Material Center [US]

NADO National Association of Development Organizations [US]

NADOA National Association of Division Order Analysts [US]

NADP National Atmospheric Deposition Program [US]

NADP β-Nicotinamide Adenine Dinucleotide Phosphate [also β-NADP] [Biochemistry]

NADP Northern Alberta Dairy Pool [Canada]

α-NADP α-Nicotinamide Adenine Dinucleotide Phosphate [Biochemistry]

β-NADP β-Nicotinamide Adenine Dinucleotide Phosphate [also NADP] [Biochemistry]

3'-NADP β-Nicotinamide Adenine Dinucleotide 3'-Phosphate [Biochemistry]

NADPH Nicotinamide Adenine Dinucleotide Phosphate, Reduced Form [also β-NADPH] [Biochemistry]

α-NADPH α-Nicotinamide Adenine Dinucleotide Phosphate, Reduced Form [Biochemistry]

β-NADPH β-Nicotinamide Adenine Dinucleotide Phosphate, Reduced Form [also NADPH] [Biochemistry]

3'-NADPH β-Nicotinamide Adenine Dinucleotide 3'-Phosphate, Reduced Form [Biochemistry]

NADPH-FMN Nicotinamide Adenine Dinucleotide Phosphate, Reduced Form–Flavin Mononucleotide (Oxidoreductase) [Biochemistry]

NADTP National Association of Desktop Publishers [US]

Na5DTPA Pentasodium Diethylenetriaminepentaacetic Acid

NADW North Atlantic Deep Water

NADWARN Natural Disaster Warning System [US]

Nadezhn Dolgovech Mash Sooruzh Nadezhnost i Dolgovechnost Mashin i Sooruzhenii [Russian Journal]

NAE National Academy of Engineering [US]

NAE National Aeronautical Establishment [of National Research Council of Canada]

NAE Not Above, or Equal

NAE Nursery Association Executives [now Nursery Association Executives of North America, US]

NAEB National Association of Educational Broadcasters [US]

NAEC National Advisory Eye Council [US]

NAEC National Aeronautical Establishment of Canada [of National Research Council of Canada]

NAEC National Aerospace Education Council [US]

NAEC National Agricultural Engineering Corporation [PR China]

NAEC National Association of Elevator Contractors [US]

NAEC National Association of Engineering Companies [US]

NAEC National Association of Exhibition Contractors [UK]

NAEC Naval Air Engineering Center [of US Navy at Philadelphia, Pennsylvania, US]

NAE(CAN) National Aeronautical Establishment (Canada) [of National Research Council of Canada]

NAECON National Aerospace Electronics Conference [of Institute of Electrical and Electronic Engineers, US]

NAED National Association of Electrical Distributors [US]

NAEDA North American Equipment Dealers Association [US]

NAEDS National Association of Educational Data Systems

Na$_2$-EDTA Ethylenediaminedtetraacetic Acid, Disodium Salt

Na$_4$-EDTA Ethylenediaminedtetraacetic Acid, Totally Deprotonated Form

NAEE National Association for Environmental Education [UK]

NAEE National Association for Environmental Education [Name Changed to North American Association of Environmental Education, US]

NAEE North American Association of Environmental Education [US]

NAEF North American Environment Fund

NAEGIS NATO Airborne Early Warning/Ground Environment Integrated Segment

NAEM National Association for Environmental Management [US]

NAEM National Association of Explosion Managers [US]

NAEMC National Association of Export Management Companies [now National Association of Export Companies, US]

NAENA Nursery Association Executives of North America [US]

NAEP National Assessment of Educational Progress [US]

NAEP National Association of Environmental Professionals [US]

NAES Naval Air Experimental Station [US Navy]

NAESA National Association of Elevator Safety Authorities [US]

NAESCO National Association of Energy Service Companies [also NAESC, US]

NAET National Association of Educational Technicians

NaEtO Sodium Ethoxide (or Sodium Ethylate)

NAEW NATO Airborne Early Warning

NAEWCSPMO NATO Airborne Early Warning and Control System Program Management Organization

NAF Naval Aircraft Factory [Philadelphia, Pennsylvania, US]

NAF Naval Air Facility [of US Navy]

NAF Naval Avionics Facility [of US Navy]

NAF Netherlands Atomic Forum

NAF Network Access Facility

N Af North Africa(n)

NAf Netherlands Antillean Florin (or Guilder) [also NAf; Currency]

NaF Sodium Fluoride (Flux)

NAFA National Aboriginal Forestry Association [Canada]

NAFA National Association of Fleet Administrators [US]

NAFAG NATO Air Force Armaments Group

NAFAX National Facsimile Network Circuit

NAFB National Association of Farm Boardcasters [US]

NAFB&AE National Association of Farriers, Blacksmiths and Agricultural Engineers [UK]

NAFC National Anti-Fluoridation Campaign [UK]

NAFC National Association of Food Chains [US]

NAFC National Average Fuel Consumption

NAFC North-American Forestry Commission [US]

NAFC Northwest Atlantic Fisheries Center [Canada]

NAFCC Northern Aquatic Food Chain Contaminants (Database) [Canada]

NAFCD National Association of Floor Covering Distributors [US]

NAFCILLIN 6-(2-Ethoxy-1-Naphthamido)penicillin [also Nafcillin] [Biochemistry]

NAFCO North Atlantic Fisheries Consultative Committee

NAFCO Northwest Atlantic Fishery Consultative Organization

NAFD Northern Alberta Forest District [Canada]

NAFEC National Administrative Facilities Experimental Center [US]

NAFEC National Aviation Facilities Experimental Center [US]

NAFED National Association of Fire Equipment Distributors [US]

NAFEM National Association of Food Equipment Manufacturers [US]

NAFEO National Association for Equal Opportunity in Higher Education [US]

n-α-Fe$_2$O$_3$ Nanostructured Alpha Iron (III) Oxide (or Nanostructured Hematite)

NAFFP National Association of Frozen Food Producers [UK]

NAFI National Association of Fire Investigators [US]

NAFI National Association of Forest Industries

NAFI Naval Avionics Facility, Indianapolis [US Navy]

NAFIC National Alcohol Fuels Information Center [US]

NAFIPS North American Fuzzy Information Processing Society [US]

NAFO National Association of Fire Officers [UK]

NAFO Northwest Atlantic Fisheries Organization [Canada]

NaFo Sodium Formate

NAFP National Academy for Fire Prevention [of US Department of Commerce] [Now Part of Federal Emergency Management Agency]

N Afr North Africa(n)

NAFRD National Association of Fleet Resale Dealers [US]

NAFS National Association of Fastener Stockholders [US]

NAFSA National Association for Foreign Student Affairs [US]

NAFSA National Fire Services Association [UK]

NAFSWMA National Association of Flood and Storm Water Management Agencies [US]

NAFTA North American Free Trade Act

NAFTA North American Free Trade Agreement [Between Canada, the USA and Mexico]

NAFTA North Atlantic Free Trade Area

NAFTC National Association of Freight Transportation Consultants [US]

NAFTC North American Forging Technology Conference

NAFTZ National Association of Foreign Trade Zones [US]

NAFV National Association of Federal Veterinarians [US]

NAFWR National Association of Warehousemen and Removers [UK]

NAG N-Acetyl Glucosamine [Biochemistry]

NAG NASA Grant [Designator]

NAG National Academy of Geosciences [US]

NAG National Association of Goldsmiths [UK and Ireland]

NAG National Gallery of Canada

NAG Non-Attaching Gas

NAG Numerical Algorithms Group

n-Ag Nanocrystalline Silver

NAGA North American Game Bird Association [US]

NAGARA National Association of Government Archives and Records Administrators [US]

NAGARD NATO Advisory Group for Aeronautical Research and Development

NAGC National Association of Government Communicators [US]

NAGCD National Association of Glass Container Distributors [now National Association of Container Distributors, US]

NAGDM National Association of Garage Door Manufacturers [US]

NAGE National Association of Government Engineers [US]

NAGGU North American Geology and Geophysics University [at University of South Dakota, Vermillion, US]

NAGHSR National Association of Governors Highway Safety Representatives [US]

NAGI&QAP National Association of Government Inspectors and Quality Assurance Personnel [US]

NAGIS Natal/Kwa-Zulu Association for Geographical Information Systems [South Africa]

NAGLO National Association of Governmental Labor Officials [US]

NAGLSTP National Association of Gay and Lesbian Scientists and Technical Professionals [US]

NAGMC North Atlantic Council and Military Committee [of NATO]

NAGMR National Association of General Merchandise Representatives [US]

Nagoya Math J Nagoya Mathematical Journal [of Nagoya University, Japan]

NAGT National Association of Geoscience Teachers [US]

NAH Nordic Association for Hydrology [Sweden]

NaH Sodium Hydride

NAHA National Association for Holistic Aromatherapy [US]

NAHAD National Association of Hose and Accessories Distributors [US]

NAHB National Association of Home Builders [Washington, DC, US]

NAHBO National Association of Hospital Broadcasting Organizations [UK]

Na$_x$Hg$_y$ Sodium Amalgam [General Formula]

Na$_2$HNTA Nitrilotriacetic Acid, Disodium Salt

NAHR National Association of Housing and Redevelopment [US]

NAHRI National Animal Husbandry Research Institute [Denmark]

NAHU North American Honeywell Users Association [US]

NAI N-Acetylimidazole

NAI National Aerospace Institute [Bangalore, India]

NAI National Association of Interpretation [US]

NAI New Alchemy Institute [US]

NAI Northrop Aeronautical School [US]

NaI Sodium Iodide

NAIC National Astronomy and Ionospheric Center [of Cornell University, Ithaca, New York, US]

NAICAA North American Initiative for Copper Architectural Applications (Campaign) [of Copper Development Association, US, and Canadian Copper and Brass Development Association]

NAICC National Association of Independent Computer Companies [US]

NAICU National Association of Independent Colleges and Universities [US]

NAID National Association of Industrial Distributors [UK]

NAID National Association of Installation Developers [US]

NAIDA National Agricultural and Industrial Development Association [UK]

NAIDM National Association of Insecticide and Disinfectant Manufacturers [US]

NAIEC National Association for Industry Education Cooperation [US]

NAIF Navigation and Ancillary Information Facility [NASA Planetary Data System]

NAIG Nippon Atomic Industry Group [Japan]

NAILD National Association of Independent Lighting Distributors [US]

NAILS NCAR Airborne Infrared Lidar System [of National Center for Atmospheric Research, Boulder, Colorado, US]

NAIMA North American Insulation Manufacturers Association

NAIN National Antimicrobial Information Network [of US Environmental Protection Agency]

NAIOP Navigational Aid Inoperative for Parts

NAIPRC Netherlands Automatic Information Processing Research Center

NAIR National Institute for Advanced Interdisciplinary Research [of Agency of Industrial Science and Technology, Ibaraki, Japan]

NAIRD National Association of Independent Record Distributors and Manufacturers [US]

NAIRO National Association of Intergroup Relations Officers [US]

NAIT National Association of Industrial Technology [US]

NAIT Northern Alberta Institute of Technology [Edmonton, Canada]

NaI(Th) Thorium-Doped Sodium Iodide

NaI(Tl) Thallium-Doped Sodium Iodide

NAITTE National Association of Industrial and Technical Teacher Educators [US]

NAIWC National Association of Inland Waterway Carriers [UK]

NaIX Sodium Isopropyl Xanthogenate

NAJ National Academy of Japan

NAJ National Association of Jewelers [US]

NAK Negative Acknowledge(ment) (Character) [also NACK]

NAK Node Access Kit

NaK Sodium-Potassium Alloy

Na/K Sodium/Potassium (Ratio)

NAL National Accelerator Laboratory [of US Department of Energy]

NAL National Acoustics Laboratory

NAL National Aeronautical Laboratory [India]

NAL National Aerospace Laboratories [Bangalore, India]

NAL National Aerospace Laboratory [Japan]

NAL National Agricultural Library [of US Department of Agriculture]

NAL National Airlines [US]

NAL Naval Aeronautical Laboratory [US Navy]

NAL New Assembly Language

NALC National Association of Litho Clubs [US]

NALC North American Landscape Characterization [Joint Project of the US Environmental Protection Agency and the US Geological Survey]

NALCD National Agricultural Library and Center for Documentation [Hungary]

Nalcom Naval Logistics Command [of Royal Singapore Navy]

NALF North American Loon Fund [US]

n-AlGaAs N-Type Aluminum Gallium Arsenide (Semiconductor)

n$^+$-AlGaAs N-Plus-Type Aluminum Gallium Arsenide (Semiconductor)

n-AlGaAs/GaAs N-Type Aluminum Gallium Arsenide/Gallium Arsenide (Superlattice)

n-AlGaAs/GaAs SL N-Type Aluminum Gallium Arsenide/Gallium Arsenide Superlattice

NALGHW National Association of Local Governments on Hazardous Waste [US]

NALGO National and Local Government Officers' Association [UK]

NALHC North American Log Homes Council [US]

NALM National Association of Lift Makers [UK]

NALMCO National Association of Lighting Management Companies [US]

NALMS North American Lake Management Society [US]

n-Al$_2$O$_3$ Nanocrystalline Aluminum Oxide (or Nanocrystalline Alumina)

NALPM National Association of Lithographic Plate Manufacturers [US]

NALR National Association of Lighting Representatives [US]

NAM N-Acetylmethionine

NAM N-Acetylmuramic Acid

NAM N-(9-Acridinyl)maleimide

NAM National Air Museum [of Smithsonian Institution, Washington, DC, US]

NAM National Association of Manufacturers [US]

NAM National Astronomy Meeting [UK]

NAM National Atomic Museum [Albuquerque, New Mexico, US]

NAM National Aviation Museum [US]

NAM Nautical Air Mile

NAM Navy Aerosol Model [of US Navy]

NAM Neutron-Absorbing Material

NAM Network Access Machine

NAM Network Access Method

NAM Nonaligned Movement

NAM Non-Aqueous Media

NAM Northrup-Allison-McCammon (Calculation)

NAM Number Assignment Module

N Am North America(n)

Nam Vietnam [In US Linguistic Usage]

NAMA National Agri-Marketing Association [US]

NAMA National Association of Mathematics Adsvisors [UK]

NAMA National Automatic Merchandising Association [US]

NAMA North American Mycological Association [US]

NAMAS National Measurement Accreditation Service [of Civil Aviation Authority, UK]

NAMB National Agricultural Marketing Board

NAMBO National Association of Motor Bus Operators [US]

NaMBT Sodium-2-Mercaptobenzothiazole

NAMC Naval Air Material Center [US Navy]

NAMC Naval Aviation Medical Center [of US Navy at Pensacola, Florida, US]

NAMCC National Advanced Materials Committee of China [PR China]

NAME National Association of Marine Enginebuilders [UK]

NAME National Association of Marine Engineers [Canada]

NAME National Association of Maritime Educators [US]

NAME National Association of Media Educators [US]

NAME National Association of Name Plate Manufacturers [US]

NAME Nitroarginine Methyl Ester [Biochemistry]

D-NAME Nω-Nitro-D-Arginine Methyl Ester [Biochemistry]

L-NAME Nω-Nitro-L-Arginine Methyl Ester [Biochemistry]

NaMeO Sodium Methoxide (or Sodium Methylate)

NAMEPA National Association of Minority Engineering Program Administrators [US]

N Amer North America(n)

NAMES National Association of Medical Equipment Suppliers [US]

NAMES NAVDAC (Navigation Data Assimilation Computer) Assembly, Monitor, Executive System

NAMF National Association of Metal Finishers [US]

NAMF North American Multifrequency (Signal)

NAMFI NATO Missile Firing Installation

NAMG Narrow-Angle Mars Gate [of NASA]

NAMH National Association for Mental Health [US]

NAMHO National Association of Mining History Organizations [UK]

NAMI National Association of Malleable Iron Founders [UK]

NAMI Naval Aerospace Medical Institute

Na-4-Mica Sodium-4-Mica [also Na-4-mica] [*Note:* A Synthetic Mica with 2 Sodium Ions Replacing the Potassium Ions]

N Am Ind North American Indian(s)

NAML Naval Aircraft Materials Laboratory [UK]

NAMM National Association of Margarine Manufacturers [US]

NAMM National Association of Mirror Manufacturers [US]

NAMM Number Average Molar Mass

NAMMA NATO Multi-Role (Combat Aircraft Development and Production) Management Agency

NAMMC Natural Asphalt Mine Owners and Manufacturers Council [UK]

NAMMO NATO Multi-Role (Combat Aircraft Development and Production) Management Organization

NAMO National Agricultural Marketing Officials [US]

NAMO National Association of Marketing Officers [US]

NAMOA National Association of Miscellaneous, Ornamental and Architectural Products Contractors [US]

NAMPP National Advanced Materials Program Plan [US]

NAMPS Narrow-Band Analog Mobile Phone Service

NAMPS National Association of Marine Products and Services [US]

NAMPUS National Association of Master Plumbers of the United States

NAMRAD Non-Atomic Military Research and Development

NAMRC North American Manufacturing Research Conference

NAMRI North American Manufacturing Research Institute [of Society of Manufacturing Engineers, US]

NAMRI/SME North American Manufacturing Research Institute of the Society of Manufacturing Engineers [US]

NAMRU Naval Medical Research Unit [of US Navy]

NAMS National Air Monitoring Service [US]

NAMS National Association of Marine Services [US]

NAMS National Association of Marine Surveyors [US]

NAMS Network Analysis and Management System

NAMS North American Membrane Society [US]

NAMS Notices of the American Mathematical Society [US]

NAMSA NATO Maintenance and Supply Agency

NAMSO NATO Maintenance and Supply Organization

NAMTA National Art Materials Trade Association [US]

NAMTC Naval Air Missile Test Center [US Navy]

NAMTS Nippon Automatic Mobile Telephone System [Japan]

NAMUR Normenarbeitsgemeinschaft für Meß–und Regelungstechnik in der Chemischen Industrie [Standards Working Group for Measurement Technology and Control Engineering in the Chemical Industry, Germany]

NAMW Number-Average Molecular Weight [Polymer Science]

NAN Neuraminidase [Biochemistry]

NAN Network Access Node

NAN Network Application Node

NAN-190 1-(2-Methoxyphenyl)-4-[4-(2-Phthalimdo)butyl]-piperazine

N/A/N Niobium-Alumina-Niobium (Trilayer)

NANA N-Acetylneuraminic Acid

NANA North American Newspaper Alliance

NANBA North American National Broadcasters Association

NANCO National Association of Noise Control Officials [US]

NANCRFUG North American NCR (National Cash Register) Financial Users Group [US]

NAND Not-And

NANEAP North Africa, Near East, Asia and Pacific (Regions)

NANEP Navy Air Navigation Electronics Project [US Navy]

NANFA North American Native Fishes Association [US]

NANFSM National Association of Nonferrous Scrap Metal Merchants [UK]

NANI National Academy of Nannies, Inc. [Denver, Colorado, US]

Na-23 NMR Sodium-23 Nuclear Magnetic Resonance [also ^{23}Na NMR]

NANO International Conference on Nanostructured Materials [of National Institute for Standards and Technology, US] [also Nano]

nano- SI prefix representing 10^{-9}

NaNO Sodionitrate

Nanocryst Nanocrystal(line) [also nanocryst]

NanoFAB Center Center for Nanostructure Materials and Quantum Device Fabrication [of Texas A&M University, College Station, US]

Nanostruct Mater Nanostructured Materials [Journal published in the US]

Nanot Nanomaterials and Technology Program [of Institute for Materials Research, Singapore] [also nanot]

Nanotechnol Nanotechnologist; Nanotechnology [also Nanotech]

Nanotechnol Nanotechnology [Publication of the Institute of Physics, UK]

NANS National Automated Nesting System

NANS Nevada Academy of Natural Sciences [US]

NANU National Association of NIDS (National Investor Data Service) Users [US]

NANWEP Navy Numerical Weather Prediction [US]

NAO National Astronomical Observatory [Tokyo, Japan]

NAO Natural Active Orbital

NAO Neodymium Aluminum Oxide [NdAlO$_3$]

NAO North Atlantic Ocean

NAO North Atlantic Oscillation

NaOAc Sodium Acetate

NaOBu Sodium Butoxide (or Sodium Butylate)

NaOCl Sodium Hypochlorite

NaOCN Sodium Cyanate

NaOD Sodium Deuteroxide

NaOEt Sodium Ethoxide (or Sodium Ethylate)

NAOGE National Association of Government Engineers

NaOH Sodium Hydroxide

NAOHSM National Association of Oil Heating Service Managers [US]

NAOHN National Association of Occupational Health Nurses

NAOJ National Astronomical Observatory of Japan [Tokyo]

NaOMe Sodium Methoxide (or Sodium Methylate)

NAOP National Association for Olmsted Parks [US]

NAOP National Association of Operative Plasterers [UK]

NaOPr Sodium Propoxide (or Sodium Propylate)

NaOPr^i Sodium Isopropoxide

NAOS North Atlantic Ocean Station

NAOSMM National Association of Scientific Material Managers [US]

NAP National Academy Press [Washington, DC, US]

NAP National Afforestation Program [Australia]

NAP National Airport Plan [of Federal Aviation Administration, US]

NAP National Archives of Pakistan [Islamabad]

NAP Navigation Analysis Program

NAP Network Access Pricing

NAP Network Access Protocol

NAP Niger Agricultural Project [Nigeria]

NAP 4-Nitro Amino Phenol

NAP Noise Analysis Program

NA-P Nonabrasive-Polishing

NAPA N-Acetyl-p-Aminophenol

NAPA N-Acetylprocainamide

NAPA National Agricultural Plastics Association [US]

NAPA National Asphalt Pavement Association [US]

NAPA National Association of Parks Administrators [UK]

NAPA National Association of Purchasing Agents [now National Association of Purchasing Management, US]

NAPA National Automotive Parts Association [US]

NAPA North Atlantic Ports Association [US]

NAPA Numerical Analytical Propagator Algorithm

NAPAEO National Association of Principal Agricultural Education Officers [UK]

NAPAN National Association for the Prevention of Addiction to Narcotics [US]

NAPAP N-Acetyl-p-Aminophenol

NAPAP 2-Naphthalenesulfonylglycyl)-4-Amidinophenyl-alanine piperidide [Biochemistry]

NAPAP National Acid(ic) Precipitation Assessment Program [US]

α-NAPAP Nα-(2-Naphthalenesulfonylglycyl)-4-Amidino-DL-Phenylalaninepiperidide [Biochemistry]

NAPB National Agricultural Products Board [Tanzania]

NaPb Sodium Lead (Spin Glass)

NAPC National Association of Plumbing Contractors [now National Association of Plumbing, Heating and Cooling Contractors, US]

NAPCA National Air Pollution Control Administration [US]

NAPCA National Association of Pipe Coating Applicators [US]

NAPCA National Association of Professional Contract Administrators [now National Contract Management Association, US]

NAPCCVC National Assessment of the Potential Consequences of Climate Variability and Change [US]

NAPCTAC National Air Pollution Control Techniques Advisory Committee [US]

NAPD National Association of Pharmaceutical Distributors [UK]

NAPD National Association of Plastic Distributors [US]

NAPD National Association of Precollege Directors [US]

NAPE N-Acyl Phosphatidyl Ethanolamine [Biochemistry]

NAPE National Association of Power Engineers [US]

NAPE National Association of Professional Engravers [US]

NAPEP National Association of Planners, Estimators and Progressmen [US]

NAPET National Association of Photo Equipment Technicians [US]

NAPF National Association of Plastic Fabricators [now Decorative Laminate Products Association, US]

NAPH Normalized Auger Peak Height [Auger Spectroscopy]

NAPHCC National Association of Plumbing, Heating and Cooling Contractors [US]

NAPH&MSC National Association of Plumbing, Heating and Mechanical Service Contractors [UK]

NAPIM National Association of Printing Ink Manufacturers [US]

NAPL National Air Photography Library [US]

NAPL National Association of Photolithographers [US]

NAPL National Association of Printers and Lithographers [US]

NAPL Nonaqueous Phase Liquid

NAPLIB National Association of Aerial Photographic Libraries [UK]

NAPLPS North American Presentation Level Protocol Syntax

NAPM National Association of Paper Merchants [UK]

NAPM National Association of Pattern Manufacturers [US]

NAPM National Association of Pharmaceutical Manufacturers [US]

NAPM National Association of Photographic Manufacturers [US]

NAPM National Association of Punch Manufacturers [US]

NAPM National Association of Purchasing Management [US]

NAPME N-Aminophenol Methylether

NAPMECA National Association of Postgraduate Education Center Administrators [UK]

NAPMM National Association of Produce Market Managers [US]

NAPNE National Association for Practical Nurse Education [US]

NAPNM National Association of Pipe Nipple Manufacturers [US]

NAPO National Association of Professional Organizers [US]

NaPO Sodiophosphate

NAPOL North Atlantic Policy Working Group [of European Civil Aviation Conference, France]

NAPP National Aerial Photography Program [of US Geological Survey]

NAPP National Aerospace Productivity Program [US]

nAPPA n-Amyl Phenylphosphonate [also n-APPA]

NAPPE Network Analysis Program using Parameter Extractions

NAPR NATO Armaments Planning Review

NAPRE National Association of Practical Refrigerating Engineers [US]

NaPrO Sodium Propoxide (or Sodium Propylate)

NAPS National Air Pollution Surveillance Network [Canada]

NAPS National Association of Buying Services [US]

NAPS National Auxiliary Publications Service [US]

NAPS Night Aerial Photographic System

NAPS Nimbus Automatic Programming System

NAPSA National Appliance Parts Suppliers Association [US]

NAPSIC North American Power Systems Interconnection Committee

NAPSS Numerical Analysis Problem Solving System

NaPSS Sodium Polystyrene Sulfonate

NAPT Nordic Association of Plumbers and Tinsmiths [Sweden]

NAPU Nuclear Auxiliary Power Unit

NAPUS Nuclear Auxiliary Power Unit System

NAPVA Naphthalene Labelled Polyvinyl Alcohol

NAPVO National Association of Passenger Vessel Owners [US]

NAPWPT National Association of Professional Word Processing Technicians [US]

NAQP National Association of Quick Printers [US]

NAQP-CUG National Association of Quick Printers–Computer Users Group [US]

NAQP-MSG National Association of Quick Printers–Management Systems Group [now NAQP-CUG]

NAR National Association of Rocketry

NAR Naval Air Reserve [US]

NAR Net Advertising Revenue

NAR Net Assimilation Rate

NAR Neutron Autoradiography

NAR Nitrogen-Bearing Alkali Resistant (Glass)

NAR Nonadiabatic Resonance (Theory) [Physics]

NAR Nuclear Acoustic Resonance

NAR Nucleic Acids Research [Journal published by Oxford University Press, UK]

NAR Numerical Analysis Research

Nar Nocardia argentinensis [Microbiology]

NARA National Archives and Records Administration [US]

NARAP National Acid Rain Assessment Program [US]

NARAS National Academy of Recording Arts and Sciences [US]

NARATE Navy Radar Automatic Test Equipment

NARB National Advertising Review Board [US]

NARBA North American Regional Broadcasting Agreement

NARBW National Association of Railway Business Women [US]

NARC National Agricultural Research Center [US]

NARC National Association for Retarded Children [US]

NARC National Association of Regional Councils [US]

NARC Non-Automatic Relay Center

NARCISSE Network of Art Research Computer Image Systems in Europe

NARDIS Navy Automated Research and Development Information System [US Navy]

NARE North Atlantic Regional Experiment

NAREC Naval Research Electronic Computer

NAREFA North Atlantic Reference Fares [of European Civil Aviation Conference, France]

NAREL National Air and Radiation Environmental Laboratory [US]

NAREMCO National Records Management Council [US]

NARF Naval Air Rework Facility

NARF Nuclear Aerospace Research Facility [of US Air Force]

NARG Navigation Aids Research Group [of International Civil Aviation Organization]

NARI Nanjing Automation Research Institute [Nanjing, PR China]

NARI Natal Agricultural Research Institute [South Africa]

NARI National Association of Recycling Industries [now Institute of Scrap Recycling Industries, US]

NARI National Association of the Remodeling Industry [US]

NARI Nuclear Aerospace Research Institute [of US Air Force]

NARI Met Rep NARI Metals Report [Published by the National Association of Recycling Industries, US]

NARL National Aerospace Research Laboratory [US]

NARL Naval Aeronautical Rocket Laboratory [of US Navy]

NARL Naval Arctic Research Laboratory [of US Navy]

NARM National Association of Relay Manufacturers [US]

NARMC National Association of Regional Media Centers [US]

NARO National Association of Royalty Owners [US]

NARO North American Regional Office

NARP National Association of Railroad Passengers [US]

NARP National Association of Registered Plans [US]

NARP Nitric Oxide–Ammonia Rectangular Pulse (Technique)

NARPV National Association for Remotely Piloted Vehicles [now Association for Unmanned Vehicle Systems, US]

NARS National Archives and Records Services [of US General Services Administration]

NARS National Association of Radiator Specialists [UK]

NARS National Association of Radiotelephone Systems [US]

NARS National Association of Rail Shippers [US]

NARS North Atlantic Radio System

NARSA National Automotive Radiator Service Association [US]

NARS-A1 National Archives and Record Service–Automation 1 [US]

NARSC National Association of Reinforced Steel Contractors [US]

NARST National Association for Research in Science Teaching [US]

NARTB National Association of Radio and Television Broadcasters [US]

NARTE National Association of Radio and Telecommunication Engineers [US]

NARTM National Association of Rope and Twine Merchants [UK]

NARTS National Association for Radio Telephone Systems [US]

NARTS Naval Aeronautics Test Station [US]

NARTS Naval Air Rocket Test Station [Dover, New Jersey, US]

NARU North Australia Research Unit [of Australian National University, Canberra]

NARUC National Association of Regulatory Utility Commissioners [US]

NARUCE National Association of Regulatory Utility Commission Engineers [US]

NARVRE National Association of Retired and Veteran Railroad Employees [US]

NARWA Nordic Agricultural Research Workers Association [Norway]

NAS Nanocluster-Assembled Solid

NAS National Academy of Sciences [Allahabad, India]

NAS National Academy of Sciences [Washington, DC, US]

NAS National Aerospace Standard

NAS National Aircraft Standards [US]

NAS National Airspace System [of Federal Aviation Administration, US]

NAS National Aquarium Society [US]

NAS National Association of Shopfitters [UK]

NAS National Astrological Society [US]

NAS National Audubon Society [US]

NAS National Avionics Society [US]

NAS Naval Air Station [US]

NAS Nebrasca Academy of Science [Lincoln, US]

NAS Network Access Server

NAS Network Access System

NAS Network Application Support [of Digital Equipment Corporation, US]

NAS Noise Abatement Society [UK]

NAS Nominal Aggregate Signal

NAS Norwegian Academy of Sciences [Oslo]

NAS Nucleophilic Aliphatic Substitution [Chemistry]

NAS Sodium Aluminosilicate

NAS Nursing Advisory Service [US]

NASA 2-Naphthylamine-1-Sulfonic Acid

NASA National Aeronautics and Space Administration [US]

NASA National Appliance Service Association [US]

NASA National Association of State Archeologists [US]

NASA North Atlantic Seafood Association

NASAA National Association for Sustainable Agriculture in Australia

NASA ADC National Aeronautics and Space Administration–Astronomical Data Center [at Goddard Space Flight Center, Greenbelt, Maryland, US]

NASA ADF National Aeronautics and Space Administration–Astrophysics Data Facility [at Goddard Space Flight Center, Greenbelt, Maryland, US]

NASA ADS National Aeronautics and Space Administration–Astrophysics Data System

NASA/AITP National Aeronautics and Space Administration/Aerospace Industry Technology Program [US]

NASA-ARC National Aeronautics and Space Administration–Ames Research Center [Mountain View, California, US]

NASA-ASEE National Aeronautics and Space Administration/American Society for Engineering Education (Fellowship Program)

NASA/BCAC National Aeronautics and Space Administration/Boeing Commercial Airplanes Corporation (Project) [US]

NASAC National Association of Scientific Angling Clubs [US]

NASA CASI National Aeronautics and Space Administration–Center for Aero-Space Information [Hanover, Maryland, US] [also NASA-CASI]

NASA CASI STI National Aeronautics and Space Administration–Center for Aero-Space Information Scientific and Technical Information (Service)

NASA Conf Publ National Aeronautics and Space Administration Conference Publication [also NASA CP]

NASA CR National Aeronautics and Space Administration Contractor Report [also NASA-CR, or NASA Contr Rep]

NASACU National Association of State Approved Colleges and Universities [US]

NASAD National Association of Sport Aircraft Designers [US]

NASA-DFRC National Aeronautics and Space Administration–Dryden Flight Research Center [Edwards, California, US]

NASA/DOD National Aeronautics and Space Administration/Department of Defense (Conference) [also NASA/DoD]

NASA DSN National Aeronautics and Space Administration–Deep Space Network [US]

NASA-ED National Aeronautics and Space Administration–Education Division [US]

NASA FOIA National Aeronautics and Space Administration Freedom of Information Act [US]

NASAGA North American Simulation and Gaming Association [US]

NASA/GAS National Aeronautics and Space Administration Get-Away Special (Program) [also NASA-GAS]

NASA GILS National Aeronautics and Space Administration Government Information Locator Service

NASA-GISS National Aeronautics and Space Administration–Goddard Institute for Space Studies [at New York City, US] [also NASA/GISS]

NASA-GSFC National Aeronautics and Space Administration–Goddard Space Flight Center [at Greenbelt, Maryl and, US]

NASA HQ National Aeronautics and Space Administration Headquarters [Washington, DC, US]

NASA-IVVF National Aeronautics and Space Administration–I ndependent Validation and Verification Facility [at Fairmont, West Virginia, US]

NASA-JPL National Aeronautics and Space Administration–Jet Propulsion Laboratory [Pasadena, California, US] [also NASA/JPL]

NASA-JSC National Aeronautics and Space Administration–Johnson Space Center [Houston, Texas, US]

NASA-KSC National Aeronautics and Space Administration–Kennedy Space Center [at Cape Canaveral, Florida, US]

NASA-LDEF National Aeronautics and Space Administration–Long-Duration Exposure Facility [US]

NASA-LeRC National Aeronautics and Space Administration–Lewis Research Center [at Cleveland, Ohio, US]

NASA-LRC National Aeronautics and Space Administration–Langley Research Center [at Hampton, Virginia, US]

NASA MDS National Aeronautics and Space Administration Mission Definition Study

NASA-MFA National Aeronautics and Space Administration–Moffett Federal Airfield [at Mountain View, California, US]

NASA-MPP National Aeronautics and Space Administration–Massively Parallel Processor

NASA-MSFC National Aeronautics and Space Administration–Marshall Space Flight Center [Huntsville, Alabama, US]

NASA-NASP National Aeronautics and Space Administration–National Aerospace Plane Program [also NASA/NASP]

NASA-NBSIR National Aeronautics and Space Administration–National Bureau of Standards Information Report [also NASA NBSIR]

NASA-NRC National Aeronautics and Space Administration–National Research Council (Program) [US]

NASAO National Association of State Aviation Officials [US]

NASA-OAET National Aeronautics and Space Administration–Office of Aeronautics, Exploration and Technology [US]

NASA-OAST National Aeronautics and Space Administration–Office of Aeronautics and Space Technology

NASAOCARE National Association of State Aviation Officials Center for Aviation Research and Education [US]

NASA-OSF National Aeronautics and Space Administration–Office of Space Flight

NASA-OSS National Aeronautics and Space Administration–Office of Space Science

NASA-OSSA National Aeronautics and Space Administration–Office of Space Science and Applications

NASA-OSTA National Aeronautics and Space Administration–Office of Space and Terrestrial Applications

NASAP Network Analysis and Systems Application Program

NASA PDS National Aeronautics and Space Administration–Planetary Data System [at Jet Propulsion Laboratory, Caltech, Pasadena, California, US]

NASA PR National Aeronautics and Space Administration Procurement Regulation

NASA Ref Publ National Aeronautics and Space Administration Reference Publication [also NASA RP]

NASARR North American Search and Range (or Ranging) Radar

NASA SEC National Aeronautics and Space Administration Space Engineering Center

NASA SERC National Aeronautics and Space Administration Space Engineering Research Center [of University of Cincinnati, Ohio, US]

NASA/SETI National Aeronautics and Space Administration Search for Extraterrestrial Intelligence (Program)

NASA SP National Aeronautics and Space Administration Special Publication [also NASA Spec Publ]

NASA-SSC National Aeronautics and Space Administration–Stennis Space Center [Mississippi, US]

NASA Spec Publ National Aeronautics and Space Administration Special Publication [NASA SP]

NASA-STAR National Aeronautics and Space Administration Scientific and Technical Aerospace Reports

NASA STI National Aeronautics and Space Administration Scientific and Technical Information (Program Office)

NASA TB National Aeronautics and Space Administration Technical Briefs [also NASA Tech Briefs]

NASA TEES National Aeronautics and Space Administration Texas Engineering Experiment Station [US]

NASA TM National Aeronautics and Space Administration Technical Memorandum [also NASA Tech Memo]

NASA TN National Aeronautics and Space Administration Technical Note [also NASA TN]

NASA TP National Aeronautics and Space Administration Technical Paper [also NASA Tech Pap]

NASA TR National Aeronautics and Space Administration Technical Report [also NASA Tech Rep]

NASA TRS National Aeronautics and Space Administration Technical Report Server [also NASA Tech Rep Server]

NASA TV National Aeronautics and Space Administration Television [US]

NASA-WFF National Aeronautics and Space Administration–Wallops Flight Facility [at Wallops Island, Virginia, US]

NASA-WSTF National Aeronautics and Space Administration–White Sands Test Facility [at White Sands, New Mexico, US]

NASBA National Automobile Safety Belt Association [UK]

NASBLA National Association of State Boating Law Administrators [US]

NASC National Aeronautics and Space Council [US]

NASC National AIS (Aeronautical Information Service) System Center [US]

NASC National Association of Scaffolding Contractors [UK]

NASC Naval Air Systems Command [US]

nasc necessary and sufficient condition [Mathematics]

NASCA National Association of State Cable Agencies [US]

NASCAR National Association for Stock-Car Auto Racing [US]

NASCAS NAS/NRC (National Academy of Sciences/National Research Council) Committee on Atmospheric Sciences [US]

NASCD National Association for Sickle-Cell Disease [US]

NaSCN Sodium Thiocyanate

NASCO National Academy of Sciences Committee on Oceanography [US]

NASCO North Atlantic Salmon Conservation Organization [Scotland]

NASCOE National Association of ASCS (Agricultural Stabilization and Conservation Service) County Office Employees [US]

NASCOM National Aeronautics and Space Administration Communications Network [also Nascom]

NASCOM II National Aeronautics and Space Administration Communications Network II (Upgraded Version)

NASCP North American Society for Corporate Planning

NASD National Association of Securities Dealers [US]

NASD National Association of Service Dealers [US]

NASDA National Association of State Departments of Agriculture [US]

NASDA National Association of State Development Agencies [US]

NASDA National Space Development Agency [Japan]

NASDAQ National Association of Securities Dealers Automated Quotations (System) [US]

NASDU National Amalgamated Stevedores and Dockers Union [UK]

NASEC National Applied Software Engineering Center

NASF National Association of State Foresters [US]

NASFCA National Automatic Sprinkler and Fire Control Association [US]

NASFM National Association of Store Fixture Manufacturers [US]

NASFW National Association of Solid Fuel Wholesalers [UK]

NASG North American Strawberry Growers Association [US]

NaSH Sodium Hydrosulfide

NASHA National Association for Speech and Hearing Action

NASHS Northwest Association of Secondary and Higher Schools [US]

NASI National Academy of Sciences of India [Allahabad, India]

NASI NetWare Asynchronous Services Interface

NASIC Northeast Academic Science Information Center [US]

NASICON Sodium Zirconium Silicon Phosphate (Ionic Conductor)

NaSiON Sodium Silicon Oxynitride

NASIRC National Aeronautics and Space Administration Automated Systems Internet Response Capability

NASIS National Aeronautics and Space Administration Aerospace Safety Information System

NASIS National Association for State Information Systems [US]

NASIS National Automated Sourcing Information System [Canada]

NASKER National Aeronautics and Space Administration Ames Kernel (Benchmark) [of NASA Ames Research Center, Mountain View, California, US]

NASL Naval Applied Sciences Laboratory [US]

NASLR National Association of State Land Reclaimationists [US]

NASM National Air and Space Museum [of Smithsonian Institution, Washington, DC, US]

NASN National Air Sampling Network [US]

NASN National Air Surveillance Network [US]

NaSn Sodium-Tin (Compound)

NASNI Naval Air Station at San Nicolas Island [California, US]

NAS/NRC National Academy of Sciences/National Research Council (Program) [US]

NAS/NRC/NAE National Academy of Sciences/National Research Council/National Academy of Engineering [US]

NaSO Sodiosulfate

NASOH North American Society for Oceanic History [US]

NASP NASP Workshop Publication [of NASA National Aerospace Plane Program]

NASP National Academy of Sciences in Panama

NASP National Aerospace Plane (Program) [of NASA]

NASP National Airport System Plan

NA-SP Nonabrasive Slightly Polished; Nonabrasive-Slight Polishing

NASPA National Society of Public Accountants

NASPD National Association of Steel Pipe Distributors [US]

NADPM National Association of Seed Potato Merchants [UK]

NASPO National Association of State Purchasing Officials [US]

NASQAN National Stream Quality Accounting Network [US]

NASR National Association of Solvent Recyclers [US]

NASRCP National Association for State River Conservation Programs [US]

NASS National Association of Steel Stockholders [UK]

NASSC National Air and Sea Systems Command [now Naval Sea Systems Command, US]

NASSM National Association of Scissors and Shears Manufacturers [US]

NASSP National Association of Secondary School Principals [US]

NASSTIE National Association of State Supervisors of Trade and Industrial Education [US]

NASSTRAC National Small Shipments Traffic Conference [US]

NASTD National Association of State Telecommunications Directors [US]

NaSt Sodium Stearate

NASTRAN NASA Structural Analysis (Software)

NASTS National Association for Science, Technology and Society [US]

NASU National Association of State Universities [US]

NASULGC National Association of State Universities and Land-Grant Colleges [US]

NASUS National Academy of Sciences of the United States [Washington, DC]

NASW National Association of Science Writers [US]

NASWF Naval Air Special Weapons Facility [US]

NAT N-Acryloyl-Tris(hydroxymethyl)aminomethane

NAT Network Address Translators

NAT Node Attached Table

NAT Normal Allowed Time

Nat Nation(al) [also nat]

Nat Native [also nat]

Nat Nature; Natural [also nat]

N At North Atlantic [alsdo N Atl]

NATA National Air Transportation Association [US]

NATA National Association of Testing Authorities [Australia]

NATA National Automobile Transporters Association [US]

NATA North American Telecommunications Association [US]

NATA North American Telephone Association

NATA Northern Air Transport Association

Nat Areas J Natural Areas Journal [US]

NATAS National Academy of Television Arts and Sciences [US]

NATAS NOAA (National Oceanic and Atmospheric Administration)/AVHRR (Advanced Very-High-Resolution Radiometer) Transcription and Archive System

NATAS North American Thermal Analysis Society [US]

NATB National Automobile Theft Bureau [US]

Nat Biotechnol Nature Biotechnology [US]

Nat Bur Stand National Bureau of Standards [now National Institute of Standards and Technology, Gaithersburg, Maryland, US]

NATC Naval Air Test Center [of US Navy at Patuxent River, Maryland, US]

NATCC National Air Transport Coordinating Committee [US]

NATCHEM National Atmospheric Chemistry Database [Canada]

NATCOL Natural Food Colors Association [Switzerland]

NATCOM National Communications Symposium [US]

NATCS National Air Traffic Control Service [of Board of Trade, UK]

NATD National Association of Test Directors [US]

NATD National Association of Tool Dealers [UK]

NATE Neutral Atmosphere Temperature Experiment

NATEC Naval Air Technical Evaluation Center [UK]

NATEF National Automotive Technicians Education Foundation [US]

NATEL Nortronics Automatic Test Equipment Language

NATES National Analysis of Trends in Emergency Systems [of Environmental Protection Service, Canada]

NATESA National Alliance of Television and Electronic Service Associations [US]

NATF Naval Air Test Facility [US]

Nat Foods Merch Natural Foods Merchanizer [US Journal]

Nat Gas Ind Technol Natural Gas Industrial Technology [Canadian Publication]

Nat Hazards Natural Hazards [Journal published in the Netherlands]

Nat Health Natural Health [Canadian Journal]

Nat Health Prod Natural Health Products [Canadian Journal]

NATHERS National House Energy Rating Scheme

Nat Hist Natural Historian; Natural History

Nat Hist Natural History [US Publication]

Nat Hist Natural History Magazine [US]

NATIE National Association for Trade and Industrial Education [US]

NATII National Association of Trade and Industrial Instructors [US]

Nat Immun Natural Immunity [International Journal]

NATIS National Information System [of UNESCO]

NATL National Agricultural Transportation League [US]

NATL Naval Aeronautical Turbine Laboratory [US]

N Atl North Atlantic [also N At]

NaTl Sodium Thallium (Spin Glass)

natl national [also natl, or nat'l]

Natl Acad Sci Lett National Academy of Science Letters [Published by the National Academy of Sciences, India]

Natl Bur Stand Build Sci Ser National Bureau of Standards Building Science Series [US]

Natl Bur Stand Monogr National Bureau of Standards Monograph [US]

Natl Bur Stand Res Rep National Bureau of Standards Research Report [US]

Natl Bur Stand Spec Publ National Bureau of Standards Special Publication [US]

Natl Bur Stand Tech Note National Bureau of Standards Technical Note [US]

Natl Bur Stand Tech Rep National Bureau of Standards Technical Report [US]

Natl Bur Stand Update National Bureau of Standards Update [US]

Natl Contract Manage J National Contract Management Journal [of National Contract Management Association, US]

Natl Electron Rev National Electronics Review [UK]

Natl Eng National Engineer [Journal of the National Association of Power Engineers, US]

Natl Geogr Mag National Geographic Magazine [of National Geographic Society, US]

Natl Inst Stand Technol Build Sci Ser National Institute of Standards and Technology Building Science Series [US]

Natl Inst Stand Technol Monogr National Institute of Standards and Technology Monograph [US]

Natl Inst Stand Technol Res Rep National Institute of Standards and Technology Research Report [US]

Natl Inst Stand Technol Spec Publ National Institute of Standards and Technology Special Publication [US]

Natl Inst Stand Technol Tech Note National Institute of Standards and Technology Technical Note [US]

Natl Inst Stand Technol Tech Rep National Institute of Standards and Technology Technical Report [US]

Natl Inst Stand Technol Update National Institute of Standards and Technology Update [US]

Natl Lucht- en Ruimtevaartlab Rep Nationaal Lucht- en Ruimtevaartlaboratorium Report [Report of the Dutch National Laboratory of Aeronautics and Astronautics]

Natl Mon National Monument [also Nat'l Mon]

Natl Museum Can Bull National Museum of Canada Bulletin

Natl News Rep National News Report [of Sierra Club, US]

nat log natural logarithm [Symbol]

Natl Park National Park [also Nat'l Park, or Nat Park]

Natl Phys Lab Rep National Physical Laboratory Reports [UK]

Natl Symp National Symposium

Natl Tech Rep National Technical Report [Published by Matsushita Electric Industries Co. Ltd., Osaka, Japan]

Natl Water Cond National Water Conditions [Publication of the US Department of the Interior]

Natl Weather Dig National Weather Digest [Publication of the National Weather Association, US]

NATM New Austrian Tunneling Method

NATMAC National Air Traffic Management Advisory Committee [of Civil Aviation Authority, UK]

NATMAP Division of National Mapping [Now Australian Surveying and Land Information Group]

NatMIS National Marine (or Maritime) Information System [of Environmental Resources Information Network] [also NATMIS]

NATMM National Association of Textile Machinery Manufacturers [now American Textile Machinery Association, US]

Nat Mon National Monument

Nat mur Sodium Chloride (Tissue Salt)

NATO North Atlantic Treaty Organization [Brussels, Belgium]

NATOA National Association of Telecommunications Officers and Advisors [US]

NATO Adv Study Inst Ser A NATO Advanced Study Institute, Series A

NATO Adv Study Inst Ser B NATO Advanced Study Institute, Series B

NATO Adv Study Inst Ser B Phys NATO Advanced Study Institute, Series B: Physics

NATO Adv Study Inst Ser C NATO Advanced Study Institute, Series C

NATO Adv Study Inst Ser D NATO Advanced Study Institute, Series D

NATO Adv Study Inst Ser E NATO Advanced Study Institute, Series E

NATO Adv Study Inst Ser E Appl Sci NATO Advanced Study Institute, Series E: Applied Sciences

NATO-AGARD NATO Advisory Group for Aerospace Research and Development

NATO-AGARD CP NATO Advisory Group for Aerospace Research and Development Conference Paper

NATO ASI North Atlantic Treaty Organization–Advanced Study Institute

NATO ASI Ser A NATO ASI , Series A

NATO ASI Ser B NATO ASI , Series B

NATO ASI Ser B Phys NATO ASI , Series B: Physics

NATO ASI Ser C NATO ASI , Series C

NATO ASI Ser D NATO ASI , Series D

NATO ASI Ser E NATO ASI , Series E

NATO ASI Ser E Appl Sci NATO ASI , Series E: Applied Sciences

NATO Conf Ser NATO Conference Series

NATO CRG NATO Collaborative Research Grants (Program)

NATO HQ NATO Headquarters [Brussels, Belgium]

NATO-PCO North Atlantic Treaty Organizatio Publication Coordination Office [Belgium]

Nat Ord Natural Order [also nat ord]

NATO SAD NATO Scientific Affairs Division

NATO Sci Environ Aff Newsl NATO Scientific and Environmental Affairs Newsletter

Nat Park National Park [also Natl Park, or Nat'l Park]

Nat phos Sodium Phosphate (Tissue Salt)

Nat Prod Rep Natural Products Reports [of Royal Society of Chemistry, UK]

NATPS National Association of Trade Protection Societies [UK]

NATR National Association of Technical Research [France]

Nat Resour Forum Natural Resources Forum [Journal published in the US]

Nat Rubber Dev Natural Rubber Development [Publication of the Malaysian Rubber Bureau]

Nat Rubber News Natural Rubber News [Publication of the Malaysian Rubber Bureau]

NATS National Activity to Test Software

NATS National Air Traffic Services [UK]

NATS National Association of Temporary Services [US]

NATS Naval Air Transport Service [US]

NATS Nordisk Avisteknisk Samarbetsnamnd [Nordic Joint Technical Press Board, Sweden]

Nat Sci Natural Science; Natural Scientist

NATSIM Network Advanced Training Simulator

NATSOPA National Society of Operative Printers and Assistants [now Society of Graphical and Allied Trades, UK]

NATSPG North Atlantic Systems Planning Group [also NAT/SPG] [of International Civil Aviation Organization]

NATSU Nominated Air Traffic Service Unit

Nat sulph Sodium Sulfphate (Tissue Salt)

NATSURV National Survey

Nat Symp National Symposium

NATT Northern Australia Tropical Transect

NATTA Network of Alternative Technology and Technology Assessment [UK]

NATTA North American Trackless Trolley Association [US]

Nat Tech Inf Serv National Technical Information Service [of US Department of Commerce]

NAT/TFG North Atlantic Traffic Forecasting Group

NATTS National Association of Trade and Technical Schools [US]

NATTS Naval Air Test Turbine Station [US Navy]

NATU Naval Aircraft Torpedo Unit [US Navy]

NATURA Fundación Ecuatoriana para la Conservación de la Naturaleza [Ecuadorean Foundation for Nature Conservation]

Naturarzt Der Naturarzt [German Journal on Naturopathy]

Nature, Int Wkly J Sci Nature–International Weekly Journal of Science [Published in the US]

Natürl Z Mensch Umw Natürlich: Zeitschrift für Mensch und Umwelt [German Journal on Man and His Natural Environment]

Naturwissenschaften Die Naturwissenschaften [German Journal on Natural Sciences]

NATVAS National Academy of Television Arts and Sciences [US]

Nat Way The Natural Way [Journal published in the US]

NAU Network Addressable Unit

NAU Network Administration Utilities

NAU Northern Arizona University [Flagstaff, US]

n-Au Nanocrystalline Gold

Nauchno-Tekh Inf 1 Nauchno-Tekhnicheskaya Informatisaya, Seriya 1 [Russian Journal on Scientific and Technical Information Processing, Series 1]

Nauchno-Tekh Inf 2 Nauchno-Tekhnicheskaya Informatisaya, Seriya 2 [Russian Journal on Automatic Documentation and Mathematical Linguistics, Series 2]

NAUG National Appleworkers Users Group [US]

Nauk Inf Naukovedenie i Informatika [Ukrainian Journal on Computer Science and Technology]

NAUOIS Netherlands Association of Users of On-Line Information Systems

NAUOR Northern Arizona University Organized Research [US]

NAUS National Airspace Utilization Study

NAUS National Association of Urban Studies [UK]

N Austr North Australia(n)

Naut Nautical; Nautics [also Naut, or naut]

Nautilus Bull Nautilus Bulletin [Publication of Nautilus Institute for Security and Sustainable Development, US]

Naut Mi Nautical Mile [also naut mi]

NAUW National Association of University Women [US]

NAV Net Asset Value

Nav Navigable; Navigate; Navigation(al); Navigator [also nav]

nav naval

NAVA National Audio Visual Association [US]

NAVAC National Audio-Visual Aids Center [UK]

Navaglobe Global Navigation System [also navaglobe]

NAVAID Air Navigation Facility

NAVAID Navigation(al) Aid [also Navaid, or navaid]

NAVAIDS Navigation(al) Aids [also Navaids, or navaids]

NAVAIR Naval Air Systems Command [US Navy]

NAVAPI North American Voltage and Phase Indicator

NAVAR Navigation Air Radar

NAVAR Navigation and Ranging [also Navar, or navar]

NAVAR Radar Air Navigation and Control System

Nav Arch Naval Architect(ure)

Navarho Navigation Aid, Rho Radio Navigation System [also navarho]

Navascope NAVAR Airborne Radarscope [also navascope]

Navascreen NAVAR Ground Screen [also navascreen]

Nav Av Naval Aviation; Naval Aviator

NAVC National Audiovisual Center

NAVCAD Naval Aviation Cadet (Program) [of US Navy]

NAVCM Navigation Countermeasures

NAVCOM Naval Communication

NAVCOMMSTA Naval Communications Station [US Navy]

NAVDAC Navigation Data Assimilation Center

NAVDAC Navigation Data Assimilation Computer

NAVDAD Navigationally Derived Air Data

NAVDAS Navigation Data Acquisition System

Nav Data Navigational Data [also nav data]

NAVDOCKS Navy Yards and Docks Bureau [US Navy]

NAV-DSN Navigation–Deep Space Network

NavE Naval Engineer [also NavEng]

NAVEAM Navigational Warning, Eastern Atlantic and Mediterranean

NAVELECSYSCOM Naval Electronics System Command [also NAVELEX] [US Navy]

NavE Naval Engineer [also Nav Eng, or NavEng]

Nav Eng Naval Engineer(ing)

Nav Eng J Naval Engineers Journal [Publication of the American Society of Naval Engineers]

NAVEX Navigation Experiment Package [of NASA]

NAVFEC Naval Facilities Engineering Command [US Navy] [also NAVFAC]

NAVHARS Navigation Heading and Attitude Reference System

Navig Navigation; Navigator [also navig]

Navig J Inst Navig Navigation, Journal of the Institute of Navigation [US]

NAVIRO Navy Industrial Relation Office [US Navy]

NAVLAB Naval Laboratory [also Navlab]

NAVLAB Navigation Laboratory

NAVMAT Office of Naval Materiel [US Navy]

NAVMC Navy Marine Corps [US Navy]

NAVMED Naval Aerospace Medical Institute [US Navy]

NAVMINDEFLAB Navy Mine Defense Laboratory [US Navy]

NAVOCEANO Naval Oceanographic Office [US Navy]

NAVORD Naval Ordnance [also NavOrd]

NAVORD Naval Ordnance Laboratory [Silver Spring, Maryland, US]

NAVORD Rep Naval Ordnance Report [of US Naval Ordnance Laboratory, Silver Spring, Maryland]

NAVPERS (Bureau of) Naval Personnel [US Navy]

NAVPHOTOCEN Naval Photographic Center [US Navy]

NAVPOOL Navigation Parameter Common Pool

NAVPS Net Asset Value per Share

Nav Res Naval Research(er) [also nav res]

Nav Res Logist Naval Research Logistics [US Publication]

Nav Res Rev Naval Research Reviews [Published by US Department of the Navy, of Office of Naval Research]

NAVS Navigation System [also NAV S, or Nav S]

NAVSAT Navigation(al) Satellite (System) [also NavSat]

NAVSEA Naval Sea Systems Command [US Navy]

NAVSEC Naval Ship Engineering Center [US Navy]

NAVSEP (Specialist Panel) Navigation and Separation of Aircrafts

NAVSHIPS Naval Ship Systems Command [US Navy]

NAVSPASUR Naval Space Surveillance System [US Navy]

NAVSTAR Navigation Satellite Timing and Ranging System [Global System of US Navigation Satellites]

NAVSTAR Navigation System using Timing and Ranging

NAVSTAR/GPS Navigation System using Timing and Ranging/Global Positioning System

NAVSUP Naval Supply Systems Command [US Navy]

NAVSWC Naval Surface Weapons Center [of US Navy at Dahlgren, Virginia]

NAVTEC National Association of Vocational Technical Education Communicators [US]

NAVTRADEVCEN Naval Training Device Center [US Navy]

NAVTRI Navigational Triangle

NAVWEPS Bureau of Naval Weapons [of US Navy] [also NAVWeps]

Navy Domest Technol Transf Fact Sheet Navy Domestic Technology Transfer Fact Sheet [US Navy]

Navy Int Navy International [UK]

Navy Technol Transfer Fact Sheet Navy Technology Transfer Fact Sheet [Publication of the US Navy]

NAW National Association of Wholesalers [US]

NAW Non Acid Washed

N/AW Night/Adverse Weather (Evaluator)

NAWAPA North American Water and Power Alliance [US/Canada]

NAWAS National Warning System [US]

NAWC National Association of Water Companies [US]

NAWC Naval Air Warfare Center [of US Army at China Lake, California and Trenton, New Jersey]

NAWC-AD Naval Air Warfare Center Aircraft Division [of US Army at Warminster, Pennsylvania]

NAWC-ADWAR Naval Air Warfare Center Aircraft Division, Warminister [of US Army in Pennsylvania]

NAWC-WD Naval Air Warfare Center Weapons Division [of US Army at China Lake, California]

NAWC-WPNS Naval Air Warfare Center Weapons Division [of US Army at China Lake, California]

NAWDAC National Association for Women Deans, Administrators and Counselors [US]

NAWDC National Association of Waste Disposal Contractors [UK]

NAWDEX National Water Data Exchange [of United States Geological Survey]

NAWF North American Wildlife Foundation [US]

NAWG National Association of Wheat Growers [US]

NAWGF National Association of Wheat Growers Foundation [US]

NAWHSL National Association of Women Highway Safety Leaders [US]

NAWIC National Association of Women in Construction [US]

NAWID National Association of Writing Instrument Distributors [US]

NAWK National Association of Warehouse Keepers [UK]

NAWLA North American Wholesale Lumber Association [US]

NAWLT Nitric Acid Weight Loss Test

NAWM National Association of Wool Manufacturers [US]

NAWMP National Association of Waste Material Producers [US]

NAWMP North American Waterfowl Management Plan [Canada]

NAWPF North American Wildlife Park Foundation [US]

NAWPU National Association of Water Power Users [UK]

NAWQA National Water Quality Assessment [of US Geological Survey/Biological Resources Division]

NAWS Negotiate About Window Size

NAWS North American Wolf Society [US]

NAWTWPC Netherlands Association on Wastewater Treatment and Water Pollution Control

NB Naphthol Blue

NB Naphthyl Butyrate

NB Narrowband (Transmission) [also N/B]

NB Narrow Beam

NB National Battlefield [US]

NB Naval Base

NB Navigation Base

NB Nebraska [US]

NB Needle Bearing

NB Needle Roller Bearing, Drawn Cup, Full Complement, Without Inner Ring, Metric-Dimensioned [Symbol]

NB Negri Bossi (Injection Molding Machine)

NB Nernst Body [Thermodynamics]

NB Nerve Bundle [Anatomy]

NB Neutral Beam

NB Neutron Beam

NB Neutron Bomb

NB Neutron Booster

NB New Brunswick [Canada]

NB Nissl Body (or Bodies) [Cytology]

NB Nitrobenzene [also Nbz]

NB Nitrogen Base

NB No-Bake (Binder) [Metallurgy]

NB No Bias (Relay)

NB Nonbonding

NB Nonbridging

NB Nopol Benzyl

NB North Borneo

NB Northbound

NB North Britain

NB *(Nota Bene)* – Note Well; Take Notice [also nb]

NB Notch Brittleness [Metallurgy]

NB Nucleate Boiling [Chemical Engineering]

NB Sodium Borate

α-NB α-Naphthyl Butyrate

N$_s$(B) Surface Density of B Atoms [Symbol]

N/B Narrowband (Transmission) [also NB]

Nb Niobium [Symbol]

Nb^{4+} Tetravalent Niobium Ion [Symbol]

Nb^{5+} Pentavalent Niobium Ion [Symbol]

Nb-90 Niobium-90 [also ^{90}Nb, or Nb^{90}]

Nb-91 Niobium-91 [also ^{91}Nb, or Nb^{91}]

Nb-92 Niobium-92 [also ^{92}Nb, or Nb^{92}]

Nb-93 Niobium-93 [also ^{93}Nb, or Nb^{93}]

Nb-94 Niobium-94 [also ^{94}Nb, or Nb^{94}]

Nb-95 Niobium-95 [also ^{95}Nb, or Nb^{95}]

Nb-96 Niobium-96 [also ^{96}Nb, or Nb^{96}]

Nb-97 Niobium-97 [also ^{97}Nb, or Nb^{97}]

Nb-99 Niobium-99 [also ^{99}Nb, or Nb^{99}]

nB noncoherent (grain) boundary (in materials science) [Symbol]

NBA γ-(2-Naphthyl)butyric Acid

NBA Narrowband Allocation

NBA Narrow-Beam Absorption

NBA National Bankers Association [US]

NBA National Benzol and Allied Products Association [UK]

NBA National Board of Aviation [Finland]

NBA National Brassfounders Association [UK]

NBA National Buffalo Association [US]

NBA National Building Agency

NBA National Business Association [UK]

NBA N-Bromoacetamide

NBA N-Butylalcohol

NBA Net Book Agreement

NBA Neuromuscular Blocking Agent [Biochemistry]

NBA Nitrobenzaldehyde

NBA Nitrobenzoic Acid

nBA n-Butylacrylate

NBAA National Business Aircraft Association [US]

NBAC National Biotechnology Advisory Committee [Canada]

NBACSTT New Brunswick Association of Certified Survey Technicians and Technologists [Canada]

Nb-Al Niobium-Aluminum (Alloy System)

NbAlCrY Niobium Aluminum Chromium Yttrium (Alloy)

Nb$_3$Al/Nb Triniobium Aluminide Reinforced Niobium (Composite)

Nb/Al$_2$O$_3$ Niobium/Aluminum Oxide (Interface)

Nb-Al$_2$O$_3$ Niobium Reinforced Aluminum Oxide

Nb-Al$_2$O$_3$ Niobium-Aluminum Oxide (System)

Nb-Al-Ti Niobium-Aluminum-Titanium (Alloy)

NBAR National Bureau of Agricultural Research [PR China]

NBARN New Brunswick Association of Registered Nurses [Canada]

n-BaTiO$_3$ Nanocrystalline Barium Titanate

NBB n-Butylboronate

NBBI National Board of Boiler and Pressure Vessel Inspectors [US]

NBBI Netherlands Organization for Libraries and Information Services

NBBPVI National Board of Boiler and Pressure Vessel Inspectors [US]

NBC Narrowband Channel

NBC Narrowband Conducted

NBC National Book Committee [US]

NBC National Book Council [UK]

NBC National Broadcasting Company [US]

NBC National Boiler Council [US]

NBC National Building Code [US]

NBC Nonbleeding Cable

NBC Norwegian Bulk Carrier

NBC Nuclear, Biological and Chemical (Protective Clothing)

NbC Niobium Carbide

4NB15C5 4'-Nitrobenzo-15-Crown-5

NBCC National Building Code of Canada

NBCC New Brunswick Community College [Canada]

NBCCA Northern British Columbia Construction Association [Canada]

NBCCO Neodymium Barium Calcium Copper Oxide (Superconductor)

NBCD Natural Binary-Coded Decimal

NBCFAE National Black Coalition of Federal Aviation Employees [US]

NBCFO Neodymium Barium Copper Iron Oxide (Superconductor)

NbCN Niobium Carbonitride

NBCNO Neodymium Barium Copper Niobium Oxide (Superconductor)

NBCO Neodymium Barium Copper Oxide (Superconductor)

NbCpCl$_4$ Cyclopentadienylniobium Tetrachloride

Nb-Cr-Ti Niobium-Chromium-Titanium (Alloy System)

NBCTO Neodymium Barium Copper Titanium Oxide (Superconductor)

Nb-Cu Niobium-Copper (Alloy System)

NBCV Narrowband Coherent Video

NBCW Nuclear, Biological and Chemical Warfare

NBD Negative Binomial Distribution

NBD Nitrobenzoxadiazole

NBD Nitro Blue Diformazan

NBDA National Bicycle Dealers Association [US]

NBD Aziridine 7-Chloro-4-Nitrobenzo-2-Oxa-1,3-Diazole Aziridine [also NBD aziridine]

NBDC National Broadcast Development Committee [UK]

NBDC New Brunswick Development Corporation [Canada]

NBD-Cl 4-Chloro-7-Nitrobenzofuran; 7-Chloro-4-Nitrobenzofuran [also NBD Chloride, or NBD chloride]

NBD-F 4-Fluoro-7-Nitrobenzofuran; 7-Fluoro-4-Nitrobenzofuran [also NBD Fluoride, or NBD fluoride]

NBDI N,N'-Diisopropyl-O-(4-Nitrobenzene)isourea; O-(4-(Nitrobenzyl)-N,N'-Diisopropylisourea

NBDMO N-Bromo-4,4-Dimethyl-2-Oxazolidinone

Nb(DPM)$_5$ Niobium Dipivaloylmethanoate [also Nb(dpm)$_5$]

NBDL Narrowband Data Line

NBD-Taurine N-(7-Nitrobenz-2Oxa-1,3-Diazol-4-yl)taurine [also NBD taurine]

NBE N-Bromoacetylethylenediamine

NBE Nitrogen-Bound Exciton (Spectrum) [also N-BE] [Solid-State Physics]

NBE Non-Bonding Electrons [Physical Chemistry]

NBE Not Below, or Equal

N-BE Nitrogen-Bound Exciton (Spectrum) [also NBE] [Solid-State Physics]

Nb:EAM Niobium Embedded-Atom Method [Materials Science]

NBEAP New Brunswick Exploration Assistance Program [Canada]

NBECN New Brunswick Education Computer Network [Canada]

NBEP National Biomass Ethanol Program [of Natural Resources Canada]

NBEPC New Brunswick Electric Power Commission [Canada]

NBER National Bureau of Economic Research [US]

NBER National Bureau of Engineers Registration

NBETF Neutral Beam Engineering Test Facility

Nb(EtO)$_5$ Niobium(V) Ethoxide (or Niobium(V) Ethylate)

NBF Neutral Buoyancy Facility

Nb(f) Niobium Fiber

NBFA National Business Forms Association [US]

NBFA New Brunswick Forest Authority [Canada]

NBFFO National Board of Fur Farm Organizations [US]

NBFL New Brunswick Federation of Labour [Canada]

NBFM Narrowband Frequency Modulation

NBFN New Brunswick Federation of Naturalists [Canada]

NBFU National Board of Fire Underwriters [US]

NBG Near Band Gap [Solid-State Physics]

NBGQA National Building Granite Quarries Association [US]

NBH Network Busy Hour

NBH Neutral-Beam Heating

NbH Niobium Hydride

NBHA N-Benzylheptadecylamine

NBHA O-(4-Nitrobenzyl)hydroxylamine

NBHC New Brunswick Hydro Commission [Canada]

Nb-Hf Niobium-Hafnium (Alloy)

Nb-HSLA Niobium-Bearing High-Strength Low Alloy (Steel)

NBI Neutral Beam Injection

NBI Niels Bohr Institute [Copenhagen, Denmark]

NBI Nothing But Initials

NBII National Biological Information Infrastructure (Program) [of US Geological Survey/Biological Resources Division]

NBIN National Biodiversity Information Network [Indonesia]

NBIOME Northern Biosphere Observation and Modeling Experiment

NBIS New Brunswick Information Service [Canada]

NBIS Northern Biosphere Information System

NBIT New Bedford Institute of Technology [Massachusetts, US]

NBIT New Brunswick Institute of Technology [Canada]

NBJ Nanobridge Junction [Electronics]

NBL Narrowband Linear

NBL Naval Biological Laboratory [of US Navy]

NBL Neuroblastoma [Medicine]

NBL New Brunswick Laboratory [of Agricultural Extension Center, Canada]

NBL Nitrile-Butadiene Latex

NBLE Nearly Best Linear Estimator

NBLS New Brunswick Land Surveyors [Canada]

NBM National Buildings Museum [US]

NBM Needle Roller Bearing, Drawn Cup, Full Complement, Closed End, Without Inner Ring, Metric-Dimensioned [Symbol]

NBM Neodymium Barium Manganate

NBM Non-Book Material

nBMA n-Butyl Methacrylate

Nb-93 MAS Niobium-93 Magic Angle Spinning [also ^{93}Nb MAS]

NBMB N Binary Digits–M Binary Digits

NBMCR Non-Book Materials Cataloguing Rules

NBMDA National Building Material Distributors Association [US]

NBMDA New Brunswick Mineral Development Agreement [also NB MDA, Canada]

NBME National Board of Medical Examiners [US]

NBMG Nevada Bureau of Mines and Geology [US]

NBMN Nitrobenzal Malononitrile

m-NBMN m-Nitrobenzal-Malononitrile

NBMO Nonbonding Molecular Orbital [Physical Chemistry]

Nb-Mo Niobium-Molybdenum (Alloy System)

Nb/MoSi$_2$ Niobium-Fiber-Reinforced Molybdenum Silicide (Composite)

NBMPR 4-Nitrobenzyl-6-Thioinosine [Biochemistry]

MBMPR 6-(4-Nitrobenzylmercapto)purin-9-β-D-Ribofuranoside [Biochemistry]

NBMPR Nitrobenzylthioribofuranosylpurine

NBN New Biological Nomenclature

NbN Niobium(I) Nitride (Superconductor)

n-BN N-Type Boron Nitride

NBNA National Benchmark Network for Agrometeorology

Nb(NEt$_2$)$_4$ Tetrakis(dimethylamido)niobium

Nb-93 NMR Niobium-93 Nuclear Magnetic Resonance [also ^{93}Nb NMR]

NBO Network Building Out; Network Buildout

NBO Nonbonding Orbital [Physical Chemistry]

NBO Nonbridging Oxygen [Physics]

NbO Niobium(II) Oxide (or Niobium Monoxide)

NbO$_x$ Niobium Oxides (e.g., NbO, NbO$_2$, or Nb$_2$O$_5$)

Nb(OEt)$_5$ Niobium(V) Ethoxide (or Niobium(V) Ethylate)

NBOHC Nonbridging Oxygen-Hole Center [Physics]

NBOC Network Building Out Capacitor

N-Bomb Neutron Bomb [also N-bomb]

Nb(OPr^i)$_5$ Niobium (V) Isopropoxide

NBOR Network Building Out Resistor

NBP National Battlefield Park [US]

NBP National Blood Policy [of US Department of Health, Education and Welfare]

NBP Normal Boiling Point [also nbp]

NBP Nu-Bit Inc. (Heat Treating) Process [Metallurgy]

NBPA National Bark Producers Association [now National Bark and Soil Producers Association, US]

NBPA New Brunswick Potato Agency [Canada]

NBPC Normal Bonded-Phase Chromatography

N-BPHA N-Benzyl-N-Phenyl Hydroxylamine

NBPM Narrowband Phase Modulation

NBPS Neutral Beam Power Systems

NBQX 2,3-Dihydroxy-6-Nitro-7-Sulfamoylbenzo(f) quinoxaline

NBR Narrowband Radiated

NBR National Board of Roads and Water [Finland]

NBR Nitrile-Butadiene Rubber

NBR Nonbonding Resonance

Nbr Number [also nbr]

Nb-Re Niobium-Rhenium (Alloy System)

NBREP National Blood Resources Education Program [US]

NBRF National Biomedical Research Foundation [Silver Spring, Maryland, US]

NBRI National Building Research Institute [South Africa]

NBRPC New Brunswick Research and Productivity Council [Canada]

NBRT National Board for Respiration Therapy [US]

N Bruns New Brunswick [Canada]

NBS Narrowband Socket

NBS National Battlefield Site [US]

NBS National Biological Service [US]

NBS National Bureau of Standards [now National Institute of Standards and Technology, US]

NBS National Bureau of Standards [of Ministry of Economic Affairs, Taiwan]

NBS Natural Black Slate

NBS N-Bromosuccinimide

NBS New British Standard(s) [UK]

NBS Nickel-Bonded Steel

NBS Nitrobenzenesulfenyl

NBS Nordiska Byggforskningsorgans Samarbetsgrupp [Nordic Building Research Cooperation Group, Norway]

NBS Numeric Backspace (Character)

Nb-S Niobium-Vanadium (Alloy System)

NBSAC National Boating Safety Advisory Council [US]

NBS Build Sci Ser NBS Building Science Series [Published by the US National Bureau of Standards]

NBSC New Brunswick Safety Council [Canada]

NBSC 2-Nitrobenzenesulfenyl Chloride

NBSCETT New Brunswick Society of Certified Engineering Technicians and Technologists [Canada]

NBS Circ National Bureau of Standards Circular [US]

NBSCS Nickel-Based Single Crystal Superalloy

NBSD Night Bombardment Short Distance

NBSFS National Bureau of Standards Frequency Standard [now National Institute of Standards and Technology Frequency Standard, US]

NBS/ICST National Bureau of Standards/Institute for Computer Sciences and Technology [now NIST/ICST, US]

Nb/Si Niobium on Silicon Substrate

NbSi$_2$/Nb Niobium Disilicide Reinforced Niobium (Composite)

Nb$_5$Si$_3$/Nb Pentaniobium Trisilicide Reinforced Niobium (Composite)

NBSIR National Bureau of Standards Information Report [now NISTIR]

NBSL National Biological Standards Laboratory [Australia]

NBS Monogr NBS Monograph [Published by the US National Bureau of Standards]

NBSPA National Bark and Soil Producers Association [US]

NBSR National Bureau of Standards Reactor [now NIST Research Reactor, Gaithersburg, Maryland, US]

NBS Res Rep NBS Research Report [Published by the US National Bureau of Standards] [now NIST Res Rep]

NBS-SIS National Bureau of Standards–Standard Information Services [now NIST-SIS]

NBS SP NBS Special Publication [Published by the US National Bureau of Standards] [also NBS Spec Publ] [now NIST SP]

NBS-SRM National Bureau of Standards–Standard Reference Material [now NIST-SRM]

NBS Tech Note NBS Technical Note [Published by the US National Bureau of Standards] [now NIST Tech Note]

NBS Tech Rep NBS Technical Report [Published by the US National Bureau of Standards] [now NIST Tech Rep]

NBS Update NBS Update [Published by the US National Bureau of Standards] [now NIST Update]

NBSV Narrowband Secure Voice

NBT Narrow-Beam Transducer

NBT Neutral Buoyancy Trainer

NBT 4-Nitro Blue Tetrazolium; Nitro Blue Tetrazolium

NBT Null-Balance Transmissometer

N-B-T Naepaine–Benzocaine–Tetracaine [Pharmacology]

NBTA New Brunswick Teachers Association [Canada]

Nb-Ta Niobium-Tantalum (Alloy System)

NBTB New Brunswick Travel Burau [Canada]

NBTD National Board for Technical Development [Sweden]

NBTDR Narrowband Time-Domain Reflectometry

NbTe Niobium Telluride (Semiconductor)

NB Tel New Brunswick Telephone Company [Canada]

NBTF National Biomedical Tracer Facility

NBTG Nitrobenzylthioguanosine [Biochemistry]

NBTGR S-(4-Nitrobenzyl)-6-Thioguanine Ribofuranoside [Biochemistry]

NBTGR 4-Nitrobenzyl-6-Thioguanosine [Biochemistry]

NbTi Niobium Titanium (Compound)

Nb-Ti Niobium-Titanium (Alloy System)

NBTL National Battery Test Laboratory [of Argonne National Laboratory, Illinois, US]

NBTL Naval Boiler and Turbine Laboratory [US Navy]

NBTR Narrowband Tape Recorder

NBU National Biodiversity Unit

Nb-U Niobium-Uranium (Alloy System)

N BUTT American Buttress Thread [Symbol]

NBV Net Book Value

Nb-V Niobium-Vanadium (Alloy System)

NBVA National Bulk Vendors Association [US]

Nb-W Niobium-Tungsten (Alloy System)

NBWA National Blacksmiths and Welders Association [US]

N by E North by East

N by NE North by Northeast

N by NW North by Northwest

N by W North by West

NBz Nitrobenzene [also NB]

Nb-Zr Niobium-Zirconium (Alloy System)

NC American National Coarse Thread [Symbol]

NC Nano-Cluster(s) [Materials Science]

NC Nanocomposite

NC Nanocrystal(line); Nanocrystallite; Nanocrystallization [also nc]

NC Nasal Cavity [Anatomy]

NC (American) National Coarse (Screw) Thread

NC Natta Catalyst [Chemistry]

NC Natural Circulation

NC Natural Convection

NC Navy Curtiss [US Navy Aircraft Symbol]

NC Near-Coincidence (Grain Boundary) [Materials Science]

NC Nearly Commensurate (Phase) [Materials Science]

NC Nearly Continuous

NC Nerve Cell (or Neuron) [Anatomy]

NC Nestell-Christy (Method)

NC Network Computer

NC Network Congestion

NC Network Connect

NC Network Control

NC Neutron Capture

NC Neutron Content

NC Neutron Contrast

NC Neutron Counter

NC Neutron Cycle [Nuclear Reactors]

NC Neutron-Radiographic Contrast

NC Neural Computation (Symposium)

NC Neural Computing

NC New Caledonia

NC New Caledonia (French) [ISO Code]

NC Newton-Cotes (Formula) [Mathematics]

NC Nickel Chrome

NC Nitrocellulose

NC Nitrocompound

NC Nitrogen Cycle [Ecology]

NC No Carry (Flag) [Computers]

NC No Change

NC No Charge [also N/C]

NC No Circuit

NC No Coil

NC No Comment

NC No Connection [also nc]

NC No Credit [also N/C]

NC Noctilucent Cloud [Meteorology]

NC Noise Criterion (or Criteria)

NC Nomarski Contrast [Microscopy]

NC Nominal Correction

NC Nominating Committee

NC Noncarboxylate

NC Nonchemical

NC Noncoherence; Noncoherent

NC Noncombustible

NC Noncondensable

NC Nonconducting; Nonconductor

NC Nonconformance; Nonconforming

NC Non-Contact

NC Noncorrodible

NC Non-Crystalline

NC Nor-Cal (Products Inc.) [US]

NC Nordic Council [Denmark, Finland, Iceland, Norway and Sweden]

NC Normal Cooling

NC Normally Closed (Contact) [also N/C]

NC North Carolina [US]

NC Nose Cone [also N/C]

NC Not Calculated

NC Notochord [Zoology]

NC Nuclear Ceramics

NC Nuclear Capability

NC Nuclear Chemist(ry)

NC Nuclear Converter [Nuclear Engineering]

NC Numbering Counter

NC Numerical Calculus

NC Numerical(ly) Control(led) [also N/C]

NC Nurse Corps

NC Nutation Cone

NC Nutrient Cycle [Ecology]

NC1 Nominal Correction 1 [Aerospace Phasing Maneuver] [also NC-I]

N2OC6 Naphtho-20-Crown-6

N/C No Charge [also NC]

N/C No Coverage

N/C No Credit [also NC]

N/C Normally Closed (Contact) [also NC]

N/C Nose Cone [also NC]

N/C Not Connected [also n/c]

N/C Not Critical

N/C Numerical Control [also NC]

Nc Naphthalocyanine

1,2-Nc 1,2-Naphthalocyanine

2,3-Nc 2,3-Naphthalocyanine

nC nanocoulomb [Unit]

nc nanocrystalline

nc no connection [also NC]

nc- nanocrystalline [Usually Followed by a Chemical Element, or Compound, e.g., nc-Ni is Nanocrystalline Nickel, etc.]

.nc New Caledonia (French) [Country Code/Domain Name]

NCA Naphthalene Carbonic Acid

NCA 1-Naphthalenecarboxylic Acid
NCA NASA Cooperative Agreement
NCA National Canners Association [US]
NCA National Capitol Area [US]
NCA National Caving Association [UK]
NCA National Chiropractic Association [US]
NCA National Coal Association [US]
NCA National Committee on Agrometeorology [US]
NCA National Communications Association [US]
NCA National Composition Association [US]
NCA National Computer Association [US]
NCA National Confectioners Association [US]
NCA National Constructors Association [US]
NCA National Cosmetology Association [US]
NCA National Council of Aviculture [UK]
NCA National Council on Alcoholism [US]
NCA Nature Conservancy Agency [UK]
NCA Naval Communications Annex
NCA Network Communications Adapter
NCA Network Computing Architecture
NCA Neutralized Current Acid (Waste)
NCA New Communities Administration
NCA Nickel Calcium Aluminate
NCA Noise Control Act [US]
NCA Non-Column Approximation [Electron Microscopy]
NCA Noncommutative Algebra
NCA Noncorrodible Aluminum
NCA Non-Crossing (Diagram) Approximation
NCA Nonspecific Cross-Reacting Antigen [Immunology]
NCA Northwest Computing Association [US]
NCAB National Cancer Advisory Board [US]
NCAC National Commission for Automatic Control
NCAC National Council of Acoustical Consultants [US]
NCAC Nordic Customs Administrative Council [Finland]
NCAC Nova Scotia Agricultural College [Truro, Canada]
NCACC National Civil Aviation Consultative Committee [UK]
NCACSS North Central Association of Colleges and Secondary Schools [US]
NCAE National Council for Adult Education [New Zealand]
NCAEI National Conference of Applications of Electrical Insulation [US]
NCAER National Council of Applied Economic Research [India]
NCAES National Center for Analysis of Energy Systems [US]
n-CaF$_2$ Nanostructured Calcium Fluoride
NC-AFM Non-Contact Atom Force Microscope [also NCAFM]
NCAIR National Center for Automated Information Retrieval [US]
N Cal New Caledonia(n)
NCAM National Congress on Applied Mechanics [of American Society of Mechanical Engineers, US]
NCAM Network Communications Access Method
N-CAM Neural Cell Adhesion Molecule
NCAMR Nordic Council for Arctic Medical Research [Finland]

NCAP Nonlinear Circuit Analysis Program
ncAPB Nonconservative Antiphase Boundary [Materials Science]
NCAPC National Center for Air Pollution Control [of Public Health Service, US]
NCAR National Center for Atmospheric Research [of National Science Foundation, at Boulder, Colorado, US]
NCAR National Center for Atmospheric Research [New Zealand]
NCAR National Committee for Antarctic Research
NCAR National Conference on the Administration of Research
NCAR National Conference on the Advancement of Research [US]
N Car North Carolina [US]
NCARB National Council of Architectural Registration Boards [US]
NCAS National Council of American Shipbuilders [US]
NCASI National Council (of the Paper Industry) for Air and Stream Improvement [US]
NCASI Tech Bull NCASI Technical Bulletin [US]
NCASI Tech Rev Index NCASI Technical Review Index [US]
NCAT National Center for Appropriate Technology [of US Department of Energy]
NCAT Non-Cumulated Aging Twins [Metallurgy]
NCATU North Carolina A&T University [Greensboro, US]
NCAVAE National Committee for Audio-Visual Aids in Education [UK]
NCAW Neutralized Current Acid Waste
NCB National Cargo Bureau [US]
NCB National Coal Board [UK]
NCB National Conservation Bureau [US] [Abolished]
NCB National Council for Cement and Building Materials [India]
NCB Naval Communications Board
NCB Near-Coincidence (Grain) Boundary) [Materials Science]
NCB Network Control Block
NCB Nitrochlorobenzene
NCBA National Cattle Breeders Association [UK]
NCBA National Cooperative Business Association [US]
NCBP National Contaminants Biomonitoring Program [US]
NCBR Near Commercial Breeder Reactor
NCBR Nitride-Cooled Breeder Reactor
NCBR Nordic Committee on Building Regulations [Finland]
NCBI National Center for Biotechnology Information [of National Library of Medicine, Bethesda, Maryland, US]
NCBI National Cotton Batting Institute [US]
NCBMP National Council of Building Material Producers [UK]
NCBTA Nordic Cooperative of Brick and Tilemakers Associations [Sweden]
N-CBZ N-Carbobenzoxy
NCC NASA Class Code
NCC National Chamber of Commerce [Bolivia]
NCC National Climate Center [Australia]

NCC National Computer Center [of US Environmental Protection Agency]

NCC National Computer Center [Manchester, UK]

NCC National Computer Conference [US]

NCC National Consultative Committee

NCC National Consultative Committee [of Australian Nature Conservancy Agency]

NCC National Consultative Council

NCC National Controls Corporation [US]

NCC National Coordinating Center

NCC National Cotton Council [US]

NCC National Curriculum Council [UK]

NCC Naturally Commutated Cycloconverter

NCC Nature Conservancy Council [UK]

NCC Nature Conservation Council [of New South Wales, Australia]

NCC Net Charge Compensation (Model)

NCC Netherlands Conference Center [The Hague]

NCC Network Computer Center

NCC Network Control Center

NCC Network Control Center [NASA Goddard Space Flight Center, Greenbelt, Maryland, US]

NCC Network Coordination Center

NCC Neurocirculatory Collapse [Medicine]

NCC Nickel-Phosphorus-Based Ceramic Composite

NCC Nitrogen-Carbon Cycle [Ecology]

NCC Nominal Corrective Combination (Maneuver) [Aerospace]

NCC Normally Closed Contact

NCC North Carolina College [Durham, US]

NCC Nutrition Coordinating Committee [US]

nc-C Nanocrystalline Carbon

NCCA National Chemical Credit Association [US]

NCCA National Coil Coaters Association [US]

NCCA National Concrete Contractors Association [now American Society for Concrete Construction, US]

NCCA National Cotton Council of America [US]

NCCAM National Center for Complementary and Alternative Medicine [of National Institutes of Health, Bethesda, Maryland, US]

NCCAS National Committee for Climate Change and Atmospheric Sciences [Australia]

NCCAT National Committee for Clear-Air Turbulence [US]

NCCAW National Consultative Committee on Animal Welfare

NCC-BN Nickel-Phosphorus-Based Boron Nitride

NCCC Nebraska Consolidated Communications Corporation [US]

NCCCC Naval Command, Control and Communications Center [US Navy]

NCCCD National Center for Computer Crime Data [Los Angeles, California, US]

NCCChE National Certification Commission in Chemistry and Chemical Engineering [US]

NCCD North Carolina College at Durham [US]

NCCDC National Computer Crime Data Center [now National Center for Computer Crime Data, US]

NCCDPC NATO Command, Control and Data Processing Committee

NCCE National Coalition for Consumer Education [US]

NCCEM National Coordinating Council on Emergency Management [US]

NCCF Network Communications Control Facility

NCCFN National Coordinating Committee on Food and Nutrition [Philippines]

NCCFO Neodymium Cerium Copper Iron Oxide (Superconductor)

NCC-HBN Nickel-Phosphorus-Based Hexagonal Boron Nitride [also NCC-hBN]

NCCHS National Conference on the Challenge of Health and Safety [Canada]

NCCHTS National Coordinating Center for Health Technology Assessment [UK]

NCCI Northamptonshire Chamber of Commerce and Industry [UK]

NCCIA North Carolina Crop Improvement Association [US]

NCCIS NATO Command Control and Information System

NCCK National Consultative Committee on Kangaroos [of Australian Nature Conservancy Agency]

NCCL National Council of Canadian Labour

NCCL National Council of Coal Lessors [US]

NCCLS National Committee for Clinical Laboratory Standards [US]

nC/cm nanocoulomb(s) per centimeter [also nC cm^{-1}]

nC/cm² nanocoulomb(s) per square centimeter [also nC cm^{-2}]

NC/CNC Numerical Control/Computer Numerical Control

NCCNSW Nature Conservation Council of New South Wales [Australia]

NCCFO Neodymium Cerium Copper Iron Oxide (Superconductor)

NCCN National Council of Catholic Nurses [US]

NCCNO Neodymium Cerium Copper Nickel Oxide (Superconductor)

NCCNO Neodymium Calcium Copper Niobium Oxide (Superconductor)

NCCO Neodymium Cerium Copper Oxide (Superconductor)

NCCOSC Naval Command, Control and Ocean Surveillance Center [San Diego, California, US]

NCCP Natural Communities Conservation Planning (Program) [California, US]

NCCPB National Council of Commercial Plant Breeders [US]

NCCPG National Council for the Conservation of Plants and Gardens [UK]

NCCr National Committee on Crystallography [US]

NCCS National Coalition for Cancer Survivorship [US]

NCCS Numerical Control Computer Sciences

NCC-SiC Nickel-Phosphorus-Based Silicon Carbide

NCCSS National Central Conference on Summer Schools [US]

NCCU National Chung Cheng University [Chia-Yi, Taiwan]

NCCU National Conference of Canadian Universities

NCCU North Carolina Central University [Durham, US]

NCCZO Neodymium Cerium Copper Zinc Oxide (Superconductor)

NCD Negotiated Critical Dates

NCD Network Computing Device

NCD Network Cryptographic Device

NCD Nicotinamide Cytosine Dinucleotide [Biochemistry]

NCDA National Cooperative Development Association [India]

NCDC National Climatic Data Center [US]

NCDC National Communicable Disease Center [now Centers for Disease Control and Prevention, Atlanta, Georgia, US]

NCDC New Community Development Corporation

NCDC 2-Nitro-4-Carboxyphenyl N,N-Diphenylcarbamate

NCDDG National Cooperative Drug Discovery Group [US]

NCDEAS National Committee of Deans of Engineering and Applied Science [Canada]

NCDF New Crop Development Fund [Canada]

n-CdHgTe N-Type Cadmium Mercury Telluride (Semiconductor)

NCDM National Conference on Disaster Management [Canada]

NCDRH National Center for Devices and Radiological Health

NCDS NASA Climate Data System

NCDS National Centre for Development Studies [of Australian National University, Canberra]

n-CdSe N-Type Cadmium Selenide (Semiconductor)

NCDWR North Carolina Department of Water Resources [US]

NCE National College of Education [Evanston, Illinois, US]

NCE National Committee for the Environment

NCE Networks of Centers of Excellence (Program) [Canada]

NCE Network Connection Element

NCE Network of Centres of Excellence [Canada]

NCE Newark College of Engineering [New Jersey, US]

NCE Noise Control Engineering

NCE Nordic Council for Ecology

NCE Nuclear Chemical Engineer(ing)

NCEA National Center for Energy Analysis [of Brookhaven National Laboratory, Upton, New York, US

NCEA National Center for Environmental Assessment [of US Environmental Protection Agency]

NCEA N-(1-Carboxyethyl)alanine [Biochemistry]

NCEA North Central Electric Association [US]

NCEB National Council for Environmental Balance [US]

NCEB NATO Communications Electronic Board

NCEC National Chemical Emergency Center [of United Kingdom Atomic Energy Authority]

NCEC National Coast Export Company [US]

NCEC National Construction Employers Council [US]

NCEC North Coast Environment Center

NCECS North Carolina Educational Computing Services [US]

NCED Northern California Earthquake Data [of US Geological Survey, Menlo Park, California]

NCEE National Congress on Engineering Education [US]

NCEE National Council of Engineering Examiners [US]

NCEES National Council of Examiners for Engineering and Surveying [US]

NCEFR National Council of Erectors, Fabricators and Riggers [US]

NCEFT National Commission on Electronic Funds Transfer

NCEL Naval Civil Engineering Laboratory [of US Navy]

NCEM National Center for Electron Microscopy [of Lawrence Berkeley National Laboratory at University of California at Berkeley, US]

NCEMT National Center for Excellence in Metalworking Technology [Johnstown, Pennsylvania, US]

NCEN National Commission on Egg Nutrition [US]

n-CeO$_2$ Nanocrystalline Cerium Oxide

NCEP National Center for Environmental Prediction [US]

NCEP National Cholesterol Education Program [US]

NCEPI National Center for Environmental Publications and Information [of US Environmental Protection Agency]

NCERQA National Center for Environmental Research and Quality Assurance [of US Environmental Protection Agency]

NCERT National Council of Educational Research and Training [India]

NCES National Center for Educational Statistics [US]

NCET National Council for Educational Technology [US]

NCF Narrow-Cut Filter

NCF National Cancer Foundation [US]

NCF National Clayware Federation [UK]

NCF National Communications Forum

NCF Neochemical Food

NCF Netware Command File

NCF Nominal Characteristics File

NCF Non-Consolidating Feeder

NCF Non-Crimped Fiber

NCF Non-Crystalline Film

NCFAE National Council of Forestry Association Executives [US]

NCFC National Council of Farmer Cooperatives [US]

NCFCA National Congress of Floor Covering Associations [US]

1,2-NcFe Iron 1,2-Naphthalocyanine

2,3-NcFe Iron 2,3-Naphthalocyanine

2,3-NcFe(pyz)$_2$ Iron 2,3-Naphthalocyanine Bis(pyrazine)

2,3-NcFe(tz)$_2$ Iron 2,3-Naphthalocyanine Bis(triazole)

NCFES North Central Forest Experiment Station [of US Department of Agriculture–Forest Service at Carbondale, Illinois]

NCFI National Cold Fusion Institute [US]

NCFM National Commission on Food Marketing [US]

NCFM National Committee on Fluid Mechanics

NCFMF National Committee for Fluid Mechanic Films [US]

NCFP National Conference on Fluid Power [US]

NCG Nanochannel Glass

NCG National Contractors Group [UK]

NCG Negatively Curved Graphite

NCG Network Control Group

NCG Nickel-Coated Graphite

NCG Non-Condensable Gas

NCG North Central Gyre [Oceanography]

NCG Nuclear Criteria Group

NCGA National Computer Graphics Association [US]

NCGA National Corn Growers Association [US]

NCGA National Cotton Ginners Association [US]

NCGCP National Climate and Global Change Program [US]

NCGDME National Consortium for Graduate Degrees for Minorities in Engineering [Notre Dame, Indiana, US]

NCGE National Council for Geographic Education [US]

NCGF Nickel-Coated Graphite Fiber

NCGG National Committee for Geophysics and Geodesy [Pakistan]

NCGIA National Center for Information and Analysis [US]

NCGM (International Conference on) Nano-Clusters and Granular Materials

NCGR National Council for Geocosmic Research [US]

NCGR National Council on Gene Resources [US]

NCGR Res J NCGR Research Journal [Publication of the National Council for Geocosmic Research, US]

NCGS Nuclear Criteria Group Secretary

NCGT National Council of Geography Teachers [US]

NCH National Center for Homeopathy [US]

NCH National Center for Horticulture [US]

NCH Network Connection Handler

NCH Nylon 6-Clay Hybrid

N Ch Normal Charge

NCHEML National Chemical Laboratory [of Ministry of Technology, UK]

NCHGR National Center for Human Genome Research [US]

NCHI National Council of the Housing Industry [US]

NCHMT National Capital Historical Museum of Transportation [US]

NCHP Nylon 6-Clay Hybrid with Silicate Layers of Saponite

NCHPTWA National Clearinghouse for Periodical Abstracts Service [US]

NCHRP National Cooperative Highway Research Program [of American Association of State Highway Officials, US]

NCHS National Center for Health Statistics [US]

NCHSE National Center for Human Settlements and Environment [India]

NCHU National Chung-Hsing University [Taiwan]

NCHVRFE National College for Heating, Ventilating, Refrigeration and Fan Engineering [US]

NCI National Cancer Institute [of National Institutes of Health, Bethesda, Maryland, US]

NCI National Cancer Institute (of Canada)

NCI National Computer Institute [US]

NCI National Computing Industries

NCI Naval Counter-Intelligence

NCI Navigation Control Indicator

NCI Negative (Ion) Chemical Ionization

NCI Net Carried Interest

NCI Netherlands Center for Informatics

NCI (Office of) New Concepts and Initiatives [US Air Force]

NCI Nodular Cast Iron [Metallurgy]

NCI Non-Coded Information

NCI Noncompetitive Inhibitor

NCI Northeast Computer Institute [US]

Nci Neisseria cinerea [Microbiology]

nCi nanocurie [Unit]

NCIA National Cavity Insulation Association [UK]

NCIAAA National Clearinghouse for Information on Alcohol Abuse and Alcoholism [of National Institutes of Health, Bethesda, Maryland, US]

NCIC National Cancer Institute of Canada

NCIC National Construction Industry Council [US]

NCIC National Crime Information Center [of Federal Bureau of Investigations in Washington, DC, US]

NCID National Center for Infectious Diseases [of Centers for Disease Control and Prevention, Atlanta, Georgia, US]

NCIDQ National Council for Interior Design Qualification [US]

NCI/GAB National Cancer Institute/Grants Administration Branch [of National Institutes of Health, Bethesda, Maryland, US]

NCIGBP National Committee for the International Geosphere/Biosphere Program [US]

NCIH National Conference on Industrial Hydraulics [US]

NCIO National Congress of Inventors Organizations [US]

NCIO Bull NCIO Bulletin [US]

NCIO Newsl NCIO Newsletter [US]

NCIPT National Institute for Integrated Photonic Technology [of Defense Advanced Research Projects Agency, US]

NCIS National Chemical Information Symposium [of American Chemical Society–Division of Chemical Information, US]

NCISC Naval Counter-Intelligence Support Center [US]

NCISE National Center for Improving Science Education [US]

NCIT National Council on Inland Transport [UK]

NCITO National Council on Industry Trading Organizations [UK]

NCIU Network Common Interface Unit

NCKU National Cheng Kung University [Tainan, Taiwan]

NCL National Central Library [Now Part of British Library]

NCL National Chemical Laboratory [Pune, India]

NCL National Communications Laboratory [US]

NCL Network Control Language

NCL Node Compatibility List

NCLA National Council of Local Administrators [US]

NCLI National Chemical Laboratory for Industry [of Tukuba Research Center, Japan]

NCLIS National Commission on Libraries and Information Science [US]

NCLT Night Carrier Landing Trainer

NCM Nanocrystalline Material

NCM National Commission of Mathematics [Portugal]

NCM Neodymium Calcium Manganate

NCM Network Control Module

NCM Nordic Council of Ministers [Scandinavia]

NCM Nordic Council on Medicine [Sweden]

N/cm² Newton(s) per square centimeter [also N cm^{-2}]

n/cm² neutron(s) per square centimeter [also n cm^{-2}]

NCMA National Catalogue Managers Association [US]

NCMA National Ceramic Manufacturers Association [US]

NCMA National Concrete Masonry Association [US]

NCMA National Contract Management Association [US]

NCMA National Critical Materials Act [US]

NCMB Nordic Council for Marine Biology [Norway]

NCMC National Coalition for Marine Conservation [US]

NCMC National Critical Materials Council [US]

NCMCE National Council of Minority Consulting Engineers [now National Association of Black Consulting Engineers, US]

NCME National Council on Measurement in Education [US]

NC Med J North Carolina Medical Journal [US]

NCML Naval Chemical and Metallurgical Laboratory [of Naval Dockyard, Bombay, India]

NCMET Non-Closed Shell Many-Electron Theory [Physics]

NCMRD National Center for Management Research and Development

NCMRED National Council on Marine Resources and Engineering Development [US]

NCMS National Center for Manufacturing Sciences [US]

NCMS National Classification Management Society [US]

n/cm²s neutron(s) per square centimeter second [also $n \, cm^{-2} \, s^{-1}$, or n/cm²sec]

n/cm²/s neutron(s) per square centimeter per second [also $n \, cm^{-2} \, s^{-1}$, or n/cm²/sec]

n/cm²/s/Å neutron(s) per square centimeter per second per angstrom [also $n \, cm^{-2} \, s^{-1}, Å^{-1}$, or n/cm²/sec/Å]

NCMSB Neutron and Consensed Matter Science Branch [of Atomic Energy of Canada Limited Research, Chalk River, Ontario, Canada]

NCMT Numerical Control for Machine Tools

NCMT Numerical Controlled Machine Tool

NCN Network Control Node

NCN Nixdorf Communications Network [Germany]

n-CN N-Type Carbon Nitride

NCNA New China News Agency

NCNB North Carolina News Bureau [US]

NCNM National College of Naturopathic Medicine [US]

NCNO Neodymium Calcium Nickel Oxide

NCNO Neodymium Cerium Nickel Oxide

NCO National Change of Address

NCO National Coordination Office (for High Performance Computing and Communications) [US]

NCO Neodymium Copper Oxide (Superconductor)

NCO Network Control Office

NCO Non-Commissioned Officer

NCO North Canadian Oils

Nco Nocardia corallina [Microbiology]

N₂/CO Nitrogen/Carbon Monoxide (Ratio)

N₂/CO₂ Nitrogen/Carbon Dioxide (Ratio)

n-Co Nanocrystalline Cobalt

NCOI National Council for the Omnibus Industry [UK]

NCOLUG North Carolina On-Line User Group [US]

NCOR National Committee for Oceanographic Research [Pakistan]

NCOS National Committee for Oceanic Sciences

NCOS Network Computer Operating System

NCOS Non-Concurrent Operating System

NCP National (Oil and Hazardous Substances Pollution) Contingency Plan [US]

NCP NetWare Core Protocol

NCP Network Control Point

NCP Network Control Processor

NCP Network Control Program

NCP Network Control Protocol

NCP Non-Central Potential

NCP Non-Consultative Party

NCP North Celestial Pole [Astronomy]

NCP Not Copy Protected

NCPA National Center for Physical Acoustics [of University of Mississippi, University, US]

NCPA National Cottonseed Products Association [US]

NCPA Nested Coherent Potential Approximation [Physics]

NCPA/UM National Center for Physical Acoustics/University of Mississippi [University, US]

NCPC Northern Canada Power Commission

NCPCA National Committee for the Prevention of Child Abuse [US]

NCPCC National Clearinghouse Poison Control Center [US]

NCPD National Council for Population and Development [US]

NCPDM National Council of Physical Distribution Management [UK]

NCPF National Council on Private Forests [of American Forestry Association, US]

NCPGG National Center for Petroleum Geology and Geophysics [Australia]

NCPI National Clay Pipe Institute [US]

NCP/L Network Control Program/Local [also NCP-L]

NCP/LR Network Control Program/Local-Remote [also NCP-L/R]

NCPM National Clay Pot Manufacturers [US]

NCPMA Noise Control Products and Materials Association [also NCP&MA, US]

NCPO National Climate Project Office [of National Oceanic and Atmospheric Administration, US]

NCPO Neodymium Copper Phosphorus Oxide

NCPO Nordic Council for Physical Oceanography [Sweden]

NCPPL Numerical Control Part Programming Language

NCPR National Congress of Petroleum Retailers [now Service Station Dealers of America, US]

NCP/R Network Control Program/Remote [also NCP-R]

NCPS National Commission on Product Safety [US]

NCPTWA National Clearinghouse for Periodical Title Word Abbreviations

NCPUA National Committee on Pesticide Use in Agriculture [Canada]

NCPUA National Committee on Pesticide Use in America [US]

NCP/VS Network Control Program/Virtual Storage [also NCP-VS]

NCPWB National Certified Pipe Welding Bureau [US]

NCPWI National Council on Public Works Improvement [US]

NCQR National Council for Quality and Reliability [UK]

NCR National Capital Region [Canada]

NCR National Cash Register (Corporation) [US]

NCR Natural Circulation (Nuclear) Reactor

NCR No Calibration Required

NCR No Carbon Required (Paper)

NCR No Circuit Request

NCR Nonconformance(/Failure) Report

NCR Non-Conforming Reports

NCR Northern Capital Region

NC/R No Change, or Reset

NCRA National Cellular Resellers Association [US]

NCRA National Cooperative Refinery Association [US]

NCRC National Cave Rescue Commission [US]

NCR-DNA NCR (Corporation) Distributed Network Architecture

NCRDS National Coal Resources Data System [of United States Geological Survey]

NCRE Naval Construction Research Establishment [UK]

NCRH National Center for Radiological Health [of Public Health Society, US]

NCRI National Consumer Research Institute [US]

NCRL National Chemical Research Laboratory [South Africa]

NCR MISSION NCR Corporation Manufacturing Information System Support Integrated On-Line [of NCR Corporation, US]

NCRP National Committee on Radiation Protection (and Measurement) [of National Institutes of Health, Bethesda, Maryland, US]

NCRPM National Council on Radiation Protection and Measurements [of National Institutes of Health, Bethesda, Maryland, US]

NCRR National Center for Research Resources [of National Institutes of Health, Bethesda, Maryland, US]

NCRSR National Commercial Refrigeration Sales Association [US]

NCRT National College of Rubber Technology [UK]

NCRUCE National Conference of Regulatory Utility Commission Engineers [US]

NCRV National Committee for Radiation Victims

NCRW Neutralized Cladding Removal Waste [Nuclear Engineering]

NCS National Communications System [of US Department of Defense]

NCS National Computer Systems

NCS National Conservation Strategy

NCS National Conservation Strategy [of Environment Australia]

NCS National Corrosion Service

NCS Natural Color System

NCS Naval Communication Station

NCS Naval Control of Shipping

NCS N-Chlorosuccinimide

NCS Net Control Station [Military]

NCS Netherlands Computer Society

NCS Network Cabling Specialist

NCS Network Controller Software

NCS Network Control Station

NCS Network Control System

NCS Network Computing System

NCS Network Coordination Station

NCS No Checking Signal

NCS No-Crimp Structure

NCS Noncoherent Scattering [Atomic Physics]

NCS Non-Coincidence (Lattice) Site [Materials Science]

NCS Nonconventional System

NCS Noncrystalline Solid

NCS North Carolina Section [of Materials Research Society, US]

NCS North Carolina State [US]

NCS Numerical Category Scaling

NCS Numerical Control Society [now Association for Integrated Manufacturing Technology, US]

NCS Numerical Control System

NCS Numeric Character Set

NCS Nutation Control System

N_2CS_3 Disodium Monocalcium Trisilicate [$2Na_2O$ CaO $3SiO_2$, or $Na_2CaSi_3O_9$]

NC/S No Change, or Set

NCSA National Center for Supercomputing Applications [Champaign, Illinois, US]

NCSA National Conservation Strategy for Australia [of Environment Australia]

NCSA National Construction Software Association [US]

NCSA National Crushed Stone Association [US]

NCSA N-Carbonylsulfamoylchloride

NCSAC National Conservation Strategy Advisory Council [Australia]

NCSAG Nuclear Cross-Section Advisory Group

NCSBCS National Conference of States on Building Codes and Standards [US]

NCSBEE National Council of State Board of Engineering Examiners [now National Council of Engineering Examiners [US]

NCSC National Computer Security Center [US]

NCSC National Council of Schoolhouse Construction

NCSC North Carolina State College [Raleigh, US]

NCSC North Carolina Supercomputer Center [US]

NCSC Nuclear Criticality Safety Committee [US]

NCSCR North Carolina State College Reactor [Raleigh, US]

NCSD Nuclear Criticality Safety Division [of American Nuclear Society, US]

NCSE National Center for Science Education [US]

NCSE North Carolina Society of Engineers [US]

nc-Se Nanocrystalline Selenium

NCSESA National Committee on Science Education Standards and Assessment [of National Academy of Sciences, US]

NCSI National Council for Stream Improvement [US]

NCSI Network Communications Services Interface

nc-Si Nanocrystalline Silicon

nc-Si:F Fluorinated Nanocrystalline Silicon

nc-Si:H Hydrogenated Nanocrystalline Silicon

NCS/ICS Network Controller Software/Integrated Controller Software

NCSL National Conference of Standards Laboratories [US]

NCSL National Conference of State Legislatures [US]

NCSL Near Coincidence Site Lattice [also nCSL] [Materials Science]

NCSL Newsl NCSL Newsletter [Published by the National Conference of Standards Laboratories, US]

NCSM National Council of Supervisors of Mathematics [US]

NCSN National Computer Service Network [US]

NCSP Nordic Committee on Salaries and Personnel [Denmark]

NCSP Northern Cod Science Program [of Department of Fisheries and Oceans, Canada]

NCSPA National Corrugated Steel Pipe Association [US]

NCSR National Center for Systems Reliability [UK]

NCSS National Cactus and Succulent Society [UK]

NCST National Conference on School Transportation [US]

NCST National Council for Science and Technology [Kenya]

NCSTD National Center for Scientific and Technical Documentation [Netherlands]

NCSTRC North Carolina Science and Technology Research Center [of University of North Carolina, US] [also NC/STRC]

NCSTS National Conference of State Transportation Specialists [US]

NCSTTO National Council for the Supply and Training of Teachers Overseas [UK]

NCSU North Carolina State University [Raleigh, US]

NCT Nagaoka College of Technology [Niigata, Japan]

NCT National Center of Tribology [UK]

NCT National Chamber of Trade [UK]

NCT Network Control and Timing

NCT Night Closing Trunks

NCT Nitrocellulose, Tubular

NCT Noncontact Tonometer [Medicine]

NCT Nosé Constant Temperature

NCT Sodium Chromium Titanate

NCTA National Cable Television Association [US]

NCTA National Council for Technological Awards [UK]

NCTE National Council for Textile Education [US]

NCTE National Council of Teachers of English

NCTE Network Channel Termination Equipment

nc-TiN Nanocrystalline Titanium Nitride

NCTM National Council of Teachers of Mathematics [US]

NCTN NASA Commercial Technology Network [of NASA Headquarters, Washington, DC, US]

NCTS National Council of Technical Schools

NCTS Northeast Corridor Transportation System [US]

NCTSI National Council of Technical Service Industries [now Contract Services Association, US]

NCTT Norwegian Council for Technical Terminology [Norway]

NCTTA National Competitiveness Technology Transfer Act [US]

NCTU National Carpet Trades Union [UK]

NCTU National Chiao Tung University [Hsinchu, Taiwan]

NCU National Central University [Chung-Li, Taiwan]

NCU National Communications Union [UK]

NCU Navigation Computer Unit

NCU North Carolina University [US]

n-Cu Nanocrystalline Copper

NCUA National Credit Union Administration

NCUAS Northwest College and University Association for Science [Richland, Washington (State), US]

NCUC Nuclear Chemistry Users Committee

n-Cu(Co) Cobalt-Doped Nanocrystalline Copper

NCUES National Center for Urban Environmental Studies [US]

NCUG Nevada COBOL Users Group [US]

NCUGAE National Computer User Group in Agricultural Education [UK]

NCUR National Committee for Utilities Radio [now Utilities Telecommunications Council, US]

NCURA National Council of University Research Administrators [US]

NCUT National Center for Upgrading Technology [of Canada Center for Mineral and Energy Technology–Western Research Center, Natural Resources Canada]

NCTU National Chiao-Tung University of Education [Hsinchu, Taiwan]

NCUE National Changhua University of Education [Changhua, Taiwan]

NCUTLO National Committee on Uniform Traffic Laws and Ordinances [US]

NCV Nerve Condition Velocity

NCV Net Calorific Value [also ncv]

NCV No Commercial Value

NCVB Norfolk Convention and Visitors Bureau [Virginia, US]

NCVDG National Cooperative Vaccine Development Group [US]

nc-VN Nanocrystalline Vanadium Nitride

NCVO National Council of Voluntary Organizations [UK]

NCVQ National Council for Vocational Qualification [UK]

NCW No Change in Weather

NCW Non-Communist World

NCWCC North Central Weed Control Committee [US]

NCWCD National Commission for Wildlife Conservation and Development [Saudi Arabia]

NCWQ National Commission on Water Quality [US]

NCWR Nordic Council for Wildlife Research [Denmark]

NCWM National Conference on Weights and Measures [US]

NCWM National Council on Women in Medicine [US]

nc-WN Nanocrystalline Tungsten Nitride

ND NASA Document

ND Natural Draft

ND Naturopathic Doctor

ND Negative Declaration

ND Network Design

ND Neutral Density (Filter)

ND Neutral Direction

ND Neutron Density

ND Neutron Detection; Neutron Detector

ND Neutron Diffraction

ND Neutron Diffractometer; Neutron Diffractometry

ND New Deal

ND New Delhi [India]

ND New Drug

ND N-Nitroso Diethanol Amine

ND Nitrosodurene

ND No Data

ND No Date [also nd]

ND No Delay

ND No Detect

ND Noise Dampening

ND Nomarski DIC (Differential Interference Contrast) [Microscopy]

ND Nominal Dimension

ND Non-Delay

ND Nondenatured; Nondenaturing

ND Nondirectional (Antenna)

ND Nondispersive

ND None Detected

ND Normal Direction

ND Normal Distribution

ND Normal Duty

ND North Dakota [US]

ND Notch-Ductile; Notch Ductility [Metallurgy]

ND Not Detectable; Not Detected [also nd]

ND Not Determinable; Not Determined [also N/D, nd, or n/d]

ND Not Done

ND Not Divulged [also nd]

ND Notre Dame [Indiana, US]

ND Nuclear Data

ND Nucleotidase [Biochemistry]

ND Number of Document(s)

ND Numerical Data

5'-ND 5'-Nucleotidase [Biochemistry]

N(D) Particle Size Distribution Function [Symbol]

N/D Need Date

N/D Not Determinable; Not Determined [also ND, nd, or n/d]

Nd Needle [also nd]

Nd Neodymium [Symbol]

Nd^{3+} Neodymium Ion [also Sm^{+++}] [Symbol]

Nd II Neodymium II [Symbol]

Nd-123 $NdBa_2Cu_3O_{7-x}$ [A Neodymium Barium Copper Oxide Superconductor]

Nd-138 Neodymium [also ^{138}Nd, or Nd^{138}]

Nd-139 Neodymium [also ^{139}Nd, or Nd^{139}]

Nd-140 Neodymium [also ^{140}Nd, or Nd^{140}]

Nd-141 Neodymium [also ^{141}Nd, or Nd^{141}]

Nd-142 Neodymium [also ^{142}Nd, or Nd^{142}]

Nd-143 Neodymium [also ^{143}Nd, or Nd^{143}]

Nd-144 Neodymium [also ^{144}Nd, or Nd^{144}]

Nd-145 Neodymium [also ^{145}Nd, or Nd^{145}]

Nd-146 Neodymium [also ^{146}Nd, or Nd^{146}]

Nd-147 Neodymium [also ^{147}Nd, or Nd^{147}]

Nd-148 Neodymium [also ^{148}Nd, or Nd^{148}]

Nd-149 Neodymium [also ^{149}Nd, or Nd^{149}]

Nd-150 Neodymium [also ^{150}Nd, or Nd^{150}]

Nd-151 Neodymium [also ^{151}Nd, or Nd^{151}]

Nd-211 Nd_2BaCuO_{10} [A Neodymium Barium Copper Oxide Superconductor]

Nd-422 $Nd_4Ba_2Cu_3O_{10}$ [A Neodymium Barium Copper Oxide Superconductor]

Nd-2111 $Nd_2BaCeCuO_4$ [A Neodymium Barium Cerium Copper Oxide Superconductor]

nD noncoherent dislocation (in materials science) [Symbol]

nd no date [also ND]

nd not detectable; not detected [also ND]

nd not determinable; not determined [also ND, N/D, or n/d]

nd not divulged [also ND]

n-d n-dimensional (space) [Mathematics]

12N46D 1,2-Naphthoquinone-4,6-Disulfonate

NDA Naphthaline Dicarboxylic Acid

NDA National Dairymen's Association [UK]

NDA National Defense Act [US]

NDA National Defense Academy [Yokosuka, Japan]

NDA New Drug Application (Program) [US]

NDA Nondestructive Analysis

NDA Nondestructive Assay

NDAB Numerical Data Advisory Board

NDAC National Defense Advisory Committee [US]

NDAC Not Data Accepted [General-Purpose Interface Bus Line]

NDAC Nuclear Defense Affairs Committee [of NATO]

$Nd(Ac)_3$ Neodymium (III) Acetate

$Nd(ACAC)_3$ Neodymium Acetylacetonate [also $Nd(AcAc)_3$, or $Nd(acac)_3$]

NdAD Nicotinamide Deoxyadenosine Dinucleotide [Biochemistry]

NDADS NSSDC (National Space Science Data Center) Data Archive Distribution System [of NASA Goddard Space Flight Center, Greenbelt, Maryland, US]

N Dak North Dakota [US]

Ndap Nitrilodiaceticpropionic Acid

NDAR National Directory of Australian Resources [of National Resource Information Center]

NDB Nautical Directional Beacon

NDB Nondirectional (Radio) Beacon [also NdB]

NDB Nucleic Acid Database [Biochemistry]

NDB Numeric Database

NDBA N-Nitroso Dibutylamine

NDBC National Data Buoy Center [of National Oceanic and Atmospheric Administration, US]

NdBCO Neodymium Barium Copper Oxide (Superconductor)

NDB/L Nondirectional Beacon/Locator

NDBMS Network Database Management System

NDBP National Data Buoy Program [of National Oceanic and Atmospheric Administration, US]

NDBP Di-n-Butylphosphate

Nd-BR Butadiene Rubber Based on Neodymium Catalyst

NDC National Dairy Council [Rosemont, Illinois, US]

NDC National Data Center

NDC National Data Communication

NDC National Defense College [of Department of National Defense at Kingston, Ontario, Canada]

NDC National Design Council

NDC National Development Corporation [Tanzania]

NDC National Dome Council [US]

NDC NATO Defense College

NDC Negative Differential Conductance; Negative Differential Conductivity

NDC Network Diagnostic Control

NDC Nondestructive Characterization

NDC NORAD (North American Aerospace Defense Command) Direction Center

NDC Normalized Device Coordinate

NDCA National Drilling Contractors Association [US]

NDCA Nuclear Development Corporation of America

NDCAA National Dutch Civil Aviation Authority

NDCB Nitrodichlorobenzene

NdCCO Neodymium Cerium Copper Oxide (Superconductor)

NDCDB National Digital Cartographic Data Base [US]

NDCEE National Defense Center for Environmental Excellence [US]

Nd-Co Neodymium-Cobalt (Alloy System)

Nd(Cp)$_3$ Tris(cyclopentadienyl)neodymium

NDCR Noise Distortion Clearance Range

NDCT Natural Draft Cooling Tower [Nuclear Engineering]

NDD Neutron Density Distribution

NDDK Network Device Development Kit

NDDN National Dry Deposition Network [US]

NDDO Neglect of Diatomic Differential Overlap [Physics]

NDDRP Natural Disaster and Drought Relief Policies

NDE Near-Death Experience

NDE Nondestructive Evaluation [also NDEx]

NDE Nondestructive Examination

NDE Nonlinear Dielectric Effect

NDE Nonlinear Differential Equation

Nde Neisseria denitrificans [Microbiology]

NDEA National Defense Education Act [US]

NDEA National Display Equipment Association [UK]

NDEA N-Nitrosodiethylamine

NDE/CAD Nondestructive Evaluation/Computer-Aided Design

NDEF Not to be Defined

NDELA N-Nitrosodiethanolamine

NDES Normal Digital Echo Suppressor

NDEx Nondestructive Examination [also NDE]

NDF National Development Foundation [South Africa]

NDF National Drilling Federation [US]

NDF Neutral Density (Light) Filter

NDF Neutral Detergent Fiber

NDF No Defect Found

NDF Nonlinear Distortion Factor

NDF Neutral-Density Factor

NDF Neutral-Density Filter [Optics]

NdFe Neodymium Iron (Alloy)

NdFeB Neodymium Iron Boron (Magnet Alloy)

NDFTA National Dried Fruit Trades Association [UK]

N^7-dGTP 7-Deaza-2'-Deoxyguanosine 5'-Triphosphate [Biochemistry]

NDGA Nordihydroguaiaretic Acid

NDGFD North Dakota Game and Fish Department [US]

Nd:Glass Neodymium-Doped Glass (Laser) [also Nd-Glass, Nd:glass, or Nd^{3+}:Glass]

NDGRAV The Nickel and Dime Gravity Meeting [Syracuse, New York, US]

NDGS North Dakota Geological Survey [US]

NDHA National District Heating Association [US]

NDHD North Dakota Health Department [US]

Nd(HFAC)$_3$ Neodymium(III) Hexafluoroacetylacetonate [also Nd(hfac)$_3$]

NDHQ National Defense Headquarters [of Department of National Defense, Ottawa, Canada]

NDI 1,5-Naphthalene Diisocyanate

NDI Neglect of Differential Integrals

NDI Neglect of Difficult Integrals

NDI Nickel Development Institute [Sidcup, UK]

NDI Nondestructive Inspection

NDI Non-Development Item

NDI Numerical Designation Index

NDIR Nondispersive Infrared (Absorption)

NDIS Network Device Interface Specification

NDIS Network Driver Interface Specification

NDL National Diet Library [Tokyo, Japan]

NDL Network Data Language

NDL Network Definition Language

NDL Nuclear Defense Laboratory [of US Army]

Ndl Needle [also ndl]

NDLC Network Data Link Control

NDLC Nitrogen-Containing Diamond-Like Carbon

Nd:LiYF$_4$ Neodymium-Doped Lithium Yttrium Fluoride (Laser) [also Nd^{3+}:LiYF$_4$]

NDLM Nondestructive Laser Mapping

Nd:LNO Neodymium-Doped Lithium Niobate [also Nd^{3+}:LNO]

Ndls Needles [also ndls]

NDM Negative Differential Mobility

NDM Nominal Disconnected Mode

NDM Normal Disconnected Mode

NDMA National Dimension Manufacturers Association [US]

NDMA Nitrosodimethylamine

NDMA Nondispersive Microanalysis

NDMC National Defense Medical Center [of Department of National Defense, Canada]

NDMS Network Design and Management System

NDN Non-Delivery Notice

NdN Neodymium Nitride

nDNA Native Deoxyribonucleic Acid [Biochemistry]

N-DNP N-Dinitrophenyl

NDO Neglect of Differential Overlap [Physics]

NDO Network Development Office [of Library of Congress, Washington, DC, US]

N-D-O Nitrogen-Deuterium-Oxygen (Bonds)

NDOA N-Nitroso-2-(3-3',7'-Dimethyl-2',6'-Octadienyl)-2-Aminoethanol

NDOP Di-n-Octylphosphate

NdOPr^i Neodymium Isopropoxide

NDOS New Disk Operating System

NDoT N-Diethyl-o-Toluidine [also Ndot]

NDP National Drought Policy

NDP Net Domestic Product

NDP Neutron Depth Profiling

NDP 1-(4-Nitrobenzyl)-4-(4-Diethylaminophenylazo)pyridinium

NDP Normal Diametral Pitch (of Gears)

NDP Nuclear Disarmament Party [Australia]

NDP Nucleoside Diphosphatase [Biochemistry]

NDP Numeric Data Package

NDP Numeric Data Processor

NdP Neodymium Phosphate (Semiconductor)

NDPA National Decorating Products Association [US]

NDPA N-Nitroso Diphenylamine

NDPA N-Nitroso Dipropylamine

Ndpa Nitrilodipropionicacetic Acid

NDPK Nucleoside 5'-Diphosphate Kinase [Biochemistry]

NDPP 4-[{4-(Diethylamino)phenyl}azo]-1-[(4Nitrophenyl)methyl]pyridinium Bromide

nDPPA n-Decyl Phenylphosphonate [also n-DPPA]

NDPRB National Dairy Promotion and Research Board [Arlington, Virginia, US]

NDPS National Data Processing Service [UK]

NDPS National Data Processing System [US]

NDR National Drivers Register [US]

NDR Negative Differential Resistance

NDR Network Data Reduction

NDR Network Data Representation

NDR Nil-Ductility Region [Metallurgy]

NDR Nondestructive Read(out)

NDR Norddeutscher Rundfunk [Broadcasting Station in Northern Germany]

NDRA Natural Disaster Relief Arrangements

NDRC National Defense Research Committee [US]

NDRC National Defense Research Council [US]

NDR/DW Nondestructive Read/Destructive Write

NDRE Norwegian Defense Research Establishment

NDRF National Defense Reserve Fleet [of US Maritime Administration]

NDRI National Dairy Research Institute [India]

NDRO Nondestructive Readout

NDS National Design Specification

NDS NetWare Directory Service

NDS Network Documentation System

NDS Neutron Dispersive System

NDS Neutron-Doped Silicon

NDS New Drug Submission

NDS Nondispersive Spectrometry

NDS Nonparametric Detection Scheme

NDS Nordic Demographic Society [Norway]

NDS North Dakota State [US]

NDS Novell NetWare Directory Services

NDS Nuclear Data Sheet

NDSA Napthalene Disulfonic Acid

NDSA National Dam Safety Act [US]

NDSA Northern Development Subsidiary Agreement [Canada]

NDSB Narcotic Drugs Supervisory Body [of United Nations]

NDSC Network for the Detection of Stratospheric Change

NDSC North Dakota Safety Council [US]

NDSEF National Defense Science and Engineering Fellowship

NDSLC North Dakota State Library Commission [US]

NDSPE North Dakota Society of Professional Engineers [US]

NDSS National Down's Syndrome Society [US]

NDST (International Symposium on) New Diamond Science and Technology

NDSU North Dakota State University [Fargo, US]

NDT Nephrology, Dialysis, Transplantation [Journal published by Oxford University Press, UK]

NDT Net Data Throughout

NDT Network Description Template

NDT Neurodevelopmental Treatment [Medicine]

NDT Newfoundland Daylight Time [Canada]

NDT Nil-Ductility Temperature [Metallurgy]

NDT Nil-Ductility Transition [Metallurgy]

NDT Nondestructive Test(ing)

NDTA National Defense Transportation Association [US]

NDTC Nondestructive Testing Center

NDTE Nondestructive Testing Equipment

NDT&E Nondestructive Testing and Evaluation

NDT&E Int Nondestructive Testing and Evaluation International [Journal]

NDT-E Nondestructive Testing–Economical

NDTF Nondestructive Test Facility

Nd(TFAC)$_3$ Neodymium(III) Trifluoroacetylacetonate [also Nd(tfac)$_3$]

NDT Int Nondestructive Testing International [Journal published in the UK]

NDTL Nondestructive Test Laboratory

Nd(TMHD)$_3$ Neodymium(III) Tris(2,2,6,6-Tetramethyl-3,5-Heptanedionate) (also Nd(tmhd)$_3$]

NDTT Nil Ductility Transition Temperature [Metallurgy]

NDTT Nil Ductility Test Temperature [Metallurgy]

NDU National Defense University [of US Department of Defense at Washington, DC]

NDU Nôtre Dame University [Paris, France]

NDUV Nondispersive Ultraviolet

NDVI Normalized Difference Vegetation Index [Ecology]

n-d VRHC N-Dimensional Variable-Range Hopping Conduction [Physics]

NDWCS National Drilling and Well Control School [of University of New South Wales, Sydney, Australia]

NDWP National Demonstration Water Project [now National Water Project, US]

.NDX Index [File Name Extension]

Nd:YAG Neodymium-Doped Yttrium Aluminum Garnet (Laser) [also Nd-YAG, or Nd^{3+}:YAG]

Nd:YAP Neodymium-Doped Yttrium Aluminum Perovskite (Laser) [also Nd-YAP, or Nd^{3+}:YAP]

Nd:YLF Neodymium-Doped Yttrium Lithium Fluoride (Laser) [also Nd-YLF, or Nd^{3+}:YLF]

Nd:YVO$_4$ Neodymium-Doped Yttrium Vanadate (Laser) [also Nd-YVO$_4$, or Nd^{3+}:YVO$_4$]

NE Monomethyl Ester of 5-Norbornene-2,3-Dicarboxylic Acid

NE Nanoelectronic(s)

NE Nanoengineered; Nanoengineering

NE National Estate [Australia]

NE Naval Engineer(ing)

NE Near (Jump) [Microsoft-Disk-Operating System]

NE Nebraska [US]

NE Negative Earnings

NE Nernst-Einstein (Approach) [Physics]

NE Nernst Equation [Physical Chemistry]

NE Nerve Ending [Anatomy]

NE Net Energy

NE Neutral Emission

NE Neutral Equilibrium [Physics]

NE Neutralization Equivalent [Chemistry]

NE Neutron Energy

NE New Edition

NE New England (States) [Maine, New Hampshire, Vermont, Massachusetts, Rhode Island and Connecticut]

NE New Resource Standard [Software Engineering Standard]

NE Niger [ISO Code]

NE No Effect

NE Nonelastic Elongation

NE Non-Embrittled (Material)

NE Non-Equilibrium

NE Norbornene Monomethyl Ester Carboxylic Acid

NE Norepinephrine [Biochemistry]

NE Normal Equation

NE North-East(ern) [ne]

NE Nose Ejection

NE Notch Effect [Mechanics]

NE Not Equal to

NE Nuclear Emulsion

NE Nuclear Energy

NE Nuclear Engine

NE Nuclear Engineer(ing)

NE Nuclear Explosive

NE Nueva Esparta [Venezuela]

NE Numerical Engineer(ing)

N/E New Edition

N(E) Density of States (in Solid-State Physics) [Symbol]

N(E) Electron Energy Distribution [Symbol]

N(E$_F$) Density of States at Fermi Level [Symbol]

N$_\uparrow$(E$_F$) Fermi Density of States of Up Band [Symbol]

N$_\downarrow$(E$_F$) Fermi Density of States of Down Band [Symbol]

N(E$_i$) Ion Density Distribution [Symbol]

N$_m$/E$_0$ Normalized Media Noise (in Electronics) [Symbol]

Ne Neon [Symbol]

Ne Neuron [Anatomy]

Ne Normal Erosion Resistance [Symbol]

Ne-19 Neon-19 [also ^{19}Ne, or Ne^{19}]

Ne-20 Neon-20 [also ^{20}Ne, Ne^{20}, or Ne]

Ne-21 Neon-21 [also ^{21}Ne, or Ne^{21}]

Ne-22 Neon-22 [also ^{22}Ne, or Ne^{20}]

Ne-23 Neon-23 [also ^{23}Ne, or Ne^{21}]

n(E) density of states (in solid-states physics) [Symbol]

n(E$_F$) Fermi density of states (in solid-state physics) [Symbol]

NEA National Education Association [US]

NEA National Electronic Association [US]

NEA National Endowment for the Arts [US]

NEA National Energy Act [US]

NEA National Environmental Act [US]

NEA National Erectors Association [US]

NEA National Exhibitors Association [UK]

NEA Negative Electron Affinity [Electronics]

NEA N-Ethylaniline [also Nea]

NEA Noise Equivalent Angle

NEA North Eastern Airlines [US]

NEA Nuclear Energy Agency [of Organization for Economic Cooperation and Development, France]

NEA Nuclear Engineering Associates

Nea N-ethylaniline [also NEA]

NEAA Non-Essential Amino Acid [Biochemistry]

NEA Act Rep NEA Activity Report [of Organization for Economic Cooperation and Development–Nuclear Energy Agency, France]

NEAC National Energy Advisory Committee [Australia]

NEAC Nippon Electric Automatic Computer

NEACP National Emergency Airborne Command Post [of US Department of Defense]

NEACSS New England Association of Colleges and Secondary Schools [US]

NEADAI National Education Association–Department of Audiovisual Instruction

NEAFC Northeast Atlantic Fisheries Commission [UK]

NEA J National Education Association Journal [US]

Neal N-Diethylaniline

NEANDC Nuclear Energy Agency–Nuclear Data Committee [France]

NEA Newsl NEA Newsletter [of Organization for Economic Cooperation and Development's Nuclear Energy Agency, France]

NEAP National Energy Audit Program

NEAP National Environmental Action Plan

NEAR National Emergency Alarm Repeater [US]

NEAR Near-Earth Asteroid Rendezvous [Astronomy]

NEARA New England Antiquities Research Association [US]

NEA Rep NEA Reporter [Publication of the National Education Association, US]

Near-IR Near Infrared [also near-IR]

NEAS Near East Archeological Society [US]

NEAT NCR (National Cash Register) Electronic Autocoding Technique

NEB National Economic Board [South Korea]

NEB National Energy Board [of Natural Resources Canada]

NEB National Enterprise Board [UK]

NEB Nebraska [US]

NEB Net Environmental Benefit(s)

NEB Noise Equivalent Bandwidth

NEB Nuclear, Electronic, Biological (Warfare)

Neb Nebraska [US]

neb *(nebula)* – spray (solution) [Medical Prescriptions]

NEBA NASA Employees Benefit Association

NEBAM Nederlandse Bevrachting en Agentur Maatschappij BV [Dutch Freight Agency located at Velsen]

NEBB National Environmental Balancing Bureau [US]

NEBCO Neodymium Erbium Barium Copper Oxide (Superconductor)

NEBC North East Biotechnology Center [UK]

Nebr Nebraska [US]

NEBS New Exporters to Border States (Program) [of Department of Foreign Affairs and International Trade, Canada]

NEBSS National Examinations Board in Supervisory Studies

NEBULA Natural Electronic Business Users Language

NE by E Northeast by East

NE by N Northeast by North

NEC National Economic Council [UK, Pakistan and Philippines]

NEC National Electric Code [US]

NEC National Electronics Conference [US]

NEC National Electronics Council [London, UK]

NEC National Electrostatics Corporation [Middleton, Wisconsin, US]

NEC National Emergency Council

NEC National Engineering Consortium [US]

NEC Nippon Electric Company [Tokyo, Japan]

NEC Nitrogen Engineering Corporation [US]

NEC North Equatorial Current [Oceanography]

NEC Northern European Country

NEC Not Elsewhere Classified

NEC Not Explicitly Covered

NEC Nuclear Energy Center

NEC Numeric Electromagnetic Code

NECA 5'-(N-Ethylcarboxamido)adenosine [Biochemistry]

NECA National Electrical Contractors Association [US]

NECA National Exchange Carrier Association [US]

NECA Near East College Association [US]

Necar New Electric Car

NECCA National Educational Closed Circuit Association [now Educational Television Association, UK]

NECEA National Engineering Construction Employers Association [UK]

NECG National Executive Committee on Guidance

NECH New England Center for Headache [at Greenwich Hospital, Connecticut, US]

NECIES Northeast Coast Institution of Engineers and Shipbuilders [UK]

NECNP New England Coalition on Nuclear Pollution [US]

NECOS Network Coordinating Station

NECPA National Emergency Command Post Afloat [of US Department of Defense]

NEC PC Nippon Electric Company Personal Computer

NECPUC New England Conference of Public Utility Commissioners [US]

N-ECR Nitrogen-Based Electron-Cyclotron Resonance [also NECR]

NEC Res Dev NEC Research and Development [Published by Nippon Electric Co. Ltd, Tokyo, Japan]

NECS National Electric Code Standards [US]

NECS Nationwide Educational Computer Service [US]

NECSP National Enhanced Cancer Surveillance Program [Canada]

NECSS Nuclear Energy Center Site Survey

NECTA National Electrical Contractors Trade Association [US]

NEC Tech J NEC Technical Journal [Published by Nippon Electric Co. Ltd, Tokyo, Japan]

NED NASA/IPAC (Infrared Processing and Analysis Center) Extragalactic Database [at Jet Propulsion Laboratory, California Institute of Technology, Pasadena, US]

NED Neutral Emission Detection; Neutral Emission Detector

NED New Editor

NED New English Dictionary

NED Normal Energy Distribution

NED Nuclear Engineering and Design

NEDA National Economic Development Authority [Philippines]

NEDA National Electronic Distributors Association [US]

NEDA National Electronics Development Association [New Zealand]

NEDA National Environmental Development Association [US]

NEDA National Exhaust Distributors Association [US]

NEDA/Ground National Environmental Development Association Groundwater Project [US]

NEDA/USA National Exhaust Distributors Association/Undercar Specialists Association [US]

NEDC National Economic Development Council [UK]

NEDDC N-[Naphthyl(1)]-Ethylenediamine

NEDN Naval Environmental Data Network [US Navy]

NEDO National Economic Development Office [of National Economic Development Council, UK]

NEDO New Energy (and Industrial Technology) Development Organization [of Ministry of International Trade and Industry, Ibaraki, Japan]

NEDRES National Environmental Data Referral System [US]

NEDS National Emissions Data System [of US Environmental Protection Agency]

NEDSA Non-Erasing Deterministic Stack Automation

Ned Tijdschr Natuurkd A Nederlands Tijdschrift voor Natuurkunde A [Dutch Journal of Physics, Section A; published by Nederlandse Natuurkundige Vereniging, Amsterdam, Netherlands]

Ned Tijdschr Natuurkd B Nederlands Tijdschrift voor Natuurkunde B [Dutch Journal of Physics, Section B; published by Nederlandse Natuurkundige Vereniging, Amsterdam, Netherlands]

NEDU Navy Experimental Diving Unit

NEE Naphthylethyl Ether

NEEB North East Engineering Bureau [UK]

NEEB Northeastern Electricity Board [UK]

NEED New Employment Expansion and Development (Program)

Need Needle [also need]

NEEDS National Emergency Equipment Data System [of Environment Canada]

NEEDS National Engineering Education Delivery System [of National Science Foundation, US]

NEEDS New England Educational Data System [US]

NEEDS Nonlinear Evolution Equations and Dynamical Systems (Conference)

NEELS National Emergency Equipment Locator System [of Environmental Protection Service, Canada]

NEEP New England Economic Project [US]

NEEP Nuclear Electronic Effects Program [US]

NEERI National Environmental Engineering Research Institute [India]

NEETU National Engineering and Electrical Trades Union [UK]

NEF National Energy Foundation [US]

NEF (American) National Extra-Fine Thread [Symbol]

NEF New England Section Fall Meeting

NEF Noise Exposure Forecast

NEFA Non-Esterified Fatty Acid

NEFA North East Forest Alliance

NEFC National Engineering Foundation Conference [US]

NEFC Near East Forest Commission

NEFD Noise Equivalent Flux Density

NEFES Northeastern Forest Experiment Station [of US Department of Agriculture at Chestnut Hill, Pennsylvania]

NEFO National Electronics Facilities Organization [US]

NEFSG Northeastern Forest Soils Group [Canada and US]

NEFTIC Northeastern Forest Tree Improvement Conference [US]

NEG Nitride Extrinsic Gettering [also N-EG]

NEG Non-Euclidean Geometry

NEG Nonevaporable Getter

N-EG Nitride Extrinsic Gettering [also NEG]

Neg Negate; Negation [also neg]

neg negative

Negatron Negative-Resistance Screen-Grid Tube [also negatron]

NEGB North Eastern Gas Board [UK]

NEGB Non-Equilibrium Grain Boundary [Materials Science]

Negistor Negative Resistor [also negistor]

NEGOR Net Current Gas-Oil Ratio

NEGP National Environmental Goals Program [of US Environmental Protection Agency]

NEGP National Estate Grants Program [of Australian Heritage Commission]

NEHA National Environmental Health Association [US]

NeHe Neon-Helium (Laser)

NEHTP New England Headache Treatment Program [US]

NEI 1-(1-Naphthyl)ethylisocyanate

NEI National Eye Institute [of National Institutes of Health, Bethesda, Maryland, US]

NEI Negative-Ion with Electron-Impact Ionization [Physics]

NEI Netherlands East Indies

NEI Netherlands Economics Institute

NEI Noise Equivalent Irradiance

NEI Noise Exposure Index

NEI Nordic Energy Index [Denmark]

NEI Not Elsewhere Included [also nei]

NEI Not Elsewhere Indicated [also nei]

NEI Nuclear Engineering International

NEIC National Earthquake Information Center [of US Geological Survey at Golden, Colorado]

NEIC National Energy Information Center [US]

NEIC National Engineering Information Center [US]

NEII National Elevator Industry, Inc. [US]

NEIMME North of England Institute of Mining and Mechanical Engineers [UK]

NEIS National Earthquake Information Service [of US Geological Survey]

NEIS National Electrical Industries Show

NEIS National Engineering Information System

NEISS National Electronic Injury Surveillance System [US]

NEITDO New Energy and Industrial Technology Development Organization [Japan]

NEJM New England Journal of Medicine [of Massachusetts Medical Society, US]

NEK 1-Naphthyl Ethyl Ketone

NEKDA New England Kiln Drying Association [US]

NEL National Electronics Council [UK]

NEL National Engineering Laboratory [Glasgow, UK]

NEL National Epilepsy League [US]

NEL Naval Electronics Laboratory [of US Navy]

NEL Nuclear Electronics Laboratory [US]

NELA National Electric Light Association [now Edison Electric Institute, US]

NELA Northeastern Loggers Association [US]

NELAT Naval Electronics Laboratory Assembly Tester

NELC Naval Electronics Laboratory Center [US Navy]

NELCON National Electronics Conference [of Institute of Electrical and Electronics Engineers, US]

NELCON NZ National Electronics Conference, New Zealand

NELIA Nuclear Energy Liability Insurance Association

NELIAC Navy Electronics Laboratory International Algorithmic Compiler [US Navy]

NELINET New England Library Information Network [US]

NELMA Northeastern Lumber Manufacturers Association [US]

NELPA Northwest Electric Light and Power Association

NELPIA Nuclear Energy Liability Property Insurance Association

NEL Rep NEL Report [National Engineering Laboratory, Glasgow, UK]

NEM NAAQO (National Ambient Air Quality Objectives) Exposure Model

NEM Nanoengineered Material

NEM N-Ethylmaleimide

NEM Nitrogen Ethylmorpholine

NEM Not Elsewhere Mentioned [also nem]

NEMA National Egg Marketing Association [UK]

NEMA National Electrical Manufacturers Association [US]

NEMA National Emergency Management Association [US]

NEMA/EEMAC National Electrical Manufacturers Association/Electrical and Electronic Manufacturers Association [US]

NEMAG Negative Effective Mass Amplifiers and Generators

NEMC New England Medical Center [Boston, Massachusetts, US]

NEMD Nonequilibrium Molecular Dynamics (Method) [Physics]

NEMRL New England Marine Research Laboratory [US]

NEMI National Elevator Manufacturing Industry [now National Elevator Industry, Inc., US]

NEMI National Energy Management Institute [US]

NEMJ (Alliance for) National Excellence in Materials Joining (Program) [US]

NEMJET NEMJ Education and Training [US]

NEMO Naval Experimental Manned Observatory

NEMO Non-Empirical Molecular Orbitals [Physical Chemistry]

NEMO Not Emanating from Main Office [also nemo]

NEMP Nuclear Electromagnetic Pulse

NEMPET Northeast Microbial Physiologists, Ecologists and Taxonomists [US]

NEMRA National Electrical Manufacturers Representatives Association [US]

NEMRIP New England Marine Resources Information Program [US]

NEMS National Environment Management Strategies (Program)

NEMS Nimbus E Microwave Spectrometer

NENA N-(2-Nitroxy)nitraminoethane

NEngl New England [US]

N Engl Northern England

N Engl J Med New England Journal of Medicine [of Massachusetts Medical Society, US]

NENO Dinitrodi(β-Nitroethyl)oxamide

NENP Nickel(II) (Ethylenediamine) Nitrite Perchlorate

NEO Naval Engineering Officer

NEO New Exports to Overseas (Program) [of Department of Foreign Affairs and International Trade, Canada]

NEO Neoplastic Effects (of Hazardous Materials)

NEO Net Electrical Output

NEO Nuclear Energy Organization [Cairo, Egypt]

NEOCON National Exposition of Contract Interior Furnishings [of Institute of Business Designers, US]

NEODA National Edible Oil Distributors Association [UK]

NEODAT Inter-Institutional Database of Fish Biodiversity in the Neotropics

NEOF No Evidence of Failure

NEOF Nordic Engineer Officers Federation [Sweden]

NEOME Network of Excellence on Organic Materials for Electronics [of Esprit, Europe]

Neorg Mater Neorganicheskie Materialy [Publication on Inorganic Materials of the Academy of Sciences of the USSR]

neo-STX neo-Saxitoxin [Biochemistry]

NEP National Emphasis Program

NEP National Energy Plan

NEP National Energy Policy (Plan) [of US Department of Energy]

NEP National Energy Program

NEP National Estuary Program [of US Environmental Protection Agency]

NEP National Extension Programs

NEP Net Electrical Power

NEP N-Ethyl-2-Pyrrolidinone

NEP Network Entry Point

NEP Never-Ending Program

NEP Noise Equivalent Power

NEP Nominal Entry Point

NEP Normal Entry Point

NEP Nonlinear Electrooptic Property

NEPA National Environmental Policy Act [US]

NEPA National Environmental Policy Administration [US]

NEPA National Environmental Protection Act

NEPA National Environment Protection Authority [Australia]

NEPA Nigerian Electric Power Authority

NEPA Nuclear Energy for Propulsion of Aircraft (Project) [of Fairshield Engine and Airplane Corporation, Oak Ridge, Tennessee, US]

NEPATS National Environmental Policy Administration Tracking System [US]

NEPB National Environmental Protection Board [Sweden]

NEPC National Environment Protection Council [Australia]

NePC Neon Photoconductor [also Ne-PC]

NEPCON National Electronic Packaging and Production Conference [US]

NEPCON WEST National Electronic Packaging and Production Conference West [US]

NEPD Noise Equivalent Power Density

NEPE National Emergency Planning Establishment

Neph Nephelometric; Nephelometry [also neph]

NEPHIS Nested Phrase-Indexing System

Nephrol Nephrological; Nephrologist; Nephrology [also nephrol]

Nephrol Dial Transplant Nephrology, Dialysis, Transplantation [Journal published by Oxford University Press, UK]

NEPI National Environmental Policy Institute [Washington, DC, US]

NEPIA Nuclear Energy Property Insurance Association

NEPIS National Environmental Publications Information System [of US Environmental Protection Agency]

NEPIS N-Ethyl-5-Phenylisoxazolium-3'-Sulfonate

NEPM National Environment Protection Measure

NEPMA National Engine Parts Manufacturers Association [US]

NEP&ME National Exposition of Power and Mechanical Engineering

NEPMs National Environment Protection Measures

NEPP National Environment Policy Plan

NEPP National Environment Protection Plan

NEPPS National Environmental Performance Partnership System [Joint US Federal/ State Environmental Project]

NEPR NATO Electronic Parts Recommendation

NEPTUNE Northeastern Electronic Peak Tracing Unit and Numerical Evaluator

NEPZ Nantze Export Processing Zone [Taiwan]

NEQ Not Equal

NEQ Non-Equivalence (Function)

NER National Eagle Repository [US]

NER National Engineers Register

NER Nederlands Elektronica– en Radiogenootschap [Netherlands Electronics and Radio Cooperative]

NER Nepalese Rupee [Currency]

NER Net Energy Ratio

n(E,r) local density of states (in solid-state physics) [Symbol]

NERA National Economic Research Association [US]

NERAC New England Research Application Center [US]

NERBA New England Road Builders Association [US]

NERBC New England River Basins Commission [US]

NERC National Electric Reliability Council [US]

NERC National Environmental Research Center [US]

NERC National Environmental Research Council [US]

NERC National Electronics Research Council [now National Electronics Council, UK]

NERC Natural Environmental Research Council [UK]

NERC New England Regional Commission [US]

NERC New England Research Center [US]

NERC New En-Route Center [of Naval Air Transport Service, UK]

NERC Nigeria Educational Research Council

NERC North American Electric Reliability Council [US]

NERC Nuclear Energy Research Center [Belgium]

NERD&D National Energy Research Development and Demonstration [Australia]

NERDDC National Energy Research Development and Demonstration Council [Australia]

NERDOP National Energy Research Development and Demonstration Program [of Australian Atomic Energy Commission]

NEREM Northeast Electronics Research and Engineering Meeting [US]

ne rep *(ne repetere)* – do not repeat [Medical Prescriptions]

NERF Naval Electromagnetic Radiation Facility [of US Navy]

NERFET Negative Resistance Field-Effect Transistor

NERHL Northeastern Radiological Health Laboratory [US]

NERIS National Energy Referral Information System [of Environment Information Center, US]

NERL National Exposure Research Laboratory [US]

NERL Nuclear Engineering Research Laboratory [US]

NERO Na (Sodium) Experimental Reactor of O (Zero) Energy

NERO National Energy Resources Organization [US]

NERO Near Earth Rescue Operation [of NASA]

NERO Near East Regional Office [of Food and Agricultural Organization]

NERP National Environmental Research Park [US]

NE&RS (Department of) Nuclear Engineering and Radiation Science

NERSC National Energy Research Supercomputer Center [of Lawrence Livermore Laboratories, Livermore, California, US]

NERV Nuclear Emulsion Recovery Vehicle

NERVA Nuclear Engine for Rocket Vehicle Applications

NES National Education Supercomputer

NES National Employment Service

NES National Energy Strategy [US]

NES National Engineering Service

NES National Environment Secretariat

NES National Ergonomic Society [Sweden]

NES National Estimating Society [US]

NES Near-End Suppressor

NES Noise Equivalent Signal

NES Non English Speaking

NES Non-Equilibrium State

NES Not Elsewhere Specified [also nes]

NES Nuclear Engineering School [Tokai, Japan]

NES Numerical Engineering Association [UK]

NESA National Electric Sign Association [US]

NESA National Energy Specialist Association [US]

NESAC-BIO NIH (National Institutes of Health) Environmental Surface Analysis Center for the Study of Biological Problems [Bethesda, Maryland, US]

NESA J NESA Journal [Published by the National Energy Specialists Association, US]

NESB Non English Speaking Background

NESBAC Northeast Shetland Basin Area Communications [UK]

NESBIC Netherlands Students Bureau for International Cooperation

NESC National Electrical Safety Code [US]

NESC National Environmental Satellite Center [of Environmental Science Services Administration, US]

NESC National Environmental Supercomputing Center [of US Environmental Protection Agency]

NESC Naval Electronics Systems Command [US Navy]

NESC Network for Evaluating (or for the Evaluation) Steel Components [of Institute for Advanced Materials, Netherlands]

NESC Northeastern State College [Tahlequah, Oklahoma, US]

NESC Nuclear Engineering and Science Conference

NESCA National Environmental Systems Contractors Association [now Air Conditioning Contractors of America, US]

NESCC Naval Electronics Systems Command Headquarters [US Navy]

NESCOM New Standards Projects Committee [of Institute of Electrical and Electronics Engineers, US]

NESDA National Electronic Sales and Service Dealers Association [US]

NESDA National Equipment Servicing Dealers Association [US]

NESDA Northeast Scotland Development Authority

NESDD NOAA Earth Science Data Directory [of National Oceanic and Atmospheric Administration, US]

NESDIS National Environmental Satellite Data and Information Service [of National Oceanic and Atmospheric Administration, US]

NESEM New England Society of Electron Microscopy [US]

NESHAP National Emission Standards for Hazardous Air Pollutants [US]

NESHP National Emission Standards for Hazardous Pollutants [US]

NESI Nonequilibrium Surface Ionization

NESP National Environment Studies Project [US]

NESR Natural Environment Support Room

NESRF Northern Environmental Studies Revolving Fund [Canada]

NESS National Environmental Satellite Service [US]

NESS Network and Evaluation Simulation System

NESSEC Naval Electronic Systems Security Engineering Center [US]

NESSUS Numerical Evaluation of Stochastic Structures Under Stress (Computer Code)

NEST National Ecological Surveys Team [of US Geological Survey/Biological Resources Division]

NEST Naval Experimental Satellite Terminal

NEST Novell Embedded Systems Technology

NESTA National Earth Science Teachers Association [US]

NESTEV Naval Electronics System Test and Evaluation

NESTOR Neutron Source Thermal Reactor

NESTS Non-Electric Stimulus Transfer System

NESW Non-Essential Service Water

NET Nachrichten Elektronik + Telematik [German Journal on Communication, Electronics and Telematics]

NET National Educational Television [US]

NET Negative Entropy Trap

NET Net Equivalent Temperature

NET Network Equipment Technology

NET Next European Torus (Nuclear Fusion Device)

NET Nitroso Ester Terpolymer

NET Noise Equivalent Temperature [Thermodynamics]

NET Non-Equilibrium Thermodynamics

.NET Networks [Internet Domain Name] [also .net]

Net Internet

Net Network [also net]

NETA National Electrical Testing Association [now International Electrical Testing Association, US]

NETA National Environmental Training Association [US]

NETA North Eastern Traders Association [UK]

NETAC Nuclear Energy Trade Associations Conference [UK]

NETA Newsl NETA Newsletter [Published by the National Environmental Training Association, US]

NetBEUI NetBIOS Extended User Interface

NetBIOS Network Basic Input/Output System [also NETBIOS]

NETC National Emergency Transportation Center [US]

NetCDF Network Common Data Format [also netCDF]

NETCOM Network Communications

NET/COM Networking and Communications Exposition [Canada]

Netcon Networking and Connectivity Show [Canada]

NETD Noise Equivalent Temperature Difference [Thermodynamics]

NETE Naval Engineering Test Establishment

NETEC Network Technical Support Group

NETFS National Educational Television Film Service [US]

Neth Netherlands

1,N⁶-Etheno-NAD $1,N^6$-Etheno-NAD Nicotinamide $1,N^6$-Ethenoadenine Dinucleotide [Biochemistry]

1,N⁶-Etheno-NADP $1,N^6$-Etheno-NADP Nicotinamide $1,N^6$-Ethenonicotinamide Adenine Dinucleotide Phosphate [Biochemistry]

NET/ITER Next European Torus/International Thermonuclear Experimental Reactor

NET IVHU Next European Torus/In-Vessel Handling Unit

NETL National Export Traffic League [US]

NETMON Network Monitor

NETMUX Network Multiplexer

NETR Nuclear Engineering Test Reactor [US Air Force]

NETSET Network Synthesis and Evaluation Technique

NETT New England Telephone and Telegraph Company [US]

Netw Network [also netw]

Netw, Comput Neural Syst Network: Computation in Neural Systems [Journal of the Institute of Physics, UK]

Netw Manage Networking Management [US Publication]

Netw Newsl Network Newsletter [Publication of the Environmental and Societal Impacts Group of the National Center for Atmospheric Research, US]

NETWORKS Polymer Networks Group Meeting

NEU Northeastern University [Boston, Massachusetts, US]

Neu Neuraminic Acid

NEUDATA Neutron Data Under Direct Access (Database) [of Organization for Economic Cooperation and Development, France]

Neue Bergbautech Neue Bergbautechnik [German Journal on Modern Mining Engineering]

Neues Jahrb Geol Paleont Neues Jahrbuch für Geologie und Paleontologie [German Yearbook on Geology and Paleontology]

Neues Jahrb Mineral Neues Jahrbuch für Mineralogie [German Yearbook on Mineralogy]

Neues Jahrb Mineral Monatsh Neues Jahrbuch für Mineralogie, Monatshefte [German Yearbook on Mineralogy, Monthly]

Neue Tech Neue Technik [Swiss Journal on New Technology]

Neue Verpack Neue Verpackung [German Journal on Modern Packaging]

NEUG National Epson Users Group [US]

NeuNAc N-Acetyl-D-Neuraminic Acid

Neural Comput Neural Computation [Journal published by MIT Press, Cambridge, US]

Neural Netw Neural Networks [Journal published in the US]

Neurobiol Neurobiological; Neurobiologist; Neurobiology

Neurobiol Aging Neurobiology of Aging (Journal)

Neurobiol Disease Neurobiology of Disease [Journal published in the US]

Neurobiol Learn Mem Neurobiology of Learning and Memory [Journal published in the US]

Neuroendocrinol Neuroendocrinology [International Journal]

Neuroepidemiol Neuroepidemiology [International Journal]

Neuroimmunomod Neuroimmunomodulation [International Journal]

Neurol Neurological; Neurologist; Neurology [also neurol]

Neuropharmacol Neuropharmacological; Neuropharmacologist; Neuropharmacology

Neuropharmacol Neuropharmacology [International Journal]

Neuropsychobiol Neuropsychobiology [International Journal]

Neuropsychol Neuropsychological; Neuropsychologist; Neuropsychology

Neuropsychol Neuropsychologica [International Journal]

Neurosci Neuroscience; Neuroscientist

Neurosci Neuroscience [International Journal]

NE USA North Eastern United States of America [also NE-USA]

neut neuter

neut neutral

Neutn Neutralization [also neutn]

NEUTR Neutral Conductor [Electricity]

NEUTRINO International Conference on Neutrino Physics and Astrophysics

NEV Net Energy Value

NEV Non-Equivalent (Gate)

Nev Nevada [US]

NEVE Non-Empirical Valence-Electron

NEW National Educators Workshop [of Materials Education Council, US]

NEW National Engineers Week [US]

NEW Nuclear Energy Women [US]

.NEW New Information [File Name Extension]

NEWA National Electric Wholesalers Association [now National Association of Electrical Distributors, US]

NEWA New England Waterworks Association [US]

NEWAC NATO Electronic Warfare Advisory Committee

New Astron New Astronomy [Published in the Netherlands]

New Electron New Electronics [Journal published in the UK]

New Engl J Med New England Journal of Medicine [of Massachusetts Medical Society, US]

Newf Newfoundland [Canada] [also Newfld]

New Gener Comput New Generation Computing [Journal published in Japan]

New Int New Internationalist [US Environmental Journal]

New J Chem New Journal of Chemistry [Published by the Royal Society of Chemistry, UK]

New M New Mexico [US]

New Mat Int New Materials International [Journal published in the UK]

New Mat/Jpn New Materials/Japan [Journal published in the UK]

New Mat/Korea New Materials/Korea [Journal published in the UK]

New Mat World New Materials World [Journal published in the UK]

New Mexico Tech New Mexico Institute of Mining and Technology [Socorro, US]

New Phys New Physics [Journal of the Korean Physical Society; published by the Korea Institute for Industrial Economics and Technology, South Korea]

New Phys (Korean Phys Soc) New Physics (Korean Physical Society) [Published by the Korea Institute for Industrial Economics and Technology, South Korea]

NEWRADS Nuclear Explosion Warning and Radiological Data System

NEWS (International Symposium on) Nuclear Electro-Weak Spectroscopy

NEWS Naval Electronic Warfare Simulator

NEWS NetWare Early Warning System

NEWS Network Extensible Window System

News Newspaper [also news]

New Sci New Scientist [Journal published in the UK]

News Govt Ind Res Inst, Osaka News of the Government Industrial Research Institute, Osaka [Japan]

Newsl Newsletter [also newsl]

Newsl NEA Data Bank Newsletter of the NEA (Nuclear Energy Agency) Data Bank [Published by Organization for Economic Cooperation and Development, France]

Newsl Sust Devel Newsletter: Sustainable Development [Publication of the President's Council on Sustainable Development, US]

Newsl Oil Spill Prof Newsletter for Oil Spill Professionals [US Publication]

News Nisshin Steel News from Nisshin Steel [Published by Nisshin Steel Corporation, Tokyo, Japan]

News Physiol Sci News in Physiological Sciences [Publication of the International Union of Physiological Sciences, France]

News Rohde Schwarz News from Rohde and Schwarz [German Newsletter of Rohde & Schwarz, Munich]

New Technol Jpn New Technology Japan (Bulletin) [also New Technol (Jpn)]

NEW Update National Educators Workshop:Update [of Materials Education Council, US]

NEWWA New England Water Works Association [US]

Nex Nynaes Energy Chemicals Complex [Sweden]

NEXAFS Near-Edge X-Ray Absorption Fine Structure

NEXRAD Next Generation Weather Radar

NEXCO National Association of Export Companies [US]

NEXPRI Nederlands Expertise Centruum voor Ruimtelijke Informatiererwerkig [Netherlands Expertise Center for Land-Use Information]

NEXT Near-End Crosstalk

NEY No Electron Yield

NF American National Fine Thread [Symbol]

NF National Forest [US]

NF National Forge Company [Irvine, Pennsylvania, US]

NF National Formulary [Grade of Chemicals]

NF Natural Fiber

NF Natural Flood

NF Natural Frequency

NF Near-Fully

NF Near Face

NF Near Field

NF Nearly Ferromagnetic (Model) [Solid-State Physics]

NF Nerve Fiber [Anatomy]

NF Neutrofibromatosis [Medicine]

NF Neutral Fiber [Mechanics]

NF Neutral Filament

NF Neutral Filter

NF Newfoundland [Canada]

NF Newtonian Fluid [Fluid Mechanics]

NF Network Flow

NF Neutron Filter

NF Newfoundland [Canada]

NF Nickel Ferrite

NF Night Fighter

NF Nitride Fiber

NF Nitrogen Fixation [Chemical Engineering]

NF Nobel Foundation [Stockholm, Sweden]

NF No Failure [also nf]

NF No Font

NF No Fracture [also nf]

NF No Funds [also nf]

NF Noise Factor

NF Noise Figure

NF Noise Frequency

NF Nonfaceted (Eutectics) [also nf] [Metallurgy]

NF Nonferrous

NF Nonflammable

NF Non-Food

NF Norfolk Island [ISO Code]

NF Normal Factor (Concentration)

NF Normal Fluid [Cryogenics]

NF Normal Force

NF Normal Freezing

NF Normalized Function [Mathematics]

NF Nose Fuse (Bomb)

NF Notch Factor [Mechanics]

NF Not Found

NF Nozzle Flow

NF Nuclear Fission

NF Nuclear Force

NF Nuclear Fuel

NF Nuclear Fusion

NF$_o$ Overall Noise Figure [Semiconductor Symbol]

N/F No Funds

Nf Nerve Fiber [Anatomy]

nF nanofarad [also nf]

.nf Norfolk Island [Country Code/Domain Name]

NFA Nanocrystalline Forming Ability [Materials Science]

NFA National Farmers Association [Ireland]

NFA National Fertilizers Association [US]

NFA National Food Authority

NFA National Founders Association [now American Cast Metals Association, US]

NFA Network File Access

NFA Nonferrous Alloy(s)

NFA Non-Food Agriculture

NFA Norsk Forening for Automatisering [Norwegian Automation Association]

NFA Northern Flood Agreement [Canada]

NFA Northwest Forestry Association [US]

NFA Numerical Functional Analysis

NFAA National Federation of Advertising Agencies [now International Federation of Advertising Agencies, US]

NFAA Northern Federation of Advertisers Associations [Sweden]

NFAC Naval Facilities Engineering Command Headquarters [US Navy]

NFAH National Foundation for the Arts and the Humanities

NFAIS National Federation of Abstracting and Information Services [US]

NFAIS National Federation of Abstracting and Indexing Services [Name Changed to National Federation of Abstracting and Information Services, US]

NFAIS Bull NFAIS Bulletin [US]

NFAIS Newsl NFAIS Bulletin [US]

NFAM Network File Access Method

NFAP National Fisheries Adjustment Program

NFAP Network File Access Protocol

NFB National Film Board [Canada]

NFB Negative Feedback

NFB Negative Feeder-Booster

NFB Neodymium Iron Boride

NFBA National Frame Builders Association [US]

NFBC National Film Board of Canada

NFBPM National Federation of Builders and Plumbers Merchants [UK]

NFBS Nordiske Forskningsbibliotekens Samarbejdskomite [Nordic Research Libraries Cooperative Committee–A Joint Scandinavian Committee]

NFBW National Federation of Building Workers [Israel]

NFBWW Nordic Federation of Building and Wood Workers [Sweden]

NFC National Fertilizer Corporation [Japan]

NFC National Fire Code [of National Fire Protection Association, US]

NFC National Forensic Center [US]

NFC National Freight Consortium [UK]

NFC No Further Consequences [also nfc]

NFC Northern Fisheries Committee

NFC Not Favorably Considered [also nfc]

NFC Nuclear Fuel Cycle

NFCA Nonfuel Core Array

NFCB National Federation of Community Broadcasters

NFCC National Farm Chemurgic Council [US] [Defunct]

NFCGA National Federation of Constructional Glass Associations [UK]

NFCI National Federation of Clay Industries [UK]

NFCR National Foundation for Cancer Research [US]

NFCRS National Fisheries Containment Research Center [US]

NFCS National Federation of Construction Supervisors [UK]

NFCS Nuclear Forces Communications Satellite

NFCSIT National Federation of Cold Storage and Ice Trades [also NFCIT, UK]

NFCTA National Federation of Corn Trade Associations [UK]

NFCTC National Foundry Craft Training Center [UK]

NFD Nuclear Fluid Dynamic(s) [Nuclear Physics]

Nfd Newfoundland [Canada]

NFDA National Fastener Distributors Association [US]

NFDC National Federation of Demolition Contractors [UK]

NFDC National Fertilizer Development Center [of Tennessee Valley Authority at Muscle Shoals, US]

NFDC National Flight Data Center [US]

NFDM Non-Fat Dry Milk

NFDPS National Flight Data Processing System [of International Civil Aviation Organization]

NFE National Faculty Exchange [US]

NFE Nearly-Free Electron

NFE Network Front-End

NFE Nitrogen-Free Extract

NFE Nuclear Fuel Element

n-Fe Nanocrystalline Iron

NFEA National Federated Electrical Association [UK]

NFEAS National Federation of Engineers, Architects and Surveyors [Israel]

NFEC National Food and Energy Council [US]

NFEC Naval Facilities Engineering Command [of US Navy in Alexandria, Virginia, US]

NFECC Naval Facilities Engineering Command Center [of US Navy in Alexandria, Virginia, US]

NFEGI National Federation of Engineering and General Ironfounders [UK]

nfer nonferrous

NFERF National Fisheries Education and Research Foundation [US]

NFES Naval Facilities Engineering Service [US]

NFES Northeastern Forest Experiment Station [of US Department of Agriculture at Chestnut Hill, Pennsylvania]

NFE-sp Nearly-Free Electron sp (Band) [Symbol]

NFET N-Channel (Junction) Field-Effect Transistor [also N-FET]

NFET Negative-Channel Field-Effect Transistor [also N-FET]

NFE-TB Nearly-Free Electron Tight-Binding [Physics]

NFETM National Federation of Engineers Tool Manufacturers [UK]

NFF National Farmers' Federation

NFF No Fault Found

NF-F Nonfaceted-Faceted (Eutectics) [also nf-f] [Metallurgy]

NFFAW Newfoundland Fishermen, Food and Allied Workers Union [Canada]

NFFE National Federation of Federal Employees [US]

NFFS National Ferrous Founders Society [US]

NFFS Nonferrous Founders Society

NFFTU National Federation of Furniture Trade Unions [UK]

NFFWU Nordic Federation of Factory Workers Unions [Sweden]

NFGC National Federation of Grain Cooperatives [US]

NFHA National Federation of Housing Associations [UK]

NFHC National Federation of Housing Co-operatives [UK]

NFI National Federation of Ironmongers [UK]

NFI National Fisheries Institute [US]

NFI National Forest Inventory [Australia]

NFI National Forestry Institute [Canada]

NFI Non-Food Item

NFIA National Feed Ingredients Association [US]

NFIB National Federation of Independent Businesses [US]

NFIC National Fisheries Industry Council

NFIMA Nonferrous Ingot Metal Institute [US]

NFIP National Foundation for Infantile Paralysis [US]

NFISM National Federation of Iron and Steel Merchants [UK]

NFK Norsk Korrsjontknisk Forening [Norwegian Corrosion Research Association]

NFL National Fire Laboratory [of National Research Council of Canada]

NFL Near-Fully Lamellar (Microstructure) [Metallurgy]

NFL Newfoundland Federation of Labour [Canada]

NFL Non-Fermi Liquid

Nfl Newfoundland [Canada]

NFLCP National Federation of Local Cable Programmers [US]

Nfld Newfoundland [Canada]

NFLPA National Freelance Photographers Association [US]

NFLPN National Federation of Licensed Practical Nurses [US]

NFM Narrow(band) Frequency Modulation

NFM Near-Field Micropotential

NFM Newtonian Fluid Mechanics

NFM Non-Fat Milk

NFM Nonferromagnetic

NFM Nonferrous Material

NFM Nonferrous Metal

NFM Nonferrous Metallurgy

NFM Nuclear Ferromagnetism

NFMA National Facility Management Association [US]

NFMA National Fireplace Makers Association [UK]

NFMA Needleloom Felt Manufacturers Association [UK]

NFMA Norwegian Furniture Manufacturers Association

NFMC National Federation of Milk Cooperatives [France]

NFMC National Food Marketing Commission [US]

NFMD National Foundation for Muscular Dystrophy [US]

NFMES National Fund for Minority Engineering Students

NFMM Nonferromagnetic Material

NFMOA National Fish Meal and Oil Association [US]

N-FMOC N-Fluoroenylmethoxycarbonyl

NFMP National Federation of Master Painters and Decorators [UK]

NFMPC Nonferrous Metal Producers Committee [US]

NFMS Non-Fat Milk Solids

NFMSAEG Naval Fleet Missile System Analysis and Evaluation Group [US Navy]

NF-NF Nonfaceted-Nonfaceted (Eutectics) [also nf-nf] [Metallurgy]

NFO National Farmers Organization [US]

NFO National Freight Organization

NFO Nickel Ferrite [$NiFe_2O_4$]

NF_3-O_2 Nitrogen Trifluoride-Oxygen (Mixture)

NFOC National Fireworks Ordnance Corporation [US]

NFOV Narrow Field of View

NFOV Near Field of View

NFOVR Narrow Field-of-View Radiometer

NFP Nationales Forschungsprogramm [National Research Program, Germany]

NFP National Focal Point

NFP National Fusion Program [Canada]

NFP Nuclear Fuel Processing

NFPA National Fire Prevention Association [Canada]

NFPA National Fire Protection Association [US]

NFPA National Flaxseed Processors Association [US]

NFPA National Flexible Packaging Association [now Flexible Packaging Association, US]

NFPA National Fluid Power Association [US]

NFPA National Food Processors Association [US]

NFPA National Forest Products Association [US]

NFPA HR National Fire Protection Association Hazard Rating [US]

NFPA/JIC National Fire Protection Association/Joint Industry Council (Standard) [US]

NFPCA National Fire Prevention and Control Administration [US]

NFPDC National Federation of Painting and Decorating Contractors [UK]

NFPDE National Federation of Plumbers and Domestic Engineers [UK]

NFPEDA National Farm and Power Equipment Dealers Association [US]

NFPMC National Farm Products Marketing Council

NFPRF National Fire Prevention Research Foundation [US]

NFPS National Forest Policy Statement [Australia]

NFPW National Federation of Professional Workers [UK]

NFQ Night Frequency [also Nfq]

NFR National Forskningsrådet [National Science Research Council, Sweden]

NFR Negative Flux Rate

NFR No Further Requirement

NFR Nuclear Fuel Reprocessing

NFRC National Federation of Roofing Contractors [UK]

NFRC National Freight Rail Corporation

NFRC Northern Forest Research Center [of Canadian Forestry Service]

NFRCC Nordic Forest Research Cooperation Committee [Norway]

NFRO National Fire Research Office [Australia]

NFS NASA FAR (Federal Acquisition Regulation) Supplement

NFS National Forests Strategy

NFS National Forest System [US]

NFS Network File Service

NFS Network File System [Internet]

NFS N-Fluorosuccinimide

NFS Not for Sale

NFS Nozzle Flow Sensor

NFS Nuclear Fuel Services [US]

NFSA National Fertilizer Solutions Association [US]

NFSA National Fire Sprinkler Association [US]

NFSA Navy Field Safety Association [now Association of Navy Safety Professionals, US]

NFSAIS National Federation of Science Abstracting and Indexing Services [US]

NFSC Nuclear Fuel Services Corporation [US]

NFSEM Near-Field Scanning Electron Microscope; Near-Field Scanning Electron Microscopy

NFSOM Near-Field Scanning Optical Microscope; Near-Field Scanning Optical Microscopy

NFSP Nonflight Switch Panel

NFSP Nuclear Fuel Services Plant

NFSR National Federation for Scientific Research [Belgium]

NFSS National Federation of Sea Schools [UK]

NFSWMM National Federation of Scale and Weighing Machine Manufacturers [UK]

NFT National Federation of Textiles [US]

NFT Navigation Flight Test

NFT Neodymium-Iron-Titanium (Compound)

NFT Networks File Transfer

NFT Nutrient Film Technique [Agriculture]

nft³ normal cubic feet [Unit]

NFTA National Freight Transportation Association [US]

NFTA Nitrogen Fixing Tree Association [US]

NFTB Niagara Frontier Tariff Bureau [Canada]

NFTC National Foreign Trade Council [US]

NFTC National Furniture Traffic Conference [US]

NFTMS National Federation of Terrazzo Marble and Mosaic Specialists [UK]

NFTS New Facing Targets Sputtering (Apparatus)

NFTSA National Film, Telvision and Sound Archives [Canada]

NFU National Farmers Union [Canada and UK]

NFUS National Farmers Union of Scotland

NFVT National Federation of Vehicle Trade [UK]

NFWFL National Fish and Wildlife Forensic Laboratory [Oregon, US]

NFW Nonfuel Wasting

NFWA National Farm Workers Association [US]

NFWC National Fire Waste Council

NFWO National Fond voor Wetenschappelijk Onderzoek [National Foundation for Scientific Research, Brussels, Belgium]

NFWMP Nuclear Fuel Waste Management Program [Canada]

NFY Notify [also Nfy, or nfy]

NFZ 5-Nitro-2-Furfurolsemicarbazone

NFZ No-Fly Zone

NG Nanoglass

NG Narrow Ga(u)ge [Civil Engineering]

NG Nasogastric [Medicine]

NG National Guard [US]

NG Natural Gas

NG Natural Gasoline

NG Near Gamma (Microstructure) [Metallurgy]

NG Negative Sign (Flag) [Computers]

NG Neodymium Gallate

NG Neodymium Glass

NG New Generation

NG New Guinea

NG Next Generation

NG Nigeria [ISO Code]

NG Night Glasses

NG Nitrogen Gage

NG Nitroglycerin

NG Noble Gas

NG No Good [also ng]

NG Noise Generator

NG Non-Government(al)

NG Noxious Gas

NG Nuclear Geophysicist; Nuclear Geophysics

NG Nuclear Grade (Material)

9G- Ghana [Civil Aircraft Marking]

N&G Navigation and Guidance

ng nanogram [Unit]

.ng Nigeria [Country Code/Domain Name]

NGA National Glass Association [US]

NGA National Graphical Association [UK]

NGA National Guard Association [US]

NGA Northwest Geophysical Associates [US]

NGA Notice of Grant Award

NGAA Natural Gas Association of America [US]

n-GaAs N-Type Gallium Arsenide (Semiconductor)

n⁺-GaAs N-Plus-Type Gallium Arsenide (Semiconductor)

NGAC National Greenhouse Advisory Committee

NGAM Noble Gas Activity Monitor

n-GaN N-Type Gallium Nitride (Semiconductor)

n-GaN:Si Silicon-Doped N-Type Gallium Nitride (Semiconductor)

n-GaP N-Type Gallium Phosphide (Semiconductor)

NGAUS National Guard Association of the United States

NGB National Guard Bureau [of US Department of the Army]

NGC National Gallery of Canada [Ottawa]

NGC New General Catalog [Johan L. Dreyer's Catalog of Star Clusters, Nebulae and Galaxies; Listing Consists of the Letters "NGC" Followed by a Numeral, e.g., NGC 5139 Refers to the ω-Centauri Cluster] [Astronomy]

NGC Noble Gas Compound

NGC Nordic Geodetic Commission [Denmark]

NGC North Georgia College [Dahlonega, US]

NGc Nitroglycol

NGCC National Guard Computer Center [US]

NGCC Newport and Gwent Chamber of Commerce [UK]

NGCS New Generation Computer System

NGCT North Gloucestershire College of Technology [Cheltenham, UK]

NGD Nicotinamide Guanine Dinucleotide [Biochemistry]

NGDA National Glass Dealers Association [US]

NGDB National Geodetic Data Base [US]

NGDC National Geophysical Data Center [of National Oceanic and Atmospheric Administration, US]

ng/dL nanogram per deciliter [also ng/dl]

NGE Natural Gas Extraction

NGE Neighboring Group Effect

NGE Not Greater, or Equal

n-Ge N-Type Germanium (Semiconductor)

NGEECO National and German Electrical and Electronic Services Company [of Siemens AG in Kuwait]

NG-EGDN Nitroglycerine Ethylene Glycol Dinitrate

N Ger Northern Germany; North German

NGF Naval Gun Fire

NGF Nerve Growth Factor [Biochemistry]

NGFA National Grain and Feed Association [US]

NGF-α Nerve Growth Factor-α [Biochemistry]

NGF-β Nerve Growth Factor-β [Biochemistry]

NGF-2.5S Nerve Growth Factor-2.5S [Biochemistry]

NGF-7S Nerve Growth Factor-7S [Biochemistry]

NGFSP Natural Gas Fuelling Station Program [of Natural Resources Canada]

NGFTB Nordic Group for Forest Tree Breeding [Norway]

NGG Normal Grain Growth [Metallurgy]

ng/g nanogram per gram [also ng g^{-1}]

NGGLT Natural Gas and Gas Liquids Tax

NGH National Guild of Hypnosis [Canada]

NGH2 p-Nitrophenylaminoglyoxime

NGI Next-Generation Internet

NGI Niederländische Gesellschaft für Informatik [Netherlands Society for Informatics]

NGI Non-Government Institution

NGILI National Geography Instructional Leadership Institute [US]

NGI Mag NGI Magazine [Magazine on Informatics published by the Niederländische Gesellschaft für Informatik]

NGISTC National Geological Information System Technology Center [Canada]

NGK New Guinea Kina [Currency of Papua New Guinea]

NGL Natural Gas Liquid(s)

NGL Neodymium-Doped Glass Laser

NGL Neon Glow Lamp

NGM Nested Grid Model

NGM Neutron-Gamma Monte Carlo [Physics]

NGM Nitrogen Generation Module

NGM No Graphics Mouse

NGMA National Gas Measurement Association [US]

NGMA National Geoscience Mapping Accord

NGMA National Greenhouse Manufacturers Association [US]

NGMC National Grid Management Council

NGMDB National Geologic Map Data Base [of US Geological Survey]

ng/mL nanogram(s) per milliliter [also ng mL^{-1}, or ng/ml]

NGN Nigerian Naira [Currency]

NGO National Gas Outlet Thread [Symbol]

NGO Neodymium-Gallium Oxide

NGO Nongovernmental Organization

NGOP Natural Gas Oxypyrolysis (Process) [Chemical Engineering]

NGOR Net Gas-Oil Ratio

NGOTP Natural Gas and Oil Technology Partnership [US]

NGP Network Graphics Protocol

NGPA National Gas Policy Act

NGPA Natural Gas Policy Act [US]

NGPA Natural Gas Processors Association [US]

ng/Pa·s·m² nanogram(s) per pascal second square meter [also ng Pa^{-1} s^{-1} m^{-2}]

NGPSA Natural Gas Processors Suppliers Association [now Gas Processors Suppliers Association, US]

NGR Neutral Grounding Resistor

NGR Nongrain-Raising Stain [Wood Treatment]

NGR Nuclear Gamma-Ray Resonance

n-Gr Nanocrystalline Graphite

NGRI National Geophysical Research Institute [India]

NGRR Nuclear Gamma-Ray Resonance

NGRS Narrow Gauge Railway Society

NGRS National Greenhouse Response Strategy [Australia]

NGRT Next-Generation Gamma-Ray Telescope [of NASA Goddard Space Flight Center, Greenbelt, Maryland, US]

NGS National Gas Straight Thread [Symbol]

NGS National Geographic Society [US]

NGS National Geodetic Survey [of National Oceanic and Atmospheric Administration, US]

NGS New Generation System

NGS Nominal Guidance Scheme

NGS Normal Goat Serum [Immunochemistry]

NGS Nuclear Generating Station

NGS Numerical Geometry System

NGSA Natural Gas Supply Association [US]

NGSC National Geological Survey of China

NGSC National Greenhouse Steering Committee [Australia]

NGSC Natural Gas Supply Committee [now Natural Gas Supply Association, US]

NGSDC National Geophysical and Solar Terrestrial Data Center [US]

NGSEF National Geographic Society Education Foundation [US]

NGSF Noble Gas Storage Facility

NGSP Next Generation Signal Processing; Next Generation Signal Processor

NGST Next Generation Space Telescope [of NASA Goddard Space Flight Center, Greenbelt, Maryland, US]

NGT NASA Ground Terminal

NGT National Gas Taper Thread [Symbol]

NGT Next Generation Trainer [of US Air Force]

NGT Nonsymmetrical Growth Theory [Metallurgy]

Ngt Night [also ngt]

ngtb negotiable

NGTE National Gas Turbine Establishment [UK]

ng TE/m³ nanogram toxicity-equivalent per cubic meter [also ng TE m^{-3}]

NGU Dinitroglycoluril [Biochemistry]

NGU Nongonococcal Urethritis [Medicine]

NGU Norges Geologiske Undersøkelse [Geological Survey of Norway]

NGu Nitroguanidine [Biochemistry]

NGu N-Methyl Guanidine [Biochemistry]

N Guin New Guinea

NGV National Gallery of Victoria [Melbourne, Australia]

NGV Natural Gas for Vehicles (Program)

NGV Natural Gas Vehicle

NGV Next Generation Vehicle

NGVD National Geodetic Vertical Datum [US]

NGVP Natural Gas Vehicle Program [of Natural Resources Canada]

NGWA National Ground Water Association [US]

NH American National Hose Coupling and Fire-Hose Coupling Threads [Symbol]

NH Nabarro-Herring (Creep) [also N-H] [Metallurgy]

NH Needle Roller Bearing, Drawn Cup, With Cage, Without Inner Ring, Metric-Dimensioned [Symbol]

NH Net Head [Fluid Mechanics]

NH Net Heat

NH New Hampshire [US]

NH New Haven [Connecticut, US]

NH New Hebrides [now Vanuatu]

NH Nominal Height

NH Non-Hazardous

NH Nonhygroscopic

NH Norsk Hydro [Norway]

NH Northern Hemisphere

NH North-Holland (Division of Elsevier Science Publishers BV) [Amsterdam, Netherlands]

NH^{2-} Imide (Ion) [Symbol]

NH$_2$ Amino (Radical) [Symbol]

NH$_2^-$ Amide (Ion) [Symbol]

NH$_3$ Ammonia

NH$_{3(aq)}$ Aqueous Ammonia

^{13}NH$_3$ Ammonia based on Nitrogen-13

NH Ammonium (Ion) [Symbol]

N$_2$H$_3^-$ Hydrazide (Ion) [Symbol]

N$_2$H$_4$ Hydrazine

N$_2$/H$_2$ Nitrogen/Hydrogen (Mixture)

N-H Nabarro-Herring (Creep) [also NH] [Metallurgy]

N7/h6 (Accurate) Locational Transition Fit; Basic Shaft System [ISO Symbol]

nH nanohenry [also nh]

NHA National Housing Act [Canada]

NHA National Housing Agency [US]

NHA National Hydropower Association [US]

NHA Newton-Horner Algorithm

NHA Next Higher Assembly

NHA Northwest Hardwood Association [US]

NHAM National Hose Assemblies Manufacturers [US]

NHAMA National Hose Assemblies Manufacturers Association [US]

NHANES National Health and Nutrition Examination Survey [US]

NHAP National High Altitude Photography [US]

NHATS National Human Adipose Tissue Survey [US]

NHAW North American Heating and Air Conditioning Wholesalers Association [US]

NHB 4-Hydroxy-3-Nitrobenzoic Acid

NHB NASA Handbook

NHB National Harbours Board

NHB Non-Hydrogen Bonded

NH$_4$Br Ammonium Bromide

NHC Nabarro-Herring-Coble (Creep) [Metallurgy]

NHC National Health Council [US]

NHC National Housing Center [of National Association of Home Builders, US]

NHC National Hurricane Center [Miami, Florida, US]

NHC Northern Horizon Circle [Astronomy]

NHC Northwest Horticultural Council [US]

NHC Numatec Hanford Inc. [Washington State, US]

NH$_4$Cl Ammonium Chloride

NH Crp Nabarro-Herring Creep [Metallurgy]

NHD Nevada Highway Department [US]

NHDA National Huntington's Disease Association [US]

NHDV Normalized Hydrodynamic Voltammograms

NHE Normal Hydrogen Electrode

Nhe Neisseria mucosa [Microbiology]

NHEA New Hampshire Education Association [US]

NHEC Niger Higher Education Center

NHELP New Hitachi Effective Library for Programming

NHF National Headache Foundation [US]

NHF National Health Federation [US]

NHF Naval Historical Foundation [US]

NH$_4$F Ammonium Fluoride

n-HgCdTe N-Type Mercury Cadmium Telluride (Semiconductor)

NHGRI National Human Genome Research Institute [of National Institutes of Health, Bethesda, Maryland, US]

NHHS New Hampshire Historical Society [US]

NHI National Health Institute

NHI National Health Insurance

NHI National Heart Institute [US]

NH$_4$I Ammonium Iodide

NHICH National Health and Information Clearinghouse [US]

NHK Nippon Hoso Kyokai [Tokyo, Japan]

NHK Lab Note NHK Laboratories Note [Published by Nippon Hoso Kyokai, Tokyo, Japan]

NHK Tech J NHK Technical Journal [Published by Nippon Hoso Kyokai, Tokyo, Japan]

NHL Neon Helium Laser

NHL Nuclear Historic Landmark [US]

NHLA National Hardwood Lumber Association [US]

NHLBI National Heart, Lung and Blood Institute [of National Institutes of Health, Bethesda, Maryland, US]

NHLI National Heart and Lung Institute [US]

NHM Natural History Museum [London, UK]

NHM (*Naturhistorisches Museum*) – German for "Natural History Museum"

NHM Needle Roller Bearing, Drawn Cup, With Cage, Closed End, Without Inner Ring, Metric-Dimensioned [Symbol]

NHM No Hot Metal [Metallurgy]

NHMA National Housewares Manufacturers Association [US]

NHMA New Hampshire Medical Association [US]

NHMFL National High Magnetic Field Laboratory [of Florida State University, Tallahassee, US]

NHMFL National High Magnetic Field Laboratory [of Los Alamos National Laboratory, New Mexico, US]

NHMO NATO Hawk Management Office

NHMRC National Health and Medical Research Council [Australia] [also NH&MRC]

NHNP N-(2-Hydroxy-3-(1-Naphthaloxy)propyl)

NHNP-E N-(2-Hydroxy-3-(1-Naphthaloxy)propyl) ethylenediamine

NH₄OH Ammonium Hydroxide

NHP National Historic Park [US]

NHP Natural History Press [Garden City, US]

NHP Network Host Protocol

NHP N-Hydroxyphthalimide (Ester)

NHP Nominal Horsepower [also nhp]

NHP Numeric Handprinting

NHPC National Historical Publication Commission

NHPLO NATO Hawk Production and Logistics Organization

N-HPM Nitrogen High-Pressure Melting

NHPMA Northern Hardwood and Pinewood Manufacturers Association [US]

NHPMo Ammonium 12-Molybdatophosphate

NHPRS National Hydroelectric Power Resources Study [of US Atomic Energy Commission]

NHPS New Hampshire Pharmaceutical Association [US]

NHPS New Hampshire Public Service [US]

NHPW Ammonium 12-Tungstatophosphate

NHQ Naphthohydroquinone

NHRA National Hot Rod Association [US]

NHRAIC Natural Hazards Research and Applications Information Center [US]

NHRC National Hydrology Research Center [Canada]

NHRDP National Health Research and Development Program [Canada]

NHRE National Hail Research Experiment [US]

NHRI National Hydrology Research Institute [Canada]

NHRL National Hurricane Research Laboratory [US]

NHRP National Hurricane Research Project [US]

NHRP Next Hop Resolution Protocol

NHS National Health Service [UK]

NHS National Health Survey [US]

NHS National Historic Site [US]

NHS National Housing Strategy

NHS N-Hydroxysuccinimide (Ester)

NHS N-Hydroxysuccinimidyl

NHS Nickel Hafnium Silicide

NHS-ABG 4-Azidobenzoylglycine N-Succinimidyl Ester

NHS-Biotin (+)-Biotin-N-Hydroxysuccinimidester

NHSC National Health Service Corps [US]

NHSC National Home Study Council [US]

NHSC New Hampshire Safety Council [US]

NH₃/SiH₄ Ammonia-to-Silane (Ratio)

NHSiW Ammonium 12-Tungstatosilicate

NHSPEI Natural History Society of Prince Edward Island [Canada]

NHSTA National Highway Safety Transportation Administration [US]

NHSW Neue Hamburger Stahlwerke [German Steel Works located at Hamburg]

NHT Nanohardness Test(er)

NHT Natural Heritage Trust [Australia]

NHT Numerical Heat Transfer

NHTPC National Housing and Town Planning Council [UK]

NH Tpk New Hampshire Turnpike [US]

NHTSA National Highway Traffic Safety Administration [US]

NHV Net Heating Value

NHW Non-Hazardous Waste

NHWA National Heating and Air Conditioning Wholesalers [now North American Heating and Air Conditioning Wholesalers Association, US]

NHWMA National Hydraulic Woodsplitter Manufacturers Association [now Logsplitters Manufacturers Association, US]

NHWU Non-Heatset Web Unit [of Printing Industries Association, US]

NI Nanoindentation; Nanoindenter [Mechanical Testing]

NI Natural Immunity [Immunology]

NI Nautical Institute [UK]

NI Navigation(al) Instrument

NI Negative Image

NI Negative Input

NI Negative-Intrinsic (Junction) [also N-I, ni, or n-i]

NI Negative Ion

NI Nerve Impulse

NI Network Integration

NI Network Interconnect(ion)

NI Network Interface

NI Neutron Interference

NI Nicaragua [ISO Code]

NI Nitrogen Intake

NI Noise Index

NI Noninductive

NI Noninhibit [Computers]

NI Non-Ionic (Detergent)

NI Norfolk Island [Australia]

NI Normal Interstitial [Crystallography]

NI Northern Ireland

NI Northern Iowa [US]

NI Not Included [also ni]

NI Nozzle Inlet

NI Nuclear Instrument(ation)

NI Numerical Index

NI Numerical Integration

NI Numeric Information

NI_C Chemical-Vapor-Deposited Nitride

N/I Noise to Interference (Ratio)

N/I Non-Interlaced

Ni Nickel [Symbol]

Ni Nitrate [Abbreviation]

Ni²⁺ Divalent Nickel Ion [also Ni⁺⁺] [Symbol]

Ni⁴⁺ Tetravalent Nickel Ion [Symbol]

Ni-56 Nickel-58 [also ⁵⁶Ni, or Ni⁵⁶]

Ni-57 Nickel-58 [also ⁵⁷Ni, or Ni⁵⁷]

Ni-58 Nickel-58 [also ⁵⁸Ni, or Ni⁵⁸]

Ni-59 Nickel-58 [also ⁵⁹Ni, or Ni⁵⁹]

Ni-60 Nickel-58 [also ⁶⁰Ni, or Ni⁶⁰]

Ni-61 Nickel-58 [also ⁶¹Ni, or Ni⁶¹]

Ni-62 Nickel-58 [also ⁶²Ni, or Ni⁶²]

Ni-63 Nickel-58 [also ^{63}Ni, or Ni^{63}]

Ni-64 Nickel-58 [also ^{64}Ni, or Ni^{64}]

Ni-65 Nickel-58 [also ^{65}Ni, or Ni^{65}]

Ni-66 Nickel-58 [also ^{66}Ni, or Ni^{66}]

.ni Nicaragua [Country Code/Domain Name]

NIA National Ice Association [US]

NIA National Institute on Aging [of National Institutes of Health, Bethesda, Maryland, US]

NIA Needle Roller Bearing, With Cage, Machined Ring, Lubrication Groove and Hole in Outer Diameter, Inch-Dimensioned [Symbol]

NIA Next Instruction Address

NIA Nova Interface Adapter

NIAA Nursery Industry Association of Australia

NIAAA National Institute on Alcohol Abuse and Alcoholism [of National Institutes of Health, Bethesda, Maryland, US]

NIAB National Institute of Agricultural Botany [UK]

NIAC National Industry Advisory Committee [of Federal Communications Commission, US]

NIAC National Information and Analysis Center [US]

NIAC Northern Ireland Automation Center

NIAC Nuclear Insurance Association of Canada

Ni(Ac)$_2$ Nickel(II) Acetate

Ni(ACAC)$_2$ Nickel(II) Acetylacetonate [also Ni(AcAc)$_2$, or Ni(acac)$_2$]

[Ni(ACAC)$_2$]$_3$ Trimeric Nickel(II) Acetylacetonate [also [Ni(AcAc)$_2$]$_3$, or [Ni(acac)$_2$]$_3$]

NIADA National Independent Automobile Dealers Association [US]

NIAG NATO Industrial Advisory Group

NIAE National Institute of Agricultural Engineering [UK]

NIAE National Institute for Architectural Education [US]

NIAG NATO Industrial Advisory Group

Ni-Ag Nickel-Silver (Alloy System)

Niag Front Niagara Frontier

NIAID National Institute of Allergy and Infectious Diseases [of National Institutes of Health, Bethesda, Maryland, US]

NIAL National Institute of Arts and Letters [US]

NIAL Nested Interactive Array Language

NiAl Nickel Aluminide (Coating)

Ni$_3$Al Trinickel Aluminide

Ni-Al Nickel-Aluminum (Alloy System)

Ni-Al/Al$_2$O$_3$ Alumina-Reinforced Nickel-Aluminum (Composite)

Ni$_3$Al(B) Boron-Doped Trinickel Aluminide

NiAl(C) Carbon-Doped Nickel Aluminide

Ni$_3$Al(C) Carbon-Doped Trinickel Aluminide

NiAl(Cr) Chromium-Doped Nickel Aluminide

NiAl/Cr Chromium Reinforced Nickel Aluminide (Composite)

NiAl/Cr$_f$ Chromium-Fiber Reinforced Nickel Aluminide (Composite) [also NiAl/Cr(f)]

NiAl/Cr$_p$ Chromium Particulate Reinforced Nickel Aluminide (Composite) [also NiAl/Cr(p)]

NiAlGaAs Nickel Aluminum Gallium Arsenide (Semiconductor)

NiAl(Hf) Hafnium-Doped Nickel Aluminide

Ni-Al-Mn Nickel-Aluminum-Manganese (Alloy System)

NiAl/Mo Molybdenum Reinforced Nickel Aluminide (Composite)

NiAl/Mo$_f$ Molybdenum-Fiber Reinforced Nickel Aluminide (Composite) [also NiAl/Mo(f)]

NiAl/Mo$_p$ Molybdenum Particulate Reinforced Nickel Aluminide (Composite) [also NiAl/Mo(p)]

NiAl(Mo) Molybdenum-Doped Nickel Aluminide

Ni-Al-Mo Nickel-Aluminum-Molybdenum (Alloy)

NiAl(N) Nitrogen-Doped Nickel Aluminide

Ni-Al$_2$O$_3$ Alumina-Dispersed Nickel

NiAlSi Nickel-Aluminum-Silicon (Alloy)

NiAl(Ti) Titanium-Doped Nickel Aluminide

NiAlZr Nickel-Aluminum-Silicon (Alloy)

NIAM Netherlands Institute for Audiovisual Media

NIAE National Institute of Agricultural Engineering [US]

NIAMS National Institute of Arthritis and Musculoskeletal and Skin Diseases [of National Institutes of Health, Bethesda, Maryland, US]

NIAR National Institute for Aviation Research [of Wichita State University, Kansas, US]

NIAR Norwegian Institute for Air Research

NIAS Nagasaki Institute of Applied Science [Japan]

NiAs Nickel Arsenide (Semiconductor)

NIATM New International Association for Testing Materials

Ni-Au Nickel-Gold (Alloy System)

NIB 4-Hydroxy-3-Iodo-5-Nitrobenzoic Acid

NIB Negative Impedance Booster

NIB Needle Roller Bearing, Drawn Cup, Full Complement, Without Inner Ring, Inch-Dimensioned [Symbol]

NIB Node Initialization Block

NIB Non-Interference Basis

NIB Nordic Investment Bank

NIB Normeninformationsbank [German Databank on Standards of DIN Deutsches Institut für Normung]

NiB Nickel Boride

Ni-B Nickel-Boron (Alloy)

NIBA National Industrial Belting Association [US]

NIBA Northern Ireland Bankers Association [UK]

Ni/BCSCO Nickel on Bismuth Strontium Calcium Yttrium Copper Oxide (Superconductor)

NiBe Nickel Beryllide

Ni-Be Nickel-Beryllium (Alloy System)

NIBESA National Independent Bank Equipment and Systems Association [US]

NIBH National Institute of Bioscience and Human-Technology [of Agency of Industrial Science and Technology, Ibaraki, Japan]

NIBHT National Institute of Bioscience and Human Technology [of Agency of Industrial Science and Technology, Ibaraki, Japan]

NIBID National Investment Bank for Industrial Development [Greece]

NIBM Needle Roller Bearing, Drawn Cup, Full Complement, Closed End, Without Inner Ring, Inch-Dimensioned [Symbol]

NIBMA Northern Ireland Builders Merchants Association [UK]

NIBR Norsk Institut for By– og Regionsforskning [Norwegian Institute for Urban and Regional Research]

Ni-BR Butadiene Rubber Based on Nickel Catalyst

NIBS National Institute of Building Sciences [US]

NIBS Nippon Institute for Biological Science [Japan]

NIBSC National Institute for Biological Standards and Control [UK]

Ni-B-Si Nickel-Boron-Silicon (Alloy)

NIBTN Nitroisobutylglycerol Trinitrate

NIC National Indicational Center

NIC National Industrial Council [US]

NIC National Inspection Council

NIC National Institute of Corrections

NIC National Interfraternity Confederation [US]

NIC Near-Infrared Camera

NIC Nearly Instantaneous Companding (Processor)

NIC Negative Impedance Converter

NIC Network Information Center [Internet]

NIC Network Interface Card

NIC Network Interface Control

NIC Newly Industrialized Countries

NIC Newsprint Information Committee [US]

NIC Nicaraguan Cordoba [Currency]

NIC Nineteen-Hundred Indexing and Cataloging

NIC Noise Isolation Class

NIC Nomarski Interference Contrast [Microscopy]

NIC Non-Ionic Compound

NIC Nonisothermal Calorimeter

NIC Not in Contact

NIC Nuclear Industry Consortium [Belgium]

NIC Numeric Intensive Computing

Ni-C Nickel-Carbon (System)

Nic Nicaragua(n)

NICA National Insulation Contractors Association [US]

NI-CAD Nickel-Cadmium (Battery) [also NICAD, Ni-Cad, or nicad]

NiCAlON Nickel Carbon Aluminum Oxynitride

NICAM Near-Instantaneous Companded Audio Multiplex

NICAP National Investigations Committee on Aerial Phenomena [US]

Nicar Nicaragua(n)

NICB National Industrial Conference Board [US]

NICC National Industries Computer Committee [UK]

NICCI Northern Ireland Chamber of Commerce and Industry [UK]

NICD National Institute of Cleaning and Dyeing [US]

NiCd Nickel Cadmium

Ni-Cd Nickel-Cadmium (Alloy System)

NICE National Institute for Computers in Engineering [Rockville, Maryland, US]

NICE National Institute of Ceramic Engineers [of American Ceramic Society, US]

NICE Normal Input-Output Control Executive

NICE³ National Industrial Competitiveness through Environment, Energy and Economics [Joint Program of the US Department of Energy, US Department of Commerce and US Environmental Protection Agency]

NICEIC National Inspection Council for Electrical Installation Contracting [UK]

NICEM National Information Center for Educational Materials [of University of Southern California, Los Angeles, US]

NICEM National Information Center for Educational Media [US]

NICET National Institute for Certification in Engineering Technologies [US]

NICHD National Institute of Child Health and Human Development [of National Institutes of Health, Bethesda, Maryland, US]

NICI Negative Ion Chemical Ionization

NICIA Northern Ireland Coal Importers Association

Nickel Top Nickel Topics [Journal published in the US]

Nickel TPP 5,10,15,20-Tetraphenyl-21H,23H-Porphine Nickel(II) [also nickel TPP]

NiCl Nickel Chloride

NICMOS Near-Infrared Camera and Multi-Object Spectrometer [also NIC-MOS]

$[Ni(CN)_4]^{2-}$ Tetracyanonickelate(II) Ion

NICNAS National Industrial Chemicals Notification and Assessment Scheme [Australia]

$Ni(Co)_2$ Bis(cyclooctadiene)nickel

Ni-Co Nickel-Cobalt (Alloy System)

NICOA National Independent Coal Operators Association [US]

NiCoCrAlY Nickel-Cobalt-Chromium-AluminumYttrium (Coating)

NICOL Nineteen-Hundred Commercial Language

NICOLAS Network Information Center On-Line Aid System [of NASA]

$Ni(Cp)_2$ Bis(cyclopentadienyl)nickel (or Nickelocene)

Ni-Cr Nickel-Chromium (Alloy System)

NiCrAl Nickel-Chromium-Aluminum (Coating)

Ni-Cr-Al Nickel-Chromium-Aluminum (Alloy System)

$NiCr-Al_2O_3$ Alumina Fiber Reinforced Nickel-Chromium (Composite)

NiCrAlY Nickel-Chromium-Aluminum Yttrium (Coating)

NiCrCoY Nickel-Chromium-Cobalt-Yttrium (Coating)

NiCrFeSiB Nickel-Chromium-Iron-Silicon-Boron (Coating)

NICS NATO Integrated Communications System

NICS Network Integrity Control System

NICSEM National Information Center for Special Educational Materials [of University of Southern California, Los Angeles, US]

NICSMA NATO Integrated Communications System Management Agency

NICSO NATO Integrated Communications Systems Organization [now NATO Communications and Information Systems Organization]

Ni-Cu Nickel-Copper (Alloy System)

NICUFO National Investigations Committee on Unidentified Flying Objects [US]

NID Naval Intelligence Department

NID Network Inward Dialing

NID New Interactive Display

NID Next Identification

NID Nuclear Instruments and Detectors [Institute of Electrical and Electronics Engineers Nuclear and Plasma Sciences Society Committee, US]

NIDA National Industrial Distributors Association [now International Distribution Association, US]

NIDA National Institute on Drug Abuse [of National Institutes of Health, Bethesda, Maryland, US]

NIDA N-Nitrosoiminodiacetic Acid

NIDA Northeastern Industrial Development Association [US]

NIDA Numerically Integrated Differential Analyzer

NIDB Nigerian Industrial Development Bank

NiDBC Nickel Dibutyl Dithiocarbamate

NIDC Nepal Industrial Development Corporation

NIDC Northern Ireland Development Council

NIDCD National Institute on Deafness and Other Communication Disorders [of National Institutes of Health, Bethesda, Maryland, US]

NIDCR National Institute of Dental and Craniofacial Research [of National Institutes of Health, Bethesda, Maryland, US]

NIDDK National Institute of Diabetes and Digestive and Kidney Diseases [of National Institutes of Health, Bethesda, Maryland, US]

NIDDM Non-Insulin-Dependent Diabetes Mellitus [also known as Type II Diabetes Mellitus]

NIDER Netherlands Institute for Documentation and Registration

NiDI Nickel Development Institute [Toronto, Canada]

NiDMC Nickel Dimethyl Dithiocarbamate

NIDOC National Information and Documentation Center [Cairo, Egypt]

NIDP Negative Imaginary Decoupling Potential

NIDR National Institute of Dental Research [of US Public Health Services]

NIDS National Investor Data Service [US]

NIDS Network Interface Data System

NIE National Index of Ecosystems [Defunct] [Now through the Environmental Resources Information Network and the Australian Nature Conservancy Agency]

NIE National Institute for the Environment

NIE National Institute of Education [US and Singapore]

NIE Newly Industrialized Economies

NIE Not Included Elsewhere [also nie]

NiEDTA Nickel Ethylenediaminetetraacetate [also NiEdta]

NIEE National Institute for Engineering Ethics

NIEF National Ironfounding Employers Federation [UK]

NIEHS National Institute of Environmental Health Sciences [of National Institutes of Health, Bethesda, Maryland, US]

NIEM Network for Industrial Environmental Management

NIEO New International Economic Order

NIER National Industrial Equipment Reserve

NIER National Institute for Educational Research [Japan]

NIES National Industry Extension Service

NIES National Institute for Environmental Studies [Japan]

NIES National Institute of Environmental Science [Japan]

NIESR National Institute for Economic and Social Research

Nieuw Arch Wisk Nieuw Archief voor Wiskunde [New Archive for Mathematics; of Wiskundig Genootschap te Amsterdam, Netherlands]

NIF National Inventors Foundation [US]

NIF Network Information File

NIF Noise Improvement Factor (for Pulse Modulation)

NIF Note Issuance Facility

Ni-Fe Nickel-Iron (Alloy System)

Nife Nickel-Iron (Core) [Geology]

NIFES National Industrial Fuel Efficiency Service [UK]

NiFeO Nickel Iron Oxygen (Alloy)

NiFe/Wp Tungsten Particulate Reinforced Nickel-Iron (Composite) [also NiFe/W(p), or NiFe/W$_p$]

NIFFT National Institute of Foundry and Forge Technology [Ranchi, India]

NIFHA Northern Ireland Federation of Housing Association [UK]

NIFO Next-In, First-Out [also nifo]

NIFP National Institute for Fresh Produce [UK]

NIFS National Institute for Farm Safety [US]

NIFTE Neon Indicating Functional Test Equipment

NIFTP Network-Independent File Transfer Protocol

NIG Nordic Industrial Group

Nig Nigeria(n)

Ni-Ga Nickel-Gallium (Alloy System)

NIGC National Iranian Gas Company

NIGDA National Industrial Glove Distributors Association [US]

Ni-Ge Nickel-Germanium (Alloy System)

n-IGFET N-Channel Insulated-Gate Field-Effect Transistor

NIGMS National Institute of General Medical Sciences [of National Institutes of Health, Bethesda, Maryland, US]

NIGP National Institute of Governmental Purchasing [US]

NIGP National Institute of Government Purchasing [Canada]

NIGTA Northern Ireland Grain Trade Association [UK]

Nigu Nitroguanidine [Biochemistry]

NIH National Information Highway [US]

NIH National Institute of Hardware [UK]

NIH National Institutes of Health [Bethesda, Maryland, US]

NIH Needle Roller Bearing, Drawn Cup, With Cage, Without Inner Ring, Inch-Dimensioned [Symbol]

NIH Not Invented Here

NIH Consens Devel Conf Consens Statem National Institutes of Health Consensus Development Conference Consensus Statement [of National Library of Medicine, Bethesda, Maryland, US]

NIH Consens Devel Conf Summ National Institutes of Health Consensus Development Conference Summary [of National Library of Medicine, Bethesda, Maryland, US]

NIH Consens Statem National Institutes of Health Consensus Statement [of Office of Medical Applications of Research, Bethesda, Maryland, US]

NIH/DOE National Institutes of Health/Department of Energy (Program) [US]

Ni(HFAC)₂ Nickel Hexafluoroacetylacetonate [also Ni(hfac)₂]

NIHL Noise Induced Hearing Loss [Medicine]

NIHM Needle Roller Bearing, Drawn Cup, With Cage, Closed End, Without Inner Ring, Inch-Dimensioned [Symbol]

NIH/NIAID National Institutes of Health/National Institute of Allergy and Infectious Diseases (Standard) [US]

NIH:OVCAR-8 National Institutes of Health Ovarian Carcinoma Cells

NIHSA Newfoundland Industrial Health and Safety Association [Canada]

NIH TAC National Institutes of Health Technology Assessment Conference [US]

NII National Information Infrastructure [US]

NII Negative Impedance Inverter

NII Neutron Image Intensifier

NII Normalized Integrated Intensities

NII Nuclear Installations Inspectorate [UK]

NIICU National Institute of Independent Colleges and Universities [US]

NIIDST National Institute for Information and Documentation in Science and Technology [Romania]

Ni-In Nickel-Indium (Alloy System)

NIIT National, International and Intercontinental Telecommunications Network

NIJSI (Vereniging der) Nederlandse Ijzer-en Staalproducerende Industrie [Dutch Federation of the Iron and Steel Industry]

NiKα Nickel K-Alpha (Radiation) [also NiK$_\alpha$]

NIKHEF Nederlands Institut voor Kern– en Hochenergiefysica [Netherlands Institute for Nuclear and High-Energy Physics]

Nikkei Electron Nikkei Electronics [Journal published in Japan]

NIL Nano-Imprint Lithography

NIL Nitrogen Inerting Line

NIL Nothing [Amateur Radio]

NiL Nickel Ligand

L-NIL L-N^6-(1-Iminoethyl)lysine [Biochemistry]

Ni-La Nickel-Lanthanum (Alloy System)

NiLα Nickel L-Alpha (Radiation) [also NiL$_\alpha$]

NILG Noninteracting Lattice Gas [Physics]

NILI Northern Interior Lumber Industries [of British Columbia, Canada]

NILO Nitrogen Implantation Local Oxidation

NILPT National Institute for Low-Power Television [US]

NILU Norsk Institut for Luftforskning [Norwegian Institute for Air Research]

NILUG National Independent Lynx User Group [UK]

NIM National Institute of Metallurgy [South Africa]

NIM (Kikuchi) Natural Iteration Method [Crystallography]

NIM Needle Roller and Cage Assembly, Inch-Dimensioned [Symbol]

NIM Network Installation Management

NIM Network Interface Machine

NIM Network Interface Monitor

NIM Neutron Interactive Materials (Program) [US]

NIM Newspapers in Microform

NIM Normal-Incidence Monochromator

NIM Nuclear Instrument(ation) Module

NIMA National Imagery and Mapping Agency [US Military]

NIMA Northern Ireland Ministry of Agriculture

NIMAT Newfoundland Institute for Management Advancement and Training [Canada]

NIMBY Not in My Backyard

NIMC National Imagery and Mapping Center [US Military]

NIMC National Institute of Materials and Chemical Research [of Agency of Industrial Science and Technology, Ministry of International Trade and Industry, Ibaraki, Japan]

NIMCR National Institute of Materials and Chemical Research [Tsukuba, Japan]

NIMD National Institute of Management Development [Egypt]

NIMESULIDE N-(4-Nitro-2-Phenoxyphenyl)methane sulfonamide [also Nimesulide, or nimesulide]

Ni-Mg Nickel-Magnesium (Alloy System)

Ni/MgO Nickel-Reinforced Magnesia (Composite)

NIMH National Institute of Mental Health [of National Institutes of Health, Bethesda, Maryland, US]

NiMH Nickel-Metal Hydride (Battery) [also Ni-MH]

NIMIS National Instructional Materials Information System [of University of Southern California, Los Angeles, US]

NIMMS Nineteen-Hundred Integrated Modular Management System

Ni-Mn Nickel-Manganese (Alloy System)

Ni-Mo Nickel-Molybdenum (Alloy System)

NiMoB Nickel Molybdenum Boride (Glass)

NiMoP Nickel Molybdenum Phosphide (Glass)

NIMP New and Improved Materials and Processes

NIMPA Northern Ireland Master Painters Association

NIMPA Northern Ireland Master Plumbers Association

NIMPHE Nuclear Isotope Monopropellant Hydrazine Engine

NIMR National Institute for Materials Research [South Africa]

NIMR National Institute for Medical Research [UK]

NIMSCO NODC (National Oceanographic Data Center) Index for Instrument Measured Subsurface Current Observation [US]

NIMT National Institute of Magnesium Technology [Canada]

Ni(MTC)₂ Nickel Monothiocarbamate [also Ni(mtc)₂]

[Ni(MTC)₂]₆ Hexameric Nickel Monothiocarbamate [also [Ni(mtc)₂]₆]

NIN National Information Network [US]

NINA National Institute Northern Accelerator

n-InGaAs N-Type Indium Gallium Arsenide (Semiconductor)

n⁺-InGaAs N-Plus-Type Indium Gallium Arsenide (Semiconductor)

n-InAs n-Type Indium Arsenide (Semiconductor)

Ni-Nb Nickel-Niobium (Alloy System)

NINCDS National Institute of Neurological and Communicative Disorders and Stroke [US]

NINDS National Institute of Neurological Disorders and Stroke [of National Institutes of Health, Bethesda, Maryland, US]

NINES Norfolk Information Exchange Scheme [UK]

n-InGaAs N-Type Indium Gallium Arsenide (Semiconductor)

NINIA Nephelometric Inhibition Immunoassay

Ni/Ni$_3$Nb Nickel-Reinforced Nickel Niobide

Ni-NiO Nickel-Nickel Monoxide (Cell)

NINP Norfolk Island National Park [of Australian Nature Conservancy Agency]

n-InP n-Type Indium Phosphide (Semiconductor)

NINR National Institute of Nursing Research [of National Institutes of Health, Bethesda, Maryland, US]

n-InSb n-Type Indium Antimonide (Semiconductor)

NINST Non-Instrument Runway

NINT Nearest Integer

NIO National Institute of Oceanography [UK]

NIO Native Input/Output

NiO Nickel(II) Oxide (or Nickel Monoxide)

NiO-Al$_2$O$_3$ Nickel Monoxide-Aluminum Oxide (System)

Niobium Tech Rep Niobium Technical Report

NIOC National Iranian Oil Company

NIOD Network Inward and Outward Dialing

NiOEP 2,3,7,8,12,13,17,18-Octaethylporphine Nickel(II)

NiO-Fe$_2$O$_3$ Nickel Monoxide-Ferric Oxide (System)

NIOGEMS National Indian Oil and Gas Evaluation System [of US Geological Survey]

NiO$_x$H$_y$ Nickel Hydroxide [General Formula]

Ni(OH)$_2$ Nickel(II) Hydroxide

NIOP National Institute of Oilseed Products [US]

NIOS National Institute for Occupational Safety

NIOSH National Institute for Occupational Safety and Health [US]

NIOSH/MSHA National Institute for Occupational Safety and Health/Mine Safety and Health Administration [US]

NIOSH REL National Institute for Occupational Safety and Health's Recommended Exposure Limit

NIOSHTIC NIOSH Technical Information Center [US]

NiO-TiO$_2$ Nickel Oxide-Titanium Dioxide (System)

NiO-Y$_2$O$_3$ Nickel Oxide-Yttrium Oxide (System)

NiO-ZrO$_2$ Nickel Oxide-Zirconium Oxide (System)

NIP 4-Hydroxy-3-Iodo-5-Nitrophenylacetic Acid

NIP Naval Intelligence Professionals [US]

NIP Negative (Layer)–Intrinsic (Layer)–Positive (Layer) (Transistor) [also N-I-P, nip, or n-i-p]

NIP Network Input Processor

NIP Network Interface Processor

NIP Nitroindene Polymer

NIP Non-Impact Printer

NIP Normal Incidence Pyrheliometer

NIP Not in Practice

NIP N-Type-Intrinsic-P-Type (Transistor) [also N-I-P, nip, or n-i-p]

NIP Nucleus Initialization Procedure [Computers]

Ni(P) Phosphorus-Doped Nickel

Ni-P Nickel-Phosphorus (Alloy System)

Nip Nipple [also nip]

n-i-p negative (layer)–intrinsic (layer)–positive (layer) (transistor) [also NIP, N-I-P, or nip]

N$_1$I/ℓ1 Alternating-Current Excitation [Symbol]

NIPA National Income and Product Accounts [of US Department of Commerce]

NIPA Notice of Initiation of Procurement Action

NIPAAm N-Isopropyl Acrylamide [also NIPAA, or NIPAM]

NIPAl N-Isopropylaniline [also NiPAl]

NiPc Nickel Phthalocyanine

NIPCC National Industrial Pollution Control Council

Ni-Pd Nickel-Palladium (Alloy System)

NIPDWS National Interim Primary Drinking Water Standards

NIPER National Institute for Petroleum and Energy Research [US]

NIPERA Nickel Producers Environmental Research Association [also NiPERA]

Ni-P-HBN Nickel-Phosphorus-Hexagonal Boron Nitride [also Ni-P-hBN]

NIPHLE National Institute of Packaging, Handling and Logistics Engineers [US]

NIPI N-Doped Intrinsic P-Doped Intrinsic (Transistor) [also N-I-P-I, nipi, or n-i-p-i]

NIPI Negative-Intrinsic-Positive-Intrinsic (Transistor) [also N-I-P-I, nipi, or n-i-p-i]

NIPO Negative Input–Positive Output

NiPO Nickelophosphate

Nippon Kinz Gakk Nippon Kinzoku Gakkaishu [Japan]

Nippon Kokan Tech Bull Nippon Kokan Technical Report [Kawasaki, Japan]

Nippon Kokan Tech Rep Nippon Kokan Technical Report [Kawasaki, Japan]

Nippon Stainless Tech Rep Nippon Stainless Technical Report [Tokyo, Japan]

Nippon Steel Tech Rep Nippon Steel Technical Report [Tokyo, Japan]

Nippon Steel Tech Rep (Overseas) Nippon Steel Technical Report (Overseas) [Tokyo, Japan]

Nippon Tungsten Rev Nippon Tungsten Review [Fukuoka, Japan]

NIPR National Institute for Personnel Research [South Africa]

NIPR National Institute for Polar Research [Japan]

NIPS Naval Intelligence Processing System [of US Navy]

NIPS Network Inputs/Outputs per Second

NIPS NMCS (National Military Command System) Information Processing System [US]

NIPSA Northern Ireland Public Service Alliance

Ni-P-SiC Nickel-Phosphorus-Silicon Carbide

NIPTS Noise-Induced Permanent Threshold Shift

NIQOA Northern Ireland Quarry Owners Association

NIR Near Infrared

NIR Near Infrared Diffuse Reflection

NIR Needle Roller Bearing Inner Ring, Lubrication Groove and Hole in Bore, Inch-Dimensioned [Symbol]

NIR Negative-Impedance Repeater

NIR Network Information Retrieval

NIR Neutral Impact Radiation

NIR Next Instruction Register

NIR Nickel-Iron Refinery [of Falconbridge Ltd., Sudbury, Ontario, Canada]

NIR Non-Inductive Resistor

NIR Non-Isothermal Reaction

NIR-1 9-Cyano-N,N,N'-Triethylpyronine-N'Caproic Acid N^4(Maleimidoethyl)piperazide Chloride

NIR-2 9-Cyano-N,N,N'-Triethylpyronine-N'Caproic Acid N-Hydroxysuccinimide Ester Chloride

Nir Nicotinamideribose [Biochemistry]

NIRA National Industrial Recovery Act [US]

NIRA National Iridology Research Association [US]

NIRA Near Infrared Analysis

NIRA Near Infrared Reflectance Analysis

NIRB Nuclear Insurance Rating Bureau [US]

NIRD National Institute for Research in Dairying [UK]

NIRE National Institute for Resources and Environment [of Agency of Industrial Science and Technology, Japan]

N Ire Northern Ireland

NIRF National Irrigation Research Fund

Ni-Rh Nickel-Rhodium (Alloy System)

NIRI National Information Research Institute [US]

NIRIM National Institute for Research in Inorganic Materials [Ibaraki, Japan]

NIRIN National Industrial Research Institute of Nagoya [of Agency of Industrial Science and Technology, Japan]

NIRMA Nuclear Information and Records Management Association [US]

NIRMS Noble Gas Ion Reflection Mass Spectroscopy

NIRNS National Institute for Research in Nuclear Science [UK]

NIRO Nike-Iroquois Rocket

NIROM National Institute for Research in Organic Materials [Tsukuba, Japan]

NIROS Nixdorf Real-Time Operating System

NIRR National Institute for Road Research [South Africa]

NIRRD National Institute for Rocket Research and Development [South Africa]

NIRS National Institute for Radiological Sciences [Chiba, Japan]

NIRS Near-Infrared Spectrophotometer; Near-Infrared Spectrophotometry

NIRS Near-Infrared Spectroradiometer; Near-Infrared Spectroradiometry

NIRS Near-Infrared Spectroscope; Near-Infrared Spectroscopy

NIRS Neutral Impact Radiation Spectroscopy

NIRT National Iranian Radio and Television

NIRTS New Integrated Range Timing System

NIRTU Northern Ireland Industrial Research and Technology Unit

NIS National Information System [US]

NIS National Institute of Science [US]

NIS NATO Identification System

NIS Negative Ion Spectroscopy

NIS Netherlands Information Service

NIS Network Information Services [Internet]

NIS Network Interface System

NIS Neutron Image Storage

NIS Neutron Inelastic Scattering

NIS Neutron Instrumentation System

NIS Newly Independent States (of the Former Soviet Union)

NIS Nordiska Ingenjorssamfundet [Nordic Engineers Association, Sweden]

NIS Normal-Insulator-Superconductor (Junction)

NIS Not in Stock

NIS NRC (National Research Council) Information System [US]

NIS Nuclear-Induced Sensitization

NIS Nuclear Instrumentation System

NiS Nickel Sulfide (Semiconductor)

Ni-S Nickel-Sulfur (System)

NISA National Independent Sign Association [now Independent Signcrafters of America, US]

NISA National Industrial Sand Association [US]

NISA National Industrial Service Association [now Electrical Apparatus Service Association, US]

NISARC National Information Storage and Retrieval Center [US]

NiSb Nickel Antimonide (Semiconductor)

Ni-Sb Nickel Antimony (Alloy System)

NiSbV Nickel Antimony Vanadium (Compound)

NISC National Industrial Space Committee [US]

NISD National Association of Steel Detailing [US]

N-ISDN Narrowband ISDN (Integrated Services Digital Network)

N-ISDN National ISDN (Integrated Services Digital Network)

NiSe Nickel Selenide (Semiconductor)

NISEC Negative Ion Secondary Emission Coefficient

NISEE National Information Service for Earthquake Engineering [US]

NISI National Institute of Sciences of India

NISI Northwest Instrument System, Inc. [US]

Ni-Si Nickel-Silicon (Alloy System)

NiSi$_2$ Nickel Disilicide

Ni-SiC Nickel-Based Silicon Carbide

Ni-Sil Nickel-Silver

NiSi$_2$/p-Si Nickel Disilicide on P-Type Silicon (Substrate)

NiSi$_2$/Si Nickel Disilicide/Silicon (Substrate)

NiSiTi Nickel-Silicon-Titanium (Structure)

NiSiZr Nickel-Silicon-Zirconium (Compound)

NISL Network Interface Sublayer

NISLAPP National Institute for Science, Law and Public Policy [US]

NISO National Individual Standing Offer [Canada]

NISO National Industrial Safety Organization [Ireland]

NISO National Information Standards Organization [US]

NiSO Nickelosulfate

NISP Networked Information Services Project

NISPA National Information System for Physics and Astronomy [US]

NISREC Northern Ireland Semiconductor Research Center [of Queen's University, Belfast]

NiSSb Nickel Sulfur Antimonide (Structure)

NISSD Nautilus Institute for Security and Sustainable Development [US]

Nisshin Steel Tech Rep Nisshin Steel Technical Report [Tokyo, Japan]

NIST National Information System for Science and Technology [Japan]

NIST National Institute of Science and Technology [Philippines]

NIST National Institute of Standards and Technology [Offices located at Gaithersburg, Maryland and Boulder, Colorado,US] [formerly National Bureau of Standards]

Ni St Nickel Steel [also NiSt]

NIST/ACerS Conf National Institute of Standards and Technology/American Ceramic Society Conference

NIST ATP National Institute of Standards and Technology/Advanced Technology Program [also NIST/ATP, US]

NIST-BD National Institute of Standards and Technology Biotechnology Division [Gaithersburg, Martyland, US]

NIST Build Sci Ser NIST Building Science Series [Published by the National Institute for Standards and Technology, US]

NIST-CD National Institute of Standards and Technology Ceramics Division [Gaithersburg, Martyland, US]

NISTCERAM National Institute of Standards and Technology Structural Ceramics Database [US]

NIST-EMFD National Institute of Standards and Technology Electromagnetic Fields Division [Boulder, Colorado, US]

NIST/EPA/MSDC NIST (National Institute of Standards and Technology)/EPA (Environmental Protection Agency) Mass Spectral Database on CD-ROM [US]

NISTFS National Institute of Standards and Technology Frequency Standard [formerly NBSFS, US]

NIST/ICST National Institute of Standards and Technology/Institute for Computer Sciences and Technology [formerly NBS/ICST, US]

NISTIR National Institute of Standards and Technology Information Report [formerly NBSIR]

NIST-MD National Institute of Standards and Technology Metallurgy Division [Gaithersburg, Maryland, US]

NIST Monogr NIST Monograph [Published by the National Institute for Standards and Technology, US]

NIST MSEL NIST Materials Science and Engineering Laboratory [of National Institute for Standards and Technology, US]

NIST NSE NIST Neutron Spin Echo Spectrometer [of National Institute of Standards and Technology, US]

NIST-PMD National Institute of Standards and Technology Process Measurements Division [Gaithersburg, Maryland, US]

NIST-RD National Institute of Standards and Technology Reactor Division [Gaithersburg, Maryland, US]

NIST Res Rep NIST Research Report [Published by the National Institute for Standards and Technology, US]

NIST-RRD National Institute of Standards and Technology Reactor Radiation Division [Gaithersburg, Maryland, US]

NIST-SIS National Institute of Standards and Technology–Standard Information Services [formerly NBS-SIS]

NIST SP NIST Special Publication [Published by the National Institute for Standards and Technology, US] [also NIST Spec Publ]

NIST-SRM National Institute of Standards and Technology–Standard Reference Material [National Bureau of Standards–Standard Reference Material, US]

NIST Tech Note NIST Technical Note [Published by the National Institute for Standards and Technology, US]

NIST Tech Rep NIST Technical Report [Published by the National Institute for Standards and Technology, US]

NIST Update NIST Update [Published by the National Institute for Standards and Technology, US]

NISU Northeastern Illinois State University [Chicago, US]

NIT Nagoya Institute of Technology [Japan]

NIT Nanking Institute of Technology [PR China]

NIT Net Ingot Ton [Metallurgy]

NIT Nevada Institute of Technology [US]

NIT New Industrial Technology

NIT New Information Technology

NIT Nippon Institute of Technology [Saitama, Japan]

NIT Non-Intelligent Terminal

NIT Nonlinear Inertialess Three-Pole

NIT Northrop Institute of Technology [Inglewood, California, US]

NIT Norwegian Institute of Technology [Trondheim]

NIT Numerical Indicator Tube [Electronics]

Nit Nitrate [also nit]

NITC National Information Technology Center [US]

NITC National Information Transfer Center [of UNESCO]

Ni$_3$(TCNQ)$_2$ Nickel(II) Tetracyanoquinodimethane

NITD Nebraska Information and Tourism Division [US]

NITEP National Incinerator Testing and Evaluation Program [Canada]

NiTe Nickel Telluride (Semiconductor)

Ni(TFAC)$_2$ Nickel Trifluoroacetylacetonate [also Ni(tfac)$_2$]

NITG Nederlands Instituut voor Toegepast Geowetenschap [Netherlands Institute of Applied Geoscience]

Ni-Ti Nickel-Titanium (Alloy System)

Ni-Ti/Al$_2$O$_3$ Alumina Reinforced Nickel-Titanium (Composite)

Ni-TiC Titanium Carbide Dispersed Nickel

Ni-Ti-Cu Nickel-Titanium-Copper (Shape Memory Alloy)

Ni-Ti-Fe Nickel-Titanium-Iron (Shape Memory Alloy)

Ni-TiN Titanium Nitride Dispersed Nickel

Ni-Ti-Nb Nickel-Titanium-Niobium (Shape Memory Alloy)

NITL National Industrial Transportation League [US]

NITP National Industrial Training Program

NITP National Institute for Theoretical Physics [at University of Adelaide, Australia]

NiTPP 5,10,15,20-Nickel Tetraphenylporphyrin

NITR National Institute for Telecommunications Research [South Africa]

NITR 2(R) 4,4,5,5-Tetramethyl-4,5-Dihydro-1H-Imidazolyl-1-Oxy-3 Oxide [*Note:* R is an Ethyl, n-Propyl, or Isopropyl]

Nitr Nitrate [also nitr]

NitRem Nitrate Removal (Process)

Nitro BT Nitro Blue Tetrazolium

NITROMIDE 3,5-Dinitrobenzamide [also Nitromide, or nitromide]

Nitro-PAPS 2-(5-Nitro-2-Pyridylazo)-5-(N-Propyl-N-Sulfopropylamino)phenol

NITROS Nitrostarch [also Nitros, or nitros] [Explosive]

Nitroso-PSAP 2-Nitroso-5-(N-Propyl-3-Sulfopropylamino) phenol

NITTA Northern Ireland Timber Trade Association

NIU Network Interface Unit

NIU Northern Illinois University [De Kalb, Illinois, US]

NIUW National Institute for Urban Wildlife [US]

NIV Negative Impedance Inverter

NIVA Norwegian Institute for Water Research

NIW Naval Inshore Warfare

NIW Niedersächsisches Institut für Wirtschaftsforschung [Lower Saxony Institute for Economic Research, Germany]

Ni-W Nickel-Tungsten (Alloy System)

NIWA National Institute of Water and Atmospheric Research Ltd. [New Zealand]

NIWMMA Northern Ireland Wholesalers, Merchants and Manufacturers Association

NIWR National Institute for Water Research [South Africa]

NIWR National Institute for Water Resources

NIX NASA Image eXchange

NiX Amorphous Binary Nickel Alloy [*X* Represents One, or More Metalloids]

NICSW Normal Incidence X-Ray Standing Wave

NIY Needle Roller Bearing, Drawn Cup, Full Complement, Rollers Retained by Lubricant, Without Inner Ring, Inch-Dimensioned [Symbol]

Ni/Y-123 Nickel on Yttrium Barium Copper Oxide (Superconductor)

Ni/YBCO Nickel on Yttrium Barium Copper Oxide (Superconductor)

NIYM Needle Roller Bearing, Drawn Cup, Full Complement, Rollers Retained by Lubricant, Closed End, Without Inner Ring, Inch-Dimensioned [Symbol]

NIZC National Industrial Zoning Committee [US]

NIZEMI NIeder-ZEntrifugal-MIkroskop [Slow-Rotating Centrifugal Microscope; of DLR-German Aerospace Research Establishment]

Ni-Zn Nickel-Zinc (Alloy System)

Ni-Zr Nickel-Zirconium (Alloy System)

NJ Network Junction

NJ New Jersey [US]

9J- Zambia [Civil Aircraft Marking]

NJAC National Joint Advisory Council [US]

NJB National Job Bank

NJC National Joint Committee

NJC National Joint Council

NJC Naval Joining Center [of US Navy]

NJC New Journal of Chemistry [Published by the Royal Society of Chemistry, UK]

NJCBI National Joint Council for the Building Industry [UK]

NJCC National Joint Computer Committee [now American Federation of Information Processing Societies, US]

NJCC National Joint Consultative Committee

NJCEC NATO Joint Communications and Electronic Committee

NJCM New Jersey College of Medicine [Newark, US]

NJCMD New Jersey College of Medicine and Dentistry [Newark, US]

NJCL Network Job Control Language

NJCST New Jersey Commission on Science and Technology [US]

NJE Network Job Entry

NJEH New Jersey Extraordinarily Hazardous Substances List [US]

NJES Nozzle Joint Environmental Simulators

n-JFET N-Channel Junction Field-Effect Transistor

NJGFE Nordic Joint Group for Forest Entomology [Sweden]

NJHS New Jersey Historical Society [US]

NJI Network Job Interface

NJIC National Joint Industrial Council [UK]

NJIT New Jersey Institute of Technology [Newark, US]

NJMS New Jersey Medical School [Newark, US]

NJP Network Job Processing

NJPA New Jersey Pharmaceutical Association [US]

NJPA New Jersey Port Authority [US]

NJPMB Navy Jet-Propelled Missile Board [of US Navy]

NJS Noise Jammer Simulator

NJ SCOE New Jersey Sematech Center of Excellence [US]

NJSPE New Jersey Society of Professional Engineers [US]

NJUG National Joint Utilities Group [US]

NK Natural Killer (Cell) [Biochemistry]

NK *(Nihon-Koku)* – Official Name of "Japan"

NK Nippon Kokakukdishi [Japan]

NK North Korea(n)

9K- Kuwait [Civil Aircraft Marking]

n(k) momentum distribution function [Symbol]

NKA Nordisk Kontaktorgan for Atomenergisporgsmal [Nordic Liaison Committee for Atomic Energy, Denmark]

NKα Nitrogen K-Alpha (Radiation) [also Nk_α]

NKAF North Korean Air Force

NK-AP Nippon Kokan Arc Process [Metallurgy]

NKBA National Kitchen and Bath Association [US]

NKCF Natural Killer Cytotoxic Factor [Biochemistry]

NKF National Kidney Foundation [US]

NKF Nationalkommittén för Fysik [National Committee for Physics, Sweden]

N/kg Netwon(s) per kilogram [also N kg^{-1}]

NKHA National Kerosene Heater Association [US]

NKI Nippon Kinoko Institute [Japan]

NKIT National Kaohsiung Institute of Technology [Taiwan]

NKK Nationalkommittén för Kemi [National Committee for Chemistry, Sweden]

NKK Nippon Kokan KK (Steel Company) [Kawasaki, Japan]

NKK-CAL Nippon Kokan–Continuous Annealing Line [Metallurgy]

NKK-RQ Nippon Kokan–Roll Quench (Process) [Metallurgy]

NKK Tech Rep NKK Technical Report [Published by Nippon Kokan KK, Kawasaki, Japan]

NKK Tech Rev NKK Technical Review [Published by Nippon Kokan KK, Kawasaki, Japan]

NKK-RQ Nippon Kokan–Water Quench (Process) [Metallurgy]

NKMU National Kangaroo Monitoring Unit [of Australian Nature Conservancy Agency]

NKPA North Korean People's Army

NKr Norwegian Krone [Currency] [also NKR, or Nkr]

NKSF Natural Killer Cell Stimulatory Factor [Biochemistry]

NKUDIC National Kidney and Urologic Diseases Information Clearinghouse [of National Institute of Diabetes and Digestive and Kidney Diseases, National Institutes of Health, Bethesda, Maryland, US]

NKW North Korean Won [also NK W] [Currency]

NL Nanolaminate

NL National Lakeshore [US]

NL National Landmark [US]

NL National Library [Ottawa, Canada]

NL Natural Language

NL Navy League

NL Navy List

NL Nearly Lamellar (Microstructure) [Metallurgy]

NL Nearly Localized (Model) [Solid-State Physics]

NL Negative Logic

NL Neodymium Laser

NL Netherlands [ISO Code]

NL Network Layer

NL Neutron Log [Mineral Exploration]

NL Newfoundland and Labrador [Canada]

NL New Line (Character) [Data Communications]

NL New London [Connecticut, US]

NL Newtonian Liquid [Fluid Mechanics]

NL Nodal Line

NL Noise Level

NL Noise Limitation; Noise Limiter

NL No Limit

NL No-Load (Voltage)

NL Non-Linear(ity)

NL Non-Luminous

NL Northern Lights

NL Not Listed

NL Non-Loaded

NL Nuclear Laser

NL Nuevo Léon [Mexico]

NL Null Line

N/L Navigation/Localizer

N(λ) Noise at Wavelength λ [Symbol]

9L- Sierra Leone [Civil Aircraft Marking]

nl *(non licet)* – it is not permitted

nl *(non liquet)* – it is not clear

.nl Netherlands [Country Code/Domain Name]

n(λ) refractive index as a function of wavelength [Symbol]

NLA National Librarians Association [US]

NLA National Library Act [New Zealand]

NLA National Library of Australia [Canberra, Australian Capital Territory]

NLA National Lime Association [US]

NLA Newfoundland Library Association [Canada]

NLA (Study Group on) New Larger Aeroplanes

NLA Next Lower Assembly

NLA Nonlinear Acoustic(s)

NLA Nonlinear Analysis

NLA Normalized Load Access

NLA Northwestern Lumbermen's Association [US]

NLA Numerical Linear Algebra

Nla Neisseria lactamica [Microbiology]

Nla III Neisseria lactamica III (Three) [Microbiology]

Nla IV Neisseria lactamica IV (Four) [Microbiology]

NLAC National Landcare Advisory Committee [Australia]

NLAC National Library Advisory Committee [Canada]

NLAE National Laboratory for the Advancement of Education

N Lat North Latitude

NLB National Lighting Bureau [US]

NLB Nonlinear Buckling [Mechanics]

NLBA National Lead Burning Association [US]

NLBA Nonlinear Buckling Analysis [Mechanics]

NLBMDA National Lumber and Building Material Dealers Association [US]

NLC National Library of Canada [Ottawa]

NLC New Line Character

NLC Nonlinear Control

NLC Northern Land Council

NLC Notched-Link Chain

NLCIF National Light Castings Ironfounders Federation [UK]

NLCS Nordic Leather Chemists Society [Denmark]

NLD Necrobiosis Lipoidica Diabeticorum [Dermatology]

NLD Nonlinear Distortion

NLD Nonlinear Dynamics

NLD Nonlinear Dynamics (International Conference)

NLD Nonlocal Density [Physics]

NLDA Nonlocal Density Approximation [Physics]

NLDC Newfoundland and Labrador Development Corporation [Canada]

NLDF Nonlocal Density–Functional [Physics]

NLDFT Nonlocal Density Functional Theory [Physics]

NLDTS Nonlinear Discrete-Time System

NLE National Lighting Exposition

NLE National Livestock Exchange [US]

NLE Nonlinear Elasticity

NLE Nonlinear Equation

NLE Nonlinear Estimation

NLE Nonlocalized Electron [Physics]

NLE Nontopological Localized Excitation

NLE Not Less, or Equal

Nle Norleucine; Norleucyl [Biochemistry]

NLEA National Lumber Exporters Association [US]

NLEE Nonlinear Evolution Equation

NLEF Nonlinear Electric Field

NLEFM Nonlinear Elastic Fracture Mechanics

NLF Nonlinearity Factor

NLF Normal Load Factor [Electrical Engineering]

NLfB Niedersächsisches Landesamt für Bodenforschung [Lower Saxony State Office for Soil Research, Hannover, Germany]

NLG Netherlands Guilder [Currency]

NLG Nose Landing Gear

NLG Sodium Lanthanum Germanate

NLGA National Lumber Grades Authority [US]

NLGDB National Local Government Boundary Database [Australia]

NLGERN National Local Government Environmental Resource Network [Australia]

NLGI National Lubricating Grease Institute [US]

NLGLP National Laboratory Gene Library Project [Joint Project of Los Alamos National Laboratory and Lawrence Livermore National Laboratory, US]

NLH Nonlinear Harmonics

nL/h nanoliter per hour [also nL h^{-1}, or nl/h]

NLHEP National Laboratory for High Energy Physics [US]

NLI National Lead Industries

NLI National Lifeboat Institution [UK]

NLI National Localities Index

NLI Natural Language Interface

NLI Nonlinear Interpolating

NLIF Nonlinear Interference Filter

NLIMT Newfoundland and Labrador Institute of Marine Technology [Canada]

NLJP Negligible Liquid Junction Potential [Physical Chemistry]

NLL National Lending Library [US]

NLL Nationaal Luchtvaart Laboratorium [National Aeronautical Laboratory, Netherlands] [now Nationaal Lucht– en Ruimtevaartlaboratorium]

NLL Neodymium Liquid Laser

NLLS Nonlinear Least-Squares (Fitting) [also NLLSQ] [Statistics]

NLLSA Nonlinear Least Squares Analysis [Statistics]

NLLSQ Nonlinear Least-Squares (Fitting) [also NLLS] [Statistics]

NLLST National Lending Library for Science and Technology [now British Library Lending Division]

NLM National Library of Medicine [of National Institutes of Health, Bethesda, Maryland, US]

NLM NetWare Loadable Module

NLM Noise Level Monitor

NLMA National Lumber Manufacturers Association [now National Forest Products Association, US]

NLMB National Livestock and Meat Board [US]

NLMC Nordic Labor Market Committee [Denmark]

NLMF Nonlinear Magnetic Field

NLMI Newfoundland and Labrador Marine Institute [Canada]

NLMO Nonlocalized Molecular Orbital [Physical Chemistry]

NLN National League for Nursing [US]

NLN Sodium Lanthanum Niobate

NLNP National Library Network Program [US]

NLNZ National Library of New Zealand

NLO National Liaison Officer

NLO Nonlinear Optic(al); Nonlinear Optics

NLO Nonlinear Oscillation; Nonlinear Oscillator

NLO Nonlocalized Orbital [Physics]

NLOD Nonlinear Optical Device

NLOGF National Lubricating Oil and Grease Federation [UK]

NLOS Natural Language Operating System

N-LOS Non-Line of Sight [US Army]

NLP National Landcare Plan [Australia]

NLP National Landcare Program [Australia]

NLP Natural-Language Processing

NLP Nonlinear Parameter

NLP Nonlinear Programming

NLP Sodium Lithium Phosphate

NlP Netherlands Patent

NLPGA National Liquefied Petroleum Gas Association [now National Propane Gas Association, US]

NLQ Near Letter-Quality (Printing)

NLR Nationaal Lucht– en Ruimtevaartlaboratorium [National Aerospace Laboratory, Netherlands]

NLR Noise Load Ratio

NLR No-Load Ratio

NLR Nonlinear Regression [Statistics]

NLR Nonlinear Resistance; Nonlinear Resistor

NLR Nonlinear Resonance

NLRA National Labor Relations Act [US]

NLRB National Labor Relations Board [US]

NLRSDA National Land Remote Sensing Data Archive [of US Geological Survey]

NLS National Language Support

NLS National Library of Sciences [Bethesda, Maryland, US]

NLS National Library of Scotland [of British Library in Edinburgh]

NLS National Library Service [Wellington, New Zealand]

NLS N-Lauroylsarcosine [Biochemistry]

NLS Noise-Level Scattering

NLS No-Load Speed

NLS Nonlinear Schroedinger [Quantum Mechanics]

NLS Nonlinear Spectroscopy

NLS Nonlinear System

NLSA National Locksmith Suppliers Association [US]

NLSA Nonlinear Stress Analysis

NLSB National Land Survey Board [Sweden]

NLSC Northeast Lousiana State College [Monroe, US]

NLSE Nonlinear Schroedinger Wave Equation [Quantum Mechanics]

NLSFUNC National Language Support Function (Command) [also nlsfunc]

NLSLS National Library of Scotland Lending Services [Edinburgh]

NLSM Nonlinear Sigma Model

NLSMB National Livestock and Meat Board [US]

NLSP Network Layer Security Protocol

NLSPA National Livestock Producers Association [US]

NLT Non Light-Tight

NLT Non-Linear Time Sequence

NLT Not Later Than

NLT Not Less Than

NLT Not Lower Than

NLT Net Long Ton [Unit]

NLTDO Newfoundland and Labrador Tourist Development Office [Canada]

NLTE Nonlocal Thermodynamic Equilibrium

NLTS Nonlinear Transition Shift [Solid-State Physics]

NLU Natural-Language Understanding

NLUA Naturkundig Laboratorium der Universiteit van Amsterdam [Scientific Laboratory of the University of Amsterdam, Netherlands]

NLUPP Northern Land Use Planning Program [Canada]

NLUS Navy League of the United States

NLV National Language Version

NLV Non-Linear Vibration [Mechanics]

NLW National Library Week [US]

NLW National Library of Wales [of British Library in Aberystwyth]

NLW Natural Linewidth [Spectroscopy]

NLW Nonlinear Wave

NLWP Nonlinear Wave Propagation

NLWM Nonlinear Wave Mixing

NM Nanocrystalline Material

NM Nanomaterial

NM Nanostructured Material

NM (1-Naphthyl)methanol

NM National Monument [US]

NM National Museum

NM Natural Magnet

NM Nautical Mile [also nm, NMI, or nmi]

NM Nearly Metamagnetic (Model) [Solid-State Physics]

NM Needle Roller and Cage Assembly, Metric-Dimensioned [Symbol]

NM Negative Medium (Error)

NM Negative Misfit (Alloy) [Metallurgy]

NM Network Management; Network Manager

NM Network Manager [of NASA Goddard Space Flight Center, Greenbelt, Maryland, US]

NM Network Model

NM Neutron Monochromation; Neutron Monochromator

NM New Mexico [US]

NM New Moon [Astronomy]

NM Newtonian Mechanics

NM Newtonian Model [Physics]

NM Nilsson Model

NM Nitromethane

NM Nitrosomesitylene

NM Noble Metal

NM Noise Margin

NM Noise Meter

NM Nomarski Microscope

NM No Measurement

NM Nonmagnetic

NM Nonmaskable

NM Nonmetal(lic)

NM Normal Magnification

NM Normal Mode [Computers]

NM Not Meaningful

NM Not Measured [also nm]

NM Nuclear Magnetism

NM Nuclear Magneton [Unit]

NM Nuclear Material

NM Nuclear Matter

NM Nuclear Medicine

NM Nuclear Membrane [Cytology]

NM Nuclear Metallurgist; Nuclear Metallurgy

NM Nuclear Method

NM Nuclear Model

NM Nuclear Moment

NM Numerical Mathematics

(NM)² Novel Magnetic Nanodevices and (Artificially Layered) Materials [of Esprit, Europe]

N/M No Mark; Not Marked

N/M Number per Section Area [Symbol]

9M- Malaysia [Civil Aircraft Marking]

Nm Newton-meter (of torque) [also N-m, or N·m]

N/m Newton(s) per meter [also N m^{-1}]

N/m² Newton(s) per square meter [also N m^{-2}]

N/m³ Newton(s) per cubic meter [also N m^{-3}]

nM nanomole [also nM]

nm nanometer [Unit]

nm nautical mile [also NM, NMI, or nmi]

nm not measured [also NM]

nm nuclear magneton [Unit]

nm^{-1} one per nanometer [also 1/nm]

nm² square nanometer [also sq nm]

nm^{-2} one per square nanometer [also 1/nm²]

nm³ cubic nanometer [also cu nm]

nm³ normal cubic meter [Unit]

nm^{-3} one per cubic nanometer [also $1/nm^3$]

nm^4 meter to the fourth (power) [Unit]

nm- nanometer [Followed by an Element or Compound, e.g., nm-TiAl is Nanometer Titanium Aluminide]

n/m^2 neutron(s) per square meter [also $n\ m^{-2}$]

NMA Nadic Methyl Anhydride

NMA National Management Association [US]

NMA National Medical Association [US]

NMA National Microfilm Association [now Association for Information and Image Management, US]

NMA National Microform Association [US]

NMA National Micrographics Association [now Association for Information and Image Management, US]

NMA National Mortgage Association [US]

NMA National Museum of Australia

NMA N-Methylolacrylamide

NMA N-Methyl Acridinium

NMA Network Management Agent

NMA Non-Marine Association [of Lloyd's Underwriters, UK]

NMAA National Machine Accountants Association [now Data Processing Management Association, US]

NMAA National Materials Advancement Award [of Federation of Materials Societies, US]

NMAB National Materials Advisory Board [US]

NMAB Newsl National Materials Advisory Board Newsletter [US]

Nmal N-Dimethylaniline

NMAP National Metric Advisory Panel [US]

N-15 MASS NMR Nitrogen-15 Magic Angle Sample Spinning Nuclear Magnetic Resonance [also ^{15}N MASS NMR]

NMB Nonmetal Bromide

NMC National Meteorological Center [Suitland, Maryland, US]

NMC National Meteorological Center [of World Meteorological Organization]

NMC National Museum of Canada [Ottawa]

NMC Naval Materials Command [of US Navy]

NMC Naval Medical Center [of National Institutes of Health, Bethesda, Maryland, US]

NMC Naval Medical Command [of US Navy]

NMC Naval Missile Center [of US Navy]

NMC Nebraska Medical Center [of University of Nebraska, Omaha, US]

NMC Network Management Center

NMC Network Measurement Center

NMC Nissan Motor Company Limited [Japan]

NMC Nonmetal Carbide

NMC Nonmetallic Cable

NMC Northern Montana College [Havre, US]

NMC NSSDC (National Space Science Data Center) Master Catalog [of NASA]

NMC Numeric Meteorological Center [of National Oceanic and Atmospheric Administration, US]

NMCC National Military Command Center [US]

NMCC Network Maintenance/Monitoring Control Center

NMCC Network Management Control Center

NMCC New Metals and Chemicals Corporation [Tokyo, Japan]

NMCL Naval Missile Center Laboratory [of US Navy]

NMCl Nonmetal Chloride

NMCPP New Mexico Center for Particle Physics [of University of New Mexico at Albuquerque, US]

NMCS National Military Command System [US]

NMCS Nuclear Materials Control System

NMCSSC National Military Command System Support Center

NMD NASA Master Directory

NMD National Mapping Division [of US Geological Survey]

NMD Naval Mine Depot [Yorktown, Virginia, US]

NMDA National Marine Distributors Association [US]

NMDA National Metal Decorators Association [US]

NMDA N-Methyl-D-Aspartate; N-Methyl D-Aspartic Acid [Biochemistry]

NMDC National Mineral Development Corporation [India]

Nm/deg Newtonmeter per degree [also $Nm\ deg^{-1}$]

NMDPP Neomethyldiphenylphosphine

NMDSC Naval Medical Data Service Center [of US Navy]

NME Naphthylmethyl Ether

NME National Military Establishment

NME Noise-Measuring Equipment

NMe Nitromethane

NMe_2 Dimethylamino-

NMEA National Marine Educators Association [US]

NMEA National Marine Electronics Association [US]

N^6-Me-DMT-dA 5'-O-(4,4'-Dimethoxytrityl)-N^6-Methyl-2'-Deoxyadenosine [Biochemistry]

NMEL Navy Marine Engineering Laboratory [US]

NMEL Nuclear Mechano-Electronic Laboratory

NMERI National Mechanical Engineering Research Institute [South Africa]

NMESFET Negative Metal-Gate Schottky Field-Effect Transistor

N-Me-trdtrd N-Methyl-trimethylenediamine-N,N',N'-Triacetate

N Mex New Mexico [US]

NMF National Microelectronics Facility [Canada]

NMF Neutron Multiplication Factor [Nuclear Reactors]

NMF New Master File

NMF Nonmetal Fluoride

NMF Norsk Matematisk Forening [Norwegian Mathematical Association]

NMFA National Mineral Feed Association [now National Feed Ingredients Association, US]

NMFC National Motor Freight Classification

NMFC New Mexico Film Center [Santa Fe, US]

NMFS National Marine Fisheries Service [of National Oceanic and Atmospheric Administration, US]

NMFTA National Motor Freight Traffic Association [US]

NMFWA National Military Fish and Wildlife Association [US]

n-$MgAl_2O_4$ Nanocrystalline Magnesium Aluminate

NMH Nautical Miles per Hour [also nmh]

NMH No-Mar Hammer

NMH Nonmetal Hydride

nm/h nanometer(s) per hour [also nm h⁻¹, or nm/hr]

NMHC National Materials Handling Center [UK]

NMHC Non-Methane Hydrocarbon(s)

NMHF National Manufacturered Housing Federation [US]

NMHS National Maritime Historical Society [US]

NMHT National Museum of History and Technology [of Smithsonian Institution, Washington, DC, US]

NMHU New Mexico Highlands University [Las Vegas, US]

NMHWMS New Mexico Hazardous Waste Management Society [US]

NMI NASA Management Institute [US]

NMI NASA Management Instruction

NMI National Maglev Initiative [US]

NMI National Museum of Ireland

NMI Nautical Mile [also nmi, NM, or nm]

NMI Neutrinos at the Main Injector (Program) [of Fermilab, Batavia, Illinois, US]

NMI No Middle Initial

NMI Nonmagnetic Inclusion [Metallurgy]

NMI Non-Maskable Interrupt

NMIA National Military Intelligence Association [US]

NMI/in Nautical Mile(s) per Inch [also nmi/in]

N Mile Nautical Mile [also n mile]

NMIMT New Mexico Institute of Mining and Technology (or New Mexico Tech) [Socorro, US]

N/min Newton(s) per minute [also N min⁻¹]

NMIP Northern Mining Information Program [Canada]

NMIRI Nagoya Municipal Industrial Research Institute [Nagoya, Japan]

NMIS Nuclear Materials Information System

NMIS Nuclear Materials Inventory System

NMIZT N-Methylimidazoline-2-Thione

NMK Naphthylmethyl Ketone

NMKL Nordisk Metodikkomite for Livsmedel [Nordic Committee on Food Analysis, Finland]

NML Narragansett Marine Laboratory [of University of Rhode Island in Saunderstown, US]

NML National Magnet Laboratory [of Massachusetts Institute of Technology, Cambridge, US]

NML National Measurement Laboratory [of Commonwealth Scientific and Industrial Research Organization, Australia]

NML National Metallurgical Laboratory [of Council for Scientific and Industrial Research at Madras and Jamshedpur, India]

NML National Mining Law [US]

NML Nuclear Magnetism Log

nml normal

NML Tech J NML Technical Journal [Published by the National Metallurgical Laboratory, India]

NMM National Museum of Man [of National Museum of Canada, Ottawa]

NMM Nautical Miles per Minute [also nmm]

NMM Net Metal Mandrel

NMM NetWare Management Map

NMM Network Measurement Machine

NMM N-Methylmorpholine

NMM Nonmagnetic Material

NMM Nonmetal(lic) Material

NMM Nuclear Magnetic Moment

NMM Nuclear Materials Management

Nmm Newton-millimeter (of torque) [also N-mm, or N·mm]

Nm/m Newton-meter per meter [also N·m/m, or Nm m⁻¹]

N/mm Newton(s) per millimeter [also N mm⁻¹]

N/m²√m Netwon(s) per meter squared square root of meter [also N m⁻² m^{1/2}]

N/μm Newton(s) per micrometer [also N μm⁻¹]

NMMA National Marine Manufacturers Association [US]

NMMA N^G-Methyl-L-Arginine [Biochemistry]

NMMD Nuclear Materials Management Department

NMMD Nuclear Materials Management Division [of Ontario Hydro, Toronto, Canada]

NMMI New Mexico Military Institute [Roswell, New Mexico, US]

nm/min nanometer(s) per minute [also nm min⁻¹]

nm²/mN square nanometer(s) per millinewton [also nm² mN⁻¹]

NM/MO Nonmagnetic (Material)/Magnetooptic (Material)

NMMSS Nuclear Materials Management and Safeguards System

NMMW Near Millimeter Wave System

NMN Nickel-Molybdenum-Nickel

NMN β-Nicotinamide Mononucleotide [also β-NMN] [Biochemistry]

NMN Nonmetal Nitride

α-NMN α-Nicotinamide Mononucleotide [Biochemistry]

β-NMN β-Nicotinamide Mononucleotide [also NMN] [Biochemistry]

NMNH National Museum of National History [US]

NMNH β-Nicotinamide Mononucleotide, Reduced Form [also β-NMNH] [Biochemistry]

β-NMNH β-Nicotinamide Mononucleotide, Reduced Form [also NMNH] [Biochemistry]

NMNS National Museum of Natural Sciences [of National Museum of Canada, Ottawa]

NMO 4-Methylmorpholine N-Oxide

NMO N-Methylmorpholine Oxide

NMO Nonmetal Oxide

NMO Normal Manual Operation

NMO Normal Moveout

NMOA National Mobile Office Association [now Mobile Modular Office Association, US]

NMOC National Marketing Organization Committee

NMOC National Meteorological Operations Center

NMOG Non-Methane Organic Gas(es)

nmol nanomole [Unit]

nmol/cm² nanomole(s) per square centimeter [also nmol cm⁻²]

nmol/cm³ nanomole(s) per cubic centimeter [also nmol cm⁻³]

nmol/cm²s nanomole(s) per square centimeter-second [also nmol cm⁻² s⁻¹]

nmol/L nanomole(s) per liter [also nmol/l, or nmol L^{-1}]

NMOR N-Nitroso Morpholine

NMOS N-Channel Metal-Oxide Silicon [also N MOS, N-MOS, or n-MOS]

NMOS Negative Metal-Oxide Semiconductor [also N MOS, N-MOS, or n-MOS]

NMOS N-Type Metal-Oxide Semiconductor [also N MOS, N-MOS, or n-MOS]

NMOS FET N-Channel Metal-Oxide Semiconductor Field-Effect Transistor

n-MOSFET N-Channel Metal-Oxide Semiconductor Field-Effect Transistor

NMOS LSI N-Type Metal-Oxide Semiconductor Large-Scale Integration

NMOS RAM N-Type Metal-Oxide Semiconductor–Random-Access Memory

NMot N-Dimethyl-o-Toluidine

NMP Nanomechanical Probe [Mechanical Testing]

NMP National Mapping Program [of US Geological Survey]

NMP National Meter Programming

NMP National Memorial Park [US]

NMP National Military Park [US]

NMP National Museum of Pakistan [Karachi]

NMP National Museum Policy [Canada]

NMP Navigational Microfilm Projector

NMP Net Material Product [Poland]

NMP Network Management Protocol

NMP N-Methylpenazine

NMP N-Methylpyrrolidone

NMP N-Methyl-2-Pyrrolidinone; 1-Methyl-2-Pyrrolidinone

NMP Nonmetal Phosphide

NMP Normenausschuß Materialprüfung [Standards Committee on Materials Testing; of DIN Deutsches Institut für Normung, Germany]

NMP North Magnetic Pole

NMPC National Minority Purchasing Council

NMPDN National Materials Property Data Network

NMPF National Milk Producers Federation [US]

NMPH Nautical Miles per Hour [also nmph]

NMPip N-Methyl Piperidine [also NMPIP]

NMPK Nucleoside Monophosphate Kinase [Biochemistry]

NMPld N-Methylpyrrolidine

NMPM Nautical Miles per Minute [also nmpm]

NMPP 4-(4'-Nitro-2'-Methylsulfonylphenylazo)phenyl Phosphate

NMPS Nautical Miles per Second [also nmps]

NMR National Military Representative [of NATO]

NMR National Missile Range [US]

NMR Negative Magnetoresistance

NMR Nordisk Ministerråd [Nordic Minister's Council, Denmark]

NMR Normal-Mode Rejection [Electronics]

NMR Nuclear Magnetic Resonance

NMRA National Marine Representatives Association [US]

NMRA National Mine Rescue Association [US]

NMRA National Model Railroad Association [US]

NMRAS Nuclear Materials Report and Analysis System

NMRC National Microelectronics Research Center [Cork, Ireland]

NMR Centre Centre for Nuclear Magnetic Resonance [UK]

NMRD Nuclear Magnetic Relaxation Dispersion

NMRF Nuclear Magnetic Resonance Flowmeter

NMRG Nanomaterials Research Group [Canada]

NMRI Nuclear Magnetic Resonance Imaging

NMRL Nuclear Magnetic Resonance Laboratory [of University of Kansas, Lawrence, US]

NMRN National Meteorological Rocket Network [US]

NMR-ON Nuclear Magnetic Resonance on Oriented Nuclei [also NMRON]

NMRR Normal-Mode Rejection Ratio [Electronics]

NMRS Nuclear Magnetic Relaxation Spectroscopy

NMRS Nuclear Magnetic Resonance Spectrometer; Nuclear Magnetic Resonance Spectrometry

NMRS Nuclear Magnetic Resonance Spectroscopy

nmrs numerous

NMRT Nuclear Magnetic Resonance Tomography

NMS National Marine Sanctuary [US]

NMS National Master Specification [US]

NMS National Museum of Science [Tokyo, Japan]

NMS Nautical Miles per Second [also nms]

NMS Naval Meteorological Service [of US Navy]

NMS Networked Messaging System

NMS Network Management Services

NMS Network Management Signal

NMS Network Management System

NMS Neutral Meson Spectrometer

NMS Neutron Magnetic Scattering

NMS Neutron Monitoring System

NMS New Mexico Section [of Materials Research Society, US]

NMS Noise Measuring Set

NMS Nonmetal Sulfide

NMS Nordic Metalworkers Secretariat [Sweden]

NMS Normal Market Size [UK Stock Exchange]

NMS Norsk Metallurgisk Selskap [Norwegian Metallurgical Society]

NMS Nuclear Medical Science

Nm/s Newton-meter(s) per second [also Nm/sec, or Nm s^{-1}]

nm/s nanometer(s) per second [also nm s^{-1}, or nm/sec]

NMSA National Metal Spinners Association [US]

NMSA National Moving and Storage Association US]

NMSC National Merit Scholarship Corporation [Evanston, Illinois, US]

NMSC Nonferrous Metals Society of China

NMSC Nonmartensitic Structural Component [Metallurgy]

NMSC Northwest Missouri State College [Maryville, US]

NMSD Next Most Significant Digit

NMSE National Agenda in Materials Science and Engineering [of Materials Research Society, US]

Nm/sec Newton-meter(s) per second [also Nm/s, or Nm s^{-1}]

nm/sec nanometer(s) per second [also nm/s, or nm s^{-1}]

NMSF Northeast Marine Support Facility [of National Oceanic and Atmospheric Administration at Woods Hole, Massachusetts, US]

NMSI National Maximum Speed Limit

NMSO National Master Standing Offer [Canada]

N-MSOC N-Methylsulfonylethyloxycarbonyl

NMSP National Merit Scholarship Program [of National Merit Scholarship Corporation, Evanston, Illinois, US]

NMSS National Multiple Sclerosis Society [US]

NMSS National Multipurpose Space Station

NMSS (Office of) Nuclear Material Safety and Safeguards [US]

NMST National Museum of Science and Technology [of National Museum of Canada, Ottawa]

NMST New Materials System Test

NMSU New Mexico State University [University Park, US]

NMT New Mexico Tech (or New Mexico Institute of Mining and Technology) [Socorro, US]

NMT N-Methyltransferase [Biochemistry]

NMT Nordic Mobile Telephone System

NMT Notification of Master Tool

NMT Not More Than

NMT Nuclear Magnetic Tomograph(y)

NMT Nuclear Medicine Technologist; Nuclear Medicine Technology

NMTA National Metal Traders Association [now American Association of Industrial Management, US]

NMTA Northwest Marine Trade Association [US]

NMtAl N-Methylaniline

NMTAS National Milk Testing and Advisory Service [UK]

NMTBA National Machine Tool Builders Association [US]

NMTC National Mechanical Trade Council [US]

NMTD Nuclear Materials Technology Division [of Los Alamos National Laboratory, New Mexico, US]

NMTDC Nonferrous Materials Technology Development Center [Hyderabad, India]

NMTF National Metal Traders Federation [UK]

NMTFA National Master Tile Fixers Association [UK]

nm-TiAl Nanometer Titanium Aluminide

NMTP Nonmartensitic Transformation Product [Metallurgy]

NMU National Maritime Union [US]

NMU Nitrosomethylurethane

NMU Northern Michigan University [Marquette, US]

NMV National Museum of (the State of) Victoria [Melbourne, Australia]

NMV Nitrogen Manual Valve

NMVS Neutron Magneto-Vibrational Scattering

NMW National Museum of Wales [Cardiff, UK]

NMWA National Mineral Wool Association [US]

NMWC New Mexico Western College [Silver City, US]

NMWCO Nominal Molecular Weight Cut-Off

NMWL Nominal Molecular Weight Limit

NMWP National Mixed Waste Program [US]

NMWQL National Marine Water Quality Laboratory [of US Environmental Protection Agency]

NN α-Naphthonitrile

NN Nearest Neighbor (Atoms) [also nn] [Crystallography]

NN Near-Net (Shape)

NN Negative-Negative (Junction) [also N-N, nn, or n-n]

NN Neodymium Nickelate [$NdNiO_3$]

NN Net Net [also nn]

NN Network Node

NN Neural Net(work)

NN Neural Networks (Council) [of Institute of Electrical and Electronics Engineers, US]

NN Newport News [Virginia, US]

NN Nitronaphthalene

NN No News [Internet]

NN+ Negative-Negative Plus (Junction) [also N-N+, nn+, or n-n+]

N/N$_f$ Number of Cycles to Failure [Symbol]

N-N Neutron-Neutron (Logging)

–N=N– Azo Compound [Formula]

9N- Nepal [Civil Aircraft Marking]

nN nanonewton [Unit]

n/N cycle ratio (i.e., ratio of number of (stress) cycles (endured) to number of cycles to failure [Symbol]

n$_i$/n mole fraction (i.e., ratio of the number of moles of component to the total number of moles)

NNA New Network Architecture

NNAG NATO Naval Armaments Group

NNAPB Nearest-Neighbor Antiphase Boundary [Solid-State Physics]

NNC National Network Congestion Signal

NNC National Nomination Committee

NNC National Nuclear Center

NNC Non-Noise Certified Aircraft

NNCC National Neutron Cross-Section Center [US]

NNCIAWPRC Netherlands National Committee of the International Association on Water Pollution Research and Control

NNCIAWPRC Norwegian National Committee of the International Association on Water Pollution Research and Control

NNC RK National Nuclear Center, Republic of Kazakhstan

NNCSC National Neutron Cross-Section Center [US]

NNCSC National Neutron Cross-Section Committee [US]

NNCT Nagano National College of Technology [Nagano-shi, Japan]

NNCT Niihama National College of Technology [Japan]

NND National Network Dialing

NND National Number Dialing

NND Nearest Neighbor Distance [Crystallography]

NND Normalized Normal Distribution

NNDO Neglect of Nonbonded Differential Overlap [Physics]

NNDTC National Nondestructive Testing Center [UK]

NNE North-Northeast; North by Northeast

NNEC National Nuclear Energy Commission [Brazil]

NNEDDC N-[Naphthyl(1)]-Ethylenediamine

NN EPI Nearest Neighbor Effective Pair Interaction [Physics]

NNF National Nanofabrication Facility [of Cornell University, Ithaca, New York, US]

NNF National Neurofibromatosis Foundation [US]

NNF Non-Newtonian Fluid [Fluid Mechanics]

NNFM Non-Newtonian Fluid Mechanics

NNGu N,N-Dimethyl Guanidine [Biochemistry]

NN'Gu N,N'-Dimethyl Guanidine [Biochemistry]

NNGA Northern Nut Growers Association [Canada]

NNH Nearest Neighbor Histogram [Crystallography]

NNH Nearest Neighbor Hopping [Crystallography]

NNHS Newfoundland National History Society [Canada]

NNI Nederlands Normalisatie-Instituut [Netherlands Standards Institute]

NNI Net National Income

NNI Network to Network Interface

NNI Noise and Number Index

NNI Non-Nuclear Instrumentation

NNI Non-Numeric Information

NNI Norsk Nobel Institut [Norwegian Nobel Institute, Oslo]

NNI Norwegian Nobel Institute [Oslo]

NNI-I NASA Standard Initiator-Type I [formerly SMSI (Standard Manned Spaceflight Initiator)]

n-Ni Nanocrystalline Nickel

NNIC Normalized Noise Isolation Class

n-NiTi Nanostructured Nickel Titanium

NNL Non-Newtonian Liquid [Fluid Mechanics]

NNLM National Network of Libraries of Medicine [US]

NNLS Nonnegative Least Squares (Algorithm)

NNMR Nutation Nuclear Magnetic Resonance

NNN Next Nearest Neighbor (Atoms) [also nnn] [Crystallography]

NNNAPB Next-Nearest-Neighbor Antiphase Boundary [Solid-State Physics]

NNNDO Neglect of Non-Neighbor Differential Overlap [Physics]

NNN EPI Next Nearest Neighbor Effective Pair Interaction [Physics]

NNNN Next-Next Nearest Neighbor (Atoms) [Crystallography]

NNO Neodymium Nickel Oxide

NNOA National Naval Officers Association [US]

NNOC Nigerian National Oil Company

NNP Net National Product

NNP Net Naturalization Potential

NNPT Nuclear Nonproliferation Treaty

NNR National Number Routed

NNR Nearest-Neighbor Relation(ship) [Crystallography]

NNR Normalized Noise Reduction

NNRI National Nutrition Research Institute [South Africa]

NNRIS National Natural Resources Information System

NNS European Conference on Near-Net Shape Manufacturing of Metal Parts

NNS Near-Net Shape

NNS Newport News Shipbuilding [Virginia, US]

NNSA National Nitrogen Solutions Association [now National Fertilizer Solutions Association, US]

NNSA National Nurses Society on Alcoholism [of National Council on Alcoholism, US]

NNSF National Natural Science Foundation [PR China]

NNSFC National Natural Science Foundation of China

NNSS Navy Navigational Satellite System [of US Navy]

NNTP Network News Transfer Protocol [Internet]

NNTP 2-Nitrimino-5-Nitroxy-1,2-Diazacyclohexane

NNUN National Nanofabrication Users Network

NNV Nederlandse Natuurkundige Vereniging [Netherlands Physical Society]

NNW North-Northwest; North by Northwest

NNWS Non-Nuclear Weapons State(s)

NNWSI Nevada Nuclear Waste Storage Investigations [of US Department of Energy]

NO Natural Orbital

NO Naval Office(r)

NO Naval Ordnance

NO Negative Output

NO Neutron Optics

NO New Orleans [Louisiana, US]

NO Nitrate

NO Nitric Oxide

NO Nitrided Oxide

NO Nitrosyl (Radical) [Symbol]

NO No Overshoot

NO No Overshoot [Magnetic Recording]

NO Normally Open (Contact) [also N/O]

NO Normal Operation

NO Norway [ISO Code]

NO Not for Off -KSC (Kennedy Space Center) Distribution [NASA]

NO Not Observed [also no]

NO Nozzle Outlet

NO Nuclear Orientation

NO$^-$ Nitroso (Radical) [Symbol]

$^+$NO Nitrosonium Ion [Symbol]

NO$_2$ Nitrogen Dioxide

NO Nitryl (or Nitronium) (Ion) [Symbol]

NO Nitrite (Ion) [Symbol]

$-$NO$_2$ Nitro Group [Symbol]

NO Nitrate Ion [Symbol]

N$_2$O Nitrous Oxide

N$_2$O$_4$ Nitrogen Tetroxide

N$_2$O$_5$ Nitrogen Pentoxide (or Nitric Anhydride)

NO$_x$ Nitrogen Oxides (i.e., Nitric Oxide, Nitrogen Dioxide, Nitrogen Peroxide, Nitrous Oxide, Nitrogen Trioxide, Nitrogen Pentoxide, or Trinitrogen Textroxide [also NOX, or Nox]

NO$_y$ Reactive Nitrogen Species (or Total Active Nitrogen) [also Noy]

N/O In the Name of

N/O Nitrogen-to-Oxygen (Ratio)

N/O Normally Open (Contact) [also NO]

N$_2$/O$_2$ Nitrogen/Oxygen (Atmosphere)

9O- Congo [Civil Aircraft Marking]

No Nobelium [Symbol]

No North(ern) [also no]

No *(Numero)* – Number [also no]

No-251 Nobelium-251 [also ^{251}No, or No^{251}]

No-252 Nobelium-252 [also ^{252}No, or No^{252}]

No-253 Nobelium-253 [also ^{253}No, or No^{253}]

No-254 Nobelium-254 [also ^{254}No, or No^{254}]

No-255 Nobelium-255 [also ^{255}No, or No^{255}]

No-256 Nobelium-256 [also ^{256}No, or No^{256}]

No-257 Nobelium-257 [also ^{257}No, or No^{257}]

No-258 Nobelium-258 [also ^{258}No, or No^{258}]

No-259 Nobelium-259 [also ^{259}No, or No^{259}]

.no Norway [Country Code/Domain Name]

nΩ nanoohm [Unit]

n(Ω) refractive index as a function of frequency [Symbol]

NOA National Oceanographic Association [US]

NOA New Obligational Authority

NOA n-Octylamine

NOA Notice of Availability

NOAA National Oceanic and Atmospheric Administration [of US Department of Commerce]

NOAA ACCP National Oceanic and Atmospheric Administration–Atlantic Climate Change Program [US]

NOAA/AFSC National Oceanic and Atmospheric Administration/Alaska Fisheries Science Center [US]

NOAA-AOML National Oceanic and Atmospheric Administration–Atlantic Oceanographic and Meteorological Laboratories [US]

NOAA-AVHRR National Oceanic and Atmospheric Administration/Advanced Very-High-Resolution Radiometer [US]

NOAA COP National Oceanic and Atmospheric Administration–Coastal Ocean Program [US]

NOAA Corps National Oceanic and Atmospheric Administration Corps [US]

NOAA/ERL National Oceanic and Atmospheric Administration/Environmental Research Laboratories [US]

NOAA FOCI National Oceanic and Atmospheric Administration Fisheries-Oceanography Coordinated Investigations (Program) [of Marine Environmental Laboratory, US]

NOAA-FSL National Oceanic and Atmospheric Administration–Forecast Systems Laboratories [US]

NOAA-GFDL National Oceanic and Atmospheric Administration–Geophysical Fluid Dynamics Laboratory [at Princeton, New Jersey, US]

NOAA-HQ National Oceanic and Atmospheric Administration Headquarters [Rockville, Maryland, US]

NOAA/NDC National Oceanic and Atmospheric Administration–National Data Center [US]

NOAA/NESDIS National Oceanic and Atmospheric Administration–National Environmental Satellite, Data and Information Service [US]

NOAA-NIC National Oceanic and Atmospheric Administration–Network Information Center [US]

NOAA-NOPC National Oceanic and Atmospheric Administration–Ocean Products Center [US]

NOAA-NSGO National Oceanic and Atmospheric Administration/National Sea Grant Office [US]

NOAA OAC National Oceanic and Atmospheric Administration–Office of Aeronautical Charting [US]

NOAA OAR National Oceanic and Atmospheric Administration–Office of Oceanic and Atmospheric Research [US]

NOAA OGP National Oceanic and Atmospheric Administration–Office of Global Programs [US]

NOAA ORA National Oceanic and Atmospheric Administration–Office of Research and Applications [US]

NOAA ORTA National Oceanic and Atmospheric Administration–Office of Research and Technology Applications [US]

NOAA-PMEL National Oceanic and Atmospheric Administration–Pacific Marine Environmental Laboratory [US]

NOAASIS National Oceanic and Atmospheric Administration Satellite Information System [US]

NOAC N^4-Octadecylcytosine β-D-Arabinofuranoside [Biochemistry]

NOAC Nordic Accelerator-Based-Research Committee [Denmark]

NOAc 2-Nitro Acetic Acid

NO2-Acac Nitro-Acetylacetonate [also NO2-acac]

NOAH National Office of Animal Health [UK]

No Amer North America(n)

NOAO National Optical Astronomy Observatories [Tucson, Arizona, US]

NOAS n-Octylaminesulfate

NOASR Netherlands Organization for Applied Scientific Research

NOB Naval Operating Base

NOBChCE National Organization of Black Chemists and Chemical Engineers [US]

NOBIN Nederlands Orgaan ter Bevordering van de Informatieverzorging [Netherlands Organization for Information Policy]

NÖBL *(Nichtöffentlicher beweglicher Landfunkdienst)* – German Private Land Mobile-Radio Service

NOBO Nonobjecting Beneficial Owner

NOBT New Orleans Board of Trade [US]

NO$_2$Bz Nitrobenzene

NOC National Occupational Classification [Canada]

NOC National On-Line Circuit [US]

NOC National Organizing Committee

NOC Naval Oceanographic Command [US]

NOC Network Operations Center [Internet]

NOC Network Operation(s) Control

NOC Non-Oxide Ceramic(s)

NOC Normally Open Contact

NOC Notation of Content

NOC Not Otherwise Classified [also noc]

NOC Not Otherwise Coded [also noc]

NOC-5 3-(Aminopropyl)-1-Hydroxy-3-Isopropyl-2-Oxo-1-Triazene

NOC-7 1-Hydroxy-3-Methyl-3-(Methylaminopropyl)-2-Oxo-1-Triazene

NOC-12 3-Ethyl-3-(Ethylaminoethyl)-1-Hydroxy-2-Oxo-1-Triazene

NOC-18 3,3-bis(Aminoethyl)-1-Hydroxy-2-Oxo-1-Triazene

nOcA n-Octanoic Acid

NOCC Network Operations Control Center

NOCC Network Operations Control Center [of NASA Goddard Space Flight Center, Greenbelt, Maryland, US]

NOCI Nederlandse Organisatie voor Chemische Informatie [Netherlands Organization for Chemical Information]

NOCO Noise Correlation

NOCP Network Operator Control Program

NOCSAE National Operating Committee on Standards for Athletic Equipment [US]

NOCT Nominal Operating Cell Temperature

NOD Naval Ordnance Department [UK]

NOD Navajo Ordnance Depot [Flagstaff, Arizona, US]

NOD Network Outward Dialing

NOD Night Observation Device

NOD Norsk Oseanografisk Datacenter [Norwegian Oceanographic Data Center]

NOD Notch Opening Displacement (Specimen) [Mechanical Testing]

NOD Notice of Deficiency

/nod no default library search option [MS-DOS Linker]

NODA N-Octyl-N-Decyl Adipate

NODA Northern Ontario Development Agreement [Canada]

NODAC Naval Ordnance Data Automation Center [of US Navy]

NODAL Network-Oriented Data Acquisition Language

NODAN Noise-Operated Device for Antinoise

NODC National Oceanographic Data Center [US]

NODC Non-Oil Developing Countries

NODE National Operational Display Equipment

NODF Nearest-Neighbor Orientation Distribution Function [Crystallography]

NODIS NSSDC (National Space Science Data Center) Online Data and Information System

NODS NASA Ocean Data System

NODUS Nondestructive Ultrasensitive Single Atomic Layer Surface Spectroscopy

NOE No Observable Effect

NOE Not Otherwise Enumerated [also noe]

NOE Nuclear Overhauser Effect [Physics]

NOE Nuclear Overhauser Enhancement (Effect) [Physics]

NOEB NATO Oil Executive Board

NOEB-E NATO Oil Executive Board–East

NOEB-W NATO Oil Executive Board–West

NOE-DIFF Nuclear Overhauser Enhancement Difference (Spectroscopy)

NOECOSS Nuclear Overhauser Enhancement Correlation (with) Shift Scaling [Physics]

NOECOSS Nuclear Overhauser Enhancement Correlation (with) Shift Slicing [Physics]

NOEL No Effect Level

NOEL Nonobservable Effect Level; No Observable Effect Level

NoEMI No Electromagnetic Interference

NOES National Occupational Exposure Survey [US]

NOES National Operational Environmental Satellite Services [US]

NOESS National Operational Environmental Satellite System

NOESY Nuclear Overhauser Enhancement (and Exchange) Spectroscopy [Physics]

NOET Nitroethane

NOEU Naval Ordnance Experimental Unit [of US Navy]

NOF National Optical Font

NOF NCR (National Cash Register) Optical Font

NOF Network Operations and Facilities

NOF (International) NOTAM (Notice to Airmen) Office

NOFA National Organic Farmers Association [US]

N of Eng North of England

NOFI National Oil Fuel Institute [US]

NOFMA National Oak Flooring Manufacturers Association [US]

NOG Nitrosoguanidine [also NsoGu] [Biochemistry]

NOG Non-Oxide Glass

/nog no group-association option [MS-DOS Linker]

NOGAD Noise-Operated Gain Adjusting Device

NOGAP Northern Oil and Gas Action Program [of Environment Canada]

NOGMA National Ornamental Glass Manufacturers Association [US]

NOGS New Orleans Geological Society [Louisiana, US]

NOHC National Open Hearth Committee [US]

NOHP Not Otherwise Herein Provided [also nohp]

NOHS National Occupational Hazard Survey [of National Institute for Occupational Safety and Health, US]

NOHSC National Occupational Health and Safety Commission [Australia]

NOI National Optics Institute [Quebec, Canada]

NOI Node Operator Interface

NOI Notice of Inquiry

NOI Notice of Intent [also NoI]

NOI Not Otherwise Indexed [also noi]

/noi no ignore/preserve lowercase option [MS-DOS Linker]

NOIA National Ocean Industries Association [US]

NOIA Newfoundland Ocean Industries Association [Canada]

NOIBN Not Otherwise Indicated by Name [also noibn]

NOIBN Not Otherwise Indicated by Number [also noibn]

No Ire Northern Ireland

NOISE National Organization to Insure a Sound-Controlled Environment [US]

Noise Control Eng J Noise Control Engineering Journal [Published by Institute of Noise Control Engineering, US]

Noise Vib Control Worldw Noise and Vibration Control Worldwide [Journal published in the UK]

NOJC National Oil Jobbers Council [now Petroleum Marketers Association of America, US]

NOK Norwegian Krone [Currency]

NOL National Ordnance Laboratory [US]

NOL Naval Ordnance Laboratory [now Naval Surface Weapons Laboratory, US]

NOL Net Operating Loss

NOL Normal Operating Loss

NOLC Naval Ordnance Laboratory at Corona [California, US]

NOLMO Nonorthogonal, Strictly Local Molecular Orbital [Physical Chemistry]

NOLS National Online Library System [of US Environmental Protection Agency]

NOLSS Northwest Ontario Lake Size Series (Database) [Canada]

NOM Network Output Multiplexer

NOM Nitromethane

.NOM Individual with Personal Site [Internet Domain Name] [also .nom]

Nom Nomenclature [also nom]

nom nominal

nΩm nanoohm meter [also nΩ-m]

N1m Peak-Amplitude Auditory-Evoked Brain Magnetic Field with Amplitude of 100 Femtotesla Observed in Magnetoencephalography of the Auditory Nerve [Symbol]

NOMA National Office Management Association [US]

NOMA National Organization of Minority Architects [US]

NOMAD Naval Oceanographic Meteorological Automatic Device

NOMAD Neutrino Oscillation MAgnetic Detector (Experiment) [of CERN–European Laboratory for Particle Physics, Geneva, Switzerland]

NOMAD Nominal Michigan Algorithmic Decoder

NOMDA National Office Machine Dealers Association [US]

Nomen Nomenclature [also nomen]

noml nominal

NOMMA National Ornamental and Miscellaneous Metals Association [US]

NOMP Niskayuna Ordnance Modification Plant [Schenectady, New York, US]

nom prop *(nomen proprium)* – by its proper name; name of drug on label [Medical Prescriptions]

NOMSA National Office Machine Service Association [US]

NOMSS National Operational Meteorological Satellite System [US]

Nom Std Nominal Standard [also nom std]

NOMTF Naval Ordnance Missile Test Facility [White Sands Proving Ground, New Mexico, US]

Nonadd Nonadditivity [also nonadd]

Noncoho Noncoherent Oscillator [also noncoho] [Radar]

Non-Com Non-Commissioned (Officer) [also Noncom, noncom, or non-com]

Non Cond Non Condensing [also non cond]

nondestr nondestructive

Nondestr Test Inst Nondestructive Testing Institute

Non-Destr Test, Aust Non-Destructive Testing, Australia [Journal]

Non-Destr Test, China Non-Destructive Testing, China [Journal]

Nondestr Test Commun Nondestructive Testing Communications [UK]

nonferr nonferrous

Non Ferr Met World Non Ferrous Metal World [Journal published in Germany]

Nonferr Met Nonferrous Metals [Journal published in PR China]

Nonlin Nonlinear(ity) [also nonlin]

Nonlinear Nonlinearity [Journal of the Institute of Physics, UK]

Nonlinear Anal Theory Methods Appl Nonlinear Analysis Theory, Methods and Applications [Journal published in the UK]

Nonlinear Sci Today Nonlinear Science Today (included in Journal of Nonlinear Science) [Journal published in Germany and the US]

Nonlinear Vib Probl Nonlinear Vibration Problems [Journal published in Poland]

NONP Non-Precision Approach Runway

N-on-P Negative-on-Positive [also n-on-p]

non rep *(non repetere)* – do not repeat [Medical Prescriptions]

non seq *(non sequitur)* – it does not follow

Non Std Non Standard [also non std]

Nonwovens Ind Nonwovens Industry [US Periodical]

Nonwovens Rep Int Nonwovens Report International [UK]

NOO Naval Oceanographic Office [of US Navy]

NOOP No Operation (Instruction) [also NO-OP, NO OP, NOP, or no op]

NOP National Oceanographic Program [US]

NOP Nebraska Ordnance Plant [Wahoo, Nebraska, US]

NOP No Operation (Instruction)

NOP Not Otherwise Provided (For) [also nop]

NOPA National Office Products Association [US]

NOPAC Network Online Public Access Catalog [Internet]

NOPC NOAA Ocean Products Center [of National Oceanic and Atmospheric Administration, US]

NOPHN National Organization for Public Health Nursing [US]

NOPI Negative Output–Positive Input

NOPL New Orleans Public Library [US]

NOP-N Nordiska Publikeringsnamnden for Naturvetenskap [Nordic Publishing Board in Science, Finland]

NOPOL No Pollution [also No Pol]

NOPON P-Bis[2,5(5-α-Naphthyloxazolyl)]benzene

NOPP National Operating Permits Program [US]

nOPPA n-Octyl Phenylphosphonate [also n-OPPA]

NOPT No Procedure Turn [also NoPT]

NOR Naval Ordnance Research [University of Minnesota, Minneapolis, US]

NOR Negative Or

NOR Nitrogen Oxide Reduction

NOR Northrup Flight Strip, New Mexico [NASA Deorbiting Site]

NOR Not and Or

NOR Not Or

NOR-1 (±)-[(E)-2-Hydroxyimino]-6-Methoxy-4-Methyl-5-Nitro-3-Hexenamide

NOR-3 (±)-(E)-4-Ethyl-2-[(E)-Hydroxyimino]-5-Nitro-3-Hexenamide

NOR-4 (±)-(E)-4-Ethyl-2-[(Z)-Hydroxyimino]-5Nitro-3-Hexen-1-ylnicotinamide

Nor North(ern) [also nor]

Nor Norway; Norwegian [also Nor]

nor normal

NORAC No Radio Contact

NORAD North American Aerospace Defense Command

NORAD North American Air Defense Command [Name Changed to North American Aerospace Defense Command]

NORAD Norwegian Agency for International Development

NORAD-COC North American Aerospace Defense–Combat Operations Center

NORAMET North American Collaboration on Metrology Standards and Services

NORAPS Naval Operations Regional Atmospheric Prediction System (Model) [of US Navy]

NORC National Opinion Research Center [of University of Chicago, Illinois, US]

NORC Naval Ordnance Research Calculator

NORC Naval Ordnance Research Computer

NORCUS Northwest College and University Association for Science [US]

NORD National Organization for Rare Disorders [US]

NORD Noise(-Modulated) Off-Resonance Decoupling

NORDA Naval Ocean Research and Development Activity [US]

NORDCO Newfoundland Oceans Research and Development Corporation [Canada]

NORDDOK Nordic Committee on Information and Documentation

NORDEK Nordic Economic Union [of Norway, Sweden, Finland and Denmark] [Defunct]

NORDEL Organization for Nordic Electrical Cooperation [Norway]

NORDFORSK Nordiska Forskningsdelegationen [Nordic Research Council] [*Note:* A Joint Scandinavian Council]

NORDICOM Nordic Documentation Center for Mass Communication Research [Denmark and Sweden]

Nordic J Build Phys Nordic Journal of Building Physics [of Royal Institute of Technology, Stockholm, Sweden]

Nordic Pulp Paper Res J Nordic Pulp and Paper Research Journal [Sweden]

NORDINFO Nordic Council for Scientific Information and Research Libraries [Finland]

NORDITA Nordic Institute for Theoretical Atomic Physics [Copenhagen, Denmark]

NORDO No Radio

NORDPLAN Nordic Institute for Studies in Urban and Regional Planning [Sweden]

NORDPOST Nordic Postal Union Conference [of Nordic Postal Union, Denmark]

NORDSAT Nordic (Scandinavian) Broadcast Satellite

Norelco Rep Norelco Report [Published by Norelco, Eindhoven, Netherlands]

NORGEN Network Operations Report Generator

Nor Geol Tidsskr Norsk Geologisk Tidsskrift [Norwegian Geological Journal]

NORIANE Normes et Règlements–Informations Automatisées Accessibles en Ligne [Standards and Regulations–On-Line Automated Information; of Association Française de Normalisation, France]

NORINDOK Norsk Komite for Informasjon og Dokumentasjon [Norwegian Committee for Information and Dokumentation]

NORM Naturally Occurring Radioactive Material

NORM Northwest Regional Meeting [of American Chemical Society, US]

NORM Not Operationally Ready Maintenance

Norm Normal(ization) [also norm]

NORMAC Northern Prawn Fishery Management Committee [Australia]

NORMARC Norwegian MARC (Machine-Readable Catalogue)

Nor Mat Tidsskr Nordisk Matematisk Tidsskrift [Nordic Mathematical Journal of the Norwegian Mathematical Association]

NORP New Oil Reference Pricing

NORPAT Northern Patrol [of Canadian Armed Forces]

NORPAX North Pacific Experiment

NORPHOS trans-Bicyclo[2.2.1]hept-5-en-2,3-Diylbis (diphenylphosphine)

Nor Polarinst Skr Norsk Polarinstitutt Skrifter [Publication of the Norwegian Polar Institute]

Nor Polarinst Temakart Norsk Polarinstitutt Temakart [Publication of the Norwegian Polar Institute]

NORS National Organic Reconnaissance Survey [US]

NORS Not Operationally Ready Supply

NORSAR Norwegian Seismic Array

NORSAT Norwegian Domestic Satellite

NORTHAG Northern Army Group Central Europe [of NATO]

Northamts Northamptonshire [UK]

North East Coast Inst Eng Shipbuild, Trans North East Coast Institution of Engineers and Shipbuilders, Transactions [UK]

Northern J Appl Forestry Northern Journal of Applied Forestry [Published by the Society of American Foresters]

NorthNet North Star Computers Network [US]

Northumb Northumberland [UK]

NORVEN Comision Venezolana de Normas Industriales [Venezuelan Commision for Industrial Standards]

Norw Norway; Norwegian

NorwP Norwegian Patent

Norw Shipp News Norwegian Shipping News [Published in Oslo, Norway]

NOS National Ocean Service [of National Oceanic and Atmospheric Administration, US]

NOS National Ocean Survey [of National Oceanic and Atmospheric Administration, US]

NOS Naval Ordnance Station [of US Navy]

NOS Network Operating System

NOS Night Observation System

NOS Nitric Oxide Synthetase [Biochemistry]

NOS Not Otherwise Specified [also nos]

Nos Nosing [also nos]

Nos Numbers [also nos]

NOSA National Occupational Safety Association [South Africa]

NOSA National One-Write Systems Association [US]

NOsartacn 9-Nitro-1,4,7,11,14,19-Hexaazatricyclo [7.7.4.2^{4,14}]docosane

NOS/BE Network Operating System/Batch Environment [also NOS-BE]

NOSC Naval Ocean Systems Center [of US Department of Defense in San Diego, California]

NOSCA New Optical Sensor Concept for Aeronautics

NOSFER Nouveau Systeme Fondamental pour la Determination des Equivalents de Reference [New Master System for the Determination of Reference Equivalents]

NOSIE NODC (National Oceanographic Data Center) Ocean Science Information Exchange [US]

NOSIG No Significant Change

N$_2$O/SiH$_4$ Nitrous Oxide/Silane (Gas Mixture)

NOSL Night/Day Optical Survey of (Thunderstorm) Lightning

NOSMO Norden Optics Setting, Mechanized Operation

NOSP Network Operations Support Plan

NOSP Network Operating Support Program

NOSS National Oceanic Satellite System

NOSS National Oceanic Survey Satellite

NOSS National Orbiting Space Station

NOSS Nimbus Operational Satellite System

NOST NASA/(Science) Office of Standards and Technology

Not Nocardia otidis-caviarum [Microbiology]

NOTAEI National Old Timers Association of the Energy Industry [US]

Notab Med Notabene Medici [Italian Journal on Medicine]

NOTAL Not Sent to All Addresses

NOTAM Notice to Airmen

NOTAR No Tail Rotor (Anti-Torque) System [Aeronautics]

NOTCC New Orleans Tourist and Convention Commission [US]

Note Recens Not Note Recensioni e Notizie [Italian Publication on Post and Telecommunications; published by Istituto Superiore Poste e Telecomunicazioni]

Note Tech-Off Nat Etudes Rech Aérosp Note Technique–Office National d'Etudes et de Recherches Aérospatiales [Technical Notes (on Aerospace Research) of Office National d'Etudes et de Recherches Aérospatiales, France]

NOTIS Network Operations Trouble Information System

NOTIS Northwestern On-Line Totally Integrated System [of Northwestern University, Evanston, Illinois, US]

Not Ist Autom/ Univ Roma Notiziario dell'Istituto di Automatica dell'Universita di Roma [Notes of the Institute of Automation of the University of Rome, Italy]

Notre Dame J Form Log Notre Dame Journal of Formal Logic [Published by the University of Notre Dame, US]

Notre Dame Sci Quart Notre Dame Science Quarterly [Published by the University of Notre Dame, US]

NOTRTR National Organization of Test, Research and Training Reactors [US]

NOTS Naval Ordnance Test Station [of US Navy at China Lake, California, US]

Notts Nottinghamshire [UK]

NOTU Naval Operational Training Unit [of US Navy]

notwg notwithstanding

NOV Notice of Violation

Nov November

NOVAC Northern Virginia Astronomy Club [US]

NOVAM Naval Oceanic Vertical Aerosol Model

NOVEL Nuclear (Spin) Orientation Via Electron (Spin) Locking [Physics]

Novosti Tselul-Khart Prom Novosti v Teslulozno-Khartienata Promishlennost [Bulgarian Journal]

NOVRAM Nonvolatile Random-Access Memory

NOVS National Office of Vital Statistics [US]

NOW Negotiable Order of Withdrawal [US]

NOW (Europhysics) Neutrino Oscillation Workshop

NO$_2$X Nitryl Halide (X Represents a Halogen, e.g., Bromine, Chlorine, etc.)

NOx Nitrogen Oxides (i.e., Nitric Oxide, Nitrogen Dioxide, Nitrogen Peroxide, Nitrous Oxide, Nitrogen Trioxide, Nitrogen Pentoxide, or Trinitrogen Textroxide) [also NOX, or No$_x$]

NOy Reactive Nitrogen Species (or Total Active Nitrogen) [also No$_y$]

NOYB None Of Your Business [Internet Jargon]

NOZ No Operating Zone

Noz Nozzle [also noz]

NP Nameplate [also N/P]

NP Nanyang Polytechnic [Singapore]

NP Nanoparticle(s) [Materials Science]

NP Nanophase [Materials Science]

NP Nanopowder [Materials Science]

NP Napalm

NP Napo-Pastaza [Ecuador]

NP National Park

NP National Preserve [US]

NP (American) National Pipe Thread

NP Natural Polymer

NP Natural Product

NP Naval Publication

NP N-Doped P-Doped (Junction) [also N-P, np, or n-p]

NP Néel Point [Solid-State Physics]

NP Negative Pole

NP Negative-Positive (Semiconductor) [also N-P, np, or n-p]

NP Neopentyl

NP Nepal [ISO Code]

NP Nernst-Planck (Equation) [Physics]

NP Net Proceeds

NP Network Polymer

NP Network Printer

NP Network Program

NP Network Protocol

NP Neumann's Principle [Crystallography]

NP Neuropeptide [Biochemistry]

NP Neuropsychiatric; Neuropsychiatry

NP Neutral Plane [Electrical Engineering]

NP Neutral Point

NP Neutrino Patents [Particle Physics]

NP Neutron Polarization

NP Newman Projections [Physics]

NP New Paragraph

NP New Pattern

NP New Providence [UK]

NP Newtonian Polynomial

NP Nickel-Plated; Nickel Plating

NP Nickel Powder

NP Nicol Prism [Optics]

NP Nitroparaffin

NP Nitropentaerythritol

NP Nitropyrene

NP Nitroso Polymer

NP Nobel Prize

NP Nodal Plane

NP Nodal Point

NP Noise Power

NP *(Nomen Proprium)* – By Its Proper Name; Name of Drug on Label [Medical Prescriptions]

NP Nondeterministic Polynomial

NP Non-Penetrating

NP Non-Persistent

NP Nonphysical

NP Nonpolar

NP Non-Priority

NP Nonpurgeable

NP Nonylphenoxy-

NP No Pagination

NP No Parity

NP No Phonon (Peak) [Solid-State Physics]

NP No Pin

NP No Place (of Publication)

NP No Print

NP Normal Phase

NP Norsk Polarinstitutt [Norwegian Polar Institute, Oslo]

NP Northern Pine

NP North Pacific

NP North Pole

NP Notary Public

NP Not Provably (Solvable Problem)

NP Not Provided [also np]

NP Nozzle Performance

NP Nuclear Paramagnetism

NP Nuclear Physicist; Nuclear Physics

NP Nuclear Pile

NP Nuclear Polarization

NP Nuclear Power

NP Nuclear Process

NP Nuclear Property

NP Nuclear Propulsion

NP Nucleoprotein [Biochemistry]

NP Nucleoside Phosphorylase [Biochemistry]

NP Nuevo Peso [Currency of Uruguay]

NP Nuissance Particulates

NP Sodium Phosphate [$NaPO_3$]

2NP 2-Nitropropane

N$ New Uruguayan Peso [Currency]

N/P Nameplate [also NP]

N/P Nitrogen-to-Phosphorus (Ratio)

N/P No-Flash containing Potassium Sulfate [British Propellant]

N/P Not Provided [also n/p]

Np Current (Gas) Nitriding Potential [Symbol]

Np Neper [Unit]

Np Neptunium [Symbol]

Np Number of Primary Turns (of Transformers) [Symbol]

Np-231 Neptunium-231 [also ^{231}Np, or Np^{231}]

Np-232 Neptunium-232 [also ^{232}Np, or Np^{232}]

Np-233 Neptunium-233 [also ^{233}Np, or Np^{233}]

Np-234 Neptunium-234 [also ^{234}Np, or Np^{234}]

Np-235 Neptunium-235 [also ^{235}Np, or Np^{235}]

Np-236 Neptunium-236 [also ^{236}Np, or Np^{236}]

Np-237 Neptunium-237 [also ^{237}Np, Np^{237}, or Np]

Np-238 Neptunium-238 [also ^{238}Np, or Np^{238}]

Np-239 Neptunium-239 [also ^{239}Np, or Np^{239}]

Np-240 Neptunium-240 [also ^{240}Np, or Np^{240}]

Np-241 Neptunium-241 [also ^{241}Np, or Np^{241}]

1Np 1-Napthalenol

2Np 2-Napthalenol

nP noncoherent perfect (crystal) lattice [Symbol]

np nonprimitive (unit cell) [Crystallography]

.np Nepal [Country Code/Domain Name]

n-p n-plus-type layer/p-type layer (Semiconductor Junction)

n$^+$-p n-plus-type p-type (semiconductor device) [also n$^+$p]

n/p neutron/proton (ratio) [Symbol]

n.p. *(nomen proprium)* – drug name (written on the container label)

n(p) momentum distribution [Symbol]

∇P Pressure Gradient [Symbol]

NPA N-(1-Naphthyl)phthalaminic Acid

NPA β-(1-Naphthoyl)propionic Acid; β-(2-Naphthoyl)propionic Acid

NPA Narcissus Pseudonarcissus Agglutinin [Immunology]

NPA National Parks Association

NPA National Particleboard Association [US]
NPA National Payphone Association [US]
NPA National Petroleum Association [US]
NPA National Pharmaceutical Association [UK]
NPA National Pilots Association [US]
NPA National Planning Association [US]
NPA National Production Authority [US]
NPA National Productivity Award [Canada]
NPA Newspaper Publishers Association [UK]
NPA Nebraska Pharmaceutical Association [US]
NPA Neopentyl Alcohol
NPA Network Printer Alliance
NPA Normal Pressure Angle (of Gears)
NPA Notice of Proposed Amendment
NPA Numbering Plan Area
NPA Numerical Production Analysis
nPA n-Propylamine [also n-PA, or nPa]
nPa nanopascal [Unit]
NPABC National Public Affairs Broadcasting Center [Washington, DC, US]
NPAC National Petroleum Advisory Committee [Australia]
NPAC National Pipeline Agency of Canada
NPAC National Plantations Advisory Committee
NPAC National Project in Agricultural Communications [US]
NPAC Northeast Parallel Architectures Center [of Syracuse University, New York State, US]
N Pac North Pacific
NPAI Network Protocol Address Information
1NpAm 1-Naphthalenamine
2NpAm 2-Naphthalenamine
NPANSW National Parks Association of New South Wales [Australia]
NPAQ National Parks Association of Queensland [Australia]
NPAR Nuclear Plant Aging Research
NPB National Plant Board [US]
NPBA National Paper Box Association [now National Paperbox and Packaging Association, US]
n-PbSe Nanophase Lead Fluoride
NPBH N-Phenyl-N-Benzoyl Hydroxylamine
n-PbSe N-Type Lead Selenide
n-PbTe N-Type Lead Telluride
NPC Nanophase Ceramics
NPC NASA Publication Control
NPC National Panhellenic Conference
NPC National Patent Council [US]
NPC National Peach Council [US]
NPC National Peanut Council [US]
NPC National Petrochemical Corporation [Iran and Thailand]
NPC National Petroleum Council [US]
NPC National Pharmaceutical Council [US]
NPC National Philatelic Center [Canada]
NPC National Plumbing Code [US]
NPC National Population Council

NPC National Ports Council [UK]
NPC National Postal Museum
NPC National Potato Council [US]
NPC National Power Conference
NPC National Preservation Conference [US]
NPC NATO Pipeline Committee
NPC Natural Products Chemistry
NPC Naval Photographic Center [of US Navy]
NPC Near-Field Photoconductivity
NPC Niagara Parks Commission [Canada]
NPC Niobium Products Company [Duesseldorf, Germany]
NPC Nitrogen Purge Control
NPC Normal-Phase Chromatograph(y)
NPC Normal-Pressure Chromatograph(y)
NPC North Pacific Current [Oceanography]
NPCA National Paint and Coatings Association [US]
NPCA National Parks and Conservation Agency [US]
NPCA National Parks and Conservation Association [US]
NPCA National Pest Control Association [US]
NPCA National Precast Concrete Association [US]
NPCC Northeast Power Coordinating Council [US]
NPCF National Pollution Control Federation [US]
NPCS (International Seminar on) Nonlinear Phenomena in Complex Systems
NPCT Nuclear Process Components Test
NPCTF Nuclear Process Components Test(ing) Facility [of Ontario Hydro Research Division, Toronto, Canada]
NPD NASA Policy Directive
NPD National Paint Distributors [US]
NPD National Power Demonstration
NPD Network Protective Device
NPD Neutron Powder Diffraction
NPD Neutron Powder Diffractometer
NPD Nitrogen-Phosphorus Detector
NPD Nominal Percent Defective
NPD Normal Photoelectron Diffraction
NPD North Polar Distance [Astronomy]
NPD Nuclear Physics Division [of Engineering and Physical Sciences Research Council, UK]
NPD Nuclear Power Demonstration (Reactor) [Rolphton, Ontario, Canada]
NPD Nuclear Power Development
n-Pd Nanocrystalline Palladium
NPDA Nitrophenylenediamine
NPDC North Portage Development Corporation [Manitoba]
NPDES National Pollutant (or Pollution) Discharge Elimination System (Permit) [of US Environmental Protection Agency]
NPDES Nuclear Pollution Discharge Elimination Specification
NPDES Nuclear Pollution Discharge Elimination System
NPDL Nitrogen-Pumped Dye Laser
NPDP National Professional Development Program
NPDPP p-Nitrophenyl Diphenyl Phosphate
NPDS Nuclear Particle Detection Subsystem

NPDS Nuclear Particle Detection System

NPDWR National Primary Drinking Water Regulations [US]

NPE National Plastics Exposition and Conference [US]

NPE Natural Parity Exchange

NPE Neopentyl Ethane

NPE No-Phonon Emission (Diode)

NPE Nuclear Power Engineering

NPE Nuclear Power Engineering Committee [of Institute of Electrical and Electronics Engineers–Power Engineering Society, US]

NPEA National Printing Equipment Association [US]

NPEC Nuclear Power Engineering Committee [of Institute of Electrical and Electronic Engineers–Power Engineering Society, US]

NPEE Nuclear Photoelectric Effect

NPEF New Product Evaluation Form

NPEGE Nonyl Phenyl Eicosa-Ethylene Glycol Ether

NPE/PES Nuclear Power Engineering/Power Engineering Society [Institute of Electrical and Electronics Engineers Committee, US]

NPES National Printing Equipment Show [US]

NPESA National Printing Equipment and Supply Association [US]

NPF National Park Foundation [US]

NPF National Parkinson Foundation [US]

NPF Naval Powder Factory [now Naval Propellant Plant]

NPF Non-Planar Family [Crystallography]

NPF Normal Power Factor

NPF Nuclear Photofission

NPF Nuclear Power Facility

NPF Nuclear Problems Forum

NPFauna National Park Fauna (Database) [of US Geological Survey/Biological Resources Division]

NPFlora National Park Flora (Database) [of US Geological Survey/Biological Resources Division]

NPFO National Power Field Office [US Army]

NPFSC North Pacific Fur Seal Commission

NPFVAS Northern Prawn Fishery Voluntary Assistance Scheme [Australia]

NPG NASA Procedures and Guidelines

NPG Naval Proving Ground [Dahlgren, Virginia, US]

NPG Neopentyl Glycol

NPG Nuclear Planning Group [of NATO]

NPG Nuclear Power Group [UK]

NPGA National Propane Gas Association [US]

NPGPA Non-Powder Gun Products Association [US]

NPGS Nanopattern Generator Systems (Software)

NPGS Naval Post Graduate School [Monterey, California, US]

NPGS Nuclear Power Generating Station

NPGTC National Prairie Grouse Technical Council [US]

NPH Nitrophenylhydrazine

NPH Normal Paraffin Hydrocarbon(s)

NPHA Nitrosophenyl Hydroxylamine

n-phase nanocrystalline phase [Materials Science]

NPHB Nonphotochemical Hole Burning

NPI National Petroleum Institute

NPI National Pollution (or Pollutant) Inventory [of Commonwealth Environment Protection Agency, Australia]

NPI National Purchasing Institute [US]

NPI Net Profits Interest

NPI Network Printer Interface

NPI Nordic Productivity Institute [Norway]

NPI Norwegian Polar Institute [Oslo]

NPI Northampton Polytechnic Institute [UK]

NPI Nuclear Propulsion Initiative

NPIA National Photography Instructors Association [US]

NPIAS National Plan of Integrated Airport Systems [US]

NPIAW National Photographic Index of Australian Wildlife [of Australian Museum]

NPIN N-Doped P-Doped Intrinsic N-Doped (Transistor) [also N-P-I-N, npin, or n-p-i-n]

NPIN Negative-Positive-Intrinsic (Transistor) [also N-P-I-N, npin, or n-p-i-n]

NPIP National Poultry Improvement Plan [US]

NPIP N-Doped P-Doped Intrinsic P-Doped (Transistor) [also N-P-I-P, npip, or n-p-i-p]

NPIP Negative-Positive-Intrinsic (Transistor) [also N-P-I-P, npip, or n-p-i-p]

NPip N-Nitrosopiperidine [also NPIP]

NPIRI National Printing Ink Research Institute [US]

NPIS National Physics Information System [of American Institute of Physics, US]

NPIU Network Processing and Interface Unit

NPK Neuropeptide K [Biochemistry]

NPK Nitrogen, Phosphorous and Potassium (Fertilizer)

N:P:K Nitrogen-Phosphorous-Potassium Ratio (of Fertilizer)

N:P:K:Ca:Mg Nitrogen-Phosphorous-Potassium-Calcium-Magnesium Ratio (of Fertilizers)

NPkwy National Parkway [US]

NPL National Physical Laboratory [New Delhi, India]

NPL National Physical Laboratory [Teddington, Middlesex, UK]

NPL National Physics Laboratory [of University of Illinois at Urbana-Champaign, US]

NPL National Physics Laboratory [Guildford, Surrey, UK]

NPL National Priorities List [US]

NPL Natural Process Limits

NPL New Programming Language

NPL Nonprocedural Language

NPL Normal Power Level

NPL Nuclear Physics Laboratory [of Queen's University, Kingston, Ontario, Canada]

NPLC National Pedigree Livestock Council [US]

NPLC Nebraska Public Library Commission [US]

NPLC Non-Periodic Layer Crystal(s)

NPLC Normal Phase Liquid Chromatography

NPL Conf National Physical Laboratory Conferences [Published by Her Majesty's Stationery Office, UK]

NPL/CSIR National Physical Laboratory/Council for Scientific and Industrial Research [New Delhi, India]

NPLE Negligible Partitioning Local Equilibrium (Mode of Growth) [Metallurgy]

NPLE No Partitioning, Local Equilibrium (Mechanism) [Metallurgy]

NPLG Navy Program Language Group [of US Navy]

NPLO NATO Production and Logistics Organization

NP LTM National Park Long-Term Monitoring [US]

NPM Nanophase Material

NPM National Postal Museum [Canada]

NPM Neodymium Lead Manganate

NPM Non-Polar Molecule

NPM Non-Priority Mail

NPM Nuclear Paramagnetism

NPMA National Property Management Association [US]

NPME Nuclear Photomagnetic Effect

NPMI Nordic Pool for Marine Insurance [Sweden]

NPMK National Plan of Management of Kangaroos [Australia]

NPMP National Pesticide Monitoring Program [US]

NPMR National Premium Manufacturers Representatives [now Incentive Manufacturers Representatives Association, US]

NPMR Native Pasture Management Research [Australia]

NPMS Neutron Program for Materials Science [of Chalk River Laboratories, Atomic Energy of Canada Limited at Chalk River, Ontario, Canada]

NPN NASA Part Number

NPN N-Doped P-Doped N-Doped (Transistor) [also N-P-N, npn, or n-p-n]

NPN Negative-Positive-Negative (Transistor) [also N-P-N, npn, or n-p-n]

NPN Nonprotein Nitrogen [Biochemistry]

NPN N-Propyl Nitrate

NPNE Non-Polluting Noiseless Engine

n-p-n AlGaAs N-Doped P-Doped N-Doped Aluminum Gallium Arsenide

n-p-n AlGaAs/GaAs HBT N-Doped P-Doped N-Doped Aluminum Gallium Arsenide/Gallium Arsenide Heterojunction Bipolar Transistor

NPNP Negative-Positive-Negative-Positive (Transistor) [also N-P-N-P, n-p-n-p, or npnp]

n-p-n Si-HET N-Doped P-Doped N-Doped Silicon Heterojunction Bipolar Transistor

NPO 1-Naphthyl-5-Phenyloxazole

NPO Negative Positive 0 (Zero)

NPO *(nil per oram)* – nothing by the mouth [Medicine]

NPO Nonpenetrating Orbit

NPO Non-Profit Organization

NPO North Pacific Ocean

NPO Nuclear Power Operator

a-**NPO** 2-(1-Naphthalenyl)-5-Phenyl Oxazole; 2-(1-Naphthyl)-5-Phenyloxazole

NpO$_2$ Neptunyl (Radical) [Symbol]

NPOC Nonpurgeable Organic Carbon

NPOP NASA Polar Orbiting Platform

np, or d no place, or date

NPP National Pesticides Policy

NPP National Pretreatment Program

NPP Naval Propellant Plant [Indian Head, Maryland, US] [formerly Naval Powder Factory]

NPP Neopentyl Pentane

NPP Net Primary Production

NPP Network Protocol Processor

NPP New Policy Proposal

NPP Nitrophenyl Phosphate

NPP N-(4-Nitrophenyl-(L)-Prolinol

NPP Normal-Pulse Polarography

NPP Norm-Conserving Pseudopotential [Physics]

NPP Notice(s) of Proposed Procurement [of Open Bidding Services, Canada]

NPP Nuclear Power Plant

NPPA National Press Photographers Association [US]

NPPA Northwest Pulp and Paper Association [US]

NP&PA National Paperbox and Packaging Association [US]

NPPAC National and Provincial Parks Association of Canada

NPPB National Potato Promotion Board [US]

NPPC National Pork Producers Council [US]

NPPC Negative Persistent Photoconductivity

NPPC Northwest Power Planning Council [Portland, Oregon, US]

NPPCI Nuclear Power Plant Control Instrumentation (Working Group)

NPPD Nebraska Public Power District [US]

NPPD 5-(4-Nitrophenyl)-2,4-Pentadienal

NPPP NASA Parts and Packaging Program

NPPS National Plants Preservation Society

NPPSO Naval Publications and Printing Service Office [of US Navy]

NPR National Performance Rewiew [US]

NPR National Police Reserve [Japan]

NPR National Public Radio

NPR Naval Petroleum Reserves [US]

NPR Near-Pristine Rivers

NPR Nepalese Rupee [Currency of Nepal]

NPR Net Profit Royalty

NPR (Office of) New Production Reactors [of US Department of Energy at Hanford, Washington State]

NPR Negative Poisson Ratio

NPR Noise Pollution Ratio

NPR Noise/Power Ratio

NPR Noise Preferential Route

NPR Nozzle Pressure Ratio

NPR Nuclear Power Reactor

NPR Numeric Position Readout

NPRA National Personal Robot Association [US]

NPRA National Petroleum Refiners Association [US]

NPRA Newspaper Personnel Relations Association [US]

NPRC Nuclear Power Range Channel

NPRCA National Petroleum Radio Coordinating Association [US]

NPRCG Nuclear Public Relations Contact Group [Italy]

NPRD Nonelectronic Parts Reliability Data

NPRDS Nuclear Plant Reliability Data System

NPRF Northrop Pulse Radiation Facility [US]

NPRGCC National Program for Responding to Greenhouse Climate Change

NPRI National Pollution Release Inventory [Canada]

NPRL National Physical Research Laboratory [of Council for Scientific and Industrial Research, South Africa]

NPRL Nonprocedural Referencing Language

NPRM Notice of Proposed Rule Making

N-proCT N-proCalcitonin [Biochemistry]

N-Propsal N-Propylsalicylidenamine

NPRP Northern Pecan Research Program [Canada and US]

NPRPM National Program for River Process and Management

NPRTZ Nonpolarized Return-to-Zero Recording [also NPRZ]

NPRV Nitrogen Pressure Relief Valve

NPS NASA Planning Studies

NPS National Park Service [of US Department of the Interior]

NPS National Parks Service [Canada]

NPS National Park System [US]

NPS National Petroleum Show

NPS Naval Postgraduate School [of US Navy at Monterey, California]

NPS Negative Pressure Silder [Disk Drives]

NPS Netherlands Physical Society

NPS Network Processing Supervisor

NPS Network Processor System

NPS Nevada Petroleum Society [US]

NPS Nitrided Pressureless Sintering

NPS o-Nitrophenylsulfenyl [also Nps]

NPS Nominal Pipe Size

NPS Nonpoint Source (Pollution Control)

NPS Northern Petro-Search

NPS Novell Productivity Specialist

NPS Nuclear and Plasma Sciences (Society) [of Institute of Electrical and Electronics Engineers, US]

NPS Nuclear Power Station

NPS Numerical Plotting System

Nps o-Nitrophenylsulfenyl

NPSA New Program Status Area

NPSC American National Standard Straight Pipe Thread in Pipe Couplings [Symbol]

NPSC National Postage Stamp Collection [of Smithsonian Institution, Washington, DC, US]

NPS-Cl 2-Nitrophenylsulfenyl Chloride

NPSD Neutron Power Spectral Density

NPSF Dryseal American National Standard Fuel Internal Straight Pipe Thread [Symbol]

NPSH American National Standard Straight Pipe Thread for Loose-Fitting Mechanical Joints for Hose Couplings [Symbol]

NPSH Net Positive Suction Head [Mechanical Engineering]

NPSH Niagara Parks School of Horticulture [Canada]

NPSHA Net Positive Suction Head Available [Mechanical Engineering]

NPSHR Net Positive Suction Head Required [Mechanical Engineering]

NPSI Dryseal American National Standard Intermediate Internal Straight Pipe Thread [Symbol]

NPSI NCP (Network Control Program) Packet Switching Interface

NPSL American National Standard Straight Pipe Thread for Loose-Fitting Mechanical Joints with Locknuts [Symbol]

NPSM American National Standard Straight Pipe Thread for Free-Fitting Mechanical Joints [Symbol]

NPSP Net Positive Static Pressure [Mechanical Engineering]

NPSP Niagara Project Simulation Program [US]

NPSP N-(Phenylseleno)phthalimide

NPSRA Nuclear-Powered Ship Research Association [Japan]

NPSS Nuclear and Plasma Sciences Society [of Institute of Electrical and Electronics Engineers, US]

NPSU Nuclear Plant Safety Unit

NPSWL New Program Status Word Location

NPT American National Standard Taper Pipe Thread [Symbol]

NPT National Pipe Thread

NPT Network Planning Technique

NPT Nonproliferation Treaty

NPT Normal Pressure and Temperature

NPT Nosé Constant-Pressure [*Note:* "P" stands for Pressure and "T" for Temperature] [Materials Science]

NPT Nuclear Particle Track

NPT Nuclear Non-Proliferation Treaty

NPTA National Paper Trade Association [US]

n-Pt Nanocrystalline Platinum

1NptA 1-Naphthoic Acid

2NptA 2-Naphthoic Acid

1Npta 1-Naphthalenecarboxylic Acid

2Npta 2-Naphthalenecarboxylic Acid

NPTC National Private Truck Council [US]

NPTEC National Power Technology and Environmental Center [UK]

NPTF National Pipe Thread, Female

NPTF Dryseal American National Standard Taper Pipe Thread

NPTF Nuclear Proof Test Facility

NPTM National Pipe Thread, Male

NPT-MD Normal Pressure and Temperature (or Constant Pressure) Molecular Dynamics [also N-T-P-MD]

NPTN National Pesticide Telecommunications Network [of US Environmental Protection Agency]

NPTN National Public Telecomputing Network [US]

NPTO National Petroleum Technology Office [of US Department of Energy]

NPTR American National Standard Taper Thread for Railing Joints [Symbol]

NPTS Noise Parameter Test System

NPU National Pharmaceutical Union [UK]

NPU Natural Processing Unit

NPU Naval Parachute Unit [of US Navy]

NPU Network Planning Unit

NPU Network Processing Unit

NPU Nitrogen Purge Unit

NPU Nordic Postal Union [Denmark]

NPu Normenausschuss Pulvermetallurgie [Standards Committee for Powder Metallurgy; of DIN Deutsches Institut für Normung, Germany]

NPUG National Prime (Computer System) User Group [US]

NPV Net Present Value

NPV Nitrogen Pressure Valve

NPV No Par Value

NPV Normal Pulse Voltammetry

NPV Nuclear Polyhedrosis Virus

NPVA Net Present Value Analysis

NPVE Neopentyl Vinyl Ether

NPVLA National Paint, Varnish and Lacquer Association [now National Paint and Coatings Association, US]

NPW National Parks and Wildlife (Act) [Australia]

NPW Act National Parks and Wildlife Act [Australia]

NPWC National Parks and Wildlife Conservation (Act) [Australia]

NPWC Act National Parks and Wildlife Conservation Act [Australia]

NPWRC Northern Prairie Wildlife Research Center [of US Geological Survey/Biological Resources Division at Jamestown, North Dakota]

NPWS National Parks and Wildlife Service [New South Wales, Australia]

NPX Numeric Processor Extension

NPY Neuropeptide Y [Biochemistry]

NPY-KLH Neuropeptide Y–Keyhole Limpet Hemocyanin (Immunogen)

NPYR N-Nitrosopyrrolidine [also NPYRR, Npyr, or Npyrr]

NQ 1,4-Naphthoquinone

NQ Nitroguanidine Flashless (Propellant)

NQ Nitroquinoline

NQ North Queensland [Australia]

NQ Numerical Quadrature

12NQ 1,2-Naphthoquinone

14NQ 1,4-Naphthoquinone

9Q- Congo [Civil Aircraft Marking]

NQA National Quality Award [US]

NQAA Nuclear Quality Assurance Agency [US]

NQCC North Queensland Conservation Council [Australia]

NQCC Nuclear Quadrupole Coupling Constant [Physics]

NQCRRP North Queensland Community Rainforest Reforestation Program [Australia]

NQd Non-Quaded (Cable)

NQDR Nuclear Quadrupole Double Resonance [Physics]

NQHFS Nuclear Quadrupole Hyperfine Structure [Physics]

NQI Nuclear Quadrupole Interaction [Physics]

NQIC National Quality Information Center [UK]

NQM Nonrelativistic Quantum Mechanics

NQM Nuclear Quadrupole Moment [Physics]

NQNC North Queensland Naturalists Club [Australia]

NQO Nitroquinoline Oxide

NQPC National Quartz Producers Council [US]

NQR Nuclear Quadrupole Resonance [Physics]

NQRF Nuclear Quadrupole Resonance Frequency [Physics]

NQRS Nuclear Quadrupole Resonance Spectroscopy [Physics]

NQS Network Queing System

12NQ4S 1,2-Naphthoquinone-4-Sulfonate

NR American National Thread with a 0.108p to 0.144p Controlled Root Radius [Symbol]

NR National River [US]

NR Natural Resource

NR Natural Rubber

NR Nature Reserve

NR Nauru [ISO Code]

NR Naval Reactor

NR Naval Reactor Program [of US Navy]

NR Naval Reserve [of US Navy]

NR Navigational Radar

NR Navigation Room

NR Navy Regulations

NR Needle Roller Bearing Inner Ring, Lubrication Groove and Hole in Bore, Metric-Dimensioned [Symbol]

NR Negative Resistance

NR Nepalese Rupee [Currency of Nepal]

NR Neural Regeneration and Recovery (Network) [Canada]

NR Neutral Red [Organic Chemistry]

NR Neutron Radiation

NR Neutron Radiograph(y)

NR Neutron Reflectivity

NR Neutron Reflectometry

NR Newton-Raphson (Formula) [Mathematics]

NR Newton's Rings [Optics]

NR Nickel Rhenate

NR Nitrile Resin

NR Nitrile Rubber

NR Noise Radiator

NR Noise Rating

NR Noise Ratio

NR Noise Reduction

NR Nonradiative

NR Nonradioactive

NR Nonreactive

NR Nonrecoverable [Computers]

NR Non-Reflecting

NR Non-Relativistic

NR Non-Resident [Computers]

NR Non-Return (Valve)

NR Non-Rotating

NR Nordheim's Rule [Solid-State Physics]

NR No Refill [Medical Prescriptions]

NR Nose Radius [Cutting Tools]

NR Not Recommended [also nr]

NR Not Reported [also nr]

NR Not Required [also nr]

NR Nuclear Reaction; Nuclear Reactor

NR Nuclear Resonance

NR Nuclear Rocket

NR Nucleophilic Reagent [Physical Chemistry]

NR Number [also Nr, or nr]

NR Number of Report

NR Nyquist Rate [Communication]

NR$_o$ Output Noise Ratio [Semiconductor Symbol]

N(R) Receive-Sequence Number

N(r) Atmospheric Molecular Number Density [Symbol]

N(r) Number of Cloud Particles per Unit Volume [Symbol]

N$_i$(r) Average Number of Atoms of Type *i* per Unit Volume [Symbol]

nr near

nr not reported [also NR]

nr not required [also NR]

nr number [also NR, or Nr]

n(r) particle size distribution function [Symbol]

.nr Nauru [Country Code/Domain Name]

NRA NASA Research Announcement

NRA National Railway Association [UK]

NRA National Recovery Administration [US] [Obsolete]

NRA National Rehabilitation Association [US]

NRA National Reclamation Association [now National Water Resources Association, US]

NRA National Recreational Area [US]

NRA National Registration Authority (for Agricultural and Veterinary Chemicals)

NRA National Renderer Association [US]

NRA National Rifle Association [UK and US]

NRA National Rivers Authority [UK]

NRA Natural Resource Accounting

NRA Naval Radio Activity

NRA Near Real-Time Advisory [also nra]

NRA Network Resolution Area

NRA Non-Relativistic Approximation [Physics]

NRA Nuclear Reaction Analysis

NRA Nuclear Resonance Absorption

NRAB National Railroad Adjustment Board [US]

NRAC Natural Resources Audit Council [New South Wales, Australia]

NRAD Neutron Radiography (Reactor)

NRaD NCCOSC (Naval Command, Control and Ocean Surveillance Center) RDT&E (Research, Development, Test(ing) and Evaluation) (Division) [US]

NRAL Nuffield Radio Astronomy Laboratories [of University of Manchester at Cheshire, UK]

NRAO National Radio Astronomy Observatory [West Virginia, US]

NRAS Neutron Resonance Absorption Spectroscopy

NR Assay Neutral Red Assay

NRB Natural Resources Board [US]

NRB Nuclear Reaction Broadening [Physics]

NRB Nuclear Reactors Branch [of US Department of Energy]

NRB Nuclear Resonance Broadening [Physics]

NRBA National Registered Builders Association [UK]

NRBP Natural Resource-Based Products

NRC National Rail Corporation [US]

NRC National Recyclers Coalition [US]

NRC National Referral Center [of Library of Congress, Washington, DC, US]

NRC National Remodelers Council [of National Association of Home Builders, US]

NRC National Replacement Character

NRC National Research Center

NRC National Research Center [Dokki-Cairo, Egypt]

NRC National Research Council [Ottawa, Canada]

NRC National Research Council [Washington, DC, US]

NRC National Research Council [Bangkok, Thailand]

NRC National Research Council [Tokyo, Japan]

NRC National Resources Council [US]

NRC National Response Center [US]

NRC National Rocket Club [now National Space Club, US]

NRC Nation's Report Card [US]

NRC Natural Resources Committee [US]

NRC Newfoundland Safety Council [Canada]

NRC Niger River Commission [Nigeria]

NRC Nissan Research Center [Japan]

NRC Noble Research Center [US]

NRC Noise Rating Curve [Acoustics]

NRC Noise Reduction Coefficient [Acoustics]

NRC Nonrecurring Change

NRC Nonrecurring Charge [Telecommunications]

NRC Nonrecurring Connection

NRC Nonrecurring Costs

NRC Noranda Research Center [Canada]

NRC Notch Root Contraction

NRC Nuclear Reactor Core

NRC Nuclear Reactor Control

NRC Nuclear Recycling Center

NRC Nuclear Regulatory Commission [Washington, DC, US]

NRC Nuclear Research Center

NRCA National Research Council Associate [US]

NRCA National Resources Council of America [US]

NRCA National Roofing Contractors Association [US]

NRCAN (Department of) Natural Resources Canada [also NRCan] [formerly Energy Mines and Resources Canada]

NRC-ARL National Research Council–Army Research Laboratory (Project) [US]

NRCC National Research Council of Canada [Ottawa, Ontario]

NRC(Can) National Research Council (of Canada)

NRC CAST National Research Council–Cooperation in Applied Science and Technology (with the Newly Independent States of the Former Soviet Union) (Program) [US]

NRCC-IAR National Research Council of Canada–Institute for Aerospace Research [Ottawa]

NRCCL Norwegian Research Center for Computers and Law

NRCC LINAC National Research Council of Canada Linear Accelerator

NRCC/NAE National Research Council of Canada/National Aeronautical Establishment

NRC Compd Nitrocellulose, Rosins and China Wood Oil in Methyl Acetate [Ammunition]

NRCD National Reprographic Center for Documentation [UK]

NRC IRAP National Research Council/Industrial Research Assistance Program [Canada]

NRCJ National Research Council of Japan

NRCL Natural Resources Conservation League

NRCLSE National Resource for Computers in Life Science Education [University of Washington, Seattle, US]

NRCLV Natural Resources Conservation League of Victoria [Australia]

NRCMAI National Railroad Construction and Maintenance Association, Inc. [US]

NRCMF National Referral Center Master File [of Library of Congress, Washington, DC, US]

NRC-NAS National Research Council–National Academy of Sciences (Program) [US]

NRC-NAS National Research Council–Naval Postgraduate School (Associateship) [US]

NRC-NASA National Research Council–National Aeronautics and Space Administration [US]

NRC-NRaD National Research Council–NCCOSC (Naval Command, Control and Ocean Surveillance Center) RDT&E (Research, Development, Test(ing) and Evaluation) (Division)

NRC-NRL National Research Council–Naval Research Laboratory (Program) [Washington, DC, US] [also NRC/NRL]

NRCP National Rainforest Conservation Program [of Environment Australia]

NRCP National Research Council of the Philippines

NRCR Nuclear Reactor Control Rod

NRCS Natural Resource Conservation Service [of US Department of Agriculture, Lincoln, Nebraska]

NRCS Nuclear Reactor Control System

NRCS National Replacement Character Set

NRCSA National Registration Center for Study Abroad [US]

NRCST National Referral Center for Science and Technology [of Library of Congress, Washington, DC, US]

NRCT National Research Council of Thailand [Bangkok, Thailand]

NRD Natural Resources Damages

NRD Natural Resources Division [of United Nations Economic Commission for Africa]

NRD Negative Resistance Diode

NRD Network Resource Dictionary

NRD Nonradiative Dielectric

NRDB Natural Rubber Development Association [Malaysia]

NRDC National Research Development Corporation [now British Technology Group, UK]

NRDC National Research Development Council [US]

NRDC Natick Research and Development Center [of US Army at Natick, Massachusetts]

NRDC Natural Resources Defense Council [US]

NRDC Northboro Research and Development Center [Iowa, US]

NRDCA National Roof Deck Contractors Association [US]

NRDF Nonrecursive Digital File

NRDL Naval Radiological Defense Laboratory [of US Navy]

NRDS Nuclear Rocket Detection System

NRDS Nuclear Rocket Development Station [of NASA in Nevada, US]

NRE Naval Research Establishment [now Defense Research Establishment Atlantic, Canada]

NRE Non-Recurring Engineering (Charges)

NRe Nepalese Rupee [Currency of Nepal]

n-Re Nanocrystalline Rhenium [Symbol]

NRECA National Rural Electric Cooperative Association [US]

NRECAIF National Rural Electric Cooperative Association, International Foundation

NRECOIL Nuclear Recoil Spectrometry

NREL National Renewable Energy Laboratory [of US Department of Energy at Golden, Colorado, US] [formerly Solar Energy Research Institute]

NREL Tech Rep National Renewable Energy Laboratory Technical Report [also NREL TR]

NREM Non-Rapid Eye Movement (Sleep) [also known as S-sleep]

NREN National Research and Education Network [US]

NREP National Registry of Environmental Professionals [US]

NRET Nonradiative Energy Transfer

NRF Napkin Ring Fracture [Mechanics]

NRF National Roofing Foundation [US]

NRF Natural Resources Forum [of United Nations]

NRF Naval Reactor Facility [US]

NRF Nelson-Riley Function [X-Ray Diffraction]

NRF Neutron Radiography Facility

NRF Non-Reflecting Film

NRF Nuclear Reactor Fuel

NRF Nuclear Research Facility

NRF Nuclear Resonance Frequency

NRFC National Rail Freight Corporation

NRFC National Railroad Freight Committee [US]

NRFD Not Ready for Data [General-Purpose Interface Bus Line]

NRFE Nuclear-Reactor Fuel Element(s)

NRFR Nuclear-Reactor Fuel Rod(s)

NRFSA Navy Radio Frequency Spectrum Activity

NRG Neutron Radiography (Facility)

NRG (University of) Newcastle Research Group [UK]

NRG Nuklearrohr-Gesellschaft mbH [German Manufacturer of Nuclear Tubing; of Kraftwerk Union AG]

NRGBT Normal Relative Grain Boundary Thickness [Materials Science]

NRGSM National Research Group for the Structure of Matter [Italy]

n-Rh Nanocrystalline Rhodium

NRHA National Retail Hardware Association [US]

NRHC National Rivers and Harbours Congress [US]

NRHM National Railway Historical Society [US]

NRHP National Register of Historic Places [US]

NRHQ Northern Region Headquarters

NRI National Resources Institute [of University of Maryland, College Park, US]

NRI Natural Resources Inventory [US]

NRI Natural Resources Institute [UK]

NRI Nebraska Research Institute [US]

NRI Net Radio Interface

NRI Net Revenue Interest

NRI Nomura Research Institute [Japan]

NRI Nonrecurring Installation Charge [Telecommunications]

NRI Nonrecurring Investment

NRI Nuclear Resonance Induction

NRIA National Railroad Intermodal Association [US]

NRIC National Reserves Investigation Committee [UK]

NRIC National Resource Information Center [Australia]

NRIC National Response and Information Center

NRIM National Research Institute for Metals [Tokyo, Japan]

NRIMS National Research Institute for Mathematical Sciences [South Africa]

NRIN National Research Institute of Nagoya [Japan]

NRIP Number of Rejected Initial Pickups

NRIS National Resource Information System (Database) [Queensland, Australia]

NRJ Non-Reciprocal Junction

NRL National Reference Library

NRL Naval Research Laboratory [of US Navy at Washington, DC, US]

NRL Nelson Research Laboratory [UK]

NRL Network Restructuring Language

NRL Nuclear Reactor Laboratory [US]

NRL Nuclear Reactor Lattice

NRL Nuclear Resonance Level

NRL Nutrition Research Laboratory [India]

NRLA Northeastern Retail Lumber Association [US]

NRLC National Railway Labor Conference [US]

NRL-CD Naval Research Laboratory, Chemistry Division [of US Navy at Washington, DC, US]

NRLDA National Retail Lumber Dealers Association [now National Lumber and Building Material Dealers Association, US]

NRLF The Northern Regional Library Facility [US]

NRLH Nonrigid Rotation Large-Amplitude (Internal Motion) Hamiltonian [Physics]

NRLM National Research Laboratory of Metrology [Ibaraki, Japan]

NRLM Nuclear Reactor Loading Machine

NRL-NRC Naval Research Laboratory–National Research Council [Washington, DC, US]

NRLSI National Reference Library for Science and Invention [UK]

NRM Natural Remanent Magnetization

NRM Nonrecurring Maintenance

NRM Nonrelativistic Mechanics

NRM Normal Response Mode

NRM Northern Rocky Mountains [US/Canada]

NRM Nuclear Reactor Material

NRM Nuclear Resonance Magnetometer

nrm normalize(d)

NRMA National Reloading Manufacturers Association [US]

NRMA National Retail Merchants Association [US]

NRMA National Roads and Motoring Association

NRMA Nuclear Records Management Association [US]

NRMCA National Ready Mixed Concrete Association [US]

NRMDP Natural Resource Management and Development Project

NRMEC North American Rockwell Microelectronics Company [US]

nrml normal

NRMLS New England Regional Medical Library Service [US]

NRMM National Register of Microform Masters [of Library of Congress, Washington, DC, US]

NRMPI Nonresonant Multiphoton Ionization

NRMR Netherlands Research and Materials Testing Reactor

NRMRL National Risk Management Research Laboratory [of US Environmental Protection Agency at Cincinnati, Ohio, US]

NRMS Natural Resource Management Strategy [of Murray-Darling Basin Commission, Australia]

NRMS Network Reference and Monitor Station

NRMS Neutralization Reionization Mass Spectrometry

NRMS Nominal Root Mean Square

NRN Negative Run Number

NRN Noise Rating Number [Acoustics]

NRN No Reply Necessary

NRO Naval Reserve Officer

NROP New River Ordnance Plant [Redford, Virginia, US]

NROS Naval Reserve Officers School [US]

N-ROSS Navy Remote Ocean Sensing System

NROTC Naval Reserve Officers Training Corps [of US Navy]

NRP New River Project [Assessment of Chemical Pollution of the New River near Mexicali, Mexico]

NRP Normal Rated Power

NRP Nuclear-Reactor Physics

NRPA National Recreation and Parks Association [US]

NRPB National Radiological Protection Board [UK]

NRPDA National Radio Parts Distributors Association [now National Electronic Distributors Association, US]

NRPMP National Rivers Processes and Management Program [of Commonwealth Environmental Protection Agency]

NRPRA Natural Rubber Producers Research Association [UK]

NRPS Non-Reciprocal Phase Shift

NRPU Northern Redwood Purchase Unit [US]

NRQM Nonrelativistic Quantum Mechanics

NRR Net Reproduction Rate

NRR Noise Reduction Rating [Acoustics]

NRR Noise Reduction Regulations [US]

NRR Nonreactive Resistor

NRR Non-Reciprocal Reflectivity

NRR Nonrenewable Resource

NRR Nonrepeatable Runout

NRR Nonreturn-to-Reference (Recording)

NRR Non-Reversing Relay

NRR Nuclear Reactor Reflector

NRR (Office of) Nuclear Reactor Regulation [US]

NRRA National Resource Recovery Association

NRRA Nuclear Resonance Reaction Analysis

NRRDC National Resources Research and Development Corporation

NRRFSS National Research and Resource Facility for Submicron Structures [of Cornell University, Ithaca, New York, US]

NRRI National Regulatory Research Institute [US]

NRRL Northern Regional Research Laboratory [US]

NRRO Nonrepeatable Runout

NRRR Nonreturn-to-Reference Recording

NRRS Naval Radio Research Station

NRS National Residues Survey

NRS National Reserves System [Australia]

NRS Naval Radio Station

NRS Naval Rocket Society [US]

NRS Navy Records Society [UK]

NRS New Regeneration System (Process)

NRS Nickel Recovery System

NRS Nonconformance Reporting System

NRS Nonrising Stem

NRS Normal Raman Spectrometry; Normal Raman Spectroscopy

NRS North Radiography Station [of Argonne National Laboratory–West, Idaho Falls, US]

NRS Nuclear Radiation Spectroscopy

NRS Nuclear Reaction Spectrometry

NRS Nuclear Reactor Shielding

NRSA National Rubber Shippers Association [US]

NRSA National Rural Studies Association [UK]

NRSC National Remote Sensing Center [UK]

NRSCP National Reserves System Cooperative Program [of Australian Nature Conservancy Agency]

NRSE Neutron Resonant Spin Echo (Technique) [Physics]

NRSE New and Renewable Sources of Energy

NRSEP National Roster of Scientific and Engineering Personnel

NRSR Nuclear Reactor Shim Rod

NRST Nonreactive Solute Transport

NRSTP National Register of Scientific and Technical Personnel

NRSW Nuclear River Service Water

NRT National Response Team

NRT Near Real-Time [also NR/T]

NRT Net Register(ed) Ton(nage)

NRT Neutron Radiographic Test(ing); Neutron Radiography

NRT Nonradiating Target

NR/T Near Real-Time [also NRT]

n(r,t) free-electron concentration [Symbol]

NRTC National Road Transport Commission

NR Technol Natural Rubber Technology [Publication of the Malaysian Rubber Producers Research Association]

NRTEE National Round Table on the Environment and the Economy [Canada]

NRTL National Recognized Testing Laboratory [US]

NRTS National Reactor Test Station [of Idaho National Engineering Laboratory, Idaho Falls, US]

NRTS Not Repairable at This Station

NRTSC Naval Reconnaissance and Technical Support Center [of US Navy]

NRU National Research Universal (Reactor) [of Atomic Energy of Canada Limited, Chalk River, Ontario, Canada]

Nru Nocardia rubra [Microbiology]

n-Ru Nanocrystalline Ruthenium

NRV Natural Resource Valuation

NRV Net Recovery Value

NRW Nuclear Radwaste

nrw narrow

NRWG Neutron Radiography Working Group [of American Society for Testing and Materials, US]

NRWO Nuclear Radwaste Operator

NRX National Research Experimental (Reactor) [of Atomic Energy of Canada Limited, Chalk River, Ontario, Canada]

NRX NERVA (Nuclear Engine for Rocket Vehicle Applications) Reactor Experiment [of US Atomic Energy Commission–Nuclear Rocket Development Station, Nevada] [Completed]

NRX Nuclear Reactor Experiment

NRX-CX NERVA (Nuclear Engine for Rocket Vehicle Applications) Reactor Critical Assembly [of US Atomic Energy Commission]

NRX-CX Nuclear-Engine Reactor Critical Assembly [of US Atomic Energy Commission]

NRZ Non-Return-to-Zero (Recording)

NRZ1 Non-Return-to-Zero Change on One [also NRZ-1]

NRZC Non-Return-to-Zero Change Recording [also NRZ-C, NRZ/C, or NRZ(C)]

NRZI Non-Return-to-Zero Indicator

NRZI Non-Return-to-Zero Inverted (Recording)

NRZI Non-Return-to-Zero I (One)

NRZL Non-Return-to-Zero Level [also NRZ-L]

NRZM Non-Return-to-Zero Mark Recording [also NRZ-M, or NRZ/M]

NS American National Thread–Standard

NS Nanostructure(d); Nanostructural

NS Nassi-Schneiderman (Chart)

NS National Seashore [US]

NS National Semiconductor [US Company]

NS National Site [US]

NS National Special (Screw) Thread

NS National Standard

NS Natural Science

NS Natural Selection [Darwinism]

NS Naval Service [UK]
NS Naval Ship
NS Naval Station
NS Navigation Satellite
NS Navigation System
NS Near Side
NS Near-Surface (Discontinuity) [Materials Science]
NS Nervous System [Anatomy]
NS NetScape Browser [Internet]
NS Net Shape
NS Network Security
NS Network Supervisor
NS Neural System
NS Neuroelectric Society [US]
NS Neutron Scattering
NS Neutron Source
NS Neutron Spectrometer; Neutron Spectrometry
NS Neutron Spectroscope; Neutron Spectroscopy
NS Neutron Spectrum
NS Neutron Star [Astronomy]
NS New Process Standard [Software Engineering Standard]
NS New Series
NS New Style (Calendar) [Astronomy]
NS New System
NS Nickel Silicate
NS Nickel Silver
NS Nickel Steel
NS Night Sky
NS Niobium Selenide
NS Nitrate of Sodium
NS Nitrile Synthesis
NS Nitrostarch [Explosive]
NS Noise Sensitivity
NS Noise Spectrum
NS Nominal Size
NS Nonsequenced
NS Non-Specific
NS Non-Staining
NS Non-Sterile
NS Non Stop
NS Non-Summing
NS Normal State
NS Normal Spectrum
NS Norte de Santander [Colombia]
NS North Sea
NS North-South
NS North Star
NS Notch Strength [Metallurgy]
NS Not Scheduled [also ns]
NS Not Specified [also ns]
NS Not Sufficient [also ns]
NS Not Switchable
NS Nova Scotia(n) [Canada]

NS N-Sulfo- [Chemical Prefix]
NS Nuclear Safety
NS Nuclear Scattering
NS Nuclear Science; Nuclear Scientist
NS Nuclear(-Powered) Ship
NS Nuclear Shuttle
NS Nuclear Spectroscopy
NS Nuclear Spin [Nuclear Physics]
NS Nuclear Submarine
NS Nuclear System
NS Nucleophilic Substitution [Physical Chemistry]
NS Number System
NS Numerical Stability
NS Nutrition Society
NS Sodium Silicate [$Na_2O \cdot SiO_2$, or Na_2SiO_3]
N/S Not Sufficient [also n/s]
N-S Normal-to-Superconductivity (Transition) [Solid-State Physics]
N:S Normal-Superconductor Contact Point
N(S) Send-Sequence Number
Ns Newton-second [also N·m]
Ns Nimbostratus (Cloud)
Ns Number of Secondary Turns (of Transformers) [Symbol]
N/s Newton per second [also N s^{-1}]
nS nanosiemens [Unit]
ns nanosecond [also nsec]
ns not scheduled [also NS]
ns not specified [also NS]
ns not sufficient [also NS]
n/s neutrons per second [also n/sec]
NSA Naphthalenesulfonic Acid
NSA National Safflower Association [US]
NSA National Sawmilling Association [UK]
NSA National Security Agency [US]
NSA National Service (Armed Forces) Act [US]
NSA National Sheep Association [UK]
NSA National Shellfisheries Association [US]
NSA National Shipping Authority [US]
NSA National Slab Association [US]
NSA National Standards Association [US]
NSA National Stereoscopic Association [US]
NSA National Stone Association [US]
NSA National Student Association [US]
NSA National Sunflower Association [US]
NSA (Vereniging) Natuurkunde en Sterrenkunde Studenten Amsterdam [Amsterdam Association of Physics and Astronomy Students, of University of Amsterdam, Netherlands]
NSA Naval Small Arms
NSA Netherlands Society for Automation
NSA New Stone Age [Anthropology]
NSA Nitride Self-Aligned
NSA Nonenyl Succinic Anhydride
NSA Nonsequenced Acknowledgement

NSA Non-Stereospecific Addition [Organic Chemistry]

NSA North Slope of Alaska

NSA Norwegian Standards Association

NSA No Suitable Applicant [also nsa]

NSA Not Seasonally Adjusted [also nsa]

NSA Nuclear Science Abstracts [Joint Publication of the US Department of Energy's Energy Research and Development Administration and Oak Ridge National Laboratory]

NSA Nuclear Stock Association (Limited) [UK]

NSA Nuclear Suppliers Association [US]

NSAA National Sulfuric Acid Association [UK]

NSAA Nova Scotia Association of Architecture [Canada]

NSAC Norwegian Society for Automatic Control

NSAC Nova Scotia Agricultural College [Truro, Canada]

NSAC Nuclear Safety Analysis Center

NSA/CSS National Security Agency/Central Security Service [US]

NSAE National Society of Architectural Engineers [US]

NSaF National Sanitation Foundation

NSAI National Standards Authority of Ireland

NSAI Nonsteroidal Anti-Inflammatory

NSAID Nonsteroidal Anti-Inflammatory Drug

NSA Ltd Nuclear Stock Association Limited [UK]

NSAP Network Service Access Point

NSAPI Netscape Server Application Program Interface

NSA POLY National Security Agency Polygraph [US]

NSAS National Smoke Abatement Society [now National Society for Clean Air (and Environmental Protection), UK]

NSAS Neutron Small-Angle Scattering

NSB National Savings Bank [UK]

NSB National Science Board [US]

NSB Naval Standardization Board [of US Navy]

NSB Network of Small Businesses [US]

NSB Network Systems Branch [of Division of Computer Research and Technology, US]

NSB Nuclear Standards Board

NSBA National School Boards Association [US]

NSBA National Silica Brickmakers Association [UK]

NSBE National Society of Black Engineers [US]

NSBET National Society of Biomedical Equipment Technicians [US]

NSBP National Society of Black Physicists [US]

NSBS Nova Scotia Bird Society [Canada]

NSBU National Small Business United [US]

NSC Naphthalene Sulfochloride

NSC National Safety Council [US]

NSC National Science Council [Taiwan]

NSC National Security Council [US]

NSC National Space Council [Washington, DC, US]

NSC National Steel Corporation [US]

NSC NATO Science Committee

NSC NATO Supply Center

NSC Naval Space Command [US]

NSC Near-Site Coincidence (Theory) [Physics]

NSC Near-Surface Chemistry

NSC Network Switching Center

NSC Neutron Scattering Center [of Los Alamos National Laboratory, New Mexico, US]

NSC Newark State College [Union, New Jersey, US]

NSC Nippon Steel Corporation [Japan]

NSC Nippon Steel Corporation Process [Metallurgy]

NSC Nodal Switching Center

NSC Noise Suppression Circuit

NSC Nonstoichiometric Compound

NSC Northern Montana College [Havre, Montana, US]

NSC Northern State College [Aberdeen, South Dakota, US]

NSC Northwestern State College [Alva, Oklahoma, US]

NSC Norwegian Space Center

NSC Nuclear Services Corporation [US]

NSC Numerical Sequence Code

NSC Numeric Space Character

NSC Nuclear Science Center [of New Delhi, India]

NSCA National Society for Clean Air (and Environmental Protection) [UK]

NSCA National Sound and Communications Association [US]

NSCAD Nova Scotia College of Art and Design [Canada]

NSCAT NASA Scatterometer

NSCB Nordic Society for Cell Biology [Denmark]

NSCC Nuclear Services Closed Cooling

NSCCEC National School Curriculum Center for Educational Computing [Tappan, New York, US]

NSCD National Spatial Cadastral Database [US]

NSCIC Nova Scotia Communication and Information Center [Canada]

NSCL National Superconducting Cyclotron Laboratory [of Michigan State University, East Lansing, US]

NSCL Northwestern State College of Louisiana [Natchitoches, US]

ns/cm nanosecond per centimeter [also ns cm^{-1}]

NSCNO Neodymium Strontium Copper Niobium Oxide (Superconductor)

NSCO Neodymium Strontium Copper Oxide (Superconductor)

NSCP National Soil Conservation Program

NSCR Nova Scotia Cancer Registry [Canada]

NSCR Nuclear Science Center Reactor [of Texas A&M University, College Station, US]

NSCS National Soil Conservation Strategy

NSCS North Sea Continental Shelf

NSCS North Star Computer (System) Society [US]

NSCUFA Nova Scotia Council of University Faculty Associations [Canada]

NSD Nairobi Sheep Disease

NSD NASA Standard Detonator

NSD National Security and Defense [US]

NSD Network Status Display

NSDA National Space Development Agency [Japan]

NSDB National Science Development Board [Philippines]

NSDC Northern Shipowners Defense Council [Norway]

NSDC Nova Scotia Design Craftsmen [Canada]

NSDD National Security Division Directive [US]

NSDI National Spatial Data Infrastructure [of US Geological Survey]

ns/div nanosecond(s) per (horizontal) division [Oscilloscope Unit]

NSDJA National Sash and Door Jobbers Association [US]

NSDM National Security Decision Memorandum

NSDO National Seed Development Organization [UK]

NSDOF Nova Scotia Department of Fisheries [Canada]

NSDU Network Service Data Unit

NSDV Nairobi Sheep Disease Virus

NSE National Student Exchange [US]

NSE Navier-Stokes Equation [Fluid Mechanics]

NSE Neuron Specific Enolase [Biochemistry]

NSE Neutron Spin Echo (Technique) [Physics]

NSE Nonlinear Schroedinger Equation [Quantum Mechanics]

NSE Normal Spectral Emittance

NSE Nottingham Society of Engineers [UK]

NSE Nuclear Science and Engineering

NSE Nuclear Solid Effect

NSE Nuclear Spin Echo [Physics]

NS&E Natural Sciences and Engineering

NSEAD National Society for Education in Art and Design [UK]

NSEC Naval Ship Engineering Center [of US Navy at Hyattsville, Maryland, US]

nsec nanosecond [also ns]

n/sec neutrons per second [also n/s]

NSEF Navy Security Engineering Facility [of US Navy]

NSEH Neutron Scattering Experimental Hall [of Los Alamos National Laboratory, New Mexico, US]

NSEIP Norwegian Society for Electronic Information Processing

NSEM Near-Field Scanning Electron Microscope; Near-Field Scanning Electron Microscopy

NSEMA National Spray Equipment Manufacturers Association [US]

NSEO Nuclear Spin Electron Orbit [Physics]

NSEP National Security and Emergency Preparedness [US]

NSEPB National Swedish Environment Protection Board

NSERC Natural Sciences and Engineering Research Council [Ottawa, Ontario, Canada]

NSERCC Natural Sciences and Engineering Research Council of Canada [Ottawa]

NSERC BEP NSERC Bilateral Exchange Program [Canada]

NSERC CG NSERC Conference Grant [Canada]

NSERC EG NSERC Equipment Grant [Canada]

NSERC IG NSERC Infrastructure Grant [Canada]

NSERC IRC NSERC Industrial Research Chairs (Program) [Canada]

NSERC-OHT NSERC-Ontario Hydro Technologies (Research Chair) [Canada]

NSERC PRAI NSERC Project Research Applicable in Industry (Grant) [Canada]

NSERC SG NSERC Strategic Grant [Canada]

NSERC VFCGL NSERC Visiting Fellowships in Canadian Government Laboratories [Canada]

NSES National Society of Electrotypers and Stereotypers [UK]

NSES Neutron Spin Echo Spectroscopy

NSESD National Strategy and Ecologically Sustainable Development

NSF National Sanitation Foundation

NSF National Science Foundation [now NSF International, Washington, DC, US]

NSF Naval Supersonic Facility [of US Navy]

NSF Neil Squire Foundation [Canada]

NSF Neutron Scattering Factor

NSF Norges Standardiseringsforbund [Norwegian Standards Association]

NSF Not Sufficient Funds [also nsf]

NSF Nuclear Science Foundation

NSF Nuclear Structure Facility [UK]

NSF Nyquist Sampling Frequency

NSFA National Science Foundation Act [US]

NSF-ARI National Science Foundation–Academic Research Infrastructure (Program) [US]

NSF-ASC National Science Foundation–Associate in Science (Program) [US]

NSFC National (or Natural) Science Foundation of China

NSF-CNRS (US) National Science Foundation/(French) Centre National de Recherche Scientifique (Project)

NSFD Note of Structural and Functional Deficiency

NSF-DMR National Science Foundation Division for Materials Research [also NSF DMR, US]

NSF-DMR-MRG National Science Foundation Division for Materials Research Materials Research Grant [US]

NSF-DMR-PYI National Science Foundation Division for Materials Research Presidential Young Investigator (Program) [US]

NSF-DPP National Science Foundation Division of Polar Programs [also NSF DPP, US]

NSF-EPIC National Science Foundation Edison Polymer Innovation Corporation [at Case Western Reserve University, Cleveland, Ohio, US] [also NSF/EPIC]

NSF-ERC National Science Foundation–Engineering Research Centers (Program) [US]

NSF International National Science Foundation International [formerly NSF, US]

NSFL National Science Film Library

NSFL National Smokeless Fuels Limited [of National Coal Board, UK]

NSFL Nova Scotia Federation of Labour [Canada]

NSF-MRC National Science Foundation–Materials Research Centers (Program) [also NSF/MRC, NSF MRC, US]

NSF-MRG National Science Foundation–Materials Research Grant [also NSF/MRG, NSF MRG, US]

NSF-MRL National Science Foundation–Materials Research Laboratories (Program) [also NSF/MRL, or NSF MRL, US]

NSF/MRSEC National Science Foundation/Materials Research Science and Engineering Center (Program) [US]

NSF-NATO National Science Foundation–North Atlantic Treaty Organization (Joint Fellowship Program) [US]

NSFnet National Science Foundation (Computer) Network [US] [also NSFNet]

NSF-OIS National Science Foundation–Office of Information Systems [US]

NSF-OPAS National Science Foundation–Organizational Prior Approval System [US]

NSF-PYI National Science Foundation–Presidential Young Investigator (Program) [US]

NSF-PYI-DMR National Science Foundation–Presidential Young Investigator (Program), Division for Materials Research [US]

NSF-REU National Science Foundation–Research Experiences for Undergraduates (Program) [US]

NSF-REU-DMR National Science Foundation–Research Experiences for Undergraduates (Program)–Division for Materials Research [US]

NSFRC National Soil and Fertilizer Research Committee [US]

NSFS Net Suction Fracture Strength

NSF-STC National Science Foundation–Science and Technology Centers [also NSF/STC, NSF STC, NSF-S&TC, NSF/S&TC, or NSF S&TC, US]

NSF-STCS National Science Foundation–Science and Technology Center for Superconductivity [also NSF/STCS, NSF STCS, NSF-S&TCS, NSF/S&TCS, or NSF S&TCS, US]

NSF-STIA National Science Foundation–(Directorate for) Scientific, Technological and International Affairs [US]

NSG NASA Grant

NSG National Sea Grant [of National Oceanic and Atmospheric Administration, US]

NSG Naval Security Group [of US Navy]

NSG Newspaper Systems Group [US]

NSGA National Sand and Gravel Association [now National Aggregates Association, US]

NSGC Naval Security Group Command [of US Navy]

NSGCP National Sea Grant College Program [of National Oceanic and Atmospheric Administration, US]

NSGD National Sea Grant Depository [of National Oceanic and Atmospheric Administration, US]

NSGO National Sea Grant Office [of National Oceanic and Atmospheric Administration, US]

NSGP National Sea Grant Program [of National Oceanic and Atmospheric Administration, US]

NSGPMA National Salt-Glazed Pipe Manufacturers Association [UK]

NSGT NASA Ground Terminal

(N,Σ,H$_s$) Number of Particles, Stress Tensor and Generalized Enthalpy [Symbol]

NSHB North of Scotland Hydro-Electric Board [Scotland]

NSHC National Solar Heating and Cooling [now Conservation and Renewable Energy Inquiry and Referral Service, US]

NSHC North Sea Hydrographic Commission [Sweden]

NSHD Nevada State Highway Department [US]

NSHP National Small Hydropower Program [of US Department of Energy]

NSHS Nebraska State Historical Society [US]

NSHS Nevada State Historical Society [US]

NSI NASA Science Internet

NSI National Security Information

NSI National Space Institute

NSI National Supervisory Inspectorate [UK]

NSI Netherlands Society for Informatics

NSI Negative Surface Ionization

NSI Next Sequential Instruction

NSI Nonlinear Surface Impedance

NSI Nonsatellite Identification

NSI Nonsequenced Information

NSI Nonstandard Item

Nsi Neisseria sicca [Microbiology]

n-Si Nanocrystalline Silicon

n-Si N-Type Silicon

NSIA National Sailing Industry Association [US]

NSIA National Security Industrial Association [US]

NSIC Nuclear Safety Information Center [Oak Ridge National Laboratory, Tennessee, US]

NSIC National Storage Industry Consortium

NSIDC National Snow and Ice Data Center [US]

NSIF National Swine Improvement Federation [US]

NSIF Near Space Instrumentation Facility [US]

n-Si:H Hydrogenated Nanocrystalline Silicon

NSI-I NASA Standard Initiator-Type I [formerly SMSI (Standard Manned Spaceflight Initiator)]

n-Si:P Phosphorus-Doped N-Type Silicon

n-Si:Pt Platinum-Doped N-Type Silicon

NSIS Nova Scotia Information Service [Canada]

NSIT Nova Scotia Institute of Technology [Canada]

NSJ Nuclear Society of Japan

NSLOOKUP Name Server Lookup [Unix Operating System]

NSL National Science Library [now Canada Institute for Scientific and Technical Information, Canada]

NSL National Science Laboratories [US]

NSL National Standards Laboratory

NSL Nebraska State Library [Lincoln, US]

NSL Net Switching Loss

NSL Nevada State Library [Carson City, US]

NSL Northrop Space Laboratories [US]

NSL Naval Supersonic Laboratory [of US Navy]

NSLA Nova Scotia Library Association [Canada]

NSLR Nuclear Spin-Lattice Relaxation

NSLRB Nova Scotia Labour Relations Board [Canada]

NSLRSDA National Satellite Land Remote Sensing Data Archive [of US Geological Survey/National Mapping Division]

NSLS National Synchrotron Light Source [of Brookhaven National Laboratory, Upton, New York, US]

NSLS-BNL National Synchrotron Light Source of Brookhaven National Laboratory [Upton, New York, US]

NSM National Society Member(s)

NSM Neodymium Strontium Manganate

NSM Netscape Server Manager

NSM Network Security Module

NSM Network Status Monitor

NSM Network Storage Module

NSM New Smoking Material (Tobacco Substitute)

NSM Nordic Semiconductor Meeting [Scandinavia]

NSM Nuclear Shell Modell

Ns/m² Newton second(s) per square meter [also N s/m², N·s/m², or Ns m^{-2}]

NSMA National Scale Men's Association [now International Society of Weighing and Measurement, US]

N-SmA Nematic-Smectic A (Phase) Transition (of Liquid Crystal)

N-SmA₁ Nematic-Smectic A$_1$ (Phase) Transition (of Liquid Crystal)

NSMB Nuclear Standards Management Board [of American National Standards Institute, US]

NSMDA Nova Scotia Mineral Development Agreement [Canada]

NSMO Neodymium Strontium Manganese Oxide

NSMP National Society of Master Patternmakers [UK]

NSMPA National Screw Machine Products Association [US]

NSMR National Society for Medical Research [US]

NSMRSE National Study of Mathematics Requirements for Scientists and Engineers [US]

NSMS National Safety Management Society [US]

NSN NASA Science Network

NSN National Stock Number [US]

NSN Nanostructured Semiconductor Network

NSN Sodium Strontium Niobate

NSNO Neodymium Strontium Nickel Oxide

n-Sn Nanocrystalline Tin

n-SnO₂ N-Type Stannic Oxide

NSO NASA Support Operation

NSO National Solar Observatory [US]

Nso Nitroso [Chemistry]

NSOEA National Stationery and Office Equipment Association [US]

NsoGu Nitrosoguanidine [also NOG] [Biochemistry]

NSOM Near-Field Scanning Optical Microscope; Near-Field Scanning Optical Microscopy

NSOP National Second Opinion Program [US]

NSOT Not-Back-Twinning Second Order Twin [Metallurgy]

NSP NASA Support Plan

NSP National Society of Painters [UK]

NSP National Space Plan [Argentina]

NSP Network Services Protocol

NSP Network Services Provider

NSP Network Signal Processor

NSP Network Support Plan

NSP Network Support Processor

NSP Non-Spin Polarized (Photoemission)

NSP Nonstandard Part

NSP Nuclear Spin Polarization

NSPA National Society of Public Accountants [US]

NSPA National Soybean Processors Association [US]

NSPA Nova Scotia Pharmaceutical Association [Canada]

NSPAC National Standards Policy Advisory Committee [US]

NSPAR Non-Standard Parts Approval Request

NSPB National Society for the Prevention of Blindness

NSPC National Sound-Program Center [US]

NSPC National Standard Plumbing Code [US]

NSPC Nova Scotia Power Commission [Canada]

NSPC Nova Scotia Power Corporation [Canada]

NSPCC National Standard Plumbing Code Committee [US]

NSPCRA National Spotted Poland China Record Association [US] [*Note:* Spotted Poland China Refers to a Breed of Swine]

NSPE National Society of Professional Engineers [US]

NSPFEA National Spray Painting and Finishing Equipment Association [now National Spray Equipment Manufacturers Association, US]

NSPI National Society for Performance and Instruction [US]

NSPI National Society for Programmed Instruction [US]

NSPIE National Society for the Promotion of Industrial Education [US]

NSPOL Non-Scheduled Operations Policy

NSPP Nuclear Safety Pilot Plant [of Oak Ridge National Laboratory, Tennessee, US]

NSPRI Nigeria Store Product Research Institute

NSPS National Society of Professional Surveyors [of American Congress on Surveying and Mapping, US]

NSPS New Source Performance Standards

NSPV Number of Scans per Vehicle

NSQ Nuclear Spin Quenching

NSQCRE National Symposium on Quality Control and Reliability in Electronics [US]

NSR Natural Sports Rub [Natural Medicine]

NSR Near-Surface Region

NSR Neutron Source Reactor [of Brookhaven National Laboratory, Upton, New York, US]

NSR New Source Review

NSR Nitrile-Silicone Rubber

NSR Nominal Slow Rate

NSR Nordiska Skogsarbetsstudiernas [Nordic Research Council on Forest Operations, Finland]

NSR Normal Slow Rate (Maneuver)

NSR No Sewing Required (Seam)

NSR Notch Strength Ratio [Mechanics]

NSR Notch Stress Rupture [Mechanics]

NSR Nuclear Spin Relaxation [Physics]

NSR Nuclear Structure References (Database) [of Oak Ridge National Laboratory, Tennessee, US]

NSR1 First Coelliptic Maneuver [Aerospace]

NSR2 Second Coelliptic Maneuver [Aerospace]

NSRA National Service Robot Association [US]

NSRA National Small-Bore Rifle Association [UK]

NSRB National Security Resources Board [US]

NSRC National Stereophonic Radio Committee [US]

NSRC National Supercomputing Research Center [Singapore]

NSRC Natural Science Research Council

NSRC Neosynthesis Research Center [Sri Lanka]

NSRDA National Reference Standards Data System [of National Institute of Standards and Technology, Gaithersburg, Maryland, US]

NSRDB National Scientific Research and Development Board [Pakistan]

NSRDB Nova Scotia Resources Development Board [Canada]

NSRDC Naval Ship Research and Development Center [of US Navy]

NSRDS National Standard Reference Data Series [of National Institute for Standards and Technology, Gaithersburg, Maryland, US]

NSRDS National Standard Reference Data System [of Office of Standard Reference Data–National Institute for Standards and Technology, Gaithersburg, Maryland, US] [also NS-RDS]

NSRDS-NBS National Standard Reference Data System–National Burau of Standards [now NSRDS-NIST]

NSRDS-NIST National Standard Reference Data System–National Institute for Standards and Technology [US]

NSRF Nova Scotia Research Foundation [also NSRFI, Halifax, Canada]

NSRFC Nova Scotia Research Foundation Corporation [Halifax, Canada]

NSRFI National Symposium on Radio Frequency Interference

NSRFI Nova Scotia Research Foundation, Inc. [also NSRF, Halifax, Canada]

NSRG Northern Science Research Group [Canada]

NSRI Nelspruit Subtropical Research Institute [South Africa]

NSRM National Strategy for Rangeland Management [Australia]

NSRP Nordic Society for Radiation Protection [Finland]

NSRQC National Symposium on Reliability and Quality Control

NSRR Nuclear Safety Research Reactor

NSRS NASA Safety Reporting System

NSRS Naval Supply Radio Station

NSRS NIR (Near-Infrared) Sphere Radiometer System

NSRT Near Surface Reference Temperature

NSRW Nuclear Service Raw Water

NSRWP Nuclear Service Raw Water Pump

NSS National Search and Rescue Secretariat [Canada]

NSS National Space Society [US]

NSS National Speleological Society [US]

NSS National Standards System

NSS Navy Secondary Standards [of US Navy]

NSS Network Supervisor System

NSS Network Synchronization Subsystem

NSS Nitrogen Supply Subsystem

NSS Nodal Switching System

NSS Nonequilibrium Solvation State

NSS Non-Sea Salt Sulfate

NSS Nonspatial Statistics

NSS Nonsteady-State

NSS Nordic Statistical Secretariat [Denmark]

NSS Normal Salt Spray (Test)

NSS Nuclear Steam System

NSS Number (Density) of Surface States [Physics]

NSSA National Sanitary Supply Association [now International Sanitary Supply Association, US]

NSSA National Science Supervisors Association [US]

NSSA Neutron Scattering Society of America

NSSA Nova Scotia Salmon Association [Canada]

NSSANS Near-Surface Small-Angle Neutron Scattering

NSSC NASA Standard Spacecraft Computer

NSSC Naval Sea Systems Command [of US Navy]

NSSC Naval Supply Systems Command [of US Navy]

NSSC Nova Scotia Safety Council [Canada]

NSSC Nuclear Safety Standards Commission [Germany]

NSSCC National Space Surveillance Control Center [US]

NSSDC National Space Science Data Center [of NASA Goddard Space Flight Center, Greenbelt, Maryland, US]

NSSDRI Nanjing Solid-State Devices Research Institute [Nanjing, PR China]

NSSE National Society for the Study of Education [US]

NSSEA National School Supply and Equipment Association [US]

NSSF National Shooting Sportsman Foundation [US]

NSSF Near Surface Storage Facility [US]

NSSFC National Severe Storms Forecast Center [US]

NSSL National Seed Storage Laboratory [US]

NSSL National Severe Storms Laboratory [of National Oceanic and Atmospheric Administration, US]

NSSL/NOAA National Severe Storms Laboratory of the National Oceanic and Atmospheric Administration [US]

NSSM National Security Study Memorandum [US]

NSSMS NATO Seasparrow Surface Missile System

NSSN National Standard Systems Network [US]

NSSP National Student Safety Program [US]

NSSS National Sewage Sludge Survey [US]

NSSS National Space Surveillance System [US]

NSSS Nuclear Steam Supply System

NSSTS Near-Surface Salinity and Temperature System

NST National Scenic Trail [US]

NST National Security Technology

NST Net Shape Technology

NST Network Support Team

NST Newfoundland Standard Time [Canada]

NST Nil Strength Temperature [Metallurgy]

NST Non-Slip Tread

NST Nuclear Science and Technology

NST Nuclear Spin Tomograph(y)

NS&T National Status and Trends [US]

NSTA National Safe Transit Association [US]

NSTA National School Transportation Association [US]

NSTA National Science Teachers Association [US]

N Staff North Staffordshire [UK]

NSTAG National Science and Technology Analysis Group

NSTAR N* Physics and Nonperturbative Quantum Chromodynamics

NSTB National Safety Transportation Board [US]

NSTB National Science and Technology Board [Singapore]

NSTB Nova Scotia Travel Bureau [Canada]

NSTC National Science and Technology Center

NSTC National Science and Technology Council [US]

NSTC Nova Scotia Technical College [Halifax, Canada]

NSTD Nanometer Science and Technology Division [of American Vacuum Society, US]

NSTD Nested [also Nstd, or nstd] [Computers]

NSTF National Scholarship Trust Fund [Pittsburgh, Pennsylvania, US]

NSTF Near Surface Test Facility [US]

NSTIC Naval Science and Technology Information Center [now Defense Research Information Center, UK]

NSTIC Naval Scientific and Technical Information Center [of US Navy]

NSTL National Software Testing Laboratories

NSTL National Space Technology Laboratories [of NASA] [formerly Mississippi Test Facility

NSTN Nonstandard Telephone Number

NSTP National Skills Training Program

NSTP Nuffield Science Teaching Project [US]

NSTPC Nova Scotia Tidal Power Corporation [Canada]

NSTS National Space Transportation System [US]

NSTS Nuclear Science Technology Section [of Economic Commission for Europe]

NSTW National Science and Technology Week [of National Science Foundation, US]

NSU Network Service Unit

NSU Nicholls State University [US]

NSU Nonspecific Urethritis [Medicine]

NSU Norfolk State University [Virginia, US]

NSU Novosibirsk State University [Russia]

NSUA Nigerian Students Union in the Americas [US]

NSug Nitrosugar

NSV Net Sales Value

NSV Nonautomatic Self-Verification

NSV Nuclear Service Vessel

NSW National Software Works [US]

NSW Neutron Spin Wave [Physics]

NSW New South Wales [Australia]

NSW Nonlinear Sinusoidal Wave

NSW NSP (Network Signal Processor) Status Word

NSWA National Small Woods Association [UK]

NSWA National Stripper Well Association [US]

NSWC Naval Surface Warfare Center [of US Navy at Carderock]

NSWC Naval Surface Weapons Center [of US Navy at Dahlgren, Virginia]

NSWC-DD Naval Surface Weapons Center–Dalgren Division [US]

NSWC-DPB Naval Surface Warfare Center–Detonation Physics Branch [of US Navy at Silver Spring, Maryland]

NSWC TR Naval Surface Weapons Center Technical Report [also NSWC Tech Rep]

NSWEPA New South Wales Environment Protection Authority [Australia]

NSWFC New South Wales Forestry Commission [Australia]

NSWIT New South Wales Institute of Technology [Australia]

NSWL Naval Surface Weapons Laboratory [of US Navy at Silver Spring, Maryland]

NSWMA National Solid Waste Management Association [US]

NSWNPWS New South Wales National Parks and Wildlife Service [Australia]

NSWP Non-Soviet Warsaw Pact

NSWS National Surface Water Survey [US]

NSWS Neutron Spin-Wave Scattering [Physics]

NSYF Natural Science for Youth Foundation [US]

NSYU National Sun Yat-Sen University [Kaohsiung, Taiwan]

NT Nanotechnology

NT Narrow(er) Term

NT Naval Threat

NT Neap Tide

NT Neél Temperature [Solid-State Physics]

NT Negative Temperature

NT Negentropy Theory [Information Technology]

NT Neon Tube

NT Neotetrazolium Chloride

NT Nephelometric Turbidity

NT Nerve Tissue (or Nervous Tissue) [Anatomy]

NT Net Ton

NT Network Termination; Network Terminator

NT Network Theory [Electrical Engineering]

NT Neumann's Triangle

NT Neurotoxin [Biochemistry]

NT Neurotransmitter [Medicine]

NT Neurotrophin [Biochemistry]

NT Neutron Temperature

NT Neutron Thermalization

NT Neutron Topography

NT New Product Standard [Software Engineering Standard]

NT New Technology

NT Newtonian Telescope

NT Night Tracer

NT Nitrate-Trotyl [French Explosive]

NT Nitrotoluene

NT Noise Thermometer

NT Nome Time [Greenwich Mean Time +11:00]

NT Nonthermal

NT Non-Tight

NT Non-Tilting (Drum Concrete Mixer)

NT Normalized and Tempered; Normalizing and Tempering [Metallurgy]

NT Normal Temperature

NT Normal Tension

NT Northern Tablelands [New South Wales, Australia]

NT Northern Territory [Australian State]

NT Northwest Territories [Canada] [also NWT]

NT Nortriptyline

NT Nose Temperature

NT Not Tested [also nt]

NT No Transmission

NT Novel Technology

NT Nozzle Throat

NT Nuclear Technique

NT Nuclear Technologist; Nuclear Technology

NT Nuclear Track

NT Numbering Transmitter

NT Number Theory

NT Nyquist's Theorem [Electrical Engineering]

NT Saudi-Arabian Irak [ISO Code]

NT-3 Neurotrophin-3 [Biochemistry]

NT-4 Neurotrophin-4 [Biochemistry]

2-NT 2-Nitrotoluene

4-NT 4-Nitrotoluene

Nt Neurofilament

Nt Newton [also nt]

Nt Niton [Chemistry]

nT nanotesla [Unit]

n(T) (charge) carrier concentration [Symbol]

nt net

nt neuter

nt not tested [also NT]

.nt Saudi-Arabian Irak [Country Code/Domain Name]

$n_m(t)$ number of (polymer) molecules of size m at time t [Symbol]

$\nabla\theta$ phase gradient [Symbol]

NTA National Technical Association [US]

NTA National Telecommunications Agency [US]

NTA National Translators Association [US]

NTA National Transportation Act [Canada]

NTA National Transportation Association [Canada]

NTA National Tuberculosis Association [US]

NTA Nitrilotriacetate; Nitrilotriacetic Acid [also Nta]

NTA Northern Textile Association [US]

NTA Norwegian Travel Association

NTAG National Technical Advisory Group

NTAG Network Technical Architecture Group [of Network Development Office, Library of Congress, Washington, DC, US]

NTAN Nitrilotriacetonitrile

NTAOCH Notice to Air Operator Certificate Holders [of Civil Aviation Authority, UK]

NTAP Northern Technology Assistance Program [Canada]

NTAS New Technology Advanced Server

NTB National Training Board [Canada]

NTB Neu-Technikum Buchs [Buchs College of Technology, Switzerland]

NTB Nonorthogonal Tight Binding (Theory) [Physics]

NTB Non-Tariff Barrier

N-t-B Nitroso-tert-Butane [also NTB]

Ntba N-tert-Butylaniline [also N-t-BA]

NTBI Non-Transferrin-Bound Iron [Biochemistry]

N-t-BOC N-tert-Butoxycarbonyl

NTBT Nitrilotris(methylene-2-Benzothiazole)

NTBT Nuclear Test Ban Treaty [US]

NTC National Telecommunications Conference [of Institute of Electrical and Electronics Engineers, US]

NTC National Telemetering Conference

NTC National Television Center [US]

NTC National Tick Collection [US]

NTC National Trade Center

NTC National Training Center [US Army]

NTC National Transformers Committee [US]

NTC National Translation Center [US]

NTC National Turfgrass Council [UK]

NTC Naval Telecommunications Command [of US Navy]

NTC Navy Training Center [of US Navy at Charleston, South Carolina]

NTC Negative Temperature Coefficient (of Resistance)

NTC Negative Transconductance

NTC Neotetrazolium Chloride

NTC Network Transmission Committee [of Video Transmission Engineering Advisory Committee, US]

NTC Nonthermal Continuum [Physics]

NTC Noranda Technology Center [Canada]

NTC Norwegian Trade Council

NTCA National Telephone Cooperative Association [US]

NTCA National Tile Contractors Association [US]

NTCB 2-Nitro-5-Thiocyanatobenzoic Acid

NTCR Negative Temperature Coefficient of Resistivity

NTCS Naval Threat/Countermeasure Simulator

NTD National Topographic Database [Canada]

NTD Naval Torpedo Depot

NTD Neutron Thermodiffractometry

NTD Neutron Transmutation Doping

NTD Nuclear Triode Detector

NT$ New Taiwanese Dollar [Currency of Taiwan] [also NTD]

NTDA National Trade Development Association [UK]

NTDB National Topographic Database [Canada]

NTDB National Trade Data Bank [of STATUSA, Department of Commerce]

NTDC National Tyre Distributors Association [UK]

NTDE Naval Tactical Display Emulator

NTDP New Technology Demonstration Program [US]

NTDPMA National Tool, Die and Precision Machining Association [US]

NTDRA National Tire Dealers and Retreaders Association [US]

NTDRA Dealer News NTDRA Dealer News [Published by the National Tire Dealers and Retreaders Association, US]

NTDRP Northern Territory Drought Relief Policy [Australia]

NTDS National Tactical Data System

NTDS Naval Tactical Data System [of US Navy]

NTE Navy Teletypewriter Exchange [of US Navy]

NTE Network Terminal Equipment

NTE Network Terminating Equipment

NTE Northern Telecom, Electronics [Canada]

NTE Not to Exceed

NTEA National Time Equipment Association [US]

NTEA National Truck Equipment Association [US]

NTEP New Technology Employment Program

NTETA National Traction Engine and Tractor Association [UK]

NTF National Tidal Facility

NTF National Transfer Format [Computers]

NTF National Turkey Federation [US]

NTF Nordic Transport Workers Federation [Norway]

NTF No Trouble Found

NTF Nuclear Test Facility

NTF Number Type Flag

NTFS New Technology File System

NTG Nachrichtentechnische Gesellschaft [German Communications Society]

NTG NATO (North Atlantic Treaty Organization) Training Group

NTG Nuclear Turbogenerator

NTH Norges Tekniske Högskole [Norwegian Institute of Technology, Trondheim]

9th ninth

NTHM Net Tonne of Hot Metal [also nthm]

NTHP National Trust for Historic Preservation [US]

nthrn northern

NTHU National Tsing Hua University [Hsinchu, Taiwan]

NTI National Technology Initiative [US]

NTI Negative Thermionic (Mass Spectrometry)

NTI Noise Transmission Impairment

NTIA National Telecommunications and Information Agency [of US Department of Commerce]

NTIAC Nondestructive Testing Information Analysis Center [of Texas Research Institute Austin, Inc., US]

NTIAM National Testing Institute for Agricultural Machinery [Sweden]

NTICL National Technical Information Center and Library [Hungary]

NTID National Technical Institute for the Deaf [US]

n-TiO$_2$ Nanocrystalline (or Nanostructured) Titanium Dioxide

NTIRA National Trucking Industrial Relations Association [US]

NTIS National Technical Information Service [of US Department of Commerce]

NTIS Chem NTIS Chemistry [Publication of the National Technical Information Service, US]

NTIS Comput Control Inform Theory NTIS Computers, Control and Information Theory [Publication of the National Technical Information Service, US]

NTIS Environ Pollut Control NTIS Environmental Pollution Control [Publication of the National Technical Information Service, US]

NTIS Mater Sci NTIS Materials Sciences [Publication of the National Technical Information Service, US]

NTIS Rep NTIS Report [Publication of the National Technical Information Service, US]

NTISS National Telecommunications and Information System Security

NTISSC National Telecommunications and Information System Security Committee [US]

NTIT National Taiwan Institute of Technology [Taipei]

NTK Net Ton Kilometer [Unit]

NTL Noncontact Temperature Monitor

NTL Nonthreshold Logic

NTLIS Northern Territory Land Information System [Australia]

NTLS National Truck Leasing System [US]

NTM National Topographical Map [Canada]

NTM Non-Transition Metal

NTM Normal-to-Metal (Dimension)

NTM Northern Territory (Art Gallery and) Museum [Darwin, Australia]

NTMA National Terrazzo and Mosaic Association [US]

NTMA National Tooling and Machining Association [US]

NTMP Nitrilo-Tris-Methylene Phosphoric Acid

NTMS National Topographic Mapping Series [US]

NTMVSA National Traffic and Motor Vehicle Safety Act [US]

NTN National Telecommunications Network

NTN National Trends Network [US]

NTNO Neodymium Thorium Nickel Oxide

NTNT New Technology National Tescope [Tucson, Arizona, US]

NTNU National Taiwan Normal University [Taipei]

NTO Natural Transition Orbitals [Physics]

NTO Neodymium-Titanium Oxide (or Neodymium Titanate) [Nd$_2$Ti$_2$O$_7$]

NTO Network Terminal Option

NTO Nitrogen Tetroxide

NTOC Number to Character (Conversion)

NTOGA North Texas Oil and Gas Association [US]

NTOTC National Training and Operational Technology Center

NTOU National Taiwan Ocean University [Keelung, Taiwan]

NTP National Toxicology Program [US]

NTP National Tree Program [of Australian Nature Conservancy Agency]

NTP Network Terminal Protocol

NTP Network Terminating Point

NTP Network Test Panel

NTP Network Time Protocol [Internet]

NTP Network Termination Processor

NTP Nitrilo-Tris-Methylene Phosphoric Acid

NTP Normal Temperature and Pressure [also ntp]

NTP No Title Page

NTP Notice To Proceed

NTP Nuclear Test Plant

NTP Sodium Titanium Orthophosphate (Ceramics) [$NaTi_2(PO_4)_3$]

Ntp Nitrilotripropionic Acid

NTPA National Toxicology Program Carcinogens List [US]

NTPC National Technical Planning Committee [Sudan]

NTPC National Technical Processing Center [US]

NTPC National Thermal Power Corporation [India]

NTPCC National Tree Program Coordinating Committee [Australia] [Defunct]

NTPF Number of Terminals per Failure

NTPG National Textile Processors Guild [US]

NTPP Sodium Tripolyphosphate

NTPT National Toxicology Program Technical Reports List [US]

NTR Next Task Register

NTR Nonthermal Radiation [Physics]

NTR Nuclear Test Reactor

NTR Nucledyne Training Reactor [US]

NTRA National Tyre Recycling Association [UK]

NTRAS New Technology Remote Access Services

NTRL National Telecommunications Research Laboratory [South Africa]

Ntrm Nitramide

NTRMA National Tile Roofing Manufacturers Association [US]

NTRR Nonthermal Radio Radiation

NTRS NASA Technical Report Server

Ntrtn Nutrition [also ntrtn]

NTS National Topographic (Map) System [Canada]

NTS National Training Survey

NTS Naval Test Station

NTS Naval Torpedo Station [Newport, Rhode Island, US]

NTS Navigation Technology Satellite

NTS Navigation Technology System

NTS Navy Transit Satellite [of US Navy]

NTS Near Term Schedule

NTS Negative Torque Signal

NTS Nevada Test Site [of NASA and US Department of Energy]

NTS Non-Traffic Sensitive

NTS Normal Tetrahedral Structure

NTS North Texas Section [of Materials Research Society, US]

NTS Notch Tensile Strength [Mechanics]

NTS Not to Scale

NTS NT Server [Microsoft Windows]

NTSA National Technical Services Association [US]

NTSA NetWare Telephony Services Architecture

NTSB National Transportation Safety Board [of US Department of Transportation]

NTSC National Television Standard Code [US]

NTSC National Television Standards Committee [US]

NTSC National Television System Committee [US]

NTSC National Thermal Spray Conference (and Exposition) [US]

NTSC North Texas State College [Denton, US]

NTSD Normal Theory Sampling Distribution

NTSEA National Trade Show Exhibitors Association [now International Exhibitors Association, US]

NTSK Nordiska Tele-Satelit Kommitton [Nordic Tele-Satellite Committee, Norway]

NTSL Nonintegrated Two-Stage Liquid

NTSN National Threatened Species Network [Australia]

NTSO NASA Test Support Office

NTSU North Texas State University [Denton, US]

NTT New Technology Telescope [of European Southern Observatory, Germany]

NTT Nippon Telegraph and Telephone Corporation [Tokyo, Japan]

NTTC National Tank Truck Carriers [US]

NTTC National Technology Transfer Center [Washington, DC, US]

NTTC Nippon Telegraph and Telephone Corporation [Japan]

NTTCIW National Technical Task Committee on Industrial Wastes [US]

NTTF Network Test and Training Facility

NTT LSI Labs NTT Large-Scale Integration Laboratories [of Nippon Telegraph and Telephone Corporation at Atsugi, Japan]

NTTP Network Test and Termination Point

NTT R&D Nippon Telegraph and Telephone Research and Development [Tokyo, Japan]

NTT Rev Nippon Telephone and Telegraph Review [Tokyo, Japan]

NTU Nanyang Technological University [PR China]

NTU Nanying Technological University [Singapore]

NTU National Taiwan University [Taipei]

NTU National Technical University

NTU National Technological University [Fort Collins, Colorado, US]

NTU Naphthylthiourea

NTU Nephelometric Turbidity Unit

NTU Network Terminating (or Termination) Unit

NTU Nishi-Tokyo University [Yamanashi, Japan]

NTU Northern Territory University [Australia]

NTU Number of Transfer Units

NTUA National Technical University of Athens [Greece]

NTV Neutron Television (System) [also N-TV]

NTV Network Television

NTV Nonlinear Thickness Variation

NTW Nonpressure Thermit Welding

NTW NT Workstation [Microsoft Windows]

NTWK Network [also Ntwk, or ntwk]

NTWS Nontrack While Scan

Nt Wt Net Weight [also nt wt]

NTX Naval Telecommunications System [of US Navy]

NTZ Nachrichtentechnische Zeitung [German Journal on Communication Engineering]

NTZ No-Transgression Zone

NTZ Arch Nachrichtentechnische Zeitung Archiv [German Archive of Communication Engineering]

NU Nagasaki University [Japan]

NU Nagoya University [Japan]

NU Name Unknown [also nu]

NU Nigata University [Japan]

NU Nihon University [Tokyo, Japan]

NU Niue [ISO Code]

NU Nonuniform

NU Northeastern University [Boston, Massachusetts, US]

NU Northwestern University [Evanston, Illinois, US]

NU No Umbra (Technique)

NU Number Unobtainable (Tone)

Nu Nucleolus [Cytology]

Nu Numeral [also nu]

Nu Nusselt Number (of Forced Convection) [Symbol]

Nu$_m$ Nusselt Number (of Mass Transfer) [Symbol]

.nu Niue [Country Code/Domain Name]

NUA Network User Address

NUA Network User Association [US]

NUANS Newly Upgraded Automated Name Search (Database)

NUAP National Union of Airline Pilots [Israel]

NUAW National Union of Agricultural Workers [UK]

NUBA National UHF (Ultrahigh Frequency) Broadcasters Association [US]

NUC Nailed-Up Connection

NUC National Union Catalog [of Library of Congress, Washington, DC, US]

NUC Nipissing University College [Ontario, Canada]

NUC Nucleation of Ice Crystals

nuc nuclear

NUCA National Utility Contractors Association [US]

Nucaps National Union of Civil and Public Servants [UK]

NUCDF National Urea Cycle Disorder Foundation [US]

NucE Nuclear Engineer [also Nuc Eng, or NucEng]

NUCEA National University Continuing Education Association [US]

NucEng Nuclear Engineer [also NucEng, or NucE]

NUCGIW National Union of Ceramics and Glass Industry Workers [Israel]

nucl nuclear

Nucleic Acids Res Nucleic Acids Research [Journal published by Oxford University Press, UK]

Nucl Act Nuclear Active [South African Publication]

Nucl Appl Nuclear Applications (Journal)

Nucl Austral Bull Nuclear Australia Bulletin [Published by the Australian Nuclear Society]

Nucl Chem Nuclear Chemist(ry)

Nucl Chem Eng Nuclear Chemical Engineer(ing)

Nucl Chem Waste Manage Nuclear and Chemical Waste Management [Journal published in the US]

Nucl Data Sheets Nuclear Data Sheets [Published in the US]

Nucl Energy Nuclear Energy [Journal of the British Nuclear Energy Society]

Nucl Energy (J Br Nucl Energy Soc) Nuclear Energy (Journal of the British Nuclear Energy Society) [UK]

NuclE Nuclear Engineer [also Nucl Eng, or NuclEng]

Nucl Eng Nuclear Engineer(ing)

Nucl Eng Nuclear Engineering [Journal of the Institution of Nuclear Engineers, UK]

Nucl Eng Des Nuclear Engineering and Design [Journal published in Switzerland]

Nucl Eng Des/Fusion Nuclear Engineering and Design/Fusion [Journal published in Switzerland]

Nucl Eng Int Nuclear Engineering International [Journal published in the UK]

NUCLENOR Controles Nucleares del Norte [Spain]

Nucleo Nucleonic(s) [also nucleo, Nucleon, or nucleon]

Nucl Eur Nuclear Europe [Swiss Publication]

NUCLEX International Nuclear Industrial Fair and Technical Meeting

Nucl Fusion Nuclear Fusion [Published by the International Atomic Energy Agency, Austria]

Nucl Fusion Plasma Phys Nuclear Fusion and Plasma Physics [Published in PR China]

Nucl Ind Nuclear Industry [Publication of the US Council for Energy Awareness]

Nucl Instrum Methods Nuclear Instruments and Methods [Now split into Sections A and B] [Journal published in the Netherlands]

Nucl Instrum Methods A Nuclear Instruments and Methods A [Journal published in the Netherlands]

Nucl Instrum Methods B Nuclear Instruments and Methods B [Journal published in the Netherlands]

Nucl Instrum Methods Phys Res Nuclear Instruments and Methods in Physics Research [Journal published in the Netherlands]

Nucl Instrum Methods Phys Res A, Accel Spectrom Detect Assoc Equip Nuclear Instruments and Methods in Physics Research, Section A: Accelerators, Spectrometers, Detectors and Associated Equipment [Journal published in the Netherlands]

Nucl Instrum Methods Phys Res B, Beam Interact Mater At Nuclear Instruments and Methods in Physics Research, Section A, Beam Interactions with Materials and Atoms [Journal published in the Netherlands]

NUCLIT Nucleare Italiana [Italian Nuclear Corporation]

Nucl Mat Nuclear Material

Nucl Med Nuclear Medicine

Nucl Med Biol Nuclear Medicine and Biology (Journal)

Nucl Met Nuclear Metallurgist; Nuclear Metallurgy

Nucl News Nuclear News [Publication of the American Nuclear Society]

Nucl Phys Nuclear Physicist; Nuclear Physics

Nucl Phys A Nuclear Physics A [Journal published in the Netherlands]

Nucl Phys B, Part Phys Nuclear Physics B, Particle Physics [Journal published in the Netherlands]

Nucl Phys B, Proc Suppl Nuclear Physics B, Proceedings Supplements [Journal published in the Netherlands]

Nucl Prof Nuclear Professional [Publication of the Institute of Nuclear Power Operations, US]

Nucl Saf Nuclear Safety [Publication of the Oak Ridge National Laboratory, Tennessee, US]

Nucl Sci Nuclear Science; Nuclear Scientist

Nucl Sci Appl A Nuclear Science and Applications, Section A [Journal published in Switzerland]

Nucl Sci Appl B Nuclear Science and Applications, Section B [Journal published in Switzerland]

Nucl Sci Eng Nuclear Science and Engineering [Journal of the American Nuclear Society]

Nucl Sci J Nuclear Science Journal [Published by the Atomic Energy Council, Taiwan]

Nucl Tech Nuclear Techniques [Published in PR China]

Nucl Technol Nuclear Technologist; Nuclear Technology

Nucl Technol Nuclear Technology [Publication of the American Nuclear Society, US]

Nucl Technol Nuclear Technology [Journal published in Switzerland]

Nucl Technol/Fusion Nuclear Technology/Fusion [Published by the American Nuclear Society, US]

Nucl Tracks Radiat Meas Nuclear Tracks and Radiation Measurements [Journal published in the UK]

NUDAC Nuclear Data Center

NUDA&GO National Union of Domestic Appliance and General Operatives [UK]

NUDET Nuclear Detection

NUDET Nuclear Detonation

NUDETS Nuclear Detection System

NUDP National Urban Development Program

NUDW National Union of Diamond Workers [Israel]

NUEA National University Extension Association [now National University Continuing Education Association, US]

NUFFIC Netherlands University Foundation for International Cooperation

NUFGW National Union of Flint Glass Workers [UK]

NUG Numerical Algorithms Group

NUGMW National Union of General and Municipal Workers [UK]

NUI National University of Ireland

NUI Network User Identification

NUI Network User Interface

NUI Notebook User Interface

NUIC National University of Ireland at Cork

NUID National University of Ireland at Dublin

NUIG National University of Ireland at Galway

NUJ National Union of Journalists [UK]

Nukl Tehnol Nuklearna Tehnologija [Yugoslavian Publication on Nuclear Technology]

NUL National University of Lesotho [Roma]

NUL Non-GSE (Ground Support Equipment) Utilization List

NUL Null (Character) [Data Communications]

NULACE Nuclear Liquid Air Cycle Engine

NULMW National Union of Lock and Metal Workers [UK]

NULW National Union of Leather Workers and Allied Trades [UK]

NUM National Union of Manufacturers [UK]

NUM National Union of Mineworkers [UK]

NUM National University of Malaysia [Kampong Bangi]

NUM National University of Mexico

Num Number(s) [also num]

Num Numeral [also num]

Num Numeric(s) [also num]

NUMA Non-Uniform Memory Access

NUMAS National Union of Manufacturers Advisory Service [UK]

NUMAST National Union of Marine Transport Officers [UK]

NUME Numerical Methods in Engineering

NUMEC Nuclear Materials and Equipment Corporation

Numer Numeric(al) [also numer]

Numer Anal Numerical Analysis

Numer Eng Numerical Engineering [UK]

Numer Funct Anal Optim Numerical Functional Analysis and Optimization [Published in the US]

Numer Heat Transf Numerical Heat Transfer [Journal published in the US]

Numer Heat Transf A Numerical Heat Transfer A [Journal published in the US]

Numer Heat Transf B Numerical Heat Transfer B [Journal published in the US]

Numer Math Numerische Mathematik [German Publication on Numerical Mathematics]

Numer Methods Partial Diff Equations Numerical Methods for Partial Differential Equations [Journal published in the US]

NUMETA Numerical Methods in Engineering–Theory and Applications

NUMIFORM Numeral Methods in Industrial Forming Processes (Conference)

NUMIS Northwestern University Multislice and Imaging System [US]

Num Numismatic(s)

NUMRC Northwestern University Materials Research Center [Evanston, Illinois, US] [also NU MRC, US]

NUMS National Union of Moroccan Students

NUMS Northwestern University Medical School [Evanston, Illinois, US]

NUMS Nuclear Materials Security

NUMSA Nordic Union of Motor Schools Associations [Denmark]

NUMSTR Numeric String (Conversion)

NUOS Naval Underwater Ordnance Station [of US Navy]

Nuovo Cimento A Nuovo Cimento, Section A [Physics Journal published by Società Italiana de Fisica, Italy]

Nuovo Cimento B Nuovo Cimento, Section B [Physics Journal published by Società Italiana de Fisica, Italy]

Nuovo Cimento C Nuovo Cimento, Section C [Physics Journal published by Società Italiana de Fisica, Italy]

Nuovo Cimento D Nuovo Cimento, Section D [Physics Journal published by Società Italiana de Fisica, Italy]

Nuovo Sagg Nuovo Saggiatore [Physics Journal published by Società Italiana de Fisica, Italy]

NUP Nuevo Peso [Currency of Uruguay]

NUPAD Nuclear-Powered Active Detection System

NUPBPW National Union of Printing, Bookbinding and Paper Workers [now Society of Graphical and Allied Trades, UK]

NUPE National Union of Public Employees [UK]

NUPPSCO Nuclear Power Plant Standards Committee

NUR National Union of Railwaymen [UK]

NUR Net Unduplicated Research

NUR Neutron Radio Facility [of Malaysian Institute of Nuclear Technology and Research, Kajang]

NUR-II Neutron Radio Facility II [of Malaysian Institute of Nuclear Technology and Research, Kajang]

NURBS Nonuniform Rational B-Spline [Computers]

NURC National Undersea Research Centers [of National Oceanic and Atmospheric Administration, US]

NuRd Neutral Red

NURDC Naval Undersea Research and Development Center [of US Navy at San Diego, California]

NURE National Uranium Resource Evaluation (Program) [US]

Nurs Mirror Midw J Nursing Mirror and Midwives Journal [US]

Nurs Outlook Nursing Outlook [US Publication]

Nurs Res Nursing Research [US Journal]

NURP National Undersea Research Program [of National Oceanic and Atmospheric Administration, US]

NURTIW National Union of Rubber and Tire Industries Workers [Israel]

NUS National Union of Seamen [UK]

NUS National Union of Students [UK]

NUS National University of Singapore

NUS Nuclear Utility Service

NUSAS National Union of South African Students

NUSC Naval Underwater Systems Center [of US Navy]

NUSD Nagasaki University School of Dentistry [Japan]

NUSL Naval Underwater Sound Laboratory [of US Navy]

NUSMWC National Union of Sheetmetal Workers and Coppersmiths [UK]

NUSS Nuclear Physics Summer School and Symposium [of Seoul National University, South Korea]

NUSS Nuclear Safety Standard(s)

NUT National Union of Teachers [UK]

NUT Northeastern University of Technology [Shenyang, PR China]

NUT Number Unobtainable Tone

Nut Nutrient [also nut]

NUTEC Norwegian Underwater Technology Center

NUTGLW National Union of Textile, Garment and Leather Workers [Israel]

NUTP National Uranium Tailings Program [of Canada Center for Mineral and Energy Technology, Natural Resources Canada]

NUTPE National Union of Technicians and Practical Engineers [Israel]

Nutr Action Health Nutrition Action Health (Newsletter) [of Center for Science in the Public Interest, US]

Nutr Cancer Nutrition and Cancer (Journal) [US]

Nutr News Nutrition News [US]

Nutr Q Nutrition Quarterly [Canada]

Nutr Rep Nutrition Reports [US]

Nutr Rep Int Nutrition Reports International

Nutr Res Nutrition Research [International Journal]

Nutr Rev Nutrition Reviews [US]

NUTWTC National Union of Transport Workers and Transport Cooperatives [Israel]

NUV Near-Ultraviolet

NUVB National Union of Vehicle Builders [UK]

NUWC Naval Undersea Warfare Center [of US Navy]

NUWES Naval Undersea Warfare Engineering Station [of US Navy]

NV Natural Vibration

NV Needle Valve

NV Negative Very-Small (Error)

NV Neurotropic Virus

NV Nevada [US]

NV New Version

NV Night Vision

NV Nile Valley [Africa]

NV Noise Voltage

NV Nominal Value

NV Nonvolatile; Nonvolatility

NV No Overflow (Flag) [Computers]

NV Nozzle Velocity

NV No Value [also nv]

NV Number and Volume (Formulation) [Physics]

N/V Electron Concentration [Symbol]

N/V Number per Unit Volume [Symbol]

nV nanovolt [Unit]

nV noncoherent vacancy (in materials science) [Symbol]

nV$_{rms}$ nanovolt, root-mean-square [Unit]

NVA Norske Videnskaps Akademi [Norwegian Academy of Sciences, Oslo]

nVA n-Valeric Acid

Nva Norvaline; Norvalyl [Biochemistry]

NVACP Neighborhood Voluntary Associations and Consumer Protection [of Organization of American States]

NVAFB North Vandenberg Air Force Base [California, US]

NVAO Norske Videnskaps-Akademi i Oslo [Norwegian Academy of Sciences at Oslo]

NVB Navigational Base

NVBO Natural Valence-Band Offset [Solid-State Physics]

NVBF Nordiska Vetenskapliga Bibliothekarieforbundet [Scandinavian Federation of Scientific Research Libraries, Denmark]

NVC Normal Valence Compound

NVC N-Vinylcarbazole

NVCA National Van Conversion Association [US]

NVCASE National Voluntary Conformity Assessment System Evaluation (Program)

NVD Nausea, Vomiting and Diarrhea [also N,V,D]

NVE Neurotropic-Virus Encephalitis [Medicine]

NVE Nonvolatile Electronics Inc. [Eden Prairie, Minnesota, US]

NVE Number, Volume and Energy (Formulation) [Physics]

(N,V,E) Number of Particles, Volume of System and Internal Energy

–ve negative

NVEOD Night Vision and Electrooptics Directorate

NVF Nordisk Vejteknisk Forbund [Nordic Road Association, Denmark]

NVFR Naturvetenskapliga Forskningsrådet [Scientific Research Council, Denmark]

NVFRAM Nonvolatile Ferroelectric Random-Access Memory

NVG Night Vision Goggles

NVGA National Vocational Guidance Association [US]

NVH Noise, Vibration and Harshness (Characteristics)

nV√Hz nanovolt square root hertz [also nV Hz$^{1/2}$]

NVIC National Vaccine Information Center [Vienna, Virginia, US]

NVLA National Vehicle Leasing Association [US]

NVLAP National Voluntary Laboratory Accreditation Program [of National Institute of Standards and Technology, Gaithersburg, Maryland, US]

NVM Nonvolatile Matter

NVM Nonvolatile Memory

NVMA National Veterinary Medical Association [UK]

NVOCC Non-Vessel Operation Common Carrier [US]

NVOC Chloiride 6-Nitroveratryl Chloroformate

NVOD Near-Video On Demand

NVP Nominal Velocity of Propagation

NVPOWG NASA/VAFB (Vandenberg Air Force Base) Payload Operations Working Group [US]

NVR Nonvolatile Residue

NVR No Verification Required

NVR No Voltage Release

NVRAM Nonvolatile Random-Access Memory [also NV RAM]

NVRS National Vegetable Research Station [UK]

NVS Nationale Verzchrstudie [National Consumption Study, Germany]

NVS National Vacuum Symposium [of American Vacuum Society, US]

NVS National Vegetable Society [UK]

NVS Night Vision System

NVSAVS National Vacuum Symposium of the American Vacuum Society [US]

NVSD Night Vision System Development

NVT Network Virtual Terminal

NVT Nosé Constant-Temperature [*Note:* "V" stands for Volume and "T" for Temperature] [Materials Science]

NVT Novell Virtual Terminal

NW Narrow Well

NW NASA Waiver

NW Net Weight [also nw]

NW Network [also N/W]

NW Neville-Winter (Acid) [Organic Chemistry]

NW Niels and Wadsworth (Ductile-to-Brittle Transition Temperature Estimation) [Metallurgy]

NW Nishiyama-Wassermann (Orientation Relationship) [also N-W] [Crystallography]

NW Nonavailable Water [Ecology]

NW Nonwoven

NW North Wales [UK]

NW Northwest(ern)

NW Northwest Orient Airlines [US]

NW Nose Wheel

NW Now [Amateur Radio]

NW Nuclear Waste

NW Nuclear Weapon

N/W Network [also NW]

N-W Nishiyama-Wassermann (Orientation Relationship) [also NW] [Crystallography]

nW nanowatt [Unit]

n-W Nanocrystalline Tungsten

NWA National Waterfowl Alliance [US]

NWA National Weather Association [US]

NWA Neville-Winter Acid [Organic Chemistry]

NWA North West Africa

NWA Northwest Airlines [US]

NWA Norwegian Water Association

NWAC National Weather Analysis Center [US]

NW-Acid Neville-Winter Acid [also NW acid] [Organic Chemistry]

NWAG Naval Warfare Analysis Group [of US Navy]

NWAHACA National Warm Air Heating and Air Conditioning Association

N Wales North Wales [UK]

NWASC Northwest Association of School and Colleges [US]

NWB National Wiring Bureau [US] [Defunct]

nWb nanoweber [Unit]

NW by N Northwest by North

NW by W Northwest by West

NWC National War College [Washington, DC, US]

NWC National Water Center [US]

NWC National Water Commission [US]

NWC National Water Conditions [Report of United States Geological Survey/Water Resources Division]

NWC National Waterfowl Council [US]

NWC National Watershed Congress [US]

NWC National Waterways Conference [US]

NWC Naval War College [Newport, Rhode Island, US]

NWC Naval Weapons Center [of US Navy]

NWC Net Weekly Circulation

NWCC National Water Companies Conference [now National Association of Water Companies, US]

NWCC National Weed Committee of Canada

NWCC National Women's Consultative Council

NWCF Northwest Citizens Forum [US]

NWCP Noxious Weeds Control Program [Northern Territory, Australia]

NWD Network Wide Directory

NWDS National Wetlands Data System [US]

NWDS Number of Words

NWDSEN Number of Words per Entry

NWEA National Wood Energy Association [US]

NWEB North Western Electricity Board [UK]

NWEF Naval Weapons Evaluation Facilities [of US Navy]

NWEP Nuclear Weapons Effects Panel [US]

NWF National Wildlife Federation [US]

NWF Nuclear Waste Fund [US]

NWF Numerical Weather Forecasting

NWFA National Wood Flooring Association [US]

NWFA Northwest Farm Managers Association [US]

NWFCCC National Wildlife Federation Corporate Conservation Council [US]

NWG National Wire Ga(u)ge

NWG National Working Group

NWGA National Wholesalers and Grocers Alliance [Ireland]

NWGA National Wool Growers Association [US]

NWGCM National Working Group on Coastal Management

NWGS Northwest Geological Society

NWH Normal Working Hours

NWHA National Wholesale Hardware Association [US]

NWHC National Wildlife Health Center [of US Geological Survey/Biological Resources Division at Madison, Wisconsin]

NWHF National Wildlife Health Foundation [US]

NWHN National Women's Health Network [US]

NWI National Wetlands Inventory [of US Fish and Wildlife Service]

NWI National Wilderness Inventory [of Australian Heritage Commission]

NWI Net Working Interest

NWIC National Wheat Improvement Committee [US]

NWIS Network of (Minority) Women in Science [US]

NWI National Wetlands Inventory (Program) [US]

Nwk Network [also nwk]

NWL Naval Weapons Laboratory [of US Navy]

NWLDYA National Wholesale Lumber Distributing Yard Association [now Hardwood Distributors Association, US]

NWM Network Management

NWM Nuclear Waste Management

NWMA National Woodwork Manufacturing Association [US]

NWMA Northern Woods Logging Association [US]

NWMA Northwest Mining Association [US]

NWMSC Northwest Missouri State College [Maryville, US]

NWN National Wireless Network

NWNET Northwest Net [Bellevue, Washington State, US]

NWO Nederlandse Organisatie voor Wetenschappelijk Onderzoek [Netherlands Organization for Scientific Research, The Hague, Netherlands]

NWO Nonwoven Oriented

NWOA National Woodland Owners Association [US]

NWP National Water Project [US]

NWP National Weather Prediction [US]

NWP National Writing Project [US]

NWP Naval Weapons Plant [US]

NWP North West Passage [Canadian Arctic]

NWP Numerical Weather Prediction

NWPA National Wildlife Protection Act [Australia]

NWPA Nuclear Waste Policy Act [US]

NWPC Nordic Wood Preservation Council [Finland]

NWPCA National Wooden Pallet and Container Association [US]

NWPF National Water Purification Foundation [US]

NWPFC Northwest Pacific Fisheries Commission

NWPU National Wildlife Protection Unit [of Australian Nature Conservancy Agency]

NWQL National Water Quality Laboratory [West Kingston, Rhode Island, US]

NWQL National Water Quality Laboratory [Canada]

NWQMS National Water Quality Management Strategy [Australia]

NWR National Wildlife Refuges [of US Fish and Wildlife Service]

NWRA National Water Resources Association [US]

NWRA National Wheel and Rim Association [US]

NWRA National Wildlife Refuge Association [US]

NWRC National Weather Records Center [of Environmental Science Services Administration, US]

NWRC National Wetlands Research Center [of US Geological Survey/Biological Resources Division at Lafayette, Madison, Wisconsin]

NWRC National Wildflower Research Center [US]

NWRC National Wildlife Research Center [Canada]

NWRF Naval Weather Research Facility [of US Navy]

NWRI National Water Research Institute

NWRT National Wildlife Rescue Team [US]

NWRV Nord– und Westdeutscher Rundfunkverband [Association of Broadcasting Stations of Northern and Western Germany]

NWS National Weather Service [of National Oceanic and Atmospheric Administration, US]

NWS NetWare Web Server

NWS Neutral Waste System

NWS North Warning System [US/Canada]

NWS Nose-Wheel Steering

NWS Nuclear Waste Storage

NWS Nuclear Weapons State

NWSA National Welding Supply Association [US]

NWSA Nose Wheel Steering Amplifier

NWSC National Weather Satellite Center [US]

NWSC National Women Student's Coalition [US]

NWSC Northwestern State College [Alva, Oklahoma, US]

NWSCL Northwestern State College of Louisiana [Natchitoches, US]

NWSD Naval Weather Service Division [of US Navy]

NWSFO National Weather Service Forecast Office [of National Oceanic and Atmospheric Administration, US]

NWSI New World Services Inc. [US]

NWSIA National Water Supply Improvement Association [US]

NWSIAH North Western Society of Industrial Archeology [UK]

NWS-OH National Weather Service–Office of Hydrology [US]

Nwspr Newsprint [also nwspr]

NWSRFS National Weather Service River Forecast System [US]

NWS RTQC North Warning System/Real-Time Quality Control

NWSSG Nuclear Weapons System Satellite Group [US]

NWT Nationaler Wettbewerb für die besten Tonaufnahmen [National Competition for the Best Sound Recordings, Germany]

NWT Non-Waste Technology

NWT Nonwatertight

NWT Northwest Territories [Canada]

N Wt Net Weight [also n wt]

NWTC National Wetlands Technical Council [US]

NWTEC National Wool Textile Export Corporation [UK]

NWTFL Northwest Territories Federation of Labour [Canada]

NWTI National Wood Tank Institute [US]

NWTRB Nuclear Waste Technology Review Board [US]

NWTRNA Northwest Territories Registered Nurses Association [Canada]

NWTS National Waste Terminal Storage (Program)

NWU Nara Women's University [Nara, Japan]

NWU National Workers Union (of Jamaica)

NWU Northwestern University [Evanston, Illinois, US]

NWUMRC Northwestern University Materials Research Center [US]

NWUS National Workers Union of the Seychelles

NW USA North Western United States of America [also NW-USA]

NWW Nose Wheel Well

NWWA National Water Well Association [US]

NWWDA National Wood Window and Door Association [US]

NX Normal to X-Axis

$N_d(x)$ (Charge) Carrier Concentration Profile [Symbol]

$n(x)$ mean amount of substance of entity "x" [Symbol]

NXDO Nike X Development Office [US]

9XR- Rwanda [Civil Aircraft Marking]

nxt next

NY Needle Roller Bearing, Drawn Cup, Full Complement, Rollers Retained by Lubricant, Without Inner Ring, Metric-Dimensioned [Symbol]

NY New York [US]

NY Normal to Y-Axis

NY Northern Yemen

NY Nuclear Yield [Nuclear Weapons]

9Y- Trinidad and Tobago [Civil Aircraft Marking]

Ny Nylon

Ny4 Nylon 4

Ny6 Nylon 6

Ny6,6 Nylon 6,6

Ny6,9 Nylon 6,9

Ny6,10 Nylon 6,10

Ny6,12 Nylon 6,12

Ny9 Nylon 9

Ny11 Nylon 11

Ny12 Nylon 12

Ny46 Nylon 46

NYA New York Aquarium [Coney Island, Brooklyn, US]

NZA New Zealand Army

NYAB Neodymium-Yttrium-Aluminum Borate (Crystal)

NYADS New York Air Defense Sector [US]

NYAM New York Academy of Medicine [US]

NYANG New York Army National Guard [US]

NYAP New York Assembly Program [US]

NYARTCC New York Air Route Traffic Control Center [US]

NYAS New York Academy of Science [US]

NYBCO Neodymium Yttrium Barium Copper Oxide (Superconductor)

NYBC New York Blood Center [US]

NYBG New York Botanical Garden [Bronx Park, US]

NYBPE New York Business Press Editors [US]

NYBT New York Board of Trade [US]

NYC Neighborhood Youth Corps [US]

NYC New York Central (Railroad) [US]

NYC New York City [US]

NYCBAN New York Center Beacon Alphanumerics [of Federal Aviation Administration, US]

NYCDH New York City Department of Health [US]

NYCE New York Cotton Exchange [also NYCTN, US]

NYCCC New York City Community College [US]

NYCS New York Cipher Society [US]

NYCSE New York Coffee and Sugar Exchange [US]

NYCTC New York City Technical College [of City University of New York, US]

NYCTN New York Cotton Exchange [also NYCE, US]

NYCX New York Commodity Exchange [US]

NYD Navy Yard

NYD Not Yet Detected [also nyd]

NYD Not Yet Determined [also nyd]

NYES New York Electrical Society [US]

NYES New York Entomological Society [US]

NYFD New York Fire Department [US]

NYFE New York Futures Exchange [US]

NYGS New York Graphic Society [US]

NYH New York Hospital [New York City, US]

NYHS New York Historical Society [US]

NYI National Young Investigator

NYI NSF Young Investigator (Awards Program) [of National Science Foundation, US]

NYIBOR New York Interbank Offered Rate [also Nyibor]

NYID Not-Yet-Invented Device

NYIT New York Institute of Technology [US]

Nyl Nylon [also nyl]

NYLSMA New York Lamp and Shade Manufacturers Association [now Eastern Lamp and Lighting Association, US]

NYM Needle Roller Bearing, Drawn Cup, Full Complement, Rollers Retained by Lubricant, Closed End, Without Inner Ring, Metric-Dimensioned [Symbol]

NYMC New York Medical College [Valhalla, New York, US]

NYME New York Metal Exchange [US]

NYMEX New York Mercantile Exchange [US]

NYMH New York Memorial Hospital [US]

NYMRRLA New York Metropolitan Reference and Research Library Agency [US]

NYNS New York Naval Shipyard [of US Navy]

NYO New Yellow-Orange (Fluorescence)

NYO New York Operations Office [US]

NYO Not Yet Operating [also nyo]

n-Y$_2$O$_3$ Nanocrystalline Yttria (or Yttrium Dioxide)

NYOD New York Ordnance District [US]

NYP Not Yet Published [also nyp]

NYPA New York Port Authority [US]

NYPA New York Power Authority [US]

NYPC New York Paint Club [US]

NYPC New York Pigment Club [US]

NYPD New York Police Department [US]

NYPL New York Public Library [US]

NYPP New York Power Pool [US]

NY Press New York Press [US]

NYPS National Yellow Pages Service [US]

NYPSC New York Public Service Commission [US]

NYS New York State [US]

NYSA New York Shipping Association [US]

NYSAA New York State Archeological Association [US]

NYSAES New York State Agricultural Experiment Station [Geneva, US]

NYSASS New York State Association of Service Stations [US]

NYSCA New York State College of Agriculture [US]

NYSCALS New York State College of Agriculture and Life Sciences [US]

NYSCAMP New York State Center for Advanced Materials Processing [of Clarkson University, Potsdam, US]

NYSCC New York State College of Ceramics [of Alfred University, US]

NYSDC New York State Department of Commerce [US] [also NYSDOC]

NYSDEC New York State Department of Environmental Conservation [US]

NYSDH New York State Department of Health [Albany, US]

NYSDOC New York State Department of Commerce [US] [also NYSDC]

NYSDOT New York State Department of Transportation [US]

NYSE New York Stock Exchange [US]

NYSEG New York State Electric and Gas [US]

NYSEM New York Society of Electron Microscopy [US]

NYSERDA New York State Energy Research and Development Authority [US]

NYSFTCA New York State Fruit Testing Cooperative Association [US]

NYSHS New York State Historical Society [US]

NYSILL New York State Interlibrary Loan System [US]

NYSIS New York State Institute of Science [US]

NYSIS New York State Institute for Superconductivity [of State University of New York at Buffalo, US]

NYS J Med New York State Journal of Medicine

NYSPA New York State Pharmaceutical Association [US]

NYSPA New York State Power Authority [US]

NYSSIM New York State Society of Industrial Medicine [US]

NYSSPE New York State Society of Professional Engineers [US]

NYSSTF New York State Science and Technology Foundation [Albany, US]

NYST New York State Thruway [US]

NYSTA New York State Thruway Association [US]

NYSU New York State University [US]

Ny Tek Ny Teknik [Swedish Publication on New Technology]

NYTIS New York Times Information Service [US]

NYU New York University [New York City, US]

NYUMC New York University Medical Center [New York City, US]

NYU/IBM New York University/International Business Machines Corporation (Ultracomputer)

NYUIMS New York University Institute of Mathematical Sciences [New York City, US]

NYU Press New York University Press [New York City, US]

NYUST National Yunlin University of Science and Technology [Toulin, Taiwan]

NYZG New York Zoological Garden [Bronx Park, US]

NYZP New York Zoological Park [New York City, US]

NYZS New York Zoological Society [US]

NZ Neutral Zone

NZ New Zealand

NZ New Zealand [ISO Code]

NZ Noise Zone

NZ Normal to Z-Axis

NZ No Zero (Flag) [Computers]

.nz New Zealand [Country Code/Domain Name]

NZA Nickel Zirconium Aluminide

NZAB New Zealand Association of Bacteriologists

NZACT New Zealand Association of Chemistry Teachers

NZAEI New Zealand Agricultural Engineering Institute

NZAOC New Zealand Army Ordnance Corps

NZAP New Zealand Antarctic Program

NZAS New Zealand Academy of Sciences

NZASc New Zealand Association of Scientists

NZB Non Zero Binary

NZBS New Zealand Broadcasting Service

NZCER New Zealand Council for Educational Research

NZCWPRC New Zealand Committee for Water Pollution Research and Control

NZ$ New Zealand Dollar [Currency of Cooke Islands, Tokelau and New Zealand] [also NZD]

NZDA New Zealand Department of Agriculture

NZDA New Zealand Dietetic Association

NZDCS New Zealand Department of Census and Statistics

NZDLS New Zealand Department of Lands and Survey

NZDSIR New Zealand Department of Scientific and Industrial Research

NZE Nuclear Zeeman Effect [Spectroscopy]

N Zeal New Zeland

NZEI New Zealand Electronics Institute

NZ Electron Rev New Zealand Electronics Review [Published by the National Electronics Development Association]

NZ Energy J New Zealand Energy Journal

NZ Eng New Zealand Engineering [Journal of the New Zealand Institution of Professional Engineers]

NZES New Zealand Ecological Society

NZF Nickel Zinc Ferrate

NZFMRA New Zealand Fertilizer Manufacturers Research Association

NZFRI New Zealand Forest Research Institute

NZFS New Zealand Forest Service

NZGA New Zealand Grassland Association

NZGenS New Zealand Genetical Society

NZGS New Zealand Geographical Society

NZIAS New Zealand Institute of Agricultural Science

NZIC New Zealand Institute of Chemistry

NZIE New Zealand Institute of Engineers

NZIE Trans Civ Eng NZIE Transactions, Civil Engineering Section [Published by the New Zealand Institution of Engineers]

NZIE Trans Electr/Mech/Chem Eng NZIE Transactions, Electrical/Mechanical/Chemical Engineering Section [Published by the New Zealand Institution of Engineers]

NZIF New Zealand Institute of Foresters

NZIIA New Zealand Institute of International Affairs

NZIIRD New Zealand Institute of Industrial Research and Development [Lower Hutt]

NZInstW New Zealand Institute of Welding [also NZIP]

NZIP New Zealand Institute of Physics

NZIPE New Zealand Institution of Professional Engineers

NZIRE New Zealand Institute of Refrigeration Engineers

NZIW New Zealand Institute of Welding [also NZInstW]

NZIW Bull NZIW Bulletin [Published by the New Zealand Institute of Welding]

NZJCB New Zealand Joint Communication Board

NZ J Dairy Sci Technol New Zealand Journal of Dairy Science and Technology

NZ J Forestry Sci New Zealand Journal of Forestry Science

NZLA New Zealand Library Association

NZ Med J New Zealand Medical Journal

NZMS New Zealand Meteorological Service

NZNAC New Zealand National Airways Corporation

NZNCIAWPRC New Zealand National Committee of the International Association on Water Pollution Research and Control

NZNCOR New Zealand National Committee on Oceanic Research

NZNEDA New Zealand National Electronics Development Association

NZNRAC New Zealand National Research Advisory Council

n-ZnS N-Type Zinc Sulfide

NZP National Zoological Park [of Smithsonian Institution, Washington, DC, US]

NZP Sodium Zirconium Phosphate (Ceramics) [$NaZr_2(PO_4)_3$]

NZPO New Zealand Post Office

n-ZrO$_2$ Nanocrystalline Zirconia (or Zirconium Dioxide)

NZS Nickel Zirconium Silicide

NZSA New Zealand Statistical Association

NZSAP New Zealand Society of Animal Production

NZSB New Zealand Soil Bureau

NZSCA New Zealand Soil Conservation Association

NZSI New Zealand Standards Institute [now Standards Association of New Zealand]

NZSLO New Zealand Scientific Liaison Office

NZSSS New Zealand Society of Soil Science

NZT New Zealand Time [Greenwich Mean Time −12:00]

NZT Non-Zero Transfer

NZT Sodium Zirconium Phosphate

NZ-UKCCI New Zealand-United Kingdom Chamber of Commerce and Industry

NZVA New Zealand Veterinary Association

NZW Sodium Zirconium Tungstate

NZWAA New Zealand Weed Control Conference

NZWC New Zealand Wool Commission

O   Annealed, Recrystallized (Wrought Products) [Basic Temper Designation for Aluminum and Magnesium Alloys]

O   Blood Group "O" [Contains Red Blood Cells with neither Substance "A" nor "B" on Surface]

O   Center of Symmetry at the Octahedral Site [Crystallographic Symbol Appended to Space Group Denotation]

O   Circumcenter (of a Triangle) [Symbol]

O   Longitudinal Compression Compliance [Symbol]

O   Oahu [Hawaiian Islands]

O   Oak

O   Oakland [California, US]

O   Object; Objective; Objectivity

O   Oblate(ness)

O   Obligate [Biology]

O   Observe(r); Observation; Observatory

O   Observation Aircraft [US Navy Symbol]

O   Obstruct(ion)

O   Occlude(d); Occlusion

O   Occupation(al)

O   Occur(ence)

O   Ocean(ic)

O   Oceania(n)

O   Octa-

O   Octagon(al)

O   Octagonal [Crystallographic Symmetry Group Symbol]

O   Octahedral; Octahedron

O   Octane

O   *(Octarius)* – pint [Medical Prescriptions]

O   Octave

O   October

O   Octopus

O   Oculomotor [Anatomy]

O   Octupole [Physics]

O   Oder (River) [Central Europe]

O   Odessa [Ukraine]

O   Oedogonium [Genus of Algae]

O   Oestrus [Genus of Flies]

O   Office(r)

O   Official

O   Offset

O   Ohio [US]

O   Oklahoma [US]

O   Oil; Oiled; Oiling

O   Oiler

O   Oil-Hardening Type Cold-Work Tool Steels [AISI-SAE Symbol]

O   Old

O   Olefin(ic)

O   Olfactory

O   Olympia [Washington State, US]

O   Omaha [Nebraska, US]

O   Oman

O   (Upper-case) Omicron [Greek Alphabet]

O   Onchoceriasis [Medicine]

O   Omasum (of Ruminant Stomach) [Zoology]

O   Omnivore; Omnivorous

O   Oncorhynchus [Genus of Fish Including the Salmon]

O   On-Line

O   Ontario [Canada]

O   Opacify; Opacifier; Opacity

O   Opal

O   Opalescence; Opalescent

O   Opaque

O   Open(ing)

O   Operand

O   Operate(d); Operating

O   Operating Number [Symbol]

O   Operation; Operator

O   Opinion

O   Opium

O   Oppose; Opposite; Opposition

O   Optic(al); Optician; Optics

O   Optimal; Optimize; Optimization; Optimum

O   Option(al)

O   Opsonin [Immunology]

O   Oral(ly)

O   Orange

O   Orbit(al); Orbiter

O   Order(ing)

O   Ordered (Phase) [Symbol]

O   Ordinary

O   Ordnance

O   Ordovician [Geological Period]

O   Ore

O   Oregon [US]

O   Organ(ic); Organics

O   Organism

O   Orient(ation)

O   Orifice

O   Origin(al)

O   Origin (of Coordinates) [Symbol]

O   Orinoco (River) [Venezuela]

O   Orlando [Florida, US]

O   Orthohedron

O   Orthohedral (Site) [Symbol]

O   Orthogonal(ity)

O   Orthogonal Group (in Mathematics) [Symbol]

O   Orthorhombic (Crystal System)

O   Osaka [Japan]

O   Osazone [Biochemistry]

O   Oscar [Phonetic Alphabet]

O   Oscillate; Oscillating; Oscillation; Oscillator(y)

O   Oscillograph

O   Oscillometer; Oscillometry

O   Oscilloscope

O   Oslo [Norway]

O   Osmometer; Osmometry

O   Osmosis; Osmotic

O   O-Shell (of Electron) [Symbol]

O   Osseous [Anatomy]

O   Ossification; Ossify [Medicine]

O   Osteocyte [Medicine]

O   Osteon [Medicine]

O   Osteoporosis [Medicine]

O   Other Engineering (Technology) Discipline [Discipline Category Abbreviation]

O   Ottawa [Canada]

O   Outdoor(s)

O   Outflow

O   Outgas(sing)

O   Output

O   Outside Diameter (of Bevel Gears, Helical Gears, etc.) [Symbol]

O   Oval(ity)

O   Ovaloid

O   Ovarian; Ovary [Anatomy]

O   Oven

O   Overall

O   Overall Readability [Symbol]

O   Overflow

O   Overhead

O   Overtone

O   Oviduct [Anatomy]

O   Ovulate; Ovulation

O   Ovule [Botany]

O   Ovum [Cytology]

O   Oxazolone

O   Oxidant; Oxide; Oxidizable; Oxidize(r); Oxidizing; Oxidation

O   Oxychloride Type (Grinding Wheel) Bond

O   Oxygen [Symbol]

O   Oxygenate; Oxygenation; Oxigenator

O   Oxytocic [Medicine]

O   Oyster

O   Ozone

O   Ozonide

O   Ozonization; Ozonizer

O   Ozonolysis

O   Ozonosphere

O   Pint [Medical Prescriptions]

O   Star Group of Spectral Type O [Surface Temperature 22,200°C, or 40,000°F] [Letter Designation]

O   Yarn Over [Woven Fabrics]

/O   Object Directories Option [Turbo Pascal]

O·   Oxygen Radical [Symbol]

O″   Loss Compliance [Symbol]

O^+ Special Orthogonal Group [Symbol]

O_2 Molecular Oxygen [Symbol]

$^{16}O_2$ Molecular Oxygen-16 [Symbol]

$^{18}O_2$ Molecular Oxygen-18 [Symbol]

O^{2-} Oxygen Ion [also O^{--}] [Symbol]

O_2^- Charged Oxygen Molecule [Symbol]

O_3 Ozone [Symbol]

O_3G Ozone Ion [Symbol]

O_{95} Apparent Opening Size (for Geotextiles) [Symbol]

O_B Bridging Oxygen [Symbol]

O_e Octahedral Slip in Easy Glide Configuration (in Metallurgy) [Symbol]

O_{NB} Non-Bridging Oxygen [Symbol]

O_u Octahedral Slip in Uneasy Glide Configuration (in Metallurgy) [Symbol]

O II   Singly Ionized Oxygen

O III   Doubly Ionized Oxygen

O-1   Ensign [US Navy and Coast Guard Grade]

O-1   Second Lieutenant [US Army, Air Force and Marine Corps Grade]

O-2   First Lieutenant [US Army, Air Force and Marine Corps Grade]

O-2   First Lieutenant, Junior Grade [US Navy and Coast Guard Grade]

O-3   Captain [US Army, Air Force and Marine Corps Grade]

O-3   Lieutenant [US Navy and Coast Guard Grade]

O-4   Major [US Army, Air Force and Marine Corps Grade]

O-4   Lieutenant Commander [US Navy and Coast Guard Grade]b

O-5   Lieutenant Colonel [US Army, Air Force and Marine Corps Grade]

O-5   Commander [US Navy and Coast Guard Grade]

O-6   Colonel [US Army, Air Force and Marine Corps Grade]

O-6   Captain [US Navy and Coast Guard Grade]

O-7   Brigadier General [US Army, Air Force and Marine Corps Grade]

O-7   Rear Admiral, Lower Half [US Navy and Coast Guard Grade]

O-8   Major General [US Army, Air Force and Marine Corps Grade]

O-8 Rear Admiral, Upper Half [US Navy and Coast Guard Grade]

O-9 Lieutenant General [US Army, Air Force and Marine Corps Grade]

O-9 Vice Admiral [US Navy and Coast Guard Grade]

O-10 General [US Army, Air Force and Marine Corps Grade]

O-10 Admiral [US Navy and Coast Guard Grade]

O-11 General of the Air Force [US Air Force Grade]

O-11 General of the Army [US Army Grade]

O-11 Admiral of the Fleet [US Navy Grade]

O-14 Oxygen-14 [also ^{14}O, or O^{14}]

O-15 Oxygen-15 [also ^{15}O, or O^{15}]

O-16 Oxygen-16 [also ^{16}O, O^{16}, or O]

O-17 Oxygen-17 [also ^{17}O, or O^{17}]

O-18 Oxygen-18 [also ^{18}O, or O^{18}]

O-19 Oxygen-19 [also ^{19}O, or O^{19}]

o normal stress [Symbol]

o occupy

o occupancy (of crystal lattice site) [Symbol]

o oral(ly)

o orange

o ordinary

o ortho-

o orthorhombic (crystal system) [Symbol]

o ovarian; ovary [Anatomy]

o ovulate; ovulation

.o object file [File Name Extension]

o- ortho- [Chemistry]

$\bar{o}$ omega [English Equivalent]

/o overlay interrupt option [MS-DOS Linker]

o (lower-case) omicron [Greek Alphabet]

Ω Activation Volume [Symbol]

Ω Angular Velocity of Precession [Symbol]

Ω Atomic Volume [Symbol]

Ω Average Phonon Frequency [Symbol]

Ω Average Volume per Atom [Symbol]

Ω Brocard Point of a Triangle [Symbol]

Ω Bulk Free Energy [Symbol]

Ω Cracking Coefficient [Symbol]

Ω Degree of Orientation (in Quantitative Metallography) [Symbol]

Ω Frequency [Symbol]

Ω Fundamental Frequency [Symbol]

Ω Grand Potential [Symbol]

Ω Magnetic Potential [Symbol]

Ω Modulating Frequency [Symbol]

Ω Molar Volume [Symbol]

Ω Molecular Volume [Symbol]

Ω Ohm [Unit]

Ω (Upper-case) Omega [Greek Alphabet]

Ω Omega Particle (or Omega Hyperon) [Symbol]

Ω Plasma Frequency [Symbol]

Ω Reflection Coefficient [Symbol]

Ω Relaxation Frequency [Symbol]

Ω Solid Angle [Symbol]

Ω Spherical Image [Symbol]

Ω Supersaturation [Symbol]

Ω Thermodynamic Work Potential [Symbol]

Ω Volume per Ion Pair [Symbol]

Ω Wavefunction [Symbol]

[Ω] Relative Molecular Magnetic Rotatory Power with Respect to Water [Symbol]

Ω^- Omega-Minus Particle [Symbol]

$\overline{\Omega}^-$ Omega-Minus Antiparticle [Symbol]

Ω_0 Atomic Volume of Pure Solvent [Symbol]

Ω^{-1} Reciprocal Ohm (or One per Ohm) [also 1/Ω]

Ω_a Atomic Volume of Solute in α Phase [Symbol]

Ω_{pl} Fractional (or Percentage) Amount of Planar Orientation (in Quantitative Metallography) [Symbol]

Ω_S Atomic Volume of Solute(s) [Symbol]

ω angle subtended by image (in optical instrument) [Symbol]

ω angular frequency; angular velocity; circular frequency [Symbol]

ω conical (a modifier for radiometric properties and quantities) [Symbol]

ω disclination (in metallurgy) [Symbol]

ω dispersion [Symbol]

ω dispersive power (of a lens, etc.) [Symbol]

ω ellipticity (of reflected light) (in ellipsometry) [Symbol]

ω energy transfer [Symbol]

ω excitation energy [Symbol]

ω first transfinite ordinal, order type of the set of all positive integers in natural order [Symbol]

ω frequency of an alternating-current magnetic field [Symbol]

ω growth rate [Symbol]

ω imaginary root of unity (usually an imaginary cube root, $\frac{1}{2}(-1+I\sqrt{3})$) [Symbol]

ω Larmor frequency [Symbol]

ω molar ratio [Symbol]

ω (lower-case) omega [Greek Alphabet]

ω omega phase (of a material) [Symbol]

ω order type of the set of positive integers [Symbol]

ω ordinary index of refraction (of a uniaxial crystal) [Symbol]

ω pulsatance (or periodic time) [Symbol]

ω solid angle [Symbol]

ω symmetry translation (in crystallography) [Symbol]

ω turn angle (of superlattice) [Symbol]

ω vibration frequency [Symbol]

ω wavenumber [Symbol]

[ω] specific magnetic rotatory power [Symbol]

ω_0 angle subtended by object (in optical instrument) [Symbol]

ω_0 characteristic angular frequency [Symbol]

ω_0 coupling frequency [Symbol]

ω_0 initial angular velocity [Symbol]

ω_0 Larmor frequency [also ω_B]

ω_c critical (angular) frequency [Symbol]

ω_D Debye frequency [Symbol]

ω_d angular frequency of natural vibrations [Symbol]

ω_d damped natural frequency [Symbol]

ω_F angular frequency of fundamental vibration [Symbol]

ω_f final angular velocity [Symbol]

ω_i idler frequency [Symbol]

ω_L Larmor frequency [Symbol]

ω_l longitudinal optical phonon frequency [Symbol]

ω_m maximum (angular) frequency [Symbol]

ω_m modulation frequency [Symbol]

ω_μ accepting phonon [Symbol]

ω_n normal angular frequency [Symbol]

ω_n undamped natural frequency [Symbol]

ω_o angle subtended by object (in optical instrument) [Symbol]

ω_p phonon frequency [Symbol]

ω_p plasma frequency [Symbol]

ω_p pump frequency [Symbol]

ω_Q quadrupole frequency [Symbol]

ω_R resonant frequency [Symbol]

ω_s signal frequency [Symbol]

ω_{sr} series resonance frequency [Symbol]

ω_t transverse optical phonon frequency

ω_{tr} transformation disclination (jn metallurgy) [Symbol]

ω_x angular velocity component in x-direction [Symbol]

ω_y angular velocity component in y-direction [Symbol]

ω_z angular velocity component in z-direction [Symbol]

ω_z total rotation [Symbol]

$\omega(783)$ omega meson having a mass of approximately 783 MeV [Particle Physics]

OA Objective Analysis

OA Objective Aperture

OA Oceanic Airlines

OA Octyl Alcohol

OA Off-Axis (Loading) [Mechanics]

OA Office Automation

OA Office of Administration [US]

OA Office of the Administrator [of US Environmental Protection Agency]

OA Office of Applications [now Office of Space and Terrestrial Applications, NASA]

OA Officers' Association [UK]

OA Oil-Immersed Air-Cooled (Transformer)

OA Oil-to-Air (Ratio)

OA Okadaic Acid

OA Oleic Acid

OA Omniantenna

OA One-Atom (Theory) [Physics]

OA Open Arc

OA Operand Address

OA Operating Assembly

OA Operating Authorization

OA Operational Aft [Aerospace]

OA Operational Amplifier

OA Operations Agent

OA Operations Analysis; Operations Analyzer

OA Optical Absorption

OA Optical Activity; Optically Active

OA Optic(al) Axis

OA Optical Alignment

OA Optical Analysis

OA Opto-Acoustic(s) [also O-A]

OA Orbital Approximation [Physics]

OA Orbital Assembly [Aserospace]

OA Ordnance Artificer

OA Organic Acid

OA Organoaluminum

OA Orsat Analysis; Orsat Analyzer [Chemistry]

OA Outdoor Air

OA Output Amplifier

OA Output Axis

OA Outside Air

OA Overaged; Overaging [Metallurgy]

OA Overall [also O/A]

OA Overflow Area

OA Oxide Accumulation

OA Oxidizing Agent

OA Oxyacetylene

OA Oxygen Absorbed; Oxygen Absorption

O/A On Account [also o/a]

O/A On Application

O/A On, or About [also o/a]

O/A Outer Anchorage

O/A Overall [also OA]

O-A (Coupled) Oceanographic-Atmospheric

O-A Opto-Acoustic(s) [also OA]

Oa 1-Octanamine

o/a on account [also O/A]

o/a on, or about [also O/A]

ω/a sharpness of resonance [Symbol]

OAA Observatorio Astronómica de Argentina [Astronomical Observatory of Argentina in Cordoba at Bosque Alegre]

OAA Ontario Association of Architects [Canada]

OAA Orbiter Access Arm [of NASA]

OAA Orbiter Alternate Airfield [of NASA]

OAA Organisation pour l'Alimentation et l'Agriculture [Food and Agricultural Organization; of United Nations]

OAA Orient Airlines Association [Philippines]

OAA Outdoor Advertising Association [UK]

OAA Oxalic Acid Anodizing [Metallurgy]

OAA Oxaloacetic Acid

OAAA Outdoor Advertising Association of America [US]

oAaBA o-Acetamidobenzoic Acid

OAAC Outdoor Advertising Association of Canada

OAAS Ontario Association of Agricultural Societies [Canada]

OAATM Office of the Assistant for Automation [US]

OAAU Orthogonal Array Arithmetic Unit

OAB Ocean Affairs Board [of National Academy of Sciences, US]

OAB One-to-All Broadcast

OAB Osservatorio Astronomico di Brera [Brera Astronomical Observatory, Milan, Italy]

OAC Oceanic Area Control

OAC Office of Aeronautical Charting [of National Ocean Service, National Oceanic and Atmospheric Administration, US]

OAC On Approved Credit [also oac]

OAC Ontario Academic Course [Canada]

OAC Ontario Agricultural College [of University of Guelph, Canada]

OAC Ontario Arts Council [Canada]

OAC Operations Advisory Committee

OAC Operations Analysis Center

OAC Opticians Association of Canada

OAC Optimally Adaptive Control

OAC Ordnance Ammunition Command [Joliet, Illinois, US]

OAC Organoaluminum Compound

OAC Overseas Automotive Club [US]

OAc Acetoxy-

Oac Acetate

oAcBA o-Acetylbenzoic Acid

OACC Oceanic Area Control Center

OACETT Ontario Association of Certified Engineering Technicians and Technologists [Canada]

OACETT/ACSTTO Ontario Association of Certified Engineering Technicians and Technologists/Association of Certified Survey Technicians and Technologists of Ontario [Canada]

OACF Orientational Autocorrelation Function

OACI Ontario Academic Courses Institute [Canada]

OACI Organisation de l'Aviation Civil Internationale [International Civil Aviation Organization; of United Nations]

OACSU Off-Air Call Setup

OACU Office of Animal Care and Use [of Office of Intramural Research, National Institutes of Health, US]

OACUL Ontario Association of College and University Libraries [now Ontario College and University Library Association, Canada]

OAD One Atom Detection

OAD Open Architecture Driver

OAD Operational Availability Date

OAD Operations Analysis Division

OAD Optical Activity Detector

OAD Optoacoustic Device

OADA Ontario Automobile Dealers Association [Canada]

OAE Opto-Acoustic Effect [Physics]

OAE Orbital Analysis Engineer

OAEAO Ontario Association of Education Administration Officials [Canada]

OAEM Ontario Approved Educational Microcomputers [Canada]

OAEP Office of Atomic Energy for Peace [Bangkok, Thailand]

OAET Office of Aeronautics, Exploration and Technology [of NASA]

OAF Open-Arc Furnace [Metallurgy]

OAF Origin Address Field

OAFD Orbiter Air Flight Deck [NASA]

OAFH Office of Air Force History [of US Air Force at Bolling Air Force Base, Washington, DC]

OAFTO Orbiter Atmospheric Flight Test Office [of NASA]

OAFU Observers Advanced Flying Unit

OAG Official Airline Guide

OAG Online Airline Guide

OAGS Ontario Association of General Surgeons [Canada]

OAHI Ontario Association of Home Inspectors

OAHRC Ontario Agricultural Human Resource Committee [Canada]

OA&HS Oxfordshire Architectural and Historical Society [UK]

OAI Ohio (State) Aerospace Institute [US]

OAI Open Applications Interface

OAI Outside Air Intake

OAIDE Operational Assistance and Instructive Data Equipment

OAII Ocean-Atmosphere-Ice Interaction

OAK Organization for the Advancement of Knowledge [US]

Oak Ridge Natl Lab Met Ceram Tech Rep Oak Ridge National Laboratory Metals and Ceramics Technical Report [Tennessee, US]

OAL Optically Active Ligand

OAL Overall Level

OALC Ogden Air Logistics Center [of Hill Air Force Base, Utah, US] [also O-ALC]

OALS Observer Air Lock System

OALS Office of Arid Lands Studies [of University of Arizona, Tucson, US]

OALT Ontario Association of Library Technicians [Canada]

OAM Observatorio Astronómica de Madrid [Astronomical Observatory of Madrid, Spain]

OAM Office of Aerospace Medicine [US]

OAM Office of Alternative Medicine [of National Institutes of Health, Bethesda, Maryland, US]

OAM Ontario Agricultural Museum [Canada]

OAM Operation and Maintenance

OAM Orbital Angular Momentum

OAM Oscillator Activity Monitor

OAMA Office Automation Management Association [US]

OAME Orbital Altitude Maneuvering Electronics

OAMP Optical Analog Matrix Processing

OAMS Orbit(al) Attitude (and) Maneuver(ing) System

OAN Observatorio Astronomico Naciónal [National Astronomical Observatory, Alcala de Henares, Madrid, Spain]

OANA Organization of Asian News Agencies

OAND Ontario Association of Naturopathic Doctors [Canada]

O and C Operations and Checkout [also O&C]

O Antigen Somatic Antigen [also O antigen] [Microbiology]

OAO Office of Aircraft Operations [of National Oceanic and Atmospheric Administration, US]

OAO Orbiting Astronomical Observatory [Series of 4 US Scientific Satellites Launched by NASA between 1966 and 1972]

OAO Orthogonalized Atomic Orbital [Physics]

OAP Office of the Assistant to the President [US]

OAP Ontario Apprenticeship Program [Canada]

OAP Optical Array Probe

OAP Optical Atom Probe

OAP Orthogonal Array Processor

OAP Osservatorio Astronomico di Palermo [Astronomical Observatory of Palermo, Italy]

OAPC Office of Alien Property Custodian

OAPEC Organization of Arab Petroleum Exporting Countries [Kuwait]

OAQ Order of Architects of Quebec [Canada]

OAQ Outdoor Air Quality

OAQPS Office of Air Quality Planning and Standards [of US Environmental Protection Agency]

OAQS On-Line Associative Query System [of University of Illinois, US]

OAR Office of Aerospace Research [of US Air Force]

OAR Office of AIDS (Acquired Immunodeficiency Syndrome) Research [of National Institutes of Health, Bethesda, Maryland, US]

OAR Office of Air and Radiation [of US Environmental Protection Agency]

OAR Office of Atmospheric Research [of National Oceanic and Atmospheric Administration, US]

OAR Office of Oceanic and Atmospheric Research [of National Oceanic and Atmospheric Administration, US]

OAR Operator Authorization Record

O₂/Ar (Molecular) Oxygen/Argon (Gas Mixture)

OARAC Office of Aerospace Research Automatic Computer

OAR Office of AIDS (Acquired Immunodeficiency Syndrome) Research Advisory Council [US]

OARC Ordinary Administrative Radio Conference

OARE Orbital Acceleration Research Experiment [of NASA]

OARM Office of Administration and Resources Management [of US Environmental Protection Agency]

OARP Office of Advanced Research Programs [Washington, DC, US]

OARS Office Automation Reporting Service

OART Office of Advanced Research and Technology [of NASA]

OAS Obstacle Assessment Surface

OAS Office Automation System

OAS Office of the Assistant Secretary [US]

OAS Ohio Academy of Science [US]

OAS One-to-All Scatter

OAS Optically Active Substance

OAS Optoacoustic Spectrometry

OAS Optoacoustic Spectroscopy

OAS Orbit Adjust Subsystem

OAS Orbiter Aeroflight Simulator [of NASA]

OAS Orbiter Atmospheric Simulator [of NASA]

OAS Orbiter Avionics System [of NASA]

OAS Organization of American States

OASC One-Atom State Self-Consistency (Model) [Physics]

OASCB Orbiter Avionics Software Control Board [of NASA]

OASD Office of the Assistant Secretary of Defense [US]

OASDHI Old-Age, Survivors and Disability Health Insurance [US]

OASDI Old-Age, Survivors and Disability Insurance [US]

OASF Orbiting Astronomical Support Facility

OASH Office of the Assistant Secretary for Housing [US]

OASI Office Automation Society International [US]

OASI Old-Age and Survivors Insurance [US]

OASIS Observations At Several Interacting Scales

OASIS Ocean All Source Information System

OASIS Oceanic and Atmospheric Scientific Information System [of National Oceanic and Atmospheric Administration, US]

OASIS Oceanic Area System Improvement Study

OASIS Office Automation Services and Information Systems (Directorate) [of Government Services Canada]

OASM Office of Aerospace Medicine [US]

OASM Ohm, Ampere, Second, Meter (System) [also oasm]

OASPL Overall Sound Pressure Level

OASR Office of Aeronautical and Space Research [now Office of Advanced Research Programs, US]

OAST Office of Aeronautics and Space Technology [of NASA]

OASTT Office of Aeronautics and Space Transportation Technology [of NASA]

OASV Orbital Assembly Support Vehicle

OASYS Office Automation System

OAT Office for Advanced Technology [of US Air Force]

OAT Operational Acceptance Test

OAT Operational Air Traffic

OAT Outside Air Temperature

OAT Overall Test

OATP Operational Acceptance Test Procedure

OATRU Organic and Associated Terrain Research Unit [Canada]

OATS Original Article Tearsheet Service

OATUU Organization of African Trade Union Unity

OAU Operator Assistance Unit

OAU Organization of African Unity [of United Nations Development Program]

1/au one per atomic unit [also au⁻¹]

1/au² one per atomic unit squared [also 1/au⁻²]

1/au³ one per atomic unit [also au^{-3}]

OAW Oxyacetylene Welding

ÖAW Österreichische Akademie der Wissenschaften [Austrian Academy of Science, Vienna]

oAxBA o-Acetoxybenzoic Acid

OAZ Oxidation-Affected Zone [Metallurgy]

OB Observation Balloon

OB Obstetric(ian); Obstetrics

OB Obtuse Bisectrix [Crystallography]

OB Octal-to-Binary [also O/B]

OB Octave Band

OB Office of Budget [of US Department of the Interior]

OB Oil Bearing

OB Oil-Break [Electrical Energineering]

OB Oil Burner

OB Old Boy [Amateur Radio]

OB Olivine-Bearing (Mineral)

OB On-Board [also O/B]

OB Operational Base [also O/B]

OB Optical Bench

OB Ordnance Board [UK]

OB Organization Blocks

OB Osteoblast [Anatomy]

OB Outboard

OB Outbound

OB Output Buffer

OB Output Bus

OB Outside Broadcast (Van)

OB Overbending

OB Overburden [Mining]

OB 4,4'-Oxybis(benzenesulfonic Acid Hydrazide)

OB Oxygen Balance

OB Oxygen-Bridging (Mechanism)

OB- Peru [Civil Aircraft Marking]

O/B Onboard [also OB]

O/B Operational Base [also OB]

O-B Octal-to-Binary [also OB]

Ob Obstetric(ian); Obstetrics

ob *(obiit)* – died

ob *(obiter)* – incidentally

ob obsolete

OBA Octave-Band Analyzer

OBA Oil Burning Apparatus

OBA Oxygen Breathing Apparatus

oBA o-Bromoaniline

OBAA 3-(4-Octadecylbenzoyl)acrylic Acid

OBAA Oil Burning Apparatus Association [UK]

OBAP Office Belge pour l'Accroissement de la Productivité [Belgian Office for the Improvement of Productivity]

OBATA Ontario Biological Aeration Tillage Association [Canada]

OBAWS Onboard Aircraft Weighing System

ÖBB Österreichische Bundesbahnen [Austrian Railways]

oBBa o-Bromobenzoic Acid

OBC Octal-to-Binary Conversion

OBC On-Board Computer

OBC Optical Bar Code

OBCE Office Belge du Commerce Exterieur [Belgian Office of Foreign Trade]

OBCO Onboard Checkout (Instrumentation)

OBCR Optical Bar Code Reader

OBCS Onboard Checkout Subsystem

OBD Omnibearing Distance Navigation

OBD On Board Diagnostics; On-Board Diagnostic System [Automotive Powertrain]

OBD Open Blade Damper

OBD Ordnance Base Depot

OBD II On-Board Diagnostic System II [Automotive Powertrain]

OBd Ordnance Board

OBDD Ordered Bicontinuous Double Diamond (Morphology) [Materials Science]

OBE Onboard Electronics

OBE Operating Basis Earthquake

OBE Operating Basis Event

OBE Order of the British Empire

OBE Overcome by Events

OBEA Ontario Business Education Association [Canada]

o-BeCr$_2$O$_4$ Orthogonal Beryllium Chromate

Oberfläche Surf Oberfläche Surface [Journal on Surface Science and Technology; published in Switzerland]

OBES Office of Basic Energy Science [of US Department of Energy]

OBES Ohio Bureau of Employment Services [US]

OBES Orthonormal Basis of an Error Space

OBES-DMS Office of Basic Energy Science–Department of Materials Science [of US Department of Energy]

OBESSI Organizing Bureau of European School Student Unions [Sweden]

OBEX Object Exchange [Word Processing]

OBF One-Bar Function

OBFS Organization of Biological Field Stations [US]

OBGS Orbital Bombardment Guidance System

OB-GYN Obstetrics and Gynecology; Physician in Obstetrics and Gynecology [also Ob-Gyn, or ob-gyn]

OBH Office Busy Hour

OBI Office of Basic Instrumentation

OBI Omnibearing Indicator

OBI Ontario Bus Industries [Canada]

OBI Open-Back Inclinable (Press)

OBIC Optical Beam Induced Current

OBIFCO On-Board In-Flight Checkout

OBIP Ontario Business Incentive Program [Canada]

.OBJ Object (File) [File Name Extension)

Obj Object(ive) [also obj]

OBK Oppenheimer-Brinkmann-Kramers (Approximation) [Physics]

obl oblique

obl oblong

Obl Ext Abd Obliquus Externus Abdominis [also obl ext abd] [Anatomy]

Obl Int Abd Obliquus Internus Abdominis [also obl int abd] [Anatomy]

oblat *(oblatum)* – a cachet [Medical Prescriptions]

OBM Oil Bulk Modulus

OBM Ontario Basic Mapping [Canada]

OBM Oxygen-Bottom Blown Maxhuette (Steelmaking Process) [Metallurgy]

OBN Out-of-Band Noise

OBN Office Balancing Network

OBNA Only But Not All

OBO Oil Bulk-Ore

OBO Or Best Offer [also obo]

O/B/O Ore/Bulk/Oil (Carrier)

OBOC Ore/Bulk/Oil Carrier

Obogashch Rud Obogashchenie Rud [Russian Publication]

OBOS Ore/Bulk/Oil Ship

OBP On-Board Processing

oBP o-Bromophenol

OBR Ohio Board of Regents [US]

OBR Outboard Recorder

Obrobka Plast Obrobka Plastyczna [Polish Publication]

OBS Obstetrician; Obstetrics

OBS Omnibearing Selector

OBS On-Board Simulation (System)

OBS Open-Back Stationary (Press)

OBS Open Bidding Service [of Public Works and Government Services Canada]

OBS Operational Bioinstrumentation System

OBS Operational Biomedical Sensors

OBS Operational Biomedical System

Obs Observation; Observatory; Observe(r) [also obs]

Obs Obsolescence; Obsolete [also obs]

Obs Obstacle [also obs]

obs obsolete

obsc obscure

Obsn Observation [also obsn]

obsol obsolete

OBSPL Octave-Band Sound Pressure Level

OBSS Ocean Bottom Scanning Sonar

Obst Obstetrician; Obstetrics

Obst Obstruction [also obst, Obstn, or obstn]

Obsy Observatory [also obsy]

OBT Office of the Building Technologies [of US Department of Energy]

OBT One Billion Trees (Program) [of Australian National Conservancy Agency]

obt obtain

obtd obtained [also obt'd]

obtg obtaining

OBU Offshore Banking Unit

OBu^t Tertiary Butoxy (or tert-Butyl Ester)

OBV Oxidizer Bleed Valve

OB Van Outside-Broadcast Van [also OB van]

OBzh Benzhydryl Ester (or Diphenylmethoxy)

OBzl Benzyl Ester (or Benzoxy)

OC Oberlin College [Ohio, US]

OC Object Code

OC Ocean Current

OC OCLC (Online Computer Library Catalog) Number [US]

OC Octal Code

OC Off-Center

OC Office of Communications [of National Institutes of Health, Bethesda, Maryland, US]

OC Officer Commanding

OC Official Classification

OC Oil Chemist(ry)

OC Oil Color

OC Oil Content

OC Oil-Cooled; Oil Cooler; Oil Cooling

OC Oil Country

OC Oklahoma City [US]

OC Olfactory Cell [Anatomy]

OC Oligocyclic (Stress)

OC On Center

OC On-Condition

OC Open Cell

OC Open Charter

OC Open Circuit [also O/C]

OC Open Coil (Annealing) [Metallurgy]

OC Open Collector

OC Open Commission (Meeting)

OC Open Controller

OC Open Cup

OC Open Cycle [Thermodynamics]

OC Operating Characteristic (Curve) [Statistics]

OC Operating Cost

OC Operating Curve

OC Operating Cycle

OC Operational Calculus

OC Operational Characteristic (Curve) [Statistics]

OC Operational Computer

OC Operational Control

OC Operation Cost

OC Operations Control(ler)

OC Operations Coordinator

OC Operator Command

OC Opportunity Cost

OC Optical Carrier

OC Optical Cavity

OC Optical Center

OC Optical Character

OC Optical Coating

OC Optical Code

OC Optical Communications

OC Optical Compensator

OC Optical Computer

OC Optical Constant
OC Optical Coupler
OC Optical Crystal
OC Optimal Control
OC Optimization Control
OC Orbital Check
OC Order Cost
OC Order-in-Council
OC Ordnance Committee
OC Ordnance Corps
OC Organic Carbon [also oc]
OC Organic Chemist(ry)
OC Organic Coating
OC Organic Compound
OC Organizing Committee
OC Organochlorine
OC Organocopper
OC Orthochromatic (Filter)
OC Oscillating Current
OC Osteocyte [Anatomy]
OC Otto Cycle [Thermodynamics]
OC Outlet Contact
OC Outside Circumference
OC Ovarian Cancer [Medicine]
OC Over Center
OC Over-Compounded [Electrical Engineering]
OC Overcorrected; Overcorrection
OC Overcritical
OC Overcurrent
OC Oxidation Catalysis
OC Oxide Catalyst
OC Oxide Cathode
OC Oxide Ceramic(s)
OC Oxide Coated; Oxide Coating
OC Oxycellulose
OC Oxygen Cell
OC Oxygen Chemisorption
OC Oxygen Consumed; Oxygen Consumption
OC Oxygen Cutting
OC Ozawa and Chen (Equation [Materials Science]
O&C Operation and Checkout
O&C Operations and Checkout (Building) [formerly Manned Spacecraft Operations Building, NASA]
O&C Operations and Control
O/C Officer in Charge
O/C Open Circuit [also OC]
O/C Operations Critical
O/C Overcharge
Oc Occurrence [also oc]
Oc Octahedral Site (in Crystallography) [Symbol]
oC o-Cresol
oC Pearson symbol for base-centered space lattice in orthorhombic crystal system (this symbol is followed by the number of atoms per unit cell, e.g. oC4, oC8, etc.)

OCA Obstacle Clearance Altitude
OCA Oceanic Control Area
OCA Operational Control Authority
OCA Organisación dos Cooperativos Americanos [of Organization of the Cooperatives of America, Peru]
OCA Organization of the Cooperatives of America [Peru]
oCa o-Chloroaniline
OCAC Overseas Chinese Affairs Commission [Taiwan]
OCAD Occupational and Career Analysis Development
OCAD Optical Character and Detect
OCAFF Office of the Chief, Army Field Forces [Fort Monroe, Virginia, US]
OCAL On-Line Cryptanalytic Aid Language
OCALC Oklahoma City Air Logistics Center [of US Air Force] [also OC-ALC]
OCAM Ontario Center for Advanced Manufacturing [Canada]
OCAM Organisation Commune Africaine et Malgache [African and Malagasy Common Organization] [now OCAMM]
OCAMC Oxford Center for Advanced Materials and Composites [of University of Oxford, UK]
OCAMM Organisation Commune Africaine, Malgache et Mauricienne [African, Malagasy and Mauretanian Common Organization] [formerly OCAM]
OCAMS Orbital Correspondence Analysis in Maximum Symmetry
OCAP Ontario Career Action Program [Canada]
OCAPT Ontario Center for Automotive Parts Technology [St. Catharines, Ontario, Canada]
OCAS On-Line Cryptanalytic Aid System
OCAS Organization of Central American States
OCB Office of Computational Bioscience [of Division of Computer Research and Technology, US]
OCB Offshore Certification Bureau [UK]
OCB Oil Circuit Breaker; Oil-Operated Circuit Breaker
OCB Outgoing Calls Barred
OCBA O-Chlorobenzoic Acid [also oCBA]
OCBC O-Chlorobenzyl Chloride [also oCBC]
OcB/L Ocean Bill of Lading
OCBN O-Chlorobenzonitrile [also oCBN]
OCC Office of Contract Compliance
OCC Office of the Comptroller of the Currency [US]
OCC Ohno Continuous Casting [Metallurgy]
OCC Oklahoma Corporation Commission [Oklahoma City, US]
OCC Old Corrugated Cardboard; Old Corrugated Containers [Recycling]
OCC Olympus Castable (Glass) Ceramic
OCC Ontario Chamber of Commerce [Canada]
OCC Ontario Corn Committee [Canada]
OCC Open-Chain Compound
OCC Operational Computer Complex
OCC Operations Control Center
OCC Operator Control Command
OCC Options Clearing Corporation
OCC Organocopper Compound

OCC Other Common Carrier [Telecommunications]

OCC Output Control Character

Occ Occulting [also occ]

Occ Occupation [also occ]

OCCA Oil and Colour Chemists Association [UK],

OCCAA Oil and Colour Chemists Association of Australia

occas occasional(ly)

OCCC Office of the Chief of the Chemical Corps [US Department of the Army]

OCCC Ontario Cereal Crops Committee [Canada]

OCCE Operational Clothing and Combat Equipment

OCCG Operations Center Coordinator Group

OCCI Oldham Chamber of Commerce and Industry [UK]

Occpn Occupation [also occpn]

Occr Ovzdusi Ochrona Ovzdusi [Polish Journal]

Occr Powietrza Ochrona Powietrza [Polish Journal]

Occr Przed Koroz Ochrona Przed Korozja [Polish Journal]

OCCS Office of Computer and Communications Systems [of National Library of Medicine, Bethesda, Maryland, US]

occsly occasionally

Occup Occupation(al) [also occup]

Occup Environ Med Occupational and Environmental Medicine [Published in the US]

Occup Health Occupational Health [Journal published in the UK]

Occup Psychol Occupational Psychology [Journal published in the US]

Occup Saf Health Admin Sub Serv Occupational Safety and Health Administration' Subscription Serv\ice [US]

OC Curve Operating Characteristic Curve; Operational Characteristic Curve [also OC curve] [Statistics]

OCCWS Office, Chief of the Chemical Warfare Service

OCD Ocean Chemistry Division [of National Oceanic and Atmospheric Administration–Atlantic Oceanographic and Meteorological Laboratories, US]

OCD Office of Civil Defense [formerly Civil Defense Administration; later Defense Civil Preparedness Agency; now Federal Emergency Management Agency, US]

OCD Operational Capability Development

OCD Other Checkable Deposits [US]

OCDM Office of Civil Defense Mobilization [US]

OCDP Operations Capability Development Plan

OCDR Orbiter Critical Design Review [of NASA]

OCDRE Organic-Cooled Deuterium-(Moderated) Reactor Experiment [of Whiteshell Nuclear Research Establishment, Manitoba, Canada]

OCdt Officer Cadet [Canadian Forces]

OCdt Officer Commandant [also O Cdt]

OCDU Optics Coupling Data Unit; Optics Coupling Display Unit

OCE Ocean Color Experiment

OCE Ocean Covered Earth

OCE Office of the Chief of Engineers [US Department of the Army]

OCE One-Center Expansion

OCE Open Collaborative Environment

OCE Oregon College of Education [Monmouth, US]

OCEAN Oceanographic Coordination Evaluation and Analysis Network

OCEAN Organisation de la Communauté Européenne des Avitailleurs des Naivres [Ship Suppliers Organization of the European Community, Netherlands]

OceanE Ocean Engineer [also Ocean Eng, or OceanEng]

Ocean Eng Ocean Engineer(ing)

Ocean Eng Ocean Engineering [Journal published in the UK]

OCEANIC Ocean Network Information Center [US]

Oceanogr Oceanographer; Oceanographic; Oceanography [also Oceanog, oceanog, oceanogr]

Oceanol Oceanological; Oceanologist; Oceanology [also oceanol]

Oceanol Oceanology [Journal published in Russia]

Oceanol Acta Oceanologica Acta [Journal published in France]

Oceanol Limnol Sin Oceanologia et Limnologia Sinica [Chinese Journal]

OCED Organisation de Coopération Economique et de la Développement [Organization for Economic Cooperation and Development, France]

OCEE Office for Central Europe and Eurasia [of National Research Council, Washington, DC, US]

OCEE Organisation de Coopération Economique Européenne [Organization for European Economic Cooperation] [now Organisation de Coopération Economique et de la Développement, France]

OCETA Ontario Center for Environmental Technology Advancement [Canada]

OCF Objects Components Framework

OCF Onboard Computational Facility

OCF Ontario Cancer Foundation [Canada]

OCF Open Channel Flowmeter

OCF Orbiter Computational Facilities [of NASA]

OCF Orientation Coherence Function [Crystallography]

OCF Orientation(al) Correlation Function

OCF Owens-Corning Fiberglass Corporation [Granville, Ohio, US]

OCF Oxygen-Coal-Flux (Injection) [Metallurgy]

OCFMFPT Ontario Center for Farm Machinery and Food Processing Technologies [Canada]

OCFNT Occluded Front [Meteorology]

OCG Optimal Code Generation

OCG Optical Crown Glass

OCGS Ontario Council on Graduate Studies [Canada]

OCH Obstacle Clearance Height [Navigation]

OCH Office for Communication in the Humanities [UK]

OCH Orbiter Common Hardware [of NASA]

O_2/C_2H_2 Oxygen/Acetylene (Gas Mixture)

OCHC Operator Call Handling Center

OCH_2CN Cyanomethyl Ester (or Cyanomethoxy)

$O_2/CHF_3/SF_6$ Trifluoromethane/Argon/Sulfur Hexafluoride (Gas Mixture)

OCHP Office of Children's Health Protection [of US Environmental Protection Agency]

OCH(Ph)CN O-Cyanobenzyl

Ochr Powietrza Ochrona Powietrza [Polish Journal]

Ochr Przed Koroz Ochrona Przed Korozja [Polish Journal]

OCI Office of Computer Information [US]

OCI Office of the Coordinator of Information

OCI Ontario Cancer Institute [Canada]

OCI Operator Control Interface

OCI Optically-Coupled Isolator

OCI Oxide Control and Indication

OCIA Organic Crop Improvement Association [Canada and US]

OCIMF Oil Companies International Marine Forum [UK]

Ω **circ-mil/ft** ohm circular-mil per foot [Unit]

OCIS Optics Classification and Indexing Scheme [of Optical Society of America]

OCIS OSHA (Occupational Safety and Health Administration) Computerized Information System [US]

OCL Obstacle (or Obstruction) Clearance Limit [Navigation]

OCL Operational Control

OCL Operational Control Level

OCL Operation Control Language

OCL Operator Control Language

OCL Overall Connection Loss

OCL Overcorrected Lens

OCLC Ohio College Library Center [now On-Line Computer Library Center]

OCLC On-Line Computer Library Center [Dublin, Ohio, US]

OCLC On-Line Computer Library Catalog

OCLC Micro On-Line Computer Library Center Micro [US]

OCLC World Cat On-Line Computer Library Center World Catalog [US]

OCLI Optical Coating Laboratory, Inc. [US]

Ocln Occlusion [also ocln]

OClO Chlorine Dioxide

OCM OLR (Outgoing Longwave Radiation) Climatological Minimum

OCM Ontario Center for Microelectronics [Canada]

OCM Opencast Mining

OCM Optical Countermeasures

OCM Organic Content Monitor

OCM Oscillator and Clock Module

OCM Oxygen Coupling of Methane (Process)

Ω**cm** ohm-centimeter [also ohm cm, ohm-cm, $\Omega \cdot$cm, or Ω-cm]

$\Omega^{-1} \cdot$**cm**$^{-1}$ reciprocal ohm centimeter [also $(\Omega \cdot$cm$)^{-1}$, $(\Omega$-cm$)^{-1}$, (ohm-cm)$^{-1}$, 1/$\Omega \cdot$cm, or 1/ohm-cm]

Ω**cm²** ohm-square centimeter [also $\Omega \cdot$cm², or Ω-cm²]

Ω**/cm** ohm(s) per centimeter [also Ω cm^{-1}]

$(\Omega \cdot$**cm**$)^{-1}$ reciprocal ohm centimeter [also $(\Omega \cdot$cm$)^{-1}$, $(\Omega$-cm$)^{-1}$, $\Omega^{-1} \cdot$cm^{-1}, (ohm-cm)$^{-1}$, 1/$\Omega \cdot$cm, or 1/ohm-cm]

1/cm one per centimeter [also cm^{-1}]

1/cm² one per square centimeter [also cm^{-2}]

1/cm³ one per cubic centimeter [also cm^{-3}]

OCMA Oil Companies Material Association

OCMG Oxygen-Controlled Melt-Growth (Method) [Metallurgy]

Ω**-cmil/ft** ohm-circular mil(s) per foot [Unit]

OCMR Ontario Center for Materials Research [now Materials and Manufacturing Ontario, Canada]

OCMS Optional Calling Measured Service

1/cm·T one per centimeter tesla [also cm^{-1} T^{-1}]

OCN Office of the Commissioner of Namibia

OCN Order Control Number

OCN$^-$ Cyanate (Ion) [Symbol]

ocnl occasional

OCNO Office of the Chief of Naval Operations [US]

OCO Office, Chief of Ordnance [US Department of the Army]

OCO OMS (Orbital Maneuvering System) Cutoff

OCO Open-Close-Open (Contact)

OCO Operation Capability Objectives

O$_2$-CO Oxygen-to-Carbon Monoxide Ratio

O$_2$-CO$_2$ Oxygen-to-Carbon Dioxide Ratio

O$_2$-CO$_2$ Oxygen–Carbon Dioxide Cycle [Ecology]

Ω**/°C/Ω** ohm per degree Celsius per ohm [also Ω °C^{-1} Ω^{-1}]

OCO-ORDTA Office, Chief of Ordnance–Artillery Ammunition Branch [US]

OCP Obstacle Clearance Panel [of International Civil Aviation Organization]

OCP Office of Commercial Programs

OCP Onchocerciasis Control Program [of World Health Organization]

OCP One-Component Plasma (System)

OCP Ontario College of Pharmacy [of University of Toronto, Canada]

OCP Open Circuit Potential [Physical Chemistry]

OCP Operating Control Procedure

OCP Operational Checkout Procedure

OCP Orbital Combustion Process

OCP Order Code Processor

OCP Output Control Program

OCP Output Control Pulse(s)

OCP Overload Control Process

OCP Oxford Concordance Project [of Oxford University, UK]

oCP o-Chlorophenol

OCPA O-Chlorophenylacetic Acid

OCPC Ontario Crop Protection Committee [Canada]

OCPCA Oil and Chemical Plant Constructors Association [UK]

OCPDB Organic Chemical Producers Database [of US Environmental Protection Agency]

oCPHAAA o-Carboxyphenylhydrazo Acetoacetanilide [also o-CPHAAA]

OCPIP O-Chlorophenol Indophenol [also oCPIP]

OCPT 2-Chloro-p-Amino Toluene

OCPV Open-Circuit Photovoltage

OCR Office of Coal Research [of US Department of the Interior]

OCR Oil Circuit Recloser

OCR Open-Cycle Reactor (System)

OCR Operational Control Room

OCR Operations Change Request

OCR Optical-Character Reader; Optical-Character Reading [also ocr]

OCR Optical-Character Recognition [also ocr]

OCR Organic-Cooled (Nuclear) Reactor

OCR Organization for the Collaboration of Railways [Poland]

OCR Output Control Register

OCR Overconsolidation Ratio [Geology]

OCR Overcurrent Relay

OCR Overhaul Component Requirement

Ocr Occurrence [also ocr]

OCRA Office Communications Research Association

OCR-A Optical-Character Recognition–Font A [also OCRA]

OCR-B Optical-Character Recognition–Font B [also OCRB]

OCRBI Organization for Cooperation in the Roller Bearings Industry [Poland]

OCRD Ocean Climate Research Division [of National Oceanic and Atmospheric Administration–Pacific Marine Environmental Laboratory, US]

OCRD Office of the Chief of Research and Development [US Army]

OCRE Office of Conservation and Renewable Energy [of US Department of Defense]

OCRI Ottawa-Carleton Research Institute [Canada]

OCRM Orbiter Crash and Rescue Manuals [of NASA]

OCRPI Office Central de Repartition des Produits Industriels [Central Office of Distribution of Industrial Products, France]

OCRR Ontario Center for Resource Recovery [Canada]

OCRS Office of Computing Resources and Services [of Division of Computer Research and Technology, US]

OCRS Ontario Center for Remote Sensing [Canada]

OCRS Open-Cycle (Nuclear) Reactor System

OCRS Optical-Character Recognition System

OCRS Organisation Commune de les Régions Saharienne [Common Organization of the Saharian Regions]

OCRUA Optical-Character Recognition Users Association [now Recognition Technologies Users Association, US]

OCR Wand Optical-Character Reader Wand

OCRWM Office of Civilian Radioactive Waste Management [of US Department of Energy]

OCS Obstacle Clearance Surface

OCS O-Carbonyl Sulfide

OCS Office Communications System

OCS Office of Commodity Standards [of National Institute for Standards and Technology, Gaithersburg, Maryland, US]

OCS Office of Communication Systems [of US Air Force]

OCS Office of Community Services [of US Department of Health and Human Services]

OCS Office of the Chief Scientist [Australia]

OCS Officer Candidate School

OCS Offshore Constitutional Settlement

OCS Onboard Checkout System

OCS On-Card Sequencer

OCS Ontario Science Center [Canada]

OCS Operations Control System

OCS Opposed Cathode System

OCS Optical-Character Scanner

OCS Optical Communication System

OCS Optical Contact Sensor

OCS Oriental Ceramic Society [UK]

OCS Outer Continental Shelf

OCS Output Control Subsystem

OCS Output Control System

OCS Overload Control Subsystem; Overload Control System

OCS Overseas Communications Service [India]

OCSA Ontario Council of Safety Associations [Canada]

OCS BBS Outer Continental Shelf Bulletin Board System

OCSE Office of Child Support Enforcement [of US Department of Health and Human Services]

OCSO Office of the Chief Signal Officer [US Department of the Army]

OCST Office of Commercial Space Transportation [of US Department of Transportation]

OCT O-Chlorotoluene

OCT Office of Critical Tables [of National Academy of Sciences/National Research Council, US]

OCT Office of the Chief of Transportation [US Department of the Army]

OCT Ontario College of Teachers [Canada]

OCT Operational Cycle Time

OCT Optimal Control Theory

OCT Ornithine Carbamyl Transferase [Biochemistry]

OCT Overseas Countries and Territories

Oct Octagon(al) [also oct]

Oct Octahedral [also oct]

Oct Octal [also oct]

Oct Octane [also oct]

Oct Octave

Oct October

oct octavo [also 8vo] [Printing]

OCTA Oceanic Control Area

Octahdr Octahedral; Octahedron [also octahdr]

OCT-BIN Octal-to-Binary (Conversion) [also Oct-Bin]

OCTC O-Chlorobenzotrichloride

OCT-DEC Octal-to-Decimal (Conversion) [also Oct-Dec]

OCTENAR Octane Extractive Distillation of Aromatics (Process)

OCTG Oil Country Tubular Goods

Oct No Octane Number [also oct no]

OCTR Office of Computational and Technology Research [of US Department of Energy]

OCTRF Ontario Cancer Treatment and Research Foundation [Canada]

OCTS Ocean Color and Temperature Scanner [of Advanced Earth Observing Satellite, National Space Development Agency of Japan]

OCT$ Octal-Value String [Programming]

OCTV Open Circuit Television

OCU Oklahoma City University [US]

OCU Ontario Council of Universities [Canada]

OCU Operational Control Unit

OCU Operational Conversion Unit

OCU Osaka City University [Japan]

OCUA Ontario Council on University Affairs [Canada]

OCUB Osmium Collidine Uranylenbloc

OCUFA Ontario Confederation of University Faculty Association [Canada]

OCUL Ontario Council on University Libraries [Canada]

ocul *(oculus)* – eye [Medical Prescriptions]

OCULA Ontario College and University Library Association [Canada]

oculent *(oculentum)* – eye ointment [Medical Prescriptions]

OCV Open-Circuit Voltage

OCX OLE (Object Linking and Embedding) Custom Control [Word Processing]

OCZM Office of Coastal Zone Management

OD Doctor of Optometry

OD Object Distance [Optics]

OD Ocean Drilling

OD Octal-to-Decimal [also O-D]

OD *(Oculus dexter)* – (in the) right eye [also o.d.] [Medical Prescriptions]

OD Office of the Director [of National Institutes of Health, Bethesda, Maryland, US]

OD Officer of the Day

OD Oligodynamic(s)

OD Olive Drab

OD On Deck [also od]

OD On Demand

OD Operating Data

OD Operational Downlink

OD Operational Downlist

OD Operations Directive

OD Opthalmological Doctor

OD Optical Density; Optically Dense

OD Optical Detection; Optical Detector

OD Optical Disk

OD *(Optometriae Doctor)* – Doctor of Optometry; Optometric Doctor

OD Order-Disorder (Transition) [also O-D] [Solid-State Physics]

OD Ordered Domain [Solid-State Physics]

OD Ordnance Department [of US Department of the Army]

OD Ordnance Depot

OD Ore Dressing

OD Orthodeuterium

OD Outer Diameter [also O/D, or od]

OD Output Data

OD Output Disable

OD Outside Diameter [also O/D]

OD Outside Dimension

OD Oven-Dried (Wood)

OD Overdesign(ed); Overdesigning

OD Overdosage; Overdose [also od]

OD Overdraft; Overdrawn

OD Overload Detection

OD Oxidation Defect

OD Oxygen Demand

OD Oxygen Drain

OD Ozone Depleting; Ozone Depletion

OD- Lebanon [Civil Aircraft Marking]

O/D On Demand

O/D On Dock

O/D Origin/Destination

O/D Outside Diameter [also OD]

O/D Overdraft

O-D Octal-to-Decimal [also OD]

O-D Order-Disorder (Transition) [also OD] [Solid-State Physics]

O-D Origin and Destination (Survey)

1D One-Dimensional [also 1-D]

od *(oculus dexter)* – (in the) right eye [also OD] [Medical Prescriptions]

od *(omni die)* – every day [Medical Prescriptions]

od on deck [also OD]

od outer diameter; outside diameter [also OD]

od overdose

ODA Octadecyl-Bonded Alumina

ODA Octanediamine

ODA O-Dianisidine

ODA Office Data Architecture

ODA Office Document Architecture

ODA Official Development Assistance

ODA Ontario Dental Association [Canada]

ODA Open Document Architecture

ODA Operational Data Analysis

ODA Operational Design and Analysis

ODA Optimum Daily Allowance

ODA Overseas Development Agency [UK]

ODA 4,4'-Oxydianiline

18ODA 1,8-Octanediamine [also 18oda]

1DA One-Dimensional Analysis [also 1-DA]

ODAA α-Octadecylacrylic Acid

ODAL Octadecenal

13-ODAL 13-Octadecenal

ODALS Omnidirectional Approach Lighting System

ODA/PMDA 4,4'-Oxydianiline/Pyromellitic Dianhydride

ODAPI Omnidirectional Approach Path Indicator

ODAPI (Borlund) Open Database Application Programming Interface

ODAPS Oceanic Display and Planning System [US]

ODAS Ocean Data Acquisition Systems

ODB Operational Data Book

ODB Operations Data Base

ODB Output Data Buffer

ODB Output to Display Buffer

ODBA Oregon Dairy Breeders Association [US]

ODBC Object-Oriented Database Connectivity

ODBC Open Data Base Connectivity

ODBMS Object-Oriented Database Management System

ODC Octal-to-Decimal Conversion

ODC Old Dominion College [Norfolk, Virginia, US]

ODC Ontario Development Corporation [Canada]

ODC Optimized Double Configuration

ODC Orientationally-Disordered Crystal

ODC Ornithine Decarboxylase [Biochemistry]

ODC Other Direct Costs

ODC Output Data Control

ODC Ozone-Depleting Chemical

ODC Ozone-Depleting Compound

ODCC Oxford and District Chamber of Commerce [UK]

1D-CDP One-Dimensional Circular Diffraction Pattern [also ODCDP]

ODCDR Orbiter Delta Critical Design Review [of NASA]

ODCF One-Dimensional Compressible Flow [also 1DCF]

ODCP One-Digit Code Point

ODCR Optically-Detected Cyclotron Resonance

ODCSOPS Office of the Deputy Chief of Staff for Military Operations [of US Army]

ODD One-Dimensionally Disordered; One Dimensional Disorder [Solid-State Physics]

ODD Operator Distance Dialing

ODD Optical Data Digitizer

ODD Optical Data Disk

ODD Optical Digital Disk

ODD Ouchterlony Double Diffusion (Assay)

ODDES Office of the Deputy Director for Extramural Science [US]

ODDF Orientation Difference Distribution Function [Crystallography]

1D DOS One-Dimensional Density of States [also 1D-DOS, or ODDOS]

ODDRE Office of the Director of Defense Research and Engineering [of US Department of Defense]

ODDS Operational Data Delivery Services

ODE On-Line Data Entry

ODE Ordinary Differential Equation

ODE Office de Documentation Economique [Office for Economic Documentation, France]

ODE Opposite Drive-End Entry

ODE Optical Doppler Effect

ODE Ordinary Differential Equation

ODE Oxygen Defect Electron

ODECA Organisation des Etats Centrafricaine [Organization of Central African States]

ODECA Organización dos Estados Centramericanos [Organization of Central American States]

ODEFA Organisation de Défense de l'Environnement et de la Faune Africaine [Organization for the Defense of the African Environment and Fauna] [of United Nations]

1-DEG One-Dimensional Electron Gas

ODENDOR Optically Detected Electron Nuclear Double Resonance [also OD-ENDOR]

ODEPA 4-[Bis(1-Aziridinyl)phosphinyl]morpholine

ODES Optical Discrimination Evaluation Study

ODESR Optically Detected Electron Spin Resonance [also OD-ESR]

ODESSA Ocean Data Environmental Science Services Acquisition

ODESY On-Line Data Entry System

ODETTE Organization for Data Exchange and Teletransmission in Europe

ODF One-Dimensional Flow [also 1DF]

ODF Orbit Determination Facility

ODF Orientation Density Function

ODF Orientation Distribution Function [Crystallography]

ODF Original Data File

1DF One-Dimensional Flow [also ODF]

ODFW Oregon Department of Fish and Wildlife [US]

ODG Off-Line Data Generator

O(δg) Orientation Correlation Function [Physics]

1D-HAF One-Dimensional Heissenberg Antiferromagnet [Solid-State Physics]

ODI Open Datalink Interface

ODI Open Device Interconnect

ODI Overseas Development Institute [UK]

ODIF Open Document Interchange Format

ODII Optically Detected Impact Ionization

ODIM Optimized Diatomics-in-Molecules (Method)

ODIN On-Line Documentation and Information Network [Germany]

ODIN Optical Design Integration

ODIN Orbital Design Integration (System)

ODIS Oceanographic Data Information System

ODL Object Definition Language

ODL Object Description Language

1DL One-Dimensional (Crystal) Lattice

ODLOS Optical Deep-Level Optical Spectroscopy

ODLRO Off-Diagonal Long-Range Order [Solid-State Physics]

ODLTS Optical Deep-Level Thermal Spectroscopy

ODM Object Data Manager

ODM Operational Data Messages

ODM Optical Diffractometer; Optical Diffractometry

ODM Optimized Distribution Model

ODM Orbital Determination Module

ODM Overseas Development Ministry [UK]

ODMA Open Document Management API (Application Program Interface) (Support)

ODMR Optical Detection of Magnetic Resonance

ODMR Optical Double Magnetic Resonance

ODMR Optically Detected Magnetic Resonance (Spectroscopy)

ODMS Object-Oriented Database Management System

ODN Octadecaneuropeptide [Biochemistry]

ODN Open Data Network

ODN Own Doppler Nullifier

ODNRI Overseas Development Natural Resources Institute [UK]

ODNMR Optically-Detected Nuclear Magnetic Resonance

1D-NMR One-Dimensional Nuclear Magnetic Resonance [also 1DNMR]

1D-NOE One-Dimensional Nuclear Overhauser Effect [also 1DNOE]

ODNR Ohio Department of Natural Resources [US]

ODNRI Overseas Development Natural Resources Institute [UK]

ODO One-Dimensionally Ordered; One Dimensional Order [Solid-State Physics]

1DOF One Degree Of Freedom

ODOP Offset Doppler

ODOP Orbital Doppler System

odorl odorless

ODP Ocean Drilling Platform

ODP Ocean Drilling Program [US]

ODP Ocean Drilling Project

ODP Office of Defense Programs [of US Department of Energy]

ODP Office of Disease Prevention [of National Institutes of Health, Bethesda, Maryland, US]

ODP Oil Diffusion Pump

ODP Open Distributed Processing

ODP Operational Display Procedure

ODP Optical Data Processing; Optical Data Processor

ODP Optical Difference of Path [also odp]

ODP Original Document Processing

ODP Ozone-Depleting Potential; Ozone-Depletion Potential

ODPA Octylated Diphenyl Amine

ODPAS Optically Detected Photoacoustic Spectroscopy

ODPCS Oceanographic Data Processing and Control System

ODPEX Offshore Drilling and Production Exhibition

ODPINS Office of Defense Programs, Intelligence and National Security [of US Department of Energy]

1DPM One-Dimensional Plasma Model [Physics]

ODPN 3,3'-Oxydipropionitrile

ODQ 1H-[1,2,4]Oxadiazolo-[4,3-a]Qunioxalin-1-One

ODR Offshore (Oil) Drilling Rig

ODR Omnidirectional Range

ODR Operator Data Register

ODR Optical Data Recognition

ODR Optical Document Reader; Optical Document Reading

ODR Optical Document Recognition

ODR Output Data Redundancy

ODR Output Definition Register

ODR Oxygen Dissolution Reaction\

ODRAN Operational Drawing Revision Advance Notice

ODRN Orbiting Data Relay Network

ODRS Ogasawara Downrange Station [of National Space Development Agency of Japan]

ODS Octadecasilica

ODS Octadecylsilyl

ODS Octadecyltrimethyloxysilane

ODS (Microsoft) Open Data Services

ODS Operational Data Store

ODS Optical Disk Storage

ODS Orientation Distribution Function

ODS Oxide Dispersion-Strengthened; Oxide Dispersion Strengthening [Metallurgy]

ODS Oxygen Diffusion Size

ODS Ozone-Depleting Substance(s)

ODSA Ontario Dental Society of Anaesthesia [Canada]

ODSA Overseas Development Service Association [UK]

ODSC Oxide-Dispersion-Strengthened Copper

1DSE One-Dimensional Schroedinger Equation [Quantum Mechanics]

ODSI Open Directory Service Interface

ODSS Ocean Dumping Surveillance System

ODT Octadecanethiol

ODT Octal Debugging Technique

ODT Office of Defense Transportation [US]

ODT On-Line Debugging Technique

ODT Open Desktop

ODT Optical Data Transmission

ODT Optical Disk Technology

ODT Order-Disorder Transformation

ODT Order-Disorder Transition [Solid-State Physics]

ODT Outside Diameter Tube

1D t-J One-Dimensional Time-Antiferromagnetism Exchange (Model) [Solid-State Physics]

ODTS Octadecyltrichlorosilane

ODU Old Dominion University [Norfolk, Virginia, US]

ODU Output Display Unit

ODVAR Orbit Determination and Vehicle Attitude Reference

1D VRH One-Dimensional Variable Range Hopping [Physics]

ODW Office of Drinking Water [of US Environmental Agency]

ODW Organic Dry Weight

ODW Omega Drop Windsonde System

OE Oceanic Engineering

OE Oceanic Engineering (Society) [of Institute of Electrical and Electronics Engineers, US]

OE Office Equipment

OE Office Environment

OE Office of Education [US]

OE Oil-Extended (Rubber)

OE Oil Extraction

OE Old English

OE Omissions Excepted [also oe]

OE Onsager Equation [Physical Chemistry]

OE Open End (V-Belt)

OE Operating Engineer

OE Optical Electronics

OE Optical Element

OE Optical Emission; Optical Emitter

OE Optical Encoder

OE Opto-Electronic(s) [also O-E]

OE Orbital Electron

OE Orbital Element [Physics]

OE Order of Engineers

OE Ordnance Engineer(ing)

OE Original Equipment

OE Otto Engine

OE Outer Electron

OE Output Enable

OE Overhauser Effect [Atomic Physics]

OE Overrun Error [Computers]

OE Own Exchange [also O/E]

OE Oxidation Exposure

OE Oxide Electrode

OE Oxygen Electrode

OE Ozone Engineer(ing)

O/E Own Exchange

O-E Opto-Electronic(s) [also OE]

OE- Austria [Civil Aircraft Marking]

Oe Oersted [Unit]

Oe^{-1} One per Oersted [also 1/Oe]

Oe^2 Square Oersted [Unit]

oe odd-even (nuclei) (i.e., atomic nuclei whose spin is equal to an odd multiple of $\frac{1}{2} \times \hbar$, $\hbar$ = Planck's constant ÷ 2π) [Nuclear Physics]

oe omissions excepted [also OE]

OEA Office of External Affairs [US]

OEA Ophthalmic Exhibitors Associastion [UK]

OEA Optometric Editors Association [US]

OEA *(Organisação dos Estados Americanos)* – Portuguese for "Organization of American States"

OEA *(Organisación dos Estados Americanos)* – Spanish for "Organization of American States"

OEA Organization of European Aluminum Smelters

OEA Oxygen-Enriched Atmosphere

OEA-CNEA Organisación des Estados Americanos—Comisión Nacional de Energia Atomica [Organization of American States–National Atomic Energy Commission of Argentina]

OEAP Operational Error Analysis Program

OEAS Office of Essential Air Service [of US Department of Transportation]

OEAS Orbital Emergency Arresting System

OEAW Oesterreichische Akademie der Wissenschaften [Austrian Academy of Sciences]

OEB Oscillatory Electric Birefringence

OEB Ontario Energy Board [Canada]

OEBR Oil-Extended Butadiene Rubber [also OE-BR]

OEC Office of Economic Competitiveness [of US Department of Energy]

OEC Office of Energy Conservation [of US Department of Energy]

OEC Ontario Economic Council [Canada]

OEC Ontario Energy Corporation [Canada]

OEC Optoelectronic Conference

OEC Organisación dos Estados Centroamericanos [Organization of Central American States]

OECA Office of Enforcement and Compliance Assurance [of US Environmental Protection Agency]

OECA Ontario Educational Communications Authority [Canada]

OECD Organization for Economic Cooperation and Development [formerly Organization for European Economic Cooperation, France]

OECD Econ Outlook OECD Economic Outlook [of Organization for Economic Cooperation and Development]

OECD/MEI OECD Main Economic Indicators

OECD-NEA OECD Nuclear Energy Agency

OECD/NIA OECD National Income Accounts

OECD/SIDS OECD Screening Information Data Set

OECO Outboard Engine Cutoff

OECUT Office of Energy Conversion and Utilization Technology [of US Department of Energy]

OED Optoelectronic Device

OED Orbiting Energy Depot

OED Oxidation Enhanced Diffusion

OED Oxford English Dictionary

OED Oxygen-Enhanced Diffusion

OEDC Ontario Engineering Design Competition [Canada]

OEDF 1-Hydroxyethylidenediphosphonic Acid

OEDM One-Electron Diatomic Molecule

OEDSF Onboard Experimental Data Support Facility

OEE Optoelectronic Effect

OEEC Organization for European Economic Cooperation [now Organization for Economic Coperation and Development, France]

OE-EPDM Oil-Extended Ethylene-Propylene Diene Monomer

OEERE Office of Energy Efficiency and Renewable Energy [of US Department of Energy]

OE-E-SBR Oil-Extended Emulsion Styrene Butadiene Rubber

OEETD Office of Environmental Engineering and Technology Demonstration

OEF Oceanic Educational Foundation [US]

OEF Organization of Employers Federations [UK]

OEF Origin Element Field

OE/FIBERS Optics and Electrooptics Fibers (Conference)

OEG Open-End (Wave-)Guide

OEG Operations Evaluation Group

OEG Outdoor Ethics Guild [US]

OEGAI Oesterreichische Gesellschaft für Artificial Intelligence [Austrian Society for Artificial Intelligence]

OEGAI J OEGAI Journal [Published by the Oesterreichische Gesellschaft für Artificial Intelligence, Austria]

OEH Ordnance Engineering Handbook

OEHO Ordnance Engineering Handbook Office [of Duke University, Durham, North Carolina, US]

OEIAZ Oesterreichische Ingenieur– und Architekten-Zeitschrift [Austrian Journal for Engineers and Architects]

OEIC Optoelectronic Integrated Circuit

OEIZ Oesterreichische Ingenieur-Zeitschrift [Austrian Journal for Engineers]

OEJ Office of Environmental Justice [of US Environmental Protection Agency]

Oe/K Oersted(s) per Kelvin [also Oe K^{-1}]

OEL Occupational Exposure Limit

OEL Ontario Electrical League [Canada]

OE LASE Optics, Electro-Optics and Laser Applications in Science and Engineering [of International Society for Photo-Optical Engineering, US] [also OE/LASE, US]

Oelhydraul Pneum Oelhydraulik und Pneumatik [German Journal on Oil Hydraulics and Pneumatics]

OEM Original Equipment Manufacturer

OEM Original Equipment Market

OEM Other Equipment Manufacturer

OEMI Office Equipment Manufacturers Institute

Oe/min Oersted(s) per minute [also Oe min^{-1}]

OEMO One-Electron Molecular Orbital

OEMS Open-Site EMI (Electromagnetic Interference) Measurement System

OEMS Optical Emission under Mechanical Stress

OEMSA Optical Equipment Manufacturers and Suppliers Association [UK]

OEMTC Ohio Edison Materials Technology Center [US]

OEMWM Office of Environmental Management and Waste Management [of US Department of Energy]

OEN Ontario Environment Network [Canada]

Oe/nm Oersted(s) per nanometer [also Oe nm^{-1}]

OE-NR Oil Extended Nitrile Rubber

OEO Office of Economic Opportunity [US]

OEO Office of Equal Opportunity [of National Institutes of Health, Bethesda, Maryland, US]

OE/OEM Original Equipment/Original Equipment Manufacturer

OENT Oficina Espaniola Nacional de Turistos [Spanish National Tourist Office]

OEOS Office of Earth Observation Systems [of National Space Development Agency of Japan]

OEP 2,3,7,8,12,13,17,18-Octaethylporphinate; 2,3,7,8,12,13,17,18-Octaethylporphine

OEP Odd-Even Predominance

OEP Office Equipment and Products [Published in Japan]

OEP Office of Emergency Planning [US]

OEP Office of Emergency Preparedness [US]

OEP Operand Execution Pipeline

OEP Optimized Effective Potential

OEP Optoelectronic Packaging

OEP Overall Economic Perspective

O&EP Occupational and Environmental Protection

OEPA Ohio Environmental Protection Agency [US]

OEPER Office of Environmental Processes and Effects Research [US].

OEPP Office of Environmental Policy and Planning [Thailand]

OEQ Order of Engineers of Quebec [Canada]

OEQO Ontario Effluent Quality Objective [Canada]

OER Office of Energy Research [of US Department of Energy]

OER Office of Exploratory Research [US]

OER Office of Extramural Research [of National Institutes of Health, Bethesda, Maryland, US]

OER Oxygen Enhancement Ratio

OER Oxygen Evolution Reaction

OER/BES Office of Exploratory Research/Basic Energy Sciences [of US Department of Energy]

OERC Optimum Earth Reentry Corridor

OERD Ocean Environment Research Division [of National Oceanic and Atmospheric Administration– Pacific Marine Environmental Laboratory, US]

OERD Office of Energy Research and Development [of Natural Resources Canada]

OEERE Office of Energy Efficiency and Renewable Energy [of US Department of Energy] .

OERN Organisation Européenne pour la Recherche Nucléaire [European Organization for Nuclear Research]

OERS Organisation des Etats de la Rivière Sénégal [Organization of Senegal River States] [Now Defunct; Formerly Included Guinea, Mali, Mauritania and Senegal]

OERWM Office of Environmental Restoration and Waste Management [of US Department of Energy]

OES Oceanic Engineering Society [of Institute of Electrical and Electronics Engineers, US]

OES Office of Earth Science [of NASA]

OES Office of Emergency Services [US]

OES Office of Employment Security [of Employment and Training Administration, US]

OES Office of Endangered Species

OES Offshore Engineering Society [UK]

OES Oklahoma Extension Society [US]

OES Optical Emission Spectrograph(y); Optical Emission Spectrometer; Optical Emission Spectrometry; Optical Emission Spectroscopy

OES Orbital Escape System [of NASA]

OES Orbiter Emergency Site [of NASA]

OES Order Entry System

OES Outgoing Echo Suppressor

OES Overseas Exhibition Service [UK]

Oe/s Oersted(s) per second [also Oe/sec, or Oe s^{-1}]

OESBR Oil-Extended Styrene Butadiene Rubber [also OE-SBR]

OESC Organisation pour l'Education, la Science et la Culture Intellectuelle [United Nations Educational, Scientific and Cultural Organization]

OESD Opto-Electronic Semiconductor Device

OESE One-Electron Schroedinger Equation [Quantum Mechanics]

Oe/sec Oersted(s) per second [also Oe/s, or Oe s^{-1}]

OESLA Office of Engineering Standards Liaison and Analysis [US]

OESS Orbiter/ET (External Tank) Separation Subsystem [of NASA]

Oesterr Ing Archit-Z Oesterreichische Ingenieur- und Architekten-Zeitschrift [Austrian Journal for Engineers and Architects]

Oesterr Ing Z Oesterreichische Ingenieur-Zeitschrift [Austrian Journal for Engineers]

Oesterr Z Elektrwirtsch Oestereichische Zeitschrift für die Elektrizitätswirtschaft [Austrian Journal for the Power Supply Industry]

OET Octachloro Endomethylene Tetrahydroindan

OET Organisación para el Estudios Tropicos [Tropical Studies Organization, University of San José, Costa Rica]

O/ET Orbiter/External Tank [of NASA]

OEt Ethyl Ester (or Ethoxy)

OETB Offshore Energy Technology Board

OE-TECHNOLOGY Applications Symposium on Optics, Electro-Optics and Lasers in Industry

OEU Osaka Electro-Communication University [Japan]

OEW Operating Empty Weight; Operational Empty Weight

OEWG Open-End Waveguide

OEX Orbiter Experiments [of NASA]

oExA o-Ethoxyaniline

OEZE Oesterreichische Zeitschrift für die Elektrizitätswirtschaft [Austrian Journal for the Power Supply Industry]

OF Objective Function

OF Ocean Floor

OF Oil Feed

OF Oil-Filled; Oil Filling

OF Oil Filter(ing)

OF Oil Furnace

OF Old French [also OFr]

OF Operational Fix(ed)

OF Operational Forward [Aerospace]

OF Optical Fiber

OF Optical Filter

OF Optical Flat

OF Optical Fluorescence

OF Orbital Flight [also O/F]

OF Ordered Flat

OF Organofunctional (Silanes)

OF Orientation Factor

OF Oriented Film [Electronics]

OF Oscillator Frequency

OF Outflow

OF Output Factor

OF Outside Face

OF Overflow [also O/F]

OF Overflow Flag [Computers]

OF Oxide Film

OF Oxidizing Flame

OF Oxyfuel

OF Oxygen Fill

OF Oxygen-Free

O/F Orbital Flight [also OF]

O/F Overflow [also OF]

O/F Oxidizer-to-Fuel (Ratio) [Propellants]

oF Pearson symbol for face-centered space lattice in orthorhombic crystal system (this symbol is followed by the number of atoms per unit cell, e.g. oF24, etc.)

OFA Oil-Immersed Forced Air-Cooled (Transformer)

OFA Ontario Federation of Agriculture [Canada]

OFA Other Federal Agencies

OFA Oxygenated Fuels Association [US]

OfA Offensive Arms

oFa o-Fluoroaniline

OFAAP Ontario Farm Adjustment Assistance Program [Canada]

OFA Oesterreichische Forschungsgesellschaft für Atomenergie [Austrian Atomic Energy Research Company]

OFAGE Orthogonal Field-Alternating Gel Electrophoresis

OFAH Ontario Federation of Anglers and Hunters [Canada]

OFAP Observational Facilities Advisory Panel

OFAR Office of Foreign Agricultural Relations [US]

OFB Operational Facilities Branch [of NASA]

OFB Oscillatory Flow Birefringence

OFB Output Feedback

oFBA o-Fluorobenzoic Acid

OFC Oil-Filled Cable

OFC Operational Flight Control

OFC Optical Fiber Communication

OFC Optical Fiber (Communications) Conference

OFC Oregon Fish Commission [US]

OFC Organic Fruit Cooperative [UK]

OFC Overfeed Combustion

OFC Oxyfuel Gas Cutting

OFC Oxygen-Free Copper [also OF Cu]

Ofc Office [also ofc]

OFCA Ontario Federation of Construction Associations [Canada]

OFC-A Oxyacetylene Cutting

OFCC Office of Federal Contract Compliance [US]

OFCC Ontario Forage Crops Committee [Canada]

OFCCP Office of Federal Contract Compliance Programs [of US Employment Standards Administration]

OFCE Office Français du Commerce Extérieur [French Office for Foreign Trade]

OFCF Overseas Farmers Cooperative Federation [UK]

OFC-H Oxyhydrogen Cutting

OFC-N Oxynatural Gas Cutting

OFC-P Oxypropane Cutting

Ofcr Officer [also ofcr]

OFCRC Ontario Field Crops Research Committee [Canada]

OF Cu Oxygen-Free Copper [also OFC]

OFD Ordnance Field Depot

OFDC Ontario Film Development Corporation [Canada]

OFDM Optical Frequency Division Multiplex(ing)

OFDM Orthogonal Frequency Division Multiplex(ing)

OFDS Orbiter Flight Dynamics Simulator [of NASA]

OFDS Oxygen Fluid Distribution System

OFE Oxygen-Free Electrolytic (or Electronic, or Electrical) (Copper)

OFE Cu Oxygen-Free Electrolytic (or Electronic, or Electrical) Copper

OFEN Office Fédéral de l'Energie [Swiss Federal Office of Energy]

OFES Office Fédéral de l'Education et de la Science [Swiss Federal Office for Science and Education, Berne]

OFES Office of Fusion Energy Sciences [of US Department of Energy]

OFF Open Force Field (Method) [Physical Chemistry]

Off Office; Officer; Official [also off]

offen offensive

Off Environ Office Environment [Published in the UK]

Off Equip Index Office Equipment Index [Published in the UK]

Off Equip Methods Office Equipment and Methods [Canadian Publication]

Off Equip Prod Office Equipment and Products [Published in Japan]

Off Equip News Office Equipment News [Published in the UK]

Off Gaz Official Gazette [US]

Off Gaz US Pat Official Gazette, US Patent and Trademarks Office

Off Gaz US Pat Trademks Off Pat Official Gazette of the United States Patent and Trademarks Office, Patents

Off Gaz US Pat Trademks Off Tradem Official Gazette of the United States Patent and Trademarks Office, Trademarks

Off Home Office at Home [Published in the UK]

offic official

Off Inf Manage Int Office and Information Management International [Publication of the Institute of Administrative Management, UK]

Off J Eur Comm Official Journal of the European Communities [Luxembourg]

Off J Pat Official Journal, Patents [UK]

Off Mag Office Magazine [UK]

Off Manage Office Management [Published in Germany]

Off Prem Off Premises [also off prem]

Off Prod News Office Products News [UK]

Off Prod News (Australia) Office Products News (Australia) [Published in Sydney, Australia]

Off Prod News (UK) Office Products News (UK)

Offr Officer [also offr]

Offset Print Reprogr Offset Printing and Reprographics [Journal published in the UK]

Offshore Eng Offshore Engineer [Journal published in the UK]

Offshore Res Focus Offshore Research Focus [Journal published in the UK]

Off Syst Office System [also off syst]

Off Syst Res J Office Systems Research Journal [Published by the Office Systems Research Association, US]

Off Technol People Office Technology and People [Published in the UK]

Off World News Office World News [Published in the US]

OFG Optical Flint Glass

OFG Optical Frequency Generator

OFG Organic Functional Group

OFG Oxyfuel Gas

OFHA Oilfield Haulers Association [US]

OFHC Oxygen-Free High-Conductivity (Copper)

OFHC Cu Oxygen-Free High-Conductivity Copper

OFI Office Français d'Importation [French Importation Office]

OFI Operational Flight Instrumentation

OFI Optical Fiber Identifier

OFI Oxford Forestry Institute [UK]

OFIA Ontario Forest Industries Association [Canada]

OFIA Optical Frame Importers Association [UK]

OFIComex Office Français d'Importation et du Commerce Extérieur [French Office for Importation and Foreign Trade]

OFIS Office Information System

OFIS Operational Flight Information Service [of International Civil Aviation Organization]

OFIX Office of the Future Information Exchange [UK]

OFK Official Flight Kit

OFK Optical Flight Kit

OFL Ontario Federation of Labour [Canada]

OFL Optical Fault Locator

Ofl Overflow [also ofl]

OFLINPS Open Frame Linear Power Supply [also OF LIN PS]

Oflo Overflow [also oflo]

OFLP Oxygen-Free Low-Phosphorus (Copper)

OFLPC Oxygen-Free Low-Phosphorus Copper

OFLS Off-Line System

Oflw Overflow [also oflw]

OFM Office of Financial Management [US]

OFM Ordnance Field Manual

OFMP Organization of Facility Managers and Planners [of Operations Management Education and Research Foundation, US]

oFmP o-Formylphenol

OFMT Output Format for Numbers

OFN Organization for Flora Neotropica [US]

OFNPS Outstate Facility Network Planning System

OFO Office of Flight Operations [of NASA]

OFO Orbiting Frog Otolith (Satellite)

OFO-1 First Orbiting Frog Otolith (Satellite)

OFP Occluded Frontal Passage

OFP Operating Force Plan

OFP Operational Flight Profile

OFP Operational Flight Program

OFP Orbiter Flight Program [of NASA]

OFP Oscilloscope Face Plane

oFP o-Fluorophenol

OFPA Ontario Food Protection Association [Canada]

OFPANA Organic Foods Production Association of North America

OFPP Office of Federal Procurement Policy [US]

OFPS Office of Field Project Support [of University Corporation for Atmospheric Research, US]

OFPS Open Frame Power Supply

OFPU Optical Fiber Production Unit

OFQ Office du Film du Québec [Quebec Film Office, Canada]

OFR Oil-Resistant, Flame-Retardant (Cable)

OFR On-Frequency Repeater

OFR Organic Free Radical

OFR Overfrequency Relay

OFr Old French [also OF]

OFRC Oregon Forest Research Center [of Oregon State University, Corvallis, US]

Of Rdg Of Reading [also of rdg] [Measurement Technology]

OFRS Office Français de Recherches Sousmarines [French Submarine Research Office]

OFS Object File System

OFS Oil from Sludge (Process)

OFS Operational Fixed (Microwave) Service

OFS Operations and Flight Support [also O&FS]

OFS Optical-Fiber Sensor

OFS (International Conference on) Optical-Fiber Sensors

OFS Orange Free State [South Africa]

OFS Orbital Flight System

OFS Orbiter Flight System [of NASA]

OFS Orbiter Functional Simulator

OFS Ordnance Field Service

OFS Output Field Separator

OFS Oxygen-Free (Copper) with Silver

O&FS Operations and Flight Support [also OFS]

Ofs Offset (of Object) [Pascal Function]

OFSD Operating Flight Strength Diagram

OFSH Ovine Follicle Stimulation Hormone [Biochemistry]

Ofshr Offshore [also ofshr]

OFSMPS Open-Frame Switch Mode Power Supply [also OF SMPS]

OFSO Ordnance Field Safety Officer [Jeffersonville, Indiana, US]

OFT Orbital Flight Test

OFT Orbiter Flight Test [of NASA]

OFT Office of Fair Trading [UK]

OFT Optical Fiber Thermometry

OFT Optical Fiber Tube

OFT Optimum Flight Trajectory

OFT Orbital Flight Test

oft often

OFTDA Office of Flight Tracking and Data Acquisition [of NASA]

OFTDS Orbital Flight Test Data System

OFTECH Internationale Ausstellung für Oberflächentechnik [International Exhibition for Surface Engineering, Germany]

OFTEL Office of Telecommunications [UK]

OFTF Optical-Fiber Transfer Function

OFTM On-Orbit Flight Technique Meeting

OFTR Orbital Flight Test Requirement(s)

OFTS Optical Fiber Transmission System

OFTTC Ontario Film and Television Tax Credit [Canada]

OFU Ofenbau-Union GmbH [German Manufacturer of Industrial Furnaces for the Iron, Steel and Ceramic Industries]

OFV Overflow Valve

OFVGA Ontario Fruit and Vegetable Growers Association [Canada]

OFW Outstanding Florida Waters [US]

OFW Oxyfuel Gas Welding

OFX Open Financial Exchange

OFXLP Oxygen-Free Extra-Low-Phosphorus (Copper)

OFXLPC Oxygen-Free Extra-Low-Phosphorus Copper

OFZ Obstacle-Free Zone

OG Oil Gas

OG Onia-Gegi (Process)

OG Open Grain (Wood)

OG Operating Grant

OG Optical Generation

OG Optical Glass

OG Orbital Grinder; Orbital Grinding

OG Orbitron Gauge

OG Organic Glass

OG Or Gate

OG Oriented Growth (Theory) [Metallurgy]

OG Original Gum [also og]

OG Outer Gimbal

OG Outgas(sing)

OG Outgoing

OG Oxide Glass

OG Oxides, Globular [Metallurgy]

OG Oxygen Gage

OG Oxygen (Converter) Gas (Steelmaking Process) [Metallurgy]

OG Ozone Generator

O/G Outgoing [also o/g]

Og Ouguiya [Currency of Mauritania]

og ogive; ogival [Architecture]

Ω/G Ohm(s) per Gauss [also Ω G^{-1}]

0G Zero Gravity (or Zero Weight) [also ZG]

OGA Open-Graded (Mineral) Aggregate

OGA Organic Growers Association [UK]

OGA Outer Gimbal Angle

OGA Outer Gimbal Axis

OGB Oil and Gas Board [Alabama, US]

OGB Original Grain Boundary [Materials Science]

ÖGB Österreichischer Gewerkschaftsbund [Federation of Austrian Trade Unions]

OGC Office of the General Council [of National Institutes of Health, Bethesda, Maryland, US]

OGC Oregon Game Commission [US]

OGC Oregon Graduate Center [Beaverton, Oregon, US]

OGCA Ontario General Contractors Association [Canada]

OGCM Ocean General Circulation Model [Meteorology]

OGDC Office of Geographic Data Coordination [of Victoria State, Australia]

OGDC Oil and Gas Development Corporation [Bangladesh]

OGDD Outgoing/Delay Dial

OGDI Oesterreichische Gesellschaft für Dokumentation und Information [Austrian Society for Documentation and Information]

OGE Operating Ground Equipment; Operational Ground Equipment

OGE Optogalvanic Effect [Physics]

OGE Out of Ground Effect

ÖGEM Österreichische Gesellschaft für Elektronenmikroskopie [Austrian Society for Electron Microscopy]

OGI Oculogyral Illusion [Medicine]

OGI Oregon Graduate Institute (of Science and Technology) [Beaverton, US]

ÖGI Österreichisches Gießerei-Institut [Austrian Foundry Institute]

OGID Outgoing/Immediate Dial

OGIP Original Gas-in-Place (Process)

OGIST Oregon Graduate Institute of Science and Technology [Beaverton, US]

OGJ Oil and Gas Journal [US]

OGJ Outgoing Junction [Telecommunications]

OGM Office of Guided Missiles [US]

OGM Oriented Gas Model

OGM Outgoing Message

OGMC Ordnance Guided Missile Center [US]

OGMS Open Gradient Magnetic Separator

OGMS Ordnance Guided Missiles School [at Redstone Arsenal, Huntsville, Alabama, US]

OGMT Orbiter Greenwich Mean Time [NASA]

OGO Orbiting Geophysical Observatory [Series of NASA Satellites]

OGP Office of Global Programs [of National Oceanic and Atmospheric Administration, US]

OGP Outgoing Message Process

1/GPa one per gigapascal [also GPa^{-1}]

OGR ORNL (Oak Ridge National Laboratory) Graphite Reactor [of US Department of Energy]

OGR Outgoing Repeater

OGR Oxygen-Gas Recovery System

OGRC Office of Grants and Research Contracts [US]

OGS Oklahoma Geological Society [US]

OGS Ontario Geological Society [Canada]

OGS Ontario Geological Survey [Canada]

OGS Ontario Graduate Scholarship [Canada]

OGS Optogalvanic Spectroscopy

OGS Osservatorio Geofisico Sperimentale [Experimental Geophysics Observatory, Trieste, Italy]

OGS Outer Glideslope

OGS Oxygen Generation System

1/G·s One per Gauss per second [also $G^{-1}s^{-1}$]

OGST Overthread Guide Sleeve Tool

OGT Outgoing Toll

OGT Outgoing Trunk

OGTT Oral Glucose Tolerance Test [Medicine]

OGU Outgoing Unit

OGV Outlet Guide Vane

OGV Oxygen Gage Valve

ÖGV Österreichische Gesellschaft für Vakuumtechnik [Austrian Vacuum Society]

OGWS Outgoing/Wink Start

OGWDW Office of Ground Water and Drinking Water [of US Environmental Protection Agency]

OGY Ott-Grebogi-Yorke (Method) [Solid-State Physics]

OH Occupational Health

OH Octahedron

OH Octal-to-Hexadecimal [also O-H]

OH Off Hook [Communications]

OH Office of Hydrology

OH Ohio [US]

OH Ohmic Heat(ing)

OH Oil-Hardened; Oil Hardening [Metallurgy]

OH Oil Hole

OH Oil Hydraulic(s)

OH Olduvai Habiline [Ancient Human Race in East Africa]

OH On Hand

OH Ontario Hydro [Now Subdivided into Ontario Hydro Services and Ontario Power Generation] [Canada]

OH Open Hearth (Furnace) [Metallurgy]

OH Operational Hardware

OH Optical Harmonic [Solid-State Physics]

OH Optical Harness

OH Optical Holograph(y)

OH Order Hardening [Metallurgy]

OH Orthohydrogen

OH Outside Height

OH Oval Head (Screw)

OH Overhaul

OH Overhead

OH Overheat(ing)

OH- Finland [Civil Aircraft Marking]

OH· Hydroxyl Radical [Symbol]

OH⁻ Hydroxide [Symbol]

OH⁻ Hydroxyl Group [Symbol]

O&H Oxygen and Hydrogen [also OH]

O-H Octal-to-Hexadecimal [also OH]

oh *(omni hora)* – every hour [Medical Prescriptions]

1/h one per hour [also h^{-1}]

oh2 *(omni hora duo)* – every two hours [Medical Prescriptions]

oh3 *(omni hora tres)* – every three hours [Medical Prescription]

OHA Office of Hearings and Appeals [of US Department of Energy]

OHA Oklahoma Heritage Association [US]

OHA Ontario Herbalist Association [Canada]

OHA Ontario Horticultural Association [Canada]

OHA Operational Hazard Analysis

OHA Orbital Height Adjustment (Maneuver)

OHA Outside Helix Angle

OHA Oxygen Hemoglobin Affinity

OHAc Hydroxy Acetic Acid (or Glycolic Acid)

OHADOE Office of Hearings and Appeals, Department of Energy [US]

OHADOI Office of Hearings and Appeals, Department of the Interior [US]

100A1 One Hundred A One [UK Publication]

OHBG Oregon Highland Bentgrass Commission [US]

OH14BQ Hydroxy-1,4-Benzoquinone

OHC Occupational Health Center

OHC Onboard Hard Copier

OHC Ontario Housing Corporation [Canada]

OHC Ontartio Hypnosis Center [Canada]

OHC Open Hole Compression

OHC Optics Hand Controller

OHC Overhead Camshaft

OHC Oxygen-Associated Hole Center [Solid-State Physics]

OHClBQ 2,5-Dihydroxy-3,6-Dichloro-1,4-Benzoquinone

OHD Oil-Hydraulic Design

OHD Optical-Heterodyne Detect(ion)

OHD Over-the-Horizon Detection; Over-the-Horizon Detector [Navigation]

OHDDD Ontario Hydro Design and Development Division [Canada]

OHDETS Over-Horizon Detection System

OH/D&D-G Ontario Hydro/Design and Development Division–Generation [Canada]

OH-DPAT Hydroxy-2-(Di-n-Propylamino)tetralin

(±)-7-OH-DPAT (±)-7-Hydroxy-2-(Di-n-Propylamino)tetralin

(±)-8-OH-DPAT (±)-8-Hydroxy-2-(Di-n-Propylamino)tetralin

OHD-RIKES Optical-Heterodyne Detected Raman-Induced Kerr Effect Spectroscopy

OHDS Office of Human Development Services [of US Department of Health and Human Services]

OHE Office of Hanford Environment [of Pacific Northwest National Laboratory, Richland, Washington State, US]

OHEA Office of Health and Environmental Assessment [of US Environmental Protection Agency]

OHER Office of Health and Environmental Research [of US Department of Energy]

OHF Occupational Health Facility

OHF Open-Hearth Furnace [Metallurgy]

5-OH-2FPTSC 5-Hydroxy-2-Formylpyridine Thiosemicarbazone

OHG Offene Handelsgesellschaft [General Partnership; Corporate Form in Central Europe] [also oHG]

OHGVT Orbiter Horizontal Ground Vibration Test [of NASA]

OHHCPA Ontario Home Health Care Providers Association [Canada]

OHHS Off-Highway Haulage System,

OHI Optical Holographic Interferometry

OHI Organisation Hydogaphique International [International Hydrographical Organization, Monaco]

OHIONET Ohio Network [US]

OHIP Ontario Health Insurance Plan [Canada]

O₂-HIP Oxygen Hot Isostatic Pressing [Metallurgy]

OHL Occupational Health Laboratory [of Ontario Ministry of Labour, Canada]

OHL Overhead Line

OHM Octadecylhydrogenmaleate (Crystal)

OHM Ohmmeter

OHM Oil and Hazardous Materials

ohm cm ohm-centimeter [also ohm-cm, Ωcm, $\Omega \cdot$cm, or Ω-cm]

ohm⁻¹·cm⁻¹ reciprocal ohm centimeter [also $(ohm \cdot cm)^{-1}$, $(\Omega\text{-cm})^{-1}$, $(ohm\text{-cm})^{-1}$, $1/\Omega \cdot cm$, or $1/ohm\text{-}cm$]

ohm cm² ohm-square centimeter [also $\Omega \cdot cm^2$, or $\Omega\text{-}cm^2$]

ohm/cm ohm(s) per centimeter [also ohm cm⁻¹, Ω/cm, or Ω cm⁻¹]

(ohm-cm)⁻¹ reciprocal ohm-centimeter [also $(\Omega \cdot cm)^{-1}$, $1/\Omega \cdot cm$, $\Omega^{-1}cm^{-1}$, $1/ohm\text{-}cm$, or $ohm^{-1} cm^{-1}$]

OHMM Ohmmeter

ohm m ohm-meter [also ohm-m, Ωm, $\Omega \cdot$m, or Ωm]

ohm⁻¹·m⁻¹ reciprocal ohm meter [also $(\Omega \cdot m)^{-1}$, $(\Omega\text{-}m)^{-1}$, $(ohm\text{-}m)^{-1}$, $1/\Omega \cdot m$, or $1/ohm\text{-}m$]

ohm/m ohm(s) per meter [also Ω m⁻¹]

ohm-m² ohm-square meter [also $\Omega \cdot m^2$, or $\Omega\text{-}m^2$]

(ohm-m)⁻¹ reciprocal ohm centimeter [also $(\Omega \cdot m)^{-1}$, $(\Omega\text{-}m)^{-1}$, $\Omega^{-1} \cdot m^{-1}$, $(ohm\text{-}m)^{-1}$, $1/\Omega \cdot m$, or $1/ohm\text{-}m$]

OHMPA Ontario Hot Mix Producers Association [Canada]

OHMR Office of Hazardous Materials Regulations

OHMS On Her/His Majesty's Service

OHMT Oil and Hazardous Materials Treatment (Regulation)

OHM-TADS Oil and Hazardous Materials–Technical Assistance Data System [of US Environmental Protection Agency]

OH/NMMD Ontario Hydro/Nuclear Materials Management Department [Canada]

OHP Operational Hydrology Program [of World Health Organization]

OHP Outer Helmholtz Plane [Physics]

OHP Overhead Projector

OHP Oxygen at High Pressure

OHPDA Occupational Hygiene Product Distributors Association [UK]

OHQ Hydroxyquinoline

2OHQ 2-Hydroxyquinoline

3OHQ 3-Hydroxyquinoline

4OHQ 4-Hydroxyquinoline

5OHQ 5-Hydroxyquinoline

6OHQ 6-Hydroxyquinoline

7OHQ 7-Hydroxyquinoline

8OHQ 8-Hydroxyquinoline

OHR Office of Health Research

OHR Office of Human Resources

OHR Overhead Railway

OHRD Ontario Hydro Research Division [also OH-RD, Canada]

OHRD/TSTA Ontario Hydro Research Division/Tritium Systems Test Assembly [Canada]

OHRE Office of Human Resources and Education [of NASA]

OHRM Office of Human Resources Management [US]

OHS Occupational Health and Safety

OHS Octadecylhydrogensuccinate (Crystal)

OHS Off-Highway Haulage System

OHS Off-Hook Service [Communications]

OHS Office of Highway Safety [US]

OHS Ohio Historical Society [US]

OHS Oil-Hydraulic System

OHS Oklahoma Historical Society [US]

OHS Ontario Historical Society [Canada]

OHS Ontario Homeopathic Society [Canada]

OHS Ontario Humane Society [Canada]

OHS Ontario Hydro Services [Formerly Ontario Hydro] [Canada]

OHS Open-Hearth Steel

OHS Oregon Historical Society [US]

OH&S Occupational Health and Safety

OHSA Occupational Health and Safety Act [Canada]

OHSB Occupational Health and Safety Branch [Canada]

OHSC Occupational Health and Safety Commission [Canada]

OHSD Occupational Health and Safety Division [of Ministry of Labour, Canada]

OHSGT Office of High-Speed Ground Transportation [of US Department of Transportation]

OHST Occupational Health and Safety Technologist

OHSU Oregon Health Sciences University [Portland, US]

OHT Ontario Hydro Technologies [Canada] [Dissolved]

OHT Open-Hole Tension

OHT Oxygen at High Temperature

OH&T Oil-Hardened and Tempered; Oil Hardening and Tempering [Metallurgy]

OHTB Ontario Highway and Transport Board [Canada]

OHTDC Ontario Hydro Tritium Dispersion Code [Canada]

OHTE Ohmically-Heated Toroidal Experiment

OHV Overhead Valve (Engine)

OHW Oxyhydrogen Welding

OHWM Office of Hazardous Waste Management [of Battelle Northwest, Richland, Washington, US]

OHWM Ordinary High Water Mark

1/Hz One per Hertz [also Hz^{-1}]

OI Object Interface

OI Octahedral Interstice [Crystallography]

OI Office Information

OI Oil-Immersed; Oil Immersion

OI Oil-Impregnated; Oil Impregnation

OI Oil-Insulated; Oil Insulation

OI Operational Instrumentation

OI Operation Interface

OI Operations Instruction

OI Opportunistic Infection [Medicine]

OI Optical Image

OI Optical Inactivity; Optically Inactive

OI Optical Indicatrix

OI Optical Information

OI Optical Instrument(ation)

OI Optical Interference

OI Optical Isomer

OI Opto-Isolator [Electronics]

OI Orkney Islands [Scotland]

OI Output Inhibit

OI Oxygen Index

O.i. One Pint [Medical Prescriptions]

oI Pearson symbol for body-centered space lattice in orthorhombic crystal system (this symbol is followed by the number of atoms per unit cell, e.g. oI6, oI14, etc.)

OIA Observatorio Ionosférica Arecibo [Arecibo Ionospheric Observatory, Costa Rica]

OIA Ocean Industries Association [US]

OIA Office Information Architecture

OIA Office of International Activities [of US Environmental Protection Agency]

OIA Office of International Affairs [of National Research Council, US]

OIA Office of Intramural Affairs [of Office of Intramural Research, National Institutes of Health, Bethesda, Maryland, US]

OIA Orbiter Interface Adapter [of NASA]

oIa o-Iodoaniline

OIB Operating Impedance Bridge

OIB Operation Instruction Block [Computers]

OIB Operations Integration Branch [of NASA]

OIB Orbiter Interface Box [of NASA]

oIBA o-Iodobenzoic Acid

OIBF Oesterreichisches Institut für Bibliotheksforschung, Dokumentations- und Informationswesen [Austrian Institute for Librarianship, Documentation and Information Science]

OIC Ocean Information Center

OIC Oceanographic Instrumentation Center [of US Navy]

OIC Officer in Charge

OIC Oil-Insulated Cable

OIC Oh I See [Internet Jargon]

OIC On-Line Instrumentation Coordinator

OIC Ontario International Corporation [Canada]

OIC Operations Instrumentation Coordinator

OIC Optical Integrated Circuit

OIC Orbital Integrated Checkout

OIC Orbiter Integrated Checkout

OIC Order-in-Council

OIC Organo-Iron Compound

OIC Oxide-Induced Closure

OICA Ontario Institute of Chartered Accountants [Canada]

OICAP Ocean Industries Capital Assistance Program

OICC Ontario Institute of Chartered Cartographers [Canada]

OICD Office of International Cooperation and Development

OICETS Optical Interorbit Communication Test Satellite [of National Space Development Agency of Japan]

OICP Office of International Communications Policy [of International Communication Agency, US]

OICR Ontario Institute for Computer Research [Canada]

O-ICTS Optical-Isothermal Capacitance Transient Spectroscopy

ÖID Österreichischer Informationsdienst [Austrian Information Service]

OIDA Optoelectronics Industry Development Association [US]

OIDC Ontario Industrial Development Council [Canada]

OIDI Optically Isolated Digital Input

OIDL Object Interface Definition Language

OIDO Ocean Industries Development Office

OIDPS Overseas Intelligence Data Processing System

OIE Office International des Epizooties [International Office of Epizootics, France]

OIE Organisation Internationale des Employeurs [International Organization of Employers]

OIES Office for Interplanetary Earth Studies [Boulder, Colorado, US]

OIF Optical Interference Filter

OIFC Oil-Insulated, Fan-Cooled (Transformer)

OIFIG Official Irish FORTH (Computer Language) Interest Group

OIFOC Oil-Immersed, Forced-Oil-Cooled (Transformer)

OIG Office of the Inspector General [US Department of the Army]

OIG Optically Isolated Gate [Electronics]

ÖIG-Kunststoff Österreichische Interessengemeinschaft zur Förderung der Kunststoffberufe [Austrian Association for the Advancement of the Plastics Industries]

OII Office of Industrial Innovation [of Department of Regional Industrial Expansion, Canada]

OII Office of Invention and Innovation [of National Institute for Standards and Technology, Gaithersburg, Maryland, US]

OII Oil Investment Institute [US]

OII Operations Integration Instruction

OII Oxygen Ion Implantation; Oxygen Ion Implanted

OIIT Ontario Institute of Information Technology [Canada]

OIL Orange Indicating Lamp

Oil Gas J Oil and Gas Journal [US]

Oilman Wkly Newsl Oilman Weekly Newsletter [UK]

OILPOL International Convention for the Prevention of Pollution of the Sea by Oil

Oil Turp Oil of Turpentine [also oil turp]

OIM Offshore Installation Manager

OIM Orientation Imaging Microscopy

OIML Organisation Internationale de Métrologie Légale [International Organization for Legal Metrology, France]

OIN Organisation Internationale de la Normalisation [International Standards Organization, Switzerland]

Ω/**in** ohm(s) per inch [also Ω in^{-1}]

OINC Officer in Charge [also OinC]

OIO Oil Industry Outreach [of National Technology Transfer Center, US]

OIP Obligate Intracellular Parasite [Biology]

OIP Office of Industrial Programs [of US Department of Energy]

OIP Office of Information Programs [of National Institute of Standards and Technology, Gaithersburg, Maryland, US]

OIP Office of International Programs [Canada]

OIP Oil-in-Place [Oil Exploration]

OIP Ontario Institute of Painters [Canada]

OIP Operating Internal Pressure

OIP Operations Improvement Program

OIP Optical Image Processing

OIP Optical Information Processing

OIP Optimized Inner Projection (Technique)

oIP o-Iodophenol

OIPN l'Office International pour la Protection de la Nature [International Office for the Protection of Nature]

OIR Office of Intramural Research [of National Institutes of Health, Bethesda, Maryland, US]

OIR Operations Integration Review

OIR Organisation Internationale de la Radiodiffusion [International Broadcasting Organization] [now Organisation Internationale de Radiodiffusion et Télévision]

OIRCA Ontario Industrial Roofing Contractors Association [Canada]

OIRM Office of Information Resources Management [of Office of the Director, National Institutes of Health, Bethesda, Maryland, US]

OIRS Orbit Information Retrieval Services

OIRT Organisation Internationale de Radiodiffusion et Télévision [International Radio and Television Organization]

OIS Object Identification System

OIS Obstacle Identification Surface

OIS Office Information System

OIS Office of Information Services [of Federal Aviation Administration, US]

OIS Office of Information Systems [of National Science Foundation, US]

OIS Office of Investigatory Services [US]

OIS Ontario Investment Service [of Ministry of Economic Development, Trade and Tourism, Canada]

OIS Operational Intercommunication System

OIS Optical Imaging System

OIS Optical Imaging Systems, Inc. [US]

OIS Optical Information System

OIS Optics and Instrumentation Software

OIS Orbiter Instrumentation Systems [of NASA]

OISA Office of International Science Activities

OISA Office of International Scientific Affairs [US]

OISC Oil-Insulated and Self-Cooled (Transformer)

OISCA Organization for Industrial, Spiritual and Cultural Advancement

OIS-D Operational Intercommunications System–Digital

OISE Ontario Institute for Studies in Education [Canada]

OISR Open Item Status Report

OISSP Office of Interim Space Station Program [of NASA]

OIT Office of Industrial Technology [of US Department of Energy]

OIT Ohio Institute of Technology [Columbus, US]

OIT Orbiter Integrated Test [of NASA]

OIT Organisation Internationale du Travail [International Labour Organization; of United Nations]

OIT Osaka Institute of Technology [Japan]

OITC Ontario Innovation Tax Credit [of Ontario Ministry of Finance, Canada]

OITC Ontario International Trade Corporation [of Ministry of Economic Development, Trade and Tourism, Canada]

OITL Outdoor-Indoor (Sound) Transmission Loss

OITT Outpulse Identifier Trunk Test

OIU Operator Interface Unit

OIV Oxidizer Isolation Valve [Aerospace]

OIVS Orbiter Interface Verification Set [of NASA]

OIWC Oil-Immersed, Water-Cooled (Transformer)

OJ Official Journal (of the European Communities)

OJ Originating Junctor [Electrical Engineering]

O.j. One Pint [Medical Prescription]

OJCS Office of Joint Chiefs of Staff [US]

OJEC Official Journal of the European Communities

OJLC Optimizing Joint Load Control

OJPS Online Journal Publishing Service [of American Institute of Physics, US]

OJT On-the-Job Training

OK All Right

OK Oklahoma [US]

OK Onuki-Kawasagi (Theory) [Physics]

OK- Czechoslovakia [Civil Aircraft Marking]

1/K one per kelvin [also K^{-1}]

$\omega(k)$ frequency-wavevector dispersion [Symbol]

OKA Out-of-Kilter (Algorithm)

OKα Oxygen K-Alpha (Radiation) [also OK$_\alpha$]

oka otherwise known as

OK'd Okayed (or [Approved])

OKE Optical Kerr Effect

OKG Ornithine Alpha-Ketoglutarate [Biochemistry]

OKITAI Okinawa–Taiwan Submarine Cable [Japan/Taiwan]

Oki Tech Rev Oki Technical Review [Published by Oki Electric Industry Co. Ltd., Tokyo, Japan]

OKL On Key Label

Okla Oklahoma [US]

ÖKS Arbeitsgemeinschaft Ölkatastrophenschutz e.V. [Oil Disaster Prevention Association, Germany]

OL Objective Lens

OL Object Language [Computers]

OL Obstruction Light

OL Occipital Lobe (of Brain) [Anatomy]

OL Ocean Liner

OL *(Oculus laevus)* – (in the) left eye [also ol] [Medical Prescriptions]

OL Office of Labor [US]

OL Ohm's Law

OL Oilless (Bearing)

OL Oil-Miscible Liquid

OL Olfactory Lobe [Anatomy]

OL On-Line

OL On-Load [Nuclear Reactor]

OL Only Loadable [Computers]

OL Open Loop

OL Operating Level (of Relays)

OL Operating License

OL Operating Line [Chemical Engineering]

OL Operating Location

OL Operational Left [Aerospace]

OL Ophthalmic Laser

OL Optical Length

OL Optics Letters [Journal of the Optical Society of America]

OL Ordinary Luminescence

OL Ordnance Lieutenant (Navy)

OL Organolithium

OL Ortho-Laterals [also known as Ortho-Lateral Cracks] [Fractography]

OL Other Line

OL Output Latch

OL Overflow Level

OL Overhead Line

OL Overlap

OL Overlay

OL Overlayer

OL Overload [also O/L]

OL Ozone Layer

O/L Operations/Logistics

O/L Overload [also OL]

ol *(oculus laevus)* – (in the) left eye [also OL] [Medical Prescriptions]

OLA On-Line Activities

OLA Ontario Library Association [Canada]

OLABS Offshore Labrador Biological Studies [Canada]

OLAC On-Line Accelerated Cooling [Metallurgy]

OLAC On-Line Audiovisual Catalogers [US]

OLADE Organisación Latinoamericana de Energia [Latin-American Energy Organization] [also OLAE]

OLAP On-Line Analytical Processing

OLAR Office of Laboratory Animal Research [of Office of Extramural Research, National Institutes of Health, Bethesda, Maryland, US]

OLAS On-Line Acquisitions System [US]

OLB On-Line Batch

OLB Outer Lead Bond [Tape Automated Bonding]

OLBI On-Line Business Information

OLBM Orbital-Launched Ballistic Missile

OLC Onion-Like Carbon

OLC On-Line Computer

OLC Ontario Land Corporation [Canada]

OLC Ontario Library Council [Canada]

OLC Open-Loop Control

OLC Organolithium Compound

OLC Outgoing Line Circuit

OLCP On-Line Complex Processing

OLCR On-Line Character Recognition

OLCr Ordnance Lieutenant-Commander (Navy)

OLCS Open-Loop Control System

OLD On-Line Debugging

OLD Open-Loop Damping

OLD Open-Loop Drive

.OLD Old Version [File Name Extension] [also .old]

OLDB On-Line Data Bank; On-Line Database

OLDI On-Line-Data Interchange

OLDS Open-Loop Drive System

OLE (Microsoft) Object Linking and Embedding

OLE Office of Library Education [of American Library Association, US]

OLE On-Line Enquiry

OLE Ontario Land Economist [Canada]

OLE Optical Laser Excimer

OLED Organic Light-Emitting Diode

Olefining Olefin Refining (Process)

OLERT On-Line Executive for Real Time

OLEX Ontario Livestock Exchange [Canada]

OLF One-Hundred Linear Feet

OLF Orbiter Landing Facility [now Shuttle Landing Facility, NASA]

OLF Orbital Launch Facility

OLF Outlying Field [US]

OLFA On-Line File Access

OLG Office of Local Government [Australia]

OLGA On-Line Guitar Archive

OLI On-Line Information

OLI Optical (Phone) Line Interface [of AT&T Corporation, US]

OLIF Orbiter Landing Instrumentation Facilities [of NASA]

OLIFLM On-Line Image Forming Light Modulator

Oligo d(pA)$_n$ Oligodeoxyadenylic Acid [n = 2 to 9] [Biochemistry]

Oligo d(pC)$_n$ Oligodeoxycytidylic Acid [n = 3 to 36] [Biochemistry]

Oligo d(pT)$_n$ Oligodeoxythymidylic Acid [n = 2 to 24] [Biochemistry]

OLIP On-Line Image Processing

OLIP On-Line Instrument Package

OLIPaC Office of Land Information Policy and Coordination [of Department of Conservation and Land Management, New South Wales, Australia]

OLIPSE Optizone Liquid-Phase Sintering Experiment

OLIS On-Line Information System

OLIVES Optical Interconnections for VLSI (Very-Large-Scale Integration) and Electronic Systems [ESPRIT Program of the European Community]

OLL On-Line Library

OLLRC Ontario Laser and Light-Wave Research Center [Canada]

OLLT Office of Libraries and Learning Technologies [US]

OLM Office of Laboratory Management [of US Department of Energy]

OLM On-Line Monitor

OLM Optical Light Microscopy

OLMA Ontario Lumber Manufacturers Association [Canada]

OLMC On-Line Microcomputer

OLMC Output Logic Macrocell

OLMS Office of Labor-Management Standards [of US Department of Labor]

OLMS Osborne Laboratories of Marine Sciences [Brooklyn, New York, US]

OLMSA Office of Life and Microgravity Sciences and Applications [of NASA]

OLO On-Line Operation

OLO Optics Letters Online [Publication of the Optical Society of America]

OLO Orbital Launch Operations

OLOC Ontological Locality

OLOE On-Line Order Entry

OLOGS Open-Loop Oxygen-Generating System

OLOW Orbiter Lift-Off Weight [NASA]

OLP Objective Lens Power Supply

OLP On-Line Processing; On-Line Processor

OLP Off-Line Program(ming)

OLP Oxygen Lance Powder (Steelmaking Converter) [Metallurgy]

OLP Oxygen/Lime/Powder (Steelmaking Process) [Metallurgy]

OLPA Office of Legislative and Policy Analysis [of National Institutes of Health, Bethesda, Maryland, US]

OLPARS On-Line Pattern Analysis and Recognition System

OLP/LBE Oxygen Lance Powder/Lance-Bubbling-Equilibrium (Steelmaking Converter) [Metallurgy]

OLPS On-Line Programming System

OLR Objective Loudness Ratio

OLR Off-Line Recovery

OLR Off Load Routes

OLR On-Line Review [UK Publication]

OLR On-Load Refuelling

OLR Outgoing Longwave Radiation

OLRA Ontario Labour Relations Act [Canada]

OLRB Ontario Labour Relations Board [Canada]

OLRC Office of the Law Revision Counsel [of US House of Representatives]

O/L-RC Overload-Reverse Current

OLRT On-Line Real-Time (System)

OLS On-Line Search

OLS On-Line Storage

OLS On-Line System

OLS Ontario Land Surveyor [Canada]

OLS Open Loop System

OLS Operational Line-Scan System [US Department of Defense, Defense Meteorological Satellite Program]

OLS Optical Line Scanner

OLS Ordinary Least Squares

OLS Over Larger Scale

OLSA Office of Local Section Activities [of American Chemical Society, US]

OLSA Orbiter/LPS (Launch Processing System) Signal Adapter [of NASA]

OLSA Ontario Land Surveyors Association [Canada]

OL'SAM On-Line Search Assistance Machine [of Franklin Institute, Philadelphia, Pennsylvania, US]

OLSC On-Line Scientific Computer

OLSDMS On-Line Strain and Damage Measurement System

OLSE Ordinary Least-Squares Estimator

OLS-L Operational Line-Scan System, Broadband Visible Channel (0.4-1.1μm) [US Department of Defense–Defense Meteorological Satellite Program]

OLSP On-Line Service Provider

OLSP Orbiter Logistics Support Plan [of NASA]

OLSS On-Line Software System

OLSS On-Line Support Software

OLS-T Operational Line-Scan System, Broadband Infrared Channel (10.2-12.8 μm) [US Department of Defense–Defense Meteorological Satellite Program]

OLSUS On-Line System Use Statistics

OLT Office of Learning Technologies [Canada]

OLT On-Line Test

OLTE On-Line Test Equipment

OLTE Optical Line Terminating Equipment

OLTEP On-Line Test Executive Program

OLTP Off-Line Tape Preparation

OLTP On-Line Transaction Processing

OLTS On-Line Test Section

OLTS On-Line Test System

OLTT On-Line Test Terminal

OLUC On-Line Union Catalog [of On-Line Computer Library Center, US]

OLUD On-Line Update

OLUG Office Landscape Users Group [now Office Planners and Users Group, US]

OLUHO Okinawa-Luzon-Hong Kong Submarine Cable [Japan/Philippines/Hong Kong]

OLUM On-Line Update Module

OLV Open-Frame Low Voltage

OLVP Office of Launch Vehicle Programs [of NASA]

OLVP Office of Launch Vehicles and Propulsion [of NASA]

OLWM Ordinary Low Water Mark

OM Object Management; Object Manager [Computers]

OM Object Module

OM Office Management

OM Office of Management [of National Institutes of Health, Bethesda, Maryland, US]

OM Office Manager

OM Office Model

OM Offshore Mechanics

OM Old Man [Radio Amateur Abbreviation Refering to Any Male, Regardless of Age]

OM Oman [ISO Code]

OM Operational Mid [Aerospace]

OM Operations Management; Operations Manager

OM Operations Maintenance

OM Optical Master

OM Optical Material

OM Optical Measurement

OM Optical Medium

OM Optical Metallograph(y)

OM Optical Method

OM Optomagnetic(s)

OM Orbital Maneuvering-Engine

OM Orbiter Main-Engine [of NASA]

OM Order of Merit

OM Organic Matrix

OM Organic Matter

OM Organomagnesium

OM Organometallic(s)

OM Orifice Meter

OM Outer Marker [Aeronautics]

OM Output Module

OM Ozone Mapper; Ozone Mapping

O&M Operation (or Operating) and Maintenance

O&M Organization and Method

O/M Oxygen-to-Metal (Ratio)

om *(omni mane)* – every morning [Medical Prescriptions]

.om Oman [Country Code/Domain Name]

Ωm ohm-centimeter [also ohm cm, ohm-cm, $\Omega \cdot$cm, or Ω-cm]

Ωm² ohm-square meter [also $\Omega \cdot$ m², or Ω-m²]

$\Omega^{-1} \cdot$ m^{-1} reciprocal ohm meter [also $(\Omega \cdot$m$)^{-1}$, $(\Omega$-m$)^{-1}$, (ohm-m)$^{-1}$, 1/$\Omega \cdot$m, or 1/ohm-m]

Ω/m ohm(s) per meter [also Ω m^{-1}]

$(\Omega$-m$)^{-1}$ reciprocal ohm meter [also $(\Omega \cdot$m$)^{-1}$, $(\Omega$-m$)^{-1}$, $\Omega^{-1} \cdot$m^{-1}, (ohm-m)$^{-1}$, 1/$\Omega \cdot$m, or 1/ohm-m]

OMA Object Management Architecture [Computer]

OMA Omaha, Nebraska [Meteorological Station Identifier]

OMA Ontario Medical Association [Canada]

OMA Ontario Mining Act [Canada]

OMA Ontario Mining Association [Canada]

OMA Ontario Ministry of Agriculture [Canada]

OMA Ontario Museums Association [Canada]

OMA Operations Maintenance Area

OMA Operations Monitor Alarm

OMA Optical Manufacturers Assiciation [US]

OMA Optical Multichannel Analyzer

OMA Orbiter Maintenance Area [of NASA]

OMA Orderly Marketing Agreement

OMA Overseas Mining Association [UK]

O&M,A Operation and Maintenance, Army

OMAC Occupational Medical Association of Canada

OMAC On-Line Manufacturing Control

OMA CCD Optical Multichannel Analyzer Charge-Coupled Device

OMAE (International Conference on) Offshore Mechanics and Arctic Engineering

OMAE Proc Proceedings of the International Conference on Offshore Mechanics and Arctic Engineering

OMAF Ontario Ministry of Agriculture and Food [now OMAFRA]

OMAFRA Ontario Ministry of Agriculture and Food and Rural Affairs [Canada]

OMAP Object Module Assembly Program

OMAR Office of Medical Applications of Research [of Office of Disease Prevention, National Institutes of Health, Bethesda, Maryland, US]

OMASD Office of Management Appraisal and Systems Development [US]

OMAT Office of Manpower, Automation and Training

OMB Office of Management and Budget [of United States Congress, Washington, DC]

OMB Outer Marker Beacon [Aeronautics]

oMBA o-Methylbenzoic Acid

OMBE Office of Minority Business Enterprises [now Minority Business Development Agency]

OMBE Organic Molecular Beam Epitaxy

OMBI Observation, Measurement, Balancing and Installation

OMBKE Orságos Magyar Bányászati es Kohászati Egyesület [Hungarian Mining and Metallurgical Society]

OMBUU Orbiter Midbody Umbilical Unit [of NASA]

OMC Open Market Committee [US]

OMC Optimum Moisture Content (of Soil)

OMC Orbiter Maintenance and Checkout [of NASA]

OMC Organic Matrix Composite

OMC Organisation Mondiale du Commerce [World Trade Organization; of United Nations]

OMC Organomagnesium Compound

OMC Organometallic Chemistry

OMC Organometallic Compound

OMC Orientation-Memory-Concentration (Test) [Alzheimer's Disease]

OMC Oxidized Microcrystalline (Wax)

O&MC Operation and Maintenance Cost

OMCA Ontario Motor Coach Association [Canada]

OMCF Operations and Maintenance Control File

OMCF Orbiter Maintenance and Checkout Facility [of NASA]

OMCOS (Symposium on) Organometallic Chemistry Directed Towards Organic Synthesis [of International Union of Pure and Applied Chemistry, UK]

OMCU Ontario Ministry of Colleges and Universities [Canada]

OMCVD Organometallic Chemical Vapor Deposition

OMD Open Macro Definition

OMD Operations and Maintenance Documentation

OMD Orbiter Mating Device [of NASA]

OMDP Ocean Margin Drilling Project

OMDR Operations and Maintainability Data Record; Operations and Maintenance Data Record

OME Office of Management Engineers

OME Office of Minerals Exploration [US]

OME Ontario Ministry of Education [Canada]

OME Ontario Ministry of Energy [Canada]

OME Ontario Ministry of the Environment [Canada]

OME Open Messaging Environment (Protocol)

OME Optimum Mineral Extraction

OME Orbiter Main Engine [of NASA]

OME Ordnance Mechanical Engineer(ing)

OMe Methyl Ester (or Methoxy)

O-Me O-Methyl Diisopropyl

OMEA Ontario Municipal Electric Association [Canada]

OMEDT Ontario Ministry of Economic Development and Trade [Canada]

OMEF Office Machines and Equipment Federation

OMEGA Octanoyl-N-Methylglucamide

OMEP Ontario Mineral Exploration Program [Canada]

OMEP Office of Marine and Estuarine Protection [of US Environmental Protection Agency]

OMER Operations Management Education and Research Foundation [US]

OMEST Ontario Ministry of Energy, Science and Technology [Canada]

OMET Ontario Ministry of Education and Training [Canada]

OMET Orbiter Mission Elapsed Time [NASA]

OMETA Ordnance Management Engineering Training Agency [US]

OMEWG Orbiter Maintenance Engineering Working Group [of NASA]

OMF Object Module File

OMF (Microsoft) Object Module Format

OMF Office of Marketing Facilities [US]

OMF Old Master File

OMF Ontario Ministry of Finance [Canada]

OMF Open Media Framework

OMF Open Message Format

OMFS Office Maser Frequency Supply

OMG Object Management Group

OMG Ocean Mapping Group [of University of New Brunswick, Fredericton, Canada]

OMG Oesterreichische Mathematische Gesellschaft [Austrian Mathematical Society]

OMG Oesterreichische Mineralogische Gesellschaft [Austrian Mineralogical Society]

OMGUS Office of Military Government of the United States

OMH Ontario Ministry of Health [Canada]

OMH-1 Sodium Diethyldihydroaluminate

omh *(omni hora)* – every hour [Medical Prescriptions]

OMHF Ontario Mental Health Foundation [Canada]

OMHSA Ontario Municipal Health and Safety Association [Canada]

OMI O-Methyl Isourea

OMI (Lotus) Open Messaging Interface

OMI Operations and Maintenance Instruction

OMI Oregon Metals Initiative [US]

OMIA Online Mendelian Inheritance in Animals [Animal Genetics Database of the Australian National Geoscience Information Service]

OMIBAC Ordinal Memory Inspecting Binary Automatic Computer

OMIM Online Mendelian Inheritance in Man [Human Genetics Database of the National Center for Biotechnology Information, US]

1/min one per minute [also min^{-1}]

OMIS Oil Market Information System

OMIS Operational (or Operations) Management Information System

OMIS Optical Microscope Inspection System

OMISS Operation and Maintenance Instruction Summary Sheet(s)

OMiU O-Methyliso-Urea

OML Object Manipulation Language

OML Ontario Ministry of Labour [Canada]

OML Ontario Motor League [Canada]

OML Ordnance Missile Laboratories [US]

OML Outer (or Outside) Mold Line [Materials Science]

OML Outgoing Matching Loss

OMM Operation and Maintenance Manual

OMM Organisation Météorologique Mondiale [World Meteorological Organization]

OM&M Operation, Maintenance and Management

omm *(omni mane)* – every morning [Medical Prescriptions]

1/mm one per millimeter [also mm^{-1}]

1/mm² square millimeter [also mm^{-2}]

1/mm³ cubic millimeter [also mm^{-3}]

om man *(omni mane)* – every morning [Medical Prescriptions]

OMMB Ontario Milk Marketing Board [Canada]

OMMH Orbiter Maintenance Man-Hours [NASA]

Ω·mm²/m ohm-square millimeter per meter [also Ω·mm² m^{-1}, or ohm-mm²/m]

1/mm/μs² one per millimeter per microsecond squared [also 1/mm μsec², mm^{-1} μs^{-2}, or mm^{-1} μsec^{-2}]

OMMSQA Office of Modeling, Monitoring Systems and Quality Assurance [US]

omn *(omni nocte)* – every night [Medical Prescriptions]

OMNDM Ontario Ministry of Northern Development and Mines [Canada]

Omni Omnirange [also omni]

omni omnidirectional

OMNITAB Omnibus Programs, Handling of Tabular Numerical Operations [Programming Language]

OMNITENNA Omnirange Antenna [also Omnitenna]

OMNIX Onyx Microcomputer Unix (System)

om noct *(omni nocte)* – every night [Medical Prescriptions]

OMNR Ontario Ministry of Natural Resources [Canada]

OMNS Open Network Management System

1/mol one per mole [also mol^{-1}]

OMP 2,3,7,8,12,13,17,18-Octamethylprophinate

OMP Office of Metric Programs [US]

OMP Olefin Metathesis Polymerization

OMP Operations and Maintenance Plan

OMP Optical Mark Printer

OMP Orotidine 5'-Monophosphate

OMP Ozone Mapping Project [of US Environmental Protection Agency]

5'-OMP Orotidine-5'-Monophosphate

oMP o-Methoxyphenol

OMPA Octamethyl Pyrophosphoramide [also OMPHA, ompa, or ompha]

iso-OMPA Tetraisopropyl Pyrophosphoramide

1/MPa One per Megapascal [also MPa^{-1}]

OMPR Operational Maintainability Problem Reporting

OMPR Optical Mark Page Reader

OMPRA One-Man Propulsion Research Apparatus [of NASA]

OMPT Observed Mass Point Trajectory

OMR Omani Rial [Currency of Oman]

OMR Operations and Maintenance Requirements

OMR Operations Management Room

OMR Optical-Mark Reader; Optical-Mark Reading

OMR Optical-Mark Recognition

OMR Optomagnetic Recorder; Optomagnetic Recording

OMR Orbiter Management Review [of NASA]

OMR Organic-Moderated (Nuclear) Reactor [US]

OMR Organometallic Reagent

OM&R Operation, Maintenance and Replacement

OMRB Operating Material Review Board

OMRC Operational Maintenance Requirements Catalog

OMRCA Organic-Moderated Reactor Critical Assembly [US]

OMRE Organic-Moderated Reactor Experiment [of Idaho National Engineering Laboratory, US Department of Energy at Idaho Falls]

OMRF Orbiter Maintenance and Refurbishment Facility [of NASA]

OMRF Orbiter Modification and Refurbishment Facility [of NASA]

OMRO Ordnance Material Research Office [US]

OMRP Operations and Maintenance Requirements Plan

OMRR Ordnance Material Research Reactor [US]

OMRS Operational Maintainability Reporting System

OMRS Operational Maintenance Requirements and Specification

OMRS Operations and Maintenance Requirements (and) Specification

OMRSD Operational Maintainability Reporting Systems Document

OMRSD Operational Maintenance Requirements and Specification Document

OMRSD Operations and Maintenance Requirements and Specifications Documentation

OMRV Operational Maneuvering Reentry Vehicle

OMS Octahedral Molecular Sieve

OMS Office Management System

OMS Office of Management Services [US and Canada]

OMS Operational Meteorological Satellite [of NASA]

OMS Optical Multichannel Spectroscopy

OMS Optoelectronic Measuring System

OMS Orbital Maneuvering Subsystem

OMS Orbital Maneuvering System (Engine) [of NASA Space Shuttle]

OMS Organisation Mondiale de la Santé [World Health Organization]

OMS Output per Manshift

OMS Output Multiplex Synchronizer

OMS Ovonic Memory Switch

ÖMS Österreichische Mathematische Gesellschaft [Austrian Mathematical Society]

1/m²·s one per square meter second [also $m^{-2}s^{-1}$]

OMSA Offshore Marine Service Association [US]

OMSF Office of Manned Space Flight [now Office of Space Transportation Systems] [of NASA]

OMSP Operational Maintenance Support Plan

OMSP Organically Modified Sodium Bentonite

OMT Ontario Ministry of Transportation [Canada]

OMT Orthomode Transducer

1/m·T one per meter tesla [also $m^{-1} T^{-1}$]

OMTF Ordnance Missile Test Facility [White Sands, New Mexico, US]

OMTNS Over Mountains

OMTRI Osaka Municipal Technical Research Institute [Japan]

OMTS Organically Modified Mica-Type Silicate

OMTS Organizational Maintenance Test Station

OMTS-A Organically Modified Mica-Type Silicate, Acid-Exchanged

OMTS-B Organically Modified Mica-Type Silicate, Base-Exchanged

OMU N'-Cyclooctyl-N,N-Dimethylurea

OMU Optical Measuring Unit

OMU Osaka Medical University [Japan]

OMUA Office Machinery Users Association [now Institute of Administrative Management, UK]

OMV Obligatory Minimum Valency

OMV Orbital Maneuvering Vehicle

OMV Oxygen Manual Valve

OMVPE Organometallic Vapor-Phase Epitaxy

OMVS Organisation pour la Mise en Valeur du Sénégal [Organization for the Development of the Senegal River]

oMxA o-Methoxyaniline

oMxBA o-Methoxybenzoic Acid

oMxCa o-Methoxycarbonylaniline

ON Octane Number

ON Odd Number; Odd-Numbered

ON Oesterreichisches Normeninstitut [Austrian Standards Institute] [also ÖN]

ON Ontario [Canada]

ON Optic Nerve [Anatomy]

ON Order Number

ON Oriented Nucleation [Metallurgy]

ON Oxidation Number [Chemistry]

ÖN Österreichisches Normeninstitut [Austrian Standards Institute] [also ON]

O(N) Order N Method (of Atoms)

O₂/N₂ Oxygen/Nitrogen (Atmosphere)

on *(omni nocte)* – every night [Medical Prescriptions]

ONA Office of National Assessments

ONA Ontario Nurses Association [Canada]

ONA Open Network Architecture

ÖNA Österreichischer Normenausschuss [Austrian Standards Committee]

ONAL Off-Network Access Line

ONAM Observatorio Nacional Astrofisica de Mexico [National Astrophysical Observatory of Mexico at Tonanzintla]

ONAM Observatorio Nacional Astronómica de Mexico [National Astronomical Observatory of Mexico at Tacubaya]

ONAT Off-Network Access Trunk

O-NAV Onboard Navigation [also O-Nav]

ONB Oesterreichische Nationalbibliothek [National Library of Austria]

ONB O-Nitrobiphenyl

ÖNB Österreichischer Naturschutzbund [Austrian Nature Protection League]

oNBAAP 4-N-(o'-Nitrobenzylidene)aminoantipyrine [also o-NBAAP]

Onbd Onboard

ONBT Orbiter Neutral Buoyancy Trainer [of NASA]

ONBzl p-Nitrobenzyloxy

ONC Open Network Computing

ONC Operational Navigation Chart(s)

ONC Ordinary National Certificate [UK]

Oncol Oncological; Oncologist; Oncology [also oncol]

Oncol Oncology [International Journal]

ONCT Oita National College of Technology [Japan]

OND Operator Need Date

OND Ordinary National Diploma [UK]

ONDB Online Database

ONDE Office of Nondestructive Evaluation [of National Institute of Standards and Technology, Gaithersburg, Maryland, US]

Onde Electr L'Onde Electrique [French Journal on Electric Waves]

ONDS Open Network Distribution Services

ONE Open Network Environment [Internet]

ONE Optimum Network Executive

ONERA Office National d'Etudes et de Recherches Aérospatiales [National Office for Aerospace Studies and Research, Toulouse, France]

ONERA-CERT Office National d'Etudes et de Recherches Aérospatiales–Centre d'Etudes et de Recherches de Toulouse [National Office for Aerospace Research of the Toulouse Research and Study Center, France]

ONF Office National des Forêts [National Forestry Office, France]

ONF Ozark National Forest [Arkansas, US]

ONGA Ontario Natural Gas Association [Canada]

ONGA Overseas Number Group Analysis

ONGC Oil and Natural Gas Commission [India]

ONGC Oil and Natural Gas Corporation Limited [India]

ONI Office of Naval Intelligence [of US Navy]

ONI Operator Number Identification [also ONID]

ONIA Office National Industriel de l'Azote [National Office of the Nitrogen Industry, France]

ONID Operator Number Identification [also ONI]

ONISEP Office National d'Information sur les Enseignements et les Professions [National Information Office on Education and the Professions, France]

Online Bus Inf Online Business Information [UK Journal]

Online Libr Microcomput Online Libraries and Microcomputers [US Journal]

Online Rev Online Review [UK]

ONLS On-Line System

ONM Office National Météorologique [National Meteorological Office, France]

1/nm one per nanometer [also nm^{-1}]

$\Omega^{-1}nm^{-1}$ one per ohm nanometer [also $1/\Omega \cdot nm$]

O-17 NMR Oxygen-17 Nuclear Magnetic Resonance [also ^{17}O NMR]

Onnes Comm Phys Lab Leiden Onnes Communications of the Physical Laboratory of Leiden [Netherlands]

ONO Or Near Offer [also ono]

ONO Oxide-Nitride-Oxide (Structure)

ONO Oxide/Reoxidized Nitrided Oxide (Film)

oNOa o-Nitroaniline [also ONOA]

oNOBA o-Nitrobenzoic Acid [also ONOBA]

oNPOE o-(Nitrophenyl)octylether [also o-NPOE]

ONOP O-Nitrophenol [also oNOP]

ÖNORM Österreichische Norm [Austrian Standard]

ONP Old Newspapers [Recycling]

ONP O-Nitrophenol [also oNP]

ONP p-Nitrophenyl Ester

ONP Optical Nuclear Polarization

Onp p-Nitrophenyl Ester (or p-Nitrophenoxy)

ONPG o-Nitrophenyl β-D-Galactopyranoside [Biochemistry]

ONPI Office of News and Public Information [US]

ONQ Order of Nurses of Quebec [Canada]

ONR Office of Naval Research [of US Navy at Arlington, Virginia]

ONR-ASEE Office of Naval Research–American Society of Engineering Education (Program) [US]

ONR/DARPA Office of Naval Research/Defense Advanced Research Projects Agency (Program) [US]

ONR/DOD Office of Naval Research of the Department of Defense [of US Navy at Arlington, Virginia]

ONRI Osaka National Research Institute [of Agency of Industrial Science and Technology, Japan]

ONR-MD Office of Naval Research–Materials Division [of US Navy at Arlington, Virginia]

ONRR Office of Nuclear Regulatory Research [of Nuclear Regulatory Commission, US]

ONRR-NRC Office of Nuclear Regulatory Research of the Nuclear Regulatory Commission [US]

ONR/SDIO Office of Naval Research/Strategic Defense Initiative Organization (Program) [US]

ONS Omega Navigation System

ONS Overseas News Service [UK]

ONSHR On Shore

ONSPS Office of Nuclear Safety Policy and Standards [of US Department of Energy]

ONST Office of National Security Technology [of Battelle Northwest, Richland, Washington, US]

ONT Office National du Tourisme [National Tourism Office, France]

ONT Office of Naval Technology [US]

Ont Ontario [Canada]

ONTAP On-Line Training and Practice [also ONTAP]

Ontario Technol Ontario Technologist [Canadian Journal]

ONT/ASEE Office of Naval Technology/American Society for Engineering Education (Fellowship) [US]

ONTC Ontario Northland Transportation Commission [Canada]

ONTERIS Ontario Educational Research Information System [Canada]

ONU Ohio Northern University [Ada, US]

ONU Optical Network Unit

ONU Organisation des Nations Unies [United Nations Organization]

ONUESC Organisation des Nations Unies pour l'Education, la Science et la Culture Intellectuelle [United Nations Educational, Scientific and Cultural Organization]

ONULP Ontario New Universities Library Project [Canada]

ONWARD Organization of Northwest Authorities for Rationalized Design [US]

ONWI Office of Nuclear Waste Isolation [of US Department of Energy]

OO Object-Oriented (Computer Programming)

OO Oceanographic Office

OO Ocean Outlook [US]

OO Optical Orientation

OO Orbital Oscillation

OO Orbital Overlap [Physics]

OO Ordnance Office(r)

OO- Belgium [Civil Aircraft Marking]

O/O Ore/Oil (Carrier) [also OO]

oo odd-odd (nuclei) (i.e., atomic nuclei whose spin is equal to an even multiple of $\frac{1}{2} \times \hbar$, $\hbar$ = Planck's constant $\div$ 2π) [Nuclear Physics]

1/Ω one per ohm [also Ω^{-1}]

OOA Object-Oriented Analysis

OOA Offshore Operators Association [UK]

OOA Ontario Opticians Association [Canada]

OOAD Object-Oriented Application Design; Object-Oriented Application Development

OOB Out-of-Balance (Current)

OOB Out of Band

OOC Object-Oriented Computing

OOC Ore/Oil Carrier [also O/OC]

OOC On-Off Control

$1/\omega C$ Reactance [Electrical Engineering]

OOCL Oriental Overseas Container Line, Limited [Canada]

$1/\Omega \cdot cm$ one over ohm-centimeter [also $\Omega \cdot cm^{-1}$, $(ohm\text{-}cm)^{-1}$, or 1/ohm-cm]

OOD Object-Oriented Design

OOD Opposite Oriented Diffusion

OOD Orbiter On-Dock [NASA]

OOD Out-of-Detent

OODB Object-Oriented Database

OODBMS Object-Oriented Database Management System [also OODMS]

OODP Out-of-Detent Pitch

OODR Optical-Optical Double Resonance (Spectroscopy)

OODR Out-of-Detent Roll

$1/Oe$ One per Oersted [also Oe^{-1}]

OOG Office of Oil and Gas [of US Department of the Interior, Washington, DC]

OOG Olive Oil Group [US]

OOG Onshore Operators Group [UK]

OOG Out of Gauge

OOIP Original Oil-in-Place [Oil Exploration]

OOK On-Off Keying [Communications]

OOL Object-Oriented Language

OOL Operator-Oriented Language

OOL Out-of-Line Errors

OOLR Overall Objective Loudness Rating

OOLUG Oklahoma On-Line Users Group [US]

$1/\Omega \cdot m$ one over ohm-meter [also Ωm^{-1}, $(ohm\ m)^{-1}$, or 1/ohm-m]

OOMM Organizational Operations and Maintenance Manual

OOO Out of Order [also ooo]

OOOS Object-Oriented Operating System

OOP Object-Oriented Programming

OOP Oblique Orthographic Projection

OOP Out of Phase [Physics]

OOP Out-of-Plane (Failure)

OOP Out of Position

OOPL Object-Oriented Programming Language

OOPS Off-Line Operating Simulator

OOPS Object Oriented Programming and Systems

OOPS Oxide One Pot Synthesis (Process)

OOR Office of Ordnance Research [of Duke University, Durham, North Carolina, US]

OOR Operator Override

OOR Out-of-Round(ness)

OORC Ontario Ornamental Research Committee [Canada]

OOS Object-Oriented Systems

OOS Off-line Operating Simulator

OOS On-Orbit Station

OOS Operational Operating System

OOS Orbit-to-Orbit Shuttle

OOS Orbit-to-Orbit Stage

OOS Ore/Oil Ship [also O/OS]

OOS Out of Service [also oos]

OOSA Office of Outer Space Affairs [of United Nations]

OOSDP On-Orbit Station Distribution Panel

OOT Object-Oriented Technology

OOT Out of Tolerance

OOT Out of True

OOU Out of Use [also oou]

OOUI Object-Oriented User Interface

OOUG Oregon On-Line Users Group [US]

OOW Oklahoma Ordnance Works [Pryor, US]

OP Object Plane [Optics]

OP Object Program

OP Object-Oriented Programming

OP Obligate Parasite [Biology]

OP Observation Plane

OP Observation Post

OP Observatoire de Paris [Paris Observatory, France]

OP Ocean Physicist; Ocean Physics

OP Octylphenoxy-

OP Odd Parity

OP Office of Policy [of US Environmental Protection Agency]

OP Office Processor

OP Oil Paint

OP Oil Paper

OP Oil Pollution

OP Oil Pressure

OP Oil Pump

OP Olber's Paradox [Astronomy]

OP Old Process (Patenting) [Metallurgy]

OP Open [also op]

OP Open Point (Ammunition)

OP Open Pore

OP Operand [also Op, or op]

OP Operating Potential

OP Operating Pressure

OP Operating Procedure

OP Operation [Computers/Communications]

OP Operational Priority

OP Operational Process

OP Operation Part

OP Operations Plan

OP Operator('s) Panel

OP Oppenheimer-Phillips (Process) [also O-P] [Nuclear Physics]

OP Optical Path

OP Optical Physicist; Optical Physics

OP Optical Processing

OP Optical Properties

OP Optical Pumping

OP Optical Purity; Optically Pure

OP Optical Pyrometer

OP Oral Pathologist; Oral Pathology

OP Orbital Polarization [Physics]

OP Ordered Polymer

OP Order Parameter [Statistical Mechanics]

OP Order Processing

OP Organic Polymer

OP Organophosphorus

OP Oriented Polymer

OP Oriented Polypropylene

OP Orthogonal Polynomial

OP Orthographic Projection

OP Orthopositronium

OP Osmotic Pressure

OP Osteogenic Protein [Biochemistry]

OP Osteopontin [Biochemistry]

OP Outlet Pressure

OP Out-of-Phase [Physics]

OP Out of Print [also op]

OP Outpatient

OP Output

OP Output Processor

OP Overpotential

OP Overpressure [also O/P]

OP Over Proof [Spirits]

OP Oxidation Potential [Physical Chemistry]

OP Oxidation Protection

OP Oxygen Point [also op]

OP Oxygen Purge; Oxygen Purging

OP Oxygen Pressure Process

1P One Pole (or Single Pole)

O&P Operations and Procedures

O/P Output [also o/p, OP, or op]

O/P Overpressure [also OP]

O-P Oppenheimer-Phillips (Process) [also OP] [Nuclear Physics]

Op Operand [also OP, or op]

Op Operate; Operator; Operation(al) [also op]

Op Optic(al); Optics [also op]

oP Pearson symbol for primitive (simple) space lattice in orthorhombic crystal system (this symbol is followed by the number of atoms per unit cell, e.g. oP4, oP16, etc.)

op open

op out of print [also OP]

op over proof [Spirits]

OPA Octadecylphosphonic Acid

OPA Office of Planning and Analysis [US]

OPA Office of Price Administration [US]

OPA Office of Program Analysis [of US Department of Energy]

OPA Office of Public Affairs [of Materials Research Society, US]

OPA Oil Pollution Act [US]

OPA Ontario Pharmacists Association [Canada]

OPA Ontario Podiatry Association [Canada]

OPA Ontario Psychological Association [Canada]

OPA Operations Planning Analysis

OPA Operator Priority Access

OPA O-Phthalaldehyde

OPA O-Phthalic Acid

OPA Optical Parametric Amplifier

OPA Optical Publishing Association [US]

OPA Optoelectronic Pulse Amplifier

ÖPA Österreichisches Patentamt [Austrian Patent Office]

1/Pa One over Pascal [also Pa^{-1}]

opa opaque

OPAC Online Public Access Catalog [Internet]

OPAC (Working Group on) Operation of Aircrafts [of European Civil Aviation Conference, France]

OPAC Operations Planning Advisory Committee

OPADEC Optical Particle Decoy

OPA/FMOCCl Orthophthalaldehyde/Fluorenylmethoxycarbonyl Chloride

OPAIT Ontario Program for the Advancement of Industrial Technology [Canada]

OPAL Omni-Purpose Apparatus at LEP ((Large Electron-Positron) Collider) (Experiment) [of CERN–European Laboratory for Particle Physics, Geneva, Switzerland]

OPAL Operational Performance Analysis Language

OPAL Order Processing Automated Line

OPAMP Operational Amplifier [also OP AMP, opamp, op-amp, or op amp]

OPAL Optical Platform Alignment Linkage [Aerospace]

OPAN Oxy-bis(phthalic Anhydride)

OPANAL Organismo para el Prohibición de los Armas Nuclear en América Latina [Agency for the Prohibition of Nuclear Weapons in Latin America (and the Caribbean)]

OPAP Ontario Prospectors Assistance Program [of Ontario Ministry of Northern Development and Mines, Canada]

OPAP Phenylazophenyl Ester

OPAS Operational Assignment

OPAS Organizational Prior Approval System [of National Science Foundation, US]

1/Pa·s One per pascal-second [also $Pa^{-1}·s^{-1}$]

OPASTCO Organization for the Protection and Advancement of Small Telephone Companies [US]

OPB Oxidizer Preburner [Aerospace]

OPBA Ontario Public Buyers Association [Canada]

OPBOV Oxidizer Preburner Oxidizer Valve [Aerospace]

OPC Ocean Products Center [US]

OPC Oilfield Production Consultants (Limited) [UK]

OPC Oligonucleotide Purification Cartridge

OPC On-Line Plotter Controller

OPC Ontario Press Council [Canada]

OPC OpenGL (Three-Dimensional Graphics Library) Performance Characterization (Subcommittee)

OPC Open Performance Characterization

OPC Operational Control [Aeronautics]

OPC Operation Code

OPC Optical Phase Conjugation

OPC Optical Photoconductor

OPC Optional Calling

OPC Ordinary Portland Cement

OPC Ordnance Procurement Center [New York, US]

OPC Organic Polymer Chemistry

OPC Organic Photoconducting; Organic Photoconductor

OPC Organophosphorus Compound

OPC Other Project Costs

OPCA Ornamental Plant Collectors Association

OPCC Optical Product Code Council [US]

OPCD Operational Planning and Coordination Directorate [of Secretary of State, Canada]

op cit *(opere citato)* – in the work cited

opcl optical

OPCM Operative Plasterers and Cement Masons International

OPCODE Operation(al) Code [also opcode, OP-CODE, op-code, OP CODE, or op code]

OPCOM Operations Communications

OpCom Operating Committee [of Institute of Electrical and Electronics Engineers, US]

OPCON Optimizing Control

OPCP Perchlorophenyl Ester

OPCR One-Pass Cold-Rolled (Material) [Metallurgy]

OPCT Osaka Prefectural College of Technology [Japan]

OPCW Organization for the Prevention of Chemical Weapons [Netherlands]

OPD Ocean Physics Division [of Defense Research Establishment Pacific, British Columbia, Canada]

OPD One per Desk

OPD Operand [also Opd, or opd]

OPD Operational Programming Department

OPD Operations Department

OPD O-Phenylenediamine

OPD Optical Particle Detector

OPD Optical Path Difference

OPD Outpatient Department

OPD Over Pin Diameter

OPDAC Optical Data Converter

OPDAR Optical Detection and Ranging [also Opdar, or opdar]

OPDAR Optical Radar [also Opdar, or opdar]

OPDEVFOR Operations Development Forces (Navy) [US]

OPDF One-Particle Distribution Function [Physics]

OPDN Oxydipropionitrile

OPDP Officer Professional Development Program

OPDP O-Phenylenebis(diphenylphosphine) [also opdp]

OPE Office of Planning and Environment [US]

OPE Operations Project Engineer

OPE Optimized Processing Element

OPE Other Project Element

OPE Oxidized Polyethylene

OPE Oxygen Plasma Etching

OPEC Organization of Petroleum Exporting Countries [Austria]

OPECNA Organization of Petroleum Exporting Countries News Agency [Austria]

OPEDA Organization of Professional Employees of the Department of Agriculture [US]

OPEDA Outdoor Power Equipment Distributors Association [US]

OPEG OPEC (Organization of Petroleum Exporting Countries) Petroleum Emergency Group

OPEI Outdoor Power Equipment Institute [US]

OPEIU Office and Professional Employees International Union

OPEN Ocean Production Enhancement Network [of Dalhousie University, Nova Scotia, Canada]

OPEN Organisation des Producteurs d'Energie Nucléaire [Organization of Nuclear Energy Producers, France]

OPEN Origins of Plasma in the Earth's Neighborhood (Satellite) [of NASA]

OpenGL Three-Dimensional Graphics Library

Open Syst Softw Open Systems and Software [Journal published in the UK]

OPEP Orbital Plane Experimental Package [of NASA]

OPEPR Optical Perturbation Electron Paramagnetic Resonance

Oper Operate; Operation(al); Operator [also oper]

OPERA Office of Policy for Extramural Research Administration [of National Institutes of Health, Bethesda, Maryland, US]

OPERA Out-of-Pile Expulsion and Reentry Apparatus

OpeRA On-Line Periodicals Research Area [of Institute of Electrical and Electronics Engineers, US]

OPERATORS Optimization Program for Economical Remote Trunk Arrangement and TSPS (Traffic Service Positions System) Operator Arrangements

Operg Operating [also operg]

Oper Dent Operative Dentistry [Journal]

Oper Dent Suppl Operative Dentistry Supplement [Journal]

Oper Geogr Operational Geographer [Journal of the Canadian Association of Geographers]

Oper Manage Rev Operations Management Review [US]

Oper Res Operations Research(er)

Oper Res Operations Research [Publication of the Operations Research Society of America]

Oper Res Lett Operations Research Letters [Publication of the Operations Research Society of America]

Oper Syst Operating System [also oper syst]

Oper Syst Netw Operating Systems and Networks [UK Journal]

Oper Syst Rev Operating Systems Review [Published by the Association for Computing Machinery, US]

OPEX Operational and Executive

OPF Operation Plan Framework

OPF Orbiter Processing Facility [of NASA]

OPF Osmium Potassium Ferrocyanide

OPFC Orbiter Preflight Checklist [of NASA]

OPFET Optical Field-Effect Transistor [also OpFET]

OPFIR Optically Pumped Far Infrared

OPFM Outlet Plenum Feature Model

OPG Oil-Pressure Gauge

OPG One-Pump Gradient (Furnace)

OPG Ontario Power Generation [Superseded Ontario Hydro, Canada]

OPG Optical Precipitation Gauge

OPG Oxalate, Peroxide and Gluconic Acid

OPG Oxypolygelatin

OPGUID Optimum Guidance (Technique)

OPh Phenyl Ester

1PH One Phase (or Single Phase)

OPHA Ontario Public Health Association [Canada]

oPhA o-Phthalic Acid

oPha o-Phenylaniline

oPhBA o-Phenylbenzoic Acid

oPhDA o-Phenylenediamine

OPHF Orbital Polarized Hartree-Fock [Quantum Mechanics]

oPhP o-Phenylphenol

Op Hrs Operation Hours [also op hrs]

opht ophthalmic

Ophthalm Physiol Opt Ophthalmic and Physiological Optics [Journal published in the UK]

Ophthalm Res Ophthalmic Research [International Journal]

Ophthalmol Opthalmological; Opthalmologist; Opthalmology [also opthalmol]

Ophthalmol Ophthalmologica [International Journal]

OPI Office of Planning and Integration [of US Department of Energy, Richland Operations, Washington State]

OPI Office of Public Information [of US Department of Justice]

OPI Ontario Petroleum Institute [Canada]

OPI Open Prepress Interface

OPI Orbiter Payload Interrogator [of NASA]

OP&I Office of Patents and Inventions [US]

OPIC Overseas Private Investment Corporation [of International Development Cooperation Agency, US]

OPIC Overseas Private Investment Council

OPIDF Operational Planning Identification File

OPIM Order Processing and Inventory Monitoring

OPIS Orbiter Prime Item Specification [of NASA]

OPIT Oxide Powder-In-Tube (Process) [Materials Science]

Opk Optokinetic(s) [also opk]

OPL Official Publications Library

OPL Open Problem List

OPL Operational [also Opl, or opl]

OPL Optically Pumped Laser

OPL Outer Plexiform Layer

OPLA Offshore Pollution Liability Agreement

OPLA Ontario Public Library Association [Canada]

OPLAC Ontario Public Libraries Advisory Committee [Canada]

OPLB Ontario Public Libraries Board [Canada]

OPLC Ontario Provincial Libraries Council [Canada]

OPLC Overpressure Layer Chromatography

OPLE Omega Position Location Experiment [of NASA]

OPLF Orbiter Processing and Landing Facility [of NASA]

OPLIN Ontario Public Libraries Information Network [Canada]

OPM Observatoire de Paris-Meudon [Paris-Meudon Observatory, France]

OPM Office of Personnel Management [of Federal Bureau of Investigations, US]

OPM Open-Pit Mining

OPM Operations per Minute [also opm]

OPM Operator Programming Method

OPM Orbits per Minute [also opm] [Sanding Equipment]

OPM Ordnance Proof Manual

OPM Output Position Map

OPMA Office Products Manufacturers Association [US]

OPMA Open Pit Mining Association [US]

OPMA Oriented Polypropylene-Film Manufacturers Assiciation

OPMA Overseas Press and Media Association [UK]

OpMan Operations Manual [of Institute of Electrical and Electronics Engineers, US]

OPMET Operational Meteorological Information

OPMG Office of the Provost Marshall General [US Army]

OPMS Optical Pressure Measurement System

OPN One-Port Network

OPN Opcode (Operation Code) Parts Number

OPN Optics and Photonic News [Magazine of the Optical Society of America]

Opn Open(ing) [also opn]

Opn Operation [also opn]

OPNAV Office of the Chief of Naval Operations [also OpNav]

OPND Operand [also Opnd, or opnd]

Opng Opening [also opng]

OPNS Operational Phase

OPO ODA–PMDA–ODA (4,4′-Oxydianiline– Pyromellitic Dianhydride–4,4′-Oxydianiline)

OPO Optic(al) Parametric Oscillator

OPO Orbiter Project(s) Office [of NASA]

OPOL Offshore Pollution Liability Agreement

OPOL Offshore Pollution Liability Association [UK]

OPOL Optimization-Oriented Language

OPOV Oxidizer Preburner Oxidizer Valve [Aerospace]

OPP Octal Print Punch

OPP 2,3,7,8,12,13,17,18-Octapropylporphinate

OPP Office of Pesticide Programs [of US Environmental Protection Agency]

OPP Office of Plans and Policy [of Federal Communications Commission, US]

OPP Office of Polar Programs [US]

OPP O-Phenylphenol

OPP Optical Precipitate Profiler

OPP Oriented Polypropylene

OPP Out-of-Phase Particles (of Intermetallics)

Opp Opportunity [aloso opp]

Opp Opposition [also opp]

opp opposite(ly)

OPPA Octylpyrophosphoric Acid

OPP CE Opposite Commutator End [also OPPCE]

OPPD Omaha Public Power District [Nebraska, US]

OPPMSA Ontario Pulp and Paper Makers Safety Association [Canada]

OPPOSIT Optimization of a Production Process by an Ordered Simulation and Iteration Technique

OPP PE Opposite Pulley End

OPPAR Orbiter Project(s) Parts Authorization Request [of NASA]

OPPL Orbiter Project Parts List [of NASA]

OPPS Overpressurization Protection Switch

OPPS Overpressurization Protection System

OPPT Office of Pollution Prevention and Toxics [of US Environmental Protection Agency]

OPPTS Office of Prevention, Pesticides and Toxic Substances [of US Environmental Protection Agency]

OPR Office of Population Research [of Princeton University, New Jersey, US]

OPR Office of Primary Responsibility

OPR Offsite Procurement Request

OPR Oil Production Rate

OPR Operational Preference

OPR Operations Planning Review

OPR Optical Page Reader; Optical Page Reading

OPR Optical Pattern Recognition

OPR Orbiter/Payload Recorder [of NASA]

OPR Oxygen Pressure Regulator

OPr Propionate

Opr Operation(al); Operative; Operator [also opr]

OPRA Office Products Representatives Association [US]

OPRB Oklahoma Planning and Resources Board [US]

OPRC Oil Pollution Preparedness, Response and Cooperation Convention

OP Register Operation Register [also op register]

OPRIS Ohio Project for Research in Information Services [US]

o Prof *(emeritus ordentlicher Professor)* – Central European Academic Title equivalent to a (Full) Professor Emeritus

OPRR Office for Protection from Research Risks [of Office of Extramural Research, National Institutes of Health, Bethesda, Maryland, US]

OPRU Oil Pollution Research Unit [Now Part of Field Studies Council, UK]

OPRV Oxygen Pressure Relief Valve

OPS Office of Pipeline Safety [of US Department of Transportation]

OPS Office of Program Support [of United States Geological Survey]

OPS Office Publishing System

OPS Off-Premise Station

OPS Offshore Power Systems

OPS Oil Pressure Switch

OPS On-Line Process Synthesizer

OPS Open Profiling Standard

OPS Operational Paging System

OPS Operational Performance Standards

OPS Operational Protection System

OPS Operational Sequence

OPS Operations [Computer/Communications]

OPS Operation(s) Sequence

OPS Operations per Second

OPS Operator's Subsystem

OPS Optical Proximity Sensor

OPS Orbiter Project Schedule [of NASA]

OPS Oriented Polystyrene

OPS Overhead Positioning System

OPS Oxidation Protection System

OPS Oxygen Purge System

Ops Operations; Operators [also ops]

o-Ps Orthopositronium [Particle Physics]

OPSA N,N'-Diethyl 4-Morpholinylphosphonothioc Diamide

OPSB Orbiter Processing Support Building [of NASA]

OPSCAN Optical Scanning

OPSCON Operations Control

OPSEC Operations Security

OPSF Orbital Propellant Storage Facility

OPSK Optimum Phase-Shift Keyed; Optimum Phase-Shift Keying

OPSKS Optimum Phase-Shift Keyed Signal

OPSP Office of Product Standards Policy

OPSK Operations Panel [of International Civil Aviation Organization]

OPSR Office of Pipeline Safety Regulations [of US Department of Transportation]

OPSR Operations Supervisor [Ground Spacecraft Tracking and Data Network Site]

OPSS Orbital Propellant Storage Subsystem

OPSWL Old Program Status Word Location

OPSYS Operating System

OPT Open Protocol Technology

OPT Operational Pressure Transducer

OPT Ophthaldialdehyde

.OPT Options [File Name Extension] [also .opt]

ÖPT Österreichische Post− und Telegraphenverwaltung [Austrian Post and Telegraph Administration]

Opt Optical; Optician; Optics [also opt]

Opt Optimum [also opt]

Opt Option(al) [also opt]

opt optimal; optinum

opt *(optimus)* – best; optimal

OPTA O-Phthaldialdehyde

OPTA Optimal Performance Theoretically Attainable

Optacon Optical/Tactile Converter

OPTAR Optical Automatic Ranging

Opt Acta Optica Acta [US]

Opt Appl Optica Applicata [Applied Optics Journal of Wroclaw Technical University Press, Poland]

OPTB Operational Program Time Base

OPTCON Trade Show and Exhibition on Optics, Photonics and Laser Applications

Opt Commun Optics Communications [Journal published in the Netherlands]

Opt Electron Microsc Optical and Electron Microscopy [Journal published in the UK]

OptE Optical Engineer [also Opt Eng, or OptEng]

Opt Eng Optical Engineer(ing)

Opt Eng Optical Engineering [Journal of the Society of Photooptical Instrumentation Engineers, US]

Opt Eng Rep Optical Engineering Report [Published by the Society of Photooptical Instrumentation Engineers, US]

Opt Expr Optics Express [Journal of the Optical Society of America]

OPTI Office of Productivity, Technology and Innovation [US]

OPTIC Ophthalmological Products Trade Conference [UK]

Opticon Optical Tactical Converter

OPTICIMP Optimization of Computer Integrated Materials Processing

OPTIM Order Point Technique for Inventory Management

Optim Optimal; Optimum [also optim]

OPTIMA Organization for the Phyto-Taxonomic Investigation of the Mediterranean Area [Germany]

OPTIMATR Optical Properties of Materials (Computation Software)

Optim Control Appl Methods Optimal Control Applications and Methods [Journal published in the UK]

Opt Inf Optical Information [also opt inf]

Opt Inf Syst Optical Information Systems [Journal published in the US]

Opt Lasers Eng Optics and Lasers in Engineering [Journal published in the UK]

Opt Laser Technol Optics and Laser Technology [Journal published in the UK]

OPTLC Overpressure Thin-Layer Chromatography

Opt Lett Optics Letters [Journal of the Optical Society of America]

Opt Mater Optical Materials [Journal published in the Netherlands]

Opt Mekh Prom Optiko-Mekhanicheskaya Promyshlennost [Russian Journal on Optical Technology]

Optn Optician [also optn]

Opt News Optical News [Published by the Optical Society of America, US]

OPTO International Optics and Optoelectronics Conference [France]

Optoelectron Optoelectronic(s) [also optoelectron]

Optoelectron, Devices Technol Optoelectronics–Devices and Technologies [Journal published in Japan]

Optoelectron Instrum Data Process Optoelectronics, Instrumentation and Data Processing [Journal published in the US]

Optoelektron Poluprovodn Tekh Optoelektronika i Poluprovodnikovaya Tekhnika [Ukrainian Journal on Optoelectronics and Related Subjects]

Optom Optometrical; Optometrist; Optometry [also optom]

Optom Vis Sci Optometry and Vision Science [Publication of the American Academy for Optometry, US]

Opt Phys Optical Physicist; Optical Physics

Opt Prop Optical Properties [also opt prop]

Opt Pura Apl Optica Pura y Aplicada [Spanish Journal on Pure and Applied Optics]

Opt Quantum Electron Optical and Quantum Electronics [UK Journal]

Optronics Optoelectronics [also optronics]

OPTS Office of Pesticide Programs and Toxic Substances [of US Environmental Protection Agency]

OPTS On-Line Peripheral Test System

OPTS Optical Sight

Opts Operations [also opts]

Opt Spectra Optical Spectra [Journal published in the US]

Opt Spectrosc Optics and Spectroscopy [Translation of: *Optika i Spektroskopiya (USSR)*; Joint Publication of MAIK Nauka, Russia and the American Institute of Physics, US]

Opt Spectry Optical Spectroscopy [Journal published in the US]

Opt Spektrosk Optika i Spektroskopiya [Russian Journal on Optics and Spectroscopy]

Opt Technol Optical Technologist; Optical Technology

OPTUL Optical Pulse Transmission Using Laser

OPU Operations Priority Unit

OPU Operations Processor Unit

OPU Osaka Prefecture University [Japan]

OPUG Office Planners and Users Group [US]

OPUR Object Program Utility Routines

OPUS Octal Program Updating System

OPUS Open Program Unified Study

OPUS Organization of Professional Users of Statistics [UK]

OPV Oral Polio Vaccine

OPV Orthopoxvirus

OPW Operating Weight

OPW Orthogonal(ized) Plane Wave [Physics]

OPWBP Office of Pension and Welfare Benefits Programs [of US Department of Labor]

OPX Off-Premise Extension

oPxBA o-Phenoxybenzoic Acid

OPZ Oesterreichisches Produktivitätszentrum [Austrian Productivity Center]

OQ Oil-Quench(ed); Oil Quenching [Metallurgy]

1Q-AFM Collinear One-Q Spin State–Antiferromagnetic Moment Alignment

1Q-FM Collinear One-Q Spin State–Ferromagnetic Moment Alignment

OQL Object Query Language

OQL On-Line Query Language

OQMG Office of the Quartermaster General [of US Department of the Army]

O-QPSK Offset Quaternary Phase Shift Keying [also OQPSK]

OQ&T Oil-Quenched and Tempered; Oil Quenching and Tempering [also OQ & T] [Metallurgy]

OR Observation Request

OR Occurence Report

OR Ocean Rescue

OR Odds Ratio

OR Office of Reinvention [of US Environmental Protection Agency]

OR Offset Ratio

OR Ohio River [US]

OR Ohmic Resistance

OR Oil Resistance; Oil-Resistant

OR Oil Rig

OR Oleoresin(ous)

OR Omani Rial [Currency of Oman]

OR Omni(directional) Radio Range

OR Operating Reactor

OR Operating Room

OR Operational Readiness

OR Operational Requirements

OR Operational Research

OR Operational Right [Aerospace]

OR Operations Requirement

OR Operations Research

OR Operations Review

OR Operations Room

OR Operator [also, or or or]

OR Optical Receiver

OR Optical Recorder; Optical Recording

OR Optical Resonance

OR Orange River [Southern Africa]

OR Oregon [US]

OR Ore Refining

OR Organizational Research

OR Orientation

OR Orientation Ratio

OR Orientation Relationship [Crystallography]

OR Orientation Response

OR O-Ring

OR Ostwald Ripening [Chemistry]

OR Ottawa River [Canada]

OR Outer Roll

OR Output Ready

OR Output Register

OR Outrigger

OR Outside Radius

OR Overhaul and Repair [also O&R]

OR Overload Relay

OR Overrange [Data Communication]

OR Override; Overriding

OR Overrun(ning)

OR Owner's Risk [also or]

OR Oxidation-Reduction (Potential) [Chemistry]

OR Oxidation Resistance; Oxidation-Resistant

OR Oxide Removal

OR Oxygen Reduced; Oxygen Reduction

OR Oxygen Relief [also O/R]

OR2000 Ocean Rescue 2000

O&R Ocean and Rail

O&R Overhaul and Repair [also OR]

O/R On Request

O/R Originator/Recipient

O/R Outside Radius

O/R Oxygen Relief [also OR]

or orange

$\Omega(r)$ Radial Distribution Function [Symbol]

ORA Office des Renseignements Agricoles [Office for Agricultural Information]

ORA Office of Research Analysis [of US Air Force]

ORA Office of Research and Applications [of National Oceanic and Atmospheric Administration, US]

ORA Output Reference Axis

ORACLE Oak Ridge Automatic Computer and Logical Engine [US]

ORACLE On-Line Inquiry and Report Generator (Program) [UNIX Database Program]

ORACLE Optical Recognition of Announcements by Code(d) Line Electronics

ORAE Operational Research (and) Analysis Establishment [Canada]

ORAN Orbital Analysis

Orange FS Orange Free State [South Africa]

ORAPA Ontario Retail Accident Prevention Association [Canada]

ORATE Ordered Random Access Talking Equipment

ORAU Oak Ridge Associated Universities [Tennessee, US]

ORAU-NSLS Oak Ridge Associated Universities–National Synchrotron Light Source [Tennessee, US]

ORB Object Request Broker

ORB Observatoire Royal de Belgique [Royal Belgian Observatory, Uccle]

ORB Omnidirectional Radio Beacon

ORB Omnidirectional Research Beacon

ORB Operational Research Branch [US and Canada]

ORB Owner's Risk of Breakage [also orb]

Orb Orbit(al); Orbiter [also orb]

ORBA Ontario Road Builders Association [Canada]

ORBID On-Line Retrieval of Bibliographic Data

ORBIS Orbiting Radio Beacon Ionospheric Satellite [of NASA]

ORBIT On-Line Real-Time Branch Information [Computer Language]

ORBIT On-Line Retrieval of Bibliographic Information, Time-Shared

ORBIT ORACLE (Oak Ridge Automatic Computer and Logical Engine) Binary Internal Translator [of Oak Ridge National Laboratory, Tennessee, US]

ORB 1-g Orbiter One-g (Trainer) [of NASA]

ORC Occidental Research Corporation

ORC Occupational Research Center [UK]

ORC Office of Radiation Control [US]

ORC Officers' Reserve Corps

ORC On-Line Reactivity Computers

ORC Operational Research Committee [Canada]

ORC Operations Research Center [US]

ORC Ordnance Rocket Center

ORC Organized Reserve Corps [US]

ORC Orthogonal Row Computer

ORC Outokumpu Research Center [at Pori, Finland]

ORC Overrunning Clutch

ORC Oxidation Reduction Cycle [Chemistry]

ORCA Oficina Regional Centroamericana [Regional Office for Central America; of International Union for Conservation of Nature and Natural Resources]

ORCA Ontario Royal Commission on Asbestos [Canada]

ORCAL Orange County Manufacturing and Metalworking Conference and Exposition [US]

ORCB Order of Railway Conductors and Brakemen [Now Part of United Transportation Union, US]

ORCHIS Oak Ridge Computerized Hierarchical Information System [of US Department of Energy]

ORCMT Oak Ridge Center for Manufacturing Technology [US]

ORCO Organization Committee [of International Standards Organization]

ORCS Organic Reactions Catalysis Society [US]

ORD Office of the Chief of Ordnance [US]

ORD Office of Rare Diseases [of National Institutes of Health, Bethesda, Maryland, US]

ORD Office of Research and Development [of US Environmental Protection Agency]

ORD Office of Research and Development [of National Space Development Agency of Japan]

ORD Once-Run Distillate [Chemical Engineering]

ORD Operational Readiness Data

ORD Operational Readiness Date

ORD Operational Ready Data

ORD Operational Requirements Document

ORD Operational Research Division

ORD Optical Rotary Dispersion

ORD Orbital Requirements Document

ORD Owner's Risk of Damage

ORD Oxidation-Retarded Diffusion

Ord Order [also ord]

Ord Ordinal Number [Pascal Function]

Ord Ordnance [also ORD, or ord]

ord ordinary

ORDA Office of Recombinant DNA (Deoxyribonucleic Acid) Activities [US]

ORDAmmDept Ordnance Ammunition Department [US]

ORDB Ordnance Board [also OrdB]

ORDC Orbiter Data Reduction Center [of NASA]

ORDCF Ontario Research and Development Challenge Fund [of Ministry of Energy, Science and Technology, Canada]

ORDE Optical Rotating Disk Electrode

OrdE Ordnance Engineer [also Ord Eng, or OrdEng]

ORDEF Ontario Research Development Foundation [Canada]

Ord Eng Ordnance Engineer(ing)

ORDIR Omnirange Digital Radar [also Ordir, or ordir]

Ordn Ordnance [also ordn]

ORDP Ordnance Pamphlet [also OrdP]

Ord Sgt Ordnance Sergeant

ORDVAC Ordnance Variable Automatic Computer

ORE Occupational Radiation Exposure

ORE Ocean Research Equipment

ORE Ocean Resources Engineer(ing)

ORE Office de Recherche et de l'Expérience [Research and Testing Office; Union Internationale des Chemins de Fer] [Railroad Engineering]

ORE On-Orbit Repair Experiment

ORE Operational Research Establishment [now Defense Operational Research Establishment, Canada]

ORE Overall Reference Equivalent

Ore Oregon [US] [also Oreg]

Ore Geol Rev Ore Geology Review [Journal published in the Netherlands]

ORELA Oak Ridge Electron Linear Accelerator [of US Department of Energy, Tennessee, US]

OREM Objective Reference Equivalent Meter

OREO Orbiting Radio Emission Observatory [of NASA]

OREP Optical Repeater Equipment

Ore Res Ore Reserve [also ore res]

ORES Online Resources for Earth Scientists

ORF Oberhausen Rotor (Steelmaking) Furnace [Metallurgy]

ORF Oklahoma Research Foundation [Oklahoma City, US]

ORF Ontario Research Foundation [now Ontario Research and Technology Foundation, Canada]

ORF Optical Rangefinder

ORF Ortho Research Foundation [US]

ORF Owner's Risk of Fire

ÖRF Österreichischer Rundfunk [Austrian Broadcasting Corporation]

Orf Orifice [also orf]

ORFEDA Ontario Regional Farm Equipment Dealers Association [Canada]

ORFM Outlet Region Feature Model

ORG Operations Research Group

ORG Optical Rain Gauge

ORG O-Ring Gasket

ORG Orthogonal Row Computer

.ORG Non-Profit Organizations [Internet Domain Name] [also .org]

Org Organ(ic) [also org]

Org Organization [also org]

Org Origin(al) [also org]

Org Origination [also org]

org organized

ORGALIME Organisme de Liaison des Industries Métalliques Européennes [Liaison Group for the European Metal Industries, Belgium]

Organ Organization [also organ]

Organomet Organometallics [Journal of the American Chemical Society]

ORGATEC International Office Trade Fair [Cologne, Germany]

Org Chem Organic Chemist(ry)

Org Chem Organic Chemistry [Journal of the Chinese Chemical Society]

ORGDP Oak Ridge Gaseous Diffusion Plant [of Oak Ridge National Laboratory, Tennessee, US]

Org Geochem Organic Geochemistry [International Journal]

Orgn Organization [also orgn]

ORGSC Oregon Ryegrass Growers Seed Commission [US]

ORHF Open-Shell-Restricted Hartree-Fock [Quantum Mechanics]

ORI Ocean Research Institute [Japan]

ORI Online Retrieval Interface

ORI Operational Readiness Inspection

ORI Optical Research Institute

ORI Overriding Royalty Interest

.ORI Original [File Name Extension] [also .ori]

ORIA Office of Regulatory and Information Affairs

ORIC Oak Ridge Isochronous Cyclotron [US]

ORICAT Original Cataloguing System

Oride Override

OR&IE Operations Research and Industrial Engineering

Orient Orientation [also orient]

Orient Med Orientación Medica [Spanish Medical Journal]

Orig Origin(al) [also orig]

orig original(ly)

Orig Life Origins of Life [Published in the Netherlands]

ORINS Oak Ridge Institute of Nuclear Studies [now Oak Ridge Associated Universities, Tennessee, US]

ORION On-Line Retrieval of Information Over a Network

ORISE Oak Ridge Institute for Science and Education [of Oak Ridge National Laboratory, Tennessee, US]

ORIT Operational Readiness Inspection Team

ORIT Organisación Regional Interamericana de Trabajadores [Regional Organization of Interamerican Workers]

ORJ O-Ring Joint

ORKID Orientation Determination from Kikuchi Diagrams [Crystallography]

ORL Orbital Research Laboratory [of NASA]

ORL Ordnance Research Laboratory [US]

ORL Journal of Oto-Rhino-Laryngology and Its Related Specialties [International Journal]

ORL Oto-Rhino-Laryngologist; Oto-Rhino-Laryngology [Medicine]

ORL Owner's Risk of Leakage

orl oral [Toxicology]

ORLA Optimum Repair Level Analysis

ORLY Overload Relay

ORM Other Regulated Materials [of US Department of Transportation]

ORM Overlapping Resolution Map [Chromatography]

ORMAK Oak Ridge Tokamak [of Oak Ridge National Laboratory, Tennessee, US]

ORMH Office of Research on Minority Health [of National Institutes of Health, Bethesda, Maryland, US]

ORMOCERS Organically Modified Ceramics [also Ormocers, or ormocers]

ORMOSIL Organically Modified Silicate [also Ormosil, or ormosil]

Orn Ornament [also orn]

Orn Ornithological; Ornithologist; Ornithology [also orn]

Orn Ornithine; Ornithyl [Biochemistry]

orn orange

Ornam Ornament(al) [also ornam]

Ornam/Misc Met Fabr Ornamental/Miscellaneous Metal Fabricator [US]

Ornith Ornithological; Ornithologist; Ornithology [also ornith]

ORNL Oak Ridge National Laboratory [of US Department of Energy in Tennessee]

ORNL/FMP ORNL/Fossil-Energy Materials Proceedings [of Oak Ridge National Laboratory, Tennessee]

ORNL Mater Res Newsl ORNL Materials Research Newsletter [Published by the Oak Ridge National Laboratory, Tennessee, US]

ORNL Rev ORNL Review [Published by the Oak Ridge National Laboratory, Tennessee, US]

ORNL TM ORNL Technical Memorandum [Published by Oak Ridge National Laboratory, Tennessee, US] [also ORNL Tech Memo]

OR/NOR Or/Not, or (Gate)

ORO Operations Research Office [of Johns Hopkins University, Baltimore, Maryland, US]

OROM Optical Read-Only Memory

ORP OFS (Orbiter Flight System) Retransmission Processor [NASA]

ORPA Optimized Random Phase Approximation [Materials Science]

ORP Operational Readiness Platform

ORP Optional Response Poll

ORP Oxidation-Reduction Potential [Physical Chemistry]

ORP Oxidation-Reduction Probe

ORPM Office of Research Program Management [US]

ORR Oak Ridge Research Reactor [of Oak Ridge National Laboratory, Tennessee, US]

ORR Oak Ridge Reservation [Tennessee, US]

ORR Operational Readiness Review

ORR Operations Requirements Review

ORR Oxygen Reduction Reaction

ORRAS Optical Research Radiometrical Analysis System

ORRMIS Oak Ridge Regional Modeling Information System [US]

ORRPB Ottawa River Regulation Planning Board [Canada]

ORRRC Outdoor Recreation Resources Review Commission [US]

ORS Object Recognition Systems

ORS Octahedral Research Satellite [of NASA]

ORS Office of Research Services [of National Institutes of Health, Bethesda, Maryland, US]

ORS Oil Reclamation System

ORS On-Site (Oil) Reclamation System

ORS Operational Research Section [UK]

ORS Operational Research Society [UK]

ORS Oral Rehydration Solution [Medicine]

ORS Orbital Refueling System

ORS O-Ring Seal

ORS Orroval Valley, Australia [Space Tracking and Data Network]

ORS Orthopedic Research Society [US]

ORS Output Record Separator

ORSA Operations Research Society of America [US]

ORSA J Comput ORSA Journal of Computing [Published by the Operations Research Society of America, US]

ORSA/TIMS Operations Research Society of America/The Institute of Management Sciences [Baltimore, Maryland, US]

ORSA/TIMS Bull ORSA/TIMS Bulletin [Published by the Operations Research Society of America/The Institute of Management Sciences, Baltimore, Maryland, US]

ORS-BR O-Ring Seal, Braze Type

ORS-BT O-Ring Seal, Bite Type

ORSDI Oak Ridge Selective Dissemination of Information [US]

ORSI Operational Research Society of India

ORS(India) Operational Research Society (India Section)

ORSJ Operations Research Society of Japan

ORSoc Operational Research Society [UK]

ORSORT Oak Ridge School of Reactor Technology [of US Department of Energy in Tennessee]

ORT Operational Readiness Test

ORT Orbit Readiness Test

ORT Overland Radar Technology

ORTA Office of Research and Technology Applications [of National Oceanic and Atmospheric Administration, US]

ORTA Office of Research and Technology Assessment [US]

ORTAC (Ortungssystem im TACAN-Frequenzband) – German for "Navigation System in the TACAN (Tactical Air Navigation System) Frequency Range"

ORTAC-M (Ortungssystem im TACAN-Frequenzband, Monopuls) – German for "Navigation System in the TACAN (Tactical Air Navigation System) Frequency Range, Monopulse"

ORTAG Operations Research Technical Assistance Group [of US Army]

ORTDC Oregon Resource and Technology Development Corporation [US]

ORTECH Ontario Research and Technology Foundation [formerly ORF, Canada]

ORTEP Oak Ridge Thermal Ellipsoid Program [of Oak Ridge National Laboratory, Tennessee, US]

ORTF Office de la Radiodiffusion et Télévision Francaise [French Office for Broadcasting and Television]

Orth Orthopedic(s) [also orth]

Ortho Orthogon(al) [also ortho]

Orthog Orthogon(al) [also orthog]

Orthop Orthopedic(s) [also orthop]

Orthop Clinics N Am Orthopedic Clinics in North America [Journal]

ORTS Optional Residential Telephone Service

ORTUEL Organized Reserve Training Unit, Electronics

ORU Orbital Replaceable Unit

ORU Oxygenate Removal Unit

ORV Ocean Range Vessel

ORV Off-(the-)Road Vehicle

ORV Orbital Rescue Vehicle

ORV Orthoreovirus

ORVR Onboard Refuelling Vapor Recovery [Automobiles]

ORWH Office of Research on Women's Health [of National Institutes of Health, Bethesda, Maryland, US]

ORZ Optimized Rouse-Zimm (Equation) [Chemistry]

OS Object Space [Optics]

OS Oblique Sounding

OS Oceanic Society [US]

OS Octahedral Site [also os] [Crystallography]

OS (Oculus sinister) – (in the) left eye [also os] [Medical Prescriptions]

OS Odd Symmetric

OS Office of Supply [US]

OS Office of the Secretary [US]

OS Office System

OS Offshore Steel

OS Ohio State [US]

OS Oil Sample; Oil Sampling

OS Oil Sand

OS Oil Separation; Oil Separator

OS Oil Shale

OS Oil Solubility; Oil-Soluble

OS Oil Spill

OS Oil Stain

OS Oil Switch

OS Oil Sump

OS Old Style (Calendar) [Astronomy]

OS Oligosaccharide

OS Omega System

OS Omnibus Society [UK]

OS Oncostatin [Immunochemistry]

OS One-Shot (Multivibrator)

OS On-Orbit Station

OS On-Site

OS "ON" Switch

OS Open Stoping [Mining]

OS Open System

OS Operating Stress

OS Operating System [also O/S]

OS Operating (or Operational) Software

OS Operational Sequence

OS Operations System

OS Optical Scan(ner); Optical Scanning

OS Optic(al) Sensor

OS Optical Spectrograph(y)

OS Optical Spectrometer; Optical Spectrometry

OS Optical Spectroscope; Optical Spectroscopy

OS Optical Spectrum

OS Optical System

OS Optics Subsystem

OS Oral Surgeon; Oral Surgery

OS Orbiter CEI (Compliance Evaluation Inspection) Specification [NASA]

OS Ordered Scattering

OS Ordered Structure [Solid-State Physics]

OS Ordinary Seaman

OS Ordnance Society [UK]

OS Ordnance Survey

OS Oregon State [US]

OS Organic Semiconductor

OS Organic Synthesis

OS Organosilane

OS Organosilicon

OS Organosulfur

OS Orthogonal System [Mathematics]

OS Oscillation; Oscillator [also Os]

OS Osseous System [Anatomy]

OS Osteosarcoma [Medicine]

OS Outer Shell (Electron)

OS Outer Space

OS Out of Service

OS Out of Stock [also O/S, or o/s]

OS Output System

OS Outsize(d)

OS Outstanding [also O/S]

OS Overhauser Shift [Physics]

OS Overshoot(ing)

OS Oversize(d)

OS Overstress(ing)

OS Oxygen Saturated; Oxygen Saturation

OS Oxygen-Saturated (Copper)

OS Oxidation-Sulfidation (Reaction)

OS Oxides, Streaky [Metallurgy]

OS Oxygen Sensor

OS Oxygen Storage

OS Ozone Science; Ozone Scientist

OS/2 Operating System Two [of IBM Corporation, US]

ÖS Östereicher Schilling [Austrian Shilling]

O/S Operating System [also OS]

O/S Out of Stock [also o/s]

O/S Outstanding [also OS]

Os Osmium [Symbol]

Os Ostrogradsky Number [Symbol]

Os-184 Osmium-184 [also ^{184}Os, or Os^{184}]

Os-186 Osmium-186 [also ^{186}Os, or Os^{186}]

Os-187 Osmium-187 [also ^{187}Os, or Os^{187}]

Os-188 Osmium-188 [also ^{188}Os, or Os^{188}]

Os-189 Osmium-189 [also ^{189}Os, or Os^{189}]

Os-190 Osmium-190 [also ^{190}Os, or Os^{190}]

Os-191 Osmium-191 [also ^{191}Os, or Os^{191}]

Os-192 Osmium-192 [also ^{192}Os, or Os^{192}]

Os-193 Osmium-193 [also ^{193}Os, or Os^{193}]

Os-194 Osmium-194 [also ^{194}Os, or Os^{194}]

os octahedral site [also OS] [Crystallography]

os (oculus sinister) – (in the) left eye [also OS] [Medical Prescriptions]

1/s one per second [also 1/sec, 1/sec, or s^{-1}]

1/s^2 one per second squared [also 1/sec^2, sec^{-2}, or s^{-2}]

Ω/□ ohm(s) per square [Unit of Sheet Resistance]

OSA Office of Safety Assessment [US]

OSA Office of Student Affairs

OSA Office of the Secretary of the Army [US]

OSA Office Secret Act [UK]

OSA Office System Architecture

OSA Old Stone Age [Anthropology]

OSA Olefin-Modified Styrene-Acrylonitrile

OSA On-Stream Analysis

OSA Open Scripting Architecture

OSA Open Systems Architecture

OSA Operational Support Area

OSA Optical Society of America [US]

OSA Organic Soil Association [South Africa]

OSA Outer Soundings Array

OSA Output Spooling Area

OSAC Open Space Action Committee [now Open Space Institute, US]

OSAC Oxford System of Automated Cartography [Geographic Information System]

OSAF Office of the Secretary of the Air Force [US]

OSAF Origin Subarea Field

OSAHS Ohio State Archeological and Historical Society [US] [also OSA&HS]

OSAM Overflow Sequential Access Method

OSAIS Oil Spillage Analytical Information Service [of Laboratory of the Government Chemists, UK]

Osaka Math J Osaka Mathematical Journal [Japan]

OSAM Overflow Sequential Access Method

OSAP Ontario Student Assistance Program [Canada]

OSAR Operational Safety Analysis Report

OSA/SPIE Optical Society of America/International Society for Optical Engineering (Task Force)

OSAT Office for the Study of Automotive Transportation [of University of Michigan Transportation Research Institute, Ann Arbor, US]

OSA-UCS Optical Society of America–Uniform Color Scales

OSB Operational Status Bit

OSB Operations and Services Building [of Pacific Northwest National Laboratory, Richland, Washington, US]

OSB Oriented Strand Board [Building Trades]

OSBP Ontario Special Bursary Plan [Canada]

OSC Ocean Science Committee [of Ocean Affairs Board, National Academy of Sciences, US]

OSC Ocean Sciences Center [of Memorial University of Newfoundland, St. John's, Canada]

OSC Offshore Survival Center [UK]

OSC Ohio Supercomputer Center [of Ohio State University, Columbus, US]

OSC On-Site Safety Committee

OSC On Surface Cracking

OSC Ontario Science Center [Toronto, Canada]

OSC Ontario Securities Commission [Canada]

OSC Operator Services Complex

OSC Optically Sensitive Controller

OSC Optical Sciences Center [of University of Arizona, Tucson, US]

OSC Orbit-Spin Coupling [Physics]

OSC Oregon State College [Corvallis, US]

OSC Organosilicon Compound

OSC Organosulfur Compound

OSC Orthosubstitution Compound

OSC Osteoporosis Society of Canada

OSC Own Ship's Course

OSC Oxygen-Storage Capacity

Osc Oscillation; Oscillator [also osc]

OSCA Optical Sensors Collaborative Association [UK]

OSCAA Oil Spill Control Association of America [now Spill Control Association of America, US]

OSCAR Occupational Safety Climate Assessment Report [of National Safety Council, US]

OSCAR Optimum Systems Covariance Analysis Results

OSCAR Orbiting Satellite Carrying Amateur Radio [of (Radio) Amateur Satellite Corporation, US] [also Oscar]

OSCAR Order Status Control and Reporting

OSCAS Office of Statistical Coordination and Standards [Philippines]

OSCC Ohio Super-Computer Center [of Ohio State University, Columbus, US]

OSCE Office of the Special Counsel on Ethics [of National Institutes of Health, Bethesda, Maryland, US]

OSCE Organization for Security and Cooperation in Europe [Czech Republic]

OSCEF Optical Scatter and Contamination Effects Facility

OSCG Oscillograph [also Oscg, or oscg]

OSCIL Oscillator [also Oscil, or oscil]

OSCL Operating System Control Language

OSC-MULT Oscillator-Multiplier [also Osc-Mult, or oscmult]

OSCOM Oslo Commission for Marine Pollution Control

OSC OUT Oscillator Out [also OSCOUT]

Os(Cp)$_2$ Bis(cyclopentadienyl)osmium (or Osmocene)

OSCRL Operating System Command and Response Language

OS-Cu Oxygen-Saturated Copper

OSD Office of Systems Development [of National Oceanic and Atmospheric Administration, US]

OSD Office of the Secretary of Defense [US]

OSD Office Systems Design

OSD On-Line System Driver

OSD Online Systems Department [of Fermilab, Batavia, Illinois, US]

OSD On-Screen Digital

OSD Operational Sequence Diagram

OSD Operational Systems Development

OSD Optical Scanning Device

OSD Organosilicon Device

OS&D Over, Short and Damage Report

OSDA Optical Shallow Defect Analyzer

OSDH Orbiter System Definition Handbook [of NASA]

OSDM Optical Space-Division Multiplexing

Os-DMEDA Osmium Dimethylethylenediamine

OSDP On-Site Data Processor

OSDR Oil Slick Detection Radar

OSDS Operating System for Distributed Switching

OSE Oceanic Society Expeditions [US]

OSE Operating Support Equipment

OSE Operational Support Equipment

OSE Optical Science and Engineering

OSE Orbiter Support Equipment [of NASA]

OSE Overall Station Efficiency

OSE Ozone Science and Engineering

OS&E Ozone: Science and Engineering (Journal) [of International Ozone Association]

OS/E Operating System/Environment

OSEC Office Suisse d'Expansion Commerciale [Swiss Commercial Expansion Office]

1/sec one per second [also 1/s, s^{-1}, or 1/sec]

1/sec^2 one per second squared [also 1/s^2, sec^{-2}, or s^{-2}]

OSEE Optically-Simulated Electron Emission

OSEOS Operational Synchronous Earth Observatory Satellite

OSEP Office of Scientific and Engineering Personnel [of National Research Council, US]

OSERP Oil Sands Environmental Research Program [Canada]

OSERS Office of Special Education and Rehabilitation Services [of US Department of Education]

OSETI Office of Science Education and Technical Information [of US Department of Energy]

OSF Office of Space Flight [of NASA]

OSF Ohio Section Fall Meeting [of American Physical Society, US]

OSF One Hundred Square Feet [also osf]

OSF Open Software Foundation [US Consortium of Digital Equipment Corporation, IBM Corporation, and Hewlett Packard]

OSF Ordnance Storage Facility

OSF Osteoblast-Specific Factor [Biochemistry]

OSF Oxidation(-Induced) Stacking Fault [Materials Science]

OSFD Office of Space Flight Development [now Office of Space Flight Programs]

OSF/DCE Open Software Foundation/Distributed Computing Environment

OSFM Office of Spacecraft and Flight Missions [of NASA]

OSFP Office of Space Flight Programs [of NASA]

OSG Office of the Secretary-General [of United Nations]

OSG Office of the Surgeon General [of US Department of the Army]

1/s·G one per second gauss [also $s^{-1} G^{-1}$]

OSGP Ontario Study Grant Plant [Canada]

OSH Occupational Safety and Health

OSH (National) Occupational Safety and Health Conference [Canada]

OSH Oil Sands and Heavy Oil

OSHA Occupational Safety and Health Act [US]

OSHA Occupational Safety and Health Administration [of US Department of Labor]

OSH Act Occupational Safety and Health Act [US]

OSHA HAZWOPER Occupational Safety and Health Administration–Hazardous Waste Operations (Regulations) [US]

OSHA HCS Occupational Safety and Health Administration Hazard Communication Standard

OSHA PEL Occupational Safety and Health Administration's Permissible Exposure Limit

OSHA STD Occupational Safety and Health Administration Standard [US]

OSHB One-Sided Height Balanced

OSHC Oregon State Highway Commission [US]

OSHO Oil Sands and Heavy Oil

OSHRC Occupational Safety and Health Review Commission [US]

OSH-ROM Occupational Safety and Health Databases on CD-ROM

OSHSA Occupational Safety and Health Standards Act [US]

OSI Office of Scientific Information [of National Institute of Mental Health, Bethesda, Maryland, US]

OSI Office of Scientific Intelligence

OSI On-Site Inspection

OSI Open Space Institute [US]

OSI Open Systems Interconnect(ion)

OSI Open Systems Interface

OSI Operating System Interface

OSI Optical Sciences Institute [of Lawrence Livermore National Laboratory at University of California at Berkeley, US]

OSI Optical Society of India

OSI Optical Storage International [US]

OSI Ounces per Square Inch [also osi, oz/in², or oz/sq in]

O/Si Oxygen-to-Silicon (Ratio)

OSIC Optimization of Subcarrier Information Capacity

OSIL Operating System Implementation Language

(OSiR₂)ₙ Polysiloxane [General Formula; R Represents Hydrogen, or an Alkyl, or Aryl]

OSIRIS On-Line Search Information Retrieval Information Storage [US Navy]

OSIRIS Optical Spectrograph and Infrared Imaging System

Os-Ir-Ru Osmium-Iriodium-Ruthenium (Alloy System)

OSIS Office of Science Information Service [of National Science Foundation, US]

OSIS Oil Slug Injection System

OSIS Optoelectronic Surface Inspection System

OSIS Oregon Species Information System

OSJ Optical Society of Japan

OSK Optical Society of Korea

OSL Observed Significance Level

OSL Ontario Safety League [Canada]

OSL Operating System Language

OSL Optically Stimulated Luminescence

OSL Outstanding Leg

OSLM Operations Shop/Lab Manager

OSLO Other Six Leases Operation [Canada]

OSLP Ontario Student Loans Plan [Canada]

OSM Office of Surface Mining [of US Department of the Interior]

OSM Off-Screen Model

OSM Oil-Sands Mining

OSM Oncostatin M [Immunochemistry]

OSM One-Shot Molding [Materials Science]

OSM On-Screen Manager

OSM Operating System Monitor

OSM Option Select Mode

OSM Orbital Service Module

OSM Ordnance Safety Manual

OSM Oregon Steel Mills, Inc. [Portland, US]

OSM Organisation Syndicale Mondiale [World Federation of Trade Unions]

Osm Osmole [Unit]

OSMA Occidental Society of Metempiric Analysis [US]

OSMA Orthopedic Surgical Manufacturers Association [US]

OSME Ornithological Society of the Middle East [UK]

OS/MFT Operating System/Multiprogramming a Fixed Number of Tasks

Osm/kg Osmole per kilogram [Unit]

OSMP Operational Support Maintenance Plan

OSMRE Office of Surface Mining, Reclamation and Enforcement [of US Department of the Interior]

OSMT Ontario Society of Medical Technologists [Canada]

OSMV One-Shot Multivibrator

OS/MVS Operating System/Multiprogramming with Virtual Storage

OS/MVT Operating System/Multiprogramming a Variable Number of Tasks

OSN Output Sequence Number

OSNI Ordinance Survey of Northern Ireland

OSNZ Ornithological Society of New Zealand

OSO Ocean Systems Operation

OSO Offshore Supplies Office

OSO Orbital (or Orbiting) Solar Observatory [Series of NASA Satellites]

OSO Orbiting Satellite Observer

OSO Orbiting Scientific Observatory

OSO Origination Screening Office

O/S/O Ore/Slurry/Oil (Carrier) [also OSO]

O₂–SO₂ Oxygen–Sulfur Dioxide (Gas Mixture)

O₂–SO₃ Oxygen–Sulfur Trioxide (Gas Mixture)

OSOC Ore/Slurry/Oil Carrier

OSOP Orbiter Systems Operating Procedures [of NASA]

OSOS Ore/Slurry/Oil Ship

O₂–SO₂–SO₃ Oxygen–Sulfur Dioxide–Sulfur Trioxide (Gas Mixture)

OSP Office of Scientific Personnel [of National Research Council–National Academy of Sciences, US]

OSP Office of Scientific Projects [of Defense Advanced Research Projects Agency, US]

OSP Office of Statistical Policy

OSP Oligo Selection Program

OSP On-Screen Programming

OSP Operations Support Plan

OSP Operator Service Provider [Telecommunications]

OSP Optical Sensor Package

OSP Optical Signal Processing

OSP Optical Storage Processor

OSP Outside Plant

OSPCA Ontario Society for the Prevention of Cruelty to Animals [Canada]

Os-Pd Osmium-Palladium (Alloy)

OSPES Outer-Shell Photoelectron Spectroscopy

OSPF Open Shortest-Path First [Internet]

OSPI Octanorm Service Partner International (Convention)

OSPM Office of Special Programs in Materials [of National Science Foundation–Division of Materials Science, US]

Os(PMCp)$_2$ Bis(pentamethylcyclopentadienyl)osmium (or Decamethylosmocene)

OSPPD Office of Strategic Planning and Policy Development [of Employment and Training Administration, US]

OSPRE Office of Science, Planning and Regulation Evaluation [US]

OSPS Operator Services Position System

OSPTT Office of Science Policy and Technology Transfer [of National Institutes of Health, Bethesda, Maryland, US]

OsPu Osmium Plutonium (Compound)

Ω/sq Ohm per square [Unit of Sheet Resistance]

Ω/□ Ohm per square [Unit of Sheet Resistance]

OSQL Object Structured Query Language

OSR OEM (Original Equipment Manufacturer) Service Release

OSR Office for Scientific Research [Indonesia]

OSR Office of Scientific Research [of US Air Force]

OSR Oil/Steam Ratio

OSR Operational Scanning Recognition

OSR Operations Support Room

OSR Operand Storage Register

OSR Optical Scanning Recognition

OSR Optical Solar Reflector

OSR Output Shift Register

OSR Over-the-Shoulder Rating

OSRA Ocean Science-Related Acronyms [of National Oceanic and Atmospheric Administration, US]

OSRA Office Systems Research Association [US]

OSRD Office of Scientific Research and Development [US]

OSRD Office of Standard Reference Data [of National Institute for Standards and Technology, Gaithersburg, Maryland, US]

OSRMD Office of Scientific Research, Mechanics Division [of US Air Force]

OSRR Office of Site Remediation and Restoration [of US Environmental Protection Agency]

OSRT Oil Spill Response Planning

OSRS Off-Resonance Stimulated Raman Scattering

OSRT Oil Spill Response Team

Os-Ru Osmium-Ruthenium (Alloy)

OSS Ocean Science and Surveys [Canada]

OSS Ocean Surveillance Satellite

OSS OEX (Orbiter Experiments) Support System [of NASA]

OSS Office of Satellite Systems [of National Space Development Agency of Japan]

OSS Office of Space Science [of NASA]

OSS Office of Statistical Standards [US]

OSS Office of the Supervising Scientist [Alligator Rivers, Australia]

OSS Office of Strategic Services [US]

OSS Operational Storage Site

OSS Operation Support System

OSS Operating System Supervisor

OSS Operating System Support

OSS Operator Service Switch [Telecommunications]

OSS OPTCON (Trade Show and Exhibition on Optics, Photonics and Laser Applications) Solution Series

OSS Optics Subsystems

OSS Orbital Space Station

OSS Orbiting Space Station

OSS Organic Solid State

O.ss. ½ Pint [Medical Prescription]

OSSA Office for Space Science and Applications [of NASA]

OSSA Office of Space Science and Applications [of Commonwealth Scientific and Indus-trial Research Organization]

OSSD Ontario Secondary School Diploma [Canada]

OSSE Observing System Simulation Experiment

OSSE Oriented Scintillation Spectrometer Experiment [at Gamma-Ray Observatory of NASA]

OSSGD Ontario Secondary School Graduation Diploma [Canada]

OSSHGD Ontario Secondary School Honour Graduation Diploma [Canada]

OSSL Operating System Simulation Language

OSSRH Orbiter Subsystem(s) Requirements Handbook [of NASA]

OSSS Orbital Space Station System [of NASA]

OSSS Orbiting Space Station Study [of NASA]

OSSU Operator Services Switching Unit

OST Office of Science and Technology

OST Office of Science and Technology [of US Department of Energy]

OST Office of Science and Technology [of US Environmental Protection Agency]

OST Office of Science and Technology [of Presidential Scientific Advisory Committee, US]

OST Office of Standards and Technology [of NASA]

OST Offshore Storage and Treatment

OST On-Shift Test

OST Operating System Toolbox

OST Operating System Trap

OST Operational System Test

OST Operations Support Team [of NASA Goddard Space Flight Center, Greenbelt, Maryland, US]

OST Orbiter Support Trolley [of NASA]

OST Originating Station Treatment

OST Outer Space Treaty [of United Nations Committee for Peaceful Uses of Outer Space]

1/s·T one per second tesla [also $s^{-1}T^{-1}$]

OSTA Office of Space and Terrestrial Applications [of NASA]

OSTAC Ocean Science and Technology Advisory Committee [of National Sailing Industry Association, US]

OSTC Office of Science and Technology Centers [of National Science Foundation, US]

OSTDS Office of Space Tracking and Data Systems [of NASA]

OSTE One-Step Temper Embrittlement [Metallurgy]

OSTEST Operating System Test

OSTF Ordnance Survey Transfer Format

OSTI Office of Science and Technology Infrastructure [of National Science Foundation, US]

OSTI Office of Scientific and Technical Information [of US Department of Energy]

OSTI Organization for Social and Technological Innovation [US]

OSTL Operating System Table Loader

Os-TMEDA Osmium Tetramethylethylenediamine

OSTNPS Operator Services Traffic Network Planning System

OSTO Office of Space Transportation Operations [of NASA]

OSTP Office of Science and Technical Policy [of National Research Council, US]

OSTP (White House) Office of Science and Technology Policy [US]

OSTP Office of Scientific and Technical Personnel [of Organization for Economic Cooperation and Development, France]

OSTP Orbiting System Test Plan

OSTR Oregon State (University) TRIGA (Training Reactor, Isotope-Production, General Atomics) Reactor [Corvallis, US]

OSTRE Objective Sidetone Reference Equivalent

OSTS Office of Space Transportation Systems [of National Space Development Agency of Japan]

OSTS Office of State Technical Service [US]

OSTS Official Seed Testing Station [UK]

ÖSTV Österreichischer Stahlbauverband [Austrian Steel Construction Association]

1s&2s Ones and Twos [Combination of Hardwood Grades of Firsts and Seconds] [Lumber]

OSU Ohio State University [Columbus, US]

OSU Oklahoma State University (of Agriculture and Applied Science) [Stillwater, US]

OSU Oregon State University [Corvallis, US]

OSU Osaka Sangyo University [Japan]

OSu N-Succinimide Ester

OSUCM Ohio State University College of Medicine [Columbus, US]

OSUHRF Ohio State University Heat Release Calorimeter [US]

OSUL Ohio State University Libraries [Columbus, US]

OSUR Ohio State University Reactor [Columbus, US]

OSUS Office of Space Utilization Systems [of National Space Development Agency of Japan]

OSURF Ohio State University Research Foundation [Columbus, US]

OSV Ocean Station Vessel

OSV On-Site Verification

OSV Orbital Support Vehicle

OSW Oblique Shock Wave [Fluid Mechanics]

OSW Office of Saline Water [of US Department of the Interior]

OSW Office of Solid Waste [of US Environmental Protection Agency]

OSWAC Ordnance Special Weapons Ammunition Command [Picatinny Arsenal, Dover, New Jersey, US]

OSWER Office of Solid Waste and Emergency Response [of US Environmental Protection Agency]

OSWS Operating System Workstation

OSWV Osteryoung Square Wave Velotammetry

OS&Y Outside Screw and Yoke

OT Object Technology

OT Obstacle Twin [Metallurgy]

OT Occupational Therapist; Occupational Therapy

OT Ocean Trench

OT Odor Threshold

OT Office of Telecommunications [US]

OT Office of Transportation [US]

OT Office Technology

OT Oil Tanker

OT Oil-Temper(ed); Oil Tempering [Metallurgy]

OT Oil-Tight

OT Old Tuberculin [Immunology]

OT On-Time

OT On Truck [also ot]

OT Opening Time

OT Open Top

OT Operating Temperature

OT Operating Time

OT Operational-Tank [NASA]

OT Operational Test

OT Operational Trajectory

OT Optical Technologist; Optical Technology

OT Optical Test(ing)

OT Optical Tooling

OT Optical Tracker; Optical Tracking

OT Optical Transform

OT Optimization Theory [Mathematics]

OT Orbital Theory [Physics]

OT Ordinary Temperature

OT Organization Theory

OT Organotin

OT Orthogonal Transformation [Mathematics]

OT Orthogonal Trees

OT Orthotolidine; O-Tolidine

OT Osseous Tissue [Anatomy]

OT (Negative) Outer Tetrahedron [Crystallography]

OT Output Terminal

OT Out Temperature

OT Overall Test

OT Overtemperature [also O/T, O/TEMP, or O/Temp]

OT Overtime

OT Overtone

OT Overturn(ing)

OT Oxidizer Tank [Aerospace]

O&T Organization and Training [of US Department of the Army]

O/T On Truck [also o/t]

O/T Orthorhombic-to-Tetragonal (Transformation) [also O-T] [Metallurgy]

O/T Other Times

O/T Overtemperature [also OT, O/TEMP, or O/Temp]

O-T Orthorhombic-to-Tetragonal (Transformation) [also O/T] [Metallurgy]

1/T Bandwidth [also T^{-1}]

OTA Office of Technology Assessment [of US Congress, Washington, DC]

OTA Official Test Aerosol

OTA Omaha Transit Authority [Nebraska, US]

OTA Ontario Trucking Association [Canada]

OTA Open Test Assembly

OTA Operational Trans(con)ductance Amplifier

OTA Operation-Triggered Architecture

OTA Optical Telescope Assembly

OTA Organic Trade Association [US]

OTA Organisation Mondiale du Tourisme et de l'Automobile [World Organization of Tourism and Automobiles, France]

OTA Orthotolidine-Arsenite

OTAC Ordnance Tank-Automotive Command [Detroit, Michigan, US]

OTAF Office of Technology Assessment and Forecasts [of US Patent and Trademark Office]

OTAN Organisation du Traité de l'Atlantique du Nord [North Atlantic Treaty Organization]

OTANA Organic Trade Association of North America

OTANS Offshore Trade Association of Nova Scotia [Canada]

OTANY Oil Trades Association of New York [US]

OTB Orbiting Tanker Base

Otbd Outboard [also otbd]

OTBG O-Tolylbiguanide [Biochemistry]

Otbor Pereda Inf Otbor i Peredacha Informatsii [Ukrainian Journal]

OTC Ocean Thermal Conversion (Process)

OTC Office of Technical Coordination [of US Department of Energy]

OTC Office of Telecommunications

OTC Office Technology Coordinator

OTC Officers Training Corps

OTC Offshore Technology Conference

OTC Off -Track Capability (Test) [Magnetic Recording Media]

OTC Once-Through Cooling

OTC Operating Telephone Company

OTC Operational Test Center

OTC Orbiter Test Conductor

OTC Ordnance Training Command [of Aberdeen Proving Ground, Maryland, US]

OTC Organotin Compound

OTC Originating Toll Center

OTC Ornithine Transcarbamylase [Biochemistry]

OTC Osaka Titanium Company Limited [Japan]

OTC Overhead Travelling Crane

OTC Overseas Telecommunications Commission [Australia]

OTC Over-the-Counter

OTC Ownership Transfer Corporation

OTC Oxytetracycline

OTCA Overseas Technical Cooperation Agency [Japan]

OTCCC Open-Type Control Circuit Contact

OTCM Ordnance Technical Committee Minutes

OTCP 2,4,5-Trichlorophenyl (Ester) [also Otcp]

OTCR Office of Technical Cooperation and Research [US]

OTCS Okinawa Tracking and Communication Station [of National Space Development Agency of Japan]

OTD Office of Technology Development [of Office of Environmental Management and Waste Management , US Department of Energy]

OTD Operational Technical Documentation

OTD Optical Time Domain

OTD Orbiter Test Director [of NASA]

OTD Original Transmission Density

OTDA Office of Tracking and Data Acquisition [now Office of Space Tracking and Data Systems]

OTDR Optical Time Domain Reflectometer; Optical Time Domain Reflectometry

OTDT Over Temperature Delta T

OTE Octadecyltriethoxysilane

OTE Operational Test and Evaluation [also OT&E]

OTE Operational Test Equipment

OT&E Operational Test and Evaluation [also OTE]

OTEC Ocean Thermal-Energy Conversion (System)

OTECH Oceaneering Technology Integration [US]

O/Temp Overtemperature [also Otemp, OT, or ot]

OTES Optical Technology Experiment System

OTF Ontario Teachers Federation [Canada]

OTF Ontario Technology Fund [Canada]

OTF Open Token Foundation

OTF Optical Transfer Function

OTFA Office of Technical and Financial Assistance [US]

OTG Option Table Generator

OTGH2 O-Tolylaminoglyoxime

OTH Over-the-Horizon (Radar)

OTH-B Over-the-Horizon Backscatter (Radar) [also OTH/B]

OTHR Over-the-Horizon Radar

OTI Office of Technology Integration [US]

OTI Opaque Thermal Insulation

OTI Oregon Travel Information [US]

ω-Ti Omega Titanium [Symbol]

OTIF Organisation Intergouvernementale pour les Transports Internationaux Ferroviaire [Intergovernmental Organization for International Carriage by Rail, Switzerland]

OTIA Ordnance Technical Intelligence Agency [Arlington, Virginia, US]

OTIO Ordnance Technical Intelligence Office [US]

OTIP Ontario Training Incentive Program [Canada]

OTIS Ordnance Technical Intelligence Service [Aberdeen Proving Ground, Maryland, US]

OTIU Overseas Technical Information Unit

OTMA Ottawa Texture Measurement System [Canada]

OTJ Off-the-Job

OTJ On-the-Job

OTK Oxidizer Tank [Aerospace]

Otkryt Isobret, Otkrytiya, Isobreteniya [Russian Technical Journal]

Otkryt Isobret, Prom Obraz, Tobar Znaki Otkrytiya, Isobreteniya, Promyshlennye Obraztsy, Tobarnye Znaki [Russian Technical Journal]

OTL On-Line Task Loader

OTL Optoelectronics Technology Research Laboratory [Japan]

OTL Order Trunk Line

OTL Ordering Tie Line

OTL Ordnance Test Laboratory

OTL Overhead Transmission Line

OTLC Orbiter Timeline Constraints [NASA]

OTLCA Overhead Transmission Line Contractors Association [UK]

Otlk Outlook [also otlk]

OTLP 0 (Zero) dBm Transmission Level Point

Otlt Outlet [also otlt]

OTM Office of Telecommunications Management [US]

OTM Office of Transportation Materials [of US Department of Energy]

OTM Orbit Trim Maneuver

OTMA Office Technology Management Association [US]

OTMJ Outgoing Trunk Message Junction

OTN Operational Teletype Network

OTO One-Time-Only

OTOH On The Other Hand [Internet Jargon]

Otolaryngol Head Neck Surgery Otolaryngology: Head and Neck Surgery [US Journal]

O to O Out to Out

O/T OR On Truck, or Railway [also o/t or]

OTOS Orbit-to-Orbit Stage

OTOS N-Oxidiethylene Dithiocarbamyl-N'-Oxydiethylene Sulfenamide

OTP Office of Technical Performance [of Jefferson Laboratory, Newport News, Virginia, US]

OTP Office of Technology Policy [of US Department of Commerce]

OTP Office of Telecommunications Policy [of US White House, Washington, DC]

OTP One-Time Programmable

OTP On Top

OTP Operability Test Procedures

OTP Operational Test Procedures

OTP Operations Test Plan

OTP Operations Turnaround Plan

OTP Ortho-Terphenyl; O-Terphenyl

OTP Overtemperature Protection

OTP Ozone Trends Panel

OTP-EPROM One-Time Programmable–Electrically Programmable Read-Only Memory

OTPP Ocean Thermal Power Plant

OTR Occupational Therapsit, Registered

OTR Office des Transports par Route [Office for Road Transport, Belgium]

OTR Operating Time Record

OTR Optical Tracking

OTR Optimum Time Recording

OTR Organic Test Reactor [of Whiteshell Nuclear Research Establishment, Manitoba, Canada]

OTR Outer [also Otr, or otr]

OTR Over-the-Road (Vehicle)

OTRAC Oscillogram Trace Reader

OTRACO Oceanic Trading Company [also Otraco]

OTRC Offshore Technology Research Center [of Texas A&M University, College Station, US]

OTRG Office Technology Research Group [US]

OTS n-Octadecyltrichlorosilane

OTS Office of Technical Services [of US Department of Commerce]

OTS Office of Technological Services

OTS Office of Toxic Substances [of US Environmental Protection Agency]

OTS Officers' Training School

OTS Off-the-Shelf (Computer System)

OTS Optical Technology Satellite

OTS Optical Transport System

OTS Orbital Technology Satellite

OTS Orbital Test Satellite [of European Space Agency]

OTS Orbital Transport System

OTS Orbiter Test Conductor

OTS Organization for Tropical Studies [Costa Rica]

OTS Organized Track Structure

OTS O-Toluenesulfonamide

OTS Out of Service

OTS Ovonic Threshold Switch [Electronics]

OTS Own Time Switch

OTS Oxford Text System [of Oxford University Press, UK]

OTSAA Officer Training School Alumni Association [US]

OTSG Once-Through Steam Generator

OTSGS Once-Through Steam Generating System

OTSR Once-Through Superheat Reactor

OTSR Optimum Track Ship Routing

OTSS Off-the-Shelf System

OTS/USDOC Office of Technical Services/United States Department of Commerce [also OTS/USDC]

OTT Observed Time of Transit [Astronomy]

OTT Office of Technology Transfer [of Office of Science Policy and Technology Transfer, National Institutes of Health, Bethesda, Maryland, US]

OTT Office of Technology Transfer [of CANMET–Canada Center for Mineral and Energy Technology, Ottawa]

OTT Office of Transportation Technologies [of US Department of Energy]

OTT One-Time Tape

OTT One-Time Transfer

OTT Optional Team Targeting

OTT Over the Top [Internet]

OTTBS N-Oxidiethylene Dithiocarbamyl-N'-tertButyl Sulfenamide

OTTLE Optically Transparent Thin Layer Electrode

OTRS Office of Technology Transfer and Regulatory Support

OTTS Organization of Teachers of Transport Studies [UK]

OTTS Outgoing Trunk Testing System

OTU Office of Technology Utilization [of NASA]

OTU Office of Technology Utilization [of US Department of Energy]

OTU Operational Taxonometric Unit

OTU Operational Test Unit

OTU Operational Training Unit [of Canadian Forces]

OTV Operational Television

OTV Orbital Transfer Vehicle

OTV Orbiter Transfer Vehicle [of NASA]

OU Oakland University [Rochester, Michigan, US]

OU Object Unit

OU Odense University [Denmark]

OU Ohio University [Athens, US]

OU Okayama University [Japan]

OU Open University

OU Operable Unit

OU Operation Unit

OU Osaka University [Japan]

OU Osmania University [Hyderabad, India]

OU Oxford University [UK]

OU Out of Use [also ou]

O&U Operations and Utilization

O/U Oxygen-to-Uranium (Ratio)

ou *(oculus uno)* – in each eye [Medical Prescription]

OUA Organisation de l'Unité Africaine [Organization of African Unity]

OUAC Ontario University Application Center [Canada]

OUAS Oxford University Archeological Society [UK]

OUB Occasional-Use Bands

OUC Okanagan University College [Kelowna, British Columbia, Canada]

1UC One-Unit-Cell [Crystallography]

OUCC Ontario Universities Computing Conference [Canada]

OUCA Ontario University Council on Admissions [Canada]

OUCC Observatorio de la Universidad Católica de Chile [Observatory of the Catholic University of Chile, Santiago]

OUD Office of Undergraduate Development [of University of California at Irvine, US]

OUF Oxygen Utilization Factor

OUG On-Line Users Group

OUG/I On-Line Users Group/Ireland

OUL Omaha University Library [Nebraska, US]

OULCS Ontario University Libraries Cooperation System [Canada]

OUN Norman, Oklahoma [Meteorological Station Designator]

OUO Official Use Only

OUP OFS (Orbiter Flight System) Uplink Processor [of NASA]

OUP Oxford University Press [UK]

OUPID Ontario University Program for Instructional Development [Canada]

OURS Open Users Recommended Solutions

OUS Okayama University of Science [Japan]

OUSEP Office of University and Science Education Programs [of US Department of Energy]

OUST Office of Underground Storage Tanks [of US Environmental Protection Agency]

OUT Orbiter Utilities Tray [of NASA]

OUT Send Byte to Machine Output Port (Statement) [Programming]

.OUT Outlines [File Name Extension] [also .out]

Out Outgoing [also out]

Out Outlet [also out]

Out Output [also out]

OUTA Ontario Urban Transit Association [Canada]

Outbd Outboard [also outbd]

OUTLIM Output Limiting Facility

OUTLN Outline [also Outln, or outln]

Outlook Res Libr Outlook on Research Libraries [Journal published in the UK]

OUTPOS Output Position [BASIC Programming]

OUTRAN Output Translator

OUTS Output String [Computers]

OUTWATS Outgoing Wide-Area Telephone Service

OV Observed Value

OV Oil of Vitriol

OV Orbital Velocity [Astronomy]

OV Orbiter Vehicle; Orbiting Vehicle

OV Overflow (Flag) [Computers]

OV Overvoltage [also O/V]

OV Oxygen Vent

Ov Ovarian; Ovary [also ov]

Ω/V ohms per volt [Unit]

.OV1 Overlay File 1 [DOS File Name Extension]

.OV2 Overlay File 2 [DOS File Name Extension]

OVA Offshore Valve Association [US]

OVA Ontario Veterinary Association [Canada]

OVA Organic Vapor Analyzer

OVA Ovalbumin [Biochemistry]

Ovako Steel Tech Rep Ovako Steel Technical Report [Sweden]

OVAL Object-Based Virtual Application Language

Ovbd Overboard [also ovbd]

OVC Ontario Veterinary College [Guelph, Canada]

OVC Optimized Valence Configuration

OVC Overcast [also Ovc, or ovc]

OVCAR-8 Ovarian Carcinoma Cells (As Defined by the National Institutes of Health, Bethesda, Maryland, US)

OVCO Operational Voice Communication Office

OVD Optically Variable Device

OVD Optical Video Disk

OVD Outside Vapor Deposition

ÖVE Österreichischer Verein der Elektrotechniker [Austrian Association of Electrical Engineers]

OVERS Orbital Vehicle Reentry Simulator

Overseas Devt Nat Resourc Inst Newsl Overseas Development Natural Resources Institute Newsletter [UK]

OvF Overhead Fire

Ovf Overfill [also ovf]

OVFF Orbital Valence Force Field [Physics]

Ovfl Overflow [also Ovflo]

OVGF Outer Valence Green's Function

Ov H Oval Head (Screw)

Ovhd Overhead [also ovhd]

Ovhl Overhaul [also ovhl]

Ovht Overheat [also ovht]

OVI Operational Validation Inspection

OVID On-Line Visual Display

.OVL Program Overlay [File Name Extension] [also .ovl]

OVLBI Orbital Very Long Baseline Interferometer; Orbital Very Long Baseline Interferometry

Ovld Overload [also ovld]

Ov Lig Ovarian Ligament [also ov lig] [Anatomy]

Ovlp Overlap [also ovlp]

OVOS Optimized Virtual Orbital Space [Physics]

OVP Overvoltage Protection

OVPO Outside Vapor-Phase Oxidation

OVPU Overvoltage Protection Unit

.OVR Program Overlay [File Name Extension]

Ovrd Override [also ovrd]

Ovrhd Overhead [also ovrhd]

Ovrn Overrun [also ovrn]

Ovrng Overrunning [also ovrng]

OVRO Owens Valley Radio Observatory [California, US]

OVRSTK Overstrike [also Ovrstk, or ovrstk]

OVS Operational Voice System

OVS OSHA (Occupational Safety and Health Administration) Versatile Sampler [US]

O/V-U/V Overvoltage/Undervoltage

OVV Overvoltage [ovv]

OW Office of Water [of US Environmental Protection Agency]

OW Oil-Immersed Water-Cooled (Transformer)

OW Oil Well

OW Open-Window (Unit) [Acoustics]

OW Open Wire

OW Operating Weight

OW Optical Window

OW Order Wire

OW Overwrite [also Ow]

O/W Oil-in-Water (Emulsion)

OWAEC Organization for West African Economic Cooperation

O Wave Ordinary Wave (Component) [also o wave] [Geophysics]

OWC Ocean Water Circulation

OWC Oil/Water Contact

OWC Oil Well Cement

OWC Ordnance Weapons Command [Rock Island, Illinois, US]

OWC Ozone World Congress [of International Ozone Association]

OWCP Office of Workers Compensation Programs [of US Employment Standards Administration]

OWD One-Way Doppler [Physics]

OWDE One-Way Doppler Extraction [Physics]

OWE Ontario Waste-Material Exchange [Canada]

OWE Operating Weight Empty

OWF One-Way Function

OWF Optimal Work Function

OWF Optimum Working Facility

OWF Optimum Working Frequency [Communications]

OWFEA Ontario Wholesale Farm Equipment Association [Canada]

OWG Oil, Water, Gas

OWG Open Waveguide

OWG Optical Waveguide

OWI Office of War Information

OWI Office of Waste Isolation [US]

Ω-W/K² Ohm-Watt per Kelvin Squared [also Ω-W K^{-2}, or Ω-W/(K)²]

OWL Object Windows Library

OWL On-Line Without Limitation

OWL Open Windows Library

OWL Optimal Waste Loading

OWM Office of Wastewater Management [of US Environmental Protection Agency]

OWM Office of Weights and Measures [of National Institute for Standards and Technology, Gaithersburg, Maryland, US]

OWM Optical Waveguide Microscopy

OWMA Ontario Waste Management Association [Canada]

OWMC Ontario Waste Management Corporation [Canada]

OWP Office of Water Policy [US]

OWP Oil Well Pump(er)

OWPR Ocean Wave Profile Recorder

OWPS Office of Water Planning and Standards [of US Environmental Protection Agency]

OWPS Offshore Windpower System

OWR Omega West Reactor [of Los Alamos Scientific Laboratory, New Mexico, US]

OWRC Ontario Water Resources Commission [Canada]

OWRR Office of Water Resources Research [of United States Department of the Interior]

OWRS Office of Water Regulations and Standards [of US Environmental Protection Agency]

OWRT Office of Water Research and Technology [US]

OWS Ocean Weather Ship

OWS Ocean Weather Station

OWS Oily Waste System

OWS Operational Weather Support

OWS Operator's Workstation

OWS Orbital Workshop (Station) [of NASA]

OWSME One-Way Shape Memory Effect [Metallurgy]

OWTD Office of Waste Technology Development [US]

OWTU Oilfields Workers Trade Union [Trinidad]

OW Unit Open-Window Unit [also OW unit] [Acoustics]

OWW Organic Wash Waste

OX$_C$ Chemical-Vapor-Deposited Oxide

OX$_N$ Native Oxide

OX$_T$ Thermal Oxide

Ox Odd Oxygen [i.e., O and O_3]

Ox Oxalic Acid

Ox Oxidation; Oxidizer [also ox]

Ox Oxidant [Symbol]

Ox Oxithiin

ox oxalate (group) [$C_2O_4)^{2-}$]

ox oxalic

Oxa-2 3,6-Dioxaoctane-1,8-Diamine

Oxa-4 3,6,9,12-Tetraoxatetradecane-1,14-Diamine

Oxa-6 3,6,9,12,15,18-Hexaoxaeicosane-1,20-Diamine

OxAc Oxaloacetic Acid

Oxac Oxalacetate

Oxal Oxalate; Oxalic [also oxal]

Oxam Oxamide

Oxan Oxanilide

OXD Oxidative Dehydrogenation (Process)

Oxd Oxide; Oxidize(d); Oxidizer [also oxd]

ox'd oxidized [also oxd]

Oxdzr Oxidizer [also oxdzr]

Oxfam Oxford Committee for Famine Relief [UK and Canada]

Oxid Oxidize(r); Oxidation [also oxid]

Oxid Met Oxidation of Metals [Journal published in the US]

Oxidn Oxidation [also oxidn]

Oxil 8-Hydroxyquinoline

OXMAT Oxidizing Material [also Ox Mat, or oxmat]

Oxon Oxfordshire [UK]

Oxon *(Oxoniensis)* – of Oxford University [UK]

OXS Oxygen Sensor

OXSEN Oxide Spin Electronics Network [of European Community]

Oxy Oxygen [also oxy]

OY Optical Yield

OY- Denmark [Civil Aircraft Marking]

OYAP Ontario Youth Apprenticeship Program [of Ministry of Education and Training, Canada]

OYI Outstanding Young Investigator (Award) [of Materials Research Society, US]

OYP Office of Youth Programs [of US Department of the Interior]

OZ Optical-Grade Fused Quartz

OZ Organic Zinc Coating

OZ Organozinc

Oz Ozone

oz ounce [Unit]

oz AI Ounce(s) of Active Ingredient [also oz ai]

oz AI/1000 ft ounce(s) of active ingredient per 1000 feet of row [also oz ai/1000 ft]

oz AI/1000 ft^3 ounce(s) of active ingredient per 1000 cubic feet of row [also oz Ai/1000 ft^3]

oz ap ounze, apothecaries' [also oz apoth]

oz av ounce, avoirdupois [Unit]

OZC Organozinc Compound

oz/cu in ounces per cubic inch [oz/in^3]

OZD Observed Zenith Distance [Astronomy]

ozf ounce-force [Unit]

ozf-in ounce-inch [Unit]

oz fl ounze, fluid [Unit]

oz ft ounce-foot [also oz-ft]

oz/ft^2 ounce(s) per foot [also oz/sq ft]

oz/gal ounce(s) per gallon [Unit]

OZI Okubo-Zweig-Iizuha (Quark Model) [Particle Physics]

oz-in ounce-inch [also oz in]

oz/in^2 ounce(s) per square inch [also oz/sq in, OSI, or osi]

oz/in^3 ounce(s) per cubic inch [also oz/cu in]

oz-in/amp ounce-inch(es) per ampere [also oz in/amp]

oz-in/V ounce-inch(es) per volt [also oz in/V]

oz-in-sec^2 ounce-inch-second squared [also oz-ins^2]

oz-in-sec/rad ounce-inch-second(s) per radian [also oz in-sec/rad]

Ozone Sci Eng Ozone Science and Engineering [Journal of the International Ozone Association, US]

oz/sq ft ounce(s) per foot [also oz/ft^2]

oz/sq in ounce(s) per square inch [also oz/in^2, osi, or OSI]

oz/sq yd ounces per square yard [also oz/yd^2]

oz t ounce, troy [also oz tr]

oz/t ounce(s) per ton [Unit]

oz/1000 ft ounce(s) (of chemical) per 1000 feet of row [Unit]

oz/1000 ft³ ounce(s) (of chemical, or spray) per 1000 cubic feet of row [Unit]

OZTC(HA) Organization and Training Center (Heavy Artillery) [UK]

oz tr ounze, troy [also oz t]

oz/y ounce(s) per year [also oz/yr]

oz/yd ounce(s) per yard [Unit]

oz/yd² ounce(s) per square yard [also oz/sq yd]

oz/yr ounce(s) per year [also oz/y]

P

P   Absolute Pressure [Symbol]

P   Acoustic Power [Symbol]

P   Applied Load [Symbol]

P   Atmospheric Pressure (at Height $h > 0$) [Symbol]

P   Average (Electric) Power (or Actual Power, Real Power, or True Power) [Symbol]

P   Circular Pitch (of Worm Gears) [Symbol]

P   Depth of Penetration [Symbol]

P   Deviating Power (of Prisms) [Symbol]

P   Diametral Pitch (of Gears, Splines, etc.) [Symbol]

P   Dielectric Polarization [Symbol]

P   Effective Pair Production Content (in Physics) [Symbol]

P   Electric Dipole Moment [Symbol]

P   Electric Polarization [Symbol]

P   Equivalent Dynamic Load (on Rolling Bearings) [Symbol]

P   Earth Pressure (i.e. Force Exerted by Soil) [Symbol]

P   Induced (Electric) Dipole Moment [Symbol]

P   Larson-Miller Parameter [Symbol]

P   Lead (II) Oxide [Ceramics]

P   Lidar Pointing Mode [Symbol]

P   Load (Range) [Symbol]

P   Molar Polarization [Symbol]

P   Mold Steel [AISI-SAE Symbol]

P   Multiplicity Factor [Symbol]

P   Near Point (of the Eye) [Symbol]

P   Number of Phases (of a Material) [Symbol]

P   Number of Grain-Boundary Intersections (in Quantitative Metallography) [Symbol]

P   Number of Intersections with Perimeter (for 2-D Planar Figures), or with Surfaces (for 3-D Grains and Particles) Made by Constant Length of Random Short Test Lines ("l") (in Quantitative Metallography) [Symbol]

P   Number of Point Elements, or Test Points (in Quantitative Metallography) [Symbol]

P   Off-Zenith Direction [Symbol]

P   Pacific (Ocean)

P   Pacific [US Geographic Region]

P   Pack(age)

P   (Data) Packet

P   Packing

P   Pad(ding)

P   Paddle

P   Page

P   Pagination

P   Paging

P   Paint(ing); Painter

P   Pair(ing)

P   Pakistan(i)

P   Palate; Palatine [Anatomy]

P   Paleozoic [Geological Era]

P   Palermo [Sicily, Italy]

P   Palestine(an)

P   Palette; Pallet(ing); Palletize(r)

P   Palisades (of Avian Eggs) [Zoology]

P   Pan

P   Panama(nian)

P   Pancreas; Pancreatic [Anatomy]

P   Pandornia [Genus of Algae]

P   Panel(ing)

P   Panorama; Panoramic

P   Pantograph(ic)

P   Papa [Phonetic Alphabet]

P   Paper

P   Parabola; Parabolic; Paraboloid(al)

P   Parachute

P   Paradox

P   Paraelectric(ity)

P   Paraffin

P   Paragraph

P   Paraguay(an)

P   Parallax

P   Parallel(ism)

P   Parallels

P   Paramaribo [Suriname]

P   Paramecium [Genus of Protozoans]

P   Parameter(ization); Parametric

P   Paraná (River) [South America]

P   Parascaris [Genus of Nematodes]

P   Parasite; Parasitic; Parasitism

P   Parenchyma [Botany]

P   Parent

P   Parental (Generation) [Genetics]

P   Paris [France]

P   Parison

P   Parity

P   Parity (or Space Reflection Symmetry) [Particle Physics]

P   Park

P   Park Gear [Symbol]

P   Parking
P   Part(ial)
P   Partial Detonation [Symbol]
P   Partial-Discharge (Corona) Power Loss [Symbol]
P   Partial Pressure [Symbol]
P   Part(ing)
P   Particle
P   Particulate
P   Partition(ing)
P   Pasadena [California, US]
P   Pass(ing)
P   Passage
P   Passenger
P   Passenger Car (or Vehicle) [Symbol]
P   Passivate; Passivation; Passivator
P   Passive; Passivity
P   Paste
P   Pasteurize(d); Pasteurization; Pasteurizer
P   Pasteurella [Genus of Microorganisms]
P   Pataca [Currency of Macau]
P   Patch(ing)
P   Patella [Anatomy]
P   Patent(ed); Patentee
P   Patented; Patenting [Metallurgy]
P   Path
P   Pathogen(ic); Pathogenicity [Medicine]
P   Pathogenesis [Medicine]
P   Patrol Aircraft [US Navy Symbol]
P   Pattern
P   Pause
P   Pavement
P   Payload
P   Peace River (Bitumen) [Alberta Oil Sands]
P   Peak
P   Peak Height [Symbol]
P   Pear
P   Pearl
P   Pearlite [Symbol]
P   Pebble
P   Peclet Number (in Chemical Engineering) [Symbol]
P   Pedestral
P   Pediastrum [Genus of Algae]
P   Pediatrician; Pediatrics
P   Pediculus [Genus of Lice]
P   Pedologist; Pedology
P   Peel(ing)
P   Peen(ing)
P   Peg
P   Peking [PR China]
P   Pellet(izing)
P   Peltier Coefficient [Symbol]
P   Pelvis [Anatomy]
P   Pemphigus [Medicine]

P   Pen
P   Pencil
P   Pendant
P   Pendulum
P   Penetrant; Penetrate; Penetration; Penetration
P   Penicillin [Microbiology]
P   Penicillium [Genus of Fungi]
P   Penile; Penis [Anatomy]
P   Pennsylvania [US]
P   Pennsylvanian [Geological Period]
P   Penta-
P   Pentagon(al)
P   Pentane
P   Pentode
P   Pentose [Biochemistry]
P   Pepsin; Pepsinogen [Biochemistry]
P   Peotic [Medicine]
P   Peptide; Peptize; Peptization [Biochemistry]
P   Percent(age)
P   Percent Linear Thermal Expansion [Symbol]
P   Perception
P   Percolate; Percolation; Percolator
P   Percussion
P   Perforate(d); Perforation
P   Perform(ance)
P   Performance Index [Symbol]
P   Perfusion
P   Periapsis [Astronomy]
P   Peridinium [Genus of Algae]
P   Perigee
P   Perihelion
P   Perimeter
P   Period; Periodic(ity)
P   Period of Oscillation [Symbol]
P   Periosteum [Anatomy]
P   Peripheral; Periphery
P   Periscope
P   Peristalsis; Peristaltic [Medicine]
P   Peritectic (Reaction) [Symbol]
P   Peritoneum [Anatomy]
P   Permanence; Permanent
P   Permeability; Permeable
P   Permeance [Symbol]
P   Permeation
P   Permian [Geological Period]
P   Permutation
P   Perovskite [Mineral]
P   Persia(n)
P   Persistence; Persistent
P   Person(al)
P   Personnel
P   Perspective
P   Perth [Australia]

P   Perturbation

P   Pertussis (or Whooping Cough) [Medicine]

P   Peru(vian)

P   Peta- [SI Prefix]

P   Petal [Botany]

P   Petrologist; Petrology

P   Petroleum

P   Petroleum Engineering (Technology) [Discipline Category Abbreviation]

P   Pewter

P   Phagocyte; Phagocytic [Biology]

P   Phalanx [Anatomy]

P   Phantom

P   Pharmacopoeia

P   Pharyngeal; Pharynx [Anatomy]

P   Pharyngitis [Medicine]

P   Phase

P   Phaser [Electronics]

P   Phasor [Physics]

P   Phenolic(s)

P   Phenomenon

P   Philadelphia [Pennsylvania, US]

P   Philippines

P   Philippine Peso [Currency]

P   Phoenix [Arizona, US]

P   Phonon [Physics]

P   Phosphor [Symbol]

P   Phosphorazo-

P   Phosphoresce(nce); Phosphorescent

P   Phosphorus [Symbol]

P   Photograph(ic); Photographer; Photography

P   Photometer; Photometric; Photometry

P   Photosynthesis; Photosynthetic

P   Physic(al); Physicist; Physics

P   Picea [Genus of Coniferous Trees Including the Spruces]

P   Pick(ing)

P   Pickle; Pickling

P   Pictorial

P   Picture

P   Pierce; Piercing

P   Pigment(ation)

P   Pile

P   Pill

P   Pillar

P   Pillow

P   Pilot(age); Piloting

P   Pin

P   Pincer

P   Pinch(ing)

P   Pine

P   Pinion

P   Pink

P   Pinning [Solid-State Physics]

P   Pinnularia [Genus of Algae]

P   Pinus [Genus of Coniferous Trees Including the Pines]

P   Pipe

P   Piperazine Dinitrate

P   Pipet(te)

P   Piraeus [Greece]

P   Pistil [Botany]

P   Piston

P   Pit(ter); Pitting

P   Pitting [Metallurgy]

P   Pitch

P   Pitch (of Threads, etc.) [Symbol]

P   Pitch (of Weld) [Symbol]

P   Pith [Botany]

P   Pittsburgh [Pennsylvania, US]

P   Pituitary [Anatomy]

P   Pivot(ed)

P   Placebo [Medicine]

P   Placenta

P   Plain

P   Plan(ning)

P   Planar

P   Plane

P   Planed [Lumber]

P   Planer

P   Planet(ary)

P   Planetarium

P   Planing

P   Plankton [Ecology]

P   Plant

P   Plant(ing); Planter

P   Plaque

P   Plasma

P   Plasmodium [Microbiology]

P   Plaster(er)

P   Plastic(s)

P   Plast(icity)

P   Plasticizer; Plastisol

P   Plastron [Zoology]

P   Plate

P   Platelet [Medicine]

P   Plateau

P   Platen

P   Plater; Plating

P   Play

P   Pleistocene [Geological Period]

P   Pleura(l) [Anatomy]

P   Pliers

P   Plot(ter)

P   Plug

P   Plug Chamfer (of Screw Taps) [Symbol]

P   Plunger

P   Ply

P   Plywood

P   Pneumatic(s)

P   Pneumocystis [Medicine]

P   Pneumonia [Medicine]

P   Pocket

P   Point

P   Pointer

P   Poise [Unit]

P   Poison(ing); Poisonous

P   Poland; Polish

P   Polar(ity); Polarizability; Polarization; Polarize(r)

P   Polarization [Symbol]

P   Pole

P   Polenske Number (in Chemistry) [Symbol]

P   Polhode [Mechanics]

P   Polish(ing)

P   Poll(ing)

P   Pollen; Pollinate; Pollination [Botany]

P   Pollutant; Pollute(r); Polluting; Pollution

P   Polymer(ic); Polymerization

P   Polynesia(n)

P   Polynomial

P   Polyp

P   Polypeptide [Biochemistry]

P   Pond

P   Pool

P   Poona [India]

P   Poor

P   Pop(ping)

P   Poplar

P   Poppet

P   Population [also p]

P   Populus [Genus of Deciduous Trees Including Poplars and Aspens]

P   Pore; Porosity; Porous

P   Porod Constant [Symbol]

P   Porifera [Zoology]

P   Port(al)

P   Portability; Portable

P   Portland [Oregon, US]

P   Portrait (Font Orientation)

P   Portugal; Portuguese

P   Position(al)

P   Positive; Positivity

P   Positron(ium)

P   Post

P   Potassium Sulfate Containing Propellant

P   Pot

P   Potting

P   Potential

P   Potential Vorticity [Symbol]

P   Potter (Orientation Relationship) [Crystallography]

P   Potter(y)

P   Pour(ing)

P   Powder

P   Power

P   Poynting Vector [Symbol]

P   P-Phase (in Metallurgy) [Symbol]

P   Practical; Practice

P   Practitioner

P   Prague [Czech Republic]

P   Precaution

P   Precession

P   Precipitate; Precipitation; Precipitator

P   Precipitin (Test) [Immunology]

P   Precise; Precision

P   Precursor

P   Predict(ion); Predictor

P   Preference; Preferential

P   Preload(ing)

P   Premolar [Dentistry]

P   Prerequisite

P   Prescription

P   Present(ation)

P   Preservation; Preservative

P   President(ial)

P   Press(ing)

P   Pressure [also p]

P   Pretoria [South Africa]

P   Prevent(ion)

P   Primary

P   Primary Principal Point [Optics]

P   Primary Wave [Seismology]

P   Primate

P   Prime

P   Prime Wave [Geology]

P   Primer

P   Primitive (Simple) Space Lattice in Triclinic (Anorthic), Monoclinic, Orthorhombic, Tetragonal, Hexagonal and Cubic Crystal Systems [Hermann-Mauguin Symbol]

P   Primitive (Phase) [Symbol]

P   Principal

P   Principal Line [Spectroscopy]

P   Principal Point [Optics]

P   Principle

P   Print(er); Printing

P   Prior(ity)

P   Prism(atic)

P   Prismatic Effect (of a Decentered Lens) [Symbol]

P   Private

P   Probability [Symbol]

P   Probability; Probable

P   Probe; Probing

P   Problem; Problematic(s)

P   Proboscidea [Order of Mammals Including Elephants]

P   Procedure

P   Proceed(ing)

P   Process(ing); Processibility; Processor

P   Process Energy [Symbol]

P   Prod

P   Produce(r); Product(ion); Productive; Productivity

P   Profession(al)

P   Professor

P   Profile; Profiling

P   Progesterone [Biochemistry]

P   Prognosis; Prognostic

P   Program

P   Programmer; Programming

P   Progress(ion); Progressive

P   Prohibited; Prohibition

P   Project

P   Projection; Projector

P   Projectile

P   Proliferation

P   (−)-Proline; Prolyl [Biochemistry]

P   Prominence

P   Prompt(ness)

P   Prong

P   Proof(ing)

P   Prop

P   Propagate; Propagating; Propagation

P   Propagation Constant [Symbol]

P   Propanol

P   Propeller

P   Property

P   Prophylactic; Prophylaxis [Medicine]

P   Proportion(al); Proportioning

P   Propulsion

P   Prospect(ing); Prospector

P   Protect(ion); Protective; Protector

P   Proterozoic [Geological Era]

P   Protein [Biochemistry]

P   Proteus [Genus of Bacteria]

P   Protherombin [Biochemistry]

P   Protococcus [Genus of Algae]

P   Protocol

P   Proton [also p, or H^{1+}]

P   Protoplasm; Protoplast [Cytology]

P   Prototype

P   Protozoa(n) [Zoology]

P   Protract(or)

P   Protrusion

P   Protuberance

P   Province

P   Providence [Rhode Island, US]

P   Provision(al)

P   Proximity

P   Prunus [Genus of Plants including Fruit Trees, such as Cherries, Peaches, Plums, etc.]

P   P-Shell (of Electron) [Symbol]

P   Psychological; Psychologist; Psychology

P   Psychosis; Psychotic(s)

P   Psychrometer; Psychrometric; Psychrometry

P   Puddle

P   Puddling [Metallurgy]

P   Pulley

P   Pula [Currency of Botswana]

P   Pulix [Genus of Fleas]

P   Pull(ing)

P   Pulp

P   Pulper; Pulping

P   Pulsatance [Physics]

P   Pulsating; Pulsation; Pulsator

P   Pulse

P   Pulse Discharge Power Loss [Symbol]

P   Pulse Modulation [Symbol]

P   Pultrusion

P   Pulverization; Pulverizer

P   Pump(ing)

P   Punch(ing)

P   Puncture(d)

P   Pupa [Entomology]

P   Pupil (of the Eye) [Anatomy]

P   Purchase(r); Purchasing

P   Pure; Purification; Purify(ing); Purity

P   Purine [Biochemistry]

P   Purl [Hydrology]

P   Pusan [South Korea]

P   Push(ing)

P   Putty

P   Pyongyang [North Korea]

P   Pylon

P   Pyramid(al)

P   Pyrenees (Mountains) [France/Spain]

P   Pyrheliometer

P   Pyridine

P   Pyrolysis; Pyrolytic

P   Pyrometer; Pyrometric; Pyrometry

P   Radiant Power (or Radiant Flux) [Symbol]

P   Reflection Coefficient [Symbol]

P   Single Paper Covered (Magnet Wire) [Symbol]

P   Soft Pad (Launch Environment) [USDOD Symbol]

P   Space-Parity (in Quantum Mechanics) [Symbol]

P   Spin Polarization [Symbol]

P   Thickness of Superconductor Outer Layer [Symbol]

P   Total Load [Symbol]

P   Total Pressure [Symbol]

/P   Prompt Option [MS-DOS Shell]

P-   Nonmagnetic Doublet Spectra [Letter Symbol Followed by a Number, e.g., P-1, P-2, etc.]

-P   Photo-Reconnaissance [US Navy Suffix]

[P]   Parachor [Chemistry]

$\bar{P}$ Mean Pressure [Symbol]

P*   Peak (or Inflection) Point Load [Symbol]

P′   First Principal Point [Optics]

P′   (Earthquake) Push Wave that has Passed through the Earth's Core [Symbol] [also PKP]

P′   Quasi-Peritectic (in Metallurgy) [Symbol]

P″   Secondary Principal Point [Optics]

P$_0$ Absolute Pressure [Symbol]

P$_0$ Atmospheric Pressure at height $h = 0$ [Symbol]

P$_0$ Earth Pressure at Rest (i.e. Force Exerted by Soil at Rest) [Symbol]

P$_0$ Equivalent Static Load (on Rolling Bearings) [Symbol]

P$_0$ Outside Pressure [Symbol]

P$_0$ Polarizer [Symbol]

P$_0$ Pressure at Ice Point (0°C) [Symbol]

P$_0$ Radiant Energy (or Power) of Incident Beam (of Electromagnetic Radiation) [Symbol]

P$_0$ Standard Atmospheric Pressure [Symbol]

P$_0$ Static Pressure (in a Fluid Flow) [Symbol]

P$_I$ Mode I Imposed Load (in Mechanical Testing) [Symbol]

P$_{II}$ Mode II Imposed Load (in Mechanical Testing) [Symbol]

P$_1$, P$_2$, P$_3$, etc.   First, Second , Third, etc., Parental Generation (in Genetics) [Symbol]

P^{3-} Phosphide [Symbol]

P$_4$ Molecular Phosphorus [Symbol]

P^{5+} Pentavalent Phosphorus Ion [Symbol]

P$_A$ Number of Point Features per Total Test Area (in Quantitative Metallography) [Symbol]

P$_a$ Active Earth (or Soil) Pressure [Symbol]

P$_a$ Allowable Pile Bearing Load (in Civil Engineering) [Symbol]

P$_a$ Alternating Load [Symbol]

P$_a$ Apparent (Electric) Power [Symbol]

P$_a$ (Partial-Discharge) Apparent (Electric) Power Loss [Symbol]

P$_a$ Loading Amplitude (in Fatigue) [Symbol]

P$_a$ Pressure at Anode [Symbol]

$\overline{P}_a$ Average Pressure [Symbol]

P$_\alpha$ Number of Test Points Falling in Alpha Phase (in Quantitative Metallography) [Symbol]

P$_{a(B;f)}$ Specific Apparent (Electric) Power [Symbol]

P$_{abs}$ Absolute Pressure [Symbol]

P$_{amb}$ Ambient (Barometric) Pressure [Symbol]

P$_{Ar}$ Pressure of Argon [Symbol]

P$_{atm}$ Atmospheric Pressure [Symbol]

P$_{AVG}$ Average Power [also P$_{avg}$] [Symbol]

P$_{BE}$ Power Input (Direct-Current) to Base, Common Emitter [Semiconductor Symbol]

P$_c$ (Total) Alternating-Current Core Loss [Symbol]

P$_c$ Carrier Power (of Transmitter Stations) [Symbol]

P$_c$ Cavitation Pressure [Symbol]

P$_c$ Chamber Pressure [Symbol]

P$_c$ Continuous (Geophysical) Micropulsation [Symbol]

P$_c$ Critcal Pressure [Symbol]

P$_c$ (Total) Load Carried by a Fiber-Reinforced Composite [Symbol]

P$_c$ Microwave Absorption [Symbol]

P$_c$ Peak Count (in Surface Roughness Determination) [Symbol]

P$_c$ Pressure at Cathode [Symbol]

P$_{CB}$ Power Input (Direct-Current) to Collector, Common Base [Semiconductor Symbol]

P$_{c(B;f)}$ Specific Alternating-Current Core Loss [Symbol]

P$_{cΔ}$ Incremental Alternating-Current Core Loss [Symbol]

P$_{CE}$ Power Input (Direct-Current) to Collector, Common Emitter [Semiconductor Symbol]

P$_{cr}$ Critial Load (on a Column) [Symbol]

P$_D$ Power (in Drilling) [Symbol]

P$_D$ Power Dissipation [Semiconductor Symbol]

P$_d$ Maximum Draw Load (in Draw Fracture Test) [Symbol]

P$_{Δe}$ Incremental AC (Alternating Current) Eddy-Current Core Loss [Symbol]

P$_{Δh}$ Incremental Hysteresis Loss [Symbol]

P$_{D(max)}$ Maximum Power Dissipation (Rating) [Semiconductor Symbol]

P$_{dyn}$ Dynamic Pressure [Symbol]

P$_e$ Electronic Polarization [Symbol]

P$_e$ Equilibrium Pressure [Symbol]

P$_e$ Normal AC (Alternating Current) Eddy-Current Core Loss [Symbol]

P$_e$ Power per Electron [Symbol]

P   Peclet Number (in Chemical Engineering) [Symbol]

P$_{EB}$ Power Input (Direct-Current) to Emitter, Common Base [Semiconductor Symbol]

P$_{elec}$ Electrical Power [Symbol]

P$_F$ Forward Power Dissipation, Direct Current (Without Alternating Component) [Semiconductor Symbol]

P$_f$ Maximum Fracture Load (in Draw Fracture Test) [Symbol]

P$_f$ Precipitation Finish [Symbol]

P$_f$ Load Carried by Fibers of Fiber-Reinforced Composite [Symbol]

P$_{F(AV)}$ Forward Power Dissipation, DirectCurrent (With Alternating Component) [Semiconductor Symbol]

P$_{FM}$ Maximum (Peak) Total Forward Power Dissipation [Semiconductor Symbol]

P$_G$ Gate Power [Semiconductor Symbol]

P$_G$ Growth Probability (in Crystallization) [Symbol]

P$_g$ Equilibrium Partial Pressure [Symbol]

P$_g$ Gauge Pressure [also P$_{ga}$] [Symbol]

P$_H$ Hugoniot Shock Pressure (in Mechanics) [Symbol]

P$_h$ Normal Hysteresis Loss [Symbol]

P$_i$ Inorganic Phosphate [Symbol]

P$_i$ Input Power [Symbol]

P$_i$ Interplanar Interaction [Symbol]

P$_i$ Ionic Polarization [Symbol]

P$_i$ Pressure of Species "i" [Symbol]

P$_i$ Probability that Probe Ion Remains Ionized after Interaction With an Species "i" Atom (in Ion Spectroscopy Scattering) [Symbol]

P$_{IB}$ Common-Base Large-Signal Input Power [Semiconductor Symbol]

P$_{IC}$ Common-Collector Large-Signal Input Power [Semiconductor Symbol]

P$_{IE}$ Common-Emitter Large-Signal Input Power [Semiconductor Symbol]

P$_{in}$ Entrance Pressure [Symbol]

P$_{in}$ Input Power [Symbol]

P$_{ind}$ Induced Polarization [Symbol]

P$_L$ Load Power [Symbol]

P$_L$ Number of Point Intersections per Unit Length of Test Line (in Quantitative Metallography) [Symbol]

P$_{LM}$ Larson-Miller Parameter [Symbol]

P$_m$ Load Carried by Matrix of Fiber-Reinforced Composite [Symbol]

P$_m$ Magnetic Scattering Length [Symbol]

P$_m$ Mean Load [Symbol]

P$_{max}$ Maximum (or Peak) Load [Symbol]

P$_{max}$ Maximum (or Peak) Pressure [Symbol]

P$_{mech}$ Mechanical Power [Symbol]

P$_{min}$ Minimum Load [Symbol]

P$_{min}$ Minimum Pressure [Symbol]

P$_{nom}$ Nominal Power [Symbol]

P$_N$ Nucleation Probability (in Crystallization) [Symbol]

P$_n$ Negative Pressure [Symbol]

P$_n$ Parity Bit [n = 1, 2, 3, etc. is the Position Number)

P$_{n,r}$ Number n(n − 1) $\cdots$ (n − r + 1), of Permutations, or Arrangements of n Distinct Objects Taken r at a Time without Repetition [also $_nP_r$, or P]

P$_o$ Organic Phosphate [Symbol]

P$_o$ Orientation Polarization [Symbol]

P$_o$ Outside Pressure [Symbol]

P$_o$ Output Power [Symbol]

Po$_2$ Pressure of Oxygen [Symbol]

P$_{OB}$ Common-Base Large-Signal Output Power [Semiconductor Symbol]

P$_{OC}$ Common-Collector Large-Signal Outut Power [Semiconductor Symbol]

P$_{OE}$ Common-Emitter Large-Signal Output Power [Semiconductor Symbol]

P$_{out}$ Exit Pressure [Symbol]

P$_{out}$ Output Power [Symbol]

P$_P$ Point Count, or Point Fraction (i.e., Number of Point Elements per Total Number of Test Points) (in Quantitative Metallography) [Symbol]

P$_P$ Pitot Pressure [Symbol]

P$_P$ Pump Power [Symbol]

P$_p$ Bearing Capacity (of a Pile) [Symbol]

P$_p$ Passive Earth (or Soil) Pressure [Symbol]

P$_p$ Perforation Probability (in Corrosion Science) [Symbol]

P$_p$ Positive Pressure [Symbol]

P$_{part}$ Partial Pressure [Symbol]

P$_Q$ Maximum Load (or Load Corresponding to Intersection of 2% Crack Growth Secant Line with the Plot) (in Plane-Strain Fracture Toughness Test) [Symbol]

P$_q$ Reactive (Quadrature) Power [Symbol]

P$_R$ Reverse Power Dissipation, Direct-Current (Without Alternating Component) [Semiconductor Symbol]

P$_r$ Remanent Polarization [Symbol]

P$_r$ Residual Core Loss [Symbol]

P$_{R(AV)}$ Reverse Power Dissipation, Direct-Current (With Alternating Component) [Semiconductor Symbol]

P$_{RM}$ Maximum (Peak) Total Reverse Power Dissipation [Semiconductor Symbol]

P$_S$ Signal Strength [Symbol]

P$_s$ Load on the Sample [Symbol]

P$_s$ Precipitation Start [Symbol]

P$_s$ Probability of Survival [Symbol]

P$_s$ Saturation Polarization [Symbol]

P$_s$ (Voltage) Source Power [Symbol]

P$_s$ Spontaneous Polarization [Symbol]

P$_s$ Static Pressure [Symbol]

P$_s$ Stiffness Performance Index [Symbol]

P$_s$ Stray Radiant Power (in Molecular Spectroscopy) [Symbol]

P$_{SN}$ Shot-Noise Power [Symbol]

P$_T$ Texture Related Permeability [Symbol]

P$_T$ Total Nonreactive Power Input to All Terminals [Semiconductor Symbol]

P$_T$ Total Number of Test Points (in Quantitative Metallography) [Symbol]

P$_T$ Trapping Probability (of Particles) [Symbol]

P$_t$ (Geophysical) Micropulsation consisting of Trains of Damped Oscillations [Symbol]

P$_t$ Pressure at Temperature t [Symbol]

P$_t$ Total Detected Radiant Power (in Molecular Spectroscopy) [Symbol]

P$_{th}$ Threshold Pressure [Symbol]

P$_{tot}$ Total Pressure [Symbol]

P$_u$ Unit Power (in Machining) [Symbol]

P$_v$ Isochoric Pressure [Symbol]

P$_v$ Vapor Pressure [Symbol]

P$_v$ Vertical Stress [Symbol]

P$_W$ Working Pressure [Symbol]

P$_{WF}$ Water-Feed Pump [Symbol]

P$_w$ Plastic Limit (for Rock, or Soil) [Symbol]

P$_w$ Winding Loss (or Copper Loss) [Symbol]

P$_x$ Load in Horizontal Direction [Symbol]

P$_Y$ Limit Load at Yield Point [Symbol]

P$_y$ Hydrostatic Loading [Symbol]

P$_y$ Load in Vertical Direction [Symbol]

P$_Z$ Metallostatic Pressure at Depth "Z" below Surface [Symbol]

P$_Z$ Zener Pinning Pressure [Symbol]

P$_z$ rms (root-mean-square) exciting power [Symbol]

P$_{z(B;f)}$ Specific Exciting Power [Symbol]

P0 Pulsed Carrier without Modulation for Information Carriage [Symbol]

P1 Petty Officer, First Class

P1 Pulse Modulation–Telegraphy without Modulation [Symbol]

P1D Telegraphy by On-Off Keying of Pulsed Carrier without Audio Frequency Modulation [Symbol] [also P1d]

P2 Petty Officer, Second Class

P2 Pollution Prevention [also P^2, or PP]

P2D Telegraphy by Modulation of the Pulse Amplitude [also P2d] [Symbol]

P2E Telegraphy by Modulation of the Pulse Width [also P2e] [Symbol]

P2F Telegraphy by Modulation of the Pulse Phase, or Position [also P2f] [Symbol]

P3 Bis[1-(Diphenylphosphino)propyl]phenylphosphine

P3 2,2',2"-Phosphinidynetris[ethyl(diphenyl)phosphine

P3 Platform for Privacy Preferences [Internet]

P3 Portable Plotting Package [also P^3, or PPP]

P3D Telephony by Modulation of the Pulse Amplitude [also P3d] [Symbol]

P3E Telephony by Modulation of the Pulse Width [also P3e] [Symbol]

P3F Telegraphy by Modulation of the Pulse Phase, or Position [also P3f] [Symbol]

P3G Telephony–Code Modulated Pulses; After Sampling and Quantization [also P3g] [Symbol]

P4 1,1,4.7,10,10-Hexaphenyl-1,4,7,10-Tetraphosphadecane

P9 Pulse Modulation–Composite Emissions (or Transmissions) and Cases Not Covered by P1 to P3G [Symbol]

P II Phosphorus II [Symbol]

P III Phosphorus III [Symbol]

P-29 Phosphorus-29 [also ^{29}P, or P^{29}]

P-30 Phosphorus-30 [also ^{30}P, or P^{30}]

P-31 Phosphorus-31 [also ^{31}P, P^{31}, or P]

P-32 Phosphorus-32 [also ^{32}P, or P^{32}]

P-33 Phosphorus-33 [also ^{33}P, or P^{33}]

P-34 Phosphorus-34 [also ^{34}P, or P^{34}]

p angular momentum of an orbital electron [Symbol]

p atmospheric pressure (at height $h > 0$) [Symbol]

p bearing pressure [Symbol]

p circular pitch (of gears, splines, etc.) [Symbol]

p contact pressure (for soil, or rock) [Symbol]

p earth pressure (i.e. pressure exerted by soil) [Symbol]

p electric dipole moment [Symbol]

p electron holes (in extrinsic semiconductors) [Symbol]

p fraction defective (of a sample) (in statistics) [Symbol]

p half the latus rectum of a conic [Symbol]

p length of a perpendicular from the origin to a line, or plane (in geometry) [Symbol]

p magnetic pole strength [Symbol]

p modulation periodicity [Symbol]

p momentum [Symbol]

p number of (electron) holes per cubic meter (of a semiconductor) [Symbol]

p number of independent variables (in statistics) [Symbol]

p operator ($py = dy/dt$)

p page [also P]

p paging (command) [Edlin MS-DOS Line Editor]

p para- [Chemistry]

p parallel

p parameter of a parabola [Symbol]

p part [also P]

p partially coherent (defect) [Metallurgy]

p particle

p particulate

p past

p Peclet number (in Chemical Engineering) [Symbol]

p pence [Plural of "Penny"]

p penetration (of metal-shearing punches) [Symbol]

p penny

p (per) – by

p percent of interest [Symbol]

p perimeter [Symbol]

p peseta [Currency of Spain]

p peso [Currency of Several Spanish-Speaking Countries]

p pi [English Equilvalent]

p pico- [SI Prefix]

p pint [Unit]

p pitch [Symbol]

p pitch of thread [Symbol]

p planar

p plate (or anode) [Semiconductor Symbol]

p p-layer [Symbol]

p pole distance [Symbol]

p p-orbital [Symbol]

p polycrystalline

p polyphase

p poor

p population [also P]

p population (in statistics) [Symbol]

p porous; porosity

p positive

p poundal [Unit]

p (total) pressure [Symbol]

p primary

p prime integer [Symbol]

p primitive (unit cell) [Crystallography]

p probability (in statistics) [Symbol]

p protein [Biochemistry]

p proton [Symbol]

p pula [Currency of Botswana]

p purl [Woven Fabrics]

p pyrochlore [Mineral]

p reflection coefficient [Symbol]

p scattering factor of conduction electrons [Symbol]

p sound pressure [Symbol]

p space frequency vector [Symbol]

p space inversion [Symbol]

p (normal) stress (within a soil mass) [Symbol]

p ultimate unit load (on a column) [Symbol]

p volume fraction [Symbol]

p- para- [Chemistry]

p- polycrystalline [Usually Followed by a Chemical Element, or Compound, e.g., $p\text{-}CaF_2$, etc.]

p- p-type [Electronics]

/p pause option [MS-DOS Linker]

p% percentage population (in statistics) [Symbol]

p̄ antiproton [Symbol]

$\overline{p}$ average fraction defective (of a sample) (in statistics) [Symbol]

$\hat{p}$ momentum [Symbol]

$p*$ probability that thermal motion will cause an ion to jump across a barrier, or a vacancy to move [Symbol]

$p*$ saturation point pressure [Symbol]

$p*$ sound radiation pressure [Symbol]

$p*$ switched polarization (in physics) [Symbol]

$p\char`\^$ nonswitched polarization (in physics) [Symbol]

p' true fraction defective (of a sample) [Symbol]

p^+ p-plus-type (semiconductor) layer [Symbol]

p^- antiproton [Symbol]

p^0 proton [Symbol]

p_0 earth pressure at rest (i.e. pressure exerted by soil at rest) [Symbol]

p_0 non-zero impurity concentration (in solid-state physics) [Symbol]

p_0 pressure (exerted by a gas) at ice point (0°C) [Symbol]

p_0 standard (atmospheric) pressure (101.325 kPa) [Symbol]

p_0 static pressure of a fluid stream [Symbol]

p_0^c critical impurity concentration (in semiconductor) [Symbol]

p_1 accetable quality level [Symbol]

p_1 initial pressure [Symbol]

p_1 vapor pressure of the solvent [Symbol]

p_2 final pressure [Symbol]

p_2 limiting quality level [Symbol]

p_2 lot tolerance fraction defective [Symbol]

p_2 vapor pressure of the solute [Symbol]

p_{100} pressure exerted by a gas at the steam point (100°C) [Symbol]

p_A active earth pressure (i.e. active pressure exerted by soil) [Symbol]

p_A partial pressure of component A [Symbol]

p_a adsorption pressure [Symbol]

p_a allowable soil (or bearing) pressure [Symbol]

p_a probability of adsorption [Symbol]

p_{amb} ambient pressure [Symbol]

p_{Ar} partial pressure of argon [Symbol]

p_B partial pressure of component B [Symbol]

p_{BE} instantaneous total power input to base, common emitter [Semiconductor Symbol]

p_c (unconfined, or uniaxial) compressive strength (of soil, or rock) [Symbol]

p_c percolation threshold [Symbol]

p_c critical concentration [Symbol]

p_c critical (specific) pressure [Symbol]

p_c percolation temperature [Symbol]

p_{CB} instantaneous total power input to collector, common emitter [Semiconductor Symbol]

p_{CE} instantaneous total power input to collector, common emitter [Semiconductor Symbol]

p_{CH_4} partial pressure of methane [Symbol]

p_{CO} partial pressure of carbon monoxide [Symbol]

p_{CO_2} partial pressure of carbon dioxide [Symbol]

p_d desorption pressure [Symbol]

p_e electronic polarization [Symbol]

p_e overpressure [Symbol]

p_e preconsolidation pressure (or prestress) (of soil) [Symbol]

p_{EB} instantaneous total power input to emitter, common base [Semiconductor Symbol]

p_F Fermi momentum [Symbol]

p_F instantaneous total forward power dissipation [Semiconductor Symbol]

p_{H_2} partial pressure of hydrogen [Symbol]

p_{H_2O} partial pressure of water [Symbol]

p_{H_2S} partial pressure of hydrogen sulfide [Symbol]

p_i electric charge density [Symbol]

p_i ionic polarization [Symbol]

p_i partial pressure of component (or species) "i" [Symbol]

p_{ib} common-base small-signal input power [Semiconductor Symbol]

p_{ic} common-collector small-signal input power [Semiconductor Symbol]

p_{ie} common-emitter small-signal input power [Semiconductor Symbol]

p_k kink concentration (in electron microscopy) [Symbol]

p_m magnetic dipole moment [Symbol]

p_{NO_2} partial pressure of nitrogen dioxide [Symbol]

$p_{N_2O_4}$ partial pressure of nitrogen tetroxide [Symbol]

p_o efficiency (of collecting system in electron microscopy) [Symbol]

p_{O_2} partial pressure of oxygen [Symbol]

p_{ob} common-base small-signal output power [Semiconductor Symbol]

p_{oc} common-collector small-signal output power [Semiconductor Symbol]

p_{oe} common-emitter small-signal output power [Semiconductor Symbol]

p_p passive earth pressure (i.e. passive pressure exerted by soil) [Symbol]

p_{part} partial pressure [Symbol]

p_R instantaneous total reverse power dissipation [Semiconductor Symbol]

p_R (Proctor) penetration resistance (of soil) [Symbol]

p_r reduced pressure [Symbol]

p_r residual (total) pressure [Symbol]

p_{rg} residual gas pressure [Symbol]

$p_{r,vap}$ residual vapor pressure [Symbol]

p_s saturation vapor pressure [Symbol]

p_{S_2} partial pressure of sulfur [Symbol]

p_{SiO} partial pressure of silicon oxide [Symbol]

p_{SO_2} partial pressure of sulfur dioxide [Symbol]

p_T nonreactive power input, instantaneous total, to all terminals [Semiconductor Symbol]

p_t pressure exerted by a gas at temperature t [Symbol]

p_{tot} total pressure [Symbol]

p_{ult} ultimate pressure [Symbol]

p_v vapor pressure [Symbol]

p_w saturated vapor pressure at wet-bulb temperature t_w [Symbol]

p_x p-orbital along x-axis [Symbol]

p_y p-orbital along y-axis [Symbol]

p_z p-orbital along z-axis [Symbol]

Φ Cartesian Tensor [Symbol]

Φ Complex Faraday Effect [Symbol]

Φ Contour Angle [Symbol]

Φ Crack Inclination (or Slant) Angle (in Mechanics) [Symbol]

Φ Critical Concentration [Symbol]

Φ Energy Density [Symbol]

Φ Fiber Volume Concentration [Symbol]

Φ Fluence [Symbol]

Φ Function [Symbol]

Φ Heat Flux [Symbol]

Φ Inclination Angle [Symbol]

Φ Integral Neutron Fluence [Symbol]

Φ Luminous Flux [Symbol]

Φ Magnetic Flux [Symbol]

Φ Magnetic Scalar Potential [Symbol

Φ Magnetostatic Scalar Potential [Symbol]

Φ Multigroup Particle Fluence (in Radiography, etc.) [Symbol]

Φ Orientation Angle (in Crystallography) [Symbol]

Φ Phase [Symbol]

Φ Particle Fluence (in Dosimetry, etc.) [Symbol]

Φ Peak Splitting Parameter [Symbol]

Φ Permeability (in Fluid Mechanics) [Symbol]

Φ Perturbation [Symbol]

Φ (Upper-case) Phi [Greek Alphabet]

Φ Potential [Symbol]

Φ Potential Energy [Symbol]

Φ Power Density [Symbol]

Φ Power Dissipation [Symbol]

Φ Radiant Flux (or Radiant Power) [Symbol]

Φ Frequency Ratio (or Relative Frequency) (in Statistics) [Symbol]

Φ Scalar Potential [Symbol]

Φ Thermodynamic Factor [Symbol]

Φ Twist Angle [Symbol]

Φ Viscoelastic Loss Function [Symbol]

Φ Work Function [Solid-State Physics] [Symbol]

Φ_0 Equivalent 2200 m/s Fluence [Symbol]

Φ_0 Magnetic Flux Quantum (2.06735×10^{-15} Wb) [Symbol]

Φ_c Critical Concentration [Symbol]

Φ_e Radiant Flux (or Radiant Power) [Symbol]

ϕ angle [Symbol]

ϕ angle of eccentricity (in astronomy) [Symbol]

ϕ angle of friction [Symbol]

ϕ angle of incidence [Symbol]

ϕ angle of longitude (in spherical polar coordinate system) [Symbol]

ϕ angle of obliquity (of soil, or rock solids) [Symbol]

ϕ angle of refraction [Symbol]

ϕ angle of twist [Symbol]

ϕ Compton photon scattering angle [Symbol]

ϕ dedendum angle (of milled bevel gears) [Symbol]

ϕ (displacement) phase angle [Symbol]

ϕ effective tunneling barrier height [Symbol]

ϕ electrode potential [Symbol]

ϕ electrochemical potential [Symbol]

ϕ entropy [Symbol]

ϕ electronic exit work function [Symbol]

ϕ fluidity [Symbol]

ϕ function [Symbol]

ϕ Gaussian error function [Symbol]

ϕ geographical latitude (in astronomy) [Symbol]

ϕ golden ratio (i.e., $(\sqrt{5} + 1) \times \frac{1}{2} = 1.618034...$) [Symbol]

ϕ helix angle (of screw threads) [Symbol]

ϕ lagging angle (in electrical engineering) [Symbol]

ϕ loss angle (in electronics) [Symbol]

ϕ magnetic flux [Symbol]

ϕ maxwell [Symbol]

ϕ packing density (of polydisperse particle systems) [Symbol]

ϕ (particle) number density (or concentration) [Symbol]

ϕ phase angle [Symbol]

ϕ phase difference [Symbol]

ϕ phase factor [Symbol]

ϕ phenyl (C_6H_5) [Symbol]

ϕ (lower-case) phi [Greek Alphabet]

ϕ photoelectric work function [Symbol]

ϕ porosity [Symbol]

ϕ pressure angle (of gears and splines) [Symbol]

ϕ proton fluence [Symbol]

ϕ rise angle (of cams) [Symbol]

ϕ shear angle [Symbol]

ϕ shear strain [Symbol]

ϕ solidification constant [Symbol]

ϕ surface coverage [Symbol]

ϕ third coordinate (usually) in spherical coordinate systems [Symbol]

ϕ Thiele modulus [Symbol]

ϕ total magnetic flux [Symbol]

ϕ velocity potential [Symbol]

ϕ volume fraction [Symbol]

ϕ volumetric strain [Symbol]

ϕ^0 standard electrode potential at 298K (25°C) [Symbol]

ϕ_0 amplitude of transmitted beam [Symbol]

ϕ_0 flux quantum [Symbol]

ϕ_0 fundamental flux quantum [Symbol]

ϕ_0 inner potential [Symbol]

ϕ_1 magnetic flux linking of first turn (of electric machinery) [Symbol]

ϕ_2 magnetic flux linking of second turn (of electric machinery) [Symbol]

ϕ_B Schottky barrier height (in solid-state physics) [also ϕ_b] [Symbol]

ϕ_B^{cv} Schottky barrier height as per capacitance-voltage measurements (in solid-state physics) [Symbol]

ϕ_B^{IV} Schottky barrier height as per current-voltage measurements (in solid-state physics) [Symbol]

ϕ_{B0} zero bias Schottky barrier height (in solid-state physics) [also ϕ_{b0}] [Symbol]

ϕ_{Bn} n-type Schottky barrier height (in solid-state physics) [Symbol]

ϕ_{Bp} p-type Schottky barrier height (in solid-state physics) [Symbol]

ϕ_{BR} (effective) Schottky barrier height after reverse bias annealing at temperature T (in solid-state physics) [also ϕ_{br}] [Symbol]

ϕ_{BZ} Schottky barrier height after zero bias annealing at temperature T (in solid-state physics) [also ϕ_{bz}] [Symbol]

ϕ^{CV} Schottky barrier height as per capacitance-voltage measurements (in solid-state physics) [Symbol]

$\phi_\Delta{}^*$ angle of incidence for which Δ-sensitivity is a maximum (in ellipsometry) [Symbol]

ϕ_e radiant flux (or radiant power) [Symbol]

ϕ_g amplitude of diffracted, or scattered beam [Symbol]

ϕ_i angle subtended by image (of optical instrument) [Symbol]

ϕ_i thermodynamic potential of component i [Symbol]

ϕ^{IV} Schottky barrier height as per current-voltage measurements (in solid-state physics) [Symbol]

ϕ_K Kerr function (in physics) [Symbol]

ϕ_k pseudo-eigenfunction [Symbol]

ϕ_m mutual flux (of transformers) [Symbol]

ϕ_m workfunction of a metal [Symbol]

ϕ_N magnetic flux linking of N-th turn (of electric machinery) [Symbol]

ϕ_n pair interaction [Symbol]

ϕ_n Schottky barrier height for n-type semiconductors [Symbol]

ϕ_o angle subtended by object (in optical instrument) [Symbol]

ϕ_p magnetic flux in primary coil (of transformers) [Symbol]

ϕ_p Schottky barrier height for p-type semiconductors [Symbol]

$\phi_\psi{}^*$ angle of incidence for which ψ-sensitivity is a maximum (in ellipsometry) [Symbol]

ϕ_s magnetic flux in secondary coil (of transformers) [Symbol]

φ argument of a complex number (e.g., $\varphi = \text{arc } z$; $z = a + jb = re^{j\varphi}$) [Symbol]

φ correlation function [Symbol]

φ electrostatic potential [Symbol]

φ (sample) frequency (in statistics) [Symbol]

φ incidence angle [Symbol]

φ Lagrangian multiplier (in mathematics) [Symbol]

φ misorientation angle (in crystallography) [Symbol]

φ phase factor [Symbol]

φ phi (variant) [Greek Alphabet]

φ thermodynamic factor [Symbol]

φ Schottky energy barrier (in solid-state physics) [Symbol]

$\varphi^{(l)}$ Bloch wave amplitude (in physics) [Symbol]

φ_A volume fraction of substance "A" [Symbol]

φ_c critical incidence angle [Symbol]

Π Osmotic Pressure [Symbol]

Π (Upper-case) Pi [Greek Alphabet]

Π Product (of Terms) (in Mathematics) [Symbol]

Π Elastic Potential [Symbol]

π bonding pi (molecular) orbital (in physical chemistry) [Symbol]

π electrolytic polarization [Symbol]

π integration period (in spectroscopy) [Symbol]

π osmotic pressure [Symbol]

π Peltier heat [Symbol]

π (lower-case) pi [Greek Alphabet]

π pi bond [Symbol]

π pi electron [Symbol]

π Pilling-Bedworth ratio (in corrosion science) [Symbol]

π pi meson (in particle physics) [Symbol]

π pi orbital [Symbol]

π reflectance [Symbol]

π ratio of a circle's circumference to its diameter (3.14159265...) [Symbol]

π overvoltage [Symbol]

π^* antibonding pi orbital (in physical chemistry) [Symbol]

π^+ pi-plus meson (i.e., positively charged pion) [Symbol]

π^- pi-minus meson (i.e., negatively charged pion) [Symbol]

π^0 pi-zero meson (i.e., neutral pion) [Symbol]

Ψ Airy Stress Function (in Mechanics) [Symbol]

Ψ Amplitude [Symbol]

Ψ Angle [Symbol]

Ψ Angle of Obliquity (of Soil, or Rock Solids) [Symbol]

Ψ Angle of Rotation [Symbol]

Ψ Dihedral Angle [Symbol]

Ψ Electric Flux [Symbol]

Ψ Gauss Sum Function [Symbol]

Ψ Ground State (of Energy) [Symbol]

Ψ Hess Angle (in Physics) [Symbol]

Ψ Luminous Flux [Symbol]

Ψ Macroscopic Quantum Mechanical Wave Function, or Order Parameter [Symbol]

Ψ Mixity (i.e., Ratio of Shear Stress and Opening-Mode Crack-Tip Stress) [Symbol]

Ψ One-Fold Symmetry Distribution of Second Harmonic Field (in Physics) [Symbol]

Ψ Profile Structure Factor (in Fracture Analysis) [Symbol]

Ψ (Upper-case) Psi [Greek Alphabet]

Ψ Reduction in Cross-section(al) Area (in Mechanics) [Symbol]

Ψ Wave Function (in Physics) [Symbol]

Ψ_0 Ground State (of Energy) [Symbol]

Ψ_a Antisymmetric Wave Function (in Physics) [Symbol]

Ψ_s Symmetric Wave Function (in Physics) [Symbol]

ψ angle [Symbol]

ψ angle between normal to diffracting lattice planes and sample surface [Symbol]

ψ angle of rotation of x-ray diffraction sample [Symbol]

ψ atomic interaction distance [Symbol]

ψ azimuthal angle [Symbol]

ψ column vector (in transmission electron microscope) [Symbol]

ψ displacement current [Symbol]

ψ eigenfunction [Symbol]

ψ ellipsometric parameter (in pysics) [Symbol]

ψ helix angle (of helical gears and worm gears) [Symbol]

ψ molar ratio [Symbol]

ψ molecular orbital (in physical chemistry) [Symbol]

ψ permittivity (of geotextiles) [Symbol]

ψ phase angle [Symbol]

ψ phase difference [Symbol]

ψ (lower-case) psi [Greek Alphabet]

ψ psi function (or free energy function) [Symbol]

ψ separation ratio [Symbol]

ψ shear strain rate (of fluids) [Symbol]

ψ specific damping capacity [Symbol]

ψ spiral angle [Symbol]

ψ stream function (in fluid mechanics) [Symbol]

ψ stress intensity phase angle (in mechanics) [Symbol]

ψ superconducting order parameter [Symbol]

ψ traction ratio, mole mixity (in fracture mechanics) [Symbol]

ψ (Schroedinger) wave function [Symbol]

ψ wetting exponent [Symbol]

ψ^2 probability density [Symbol]

ψ^* complex conjugate of (Schroedinger) wave function [Symbol]

ψ_+ bonding (molecular) orbital (in physical chemistry) [Symbol]

ψ_- antibonding (molecular) orbital (in physical chemistry) [Symbol]

ψ_B Bloch wave (or eigensolution) (in physics) [Symbol]

ψ_k eigenfunction [Symbol]

ψ_k crystal wavefunction (in physics) [Symbol]

$\psi_{n,l,m}$ wave function (in physics) [Symbol]

$\mathcal{P}$ Dielectric Polarization [Symbol]

$\mathcal{P}$ Haldane's Field Momentum Functional (in Physics) [Symbol]

$\mathcal{P}$ (Magnetic) Permeance [Symbol]

$\mathcal{P}$ Probability (in Statistics) [Symbol]

$\mathcal{P}_r$ Remanent Polarization [Symbol]

$\wp$ Electric Polarization [Symbol]

$\wp$ Weierstrass (in Mathematics) [Symbol]

$ Pataca [Currency of Macau]

£ Pound Sterling [Currency of Northern Ireland and the UK]

1° Primary (Alcohol, Amine, etc.) [Symbol]

3-Ps The Three Presidents of the Institute of Electrical and Electronics Engineers [US]

PA Highest Priority

PA Pack Area [Computers]

PA Pack Artillery

PA Pad Abort

PA Paging and Area Warning

PA Paintmakers Association [UK]

PA Pair Association (Method) [Physics]

PA Paleolithic Age [Anthropology]

PA Palladium Aluminate

PA Palmitic Acid

PA Palo Alto [Texas, US]

PA Panama [ISO Code]

PA Pan America(n)

PA Panel Absorber

PA Paper Advance

PA Parallel Adder

PA Parallel Algorithm

PA Parametric Amplifier

PA Parasitic Antenna

PA Particle Acceleration; Particle Accelerator

PA Particle Atlas

PA Particular Average [also pa]

PA Passenger Address (System)

PA Patent Application

PA Peak Action

PA Peak-Aged; Peak Aging [Metallurgy]

PA Peak Amplitude

PA Pending Availability

PA Pendulous Axis

PA Pennsylvania [US]

PA Performance Analysis

PA Performance Assessment

PA Perigee Altitude

PA Permanently Associated

PA Pernicious Anemia

PA Personal Assistant

PA Perturbation Analysis [Physics]

PA Petroleum Abstracts [of University of Tulsa, Oklahoma, US]

PA Pharmacy Act [UK]

PA Phase Angle [Physics]

PA Phenolic Acid

PA Phenylalanine [also P-A] [Biochemistry]

PA 1-Phenylallyl [Biochemistry]

PA Phosphoric Acid

PA Photoacoustic; Photoacoustic(s)

PA Photoaddition

PA Photoadsorption

PA Photoinduced Absorption

PA Phthalic Acid
PA Phthalic Anhydride
PA Physical Acoustics
PA Physical Adsorption
PA Physical Aging
PA Physical Analysis
PA Physician's Assistant
PA Physics Abstracts [of INSPEC–Information Services Physics, Electrical and Electronics, and Computers and Control, UK]
PA Picatinny Arsenal [Dover, New Jersey, US]
PA Picolinic Acid
PA Picric Acid
PA Pilotless Aircraft
PA Piper Aircraft (Corporation) [US]
PA Piscis Australis [Astronomy]
PA Planning Approach
PA Plan of Action
PA Plant Assessment
PA Planar Anisotropy
PA Plasma Acceleration; Plasma Accelerator
PA Plasma-Activated; Plasma Activation
PA Plasma Arc
PA Plasma-Assisted
PA Plasminogen Activator [Biochemistry]
PA Plastic Anisotropy
PA Pneumatic Analyzer
PA Point of Aim
PA Polar Air [Meteorology]
PA Polar Axis
PA Polarization Analyzer
PA Pollock and Argon [Metallurgy]
PA Polyacetylene
PA Polyaddition; Polyadduct
PA Polyallomer
PA Polyamide
PA Polyaniline
PA Polyatomic
PA Positron Annihilation [Physics]
PA Postacceleration [Electronics]
PA Posterior-Anterior (Projection) [X-Ray Technology]
PA Post-Secondary Accreditation
PA Power Amplifier
PA Power of Attorney [also [P/A]
PA Preacceleration
PA Prealloyed; Prealloying
PA Preamplifier
PA Precision Approach
PA Preferential Adsorption
PA Preplaced Aggregate [Civil Engineering]
PA Press Agent
PA Press Association [UK]
PA Pressure Accumulator
PA Pressure Angle (of Gears)

PA Principal Axis
PA Principle of Adding
PA Priority Area
PA Privacy Act
PA Private Account
PA Probability of Acceptance
PA Process Allocator
PA Process Analysis; Process Analyzer
PA Process Annealed; Process Annealing [Metallurgy]
PA Process Automation
PA Producers Association [UK]
PA Product Analysis
PA Product Assurance
PA Production Adjustment
PA Production Analysis; Production Analyst
PA Professional Archeologist
PA Program Access
PA Program Action
PA Program Address
PA Program Announcement
PA Program Analysis
PA Program Application
PA Program Attention (Key)
PA Program Authorization
PA Programmatic Agreement
PA Propionic Acid
PA Propylamine
PA Propulsion Assisted
PA Prostate-Specific Antigen (Test) [Medicine]
PA Protected Area
PA Proton Accelerator
PA Proton Affinity
PA Pseudomonas Aeruginosa [Microbiology]
PA Psychoacoustics
PA Psychological Abstracts [of American Psychological Society, US]
PA Public Address (System)
PA Public Administration; Public Administrator
PA Public Affairs
PA Public Assistance
PA Publishers Association [UK]
PA Pulmonary Artery [Anatomy]
PA Pulse Amplifier
PA Pulse Analysis; Pulse Analyzer
PA Purchasing Agent
PA Puromycin Aminonucleoside [Microbiology]
PA Pyroelectric Anemometer
PA4 Polyamide 4 (or Nylon 4)
PA6 Polyamide 6 (or Nylon 6)
PA6T Polyamide 6T (or Nylon 6T)
PA9 Polyamide 9 (or Nylon 9)
PA11 Polyamide 11 (or Nylon 11)
PA12 Polyamide 12 (or Nylon 12)
PA46 Polyamide 46 (or Nylon 46)

PA-I Pseudomonas Aeruginosa Lectin I [Biochemistry]

PA-II Pseudomonas Aeruginosa Lectin II [Biochemistry]

P4A Pyridine-4-Aldehyde [also P4a]

P&A Personnel and Administration [of US Department of the Army]

P&A Planning and Analysis

P&A Plugged and Abandoned [also P and A, p and a, or p&a] [Oil and Gas Industry]

P/A Power of Attorney [also PA]

P/A Mean Pressure under Indenter (i.e., Applied Indentation Load per Contact Area) [Mechanical Testing]

P/A Problem Analysis

P-A Peak-to-Average (Ratio) [also P/A]

P-A Phenylalanine [also P-A] [Biochemistry]

P(A$_B$) Probability of A Atoms Occupying Surface B [Symbol]

Pa Pascal [Unit]

Pa Paste [also pa]

Pa Protactinium [Symbol]

Pa Pennsylvania [US]

Pa-225 Protactinium-225 [also ^{225}Pa, or Pa^{225}]

Pa-226 Protactinium-226 [also ^{226}Pa, or Pa^{226}]

Pa-227 Protactinium-227 [also ^{227}Pa, or Pa^{227}]

Pa-228 Protactinium-228 [also ^{228}Pa, or Pa^{228}]

Pa-229 Protactinium-229 [also ^{229}Pa, or Pa^{229}]

Pa-230 Protactinium-230 [also ^{230}Pa, or Pa^{230}]

Pa-231 Protactinium-231 [also ^{231}Pa, or Pa^{231}]

Pa-232 Protactinium-232 [also ^{232}Pa, or Pa^{232}]

Pa-233 Protactinium-233 [also ^{233}Pa, or Pa^{233}]

Pa-234 Protactinium-234 [also ^{234}Pa, Pa^{234}, UX$_2$, or UZ]

Pa-235 Protactinium-235 [also ^{235}Pa, or Pa^{235}]

Pa^{-1} One over Pascal [also 1/Pa]

pA picoampere [Unit]

pA negative logarithm of the acid dissociation constant (for weak acids)

pa particular average [also PA]

pa *(per annum)* – each year

pa *(pro analysi)* – by analysis

.pa Panama [Country Code/Domain Name]

PAA Pad Area Array

PAA Pan American World Airways System

PAA Paper Agents Association [UK]

PAA Paraazoxyanisole

PAA P-Azoxyanisole

PAA Phase Antenna Array

PAA Phenylacetic Acid

PAA Phosphoric Acid Aluminum (Treatment) [Metallurgy]

PAA Phosphoric Acid Anodizing [Metallurgy]

PAA Photon Activation Analysis

PAA Phthalic Acid Anhydride

PAA Physiological Amino Acid

PAA Picramic Acid

PAA Polyacetic Acid

PAA Polyacrylacetylene

PAA Polyacrylamide

PAA Polyacrylic Acid

PAA Polyarylacetylene

PAA Polyamic Acid

PAA Population Association of America [US]

PAA Potato Association of America [US]

PAA Primary Aromatic Amine(s)

PAA Public Affairs Office(r)

PAA Pulse-Amplitude Analyzer

PAAAC Pan-American Agricultural Aviation Center

PAABA P-Acetamidobenzoic Acid [also pAaBA]

PAABS Pan-American Association of Biochemical Societies [US]

PA-ABS Polyamide/Acrylonitrile-Butadiene Styrene (Blend) [also PA/ABS]

PAAC Program Analysis Adaptable Control

PAA-co-AA Poly(acrylamide-co-Acrylic Acid)

PAA-co-MA Poly(acrylic Acid-co-Maleic Acid)

PAAI Poster Advertising Association of Ireland

PAAM Polyacrylamide

PAA-PS Polyacrylic Acid–Polysulfone

PAAS Pakistan Association for the Advancement of Science

Pa/at% Pascal(s) per atomic percent [Unit]

PAB Aminopentyl Benzimidazole

PAB P-Aminobenzoic Acid

PAB P-Aminobenzyl

PAB Panamanian Balboa [Currency]

PAB Personal Address Book

PAB Polymer Alloys and Blends

PAB Primary Application Block

PAB Public Affairs Bureau [of Environment Australia]

PAB Publications Activities Board [of Institute of Electrical and Electronics Engineers, US]

PAB Pulse Adsorption Bed

P(A|B) Conditional Probability of Events A and B [also $\mathcal{P}$(A|B)]

P($a_p \beta_p$) Preisach Distribution Function [*Note*: α and β Represent Characteristic Dipole Parameters] [Solid-State Physics]

PABA P-Aminobenzoic Acid; Para-Aminobenzoic Acid [also pABA]

PABA Sodium Sodium-P-Aminobenzoate [also PABA sodium]

PABD 3,5-Dimethyl-N-[3,5-Dimethyl-1H-Pyrazol-1-ylmethyl]-N-Phenyl 1H-Pyrazole-1-Methanamine [also pabd]

PABLA Problem Analysis By Logical Approach

PABN Polyaminobenzonitrile

PABS Pan-American Biodeterioration Society [US]

PABs Polymer Alloys and Blends

PABST Primary Adhesively Bonded Structures Technology

PABX Private Area Exchange

PABX Private Automatic Branch Exchange [also pabx]

PAC Package Assembly Circuit

PAC Packaging Association of Canada

PAC Pan American College [Edinburg, Texas, US]

PAC Parachute and Cable (Rocket)

PAC Particle Accelerator Conference [of American Physical Society, Division of Physics of Beams]

PAC Pedagogic Automatic Computer

PAC Peripheral Autonomous Control

PAC Permanent Accomodation Complex

PAC Permanent Agricultural Committee [of International Labour Organization]

PAC Personal Analog Computer

PAC Perturbed Angular Correlation (of Gamma Radiation)

PAC Pharmaceutical Advertising Council [US]

PAC Physician's Assistant, Certified

PAC Pilotless Aircraft

PAC Plasma-Arc Cutting

PAC Political Action Committee

PAC Polled Access Circuit

PAC Pollution Abatement Control

PAC Polyalkene Carbonate

PAC Poly-Aluminum Chloride

PAC Polycyclic Aromatic Hydrocarbon

PAC Poly(dimethyldiallylammonium Chloride)

PAC Portable Air Compressor

PAC Powder Alloy Corporation [US]

PAC Preauthorized Checking

PAC President's Advisory Council [US]

PAC Pressure Alpha Center

PAC Probe Aerodynamic Center

PAC Problem Action Center

PAC Process and Control

PAC Process Automation Center [of Texas Instruments, US]

PAC Process Automation Computer

PAC Professional Activities Committee [of Institute of Electrical and Electronics Engineers, US]

PAC Program Address Counter

PAC Program Assembly Card

PAC Program Authorized Credentials

PAC Programmable Automatic Comparator

PAC Public Access Catalog [US]

PAC Public Archives Canada

PAC Public Archives Commission

Pac Pacific

Pac Phenylaminocarbonyl

PACA Picture Agency Council of America [US]

PACA Polyamide Carboxylic Acid

PACAF Pacific Air Force [at Hickam Air Force Base, Oahu, Hawaii, US]

PACAN Program for Accreditation of Laboratories in Canada

PACAP Adenylate Cyclase Activating Polypeptide [Biochemistry]

PACAS Personnel Access Control Accountability System

PACB Pan-American Coffee Bureau

PACBA P-Acetylbenzoic Acid [also pAcBA]

PACC Pesiticides and Agricultural Chemicals Committee

PACC Problem Action Control Center

PACC Product Administration and Contract Control

PACCT PERT (Production Evaluation and Review Technique) and Cost Correlation Technique

PACD Plan of Action to Combat Desertification

PACE Pacific Agricultural Cooperative for Export [US]

PACE Packaged CRAM (Card Random Access Memory) Executive

PACE Passive Attitude Control Experiment

PACE Petroleum Association for the Conservation of the Environment [Canada]

PACE Philippine Association of Civil Engineers

PACE Physics and Chemistry Experiment

PACE Planetary Association for Clean Energy [Canada]

PACE Plant Acquisition and Construction Equipment

PACE Plant and Capital Equipment

PACE Pollution and Abatement Control Expenditures

PACE Precision Analog Computing Equipment

PACE Preflight Acceptance Checkout Equipment

PACE Prelaunch Automatic Checkout Equipment

PACE Priority Access Control Enabled

PACE Process Automation and Control Executive

PACE Professional Activities Committee for Engineers [of Institute of Electrical and Electronics Engineers, US]

PACE Professional Association of Consulting Engineers

PACE Programmed Automatic Communications Equipment

PACE Programming Analysis Consulting Education

PACE Projects to Advance Creativity in Education

PACE Protection Against Corrosion/Erosion

PACE Pulsed Analog-to-Digital Converter and Encoder

PACED Program for Advanced Concepts in Electronic Design

PACE-LV Preflight Acceptance Checkout Equipment for Launch Vehicle [also PACE/LV]

PACER Portfolio Analysis, Control, Evaluation and Reporting

PACER Process Assembly Case Evaluator Routine

PACER Program-Assisted Console Evaluation and Review

PACER Programmed Automatic Communications Equipment

PACER Program of Active Cooling Effects and Requirements

PACES Particle Accelerator Control Expert System

PACE-SC Preflight Acceptance Checkout Equipment for Spacecraft [also PACE-S/C]

PACF Periodic Autocorrelation Function

PACFORNET Pacific Coast Forest Research (Information) Network

PACGSR Pan-American Center for Geographical Studies and Research [Ecuador]

PACHG Program Advisory Committee on the Human Genome [of US Department of Health and Human Services]

PACIA Particle Counting Immunoassay

PACIA Particle Agglutination Counter Immunoassay

Pacif Pacific

Pac J Math Pacific Journal of Mathematics [of University of California Press, Berkeley, US]

Packag Packaging [also packag]

Packag Packaging [Journal published in the UK and the US]

Packag (Denver) Packaging (Denver) [Journal published in Denver, Colorado, US]

Packag (UK) Packaging (UK) [Published in the UK]

Packag News (Australia) Packaging News (Australia)

Packag Newsl Packaging Newsletter [US]

Packag News (UK) Packaging News (UK)

Packag Technol Sci Packaging Technology and Science [Journal published in the UK]

Packag Week Packaging Week [Published in the UK]

Packetnet Packet-Switched (Data) Network

Packg Packaging [also packg]

PACLR Picatinny Arsenal Chemical Laboratory Report [Dover, New Jersey, US]

PACM P-Aminocyclohexyl Methane

PACM Pulse Amplitude Code Modulation

Pac O Pacific Ocean

PACOL Paraffin Conversion, Linear (Process)

PACOPA 2-[{(Phenylamino)carbonyl}oxy]propanoic Acid

PACOR Passive Correlation and Ranging

PACOSS Passive and Active Control Space Structures

Pac Pro Internationale Messe Packmittelproduktion Maschinen–Materialien–Verfahren [International Fair for Packaging Machinery, Materials and Techniques, Germany]

PAC(R) Parachute and Cable (Rocket)

PACRA Pottery and Ceramics Research Association [New Zealand]

PacRim International Meeting of Pacific Rim Ceramic Societies

PACS Pacific Area Communications System

PACS Perturbed-Angular Correlation Spectroscopy

PACS Physics and Astronomy Classification Scheme (Index) [of American Institute of Physics, US]

PACS Picture Archiving and Communication System

PACS Pointing and Attitude Control System

PACS Project Analysis and Control System

PACS Public Access Computer System

PACSC Pesticides and Agricultural Chemicals Standing Committee

PACS-L Public Access Computer Systems List [Internet]

PACT Pay Actual Computer Time

PACT Plasma Arc Centrifugal Treatment (System)

PACT Powdered Activated Carbon Treatment

PACT Production Analysis Control Technique

PACT Program for Advancement of Commercial Technology

PACT Program for Automatic Coding Techniques

PACT Programmable Asynchronous Clustered Teleprocessing

PACT Programmed Analysis Computer Transfer

PACT Programmed Automatic Circuit Tester

PACTG Pediatric AIDS Clinical Trials Group

PACV Patrol Air Cushion Vehicle

PACVD Plasma-Activated Chemical Vapor Deposition; Plasma-Assisted Chemical Vapor Deposition [also PA-CVD]

PACX Private Automatic Computer Exchange

PAD (Data) Packet Assembler/Disassembler; (Data) Packet Assembly/Disassembly

PAD Pattern Analysis Diffractometer [X-Ray Diffraction]

PAD Peripheral Artery Disease

PAD Percutaneous Device

PAD Perturbed Angular Distribution

PAD Photoemission Angular Distribution

PAD Pitless Adapter Division [of Water Systems Council, US]

PAD Pixel Access Definition

PAD Plastics Analysis Division [of Society of Plastics Engineers, US]

PAD Polar Angle Distribution

PAD Poly-Aperture Device

PAD Positioning Arm Disk

PAD Postacceleration Detector

PAD Post-Activation Diffusion

PAD Post-Alloy Diffused (Transistor)

PAD Power Amplifier Device

PAD Preliminary Advisory Data

PAD Program Approval Document

PAD Propellant-Acquisition Device

PAD Propellant-Actuated Device

PAD Pulsed Amperometric Detector

PADA 2-(5-Bromo-2-Pyridylazo)-5-Diethylaminophenol

PADA Private Automobile Dealers Association [Canada]

PADA Pyridine-2-Azo-p-Dimethylaniline

PADAP 2-(2-Pyridylazo)-5-Diethylaminophenol

PADAR Passive Detection and Ranging [also Padar, or padar]

PADD Petroleum Administration for Defense Districts

PADE Pad Automatic Data Equipment

PADF Pan-American Development Foundation [Washington, DC, US]

PADIA Patrol Diagnosis

PADIRT Platform for Atmospheric Data in Real-Time [of Bedford Institute of Oceanography, Dartmouth, Nova Scotia, Canada]

PADIS Pan-African Documentation and Information System [Ethiopia]

PADL Part and Assembly Description Language

PADL Parts and Design Language

PADL Pilotless Aircraft Development Laboratory [US Navy]

PADLA Programmable Asynchronous Dual Line Adapter

PADLOC Passive-Active Detection and Location [also Padloc, or padloc]

PADLOC Passive Detection and Location of Countermeasures [also Padloc, or padloc]

PADLOCC Passive-Active Detection and Location Countermeasures

PADM Portable Audio Data Modem

PADRE Patient Automatic Data Recording Equipment

PADRE Portable Automatic Data Recording Equipment

PADS Passive-Active Data Simulation

PADS Pen Application Development System

PADS Performance Analysis and Design Synthesis

PADS Performance Analysis Display System

PADS Plasma-Assisted Dry Soldering

PADS Port and Airport Development Strategy

PADS Precision Aerial Display System

PADT Post-Alloy Diffused Transistor

PADU Protected Areas Data Unit [cf World Conservation Monitoring Center]

PAE Photoacoustic Effect

PAE Photo-Induced Absorption Excitation

PAE Plasma Accelerator Engine

PAE Polyarylene Ether

PAE Polyaryl Ether

PAE Post-Accident Environment [Nuclear Engineering]

PAE Power Added Efficiency (Transistor)

PAE Preventive Action Engineer

PAE Problem Assessment Engineer

PAE Products–Activities–End Products (Analysis) [Battelle Memorial Institute, Columbus, Ohio, US]

PAE² Particle Atlas, Electronic Edition [of MicroDataware, US]

p ae *(partes aequales)* – equal parts [Medical Prescriptions]

PAEA Pakistan Atomic Energy Authority

PAEC Pakistan Atomic Energy Commission [Karachi]

PAECI Pan-American Association of Educational Credit Institutions [Colombia]

PA-EPDM Polyamide/Ethylene-Propylene-Diene Monomer (Blend) [also PA/EPDM]

PAEK Polyarylether Ketone

PAEM Program Analysis and Evaluation Model

PAES Positron Annihilation Induced Auger Electron Spectroscopy

PAES Proton-Induced Auger Electron Spectroscopy

PAF Pacific Air Forces [at Hickam Air Force Base, Oahu, Hawaii, US]

PAF Page Address Field

PAF Payload Attach Fitting

PAF Peak Annual Funding

PAF Peripheral Address Field

PAF Picric Acid Formaldehyde

PAF Plastic After-Flow

PAF Platelet-Activating Factor [Biochemistry]

PAF Poly(dimethyldiallylammonium Fluoride)

PAF Pre-Atomized Fuel

PAF Pulsed Accelerated Flow (Spectroscopy)

PAF(C$_{16}$) L-α-Phosphatidylcholine, β-Acetyl-γ-O-Hexadecyl (Platelet-Activating Factor) [Biochemistry]

PAF(C$_{16}$) L-α-Phosphatidylcholine, β-Acetyl-γ-O-Octadecyl (Platelet-Activating Factor) [Biochemistry]

PAFA Permissible Additives to Food Act [US]

PAFB Patrick Air Force Base [Florida, US]

PAFC Phase-Locked Automatic Frequency Control

PAFC Philippine-American Financial Commission

PAFC Pairwise Additive Function Counterpoise (Method)

PAFC Phosphoric Acid Fuel Cell

PAFLU Philippine Association of Free Labour Unions

PA-FTIR Plasma-Assisted Fourier Transform Infrared (Spectroscopy)

PAG Photo Acid Generator

PAG Plant Advisory Group [of International Union for Conservation of Nature and Natural Resources and World Wildlife Fund]

PAG Plasma-Arc-Based Gasifier

PAG Polyalkylene Glycol

PAG Prealbumin Globulin [Biochemistry]

PAG Protein A Colloidal Gold [Biochemistry]

PAG Protein Advisory Group [of United Nations]

pAg Protein A Gold [Biochemistry]

PAGB Prior Austenite Grain Boundary [Materials Science]

PAGB Proprietary Association of Great Britain [UK]

PAGC Power Amplifier Gain Control [also pagc]

PAGE PERT (Production Evaluation and Review Technique) Automated Graphical Extension

PAGE Polyacrylamide Gel Electrophoresis

PAGEOS Passive Geodetic Earth Orbiting Satellite [of NASA]

PAGEOPH Pure and Applied Geophysics [Journal published in Germany]

PAGES Past Global Changes

PAGES Program Affinity Grouping and Evaluation System

PAGICEP Petroleum and Gas Industry Communications Emergency Plan [US]

PAGIF Polyacrylamide Isoelectric Focusing

PAGS Prior Austenite Grain Size [Metallography]

PAGSMBE Plasma-Assisted, Gas-Source Molecular Beam Epitaxy

PAGT Provincial Association of Geography Teachers [Canada]

PAH P-Aminohippurate; Para-Aminohippurate [Biochemistry]

PAH Pan-American Highway

PAH Payload Accommodations Handbook [of NASA]

PAH Polyaromatic Hydrocarbon(s); Polycyclic Aromatic Hydrocarbon(s); Polynuclear Aromatic Hydrocarbon(s)

PAHA P-Aminohippuric Acid

PAHC Pan-American Highway Congress [US]

PAHO Pan-American Health Organization [of Organization of American States]

PAHP Plasma Assisted Hot Pressing [Metallurgy]

PAHR Post-Accident Heat Removal [Nuclear Engineering]

PA-HT Pollock and Argon–High Tension [Metallurgy]

pA/$\sqrt{\text{Hz}}$ picoampere per square root of hertz [also pA Hz^{-1}]

PAI Passive Acquired Immunity

PAI Personnel Accrediation Institute [US]

PAI Plasminogen Activator Inhibitor [Biochemistry]

PAI Polyamideimide

PAI Poly(dimethyldiallylammonium Iodide)

PAI Process Analytical Instrumentation

PAI Programmer Appraisal Instrument

PAIC Public Address Intercom System

PAID Pan-African Institute for Development [Cameroon]

PAID Personnel and Accounting Integrated Data System

PAIGH Pan-American Institute of Geography and History [of Organization of American States]

PAIH Public-Access Internet Host

PAIMEG Pan-American Institute of Mining Engineering and Geology

Paint Painting [also paint]

Paint Resin Paint and Resin [Journal published in the US]

PAIP Programme d'Accélération des Investissements Privés [Private Investment Expedition Program, Canada]

PAIP Public Affairs and Information Program [of Atomic Industries Forum]

PAIR Performance and Integration Retrofit

PAIR Precision Approach Interferometer Radar

PAIRS Pushbroom Airborne Infrared Remote Sensor

PAIS Public-Access Internet Site

PAIS Public Affairs Information Service [US]

PAIS FLI Public Affairs Information Service–Foreign Language Index [US]

PAIT Program for Advancement of Industrial Technology [Canada]

PAJ Plasma-Assisted Joining

PAK Personalausbildungskommission [Personnel Training Commission; of Verband Schweizerischer Kunststoff– Press– und Spritzwerke, Switzerland]

PAK Polyester Alkyd

.PAK Packed (Compressed) [File Name Extension]

Pak Pakistan(i)

Pak *(Panzerabwehrkanone)* – German for "Antitank Gun" [also PAK]

PAKEX Packaging Exhibition

Pak J Sci Pakistan Journal of Science [of Pakistan Association for the Advancement of Science]

Pak J Sci Res Pakistan Journal of Scientific Research [of Pakistan Association for the Advancement of Science]

PAKNET Pakistan Information Network

PAKOE Panhellenic Center for Environmental Studies

Pak Re Pakistan Rupee [Currency]

PAKTUS PRC (Inc.)'s Adaptive Knowledge-Based Text Understanding System

PAL PAM (Programmable Algorithm Machine) Assembly Language

PAL Paradox Application Language

PAL Partition Allocation Utility

PAL Pedagogic Algorithmic Language

PAL Performance and Reliability

PAL Permanent Artificial Lighting

PAL Permissive Action Link

PAL Phase Alternation Line (Television System)

PAL Phase Attenuation by Line

PAL Philippine Air Lines

PAL Phillips Air Liquefier

PAL Polyaniline

PAL Positron Annihilation Lifetime (Spectroscopy)

PAL Process Assembly Language

PAL Production and Application of Light [Institute of Electrical and Electronic Engineers–Industry Applications Society Committee]

PAL Programmable Array Logic

PAL Programmed Application Library

PAL Programmer Assistance and Liaison

PAL Programming Assembly Language

PAL Prototype Application Loop

PAL Psychoacoustic Laboratory

PAL Pyridoxalphosphate

.PAL Palette [File Name Extension] [also .pal]

Pal Palestine; Palestinian

PALACE Profiling Autonomous Lagrangian Circulation Explorer(float)

Palaeogeogr Palaeoclimatol Palaeoecol Palaeogeography, Palaeoclimatology, Palaeoecology [International Journal]

Palaeontol Palaeontology [Journal published by the Natural History Museum, London, UK]

PAL APA Indonesian Communications Satellite

PALASM Programmable Array Logic Assembler

PALC Plasma-Addressed Liquid Crystal (Display)

PAL-D Phase Alternation Line–Delay [also PAL D]

Paleobiol Paleobiological; Paleobiologist; Paleobiology [International Journal]

Paleobiol Paleobiology [International Journal]

Paleontol Paleontological; Paleontologist; Paleontology [also Paleon, Paleont, paleon, paleont or paleontol]

Paleontol J Paleontological Journal [Published in Russia]

PALERMO Planning and Learning for Resources-Oriented Data Management and Organization [of University of Ottawa, Canada]

p-AlGaAs P-Type Aluminum Gallium Arsenide (Semiconductor)

p⁺-AlGaAs P-Plus-Type Aluminum Gallium Arsenide (Semiconductor)

PALINET Pennsylvania Area Library Network [US]

PALLAS Phased Automation of the Hellenic ATC (Air Traffic Control) System

palliat palliative

Palliat Med Palliative Medicine (Journal) [US]

PALM Philips Automated Laboratory Management

PALM Precision Altitude and Landing Monitor

PALS Photo Area and Location System

PALS Positron Annihilation Lifetime Spectroscopy

PALS Precision Approach and Landing System

PALS Principles of the Alphabet Literacy System

PAL-S Phase Alternation Line–Simple [also PAL S]

PA-LT Pollock and Argon–Low Tension [Metallurgy]

Palynol Palynological; Palynologist; Palynology

PAM Payload Assist Module [of NASA]

PAM Peptidyl α-Amidating Monoxygenase [Biochemistry]

PAM Periodic Anderson Model (for Heavy Fermion Systems) [Solid-State Physics]

PAM Peripheral Adapter Module

PAM Phenylacetamidomethyl

PAM Photoacoustic Microscopy

PAM Plan-Applier Mechanism

PAM Plasma-Arc Machining

PAM Plasma-Arc Melting

PAM Plasma Assisted (or Aided) Machining

PAM Pole Amplitude Modulation

PAM Polyazomethine

PAM Portable Activity Monitor

PAM Portable Alpha Monitor

PAM Portable Automated Mesonet

PAM Post-Accident Monitoring [Nuclear Engineering]

PAM Pozzolan Aggregate Mixture [Civil Engineering]

PAM Primary Access Method

PAM Procaine Benzylpenicillin [Pharmacology]

PAM Process Automatic Monitor

PAM Process Automation Monitor

PAM Programmable Algorithmic Machine

PAM Pulse Amplifier Modulation

PAM Pulse-Amplitude Modulation

PAM 2-Pyridine Aldoxime Methiodide [also 2-PAM]

PAM Pyridoxamine Phosphate

L-PAM L-Phenylalanine Mustard [Biochemistry]

2-PAM 2-Formyl-1-Methyl-Pyridinium Chloride Oxime; Pralidoxime [Biochemistry]

2-PAM 2-Pyridine-2-Aldoxime-N-Methyl (Iodide) [Biochemistry]

Pam Pamphlet [also pam]

Pa√m Pascal(s) square root of meter [also Pa·m$^{1/2}$]

PAMA Pan-American Medical Association

PAMA Professional Aviation Maintenance Association [US]

PAMA Pulse-Address Multiple Access

PAM-A Payload Assist Module, Atlas-Centaur Class Spacecraft

PAMAC Parts and Materials Accountability Control

PAMBA P-Aminomethylbenzoic Acid

PAMC Panhandle A&M (Agricultural and Mechanical) College [Goodwell, Oklahoma, US]

PAMC Provisional Acceptable Means of Compliance

PAMCA Plasma-Arc-Melting/Centrifugal Atomization [Metallurgy]

2-PAM Chloride 2-Pyridine-2-Aldoxime-N-Methyl Chloride

PAMCO Pittsburgh and Midway Coal Mining Company (Process)

PAMD Periodic Acid Mixed Diamine

PAM/D Process Automation Monitor/Disk

PAM-D Payload Assist Module, Delta Class Spacecraft

PAMEE Philippine Association of Mechanical and Electrical Engineers

PAMELA Plan-Applier Mechanism for English Language Analysis

PAMERAR Program for Assessment and Mitigation of Earthquake Risk in the Arab Region [of UNESCO]

PAMETRADA Parsons and Marine Engineering Turbine Research and Development Association [UK]

PAM-FM Pulse Amplitude Modulation–Frequency Modulation

PAMI Prairie Agricultural Machinery Institute [Saskatchewan, Canada]

PAML Program Authorized Materials List

PAMM (British Ceramic) Plant and Machinery Manufacturers Association [UK]

PAMOCVD Plasma-Assisted Metal-Organic Chemical Vapor Deposition

PAMP Plasma-Assisted Materials Processing

PAMPER Practical Application of Midpoints for Exponential Regression

PAMPS Poly(acrylamido-Methylpropane Sulfonic Acid)

PAMS Pad Abort Measuring System

PAMS Parallel Application Management System

PAMS Permanent Air Monitoring System

PAMS Photochemical Air Monitoring System

PAMS Portable AFT (Adiabatic Flame Temperature) Mass Spectrometer

PAMS Portable Automated Mesonet Stations

PAMS Precision Abrasion Mass Spectrometry

PAMS Preselected Alternate Master/Slave

PAMS Proceedings of the American Mathematical Society

Pa·m³/s Pascal cubic meter(s) per second [also Pa·m³/sec, or Pa·m³ s^{-1}]

Pa·m³/sec Pascal cubic meter(s) per second [also Pa·m³/s, or Pa·m³ s^{-1}]

PAMSA P-Aminoethylsalicylic Acid

PAN International Aircraft Distress Signal

PAN Peroxyacetyl Nitrate

PAN Pesticides Action Network [Belgium]

PAN Phenylazonaphthol

PAN N-Phenyl-α-Naphthyl Amine

PAN Phthalic Anhydride

PAN Polled Access Network

PAN Polskiej Akademii Nauk [Polish Academy of Sciences, Warsaw]

PAN Polyacrylonitrile

PAN Positional Alcohol Nystagmus [Medicine]

PAN 1-(2-Pyridylazo)-2-Naphthol

PAn Polyaniline

Pan Panama(nian)

Pan Panorama [also pan]

Pan Panchromic [also pan]

Pan Panel [also pan]

PANA Pan-African News Agency

PANACEA Package for Analysis of Networks of Asynchronous Computers with Extended Asymptotics

PANAFTEL Pan-African Telecommunications Network

PANAGRA Pan-American-Grace Airways

PANAIR Pan-American (World) Airways

PANAM Pan-American (World) Airways

PANAR Panoramic Radar [also Panar, or panar]

PANASH Paleoclimates of the Northern and Southern Hemisphere

PAN-CF Polyacrylonitrile-Based Carbon Fiber

PAN-co-BD Poly(acrylonitrile-co-Butadiene)

PAN-co-B-co-AA Poly(acrylonitrile-co-Butadiene-co-Acrylic Acid)

PAN-co-B-co-S Poly(acrylonitrile-co-Butadiene-co-Styrene)

PAN-CSA Polyaniline Doped with Camphor Sulfonic Acid

PANDA Panoramic Adapter [also Panda, or panda]

PANDA Positive And Negative Ions Deposition Apparatus

PANDA Prestel Advanced Network Design Architecture

PANDA Programmers Analysis and Development Aid

P and A Plugged and Abanoned [also p and a, P&A, or p&a] [Oil and Gas Industry]

P and L Profit and Loss (Statement) [also P&L, or P/L]

PANDORA Prototyping a Navigation Database of Road Network Attributes

P and S Plugged and Suspended [also p and s, P&S, or p&s] [Oil and Gas Industry]

PANE Performance Analysis of Electrical Networks [Computer Program]

PANG Polskiej Akademii Nauk w Gdansku [Polish Academy of Sciences at Gdansk, Poland]

PANJU Pan-African Union of Journalists

PANI Polyaniline [also Pani]

PANIC Particles And Nuclei–International Conference [Physics]

PANIC Potential and Needs, Investments and Capabilities

PANI-EB Emeraldine Base Polyaniline

PAN-PVC Polyacrylonitrile–Polyvinyl Chloride Copolymer

PANS Potentially Attractive New Services

PANS Procedures for Air Navigation Services

PANS Public Archives of Nova Scotia [Canada]

PANSDOC Pakistan National Scientific and Technical Documentation Center

PANSMET Procedures for Air Navigation Services–Meteorology

PANS/OPS Procedures for Air Navigation Services/Aircraft Operations

PANS/RAC Procedures for Air Navigation Services/Rules of the Air Traffic Services

Pant Pantograph [also pant]

PANTHEON Public Access by New Technology to Highly Elaborate On-Line Networks

PANTS Public Acceptance of New Technologies

PAO Personnel Administration Office

PAO Phenylarsine Oxide

PAO Polyalkaline Oxide

PAO Poly-Alpha-Olefin (Type Lubricant)

PAO Public Affairs Office(r)

PAO Pulsed Avalanche Diode Oscillator

PAP 1,4-Bis(2-Pyridylamino)phthalazine

PAP (Data) Packet-Level Procedure

PAP P-Aminophenylalanine [Biochemistry]

PAP Paraazoxyphenetole

PAP Password Authentication Protocol

PAP Payload Activity Planner

PAP Peroxidase-Anti-Peroxidase [Biochemistry]

PAP PET (Polyethylene Terephthalate) Inner Sheath–Aluminum Intermediate Sheath–PET (Polyethylene Terephthalate) Outer Sheath (Cable) [also pap]

PAP Phase Advance Pulse

PAP 3'-Phosphoadenosine 5'-Phosphate [Biochemistry]

PAP Photon-Assisted Processing

PAP Placental Anticoagulant Protein [Biochemistry]

PAP Plasma-Arc Processing

PAP Plasma-Assisted Pressing

PAP Plasma-Assisted Processing

PAP Pouchou and Pichoir (Model) [Physics]

PAP Prealloyed Powder [Powder Metallurgy]

PAP Preapproved Payment

PAP Preauthorized Payment

PAP Primary Atypical Pneumonia [Medicine]

PAP Printer Access Protocol

PAP Product Assurance Plan

PAP Prospectors Assistance Program [Canada]

PAP Prostatic Acid Phosphatase [Biochemistry]

Pap Papanicolaou (Smear Test) [Medicine]

Pap Paper [also pap]

pAP p-Acetylphenol

PAPA Parallax Aircraft Parking Aid

PAPA Programmer and Probability Analyzer

pAPAO p-Aminophenylarsine Oxide

PA-PAR Polyamide/Polyarylate (Blend) [also PA/PAR]

PA-PC Polyamide/Polycarbonate (Blend) [also PA/PC]

PAPD Periodate Dimethylphenylenediamine

PAPE Photoactive Pigment Electrophotography

PAPERCHEM Paper Chemistry (Database) [of Institute of Paper Chemistry, US] [also PaperChem]

Paper Conservator The Paper Conservator [Publication of the Institute of Paper Conservation, UK]

Paper Conserv News Paper Conservation News [Publication of the Institute of Paper Conservation, UK]

Paper Film Foil Conv Paper, Film and Foil Converter [Journal published in the US]

Paperi Puu Paperi ja Puu [Finnish Journal on Pulp and Paper]

Paper Technol Paper Technology [Journal of the Paper Industries Research Association, US]

Paper Twine J Paper and Twine Journal [US]

Paper Yearb Paper Yearbook [US]

Paper Film Foil Convert Paper, Film and Foil Converter [Journal published in the US]

PAPI Polymethylene Polyphenyl Isocyanate

PAPI Precision Approach Path Indicator

Papier Das Papier [German Publication on Paper and Papermaking]

Papier Carton Cellulose Papier, Carton et Cellulose [French Publication on Paper, Cardboard and Cellulose]

Papier-Kunstst-Verarb Papier- und Kunststoff-Verarbeiter [German Journal on Paper and Plastics Processing]

Papiermacher Der Papiermacher [German Papermaking Journal]

Papir Celuloza Papir a Celuloza [Czech Publication on Paper and Cellulose]

PAPL Plant Air Pollution Laboratory [of US Department of Agriculture at Beltsville, Maryland]

PAPM Pulse Amplitude Phase Modulation

PAP4ME N,N'-Bis(4-Methyl(-2-Pyridyl] 1,4-Phthalazinediamine

pAPMSGF (p-Amidinophenyl)methylsulfonylfluoride [also p-APMSF]

PAPP P-Aminopropiophenone

PAPP Pre-Approved Payment Plan

PA-PPE Polyamide/Polyphenylene Ether (Blend) [also PA/PPE]

PA-PPO Polyamide/Polyphenylene Oxide (Blend) [also PA/PPO]

Pap Puu Paperi ja Puu [Finnish Journal on Pulp and Paper]

PAPR Powered Air-Purifying Respirator

Papreg Paper Impregnated with Thermosetting Plastic(s) [also papreg]

PAPRICAN Pulp and Paper Research Institute of Canada

PAPS Periodic Acid Phenylhydrazine Schiff [Chemistry]

PAPS Periodic Armaments Planning System

PAPS 3'-Phosphoadenosine-5'-Phosphosulfate [Biochemistry]

PAPS Phosphoadenosine Phosphosulfate [Biochemistry]

PAPS Physics Auxiliary Publication Service [of American Institute of Physics, US]

2'-PAPS 2'-Phosphoadenosine-5'-Phosphosulfate [Biochemistry]

PAPTC Paper Tape Controller

PAPU Pan-African Postal Union [Tanzania]

PAPVD Plasma-Assisted Physical Vapor Deposition [also PA-PVD]

PAR Page Address Register

PAR Pakistan Rupee [Currency]

PAR Peak Accelerometer Recorder

PAR Peak-to-Average Ratio

PAR Pennsylvania Advanced Reactor [US]

PAR Performance Analysis and Review

PAR Performance Appraisal Report

PAR Perimeter Acquisition Radar

PAR Personal Animation Recorder

PAR Photosynthetically Active Radiation [Remote Sensing]

PAR Plasma-Arc Reduction [Metallurgy]

PAR Plasma-Arc Remelting [Metallurgy]

PAR Planning Action Request

PAR Platelet Aggregation Reagent [Biochemistry]

PAR Polyarylate

PAR Positron Accumulator Ring [of (US) Argonne National Laboratory's Advanced Photon Source Linear Accelerator]

PAR Precision Approach Radar

PAR Preferential Arrival Route

PAR Princeton Applied Research [of Princeton University, New Jersey, US]

PAR Problem Accountability Record

PAR Problem Action Record

PAR Problem Action Request

PAR Processor Address Register

PAR Product Acceptance Review

PAR Production Automated Riveting

PAR Professional Abstracts Registry

PAR Program Address Register

PAR Program Appraisal and Review

PAR Progressive Aircraft Rework

PAR Project Appropriation Request

PAR Project Authorization Request

PAR Pulse Acquisition Radar

PAR Purchasing Approval Request

PAR 4-(2-Pyridylazo)resorcinol

P/AR Peak-to-Average Ratio

Par Parabola; Parabolic [also par]

Par Paragraph [also par]

Par Parallel(ism) [also par]

Par Parameter; Parametric [also par]

Par Parenthesis [also par]

Par Parity [also par]

PARA Problem Analysis and Recommended Action

PARA Professional Audio-Video Retailers Association [US]

PARA Polyaryl Amide

Para Parabolic [also para]

Para Paragraph [also para]

Para Paraguay(an)

PARACHUTE PPP (Point-to-Point Protocol) and ARA (AppleTalk Remote Access) Central High-Speed User Telecommuting Engine

PARADE Passive-Active Ranging and Determination

paral parallel

Parallel Comput Parallel Computing [Journal published in the Netherlands]

Param Parameter [also param]

ParamCount Parameter Count [Pascal Function]

PARAMI Parsons Active Ring Around Miss Indicator

PARAMP Parametric Amplifier [also Paramp or paramp]

ParamStr Parameter String [Pascal Function]

PARASEV Paraglider Search Vehicle

PARASYN Parametric Synthesis

PARC Palo Alto Research Center [of Xerox Corporation, California, US]

PARC Progressive Aircraft Reconditioning Cycle

PARC Public Access Resource Center

PARC Public Archives Records Center [Ottawa, Canada]

PARCA Pan-American Railway Congress Association

PARCC Precision, Accuracy, Representativeness, Completeness, Comparability

Parch Parchment [also parch]

PARCOM Paris Commission (for Marine Pollution Control) [UK]

PARCOS (Tropical Workshop on) Particle Physics and Cosmology [San Juan, Puerto Rico, US]

PARCS Perimeter Acquisition Radar Characterization System

PARCS Pesticide Analysis Retrieval and Control System [of US Environmental Protection Agency]

PA-RISC Precision Architecture-Reduced Instruction Set Computing

PARD Pakistan Academy for Rural Development

PARD Partnership Agreement on Rural Development [Canadian Federal/Provincial Program]

PARD Parts Application Reliability Data

PARD Periodic and Random Deviation(s)

PARD Precision Annotated Retrieval Display System

PARDOP Passive Ranging Doppler

PARE Physical Ability Requirement Evaluation

Paren Parenthesis [also paren]

PARICOM Paris Commission (for Marine Pollution Control) [UK]

PARIF Program for Automation Retrieval Improvement by Feedback [of EURATOM]

PARIS Pulse Analysis-Recording Information System

PARIS Portable Automated Remote Inspection System

PARISCOS Paris Cosmology Colloquium [France]

PARL Palo Alto Research Laboratory [of Xerox Corporation in California, US]

PARL Prince Albert Radar Laboratory [of Communications Research Center, Canada]

Parl Parliament(ary) [also parl]

parl parallel [also parl]

PARL-PR Parallel Pair [also Parl-Pr, or parl-pr]

PARM Program Analysis for Resource Management

Parm Parameter [also parm]

PARMA Public Agency Risk Managers Association [US]

Par Num Paragraph Number

PAROS Passive Ranging of Submarines

PAR/PA Polyarylate/Polyamide (Blend) [also PAR/PA]

PARPES Polarization-Dependent Angle-Resolved Photoemission Spectra

PARR Pakistan Atomic Research Reactor [Islamabad]

PARR Post-Accident Radioactivity Removal [Nuclear Engineering]

PARR Procurement Authorization and Receiving Report

PARR-1 Pakistan Atomic Research Reactor 1 [Islamabad]

PARR-1 NRF Pakistan Atomic Research Reactor-1–Nuclear Research Facility [Islamabad]

PARS Photoacoustic Raman Spectroscopy

PARS Portable Analyzer for Residual Stresses

PARS Pressure-Assisted Reaction Sintering (Technique) [Metallurgy]

PARS Private Advanced Radio Service

PARS Property Accountability Record System

Pars Paragraphs [also pars]

PARSAC Particle Size Analog Computer

PARSAVAL Pattern Recognition System Application Evaluation [also Parsaval]

parsec parallax-second [Astronomical Unit Equal to 3.258 Light Years, or 3.08572×10^{13} Kilometers]

PARSEV Paraglider Search Vehicle

Parsyn Parametric Synthesis [also parsyn]

PART Parts Allocation Requirement Technique

Part Particle [also part]

part particular(ly)

PARTAC Precision Askania Range Target Acquisition

Part Accel Particle Accelerators [Journal published in the UK]

Part Charact Particle Characterization [Journal published in Germany]

PARTEI Purchasing Agents of the Radio, Television and Electronics Industry

PARTEQ Partners in Technology at Queen's [also Parteq] [of Queen's University, Kingston, Ontario, Canada]

Partic Phys Particle Physicist; Particle Physics

Partn Partition [also partn]

PARTNER Proof of Analog Results Through a Numerical Equivalent Routine

Part Part Syst Charact Particle and Particle Systems Characterization [Journal published in the US and Germany]

Part Phys Particle Physicist; Particle Physics

Part Phys On-Line Rev Particle Physics On-line Review [Joint Publication of Eagle Intermedia Publishing Ltd., UK and the University of Oldenburg, Germany]

Part Sci Technol Particulate Science and Technology [Journal of the Fine Particle Society, US]

Part World Particle World [Journal published in the UK]

PARUPS Polarization-Dependent Angle-Resolved Ultraviolet Photoemission Spectroscopy

Parv Paravane

parv *(parva)* – small [Medical Prescription]

parvo parvovirus

PAS Pakistan Academy of Sciences [Islamabad, Pakistan]

PAS P-Aminosalicylic Acid; Para-Aminosalicylic Acid

PAS Particle Accelerator School [of Fermilab, Batavia, Illinois, US]

PAS Patent Applicant Service [of International Patent Documentation Center, Austria]

PAS Payload Accommodation Studies

PAS Perigee/Apogee Impulse System

PAS Periodic Acid-Schiff [Chemistry]

PAS Perthshire Agricultural Society [UK]

PAS Phase Address System

PAS Phenylaminosalicylate

PAS Philadelphia Astronautical Society [US]

PAS Phosphorus–Arsenic–Strychnine (Tonic) [Pharmacology]

PAS Photoacoustic Spectroscopy

PAS Physics Academic Software [of American Institute of Physics, US]

PAS Plasma Activated (or Assisted) Sintering

PAS Plasma Arc Spraying

PAS Polish Academy of Sciences [Warsaw, Poland]

PAS Polyacrylsulfone

PAS Polyarylsulfone

PAS Poly(p-Amino Styrene)

PAS Positron Annihilation Spectroscopy

PAS Pressure-Assisted Sintering [Metallurgy]
PAS Primary Alert System
PAS Primary Ascent System
PAS Principal Axis System
PAS Privacy Act Statement
PAS Process Analysis Services
PAS Professor of Air Science
PAS Program Activity Structure
PAS Program Address Storage
PAS Protocol Analysis System
PAS Public Address System
PAS Pulse-Assisted Sintering [Metallurgy]
PAS Pulsed Atom Site
PAS Pyrotechnic Arming Switch
.PAS PASCAL Source Code [File Name Extension]
PAs Phosphorus Arsenide
Pa·s Pascal second [also Pa-s]
$Pa^{-1} \cdot s^{-1}$ One per pascal-second [also 1/Pa·s]
PASA P-Aminosalicylic Acid; Para-Aminosalicylic Acid
PASA Poly(amide Sulfonamide)
PASB Pan-American Sanitary Bureau
PASC Pan-American Standards Committee
PASC Pacific Area Standards Congress [US]
PASC Planning Advisory Subcommittee
PASC Portable Applications Standards Committee
PASCA Positron Annihilation Spectroscopy for Chemical Analysis
PASCAL Philips Automatic Sequence Calculator
PASCOS (International Symposium on) Particles, Strings and Cosmology
PASE Power-Assisted Storage Equipment
PASEM Particle Analysis Scanning Electron Microscopy
PASEM Program of Assistance to Solar Equipment Manufacturers
PASF Photographic Art and Science Foundation [US]
PASH Polycyclic Aromatic Sulfur Heterocycle(s)
PA/SI Preliminary Assessment/Site Inspection
p-a-Si:H P-Type Hydrogenated Amorphous Silicon
p-a-SiC:H P-Type Hydrogenated Amorphous Silicon Carbide
p-a-SiO:H P-Type Hydrogenated Amorphous Silicon Monoxide
PASL Physical Activity Sciences Laboratory [Canada]
PASLA Programmable Asynchronous Line Adapter
PASLIB Pakistan Association of Special Libraries
PA-SM Periodic Acid-Silver Methenamine
Pa·s/m Pascal-second(s) per meter [also Pa·s m^{-1}]
Pa·s/m² Pascal-second(s) per square meter [also Pa·s m^{-2}]
Pa·s/m³ Pascal-second(s) per cubic meter [also Pa·s m^{-3}]
PASO Pan-American Sanitary Organization
PASOS Paperless Shop-Order System
PASP Photovoltaic Array Space Power
PASS Performance Analysis Software System
PASS Performance Assessment Scientific Support
PASS Photo-Access Security System

PASS Planning and Scheduling System
PASS Planning and Specification Software
PASS Prenotification Analysis Support System
PASS Primary Avionics Software System
PASS Private Automatic Switching System
PASS Procurement and Acquisition Support System
PASS Production Automated Scheduling System
PASS Program Aid Software System
PASS Program Alternative Simulation System
PASS Programmed Access Security System
PASS Purchasing Activities Support System
Pass Passage [also pass]
Pass Passenger [also pass]
pass (passim) – throughout
pass passive
PASSIM President's Advisory Staff on Scientific Information Management [US]
PAS Sodium Sodium-P-Aminosalicylate [also PAS sodium]
PAST Process Accessible Segment Table
PAS&T Particle Accelerator Science and Technology [Society of the Institute of Electrical and Electronics Engineers, US]
past (pastillus) – pastille; paste [Medical Prescriptions]
PAT Parametric Artificial Talker
PAT Paroxysmal Atrial Tachycardia [Medicine]
PAT Passive Angle Tracking
PAT Peripheral Allocation Table
PAT Personalized Array Translator
PAT Petroleum Authority of Thailand
PAT 1-Phenyl-5-Amino Tetrazole
PAT Phosphoramidothioic Acid
PAT Picric Acid Turbidity
PAT Platoon Antitank
P3AT Poly(3-Alkyl Thiophene)
PAT Polyaminotriazole
PAT Polymers for Advanced Technologies [International Journal]
PAT Positron Annihilation Technique
PAT Power Ascension Testing
PAT Prediction Analysis Technique
PAT Printer Action Table
PAT Problem Action Team
PAT Production Acceptance Test
PAT Professional Association of Teachers [UK]
PAT Proficiency Analytical Testing
PAT Program Attitude Test
PAT Programmable Automatic Tester
PAT Programmer Aptitude Test
PAT Propylaminohydroxy Tetrahydronaphthalene
.PAT Patch [File Name Extension]
.PAT Pattern [File Name Extension]
7-PAT 2-N,-Di[2,3(n)-]Propylamino-7-Hydroxy-1,2,3,4,-Tetra-hydronaphthalene
Pat Patent [also pat]

Pat Pattern [also pat]

pat patent(ed)

PATA Pneumatic All-Terrain Amphibian

Pat Appl Patent Application

PATBX Private Automatic Telex Branch Exchange

PATC Professional, Administrative, Technical and Clerical

PATC Professional Air Traffic Controller

PATCA Phase-Lock Automatic Tuned Circuit Adjustment

PATCA Professional and Technical Consultants Association [US]

PATCO Professional Air Traffic Controllers Organization [US]

patd patented [also pat'd]

PATE Programmed Automatic Test Equipment

Patentbl Patentblatt [Official Gazette of the German Patent Office]

PATH Performance Analysis and Test Histories

PATH Physiotherapy Active Treatment towards Health

PATH Port Authority Trans-Hudson Corporation [New York City, US]

Path Pathological; Pathologist; Pathology [also path]

PATHFINDER Pathological (Element) Finder

Pathobiol Pathobiology; Pathobiological; Pathobiology

Pathobiol Pathobiology [International Journal]

Pathol Pathological; Pathologist; Pathology [also pathol]

Pathol Biol Pathological Biology [US Journal]

Pathol Oncol Res Pathological Oncology Research [Journal published in Hungary]

PATI Passive Airborne Time-Difference Intercept

PATLAW Patent Law (Database) [of Bureau of National Affairs, US]

Patman 6-Palmitoyl-2{[(Trimethylammonio)ethyl]-methylamino}naphthalene Chloride

PATMI Power Actuated Tool Manufacturers Institute [US]

PATN Pattern Analysis Package [Software]

PATN Promotional Port Access Telephone Number

Patn Pattern [also patn]

PATO Partial Acceptance and Takeover

Pat Off Patent Office [also PatOff]

Pat Off Rec Patent Office Record [of Public Works and Government Services Canada]

PATOLIS Patent On-Line Information System

PATOS Patentdatenbank [Patent Data Base, Germany]

PATP Preliminary Authority To Proceed

P-32 ATP Phosphorus-32 Adenosine-5'-Triphosphate [also [also ^{32}P-ATP] [Biochemistry]

pat pend patent pending

PA Tpk Pennsylvania Turnpike [also Pa Tpk, US]

PATR Polarized Attenuated Total Reflectance (Technique)

PATRA Packaging and Allied Trades Research Association [now Printing and Packaging Industries Research Association, UK]

PATRIC Pattern Recognition and Information Correlation

PATRIC Pattern Recognition Interpretation and Correlation

PATRICIA Practical Algorithm To Receive Information Coded In Alphanumeric

Patricia Seybold's Off Comput Res Partricia Seybold's Office Computing Report [of Patricia Seybold's Office Computing Group, US]

PATROL Program for Administrative Traffic Reports On-Line

PATS Passive Anti-Theft System

PATS Precision Altimeter Techniques Study

PATS Program for Analysis of Time Series

2PATSC 2-Pyridinaldehyde Thiosemicarbazone

PATSEARCH Patent Search System [UK]

PATSY Programmer's Automatic Testing System

PATT Project for the Analysis of Technology Transfer

Patt Pattern [also patt]

PATTERN Planning Assistance Through Technical Evaluation of Relevance Numbers [of Operations Research Society of America, US]

Pattern Recognit Pattern Recognition [Journal published in the UK]

Pattern Recognit Lett Pattern Recognition Letters [Published in the Netherlands]

PATTI Pneumatic Adhesion Tensile Testing Instrument

PATU Pan-African Telecommunications Union [Zaire]

PATWAS Pilot's Automatic Telephone Weather Answering Service

PATX Private Automatic Telex Exchange

PAU Pan-American Union

PAU Pan American University [Edinburg. Texas, US]

PAU Pattern Articulation Unit

PAU Pilotless Aircraft Unit

PAU Polska Akademja Umiejętności [Polish Academy of Science and Letters]

PAU Power and Alarm Unit

PAU Precision Approach–UNICOM (Universal Integrated Communications)

PAU Probe Aerodynamic Upper

PAu Piscis Australis [Astronomy]

PAUJ Pan-African Union of Journalists

PAUL Parallel-Axis Ultraprecision Lathe

PAV Phase Angle Voltmeter

PAV Position and Velocity

PAV Pressure Actuated Valve

PAV Pressure Aging Vessel

Pav Paving [also pav]

PAVC Pan-American Veterinary Congress

PAVC Plasma Arc Vitreous Ceramic (Process)

PAVD Plasma-Assisted Vapor Deposition

PAVE Philippine Association for Vocational Education

PAVE Position and Velocity Extraction

PAVT Position and Velocity Tracking

PAW Plasma-Arc Welding

PAW Power-All-the-Way [also paw]

PAWA Power and Water Authority [Northern Territory, Australia]

PAWC Pan-African Workers Congress

PAWOS Portable Automatic Weather Observable Station

PAWRS Private Aviation Weather Research Station

PAWS Paging and Area Warning System

PAWS Parts Analysis Web System [of NASA Goddard Space Flight Center, Greenbelt, Maryland, US]

PAWS Phased Array Warning System

PAWS Portable Acoustic Wave Sensor

PAWS Private Aviation Weather Station

PAWS Programmable (or Programmed) Automatic Welding System

PAWWSWR Portuguese Association on Water, Wastewater and Solid Wastes Research

PAX Parallel Architecture Extended

PAX Passengers

PAX Photoemission of Adsorbed Xenon

PAX Physical Address Extension

PAX (Unix) Portable Archive Exchange

PAX Private Area Exchange

PAX Private Automatic Exchange [also pax]

pAxBA p-Acetoxybenzoic Acid

PAYCOM Payload Command (Coordinator)

PAYCOM Payload Commander

PAYDAT Payload Data

PAYE Pay As You Earn (System)

Payll Payroll [also payll]

PAYT Pay-As-You-Throw (Program) [of US Environmental Protection Agency]

Payt Payment [also payt]

PB Lead Borate

PB Page Buffer

PB Paint Brush

PB Paired Bond

PB Pancake-to-Brush (Transition) [Electrical Engineering]

PB Parallel Band

PB Parallel Beam

PB Parity Bit

PB Particle Beam

PB Passband

PB Pathogenic Bacteria [Microbiology]

PB Payback (Period)

PB Peak Broadening [Spectroscopy]

PB Periodic Boundary [Computers]

PB Peripheral Buffer

PB Petabyte [Unit]

PB Phase Boundary [Materials Science]

PB Phenobarbital [Pharmacology]

PB Phonetically Balanced

PB Phosphor Bronze

PB Pilling-Bedworth (Ratio) [also P-B] [Corrosion Science]

PB Pilot Balloon

PB Pilotless Bomber

PB Pine Bluff [Arkansas, US]

PB Pipe Break

PB Piperonyl Butoxide

PB Plackett-Burman (Design) [also P-B]

PB Plain Back (Adhesive)

PB Planning Board

PB Playback

PB Plotboard

PB Plugboard

PB Poisson-Boltzmann (Theory) [Physics]

PB Polar Bond

PB Polyblend

PB Polybutadiene

PB Polybutylene

PB Polymer Blend

PB Positive Big (Error)

PB Potential Barrier [Physics]

PB Power Balance

PB Power Block

PB Pre-Bloom (Application) [Pesticides]

PB Preburner [also P/B]

PB Precipitation Body

PB Projectile Bremsstrahlung

PB Proportional Band

PB Publications Board [of Office of Technical Services, US]

PB Pull Box

PB Pulsed Beam

PB Purple–Blue

PB Pushbutton [also P/B]

P(B_A) Probability of B Atoms Occupying Surface A [Symbol]

P(B_{hf}) Hyperfine Field Distribution (in Solid-State Physics) [Symbol]

P/B Parts per Billion [also ppb]

P/B Peak-to-Background (Ratio) [also P-B]

P/B Phosphorus-to-Boron (Ratio)

P/B Preburner [also PB]

P/B Pushbutton [also PB]

P-B Pilling-Bedworth (Ratio) [also PB] [Corrosion Science]

P-B Plackett-Burman (Design) [also PB]

Pb *(Plumbum)* – Lead [Symbol]

Pb Ptychodiscus brevis [A Marine Dinoflagellate]

Pb^{2+} Divalent Lead Ion [also Pb^{++}] [Symbol]

Pb^{4+} Tetravalent Lead Ion [Symbol]

Pb II Lead II [Symbol]

Pb-198 Lead-198 [also ^{198}Pb, or Pb^{198}]

Pb-199 Lead-199 [also ^{199}Pb, or Pb^{199}]

Pb-200 Lead-200 [also ^{200}Pb, or Pb^{200}]

Pb-201 Lead-201 [also ^{201}Pb, or Pb^{201}]

Pb-202 Lead-202 [also ^{202}Pb, or Pb^{202}]

Pb-203 Lead-203 [also ^{203}Pb, or Pb^{203}]

Pb-204 Lead-204 [also ^{204}Pb, or Pb^{204}]

Pb-206 Lead-206 [also ^{206}Pb, Pb^{206}, or RaG]

Pb-207 Lead-207 [also ^{207}Pb, Pb^{207}, or Pb]

Pb-208 Lead-208 [also ^{208}Pb, Pb^{208}, or ThD]

Pb-209 Lead-209 [also ^{209}Pb, or Pb^{209}]

Pb-210 Lead-210 [also ^{210}Pb, Pb^{210}, or RaD]

Pb-211 Lead-211 [also ^{211}Pb, Pb^{211}, or AcB]

Pb-212 Lead-212 [also ^{212}Pb, Pb^{212}, or ThB]

Pb-214 Lead-214 [also ^{214}Pb, Pb^{214}, or RaB]

Pb-1212 $PbSr_2CaCu_2O_y$ [A Lead Strontium Calcium Copper Oxide Superconductor]

Pb-2212 $Pb_2Sr_2CaCu_3O_y$ [A Lead Strontium Calcium Copper Oxide Superconductor]

Pb-2223 $Pb_2Sr_2Ca_2Cu_3O_{10}$ [A Lead Strontium Calcium Copper Oxide Superconductor]

pB negative logarithm of the base dissociation constant (for weak bases) [Symbol]

pB partially coherent (grain) boundary (in materials science) [Symbol]

p(B) hyperfine field distribution function (in solid-state physics) [Symbol]

$B Bolivian Peso [Currency] [also $b]

$\pi\beta$ Pion Beta (Experiment) [of Paul Scherrer Institute, Villigen, Switzerland]

PBA Permanent Budget Account

PBA γ-Phenylbutyric Acid

PBA Plastic Bag Association [US]

PBA Polybutylene Acrylate

PBA Printed Board Assembly

PBA Prismatic Beta Spectrometer

PBA Proportional Band Adjustment

PBA Provincia de Buenos Aires [Province of Buenos Aires, Argentina]

PBA Public Buildings Administration [US]

PBA Pyranyl Benzyl Adenine [Biochemistry]

PBA Pyrene Butyric Acid

P(B|A) Conditional Probability of Events B and A [also $\mathcal{P}$(B|A)] [Statistics]

pBA p-Bromoaniline [also pBa]

PBAOH^{4-} 2-Hydroxy-1,3-Propylenebis(oxamato) (Anion) [alsp pbaOH^{4-}]

PBAA Polybutadiene Acrylic Acid

Pb(Ac)$_2$ Lead(II) Acetate [also Pb(ac)$_2$]

Pb(Ac)$_4$ Lead(IV) Acetate [also Pb(ac)$_4$]

Pb(ACAC)$_2$ Lead(II) Acetylacetonate [also Pb(AcAc)$_2$, or Pb(acac)$_2$]

PBAH Polycyclic Benzenoid Aromatic Hydrocarbon

PBAPS Pipe Break Air Piping System

PBABS Pipe Break Automatic Protective System

PBAN Polybutadiene Acrylonitrile

PBB Polybrominated Biphenyl

PBBA P-Bromobenzoic Acid [also pBBA]

Pb(B$_1$B$_2$)O$_3$ Complex Lead Based Perovskite [General Formula]

PBBE Polybrominated Biphenyl Ether

PBBI Poly(butadiene Bisimide)

Pb:BiSCO Lead-Substituted Bismuth Strontium Copper Oxide (Superconductor) [also Pb:BISCO, or Pb:BSCO]

PBBO 2-(4-Biphenylyl)-6-Phenylbenzoxazole

PBBO Phenylbiphenylylbenzoxazole

Pb-Bronze Lead Bronze [also Pb bronze]

PBBS Projected Bulk-Based Structure

Pb:BSCO Lead-Substituted Bismuth Strontium Copper Oxide (Superconductor) [also Pb:BISCO, or Pb:BiSCO]

PBC Periodic Bond Chain

PBC Periodic (Grain) Boundary Condition(s) [Materials Science]

PBC Peripheral Buffer Computer

PBC Photobiochemical; Photobiochemist(ry)

PBC Program Booking Center

Pb-Ca Lead-Calcium (Alloy System)

PBCCO Lead-Barium-Calcium-Copper Oxide (Superconductor)

PBCCO Praseodymium-Barium-Calcium-Copper Oxide (Superconductor)

PBCFO Praseodymium Barium Copper Iron Oxide (Superconductor)

PBCGO Praseodymium Barium Copper Gallium Oxide (Superconductor)

PBCNO Praseodymium Barium Copper Niobium Oxide (Superconductor)

PBCMR Packed Bed Catalytic Membrane Reactor

PBCO Praseodymium Barium Copper Oxide (Superconductor)

PBCTO Praseodymium Barium Copper Tantalum Oxide (Superconductor)

PBCS Post-Boost Control System

PBCT Proposed Boundary Crossing Time

Pb-Cu Lead-Copper (Alloy System)

PbCV Paramecium bursaria Chlorella Virus

PbCV-1 Paramecium bursaria Chlorella Virus, Type 1

PBD Parallel Blade Damper

PBD Payload Bay Door

PBD Phase Boundary Dislocation [Solid-State Physics]

PBD 2-Phenyl-5-(4-Biphenylyl)-1,3,4-Oxadiazole

PBD Polybutadiene

Pbd Paperboard [also pbd]

PbDEC Lead Diethyl Dithiocarbamate

PBDF Payload Bay Door Forward

PBDG Pushbutton Data Generator

PBDI Position Bearing and Distance Indicator

PBDM Payload Bay Door Mechanism

PbDMC Lead Dimethyl Dithiocarbamate

Pbd Pkg Paperboard Packaging [Journal published in the US]

PBD-PS Polybutadiene/Polystyrene (Blend) [also PBD/PS]

PBDS Photothermal Beam Deflection Spectroscopy

PBE Prompt Burst Experiments

PBE Prompt By Example [Computers]

PBE Proton Balance Equation [Physics]

PBE Pulsed Bridge Element

PBEA Paint, Body and Equipment Association [US]

PBEC Pacific Basin Economic Committee

PBEIST Planning Board for European Inland Surface Transport

Pb(Et)$_4$ Tetraethyllead

PbEuSe Lead Europium Selenide (Semiconductor)

PbEuSeTe Lead Europium Selenide Telluride (Laser)

PbEuTe Lead Europium Telluride (Semiconductor)

PBF Potential Benefit Factor

PBF Power Burst Facility [of Idaho National Engineering Laboratory, Idaho Falls, US]

PBFA Particle Beam Fusion Accelerator [of US Department of Energy at Sandia National Laboratories, Albuquerque, New Mexico, US]

PBFA Provincial Booksellers Fair Association [UK]

PbFe Lead Ferrite

PBG Photonic Band Gap [Physics]

PBG Polybenzyl Glutamate [Biochemistry]

PBG Positive Pressure Breathing under G [Aerospace]

PBGA Plastic Ball Grid Array

PBGC Permian Basin Graduate Center [at Midland, Texas, US]

Pb/Ge Lead/Germanium (Multilayer)

Pb-Gr Lead-Graphite (Symbol)

PBH Planar-Buried Heterostructure [Electronics]

PBH Primordial Black Hole [Astronomy]

PBHA Poly(biphenyloxyhexyl Acrylate)

Pb(HFAC)$_2$ Lead(II) Hexafluoroacetylacetonate [also Pb(hfac)$_2$]

PBHP Pounds per Brake Horsepower [also pbhp]

PB-HTGR Peach Bottom High-Temperature Gas-Cooled (Nuclear) Reactor

PBI Paper Bag Institute [US]

PBI Plant Biological Institute

PBI Plant Biotechnology Institute [Canada]

PBI Plumbing Brass Institute [now Plumbing Manufacturers Institute, US]

PBI Polybenzimidazole

PBI Process Branch Indicator

PBI Protein-Bound Iodine (Test) [Medicine]

PBI Push Button Indicator

PBIB Partially Balanced Incomplete Block Design [Statistics]

PBIC Programmable Buffer Interface Card

PBIM Programmable Buffer Interface Module

Pb-In Lead-Indium (Alloy System)

PBIPA Polybenzimidazo Benzophenanthroline

Pb-I-Pb Lead–Insulator–Lead [Solid-State Physics]

PBIT Parity Bit [also pbit]

PBJ Paper-Braided Jute

PBK Payload Bay Kit

ΦBK Phi Beta Kappa (Student Society)

PBL Payload Bay Liner

PBL Planetary Boundary Layer [Meteorology]

PBL Public Broadcast Laboratory

PbLα Lead L-Alpha (Radiation) [also PbL$_\alpha$]

pbl probable

PBLG Poly(benzyl-L-Glutamate) [Biochemistry]

PBM Patrol Bomber Martin [Flying Boat]

PBM Program Business Management

PBM Permanent Benchmark

PBM Planetary Ball Mill(ing)

PBM Plastic Bonded Magnet

PBM Prikladnaya Biokhimiya i Microbiologiya [Russian Journal on Applied Biochemistry and Microbiology]

PBMA Poly(butyl Methacrylate)

PbMα Lead M-Alpha (Radiation) [also PbM$_\alpha$]

PBMC Peripheral Blood Mononuclear Cells [Cytology]

Pb5MC Lead Pentamethylene Dithiocarbamate

PbMgNbO$_3$ Lead-Magnesium Niobate

PbMnTe Lead Manganese Telluride (Semiconductor)

PBMO Praseodymium Barium Manganese Oxide (Superconductor)

PBN Butylphenylnitrone

PBN Lead Barium Niobate (Ceramics)

PBN Phenyl-N-tert-Butylnitrone

PBN N-Phenyl-β-Naphthylamine

PBN Physical Block Number

PBN Polybenzonitrile

PBN Polybutylene Naphthalate

PBN Pyrolytic Boron Nitride

PBNA N-Phenyl-β-Naphthyl Amine

PBNA Polytechnisch Bureau Nederland, Arnhem [Netherlands Polytechnical Bureau, Arnhem]

Pb-Na Lead-Sodium (Alloy System)

PBNS Polonium-Beryllium Neutron Source

PBO 5-Phenyl-2-[(1,1'-Biphenyl)-4-yl]oxazole

PBO Piperonyl Butoxide

PBO Poly(cis-Benzoxazole)

PBO Poly p-Phenylene Benzobioxazole

PbO Lead Monoxide

PbO$_x$ Lead Oxides (e.g., Lead Monoxide, Lead Dioxide, Lead Suboxide, or Lead Tetroxide) [also PbOx]

Pb(OAc)$_2$ Lead (II) Acetate

Pb(OAc)$_4$ Lead (IV) Acetate

PBO/PPA Poly(cis-Benzoxazole)/Polyphosphoric Acid

PBOS Planning Board for Ocean Shipping [of NATO]

PbO-SiO$_2$ Lead Monoxide-Silicon Dioxide (System)

PBOT Philadelphia Board of Trade [US]

PBOW Plum Brook Ordnance Works [Sandusky, Ohio, US]

PbOx Lead Oxides (e.g., Lead Monoxide, Lead Dioxide, Lead Suboxide, or Lead Tetroxide) [also PbO$_x$]

PBP Pentagonal-Bipyramidal [Crystallography]

PBP Performance Based Pay

PBP Preheating Blistering Porosity [Metallurgy]

PBP Photobiophysical; Photobiophysicist; Photobiophysics

PBP Poly(benzoyl-1,4-Phenylene)

PBP Proton-Blocking Pattern [Physics]

PBP Pushbutton Panel

pBP p-Bromophenol

PBPB Pyridinium Bromide Perbromide

PbPc Lead Phthalocyanine

PBPP Poly(bisphenoxyphosphazene)

PBPS Post-Boost Propulsion System

PBQS Parallel Batch Queuing System

PBR Paint Bake Response

PBR Pebble Bed Reactor [of Oak Ridge National Laboratory, Tennessee, US]

PBR Permit by Rule

PBR Plum Brook Reactor [of NASA]

PBR Pole Broken

PBR Polished-Bore Receptacle

PBR Professional Board for Radiography [South Africa]

PBR Publications Board Report [of Office of Technical Services, US]

PBR Pyridine Butadiene Rubber

PBRE Pebble Bed Reactor Experiment [of Oak Ridge National Laboratory, Tennessee, US]

PBRept Publications Board Report [of Office of Technical Services, US]

PBRF Plant Breeding Research Forum [US]

PBRF Plum Brook Reactor Facility [of NASA]

P-Bronze Phosphor Bronze [also P-bronze]

PbrS Polybromostyrene

PBS Pacific Biological Station [Canada]

PBS Palladium-Barium Sulfate

PBS Phosphate-Buffered Saline

PBS Phosphate Buffer Solution

PBS Picture Building System

PBS Polybutadiene Styrene

PBS Poly(butene-1-Sulfone)

PBS Poly-Si(licon) Back Seal

PBS Prefabricated Bituminous (Road) Surfacing

PBS Pressure Boundary Subsystem

PBS Project Breakdown Structure

PBS Projected Band Structure [Solid-State Physics]

PBS Protective Breathing System

PBS Public Buildings Service [of General Services Administration, US]

PBS Public Broadcasting Service

PbS Lead Sulfide (Semiconductor)

PbSb Lead Antimonide (Semiconductor)

Pb-Sb Lead Antimony (Alloy System)

PBS-BSA Phosphate Buffered Saline with Bovine Serum Albumin [Biochemistry]

PBSC Penicillin–Bacitracin–Streptomycin–(Sodium) Caprylate [Pharmacology]

PbSe Lead Selenide (Semiconductor)

Pb/Si Lead on Silicon (Substrate)

Pb-Sn Lead-Tin (Alloy System)

PbSnMnTe Lead Tin Manganese Telluride (Semiconductor)

PbSnTe Lead Tin Telluride (Semiconductor)

PbSO Plumbosulfate

PBSS Planets Beyond the Solar System (Program) [of Space Telescope Science Institute, US]

PBS-T Phosphate Buffered Saline with Tween 20 [Note: *Tween 20* is a Trademark for Polyoxyethylene (20) Sorbitan Monolaurate] [Biochemistry]

PBSTA Push Button Station

PBT Permeable-Base Transistor

PBT Plant Biotechnology Institute [of National Research Council of Canada]

PBT Polybutylene Terephthalate [also PBTP]

PBT Poly(-3-Hydroxybutyrate)

PBT Poly(p-Phenylene Benzobisthiazole)

PbTe Lead Telluride (Semiconductor)

PBTF Polybromotrifluoroethylene

PBTF Pump Bearing Test Facility [of Liquid Metal Engineering Center, US]

Pb(THD)$_2$ Lead(II) 2,2,6,6-Tetramethyl-3,5-Heptanedionate [also Pb(thd)$_2$]

Pb(THD)$_4$ Lead(IV) 2,2,6,6-Tetramethyl-3,5-Heptanedionate [also Pb(thd)$_4$]

PBT-LDPE Polybutylene Terephthalate/Low-Density Polyethylene (Blend) [also PBT/LDPE]

PBTMA P-Phenylene Bis(trimellitate Anhydride)

Pb(TMHD)$_2$ Lead(II) Bis(2,2,6,6-Tetramethyl-3,5-Heptanedionate) [also Pb(tmhd)$_2$]

PBTP Polybutylene Terephthalate [also PBT]

PBT-PAR Polybutylene Terephthalate/Polyarylate (Blend) [also PBT/PAR]

PBT-PC Polycarbonate/Polybutylene Terephthalate (Blend) [also PC/PBT]

PBT-PET Polybutylene Terephthalate/Polyethylene Terephthalate (Blend) [also PBT/PET]

PBT-PES Polybutylene Terephthalate/Polyether Sulfone (Blend) [also PBT/PES]

PBT-PPE PolybutyleneTerephthalate/Polyphenylene Ether (Blend) [also PBT/PPE]

PBT-PPO PolybutyleneTerephthalate/Polyphenylene Oxide (Blend) [also PBT/PPO]

PbTx-2 Brevitoxin 2 [from Ptychodiscus brevis]

PbTx-3 Brevitoxin 3 [from Ptychodiscus brevis]

PbTx-9 Brevitoxin 9 [from Ptychodiscus brevis]

PBU Phenylbenzylurethane

PBU Pushbutton Unit

PBV Post-Boost Vehicle

PBW Parts by Weight

PBW Proportional Band Width

PBWA Plasma-Beat-Wave Accelerator

PBX Plastic Bonded Explosive

PBX Private Branch (Telephone) Exchange [also pbx]

PBXRD Parallel Beam X-Ray Diffraction

PBZ Physiological and Biochemical Zoology [Journal published by University of Chicago Press, US]

PBZ Pyribenzamine

PBzMA Poly(benzylmethacrylate)

Pb-Zn Lead-Zinc (Alloy System)

PBZT Polyphenylene Benzalthiazole

PBZT Poly(p-Phenylene Benzobisthiazole)

PBZT Poly(trans-Benzothiazole)

PBZT/PA Poly(trans-Benzothiazole)/Polyamide

PBZT/PPA Poly(trans-Benzothiazole)/Polyphosphoric Acid

PC Lowest Priority

PC Pace College [New York City, US]

PC Palmer Configuration [Physics]

PC Panama Canal

PC Panama City

PC Paper Chromatograph(y)

PC Paracrystalline; Paracrystallinity

PC Parallel Circuit
PC Parallel Communication
PC Parallel Computer; Parallel Computing
PC Parallel Connection
PC Parallel Cube
PC Parametric Cubic
PC Parenchyma Cells [Botany]
PC Parity Check(er) [Computers]
PC Partial Combustion
PC Participation Certificate
PC Particle Characterization
PC Particle Collision
PC Particulate Composite
PC Partition Chromatography
PC Partition Coefficient [Chemistry]
PC Parts Catalogue
PC Passenger Car
PC Patent Classification
PC Path Control
PC Patrol Subchaser, Steel-Type [US Navy Symbol]
PC Peace Corps [US]
PC Peg Count [Telecommunications]
PC Peltier Coefficient [Physics]
PC Pemphigus Contageosus [Medicine]
PC Pentagon Column
PC Perfect Crystal
PC Performance Coefficient
PC Periodic-Chaotic (Sequence) [Physics]
PC Peripheral Collision
PC Peripheral Controller
PC Perkin Condensation [Chemistry]
PC Permeability Coefficient [Fluid Mechanics]
PC Personal Chromatograph
PC Personal Computer
PC Petersen Conductivity
PC Petro Canada
PC Petrochemical(s); Petrochemist(ry)
PC Petroleum Chemical(s)
PC Petroleum Coke
PC Petty Cash [also P/C, or pc]
PC Phase Change [Physics]
PC Phase Code
PC Phase Coherence; Phase-Coherent [Physics]
PC Phase Conjugation [Physics]
PC Phase Constant [Electrical Engineering]
PC Phase Contrast [Microscopy]
PC Phase Control [Electronics]
PC Phase Conversion; Phase Converter
PC Phase Correction [Telecommunications]
PC Phase Current
PC Phenol Coefficient [Chemistry]
PC Pheochromocytoma [Medicine]
PC Phosphatidylcholine [Biochemistry]

PC Phosphocreatine [Biochemistry]
PC Photocarrier
PC Photocatalysis; Photocatalyst
PC Photocathode
PC Photocell
PC Photochemical; Photochemist(ry)
PC Photochromatic(ity)
PC Photochromic; Photochromism
PC Photocomposing; Photocomposition
PC Photoconductance; Photoconduction; Photoconductive; Photoconductivity; Photoconductor
PC Photoconductive Cell
PC Photocopier; Photocopy(ing)
PC Photocontrol
PC Photocoupler [Electronics]
PC Photocurrent [Physical Chemistry]
PC Photoelectric Control
PC Photoemissive Converter
PC Photographic Camera
PC Phthalocyanine [also Pc]
PC Physical Ceramics
PC Physical Chemist(ry)
PC Physical Constant
PC Physics Computing (Conference)
PC Pie Chart
PC Piezochemical; Piezochemist(ry)
PC Pilot Chute
PC Pin Cluster
PC Pitch Circle (of Gears) [also pc]
PC Pitch Control [also P/C]
PC Planck's Constant [Quantum Mechanics]
PC Plane Change
PC Plasma Cell [Cytology]
PC Plasma Chemist(ry)
PC Plasma Chromatography
PC Plastocyanine
PC Plenum Chamber
PC Pneumatic Circuit
PC Pneumatic Control(ler)
PC Pneumococcal; Pneumococcus [Microbiology]
PC Pocket Calculator
PC Point Cathode
PC Point Contact
PC Point of Curvature
PC Polar Cap
PC Polar Coordinate
PC Polar Crane
PC Polar Crystal
PC Polhode Cone [Mechanics]
PC Police Constable
PC Politechnika Czestochowska [Czechstochowa Polytechnic Institute, Poland]
PC Pollution Control
PC Polycarbonate

PC Polycarbosilane
PC Polychloroprene
PC Polychromatic; Polychromatism
PC Polycondensate; Polycondensation
PC Polycrystal(line); Polycrystallinity
PC Polymer Chain
PC Polymer Chemist(ry)
PC Polymer Composite
PC Polymerizable Complex
PC Population Commission [of United Nations]
PC Portable Computer
PC Port Control
PC Portland Cement
PC Position Control
PC Positive Charge
PC Positive Crystal
PC Positron Camera
PC Postcard [also pc]
PC *(Post cibum)* – After Food, or After Meals [Medical Prescriptions]
PC Post-Combustion
PC Power Coefficient [Aerodynamics]
PC Power Component
PC Power Conditioning
PC Power Control
PC Power Conversion; Power Converter
PC Practical Computing
PC Precision Cast(ing) [Metallurgy]
PC Pre-Emphasis Circuit
PC Preferential Combustion
PC Preparatory Commission
PC Pressure Cable
PC Pressure Cap
PC Pressure Circuit
PC Pressure Control
PC Pressure Cooker
PC Prestressed Concrete
PC Primary Cell
PC Primary Color
PC Primary Consumer [Ecology]
PC Print Chain
PC Printed Circuit
PC Process Chart
PC Process Control(ler)
PC Processing Center
PC Processor Control(ler)
PC Product Code
PC Product Cycle
PC Production Center
PC Production Control
PC Production Cycle
PC Professional Communication (Group) [of Institute of Electrical and Electronics Engineers, US]
PC Professional Computing

PC Professional Consultant
PC Professional Corporation
PC Professors of Curriculum [US]
PC Progenitor Cryptocide
PC Program Controller
PC Program Coordination
PC Program Counter
PC Programmable Controller [also pc]
PC Project Control
PC Proportional Control
PC Proportional Counter
PC Propylene Carbonate
PC Protective Clothing
PC Protocol Converter
PC Proton Count(er); Proton Counting
PC Provisional Costs
PC Pseudo Code [Computers]
PC Psychrometric Chart [Thermodynamics]
PC Publications Committee
PC Pulmocutaneous [also pc] [Zoology]
PC Pulmonary Circulation [Medicine]
PC Pulp Cavity [Dentistry]
PC Pulsating Current
PC Pulse Circuit
PC Pulse Compression
PC Pulse Control(ler)
PC Pulsed Cathode
PC Pulsed Current
PC Pulverized Coal
PC Punched Card
PC Pure Culture [Microbiology]
PC Pyrocellulose
PC Pyrometric Cone
PC Pyrurate Carboxylase [Biochemistry]
PC Single-Paper, Single-Cotton (Electric Wire)
PC-104 Personal Computer 104 (Interface)
P3C [2-((Diphenylphosphino)methyl)-2-Methyl-1,3-Propanediyl]bis(diphenylphosphine)
P&C Process and Control
P/C Petty Cash [also PC, or pc]
P/C Pitch Control [also PC]
P/C Price Current
P-C Plastic-Carbon (Replica)
P-C Processor-Controller
P-C Pulse Counter
Pc Column Parameter [ANSI Escape Sequence Code]
Pc Phthalocyanine [also PC]
Pc Picture [also pc]
Pc Piece [also pc]
Pc Pneumocystis carinii (Protozoa) [Microbiology]
Pc Price [also pc]
pC p-Cresol
pC picocoulomb [Unit]
pC picocurie [Unit]

pc parsec [Unit]

pc per cent

pc plastic strain in tension, creep in compression [Mechanical Testing]

pc *(post cibum)* – after food, or after meals [Medical Prescriptions]

pc pulmocutaneous [also PC] [Zoology]

£C Cyprus Pound [Currency]

PCA Packaging Council of Australia

PCA Parallel Computer Architecture

PCA Partnership and Cooperation Agreement (of European Community with former Soviet States)

PCA Pennsylvania Coal Association [US]

PCA Perchloric Acid

PCA Performance and Coverage Analyzer

PCA Permanent Court of Arbitration

PCA Personal Computer Analyzer

PCA Personal Computers Association

PCA Phaseolus Coccineus Agglutinin [Immunology]

PCA Philadelphia College of Art [Pennsylvania, US]

PCA Photocurrent Amplifier

PCA Physical Configuration Audit

PCA Pitch Control Assembly

PCA Pneumatic Control Assembly

PCA Point of Closest Approach

PCA Polar Cap Absorption; Pole-Cap Absorption

PCA Polycaproamide

PCA Polycrystalline Alumina

PCA Pool Critical Assembly [of Oak Ridge National Laboratory, Tennessee, US]

PCA Portland Cement Association [US]

PCA Positive Control Area [Air Traffic Control]

PCA Posterior Cerebral Artery [Anatomy]

PCA Power Control(ler) Assembly

PCA Precision Cleaning Agent [Chemistry]

PCA Preconditioned Air System

PCA Prestressed Concrete Association [UK]

PCA Principal Component Analysis

PCA Printed Circuit Assembly

PCA Printed Communications Adapter

PCA Private Communications Association [now National Communications Association, US]

PCA Process(ing) Control Agent [Chemical Engineering]

PCA Producers Commission Association [US]

PCA Production Credit Association [US]

PCA Programmable Communications Adapter

PCA Programmable Counter Array

PCA Protective Clothing Arrangement

PCA Protective Connecting Arrangement

PCA Protein Crystallization Apparatus [Biochemistry]

PCA Pulp Chemicals Association [US]

PCA Pulse Code Adapter

PCA Pulse Counter Adapter

PCA Pyrotechnic Control Assembly

PCA Pyrrolidonecarboxylic Acid

pCa p-Chloroaniline

PCaBCO Praseodymium-Calcium-Barium-Copper Oxide (Superconductor)

PC-ABS Polycarbonate/Acrylonitrile-Butadiene-Styrene (Blend) [also PC/ABS]

PCAC Partially Conserved Axial Current; Partially Conserved Axial-Vector Current

PCACIAS Personal Computer Automated Calibration Interval Analysis System

PCAD Pointing Control and Aspect Determination [Aerospace]

P-CAD Personal (Computer) Computer-Aided Design

p-CaF$_2$ Polycrystalline Calcium Fluoride

PCAM Partitioned Content Addressable Memory

PCAM Probe Card Assembly Machine

PCAM Punched Card Accounting Machine

PCAO Pollution Control Association of Ontario [Canada]

PCAP Payload Crew Activity Plan

Pcar Pneumocystis carinii [Microbiology]

PCA-PMPIPA Polycaproamide Poly-m-Phenyleneiso-phthalamide

PCA-PPABIA Polycaproamide Poly-p-Amidobenzimidazole

PCAS Primary Central Alarm Station

PCASP Passive Cavity Aerosol Spectrometer

PCAST President's Council of Advisors on Science and Technology [US]

PCAT Pharmacy College Admission Test

PC AT Personal Computer Advanced Technology [also PC-AT]

PC ATE Personal Computer Advanced Technology Enhanced [also PC-ATE]

PC ATX Personal Computer Advanced Technology Expanded [also PC-ATX]

PCB Page Control Block

PCB Pentylcyanobiphenyl

PCB Physics, Chemistry and Biology

PCB (Certificat d'Etudes) Physiques, Chimiques, Biologiques [Diploma of Physical, Chemical and Biological Studies, France]

PCB Plenum Chamber Burning

PCB Polychlorinated Biphenyl

PCB Polychlorobenzene

PCB Polychlorobiphenyl

PCB Power Circuit Breaker

PCB Power Control Box

PCB Printed Circuit Board

PCB Process Control Block

PCB Processor Command Bus

PCB Product Configuration Baseline

PCB Program Communication Block

PCB Program Control Block

PCB Public Coin Box [Telecommunications]

PCBA P-Chlorobenzoic Acid [also pCBA]

PCBC Partially Conserved Baryon Current

PCBC P-Chlorobenzyl Chloride

PCBC Plain Cipher Block Chaining

PCBC Propagating Cipher Block Chaining

PCBCR Phenyl-C_{61}-Butyric Acid Cholestryl Ester

PCBE Pulse Chemical Beam Epitaxy

PCBM (1-(3-Methoxycarbonyl)propyl)-1-Phenyl[6,6] C_{61}

PCBN P-Chlorobenzonitrile

PCBN Polycrystalline Cubic Boron Nitride [also PcBN]

PCBPS Printed Circuit Board Power Supply [also PCB PS]

PCB/PWB Printed Circuit Board/Printed Wiring Board

PCBS Permanent Committee on Biological Standards

PCBS Positive Control Bombardment System

PCBT Polycarbonate-Butylate Terephthalate

PCBT Pressure-Cooker Bias Test

PCBTF P-Chlorobenzotrifluoride

PC Bus Softw Personal Computer Business Software [Journal published in the UK]

PCC (Launch) Pad Control Center

PCC Panama Canal Commission [US]

PCC Parametric Channel Controller

PCC Parklawn Computer Center [of Public Health Service/Food and Drug Administration, US]

PCC Partial Crystal Control

PCC Participating Clinical Center

PCC Payload Control and Checkout

PCC Peripheral Control Computer

PCC Personal Code Calling

PCC Petroleum Compensation Charge

PCC Philadelphia Chamber of Commerce [US]

PCC Photoconductive Cell

PCC Point of Common Coupling

PCC Point of Compound Curvature

PCC Poison Control Center

PCC Pole-Changing Control [Electrical Engineering]

PCC Polychlorinated Camphene

PCC Polymers/Ceramics/Composites (Database)

PCC Portland Cement Concrete

PCC Power Control Center

PCC Precipitated Calcium Carbonate

PCC Precision Castparts Corporation [US]

PCC Processor Control Console

PCC Producers Council of Canada

PCC Production Control Center

PCC Professional Conduct Committee [of General Medical Council, UK]

PCC Program Control Counter

PCC Program-Controlled Computer

PCC Program Coordinating Center

PCC Pyridinium Chlorochromate

P/C/C Polymers/Ceramics/Composites (Alert) [Joint Publication of ASM International US, and Institute of Materials, UK]

PC-CAD Personal-Computer Computer-Aided Design [also PC CAD]

P/C/C Alert Polymers/Ceramics/Composites Alert [Joint Publication of ASM International, US, and the Institute of Materials, UK]

PCCB Program Configuration Control Board

PCCC Pakistan Central Cotton Committee

PCCC Phoenix Conference on Computers and Communications [US]

PCCD Peristaltic Charge-Coupled Device

PCCE Payload Common Communication Equipment

PCCE Proton Coupling Constant Extraction

PCCI Peterborough Chamber of Commerce and Industry [UK]

PCCM Private Circuit Control Module

PCCM Program Change Control Management

PCCN Provisioning Contract Control Number

PCCNO Praseodymium Calcium Copper Niobium Oxide (Superconductor)

PCCO Praseodymium Cerium Copper Oxide (Superconductor)

PcCo Cobalt Phthalocyanine

PcCo [29H,31H-Phthalocyanato(2−)N^{29},N^{30},N^{31},N^{32}]cobalt

PcCo(t-bupy)$_2$ Cobalt Phthalocyanine Bis(4-tert-Butylpyridine)

PcCo(py)$_2$ Cobalt Phthalocyanine Bis(pyridine)

PcCo(pyz)$_2$ Cobalt Phthalocyanine Bis(pyrazine)

PCCP Precalculated Collision Probability

PCCP Preliminary Contract Change Proposal

PCCS Photographic Camera Control System

PCCS Photon-Capture Cross-Section (Spectrum) [Physics]

PCCS Positive Control Communications System

PCCS Processor Common Communications System

PCD Parametric Control Device

PCD Photo Compact Disk

PCD Photoconductive Decay

PCD Photoconductive Device

PCD Physical Chemistry Division [of National Chemical Laboratory, Pune, India]

PCD Pincushion Distortion

PCD Pitch Circle Diameter (of gears) [also pcd]

PCD Planned Continuation Date

PCD Polycarbodiimide

PCD Polycrystalline Diamond

PCD Port Control Diagnostic

PCD Power Circle Diagram

PCD Power Control and Distribution

PCD Power Control Device

PCD Process Control Division

PCD Production Common Digitizer

PCD Procurement and Contracts Division [of NASA]

PCD Procurement Control Document

PCDC Punched Card Data Processing

PCDDS Private Circuit Digital Data Service

PC-DOS Personal Computer–Disk Operating System [also PC DOS, or PCDOS]

PCDP Punched Card Data Processing

PCD-ROM Paper/Compact-Disk Read-Only Memory

PCDS Pilot Climate Data System [of NASA Goddard Space Flight Center, Greenbelt, Maryland, US]

p-CdTe P-Type Cadmium Telluride (Semiconductor)

PCDU Payload Command Decoder Unit

PCE Papillary Cystadenomas of the Epididymis [Medicine]

PCE Perchloroethylene

PCE Peripheral Controller Enclosure

PCE Photocatalytic Effect

PCE Photoconductive Effect

PCE Power Conditioning Electronics

PCE Program Cost Estimate

PCE Project Coordination Center Europe

PCE Punched-Card Equipment

PCE Pyrocarbonic Acid Diethyl Ester

PCE Pyrometric Cone Equivalent

PC&E Pittsburgh Conference and Exhibition [Pennsylvania, US]

PCEA Pacific Coast Electric Association [US]

PCEAA Professional Construction Estimators Association of America [US]

PCEB PCI (Peripheral Component Interconnect) to EISA (Extended Industry Standard Architecture) Bridge

PCEH President's Committee on Employment of the Handicapped [US]

PCEK Pappas, Carter, Evans and Koop (Theory) [Physics]

PCEM Phase Contrast Electron Microscopy

PCEM Process Chain Evaluation Model

PCEM Propulsion Contamination Effects Module

PCES Phase Change Energy Storage

PCETF Power Conversion Equipment Test Facility

PCEVE Polychloroethyl Vinyl Ether

PCEVE Poly(chloroethyl Vinyl Ether)

PCF Lead Chromium Fluoride

PCF Pair Correlation Function [Physics]

PCF Parabolic Cylinder Function

PCF Partial Correlation Function [Statistics]

PCF Payload Control Facility

PCF Peripheral Control Facility

PCF Pitch-Based Carbon Fiber

PCF Planning Coordination Facility

PCF Point Characteristic Function [Optics]

PCF Potential Controlled Flotation

PCF Pounds per Cubic Foot [also pcf, or lb/ft^3]

PCF Power Cathode Follower

PCF Primary Crystallization Field

PCF Printed Circuit Fabrication

PCF Program Control Facility

PCF Pulse-to-Cycle Fraction

PCFA Precast Concrete Frame Association [UK]

PcFe Iron Phthalocyanine

PcFe(bpe)$_2$ Iron Phthalocyanine Bis(4-Pyridyl)ethylene

PcFe(bpy)$_2$ Iron Phthalocyanine Bis(4,4'-Bipyridine)

PcFe(t-bupy)$_2$ Iron Phthalocyanine Bis(4-tert-Butylpyridine)

PcFe(Clpyz)$_2$ Iron Phthalocyanine Bis(chloropyrazine)

PcFe(dib)$_2$ Iron Phthalocyanine Bis(1,4-Diisocyanobenzene)

PcFe(me$_4$dib)$_2$ Iron Phthalocyanine Bis(Tetramethyl Diisocyanobenzene)

PcFe(etpyz)$_2$ Iron Phthalocyanine Bis(ethylpyrazine)

PcFe(mepyz)$_2$ Iron Phthalocyanine Bis(methylpyrazine)

PcFe(me$_2$pyz)$_2$ Iron Phthalocyanine Bis(dimethylpyrazine)

PcFe(py)$_2$ Iron Phthalocyanine Bis(pyridine)

PcFe(pyz)$_2$ Iron Phthalocyanine Bis(pyrazine)

PcFe(tz)$_2$ Iron Phthalocyanine Bis(triazole)

PCFIA Particle Concentration Fluorescence Immunoassay

PCFS Pin, Clip and Fastener Services [US]

PCG Phonocardiogram; Phonocardiograph(y)

PCG Phosphate Crown Glass

PCG Plains Cotton Growers [US]

PCG Planning and Control Guide

PCG Programmable Character Generator

PCG Protein Crystal Growth (Experiment) [NASA Space Shuttle]

PCGN Permanent Committee on Geographical Names [UK]

PCGOAL Personal Computer Ground Operations Aerospace Language [of NASA]

PCGS Protein Crystal Growth System

PCH Polycyclohexane

PCH Proton Channeling [Physics]

PC(H) Patrol Craft, Hydrofoil [US Navy Symbol]

Pch Punch [also pch]

PCHDMT Polycyclohexane Dimethylene Terephthalate; Poly(1,4-Cyclohexylenedimethylene)terephthalate

PCHIS Population Clearinghouse and Information System [of Economic and Social Commission for Asia and the Pacific, United Nations]

PCHK Parity Check [Computers]

PCHM Plasma Cold Hearth Melting (Process) [Metallurgy]

PCHMA Poly(cyclohexyl Methacrylate)

PCHR Plasma Cold-Hearth Refining (Process) [Metallurgy]

PCI Panel Call Indicator

PCI Pantone Color Institute [US]

PCI Pattern Correspondence Index

PCI Pavement Condition Index [Civil Engineering]

PCI Pellet-Cladding Interaction [Nuclear Engineering]

PCI Peripheral Command Indicator

PCI Peripheral Component Interconnect

PCI Peripheral Component Interface

PCI Peripheral Controller Interface

PCI Peripherals Connection Interface (Accelerator)

PCI Personal Computer Interface

PCI Pilot Controller Integration

PCI Plant Control Interface

PCI Portable Cesium Irradiator [of US Department of Energy]

PCI Positive Chemical Ionization

PCI Postcollision Interaction [Physics]

PCI Powder Coating Institute [US]

PCI Power Conversion International (Congress)

PCI Precision Cascade Impactor

PCI Prestressed Concrete Institute [US]

PCI Procedure Change Unit

PCI Process Capability Index

PCI Process Consultants, Inc. [US]

PCI Process Control Interface

PCI Product Configuration Identification

PCI Program Check Interruption

PCI Program Control Input

PCI Program-Controlled Interruption

PCI Programmable Communications Interface

PCI Programmable Controller Interface

PCI Programmed Control Interrupt

PCI Pulverized Coal Injection (Technology) [Metallurgy]

pCi picocurie [Unit]

PCIA Personal Communications Industry Association [US]

PCIAC Petro-Canada International Assistance Corporation

PCIC Petroleum and Chemical Industry Conference

PCIC Printed Circuit-Card Interrupt Controller

PCIF Personal Computer Interface

PCIF Printed Circuit Interconnection Federation [UK]

PCIF Programmable Controller Interface Facility

PCIFC Permanent Commission of the International Fisheries Convention

PCI-GBIP Personal Computer Interface–General-Purpose Interface Bus

PCIJ Prestressed Concrete Institute Journal [US]

PCIL Pilot-Controlled Instrument Landing

pCi/L picocurie per liter [also pCi/l also pCi L^{-1}]

PCILO Perturbative Configuration Interaction Based on Localized Orbitals [Physics]

PCILOCC Perturbative Configuration Interaction Using Localized Orbitals for Crystal Calculation [Physics]

PCIM Power Conversion and Intelligent Motion

PCIN Program Change Identification Number

PCIN Program Change Integration

PC-I/O Program Controlled Input/Output

PCIOS Processor Common Input/Output System

PCIP Personal Computer Instrument Product

PCIS Passive Cycle Isolation System [Nuclear Engineering]

PCIS Personal Computer Information Service

PCIS Primary Containment Isolation System [Nuclear Engineering]

PCIS Project Control Information System

PC IT Personal Computer Information Technology [also PC-IT]

PCIV Prestressed Cast Iron (Pressure) Vessel

PCIZC Permanent Committee of International Zoological Congresses

PCJC Pakistan Central Jute Committee

PCJCP Phase-Corrected Coupled J Cross-Polarization (Spectra)

PCK Processor-Controlled Keying

.PCK Pick File [Turbo Pascal File Extension]

PCL Pallet Coolant Loop [Nuclear Reactors]

PCL Parallel Communications Link

PCL Permissible Contamination Limit(s)

PCL Phagocytic Cell Labeling [Microbiology]

PCL Physical Chemistry Laboratory [of University of Oxford, UK]

PCL Pilot Controlled Lighting

PCL Planning and Conservation League [Sacramento, California, US]

PCL Polycaprolactone

PCL Polytechnic of Central London [UK]

PCL Poly(ε-Caprolactone)

PCL Power Control List

PCL Primary Coolant Line [Nuclear Reactors]

PCL Printed Circuit Lamp

PCL Printer Command Language

PCL Printer Control Language

PCL Process Control Language

PCL Programmasyst Control Language

PCL Programming Check List

PCL Polytechnic of Central London [UK]

PCL Primary Coolant Loop [Nuclear Reactors]

PCLA Process Control Language

PCLASS Processed CLASS (Cross-Chain Loran Atmospheric Sounding System)

PCLC Poly(ε-Caprolactone) Nano-Composite

PCLC Preparative Column Liquid Chromatography

PCL-co-AA Poly(ε-Caprolactone)-co-Acrylic Acid

PCL-co-DMVPA Poly(ε-Caprolactone)-co-Dimethyl Vinylphosphonic Acid

PCL-co-VPA Poly(ε-Caprolactone)-co-Vinylphosphonic Acid

PCLDI Prototype Closed Loop Development Installation [Nuclear Reactors]

PC LNP Personal Computer Local Network Program

PCLO Printed Circuit (Board) Layout

PCL-PC Polycaprolactone/Polycarbonate (Blend) [also PCL/PC]

PCLR Parallel Communications Link Receiver

PCLS Prototype Closed Loop System [Nuclear Reactors]

PCLT Prototype Closed Loop Test [Nuclear Reactors]

PCM Partial Charge Model

PCM Passive Countermeasures

PCM Personal Computer Manufacturer

PCM Phase-Change Material

PCM Phase-Contrast Microscope; Phase-Contrast Microscopy

PCM Pitch Control Motor

PCM Plasma Cell Mastitis [Medicine]

PCM Plug Compatible Machine

PCM Plug Compatible Mainframe

PCM Plug Compatible Memory

PCM Plug Compatible Module

PCM Plug Control Module

PCM Polymer-Coated Material

PCM Port Command Module

PCM Power Control Mission

PCM Power Cooling Mismatch

PCM Powertrain Control Module [Automobiles]

PCM Printer Cartridge Metric

PCM Process Control Module

PCM Process Control Monitor

PCM Profiling Current Meter

PCM Pulse-Code Modulation; Pulse Code Modulator

PCM Pulse-Counting Method

PCM Pulse-Count Modulation

PCM Punched Card Machine

PC²M Plasma Chemical Cleaning Module

PCMA Paired Carrier Multiple Access

PCMA Personal Computer Management Association [US]

PCMA Professional Convention Management Association [US]

PC Mag Personal Computer Magazine [US]

PC Mag (UK Ed) Personal Computer Magazine (UK Edition)

PCMB P-Chloromercuribenzoate

PCMB P-Chloromercuribenzoic Acid

PCMB Sodium P-Chloromercuribenzoate

pCMBS p-Chloromercuribenzenesulfonic Acid

PCMC PCI (Peripheral Component Interconnect/Interface) Cache, Memory Controller

PCMC Preparatory Commission for Metric Conversion

PCMCIA Personal Computer Manufacturers Communications Interface Adapter

PCMCIA Personal Computer Memory Card International Association

PCMD Pulse Code Modulation, Digital

PCME Pulse Code Modulation Event

PCM-FM Pulse Code Modulation–Frequency Modulation

PCMGB Proportionate Control Mixing Gas Burner

PCMH Professional Certified in Materials Handling

PCMI Pellet-Cladding Mechanical Interaction [Nuclear Engineering]

PCMI Photochemical Machining Institute [US]

PCMI Photochromic Micro-Image

PCMIA Pittsburgh Coal Mining Institute of America [Pennsylvania, US]

PCMIM Personal Computer Media Interface Module

Pc Mk Piece Mark

PCMM Plug Compatible Mainframe Manufacturer

PCMM Professional Certified in Materials Management

PCMMU Pulse-Code Modulation Master Unit

PCMO Praseodymium Calcium Manganese Oxide (Superconductor)

PCM/PAM Pulse Code Modulation/Pulse Amplitude Modulation

PCMR President's Committee on Mental Retardation [US]

PCMS P-Chloromercurphenylsulfonic Acid

PCMS Plasma-Controlled Magnetron Sputtering

PCMS Polycyclomethylsilazane

PCMS Pulse-Code Modulation Shared

PCMS Punched Card Machine System

PCMX P-Chlorodimethylphenol

PCN Lead Cobalt Niobate

PCN Page Change Notice

PCN Pavement Classification Number [Civil Engineering]

PCN Personal Communications Network

PCN Personal Computer Network

PCN Personal Computer News [Journal published in the UK]

PCN Personal Computer Network

PCN Program Change Notice

PCN Program Control Number

PCN Publication Code Number

pC/N picocoulomb per Newton [also pC N^{-1}]

PCNA Proliferating Cell Nuclear Antigen [Immunology]

pCNa p-Cyanoaniline

PCNB Pentachloronitrobenzene

pCNBA p-Cyanobenzoic Acid

PC Net Personal Computer Network

PCNFS Personal Computer Network File System

pCNP p-Cyanophenol

PCO Peak Conservation Organization

PCO Photocatalytic Oxidation

PCO Plant Control Office

PCO Point of Control and Observation

PCO Post-Checkout (Operations)

PCO Praseodymium Cerium Oxide (Superconductor)

PCO Praseodymium Copper Oxide (Superconductor)

PCO Process Configuration and Operation

PCO Procuring Contracting Officer

PCO Procyanidolic Oligomer

PCO Professional Conference Organizer

PCO Program Controlled Output

PCO Program Control Office

PCO Publication Coordination Office [of NATO]

PCOM Philadelphia College of Osteopathic Medicine [Pennsylvania, US]

PCONA P-Chloro-O-Nitroaniline

PCOPY (Screen) Page Copy (Command)

PCOS Primary Communications Oriented System

PCOS Process Control Operating System

P$_n$(cos θ) Legendre polynomial (in mathematics) [Symbol]

PCOT Payload Center Operations Team

PCOT P-Chloro-o-Aminotoluene

PCQ Production Control Quantometer

PCP Packet Control Panel

PCP Parachlorobisphenyl

PCP Parallel Cascade Processor

PCP Parallel Circular Plate

PCP P-Chlorobisphenyl

PCP Pentachlorophenol

PCP Pericyclic Process [Biology]

PCP Personal Communication Services

PCP Personal Communication System

PCP Personal Conferencing Specification

PCP Pest(icide) Control and Prevention (Act) [Canada]

PCP Phencyclidine; Phenylcyclidine [Biochemistry]

PCP Photochemical Processing

PCP Photon-Coupled Pair

PCP Planning Control Sheet

PCP Plug Compatible Peripheral

PCP Pneumocystis carinii Pneumonia [Medicine]

PCP Pneumococcal Pneumonia [Medicine]

PCP Polychloroprene

PCP Polyconic Projection [Cartography]

PCP Port Call Processing

PCP Post-Construction Permit

PCP Power Control Panel

PCP Primary Control Program

PCP Primary Coolant Pump

PCP Primary Cross-Connection Point

PCP Process Control Plan

PCP Process Control Processor

PCP Processor Control Program

PCP Program Change Proposal

PCP Programmable Communications Processor

PCP Project Change Proposal

PCP Project Control Plan

PCP Project Cost Plan

PCP Pulse Comparator

PCP Punched Card Punch

Pcp Perchlorophenyl

pCP p-Chlorophenol

pCp Cytidine Biphosphate; Cytidine-3',5'-Biphosphate [Biochemistry]

PCPA P-Chlorophenylacetic Acid

PCPA P-Chlorophenylalanine [Biochemistry]

PCPA Pest Control Products Act [Canada]

PC-PA Polycarbonate/Polyamide (Blend) [also PC/PA]

PCPB P-Chlorophenylbenzene

PCPBMA Pacific Coast Paper Box Manufacturers Association [US]

PCPBS P-Chlorophenylbenzenesulfonate; P-Chlorophenylbenzenesulfonic Ester

PCP-BSA Phenylcyclidene–Bovine Serum Albumin [Immunology]

PC-PBT Polycarbonate/Polybutylene Terephthalate (Blend) [also PC/PBT]

PCPCI Power Conversion Products Council International [US]

PC-PE Polycarbonate/Polyester (Blend) [also PC/PE]

PC Perspect Newsl PC (Personal Computer) Perspectives Newsletter [US]

PC-PET Polycarbonate/Polyethylene Terephthalate (Blend) [also PC/PET]

PCPIP Paris Convention for the Protection of Industrial Property

PCPM Pennsylvania College of Podiatric Medicine [Philadelphia, US]

Pcpn Precipitation [also pcpn]

PCPS Philadelphia College of Pharmacy and Science [Pennsylvania, US]

PCPS Private Carrier Paging System

PCPS Prodat Communication and Processing System

PCR Page Control Register

PCR Partial Combustion Reactor

PCR Pass Card Reader

PCR Payload Changeout Room

PCR Payload Checkout Room

PCR Payload Control Room

PCR Peer Code Review

PCR Phase-Change Record(ing)

PCR Phase-Controlled Rectifier

PCR Photochemical Reaction

PCR Photoconductive Relay

PCR Photoconductive Resonance

PCR Pickled and Cold Rolled; Pickling and Cold Rolling [Metallurgy]

PCR Polymerase Chain Reaction [Biochemistry]

PCR Polymer Chain Reaction

PCR Post-Column Reaction

PCR Postcombustion Ratio

PCR Power Change Request

PCR Power Control Rod [Nuclear Reactors]

PCR Precision Control Relay

PCR Preconsolidated Roll (Product) [Metallurgy]

PCR Principal Component Regression

PCR Print Command Register

PCR Procedure Change Request

PCR Program Change Request

PCR Program-Controlled Request

PCR Program Control Register

PCR Project Control Room

PCR Publication Change Request

PCR Punched Card Reader

PCRA Poland-China Record Association [*Note:* Poland China Refers to a Breed of Swine]

PCRB Program Change Review Board

PCRC Primary Communications Research Center [UK]

PCRCA Pickled, Cold Rolled and Close Annealed; Pickling, Cold Rolling and Close Annealing [Metallurgy]

PCR/DNA Polymerase Chain Reaction/Deoxyribonucleic Acid (Technique) [Biochemistry]

PCRL Phoenix Corporation Research Laboratories [of Motorola Inc., Tempe, Arizona, US]

PCRM Physicians Committee for Responsible Medicine

PCRNB President's Council on Recreation and Natural Beauty [US] [Defunct]

PCRS Primary Control Rod System [Nuclear Reactors]

PcRu(dib)$_2$ Ruthenium Phthalocyanine-Bis(1,4-Diisocyanobenzene)

PcRu(me$_4$dib)$_2$ Ruthenium Phthalocyanine Bis(Tetramethyl Diisocyanobenzene)

PcRu(tz)$_2$ Ruthenium Phthalocyanine Bis(pyridine)

PCRV Prestressed Concrete Reactor Vessel

PCS PAL (Phase-Alternation Line) Contour Sharpness

PCS Paracrystalline Substance

PCS Parts Collection Survey [US]

PCS Passenger Car Study (Database) [of University of Calgary, Alberta, Canada]

PCS Patchable Control Store

PCS Patent Classification Service [of International Patent Documentation Center, Austria]

PCS Payload Checkout System

PCS Payload Control Supervisor

PCS Penetrable-Concentric-Shell (Model) [Physics]

PCS Peripheral Computer System

PCS Personal Communications Service

PCS Personal Communication System

PCS Personal Conferencing Specification

PCS Personal Computer System

PCS Personal Computing Society [US]

PCS Phono-Cardio-Scan (System) [Medicine]

PCS Photochemical Smog [Meteorology]

PCS Photon Correlation Spectroscopy

PCS Planning Control Sheet

PCS Plasma Chemical Synthesis

PCS Plastic-Clad Silica (Fiber)

PCS Plastic-Clad Silicon

PCS Plastic-Coated Silica (Fiber)

PCS Pneumatic Control System

PCS Pointing Control System

PCS Polycarbosilane

PCS Portable Computer System

PCS Port Command Store

PCS Port Control System

PCS Position Control System [Numerical Control]

PCS Potash Corporation of Saskatchewan [Canada]

PCS Power Conditioning System

PCS Power Conversion System

PCS Preferred Character Set

PCS Premium Coal Sample (Facility) [of Argonne National Laboratory, Illinois, US]

PCS Pressure Control System

PCS Primary Coolant System

PCS Print Contrast Scale

PCS Print Contrast Signal

PCS Print Contrast System

PCS Procedure Completion Sheet

PCS Process Control System

PCS Professional Communications Society [of Institute of Electrical and Electronics Engineers, US]

PCS Program Center Store

PCS Program Counter Store

PCS Programmable Communications Subsystem

PCS Programmable Communications System

PCS Project Control System

PCS Protonated Carbon Suppression

PCS Punch(ed) Card System

PCS Patrol Subchaser, Wooden-Type [US Navy Symbol]

Pcs Pieces [also pcs]

PCS$_p$ Print Contrast Signal at Point "p" [Symbol]

PCSA Personal Computing Systems Architecture

PCSA Power Crane and Shovel Association [US]

PC-SAN Polycarbonate/Styrene-Acrylonitrile (Blend) [also PC/SAN]

PCSAS Policy Committee for Scientific Agricultural Societies [US]

PCSB Pulse-Code Scanning Beam

PCSC Pacific Coast Science Center [US]

PCSC Power Conditioning, Switching and Control

PCSD Planar Coincidence Site Density [Crystallography]

PCSD President's Council on Sustainable Development [US]

PCSD Printer Control Sequence Description

PCSFSK Phase Comparison Sinusoidal Frequency Shift Keying

pcs/hr (work)pieces per hour [Unit]

PCSI Annual Conference on the Physics and Chemistry of Semiconductor Interfaces

PCSIG Personal Computer Software Interest Group [US]

[PcSiO]$_n$ Silicon Oxide Phthalocyanine Complex [General Formula]

PCSIR Pakistan Council of Scientific and Industrial Research

PCSIR Labs Pakistan Council of Scientific and Industrial Research Laboratories [Lahore]

PC-SMA Polycarbonate/Styrene Maleic Anhydride (Blend) [also PC/SMA]

pcs/min (work)pieces per minute [Unit]

PCSMQC Polish Committee for Standardization, Measures and Quality Control

PCSP Polar Continental Shelf Project [Canada]

PCSP Program Communications Support Program

PCSS Photoconductive Semiconductor Switch

PCT (International) Patent Cooperation Treaty

PCT P-Chlorotoluene

PCT Peak Can Temperature [Nuclear Engineering]

PCT Perfect Crystal Technology

PCT Peripheral Control Terminal

PCT Phosphorochloridothioate

PCT Photochemical Transfer

PCT Photon-Coupled Transistor

PCT Planning and Control Techniques

PCT Plasmacytoma [Medicine]

PCT Polychlorinated Triphenyl

PCT Polycyclohexane (Dimethylene) Terephthalate

PCT Portable Camera-Transmitter

PCT Pressure–Composition–Temperature [also P-C-T]

PCT Pressure-Cooker Test

PCT Private Communications Technology

PCT Product Consistency Test

PCT Program Control Table

PCT Pulse Count [also pct]

.PCT Picture [File Name Extension] [also .pct]

pct percent(age)

PCTA Polymer of Cyclohexanedimethanol Terephthalic Acid

PCTC P-Chlorobenzotrichloride

PCTE Portable Commercial Test Equipment

PCTEM Phase-Contrast Transmission Electron Microscopy

PCTF Phase-Contrast Transfer Function

PCTF Plant Component Test Facility

PCTF Premier's Council Technology Fund [Canadian Provinces]

PCTFE Polychlorotrifluoroethylene

PCTG Poly(1,4-Cyclohexylene Dimethylene Terephthalate)-glycol

PCT-G Glycol-Modified Polycyclohexyl Terephthalate

PCT-GF Plasmacytoma Growth Factor [Biochemistry]

PCTIS Preston Commercial and Technical Information Service [UK]

PCTM Pulse Count Modulation

PCTO Payload Cost Tradeoff Optimization

PCTP Pentachlorothiophenol [also 5CTP]

PC-TPUR Polycarbonate/Thermoplastic Polyurethane (Blend) [also PC/TPUR]

PCTR (Launch) Pad Connection Terminal Room

PCTR Physical Constant Test Reactor [of Hanford Test Reactor, Washington State, US]

PCTS Philadelphia College of Textiles and Science [Pennsylvania, US]

PCTS Photocapacitance Transient Spectroscopy

PC-TV Personal Computer–Television (Interface)

pct vac percent of vacuum [also % vac]

PCU Paging Control Unit

PCU Passenger Car Unit

PCU Payload Checkout Unit

PCU Peripheral Control Unit

PCU Physical Control Unit

PCU Pound Centigrade Unit [also Pcu, or pcu]

PCU Power and Controller Unit

PCU Power Control Unit

PCU Power Conversion Unit

PCU Pressure Control Unit

PCU Process Control Unit

PCU Processor Control Unit

PCU Program Control Unit

PCU Progress Control Unit

PCU Punched Card Unit

PCUA Power Controller Unit Assembly

PCUA PROFIT (Computer System) Control Users Association [US]

PCUN Population Commission of the United Nations

PCYS Polycrystal Yield Surface

PCV Packed Cell Volume

PCV Plano-Concave (Lens)

PCV Pollution Control Valve

PCV Positive Crankcase Ventilation [Automobiles]

PCV Precheck Verification

PCV Pressure Control Valve

PCV Primary Containment Vessel [Nuclear Reactors]

PCV Purge Control Valve

PCVB Pyro Continuity Verification Box

PCVD Plasma-Assisted Chemical Vapor Deposition; Plasma Chemical Vapor Deposition [also P-CVD]

PCVL Pilot Controlled Visual Landing

PCW Program Control Word

PCW Pulsed Continuous Wave

PCWCA Poured Concrete Wall Contractors Association [US]

PCWP Pulmonary Capillary Wedge Pressure [Medicine]

PCWPC Permanent Committee of the World Petroleum Congress

PCX 3,5-Dinitro-3,5-Diazapiperidinium Nitrate [Symbol]

PCX Plano-Convex (Lens)

PCX Plasma Confinement Experiment [Physics]

.PCX Picture Image [also .pcx] [File Name Extension]

PC XT Personal Computer Expanded Technology [also PC-XT]

PCy$_3$ Tricyclohexyl Phosphine

PCYL Plano-Cylindrical (Lens)

PCZ Panama Canal Zone

PCZ Physical Control Zone

PCZ Positive Control Zone

PD Pharmaceutical Doctor; Doctor of Pharmacy

PD Packing Density

PD Pallet Decoupler

PD Pancreatic Duct [Anatomy]

PD Parabolic Dish

PD Para-Deuterium

PD Parasite Drag [Fluid Mechanics]

PD Parallel Data

PD Parkinson's Disease [Medicine]

PD Partial Depletion

PD Partial Derivative

PD Partial Detonation

PD Partial Discharge

PD Partial Dislocation [Materials Science]

PD Particle Density

PD Particle Detector; Particle Detector

PD Particle Dynamics

PD Partly Deformed [also pd]

PD Passive Detection

PD Pavement Design [Civil Engineering]

PD Peak Demand

PD Pediatric Dentist(ry)

PD Pen Down

PD Penning Deexcitation [Physics]

PD Per Day

PD Perfect Dielectric

PD Periodic Damping [Physics]

PD Periodic Dryer; Periodic Drying

PD Peripheral Device

PD Peritoneal Dialysis [Medicine]

PD Permanent Dipole

PD Permissable Dosage

PD Personal Data

PD Perspective Drawing

PD Petri Dish
PD Phase-Change Dual
PD Phase Difference
PD Phase Diagram
PD Phenyldichloroarsine [Symbol]
PD Photodarkening (Effect)
PD Photodegradable (Plastics)
PD Photodesorption
PD Photodetection; Photodetector
PD Photodiffusion
PD Photodiode
PD Photodisintegration [Nuclear Physics]
PD Photodissociation [Physical Chemistry]
PD Photodynamic(s)
PD Photoelectron Diffraction
PD Photographic Density
PD Photographic Development
PD Photon Detection; Photon Detector [Physics]
PD Physical Distribution
PD Physics Department
PD Physics Division
PD Pictorial Display
PD Pipe Diameter
PD Pitch Diameter (of Gears)
PD Pitch Down
PD Pit Depth [Corrosion]
PD Placer Deposit [Mining]
PD Planar Defect [Crystallography]
PD Plane Deformation
PD Plane Disagreement
PD Plant Design(er)
PD Plasma-Deposited; Plasma Deposition
PD Plasma Doping
PD Plastic Deformation
PD Planned Derating
PD Plasma Display
PD Point Defect [Crystallography]
PD Point Detonating (Fuse)
PD Polar Distance [Astronomy]
PD Policy Directive
PD Police Department
PD Polydisperse; Polydispersity
PD Polymer Degradation
PD Population Density [Ecology]
PD Population Distribution [Statistics]
PD Port Dues
PD Positive Displacement [Mechanical Engineering]
PD Postal District
PD Potassium Dichromate
PD Potential Difference [also pd]
PD Powder Diffraction [Crystallography]
PD Powder Diffractometer; Powder Diffractometry
PD Power Density

PD Power Diode
PD Power Dissipation [also pd]
PD Power Distribution
PD Power Doubler
PD Power Down
PD Preferred Direction [Metallurgy]
PD Pre-Deformation [Mechanics]
PD Preisach Dipole [Solid-State Physics]
PD Preliminary Design
PD Pressed Direction [Metallurgy]
PD Pressure Demand (Respirator)
PD Pressure Differential
PD Pressure Distillate; Pressure Distillation
PD Prime Driver
PD Principal Direction
PD Principal Distance
PD Printer's Devil
PD Priority Directive
PD Prism Diopter
PD Probability Density [Quantum Mechanics]
PD Probability Distribution [Statistics]
PD Probability of Detection
PD Problem Definition
PD Problem Determination
PD Procedure Division [Computers]
PD Process Data
PD Process Design(er)
PD Procurement Division
PD Procurement Document
PD Procurement Drawing
PD Product Data
PD Product Development
PD Production Department
PD Professional Development
PD Professional Diploma
PD Profile Drag [Fluid Mechanics]
PD Program Development
PD Program Directive
PD Programmable Device
PD Project Development
PD Project Directive
PD Projected Display
PD Project Documentation
PD Propellant Dispersal; Propellant Dispersion
PD Propulsion Directorate [of Phillips Laboratory, Edwards Air Force Base, California, US]
PD Propanediamine
PD Proportional Derivative
PD Proportional Differential (Controller)
PD Proprietary Drug [Pharmacology]
PD Propulsion Directorate
PD Proton-Donor [Physics]
PD Proximity Detector
PD Publication Date

PD Public Domain
PD Pulse Delay
PD Pulse Doppler
PD Pulse Driver
PD Pulse Duration
P&D Pickup and Delivery
Pd Palladium [Symbol]
Pd Period [also pd]
Pd^{2+} Divalent Palladium Ion [also Pd^{++}] [Symbol]
Pd^{4+} Tetravalent Palladium Ion [Symbol]
Pd-100 Palladium-100 [also ^{100}Pd, or Pd^{100}]
Pd-101 Palladium-101 [also ^{101}Pd, or Pd^{101}]
Pd-102 Palladium-102 [also ^{102}Pd, or Pd^{102}]
Pd-103 Palladium-103 [also ^{103}Pd, or Pd^{103}]
Pd-104 Palladium-104 [also ^{104}Pd, or Pd^{104}]
Pd-105 Palladium-105 [also ^{105}Pd, or Pd^{105}]
Pd-106 Palladium-106 [also ^{106}Pd, or Pd^{106}]
Pd-107 Palladium-107 [also ^{107}Pd, or Pd^{107}]
Pd-108 Palladium-108 [also ^{108}Pd, or Pd^{108}]
Pd-109 Palladium-109 [also ^{109}Pd, or Pd^{109}]
Pd-110 Palladium-110 [also ^{110}Pd, or Pd^{110}]
Pd-111 Palladium-111 [also ^{111}Pd, or Pd^{111}]
Pd-112 Palladium-112 [also ^{112}Pd, or Pd^{112}]
pD partially coherent dislocation (in materials science) [Symbol]
pd paid
pd partly deformed [also PD]
pd *(per diem)* – per day; for each day
PDA Packaging Distributors Association [UK]
PDA Parenteral Drug Association [US]
PDA Partial Discharge Analyzer
PDA Percent Defective Allowable
PDA Personal Digital Assistant [Telecommunications]
PDA P-Phenylene Diamine
PDA Photodiode Array
PDA Photon Diffusion Approximation [Physics]
PDA Physical Device Address
PDA Piperazine Diacrylamide
PDA Polydiacetylene
PDA Post-Deflection Acceleration [also pda]
PDA Post-Deposition Anneal(ing) [Metallurgy]
PDA Precision Drive Axis
PDA Predocketed Application
PDA Preliminary Design Approval
PDA Preliminary Design Authorization
PDA Probability Discrete Automaton
PDA Probability Distribution Analyzer
PDA Problem Determination Aid
PDA Professional Designers Association [UK]
PDA Propellant Drain Area
PDA Proposed Development Approach
PDA Propylenediamine
PDA Prospectors and Developers Association [Canada]
PDA Pulse Discrimination Analysis

PDA Pulse Height Distribution Analysis
PDA Pump Distributors Association [UK]
PDA Pump Drive Assembly
13PDA 1,3-Propanediamine [also 13Pda]
PD3A 1,2-Propylenediaminetriacetate [also pd3a]
PDAC Prospectors and Developers Association of Canada
Pd(Ac)$_2$ Palladium(II) Acetate
Pd(ACAC)$_2$ Palladium(II) Acetylacetonate [also Pd(acac)$_2$, or Pd(acac)$_2$]
PDAF Provincial Department of Agriculture and Forestry [Taiwan]
Pd-Ag Palladium-Silver (Alloy)
PDAID Problem Determination Aid
PdAl Palladium Aluminide (Compound)
Pd-Al Palladium-Aluminum (Alloy System)
PDA LB Polydiacetylene Langmuir-Blodgett Film [Physical Chemistry]
Pd-Al$_2$O$_3$ Palladium-Aluminum Oxide (System)
PDAP Polydiallylphthalate
PDAPS Pollution Detection and Prevention System
PDAR Preferential Departure and Arrival Route [Aeronautics]
PDAR Program Description and Requirements
PDA Rec PDA Recorder [Publication of the Prospectors and Developers Association of Canada]
PDAT Phosphorodiamidothioc Acid
Pd-Au Palladium-Gold (Alloy System)
PDB P-Dichlorobenzene; Paradichlorobenzene
PDB Performance Data Book
PDB Physical Database
PDB Power Distribution Box
PDB Process Descriptor Base
PDB Protein Databank [of Brookhaven National Laboratory, Upton, New York, US]
PDB Public Debt Bureau [US]
PDBG Polydibutylgermanium
p-DBR P-Type Distributed Bragg Reflector (Mirror)
PDBS Polydibutyl Stannate
PDBTA Poly(diaminobenzamide Terephthalate)
PDBu Phorbol-12,13-Dibutyrate
PDC Package Design Council [now Package Design Council International, US]
PDC Paper Distribution Council [US]
PDC Parallel Data Controller
PDC Performance Data Computer
PDC Permanent Data Call
PDC Photonuclear Data Center [of National Institute for Standards and Technology, Gaithersburg, Maryland, US]
PDC Polycrystalline Diamond Compact
PDC Polydisperse Colloid [Chemistry]
PDC Power Distribution Control
PDC Power Duration Curve
PDC Predefined Command
PDC Premission Documentation Change
PDC Primary Domain Controller

PDC Printing Density Controller

PDC Probability of Detection and Conversion

PDC Problem Domain Component [Computer Programming]

PDC Procurement Document Change

PDC Pyridinium Dichromate

PDC Pyrrolidinedithiocarbamate

PDC Single-Paper, Double-Cotton (Electric Wire)

PDCA Painting and Decorating Contractors of America [US]

PDCE Paramagnetic Design and Cost Effectiveness

PDCI Package Design Council International [US]

PDCLC Pictorial Display Course Line Computer

PD&Co Parke, Davis and Company Limited [Canadian Pharmaceutical Manufacturer]

Pd-Co Palladium-Cobalt (Alloy System)

PDCP Polydicyclopentadiene [also PDCPD]

pDCP Dicresylphospohate [also p-DCP]

PDCR Polished and Double-Stage Cold Rolled; Polishing and Double Stage Cold Rolling [Metallurgy]

PDCR Proprietary Data Control Record

PDCS Performance Data Computer System

PDCS Power Distribution and Control Subsystem; Power Distribution and Control System

Pd-Cu Palladium-Copper (Alloy System)

PDD Partial Discharge Detector

PDD Past Due Date

PDD Physical Damage Division

PDD Physical Data Description

PDD Physical Device Driver

PDD Portable Digital Document

PDD Post Dialing Delay

PDD Precision Digital Dilatometry

PDD Probability Density Distribution [Statistics]

PDD Program Design Data

PDD Projected Data Display

PDD Projected Decision Date

PDD Prospective Decision Date

PDD Pulse Delay Device

PDD PWB (Printed Wiring Board) Design Device

PDDA Polydiallyldimethylammonium Chloride

PDDF Portable Document Delivery Format

PDDI Product Data Definition Interface

Pd(DIPHOS)$_2$ Bis[1,2-Bis(diphenylphosphino)ethane] Palladium

PDDP Product Design and Development Program

PDE Partial Differential Equation [Mathematics]

PDE PEG (Polyethylene Glycol) Diglycidyl Ether

PDE Phosphodiesterase [Biochemistry]

PDE Photon Drag Effect [Physics]

PDE Position-Determining Equipment

PDE Problem Domain Expert [Computer Programming]

PDE 2-(2-Pyridyldithio)ethanol

PDEA Phenyldiethanolamine

PDEAS Phenyl Diethanolamine Succinate

PD-ECGF Platelet-Derived Endothelial Cell Growth Factor [Biochemistry]

PDED Partial Double Error Detecting

PDELS Parallel-Detection Electron Loss Spectrometer

PDEP Pennsylvania Department of Environmental Protection [US]

PDES Preliminary Draft Environmental Statement

PDES Product Data Exchange Specification

PDES Product Definition Exchange Specification

PDES/STEP Product Data Exchange Specification/Standard for the Exchange of Product Model Data

PDF Package Definition File

PDF Pair Distribution Function [Physics]

PDF Parkinson's Disease Foundation [US]

PDF Plant Design Factor

PDF Pocket Dosimeter, FEMA (Federal Emergency Management Agency)

PDF Point Detonating Fuse

PDF Pole Density Function

PDF Portable Data File

PDF Portable Document Format (File) [also pdf]

PDF Portable Document File

PDF Post-Detection Filter

PDF Postdoctoral Fellow

PDF Powder Diffraction File [of International Center for Diffraction Data, Joint Committee on Powder Diffraction Standards at Swarthmore, Pennsylvania, US]

PDF Power Density Function

PDF Preisach Distribution Function [Solid-State Physics]

PDF Principal's Development Fund [of Queen's University, Canada]

PDF Printer Description File

PDF Probability Density Function [also pdf] [Statistics]

PDF Probability Distribution Function [also pdf] [Statistics]

PDF Processor Defined Function

PDF Product Development Fund [of Advanced Systems Institute, Vancouver, British Columbia, Canada]

PDF Program Development Facility

PDF Protected Difference Fat

PDF-1 Powder Diffraction File 1 (Database) [of International Center for Diffraction Data, Joint Committee on Powder Diffraction Standards at Swarthmore, Pennsylvania, US]

PDF-2 Powder Diffraction File 2 (Dasebase) [International Center for Diffraction Data, Joint Committee on Powder Diffraction Standards at Swarthmore, Pennsylvania, US]

.PDF Portable Document Format [File Name Extension] [also .pdf]

.PDF Printer Description (File) [Borlund Lotus File Name Extension]

PDFD Predemonstration Fusion Device

PDG Precision Drop Glider

PDG Pregnanediol Glucoronide [Biochemistry]

Pd-Ga Palladium-Gallium (Alloy System)

Pd/GaAs Palladium on Gallium Arsenide (Substrate)

PDGDL Plasma Dynamics and Gaseous Discharge Laboratory [US]

PdGe Palladium Germanide

Pd/Ge Palladium on Germanium (Substrate)

PDGF Platelet-Derived Growth Factor [Biochemistry]

PDGF Power Data Grapple Fixture

PDGF-AA Platelet-Derived Growth Factor AA [Biochemistry]

PDGF-AB Platelet-Derived Growth Factor AB [Biochemistry]

PDGF-BB Platelet-Derived Growth Factor BB [Biochemistry]

PDGFR Platelet-Derived Growth Factor Receptor [also PDGF-R] [Biochemistry]

PDGS Precision Delivery Glide System

PDGS Product Design Graphics System

PDH Philibert-Duncumb-Heinrich (Equation) [Physics]

PDH Planned Derated Hours

PDH Plesiosynchronous Digital Hierarchy

PDH Phosphate Dehydrogenase [Biochemistry]

PDH Pocket Dosimeter–High

PdH Palladium Hydride

Pd(HFAC)$_2$ Palladium Bis(1,1,1,5,5,5-Hexafluoro-2,4-Acetylacetonate) [also Pd(hfac)$_2$]

PDHS Poly(di-n-Hexylsilane)

PDI Paul-Drude-Institut (für Festkörperelektronik) [Paul Drude Institute (for Solid-State Electronics), Berlin, Germany]

PDI Payload Data Interleaver

PDI Perfect Digital Invariant

PDI Periodic Domain Inverted (Crystal)

PDI Performance Demonstration Initiatives

PDI Pharmacy Dedication Information

PDI Phenylene Diisocyanate

PDI Pictorial Deviation Indicator

PDI Picture Description Instruction

PDI Plastic Drum Institute [US]

PDI Plastic Dual In-Line (Package) [Electronics]

PDI Plumbing and Drainage Institute [US]

PDI Polydispersity Index [Chemistry]

PDI Positive Displacement Injector

1,3-PDI 1,3-Phenylene Diisocyanate

PDIAL Public Dialup Internet Access List [Internet]

PDII Pusat Dokumentasi dan Informasi Ilmiah [Scientific Documentation and Information Center, Indonesia]

Pd-In Palladium-Indium (Alloy System)

PDIO Parallel Digital Input/Output

PDIO Photodiode [also pdio]

PDIP Plastic Dual In-Line Package [Electronics]

PDIP Plastic Dual In-Line Pinout [Electronics]

PDIPP Polydiisopropyl Vinylphenol

PDIR Peripheral Data Set Information Record

PDIS Pressure Differential Switch

PDK 1-Phospho-2,3-Diketo-5-S-Methylthiopentose [Biochemistry]

ΦΔK Phi Delta Kappa (Student Society)

PDL Page-Description Language

PDL Particle Physics Laboratory [US]

PDL Particle Dynamics Laboratory [Canada]

PDL Photodissociation Dye Laser

PDL Plastics Design Library [Division of William Andrews Inc., Norwich, New York, US]

PDL Pocket Dosimeter–Low

PDL Polarization Diversity Lidar

PDL Print Definition Language

PDL Procedure Definition Language

PDL Procedure Distribution List

PDL Process Description Language

PDL Program Description Language

PDL Program Design Language

PDL Program Development Language

PDL Programmable Data Language

PDL Pulse Delta Modulation

PDL Pulse Duration Modulation

PDL Pulse Dye Laser

pdl poundal [Unit]

PdLα Palladium L-Alpha (Radiation) [also PdL$_\alpha$]

PDLC Polymer-Dispersed Liquid Crystal

PDLCD Polymer-Dispersed Liquid Crystal Display

PDLE Process Development Laboratory East [of US Department of Energy]

pdl-ft foot-poundal [Unit]

pdl/ft^2 poundal per square foot [also pdl/sq ft]

pdl·sec/ft^2 poundal-second per square foot [also pdl·sec/sq ft]

pdl·sec/sq ft poundal-second per square foot [also pdl·sec/ft^2]

pdl·s/ft^2 poundal-second per square foot [also pdl·s/sq ft]

pdl sq ft poundal per square foot [also pdl/ft^2]

pdl·s/sq ft poundal-second per square foot [also pdl·s/ft^2]

PD-LSF Potential-Derived (Point Charge) Least-Squares Fitting (Model)

P-DLTS Photo-Deep Level Transient Spectroscopy [also photo-DLTS]

PDLW Process Development Laboratory West [of US Department of Energy]

P-EPR Photo-Electron Paramagnetic Resonance [also photo-EPR]

PDM Perchlorodiphenylmethyl (Radical)

PDM Physical Distribution Management

PDM Pilot Decision Making

PDM Polarization Division Multiplex(ing)

PDM Polynomial Discriminant Method

PDM Positive Displacement Meter

PDM Practical Data Manager

PDM Predictive Maintenance

PDM Processor Data Monitor

PDM Product Data Management; Product Data Manager

PDM Project Development Methodology

PDM Protected Difference Milk

PDM Pulse Delta Modulation

PDM Pulse Duration Modulation

PDMA 1,2-Phenylenebis(dimethylarsine) [also pdma]

PDMA Plumbing and Drainage Manufacturers Association [now Plumbing and Drainage Institute, US]

PDMA Polyamic Acid and Polymellitic Dianhydride

PDMA Product Development and Management Association

PDMA-ODA Polyamic Acid and Polymellitic Dianhydride–4-Oxydianiline

PDME Pendant Drop Melt Extraction (Process) [Powder Metallurgy]

PDME Precision Distance-Measuring Equipment [also P-DME]

PDMEA Phosphoric Acid Mono[2-(Dimethylamino)ethyl]-ester

PDM-ECD Photodetachment-Modulated Electron Capture Detector

PDM-FM Pulse Duration Modulation–Frequency Modulation

PDM-FM-FM Pulse Duration Modulation–Frequency Modulation–Frequency Modulation (System)

PDMG Platinum-Doped Melt Growth (Process) [Metallurgy]

Pd-Mg Palladium-Magnesium (Alloy System)

Pd-MgO Palladium-Magnesium Oxide (System)

PDMHS Peak District Mines Historical Society [UK]

PDMP 1,2-Phenylenebis(dimethylphosphine) [also pdmp]

PDMP Product Development Management Program

PDM-PAM Pulse Duration Modulation–Pulse Amplitude Modulation (System)

PDMS Plasma-Desorption Mass Spectrometry; Plasma-Desorption Mass Spectroscopy

PDMS Polydimethylsiloxane; Poly-2-Dimethylsiloxane

PDMS-PS-PDMS Polydimethylsiloxane-Polystyrene-Polydimethylsiloxane (Copolymer)

PDN Public Data Network

Pdn Production [also pdn]

PDNA Propylene-1,2-Dinitramine

Pd/n-GaAs Palladium on N-Type Gallium Arsenide (Substrate)

PdnHS Poly(di-n-Hexyl Silane)

PdNi Palladium Nickel

Pd-Ni Palladium-Nickel (Alloy System)

Pd/n-InP Palladium on N-Type Indium Phosphide (Substrate)

Pd-NiO Palladium-Nickel Oxide (System)

PDO Portable Distributed Objects

PDO Program Directive Operations

PdO Palladium(II) Oxide (or Palladium Monoxide)

PdOEP 2,3,7,8,12,13,17,18-Octaethylporphine Palladium(II)

PDOS Partial Density of States [Solid-State Physics]

PDOS Phonon Density of States [Solid-State Physics]

PDP Pacific Data Products, Inc. [US]

PDP Parallel Data Processing

PDP Pasture Development Plan [South Australia]

PDP Plasma Diagnostics Package

PDP Plasma Display Panel

PDP Positive Displacement Pump

PDP Postinsertion Deorbit Preparation

PDP Powder Diffraction Program [of Joint Committee on Powder Diffraction Standards, US]

PDP Preliminary Definition Plan

PDP Procedure Definition Processor

PDP Process Development Pile [US]

PDP Procurement Data Package

PDP Product Development Program

PDP Productivity Development Program [of New York State Science and Technology Foundation, US]

PDP Professional Data Processor

PDP Program Definition Phase

PDP Program Development Plan

PDP Programmable Data Processor; Programmed Data Processor

PDP Programmable Digital Processor

PDP Project Definition Phase

PDPC Position Display Parallax-Corrected

Pd-P Palladium-Phosphorus (System)

Pd-PEI Palladium-Polyethylenimine

PD/PGM Power Down/Program

PDPI Pacific Data Products, Inc. [US]

PDPI Primitive Discretized Path Integral

PDPI Proportional Differential Proportional Integral (Controller)

PDPS Parts Data Processing System

Pd-Pt Palladium-Platinum (Alloy System)

PDQ Physician Data Query

PDQ Pretty Damned Quick [Computer Jargon]

PDQ Product Demand Quotation

PDQ Programmed Data Quantizer

PDR Page Data Register

PDR Parallel Downflow Rinse

PDR Peninsula Development Road [Australia]

PDR Performance Data Rate

PDR Periscope Depth Range

PDR Physician's Desk Reference

PDR Pilot's Display Recorder

PDR Policy Development and Research

PDR Potential-Drop Ratio

PDR Power Directional Relay

PDR Precision Depth Recorder

PDR Predetermined Route

PDR Preferential Departure Route

PDR Preferential Diffuse Reflection

PDR Preliminary Data Report

PDR Preliminary Data Requirements

PDR Preliminary Design Review

PDR Pressurized Deuterium Reactor

PDR Priority Data Reduction

PDR Processed Data Recorder

PDR Processing Date Rate

PDR Processing, Distribution and Retailing

PDR Production, Distribution and Retailing

PDR Program Design Review

PDR Program Director's Review
PDR Program Discrepancy Report
PDR Program Drum Recording
PDR Property Disposal Request
PDR Public Document Room
Pdr Pounder [also pdr]
Pdr Powder [also pdr]
PDRD Procurement Data Requirements Document
Pd-Rh Palladium-Rhodium (Alloy System)
PDRL Procurement Data Requirements List
PDRM Payload Deployment and Retrieval Mechanism
PDRP Program Data Requirement Plan
PDRS Payload Data and Retrieval System
PDRS Payload Deployment and Retrieval System
PDRS Pulse Doppler Radar Simulation
PD&RS Payload Deployment and Retrieval Subsystem
PDRSS Payload Development and Retrieval System Simulator
PDRSTA Payload Deployment and Retrieval System Test Article
Pd-Ru Palladium-Ruthenium (Alloy System)
PDS Package Data System
PDS Packet Driver Specification
PDS Parallel Data Storage
PDS Partitioned Data Set
PDS Payload Data Subsystem [of NASA Mars Observer]
PDS Penultimate Digit Storage
PDS Periodicals Data System
PDS Personal Data Sheet
PDS Personal Data System
PDS Petroleum Data System [of University of Oklahoma, Norman, US]
PDS Photodigital Store
PDS Photodischarge Spectroscopy
PDS Photothermal Deflection Spectroscopy
PDS Planar Density of States [Physics]
PDS Planetary Data System [of NASA at Jet Propulsion Laboratory, Caltech, Padadena, California, US]
PDS Plasma Deposition System
PDS Portable Document Software
PDS Power Density Spectrum
PDS Power Distribution Subsystem
PDS Power Distribution System
PDS Predocketed Special Report
PDS Premises Distribution System
PDS Private Database Service
PDS Problem Data System
PDS Procedures Development Simulator
PDS Processor Direct Slot
PDS Professional Development System
PDS Program Data Set
PDS Program Data Source
PDS Programmable Data Station
PDS Programmable Device Support

PDS Programmable Distribution System
PDS Propellant Dispersion System
2-PDS Di-(2-Pyridyl)disulfide
PD/S Problem Definition/Solution
PdS Palladium(II) Sulfide
PDSAR Public Document Status of Assessment Report
PDSC Parabolic Dish Solar Collector
PDSC Pressure Differential Scanning Calorimetry
PDSCWFRS Parabolic Dish Solar Collector Waste Fluid Reduction System
Pd-Si Palladium-Silicon (Alloy System)
Pd/SiC Silicon Carbide Reinforced Palladium (Composite)
PDSO Polydimethoxysilane
PDSPI Polyurethane Division, Society of the Plastics Industry [US]
PDSMS Point-Defense Surface Missile System
PDSS Payload Data Services System
PDSS Post Development and Software Support
PDT Pacific Daylight Time [Greenwich Mean Time +09:00] [also pdt]
PDT Performance Diagnostic Tool
PDT Photodynamic Therapy
PDT Physical Device Table
PDT Picture Description Test
PDT Plasma Diffusion Treatment
PDT Plate Dent Test
PDT Preferential Diffuse Transmission
PDT Product [also Pdt, or pdt]
PDT Programmable Data Terminal
PDT Programmable Drive Table
PDT 3-(2-Pyridyl)-5,6-Diphenyl-1,2,4-Triazine
PDTA 1,2-Propanediaminetetraacetate
PD3A 1,2-Propylenediaminetriacetate [also pd3a]
1,3-PD3A 1,3-Propanediamine-N,N,N'-Triacetate [also 1,3-pd3a]
PDTC Pyrrolidinedithiocarbamate
Pd(TFA)$_2$ Palladium(II) Trifluoroacetate
PDU Pilot('s) Display Unit
PDU Plug Distribution Unit
PDU Power Distribution Unit
PDU Power Drive Unit
PDU Pressure Distribution Unit
PDU Process Development Unit
PDU Product Distribution Unit
PDU Programmable Delay Unit
PDU Protocol Data Unit [Internet]
PDU Pulse Detection Unit
PDV Premodulation Processor Deep-Space Voice
PDV Pressure Disconnect Valve
PdV Palladium Vanadide
Pd-V Palladium-Vanadium (Alloy System)
P-ΔV Load–Differential Voltage (Diagram)
PDW Partially Delactosed Whey
PDW Planar Dielectric Waveguide

Pd-W Palladium-Tungsten (Alloy System)

PDWP Partially Delactosed Whey Powder

PDWS Primary Drinking Water Standards

PDX Poloidal Divertor Experiment

PDX Private Digital Exchange

.PDX Paradox (Files) [Borland File Name Extension]

.PDX Printer Description Extension [File Name Extension] [also .pdx]

pdXi Process Data Exchange Institute [of American Institute of Chemical Engineers, US]

Pd-Zn Palladium-Zinc (Alloy System)

Pd-ZnO Palladium-Zinc Oxide (System)

PE Pacific Electric (Railroad) [US]

PE Packaging Engineer(ing)

PE Page-End (Character)

PE Pairing Energy [Nuclear Physics]

PE Paloecological; Paleoecologist; Paleoecology

PE Paper Electrophoresis

PE Paraelectric(ity)

PE Parallel Element [Computers]

PE Parity (Checking) Enable

PE Parity Error

PE Parity Even (Flag)

PE Partial Erasure [Magnetic Media]

PE Particle Emission

PE Peltier Effect [Physics]

PE Penicillininase Enzyme [Biochemistry]

PE Pentaerythritol

PE Performance Evaluation

PE Peripheral Enable

PE Peripheral Equipment

PE Permanent Echo [Radar]

PE Peroxide Effect

PE Personal Equation

PE Peru [ISO Code]

PE Petroleum Engineer(ing)

PE Petroleum Ether

PE Pharmaceutical Engineer(ing)

PE Phase Encoded; Phase Encoder; Phase Encoding

PE Phosphatidyl Ethanolamine [Biochemistry]

PE Photoeffect

PE Photoejection

PE Photoelastic(ity)

PE Photoelectric(ity)

PE Photoelectrolysis; Photoelectrolytic

PE Photoelectron(ic)

PE Photoelectron Emission

PE Photoelement

PE Photoellipsometer; Photoellipsometric; Photoellipsometry

PE Photoemission; Photoemissivity; Photoemitter

PE Photoengraving

PE Photoetched; Photoetching

PE Photoexcitation

PE Photographic Effect

PE Photographic Engineer(ing)

PE Photogrammetric Engineer(ing)

PE Photon Echo [Physics]

PE R-Phycoerythrin [Biochemistry]

PE Physical Examination

PE Physics Education [of Institute of Physics Publishing, UK]

PE Picture Element [also Pixel, or pixel]

PE Piezoelectric(ity)

PE Pinch Effect

PE Pinion End

PE Piston Engine

PE Planning Engineer(ing)

PE Plant Efficiency

PE Plant Engineer(ing)

PE Plasma Engine

PE Plasma Enhanced; Plasma Enhancement

PE Plasma-Etched; Plasma Etching

PE Plastic Engine

PE Plastic Explosive

PE Plastics Engineer(ing)

PE Plastoelastic(ity)

PE Platinum Electrode

PE Plumbing Engineer(ing)

PE Pockels Effect [Optics]

PE Polar Effect

PE Polluting Emission

PE Polyelectrolyte

PE Polyelectron

PE Polyester

PE Polyether

PE Polyethylene

PE Polymer Engineer(ing)

PE Population Equivalent

PE Porcelain Enamel(ing)

PE Positional Efficiency

PE Position Error

PE Positive Electron

PE Positron Emission [Particle Physics]

PE Post Exchange

PE Potential Energy

PE Powdered Extract

PE Power Engineer(ing)

PE Power Engineering (Society) [of Institute of Electrical and Electronics Engineers, US]

PE Preemption [Telecommunications]

PE Precision Engineer(ing)

PE Pressure Effect [Spectroscopy]

PE Primary Electron

PE Primary Energy

PE Prince Edward Island [Canada]

PE Printer Emulation

PE Printer's Error [also pe]

PE Priority Encoder

PE Probable Error [Statistics]

PE Process Engineer(ing)

PE Processing Element

PE Processor Element

PE Production Engineer(ing)

PE Professional Ecologist

PE Professional Engineer

PE Professional Engraver

PE Professional Estimator

PE Professor Emeritus

PE Programming Environment

PE Project Engineer

PE Propeller Efficiency

PE Protect Enable

PE Protein Engineering

PE Protoenstatite [Mineral]

PE Proton Emission

PE Proton Exchange

PE Proximity Effect [Electrical Engineering]

PE Pseudoelastic(ity)

PE Pulley End

PE Pulse-Echo (Ultrasonic Testing) [also P/E]

PE Pyroelectric(ity)

P(E) Energy Barrier Distribution [Symbol]

P(E) Polarization as a Function of Electric Field [Symbol]

P(E) Probability of an Event [Symbol]

P/E Precipitation-Evaporation (Ratio) [also P-E]

P/E Price/Earnings (Ratio) [also p/e]

P/E Pulse-Echo (Ultrasonic) Testing [also PE]

P-E Pivalic Acid–Ethanol

P-E (Dielectric) Polarization versus Electric Field (Curve)

P-E Precipitation-Evaporation (Ratio) [also P/E]

Pe Péclet (Diffusion) Number [Symbol]

Pe Pellet [also pe]

Pe Pentyl

Pe Perovskite [Mineral]

Pe Petroleum Ether

Pe$_s$ Surface Péclet (Diffusion) Number [Symbol]

.pe Peru [Country Code/Domain Name]

Φ(E) Incident-Energy Fluence Spectrum [Symbol]

Φ$_{eq}$(E$_0$) Equivalent Monoenergetic Neutron Fluence [Symbol]

£E Egyptian Pound [Currency]

PEA Pennsylvania Electric Association [US]

PEA Phenylethanolamine

PEA α-Phenylethyl Alcohol

PEA Phenylethylamine

PEA Phosphatidyl Ethanolamine

PEA Photoelectric Absorption

PEA Pocket Ethernet Adapter

PEA Polyethylacrylate

PEA Positive Electron Affinity [Electronics]

PEA Push-Effective Address

2-PEA 1-(2-Pyridyl)ethanamide [also 2-Pea, or 2-pea]

PEAA Photoelectric Absorption Analysis

PEAB Professional Engineers Appointments Bureau [UK]

PEAC Power Electronics Applications Center [of Electric Power Research Institute, US]

PEAC Program Evaluation and Audit Committee

n-PeAc n-Pentyl Acetate

PEACER Petroleum Employers Advisory Council on Employee Relations

PEACESAT Pan-Pacific Editing and Communication Experiment by Satellite [of University of Hawaii, US]

PEACU Plastic Energy Absorption in Compression Unit

PEAP Pentaerythritol Diacetate Dipropionate

PEAP Personal EGRESS (Emergency Global Rescue, Escape and Survival System) Air Pack

PEAR Plasma Extended Arc Reactor [Canada]

PEARL Package for Efficient Access to Representations in LISP (List Processing)

PEARL Performance Evaluation of Amplifiers from a Remote Location

PEARL Periodical Enquiry, Acquisition and Registration Locally [UK]

PEARL Process, Experiment and Automation Real-Time Language

PEARL Programmed Editor and Automated Resources for Learning

PEAS Pacific's Electric Acquisition Service [of Pacific University, Forest Grove, Oregon, US]

PEB Performance Evaluation Board

PEB α-Phenylethyl Bromide

PEB Potential Energy Barrier

PEB Proposal Evaluation Board

PEB Pulse Electron Beam

PEBA Polyether Block Amide; Polyester Block Amide

PEBC S-Propyl-N-Ethyl-N-Butyl Monothiocarbamate

PEBCO Praseodymium Erbium Barium Copper Oxide (Superconductor)

PE-b-PEE Polyethylene-Poly(ethylethylene) Diblock Copolymer

n-PeBr n-Pentyl Bromide

PEC Packaged Electronic Circuit

PEC Panel Electronic Circuit

PEC Paper Electrochromatography

PEC Passing Electric (DC) Current

PEC Petroleum Energy Center

PEC $(-)$-α-Phenylethyl Chloride

PEC Photoelectric Cell

PEC Photoelectric Colorimeter; Photoelectric Colorimetry

PEC Photoelectrochemical; Photoelectrochemist(ry)

PEC Photoelectrochemical Cell

PEC Photoemissive Cell

PEC Plant Equipment Code

PEC Platform Electronic Cards

PEC Polyester Carbonate

PEC Polyphenylene Ether Copolymer

PEC Position Error Correction

PEC Power Electronics Council [now IEEE (Institute of Electrical and Electronics Engineers) Power Electronics Society, US]

PEC President's Emergency Council [US]

PEC Previous Element Coding

PEC Primary Environment Care

PEC Program Exception Code

PEC Proposal Evaluation Committee

PEC Public Extension Circuit

4-PEC S-[2-(4-Pyridinyl)ethyl]-L-Cysteine [Biochemistry]

Pec Pectoralis [also pec] [Anatomy]

PECAN Pulse Envelope Correlation Air Navigation

PECBI Professional Engineers Conference Board for Industry

PECC Petroleum Exploration Computer Consultants [UK]

PECC Product Engineering Control Center

PECH Poly(epichlorohydrin)

PECG Phonoelectrocardiograph(y)

PECL Positive Electron Coupled Logic

n-PeCl n-Pentyl Chloride

PECM Passive Electronic Countermeasures

PECM Provisional Extracellular Matrix [Biology]

PECM Pulse Electrochemical Machining

PE-co-AA Poly(ethylene-co-Acrylic Acid)

PE-co-EA Poly(ethylene-co-Ethyl Acrylate)

PE-co-MA Poly(ethylene-co-Maleic Anhydride)

PE-co-MAA Poly(ethylene-co-Methacrylic Acid)

PE-co-VA Poly(ethylene-co-Vinyl Acetate)

PECOSY Primitive Exclusive Correlated Spectroscopy [also PE-COSY]

PECP Processing and Environmental Catalysis Program [of Energy Technology Branch, Natural Resources Canada]

PECS Portable Environmental Control System

PECTFE Polyethylene Chlorotrifluoroethylene [also PE-CTFE]

PECUS Personal Engineering Computer Users Society [US]

PECVD Plasma-Enhanced Chemical Vapor Deposition [also PE-CVD]

PECVD-W$_2$N Plasma-Enhanced Chemical Vapor Deposited Ditungsten Nitride

PECVI Plasma-Enhanced Chemical Vapor Infiltration [also PE-CVI]

PED Personnel Equipment Data

PED Photoelectr(on)ic Device

PED Photoelectron Diffraction

PED Polymer Engineering Directorate [UK]

PED Potential Energy Distribution

PED Pulse Electric Discharge; Pulse Electric Discharging

PED Pulsed Electrodeposition

PED Pulse Edge Discrimination

PED Pyroelectric Detector

Ped Pedal [also ped]

Ped Pedestal [also ped]

Ped Pediatric(ian); Pediatrics

Ped Pediatrics [US Journal]

Ped Pedological; Pedologist; Pedology

pedalfer soil with aluminum and iron accumulation [Geology]

Ped Clin North Am Pediatric Clinics of North America

PEDCUG Planning Engineers Desktop Computer Users Group [US]

PedD *(Paedagogiae Doctor)* – Doctor of Pedagogy

Pediatr Pediatric(ian); Pediatrics

Pediatr Pediatrics [of American Academy of Pediatrics]

Pediatr Clin North Am Pediatric Clinics of North America (Journal)

Pediatr Neurosurg Pediatric Neurosurgery [International Journal]

Pediatr Res Pediatric Research [US Journal]

PEDIN Petroleum Exploration Data Index

PEDN Planned Event Discrepancy Notification

Ped Neurosurg Pediatric Neurosurgery [International Journal]

pedocal soil with calcium carbonate accumulation [Geology]

PEDODSDTF Pyrazino Ethylenedioxodiselenadithiafulvalene

PEDRA Parallel Electrical Debye Response Analysis (Method) [Physics]

Ped Res Pediatric Research [US Journal]

PEDRO Pneumatic Energy Detector with Remote Optics

PEDS Peltier Effect Diffusion Separation

PEDS Precise Engineering and Deformation Survey [Canada]

PEDS Protective Equipment Decontamination Section [of Institute of Electrical and Electronics Engineers, US]

PEDT Polyethyldioxythiophene

PEE Photoelastic Effect

PEE Photoelectric Effect

PEE Photoelectric Ellipsometer; Photoelectric Ellipsometry

PEE Photoelectric Emission

PEE Photoelectron Emission

PEE Piezoelectric Effect

PEE Poly(ethylethylene)

PEE Pyroelectric Effect

P&EE Proving and Experimental Establishment [UK]

PEEA N-Phenyl-N-Ethyl Ethanolamine

PE/ED&PG Power Engineering/Energy Development and Power Generation [Institute of Electrical and Electronics Engineers–Power Engineering Society Committee, US]

PEEE Photostimulated Exoelectron Emission

PEEK Polyetherether Ketone

PEEKK Polyetherether Ketone Ketone

PEEKKK Polyetherether Ketone Ketone Ketone

PEEK-PES Polyetheretherketone/Polyether Sulfone (Blend) [also PEEK/PES]

PEEK-PESV Polyetheretherketone/Polyether Sulfone (Blend) [also PEEK/PESV]

PEEL Programmable Electrically Erasable Logic

PEELS Parallel Electron-Energy Loss Spectrometer; Parallel Electron Energy-Loss Spectrometry

PEEM Panel of Experts on Environmental Management (for Vector Control)

PEEM Phosphorescence Emission Excitation Matrix

PEEM Photoelectron Emission Microscopy

PEEM Photoelectron Emission Electron Microscopy

PEEM Photoemission Electron Microscopy

PE/EM Power Engineering/Electric Machinery [Institute of Electrical and Electronics Engineers–Power Engineering Committee, US]

PEEP Pilot's Electronic Eye-Level Presentation

PEEP Positive End-Expiratory Pressure

PEER Partnerships for Environmental Education and Research [US]

PEER Performance Evaluation and Expenditure Review

PEER Professional Engineers Employment Registry [US]

PEES Polyetherether Sulfone

PEES-co-ES Polyetherether Sulfone-co-Ethersulfone

PEF Packaging Education Foundation [US]

PEF Photoelectric Fluorometer

PEF Physical Electronics Facility

PEF Potential Energy Function

PEFMA Polyethylene Foam Manufacturers Association [UK]

PEFO Payload Effects Follow-on (Study)

PEFR Peak Expiratory Flow Rate

PEG Penning (Ionization) Gauge

PEG Petrochemical Energy Group [US]

PEG Pneumoencephalography [Medicine]

PEG Polycrystalline (Silicon) Extrinsic Gettering

PEG Polyethylene Glycol [also PG]

PEG Powered Explicit Guidance

PEG Project Execution Guidelines

PEGA Precision Engineering Grinding Apparatus

PEGAD Permission Granted to Add

PEGASUS Precision Engineering Grinding Apparatus for Superfinishing Ultrahard Surfaces

PEG-DMSO Polyethylene Glycol–Dimethyl Sulfoxide (Solution)

PEG-HRP Polyethylene Glycol–Horseradish Peroxidase [Biochemistry]

PEG-Lipase Polyethylene Glycol Lipase [Biochemistry]

PEG-Protein Polyethylene Glycol Protein [Biochemistry]

PEG-SOD Polyethylene Glycol–Superoxide Dismutase [Biochemistry]

PEG-Subtilisin Polyethylene Glycol–Subtilisin [Biochemistry]

PEH Polyphenylene Ether Homopolymer

PEHA Pentaethylenehexamine

PEI Paul-Ehrlich-Institut–Bundesamt für Sera und Impfstoffe [Paul Ehrlich Institute–Federal Office for Sera and Vaccines, Germany]

PEI Petroleum Equipment Institute [US]

PEI Philips Electronic Instruments Company

PEI Polyetherimide

PEI Polyethylene Imide

PEI Polyethylenimine

PEI Porcelain Enamel Institute [US]

PEI Power Engineering Institution [of Chinese Mechanical Engineering Society]

PEI Preliminary Engineering Inspection

PEI Prince Edward Island [Canada]

PEIC Periodic Error Integrating Controller

PE/IC Power Engineering/Insulated Conductors [Institute of Electrical and Electronics Engineers–Power Engineering Committee, US]

PEIEC Prince Edward Island Energy Corporation [Canada]

PEIFL Prince Edward Island Federation of Labour [Canada]

PEI-GTA Polyethyleneimine–Activated Glutaraldehyde

PEILS Prince Edward Island Land Surveyors [Canada]

PEIM Polyester Imide

Peine Salzgitter Ber Peine Salzgitter Berichte [Peine Salzgitter Report; of Peine Salzgitter AG, Germany]

PEIR Photoelectric Infrared Radiation

PEIR Project Equipment Inspection Record

PEIS Programmatic Environmental Impact Statement [of Argonne National Laboratory, Illinois, US on Depleted Uranium Hexafluoride]

PEISC Prince Edward Island Safety Council [Canada]

PEITB Prince Edward Island Travel Burau [Canada]

PEK Phase-Exchange Keying

PEK Polyether Ketone

PEKEKK Polyether Ketone Ether Ketone Ketone

PEKK Polyether Ketone Ketone

PEL Permissible Exposure Limit [of US Occupational Safety and Health Administration]

PEL Personnel Exposure Limit

PEL Physics Express Letters [of Institute of Physics Publishing, UK]

PEL Picture Element [also Pixel, or pixel]

PEL Power Electronics (Council) [of Institute of Electrical and Electronics Engineers, US]

PEL Precision Elastic Limit

PEL Pressure Exposure Limit

PELA P.E. LaMoreaux and Associates Inc. [Oak Ridge, Tennessee, US]

PEL/ET Power Electronics/Electronic Transformers [Institute of Electrical and Electronics Engineers– Power Engineering Committee, US]

PELS Proton Energy Loss Spectroscopy

PELSS Precision Emitter Location Strike System

PELTP Personnel Licensing and Training Panel [of International Civil Aviation Organization]

PEM Particle Environment Monitor

PEM Photoelastic Modulator

PEM Photoelectric Microscope; Photoelectric Microscopy

PEM Photoelectromagnetic(s)

PEM Photoelectron Microscopy

PEM Piezoelastic Modulation

PEM Piezoelectric Material

PEM Piezoelectric Microphone

PEM Plant Engineering and Maintenance [Canadian Journal]

PEM Plasma Electrolyte Membrane (Fuel Cell)

PEM Plastic-Envelope Method

PEM Privacy Enhanced Mail [Internet]

PEM Process Engineering Magazine [Germany]

PEM Process Execution Module

PEM Processor Element Memory

PEM Processing Element Memory

PEM Procurement Equipment and Missiles [of US Army]

PEM Production Engineering Measure

PEM Program Execution Monitor

PEM Proto-Environmental Model

PEM Proton Exchange Membrane

PEM Pulse Electromagnetic

PEMA 2-Phenyl-2-Ethylmalonamide

PEMA Polyethyl Methacrylate

PEMA Process Equipment Manufacturers Association [US]

PEMA Procurement Equipment and Missiles–Army [US]

PEMAC Plant Engineering and Maintenance Association of Canada

PEMAC Professional Engineers Manpower Assessment Committee [Canada]

Pemb Pembrokeshire [Wales]

PEMC Photoelectromagnetic Cell

PEMCSCF Pair-Excitation Multiconfiguration Self-Consistent Field [Physics]

PEMD Program for Export Market Development [of Export Programs Division, Department of Foreign Affairs and International Trade, Canada]

PEME Photoelectromagnetic Effect

Pemex Petróleos Mexicanos [Mexican Petroleum and Gas Corporation]

PEMF Photoelectromotive Force

PEMFC Polymer Electrolyte Membrane Fuel Cell

PEMM Dimethyl Phosphatidyl Ethanolamine

PE-MOCVD Plasma-Enhanced Metalorganic Chemical Vapor Deposition

PEMS Pesticide Enforcement Management System [of US Environmental Protection Agency]

PEMS Policy and Expenditure Management System

PEMS Porcelain-Enamelled Metal Substate

PEMV Pea Enation Mosaic Virus

PEN Peninsula Environmental Network [Australia]

PEN Polyethylene Naphthalene

Pen Penicillamine; Penicillamyl [Biochemistry]

Pen Penicillin [also PEN] [Biochemistry]

Pen Peninsula [also pen]

Pen Pentane

PENA Primary Emission Neutron Activation

PEN-B Penicillin, Benzalthine Salt [also Pen-A] [Biochemistry]

PENBAL Peninsular Spain-Balearic Islands Submarine Cable [Spain]

PENCAN Peninsular Spain-Canary Islands Submarine Cable [Spain]

PENCE Protein Engineering Network of Centres of Excellence [Canada]

PENCIL Pictorial Encoding Language

Pend Pendulum [also pend]

P-ENDOR Pulsed Electron Nuclear Double Resonance

P/E News Petroleum Energy Business News Index [of American Petroleum Institute, US]

PEN-F Penicillin F [also Pen-F] [Biochemistry]

PEN-G Penicillin G; Benzylpenicillin [also Pen-G] [Biochemistry]

PEng Professional Engineer

Penin Peninsula [also penin]

PEN-K Penicillin, K (Potassium) Salt [also Pen-K] [Biochemistry]

Penn Pennsylvania [US]

Penn University of Pennsylvania [Philadelphia, US]

PEN-NA Penicillin, Sodium Salt [also Pen-Na] [Biochemistry]

Penna Pennsylvania [US]

Penn-MRL University of Pennsylvania, Materials Research Laboratory [Philadelphia, US]

PENNSTAC Penn(sylvania) State University Automatic Computer

Penn State Pennsylvania State University [University Park, US]

PEN-O Penicillin O [also Pen-O] [Biochemistry]

PEN-P Penicillin, Procain Salt [also Pen-P] [Biochemistry]

PE/NPE Power Engineering/Nuclear Power Engineering [Institute of Electrical and Electronics Engineers– Power Engineering Committee, US]

PENRAD Penetration Radar

PEN SDK Pen Computing Software Development Kit

PENT Poly(Ethylene-Naphthalate-Terephthalate)

PENT Polyethylene Notch Tensile Test

Pent Pentagon(al) [also pent]

Pent Pentagon Building [Washington, DC, US]

Pent Pentode [also pent]

Pent Pentolite [High Explosive]

PEN-V Penicillin V; Phenoxymethyl Penicillin [also Pen-V] [Biochemistry]

PEO Patrol Emergency Officer

PEO Piezoelectric Oscillator

PEO Polyethylene Oxide

PEO Polymer Electrolyte

PEO Professional Engineers of Ontario [Canada]

2PeoA trans-2-Pentenoic Acid

3PeoA trans-3-Pentenoic Acid

PEOC-Cl 2-Triphenylphosphonioethyl Chloroformate Chloride

PEOE Partial Equalization of Orbital Electronegativity

n-PeOH n-Pentyl Alcohol

PEO-PBMA Polyethylene Oxide/Polybutyl Methacrylate

PEO-PC Polyethylene Oxide/Polycarbonate

PEO-PET Polyethylene Oxide/Polyethylene Terephthalate

PEO-PMA Polyethylene Oxide/Polymethyl Acrylate

PEOS Propulsion and Electrical Operating System

PEO-SDS Polyethylene Oxide/Sodium Dodecylsulfate (Solution)

PEOX Polyethyl Oxazoline [also PEOx]

PEP (Data) Packet Exchange Protocol

PEP Packetized Ensemble Protocol

PEP Parametric Element Processor

PEP Partitioned Emulation Program(ming)

PEP Pauli Exclusion Principle [Quantum Mechanics]

PEP Peak Envelope Power

PEP People for Energy Progress [US]

PEP Phosphoenolpyruvate; Phosphoenolpyruvic Acid

PEP Photoelectric Photometry

PEP Photoelectric Potential

PEP Piezoelectric Polymer

PEP Planar Epitaxial Passivated (Transistor)

PEP Planar Epitaxial Technology with Passivation

PEP Planetary Ephemeris Program [US]

PEP Plastics Education Foundation [of Society of Plastics Engineers, US]

PEP Political and Economic Planning

PEP Polyethylene Polyamine

PEP Poly(ethylene Propylene)

PEP Polymer Engineering Program

PEP Porsche Experimental Prototype

PEP Positron Electron Photon

PEP Power Extension Package

PEP Program Evaluation Procedure

PEP Program to Enhance Productivity

PEP Programmed Emulation Partition

PEP Propulsion and Energetics Panel [of NATO Advisory Group for Aerospace Research and Development]

PEP Proton-Electron-Positron (Beam)

PEP Public Education Program [of American Nuclear Society, US]

4-PEP S-[2-(4-Pyridyl)ethyl]-DL-Penicillamine

PeP Proton-Electron-Proton (Reaction) [also PEP] [Nuclear Physics]

PEPA Phenylthynyl-Terminated BPDA-PDA (Poly(Bisphenyldianhydride-p-Phenylenediamine)) (Oligomer)

PEPAG Physical Electronics and Physical Acoustics Group

PEPAT Protocol on Environment Protection of the Antarctic Treaty

PEP-b-PEE Poly(ethylenepropylene)-Poly(ethylethylene) Diblock Copolymer

PEPC Phosphoenol Pyruvate Carboxy

PEPCK Phosphoenolpyruvate Carboxykinase [Biochemistry]

PEPE Parallel Element-Processing Ensemble

PE-PEE Polyethylene-Poly(ethylethylene) Diblock Copolymer

PEPEOA N,N-Bis[2-(Diphenylphosphino)ethyl] 2-Diphenylphosphinyl Ethanamine

PEPICO Photoelectron-Photo-Ion Coincidence

PEPIPICO Photoelectron-Photo-Ion Photo-Ion Coincidence

PEP-L Partitioned Emulation Programming–Local

PEP-LR Partitioned Emulation Programming–Local/Remote

PEPP Planetary Entry Parachute Program

PEPP Professional Engineers in Private Practice

PEP-PEE Poly(ethylenepropylene)/Poly(ethylethylene) Diblock Copolymer

PEPPI Program to Encorage Product and Process Innovation

PEPR Precision Encoding and Pattern Recognition

PEPR Precision Encoding and Pattern Recognition Device [of Massachusetts Institute of Technology, Cambridge, US]

PE/PSC Power Engineering/Power System Communications [Institute of Electrical and Electronics Engineers–Power Engineering Committee, US]

PE/PSR Power Engineering/Power System Relaying [Institute of Electrical and Electronics Engineers– Power Engineering Committee, US]

PEPVD Plasma-Enhanced Physical Vapor Deposition [also PI-PVD]

PER Diammonium Peroxydisulfate

PER Path Extension Ratio

PER Perchloroethylene

PER Pollution Emission Register

PER Polluting Emissions Register [of European Community]

PER Preliminary Engineering Report

PER Price-Earnings Ratio

PER Process Energy Requirement

PER Program Event Recording

PER Program Execution Request

PER Preliminary Engineering Report

PER Production Engineering and Research

PER Protein Efficiency Ratio

PER Pseudoequilibrium Ratio

PER Public Environment Report

Per Performance [also per]

Per Pericardium [Medicine]

Per Perigee [Astronomy]

Per Period(ical) [also per]

Per Persia(n)

Per Person [also per]

PERA Planning and Engineering for Repairs and Alterations

PERA Production Engineering Research Association [UK]

PERA Production Engine Remanufacturers Association [US]

PERA Project Engineering Research Association [UK]

Perc Percussion [also perc]

Perc Dr Percussion Drilling

Percn Percussion [also percn]

PERCOS Performance Coding System

PERD Panel on Energy Research and Development [Canada]

PERD Program of Energy Research and Development [of Natural Resources Canada]

PEREF Propellant Engine Research Environmental Facility

Perem Zvezdy Peremennye Zvezdy [Russian Journal on Astronomy]

PERF Petroleum Environmental Research Forum

Perf Perfection [also perf]

Perf Perforation [also perf]

Perf Perform(ance) [also perf]

perf perfect

perf perforate(d)

PERFECT Productivity Erosion Runoff Functions to Evaluate Conservation Techniques

Perfn Perforation [also perfn]

Perform Performance [also perform]

Perform Chem Performance Chemicals [Journal published in the US]

Perform Eval Performance Evaluation [Journal published in the Netherlands]

Perform Eval Rev Performance Evaluation Review [Published by the Association for Computing Machinery, US]

Perfs Perforations [also perfs]

PERGO Project Evaluation and Review with Graphic Output

perh perhaps

PERI Photoengravers Research Institute [US]

PERI Protein Engineering Research Institute [Japan]

Perif Peripheral [also perif]

Period Periodical [also period]

Periodic EMF Periodic Electromotive Force [also periodic emf]

Period Math Hung Periodica Mathematica Hungarica [Hungarian Mathematical Periodical]

Period Polytech Periodica Polytechnica [Hungarian Polytechnical Periodical]

Period Polytech, Chem Eng Periodica Polytechnica, Chemical Engineering [Hungarian Periodical on Chemical Engineering]

Period Polytech, Electr Eng Periodica Polytechnica, Electrical Engineering [Hungarian Periodical on Electrical Engineering]

Period Polytech, Mech Eng Periodica Polytechnica, Mechanical Engineering [Hungarian Periodical on Mechanical Engineering]

Perkin-Elmer Tech News Perkin-Elmer Technical News [Published by Perkin-Elmer Corporation, Danbury, Connecticut, US]

Perkin Trans Perkin Transactions [of Royal Society of Chemistry, UK]

Perkin Trans I Perkin Transactions I [of Royal Society of Chemistry, UK]

Perkin Trans II Perkin Transactions II [of Royal Society of Chemistry, UK]

PERL Perkin Elmer Robot Language

PERL Practical Extraction and Report(ing) Language [also perl] [Internet]

PERL Production Engineering Research Laboratory [of Hitachi Ltd., Yokohama, Japan]

PERM Program Evaluation for Repetitive Manufacture

Perm Permanance [also perm]

Perm Permeability [also perm]

perm permanent(ly)

perm-dyn permanent dynamic

Perminvar Permeability Invariant (Material)

permly permanently

Permn Permission [also permn]

PERP Process Evaluation/Research Planning

perp perpendicular

per pro (per procura) – by proxy

Pers Persia(n)

Pers Persist(ence) [also pers]

Pers Person(al) [also pers]

Pers Personnel [also pers]

Pers Comput (Germany) Personal Computer (Germany)

Pers Comput (USA) Personal Computing (USA) [US Magazine]

Pers Comput Mag Personal Computer Magazine [UK]

Pers Comput World Personal Computer World [Journal published in the UK]

Pers Manage Personnel Management [Journal published in the UK]

PERT Performance Evaluation (and) Review Technique

PERT Pollution Emergency Response Team [Canada]

PERT Production and Evaluation Review Technique

PERT Program Evaluation and Review Technique

PERT Program Evaluation Research Task

Pert Pertussis (or Whooping Cough) [Medicine]

pert pertaining

PERTCO Program Evaluation and Review Technique with Cost

pertg pertaining

PERU Production Equipment Records Unit

Peruv Peruvian

PES Paraelectric Substance

PES Partie Ecologiste Suisse [Swiss Green Party]

PES Peruvian Sol [Currency of Peru]

PES Petroleum Engineer Search

PES Petroleum Exploration Society [UK]

PES Photoelectric Scanning

PES Photoelectron Spectroscopy

PES Photoelectron Spectrum

PES Photoemission Spectroscopy

PES Photoemission Spectrum

PES Piezoelectric Semiconductor

PES Piezoelectric Sensor

PES Plasma Emission Spectroscopy

PES Polyether Sulfone

PES Positioning Error Signal

PES Positron Emission Spectroscopy

PES Potential Energy Surface

PES Power Electronics Society [of Institute of Electrical and Electronics Engineers, US]

PES Power Engineering Society [of Institute of Electrical and Electronics Engineers, US]

PES Primary Earnings per Share

PES Processor Enhancement Socket

PES Program Execution System

PES Programmer Electronic Switch

PES Public Enquiry System

PES Pyroelectric Substance

PESA Petroleum Equipment Suppliers Association [US]

PESA Petroleum Exploration Society of Australia

PESC Passivated Emitter Solar Cell

PESC Power Electronics Specialists Conference

PESDA Printing Equipment Supply Dealers Association

PESGB Petroleum Exploration Society of Great Britain

PESICO Photoelectron Secondary Ion Coincidence

PESIS Photoelectron Spectroscopy of Inner Shell(s)

PESM Photoelectron Spectromicroscopy

PESOS Photoelectron Spectroscopy of Outer Shell(s)

PESP Pesticide Environmental Stewardship Program [of US Environmental Protection Agency]

PE/SPD Power Engineering/Surge-Protective Devices [Institute of Electrical and Electronics Engineers– Power Engineering Society, US]

PESR Pressurized Electroslag Remelting [Metallurgy]

PEST Parameter Estimation by Sequential Testing

PEST People for Environmentally Sustainable Transport

PESTAB Pesticides Abstracts (Database) [of US Environmental Protection Agency]

PESTDOC Pesticide Documentation (Database) [UK]

PESTF Proton Event Start Forecast

PE/SUB Power Engineering/Substations [Institute of Electrical and Electronics Engineers–Power Engineering Committee, US]

PESV Pressurized Electroslag Remelting [Metallurgy]

PE/SWG Power Engineering/Switchgear [Institute of Electrical and Electronics Engineers–Power Engineering Committee, US]

PET Lead Europium Telluride

PET Patterned Epitaxial Technology

PET Pentaerythritol

PET Peripheral Equipment Tester

PET Permeation Enhancement Technology

PET Personal Electronic Transaction Computer

PET Personal Electronic Transactor

PET Phase Elapsed Time

PET Photoelectric Transition

PET Photoelectric Tube

PET Photographic Equipment Technician

PET Physical Equipment Table

PET Piezoelectric Transducer

PET Pilot-Line Experiment Technology [of US Army]

PET Plasma Edge Technique

PET Plastic Engine Technology

PET Point of Equal Time

PET Polyethylene Terephthalate [also PETP]

PET Portable Executive Telephone

PET Position-Event-Time

PET Positron Emission Tomography

PET Precision End Trimmed

PET Print Enhancement Technology

PET Production Environmental Testing

PET Program Evaluator and Tester

Pet Petroleum [also pet]

PEt₃ Triethyl Phosphine

PETA Pentaerythritol Tetraacetate

PETA Pentaerythritol Triacrylate

PETA Performance Evaluation and Trend Analysis

PETAN Perkin-Elmer Thermal Analysis Newsletter [US]

PETB Preflight Test Bus

PETC Pittsburgh Energy Technology Center [of US Department of Energy in Pennsylvania]

PE/TC Power Engineering/Technical Council [Institute of Electrical and Electronics Engineers–Power Engineering Committee, US]

PETCO Plastic Engine Technology Corporation

PE/T&D Power Engineering/Transmission and Distribution [Institute of Electrical and Electronics Engineers–Power Engineering Committee, US]

PETE Partnership for Environmental Technology Education [US]

PETE Pneumatic End-to-End

PetE Petroleum Engineer [also Pet Eng, or PetEng]

Pet Eng Petroleum Engineer(ing)

Pet Eng Int Petroleum Engineering International (Journal)

PETFE Polyethylene Tetrafluoroethylene; Polyethylene-co-Tetrafluoroethylene [also PE-TFE]

PETG Glycol-Modified Polyethylene Terephthalate; Polyethylene Terephthalate Glycol [also PET-G]

Pet Geol Petroleum Geologist; Petroleum Geology

Pet Intell Wkly Petroleum Intelligence Weekly [US]

PETMA Portable Electric Tool Manufacturers Association [UK]

Pet Manage Petroleum Management [Journal published in the US]

PETN Pentaerythritol Tetranitrate

PETP Polyethylene Terephthalate [also PET]

PET-PC Polyethylene Terephthalate/Polycarbonate (Blend) [also PET/PC]

PE/TR Power Engineering/Transformers [Institute of Electrical and Electronics Engineers–Power Engineering Committee, US]

Petr Petrol(eum) [also petr]

Petr Petrological; Petrologist; Petrology [also petr]

PETRA Positron-Elektron-Tandem-Ringbeschleuniger-Anlage [Positron Electron Tandem Ring Accelerator, Germany]

Petr Eth Petroleum Ether [also petr eth]

Petrin Pentaerythritol Trinitrate

Petrin Acr Pentaerythritol Trinitrate Acrylate

Petro Petroleum [also petro]

Petróbrás Petróleos Brasileiros [Brazilian National Petroleum Company]

PetroCan Petroleum Company of Canada

PetroE Petroleum Engineer [also Petro Eng, or PetroEng]

Petro Eng Petroleum Engineer(ing)

PETROEX Petroleum Products Exchange Data Clearinghouse [US]

Petrol Petroleum [also petrol]

Petrol Petrology [Journal Published in Russia]

Petroleum X Petroleum Exchange [US Organization]

PETROMIN General Petroleum and Minerals Organization [also Petromin, Saudi Arabia]

Petro-PC Microcomputers in Petroleum Exploration and Production

PETS Pacific Electronics Trade Show

PETS Payload Environmental Transportation System

PETS Payload Experiment Test System [of NASA Kennedy Space Center]

PETS Pentaerythritol Tetrastearate

PETS Plastic Education and Troubleshooting System

PETS POCC (Payload Operations Control Center) Experiments Timeline System [NASA Ground Data Systems Division and Spacelab]

PET Scan Positron Emission Tomography Scan [also PET scan]

PETT Portable Ethernet Transceiver Tester

PETX Tetra(nitraminomethyl)methane [Symbol]

PEU Port Expander Unit

PEV Peak Envelope Voltage

PEV Photoelectric Voltage

PeV Petaelectronvolt [Unit]

PEW Percussion Welding

PEW Polyurethane Enamelled (Electric) Copper Wire

PEX PHIGS (Programmers' Hierarchical Interactive Graphics System) Extension to X-Windows System

PEX Pituitary Extract [Medicine]

PEX Projectable Excitation

p ex *(par exempli)* – for example

pExA p-Ethoxyaniline

PEXAFS Photoemission Extended X-Ray Absorption Fine Structure

PEY Photoelectron Yield

PEYS Photoelectron Yield Spectroscopy

PF Packing Factor [Metallurgy]

PF Packing Fraction [Nuclear Physics]

PF Page Footing

PF Page Formatter

PF Parachute Facility

PF Parallel Flow [Electrical Engineering]

PF Parasitic Fluorescence

PF Parkinson Foundation

PF Part Family

PF Partition Function [Physics]

PF Path Function

PF Payload Forward

PF Payload Function

PF Peak Factor

PF Penetration Fracture (Test) [Mechanics]

PF Perchlorylfluoride

PF Percussion Fuse

PF Perfect Fluid

PF Periodic Function [Mathematics]

PF Permafrost

PF Permeability Factor

PF Personal Freezer

PF Phase Factor [also pf]

PF Phase Focusing

PF Phenol-Formaldehyde

PF Photofinish(ing)

PF Photofission

PF Photofluorograph(ic); Photofluorography

PF Photographic Film

PF Photon Factory [of National Laboratory for High-Energy Physics, US]

PF Picrylfluoride

PF Picture Frequency

PF Pinning Force [Solid-State Physics]

PF Plane Front

PF Plasma Frequency

PF Platelet Factor [Immunochemistry]

PF Plastic Flow

PF Plug Flow (Transport)

PF Pneurop Flange

PF Point Foundation [US]

PF Point of Frog [Civil Engineering]

PF Poiseuille Flow [Fluid Mechanics]

PF Polar Front [Meteorology]

PF Polarization Factor

PF Polarizing Filter [Optics]

PF Pole Figure [Crystallography]

PF Poloidal Field [Physics]

PF Polygonal Ferrite [Metallurgy]

PF Polymer(ic) Foam

PF Polynesia (French) [ISO Code]

PF Polynésie Française [French Polynesia]

PF Poole-Frenkel (Effect) [Physics]

PF Position Finder; Position Finding

PF Potential Flow

PF Potential Function

PF Powder Forged; Powder Forging [also P/F]

PF Powered Flight

PF Power Factor [also pf]

PF Powder Filter

PF Power Frequency

PF Preflight

PF Pressure Feed

PF Pressure Filter

PF Primary Fuel

PF Prime Function

PF Principal Focus

PF Printing Foil

PF Probability Factor

PF Probability Function

PF Probability of Failure

PF Program Function

PF Programmable Function

PF Protection Factor

PF Pulse Frequency

PF Pulverized Fuel

PF Punch Off [Data Communications]

PF4 Platelet Factor 4 [Immunochemistry]

P&F Scholarships and (Postdoctoral) Fellowships

P/F Poll/Final (Bit) [Data Communications]

P/F Powder Forged; Powder Forging [also PF]

Pf Profile [also pf]

pF picofarad [also pf] [Unit]

pf pfennig [German Monetary Unit; 1/100 of a Deutschmark] [also pfg]

.pf French Polynesia [Country Code/Domain Name]

PFA Palmdale Final Assembly [California, US]

PFA Perfluoroalkoxy (Polymer)

PFA Perfluoroalkoxy Alkene

PFA Pesticide Formulators Association [now Pesticide Producers Association, US]

PFA Polyfluoroalkoxy

PFA Polyfurfuryl Alcohol

PFA Polyurethane Foam Association [US]

PFA Popular Flying Association [UK]

PFA Power Fastenings Association [UK]

PFA Press Foundation of Asia

PFA Presse Française Associée [French Associated Press]

PFA Program and File Analysis

PFA Provincial Forestry Administration [Taiwan]

PFA Pulse Flip Angle

PFA Pulverized Fuel Ash

pFA p-Fluoroaniline

Pfab Prefabrication [also pfab]

PFB Payload Feedback

PFB Payload Forward Bus

PFB Position Feedback

PFB Praseodymium Iron Boron (Compound)

PFB Preformed Beams [Sonar]

PFB Pressure Fed Booster

PFA Pressurized Fluidized Bed

PFB Primary Fore-Brain [also pfb]

PFB Printer Font Binary (File)

PFB Provincial Food Bureau [Taiwan]

PFB Provisional Frequency Board

.PFB Printer Font Binary [File Name Extension] [also .pfb]

PFBA Polyperfluorobutylacrylate

pFBA p-Fluorobenzoic Acid

PFBC Pressurized Fluidized-Bed Combustion

PFBHA O-(Pentafluorobenzyl)hydroxylamine

PFBR Prototype Fast Breeder Reactor

PFBT Performance Functional Board Tester

PFC Pacific Forestry Center [Canada]

PFC Pack Feed and Converter

PFC Park and Forest Commission [Connecticut, US]

PFC Parkinson Foundation of Canada

PFC Part Failure Code

PFC Perfluorocarbon

PFC Perfluorocompound

PFC Performance Flight Certification

PFC Phase Frequency Characteristics

PFC Phenylacetamidotrifluoromethylcoumarin

PFC Photofinish Camera

PFC Planar Flow-Casting [Metallurgy]

PFC Plasma Fusion Center [of Massachusetts Institute of Technology, Cambridge, US]

PFC Polymerization-Filled Composite

PFC Postflight Checklist

PFC Power Factor Corrector

PFC Preliminary Flight Certification

PFC Priority Foreign Country

PFC Private First Class [also Pfc]

PFC Professional Fee Costing

PFC Pulse-Flow Coulometry

PFCB Perfluorocyclobutane

Pfce Performance [also pfce]

PFCF Payload Flight Control Facility

PFCF Producer Fixed Capital Formation

pF/cm² picofarad(s) per square centimeter [also pF cm^{-2}]

PFCS Primary Flight Control System

PFCU Power Flying Control Unit

PFD Personal Flotation Device

PFD Planned Flight Data

PFD Power Flux Density

PFD Primary Flash Distillate

PFD Primary Flight Display

pfd preferred

PFDC Pole Figure Data Collection [Crystallography]

PFDF Pacific Fisheries Development Foundation [US]

PFDF Payload Flight Data File

PFDME Perfluoronated Dimethoxyethane

PFDMM Perfluoronated Dimethoxymethane

PFDMS Pulsed Field Desorption Mass Spectrometer

PFE Path Following Error [Aeronautics]

PFE Paramagnetic Faraday Effect

PFE Partners for the Environment [of US Environmental Protection Agency]

PFE Perfluoroethylene

PFE Photoferroelectric(ity)

PFE Plenum Fill Experiment

PFE Process Fuel Equivalent [Metallurgy]

PFE Programmer's File Editor

(PFE)₁ Process Fuel Equivalent of Mining a Raw Material [Metallurgy]

(PFE)₂ Process Fuel Equivalent of Ore Beneficiation [Metallurgy]

(PFE)₃ Process Fuel Equivalent of Smelting or Extracting Metal [Metallurgy]

(PFE)₄ Process Fuel Equivalent of Refining the Extracted Metal [Metallurgy]

(PFE)ₚ Process Fuel Equivalent of All Means Used for Pollution Abatement during Mining, Beneficication, Smelting, Extraction and Refining [Metallurgy]

PFEA Polish Federation of Engineering Associations

PFEE Photoferroelectric Effect

PFEP Polyfluoroethylenepropylene

PFEP Programmable Front-End Processor

PFES Proposed Final Environmental Statement

PFET P-Channel (Junction) Field-Effect Transistor [also P-FET]

PFET Photo-Field-Effect Transistor

PFET Positive-Channel Field-Effect Transistor [also P-FET]

PFF Lead Iron Fluoride [$Pb_3(FeF_6)_2$]

PFF Precast Flooring Federation [UK]

PFF Presidential Faculty Fellows (Program) [US]

PFG Conference on Particles, Fields and Gravitation

PFG Pulsed-Field Gel Electrophoresis

Pfg Pfennig [German Monetary Unit; 1/100 of a Deutschmark] [also pfg, or pf]

PFGC Parameters from Group Contribution (Equation of State) [Physics]

PFGNMR Pulsed Field Gradient (Spin Echo) Nuclear Magnetic Resonance

PFGSE Pulsed Field Gradient Spin Echo

PFH Pressurized Fluidized-Bed Hydroretorting (Process)

PFI Pacific Forest Industry

PFI Pack File Indexer

PFI Parapsychology Foundation Inc. [US]

PFI Pipe Fabrication Institute [US]

PFI Power Factor Indicator

PFI Pulsed Field Ionization

PFIR Part Fill-In and Ram

PFIX Power Failure Interrupt

PFK Payload Function Key

PFK Perfluoro Kerosine

PFK Phosphofructokinase [Biochemistry]

PFK Plastic Fluted Knob

PFK Program(med) Function Key

PFL Primary Freon Loop

PFL Propulsion Field Laboratory [US]

PflM Pseudomonas fluorescens [Microbiology]

PFM Parallel Field Magnetization

PFM Power Factor Meter

PFM Printer Font Metrics

PFM Process Facility Modification

PFM Pulse-Frequency Modulation

.PFM Printer Font Metrics [Windows File Name Extension] [also .pfm]

Pfm Platform

PFMC Pacific Fisheries Management Council [US]

PF Meter Power Factor Meter [also PF meter]

pFmP p-Formylphenol

PFMS Pyrolysis Field Ionization Mass Spectrometry

PFMVE Perfluoromethylvinylether

PFN Permanent File Name

PFN Pulse Forming Network

PFO Pasadena Field Office [of US Geological Survey at Pasadena, California]

PFO Perfluorooctanoate

PFOMA Perfluorooctyl Methacrylate

PFP Perfluorophenyl

PFP Perfluoropropylene

PFP Performic Acid Phosphotungsite

PFP P-Fluorophenylalanine [Biochemistry]

PFP Plastic-Faced Plaster

PFP Plutonium Finishing Plant [of US Department of Energy at Hanford, Washington State]

PFP Post Flight Processor

PFP Program File Processor

PFP Program Financial Plan

PFP Programmable Function Panel

pFP p-Fluorophenol

PFPA Pentafluoropropanoic Acid Anhydride

PFPA Perfluorophenyl Azide

PFPA Pitch Fiber Pipe Association [US]

PFPD Pulsed Flame Photometric Detector

PFPE Perfluoropolyether

PFPE Perfluoropolyalkylether

PFPEO Perfluoropolyethylene Oxide [also PFPEO]

PFPOE Perfluoronated Polyperfluoro-Oxyethylene

PFPOH 2,2,3,3,3-Pentafluoropropanol

PFPOM Perfluoronated Polyperfluoro-Oxymethylene

PFPP tris(2,2,3,3,3-Pentafluoropropyl)phosphate

PFR Parts Failure Rate

PFR Permitted Flying Route

PFR Plug Flow Reactor

PFR Polynesian Franc [also PFr; Currency of French Polynesia and Tahiti]

PFR Portable Foot Restraint

PFR Power Fail Recovery

PFR Power Fail Restart [also PF/R]

PFR Programmed Film Reader

PFR Prototype Fast Reactor

PFR Pulse Frequency

PF/R Power Fail/Restart [also PFR]

P/FR Problem/Failure Report [of NASA Jet Propulsion Laboratory, Pasadena, California, US]

Pfr Perforator [also pfr]

PFRA Prairie Farm Rehabilitation Act [Canada]

PFRA Prairie Farm Rehabilitation Administration [Canada]

PFRS Portable Field Recording System

PFRT Preliminary Flight Rating Test

PFR/UK Prototype Fast Reactor/United Kingdom

PFRX Preflight Relmat (X = 1, 2, 3 ...9, or A, B, C...Z) [of NASA]

PFS Percent Full Scale

PFS Peripheral Fixed Shim

PFS Primary Flight System

PFS Primary Frequency Supply

PFS Programmable Frequency Source

PFS Propellant Field System

PFS Pulse Fusion Surfacing

P&FS Particles and Fields Subsatellite

PFSEPR Pulse Field-Sweep Electron Paramagnetic Resonance

PFSI Perfluorosulfonate Ionomer

PFT Paper, Flat Tape

PFT Partial Fourier Transform

PFT Pulse Fourier Transformation

PFTA Payload Flight Test Article

PFTBA Perfluorotri-N-Butylamine

PF-3MO Perfluorotrimethylene Oxide

PFTP Pentafluorothiophenol

PFU Plaque Forming Units [alsp pfu] [Virology]

P4VP Poly(4-Vinylpyridine)

PFZ Precipitate-Free Zone; Precipitation-Free Zone [Metallurgy]

PG Papua–New Guinea [ISO Code]

PG Parental Guidance [Motion-Picture Rating]

PG Parity Generator

PG Pattern Generation; Pattern Generator

PG Pentaglycerine

PG Perfect Gas

PG Periodic Group

PG Permanent Glow

PG Persian Gulf

PG Petrogenesis; Petrogenic; Petrogeny

PG Pharmacopeia Germanica

PG Phase Gradient

PG Phosphatidylglycerol

PG Phosphoglycerate

PG Photogrammetric; Photogrammetry

PG Photon Gas [Statistical Mechanics]

PG Physical Geography

PG Picture Generation; Picture Generator

PG Pin Grid [Electronics]

PG Pituitary Gland [Anatomy]

PG Pivot Gun

PG Planar Graph

PG Planetary Gear

PG Planning Group

PG Plate Glass

PG Pospiech and Gryziecki (Model) [Metallurgy]

PG Point Group [Crystallography]

PG Politechnika Gdańska [Gdansk Polytechnic Institute, Poland]

PG Polyethylene Glycol [also PEG]

PG Polypropylene Glycol [also PPG]

PG Port Group

PG Postgraduate

PG Potential Gradient

PG Power Gain

PG Power Gate

PG Power Generation; Power Generator

PG Pressure Gauge

PG Procedures and Guidelines

PG Producer Gas

PG Product Group

PG Professional Geologist

PG Professional Geophysicist

PG Program Generation; Program Generator

PG Program Generic

PG Project Group

PG Projective Geometry

PG Prompt Gamma Ray [Physics]

PG Prostaglandin [Biochemistry]

PG Prostate Gland [Anatomy]

PG Prosthetic Group [Biochemistry]

PG Protective Ground

PG Proton Glass

PG Proton Gradiometer

PG Proving Ground

PG Pseudogel

PG Pulse Generation; Pulse Generator

PG Pumped-Storage Generation

PG Purple with Green

PG Pyrolytic Graphite

2-PG D-2-Phosphoglyceric Acid

3-PG 3-Phosphoglycerate

6-PG 6-Phosphogluconate

Pg Page [also pg]

Pg Petagram [Unit]

Pg Portuguese; Portugal

pg picogram [Unit]

.pg Papua–New Guinea [Country Code/Domain Name]

PGA Pad Grid Array

PGA Penicillin G Ammonium

PGA Phosphoglycerate

PGA Phosphoglyceric Acid

PGA Pin Grid Array

PGA Polyglycolic Acid; Poly(L-Glycolic) Acid

PGA Power Generating Assembly

PGA Pressure Garment Assembly

PGA Proceedings of the Geological Association [London, UK]

PGA Professional Graphics Adapter

PGA Programmable Gain Amplifier

PGA Programmable Gate Array

PGA Propane Gas Association

PGA Pteroylglutamic Acid [Biochemistry]

PGA Purchased Gas Adjustment

PGA1 Prostaglandin A1 [Biochemistry]

PGA2 Prostaglandin A2 [Biochemistry]

3-PGA 3-Phosphoglycerate

3-PGA 3-Phosphoglyceric Acid

PGAA Pin Grid Array Adapter

PGAA Prompt Gamma(-Ray) Activation Analysis

p-GaAs P-Type Gallium Arsenide (Semiconductor)

p⁺-GaAs P-Plus Type Gallium Arsenide (Semiconductor)

p-GaAsSb P-Type Gallium Arsenide Antimonide (Semiconductor)

PGAC Public and Governmental Affairs Committee [of Minerals, Metals and Materials Society of American Institute of Mining, Metallurgical and Petroleum Engineers, US]

p-GaInP P-Type Gallium Indium Phosphide (Semiconductor)

PGAL Proof Gallons [also PGal]

p-GaN P-Type Gallium Nitride (Semiconductor)

p-GaN:Mg Magnesium-Doped P-Type Gallium Nitride (Semiconductor)

-GaP P-Type Gallium Phosphide (Semiconductor)

p-GaSb P-Type Gallium Antimonide (Semiconductor)

p-GaSe P-Type Gallium Selenide (Semiconductor)

PG B_1 Prostaglandin B_1 [Biochemistry]

PG B_2 Prostaglandin B_2 [Biochemistry]

PG B_1-BSA Prostaglandin B_1–Bovine Serum Albumin (Immunogen) [Biochemistry]

PGBD Primary Grain Boundary Dislocation [Materials Science]

Pgbd Plugboard

PGC Pacific Geoscience Center [of Natural Resources Canada]

PGC Parameter Generation and Controls

PGC Port Group Control

PGC Post-Graduate Certificate

PGC Power Generation Committee [of IEEE (Institute of Electrical and Electronics Engineers) Power Engineering Society, US]

PGC Professional Graphics Controller

PGC Program Group Control

PGC Pulsed Gas Chromatography

PGC Pyrolysis Gas Chromatography

PGCA Patent Glazing Contractors Association [UK]

PGCC Presidential Green Chemistry Challenge [US Government/Chemical Industry Program]

PGC/MS Pyrolysis Gas Chromatography/Mass Spectroscopy

PGCOA Pennsylvania Grade Crude Oil Association [US]

PGCR Portable Gas-Cooled Reactor [US]

PGD Phosphogluconate Dehydrogenase [Biochemistry]

PGD Planar Gas Discharge

PGD Professional Graphics Display

PGD Provinciale Gezondheidsdienst voor Dieren [Provincial Animal Health Service, Netherlands]

PG D_2 Prostaglandin D_2 [Biochemistry]

PGDCS Power Generation, Distribution and Control Subsystem; Power Generation, Distribution and Control System

PGDN Page Down [also PgDn]

PGE Page [also Pge, or pge]

PGE Pacific Gas and Electric Company [San Francisco, California, US]

PGE Petroleum and Geosystems Engineering

PGE Phenyl Glycidyl Ether

PGE Platinum-Group Element

PGE Primary Ground Electrode

PG&E Pacific Gas and Electric Company [San Franscisco, California, US]

PG E_1 Prostaglandin E_1 [Biochemistry]

PG E_2 Prostaglandin E_2 [Biochemistry]

Pge Page [also pge]

Pge Purge [also pge]

p-Ge P-Type Germanium (Semiconductor)

PG E_1-BSA Prostaglandin E_1–Bovine Serum Albumin (Immunogen) [Biochemistry]

PG E_2-BSA Prostaglandin E_2–Bovine Serum Albumin (Immunogen) [Biochemistry]

PGEC Professional Group on Electronic Computers

PGEN Program Generator

PGeol Professional Geologist

PGEWS Professional Group on Engineering Writing and Speech [of Institute of Electrical and Electronics Engineers, US]

PGF Lead Gallium Fluoride

PGF Plant Growth Facility [of Pacific Northwest National Laboratory, Richland, Washington, US]

PG $F_{1\alpha}$ Prostaglandin $F_{1\alpha}$ [Biochemistry]

PG $F_{2\alpha}$ Prostaglandin $F_{2\alpha}$ [Biochemistry]

PG $F_{1\alpha}$-BSA Prostaglandin $F_{1\alpha}$–Bovine Serum Albumin (Immunogen) [Biochemistry]

PG $F_{2\alpha}$-BSA Prostaglandin $F_{2\alpha}$–Bovine Serum Albumin (Immunogen) [Biochemistry]

PGH Paragraph [also Pgh, or pgh]

PGH Patrol Gunboat, Hydrofoil

PGH Philadelphia General Hospital [Pennsylvania, US]

PGH Port Group Highway

Pgh Pittsburgh [Pennsylvania, US]

PGHM Payload Ground Handling Mechanism

PGHTS Port Group Highway Time Slot

PGI Parameter Group Identifier

PGI Phosphoglucoisomerase [Biochemistry]

PGI Port Group Interface

PGI Programmable Graphics Interface

PGI Pyrotechnics Guild International [US]

PGI$_2$ Prostaglandin I_2 [Biochemistry]

PGK Papua New Guinea Kina [Currency]

PGK Penicillin G Potassium [Biochemistry]

PGK Phosphoglycerate Kinase [Biochemistry]

PGK Phosphoglyceric Phosphokinase [Biochemistry]

PGK Program Check

PGL Precursor Geometry Limited (Reaction)

.PGL Graphics [Hewlett-Packard File Name Extension]

PGLIN Page and Line

pGlu Pyroglutamic Acid; Pyroglutamyl- [Biochemistry]

PGM Paleogeomorphological; Paleogeomorphologist; Paleogeomorphology

PGM Phosphoglucomutase [Biochemistry]

PGM Phosphoglycerate Mutase [Biochemistry]

PGM Platinum-Group Metal(s)

PGM Precision-Guided Munition

Pgm Program(mable) [also pgm]

P-GMAW Pulsed(-Current) Gas Metal-Arc Welding

PGMC Primary Glass Manufacturers Council [US]

PGMS Precision-Guided Munitions

PGMTT Professional Group on Microwave Theory and Techniques [US]

PGNAA Prompt Gamma Neutron Activation Analysis [Physics]

PGNCS Primary Guidance and Navigation Control Subsystem; Primary Guidance and Navigation Control System

PGNS Primary Guidance and Navigation System

PG NUM (New) Page Number [also Pg Num] [Computers]

PGOC Payload Ground Operations Contract(or)

PgOH Propargyl Alcohol

p-gon p-sided polygon [*Note:* "p" Indicates the Number of Line Segments, e.g. in a Tetragon p Equals 4, and in a Hexagon 6]

PGOR Payload Ground Operation Requirements

PGORS Payload Ground Operations Requirements Study

PGOS Petroleum, Gas and Oil Shale

PGP Pretty Good Privacy [Internet]

PGP Programmable Graphics Processor

.PGP ProGram Parameter [AutoCAD File Name Extension]

PGR (Corrosion) Pit Generation Rate

PGR Planning and Ground Rule [NASA Spacelab]

PGR Plant Genetic Resources

PGR Precision Graph(ic) Recorder

PGR Prompt Gamma Rays

PGR Psychogalvanic Reaction; Psychogalvanic Reflex; Psychogalvanic Response

PGRA Plant Genetic Resources for Agriculture

PGRC Plant Genetic Resource Center

PGRF Pacific Gamefish Research Foundation [US]

PGRF Pulse Group Repetition Frequency

PGRO Processors and Growers Research Organization [UK]

PGRSA Plant Growth Regulator Society of America [US]

PGRT Petroleum and Gas Revenue Tax

PGRWG Payload Ground Requirements Working Group [of NASA]

PGRWG Plant Growth Regulator Working Group [now Plant Growth Regulator Society of America, US]

PGS PCM (Pulse-Code Modulation) Gateway Services

PGS Pennsylvania Geological Survey [US]

PGS Plane Grating Spectrograph(y)

PGS Postgraduate Scholarship

PGS Power Generation Satellite

PGS Power Generation Subsystem; Power Generation System

PGS Power Generator Section

PGS Product Generation System

PGS Program Generation System

PGSE Payload Ground Support Equipment

PGSE Pulse(d)-Gradient Spin-Echo [Physics]

PGT Lead Germanium Telluride

PGT Page Table [also PgT]

PGT Planetary Gear Train

PGT Princeton Gamma-Tech, Inc. [New Jersey, US]

PGTN Pentagycerin Trinitrate

PGU Plant Growth Unit

PGU Power Generator Unit

PGU Pressure Gas Umbilical

PGUP Page Up [also PgUp]

PGW Periodic Gravity Wave

PGW Pressure Gas Welding

PGWG Particles and Gases Working Group

PGY Postgraduate Year

PH *(Pädagogische Hochschule)* – German for "University of Education", or "College of Education"

PH Para-Hydrogen

PH Page Heading

PH Peak Height [also P/H]

PH Pearl Harbor [Oahu, Hawaii]

PH Pearson's Handbook [Crystallography]

PH Peierls-Hubbard (Model) [Solid-State Physics]

PH Peltier Heat [Physics]

PH Performance History

PH Period Hours

PH Personal Hygiene

PH Phasemeter

PH Philippines [ISO Code]

PH Photoelectron Holography

PH Pin Hole

PH Portable Handtruck

PH Positive Hole

PH Post-Harvest (Application) [Pesticides]

PH Power House

PH Pratt-Hayford (System) [Geodesy]

PH Precipitation Hardenable; Precipitation Hardened; Precipitation Hardening [Metallurgy]

PH Pressure Head

PH Professional Hydrologist

PH Public Health

PH Public Holidays

PH Pulse Height

PH- Netherlands [Civil Aircraft Marking]

P(H) Hyperfine Field Distribution [also p(H)] [Solid-State Physics]

P&H Postage and Handling

P/H Parts per Hundred [also PPH, or pph]

P/H Peak Height [also PH]

P-H Pentagonal-Hexagonal (Carbon Ring)

P-H Phosphorus-Hydrogen (System)

Ph Hydrostatic Pressure

Ph Phantom [also ph]

Ph Pharmacy [also ph]

Ph Phase [also ph]

Ph Phenol(ic) [also ph]

Ph 1,10-Phenanthroline

Ph Phenol

Ph Phenyl

Ph Phone [also ph]

Ph Phosphorus [Abbreviation; Chemical Symbol: P]

Ph Phthalimide

Ph Phytophthora [Genus of Fungi]

P7/h6 Locational Interference Fit; Basic Shaft System [ISO Symbol]

P-h Indentation Load-Indentation Depth (Hysteresis Loop) [Mechanical Testing]

pH picohenry [also ph] [Unit]

pH (potentia hydrogenii) – negative logarithm of hydrogen-ion activity; represents the acidity, or alkalinity of aqueous solutions

pH$_0$ pH value at time $T = 0$ [Symbol]

pH$_T$ pH value after Time T [Symbol]

ph phi [English Equilvalent]

ph phot [Unit]

ph photon

.ph Phillipines [Country Code/Domain Name]

PHA Periodic Hexagonal Array (Model) [Materials Science]

PHA Phaseolus Vulgaris Agglutinin [Immunology]

PHA Phytohemagglutinin [Biochemistry]

PHA Pivalohydroxamic Acid Ester

PHA Polyhydroxyalkanoate

PHA Precipitation-Hardening Alloy [Metallurgy]

PHA Prelimiary Hazard Analysis

PHA Process Hazards Analysis

PHA Public Health Alliance [UK]

PHA Public Health Association [Australia]

PHA Public Housing Authority [US]

PHA Pulse-Height Analysis; Pulse Height Analyzer

PHABSIM Physical Habitat Simulation (Model)

PhAc Phenylacetate; Phenylacetic Acid

PHACOMP Phase Computation [Solid-State Physics]

PHA-E Phaseolus Vulgaris Agglutinin–Erythroagglutinin [Immunology]

PHA-L Phaseolus Vulgaris Agglutinin–Leucoagglutinin [Immunology]

PHA-M Phaseolus Vulgaris Agglutinin–Mucoprotein [Immunology]

Ph$_3$[12]aneAs$_3$ 1,5,9-Triphenyl 1,5,9-Triarsacyclododecane

Phant Phantastron [also phant] [Electronics]

PHA-P Phaseolus Vulgaris Agglutinin–Phytohemagglutinin [Immunology]

Phar Pharmaceutical; Pharmacy; Pharmacopoeia [also phar]

PHARB Public Health Applications Research Branch [of National Cancer Institute, Bethesda, Maryland, US]

PharD Pharmaceutical Doctor; Doctor of Pharmacy

PHARE Program for Harmonized Air Traffic Management Research in Europe [of Eurocontrol]

Pharm Pharmaceutical(s); Pharmacist; Pharmacy; Pharmacopoeia [also pharm]

Pharmacol Pharmacological; Pharmacologist; Pharmacology [also pharmacol]

Pharmacol Pharmacology [International Journal]

Pharmacol Biochem Behav Pharmacology, Biochemistry and Behavior (Journal)

Pharmacol Res Pharmacological Research [Journal published in the US]

Pharmacol Rev Pharmacological Reviews [Journal published in the US]

Pharm Acta Helv Pharmaceutica Acta Helvetica [Switzerland]

Pharma Int Pharma International [Journal published in the US]

PharmD Pharmaceutical Doctor; Doctor of Pharmacy

PharmE Pharmaceutical Engineer [also Pharm Eng, or PharmEng]

Pharm Eng Pharmaceutical Engineer(ing)

Pharm Eng Pharmaceutical Engineering [Journal of the International Society of Pharmaceutical Engineers]

Pharm Ind Pharmazeutische Industrie [German Journal on the Pharmaceutical Industry]

Pharm Pat Pharmaceutical Patent [also PharmPat]

Pharm Res Pharmaceutical Research [Journal of the American Association of Pharmaceutical Scientists]

Pharm Tech Jpn Pharmaceutical Technology in Japan [Journal]

Pharm Technol Pharmaceutical Technology

Pharm Technol Pharmaceutical Technology (Journal)

PHAROS Plan Handling and Radar Operating System

PHASEQCD Phase Transitions in Quantum Chromodynamics (Workshop) [at Brookhaven National Laboratory, Upton, New York, US]

Phase Trans Phase Transformations (Journal)

Phase Transit Phase Transitions [Journal published in the UK]

PHASS Precipitation-Hardenable Austenitic Stainless Steel

PHAXSCAN Philips High-Speed Automatic X-Ray Analysis System

PHB Photochemical Hole Burning (Experiments) [Physics]

PHB p-Hydroxybenzoic Acid

PHB Poly(hydroxybutyrate); Poly-3-Hydroxybutyrate; D(-)-3-Hydroxybutyric Acid

PHB Primary Hind-Brain [also phb]

PhB (Philosophiae Baccalaureus) – Bachelor of Philosophy

PHB A 3-Ketothiolase [also phb a] [Biochemistry]

PHB B NADPH-Linked Acetoacetyl-CoA Reductase [also phb b] [Biochemistry]

PHB C Polyhydroxyalkanoate Polymerase [phb c] [Biochemistry]

PHBA Polyhydroxybenzoic Acid

P(3HB-co-4HB) Copolymer of 3- and 4-Hydroxybutyrate

P(3HB-co-3HP) Copolymer of 3-Hydroxybutyrate and 3-Hydroxypropionate

P(3HB-co-3HV) Copolymer of 3-Hydroxybutyrate and 3-Hydroxyvalerate

PHB-co-PHV Poly(3-Hydroxybutyric Acid-co-3-Hydroxyvaleric Acid

PHBE p-Hydroxybenzoic Ester

PHBHV Poly(3-Hydroxybutyric Acid-co-3-Hydroxyvaleric Acid)

PhBr Pharmacopeia Britannica

PhBr Phenyl Bromide

Ph₃BrGe Triphenylbromogermane

Ph Brz Phosphor Bronze

PhBU Phenylbenzylurethane

PHBV Poly(3-Hydroxybutyrate Valerate)

Ph Bz Phosphor Bronze

PHC Petroleum Hydrocarbon

PHC Physical Hydrogen Cracking

PHC Primary Health Care

PHC Public Health Committee

PhC Pharmaceutical Chemist

Phc Phenanthrenocyanine

PhcFe Iron Phenanthrenocyanine

PhCl Phenyl Chloride

Ph₃ClGe Triphenylchlorogermane

[PhcFe(dib)]ₙ Iron Phenanthrenocyanine 1,4-Diisocyanobenzene

PHCT Perturbed Hard-Chain Theory [Physics]

PHCVD Photochemical Chemical Vapor Deposition

PHCVD Photosensitized Chemical Vapor Deposition

PHD Parallel Head Disk

PHD Physical Hydrology Division [of National Hydrology Research Institute, Canada]

PHD Plastohydrodynamic(s)

PHD Pulse Height Discriminator

PhD *(Philosophiae Doctor)* – Doctor of Philosophy

PhD Photoelectron Diffraction

PhDChE Doctor of Philosophy in Chemical Engineering [also PhD ChE]

PhDEE Doctor of Philosophy in Electrical Engineering [also PhD EE]

PhD(Hons) Doctor of Philosophy (Honours)

PHE Packaging and Handling Engineer

PHE Plastic High Explosive

PHE Plate Heat Exchange

PHE Poly(β-Hydroxyether)

PHE Public Health Engineer

Phe (−)-Phenylalanine; Phenylalanyl [Biochemistry]

phE Photon Emission

PheAc Phenoxyacetyl

Phe-Ala Phenylalanine-Alanine [Biochemistry]

PHEEM Photoemission Electron Microscopy

PHEMA Polyhydroxyethyl Methacrylate

P-HEMT Pseudomorphic High-Electron Mobility Transistor [also PHEMT, or pHEMT]

PHEN Phenanthroline Monohydrate

Phen 1,10-Phenanthroline [also phen]

Phen Phenol(ic) [also phen]

Phen Phenomenological; Phenomenology [also phen]

Phen Phenomenon; Phenomena

PHENO Recent Developments in Phenomenology (Conference)

PHENO Precise Hybrid Elements for Nonlinear Operation

Phenyl-PAS Phenyl-p-Aminosalicylate

4-Phenyl-TAD 4-Phenyl-(δ)1-1,2,4-Triazoline-3,5-Dione

P&HEP Plasma and High-Energy Physics [Group of Institute Electrical and Electronics Engineers, US]

PHERMEX Pulsed High-Energy Radiographic Machine Emitting X-Rays [of Los Alamos Scientific Laboratory, New Mexico, US]

PHEXIC Polyhexylisocyanate

PHF Payload Handling Fixture

PHF Personal Hygiene Facility

PHF Plug Handling Fixture

PHF Poly(hydrogen Fluoride)

PhF Phenyl Fluoride

PhF Physics of Fluids [Journal of the American Institute of Physics, US]

PhG Graduate in Pharmacy

Ph₂Ge Diphenylgermane

Ph₃Ge Triphenylgermane

Ph₄Ge Tetraphenylgermane

PhGer Pharmacopeia Germanica

Ph H Phillips Head (Screw) [also PhH]

PH₃/H₂ Phosphine/Hydrogen (Gas Mixture)

PHHC Phenylhydrazine Hydrochloride

PH₃/He Phosphine/Helium (Gas Mixture)

PhHgAc Phenylmercury Acetate

PhHgBr Phenylmercury(II) Bromide

PhHgCl Phenylmercury(II) Chloride

PHI Pattern Hit Initiated

PHI Perkin-Elmer, Physical Electronics (Division) [Eden Prairie, Minnesota, US]

PHI Phosphohexose Isomerase [Biochemistry]

PHI Position and Homing Indicator

PHI Public Health Inspector

PhI Phenyl Iodide

phib amphibian

PHI-BLAST Pattern Hit Initiated Basic Local Alignment Search Tools [of National Center for Biological Information, US]

PHICs Poorest and Most Heavily Indebted Countries

PHIG Programmer's Hierarchical Interactive Graphics

PHIGS Programmer's Hierarchical Interactive Graphics System

PHIL Predictive Hypsometry Incorporating Lag (Model) [Earth Sciences]

Phil Philippine(s)

Phil Philosophical; Philosopher; Philosophy [also phil]

Phila Philadelphia [Pennsylvania, US]

Philcite Philippine Center for International Trade and Exhibition [Manila]

Philips Electron Opt Bull Philips Electron Optics Bulletin [Published by Philips-Norelco, Eindhoven, Netherlands]

Philips J Res Philips Journal of Research [N.V. Philips, Eindhoven, Netherlands]

Philips J Res Suppl Philips Journal of Research, Supplement [N.V. Philips, Eindhoven, Netherlands]

Philips Res Rep Philips Research Report [N.V. Philips, Eindhoven, Netherlands]

Philips Res Rep Suppl Philips Research Report, Supplement [N.V. Philips, Eindhoven, Netherlands]

Philips Tech Rev Philips Technical Review [N.V. Philips, Eindhoven, Netherlands]

Philips Telecommun Data Syst Rev Philips Telecommunication and Data Systems Review [N.V. Philips, Eindhoven, Netherlands]

Philips Weld Rep Philips Welding Reporter [N.V. Philips, Eindhoven, Netherlands]

Phil Labs Philadelphia Laboratories Inc. [US Pharmaceutical Manufacturer]

PhilM Master of Philosophy

Philos Philosopher; Philosophical; Philosophy [also philos]

Philos Mag Philosophical Magazine [UK]

Philos Mag A, Phys Condens Matter Defects Mech Prop Philosophical Magazine A, Physics of Condensed Matter, Defects and Mechanical Properties [UK]

Philos Mag B, Phys Condens Matter Electron Opt Magn Prop Philosophical Magazine B, Physics of Condensed Matter, Electronic, Optical and Magnetic Properties [UK]

Philos Mag Lett Philosophical Magazine Letters [UK]

Philos Sci Philosophy of Science [Journal of the Philosophy of Science Association, US]

Philos Trans R Soc A Philosophical Transactions of the Royal Society, Series A [UK]

Philos Trans R Soc London A Philosophical Transactions of the Royal Society of London, Series A [UK]

Philos Trans R Soc London A, Math Phys Sci Philosophical Transactions of the Royal Society of London, Series A, Mathematical and Physical Sciences [UK]

PHILQA Philips Question Answering System [of Philips Research Laboratories, Eindhoven, Netherlands]

PHILSIN Philippines-Singapore Submarine Cable

Phil Trans R Soc Lond Philosophical Transactions of the Royal Society of London [UK]

PHIN Position and Homing Inertial Navigator

PHIP Para-Hydrogen-Induced Polarization

pH/ISE pH Value/Ion-Selective Electrode [Chemistry]

Phisyl Cl (1,3-Dihydro-1-Oxo-2H-Isoindol-2-yl)-4-Benzenesulfonyl Chloride

PHK Personal Hygiene Kit

PHL Packaging, Handling and Logistics

PHL Bull PHL (Packaging, Handling and Logistics) Bulletin [of National Institute of Packaging, Handling and Logistic Engineers, US]

PhL (Philosophiae Licentiatus) – Licentiate in Philosophy

PHLAG Philips Load and Go

PhLi Phenyllithium

PHLW Pressurized Heavy and Light Water (Reactor)

PHLODOT Phase Lock Doppler Tracking

PHLX Philadelphia Stock Exchange [Pennsylvania, US]

PHM Patrol Hydrofoil Missile(ship)

PHM Per Hundred Million [also phm]

PHM Phasemeter

PHM Phase Modulation

PhM (Philosophiae Magister) – Master of Philosophy

PHMC Probe Heater Motor Controller

PhMgBr Phenylmagnesium Bromide

PhMgCl Phenylmagnesium Chloride

PHMSO Polyhydridomethylsiloxane

PHMSS Precipitation-Hardenable Martensitic Stainless Steel

pH/mV pH value versus millivolt [also pH mV^{-1}]

pH/mV/°C pH value versus millivolt per degree centigrade (or celsius) [also pH mV^{-1}°C^{-1}]

PHN Public Health Nurse; Public Health Nursing

Phnz Phenazine

PHO Polyhydroxyoctanoate

.PHO Phone (or Photo) List [File Name Extension]

Ph$_2$O Diphenyl Oxide (or Diphenyl Ether)

PHOBIA Photon Counting Cameras for Bioluminescence and Autoradiography (Project) [of European Research Coordination Agency]

PHODEC Photometric Determination of Equilibrium Constants

Phoen Phoenicia(n)

PHOENIX Plasma Heating Obtained by Energetic Neutral Injection Experiment

Phoenix Q Phoenix Quarterly [of Institute of Scrap Recycling Industries, US]

Phoenix: Voice Scrap Ind Phoenix: Voice of the Scrap Industries [US]

PHOFEX Photofragment Excitation

PHOFRY Photofragment Yield (Spectroscopy)

Phon Phonetic(s) [also phon]

Phon Phonological; Phonology [also phon]

Phonet Phonetic(s) [also phonet]

Phonet Phonetica [International Journal]

Phono Phonograph(y) [also phono]

Phos Phosphate [also phos]

PHOSCHEM Phosphate Chemicals Export Association [US]

Phot Photographer; Photographic; Photography [also phot]

PHOTAC Phototypesetting and Composing

PHOTHO Phosphate Glass Solidification of Thorium-Bearing Reprocessing Waste (Process) [Nuclear Engineering]

Photo Photograph(ic) [also photo]

Photobiochem Photobiophys Photobiochemistry and Photobiophysics [Journal published in the Netherlands]

Photochem Photobiol Photochemistry and Photobiology [Journal]

Photo-DLTS Photo-Deep Level Transient Spectroscopy [also photo-DLTS]

Photo-EPR Photo-Electron Paramagnetic Resonance [also photo-EPR]

Photog Photographical; Photographer; Photography [also photog, Photogr, or photogr]

Photogramm Photogrammetric; Photogrammetry [also photogramm]

Photogramm Eng Remote Sens Photogrammetric Engineering and Remote Sensing [Journal of the American Society of Photogrammetry, US]

Photogramm Rec Photogrammetric Record [Journal of the Photogrammetry Society, UK]

Photogr Sci Eng Photographic Science and Engineering (Journal)

Photom Photometric; Photometrist; Photometry [also photom]

PHOTON International Conference on the Structure and Interactions of the Photon

Photophys Photophysical; Photophysicist; Photophysics [also photophys]

Photosynth Res Photosynthesis Research [Journal published in the Netherlands]

PHOXIM Phenylglyoxylonitriloxime

PHP Parahexaphenyl (Oligomer)

PHP Parts, Hybrids, and Packaging [Group of Institute of Electrical and Electronics Engineers Group]

PHP Payload Handling Panel

PHP P-Hexaphenyl (Oligomer)

PHP Philippine Peso [Currency]

PHP Pound(s) per Horsepower [also php, lb/HP, or lb/hp]

PHP Primary Hydrogen Porosity

PHP Pulsed Holography Development [of Sandia National Laboratories, Albuquerque, New Mexico, US]

PHP Pump Horsepower

Ph$_2$P Diphenylphosphine

Ph$_3$P Triphenylphosphine

P(3HP) D(-)-3-Hydroxypropionic Acid

PHPA Perlindungan Hutan dan Pelestrian Alam [Directorate General of Forest Protection and Nature Conservation, Indonesian Ministry of Forestry]

PhPhA Phenylphosphonic Acid

Phpht Phenolphthalein [also phpht]

PHPM Pulsed (Electron-Beam) High-Pressure Mass Spectrometry

Ph$_3$PO Triphenylphosphine Oxide

PhPy 2-Phenyl Pyridine [also phpy]

P5HQ 1-Propyl-5-Hydroxyquinoline

P6HQ 1-Propyl-5-Hydroxyquinoline

PHR Payload Hazardous Report

PHR Peak Height Ratio

PHR Phorbol [also Phr]

PHR Physical Record

PHR Pound-Force per Hour [Unit]

PHR Process Hazards Review

PHR Process Heat Reactor Program

PHR Pulse-Height Resolution

Phr Phorbol [also PHR]

Phr Phrase [also phr]

phr parts per hundred resin [Unit]

phr parts per hundred rubber [Unit]

PHRAN Phrasal Analyzer

PHRED Phrasal English Diction

PHRI Public Health Research Institute [New York City, US]

Phrm Pharmaceutical [also phrm]

PHS Pan-Head Steel (Screw)

PHS Part-Handling System

PHS Payload Handling Station

PHS Personal Handy-Phone System

PHS Plastics Historical Society [UK]

PHS Polystyrene Hydroxystyrene

PHS Postal History Society [UK]

PHS Potential Hypersurface

PHS Printing Historical Society [UK]

PHS Public Health Service [of US Department of Health and Human Services, Washington, DC]

PHS Public Health Statement

PHS Pulse Height Selection

pH(S) pH (Value) Standard [Chemistry]

Ph-Salt Ethylenediamine Dinitrate [also Ph-salt]

PHSE Phase [also Phse, or phse]

PHSF Payload Hazardous Servicing Facility

PhSH Phenylmercaptan

PHSI Positive Ion Hyperthermal Surface Ionization

(PhSiO$_{1.5}$)n Polyphenyl Silsesquioxane

PHSPS Preservation, Handling, Storage, Packaging and Shipping

PHSS Precipitation Hardenable (or Hardening) Stainless Steel

PH Steel Precipitation Hardenable (or Hardening) Steel

PHT Phototube

PHT Plasma Hydrogen Treatment

PHT Polyhexylthiophene

PHT Precipitation Heat-Treated; Precipitation Heat Treatment [Metallurgy]

PHT Pyrrolidone Hydrotribromide

PhT Phthalic Acid

Pht Phenetole

Pht Phthalyl

Ph(t) Time-Dependent Hydrostatic Pressure

PHTC Pulse Height to Time Converter

PHTD PNNL HWVP Technology Development [of Pacific Northwest National Laboratory–Hanford Waste Vitrification Plant, US Department of Energy]

PHTP Perhydrotriphenylene

PHTS Primary Heat Transport System

PhUS Pharmacopeia of the United States

P(3HV) D(-)-3-Hydroxyvaleric Acid

PHW Pressurized Heavy Water

PHWR Pressurized Heavy Water (Nuclear) Reactor

Phycol Phycologia [International Journal on Algae]

PHYREC (Current) Physical Record Number [BASIC Programming]

Phys Physical; Physicist; Physics [also phys]

Phys Physician [also phys]

Phys Physiological; Physiologist; Physiology [also phys]

Phys Abstr Physics Abstracts [of INSPEC–Information Services Physics, Electrical and Electronics, and Computers and Control, UK]

Phys At Nucl Physics of Atomic Nuclei [Translation of: *Fizika Atomnoya Yadra*; jointly published by American Institute of Physics, US and MAIK Nauka, Russia]

PHYSBE Physiological Simulation Benchmark Experiment

Phys Bl Physikalische Blätter [German Physics Publication]

Phys Briefs Physics Briefs [Published by the American Institute of Physics, US]

Phys Bull Physics Bulletin [of Institute of Physics, UK]

Phys Can Physics in Canada [Canadian Journal]

Phys Chem Physical Chemist(ry)

Phys Chem Earth Physics and Chemistry of the Earth [Journal published in the Netherlands]

Phys Chem Earth Sci Physics, Chemistry and Earth Science [International Journal]

Phys Chem Glasses Physics and Chemistry of Glasses [Journal of the Society of Glass Technology, UK]

Phys Chem Liq Physics and Chemistry of Liquids [Journal published in the UK]

Phys Chem Mater Treat Physics and Chemistry of Materials Treatment [Translation of: *Fizika i Khimiya Obrabotki Materialov (USSR)*; published in the UK]

Phys Chem Mech Surf Physics, Chemistry and Mechanics of Surfaces [Journal published in the UK]

Phys Chem Miner Physics and Chemistry of Minerals [Journal published in Germany]

Phys Cer Physical Ceramics

Phys Daten Physik Daten [Data Sheets on Physics; published by Fachinformationszentrum Energie–Physik–Mathematik, Germany]

Phys-Dokl Physics-Doklady [Russian Translation Journal jointly published by the American Institute of Physics, US and MAIK Nauka, Russia]

Phys Earth Planet Inter Physics of the Earth and Planetary Interiors [Journal published in the Netherlands]

Phys Educ (India) Physics Education (India) [Published by the University of Poona, India]

Phys Educ (UK) Physics Education [Journal of the Institute of Physics, UK]

Phys Educ News Physics Education News [Publication of the American Institute of Physics, US]

Phys Educ Univ Physics Education in University [Publication of the Physical Society of Japan]

Phys Energ Fortis Phys Nucl Physica Energiae Fortis et Physica Nuclearis [Journal on High-Energy and Nuclear Physics; published in PR China]

Phys Expr Lett Physics Express Letters [Journal of the Institute of Physics, UK]

Phys Fluids Physics of Fluids [Journal of the American Institute of Physics, US]

Phys Fluids A, Fluid Dyn Physics of Fluids A, Fluid Dynamics [Journal of the American Institute of Physics, US]

Phys Fluids B, Plasma Phys Physics of Fluids B, Plasma Physics [Journal of the American Institute of Physics, US]

Phys Geogr Physical Geographer; Physical Geography

Physic Physician [also physic]

Physicochem Hydrodyn Physicochemical Hydrodynamics [Journal published in the US]

Physio Physiotherapy; Physiotherapist [also physio]

Physiol Physiological; Physiologist; Physiology [also physiol]

Physiol Biochem Zool Physiological and Biochemical Zoology [Journal published by University of Chicago Press, US]

Physiol Ent Physiological Entomology [Publication of the Royal Entomological Society of London, UK] [also Physiol Entomol]

Physiol Meas Physiological Measurement [Journal of Institute of Physics, UK]

Physiol Plantarum Physiologia Plantarum [Journal on Plant Physiology; published in Denmark]

Physiol Rev Physiological Reviews [of American Physiological Society, US]

Phys Lett Physics Letters [Published in the Netherlands]

Phys Lett A Physics Letter A [Published in the Netherlands]

Phys Lett B Physics Letter B [Published in the Netherlands]

Phys Lett C Physics Letter C [Published in the Netherlands]

Phys Mag Physicalia Magazine [of Belgian Physical Society]

Phys Med Physical Medicine [also phys med]

Phys Med Biol Physics in Medicine and Biology [Journal of the Institute of Physics, UK]

Phys Met Physical Metallurgist; Physical Metallurgy

Phys Met Physics of Metals [Translation of: *Metallofizika (USSR)*; published in the UK]

Phys Met Metallogr Physics of Metals and Metallography [Translation of: *Fizika Metallov i Metallovedenie (USSR)*; published in the UK]

Phys Mineral Physical Mineralogist; Physical Mineralogy

Phys News Update Physics News Update [of American Institute of Physics, US]

Phys Oceanogr Physical Oceanographer; Physical Oceanography

Phys Org Chem Physical Organic Chemist(ry)

Phys Pap Physics Papers [Journal published in the Netherlands]

Phys Part Nucl Physics of Particles and Nuclei [Translation of: *Fizika Elementarnykh Chastits i Atomnoya Yadra*; jointly published by American Institute of Physics, US and MAIK Nauka, Russia]

Phys Plasma Physics of Plasma [Journal of the American Institute of Physics, US]

Phys Proc Physical Processing [also phys proc]

Phys Prop Physical Properties [also phys prop]

Phys Rep Physics Reports [Published in the Netherlands]

Phys Rep A Physics Reports A [Published in the Netherlands]

Phys Rep B Physics Reports B [Published in the Netherlands]

Phys Rep C Physics Reports C [Published in the Netherlands]

Phys Rev Physical Review [also phys rev]

Phys Rev Physical Review [of American Physical Society, US]

Phys Rev A Physical Review A [of American Physical Society, US]

Phys Rev A, At Mol Opt Phys Physical Review A, Atomic, Molecular and Optical Physics [of American Physical Society, US]

Phys Rev Abstr Physical Review, Abstracts [of American Institute of Physics, US]

Phys Rev A, Stat Phys Plasmas Fluids Relat Interdiscip Top Physical Review A, Statistical Physics, Plasmas, Fluids and Related Interdisciplinary Topics [of American Physical Society, US]

Phys Rev B Physical Review B [of American Physical Society, US]

Phys Rev B, Condens Matter Physical Review B, Condensed Matter [of American Physical Society, US]

Phys Rev C Physical Review C [of American Physical Society, US]

Phys Rev C, Nucl Phys Physical Review C, Nuclear Physics [of American Physical Society, US]

Phys Rev D Physical Review D [of American Physical Society, US]

Phys Rev D, Part Fields Physical Review D, Particles and Fields [of American Physical Society, US]

Phys Rev E Physical Review E [of American Physical Society, US]

Phys Rev Lett Physical Review Letters [of American Physical Society, US]

Phys Rev Spec Top Acc Beams Physical Review Special Topics–Accelerators and Beams [of American Physical Society, US]

PhysS Physical Society [also PhysSoc]

Phys Sci Physical Science; Physical Scientist

Phys Scr Physica Scripta [Physics Journal of the Royal Swedish Academy of Sciences]

Phys Scr Vol T Physica Scripta Volume T [Physics Journal of the Royal Swedish Academy of Sciences]

PhysSoc Physical Society [also PhysS]

Phys Solid State Physics of the Solid State [Translation of: *Fizika Tverdogo Tela*; jointly published by the American Institute of Physics, US ans Maik Nauka, Russia]

Phys Status Solidi A Physica Status Solidi A [Journal of SolidState Physics, Part A; published in Germany]

Phys Status Solidi B Physica Status Solidi B [Journal of Solid-State Physics, Part B; published in Germany]

Phys Teach Physics Teaching [Publication of the Korean Physical Society, South Korea]

Phys Technol Physics in Technology [Journal of the Institute of Physics, UK]

Phys Test Chem Anal, Chem Anal Physical Testing and Chemical Analysis, Chemical Analysis [Published in PR China]

Phys Today Physics Today [Journal of the American Institute of Physics, US]

Phys World Physics World [Magazine of the Institute of Physics, UK]

Phytochem Phytochemical; Phytochemist(ry) [also phytochem]

Phytochem Phytochemistry [Journal of the Phytochemical Society of Europe]

Phz Phthalazine

PI International Conference on Path-Integrals from peV to TeV [Physics]

PI Pacific Islands

PI Packaging Institute

PI Packing Index [Crystallography]

PI Paper Insulation

PI Parainfluenza [Medicine]

PI Parallel Input

PI Parallel Interface

PI Parameter Identifier

PI Parent Ion

PI Particular Integral

PI Passive Immunity

PI Path Integral

PI Payload Interrogator

PI Pen and Ink

PI Performance Index [Industrial Engineering]

PI Perlite Institute [US]

PI Phase Inverter

PI Philippine Islands

PI Phosphatidyl Inositol [Biochemistry]

PI Photographic Illumination

PI Photoionization

PI Physical Interpretation

PI Pig Iron [Metallurgy]

PI Pilotless Interceptor

PI Pin Insulator

PI Pitting Index [Metallurgy]

PI Planetary Index

PI Plant Industry (Division) [of Commonwealth Scientific and Industrial Research Organization]

PI Plasma Induced

PI Plasticity Index (for Rock, or Soil)

PI Plastics Institute [UK]

PI Plastoinelastic(ity)

PI Point Initiating

PI Point Insulating

PI Point of Intersection

PI Poisoning Intermediate

PI Polarization Interferometer

PI Polyimide

PI Polyisoprene; 1,4-Polyisoprene

PI Polymer Impregnation

PI Population Inversion [Atomic Physics]

PI Position Indicator

PI Positive Input

PI Positive Ion

PI Pratt Institute [Brooklyn, New York, US]

PI Preionization

PI Preliminary Investigation

PI Pressure Indication; Pressure Indicator

PI Principal Investigator

PI Priority Inhibit

PI Priority Interrupt

PI Pristine Ice

PI Process Image

PI Process Integration

PI Processor Interface
PI Procurement Item
PI Product Innovation
PI Productivity Index [Oil Industry]
PI Program Indicator
PI Program Interface
PI Program Interruption
PI Program Introduction
PI Programmed Instruction
PI Programmed Interrupt
PI Propidium Iodide
PI Proportional Integral (Controller)
PI Propyl Isome
PI Public Information
PI Public Involvement
PI Pulse Interval
PI3 Parainfluenza (Virus), Type 3
PI- Phillippines [Civil Aircraft Marking]
P&I Performance and Interface (Specification)
P&I Piping and Instrumentation
P&I Protection and Indemnity (Insurance)
Pi Inorganic Phosphate
p-i p-type–intrinsic (junction) [Electronics]
PIA Pacific Institute of Aromatherapy [San Rafael, California, US]
PIA Pakistan International Airlines
PIA Partitioning Industry Association [UK]
PIA Peoria, Illinois [Meteorological Station Identifier]
PIA Peripheral Interface Adapter
PIA Petroleum Incentives Administration
PIA Petroleum Industry Association [UK]
PIA Pharmaceutical Industries Association [of European Free Trade Association]
PIA N-(2-Phenylisopropyl)adenosine [Biochemistry]
PIA Photoinduced Absorption
PIA Photoinduced Anisotropy
PIA Pilots International Association [US]
PIA Plastics Industry Association
PIA Plastics Institute of America [US]
PIA Post-Irradiation Annealing [Metallurgy]
PIA Pre-Installation Acceptance
PIA Pressure-Induced Amorphization [Metallurgy]
PIA Printing Industries of America [US]
PIA Printing Industries Association [US]
PIA Program Implementation Agency
PIA Project Impact Analysis
PIA Pulse-Interval Analyzer
pIa p-Iodoaniline
(R)-PIA (R)-N6-(1-Methyl-2-Phenylethyl)adenosine [Biochemistry]
(S)-PIA (S)-N6-(1-Methyl-2-Phenylethyl)adenosine [Biochemistry]
PIAC Peak Instantaneous Airborne Counts
PIAC Petroleum Industry Advisory Committee [UK]

PIAD Plastics in Automotive Design [of Society of Manufacturing Engineers, US]
PIAM Petroleum Industry Application of Microcomputers
PIANC Permanent International Association of Navigation Congresses [Belgium]
PIANEC Planning of the Implementation of an Improved AFTN/AFS (Aeronautical Fixed Telecommunications Network/Aeronautical Fixed Service) Network [of International Civil Aviation Organization]
PIAPACS Psychophysiological Information Acquisition, Processing and Control System
PIAR Project Impact Analysis Report
PIARC Permanent International Association of Road Congresses [France]
PIAS Photon-Counting (Two-Dimensional) Image Acquisition System
PIAT Projector Infantry Antitank
PIB Partially Ionized Beam (Deposition Technique)
PIB Petroleum Information Board
PIB Petroleum Information Bureau [UK]
PIB Polar Ionosphere Beacon
PIB Polyisobutene
PIB Polyisobutylene
PIB Polytechnic Institute of Brooklyn [New York State, US]
PIB Potential-Induced Breathing (Model)
PIB Preflight Information Bulletin
PIB Processor Interface Buffer
PIB Programmable Input Buffer
PIB Public Investments Board [India]
PIB Publishing Information Bulletin
PIB Pyrotechnic Installation Building
PIBA P-Iodobenzoic Acid [also pIBA]
PIBAC Permanent International Bureau of Analytical Chemistry
PIBAL Pilot-Balloon (Observation) [also Pibal or pibal]
PIBAL Polytechnic Institute of Brooklyn, Aerodynamics Laboratory [US]
PIBD Program for International Business Development [of Department of Foreign Affairs and International Trade, Canada]
PIB-co-MA Polyisobutylene-co-Maleic Anhydride
PIBMM Permanent International Bureau of Motorcycle Manufacturers [now International Motorcycle Manufacturers Association, France]
PIBMRI Polytechnic Institute of Brooklyn, Microwave Research Institute [US]
PIBUC Pilot Backup Control
PIBS Polar Ionosphere Beacon Satellite [of NASA]
PIC Pacific Island Country
PIC Paired Ion Chromatography
PIC Paper-Insulated Cable
PIC Particle-in-Cell
PIC Particle-Induced Conductivity
PIC Payload Integration Center
PIC Payload Integration Committee
PIC Personal Identification Code

PIC   Personal Information Carrier

PIC   Personal Intelligent Communicator

PIC   Phenyl Isocyanate

PIC   Photographic Industries Council [US]

PIC   Photographic Interpretation Center

PIC   Picrinic Acid

PIC   Pilot in Command

PIC   Plastic-Insulated Cable

PIC   Plastic-Insulated Conductor

PIC   Plasticity Induced Closure

PIC   Platform for Internet Content

PIC   Polyethylene-Insulated Cable

PIC   Polyethylene-Insulated Conductor

PIC   Polymer Impregnated Concrete

PIC   Portable Inventory Collector (Software Program)

PIC   Position Independent Code

PIC   Positive Impedance Converter

PIC   Potential Icing Category

PIC   Power Information Center

PIC   Preimpregnated Cable

PIC   Preinstallation Checkout

PIC   Pressure-Impregnation Carbonization (Process)

PIC   Pressure Indication and Control [also PI&C]

PIC   Pressure Infiltration Casting [Metallurgy]

PIC   Prices and Income Commission

PIC   Primary Inter-LATA (Local Access and Transport Area) Carrier

PIC   Prior Informed Consent

PIC   Priority Interrupt Control(ler)

PIC   Process Interface Control

PIC   Processor Interconnection Channel

PIC   Product of Incomplete Combustion

PIC   Program Indicator Code

PIC   Program Information Code

PIC   Program Interrupt Control(ler)

PIC   Programmable Interrupt Controller

PIC   Programmable Interval Clock

PIC   Pyrotechnic Ignition Control

PIC   Pyrotechnic Initiator Capacitor

PIC   Pyrotechnic Initiator Controller

PI&C   Pressure Indication and Control [also PIC]

.PIC   Picture (or Graphics) [Lotus 1-2-3 File Extension]

2-Pic   2-Picolinylamine [also 2-pic]

PICA   Power Industry Computer Applications

PICA   Project for Integrated Catalogue Automation

PICA   Public Interest Computing Association [US]

PICAA   Permanent International Committee of Agricultural Associations

PICAC   Power Industry Computer Applications Conference

PICAN   Probabilistic Integrated Composite Analyzer (Computer Code)

PICAO   Provisional International Civil Aviation Organization

Pic Arsn   Picatinny Arsenal [of US Army at Dover, New Jersey, US]

PICB   Peripheral Interface Control Bus

PICC   Philippine International Convention Center [Manila, Philippines]

PICC   Provisional International Computation Center

PICED   Photoinduced Changes (of the) Energy Distribution

PICG   Petrolite Industrial Chemicals Group [US]

PICGC   Permanent International Committee for Genetic Congresses

PICI   Positive Ion Chemical Ionization

PICLS   Purdue International and Computational Learning System [of Purdue University, West Lafayette, Indiana, US]

PICM   (Laboratoire de) Physique des Interfaces et Couche Minces [Interface and Thin-Film Physics Laboratory; Ecole Polytechnique, Palaiseau, France] [also Lab PCM]

PICOP   Paper Industries Corporation of the Philippines

PICP   Program Interface Control Plan

PICPSA   Permanent International Commission for the Proof of Small Arms [Belgium]

PICRS   Program Information Control and Retrieval System

PICRS   Program Information Coordination and Review Service

PICS   Partners in Customer Satisfaction [of Pontiac/General Motors Corporation, US]

PICS   Personnel Information Commission System

PICS   Personnel Information Communication System

PICS   Philippine Instrumentation and Control Society

PICS   Photo Index and Cataloging System

PICS   Planetary Integrated Camera-Spectrometer

PICS   Platform for Internet Content Selection

PICS   Plug-In Control System

PICS   Plug-In Inventory Control System

PICS   Predefined Input Control Sequence

PICS   Production Information and Control System

PICS   Programmable Industrial Control Simulation

PICS/DCPR   Plug-In Inventory Control System/Detailed Continuing Property Record

PICT   Prague Institute of Chemical Technology [Prague, Czech Republic]

Pict   Picture [also pict]

PICTEL   Picture Telephone [US]

PICTS   Photoinduced Current Transient Spectroscopy

PICU   Parallel Instruction Control Unit

PICUTP   Permanent International Committee of Underground Town Planning

PICVD   Plasma-Induced Chemical Vapor Deposition [also PI-CVD]

PICXAM   Pacific-International Congress on X-Ray Analytical Methods

PID   Parallel Interface Device

PID   Particle-Induced Desorption

PID   Payload Insertion Device

PID   Pelvic Inflammatory Disease [Medicine]

PID   Peripheral Interface Device

PID   Personal Identification Device

PID Photoinduced Desorption

PID Photo-Ion Detector

PID Photo-Ionization Detector

PID Planning and Integration Division [US]

PID Port Identification

PID Process Identification Number

PID Product Identification

PID Product Identifier

PID Program Information Department

PID Program Information Document

PID Program Introduction Document

PID Prolapsed Intervertebral Disk [Medicine]

PID Proportional Integral Derivation; Proportional Integral Derivative; Proportional Integral Differential (Controller)

PID Pseudo Interrupt Device

PID Public Information Division [of US Army]

P&ID Piping and Instrument(ation) Diagram

P&ID Piping and Instrument(ation) Drawing

P&ID Process and Instrument(ation) Diagram

PIDA Payload Installation and Deployment Aid

PIDA Pharmaceutical Industry Development Assistance

PIDA Phenylindane Dicarboxylic Acid

PIDA N-(Phosphonomethyl)iminodiacetic Acid

PIDAS Perimeter Intrusions Detection Analysis System

PIDB Peripheral Interface Data Bus

PIDC Pakistan Industrial Development Corporation

PIDC Photo-Induced Discharge Characteristics

PIDCOM Process Instruments Digital Communications System

PIDE Pakistan Institute of Development Economics

PIDFTMS Particle-Induced Desorption Fourier Transform Mass Spectrometry

PIDI Philippines Invention Development Institute

PIDP Programmable Indicator Data Processor

PIDS Parameter Inventory Display System [of National Oceanographic Data Center, US]

PIDS Personal Inspirable Dust Spectrometer

PIDS Portable Image Display System

PIE Pacific Islands Ecosystems (Database) [of US Fish and Wildlife Service]

PIE Parallel Instruction Execution

PIE Parallel Interface Element

PIE Payload Integration Equipment

PIE Peripheral Interface Element

PIE Photoionization Efficiency (Spectroscopy)

PIE Plug-In Electronics

PIE Positive Ion Emission

PIE Post-Infective Encephalitis [Medicine]

PIE Post-Irradiation Examination

PIE Pulse Interference Elimination

PIE Pulse Interference Emitting

PIEA Petroleum Industry Electrical Association [now Energy Telecommunications and Electrical Association, US]

PIECE Petroleum Institute Environmental Conservation Executive [Australia]

PIECP Preliminary Impact Engineering Change Proposal

PIER Procedures for Internet/Enterprise Renumbering

PIERC Pacific Island Ecosystems Research Center [of US Geological Survey/Biological Resources Division at Honolulu, Hawaii]

PIES Penning Ionization Electron Spectroscopy

PIES Pollution Prevention Information Exchange System

PIF Payload Integration Facility

PIF Picture Interchange Format File

PIF Presso l'Istituto d Fisica [Institute of Physics Press, Turin, Italy]

PIF Program Information File

PIFA Particle Immuno-Fluorescence Assay

PIFA Packaging and Industrial Films Association [UK]

PIFC Pakistan Industrial Finance Corporation

PIFEX Programmable Image Feature Extractor

PIFIMS Photon-Induced Field Ionization Mass Spectrometry

PIG Penning Ion(ization) Gauge

PIG Percentage Initiation by Grit [Explosives]

PIG Philips Ion(ization) Gauge

PIG Pirani (Thermal Conductivity) Gauge

PIGA Pendulous Integrating Gyro Accelerometer

Pigm Pigment [also pigm]

PIGME Particle-Induced Gamma-Ray Emission

PIGME Programmed Inert-Gas Multi-Electrode (Welding)

Pigm Resin Technol Pigment and Resin Technology [Journal published in the UK]

PIH Payload Integration Hardware

PIH β-Phenyl Isopropylhydrazine

PIH Pyridoxal Isonicotinoyl Hydrazone

PII Plasma Ion Implantation

PII Positive Immittance Inverter

PII Positive Impedance Inverter

PII Program Integrated Information

PIIF Pakistan International Industrial Fair

PIII Plasma Immersion Ion Implantation

PIIN Procurement Instrument Identification Number

PI/INT'L Packaging Institute International [US]

PIITS Philips International Institute of Technological Studies [Eindhoven, Netherlands]

PIK Potsdam-Institut für Klimafolgenforschung e.V. [Potsdam Institute for Climate Change Research, Germany]

PIKES Photo-Interpretation Keys Expert System

PIL Percentage Increase in Loss

PIL Pitt Interpretive Language

PIL Precision-In-Line

PIL Precision and Intelligence Laboratory [of Tokyo Institute of Technology, Japan]

PIL Procurement Instrument Identification

Pil Pilot(age) [also pil]

pil *(pilula)* – pill [Medical Prescriptions]

PILA Power Industry Laboratory Association [US]

PILC Paper-Insulated Lead-Covered (Cable)

PILE Product Inventory Level Estimator

PILL Programmed Instruction Language Learning

PILOT Panel on Instrumentation for Large Optical Telescopes [UK]

PILOT Permutation Indexed Literature of Technology

PILOT Piloted Low-Speed Test

PILOT Programmed Inquiry Learning, or Teaching

PilotACE Pilot Automatic Computing Engine

PILP Program for Industrial Laboratory (or Industry/Laboratory) Projects [of National Research Council of Canada]

PILPS Project for Intercomparison of Land Surface Parameterization Schemes [of Working Group on Numerical Experimentation]

PIM Passive Intermodulation Measurement

PIM Peak Integration Method

PIM Peripheral Interface Module

PIM Personal Information Manager

PIM Phase Integral Method [Quantum Mechanics]

PIM Pilot Machine

PIM Powder Injection Molding

PIM Precision Indicator of the Meridian

PIM Primary Interface Module

PIM Processor Interface Module

PIM Product Innovation Management

PIM Production Inventory Management; Production Inventory Manager

PIM Program for Information Managers [now Association of Information Managers, US]

PIM Program Integration Manual

PIM Protocol-Independent Multicast

PIM Pulse Interval Modulation

P&IM Production and Inventory Management (Review) [Publication of American Production and Inventory Control Society, US]

PIMA Paper Industry Management Association [US]

PIMA Prairie Implement Manufacturers Association [Canada]

PIMA Polyisocyanurate Insulation Manufacturers Association [US]

PIMC Path-Integral Monte Carlo (Method) [Physics]

PIME Phthalimidomalonic Ester

PIMERS Periodical Immersion and Emersion Raman Spectroscopy

PIMISS Pennsylvania Interagency Management Information Support System [US]

P&IM Rev Production Inventory Management Reviews [of American Production and Inventory Control Society, US]

P&IM Rev APICS News P & IM (Production and Inventory Management) Review and APICS News [Published by the American Production and Inventory Control Society, US]

PIMS Payload Information Management System

PIMS Personnel Information Management System [of NASA]

PIMS (Conference on) Polymers in Medicine and Surgery [US]

PIMS Precision Ion Milling System

PIMS Principal Investigator Microgravity Services

PIMS Programmable Implantable Medicine System

PIN Pacific Island Network

PIN Personal Identification Name

PIN Personal Identification Number

PIN Personal Identifier Number

PIN Personal Identity Number

PIN Piece Identification Number

PIN Plant Information Network

PIN P-N Junction with Isolation Region (Diode)

PIN Police Information Network

PIN Position Indicator

PIN Positive-Intrinsic-Negative (Transistor) [also P-I-N, pin, or p-i-n]

PIN Power Input

PIN Private Intelligent (Data) Networker

PIN Process Identification Number

PIN Product Identification Number [Part of Tranportation and Dangerous Goods Regulation]

PIN Program Identification Number

PIN P-Type-Intrinsic-N-Type (Transistor) [also P-I-N, pin, or p-i-n]

PINAC Permanent International Association of Navigation Congresses

PINCCA Price Index Numbers for Current Cost Accounting

Pinacol 2,3-Dimethyl-2,3-Butanediol [also pinacol]

Pinacolone Methyl tert-Butyl Ketone [also pinacolone]

PIND Particle Impact Noise Detection; Particle Impact Noise Detector

PINE Program for Internet News and E-mail

PINFET Positive-Intrinsic-Negative Field-Effect Transistor

PING Packet Internet Groper

p-InGaAs P-Type Indium-Gallium Arsenide (Semiconductor)

p$^+$-InGaAs P-Plus-Type Indium-Gallium Arsenide (Semiconductor)

PINI Plug-In Neutral Injector

PINO Positive Input–Negative Output

p-InP P-Type Indium Phosphide (Semiconductor)

p-InP/Au P-Type Indium Phosphide on Gold (Substrate)

p-InP/Au MIS P-Type Indium Phosphide on Gold Metal–Interface Layer—Semiconductor (Diode)

PINPD Positive-Intrinsic-Negative Photodiode

PINS Portable Inertial Navigation System

PINS Precise Integrated Navigation System

PINS Precision Integrated Navigation System

PINS Precision's Improved Nesting System

PINSAC PINS (Precision Integrated Navigation System) Alignment Console

PINSCH Periodic-Index Separate-Confinement Heterostructure [Electronics]

PINSCH QW Periodic-Index Separate-Confinement Heterostructure Quantum-Well (Laser) [Electronics]

PINSTECH Pakistan Institute of Technology [Islamabad]

PINT Priority Interrupt Controller

PINT Purdue Interpretive Programming and Operating System [of Purdue University, Lafayette, Indiana, US]

PINTEC Plastics Institute National Technical Conference [UK]

PINY Polytechnic Institute of New York [Brooklyn, US]

PIO Paired Interacting Orbitals [Physics]

PIO Parallel Input/Output

PIO Payload Integration Plan

PIO Peripheral Input/Output

PIO Photocomposition Input Option

PIO Pilot-Induced Oscillation

PIO Port Information Office [of International Association of Independent Tankers Owners, Norway]

PIO Position Iterative Operation

PIO Precision Iterative Operation

PIO Private Input/Output

PIO Processor Input/Output

PIO Programmed Input/Output

PIO Public Information Office

PIOCS Parallel Input/Output Control System

PIOCS Physical Input/Output Control System

pion pi meson (i.e. π^+, π^-, or π^0 meson) [also π meson] [Particle Physics]

PIOS Penning Ionization Optical Spectroscopy

PIOS Planning Information Overlay System [Geographic Information System]

PIOSA Pan-Indian Ocean Science Association

PIOTA Post Irradiation Open Test Assembly

PIOTA Proximity Instrumented Open Test Assembly

PIOU Parallel Input/Output Unit

PIP N,N'-Bis(4-Aminophenyl)piperazine

PIP Part Information Program

PIP Path Independent Protocol

PIP Payload Integration Plan

PIP Payload Interface Plan

PIP Peripheral Interchange Program

PIP Personal Identification Project

PIP Petroleum Incentives Program [Canada]

PIP Picture-in-Picture

PIP Pipeline Image Processor

PIP Plant-in-Place

PIP Plant Instrumentation Program

PIP Plug-in Programmer

PIP Polar Information Program [of National Science Foundation, US]

PIP Pollution Information Project [of National Research Council of Canada]

PIP Polymer Impregnation and Pyrolysis (Technique)

PIP Position Indicating Probe

PIP Positive Incentive Program

PIP Poured-in-Place

PIP Precision Ion Polishing (System)

PIP Predicted Impact Point

PIP Primary Indicating Position

PIP Probabilistic Information Processing

PIP Problem Input Preparation

PIP Problem Isolation Procedure

PIP Product Improvement Plan

PIP Product Improvement Program [US]

PIP Production Instrumentation Package

PIP Programmable Integrated Processor

PIP Programmable Interconnect Point

PIP Programmed Individual Presentation

PIP Project Implementation Plan

PIP Project on Information Processing

PIP Prototype Inlet Piping

PIP Public Involvement Plan

PIP Pulsed Integrating Pendulum

Pip Piperidine

pIP p-Iodophenol

PIPA Pacific Industrial Property Association [US]

PIPA Pulsed Integrating Pendulum Accelerometer

PIPA Pulsed Integrating Pendulum Assembly

PIPE Points, Income, Personnel Expense

PIPECO Photo-Ion Photo-Electron Coincidence

Pipeline Gas J Pipeline and Gas Journal [US]

Pipe Line Ind Pipe Line Industry [Journal published in the US]

PIPER Pulsed Intense Plasma for Exploratory Research

PIPES 1,4-Piperazinebis(ethanesulfonic Acid)

Pipes Pipelines Int Pipes and Pipelines International [Journal published in the UK]

Pipmtc Piperidinemonothiocarbamate

PIPO Parallel-In, Parallel-Out (Shift Register)

PIPO Parallel Input/Parallel Output

PIPOR Program of International Polar Oceans Research

PIPR Plant-in-Place Records

PIPRO Program Development for the Analysis and Assessment of the Pipeline Segments [Part of Panel on Energy Research and Development (Program), Canada]

PIPS Passivated Implanted Planar Silicon

PIPS Pattern Information Processing System

PIPS Payload Inspection and Processing System

PIPS Precision Ion Polishing System

PIPS Pulsed Integrating Pendulums

PIPSYL 4-Iodobenzenesulfonyl (Chloride) [also Pipsyl]

PIPUB Pseudo-Isotropic Pseudo-Umklapp Backscattering (Theory) [Physics]

PIPVD Plasma-Induced Physical Vapor Deposition [also PI-PVD]

Pipz Piperazine

PIQ Parallel Instruction Queue

PIQMC Path Integral Quantum Monte Carlo (Method) [Physics]

PIQSY Probe for the International Quiet Solar Year

PIR Peak-Intensity Recording

PIR Petrolite Irradiation Reactor [US]

PIR Photon-Induced X-Ray

PIR Polyisocyanurate

PIR Precision Infrared Radiometer

PIR Pressure-Arming Impact Rocket

PIR Program Incident Report

PIR Protein Identification Resource [of United Nations International Center for Genetic Engineering and Biotechnology]

PIR Protocol Independent Routing

PIRA Paper Industries Research Association [US]

PIRA Printing and Packaging Industries Research Association [UK]

PIRA Pbd Abstr PIRA Paperboard Abstracts [of Paper Industries Research Association, US]

PIRA Printing Abstr PIRA Printing Abstracts [of Paper Industries Research Association, US]

PIRD Program Instrumentation Requirement Document

PIREP Pilot Report [also Pirep, or pirep]

PIRETS Pittsburgh Retrieval System [of University of Pittsburgh, Pennsylvania, US]

PIRG Public Interest Research Group [US]

PIRI Paint Industries Research Institute [South Africa]

PIRINC Petroleum Industry Research Foundation, Inc. [US]

PIRN Preliminary Interface Revision Notice

PIROGAS Plasma Injection of Reducing Overheated Gas System (for Cowper Heating) [also Pirogas] [Metallurgy]

PIRS Personal Information Retrieval System

PIRS Pulsed Infrared Stimulation

PIRT Physical Interpretations of Relativity Theory

PIRT Precision Infrared Tracking

PIRT Precision Infrared Triangulation

PIRT Public Information Retrieval Terminal

PIRV Programmed Interrupt Request Vector

PIS Patent Inventor Service [of International Patent Documentation Center, Austria]

PIS Penning Ionization Spectroscopy

PIS Postal Inspection Service [US]

PISA Portable Information System Architecture

PISA Public Interest Satellite Association

PISAB Pulse Interference Suppression and Blanking

PISC Petroleum Industry Security Council [US]

PISCES Parallel Implementation of Scientific Computing Environments [also Pisces]

PISE Plasma and Ion Surface Engineering (Techniques)

PISE/EUROPE European Joint Committee of PISE (Plasma and Ion Surface Engineering)

PISH Program Instrumentation Summary Handbook

Pis'ma Astron Zh Pis'ma v Astronomischeskii Zhurnal [Soviet Astronomy Letters]

Pis'ma Zh Eksp Teor Fiz Pis'ma v Zhurnal Eksperimental'noi i Teoreticheskoi Fiziki [Soviet Physics, Journal of Experimental and Theoretical Physics Letters]

Pis'ma Zh Tekh Fiz Pis'ma v Zhurnal Tekhnicheskoi Fiziki [Soviet Technical Physics Letters]

PISO Parallel-In, Serial-Out (Shift Register)

PISO Parallel Input/Serial Output

PISO Polyimidesulfone

PISP (International) Polar Ice Sheet Program

PIST Pohang Institute of Science and Technology [South Korea]

PISW Program Interrupt Status Word

PIT Peripheral Input Tape

PIT Plastics in Telecommunications (Conference) [UK]

PIT Powder-in-Tube (Technique) [Superconductor Science]

PIT Pre-Installation Test

PIT Principal, Interest and Taxes

PIT Printing and Information Technology (Division) [of Printing and Packaging Industries Research Association, UK]

PIT Processing Index Terms

PIT Program Instruction Tape

PIT Programmable Interval Timer

PIT Protection Identification Key

PIT Przemyslowy Instytut Telekomunikacji [Telecommunications Research Institute, Warsaw, Poland]

PITA Pain In The "Ass" [Computer Jargon]

PITA Parison Inflation Thinning Analysis

PITAC Pakistan Industrial Technical Assistance Center

PITB Petroleum Industry Training Board [UK]

PITC Phenyl Isothiocyanate

Pitch-CF Pitch-Based Carbon Fiber [also pitch-CF]

PITCOM Parliamentary Information Technology Committee [UK]

PITG Payload Integration Task Group

PITI Principal, Interest, Taxes and Insurance

PITL Photoinduced Thermoluminescence

PI-TOF-MS Photoionization Time-of-Flight Mass Spectrum

PITS Payload Integration Test Set

PITS Photo-Induced Transient Spectroscopy

PITS Passive Intercept Tracking System

PITS Petroleum Industry Training Service

Pitt Pittsburgh [Pennsylvania, US]

PITTCON Pittsburgh Conference (and Exposition) [also PITT-CON, or Pittcon, US]

Pitture Vern Pitture e Vernici [Italian Journal on Paints, Lacquers and Varnishes]

PIU Path Information Unit

PIU Plug-In Unit

PIU Power Interface Unit

PIU Process Interface Unit

PIU Programmable Interface Unit

PIU Pyrotechnic Initiator Unit

PIUMP Plug-In Unit Mounting Panel

PIUS Process Inherent Ultimate(ly) Safe (Reactor)

PIV Parainfluenza Virus

PIV Particle Image Velocimeter; Particle Image Velocimetry

PIV Peak Inverse Voltage [also piv]

PIV Product Inspection Verification

PIV$_{sust}$ Sustained Peak Inverse Voltage

PIW Program Interrupt Word

PIX Particle-Induced X-Ray

PIX Photo-Induced X-Ray

PIX Picture [also pix]

PIXE Particle-Induced X-Ray Emission

PIXE Particle-Induced X-Ray Excitation

PIXE Photon-Induced X-Ray Emission

PIXE Proton-Induced X-Ray Emission

PIXEL International Pixel Detector Workshop

Pixel Picture Element [also pixel, or PEL]

Pixel, Computer Graph CAD/CAM Image Process Pixel, Computer Graphics, Computer-Aided Design/Computer-Aided Manufacturing, Image Processing [Journal published in Italy]

PIXES Particle-Induced X-Ray Emission Spectroscopy

PIXI Portable Image X-Ray Intensifier

PIXI Professional Industrial X-Ray Imaging

PJ petajoule [Unit]

PJ Photojunction [Electronics]

PJ Plasma Jet

PJ Pulse-Jet (Engine)

PJ- Netherlands West Indies [Civil Aircraft Marking]

pJ picojoule [Unit]

PJA Pakistan Jute Association

PJA Pipe Jacking Association [UK]

PJE Parachute Jumping Exercise

PJE Pulse-Jet Engine

PJM Pennsylvania–New Jersey–Maryland [US]

PJM Plasma-Jet Machining

pJ/m picojoule per meter [also pJ m^{-1}]

PJMA Pakistan Jute Mills Association

PJT Prices Justification Tribunal [Australia]

PK Pakistan [ISO Code]

PK Peach-Koehler (Equation) [Metallurgy]

PK Peng-Robinson (Equation of State) [Physical Chemistry]

PK Phenyl Ketone

PK Politechnika Krakówska [Krakow Polytechnic Institute, Poland]

PK Polymerization Kinetics

PK Printer-Keyboard

PK Protein Kinase [Biochemistry]

PK Psychokinesis; Psychokinetic [Medicine]

PK Pyruvate Kinase [Biochemistry]

PK- Indonesia [Civil Aircraft Marking]

Pk Pack [also pk]

Pk Park [also pk]

Pk Peak [also pk]

pK *(potentia kalium)* – negative logarithm of the equilibrium constant K for a specified reaction; used to express dissociation extent of weak acids and bases]

pK$_a$ negative logarithm of the acid dissociation constant [Symbol]

pK$_{a1}$ first acid dissociation constant [Symbol]

pK$_{a2}$ second acid dissociation constant [Symbol]

pK$_b$ negative logarithm of the base dissociation constant [Symbol]

pK$_{HIn}$ negative logarithm of the dissociation constant of the indicator acid [Symbol]

pk peck [Unit]

.pk Pakistan [Country Code/Domain Name]

$\phi(k)$ phase shift of photoelectron [Symbol]

PKA Primary Knock-on Atom [Solid-State Physics]

pKa acid dissociation constant [Symbol]

pKa$_1$ first acid dissociation constant [Symbol]

pKa$_2$ second acid dissociation constant [Symbol]

PKα Phosphorus K-Alpha (Radiation) [also Pk$_\alpha$]

Pk Arty Park Artillery

pKb base dissociation constant [Symbol]

PKC Problem-Knowledge Coupler

PKC Protein Kinase C [Biochemistry]

PKC α Protein Kinase C Alpha [Biochemistry]

PKC β_1 Protein Kinase C Beta 1 [Biochemistry]

PKC β_2 Protein Kinase C Beta 2 [Biochemistry]

PKC γ Protein Kinase C Gamma [Biochemistry]

PKC δ Protein Kinase C Delta [Biochemistry]

PKC ε Protein Kinase C Epsilon [Biochemistry]

PKC ζ Protein Kinase C Zeta [Biochemistry]

PKC η Protein Kinase C Eta [Biochemistry]

PKD Partially Knocked Down

PKD Polycystic Kidney Disease [Medicine]

PKD Programmable Keyboard (and) Display

PKE Polar Kerr Effect [Physics]

Pkg Package; Packaging; Packing [also pkg]

Pkg Packaging [Journal published in the UK and the US]

Pkg (Denver) Packaging (Denver) [Journal published in Denver, Colorado, US]

Pkg News (Australia) Packaging News (Australia)

Pkg Newsl Packaging Newsletter [US]

Pkg News (UK) Packaging News (UK) [Published in the UK]

Pkg Technol Sci Packaging Technology and Science [Journal published in the UK and the US]

Pkg (UK) Packaging (UK) [Journal published in the UK]

Pkg Week Packaging Week [Published in the UK]

Pk How Pack Howitzer

PKI Public Key Infrastructure

PK/LDH Pyruvate Kinase/Lactic Dehydrogenase [Biochemistry]

PKM Perigee Kick Motor [Aerospace]

PKO Primary Knock-On (Atom) [Solid-State Physics]

PKP Professional Kinesiology Practice

PKP Pseudo-Kikuchi Pattern [Crystallography]

PKP (Earthquake) Push Wave that has Passed through the Earth's Core [Symbol] [also P']

ΦΚΦ Phi Kappa Phi (Student Society)

PK/PK Peak-to-Peak [also PK-PK, Pk/Pk, or Pk/Pk]

PKR Pakistan Rupee [Currency]

Pkr Packer [also pkr]

PKSE Primary Kinetic Salt Effect [Chemistry]

PKT Polymer Kinetic Theory [Chemical Engineering]

Pkt Packet [also pkt]

Pkt Pocket [also pkt]

PKU Peking University [PR China]

PKU Phenylketonuria [Medicine]

PK/VAL Peak-to-Valley (Ratio) [also Pk/Val]

Pkwy Parkway [also pkwy]

PKZ Pectin–Kaolin–Zinc Phenolsulfone (Compound)

PL Palisade Layer [Botany]

PL Panoramic Lens

PL Panleucopenia [Medicine]

PL Paper Laminate

PL Parallel Load

PL Parietal Lobe (of Cerebral Cortex) [Anatomy]

PL Partial Loss [also PL]

PL Parting Line [Metallurgy]

PL Parts List [also P/L]

PL Passivation Layer

PL Patent Law(yer)

PL Path Length

PL Payload [also P/L]

PL Peak Level

PL Peak Load

PL Pentane Lamp

PL Perception of Light

PL Phantom Line

PL Phase Line

PL Phillips Laboratory [of US Air Force at Edwards Air Force Base, California]

PL Phospholipid [Biochemistry]

PL Photographic Lamp

PL Photolithography

PL Photoluminescence; Photoluminescent

PL Physicial Limnologist; Physical Limnology

PL Physical Line

PL Physical Link

PL Pigment Layer (of the Eye) [Anatomy]

PL Pinned Layer (Magnetization) [Solid-State Physics]

PL Piobert-Lüders (Band) [Metallurgy]

PL Piping Loads

PL Plasticity Limit

PL Plastic Limit [Geology]

PL Plus Sign (Flag) [Computers]

PL Plumb Line

PL Poland [ISO Code]

PL Polarized Light

PL Politechnika Lodzka [Lodz Polytechnic Institute, Poland]

PL Polymer Laboratories [US]

PL Polymer Laboratory

PL Portable Low-Power Plant [US]

PL Portevin-Le Chatelier (Effect) [also P-L] [Metallurgy]

PL Port of London [UK]

PL Port Louis [Mauritius]

PL Position Line

PL Positive Logic

PL Postlanding [also P/L]

PL Power Law (Creep) [also P-L] [Metallurgy]

PL Power Level

PL Power Loading

PL Prelaunch

PL Pressurizer Level

PL Private Line

PL Procedural Language

PL Production Language

PL Production Level

PL Professional Librarian

PL Program Library

PL Program Logic

PL Programming Language

PL Programming Logics

PL Prolactin [Biochemistry]

PL Proportional(ity) Limit [Mechanics]

PL Protective Layer

PL Proximity Log

PL Public Law

PL Public Library; Public Librarian

PL Pulsed Laser

PL Steel Plate [AISI/AISC Designation]

.PL Pearl Language [File Name Extension] [also .pl]

PL/1 Programming Language 1 [also PL/I]

PL/9900 Programming Language 9900

P&L Profit and Loss (Statement) [also P/L, or P and L]

P/L Parts List [also PL]

P/L Payload [also PL]

P/L Postlanding [also PL]

P/L Purchased Labor

P-L Portevin-Le Chatelier (Effect) [also PL] [Metallurgy]

P-L Power-Law (Creep) [also PL] [Metallurgy]

Pl Line Parameter [ANSI Escape Sequence Code]

Pl Pile [also pl]

Pl Place [also pl]

Pl Placenta

Pl Plant [also pl]

Pl Plate [also pl]

Pl Platinum [Abbreviation; Symbol: Pt]

Pl Plug [also pl]

Pl Plus [also pl]

Pl Poiseuille [Unit]

Pl Pulse [also pl]

pl partial loss [also PL]

pl plural

.pl Poland [Country Code/Domain Name]

$\phi_\lambda(\lambda)$ Color Stimulus Function [Symbol]

PLA Pakistan Librarians Association

PLA Parachute Location Aid

PLA Passive Localization Assistant

PLA Peak Load Autotransformer

PLA Phase Locked Arrays

PLA Photo-Induced Luminescence Excitation

PLA Physiological Learning Aptitude

PLA Plain Language Address

PLA Platteville, Wisconsin [Meteorological Station Identifier]

PLA Polish Librarians Association

PLA Pollutant Load Allocation [of US Environmental Protection Agency]

PLA Polylactic Acid; Poly(L-Lactic) Acid [Biochemistry]

PLA Polylalamine [Biochemistry]

PLA Port of London Authority [UK]

PLA Power Lever Angle

PLA Practise Low Approach [Aeronautics]

PLA Print Load Analyzer

PLA Product Liability Act

PLA Programmable Line Adapter

PLA Programmable Logic Array; Programmed Logic Array

PLA Progressive Localization Assistant

PLA Proton Linear Accelerator

PLA Public Library Association [US]

PLA Pulsed Laser Ablation

PLA Pulsed Laser Anneal(ing) [Metallurgy]

PLA Pulverized Limestone Association [US]

DL-PLA DL-Polylactic Acid [Biochemistry]

L-PLA L-Polylactic Acid [Biochemistry]

PLAAR Packaged Liquid Air-Augmented Rocket

PLAK PLM (Pressurized Logistics Module) Late Access Kit

PLACE Position Location Aircraft Communications Equipment

PLACE Post-Landsat Advanced Concept Evaluation [Note: "Landsat" Refers to a Series of NASA Earth Rersources Technology Satellites]

PLACE Programming Language for Automatic Checkout Equipment

Placid Plombo Acido (Lead Recovery Process) [Metallurgy]

PLACIR Plated Circuit Board [also placir]

PLACO Planning Committee [of International Standards Organization]

PLAD Parachute Low-Altitude Delivery

PLAD Plasma Diode

PLAD Plasma Doping

PLAD Process for Lilly Application Development [of Eli Lilly & Co., Indianapolis, US]

PLADS Parachute Low-Altitude Delivery System

PLAGA Poly(lactic Acid-Glycolic Acid) [Biochemistry]

PLAGE Überparteiliche Plattform Gegen Atomgefahren und Zukunftswerkstatt Energie [Non-Partisan Platform against Nuclear Hazards and Workshop on Future Energy Sources, Germany]

PLAL Poly(lactic Acid-co-Lysine) [Biochemistry]

PLAN Parts Logistics Analysis Network

PLAN Personal Local Area Network

PLAN Program Language Analyzer

PLAN Programming Language Nineteen-Hundred

PLANES Language Enquiry System

PLANET Planned Logistics Analysis and Evaluation Technique

PLANET Private Local Area Network [UK]

planet planetary

Planet Rep Planetary Report [of Planetary Society, US]

Planet Space Sci Planetary and Space Science [Journal published in the UK]

Planistor Planar Resistor [also planistor]

PLANIT Programming Language for Interactive Teaching

PLANN Plant Location Assistance Nationwide Network [US]

PLANOX Planar Oxide [also Planox]

PLANS Program Logistics and Network Scheduling System

Planseeber Pulvermetall Planseeberichte für Pulvermetallurgie [Plansee Reports for Powder Metallurgy; published by Plansee, Reutte, Austria]

Plansee Proc Plansee Conference Proceedings

Plant Cell Plant Cell [Journal of the American Society of Plant Physiologists]

Plant Cell Physiol Plant and Cell Physiology [Journal of the Japanese Society of Plant Physiologists]

Plant Cell Rep Plant Cell Reports [Germany and US]

Plant Cell Tissue Organ Culture Plant Cell, Tissue and Organ Culture [Journal published in the Netherlands]

PlantE Plant Engineer [also Plant Eng or PlantEng]

Plant Eng Plant Engineer(ing)

Plant Eng Plant Engineer [Journal of the Institution of Plant Engineer, UK]

Plant Eng Plant Engineering [US Journal]

Planters Bull Planters Bulletin [Published by the Rubber Research Institute of Malaysia]

Plant Geogr Plant Geographer; Plant Geography

Plant Growth Regul Plant Growth Regulation [Journal published in the Netherlands]

Plant Mol Biol Plant Molecular Biology [Journal of the International Society for Plant Molecular Biology, US]

Plant Mol Biol Rep Plant Molecular Biology Reporter [Journal of the International Society for Plant Molecular Biology, US]

Plant Oper Prog Plant Operation and Progress [Journal of the American Institute of Chemical Engineers]

Plant Physiol Plant Physiology [Journal of the American Society of Plant Physiologists]

Plant Physiol Biochem Plant Physiology and Biochemistry [Journal of the Plant Physiology and Biochemistry Society, India]

Plant Sci Plant Science [Journal published in Ireland]

Plant Sci Bull Plant Science Bulletin [Published by the Botanical Society of America]

Plants Mach Plants and Machinery [Journal published in the US]

PLAP PCR (Payload Changeout Room) Late Access Platform [Aerospace]

PLAP Placental Alkaline Phosphatase [Biochemistry]

PLAP Pulsed-Laser Atom Probe (Technique)

PLA/PGA Polylactic Acid/Polyglycolic Acid (Copolymer)

PLAQ Planned Quantity

PLAS Private Line Assured Service

PLAS Program Logical Address Space

Plas Plaster [also plas]

Plas Plastic(s) [also plas]

PLASI Pulse Light Approach Indicator System

Plasma Chem Plasma Process Plasma Chemistry and Plasma Processing [Journal published in the US]

Plasma Phys Control Fusion Plasma Physics and Controlled Fusion [Journal of the Institute of Physics, UK]

Plasma Phys Rep Plasma Physics Reports [Russian Translation Journal jointly published by the American Institute of Physics, US and MAIK Nauka, Russia]

PLASMEX International Plastics Machinery, Equipment and Materials Exhibition

Plast Plastic(s) [also plast]

Plast Plastic(ity) [also plast]

Plast Age Plastics Age [Journal published in Japan]

Plast Bldg Contr Plastics in Building Construction [Journal published in the UK]

Plast Bus Plastics Business [Canadian Journal]

Plast Compounding Plastics Compounding [Journal published in the US]

Plast Des Forum Plastics Design Forum [Journal published in the US]

PLASTEC Plastics Technical Evaluation Center [US Army]

PLASTEC Plastics Technology Information Analysis Center

Plaste Kautsch Plaste und Kautschuk [German Journal on Plastics and Rubber]

Plast Eng Plastics Engineering [Journal of the Society of Plastics Engineers, US]

Plast Eng Plastics in Engineering [Journal published in the UK]

Plasteuropac Organization of European Manufacturers of Plastic Packaging Materials

Plasteurotec Organization of European Manufacturers of Technical Plastic Parts

PlastForum Scand PlastForum Scandinavia [Plastics Journal published in Sweden]

Plast Ind News Plastics Industry News [Japan]

Plast Rubb Wkly Plastics and Rubber Weekly

PlasTIPS Plastics Training and Information to Promote Safety

Plast Massy Plasticheskie Massy [Russian Publication on Moldable and Plastic Masses]

Plast Mod Elastom Plastiques Modernes et Elastomères [French Journal on Advanced Plastics and Elastomers]

Plast News Plastics News [Published by the Plastics Institute of Australia]

Plast News Briefs Plastics News Briefs [Published by the Society of the Plastics Industry, US]

Plast Reconstr Surg Plastic and Reconstructive Surgery [US Journal]

Plast Retail Packag Bull Plastics in Retail Packaging Bulletin [Switzerland]

Plast Rubber Int Plastics and Rubber International [Published by the Plastics and Rubber Institute, UK]

Plast Rubber Process Appl Plastics and Rubber Processing and Applications [Published by the Plastics and Rubber Institute, UK]

Plast Rubber Wkly Plastics and Rubber Weekly [UK]

Plast S Afr Plastics Southern Africa [Journal]

Plast Technol Plastics Technology [Journal published in the US]

Plast Today Plastics Today [Journal published in the UK]

Plast World Plastics World [Journal published in the US]

Plasty Kauč Plasty a Kaučuk [Czech Journal on Plastics and Rubber]

PLAT Pilot Landing Aid Television

Plat Plateau [also plat]

Plat Platform [also plat]

Plat Plating [also plat]

Plat Platoon [also plat]

Platf Platform [also platf]

Platinum Met Rev Platinum Metals Review [UK]

PLATO Programmed Learning AutomaticTeaching Organization [US]

PLATO Programmed Logic for Automated Teaching Operations; Programmed Logic for Automatic Teaching Operations

PLATR Pawling Lattice Test Rig [US]

Plat Sgt Platoon Sergeant [also PlatSgt]

PLATSIM Linear Simulation and Analysis Package for Large-Order Flexible System [of NASA]

Plat Surf Fin Plating and Surface Finishing [Journal of the American Electroplaters and Surface Finishers Society, US]

PLB Payload Bay

PLB Personal Locator Beacon

PLB Picture Level Benchmark

PLBD Payload Bay Door(s)

Plbg Plumbing [also plbg]

Plbk Playback [also plbk]

PLBR Prototype Large Breeder Reactor

PLC Pacific Logging Congress [US]

PLC Pattern Length Coding

PLC Percent Linear Change

PLC Photoluminescence Excitation

PLC Plague Locust Commission

PLC Plasma Chromatography

PLC Pneumatic Logic Circuit

PLC Polymer Liquid Crystal

PLC Portevin-Le Chatelier (Effect) [Metallurgy]

PLC Power-Law Creep [Metallurgy]

PLC Power-Line Carrier

PLC Power-Line Conditioner

PLC Power-Line Cycle

PLC Preparative Layer Chromatography

PLC Prime Level Code

PLC Process Liquid Chromatography

PLC Programmable Line Controller

PLC Programmable Logic Control(ler); Programmed Logic Control(ler)

PLC Programming Language Committee [of CODASYL–Conference on Data System Languages]

PLC Public Limited Company [also plc]

PLCA Parallel Line Communication Adapter

PLCA Pipeline Contractors Association [US]

PLCAC Pipeline Contractors Association of Canada

PLCB Pseudo-Line Control Block

PLCC Plastic Leaded Chip Carrier (Package) [Electronics]

PLCC Programmable Logic Control Circuit

PLCD Product Liability Common Defense

PLC I/O Programmable Logic Controller Input/ Output (Bus)

PLCO Praseodymium Lanthanum Copper Oxide (Superconductor)

PL Crp Power Law Creep [Metallurgy]

PLCU Programmable Logic Control Unit

PLC/VDT Programmable Logic Controller/Visual Display Terminal

PLD Partial Line Down

PLD Periodic (Crystal) Lattice Distortion

PLD Phase-Locked Demodulator

PLD Program Listing Document

PLD Programmable Logic Device

PLD Pulsed-Laser Deposition

PLD Pulse-Length Discrimination; Pulse-Length Discriminator

Pld Pyrrolidine

PLDI Payload Data Interleaver

PLDM Payload Data Management

PLDMR Photoluminescence-Detected Magnetic Resonance

PLDS Payload Data Support

PLDS Pilot Land Data System [of NASA]

PLDT Philippines Long Distance Telephone Company

PLDTS Propellant Loading Data Transmission System

PLE Partitioning Local Equilibrium (Mode of Growth) [Metallurgy]

PLE Photoluminescence Excitation (Spectrum, or Spectroscopy)

PLE Pulsed Laser Etching

PLEA Pacific Lumber Exporters Association [US]

£Leb Lebanese Pound [Currency]

PLEED Polarized Low-Energy Electron Diffraction

PLEM Pipeline and Manifold

PLENCH Pliers and Wrench (Combination) [also Plench]

PLES Photoluminescence Excitation Spectroscopy

PLF Page Length Field

PLF Parachute Landing Fall

PLF Plant Load Factor

PLF Plasma Ladle Furnace [Metallurgy]

PLFA Phospholipid Ester-Linked Fatty Acid [Biochemistry]

PLFA Pulsed Laser Flash Absorption

PLFT Lead Lithium Iron Tungstate

PLGA Poly(DL-Lactic-*co*-Glycolic Acid) Copolymer

PlGF Placenta Insulin-Like Growth Factor [Biochemistry]

Pl Gl Plate Glass

PLGSS Payload Ground Support Systems

PLH Payload Handling

PLH Pituitary Luteinizing Hormone [Biochemistry]

PLHC Poeciliopsis Lucida Hepatoma Cell [Medicine]

PLI Payload Interrogator

PLI Private Line Interface

PLIANT Procedural Language Implementing Analog Techniques

PLIB Pacific Lumber Inspection Bureau [US]

PLIB Program Library

PLIF Planar Laser Induced Fluorescence

PLIM Post-Launch Information Message

PLISN Provisioning List Item Sequence Number

PLJ Permanent Loop Junctor [Electrical Engineering]

PLL Peripheral Light Loss

PLL Phase-Locked Loop

PLL Pollution Legal Liability

PLL Prelinked Library

.PLL Prelinked Library [Clipper File Name Extension] [also .pll]

PLLA Poly(L-Lactic Acid) [Biochemistry]

PLL CMOS Phase-Locked Loop Complementary Metal-Oxide Semiconductor [also PLL C-MOS]

PLM Passive Line Monitor

PLM Payload Management

PLM Payload Monitoring

PLM Polarized Light Microscope; Polarized Light Microscopy

PLM Precision Laser Machining

PLM Pressurized Logistics Module

PLM Pulsed Laser Melting (Technique)

PLM Pulse-Length Modulation

PL/M Programming Language for Microcomputers (or Microprocessors)

Plmbg Plumbing [also plmbg]

Plmbr Plumber [also plmbr]

PLMS Program Logistics Master Schedule

PLN Flight Plan

PLN Plain (Sawed Lumber) [also pln]

PLN Program Logic Network

PLNCIAWPRC Polish National Committee of the International Association on Water Pollution Research and Control

plnd planned

Plng Planning [also plng]

PLO Pacific Launch Operations [of NASA]

PLO Phase-Locked Oscillator

PLO Process Limit Option

PLO Program Line Organization

PLOCAP Post Loss-of-Coolant Accident Protection [Nuclear Reactors]

PLOD Planetary Orbit Determination

PLOHI Pulsed-Laser Optical Holographic Interferometry

PLOO Pacific Launch Operations Office [of NASA]

PLOP Pressure Line of Position

Plotg Plotting [also plotg]

PLP Partners for Livable Places [US]

PLP Pattern Learning Parser

PLP Periodate-Lysine-Paraformalin [Biochemistry]

PLP Portable Low-Power Plant [US]

PLP Presentation Level Protocol

PLP Procedural Language Processor

PLP Programming Language Processor

PLP Pulsed Laser Photolysis

PLPA Pageable Link-Pack Area

PLPA Permissible Low-Pressure Alarm

Pl Path Plant Pathologist; Plant Pathology

P-LPD Load versus Load-Point Displacement (Curve) [Mechanics]

p-LPE P-Type Liquid Phase Epitaxy (Film)

Pl Phys Plasma Physicist; Plasma Physics

PLP-LPLA Pulsed Laser Photolysis Long-Path Laser Absorption

PLP-RF Pulsed Laser Photolysis Resonance Fluorescence

PL-Process Phönix-Lanzen Process [also PL-process] [Metallurgy]

PLPS Propellant Loading and Pressurization System

PLQ Photoluminescence Quenching Spectroscopy

PLR Program Lock-In Register

PLR Pulse Link Relay

PLR Pulse Link Repeater

PLR Pupil Light Reflex

PLRACTA Position Location Reporting and Control Tactical Aircraft

PLRD Payload Requirements Document

PLRS Position Location Reporting System

Plrty Polarity [also prlty]

PLRV Payload Launch Readiness Verification

PLRV Potato Leaf Roll Virus

PLS Partial Least Squares (Method)

PLS Payload Systems

PLS Physical Signalling

PLS Plugging Switch

PLS Polystyrene Latex Spheres

PLS Positive Lubrication System

PLS Positron Lifetime Spectroscopy

PLS Post Landing and Safing

PLS Preliminary Landing Site

PLS Primary Landing Site

PLS Primary Link Station

PLS Private-Line Service

PLS Professional Land Surveyor

PLS Programmable Limit Switch

PLS Propellant Loading System

PLS Publishers Licensing Society [UK]

PLS Pure Live Seed

PLSA Partner Launch Site Agreement [NASA/Canadian Space Agency]

PLSA Pulsed Laser Spectrum Analyzer

PLSL Propellants and Life Support Laboratory

PL/SNSR Payload Sensor

PLSP Payload Signal Processor

PLSS Portable Life-Support Subsystem; Portable Life-Support System

PLSS Precision Location Strike System

PLSS Prelaunch Status Simulator

PLSS Primary Life Support Subsystem

Plst Plastering [also plst]

PL-STA Polymer Laboratories' Simultaneous Thermal Analyzer

Plstg Plastering [also plstg]

Plstc Plastic [also plstc]

PLT Lanthanum-Doped Lead Titanate

PLT Lead Lanthanum Titanate (Ceramics)

PLT Payload Test

PLT Polyphenothiazine

PLT Power-Line Transient

PLT Princeton Large Torus [New Jersey, US]

PLT Program Library Tape

PLT Programming-Language Theory

PLT Progressive Lowering of Temperature

PLT Psittacosis-Lymphogranuloma-Trachoma [Medicine]

Plt Pilot [also plt]

Plt Plant [also plt]

Plt Plate [also plt]

ΦΛΘ Phi Lambda Theta (Student Society)

PLTC Power-Limited Tray Cable

PLTF Purple Loosestrife Task Force [US]

Pltg Plating [also pltg]

Pltlt Platelet [also pltlt]

Pltlts Platelets [also pltlts]

PLT-PLZT Lead Lanthanum Titanate–Lead Lanthanum Zirconate Titanate (Ceramics)

PLTS Precision Laser Tracking System

Plts Plates [also plts]

PLTTY Private Line Teletypewriter Service

PLTX Plectreurys Tristes (Spider) Toxin

PLTX-II Plectreurys Tristes (Spider) Toxin II

PLU Partial Line Up

PLU Price Loop-Up (of a Cash Register)

PLU Primary Logical Unit

ΦΛΥ Phi Lambda Upsilon (Student Society)

PLUM Planning Land Use Model

PLUM Payload Umbilical Mast

PLUM Priority Low-Use Minimal

PLUOT Parts Listing and Used On Technique

PLUS Program Language for User Systems

PLUS Program Library Update System

PLUTO Pipeline Under the Ocean

PLUTO Plutonium Loop Testing Reactor

PLV Pitch Line Velocity (of Cam Mechanism)

PLV Postlanding Vent

PLV Pressure Limiting Valve

PLV Production Level Video

PLVC Postlanding Vent Control

PLW Pressurized Light Water (Reactor)

PLWC Profiles of Liquid Water Content

PLWR Pressurized Light Water Reactor

PLX Picatinny Liquid Explosives [Picatinny Arsenal, US Army at Dover, New Jersey, US]

ΦΛΥ Phi Lambda Upsilon (Student Society)

Plywd Plywood [also plywd]

PLZ Lead-Lanthanum Zirconate (or Lanthanum-Doped Lead Zirconate)

PLZ Polish Zloty [Currency]

PLZ *(Postleitzahl)* – German for "Postal Code," or "Zip Code" [also Plz]

PLZ Programming Language Zilog

PLZT Lead Lanthanum Zirconate Titanate (or Lanthanum-Doped Lead Zirconate Titanate)

PLZT Polycrystalline Lead Zirconium Titanate

PM Package Monitor

PM Packing Material

PM PageMaker (Desk Publishing Software)

PM Paleomagnetic; Paleomagnetism

PM Palmgren-Miner (Rule) [Physics]

PM Panel Meter

PM Parachute Mine

PM Paramagnet(ic); Paramagnetism

PM Parametric Monitor

PM Parent Material

PM Parent Metal

PM Partial Melt (Process) [Materials Science]

PM Particle Mechanics

PM Particulate Material

PM Particulate Matter

PM Patternmaker; Patternmaking [Metallurgy]

PM Pavement Material

PM Payload Management

PM Payload Midbody

PM Paymaster

PM Pebble Mill

PM Pensky-Martens (Flash Point)

PM Performance Management

PM Performance Monitor(ing) [Radar]

PM Permanent Magnet(ism) [also P-M, pm, or p-m]

PM Permanent Material [Geology]

PM Permanent Mold [Metallurgy]

PM *(Per Mille)* – Per Thousand

PM Perovskite Manganite

PM Perpendicular (Recording) Medium

PM Perpetual Motion

PM Pesticide Manual

PM Petroleos Mexicanos [Mexican Petroleum Agency]

PM Petroleum Management

PM Phasemeter

PM Phase Mixture

PM Phase Modulated; Phase Modulation [also pm, or p-m]

PM Photographic Material

PM Photomagnetic; Photomagnetism

PM Photomechanical; Photomechanics

PM Photomicrograph(y)

PM Photomicroscopy

PM Photomodulation

PM Photomultiplier

PM Physical Measurement

PM Physical Medicine

PM Physical Metallurgist; Physical Metallurgy

PM Physical Meteorologist; Physical Meteorology

PM Physical Metrologist; Physical Metrology

PM Physical Modelling

PM Piezomagnetic; Piezomagnetism

PM Piezomagnetometer

PM Pia Mater (of the Brain) [Anatomy]

PM Pilot Motor

PM Placer Mine; Placer Mining

PM Planar (or Plane) Matching (Model) [Crystallography]

PM Planetary Mission

PM Planetary Motion

PM Plant Mix

PM Plasma Membrane [Cytology]

PM Plastic Memory

PM Platinum(-Group) Metal

PM Plugging Meter

PM Poincaré Map

PM Poisonous Material

PM Polarization Microscope; Polarization Microscopy

PM Polarization Modulation

PM Polarizing Material

PM Police Magistrate

PM Poliomyelitis [Medicine]

PM Polymeric Material

PM Polymethylene

PM Pony Motor

PM Porous Medium

PM Portable Medium-Power Plant [US]

PM Port Moresby [Papua-New Guinea]

PM Positive Medium (Error)

PM Postmaster

PM *(Post Meridiem)* – Afternoon [also pm]

PM Post Mortem

PM Potential Motion

PM Potentiometer

PM Pounds per Minute [also lbs/min]

PM Powdered Medium

PM Powder Magnet

PM Powder Metal

PM Powder Metallurgical; Powder Metallurgy [also P/M]

PM Powder Method [Solid-State Physics]

PM Power Management

PM Precious Metal

PM Precision Measurement

PM Predictive Modelling

PM Preisach (Hysteresis) Model [Solid-State Physics]

PM Premium [also Pm, or pm]

PM Presentation Manager [Computers]

PM Pressure Melting

PM Pressure Measurement

PM Preventive Maintenance

PM Prikladnaya Mekhanika [Ukrainian Journal on Applied Mechanics]

PM Primary Memory

PM Primary Mineral

PM Prime Minister

PM Prime Mover

PM Priority Mail

PM Probability Mass

PM Procedures Manual

PM Process Management; Process Manager

PM Process Metallurgist; Process Metallurgy

PM Processing Machinery

PM Process Monitor

PM Processor Module

PM Production Management; Production Manager

PM Product Management; Product Manager

PM Professional Management; Professional Manager

PM Program Manager; Program Management

PM Program Milestone

PM Projectile Motion

PM Project Management; Project Manager

PM Project Mohole [Geology]

PM Proton Magnetometer

PM Provost Marshall

PM Pulmonary Mechanics

PM Pulsating Mixing (Process) [Metallurgy]

PM Pulse Modulation; Pulse Modulator

PM Purchasing Management; Purchasing Manager

PM Purple Membrane (Domains)

PM Putzeiko Method [Physics]

PM Pyrometallurgical; Pyrometallurgist; Pyrometallurgy

PM Saint Pierre and Miquelon [ISO Code]

PM World Conference on Powder Metallurgy

PM² Powder Metallurgy and Particulate Materials

PM-1 Portable Medium-Power Reactor 1 [US]

P&M Processes and Materials

P/M Parts per Million [also PPM, or ppm]

P/M Polycrystalline (Material) on Microcrystalline (Material)

P/M Powder Metallurgical; Powder Metallurgy [also PM]

P-M Parent Phase-to-Martensite (Transformation) [also P/M] [Metallurgy]

P-M Permanent Magnet [also PM]

P-M Phase Modulation [also p-m]

Pm Petroleum [also pm]

Pm Promethium [Symbol]

Pm-141 Promethium-141 [also ^{141}Pm, or Pm^{141}]

Pm-142 Promethium-142 [also ^{142}Pm, or Pm^{142}]

Pm-143 Promethium-143 [also ^{143}Pm, or Pm^{143}]

Pm-144 Promethium-144 [also ^{144}Pm, or Pm^{144}]

Pm-145 Promethium-145 [also ^{145}Pm, or Pm^{145}]

Pm-146 Promethium-146 [also ^{146}Pm, or Pm^{146}]

Pm-147 Promethium-147 [also ^{147}Pm, Pm^{147} or Pm]

Pm-148 Promethium-148 [also ^{148}Pm, or Pm^{148}]

Pm-149 Promethium-149 [also ^{149}Pm, or Pm^{149}]

Pm-150 Promethium-150 [also ^{150}Pm, or Pm^{150}]

Pm-151 Promethium-151 [also ^{151}Pm, or Pm^{151}]

pm picometer [Unit]

pm *(post meridiem)* – afternoon [also PM]

pm *(post mortem)* – after death

pm premium

pm premolar (tooth)

.pm Saint Pierre and Miquelon [Country Code/Domain Name]

£M Maltese Pound [Currency of Malta]

PMA Ammonium Phosphomolybdate

PMA Pacific Maritime Association [US]

PMA Pakistan Medical Association

PMA Parts Manufacturing Authority

PMA Pencil Makers Association [US]

PMA Performance Monitor Annunciator

PMA Permanent Magnet Alloy(s)

PMA Permanent Magnet Association

PMA Permanent Management Arrangements

PMA Perpendicular Magnetic Anisotropy

PMA Petroleum Monitoring Agency

PMA Pharmaceutical Manufacturers Association [US]

PMA Phenylmercuric Acetate (or Phenyl Mercury Acetate)

PMA Phorbol Myristate Acetate; Phorbol-12-Myristate-13-Acetate

PMA Phosphomolybdic Acid

PMA Photo Marketing Association International [US]

PMA Physical Medium Attachment

PMA Physical Memory Address

PMA Plane Mirror Analyzer

PMA P-Methoxyamphetamine

PMA Polymeth(yl)acrylate

PMA Polyurethane Manufacturers Association [US]

PMA Positive Misfit Alloy

PMA Post-Metallization Anneal(ing) [Metallurgy]

PMA Powder Metallurgy Association [Taiwan]

PMA Precision Measurements Association [US]

PMA Precision Metalforming Association [US]

PMA Premarket Approval

PMA Pressurized Mating Adapter

PMA Prime Macro-Assembler

PMA Priority Memory Access

PMA Priority Memory Address

PMA Produce Marketing Association [US]

PMA Production and Marketing Administration [US]

PMA Production Monitoring Analysis

PMA Progressive Muscular Atrophy [Medicine]

PMA Protected Memory Address

PMA Protective Metal Alloy

PMA Protective Metal Alloys, Inc. [San Diego, California, US]

PMA Pyromellitic Acid

pMA p-Methylaniline

PM-A Polishing Medium–Abrasive

PM-A Preventive Maintenance–Level A

PMAA Petroleum Marketers Association of America [US]

PMAC Pharmaceutical Manufacturers Association of Canada

PMAC Phenylmercuric Acetate

PMAC Purchasing Management Association of Canada

PMAD Performance Monitor Annunciation Driver

PMAD Polymer Modifiers and Additives Division

PMAF Pharmaceutical Manufacturers Association Foundation [US]

PMAI Powder Metallurgy Association of India

P-Mail Paper Mail [also p-mail]

PMAI Newsl Powder Metallurgy Association of India Newsletter

PMAN Polymethacrylonitrile; Methacrylonitrile/Divinylbenzene, Polymeric Form

PMAP Performance Monitor Annunciation Panel

PMAP Physical, or Logical Coordinate Mapping (Function) [Microsoft-Disk-Operating System]

pMAP Professional Map Analysis Package [Software]

PMAR Page Map Address Register

P-31 MAS NMR Phosphorus-31 Magic-Angle Spinning Nuclear Magnetic Resonance [also ^{31}P MAS NMR]

PMASA Powder Metallurgy Association of South Africa

P-31 MASS NMR Phosphorus-31 Magic-Angle Sample Spinning Nuclear Magnetic Resonance [also ^{31}P MASS NMR]

PMAT Page Map Address Table

PMATA Paper Makers Allied Trades Association [UK]

PMB Pentamethylbenzene

PMB PHARE (Program for Harmonized Air Traffic Management Research in Europe) Management Board [of Eurocontrol]

PMB Physical Metallurgy Branch [of Naval Research Laboratory, Washington, DC, US]

PMB Pilot Make Busy

PMB Plant Molecular Biology

PMB Plastic Media Blasting

PMB Print Measures Bureau [Canada]

PMB PROM (Programmable Read-Only Memory) Memory Board

Pmb Pentamethylbenzyl

PM-B Preventive Maintenance–Level B

PMBA Plant Molecular Biology Association [now International Society for Plant Molecular Biology, US]

pMBA p-Methylbenzoic Acid

pMBAAP 4-N-(4'-Methylbenzylidene)aminoantipyrine

PMBC Pilot Make Busy Call

PMBC Plywood Manufacturers of British Columbia [Canada]

PMBX Private Manual Branch Exchange [also pmbx]

PMC Pacific Marine Center [of National Oceanic and Atmospheric Administration at Seattle, Washington, US]

PMC Payload Monitoring and Control

PMC Pennsylvania Military College [Chester, US]

PMC Phenylacetamidocoumarin

PMC Photomechanochemist(ry)

PMC Planning Ministers Council

PMC Plutonium-Molybdenum Cermet

PMC Polymer-Matrix Ceramic

PMC Polymer-Matrix Composite

PMC Polymer-Metal Composite

PMC Polymer-Modified Cement

PMC Polymer-Reinforced Matrix Composite

PMC Postmanufacturing Checkout

PMC Post Master's Certificate

PMC Post Ministerial Conference

PMC Pressure-Modulator Cell

PMC Private Medical Communication

PMC Private Meter Check

PMC Process Monitoring and Control

PMC Procurement Method Code

PMC Programmable Machine Controller

PMC Program Management Control

PMC Program Marginal Checking

PMC Project Management Corporation

PMC Pseudo Machine Code

PM-C Preventive Maintenance–Level C

PMCB Pentamethyl Chlorobenzene

PMCC Pensky-Martens Closed Cup (Flash Point Tester)

PMCG Pyrrolidylmethyl Cyclopentylphenylglycolate

PMCN Lead Magnesium Cadmium Niobate

pMCP p-Methoxycarbonylphenol

PMCS Pulse-Modulated Communications System

p-μc-Si P-Type Microcrystalline Silicon

p-μc-SiC P-Type Microcrystalline Silicon Carbide

p-μc-SiC:H P-Type Microcrystalline Hydrogenated Silicon Carbide

p-μc-Si:H P-Type Hydrogenated Microcrystalline Silicon

PMD Packet Mode Data

PMD Palmdale, California [Tactical Air Navigation Station]

PMD Paramagnetic Dispersion

PMD Parts Manufacturing Division

PMD Point of Maximum Definition

PMD Pollution Measurement Division [of Environmental Technology Center, Canada]

PMD Post Mortem Dump [Computers]

PMD Process Measurements Division [of National Institute of Standards and Technology, Gaithersburg, Maryland, US]

PMD Program Module Dictionary

PMD Project Management Division

PM&D Power Management and Distribution

PMDA Photographic Manufacturers and Distributors Association [US]

PMDA Plastics Machinery Distributors Association [UK]

PMDA Pyromellitic Acid Dianhydride

PMDA Pyromellitic Dianhydride

PMDA-ODA Pyromellitic Dianhydride/Oxydianiline (Polymer) [also PMDA/ODA]

PMDA-ODA/DABSA Pyromellitic Dianhydride/Oxybis(aniline)/2,5-Diaminobenzenesulfonic Acid

PMDBD 3,3,6,9,9-Pentamethyl-2,10-Diazabicyclo-[4.4.0]dec-1-ene

PMDI Polydiphenylmethanediisocyanate

PMDI Polymethylene Diisocyanate

PMDI Polymethylene Diphenylene Isocyanate

PMDL Palmdale, California [Tactical Air Navigation Station]

PMDL Photonic Materials and Devices Laboratory [of University of Southern California, Los Angeles, US]

PMDR Phosphorescence Microwave Double Resonance (Spectroscopy)

PMDS Plant Management, Maintenance/Design Engineering Show [Canada]

PMDS Power Management and Distribution System

PMDTA Pentamethylenediaminetetraacetate [also pmdta]

PMDTA Pentamethyl Diethylene Triamine

PME Pacific Materials Exchange

PME Photomagnetic Effect

PME Photomagnetoelectric(ity)

PME Photomagnetoelectric Effect

PME Photomechanical Effect

PME Precision Measuring Equipment

PME Process and Manufacturing Engineering

PME Process Mechanical Engineering

PME Processor Memory Enhancement

PME Protective Multiple Earthing

PMe$_3$ Trimethyl Phosphine

PMEA Powder Metallurgy Equipment Association [US]

PMEE Prime Mission Electronic Equipment

PMEF Petroleum Marketing Education Foundation [US]

PMEG Perforated Metal Export Group [UK]

PMEL Pacific Marine Environmental Laboratory [of National Oceanic and Atmospheric Administration, US]

PMEL Precision Measuring (or Measurements) Equipment Laboratory

PM-ENDOR Polarization Modulated (Radio-Frequency Fields) in Electron Nuclear Double Resonance

PMEP Principal Metal ESCA (Electron Spectroscopy for Chemical Analysis) Peak

PMET Poly(L-Methionine) [Biochemistry]

PMF Performance Monitor(ing) Function

PMF Perigee Motor Firing

PMF Potential of Mean Force

PMF Probable Maximum Flood

PMF Probable Maximum Flow

PMF Program Management Facility

PMF Pulsed Magnetic Field

PMFC Pacific Marine Fisheries Commission [US]

PMFC Polymeric Matrix Fibrous Composite

PMFG Pulsed Magnetic Field Gradient

PMFJI Pardon Me For Jumping In [Internet Jargon]

PM-FM Paramagnetic-Ferromagnetic (Transition) [Solid-State Physics]

PMF-NMR Pulsed Magnetic Field Nuclear Magnetic Resonance

PMFT Pulsed Magnetic Field Treatment

PMFTIRAS Polarization Modulated Fourier Transform Infrared Reflection-Absorption Spectroscopy [also PMFTIRRAS]

PMG Pacific Marine Geology [of US Geological Survey]

PMG Paymaster General

PMG Photographic Materials Specialty Group [of American Institute for Conservation, US]

PMG Poly(N-Methacryloyl-D-Glucosamine [Biochemistry]

PMG Portfolio Marine Group [of Environment Australia]

PMG Postmaster General

PMG Prediction Marker Generator

PMgr Professional Manager

PMGS Predictable Model Guidance Scheme

PMH Phenylmercuric Hydroxide

PMH Probable Maximum Hurricane

PMH Production per Man-Hour

PMHL Preferred Measurement Hardware List

PMHP P-Methane Hydroperoxide

PMHS Polymethylhydrosiloxane

PM-HSS Powder Metallurgy High-Speed Steel

PMI Phosphomannose Isomerase [Biochemistry]

PMI Plumbing Manufacturers Institute [US]

PMI Poly-Meta-Phenyleneisophthalamide

PMI Positive Metal Identification

PMI Postmaintenance Inspection

PMI Powder Metallurgy International [Journal published in Germany]

PMI Precious Metals Institute

PMI Preventive Maintenance Inspection

PMI Preventive Maintenance Instruction

PMI Princeton Materials Institute [of Princeton University, New Jersey, US]

PMI Principal Maintenance Inspector

PMI Private Memory Interconnect

PMI Private Memory Interface

PMI Project Management Institute

PMI Pseudo Matrix Isolation [Physics]

PMIA Parallel Multiplexer Interface Adapter

PMIA Precious Metals Industry Association [US]

P/MIA Powder Metallurgy Industries Association [also PMIA]

PMIAA Powder Metallurgy Industries Association of Australia

PMIC Parallel Multiple Incremental Computer

PMIC Payload Mission Integration Contractor

PMIC Precious Metals Institute Conference

PMIF Powder Metal Industries Federation

PMIM Powder Metal Injection Molding

PMIMRS Potential Modulation Induced Microwave Reflectance Spectroscopy

PMIP Paleoclimatological Model Intercomparison Project [Geology]

PMIP Post-Maintenance Inspection Pilot

PMIR Pressure Modulator Infrared Radiometer [of NASA Mars Observer]

PMIR Program Manager's Integration Review

PM-IRRAS Polarization Modulated Infrared Reflection-Absorption Spectroscopy

PMI Powder Metall Int PMI Powder Metallurgy International [Journal published in Germany]

PMIS Personnel Management Information System

PMIS Plant Management Information System

PMIS Precision Mechanisms in Sodium

MIS Project Management Information System

p-MIS P-Type Metal–Interface Layer–Semiconductor (Junction)

PMJI Pardon My Jumping In [Internet Jargon]

Pmk Postmark [also pmk]

PML Phoenix Memorial Laboratory [of University of Michigan, Ann Arbor, US]

PML Physical Memory Level

PML Plymouth Marine Laboratory [UK]

PML Polymer Microdevice Laboratory

PML Polymer Multilayer (Coating)

PML Powder Metallurgy Laboratory [of Max-Planck Institut, Stuttgart, Germany]

PML Preliminary Materials List

PML Programmable Macro Logic

PML Progressive Multifocal Leukoencephalopathy [Medicine]

PML Pulvermetallurgisches Laboratorium [Powder Metallurgy Laboratory; of Max-Planck Institute (for Metals Research), Stuttgart, Germany]

PMLC Programmed Multi-Line Controller

PMM Paramagnetic Material

PMM Parametric Modelling

PMM Permanent-Magnet Motor

PMM Perpetual Motion Machine [Physics]

PMM Polymer Matrix Metal

PMM Polymethylmethacrylate

PMM Pool Maintenance Module

PMM Prikladnaya Matematika i Mekhanika [Journal on Applied Mathematics and Mechanics; published by the Academy of Sciences of the USSR]

PMM Probability Matching Method

PMM Programmable Microcomputer Module

PMM Programmable Multimeter

PMM Property Management Manual

PMM Pulse Mode Multiplex

PMMA Paper Machinery Makers Association

PMMA Polymethyl Methacrylate

PMMA Polymethyl Methacrylate Association [of Conseil Européen des Fédérations de l'Industrie Chimique, Belgium]

PMMB Parallel Memory to Memory Bus

PMMC Particulate Metal Matrix Composite

PMMC Permanent Magnetic Movable Coil

PMMI Packaging Machinery Manufacturers Institute [US]

PM MMC Powder Metallurgy-Based Metal-Matrix Composite [also P/M MMC]

PMMT Procurement, Materials Management and Transportation Department [of Alcoa Corporation, US]

PMMU Paged Memory Management Unit

PMN Lead Magnesium Niobate (Ceramics)

PMN Lead Manganese Niobate (Ceramics)

PMN 1,2,3,4,5-Pentamethyl-6-Nitroso Benzene

PMN Polymorphonuclear Neutrophil [Medicine]

PMN Premanufacture Notice (or Notification)

PMN Program Management Network

PMn Pesticide Manual n-th Edition

P_6MN_3 Lead Magnesium Niobate [$6PbO \cdot MgO \cdot 3Nb_2O_5$]

PMNBT Methylated Poly(nonyl Bithiazole)

PMNH Peabody Museum of Natural History [New Haven, Connecticut, US]

PMNP Platform-Mounted Nuclear Plant

PMN-PT Lead Magnesium Niobate–Lead Titanate (Ceramics)

PMN-PZT Lead Magnesium Niobate–Lead Zirconate Titanate (Ceramics)

PMNS Provincial Museum of Nova Scotia [Halifax, Canada]

PMNT Lead Magnesium Niobate Titanate

PMO Perturbational Molecular Orbital [Physical Chemistry]

PMO Phenyl Mercury Oleate

PMO Plasma Melt Overflow

PMO Polymethylene Oxide

PMO Program Management Office [of Defense Data Network, US Department of Defense]

PM&O Plant Maintenance and Operation

PMOKE Polar Magnetooptical Kerr Effect

pmol picomole [Unit]

pmol/μL picomole per microliter [also pmol μL^{-1}]

PMOM Performance Management Operations Manual

PMON Performance Management Operations Network

PMOS P-Channel Metal-Oxide Semiconductor [also P MOS, P-MOS, or p-MOS]

PMOS Positive Metal-Oxide Semiconductor [also P MOS, P-MOS, or p-MOS]

PMOS P-Type Metal-Oxide Semiconductor [also P MOS, P-MOS, or p-MOS]

p-MoS$_2$ P-Type Molybdenum Disulfide

p-MOSFET P-Channel Metal-Oxide Semiconductor Field-Effect Transistor [also PMOSFET]

p-MOSFET P-Type Metal-Oxide Semiconductor Field-Effect Transistor [also PMOSFET]

Ψ-MOSFET Pseudo Metal-Oxide Semiconductor Field-Effect Transistor

PMP O,O-Dimethyl S-(Phthalimidomethyl)-phosphorodithionate

PMP International Conference on Processing Materials for Properties

PMP Participative Management Program [of Motorola Corporation, US]

PMP Parts-Material-Packaging

PMP Parts, Materials and Processes [also PM&P]

PMP 1,2,2,6,6-Pentamethylpiperidine

PMP Performance Management Package

PMP Phased Manufacturing Program

PMP Phenylmethylpyrazolone

PMP Photomechanical Process

PMP P-Methoxyphenol

PMP (±)-2-(P-Methoxyphenoxy)propionic Acid

PMP Polymethylpentene

PMP Poly(4-Methyl-1-Pentene)

PMP Pontryagin's Maximum Principle [Mathematics]

PMP Portable Medium-Powder Plant [US]

PMP Premodulation Processor

PMP Premolded Plastic(s)

PMP Pressure Measurement Package

PMP Preventive Maintenance Plan

PMP Probable Maximum Precipitation [Meteorology]

PMP Program Management Plan

PMP Project Management Plan

PMP Project Master Plan

P4M1P Poly(4-Methyl-1-Pentene)

PM&P Parts, Materials and Processes [also PMP]

Pmp Pump [also pmp]

pMP p-Methoxyphenol

PMPA Permanent Magnet Producers Association [now Magnetic Material Producers Association, US]

PMPA Powder Metallurgy Parts Association [also P/MPA, US]

PMPE Punch Memory Parity Error

PMPIPA Poly-m-Phenyleneisophthalamide

PMPL Preferred Mechanical Parts List

PMPMA Plastic and Metal Products Manufacturers Association [US]

PMPO Pulse-Modulated Power Oscillator

PMPPIC Polymethylene–Polyphenyl Isocyanate

PMPPOH 4-Propionyl-3-Methyl-1-Phenyl 2-Pyrazoline-5-one

PMPRB Program to Manage Pesticide Residues in Beef [Western Australia]

PMPS Poly(2-Methyl-1-Pentene Sulfone)

PMPS Polymethylphenylsilane

PMPSPT Polarization Moeller-Plesset Static Perturbation Theory [Physics]

PMR Pacific Missile Range [of US Navy at Point Mugu, California, US]

PMR Paramagnetic Relaxation

PMR Paramagnetic Resonance

PMR Performance Measurement Report

PMR Polymerizable Monomer Reactant; Polymerized Monomeric Reactant

PMR Portable Multiparameter Radar

PMR Positive Magnetoresistance [Solid-State Physics]

PMR Post Mortem Review

PMR Powder Magnetoresistance [Solid-State Physics]

PMR Precious Metals Recovery

PMR Pressure Modulator Radiometer

PMR Private Mobile Radio

PMR Problem Management Report

PMR Program Management Review

PMR Program Manager's Review

PMR Projection Microradiography

PMR Proton Magnetic Resonance

PM-Re Powder Metallurgical Rhenium

PMRI Powder Metallurgy Reseach Institute [PR China]

PMRI Prairie Masonry Research Institute [Canada]

PMRI Preventive Medicine Research Institute [US]

PMRL Physical Metallurgy Research Laboratories [of Canada Center for Mineral and Energy Technology, Natural Resources Canada]

PMRL Pulp Manufacturers Research League

PMRS Plasma-Melted Rapidly Solidified (Material)

PMRS Potential Modulated Reflectance Spectroscopy

PMS Lead Molybdenum Sulfide (Semiconductor)

PMS 5-Methyl-Phenazinium Methylsulfate

PMS P-Methylstyrene

PMS Paramagnetic Salt

PMS Paramethylpentene

PMS Particle Measurement Systems, Inc. [US]

PMS Particle Measuring System

PMS Performance Management System

PMS Performance Measurement System

PMS Performance Monitoring System

PMS Permanent Measurement System

PMS Phenazine Methosulfate

PMS Physical and Mathematical Sciences

PMS Physical Metallurgy Section [of Indira Gandhi Centre for Atomic Research, Kalpakkam, India]

PMS Picturephone Meeting Service [US]

PMS Planar Magnetron Sputtering

PMS Plant Monitoring System

PMS Policy Management System

PMS Polysciences Multiple Stain [Biology]

PMS Pre-Main-Sequence

PMS Premenstrual Stress

PMS Premenstrual Syndrome

PMS Probable Maximum Surge

PMS Processor Memory Switch

PMS Product Monitoring System

PMS Program Management System

PMS Progress in Materials Science [Journal published in the UK]

PMS Project Map System

PMS Public Message Service [US]

PMS Pulsed (Time-of-Flight) Mass Spectrometer

PMS-C Premenstrual Syndrome, Craving

PMS-D Premenstrual Syndrome, Depression

PMSF Phenylmethanesulfonyl Fluoride

PMSF Phenylmethylsulfonyl Fluoride

pMSfa p-Methylsulfonylaniline

PMS-H Premenstrual Syndrome, Heavy

PMSO Polydimethylsilsesquioxane

PMS-P Premenstrual Syndrome, Pain

pMSP p-Methylthiophenol

pMsP p-Methylsulfonyl Phenol

PMS-PVD Planar Magnetron Sputtering Physical Vapor Deposition

PMSRP Physical and Mathematical Sciences Research Paper

PMSX Processor Memory Switch Matrix

PMT Lead Manganese Telluride (Semiconductor)

PMT Phenylmercaptotetrazole

PMT Photomechanical Transfer

PMT Photomultiplier Tube

PMT Planning/Management Team

PMT Precious-Metal Tip

PMT Premenstrual Tension

PMT Production Monitoring Test

PMT Program Master Tape

PMT Pulsed Magnetic Treatment (Technology)

Pmt Payment [also pmt]

PMTA Phosphomolybdic and Phosphotungstic Acid Mixture

pMTa p-Methylthioaniline

PMTC Pacific Missile Test Center [US]

PMTC Phenomenological Martensite Transformation Concept

PMTC Private Motor Truck Council [US]

PMTCC Private Motor Truck Council of Canada

PM Tec World Congress on Powder Metallurgy and Particulate Materials [of Metal Powder Industries Federation, US]

PM²TEC International Conference (and Exhibition) on Powder Metallurgy and Particulate Materials

PMTHP Project Mercury Technical History Program [of NASA]

PM/TMDE Project Manager for Test, Measurement and Diagnostic Equipment [US Army]

PM304SS Powder Metallurgy AISI-SAE 304 Stainless Steel

PM316SS Powder Metallurgy AISI-SAE 316 Stainless Steel

1-PMTR 1-Phospho-5-S-Methylthio-D-Ribofuranoside [Biochemistry]

1-PMT-Ru 1-Phospho-5-S-Methylthioribulose [Biochemistry]

P/M Technol Newsl P/M Technology Newsletter [Published by the Metal Powder Industries Federation, US]

PMU Parametric Measurement Unit

PMU PCM (Pulse-Code Modulation) Master Unit

PMU Portable Memory Unit

PMU Precision Measurement Unit

PMU Pressure Measuring Unit

PMU Processor Management Unit

PMU Pulse Modulation Unit

PMUX Programmable Multiplex(er)

PMV Poliomyelitis Virus

pm/V picometer per volt [also pm V^{-1}]

PMVE Perfluoromethyl Vinyl Ether

PMVI Periodic Motor Vehicle Inspection

PMW Lead Magnesium Tungstate

PMW Project Manager Workbench

PMWC Powder Metallurgy World Congress

PMX Packet Multiplexer

PMX Physical Modeling Extension

PMX Private Manual Exchange [also pmx]

PMX Protected Message Exchange

pMxA p-Methoxyaniline

pMxBA p-Methoxybenzoic Acid

pMxCa p-Methoxycarbonylaniline

PMZ Partially Melted Zone

PMZ Postmenopausal Zest

PN Lead Niobate

PN Packet Network

PN Paging Network

PN Paranodal; Paranode

PN Partition [also pn] [Building Trades]

PN Partition Noise [Electronics]

PN Partition Number

PN Part Number [also P/N]

PN Passive Network

PN Pass Number

PN Patent Number

PN P-Doped N-Doped (Junction) [also P-N, pn, or p-n]

PN Peierls-Nabarro (Approach) [Solid-State Physics]

PN Pemphigus Neonatorum [Medicine]

PN Perceived Noise Level

PN Percussion Noise

PN Perfect Number

PN Performance Number [Aviation Gasoline]

PN Periodic Number

PN Phase Name

PN Photonegative; Photonegativity

PN Photoneutron

PN Photonic Network

PN Photonuclear

PN Pitcairn [ISO Code]

PN Planetary Nebula [Astronomy]

PN Plasma Nitrided; Plasma Nitriding

PN Plug Nozzle

PN Pneumatic Nebulizer

PN Point of No Return

PN Polish Notation

PN Polymer Network

PN Porto Novo [Benin]

PN Positive-Negative (Semiconductor) [also P-N, pn, or p-n]

PN Practical Nurse

PN Praseodymium Nickelate

PN Preliminary Notification

PN Prime Number

PN Prior Notice

PN Processing Node

PN Product Name

PN Professional Nurse

PN Programmable Network

PN Project Number

PN Promissory Note [also P/N, pn, or p/n]

PN Prompt Neutron

PN Propenenitrile

PN Proportional Navigation

PN 1-(n-Propyl)naphthalene

PN Propyl Nitrate

PN Pseudonoise; Pseudorandom Noise

PN Pseudorandom Numbers

PN Psychoneurotic; Psychoneurosis

PN Public Network

PN Punch On

PN Punch Only

1,2-PN 1,2-Propanediamine

1,3-PN 1,3-Propanediamine

P/N Park/Neutral [Automobiles]

P/N Part Number [also PN]

P/N Print plus Negative (Film)

P/N Promissory Note [also p/n, PN, or pn]

Pn Numeric Parameter [ANSI Escape Sequence Code]

Pn Propylene Diamine [also pn]

p-n p-plus-type layer/n-type layer (semiconductor junction)

p$^+$-n p-plus-type/n-type (semiconductor device) [also p$^+$n]

ϕ(n) Euler function (i.e., number of positive integers not exceeding and prime to the positive integer n) [Symbol]

PNA Nucleic Acids with Peptide Backbones [Biochemistry]

PNA Pacific North America

PNA Packet Network Adapter

PNA Peanut (*Arachis hypogaea*) Agglutinin

PNA Pentanitroaniline

PNA Pentose Nucleic Acid [Biochemistry]

PNA Polynuclear Aromatic Hydrocarbon(s)

PNA Programmable Network Access

PNA Project Network Analysis

pNA p-Nitroanaline

pNa (*potentia natrium*) – logarithm of the sodium-ion concentration in a solution [also PNa]

PNAD Personnel Nuclear Accident Dosimeter

PNAF Potential Network Access Facility

PNAS Proceedings of the National Academy of Sciences [US]

PNB Pernigraniline Base

PNB P-Nitrobenzoate

PNBA P-Nitrobenzyloxyamine

PNBC Pacific Northwest Bibliographic Center [Canada and US]

PNBH P-Nitrobenzaldehyde Hydrazone

PNBN Polynitrobenzonitrile

PNBPA P-Nitrobenzyl-N-(n-Propyl)amine

PNBS Pyridinium 3-Nitrobenzenesulfonate

pNBST 1-[(p-Nitro)benzenesulfonyl]-1H-1,2,4-Triazole [also p-NBST]

PNBT Poly(nonylbithiazole)

PNC Parity Non-Conservation [Quantum Mechanics]

PNC Phosphonitrilic Chloride

PNC Photonitrosation of Cyclohexane (Process)

PNC Police National Computer [UK]

PNC Polynuclear Compound

PNC Power Reactor and Nuclear Fuel Development Corporation [Japan]

PNC Programmed Numerical Control

PNC PRIMENET (Prime Network Software Package) Node Controller

PNCC Partial Network Control Center

PNCFN Permanent Nordic Committee on Food and Nutrition [Denmark]

Pnch Punch [also pnch]

PNCIAWPRC Philippine National Committee of the International Association on Water Pollution Research and Control

PNCIAWPRC Portuguese National Committee of the International Association on Water Pollution Research and Control

PNCMI Polarized Neutrons for Condensed Matter Investigations (Symposium)

PNCO Praseodymium Neodymium Copper Oxide (Superconductor)

PNCS Pakistani National Conservation Strategy

PND Pictorial Navigation Display

PND Polarized Neutron Diffraction

PND Positive-Negative Diode

PND Program Network Diagram

PND Proton Noise Decoupling

PNdB Perceived Noise Level in Decibels

PNDC Parallel Network Digital Computer

PNDO Partial Neglect of Differential Overlap [Physics]

PNDPhEth Pentanitrodiphenylether

PNDPhEtl Pentanitrodiphenylethanol

PNDPhSfo Pentanitrodiphenylsulfone

PNE Pacific National Exhibition [Vancouver, Canada]

PNE Peaceful Nuclear Explosion(s)

PNE Poblacht Na h'Eireann [Republic of Ireland]

PNEA Programma Nazionale di Ricerche in Antartide [National Antarctic Research Program, Italy]

PNEC Proceedings of the National Electronics Conference

PNEG Polycrystalline Nitride Extrinsic Gettering [also PN-EG]

Pneu Pneumatic(s) [also pneu]

Pneu Pneumococcus [also pneu] [Microbiology]

PNEUROP European Committee of Manufacturers of Compressors, Vacuum Pumps and Pneumatic Tools [UK]

PNF Partial Neutralization Feed

PNF Phosphonitrilic Fluoroelastomer

PNF Pro-Nectin F [Biochemistry]

PNF Proprioceptive Neuromuscular Friction [Medicine]

PNFI Petawawa National Forestry Institute [of Natural Resources Canada]

PNG Papua–New Guinea

PNG Percent Normalized Gradient (of Pavement Surface) [Civil Engineering]

PNG (*Persona Non Grata*) – Undesirable Person

PNG Petroleum and Natural Gas

PNG Polymer Network Group

PNG Portable Network Graphics

PNGase F Peptide N-Glycosidase F [Biochemistry]

PNGCB Petroleum and Natural Gas Conservation Board [Alberta, Canada]

PNGCS Primary Navigation, Guidance, and Control System

PNGFA Pacific Northwest Grain and Feed Association [US]

PNGV Partnership for Next Generation Vehicle [US]

Pn H Pan Head (Screw) [also PnH]

PNI Pharmaceutical News Index [US]

PNI Pictorial Navigation Indicator

PNI Pohnpei, Federated State of Micronesia [Meteorological Station Designator]

P-NID Precedence Network In-Dialing

PNIMIC Pilot NATO Insensitive Munitions Information Center

PNIN P-Doped, N-Doped, Intrinsic, N-Doped (Transistor) [also P-N-I-N, p-n-i-n, or pnin]

PNIN Positive-Negative-Intrinsic-Negative (Transistor) [also P-N-I-N, p-n-i-n, or pnin]

PNIP P-Doped, N-Doped, Intrinsic, P-Doped (Transistor) [also P-N-I-P, p-n-i-p, or pnip]

PNIP Positive-Negative-Intrinsic-Positive (Transistor) [also P-N-I-P, p-n-i-p, or pnip]

PNIPAAM Poly(N-Isopropyl Acrylamide)

PNL Pacific Naval Laboratory [now Defense Research Establishment Pacific, Canada]

PNL Pacific Northwest (National) Laboratory [now Pacific Northwest National Laboratory]

PNL Perceived Noise Level

PNL Programme National de Laboratoire [National Laboratory Program, France]

Pnl Panel [also Pnl, or pnl]

PNLA Pacific Northwest Library Association [US]

Pnlbd Panelboard [also pnlbd]

PNLT Pippard Non-Local Theory [Physics]

PNM Pulse-Number Modulation

PNN Lead Nickel Niobate (Ceramics)

p⁺nn⁺ p-plus layer, n-layer, n-plus layer (magnetodiode)
p^+nn^+ p-plus layer, n-layer, n-plus layer (magnetodiode)

p⁺n⁻n⁺ p-plus layer, n-minus layer, n-plus layer (semiconductor device)
$p^+n^-n^+$ p-plus layer, n-minus layer, n-plus layer (semiconductor device)

PNN-PT-PZ Lead Nickel Niobate–Lead Titanate–Lead Zirconate (Ceramics)

PNN-PZT Lead Nickel Niobate–Lead Zirconate Titanate (Ceramics)

PNMR Proton Nuclear Magnetic Resonance

PNMR Pulsed Nuclear Magnetic Resonance

P-31 NMR Phosphorus-31 Nuclear Magnetic Resonance [also ^{31}P NMR]

PNMT Phenylethanolamine N-Methyltransferase [Biochemistry]

PNNI Private Network to Network Interface

PNNL Pacific Northwest National Laboratories [of Battelle Memorial Institute, Richland, Washington, US]

PNO Pair Natural Orbitals [Physics]

PNO Praseodymium Nickel Oxide (Compound)

PNO PROFIBUS (Process Field Bus) Nutzerorganisation [Profibus User Organization]

PNO Pseudonatural Orbital [Physics]

P-No Welding Procedure-Group Number

pNOa p-Nitroaniline

pNOBA p-Nitrobenzoic Acid

PNOC Philippines National Oil Company

PNOCI Pair Natural Orbital Configuration Interaction [also PNO-CI]

PNOCEPA Pseudonatural Orbital Coupled Electron Pair Approximation [also PNO-CEPA]

PNOCID Pair Natural Orbital Configuration Interaction (with) Double (Excitation) [also PNO-CID]

pNOP p-Nitrophenol

PNP P-Doped, N-Doped, P-Doped (Transistor) [also P-N-P, pnp, or p-n-p]

PNP Pick-and-Place Robot

PNP Plug and Play (Capability) [Computers]

PNP Positive-Negative-Positive (Transistor) [also P-N-P, pnp, or p-n-p]

PNP Precision Navigation Processor

PNP Prenegotiation Position

PNPDPP P-Nitrophenyl Diphenyl Phosphate

PNPF Piqua Nuclear Power Facility [Piqua, Ohio, US] [Shutdown]

PNPG P-Nitrophenylglycerine [Biochemistry]

PNPG α-P-Nitrophenylglycerol [Biochemistry]

PNPG1 P-Nitrophenyl-α-D-Glycoside [Biochemistry]

PNPG3 P-Nitrophenyl-α-D-Maltotriose [Biochemistry]

PNPG7 P-Nitrophenyl-α-D-Maltoheptaoside [Biochemistry]

PNPGB P-Nitrophenyl-p-Guanidinobenzoate [Biochemistry]

PNPN P-Doped, N-Doped, P-Doped, N-Doped (Transistor) [also P-N-P-N, p-n-p-n or pnpn]

PNPN Positive-Negative-Positive-Negative (Transistor) [also P-N-P-N, p-n-p-n or pnpn] [Biochemistry]

pNPP p-Nitrophenyl Phosphate

PNPS Plant Nitrogen Purge System

PNPS Plant Nuclear Protection System

PNR Packaged Nuclear Reactor

PNR Photonuclear Reaction

PNR Pittsburgh Naval Reactors Office [of US Department of Energy]

PNR Point of No Return

PNR Polarized Neutron Reflectometry

PNR Prior Notice Required

PNR Programme National de Recherche [National Research Program; of Ecole Polytechnique Fédérale de Lausanne, Switzerland]

P(n,r) Number of r-tuples , without duplication, that can be formed from a set of n elements [Symbol]

PNRI Philippine Nuclear Research Institute [Quezon City, The Philippines]

PNRP Pollution and Natural Resources Program [of International Union for Conservation of Nature and Natural Resources]

PNS Peripheral Nervous System

PNS Pulsed Neutron Source

PNSC Packet Network Service Center

PNSC Philippines National Science Society

PNSRC Plant Nuclear Safety Review Committee

PNT Lead-Neodymium Titanate (or Neodymium-Doped Lead Titanate)

PNT Paint [also Pnt, or pnt]

PNT Plasma Nitriding

PNT Project Network Techniques

p-NT p-Nitrotoluene

PNTA P-Nitrophenylthioacetate

PNTD Personnel Neutron Threshold Detector

PNU Pusan National University [South Korea]

PNUMA Chilean National Union for the Environment

PNVS Pilot Night Vision System

PNW Pacific Northwest

PNWD Pacific Northwest Division [of Battelle Memorial Institute, Columbus, Ohio, US]

PNWL Pacific Northwest Laboratory [now Pacific Northwest National Laboratory]

PNX Private Network Exchange

PNZ Lead-Neodymium Zirconate (or Neodymium-Doped Lead Zirconate)

PNZST Lead Niobium Zirconium Tin Titanate (Ceramics)

PNZT Lead-Neodymium Zirconate Titanate (or Neodymium-Doped Lead Zirconate Titanate)

PO Palm Oil

PO Palomar Observatory [of California Institute of Technology located on Mount Palomar near San Diego, California, US]

PO Paraffin Oil

PO Parallel Output

PO Paramagnetic Oscillator

PO Parity Odd (Flag) [Computers]

PO Parity Operator

PO Parking Orbit

PO Particle Orbit

PO Part Of

PO Patent Office

PO Patents Office [UK]

PO Periodic Orbit

PO Personnel Office

PO Personnel Office [NASA Kennedy Space Center Directorate, Florida, US]

PO Petty Officer

PO Phase Oscillation

PO Phosphorus Oxide(s) [e.g., Phosphorus Monoxide, Phosphorus Trioxide, Phosphorus Pentoxide, Phosphorus Suboxide, etc.] [also POx, or Po$_x$]

PO Phosphoryl (Radical) [Symbol]

PO Physical Oceanographer; Physical Oceanography

PO Physical Optics

PO Pick-Off

PO Pierce Oscillator

PO Pilot Officer

PO Planetary Orbit

PO Plasma Oscillation

PO Polarity [also Po, or po]

PO Polar Orbit(er)

PO Polymerizable Oligomer

PO Polyolefin

PO Portugal [Keyboard Code] [also po]

PO Positional Order(ing) [Physics]

PO Positive Output

PO Postal Order [also po]

PO Post-Orbit

PO Post Office

PO Pour-On

PO Power Oscillator

PO Power Output

PO Preferred Orientation [Crystallography]

PO Pressure Oxidation

PO Primary Office [Data Communications]

PO Printout

PO Production Order

PO Process Optimization

PO Program Objective

PO Project Office(r)

PO Propylene Oxide (Homopolymer)

PO Publicly Owned; Public Ownership

PO Pulse Output

PO Punch On

PO Purchase Order

PO Pure Oxide

PO Push-Out

PO$_x$ Phoshorus Oxides [e.g., Phosphorus Monoxide, Phosphorus Trioxide, Phosphorus Pentoxide, Phosphorus Suboxide, etc.] [also POx]

PO$^-$ Hypophosphite (Ion) [Symbol]

PO$^-$ Phosphite (Ion) [Symbol]

PO$_4$ Phosphorus Tetrahedron [Symbol]

PO$^-$ Orthophosphate (Ion) [Symbol]

P(O$_2$) Oxygen Pressure [Symbol]

P(ω) Frequency-Dependent Polarization [Symbol]

P(ω) Power Spectrum [Symbol]

P&O Paints and Oils

P&O Pickled and Oiled

P&O Plans and Operations [of US Department of the Army]

P/O Phosphorus-to-Oxygen (Ratio)

Po Peso [Currency of Cuba]

Po Polonium [Symbol]

Po Potash

Po-200 Polonium-200 [also ^{200}Po, or Po^{200}]

Po-207 Polonium-207 [also ^{207}Po, or Po^{207}]

Po-208 Polonium-208 [also ^{208}Po, or Po^{208}]

Po-209 Polonium-209 [also ^{209}Po, or Po^{209}]

Po-210 Polonium-210 [also ^{210}Po, Po^{210}, RaF or Po]

Po-211 Polonium-211 [also ^{211}Po, Po^{211}, AcC′ or AcC$_1$]

Po-212 Polonium-212 [also ^{212}Po, Po^{212}, ThC′ or ThC$_1$]

Po-213 Polonium-213 [also ^{213}Po, or Po^{213}]

Po-214 Polonium-214 [also ^{214}Po, Po^{214}, RaC′ or RaC$_1$]

Po-215 Polonium-215 [also ^{215}Po, Po^{215}, or AcA]

Po-216 Polonium-216 [also ^{216}Po, Po^{216}, or ThA]

Po-217 Polonium-217 [also ^{217}Po, or Po^{217}]

Po-218 Polonium-218 [also ^{218}Po, Po^{218}, or RaA]

pO$_2$ partial pressure of oxygen [also p(O$_2$)]

po *(per os)* – by way of the mouth [Medical Prescriptions]

pΩ picoohm [Unit]

φ(ω) phase angle [Symbol]

POA Pay on Arrival

POA Pilot-Operated Absolute System

POA Plan of Action

POA Point of Action

POA Post-Oxidation Annealing [Metallurgy]

POA Provisional Operating Authorization

POAV Pi (π) Orbital Axis Vector (Method) [Physical Chemistry]

POB Peripheral Order Buffer

POB Persons on Board

POB Point of Banking

POB Polyoxybutylene Glycol

POB Port of Benton [US]

POB Post Office Box [also PO Box]

POB Procurement Opportunities Board [of Supply and Services Canada] [Now Replaced by Open Bidding Service, Canada]

POB Push-Out Base (Transistor)

POBATO Propellant on Board at Takeoff

POBN Pyridyloxide Butylnitrone

POB-N Polyoxybutylene Glycol Nylon-12 (Copolymer)

4-POBN 4-Pyridyl 1-Oxide Butylnitrone

PO Box Post Office Box [also POB]

POBT Polybutylene Terephthalate

POC Metal Powder Cutting

POC 4-Methyl-2,6,7-Trioxa-1-Phosphabicyclo[2.2.2]octane

POC Payload Operations Center

POC Payload Operations Control

POC Physical Organic Chemistry

POC Point Of Contact

POC Port of Call

POC Process Operator Console

POC Procurement Outlook Conference

POC Product(s) of Combustion

POC Proof-Of-Concept

POC Purchase Order Closeout

POC Purgeable Organic Carbon

Poc Cyclopentyloxycarbonyl

POCC Payload Operations Control Center [at NASA Marshall Space Flight Center, Greenbelt, Maryland, US]

POCH Post Office Central Headquarters [London, UK]

POCL Photooxygenation Chemiluminescence

POCN Purchase Order Change Notice

POCP Program Objectives Change Proposal

POCS Patent Office Classification System [US]

POCS Photo-Optical Control System

POCSAG Post Office Code Standardization Advisory Group [UK]

POCW Park Operations and Capital Works [of Australian Nature Conservancy Agency]

POD Payload Operations Director

POD Payload Operations Division [of NASA Johnson Space Center, Houston, Texas, US]

POD Pay on Delivery [also pod]

POD Philadelphia Ordnance District [Pennsylvania, US]

POD Pin-On-Disk (Abrasion Test Apparatus)

POD Pittsburgh Ordnance District [Pennsylvania, US]

POD Plain Old DOS (Disk-Operating System)

POD Podbielniak Analysis [Chemistry]

POD Point of Origin Device

POD Poly(p-Phenylene-1,3,4-Oxadiazole)

POD Post Office Department [US]

POD Pounds-out-the-Door

POD Power On Diagnostics (Project)

POD Preflight Operations Division

POD Principle of Determinism

POD Probability of Detection

POD Proximity Optical Device

POD Pueblo Ordnance Depot [Colorado, US]

PoD Parsimony-of-Dichotomy

PODA Priority-Oriented Demand Assignment

PODAF Power Density, Assigned Frequency Band

POD/CL Probability of Detection/Confidence Level

PODEM Path-Oriented Decision Making

PODS Post-Operative Destruct System

Podst Sterow Podstawy Sterowania [Polish Journal on Automation and Control]

POE Petroleum Operations Engineer

POE Piezooptical Effect

POE Poly(octadecylethylene)

POE Poly(oxyethylene)

POE Port of Embarkation

POE Port of Entry

POE Power Open Environment

POE Property Owning Expenses

POED Post Office Engineering Department [UK]

POEM Advanced Technology Center for Photonic and Optoelectronic Materials [Princeton University, New Jersey, US]

POEM Polar Orbit Earth-Observation Mission [of European Space Agency]

POEMS Plasma Optical Emission Mass Spectrometer

POEMS Polyoxyethylene Monostearate

POEMS Positron-Electron Magnetic Spectrometer

P(OEt)$_3$ Triethyl Phosphite

POEOP Polyoxyethyleneoxypropylene

POES Polar-Orbiting Environmental Satellite

POETRI Program on Exchange and Transfer of Information [of United Nations]

POEU Post Office Engineering Union [UK]

POF DL-α-Liproic Acid [Biochemistry]

POF Planned Outage Factor

POF Plastic Optic Fiber

POF Point of Failure

POF Polymer Optical Fiber

POF Product Operator Formalism

POF Pulsed Optical Feedback

POF Pyruvate Oxidation Factor [Biochemistry]

POFA Probability of False Alarms

POGG Petroleum Organic Geochemical Group [of Woods Hole Oceanographic Institute, Massachusetts, US]

POGO Pilot Operational Equipment

POGO Polar Orbit(ing) Geophysical Observatory [US]

POGO Programmer-Oriented Graphic Operation

POH Pilot's Operating Handbook

POH Planned Outage Hours

POH Power-On Hours

pOH negative logarithm of hydroxyl group [OH$^-$]

pOHBA p-Hydroxybenzoic Acid

POI Parking Orbit Injection

POI Plane of Incidence

POI Point of Interface

POI Principle of Indifference

POI Product of Inertia

POI Program of Instruction

POI Propellant Quantity Indicator

POIC Payload Operations Integration Center

POIL Power Density Imbalance Limit

POINTER Particle Orientation Interferometer

POIS Project-Oriented Information System

POIS Prototype On-Line Instrument System

Pois Poison(ous) [also pois]

POISE Panel on In-Flight Scientific Experiments

POISE Photosynthetic Oxygenation Illuminated by Solar Energy

POISE Pre-Operational Inspection Services Engineering

Pokr Prásk Metal Pokrody Práškové Metalurgie [Czech Publication on Metallurgy]

POL Pacific Oceanographic Laboratory

POL Pair Orthogonalized Lowdin [Physics]

POL Petrol, Oil and Lubricants [UK]

POL Petroleum, Oil and Lubricants

POL Problem-Oriented Language

POL Procedure-Oriented Language

POL Process-Oriented Language

POL Program-Oriented Language

POL Provisional Operating License

Pol Poland; Pole; Polish

Pol Polarization [also pol]

Pol Pole [also pol]

Pol Polish(ing) [also poll]

Pol Akad Nauk Pr Kom Nauk Tech Metal Fiz Metal Stopow Polska Akademia Nauk Prace Komisji Nauk Technicznych Metalurgia Fizyka Metali i Stopow [Publication of the Polish Academy of Science on Metallurgical Engineering, Physics and Materials Related Subjects]

POLANG Polarization Angle

POLAR Production Order Location and Reporting

Pol Arch Weterynar Polskie Archiwum Weterynaryine [Polish Veterinary Archives]

Polar Global Geospace Science Polar Spacecraft [of NASA]

Polar Res Polar Research [Journal published in Norway]

POL-CI Polarization Configuration Interaction

POL-CM Polarization Configuration Mixing

POLDER POLarimètre à DEuton de Recul

POLDER Polarization and Directionality of Earth Reflectances (Experiment)

Pol Eng Polish Engineering (Journal)

POLES Polar Exchange at the Sea Surface

POLEX Polar Experiment

POLGEN Problem-Oriented Language Generator

Pol Geogr Political Geographer; Political Geography

Polim Tworz Wielk Polimery Tworzywa Wielkoczasteczkowe [Polish Journal on Polymer Science and Engineering]

POLINEAR Polar and Linear (Recorder) [also polinear]

Polio Poliomyelitis (or Infantile Paralysis) [also polio] [Medicine]

Poliplasti Plast Rinf Poliplasti e Plastici Rinforzati [Italian Journal on Synthetic Resins and Reinforced Plastics]

POLLS Parliamentary On-Line Library Study [of United Kingdom Atomic Energy Authority]

Pollut Pollution [also pollut]

Pollut Atmos Pollution Atmospherique [French Journal on Air Pollution]

Pollut Eng Pollution Engineering [Journal published in the US]

Pol Sci Tech Data Polish Scientific and Technical Database

Pol Tech Rev Polish Technical Review [Published in Poland]

Pol Technol News Polish Technological News [Published in Poland]

Poly Polyethylene [also poly]

Poly Polymer [also poly]

poly polycrystalline

Poly(A) Polyadenylic Acid; Polyadenylate [also poly-A] [Biochemistry]

Poly(A,C) Polyadenylic-Cytidylic Acid [Biochemistry]

Poly(A,C,G) Polyadenylic-Cytidylic-Guanylic Acid [Biochemistry]

Poly(A,C,U) Polyadenylic-Cytidylic-Uridylic Acid [Biochemistry]

Poly(A)–p(dT)$_{10}$ Polyadenylic Acid–Decathymidylic Acid [Biochemistry]

Poly(A)–p(dT)$_{12}$ Polyadenylic Acid–Dodecathymidylic Acid [Biochemistry]

Poly(A,G) Polyadenylic-Guanylic Acid [Biochemistry]

Poly(A,G,U) Polyadenylic-Guanylic-Uridylic Acid [Biochemistry]

Poly(A)–Poly(U) Polyadenylic–Polyuridylic Acid [Biochemistry]

Poly(A)–Poly(U)–Poly(U) Polyadenylic–Polyuridylic–Polyuridylic Acid [Biochemistry]

Poly(A,U) Polyadenylic-Uridylic Acid [Biochemistry]

Poly-BTO Polycrystalline Barium Titanate [also poly-BTO]

Poly(C) Polycytidylic Acid [also poly-C] [Biochemistry]

Poly(C)–(dG)$_{12-18}$ Polycytidylic Acid–Oligodeoxyguanylic Acid [Biochemistry]

POLYCHAR (International Conference on) Polymer Characterization

Poly(C,I) Polycytidylic-Inosinic Acid [Biochemistry]

Polycide Poly-Silicon/Silicide [also polycide]

Poly(C)–Poly(G) Polycytidylic–Polyguanylic Acid [Biochemistry]

Poly(C,U) Polycytidylic-Uridylic Acid [Biochemistry]

Poly(dA) Polydeoxyadenylic Acid [Biochemistry]

Poly(dA-dC)–Poly(dG-dT) Poly(deoxyadenylic-Deoxycytidylic)–Poly(deoxyguanylic-Thymidylic Acid) [Biochemistry]

Poly(dA)–(dT)$_{10}$ Polydeoxyadenylic Acid-Decathymidylic Acid [Biochemistry]

Poly(dA-dT)–Poly(dA-dT) Polydeoxyadenylic–Thymidylic Acid [Biochemistry]

Poly(dA)–Poly(dT) Polydeoxyadenylic Acid–Polythymidylic Acid [Biochemistry]

Poly(dC) Polydeoxycytidylic Acid [Biochemistry]

Poly(dC,dT) Polydeoxycytidylic-Thymidylic Acid [Biochemistry]

Poly(dI-dC)–Poly(dI-dC) Polydeoxyinosinic–Deoxycytidylic Acid [Biochemistry]

Poly(dG) Polydeoxyguanylic Acid [Biochemistry]

Poly(dG,dC)–Poly(dG-dC) Polydeoxyguanylic–Deoxycytidylic Acid [Biochemistry]

Poly(dG)–Poly(dC) Polydeoxyguanylic–Polydeoxycytidylic Acid [Biochemistry]

Poly(dI) Polydeoxyinosinic Acid [Biochemistry]

Poly(dT) Polydeoxythymidylic Acid [Biochemistry]

POLYDOC Polytechnical Documentation

Poly(G) Polyguanylic Acid [also poly-G] [Biochemistry]

Poly(G,U) Polyguanylic-Uridylic Acid [Biochemistry]

Poly-HEME Poly(2-Hydroxyethyl Methacrylate)

Poly(I) Polyinosinic Acid [also poly-I] [Biochemistry]

Poly I:C Polyinosinic Acid–Polycytidylic Acid [Biochemistry]

Poly(I)–Poly(C) Polyinosinic–Polycytidylic Acid [Biochemistry]

Poly(I,U) Polyinosinic-Uridylic Acid [Biochemistry]

Polym Polymer [also polym]

Polym Adv Technol Polymers for Advanced Technologies [International Journal]

POLYMAT Polymeric Materials Databank [of Deutsches Kunststoffinstitut, Darmstadt, Germany]

Polym Bull Polymer Bulletin [Published in Germany and the US]

Polym Chem Polymer Chemist(ry)

Polym Commun Polymer Communications [UK]

Polym Commun (China) Polymer Communications (China) [Published by the Chinese Chemical Society, PR China]

Polym Commun (Engl Ed) Polymer Communications (English Edition) [Published by the Chinese Chemical Society]

Polym Compos Polymer Composites [Journal of the Society of Plastics Engineers, US]

Polym Cont Polymer Contents [Published in the US]

Polym Degrad Stabil Polymer Degradation and Stability [Journal published in the UK] [also Polym Deg Stab]

Polym Dig Polymer Digest [Japan]

POLYMEG Polytetramethylene Ether Glycol

PolymE Polymer Engineer [also Polym Eng, or PolymEng]

Polym Eng Polymer Engineer(ing)

Polym Eng Sci Polymer Engineering and Science [Journal of the Society of Plastics Engineers, US]

Polym Gels Netw Polymer Gels and Networks [Journal pblished in UK]

Polym J Polymer Journal [Published by the Society of Polymer Science, Japan]

Polym J (Tokyo) Polymer Journal (Tokyo) [Published by the Society of Polymer Science, Japan]

Polym News Polymer News [UK]

polymorph polymorphous

Polym Paint Col J Polymers Paint Colour Journal [UK]

Polym Phys Polymer Physicist; Polymer Physics

Polym Plast Technol Eng Polymer-Plastics Technology and Engineering [Journal published in the US]

Polym Prepr Polymer Preprints [of American Chemical Society, Division of Polymer Chemistry, US] [also Polym Prpts]

Polym Prepr Am Chem Soc Polymer Preprints of the American Chemical Society [US]

Polym Proc Eng Polymer Process Engineering [Journal published in the US]

Polym Prpts Polymer Preprints [of the American Chemical Society, US] [also Polym Prepr]

Polym Rubb Asia Polymers and Rubber Asia [Journal published in the UK]

Polym Sci Polymer Science; Polymer Scientist

Polym Sci USSR Polymer Science USSR [Translation of: *Vysokomolekulyarnye Soedineniya, Seriya A*; published in the UK]

Polym Test Polymer Testing [Journal published in the UK]

Polyol Polyhydric Alcohol [also polyol]

Polysar Prog Polysar Progress [Publication of Polysar Ltd., Canada]

POLYSE Polycrystalline Semiconductor Conference

poly-Si Polycrystalline Silicon

poly-SiC Polycrystalline Silicon Carbide

poly-Si/Si Polycrystalline Silicon on Silicon

poly-Si TFT Polycrystalline Silicon Thin-Film Transistor

Poly(T) Polythymidylic Acid [also poly-T] [Biochemistry]

Polytech Polytechnic(al); Polytechnics [also polytech]

poly-TiN Polycrystalline Titanium Nitride
POLYTRAN Polytranslation Analysis and Programming
Poly(U) Polyuridylic Acid [also poly-U] [Biochemistry]
POM Pallet-Only Mode
POM Pivaloyloxymethyl Chloride
POM Plate Orifice Meter
POM Polarized-Light Optical Microscopy
POM Polycyclic Organic Matter
POM Polymethylene Oxide
POM Polyoxymethylene
POM Pool Operational Module
POM Printer Output Microfilm
POM Printout Microfilm
PoM Plan of Management
POM₃ Poly(oxytrimethylene)
POM₄ Poly(oxytetramethylene)
P/OM Productions and Operations Management
POMBE Pulsed Organometallic Beam Epitaxy
POM-Cl Pivaloyloxymethylchloride
Pomiary Autom Kontrola Pomiary Automatyka Kontrola [Polish Journal on Automation and Control]
POMM Preliminary Operating and Maintenance Manual
POMMIE Phase Oscillations to Maximize Editing
POMS Panel on Operational Meteorological Satellites
POMSA Post Office Management Staff Association [UK]
POMSEE Performance, Operational and Maintenance Standards for Electronic Equipment
POMT Planning Operations Management Team
PON Phosphorus Oxynitride
PONA Paraffins, Olefins, Naphthenes and Aromatics
pond *(ponderosus)* – heavy [Medical Prescriptions]
PONI Positive Output, Negative Input
POO Post Office Order
POP Pantex Ordnance Plant [Amarillo, Texas, US]
POP Particle-Oriented Paper
POP Percentage of Precipitation
POP Permanent Operating Permit
POP Perpendicular Ocean Platform
POP Perpendicular-to-Orbit Plane
POP Plain Old Paper (Publication)
POP PMDA–ODA–PMDA (Pyromellitic Dianhydride–4,4'-Oxydianiline– Pyromellitic Dianhydride)
POP Point of Presence [Internet]
POP Point-of-Problem
POP Point of Purchase (Advertising)
POP Polar Orbiting Platform
POP Polymer Outreach Program [US]
POP Pop from Stack [Computers]
POP Post Office Protocol [Internet]
POP Power On/Off Protection
POP Preburner Oxidizer Pump
POP Preflight Operations Procedures
POP Prelaunch Operations Plan
POP Pressurizer Overpressure Protection System

POP Printing-Out Paper
POP Progesterone Only Pill [Pharmacology]
POP Program Operating Plan
POP Proof of Principle
Pop Popular(ity) [also pop]
Pop Populate(d); Population [also pop]
POPA Panel on Public Affairs [of American Physical Society, US]
POPA Payload Ordnance Processing Area
POPA Patent Office Professional Association [US]
POPA Pop All Registers [Computers]
POPAE Protons on Protons and Electrons
POPAI Point-of-Purchase Advertising Institute [US]
POPC Petroleum Open Projects Consortium [US]
POPDA Polyoxypropylene Diamine
POPF Pop Flags [Computers]
POPI Post Office Position Indicator
popLA Los Alamos Software for Pole Figure/Inverse Pole Figure Generation [of Los Alamos National Laboratory, New Mexico, US]
POPLINE Population On-Line [Joint US Databank of Johns Hopkins University, Columbia University and Princeton University]
POPO Push-On, Pull-Off
POPOP 1,4-bis-(5-Phenyloxazol-2-yl)benzene
POPOP 2,2'-p-Phenylene-Bis(5-phenyloxazole)
POPOP P-bis[2-(5-Phenyloxazolyl)]benzene
POPOP Phenyl-Oxazolylphenyloxazolylphenyl
POPOP Phenyloxazolyl Phenyl Oxazolylphenyl
Pop Plast Popular Plastics [Journal published in India]
Pop Plast Packag Popular Plastics and Packaging [Journal published in India]
POPR Prototype Organic Power Reactor
POPS Pantograph Optical Projection System
POPS Performance-Oriented Packing Standards
POPS Process Operating System
POPS Program for Operator Scheduling
POPs Points of Presence [Internet]
POPSI Precipitation and Off-Path Scattered Interference
POPSI Postulate-Based Permuted Subject Index
POPSO Piperazine-N,N'-bis(2-Hydroxypropanesulfonic Acid); β,β'-Dihydroxy 1,4-Piperazinedipropanesulfonic Acid
POPU Push Over Pull Up
POPUS Post Office Processing Utility Subsystem
POR Pacific Ocean Region
POR Pathology Oncology Research [Journal published in Hungary]
POR Payable on Receipt
POR Pay on Return [also por]
POR Point of Resolution
POR Portrait (Font Orientation)
POR Power-On-Reset
POR Problem-Oriented Routine
POR Project Office Room

POR Purchase Order Request

Por Porosity [also por]

PORACC Principles of Radiation and Contamination Control

PORB Production Operations Review Board

PORC Petroleum Open Projects Consortium

Porc Porcelain [also porc]

PORCN Production Order Records Change Notice

PORD Performance and Operations Requirements Document

Porist Met Mashin Poristye Metally v Mashinostroenii [Russian Journal on Porous Metals in Mechanical Engineering]

PORM Plus, or Minus [also PorM]

Poroshk Metall Poroshkovaya Metallurgiya [Journal on Powder Metallurgy and Metal Ceramics; published in the Ukraine]

PORP Porpoise (Motion)

Porp Porpoise

PORR Preliminary Operations Requirements Review

PORS Portable Oil Reclamation System

por-Si Porous Silicon

PORT Photo-Optical Recorder Tracker

Port Portability; Portable [also port]

Port Portugal; Portuguese

PORTEX International Port Exhibition featuring Harbour Construction, Harbour Technology and Harbour Organization

Port Guin Portuguese Guinea

Port Math Portugaliae Mathematica [Mathematical Journal; published by Sociedade Portuguesa de Matemática, Portugal]

Port Phys Portugaliae Physica [Physics Journal; published by Sociedade Portuguesa de Fisica, Portugal]

PORV Pilot-Operated Relief Valve

PORV Power-Operated Relief Valve

POS (Current) Cursor Position (Function) [Programming]

POS Pacific Ocean Ship

POS Patent Office Society [now Patent and Trademark Office Society, US]

POS Plant Operating System

POS Point of Sale (System) [also pos]

POS Porcelain-on-Steel (Substrate)

POS Portable Operating System

POS Portable Oxygen System

POS Primary Operating System

POS Product of Sums (Form)

POS Professional Operating System

POS Programmable Object Select

Pos Position [also pos]

Pos Search for Substring in String [Pascal Function]

pos positive

POSA Preliminary Operating Safety Analysis

POSAP Position Sensitive Atom Probe

POSB Post Office Savings Bank

POSC Petroleum Open Software Corporation [US]

POSCO Pohang Iron and Steel Company Limited [Pohang, South Korea] [also Posco]

POSCO Tech Res Lab Pohang Iron and Steel Company Limited Technical Research Laboratory [Pohang, South Korea]

POSH Permuted on Subject Headings

POSH Principles of Occupational Safety and Health [of National Safety Council, US]

POSI Parallel-Out, Serial-In

positron positive electron

POSIX Portable Operating Systems Interface for Computer Environments; Portable Operating System Interface for Unix [also Posix]

POS/MV Marine Position and Orientation Vessel

Posn Position [also posn]

POSS Passive Optical Satellite Surveillance

POSS Photo-Optical Surveillance Subsystem

POSS Plant Occurrence and Status System

POSS Polar Orbiting Satellite System [of University of Michigan, US]

POSS Polyhedral Oligosilsequioxane

Poss Possession [also poss]

Poss Possibility [also poss]

poss possibe; possibly

POSSUM Polar Orbiting Satellite System–University of Michigan [US]

POST Payload Operations Support Team

POST Polymer Science and Technology (Database) [of Chemical Abstracts Service of the American Chemical Society, US]

POST Power-On Self Test

POST Program to Optimize Simulated Trajectories

Post Posting [also post]

post positive

POSTEC Program of Research in Powder Science and Technology [Norway]

POSTECH Pohang Institute of Science and Technology [South Korea]

POST-EM Post-Emergence [also Post-Em, or postem]

Postepy Astron Postepy Astronomii [Polish Journal on Astronomy]

Postepy Fiz Postepy Fizyki [Polish Journal on Physics]

POSTER Post-Strike Emergency Reporting

PostGrad Post-Graduate

PostGradDip Post-Graduate Diploma

Postgrad Med Postgraduate Medicine [Journal]

POSTNET Postal Numeric Encoding Technique [of US Postal Service]

Post-op Post-Operative [also post-op] [Medicine]

Postp Postprocessor

Post-RTA Post-Rapid Thermal Anneal(ing) [Metallurgy]

POT Paint on Targent

POT Plain Old Telephone

POT Poly(3-Octylthiophene)

POT Poly(o-Toluidine)

p(O₂)-T oxygen partial pressure–temperature (diagram)

Pot Potential [also pot]

Pot Potentiometer [also pot]

Pot Pottery [also pot]

pot potable

PO/TBA Propylene Oxide/Tertiary Butyl Alcohol

POTC PERT (Program Evaluation and Review Technique) Orientation and Training Center [of US Department of Defense]

POT-CSA Poly(O-Toluidine) doped with Camphor Sulfonic Acid

POTF Polychromatic Optical Thickness Fringes

Potier EMF Potier Electromotive Force

POTS Plain Old Telephone Service; Plain Old Telephone System

POTW Publicly-Owned Treatment Works

Pot W Potable Water

POU Point of Use

Poultry Sci Poultry Science (Journal) [US]

POUNC Post Office Users National Council [UK]

POUT Power Output

POV Peak Operating Voltage

POV Pneumatic(ally) Operated Valve

Poverkhnost' Fiz, Khim, Mekh Poverkhnost' Fizika, Khimiya, Mekhanika [Russian Journal on Surface Physics, Chemistry and Mechanics]

Poverkhnost Ionizat Poverkhnostnaya Ionizatsiya [Russian Journal on Surface Ionization]

Powd Powder(ed) [also powd]

Powder Bulk Eng Powder and Bulk Engineering [Journal published in the US]

Powder Diffr Powder Diffraction [Journal published in the US]

Powder Handl Process Powder Handling and Processing [Journal published in Germany]

Powder Met Powder Metallurgy [Journal of the The Materials Society, UK]

Powder Met Def Technol Powder Metallurgy in Defense Technology [Journal of the Metal Powder Industries Federation, US]

Powder Met Int Powder Metallurgy International [Journal published in Germany]

Powder Met Sci Technol Powder Metallurgy Science and Technology [Journal published in India]

Powder Met Technol Powder Metallurgy Technology [Journal published in the PR China]

Powder Sci Eng Powder Science and Engineering [Journal published in the US]

Powder Technol Powder Technology [Journal published in Switzerland]

POWER Performance Optimization with Enhanced RISC (Reduced Instruction Set Computer) [also Power]

Power Convers Intell Motion Power Conversion and Intelligent Motion [Journal published in the US]

Power Eng Power Engineering [Journal published in the US]

Power Eng (USSR Acad Sci) Power Engineering (USSR Academy of Sciences) [Translation of: *Izvestiya Akademii Nauk SSSR, Energetika i Transport*; published in the US]

Power Eng J Power Engineering Journal [Published by the Institution of Electrical Engineers, UK]

Power Int Power International [Journal published in the UK]

Power Int Ed Power International Edition [Journal published in the US]

PowerPC Performance Optimization with Enhanced RISC (Reduced Instruction Set Computer)– Performance Computing

Power Transm Des Power Transmission Design [Journal published in the US]

Power Works Eng Power and Works Engineering [Journal published in the UK]

Powloki Ochr Powloki Ochronne [Polish Materials Journal]

POWP Page-Oriented Word Processor

POWS Project Operating Work Statement

POWS Pyrotechnic Outside Warning System

POWU Post Office Workers Union [UK]

POWU Post Office Work Unit

POV Peak Operating Voltage

POWTECH International Power and Bulk Solids Technology Exhibition and Conference

POWWER Power of World Wide Energy Resources

POX Partial Oxidation

POX Purgeable Organic Halogen

POx Phosphorus Oxide(s) [e.g., Phosphorus Monoxide, Phosphorus Trioxide, Phosphorus Pentoxide, Phosphorus Suboxide, etc.] [also Po_x]

POXRD Preferred-Orientation X-Ray Diffraction

PP Pacific Press [US]

PP Page Printer

PP Pair Production [Physics]

PP Panel Point [Civil Engineering]

PP Parallel Port

PP Parallel Processing; Parallel Processor

PP Para-Positronium [Particle Physics]

PP Parcel Post

PP Partial Pressure

PP Participatory Program [Part of the Australian Biological Resources Study of the Australian Nature Conservation Agency]

PP Particle Physicist; Particle Physics

PP Particle Production

PP Part Program(mer)

PP Parts Per [also pp]

PP Passivation Potential [Physical Chemistry]

PP Pauli Principle [Quantum Mechanics]

PP Peak-to-Peak [also P/P, P-P, pp, p/p or p-p]

PP Peak Point

PP Percussion Primer

PP Perfect Pairing

PP Peripheral Processor

PP Peristaltic Pump

PP Permanent Press (Fabrics)

PP Personal Protection

PP Person-to-Person

PP Perspective Projection
PP Phase Plate [Optics]
PP Philippine Peso [Currency]
PP Phosphoprotein [Biochemistry]
PP Phosphoric Acid Bis(2-Ethylhexyl)ester
PP Photographic Plate
PP Photographic Printer; Photographic Printing
PP Photographic Projection
PP Photophysical; Photophysics
PP Photopolarimeter; Photopolarimetry
PP Photopositive; Photopositivity
PP Photoplotter
PP Photoproton
PP Physical Process(ing)
PP Physical Properties
PP Picric Powder
PP Picture Plane
PP Piezopolymer
PP Pilotless Plane
PP Pilot Plant
PP Pilot Punch
PP Piston Phone
PP Plane Parallel
PP Plane Polarization; Plane-Polarized
PP Planning Package
PP Plasma Physicist; Plasma Physics
PP Plasma Polymerization; Plasma-Polymerized
PP Plasma Processing
PP Plastics Processing
PP Pleuropneumonia [Medicine]
PP Plotter Pen
PP Polarity Practitioner
PP Polarization Pyrometer
PP Polar Projection
PP Pole Piece
PP Pollution Prevention [also P2, or P^2]
PP Polymer Physicist; Polymer Physics
PP Polymer Processing
PP Polyphenylene
PP Polyphosphazene
PP Polyplophrein [Biochemistry]
PP Polypropylene
PP Polypyrrole
PP Ponderosa Pine
PP Population Parameter
PP Porous Plug
PP Port-au-Prince [Haiti]
PP Positive Pole
PP Positive-Positive (Semiconductor) [also P-P, pp, or p-p]
PP Post(age) Paid [also pp]
PP Postpartum [Medicine]
PP Postprocessor
PP Pounds of Pressure

PP Pour Plate (Culture) [Microbiology]
PP Pour Point
PP Powder Processing
PP Powder Product
PP Power Pack
PP Power Plant
PP Power Pole
PP Prepaid
PP Preprocessor
PP Present Position
PP Pressure Permeation
PP Pressure Pin
PP Pressureproof
PP Principal Plane [Optics]
PP Principal Point [Optics]
PP Printer's Points
PP Printing Paper
PP Print Position
PP Print/Punch
PP Proceedings Planned
PP Process Parallelism
PP Process Physicist; Process Physics
PP Production Plan(ning)
PP Professional Pedicurist
PP Professional Photographer
PP Professional Programmer
PP Professional Purchaser
PP Program Performance
PP Program Preparation
PP Program Product
PP Proof Paper (Firing Report)
PP Propeller Pitch
PP Propylene Plastic
PP Proton-Proton (Reaction)
PP Provincial Park [Canada]
PP Pseudoplastic(ity)
PP Pseudopotential [Physics]
PP Pseudo Program
PP Pulp and Paper [also P&P]
PP Pulse Plated; Pulse Plating
PP Pump Power
PP *(Punctum Proximum)* – Near Point [also pp]
PP Pusher Propeller
PP Push-Pull [also P-P, pp, or p-p]
PP (Earthquake) Push Wave Reflected once from the Earth's Surface [also PR$_1$] [Symbol]
PP Pyropheophorbide
PP Pyrophosphate
PP Pyroplastic(ity)
PP- Brazil [Civil Aircraft Marking]
P$_c$P (Earthquake) Push Wave Partially Reflected from Core [Symbol]
P3P Platform for Privacy Preferences Project [also P^{3}P]
P&P Pick and Place (Robot)

P&P Policies and Procedures Manual [of Institute of Electrical and Electronics Engineers, US]

P&P Post(age) and Packing [also p&p]

P&P Pulp and Paper

P/P Patch Panel

P/P Point-to-Point

P/P Printer/Plotter

P-P Peak-to-Peak [also P/P, PP, p/p, p-p or pp]

P-P Push-Pull [also PP, pp, or p-p]

P-5-P Pyridoxal-5-Phosphate

pP partially coherent perfect (crystal) lattice [Symbol]

pp pages

pp parts per [also PP]

pp *(per procurationem)* – on behalf of; per procura

pp peak to peak [also PP, P/P, P-P, p/p, or p-p]

pp plastic strain in tension and compression [Mechanical Testing]

pp post(age) paid [also PP]

pp powder supply [Military]

pp prepaid

p-p plate-to-plate

p_i/p ratio of partial pressure of component i to the total pressure [Symbol]

PPA Paired Phonon Analysis [Solid-State Physics]

PPA Pakistan Press Association

PPA Partial Pressure Analyzer

PPA Peierls Pinning Approximation [Solid-State Physics]

PPA Periodical Publishers Association [UK]

PPA Pesticide Producers Association [US]

PPA Phenylpropanolamine

PPA (+)-α-Phenylpropionamide

PPA Phenylpyruvic Acid

PPA Photo-Peak Analysis

PPA Piezoelectric Photoacoustic (Measurement)

PPA Pipeline Protection Association [UK]

PPA Pixel Processing Accelerator

PPA Pollution Prevention Act [US]

PPA Polymer Processors Association

PPA Poly-(P-Phenylacetylene)

PPA Polyphosphoric Acid

PPA Polyphthalamide

PPA P-Phthalic Acid

PPA Preliminary Pile (Reactor) Assembly [US]

PPA Preliminary Project Approval

PPA Pressure-Plate Anemometer

PPA Princeton Particle Accelerator [of Princeton University, New Jersey, US]

PPA Princeton–Pennsylvania (Proton) Accelerator [US] [Shutdown]

PPA Process Plant Association [UK]

PPA Professional Photographers of America

PPA Professional Programmers Association

PPA Propionic Acid

PPA Ptilota Plumosa Agglutinin [Immunology]

PPA Pulsed Plasma Accelerator

PPA N'-(2-Pyridylmethyl) 2-Pyridinecarboximidamide [also ppa]

PPA Shake the Bottle First [Medical Prescriptions]

pPa picopascal [Unit]

ppa *(per procura)* – by proxy

PPAA Personal Protective Armor Association [US]

PPAAR Princeton University/Pennsylvania University Army Avionics Research [US]

PPABC Professional Photographers Association of British Columbia [Canada]

PPABIA Poly-P-Amidobenzimidazole

P2AD Pollution Prevention Assistance Division [of Georgia Department of Natural Resources, US]

PP/Al Aluminum Particulate Dispersed in Polypropylene Matrix

PPARC Particle Physics and Astronomy Research Council [UK]

PPB 1,4-Diphenyl-1,3-Butadiene

PPB Part Period Balancing

PPB Parts per Billion [also ppb, or P/B]

PPB Pin-Point Bombardment [also PPB't]

PPB Plant Population Biology

PPB Powder Particle Boundary

PPB Pressure Positive Breathing

PPB Prior Particle Boundary [Materials Science]

PPB Prior Powder Boundary [Materials Science]

PPB Professional Public Buyer

PPB Program Performance Baseline

PPB PROM (Programmable Read-Only Memory) Programmer Board

PPB Pyrethrins Piperonyl Butoxide

ppb parts per billion [also PPB]

PPBA Parts per Billion Atomic [also ppba]

PPBC Plant Pathogenic Bacteria Committee [Hungary]

PPBE Passenger Protective Breathing Equipment

PPBF Pan-American Pharmaceutical and Biochemical Federation

PPBS Planning, Programming and Budgeting System

PPC Chlorinated Polypropylene

PPC Paperboard Packaging Council [US]

PPC Paper Partition Chromatography

PPC Parallel-Plate Capacitor

PPC Peak Power Control

PPC Penetrant Protective Carbon

PPC Persistent Photoconductivity

PPC Phased Provisioning Code

PPC Piperidine Pentamethylene Dithiocarbamate

PPC Plane Polar Coordinate(s)

PPC Plasma Physics Conference [of International Atomic Energy Agency]

PPC Pollable Protocol Converter

PPC Pollution Prevention Center [Sarnia, Ontario, Canada]

PPC Pour Plate Culture [Microbiology]

PPC Polyphthalate Carbonate

PPC Polypropylene Carbonate

PPC (International Workshop on) Positron and Positronium Chemistry

PPC Preprocessing Center

PPC Print Position Counter

PPC Production Planning and Control

PPC Project Planning Center

PPC Pro-Personal Computer

PPC Pulsed Power Circuit

PPCC Pittsburgh Poison Control Center [Pennsylvania, US]

PPCO Polypropylene Copolymer

PPCO$_2$ Partial Pressure CO$_2$ (Carbon Dioxide)

PPCP Pulse-Corona-Induced Plasma Chemical Process

p pct porosity in percent

PPCS Person-to-Person, Collect and Special Instruction

PPCVD Particle-Precipitation-Aided Chemical Vapor Deposition

PPD 3,6-Bis(1-Pyrazolyl)pyridazine

PPD 2,5-Diphenyl-1,3,4-Oxadiazole

PPD Particle Physics Division [of Fermilab, Batavia, Illinois, US]

PPD Past Due Date

PPD Phorbol-12,13-Didecanoate

PPD P-I-N (Positive-Intrinsic-Negative) Photodiode

PPD Plant Protection Division [Now Part of Plant Protection and Quarantine, Animal and Plant Health Inspection Service, US Department of Agriculture]

PPD Plastic Product Design

PPD Port Protection Device

PPD Pour-Point Depressant

PPD P-Phenylenediamine

PPD Project Planning Document

PPD Pulse-Type Phase Detector

PPD Pure Phase Detection

PPD Purified Protein Derivative [Biochemistry]

.PPD PostScript Printer Description [File Name Extension] [also .ppd]

ppd post(age) paid

ppd prepaid

PPDA P-Phenylene Diamine

PPDA Phenyl Phosphorodiamidate

PPDC Phenylphosphorodichloridate

PPDC Programming Panels and Decoding Circuits [also PP-DC]

PPDD Plan Position Data Display

PPDD Preliminary Project Design Description

p,p'-DDD p,p'-Dichlorodiphenyldichloroethane

p,p'-DDE p,p'-Dichlorodiphenyldichloroethene

p,p'-DDT p,p'-Dichlorodiphenyl-1,1,1-Trichloroethane

PPDF Partial Pair Distribution Function

PPDM Public Petroleum Data Model Association

PPDP/S Pariser-Parr-Del Bene-Pople/Segal (Calculations) [Physics]

PPDR Pilot Performance Description Record

PPDS Personal Printer Data Stream

PPDS Physical Property Data Service [of Institution of Chemical Engineers, UK]

PPD-T Poly-p-Phenyleneterephthalamide

PPD-T Poly p-Phenyleneterephthalate

PPDTS 2-(5,6-Bis(4-Sulfophenyl)-1,2,4-Triazin-3yl)-4-(4-Sulfophenyl)pyridine

PPE Personal Protective Equipment

PPE Phase Partitioning Experiment [of NASA]

PPE Photopyroelectric(ity)

PPE Poly-M-Phenoxylene

PPE Polyphenylene Ether

PPE Polyphenylether

PPE Polyphosphate Ester

PPE Power Plant Engineer

PPE Precise Parameter Estimation

PPE Premodulation Processing Equipment

PPE Problem Program Efficiency

PPE Problem Program Evaluator

PPE Property, Plant and Equipment

PPE-HIPS Polyphenylene Ether/High-Impact Polystyrene (Blend) [also PPE/HIPS]

PPEM Parallel Plate Electron Multiplier

PPEMA Portable Power Equipment Manufacturers Association [US]

PPEP Plasma Physics and Environmental Perturbation

PPE-PA Polyphenylene Ether/Polyamide (Blend) [also PPE/PA]

PP-EPDM Polypropylene/Ethylene-Propylene-Diene Monomer (Blend) [also PP/EPDM]

PPES Poly(1,4-Phenylene Ether Sulfone)

PP-FILM Plasma-Polymerized Film

PPF Payload Processing Facility [US Air Force]

PPF Pitch Precursor Factor

PPF (Blood) Plasma Protein Factor [Biochemistry]

PPF Pre-Printed Form

PPFA Plastic Pipe and Fittings Association [US]

PPFM Perovskite/Pyrochlore Phase Formation Model [Materials Science]

PPFRT Prototype Preliminary Flight Rating Test

PPFS Pre-Printed Forms System

PPG Pacific Proving Grounds [of US Atomic Energy Commission]

PPG Palladium-Pirani Gauge

PPG Performance Partnership Grants (for State and Tribal Environmental Programs) [of US Environmental Protection Agency]

PPG Polypropylene Glycol [also PG]

PPG Preprinted Gothics

PPG Priority Planning Grid

PPG Program Pulse Generator

PPG Propulsion and Power Generator

ppg pounds per gallon [Unit]

PPGA Peripheral Pin Grid Array

PPGA Plastic Pin Grid Array

PPGA Professional Plant Growers Association [US]

PPG Prod PPG Products [Publication of PPG Industries Inc., Pittsburgh, Pennsylvania, US]

PP/Gr Graphite Filled Polypropylene

PPH Parts per Hundred [also pph, or P/H]

PPH Peak-to-Peak Height [also pph]

PPH Phosphopyruvate Hydratase [Biochemistry]

PPH Postpartum Hemorrhage [Medicine]

PPH Pounds per Hour [also lbs/hr]

PPH Pulses per Hour [alsp pph]

PPh₃ Triphenyl Phosphine

PPh Tetraphenylphosphonium (Ion)

pph parts per hundred [also PPH, or P/H]

pPhA p-Phenylaniline

pPhDA p-Phenylenediamine

PPHM Parts per Hundred Million [also pphm]

PPHMDSO Plasma-Polymerized Hexamethyldisiloxane [also PP-HMDSO]

pPhP p-Phenylphenol

PPHS Panhandle-Plains Historical Society [US]

PPHT Phenylethyl Propylamino Hydroxytetralin

(±)-PPHT (±)-2-(N-Phenylethyl-N-Propylamino)-5-Hydroxytetralin

PPI Parcel Post Insured

PPI Parts per Inch [also ppi]

PPI Personnel Protection Indicators

PPI Pictorial Position Indicator

PPI Planar Plug-In

PPI Plan Position Indication; Plan Position Indicator

PPI Plastic Pipe Institute [US]

PPI Plastic Pronged Knob

PPI Points per Inch [also ppi]

PPI Policy Proof of Interest [Marine Insurance]

PPI Polymeric Polyisocyanate

PPI Pores per Inch [also ppi]

PPI Potash and Phosphate Institute [US]

PPI Precise Pixel Interpolation

PPI Pre-Plant Incorporated [also ppi]

PPI Programmable Peripheral Interface

PPI Pulse Position Indicator

P&PI Pulp and Paper Industry

PPi Inorganic Pyrophosphatase [Biochemistry]

PPIA Programmable Peripheral Interface Adapter

PPIB Programmable Peripheral Interface Board

PPIC Pollution Prevention Information Clearinghouse

PPIC Polyphenyl Isocyanate

PPIC Potash and Phosphate Institute of Canada

P2ID (International Symposium on) Plasma Process-Induced Damage

PPIL Priced Provisioned Item List

PPINICI Pulsed Positive Ion Negative Ion Chemical Ionization

PPIRB Pulp and Paper Industrial Relations Bureau

PPITB Plastics Processing Industry Training Board [UK]

PPITB Plast Bull PPITB Plastics Bulletin [Published by the Plastics Processing Industry Training Board, UK]

PPIU Programmable Peripheral Interface Unit

PPK n-Pentyl Phenyl Ketone

PPL Pacific Power and Light Company [Oregon, US]

PPL Parallel-Plate Laser

PPL Periodic Payment (Software) License

PPL 1-Phenyl 1H-Phosphole

PPL Pig Pancrease Lipase [Biochemistry]

PPL Plane-Polarized Light

PPL Planning Parts List

PPL Plasma Physics Laboratory [of Princeton University, New Jersey, US]

PPL Plasma Polymer Layer

PPL Plasma Processing Laboratory [of Drexel University, Philadelphia, Pennsylvania, US]

PPL Polarized Photoluminescence Excitation

PPL Polymorphic Programming Language

PPL Preferred Parts List

PPL Priced Parts List

PPL Private Pilot's License

PPL Programmable Peripheral Interface

PPL Providence Public Library [Rhode Island, US]

PPL Provisioning Parts List

P/PL Primary Payload

PPLA Professional Photographic Laboratories Association [UK]

PplA Propiolic Acid

PPL(A) Private Pilot's License (Aeroplanes) [UK]

PPL(AS) Private Pilot's License (Airships) [UK]

PPL(B) Private Pilot's License (Balloons) [UK]

PPLC Personal Programmable Logic Controller [also P²LC]

PPL(G) Private Pilot's License (Gyroplanes) [UK]

PPL(H) Private Pilot's License (Helicopters) [UK]

PP-LIF Pump-Probe-Laser-Induced Fluorescence

PPLO Pleuropneumonia-Like Organism [Microbiology]

PPM (2S-cis)-4-(Diphenylphosphino)-2-[(Diphenylphosphino)-methyl]pyrrolidine

PPM Pages per Minute [also ppm]

PPM Palladium Platinum Manganate

PPM Parallel Plate Micrometer

PPM Parts per Million [also ppm]

PPM Parts per Minute [also ppm]

PPM Path Probability Method

PPM Peak Program Meter

PPM Periodic Permanent Magnet

PPM Periodic Pulse Metering

PPM Phase Probability Method [Materials Science]

PPM Planned Preventive Maintenance

PPM Plasma Process Monitor

PPM Press Piercing Mill [Metallurgy]

PPM Previous Processor Mode

PPM Prioritätsprogramm für Materialforschung [Priority Program on Materials Research of the Swiss Government]

PPM Product Physical Metallurgy (Committee) [of Iron and Steel Society–Mechanical Working and Steel Processing (Division), US]

PPM Pulse-Phase Modulation

PPM Pulse-Position Modulation [also ppm]

PPM Pulses per Minute [also ppm]

ppm parts per million [also PPM]

ppm parts per minute [also PPM]

PPMA Petrol Pump Manufacturers Association [UK]

PPMA 1-Phenyl-2-Palmitoylamino-3-Morpholino-1-Propanol

PPMA Poly(phenylmethacrylate)

PPMA Polypropyl Methacrylate

PPMA Precision Potentiometer Manufacturers Association [US]

PPMA Processing and Packaging Machinery Association [UK]

PPMA Pulp and Paper Machinery Association [now Pulp and Paper Machinery Manufacturers Association, US]

(±)-PPMA DL-threo-1-Phenyl-2-Palmitoylamino-3-Morpholino-1-Propanol

ppm/°C parts per million per degree celsius [Unit]

PPMD Programmed Permanent Memory Device

PPME Pacific Plate Motion Experiment [Geology]

PPMF Pulp and Paper Manufacturers of Australia

PPMFA Pulp and Paper Manufacturing Federation of Australia

PPMIS Personal Property Management Information System

ppm/K parts per million per kelvin [Unit]

PPMMA Pulp and Paper Machinery Manufacturers Association [US]

PPM/MPM Press Piercing Mill/Multistand Pipe Mill [Metallurgy]

PPMO Praseodymium Lead Manganese Oxide (Superconductor)

PPMS Physical Property Measurement System

PPMS Plastic Pipe Manufacturers Society [UK]

PPMS Program Performance Measurement System

ppmv parts per million by volume

PPMW Primary Plant Mineralized Water

ppm-wt parts per million by weight [Unit]

PPN Bis(triphenylphosphine)iminium

PPN Poly(p-Naphthalene)

p⁺pn⁺ p-plus layer, p-layer, n-plus layer (magnetodiode)

PPNA Pre-Proto-Neolithic A [Archeology]

PPNB Pre-Proto-Neolithic B [Archeology]

PPO 2,5-Diphenyloxazole

PPO Diphenyl-p-Vinylphenylphosphine Oxide

PPO Phenylphenyloxazole

PPO Pollution Prevention Office

PPO Polyphenoloxidase [Biochemistry]

PPO Polyphenylene Oxide

PPO Polypropylene Oxide

PPO Precedence Partition and Outdegree

PPO Preferred (Health Care) Provider Organization [US]

PPO Prior Permission Only

PPO Professional Photographers of Ontario [Canada]

PPO₂ Partial Pressure of O_2 (Oxygen)

PPOA Power Plant Operations Analysis

PpOBA Poly(p-Oxybenzoate) [also P(pOBA)]

PPOC Professional Photographers of Canada

P POD Priority Property Operation Disk [also PPOD]

PP of A Professional Photographers of America [US]

P-Pol Plane Polarization [also p-pol]

PPO-PA Polyphenylene Oxide/Polyamide (Blend) [also PPO/PA]

PPO-PBT Polyphenylene Oxide/Polybutylene Terephthalate (Blend) [also PPO/PBT]

PPO-PS Polyphenylene Oxide/Polystyrene (Blend) [also PPO/PS]

PPOX Polypropylene Oxide

PPP Parallel Pattern Processor

PPP Parallel-Push-Pull (Amplifier) [also ppp]

PPP Parcel Processing Plant

PPP Pariser-Parr-Pople (Method) [Physics]

PPP Payload Patch Panel

PPP Peak Pulse Power

PPP 2,2',2"-Phosphinidynetris[ethyl(diphenyl)phosphine [also P3]

PPP Plastic Protective Plate

PPP Point-to-Point Protocol [Internet]

PPP Polluter Pays Principle

PPP Pollution Prevention Pays (Program) [also 3P, or P³, Canada]

PPP Pollution Prevention Plan [US]

PPP Poly(paraphenylene)

PPP Poly-p-Phenylene; Poly-1,4-Phenylene

PPP Portable Plotting Package [also P3, or P³]

PPP Prairie Plan Project [US]

PPP Programmed Potentiostatic Pulse

PPP Progress Payment Program [of Canadian Commercial Corporation]

PPP Public Policy Program

PPP (Earthquake) Push Wave Reflected twice from the Earth's Surface [Symbol] [also PR₂]

PPPA Poison Prevention Packaging Act [US]

PPPA Push Pull Power Amplifier

PP/PA Polypropylene/Polyallomer (Blend)

PPPB Pusat Penelitian dan Pengembangan Biologi [Research and Development Center for Biology, Indonesia]

PPPEA Pulp, Paper and Paperboard Export Association [US]

PPPI Precision Plan Position Indicator

PPPL Princeton Plasma Physics Laboratory [Princeton University, New Jersey, US]

PPPL-TFTR Princeton Plasma Physics Laboratory–Tokamak Fusion Test Reactor [of Princeton University, New Jersey, US]

PPPO Pusat Penelitian dan Pengembangan Oseanologi [Research and Development Center for Oceanology, Indonesia]

PPPP Pollution Prevention Pledge Program [also P⁴]

PPPRF Pan-Pacific Public Relations Federation [Thailand]

PPPS Pulse Pairs per Second [also ppps]

PPQ Plant Protection and Quarantine [of Animal and Plant Health Inspection Service, US Department of Agriculture]

PPQ Polyphenylquinoxaline

PPQ Possible Parliamentary Question

ppq parts per quadrillion [also PPQ, or P/Q]

PPQA Pageable Partition Queue Area

PPQ-APHIS-USDA Plant Protection and Quarantine (Division) of the Animal and Plant Health Inspection Service of the US Department of Agriculture

PPR Payload Preparation Room

PPR Photoplastic Recording

PPR Photopolarimeter Radiometer

PPR Polypropylene

PPR Positive Pressure Respirator [Medicine]

PPR Prior Permission Required

PPR Pulses per Revolution

Ppr Paper [also ppr]

PPRA Precedence Partition and Random Assignment

PPRA Pulp and Paper Research Institute [Montreal, Canada]

Pprbd Paperboard [also pprbd]

PPRC Pollution Prevention Resource Center [US]

PPRF Pulse Pair Repetition Frequency

PPRI Pulp and Paper Research Institute [US]

PPRIC Pulp and Paper Research Institute of Canada [Montreal, Quebec] [also PPRI(Can)]

pp/rms peak-to-peak/root-mean square (voltage)

PPS (Data) Packets per Second

PPS Page Printing System

PPS Page Processing System

PPS Parallel Processing System

PPS Parcel Processing System

PPS Pest Population Suppression

PPS Philadelphia Programming Society [US]

PPS Phosphorous Propellant System

PPS Physical Protection Section (Equipment)

PPS Plant Protection System

PPS Pneumatic Power Subsystem

PPS Point-to-Point System [Numerical Control]

PPS Policies and Procedures

PPS Polymer Processing Society

PPS Polyphenylene Sulfide

PPS Positive Pressure Slider [Disk Drives]

PPS *(Post Postscriptum)* – Additional Postscript; Second Postscript [also pps]

PPS Pound(s) per Square-Inch [also pps]

PPS Power Personal System

PPS Power and Pyro Subsystem

PPS Primary Power System

PPS Primary Propulsion System

PPS Printer/Plotter System

PPS Product Processing System

PPS Programmed Processor System

PPS Provisioning Performance Schedule

PPS Pulp and Paper Science

PPS Pulses per Second [also pps]

PPS 3-(1-Pyridino)-1-Propanesulfonate

PPSA Poly(phenylene Sulfide Amide)

PPSA Pulp and Paper Safety Association [Canada]

PPSC Privacy Protection Study Commission [US]

PPSC Processor Program State Control

PPSE Polyphosphoric Acid Trimethylsilyl Ester

PPSi Porous Polycrystalline Silicon

p/p⁺-Si P-Type/P-Plus-Type Silicon

PPSN Present Position

PPSN Public Packet-Switching Network

PPSP Power Plant Siting Program [of Maryland Department of Natural Resources, US]

PPSS Public Packet Switching Service [UK]

PPSU Polyphenylene Sulfone

PPSU Programmable Power Supply Unit

PPT Partial Prothromboplastin Time [Biochemistry]

PPT Particle Processing Technology

PPT Pattern Processing Technology

PPT Pharmaceutical Production Technology

PPT Phenylalanine-Pyruvate Transaminase [Biochemistry]

PPT 4-Phenyl-1-(Phenylamino)-2-(1H)Pyrimidinethione

PPT Plasma-Polymerized Thiophene

PPT PMDA (Pyromellitic Dianhydride)–PPD (p-Phenylenediamine)–TAB (Tetraaminobiphenyl) (Film)

PPT Polymer Production Technology

PPT Polypropylene Terephthalate

PPT Priorities and Protection Team [of Environment Australia Environmental Resources Information Network]

PPT Process Page Table

PPT Processing Program Table

PPT Pseudopotential Theory [Physics]

PPT Punched Paper Tape

.PPT Powerpoint File [File Name Extension] [also .ppt]

Ppt Precipitate [also ppt]

ppt parts per thousand [also PPT]

ppt parts per trillion [also PPT]

ppt precipitate(d) [Medical Prescriptions]

PPTA Polyphenylene Terephthalate

PPTA Poly-p-Phenylene Terephthalamide

PPTA-co-XTA Poly(phenylene Terephthalamide-co-Crosslinkable Terephthalic Acid)

p'p't'd precipitated [also pptd]

Ppte Precipitate [also ppte]

Pptg Precipitating [also pptg]

PPT-H 4-Phenyl-1-(Phenylamino) 2(1H)-Pyrimidinethione

Pptn Precipitation [also pptn]

PPTP Point-to-Point Tunneling Protocol [Internet]

PPTP Poly-P-Phenyleneterephtalamide

PPTS Pyridinium p-Toluenesulfonate

PPTT Parts per Ten Thousand [also pptt]

Ppty Property [also ppty]

PPU Peripheral Processing Unit

PPU Preprocessor Utility

PPU Probe Processing Unit

PPUG Professional Plant Users Group [UK]

P-PULSE Position Pulse [also P-Pulse] [Computers]

PPV Paidar-Pope-Vitek (Model) [Solid-State Physics]

PPV Polarized Platen Viewer

PPV Poly(p-Phenylene Vinylene)

PPV Positive Pressure Ventilation [Medicine]

PPV Primary Pressure Vessel

PPVD Plasma Physical Vapor Deposition

PPW Plane-Polarized Wave

PPW Plasma Powder Welding

PPWB Prairie Provinces Water Board [Canada]

PPWC Pulp and Paper Workers of Canada [now Pulp, Paper and Woodworkers of Canada]

PPWC Pulp, Paper and Woodworkers of Canada

PPX Packet Protocol Extension

PPX Poly(P-Xylylene)

PPX Private Packet Exchange

pPxBA p-Phenoxybenzoic Acid

PPY Palladium Poly-Yne

PPy Polypyrrole

4-PPy 4-Phenyl Pyridine

PPy-PSS Polypyrrole-Polystyrenesulfonate (Composite Polymer)

PQ Permeability Quotient

PQ Physical Quantity

PQ Plastoquinone

PQ Precision Qualifiable (Toolholder)

PQ Previous Question [also pq]

PQ Province of Quebec [Canada]

P(Q) Particle Form Factor [Symbol]

pq previous question [also PQ]

PQA Pharmaceutical Quality Assurance

PQA Plant Quarantine Act [US]

PQA Polymerizable Quaternary Ammonium

PQA Protected Queue Area

PQAA Province of Quebec Association of Architects [Canada]

PQAD Plant Quality Assurance Director

PQAMS Polymerizable Quaternary Ammonium Modified Silicate

PQC Pharmaceutical Quality Control

PQC Product Quality Control

PQC Production Quality Control

pQCD Perturbation Quantum Chromodynamics [Physics]

PQD Plant Quarantine Division [Now Part of Plant Protection and Quarantine–Animal and Plant Health Inspection Service, US Department of Agriculture]

PQFP Plastic Quad Flat Pack [Electronics]

PQFT International Conference on Problems of Quantum Field Theory [Physics]

PQGS Propellant Quantity Gauging System

PQINF (International Workshop on the) Physics of Quantum Information

PQL Practical Quantification (or Quantitation) Limit

PQLI Physical Quality of Life Index

PQM Proto-Qualification Model

PQMP Pressure-Quenched Metastable Phase [Metallurgy]

PQN Principal Quantum Number

PQQ Pyrroloquinolinequinone Tricarboxylic Acid

PQR Procedure Qualification Record [Welding]

PQRG Physics Quick Reference Guide [of American Institute of Physics, US]

PQRN Periodic Quantum Resistor Network

PQSCHA Pure-Quantum Self-Consistent Harmonic Approximation [Physics]

Pquin 2-(2'-Pyridyl)quinoline [also pquin]

p-QWMOSFET p-Channel Quantum-Well Metal-Oxide Semiconductor Field-Effect Transistor

PR Pacific Rim

PR Pakistan Rupee [Currency]

PR Paper Roll

PR Paper-Tape Reader

PR Parabolic Reflector

PR Parallel Resonant (Circuit)

PR Paraguay River [South America]

PR Paramagnetic Resonance [also PMR]

PR Park Ranger

PR Pathogenicity Related (Protein)

PR Pattern Recognition

PR Pawling Research Reactor [New York, US]

PR Payroll

PR Penicillium Roqueforti (Toxin)

PR Performance Report

PR Periodic Reverse (Current Plating)

PR Permanent Resident

PR (Magnetic) Permeability Ratio

PR Personal Robot

PR Pesticide Registration

PR Pesticide Residue

PR Petroleum Refining

PR Phase Rule [Physical Chemistry]

PR Phenol Red

PR Philippine Reactor

PR Photographic Radiometry

PR Photographic Reconnaissance

PR Photographic Record(ing)

PR Photoreactivation

PR Photo-Realism

PR Photoreflectance (Spectroscopy)

PR Photorefraction; Photorefractive

PR Photoresist(ance); Photoresistor

PR Photo-Roentgen (Unit)

PR Physical Reader

PR Physical Record

PR Physics Review [Journal Series of American Physical Society, US]

PR Piezoresistance; Piezoresistive; Piezoresistor

PR Piston Rod

PR Pitch Ratio

PR Planetary Radar

PR Planetary Reactor

PR Plasma Reactor

PR Poisson's Ratio [Mechanics]

PR Polarity Reversal; Polarity Reverser

PR Polarization Resistance

PR Polarized Reflection

PR Polar Research

PR Polymerization Reaction; Polymerization Reactor

PR Pool Reactor

PR Position Record

PR Power Reactant; Power Reactor

PR P-Polarized Reflectance

PR Preamplifier

PR Precipitation Radar

PR Precipitation Rate

PR Preliminary Record

PR Premature Release

PR Press Release

PR Pressure Ratio

PR Pressure Regulator

PR Primary Radar

PR Primary Radiation

PR Prince Rupert [British Columbia, Canada]

PR Principal Ray

PR Prism Rotator

PR Private Renter

PR Problem Report

PR Procurement Regulation

PR Procurement Requisition

PR Production Rate

PR Production Record

PR Productivity Ratio

PR Product Removal

PR Program Register

PR Program Requirement

PR Progress Record

PR Progress Report

PR Project Report

PR Proportional Representation

PR Pseudorandom

PR Public Relations

PR Puerto Rican; Puerto Rico

PR Puerto Rico [ISO Code]

PR Pulfrich Refractometer [Optics]

PR Pulse Rate

PR Pulse Ratio

PR Pulse Radiolysis

PR Pulse Reversal (Plating)

PR *(Punctum Remotum)* – Far Point [also pr] [Optics]

PR Purchase Request

PR Purchase Requisition

PR Pure Rubber

PR Purplish Red

PR Push Rod

PR$_1$ (Earthquake) Push Wave Reflected once from the Earth's Surface [Symbol] [also PP]

PR$_2$ (Earthquake) Push Wave Reflected twice from the Earth's Surface [Symbol] [also PPP]

PR-1 Print Register 1

P&R Performance and Resources

P&R Preview and Relocation (Station) [Electron Microscopy]

P-R Interval Between Start of P Wave and Start of QRS Complex (on Electrocardiogram) [Medicine]

P(R) Packet Receive-Sequence Number [Data Communications]

P(R) Power Incident upon Receiver

P-R Parent-to-R-Phase (Transformation) [also P/R] [Metallurgy]

P-R Parent Phase-to-R-Phase (Transformation) [also P/M] [Metallurgy]

Pr Pair [also pr]

Pr Prandtl Number [Symbol]

Pr Praseodymium [Symbol]

Pr Prefix [also pr]

Pr Prescription Required

Pr Pressure [also pr]

Pr Price [also pr]

Pr Prince [also pr]

Pr Print(er) [also pr]

Pr Prism [also pr]

Pr Probability [Symbol]

Pr Profession(al) [also pr]

Pr Propanoic Acid

Pr Propyl (Alcohol)

Pr^{3+} Trivalent Praseodymium Ion [also Pr^{+++}] [Symbol]

Pr^{4+} Tetravalent Praseodymium Ion [Symbol]

Pr$_M$ Magnetic Prandtl Number [Symbol]

Pr$_m$ Prandtl Number (of Fluid Diffusion) [Symbol]

Pr$_{turb}$ Turbulent Prandtl Number [Symbol]

Pr II Praseodymium II [Symbol]

Pr-123 Praseodymium Barium Copper Oxide (Superconductor) [PrBa$_2$Cu$_3$O$_7$] [also Pr123]

Pr-137 Praseodymium [also ^{137}Pr, or Pr^{137}]

Pr-138 Praseodymium [also ^{138}Pr, or Pr^{138}]

Pr-139 Praseodymium [also ^{139}Pr, or Pr^{139}]

Pr-140 Praseodymium [also ^{140}Pr, or Pr^{140}]

Pr-141 Praseodymium [also ^{141}Pr, Pr^{141}, or Pr]

Pr-142 Praseodymium [also ^{142}Pr, or Pr^{142}]

Pr-143 Praseodymium [also ^{143}Pr, or Pr^{143}]

Pr-144 Praseodymium [also ^{144}Pr, or Pr^{144}]

Pr-146 Praseodymium [also ^{146}Pr, or Pr^{146}]

P(r) Poisson Distribution (in Statistics) [Symbol]

P$_{nl}$(r) Radial Probability Density [Symbol]

pR (earthquake) push wave reflected at surface on observing station [Symbol]

pr present

pr primary

pr printed

.pr Puerto Rico [Country Code/Domain Name]

p(r) distribution function [Symbol]

p.r. *(per rectum)* – by the rectum [Medicine]

$\phi(R)$ pair potential (in physics) [Symbol]

$\phi(r)$ radial distribution function [Symbol]

$\Phi(r)$ pair potential (in physics) [Symbol]

$\Psi(r)$ Bloch Function (for Electron in Periodic Lattice) [Symbol]

$\Psi(r)$ Eigenfunction [Symbol]

$\psi_k(r)$ plane wave [Symbol]

$\psi_{nlm}(r)$ Schroedinger wavefunction (in quantum mechanics) [Symbol]

PRA Paint Research Association [UK]

PRA Peace Research Abstracts [of International Peace Research Association, Netherlands]

PRA Peak Recording Accelerograph

PRA Pendulous Reference Axis

PRA Permanent Resident Alien [US]

PRA Perturbed Rotating Atom Approximation [Physics]

PRA Pesticide Residue Analysis

PRA Photochemical Research Associates

PRA Physical Review A [Journal of the American Physical Society, US]

PRA Port Rear Access

PRA Prairie Rail Authority

PRA Precision Axis

PRA Preliminary Reserve Analysis

PRA President of the Royal Academy

PRA Primary Recoil Atom

PRA Prime Rate Access

PRA Print Alphamerically

PRA Probabilistic Risk Assessment

PRA Production Reader Assembly

PRA Program Reader Assembly

PRA Progressive Retinal Atrophy [Medicine]

PRA Prompt Radiation Analysis

PRA Public Roads Administration [US]

PRAB Program Review Approval Board [Canada]

PRAC Pyrethrum Research Advisory Committee [Kenya]

Prac Practice [also prac]

PRACA Problem Reporting and Corrective Action

n-PrAc n-Propyl Acetate

Pr(Ac)$_3$ Praseodymium(III) Acetate

Pr(ACAC)$_3$ Praseodymium(III) Acetylacetonate [also Pr(AcAc)$_3$, or Pr(acac)$_3$]

PRACSA Public Remote Access Computer Standards Association

pract practical

Pract Archit Mag Practicing Architects Magazine [Published by the Society of American Registered Architects, US]

Pract Comput Practical Computing [Journal published in the UK]

Pract Electron Practical Electronics [Journal published in the UK]

Pract Wirel Practical Wireless [Journal published in the UK]

PRADOR PRF (Pulse Repetition Frequency) Ranging Doppler Radar

PRAGB Paint Research Association of Great Britain

Prague Bull Math Linguist Prague Bulletin of Mathematical Linguistics [Czech Republic]

PRAI Phosphoribosyl Anthranilate Isomerase [Biochemistry]

PRAI Project Research Applicable in Industry (Grant) [of Natural Sciences and Engineering Research Council, Canada]

PRAIS Pesticide Residue Analysis Information Service [UK]

Prakt Metallogr Praktische Metallographie [German Publication on Practical Metallography]

Pr Alc Propyl Alcohol [also pr alc]

PRAM Parallel Random-Access Machine

PRAM Parameter Random Access Memory

PRAMS Programmable Risk Assessment Model System

PRARE Precise Range and Rate Equipment

PRAREE Precise Range and Rate Equipment–Extended Version

PRAS Planning and Reporting Accountability Structure

Prazepam 7-Chloro-1-(Cyclopropylmethyl)-1,3-Dihydro-5-Phenyl 2H-1,4-Benzodiazepin-2-one

PRB Panel Review Board

PRB Parachute Refurbishment Building

PRB Paso Robles, California [Tactical Air Navigation Station]

PRB Physical Review B (Condensed Matter) [of American Physical Society, US]

PRB Population Reference Bureau [US]

PRB Primary Resonator Block

PRB Project Review Board

Pr Bal Previous Balance

PrBCCO Praseodymium Barium Calcium Copper Oxide (Superconductor)

PrBCFO Praseodymium Barium Copper Iron Oxide (Superconductor)

PrBCGO Praseodymium Barium Copper Gallium Oxide (Superconductor)

PrBCNO Praseodymium Barium Copper Niobium Oxide (Superconductor)

PrBCO Praseodymium Barium Copper Oxide (Superconductor)

PrBCTO Praseodymium Barium Copper Tantalum Oxide (Superconductor)

PrBr Propyl Bromide

PRBR-o Physical Review B, Rapid Communications on line [of American Physical Society, US]

PRBS Pseudorandom Binary Sequence

PRBSG Pseudorandom Binary Sequence Generator [of Liquid Metal Engineering Center, US]

n-PrBz n-Propyl Benzene

PRC Pacific Rim Country

PRC Partial Response Coding

PRC People's Republic of China

PRC Periodic Reverse Current (Plating)

PRC Physical Review C (Nuclear Physics) [of American Physical Society, US]

PRC Pittsburgh Research Center [Pennsylvania, US]

PRC Planning Research Corporation

PRC Plant Record Center [of American Horticultural Society, US]

PRC Plant Research Center [of Environment Canada]

PRC Point of Reverse Curve

PRC Policy Research Centers (Program) [US]

PRC Postal Rate Commission [US]

PRC Poultry Research Center [UK]

PRC Precision (Function) [Computers]

PRC Primary Routing Center

PRC Procession Register Clock

PRC Product Research Center

PRC Programmed Rate Control

PRC Programmer's Calculator (Software)

PRC Projects Review Committee

PRC Proportional Response Controller

PRC Proton Recoil Detector

Prc Pierce [also prc]

Pr2C 1H-Pyrrole-2-Carboxylic Acid

PRCA President of the Royal Canadian Academy

PRCA Problem Reporting and Corrective Action

PRCA Public Relations Consultants Association [UK]

PRCB Program Requirements Change Board

PRCB Program Requirements Control Board

PRCB Program Review Control Board

PRCBD Program Requirements Control Board Directive

PRCBD Program Review Control Board Directive

PRCC Pacific Rim Coal Conference

PrCCNO Praseodymium Calcium Copper Niobium Oxide (Superconductor)

PrCCO Praseodymium Cerium Copper Oxide (Superconductor)

PRCF Plutonium Recycling Critical Facility

Prch Purchase [also prch]

PR China People's Republic of China

Prcht Parachute [also prcht]

PrCl Propyl Chloride

PrCN Propyl Cyanide

PrCO Praseodymium Copper Oxide (Superconductor) [also PrCoO]

Pr(Cp)₃ Tris(cyclopentadienyl)praseodymium

PRCS Precured Resin-Coated Sand [Metallurgy]

PRCS Primary Reaction Control System

PRCST Pittsburgh Regional Center for Science Teachers [Pennsylvania, US]

Prcst Precast [also prcst]

PRCVD Programmed Rate Chemical Vapor Deposition

PRD Paper-Tape Read

PRD Personal Radiation Dosimeter

PRD Pesticides Regulation Division [Now Part of Plant Protection and Quarantine–Animal and Plant Health Inspection Service, US Department of Agriculture]

PRD Pharmaceutical Research Development

PRD Physical Review D (Particles and Fields) [of American Physical Society, US]

PRD Power Range Detector

PRD Prime Radar Digitizer

PRD Printer Description

PRD Printer Driver

PRD Printer Dump

PRD Procurement Regulation Directive

PRD Procurement Requirements Document

PRD Program Requirements Data

PRD Program Requirements Document

PRD Progressive Retinal Degeneration

.PRD Printer Driver [Microsoft Word File Name Extension]

PRDA Primary Resource Development Cooperation Agreement [Canada]

PRDA Program Research and Development Agreement

PRDC Polar Research and Development Center [of US Army at Fort Belvoir, Virginia]

PRDC Power Reactor Development Company [US]

PRDDO Partial Retention of Diatomic Differential Orbitals [Physics]

PRDDO Partial Retention of Diatomic Differential Overlap [Physics]

PRDF Partial Radial Distribution Function

PRDP Power Reactor Development Program [Canada]

PRDPEC Power Reactor Development Program Evaluation Committee [of Atomic Energy of Canada Limited]

PRDS Processed Radar Display System

PRE Payload Replanning Engineer

PRE Penicillininase Receptor Enzyme [Biochemistry]

PRE Personal Rescue Enclosure

PRE Petroleum Refining Engineer

PRE Petroleum Reservoir Engineering

PRE Photoreflectance Excitation

PRE Photorefractive Effect

PRE Physical Review E [of American Physical Society, US]

PRE Pitting Resistance Equivalent [Metallurgy]

PRE Polymer Reaction Engineering

PRE Prefix [Data Communications]

PRE Preformatted [HyperText Markup Language]

PRE Preliminary Amplifier

PRE Preset (Flip-Flop Input)

PRE Processing Refabrication Experiment

PRE (Fédération Européenne des Fabricants de) Produits Refractaires [European Federation of Refractory Products Manufacturers, Switzerland]

PRE Protective Reservation Equipment

PRE Protein Relaxation Enhancement

PRe Pakistan Rupee [Currency]

PrE Printer Emulator

PREAMP Preamplifier [also Preamp, or preamp]

PREAMP Pre-Competitive Advanced Manufacturing of Electrical Products

Prec Precipitate; Precipitation [also prec]

Prec Precision [also prec]

prec preceded; preceding

Precambr Res Precambrian Research [Journal published in the Netherlands]

PRECARN Pre-Competitive Applied Research Network [Canada]

Precious Met Precious Metals [Journal published in PR China]

Precious Met News Rev Precious Metals News and Review [Published by the International Precious Metals Institute, US]

PRECIP Preserved Context Index System

Precip Precipitate [also precip]

PRECIS Pre-Coordinate Index System

Precis Precision [also precis]

Precis Eng Precision Engineering [Journal published in the UK]

Precis Met Precision Metal [Journal published in the US]

Precis Toolmak Precision Toolmaker [Journal published in the UK]

prec/m³ precipitates per cubic meter [Symbol]

PRECP Processing of Emissions by Clouds and Precipitation

Pred Predecessor (of Argument) [Pascal Function]

PREDICT Prediction of Radiation Effects by Digital Computer Techniques

Pre-Em Pre-Emergence [also pre-em]

PREF Propulsion Research Environmental Facility

Pref Preface [also pref]

Pref Prefecture [also pref]

Pref Preference [also pref]

Pref Prefix [also pref]

pref preferred

Prefab Prefabrication [also prefab]

Preflt Preflight

PreH2 3,9-Dimethyl-2,10-Undecanedione Dioxime

prel preliminary [also prelim]

PRELIMS Preliminary Matter

PRELORT Precision Long-Range Tracking Radar

PREM Preliminary Reference Earth Model

Prem Premier [also prem]

Prem Premium [also prem]

Premed Premedical (Student, or Course) [also premed]

PREMID Programmable Remote Identification (System)

Pre-Mod Premodulation [also pre-mod]

PREN Pitting Resistance Equivalency Number [Metallurgy]

PrEN Preliminary European Normal Specification

Pren N,N'-(1,2-Ethanediyl)bis(pyrrolidine-2-Carboxylic Acid) [also pren]

PRENFLO Pressurized Entrained-Flow (Process)

pre-op pre-operative

PREP Plasma Rotating Electrode Process [also P-REP] [Powder Metallurgy]

PREP Programmed Educational Package

PReP Power-Personal Computer Reference Platform

Prep Preparation; Preparatory; Prepare [also prep]

prep prepare (function) [Microsoft-Disk-Operating System]

PREPCOM Preparatory Consultation Meeting

prepd prepaid [also prep'd]

prepd prepared [also prep'd]

Prepg Preparing [also prepg]

Prepn Preparation [also prepn]

PREPP Process Experimental Pilot Plant

Prepr Preprint [also prepr]

Prepreg Preimpregnated Material [also prepreg]

Prepr Pap Am Chem, Div Fuel Chem Preprints of Papers– American Chemical Society, Division of Fuel Chemistry [US]

Pre-RTA Pre-Rapid Thermal Anneal(ing) [Metallurgy]

PRES Presentation (Font)

PRES Program Reporting and Evaluation System

Pres Presence [also pres]

Pres President [also pres]

Pres Pressure [also pres]

pres present

PRESFR Pressure Falling Rapidly

PRESRR Pressure Rising Rapidly

PRESS Pacific Range Electromagnetic Signature Studies

Press Pressurant; Pressure [also press]

Press Alt Pressure Altitude [also press alt]

PRESSAR Presentation Equipment for Slow-Scan Radar

Pressure Eng Pressure Engineering [Journal published in Japan]

PRESTO Program for Rapid Earth-to-Space Trajectory Optimization

PRESTO Program Reporting and Evaluation System for Total Operations

Prestr Concr Inst J Prestressed Concrete Institute Journal [US]

PrEth Di-n-Propyl Ether

Prev Prevention; Preventive [also prev]

prev previous(ly)

PrevAGT Previous Abnormality of Glucose Tolerance [Medicine]

Prevlv Prevalve

Prev Med Preventive Medicine [US Journal]

PREXTEND Prestel Extended [Computers]

PRF Parachute Refurbishment Facility

PRF Petroleum Research Fund [of American Chemical Society, US]

PRF Phenol-Resorcinol-Formaldehyde

PRF Plastics Recycling Foundation [US]

PRF Plutonium Reclamation Facility [US]

PRF Plywood Research Foundation [US]

PRF Potential Requirements File

PRF Precision Roll Feed

PRF Processor Request Flag

PRF Publication Reference File [of US Government Printing Office]

PRF Pulse Recurrence Frequency

PRF Pulse Repetition Frequency

PRF Purdue Research Foundation [of Purdue University, West Lafayette, Indiana, US]

.PRF Preferences [Grammatik IV File Name Extension] [also .prf]

Prf Proof [also prf]

PrF Propyl Fluoride

PRFD Pulse Recurrence Frequency Discrimination

PRFCS Pattern Recognition Feedback Control System

PrFeB Praseodymium Iron Boron (Magnet Alloy)

PRFL Pressure-Fed Liquid

PRFL Puerto Rico Federation of Labour

PR Focus Physical Review Focus [Publication of American Physical Society, US]

Pr(FOD)$_3$ Praseodymium Tris(1,1,2,2,3,3,3-Heptafluoro-7,7-Dimethyl-4,6-Octanedionate); Praseodymium Tris(6,6,7,7,8,8,8-Heptafluoro-2,2-Dimethyl-3,5-Octanedionate) [also Pr(fod)$_3$]

PRFS Pulse Recurrence Frequency Stagger

PRFU Processor Ready for Use

PRG Pacific Research Group [Japan]

PRG Program Resources Guide [of Institute of Electrical and Electronics Engineers, US]

.PRG Program (File) [File Name Extension] [also .prg]

Prg Program [also prg]

PrGe Professional Geologist

PRGMR Programmer [also Prgmr]

PRH Prolactin-Releasing Hormone [Biochemistry]

PR(H) (Magnetic) Field Dependent Permeability Ratio [Symbol]

Pr(HFAC)$_3$ Praseodymium Hexafluoroacetylacetonate [also Pr(hfac)$_3$]

Pr(HFC)$_3$ PraseodymiumTris[(heptafluoropropylhydroxymethylene) camphorate] [also Pr(hfc)$_3$]

PRHQ Pacific Region Headquarters

PRI Paleontological Research Institution [of Cornell University, Ithaca, New York, US]

PRI Performance Review Institute [Warrendale, Pennsylvania, US]

PRI Petroleum Recovery Institute [Calgary, Alberta, Canada]

PRI Photo Radar Intelligence

PRI (Rubber) Plasticity Retention Index

PRI Plastics and Rubber Institute [London, UK]

PRI Primary-Rate Interface

PRI Primate Research Institute [of Kyoto University, Japan]

PRI Project Readout Indicator

PRI Psoriasis Research Institute [US]

PRI Public Relations Institute

PRI Public Research Institute

PRI Pulse Rate Indicator

PRI Pulse Repetition Interval

PrI Propyl Iodide

Pri Priority [also pri]

pri primary [also prim]

pri private

PRIBA President of the Royal Institute of British Architects

P-Rib-PP 5-Phosphorylribose 1-Pyrophosphate [Biochemistry]

Prib Sist Upr Pribory i Sistemy Upravleniya [Russian Journal on Instrumentation and Control]

Prib Tekh Eksp Pribory i Tekhnika Eksperimenta [Russian Journal on Instruments and Experimental Techniques]

PRICE Productivity Improvement and Cost Effectiveness

Price Waterhouse Rev Price Waterhouse Review [US]

PRICM Pacific Rim International Conference on Advanced Materials and Processing

PRIDE Programmed Reliability in Design

PRIDE Programmed Reliability in Design Engineering

PRIH Prolactin-Release-Inhibiting Hormone [Biochemistry]

PRI/IRL Public Research Institutes/Industrial Research Laboratories (Scheme) [UK]

Prikl Biokhim Mikrobiol Prikladnaya Biokhimiya i Microbiologiya [Russian Journal on Applied Biochemistry and Microbiology]

Prikl Mat Mekh Prikladnaya Matematika i Mekhanika [Journal on Applied Mathematics and Mechanics; published by the Academy of Sciences of the USSR]

Prikl Mekh Prikladnaya Mekhanika [Ukrainian Journal on Applied Mechanics]

prim primary [also pri]

Prim Primate

prim primary

prim primitive

prim- primary [Chemistry]

PRIMA Process Information Maps

Primatol Primatological; Primatologist; Primatology

PRIME Precision Integrator for Meteorological Echoes

PRIME Precision Recovery Including Maneuverable Entry [Aerospace]

PRIME Program Independence, Modularity and Economy

PRIME Programmed Instruction-Form Management Education

PRIMENET Prime Network Software Package

PRIMEX Pressureless Infiltration by Metal (Process) [Powder Metallurgy]

PRIMO Platform Repairs Inspection Maintenance Offshore

PRIMORDIAL Primary Order Dialling [of European Space Agency–Information Retrieval System]

PRIMOS Prime Operating System

Prin Principal [also prin]

Prin Principle [also prin]

PRI/NADCAP Performance Review Institute/National Aerospace Defense Contractors Accreditation Program [also PRI NADCAP, US]

PRINCE Parts Reliability Information Center [of NASA]

PRINCE Programmed Reinforced Instruction Necessary to Continuing Education

PRINS Process Information System

Pr Inst Elektrotech Prace Instytutu Elektrotechniki [Proceedings of the Institute of Electrical Engineering, Warsaw, Poland]

Pr Inst Lacz Prace Instytutu Lacznosci [Proceedings of the Lacznosci Institute, Warsaw, Poland]

Pr Inst Metal Zelaza Prace Instytutu Metalurgii Zelasa [Proceedings of the Zelesa Institute of Metallurgy, Gliwice, Poland]

Pr Inst Met Niezelaz Prace Instytutu Metali Niezelaznych [Proceedings of the Nonferrous Metals Institute, Warsaw, Poland]

Pr Inst Odlew Prace Instytutu Odlewnictwa [Engineering Proceedings; published in Krakow, Poland]

Pr Inst Spaw Prace Instytutu Spawalnictwa [Proceedings; published in Gliwice, Poland]

Pr Inst Tele- & Radiotech Prace Instytutu Tele- i Radiotechnicznego [Proceedings of the Institute of Television and Radio Engineering, Warsaw, Poland]

Pr Inst Technol Drewna Prace Instytutu Technologii Drewna [Proceedings of the Drewna Institute of Technology, Poland]

Print Printer; Printing [also print]

Printed Circuit Fabr Printed Circuit Fabrication [Journal published in the US]

Printed Circuit Assem Printed Circuit Assembly [Journal published in the US]

PRINTF Print with Formatting [C Programming Language]

PRINUL Puerto Rico Inter-National Undersea Laboratory [US]

PRIO (International) Peace Research Institute, Oslo [Norway]

Prio Priority [also prio]

PRIS Propeller Revolution Indicator System

PRIS Puerto Rico Information Service

PRISA Public Relations Institute of Southern Africa

PRISE Program for Integrated Shipboard Electronics

PRISM Parallel Reduced Instruction Set Multiprocessor

PRISM Personnel Record Information System [UK]

PRISM Personal Records Information System Management

PRISM Photorefractive Information Storage Materials (Program)

PRISM Polymer Reference Interaction Site Model

PRISM Power Reactor Inherently Safe Module

PRISM Power Reactor Innovative Small Module

PRISM Professional as a Resource for Instructors in Science and Mathematics [of ASM International and National Science Foundation, US]

PRISM Programmable Interactive System

PRISM Programmed Integrated System Maintenance

PRISM Program Reliability Information System for Management

PRISMA Prozesse im Schadstoffkreislauf Meer–Atmosphere [Processes in the Pollution Cycle (North) Sea–Atmosphere (Program); of Bundesministerium für Forschung und Technology, Germany]

PRK People's Republic of Korea [North Korea]

PRK Phase Reversal Keying

PRK Photorefractive Keratecomy [Cornea Surgery]

PRK Primary Rat Kidney [Biochemistry]

Prkg Parking [also prkg]

Priv Doz *(Privatdozent)* – German Academic Title of University Lecturers

.PRJ Project [Borland File Name Extension]

Pr Kom Metal-Odlew, Pol Akad Nauk–Oddz Krakow, Metal Prace Komisji Metalurgiczno-Odlewniczej, Polska Akademia Nauk–Oddzial w Krakowie, Metalurgia [Proceedings on Metallurgical Engineering of Polish Academy of Science, Krakow]

PRL Page Revision Log

PRL Periodical Requirements

PRL Philips Research Laboratories [Eindhoven, Netherlands]

PRL Physical Review Letters [Publication of American Physical Society, US]

PRL Photoreactivating Light

PRL Prairie Regional (Research) Laboratory [of National Research Council of Canada at the University of Saskatchewan in Sakatoon]

PRL Print Lister

PRL Priority Rate Limiting

PRL Psychophysiological Reasearch Laboratory [of NASA Ames Research Center, Mountain View, California, US]

prl parallel

PRLC Pittsburgh Regional Library Center [Pennsylvania, US]

PrLCO Praseodymium Lanthanum Copper Oxide (Superconductor)

Pr:LiYF$_4$ Praseodymium-Doped Lithium Yttrium Fluoride [also Pr^{3+}:LiYF$_4$]

PRL-o Physical Review Letters online [of American Physical Society, US]

Pr:LSCO Praseodymium-Doped Lanthanum Strontium Copper Oxide (Superconductor)

PRM Parameter [also Prm]

PRM Particle Rotation Method [Metallurgy]

PRM Payload Retention Mechanism

PRM Personal Radiation Monitor

PRM Posigrade Rocket Motor

PRM Power Range Monitor

PRM Presidential Review Memorandum [US]

PRM Pressure Remanent Magnetization

PRM Process Radiation Monitor

PRM Program Review Meeting

PRM Pulse Rate Modulation

P-R-M Parent Phase-R-Phase-Martensite (Transformation) [also P/M] [Metallurgy]

PRMD Private Mangement Domain [X.400 CCITT-ISO Electronic Mail Standard]

PrMet Professional Meteorology [also Prof Met]

PrMgCl n-Propylmagnesium Chloride

PRML Partial-Response Maximum-Likelihood

Prmld Premold [also prmld]

PRMMC Particulate-Reinforced Metal-Matrix Composite [also PR MMC]

PRMO Partially Restricted Molecular Orbital [Physical Chemistry]

PRN Previous Result Negative

PRN Printer [also Prn]

PRN Printer Name

PRN Print Numerically

PRN Program Release Notice

PRN Pseudorandom Noise [Communications]

PRN Pseudorandom Numbers

PRN Pulse Range (or Ranging) Navigation

prn *(pro re nata)* – as, or when required [Medical Prescriptions]

Pr Nauk Inst Chem Nieorg Metal Pierwiastkow Rzadkich Politech Wroc, Konf Prace Naukowe Instytutu Chemii Nieorgaicznej i Metalurgii Pierwiastkow Rzadkich Politechniki Wroclawskiej, Konferencje [Publication of the Wroclaw Polytechnic Institute–Inorganic Chemistry and Metallurgical Engineering, Conference Series, Poland]

Pr Nauk Inst Chem Nieorg Metal Pierwiastkow Rzadkich Politech Wroc, Monogr Prace Naukowe Instytutu Chemii Nieorgaicznej i Metalurgii Pierwiastkow Rzadkich Politechniki Wroclawskiej, Monografie [Publication of the Wroclaw Polytechnic Institute–Inorganic Chemistry and Metallurgical Engineering, Monographs, Poland]

Pr Nauk Inst Chem Nieorg Metal Pierwiastkow Rzadkich Politech Wroc, Stud Mater Prace Naukowe Instytutu Chemii Nieorgaicznej i Metalurgii Pierwiastkow Rzadkich Politechniki Wroclawskiej, Studia i Materialy [Publication of the Wroclaw Polytechnic Institute–Inorganic Chemistry and Metallurgical Engineering, Materials Science Series, Poland]

Pr Nauk Inst Cybern Tech Politech Wroc, Ser Konf Prace Naukowe Instytutu Cybernetyki Technicznej Politechniki Wroclawskiej, Seria: Konferencje [Publication of the Wroclaw Polytechnic Institute–Cybernetic Science and Engineering, Conference Series, Poland]

Pr Nauk Inst Fiz Politech Wroc, Ser Monogr Prace Naukowe Instytutu Fizyki Politechniki Wroclawskiej, Seria: Monografie [Publication of the Wroclaw Polytechnic Institute–Physics, Monograph Series, Poland]

Pr Nauk Inst Konstr Eksploat Masz Politech Wroc, Konf Prace Naukowe Instytutu Konstrukcji i Eksploatacji Maszyn Politechniki Wroclawskiej, Konferencje [Publication of the Wroclaw Polytechnic Institute– Design and Utilization of Machinery, Conference Series, Poland]

Pr Nauk Inst Konstr Eksploat Masz Politech Wroc, Monogr Prace Naukowe Instytutu Konstrukcji i Eksploatacji Maszyn Politechniki Wroclawskiej, Monografie [Publication of the Wroclaw Polytechnic Institute– Design and Utilization of Machinery, Monographs, Poland]

Pr Nauk Inst Konstr Eksploat Masz Politech Wroc, Stud Mater Prace Naukowe Instytutu Konstrukcji i Eksploatacji Maszyn Politechniki Wroclawskiej, Studia i Materialy [Publication of the Wroclaw Polytechnic Institute–Design and Utilization of Machinery, Materials Research Series, Poland]

Pr Nauk Inst Materialozn Mech Tech Politech Wroc, Konf Prace Naukowe Instytutu Materialoznawstwa i Mechaniki Technicznej Politechniki Wroclawskiej, Konferencje [Publication of the Wroclaw Polytechnic Institute–Materials and Mechanical Engineering, Conference Series, Poland]

Pr Nauk Inst Materialozn Mech Tech Politech Wroc, Monogr Prace Naukowe Instytutu Materialoznawstwa i Mechaniki Technicznej Politechniki Wroclawskiej, Monografie [Publication of the Wroclaw Polytechnic Institute–Materials and Mechanical Engineering, Monographs, Poland]

Pr Nauk Inst Materialozn Mech Tech Politech Wroc, Stud Mater Prace Naukowe Instytutu Materialoznawstwa i Mechaniki Technicznej Politechniki Wroclawskiej, Studia i Materialy [Publication of the Wroclaw Polytechnic Institute–Materials and Mechanical Engineering, Materials Research Series, Poland]

Pr Nauk Inst Metrol Elektr Politech Wroc, Ser Konf Prace Naukowe Instytutu Metrologii Elektrycznej Politechniki Wroclawskiej, Seria: Konferencje [Publication of the Wroclaw Polytechnic Institute– Electrical Metrology, Conference Series, Poland]

Pr Nauk Inst Technol Budowy Masz Politech Wroc, Konf Prace Naukowe Instytutu Technologii Budowy Maszyn Politechniki Wroclawskiej, Konferencje [Publication of the Wroclaw Polytechnic Institute– Machine Design and Engineering, Conference Series, Poland]

Pr Nauk Inst Technol Budowy Masz Politech Wroc, Monogr Prace Naukowe Instytutu Technologii Budowy Maszyn Politechniki Wroclawskiej, Monografie [Publication of the Wroclaw Polytechnic Institute–Machine Design and Engineering, Monographs, Poland]

Pr Nauk Inst Technol Budowy Masz Politech Wroc, Stud Mater Prace Naukowe Instytutu Technologii Budowy Maszyn Politechniki Wroclawskiej, Studia i Materialy [Publication of the Wroclaw Polytechnic Institute– Machine Design and Engineering, Materials Research Series, Poland]

Pr Nauk Inst Technol Elektron Politech Wroc, Ser Konf Prace Naukowe Instytutu Technologii Elektronowej Politechniki Wroclawskiej, Seria: Konferencje [Publication of the Wroclaw Polytechnic Institute– Electrical Engineering, Conference Series, Poland]

Pr Nauk Inst Technol Elektron Politech Wroc, Ser Monogr Prace Naukowe Instytutu Technologii Elektronowej Politechniki Wroclawskiej, Serie: Monografie [Publication of the Wroclaw Polytechnic Institute–Electrical Engineering, Monographs, Poland]

Pr Nauk Inst Technol Elektron Politech Wroc, Ser Stud Mater Prace Naukowe Instytutu Technologii Elektronowej Politechniki Wroclawskiej, Serie: Studia i Materialy [Publication of the Wroclaw Polytechnic Institute– Electrical Engineering, Materials Research Series, Poland]

Pr Nauk Inst Ukl Elektromasz Politech Wroc, Ser Konf Prace Naukowe Instytutu Ukladow Elektromaszynowych Politechniki Wroclawskiej, Seria: Konferencje [Publication of the Wroclaw Polytechnic Institute– Electrical Machinery, Conference Series, Poland]

Pr Nauk Inst Ukl Elektromasz Politech Wroc, Ser Monogr Prace Naukowe Instytutu Ukladow Elektromaszynowych Politechniki Wroclawskiej, Seria: Monografie [Publication of the Wroclaw Polytechnic Institute–Electrical Machinery, Monograph Series, Poland]

Pr Nauk Inst Ukl Elektromasz Politech Wroc, Stud Mater Prace Naukowe Instytutu Ukladow Elektromaszynowych Politechniki Wroclawskiej, Studia i Materialy [Publication of the Wroclaw Polytechnic Institute–Electrical Machinery, Materials Research Series, Poland]

PRNC Puerto Rico Nuclear Center [of University of Puerto Rico at Rio Piedras–Mayaguez]

PrNCO Praseodymium Neodymium Copper Oxide (Superconductor)

PRNDL Park-Reverse-Neutral-Drive-Low Gear [Automobiles]

PRNET Packet Radio Network

PRNS Puerto Rico News Service

Prntr Printer [also prntr]

PRO Photonics Research Ontario [of Ministry of Economic Development, Trade and Tourism, Canada]

PRO Print Octal

PRO Procedure and Requirements Overview (Working Group) [of International Civil Aviation Organization]

PRO Proceed

PRO Programmable Remote Operation

PRO Protriptyline [Biochemistry]

PRO Provincial Research Organization [Canada]

PRO Public Record Office [UK]

PRO Public Relations Officer

.PRO Profile [File Name Extension] [also .pro]

PrO Propyl(ene) Oxide; Propylate; Propoxide

PrO Propoxy-

Pro Professional [also pro]

Pro (−)-Proline; Prolyl [Biochemistry]

Pro Protectorate

Pro Protein

pro prolinate

ProA Protein A [Biochemistry]

Prob Probability [also prob]

Prob Problem [also prob]

prob probable; probably

Probab Probability [also probab]

Prob Eng Inform Sci Probability in the Engineering and Informational Sciences [US]

Prob Theory Rel Fields Probability Theory and Related Fields [Published in Germany]

PROBE Pilot Radiation Observation Experiment

Probl Problem [also probl]

Probl Bioniki Problemy Bioniki [Ukrainian Journal on Problems of Bionics]

Probl Control Inf Theory Problems of Control and Information Theory [Journal published in Hungary]

Probl Inf Transm Problems of Information Transmission [Translation of: *Problemy Peredachi Informatsii (USSR)*; published in the US]

Probl Kibern Problemy Kibernetiki [Russian Journal on Problems of Cybernetics]

Probl Mashinostr Nadezhn Mash Problemy Mashinostroeniya i Nadezhnosti Mashin [Russian Journal]

Probl Materialov Fiz Met Problemy Materialovedenia i Fiziki Metallov [Russian Journal on Problems of Materials Science and Metal Physics]

Probl Metall Proizv Problemy Metallurgicheskogo Proizvodstva [Russian Journal on Problems of Metallurgical Engineering]

Probl Peredachi Inf Problemy Peredachi Informatsii [Russian Journal on Problems of Information Transmission]

Probl Phys Dielectr Problems of Physics of Dielectrics [Journal published in Poland]

Probl Planetol Problems of Planetology [Journal of the International Association of Planetology, Belgium]

Probl Prochn Problemy Prochnosti [Ukrainian Journal on Strength of Materials]

Probl Spets Elektrometall Problemy Spetsial vory Elektrometallurgii [Ukrainian Journal on Problems of Electrometallurgy]

Probl Tekh Kibern Robot Problemy na Tekhnicheskata Kibernetika i Robotikata [Bulgarian Journal on Problems of Engineering Cybernetics and Robotics]

PROC Programming Computer

Proc Procedure [also Proc, or proc]

Proc Process(ing); Processor [also proc]

Proc Procedure [also proc]

Proc Proceedings [also proc]

Proc Procurement [also proc]

Proc AAAS Proceedings of the American Academy of Arts and Sciences [US]

Proc ACS Proceedings of the American Chemical Society [US]

Proc AESF Annu Tech Conf Proceedings of the American Electroplaters and Surface Finishers Society Annual Technical Conference

Proc AES Meet Proceedings of the American Electrochemical Society Meeting [US]

Proc AIAA Proceedings of the American Institute of Aeronautics and Astronautics [US]

Proc Am Assoc Mental Deficiency Proceedings of the American Association for Mental Deficiency [US]

Proc Am Chem Soc Proceedings of the American Chemical Society [US]

Proc Am Chem Soc Proceedings of the American Chemical Society [US]

Proc Am Electrochem Soc Meet Proceedings of the American Electrochemical Society Meeting [US]

Proc Am Ethnol Soc Proceedings of the American Ethnological Society [US]

Proc Am Math Soc Proceedings of the American Mathematical Society [US]

Proc Am Phil Soc Proccedings of the American Philosophical Society [Philadelphia, Pennsylvania, US]

Proc AMS Proceedings of the American Mathematical Society [US]

Proc Am Soc Biol Proceedings of the American Society of Biology [US]

Proc Am Soc Civil Eng Proceedings of the American Society of Civil Engineers [US]

Proc Am Soc Compos Proceedings of the American Society for Composites [US]

Proc Annu Meet Am Assoc Cancer Res Proceedings of the Annual Meeting of the American Association for Cancer Research [US]

Proc Annu Meet Am Soc Clin Oncol Proceedings of the Annual Meeting of the American Society for Clinical Oncology [US]

Proc Annu Meet EMSA Proceedings of the Annual Meeting of the Electron Microscopy Society of America [US]

Proc API Proceedings of the American Petroleum Institute [US]

Proc APS Proccedings of the American Philosophical Society [Philadelphia, Pennsylvania, US]

Proc ASB Proceedings of the American Society of Biology [US]

Proc ASCE Proceedings of the American Society of Civil Engineers [US]

ProCAST Professional Casting Simulation System

Proc Astron Soc Aust Proceedings of the Astronomical Society of Australia

Proc Australas Inst Min Metall Proceedings of the Australasian Institute of Mining and Metallurgy

Proc Aust Soc Sugar Cane Technol Proceedings of the Australian Society of Sugar Cane Technologists

Proc Br Ceram Soc Proceedings of the British Ceramic Society

Proc Camb Phil Soc Proceedings of the Cambridge Philosophical Society [of Cambridge University, UK]

Proc CDI Conf Proceedings of the Corrosion-Deformation Interactions Conference

Proc Chem Soc Proceedings of the Chemical Society [UK]

Proc Conf Appl Cryst Proceedings of the Conference on Applied Crystallography

Proc Cosm-Ray Res Lab Nagoya Univ Proceedings of the Cosmic-Ray Research Laboratory of Nagoya University [Japan]

Proc CSEE Proceedings of the Chinese Society of Electrical and Electronics Engineers

procd processed [also proc'd]

Proc DIMETA Proceedings of the International Conference on Diffusion in Metals and Alloys

Proc Edinb Math Soc Proceedings of the Edinburgh Mathematical Society [UK]

Proc Egypt Acad Sci Proceedings of the Egyptian Academy of Sciences [Egypt]

Proc Electrochem Soc Proceedings of the Electrochemical Society [US]

ProcE Process Engineer [also Proc Eng, or ProcEng]

Proc EMRS Proceedings of the European Materials Research Society [also Proc Eur Mater Res Soc]

Proc EMSA Mtg Proceedings of the Electron Microscopy Society of America Meeting [US]

Proc Eng Process Engineer(ing)

PrO$_2$-CeO$_2$ Praseodymium Oxide-Cerium Oxide (System)

Process Autom Process Automation [Journal published in Germany]

Process Biochem Process Biochemistry [Journal published in the US]

Proc ESSDERC Proceedings of the European Solid-State Device Research Conference

Process Eng (Australia) Process Engineering (Australia) [Journal published in Australia]

Process Eng News Process Engineering News [Australian Publication]

Process Eng (UK) Process Engineering (UK) [Journal published in the UK]

Process Ind Canada Process Industries Canada [Canadian Publication]

Process Safety Prog Process Safety Progress [Quarterly published by the American Institute of Chemical Engineers]

Process Technol Proc Process Technology Proceedings [US]

Proc EUREM Proceedings of the European Conference on Electron Microscopy

Proc EUROMAT Proceedings of the European Conference on Advanced Materials and Processes

Proc Eur Mater Res Soc Proceedings of the European Materials Research Society [also Proc EMRS]

Proc Eur Pept Symp Proceedings of the European Peptide Symposium

Proc Fac Eng Tokai Univ Proceedings of the Faculty of Engineering, Tokai University [Japan]

Proc Fac Eng Tokai Univ (Engl Ed) Proceedings of the Faculty of Engineering, Tokai University (English Edition) [Japan]

Proc Fall Mtg Inst Met Proceedings of the Fall Meeting of the Institute of Metals [Japan]

Proc Geol Assoc Proceedings of the Geological Association [London, UK]

Proc HNS Proceedings of the High-Nitrogen Steels Conference

Proc IAHR Int Symp Ice Proceedings of the International Association for Hydraulic Research, International Symposium on Ice

Proc ICM Proceedings of the International Conference on Mechanical Behaviour of Materials

Proc IECEC Proceedings of the Intersociety Energy Conversion Engineering Conference [of Institute of Electrical and Electronics Engineers, US]

Proc IEE Proceedings of the Institution of Electrical Engineers [UK]

Proc IEEE Proceedings of the Institute of Electrical and Electronics Engineers [US]

Proc IMAM Proceedings of the International Meeting on Advanced Materials [of Materials Research Society, US]

Proc IMC Proceedings of the International Microelectronic Conference

Proc IMMM Proceedings of the International Seminar on Microstructures and Mechanical Properties of New Engineering Materials

Proc Indian Acad Sci Proceedings of the Indian Academy of Sciences [India]

Proc Indian Acad Sci, Chem Sci Proceedings of the Indian Academy of Sciences, Chemical Sciences

Proc Indian Acad Sci, Earth Planet Sci Proceedings of the Indian Academy of Sciences, Earth and Planetary Sciences

Proc Indian Acad Sci, Math Sci Proceedings of the Indian Academy of Sciences, Mathematical Sciences

Proc Indian Acad Sci, Phys Proceedings of the Indian Academy of Sciences, Physical Sciences

Proc Indian Natl Sci Acad A Proceedings of the Indian National Science Academy, Part A

Proc Indian Natl Sci Acad B Proceedings of the Indian National Science Academy, Part B

Proc Ind Waste Conf Proceedings of the Industrial Waste Conference [US]

Proc Infrared Soc Jpn Proceedings of the Infrared Society of Japan [Japan]

Proc Inst Civ Eng 1 Proceedings of the Institution of Civil Engineers 1 [UK]

Proc Inst Civ Eng 2 Proceedings of the Institution of Civil Engineers 2 [UK]

Proc Inst Electr Eng Proceedings of the Institution of Electrical Engineers [UK]

Proc Inst Electrost Jpn Proceedings of the Institute of Electrostatics Japan

Proc Inst Mech Eng Proceedings of the Institution of Mechanical Engineers [UK]

Proc Inst Mech Eng A, Power Process Eng Proceedings of the Institution of Mechanical Engineers A, Power and Process Engineering [UK]

Proc Inst Mech Eng B, J Eng Manuf Proceedings of the Institution of Mechanical Engineers B, Journal of Engineering Manufacture [UK]

Proc Inst Mech Eng B, Manage Eng Manuf Proceedings of the Institution of Mechanical Engineers B, Management and Engineering Manufacture [UK]

Proc Inst Mech Eng C, J Mech Eng Sci Proceedings of the Institution of Mechanical Engineers C, Journal of Mechanical Engineering Science [UK]

Proc Inst Mech Eng D, J Automob Eng Proceedings of the Institution of Mechanical Engineers D, Journal of Automobile Engineering [UK]

Proc Inst Mech Eng D, Transp Eng Proceedings of the Institution of Mechanical Engineers D, Transport Engineering [UK]

Proc Inst Mech Eng E, J Process Mech Eng Proceedings of the Institution of Mechanical Engineers E, Journal of Process Mechanical Engineering [UK]

Proc Inst Mech Eng F, J Rail Rapid Transit Proceedings of the Institution of Mechanical Engineers E, Journal of Rail and Rapid Transit [UK]

Proc Inst Mech Eng G, J Aerosp Eng Proceedings of the Institution of Mechanical Engineers G, Journal of Aerospace Engineering [UK]

Proc Inst Mech Eng H, J Eng Med Proceedings of the Institution of Mechanical Engineers H, Journal of Engineering in Medicine [UK]

Proc Inst Railw Signal Eng Proceedings of the Institution of Railway Signal Engineers [UK]

Proc Int Conf CVD Proceedings of the International Conference on Chemical-Vapor Deposition

Proc Int Conf HVEM Proceedings of the International Conference on High-Voltage Electron Microscopy

Proc Int Conf Mech Med Biol Proceedings of the International Conference on Mechanics in Medicine and Biology [US]

Proc Int Conf Mag Proceedings of the International Conference on Magnetism

Proc Int Conf Mag Rec Media Proceedings of the International Conference on Recording Media

Proc Int Conf OMAE Proceedings of the International Conference on Offshore Mechanics and Arctic Engineering

Proc Int Conf Sci Technol New Glass Proceedings of the International Conference on the Science and Technology of New Glass

Proc Int Conf Stainl Steels Proceedings of the International Conference on Stainless Steels

Proc Int Conf Trib Proceedings of the International Conference on Tribology

Proc Int Congr Protozool Proceedings of the International Congress of Protozoology

Proc Int Joint Conf Patt Recog Proceedings of the International Joint Conference on Pattern Recognition [Japan]

Proc Int Meet Adv Mater Proceedings of the International Meeting on Advanced Materials [of Materials Research Society, US]

Proc Int Symp Appl Ferroelectr Proceedings of the International Symposium on Applications of Ferroelectrics [of Institute of Electrical and Electronics Engineers, US]

Proc Int Symp Met Mater Sci Proceedings of the International Symposium on Metallurgy and Materials Science

Proc Int Symp Plasma Chem Proceedings of the International Symposium on Plasma Chemistry

Proc Int Trib Conf Proceedings of the International Tribology Conference

Proc ISA Proceedings of the Instrument Society of America [also Proc Instr Soc Am]

Proc ISAF Proceedings of the International Symposium on Applications of Ferroelectrics [of Institute of Electrical and Electronics Engineers, US]

Proc ISS Proceedings of the Institution of Electrical Engineers [UK]

Proc ISS Proceedings of the Iron and Steel Society [of American Iron and Steel Institute, US]

Proc ISPC Proceedings of the International Symposium on Plasma Chemistry

Proc ISTFA Proceedings of the International Symposium for Testing and Failure Analysis

Proc JANNAF Proceedings of the Joint Army-Navy-NASA-Air Force (Committee) [US]

Proc Joint FEMS/ICF Proceedings of the (International) Joint Conference on Fracture of Engineering Materials and Structures/International Congress on Fracture

Proc Joint Int Waste Manage Conf Proceedings of the International Waste Management Conference

Proc Jpn Acad Poceedings of the Japanese Academy

Proc Jpn Acad A, Math Sci Proceedings of the Japanese Academy, Series A, Mathematical Sciences [Japan]

Proc Jpn Acad B, Phys Biol Sci Proceedings of the Japanese Academy, Series B, Physical and Biological Sciences [Japan]

Proc Jpn Cer Soc. Proceedings of the Japanese Ceramic Society

Proc Jpn Congr Mater Res Proceedings of the Japan Congress on Materials Research [Japan]

Proc Jpn Soc Civ Eng Proceedings of the Japan Society of Civil Engineers [Japan]

Proc KNAW Proceedings of the Koninklijke Nederlandse Akademie van Wetenschappen [Proceedings of the Royal Netherlands Academy of Sciences, Amsterdam] [also Proc Kon Ned Akad Wet]

Proc K Ned Akad Wet A Proceedings of the Koninklijke Nederlandse Akademie van Wetenschappen, Series A [Proceedings of the Royal Netherlands Academy of Sciences, Series A]

Proc K Ned Akad Wet B Proceedings of the Koninklijke Nederlandse Akademie van Wetenschappen, Series B [Proceedings of the Royal Netherlands Academy of Sciences, Series B]

Proc Korean Inst Electr Eng Proceedings of the Korean Institute of Electrical Engineers [South Korea]

Proc LAMP Proceedings of the Laser Advanced Materials Processing Conference

Proc LMS Proceedings of the London Mathematical Society [UK] [also Proc Lond Math Soc]

Proc Mil Oper Res Symp Proceedings of the Military Operations Research Symposium [of the Military Operations Research Society, US]

PROC LIB Procedure Library [also PROCLIB]

Proc MRS Proceedings of the Materials Research Society [US]

Proc MRS Int Meet Adv Mater Proceedings of the Materials Research Society's International Meeting on Advanced Materials

Proc NAS Proceedings of the National Academy of Sciences [also Proc Natl Acad Sci, US]

Proc Natl Acad Sci India A Proceedings of the National Academy of Sciences of India, Section A

Proc Natl Acad Sci India B Proceedings of the National Academy of Sciences of India, Section B

Proc Natl Comp Conf Proceedings of the National Computer Conference [Published by the American Federation of Information Processing Societies]

Proc Natl Electron Conf Proceedings of the National Electronics Conference [US]

Proc NATO Adv Res Worksh Proceedings of the NATO Advanced Research Workshop

Proc NEA Proceedings of the National Education Association [US]

Proc NEC Proceedings of the National Electronics Conference [US]

Proc Nutr Soc Proceedings of the Nutrition Society

Proc Ocean Drill Program Proceedings of the Ocean Drilling Program [US]

Proc Ocean Drill Program, Initial Rep Proceedings of the Ocean Drilling Program, Initial Reports [US]

Proc Ocean Drill Program, Sci Results Proceedings of the Ocean Drilling Program, Scientific Results [US]

PROCOM Procedures Committee [of Institute of Electrical and Electronics Engineers, US]

PROCOMP Process Computer

PROCOMP Program Compiler

PROCON Protocol Converter

Proc Pak Acad Sci Proceedings of the Pakistan Academy of Sciences

Proc Phys Soc Proceedings of the Physical Society [London, UK]

Proc Radio Club Am Proceedings of the Radio Club of America [US]

Proc Rare Earth Res Conf Proceedings of the Rare Earth Research Conference [University of Kentucky, Lexington, US]

Proc Res Inst Atmos Nagoya Univ Proceedings of the Research Institute of Atmospherics, Nagoya University [Japan]

Proc R Anthrop Inst Proceedings of the Royal Anthropological Institute [UK]

Proc R Geogr Soc Proceedings of the Royal Geographical Society [UK]

Proc R Inst GB Proceedings of the Royal Institute of Great Britain [UK]

Proc R Inst UK Proceedings of the Royal Institute of the United Kingdom

Proc R Ir Acad A Proceedings of the Royal Irish Academy, Section A

Proc R Ir Acad B Proceedings of the Royal Irish Academy, Section B

Proc R Soc Proceedings of the Royal Society [UK]

Proc R Soc Edinb Proceedings of the Royal Society of Edinburgh [UK]

Proc R Soc Edinb A Proceedings of the Royal Society of Edinburgh, Section A [UK]

Proc R Soc Edinb B Proceedings of the Royal Society of Edinburgh, Section B [UK]

Proc R Soc Lond Proceedings of the Royal Society of London [UK]

Proc R Soc Lond A Proceedings of the Royal Society of London, Series A [UK]

Proc R Soc Lond A, Math Phys Sci Proceedings of the Royal Society of London, Series A: Mathematical and Physical Sciences [UK]

Proc R Soc Lond B Proceedings of the Royal Society of London, Series B [UK]

Proc R Soc Med Proceeding of the Royal Society of Medicine [UK]

Proc R Soc Vic Proceedings of the Royal Society of Victoria [Australia]

Proc R Soc NSW Proceedings of the Royal Society of New South Wales [Australia]

PROCR Processor [also Procr, or procr]

PROCSAN Probabilistic Composite Structural Analysis (Computer Code)

Proc SEBM Proceedings of the Society for Experimental Biological Medicine [US]

Proc SID Proceedings of the Society for Information Display [US]

Proc Soc Exp Biol Med Proceedings of the Society for Experimental Biology and Medicine [US]

Proc South Wales Inst Inst Eng Proceedings of the South Wales Institute of Engineers [Australia]

Proc SPIE Proceedings of the SPIE [of Society of Photo-Optical Instrumentation Engineers, US]

Proc SPIE–Int Soc Opt Eng Proceedings of the SPIE–The International Society for Optical Engineering [US]

Proc Symp Pure Math Proceedings of the Symposium on Pure Mathematics

Proc TAPPI Coating Conf Proceedings of the Technical Association of the Pulp and Paper Industry's Coating Conference

Proc TAPPI Papermakers Conf Proceedings of the Technical Association of the Pulp and Paper Industry's Papermarkers Conference

Proc TAPPI Pulp Conf Proceedings of the Technical Association of the Pulp and Paper Industry's Pulping Conference

Proc USNI Proceedings of the United States Naval Institute

Proc USNM Proceedings of the United States National Museum

PROD Paramagnetic Resonance by Optical Detection

PROD Preferred Orientation and Displacement [Crystallography]

Prod Produce(r); Product(ion) [also prod]

prod produce(d)

PRODAC Production Advisers Consortium [UK]

PRODAC Programmed Digital Automatic Control

ProdE Production Engineer [also Prod Eng, or ProdEng]

Prod Eng Production Engineer(ing)

Prod Eng (Cleveland) Production Engineering (Cleveland) [Journal published in the US]

Prod Eng (London) Production Engineer (London) [Journal published in the UK]

Prod Eng (New York) Product Engineering (New York) [Journal published in the US]

Prod Finish (Cincinnati) Product Finishing (Cincinnati) [Journal published in the US]

Prod Finish (London) Product Finishing (London) [Journal published in the UK]

Prod Inv Manage J Production and Inventory Management Journal [Published by the American Production and Inventory Control Society, US]

Prod Inv Manage Rev Production Inventory Management Reviews [Published by the American Production and Inventory Control Society, US]

Prod J Production Journal [UK]

Prodn Production [also prodn, Prod'n, or prod'n]

Prodr Producer [also prodr]

Prod Tech Production Technologist; Production Technology

Productronica International Trade Fair for Electronics Production [Germany]

Pro/E Pro/Engineers (Solid Modeling Package)

Prof Profession(al) [also prof]

Prof Professor

PROFAC Propulsive Fluid Accumulator

ProfBTM Professional Business and Technical Management [UK]

Prof Comput (Australia) Professional Computing (Australia) [Journal published in Australia]

Prof Comput (USA) Professional Computing (USA) [Journal published in the US]

Prof Dr-Ing *(Professor Doktor-Ingenieur)* – Professor with Doctor of Engineering (Sciences) Degree

Prof Dr phil *(Professor Doctor philosopiae)* – Professor with Doctor of Philosophy Degree

Prof Dr rer nat *(Professor Doctor rerum naturalium)* – Professor with Doctor of Natural Sciences Degree

Prof Dr sc agr *(Professor Doctor scientiarum agrariarum)* – Professor with Doctor of Agricultural Sciences Degree

Prof Dr sc math *(Professor Doctor scientiarum mathematicarum)* – Professor with Doctor of Mathematical Sciences Degree

Prof Dr sc nat *(Professor Doctor scientiarum naturalium)* – Professor with Doctor of Natural Sciences Degree

Prof Dr sc techn *(Professor Doctor scientiarum technicarum)* – Professor with Doctor of Technical Sciences Degree

ProfE Professional Engineer [also ProfEng]

Prof Electron Mag Professional Electronics Magazine [Published by the International Society of Certified Electronics Technicians, US]

Prof Emer Professor Emeritus

Prof Energy Manager Professional Energy Manager [of Association of Professional Energy Managers, US]

ProfEng Professional Engineer [also Prof Eng, or ProfE]

Prof Eng Professional Engineering [Journal published in the UK]

Prof Geol Professional Geologist

Prof Geol Professional Geologist [Journal of the American Institute of Professional Geologists]

Prof Geop Professional Geophysicist

PROFI Project Financing

PROFIBUS Process Field Bus [also Profibus]

PROFILE Programmed Functional Indices for Laboratory Evaluation

Profile Mag Profile Magazine [Norway]

PROFIT Program for Financed Insurance Techniques

PROFIT Programmed Reviewing, Ordering, and Forecasting Inventory Technique

Prof Manage Professional Management [Journal published in the UK]

Prof Met Professional Meteorologist; Professional Meteorology

Prof N Professional Nurse

Prof Ped Professional Pedicurist

Prof Pharm Corp Professional Pharmaceutical Corporation

Prof Printer Professional Printer [Journal published in the UK]

PROFS Professional Office System

PROFS Program for Regional Observing and Forecasting Services [of National Oceanic and Atmospheric Administration, Forecast Systems Laboratories, US]

Profs Professors

Prof Safety Professional Safety [Journal of the American of Society of Safety Engineers, US]

Prof Stat Professional Statistician; Professional Statistics

Prof Stat Professional Statistician [Journal of the Institute of Statisticians, UK]

ProG Protein G [Biochemistry]

Prog Prognosis; Prognostic [also prog]

Prog Program(mable); Programmer; Programming [also prog]

Prog Progress(ion) [also prog]

Prog Aerosp Sci Progress in Aerospace Sciences [Journal published in the UK]

Prog Anal Spectrosc Progress in Analytical Spectroscopy [Journal published in the UK]

Prog Arch Progressive Architecture [US Journal]

Prog Biophys Mol Biol Progress in Biophysics and Molecular Biology [Journal published in the UK]

Prog Chem Fats Other Lipids Progress in the Chemistry of Fats and Other Lipids

Prog Clin Biol Res Progress in Clinical Biological Research

Prog Cntrl Programmable Control

Prog Coll Polym Sci Progress in Colloid and Polymer Science [Journal published in Germany]

Prog Cryst Growth Charact Progress in Crystal Growth and Characterization [Journal published in the UK]

PROGDEV Program Device [also PROG DEV, or Prog Dev]

Prog Energy Combust Sci Progress in Energy and Combustion Science [Published in the UK]

PROGINFO Programming Information [also PROG INFO, or Prog Info]

Prog Mater Sci Progress in Materials Science [Journal published in the UK]

Prog Metal Phys Progress in Metal Physics (Journal)

Progn Materialov Prim EVM Prognozirovanie v Materialovdedenii s Primeneniem [Russian Journal on the Progress in Materials Science and Engineering; published in Moscow]

Prog NMR Spectrosc Progress in Nuclear Magnetic Resonance Spectroscopy [Journal published in the UK] [also Prog Nucl Magn Reson Spectrosc]

Prog Nucl Energy Progress in Nuclear Energy [Journal published in the UK]

Prog Nucl Magn Reson Spectrosc Progress in Nuclear Magnetic Resonance Spectroscopy [Journal published in the UK] [also Prog NMR Spectrosc]

Prog Oceanogr Progress in Oceanography [Journal published in the UK]

Prog Org Coat Progress in Organic Coatings [Journal published in Switzerland]

Prog Paper Recycl Progress in Paper Recycling (Journal)

Prog Part Nucl Phys Progress in Particle and Nuclear Physics [Journal published in the UK]

Prog Photovolt Res Appl Progress in Photovoltaics Research and Application (Journal)

Prog Phys Org Chem Progress in Physical Organic Chemistry [Journal]

Prog Polym Sci Progress in Polymer Science [Journal published in the UK]

Prog Powder Metall Progress in Powder Metallurgy [Publication of the International Powder Metallurgy Institute]

Prog Quantum Electron Progress in Quantum Electronics [Journal published in the UK]

Progr Programmer; Programming [also progr]

Progr Progress [also progr]

Program Programming [also program]

Program Comput Softw Programming and Computer Software [Translation of: *Programmirovanie (USSR)*; published in the US]

Program J Programmer's Journal [US]

PROGRES Programmable Resistor

PROGRESS Point-Resolved Rotating Gradient Surface Coil Spectroscopy

Prog Rubber Plast Technol Progress in Rubber and Plastics Technology [Published by the Rubber and Plastics Research Association, UK]

Prog Solid State Chem Progress in Solid-State Chemistry [Journal published in the UK]

Prog Space Res Progress of Space Research [Published by the United Nations Committee on the Peaceful Uses of Outer Space]

Prog Surf Membr Sci Progress in Surface and Membrane Science (Journal)

Prog Surf Sci Progress in Surface Science [Journal published in Canada]

Prog Tech Progres Technique [French Journal on Progress in Engineering and Technology]

Prog Theor Phys Progress in Theoretical Physics [Journal of the Physical Society of Japan]

Prog Theor Phys, Suppl Progress of Theoretical Physics, Supplement [Journal of the Physical Society of Japan]

PrOH Propanol (or Propyl Alcohol)

1-PrOH 1-Propanol

2-PrOH 2-Propanol

PROHC Peroxyl Radical Oxygen Hole Center

Proj Project [also proj]

Proj Projection; Projector [also proj]

Proj Projectile [also proj]

PROJACS Project Analysis and Control System

Project XL Project Excellence and Leadership [US National Environmental Protection/Public Health Project]

Project XLC Project Excellence and Leadership for Communities [US Community-Based Environmental Project]

Projs Projectiles [also projs]

PROLA Physical Review On-Line Archive [of American Physical Society, US]

PROLAN Processed Language

PROLOG Programming in Logic [also Prolog]

PROM Pockels Readout Optical Memory

PROM Pockels Readout Optical Modulator

PROM Programmable Read-Only Memory

PROM Program, Resources and Objectives Management (System) [of US Air Force]

Prom Promontory

PROMAT Profit through Materials Technology

Prom Energ Promyshlennaya Energetika [Russian Journal on Energy Science and Engineering]

PROMIS Problem-Oriented Medical Information System

PROMIS Process Management Information System

PROMIS Project Monitoring Information System

PROMIS Project-Oriented Management Information System

Promo Promotion [also promo]

PROMPT Predicast's Overview of Markets and Terminology [US]

PROMPT Production Reviewing, Organizing and Monitoring of Performance Techniques

PROMPT Program Monitoring and Planning Technique

PROMPT Program Reporting, Organization and Management Planning Technique

PROMS Projector On-Line Monitoring System

Prom Sint Kauch Shin Rez-tekhn Iz Promyshlennost Sinteticheskogo Kauchuka Shin i Rezinotekhnicheskikh Izdelli [Russian Journal on Synthetic Resins, Plastics and Rubber]

PROMT Precision Optimized Measurement Time

PROMUX Programmable Multiplexer [also ProMux]

Pron Pronunciation [also pron]

PRONTO Program for Numerical Tools Operation

PRONTO Programmable Network Telecommunications Operating System

Pro(OH) Hydroxyproline [Biochemistry]

PROP Performance Review for Operating Programs

PROP Planetary Rocket Ocean Platform

PROP Profit Rating of Projects

Prop Propellant; Propulsion [also prop]

Prop Propeller

Prop Property [also prop]

Prop Proportion(ing) [also prop]

Prop Proposition [also prop]

Prop Proprietor; Proprietary [also prop]

prop proper

Prop Data Update Property Data Update [Journal published in the US]

Propellants Explos Pyrotech Propellants, Explosives and Pyrotechnics [Journal published in the US]

PROPHOS 1,2-Bis(diphenylphosphino)propane [also Prophos]

PROPHOS 1-Methyl-1,2-Ethanediylbis(diphenylphosphine) [also Prophos]

Propineb Zinc-1,2-Propylene-bis-Dithiocarbamate [also propineb]

Proplnt Propellant [also Proplt, proplt, or proplnt]

Proplnts Propellants [also proplnts]

Propn Propulsion [also propn]

Propoxur o-Isopropoxyphenyl-N-Nethyl Carbamate [also propodur]

Propr Proprietor [also propr]

Propr Proprietary [also propr]

ProPr^i Praseodymium Isopropoxide

Props Properties [also props]

Propty Property [also propty]

Propy Proprietary [also propy]

PROS Professional Reactor Operator Society [US]

Pros Prospectus [also pros]

PROSAT Promotional Satellite Project [of European Space Agency]

PROSEA Plant Resources of Southeast Asia

PROSPEC Public Record Office Specification [UK]

PROSPER Profit Simulation Planning and Evaluation of Risk

PROSPRO Process Systems Program

Pr Osrodka Badaw-Rosw Elektron Prozniowej Prace Osrodka Badawczo-Roswojowego Elektroniki Prozniowej [Proceedings on Electronics and Related Subjects; published in Warsaw, Poland]

PROSUS Program on Submicrometer Structures [of Cornell University, Ithaca, New York, US]

Prot Protection [also prot]

Prot Protector(ate) [also prot]

Prot Protein [also prot]

Prot Chem Protein Chemist(ry)

ProTech Prospective Technology Communication System

PROTECT Probabilities Recall Optimizing the Employment of Calibration Time

Prot Expr Purif Protein Expression and Purification [Journal published in the US] [also Protein Expr Purif]

PROTEM/CC Programa Nactional de Projetos Cooperativos em Informática [National Program for Cooperative Projects on Information Technology; of Instituto Nacional de Pesquisas Espaciais, Brazil]

pro tem *(pro tempore)* – temporarily; for the time being

PROTEUS Profile Telemetry of Upper Ocean Currents

Prot Met Protection of Metals [Translation of: *Zashchita Metallov (USSR)*; published in the US]

Prov Provision(al) [also prov]

Prov Province; Provincial [also prov]

Prov Provost [also prov]

PROVER Procurement for Minimum Total Cost Through Value Engineering and Reliability

PROVER Procurement–Value–Economy–Reliability

Provet Projektgruppe Verfassungsverträgliche Technikgestaltung [Joint Project Team on Constitutionally Compatible Technology Design; of Bundesministerium für Forschung und Technologie and Gesellschaft für Mathematik und Datenverarbeitung, Germany]

Provit Provitamin [Medicine]

PROWAY Process Data Highway

PROWL Procedure Work Log System

PROWORD Procedure Word [also Proword, or proword]

Prox Proximity [also prox]

prox *(proximo mensa)* – in the next month

PROXI Projection by Reflection Optics of Xerographic Images

Proxy ARP Proxy Address Resolution Protocol (Technique) [also proxy ARP] [Internet]

PROXYL Tetramethylpyrrolidinyloxy; Tetramethylpyrrolidine-N-Oxyl [also Proxyl, or proxyl]

PRP Peer Review Panel

PRP Personnel Reliability Program

PRP Petroleum, Rosin and Paraffin Wax [Explosives]

PRP Platelet-Rich Plasma [Medicine]

PRP Polymer Reverse Phase

PRP Potentially Responsible Party

PRP Precision Rolled Products, Inc. [US]

PRP Pseudorandom Pulse

PRP Pulse Repetition Period

Prp Preparation [also prp]

PRPF Planar Radial Peaking Factor

PRPM Primary Power Monitor

Prpn Propane [also prpn]

Prpnt Propellant [also prpnt]

PRPP 5-Phosphorylribose 1-Pyrophosphate; Phosphoribosyl Pyrophosphate [Biochemistry]

PRPQ Programming Request Price Quotation

PrPq *(Próreitoria de Pesquisas)* – Portuguese for "Office of the Deputy Vice-Chancellor of Science"

PrPq-UFMG Próreitoria de Pesquisas da Universidade Federal de Minas Gerais [Office of the Deputy Vice-Chancellor of Science of the Federal University of Minas Gerais, Belo Horizonte, Brazil]

Pr Przem Inst Elektron Prace Przemyslowego Instytutu Elektroniki [Proceedings on Electronics; published in Warsaw, Poland]

Pr Przem Inst Telekomin Prace Przemyslowego Instytutu Telekomunikacji [Proceedings on Telecommunications; published in Warsaw, Poland]

PRPS Programming Requirements Process Specification

PRR Parts Replacement Request

PRR Pawling Research Reactor [New York, US]

PRR Philippine Research Reactor [Quezon City, The Philippines]

PRR Preliminary Requirements Review

PRR Program Requirements Review

PRR Puerto Rico Reactor [of Puerto Rico Nuclear Center, University of Puerto Rico at Rio Piedras–Mayaguez]

PRR Pulse-Repetition Rate

PRR-1 Philippine Research Reactor 1 [Quezon City, The Philippines]

PRRD Proprietary Rights to Registration Data

PRRM Pulse-Repetition Rate Modulation

PRS Pacific Rocket Society [US]

PRS Paint Research Station [UK]

PRS Partial Response Signalling

PRS Pattern Recognition Society [US]

PRS Pattern Recognition System

PRS Payload Retention Subsystem

PRS Personnel Rescue Service

PRS Personnel Rescue System

PRS Plasma Reactor System

PRS Polymer Reactor System

PRS Portable Rework System

PRS Power Reactant Subsystem

PRS Power Reactant System

PRS Power Relay Satellite

PRS P-Polarized Reflectance Scattering

PRS P-Polarized Reflectance Spectroscopy

PRS Precision Ranging System

PRS Predetermined Revolutions per Second

PRS Press [also Prs, or prs]

PRS Primary Recovery Site

PRS Primary Rescue Site

PRS Printer Resource (File)

PRS Process Radiation Sampler

PRS Provisioning Requirements Statement

PRS Pulsed Radiation Source

.PRS Printer Resource (Files) [WordPerfect File Name Extension]

PrSCNO Praseodymium Strontium Copper Niobium Oxide (Superconductor)

PrSCO Praseodymium Strontium Copper Oxide (Superconductor)

PRSD Power Reactant Storage and Distribution

PRSD Power Reactant Supply and Distribution

PRSDS Power Reactant Storage and Distribution System

PrSMO Praseodymium Strontium Manganese Oxide (Superconductor)

PRSS Problem Report Squawk Sheet

prst persist

PRSTAB Physical Review Special Topics–Accelerators and Beams [of American Physical Society, US] [also PRST-AB]

PRT Payload Replanning Team

PRT Pattern Recognition Technique

PRT Personal Rapid Transit

PRT Petroleum Revenue Tax [UK]

PRT Phosphoribosyl Transferase [Biochemistry]

PRT Platinum Resistance Thermometer

PRT Polychromatic Reflection Topography

PRT Portable Radio Terminal

PRT Portable Remote Terminal

PRT Precision Radiation Thermometer
PRT Primary Ranging Test
PRT Principal Research Team
PRT Print(er) [also Prt, or prt]
PRT Production Run Tape
PRT Program Reference Table
PRT Prompt Relief Trip
PRT Proof Test(ing)
PRT Public Rapid Transit
PRT Public Rapid Transport
PRT Purto Rico Telephone Company
PRT Pulse Recurrence Time
PRT Pulse Repetition Time
Prt Print(er) [also prt]
$\varphi(\mathbf{r,t})$ electrostatic potential [Symbol]
$\Psi(\mathbf{r,t})$ Velocity Potential [Symbol]
PRTC Ports Canada
PRTDC Puerto Rico Tourist Development Corporation
Prtg Printing [also prtg]
Pr(TFAC)$_3$ Praseodymium Trifluoroacetylacetonate [also Pr(tfac)$_3$]
Pr(TFC)$_3$ Praseodymium Tris[(trifluoromethylhydroxymethylene)camphorate] [also Pr(tfc)$_3$]
Pr(THD)$_3$ Praseodymium Tris(2,2,6,6-Tetramethyl-3,5-Heptanedionate) [also Pr(thd)$_3$]
PRTLS Powered Return to Launch Site
Pr(TMHD)$_3$ Praseodymium Tris(2,2,6,6-Tetramethyl-3,5-Heptanedionate) [also Pr(tmhd)$_3$]
Prtlnd Cem Portland Cement
PRTM Prague Ring Tunneling Method
PRTM Printing Response-Time Monitor
PRTOT Prototype Real-Time Optical Tracker
PRTR Plutonium Recycle (or Recycling) Test Reactor [US]
PRTR Printer [also Prtr, or prtr]
PRTRC Printer Controller
PRTS Pseudorandom Ternary Sequence
PRTSC Print Screen [also PrtSc]
PRTSCRN Print Screen [also PrtScrn]
Prty Priority [also prty]
PRU Packet Radio Unit
PRU Physical Record Unit
PRU Programs Research Unit
PRUF Program of Research by Universities in Forestry
PRUF Program Request Under Format
Prus Prussia(n) [also Pruss]
PRV Peak Reverse Voltage
PRV Pressure Reducing Valve
PRV Pressure Reduction Valve
PRV Pressure Regulating Valve
PRV Pressure Relief Valve
PRVS Penetration Room Ventilation System
PRVT Production Reliability Verification Testing
PRW Paired Wire

PRW Paper-Tape Rewind
PRW Paper-Tape Write
PRW Percent Rated Wattage
PRWRA Puerto Rico Water Resources Authority [US]
PRX Proximity Switch
n-PrX n-Propyl Halide
PRXN Surface-Aligned Photoreaction
PRY People's Republic of Yemen
PRY Priority [also Pry, or pry]
PR Yemen People's Republic of Yemen
Pr:YIG Praseodymium-Doped Yttrium Iron Garnet
PRZ Partially Reacted Zone [Welding]
PRZ Partially Recrystallized Zone [Metallurgy]
Pr:ZBLAN Praseodymium-Doped Zirconium Tetrafluoride–Barium Difluoride–LanthanumTrifluoride– Aluminum Trifluoride–Sodium Fluoride (Laser) [also Pr^{3+}:ZBLAN]
Przegl Gorn Przeglad Gorniczny [Polish Publication]
Przegl Odlew Przeglad Odlewnictwa [Polish Engineering Journal]
Przegl Papier Przeglad Papierniczy [Polish Journal on Papermaking]
Przegl Spawal Przeglad Spawalnictwa [Polish Journal]
Przegl Wlok Przeglad Wlokienniczy [Polish Journal]
Przem Chem Przemsyl Chemiczny [Polish Journal on Chemistry]
Prz Elektrotech Przeglad Elektrotechniczny [Polish Journal on Electrical Engineering]
Prz Stat Przeglad Statystyczny [Polish Journal on Statistics]
Prz Telekomun Przeglad Telekomunikacynjy [Polish Journal on Telecommunications]
PS Packet Switching
PS Paint Spray(ing); Paint Sprayer
PS Pair Spectrometer; Pair Spectroscope
PS Particle Size
PS Paleontological Society [US]
PS Palm Society [now International Palm Society, US]
PS Parachute Subsystem
PS Parallel-to-Serial [also P-S]
PS Parity Switch
PS Partially Soluble [also ps]
PS Partially Stabilized; Partial Stabilization
PS Particle Size
PS Particle System
PS Particle Symmetry
PS Particulate System
PS Part Surface [Numerical Control]
PS Part-Through Surface-Crack (Specimen) [Mechanics]
PS Paschen Series [Spectroscopy]
PS Passenger Steamer
PS Passive Satellite
PS Payload Specialist [also P/S]
PS Payload Station
PS Payload Support
PS Peak Shift
PS Pear Shape

PS Pearson Symbol [Crystallography]
PS Percussion Shrapnel
PS Periodical System (of Elements)
PS Permanent Secretary
PS Permanent Signal
PS Permanent Stress
PS Per Second [also ps]
PS Personal System
PS Petroleum Society [of Canadian Institute of Mining and Metallurgy]
PS *(Pferdestärke)* – German for "Horsepower"
PS Pfund Series [Spectroscopy]
PS Pharmaceutical Society [UK]
PS Phase Separation
PS Phase Shift(er)
PS Phasing System
PS Phenylsilane
PS Phonon Scattering [Solid-State Physics]
PS Phonon Spectroscopy [Solid-State Physics]
PS Phosphatidylserine [Biochemistry]
PS Phosphosilicate
PS Photogrammetry Society [UK]
PS Photographic Science(s)
PS Photoscanner
PS Photosensitive; Photosensitivity; Photosensor
PS Photosynthesis; Photosynthetic
PS Physical Science(s)
PS Physical Society [UK and Japan]
PS Physical State
PS Physiological Society [UK]
PS Pitsch-Schrader (Orientation Relationship) [Crystallography]
PS Planar Slip [Crystallography]
PS Plane Site [also ps] [Crystallography]
PS Plane Strain [Mechanics]
PS Plane Stress [Mechanics]
PS Plane Survey(ing)
PS Planetary Science; Planetary Scientist
PS Planetary Society [US]
PS Planetary Spectrum
PS Planning and Scheduling
PS Plasma Screen
PS Plasma Source
PS Plasma Spray(er); Plasma Spraying
PS Plasma Stability
PS Plasma Synthesis
PS Plastic Solid
PS Plastic Surgeon; Plastic Surgery
PS Plummer-Vinson (Syndrome) [Medicine]
PS Pneumatic System
PS Point of Shipment [also P/S]
PS Point of Switch [Civil Engineering]
PS Point Source
PS Point Symmetry

PS Polarization Spectroscope; Polarization Spectroscopy
PS Pole Strength [Physics]
PS Police Sergeant
PS Polymer Science; Polymer Scientist
PS Polymer Synthesis
PS Polysaccharide
PS Polysulfone
PS Polystyrene
PS Pore Solution
PS Pore Space
PS Porous Silicon
PS Port-of-Spain [Trinidad]
PS Port Said [Egypt]
PS Port Store
PS Port Strobe
PS Position Sensor
PS Positive Small (Error)
PS Postal Service
PS Postscript [also ps]
PS Post Secondary
PS Post Status
PS Potential Scattering [Quantum Mechanics]
PS Potentiometer Synchro
PS Potentiometric Stripping
PS Power and Sound (Program)
PS Power Source
PS Power Station
PS Power Steering [also P/S]
PS Power Strip
PS Power Stroke
PS Power Supply
PS Power System
PS Presentation Services
PS Pressed and Sintered; Pressing and Sintering [Powder Metallurgy]
PS Pressure-Sensitive [also P-S]
PS Pressure Sensor
PS Pressure Sintering [Powder Metallurgy]
PS Pressure Surge
PS Pressure Switch
PS Prestress(ing)
PS Primary Standard
PS Primary Storage
PS Prime Select
PS Principal Strain [Mechanics]
PS Principal Stress [Mechanics]
PS Prismatic Spectrum
PS Private Secretary
PS Problem Solution
PS Problem Specification
PS Processor Status
PS Process Solvent
PS Process Subsystem
PS Production Schedule

PS Production Simulator
PS Production System
PS Production Stage
PS Production System
PS Product-of-Sums (Form)
PS Product Safety
PS Professional Surveyor
PS Profile Scanner
PS Programming System
PS Program Store
PS Proof Stress [Mechanics]
PS Propeller Shaft
PS Propeller Specialist
PS Proportionally Spaced; Proportional Spacing
PS Protic Solvent
PS Proton Synchrotron (Complex) [of CERN–European Laboratory for Particle Physics, Geneva, Switzerland]
PS Proximity Sensor
PS Public School
PS Puget Sound [Washington State, US]
PS Pulsating Star [Astronomy]
PS Pulse Shaper
PS Pulses per Second [also PPS, or p/s]
PS Pumped (Hydroelectric) Storage
PS Pumped System
PS Punching Station
PS PVC (Polyvinyl Chloride) Sleeve
PS Pyloric Stenosis [Medicine]
PS Reflected Earthquake Wave that has Travelled One Leg as Push Wave and the Other as Shear Wave [Symbol]
.PS PostScript [File Name Extension] [also .ps]
0.2%PS 0.2% Proof Stress [Mechanics]
PS/2 Personal System/2 [IBM Corporation Personal Computer Series]
PS/2 Programming System 2
P2S Pralidoximmethanesulfonate
P(S) Packet Send-Sequence Number [Data Communications]
P&S Plugged and Suspended [also P and S, p and s, or p&s] [Oil and Gas Industry]
P&S Port and Starboard
P&S Press-and-Sinter (Part) [Powder Metallurgy]
P/S Payload Specialist [also PS]
P/S Point of Shipment [also PS]
P/S Power Steering [also PS]
P/S Processed and Sintered; Processing and Sintering [Metallurgy]
P-S Parallel-to-Serial [also P/S]
P-S Penicillin–Sulfacetamide
P-S Pressure-Sensitive [also PS]
Ps Positronium [Particle Physics]
Ps Pseudomonadaceae; Pseudomonas (Bacteria) [Microbiology]
Ps Selective Parameter [ANSI Escape Sequence Code]
pS picosiemens [Unit]

ps partially soluble [also PS]
ps per second [also PS]
ps picosecond [also psec]
ps plane site [also PS] [Crystallography]
ps psi [English Equilvalent]
p/s pulses per second [also PS, or PPS]
ϕs angle of friction (between two rocks, or soil solids) [Symbol]
£S Sudanese Pound [Currency]
PSA 4-Ethenyl-N,N-Dimethyl Benzenesulfonamide
PSA Pacific Science Association [Hawaii, US]
PSA Parametric Semiconductor Amplifier
PSA Particle-Size Analysis; Particle Size Analyzer
PSA Particle Surface Area
PSA Passenger Shipping Association [UK]
PSA Path Selection Algorithm
PSA Payload Service Area
PSA Payload Support Avionics
PSA Petroleum Services Association
PSA Phase-Shift Analysis
PSA Philosophy of Science Association [US]
PSA Photographic Society of America [US]
PSA Phycological Society of America [US]
PSA Pisum Sativum Agglutinin [Immunology]
PSA Plataforma Solar de Almeria [Solar Platform of Almeria, Spain]
PSA Poly(sodium Acrylate)
PSA Portable Sanitation Association (International) [US]
PSA Port Storage Area
PSA Post-Sleep Activity [Medicine]
PSA Potentiometric Stripping Analysis
PSA Poultry Science Association [US]
PSA Power Servo Amplifier
PSA Power Servo Assembly
PSA Preferred Storage Area
PSA Prefix Storage Area
PSA Pre-Sleep Activity [Medicine]
PSA Pressure-Sensitive Adhesive
PSA Pressure-Sensitive Adsorption
PSA Pressure-Swing Adsorption
PSA Pressure Switch Assembly
PSA Prices Surveillance Authority
PSA Problem Specification Analyzer
PSA Product Safety Association
PSA Prostate-Specific Antigen (Test) [Medicine]
PSA Provisions Stowage Assembly
PSA Public Service Announcement
PSA Pushdown Stack Automaton
PS&A Plasma Sciences and Applications (Society) [of Institute of Electrical and Electronics Engineers US]
PSAAPG Pacific Section of the American Association of Petroleum Geologists [US]
PSAC Petroleum Services Association of Canada

PSAC Presidential Scientific Advisory Committee; President's Science Advisory Committee [US]

PSAC Public Servants Alliance of Canada

PSAD Prediction, Simulation, Adaption and Decision

PSAE Philippine Society of Agricultural Engineers

PSAG Probabilistic Systems Assessment Group

PSAL Permanent Supplementary Artificial Lighting

PSALI Permanent Supplementary Artificial Lighting Installation

PSALI Permanent Supplementary Artificial Lighting of Interiors

PSAM Partitioned Sequence Access Method

PSANS Positional Small-Angle Neutron Scattering

PSAP Public Safety Answering Point

PSAPI Presentation Space Application Programming Interface

PSAR Polarized Synthetic Aperture Radar (Antenna)

PSAR Preliminary Safety Analysis Report

PSAR Programmable Synchronous/Asynchronous Receiver

PSA-RIT Photographic Society of America–Rochester Institute of Technology (Scholarship) [US]

PS-ASA Polystyrene–Acrylonitrile-Styrene Acrylate

PSAT Preliminary Scholastic Aptitude Test

PSAT Programmable Synchronous/Asynchronous Transmitter

PSAXS Positional Small-Angle X-Ray Scattering

PSB Palacio, Solari and Biloni (Growth Velocity) [also P-S-B] [Metallurgy]

PSB Parallel System Bus

PSB Persistent Slip Band [Metallurgy]

PSB (Snow) Plough, Sweeper and Blower

PSB Polished Silica Block

PSB Productivity and Standards Board [Singapore]

PSB Program Specification Block

PSB Proton Synchrotron Booster [of CERN– European Laboratory for Particle Physics, Geneva, Switzerland]

PSBCCO Lead (Pb) Strontium Barium Calcium Copper Oxide (Superconductor)

psbl possible

PSBLS Permanent Space Based Logistics System

PSBMA Professional Services Business Management Association

P(S-b-dMMA) Symmetric Diblock Copolymer of Polystyrene and Deuterated Polymethyl Methacrylate [also P(S-b-d-MMA)]

P(S-b-MMA) Symmetric Diblock Copolymer of Polystyrene and Polymethyl Methacrylate

PS-BMWD Polystyrene of Broad Molecular Weight Distribution

P(S-b-nBMA) Poly(styrene/n-Butyl Methacrylate) Diblock Copolymer

PS-b-PDMS Polystyrene-Polydimethylsiloxane Diblock Copolymer

PS-b-PEO Polystyrene-Polyethyleneoxide Diblock Copolymer

PS-b-PI Polystyrene-Polyisoprene Diblock Copolymer

PS-b-PMMA Polystyrene-Polymethylmethacrylate Diblock Copolymer

PS-b-PVP Polystyrene-Polyvinylpyridine Diblock Copolymer

PSBR Public Sector Borrowing Requirements [UK]

PSBR Pyridine Syrene Butadiene Rubber

PSC Pacific Salmon Commission [US]

PSC Pacific Science Council

PSC Parallel Supercomputer

PSC Parallel Switch Control

PSC Parasubstitution Compound [Chemistry]

PSC Part Status Code

PSC Paterson State College [Wayne, New Jersey, US]

PSC Pembroke State College [North Carolina, US]

PSC Periodic Surveillance Committee

PSC Permanent-Split Capacitor

PSC Personal Supercomputer

PSC Personal Superconductor

PSC Peru State College [Nebraska, US]

PSC Pharmacological Society of Canada

PSC Photosensitive Cell

PSC Pittsburgh Supercomputing Center [of National Science Foundation in Pennsylvania, US]

PSC Plane Strain Compression [Mechanics]

PSC Plant Service Center

PSC Plant Simulation Code

PSC Plymouth State College [New Hampshire, US]

PSC Polar Stratospheric Cloud(s)

PSC Point Sources Committee [US]

PSC Polyelectrolyte-Surfactant Complex

PSC Porous Silicon Carbide

PSC Portland State College [Oregon, US]

PSC Power Supply Circuit

PSC Power System Communications [Institute of Electrical and Electronic Engineers–Power Engineering Society Committee, US]

PSC Print Server Command

PSC Product Service Center

PSC Program Schedule Chart

PSC Program Sequence Control

PSC Program Support Center [of US Department of Health and Human Services]

PSC Protocol Support Component

PSC Public Service Commission [US and Canada]

PSC Pulse Signaling Circuit

PSCC Power System Computation Conference

PSCC Public Service Commission of Canada [Ottawa]

PSCCO Lead Strontium Calcium Copper Oxide (Superconductor)

PSCE Partially Spin Coupled Echo [Physics]

PSCF Primary System Control Facility

PSCF Processor Storage Control Function

PSCI Plastic Shipping Container Institute [US]

PSCL Programmed Sequential Control Language

PSCL Propellant Systems Cleaning Laboratory

PS-Cl 2-Pyridine Sulfenic Chloride

PSCM Pilot Scale Ceramic Melter

PSCM Process Steering and Control Module

PSCN Preliminary Specification Change Notice

PSCNO Praseodymium Strontium Copper Niobium Oxide (Superconductor)

PSCO Lead Strontium Copper Oxide (Superconductor)

PSCO Praseodymium Strontium Copper Oxide (Superconductor)

PSCO Pennsylvania State College of Optometry [Philadelphia, US]

PS-COOH Carboxylic Acid Terminated Polystyrene

PSC/PES Power System Communications/Power Engineering Society [of Institute of Electrical and Electronics Engineers, US]

PSCR Pauling Second Crystal Rule [Solid-State Physics]

P-SCR Photoelectric Silicon-Controlled Rectifier

PSCYCO Praseodymium Strontium Calcium Yttrium Copper Oxide (Superconductor)

PSD International Position-Sensitive Detector Conference

PSD Packet Switched Data

PSD Particle-Size Distribution

PSD Passive Sampling Device

PSD Patent Search Documentation [of European Patent Office, Munich, Germany]

PSD Permanent Signal Detection

PSD Perpendicular to Spray Direction [Thermal Spraying]

PSD Phase-Sensitive Demodulator

PSD Phase-Sensitive (Proportional) Detector [also psd]

PSD Photon-Stimulated (Ion) Desorption

PSD Physical Science Division

PSD Plastic Shear Defect

PSD Polystyrene, Deuterated

PSD Pore Size Distribution

PSD Position-Sensitive Detector

PSD Positron-Sensitive Detector

PSD Post Sending Delay

PSD Power Shutdown

PSD Power Spectral Density

PSD Power Spectrum Density

PSD Power System Dynamics

PSD Presolidified Droplet [Thermal Spraying]

PSD Prevention of Significant Deterioration

PSD Printing Systems Division

PSD Product State Distribution

PSD Programmable Dispenser

PSD Program Status Double-Word

PSD Program Support Document

PSD Protective Sciences Division [of Defense Research Establishment Ottawa, Canada]

PSD Pulse Shape Discrimination

psd passed

PSDC Permanent Signal Detection Circuit

PSDC Public Switched Digital Capability

PSDC Purge Sample, Detect and Calibrate

PSDD Planetary Science Data Dictionary [of NASA Planetary Data System]

PSDD Preliminary System Design Description

PSDDS Pilot Switched Digital Data Service

PSDDS Public Switched Digital Data Service

PSDE Payload and Spacecraft Development and Experimentation

PSDF Propulsion Systems Development Facility

PSDIAD Photon-Stimulated Desorption Ion Angular Distribution

PSDMS Photon-Stimulated (Ion) Desorption Mass Spectroscopy

PSDN Packet-Switched Data Network

PSDN Public Switched Data Network

PS-DOP Polystyrene in Dioctyl Phthalate [also PS/DOP]

PSDP Payload Station Distribution Panel

PSDR Planning and Scheduling Document Record

PS-DVB Polystyrene in Divinyl Benzene [also PS/DVB]

PSDS Packet-Switched Data Service

PSE Pacific (Coast) Stock Exchange [San Francisco and Los Angeles, US]

PSE Packet Switching Exchange

PSE Payload Service (or Servicing) Equipment

PSE Payload Support Equipment

PSE Payload Systems Engineer

PSE Periodic System of (the Chemical) Elements

PSE Phytochemical Society of Europe [UK]

PSE Please [Amateur Radio]

PSE Polymer Science and Engineering

PSE Poly(sodium Ethylacrylate)

PSE Polysulfide Elastomer

PSE Portuguese Society of Engineers

PSE Post-Secondary Education

PSE Power Systems Engineering [Institute of Electrical and Electronic Engineers–Power Engineering Society Committee]

PSE Pressurized Subcritical Experiment

PSE Primary Salt Effect [Chemistry]

PSE Producer Subsidy Equivalent

PSE Programming Support Environment

psec picosecond [also ps]

PSECCO Lead Strontium Europium Cerium Copper Oxide (Superconductor)

PSECT Prototype Section

PSED Physical Sciences and Engineering Division

PSEE Photostimulated Exoelectron Emission

PSEM Philips Scanning Electron Microscope [of N.V. Philips, Eindhoven, Netherlands]

PSEP Passive Seismic Experiment Package

PSEP Polyhedral Skeletal Electron Pair (Theory) [Physics]

PSEP Polymer Science and Engineering Program

PSEPT Polyhedral Skeletal Electron Pair Theory [Physics]

PSER Poly(L-Serine) [Biochemistry]

PSERVER Print Server

PSES Passive Solar Energy System

PSES (European Meeting) From Planck Scale to Electroweak Scale [Particle Physics]

PSES Pretreatment Standards for Existing Sources

Pseud Pseudonym [also pseud]

Pseudo-RSP Pseudo-Rapid Solidification Processing [also pseudo-RSP]

PSF Partial Stacking Fault [Materials Science]

PSF Passive Solar Foundation

PSF Performance Shaping Factor

PSF Permanent Signal Finder

PSF Permanent Swap File

PSF Physical Society of Finland

PSF Plastic Stress Function

PSF Polysulfone

PSF Point Spread Function

PSF Pound-Force per Square Foot [also psf, or lbf/ft^2]

PSF Pound(s) per Square Foot [also psf, or lb/ft^2]

PSF Print(ing) Services Facility

PSF Processing and Staging Facility (for NASA Solid Rocket Booster)

PSF Processing and Storage Facility (for NASA Shuttle External Tank)

PSF Process Signal Former

PSF Progressive Space Forum [US]

PSF Provisional System Feature

PSF Pumped Storage Facility

PSFA Power System Fault Analysis

pSfBA p-Sulfobenzoic Acid

PSFC Pacific Salmon Fisheries Commission

PS/FC Power Supply/Fuel Cell

PSFEM Photostimulated Field-Emission from Metals

PSFG Permanent Service for the Fluctuation of Glaciers

PSFI Photon-Stimulated Field Ionization

Psfn Phenosafranine

pSfP p-Sulfophenol

PSFT Plane Strain Fracture Toughness

PSFT Progressive-Saturation Fourier Transform

PSG Percentage Sensitization by Grit [Explosives]

PSG Phenol Sector Group [of Conseil Européen des Fédérations de l'Industrie Chimique, Belgium]

PSG Phosphorus-Doped Silica Glass

PSG Phosphosilicate Glass

PSG Photoelastic Strain Gauge

PSG Power Subsystem Group

PSG Primate Specialists Group [of International Union for Nature and Natural Resources]

PSG Publishing Systems Group [US]

PSG Pulse Sequence Generation

PSG Pumped-Storage Generation

Psg Passage [also psg]

Psg Passing [also psg]

PSGB Pharmaceutical Society of Great Britain

PSGB Primate Society of Great Britain

PSG/BPSG Phosphosilicate Glass/Borophosphosilicate Glass

PSGC Puget Sound Governmental Conference [US]

PSGCCO Lead (Pb) Strontium Gadolinium Cerium Copper Oxide (Superconductor)

Psgr Passenger [also Psgr, or psgr]

PSH Protonated Polystyrene

PSHB Persistant Spectral Hole Burning [Physics]

PSHTM Pilot Scale High-Temperature Melter

PSI Gesellschaft für Prozeßsteuerungs– und Informationssysteme mbH [German Process Control and Information Systems Manufacturer; located at Berlin]

PSI Packetnet Systems Interface

PSI Paid Service Indication

PSI Pakistan Standards Institution

PSI Paper Stock Institute [US]

PSI Parallel-Slit Interferometer

PSI Parapsychology Sources of Information

PSI Park Scientific Instruments [Mountain View, California, US]

PSI Passive Solar Institute [US]

PSI Paul Scherrer Institute [Villigen, Switzerland]

PSI Peripheral Subsystem Interface

PSI Peripheral System Interface

PSI Permuterm Subject Index [of Institute for Scientific Information, US]

PSI Personal Sequential Interface

PSI Pharmaceutical Society of Ireland

PSI Phase-Shifting Interferometry

PSI Planned Start Installation

PSI Plan-Speed Indicator

PSI Plasma Sciences Inc. [Lorton, Virginia, US]

PSI (International Conference on) Plasma Surface Interactions (in Controlled Fusion Devices)

PSI Polysiloxane Imide

PSI Pool of Scientific Instruments (for Medical Technology) [Germany]

PSI Position Specific Iterated

PSI Positive Surface Ionization

PSI Pound-Force per Square Inch [also psi, or lbf/in^2]

PSI Pound(s) per Square Inch [also psi, or lb/in^2]

PSI Power Static Inverter

PSI Preprogrammed Self-Instruction

PSI Present Serviceability Index

PSI Pressure-Sensitive Identification

PSI Printed Subject Index

PSI Proctorial System of Instruction

PSI Professional Services Institute

PSI Project Starlight International [US]

PSI Protosynthetic Indexing

PSI Psychonomic Society, Inc. [US]

P-Si Porous Silicon [also PSi, pSi, or p-Si]

p-Si Polysilicon (or Polycrystalline Silicon)

p-Si p-Type Silicon (Semiconductor)

psi pound-force per square Inch [also PSI, or lbf/in^2]

psi pound(s) per square inch [also PSI, or lb/in^2]

psia pounds per square inch, absolute [also PSIA]

PSI-BLAST Position Specific Iterated Basic Local Alignment Search Tools [of National Center for Biological Information, US]

PSIC Passive Solar Industries Council [US]

PSIC Process Signal Interface Controller

p-SiC P-Type Silicon Carbide

PSI Center Parapsychology Sources of Information Center [Dix Hills, New York State, US]

PSID Photon-Stimulated Ionic Desorption

PSID PostScript Image Data

PSID Preliminary Safety Information Document

psid pounds per square inch, differential [also psid]

PSIF pounds per square inch force [also PSIF]

psi·ft/min pound(s) per square inch feet per minute

PSIG Pipeline Simulation Interest Group [US]

PSIG Propulsion Systems Integration Group

psig pounds per square inch, ga(u)ge [also PSIG]

PSII Plasma Source Ion Implantation

psi$\sqrt{\text{in}}$ pound(s) per square inch square root of inch [also psi·in$^{1/2}$]

PSIM Power System Instrumentation and Measurement (Committee) [of Institute of Electrical and Electronics Engineers Committee]

PSI Nucl Part Phys Newsl PSI Nuclear and Particle Physics Newsletter [Published by the Paul Scherrer Institute, Villigen, Switzerland]

PSI-RIT Photographic Society of America/Rochester Institute of Technology [US]

PSIS Pounds per Square Inch, Sealed [also psis]

PSJ Physical Society of Japan

PSK Phase-Shift Keyed; Phase-Shift Keying

PSK Program Selection Key

PSKM Phase Shift Keyed Modulation

PSL Parallel Strand Lumber

PSL Photographic Science Laboratory

PSL Photostimulated Luminescence

PSL Physical Sciences Laboratory [of Division of Computer Research and Technology, National Institutes of Health, Bethesda, Maryland, US]

PSL Physical Sciences Laboratory [of Pacific Northwest National Laboratory, US Department of Energy]

PSL Power and Signal List

PSL Pressure Seal

PSL Problem Specification Language

PSL Problem Statement Language

PSL Process Simulation Language

PSL Propellant Seal

PSL Pusat Studi Linkugan [Environmental Study Center, Indonesia]

P Sl Pipe Sleeve

PSLP Private Sector Linkages Program

PSLI Packet Switch Level Interface

PSL/PSA Problem Statement Language/Problem Specification Analyzer

PSLRB Public Service and Labour Relations Board [of New Brunswick, Canada]

PSM Parallel Slit Map

PSM Phase Sensitive Modulator

PSM Poly(sodium Methacrylate)

PSM Ponce School of Medicine [Puerto Rico]

PSM Post-Synaptic Membrane [Anatomy]

PSM Printing System Manager

PSM Process Safety Management

PSM Programming Support Monitor

PSM Propellant Storage Module

PSM Pulse Shape Modulation

PSM Pyro(technic) Substitute Monitor

P&SM Procurement and Subcontract Management

PSMA Poly(styrene-Maleic Anhydride)

PSMA Power Saw Manufacturers Association [now PPEMA, US]

PSMA Power Supply Manufacturers Association [UK]

PSMA Pressure Sensitive Manufacturers Association [UK]

PSMA Professional Services Management Association

PSMA Pyrotechnic Signal Manufacturers Association [US]

PS-co-MA Poly(styrene-co-Maleic Anhydride)

PSMBE Plasma-Source Molecular Beam Epitaxy

pSmBA p-Sulfamybenzoic Acid

PSMD Photoselective Metal Deposition

PSMDE Pseudostationary Mercury Drop Electrode

PSMMA Plastic Soft Materials Manufacturers Association [US]

PSMO Praseodymium Strontium Manganese Oxide (Superconductor)

PSMO/LAO Praseodymium Strontium Manganese Oxide on Lanthanum Aluminate

PSMR Parts Specification Management for Reliability

PSMS Permanent Section of Microbiological Standardization [of International Union of Microbiological Societies, UK]

PSMS (Lawton and Brody) Physical Self-Maintenance Skills

PSMSL Permanent Service for Mean Sea Level [UK]

PSMT Lead Tin Manganese Telluride (Semiconductor)

PSN Packet-Switched Network; Packet-Switching Network

PSN Packet Switch(ing) Node

PSN Particle Stimulated Nucleation [Metallurgy]

PSN Print Sequence Number

PSN Private Satellite Network

PSN Programmable Sampling Network

PSN Public Switched Network

Psn Position [also psn]

PSNA Phytochemical Society of North America [US]

PS-NH$_2$ Amine-Terminated Deuterated Polystyrene

PSNI Pharmaceutical Society of Northern Ireland

ps/nm/km picosecond per nanometer per kilometer [also ps nm^{-1} km^{-1}]

PSNS Perthshire Society of Natural Science [UK]

PSNS Pretreatment Standards for New Sources

PSNS Programmable Sampling Network Switch

PSNS Puget Sound Naval Shipyard [US]

PSO Methyl-p-Vinylphenyl Sulfoxide

PSO Pakistan Science Office [Islamabad]

PSO Pilot Systems Operator

PSO Polystyrene Oxide

PSO Polysulfone [also PSU]

PSO Project Science Office

PSO2 Methyl-p-Vinylphenylsulfone

PSOCG Patricia Seybold's Office Computing Group [US]

PSOI Porous Silicon-on-Insulator (Substrate)

P Sol Partly Soluble [also p sol]

PSOP Payload Systems Operating Procedures

PSP Packet Switching Processor

PSP Paper-Tape Space

PSP Paratytic Shellfish Poisoning

PSP Payload Signal Processor

PSP Payload Specialist Panel

PSP Payload Support Plan

PSP Pentagonal Small Particle

PSP Performance Shaping Parameters

PSP Permanent Sample Plot

PSP Personal Software Products (Group) [of IBM Corporation, US]

PSP Phenolsulfonphthalein [Biochemistry]

PSP Pierced Steel Planking

PSP Pitot-Static Pressure

PSP Planet Scan Platform

PSP Planned Schedule Performance

PSP Planned Standard Programming

PSP Plasma Spraying

PSP Plasmon Surface Paritons [Solid-State Physics]

PSP Polymer Science Program

PSP Polystyrylpyridine

PSP Portable Service Processor

PSP Power System Protection

PSP Precision Solar Pyranometer

PSP Preferred System Provider

PSP Presending Pause

PSP Primary Sodium Pump

PSP Programmable Signal Processor

PSP Program Segment Prefix

PSP Program Support Plan

PSP Project Schedule Plan

PSP Pseudostatic Spontaneous Potential [Petroleum Engineering]

PSPA Pressure Static Probe Assembly

PSPA Program Standards Project Authority (for Materials) [of IBM Corporation, US]

PS-PB Polystyrene/Polybutadiene (Blend) [also PS/PB]

PS-PBrS Polystyrene-Polybromostyrene (Blend) [also PS/PBrS]

PSPC Permanent South Pacific Commission

PSPC Position Sensitive Proportional Counter

PSPD Position Sensitive Proportional Detector

PS-PDMS Polystyrene/Poly-2-Dimethylsiloxane (Blend) [also PS/PDMS]

PSPDN Packet-Switched Public Data Network

PS-PE Polystyrene/Polyethylene (Blend) [also PS/PE]

PS-PEO Polystyrene/Polyethyleneoxide (Blend) [also PS/PEO]

PSPF Post-Source Pulse Focusing

PSPGV Primary Sodium Pump Guard Vessel

PS-PI Polystyrene-Polyisoprene (Diblock Copolymer) [also PS/PI]

PSPL Priced Spare Parts List

PSPL Puget Sound Power and Light Company [Montana, US]

PS-PMMA Polystyrene/Polymethyl Methacrylate (Blend) [also PS/PMMA]

PS/PMMA/dBZ Polystyrene/Polymethyl Methacrylate/ Deuterated Benzene

PS/P2ClS Polystyrene/Poly(2-Chlorostyrene)

PS/P2ClS/DBP Polystyrene/Poly(2-Chlorostyrene)/Dibutyl Phthalate

PSPMO Praseodymium Strontium Lead Manganese Oxide (Superconductor)

PSPO Public Safety Project Office [of National Research Council of Canada]

PSPP Pressure Swing Parametric Pumping

PS-PPO Polystyrene/Polyphenylene Oxide (Blend) [also PS/PPO]

PS-PPP Polystyrene/Polyparaphenylene (Blend) [also PS/PPP]

PSPS Paddle Steamer Preservation Society [UK]

PSPS Planar Silicon Photoswitch

PS-PVME Polystyrene/Polyvinyl Methyl Ether (Blend) [also PS/PVME]

PS-PVP Polystyrene/Poly(2-Vinylpyridine) (Diblock Copolymer) [also PS/P2VP]

PS-PVP Polystyrene/Poly(vinylpyridine) (Diblock Copolymer) [also PS/PVP]

PSQ Personnel Security Questionnaire

PSR Packed Snow on Runway

PSR Payload Support Room

PSR Pennsylvania State Research Reactor [of Pennsylvania State University, University Park, US]

PSR Peripheral Shim Rods [Nuclear Reactors]

PSR Photosynthesis Research

PSR Point of Safe Return

PSR Power System Relaying [Institute of Electrical and Electronics Engineers–Power Engineering Society Committee, US]

PSR Present Serviceability Rating

PSR Pre-Soak Rail

PSR Primary Surveillance Radar

PSR Processor State Register

PSR Procurement Status Report

PSR Program Status Register

PSR Program Status Review

PSR Program Support Representative

PSR Proportional Specimen Resistance (Model) [Mechanical Testing]

PSR Proton Storage Ring [Nuclear Engineering]

PSRA Pulse Surface Reaction Rate Analysis

P-SRAM Pseudo-Static Random-Access Memory

PSRB Post-Sintered Reaction-Bonded

PSRBSN Post-Sintered Reaction-Bonded Silicon Nitride

PSRC Particulate Systems Research Center [of University of Missouri at Columbia, US]

PSRD Program Support Requirements Document

PSRG Position Space Renormalization Group

PSRI Physiographical Science Research Institute [of Tokyo University, Japan]

PSRO Professional Standards Review Organizations [US]

PSROC Physical Society of the Republic of China [Taiwan]

PSRP Physical Sciences Research Paper

PSRPP Public Safety and Resource Protection Program [US]

PSRR Power Supply Rejection Ratio

PSRT PostScript Round Table

PSS Packet-Switched Service [UK]

PSS Packet-Switched System

PSS Packet Swichstream (Service) [of British Telecom, UK]

PSS Pad Safety Supervisor [Aerospace]

PSS Palomar Sky Survey

PSS Payload Specialist Station

PSS Payload Support System

PSS Personal Signaling System

PSS Phase-Sensitive System

PSS Photostationary State

PSS Physics of the Solid State [of American Institute of Physics, US]

PSS Polystyrenesulfonate

PSS Power Supply System

PSS Pressure-Sensitive Schottky (Transistor)

PSS Production Support System

PSS Program Support System

PSS Proprietary Software System

PSS Proprietary Support System

PSS Propellant Supply Subsystem

PSS Propulsion Support System

PSS Public Services Satellite

PSS PUREX (Plutonium-Uranium Reduction and Extraction) Sludge Supernatent Liquid

PSSA Parasitological Society of Southern Africa

PSSA$_x$ Partially Sulfonated Polystyrene Random Copolymer

PSSA Pharmaceutical Society of South Africa

PSSA Photogrammetric Society of South Africa

PSSA Poly(styrene Sulfonic Acid)

PSSAANDPS Permanent Secretariat of the South American Agreement on Narcotic Drugs and Psychotropic Substances [Argentina]

PSSC Packaging, Shipping, and Safety Committee [US]

PSSC Physical Sciences Study Committee [US]

PSSC Public Service Satellite Consortium [US]

PSSCC Private Sector Standards Coordinating Center

PSSCCO Lead Strontium Samarium Cerium Copper Oxide (Superconductor)

PSSD Position-Sensitive Scintillation Detector

PSSG Physical Sciences Study Group

PS-SiCl$_3$ Trichlorosilane Terminated Polystyrene

PSSN Pressure-Sintered Silicon Nitride

PSSMA Paper Shipping Sack Manufacturers Association [US]

psso pass slipped stitch over [Woven Fabrics]

PSSP Payload Specialist Station Panel

PSSP Phone Center Staffing and Sizing Program

PSSST Pakistan Society for Semiconductor Science and Technology

PSST Periodic Significant Scheduled Tasks

PSSU Patch Survey and Switching Unit

PST Lead Scandium Tantalate

PST Lead Scandium Titanate

PST Lead Tin Telluride (Semiconductor)

PST Pacific Standard Time [Greenwich Mean Time +08:00] [also pst]

PST Paired Selected Ternary; Pair-Select Ternary

PST Pakistan Standard Time [Greenwich Mean Time −05:00]

PST Partition Specification Table

PST Phase Space Theory [Physics]

PST N-[4-Phenyl-2-Thioxo-1(2H)-Pyrimidinyl]benzene-sulfonamidato-

PST Photostress Technology

PST Piezoelectric Scanning Tube

PST Plasma Separator Tube

PST Point of Spiral Tangent

PST Point-Set Topology

PST Polished Surface Technique

PST Polycrystal Scattering Topography

PST Polysynthetically Twinned; Polysynthetical Twinning [Crystallography]

PST Post-Stimulus Time

PST Precision Static Toroid

PST Pressure Sensitive Tape

PST Primary Surge Tank

PST Provincial Sales Tax [Canada]

Pst Pesticide [also pst]

Pst Providencia stuartii [Microbiology]

PstI Providencia stuartii I (One) [Microbiology]

PSTA Packaging Sciences and Technology Abstract

PSTC Pressure Sensitive Tape Council [US]

PSTC Public Switched Telephone Network

PSTF Pressure Suppression Test Facility

PSTF Proximity Sensor Test Facility

PSTF Pump Seal Test Facility

Pstg Postage [also pstg]

£Stg Pound Sterling [Northern Ireland and UK]

PSTGS Polystyrene–Triglycene Sulfate (Composite)

PST-H N-[4-Phenyl-2-Thioxo-1(2H)pyrimidinyl]-benzenesulfonamide

PSTI Petroleum Science and Technology Institute [UK]

PSTM Photon Scanning Tunneling Microscope; Photon Scanning Tunneling Microscopy

PSTN Public Service Telephone Network; Public Switched Telephone Network

PSTO Lead Scandium Tantalum Oxide (Superconductor)

PSTV Potato Spindle Tuber Viroid [Microbiology]

PSU Packet Switching Unit

PSU Path Setup

PSU Penn(sylvania) State University [University Park, US]

PSU Peripheral Switching Unit

PSU Pittsburgh State University [Pennsylvania, US]

PSU Polysulfone [also PSO]

PSU Portland State University [Oregon, US]

PSU Port Sharing Unit

PSU Port Storage Utility

PSU Power Supply Unit

PSU Power Switching Unit

PSU Primary Sampling Unit

PSU Primary Switching Unit

PSU Program Storage Unit

PSU Pseudospin Valve [Magnetic Media]

PSU Publications Services Unit [of International Union for Conservation of Nature and Natural Resources]

PSU Public Service Union

PSU-ABS Polysulfone/Acrylonitrile-Butadiene-Styrene (Blend) [also PSU/ABS]

PSUCS Penn(sylvania) State University Continental Stratus

PSU-PBT Polysulfone/Polybutylene Terephthalate (Blend) [also PSU/PBT]

PSU-PET Polysulfone/Polyethylene Terephthalate (Blend) [also PSU/PET]

PSU Press Penn(sylvania) State University Press [University Park, US]

PSUR Pennsylvania State University Reactor [University Park, US]

PSV Pair Shield Video

PSV Planetary Space Vehicle

PSV Program Status Vector

PSV Pulverization Sous Vede [Metallurgy]

PSV Pseudo-Spin Valve [Magnetic Media]

Psvt Passivate [also psvt]

PSW Pacific Southwest

PSW Partial Spectral Weight

PSW Pipe Socket Weld

PSW Plasmon Surface Wave [Solid-State Physics]

PSW Processor Status Word

PSW Program Status Word

PSW PUREX (Plutonium-Uranium Reduction and Extraction) Sludge Waste

PSWP Pacific Southwest Water Plan [US]

PSWP Plant Service Water Pump

PSWR Power Standing Wave Ratio

PSY Psychotropic Effects (of Hazardous Materials)

Psych Psychiatrical; Psychiatrist; Psychiatry [also psych]

Psych Psychological; Psychologist; Psychology [also psych]

Psychoanal Psychoanalyst; Psychoanalytical; Psychoanalysis [also psychoanal]

Psychol Psychological; Psychologist; Psychology [also psychol]

Psychol Abstr Psychological Abstracts [of American Psychological Society, US]

Psychol Bull Psychological Bulletin

Psychol Forsch Psychologische Forschung [German Journal on Psychological Research]

Psychol Med Psychological Medicine [US Journal]

Psychol Monogr Psychological Monographs

Psychol Rep Psychological Reports

Psychol Rev Psychological Reviews [of American Psychological Society, US]

Psychol Today Psychology Today [International Journal]

Psychopathol Psychopathology [International Journal]

Psychosom Med Psychosomatic Medicine [US Journal]

Psychother Psychosomat Psychotherapy and Psychosomatics [International Journal]

PsycINFO Psychological Information Database

PSYOPS Psychological Operations [also Psyops, or psyops]

£Syr Syrian Pound [Currency]

PSYWAR Psychological Warfare [also Psywar, or psywar]

PSZ Partially-Stabilized Zirconia

.ps.Z Compressed PostScript [UNIX File Name Extension]

PSZT Lead-Tin Zirconate Titanate (Ceramics)

PT Advanced Planning and Technology Office [NASA Kennedy Space Center Directorate, Florida, US]

PT Lead Titanate (Ceramics)

PT Liquid Penetrant Examination [Symbol]

PT Pacific Time [also pt]

PT Packet Terminal

PT Page Table

PT Pain Threshold

PT Paleotemperature [Geology]

PT Palladium Tube

PT Paper Tape

PT Parallel (Data) Transmission

PT Parathormone [Biochemistry]

PT Parathyroid [Anatomy]

PT Parasitic Transistor

PT Partellar Tendon

PT Particle Technology

PT Part Time [aklso P/T]

PT Part Tolerance

PT Passive Transport [Medicine]

PT Patrol Torpedo (Boat)

PT Pay Tone

PT Peaking Transformer

PT Pelton Turbine

PT Pencil Tube [Electronics]

PT Penetrant Testing

PT Pentode Transistor

PT Percussion Tube

PT Performance Test

PT Periodic Table (of Elements)

PT Perturbation Theory

PT Petroleum Technologist; Petroleum Technology

PT Phase Transformation
PT Phase Transition
PT Phase Type
PT Photoelastic Transducer
PT Photo-Technologist
PT Phototelegraphy
PT Phototherapy
PT Photothermal
PT Phototransistor
PT Phototube
PT Physical Therapist; Physical Therapy
PT Physiotherapeutical; Physiotherapist; Physiotherapy
PT Pickup Truck
PT Picture Transformation
PT Pinch Thetatron [Electronics]
PT Pine Tar
PT Pipe Tap
PT Pipe Thread
PT Pitch Trim
PT Pitot Tube
PT Planning and Technology
PT Plant Taxonomist; Plant Taxonomy
PT Plasma Technology
PT Plasma Torch
PT Plastics Technologist; Plastics Technology
PT Plate Tectonics [Geology]
PT Plutonium Titanate
PT Pneumatic Tool
PT Pocket Telephone
PT Point of Tangent
PT Politechnico di Torino [Turin Polytechnic Institute, Italy]
PT Polymerized Thiophene
PT Polymer Technologist; Polymer Technology
PT Polymorphic Transformation [Crystallography]
PT Polythiophene
PT Pomeranchuk Theorem [Particle Physics]
PT Portable Terminal
PT Portugal [ISO Code]
PT Positional Tolerance
PT Posttensioning
PT Potential Theory [Mathematics]
PT Potential Transformer
PT Potentiometric Titration
PT Powder Technology
PT Power Taps
PT Power Train
PT Power Transistor
PT Power Transmission
PT Prathet T'hai [Kingdom of Thailand]
PT Pressure–Temperature (Diagram) [also P-T]
PT Pressure Transducer
PT Primary Target (Artillery)
PT Primary Twin [Metallurgy]

PT Printer Terminal
PT Probability Theory
PT Procedure Turn [Aeronautics]
PT Processing Time
PT Proof Test
PT Propellant Transfer
PT Propeller Thrust
PT Proper Time [Relativity]
PT Property Table
PT Prothrombin Time [Medicine]
PT Prototype; Prototyping
PT P-Terphenyl
PT Public Transport
PT Pulse Timer
PT Pulse Transformer
PT Pulse-Triggered (Flip-Flop)
PT Pump Turbine
PT Punched Tape
PT- Brazil [Civil Aircraft Marking]
P(T) Temperature Dependent Polarization (Function)
P&T Posts and Timbers
P/T Part Time [also PT]
P/T Parts per Trillion [also PPT, or ppt]
P/T Pressure/Temperature
P/T Pump/Turbine
P-T Plasma Thermocouple Reactor [US]
P-T Pressure-Temperature (Diagram) [also PT]
Pt Part [also pt]
Pt Patient [also pt]
Pt Payment [also pt]
Pt Peseta [Currency of Spain]
Pt Petaton [Unit]
Pt Platinum [Symbol]
Pt Point [also pt]
Pt Pontederia [Biology]
Pt Port [also pt]
Pt Potorous tridactylis [Genus of Marsupials]
Pt Radiated Transmitter Power [Symbol]
Pt^{2+} Divalent Platinum Ion [also Pt^{++}] [Symbol]
Pt^{4+} Tetravalent Platinum Ion [Symbol]
Pt-27 Platinum 27 (NBS Platinum Reference Standard for Thermoelectric Measurements until 1972)
Pt-67 Platinum 67 (NBS Platinum Reference Standard for Thermoelectric Measurements after 1972)
Pt-190 Platinum-190 [also ^{190}Pt, or Pt^{190}]
Pt-191 Platinum-191 [also ^{191}Pt, or Pt^{191}]
Pt-192 Platinum-192 [also ^{192}Pt, or Pt^{192}]
Pt-193 Platinum-193 [also ^{193}Pt, or Pt^{193}]
Pt-194 Platinum-194 [also ^{194}Pt, or Pt^{194}]
Pt-195 Platinum-195 [also ^{195}Pt, or Pt^{195}]
Pt-196 Platinum-196 [also ^{196}Pt, or Pt^{196}]
Pt-197 Platinum-197 [also ^{197}Pt, or Pt^{197}]
Pt-198 Platinum-198 [also ^{198}Pt, or Pt^{198}]
Pt-199 Platinum-199 [also ^{199}Pt, or Pt^{199}]

P(t) (Survival) Probability as a Function of Time [Statistics]

P(τ) Relaxation Time Distribution (in Solid-State Physics) [Symbol]

pT picotesla [Unit]

p(T) displacement probability [Symbol]

p-T pressure-temperature (diagram) [also pT, or p(T)]

pt part(s) [Unit]

pt patient [also Pt]

pt pint [Unit]

pt point [Unit]

.pt Portugal [Country Code/Domain Name]

£T Turkish Pound [Currency]

£/t (UK) Pounds per ton [Unit]

Pt Peseta [Currency of Spain]

$\phi(\theta)$ Lorentz polarization factor [Symbol]

$\pi(\tau)$ relaxation time dependent distribution (in physics) [Symbol]

$\psi(\tau)$ autocorrelation function [Symbol]

PTA Parent-Teachers Association

PTA Particle Tracking Autoradiography

PTA Passenger Transport Association [South Africa]

PTA *(Pharmazeutisch-technische(r) Assistent(in))* – German for "Pharmaceutical Assistant"

PTA Phenyltrimethylammonium

PTA Phosphotungstic Acid

PTA Phthalic Anhydride

PTA Pivaloyltrifluoroacetone

PTA Planar Turbulence Amplifier

PTA Plasma Thromboplastin Antecedent [Medicine]

PTA Plasma Transferred Arc

PTA Polythionic Acid

PTA Positest Analysis

PTA Post-Turbine Augmentation

PTA Potential(ly) Toxic Area

PTA Power Transfer Assembly

PTA Preferential Trading Agreement

PTA Primary Tungsten Association [UK]

PTA Programmable Translation Array

PTA Programmed Time of Arrival

PTA Propulsion Test Article

PTA Psophocarpus Tetragonolobus Agglutinin [Immunology]

PTA Public Transport Association [UK]

PTA Pulse Torquing Assembly

PTA Purified Terephthalic Acid

Pta Peseta [Currency of Spain]

Pta *(Punta)* – Spanish for "Point" [Geography]

PTAB Photographic Technical Advisory Board [US]

PTAC Packaged Terminal Air Conditioner

$Pt(Ac)_2$ Platinum(II) Acetate

$Pt(ACAC)_2$ Platinum(II) Acetylacetonate [also $Pt(AcAc)_2$, or $Pt(acac)_2$]

PTACV Prototype Tracked Air Cushion Vehicle

P2AD Pollution Prevention Assistance Division [of Georgia Department of Natural Resources, US]

PTAH Phosphotungstic Acid Hematoxylin

PtAl Platinum Aluminide

Pt/Al_2O_3 Alumina Fiber Reinforced Platinum (Composite)

Pt/Al_2O_3 Platinum-Alumina (Catalyst)

PTAP Phenyltrimethylammonium Tribromide

P-Tape Paper Tape [also P-tape]

Ptas Pesetas [Plural of Peseta–Currency of Spain]

PTA-SA Passenger Transport Association of South Africa

P3AT Poly(3-Alkyl Thiophene)

Pt-Au Platinum-Gold (Alloy System)

PTB Patrol Torpedo Boat

PTB Payload Timing Buffer

PTB Personal Touch Banking

PTB Phosphoric Acid Tributyl Ester

PTB Physical Transaction Block

PTB Physikalisch-Technische Bundesanstalt [Federal Institute for Physical Science and Technology, Berlin, Germany]

PTBBA P-tert-Butylbenzoic Acid

PT Boat Patrol Torpedo-Boat [also PT-Boat, or PT boat]

PTBR Punched Tape Block Reader

PTC Pacific Telecommunications Conference [US]

PTC Pacific Telecommunications Council [US]

PTC Papillary Thyroid Carcinoma

PTC Passive Thermal Control

PTC Patrol Torpedo-Craft

PTC Personal Technical Certificate

PTC Personnel Transfer Capsule

PTC Phase-Transfer Catalysis

PTC Phase Transition Curve

PTC Phenyl Isothiocyanate

PTC Phenylthiocarbamide

PTC Phenylthiocarbamoyl

PTC Philadelphia Transportation Company [Pennsylvania, US]

PTC Philippine Trade Commission

PTC Photographic Training Center [UK]

PTC Plant Test Date

PTC Plasma Thromboplastin Component [Biochemistry]

PTC Plastics Technology Center

PTC Polytitanocarbosilane

PTC Portable Tele-Transaction Computer

PTC Portable Temperature Controller

PTC Portable Thermocouple Calibrator

PTC Positive Temperature Coefficient

PTC Postal Telegraph Cable

PTC Posttensioned Concrete

PTC Posttensioning Concrete

PTC Power Toggle Clamp

PTC Product Technology Center [of Cominco Ltd., Mississauga, Canada]

PTC Programmable Thermal Control(ler)

PTC Programmed Temperature Cycling

PTC Programmed Transmission Control

PTC Progressive Temperature Compensation

PTC Propionylthiocholine

PTC Pulse Time Code

Pt/C Platinum/Carbon

Ptc Pataca [Currency of Macao]

PTCA Percutaneous Transluminal Coronary Angioplasty [Medicine]

PTCD Property Table Current Date

Pt-Cd Platinum-Cadmium (Alloy System)

PTCDA Perylene Tetracarboxylic Dianhydride

PTCE Pilot Tropical Cirrus Experiment

PTCI Perturbation Treatment Configuration Interaction

P2ClS Poly(2-Chlorostyrene)

PTCMA Plastic Tanks and Cisterns Manufacturers Association [UK]

PtCo Platinum Cobalt (Alloy)

Pt(COD)Cl$_2$ Cyclooctadieneplatinum (II) Chloride [also Pt(cod)Cl$_2$]

PTCR (Launch) Pad Terminal Connection Room

PTCR Positive Temperature Coefficient of Resistivity

Pt-Cr Platinum-Chromium (Alloy System)

PTCRI Patent, Trademark and Copyright Research Institute [US]

PTCS Passive Thermal Control Section

PTCS Passive Thermal Control System

PTCS Payload Test and Checkout System

PTCS Phenyltrichlorosilane

PTCS Planning, Training and Checkout System

PTCS Propellant Tanking Computer System

Pt-Cu Platinum-Copper (Alloy System)

PTD Parallel Transfer (Disk) Drive

PTD Photothermal Deflection (Technique)

PTD Photothermal Displacement

PTD Physical Theory of Diffraction

PTD Post-Tuning Drift [Electronics]

PTD Power Transmission Design

PTD Precision Twist Drill

PTD Process Technology Division [of Iron and Steel Society, US]

PTD Provisioning Technical Documentation

ptd painted

PTDA Power Transmission Distributors Association [US]

Ptd A Pointed Arch

PTDDSS PTD (Provisioning Technical Documentation) Data Selection Sheet

PTDF Pacific Tuna Development Foundation [now Pacific Fisheries Development Foundation, US]

PT-DIV Patrol Torpedo-Boat Division [also PTDiv]

PTDL Patent and Trademark Depository Libraries [US]

PT-DOS Processor Technology Disk-Operating System [also PT DOS, or PTDOS]

PTE Page Table Entry

PTE Peculiar Test Equipment

PTE Periodic Table of Elements

PTE Photothermoelastic(ity)

PTE Portuguese Escudo [Currency]

PTE Pressure-Tolerant Electronics

PTE Professional Technical Engineer

Pte *(Pointe)* – French for "Point" [Geography]

Pte Private [Military Rank]

PTEC Plastics Technical Evaluation Center [US Army]

PT/Elektron Elektrotech PT/Elektronica Elektrotechniek [Dutch Journal on Electronics and Electrical Engineering]

PTEM Plan-View Transmission Electron Microscope; Plan-View Transmission Electron Microscopy

Pt(en)$_2$ Platinum(II) (1,2-Diaminoethane)

Pt(en)$_2$Cl$_2$ Platinum(II) (1,2-Diaminoethane) Dichloride

PTERM Physical Termination

PTF Lead Titanium Fluoride

PTF Phase Transfer Function

PTF Polymer Thick Film

PTF Polytetrafluoroethylene

PTF Problem Temporary Fix

PTF Problem Trouble Fix

PTF Programmable Transversal Filters

PTF Program Temporary Fix

PTFA Primary Transverse Field Anneal(ing) [Solid-State Physics]

PTFCE Polytrifluorochloroethylene

PTFE Polytetrafluoroethylene

P3FE Polytrifluoroethylene

Pt-Fe Platinum-Iron (Alloy System)

PTFE-F Fluoropore Polytetrafluoroethylene (Polymer)

PTFE-M Mitex Polytetrafluoroethylene (Polymer)

pTFMa p-Trifluoromethylaniline

PTFS Precision Temperature Forcing System

PTF-SAE-SHORT Dryseal SAE (Society of Automotive Engineers) Short Taper Pipe Thread [Symbol]

PTG Parathyroid Gland [Anatomy]

PTG Parsons Turbines and Generators

PTG Precise Tone Generator

PTG Prepared Town Gas

Ptg Painting [also ptg]

Ptg Petrograd [formerly St. Petersburg, Petrograd and Leningrad, now St. Petersburg, Russia]

Ptg Printing [also ptg]

PTGA Pteroyl Triglutamic Acid [Biochemistry]

PtGa Platinum Gallium (Compound)

Pt-Ga Platinum-Gallium (Alloy System)

Pt/GaAs Platinum on Gallium Arsenide (Substrate)

PTGC Parsons Turbines and Generators Canada [St. Catharines, Ontario, Canada]

PTGC Programmed-Temperature Gas Chromatography

PTGH2 P-Tolylaminoglyoxime

Ptg Std Petrograd Standard

PTH Parathyroid Hormone [Biochemistry]

PTH Phenylthiohydantoin [Biochemistry]

PTH Plated Through Hole

Pt-H Platinum-Hydrogen (Complex)

PTHF Polytetrahydrofuran

Pt(HFAC)₂ Platinum(II) Hexafluoroacetylacetonate [alsp Pt(hfac)₂]

PTHrP Parathyroid Hormone Related Peptide [also PTHRP]

PTI Packaged Ice Association [US]

PTI Pancreatic Trypsin Inhibitor [Medicine]

PTI Party Identity

PTI Philadelphia Textile Institute [Pennsylvania, US]

PTI Pipe Test Insert

PTI Plugging Temperature Indicator

PTI Post-Tensioning Institute [US]

PTI Power Tool Institute [US]

PTI Preliminary Test Information

PTI Presentation of Technical Information

PTI Pre-Turbine Injection

PTI Programmed Test Input

PTI Program Transfer Interface

PTI Public Technology, Inc. [US]

PtI Platinum(I) Iodide

PTIC Patent and Trademark Institute of Canada

PTID Property Table Issue Data

P2ID (International Symposium on) Plasma Process-Induced Damage

PTIDG Presentation of Technical Information Discussion Group [UK]

PTIO Pesticides Technical Information Office [Canada]

PTIO 2-Phenyl-4,4,5,5-Tetramethylimidazoline1-Oxyl 3-Oxide

Pt-Ir Platinum-Iridium (Alloy System)

PTIS Photothermal Ionization Spectroscopy

PTIS Program Test Input System

PTK Percentage of Perturbation Theory Kept [Physics]

PTK Plastic Tapered Knob

PTK n-Propyl m-Tolyl Ketone

PTK Protection Check

PTL Pittsburg Testing Laboratory [Pennsylvania, US]

PTL Polyphenothiazine Ladder Polymer

PTL Power Transmission Line

PTL Process and Test Language

Ptlbd Particleboard [also ptlbd]

PTLP Partial Transient Liquid Phase

PTLPB Partial Transient-Liquid-Phase Bonding

PT-LT Lead Titanate Lanthanum Titanate (System)

PTLU Pretransmission Line-Up

ptly partly

PTM Pennsylvania Test Methods [US]

PTM Perchlorotriphenylmethyl

PTM Performance Test Model

PTM (Solid-Solid) Phase Transformation in Inorganic Materials

PTM Portable Terminal Monitor

PTM (International) Powder Technology Materials (Conference)

PTM Programmable Terminal Monitor

PTM Programmable Terminal Multiplexer

PTM Programmable Timer Module

PTM Proof Test Model

PTM Pulse-Time Modulation

PTM Pulse Time Multiplex

PTMA Phosphotungstic and Phosphomolybdic Acid Mixture

PTMC Phenomenological Theory of Martensite Crystallography [Metallurgy]

PTMEG Polytetramethane Glycol

PTMEG Polytetramethylene Ether Glycol

PtMe₃I Iodotrimethylplatinum (IV)

PTML PNPN (Positive-Negative-Positive-Negative) Transistor Magnetic Logic

PtMn Platinum-Manganese (Alloy)

PtMnSb Platinum Manganese Antimonide

PTMO Polytetramethane Oxide

PTMO n-Propyltrimethoxymethane

Pt-Mo Platinum-Molybdenum (Alloy System)

PTM/OS Programmable Terminal Monitor/Operating System

PTMS Precision Torque Measuring System

PTMS P-Toluidine-M-Sulfonic Acid

PTMT Polytetramethylene Terephthalate

PTMT Preliminary Thermomechanical Treatment [Metallurgy]

PTN Personal Telephone Number

PTN Plant Test Number

PTN Procedure Turn [Aeronautics]

PTN Public Telephone Network

Ptn Partition [also ptn]

Ptn 2,4-Pentanediamine [also ptn]

Ptn Portion [also ptn]

Ptn 1,2,3-Propanetriamine

Pt-Ni Platinum-Nickel (Alloy System)

1,5-Ptnta 1,5-Pentanediaminetetraacetate

2,4-Ptnta 2,4-Pentanediaminetetraacetate

PTO Lead Titanate (or Lead Titanium Oxide) Ceramics [PbTiO₃]

PTO Participating Test Organizations

PTO Part Time Operation

PTO Patent and Trademark Office [of US Department of Commerce]

PTO Please Turn Over [also pto]

PTO Power Takeoff [also pto]

PTO Power Test Operations

PTO Public Telecommunications Operator

P(2ω) Nonlinear Polarization [Symbol]

PTOS Paper-Tape-Oriented Operating System

PTOS Patent and Trademark Office Society [US]

£/t oz (UK) Pounds per troy ounce [Unit]

PTQ Poly-P-Tolyquinoxaline

PTP 3,6-Bis(2-Pyridylthio)pyridazine

PTP Paper-Tape Punch

PTP Peripheral Target Position

PTP Personal Touch Payment

PTP Phase Transition Point

PTP Point-to-Point (Control)

PTP Preferred Target Point
PTP Proximity Test Plug
PTP Pseudothermoplastics
P3P Platform for Privacy Preferences Project [also P³P]
PTPase Protein Tyrosine Phosphatase [Biochemistry]
PTPD Property Table Previous Date
PtPdMn Platinum Palladium Manganese (Alloy)
PTPI 2-p-Tolyl Pyridinecarboxaldimine
PTPR Paper Tape Punch Reader
PTPS Plasma-Treated Porous Silicon
pTpT Phosphoryl-Thymidylyl-Thymidine [Biochemistry]
Pt-PtRh Platinum-Platinum Rhodium (Thermocouple)
PT-PZ Lead Titanate–Lead Titanate (Ceramics)
PTR Paper-Tape Reader
PTR Part Throttle Reheat
PTR Pool Test Reactor [of Chalk River Nuclear Laboratories, Ontario, Canada]
PTR Pool Training Reactor [US]
PTR Pool-Type (Nuclear) Reactor
PTR Position Track Radar
PTR Power Transformer
PTR Predominant Twin Reorientation [Metallurgy]
PTR Preliminary Test Report
PTR Pressure Tube Reactor [US]
PTR Pressurized Tube Reactor [US]
PTR Printer [also Ptr, or ptr]
PTR Problem Tracking and Reporting
PTR Processor Tape Read
PTR Program(mer) Trouble Report
PTR Progress Through the Rank
PTR Proof Test Reactor [of Knoll's Atomic Power Laboratory; of General Electric Company, Schenectady, New York, US]
PTR Pointer [also ptr]
Ptr Pointer(-Type Value) [Pascal Function]
Ptr Printer [also ptr]
PTRA Power Transmission Representatives Association [US]
PTR CMND Printer Command [also Ptr Cmnd]
PTRD Property Table Revision Data
Pt-Re Platinum-Rhenium (Alloy System)
Pt-Rh Platinum-Rhodium (Thermocouple)
PTRM Partial Thermoremanent Magnetization
Pt-Ru Platinum-Ruthenium (Alloy System)
PTS Payload Test Set
PTS Payload Timeline Summary
PTS Payload Transportation System
PTS Permanent Threshold Shift
PTS Phototypesetting
PTS Plane Transport System
PTS Pneumatic Test Set
PTS Polydiacetylene-Toluene-Sulfonate
PTS Portable Transfer System
PTS Post-Traumatic Stress
PTS Power Transient Suppressor

PTS Predicast (Inc.)'s Terminal System
PTS Pressure Tuning Spectroscopy
PTS Pressurized Thermal Shock
PTS Proceed to Send
PTS Program Test System
PTS Propellant Transfer System
PTS p-Toluenesulfonate
PTS Public Telephone System
PTS Pure Time Sharing
1-PTS 1-Phenyl Thiosemicarbazone
4-PTS 4-Phenyl Thiosemicarbazide
Pts Patients [also pts]
PTSA P-Toluenesulfonamide; P-Toluenesulfonic Acid
PTSD Post-Traumatic Stress Disorder
PTSI P-Toluenesulfonyl Isocyanate
PtSi Platinum Silicide
Pt/Si Platinum on Silicon (Substrate)
pTSiA p-Toluenesulfinic Acid
Pt/SiC Silicon Carbide Reinforced Platinum (Composite)
PtSi/Si Platinum Silicide on Silicon (Substrate)
PTSP Paper Tape Software Package
PT/SP Pressure Tube to Spool Piece
PTSR Pressure Tube Superheat Reactor [US]
PTSR Process Tool Support Requirement
PTT Party Test
PTT Platform Transmitter Terminal [of Global Positioning System]
PTT Poly(terthiophene)
PTT Postal, Telegraph and Telephone
PTT Postal, Telephone and Telegraph
PTT Postal, Telephone and Telex
PTT Precipitation–Time–Temperature (Curve) [also P-T-T]
PTT Predicted Time of Transit [Astronomy]
PTT Program Test Tape
PTT Prothrombin Time [Medicine]
PTT Push-to-Talk
PTTC Paper Tape and Transmission Code
PTTC Petroleum Technology Transfer Council [US]
PTTI Postal, Telegraph and Telephone International [Switzerland]
Pt/Ti Platinum/Titanium (Electrode)
P-32 TTP Phosphorus-32 Thymidine 5'-Triphosphate [also ³²P-TTP]
Pttrn Pattern [also pttrn]
PTU Package Transfer Unit
PTU Plumbing Trades Union [UK]
P-Tube Pneumatic Tube [also P tube]
PtUYBaO Platinum Uranium Yttrium Barium Oxide (Superconductor)
PTUC Pacific Trade Union Community [Australia]
PTUC Philippines Trade Union Council
PTV Passenger Transfer Vehicle
PTV Pathfinder Test Vehicle [NASA Mars Mission]
PTV Polythienylenevinylene

PTV Predetermined Time Value

PTV Programmed Temperature Vaporizer

PTV Projection Television

PTV Projective Television

PTV Public Television

PTVA Propulsion Test Vehicle Assembly

P2VP Poly(2-Vinylpyridine)

P2VP-PAA Poly(2-Vinylpyridine)–Polyacrylic Acid

PTW Pressure Thermit Welding

PTX Picatinny Ternary Explosive [Picatinny Arsenal, US Army at Dover, New Jersey, US]

PTX Pressure–Temperature–Concentration (Diagram) [also P-T-X]

Pty Party [also pty]

Pty Proprietary [also pty]

Pty Proprietary Company [Australia and South Africa]

PTZ Lead Titanate Zirconate

PTZ Pentylenetetrazol

PTZ Phenothiazine

PTZ Posttechnisches Zentralamt [Postal Engineering Center, Darmstadt, Germany]

PTZ Propyl 1H-Tetrazole [also ptz]

Pt-Zn Platinum-Zinc (Alloy System)

PU Pacific University [Forest Grove, Oregon, US]

PU Paris University [France]

PU Peking University [Beijing, PR China]

PU Peptic Ulcer [Medicine]

PU Peripheral Unit

PU Per Unit

PU Philips-Universität [Philips University, Marburg, Germany]

PU Physical Unit

PU Pickup

PU Polyurethane [also PUR]

PU Poona University [Undia]

PU Power Unit

PU Princeton University [New Jersey, US]

PU Probe Unit

PU Processing Unit

PU Processor Utility

PU Production Unit

PU Propellant Unit

PU Propellant Utilization

PU Propulsion Unit

PU Proteinuria [Medicine]

PU Purdue University [West Lafayette, Indiana, US]

PU Pure Ultrasonics

Pu Plutonium [Symbol]

Pu Pula [Currency of Botswana]

P-u Load versus Load-Point Displacement (Relationship) [Mechanics]

Pu Pick-up (of a combine) [also pu, or p/u] [Agriculture]

Pu-232 Plutonium-232 [also ^{232}Pu, or Pu^{232}]

Pu-233 Plutonium-233 [also ^{233}Pu, or Pu^{233}]

Pu-234 Plutonium-234 [also ^{234}Pu, or Pu^{234}]

Pu-235 Plutonium-235 [also ^{235}Pu, or Pu^{235}]

Pu-236 Plutonium-236 [also ^{236}Pu, or Pu^{236}]

Pu-237 Plutonium-237 [also ^{237}Pu, or Pu^{237}]

Pu-238 Plutonium-238 [also ^{238}Pu, or Pu^{238}]

Pu-239 Plutonium-239 [also ^{239}Pu, or Pu^{239}]

Pu-240 Plutonium-240 [also ^{240}Pu, or Pu^{240}]

Pu-241 Plutonium-241 [also ^{241}Pu, or Pu^{241}]

Pu-242 Plutonium-242 [also ^{242}Pu, or Pu^{242}]

Pu-243 Plutonium-243 [also ^{243}Pu, or Pu^{243}]

Pu-244 Plutonium-244 [also ^{244}Pu, or Pu^{244}]

Pu-245 Plutonium-245 [also ^{245}Pu, or Pu^{245}]

Pu-246 Plutonium-246 [also ^{246}Pu, or Pu^{246}]

PUAS Postal Union of the Americas and Spain [Uruguay]

PUB Public (Directory) [Internet]

PUB Public Utilities Board

PUB Publications Board [of Institute of Electrical and Electronics Engineers, US]

.PUB Publication [Ventura File Name Extension]

Pub Publication; Publisher; Publishing [also pub]

pub public

pub publish(ed)

Pubg Publishing [also pubg]

Publ Publish; Publication [also publ]

Publ Astron Inst Czech Acad Sci Publications of the Astronomical Institute of the Czech Academy of Sciences

Publ Astron Soc Jpn Publications of the Astronomical Society of Japan

Publ Astron Soc Pac Publications of the Astronomical Society of the Pacific [US]

publd published [also publ'd]

Publ Dom Astrophys Obs Publications of the Dominion Astrophysical Observatory [Canada]

Publ Dom Astrophys Obs Victoria BC Publications of the Dominion Astrophysical Observatory, Victoria, British Columbia [Canada]

Publ Earth Phys Branch Dep Energy Mines Resour Publications of the Earth Physics Branch of the Department of Energy, Mines and Resources [Canada]

Publ Elektroteh Fak Ser, Elektroenerg Publicaije Elektrotehnickog Fakulteta Serija: Elektroenergetika [Publications of the Faculty of Electrical Engineering, Series: Electrical Energy, Belgarde, Yugoslavia]

Publ Elektroteh Fak Ser, Elektron Telekomun Autom Publikacije Elektrotehnickog Fakulteta Serija: Elektronika, Telekomunikacije, Automatika [Publications of the Faculty of Electrical Engineering, Series: Electronics, Telecommunications, Automation, Belgrade, Yugoslavia]

Publ Elektroteh Fak Ser, Mat Fiz Publikacije Elektrotehnickog Fakulteta Serija: Matematika i Fizika [Publications of the Faculty of Electrical Engineering, Series: Mathematics and Physics, Belgrade, Yugoslavia]

Publg Publishing [also publg]

Publ Health Rep Public Health Reports [US]

Publ Inst Math Publications de l'Institut Mathématique [Publications of the Mathematical Institute, Serbian Academy of Sciences, Belgrade]

Publ Inst Math, Acad Serbe Sc Publications de l'Institut Mathématique, Académie Serbe des Sciences [Publications of the Mathematical Institute, Serbian Academy of Sciences, Belgrade]

Publ Inst R Meteorol Belg A Publications, Institut Royal Meteorologique de Belgique, Serie A [Publications of the Royal Belgium Institute of Meteorology, Series A]

Publ Inst R Meteorol Belg B Publications, Institut Royal Meteorologique de Belgique, Serie B [Publications of the Royal Belgium Institute of Meteorology, Series B]

Publ Korean Astron Soc Publications of the Korean Astronomical Society [South Korea]

Publ Math Publicationes Mathematicae [Mathematical Publication of the Mathematical Institute of the University of Debrecen, Hungary]

Publn Publication [also publn]

Publ Natl Astron Obs Jpn Publications of the National Astronomical Observatory of Japan

Publns Publications [also publns]

Publ Opin Q Public Opinion Quarterly [US]

Publ Tech Univ Heavy Ind Publications of the Technical University for Heavy Industry [Miscolc, Hungary]

Publ Tech Univ Heavy Ind (Miskolc) B, Metall Publications of the Technical University for Heavy Industry (Miskolc) Series B, Metallurgy [Miscolc, Hungary]

Publ Util Fortn Public Utilities Fortnightly [Published in the US]

Pubn Publication [also pubn]

PUBnet The Public Network [of National Institutes of Health, Bethesda, Maryland, US]

Pubr Publisher [also pubr]

PUC Paid-Up Capital

PUC Program Under Control

PUC Processor Utility Controller

PUC Public Utilities Commission [Canada and US]

PUC Public Utilities Committee

PUCA Public Utilities Communicators Association [US]

PUCC Pontifica Universidad Católica de Chile [Pontifical Catholic University of Chile, Santiago]

PUCE Pontifica Universidad Católica del Ecuador [Pontifical Catholic University of Ecuador, Quito]

PU/CEES Princeton University/Center for Energy and Environmental Studies [US]

PUCK Propellant Utilization Checkout Kit

PUCOT Piezoelectric Ultrasonic Composite Oscillator Technique

PUCP Physical Unit Control Point

PUCRJ Pontifica Universidade Catolica do Rio de Janeiro [Pontifical Catholic University of Rio de Janeiro, Brazil] [also PUC-RJ, PUC/RJ, or PUC-Rio]

PUCRS Pontifica Universidade Catolica do Rio do Sul [Pontifical Catholic University of Rio do Sul, Porte Alegre, Brazil] [also PUC-RS, or PUC/RS]

PUCS Propellant Utilization Control System

PUCSP Pontifica Universidade Catolica do São Paulo [Pontifical Catholic University of Sao Paulo, Brazil] [also PUC-SP, or PUC/SP]

PUD Pickup and Delivery

PUD Polyurethane Dimethacrylate

PUD Public Utility (or Utilities) District

PUF Polyurethane Foam

PUF Presses Universitaires de France [University Press of France, Paris]

PUF Pressurized Unsaturated Flow (Method)

PUFA Polyunsaturated Fatty Acid

PUFFS Passive Underwater Fire Control Feasibility Study

PUFFT Purdue University Fast FORTRAN Translator [US]

PUG PASCAL (Computer Language) Users Group [US]

PUG Penta Users Group [US]

PUG PET (Personal Electronic Transactor) Users Group

PUG Petroleum User Group

PUGS Propellant Utilization (and) Gaging System

PUJ Pontifica Universidad Javeriana [Javeriana Pontifical University, Bogota, Colombia]

PUJT Programmable Unijunction Transistor

PUL Program Update Library

PUL Princeton University Library [New Jersey, US]

PUL Pulmonary Systems Effects (of Hazardous Materials)

pul pulmonary [Anatomy]

Pull B Sw Pull Button Switch

Pulm Art Pulmonary Artery [also pulm art]

Pulm Pharmacol Therap Pulmonary Pharmacology and Therapeutics [Journal published in the US]

Pulp Paper Pulp and Paper [Journal published in the US]

Pulp Paper Can Pulp and Paper Canada [Canadian Journal]

Pulp Paper Int Pulp and Paper International [Canadian Journal]

Pulp Paper Mag Pulp and Paper Magazine [Sweden]

Pulp Paper Wk Pulp and Paper Week [Published in the US]

Pulsar Pulsed Stellar Radio Source [also pulsar]

Pulsar Pulsed Electromagnetic Device [also pulsar] [Medicine]

PULSE Prime User Library Service [of National Prime (Computer System) User Group, US]

Pulv Pulverize(r) [also pulv]

pulv (pulvis) – powder [Medical Prescriptions]

pulvd pulverized [also pulv'd]

Pulvn Pulverization [also pulvn]

PUM Philipps-Universität Marburg [Philipps University at Marburg, Germany]

PUM Processor Utility Monitor

PUMA Programmable Universal Manipulator for Assembly

PUMA Programmable Universal Mechanical Assembly

PUMP Predictive Upkeep and Maintenance Program [of Transport Canada]

PUN Phenolic Urethane No-Bake (Binder) [Foundry Engineering]

PUN Physical Unit Number

PUN Punch [also Pun, or pun]

Punc Punctuation [also punc]

PUNS Permanently Unfit for Naval Service

PuO₂ Plutonyl (Radical) [Symbol]

PUP PARC (Palo Alto Research Center) Universal Packet (protocol) [of Xerox Corporation, US]

PUP Pennsylvania University Press [of Pennsylvania State University, University Park, US]

PUP Peripheral Unit Processor

PUP Peripheral Universal Processor

PUP Plutonium Utilization Program

PUP Presses Universitaires de France [University Press of France]

PUP Princeton University Press [New Jersey, US]

PU-PDMS Polyurethane/Poly-2-Dimethylsiloxane (Copolymer)

PUPO Pull Up Push Over

PUR Polyurethane [also PU]

PUR Procurement Request

PUR Purdue University Reactor [US]

PUR 1H-Purine [Biochemistry]

Pur Purchase(r); Purchasing [also pur]

Pur Purify; Purified; Purifier [also pur]

PUR-ABS Polyurethane/Acrylonitrile-ButadieneStyrene (Blend) [also PUR/ABS]

Pur Ig Purified Immunoglobulin [Biochemistry]

PURC Public Utilities Review Commission

Purch Purchase [also purch]

Purch Supply Manage Purchasing and Supply Management [Journal of the Institute of Purchasing and Supply, UK]

PURE Protecting and Utilizing Resources for our Environment (Program)

Pure Appl Chem Pure and Applied Chemistry [Journal of the International Union of Pure and Applied Chemistry]

Pure Appl Geophys Pure and Applied Geophysics [Journal published in Germany]

PUREX Plutonium and Uranium Recovery by Extraction (Process) [also Purex]

PUREX Plutonium-Uranium Reduction and Extraction (Process) [also Purex]

PURL Permanent Uniform Resource Locator [Internet]

purp purple

PURPA Public Utilities Reform Policy Act

PURPA Public Utility Regulatory Policy Act [US]

PUR-RIM Polyurethane Reaction-Injection Molding; Reaction-Injection Molded Polyurethane [also PUR RIM]

PUR-RRIM Polyurethane Reinforced Reaction-Injection Molding; Reinforced Reaction-Injection Molded Polyurethane [also PUR RRIM]

PURSUIT Purchaser/Supplier Information Transfer

PURT Pet Urine Removal Treatment

PURV Powered Underwater Research Vehicle

PUS Pontifica Universidad de Salamanca [Pontifical University of Salamanca, Spain]

PUS Processor Upgrade Socket

PUS Processor Utility Subsystem

PUS Programming Under Stress

PUS Public Understanding of Science [Journal of Institute of Physics, UK]

PUSAS Proposed United States of America Standard

PUSH Purchase and Use of Solar Heating

PUSHA Push All Registers [Computers]

PUSH-BUTT Buttress Thread, Push-Type [also PushButt]

PUSHF Push Flags [Computers]

PUSS Pallet Utility Support Structure

PUSS Pilot's Universal Sighting System

PUSSY Programmable Universal Seaching System

PUT Phosphate Uridyl Transferase [Biochemistry]

PUT Pickup Truck

PUT Pickup Tube

PUT Pilot under Training [also PU/T]

PUT Programmable Unijunction Transistor

PUT Program Update Tape

P-32 UTP Phosphorus-32 Uridine 5'-Triphosphate [also ^{32}P-UTP]

PUV Propellant Utilization Valve

PUV Pulsed Ultraviolet Excitation

PV Papilloma Virus

PV Papovavirus

PV Par Value

PV Particle Velocity

PV Parvovirus

PV Passport Verification

PV Path Verification

PV Patrol Vessel

PV Peak Value

PV Peak Voltage

PV Peripheral Vision

PV Perovskite [Mineral]

PV Phase Velocity

PV Phase Voltage

PV Photovoltage; Photovoltaic(s)

PV Pigment Volume

PV Pilot Vessel

PV Pipe Ventilated; Pipe Ventilation

PV Plan View

PV Plasticoviscous; Plasticoviscosity

PV Plug Valve

PV Polio Virus

PV Pons Variolii (of Brain) [Anatomy]

PV Pore Volume

PV Positional Vector

PV Positive Very-Small (Error)

PV Positive Volume

PV Potato Virus

PV Power Valve

PV Poynting Vector [Physics]

PV Present Value

PV Pressure-Control Valve

PV Pressure/Velocity (Limit)

PV Pressure Vessel

PV Pressure–Volume (Diagram) [also P-V]

PV Pressure–Velocity (Diagram)

PV Prevalve

PV Prime Vertical [Geodesy]

PV Pseudoviscosity; Pseudoviscous

PV Pulmonary Vein [Anatomy]

PV Puncture Voltage

4PV 1,4-Bis[4-(3,5-Di-tert-Butylstyryl)styryl] Benzene

P-V Peak-to-Valley (Ratio) [also P/V] [Telecommunications]

P-V Load–Displacement Voltage (Diagram)

P-V Pressure–Volume (Diagram) [also PV]

Pv Perovskite [Mineral]

pV partially coherent vacancy (in materials science) [Symbol]

p.v. *(per vaginam)* – through the vagina [Medicine]

PVA Pivalic Acid

PVA Plan View Area

PVA (Blood) Plasma Volume Expander [Medicine]

PVA Polyvinyl Acetate

PVA Polyvinyl Acid

PVA Polyvinyl Alcohol

PVA Population Viability Analysis

PVA Preburner Valve Actuator

PVA Propellant Valve Actuator

PVA+ Polyvinyl Acetal Plus

PVAA Polyvinyl Acetal

PVAC Present Value of Annual Charges

PVAc Polyvinyl Acetate [also PVAC]

PVAl Polyvinyl Alcohol [also PVAL, or PVAlc]

PVAMC Prairie View A&M (Agricultural and Mechanical) College [Prairie View, Texas, US]

PVAST Propeller Vibration and Strength Analysis

PVATSC Pyruvic Acid Thiosemicarbazone

PVB Polyvinyl Bromide

PVB Polyvinyl Butyral

PVB Polyvinyl Butyryl

PVBA Polyvinyl Benzoic Acid

PV:BONUS Building Opportunities in the United States for Photovoltaics [of US Department of Energy]

PVC Permanent Virtual Circuit

PVC Photovoltaic Cell

PVC Photovoltaic Conversion

PVC Pigment Volume Concentration

PVC Polyvinylcarbazole

PVC Polyvinyl Chloride

PVC Position and Velocity Computer

PVC Pulse Voltage Converter

PVC+ Chlorinated Polyvinylchloride

PVCA Polyvinyl Chloride Acetate

PVC-ABS Polyvinyl Chloride/Acrylonitrile-Butadiene-Styrene (Blend) [also PVC/ABS]

PVCBMA Polyvinyl Chloride Belting Manufacturers Association [US]

PVCC Chlorinated Polyvinylchloride

PVC-co-VA Poly(vinyl Chloride-co-Vinyl Acetate)

PVC-co-VA-co-MA Poly(vinyl Chloride-co-Vinyl Acetate-co-Maleic Acid)

PVC-CPE Polyvinyl Chloride/Chlorinated Polyethylene (Blend) [also PVC/CPE]

PVC-CPVC Polyvinyl Chloride/Chlorinated Polyvinyl Chloride (Blend) [also PVC/CPVC]

PVC-EPDM Polyvinyl Chloride/Ethylene Propylene Diene Monomer (Blend) [also PVC/ EPDM]

PVC-EVA Polyvinyl Chloride/Ethylene-Vinyl Acetate (Blend) [also PVC/EVA]

PVCH Polyvinylcyclohexane

PVC-MBS Polyvinyl Chloride/Methacrylate-Butadiene-Styrene (Blend) [also PVC/MBS]

PVCN Polyvinyl Cinnamate

PVC-NR Polyvinyl Chloride/Nitrile Rubber (Blend) [also PVC/NR]

PVC-PAE Polyvinyl Chloride/Polyaryl Ether (Blend) [also PVC/PAE]

PVC-PMMA Polyvinyl Chloride/PolymethylMethacrylate (Blend) [also PVC/PMMA]

PVD Paravisual Director

PVD Peripheral Vascular Disease [Medicine]

PVD Photovoltaic Detector

PVD Photovoltaic Device

PVD Physical(ly) Vapor-Deposited (Coating) [Often Followed by a Chemical Symbol, or Formula, e.g., PVD TiN Represents Physically Vapor Deposited Titanium Nitride]

PVD Physical Vapor Deposition

PVD Plan Video Display

PVD Plan View Display

PVD Polyvinylidene Fluoride

PV&D Purge, Vent and Drain [also PVD]

PVDA Polyvinylidene Acetate

PVDC Polyvinyl Dichloride [also PVdC]

PVDC Polyvinylidene Chloride [also PVdC]

PVDC Polyvinylidene Dichloride

PVD/CVD Physical Vapor Deposition/Chemical Vapor Deposition

PVDF Polyvinyl Difluoride [also PVdF]

PVDF Polyvinylidene Difluoride

PVDF Polyvinylidene Fluoride [also PVdF]

PVDF-D Durapore Polyvinylidene Difluoride (Polymer)

PVE Photoviscoelastic(ity)

PVE Photovoltaic Effect

PVE Polyvinyl Ether

PVE Polyvinyl Ethyl Ether

PVE Post-Vaccinal Encephalitis [Medicine]

PVE Prosthetic Valve Endocarditis [Medicine]

+ve positive

PVER Pattern Visual Evoked Response

PVF Lead Vanadium Fluoride

PVF Polyvinyl Fluoride

PVF Polyvinyl Formal

PVF Polyvinylidene Fluoride

PVF$_2$ Poly(vinylidene Difluoride)

PVFM Polyvinylformal

PVFO Polyvinyl Formal

PVGP Program Vehicle Grant Program [US]

PVHD Peripheral Vision Horizon Device

PVHD Peripheral Vision Horizon Display

PVI Poly(vinyl Imidazole)

PVI Polyvinyl Isobutyl Ether

PVI Premature Vulcanization Inhibitor

PVI Programmable Video Interface

PVK Polyvinyl Carbazole

PVL Parameter Value Language

pvl prevail

PVLA Poly-N-p-Vinylbenzyl-D-Lactonamide [Biochemistry]

PVM Parallel Virtual Machine

PVM Particulate Volumn Monitor

PVM Passthrough Virtual Machine (Protocol)

PVM Permanent Virtual Machine

PVM Planar Vibrational Mode

PVM Polyvinyl Methyl Ether

PVM Pulse Velocity Meter

PVMA Pressure Vessel Manufacturers Association [US]

PVMaT Photovoltaic Manufacturing Technology (Project) [of US Department of Energy]

PVME Polyvinyl Methyl Ether

PVME/MA Polyvinyl Methyl Ether/Maleic Anhydride

PVOH Polyvinyl Alcohol

P Vol Pore Volume [also p vol]

PVOR Precision VOR (Very-High Frequency Omnidirectional Radio Range)

PVP Parallel Vector Processing

PVP Pipelined Vector Processor

PVP Present Value Profit

PVP Polyvinyl Propionate

PVP Polyvinylpyridine

PVP Poly(2-Vinylpyridine); Poly(4-Vinylpyridine)

PVP Polyvinylpyrrolidone

PV-P Past Vice-President

P2VP Poly(2-Vinylpyridine); Poly(4-Vinylpyridine)

PVPA Plant Variety Protection Act [US]

PVPCC Poly(4-Vinylpyridinium Chlorochromate)

PVP-co-AA Poly(1-Vinylpyrrolidone-co-Acrylic Acid)

PVP-co-BMA Poly(4-Vinylpyridine-co-Butyl Methacrylate)

PVP-co-VA Polyvinylpyrrolidone-Polyvinyl Acetate Copolymer

PVPDC Polyvinylpyridinium Dichromate

PVPDC Poly(4-Vinylpyridinium Dichromate)

PVP-dPS-PVP Polyvinylpyridine-Deuterated Polystyrene Polyvinylpyridine (Triblock Polymer)

PVPHF Poly-[4-Vinylpyridinium Poly(hydrogen Fluoride)]

PVPHP Poly(4-Vinylpyridine Hydrogentribromide)

PVPI Polyvinylpyrrolidone-Iodine (Complex) [also PVP-I]

PVPO Plant Variety Protection Office [of US Department of Agriculture]

P2VP-PAA Poly(2-Vinylpyridine)–Polyacrylic Acid

PVR Plant Variety Rights

PVR Precision Voltage Reference

PVR Process Variable Record

P&VR Pure and Vulcanized Rubber (Insulation)

PVRC Pressure Vessel Research Committee [o fWelding Research Council, US]

PVRD Purge, Vent, Repressurize and Drain

PVS Panoramic Vision System

PVS Parallel Visualization Server

PVS Performance Verification System

PVS Photovoltaic System

PVS Polyvinyl Sulfate

PVS Portable Vacuum Standards

PVS Portable Vacuum System

PVS Positronium Velocity Spectroscopy

PVS Premate Verification System Test

PVS Program Validation Services

PVSEC (International) Photovoltaics Science and Engineering Conference [of Institute of Electrical and Electronics Engineers, US]

PVSK Polyvinyl Sulfate, Potassium Salt

PVT Page View Terminal

PVT Paint and Varnish Technologist; Paint and Varnish Technology

PVT Personal Verifier Terminal

PVT Physical Vapor Transport

PVT Polyvinyl Toluene

PVT Preflight Verification Test

PVT Pressure Vessel Technology

PVT Pressure Viscosity Test

PVT Pressure–Volume–Temperature (Diagram) [also P-V-T]

PVT Proportional Variable Transmission [Automobiles]

PVT Pulsed Video Thermography

PVT Pyrotechnic Verification Test

Pvt Private [Military Rank] [alsp PVT]

PVTB Productivity Validation Test Bed

PVTD PNNL (Pacific Northwest National Laboratory) Vitrification Technology Department [of US Department of Energy]

PVTEM Plan-View Transmission Electron Microscopy

PVTI Piping and Valve Test Insert

PVTOS Physical Vapor Transport of Organic Solids

PVTOS Physical Vapor Transport of Organic Solutions

PVTX Pressure–Volume–Temperature– Composition (Diagram) [also P-V-T-X]

Pvu Proteus vulgaris [Microbiology]

Pvu I Proteus vulgaris I (One) [Microbiology]

Pvu II Proteus vulgaris II (Two) [Microbiology]

PV USA Photovoltaics for Utility Scale Application (Site) [of University of California at Davis, US]

PVW Process Validation Wafer

PVWA Planned Value of Work Accomplished

PVWS Planned Value of Work Scheduled

PVX Potato Virus X

PVY Potato Virus Y

PW Packed Weight
PW Palau [ISO Code]
PW Paraffin Wax
PW Partial Wave
PW Password
PW Peak Width
PW Permanent Way
PW Physical Weathering [Geology]
PW Piano Wire
PW Pilot Wire
PW Plain Washer
PW Plasma Wave
PW Politechnika Wroclawska [Wroclaw Polytechnic
    Institute, Poland]
PW Potential Well [Physics]
PW Powder Weight
PW Precipitable Water [Meteorology]
PW Pressure Wave
PW Primary Winding
PW Prime Western (Grade of Slab Zinc)
PW Printed Wiring
PW Private Wire
PW Professional Write (Software)
PW Progressive Wave
PW Pulsewidth
PW50 Pulsewidth (Jitter) 50 [also PW_{50}] [Magnetic Media]
P&W Pratt and Whitney (Engine)
P&W Printing and Writing
P/W Doppler (Spectroscopic) Broadening Lineshape
    Parameter [Symbol]
pW negative logarithm of the ion product constant for
    water [Symbol]
pW picowatt [also pw]
PWA Ammonium Phosphotungstate
PWA Pacific Western Airlines [Canada]
PWA Partial Wave Analysis
PWA Peak Width Analysis
PWA Performance Warehouse Association [US]
PWA Pratt and Whitney Aircraft [US]
PWA Printed Wire Assembly
PWA Private Write Area
PWA Product Work Authorization
PWA Public Works Administration [US] [Defunct]
P&WA Pratt and Whitney Aircraft [US]
PWAA Paint and Wallpaper Association of America [now
    National Decorating Products Association, US]
PWAC Present Worth of Annual Charges
PWAC Present Worth of Annual Cost
PWAD Pratt and Whitney Aircraft Division [US]
P Wave Auricular Contraction (of Electrocardiogram) [also
    P wave] [Medicine]
P Wave Pressure Wave [also P wave] [Physics]
PWB Parallel Wire Bundle
PWB Printed Wiring Board

PWB Programmer's Workbench
PWBA Plane Wave Born Approximation [Physics]
PWBA Printed Wiring Board Assembly
PWBS Program Work Breakdown Structure
PWC Paper Wood Cellulose
PWC Professional Women in Construction [US]
PWC Public Works Canada
PWC Public Works Committee [US Senate]
PWC Pulsewidth Coded
PWCB Provincial Water Conservancy Bureau [Taiwan]
pW/cm² picowatt per square centimeter [Symbol]
PWD Participatory Work Design
PWD Petroleum Warfare Development [UK]
PWD Power Distribution
PWD Present Working Directory [Internet]
PWD Print Working Directory
PWD Public Works Department [US]
PWD Pulsewidth Discriminator
Pwd Powder [also Pwdr, pwdr, or pwd]
PWE Present Working Estimate
PWE Present Worth Expenditures
PWE Pulsewidth Encoder
PWEDA Public Works and Economic Development
    Association [US]
PWF Pacific Whale Foundation [US]
PWF Present Worth Factor
PWFA Plasma Wave Field Accelerator
PWFG Primary Waveform Generator
PWG Planning Working Group [of Eurocontrol]
PWGeYbTm Lead Fluoride–Tungstic Oxide–Germanium–
    Yttrium Fluoride–Thulium Fluoride (Glass Ceramic)
PWGSC Public Works and Government Services of Canada
    [Ottawa, Ontario Canada] [Formerly Department of
    Supplies and Services]
PWH PHB Weserhütte AG [Major German Manufacturer of
    Metallurgical, Mining, Offshore and Stockyard
    Equipment, Building Machinery, Transshipping
    Installations, etc.]
PWHS Public Works Historical Society [US]
PWHT Postweld Heat Treated; Postweld Heat Treatment
PWI E.O. Paton Electric Welding Institute [Kiev, Ukraine]
PWI Permanent Way Institution [UK]
PWI Pilot Warning Indicator
PWI Pilot Warning Instrument
PWI Proximity Warning Indicator
PWIA Plane Wave Impulse Approximation [Physics]
P-WIRE Private Wire [also P-Wire]
PWL Piecewise Linear (Control System)
PWL Power Level
PWM Plated Wire Memory
PWM Pokeweed (*Phytolacca americana*) Mitogen
    [Biochemistry]
PWM Pulsewidth Modulation
PWM Pulsewidth Multiplier

PWMCSCF Partial Wave Multiconfiguration Self-Consistant Field [Physics]

PWMD Printed Wiring Master Drawing

PWM-FM Pulsewidth Modulation–Frequency Modulation

PWN Pinewood Nematode [Biology]

PWO Paul Wild Observatory [Narrabri, New South Wales, Australia]

PWOM Peat/Water/Oil Mixture

PWP Particle per Wafer Pass [Electronics]

PWP Permanent Wilting Percentage [Ecology]

PWP Personal Word Processor

PWP Plasticized White Phosphorus

PWP Programmable Weld Positioner

pWp picowatt, psophometrically weighted

PWPP Pittsfield Works Power Plant [of General Electric Company in Massachusetts, US]

pWp0/pWpP picowatt, psophometrically weighted measured at a point of zero reference level

PWPT Professional Word Processing Technician

PWQO Provincial Water Quality Objectives [Canada]

PWR Power Amplifier

PWR Pressurized Water Reactor

Pwr Power [also pwr]

PWRC Patuxent Wildlife Research Center [of US Geological Survey/Biological Resources Division at Laurel, Maryland]

PWRC FMS Patuxent Wildlife Research Center Financial Management System [of US Geological Survey/Biological Resources Division at Laurel, Maryland]

PWR DWN Power Down [also Pwr Dwn]

PWR-FLECHT Pressurized Water Reactor–Full Length Emergency Core Heat Transfer

PWRS Programmable Weapons Release System

PWS Pricing Work Statement

PWS Programmer Workstation

PWS Psychological Warfare Association [US]

PWSCS Programmable Workstation Communication Services

p-WSe₂ P-Type Tungsten Selenide (Semiconductor)

PWSiYbTm Lead Fluoride–Tungstic Oxide–Silicon–Yttrium Fluoride–Thulium Fluoride (Glass Ceramic)

PWT Propulsion Wind Tunnel

pwt pennyweight [Unit]

PWTeYbTm Lead Fluoride–Tungstic Oxide–Tellurium–Yttrium Fluoride–Thulium Fluoride (Glass Ceramic)

PWThYbTm Lead Fluoride–Tungstic Oxide–Thorium–Yttrium Fluoride–Thulium Fluoride (Glass Ceramic)

PWTUCR Pressurized Water-Thorium-Uranium Converter Reactor

PWU Power Workers Union [Canada]

PWZrYbTm Lead Fluoride–Tungstic Oxide–Zirconium–Yttrium Fluoride–Thulium Fluoride (Glass Ceramic)

PX Phantom Coil [Telecommunications]

PX Polyxylylene

PX Post Exchange [US]

PX Press (Message)

PX Pressure–Concentration (Diagram) [also P-X]

PX Private Exchange

PX Pyroxene [Mineral]

PX₃ Phosphorus Trihalide [Formula]

PX₅ Phosphorus Halide [Formula]

.PX Primary Index [Paradox File Name Extension]

P-X Pressure–Composition (Diagram) [also P-X]

Px Peroxidase [Biochemistry]

Px Pyroxene [Mineral]

P(x) Poisson Distribution [Mathematics]

P-x Load-Deflection (Curve) [Mechanics]

p(X) Rayleigh Distribution [Mathematics]

$\phi(x)$ function of x [Symbol]

$\psi(x)$ distribution function [Symbol]

PXA Primary X-Ray Analysis

PXAS Polarized X-Ray Absorption Spectroscopy

PXBDE p-Xylenebis(diaminoethyl)amine

PXML Private Exchange Master List

PXS Portable X-Ray Source

PXSTR Phototransistor [also Pxstr, or pxstr]

PXT Polyxylylene Tetrahydrothiophenium Chloride

PXT Poly(xylylidenetetrahydrothiophenium)

$\phi(x,t)$ phase of the wave function dependent on position and time [Symbol]

P(x,y) Fresnel Propagation Function [Symbol]

$\varphi(x,y)$ position-dependent phase vector (of light wave) [Symbol]

$\psi(x,y,z)$ (Schroedinger's) field scalar (in quantum mechanics) [Symbol]

PY Paraguay [ISO Code]

PY Percus-Yevick (Hard-Sphere Interaction Model) [Physical Chemistry]

PY Per Year [also py]

PY Plastic Yield(ing) [Mechanics]

PY Platonic Year [Astronomy]

PY Porphyrin [also Py] [Biochemistry]

PY Program Year

PY Publication Year

PY2 Bis[2-(2-Pyridyl)ethyl]amine

P-Y Percus-Yevick (Equation) [Physics]

P/Y Pitch/Yaw [Aerospace]

Py Permalloy

Py Porphyrin [also PY] [Biochemistry]

Py Pyrites [Mineral]

Py Pyridine [also py] [Biochemistry]

Py Pyridyl [also py]

Py Pyrochlore [Mineral]

Py Pyrometric; Pyrometer; Pyrometer

py per year [also PY]

py pyridine [Abbreviation Used in Coordination Compound Formulas]

.py Paraguay [Country Code/Domain Name]

Pyaz 4-[{2-(Hexahydro-5,5,7-Trimethyl-1H-1,4-Diazepin-1-yl)-ethyl}amino]-4-Methyl-2-Pentanol [also pyaz]

PYBCO Praseodymium Yttrium Barium Copper Oxide (Superconductor)

Pyben 2-(2-Pyridyl) Benzimidazole [also pyben]

PyBOP Benzotriazol-1-yl-Oxytripyrrolidinophosphonium Hexafluorophosphate

PyC Pyrolytic Carbon (or Pyrocarbon)

Py2C Pyridine-2-Carboxylic Acid

Py3C Pyridine-3-Carboxylic Acid

Py4C Pyridine-4-Carboxylic Acid

Pyc Pycnometer [also pyc]

PyCloP Chlorotripyrrolidinophpsphonium Hexafluorophosphate

(Py$_m$CuO)$_n$ μ-Oxo Copper (II) [Soluble Form of CuO (II) in Pyridine Solution] [also (py$_m$CuO)$_n$]

Pyd Pyridazine [also pyd]

PYDCA Pyridine-2,6-Dicarboxylate [also pydca]

PYDIEN N-(2-Pyridinylmethyl)-N'-[2-{(2-Pyridinylmethyl)-amino}ethyl] 1,2-Ethanediamine

[15]pydieneN5 3,6,9,12,18-Pentaazabicyclo[12.3.1]octadeca-2,12-diene

[16]pydieneN5 3,6,10,13,19-Pentaazabicyclo[13.3.1]-nonadeca-2,13-diene

[17]pydieneN5 3,7,10,14,20-Pentaazabicyclo[14.3.1]-eicosa- 2,14-diene

PYDIPY 1-(2-Pyridyl)-3,5-Dimethylpyrazole

PYDPT N-(2-Pyridinylmethyl)-N'-[3-{(2-Pyridinylmethyl)-amino}propyl] 1,3-Propane-diamine

PYDV Potato Yellow Dwarf Virus

Pydz Pyridazine [also pydz]

PYF Praseodymium Yttrium Ferrite

PYFM Post-Yield Fracture Mechanics

PYFV Parsnip Yellow Fleck Virus

PYG Paraguayan Guarani [Currency]

PYI Presidential Young Investigator (Award) [of National Science Foundation, US]

Pyim 2-(2-Pyridyl)imidazole [also pyim]

PYLL Potential Years of Life Lost

Pym Pyrimidine [also pym]

Pymd Pyrimidine

PYMI 2-Pyridinemethanimine

PYNAPY 2-(2-Pyridyl) 1,8-Napthylridine

PyNO Pyridine-N-Oxide

PYO Pick-Your-Own (Fruit) [Agriculture]

PyO Permalloy Oxide

Pyo Pyridine-N-Oxide

Pyph Polyphase [also pyph]

Pyr Pyrometer; Pyrometry; Pyrometry [also pyr]

Pyr Pyrethroid

Pyr Pyridine [also pyr]

Pyr Pyrrolidine

Pyram Pyramid(al) [also pyram]

Pyrmtc Pyrrolidinemonothiocarbamate

Pyro Pyrotechnic(s) [also pyro]

Pyro Pyrocellulose

Pyro Pyrocotton

Pyromet Pyrometallurgical; Pyrometallurgist; Pyrometallurgy [also pyromet]

Pyrot Pyrotechnic(s) [also pyrot]

Pyrphos 3,4-Bis(diphenylphosphiono)pyrrolidine

3-Pyr-Py 3-(Pyrrol-1-ylmethyl)-Pyridine [also 3-PYR-PY]

Pyrr Pyrrole

PYS Partial Yield Spectroscopy

PYSAL N-(2-Pyridyl)salicyclaldimine

PYSZ Partially Yttria Stabilized Zirconia

(Py)3tach 1,3,5-tris(Pyridine-2-carboxaldimino) cyclohexane [also (py)3tach]

PYTHIA 2-Methyl-2-(2-Pyridyl)-4-Carbomethoxy-1,3-Thiazoline

Pythiaz 2,5-Bis(2-Pyridyl)thiazole [also pythiaz]

(Py)3tren N'-(2-Pyridylmethylene)-N,N-Bis[2{(2Pyridyl-methylene)amino}ethyl]-1,2-Ethanediamine [also(py)3tren]

PYY Peptide YY [Biochemistry]

Pyz Pyrazine [alsp pyz]

PYZC Pyrazinecarboxylate [also Pyzc]

Pyzl Pyrazole

Pyzn Pyrazine

PZ Lead Zirconate (Ceramics)

PZ Pick-Up Zone

PZ Plastic Zone

PZ Print Zone

PZ Pyrazine [also Pz, or pz]

PZ Pyrazole

PZ- Suriname [Civil Aircraft Marking]

Pz p-Phenylazobenzyloxycarbonyl

Pz Piezo [also pz]

Pz Pyrazine [also PZ, or pz]

pz pièze [Unit]

PZA Pyrazinamide [also Pza]

PZC Point of Zero Charge [Surface Science]

PZC Potential (of) Zero Charge [also pzc]

PZC Zirconium-Modified Polycarbosilane (Polymer)

PZNPC Point of Zero Net Proton Charge [Surface Science]

PZE Piezoelectric Element

PZI Protamine Zinc Insulin [Medicine]

PZJA Protected Zone Joint Authority [Torres Strait]

PZL Pyrazole [also Pzl, or pzl]

PZMN Lead Zinc Magnesium Niobate (Ceramics)

PZMN-PT Lead Zinc Magnesium Niobate–Lead Titanate (Ceramics)

PZN Lead Zinc Niobate (Ceramics)

p-ZnGeP$_2$ P-Type Zinc Germanium Phosphide

p-ZnMgSe P-Type Zinc Magnesium Selenide (Semiconductor)

PZN-PMN-PT Lead Zinc Niobate–Lead Magnesium Niobate–Lead Titanate (Ceramics)

PZN-PT Lead Zinc Niobate–Lead Titanate (Ceramics)

p-ZnSe P-Type Zinc Selenide (Semiconductor)

p-ZnSe:N Nitrogen-Doped P-Type Zinc Selenide (Semiconductor)

p-ZnSe:P Phosphorus-Doped P-Type Zinc Selenide (Semiconductor)

PZNT Lead Zinc Niobate Titanate (Ceramics)

p-ZnTe P-Type Zinc Telluride (Semiconductor)

Pzr Pressurizer [also pzr]

PZS Plastic Zone Site [Materials Science]

PZST Lead Zorconium Tin Titanate (Ceramics)

PZT Lead Zirconate Titanate (Ceramics)

PZT Photographic Zenith Time

PZT Photographic Zenith Tube

PZT Piezoelectric Transducer

PZT Piezoelectric Translator

PZT/PMN Lead Zirconate Titanate/Lead Magnesium Niobate (Ceramics)

PZZT Point of Zero Zeta Potential [Physics]

Q   Activation Energy [Symbol]

Q   Activation Enthalpy [Symbol]

Q   Arc Length of Wormgear Tooth Measured along Root [Symbol]

Q   (2,5-Benzoquinoyl)acetyl

Q   Calibrated Leak Flow [Symbol]

Q   (Electric) Charge [Symbol]

Q   (Electric) Charge Transfer [Symbol]

Q   Concentration in a Supersaturated Solution [Symbol]

Q   Demagnetization Energy [Symbol]

Q   Discharge (of Fluid) [Symbol]

Q   Discharge (Corona) Quantity [Symbol]

Q   Drone (Aircraft, Missile, or Rocket) [USDOD Symbol]

Q   Dynamic Pressure [Symbol]

Q   Energy Factor [Symbol]

Q   Exposure Index [Symbol]

Q   Extinction Efficiency [Symbol]

Q   Flow Rate [Symbol]

Q   Glutamine; Glutaminyl [Biochemistry]

Q   Heat of Adsorption [Symbol]

Q   Heat Flow (Rate) [Symbol]

Q   Integrated Intensity [Symbol]

Q   Interference Factor (of Bevel Gears) [Symbol]

Q   Ion Dose [Symbol]

Q   Magnetooptical Coupling Constant [Symbol]

Q   Magnification Factor [Symbol]

Q   Maximum Tire Speed of 99 mph (or 160 km/h) [Rating Symbol]

Q   Metal (or Material) Removal Rate (in Machining) [Symbol]

Q   Molecular Partition Function [Symbol]

Q   Moment of (Cross-Sectional) Area [Symbol]

Q   Momentum Transfer [Symbol]

Q   Neutron Momentum Transfer [Symbol]

Q   Number of Pieces Produced [Symbol]

Q   Pitch Rate (or Angular Rate) [Symbol]

Q   Precision of (Earthquake) Location [Symbol]

Q   Qatar [(Persian) Gulf State]

Q   Q-Factor [Symbol]

Q   Q-Ratio (i.e., Ratio of Mean Length to Mean Width of Elongated Grains) (in Quantitative Metallography) [Symbol]

Q   Q-Shell (of Electron) [Symbol]

Q   (Hydrostatic) Q-Stress (in Continuum Mechanics) [Symbol]

Q   Quadrant

Q   Quadratic

Q   Quadratics

Q   Quadrature

Q   Quadrillion

Q   Quadrupole

Q   Qualification; Qualifier; Qualify

Q   Qualitative; Quality

Q   Quality Factor [Symbol]

Q   Quantitative; Quantity

Q   Quantity of Electricity [Symbol]

Q   Quantity (or Amount) of Heat [Symbol]

Q   Quantization; Quantize

Q   Quantum

Q   Quarantine

Q   Quarter(ly)

Q   Quartermaster

Q   Quartile (of Distribution)

Q   *(Quasi)* – as it were; almost

Q   Quaternary [Geological Period]

Q   Quebec [Canada]

Q   Quebec [Phonetic Alphabet]

Q   Quebrachitol,

Q   Queen

Q   Queensland [Australia]

Q   Quench(ed); Quencher; Quench(ing)

Q   Quenosine [Biochemistry]

Q   Quercus [Genus of Deciduous Trees Including the Oaks]

Q   Query

Q   Question(able); Questioned; Questioning [also q]

Q   Quetzal [Currency of Guatemala]

Q   Queue

Q   Quisce(nce); Quiescent

Q   Quinhydrone

Q   Quinoid

Q   Quinol

Q   Quintillion [also quint, or QUINT]

Q   Quinuclidine

Q   Quire [Sheets of Paper]

Q   Quito [Ecuador]

Q   Quit(ting)

Q   Quotient

Q   Radiant Energy [Symbol]

Q   Rate of Stock Removal (in Grinding) [Symbol]

Q   Reaction Quotient [Symbol]

Q Reaction Rate [Symbol]

Q Resonance [Symbol]

Q Rotation (in Crystallography) [Symbol]

Q Scattering Factor [Symbol]

Q Scattering Vector [Symbol]

Q Statical Moment (of an Area about an Axis) [Symbol]

Q Storage Factor [Symbol]

Q Strain (in Electron Microscopy) [Symbol]

Q Target and Drone (Aircraft) [US Air Force Symbol]

Q Total-Water Mixing Ratio [Symbol]

Q Transistor [Semiconductor Symbol]

Q Value of a Physical Quantity [Symbol]

Q Wave Vector [Symbol]

Q Weight [Symbol]

/Q Quiet Option [Turbo Pascal]

-Q Countermeasures (Aircraft) [US Navy Suffix]

Q_∞ Final Value of a Physical Quantity [Symbol]

Q^+ quat [Symbol]

Q_0 Initial Value of a Physical Quantity [Symbol]

Q_0 Intrinsic Q [Symbol]

Q_0 Q-Factor at Resonant Frequency [Symbol]

Q_0 Output of a Transducer [Symbol]

Q^{-1} Internal Friction Spectrum [Symbol]

Q^{-1} Inverse Quality Factor [Symbol]

Q_1 First Quartile [Symbol]

Q_1 Quantity of Heat Absorbed (in Carnot Cycle) [Symbol]

Q_1 Input of a Transducer [Symbol]

Q_1 Final Quantity of Heat [Symbol]

Q_2 Initial Quantity of Heat [Symbol]

Q_2 Quantity of Heat Rejected (in Carnot Cycle) [Symbol]

Q_2 Second Quartile [Symbol]

Q_3 Third Quartile [Symbol]

Q_4 Fourth Quartile [Symbol]

Q_a Allowable Pile Bearing Load (in Civil Engineering [Symbol]

Q_a Apparent Creep Energy [Symbol]

Q_{app} Apparent Activiation Energy [Symbol]

Q_b Grain Boundary Transport (in Materials Science) [Symbol]

Q_{BD} Dielectric Breakdown Charge [also Q_{bd}] [Symbol]

Q_c Creep Activation Energy [Symbol]

Q_c Creep Activation Energy [Symbol]

Q_c Critical Wave Vector Transfer for Total Reflection [Symbol]

Q_d Diffusion Activation Energy [Symbol]

Q_e (Electric) Charge Extraction [Symbol]

Q_e Effective Creep Energy [Symbol]

Q_e Radiant Energy [Symbol]

Q_f Charge Density [Symbol]

Q_{g1} Control-Grid Dissipation [Electronic Symbol]

Q_{g2} Screen-Grid Dissipation [Electronic Symbol]

Q_{hkl} Wave Vector of Lattice Plane Set (hkl) [Symbol]

Q_i (Electric) Charge Injection [Symbol]

Q_i Imaginary Part of Magnetooptical Coupling Constant [Symbol]

Q_i Partial-Discharge (Corona) Magnitude [Symbol]

Q_L Lattice Diffusion Activation Energy (in Materials Science) [Symbol]

Q_L Loaded Q [Symbol]

Q_m Magnetic Storage Factor [Symbol]

Q_m Mobile Charge Density [Symbol]

Q_P Pipe Diffusion Activation Energy [Symbol]

Q_p Bearing Capacity (of a Pile) [Symbol]

Q_p Net Calorific Value [Symbol]

Q_r Real Part of Magnetooptical Coupling Constant [Symbol]

Q_S Stored Charged [Semiconductor Symbol]

Q_{sc} Semiconductor Depletion Layer Charge [Symbol]

Q_{sc} Space Charge Density [Symbol]

Q_{SD} Self-Diffusion Activation Energy [Symbol]

Q_{SDW} Wave Vector of Spin Density Wave (in Solid-State Physics) [Symbol]

Q_{SS} Steady-State Creep Activation Energy (in Materials Science) [Symbol]

Q_T Amount of Material Removed per Tool Life [Symbol]

Q_T Total Electric Charge [Symbol]

Q_t Partial-Discharge Quantity (Terminal Corona Discharge) [Symbol]

Q_{tj} Partial-Discharge (Corona) Quantity of the j-th Discharge Pulse [Symbol]

Q_v Vacancy Formation Energy (in Materials Science) [Symbol]

Q_V Activation Energy for Volume Diffusion [Symbol]

Q_v Gross Calorific Value (or Higher Heating Value) [Symbol]

Q_v Volume Diffusion [Symbol]

Q_X In-Plane Wavevector [Symbol]

Q_Z Out-of-Plane Wavevector [Symbol]

2Q 2-Q Spin State (in Solid-State Physics) [Symbol]

3Q 3-Q Spin State [Solid-State Physics]

q (electric) charge (of each charge carrier in a semiconductor) [Symbol]

q complementary probability (e.g., $q = 1-p$) [Symbol]

q coordinate of the imaginary axis (of complex plane) [Symbol]

q depolarization factor [Symbol]

q diffusional cross-section of dislocation (in materials science) [Symbol]

q discharge velocity (of water through soil or rock) [Symbol]

q dynamic pressure (in Fluid Dynamics) [Symbol]

q electron charge (0.1602×10^{-18} C) [Symbol]

q fatigue notch sensitivity (value) [Symbol]

q flow factor (for journal bearings) [Symbol]

q flow rate [Symbol]

q force constant [Symbol]

q heat (entering, or leaving a thermodynamic system) [Symbol]

q heat flow [Symbol]

q heat flux [Symbol]

q   (latent) heat of fusion [Symbol]

q   heat transfer rate [Symbol]

q   index of sensitivity (of a material) (used in stress concentration calculations) [Symbol]

q   ionic charge [Symbol]

q   load(ing) [Symbol]

q   load per unit distance [Symbol]

q   moisture [Symbol]

q   notch sensitivity [Symbol]

q   number of valence electrons per atom [Symbol]

q   probability [Symbol]

q   quadrillion [also quad, or QUAD]

q   quadrupole moment [Symbol]

q   quantity

q   quantity of electricity [Symbol]

q   quantity of heat [Symbol]

q   quart [Unit]

q   quaternion (operator) [Symbol]

q   quarterly

q   quarto

q   quasi

q   question(able)

q   quintal [Unit]

q   quire [1/20 of a ream of paper]

q   quit/no save (command) [Edlin MS-DOS Line Editor]

q   quotient of geometric series [Symbol]

q   radiation [Symbol]

q   ratio factor (of gears) [Symbol]

q   safety factor (of fusion reactor) [Symbol]

q   scattering vector [Symbol]

q   specific humidity [Symbol]

q   throughput [Symbol]

q   water vapor mixing ratio [Symbol]

q   wavenumber [Symbol]

q   wave vector [Symbol]

q^* saturation mixing ratio [Symbol]

q_+ (quantity of) positive charge [Symbol]

q_- (quantity of) negative charge [Symbol]

$q_\parallel$ wave vector parallel to surface [Symbol]

$q_\perp$ wave vector parallel to perpendicular [Symbol]

q_0 point charge [Symbol]

q_1 electric charge at point 1 [Symbol]

q_1, q_2 quartile distances from median [Symbol]

q_a allowable soil (or bearing) pressure [Symbol]

q_e electron charge (in conductors) [Symbol]

q_e ultimate bearing capacity (of soil, or rock mass) [Symbol]

q_{eff} effective depolarization factor [Symbol]

q_G throughput (of a pump) [Symbol]

q_i charge on the i-th component, element, grain, etc. [Symbol]

q_L leak rate [Symbol]

q_M magnetic interaction vector [Symbol]

q_m mass flow rate [Symbol]

q_m vapor flux [Symbol]

q_N molecule flow rate [Symbol]

q_n charge of a single negative carrier (of intrinsic semiconductor) [Symbol]

q_p charge of a single positive carrier (of intrinsic semiconductor) [Symbol]

q_p heat entering, or leaving a thermodynamic system during constant-pressure process [Symbol]

q_{pV} flow rate (of a pump) [Symbol]

q_u (unconfined, or uniaxial) compressive strength (of soil, or rock) [Symbol]

q_{ult} ultimate bearing capacity (of soil, or rock mass) [Symbol]

q_V volume flow rate (of a fluid) [Symbol]

q_v molar flow rate [Symbol]

4°   Quaternary (Ammonium Salt) [Symbol]

QA   Qatar [ISO Code]

QA   Quadrupole Amplifier

QA   Qualitative Analysis

QA   Quality Analysis

QA   Quality Assessment

QA   Quality Assurance

QA   Quantas Airways [Australia]

QA   Quantitative Analysis

QA   Quantum Acoustics

QA   Quasiamorphous [also qa]

QA   Quasi-Atom

QA   Quench Aging [Metallurgy]

QA   Quench Annealing [Metallurgy]

QA   Quenching Action

QA   Quenching Agent

QA   Query Analyzer

QA   Quick Acting

QA   Quick Assembly

QA   Quinic Acid

Q&A   Questions and Answers

Q/A   Electric Charge per Unit Area [Symbol]

.qa   Qatar [Country Code/Domain Name]

QAA   Qualified Administrator Assistant

QAB   Quaternary Ammonium Base

QAC   Quality Assurance Chart

QAC   Quality Assurance Checklist

QAC   Quasistatic Adiabatic Change

QAC   Quaternary Ammonium Compound

qa-C   quasiamorphous carbon

QACAD   Quality Assurance Corrective Action Document

QADS   Quality Assurance Data System

QAE   Diethyl-(2-Hydroxypropyl)aminoethyl (Cellulose)

QAE   Quality Assurance Engineer(ing)

QAE   Quaternary Aminoethyl (Cellulose)

QAE   Quench-Age Embrittlement [Metallurgy]

QAE-C   Quaternary Aminoethyl Cellulose

QAE-C   Diethyl-(2-Hydroxypropyl)aminoethyl Cellulose

QAF   Quality of Approach Flow [Water Turbines]

QAGC Quite Automatic Gain Control [also qagc]

QAH Quasi-Acoustical Holography

QAH Quaternary Ammonium Hydroxide

QAI Quality Assurance Instruction

QAIP Quality Assurance Inspection Procedure

QAK Quick-Acting Knob

QALY Quality Adjusted Life Years

QAM Quadrature Amplitude Modulation

QAM Quality Assurance Manual

QAM Quaternary Ammonium Membrane

QAM Queued Access Method

QAM-PAM Quadrature Amplitude Modulation–Pulse Amplitude Modulation

QAMA Quebec Asbestos Mining Association [Canada]

QAMP Quality Assurance Management Plan

QANGO Quasi-Autonomous Non-Government(al) Organization

QAO Quality Assurance Office

QA&O Quality Assurance and Operations

QAP Quality Assurance Procedure

QAP Quality Assurance Program

QAP Quaternary Amine Polymer

QAP Quaternary Ammonium Polymer

QAPI Quality Assurance Program Index

QAPP Quality Assurance Program Plan

QAPP Quality Assurance Project Plan

Q Appl Math Quarterly of Applied Mathematics [Published by the American Mathematical Society, US]

QA/QC Quality Assurance/Quality Control [also QAQC]

QAR Quality Assurance Representative

QAR Quality Assurance Requirements

QAR Quality Assurance Responsible

QAR Qatari Rial [Currency of Qatar]

QAS tris(2-Phenylarsinophenyl)arsine

QAS Quality Assurance System

QAS Quaternary Ammonium Salt

QAS Question Answering System

QASK Quadrature Amplitude Shift Keying

QASL Quality Assurance Systems List

QASP Quality Assurance Surveillance Plan

QAT Quality Action Team

QAT Quality Assurance Test

8QATSC 8-Quinolinaldehyde Thiosemicarbazone

QATT Qualification for Acceptance Thermal Testing

QAVC Quiet Automatic Volume Control [also qavc]

QAVT Qualification Acceptance Vibration Test

QB Q Branch [Spectroscopy]

QB Quantum-Box (Laser)

QB Quasi-Bragg [Metallurgy]

QB Quench Bath [Metallurgy]

QB Quick-Break (Switch) [also Q-B]

QB Quick Burning (Propellant)

QB Quincke Balance

Q band Microwave Frequency Band of 36.00 to 46.00 gigahertz

QBB Quartz Beam Balance

QBE Query by Example

QBF Query by Form

QBFP Queensland Boating and Fisheries Patrol [Australia]

QBIC Query By Image Content

Q-Boat Vessel for Trapping Boats [also Q-boat]

Q-BOP Quick-Quiet Basic Oxygen (Steelmaking Process); Quiet-Quick-Quality Basic Oxygen (Steelmaking) Process [Metallurgy]

Q-BOP/OBM Quiet-Quick Basic Oxygen (Steelmaking) Process/Oxygen-Bottom Blown Maxhuette [Metallurgy]

QBS Qualifications-Based Selection

QBS Quebec Bureau of Statistics [Canada]

QBS Quick-Break Switch

QBT Quad Bus Transceiver

QBT Quadrax Biaxial Tape

Q Bull Int Assoc Agric Libr Doc Quarterly Bulletin of the International Journal of the Association of Agricultural Librarians and Documentalists [Published in the Netherlands]

QC Quadratic Convergence

QC Quality Circles

QC Quality Code

QC Quality Control

QC Quantum Count

QC Quantum-Chemical; Quantum Chemist(ry)

QC Quantum Communications

QC Quantum Condition

QC Quantum Constant

QC Quantum Coordinates [Physics]

QC Quantum Counter

QC Quantum Critical

QC Quartz Clock

QC Quartz Crystal(line)

QC Quasichemical

QC Quasi-Cleavage [Crystallography]

QC Quasi-Conductor

QC Quasicontinuous

QC Quasi Continuum

QC Queens College [of City University of New York, Flushing, US]

QC Quench Cracking [Metallurgy]

QC Quenching Circuit

QC Quenching Crack [Metallurgy]

QC Queue Control [Computers]

QC Quezon City [Philippines]

QC Quick-Change (Tool)

QC Quick Code

QC Quick-Connect

QC Quick Count

QC Quiesce-Completed [Computers]

QCA Quality Control Analysis

QCA Quasi-Chemical Analysis

QCA Quaternary Carbon Atom

QCA Quantitative Chemical Analysis

QCA Quantum-Dot Cellular Automata

QCB Queue Control Block

QCC Quadrupole Coupling Constant

QCC Quality Control Center

QCC Quality Control Chart

QCC Quality Control Circle

QCC Queensland Conservation Council [Australia]

QCD Quantum Chromodynamics [Particle Physics]

QCCEC Queensland Conservation Council and Environment Center [Australia]

QCDFR Quasichemical Defect Formation Equilibrium Reaction

QCDR Quality Control Deficiency Report

Q-CdS Quantized (or Quantum Confined) Cadmium Sulfide

QCE Quality Control Engineer(ing)

QCE Quality Control Evaluation

QCE Quantum Computing in Europe (Pathfinder Conference)

QCESC Quad Cities Engineering and Science Control

QCF Quartz-Crystal Filter

QCFF Quantum Chemical Force Field (Calculation)

QCFO Queensland Commercial Fishermen's Organization [Australia]

QCFS Quartz-Controlled Frequency Standards

QCI Quadratic Configuration Interaction

QCI Quality Conformance Inspection

QCI Quality Counsellor to Industry

QCI Queen Charlotte Islands [Canada]

QCISD Quadratic Configuration Interaction Single (and) Double (Excitation)

QCL Quasi-Classical

QCM Quantitative Computer Management

QCM International Conference on Quantum Communication, Measurement, and Computing

QCM Quartz Crystal Microbalance

QCM Quartz Crystal Microgravimetric; Quartz Crystal Microgravimetry

QCM Quasichemical Method

QCMBPT Quasi-Degenerate Many-Body Perturbation Theory [Physics]

QCN Quadrupole Chemical-Shift Nonspinning (Subroutine) [of Fourier Transform Nuclear Magnetic Resonance (Computer Program)]

QCO Quartz Crystal Oscillator

QCO Quebec Construction Office [Canada]

Q Code Q-Type Telegraphic Code [also Q code]

Q Colorado Sch Mines Quarterly–Colorado School of Mines [US]

QCOM Quartz Crystal Oscillator Microbalance

QCOMPU Quantum Computation Conference [of SPIE– International Society for Optical Engineering]

QCONF Quantity Conversion Factor

QCOP Quality Control Operating Procedure

QCOT Qualitative Crystal Orbital Theory [Physics]

QCP Quality Control Protocol

QCP Quick Connect Panel

QCPE Quantum Chemistry Program Exchange [of Indiana University, Bloomington, US]

QCPE Bull Quantum Chemistry Program Exchange Bulletin [US]

QCPSK Quaternary Coherent Phase-Shift Keying

QC/QA Quality Control/Quality Assurance

QCR Quality Control/Reliability

QCR Quartz-Crystal Resonator

QCR Quick Change Response

QCS Quasichemical Solution

QCS Query Control Station

QCS Quasicrystal Structure

QCS Quick-Change System

QCSE Quantum-Confined Stark Effect [Physics]

QCSEE Quick, Clean, Short-Haul Experimental Engine

QCSEE Quiet Clean STOL (Short Takeoff and Landing) Experimental Engine

QCT Quality Control Test

QCT Quasi-Classical Trajectory (Calculation) [Physics]

QC&T Quality Control and Test

Q Cum Index Med Quarterly Cumulative Index Medicus [Publication of American Medical Association, US]

QCW Quadrature Carrier Wave

QD Quadratic Demodulation

QD Quantum Defect [Atomic Physics]

QD Quantum Detector

QD Quantum Discontinuity

QD Quantum Dot

QD Quartile Deviation

QD Quasi-Dielectric

QD Quenching Distance

QD Querty/Dvorak

QD Quick Disassembly

QD Quick Disconnect(ion)

QD Quiet Day

QD Quill Drive

QD Quinoline Dye

Q&D Quick and Dirty

Qd Quad

qd (quaque die) – every day [Medical Prescriptions]

QDA Quantity Discount Agreement

QDA Quantum Dot Array

QDA Quasiharmonic Debye Approximation [Physics]

QDC Quality Products, Delivered on Time at a Competitive Cost

QDC Quick Dependable Communications

QDC Quick Disconnect Coupling

QDC Quinolinium Dichromate

QDE Quantum Detection Efficiency

QDEH Queensland Department of Environment and Heritage [Australia]

QDI Quantum Design Inc. [San Diego, California, US]

.QDI Quicken Dictionary [File Name Extension]

QDLEED Quasi-Dynamical Low-Energy Electron Diffraction

Q-DLTS Charge-Based Deep Level Transient Spectroscopy

QDM Quantum Defect Method [Physics]

QDOS Quasiparticle Density of States [Physics]

QDOS Quick and Dirty Operating System

QDPI Queensland Department of Primary Industries [Australia]

QDPSK Quaternary Differential Phase Shift Keying

QDR Qualification Design Review

QDR Quantum Dot Refrigerator

QDRI Qualitative Development Requirements Information [US Army]

QDS Quality Data System

qds *(quater in die sumendus)* – four times daily [Medical Prescriptions]

QDSV Quiet Day Solar Variation [Astronomy]

.QDT Quicken Data [File Name Extension]

QDVPT Quasi-Degenerate Variational Perturbation Theory [Physics]

QE Quadrant Electrometer

QE Quadrature Error

QE Quadratic Equation

QE Qualitative Evaluation

QE Quality Engineer(ing)

QE Quality Essentials

QE Quadrant Elevation

QE Quadrupole Echo (Spectroscopy)

QE Quantum Effect

QE Quantum Efficiency [Electronics]

QE Quantum Electrodynamics

QE Quantum Electronic(s)

QE Quantum Emission

QE Quantum Equivalence

QE Quartz Electrometer

QE Quasi-Elastic(ity)

QE Quasi-Equilibrium [Geology]

QE Queue Entry

QE Quiesce (Communication)

QE Quincke Effect [Physics]

QE Quinhydrone Electrode

QE$_{ext}$ External Quantum Efficiency [Electronics]

QE$_{int}$ Internal Quantum Efficiency [Electronics]

QE II Queen Elizabeth II [Ocean Liner]

QEA Quantum Electronics and Applications [Published in the US]

QEA Queue Element Area

QEAS Quantum Electronics and Applications Society [now Lasers and Electro-Optics Society of Institute of Electrical and Electronics Engineers, US]

QEC Quick Engine Change

QED Quantum Electrodynamics [Particle Physics]

QED Quantum Electronics Division [of European Physical Optical Society]

QED Quantum Epitaxial Design

QED Quenched Epitaxy Desorption

QED Quick (Text) Editor

qed *(quod erat demonstrandum)* – which was to be proved, or demonstrated [also QED]

QEDC Quebec Engineering Design Competition [Canada]

QEDG Quadrant-Controlled Electro-Dynamic Gradient (Crystal Growth) [Materials Science]

QEE Quadruple-Expansion Engine

qef *(quod erat faciendum)* – which was to be done, or constructed [also QEF]

QEGS Quasi-Elastic Gamma-Ray Scattering

QEH Queen Elizabeth Hospital [Birmingham, UK]

QEI Quality Exposition International [US]

QEI Queen Elizabeth Islands [Canada]

QEJG Quantum Electronics Joint Group [Japan]

QEL Quality Element

QELS Quantum Electronics and Laser Science (Conference) [US]

QELS Quasi-Elastic Light Scattering

QEMM Quarterdeck Expanded Memory Manager [*Note:* Quarterdeck Corporation is a US Company]

QENS Quasi-Elastic Neutron Scattering

QEO Quasi-Electron-Occupation

QEP Quality Evaluation Package

QES Quantized Electronic Structure

QES Quasi-Elastic Scattering

QET Quantitative Extraction Technique

QET Quasi-Equilibrium Theory [Geology]

QETE Quality Engineering Test Establishment [of Department of National Defense, Canada]

QETG Quantum Electrotopography

QEW Quick Early Warning (Test) (in Vitamin Therapy) [Medicine]

QF Qabel Foundation [US]

QF Quadrupole Field [Physics]

QF Quadrupole Force [Physics]

QF Quality Factor

QF Quality Form

QF Quantum-Film (Laser)

QF Quantum Fluid

QF Quantum Fluxoid

QF Quartz Fiber

QF Quasi-Fission

QF Queue Full

QF Quick Firing (Gun)

QF Quick Freezing

QFA Quick File Access

QFA Quick Firing Ammunition

Q-FAX Quick Facsimile Transmission

QFA Queen's Faculty Association [of Queen's University, Kingston, Canada]

QFC Quick-Firing Converted

QFD Quality Function Deployment

QFD Quantum Formation Diagram [Physics]

QFD Quartz-Fiber Dosimeter

QFE Quartz-Fiber Electroscope

QFE Quasi Fermi Energy [Physics]

QFE Quick-Fix Engineering

QFE Transistor, Field Effect

QFET Quasi-Free-Electron Theory [Solid-State Physics]

Q Fever Query Fever [also Q fever] [Medicine]

QFEXT (Workshop on) Quantum Field Theory under the Influence of External Conditions

QFG Quick-Firing Gun

QFI Qualified Flying Instructor

QFIRC Quick Fix Interference Reduction Capability

QFITC Quinolizino-Substituted Fluorescein Isothiocyanate

QFL Quebec Federation of Labour [Canada]

QFL Quilted Full Liner

QFLOW Quota Flow Control

QFM Quality-Factor Meter

QFM Quantized Frequency Modulation

QFM Quartz-Fiber Manometer

QFMA Quebec Furniture Manufacturers Association [Canada]

QFMA Queensland Fish Management Authority [Australia]

QFP Quad Flat Pack [Electronics]

QFS Queensland Fisheries Service [Australia]

QFSK Quadrature Frequency Shift Keying

QFT Quantitative Feedback Technique

QFT Quantum Field Theory [Quantum Mechanics]

QFTHEP International Workshop on High Energy Physics and Quantum Field Theory

QG Quantum Gravitation; Quantum Gravity

QG Meeting on Constrained Dynamics and Quantum Gravity

QG Quartermaster-General

QG Quartz Glass

QG Quenching Gas

QG Quinidine Gluconate [Biochemistry]

QGIS (International Colloquium on) Quantum Groups and Integrable Systems

QGP Quark-Gluon Plasma [Nuclear Physics]

QGPC Qatar General Petroleum Corporation

QGS Quantity Gaging System

QH Quantum Hall (Effect) [Electronics]

QH Quarter-Hard (Temper) [also ¼H] [Metallurgy]

QH Quasi-Harmonic

QH Quebec Hydro [Canada]

QH Queens Hospital [New York, US]

QH Quench-Harden(ed); Quench Hardening [Metallurgy]

QH2 2,5-Dihydroxyphenylacetyl

¼H Quarter-Hard (Temper) [also QH] [Metallurgy]

qh (quaque hora) – every hour [Medical Prescriptions] [also qqh]

q2h (quaque duo hora) – every two hours [Medical Prescriptions] [also qq2h]

q4h (quaque quater hora) – every four hours [Medical Prescriptions] [also qqqh, qqh, or qq4h]

q6h (quaque sex hora) – every six hours [Medical Prescriptions] [also qq6h]

QHA Quasiharmonic Approximation

QHAF Quantum Heisenberg Antiferromagnet [also Q-1D HAF] [Solid-State Physics]

QHC Quick Heating/Cooling

QHD Quantum Hydrodynamics [Low-Temperature Physics]

QHE Quantum Hall Effect [Electronics]

QHE Quinhydrone Electrode

QHEC Quebec Hydro-Electric Commission [Canada]

QHERB Queensland Herbarium [Australia]

Q-HIP Quick Hot Isostatic Pressing [Powder Metallurgy]

QHM Quartz Horizontal-Force Magnetometer

QHN (Symposium on) Quarks in Hadrons and Nuclei [Particle Physics]

QHR Quality History Records

qhr (quaque hora) – every hour [Medical Prescriptions] [also qh]

QHS Quantum Hyperfine Structure [Physics]

qhs (quaque hora somnus) – at bedtime [Medical Prescriptions]

QI Quadratic Interpolation

QI Qualitative Inspection

QI Quantitative Inspection

QI Quality Inspection

QI Quartz Iodide

QI Quinoline-Insoluble (Material)

QI Quinone Inhibitor

QIA Quantitative Image Analysis

QIC Quality Information using Cycle Time [Computers]

QIC Quarter-Inch Cartridge

QIC Quarter-Inch Compatibility

QICT Qingdao Institute of Chemical Technology [PR China]

qid (quater in die) – four times daily [Medical Prescriptions]

QIDC Quebec Industrial Development Corporation [Canada]

.QIF Quicken Interchange Format (or Import File) [File Name Extension]

QIFC Quebec Institute of Floor Covering [Canada]

QIL Quad-In-Line [Electronics]

QIL Quartz-Iodine Lamp

QIMA Queensland Institute of Municipal Administration [Australia]

QIMR Queensland Institute of Medical Research [Australia]

QINFO Quantum Information, Decoherence and Chaos (Conference)

QIO Queue Input/Output

QIP Quad-In-Line Package [Electronics]

QIP Quality Incentive Program

QIP Quality Improvement Process

QIRC Quebec Industrial Research Center [Canada]

QIRI Quebec Industrial Relations Institute [Canada]

QIS Quench(ing) Induced Segregation [Metallurgy]

QISAM Queued Indexed Sequential Access Method

QIT Quality Improvement Team

QIT Quality Information and Test System
QIT Quebec Iron & Titanium Company [Sorel, Canada]
QIT Quick Immersion Thermocouple
QITL Quartz-Iodine Tungsten Lamp
QITMS Quadrupole Ion Trap Mass Spectrometry
QIUF Quebec Industrial Union Federation [Canada]
QJ Quantum Jump
QJEG Quarterly Journal of Engineering Geology [of Geological Society, London, UK] [also Q J Eng Geol]
Q J Exp Psychol Quarterly Journal of Experimental Psychology
Q J Jpn Weld Soc Quarterly Journal of the Japan Welding Society
Q J Math Quarterly Journal of Mathematics [of Oxford University Press, UK]
Q J Mech Appl Math Quarterly Journal of Mechanics and Applied Mathematics [of Oxford University Press, UK]
Q J R Astron Soc Quarterly Journal of the Royal Astronomical Society [UK]
QJRMS Quarterly Journal of the Royal Meteorological Society [UK] [also Q J R Meteorol Soc]
Q J Seismol Quarterly Journal of Seismology [of the Japan Meteorological Agency]
Q J Stud Alcohol Quarterly Journal of Studies on Alcohol [US]
QL Quadrupole Lens
QL Quantification (or Quantitation) Limit
QL Quantum Leap
QL Quantum Limit
QL Quartz Lamp
QL Quartz Lens
QL Quartz Liquefier
QL Quasiline (Spectrum)
QL Quasilinear(ization)
QL Query Language
QL Queue Length
QL Quick Look
Q&L Quote and Literature
QLA Quebec Library Association [Canada]
QLAP Quick Look Analysis Program
QLCR Quadrupolar Level-Crossing Resonance
QLD Queensland [Australia] [also Qld]
QLDM Quantum Liquid Drop Model
QLDR Quick Look Data Reference
QLDS Quick Look Data Station
QLE Quantum Langevin Equation
QLEED Quantitative Low-Energy Electron Diffraction
QLF Quality Loss Function
qlfd qualified
QLIC Queensland Land Information Council [Australia]
QLIS Queensland Land Information System [Australia]
QLIT Quick Look Intermediate Tape
QLP Quasi-Living Polymerization
QLQC Quasi-Lattice Quasi-Chemical (Method)
QLR Quasi Long-Range

QLR Quick Look Report
QLRB Quebec Labour Relations Board [Canada]
QLS Quasi-Elastic Light Scattering
QLS Quebec Land Surveyor [Canada]
QLS Quick Look Station
QLSA Queue Line Sharing Adapter
Qlty Quality [also qlty]
Qly Quality [also qly]
QM Equalizer Mode
QM Q-Machine [Plasma Physics]
QM Quadrature Modulation
QM Quadrupole Moment
QM Qualification Model
QM Qualification Motor
QM Quality Management; Quality Manager
QM Quality Monitoring
QM Quantitative Microscopy
QM Quantity Meter
QM Quantum-Mechanical(ly); Quantum Mechanics
QM Quartermaster
QM Quasi-Molecule [Atomic Physics]
QM Quasi-Momentum [Physics]
QM Queensland Museum [Brisbane, Australia]
QM Quenching Medium
QM Quick-Make (Switch) [also Q-M]
QMA Quadrupole Magic-Angle Spinning (Subroutine) [of Fourier Transform Nuclear Magnetic Resonance (Computer Program)]
QMA Quadrupole Mass (Spectrum) Analysis; Quadrupole Mass (Spectrum) Analyzer
QMA Quantitative Microbeam Analysis
QMA Queen's (University) Marketing Association [Kingston, Ontario, Canada]
QMAS Quadrupole Mass Analyzer for Solids
QMATH Mathematical Results in Quantum Mechanics (Conference)
QMC Quantum Mechanical Calculations
QMC Quantum Monte Carlo (Method)
QMC Quartermaster Corps [of US Army]
QMC Quebec Management Council [Canada]
QMC Queen Mary College [of University of London, UK]
QMC Quekett Microscopical Club [UK]
QMD Quadrupole Mass Detector
QMD Quantized Magnetic Disk; Quantum Magnetic Disk
QMD Quantum Molecular Dynamics
QMDO Qualitative Material Development Objective
QME Quantistical Macroscopic Effect
QMF Quadrupole Mass Filter
QMF Query Management Facility
QMG Quartermaster General [also QM Gen]
QMG Quench and Melt Growth (Solidification) [Metallurgy]
QM Gen Quartermaster General [also QMG]
QMH Queen Mary Hospital [Toronto, Canada]
QMI Qualification Maintainability Inspection

QMI Quality Management Institute [US]

QML Qualified Manufacturers List

QMMA Quebec Metals Mining Association [Canada]

QMP Q-Machine Plasma [Plasma Physics]

QMP Quality Management Plan

QMP Quantum-Mechanical Principles

QMP Quebec Metal Powders Limited [Canada]

QMQB Quick-Make, Quick-Break (Switch)

QMR Qualitative Material Requirement

QMRE Quantum-Mechanical Resonance Energy

QMRP Quality Management Registration Program

QMS Quadrupole Mass Spectrometer; Quadrupole Mass Spectrometry

QMS Quality Management Services

QMS Quartermaster Sergeant [also QM Sgt]

QMSA Quantitative Mobility Spectrum Analysis; Quantitative Mobility Spectrum Analyzer

QM Sgt Quartermaster Sergeant [also QMS]

QMST Quark Model of Superconductivity Type [Particle Physics]

QMT Quantum Mechanical Tunneling

.QMT Quicken Memorized List [File Name Extension]

QM&TC Quality Management and Technology Center [of DuPont Inc.]

QMTI Quadrature Moving Target Indicator

QMWC Queen Mary and Westfield College [of University of London, UK]

QMZPE Quantum Mechanical Zero Point Energy

QN Qiana Nylon

QN Quadruple Node (of Grain Boundary) [also qn] [Materials Science]

QN Quantum Noise

QN Quantum Number

QN Quasi-Newton(ian)

QN Query Normalization

Qn Question [also qn]

Qn Quotation [also qn]

Qn Quinoline

QNC Queensland Naturalists' Club [Australia]

QND Quantum Non-Demolition (Measurement)

QNDE Quantitative Nondestructive Evaluation (Conference) [also Q-NDE]

QNH Air Pressure in Inches [Unit]

QNLσM Quantum Nonlinear Sigma Model [Physics]

Q/nm Momentum Transfer per nanometer [also Q nm^{-1}]

QNPC Qatar National Petroleum Company

QNPWA Queensland National Parks and Wildlife Authority [Australia]

QNPWS Queensland National Parks and Wildlife Service [Australia]

QNQ Quantum Number of Quarks [Particle Physics]

QNQ Quinoid-Nitrogen-Quinoid

QNS Quadrupole Nonspinning Pattern (Subroutine) [of Fourier Transform Nuclear Magnetic Resonance (Computer Program)]

QNS Quantity Not Sufficient

QNS Quasi-Elastic Neutron Scattering

QNS Quinary Number System

Qnsld Queensland [Australia]

QNT Quantizer [also Qnt, or qnt]

.QNX Quicken Indexes to Data [File Name Extension]

Qnty Quantity [also qnty]

Qnxl Quinoxaline

QO Quantum Optics

QO Quinoline Oxide

QO Quinoxaline Oligomer

QOA Quality on Arrival

QOD Quick-Opening Device

Q1D Quasi-One-Dimensional [also Q-1D, Quasi-1D, or quasi-1D]

Q1D EG Quasi-One-Dimensional Electron Gas [also Q-1D EG] [Physics]

Q1D HAF Quasi-One-Dimensional Heisenberg Antiferromagnet [also Q-1D HAF] [Solid-State Physics]

Q1D P Quasi-One-Dimensional Peierls (System) [also Q-1DP] [Physics]

QOF Quenching of Fluorescence

QOL Quenching of Luminescence

QOM Quartz Oscillator Microgravimetry

QOM Quenching of Orbital Moments [Physics]

QOP Quenching of Phosphorescence

QOR Quenching of Radiation

QORC Quadrature Overlapped Raised Cosine

QOSRC Quadrature Overlapped Squared Raised Cosine

QOS Quality of Service

QP tris[(2-Diphenylphosphino)phenyl]phosphine

QP Quadruple Point

QP Quantum-Physical; Quantum Physicist; Quantum Physics

QP Quarter-Phase [Physics]

QP Quartz Plate

QP Quasi-Particle [Physics]

QP Quasi Peak [Physics]

QP Quasi-Periodic [Physics]

QP Quentin Process [Metallurgy]

QP Query Processing

QP Quiescent Prominence [Astronomy]

QP Quinacridone Pigment

QP Quinoline Polymer

QP Quinoxaline Polymer

qp *(quantum placet)* – as much as you please [Medical Prescriptions] [also qpl]

QPA Quantitative Precipitin Analysis [Immunology]

QPA Quaternary Polyacrylamide

QPAM Quadrature Phase and Amplitude Modulation

QPC Quantum Point Contact

QPC Quebec Press Council [Canada]

QPD Quadrature Phase Detection

QPD Quasi-Particle Detector

QPDM Quad Pixel Dataflow Manager

QPG Quantum Phase Gate

QPL Qualified Parts List

QPL Qualified Products List [of US Military]

qpl *(quantum placet)* – as much as you please [Medical Prescriptions] [also qp]

QPM Quality Program Manager

QPM Quantized Pulse Modulation [Communications]

QPM Quark-Parton Model [Particle Physics]

QPM Quasi-Periodic Motion [Physics]

QPM Quasi Phase Matched

QPN Quebrachitol Pentanitrate

QPNC Queensland Place Name Committee [Australia]

Q-pole Quadrupole

QPP Quantized Pulse Position [Communications]

QPP Quiescent Push-Pull (Amplifier)

QPPM Quantized Pulse Position Modulation

QPQ Quench-Polish-Quench (Technique) [Metallurgy]

QPR Quadrature Partial Response

QPR Quality/Price Ratio

QPR Quarterly Progress Report

QPRD Quality Planning Requirements Document

QPRI Qualitative Operational Requirement

QPRI Queensland Pharmaceutical Research Institute [Australia]

QPRK Quadrature Partial-Response Keying

QPRS Quadrature Partial-Response Signalling

QPS Quality Planning Specification

QPS Quantitative Physical Science

QPS Quantity Planning Specification

QPSK Quadrature Phase-Shift Key

QPSK Quaternary Phase-Shift Keying

QPSL Quasiperiodic Superlattice [Solid-State Physics]

QPSX Queued Package Synchronous Exchange

QPT Quadrant Power Tilt

QPTT Quick Prothrombin Time Test [Biochemistry]

QQ Quill and Quire

QQ Quintus Quartz

qqh *(quaque hora)* – every hour [Medical Prescriptions] [also qh]

qq2h *(quaque duo hora)* – every two hours [Medical Prescriptions] [also q2h]

qq4h *(quaque quater hora)* – every four hours [Medical Prescriptions] [also qqqh, qqh, or q4h]

qq6h *(quaque sex hora)* – every six hours [Medical Prescriptions] [also q6h]

qqv *(quae vide)* – which (things) see

QR Qatar Riyal [Currency of Qatar]

QR Quadrupole Radiation

QR Quadrupole Resonance

QR Quality and Reliability

QR Quantity Received

QR Quantum Resonance

QR Quartz Resonator

QR Quasi-Random

QR Quasi-Reflection

QR Quasi-Rigid

QR Queensland Railways [Australia]

QR Queen's Regulations [UK]

QR Quick Reaction

QR Quick Release

Qr Quarter [also qr]

Q(r) Rotational Partition Function [Symbol]

qr quarter [Unit]

qr quire [1/20 of a Ream of Paper]

QRA Quality Reliability Assurance

QRA Quasi-Relativistic Approximation [Physics]

QRA Queen's Research Award [of Queen's University, Kingston, Ontario, Canada]

QRA Quick Reaction Alert

QRBHCA Quebec Road Builders and Heavy Construction Association [Canada]

QRBM Quasi-Random Band Model [Physics]

QRC Quick Reaction Capability

QRCA Qualitative Research Consultants Association [US]

QRCC Query Response Communications Console

Q Rd Quarter-Round [also 1/4 Rd]

QRDP Quality, Reliability, Durability, Performance

QRE Quick-Reaction Estimate

Q Rep Railw Tech Res Inst Quarterly Report of the Railway Technical Research Institute [Japan]

Q Rev Quarterly Reviews [Chemical Journal published in the UK]

Q Rev Lond Quarterly Reviews, London [Chemical Journal published at London, UK]

Q Rev Biophys Quarterly Review of Biophysics [Published by the International Union for Pure and Applied Biophysics]

QRGA Quadrupole Residual Gas Analyzer

QRI Qatar(i) Rial [Currency of Qatar]

QRI Qualitative Requirements Information

QRI Quaternary Research Institute [of University of Waterloo, Ontario, Canada]

QRI Quick-Reaction Integration

QRI Quick-Reaction Interceptor

QRi Qatari Rial [Currency of Qatar]

QRIA Quick-Reaction Integration Activity

QRL Quick Relocate and Link

QRN Quasi-Random Noise

QRN Quasi-Resonant Neutralization

QRP Query and Reporting Processor

QRPG Quebec Rubber and Plastics Group [Canada]

QRR Quantum Railroads

QRS Quality and Reliability

QRS Quantity Received

QRS Quasi-Random

QRS Quick Reaction Sortie

QRS Ventricular Depolarization (on Electrocardiogram) [Medicine]

Qrs Quarters [also qrs]

QRSL Quick-Reaction Space Laboratory

QRSS Quasi-Random Signal Source

QRT Queue Run Time

qrtly quarterly

qr tr troy quarter [Unit]

Qry Quarry [also qry]

QS Q-Switch(ing) (Laser)

QS Quadrisoft

QS Quadrupole Scattering

QS Quadrupole Spectrometer

QS Quadrupole Splitting

QS Quality System

QS Quantitative Spectroscopy

QS Quantitative Stereology

QS Quantity Surveyor

QS Quantum Scattering

QS Quantum Solid [Solid-State Physics]

QS Quantum State [Quantum Mechanics]

QS Quantum Statistics [Statistical Mechanics]

QS (Workshop on) Quantum Systems [Physics]

QS Quartered Square (Method)

QS Quarter Session

QS Quartz Spectrograph(y)

QS Quasielastic Scattering

QS Quasistatic

QS Quasistable

QS Quasistationary

QS Quasisteady

QS Quaternary Structure

QS Queuing System

QS Query System

QS Queue Select

QS Quiet Sun [Astronomy]

Q-S Queneau-Schumann (Continuous Oxygen Converter) [Metallurgy]

qs *(quantum sufficiat; quantum satis)* – as much as may suffice, or as much as necessary [Medical Prescriptions]

QSA Quad Synchronous Adapter

QSA Qualification Site Approval

QSA Quantitative Surface Analysis

QSA Queen's Student Association [of Queen's University, Kingston, Ontario, Canada]

QSA Signal Strength Scale [Radion Engineering]

QSA1 Hardly Perceptible, Unreadable Signal [Radio Engineering]

QSA2 Weak Signal, Readable Now and Then [Radio Engineering]

QSA3 Fairly Good Signal, Readable with Difficulty [Radio Engineering]

QSA4 Good, Readable Signal [Radio Engineering]

QSA5 Very Good, Perfectly Readable Signal [Radio Engineering]

QSAM Quadrature Sideband Amplitude Modulation

QSAM Quasi-Steady Acceleration Measurement (Experiment) [of NASA]

QSAM Queued Sequential Access Method

QSAT Quebec Standard Asbestos Test

QSATS Quiet Short-Haul Air Transportation System

QSB Quartz Spring Balance

QSC Quasi-Single-Crystal(line)

QSD Quantum Structural Diagram [Physics]

QSE Quadratic Stark Effect [Atomic Physics]

QSE Qualified Scientists and Engineers

QSE Quantum Size Effect

QSEE Quiet STOL (Short Takeoff and Landing) Experimental Engine

QSF Quantized Signal Field (Method)

QSF Quasistatic Field [Physics]

QSF Quasi-Steady Flow (of Gases)

QSF Quantum Spin Fluid

QSFS Queensland Space Frontier Society [Australia]

QSG Quantum Sine-Gordon (Method) [Physics]

QSG Quasi-Stellar Galaxy [Astronomy]

QSG Quince Seed Gun

QSGVT Quarter Scale Ground Vibration Test

Q-Ship Vessel for Trapping Boats [also Q-ship]

QSL Q-Switched Laser

QSL Qualified Source List

QSL Quebec Safety League [Canada]

QSL Queneau-Schuhmann-Lurgi (Process) [Metallurgy]

QSL Queue Search Limit

QSM Quarter-Scale Model

QSMVT Quarter-Scale Model Vibration Testing

QSNT 1-(8-Quinolinesulfonyl)-3-Nitro-1H-1,2,4-Triazole

QSO Quasi-Stellar Object [also quasar]

QSOP Quadripartite Standing Operating Procedures

QSP Quasistatic Process [Thermodynamics]

QSP Quench Spray Pump

QSPP Quebec Society for the Protection of Plants [Canada]

QSPR Quantitative Structure-Property Relationships (Computer Program) [Materials Science]

QSR Quarterly Statistical Report

QSR Quasi Short-Range

QSR Quick-Set Recording

QSR Quiet-Sun Radiation

QSRA Quiet STOL (Short Takeoff and Landing) Research Aircraft

QSRMC Quality Scheme for Ready Mixed Concrete [UK]

QSRR Quantitative Structure Retention Relationship

QSRS Quasi-Stellar Radio Source

QSS Quartz Standby System

QSS Quasistable State

QSS Quasisteady State

QSS Quasistellar (Radio) Source

QSS Quench Spray Subsystem

QSS Quindar Scanning System

QST Have You Received the General Call? [Radio Code Call]

QST Quasi-Static Tester

QST Quasistationary Theory [Physics]

QST Quenched and Self-Tempered; Quenching and Self-Tempering [Metallurgy]

QSTAGS Quadripartite Standardization Agreements

QSTM Quebec Standard Test Method (for Asbestos)

Qstnry Quasistationary [also qstnry]

QSTOL Quiet Short Takeoff and Landing

QSW Quantized Spin Wave [Solid-State Physics]

QT Quadruple Thermoplastic (Electric Wire)

QT Qualification Test(ing)

QT Qualifier Type

QT Qualitative Test

QT Quality Technologist; Quality Technology

QT Quantitative Test

QT Quantum Theory

QT Quantum Transition

QT Quantum Turbulence

QT Quartz Thermometer

QT Quartz Tube

QT Quasi-Transverse

QT Quenched and Tempered; Quenching and Tempering [also Q&T] [Metallurgy]

QT Quench(ing) Temperature

QT Quench Test

QT Quench Time

QT Queuing Time

QT Queuing Theory [Mathematics]

QT Quill Tip

QT Quotient [also Qt, or qt]

QT Quotient Theorem

QT Transistor, Tetrode

Q&T Quenched and Tempered; Quenching and Tempering [also QT] [Metallurgy]

Q-T Interval from Start of QRS Complex to Finish of T Wave (on Electrocardiogram) [Medicine]

Qt Quantity [also qt]

Q(t) Cumulative Amount [Symbol]

Q(t) Time Dependent Amplitude [Symbol]

Q(t) Translational Partition Function [Symbol]

qt quart [Unit]

QTA Quebec Trucking Association [Canada]

QTAM Queued Telecommunications Access Method

QTAM Queued Terminal Access Method

QTAT Quick Turnaround Time (Line)

QTB Quarry Tile Base

QTBM Quantum Transmitting Boundary Method

QTC Quantum Theory of Cohesion [Physics]

QTC Quicktime Conferencing

QTCL Quantum Tunneling Cyclic Loop

Qtd Quartered [also qtd]

Q-2D Quasi Two-Dimensional (Electron Gas) [also Q2D, or QTD]

QTEST Quantitative Test [also QTest, or Qtest]

QTF Quantum Theory of Field

QTF Quarry Tile Floor

QTH Queued Transaction Handling

Q-TiO₂ Q-Sized Titanium Dioxide

QTIPS Quick Turn-Around Digital Image Printer Subsystem

QTM Quantitative Television Microscope

QTM Quantum Tunneling of Magnetization (Effect)

Qto Quarto [also qto]

QTOL Quiet Takeoff and Landing

QTP Qualification Test Plan

QTP Quality Test Plan

QTR Qualification Test Report

QTR Quantum Theory of Radiation [Physics]

QTR Quarry Tile Roof

QTR Quasi-Temperature Reduction

Qtr Quarter [also qtr]

QTRC Quality Training Research Center

QTRCD Quarter Code

Qtrs Quarters [also qtrs]

QTS Quarter Turn Screw

qts quarts [Unit]

QTST Quantum Transition State Theory [Physics]

QT Steel Quenched and Tempered Steel [also QT steel, Q&T Steel, or Q&T steel]

QTVM Quantitative Television Microscope

Qty Quantity [also qty]

QTYOH Quantity on Hand

QTYOO Quantity on Order

QTYOR Quantity Ordered

Qty/Pkg Quantity per Package

QTYSH Quantity Shipped

Qtz Quartz [also qtz]

QU Qinghua University [Beijing, PR China]

QU Queen's University [Kingston, Ontario, Canada]

QU Queen's University [Belfast, Northern Ireland]

QU Transistor, Unijunction

Qu Query [also qu]

Qu Question [also qu]

qu quarterly

qu quart [Unit]

qu quasi

QUA Queen's University Archives [of Kingston, Canada]

Quad Quadrangle [also quad]

Quad Quadrant [also quad]

Quad Quadrature [also quad]

Quad Quadrilateral [also quad]

Quad Quadrillion [also quad, or q]

Quad Quadrophonic; Quadrophony [also quad]

Quad Quadruplex [also quad]

quad quadruple

QUAD MOS Quad Metal-Oxide Semiconductor [also Quad MOS, or quad MOS]

Quad Ric Sci Quaderni de la Recerca Scientifica [Scientific Research Journal published by Consiglio Nazionale delle Ricerche, Italy]

QUADS Quality Achievement Data System,

Quaest Inf Quaestiones Informaticae [Computer Science Journal published by the Computer Society of South Africa]

Qual Qualification; Qualified; Qualifier; Qualitative; Quality [also qual]

Qual Assur Quality Assurance

Qual Assur Quality Assurance [Publication of the Institute of Quality Assurance, UK]

QualE Quality Engineer [also QualEng]

Qual Eng Quality Engineer(ing)

Qual Eng Quality Engineering [Journal of the American Society for Quality Control, US]

Qual Eval Quality Evaluation [Journal published in India]

QUALHYMO Quality/Quantity Simulation Model (for Ontario Waterbeds)

Qual Prog Quality Progress [Journal of the American Society for Quality Control]

Qual Reliab Eng Int Quality and Reliability Engineering International [US]

Qual Rev Quality Review [Published by the American Society for Quality Control]

Qual Rev Prat Contrôle Ind Qualité Revue Pratique de Contrôle Industriel [French Quality Review on Industrial Control Engineering]

QUALTIS Quality Technical Information Service [of United Kingdom Atomic Energy Authority]

Qual Today Quality Today [Published in the UK]

Qual Zuverlässigk Qualität und Zuverlässigkeit [German Journal on Quality and Reliability]

qualy qualitatively

QUAM Quadrature Amplitude Modulation

Quan Quantity [also quan]

quan quantitative

QUANGO Quasi-Autonomous National Government Organization

Quant Quantitative; Quantity; Quantum [also quant]

Quant Mech Quantum Mechanic(s) [also Quant Mech]

Quant Phys Quantum Physicist; Quantum Physics

QUANTRAS Question Analysis, Transformation and Search

Quantum Opt Quantum Optics [Journal published in the UK]

Quantum Semiclass Opt Quantum and Semiclassical Optics [Journal of the Institute of Physics, UK]

quanty quantitatively

Quar Quarry [also quar]

Quar Quarter [also quar]

QUARK Quantizer, Analyzer and Record Keeper

QUARKS International Seminar on High Energy Physics

Quarry Manage Quarry Management [Published in the US]

Quart Quarter(ly) [also quart]

Quart Appl Math Quarterly Applied Mathematics [of American Mathematical Society, US]

Quart Bull Int Assoc Agric Libr Doc Quarterly Bulletin of the International Journal of the Association of Agricultural Librarians and Documentalists [Published in the Netherlands]

Quart J Eng Geol Quarterly Journal of Engineering Geology [of Geological Society, London, UK]

Quart J Jpn Weld Soc Quarterly Journal of the Japan Welding Society

Quart J Math Quarterly Journal of Mathematics [of Oxford University Press, UK]

Quart J Mech Appl Math Quarterly Journal of Mechanics and Applied Mathematics [of Oxford University Press, UK]

Quart J R Astron Soc Quarterly Journal of the Royal Astronomical Society [UK]

Quart J R Met Soc Quarterly Journal of the Royal Meteorological Society [UK]

Quart J Seismol Quarterly Journal of Seismology [of the Japan Meteorological Agency]

Quart Rep Railw Tech Res Inst Quarterly Report of the Railway Technical Research Institute [Japan]

Quart Rev Quarterly Reviews [Chemical Journal published in the UK]

Quart Rev Biophys Quarterly Review of Biophysics [Published by the International Union for Pure and Applied Biophysics]

Quasar Quasi-Stellar (Object) [also quasar]

Quasar Quasi-Stellar (Radio) Source [also quasar]

Quasi-1D Quasi-one-dimensional [also quasi-1D, Q1D, or Q-1D]

Quat Quaternary [also quat]

Quat Quaternion [Mathematics]

Quat Quaternary Ammonium Compounds

Quat Int Quaternary International [Journal]

Quat Res Quaternary Research [Journal published in US]

Quat Sci Rev Quaternary Science Reviews [UK]

QUB Queen's University, Belfast [Northern Ireland]

qubit quantum bit [Quantum Computing]

QUC Queen's University Council [Kingston, Ontario, Canada]

QUE Quetzal [Currency of Guatemala]

Que Quebec [Canada]

Queensl Queensland [Australia]

QUELD Queen's University Experiment in Liquid-Metal Diffusion [Kingston, Ontario, Canada]

Quen/Temp Quenched and Tempered [also Quen/Temp] [Metallurgy]

Ques Question [also ques]

QUESAR Quasi-Elliptical Self-Adaptive Rotation

QUESSI Queen's University Engineering Society Services Incorporation [Kingston, Ontario, Canada]

QUEST Center for Advanced Quantized Electronic Structures [of National Science Foundation Science and Technology Center at University of California at Santa Barbara, US]

QUEST Quality Electrical Systems Test

QUEST Quality, Environment and Safety Tracking

QUEST Quantitative Experimental Stress Tomography Laboratory

QUEST Quantized Electronic Structures

QUEST Queen's University Engineering Solar Team [Kingston, Ontario, Canada]

QUEST Query Evaluation and Search Technique

quest questioned

Queueing Syst Theory Appl Queueing Systems Theory and Applications [Journal published in Switzerland]

QUI Quito, Ecuador [Space Tracking and Data Network Site]

QUIBEC N-Benzylchininium Chloride

QUIC Quality (Data) Information and Control (System)

QUIC Quality Unit Inventory Control

QUICK Queen's University Interpretative Coder, Kingston [Canada]

QUICKTRAN Quick FORTRAN

QUICO Quality Improvement through Cost Optimization

QUIL Quad-In-Line (Package) [Electronics]

QUILL Queen's University Inter-Library Loan [Kingston, Ontario, Canada]

QUIMAL Queen's University Immiscible Alloys Experiment [Kingston, Ontario, Canada]

Quin Quinaldinate [also quin]

Quin-2 N-[2-{(8-(Bis(carboxymethyl)amino)-6-Methoxy-2-Quinolinyl)methoxy}-4-Methylphenyl]-N-Carboxymethyl Glycine

Quin II AM (2-(2-Amino-5-Methylphenoxy)methyl-6-Methoxy-8-Aminoquinoline Tetraacetoxy Methyl Ester [also Quin2 AM]

Quintes Int Quintessence International [Journal]

QUIP Quad-In-Line Package [Electronics]

QUIP Query Interactive Processor

QUIP Questionnaire Interpreter Program

QUIP Quick-In-Line Package [Electronics]

QUIP Quota Input Processor

QUIPS Quantimet User Interactive Processing Software [also Quips]

QUIS Queen's University Information System [Belfast, Northern Ireland]

QUIT Quantum Induced Phase Transition

QUO Quadex Users Organization [US]

QUOBIRD Queen's University On-Line Bibliographic Information Retrieval and Dissemination [Belfast, Northern Ireland]

QUODAMP Queen's University Databank on Atomic and Molecular Physics [Belfast, Northern Ireland]

Quot Quotation [also Quotn, quot, or quotn]

QUP Quality Unit Pack

QUPID Quantum Path Integral Molecular Dynamics (Method)

QUSB Queen's University School of Business [Kingston, Ontario, Canada]

QUT Queensland University of Technology [Australia]

QV Quality Verification

QV Quantum Voltage

QV Quasiviscous

QV Quench Vessel

QV Q Value [Nuclear Physics]

Q(v) Vibrational Partition Function [Symbol]

qv *(quantum vis)* – as much as you like (or will) [Medical Prescriptions]

qv *(quod vide)* – which, or whom see

QVC Quasiviscous Creep [Mechanics]

QVE Quasiviscous Effect

QVM Queen Victoria Museum [Launceston, Tasmania, Australia]

QVT Qualified Verification Testing; Quality Verification Testing

QVVT Qualified Verification Vibration Testing

QW Quantum Wave

QW Quantum Well [Electronics]

QW Quantum Wire (Laser)

QW Quarter Wave

QW Quartz Wedge

QW Quer Wave [Physics]

QWA Quarter-Wave Antenna

Q Wave Initial Downward Defection of QRS Complex Indicating Start of Ventricular Depolarization (on Electrocardiogram) [Medicine]

QWB Quantum Well Box [Electronics]

QWC Quarter Wave Coupling

QWE Quantum-Wave Equation

QWG/CD Quadripartite Working Group on Combat Development

QWH Quantum Well Heterostructure [Electronics]

QWID Quantum Well Infrared Detector [Electronics]

QWIP Quantum Well Infrared Photodetector [Electronics]

QWL Quality of Work Life

QWL Quantum-Well Laser [Electronics]

QWLD Quantum Well Laser Diode [Electronics]

QWP Quarter-Wave Plate

QWR Quantum Wire [Electronics]

QWS Quantum Well Structure [Electronics]

QWT Quarter-Wave Transformer

QWTMA Queensland Wet Tropics Management Authority [Australia]

QWW Quantum Well Wire [Electronics]

QWW-HEMT Quantum Well Wire–High-Electron-Mobility Transistor [Electronics]

QY Quantum Yield [Physical Chemistry]

Qy Quarry [also qy]

Qy Query [also qy]

Q-YAG Q-Switched (Neodymium-Doped) Yttrium Aluminum Garnet (Laser)

QZ Qualität und Zuverlässigkeit [German Journal on Industrial Quality Assurance]

QZE Quadratic Zeeman Effect [Atomic Physics]

QZP Quadruple Zooming Power

R   Airflow Resistance [Symbol]

R   Alkyl Group [Symbol]

R   Arginine; Arginyl [Biochemistry]

R   Atomic Radius [Symbol]

R   Backscatter Factor [Symbol]

R   Bend Radius [Symbol]

R   Bond Length [Symbol]

R   Bulk Resistance [Symbol]

R   Charged Groups of Stationary Phase (in Ion Chromatography) [Symbol]

R   Circumradius (of a Triangle) [Symbol]

R   Clockwise Configuration (of Stereoisomer) (e.g., (R)-(+)-sec-Butyl Chloride) [Symbol]

R   Cloud Albedo [Symbol]

R   Code Grade Rubber Type (Conductor Insulation) [Symbol]

R   Cooling Rate [Symbol]

R   Curvature (of Mirrors, Lenses, etc.) [Symbol]

R   (Dislocation) Displacement Field (in Materials Science) [Symbol]

R   Distance from Load to Central Axis (of Springs) [Symbol]

R   Far Point (of the Eye) [Symbol]

R   Fatigue Ratio (in Mechanics) [Symbol]

R   Free Radical [Symbol]

R   (Universal) Gas Constant (8.31451 J/(mol·K)) [Symbol]

R   Grain Radius (in Metallurgy) [Symbol]

R   Growth Rate (in Metallurgy) [Symbol]

R   Heating Rate [Symbol]

R   Internuclear Separation [Symbol]

R   Jumping Probability (in Thin-Film Technology) [Symbol]

R   (Crystal) Lattice Vector [Symbol]

R   Lay Approximately Radial Relative to Center of the Nominal Surface [Symbol]

R   Lineshape Parameter (in Spectroscopy) [Symbol]

R   Load Ratio (in Fatigue) [Symbol]

R   Molar Gas Constant (8.31451 J/(mol·K)) [Symbol]

R   Multiple Correlation Coefficient (in Molecular Spectroscopy) [Symbol]

R   Number of Independent Reaction Equilibria [Symbol]

R   Number of Teeth (of Cutting Tools) [Symbol]

R   Organic Group (in Chemical Formulas) [Symbol]

R   Pair Correlation Coefficient (in Physics) [Symbol]

R   Perceived Intensity of a Stimulus [Symbol]

R   Pilling-Bedworth Ratio (in Corrosion Science) [Symbol]

R   Primitive (Simple) Space Lattice in Rhombohedral Crystal Systems [Hermann-Mauguin Symbol]

R   Rabat [Morocco]

R   Race

R   Racemic; Racemization

R   Rack(ing)

R   Radar

R   Radial

R   Radial Direction [Symbol]

R   Radial Tire [Symbol]

R   Radiance; Radiant

R   Radiancy [Symbol]

R   Radiate; Radiation; Radiator; Radiative

R   Radiative Heat Transfer [Symbol]

R   Radical [Chemistry]

R   Radio

R   Radioactive; Radioactivity

R   Radiograph(ic); Radiography

R   Radioisotope

R   Radiologic(al); Radiology

R   Radiometric; Radiometer; Radiometry

R   Radionuclide

R   Radiosonde

R   (External, or Outside) Radius [Symbol]

R   Radius [Anatomy]

R   Radiusing

R   Radius of Rotation (of a Solid Body) [Symbol]

R   Radius of Curvature (of a Surface, Lens, Mirror, etc.) [Symbol]

R   Radix

R   Radula (of Mollusks) [Zoology]

R   Raft

R   Rag

R   Rail

R   Railing

R   Railroad

R   Railway [also r]

R   Rain [also r]

R   Rainbow

R   Rain Rate [Symbol]

R   Raise(d); Raising

R   Rake

R   Raleigh [North Carolina]

R   Ram [also r]

R   Rammer; Ramming [Metallurgy]

R Raman Line Intensity (in Spectroscopy) [Symbol]
R Ramp
R Rand [Currency of Namibia and South Africa]
R Random (Copolymer)
R Random Error
R Randomization; Randomize
R Range [also r]
R Range Rate (in Electronics) [Symbol]
R Rangoon [Burma]
R Rank(ing)
R Rank Correlation Coefficient [Symbol]
R Rankine [Unit]
R Rapid(ity)
R Rare Earth
R Rarefaction; Rarefied; Rarefy
R Raster
R Rat [also r]
R Ratchet
R Rate [Symbol]
R Rate of (Information) Transmission [Symbol]
R Rating
R Ratio [also r]
R Rational(ization)
R Raw
R Rawinsonde [Meteorology]
R Ray
R Rayleigh Wave [Geophysics]
R Rayleigh Number [Symbol]
R Reach
R React(ant); Reaction; Reactor
R Reaction Layer
R Reaction Rate [Symbol]
R Reactive; Reactivity
R Read(ability); Readable
R Read(er); Reading
R Reagent
R Real(ity)
R Ream(er); Reaming
R Rear
R Reason
R Réaumur [Symbol]
R Rebound
R Recalescence; Recalescent [Metallurgy]
R Recarburization; Recarburize(r) [Metallurgy]
R Receipt
R Receive(r); Reception; Receptor
R Recess(ing)
R Recharge(able)
R Recife [Brazil]
R Recipient
R Reciprocate; Reciprocating; Reciprocation
R Recirculate; Recirculation
R Reclamation; Reclaim

R Recognition; Recognize(d)
R Recoil(ing)
R Recoilless
R Recombination
R Recommend(ation)
R Reconnaissance Aircraft [US Air Force and Army Symbol]
R Reconstruct(ion)
R Record(er); Recording
R Recover(able); Recovery
R Recrystallization; Recrystallize(d)
R Rectal; Rectum [Anatomy]
R Rectangle; Rectangular(ity)
R Rectification; Rectifier; Rectify(ing)
R Recuperation; Recuperative; Recuperator
R Recur(rence)
R Recycle(d); Recycling
R Red [also r]
R Reduce(r); Reducing; Reduction
R Redundancy; Redundant
R Reed
R Reef
R Reel(ing)
R Refer(ence)
R Referee
R Referral
R Refine(d); Refinement; Refiner(y); Refining
R Reflect(ion) [also r]
R Reflectance [Symbol]
R Reflection Coefficient [Symbol]
R Reflection Efficiency [Symbol]
R Reflective; Reflectivity
R Reflectivity [Symbol]
R Reflectometer; Reflectometry
R Reflector
R Reflex
R Reflux
R Reform(ation); Reformer
R Refract(ion); Refractive; Refractivity
R Refractometer; Refractometry
R Refractoriness; Refractory
R Refrigerant; Refrigerate(d); Refrigeration; Refrigerator
R Regeneration; Regenerative; Regenerator
R Regina [Saskatchewan, Canada]
R Region(al)
R Region (in Staticstics) [Symbol]
R Register
R Registrar
R Registers (Command) [Programming]
R Registration
R Regress(ion)
R Regular(ity)
R Regulate; Regulation; Regulator
R Rehydratable; Rehydrate; Rehydration

R Reinforce(d); Reinforcement; Reinforcer
R Reject(ion)
R Relation(al); Relationship
R Relative
R Relativistic; Relativity
R Relax(ation)
R Relaxation Rate (in Mechanics) [Symbol]
R Relay
R Relative Signal Strength [Symbol]
R Release
R Reliability; Reliable
R Reliability Factor [Symbol]
R Relief
R Relieving
R Reluctance; Reluctant
R (Magnetic) Reluctance [Symbol]
R Reluctivity
R Remainder [Symbol]
R Remanence; Remanent
R Remedial; Remedy
R Remit(tance)
R Remote [also r]
R Renal
R Renormalization; Renormalize
R Rendezvous [Aerospace]
R Reno [Nevada, US]
R Repair(ing)
R Repeat(er); Repeatability; Repetitive
R Reproducibility Limit (in Statistics) [Symbol]
R Replace(ment)
R Replenish(ment)
R Replica(tion)
R Report(ing)
R Represent(ation); Representative
R Reprocess(ing)
R Reproduce(r); Reproduction
R Reproducible; Reproducibility
R Reproducibility Limit [Symbol]
R Republic
R Repulsion; Repulsive
R Request
R Require(ment)
R Requisition
R Rescue (Dump)
R Research(er); Researching
R Reserve
R Reservoir
R Reset(ting)
R Residence; Resident(ial)
R Residual
R Residual Term (in Mathematics) [Symbol]
R Residue
R Resilience; Resilient

R Resin(ous)
R Resist
R Resist(ance); Resistant
R Resistive; Resistivity; Resistor
R (DC, Ohmic, or True) Resistance [Symbol]
R (Mechanical) Resistance [Symbol]
R Resolution; Resolve(r); Resolving
R Resonance; Resonant
R Resonate; Resonator
R Resorcinol
R Resource
R Resperine [Biochemistry]
R Respiration; Respirator(y)
R Respond(ent); Response; Responsive; Responsivity
R Rest
R Restore(d); Restoration
R Restrain(er); Restraint
R Restrict(ion)
R Restricted [Motion-Picture Rating]
R Restrike; Restriking
R Result
R Resultant Force [Symbol]
R Resumé
R Resuscitation; Resuscitator
R Retain(er)
R Retard(ation)
R Retardant; Retarder
R Retention; Retentivity
R Reticle; Reticular [Optics]
R Reticulum [Biology]
R Retina(l) [Anatomy]
R Retire(ment)
R Retort(ing)
R Retract(ion)
R Retrogression; Retrogressive
R Retry
R Return
R Revenue
R Reverberate; Reverberation; Reverberatory
R Reversal; Reverse(d); Reversibility; Reversible; Reversion
R Reverse Gear [Symbol]
R Revision
R Revolution; Revolve(d); Revolving
R Revolver
R Rework(ing)
R Reykjavik [Iceland]
R Reynolds Number [Symbol]
R Rheological; Rheology
R Rheostat(ic)
R Rhine [Europe]
R Rhodes [Greece]
R Rhodopsin [Biochemistry]
R Rhodotorula [Microbiology]

R   Rhomb(ic)
R   Rhombohedral (Crystal System) [Symbol]
R   Rhombohedron
R   Rhône (River) [Switzerland/France]
R   Rhizopus [Mycology]
R   Rib(bed)
R   Ribbon
R   (−)-Ribose [Biochemistry]
R   Ribosome [Cytology]
R   Richmond [Virginia, US]
R   Rickettsia [Genus of Bacteria]
R   Rider
R   Ridge
R   Riffle
R   Rifle
R   Rift
R   Right
R   Rigid(ity)
R   Rim
R   Rimmed; Rimming [Metallurgy]
R   Ring
R   Ringing
R   Rinse
R   (Rio) – Portuguese and Spanish for "River"
R   Ripple
R   Rise
R   Rise Time (in Acoustic Emission) [Symbol]
R   River
R   Rivet(ing)
R   Riyadh [Saudi Arabia]
R   Road
R   Roast(er); Roasting
R   Robot(ic); Robotics
R   Robust [Anthropology]
R   Rochester [New York, US]
R   Rock
R   Rocker
R   Rocking
R   Rocket
R   Rockwell Hardness [Symbol]
R   Rod
R   Rodent(icide)
R   Rodentia [Order of Mammals]
R   Rodrigues Vector [Symbol]
R   Roentgen [Symbol] [formerly also r]
R   Roll
R   Roller; Rolling
R   Roof
R   Roofer; Roofing
R   Romania(n)
R   Rome [Italy]
R   Romeo [Phonetic Alphabet]
R   Room

R   Root
R   Root Opening (of Weld) [Symbol]
R   Rope
R   Rosario [Argentina]
R   Rose
R   Rosette
R   Rot(ting)
R   Rotary; Rotatable; Rotate; Rotation(al); Rotator(y)
R   Rotation Matrix (in Crystallography) [Symbol]
R   Rotary Mode
R   Rotor
R   Rotterdam [Netherlands]
R   Rouen [France]
R   Rough(ness)
R   Roughing (of Vacuum Systems)
R   Round(ness)
R   Route(r)
R   Routine
R   Routing
R   Roving
R   Row
R   Royal(ty)
R   R-Phase (in Metallurgy) [Symbol]
R   Rubber(ize)
R   Rubber Type (Grinding Wheel) Bond
R   Ruble [Currency of Russian Federation]
R   Ruby
R   Rudder
R   Rugged(ness)
R   Rule
R   Ruler
R   Ruling
R   Rumania(n)
R   Rumen [Zoology]
R   Ruminant [Zoology]
R   Run
R   Runner
R   Running
R   Runway
R   Rupee [Currency of India]
R   Rupture(d)
R   Russia(n)
R   Rust [Botany]
R   Rust(ing) [Metallurgy]
R   Rutile [Mineral]
R   R-Value (or Crack Resistance) [Symbol]
R   Rwanda(n)
R   Rydberg Constant (= $1.097373 \times 10^7 \ m^{-1}$) [Symbol]
R   Scattering Ratio [Symbol]
R   Ship Lauched (Vehicle) [USDOD Symbol]
R   Shrinkage Ratio (of Soil) [Symbol]
R   Solomon (Unit) [Symbol]
R   Spectral Resolution [Symbol]

R Spring Rate [Symbol]

R Stress Ratio (i.e., Ratio of Minimum Stress S$_{min}$ to Maximum Stress S$_{max}$) (in Fatigue) [Symbol]

R Surface Roughness Factor [Symbol]

R Thermal Resistance [Symbol]

R Thermal Shock Resistance for Severe Heat Transfer (in Ceramics) [Symbol]

R Transmission Ratio (of Worm Gears) [Symbol]

R Variance [Symbol]

R Vector Space [Symbol]

R/ *(Recipe)* – Prescription

/R Run In Memory Option [Turbo Pascal]

-R Transport (Aircraft) [US Navy Suffix]

R· Alkyl Radical (in Chemistry) [Symbol]

R· Catalyst Species, or Initiator (in Polymer Science) [Symbol]

Ṙ (Nucleus) Growth Rate (in Metallurgy) [Symbol]

°R Degrees Rankine [Symbol]

°R Degrees Réaumur [Symbol]

R% Range (of Values) [Symbol]

R% Reduction in Percent [Symbol]

R% Relaxation in Percent (in Mechanics) [Symbol]

[R] Molecular Refraction [Symbol]

R̄ Average Range (in Statistics) [Symbol]

R̄ Mean Atomic Radius [Symbol]

R$_{\parallel}$ Magnetoresistance for Field Applied Parallel to Current [Symbol]

R$_{\parallel}$ Parallel Electric Resistance [Symbol]

R$_{\parallel}$ Resistance of Magnetoresistive Elements with Magnetization Oriented Parallel to Current [Symbol]

R$_{\perp}$ Magnetoresistance for Field Applied Perpendicular to Current [Symbol]

R$_{\perp}$ Perpendicular Electrical Resistance [Symbol]

R$_{\perp}$ Resistance of Magnetoresistive Elements with Magnetization Oriented Perpendicular to Current [Symbol]

R↓ Roll Down (Function) [Computers]

R↑ Roll Up (Function) [Computers]

R$_{\square}$ Sheet Resistance [Symbol]

R$_{\infty}$ Reflectivity [Symbol]

R$_{\infty}$ Rydberg Constant (1.09737315 × 10^7 m^{-1}) [Symbol]

R° Radius in Degrees of Arc [Symbol]

R′ Radius in Minutes of Arc [Symbol]

R′ Radius of Curvature of First Surface (of Lens, Mirror, etc.) [Symbol]

R′ Thermal Shock Resiatance for Relatively Mild Heat Transfer (in Ceramics) [Symbol]

R″ Radius in Seconds of Arc [Symbol]

R″ Radius of Curvature of Second Surface (of Lens, Mirror, etc.) [Symbol]

R⁗ Thermal Shock Resiatance for Damage Resistance Parameter (in Ceramics) [Symbol]

R$_0$ Coefficient of Ordinary Hall Effect [Symbol]

R$_0$ Matrix Crack Resistance (of Composites) [Symbol]

R$_0$ (Chemical) Reducing Agent [Symbol]

R$_0$ (Ohmic) Resistance (of a Conductor) at 0°C [Symbol]

R$_0$ (Ohmic) Resistance at Reference Temperature t$_0$°C [Symbol]

R$_0$ Universal Gas Constant (8.31451 J/mol·K) [Symbol]

R$_0$ Zero-Field Resistance [Symbol]

R$_1$ Coefficient of Extraordinary Hall Effect [Symbol]

R$_1$ (Magnetic) Core Resistance [Symbol]

R^2 Coefficient of Multiple Determination (in Molecular Spectroscopy) [Symbol]

R^2 Regression Coefficient (in Statistics) [Symbol]

R^2 R-Squared (in Statistics) [Symbol]

R^2 Two-Dimensional (Vector) Space (in Mathematics) [Symbol]

R$_2$′ Rayleigh Number 2 (for Free Convection) [Symbol]

R^3 Reclaim, Reuse, Reduce [also R3, RRR, or 3R]

R^3 Reduce, Reuse, Recycle [also R3, RRR, or 3R]

R^3 Three-Dimensional (Vector) Space (in Mathematics) [Symbol]

R$_3$ Rheumatoid Arthritis (Test) [Symbol]

R^5 Reduce, Repair, Reuse, Recycle, Rethink [also R5]

R$_{10}$ Alpha-Cellulose

R$_{100}$ (Electrical) Resistance (of a Conductor) at 100°C [Symbol]

R$_A$ Atomic Radius [Symbol]

R$_A$ Resistance of Amorphous Layer [Semiconductor Symbol]

R$_A$ Coefficient of Retroreflection [Symbol]

R$_A$ Rockwell 'A' Hardness (Conical Diamond Indenter; Applied Load 60 kg) [Symbol]

R$_a$ Anode (or Plate) Resistance [Symbol]

R$_a$ Antisymmetric (Component of Dislocation) Displacement Field (in Materials Science) [Symbol]

R$_a$ Arithmetic Average (Surface) Roughness [Symbol]

R$_a$ Armature Resistance [Symbol]

R$_a$ Reduction of Area (in Tension Test) [Symbol]

R$_{AC}$ AC (Alternating-Current) Resistance [also R$_{a-c}$]

R$_B$ Base Resistance [Semiconductor Symbol]

R$_B$ Electron Backscattering Factor [Symbol]

R$_B$ Rockwell 'B' Hardness (1/16 in. Dia. Steel Sphere Indenter; Applied Load 100 kg) [Symbol]

R$_B$ Specific Gas Constant [Symbol]

R$_b$ Base Resistance [Semiconductor Symbol]

R$_b$ Reflectance Factor of a Thin Film or Sheet of Material with a Black Backing [Symbol]

R$_{b1}$ Resistance of Base 1 [Semiconductor Symbol]

R$_{b2}$ Resistance of Base 2 [Semiconductor Symbol]

R$_{Be}$ Residual Term of Bessel's Formula [Symbol]

R$_C$ Collector Resistance [Semiconductor Symbol]

R$_C$ Effective Degree of Polymerization [Symbol]

R$_C$ Rockwell 'C' Hardness (Conical Diamond Indenter; Applied Load 150 kg) [Symbol]

R$_c$ Ashcroft Empty-Core Pseudopotential Radius (in Physics) [Symbol]

R$_c$ Contact End Resistance [Symbol]

R$_c$ Critical Cooling Rate [Symbol]

R$_c$ Critical Radius [Symbol]

R_c Ion Core Radius [Symbol]

R_D Diode Resistance [Semiconductor Symbol]

R_D Direct Reset (of a Flip-Flop) [Symbol]

R_D Rockwell 'D' Hardness (Conical Diamond Indenter; Applied Load 100 kg) [Symbol]

$R_{\Delta G}$ Driving Force Ratio (in Chemical Vapor Deposition) [Symbol]

R^E Enhanced Rutherford Backscattering (Spectroscopy) [Symbol]

R_E Emitter Resistance [Semiconductor Symbol]

R_E Rockwell 'E' Hardness (1/8 in. Dia. Steel Sphere Indenter; Applied Load 100 kg) [Symbol]

R_e Contact Resistance [Symbol]

R_{eff} Effective Resistance [Symbol]

R_{eq} Equivalent (Electrical) Resistance [Symbol]

R_{eq} Equivalent Noise Resistance [Symbol]

R_{eq} Resistive Component of a Series Equivalent Circuit [Symbol]

R_{EXT} External Resistance (or Resistor) [Symbol]

R_F Feedback Resistor [Semiconductor Symbol]

R_F Fresnel Reflectivity [Symbol]

R_F Rockwell 'F' Hardness (1/16 in. Dia. Steel Sphere Indenter; Applied Load 100 kg) [Symbol]

R_f Relative Molecular Mobility [Symbol]

R_f Feedback Resistance [Semiconductor Symbol]

R_G Guinier Radius (in Solid-State Physics) [Symbol]

R_G Radius of Gyration [Symbol]

R_G Rockwell 'G' Hardness (1/16 in. Dia. Steel Sphere Indenter; Applied Load 100 kg) [Symbol]

R_g Grid-Leak Resistance [Semiconductor Symbol]

R_g Grid Resistor [Semiconductor Symbol]

R_g Radius of Gyration [Symbol]

R_{g1} Grid-Leak Resistance [Semiconductor Symbol]

R_H Hall Coefficient (or Constant) [Semiconductor Symbol]

R_H Quantized Hall Resistance (2.581281×10^4 Ω) [Symbol]

R_H Rockwell 'H' Hardness (1/8 in. Dia. Steel Sphere Indenter; Applied Load 100 kg) [Symbol]

R_I Coefficient of Luminous Intensity (of a Retroreflector) [Symbol]

R_i Internal Resistance (of Electron Tubes) [Symbol]

R_i Thermal Interfacial Resistance [Symbol]

R_{ig} Global Interface Thermal Resistance [Symbol]

R_{in} Input Resistance [Semiconductor Symbol]

R_{INT} Internal Resistance (or Resistor); Internal (Source) Resistance [Semiconductor Symbol]

R_K Rockwell 'K' Hardness (1/8 in. Dia. Steel Sphere Indenter; Applied Load 150 kg) [Symbol]

R_k Cathode Bias [Electronic Symbol]

R_k Cathode Resistor [Electronic Symbol]

R_L Coefficient of Retroreflected Luminance [Symbol]

R_L Larmor Radius (of Electrons) [Symbol]

R_L Load Resistance; Load Resistor [Symbol]

R_L Profile Roughness Parameter (in Fracture Mechanics) [Symbol]

R_L Rockwell 'L' Hardness (1/4 in. Dia. Steel Sphere Indenter; Applied Load 60 kg) [Symbol]

R_M Coefficient of Line Retroreflection [Symbol]

R_M Rockwell 'M' Hardness (1/4 in. Dia. Steel Sphere Indenter; Applied Load 100 kg) [Symbol]

R_m Moving Average (in Staticstics) [Symbol]

R_m Thermal Resistance of Mold (in Metallurgy) [Symbol]

R_{max} Maximum Peak-to-Valley Height within Sampling Length "L" (Surface Roughness) [Symbol]

R_{min} Minimum Bend Radius [Symbol]

R_N Normal-State Sheet Resistance [Symbol]

R_N Norton's Equivalent Resistance [Symbol]

R^n n-Dimensional Vector Space (in Mathematics) [Symbol]

R_n (Crystal) Growth Rate Normal to Isotherm [Symbol]

R_n Normal Resistance [Symbol]

R_{off} Off-Resistance [Symbol]

R_{on} On-Resistance [Symbol]

R_{out} Output Resistance [Semiconductor Symbol]

R_P Rockwell 'P' Hardness (1/4 in. Dia. Steel Sphere Indenter; Applied Load 150 kg) [Symbol]

R_p Alternating-Current Plate Resistance [Symbol]

R_p Complex Reflection Coefficient of Light Polarized Parallel to Incidence Plane [Symbol]

R_p Maximum Profile Height from the Mean Line within the Sampling Length (Surface Roughness) [Symbol]

R_p Parallel Equivalent Resistance [Symbol]

R_p Polarization Resistance [Symbol]

R_p Projected Range [Symbol]

R_p Pull-up Resistor [Symbol]

R_p Reflectance at Point "p" [Symbol]

R_p Reflection Coefficient for Plane Polarization [Symbol]

R_p Resistance in Primary Coil (of Electric Machinery) [Symbol]

R_ϕ Thermal Resistance at Phase Boundary (in Fluid Flow) [Symbol]

R_{pm} Maximum Profile Height R_p Determined over Five Sampling Lengths (Surface Roughness) [Symbol]

R_{pp} Peak-to-Peak Reflectivity [Symbol]

R_q Root-Mean-Square (Surface) Roughness [Symbol]

R_R Rockwell 'R' Hardness (1/2 in. Dia. Steel Sphere Indenter; Applied Load 60 kg) [Symbol]

R_{ref} Reflected Resistance [Symbol]

R_S Rockwell 'S' Hardness (1/2 in. Dia. Steel Sphere Indenter; Applied Load 100 kg) [Symbol]

R_S Series Resistance [Symbol]

R_S Spontaneous Hall Coefficient [Symbol]

R_S Surface Resistance [Symbol]

R_S Surface Roughness Parameter (in Fracture Mechanics) [Symbol]

R_s Complex Reflection Coefficient of Light Polarized Perpendicular to Incidence Plane [Symbol]

R_s Effective Shunt Resistance [Symbol]

R_s Reflection Coefficient for S-Polarization [Symbol]

R_s Resistance of Sample (or Specimen) [Symbol]

R_s Series Resistance [Symbol]

R$_s$ Sheet Resistance (of Semiconductors, or Thin Films) [Symbol]

R$_s$ Solution Resistance [Symbol]

R$_s$ (Voltage) Source Resistance [Semiconductor Symbol]

R$_s$ Spreading Resistance (of Semiconductors) [Symbol]

R$_s$ Strain Resolution (or Strain Sensitivity) [Symbol]

R$_s$ Substrate Resistance [Symbol]

R$_s$ Surface Resistance [Symbol]

R$_s$ Symmetric (Component of Dislocation) Displacement Field (in Materials Science) [Symbol]

R$_s$ Thermal Resistance of Solid(ified Material) [Symbol]

R$_{sa}$ Sheet Resistance in Alloyed Semiconductor Layer [Symbol]

R$_{sc}$ Screening Length [Symbol]

R$_{sh}$ Sheet Resistance (of Semiconductor) [Symbol]

R$_{sk}$ Sheet Resistance of Semiconductor Layer Beneath Contact [Symbol]

R$_{St}$ Residual Term of Stirling's Formula (in Mathematics) [Symbol]

R$_{su}$ Sheet Resistance in Unalloyed Semiconductor Layer [Symbol]

R$_T$ Radiancy (of a Blackbody) [Symbol]

R$_T$ Total (Electrical) Resistance [Symbol]

R$_T$ Transition-Metal Atomic Radius [Symbol]

R$_T$ Trigger Resistance [Semiconductor Symbol]

R$_t$ Maximum Peak-to-Valley Height within Assessment Length (Surface Roughness) [Symbol]

R$_t$ Resistance (of a Conductor) at t°C [Symbol]

R$_\theta$ Thermal Resistance [Semiconductor Symbol]

R$_{\theta CA}$ Thermal Resistance, Case to Ambient [Semiconductor Symbol]

R$_{Th}$ Thevenin's Equivalent Resistance [Symbol]

R$_{\theta JA}$ Thermal Resistance, Junction to Ambient [Semiconductor Symbol]

R$_{\theta JC}$ Thermal Resistance, Junction to Case [Semiconductor Symbol]

R$_{tm}$ Maximum Peak-to-Valley Height based on Average of Five Consecutive Sampling Lengths (Surface Roughness) [Symbol]

R$_u$ Ultimate Resilience [Symbol]

R$_V$ Rockwell 'V' Hardness (½ in. Dia. Steel Sphere Indenter; Applied Load 150 kg) [Symbol]

R$_w$ Liquidity Index (or Water-Plasticity Ratio) [Symbol]

R$_w$ Local Solidification Rate (in Metallurgy) [Symbol]

R$_w$ Reflectance Factor of a Thin Film or Sheet of Material with a White Backing [Symbol]

R$_w$ Winding Resistance (of Electric Machinery) [Symbol]

R$_X$ Nonmetal Atomic Radius [Symbol]

R$_x$ Resistance at Operating Temperature t$_x$ [Symbol]

R$_Z$ Zener Resistance [Semiconductor Symbol]

R$_z$ Ten-Point Height (Surface Roughness) [Symbol]

R1   Type 1 Rutile, Magneli Phase [Mineral]

R2   Type 2 Rutile, Magneli Phase [Mineral]

RIα Receptor Iα [Biochemistry]

RIβ Receptor Iβ [Biochemistry]

R17934   5-(2-Thenoyl 2-Benzimidazolecarbamic Acid Methyl Ester

3R   Martensitic Phase in Nickel-Aluminum Alloys [Metallurgy]

15R-   15R Polytype [Ramsdell Notation: 15 Refers to Number of Layers Necessary to Produce a Unit Cell, and R Refers to the Rhombohedral Symmetry, e.g., 15R-SiC]

21R-   21H Polytype [Ramsdell Notation: 21 Refers to Number of Layers Necessary to Produce a Unit Cell, and R Refers to the Rhombohedral Symmetry, e.g., 21R-SiC]

r   absolute value of a complex number [Symbol]

r   anisotropy parameter [Symbol]

r   common ratio between successive terms of a geometric progression [Symbol]

r   complex reflection coefficient [Symbol]

r   coordinate of the real axis (of complex plane) [Symbol]

r   correlation coefficient (in statistics) [Symbol]

r   distance [Symbol]

r   factorial [Symbol]

r   first coordinate in cylindrical, polar and spherical coordinate systems [Symbol]

r   (latent) heat of vaporization [Symbol]

r   inradius (of a triangle) [Symbol]

r   interatomic distance [Symbol]

r   interest rate [Symbol]

r   Lankford value (in metallurgy) [Symbol]

r   length of a moment arm [Symbol]

r   modulus (of a complex quantity) [Symbol]

r   (air-water, or air-ice) mixing ratio [Symbol]

r   particle radius (in metallurgy) [Symbol]

r   plastic strain ratio (in formability testing) [Symbol]

r   polar coordinate [Symbol]

r   radial

r   radial distance [Symbol]

r   (inside, or internal) radius [Symbol]

r   radius of curvature (in optics) [Symbol]

r   radius vector (in cylindrical polar coordinate system) [Symbol]

r   rain [also R]

r   ram [also R]

r   range [also R]

r   rare

r   rat [also R]

r   rate [also R]

r   ratio [also R]

r   reaction rate [Symbol]

r   red [also R]

r   reflection [also R]

r   reflection amplitude [Symbol]

r   refraction [also R]

r   remote [also R]

r   relative

r   relative redundancy [Symbol]

r   repeatability limit (in statistics) [Symbol]

r   replace text (command) [Edlin MS-DOS Line Editor]

r   reset

r   (internal) resistance [Symbol]

r   resolution [also R]

r   rheostat(ic) [also R]

r   rhesus negative (gene) [Symbol]

r   rho [English Equivlalent]

r   ribosomal; ribosome [Cytology]

r   right [also R]

r   river [also R]

r   rod [also R]

r   rodenticide

r   roentgen [Symbol] [now R]

r   root of equation [Symbol]

r   running index (e.g., the rth term of a sequence, or series) [Symbol]

r   space vector [Symbol]

r   specific airflow resistance [Symbol]

r   specific gas constant [Symbol]

r   square-root diffusivity [Symbol]

r   thermal resistivity [Symbol]

r   water-to-cement ratio (of a concrete or mortar) [Symbol]

$\bar{r}$ position vector (in mathematics) [Symbol]

r^* critical nucleation radius (of a particle) [Symbol]

r' Fresnel reflection coefficient [Symbol]

r_0 airflow resistivity [Symbol]

r_0 atomic radius [Symbol]

r_0 equilibrium (crystal) lattice spacing [Symbol]

r_0 equilibrium interatomic separation [Symbol]

r_A anion ionic radius [Symbol]

r_A fraction of A sites occupied by A atoms in a solid solution [Symbol]

r_a adsorption rate [Symbol]

r_a anode slope resistance (or electrode alternating-current resistance) (of a thermionic tube) [Symbol]

r_a apparent thermal resistivity [Symbol]

r_a major radius of an ellipse [Symbol]

$r_{a,b}$ absorption edge jump ratio (in physics) [Symbol]

r_b minor radius of an ellipse [Symbol]

r_{BB} interbase resistance [Semiconductor Symbol]

r_C cation ionic radius [Symbol]

r_c contact resistivity [Symbol]

r_c critical radius of a nucleus (in metallurgy) [Symbol]

$r_{CE(sat)}$ saturation resistance, collector-to-emitter [Semiconductor Symbol]

$r_{DS(on)}$ static drain-source on-state resistance [Semiconductor Symbol]

$r_{ds(on)}$ small-signal drain-source on-state resistance [Semiconductor Symbol]

r_E Eckhaus instability boundary [Symbol]

r_e classical electron radius (2.817941×10^{-15} m) [Symbol]

r_e emitter resistance (for transistors) [Semiconductor Symbol]

$r_{ele2(on)}$ small-signal emitter-emitter on-state resistance [Semiconductor Symbol]

r_F Fresnel reflection amplitude [Symbol]

r_G specific refractivity [Symbol]

r_g radius of gyration [Symbol]

r_γ plastic-zone adjustment (in mechanics) [Symbol]

r_h enthalpy recovery factor [Symbol]

r_{Hc} coercivity radius [Symbol]

r_i dynamic resistance at inflection point [Semiconductor Symbol]

r_i saturation mixing ratio (for moist air saturated with respect to ice) [Symbol]

r_L specific refraction [Symbol]

r_{min} ultimate resolution (of electron microscope) [Symbol]

r_n electron orbit radius of hydrogen atom corresponding to n-th state [Symbol]

r^p reflection coefficient for component in plane of incidence [Symbol]

r_p complex reflection coefficient for polarization [Symbol]

r_p plate (or anode) resistance [Semiconductor Symbol]

r_p pore radius [Symbol]

r_p radius of plastic zone (in fracture testing) [Symbol]

r^s reflection coefficient for component in plane of the surface [Symbol]

r_s complex reflection coefficient for s-polarization [Symbol]

r_s radius of sphere containing one electron [Symbol]

r_s surface roughness [Symbol]

r_t chip thickness ratio (in machining) [Symbol]

r_w saturation mixing ratio (for moist air saturated with respect to water) [Symbol]

r_{WS} Wigner-Seitz radius (in materials science) [Symbol]

r_{XX} intercorrelation coefficient (in statistics) [Symbol]

r_Y plastic-zone adjustment (in fracture testing) [Symbol]

P (Upper-case) Rho [Greek Alphabet]

ρ bulk density [Symbol]

ρ charge distribution [Symbol]

ρ coefficient of correlation (in statistics) [Symbol]

ρ crack tip radius of Griffith type crack [Symbol]

ρ density [Symbol]

ρ dislocation density (in materials science) [Symbol]

ρ electron density [Symbol]

ρ first coordinate in cylindrical, polar and spherical coordinate systems [Symbol]

ρ magnitude of a complex number [Symbol]

ρ mass density

ρ notch-tip radius (in fatigue cracking) [Symbol]

ρ optical density [Symbol]

ρ packing density [Symbol]

ρ proportionality factor [Symbol]

ρ radius of curvature [Symbol]

ρ (chemical) reaction constant [Symbol]

ρ reflectance [Symbol]

ρ reflection coefficient ratio (i.e., r^p/r^s) [Symbol]

ρ reflectivity [Symbol]

ρ (electrical) resistivity (or specific resistance) [Symbol]

ρ (lower-case) rho [Greek Alphabet]

ρ surface relaxivity [Symbol]

ρ volume charge density [Symbol]

ρ^* complex resistivity [Symbol]

ρ^* optical density [Symbol]

ρ' relative density [Symbol]

$\rho_\parallel$ in-plane (or parallel) resistivity [Symbol]

$\rho_\perp$ out-of-plane (or perpendicular) resistivity [Symbol]

$\rho\uparrow$ resistivity in the spin-up channels (in physics) [Symbol]

$\rho\downarrow$ resistivity in the spin-down channels (in physics) [Symbol]

$\overline{\rho}^{1/3}$ average cube root of electron density [Symbol]

$\rho\infty$ reflectivity [Symbol]

ρ_0 bulk resistivity [Symbol]

ρ_0 initial (or residual) resistivity [Symbol]

ρ_0 mean density [Symbol]

ρ_0 (electrical) resistivity of a pure metal [Symbol]

ρ_{300} room-temperature resistivity [Symbol]

ρ_A density of component, phase, solid, etc. "A" [Symbol]

ρ_a density of the alpha phase (of an alloy) [Symbol]

ρ_a (electrical) resistivity of the alpha phase (of an alloy) [Symbol]

ρ_{ab} in-plane (in a-b plane) resistivity (of superconductor) [Symbol]

ρ_{AH} anomalous Hall resistance [Symbol]

ρ_β density of the beta phase (of an alloy) [Symbol]

ρ_β (electrical) resistivity of the beta phase (of an alloy) [Symbol]

ρ_c calculated density [Symbol]

ρ_c critical density [Symbol]

ρ_c critical distribution function [Symbol]

ρ_c particle density of condensate [Symbol]

ρ_c resistivity along c-axis (of superconductor) [Symbol]

ρ_c resistivity of composite [Symbol]

ρ_c specific contact resistance [Symbol]

ρ_{ca} resistivity in metal-alloyed semiconductor interface [Symbol]

ρ_{cu} resistivity in alloyed-unalloyed semiconductor interface [Symbol]

ρ_d deformation resistivity (of metals) [Symbol]

ρ_d density of dry soil, or rock [Symbol]

ρ_d diffuse reflectance [Symbol]

ρ_d dislocation resistivity (in materials science) [Symbol]

ρ_d resistivity due to (crystal) lattice imperfections [Symbol]

ρ_e reflectance [Symbol]

ρ_e (electrical) resistivity [Symbol]

ρ_F (electrical) resistivity contribution per unit concentration of Frenkel defects (in materials science) [Symbol]

ρ_f resistivity of fiber [Symbol]

ρ_g density of gas [Symbol]

ρ_{gb} grain boundary resistivity (in materials science) [Symbol]

ρ_i electron density at atom "i" [Symbol]

ρ_i impurity resistivity (i.e., the resistivity due to impurity atoms) [Symbol]

ρ_K Kondo resistivity (in materials science) [Symbol]

ρ_l density of liquid [Symbol]

ρ_M density of metal [Symbol]

ρ_m mass density [Symbol]

ρ_m matching resistivity [Symbol]

ρ_m matrix resistivity [Symbol]

ρ_m measured density [Symbol]

ρ_{min} minimum resistivity [Symbol]

ρ_n normal-state resistivity [Symbol]

ρ_O density of oxide [Symbol]

ρ_R residual resistivity [Symbol]

ρ_r regular reflectance [Symbol]

ρ_{RT} room-temperature resistivity [also ρ_{rt}] [Symbol]

ρ_s density of sample, or specimen [Symbol]

ρ_s density of solid [Symbol]

ρ_s saturation defect density [Symbol]

ρ_s saturation resistivity [Symbol]

ρ_s spin density (in solid-state physics) [Symbol]

ρ_s superfluid density [Symbol]

ρ_{sat} density of saturated soil, or rock [Symbol]

ρ_{sub} density of submerged soil, or rock [Symbol]

ρ_T thermal resistivity (coefficient) [Symbol]

ρ_T total dislocation density (in materials science) [Symbol]

ρ_T total resistivity [Symbol]

ρ_t radius of curvature of tip of a (elliptical) crack [Symbol]

ρ_t thermal resistivity [also ρ_{th}] [Symbol]

ρ_v vapor density [Symbol]

ρ_v reflectance [Symbol]

ρ_v (electrical) resistivity contribution per unit concentration of vacancies (in materials science) [Symbol]

ρ_v valence charge density [Symbol]

ρ_w density of water [Symbol]

ρ_x density calculated from x-ray measurements [Symbol]

ρ_x resistivity coefficient (in solid solutions) [Symbol]

ρ_x resistivity at any temperature t_x [Symbol]

ρ_{xx} magnetoresistivity [Symbol]

ρ_{xy} Hall resistivity [Symbol]

ϱ (lower-case) rho (variant) [Greek Alphabet]

ϱ coordinate [Symbol]

ϱ mass concentration (density) [Symbol]

ϱ radius [Symbol]

ϱ reflectance [Symbol]

ϱ reflection factor (in electrical engineering) [Symbol]

ϱ resistivity [Symbol]

ϱ volume (or space) charge density [Symbol]

ϱ_0 standard state density of a gas [Symbol]

$\mathbb{R}$ Real Number [Symbol]

®   Registered Trademark [Symbol]

$\Re$ Real Part (of Complex Number) [also Re]

ℜ Wendt-Abraham Parameter (of Pair Distribution Function) [Symbol]

RA Activated Rosin [Soldering]

RA Cylindrical Roller Bearing, Double Row, Non-Locating, Outer Ring With Three Integral Ribs, Inner Ring Separable, Inner Ring Without Ribs [Symbol]

RA Metric Untrimmed Paper Size [Usually Designated by the Letters "RA" Followed by a Numeral, e.g., RA0, RA1, etc.]

RA Radar Altimeter [also R/A]

RA Radar Astronomy

RA Radford Arsenal [Virginia, US]

RA Radial Acceleration

RA Radial Axis

RA Radiation Absorber

RA Radiative Auger [Physics]

RA Radioactive; Radioactivity

RA Radioactive Aerosol

RA Radio Altimeter

RA Radioanalysis

RA Radioassay [Analytical Chemistry]

RA Radio Astronomer; Radio Astronomical; Radio Astronomy

RA Radio Array

RA Radioatmospheric(s)

RA Radioautograph(y)

RA Radius of Action [Navigation]

RA Radonic Association [UK]

RA Rainforest Alliance [US]

RA Random-Access (Unit)

RA Random Assembly

RA Random Arc (Evaporation)

RA Range Area

RA Rapid Access

RA Rapid Acquisition

RA Raritan Arsenal [Metuchen, New Jersey, US]

RA Rate of Application

RA Rate of Approach [Aeronautics]

RA Rational Number

RA Ravenna Arsenal [Apco, Ohio, US]

RA Reactor Argentina [Argentinian (Nuclear) Reactor]

RA Read Amplifier

RA Ready-Access

RA Rear Access

RA Rear-Admiral [also RADM, or RAdm]

RA Rear Axle

RA Recoil Atom

RA Record Address

RA Recorded Announcement

RA Recrystallization Anneal(ed); Recrystallization Annealing [Metallurgy]

RA Rectifier

RA Recycling Agent

RA Redstone Arsenal [Huntsville, Alabama, US]

RA Reducing Adapter

RA Reducing Agent

RA Reducing Atmosphere

RA Reduction of (or in) Area [Mechanical Testing]

RA Reductive Amination [Organic Chemistry]

RA Reflection-Absorption

RA Reflex Arc [Anatomy]

RA Refractories Association (of Great Britain)

RA Refractory Alloy(s)

RA Regression Analysis [Statistics]

RA Regular Army

RA Relative Accuracy

RA Relative Acidity

RA Relative Address

RA Relative Area

RA Relocation Address

RA Remedial Action

RA Remote Access

RA Renal Artery [Anatomy]

RA Renew America [US]

RA Rental Agreement

RA Repeat Attempt

RA Replacement Algorithm

RA República Argentina [Republic of Argentina]

RA Republic of Argentina

RA Research Administration; Research Administrator

RA Research Associate

RA Research Association

RA Reserve Alkalinity

RA Resonance Absorption

RA Resorcylic Acid

RA Retinoic Acid [Biochemistry]

RA Return Address

RA Return Air

RA Return Authorization

RA Rheumatoid Arthritis [Medicine]

RA Rice Association [UK]

RA Right Aft [Naval Architecture]

RA Right Angle

RA Right Ascension [Astronomy]

RA Right Auricle (of the Heart) [Anatomy]

RA Ripple Adder

RA Risk Analysis

RA Road Assist(ance)

RA Robotics and Automation

RA Robotics and Automation (Council) [of Institute of Electrical and Electronics Engineers, US]

RA Robustness Analysis [Statistics]

RA Rolled Alloys Company [Temperance, Michigan, US]

RA Root Addressable (Record)

RA Rosin-Activated; Rosin Activation [Soldering]

RA Rotating Analyzer

RA Rotation Axis [Crystallography]

RA Royal Academician

RA Royal Academy [UK]

RA Royal Arsenal [UK]

RA Royal Artillery [UK]

.RA Real Audio [File Name Extension] [also .ra]

%RA Percent Reduction in (or of) Area (in Mechanics) [Symbol]

%RA Percent Relative Accuracy [Symbol]

RA0 Metric Untrimmed Paper Size [Sheet Size: 860 × 1220 mm]

RA1 Metric Untrimmed Paper Size [Sheet Size: 610 × 860 mm]

RA2 Metric Untrimmed Paper Size [Sheet Size: 430 × 610 mm]

R(A$_B$) Reevaporation Rate of A Atoms from Surface B [Symbol]

R′(A$_B$) Reevaporation Probability of A Atoms from Surface B [Symbol]

R&A Robotics and Automation

R/A Radar Altimeter [also RA]

R/A Recorded Announcement

Ra Radium [Symbol]

Ra Range [Geography]

Ra Rayleigh Number [Symbol]

Ra Rayon

Ra Reaction Amorphization [Metallurgy]

Ra$_3$ Rayleigh Number 3 (of Combined Free and Forced Convection) [Symbol]

Ra-218 Radium-218 [also ^{218}Ra, or Ra^{218}]

Ra-219 Radium-219 [also ^{219}Ra, or Ra^{219}]

Ra-220 Radium-220 [also ^{220}Ra, or Ra^{220}]

Ra-221 Radium-221 [also ^{221}Ra, or Ra^{221}]

Ra-222 Radium-222 [also ^{222}Ra, or Ra^{222}]

Ra-223 Radium-223 [also ^{223}Ra, Ra^{223}, or AcX]

Ra-224 Radium-224 [also ^{224}Ra, Ra^{224}, or ThX]

Ra-225 Radium-225 [also ^{225}Ra, or Ra^{225}]

Ra-226 Radium-226 [also ^{226}Ra, or Ra^{226}]

Ra-227 Radium-227 [also ^{227}Ra, or Ra^{227}]

Ra-228 Radium-228 [also ^{228}Ra, Ra^{228}, or MsTh$_1$]

Ra-229 Radium-229 [also ^{229}Ra, or Ra^{229}]

Ra-230 Radium-230 [also ^{230}Ra, or Ra^{230}]

ra rat [Immunochemistry]

RAA Cylindrical Roller Bearing, Single-Row, One-Direction-Locating, Single-Ribbed Outer Ring, Double-Ribbed Inner Ring, Both Rings Separable [Symbol]

RAA Radar Aircraft Altitude (Calculator)

RAA Radioactive Activation Analysis

RAA Random-Access Array

RAA Regenerative Agriculture Association [US]

RAA Regional Airlines Association [US]

RAA Relocation Authorization Agreement

RAA Remote Access Audio

RAA Remote Axis Admittance

RAA Rice Growers' Association of Australia

RAA Royal Academy of Arts [UK]

RaA Radium A [also Po-218, ^{218}Po, or Po^{218}]

RAAB Remote Amplifier Acquisition and Advisory Box

RAAB Remote Amplifier and Adoption Box

RAAB Remote Application and Advisory Box

RAAC Radar Aircraft Altitude Calculator

RAAF Royal Australian Air Force

RAAP Residue Arithmetic Associative Processor

RAAP Resource Allocation and Planning

RAAR RAM (Random-Access Memory) Address Register

RAAS Remedial Action Assessment System

RAAS/TIS Remedial Action Assessment System/Technology Information System

RAB Cylindrical Roller Bearing, Single-Row, Non-Locating, Single-Ribbed Outer Ring, Inner Ring without Ribs, Both Rings Separable [Symbol]

RAB Rabaul, Papua–New Guinea [Meteorological Station Designator]

RAB Radio Advertising Bureau [US]

RAB Regge-Type Asymptotic Behaviour [Physics]

RAB Regional Activities Board [of Institute of Electrical and Electronics Engineers, US]

RaB Radium B [also Pb-214, ^{214}Pb, or Pb^{214}]

RABAR Raytheon Advanced Battery Acquisition Radar [also Rabar, or rabar]

RABDF Royal Association of British Dairy Farmers

RABFM Research Association of British Flour Millers

RABIN Raad van Advies voor Bibliotheekwesen en Informatieverzorging [Netherlands Council for Libraries and Information Services]

RABiTS Rolling-Assisted Biaxially-Textured Substrate [Metallurgy]

RABPCVM Research Association of British Paint, Color and Varnish Manufacturers

RABRM Research Association of British Rubber Manufacturers [now Rubber and Plastic Research Association, UK]

RABSA Rabbit Anti-Bovine Serum Albumin [Immunology]

RAB/TAB Regional Activities Committee/Technical Activities Committee [of Institute of Electrical and Electronics Engineers, US]

RAC Radioanalytical Chemistry

RAC Railway Association of Canada

RAC Random-Access Control(ler)

RAC Rapid Analysis Chromatography

RAC Real-Time Active Controller

RAC Recombinant DNA (Deoxyribonucleic Acid) Advisory Committee [of National Institutes of Health, Bethesda, Maryland, US]

RAC Rectified Alternating Current [also rac]

RAC Recurring Annual Costs

RAC Recycling Advisory Committee

RAC Reference Air Concentration(s)

RAC Reflexology Association of Canada

RAC Regional Advisory Committee

RAC Reliability Action Center

RAC Reliability Analysis Center [US]

RAC Remote Access and Control

RAC Remote Access Computer

RAC Remote Analysis Computer

RAC Remote Automatic Calibration

RAC Repairable Asset Control [Joint Project of Ford Motor Company and General Motors, US]

RAC Resource Assessment Commission [Australia]

RAC Response Action Contractor

RAC Robotics and Automation Council [of Institute of Electrical and Electronics Engineers, US]

RAC Royal Agricultural College

RAC Royal Armoured Corps

RAC Royal Automobile Club [UK]

RAC Rubber Association of Canada

RAC Rules of the Air and Air Traffic Control

RaC Radium C [also Bi-214, ^{214}Bi, or Bi^{214}]

RaC′ Radium C′ [also RaC$_1$, Po-214, ^{214}Po, or Po^{214}]

RaC″ Radium C″ [also RaC$_2$, ^{210}Tl, or Tl^{210}]

rac racemic [Chemistry]

RACC Radiation and Contamination Control

RACC Reykjavik Area Control Center [Iceland]

RACE Random-Access Computer Equipment

RACE Random-Access Control Equipment

RACE Rapid Acquisition and Computer Extraction

RACE Rapid Automatic Checkout Equipment

RACE Reactive Accelerated Cluster Erosion

RACE Regional Automatic Circuit Exchange

RACE Research and Development in Advanced Communications Technologies in Europe

RACE Research in Advanced Communication in Europe

RACE Research into Artifacts, Center for Engineering [of University of Tokyo, Japan]

RACEP Random-Access and Correlation for Extended Performance

RACES Radio Amateur Civil Emergency Service

RACF Resource Access Control Facility

RACG Radiometric Area-Correlation Guidance

RACG Reaction Alternating-Current Generator

RACI Royal Australian Chemical Institute

RACIC Remote Area Conflict Information Center [of Battelle Memorial Institute, Columbus, Ohio, US]

RACIP International Conference on Research and Communications in Physics

R-Acid 3-Hydroxy 2,7-Naphthalenedisulfonic Acid [also R-acid]

2R-Acid 2-Naphthol-3,6-Disulfonic Acid [also R-acid, 2R-Acid, or 2R-acid]

RACIP (International Conference on) Research and Communication in Physics

RACKPC Rack-Mount Personal Computer

RAC MSA Royal Automobile Club Motor Sports Association [UK]

RACON Radar Beacon [also Racon, or racon]

rac-PAF Racemic Platelet Activating Factor [Biochemistry]

RACS Remote Access Calibration System

RACS Remote Access Computing System

RACS Remote Automatic Calibration System

RACS Rotation Axis Coordinate System

RACT Reasonably Available Control Technology

RACT Reasonably Available Control Technique

RACT Remote Access Computer Technique

RACU Remote Acquisition and Command Unit

RAD Radial Detector

RAD Radiation Absorbed Dose [also rad]

RAD Radioactive Decay

RAD Radioactive Disintegration

RAD Radioactivity Detection

RAD Random-Access Data

RAD Random-Access Device

RAD Random-Access Disk

RAD Rapid-Access Data

RAD Rapid-Access Device

RAD Rapid-Access Disk

RAD Rapid Application Development

RAD Ratio Analysis Diagram

RAD Reactive Atomization and Deposition [Metallurgy]

RAD Records Arrival Date

RAD Regional Activities Department [of Institute of Electrical and Electronics Engineers, US]

RAD Relative Air Density

RAD Ribbon Against Drop

RAD Roentgen Absorbed Dose [also rad]

RAD Rural Areas Development

RaD Radiative Deexcitation

RaD Radium D [also Pb-210, ^{210}Pb, or Pb^{210}]

Rad Radar [also rad]

Rad Radial [also rad]

Rad Radiant [also rad]

Rad Radiator [also rad]

Rad Radical [also rad]

Rad Radio [also rad]

Rad Radiometer [also rad]

rad radian [Unit]

rad radiation absorbed dose [also RAD]

rad radius [Symbol]

rad roentgen absorbed dose [also RAD]

rad *(radix)* – root [Medical Prescription]

RADA Random-Access Discrete Address

RADAC Radar Analog/Digital Data and Control

RADAC Rapid Digital Automatic Computing [also radac]

RAD-ALT Radar Altimeter [also Rad-Alt, or radalt]

RADAN Radar Doppler Automatic Navigator [also Radan, or radan]

Radant Integrated Radome Antenna

RADAR Radio All-Dimension Audience Research

RADAR Radio Detection and Ranging [also Radar, or radar]

RADAR Random Access Dump and Reload

RADAR Rights, Availabilities, Distribution Analysis and Reporting

RADARSAT Radar Satellite [also Radarsat]

RADARSAT (Synthetic Aperture) Radar Observation Satellite [also Radarsat, Canada]

RADARSAT DDP Radar Observation Satellite Data Development Program [of Public Works and Government Services Canada]

RADARSAT UDP Radar Observation Satellite User Development Program [of Public Works and Government Services Canada]

RADAS Random-Access Discrete Address System

RADATA Radar Automatic Data Transmission and Assembly

RADATS Radiographic Automated Testing System

RADB Routing Arbiter Data Base

RADC Rome Air Development Center [of US Air Force Systems Command at Griffiss Air Force Base, New York]

RADCC Radiation Control Center

RADCM Radar Countermeasures

RADCON Radar Data Converter

RadCon Radiological Control

RADCOR International Symposium on Radiative Corrections

RADD Rapid-Access Data Drum

RADEFF Radiation Effects

RADEED Radial Distributions of Exterior Electron Densities

Rad Elect Eng Radio and Electronic Engineer [Journal published in the UK]

RADEM Random-Access Delta Modulation

Radex Runds Radex Rundschau [Austrian Publication] [also Radex Rdsch]

RADFAC Radiation Facility

RADFET Radiation (Sensing) Field-Effect Diode

RADHAZ Radiation Hazard

RADIAC Radioactivity Detection, Identification and Computation [also Radiac, or radiac]

RADIALS Research and Development in Information and Library Science [UK]

RADIAN Radial Detector for Ion Beam Analysis

Radiat Radiation; Radiative [also rad]

Radiat Def Met Radiatsionnye Defekty v Metallakh [Kazakhstan Journal on Radiation Effects and Defects in Metals]

Radiat Eff Radiation Effects [Journal published in the UK]

Radiat Eff Defects Solids Radiation Effects and Defects in Solids [Journal published in the UK]

Radiat Eff Express Radiation Effects Express [Journal published in the UK]

Radiat Eff Lett Radiation Effects Letters [Journal published in the UK]

Radiat Eff Lett Sect Radiation Effects Letters Section [Journal published in the UK]

Radiat Environ Biophys Radiation and Environmental Biophysics [Journal published in Germany]

Radiat Phys Chem Radiation Physics and Chemistry [Journal published in the UK]

Radiat Prot Radiation Protection [of Korean Association for Radiation Protection, South Korea]

Radiat Prot Dosim Radiation Protection Dosimetry [Journal published in the UK]

Radiat Res Radiation Research [Journal of Radiation Research Society, US]

RADINT Radar Intelligence [also Radint, or radint]

Radioact Radioactive; Radioactivity [also radioact]

Radioact Waste Manage Nucl Fuel Cycle Radioactive Waste Management and the Nuclear Fuel Cycle [Published in Switzerland]

Radiobiol Radiobiologist; Radiobiology

Radiobiol Radiobiologiia [Russian Journal on Radiobiology]

RADIOCOM Radiocommunication [also Radiocom or radiocom]

Radio Commun Radio Communication [Publication of the Radio Society of Great Britain]

Radiodiffus Telev Radiodiffusion Television [French Publication on Broadcasting and Television]

Radio-Electron Radio-Electronics [Published in the US]

Radioelectron Commun Syst Radioelectronics and Communications Systems [Translation of: *Izvestiya Vysshikh Uchebnykh Zavedenii, Radioelektronika (USSR)*; published in the US]

Radio-Electron World Radio and Electronics World [Published in the UK]

Radiol Radiological; Radiologist; Radiology

Radiophys Quantum Electron Radiophysics and Quantum Electronics [Translation of: *Izvestiya Vysshikh Uchebnykh Zavedenii, Radiofizika (USSR)*; published in the US]

Radio Sci Radio Science; Radio Scientist

Radio Sci Radio Science [Journal of the American Geophysical Union, US]

Radiotekh Elektron Radiotekhnika i Elektronika [Russian Journal of Communication Technology and Electronics]

Radiotekhnika, Kharkov Radiotekhnika, Kharkov [Journal of Radio Engineering published in Kharkov, Ukraine]

Radiotekhnika, Mosk Radiotekhnika, Moskva [Journal of Radio Engineering published in Moscow]

Radio Telev Radio and Television [Publication of Organisation Internationale de Radiodiffusion et Television]

RADIQUAD Radio Quadrangle

RADIR Random-Access Document Indexing and Retrieval

RADIST Radar Distance Indicator

RADIT Radio Teletype

RADM Random-Access Data Modulation

RADM Rear-Admiral [also RAdm, or RA]

rad/m radian per meter [also rad m^{-1}]

RADMAT Radioactive Material [also RAD MAT or radmat]

Radn Radiation [also radn]

RADNOTE Radio Note

rad/Oe/s radian per Oersted per second [also rad Oe^{-1} s^{-1}]

RADOME Radar Dome [also Radome, or radome]

RADOP Radar Operator

RADOPWEAP Radar Optical Weapons

RADOSE Radiation Dosimeter Satellite [of NASA]

RADOT Real-Time Automatic Digital Optical Tracker

RADP Rubidium Ammonium Dihydrogen Phosphate

RADPLANBD Radio Planning Board

RADRON Radar Squadron

RADPROPCAST Radio Propagation Forecast

RADS Radiation and Dosimetry Services

RADS Rocking-Curve Diffraction Analysis Software [X-Ray Diffraction]

rad/s radian(s) per second [also rad/sec also rad s^{-1}]

rad/s² radian(s) per second squared [also rad/sec² also rad s^{-2}]

RADSCAT Radiometer-Scatterometer Sensor

rad/sec radian(s) per second [also rad/s also rad s^{-1}]

rad/sec² radian(s) per square second [also rad/s² also rad s^{-2}]

RADSL Rate Adaptive Digital Subscriber Line

RADSO Radiological Survey Officer

RADTT Radio Teletype

RADU Radar Analysis and Detection Unit

RADUX Long-Distance Continuous-Wave Low-Frequency Navigation System

RADVS Radar Altimeter and Doppler Velocity Sensor

Radwaste Radioactive Waste [also radwaste]

rad/wk radiation absorbed dose per week [Radiology]

RAE Radioastronomical Explorer; Radio Astronomy Explorer [Research Satellite]

RAE Random Arc Evaporation

RAE Range Azimuth Elevation

RAE Royal Aeronautical Establishment [now Royal Aerospace Establishment, UK]

RAE Royal Aerospace Establishment [UK]

RAE Royal Aircraft Establishment [later Royal Aeronautical Establishment, UK]

RaE Radium E [also Bi-210, ^{210}Bi, or Bi^{210}]

RaE″ Radium E″ [also RaE$_2$, Tl-206, ^{206}Tl, or Tl^{206}]

RAE-A Radio Astronomy Explorer (Satellite), Series A

RAeC Royal Aero Club [UK]

RAEN Radio Amateur Emergency Network [also RAENET]

RAER Range, Azimuth, Elevation and Reproduction

RAES Remote Access Editing System

RAES Rural Afforestation and Extension Services [Australia]

RAeS Royal Aeronautical Society [also RAES, UK]

RAF Reserved Airfreight

RAF Regular Air Force

RAF Requirements Analysis Form

RAF Research Aircraft Experiment

RAF Research Aviation Facility [of National Center for Atmospheric Research, US]

RAF Royal Air Force [UK]

RAF Richard Arnowitt Fest: A Symposium on Supersymmetry and Gravitation [of Texas A&M University, College Station, US]

RaF Radium F [also Po-210, ^{210}Po, or Po^{210}]

RAFA Royal Air Force Association [UK]

RAFAX Radar Facsimile Transmission

RAFC Radar Automatic Frequency Control

RAFC Regional Area Forecast Center

RAFES Royal Air Force Educational Service [UK]

RAFHS Royal Air Force Historical Society [UK]

RAFI Rural Advancement Fund International [US]

RAFISBENQO Radio-Signal Reporting Code for Transmission-Condition Rating

Rafl Rainfall

RAFSC Royal Air Force Staff College [UK]

RAFT Radially Adjustable Facility Tube

RAFT Ramp-Assisted Foil Casting (Technique) [Electronics]

RAFT Rapamycin and FKBP12 (FK-Binding Protein 12) Target [Biochemistry]

RAFTIR Reflection/Absorption Fourier Transform Infrared Spectroscopy

RAG Rabbit Anti-Goat [Immunochemistry]

RAG Rainforest Action Group

RAG Residential Advisory Group

RAG Reusable Agena

RAG ROM (Read-Only Memory) Address Gate

RAG Row Address Generator [Computers]

RAG Ruhrkohle AG [German Energy Supplier]

RAG Runway Arresting Gear

RaG Radium G [also Pb-206, ^{206}Pb, or Pb^{206}]

RAGB Refractories Association of Great Britain

RAGS Radiatively Active Gases; Radiatively Active Gases Sites

RAGS Risk Assessment Guidance for Superfund

RAGU Radio Receiving and Generally Useful

RAH Radiation Anneal Hardening [Metallurgy]

RAHF Research Animal Holding Facility

RAHS Royal Australian Historical Society

RAI Radar Altimeter Indicator

RAI Radioactive Interference

RAI Radioactive Iodine

RAI Random-Access and Inquiry

RAI Registro Aeronautico Italiano [Italian Aeronautical Registry]

RAI Robot Automatix Inc. [US]

RAI Royal Anthropological Institute (of Great Britain and Ireland)

RAI Royal Architectural Institute [UK]

RAI Royal Archeological Institute [UK]

RAI Runway Alignment Indicator

RAIA Royal Australian Institute of Architects

RAIC Redstone Arsenal Information Center [US Army]

RAIC Royal Architectural Institute of Canada

RAID Redundant Array of Independent Disks

RAID Redundant Array of Independent Drives

RAID Redundant Array of Individual Disks

RAID Redundant Array of Inexpensive Disks

RAID Remote Access Interactive Debugger

RAIDS Rapid Availability of Information and Data for Safety

RAIL Railroad Advancement through Information and Law (Foundation) [US]

RAIL Recupero Alluminio in Forma di Lattina [Aluminum Recycling (Group), Italy]

RAIL Robot Automatix Incorporated Language

RAIL Runway Alignment Indicator Lights

Rail Railroad [also rail]

Rail Railway(s) [also rail]

Rail Int Rail International [Publication of the International Railway Congress Association]

RAILS Remote Area Instrument Landing Sensor

Railw Railway(s) [also railw]

Railw Electr Railway and Electricity [Publication of Railway Electrification Association, Japan]

Railw Eng Int Railway Engineer International [UK Publication]

Railw Gaz Int Railway Gazette International [UK Publication]

RAIN Relational Algebraic Interpreter

RAINBOW Reduced and Internally Biased Oxide Wafer (Process) [Electronics]

Rainforests LAP Rainforests Labour Adjustment Package [North Queensland, Australia]

RAIP Requester's Approval in Principle

RAIPA Royal Australian Institute of Public Administration

RAIPR Royal Australian Institute of Parks and Recreation

RAIR Reflectance Absorption Infrared (Spectrum)

RAIR Remote Access Immediate Response

RAIR Random-Access Information Retrieval

RAIRS Raman Infrared Spectroscopy

RAIRS Reflection-Absorption Infrared Spectroscopy

RAIS Redundant Arrays of Inexpensive Systems

RAISON Regional Analysis by Intelligent Systems on Microcomputer

RAJ Reverse Air Jet

RAK Read-Access Key

RAK Remote Access Key

RAL Radio Annoyance Level

RAL RAL–Deutsches Institut für Gütesicherung und Kennzeichnung e.V. [German Institute for Quality Assurance and Identification]

RAL Resorcyclic Acid Lactone [Biochemistry]

RAL Responsibility Alignment List

RAL Riverbend Acoustical Laboratory [US]

RAL Robotics and Automation Laboratory [of University of Toronto, Canada]

RAL Row-Address Latch [Computers]

RAL Runway Alignment (Beacon)

RAL Rutherford Appleton Laboratory [UK]

R(Al,Cr,Ga)$_3$(BO$_3$)$_4$ Rare-Earth Huntite Borate [Formula]

RALF Rapid-Access to Literature via Fragmentation Codes [Germany]

RALF Relocatable Assembly Language Floating Point

RALI Resource and Land Investigation

R-AlLiCu R-Phase Aluminum Lithium Copper (Alloy) [Metallurgy]

RALPH Reduction and Acquisition of Lunar Pulse Heights

RALT Radar Altimeter

RALU Register(s) and Arithmetic and Logic Unit

RALU Register Arithmetic Logic Unit

RALW Radioactive Liquid Waste

RAM Radar-Absorbent Material; Radar Absorbing Material; Radar Absorption Material

RAM Radiation Area Monitor

RAM Radioactive Material

RAM Radio Attenuation Measurement

RAM Random-Access Memory

RAM Random Anisotropy Magnet

RAM Random Anisotropy Material

RAM Random Anisotropy Model

RAM Rapid Access Manual (Airlock)

RAM Real Audio Metafile

RAM Real-Time Aerosol Monitor

RAM Reliability, Availability and Maintainability

RAM Removable Atmosphere Muffle

RAM Research and Applications Module

RAM Responsibility Assignment Matrix

RAM Right Ascension of the Meridian [Astronomy]

.RAM Real Audio Metafile [File Name Extension]

RAMA Railway Automotive Management Association [US]

RAMA Raman Microanalysis

RAMA Recap and Movement Authorization

RAMA Rheometrics Automatic Melt Analyzer [Metallurgy]

RAMA Rome Air Materiel Area [US Air Force Systems Command at Griffiss Air Force Base, New York]

RAMAC Radio Marine Associated Companies [UK]

RAMAC Random-Access-Memory Accounting Computer [also Ramac]

RAMAC Random-Access-Memory Accounting Machine [also Ramac]

RAMAC Random-Access Method of Accounting and Control [also Ramac]

RAMAR Random-Access Memory Address Register

RAMARK Radar Marker [also Ramark, or ramark]

RAM BIOS Random-Access Memory–Basic Input/Output Subsystem [also RAMBIOS, RAM-BIOS, or RAM/BIOS]

RAMC Regional Assessment Management Committee

RAMC Royal Army Medical Corps [UK]

RAMDAC Random Access Memory Digital-to-Analog Converter

RAMIO Random-Access Memory plus Input/Output

RAMIS Random-Access Management Information System

RAMIS Rapid-Access Management Information System

RAMIS Rapid Automatic Malfunction Isolation System

RAMNAC Radio Aids to Marine Navigation Application Committee [UK]

RAMONT Radiological Monitoring

RAMP Radar Modernization Project [of Transport Canada]

RAMP Radiation Airborne Measurement Program

RAMP Raman Microprobe

RAMP (R)-1-Amino-2-(Methoxymethyl)pyrrolidine

RAMP Rapid Access to Manufactured Parts

RAMP Rapid Acquisition of Manufactured Parts

RAMP Rapid Modeling Platform

RAMP Raytheon Airborne Microwave Platform

RAMP Reliability and Maintainability Program

RAMP Remote Access Maintenance Protocol [Internet]

RAMP Rural Abandoned Mine Program

RAMPART Radar Advanced Measurement(s) Program for Analysis of Reentry Techniques

RAMPI Raw Material Price Index [Italy]

RAMPS Resource Allocation and Multi-Project Scheduling

RAM/ROM Random-Access Memory/Read-Only Memory

RAMS Radiation Measurement System

RAMS Random-Access Measurement System

RAMS Random-Access Memory Store

RAMS Regional Air Monitoring Station

RAMS Regional Atmospheric Modeling System [of Colorado State University, Fort Collins, US]

Ramsar International Convention on Wetlands [Ramsar, Iran]

RAMSH Reliability, Availability, Maintainability, Safety and Human Factors

RAMT Registered Aroma-Massage Therapist

RAMTOP Random-Access Memory Top

RAN Radio Aids to Navigation

RAN Rainforest Action Network [US]

RAN Read-Around Number [Computers]

RAN Reconnaissance/Attack Navigator

RAN Regional Air Navigation Meeting [of International Civil Aviation Organization]

RAN Royal Australian Navy

RAN Russkoi Akademii Nauk [Russian Academy of Sciences]

RaN Radiative Neutralization

RANC Radar Absorption Noise and Clutter

RANCOM Random Communications Satellite

RAND Radio Detection of Neutrinos in Ice (Experiment) [of Lawrence Berkeley Laboratory, University of California at Berkeley, US]

RAND RandomChallenge [Telecommunications]

rand random

RANDAM Random-Access Nondestructive Advanced Memory

R and D Research and Development [also R&D, R+D, or RD]

RANDID Rapid Alphanumeric Digital Indicating Device

Rand J Econ Rand Journal of Economics [US]

R and R Rest and Recreation; Rest and Recuperation [and R&R]

R and S Reset and Start [also R&S]

RANK Replacement Alphanumeric Keyboard

RANN Research Applied to National Needs (Program) [of National Science Foundation, US]

rANP Rat Atrial Natriuretic Peptide [also RANP] [Biochemistry]

iso-rANP Rat Iso-Atrial Natriuretic Peptide [also iso-RANP] [Biochemistry]

RANRL Royal Australian Navy Research Laboratory

RANSELM (International Conference on) Recent Advances in Science and Engineering of Light Materials

RANTES Regulated on Activation, Normal T-Cell Expressed and Sectreted (Chemokine) [Biochemistry]

RAO Radio Astronomical (or Astronomy) Observatory

RAO Regional Accounting Office

RAOB Radiosonde Observation [also Raob, or raob]

RAOC Royal Army Ordnance Corps [UK]

RAOD Royal Army Ordnance Depot [UK]

RAOIS Repository Acquisition Optical Information System

RAOU Royal Australasian Ornithologists Union; Royal Australian Ornithologists Union

RAP ortho-Phthalate Rubidium Hydrogen (Crystal)

RAP Rapamycin-Associated Protein [Biochemistry]

RAP Rapid Application Prototyping

RAP Rational Alternative to Pseudosciences [Spain]

RAP Reactive Atmosphere Process(ing) [Metallurgy]

RAP Reclaimed Asphalt Pavement

RAP Recommended Area for Protection

RAP Reduced Air Pressure

RAP Redundancy Adjustment of Probability

RAP Regional Acceleratory Phenomenon

RAP Relational Associative Processor

RAP Reliable Acoustic Path

RAP Remedial Action Plan [of International Joint Commission, US]

RAP Remedial Action Program

RAP Resident Assembler Program

RAP Resource Allocation and Planning

RAP Resource Allocations Processor

RAP Response Analysis Program

RAP Rocket-Assisted Projectile

RAP Rubidium Acid Phthalate

RAPA Regional Office for Asia and the Pacific [of Food and Agricultural Organization]

RAPBPPI Research Association for the Paper and Board, Printing and Packaging Industries [UK]

RAPC Royal Army Pay Corps [UK]

RAPCOE Random-Access Programming and Checkout Equipment

RAPCON Radar Approach (and) Control [of US Air Force]

RAPCON Radar Approach Control Center

RAPEC Rocket-Assisted Personnel Ejection Catapult

RAPI Royal Australian Planning Institute (Inc.)

RAPID Random-Access Personnel Information Disseminator

RAPID Reactor and Plant Integrated Dynamics

RAPID Relative Address Programming Implementation Device

RAPID Remote Access Protocol Interface Device

RAPID Research in Automatic Photocomposition and Information Dissemination

RAPL Robot Arm Programming Language

RAPLOT Radar Plot(ting) [also Raplot, or raplot]

RAPO Resident Apollo Project Office [of NASA]

RAPO Road Assist Purchase Order

RAPPI Random-Access Plan Position Indicator

RAPRA Rubber and Plastic Research Association [also Rapra, UK]

Rapra Abstr Rapra Abstracts [of Rubber and Plastics Research Association, UK]

Rapra Data Rapra Database [of Rubber and Plastics Research Association, UK]

Rapra Rev Rep Rapra Review Reports [of Rubber and Plastics Research Association, UK]

RAPS Radioactive Argon Processing System

RAPS Remedial Action Programs [Canada]

RAPS Remote Access Power Support

RAPS Retrieval Analysis and Presentation System

RAPS Right Aft Propulsion System

RAPS Risk Appraisal of Program System

RAPs Recommended Areas for Protection [Tasmania, Australia]

RAPSAT Ranging and Processing Satellite

RAPT Reusable Aerospace Passenger Transport

RAPT Robot Assembly Programming Technique

RAPTUS Rapid Thorium-Uranium-Sodium Reactor

RAPTUS Rapid Thorium-Uranium System

RAPUD Revenue Analysis from Parametric Usage Descriptions

RAR Radar Arrival Route

RAR Random Anisotropy Resonance

RAR Rapid Access Recording

RAR Reasonably Assured Resource

RAR Recording, Archival and Retrieval

RAR Return Address Register

RAR ROM (Read-Only Memory) Address Register

RARA Random Access to Random Access

RARC Regional Administrative Radio Conference [of International Telecommunications Union]

RARDE Royal Armament Research and Development Establishment [UK]

RARE Ram Air Rocket Engine

RARE Rare Animal Relief Effort [US]

RARE Réseaux Associes pour la Recherche Européenne [European Associated Research Networks]

Rare Earth Inf Cent News Rare Earth Information Center News [US]

RAREF Radiation and Repair Engineering Facility

Rare Met Rare Metals [Journal published in PR China]

RAREP Radar Report [also Rarep, or rarep]

RARG Regulatory Analysis Review Group

RAROM RAM (Random-Access Memory) and ROM (Read-Only Memory)

RARP Reverse Address Resolution Protocol [Internet]

RARS Ruakura Animal Research Station [near Hamilton, New Zealand]

RArt Royal Artillery

RAS Rack Access Stands

RAS Radar Advisory Service

RAS Radioactive Sensor

RAS Radioactive Series

RAS Radioactive Source

RAS Random Access Storage (or Store)

RAS Ranking and Scaling

RAS Rapid Access Storage

RAS (Nuclear) Reactor Alarm System

RAS (Nuclear) Reactor Analysis and Safety

RAS Reader Admission System [of British Library]

RAS Receptor Analysis System

RAS Records and Analysis Subsystem

RAS Rectified Air Speed

RAS Reflectance Anisotropy Spectroscopy

RAS Reflection/Absorption Spectrometry; Reflection/Absorption Spectroscopy

RAS Regional AIS (Aeronautical Information Service) System Center

RAS Reliability, Availability and Serviceability

RAS Remote Access Service

RAS Remote Access System

RAS Remote Analysis System

RAS Replenishment at Sea

RAS Requirements Allocation Sheet

RAS Reynolds Averaged Simulation [Computerized Fluid Dynamics]

RAS Ribi Adjuvant System (Emulsion) [Immunochemistry]

RAS Romanian Academy of Sciences [Bucharest, Romania]

RAS Rome Air Service Command [US Air Force Systems Command at Griffiss Air Force Base, New York]

RAS Row Address Select [Computers]

RAS Route Accounting Subsystem

RAS Row-Address Strobe

RAS Royal Agricultural Society (of the Commonwealth) [UK]

RAS Royal Army Service Corps [UK]

RAS Royal Astronomical Society [London, UK]

RAS Rural Adjustment Scheme

RAS Russian Academy of Sciences [Moscow and St. Petersburg]

RAS Rutgers Annihilation Spectrometer [of Rutgers University, Piscataway, New Jersey, US]

RASAC Rural Assistance Scheme Advisory Committee

RASAPI Remote Access Service Application Programming Interface

RASC Royal Army Service Corps [UK]

RASC Royal Astronomical Society of Canada

RASC Royal Astronomical Society of the Commonwealth

RASE Royal Agricultural Society of England

RASELM Recent Advances in Science and Engineering of Light Metals (Conference) [of Japan Institute of Light Metals]

RASER Radio Frequency Amplification by Stimulated Emission of Radiation [also Raser, or raser]

RASH Rain Shower

RASI Reliability, Availability, Serviceability and Improvability

RASIS Reliability, Availability, Serviceability, Integrity and Security

RA Size Metric Untrimmed Paper Size [Usually Designated by the Letters "RA" Followed by a Numeral, e.g., RA0, RA1, etc.] [also RA size]

RASM Remote Access Scheduling Mailbox

RASN Rain and Snow [Meteorology]

RASofC Royal Agricultural Society of the Commonwealth [UK]

RASP Radar Applications Specialist Panel [of Eurocontrol–European Organization for the Safety of Air Navigation]

RASP Regional Assimilation and Prognosis System

RASP Remote Access Switching and Patching

RASP Remote Antenna Signal Processor [Telecommunications]

RASP Retrieval and Statistics Processing

RASPO Resident Apollo Spacecraft Program Office [of NASA]

RASS Radar Analysis Support System

RASS Radio Acoustic Sounding System

RASS Rapid Area Supply Support

RASS Research Association of Statistical Sciences [Kyushu University, Fukuoka, Japan]

RASS Rotating Acoustic Stereo-Scanner

RASSPVDN Random-Sphere Model of Simultaneous Particle and Vapor Deposition

RASSR Reliable Advanced Solid-State (Phased Array) Radar

RAST Radioallergosorbent Test [Medicine]

RAST Recover Assist, Secure and Traverse System; Recovery Assist Secure and Traverse System

RASTA Radiation Augmented Special Test Apparatus; Radiation Special Test Apparatus

RASTAC Random-Access Storage and Control

RASTAD Random-Access Storage and Display

RASTI Rapid Speech Transmission Index

R Astron Soc Quart J Royal Astronomical Society Quarterly Journal [London, UK]

RASU Rangoon Arts and Science University [Burma]

RASVY Royal Australian Survey Corps [of Australian Army]

RAT Radioactive Tracer

RAT Reactive Acidified Tailings [Mining]

RAT Reliability Assurance Test

RAT Reserve Auxiliary Transformer

RAT Rocket-Assisted Torpedo

.RAT Rating [File Name Extension]

R&AT Research and Advanced Technology

RATA Rankine Cycle Air Turboaccelerator

RATAC Radar Analog Target Acquisition Computer

RATAC Radar Target Acquisition

RATAN Radar and Television Aid to Navigation [also Ratan, or ratan]

RATAP Reactive Tailings Assessment Program [Mining]

RATC Rate-Aided Tracking Computer

RATCC Radar Air-Traffic Control Center

RATCF Radar Air-Traffic Control Facility [of US Navy]

RATD Registered Apparel and Textile Designers [UK]

RATE Remote Automatic Telemetry Equipment

RATEL Radio-Telephone

RATER Response Analysis Tester

RATFOR Rational FORTRAN

RATG Radioactive Thickness Gauge

RATG Radio-Telegraph

Ra-Th Radium-Thorium

RATIO Radio Telescope in Orbit

RATO Rocket-Assisted Takeoff [also Rato, or rato]

RATOG Rocket-Assisted Takeoff Gear

RATS Rate and Track Subsystem

RATS Reactive Acidified Tailings Study [Mining]

RATS Reactive Acid Tailings Stabilization [Mining]

RATSC Rome Air Technical Service Command [US Air Force Systems Command at Griffiss Air Force Base, New York]

RATSCAT Radar Target Scatter

RATSIMP Rational Simplification

RATSTADS Radar Tracking System Target Acquisition and Display Subsystem [of NASA]

RATT Radio Teletype

RATT Radio Teletypewriter

RAU Rand Afrikaans University [Johannesburg, South Africa]

RAU Remote Acquisition Unit

Ra/U Radium/Uranium (Ratio)

RAUIS Remote Acquisition Unit Interconnecting Station

RAV Rising Arc Voltage

RAVC Royal Army Veterinary Corps [UK]

RAVE Radar Acquisition Visual-Tracking Equipment

RAVE Rendering Acceleration Virtual Engine [Computers]

RAVEN Ranging and Velocity Navigation

RAVIR Radar Video Recording

RAW Radioactive Waste

RAW Rationalisierungsausschuss der Deutschen Wirtschaft [German Productivity and Management Committee]

RAW Read After Write

RAW Relative Atomic Weight

.RAW Raw Data [File Name Extension]

RAWARC Radar and Warning Coordination

Rawin Radar/Wind [also rawin]

Rawin Radar Wind Sounding [also rawin]

Rawin Radio/Wind [also rawin]

Rawinsonde Radio-Wind-Sounding Device [also rawinsonde]

RAWS Remote Automated Weather System [US]

Raw Mater Rep Raw Materials Report [Sweden]

RAX Remote Access (Terminal)

RAX Rural Automatic Exchange

RAXS Rapid Analyzer of X-Ray Stress

RAYDAC Raytheon Digital Automatic Computer

rayl/m rayl per meter [Unit]

RAYNET Raytheon Data Communications Network

RAZEL Range, Azimuth and Elevation

RAZON Range and Azimuth Only

RB Bait (Ready for Use) [Chemistry]

RB Cylindrical Roller Bearing, Double Row, Two-Direction-Locating, Outer Ring Without Ribs, With Two Internal Snap Rings, Inner Ring With Three Integral Ribs, Non-Separable [Symbol]

RB Radar Beacon

RB Radiation Belt

RB Radiation Biologist; Radiation Biologist

RB Radiobalance

RB Radio Beacon

RB Radiobiological; Radiobiologist; Radiobiology

RB Radio Bremen [Bremen, Germany]

RB Randomized Blocks [Statistics]

RB Rayleigh-Benard (Convection) [Physics]

RB Reaction-Bond(ed); Reaction Bonding

RB (Nuclear) Reactor Building

RB Read Back

RB Read Backward

RB Read Buffer

RB Reclamation Bureau

RB Reconnaissance Bomber (Aircraft) [US Air Force Symbol]

RB Red Brown

RB Reducing Bushing

RB Reference Block [Computers]

RB Reinforced Brick

RB Reinforcing Bar

RB Relative Basicity

RB Remazol Black

RB Remote Batch

RB Renegotiation Board

RB Reorientation Band [Metallurgy]

RB República de Bolivia [Republic of Bolivia]

RB Request Block

RB Resistance Brazing

RB Rescue Boat

RB Reset Button

RB Resonance Bonding

RB Return-to-Bias

RB Reverse Bias [Solid-State Physics]

RB Rheostat Braking

RB Rifle Bomber

RB Rifle Brigade

RB Rigid Band (Condition) [Solid-State Physics]

RB Rigid Body [Mechanics]

RB Right Boundary

RB Right Button

RB Rock Breaking

RB Rock Burst [Mining]

RB Rocket Boost(er)

RB Rollback

RB Roller Bearing

RB Rotary Beacon

RB Round Baler [Agriculture]

RB Royal Blue

RB Royaume de Belgique [Kingdom of Belgium]

RB Rubber Board [India]

RB Rubber Bond [Grinding Wheels]

RB Rule Based (Computing)

R(B) High Magnetic-Field Resistance [Symbol]

R(B$_A$) Reevaporation Rate of B Atoms from Surface A [Symbol]

R′(B$_A$) Reevaporation Probability of B Atoms from Surface A [Symbol]

R&B Ring and Ball (Method)

Rb Rubidium [Symbol]

Rb Ruble [Currency of Russia]

Rb$^+$ Rubidium Ion [Symbol]

Rb-81 Rubidium-81 [also ^{81}Rb, or Rb^{81}]

Rb-82 Rubidium-82 [also ^{82}Rb, or Rb^{82}]

Rb-83 Rubidium-83 [also ^{83}Rb, or Rb^{83}]

Rb-84 Rubidium-84 [also ^{84}Rb, or Rb^{84}]

Rb-85 Rubidium-85 [also ^{85}Rb, or Rb^{85}]

Rb-86 Rubidium-86 [also ^{86}Rb, or Rb^{86}]

Rb-87 Rubidium-87 [also ^{87}Rb, or Rb^{87}]

Rb-88 Rubidium-88 [also ^{88}Rb, or Rb^{88}]

Rb-89 Rubidium-89 [also ^{89}Rb, or Rb^{89}]

Rb-90 Rubidium-90 [also ^{90}Rb, or Rb^{90}]

Rb-91 Rubidium-91 [also ^{91}Rb, or Rb^{91}]

Rb-92 Rubidium-92 [also ^{92}Rb, or Rb^{92}]

Rb-93 Rubidium-93 [also ^{93}Rb, or Rb^{93}]

Rb-94 Rubidium-94 [also ^{94}Rb, or Rb^{94}]

rb rabbit [Immunochemistry]

ρ(B) magnetic induction dependent resistivity [Symbol]

RBA Radar Beacon Antenna

RBA Radiation Barrier Alloy

RBA Rapid Biodiversity Assessment

RBA Recovery Beacon Antenna

RBA Refined Bitumen Association [UK]

RBA Relative Bioavailability [also RBAV]

RBA Relative Byte Address

RBA Reserve Bank of Australia

RBA Reverse Bias Annealing [Solid-State Physics]

RBA Road Bitumen Association

RBAC Recycling and Reuse Business Assistance Center (Initiative) [of US Environmental Protection Agency]

RbAc Rubidium Acetate

Rb(ACAC) Rubidium Acetylacetonate [also Ru(AcAc), or Ru(acac)]

RBAI Royal Belfast Academical Institution [Northern Ireland]

RBAO Reaction-Bonded Aluminum Oxide

RbAP Rubidium Acid Phthalate

RBAS Royal Belgian Academy of Sciences

RBAV Relative Bioavailability [also RBA]

RBA/ZBA Reverse Bias Annealing/Zero Bias Annealing (Effect) [Solid-State Physics]

RBB Remazol Brilliant Blue

RbBr Rubidium Bromide

RBBS Remote Bulletin Board System

RBC Radial Block Copolymer

RBC Red/Black Control

RBC Red Blood Cell(s)

RBC Red Blood Count

RBC Rotating Beam Ceilometer

RBCA Risk-Based Corrective Action [Petroleum and Oilfield Cleanup]

RBCCW (Nuclear) Reactor Building Closed Coolant Water

Rb$_x$C$_{60}$ Rubidium-Doped C$_{60}$ Buckmisterfullerene

RbCl Rubidium Chloride

RBCO Rhenium Barium Copper Oxide (Superconductor)

RBCO Rare-Earth Barium Copper Oxide (Superconductor)

RBCP Random Block Copolymer

RBCS Remote Bar Code System

RBCU Rod Bank Coil Unit

RBCuZn Copper-Zinc Resistance-Brazing Alloy

RBCWS (Nuclear) Reactor Building Cooling Water System

RBD Randomized Block Design [Statistics]

RBD Reliability Block Diagram

RBDE Radar Bright Display Equipment

RBDV Raspberry Bushy Dwarf Virus

RBE Radiation Biological Effectiveness

RBE Red Brown Earth

RBE Relative Biological Effectiveness (of Radiation) [also rbe]

RBE Remote Batch Entry

RBEC Roller Bearing Engineers Committee [US]

RBF Relocatable Binary Format

RBF Remote Batch Facility

RBF Resonant Bar Flexture [Mechanics]

RBF Rockefeller Brothers Fund

RbF Rubidium Fluoride

RBFD Randomized Block Factorial Design [Statistics]

RBG Royal Botanical Gardens [Australia, Canada and the UK]

rBGH Recombinant Bovine Growth Hormone [also r-BGH, or R-BGH] [Biochemistry]

RBHPF (Nuclear) Reactor Building Hydrogen Purge Fan

RBI Radar Bearing Indicator

RBI Radar Blip Identification

RBI Relative Bearing Indicator

RBI Ripple-Blanking Input

RBI Roof-Bolt Inserter [Mining]

RBI Rudjer Boscovic Institute [Yugoslavia]

RbI Rubidium Iodide

RBIM Radiation Beam Intensity Modulation

RBK Right Bank [Geography]

RBL Radiation Biology Laboratory [of Smithsonian Institution, Washington, DC, US]

RBL Right Buttock Line

RBL Ruble [Currency of Russia]

RBLG Rifled Breech-Loading Gun

RBM Random Bonding Model [Materials Science]

RBM Reactive Ball Mill(ing)

RBM Real Batch Monitor

RBM Real-Time Batch Monitor

RBM Reinforced Brick Masonry

RBM Remote Batch Module

RBM Rigid Band Model [Solid-State Physics]

RBM Rock Burst Monitor [Mining]

RBM Rod Block Monitor

RBMS Remote Bridge Management Software

RBMT Retrospective Bibliographies on Magnetic Tape

RBN Radio Beacon [also RBn]

RBN Rhombohedral Boron Nitride [also r-BN, or R-BN]

RBNA Royal British Nurses Association [UK]

Rb-87 NMR Rubidium-87 Nuclear Magnetic Resonance [also ^{87}Rb NMR]

RBO Rapid Burn-Off (Zone)

RBO Ripple-Blanking Output

RBOC Regional Bell Operating Company [US]

RBOF Receiving Basin for Off-Site Fuel

RbOH Rubidium Hydoxide

RBOT Rotating Bomb Oxidation Test

RBP Registered Business Programmer

RBP Reservoir Bubble Point

RBP Remote Batch Processor

RBP Retinol Binding Protein [Biochemistry]

Rb/P Rubidium/Phosphorus (Concentration Ratio)

RBPSSD Ruud-Barrett Position Sensitive Scintillation Detector

RBR Radar Boresight Range

RBR Roll-Bend-Roll [Mechanics]

RBR Russian Breeder Reactor

Rbr Rubber [also rbr]

RBRG Reinforcing Bar Ring Ground

RBS Radar Bomb Scoring

RBS Radio Base Station

RBS Random Barrage System

RBS (Nuclear) Reactor Building Sump

RBS Remote Batch Terminal

RBS Rock Bolting and Screening (Machine) [Mining]

RBS Royal Botanical (and Horticultural) Society [UK]

RBS Rutherford Backscattering

RBS Rutherford Backscattering Spectra; Rutherford Backscattering Spectrometry; Rutherford Backscattering Spectroscopy

RBSC Reaction-Bonded Silicon Carbide [also RBSiC]

RBS-C Channeling Rutherford Backscattering Spectrometry; Rutherford Backscattering Spectrometry/ (Ion) Channeling; Rutherford Backscattering Spectrometry with Channeling [also RBS/C]

RBSCU (Nuclear) Reactor Building Sump Cooling Unit

RB-SiOC Reaction-Bonded Silicon Oxycarbide

RBSN Reaction-Bonded Silicon Nitride

RBSOA Reverse Bias Safe Operating Area

RBSS Recoverable Booster Support System

rBST Recombinant Bovine Somatotropin [also r-BST, or R-BST] [Biochemistry]

RBSWG Risk-Based Standards Working Group [US]

RBT Radial-Beam Tube [Electronics]

RBT Remote Batch Terminal

RBT Resistance Bulb Thermometer

RBT Resonant Bar Torsion [Mechanics]

RBT Rifle Bullet Test

RBT Rigid Body Translation (Mechanism)

RBT Ringback Tone

rbt rabbit [Immunochemistry]

ρ**(B,T)** Temperature and Magnetic Induction Dependent Resistivity [Symbol]

RBU Rabindra Bharati University [Calcutta, India]

RBU Radium-Beam Unit

RBU Reaktor-Brennelement Union GmbH [German Manufacturer of Nuclear Fuel Elements]

RBUPC Research in British Universities, Polytechnics and Colleges [of British Library Lending Division, UK]

RBV (Nuclear) Reactor Building Vent

RBV Remazol Brilliant Violet

RBV Return Beam Vidicon

RBWO Resonant Backward Wave Oscillator

RC Cylindrical Roller Bearing, Single-Row, Two-Direction-Locating, Double-Ribbed Outer Ring, Double-Ribbed Inner Ring, Non-Separable [Symbol]

RC Radial Chromatography

RC Radiation Chemist(ry)

RC Radiation Cooling

RC Radiation Counter

RC Radiation Curing

RC Radiochemical; Radiochemist(ry)

RC Radio Communication

RC Radio Compass

RC Radium Content

RC Radix Complement [Mathematics]

RC Ramsden Circle [Optics]

RC Random Copolymer

RC Raney Copper

RC Range Command [Aeronautics]

RC Range Control [Aeronautics]

RC Range Correction

RC Rankine Cycle [Thermodynamics]

RC Rapid City [South Dakota, US]

RC Rapid Curing (Asphalt)

RC Rate Center

RC Rate Command

RC Rate Constant [Physical Chemistry]

RC Rate Control(led)

RC Ray-Control Electrode

RC Reactance Coil

RC Reaction Center

RC Reaction Control

RC Reaction Curing

RC Read and Compute

RC Reader Code

RC Reaction Chamber

RC Reactive Component

RC Real Circuit

RC Recent Change

RC Receiver Clock

RC Recording Completing

RC Recoil Counter

RC Recovery [also Rc] [Metallurgy]

RC Recovery Controller

RC Recurring Cost(s)

RC Red Crescent [Organization Analogous to the Red Cross in Muslim Countries]

RC Red Cross

RC Reduced Charge

RC Reduction Cell [Physical Chemistry]

RC Reed College [Portland, Oregon]

RC Reference Clock

RC Reference Coil

RC Reflex Camera

RC Refractory Ceramics

RC Refrigeration Cycle

RC Regenerative Cooling

RC Regional Center

RC Registered Chiropractor

RC Reinforced Composite

RC Reinforced Concrete

RC Relaxed Contraint (Grain Interaction Model) [Materials Science]

RC Remote Calibration (Function)

RC Remote Channel

RC Remote Computer

RC Remote Control [also R/C]

RC Reply Check

RC Report on Carcinogens [of National Institute of Environmental Health Sciences, National Institutes of Health, Bethesda, Maryland, US]

RC Reproductive Cell [Biology]

RC República de Chile [Republic of Chile]

RC República de Columbia [Republic of Columbia]

RC República de Cuba [Republic of Cuba]

RC Republic of Chad

RC Research Center

RC Research Chemist

RC Research Communication

RC Research Corporation [Tucson, Arizona, US]

RC Research Council

RC Reserve Capacity

RC Reserve Corps

RC Residual Contrast

RC Resin Coating

RC Resistance-Capacitance [also R/C, or R-C]

RC Resolver Control

RC Resonant Circuit

RC Responsible Care

RC Restrained Cursor

RC   Reticulocyte [Medicine]

RC   Retrograde Condensation [Organic Chemistry]

RC   Revenue Canada [now Canada Customs and Revenue Agency]

RC   Reverse Course

RC   Reverse Current

RC   Rheostatic Control

RC   Ribbon Comminution

RC   Richmond College [UK]

RC   Riding Circuit

RC   Ring Counter

RC   Ringelmann Chart [Environmental Sciences]

RC   Ringing Code

RC   Robot Control

RC   Rocking Curve [Materials Science]

RC   Rod Cell (of the Eye) [Anatomy]

RC   Roll Control [Aeronautics]

RC   Roller Conveyor

RC   Rolling Contact [Mechanics]

RC   Rolling Curve [Geometry]

RC   Root Canal [Dentistry]

RC   Root Circle (of Gears)

RC   Rotary Converter

RC   Rotational Coefficient [Physical Chemistry]

RC   Rotation Control

RC   Rubber-Covered (Cable)

RC   Running (or Sliding) Clearance Fit [ANSI Symbol]

RC   Rushlight Club [US]

RC   Time Constant (of a Capacitor) [Symbol]

RC1 Close Sliding Fit (H5/g4) [ANSI Symbol]

RC2 Sliding Fit (H6/g5) [ANSI Symbol]

RC3 Precision Running Fit (H7/f6) [ANSI Symbol]

RC4 Close Running Fit (H8/f7) [ANSI Symbol]

RC5 Medium Running Fit (H8/e7) ([ANSI Symbol]

RC6 Medium Running Fit (H9/e8) [ANSI Symbol]

RC7 Free Running Fit (H9/d8) [ANSI Symbol]

RC8 Loose Running Fit (H10/c9) [ANSI Symbol]

RC9 Loose Running Fit (H11/...) [ANSI Symbol]

R&C Reed and Carnrick [Pharmaceutical Manufacturer]

R/C Radio Command

R/C Radio Control

R/C Range Clearance

R/C Range Control [Aeronautics]

R/C Rate of Climb [Aeronautics]

R/C Remote Control [also RC]

R/C Resorcinol-to-Carbonate (Ratio)

R/C Resorcinol-to-Catalyst (Ratio)

R-C Radial Direction–Circumferential Direction (Orientation) [Forging]

R-C Rapid-Curing (Cement)

R-C Resistance-Capacitance [also RC, or R/C]

Rc   Recovery [also RC] [Metallurgy]

R(c) (Mechanical) Resistance of Composite [Symbol]

rC_c collector-base time constant [Semiconductor Symbol]

ρc characteristic (acoustic) impedance [Symbol]

RCA Radiochemical Analysis

RCA Radio Club of America [US]

RCA Radio Corporation of America [Princeton, New Jersey, US]

RCA Reach Climb Altitude

RCA Reinforced Concrete Association [UK]

RCA Remote Control Amplifier

RCA Research Council of Alberta [Canada]

RCA Ricinus Communis Agglutinin [Immunology]

RCA Right Coronary Artery [Anatomy]

RCA Rotary Cup Atomizer [Metallurgy]

RCA Royal Canadian Academy

RCA Royal Canadian Artillery

RCA Royal College of Art [UK]

RCA_I Ricinus Communis Agglutinin with Molecular Weight of approx. 120,000 [also RCA_{120}, or RCL I + II]

RCA_{II} Ricinus Communis Agglutinin with Molecular Weight of approx. 60,000 [also RCA_{60}, or RCL (III)]

RCA_{60} Ricinus Communis Agglutinin with Molecular Weight of approx. 60,000 [also RCA_{II}, or RCL III]

RCA_{120} Ricinus Communis Agglutinin with Molecular Weight of approx. 120,000 [also RCA_I, or RCL (I + III)]

RCAA Royal Canadian Academy of Arts

RCAA Royal Cornwall Agricultural Association [UK]

RCA Broadcast News RCA Broadcast News [of the Radio Corporation of America]

RCAC Radio Corporation of America Communications [US]

RCAC Remote Computer-Access Communications Service

RCAC Royal Canadian Armored Corps [UK]

RCACA Royal Canadian Armored Corps Association [UK]

RCACC Royal Canadian Army Cadet Corps

RCA Eng RCA Engineer [Journal of the Radio Corporation of America]

RCAF Royal Canadian Air Force [now Canadian Armed Forces]

RCAFM Royal Canadian Air Force Museum

RCAG Remote Center Air/Ground

RCAG Remote Communications Air-to-Ground

RCAG Remote Control Air/Ground

RCAMC Royal Canadian Army Medical Corps

RCAMMP Research Center for Advanced Mineral and Materials Processing [of University of Western Australia, Nedslands, Perth]

RCAN Recorded Announcement

RCA Rev RCA Review [Journal of the Radio Corporation of America, Princeton, New Jersey, US]

RCARS Resonance Coherent Anti-Stokes Raman Spectroscopy

RCAST Research Center for Advanced Science and Technology [of University of Tokyo, Japan]

RCAT Radio Code Aptitude Test

RCAT Radio Controlled Airplane Target

RCAT Ridgewood College of Agricultural Technology [Canada]

RCA Tech Notes RCA Technical Notes [of the Radio Corporation of America]

RCB Randomly Connected Bubble (Model) [Metallurgy]

RCB Reactor Containment Building

RCB Regional Chief Biologist

RCB Remote Circuit Breaker

RCB Resource Control Block

RCBC Rapid Cycling Bubble Chamber

RCBC Recycling Council of British Columbia [Canada]

RCC Radiochemical Center [UK]

RCC Radio Common Carrier

RCC Radio Common Channels

RCC Random Correction Code

RCC Range Commanders Council [of US Department of Defense]

RCC Range Control Center [Aeronautics]

RCC Read Channel Continue

RCC Reactive Current Compensation

RCC Reader Common Contact

RCC Real-Time Computer Complex

RCC Recovery Control Center

RCC Reduced Crude-Oil Conversion (Process)

RCC Regional Conferences Committee [of Institute of Electrical and Electronics Engineers, US]

RCC Reinforced Carbon-Carbon

RCC Remote Center Compliance

RCC Remote Communications Center

RCC Remote Communications Complex

RCC Remote Communications Concentrator

RCC Remote Communications Console

RCC Remote Control-Rod Cluster Assembly [Nuclear Reactors]

RCC Renal Cell Carcinoma [Medicine]

RCC Rescue Control Center

RCC Rescue Coordination Center [US and Canada]

RCC Resistance-Capacitance Coupling

RCC Reusable Carbon-Carbon

RCC Richmond Chamber of Commerce [Virginia, US]

RCC Rochester Cariology Center [US]

RCC Rod Cluster Control [Nuclear Reactors]

RCC Roller-Compacted Concrete (Dam) [Civil Engineering]

RCC Rough Combustion Cutoff

RCC Routing Control Center

R&CC Recorder and Communication Control

RCCA Rough Combustion Cutoff Assembly

RCCB Remote Control Circuit Breaker

RCCB Residual Current Circuit Breaker

RCCE Revenue Canada, Customs and Excise [also RC/CE]

RCCI Rotherham Chamber of Commerce and Industry [UK]

RCCM Remote Carrier-Controlled Modem

RCCP Recorder and Communications Control Panel

RCCPLD Resistance-Capacitance Coupled

RCCRU Research Center for Carbon Recycling and Utilization [of Tokyo Institute of Technology, Japan]

RCC&S Riots, Civil Commotion and Strike [also rcc&s]

RCD Receiver-Carrier Detector

RCD Reduced Crude Desulfuration

RCD Regional Cooperation for Development

RCD Residual Current Device

RCD Route Control Digit

Rcd Record [also rcd]

rcd received [also rc'd]

RCDAMP Research Center for Dielectric and Advanced Matter Physics [of Pusan National University, South Korea]

RCDB Rubber-Covered Double-Braided (Electric Wire)

RCDC Radar Course-Directing Center

RCDC Royal College of Dentists of Canada

RCDCD Record Code

RCDF Rural Credit Development Fund [of Reserve Bank of Australia]

RCDP Record Parallel

RCDP Remote Community Demonstration Program

RCDR Recorder [also Rcdr]

RCDSO Royal College of Dental Surgeons of Ontario [Canada]

RCE Rapid Circuit Etch [Electronics]

RCE Ray-Casting Engine

RCE Refractories Consulting and Engineering GmbH [Austria]

RCE Royal Canadian Engineers

RCEA Research Council Employees Association

RCEEA Radio Communications and Electronic Engineering Association [UK]

RCEC Resources Cost-Shared Energy Conversion (Program) [of University of Moncton, New Brunswick, Canada]

RCEI Range Communications Electronics Instruction

RCEM Research Center for Energetic Materials [of New Mexico Institute of Mining and Technology, Socorro, US]

RCEM Research Center for Extreme Materials [Osaka University, Japan]

RCEME Royal Canadian Electrical and Mechanical Engineers

RCEP Royal Commission on Environmental Pollution [UK]

RCF Radial Centrifugal Force

RCF Radio Communication Failure

RCF Recall Finder

RCF Refrigeration Centrifuge

RCF Relative Centrifugal Force

RCF Remote Call Forwarding

RCF Reverse Column Flotation

RCF Rolling Contact Fatigue [Mechanics]

RCFA Royal Canadian Field Artillery

RCFC (Nuclear) Reactor Core Fan Cooling

RCFCU (Nuclear) Reactor Core Fan Cooling Unit

RCG Radioactivity Concentration Guide

RCG Reaction-Cured Glass

RCG Relative Cooling Gain

RCG Rheocardiograph(y)

RCG Rotating-Coil Gaussmeter

RCGA Royal Canadian Garrison Artillery

RCGBR Royal Commission on the Great Barrier Reef [Australia]

RCGM (Nuclear) Reactor Cover Gas Monitor

RCGP Royal College of General Practitioners [UK]

RCGS Royal Canadian Geographical Society

RCGS Royal Canadian Geological Society

RCH Royal Canadian Horse Artillery

RCH Ruhrchemie (Process) [Chemical Engineering]

R Ch Reduced Charge

RCHCS Regenerable CO (Carbon Monoxide) and Humidity Control System

R Chem Soc Spec Publ Royal Chemical Society Special Publications [UK]

R Chg Reduced Charge

RCHO Aldehyde [Formula]

RCH/RP Ruhrchemie/Rhône-Poulenc (Process)

RCHS Royal Caledonian Horticultural Society [UK]

R&CHS Railway and Canal Historical Society [UK]

RCHSA Radiation Control for Health and Safety Act [of US Food and Drug Administration]

RCI Radar Coverage Indicator

RCI Radio Canada International

RCI Rating Core Index

RCI Raycomm Industries [US]

RCI Read Channel Initialize

RCI Remote Control Interface

RCI République de la Côte-d'Ivoire [Republic of the Ivory Coast]

RCI Restricted Configuration Interaction

RCI Rogers Communications Inc. [Canada]

RCI Roof Consultants Institute [US]

RCI Routing Control Indicator

RCI Royal Canadian Institute

RCIA Royal Canadian Institute of Architects

RCIBT Research Center of Ion Beam Technology [Hosei University, Koganei, Japan]

RCIC (Nuclear) Reactor Core Isolation Cooling

RCICS (Nuclear) Reactor Core Isolation Cooling System

RCIGG Research Center of Isotropic Geochemistry and Geochronology [US]

RCIS Regional Cadet Instructor School [Canada]

RCIU Remote Computer Interface Unit

RCJ Reaction Control Jet

RCJ Roll Control Jet

RCL Radiation Counting Laboratory [US]

RCL Radiochemical Laboratory

RCL Recall (Function)

RCL Reproduction Consideration Listing

RCL Resistance-Capacitance-Inductance (Meter) [also R-C-L]

RCL Railway Conversion League

RCL Ring Calorimeter

RCL Robot Command Language

RCL Rotate Carry Left [Computers]

RCL Royal Canadian Legion

RCL Runway Center Light

RCL Runway Centerline

RCL (I + III) Ricinus Communis Agglutinin with Molecular Weight of approx. 120,000 [also RCA$_{120}$, or RCA$_I$]

RCL III Ricinus Communis Agglutinin with Molecular Weight of approx. 60,000 [also RCA$_{60}$, or RCA$_{II}$]

RCLC (Nuclear) Reactor Coolant Leakage Calculation

RCLED Resonant-Cavity Light Emitting Diode [also RC-LED]

RCLIF Rotationally Cooled Laser Induced Fluorescence

RCLM Runway Centerline Markings

RCLS Runway Centerline Lighting System

RCLWUNE Regional Commission on Land and Water Use in the Near East [of Food and Agricultural Organization]

RCM Radar Countermeasures

RCM Radioactive Convective Model

RCM Radio Countermeasures

RCM Real-Time Confocal Microscope

RCM Reliability Corporate Memory

RCM Requirements Planning Criteria and Methods of Application

RCM Royal Canadian Mint [Ottawa]

RCM Royal College of Mines [UK]

r/cm revolution(s) per centimeter [Unit]

RCMA Railroad Construction Maintenance Association [now National Railroad Construction and Maintenance Association, Inc., US]

RCMA Roof Coatings Manufacturers Association [US]

RCMA Royal Canadian Military Institute

RCMC Resolved Motion Rate Control

RCMD Rice Council for Market Development [US]

RCMI Research Centers in Minority Institutions

RCMP Royal Canadian Mounted Police

RCMS Regional Center for Mass Spectrometry

RCN Fatty Nitrile [Formula]

RCN Reactor Center Netherlands

RCN Requirements Change Notice

RCN Royal Canadian Navy [now Canadian Armed Forces]

RCN Royal College of Nursing [UK]

RCn Rotate Right through Carry n Number of Bits (Function) [Computers]

RCNDO Rydberg Complete Neglect of Differential Overlap [Physics]

RCNVR Royal Canadian Naval Volunteer Reserve

RCO Acyl Group [Formula]

RCO (Nuclear) Reactor Core

RCO Receiver Cuts Out

RCO Recycling Council of Ontario [Canada]

RCO Red Copper Oxide

RCO Remote Communications Outlet

RCO Remote Control Office

RCO Remote Control Oscillator

RCO Representative Calculating Operation

RCO Restoration Control Office

RCO Ripple Carry Output

RCO Ripple Clock Output

R$_2$CO Ketone [Formula]

RCo$_2$ Rare-Earth Cobaltate [Formula]

R-Co Rare Earth-Cobalt (Alloy)

RCOC Royal Canadian Ordnance Corps

RCOG Royal College of Obstetricians and Gynaecologists [UK]

RCOM Regional Chapter Officers Meeting [of ASM International, US]

RCONH$_2$ Amide [Formula]

RCOOH Alkyl Group, or Carboxylic Acid [Formula]

RCOOM Ionic Metal Salt [Formula]

RCOOR Ester [Formula]

RCOOTHP Tetrahydropyranyl Ester [Formula]

RCOphth Royal College of Ophthalmologists [UK]

RCOR Ketone [Formula]

RC(OR)$_3$ Orthoester [Formula]

RCO$_2$R Ester [Formula]

RCOX Acid Halide (X = Halogen) [Formula]

RCo$_{7-x}$Zr$_x$ Rare-Earth Cobalt Zirconium (Alloy)

RCP Radiological Control Program

RCP Radiation Constraints Panel

RCP Radon Contractor Proficiency

RCP Random Close-Packed; Random Close-Packing [Materials Science]

RCP Rapid Cooling Process [Metallurgy]

RCP (Nuclear) Reactor Coolant Pump

RCP Receive Clock Pulse

RCP Recognition and Control Processor

RCP Réseau à Commutation par Paquet [Packet Switched System, France]

RCP Remote Control Panel

RCP Remote Copy [Internet]

RCP Restoration Control Point

RCP Restore Cursor Position

RCP Right Circular (Light) Polarization; Right Circular (Light) Polarizer

RCP Royal College of Physicians [UK]

RCPA Rice and Corn Production Administation [Philippines]

RCPath Royal College of Pathology [UK]

RCPB (Nuclear) Reactor Coolant Pressure Boundary

RCPC Royal College of Physicians of Canada [also RCP(C)]

Rcpt Receipt [also rcpt]

Rcpt Receptacle [also rcpt]

Rcpt Recipient [also rcpt]

R-CPZ Random Carbon Precipitation Zone [Metallurgy]

RCR Reader Control Relay

RCR Recrystallization-Controlled Rolling [Metallurgy]

RCR República de Costa Rica [Republic of Costa Rica]

RCR Required Carrier Return

RCR Retrofit Configuration Record

RCR Review Comment Record

RCR Room Cavity Ratio

RCR Rotate Carry Right [Computers]

RCR Route Contingency Reserve

RCR Royal Canadian Regiment

RCR Runway Condition Reading

RCRA Refrigeration Compressor Rebuilders Association [US]

RCRA Resource Conservation and Recovery Act [of US Environmental Protection Agency]

RCRA/EPA Resource Conservation and Recovery Act/Environmental Protection Agency [US]

RCRC Reinforced Concrete Research Council [Urbana, Illinois, US]

RCRC Russian Cosmic Ray Conference [Moscow]

RCRDC Radio Components Research and Development Committee [UK]

RCRF Rei Cretariae Romanae Fautores [Society of Roman Pottery and Ceramic Archeologists, Germany]

RCRI Radiation Chemistry Research Institute [of Tokyo University, Japan]

R-CRS Report on Course [Aeronautics]

RCS Cylindrical Roller Bearing, Single-Row, Two-Direction-Locating, Double-Ribbed Outer Ring, Double-Ribbed Inner Ring, Spherical Outside Surface, Non-Separable [Symbol]

RCS Radar Cross-Section

RCS Radio Command System

RCS Radio Compass Station

RCS Radio-Controlled Solar

RCS Rainforest Conservation Society [Australia]

RCS Rapid Cycling Synchrotron

RCS Rate-Controlled Sintering [Powder Metallurgy]

RCS Reaction Control Subsystem; Reaction Control System

RCS (Nuclear) Reactor Coolant System

RCS Rearward Communications System

RCS Records Communications Switching System

RCS Reentry Control System [of NASA]

RCS Reference Climate Station

RCS Region(s) of Coherent Scattering

RCS Registration Control System

RCS Reloadable Control Storage

RCS Remote Characterization System

RCS Remote Computing Service

RCS Remote Contact Sensor

RCS Remote Control Station

RCS Remote Control System

RCS Remote Core Sampler

RCS Replacement Collision Sequence [Materials Science]

RCS Resin-Coated Sand [Metallurgy]

RCS Reverse Circulation System

RCS Reversing Color Sequence

RCS Revision Control System

RCS Rigid Container Sheet

RCS Round-Cornered Square

RCS Royal College of Science [UK]

RCS Royal College of Surgeons [UK]

RCS Royal Commonwealth Society [UK]

RCS Rural Counselling Service

RCSC Radio Components Standardization Committee [UK]

RCSC Reaction Control Subsystem Controller

RCSC Research Council on Structural Connections [US]

RCSC Royal College of Surgeons of Canada

RCSDE Reactor Coolant System Dose Equivalent

RCSHSB Red Cedar Shingle and Handsplit Shake Bureau [now Cedar Shake and Shingle Bureau, US]

RCSS Random Communication Satellite System

RCSS Resin-Coated Stainless Steel

RCT Radiation/Chemical Technician

RCT Radiation Counter Tube

RCT Randomized Clinical Trials [Medicine]

RCT Red-Cross Telegram

RCT Regimental Combat Team

RCT Relayed Coherence Transfer Spectroscopy

RCT Remote Control Terminal

RCT Research Corporation Technologies [US]

RCT Resolver Control Transformer

RCT Revenue Canada, Taxation

RCT Rework/Completion Tag [Computers]

RCT Robotic Cluster Tool

RCT Round Compact Test [Powder Metallurgy]

RCT Rubber Chemistry and Technology

Rct Receipt [also rct]

Rct Recruit [also rct]

RCTBC RARE (Rare Animal Relief Effort) Center for Tropical Bird Conservation [US]

RCTD Rate Constrained Target Detection

RCTL Rectangular Coaxial Transmission Line

RCTL Resistor-Capacitor-Transistor Logic

R-CTOD Rock Crack-Tip Opening Displacement [Fracture Mechanics]

RCTS Railway Correspondence and Travel Society [UK]

RCTSR Radio Code Test Speed on Response

RCTIG Reservoir Characterization Technical Interest Group [of Society of Petroleum Engineers of the American Institute of Mining, Metallurgical and Petroleum Engineers, US]

RCU Redundancy Control Unit

RCU Remote Control Unit

RCU Road Construction Unit

RCu$_2$ Rare-Earth Cuprate [Formula]

RCUHVEM Research Center for Ultrahigh Voltage Electron Microscopy [of Osaka University, Japan]

R$_2$CuLi Lithium Dialkylcopper (or Lithium Organocuprate) [Formula]

R-Curve Resistance Curve [also R-curve] [Fracture Mechanics]

RCV Remotely Controlled Vehicle

RCV Research Center of Virology [Canada]

Rcv Receive [also rcv]

rcvd received [also rcv'd]

RC/VP Recent Change/Verify Position

Rcvr Receiver [also rcvr]

RCVS Remote Control Video Switch

RCVS Royal College of Veterinary Surgeons [UK]

Rcvy Recovery [also rcvy]

RCW Return Control Word

RCWP Rural Clean Water Program [US]

RCWP Rubber-Covered Weatherproof (Electric Wire)

RCWV Rated Continuous Working Voltage

RCY Remaining Cycles [also Rcy]

Rcy Recovery [also rcy]

RCYC Royal Canadian Yacht Club

RD Cylindrical Roller Bearing, Double Row, Non-Locating, Outer Ring Without Ribs, Outer Ring Separable, Inner Ring With Three Integral Ribs [Symbol]

RD Radar Data

RD Radar Detection

RD Radar Display

RD Radial Direction

RD Radiation Damage

RD Radiation Detection; Radiation Detector

RD Radiation Dose

RD Radiation Dosimeter; Radiation Dosimetry

RD Radioactive Decay

RD Radioactive Disintegration

RD Random Data

RD Random Drift

RD Raster Display

RD Rate of Departure [Aeronautics]

RD Rate of Descent [also R/D] [Aeronautics]

RD Raynaud's Disease [Medicine]

RD Reaction Diffusion (Equation)

RD Reaction Dynamics

RD Reactor Division [of National Institute of Standards and Technology, Gaithersburg, Maryland, US]

RD Read [also Rd, or rd]

RD Read Data

RD Read Direct

RD Reader Only

RD Readiness Date

RD Receive(d) Data

RD Recording Demand-Meter

RD Recording Density

RD Recording Device

RD Rectangular Dense (Filament)

RD Rectifier Diode

RD Recursive Definition

RD Recycle Delay

RD Refer to Drawer

RD Reference Designator

RD Reference Dimension

RD Reference Division [of British Library]

RD Reference Diode

RD Register Drive

RD Registered Dietician

RD Relative Density

RD Reliable and Durable

RD Remedial Design

RD Remote Diagnosis

RD Remove Directory (Command) [also rd]

RD Renal Disease [Medicine]

RD Report Departing

RD República Dominicana [Dominican Republic]

RD Requirements Document

RD Research and Development [also R&D, R+D, or R and D]

RD Research Department

RD Resin Duct (in Wood)

RD Respiratory Disease [Medicine]

RD Restricted Data

RD Return Duct

RD Rigid Dynamics

RD Ringdown [Telecommunications]

RD Rolling Direction [Metallurgy]

RD Rong-Dunlop (Orientation Relationship) [Crystallography]

RD Roof Drain

RD Root Diameter (of Gears)

RD Rotating Disk

RD Rotational Delay [Computers]

RD Royal Dutch (Petroleum Company)

RD Rubber Division [of American Chemical Society, US]

RD Ruff Degradation [Organic Chemistry]

RD Rural Delivery

RD Rural District

R&D Research and Development [also RD, R+D, or R and D]

R&D Research and Development [US Publication]

R/D Rate of Descent [also RD] [Aeronautics]

R/D Refer to Drawer [also r/d]

Rd Diffuse Reflectance [Symbol]

Rd Read [also rd]

Rd Red [also rd]

Rd Road [also rd]

Rd Rod [also rd]

Rd Round [also rd]

Rd Rudder [also rd]

rd round

rd rutherford [Unit]

RDA Radial Distribution Analysis

RDA Received Data Inverted

RDA Recommended Daily Allowance; Recommended Dietary Allowance

RDA Reflectance Difference Anisotropy

RDA Reliability Design Analysis

RDA Remote Data(base) Access

RDA Resident Data Area

RDA Reverse Diels-Alder (Reaction) [Organic Chemistry]

RDA Rod Drop Accident [Nuclear Engineering]

RDA Rotary Disk Atomizer [Metallurgy]

R&DA Research and Development Associates (for Military Food and Packaging Systems) [US]

RdAc Radioactinium [also Th-227, ^{227}Th, or Th^{227}]

RDAF Remote Data Analysis Facility

RDAFED Research and Development Association for Future Electron Devices

RDAL Representation Dependent Accessing Language

RDARA Regional and Domestic Air Route Area

RDAS Royal Danish Academy of Sciences [Copenhagen, Denmark]

R-DAT Rotary Head–Digital Audio Tape

R/day Roentgen per day [formerly also r/day]

RDB Radar Decoy Balloon

RDB Raw Data Base

RDB Receive Data Buffer

RDB Received Data Noninverted

RDB Red Data Books [of International Union for Conservation of Nature and Natural Resources]

RDB Reference Database

RDB Relational Database [also RDb]

RDB Research and Development Board [US]

RDB Research Department "B" [British Propellant used in World War I]

RDB Rifle Double-Base (Propellant)

RDBA Rural Design and Building Association [UK]

Rdbl Readable [also rdbl]

RDBMS Relational Database Management System

R&D Bull. Research and Development Bulletin [of Public Works and Government Services Canada]

RDB/VAX Relational Database/Virtual Address Extension [also Rdb/VAX]

RDC Rail Diesel Car

RDC Reactive Droop Compensation

RDC Recording Doppler Comparator

RDC Reference Designator Code

RDC Regional Dissemination Center [of NASA]

RDC Relative Dielectric Constant

RDC Reliability Data Center

RDC Remote Data Collection

RDC Remote Data Concentrator

RDC Remote Diagnostic Center

RDC Request for Document Change

RDC Research and Development Center [also R&DC]

RDC Research and Development Corporation

RDC Research and Documentation Center [of the United Nations]

RDC Resources and Development Commission [Missouri, US]

RDC Rotary-Disk Contactor

RDC Royal Drawing Society [UK]

RDC Running Down Clause [Commerce]

RDC Rural District Council

R&DC Research and Development Center [also RDC]

RDCE Radio Distribution and Control Equipment

RD CHK Read Check [also Rd Chk]

RDCIS Research and Development Center for Iron and Steel [of Steel Authority of India, Ranchi]

RDCLK Received Timing Clock

RDCS Reconfiguration Data Collection System

RDD Requirements Definition Document

RDD Requisition Due Date

RDD Research and Development Department [also R&DD]

RDD Research and Development Division [also R&DD]

RDD Research, Development and Demonstration [also RD&D]

R&DD Research and Development Department [also RDD]

R&DD Research and Development Division [also RDD]

RD&D Research, Development and Demonstration [also RDD]

RDDC Rack Drawer Delivery Cart

R&D Dig Research and Development Digest [Published in the UK]

RDDM (Nuclear) Reactor Deck Development Mockup

RDDP Radar Data Development Program [Canada]

RDDP Radarsat Data Development Project [of Marine Environmental Data Service/Department of Fisheries and Oceans, Canada]

RDDT&E Research, Development, Demonstration Testing and Evaluation

RDE Radial Defect Examination

RDE Radial Doppler Effect [Physics]

RDE Reactive Deposition Epitaxy

RDE Receptor Destroying Enzyme [Biochemistry]

RDE Remote Data Entry

RDE Research and Development Effectiveness [also R&DE]

RDE Resorcinol Diglycidyl Ether

RDE Rotating Disk Electrode

R&DE Research and Development Effectiveness [also RDE]

RDEC Research, Development and Evaluation Commission [Taiwan]

RDEU Reaction Diagnostic Experimental Unit

RDF Radial Dielectric Function

RDF Radial Distribution Function

RDF Radiodiffusion Française [French Broadcasting Company]

RDF Radio Direction Finder; Radio Direction Finding

RDF Real Data Fine

RDF Record Definition Field

RDF Refuse-Derived Fuel

RDF Repeater Distribution Frame

RDF Research and Development Fund

RDF Resource Data File

RDF Resource Description Framework

RDF-1 Refuse-Derived Fuel 1 (i.e., Wastes Used as-Discarded)

RDF-2 Refuse-Derived Fuel 2 (i.e., Wastes Processed to Coarse Particle Size and Which May, or May not Have Ferrous Metals Removed)

RDF-3 Refuse-Derived Fuel 3 (i.e., Shredded Fuel Derived from Municipal Solid Waste Processed to Remove Inorganic Materials)

RDF-4 Refuse-Derived Fuel 4 (i.e., Combustible Waste Processed into Power)

RDF-5 Refuse-Derived Fuel 5 (i.e., Combustible Waste Compressed into Briquettes, Cubettes, Pellets, or Slugs)

RDF-6 Refuse-Derived Fuel 6 (i.e., Combustible Waste Processed into Liquid Fuel)

RDF-7 Refuse-Derived Fuel 7 (i.e., Combustible Waste Processed into Gaseous Fuel)

RDFA Radial Distribution Function Analysis

RDFE Radiation-Diffraction Finite Element (Computer Program) [of National Research Council of Canada]

RDG Resolver Differential Generator

Rdg Reading [also rdg]

Rdg Ridge [also rdg]

RDGE Resorcinol Diglycidyl Ether

RDH Rack Drawer Handler

RDH Recirculating Document Handler

RDH Reference Datum Height

RDH Remote Data Harvester

RDH Round-Head (Screw)

RDI Radio Doppler Inertial

RDI Recommended Daily Intake

RDI Resource Development Impact (Program) [Canada]

RDI Route Digit Indicator

RDIA Regional Development Incentives Act

RDIP Research Development Incentives Program

RDIPE Research and Development Institute of Power Engineering [Russian Federation]

RDIU Remote Device Interface Unit

RDJTF Rapid Deployment Joint Task Force

RDL Radiological Defense Laboratory [US Navy]

RDL Research and Development Laboratories [of US Air Force]

RDL Research and Development Laboratory

RDL Resistor Diode Logic

RDL Resource Description Language [of World Wide Web Consortium]

RDL Rocket Development Laboratory [US Air Force]

rdl radial

RDM Random [also rdm]

RDM Real-Time Data Manager

RDM Recording Demand Meter

RDM Relational Data Model

RDM Repoblika Demokratika Malagasy [Malagasy Democratic Republic]

RDM Research and Development Management; Research and Development Manager [also R&DM]

RDM Ricoh Document Management

Rdm Radarman

rdm random [also rdm]

R&D Manage Research and Development Management [Published in the UK]

R&D Mag Research and Development Magazine

RDMS Relational Database Management System

Rdmtr Radiometer

RDMU Range-Drift Measuring Unit

rDNA Recombinant Deoxyribonucleic Acid [also r-DNA, or R-DNA] [Biochemistry]

RDO Radio [also Rdo]

RDO Radio Read-Out

RDO Remote Data Object

R&DO Research and Development Operations [at NASA Marshall Space Flight Center, Huntsville, Alabama, US]

RDOC Reference Designation Overflow Code

RDOS Real-Time Disk-Operating System [also RDOS, or R DOS]

RDOUT Readout [also Rdout, or rdout]

RDP Radar Data Processing

RDP Rapid Development Platform

RDP Reactor Development Program

RDP Receiver and Data Processor

RDP Regional Development Program [Australia]

RDP Reliable Datagram Protocol

RDP Remote Desktop Protocol

RDP Requirements Data Plan

RDP Requirements Development Plan

RDP Research Data Program

RDP Research Data Publication

RDP Research, Development and Production [also RD&P]

RD&P Research, Development and Production [also RDP]

RDPL République Démocratique Populaire du Laos [People's Republic of Laos]

RDPS Radar Data Processing System

RDRINT Radar Intermittent

RDR Radar [also Rdr, or rdr]

RDR Radar Departure Route

RDR Raw Data Recorder

RDR Reader [also Rdr, or rdr]

RDR Risk-Driven Remediation

Rdr Rudder [also rdr]

RDRAM Rambus Dynamic Random-Access Memory

RDR XMTR Radar Transmitter [also Rdr Xmtr]

RDS Radio Data System

RDS Radio Digital System

RDS Railway Development Society [UK]

RDS Raman Difference Spectroscopy

RDS Rate-Determining Step [Chemistry]

RDS (Nuclear) Reactor Depressurization System

RDS Real Data Smooth

RDS Reflectance-Difference Spectroscopy; Reflection-Difference Spectroscopy

RDS Related Deviation Standard

RDS Relational Data Structure

RDS Relational Data System

RDS Relative Detection Sensitivity

RDS Remote Degassing Station

RDS Rendezvous and Docking Simulator

RDS Research Defense Society [UK]

RDS Residuum Desulfurization

RDS Respiratory Distress Syndrome (of Newborn) [Medicine]

RDS Review-Item Discrepancy System

RDS Rheological Dynamic Spectroscopy

RDS Rheometric Dynamic Scanning

RDS Rheometric Dynamic Spectroscopy

RDS Robust Detection Scheme

RDS Rocketdyne Digital Simulator

RDS Royal Dublin Society [US]

RDS Rural Development Service

Rds Readings

rds rounds (of ammunition)

RDSA Rheometric Dynamic Scanning Analysis

RDSAP Royal Dutch Society for Advancement of Pharmacy [Netherlands]

RD/SB Rudder Speed Brake

rds/m rounds (of ammunition) per minute

R&DSoc Research and Development Society [UK]

RDSR Receiver Data Service Request

RDSS Radar Determination Satellite System [US]

RDSS Radio Determination Satellite Services

RDSS Research and Data Support System [of National Center for Atmospheric Research]

RDSS Royal Danish Scientific Society

RDT Reactor Development and Technology

RDT Recombinant DNA (Deoxyribonucleic Acid) Technology [Biochemistry]

RDT Remote Data Transmitter

RDT Resource Definition Table

RDTC Remote Distributed Terminal Controller

RDT&E Research, Development, Test(ing) and Evaluation [also RDTE]

RDT&E Div Research, Development, Test(ing) and Evaluation Division

RdTh Radiothorium [also Th-228, ^{228}Th, or Th^{228}]

RDTL Resistor Diode Transistor Logic

RDTO Receive Data Transfer Offset

RDTP Symposium on Recent Developments in Theoretical Physics

RDTR Radiographic Dielectric Track Registration

RDTR Research Division Technical Report

Rdtr Radiator

RDU Runtime Debugging Unit

RDW Response Data Word

RDW Return Data Words

RDX Realistic Display Mixer

RDX Resolver Differential Transmitter

RDX Research Department Explosive [Cyclotrimethylene Trinitramine]

RDX Hexahydro-1,3,5-Trinitro-1,3,5-Triazine

RDY Ready [also Rdy, or rdy]

RDYMX Ready Mix [also Rdymx, or Rdy Mx]

RDZ Reinforcement-Denuded Zone [Metallurgy]

RE Cylindrical Roller Bearing, Double Row, Non-Locating, Outer Ring Without Ribs, With Two Internal Snap Rings, Inner Ring Without Ribs, Inner Ring Separable [Symbol]

RE Radar Echo

RE Radial Engine

RE Radiant Energy

RE Radiated Emission

RE Radiation Effect

RE Radiochemical Engineer(ing)

RE Radioelectronic(s)

RE Radioelement

RE Radio Engineer(ing)

RE Radiographic Equivalence
RE Railroad Engineer(ing)
RE Railway Engineer(ing)
RE Raman Effect [Optics]
RE Ram Effect
RE Ramsauer Effect [Atomic Physics]
RE Random Error
RE Raw End
RE Rare Earth
RE Rate Effect [Electronics]
RE Reaction Engine
RE Reactive Element
RE Reactive Extrusion
RE Read Error
RE Real Number
RE Reciprocating Engine
RE Recoil Electron
RE Recovery Equipment
RE Recording Ellipsometer
RE Reduced Entity
RE Reentry [also R/E]
RE Reference Electrode [Physical Chemistry]
RE Reference Equivalent
RE Refrigeration Engineer(ing)
RE Registry Editor
RE Rehbinder Effect [Physics]
RE Relative Efficiency
RE Relative Error
RE Reliability Engineer(ing)
RE Renewable Energy
RE República del Ecuador [Republic of Ecuador]
RE Research Engineer
RE Residual Elongation (of Rope)
RE Resolution Enhancement
RE Resonance Effect
RE Resonance Energy
RE Resource Economics
RE Responsible Engineer
RE Rest Energy [Relativity]
RE Reticuloendothelial (System); Reticuloendotheliosis
RE Reunion (French) [ISO Code]
RE Reverse Engineering
RE Reynolds Equation [Fluid Dynamics]
RE Richardson Equation [Electronics]
RE River Engineer(ing)
RE Rocket Engine
RE Roentgen Equivalent
RE Roothaan's Equation [Physics]
RE Rotary Encoder
RE Rotary Evaporator
RE Rotational Energy
RE Royal Engineer [UK]
RE Rubber Elastic(ity)

RE-123 Rare-Earth Barium Copper Oxide [$REBa_2Cu_3O_y$]
R&E Rare and Endangered (Species)
R&E Research and Engineering [also RE]
R/E Reentry [also RE]
$R_L(E)$ Larmor Radius for Electron with Energy E [Symbol]
Re Real Part (of Complex Number) [also $\Re$]
Re Réaumur (Temperature)
Re Reynolds Number [Symbol]
Re Rhenium [Symbol]
Re Rupee [Currency of India]
Re_M Magnetic Reynolds Number [Symbol]
Re-182 Rhenium-182 [also ^{182}Re, or Re^{182}]
Re-183 Rhenium-183 [also ^{183}Re, or Re^{183}]
Re-184 Rhenium-184 [also ^{184}Re, or Re^{184}]
Re-185 Rhenium-185 [also ^{185}Re, or Re^{185}]
Re-186 Rhenium-186 [also ^{186}Re, or Re^{186}]
Re-187 Rhenium-187 [also ^{187}Re, or Re^{187}]
Re-188 Rhenium-188 [also ^{188}Re, or Re^{188}]
Re-189 Rhenium-189 [also ^{189}Re, or Re^{182}]
Ré Réaumur [Unit]
re regarding
.re Reunion (French) [Country Code/Domain Name]
$\rho(\varepsilon)$ strain dependent resistivity [Symbol]
$\rho(\varepsilon)$ density of state at energy level ε [also $\rho_\zeta(\varepsilon)$]
REA Radar Echoing Area
REA Radiation-Enhanced Adhesion
REA Railway Electrification Association [Japan]
REA Railway Express Agency
REA Rare-Earth Alloy
REA Redwood Empire Association [US]
REA Registered Environmental Assessor
REA Request for Equitable Adjustment
REA Rhein-Emscher Armaturen GmbH [German Manufacturer of Valves and Fittings; located in Duisburg]
REA Rural Electrification Administration [of US Department of Agriculture]
REAC Rod Ejection Accident [Nuclear Reactors]
REAC Reeves Electronic Analog Computer
REAC Regulatory Enforcement and Animal Care
Reac Reactant; Reactor; Reaction; Reactive [also reac]
REACH Robotics Evaluation And Characterization
Reacq Reacquisition [also reacq]
REACT Radio Emergency Associated Communications Team [US]
REACT Real-Time Expert Analysis and Control System
REACT Register-Enforced Automated Control Technique
React Reactive; Reactivity [also react]
React Polym Reactive Polymers [Journal published in the US and the Netherlands]
React Solids Reactivity of Solids [Journal published in the Netherlands]
READ Radar Echo Augmentation Device
READ Real-Time Electronic Access and Display
READ Relative Element Address Designation

READ Remote Electronic Alphanumeric Display

READAC Remote Environmental Automatic Data Acquisition Concept

READI Resonant-Excitation Auto-Double Ionization

READI Rocket Engine Analysis and Decision Instrumentation

READLN Read Line [Statement in Pascal Programming]

RE$_{1-x}$AE$_x$MnO$_3$ Rare-Earth Alkaline-Earth Manganate [Formula]

REAL Road Emulsion Association, London [UK]

REAl$_6$TM$_6$ Rare-Earth-Aluminium-Transition Metal [Formula]

REALCOM Real-Time Communications

REAP Remote Entry Acquisition Package

REAP Removable End Access Platform

REAP Resource Efficient Agricultural Production [Canada]

REAP Roof Evaluation Accident Prevention

REAP Rural Education Access Program

REAR Reliability Engineering Analysis Report

Rear Adm Rear Admiral [also Radm]

reasm reassemble

REB Radar Evaluation Branch

REB RAMP (Rapid Modeling Platform) Equipment Builder

REBA Relativistic Electron Beam Accelerator

REBAM Rolling Element Bearing Activity Monitor

REBCO Rare-Earth Barium Copper Oxide (Superconductor) [also ReBCO]

REBE Recovery Beacon Evaluation

REBUD Rehabilitation Budgeting Program

REC Radiant Energy Conversion

REC Radiation Electron Capture

REC Radiochemical Engineering Cell

REC Radiochemical Engineering Complex

REC Regional Engineering College [Warangal, India]

REC Regional Environmental Center (for Central and Eastern Europe)

REC Registration of Engineers Committee

REC Request for Engineering Change

REC Reversible Embedment for Cytochemistry

REC Rural Electrification Conference

.REC Database Record [File Name Extension]

.REC Recorder [File Name Extension] [also .rec]

.REC Recreational, or Entertainment Activities [Internet Domain Name] [also .rec]

Rec Receipt [also rec]

Rec Receive(r) [also rec]

Rec Record(er); Recorded; Recording [also rec]

Rec Recreation [also rec]

Rec Rectifier [also rec]

RECCE Reconnaissance [UK Military] [also recce]

recd received [also rec'd]

RECE Relativistic Electron Coil

Rec Electr Commun Eng Conversazione Tohoku Univ Record of Electrical and Communication Engineering Conversazione, Tohoku University [Japan]

Rec Chem Progr Record of Chemical Progress [Journal]

RECEN Receivable Entry (Code)

Recert Recertification [also recert]

RECFM Record Format

RECGAI Research and Engineering Council of the Graphic Arts Industry [US]

Rech Aéronaut Recherche Aéronautique [French Aeronautic Research Journal]

Rech Aérosp Recherche Aérospatiale [French Space Research Journal of Office National d'Etudes et de Recherches Aérospatiales]

Rechentech Datenverarb Rechentechnik und Datenverarbeitung [German Journal on Computation and Data Processing]

Rechg Recharge [also Rechg]

Recip Reciprocal; Reciprocity; Reciprocate [also recip]

RECIPE Recomp Computer Interpretive Program Expediter

Recirc Recirculation [also recirc]

RECl$_3$ Rare-Earth (III) Chloride [Formula]

Recl Reclosing [also recl]

recl reconcile

Reclam Rev Reclamation Review [US]

Recl Trav Chim Pays-Bas Recueil des Travaux Chimiques des Pays-Bas [Selection of Dutch Chemical Studies]

recm recommend

RECMA Radio and Electronic Component Manufacturers Association

REC MARK Record Mark [also Rec Mark, or rec mark]

RECMF Radio and Electronic Components Manufacturers Federation [UK]

RE-Co Rare Earth-Cobalt (Alloy) [also RE/Co]

Recog Recognition [also Recogn, recogn, or recog]

RECOL Retrieval Command Language

Recomp Recomplement [also recomp]

RECON Remote Console

RECON Remote Control On-Line Information Service

RECON Retrospective Conversion

RECONFIG Reconfiguration [also Reconfig, or reconfig]

Reconn Reconnaisance

REconS Royal Economic Society [UK]

Recov Recovery [also recov]

RECP Relativistic Effective Core Potential [Physics]

RECP Request for Engineering Change Proposal

Recp Receptacle [also recp]

ReCp(CO)$_3$ Cyclopentadienylrhenium Tricarbonyl

Recr Recruit [Military]

Recrc Recirculate [also recrc]

Recpt Receipt [also recpt]

recryst recrystallize

reccrystd recrystallized [also recryst'd]

Recrystn Recrystallization [also recrystn]

RECS Representative Shuttle Environmental Control System [of NASA]

Rec Sec Recording Secretary

RECSTA Receiving Station

Rect Receipt [also rect]

Rect Rectangle; Rectangular [also rect]

Rect Rectification; Rectifier [also rect]

Rect Rector(y) [also rect]

Rect Rectal; Rectum [also rect]

rect rectify; rectified [also rect]

Rect Abd Rectus Abdominis [also rect abd] [Anatomy]

RECTAS Regional Center for Training in Aerial Surveys [Nigeria]

Rect Fem Rectus Femoris [also rect fem] [Anatomy]

Rect Ocul Rectus Oculis [also rect ocul] [Anatomy]

Rec Trav Chim Pays-Bas Recueil des Travaux Chimiques des Pays-Bas [Selection of Dutch Chemical Studies]

Recursos Min Recursos Minerales [Chilean Journal on Minerals]

Recv Receive [also recv]

Recvr Receiver [also recvr]

Recycl Recycling [also recycl]

RED Radial Electron Distribution

RED Radiation-Enhanced Diffusion

RED Reflection Electron Diffraction

RED Relativistic Electrodynamics

Red Reducer; Reduction [also red]

Red Reductant [Symbol]

REDA Recycling Economic Development Advocates (Initiative) [of US Environmental Protection Agency]

REDA Regional Economic Development Authority [Canada]

REDA Resonant-Excitation Double Autoionization

REDAC Real-Time Data Acquisition

REDAP Reentrant Data Processing; Reentry Data Processing

REDEX Recycle Extract Dual Extraction (Process)

REDIT Random Access Edit

REDLN Redlining [also Redln, or redln]

Redn Reduction [also redn]

REDOX Reduction-Oxidation (Reaction) [also Redox, or redox]

REDOR Rotational Echo Double Resonance

REDSOD Repetitive Explosive Device for Soil Displacement

redsh reddish

REDSTAR Resource Data Storage and Retrieval (Database)

Redtn Reduction [also redtn]

Redun Redundancy [also redun]

redupl reduplicated [also redupl]

REDZ Recent Drizzle [Meteorology]

REE Rare-Earth Element(s)

REE Reactive Element Effect [Metallurgy]

REE Reentrant; Reentry

REE Revised Enskog Equation [Physics]

Re(ε) Real Part of Dielectric Function

REECDP Renewable Energy and Energy Conservation Demonstration Program

REECo Reynolds Electrical and Engineering Company [US]

REED Radio and Electrical Engineering Division [of National Research Council of Canada]

REEDM Rocket Exhaust Effluent Diffusion Model

REEF Great Barrier Reef Database [of Great Barrier Reef Marine Park Authority, Australia]

REELM Reflected Electron Energy Loss Microscopy

REELS Reflection Electron Energy Loss Spectroscopy

REEP Range Estimating and Evaluation Procedure

REEP Regional Environment Employment Program [of Environment Australia]

REEP Regression Estimation of Event Probabilities

REEP Restricted Edge Emitting Diode

REF Radiographic Equivalence Factor

REF$_3$ Rare-Earth (III) Fluoride [Formula]

.REF Reference [File Name Extension] [also .ref]

Ref Refiner(y); Refining [also ref]

Ref Referee [also ref]

Ref Reference [also ref]

Ref Reform(ation)

Ref Refund(ing) [also ref]

Ref Refurbish(ment) [also ref]

ref referred

ref reformed

REFA REFA Verband für Arbeitsstudien und Betriebsorganisation e.V. [Association for Work Studies and Plant Organization, Germany]

refd referred [also ref'd]

RE-Fe Rare-Earth–Iron (Alloy) [also RE/Fe]

RE-Fe-B Rare Earth-Iron-Boron (Alloy) [also REFeB]

REFET Reference Field-Effect Transistor

Refg Refrigerating [also refg]

Refgn Refrigeration [also refgn]

RE(FHD)$_3$ Rare-Earth Tris(decafluoroheptanedionate)

RefIT Reinvention for Innovative Technologies

Ref L Reference Line

Refl Reflectance; Reflection; Reflector [also refl]

Refl Reflux [also refl]

REFLECS Retrieval from the Literature on Electronics and Computer Science (Service) [of Institute of Electrical and Electronics Engineers, US]

REFLEXAFS Reflection Extended X-Ray Absorption Fine Structure

REFLES Reference Librarian Enhancement System [of University of California, US]

Ref Lib Reference Library; Reference Librarian

Ref Libr Reference Librarian [Published in the US]

Ref Man Reference Manual

Refny Refinery [also refny]

REFORM Reactive Forming (Technology)

Refr Refraction; Refractive; Refractory [also refract]

REFRA Recent Freezing Rain [Meteorology]

Refract J Refractories Journal [UK]

Refract News Refractory News [of The Refractories Institute, US]

Refrgn Refrigeration [also refrgn]

Refrig Refrigerate; Refrigeration; Refrigerator [also refrig]

REFS Reflectivity Simulation Software (for Rocking Rurves) [Materials Science]

REFS Remote Entry Flexible Security

Refs References [also refs]

REFSEARCH Reference Materials Searching System [of University of California at Berkeley, US]

Refurb Refurbish [also refurb]

Ref Zh Referativnyi Zhurnal [Russian Journal with Research Papers]

Ref Zh Inf Referativnyi Zhurnal, Informatika [Russian Journal with Research Papers on Information Science and Technology]

Ref Zh Korroz Referativnyi Zhurnal, Korroziya [Russian Journal with Research Papers on Corrosion Science and Engineering]

Ref Zh Metal Referativnyi Zhurnal, Metallurgiya [Russian Journal with Research Papers on Metallurgy]

REG Radioencephalogram [Medicine]

REG Range Extender with Gain

REG Rare-Earth Garnet

REG Rheoencephalography [Medicine]

Reg Regent

Reg Regiment [also reg]

Reg Region(al) [also reg]

Reg Register; Registration; Registrar; Registry [also reg]

Reg Regularity; Regulate; Regulation; Regulator [also reg]

reg registered

reg regular(ly)

REGAL Range, Elevation, Guidance, Approach, Landing (Device) [Aeronautics]

REGAL Range and Elevation Guidance for Approach and Landing [Aeronautics]

REGAL Rigid Epoxy Glass Acrylic Laminate

regd registered [also reg'd]

REGEN Regenerative Repeater

Regen Regenerate; Regeneration; Regenerator [also regen]

Reg-Gen Registrar-General

REGIS Regional Geographical Information System

REGIS Relational General Information System

REGIS Remote Graphics Instruction Set

Regis Register [also regis]

Reg Prof Regius Professor

REGF Reference Energy Green's Function [Physics]

regl regimental

REG/MEN Regency/Mentor

RegN Registered Nurse

Regs Regulations [also regs]

Regt Regiment [also regt]

Reg TM Registered Trademark

Reg Tn Register Ton [also Reg tn, or reg tn]

RegTP Regulierungsbehörde für Telekommunikation und Post [Federal Regulatory Agency for Post and Telecommunications, Germany] [formerly BAPT]

Regul Toxicol Pharmacol Regulatory Toxicology and Pharmacology [Journal published in the US]

Re(h$_{ie}$) Real Part of Common-Emitter Small-Signal Short-Circuit Input Impedance [Semiconductor Symbol]

Re(h$_{oe}$) Real Part of Common-Emitter Small-Signal Open-Circuit Output Admittance [Semiconductor Symbol]

Rehab Rehabilitation [also rehab]

REHIS Royal Environmental Health Institute of Scotland

REHVA Representatives of the European Heating and Ventilating Association

REI Radiation Effects in Insulators (Conference)

REI Range from Entry Interface

REI Relative Erodability

REI Research-Engineering Interaction

REIC Radiation Effects Information Center [of Battelle Memorial Institute, Columbus, Ohio, US]

REIC Rare-Earth Information Center [of Iowa State University, Ames, US]

REIC Renewable Energy Information Center [US]

REIG Rare-Earth Iron Garnet

REIIF Renewable Energy Innovation Investment Fund [Australia]

REIL Runway End Identification Lights

Reinf Reinforce(ment); Reinforcing [also reinf]

Reinfd Reinforced [also reinfd]

Reinf Plast Reinforced Plastics [Published in the UK]

REINS Requirements Electronic Input System

Reinit Reinitialization; Reinitialize [also reinit]

REIT Real Estate Investment Trust [US]

Rej Reject(ion) [also rej]

REL Radiation Evaluation Loop [of Hanford Engineering Development Laboratory, Washington State, US]

REL Radio Engineering Laboratories [US]

REL Rapidly Extensible Language

REL Rate of Energy Loss

REL Recommended Exposure Limit [of National Institute for Occupational Safety and Health, US]

REL Relative Exposure Limit

REL Remington Energy Limited [US]

REL Riksforbundet for Elektrikfieringen pa Landsbygden [National Association for Rural Electrification, Sweden]

Rel Relation; Relative [also rel]

Rel Relay [also rel]

Rel Release [also rel]

Rel Relief [also rel]

Rel Relocation [also rel]

Rel Reluctance [also rel]

rel relating; relative(ly)

rel released

Relat Relation [also relat]

Relat Relativity [also relat]

Relat J Relational Journal [US]

RELCODE Relative Code

Reliab Reliability [also reliab]

Reliab Eng Reliability Engineering [Published in the UK]

Reliab Eng Syst Saf Reliability Engineering and System Safety [Published in the UK]

RELIPOSIS Research Liaison Panel on Scientific Information Services [UK]

Reln Relation [also reln]

RELP Residual Excited Linear Predictive Coding

RELSECT Relative Sector

RELT Registered Environmental Technologist

Rel U Relative Unit [also RU]

REM Radon Emanation Method

REM Random Energy Model

REM Random Entry Memory

REM Rapid Eye Movement (Sleep) [also known as D-sleep]

REM Rare-Earth Magnet

REM Rare-Earth Metal

REM Recognition Memory

REM Reflection Electron Microscopy

REM Regeln für die Bewertung und Prüfung elektrischer Maschinen [Regulations for the Assessment and Testing of Electric Machines, Germany]

REM Registered Environmental Manager

REM Release-Engage Mechanism

REM Release Engine Mechanism

REM Release Engine Module

REM Release Escape Mechanism

REM Reliability Engineering Model

REM Remark (Statement) [Computers]

REM Remote-Manual (Bolter)

REM Replacement Micrographs

REM Replica Electron Microscopy

REM Ring Error Monitor

REM Roentgen Equivalent Man [also rem]

Rem Remainder [also rem]

Rem Remark [also rem]

Rem Reminder [also rem]

Rem Removal [also rem]

rem remote [also rem]

rem roentgen equivalent man [also REM]

REMA Retroreflective Equipment Manufacturers Association [UK]

REMA Rotating Electrical Machines Association [UK]

REMAD Remote Magnetic Anomaly Detection

REMAP Record Extraction Manipulation and Print

REMARC Retrospective Machine-Readable Catalogue [US]

REMAT Research Center for Management of New Technology [Canada]

REMC Radio and Electronics Measurements Committee [UK]

REMC Resin-Encapsulated Mica Capacitor

REMDOS Remote Disk Operating System

REME Royal Electrical and Mechanical Engineers [UK]

REMEDIE Reflection Electron Microscopy and Electron Diffraction at Intermediate Energy

REMI Radiated Electromagnetic Interference

REMI Reliability Engineering and Management Institute [US]

re mist *(repetatur mistura)* – repeat the same mixture [Medical Prescriptions]

Remittance Doc Process Today Remittance and Document Processing Today [Publication of the Recognition Technologies Users Association, US]

REMO Real-Time Event Monitor

REMO Referenzmodell für sichere Informationstechniksysteme [Reference Model for Safe Information Technology Systems (Project)] [of Gesellschaft für Mathematik und Datenverarbeitung, Germany]

REMOS Real-Time Event Monitoring System

Remote Sens Environ Remote Sensing of Environment [US Journal]

Remote Sens Rev Remote Sensing Reviews [UK]

REMPI Resonance-Enhanced Multiphoton Ionization

REM-RED Reflection Electron Microscopy–Reflection Electron Diffraction

REM-RHEED Reflection Electron Microscopy–Reflection High-Energy Electron Diffraction

REMS Resource Management System

REMSA Railway Engineering Maintenance Suppliers Association [US]

REMSA (Joint) Requirements for Emergency and Safety [of European Civil Aviation Conference, France]

REMSCAN Remote Sensor Communication and Navigation

REMSTAR Remote Electronic Microfilm in Storage Transmission and Retrieval

REMSYS Replicated Microsystems (Project) [Part of Swiss MINAST Research Program]

RE-MTS Rare Earth–Muffin-Tin Sphere [Materials Science]

rem/wk roentgen equivalent man units per week

rem/yr roentgen equivalent man units per year

REN Remote Enable [Data Processing]

REN Rendezvous [Aerospace]

REN Rename (Command) [also ren] [Computers]

REN Ringer Equivalence Number

Ren Rename [also ren]

Ren Renewal [also ren]

RENbO$_4$ Rare-Earth Niobium Oxide (or Rare-Earth Niobate) [also RENO] [Formula]

REND Rendezvous [Aerospace]

Rend Acad Lincei Rendiconti Academia dei Lincei [Journal of the Academia dei Lincei, Rome, Italy]

Rend Ist Lomb Accad Lett A Rendiconti Istituto Lombardo Accademia di Scienze e Lettere A, Scienze Matematiche e Applicazioni [Italian Journal on Mathematical Sciences and their Applications]

Rend Ist Lomb Accad Lett B Rendiconti Istituto Lombardo Accademia di Scienze e Lettere B [Italian Scientific Journal]

Rend Sem Mat Rendiconti del Seminario Matematico [Mathematical Journal of the University of Padua, Italy]

RENE Rocket Engine/Nozzle Ejector

RENM Request for Next Message

RENO Rare-Earth Niobium Oxide (or Rare-Earth Niobate) [also RENbO$_4$] [Formula]

RENPAP Regional Network on (the Production, Marketing and Control of) Pesticides for Asia and the Pacific

RENPE Rare and Endangered Native Plant Exchange [US]

REN RAD Rendezvous Radar [also Ren Rad]

Rentgenostruk Anal Polikrist Rentgenostrukturnyi Analiz Polikristallov [X-Ray Structural Analysis of Polycrystalline Materials published in Elista, Ukraine]

REntSoc Royal Entomological Society [UK]

RENUM Renumber (Program Lines) (Command) [Computers]

Renum Renumbering; Renumeration [also renum]

REO Rare-Earth Orthoferrite

REO Rare-Earth Oxide(s)

REO Regenerated Electrical Output

REO Rocket Engine Operations

RE$_2$O$_3$ Rare-Earth (III) Oxide [Formula]

RE(OH)$_3$ Rare-Earth Hydroxide [Formula]

REOM Reduced Equations-of-Motion (Theory)

REON Rocket Engine Operations–Nuclear

REOS Rare-Earth Oxides [also reos]

REP Radar Evaluation Pad

REP Range Error Probable

REP Recognized Educational Program [of Institute of Electrical and Electronics Engineers, US]

REP Reentrant Processor

REP Reentry Physics Program

REP Registered Environmental Professional

REP Registry Enhancement Project [US]

REP Relativistic Effective (Core) Potential [Physics]

REP Remote Emergency Power

REP Rendezvous Evaluation Pad [of NASA]

REP Request for Proposals [also RFP]

REP Robert Esnault-Pelterie (Astronautics Price)

REP Roentgen Equivalent Physical [also rep]

REP Rotating-Electrode Powder [Powder Metallurgy]

REP Rotating-Electrode Process [Powder Metallurgy]

Rep Repair [also rep]

Rep Repeat [also rep]

Rep Repertory [also rep]

Rep Repetition [also rep]

Rep Replace(ment) [also rep]

Rep Replication [also rep]

Rep Report(er) [also rep]

Rep Representation; Representative [also rep]

Rep Republic [also rep]

Rep Repulsion [also rep]

rep repeat

rep reported

rep *(repetatur)* – repeat; let it be repeated [Medical Prescriptions]

REPA Registered Environmental Property Assessor

REPAG Repagination [also Repag, or repag]

Rep Cast Res Lab, Waseda Univ Report of the Castings Research Laboratory, Waseda University [Japan]

Rep Chiba Inst Technol Report of the Chiba Institute of Technology [Japan]

Rep Coll Eng Hosei Univ Report of the College of Engineering, Hosei University [Japan]

REPComm Recognized Educational Program Committee [of Institute of Electrical and Electronics Engineers–Educational Activities Board/Regional Activities Board, US]

Rep Conf Strength Solids Report of the Conference on the Strength of Solids

Rep Du Pont Report on Du Pont [US]

REPE Repeat while Equal

REPE Resonance Energy per Electron

Reperf Reperforator [also reperf]

Rep Fac Eng, Nagasaki Univ Reports of the Faculty of Engineering, Nagasaki University [Japan]

Rep Fac Eng, Shizuoka Univ Reports of the Faculty of Engineering, Shizuoka University [Japan]

Rep Fac Eng, Tottori Univ Reports of the Faculty of Engineering, Tottori University [Japan]

Rep Fac Eng, Yamanashi Univ Reports of the Faculty of Engineering, Yamanashi University [Japan]

Rep Fac Sci Eng, Saga Univ Reports of the Faculty of Engineering, Saga University [Japan]

Rep Fac Sci Technol Meijo Univ Reports of the Faculty of Science and Technology, Meijo University [Japan]

Rep Fukui Prefect Ceram Ind Lab Report of Fukui Prefectural Ceramic Industry Laboratory [Japan]

Rep Gov Ind Res Inst, Chugoku Reports of the Government Industrial Research Institute, Chugoku [Japan]

Rep Gov Ind Res Inst, Nagoya Reports of the Government Industrial Research Institute, Nagoya [Japan]

Rep Gov Ind Res Inst, Osaka Reports of the Government Industrial Research Institute, Osaka [Japan]

Rep Grad Sch Electron Sci Technol Shizuoka Univ Reports of the Graduate School of Electronic Science and Technology, Shizuoka University [Japan]

Rep Himeji Inst Technol Reports of Himeji Institute of Technology [Japan]

Rep Hokkaido Forest Prod Res Inst Report of the Hokkaido Forest Products Research Institute [Japan]

Rephos Rephosphorized (Steel) [also rephos]

Rep Ind Res Inst Ishikawa Report of the Industrial Research Institute of Ishikawa [Japan]

Rep Inst High Speed Mech Tohoku Univ Reports of the Institute of High Speed Mechanics, Tohoku University [Japan]

Rep Inst High Speed Mech Tohoku Univ, Ser A Reports of the Institute of High Speed Mechanics, Tohoku University, Series A [Japan]

Rep Inst High Speed Mech Tohoku Univ, Ser B Reports of the Institute of High Speed Mechanics, Tohoku University, Series B [Japan]

Rep Inst Ind Sci, Univ Tokyo Reports of the Institute of Industrial Science, University of Tokyo [Japan]

Rep Inst Phys Chem Res Reports of the Institute of Physical and Chemical Research [Japan]

Repl Replace(ment) [also repl]

Repl Replica(tion) [also repl]

REPLAB Responsive Environment Programmed Laboratory

REPM Representatives of Electronic Products Manufacturers [now Electronics Representatives Association, US]

Rep Math Phys Reports on Mathematical Physics [UK]

Re(PMCp)(CO)₃ Pentamethylcyclopentadienylrhenium Tricarbonyl

REPNE Repeat while Not Equal

REPNZ Repeat while Not Zero

REPO Remote Emergency Power Off

Repo Repossession [also repo]

Rep of China Republic of China

Rep of Ire Republic of Ireland

Rep of Korea Republic of Korea

REP-OP Repetitive Operation [also Rep-Op, or repop]

REPPAC Repetitively Pulsed Plasma Accelerator

Rep Prog Appl Chem Reports on the Progress of Applied Chemistry [of Society of Chemical Industry, US]

Rep Prog Phys Reports on Progress in Physics [of Institute of Physics, UK]

Rep Prog Polym Phys Jpn Reports on Progress in Polymer Physics in Japan [of Association for Science Documentation Information, Japan]

Repr Repressurization

Repres Theory Representation Theory [Journal of the American Mathematical Society, US]

Rep Res Inst Appl Mech Reports of Research Institute for Applied Mechanics [Japan]

Rep Res Inst Ind Sci Kyushu Univ Reports of the Research Institute of Industrial Science, Kyushu University [Japan]

Rep Res Inst Sci Technol, Nihon Univ Report of the Research Institute of Science and Technology, Nihon University [Japan]

Rep Res Inst Strength Fract Mater, Tohoku Univ Reports of the Research Institute for Strength and Fracture of Materials, Tohoku University [Japan]

Rep Res Lab, Asahi Glass Reports of the Research Laboratory, Asahi Glass [Japan]

Rep Res Lab Eng Mater Reports of the Research Laboratory of Engineering Materials [Japan]

Rep Res Lab Eng Mater, Tokyo Inst Technol Reports of the Research Laboratory of Engineering Materials, Tokyo Institute of Technology [Japan]

Rep Res Lab Surf Sci Okayama Univ Reports of the Research Laboratory for Surface Science, Okayama University [Japan]

Rep Res Nippon Inst Technol Report of Researches, Nippon Institute of Technology [Japan]

Repro Reproduction [also repro]

Repro Reprographic(s) [also repro]

Reprod Reproduction [also reprod]

Reprod Nutr Devel Reproduction, Nutrition, Development [International Journal]

REPROM Reprogrammable PROM (Programmable Read-Only Memory)

REPROM Reprogrammable Read-Only Memory

REP-QMC Relativistic Effective Potential Quantum Monte Carlo [Physics]

Repr Representation [also repr]

Repr Reprint [also repr]

Repro Reproduction [also repro]

REPROM Reprogrammable Read-Only Memory

REPS Research, Publications and Secondment [of Arctic Institute of North America, Canada]

Rep Stat Appl Res UJSE Reports of Statistical Application Research of the Union of Japanese Scientists and Engineers

Rep Stat Appl Res Union Jpn Sci Eng Reports of Statistical Application Research, Union of Japanese Scientists and Engineers [Japan]

Rept Receipt [also rept]

Rept Repeater [also rept]

Rept Report [also rept]

Rep Technol Iwate Univ Report on Technology of Iwate University [Japan]

Rep Tohoku Univ Res Inst A Report of the Tohoku University Research Institute A [Japan]

Rep Tohoku Univ Res Inst B Report of the Tohoku University Research Institute B [Japan]

Rep Tokyo Metrop Ind Technol Cent Report of the Tokyo Metropolitan Industrial Technology Center [Japan]

Rep Tokyo Metrop Ind Technol Cent Report of the Tokyo Metropolitan Industrial Technology Center [Japan]

Rep Univ Electro-Commun Reports of the University of Electro-Communications [Japan]

rep/wk roentgen equivalent physical units per week

REPZ Repeat while Zero

Req Request [also req]

Req Require(ment) [also req]

Req Requisition [also req]

req require(d)

reqd required [also req'd]

Reqmts Requirements [also reqmts]

Reqt Requirement [also reqt]

REQVER Requirements Verification

REQWQ Requisition Word Queue

RER Radiation Effects Reactor [US Air Force]

RER Refocusing Environmental Restoration

RER Residual Error Rate [Communications]

RER Rough Endoplasmic Reticulum [also rER] [Cytology]

RERA Recent Rain [Meteorology]

RERC Rare Earth Research Conference [of University of Kentucky, Lexington, US]

RE-RE Rare Earth-Rare Earth (Interaction)

REREG Re-Register

RERF Radiation Effect Research Foundation [Japan]

(RE)Rh₄B₄ Rare-Earth Rhodium Tetraboride

RERL Residual Equivalent Return Loss

RES Radar Echo Simulation

RES Radiation Effects Section [of Defense Research Establishment Ottawa, Canada]

RES Radiation Enhanced Sublimation

RES Rapid Etching System

RES Rare-Earth Salt(s)

RES Relative Electric Strength

RES Relative Emission Sensitivity

RES Relaxed Excited State

RES Remote Entry Service

RES Remote Entry Subsystem

RES Remote Entry System

RES Remote Execution Service

RES Remote Experiment Station [US]

RES República de El Salvador [Republic of El Salvador]

RES Residual Elastic Strain

RES Resistance Electroslag [Metallurgy]

RES Resistance Electroslag Surfacing [Metallurgy]

RES Restore [Data Communications]

RES Reticuloendothelial System [Anatomy]

RES Reynolds Experimental Station [Pennsylvania, US]

RES Royal Economic Society [UK]

RES Royal Entomological Society [UK]

.RES Resource [File Name Extension] [also .res]

R&Es Rare and Endangered Plants

Res Resawed; Resawing [also res]

Res Research(er) [also res]

Res Reserve

Res Reservation [also res]

Res Reservoir [also res]

Res Reset [also res]

Res Residence; Resident(ial) [also res]

Res Residue [also res]

Res Resistance

Res Resistivity; Resistor [also res]

Res Resolution [also res]

Res Resource [also res]

res reserve(d)

res resigned

RESA Reactive Electrode Submerged Arc (Process) [Metallurgy]

RESA (Scientific) Research Society of America [US]

RESA Runway End Safety Area

Res Act Civ Eng Relat Fields Kyoto Univ Research Activities in Civil Engineering and Related Fields at Kyoto University [Japan]

Res Alert Research Alert (Database)

RESAR Reference Safety Analysis Report

Res Ass Research Associate [also Res Assoc]

Res Asst Research Assistant

RESB Royal Engineers Standards Board [UK]

Res Biol Research Biologist

Res Bull Research Bulletin

Res Bull Hiroshima Inst Technol Research Bulletin of the Hiroshima Institute of Technology [Japan]

Res Bull Print Bur Research Bulletin of the Printing Bureau [Japan]

Res Bull Reg Eng Coll Warangal Research Bulletin, Regional Engineering College, Warangal [India]

RESC Rare-Earth-Doped Semiconductor

RESCAN Reflecting Satellite Communication Antenna

Res Chem Research Chemist

Res Commun Chem Pathol Pharmacol Research Communications in Chemical Pathology and Pharmacology

RESCU Radio Emergency Search Communications Unit

RESCUE Remote Emergency Salvage and Clean-Up Equipment

RESD Research and Engineering Support Division [of International Development Association]

Res Dev Research and Development [US]

Res Dev Jpn Awarded Okochi Mem Prize Research and Development in Japan Awarded the Okochi Memorial Prize [Japan]

Res Discl Research Disclosure [Published in the UK and the US]

RE-SEED Retirees Enhancing Science through Experiments and Demonstrations

Res Electrotech Lab Researches of the Electrotechnical Laboratory [Japan]

Res Eng Des Research in Engineering Design [Journal published in the US]

RESER Reentry Systems Evaluation Radar

RESG Research Engineering Standing Group [of US Department of Defense]

RESH Recent Showers [Meteorology]

RESi$_{2-x}$ Rare-Earth Silicide [Formula]

resid residual

resin resinous

Resin Rev Resin Review [US]

Res Inst Tech Phys Hung Acad Sci Yearb Research Institute for Technical Physics of the Hungarian Academy of Sciences, Yearbook

RE$_2$Si$_2$O$_7$ Rare-Earth Silicate [Formula]

Resis Resistance [also resis]

RESL Radiological and Environmental Sciences Laboratory [US]

RESL Royal Entomological Society of London [UK]

Res Lab Research Laboratory

Res Lab Precis Mach Electron Research Laboratory of Precision Machinery and Electronics [Japan]

Res Lib Research Librarian; Research Library [also Res Libr]

Res Mech Res Mechanica [Journal published in the US]

Res Mech Lett Res Mechanica Letters [US]

Res Met Research Metallurgist

RESN Recent Snow [Meteorology]

RESNA Rehabilitation Engineering Society of North America

Res News Researcher News [US]

Res Nondestr Eval Research in Nondestructive Evaluation [Publication of the American Society for Nondestructive Testing, US]

RESORS Remote Sensing On-Line Retrieval System [of Canadian Center for Remote Sensing, Canada]

RESOS N-(Resorufin-4-Carbonyl)piperidine-4-Carboxylic Acid N-Hydroxysuccinimide Ester

Resourc Resource [also resour]

Resour Conserv Resources and Conservation [Journal published in the Netherlands]

Resour Conserv Recycl Resources, Conservation and Recycling [Journal published in the Netherlands]

Resour Policy Resources Policy [Journal published in the UK]

Resour Process Resources Processing [Journal published in the US]

Resour Recovery Conserv Resource Recovery and Conservation [Journal published in the Netherlands]

Resour Shar Inf Netw Resource Sharing and Information Networks [Published in the US]

RESP Regulated Electrical Supply Package

RESP Remote Batch Station Program

Resp Respiration; Respirator(y) [also resp]

Resp Respondent [also resp]

Resp Responsibility [also resp]

resp respective(ly)

RESPA Reference System Propagator Algorithm

Res Pap Research Paper

RESPES Resonant Photoemission Spectrum

Res Phys Research Physicist

Res Prof Research Professor

Res Prog SSE Research and Progress of SSE (Solid-State Engineering) [of Nanjing Solid State Devices Research Institute, PR China]

Res Publ Research Publication

Res Q Am Assoc Health, Phys Ed Recr Research Quarterly of the American Association for Health, Physical Education and Recreation [US]

Resr Reservoir [also resr]

RESRAD Residual Radioactivity (Computer Code)

Res Rep Research Report

Res Rep Fac Eng, Gifu Univ Research Report of the Faculty of Engineering, Gifu University [Japan]

Res Rep Fac Eng, Kagoshima Univ Research Reports of the Faculty of Engineering, Kagoshima University [Japan]

Res Rep Fac Eng, Meiji Univ Research Reports of the Faculty of Engineering, Meiji University [Japan]

Res Rep Fac Eng, Mie Univ Research Reports of the Faculty of Engineering, Mie University [Japan]

Res Rep Fac Eng, Niigata Univ Research Reports of the Faculty of Engineering, Niigata University [Japan]

Res Rep Fac Eng, Tokyo Univ Research Reports of the Faculty of Engineering, Tokyo University [Japan]

Res Rep Ikutoku Tech Univ B, Sci Technol Research Reports of Ikutoku Technical University, Part B, Science and Technology [Japan]

Res Rep Inf Sci A, Math Sci Research Reports on Information Sciences, Series A, Mathematical Science [Japan]

Res Rep Inf Sci B, Oper Res Research Reports on Information Sciences, Series B, Operations Research [Japan]

Res Rep Inf Sci C, Comput Sci Research Reports on Information Sciences, Series C, Computer Science [Japan]

Res Rep Inst Inf Sci Technol, Tokyo Denki Univ Research Reports, Institute of Information Science and Technology, Tokyo Denki University [Japan]

Res Rep Kogakuin Univ Research Reports of Kogakuin University [Japan]

Res Rep Nagaoka Coll Technol Research Reports of Nagaoka College of Technology [Japan]

Res Rep Nagoya Munic Ind Res Inst Research Reports of the Nagoya Municipal Industrial Research Institute [Japan]

Res Rev Research Review [Publication of Lawrence Berkeley Laboratory, University of California at Berkeley, US]

Resrt Restart [also resrt]

RESS Radar Echo Simulation Subsystem

RESS Radar Echo Simulation Study

RESS Rapid Expansion of Supercritical Solutions (Process)

RESS Redox Energy-Storage System [of NASA]

RESSA Regional Engineering Students Societies Association [Canada]

RESSAC Réseau d'Etude du Spectre de Surface par Analyse Circulaire [Surface Spectra for Circular Analysis Network]

Res Sci Research Scientist

REST Radar Electronic Scan Technique

REST Radar Electronic Scan Test

REST Reentry Environment and Systems Technology

REST Reentry System Test

RESTA Reconnaissance, Security and Target Acquisition [US Army]

RESTBL Relative Page to External Storage Correspondence Table

Res Technol Manage Research Technology Management [Journal of Industrial Research Institute, US]

Restr Restrict(ed); Restriction [also restr]

Resvr Reservoir [also resvr]

Res Word Process Newsl Research in Word Processing Newsletter [of South Dakota School of Mines and Technology, Rapid City, US]

Resynch Resynchronize; Resynchronous [also resynch]

RET Reliable Earth Terminal

RET Renewable Energy Technology

RET Resolution Enhancement Technology

RET Revised Enskog Theory [Physics]

RET Rotational Energy Transfer

RET Rydberg Electron Transfer

RE&T Research Engineering and Test

RE-T Rare-Earth–Transition-Metal (Intermetallics) [also RE-TM, RETM, R-T ot R-TM]

Ret Retainer [also ret]

Ret Retardation [also ret]

Ret Retirement [also ret]

Ret Retract(ion) [also ret]

Ret Return [also ret]

ret retired

RETA Refrigerating Engineers and Technicians Association [US]

RETA Regional Environmental Technical Assistance [South Pacific Regional Environment Program]

RETA Retrieval of Enriched Textual Abstracts

RETAD Rapid Exchange of Tooling and Dies

RETAIN Remote Technical Assistance and Information Network

RETC Rapid Excavation and Tunneling Conference

RETC Regional Emergency Transportation Center [US]

retd retained [also ret'd]

retd retired [also ret'd]

retd returned [also ret'd]

RETEC Regional Technical Conference [of the Society of Plastics Engineers, US]

Reten Retention [also reten]

RETF Far Return [Aerospace]

RE-TM Rare-Earth–Transition Metal [also RE/TM, R-T, or R-TM]

RETMA Radio, Electronics and Television Manufacturers Association [now Electronic Industries Association, US]

RETMA Rare-Earth Transition Metal Alloy

RE-TM CMF Rare-Earth–Transition Metal Compositional Modulated Film [also RE/TM CMF]

retn retain

RetNet (Database of) Genes Causing Retinal Diseases [of University of Texas at Houston, US]

RETOS ReGIS-to-Sixels (Converter)

RETP Renewable Energy Technologies Program [of Canada Center for Mineral and Energy Technology–Energy Technology Center, Natural Resources Canada]

Retr Retract(ion) [also retr]

retr retrieve

RETRA Radio, Electrical and Television Retailers Association [UK]

RETRO Retrofire [Aerospace]

RETRO Retrofire Officer

RETRO Retro Controller [of NASA]

Retrofit Retroactive Refit [also retrofit]

RETROSPEC Retrospective Search System [of Information Services Physics, Electrical and Electronics, and Computers and Control, UK]

retrv retrieve

RETS Recent Thunderstorm [Meteorology]

RETS Reconfigurable Electrical Test Stand

RETS Reconfiguration Electrical Test Stand

RETS Renewable Energy Technologies

RETSIE Responsive Energy Technology Symposium and International Exchange

RETSPL Reference Equivalent Threshold Sound Pressure Level

REU Rectifier Enclosure Unit

REU Research Experiences for Undergraduates [of National Science Foundation, US]

REV Reentry Vehicle

Rev Revenue [also rev]

Rev Reversal [also rev]

Rev Review [also rev]

Rev Revision [also rev]

Rev Revolution [also rev]

rev reverse(d)

rev reviewed

rev revise(d)

Rev Acad Cienc Zaragoza Revista de la Academia de Ciencias Zaragoza [Review of the Zaragoza Academy of Sciences, Spain]

Rev Acoust Revue d'Acoustique [Review on Acoustics of Société Française d'Acoustique, France]

Rev Act-Metallges Review of Activities–Metallgesellschaft AG [Germany]

Rev AIBDA Revista AIBDA [Review on Agriculture of the Associación Interamericana de Bibliothecarias y Documentalistas Agricoles, Costa Rica]

Rev Alum Revue de l'Aluminium [French Aluminum Review]

Rev Aluminio Revista do Aluminio [Brazilian Aluminum Review]

Rev Anal Chem Reviews in Analytical Chemistry [UK]

Rev Anthrop Reviews in Anthropology [International Publication]

Rev ATIP Revue ATIP [French Review on Paper Industry; published by Association Technique de l'Industrie Paptière, France]

Rev Bio-Math. Revue de Bio-Mathematique [French Review on Biomathematics of the International Society of Mathematical Biology]

Rev Bras Fis Revista Brasileira de Fisica [Brazilian Physics Review; published by Sociedade Brasileira de Fisica, Brazil]

Rev Can Biol Reviews of Canadian Biology

Rev Caucho Revisto del Caucho [Spanish Rubber Review; published by Consorcia Nacional de Industriales del Caucho]

Rev Cenic Cienc Fis Revista Cenic Ciencias Fisicas [Cuban Review on Physical Sciences]

Rev Chem Eng Reviews in Chemical Engineering [US]

Rev Chim Revista de Chimie [Spanish Review on Chemistry]

Rev Chim Minér Revue de Chimie Minérale [French Review on Minerals Chemistry]

Rev Cienc Quim Revista de Ciencias Quimicas [Cuban Review of the Chemical Sciences]

Rev Coat Corros Reviews on Coatings and Corrosion [UK]

Rev Colomb Fis Revista Colombiana de Fisica [Colombian Physics Review; published by Sociedad Colombiana de Fisica, Colombia]

Rev Commun Res Lab Review of the Communications Research Laboratory [of Ministry of Post and Telecommunications, Japan]

Rev Cuba Quim Revista Cubana de Quimica [Cuban Chemistry Review]

Rev Deform Behav Mater Reviews on the Deformation Behavior of Materials [UK]

Rev E Revue E [Belgian Review on Electrical Engineering; published by Société Royale Belge des Electriciens]

Rev Electr Commun Lab Review of the Electrical Communication Laboratory [Japan]

Rev Electrotec (Argentina) Revista Electrotecnica (Argentina) [Review on Electrical Engineering; published by Asociación Argentina de Electrotechnicos y del Comite Electrotecnico Argentina]

Rev Electrotec (Spain) Revista Electrotecnica (Spain) [Review on Electrical Engineering; published by Asociación Electrotecnica Espanola]

Rev Energ Revue de l'Energie [French Review on Energy]

Rev Esp Doc Cient Revista Espanola de Documentacion Cientifica [Spanish Review on Scientific Documentation; published by Instituto de Informacion y Documentacion en Ciencia y Tecnologia]

Rev Esp Electron Revista Espanola de Electronica [Spanish Electronics Review]

Rev Esp Med Nucl Revista Espanola de Medicina Nuclear [Spanish Nuclear Medicine Review]

Rev Eur Papiers Cartons Complexes Revue Européene des Papiers, Cartons, Complexes [European Review on Paper, Cardboard, Paperboard, etc.]

Rev Fac Eng Univ Porto Revista de Faculdade de Engenharia, Universidade do Porto [Review of the Faculty of Engineering, University of Porto, Portugal]

Rev Fac Sci Univ Istanb A Review of the Faculty of Science, University of Istanbul, Series A [Turkey]

Rev Fac Sci Univ Istanb B Review of the Faculty of Science, University of Istanbul, Series B [Turkey]

Rev Fac Sci Univ Istanb C Review of the Faculty of Science, University of Istanbul, Series C [Turkey]

Rev FITCE Revue de Federation des Ingenieurs des Telecommunications de la Communauté Européenne [Review of the Federation of Telecommunication Engineers of the European Community, Brussels, Belgium]

Rev Forest Fr Revue Forestière Française [French Forestry Review]

Rev Fr Electr Revue Française de l'Electricité [French Review on Electrical Engineering and Electronics]

Rev Fr Mec Revue Française de Mécanique [French Review of Mechanics]

Rev Fr Metall Revue Française des Metallurgistes [French Metallurgy Review]

Rev Gen Caout Plast Revue Générale des Caoutchoucs et Plastiques [French General Review on Rubber and Plastics; published by Société d'Expansion Technique et Economique]

Rev Gen Chemins Fer Revue Générale des Chemins de Fer [French General Railway Review]

Rev Gen Electr Revue Générale de l'Electricité [French General Review on Electrical Engineering and Electronics; published by Société des Electriciens des Electroniciens et des Radioélectriciens]

Rev Gen Nucl Revue Générale Nucléaire [French General Review on Nuclear Energy; published by Société Française d'Energie Nucléaire]

Rev Gen Nucl (Int Ed) Revue Générale Nucleaire, (International Edition) published by Société Française d'Energie Nucleaire]

Rev Gen Therm Revue Générale de Thermique [French General Review on Thermal Engineering]

Rev Geofis Revista de Geofisica [Spanish Review of Geophysics]

Rev Geophys Reviews of Geophysics [of American Geophysical Union, US]

Rev Geophys Space Phys Reviews of Geophysics and Space Physics [of American Geophysical Union, US]

Rev HF Revue HF [Belgian Review on Radio Frequency Engineering; published by Société Belge des Ingénieurs des Télécommunications et d'Electronique]

Rev High-Temp Mater Reviews on High-Temperature Materials [UK]

Rev Hist Sci Revue d'Histoire des Sciences [French Review on the History of Science; published by Foundation for International Scientific Coordination]

Rev Iberoam Corros Prot Revista Iberoamericana de Corrosion y Proteccion [Ibero-American Review on Corrosion and Corrosion Protection published in Spain]

Rev Ing Revue de l'Ingénierie [Canadian Revue on Engineering Sciences published by Institut Canadien des Ingénieurs]

Rev Inform Autom Revista de Informatica y Automatica [Spanish Review on and Computer Science and Automation; published by Asociación Espanola de Informatica y Automatica]

Rev Inorganic Chem Reviews in Inorganic Chemistry [UK]

Rev Int Hautes Temp Réfract Revue Internationale des Hautes Temperatures et des Réfractaires [International Review of High Temperatures and Refractories; published in France]

Rev Int Metodos Numer para Calc Diseño Ing Revista Internacional de Metodos Numericos para Calcula y Diseño en Ingenieria [International Review of Numerical Methods for Engineering Calculation and Design; published in Spain]

Rev Int Syst Revue Internationale de Systematique [International Review on Systematics; published by Association Française pour le Cybernetique, Economiqie et Technique, France]

Rev Laser Eng Review of Laser Engineering [of Laser Society of Japan]

Rev Latinoam Metal Mater Revista Latinoamericana de Metalurgia y Materiales [Latin-American Review of Metallurgy and Materials Science published in Venezuela]

Rev Mat Revista di Matematica [Mathematical Review of the University of Parma, Italy]

Rev Mat Hispano-Am Revista Matemática Hispano-Americana [Spanish-American Review on Mathematics; published by the Spanish Mathematical Society]

Rev Méc Tijdschr Revue Mécanique Tijdschrift [Belgian Review on Mechanics Journal]

Rev Metal Revista de Metalurgia [Review on Metallurgy; published in Spain]

Rev Métal Revue de Métallurgie [Review on Metallurgy; published in France]

Rev Métall, Cah Inf Tech Revue de Métallurgie, Cahiers d'Informations Techniques [French Review on Metallurgy, Information Book on Techniques]

Rev Mex Astron Astrofis Revista Mexicana de Astronomia y Astrofisica [Mexican Review on Astronomy and Astrophysics; published by Instituto de Astronomia]

Rev Mex Fis Revista Mexicana de Fisica [Mexican Physics Review published by Sociedad Mexicana de Fisica]

rev/min revolutions per minute [also rev min^{-1}, RPM, or rpm]

Rev Mineral Reviews in Mineralogy [of Mineralogical Society of America]

Rev Mod Phys Reviews of Modern Physics [Joint Publication of American Physical Society and American Institute of Physics, US]

Revn Revolution [also revn]

REVOCON Remote Volume Control

Rev Odonto-Stom Revue d'Odonto-Stomatologie [French Review on Odontostomatology]

Rev Palaeobot Palynol Review of Palaeobotany and Palynology [International Journal Journal]

Rev Partic Mater Reviews in Particulate Materials [US Publication]

Rev Phys Appl Revue de Physique Appliquée [French Review of Applied Physics]

Rev Phys Appl Paris Revue de Physique Appliquée, Paris [French Review of Applied Physics]

Rev Phys Appl (Suppl J Phys) Revue de Physique Appliquée (Supplement to *Journal de Physique*) [France]

Rev Plast Mod Revista de Plasticos Modernos [Spanish Review of Advanced Plastics]

Rev Pol Acad Sci Review of the Polish Academy of Sciences

Rev Powder Metall Phys Ceram Reviews on Powder Metallurgy and Physical Ceramics [UK]

Rev Prat Metall. Revue Pratique des Metallurgistes [French Review on Practical Metallurgy]

Rev Prog Quant Nondestr Eval Review and Progress of Quantitative and Nondestructive Evaluation

Rev Radio Res Lab Review of the Radio Research Laboratories [Japan]

Rev Roum Chim Revue Roumaine de Chimie [Romanian Review of Chemistry]

Rev Roum Phys Revue Roumaine de Physique [Romanian Review of Physics]

Rev Roum Sci Tech Revue Roumaine des Sciences Techniques [Romanian Review on Technical Sciences]

Rev Roum Sci Tech-Electrotech Energ Revue Roumaine des Sciences Techniques–Série Electrotechnique et Energetique [Romanian Review of Technical Sciences–Electrical Engineering and Energetics Series]

Rev Roum Sci Tech-Mec Appl Revue Roumaine des Sciences Techniques–Série Mécanique Apliquée [Romanian Review of Technical Sciences–Applied Mechanics Series]

REVS Requirements Engineering and Validation System

REVS Rotor Entry Vehicle System

Revs Reviews [also revs]

Rev Sci Instrum Review of Scientific Instruments [of American Institute of Physics, US]

rev/sec revolutions per second [also rev/s, rev sec^{-1}, RPS, or rps]

Rev Silicon, Germanium, Tin Lead Compd Reviews on Silicon, Germanium, Tin and Lead Compounds [UK]

Rev Soc R Belge Ing Ind Revue de la Société Royale Belge des Ingéniers et des Industriels [Review of the Royal Belgian Society of Engineers and Manufacturers]

Rev Sold Revista de Soldadura [Spanish Review on Soldering; published in Madrid]

Rev Solid State Sci Review of Solid State Science [US]

Rev Soudure-Lastijdschr Revue de la Soudure-Lastijdschrift [Belgian Review on Soldering]

Rev Tec Revista Tecnica [Venezuelan Technical Review]

Rev Tech Luxemb Revue Technique Luxembourgeoise [Luxembourg Technical Review]

Rev Tech Thomson-CSF Revue Technique Thomson-CSF [Thomson-CSF Technical Review, France]

Rev Tecnol Revista Tecnologica [Cuban Technological Review]

Rev Telecomun Revista de Telecommunicación [Spanish Review on Telecommunications; published by Consejo Tecnica de Telecommunicación]

Rev Telegr Electron Revista Telegrafica Electronica [Argentinian Review on Electronic Telegraphy]

Rev Ver Revised Version [also rev ver]

Rew Rewind [also rew]

REWA Refrigeration Equipment Wholesalers Association [now Air-Conditioning and Refrigeration Wholesalers, US]

REWAS (Global Symposium on) Recycling, Waste Treatment and Clean Technology

REWS Radar Early-Warning System

REWSONIP Reconnaissance Electronic Warfare Special Operation and Naval Intelligence Processing

REWTEL Radio and Electronics World Telecommunications [UK]

REX Radiation Exposure System

REX Real-Time Executive (Routine)

REX Reduced Exoatmospheric Cross-Section

REX Relocatable Executable

REX Robotic Excavator

REX Route Extension

Re(x) Real Part of Variable "x"

REXEC Remote Execution

REXS Radio Experimental Satellite [Japan]

REXX Restructured Extended Executor (Language) [of IBM Corporation, US] [also Rexx]

RF Cylindrical Roller Bearing, Single-Row, One-Direction-Locating, Single-Ribbed Outer Ring, Double-Ribbed Inner Ring, Outer Ring Separable [Symbol]

RF Radar Frequency

RF Radial Flow

RF Radiant Flux

RF (Electromagnetic) Radiation Field

RF Radiation Filter

RF Radio Facility

RF Radio Frequency [also R-F, rf, or r-f]

RF Rainfall

RF Raised Face

RF Raised Ferrite [Metallurgy]

RF Random Function [Mathematics]

RF Range Finder; Range Finding

RF Rapid Firing

RF Rate of Flow [Fluid Mechanics]

RF Rating Factor

RF Ratio of Fronts

RF Rayleigh Function

RF Reaction Field [Physics]

RF Reaction-Formed; Reaction Forming

RF Reaction Front

RF Reactor Fuel

RF Reactive Factor [Electrical Engineering]

RF Read Forward

RF Recuperative Furnace

RF Reducing Flame

RF Reduction Factor

RF Refractory Fiber

RF Register File

RF Reinforcing Furnace (Black)

RF Relative Frequency [Quality Control]

RF Relativistic Fluid

RF Relaxation Frequency

RF Relaxation Function

RF Release Fraction

RF Reliability Factor

RF Reporting File

RF Report Footing

RF Representative Fraction

RF République Française [French Republic]

RF Repulsive Force

RF Research Fellow

RF Resilient Flooring

RF Resin Film

RF Resistance Furnace

RF Resonance Fluorescence

RF Resonant Frequency

RF Resorcinol Formaldehyde [also R-F]

RF Response Factor

RF Restoring Force

RF Retarding Field

RF Retention Factor

RF Return Flow

RF Reverse Field

RF Reverse Free

RF Rheumatic Fever [Medicine]

RF Rheumatoid Factor [Medicine]

RF Rice Flour

RF Right Forward

RF Rimfire (Ammunition)

RF Risk Factor

RF Rockefeller Foundation [New York City, US]

RF Rocket Fuel

RF Rock Fracture

RF Rodrigues-Frank (Space Parameter) [also R-F] [Metallurgy]

RF Rolling Friction

RF Roof Fan

RF Rosenbluth Formula [Physics]

RF Rotary Forging

RF Rotating Fluid

RF Rotational Field

RF Rotational Flow

RF Rough Finish(ing)

RF Roughness Factor

RF Rubber Reinforced Type (Grinding Wheel) Bond

RF Rutgers Formula [Physics]

RF Rye Flour

R-F Radio Frequency [also RF, rf, or r-f]

R-F Resorcinol Formaldehyde [also RF]

R-F Rodrigues-Frank (Space Parameter) [also RF] [Symbol]

R/F Refrigerator/Freezer

Rf Retention Factor

Rf Rifle [also rf]

Rf Roof [also rf]

Rf Rutherfordium

Rf Friction Radius [Symbol]

rf radio frequency [also RF, R-F, rf, or r-f]

rf *(regula falsi)* – rule of false position [Mathematics]

RFA Radio-Frequency Amplification; Radio-Frequency Amplifier

RFA Radio-Frequency Authorization

RFA Radiographic Fluorescence Spectral Analysis

RFA RCRA (Resource Conservation and Recovery Act) Facility Assessment [US]

RFA Recurrent Fault Analysis

RFA Redundant Force Analysis

RFA Regional Forest Agreement

RFA Remote File Access

RFA Renewable Fuels Association [US]

RFA Request for Action

RFA Request for Application

RFA Requirements Forecasting and Accounting

RFA Retarding Field Analyzer

RFA Royal Field Artillery [UK]

R-Factor Thermal Resistance Factor [also R-factor] [Microbiology]

RFAmmo Rimfire Ammunition

RFB Radio-Frequency Bridge

RFB Ready for Baseline

RFB Reason for Backlog

RFB República Federativa do Brasil [Federal Republic of Brazil]

RFB Request for Bid

RFB Reverse Feedback

RF Black Reinforcing Furnace Black [also RF black]

RFBR Russian Foundation for Basic Research

RFC   Radar Fire Control
RFC   Radial Flow Compressor
RFC   Radio Facility Chart
RFC   Radio-Frequency Cable
RFC   Radio-Frequency Chart
RFC   Radio-Frequency Choke
RFC   Radio-Frequency Current
RFC   Rate-of-Flow Controller
RFC   Reconstruction Finance Corporation
RFC   Regenerative Fuel Cell
RFC   Request for Change
RFC   Request for Comments
RFC   Request for Connection
RFC   Retirement for Cause
RFC   Reverse-Flow Compressor
RFC   Rosette Forming Cells
RF Choke   Radio-Frequency Choke [also RF choke]
RFCI   Rare Fruit Council International [US]
RFCI   Resilient Floor Covering Institute [US]
RFCP   Radio Frequency Compatibility Program
RFCP   Request for Computer Program
RFCP   Route Facility Costs Panel [of International Civil
    Aviation Organization]
RFCVD   Radio-Frequency Chemical-Vapor Deposition
RFD   Radio-Frequency Diode
RFD   Radio-Frequency Display
RFD   Ready for Data
RFD   Reentry Flight Demonstration
RFD   Request for Discussion
RFD   Requirements Formulation Document
RFD   Rural Free Delivery
RFDC   Radio-Frequency Data Collection
RFDC   Radio-Frequency Data Communications
RF/DC   Radio Frequency/Data Communications
RF/DC   Radio Frequency/Direct-Current (Magnetron)
RFDF   Radio-Frequency Direction Finding
RFDS   Radio-Frequency Discrete Saturation
RFDU   Reconfiguration and Fault Detection Unit
RFE   Radio Free Europe
RFE   Radio Frequency Emission
RFE   Reaction Field with Exclusion (Model) [Physics]
RFE   Request for Estimate
RFE   Requisition for Expenditure
rfe   fillet radius, external spline [Symbol]
R-Fe-B   Rare Earth-Iron-Boron (Alloy) [also REFe-B]
RFEC   Redox Flow-Cell Energy Conversion (System)
RFEC   Remote-Field Eddy Current
RFEI   Request for Engineering Information
RFF   Rabbit Fever Factor
RFF   Radio-Frequency Furnace
RFF   Remote-Fiber Fluorometry
RFF   (Study Group on) Rescue and Firefighting [of
    International Civil Aviation Organization]
RFF   Resources for the Future

RFF   Royal Firearms Factory [UK]
RfF   Resources for the Future, Inc. [US]
RfF Resources   RfF Resources [of Resources for the Future,
    Inc., US]
RF-Film   Radio-Frequency-Sputtered Film [also rf-film]
RFFS   Rescue and Firefighting Services
RFG   Radar Field Gradient
RFG   Radio-Frequency Generator
RFG   Rapid-Fire Gun
RFG   Receive Format Generator
RFG   Reformed Gasoline
RFG   Report Format Generator
RFG   Rifle Fine Grain (Propellant)
Rfg   Refining [also rfg]
Rfg   Roofing [also rfg]
RFH   Radio-Frequency Heating
RFI   Radio-Frequency Induction
RFI   Radio-Frequency Interference
RFI   RCRA (Resource Conservation and Recovery Act)
    Facility Investigation [US]
RFI   Ready for Installation
RFI   Remote Facility Inquiry
RFI   Remote File Inquiry
RFI   Request for Information
RFI   Resin Film Infusion (Process)
rfi   fillet radius, internal spline [Symbol]
RFICP   Radio-Frequency Inductively Coupled Plasma
RFID   Radio-Frequency Identification [also RF-ID]
RFI/EMI   Radio-Frequency Interference/Electromagnetic
    Interference
RF/IF   Radio Frequency/Intermediate Frequency
RFIM   Random Field Ising Model [Physical Chemistry]
RFIP   Research Facilities Infrastructure Program [US]
RF/IR   Radar Frequency/Infrared Frequency
RFIT   Radio-Frequency Interference Test
RFL   Radio-Frequency Laboratories
RFL   Requested Flight Level
RFL   Resorcinol-Formaldehyde Latex
rfl   reflect
Rflg   Refuelling [also rflg]
RF LHCD   Radio-Frequency Lower Hybrid Current Drive [of
    Tokamak de Varennes, Quebec, Canada]
RFLP   Restriction Fragment Length Polymorphism [Genetics]
RFM   Reactive Factor Meter
RFM   Reactive Flammable Material
RFM   Rota Flowmeter
RF Meter   Reactive-Factor Meter [also RF meter]
RFMO   Radio-Frequency Management Office
RFMS   Radio-Frequency Mass Spectrometer
RFMS   Remote File Management System
RFMWDR   Radio-Frequency Microwave Double Resonance
RFN   Red Fuming Nitric
RFNA   Red Fuming Nitric Acid
RFNM   Request for Next Message

RF NOTAM Royal Flight Notice to Airmen

RFO Radio-Frequency Oscillator

RFO Reason for Outage

RFO Request for Order

RFO Residual Fuel Oil

RFO Retarding-Field Oscillator

RF&OOA Railway Fuel and Operating Officers Association [US]

RFP Request for Procurement

RFP Request for Proposal

RFP Requirements and Formulation Phase

RFP Reverse Field Pinch [Physics]

RFP Rocky Flats Plant [US]

RFPA Request for Preliminary Application

RFPA Request for Proposal Authorization

RF PECVD Radio-Frequency Plasma-Enhanced Chemical Vapor Deposition [also rf-PECVD]

RFPG Radio-Frequency Protection Guide

RFPR Reversed Field Pinch Reactor

RFQ Radio-Frequency Quadrupole

RFQ Request for Quotation

RFR Radio-Frequency Radiation

RFR Reason for Repair

RFR Reduced Frequency Responses

RFR Reject Failure Rate

RFr Rwandan Franc [Currency]

Rfr Roofer [also rfr]

RFRL Remote Fuels Refabrication Laboratory [US]

Rfrs Roofers [also rfrs]

RFS Radio-Frequency Shift

RFS Radio-Frequency Signal

RFS Radio-Frequency Spectrometer; Radio-Frequency Spectrometry

RFS Radio-Frequency Spectroscopy

RFS Radio-Frequency Spectrum

RFS Random Filing System

RFS Ready for Service

RFS Regardless of Feature Size

RFS Regional Frequency Supplies

RFS Remote File Sharing

RFS Remote File System

RFS Renewable Fuel Sources

RFS Renormalized Forward Scattering [Physics]

RFS Request for Service

RFS Royal Forestry Society [UK]

RFS Rutherford Forward-Scattering Spectroscopy

RFSC Reaction-Formed Silicon Carbide [also RFSiC]

RFSEW Royal Forestry Society of England and Wales

RFSEWNI Royal Forestry Society of England, Wales and Northern Ireland [UK]

RFSH Rust Federal Services of Hanford, Inc. [Washington State, US]

Rfsh Refresh(ment) [also Rfshmt, rfshmt, or rfsh]

RFSiC Reaction-Formed Silicon Carbide [also RFSC]

RFSNW Rust Federal Services Northwest [Hanford, Washington, US]

RFSP Replacement Flight Strip Printer

Rf Sq Roof Squares

RF SQUID Radio-Frequency SQUID (Superconducting Quantum Interference Device) [also RF-SQUID, or rf-SQUID]

RFSTF Radio Frequency Systems Test Facility

RFSWP Request for Service Work Package

RFT Radial-Flow Turbine

RFT Radio-Frequency Transmission

RFT Ready for Training

RFT Repeat Formation Tester

RFT Reverse Form Text

RFT Revisable Form Text

RFT Rotational Frame Transformation

RFTD Radial Flow Torr Deposition

RFT DCA Reverse Form Text Document Content Architecture

RFTF Research for the Future (Program) [of Japan Society for the Promotion of Science]

RFTF Retail Fruit Trade Federation [UK]

RF-TK Radio Frequency Tracking

RFU Radio Frequency Unit

RFU Ready for Use

RFU Reference Frequency Unit

RFU Reserved For Future Use

RFW Radio Frequency Welding

RFW Request for Waiver

RFWU Rheinische Friedrich-Wilhelms-Universität [University of Bonn, Germany]

RFX Reversed Field (Pinch) Experiment [Physics]

RG Cylindrical Roller Bearing, Single-Row, Two-Direction-Locating, Double-Ribbed Outer Ring, Inner Ring With One Rib and One Snap Ring, Non-Separable [Symbol]

RG Radiation Gauge; Radiation Gauging

RG Radio Galaxy [Astronomy]

RG Radiogoniometer

RG Radiograph(er); Radiography

RG Radio Guide

RG Radioisotopic Generator

RG Rain Gauge

RG Range [also Rg, or rg]

RG Rare Gas

RG Raster Graphics

RG Rate Growth

RG Rate Gyro(scope)

RG Reactive Gas

RG Reactor Grade (Chemical)

RG Real Gas

RG Red-Green

RG Reduction Gear

RG Reference Gauge

RG Refinery Gas

RG Reflection Grating

RG Regulatory Guide

RG Relative Growth

RG Release Guard [Telecommunications]

RG Renormalization Group (Analysis) [Statistical Mechanics]

RG Report Generator

RG República de Guatemala [Republic of Guatemala]

RG Republic of Ghana

RG Reset Gate

RG Residual Gas

RG Reticulated Grating

RG Reverse Gate

RG Reverse Gear

RG Riemannian Geometry

RG Rifle Grenade

RG Ring Ground

RG Ringing Generator

RG Rio Grande [US]

RG Rolled Gold

RG Rolling Geometry (Model) [Metallurgy]

RG Rosette Gauge

RG Rubber Gel

R/G Abrasive Belt with Glue-Type Making Coat and Resin-Type Sizing Coats [Symbol]

Rg Range [also rg]

Rg Register [also rg]

RGA Rate Gyro(scope) Assembly

RGA Registered General Assignment

RGA Residual-Gas Analysis; Residual-Gas Analyzer

RGA Return Goods Authorization

RG-Acid 5-Hydroxy 2,7-Naphthalenedisulfonic Acid

RGAHS Royal Guernsey Agricultural and Horticultural Society [UK]

RGAL Rate Gyro Assembly-Left SRB (Solid-Rocket Booster) [of NASA]

RGAO Rate Gyro Assembly-Orbiter [of NASA]

RGAR Rate Gyro Assembly-Right SRB (Solid-Rocket Booster) [of NASA]

RGB Red/Green/Blue (Monitor)

RGB River Gunboat [also Rgb]

RGBM Reinforced Grouted Brick Masonry

RGBY Red/Green/Blue/Yellow (Camera)

RGC Radioactive Gas Chromatography

RGC Reconstructed Gas Chromatogram

RGC Research Grants Council [of Hong Kong University of Science and Technology]

RGCC Roll Gas Combined Cooling System (for Annealing Lines) [Metallurgy]

RGCGAI Roman-Germanic Commission of German Archeological Institutes [Germany]

RGCSP Review of the General Concept of (Aircraft) Separation Panel [of International Civil Aviation Organization]

RGD Radiation Generating Device

RGD Rayleigh-Gans-Debye (Approximation) [Physics]

RGD Rijks Geologische Dienst [Geological Survey of the Netherlands]

RGE Range [also Rge, or rge]

RGF Regional Growth Forecast

RGF Royal Gun Factory [UK]

rgh rough

Rgh Opng Rough Opening [Building Construction]

RGI Reactive Gas Infiltration

RGI Reactive Gas Injection

RGI Refocused Gradient Imaging

RGI Residual Gas Ion

RGL Report Generator Language

RGL Runway Guard Lights

R-Glass High-Strength, High-Modulus (Corrosion-Resistant) Fiberglass [also R-glass]

RGLET Rise-Time Gated Leading Edge Trigger

Rgltr Regulator [also rgltr]

RGM Recorder Group Monitor

RGM Redundant Gyro(scope) Monitor

RGM Residual Gas Molecule

RGM Römisch-Germanisches Museum [Roman-Germanic Museum, Cologne, Germany]

RGM Rounds per Gun per Minute

RGN Registered General Nurse [UK]

RGN Rio Grande do Norto [Brazil]

RGn Recoilless Gun

Rgn Region [also rgn]

RGNCD Region Code

RGO Reference Gear Oil

RGO Regular Grain Oriented (Electrical Steel) [Metallurgy]

RGO Royal Greenwich Observatory [Hurstmonceux, Sussex, UK]

RGP Rate Gyro(scope) Package

RGP Remote Graphics Processor

RGP Residual Gas Pressure [or rgp]

RGP Rolled Gold Plate

RGP Rotary Gas Pump

RGPD Range-Gated Pulse Doppler

RGPF Royal Gunpowder Factory [UK]

RGPS Radarsat Geophysical Processing System (Working Group) [of NASA]

RGQ Rapid Gas Quenching [Metallurgy]

RGR Relative Growth Rate

RGRMA Rate Gyro(scope) Redundancy Management Algorithm

RGS Radio Guidance System

RGS Rate Gyroscope System

RGS Release Guard Signal [Telecommunications]

RGS Remote Guidance System

RGS Ribbon Growth on Substrate [Solid-State Physics]

RGS Rio Grande do Sul [Brazilian State]

RGS Rocket Guidance System

RGS Royal Geographical Society [UK]

RGSA Royal Geographical Society of Australia

RGSC Ramp Generator and Signal Converter

RGT Resonance-Gate Transistor; Resonant Gate Transistor

RGT Ringgit [Currency of Malaysia]

rgt right

RGTO Rheotaxial Growth and Thermal Oxidation [Materials Science]

RGT TAB Right-Aligned Tab [also Rgt Tab]

RGU Real-Time Graphcis Unit

RGU Red/Green/Ultraviolet (Filter) [also RGUV]

RH Alkane [Symbol]

RH Heat-Resistant Rubber Type (Conductor Insulation) [Symbol]

RH Radar Horizon

RH Radiant Heat(ing)

RH Radiation Hazard

RH Radiological Health

RH Rapid Heating

RH Read Head

RH Receive Hub

RH Receiver Hopping (Mode)

RH Refrigeration Hardened; Refrigeration Hardening

RH Regenerative Heating

RH Reheater

RH Relative Humidity [also rh]

RH Remote Handling; Remotely Handled

RH Report Heading

RH República de Honduras [Republic of Honduras]

RH Request Header

RH Request/Response Header

RH République d'Haïti [Republic of Haiti]

RH Reserve-Shutdown Hours

RH Resistance Heating

RH Resonance Hybrid [Chemistry]

RH Response Header

RH Rheostat [also Rh, or rh]

RH Right-Hand(ed) [also rh]

RH Rockwell Hardness

RH Roughness Height

RH Royal Holloway [at University of London, UK]

RH Ruhrstahl-Henrichshütte/Heraeus (Degassing Process) [also R-H] [Metallurgy]

RH- Rocket Vehicle Design by Franz von Hoefft [Followed by a Roman Numeral, e.g., RH-VI] [Early Rocket Design Program concluded in 1928]

R-H Resistance–Magnetic Field (Loop)

R(H) Magnetic-Field-Dependent Resistance [Symbol]

R(H) Magnetoresistance [Symbol]

R$_{pp}$(H) Magnetic Field Dependent Peak-to-Peak Reflectivity [Symbol]

Rh Rhesus Factor [Immunology]

Rh Rhodium [Symbol]

Rh-102 Rhodium-102 [also ^{102}Rh, or Rh^{102}]

Rh-103 Rhodium-103 [also ^{103}Rh, Rh^{103}, or Rh]

Rh-104 Rhodium-104 [also ^{104}Rh, or Rh^{104}]

Rh-105 Rhodium-105 [also ^{105}Rh, or Rh^{105}]

Rh-106 Rhodium-106 [also ^{106}Rh, or Rh^{106}]

Rh-107 Rhodium-107 [also ^{107}Rh, or Rh^{107}]

Rh-108 Rhodium-187 [also ^{108}Rh, or Rh^{108}]

Rh+ Rhesus Positive [Immunology]

Rh− Rhesus Negative [Immunology]

Rh$_0$ Rhesus Factor 0 [Immunology]

R/h Roentgen per hour [also R/hr, r/hr, or R h^{-1}]

rH Oxidation-Reduction Potential [Symbol]

r(H) Hall Coefficient (Factor) [Symbol]

rh relative humidity [also RH]

rh rhesus negative [Immunology]

ρ(H) magnetic field dependent resistivity [Symbol]

RHA Regional Health Authority [of National Health Service, UK]

RHA Road Haulage Association [UK]

RHA Royal Horse Artillery [UK]

RHAB Random High-Angle (Grain) Boundary [Materials Science]

RH/AC Rapid Heating/Air Cooling (Treatment) [Metallurgy]

Rh(Ac)$_2$ Rhodium(III) Acetate

Rh(ACAC)$_3$ Rhodium(III) Acetylacetonate [also Rh(AcAc)$_3$, or Rh(acac)$_3$]

RH/ACC Rapid Heating/Accelerated Cooling (Treatment) [Metallurgy]

RHASS Royal Highland and Agricultural Society of Scotland [UK]

RHAW Radar Homing and Warning

RHB Radar Homing Beacon

RHB Reference Heat Balance

RHB Renal Hemangioblastoma [Medicine]

RHB Rivers and Harbors Board [US]

(R)-3HB D(-)-3-Hydroxybutyric Acid Random Copolymer

(R)-4HB D(-)-4-Hydroxybutyric Acid Random Copolymer

rhbhdr rhombohedral

RHBNC Royal Holloway of the British National Committee [at University of London, UK]

RHC Regional Holding Company

RHC Reheat Coil

RHC Right-Hand Circular (Polarization) [Physics]

RHC Rotation(al) Hand Controller

RhCl(PPh$_3$)$_3$ Tris(triphenylphosphine)chlororhodium (I)

RHCM Relative Humidity Control/Monitor

RHCP Right-Hand Circularly Polarized; Right-Hand Circular Polarization [Physics]

RHD Radar Horizon Distance

RHD RAMP (Rapid Modeling Platform) HVAC (Heating, Ventilation and Air Conditioning) Design

RHD Rheumatic Heart Disease [Medicine]

RHD Right-Hand Drive [also rhd]

RHD Right Hepatic Duct [Anatomy]

RHE Radiation Hazard Effects

RHE Rankine-Hugoniot Equation [Thermodynamics]

RHE Residuum Hydrodesulfurizer [Oil Refinery]

RHE Reversible Hydrogen Electrode

RHE Road Haulage Executive [UK]

RHEB Right Hand Equipment Bay

RHEED Reflection High-Energy Electron Diffraction

RHEL Rutherford High Energy Laboratory [UK]

RHELS Reflection High-Energy Loss Spectrum

Rheo Rheostat [also rheo]

Rheol Rheological; Rheology [also rheol]

Rheol Acta Rheologica Acta [Journal published in Germany]

Rheol Bull Rheology Bulletin [of Society of Rheology, US]

RHEPL Rutherford High-Energy Physics Laboratory [US]

RHEPP Repetitive High-Energy Pulsed Power (Facility) [of Sandia National Laboratories, Albuquerque, New Mexico, US]

RHET Resonant Hot Electron Transistor

Rheumatol Rheumatologist; Rheumatology

RHF Radar Height Finding

RHF Relativistic Hartree-Fock (Method) [Physics]

RHF Restricted Hartree-Fock (Theory) [Physics]

RHF Rotary Hearth Furnace [Metallurgy]

RHF/EAF Rotary Hearth Furnace/Electric-Arc Furnace [Metallurgy]

RHFEB Right-Hand Forward Equipment Bay

RHF/SRV Rotary Hearth Furnace/Smelting Reduction Vessel (Process) [Metallurgy]

RHH Reproductive Health Hazard

RHI Range-Height Indicator

RHI Reaction Hazard Index

RHI Registered Home Inspector

RHI Relative Humidity with respect to Ice

RHI Rigid Hull Inflatable (Model)

RHIC Relativistic Heavy Ion Collider [of Brookhaven National Laboratory, Upton, New York, US]

RHIC Bull Relativistic Heavy Ion Collider Bulletin [of Brookhaven National Laboratory, Upton, New York, US]

RHIC Relativistic Heavy Ion Committee

Rhino Rhinocerus [also rhino]

RHIP Radiation Health Information Project

RHIP Reaction Hot Isostatic Pressing [Metallurgy]

RHIP Relativistic Heavy Ion Physics

RHistS Royal Historical Society [UK]

RHIT Rose-Hulman Institute of Technology [Terre Haute, Indiana, US]

RHL Residual Hazards List

RHM Relative Humidity Monitor

RHM Remote Hardware Monitoring

RHM Roentgen per Hour at One Meter [also rhm]

rhmb rhombic

RHM/Ci Roentgen per Hour at One Meter per Curie [also rhm/Ci]

RHN Rockwell Hardness Number [also Rhn]

RHNC Reference Hypernetted Chain (Theory) [Physics]

RHO Rockwell-Hanford Operations [US]

RH-OB Ruhrstahl Heraeus Oxygen Blowing (Process) [Metallurgy]

Rhod Rhodesia(n)

RhoGAM Rho Goat Anti-Mouse [Biochemistry]

RHOGUI Radar Homing Guidance

Rhomb Rhombohedric; Rhombohedron [also rhomb]

Rhomb Rhombic; Rhombus [also rhomb]

RHP Reduced Hard Pressure

RHP Reheat Pressure

RHP Rice Hull Pyrolysis

RHP Right Hand Panel

(R)-3HP D(-)-3-Hydroxypropionic Acid Random Copolymer

RH-PB Ruhrstahl Heraeus Powder Blowing (Process) [Metallurgy]

RHPG Reference High-Performance Grout

RHPW Reduced Half-Peak Width [Raman Spectroscopy]

RHQ Regional Headquarters

RHR Receiver Holding Register

RHR Rejectable Hazard Rate

RHR Residual Heat Removal

RHR Roughness Height Rating

Rhr Reheater [also rhr]

R/hr Roentgen(s) per hour [also R/h, r/hr, or $R\,h^{-1}$]

RHRP Residual Heat Removal Pump

RHRSW Residual Heat Removal Service Water

Rh-Ru Rhodium-Ruthenium (Alloy System)

RHS Rectangular Hollow Section

RHS Right-Hand Side [also rhs]

RHS Rocketdyne Hybrid Simulator

RHS Royal Historical Society [Canada]

RHS Royal Horticultural Society [UK]

RHSC Right-Hand-Side Console

RHSI Royal Horticultural Society of Ireland

RhSi Rhodium Silicide (Semiconductor)

RHSP Registered Hazardous Substances Professional

RhSSb Rhodium Sulfur Antimonide

RHSV Royal Historical Society of Victoria [Canada]

RHT Radiant Heat Temperature

RHT Radiative Heat Transfer

RHT Rapid Heat Treatment [Metallurgy]

RHT Register Holding Time

RHT Reheat Temperature [Metallurgy]

RHTM Regional Highway Traffic Model [UK]

RHTRU Remotely Handled Transuranic Waste

RHTS Reactor Heat Transport System

RHU Radioisotope Heater Unit [of NASA]

(R)-3HV D(-)-3-Hydroxyvaleric Acid Random Copolymer

RHW Router Header Word

Rh^{3+}:YIG Trivalent Rhodium-Doped Yttrium Iron Garnet

Rh^{4+}:YIG Tetravalent Rhodium-Doped Yttrium Iron Garnet

RI Racial Immunity

RI Radar Indication

RI Radar Input

RI Radar Interrogation

RI Radiographic Inspection

RI Radio Inertial

RI Radio Influence

RI Radio Interference
RI Radio Interferometer
RI Radioisotope
RI Rayleigh Instability [Fluid Mechanics]
RI Range Instrumentation
RI Rayleigh Interferometer [Optics]
RI REACT (Radio Emergency Associated Communications Team) International
RI Reactive Intermediate
RI Read-In
RI Real Image
RI Receiving Inspection [also R/I]
RI Rectifier Instrument
RI Referential Integrity
RI Reflective Insulation
RI Refraction Intercept
RI Refractive Index
RI Regal Industries Inc. [Willoughby Hills, Ohio, US]
RI Relative Intensity
RI Reliability Index
RI Repeat Indication
RI Report of Investigation
RI Reproducibility Index
RI Republica Italiana [Italian Republic]
RI Republic of Indonesia
RI Research Institute
RI Resistance-Inductance
RI Resonance Ionization
RI Retention Index [Analytical Chemistry]
RI Retinal Image [Microscopy]
RI Rhode Island [US]
RI Rial [Currency of Iran]
RI Rideability Index (of Pavement) [Symbol]
RI Ring Indicator [Data Communications]
RI Risk Investigation
RI Robotics Institute [of Carnegie-Mellon University, Pittsburgh, Pennsylvania, US]
RI Robotics International [of Society of Manufacturing Engineers, US]
RI Rockefeller Institute [New York City, US]
RI Rodale International [US]
RI Roofing Industry
RI Routing Indicator
RI Ryukyu Islands [Japan]
R&I Removal and Installation
R/I Receiving Inspection [also RI]
Ri Richardson Number (of Fluid Flow) [Symbol]
r(i) relative Auger peak-to-peak height of substance "i" [e.g., r(S) is the relative Auger peak-to-peak height of sulfur] [Symbol]
RIA Radioimmunoassay [Immunology]
RIA Radio Interference Advisory
RIA Radioisotope Assay [Chemistry]
RIA Railway Industry Association [UK]

RIA Reactivity Initiated Accident [Nuclear Reactors]
RIA Recording Industry Association [US]
RIA Registered Industrial Accountant
RIA Removable Instrument Assembly [of Idaho National Engineering Laboratory, US]
RIA Research Institute of America [US]
RIA Research Institute of Atmospherics [of Nagoya University, Japan]
RIA Rigid Ion Approximation
RIA Robotics Industries Association [US]
RIA Robot Institute of America [Name Changed to Robotics Industries of America, US]
RIA Rock Island Arsenal [Rock Island, Illinois, US]
RIA Royal Irish Academy [Dublin]
RIAA Recording Industry Association of America [US]
RIAC Royal Irish Automobile Club
RIAD Research Information and Application Division
RIAEC Rhode Island Atomic Energy Commission [US]
RIAI Royal Institute of Architects of Ireland
RIAM Research Institute for Applied Mechanics [of Kyushu University, Fukuoka, Japan]
RIAS Radio in the American Sector (of Berlin)
RIAS Regulatory Impact Assessment Statement
RIAS Research Institute for Automation Systems [Prague, Czech Republic]
RIAS Research Institute of Advanced Studies [US]
RIAS Research Institute of Animal Science [South Africa]
RIAS Royal Incorporation of Architects in Scotland
RIAST Research Institute for Advanced Science and Technology [of University of Osaka Prefecture, Japan]
RIB Rapid Infiltration Basin [Civil Engineering]
RIB Recoverable Item Breakdown
RIB Request Indicator Byte
RIB Research Information Bulletin
RIB Research Internet Backbone
RIB Right Inboard
Rib D-Ribose; Ribosyl [Biochemistry]
RIBA Royal Institution of British Architects
RIBE Reactive Ion Beam Etching
RIBS Research Information Bulletins [also RIBs]
RIBS Rutherford Ion Backscattering
RIBS Rutherford Ion Backscattering Spectroscopy
RIC Radar Indicating Console
RIC Radar Input Control
RIC Railway Industry Council
RIC Rainforest Information Center [Australia]
RIC Range Instrumentation Coordination
RIC Rare-Earth Information Center [of Iowa State University, Ames, US]
RIC Read-In Counter
RIC Reconstructed Ion Chromatography
RIC Regional Engineering College
RIC Relocation Instruction Counter
RIC Research and Information Commission [of Coordinating Secretariat of National Unions of Students, Netherlands]

RIC Research Institute for Catalysis [of Hokkaido University, Japan]

RIC Resistance Inductance and Capacitance

RIC Rhode Island College [Providence, US]

RIC Rice Improvement Conference [of Provincial Department of Agriculture and Forestry, Taiwan]

RIC Rockwell International Corporation [US]

RIC Roughness Induced Closure

RIC Royal Institute of Chemistry [UK]

RI&C Reactor Instrumentation and Control [of Institute of Electrical and Electronics Engineers Nuclear and Plasma Sciences Society, US]

RICA Railway Industry Clearance Association [US]

RICA Research Institute for Consumer Affairs [UK]

RICASIP Research Information Center and Advisory Service on Information Processing [US]

Ric Autom Ricerche di Automatica [Italian Journal on Automation Research]

RICB Reactive Ionized Cluster Beam

RICC Remote Intercomputer Communications (Interface)

RICE Regional Interactions of Climate Ecosystems

RICE Rest–Ice–Compression–Elevation (Technique) [Strain Injury Treatment]

RICE Rice Information Cooperative Effort [Philippines]

RICE Rosemount Icing Detector

RICH (International Workshop on) Ring Imaging Cerenkov Detect [Physics]

RICMO Radar Input Countermeasures Officer

RICMT Radar Inputs Countermeasures Technique

RICOB Rice and Corn Board [Philippines]

RICPE Reduced Ilmenite Carbon Paste Electrode

RICS Range Instrumentation Control System

RICS Royal Institution of Chartered Surveyors [UK]

RICW Regional Engineering College, Warangal [India]

RID Rack Insertion Device

RID Radar Input Drum

RID Radial Immunodiffusion

RID Radiation-Induced Diffusion

RID Records Issue Date

RID Refractive-Index Detector

RID Reset Inhibit Drum

RID Review Item Discrepancy

RID Review Item Disposition

RIDA Rural Industrial Development Authority [Malaysia]

RIDC Rhode Island Development Council [US]

RIDD Range Instrumentation Development Division

RIDI Receiving Inspection Detail Instruction

RIDL Ridge Instrument Development Laboratory [US Navy]

RIDS Receiving Inspection Data Status Report

RIDS Records Inventory and Disposition Schedule (and File Index)

RIDS Review Item Discrepancy System

RIE Reactive Ion Etch(er); Reactive Ion Etching

RIE Research in Education (Database) [of Educational Resources Information Center, US]

RIE Research Institute of Electronics [Shizuoka University, Japan]

RIE Rhodesian Institution of Engineers [now Zimbabwe Institution of Engineers]

RIE Right Inboard Elevon [Aeronautics]

RIE Royal Institute of Engineers [UK]

RIE Royal Institution of Engineers [Netherlands]

RIEC Research Institute of Electrical Communication [Tohoku University, Sendai, Japan]

RIEF Recirculating Isoelectric Focusing

RIEI Roofing Industry Educational Institute [Denver, Colorado, US]

RIEMM Research Institute for Electric and Magnetic Materials [Sendai, Japan]

RIE/PE Plasma-Enhanced Reactive Ion Etching

RIE PECVD Reactive Ion Etching Plasma-Enhanced Chemical Vapor Deposition [also RIE-PECVD, or RIE/PECVD]

RIES Research Institute for Electronic Sciences [of Hokkaido University, Japan]

RIETCOM Regional Interagency Emergency Transportation Committee [US]

RIF Radiation Interchange Factor

RIF Radio Influence Field

RIF Radio Interference Field

RIF Rapid Infrared Forming (Technique)

RIF Reclearance in Flight

RIF Reduction in Force

RIF Relative Importance Factor

RIF Reliability Improvement Factor

RIFI Radio Interference Field Intensity

RIFI Radio-Interference-Free Instrument

RIFF Resource Interchange File Format

RIFM Research Institute for Fragrance Materials [US]

RIFP Research Institute for Fundamental Physics [Kyoto University, Japan]

RIFS Radioisotope Field Support

RI&FS Remedial Investigation and Feasibility Study [also RI/FS] [of US Environmental Protection Agency]

RI/FS Remedial Investigation/Feasibility Study

RI/FS Risk Investigation/Feasibility Study

RIFT Reactor-In-Flight Test (Vehicle) [of NASA]

RIG Rate Integration Gyro(scope)

RIG Research Initiation Grant [of Engineering Foundation, US]

RIG Transmitter [Amateur Radio]

RIGB Royal Institution of Great Britain [London, UK]

R³ IGF-I Recombinant Insulin-Like Growth Factor I (One) [Biochemistry]

RIGFET Resistive Insulated Gate Field-Effect Transistor

Right Rect Abd Right Rectus Abdominis [also right rect abd] [Anatomy]

RIGI Receiving Inspection General Instruction

RIGS Radio Inertial Guidance System

RIGS Runway Identifiers with Glide Slope

RIHANS River and Harbor Aid to Navigation System

RIHS Rhode Island Historical Society [US]

RII Receiving Inspection Instruction

RIIA Royal Institute of International Affairs [UK]

RIIC Research Institute on International Change

RIIC Rural Industries Innovation Center

RIIS Research Institute of Industrial Science [of Kyushu University, Fukuoka, Japan]

RIISOM Research Institute for Iron, Steel and Other Metals [of Tohoku University, Sendai, Japan]

RIKEN The Institute of Physical and Chemical Research [Saitama, Japan]

RIKES Raman-Induced Kerr Effect Spectroscopy

RIL Radio Interference Level

RIL Recoverable Item(s) List(s)

RIL Red Indicating Lamp

RIL Repa(i)rable Item List

RIL Representation Independent Language

rIL Rat Interleukin [Biochemistry]

rIL-2 Rat Interleukin-2 [Biochemistry]

RILEM Réunion Internationale des Laboratoires d'Essais et de Recherches sur les Matériaux et les Construction [International Union of Testing and Research Laboratories for Materials and Structures, France]

RIM Radar Input Mapper

RIM Radio Imaging Method

RIM Radio Induction Method

RIM Rapid Infrared Manufacturing

RIM Reactant Ion Monitoring

RIM Reaction Injection Molded; Reaction Injection Molding

RIM Read-In Mode

RIM Receiver Intermodulation

RIM Remote Installation and Maintenance

RIM Repository Information Model

RIM Repulsion-Induction Motor

RIM Resource Interface Module

RIM Retention Index Method [Gas Chromatography]

RIM Reversed Isostatic Molding

RIM Rigid-Ion Model

RIM Rotary Index Machine

RIMCOF R&D (Research and Development) Institute of Metals and Composites for Future Industries [of New Energy and Industrial Technology Development Organization, Japan]

RIMDM Research Institute of Mineral Dressing and Metallurgy [of Tohoku University, Sendai, Japan]

RIME RelayNet International Message Exchange

RIMES Research Institute for Metal Surfaces of High Performance [of Kawasaki Steel Corporation, Japan]

RIM PUR Reaction Injection Molding/Polyurethane [also RIM/PUR]

RIMS Radiation Intensity Measuring System

RIMS Remote Information Management System

RIMS Resonant Ionization Mass Spectrometry; Resonant Ionization Mass Spectroscopy

RIMS Retarding Ion Mass Spectrometer

RIMS Risk and Insurance Management Society US]

RIMS/TIDE Retarding Ion Mass Spectrometer/Thermal Ion Dynamics Experiment

RIMtech Research Institute for the Management of Technology

RIN Reference Indicator Number

RIN Regular Inertial Navigator

RIN Royal Institute of Navigation [London, UK]

RINA Registro Italiano Navale [Italian Naval Registry]

RINA Royal Institution of Naval Architects [UK]

RINAL Radar Inertial Altimeter

RIND Research Institute of National Defense [Sweden]

RINDO Rydberg Intermediate Neglect of Differential Overlap [Physics]

RINEX Receiver Independent Exchange Format [Computers]

Ring Ringing [also ring]

RINGS Reports and Investigations by Networked Groups of Students (Project) [Canada]

RINS Research Institute for the Natural Sciences

RINS Rotorace Inertial Navigation System

RINT Radar Intelligence

RINT Radar Intermittent

RIO Omani Rial [Currency of Oman]

RIO Real-Time Input/Output

RIO Red Iron Oxide

RIO Relocatable Input/Output

RIO Remote Input/Output

RIO Roll-In Only [Computers]

Rio Rio de Janeiro [Brazil]

RIOMETER Relative Ionospheric Opacity Meter

RIOPR Rhode Island Open Pool Reactor [US]

RIOSA Reactive Infinite Order Sudden Approximation [Physics]

RIOT RAM (Random-Access Memory) Input/Output Timer

RIOT Real-Time Input/Output Transducer

RIOT Real-Time Input/Output Translator

RIOT Retrieval of Information by On-Line Terminal

RIP Rapid Isothermal Processing [Metallurgy]

RIP Raster Image Processor

RIP Reaction Injection Pultrusion

RIP Reactive Ion Plating

RIP Reactor Instrument Penetration

RIP Receiving Inspection Plan

RIP Regulatory and Information Policy

RIP Remote Imaging Protocol

RIP *(Requiescat in pace)* – may he (or she) rest in peace

RIP Resin-In-Pulp (Process) [Chemical Engineering]

RIP Rest in Proportion (Printing)

RIP Retired In Place

RIP Ring Index Pointer

RIP Routing Information Process

RIP Routing Information Protocol [Internet]

RIP Routing Internet Protocol

RIP Rural Industrialization Program [US]

Rip Ripped [also rip] [Building Trades]

RIPA Royal Institute of Public Administration [UK]

RIPAS Regional Integrated Protected Areas System [of United Nations]

RIPCAM Real-Time Interactive Process Control and Management

RIPE Réseaux IP Européens [European Internet Protocol Network]

RIPEM Riordan's Internet Privacy Enhanced Mail

RIPHH Royal Institute of Public Health and Hygiene [UK]

RIPL Representation Independent Programming Language

RIPPLE Radioactive Isotope Powered Pulse Light Equipment

RIPPLE Radioisotope Powered Prolonged Life Equipment

RIPS Radar Impact Prediction System

RIPS Radio Isotope Power Supply

RIPS Radio Isotope Power System

RIPS Range Instrumentation Planning Study

RIPS Raster Image Processing System

RIPS Regional Institute for Population Studies [of United Nations in Ghana]

RIPS Research Information Processing System

RIPSC Research Information Processing System Center [of Agency of Industrial Science and Technology–Ministry of International Trade and Industry, Japan]

RIPV (Nuclear) Reactor Isolation Pressure Valve

RIQ Rexroth Information Quarterly

RIQS Remote Information Query System

RIR Reference Intensity Ratio (Method) [X-Ray Diffraction]

RIR Relative Index Register

RIR Reliability Investigation Request

RIR Reportable Item Report

RIR ROM (Read-Only Memory) Instruction Register

RIRA Reports and Information Retrieval Activity

RIRDC Rural Industry Research and Development Corporation [Australia]

RIRO Roll In/Roll Out [Computers]

RIRS Reflectance Infrared Spectroscopy

RIRTI Recording Infrared Tracking Instrument

RIRV Research Institute for Road Vehicles [of TNO–(Organisatie voor) Toegepast-Naturwetenschappelijk Onderzoek, Netherlands]

RIS Radar Information Service

RIS Radiation-Induced Segregation [Metallurgy]

RIS Radiology Information System

RIS Range Instrumentation Ship

RIS Raster Input Scanner

RIS Receipt Inspection Segment

RIS Recorded Information Service

RIS Redwood Inspection Service [US]

RIS (Laser-Beam Absorption) Reflector In Space [Measurement of Chloroflurocarbons, Ozone etc.]

RIS Remote Information System

RIS Reporting Identification Symbols

RIS Research Information Service

RIS Resonance Ionization Spectroscopy; Resonant Ionization Spectrometry

RIS Retail Information System

RIS Retransmission Identity Signal

RIS Revolution Indicating System

RIS Risø International Symposium [Risø, Denmark]

RIS Rotatable Initial Susceptibility

RIS Rotational Isomeric State

RIS Rowland Institute of Science [US]

RIS Rural Information System

RIS Russian Intelligence Service

RISAFMONE Radio-Signal Reporting Code for Transmission-Condition Rating

RISC Radiology Information Systems Consortium

RISC Reduced Instruction Set Chip

RISC Reduced Instruction Set Computer; Reduced Instruction Set Computing

RISC Rockwell International Science Center [Thousand Oaks, California, US]

RISD Rhode Island School of Design [Providence, US]

RI/SD Rockwell International/Space Division [US]

RISE Research Information Service for Education [of Ontario Institute for Studies in Education, Canada]

RISE Research Institute of Science and Engineering [of Ritsumeikan University, Kyoto, Japan]

RISE Research in Supersonic Environment

RISE (National Institute for) Resources in Science and Engineering [US]

RISFM Research Institute for Strength and Fracture of Materials [of Tohoku University, Sendai, Japan]

RISKAC Risk Acceptance

Risk Anal Risk Analysis [Published in the US]

Risk Manage Risk Management [Journal of the Risk Management Society, US]

RISM Radio Interface Switch Module

RISM Reference Interaction Site Model

RISM Research Institute for Scientific Measurements [Tohoku University, Sendai, Japan]

RISM Research Institute for Strength of Materials [of Xi'an Jiaotung University, Sian, PR China]

RISM Rotational Isomeric State Model

RI/SME Robotics International of the Society of Manufacturing Engineers [US]

RISO Regional Individual Standing Offer [Canada]

Risø Rep Risø Report [of Risø Laboratory, Roskilde, Denmark]

RISOP Resonant Intersubband Scattering by Optical Phonons [Solid-State Physics]

RISP Rapid-Impingement Speed Plating

RISPA Research Institute of the Sumatra Planters Association [Indonesia]

RISSP Research Institute for Solid-State Physics [of Hungarian Academy of Sciences]

RIST Radar Installed System Tester

RIST Research Institute of Science and Technology [of Nihon University, Tokyo, Japan]

RIST Research Institute of Industrial Science and Technology [Pohang, South Korea]

RIST Rep Research Institute of Industrial Science and Technology Report [Pohang, South Korea]

RIT Radar Input Test

RIT Radio Information Test

RIT Radio Network for Inter-American Telecommunications

RIT Rate of Information Transfer

RIT Raw Input Thread

RIT Receiver Incremental Tuning

RIT Remote Interactive Terminal

RIT Request for Interface Tool

RIT Resonant Interband Tunneling

RIT Rochester Institute of Technology [New York State, US]

RIT Rocket Interferometer Tracking

RIT Royal Institute of Technology [Stockholm, Sweden]

RITA Radio-Frequency Ion Thruster Assembly

RITA Rand Intelligent Terminal Agent [US]

RITA Recognition of Information Technology Achievement [UK]

RITA Reusable Interplanetary Transport Approach Vehicle

RITC Regional Information Technology Coordinators [UK]

RITC Rhodamine Isothiocyanate [Biochemistry]

RITC-Dextran Rhodamine B Isothiocyanate-Dextran [Biochemistry]

RITE Rapidata Interactive Editor

RITE Research Institute of Innovative Technology for the Earth [Kyoto, Japan]

RITENA Reunion Internacional de Técnicos de la Nutrición Animal [International Meeting of Animal Nutrition Experts, Spain]

RITL Royal Institute of Technology Libraries [Sweden]

RITP Research Institute for Technical Physics [of Hungarian Academy of Sciences, Budapest]

RIU Remote Interface Unit

RITU Research Institute of Temple University [Philadelphia, Pennsylvania, US]

RIUK Royal Institute of the United Kingdom

RIV Radio-Influence Voltage

RIV Rapid Intervention Vehicle

RIV Recirculation Isolation Valve

Riv River [also riv]

Riv Rivet [also riv]

Riv Combust Rivista dei Combustibili [Italian Review on Combustibles]

Riv Inf Rivista Informatica [Italian Computer Science Review; published by Associazione Italiana per il Calcolo Automatica] [also Rev Inform]

Riv Ital Saldatura Rivista Italiana della Saldatura [Italian Welding Review]

Riv Mecc Rivista di Meccanica [Italian Review on Mechanics]

Riv Nuevo Cimento Rivista del Nuevo Cimento [Italian Review on Physics; published by Societa Italiana di Fisica, Italy]

RIW Reliability Improvement Warranty

RIWAS Royal Isle of Wight Agricultural Society [UK]

RIX Rapid Ion Extraction

RJ Cylindrical Roller Bearing, Single-Row, One-Direction-Locating, Double-Ribbed Outer Ring, Single-Ribbed Inner Ring, Inner Ring Separable [Symbol]

RJ Ram-Jet (Engine) [Aeronautics]

RJ Rayleigh-Jeans (Equation) [Statistical Mechanics]

RJ Reaction Jet [Aeronautics]

RJ Reference Junction

RJ Right Justify (or Justification)

RJ Rio de Janeiro [Brazil]

RJ Royal Jelly [Biology]

$\rho(\mathbf{J})$ Current Density Dependent Resistivity [Symbol]

RJA&HS Royal Jersey Agricultural and Horticultural Society [formerly RJAS, UK]

RJAS Royal Jersey Agricultural Society [now RJA&HS, UK]

RJC Reaction Jet Control [Aeronautics]

RJCP Refocused J Cross-Polarization

RJD Reaction Jet Device [Aeronautics]

RJD Reaction Jet Driver [Aeronautics]

RJDA Reaction Jet Driver Aft [Aeronautics]

RJDF Reaction Jet Driver Forward [Aeronautics]

RJEC Reaction Jet Engine Control [also RJ/EC]

RJE Ramjet Engine [Aeronautics]

RJE Remote Job Entry

RJES Remote Job Entry System

RJO Remote Job Output

RJOD Reaction Jet OMS (Orbital Maneuvering System) [of NASA Space Shuttle]

RJP Cylindrical Roller Bearing, Single-Row, One-Direction-Locating, Double-Ribbed Outer Ring with One Loose Rib, Single-Ribbed Inner Ring, Both Rings Separable [Symbol]

RJP Remote Job Processing

RK Cylindrical Roller Bearing, Single-Row, Two-Direction-Locating, Outer Ring With Two Internal Snap Rings, Double-Ribbed Inner Ring, Non-Separable [Symbol]

RK Rangekeeper

RK Reaction Kinetics [Physical Chemistry]

RK Reactor Kinetics

RK Relativistic Kinematics [Relativity Theory]

RK Republic of Kazakhstan

RK Rotary Kiln

RK Runge-Kutta (Approximation Method) [Mathematics]

r(k) reflection coefficient [Symbol]

$\rho(\kappa)$ density scattering vector [Symbol]

RKD Remote Key Distribution

RKD Rocketdyne [US]

RKDW Rationalisierungs-Kuratorium der Deutschen Wirtschaft e.V. [German Productivity and Management Association]

RKHS Reducing Kernel Hilbert Space [Mathematics]

RKI Robert-Koch-Institut–Bundesinstitut für Infektionskrankheiten und nicht übertragbare Krankheiten) [Robert Koch Institute of the Federal Institute for Infectious and Non-Communicable Diseases, Germany]

RKKY Rudermann-Kittel-Kasuya-Yosida (Theory of Electron Interactions) [Physics]

RKO Range Keeper Operator

RKOC Rotary Kiln Optimization Control

RKR Ritchie, Knott and Rice Criterion [Metallurgy]

RKR Rydberg-Klein-Rees (Potential) [Physics]

RKS Redlich-Kwong-Soave (Equation of State) [Physical Chemistry]

Rkt Rocket [also rkt]

RKTL Reichskuratorium für Technik in der Landwirtschaft [German Association for Mechanization in Agriculture, Germany]

rkva reactive kilovolt-ampere [also rkVA]

Rktry Rocketry [also rktry]

RKW Rationalisierungskuratorium der Deutschen Wirtschaft [German Productivity and Management Association]

RL Radiation Laboratory

RL Radiation Lobe [Electrical Engineering]

RL Radio Link

RL Radiolocation; Radiolocator

RL Radioluminescence; Radioluminescent

RL Radioluminescent Light

RL Raman Line [Spectroscopy]

RL Random Length [also rl]

RL Reaction Layer

RL Reactor Licensing

RL Receive Leg

RL Reciprocal Lattice [Crystallography]

RL Record Length

RL Red Lead

RL Red Light

RL Reduced Level

RL Regression Line [Statistics]

RL Reference Librarian; Reference Library

RL Reflection Loss

RL Relay Level

RL Relay Logic

RL Remote Loopback

RL Report Leaving

RL Research Laboratory

RL Research Library

RL Resistance-Inductance

RL Resistor Logic

RL Resonance Line [Spectroscopy]

RL Return Line

RL Return Loss

RL Reversed Loading [Mechanics]

RL Rhumb Line [Cartography]

RL Richland Operations Office [of US Department of Energy in Washington State, US]

RL Right Lung [Anatomy]

RL Rocket Launcher

RL Rogowski Loop

RL Rome Laboratory [of US Air Force]

RL Root Locus [Control Systems]

RL Rotate Left [Computers]

RL Royal Laboratory [UK]

RL Rubber Latex

RL Ruby Laser

RL Runway Lights

R/L Rate/Limited

R/L Remote/Local

Rl Reel [also rl]

Rl Riel [Currency of Kampuchea]

RLA Rapid Lamp Annealing

RLA Remote Line Adapter

RLA Remote Loop Adapter

RLA Repair Level Analysis

RLA Repair Line Agreement

RLA Resonant Laser Ablation

RLAC Recycling Legislation Action Coalition [US]

RLAP Rural Land Analysis Program [Canada]

RLAS Royal Lancashire Agricultural Society [UK]

RLBM Rearward Launched Ballistic Missile

RLC Radio Launch Control System

RLC Remote Load Controller

RLC Resistance-Inductance-Capacitance

RLC Resistor-Inductor-Capacitor

RLC Ripple-Load Cracking

RLC ROM (Read-Only Memory) Location Counter

RLC Rotate Left through Carry (Function) [Computers]

RLC Run Length Coding

RLCA Reaction-Limited Cluster Aggregation [Metallurgy]

RLCC Rotation Locular Countercurrent Chromatography

RLCE Request Level Change En-Route [Aeronautics]

RLCS Radio Launch Control System

RLCTS Research Laboratory and Conservation and Technical Services (Division) [of British Museum]

RLCU Reference Link Control Unit

RLD Received Line Detect

RLD Relocation Dictionary (Card)

RLD Relocation Directory

RLD Ribbon-Like Defect [Materials Science]

RLD Rijksluchtvaartdienst [Royal Dutch Aviation Administation]

RLE Run Length Encoded

RLE Research Laboratory of Electronics [of Massachusetts Institute of Technology, Cambridge, US]

RLE Roasting, Leaching, Electrowinning (Process)

RLE Run Length Encoder; Run Length Encoding

RLEA Railway Labor Executives Association [US]

RLE Curr Research Laboratory of Electronics Currents [Publication of Massachusetts Institute of Technology, Cambridge, US]

RLEM Research Laboratory of Engineering Materials [of Tokyo Institute Technology, Japan]

RLEO Request Liaison Engineering Order

RLF Retrolental Fibroplasia [Medicine]

RLF Reverse Line Feed

Rlf Relief [also rlf]

RLFCM Radioactive Liquid Fed Ceramic Melter

RLG Release Guard [Telecommunications]

RLG Research Libraries Group [US]

RLG Rifle, Large Grain (Propellant)

RLG Ring Laser Gyro(scope)

Rlg Railing [also rlg]

RLHC Research Laboratory of Hydrothermal Chemistry [of Kochi University, Japan]

RLHNC Reference-Linearized Hypernetted Chain (Theory) [Physics]

RLHTE Research Laboratory of Heat Transfer in Electronics

RLi Alkyl Lithium (or Organolithium) [Formula]

RLIN Research Libraries Information Network [US]

RLL Recorded Lithology Logging

RLL Relay Ladder Language

RLL Relocating Linking Loader

RLL Run Length Limited

RLLTA Reduced Lead, Low Toxicity Ammunition

RLM Dome-Type of Lighting Reflector

RLM Reactive-Layer Model

RLM Reflector and Lamp Manufacturers [US]

RLM Reflector and Lighting Equipment Manufacturers

RLMC Reaction-Limited Monomer Cluster

RLN Remote LAN (Local-Area Network) Node

RLn Rotate Left n (Function) [Computers]

RLNR Research Laboratory for Nuclear Reactors [of Tokyo Institute of Technology, Japan]

RLO Restoration Liaison Officer

RLO Returned Letter Office

RLOGIN Remote Login

RLOM Reflected Light Optical Microscopy

RLOP Reactor Licensing Operating Procedure

RLOS Radar Line-of-Sight [also R-LOS, or RLS]

RLP Reactive Liquid Polymer

RLPME Research Laboratory of Precision Machinery and Electronics [Yokohama, Japan]

RLQ Right Lower Quadrant (of Abdomen) [Medicine]

RLR Record Length Register

RLS Radar Line-of-Sight [also RLOS, or R-LOS]

RLS Recursive Least-Squares [Statistics]

RLS Remote Line Switch

RLS Residual Lattice Strain [Crystallography]

RLS Resonant Light Scattering

RLS Riel [Currency of Kampuchea]

RLS Rubidium Lithium Sulfate

Rls Reels [also rls]

Rls Release [also rls]

RLSD Received Line Signal Detected; Received Line Signal Detector

Rlse Release [also rlse]

RLSS Research Laboratory for Surface Science [Okayama University, Japan]

RLSS Regenerative Life Support System

RLSSC Royal Life Saving Society Canada

RLST Release Timer

RLT Radiative Lifetime

RLT Reaction Liquid Type

RLT Remote Line Test

RLT Return Line Tether

RLT Rocket, Light Tube

RLT Runway Localizer Transmitter

RLTS Radio-Linked Telemetry System

RLU Remote Line Unit

RLV Reusable Launch Vehicle

RLWL Reactor Low Water Level

RLWS Radioactive Liquid Waste System

Rlwy Railway [also Rlwy, or rlwy]

Rly Relay [also rly]

RM Cylindrical Roller Bearing, Single-Row, Non-Locating, Inner Ring without Ribs, Roolers Located by Cage, Internal Snap Rings, or End-Rings Recesses in Outer Ring, Inner Ring Separable [Symbol]

RM Racemic Mixture [Chemistry]

RM Radar Map(per)

RM Radiation Measurement

RM Radiation Monitor(ing)

RM Radiometal

RM Radiometeorograph(y)

RM Radiometer

RM Radiomicrometer

RM Radio Monitoring

RM Rainmaker; Rainmaking

RM Random Matrix [Physics]

RM Raney Method [Metallurgy]

RM Range Mark [Radar]

RM (Radiation Counting) Ratemeter

RM Raw Material

RM Reaction Mechanism

RM Reaction Milled; Reaction Milling [also RM]

RM Reactive Material

RM Reactive Metal

RM Reactive Mixing

RM Ready Mix(ed)

RM Receptance Matrix

RM Record Mark

RM Recrystallization Microstructure [Metallurgy]

RM Rectangular Mode

RM Rectangular Module (Core)

RM Rectilinear Motion

RM Recycled Material

RM Red Mud

RM Redundancy Management

RM Reed-Muller (Code)

RM Reference Material

RM Reference Method

RM Reference Mission

RM Reflecting Microscope

RM Refractory Material

RM Refractory Metal

RM Refrigerating Machine

RM Registered Midwife

RM Register Memory

RM Regular Matrix

RM Reichert-Meissl (Number) [also R-M] [Analytical Chemistry]

RM Relative Motion

RM Relativistic Mechanics

RM Relativity Mission

RM Relay Mirror

RM Reliability and Maintainability; Reliability and Maintenance

RM Remanent Magnetism

RM Remanent Magnetization

RM Remote Manipulation; Remote Manipulator

RM Remote Mobile

RM Rendezvous Maneuver

RM Republic of Malawi

RM République du Mali [Republic of Mali]

RM Repulsion Motor

RM Rescue Module

RM Research Memorandum

RM Research Metallurgist

RM Reset Mode

RM Residual Magnetism

RM Resistance Melting

RM Resistance Monitor

RM Resource Management; Resource Manager [Computers]

RM Rest Mass [Relativity]

RM Reusable Module

RM Reverse Micelle

RM Right Mantle (of Oysters)

RM Right Mid

RM Risk Management; Risk Manager

RM Robotic Motion

RM Rocket Motor

RM Rock Mechanics [Geophysics]

RM Rocky Mountains [Western North America]

RM Rod Mill(ing)

RM Rollback Module

RM Rolling Mill

RM Rosemount

RM Rotary Motion

RM Rotating Machinery

RM Rotating Machinery [Institute of Electrical and Electronic Engineers–Power Engineering Society Committee]

RM Rotational Motion

RM Royal Mail [UK]

RM Royal Marines [UK]

RM Rural Municipality

RM-1 Radiation Meteoroid Satellite

R&M Redistribution and Marketing

R&M Reliability and Maintainability; Reliability and Maintainance

R&M Reports and Memoranda

R/M Reaction Milled [also RM]

R/M Read/Mostly Memory

R-M Reichert-Meissl (Number) [also RM] [Analytical Chemistry]

Rm Ream(ing) [also rm]

Rm Room [also rm]

rm ream [Unit of quantity for paper; usually equal to 500 sheets]

RMA Radio Manufacturers Association [now Electronic Industries Association, US]

RMA Random Magnetic Anisotropy

RMA Random Multiple Access

RMA Reactive Modulation Amplifier

RMA Refractory Metals Association [US]

RMA Remote Maintenance Access

RMA Remote Manipulator Arm

RMA Retread Manufacturers Association [UK]

RMA Return Materials Authorization

RMA Return to Manufacturer Authorization

RMA Rocky Mountain Arsenal [near Denver, Colorado, US]

RMA Rosin, Mildly-Activated [Soldering]

RMA Royal Military Academy [Sandhurst, UK]

RMA Rubber Manufacturers Association [US]

RMA# Return Materials Authorization Number

RMAAS Reactivity Monitoring and Alarm System

RMAC Remote Maintenance Access Control

RMAED Rocking Micro-Area Electron Diffraction [Materials Science]

R-MAD Reactor Maintenance, Assembly, and Disassembly [of US Atomic Energy Commission]

RMAG Rocky Mountain Association of Geologists [US]

R-MAP Regional Environmental Mapping and Assessment Program [of US Environmental Protection Agency]

RMARL Rocky Mountain Analytical Research Laboratories [Golden, Colorado, US]

RMArt Royal Marine Artillery [also RMArty]

RMAS Remote Memory Administration System

RMAS Remote Monitor and Alarm System

RMATS Remote Maintenance, Administration and Traffic System

RMAX Range Maximum [also rmax]

RMB Renminbi [Currency of PR China]

RMB Right Mouse Button

RMB ROM (Read-Only Memory) Memory Board

RMBE Reactive Molecular Beam Epitaxy

RMBI Risk Management and Business Insurance

RMC Radiative Muon Capture [Physics]

RMC Ready-Mixed Concrete

RMC Redundancy Management Control

RMC Reed-Muller Code

RMC Regional Media Center

RMC Regional Meteorological Center [of World Meteorological Organization, Switzerland]

RMC Reserve Mining Company [US]

RMC River Murray Commission [Replaced by Murray-Darling Basin Commission, Australia]

RMC Rocky Mountain College [Billings, Montana, US]

RMC Rod Memory Computer

RMC Rotating Modulation Collimator

RMC Royal Marine Corps [UK]

RMC Royal Military College [UK and Canada]

RM&C (Nuclear) Reactor Monitoring and Control

RMCAO Ready-Mixed Concrete Association of Ontario [Canada]

RMCC Royal Military College of Canada

RMCD Radiation Monitoring Container/Dosimeter

RMCDE Radar Message Conversion and Distribution Equipment

RMCGP River Murray Corridor of Green Program [Australia]

RMCL Recommended Maximum Contaminant Level

RMCMI Rocky Mountain Coal Mining Institute [US]

RMCS (Nuclear) Reactor Manual Control System

RMCS Remote Maintenance Control System

RMCS Royal Military College of Science [UK]

RMD Radiation Monitoring Device

RMD Radiation Monitoring Devices, Inc. [US]

RMD Reactor Materials Division [of Atomic Energy of Canada Limited, Chalk River Nuclear Laboratories, Ontario, Canada]

RMD Rebonded-Missing-Dimer [Chemistry]

RMD Remainder (Function) [Computers]

RMD Row of Missing Dimers [Chemistry]

RMDIR Remove Directory [also RmDir, or rmdir]

Rmdr Remainder [also rmdr]

RME Relay Mirror Experiment

RME Risk Mitigation Experiment

RMEA Rubber Manufacturing Employers Association

RMe$_3$(BO$_3$)$_4$ Rare-Earth Metal Borate [Formula]

RMEE Rotating Mercury Film Electrode

RMetS Royal Meteorological Society [also RMetSoc, UK]

RMF RCS (Reaction Control System) Module Forward

RMF Reactivity Measurement Facility [of Idaho National Engineering Laboratory, Idaho Falls, US]

RMF Research and Management Foundation [of American Consulting Engineers Council, US]

RMF Resource Measurement Facility

RMF Rubidium Manganese Fluoride

RMG Redhead Modulator Gauge

RMG Right Main Gear

RMgX Alkylmagnesium Halide (or Grignard Reagent) [R Represents an Organic Group, e.g. Alkyl, or Aryl and X Represents a Halogen]

RMgBr Alkylmagnesium Bromide [Formula] [R Represents an Organic Group, e.g. Alkyl, or Aryl]

Rmh Roentgen per hour at one meter [also rmh]

RMI Rack Manufacturers Institute [US]

RMI Radio Magnetic Indicator

RMI Reliability Maturity Index

RMI Remote Mechanical Investigator

RMI Remote Method Invocation

RMI Rocky Mountain Institute [US]

RMI Route Monitoring Information

RMIC Refractory Metal-Intermetallic Composite

RMIC Research Materials Information Center [of Oak Ridge National Laboratory, Tennessee, US]

RMICBM Road Mobile Intercontinental Ballistic Missile

RMID Risk Management Implementing Directive

RMIF Retail Motor Industry Federation [UK]

R Microsc Soc, Proc Royal Microscopic Society, Proceedings [UK]

RMIN Range Minimum [also rmin]

R/min Roentgen per Minute [also r/min]

RMIS Resource Management Information System

RMIT Royal Melbourne Institute of Technology [Australia]

RMK Remark [also Rmk, or rmk]

RML Radar Mapper, Long-Range

RML Radar Microwave Link

RML Radio Microwave Link

RML Relational Machine Language

RML Rifle, Medium-Light

RML Rocky Mountain Laboratory [of US Public Health Service at Hamilton, Montana]

RMM Radar Map Matching

RMM Random Mixture Model [Materials Science]

RMM Read Mostly Memory

RMM Refractory Metals and Materials

RMM Remanence Magnetometer

RMM Remanent Moment Magnetometer

RMM Remote Maintenance Monitoring

RMM Roentgen per Minute at One Meter [also rmm]

RMMC Refractory Metals and Materials Committee [of AIME Minerals, Metals and Materials Society, US]

RMMC Rocky Mountain Mapping Center [of US Geological Survey]

RMMC Rocky Mountain Mathematics Consortium [of Department of Mathematics, Arizona State University, Tempe, US]

RMMLF Rocky Mountain Mineral Law Foundation [US]

RM/MS&C Redundancy Management/Moding, Sequencing and Control

RMMU Removable Media Memory Unit

RMN Registed Mental Nurse [UK]

RMN Return Material Number

rmn remain

RMNPP Robert Moses Niagara Power Project [of New York Power Authority, US]

RMO Radio Material Officer

RMO Red Mercury (or Mercuric) Oxide

RMOC Regional Municipality of Ottawa-Carleton [Canada]

RMOJ Rocky Mountain Oil Journal [US]

RMON Remote Monitor(ing)

RMON Resident Monitor

RMOS Refractory Metal-Oxide Semiconductor

RMP Radon Measurement Proficiency

RMP Rapid Melt and Resolidification

RMP Reentry Measurement Program

RMP Remote Maintenance Processor

RMP Reviews of Modern Physics [Joint Publication of American Institute of Physics and American Physical Society, US]

RMP Risk Management Plan

RMP Royalty Management Program

RMPI Remote Memory Port Interface

RMPI Resonant Multiphoton Ionization

RMRB Reactor Materials Research Branch [of Atomic Energy of Canada Limited, Chalk River Nuclear Laboratories, Ontario, Canada]

RMRC Resolved Motion Rate Control

RMRS Repeatable Maintenance and Recall System

RMRS Remote Meter Resetting System

RMS Radar Mapper, Short-Range

RMS Radian Means per Second [Unit]

RMS Radiation Monitoring System

RMS Railway Mail Service

RMS Random Motion Simulator

RMS (Nuclear) Reactor Monitor System

RMS Record Management Service(s)

RMS Record Management System

RMS Recovery Management Support

RMS Red Mercury (or Mercuric) Sulfide

RMS Redundancy Management System

RMS Refractory Metal Silicide

RMS Regulatory Manpower System

RMS Reliability, Maintainability and Supportability (Committee) [of Society of Automotive Engineers, US]

RMS Remote Maintenance System

RMS Remote Manipulator Subsystem; Remote Manipulator System [of NASA]

RMS Resource Management Support

RMS Resource Management System

RMS Revised Management Scheme

RMS Rheometrics Mechanical Spectrometer

RMS Risk Management Society [US]

RMS Romanian Measurement Society

RMS Root Mean Square [also rms]

RMS Rotating Mass Storage

RMS Royal Mail Ship

RMS Royal Medical Society [UK]

RMS Royal Meteorological Society [UK]

RMS Royal Microscopical Society [UK]

RMSA Reference Mean Spherical Approximation [Physics]

RMSA Rescaled Mean Spherical Approximation [Physics]

RMSCC Rock Mechanics and Strata Control Committee [of Canadian Institute of Mining and Metallurgy]

RMSD Root Mean Square Difference

RMSE Root Mean Square Error [Statistics]

RMSF Rocky Mountain Spotted Fever

RMSIP Reactive Magnetron Sputter Ion Plating

RMSO Regional Master Standing Offer [Canada]

RMSR Root-Mean-Square Radius

RMSDEV Root-Mean-Square Deviation [Statistics]

R&MSG Robotics and Mechanical Systems Group [US]

RMSS Root-Mean-Square Strain

RMSVP Remote Manipulator Subsystem Verification Plan [of NASA]

RMT (National Union of) Rail, Maritime and Transport Workers [UK]

RMT Random Matrix Theory [Physics]

RMT Registered Massage Therapist

RMT Remote Terminal

RMT Reverse Martensitic Transformation [Metallurgy]

rmt remote

RMTB Reconfiguration Maximum Theoretical Bandwidth

rmte remote [also rmte]

RMTP Regional Marine Turtle (Management and Conservation) Program [Australia]

R-MTS Rare Earth Muffin-Tin Sphere [Materials Science]

RM/TST Random Matrix/Transition State Theory [Physics]

RMU Reload and Maintenance Unit

RMU Remote Maneuvering Unit

RMU Remote Multiplexer Unit

RMV Reentry Measurement Vehicle

RMV Remote Maintenance Vehicle

RMV Ryegrass Mosaic Virus

Rmv Remove [also rmv]

RMVSO Road and Motor Vehicle Safety Office [Canada]

RMW Radioactive Mixed Waste

RMW Read-Modify-Write

RMWAA Roadmasters and Maintenance of Way Association of America [US]

RMWS (Nuclear) Reactor Make-Up Water System

RMX Remote (Data) Multiplexer

RN Cylindrical Roller Bearing, Single-Row, Non-Locating, Outer Ring without Ribs, Double-Ribbed Inner Ring, Outer Ring Separable [Symbol]

RN Radar Navigation

RN Radio Navigation

RN Radionuclide

RN Random Network (Theory)

RN Random Noise

RN Random Number

RN Raney Nickel

RN Rating Number

RN Read News [Internet]

RN Real Number

RN Reception Node

RN Recipient Name

RN Recoil Nucleus

RN Record Number

RN Reference Noise

RN Release Notice

RN Registered Nurse

RN Registry Number

RN Removable Needle

RN Report Number

RN République du Niger [Republic of Niger]

RN Research Note

RN Resonance Neutralization

RN Retinal Neuron [Anatomy]

RN Reynolds Number [Fluid Dynamics]

RN Rio Negro [Argentina and Uruguay]

RN Round Nose

RN Royal Navy [UK]

R₃N R_3N Tertiary Amine [Formula]

Rn Range [also rn]

Rn Radon [Symbol]

Rn-204 Radon-204 [also ^{204}Rn, or Rn^{204}]

Rn-205 Radon-205 [also ^{205}Rn, or Rn^{205}]

Rn-206 Radon-206 [also ^{206}Rn, or Rn^{206}]

Rn-207 Radon-207 [also ^{207}Rn, or Rn^{207}]

Rn-208 Radon-208 [also ^{208}Rn, or Rn^{208}]

Rn-209 Radon-209 [also ^{209}Rn, or Rn^{209}]

Rn-210 Radon-210 [also ^{210}Rn, or Rn^{210}]

Rn-211 Radon-211 [also ^{211}Rn, or Rn^{211}]

Rn-212 Radon-212 [also ^{212}Rn, or Rn^{212}]

Rn-215 Radon-215 [also ^{215}Rn, or Rn^{215}]

Rn-216 Radon-216 [also ^{216}Rn, or Rn^{216}]

Rn-217 Radon-217 [also ^{217}Rn, or Rn^{217}]

Rn-218 Radon-218 [also ^{218}Rn, or Rn^{218}]

Rn-219 Radon-219 [also ^{219}Rn, Rn^{219}, or An]

Rn-220 Radon-220 [also ^{220}Rn, Rn^{220}, or Tn]

Rn-221 Radon-221 [also ^{221}Rn, or Rn^{221}]

Rn-222 Radon-222 [also ^{222}Rn, Rn^{222}, or Rn]

Rn-224 Radon-224 [also ^{224}Rn, or Rn^{224}]

RNA Radio Navigational Aids

RNA Red Nitric Acid

RNA Registered Nurses Association

RNA Ribonucleic Acid [Biochemistry]

RNA Royal Naval Association [UK]

RNAA Radioactive Neutron Activation Analysis

RNAA Royal Norfolk Agricultural Association [UK]

RNAAS Royal Netherlands Academy of Arts and Sciences

RNABC Registered Nurses Association of British Columbia [Canada]

RNAC Remote Network Access Controller

RNA-DNA Ribonucleic Acid/Deoxyribonucleic Acid [Biochemistry]

RNAM Regional Network for Agricultural Machinery [Philippines]

RNANS Registered Nurses Association of Nova Scotia [Canada]

RNAO Registered Nurses Association of Ontario [Canada]

RNAS Royal Northern Agricultural Society [UK]

RNase Ribonuclease [Biochemistry]

RNase I Ribonuclease I (or Ribonuclease A) [Biochemistry]

RNase II Ribonuclease II (or Ribonuclease T₂) [Biochemistry]

RNase A Ribonuclease A (or Ribonuclease I) [Biochemistry]

RNase B Ribonuclease B [Biochemistry]

RNase C Ribonuclease C [Biochemistry]

RNase H Ribonuclease H [Biochemistry]

RNase S Ribonuclease S [Protease-Modified RNase A] [Biochemistry]

RNase T₁ Ribonuclease T_1 (or Guanyloribonuclease) [Biochemistry]

RNase T₂ Ribonuclease T_2 (or Ribonuclease II) [Biochemistry]

RNase U₁ Ribonuclease U_1 [Biochemistry]

RNase U₂ Ribonuclease U_2 (or Puryloribonuclease) [Biochemistry]

RNAV Radial/Area Navigation [also Rnav]

RNAV Radio Navigation (System) [also R/NAV]

RNAV/RNPC Radial/Area Navigation/Required Navigation Performance Capability

RNB Radionuclide Battery

RNBWS Royal Navy Bird Watching Society [UK]

RNC Request Next Character

RNC Royal Naval College [Greenwich, UK]

RNC Royal Norwegian Council

RNCS Royal Netherlands Chemical Society

RNCSIR Royal Norwegian Council for Scientific and Industrial Research [of Norwegian Institute of Technology, Trondheim, Norway]

RND Random [also Rnd]

RND Random Number (Function) [Programming]

Rnd Round [also rnd]

RNase Ribonuclease [Biochemistry]

RNCP Rural Nature Conservation Program

RNDB Road Network Data Base [of Victoria State, Australia]

RNDZ Rendezvous [also Rndz]

RNE Register of the National Estate [of Australian Heritage Commission]

RNF Receiver Noise Figure

RNFP Radar Not Functioning Properly

RNG Radio Range

RNG Random Nucleation and Growth (of Crystallites)

RNG Random Number Generator

RNH₂ RNH_2 Primary Amine [Formula]

R₂NH R_2NH Secondary Amine [Formula]

Rng Range

Rng Rt Range Rate [Electronics]

RNI Radio Nordsee International [Radio North Sea International]

RNI Recommended Nutritional Intake

RNiO₃ $RNiO_3$ Rare-Earth Nickelate [Formula]

RNIT Radio Noise Interference Test

RNL Risø National Laboratory [Roskilde, Denmark]

RNLI Royal National Lifeboat Institution [UK]

RNMH Registered Nurse for the Mentally Handicapped [UK]

RNMP Replacement of the Nautical Mile Panel [of International Civil Aviation Organization]

RNO Reoxidized Nitrided Oxide [Metallurgy]

RNODC Responsible National Oceanographic Data Center

RNP Regional Network Provider

RNP Remote Network Processor

RNP Ribonucleoprotein (Particle) [Biochemistry]

RNPC Required Navigation Performance Capability

RNPDL Risley Nuclear Power Development Laboratories [of UK Atomic Energy Authority]

RNPL Royal Naval Physiological Laboratory [Alverstoke, UK]

RNPRC Royal Naval Personnel Research Committee [UK]

RNR Receive Not Ready

RNR Resonant Nuclear Reaction (Technique) [Physics]

RNR Royal Navy Reserve [UK]

RNRF Renewable Natural Resources Foundation [US]

RNS Cylindrical Roller Bearing, Single-Row, Non-Locating, Outer Ring without Ribs, Double-Ribbed Inner Ring, Outer Ring Separable, Spherical Outside Surface [Symbol]

RNS Residue Number System [Mathematics]

RNSA Royal Naval Sailing Association [UK]

RNSS Royal Naval Scientific Service [UK]

RNSS Royal Norwegian Scientific Society

RNSYS Royal Nova Scotia Yacht Squadron [Canada]

RNT Radionuclide Technique

RNT Relative Normal-Momentum Transfer

RNTB Russian Network Test Bed

RNTF Royal Navy Torpedo Factory [UK]

RNU Cylindrical Roller Bearing, Single-Row, Non-Locating, Outer Ring without Ribs, Inner Ring without Ribs, Both Rings Separable [Symbol]

RNV Radio Noise Voltage

RNVR Royal Navy Volunteer Reserve [UK]

RNVSR Royal Navy Volunteer Supplement Reserve [UK]

RNWC Royal Naval War College [UK]

RNZA Royal New Zealand Army

RNZArt Royal New Zealand Artillery [also RNZArty]

RNZAS Royal New Zealand Astronomical Society

RNZIH Royal New Zealand Institute of Horticulture

RNZN Royal New Zealand Navy

RO Alkoxy (Radical) [Symbol]

RO Omani Ryal (or Rial) [Currency of Oman]

RO Radar Operator

RO Radiation Oscillator

RO Radio-Opaque

RO Radio Operator

RO Radio-Oscillator

RO Random Order

RO Random Orientation [Crystallography]

RO Range Operations

RO Read Only [also R/O, or R-O]

RO Reader Only

RO Readout

RO Real Object

RO Receive Only [also R/O]

RO Record Zero

RO Recovery Operations

RO Reddish Orange

RO Red-Orange

RO Red-Out [also R-O] [Aviation Medicine]

RO Reference Oscillator

RO Regional Office [Data Communications]

RO Register Output

RO Relative (Record) Organization

RO Repair Order

RO Report Over

RO Research Observatory

RO Reserve Officer

RO Reverse Osmosis [Chemical Engineering]

RO Roll Off

RO Roll On

RO Roll-Over [also R-O] [Mining]

RO Romania [ISO Code]

RO Rough Opening [Building Trades]

RO Routine Order

RO Royal Observatory [UK]

RO Runoff

RO Runout

RO Ruthenium Oxide

RO# Repair Order Number

R/O Read Only [also R-O, or RO]

R/O Receive Only [also RO]

R/O Rollout

R-O Red-Out [also RO] [Aviation Medicine]

R-O Rhombic-to-Orthorhombic (Transformation) [also R/O] [Metallurgy]

R-O Roll-Over [also RO] [Mining]

R(O^{2-}) Atomic Radius of Oxygen [Symbol]

R(ω) Reflectance [Function]

.ro Romania [Country Code/Domain Name]

ROA Raman Optical Activity

ROA Receive on Account

ROA Regional Office for Africa [of United Nations Environmental Program]

ROA Research Opportunity Awards [of National Science Foundation, US]

ROA Reserve Officers Association [US]

ROA Return on Assets

R&OA Reliability and Quality Assurance

ROAD Australian Road Research Board Database

ROAD Re-Organization of Army Division [US]

ROADRES Road Research in Australia Database

ROAP Removable Overhead Access Platform

ROAP Reorganization, Office of the Assistant to the President [US]

ROAR Royal Optimizing Assembly Routine

ROAT Radio Operator's Aptitude Test

ROAUS Reserve Officers Association of the United States (of America)

ROB (Alkyl) Brosylate [Formula]

ROB Radar Order of Battle

ROB Right Outboard

ROB Remaining on Board

ROB Remote Order Buffer

ROB Reorder Buffer [of Intel Corporation, US]

ROB Research Operations Building [of Pacific Northwest National Laboratory, Richland, Washington, US]

ROB Resonatorless Optical Bistability

ROB Robotic Operating Buddy

ROBEX Regional OPMET (Operational Meteorological Information) Bulletin Exchange Scheme

ROBIC Robotic Integrated Cell

ROBIN Remote On-Line Business Information Network

ROBO Rocket Bomber

ROBOMB Robot Bomb [also Robomb, or robomb]

Robot Robotic(s) [also robot]

Robot Auton Syst Robotica and Autonomous Systems [Journal published in the Netherlands]

Robot Comput-Integr Manuf Robotics and Computer-Integrated Manufacturing [Published in the UK]

Robot Q Robotics Quarterly [Published by Robotics International of the Society of Manufacturing Engineers, US]

Robot Today Robotics Today [Published by Robotics International of the Society of Manufacturing Engineers, US]

Robot World Robotics World [Published in the US]

ROC Range Operations Conference

ROC Rapid Omnidirectional Compaction [Powder Metallurgy]

ROC Rapid Omnidirectional Consolidation [Powder Metallurgy]

ROC Rate of Change

ROC Rate of Climb

ROC Rate of Convergence

ROC Receiver Operating Characteristic

ROC Record of Comments

ROC Recurrent Ovarian Cancer [Medicine]

ROC Relative Operating Characteristic (Curve) [Statistics]

ROC Repair Order Complete

ROC Request of Change

ROC Required Operation Capability

ROC Relative Operating Characteristic [Statistics]

ROC Remote Operator Console

ROC Republic of China

ROC Required Operational Capacity

ROC Reserve Officer Candidate (Program) [of US Navy]

ROC Return on Capital [also roc]

ROC Reusable Orbital Carrier

ROC Ruthenium-on-Carbon

RoC Report on Carcinogens [of National Institute of Environmental Health Sciences, National Institutes of Health, Bethesda, Maryland, US]

roc reciprocal ohm centimeter [Unit]

ROCAP Regional Office for Central America and Panama

ROCAPPI Research on Computer Applications for the Printing and Publishing Industries

ROCC Regional Operation Control Center

ROCC Rensselaer Operations and Control Center [of Rensselaer Polytechnic Institute, Troy, New York State, US]

ROCE Return on Capital Employed

ROCEW Role of Clouds, Energy and Water (Program) [of National Science Foundation, US]

ROCI Rate-of-Climb Indicator

Rock Prod Rock Products [Journal published in the US]

ROCKET Rand's Omnibus Calculator of the Kinetics of Earth Trajectories

Rock Mech Rock Eng Rock Mechanics and Rock Engineering [Journal published in Germany]

ROCKOON Rocket-Balloon System [also Rockoon, or rockoon]

Rocky Mt J Math Rocky Mountain Journal of Mathematics [Published by the Rocky Mountain Mathematics Consortium, US]

ROCP Radar Out of Commission for Parts

ROCR Reimbursable Obligation and Cost Reporting System

ROCR Remote Optical Character Recognition

ROD Rate of Descent

ROD Recorder on Demand

ROD Record of Decision

ROD Release Order Directive

ROD Remote(ly) Operated Door

ROD Repair on Demand

ROD Required Operational Data

ROD Rochester Ordnance District [Now Part of New York Ordnance District, US]

ROD Rossford Ordnance Depot [Toledo, Ohio, US]

ROD Royal Ordnance Depot [UK]

RODATA Registered Organizations Databank

Rod, Wire Fastener Rod, Wire and Fastener [Journal published in the US]

ROE Reflector Orbital Equipment

ROE Return on Equity

Roentgen-Bl Prax Klin Roentgen-Blätter: Praxis und Klinik [German Publication on Medical Radiology]

ROESY Rotating Frame Overhauser Effect Spectroscopy

ROF Rate of Flow [Fluid Mechanics]

ROF Remote Operator Facility

ROF Royal Ordnance Factory [UK]

R of A Reduction of Area [Mechanical Testing]

ROFL Rolling on the Floor Laughing [also ROTFL] [Internet Jargon]

ROFOR Route Forecast

ROFT Radar Off-Target

ROG Radius of Gyration

ROG Reactive Organic Gas(es)

ROG Receipt of Goods

ROG Rise-off-Ground (Model Airplane)

ROG ROme Gravitational Wave (Experiment) [of Instituto Nazionale de Fisico Nucleare, Modena, Italy]

ROH Alcohol [Formula]

ROH Receiver Off-Hook [Telecommunications]

ROHF Restricted Open-Shell Hartree-Fock [Physics]

ROI Range Operations Instruction

ROI Region of Interest

ROI Report of Incidents

ROI Republic of Ireland

ROI Return on Investment [also RoI]

ROIC Return of Invested Capital

ROIP Residual Oil In-Place

ROIS Radio Operational Intercom System

ROJ Range on Jamming

ROK Potassium Alkoxide [Formula]

ROK Republic of Korea [South Korea]

ROK Run-of-Kiln (Quicklime)

ROL Romanian Leu [Currency]

ROL Rotate Left [Computers]

ROLAC Regional Office for Latin America and the Caribbean [of United Nations Environmental Program in Mexico]

ROLAP Relational On-Line Analytical Processing

ROLF Remotely Operated Longwall Face [Mining]

ROLR Receiving Objective Loudness Rating

ROLS Recoverable Orbital Launch System

ROLS Remote On-Line System

ROLS Request On-Line Status

ROM (Alkyl) Mesylate [Formula]

ROM Read-Only Memory [also rom]

ROM Readout Memory

ROM Reciprocal Ohm Meter [also rom]

ROM Remotely Operated

ROM Return on Market Value

ROM Rough Order of Magnitude

ROM Royal Ontario Museum [Toronto, Canada]

ROM Rule of Mixtures [Chemistry]

ROM Run-of-Mill (Ore) [Mining]

ROM Run-of-Mine (Ore) [Mining]

Rom Roman(ic); Romans; Romance

Rom Romania(n)

Rom Roman Type [also rom]

rom reciprocal ohm meter [also ROM]

ROMAC Robotic Muscle-Like Actuator; Robotic Muscle-Like Actuator Control [also RoMAc]

Roman Romania(n)

ROMBUS Reusable Orbital Module Booster and Utility Shuttle

ROM BIOS Read-Only Memory–Basic Input/Output Subsystem [also ROM-BIOS, or ROM/BIOS]

ROMCOM Romulus Communications Software [*Note:* "Romulus" is a CD-ROM System for Interlibrary Loans]

ROME Reason-Oriented Modeling Environment

ROM EPROM Read-Only Memory–Erasable Programmable Read-Only Memory [also ROM-EPROM, or ROM/EPROM]

ROMIO Read-Only Memory plus Input/Output

Rom J Chem Romanian Journal of Chemistry

ROMMIDS Remotely Operated Metallic Mine Detection System

ROMON Receive-Only Monitor

ROMOTAR Range-Only Measurement of Trajectory and Recording

ROMP Ring-Opening Metathesis Polymerization

ROM RPROM Read-Only Memory–Reprogrammable Read-Only Memory [also ROM-RPROM, or ROM/RPROM]

ROMV Return on Market Value

ROMWARE ROM (Read-Only Memory) Software [also romware]

RON Receive Only

RON Remain Overnight

RON Research Octane Number

RONa Sodium Alkoxide

RONS Read-Only Name Store

RONT Radar On Target

RONTP Receive-Only Nontyping Perforator

Roofg Roofing [also roofg]

Roof Cladding Insul Roofing, Cladding and Insulation [Published in the UK]

Roof/Siding/Insul Roofing/Siding/Insulation [Published in the UK]

RoofTech Roofing Technology Trade Show

ROOI Return on Original Investment

ROOM Real-Time Object-Oriented Modeling

ROOST Rapid Optical Ocean Surveillance Testbed

ROOST Reusable One-Stage Orbital Space Truck

ROOT Relaxation Oscillator Optically Tuned

ROP Raster Operation

ROP Rate of Penetration

ROP Read/Receive-Only Printer

ROP Record of Performance (Milk-Recording Program) [of Agriculture Canada]

ROP Record of Production

ROP Record of Purchase

ROP Recovery Operating Plan

ROP Ridgewood Ordnance Plant [Cincinnati, Ohio, US]

ROP Ring-Opening Polymerization

ROP RISC (Reduced Instruction Set Computer) Operation

ROP Riverbank Ordnance Plant [California, US]

ROP Rockford Ordnance Plant [Illinois, US]

ROP Run of Paper

Ropa Uhlie Ropa a Uhlie [Slovakian Publication]

ROPE Research on Orbital Plasma Electrodynamics

ROPE Ring of Prefetch Elements

ROPES Remote On-Line Print Executive System

ROPME Regional Organization for the Protection of the Marine Environment [Kuwait]

ROPP Receive-Only Page Printer

ROPS Range Operation Performance Summary

ROPS Roll-Over-Protective Structure [Mining]

ROPS-FOPS Roll-Over-Protective (Structure) and Falling-Object-Protective Structure [also ROPS/FOPS] [Mining]

ROR Ether [Formula]

ROR Range-Only Radar

ROR Rate of Return

ROR Rate of Rise [Leak Testing]

ROR Resonant Optical Reflection

ROR Ring-On-Ring (Loading) [Mechanical Testing]

ROR Rocket on Rotor

ROR Rotate Right [Computers]

ROR Run-of-Retort (Coke) [Metallurgy]

ROR Run-of-River (Hydrostation)

RORC Rolla Research Institute [of University of Missouri–Rolla, US]

RORD Return on Receipt of Document

RORO Roll-On/Roll-Off (Transport) [also RoRo]

ROS Raster Output Scanner

ROS Rat Osteosarcoma [Medicine]

ROS Reactive Oxygen Species

ROS Read-Only Storage

ROS Regulated Oxygen Supply

ROS Regulated Oxygen System

ROS Remote Operation System

ROS Removable Overhead Structure

ROS Renewable Oxygen Standard

ROS Rosman, North Carolina [Space Tracking and Data Network Site]

ROS Royal Observatory of Scotland [Edinburgh]

ROSAR Read-Only Storage Address Register

ROSAT Roentgen Satellite [of NASA]

ROSCOP Report of Observations Samples Collected by Oceanographic Programs [of National Oceanographic Data Center, US]

ROSDALE Representation of Structure Description Arranged Linearly

ROSDR Read-Only Storage Data Register

ROSE Remotely-Operated Special Equipment

ROSE Reporting On the State of the Environment

ROSE Residuum Oil Supercritical Extraction (Process) [Chemical Engineering]

ROSE Retrieval by On-Line Search

ROSIE Reconnaissance by Orbiting Ship Identification Equipment

ROSL Royal Overseas League [UK]

ROSO₃H Alkyl Hydrogen Sulfate [Formula]

RoSPA Royal Society for the Prevention of Accidents [also ROSPA, UK]

ROSPEC Rotating Spectrometer

ROSS Rewritable Optical Storage System

ROSTA Regional Office for Science and Technology in Africa [of UNESCO in Kenya]

ROSTENA Regional Office for Science and Technology for Europe and North America [of UNESCO]

ROSTSEA Regional Office for Science and Technology for Southeast Asia [of UNESCO]

ROT Radar On Target

ROT Rare, or Threatened (Species)

ROT Remaining Operating Time

ROT Reusable Orbital Transporter

ROT Running Object Table

ROT (Alkyl) Tosylate [Formula]

Rot Rotary; Rotate; Rotation [also rot]

rot rotation (of a vector) [Mathematics]

ROTAP Rare, or Threatened Australian Plants (Database) [of Environment Australia's Environmental Resources Information Network and Endangered Species Unit]

ROTC Reserve Officers Training Corps [of US Navy]

ROTCC Receiver Off-Hook Tone Connecting Circuit [Telecommunications]

ROTEX Robotic Extractor [of NASA]

ROTFL Rolling On The Floor Laughing [also ROFL] [Internet]

ROTH Read-Only Type Handler

RO-THP Alkyl Tetrahydropyranyl Ether [also ROThp] [Formula]

ROTI Recording Optical Tracking Instrument

ROTL Remote Office Test Lines

ROTO ROESY (Rotating Frame Overhauser Effect Spectroscopy)–TOCSY (Total Correlation Spectroscopy) [A Two-Dimensional Nuclear Magnetic Resonance Technique]

Roto Rotogravure [also roto]

ROTP Regular Officer Training Program

ROTR Receive-Only Tape Reperforator

ROTR Receive-Only Typing Reperforator

ROTR S/P Receive-Only Typing Reperforator, Series to Parallel

ROTS Rotary Out Trunk Switch

ROTs Rare, or Threatened Plants

ROTT Reorder Tone Trunks

ROU República Oriental de Uruguay [Oriental Republic of Uruguay]

ROUT Rollout [Computers]

ROV Remotely-Operated Vehicle

ROVD Remotely-Operated Volume Damper

ROVOS-C Remote Automated Color Video Observing System

ROW Radford Ordnance Works [Virginia, US]

ROW Resonant Optical Waveguide (Array)

ROW Right-of-Way [also RW] [Civil Engineering]

ROW Roll Welding

ROWAP Rotating Water Atomization Process [Powder Metallurgy]

ROWS Radar for Observing the Wave Spectrum

ROWS Robotic Winding System

Roy Royal(ty); Royalties [also roy]

Rozpr Electrotech Rozprawy Elektrotechniczne [Polish Journal on Electrical Engineering]

Rozpr Hydrotech Rozprawy Hydrotechniczne [Polish Journal on Hydraulic Engineering]

RP Cylindrical Roller Bearing, Single-Row, Two-Direction-Locating, Double-Ribbed Outer Ring with One Loose Rib, Double-Ribbed Inner Ring, Outer Ring Separable [Symbol]

RP Radiation Physicist; Radiation Physics

RP Radiation Pressure
RP Radiation Protection
RP Radiation Pyrometer
RP Radiological Protection
RP Ram Pump
RP Rapid Prototyping
RP Rayleigh Principle [Physics]
RP Reaction Principle [Mineralogy]
RP Reaction Product
RP Reactive Processing
RP Reactor Physics
RP Reactor Power
RP Reader-Printer
RP Reader Punch
RP Read Printer
RP Real Part (of a Complex Number)
RP Real Property
RP Rear Projection [Television]
RP Receive Processor
RP Reception Poor
RP Reciprocating Pump
RP Recoil Particle [Physics]
RP Recoil Proton
RP Recommended Practice
RP Record Playback
RP Record Processor
RP Recovery Phase
RP Recreation Park [Australia]
RP Red–Purple
RP Red Phosphorus
RP Reduced Propagation
RP Reference Point
RP Reference Publication
RP Reference Pulse
RP Refilling Point
RP Reflecting Power
RP Refractory Product
RP Refrigeration Plant; Refrigerating Plant
RP *(Regius Professor)* – Royal Professor
RP Reiki Practitioner [Medicine]
RP Reinforced Plastics
RP Relative Pressure
RP Remote Processor
RP Repair Period
RP Replacement Project
RP Reply Paid
RP Report Passing
RP República del Paraguay [Republic of Paraguay]
RP República del Perú [Republic of Peru]
RP República de Panamá [Republic of Panama]
RP República Portuguesa [Portuguese Republic]
RP Repurchase Agreement [US]
RP Research Paper

RP Research Physicist
RP Research Project
RP Reservoir Pressure [Geology]
RP Resolving Power
RP Restoration Priority
RP Retinitis Pigmentosa [Medicine]
RP Return Point [Data Processing]
RP Reverse(d) Phase [Chemistry]
RP Reverse Polarity
RP Reversible Pendulum
RP Rheological Properties
RP Rhône-Poulenc Group [French Chemical, Pharmaceutical and Medical Company]
RP Rio de la Plata [Argentina/Uruguay]
RP Rocket Projectile
RP Rocket Projector
RP Rocket Propellant; Rocket Propulsion
RP Rollback Process
RP Rolled Product
RP Rotary Press
RP Rotary Pump
RP Rotation Plane [Crystallography]
RP Rotor Process
RP Rough(ing) Pump
RP Round Pin
RP Rubber Processing
RP Rudder Pedals
RP Ruddlesden-Popper (Phase) [Metallurgy]
RP Rust Preventive
RP Rustproof(ing)
R&P Reserve and Process
R-P (Reverse Transformation of) R-Phase to Parent Phase [also R/P] [Metallurgy]
R/P Rocket Projector
Rp Repair [also rp]
Rp Repeater [also rp]
Rp Rupiah [Currency of Indonesia]
R-5-P Ribose-5-Phosphate [Biochemistry]
RPA Radar Performance Analyzer
RPA Radio Paging Association [UK]
RPA Random-Phase Approximation (Model) [Materials Science]
RPA Rapid Pressure Application
RPA Record and Playback Assembly
RPA Reentrant Process Allocator
RPA Registered Public Accountant
RPA Relative Percent Accuracy
RPA Remote Piloted Aircraft
RPA República Popular de Angola [People's Republic of Angola]
RPA Request for Procurement Action
RPA Request for Procurement Authorization
RPA Retarding Potential Analyzer

RPA Revue de Physique Appliquée [French Journal on Applied Physics]

RPA Robinia Pseudoacacia Agglutinin [Immunology]

rPA Recombinant Protein A [Biochemistry]

RPAA Regional Planning Association of America

RPAE Random Phase Approximation with Exchange [Materials Science]

RP-Al$_2$O$_3$ Rhône-Poulenc Aluminum Oxide [*Note:* Rhône-Poulenc is a French Chemical, Pharmaceutical and Medical Manufacturer]

RPAO Radium Plaque Adaptometer Operator

RPAR Rebuttable Presumption Against Registration

RPB Radiation Protection Bureau [of Health and Welfare Canada]

RPB Radiological Protection Board [UK]

RPB République Populaire du Benin [People's Republic of Benin]

RPBCO Rare-Earth Praseodymium Barium Copper Oxide (Superconductor)

RPC Radar Planning Chart

RPC Radial-Paper Chromatography

RPC Reactive Powder Concrete

RPC Recreational Pilot Certificate [US]

RPC Rectifier Photocell

RPC Reference Publications Committee [of National Association of Corrosion Engineers, US]

RPC Reflector Polycoat

RPC Registered Protective Circuit

RPC Regional Project Coordinator

RPC Remote Parameter Control

RPC Remote Position Control

RPC Remote Power Controller

RPC Remote Procedure Call (Protocol) [Internet]

RPC Remote Processor Controller

RPC République Populaire du Congo [People's Republic of Congo]

RPC Research and Productivity Council [New Brunswick, Canada]

RPC Reversed-Phase Chromatography

RPC Rotary Phase Converter [Electrical Engineering]

RPC Rotary Powder Compaction [Powder Metallurgy]

RPC Row Parity Check

RPCI Roswell Park Cancer Institute [Buffalo, US]

RP/CI Reinforced Plastics/Composites Institute [formerly also RPCI; now Composites Institute] [of Society of the Plastics Industry, US]

RPCRS (Nuclear) Reactor Protection Control Rod System

RPCS (Nuclear) Reactor Plant Control System

RPCVD Remote Plasma-Enhanced Chemical Vapor Deposition

RPD Radar Planning Device

RPD Radiation Protection Dosimetry

RPD Random Pulse Discrimination

RPD RAMP (Rapid Modeling Platform) Piping Design

RPD Reactor Plant Designer

RPD Reference Power-Train Design

RPD Relative Percent Difference

RPD Relativistic-Particle Dynamics

RPD Removable Partial Denture

RPD Residence-Time Probability Density [also rpd]

RPD Resistance Against Plastic Deformation [Mechanics]

RPD Retarding Potential Difference [Physics]

RPD Rocket Propulsion Department

RPD Rupture Probability Density [Mechanics]

rpd rapid

RPDH Reserve-Shutdown, Planned Derated Hours

RPE Radial Probable Error

RPE Rapid Post-Editing

RPE Registered Professional Engineer

RPE Reliability Project Engineer

RPE Remote Peripheral Equipment

RPE Remote Plasma-Enhanced (Vapor Deposition)

RPE Required Page-End (Character)

RPE Resource Planning and Evaluation

RPE Retinal Pigment Epithelium Cell [Anatomy]

RPE Rocket Propulsion Establishment [UK]

RPE Rotating Platinum Electrode

RPECVD Remote Plasma-Enhanced Chemical Vapor Deposition [also RPE-CVD]

RPED Rocking Probe Electron Diffraction

RPEMOCVD Remote Plasma-Enhanced Metal-Organic Chemical Vapor Deposition [also RPE-MOCVD]

RPES Resonant Photoemission Spectroscopy

R-PES Reactive Polyethersulfone

RPF Radiometer Performance Factor

RPF Registered Professional Forester

RPF Rigid Plastic Foam

RPF Rotational Partition Function [Statistical Mechanics]

RPFC Reversed Phase Flash Chromatography

RPFS Radio Position Fixing System

RPG Radiation Protection Guide [of US Environmental Protection Agency]

RPG Random Pulse Generator

RPG Regional Planning Group

RPG Remote Processing Gateway

RPG Report Program Generator [High-Level Programming Language]

RPG Research Project Grant

RPG Rocket-Propelled Grenade

RPG Rounds per Gun [also rpg]

rPG Recombinant Protein G [Biochemistry]

RPGPD Rounds per Gun per Diem [also rpgpd]

RPGPM Rounds per Gun per Minute [also rpgpm]

RPH Recommended Power Handling

RPH Reduced Plate Height

RPH Relative Pulse Height

RPH Remotely-Piloted Helicopter

RPH Revolutions per Hour [also rph]

RPh Registered Pharmacist

RPI Radar Precipitation Integrator

RPI Radio Press International

RPI Railway Progress Institute [US]

RPI Read Punch and Interpret

RPI Régie Publicité Industrielle [French Industrial Publicity Corporation]

RPI Relative Performance Index

RPI Rensselaer Polytechnic Institute [Troy, New York, US]

RPI Resource Policy Institute [US]

RPI Revolution per Inch (of Penetration) [also rpi]

RPI Richmond Professional Institute [Virginia, US]

RPI (Nuclear) Rod Position Indicator [Nuclear Reactors]

RPI Rose Polytechnic Institute [Terre Haute, Indiana, US]

RPI Rubber and Plastics Industry

RPI Ryerson Polytechnic Institute [Toronto, Canada]

R&PI Rubber and Plastics Industry [Institute of Electrical and Electronic Engineers–Industry Applications Society Committee]

rpi revolution per inch (of penetration) [also RPI]

RPIE Real Property Installed Equipment

RPIF Regional Planetary Image Facility [of NASA Planetary Data System]

RPIPP Reverse Phase Ion Pair Chromatography

RPIS Real Property Inventory System

RPIS Rod Position Indicator System [Nuclear Reactors]

RPL Radar Processing Language

RPL Radiation Physics Laboratory [US]

RPL Radio Photoluminescence

RPL Radiophysics Laboratory [of Commonwealth Scientific and Industrial Research Organization, Australia]

RPL Rated Power Level

RPL (Standing Working Group on) Regional Plans [of International Civil Aviation Organization]

RPL Remote Program Loader

RPL Requested Privilege Level

RPL Request Parameter List

RPL Research Programming Language

RPL Resident Programming Language

RPL Reverse Polish Logic

RPL Robot Programming Language

RPL Rocket Propulsion Laboratory [of US Air Force]

RPL Running Program Language

rplc replace

RPLS Reactor Protection Logic System

RPM Random Phase Model [Materials Science]

RPM Rapid Prototyping and Manufacturing

RPM Raster Processing Machine

RPM Rate per Minute

RPM Reinforced Plastic Mortar

RPM Relaxation Potential Model [Physics]

RPM Reliability Performance Measure

RPM Resale Price Maintenance

RPM Restricted Primitive Model

RPM Resupply Provisions Module

RPM Return-Point Memory (Effect)

RPM Rollcast Planetary Mill [Metallurgy]

RPM Rotary Piston Motor

RPM Rounds per Minute [also rpm]

R&PM Research and Program Management

rpm revolutions per minute [also RPM, rev/min, or rev min^{-1}]

RPMC Remote Performance Monitoring and Control

RPMI Revolutions-per-Minute Indicator

RPMOCVD Reduced-Pressure Metallorganic Chemical Vapor Deposition [also RP-MOICVD]

RPN Real Page Number

RPN Recherche en Précision Numérique [Numeric Precision Research; of Atmospheric Environment Service, Canada]

RPN Registered Practical Nurse

RPN Reverse Polish Notation

RPN Risk Proprietary Number

RPNA Reverse Pacific/North American

RPO Radiation Protection Officer

RPO Radiation Protection Ordinance

RPO Research Program Office [of Canada Center for Mineral and Energy Technology]

RPO Revolutions per Orbit

RPOA Recognized Private Operating Agency

RPOC Remote Payload Operations Center

RP1D Repeated Unidirectional

RPP Radar Power Programmer

RPP Radarsat Precision Processor

RPP Radar Processing Plant

RPP Reductive Pentose-Phosphate (Cycle) [Biochemistry]

RPP Reinforced Polypropylene

RPP Reinforced Pyrolytic Plastic(s)

RPP Reinforced Pyrolyzed Plastic(s)

RPP Reversible-Pitch Propeller

RPP Review of Particle Physics [Published in the European Physical Journal C]

RPP Rotary Piston Pump

RPP Rural Policy Plan [Australian Capital Territory]

RP^{2}D Rational Product and Process Design

RPPG Radar Planning and Policy Group [of National Air Traffic Services, UK]

RPPICS Relative Partial Photoionization Cross-Section

RPPROM Reprogrammable Programmable Read-Only Memory

RPQ Request for Price Quotation

RPR Ram Pressure Ratio [Aerospace]

RPR Rapid Plasma Reagin (Test) [Medicine]

RPR Rapid Power Reduction

RPR Rayonet Photochemical Reactor

RPR Read Printer

RPR Rectangular Parallelepiped Resonance (Method) [Physics]

RPR Relative Performance Rating

RPR Replacement Production Reactor

RPRA Railroad Public Relations Association [US]

RPRA Rubber and Plastics Research Association [UK]

RPRC Regional Primate Research Centers [US]

RPRINTER (NetWare) Remote Printer

RPROM Reprogrammable Read-Only Memory

Rprt Report [also rprt]

Rprt Reprint [also rprt]

RPRV Remotely-Piloted Research Vehicle

RPRWP (Nuclear) Reactor Plant River Water Pump

RPS Radar Plotting Sheet

RPS Radar Position Symbol

RPS Random Program Selection

RPS Random Pulse Sequence

RPS Reactive Polystyrene

RPS Reactive Powder Sintering [Metallurgy]

RPS (Nuclear) Reactor Protection System

RPS Real-Time Photogrammetry System

RPS Real-Time Processing System

RPS Real-Time Programming System

RPS Record (and) Playback Subsystem

RPS Record (and) Playback System

RPS Records per Sector

RPS Reduced Pressure Spraying

RPS Refrigerator-Cooled Pump System

RPS Regional Pressure Setting

RPS Regulatory Performance Summary

RPS Relative Performance Score

RPS Remote Printing System

RPS Remote Processing Service

RPS Renal Portal System [Zoology]

RPS Research and Promotion of Structures

RPS Reverse Polarization Transfer

RPS Riser Pouring System [Metallurgy]

RPS Robot Programming System

RPS Rotational Position Sensing

RPS Royal Philatelic Society [UK]

RPS Royal Photographic Society [UK]

rps revolutions per second [also RPS, rev/s, rev/sec, or rev sec^{-1}]

RPSA Rapid Pressure Swing Adsorption (Process)

RPSF Rotation, Processing and Surge Facility

RPSGB Royal Photographic Society of Great Britain

RPSI Railway Preservation Society of Ireland [UK]

RPSM Resources Planning and Scheduling Method

RPSMA Rope Paper Sack Manufacturers Association [now Paper Shipping Sack Manufacturers Association, US]

RPSMG Reactor Protective System Motor Generator

RPSP Reference Preparation for Serum Proteins [of College of American Pathologists, US]

RPT Radiation Protection Technician; Radiation Protection Technologist; Radiation Protection Technology

RPT Rapid Projectile Thory

RPT Recirculation Pump Trip

RPT Registered Physical Therapist; Registered Physiotherapist

RPT Regular Public Transport

RPT Repeat (Character)

RPT Resident Provisioning Team

RPT Reverse Polarization Transfer

RPT Royal Polytechnic Institute [UK]

RPT Rudder Pedal Transducer

Rpt Repeat [also rpt]

Rpt Report [also rpt]

RPTA Rudder Pedal Transducer Assembly

RPtAl Rare-Earth Aluminide

RPTO Radarsat Program Technical Office [Canada]

RPU Radio Phone Unit

RPU Radio Propagation Unit

RPU Real-Time Photogrammetry Unit

RPU Regional Processing Unit

RPU Ryerson Polytechnical University [Toronto, Canada]

RP1D Repeated Unidirectional

R Pulm Art Right Pulmonary Artery [also r pulm art] [Anatomy]

R Pulm Vein Right Pulmonary Vein [also r pulm vein] [Anatomy]

RPV Rabbit Papilloma Virus

RPV (Nuclear) Reactor Pressure Vessel

RPV Remote(ly) Piloted Vehicle

RPV Remote Pilotless Vehicle

RPVC Reinforced Polyvinyl Chloride

RPW Resistance Projection Welding

RPWS Radio & Plasma Wave Subsystem [of NASA]

RQ Rapidly Quenched; Rapid Quenching [Metallurgy]

RQ International Conference on Rapidly Quenched and Metastable Materials

RQ Rayleigh Quotient [Physics]

RQ Reheat Quenching [Metallurgy]

RQ Reportable Quantity

RQ Request [also Rq, or rq]

RQ Respiratory Quotient [Medicine]

R/Q Resolver/Quantizer

R(Q) Wave Vector Dependent Reflectivity [Symbol]

R$_F$(Q) Wave Vector Dependent Fresnel Reflectivity [Symbol]

RQA Recursive Queue Analyzer

R&QA Reliability and Quality Assurance

RQD Rock Quality Designation

rqd required [also rq'd]

RQE Relative Quantum Efficiency

RQHNC Reference Quadratic Hypernetted Chain (Theory) [Physics]

RQI Request for Initialization

RQI Rice Quantum Institute [of Rice Institute, Houston, Texas, US]

RQL Reference Quality Level

RQL Rejectable Quality Level

RQM Relativistic Quantum Mechanics

Rqmnt Requirement [also rqmnt]

RQMS Regimental Quartermaster Sergeant

Rqmt Requirement [also rqmt]

Rqmts Requirements [also rqmts]

RQN Radial Quantum Number

Rqnts Requirements [also Reqnts]

R(Q,ω) Final-State Resolution Function [Symbol]

RQP Request Flight Plan

RQBE Relational Query By Example

rqrd required [also rqr'd]

RQS Rate Quoting System

RQS Request Supplementary Flight Plan

RQT Relativistic Quantum Theory

Rqumt Requirement [also rqumt]

RR Cylindrical Roller Bearing, Single-Row, One-Direction-Locating, Outer Ring With Two Internal Snap Rings, Single-Ribbed Inner Ring, Inner Ring Separable [Symbol]

RR Radiation Research

RR Radio Radiation

RR Radio Range

RR Radio Regulation

RR Radius Ratio [Physical Chemistry]

RR Railroad

RR Ram Recovery [Aeronautics]

RR Range Rate [Electronics]

RR Rapid Rectilinear (Lens)

RR Raschig Ring [Chemical Engineering]

RR Rayleigh-Ritz (Method) [Mathematics]

RR Reaction Rate

RR (Nuclear) Reactor Room

RR Readout and Relay

RR Real Rational [also rr]

RR Real Reality

RR Real Root

RR Receive Ready

RR Receiving Report

RR Recoilless Rifle

RR Recovery Room [Medicine]

RR Recurrence Rate [Communications]

RR Red Record

RR Red River [Southwestern US]

RR Reformatsky Reaction [Organic Chemistry]

RR Regioselective Reaction [Organic Chemistry]

RR Register-to-Register (Operation) [Computers]

RR Regulating Rod [Nuclear Reactors]

RR Relative Reactivity

RR Relative Repression

RR Relay Rack

RR Release Report

RR Remote Relay

RR Rendezvous Radar [Aerospace]

RR Repeatable Runout

RR Repetition Rate [Communications]

RR Replanning Request

RR Report Reaching

RR République Rwandaise [Republic of Rwanda]

RR Requirements Review

RR Research Reactor

RR Research Report

RR Residual Resistance [Solid-State Physics]

RR Resonance Raman (Spectroscopy)

RR Resource Recovery

RR Respiration Rate [Medicine]

RR Retroreflection; Retroreflector [Physics]

RR Return Rate

RR Return Register

RR Retro-Rocket

RR Reversing Relay

RR Revised Report

RR Rigid Rotator (Approximation) [Physics]

RR River Regulation

RR Rocket Research

RR Roll Roofing

RR Rolls Royce

RR Rolls Royce, Inc. [UK]

RR Rosin-Rammler (Distribution) [Liquid Particle Statistics]

RR Rotate Right [Computers]

RR Round Robin

RR Routh's Rule [Mathematics]

RR Running Reverse

RR Run-of-River (Hydropower System)

RR Rural Road

RR Rural Route

RR Ruthenium Red

R_0R_e Uniaxial Reflectance (Value) [Symbol]

R&R Remove and Replace

R&R Rendezvous and Recovery

R&R Rest and Recreation; Rest and Recuperation [also R and R]

R/R Abrasive Belt with Resin-Type Making and Sizing Bond Coats [Symbol]

R/R Readout and Relay

R/R Record/Retransmit

R-R Rare Earth-Rare Earth (Interaction) [Chemistry]

Rr Antenna Radiation Resistance [Symbol]

R(r) Atomic Displacement Field [Symbol]

$R_{nl}(r)$ Radial Function [Symbol]

r/R Radius Ratio [Radius of Smaller Ion to Radius of Larger Ion]

r/R Ratio of Inside Radius to Outside Radius

rr real rational [also RR]

$\rho(r)$ (unperturbed) charge density [Symbol]

$\rho(r)$ electron density [Symbol]

$\rho_e(r)$ electron density function [Symbol]

$\rho_f(r)$ scattering length density [Symbol]

ρ/ρ_0 macroscopic density [Symbol]

ρ/ρ_0 resistivity to bulk resistivity (ratio) [Symbol]

RRA Radar Recording and Analysis

RRA Red River Arsenal [Texarkana, Texas, US]

RRA Registered Records Administrator [Medicine]

RRA Remote Record Address

RRA Resident Research Associate [of NASA]

RRA Restrictive Requirement Quality A

RRA Retrogression and Reaging (Treatment) [Metallurgy]

RRA Road Roller Association [UK]

RRA Rubber Research Association [Israel]

RR-Acid 3-Amino-5-Hydroxy 2,7-Naphthalenedisulfonic Acid; 2-Amino-8-Naphthol-3,6-Disulfonic Acid [also R-acid, 2R-Acid, or 2R-acid]

RRAE Radar Recording and Analysis Equipment

RRAG Renewable Resources Assessment Group

RRAR ROM (Read-Only Memory) Return Address Register

RRB Railroad Retirement Board [US]

RRB Reinforced Rim and Bottom (Crucible) [Metallurgy]

RRB Restrictive Requirement Quality B

RRC Remodeling and Rehabilitation Committee [now National Remodelers Council; of National Association of Home Builders, US]

RRC Richland Research Complex [of Pacific Northwest National Laboratory, Washington, US]

RRC Rollin' Rock Club [US]

RRC Rolls Royce Canada [Lachine, Quebec]

RRC Rolls Royce Club [UK]

RRC Rotate Right through Carry (Function) [Computers]

R&RC Reactors and Reactor Controls [Institute of Electrical and Electronic Engineers–Nuclear and Plasma Sciences Society Committee]

RRCC Redwood Region Conservation Council [US]

RRD Reactor Radiation Division [of National Institute of Standards and Technology, Gaithersburg, Maryland, US]

RRD Reactor Research and Development

RRD Requisition Received Date

RRD Restricted Rotational Diffusion (Theory) [Physics]

RRD Ring-Rate Decay

RRDE Road-Runner Digital Edition

RRDE Rotating Ring Disk Electrode

RRDS Relative-Record Data Set

RRE Radar Research Establishment [UK]

RRE Receive Reference Equivalent

RRE Resonant Raman Effect [Atomic Physics]

RRE Royal Radar Establishment [UK]

R&RE Radiation and Repair Engineering

RREA Rendezvous Radar Electronics Assembly

RREAC Royal Radar Establishment Automatic Computer

RREG Regional Rheoencephalography [Medicine]

RREL Risk Reduction Engineering Laboratory

RREP Resonance Raman Excitation Profile

RREU Rendezvous Radar Electronics Unit

RRF Rapid Reaction Force [of United Nations]

RRF Reiner-Rivlin Fluid

RRF Relative Response Factor [Electronics]

RRF Research Reactor Facility [of University of Missouri-Columbia, US]

RRF Reservoir Recovery Forum [US]

RRF Resonant Ring Filter

RRF Retail Research Foundation [Canada]

RRF Risk Reduction Factor

RRG Relay Rack Ground

RRG Resource Request Generator

RRGM Recursive Residue Generation Method

RRGTE Rolls Royce Gas Turbine Engines [Dorval, Quebec, Canada]

RRHO Rigid Rotator Harmonic Oscillator (Approximation) [Physics]

RRI Raman Research Institute [Bangalore, India]

RRI Range-Rate Indicator [Electronics]

RRI Rendezvous Radar Indicator

RRI Re-Route Inhibit

RRI Research Reactor Institute [of Kyoto University, Kumatori-Osaka, Japan]

RRI Rocket Research Institute [US]

RRI Rowett Research Institute [Scotland]

RRI Rubber Research Institute [Malaysia]

RRIC Rubber Research Institute of Ceylon [now Rubber Research Institute of Sri Lanka]

RRICI Rolls Royce Industries Canada Inc. [Lachine, Quebec]

RRIM Reinforced Reaction Injection Molding

RRIM Rubber Research Institute of Malaysia

RRIN Rubber Research Institute of Nigeria

RRIS Railroad Research Information Service

RRIS Record Room Interrogation System

RRIS Remote Radar Integration Station

RRISL Rubber Research Institute of Sri Lanka

RRISL Bull Rubber Research Institute of Sri Lanka Bulletin

RRIWB River Research Institute, West Bengal [India]

RRKM Reaction Rate Kinetic Molecular Theory [Physical Chemistry]

RRKM Rice-Ramsperger-Kassel-Marcus (Theory of Mononulecular Gas-Phase Reaction)

RRL Radio Relay Link

RRL Radio Research Laboratory [now Communications Research Laboratory, Japan]

RRL Reaction Rate Limited (Kinetics) [Physical Chemistry]

RRL Regional Research Laboratory [India]

RRL Registered Records Librarian [Medicine]

RRL Reynolds Research Laboratory [Tamaqua, Pennsylvania, US]

RRL Road Research Laboratory [UK]

RRL Rudder Reference Line

RRLC Redwood Region Logging Conference [US]

RR&M Risk, Reliability and Maintenance [also RRM]

RRMC Royal Roads Military College [Canada]

RRMG (Nuclear) Reactor Recirculation Motor Generator

RRN Relative-Record Number

RRn Rotate Right n (Function) [Computers]

rRNA Ribosomal Ribonucleic Acid [also R-RNA or r-RNA] [Biochemistry]

RRNS Redundant Residue Number System

RRO Responsible Reporting Office

RROS Resistive Read-Only Storage

RRP (Nuclear) Reactor Refueling Plug

RRP Reader and Reader-Printer

RRP Recommended Retail Price

RRP Relative Resolving Power [Crystallography]

RRP Rotary Reciprocating Pump

RRP Rudder Reference Plane

RRP Runway Reference Point

RRPA Relativistic Random-Phase Approximation [Physics]

RRPI Relative Rod Position Indicator [Nuclear Reactors]

RRPI Resident Required Page Index

RRPI Rotary Relative Position Indicator

RRPT Receiving Report

RRPTN Receiving Report Number

RRR Raleigh Research Reactor [of North Carolina State College, Raleigh, US]

RRR Random-Random-Random (Threefold Grain Boundary Edge) [Materials Science]

RRR Reciprocal Residual Resistance (Ratio) [Solid-State Physics]

RRR Reclaim, Reuse, Reduce [also R3, 3R or R^3]

RRR Reduce, Reuse, Recycle [also R3, 3R or R^3]

RRR Residual Resistance Ratio [Solid-State Physics]

RRR Residual Resistivity Ratio [Solid-State Physics]

RRR Review of Local Government Residential Regulations

RRR Romeo Romeo Romeo [Radiotelephony]

$R_1R_2R_3$ Biaxial Reflectance (Value) [Symbol]

RRRR Reduction, Reuse, Recycling and Recovery [also 4R]

RRRS Route Relief Requirements System

RRRV Rate of Rise of Recovery Voltage

RRRV Rate of Rise of Restriking Voltage [also rrrv]

RRS Radiation Research Society [US]

RRS Radio Range Station

RRS Radio Ranging System

RRS Radio Relay Station

RRS Radio Research Station [UK]

RRS Random-Random-Special (Threefold Grain Boundary Edge) [Materials Science]

RRS Reaction Research Society [Glendale, California, US]

RRS (Nuclear) Reactor Recirculation System

RRS (Nuclear) Reactor Refueling System

RRS (Nuclear) Reactor Regulation System

RRS Required Response Spectrum

RRS Resonance Raman Scattering (Spectrometry)

RRS Resonant Raman Scattering

RRS Restraint Release System

RRS Retransmission Request Signal

RRS Resonance Raman Scattering Spectrometry

RRS Retrograde Rocket System

RRS Rule of Reversed Stability

RRSB Rosin-Rammler-Sperling-Benett (Distribution) [Liquid Particle Statistics]

(RR′SiNR″)ᵧ Polysilazane [General Formula; R, R′, and R″ Represent Hydrogen, or Alkyls]

RRSV Rice Ragged Stunt Virus

RRT Rail Rapid Transit

RRT Regional Response Teams

RRT Registered Respiratory Therapist

RRT Relative Retention Time

RRT Rendezvous Radar Transponder

RRT Resazurin Reduction Test [Food Technology]

RRT Ring-Ring Trip

RRT Round Robin Test

RRTC Retractable Replaceable Thermocouple

RRU Radiobiological Research Unit

RRV Red River Valley [North Dakota, US]

RRV Remote Reconnaissance Vehicle

RRX Railroad Crossing

RRZ Radar Regulation Zone

RRZ Restricted Release Zone

RS Alkylthio (Radical) [Symbol]

RS Cylindrical Roller Bearing, Single-Row, One-Direction-Locating, Outer Ring With One Rib and One Internal Snap Ring, Single-Ribbed Inner Ring, Inner Ring Separable [Symbol]

RS Radar Scan(ning)

RS Radially Symmetric; Radial Symmetry

RS Radiated Susceptibility

RS Radiation Source

RS Radioisotope Scanning

RS Radio Science

RS Radioscintillation

RS Radiosensitive; Radiosensitivity

RS Radio Shack Corporation [Fort Worth, Texas, US]

RS Radio Simulator

RS Radiosonde

RS Radio Sounder; Radio Sounding

RS Radio Source [Astrophysics]

RS Radio Spectroscope

RS Radio Spectrum

RS Radio Star [Astronomy]

RS Radio-Suisse [Swiss National Radio Station]

RS Rail Steel

RS Railway Signal

RS Railway System

RS Raman Scattering [Physics]

RS Raman Shift [Physics]

RS Raman Spectrometer; Raman Spectrometry

RS Raman Spectroscopy

RS Raman Spectrum

RS Random Sample; Random Sampling

RS Random Splice

RS Ranger Spacecraft

RS Range Safety (System) [NASA Apollo Program]

RS Range Selection; Range Selector [Electronics]

RS Range Sensing; Range Sensor [Robotics]

RS Range Surveillance

RS Rapid Setting

RS Rapidly Solidified; Rapid Solidification [also R/S] [Powder Metallurgy]

RS Rat Serum (Protein) [Biochemistry]

RS Raw Stock [Paper]

RS Ray Society [UK]

RS Raster Scan(ning)

RS Rawinsonde [Meteorology]

RS Rayleigh Scattering [Physics]

RS Reaction Sintered; Reaction Sintering [Metallurgy]

RS Reactive Sputtering

RS Reader Stop [Data Communications]

RS Reading Station

RS Read-Shockley (Model) [Materials Science]

RS Real Storage

RS Recoil Spectrum

RS Recommended Standard [of Electronic Industries Association, US]

RS Recording Secretary

RS Record Separator (Character) [Data Communications]

RS Red Sea

RS Redundancy Status

RS Redundant Set

RS Redwood Seconds [Viscosity]

RS Reference Section [Geology]

RS Reference Standard

RS Reference System

RS Reed-Solomon [Computers]

RS Reference Surface

RS Refinery Sludge

RS Reflectance Spectrophotometer; Reflectance Spectrophotometry

RS Reflection Spectrometer

RS Reflection Spectrum

RS Refrigerating System

RS Refurbishment Spare

RS Register Select

RS Register-to-Storage (Operation) [Computers]

RS Registry System

RS Relative Softening

RS Relative Stability

RS Relaxation Spectrum

RS Relay Switch

RS Reliability Society [of Institute of Electrical and Electronics Engineers, US]

RS Remote Sensing

RS Remote Site

RS Remote Station

RS Republic of Singapore

RS République du Sénégal [Republic of Senegal]

RS Request to Send

RS Research Satellite

RS Research Scientist

RS Reset (Key)

RS Reset/Set [also R-S]

RS Residual Stress [Mechanics]

RS Resistance Soldering

RS Resonance Scattering [Nuclear Physics]

RS Resonance Spectrum

RS Respiratory Syncytial (Virus)

RS Respiratory System [Anatomy]

RS Restrike Starter

RS Reversed Stability

RS Revolved Section [Design Engineering]

RS Rey and Saada (Equation) [Materials Science]

RS Richter Scale [Geophysics]

RS Riemann Surface [Mathematics]

RS Right Side

RS Rigid Styrene

RS Rio Grande do Sul [Brazilian State]

RS Robot(ic) System

RS Rochelle Salt

RS Rocket Science; Rocket Scientist

RS Rocket Specialist

RS Rolled Shape

RS Rolled Steel

RS Rolling Stock

RS Rough Surface

RS Route Switching

RS Royal Society [UK]

RS Rupture Stress [Mechanics]

RS Russell-Saunders (Coupling) [Physics]

R₃S⁺ R_3S^+ Sulfonium Ion [Symbol]

RS/6000 RISC (Reduced-Instruction-Set Computer) System/6000 [of IBM Corporation, US]

RS-232 Recommended Standard 232 (Communication Interface) [of Electronic Industries Association, US]

RS-232C Recommended Standard 232C (Communication Interface) [of Electronic Industries Association, US]

RS-366 Recommended Standard 366 (Communication Interface) [of Electronic Industries Association, US]

RS-422 Recommended Standard 422 (Communication Interface) [of Electronic Industries Association, US]

RS-422A Recommended Standard 422A (Communication Interface) [of Electronic Industries Association, US]

RS-423 Recommended Standard 423 (Communication Interface) [of Electronic Industries Association, US]

RS-423A Recommended Standard 423A (Communication Interface) [of Electronic Industries Association, US]

RS-449 Recommended Standard 449 (Communication Interface) [of Electronic Industries Association, US]

RS-485 Recommended Standard 485 (Communication Interface) [of Electronic Industries Association, US]

R&S Research and Statistics

R&S Reset and Start

R&S Rohde and Schwarz (Connector) [Electronics]

R&S Rouse and Shearer (Plastometer)

R/S Range Safety

R/S Rapidly Solidified; Rapid Solidification [also RS] [Powder Metallurgy]

R/S Redundant Set

R/S Relay Set

R-S Register-to-Storage (Operation) [Computers]

R-S Reset-Set (Flip Flop) [also RS]

R-S Rubidium (Rb-87) to Strontium (Sr-87) (Ratio) [also RS] [Geochronology]

Rs Rupees [Plural of Rupee–Currency of India]

R(s) Reference Input (in Automatic Control) [Symbol]

R/s Roentgen per second [also R s^{-1}]

r/s revolution(s) per second [also RPS, or rps]

RSA Rabbit Serum Albumin [Immunology]

RSA Radar Signature Analysis

RSA Railway Signal Association [US]

RSA Railway Supply Association [US]

RSA Random Sequential Adsorption (Model)

RSA Range Standardization and Automation

RSA Rapid(ly) Solidified Alloy(s)

RSA Redstone Arsenal [Huntsville, Alabama, US]

RSA Reference Satellite A

RSA Refined Sugar Association [UK]

RSA Regional Science Association [US]

RSA Regional Studies Association [UK]

RSA Rehabilitative Services Administration [of Office of Human Development Services, US]

RSA Related Scientific Activities

RSA Remote Session Access

RSA Remote Station Alarm

RSA Repair Service Attendant

RSA Republic of South Africa

RSA Republiek van Suid Afrika [Republic of South Africa]

RSA Requirements Statement Analyzer

RSA Research Society on Alcoholism [of US National Council on Alcoholism]

RSA Rheometrics Solids Analyzer

RSA Rigid Spin Approximation [Physics]

RSA Rivest-Shamir-Aldleman (Algorithm) [Communications]

RSA Royal Scottish Academy

RSA Royal Society of Arts [UK]

RSA Royal Society of Australia [Canberra]

RSA Rubber Shippers Association [US]

RSA Rural Service Area [Telecommunications]

RSA Russian Space Agency

RSA129 Cryptographic Security Number Based on 129 Digits and Invented by R. L. Rivest, A. Shamir and L. M. Adleman [Communications]

Rsa Rhodopseudomonas sphaeroides [Microbiology]

RSAB Royal Society of Armourers and Brasiers [UK]

RSAC Radiological Safety Analysis Computer

RSAC (Nuclear) Reactor Safety Advisory Committee [Canada]

RSAC Recreational Software Advisory Council

RSAC Royal Scottish Automobile Club [UK]

R-SAC Region Student Activities Committee [of Institute of Electrical and Electronics Engineers, US]

RSACE Research and Advanced Communications for Europe

RSAED Rocking Selected Area Electron Diffraction Pattern

RSAF Royal Small Arms Factory

R-SALT Disodium 3-Hydroxy-2,7-Naphthalenedisulfonate [also R-Salt]

RSARR Republic of South Africa Research Reactor

RSAS Royal Swedish Academy of Sciences [Stockholm]

RSASA Royal South Australian Society of Arts

RSB Reactor Service Building

RSB Repair Service Bureau

RSB Reticulocyte Standard Buffer [Biochemistry]

RSB Rudder Speed Brake

RSBS Research School of Biological Sciences [of Australian National University, Canberra]

RSC Radar Set Control

RSC Range Safety Command [Aerospace]

RSC Range Safety Control [Aerospace]

RSC Rapid Spinning Cup (Process) [Powder Metallurgy]

RSC Reactor Safety Commission

RSC Redistributed Successive Collision (Model) [Physics]

RSC Reed-Solomon Code [Computers]

RSC Relative Sensitivity Coefficient

RSC Remote Services Center

RSC Remote Storage Controller

RSC Rescue Sub-Center

RSC Research School of Chemistry

RSC Reversed Sigmoidal Curve

RSC Richmond School of Commerce [Toronto, Canada]

RSC Rockwell Science Center [US]

RSC Royal Society of Canada

RSC Royal Society of Chemistry [UK]

RSC Runway Surface Condition

RSC Russian Space Center

RSC-E Russian Space Corporation-Energia

RSCESFS Rapid Scanning Constant Energy Synchronous Fluorescence Spectrometry

RSCF Relativistic Self-Consistent Field [Physics]

Rsch Research [also rsch]

RSCIE Remote Station Communication Interface Equipment

RSCN Registered Sick Children's Nurse [UK]

RSCN Remote Scientific Computing Network

RSCS Rate Stabilization and Control System

RSCS Remote Spooling (and) Communications Subsystem; Remote Spooling (and) Communications System

RSCS/PVM Remote Spooling Control Subsystem/Permanent Virtual Machine

RSCS Rod Sequence Control System [Nuclear Reactors]

RSCW Research Reactor, State College of Washington [US]

RSD Radar Signal Discrimination

RSD Reactive Skin Decontaminant

RSD Reactive Sputter Deposition

RSD Refueling Shutdown

RSD Relative Standard Deviation

RSD Remote Site Data-Processing

RSD Requirements and Specifications Document

RSD Responsible System Designer

RSD Ring System Descriptor

RSD Route Server Daemon

%RSD Percent Relative Standard Deviation

RS&D Receipt, Storage and Delivery [also RSD]

R&SD Regulatory and Safety Data

RSDA Road Surface Dressing Association [UK]

RSDC Residual Stress Data Collection

RSDC Roentgen Satellite Science Data Center [of NASA Astrophysics Data Facility, Goddard Space Flight Center, Greenbelt, Maryland, US]

RSDL Reactive Skin Decontamination Lotion

RSDP Remote Shutdown Panel

RSDP Remote Site Data Processor

RSDS Range Safety Destruct System

RSE Railway Signal Engineer(ing)

RSE Reducing Street Elbow

RSE Reference Standard Endotoxin [Microbiology]

RSE Register Signaling Equipment

RSE Renewable Sources of Energy

RSE Research Scientists and Engineers

RSE Resonant Secondary Emission

RSE Royal Society of Edinburgh [Scotland]

RSECI Restricted Single Excitation Configuration Interaction

RSEP Resource Supply Expansion Program [US]

RSES Refrigeration Service Engineers Society [US]

RSEU Remote Scanner and Encoder Unit

RSEW Resistance-Seam Welding

RSF Receiving-Safing Facility

RSF Refurbishment and Subassembly Facilities

RSF Refurbishment and Subassembly Facility [of NASA Kennedy Space Center, Florida, US]

RSF Relative (Elemental) Sensitivity Factor

RSF Remote Support Facility

RSF Rotating Stall Flutter

RSFQ Resistive Single Flux Quantum (Device)

RSFS Royal Scottish Forestry Society [UK]

RSFSR Russian Soviet Federated Socialist Republic [Defunct]

RSG Reentrant Spin Glass [Solid-State Physics]

RSG Resistance Strain Gauge

Rsg Rising [also rsg]

RSGB Radio Society of Great Britain

RSGF Real Space Green Function [Mathematics]

RSGS Royal Scottish Geographical Society

RSH Mercaptan (or Thiol) [Formula]

RSH Remote Shell

RSH Royal Society of Health [UK]

RS&H Reynolds, Smith & Hill, Inc. [US]

RSI Radarsat International (Inc.) [Canada]

RSI Radial Shear Interferometer

RSI Rationalization, Standardization and Integration

RSI (Nuclear) Reactor Siting Index

RSI Register Sender Inward

RSI Remote Sensing Imagery

RSI Repetitive Strain Injury [also Carpal Tunnel Syndrome] [Medicine]

RSI Research Studies Institute

RSI Reusable Surface Insulation [NASA Space Shuttle]

RSI Review of Scientific Instruments [of American Institute of Physics, US]

RSI Royal Sanitary Institute [UK]

RSI Royal Signals Institution [London, UK]

RSI Ryzner Saturation Index

RS&I Rules, Standards and Instructions [also RSI]

RSIC Radiation Shielding Information Center [of Oak Ridge National Laboratory, Tennessee, US]

RSIC Redstone Scientific Information Center [of US Army at Huntsville, Alabama]

RSID Resource Identification (Table)

RSigG Rocket, Signal, Green

RSigR Rocket, Signal, Red

RSIL Reinforced Silicone

RSIM Repulsion-Start Induction Motor

RSIN Radiation Safety Incident Notice

RSIO Reinforced Steel Institute of Ontario [Canada]

$(R_2SiO)_n$ Polysiloxane [General Formula; R = Hydrogen, Alkyl, or Aryl]

RSIP Receiving Site Implementation Plan

RSIS Reference, Special and Information Section [of Library Association, UK]

RSJ Resistively Shunted Junction (Model) [Electronics]

RSJ Rolled Steel Joist

RSJJ Resistivity Shunted Josephson Junction [Electronics]

RSL Radio Standards Laboratory [US]

RSL Received Signal Level

RSL Request-and-Status Link

RSL Requirements Specifications Language

RSL Resource Support List

RSL Royal Society of London [UK]

Rsl Resorcinol [Organic Chemistry]

RSLC Radar Searchlight Control

RSLS Receiver Side-Lobe Suppression

RSLS Reply-Path Side-Lobe Suppression

rslv resolve

Rslvr Resolver [also rslvr]

RSM Randall, Shao, Moeng (Reference) [Physics]

RSM Rapidly Solidified Material [Powder Metallurgy]

RSM Reciprocal Space Mapping [Physics]

RSM Real Storage Management

RSM Reflective Solar Mirror

RSM Regimental Sergeant-Major

RSM Remote Switching Module

RSM Remote System Manager

RSM Repulsion-Start Motor

RSM Research Scale Melter

RSM Resource Management System

RSM Resume [Instruments]

RSM Ringing and Signal Machine

RSM Rocket Sea-Marker

RSM Royal Scottish Museum [Edinburgh]

RSM Royal School of Mines [UK]

RSM Royal Society of Medicine [UK]

RSMA Radiological Systems Microfilm Associates [US]

RSMAS Rosenstiel School of Marine and Atmospheric Science [US]

RSME Reversible Shape Memory Effect [Metallurgy]

RSMM Redundant System Monitor Model

RSMPS Romanian Society for Mathematics and Physical Sciences

RSMW Rapid Scanning Multiple Wavelength

RSN Radiation Surveillance Network

RSN Record Sequence Number

RSN Real Soon Now [Internet]

RSN Royal Singapore Navy

RSNA Radiological Society of North America

RSNA Royal School of Naval Architects [UK]

RSNC Royal Society for Nature Conservation [UK]

RSNSW Royal Society of New South Wales [Sydney, Australia]

RSNZ Royal Society of New Zealand

RSO Radiation Safety Officier

RSO Radiological Safety Office

RSO Range Safety Officer

RSO Register Sender Outward

RSO Revenue Sharing Office

RSO$_3$H Sulfonic Acid [Formula]

RSOP Reconnaissance, Selection and Organization of Artillery Positions

RSORS Remote Sensing On-Line Retrieval System

RSOS Restricted Solid-on-Solid (Approximation) [Physics]

RSP Radial Structure Plot

RSP Radio Switch Panel

RSP Random Signal Processing

RSP Rapid(ly) Solidified Powder [Powder Metallurgy]

RSP Rapid(ly) Solidified Product [Powder Metallurgy]

RSP Rapid(ly) Soliffication Processed; Rapid Solidification Processing [Powder Metallurgy]

RSP Rapid Solidification Process(ing) [Powder Metallurgy]

RSP Reactivity Surveillance Procedures

RSP Reader/Sorter Processor

RSP Record Select Program

RSP Reflective Solar Panel

RSP Registered Safety Professional

RSP Regional Seas Program

RSP Remote Site Processor

RSP Rendezvous Station Panel

RSP Replication Selection Process

RSP Required Space (Character)

RSP Research and Surveys Program

RSP Respirable, Suspended Particulate

RSP Restoration Priority

RSP Reverse Scattering Perturbation

RSP Robot Support System

RSP Rotating Shield Plug

Rsp Respond(er) [also rsp]

RSPA Research and Special Programs Administration

RSPA Research and Special Programs Association [US]

RS&PAA Remote Sensing and Photogrammetry Association of Australia Ltd.

RSPB Royal Society for the Protection of Birds [UK]

RSPCA Royal Society for the Prevention of Cruelty to Animals [UK]

RSPD Rapid-Solidification Plasma Deposition

RSPhysSE Research School of Physical Sciences and Engineering [of Australian National University, Canberra]

RSPL Recommended Spare Parts List

RS/PM Rapidly-Solidified Powder Metallurgy (Material); Rapid Solidification/Powder Metallurgy

RS/PM Alloy Rapidly-Solidified Powder Metallurgy Alloy

RSPS Research School of Pacific Studies [of Australian National University, Canberra]

RSPS Research School of Physical Sciences [of Australian National University, Canberra]

RSPT Rayleigh-Schroedinger Perturbation Theory [Physics]

RSPT Real Storage Page Table

RSPT Report Starting Procedure Turn [Aeronautics]

RSPTn Rayleigh-Schroedinger Perturbation Theory of n-th Order [Physics]

RSPX Remote Sequenced Packet Exchange

RSR Random Signal Reject

RSR Rapid Solidification Rate (Process) [Powder Metallurgy]

RSR Rapid Stain Removal

RSR Reactor Safety Research

RSR Region Student Representative [of Institute of Electrical and Electronics Engineers, US]

RSR Republica Socialista România [Romanian Socialist Republic]

RSR Restore [also Rsr, or rsr]

RSR Rod Select Relay [Nuclear Reactors]

RSR Rotary Saturation Resonance

RSR Route Surveillance Radar

Rsr Rhodopseudomonas sphaeroides [Microbiology]

RSRA Rotor Systems Research Aircraft

RSRE Royal Signals and Radar Establishment [of Defence Research Agency, Malvern, UK]

RSRI Regional Science Research Institute [US]

RSRM Redesigned Solid Rocket Motor

RSRM Revised Solid Rocket Motor

RSRM Reusable Solid Rocket Motor

RSRP Richards Sulfur Recovery (Process) [Chemical Engineering]

RSRS Radio and Space Research Station [US]

RSS Random Selection for Service

RSS Random-Special-Special (Threefold Grain Boundary Edge) [Materials Science]

RSS Range Safety Switch

RSS Range Safety System

RSS Reactant Service System

RSS Reactants Supply System

RSS Reactor Safety Study

RSS Reactor Shutdown System

RSS Registry Structure Sheet

RSS Relaxed Static Stability

RSS Remote Sensing Society [UK]

RSS Remote Shutdown System [Nuclear Reactors]

RSS Residual Sum of Squares

RSS Resolved Shear Stress [Mechanics]

RSS Resource Scheduling Software

RSS Reusable Subsystem

RSS Ribbed Smoked Sheets [Metallurgy]

RSS Rib Structure Station

RSS Root-Sum-of-Squares [Physics]

RSS Root Sum Square (Value) [also rss]

RSS Rotating Service Structure

RSS Route Switching Subsystem

RSS Royal Scientific Society [Jordan]

RSS Royal Statistical Society [UK]

RS&S Receiving, Shipping and Storage

RSSA Railway Systems Suppliers Association [US]

RSSA Royal Scottish Society of Arts (Science and Technology)

RSSA Royal Society of South Africa

RSSF Retrievable Surface Storage Facility

RSSP Radar Systems Specialist Panel [of International Civil Aviation Organization]

RSSPO Resident Space Shuttle Project Office(r) [of NASA]

RSSR Russian Soviet Socialist Republic [USSR]

RSSS Radiographic Standard Shooting Sketch

RSST Reactive System Scanning Tool

RST Radiation Science and Technology

RST Radiometric Surface Temperature

RST Rapid Solidification Technique; Rapid Solidification Technology [Powder Metallurgy]

RST Reactive Solute Transport

RST Readability Strength Tone

RST Real Space Transfer

RST Recessed Selectromatic Terminal

RST Remote Station

RST Research Study Team

RST Reset

RST Reset-Set Trigger

RST Restore (Function) [Computers]

RST Restrained Strength [Mechanics]

RST Retained Strength [Mechanics]

RST Rotating-Frame Spin Resonance

RST Rough Surface Tester

RS-T Segment between QRS Complex and T-Wave (in Electrocardiogram) [Medicine]

Rst Reset [also rst]

Rst Restart [also rst]

rstd restricted [also rst'd]

RSTMH Royal Society of Tropical Medicine and Hygiene [UK]

RSTN Regional Seismic Test Network

RST OUT Reset Out

Rstrt Restart [also rstrt]

RSTS Resource-Sharing Time-Sharing

RSU Register Storage Unit

RSU Remote Service Unit

RSU Remote Switching Unit

RSU Rostov State University [Rostov-on-Don, Ukraine]

RSUA Royal Society of Ulster Architects [Northern Ireland]

RSV Raster Scan Video

RSV Rat Sarcoma Virus

RSV Reserve [also Rsv, or rsv] [Computers]

RSV Respiratory Syncytial Virus

RSV Revised Standard Version

RSV Rice Stripe Virus

RSV Right Subclavian Vein [Anatomy]

RSV Rous Sarcoma Virus

RSV Royal Society of Victoria [Melbourne, Australia]

RSV Run-Control Solenoid Valve

RSVP Remote System Verification Program

RSVP *(Répondez s'il vous plaît)* – French for "Please Answer," or "Please Reply" [also rsvp]

RSVP Resource Reservation Protocol

Rsvr Reservoir [also rsvr]

RSW Renormalized Spin Wave (Theory) [Solid-State Physics]

RSW Resistance-Spot Welding

RSWC Right Side up With Care

RSWF Radioactive Scrap and Waste Facility

RSWG Response Strategies Working Group [Intergovernmental Panel on Climate Change]

RSWGSC RSWG Steering Committee

RSYS Responsible System

RSWPS Repetitive Square Wave Potential Signal

RSX Realistic Sound Experience

RSX Resource-Sharing Executive

RT Cylindrical Roller Bearing, Single-Row, Two-Direction-Locating, Double-Ribbed Outer Ring, Double-Ribbed Inner Ring with One Loose Rib, Inner Ring Separ-able [Symbol]

RT Radar Technology

RT Radiant Technologies, Inc. [Albuquerque, New Mexico, US]

RT Radiant Tube

RT Radiation Temperature

RT Radiation Therapy [Medicine]

RT Radiation Thermometer; Radiation Thermometry

RT Radiative Transfer [Physics]

RT Radioactive Tracer

RT Radiographic Test(ing)

RT Radiological Technician

RT Radiological Technologist; Radiological Technology

RT Radio Technician

RT Radiotelegraph(y) [also R/T]

RT Radiotelephone; Radiotelephony [also R/T]
RT Radiotelescope
RT Radio Transformer Unit
RT Rail Transit
RT Raise Top
RT Random Test(ing)
RT Range-to-Target [also R-T]
RT Range Tracking
RT Rapid Transit
RT Rapid Traverse [Machine Tools]
RT Rate [also Rt, or rt]
RT Rated Time
RT Ratio Transformer
RT Reactance Voltage
RT Reaction Time
RT Reaction Turbine
RT (Nuclear) Reactor Trip
RT Real Time [also R/T]
RT Racapitulation Theory [Biology]
RT Receiver-Transmitter
RT Receptor Technology
RT Recognition Technology
RT Recovery Treated; Recovery Treatment [Metallurgy]
RT Recreational Therapy [Medicine]
RT Recrystallization Time [Metallurgy]
RT Rectifier Tube
RT Reduction Table
RT Reference Temperature
RT Reference Test
RT Reference Trajectory
RT Refrigerating Technician
RT Registered Technician
RT Registed Telephone
RT Register Ton
RT Register Traffic
RT Register Transfer
RT Register Translator [also R/T]
RT Regression Testing
RT Reimer-Tiemann (Synthesis) [Organic Chemistry]
RT Related Term
RT Relativity Theory
RT Relaxation Time [Physics]
RT Release Time
RT Remote Terminal
RT Renewal Theory
RT Reperforator-Terminal
RT Reperforator-Transmitter
RT République du Tchad [Republic of Chad]
RT République Togolaise [Republic of Togo]
RT Research Topic
RT Research Triangle [Durham–Chapel Hill–Raleigh, North Carolina, US]
RT Residence Time
RT Resident Time

RT Residual Term [Mathematics]
RT Residual Torque
RT Residue Theorem [Mathematics]
RT Resistance Test [also R-T]
RT Resistance Thermometer
RT Resolution Tested
RT Resolver Transformer
RT Resolving Time
RT Resonance Theory
RT Resonant Tunneling [Solid-State Physics]
RT Respiratory Therapist
RT Respiratory Tract [Anatomy]
RT Response Tone
RT Restoring Torque
RT Return [also Rt, or rt]
RT Reverberation Time [Acoustics]
RT Reverse Transcriptase [Biochemistry]
RT D-Ribosylthymine [Biochemistry]
RT Rice-Thomson (Model)
RT Ring Trip
RT RISC (Reduced Instruction Set Computer) Technology
RT Rise Time
RT Road Tar
RT Robot Trainer
RT Rocket Technology
RT Rolls Royce/Turboméca (Engine)
RT Room Temperature
RT Rooted Tree [Graph Theory]
RT Rotary Table
RT Rotary Transformer
RT Rothwarf-Taylor (Method) [Metallurgy]
RT Route [also Rt, or rt]
RT Route Treatment
RT Rubber Technology
RT Run(ning) Time
RT Ruthenium Titanate
R(T) Temperature Dependent Resistance [Function]
$\bar{R}$(T) Average Cluster Radius [Symbol]
R_H(T) Temperature Dependent Hall Resistance [Symbol]
R&T Research and Technology [also RT]
R/T Radiotelegraph(y) [also RT]
R/T Radiotelephone; Radiotelephony [also RT]
R/T Real Time [also RT]
R/T Receive/Transmit; Receiver/Transmitter [also R-T]
R/T Reflectivity/Transmission (Ratio)
R/T Register Translator [also RT]
R/T Reset, or Toggle
R-T Range-to-Target [also RT]
R-T Rare-Earth–Transition-Metal (Intermetallics) [also R-TM, RE-T, RE-TM, or RETM]
R-T Receive/Transmit; Receiver/Transmitter [also R/T]
R-T Resistance-Temperature (Curve)
R-T Resistance Test [also RT]
Rt Return [also rt]

R/t (Minimum) Bend Ratio (of Sheetmetal) (i.e. Ratio of Bend Radius to Sheet Thickness) [Symbol]

R-t Reflectivity versus Time (Curve)

R(t) Cluster Radius [Symbol]

R(t) Nucleation Rate (in Metallurgy) [Symbol]

R(t) Time Decay of Resistance [Symbol]

R(t) Time-Dependent Domain Size (in Solid-State Physics) [Symbol]

R3 Reclaim, Reuse, Reduce [also RRR, R^3, or 3R]

R3 Reduce, Reuse, Recycle [also RRR, R^3, or 3R]

rT D-Ribosylthymine [Biochemistry]

r(T) Temperature Dependent Desorption Rate [Symbol]

rt right

ρ(T) temperature dependent density [Symbol]

ρ(T) temperature dependent resistivity [Symbol]

ρ-T resistivity versus temperature (curve)

RTA Railway Tie Association [US]

RTA Rapidly Thermal Annealed; Rapid Thermal Anneal(ing) [Metallurgy]

RTA Rapid Transit Association [US]

RTA Real-Time Accumulator

RTA Reliability Test Assembly

RTA Remote Technical Assistance

RTA Remote Test Access

RTA Remote Trunk Arrangement

RTA Required Time of Arrival

RTA Roads and Transportation Association

RTA Roofing Tile Association [UK]

RTA Rubber Trade Association [US and UK]

RTAC Real-Time Adaptive Control

RTAC Regional Technical Aids Center [US]

RTAC Research and Technology Advisory Committee [US]

RTAC Roads and Transportation Association of Canada [Ottawa, Ontario, Canada]

RTAed Rapid Thermal Annealed [Metallurgy]

RTAM Remote Terminal Access Method

RTAP Real-Time Automation Program

RTA/RTP Rapidly Thermal Annealed–Rapidly Thermal Processed; Rapid Thermal Annealing–Rapid Thermal Processing [Metallurgy]

RTB Read Tape Binary

RTB Resistance Temperature Bulb

RTB Response/Throughout Bias [Electronics]

RTB Rural Telephone Bank

R(T,B) Temperature and Magnetic Induction Dependent Resistance (or Magnetoresistance) [Symbol]

RTBM Real-Time Bit Mapping

RTC Radar Tracking Center

RTC Radar Tracking Control

RTC Radio Transmission Control

RTC Railway Transport Committee

RTC Range Telemetry Center

RTC Rapid Thermal Chemical Vapor Deposition

RTC Reactor Technology Center

RTC Reader/Tape Contact

RTC Real-Time Clock

RTC Real-Time Command

RTC Real-Time Computer

RTC Real-Time Control

RTC Reference Transfer Calibrator

RTC Remote Terminal Controller

RTC Removable Top Closure

RTC Resistance Temperature Coefficient

RTC Right-to-Choose

RTC Rizac Technological College [Philippines]

RTC Room-Temperature Cure(d); Room-Temperature Curing [Chemistry]

RTC Royal Tank Corps [UK]

RTC Royal Technical College [Kenya]

RTC Rural Telephone Center

RTCA Radio Technical Commission for Aeronautics [Washington, DC, US]

RTCA Radio Technical Commission of America [US] [Defunct]

RTCC Real-Time Command Controller

RTCC Real-Time Communications Control

RTCC Real-Time Computer Center [of NASA]

RTCC Real-Time Computer Command

RTCC Real-Time Computer Complex

RTCE Rotation/Translation Control Electronics

RTCF Radial Tube Component Feeder

RTCF Real-Time Computer Facility

Rtcl Reticle [also rtcl]

RTCM Radio Technical Commission for Marines [US]

RTCM Radio Technical Commission for Maritime Services [US]

RTCMA Rubber and Thermoplastic Cable Manufacturers Association [UK]

RTCP Real-Time Communications Processor

RTCS Radarsat Topographic Correction System

RTCS Real-Time Communications System

RTCS Real-Time Composition System

RTCS Real-Time Computer System

RTCU Real-Time Control Unit

RTCVD Rapid Thermal Chemical Vapor Deposition [also RT-CVD]

RTD Range Time Decoder

RTD Rapid Thermal Diffusion

RTD Rate Damping

RTD Read Tape Decimal

RTD Real-Time Display

RTD Research and Technological Development

RTD Research and Technology Division [of US Air Force]

RTD Residence Time Distribution

RTD Resistance-Temperature Detector

RTD Resistance-Temperature Device

RTD Resistance-Thermometer Detector

RTD Resistive Temperature Device

RTD Resonant Tunneling Diode

RTD Room Temperature, Dry

rtd returned [also rt'd]

RTDC Real-Time Data Channel

RTDD Real-Time Data Distribution

RTDHS Real-Time Data Handling System

RTDL Real-Time Data Link

RTDM Real-Time Data Migration

RTDS Rapid Thermal Decomposition (of Precursors) in Solution

RTDS Real-Time Data System

RTE Radio and Telecommunication Engineer(ing)

RTE Road Transport Engineer(ing)

RTE Real-Time Execution

RTE Real-Time Executive

RTE Regenerative Turboprop Engine

RTE Remote Terminal Emulator

RTE Residual Total Elongation [Mechanics]

RTE Responsible Test Engineer

RTE Reversible Temper Embrittlement [Metallurgy]

RTE Run-Time Executive

Rte Route [also rte]

RTEB Radio Trades Examination Board [UK]

RTECS Registry of Toxic Effects of Chemical Substances [of National Institute for Occupational Safety and Health, US]

RTECS# RTECS Number [of National Institute for Occupational Safety and Health, US]

RTEL Reverse Telnet [Internet]

RTEM Radar Tracking Error Measurement

R-TEM Retractable Transmission Electron Microscope (Detector)

RTEP Rapid Transit Expansion Program [Canada]

RTEP Ryerson Test of English Proficiency [Canada]

RTES Real-Time Executive System

RTES Real-Time Expert System

RTF Radiant Tube Furnace

RTF Radiodiffusion et Télévision Française [French Broadcasting and Television Company]

RTF Radiotelephone; Radiotelephony

RTF Resistance Transfer Factor

RTF Rich Text Format [Internet]

RTF Rotary Tube Feeder

RTFM Read The Fantastic (or Flipping) Manual [Internet Jargon]

RTFOT Rolling Thin-Film Oven Test

RTG Radiant-Tube Gas (Burner)

RTG Radioisotope Thermal Generation

RTG Radioisotope Thermoelectric Generator

RTG Radiotelegraphy

RTG Range-to-Ground

RTGB Reactor Turbine General Board

RT GMR Room-Temperature Giant Magnetoresistance [also RT-GMR]

rTHD D-Ribosylthymine [Biochemistry]

RTHBT Resonant Tunneling Heterostructure Bipolar Transistor

RTHC Rotation-Translation Hand Controller

RTHE Round, Threaded End

RTHF Real-Time Hydrological Forecasting

RTHI Real-Time Holographic Interferometry

RTHS Real-Time Hybrid System

RTI Radiation Transfer Index

RTI Railway Technical Institute [Tokyo, Japan]

RTI Range-Time Indicator

RTI Rayleigh-Taylor Instability [Fluid Mechanics]

RTI Real-Time Imaging

RTI Real-Time Interface

RTI Real-Time Interferometry

RTI Referred to Input [also rti]

RTI Relative Thermal Index

RTI Research Triangle Institute [of Research Triangle Park, North Carolina, US]

RTI Royal Tropical Institute [Netherlands]

RTI Rundfunktechnisches Institut [Institute for Radio Engineering, Nuremberg, Germany]

RTIC Regional Technical Information Center [UK]

RTIF Rapid Test and Integration Facility [US Navy]

RTIF Real-Time Interface

RTIFRAMP Rapid Test and Integration Facility, Rapid Access to Manufactured Parts [US Navy]

RTIO Real-Time Input/Output

RTIO Remote Terminal Input/Output

RTIOC Real-Time Input/Output Controller [also RTI/OC]

RTIP Remote Terminal Interface Package

RTIRS Real-Time Information Retrieval System

RTITB Road Transport Industry Training Board [UK]

RTK Range Tracker

RTK Right-to-Know

RTL Radio-Isotope Transport Loop [Nuclear Engineering]

RTL Radiodiffusion et Télévision de Luxembourg [Luxembourg Radio and Television]

RTL Real-Time Language

RTL Register Transfer Language

RTL Register Transfer Level

RTL Resistor-Transistor Logic

RTL Right-To-Left

RTL Runtime Library [Computers]

RTLS Return to Launch Site

RTLP Reference Transmission Level Point

RTM Rapid Thermal Multiprocessing

RTM Rapid Tuning Magnetron

RTM Raster Tunnel Microscopy

RTM Real-Time Module

RTM Real-Time Monitor(ing)

RTM Receiver-Transmitter Modulator

RTM Recording Tachometer

RTM Register Transfer Module

RTM Research Technical Memorandum

RTM Resin-Transfer Model
RTM Resin-Transfer Molding
RTM Resonant Torque Magnetometry
RTM Response Time Module
RTM Response Time Monitor
RTM Revenue Ton-Mile
RTM Robot Time and Motion
RTM Rotational Transverse Magnetometry
RTM Runtime Manager [Computers]
R-TM Rare Earth-Transition Metal (Intermetallics) [Materials Science]
RTMA Radio and Television Manufacturers Association [now Electronic Industries Association, US]
RTMOS Real-Time Multiprogramming Operating System
RTMP Routing Table Maintenance Protocol
RTMS Real-Time Memory System
RTMS Real-Time Multiprogramming System
RTN Random Telegraph Noise
RTN Rapid Thermal Nitridation
RTN Registered Tradename
RTN Remote Terminal Network
Rtn Return [also rtn]
Rtn Routine [also rtn]
R/T NET Radio Telephone Network [also R/T Net]
RTNR Ring-Tone, No Reply
RTO Railway Transport Officer
RTO Rapid Thermal Oxidation
RTO Real-Time Operation
RTO Referred to Output [also rto]
RTO Regenerative Thermal Oxidation
RTO Remote Terminal Operator
RTO Responsible Test Organization
RTO Run-Time Organization
RTOAA Rejected Take-Off Area Available
RTOL Restricted Takeoff and Landing
RTOP Real-Time Optional Processing
RTOP Research and Technology Objectives and Plans
RTOP Research and Technology Operating Plan
RTOP Research and Technology Operations and Plans [of NASA]
RTOP Research Topic Operating Plan
RTO/RTN Rapid Thermal Oxidation/Rapid Thermal Nitridation, Nitroxide (Capacitor)
RTO/RTN/RTO Rapid Thermal Oxidation/Rapid Thermal Nitridation/Rapid Thermal Oxidation, Reoxidized Nitroxide
RTOS Real-Time Operating System
RTOS/PEARL Real-Time Operating System/Process, Experiment and Automation Real-Time Language
RTP Rapidly Thermally Processed; Rapid Thermal Process(ing); Rapid Thermal Processor [Metallurgy]
RTP Rapid Thermoset Processing
RTP Rapid Transport Protocol [Data Communications]
RTP Reactor Thermal Power
RTP Real-Time Peripheral
RTP Real-Time Process(ing)

RTP Real-Time Protocol
RTP Records Turnover Package
RTP Reference Telephonic Power
RTP Reinforced Thermoplastics
RTP Reinforced Thermoset(ting) Plastics
RTP Remote Terminal Processor
RTP Remote Transfer Point
RTP Requirement and Test Procedures
RTP Research Triangle Park [Durham, North Carolina, US]
RTP Room Temperature and Pressure
(r,Θ,ϕ) spherical polar coordinates [Symbol]
RTPB Radio Technical Planning Board [US]
RTPC Real-Time Process Control(ler)
RTPC Restrictive Trade Practices Commission
RTP CVD Rapid Thermal Processing Chemical Vapor Deposition
RTPH Round Trips per Hour
R2PI Resonant Two-Photon Ionization (Spectroscopy) [also R^{2}PI]
RTPI Royal Town Planning Institute [UK]
RTPL Room Temperature Phosphorescence in Liquid Chromatography
RTPO Room Temperature Phosphorescence
RTP P POD RTP Company's Priority Property Operation Disk
RTPS Repetitive Triangular Potential Sweep
RTPSC Real-Time Power System Control
RTP/VLP-CVD Rapid Thermal Processing/Very-Low-Pressure Chemical Vapor Deposition
RTQC Real-Time Quality Control
RTR Real-Time Radiography
RTR Real-Time Radioscopy
RTR Reinforced Thermosetting Resin
RTR Repeater Test Rack
RTR Response Time Reporting
RTR Return and Restore Status
RTR Return-to-Reference (Recording)
RTR Ribbon-to-Ribbon
Rtr Rotor [also rtr]
RTRC Radio and Television Research Council [US]
Rtrd Retard [also rtrd]
RTRI Railway Technical Research Institute [Tokyo, Japan]
Rtrn Return [also rtrn]
RTRP Reinforced Theomosetting Resin Pipe
RTS Radar Target Simulator
RTS Radar Tracking Station
RTS Range Time Signal
RTS Rapid Transmission and Storage
RTS Reactive Terminal Service
RTS (Nuclear) Reactor Trip System
RTS Ready to Send
RTS Real-Time Solution
RTS Real-Time Subroutine
RTS Real-Time Supply
RTS Real-Time System

RTS   Relaxation Time Spectrum
RTS   Relative Timed Sequence
RTS   Remote Takeover System
RTS   Remote Targeting System
RTS   Remote Testing System
RTS   Remote Tracking Station
RTS   Request-to-Send (Signal)
RTS   Requirements Tracking System
RTS   Resonant Tunneling Structure
RTS   Resource Tracking System
RTS   Return to Service
RTS   RFS (Request for Service) Tracking System
RTS   Royal Television Society [UK]
RTS   Rural Telephone System
RTSA   Retail Trading Standards Association [UK]
RTSC   Read The Source Code [Internet]
RTS/CTS   Request to Send/Clear to Send (Delay)
RTSD   Resources and Technical Services Division [of American Library Association, US]
RTSE   Real-Time Spectroscopic Ellipsometry
RTSF   Real-Time Simulation Facility
RTSN   Random Telegraph Switching Noise
RTSP   Real-Time Streaming Protocol
RTSRS   Real-Time Simulation Research System
RTSS   Real-Time Scientific System
RTST   Radio Technician Selection Test
RTSV   Rice Tungro Spherical Virus
RTT   Radiation Tracking Transducer
RTT   Radioteletype
RTT   Radioteletypewriter
RTT   Real-Time Telemetry
RTT   Real-Time Tracer
RTT   Recrystallization Time–Temperature (Curve) [Metallurgy]
RTT   Reservoir and Tube Tunnel
RTT   Round-Trip Time [Internet]
RTTDS   Real-Time Telemetry Data System
RTTI   Runtime Type Information
RTTV   Real-Time Television
RTTY   Radioteletype
RTTY   Radio Teletypewriter
RTU   Real-Time Ultrasonic System
RTU   Remote Terminal Unit
RTU   Remote Test Unit
RTU   Reserve Training Unit
RTU   Right to Use
RTUA   Recognition Technologies Users Association [US]
RTV   Real Time Video
RTV   Remote Television
RTV   Return to Vendor
RTV   Room-Temperature-Vulcanisate; Room-Temperature Vulcanization; Room-Temperature-Vulcanized; Room-Temperature Vulcanizing [Chemical Engineering]
RTV   Rotating Traveling Vehicle
Rtv   Retrieve [also rtv]

RTVD   Room-Temperature Vapor Doping
RTW   Room Temperature, Wet
RTWS   Raw Tape Write Submodule
RTX   Real-Time Executive (System)
RTX   Run Time Extension
RTZ   Return-to-Zero (Recording)
RTZL   Return-to-Zero Level
RTZM   Return-to-Zero Mark
RTZ(NP)   Nonpolarized Return-to-Zero (Recording)
RTZ(P)   Polarized Return-to-Zero (Recording)
RU   Latex Rubber Type (Conductor Insulation) [Symbol]
RU   Cylindrical Roller Bearing, Single-Row, Non-Locating, Double-Ribbed Outer Ring, Inner Ring without Ribs, Inner Ring Separable [Symbol]
RU   Are You?
RU   Relative Unit [also REL U, or Rel U]
RU   Remote Unit
RU   Reproducing Unit
RU   Request/Response Unit
RU   Request Unit
RU   Response Unit
RU   Rice University [Houston, Texas, US]
RU   Rikkyo University [Kanagawa, Japan]
RU   Ritsumeikan University [Kyoto, Japan]
RU   Rockefeller University [New York City, US]
RU   Roosevelt University [Chicago, Illinois, US]
RU   Run Unit
RU   Rutgers University [Piscataway, New Jersey, US]
RU   Russian Federation [ISO Code]
RU 486   Mifepristone [Abortion-Inducing Drug]
R&U   Repairs and Utilities
Ru   Ruthenium [Symbol]
Ru-94   Ruthenium-94 [also ^{94}Ru, or Ru^{94}]
Ru-95   Ruthenium-95 [also ^{95}Ru, or Ru^{95}]
Ru-96   Ruthenium-96 [also ^{96}Ru, or Ru^{96}]
Ru-97   Ruthenium-97 [also ^{97}Ru, or Ru^{97}]
Ru-98   Ruthenium-98 [also ^{98}Ru, or Ru^{98}]
Ru-99   Ruthenium-99 [also ^{99}Ru, or Ru^{99}]
Ru-100   Ruthenium-100 [also ^{100}Ru, or Ru^{100}]
Ru-101   Ruthenium-101 [also ^{101}Ru, or Ru^{101}]
Ru-102   Ruthenium-102 [also ^{102}Ru, or Ru^{102}]
Ru-103   Ruthenium-103 [also ^{103}Ru, or Ru^{103}]
Ru-104   Ruthenium-104 [also ^{104}Ru, or Ru^{104}]
Ru-105   Ruthenium-105 [also ^{105}Ru, or Ru^{105}]
Ru-106   Ruthenium-106 [also ^{106}Ru, or Ru^{106}]
Ru-107   Ruthenium-107 [also ^{107}Ru, or Ru^{107}]
.ru   Russian Federation [Country Code/Domain Name]
RUA   Cylindrical Roller Bearing, Single-Row, Non-Locating, Double-Ribbed Outer Ring, Inner Ring without Ribs, Inner Ring Separable, Spherical Outside Surface [Symbol]
Ru(Ac)$_3$ Ruthenium(III) Acetate
Ru(ACAC)$_3$ Ruthenium(III) Acetylacetonate [also Ru(AcAc)$_3$, or Ru(acac)$_3$]
RuAl   Ruthenium Aluminide

RUAS Royal Ulster Agricultural Society [Northern Ireland]

RUB Ruhr-Universität Bochum [University of Bochum, Germany]

Rub Rouble [Currency of Russian Federation]

Rub Rubber [also rub, Rubb, or rubb]

Rubb Board Bull Rubber Board Bulletin [India]

Rubb Chem Technol Rubber Chemistry and Technology [Journal of the American Chemical Society, US]

Rubb Dev Rubber Development [Journal of the Malaysian Natural Rubber Producers Research Association]

Rubb India Rubber India [Journal of the All-India Rubber Industries Association]

Rubb Plast Fire Flamm Bull Rubber and Plastics Fire and Flammability Bulletin [Switzerland]

Rubb Plast News Rubber and Plastics News [Journal published in the US]

Rubb Plast News II Rubber and Plastics News II [Journal published in the UK]

Rubb S Africa Rubber Southern Africa [South Africa]

Rubb Stat Bull Rubber Statistical Bulletin [Published by the International Rubber Study Group, UK]

Rubb Trends Rubber Trends [Journal published in the UK]

Rubb World Rubber World [Journal published in the US]

RUBISCO D-Ribulose 1,5-Diphosphate Carboxylase [also Rubisco] [Biochemisity]

RU Bochum Ruhr-Universität Bochum [University of Bochum, Germany]

RuBP Ribulosebiphosphate; Ribulose-1,5-Biphosphate [Biochemistry]

Ru(bpy)$_3$(C60)$_2$ Ruthenium Tris(4,4′-Bipyridine) Difulleride

RUC République Unié du Cameroun [United Republic of Cameroon]

RUCA Reversible Universal Cellular Automaton

Ru(Cp)$_2$ Bis(cyclopentadienyl)ruthenium (or Ruthenocene)

Ru-Cu Ruthenium-Copper (Alloy System)

Rud Rudder [also rud]

RUDH Reserve-Shutdown, Unplanned Derated Hours

RUDH1 Reserve Shutdown, Unplanned Derated Hours, Class 1

RUDI Restricted Use Digital Instrument

Rud Metal Zb Rudarsko Metalurski Zbornik [Yugoslavian Journal on Metallurgy]

Rudodobiv Metal Rudodobiv Metalurgiya [Bulgarian Publication on Metallurgy]

RUDP Radarsat User Development Program [of Canadian Space Agency]

RuDP D-Ribulose-1,5-Diphosphate [Biochemistry]

Rudy Met Niezel Rudy i Metale Niezelazne [Polish Metals Journal]

RUEI Ranger Uranium Environmental Inquiry [Fox Report, Australia]

RUF Refractory Users Federation [UK]

RUF Resource Utilization Factor

RUF Revolving Underwriting Facility

RUG Rijksuniversiteit te Gent [University of Ghent, Belgium]

RUG Rijksuniversiteit te Groningen [University of Groningen, Netherlands]

Rugby ball (70-Carbon-Atom-)Buckminsterfullerene Molecule

RUI Resource Use Institute [UK]

RUIN Regional Urban Information Network [US]

RUL Refractoriness under Load [also RuL] [Metallurgy]

RUL Rijksuniversiteit Leiden [University of Leiden, Netherlands]

RUL Rutgers University Library [Piscataway, New Jersey, US]

Rul D-Ribulose [Biochemistry]

RUM Ranger Uranium Mines Proprietary Limited [Australia]

RUM Remote Underwater Manipulator

RUM Remote Utilization Monitor

RUM Royal University of Malta [Valletta]

Rum Rumania(n)

RUMP Ribulose Monophosphate [Biochemistry]

RUMS Resource Use Management Subgroup [of Intergovernmental Panel on Climate Change]

RUN Rewind and Unload

RUN Royal University of Norway [Oslo]

Rundfunktech Mitt Rundfunktechnische Mitteilungen [German Communications on Radio Engineering]

RUNDH Reserve-Shutdown, Unit Derated Hours

R-Unit Roentgen Unit [also R-unit]

RUP Cylindrical Roller Bearing, Single-Row, Non-Locating, Double-Ribbed Outer Ring with One Loose Rib, Inner Ring without Ribs, Both Rings Separable [Symbol]

RUP Rockefeller University Press [New York City, US]

RUP Rutgers University Press [Piscataway, New Jersey, US]

RuP Ribulose Phosphate; Ribulose-5-Diphosphate [Biochemistry]

Ru-5-P D-Ribulose-5-Phosphate [also Ru5P] [Biochemistry]

Ru(PMCp)$_2$ Bis(pentamethylcyclopentadienyl)ruthenium (or Decamethylruthenocene)

RU Press Rockefeller University Press [New York City, US]

RU Press Rutgers University Press [Piscataway, New Jersey, US]

RUPT Interrupt

Rupt Rupture [also rupt]

RUQ Right Upper Quadrant (of Abdomen) [Medicine]

RUR Rosum's Universal Robot

rur rural

RURAL (Society) Responsible for the Use of Resources, Agriculture and Land [UK]

RURAX Rural Automatic Exchange

RUS Resonant Ultrasound Spectroscopy

Rus Russia(n)

RuSc Ruthenium-Scandium (Compound)

RUSDIC Russian Dictionary

RUSG Remote Ultrasonic Stream Gage

RUSH Remote Use of Shared Hardware

RUSI Royal United Service Institute [UK]

RuSi Ruthenium Silicide

RUSS Robotic Ultrasonic Scanning System

Russ Russia(n)

Russ Chem Rev Russian Chemical Reviews [Translation of: *Uspekhi Khimii* ; published in the UK]

Russ J Inorg Chem Russian Journal of Inorganic Chemistry [Translation of: *Zhurnal Neorganicheskoi Khimii*; published in the UK]

Russ J Phys Chem Russian Journal of Physical Chemistry [Translation of: *Zhurnal Fizicheskoi Khimii (USSR)*; published in the UK]

Russ Metall Russian Metallurgy [Translation of: *Izvestiya Akademii Nauk SSSR, Metally*; published in the US]

RUT Resource Utilization Time

Rutgers Comput Technol Law J Rutgers Computer and Technology Law Journal [of Rutgers University, Piscataway, New Jersey, US]

RUVP Research Unit on Vector Pathology [of Memorial University of Newfoundland, St. John's, Canada]

RV Cylindrical Roller Bearing, Multi-Row, Non-Locating, Double-Ribbed Outer Ring, Inner Ring Without Ribs, Both Rings Separable [Symbol]

RV Rabies Virus

RV Radar Vector(ing)

RV Radius Vector [Astronomy]

RV Random Variable [Mathematics]

RV Random Vibration

RV Rated Voltage

RV Reactive Voltage

RV Reaction Vessel

RV Reactor Vessel

RV Rear View

RV Recirculation Valve

RV Recovery Vehicle

RV Recovery Vessel

RV Recreational Vehicle

RV Redwood Viscometer

RV Reentry Vehicle

RV Relative Volume

RV Relief Valve

RV Remaining Velocity

RV Renal Vein [Anatomy]

RV Rendezvous

RV República de Venezuela [Republic of Venezuela]

RV Rescue Vessel

RV Research Vessel [also R/V]

RV Restriking Voltage

RV Retrieval Vessel

RV Return Valve

RV Revised Version

RV Rifle Volunteers

RV Right Ventricle (of the Heart) [Anatomy]

RV Right Ventricular [Anatomy]

RV Robot Vision

RV Rotational Viscometer

RV Rough Vacuum

RV Royal Vessel

RV Rubella Virus

R/V Research Vessel [also RV]

RVA Radar Vectoring Area

RVA Reactive Volt-Ampere (Meter) [also rva]

RVA Recorded Voice Announcement

RVA Relative Virtual Address

RVA Reliability Variation Analysis

r-value normal plastic anisotropy parameter [Mechanics]

r-value right value [Computers]

RVAM Reactive Volt-Ampere Meter

RVB Resonant Valence Bond

RVB Resonating-Valence Band (Model) [Solid-State Physics]

RVC Relative Velocity Computer

RVC Research Volume Cost

RVC Reticulated Vitreous Carbon

RVC Ribonucleoside Vanadyl Complex [Biochemistry]

RVCAT Research Vessel Catch

RVCF Remote Vehicle Checkout Facility

RVCI Royal Veterinary College of Ireland

RVCM Residual Vinyl Chloride Monomer

RVD Rapid Vapor Doping

RVDT Rotary Variable Differential Transducer

RVDT Rotary Variable Differential Transformer

RVE Representative Volume Element

RVF Right Ventricular Failure [Medicine]

RVI Reactive Vapor Infiltration

RVI Reverse Interrupt

RVIA Recreational Vehicle Industry of America [US]

RVIA Recreational Vehicle Inspection Association

RVIA Royal Victoria Institute of Architects [Australia]

RVIS Reactor and Vessel Instrumentation System

RVIT Rotary Variable Inductance Transformer

RVLIS (Nuclear) Reactor Vessel Water Level Indication System

RVM Reactive Voltmeter

RVN Requirements Verification Network

RVO Runway Visibility by Observer

RVP Red Veterinary Petrolatum

RVP Reid Vapor Pressure (of Gasoline)

RVP Rotary Vane Pump

RVPS Remnant Vegetation Protection Scheme [Western Australia]

RVR Refuelling Vapor Recovery

RVR Runway Visual (or Visible) Range

RVRC Runway Visual Range Center

RVRR Runway Visual Range Rollout

RVRT Runway Visual Range Touchdown

RVS Real-Time Visual Simulation

rvs reverse

RVSFC Real Voice Sensitivity Frequency Characteristic [Telephone Handsets]

RVSS (Nuclear) Reactor Vessel Support System

RVT Registered Veterinary Technician

RVT Reliability Verification Tests

RVT Resource Vector Table

RVV Runway Visibility Values

RVV Russell's Viper Venom [Medicine]

RW Moisture-Resistant Rubber Type (Conductor Insulation) [Symbol]

RW Office of Civilian Radioactive Waste Management [of US Department of Energy]

RW Radar Wave

RW Radioactive Waste

RW Radio Wave

RW Random Walk (Method) [Mathematics]

RW Random Width [also rw]

RW Raw Water (Wastewater Treatment)

RW Rayleigh Wave

RW Readily Weldable

RW Read/Write [also R/W]

RW Relative Wind

RW Resistance Welded; Resistance Welding

RW Resistance Wire

RW Response Word

RW Ringworm [Medicine]

RW Rwanda [ISO Code]

Rw Rwanda [East Africa]

R/W Read/Write [also R-W, or RW]

R/W Right-of-Way [also ROW]

R/W Runway

.rw Rwanda [Country Code/Domain Name]

RWA Rotary Wheel Atomizer

RWA Rotating Wave Approximation

RWAA Resistance Welding Alloy Association

RWAGP Rural Women's Access Grant Program

RWAS Royal Welsh Agricultural Society [UK]

R Wave Upward Deflection Following Q Wave of QRS Complex (in Electrocardiogram) [Medicine]

RWC Read, Write and Compare

RWC Read, Write and Compute

RWC Read/Write Continue

RWC Rural Water Commission [of Victoria State, Australia]

RWCC Recycled Water Coordination Committee

RWCS (Nuclear) Reactor Water Cleanup System

RWCS Report Writer Control System

RWCT Radioactive Waste Collection Tank

RWCU (Nuclear) Reactor Water Cleanup Unit

RWD Radioactive Waste Disposal

RWD Reaction With Distillation (Process)

RWD Rear Wheel Drive

RWD Rewind [also Rwd, or rwd]

RWD Right Wing Down

RWE Rheinisch-Westfälisches Elektrizitätswerk [Utility Company of the State of North Rhine-Westphalia, Germany]

RWED Read/Write Extend Delete

Rw Fr Rwandan Franc [also RWF] [Currency of Rwanda]

RWG Ridge Waveguide

RWG Roebling Wire Gauge

RWH Read/Write Head

RWI Radio Wire Integration

RWI Read, Write and Initialize

RWI Rheinisch-Westfälisches Institut für Wirtschaftsforschung [North Rhine-Westphalian Institute for Economic Research, Essen, Germany]

RWI Rockwell International [US]

R/wk Roentgen per week [formerly also r/wk]

RWM Radioactive Waste Management

RWM Read/Write Memory

RWM Rectangular Wave Modulation

RWM Rod Worth Minimizer

RWMA Resistance Welder Manufacturers Association [US]

RWMC Radioactive Waste Management Center [of US Department of Energy in Idaho]

RWMS Radioactive Waste Management Site

RWN Reacting Weight Number [Chemistry]

RWN Rydberg Wave Number [Physics]

RWND Rewind [also Rwnd, or rwnd]

RWO Eight-Wrong-Omit (Counter) [Computers]

RWP Radiation (or Radiological) Work Permit

RWP Radiation (or Radiological) Work Procedure

RWP Radio Wave Propagation

RWP Reaction Wave Polymerization

RWP (Nuclear) Reactor Work Permit

RWPP Rural Water Pricing Policy

RWR Radar Warning Receiver

RWR Read/Write Register

RWR Relative Wear Resistance

RWS Radwaste System [Radioactive Waste Cleanup]

RWS Receiver Waveform Simulation

RWS Ruano, Wadsworth and Sherby (Slip) [Metallurgy]

RWSS River Water Supply System

RWST Refueling Water Storage Tank [Nuclear Engineering]

RWT Right When Tested

RWTH Rheinisch-Westfälische Technische Hochschule [Technical University of Aachen, Germany]

RWTH Aachen Rheinisch-Westfälische Technische Hochschule Aachen [Technical University of Aachen, Germany]

RWTÜV Rheinisch-Westfälischer Technischer Überwachungsverein [Technical Inspection Association of North Rhine-Westphalia, Germany]

RWU Research Work Unit

RWUM Railway Workers Union of Malawi

RWV Remote Work Vehicle

RWW Read-While-Write (Memory)

RWY Register to Indexed Storage

Rwy Railway [also rwy]

Rwy Runway [also rwy]

RX Alkyl Halide [Formula]

RX Receive(r)

RX Receiver Controller

RX Register to Indexed Storage (Operation) [also R-X]

RX Remote Exchange

RX Report Crossing

RX Resolver-Transmitter [also R-X]

R-X Organic Halide [Formula] [*Note:* R Indicates a Hydrogen Group; X a Halogen]

R-X Register to Indexed Storage (Operation) [also RX]

R-X Resolver-Transmitter [also RX]

R/X Reactor Experiments Inc. [US]

Rx Endothermic (Generator)

Rx Medical Prescription

Rx Recrystallized; Recrystallization

R*(x) Residual Term [Numerical Integration]

$R_H(x)$ Hall Coefficient as a Function of Concentration [Symbol]

$\rho(x)$ (electrical) resistivity as a function of concentration [Symbol]

RxC Receiver Clock [Data Processing]

RXD Receive Data

RxD Receive Data from Equipment

RXP Reply and Delivery of Reply to Telegram Prepaid

$\rho(x,T)$ normal-state resistivity [Symbol]

RY Cylindrical Roller Bearing, Single-Row, Two-Direction-Locating, Outer Ring With One Rib and One Internal Snap Ring, Double-Ribbed Inner Ring, Non-Separable [Symbol]

Ry Railway [also ry]

Ry Relay [also ry]

ry rydberg [Unit]

RYA Royal Yachting Association [UK]

ry^{-1}/at one over rydberg per atom [Unit]

Ry Bn Railway Battalion [also RyBn]

RYD Real Year Dollars

RYDMR Radical-Yield-Detected Magnetic Resonance

RYESAC Ryerson Students Administraive Council [of Ryerson Polytechnic University, Toronto, Canada]

ry-fu rydberg-functional unit

Ry Jct Railway Junction [also RyJct, Ry Jn]

Rylty Royalty [also rylty]

R/yr Roentgen per year [formerly also r/yr]

Ry Pt Railway Point [also RyPt]

RYS Royal Yacht Squadron [UK]

Rys Railways [also rys]

RZ Reaction Zone

RZ Reinheitszahl (or "Purity Number") [A Measure of Hemin Content Used in Biochemistry]

RZ Republic of Zambia

RZ République de Zaïre [Republic of Zaire]

RZ Reset-to-Zero

RZ Return-to-Zero (Recording)

RZ *(Roheisenzunder)* – German for "Pig Iron Scale" [Metallurgy]

R&Z Range and Zero

$\rho(z)$ electron density profile [Symbol]

$\rho(z)$ longitudinal density distribution [Symbol]

RZL Return-to-Zero Level

RZM Remote Zero Module

RZM Return-to-Zero Mark

RZM Root Zone Method

RZ(NP) Nonpolarized Return-to-Zero (Recording)

RZ(P) Polarized Return-to-Zero (Recording)

RZS Royal Zoological Society [UK]

RZSI Royal Zoological Society of Ireland

RZSNSW Royal Zoological Society of New South Wales [Australia]

RZSS Royal Zoological Society of Scotland

RZSSA Royal Zoological Society of South Australia

S

S   Addendum (of Bevel Gears, Helical Gears, etc.) [Symbol]

S   Alternating Fatigue Strength at Zero Mean Stress (in Mechanics) [Symbol]

S   Antisubmarine Aircraft [US Navy Symbol]

S   Apparent (Center-to-Center) Spacing in Lamellar Structures (in Quantitative Metallography) [Symbol]

S   Area [Symbol]

S   Axial Loading [Symbol]

S   Basic Shaft System (of Fits) [Symbol]

S   Bias-Point Stability Factor [Semiconductor Symbol]

S   Compliance Tensor (in Mechanics) [Symbol]

S   Concentration of Solute in Saturated Solution [Symbol]

S   Contact Stiffness (of Hardness Indenter) [Symbol]

S   Counterclockwise Configuration of Stereoisomers (e.g., (S)-(+)-sec-Butyl Chloride) [Symbol]

S   Center of Symmetry at the Closed-Packed Site [Crystallographic Symbol Appended to Space Group Denotations]

S   Degree of Ordering (in Solid-State Physics) [Symbol]

S   Depth of (Eddy-Current) Penetration [Symbol]

S   Distance [Symbol]

S   Effective Scattered Neutron Content [Symbol]

S   Electron(ic) Spin (in Physics) [Symbol]

S   Engineering Science and Engineering Technology [Discipline Category Abbreviation]

S   Engineering (or Conventional) Stress [Symbol]

S   Entropy [Symbol]

S   Eshelby's Transformation Tensor (in Solid-State Physics) [Symbol]

S   Fatigue Stress Amplitude (in Mechanics) [Symbol]

S   Filament Spacing (in Superconductors) [Symbol]

S   Glass-Forming Ability [Symbol]

S   Hoop Stress (in Mechanics) [Symbol]

S   Huang-Rhys Parameter (for Electron-Phonon Coupling Strength) [Symbol]

S   Kubelka-Munk Scattering Coefficient (in Physics) [Symbol]

S   (Magnetic Core) Lamination Factor [Symbol]

S   Limit of Detectibility [Symbol]

S   Long-Range Order Parameter (in Solid-State Physics) [Symbol]

S   Loudness [Symbol]

S   Magnetic Viscosity (Coefficient) [Symbol]

S   Mean Particle Surface Area (in Quantitative Metallography) [Symbol]

S   Optical Emission (of Electrons and Electron Holes) [Symbol]

S   Overlap Matrix Element [Symbol]

S   Physical Amount of a Stimulus [Symbol]

S   Position-Dependent Phase Term (in Physics) [Symbol]

S   Poynting Vector (in Physics) [Symbol]

S   Pumping Speed [Symbol]

S   Reflection [Symbol]

S   Relaxation Coefficient [Symbol]

S   Relaxation Rate [Symbol]

S   Saccharase (or Sucrase) [Biochemistry]

S   Saccharin

S   Saccharomyces [Mycology]

S   Sacral; Sacrum [Anatomy]

S   Sacramento [California, US]

S   Saddle

S   Saddle Point (in Mathematics) [Symbol]

S   Safe Stress (in Mechanics) [Symbol]

S   Safe Tensile Strength (of Cylinders and Shells) [Symbol]

S   Safety

S   Safety Factor [Symbol]

S   Sag(ging)

S   Saha Equilibrium (in Optical Spectroscopy) [Symbol]

S   Sahara

S   Saigon [South Vietnam]

S   Saint

S   Saline; Salinity

S   Salix [Genus of Trees Including the Willows]

S   Salmon

S   Salmonella

S   Salonika [Greece]

S   Salt

S   Salzburg [Austria]

S   Sample(r); Sampling

S   Sample Space (in Statistics) [Symbol]

S   Sampling Length [Symbol]

S   Sanaa [Yemen Arab Republic]

S   Sand

S   Sander; Sanding

S   Santiago [Chile]

S   Saprophyte; Saprophytic [Botany]

S   Saponification; Saponifier; Saponify

S   Sapphire

S   Sapporo [Japan]

S   Sarajevo [Bosnia-Hercegowina]

S   Sarcoma [Medicine]

S   Sarcoptes [Genus of Mites]
S   Sardinia [Italian Island]
S   Saskatchewan [Canada]
S   Saskatoon [Saskatchewan, Canada]
S   Sassafras
S   Satellite
S   Satisfaction; Satisfactory [also s]
S   Saturate; Saturation
S   Saturation Ratio (in Meteorology) [Symbol]
S   Saturday
S   Saturn
S   Savannah [Georgia, US]
S   Save
S   Saw(ing)
S   Scalar
S   Scale(r); Scaling
S   Scaling Factor [Symbol]
S   Scandinavia(n)
S   Scan(ner); Scanning
S   Scapula [Anatomy]
S   Scarf(ing) [Metallurgy]
S   Scatter(ing)
S   Scattering Matrix [Symbol]
S   Scavenger; Scavenging
S   Scene
S   Schedule(r); Scheduling
S   Schema(ta); Schematic
S   Schistosoma [Genus of Parasitic Flatworms]
S   Schizophrenia; Schizophrenic [Psychology]
S   Schmid (Tensor Resolution) Factor (in Materials Science) [Symbol]
S   School
S   Schottky (Logic)
S   Science; Scientific; Scientist
S   Scintillation; Scintillator
S   Scope
S   Score
S   Scotland [UK]
S   Scouring
S   Scram [Nuclear Reactors]
S   Scrap
S   Scratch(ing)
S   Screen(ing)
S   Screw(ing)
S   Scribe(r)
S   Scrub(ber); Scrubbing
S   Scuttle [Building Trades]
S   Sea
S   Seal [Zoology]
S   Seal(ing); Sealer
S   Seam(er); Seaming
S   Search(ing)
S   Season
S   Seat(ing)

S   Seattle [Washington, US]
S   Second(ary) [also s]
S   Secondary Wave (or Shear Wave) [Seismology]
S   Second Piola-Kirchhoff Stress [Symbol]
S   Secret(ive)
S   Secretary
S   Secretin [Biochemistry]
S   Secretion
S   Section [also s]
S   Sector
S   Sedan
S   Sediment(ation)
S   Seed(ing)
S   Seebeck Coefficient [Symbol]
S   Segment(ation)
S   Segregation
S   Seine (River) [France]
S   Seismic(s); Seismicity
S   Seismogram; Seismograph
S   Seismologist; Seismology
S   Seizing
S   Seizure
S   Select(ion); Selective; Selectivity
S   Semen
S   Semiconductive; Semiconductivity; Semiconductor
S   Senate; Senator
S   Senegal
S   Sensation
S   Sense; Sensing
S   Sensitive; Sensitivity [also s]
S   Sensitization; Sensitizer
S   Sensor
S   Seoul [South Korea]
S   Separate; Separation; Separator
S   September
S   Septum
S   Sequence(r); Sequential
S   Sequoia
S   Serial; Series [also s]
S   (−)-Serine; Seryl [Biochemistry]
S   Serological; Serologist; Serology
S   Serotonin [Biochemistry]
S   Serrate; Serration
S   Serratia [Genus of Enterobacteria]
S   Serum
S   Server
S   Service
S   Servo
S   Session
S   Set(ting)
S   Settle(r); Settling
S   Settlement
S   Set-Up

S   Severity
S   Seville [Spain]
S   Sewage; Sewer
S   Sex; Sexual(ity)
S   Sextant
S   Seychelles
S   Shackle
S   Shade
S   Shadow
S   Shaft(ing)
S   Shake [Wood]
S   Shaker; Shaking
S   Shale
S   Shanghai [PR China]
S   Shank
S   Shank Length (of Drills) [Symbol]
S   Shape
S   Shaper [Machining]
S   Shaping
S   Shark
S   Sharp(ness)
S   Sharp Line [Spectroscopy]
S   Shatter(ing)
S   Shave
S   Shaving(s)
S   Shear [also s]
S   Shearing
S   Shearing Strength [Symbol]
S   Shear Modulus [Symbol]
S   Shear Stress [Symbol]
S   Shear Wave (or Secondary Wave) [Seismology]
S   Sheath
S   Sheathing
S   Sheave
S   Sheet(ing)
S   Shelf; Shelving
S   Shell
S   Shenyang [PR China]
S   Sherman Function (in Physics) [Symbol]
S   Shield(ing)
S   Shift(ing)
S   Shim(ming)
S   Shingle
S   Ship(ping); Shipper
S   Shock
S   Shock-Resisting Tool Steels [AISI-SAE Symbol] [Usually Followed by a Number Ranging from 1 to 7; e.g., S1]
S   Shop
S   Shore
S   Short(ness)
S   Short Transverse Direction (in Fracture Testing of Rectangular Specimens) [Symbol]
S   Shot
S   Shotting [Metallurgy]

S   Shoulder
S   Shovel(ing)
S   Show
S   Shower
S   Shrink(age)
S   Shroud
S   Shunt(ing)
S   Shut(ter)
S   Shuttle; Shuttling
S   Sicily [Italian Island]
S   Side
S   Side Signal (in Electronics) [Symbol]
S   Siemens [Unit]
S   Sierra [Phonetic Alphabet]
S   Sieve(r); Sieving
S   Sift(er); Sifting
S   Sight
S   Sigmoid(al)
S   Sign
S   Signal(ing)
S   Signal Input [Symbol]
S   Signature
S   Significance; Significant
S   Silage [Agriculture]
S   Silence; Silent
S   Silica (or Silicon Dioxide) [Ceramics]
S   Silicate
S   Silicate Type (Grinding Wheel) Bond
S   Silk
S   Sill
S   Silo
S   Silt(ing)
S   Silurian [Geological Period]
S   Silver [Abbreviation; Symbol: Ag]
S   Similar(ity)
S   Simple; Simplicity
S   Simulation; Simulator
S   Simulium [Species of Black Flies]
S   Simultaneity; Simultaneous
S   Sinai [Egyptian Peninsula]
S   Singapore(an)
S   Single [also s]
S   Single-Pole Switch [Symbol]
S   Single Silk Covered (Magnet Wire)
S   Sink(ing)
S   Sinter(ed); Sintering
S   Siphon
S   Site
S   Size
S   Sizing
S   Skim(mer)
S   Skin
S   Skip(ping)

S   Skip Distance (in Ultrasonic Inspection) [Symbol]
S   Skirt
S   Skull
S   Skunk
S   Sky
S   Slab(bing)
S   Slack
S   Slag(ging)
S   Slake
S   Slate
S   Slave
S   Sledge(hammer)
S   Sleep(ing)
S   Sleeve
S   Slice
S   Slide(r); Sliding
S   Sling
S   Slip(ping)
S   Slit(ter); Slitting
S   Slope [also s]
S   Slotted Tubular Propellant [UK]
S   Slot(ting)
S   Sludge
S   Slug
S   Slugging [Metallurgy]
S   Sluice
S   Slump (of Concrete)
S   Slurry(ing)
S   Slush(ing)
S   Small(ness)
S   Smectic (Phase) [Physical Chemistry]
S   Smelt(er); Smelting
S   Smog
S   Smoke
S   Smooth(ing); Smoothness
S   Snail
S   Snake
S   Snap(per)
S   Snow
S   Snubber
S   Soak(ing)
S   Soap
S   Society
S   Sodar; Sound Radar
S   Sofia [Bulgaria]
S   Soft(ness)
S   Soften(er); Softening
S   Softfont
S   Software
S   Soil
S   Sol [Currency of Peru]
S   Solar
S   Solder(ing)

S   Solenoid
S   Solid(ity) [also s]
S   Solidification; Solidify
S   Solidus [also s]
S   Solubility; Soluble; Solute; Solution; Solvent
S   Solventless Propellant [UK]
S   Solvolysis [Chemistry]
S   Somalia(n)
S   Sommerfeld (Lubrication) Number [Symbol]
S   Son [also s]
S   Sonar
S   Sonic(s)
S   Soot
S   Sorption
S   Sort(er); Sorting
S   Sound
S   Sounder; Sounding
S   Source [also s]
S   Source [Semiconductor Symbol]
S   South(ern)
S   South Pole (of a Magnet)
S   Space
S   Spacer; Spacing
S   (Laminated Magnetic Core) Space Factor [Symbol]
S   Spain; Spaniard; Spanish
S   Span (Width)
S   Spar
S   Spare
S   Spark(ing)
S   Special(ty)
S   Species
S   Specific; Specify; Specification
S   Specimen
S   Spectral
S   Spectrogram
S   Spectrograph(y)
S   Spectrometer; Spectrometry
S   Spectroscope; Spectroscopy
S   Spectrum
S   Speculum
S   Speed [also s]
S   Speedometer
S   Sperm(atozoon); Spermatic [Anatomy]
S   Sphere; Spherical [also s]
S   Spheroid(al)
S   Spike
S   Spill
S   Spin [also s]
S   Spindle
S   Spine
S   Spinel
S   Spinner
S   Spinneret

S   Spinning

S   Spin Operator (in Solid-State Physics) [Symbol]

S   Spin Quantum Number (in Quantum Mechanics) [Symbol]

S   Spiral

S   Spiral Type Galaxies [Classification Consisting of the Letter "S" Followed by the Numeral "0", or the Lower-Case Letters a, b, c, d, or m] [Astronomy]

S   Spirit

S   Spirochete [Microbiology]

S   Spirogyra [Genus of Algae]

S   Spleen; Splenic

S   Spline

S   Split(ter); Splitting

S   Sponge

S   Spontaneity; Spontaneous

S   Spool(ing)

S   Spoon

S   Spore [Biology]

S   Spot(ting)

S   Spray(able); Sprayer; Spraying

S   Spread(er); Spreading

S   Spreading Parameter (in Surface Tension) [Symbol]

S   Spring

S   Spring Temper (in Metallurgy) [Symbol]

S   Sprinkler

S   Sputter(ing)

S   Sputtering Rate [Symbol]

S   Sputtering Yield [Symbol]

S   Square(d); Squaring

S   Squareness (in Solid-State Physics)

S   Stab [Microbiology]

S   Stability; Stabilization; Stabilize(d); Stabilizer; Stable

S   Stachybotrys [Genus of Molds]

S   Stack(ing)

S   (Laminated Magnetic Core) Stacking Factor [Symbol]

S   Staff

S   Stage

S   Stagger(ed); Staggering

S   Stagnation

S   Stain(ing)

S   Stair

S   Stake

S   Stalk [Botany]

S   Stall(ing)

S   Stamen [Botany]

S   Stamp(ing)

S   Stand

S   Standard(ization) [also s]

S   Standard Deviation [Symbol]

S   Standard Error of Estimate [Symbol]

S   (Voltage) Standing-Wave Ratio [Symbol]

S   Stapes (of Middle Ear) [Anatomy]

S   Staphylococcus [Microbiology]

S   Staple(r)

S   Star

S   Starboard

S   Starch

S   Start(ing); Starter

S   State

S   Statement

S   Static(s)

S   Station(ary)

S   Statistic(al); Statistician; Statistics

S   Stator

S   Status

S   Stay

S   Steadiness; Steady

S   Steam

S   Steamer

S   Steel

S   Steer(ing)

S   Stellar

S   Stem

S   Step

S   Sterile; Sterility; Sterilize(d); Sterilizer; Sterilization

S   Sternum [Anatomy]

S   Stewart Number (in Physics) [Symbol]

S   Stick(ing)

S   Sticking Coefficient (in Physical Chemistry) [Symbol]

S   Stiff(ness)

S   Stiffen(ing)

S   Stigeoclonium [Genus of Algae]

S   Stigma(tic); Stigmator

S   Stimulation

S   Stipe [Biology]

S   Stir(rer); Stirring

S   Stitch(ing)

S   Stock

S   Stockholm [Sweden]

S   Stoichiometric; Stoichiometry

S   Stoke [now St] [Unit]

S   Stoker

S   Stokes (Spectral Lines)

S   Stoking [Powder Metallurgy]

S   Stomach [Anatomy]

S   Stone

S   Stop(ping)

S   Stopper

S   Stope; Stoping [Mining]

S   Storage

S   Store

S   Storm

S   Stove

S   Straight(ness)

S   Strain

S   Strainer

S   Strain-Rate Sensitivity (in Mechanics) [Symbol]
S   Strand
S   Strangeness (in Particle Physics) [Symbol]
S   Strasbourg [France]
S   Strategic; Strategy
S   Stratum [Geology]
S   Stratus [Meteorology]
S   Stream(ing)
S   Streamer
S   Strength [Symbol]
S   Streptobacillus [Microbiology]
S   Streptococcus [Microbiology]
S   Streptomyces [Microbiology]
S   Stress [Symbol]
S   Stress Tensor [Symbol]
S   Stretch(ing)
S   Striate(d); Striation
S   Strike
S   String
S   Strip
S   Stripper; Stripping
S   Strobe
S   Stroke
S   Strong [also s]
S   Strong [Particle Physics]
S   Structural; Structure
S   Structure Factor (in Solid-State Physics) [Symbol]
S   Strut
S   Stub
S   Stud
S   Student; Study
S   Stuff(ing)
S   Stuttgart [Germany]
S   Style
S   Stylus
S   S-Type Structural Steel Shape [AISI/AISC Designation]
S   Styrene
S   Subject
S   Sublimate; Sublime(d); Sublimation [also s]
S   Submarine
S   Substance
S   Substation
S   Substituent
S   Substitute; Substitution(al)
S   Substrate [also s]
S   Sucrase (or Saccharase) [Biochemistry]
S   Sucre [Capital of Bolivia]
S   Sucre [Currency of Ecuador]
S   (+)-Sucrose [Biochemistry]
S   Suction
S   Suction Specific Speed [Symbol]
S   Sudan(ese)
S   Suez [Egypt]

S   Sugar
S   Suggest(ion)
S   Sulfur [Symbol]
S   Sum [Symbol]
S   Sumatra [Indonesian Island]
S   Sum(ming); Summation
S   Summer(time)
S   Sump
S   Sun
S   Sunday
S   Superconducting; Superconductive; Superconductivity; Superconductor
S   Superior(ity)
S   Superposition(ed); Superpositioning
S   Supersaturate(d); Supersaturation
S   Supersaturation Ratio [Symbol]
S   Supersonic(s)
S   Supervision; Supervisory
S   Supplement(ary)
S   Supplier; Supply
S   Support(ing)
S   Suppress(ion); Suppressor
S   Surabaya [Java]
S   Surface [also s]
S   Surface Area [Symbol]
S   Surface Area, or Interface (of Curved Surfaces) (in Quantitative Metallography) [Symbol]
S   Surface Coverage [Symbol]
S   Surge
S   Surgeon; Surgery; Surgical
S   Suriname [South America]
S   Surround(ings)
S   Survey(ing)
S   Surveyor
S   Survival
S   Susceptible; Susceptiblity
S   Suspect
S   Suspend(ed); Suspension
S   Swage; Swaging
S   Swap(ping)
S   Swaziland [Southeast Africa]
S   Swede(n); Swedish
S   Sweep(ing)
S   Swing(ing)
S   Switchyard
S   Swiss; Switzerland
S   Switch(ed); Switching
S   Switchboard
S   Swivel(ing)
S   Sydney [Australia]
S   Symbiosis; Symbiotic [Ecology]
S   Symbol(ization)
S   Symmetrical; Symmetry
S   Symposium

S Symptom(atic)

S Synapse; Synaptic [Anatomy]

S Synchro; Synchronicity; Synchronous; Synchronization; Synchronized

S Synchrotron

S Synchroscope; Synchroscopy

S Syndrome

S Synergism

S Synergistic(s)

S Synergy

S Synonym(ous)

S Synthesis; Synthesize; Synthetic

S Syracuse [New York State, US]

S Syria(n)

S Syringe

S System

S Systematic(s)

S Systematic Error

S Systemization

S Systole [Medicine]

S Tensile Strength [Symbol]

S Thermopower (or Thermoelectric Power) [Symbol]

S Thio (Radical) [Symbol]

S (Maximum) Tire Speed of 112 mph (or 180 km/h) [Rating Symbol]

S Total Linear Stopping Power (in Physics) [Symbol]

S Total Shear Across a Section [Symbol]

S Traction [Symbol]

S Ultimate Compressive Strength of Column Material [Symbol]

S Wigner-Seitz Radius (in Physics) [Symbol]

S- Sigorsky Helicopter [Followed by a Numeral, e.g., S-76]

-S Antisubmarine Warfare Aircraft [US Navy Suffix]

S+ S-Plus [Statistical Software Package]

S$_∥$ In-Plane (Magnetic) Squareness [Symbol]

S$_∥$ Momentum Transfer Parallel to Surface [Symbol]

S$_⊥$ Momentum Transfer Perpendicular to Surface [Symbol]

S$_⊥$ Perpendicular (Magnetic) Squareness [Symbol]

S* (Magnetic) Coercivity Squareness [Symbol]

S* Effective Spin (in Physics) [Symbol]

S′ (Earthquake) Shear Wave that has Passed through the Earth's Core [Symbol] [also SKS]

S′ Stress Deviator (for Soil, Rock, etc.) [Symbol]

S° Standard Entropy [Symbol]

$\overline{S}$ Average (or Mean) Surface (Area) [Symbol]

$\overline{S}$ Entropy of Mixing [Symbol]

$\overline{S}$ Molar Entropy [Symbol]

$\overline{S}$ Upper Sum [Symbol]

$\underline{S}$ Lower Sum [Symbol]

S^0 Neutral Soliton (in Physics) [Symbol]

S$_0$ Entropy at 0°C [Symbol]

S$_0$ Equilibrium Wigner-Seitz Radius (in Physics) [Symbol]

S$_0$ Flow Stress [Symbol]

S$_0$ Interlamellar, or True (Center-to-Center) Spacing in a Lamellar Structure (in Quantitative Metallography) [Symbol]

S$_0$ Intrinsic Shear Strength (of a Rock) [Symbol]

S$_0$ Lenticular Type of Galaxy [Astronomy]

S$_0$ Spherical Stress Tensor (for Hydrostatic Stress State) [Symbol]

S^{-1} One per Siemens [or 1/S]

S$_1$ Ground Speed Out [Aeronautics]

S$_1$ Seidel Term for Axial Spherical Aberration [Optics]

S$_1$ Solid 1 [Symbol]

S$_1$ Solid Phase 1 [Symbol]

S^{2-} Sulfur Ion [also S⁻] [Symbol]

S$_2$ Double-Pole Switch [Symbol]

S$_2$ Ground Speed Back [Aeronautics]

S$_2$ Seidel Term for Coma [Optics]

S$_3$ Three-Way Switch [Symbol]

S$_3$ Seidel Term for Astigmatism [Optics]

S^4 Standard Submarine Sonar System [also SSSS, or S4]

S$_4$ Four-Way Switch [Symbol]

S$_4$ Seidel Term for Curvature of Field [Optics]

S$_5$ Seidel Term for Distortion [Optics]

S$_8$ Molecular Sulfur [Symbol]

S$_{10}$-S$_{18}$ Beta-Cellulose

S$_{18}$ Gamma-Cellulose

S$_{300}$ Room-Temperature Thermoelectric Power [Symbol]

S$_a$ Alternating Load (or Stress) [Symbol]

S$_a$ Alternating Stress Amplitude [Symbol]

S$_a$ Apparent Specific Gravity [Symbol]

S$_a$ Entropy of Activation [Symbol]

S$_a$ Loading Amplitude (in Fatigue) [Symbol]

S$_a$ Stress Amplitude [Symbol]

S$_α$ Relative Sensitivity Factor of Element $α$ [Symbol]

S$_{ax}$ Alternating Stress at Mean Stress S$_{mx}$ [Symbol]

S$_b$ Bending Stress (for Springs) [Symbol]

S$_b$ Grain Boundary Entropy (in Materials Science) [Symbol]

S$_b$ Modulus of Rupture (in Bending) [Symbol]

S$_{BET}$ Specific Brunauer-Emmett-Teller Surface Area (in Physical Chemistry) [Symbol]

S$_C$ Residual Stress of Coating [Symbol]

S$_c$ Configurational Entropy [Symbol]

S$_{CB}$ Circuit Breaker [Symbol]

S$_{col}$ Linear Collision Stopping Power (in Physics) [Symbol]

S$_{cu}$ Compressive Ultimate Strength [Symbol]

S$_{cy}$ Compressive Yield Strength [Symbol]

S$_D$ Automatic Door Switch [Symbol]

S$_D$ Direct Set (Flip-Flop) [Symbol]

S$_d$ Dimensionless Sensitivity (of a Physical Quantity to a Stimulus) [Symbol]

S$_Δ$ Delta Sensitivity (in Drude Theory) [Symbol]

S$_E$ Electrolier Switch [Symbol]

S$_E$ Surface (Area) of Electrode [Symbol]

S$_e$ Effective Saturation [Symbol]

S$_e$ Electronic Stopping Power [Symbol]

S$_e$ Endurance Limit [Symbol]

S$_{en}$ Endurance Limit (in Fatigue) [Symbol]

S$_F$ Fused Switch [Symbol]

S$_F$ Sensitivity Factor (in Spectroscopy) [Symbol]

S$_f$ Fracture Strength [Symbol]

S$_f$ Fatigue-Strength Limit [Symbol]

S$_f$ Formation Entropy [Symbol]

S$_f$ Steady-State Strain-Rate Sensitivity [Symbol]

S$_f$ Fatigue (Endurance) Limit [Symbol]

S$_G$ Entropy of Gas [also S$_g$] [Symbol]

S$_g$ Deviation Parameter [Symbol]

S$_g$ Excitation Error (in Transmission Electron Microscopy) [Symbol]

S$_h$ Solid in Horizontal Plane (in Geometric Projection) [Symbol]

S$_I$ Interaction Energy [Symbol]

S$_i$ Average Spacing [Symbol]

S$_i$ Excess Entropy [Symbol]

S$_i$ Image Distance (in Optics) [Symbol]

S$_i$ Image Size (in Radiography) [Symbol]

S$_i$ Instantaneous Strain-Rate Sensitivity [Symbol]

S$_{it}$ Torsional Stress due to Initial Spring Tension [Symbol]

S$_K$ Key-Operated Switch [Symbol]

S$_L$ Entropy of Liquid [also S$_l$] [Symbol]

S$_L$ Load Range (in Fatigue Loading) [Symbol]

S$_m$ Bulk Specific Gravity (or Specific Mass Gravity) [Symbol]

S$_m$ Mean Stress (or Steady-Stress Component) [Symbol]

S$_m$ (Fatigue, or Yield) Strength of Material [Symbol]

S$_m$ Semenov Number (in Reaction Kinetics) [Symbol]

S$_{max}$ Maximum Stress [Symbol]

S$_{MC}$ Momentary Contact Switch [Symbol]

S$_{min}$ Minimum Stress [Symbol]

S$_{mx}$ Mean Stress at Alternating Stress S$_{ax}$ [Symbol]

S^n N-Space (in Mathematics) [Symbol]

S$_N$ Fatigue Strength at N Cycles [Symbol]

S$_n$ Space of n Dimensions [Symbol]

S$_o$ Object Distance (in Optics) [Symbol]

S$_o$ Object Size (in Radiography) [Symbol]

S$_p$ Interparticle Separation [Symbol]

S$_P$ Switch and Pilot Lamp [Symbol]

S$_p$ Solid in Profile Plane (in Geometric Projection) [Symbol]

S$_p$ Stopping Power (in Physics) [Symbol]

S$_\psi$ Psi Sensitivity (in Drude Theory) [Symbol]

S$_r$ Percent (or Degree of) Saturation (of a Soil, or Rock Mass) [Symbol]

S$_r$ Stress Range [Symbol]

S$_r$ Strouhal Number (in Mechanics) [Symbol]

S$_{rad}$ Linear Radiative Stopping Power [Symbol]

S$_{RC}$ Romte-Control Switch [Symbol]

S^s Surface Entropy [Symbol]

S$^\sigma$ Interfacial Entropy (in Materials Science) [Symbol]

S$_s$ Modulus of Rupture (in Torsion) [Symbol]

S$_s$ Specific Gravity (of Soilds) [Symbol]

S$_{SI}$ Specific Internal Surface Area [Symbol]

S$_{su}$ Shear Ultimate Strength [Symbol]

S$_{sy}$ Shear Yield Strength [Symbol]

S$_T$ Entropy at Temperature T [Symbol]

S$_T$ Total Spin [Symbol]

S$_t$ Degree of Sensitivity (of Clay) [Symbol]

S$_t$ Remodeling Sensitivity (or Sensitivity Ratio of Soil) [Symbol]

S$_t$ Tensile Strength [Symbol]

S$_t$ Torsional Stress (for Springs) [Symbol]

S$_u$ Ultimate (Tensile) Strength [Symbol]

S$_V$ Entropy of Vapor [also S$_v$] [Symbol]

S$_V$ Surface Area per Unit Volume [Symbol]

S$_V$ Surface, or Interface Area Divided by Total Test Volume (or Surface-to-Volume Ratio) (in Quantitative Metallography) [Symbol]

S$_v$ Grain Boundary Area (in Materials Science) [Symbol]

S$_v$ Solid in Vertical Plane (in Geometric Projection) [Symbol]

S$_v$ Viscosity Coefficient [Symbol]

S$_{WCB}$ Weatherproof Circuit Breaker [Symbol]

S$_{WF}$ Weatherproof Fused Switch [Symbol]

S$_{WP}$ Weatherproof Switch [Symbol]

S$_x$ Normal (Compressive, or Tensile) Stress in X-Direction [Symbol]

S$_x$ Relative Sensitivity Factor of Element x [Symbol]

S$_{xy}$ Shear Stress in Y-Z Plane with X-Axis Perpendicular and Y-Axis Parallel [Symbol]

S$_{xz}$ Shear Stress in Y-Z Plane with X-Axis Perpendicular and Z-Axis Parallel [Symbol]

S$_Y$ Yield Strength [Symbol]

S$_y$ (Pure) Axial Loading [Symbol]

S$_y$ Normal (Compressive, or Tensile) Stress in Y-Direction [Symbol]

S$_y$ Yield Strength [Symbol]

S$_{yx}$ Shear Stress in X-Z Plane with Y-Axis Perpendicular and X-Axis Parallel [Symbol]

S$_{yz}$ Shear Stress in X-Z Plane with Y-Axis Perpendicular and Z-Axis Parallel [Symbol]

S$_z$ Normal (Compressive, or Tensile) Stress in Z-Direction [Symbol]

S$_{zx}$ Shear Stress in X-Y Plane with Z-Axis Perpendicular and X-Axis Parallel [Symbol]

S$_{zy}$ Shear Stress in X-Y Plane with Z-Axis Perpendicular and Y-Axis Parallel [Symbol]

S2 Shade Two [Safety Glasses Designation]

S3 Shade Three [Safety Glasses Designation]

S4 Standard Submarine Sonar System [also SSSS, or S^4]

S5 Shade Five [Safety Glasses Designation]

S$_N$1 Substitution Nucleophilic Unimolecular Reaction [Symbol]

S$_N$2 Substitution Nucleophilic Bimolecular Reaction [Symbol]

S-31 Sulfur-31 [also ^{31}S, or S^{31}]

S-32 Sulfur-32 [also ^{32}S, S^{32}, or S]

S-33 Sulfur-33 [also ^{33}S, or S^{33}]

S-34 Sulfur-34 [also ^{34}S, or S^{34}]

S-35 Sulfur-35 [also ^{35}S, or S^{35}]

S-36 Sulfur-36 [also ^{36}S, or S^{36}]

S-37 Sulfur-37 [also ^{37}S, or S^{37}]

2S 2-Sulfate

4S 4-Sulfate

6S 6-Sulfate

12S4 1,4,7,10-Tetrathiacyclododecane

14S4 1,4,8,11-Tetrathiacyclotetradecane

15S5 1,4,7,10,13-Pentathiacyclopentadecane

18S6 1,4,7,10,13,16-Hexathiacyclooctadecane

24S6 1,5,9,13,17,21-Hexathiacyclotetracosane

s actual space width (for involute splines) [Symbol]

s addendum at small end of milled bevel gear teeth [Symbol]

s adsorption probability [Symbol]

s arc length (of a curve) [Symbol]

s area [Symbol]

s bending stress [Symbol]

s condensation (in sound) [Symbol]

s deformation (of springs) [Symbol]

s displacement [Symbol]

s distance [Symbol]

s dry static energy [Symbol]

s engineering (or conventional) stress [Symbol]

s entropy [Symbol]

s excitation error (in transmission electron microscopy) [Symbol]

s hydraulic gradient [Symbol]

s Laplace transform operator $s = \sigma + j\omega$ [Symbol]

s lift (height) [Symbol]

s linear distance [Symbol]

s maximum tensile stress in line-loaded compression testing [Symbol]

s nominal stress [Symbol]

s normal stress [Symbol]

s path length [Symbol]

s sagitta (of an arc) [Symbol]

s sample standard deviation [Symbol]

s satisfaction; satisfactory [also S]

s scattering

s scruple [Unit]

s search text (command) [Edlin MS-DOS Line Editor]

s second [also sec, or ″] [Unit of Time and Plane Angle]

s second(ary) [also s]

s section [also S]

s sedimentation coefficient [Symbol]

s see

s segregation factor (in metallurgy) [Symbol]

s semiperimeter of a triangle [Symbol]

s sensitivity [also S]

s series [also S]

s set

s shear [also S]

s shear strength [Symbol]

s side [Symbol]

s sideways-type quark (in particle physics) [Symbol]

s sigma [English Equilvalent]

s *(signa)* – label [Medical Prescriptions]

s signed

s single [also S]

s slant height of a geometric solid [Symbol]

s slope [also S]

s small

s soft

s sol

s solid [also S]

s solid (state) [Chemistry]

s solidus [also S]

s soluble

s son [also S]

s s-orbital (in physics) [Symbol]

s source [also S]

s source [Semiconductor Symbol]

s spacing [Symbol]

s span width [Symbol]

s specific entropy [Symbol]

s specific heat [Symbol]

s speed [also S]

s sphere; spherical [also S]

s spin [also S]

s spin quantum number (in quantum mechanics) [Symbol]

s stable

s standard [also S]

s standard deviation [Symbol]

s standing-wave ratio [Symbol]

s static

s steel [also S]

s step free energy [Symbol]

s stere [Unit]

s sticking coefficient (in physical chemistry) [Symbol]

s strange quark (in particle physics) [Symbol]

s stress [Symbol]

s stretch (of springs) [Symbol]

s strong

s sublimation; sublime(d); sublimes [also S]

s substrate [also S]

s sum [also S] [Symbol]

s surface [also S]

s surface area [Symbol]

s symmetrical [also S]

s synchronous [also S]

s$_0$ nearest neighbor distance (in crystallography) [Symbol]

s^{-1} one per second [also sec^{-1}, 1/s, or 1/sec]

s$_1$ entropy of phase 1 [Symbol]

s$_1$ specific heat of phase 1 [Symbol]

s^{-2} one per second squared [also sec^{-2}, 1/s^2 or 1/sec^2]

s^2 (unbiased) sample variance (in statistics) [Symbol]

s^2 square second [also sq s]

$\hat{s}^2$ biased sample variance (in statistics) [Symbol]

s^E molar excess entropy [Symbol]

s_E effort distance (for simple machines) [Symbol]

s^H compliance at constant magnetic field [Symbol]

s_k sum of the k-th powers of the roots of a polynomial equation [Symbol]

$s_k(n)$ sum of the k-th powers of the first n positive integers (e.g., $s_k(n) = 1^k + 2^k + \cdots + n^k$) [Symbol]

s_m mean stress (or steady-stress component) [Symbol]

s_{max} maximum stress [Symbol]

s_{min} minimum stress [Symbol]

s_n sum of first n terms of a sequence, series, or progression [Symbol]

s_R reproducibility standard deviation [Symbol]

s_R resistance force (for simple machines) [Symbol]

s_r repeatability standard deviation [Symbol]

s_{tr} transformation shear (in metallurgy) [Symbol]

s_v variable stress [Symbol]

s_w allowable working stress [Symbol]

s_x strain component in x-direction [Symbol]

s_y strain component in y-direction [Symbol]

s_z intrinsic spin (in physics) [Symbol]

s_z strain component in z-direction [Symbol]

Σ Areal Density (of Polymer Chains) [Symbol]

Σ Flexural Stiffness (in Mechanics) [Symbol]

Σ Interface (in Materials Science) [Symbol]

Σ Interfacial (or Surface) Free Energy [Symbol]

Σ Reciprocal Density of Coincidence Grain Boundary Sites (in Materials Science) [Symbol]

Σ Scattering Cross-Section (in Physics) [Symbol]

Σ Shaft Angle (of Bevel Gears) [Symbol]

Σ (Upper-case) Sigma [Greek Alphabet]

Σ Sigma Particle (or Sigma Hyperon) (in Particle Physics) [Symbol]

Σ Stress Tensor [Symbol]

Σ Summation (or Sum of Terms) [Symbol]

Σ^+ Sigma-Plus Particle (i.e., Positively Charged Sigma Hyperon) (in Particle Physics) [Symbol]

$\overline{\Sigma}^+$ Sigma-Plus Antiparticle (i.e., Positively Charged Sigma Antiparticle) (in Particle Physics) [Symbol]

Σ^- Sigma-Minus Particle (i.e., Negatively Charged Sigma Hyperon) (in Particle Physics) [Symbol]

$\overline{\Sigma}^-$ Sigma-Minus Antiparticle (i.e., Negatively Charged Sigma Antiparticle) (in Particle Physics) [Symbol]

Σ^0 Sigma-Zero Particle (i.e., Neutral Sigma Hyperon) (in Particle Physics) [Symbol]

$\overline{\Sigma}^0$ Sigma-Zero Antiparticle (i.e., Neutral Sigma Antiparticle) (in Particle Physics) [Symbol]

Σ_a True Linear Absorption Coefficient [Symbol]

Σ_s Step Free Energy [Symbol]

Σ_t Linear Absorption Coefficient [Symbol]

σ bonding sigma (molecular) orbital (in physics) [Symbol]

σ Cauchy stress tensor [Symbol]

σ cavitation number (in fluid mechanics) [Symbol]

σ Debye-Waller factor (in solid-state physics) [Symbol]

σ deviator stress (for soil, or rock) [Symbol]

σ electrical conductivity [Symbol]

σ electron spin [Symbol]

σ (tensile, or compressive) engineering (or conventional) stress [Symbol]

σ flexural strength [Symbol]

σ flow stress [Symbol]

σ helix angle (of helical gears) [Symbol]

σ interfacial energy [Symbol]

σ ionic strength (in physical chemistry) [Symbol]

σ ionization cross-section (in physics) [Symbol]

σ loss angle [Symbol]

σ magnetic leakage coefficient [Symbol]

σ (mass) magnetization [Symbol]

σ mean random, or intercept (center-to-center) spacing in a lamellar structure (in quantitative metallography) [Symbol]

σ mean stress (or steady-stress component) [Symbol]

σ molecular diameter [Symbol]

σ nondimensional grafting density (of polymer chains) [Symbol]

σ (simple) normal (tensile, or compressive) stress [Symbol]

σ nuclear cross-section [Symbol]

σ propagation constant (in electrical engineering) [Symbol]

σ radius of torsion [Symbol]

σ root-mean-square roughness [Symbol]

σ scattering cross-section [Symbol]

σ short-range order parameter (in solid solutions) [Symbol]

σ (lower-case) sigma [Greek Alphabet]

σ sigma bond (in physical chemistry) [Symbol]

σ sigma phase (e.g., in iron-chromium alloys and complex body-centered crystalline materials) [Symbol]

σ sigma orbital (in physical chemistry) [Symbol]

σ smectic susceptibility (of liquid crystals) [Symbol]

σ specific heat of electricity [Symbol]

σ specific conductance [Symbol]

σ spin (of particle) [Symbol]

σ spin quantum number [Symbol]

σ (population) standard deviation (of a statistical distribution) [Symbol]

σ Stefan-Boltzmann constant (5.6705×10^{-8} W/(m²·K⁴)) [Symbol]

σ stereo angle (in electron stereomicroscopy) [Symbol]

σ stress [Symbol]

σ (chemical) substituent constant [Symbol]

σ supersaturation factor (in physical chemistry) [Symbol]

σ surface charge density [Symbol]

σ surface energy per unit area [Symbol]

σ surface stress [Symbol]

σ surface tension [Symbol]

σ (simple) tensile stress [Symbol]

σ Thomson coefficient (in physics) [Symbol]

σ true stress [Symbol]

σ traction [Symbol]

σ (statistical) variance [Symbol]

$\bar{\sigma}$ effective stress (or effective pressure) (in a soil mass) [Symbol]

$\bar{\sigma}$ mean strength [Symbol]

$\bar{\sigma}$ spectrum averaged cross section (in radiography and dosimetry) [Symbol]

$\sigma_{\parallel}$ (electrical) conductivity in (or along) composite fiber direction [Symbol]

$\sigma_{\parallel}$ longitudinal (electrical) conductivity [Symbol]

$\sigma_{\perp}$ (electrical) conductivity across composite fiber direction [Symbol]

$\sigma_{\perp}$ transverse (electrical) conductivity [Symbol]

σ^* antibonding sigma orbital (in physics) [Symbol]

σ^* long-range hardening stress [Symbol]

σ' combined normal stress [Symbol]

σ_0 constant applied stress (in viscoelastic polymers) [Symbol]

σ_0 critical shear stress [Symbol]

σ_0 extinction length (in electron microscopy) [Symbol]

σ_0 flow stress [Symbol]

σ_0 friction stress [Symbol]

σ_0 initial stress [Symbol]

σ_0 short-range stress field [Symbol]

σ_0 uniform stress [Symbol]

σ_0 yield stress [Symbol]

$\sigma_{0.2}$ macroscopic yield strength [Symbol]

σ_1 maximum stress [Symbol]

σ_1 major principal stress (within a body) [Symbol]

σ^2 (statistical) population variance [Symbol]

σ_2 minimum stress [Symbol]

σ_2 intermediate principal stress (within a body) [Symbol]

σ_3 minor principal stress (within a body) [Symbol]

σ_a annealed-state (dark) conductivity [Symbol]

σ_a applied stress [Symbol]

σ_a stress amplitude (in fatigue) [Symbol]

σ_a meridional stress (of rotationally symmetric bodies) [Symbol]

σ_{ab} in-plane conductivity (of superconductors) [Symbol]

$\sigma_{a,m}$ residual axial stress in (composite) matrix [Symbol]

σ_b back stress [Symbol]

σ_b bending stress [Symbol]

σ_b bulk conductivity [Symbol]

σ_c calculated stress [Symbol]

σ_c capture thermal neutron cross section [Symbol]

σ_c crack strength [Symbol]

σ_c crazing stress [Symbol]

σ_c critical stress (for crack propagation) [Symbol]

σ_c isostress of composite [Symbol]

σ_c tensile stress to be applied to produce critical resolved shear stress [Symbol]

σ_c^u ultimate tensile strength of composite [Symbol]

σ_{CRSS} critical resolved shear stress [also σ_{crss}] [Symbol]

σ_{cu} composite ultimate strength [Symbol]

σ_{cy} cyclic normal stress [Symbol]

σ_{cy} yield strength of composite [Symbol]

σ_d dark conductivity [Symbol]

σ_d design stress [Symbol]

σ_{dc} direct-current conductivity [Symbol]

σ_e effective (electrical) conductivity [Symbol]

σ_e effective stress (or local stress) [Symbol]

σ_e electrical (or electronic) conductivity [Symbol]

σ_e scattering cross-section for electrons [Symbol]

σ_e shear stress [Symbol]

σ_e Thomson cross section (0.665246×10^{-28} m²) [Symbol]

σ_{eff} effective stress [Symbol]

σ_f fiber tensile strength (of composites) [Symbol]

σ_f film stress (i.e., stress in thick and thin films) [Symbol]

σ_f fission thermal neutron cross section [Symbol]

σ_f flow stress [Symbol]

σ_f (tensile) fracture strength [Symbol]

σ_f fracture stress [Symbol]

σ_f friction(al) stress [Symbol]

σ_f isostress of composite fibers [Symbol]

σ_f' fatigue-strength coefficient [Symbol]

σ_f^u ultimate tensile strength of composite fibers [Symbol]

$\bar{\sigma}_f$ average fiber stress (of composites) [Symbol]

σ_{fs} flexural strength [σ_{fs}] [Symbol]

σ_{fu} fiber ultimate strength (of composites) [Symbol]

σ_f fiber stress (of fiber-reinforced composites) [Symbol]

σ_f fiber tensile strength (of composite) [Symbol]

σ_f fluid conductivity [Symbol]

σ_f fracture stress [Symbol]

σ_f (overal applied) stress at failure [Symbol]

σ_{fs} flexural strength [Symbol]

σ_g grain boundary energy (in materials science) [Symbol]

σ_{gb} grain boundary conductivity (in materials science) [Symbol]

σ_{gm} standard deviation of geometric mean diameter (of a liquid drop) [Symbol]

σ_H Thoma cavitation parameter (in mechanical engineering) [Symbol]

σ_h (electron) hole capture cross-section [Symbol]

σ_h hydrostatic pressure [σ_{hyd}] [Symbol]

σ_i Debye-Waller type interfacial roughness factor (in solid-state physics) [Symbol]

σ_i interfacial surface energy (in material science) [Symbol]

σ_i internal stress [Symbol]

σ_i ionic conductivity [Symbol]

σ_i multigroup cross section (in radiography) [Symbol]

σ_L longitudinal stress (of composite) [Symbol]

σ_{LV} liquid/vapor Interfacial tension (in physical chemistry) [Symbol]

σ_{LYS} lower yield stress [also σ_{lys}] [Symbol]

σ_m isostress of (composite) matrix [Symbol]

σ_m (electrical) conductivity of (composite) matrix [Symbol]

σ_m maximum stress (at crack tip) [Symbol]

σ_m mean stress (at crack tip) [Symbol]

σ_m^u ultimate tensile strength of (composite) matrix [Symbol]

σ_{mr} modulus of rupture [Symbol]

σ_N nominal (net-section) stress (in fracture testing) [Symbol]

σ_n normal stress [Symbol]

σ_O Orowan stress (in metallurgy) [Symbol]

σ_o overgrowth surface energy (in metallurgy) [Symbol]

σ_{ol} overlap grafting density (of polymer chains) [Symbol]

σ_P Peierls stress (in solid-state physics) [Symbol]

σ_p cavitation number (in fluid mechanics) [Symbol]

σ_p particle conductivity [Symbol]

σ_p peak stress [Symbol]

σ_p plastic flow stress [Symbol]

σ_p^u ultimate tensile strength of (composite) particles [Symbol]

σ_{ph} photoconductivity [Symbol]

σ_{ppc} persistent-photoconductivity-state (dark) conductivity [Symbol]

σ_r radial stress (component) [Symbol]

σ_r range of stress [Symbol]

σ_r resolved shear stress [Symbol]

σ_{red} reduced (electrical) conductivity [Symbol]

$\sigma_{r,m}$ residual radial stress in (composite) matrix [Symbol]

$\sigma_{r\theta}$ stress in polar coordinate system [Symbol]

σ_s Debye-Waller type surface roughness factor (in solid-state physics) [Symbol]

σ_s saturation stress [Symbol]

σ_s sharp-notch strength (in mechanical testing) [Symbol]

σ_s sintering stress (in metallurgy) [Symbol]

σ_s (substrate) surface energy [Symbol]

σ_{SL} solid/liquid interfacial tension (in physical chemistry) [Symbol]

σ_{ss} steady-state stress [Symbol]

σ_{SV} solid/vapor interfacial tension (in physical chemistry) [Symbol]

σ_T longitudinal stress (in composites) [Symbol]

σ_T thermal stress [Symbol]

σ_T true stress [also σ_t, or σ_{tr}] [Symbol]

σ_t axial transition stress [Symbol]

σ_t torsional stress [Symbol]

σ_t total (electrical) conductivity [Symbol]

σ_t true stress [also σ_T, or σ_{tr}] [Symbol]

σ_θ hoop stress (or stress in circumferential direction) [Symbol]

σ_{th} threshold stress (for stress-corrosion cracking) [Symbol]

σ_{tr} true stress [also σ_T, or σ_t] [Symbol]

σ_{TS} tensile strength [also σ_{ts}] [Symbol]

σ_φ tangential stress (of rotationally symmetric bodies) [Symbol]

σ_u (monotonic) ultimate tensile strength [Symbol]

σ_{uc} cyclic ultimate tensile strength [Symbol]

σ_{UTS} ultimate tensile strength [also σ_{uts}] [Symbol]

σ_{UYS} upper yield stress [also σ_{uys}] [Symbol]

σ_w Bloch interfacial wall energy (in solid-state physics) [Symbol]

σ_w working stress (or safe stress) [Symbol]

σ_x normal stress in x-direction [Symbol]

σ_x estimate of process standard deviation [Symbol]

$\sigma_{\bar{x}}$ standard deviation of $\bar{x}$ [Symbol]

σ_{xx} stress along x-x axis [Symbol]

σ_{xy} shear stress in Y-Z plane with x-axis perpendicular and y-axis parallel [Symbol]

σ_{xz} shear stress in Y-Z plane with x-axis perpendicular and z-axis parallel [Symbol]

σ_Y (effective) yield strength [also σ_y] [Symbol]

σ_y normal stress in y-direction [Symbol]

σ_y uniaxial yield stress [Symbol]

σ_y (monotonic) yield strength [Symbol]

σ_y yield stress [Symbol]

$\bar{\sigma}_y$ average stress in y-direction [Symbol]

σ_{yc} cyclic yield strength [Symbol]

σ_{YS} (0.2% offset) yield strength [also σ_{ys}] [Symbol]

σ_{ys} yield stress [Symbol]

σ_{yx} shear stress in X-Z plane with y-axis perpendicular and x-axis parallel [Symbol]

σ_{yy} (normal) stress along y-y axis [Symbol]

σ_{yz} shear stress in X-Z plane with y-axis perpendicular and z-axis parallel [Symbol]

σ_z axial stress (or stress in axial direction) [Symbol]

σ_z normal stress in z-direction [Symbol]

σ_{zx} shear stress in X-Y plane with z-axis perpendicular and x-axis parallel [Symbol]

σ_{zy} shear stress in X-Y plane with z-axis perpendicular and y-axis parallel [Symbol]

σ_{zw} rotation pseudostress [Symbol]

σ_{zz} stress along Z-Z axis [Symbol]

γ (lowercase) sigma (lunate) [Symbol]

γ superfluid stiffness (in physics) [Symbol]

2° Secondary (Alcohol, Amine, etc.) [Symbol]

SA Normal (or Ordinary) Spiral Type Galaxy [Astronomy]

SA Sachs Average

SA Sacrificial Anode [Physical Chemistry]

SA Safe Association [US]

SA Safety Altitude

SA Safety Analysis

SA Safing Area

SA Sail Area

SA Salt Added; Salt Addition

SA Samarium Aluminate

SA Sample Analysis

SA Sample Average

SA San Antonio [Texas, US]

SA Saturated Air [Meteorology]

SA Saturated Atmosphere [Meteorology]

SA Saudi Arabia(n)

SA Saudi Arabia [ISO Code]

SA Scattering Amplitude [Quantum Mechanics]

SA Science Abstracts [of INSPEC–Information Services Physics, Electrical and Electronics, and Computers and Control, UK]

SA Scientific Area

SA Scientific Authority

SA Screw Axis [Crystallography]

SA Screen Analysis

SA Seaman Apprentice

SA Secretary of the Army [US]

SA Sedimentation Analysis; Sedimentation Analyzer

SA Selected Area (Diffraction)

SA Selective Area

SA Selective Availability

SA Self-Absorption

SA Self-Acting

SA Self-Activated; Self-Activation

SA Self-Aligned; Self-Aligning; Self-Alignment

SA Self-Aligning Roller Bearing, Single-Row, Angular-Contact, Outer-Ring Raceway Spherical [Symbol]

SA Self-Assembled; Self-Assembly

SA Self-Propelled Artillery

SA Semi-Annual

SA Semianthracite

SA Semi-Autogenous

SA Semi-Automatic

SA Sense Amplifier

SA Sequential Access [Computers]

SA Sequential Analysis [Statistics]

SA (Blood) Serum Albumin [Biochemistry]

SA Service d'Aéronomie [Aeronomy Service, France]

SA Service Assistant

SA Sexual Assault

SA Shaft Alley [Naval Architecture]

SA Shaft Angle

SA Shear Area [Mechanics]

SA Ship Abstracts [Joint Scandinavian Publication]

SA Shock Absorber; Shock-Absorbing; Shock Absorption

SA Siegbahn-Avogadro (Constant) [Spectroscopy]

SA Signal Amplitude

SA Signal Analysis; Signal Analyzer

SA Signature Analysis; Signature Analyzer

SA Silicon-Asbestos

SA Simple Approach

SA Single Access

SA Single-Acting; Single-Action

SA Single Aircraft

SA Single Armored (Cable)

SA Sinoatrial (Node) [also S-A] [Anatomy]

SA Sinoauricular (Node) [also S-A] [Anatomy]

SA Sintered Alumina

SA Site Activation [also S/A] [Metallurgy]

SA (Particle) Size Analysis

SA Slip Activation

SA Slow Acting

SA Slow Anneal(ed); Slow Annealing [Metallurgy]

SA Small Arms

SA Snubber Adapter

SA (Sociedad Anónima) – Spanish for "Joint Stock Company"

SA (Sociedade Anônima) – Portuguese for "Joint Stock Company"

SA (Società Anonima) – Italian for "Joint Stock Company"

SA (Société Anonyme) – French for "Incorporated Company"

SA Soil Association [UK]

SA Solar Array

SA Solution Annealed; Solution Annealing [Metallurgy]

SA Sorbic Acid

SA Sound Analysis; Sound Analyzer

SA Source Address [Computers]

SA South Africa(n)

SA South America(n)

SA South Asia(n)

SA South Atlantic

SA South Atlantic [US Geographic Region]

SA South Australia(n)

SA Space Act [US]

SA Space Age

SA Space Agency

SA Spark Arrester

SA Special Application

SA Special Area

SA Spectral Analysis; Spectral Analyzer; Spectroanalysis

SA Spectrographic Analysis

SA Spectrum Analysis; Spectrum Analyzer

SA Speech Analysis; Speech Analyzer

SA Spherical Aberration

SA Spherical Agglomeration [Metallurgy]

SA Spin Axis [Physics]

SA Springfield Armory [Massachusetts, US]

SA Stability Analysis

SA Stability Augmentation

SA Standard Addition

SA Standards Association [of Institute of Electrical and Electronics Engineers, US]

SA Star Alliance [Aeronautics]

SA State Analysis

SA Static Approximation

SA Statistical Analysis

SA Stearic Acid

SA Steered Arc (Evaporation)

SA Steam Accumulator

SA Stellar Aberration

SA Stellar Atmosphere [Astronomy]

SA Stereospecific Addition [Organic Chemistry]

SA Stochastic Approximation

SA Stone Age

SA Storage Allocation [Computers]

SA Strain Aging [Metallurgy]

SA Strain Analysis

SA Stress Analysis; Stress Analyzer

SA Stress Annealed; Stress Annealing [Metallurgy]

SA Stress Axis [Mechanics]

SA Stress-Induced (Magnetic) Anisotropy [Solid-State Physics]

SA Strong Acid

SA Structural Adhesive

SA Structural Analysis; Structure Analysis

SA Stuck-at (Error) [also sa] [Computers]

SA Subaccount

SA Subassembly

SA Subatomic

SA Subclavian Artery [Anatomy]

SA Subject to Approval [also S/A, or sa]

SA Submerged Arc [Metallurgy]

SA Substituted Amide

SA Successive Approximation [Computers]

SA Succinic Anhydride

SA Sucrose Acetate

SA Sudden Approximation [Physics]

SA Sulfonamide

SA Sulfur Asphalt

SA Sulfuric Acid

SA Superacid

SA Superalloy

SA Supplemental Agreement

SA Support Area

SA Surface-Active; Surface Activity

SA Surface Adsorption

SA Surface Analysis; Surface Analyzer

SA Surface Area

SA Surge Arrester

SA Svenska Akademien [Swedish Academy]

SA Symbolic Address

SA Symbolic Assembler

SA System Acquisition

SA System Administrator

SA System Analysis; System Analyzer

SA Systems Analysis; Systems Analyst

SA Systems Approach

SA 0 Stuck-at-Zero (Error) [also SA0, or sa 0]

SA 1 Stuck-at-One (Error) [also SA1, or sa 1]

S&A Safe and Arm (Device) [also S/A]

S&A Science and Application

S+A Dedendum (of a Milled Bevel Gear) [Symbol]

S/A Safe and Arm [also S&A]

S/A Scheduled/Actual

S/A Site Activation [also SA] [Metallurgy]

S/A Spacecraft Adapter

S/A Subassembly [also SA]

S/A Subject to Approval [also SA, or sa]

S-A Sinoatrial (Node) [also SA] [Anatomy]

S-A Sinoauricular (Node) [also SA] [Anatomy]

S-A Succinonitrile-Acetone

Sa Barred Spiral Galaxy with Large Nucleus having Tightly-Wound Arms [Astronomy]

Sa Salt

Sa Sapphirine [Mineral]

Sa Saturday

Sa *(Sierra)* – Spanish for "Mountain Range", or "High Range of Hills"

sa *(secundum artum)* – according to art [Medical Prescriptions]

sa *(sine anno)* – without year (or date)

sa subject to approval [also SA, or S/A]

.sa Saudi Arabia [Country Code/Domain Name]

SAA Safety Assurance Analysis

SAA Satellite Active Archive [of National Oceanic and Atmospheric Administration, US]

SAA Small Arms Ammunition

SAA Sociedad Argentina de Agronomia [Argentinian Agricultural Society]

SAA Society for American Archeology [US]

SAA Society of American Archivists [US]

SAA Society of Architectural Administrators [US]

SAA Society of Automotive Analysts [US]

SAA Solution Annealed and Aged; Solution Annealing and Aging [Metallurgy]

SAA South African Airways

SAA South Atlantic Anomaly [Oceanography]

SAA South Australian Agriculture Department [Adelaide]

SAA Space Act Agreement [US]

SAA Standards Association of Australia

SAA Storage Accounting Area

SAA Student Alumni Association

SAA Succinic Acid Anhydride

SAA Suffolk Agricultural Society [UK]

SAA 2-Sulfoaniline-N,N-Diacetic Acid [also Saa]

SAA Sulfuric Acid Anodizing [Metallurgy]

SAA Sulfuric Anodized Aluminum (Alloy) [Metallurgy]

SAA Superconductor Applications Association [US]

SAA Supima Association of America [US]

SAA Surface Active Agent

SAA Surface Area Analysis; Surface Area Analyzer

SAA Systems Application Architecture [Computers]

SA A Science Abstracts, Section A: Physics Abstracts [of INSPEC–Information Services Physics, Electrical and Electronics, and Computers and Control, UK]

Saa 2-Sulfoaniline-N,N-Diacetic Acid

SAAC Schedule Allocation and Control

SAAC Scientific Advisor to the Army Council [UK]

SAAC Space Applications Advisory Committee [US]

SAACB South African Association of Clinical Biochemists

SAACE South African Association of Consulting Engineers

SAACI Salesmen's Association of the American Chemical Industry [now Sales Association of the Chemical Industry, US]

SAAD Small Arms Ammunition Depot

SAAEB South African Atomic Energy Board

SAAI South African Acoustics Institute

SAAL Single-Axis Acoustic Levitator

SAALC San Antonio Air Logistics Center [Texas, US]

SAAM Special Air Force Airlift Mission [US]

SAAMA San Antonio Air Material Area [Texas, US]

SAAME South African Association of Municipal Employees

SAAO South African Astronomical Observatory

SAAP Saturn/Apollo Application Program [of NASA]

sAAPFpna Succinyl-L-Alanyl-L-Alanyl-L-Propyl-L-Phenylalanine-p-Nitroanilide [Biochemistry]

SAAPMB South African Association of Physicists in Medicine and Biology

SAAPS Styryl Aminoethylaminopropyltrimethoxysilane

SAAR Seasonally Adjusted at Annual Rates

SAARC South Asian Association for Regional Cooperation

SAAS Science Achievement Awards for Students

SAAS Shuttle Aerosurface Actuator Simulation [of NASA]

SAAS South African Association for the Advancement of Science

SAAS Southern Association of Agricultural Scientists [US]

SAAT Society of Architectural and Associated Technicians [now British Institute of Architectural Technicians, UK]

SAATC Southern African Air Transport Council

SAAU South African Agricultural Union

SAAUW South African Association of University Women

SAAWK Suid-Afrikaanse Akademie vir Wetenskap en Kunst [South African Academy of Science and Arts, Pretoria]

SAB Ordinary, or Barred Spiral Type Galaxy [Astronomy]

SAB Sabena Belgian World Airlines

SAB Science Advisory Board [of US Environmental Protection Agency]

SAB Scientific Advisory Board

SAB Secondary Application Block

SAB Shuttle Avionics Breadboard [of NASA]

SAB Silicon-Aluminum Bronze

SAB Sociedad Argentina de Biología [Argentinian Society for Biology]

SAB Society for Applied Bacteriology [UK]

SAB Society of American Bacteriologists [US]

SAB Solar Alignment Bay

SAB Space Applications Board

SAB Spacecraft Assembly Building

SAB Stack Address Block

SAB Storage and Assembly Building

SAB Surface Activated Bonding

SAB System Advisory Board

SA B Science Abstracts, Section B: Electrical Engineering Abstracts [of INSPEC–Information Services Physics, Electrical and Electronics, and Computers and Control, UK]

SAb Serum Antibody [Immunology]

SABA Spherical Agglomeration-Bacterial Adsorption (Process)

SABAP Southern African Bird Atlas Project [South Africa]

SABAR Strong Acid by Azeotropic Rectification (Process)

SABC South African Broadcasting Corporation

SABE Society for Automation in Business Education [US]

SABE Sub-Acute Bacterial Endocarditis [Medicine]

SABIC Saudi Arabia Basic Industrial Corporation [Riyadh] [also Sabic]

SABIR Semi-Automatic Bibliographic Information Retrieval

SABIT Special American Business Internship Training (Program) [US]

SABMIS Seaborne (or Sea-Based) Antiballistic Missile Intercept System [US Navy]

SABO Sense Amplifier Blocking Oscillator

SABR Symbolic Assembler for Binary Relocatable Programs

SAB(r) Ordinary, or Barred Spiral Type Galaxy, r-shaped [Astronomy]

SAB(rs) Ordinary, or Barred Spiral Type Galaxy, r-shaped, s-shaped [Astronomy]

SABRAO Society for the Advancement of Breeding Research in Asia and Oceania [Japan]

SABRE Sales and Business Reservations done Electronically

SABRE Secure Airborne Radar Equipment

SABRE Self-Aligning Boost and Reentry System

SABRE Store Access Bus Recording Equipment

SABRITA South Africa-Britain Trade Association

SABRU South African Bird Ringing Unit

SABS South African Bureau of Standards

SABS Stanford Automated Bibliographic System [of Stanford University, California, US]

SAB(s) Ordinary, or Barred Spiral Type Galaxy, s-shaped [Astronomy]

SAC Safety Air Cushion

SAC Salicylamide Compound [Pharmacology]

SAC Sample Analysis Chamber

SAC Sapphire Anvil Cell

SAC Satelite de Aplicaciónes Cientificas [Scientific Applications Satellite; Argentinian Satellite Projects with International Cooperation]

SAC Satellite Access Controller [of NASA, Apollo Project]

SAC Semi-Automatic Coding

SAC Science Advisory Committee [US]

SAC Scientific Advisory Committee

SAC Scientific Advisory Committee [of International Atomic Energy Agency]

SAC Scientific Advisory Council [UK]

SAC Service Area Computer

SAC Serving Area Concept

SAC Shipbuilding Advisory Council [UK]

SAC Short-Run Average Cost

SAC Shuttle Action Center [of NASA]

SAC Single Attachment Concentrator

SAC Soaring Association of Canada

SAC Sociedad Agronómica de Chile [Chilean Agronomical Society]

SAC Société de l'Automation et du Contrôle [Society for Automation and Control, Rabat, Morocco]

SAC Society for Analytical Chemistry [UK]

SAC Society for Analytical Cytology [US]

SAC Society for Automation and Control [Rabat, Morocco]

SAC Society of Applied Chemistry [US]

SAC Sound Absorption Coefficient

SAC Special Area Code

SAC Specific Absorption Coefficient (Wavelength)

SAC St. Andrews College [Canada]

SAC Statistical Advisory Committee [of Food and Agricultural Organization]

SAC Storage Access Channel

SAC Storage Address Computer

SAC Store Access Control(ler)

SAC Store and Clear Accumulator

SAC Strategic Advisory Committee [US]

SAC Strategic Air Command [of US Air Force]

SAC Structural Adjustment Committee

SAC Student Activities Committee [of Institute of Electrical and Electronics Engineers–Regional Activities Board, US]

SAC Student Affairs Committee [of ASM International, US]

SAC Sugar Association of the Caribbean [Trinidad and Tobago]

SAC Sulfacetamide

SAC Support Action Center

SAC Supreme Allied Commander [of NATO]

SAC Surface-Area-Center (Hardness Test)

SAC Symmetry Adapted Cluster

SAC Synchronous Astro-Compass

S/AC Stabilization and Attitude Control

Sac Streptomyces achromogenes [Microbiology]

Sac I Streptomyces achromogenes I (One) [Microbiology]

Sac II Streptomyces achromogenes II (Two) [Microbiology]

SACA South Africa Council for Automation

SAC-A Satelite de Aplicaciónes Cientificas A [Scientific Applications Satellite A, of Comisión Nacional de Actividades Espaciales, Argentina]

SACAC South African Council for Automation and Computation

SACANGO South African Committee on Air Navigation and Ground Operation

SAC-B Satelite de Aplicaciónes Cientificas B [Scientific Applications Satellite B; Joint Project of NASA and Comisión Nacional de Actividades Espaciales, Argentina]

SACBD Scientific Advisory Committee on Biological Diversity

SACC Sexual Assault Care Center

SACC Society of Air Cargo Correspondents [UK]

SAC-C Satelite de Aplicaciónes Cientificas C [Scientific Applications Satellite C; Project of Comisión Nacional de Actividades Espaciales, Argentina]

SACCEI Strategic Air Command Communications Electronics Instruction [of NATO]

SACCH Slow Associated Control Channel [Telecommunications]

SACCI Symmetry Adapted Cluster Configuration Interaction [also SAC-CI]

SACCOMNET Strategic Air Command Communications Network [of NATO]

SACCS Strategic Air Command Communications System [of NATO]

SACCS Strategic Air Command Control System [of NATO]

SACD Spheroidized-Annealed (and) Cold-Drawn [Metallurgy]

SACDA South African Copper Development Association

SACDA System Analysis Control and Design Activity

SACDIN Strategic Air Command Digital Information [of NATO]

SACE Saskatchewan Association for Computers in Education [Canada]

SACE Science Alliance Center for Excellence (Grant) [of University of Tennessee, Tullahoma, US]

SACEP South Asia Cooperative Environment Program

SACEUR Supreme Allied Commander Europe [of NATO]

SACG Synchronous Alternating-Current Generator

SACH Solid Angle Cushioned Heel (Artifical Foot)

SACI Sales Association of the Chemical Industry [US]

SACI South African Chemical Institute

SACI South African Corrosion Institute

S-Acid 4-Amino-5-Hydroxy-1-Naphthalenesulfonic Acid; 4-Amino 1,5-Naphthalenedisulfonic Acid [also S-acid]

S-Acid 1-Amino-Naphthol-4-Sulfonic Acid [also S-acid]

SACK Scientific Advisory Committee on Kangaroos [of Australian Nature Conservancy Agency]

SACL South African Conferation of Labour

SACL Space and Component Log

SACLANT Supreme Allied Commander Atlantic [of NATO]

SACM Single Adiabatic Channel Model

SACM Smart Acoustic Current Meter

SACM Statistical Adiabatic Channel Model

SACMA Suppliers of Advanced Composite Materials Association [US]

SACMA/ETAC Suppliers of Advanced Composite Materials Association/European Trade Association for Composite Materials

SACMAPS Selective Automatic Computational Matching and Positioning System

SACNAS Society for Advancement of Chicanos and Native Americans in Science [US]

SACNET Secure Automatic Communications Network

SACO Sensitive Applications Control Office

SACO Swedish Confederation of Professional Associations

SACOM Ships Advanced Communications Operational Model

SACON Shock-Absorbing Foamed Concrete

SACP Selected-Area (Electron) Channeling Pattern [also sacp]

SACP Society for Analytical Chemists of Pittsburgh [Pennsylvania, US]

SACPA South African Cement Producers Association

SACPD Scientific Advisory Committee on Panic Disorder [of National Institute of Mental Health, Bethesda, Maryland, US]

SACRS South Australian Center for Remote Sensing

SACS Small Area Census Studies

SACS Software Avionics Command Support

SACS Southern Association of Colleges and Schools [US]

SACS STU (Secure Telephone Unit) Access Control System

SACS Swedish Agro Cooperative Services

SACS Systems Software Avionics Command Support

SACSAC Dithioacetylacetone

SACSIR South African Council for Scientific and Industrial Research

SACSS Southern Association of Colleges and Secondary School [US]

SACTTYNET Strategic Air Command Teletype Network [of NATO]

SACTU South African Congress of Trade Unions [Zambia]

SACU Scottish Auto-Cycle Union

SACVB San Antonio Convention and Visitors Bureau [Texas, US]

SACVT Society of Air Cushion Vehicle Technicians

SAD Safety Assurance Diagram

SAD Scientific Affairs Division [of NATO]

SAD Seasonal Affective Disorder [Medicine]

SAD Sediment Analysis Database

SAD Selected Area (Electron) Diffraction

SAD Selective Area Diffraction

SAD Sentence Appraiser and Diagrammer

SAD Shuttle Authorized Document [of NASA]

SAD Single Atom Detection

SAD Small-Angle Diffraction

SAD Small-Angle Diffractometer

SAD Solar Array Drive

SAD Space Antennas Diversity

SAD Spacecraft Attitude Display

SAD State Aeronautics Director [US]

SAD Store Access Director

SAD Store Address Director

SAD Strong Acid Dissociables

SAD Superaerodynamic(s)

SAD Swiss Association for Documentation

SAD Systems Allocation Document

S&AD Science and Applications Directorate [of NASA]

SADA Seismic Array Data Analyzer

SADA Solar Array Drive Assembly

SADA Southern Avalon Development Association [Newfoundland, Canada]

SADA Stand-Alone Data Acquisition

SADA Standard Advanced Dewar

SADAP Simplified Automatic Data Plotter

SADAPTA Solar Array Drive and Power Transfer Assembly

SADAS Sperry Airborne Data Acquisition System

SADBU Small and Disadvantaged Business Utilization [US]

SADC Sequential Analog-Digital Computer

SADC Small Angle Data Collection

SADCC Southern African Development Coordination Conference

SADE Solar Array Drive Electronics (for Topology Experiment)

SADE Structural Assembly Demonstration Experiment

SADE Superheat Advanced Demonstration Experiment

SADEC Spin Axis Declination [Physics]

SADELM South Australian Department of Environment and Land Management

SADEP South Australian Department of Environment and Planning

SADF Semi-Automatic Document Feed

SADF South Australian Department of Fisheries

SADIC Solid-State Analog-to-Digital Computer

SADIE Scanning Analog-to-Digital Input Equipment

SADIE Semi-Automatic Decentralized Intercept Environment

SADIE Sterling and Decimal Invoicing Electronically

SADL Sterilization Assembly Development Laboratory [of NASA]

SADP 3-(4-Azidophenyldithio)propionic Acid NHydroxysuccinimide Ester

SADP Scandinavian Association of Directory Publishers [Denmark]

SADP Selected Area Diffraction Pattern

SADP System Architecture Design Package

SADPO System Analysis and Data Processing Office [of New York Public Library, US]

SADR Six-Hundred-Megacycles Air Defense Radar

SADRAC South African Defense Research Advisory Committee

SADS Silicides As Diffusion Sources

SADSAC Sampled Data Simulator and Computer

SADSAC Seiler ALGOL Digitally Simulated Analog Computer

SADSACT Self-Assigned Descriptors from Self and Cited Titles

SADT Self-Accelerating Decomposition Temperature

SADT Structured Analysis and Design Technique

SADTC SHAPE (Supreme Headquarters, Allied Powers Europe) Air Defense Technical Center [now SHAPE Technical Center; of NATO]

SADVR Symmetry-Adapted Discrete Variable Representation

SAE Selective Area Epitaxy

SAE Shaft-Angle Encoder

SAE Society of Automotive Engineers [Name Changed to Society of Automotive Engineers International, US]

SAE Society of Automotive Engineers [Australia]

SAE Space Age Engineering

SAE Spiral Aftereffect

SAE Stamped Addressed Envelope [also sae]

SAEA Soda Ash Export Association [now American Natural Soda Ash Corporation, US]

SAE/AISI Society of Automotive Engineers/American Iron and Steel Institute (Specification)

SAE/AMS Society of Automotive Engineers/Aerospace Material Specification [also SAE AMS]

SAE/ASTM Society of Automotive Engineers/American Society for Testing and Materials (Specification)

SAE Australas Society of Automotive Engineers of Australasia

SAECP Selected Area Electron Channeling Pattern (Method)

SAED Selected-Area Electron Diffraction; Selective Auger Electron Diffraction

SAEDP Selected-Area Electron Diffraction Pattern

SAEE Swiss Association for Energy Economics

SAEF Spacecraft Assembly and Encapsulation Facility [of NASA Kennedy Space Center, Florida, US]

SAEH Society for Automation in English and the Humanities

Saehkoe Electr Electron Saehkoe Electricity and Electronics [Journal published in Finland]

SAE Int'l Society of Automotive Engineers International [US]

SAE Int Congr, Tech Pap Ser Society of Automotive Engineers International Congress, Technical Paper Series [US]

sael *(sine anno et loco)* – without year and place

SAEM Scanning Auger Electron Microscope; Scanning Auger Electron Microscopy

SAEMA Suspended Access Equipment Manufacturers Association [UK]

SAEN Salmon Association of Eastern Newfoundland [Canada]

SAES Scanning Auger Electron Spectrometry (or Spectrometer)

SAES Scanning Auger Electron Spectroscopy

SAES Stand-Alone Engine Simulator

SAES Storrs Agricultural Experiment Station [of University of Connecticut, US]

SAET Society of Architectural and Engineering Technologists [Canada]

SAET Society of Automotive-Electrical Technicians [UK]

SAE TP Society of Automotive Engineers Technical Paper [also SAE Tech Pap, US]

SAEWA South African Electrical Workers Association

SAF Sample Analysis Form

SAF San Andreas Fault [California, US]

SAF Saskatchewan Agriculture and Food [Canada]

SAF Secretary of the Air Force [US]

SAF Segment Address Field

SAF Short Address Form

SAF Sigma, Aldrich and Fluka (Chemical Companies)

SAF Singapore Armed Forces

SAF Société Astronomique de France [Astronomical Society of France]

SAF Society of American Foresters [US]

SAF Space Agency Forum

SAF Spacecraft Assembly Facility [of NASA]

SAF Spin-Angle Function [Physics]

SAF Strontium Aluminum Ferrite

SAF Stuck-at-Fault (Test) [Computers]

SAF Super-Abrasion Furnace (Black)

SAF Synthetic Antiferromagnetic (Structure) [Solid-State Physics]

SAF Tolusafranine

Saf Safety [also saf]

S Af South Africa(n) [also S Afr]

SAFA Shipping and Forwarding Agent

SAFAD Swedish Agency for Administrative Development

SAFAIA Synthetic Approach to Furnace Conditions with Artificial Intelligence Analysis

SAF Black Super-Abrasion Furnace Black [also SAF black]

SAFE Safe Assessment and Facilities Establishment [of Japan Atomic Industrial Forum]

SAFE Safety and Facilities Engineering Department

SAFE San Andreas Fault Experiment [US]

SAFE Signature Analysis Using Functional Analysis

SAFE Société des Aciers Fins [French Steel Products Manufacturer]

SAFE Society for the Advancement of Fission Energy [US]

SAFE Society for the Application of Free Energy [US]

SAFE Solar Array Flight Experiment

SAFE Space and Flight Equipment

SAFE STM (Scanning Tunneling Microscope) Aligned Field Emission

SAFE Store and Forward Element

SAFE Survival and Flight Equipment Association [US]

SAFEA Space and Flight Equipment Association [US]

SAFE J SAFE Journal [of Survival and Flight Equipment Association, US]

SAFER Special Aviation Fire and Explosion Reduction

SAFER Streamlined Approach for Environmental Restoration

SAFER Stress and Fracture Evaluation of Rotors

Safety Health Safety and Health [Publication of the National Safety Council, US]

Safety Health Int Safety and Health International [Publication of the National Safety Council, US]

SAFF Store and Forward Facsimile

SAFFI Special Assembly for Fast Installation

SAFHL St. Anthony Falls Hydraulic Laboratory [of University of Minnesota, US]

SAFI Semi-Automatic Flight Inspection

SAFI Special Assembly for Fast Installation

SAFIC Société Anonyme Française de l'Industrie de Caoutchouc [French Natural Rubber Industry, Inc.]

SAFIC South Australian Fishing Industry Council

SAFIRE Simulator for Advanced Fighter Radar ECCM (Electronic Counter-Countermeasures) (Software)

SAFIRE Spectroscopy of the Atmosphere using Far-Infrared Emission

SAFISY Space Agency Forum on the International Space Year (1992)

SAFOC Semi-Automatic Flight Operations Center

S Afr South Africa(n) [also S Af]

S Afr Comput J South African Computer Journal [of Computer Society of South Africa]

SAFRI South African Forestry Research Institute

S Afr J Antarct Res South African Journal of Antarctic Research

S Afr J Chem South African Journal of Chemistry [of South African Chemical Institute]

S Afr J Libr Inf Sci South African Journal of Library and Information Science [of South African Institute for Librarianship and Information Science]

S Afr Med J South African Medical Journal

S Afr J Phys South African Journal of Physics

S Afr J Sci South African Journal of Science

S Afr Mach Tool Rev South African Machine Tool Review

S Afr Mech Eng South African Mechanical Engineer [Publication of South African Institute of Mechanical Engineers]

S-Afr Tydskr Natuurwet Tecnol Suid-Afrikaanse Tydskrif vir Natuurwetenskap en Tecnologie [South African Journal of Science and Technology]

SAFSR Society for the Advancement of Food Service Research [US]

SAFT Shortest Access First Time

SAFT Spark Analysis for Traces (Time-Resolved Spectroscope)

SAFT Synthetic Aperture Focussing Technique

SafT Safranine T

SAFTO South African Foreign Trade Organization

SAFWA Southeastern Association of Fish and Wildlife Agencies [US]

SAG Segregant-Assisted Growth

SAG Selective Area Growth

SAG Self-Aligned Gate [Electronics]

SAG Semi-Autogenous Grinding

SAG Senior Advisory Group [of International Atomic Energy Agency]

SAG Special Area Group

SAG Standard Address Generator

SAG Steroid Aid Group [UK]

SAGA Sand and Gravel Association [UK]

SAGA Short-Arc Geodetic Adjustment

SAGA Society of American Graphic Artists [US]

SAGA Statistics of Accidents in General Aviation [of European Civil Aviation Conference, France]

SAGA Studies Analysis and Gaming Agency

Sagamore Army Conf Proc Sagamore Army Conference Proceedings [of US Army]

SAGASCO South Australian Gas Company

SAGB Silk Association of Great Britain

SAGB Small-Angle Grain Boundary [Materials Science]

SAGBO Stress-Accelerated Grain Boundary Oxidation; Stress-Assisted Grain-Boundary Oxidation [Materials Science]

SAGE Semi-Automatic Ground Environment [Air Defense]

SAGE Solar-Assisted Gas Energy

SAGE Solar Atmospheric Gas Experiment

SAGE Solvent Alternatives GuidE [of Research Triangle Institute, Research Triangle Park, North Carolina, US]

SAGE Soviet (Russian) American Gallium Experiment

SAGE Strategic Advisory Group on the Environment [US]

SAGE Stratospheric Aerosol and Gas Experiment [of NASA]

SAGE Swedish Association of Graduate Engineers

SAGE I Stratospheric Aerosol and Gas Experiment I [NASA Second Applications Explorer Mission]

SAGE II Stratospheric Aerosol and Gas Experiment II [NASA Earth Radiation Budget Satellite Mission]

SAGE III Stratospheric Aerosol and Gas Experiment III [NASA Mission]

SAGEEP Symposium on the Application of Geophysics to Engineering and Environmental Problems [US]

SAGENAP Scientific Assessment Group for Experiments in Non-Accelerator Physics [of Department of Energy–Committee on High-Energy Physics, US]

SAGFET Self-Aligned Gate Field-Effect Transistor

SAGH2 p-Sulfonamidophenylaminoglyoxime

SAGIT Sectoral Advisory Committees on International Trade

SAGMOS Self-Aligning Gate Metal Oxide Semiconductor

SAGS School of Advanced Graduate Studies

SAGT Scottish Association of Geography Teachers [UK]

SAGTA School and Group Travel Association [UK]

SAGUF Schweizer Arbeitsgemeinschaft für Umweltforschung [Swiss Association for Environmental Research]

SAH Scanning Acoustical Holography

SAH Superequilibrium Atomic Hydrogen

Sah Sahara

SAHGB Society of Architectural Historians of Great Britain

SAHF Semi-Automatic Height Finder

SAHPS Solar-Energy-Assisted Heat Pump System

SAHR Society for Army Historical Research [UK]

SAHYB Simulation of Analog and Hybrid Computers

SAI Science Applications International [US]

SAI Scientific Advances, Inc. [US]

SAI Self-Annealing Implantation

SAI Società Astronomica Italiana [Italian Astronomical Society]

SAI Society of American Inventors [US]

SAI Society of Architectural and Industrial Illustrators [UK]

SAI Steel Authority of India

SAI Sub-Architectural Interface

SAI Systems Applications International [California, US]

SAIA South Australian Institute of Architects

SAIB Sucrose Acetate Isobutyrate

SAIC Science Applications International Corporation [San Diego, California, US]

SAIC Scottish Agricultural Improvement Council

SAIC Canada Science Applications International Corporation [Canada]

SAICCOR South African Pulp from Rayan Project [of Industrial Development Corporation of South Africa, Ltd.]

SAICE South African Institute of Civil Engineers [now SAICET]

SAICET South African Institute of Civil Engineering Technicians and Technologists

SAID Safety Analysis Input Data

SAID Speech Auto-Instructional Device

SAIEE South African Institute of Electrical Engineers [Marshalltown]

SAIF South African Industrial Federation

SAIF Spatial Archive Interchange Format

SAIL Sea-Air Interaction Laboratory [US]

SAIL Serial ASCII (American Standard Code for Information Interchange) Instrumentation Loops

SAIL Shuttle Avionics Integration Laboratory [of NASA Johnson Space Center, Houston, Texas, US]

SAIL Stanford Artificial Intelligence Laboratory [of Stanford University, California, US]

SAIL Stanford Artificial Intelligence Language [of Stanford University, California, US]

SAIL Systems Analysis and Integration Laboratory

SAILA Sail Assist International Liaison Associates [US]

SAILIS South African Institute of Librarianship and Information Science [Pretoria, South Africa]

SAILS Simplified Aircraft Instrument Landing System

SAILS Structured Assessment of Independent Living Skills

SAIM South African Institute of Mining and Metallurgy [also SAIMM]

SAIME South African Institute of Mechanical Engineers [Marshalltown]

SAIMM South African Institute of Mining and Metallurgy [also SAIM]

SAIMR South African Institute for Medical Research

SAIMS Selected Acquisitions Information and Management System

SAINT Satellite Interceptor

SAINT Self-Aligned Implantation for N^+-Layer Technology [Electronics]

SAINT Semi-Automatic Indexing of Natural Language

SAINT Symbolic Automatic Integrator

SAIP (The) South African Institute of Physics

SAIPA South African Institute for Public Administration

SAIRE Scaleable Agent-Based Information Retrieval Engine [of NASA]

SAIS School of Advanced International Studies [at Johns Hopkins University, Baltimore, Maryland, US]

SAIS South African Interplanetary Society

SAISAC Ship's Aircraft Inertial System Alignment Console

SAIT Samsung Advanced Institute of Technology [Suwon, South Korea]

SAIT South African Institute of Translators

SAIT South Australian Institute of Technology [Whyalla]

SAIT Southern Alberta Institute of Technology [Canada]

SAIW South African Institute of Welding

SAKI Solatron Automatic Keyboard Instructor

SAL Salsolinol Hydrobromide

SAL Saskatchewan Accelerator Laboratory [Canada]

SAL Satellite Applications Laboratory [of Office of Research and Applications, National Oceanic and Atmospheric Administration in Suitland, Maryland, US]

SAL Semantic Abstraction Language

SAL Scientific Airlock

SAL Self-Averaging Limit [Solid-State Physics]

SAL Shielded Analytical Laboratory

SAL Shift Arithmetic Left

SAL Short Approach Light

SAL Shuttle Avionics Laboratory [of NASA]

SAL Soft Adjacent Layer (Technique) [Solid-State Physics]

SAL Space Astronomy Laboratory [US]

SAL Structured Assembly Language

SAL Supersonic Aerophysics Laboratory [US]

SAL Surface Air-Lifted (Mail)

SAL Surface Analysis Laboratory [of Science and Engineering Research Council, UK]

SAL Symbolic Assembly Language

SAL Systems Assembly Language

Sal Salary [also sal]

Sal Salicylic Acid

Sal Salicylidene

Sal Saline; Salinity [also sal]

Sal Salinometer [also sal]

Sal Streptomyces albus [Microbiology]

SALA South African Library Association

SalBzen N-Salicylidene-N'-Benzylethylenediamine

SALC Sacramento Air Logistics Center [California, US]

SALD Selected-Area Laser Deposition; Selective Area Laser Deposition

Sald Salicylaldehyde

SALDIEN Bis(salicylidene)diethylene Triamine

SALDMPHEN N,N'-Bis(salicylidene)-4,5-Dimethylphenylenediamine Dianion [also saldmphen]

SALDVI Selective Area Laser Deposition Vapor Infiltration

SALE Safeguards Analytical Laboratory Evaluation

SALE Simple Algebraic Language for Engineers

SALE Symposium on Atomic Layer Epitaxy

SalEen N-Salicylidene-N'-Ethylethylenediamine

SALEN Ethylenebis(salicylcimine)

SALEN N,N'-Bis(salicylidene)ethylenediamine [also salen]

SAL2EN N,N'-Bis(salicylidene)ethylenediamine

SALGLY Salicylidene Glycine

SALIC South Australian Land Information Council

SALICIDE Self-Aligned Silicide (Technology) [also Salicide, or salicide]

SALINET Satellite Library Information Network [US and Canada]

SALM Single Anchor Leg Mooring

SALM Society of Airline Meteorologists [US]

SALMEEN N-Salicylidene-N'-Methylethylenediamine [also Salmeen]

SALMPHEN N,N'-Bis(salicylidene)-4-Methylphenylenediamine [also Salmphen]

SAL MR Soft Adjacent Layer Biased Magnetoresistance [Solid-State Physics]

SALOPH N,N'-Bis(salicylidene)-o-Phenylenediamine [also Saloph]

SALOPHEN Salicylidene Phenylenediamine [also Salophen]

SALORS Structural Analysis of Layered Orthotropic Ring-Stiffened Shells [Physics]

SALPA N-(3-Hydroxypropyl)salicylaldimine [also Salpa]

SALPD N,N'-Bis(salicylidene)propylenediamine [also salpd]

SALPHEN N,N'-Bis(salicylidene)-o-Phenylenediamine Anion [also salphen]

SAL2PHEN N,N'-Bis(salicylidene)-o-Phenylenediamine [also salphen]

SALPN Salicylidene Propanediamine [also Salpn]

SAL2PROP N,N'-Bis(salicylidene)propylenediamine [also Sal2Prop]

SALPS 2-(Salicylideneamino)phenyl Disulfide [also Salps]

SALR Saturated Adiabatic Lapse Rate [Meteorology]

SALS Separate Access Landing System [US]

SALS Ship Aircraft Locating System

SALS Short Approach Light System

SALS Single Anchor Leg Storage

SALS Small-Angle Light Scattering

SALS Solid-State Acoustoelectric Light Scanner

SALSF Short Approach Light System with Sequenced Flashing Lights

SALT Saskatchewan Association of Library Technicians [Canada]

SALT Society for Applied Learning Technology [US]

SALT Strategic Arms Limitation Talks [Between USA and USSR]

SALT Strategic Arms Limitation Treaty [Between USA and USSR]

SALT Subscriber Access Line Terminal

SALT Subscriber Apparatus Line Tester

(Sal)3tach 2,2',2"-[1,3,5-Cyclohexanetriyltris-(nitrilomethylidyne)]trisphenol [also (sal)3tach]

SALTN N,N'-Bis(salicylidene) Trimethylenediamine [also Saltn]

SALTS Self-Alignment Lift-Off Technique by Selective Oxidation

SALTSC Salicylaldehyde Thiosemicarbazone [also Saltsc]

Salv El Salvador

Salv Salvage [also salv]

SAM S-Adenosyl-L-Methionine; S-(5'-Adenosyl)-L-Methionine [Biocxhemistry]

SAM Safety Activation Monitor

SAM Scalar Audio Magnetotellurics

SAM Scanning Acoustic Microscope; Scanning Acoustic Microscopy

SAM Scanning Auger Microprobe

SAM Scanning Auger Microscope; Scanning Auger Microscopy

SAM School of Aerospace Medicine [of US Air Force Aerospace Medical Center at Brooks Air Force Base, Texas, US]

SAM Scottish Association for Metals [UK]

SAM Script-Applier Mechanism

SAM Secure Access Multiport

SAM Security Accounts Manager

SAM Selective Automonitoring

SAM Selective Availability Mitigation

SAM Self-Assembled Monolayer [Physical Chemistry]

SAM Semantic Analyzing Machine

SAM Semi-Autogenous Milling

SAM Semi-Automatic Mathematics

SAM Separate Absorption and Multiplication [Electronics]

SAM Sequence Amplitude Margin

SAM Sequential-Access Memory

SAM Sequential-Access Method

SAM Serial-Access Memory

SAM Service Attitude Measurement

SAM Shuttle Attachment Manipulator [NASA]

SAM Simulation of Analog Methods

SAM Single Application Mode

SAM Sociedad Argentina de Metales [Argentinian Society for Metals]

SAM Society for Advancement of Management [US]

SAM Sort and Merge

SAM South African Museum [Capetown]

SAM South Australian Museum [Adelaide]

SAM Spatio-Angularly Multiplexed (Hologram)

SAM Special Application Module

SAM Strain Absorbent Module

SAM Strategic Aerospace Museum [of Stategic Air Command at Offutt Air Force Base, Nebraska, US]

SAM Stratospheric Aerosol Measurement (Experiment)

SAM Strong Absorption Model

SAM Subsequent Address Message

SAM Substitute Alloy Material

SAM Supply Analysis Model

SAM Surface-to-Air Missile

SAM Suspension Automatic Monitor

SAM Symbolic and Algebraic Manipulation

SAM Symantec Antivirus for the Mac(intosh) (Computer)

SAM System Activation and Monitoring

SAM System Activity Monitor

SAM Systems Adapter Module

SAM Systems Analysis Module

SAM Systems for Automated Manufacture

S Am South America(n) [also S Amer]

SAMA Saudi Arabian Monetary Authority

SAMA Scientific Apparatus Makers Association [US]

SAMA Scottish Agricultural Machinery Association

SAMA South Australia Museum, Adelaide

SAMAB Southern Appalachian Man and Biosphere Program [US]

SAMANTHA System for the Automated Management of Text in Hierarchical Arrangement

SAM-APD Separate Absorption and Multiplication Avalanche Photodetector

SAMD Surface-to-Air Missile Development [also SAM-D]

SAME Society of American Military Engineers [US]

SAMe S-Adenosyl-L-Methionine [Biochemistry]

S Amer South America(n) [also S Am]

SAMG Sociedad Argentina de Mineria y Geología [Argentinian Society of Mining and Geology]

SAMHSA Substance Abuse and Mental Health Services Administration [of US Department of Health and Human Services]

SAMI Section Access to Membership Information [of Institute of Electrical and Electronics Engineers, US]

SAMI Sensory Adaptive Machines Inc. [Canada]

SAMI Service and Maintenance Indicator (Panel)

SAMI Socially Acceptable Monitoring Instrument(s)

SAMI Southern Appalachian Mountains Initiative [of US Environmental Protection Agency]

SAMICS Solar Array Manufacturing Industry Costing Standards

S Am Ind South American Indian(s)

SAMIS Structural Analysis and Matrix Interpretative System

SAMMIE Scheduling Analysis Model for Mission Integrated Experiments [Aerospace]

SAMMIE System for Aiding Man-Machine Interaction Evaluation

SAMO Semilocalized Alternant Molecular Orbital [Physical Chemistry]

SAMO Simulated Ab Inito Molecular Orbital [Physical Chemistry]

SAMOS Satellite and Missile Observation System

SAMOS Silicon and Aluminum Metal-Oxide Semiconductor

SAMOS Stacked Gate Avalanche Injection Metal Oxide-Semiconductor

SAMP S-1-Amino-2-(Methoxymethyl)pyrrolidine

SAMP Sense Amplifier

SAMP Shuttle Automated Mass Properties [NASA]

SAMP Succinoadenosine Monophosphate [Biochemistry]

Samp Sample; Sampling [also samp]

SAMPA S-Acetylmercaptosuccinic Anhydride

SAMPE Society for the Advancement of Materials and Process Engineering [US]

SAMPE Society of Aerospace Material and Process Engineers [US]

SAMPE Conf Proc SAMPE Conference Proceedings [of Society for the Advancement of Materials and Process Engineering, US]

SAMPE J SAMPE Journal [of Society for the Advancement of Materials and Process Engineering, US]

SAMPE Q SAMPE Quarterly [of Society for the Advancement of Materials and Process Engineering, US]

SAMPE Symp Proc SAMPE Sympoisium Proceedings [of Society for the Advancement of Materials and Process Engineering, US]

SAMPEX Solar, Anomalous and Magnetospheric Particle Explorer

SAMPM Scottish Association of Milk Product Manufacturers

SAMPE Q SAMPE Quarterly [of Society for the Advancement of Materials and Process Engineering, US]

SAMR Small-Angle Magnetization Rotation (Method) [Solid-State Physics]

SAMRC South African Medical Research Council

SAMS Satellite Automonitor System

SAMS Shuttle Attachment Manipulator System [of NASA Space Shuttle]

SAMS South African Mathematical Society

SAMS Space Acceleration Measurement System [of NASA]

SAMS Statospheric and Mesospheric Sounder

SAMS Swiss Academy of Medical Sciences

SAMSA Silica and Moulding Sands Association [UK]

SAMSA Southern Africa Mathematical Sciences Association

SAMSARS Satellite-Based Maritime Search and Rescue System

SAMSAT South America/South Atlantic [also SAM SAT, or SAM/SAT]

SAMSO Space and Missile Systems Organization [of US Air Force]

SAMSOM Support-Availability Multisystem Operations Model

SAMSON Strategic Automatic Message Switching Operational Network

SAMSON System Analysis of Manned Space Operations

SAMSOR SAMSO (Space and Missile Systems Organization) Regulation [now Space Division Regulation; of US Air Force]

SAMT State-of-the-Art Medium Terminal [US Army]

SAMT Swiss Association for Materials Testing

SAMTEC Space and Missile Test Center [at Vandenberg Air Force Base, California, US]

SAMTO Space and Missile Test Organization [at Vandenberg Air Force Base, California, US]

SAN Scanning Auger Nanoprobe

SAN Science Association of Nigeria

SAN Serpska Akademija Nauka [Serbian Academy of Sciences, Belgrade]

SAN Small Area Network

SAN Standard Address Number [of Book Industry Systems Advisory Committee, US]

SAN Strontium Aluminum Niobate

SAN Styrene-Acrylonitrile (Polymer)

SAN System Area Network

san sanitary

SANA Sanierung der Atmosphäre über den neuen Bundesländern (Projekt) [Atmospheric Clean-Up in the "New" German States (Project)] [of Bundesministerium für Forschung und Technologie, Germany]

SANA Scientists Against Nuclear Arms [UK]

SANA Soyfoods Association of North America [US]

SAN-ABS Styrene Acrylonitrile/Acrylonitrile-Butadiene-Styrene (Blend) [also SAN/ABS]

SANACC State Army-Navy-Air Coordinating Committee

SANAE South African National Antarctic Expedition

SANB South African National Bibliography [of Council for Scientific and Industrial Research]

SANCAR South African National Committee for Antarctic Research

SANCAR South African National Council for Antarctic Research

SANCIAWPRC South African National Committee of the International Association on Water Pollution Research and Control

SANCI South African National Committee on Illumination

SAND Shelter Analysis for New Designs

SAND Spacelab Ancillary Data

SAND Site Activation Need Date

SAND Statistical Analysis of Natural Resource Data [Norway]

s and c suspended and capable [Oil and Gas Industry]

SANDI Shear and Normal Displacement Instrument

S and L Savings and Loan (Association) [also S&L]

SANE Solar Alternatives to Nuclear Energy

SANE Standard Apple Numeric Environment [of Apple Computer Inc., US]

SanE Sanitary Engineer [also San Eng]

San Eng Sanitary Engineer(ing)

SANHS Somerset Archeological and Natural History Society [UK]

Sanit Sanitary; Sanitation [also sanit]

Sanken Tech Rep Sanken Technical Report [Published by Sanken Electric Co. Ltd., Osaka, Japan]

SANOVA Simultaneous Analysis of Variance [Statistics]

SAN-PC Styrene-Acrylonitrile/Polycarbonate (Blend) [also SAN/PC]

SAN-PVC Styrene-Acrylonitrile/Polyvinyl Chloride (Blend) [also SAN/PVC]

SANR Scaled Adiabatic Nuclear Rotation (Method) [Physics]

SANR Subject to Approval No Risk

SANS Small-Angle Neutron Scattering

SANSS Structure and Nomenclature Search System

SANTA Systematic Analog Network Testing Approach

Sanyo Tech Rev Sanyo Technical Review [Published by Sanyo Electric Co. Ltd., Osaka, Japan]

SANZ Standards Association of New Zealand

SAO Samarium Aluminum Oxide

SAO Smithsonian Astrophysical Observatory [Cambridge, Massachusetts, US]

SAO South Atlantic Ocean

SAO Spetsial'noi Astrofizicheskoi Observatorii [Special Astrophysical Observatory, North Caucasus, Russian Federation]

Sa$_x$O$_y$ Sacrificial Oxide [Oxide/Metal Composites]

SAOL Sugar Association of London [UK]

SAONC Special Astrophysical Observatory–North Caucasus [Russian Federation]

SAOS Scottish Agricultural Organization and Society

SAOUG South African On-Line User Group

SAP Sample Analysis Plan

SAP Sampling and Analysis Plan [also S&AP]

SAP Second Audio Program

SAP Semi-Armor Piercing

SAP Serum Alkaline Phosphatase [Biochemistry]

SAP Serum Amyloid P [Biochemistry]

SAP Service Access Point

SAP Service Advertising Protocol

SAP Service Assessment Pool

SAP Share Assembly Program

SAP Simple Asynchronous Protocol

SAP Sintered Aluminum Powder (Material)

SAP Sintered Aluminum Product

SAP Specific Action Potential

SAP Start of Active Profile

SAP State Assessment Panels (for the National Landcare Program) [Australia]

SAP Strain Arrestor Plate

SAP Structural Adjustment Program

SAP Structural Analysis Program

SAP Subatomic Particle

SAP Sustainable Agricultural Practices

SAP Symbolic Address Program

SAP Symbolic Assembly Program

SAP System Access Protocol

SAP Systems, Applications and Products (Company) [US]

SAP Systems Assurance Program

S&AP Sampling and Analysis Plan [also SAP]

s ap scruple, apothecaries' [Unit]

SAPA South African Press Association

SAPE Society for Professional Education

SAPE Solenoid Array Pattern Evaluator

SAPH Styrenated and Alkylated Phenol

SAPI Sales Association of the Paper Industry [US]

SAPI Speech Application Program Interface

SAPHIR Spectrometer Arrangement for Photon Induced Reactions (Experiment) [of Rheinische Friedrich-Wilhelms-Universität Bonn, Germany] [also Saphir]

SAPIR System of Automatic Processing and Indexing of Reports

SAPL Seacoast Anti-Pollution League [US]

SAPL South African Public Library

SAPLA Standing Advisory Panel on Library Automation [of Library Resources Coordinating Committee, UK]

SAPO Silicoaluminophosphate

Sapon Saponification; Saponify [also sapon]

sapond saponified [also sapond, or sapon'd]

SAPONET South African Packet-Oriented Switching Network

Sapong Saponifying [also sapong, Sapon'g, or sapon'g]

SAPP Sodium Acid Pyrophosphate

SAPROS Smart Armor Protection System

SAPT Scottish Association for Public Transport

SAPT Symmetry-Adapted Perturbation Theory [Physics]

SAPTA S-Acetylthiopropionic Acid N-Hydroxysuccinimide Ester

Sap Val Saponification Value [Chemistry]

SAPWW Spanish Association for Purification of Water and Wastewater

SAPYAL Salicyl(2-Pyridyl)aldazine [also sapyal]

SAQC Spanish Association for Quality Control

SAR Safety Analysis Report

SAR Saudi Riyal (or Rial) [Currency of Saudi Arabia]

SAR Scanning Angle Reflectometry

SAR School of American Research [Santa Fe, New Mexico, US]

SAR Search and Rescue

SAR Segment Address Register

SAR Segmentation and Reassembly

SAR Semi-Automatic Rifle

SAR Service Analysis Request

SAR Service Analysis Report

SAR Shift Arithmetic Right

SAR Simulated Acid Rain

SAR Sodium Absorption Ratio

SAR Source Address Register

SAR South African Rand [Currency of Namibia and South Africa]

SAR Specific Absorption Rate

SAR Staffing Audit and Review

SAR Statistical Application Research

SAR Storage (or Store) Address Register

SAR Street Address Record

SAR Structure-Activity Relationship [Crystallography]

SAR Student Aid Report [US]

SAR Successive-Approximation Register [Computers]

SAR Sulfuric Acid Recovery (Process)

SAR Synthetic Aperture Radar

SAR Syrian Arab Republic

SA R Saudi Riyal (or Rial) [Currency of Saudi Arabia]

SA(r) Normal (or Ordinary) Spiral Type Galaxy, r-shaped [Astronomy]

Sar Sarcosine; Sarcosyl [Biochemistry]

sar sarcophagine [Biochemistry]

SARA Sampled Aperture Receiving Array

SARA Society of American Registered Architects [US]

SARA Superfund Amendments and Reauthorization Act [US]

SARAH Search, Rescue and Homing [also Sarah or sarah]

SARARC Stable Auroral Red Arc

SARBE Search and Rescue Beacon Equipment

SARBICA Southeast Asia Regional Branch of the International Council for Archives [Malaysia]

SARBR Southern Appalachian Regional Biosphere Reserve [US]

SARC Space and Astronomy Research Center [of Council for Scientific Research, Baghdad, Iraq]

SARCCUS South African Regional Committee for the Conservation and Utilization of the Soil

SARCOM Search and Rescue Communicator

SARCUP Search and Rescue Capability Update Program

SARD Swedish Agency Research Cooperation with Developing Countries

Sard Sardinia(n)

SARDA Special Agricultural and Rural Development Act

SARDPF Synthetic Aperture Radar Data Processing Facility

SARDS Suddenly Acquired Retinal Degeneration Syndrome [Veterinary Medicine]

SAREC Search and Rescue Emergency Center

SAREF Safety Research Experiment Facility

SAREX Saccharide Extraction (Process)

SAREX (Space) Shuttle Amateur Radio Experiment

SARF South African Road Federation

SAR GPF CSCI Synthetic Aperture Radar Ground Processing Facility Computer Software Configuration Item

SARI Subduction Accelerated Research Initiative

SARI Synthetic Aperture Radar Imaging

SARIS South African Retrospective Information System

SARISA Surface Analysis by Resonance Ionization of Sputtered Atoms

SARL *(Société à Responsabilité Limitée)* – French for "Limited Liability Company"

SARL South African Radio League

SARM Set Asynchronous Response Mode

SARMCS Synthetic Aperture Radar Motion Compensation System

SAROAD Storage and Retrieval of Aerometric Data [of US Environmental Protection Agency]

SARP Safety Analysis Report for Packaging

SARP Schedule and Report Procedure

SARP Signal Automatic Radar-Data Processing System

SARP Small Aspect Ratio Particle

SARP Small Autonomous Research Package

SARP Sulfuric Acid Recovery Process

SARP Sumitomo Alkali Refining Process

SARPS Standards and Recommended Practices [of International Civil Aviation Organization]

SARS Single-Axis Reference System

SA(rs) Normal (or Ordinary) Spiral Type Galaxy, r-shaped, s-shaped [Astronomy]

SARSAT Search and Rescue Satellite

SARSAT Search and Rescue Satellite Aided Tracking System

SART Shuttle Astronaut Recruitment Program [US]

SART Stimuli Analog Refresh Table

SARTECH Search and Rescue Technician

SARTS Switched Access Remote Test System

SARUC Southeastern Association of Regulatory Utility Commissioners [US]

SAS International Conference on Small-Angle Scattering

SAS Safeguards and Security

SAS Sales Accounting System

SAS Saturated Ammonium Sulfate

SAS Scandinavian Airlines System

SAS School of Applied Science

SAS Secondary Alarm Station

SAS Security Agency Study

SAS Segment Arrival Storage

SAS Selected Applicant Service

SAS Self-Avoiding Surfaces

SAS Sensitive Application Systems

SAS Sequenced Answer Signal

SAS Serbian Academy of Sciences [Belgrade]

SAS Single Attached Station

SAS Single Audio System

SAS Slovak Academy of Sciences [Bratislava]

SAS Slovenian Academy of Sciences [Ljubljana]

SAS Small-Angle Scattering

SAS Small Astronomical Satellite [Series of US Satellites]

SAS Society for Applied Spectroscopy [US]

SAS Sodium Acid Sulfate

SAS Sodium Alkane Sulfonate

SAS Sodium Aluminosilicate

SAS Sodium Aluminum Sulfate

SAS Solar Array System

SAS Special Air Service [UK Army]

SAS Stability Augmentation Subsystem; Stability Augmentation System [Aerospace]

SAS Staffordshire Agricultural Society [UK]

SAS Statistical Analysis (Software Package)

SAS Statistical Analysis System (Software Package)

SAS Straight Alkane Sulfonate

SAS Straight Alkyl Sulfonate

SAS Stress Analyst Station

SAS Succine-Aldehyde Acid

SAS Support Amplifier Station

SAS Surface-Active Substance(s)

SAS Sverige-Amerika Stiftelsen [Sweden-America Foundation]

SAS Swedish Academy of Sciences [Stockholm]

SAS Swedish Airlines

SAS Switched Access System

SAS Syrian Academy of Sciences [Damascus]

SAS-1 First Small Astronomical Satellite [US]

S/AS Stokes/Anti-Stokes (Mode) [Spectroscopy]

SASA Serbian Academy of Science and Arts [Belgrade]

SASA Solvent-Accessible Surface Area

SASA South African Sugar Association

SASAR Segmented Aperture–Synthetic Aperture Radar

SASE Self-Addressed and Stamped Envelope

SASE Specific Application Service Element

SASE Statistical Analysis of a Series of Events

SASF Sensitive Application Systems and Facilities

SASFA South Australian Shark Fishermen's Association

SASH Symmetry-Adapted Spherical Harmonic (Function) [Physics]

SASI Shugart Associates Standard Interface

SASI Shugart Associates System Interface

SASI South African Standards Institution

SASI System on Automotive Safety Information

SASIDS Stochastic Adaptive Sequential Information Dissemination System

Sask Saskatchewan [Canada]

Sask Power Saskatchewan Power Corporation [Canada]

Sask Tel Saskatchewan Telephone Company [Canada]

SASLO South African Scientific Liaison Office

SASM Society for Applied Science and Mathematics

SASMIRA Silk and Art Silk Mills Research Association [India]

SASOL South African Oil from Coal Project [of Industrial Development Corporation of South Africa, Ltd.]

SASOL South African Coal, Oil and Gas Corporation [also Sasol]

SASPL Saturated Ammonium Sulfate Precipitation Limit

SASPRSC South African Society for Photogrammetry, Remote Sensing and Cartography

SASR Sri Aurobindo Society Research [India]

SASREG Southern Africa Sub-Regional Environment Group

SASS Subsonic Assessment

SASS Subsonic Assessment Program [of NASA Atmospheric Effects of Aviation Project]

SASS Suspended Array Surveillance System

SASSI Synthetic Amorphous Silica and Silicates Industry Association [US]

SASSY Supported Activity Supply System

SAST Scientific Assessment and Strategy Team [US]

SASTA South African Sugar Cane Technologists Association

SASTP Stand-Alone Self-Test Program

SASTU Signal Amplitude Sampler and Totalizing Unit

SASTW Second ASEAN (Association of Southeast Asian Nations) Science and Technology Week

SASY Spin ASymmetry Detector ArraY [of Brookhaven National Laboratory, Upton, New York, US]

SAT Strontium Aluminum Tantalate

SAT Samoan Tala [Currency of Western Samoa]

SAT Scholastic Aptitude Test

SAT Sheet-Feeder Action Table

SAT Shock Aversion Therapy [Medicine]

SAT Sialytransferase [Biochemistry]

SAT Society of Acoustic Technology

SAT Sodium Ammonium Thiosulfate

SAT Stabilization Assurance Test

SAT Status and Trends (Report) [US]

SAT Stepped Atomic Time

SAT Strontium Aluminum Tantalate

SAT Subatomic Theory

SAT Supervisory Audio Tone

SAT Symmetric Antitrapping

SAT Symmetric Axis Transform

SAT System Access Technique

SAT Systems Approach to Training

Sat Satellite [also sat]

Sat Saturate(d); Saturation [also sat]

Sat Saturday

Sat Saturn [also Sat]

S At South Atlantic

SATA S-Acetylthioglycolic Acid N-Hydroxysuccinimide Ester

SATA Servico Açoreano de Transportes Areos [Azores Air Transport Service]

SATA N-Succinimidyl S-Acetyl Thioacetate

SATAF Shuttle Activation Task Force [of NASA]

SATAN Satellite Automatic Tracking Antenna

SATAN Security Administrator Tool for Analyzing Networks

SATAN Sensor for Airborne Terrain Analysis

SATANAS Semi-Automatic Analog Setting

SATAR Satellite for Aerospace Research [of NASA]

SATC Schweizerische Akademie der Technischen Wissenschaften [Swiss Academy of Engineering Sciences]

SATCC Southern African Transport and Communications Commission

SATCO Senior Air Traffic Control Officer

SATCO Signal Automatic Air Traffic Control

SATCOL Satellite Network of Columbia

SATCOM Satellite Communication System [also Satcom]

SATCOM Scientific and Technical Communication Committee [US]

SATCOM Satellite Communications Agency [of US Department of Defense]

SATCOM Satellite Communications [also Satcom]

SATCOMA Satellite Communications Agency [of US Department of Defense]

satd saturated [also sat'd]

SATE Stress-Assisted Transformation Effect [Metallurgy]

Satel Satellite [also Satell, satel, or satell]

Satell Commun Satellite Communications [US Publication]

SATF Shortest Access Time First

SATF Substituted Anilines Task Force [US]

Satg Saturating [also satg, Stat'g, or sat'g]

SATIF Scientific and Technical Information Facility

SATIN SAGE (Semiautomatic Ground Environment) Air Traffic Integration

SATINFO Satellite Information Team [of National Environmental Satellite, Data and Information Service–National Oceanic and Atmospheric Administration, US] [also Satinfo]

SATIRE Semi-Automatic Technical Information Retrieval

SATIS Scientific and Technical Information Service [of National Library of New Zealand]

satisf satisfactory

SATIVA Society for Agricultural Training through Integrated Voluntary Activities [US]

SATKA Surveillance, Acquisition, Tracking and Kill Assessment [of US Department of Defense]

Satn Saturation [also satn]

SATNAV Satellite Navigation [also Satnav]

SATNUC Société pour les Applications Techniques dans le Domaine de l'Energie Nucléaire [Society for Technical Applications of Nuclear Energy, France]

SATO Self-Aligned Thick Oxide

SATO Shuttle Attached Teleoperator [of NASA]

SATO Supply and Transportation Operations

SATO Synthetic Aircraft Turbine Oil

SATRA Shoe and Allied Trades Research Association [Name Changed to SATRA Footwear Technology Center, UK] [also Satra]

SATRA Bull SATRA Bulletin [of SATRA Footwear Technology Center, UK]

SATRAC Satellite Automatic Terminal Rendezvous and Coupling

SATRO Science and Technology Regional Organization [Aberdeen, Scotland]

SATS Short Airfield and Tactical Support

SATS Shuttle Avionics Test System [of NASA]

SATS Small Applications Technology Satellite

SATS Solar Alignment Test Site

SATSAR (Study Group on) Satellite Aided Search and Rescue [of International Civil Aviation Organization]

SATSTREAM Satellite Switchstream Service [UK]

SATT Bis(salicylaldehyde)triethylenetetramine

SATT Shear-Area Transistion Temperature

SATT Strowger Automatic Toll Ticketing [US]

SATUCC Southern African Trade Union Coordination Council [Botswana]

SAU Service d'Architecture et d'Urbanisme [Architectural and Town-Planning Services, France]

SAU Signal Acquisition Unit

SAU Smallest Addressable Unit

SAU Standard Advertising Unit

SAU St. Andrew's University [UK]

SAU Strap-Around Unit

SAU System Availability Unit

Sau Staphylococcus aureus [Microbiology]

Sau3A Staphylococcus aureus 3A [Microbiology]

Sau96 Staphylococcus Aureus PS96 [Microbiology]

SAUCERS Saucer and Unexplained Celestial Events Research Society [US]

Saud Arab Saudi Arabia(n)

Saudi Aramco Saudi Arabian Oil Company

S Austr South Australia(n) [also S Austral]

SAV Space-Air Vehicle(s)

SAV Submerged Aquatic Vegetation

.SAV Saved [File Name Extension]

SA/V Surface Area-to-Volume (Ratio)

Sav Savings [also sav]

SAVA Society for Accelerator and Velocity Apparatus

SAVDM Single-Application Virtual DOS (Disk-Operating System) Machine

SAVE Society of American Value Engineers [US]

SAVE System for Automatic Value Exchange

SAVER Shuttle Avionics Verification and Evaluation [NASA]

SAVES Sizing of Aerospace Vehicle Structures

SAVITAR Sanders Associates Video Input/Output Terminal Access Resource

SAVOR Single-Actuated Voice Recorder

SAVOY Sales Forecasting System

SAVP Strabism, Amblyopia and Visual Processing (Program) [of National Eye Institute, National Institutes of Health, Bethesda, Maryland, US]

SAVS Safeguards for Area Ventilation System

SAVS Status and Verification System

SAW Schweizer Akademie der Wissenschaften [Swiss Academy of Sciences]

SAW Self-Avoiding Walk

SAW Strontium Aluminum Tungstate

SAW Submerged Arc Welding

SAW Subsidiary Agreement on Water [Canada]

SAW Surface Acoustic Wave

SAWC Special Air Warfare Center [of US Air Force]

SAWC Stress Analysis of Woven Composites (Software) [of NASA]

SAWCAA Soil and Water Conservation Association of Australia

SAWD Solid Amine Water Desorbed

SAWD Surface Acoustic Wave Device

SAWDAC Siding and Window Dealers Association of Canada

SAWE Society of Aeronautical Weight Engineers [Name Changed to Society of Allied Weight Engineers, US]

SAWE Society of Allied Weight Engineers [US]

SAWE Newsl Society of Allied Weight Engineers Newsletter [US]

SAWF Surface Acoustic Wave Filter

SAW&F South Australian Woods and Forests Department

SAWIC South African Water Information Center

SAWMA Southern African Wildlife Management Association [South Africa]

SAWMARCS Standard Aircraft Weapons Management and Release Control System

SAWO Surface Acoustic Wave Oscillator

SAWP Society of American Wood Preservers [US]

SAWRS Supplementary Aviation Weather Observatories

SAWS Small Arms Weapon System

SAWS Supply Authorization Withdrawal System

SAW-S Series Submerged Arc Welding

SAWTRI South African Wool and Textile Research Institute

SAX Selected Area X-Ray (Photoelectron Spectroscopy)

SAX Small-Angle Scattering of X-Rays

SAX Small Automatic Exchange

SAX Strong Anion Exchange(r)

SA-XAES Selected Area X-Ray Induced Auger Electron Spectroscopy

SAXPS Soft X-Ray Appearance Spectroscopy

SA-XPS Selected Area X-Ray Photoelectron Spectroscopy

SAXRD Small-Angle X-Ray Diffraction

SAXS Small-Angle X-Ray Scattering

SAXS Small-Angle X-Ray Spectroscopy

SAYE Safe As You Earn

SAYTD Sales Amount Year-to-Date

SAZ South African Rand [Currency of Namibia and South Africa]

SAZU Slovenska Akademija Znanosti i Umjetnosti [Slovenian Academy of Sciences and Fine Arts]

SB Bachelor of Science

SB Barred-Spiral-Type Galaxy [Astronomy]

SB Manufacturers' Stovebolt Standard (Screw) Thread [Symbol]

SB Salomon Islands [ISO Code]

SB Salt Bath [Metallurgy]

SB Salt Bridge [Physical Chemistry]

SB San Bernardino [California, US]

SB Sand Blast(ing)

SB Santa Barbara [California, US]

SB Saturated Boiling

SB Schiff Base [Chemistry]

SB Schotten-Baumann (Reaction) [Organic Chemistry]

SB Schottky Barrier [Electronics]

SB Schottky Behaviour [Electronics]

SB Schwinger Boson [Particle Physics]

SB *(Scientiae Baccalaureus)* – Bachelor of Science

SB Scrap Bait [Chemistry]

SB Secondary Battery

SB Sector Boundary [Astrophysics]

SB Seeman-Bohlin (X-Ray Diffraction Geometry) [Physics]

SB Selection Board

SB Self-Aligning Roller Bearing, Single-Row, Angular-Contact, Inner-Ring Raceway Spherical [Symbol]

SB Self-Bias [Electronics]

SB Serial Binary

SB Set Bit (Function)

SB Shear Band [Metallurgy]

SB Sheet Boiling

SB Shell Bullet

SB Shipping Bill

SB Shoe Bracket

SB Shoe Brake

SB Short Bill

SB Shoulder Blade [Anatomy]

SB Shunting Boundaries [Materials Science]

SB Sideband

SB Side Blowing; Side-Blown [Metallurgy]

SB Side Brazed

SB Siege Battery

SB Signaling Battery

SB Simultaneous Broadcast(ing)

SB Siphon Barometer

SB Sleeve Bearing

SB Slide Bearing

SB Slip Band [Crystallography]

SB Slow-Break

SB Slow Burning Type (Conductor Insulation) [Symbol]

SB Small Business

SB Smooth-Bore

SB Society for Biomaterials [US]

SB Solar Battery

SB Soleil-Babinet (Compensator) [Optics]

SB Sonic Barrier

SB Sonic Boom

SB Soot Blower

SB Sound Barrier

SB Sound Blaster

SB Sound Board

SB Southbound

SB Soybean

SB Space Base

SB Special Billing

SB Spectral Band

SB Spectrobolometer

SB Speed Brake

SB Splash Block

SB Split Beam (Test Specimen) [Mechanical Testing]

SB Spring Balance

SB Standard Bead [Building Trades]

SB Standby [also S/B]

SB Stanford-Binet (Intelligence Test) [Psychology]

SB State Block

SB Static Balance(r)

SB Stefan-Boltzmann (Constant) [Statistical Mechanics]

SB Stony Brook [New York State, US]

SB Stop Bath [Photography]

SB Storage Battery

SB Storage Buffer

SB Stove Bolt

SB Straight Binary

SB Strong Base

SB Stuffing Box

SB Styrene Butadiene

SB Submarine Boat

SB Sulfobetaine

SB Sun and Bauer (Grain Boundary Model) [Materials Science]

SB Supply Bulletin

SB Surface Barrier [Electronics]

SB Sweepback [Aerospace]

SB Switch Bridge [Horology]

SB Symmetric Bragg (X-Ray Analysis Geometry) [Physics]

SB Symmetry Breaking [Physics]

SB Synchronization Base

SB Synchronization Bit

SB Systematic Botany

SB3-8 N-Octyl-N,N-Dimethyl-3-Ammonio-1-Propanesulfonate (or Octyl Sulfobetaine)

SB3-10 N-Decyl-N,N-Dimethyl-3-Ammonio-1-Propane-sulfonate (or Caprylyl Sulfobetaine)

SB3-12 N-Dodecyl-N,N-Dimethyl-3-Ammonio-1-Propane-sulfonate (or Lauryl Sulfobetaine)

SB3-14 N-Tetradecyl-N,N-Dimethyl-3-Ammonio-1-Propane-sulfonate (or Myristyl Sulfobetaine)

SB3-16 N-Hexadecyl-N,N-Dimethyl-3-Ammonio-1-Propane-sulfonate (or Palmityl Sulfobetaine)

SB3-18 N-Octadecyl-N,N-Dimethyl-3-Ammonio-1-Propane-sulfonate (or Stearyl Sulfobetaine)

S/B Standby [also SB]

Sb Spiral Galaxy with Medium Nucleus having Relatively Open Arms [Astronomy]

Sb *(Stibium)* – Antimony [Symbol]

Sb^{5+} Antimony Ion [Symbol]

Sb$_4$ Molecular Antimony [Symbol]

Sb II Antimony II [Symbol]

Sb III Antimony III [Symbol]

Sb-116 Antimony-116 [also ^{116}Sb, or Sb^{118}]

Sb-117 Antimony-117 [also ^{117}Sb, or Sb^{119}]

Sb-118 Antimony-118 [also ^{118}Sb, or Sb^{118}]

Sb-119 Antimony-119 [also ^{119}Sb, or Sb^{119}]

Sb-120 Antimony-120 [also ^{120}Sb, or Sb^{120}]

Sb-121 Antimony-121 [also ^{121}Sb, or Sb^{121}]

Sb-122 Antimony-122 [also ^{122}Sb, or Sb^{122}, or Sb]

Sb-123 Antimony-123 [also ^{123}Sb, or Sb^{124}]

Sb-124 Antimony-124 [also ^{124}Sb, or Sb^{124}]

Sb-125 Antimony-125 [also ^{125}Sb, or Sb^{125}]

Sb-126 Antimony-126 [also ^{126}Sb, or Sb^{126}]

Sb-127 Antimony-127 [also ^{127}Sb, or Sb^{127}]

Sb-128 Antimony-128 [also ^{128}Sb, or Sb^{128}]

Sb-129 Antimony-129 [also ^{129}Sb, or Sb^{129}]

Sb-130 Antimony-130 [also ^{130}Sb, or Sb^{130}]

Sb-131 Antimony-131 [also ^{131}Sb, or Sb^{131}]

Sb-132 Antimony-132 [also ^{132}Sb, or Sb^{132}]

Sb-133 Antimony-133 [also ^{133}Sb, or Sb^{133}]

Sb-134 Antimony-134 [also ^{134}Sb, or Sb^{134}]

Sb-135 Antimony-135 [also ^{135}Sb, or Sb^{135}]

sb stilb [Unit]

SBA Scene Balance Algorithms

SBA School-Based Assessment

SBA School of Business Administration

SBA Scottish Biomedical Association [UK]

SBA Seat Back Assembly

SBA sec-Butyl Alcohol (or Secondary Butyl Alcohol)

SBA Security By Analysis

SBA Shared Batch Area

SBA Singapore Booksellers Association

SBA Slurry Blasting Agent

SBA Small Business Administration [US]

SBA Sociedade Brasileira de Automatica [Brazilian Society for Automation]

SBA Société Belge de l'Azote [Belgian Nitrogen Society]

SBA Société Belge de l'Azote (Process)

SBA Soybean Agglutinin [Immunology]

SBA Spina Bifida Association [US]

SBA Standard Beam Approach

SBA Statistisches Bundesamt [Federal Statistics Office, Germany]

SBA Steamboat Association [UK]

SBA Structure Borne Acoustic

SBA Styrene-Butadiene-Acrylonitrile (Copolymer) [also S/BA]

SBA Systems Builders Association [US]

SBa Barred Spiral Galaxy with Large Nucleus Having Tightly-Wound Arms [Astronomy]

SBAA Swedish Business Archives Association

SBAC Society of British Aerospace Companies

SBAFWP Standby Auxiliary Feedwater Pump

SBAMPG Société Belge d'Astronomie, de Météorologie et de Physique du Globe [Belgian Society of Astronomy, Meteorology and Geophysics]

S Band Microwave Frequency Band of 1.55 to 5.20 gigahertz [also S band]

SBAS S-Band Antenna Switch

SBAS Standard Beam Approach System [Aeronautics]

SBASI Single-Bridge Apollo Standard Initiator [NASA Apollo Program]

SBB Monothiobis(benzoylmethane)

SBB Schweizer(ische) Bundesbahn [Swiss Federal Railways]

SBB Single Beam Blanking

SBB Sociedade Botanica do Brasil [Brazilian Botanical Society]

SBB Société Belge de Biologie [Belgian Society of Biology]

SBb Barred Spiral Galaxy with Medium Nucleus Having Relatively Open Arms [Astronomy]

SBBC Styrene-Butadience Block Copolymer

SBBNF Ship and Boat Builders National Federation [UK]

Sb(BuO)₃ Antimony(III) Butoxide

SBC Santa Barbara Channel [California, US]

SBC Schmidt-Baker Camera [Astronomy]

SBC Silicon Blue Cell [also sbc]

SBC Simplified Brick Construction (Method)

SBC Single Binary Collision [Physics]

SBC Single-Board Computer [also sbc]

SBC Single Bond Correlation

SBC Small Business Computer

SBC Sonic Boom Committee [of International Civil Aviation Organization]

SBC Speed Brake Command

SBC Standard Buried Collector

SBC Steel-Bonded (Titanium) Carbide

SBC Structural Biology Center [of Argonne National Laboratory, Illinois, US]

SBC Styrene Block Copolymer

SBC Surface Boundary Condition

SBC Swiss Bank Corporation

SBC System Bus Controller

SBc Barred Spiral Galaxy with Small Nucleus Having Widely Open Arms [Astronomy]

SBCA Satellite Broadcasting and Communications Association [US]

SBCA Scottish Building Contractors Association

SBCA Sensor-Based Control Adapter

SBCC Small Business Consumer Center [Canada]

SBCC Southern Building Code Congress [now SBCCI]

SBCCI Southern Building Code Congress International [US]

SbCl₅ Antimony Pentachloride (Compound)

SbCl₅-GIC Antimony Pentachloride Graphite Intercalation Compound

SBCO Samarium Barium Copper Oxide (Superconductor)

SBCO Strontium Barium Copper Oxide (Superconductor) [also SrBCO]

SBCP Slow-Burning Cocoa Powder

SBCR Stock Balance and Consumption Report

SBCS Single-Byte Character Set

SBCS Small Business Consulting Service [of Queen's University, Kingston, Ontario, Canada]

SBCTO Samarium Barium Copper Titanium Oxide (Superconductor)

SBCU Sensor Board Control Unit

SBD Schematic Block Diagram

SBD Schottky Barrier Diode

SBD Single Button Dial

SBD Surface Barrier Diode

SBD System Block Diagram

S-BD S-Band [also S-Bd]

SBDA Science-Based Departments and Agencies

SBDB Small Business Development Bank

SBDB Small Business Development Branch [of Ministry of Industry, Trade and Technology, Canada]

SBDC Small Business Development Center

SBDC Small Business Development Corporation

SBD-Cl 4-Chloro-7-Sulfobenzofuran, Ammonium Salt

SBD-F Ammonium 7-Fluorobenzo-2-Oxa-1,3-Diazole-4-Sulfonate

SB-DIP Side-Brazed Dual-In-Line Package [Electronics]

SBE Society of Broadcast Engineers [US]

SBE Society of Business Economists [UK]

SBE Spouted Bed Electrode

SBE Subacute Bacterial Endocarditis [Medicine]

SBE Sub-Bit Encoder

SBE Supertwisted Birefringence Effect [Physics]

SBE System Buffer Element

SBEC Single Binary Elastic Collision [Physics]

SBEC Single Board (Automotive) Engine Controller

SBEC Small Business Enterprise Centers [Canada]

SBECP Small Business Equity Corporations Program

SBeO Sintered Beryllia

SBET Society of Biomedical Equipment Technicians [now National Society of Biomedical Equipment Technicians, US]

Sb(EtO)₃ Antimony(III) Ethoxide

SBF Short Backfire (Antenna)

SBF Simulated Body Fluid

SBF Sociedade Brasileira de Fisica [Brazilian Society of Physics]

SBFM Silver-Band Frequency Modulation

SBFU Standby Filter Unit

SBG Schweizer Botanische Gesellschaft [Swiss Botanical Society]

SBG Sociedade Brasiliera do Genetica [Brazilian Genetics Society]

SBG Standard Battery Grade

SBGI Society of the British Gas Industries

SBGTS Standby Gas Treatment System

SBH Schottky Barrier Height [Electronics]

SBH Sodium Borohydride

SBH Strip-Buried Heterostructure [Electronics]

SBH Switch Busy Hour

SBHC Speed Brake Hand Controller

Sbhd Subtrahend [also sbhd]

SBHSC Stony Brook Health Sciences Center [New York State, US]

SBI Single Byte Interleaved

SBI Society for Business Information [Germany]

SBI Sound Blaster Instrument

SBI Steel Boiler Institute

SBI Synchronous Backplane Interconnect

SBIG Small Business Investment Grant

SBIP Small Business Intern Program

SBIR Small Business Innovation Research [Joint Program of the US Department of Defense, the US Department of Energy and NASA]

SBIR Society for Basic Irreproducible Research

SBIR Space-Based Infrared

SBIR Storage Bus in Register

SBIR/STTR Small Business Innovation Research/Small Business Technology Transfer (Program) [US]

SBIS Small Business Information Service [of Federal Business Development Bank, Canada]

SBIS Standard Base Information System

SBITE Société Belge des Ingénieurs des Télécommunications et d'Electronique [Belgian Society of Telecommunications and Electronics Engineers]

SBITP Small Business Industry Technology Program

SBK Sinclair-Baker-Kellogg (Process)

SBK Single-Beam Klystron

Sb Kratk Soobshch Fiz AN SSSR Fiz Inst PN Lebed Sbornik Kratkie Soobshcheniya po Fisike, AN SSSR, Fizicheskii Institut PN Lebedeva [Soviet Physics–Lebedev Institute Reports; of the Akademy of Sciences of the USSR]

SBL Styrene-Butadiene Latex

SBLA Small Business Loans Act [Canada]

SBLC Standby Liquid Control

SBLG Small Blast Load Generator

s-BLM Supported Bilayer Lipid Membrane

SBLOCA Small Break Loss-of-Coolant Accident [Nuclear Reactors]

SBM Samarium Barium Manganate

SBM Secondary Bacterial Meningitis [Medicine]

SBM Separate Bombardment Mode

SBM Single Buoy Mooring

SBM Solomon-Bloembergen-Morgan (Theory) [Physics]

SBM System Balance Measure

SBMA Steel Bar Manufacturers Association [US]

SBMA Steel Bar Mills Association

SBMAC Sociedade Brasileira de Matematica Aplicada e Computacional [Brazilian Society of Applied and Computational Mathematics]

SBME Société Belge de la Microscopie Electronique [Belgian Society for Electron Microscopy]

Sb(MeO)$_3$ Antimony(III) Methoxide

SBMI School Bus Manufacturers Institute [US]

SBML Special Bauxite Mining Lease

SBML Suspended Bed-Material Load

Sb-MnSb Manganese Antimonide Reinforced Antimonide (Composite)

SBMO Samarium Barium Manganate

SBMO Sociedade Brasileira de Microondas [Brazilian Microwave Society]

SBMPL Simultaneous Binaural Midplane Localization

SBMV Southern Bean Mosaic Virus

SBN Schweizerischer Bund für Naturschutz [Swiss Nature Protection League]

SBN Sintered Boron Nitride

SBN Small Business Network [US]

SBN Standard Book Number

SBN Strontium Barium Niobate

SBN Strontium Bismuth Niobate

Sb Nauchni Tr Sbornik Nauchni Trudove [Bulgarian Scientific Journal]

SBNWM Scientific Basis of Nuclear Waste Management

SBNT Single-Breath Nitrogen Test

SBO Sideband(s) Only

SBO Soy Bean Oil

SBO System for Business Operations

SbOCl Antimony Oxychloride

Sb(OEt)$_3$ Antimony(III) Ethoxide

Sb(OMe)$_3$ Antimony(III) Methoxide

SBP San Luis Obispo (Tactical Air Navigation Station)

SBP Set Binding Protein [Biochemistry]

SBP Shore-Based Prototype

SBP Single-Strand Binding Protein [Biochemistry]

SBP Slotted-Blade Propeller

SBP Société Belge de Photogrammétrie [Belgian Society for Photogrammetry]

SBP Société Belge de Physique [Belgian Physical Society]

SBP Sonic Boom Panel

SBP Special Boiling Point (Gasoline)

SBP Stainless Ball Plunger

SBP Sulfobromophthalein

SBP Systolic Blood Pressure [Medicine]

SBPA Singapore Book Publishers Association

SBPC Sociedade Brasileira para o Progresso da Ciência [Brazilian Society for the Progress of Science]

Sb/Pd Antimony/Palladium (Heterojunction)

SBPF Sociedade Brasileira de Pesquisas Fisicas [Brasilan Society for Physical Research]

SBPIM Society of British Printing Ink Manufacturers

SBPM Society of British Paint Manufacturers

SBPO Spun-Bounded Polyolefin

SBPT Small-Modular-Weight Basic Protein Toxin

SBQ Special Bar Quality (Steel) [Metallurgy]

SBR Selectively Buried Ridge (Waveguide)

SBR Shear-Wave Back Reflectivity

SBR Signal-to-Background Ratio

SBR Society for Biological Rhythm [Puerto Rico]

SBR Society of Bead Reseachers [Canada]

SBR Soviet Breeder Reactor [USSR]

SBR Space-Based Radar

SBR Storage Buffer Register

SBR Styrene-Butadiene Rubber

.SBR Source Browser [Borland File Name Extension]

SB(r) Barred Spiral Type Galaxy, r-shaped [Astronomy]

SBRC Santa Barbara Research Center [California, US]

SBRC Solubility/Bioavailability Research Consortium [US]

SB(rs) Barred Spiral Type Galaxy, r-shaped, sshaped [Astronomy]

SBS Satellite Business System [US]

SBS Save British Science Society [UK]

SBS Serially Balanced Sequence

SBS Short-Beam Shear [Mechanics]

SBS Sick Building Syndrome [Medicine]

SBS Silicon Bidirectional Switch

SBS Silicon Bilateral Switch

SBS Single Buoy Storage

SBS Slow-Break Switch

SBS Smart Battery Specification

SBS Solid Bleached Sulfate (Softwood Pulp)

SBS Special Broadcasting Service

SBS Stimulated Brillouin Scattering [Physics]

SBS Styrene-Butadiene-Styrene (Block Copolymer) [also S-B-S]

SBS Submerged Bed Scrubber

SBS Subscript (Character)

SBS Swiss Botanical Society

S-B-S Styrene-Butadiene-Styrene (Block Copolymer) [also SBS]

SB(s) Barred Spiral Type Galaxy, s-shaped [Astronomy]

SBSC Schottky Barrier Solar Cell

SB/SDB Small Business/Small Disadvantaged Business

SBSE Scintillator Backscattered Electron

SBSED Scintillator Backscattered Electron Detector

SbSn Antimony Stannide

SBSS Semi-Bleached Sulfate Softwood (Pulp)

SBSS Short-Beam Shear Strength [Mechanics]

SBSTTA Subsidiary Body on Scientific, Technical and Technological Advice [of Convention on Biological Diversity]

SBT Self-Briefing Terminal

SBT Six-Bit Transcode

SBT South Bay Technology, Inc. [US]

SBT Southern Bell Telephone Company [US]

SBT Southern Bluefin Tuna

SBT Strontium Bismuth Tantalate

SBT Submarine Bathythermograph

SBT Surface-Barrier Transistor

SBTC Speed Brake/Thrust Controller

SBTMA Swedish Brick and Tile Manufacturers Association

Sb Tr Ts NIIB Sbornik Trudov Tsentral'nogo Nauchno Issledovatel'skogo Instituta Bumagi [Russian Scientific Journal]

Sb Tr Ts NILK Prom Sbornik Trudov Tsentral'nogo Nauchno Issledovatel'skogo i Proektnogo Instituta Lesokhimicheskoi Promyshlennosti [Russian Scientific Journal]

Sb Tr VNIIB Prom Sbornik Trudov Vsesyuznogo Nauchno Issledovatel'skogo Instituta Tsellulozno-Bumazhnoi Promyshlennosti [Russian Scientific Journal]

SBTT Small Business Technology Transfer (Program)

SBU Simon Bolivar University [Caracas, Venezuela]

SBU Station Buffer Unit

SBU System Billing Unit

SBUAM Société Belge des Urbanistes et Architectes Modernistes [Belgian Society of Urban Planners and Modern Architects]

SBUEW Polyurethane Coated with Self-Bonding Outer Coating Enamelled (Wire)

SBUV Solar Backscatter Ultraviolet (Radiometer)

SBUV/TOMS Solar Backscatter Ultraviolet/Total Ozone Mapper System

SBV Schweizerischer Bankverein [Swiss Bank Corporation]

SBV Schweizerischer Bauernverband [Swiss Farmers' Organization]

SBV Shield Building Vent

Sb Věd Pr Vys Šk Báňské Ostravé, Horn-Geol Sborník Vědeckých Prací Vysoké Školy Báňske v Ostravé, Řada Hornicko-Geologická [Journal of the Mining and Geological Engineering Department of the Technical University Ostrava, Czech Republic]

Sb Věd Pr Vys Šk Báňské Ostravé, Hutn Sborník Vědeckých Prací Vysoké Školy Báňske v Ostravé, Řada Hutnicka [Journal of the Metallurgical Engineering Department of the Technical University Ostrava, Czech Republic]

Sb Věd Pr Vys Šk Báňské Ostravé, Strojni Elektrotech Sborník Vědeckých Prací Vysoké Školy Báňske v Ostravé, Řada Strojni Elektrotechnicka [Journal of the Electrical Engineering Department of the Technical University Ostrava, Czech Republic]

SBVS Shield Building Vent System

SBW Slow-Burning, Weatherproof Type (Conductor Insulation) [Symbol]

SBWE State Board of Water Engineers [US]

SBWR Simplified Boiling Water Reactor
SBX S-Band Transponder
SBX Small Business Exchange
SBX Subsea Beacon/Transponder
S by E South by East
S by SE South by Southeast
S by SW South by Southwest
S by W South by West
SBZ Schweizer Buchzentrum [Swiss Book Center]
SBZ Surface Brillouin Zone [Solid-State Physics]
SBZS Strontium Barium Zirconium Selenide
SC Samarium Chromite
SC San Carlos [Chile]
SC San Carlos [Luzon, Philippines]
SC Sandwell College [UK]
SC Sanitary Corps
SC Santa Catharina [Brazil]
SC Santa Cruz [California, US]
SC Santa Cruz [Bolivia and Argentina]
SC Satel Conseil [Satel Council, France]
SC Satellite Communication
SC Satellite Computer
SC Saturable Core
SC Saturation Current
SC Scanning Coil
SC Scattering Coefficient
SC Scattering Constant
SC Schlumberger Configuration
SC Schultz-Charlton (Antibody-Antigen Reaction) [Immunology]
SC Scientific Committee
SC Scientific Computer; Scientific Computing
SC Scintillation Counter
SC Screen Color
SC Screened Cable
SC Screw Compressor
SC Screw Conveyor
SC Screwed and Coupled
SC Search Coil
SC Search Control
SC Season Cracking [Metallurgy]
SC Secondary Consumer [Ecology]
SC Secondary Containment [Nuclear Reactors]
SC Second Class
SC Sectional Center [Telecommunications]
SC Section Code
SC Security Committee [of NATO]
SC Security Council [of United Nations]
SC Sedimentation Constant [Physical Chemistry]
SC Seebeck Coefficient
SC Seed Crystal
SC Selector Channel
SC Selenium Cell

SC Self-Aligning Roller Bearing, Double-Row, Outer-Ring Raceway Spherical, Rollers Guided by Separate Axially Floating Guide Ring on Inner Ring [Symbol]
SC Self-Capacitance
SC Self-Centering
SC Self-Charge [Quantum Mechanics]
SC Self-Closing
SC Self-Complementing
SC Self-Consistent (Model) [Metallurgy]
SC Self-Contained [also S/C]
SC Semicircle; Semicircular
SC Semi-Closed (System)
SC Semiconductor [also sc]
SC Send Common
SC Semi-Coprecipitation (Method) [Solid-State Physics]
SC Sending Complete
SC Seneca College [Canada]
SC Sense Contact
SC Sequence Controller
SC Series Circuit
SC Series Coil
SC Service Cable
SC Service Charge [also S/C]
SC Service Code
SC Servo Control
SC Session Control
SC Set Clock
SC Sex Cell
SC Seychelles [ISO Code]
SC Shaped Charge
SC Shaping Circuit [Electrical Engineering]
SC Shaw College [Toronto, Canada]
SC Sheridan College [Canada]
SC Short Case
SC Short Circuit [also S/C]
SC Short Course
SC Shift-Control Counter
SC Shutdown Controller
SC Sickle Cell [Medicine]
SC Side Chain [Chemistry]
SC Sierra Club [San Francisco, California, US]
SC Signal Conditioner; Signal Conditioning [also S/C]
SC Signal Converter
SC Signal Corps [of US Army]
SC Sign Change
SC Silent Chain
SC Silicon Carbide
SC Silk-Covered (Electric Cable)
SC Silver-Copper (Electric Wire)
SC Simple Compression
SC Simple Cubic (Crystal) [also sc]
SC Simulation Council
SC Sine-Cosine
SC Single Column

SC Single Contact [also sc]
SC Single Crystal
SC Single-Current
SC Sioux City [Iowa, US]
SC Sister Chromatids [Cytology]
SC Slag Chemistry [Metallurgy]
SC Slag Concrete
SC Slave Clock
SC Slave Cylinder
SC Slip Casting
SC Slow Cooling; Slowly Cooled [Metallurgy]
SC Slow-Curing (Asphalt) [also S-C]
SC Small Capitals [also sc]
SC Small Computer
SC Smith College [Northampton, Massachusetts, US]
SC Society for Cryobiology [of Federation of American Societies for Experimental Biology, US]
SC Sodium Coding
SC Soft Copy
SC Soft-Cover (Format)
SC Soil Conservation [Ecology]
SC Solar Calendar
SC Solar Cell
SC Solar Cooker; Solar Cooking
SC Solid Case
SC Solid Core
SC Solidification Coefficient
SC Soller Collimator [Optics]
SC Solubility Coefficient [Physical Chemistry]
SC Solution Cast(ing)
SC Solution Chemist(ry)
SC Solution Coating (Process)
SC Solvay Council
SC Solventless Carbamite (Propellant)
SC Solventless Cordite (Explosive)
SC Sonochemistry
SC Source Code
SC South Carolina [US]
SC Space Capsule
SC Space Charge
SC Space Communications
SC Spacecraft [also S/C]
SC Spacer Coupling
SC Spatial Coherence [Physics]
SC Spacing Criterion [also S/C, sc, or s/c]
SC Spark Chamber
SC Special Committee
SC Specific Conductivity
SC Spectral Cluster
SC Spectrochemical; Spectrochemist(ry)
SC Speech Communication
SC Speed Control
SC Spherical Coordinate(s)
SC Spinal Column [Anatomy]

SC Spinal Cord [Anatomy]
SC Spindle Cell (of Frogs) [Zoology]
SC Splat-Cooled; Splat Cooling
SC Spontaneous Combustion
SC Sports Car
SC Spray(able) Concentrate
SC Spray Cooled; Spray Cooling
SC Spreading Coefficient [Thermodynamics]
SC Spring Constant
SC Squeeze Cast(ing)
SC Squirrel Cage (Motor)
SC Stab Culture [Microbiology]
SC Staggered Conformation
SC Standard Cell
SC Standard Clean
SC Standard Configuration
SC Standing Committee
SC Standing Conference
SC Star Catalog [Astronomy]
SC Star Coupler
SC Starting Current
SC Statement of Capability
SC Statement of Compatibility
SC Statistical Computing
SC Statistics Canada
SC Steam Calorimeter
SC Steam Cracking [Chemical Engineering]
SC Steel Can
SC Steel Casting
SC Steel-Cored (Electric Wire)
SC Steel Cover
SC Steering Committee
SC Step Coverage
SC Stereochemical; Stereochemist(ry)
SC Stereo-Comparator
SC Stevens and Ciesielski [Physics]
SC Sticking Coefficient [Physical Chemistry]
SC Stochastic Control
SC Stoichiometric Compound
SC Stop Cock
SC Stop-Continue (Register)
SC Storage Capacity
SC Straight-Chain (Structure)
SC Strain Compass
SC Stress Concentration [Mechanics]
SC Stress Corrosion
SC Strip Cast(ing) [Metallurgy]
SC Strip Cathode
SC Strontium Cuprate
SC Structural Ceramic(s)
SC Structural Chemist(ry)
SC Structural Clay
SC Structural Composite

SC Subcommittee
SC Subcooling
SC Subcritical(ity) [Nuclear Engineering]
SC Subculture [Biology]
SC Subcutaneous [also sc]
SC Subject Category
SC Subject Code
SC Subsystem Computer
SC Suez Canal
SC Sulfidation Corrosion
SC Supercalendered; Supercalendering [Papermaking]
SC Supercapacitance
SC Supercharge(r)
SC Supercluster [Astronomy]
SC Super-Composite
SC Supercomputer; Supercomputing
SC Superconductive; Superconductivity; Superconductor [also Sc]
SC Supercooled; Supercooling [Thermodynamics]
SC Superimposed Coding
SC Superimposed Current
SC Supervisor Call
SC Supervisory Control
SC Support Contractor
SC Suppressed Carrier
SC Supreme Court
SC Surface Channel
SC Surface Charge
SC Surface Chemist(ry)
SC Surface Combustion
SC Surface Command
SC Surface Complexation
SC Surface Composition
SC Surface Conductivity
SC Surface Crack
SC Suspension Concentrate
SC Swarthmore College [Pennsylvania, US]
SC Sweep Circuit
SC Swing Clamp
SC Switch Cell
SC Switched Capacitor
SC Switching Circuit
SC Switching Computer
SC Symbol(ic) Code
SC Symbolic Computation
SC Synchrocyclotron
SC Synchronous Communication
SC Synchronous Converter
SC Synchronous Counter
SC Systemic Circulation [Medicine]
SC Thiocarbonyl
.SC Script [File Name Extension] [also .sc]
S/C Self-Contained [also s/c]
S/C Sensor/Controller

S/C Service Charge [also SC]
S/C Short Circuit [also SC]
S/C Signal Conditioner [also SC]
S/C Software Contractor [also SC]
S/C Spacecraft [also SC]
S/C Spacing Criterion [also SC, sc, or s/c]
S/C Splitter/Combiner
S/C Stabilization and Control
S/C Strip-Chart (Recorder)
S/C Subcontractor
S-C Slow-Curing (Asphalt) [also SC]
S-C Switched-Capacitor
Sc Scale(r) [also sc]
Sc Scandium [Symbol]
Sc Scene [also sc]
Sc Schmidt Number [Symbol]
Sc Schneiderite [Explosive]
Sc Science; Scientific; Scientist [also sc]
Sc Scot(s); Scotland; Scottish
Sc Screen [also sc]
Sc Screw [also sc]
Sc Spiral Galaxy with Small Nucleus Having Widely Open Arms [Astronomy]
Sc Stratocumulus (Cloud)
Sc Superconductor [also SC]
Sc^{2+} Divalent Scandium Ion [also Sc^{++}] [Symbol]
Sc_3 Schmidt Number (of Electrochemistry) [Symbol]
Sc_{turb} Turbulent Schmidt Number [Symbol]
Sc-41 Scandium-41 [also ^{41}Sc, or Sc^{41}]
Sc-43 Scandium-43 [also ^{43}Sc, or Sc^{43}]
Sc-44 Scandium-44 [also ^{44}Sc, or Sc^{44}]
Sc-45 Scandium-45 [also ^{45}Sc, Sc^{45}, or Sc]
Sc-46 Scandium-46 [also ^{46}Sc, or Sc^{46}]
Sc-47 Scandium-47 [also ^{47}Sc, or Sc^{47}]
Sc-48 Scandium-48 [also ^{48}Sc, or Sc^{48}]
Sc-49 Scandium-49 [also ^{49}Sc, or Sc^{49}]
sc (scilicet) – namely
sc small capital letters; small capitals
sc subcutaneous [also SC]
$\sigma_0(c)$ concentration dependent critical shear stress [Symbol]
SCA Saskatchewan Construction Association [Canada]
SCA Schedule Change Authorization
SCA Scientific Computing Associates [of Yale University, New Haven, Connecticut, US]
SCA Sea Cadet Association [UK]
SCA Seacoast Artillery
SCA Secondary Communications Authorization
SCA Sectional Chambers Association [UK]
SCA Selectivity Clear Accumulator
SCA Semiconductor Alloy
SCA Sensor-Controlled Automation
SCA Sequence Control Area
SCA Sequence Control Assembly
SCA Sheep-Cell Agglutination [Immunology]

SCA Shipbuilders Council of America [US]

SCA Short Code Address

SCA Shuttle Carrier Aircraft [of NASA]

SCA Sickle-Cell Anemia [Medicine]

SCA Single-Channel Analyzer

SCA Simulated Core Assembly

SCA Simulation Control Area

SCA Sister Chromatid Exchange [Biochemistry]

SCA Smoke Control Area [UK]

SCA Smoke Control Association [US]

SCA Sneak Circuit Analysis

SCA Source Code Analyzer

SCA Southern Cotton Association [US]

SCA Spectrochemical Analysis

SCA Spinning-Cup Atomization [Powder Metallurgy]

SCA Sprayed Concrete Association [UK]

SCA Standing Committee on Agriculture

SCA Steel-Cored Aluminum (Conductor)

SCA Subcritical Assembly

SCA Subsidiary Carrier Authorization

SCA Subsidiary Communications Authorization [US]

SCA Superconductor Alloy

SCA Supersidiary Communications Authorization

SCA Supplemental Coolant Additive

SCA Surface Charge Analysis; Surface Charge Analyzer

SCA Suspended Ceiling Association [UK]

SCA Symmetrical Component Analysis

SCA Synchronous Communications Adapter

SCA System Control Area

Sca Streptomyces caespitosus [Microbiology]

SCAA Spill Control Association of America [US]

SCAA Superconductor Applications Association [US]

SCAAT Sault College of Applied Arts and Technology [Sault Ste. Marie, Ontario, Canada]

SCAC Soil Conservation Advisory Committee

Sc(Ac)$_3$ Scandium(III) Acetate [also Sc(ac)$_3$]

Sc(ACAC)$_3$ Scandium(III) Acetylacetonate [also Sc(AcAc)$_3$, or Sc(acac)$_3$]

SCAD Subsonic Cruise Armed Decoy

SCAD Surface Characterization and Depth Profiling

SCADA Supervisory Control and Data Acquisition (System)

SCADAR Scatter Detection and Ranging [also Scadar, or scadar]

SCADS Scanning Celestial Attitude Determination System

SCADS Simulation of Combined Analog Digital Systems

SCAE Society for Computer-Aided Engineering [US]

SCAF Société Centrale des Agriculteurs de France [Central Society of French Farmers]

SCAF Standing Committee of the Australian Forestry Council

SC Al Steel-Cored Aluminum (Electric Wire) [also SCAl]

SCALD Structural Computer-Aided Logic Design

SCALE Space Checkout and Launch Equipment

SCALOP Standing Committee on Antarctic Logistics and Operations [Australia]

SCAM SCSI (Small Computer Systems Interface) Configuration Automatically

SCAM Spectrum Characteristics Analysis and Measurement

SCAM Subcarrier Amplitude Modulation

SCAM Synchronous Communications Access Method

SCAMA Station Conferencing and Monitoring Arrangement [also SCAMMA]

SCAMA Switching, Conference and Monitoring Arrangement [of NASA]

SCAMP Sectionalized Carrier and Multipurpose Vehicle

SCAMP State-of-the-Art Computer-Assisted Machine-Tool Project

SCAMPS Small Computer Analytical and Mathematical Programming System

SCAMS Scanning Microwave Spectrometer

SCAN Second Career Assistance Network

SCAN Selected Current Aerospace Notices [of NASA]

SCAN Self-Correcting Automatic Navigation

SCAN Semiconductor Component Analysis Network

SCAN Small Computers in the Arts Network [US]

SCAN Stock Control and Analysis

SCAN Stock-Market Computer Answering Network [UK]

SCAN Student Career Automated Network

SCAN Supermarket Computer Answering Service

SCAN Surface Condition Analyzer

SCAN Switched Circuit Automatic Network

Scand Scandinavia(n)

Scand Audiol Scandinavian Audiology [Journal published in Sweden]

Scand Audiol Suppl Scandinavian Audiology Supplementum [Sweden]

Scand J Dent Res Scandinavian Journal of Dental Research

Scand J Forest Res Scandinavian Journal of Forest Research [Sweden]

Scand J Gastroent Scandinavian Journal of Gastroenterology

Scand J Haematol Scandinavian Journal of Haematology

Scand J Metall Scandinavian Journal of Metallurgy [Denmark/Finland]

Scand J Nephrol Scandinavian Journal of Nephrology

Scand J Plast Reconstr Surg Scandinavian Journal of Plastic and Reconstructive Surgery

Scand J Psychol Scandinavian Journal of Psychology

Scand J Rehab Med Scandinavian Journal of Rehabilitative Medicine

Scand J Stat Theory Appl Scandinavian Journal of Statistics Theory and Applications [Sweden]

SCANDOC Scandinavian Documentation Center

Scand Refrig Scandinavian Refrigeration [Danish Publication]

Scan Electron Microsc Scanning Electron Microscopy [Publication of Scanning Microscopy International, US]

Scan Electron Microsc Symp Proc Scanning Electron Microscope Symposium Proceedings [US]

Scan Microsc Scanning Microscopy [Publication of Scanning Microscopy International, US]

SCANIIR Surface Chemical Analysis by Neutral and Ionized Impact Radiation; Surface Composition Analysis by Neutral and Ion Impact Radiation

SCANNET Scandinavian Information Retrieval Network [of Nordic Council for Scientific Information and Research Libraries, Finland]

Scanning Electron Microsc Scanning Electron Microscopy [Publication of Scanning Microscopy International, US]

Scanning Microsc Scanning Microscopy [Publication of Scanning Microscopy International, US]

SCANS Scheduling and Control by Automated Network System(s)

SCANS System Checkout Automatic Network Simulator

SCANSAR Scanning Synthetic Aperture Radar

SCANSCAT Scanning Scatterometer

scans/min scans per minute [Unit]

scans/sec scans per second [Unit]

SCAN-Test Scandinavian Pulp, Paper and Board Testing Committee [Sweden]

SCAO Standing Conference of Atlantic Organizations [UK]

SCAO Spherical Cloud Atomic Orbital [Physics]

SCAP Silent Compact Auxiliary Power

SCAP States Cooperative Assistance Program [of Australian Nature Conservancy Agency]

SCAPE Self-Contained Atmospheric Protection Ensemble

SCAPE Self-Contained Atmospheric Protective Ensemble (Suit)

S CAPS Small Capitals [also s caps]

SCAR Satellite Capture and Retrieval

SCAR Scandinavian Council for Applied Research

SCAR Scientific Committee on Antarctic Research [UK]

SCAR Smoke, Clouds and Radiation

SCAR Special Committee on Antarctic Research [now Scientific Committee on Antarctic Research, UK]

SCAR Subcaliber Aircraft Rocket

SCAR Submarine Celestial Altitude Recorder

SCAR Sulphates, Clouds and Radiation

SCARA Selective Compliance Assembly Robot Arm; Selectively Compliant Assembly Robot Arm

SCAR A Sulphates, Clouds and Radiation–Atlantic [Oceanography]

SCARAB Scanner for the Radiation Budget

SCARAB Submersible Craft Assisting Recovery/ Repair and Burial

ScaRaBE Scanner for Radiation Budget Experiment

SCARABE Scanning Radiation Budget Experiment

SCAR B Smoke, Clouds and Radiation–Brazil Mission [of NASA]

SCARF Santa Cruz Acoustic Range Facility [California, US]

SCARF Side-Looking Coherent All-Range Focussed

SCARF Standing Committee of the Australian Forestry Council

SCARM Standing Committee on Agriculture and Resource Management [Australia]

SCARS Sneak Circuit Analysis Report Summary

SCARS Software Configuration Accounting and Reporting System

SCAQMD South Coast Air Quality Management District [US]

SCAS Scan String [Computers]

SCAS Science Calibration Subsystem

SCAS South Carolina Academy of Science [US]

SCAS Stability and Control Augmentation System

SCAS Subsystem Computer Application Software

SCAS Surrey County Agricultural Society [UK]

SCAT Scatter [also Scat, or scat]

SCAT School and College Ability Test

SCAT Schottky Cell Array Technology

SCAT Sequentially-Controlled Automatic Transmitter

SCAT Share Compiler-Assembler and Translator

SCAT Small Car Automatic Transit System

SCAT Space Communication and Tracking

SCAT Speed Command of Attitude and Thrust

SCAT Supersonic Commercial Air Transport

SCAT Surface-Controlled Avalanche Transistor

SCAT Systems and Components Automated Test

SCATANA Security Control of Air Traffic and Air Navigation Aids

ScATCC Scottish Air Traffic Control Center

SCATE Stromberg-Carlson Automatic Test Equipment

SCATHA Spacecraft Charging at High Altitude

SCATS Sequential Controlled Automatic Transistor Start

SCATS Simulation, Checkout and Training System

SCATS Standing Conference for the Advancement of Training and Supervision [UK]

SCATS Storage, Checkout and Transport

SCATS Systems and Components Automated Test Systems

SCATT Scatterometer [also Scatt, or scatt]

SCATT Scientific Communications and Technical Transfer System [US]

SCAUL Standing Conference of African University Libraries [Nigeria]

SCAV Soil Conservation Association of Victoria [Australia]

Scav Scavenge(r) [also scav]

SCAW Scientists Center for Animal Welfare [US]

SCAW Shielded Carbon-Arc Welding

SCB Schedule Change Board

SCB Segment Control Bit

SCB Selection Control Board

SCB Selector Control Box

SCB Selenite Cystine Broth [Biochemistry]

SCB Silacyclobutane

SCB Silicon Circuit Board

SCB Site Control Block

SCB Société Chimique Belge [Belgian Chemical Society]

SCB Society for the Conservation of Biology

SCB Software Control Board

SCB South Central Bell [US]

SCB Specification Control Board

SCB Stack Control Block

SCB Station Control Block

SCB Statistiska Centralbyråns Bibliotek [Statistical Central Bureau Library, Sweden]

SCB String Control Byte

SCB Subsystem Control Block

SCB System Control Block

ScB *(Scientiae Baccalaureus)* – Bachelor of Science

SCBA Self-Consistent Born Approximation [Physics]

SCBA Self-Contained Breathing Apparatus

SCBAF Self-Contained Breathing Apparatus with Full Facepiece

SCBC Science Council of British Columbia [Canada]

ScBC Bachelor of Science in Chemistry

ScBE Bachelor of Science in Engineering

SCBR Steam-Cooled Breeder Reactor [US]

SCBS System Control Blocks

SCC Safety Control Center

SCC Satellite Communication Concentrator

SCC Satellite Communications Controller

SCC Satellite Control Center

SCC Science Council of Canada

SCC Science Culture Canada (Program) [of Public Works and Government Services Canada and Ministry of State for Science and Technology]

SCC Scientific Coordinating Committee

SCC Sea Cadet Corps [UK]

SCC Secondary Category Code

SCC Secondary Containment Cooling [Nuclear Reactors]

SCC Sectional Classification Code

SCC Self-Consistent Charge [Physical Chemistry]

SCC Semiconductor Circuit

SCC Serial Communications Controller

SCC Serial Controller Chip

SCC Seta Closed Cup (Flash Point Tester) [Chemistry]

SCC Shock Compression of Condensed Matter [Meeting of the American Physical Society, US]

SCC Short-Circuit Current

SCC Sierra Club of Canada

SCC Sigmund Cohn Corporation [US]

SCC Signal Conversion Circuit

SCC Simulation Control Center

SCC Single Channel per Carrier

SCC Single-Conductor Cable

SCC Single Configurational Coordinate (Model) [Physics]

SCC Single-Cotton Covered (Electric Wire)

SCC Slidell Computer Complex

SCC Smooth-Conductor Cable

SCC Society of Cosmetic Chemistry [US]

SCC Soil Classification Chart

SCC Soil Conservation Committee [South Carolina, US]

SCC Somatic Cell Concentration

SCC Space Consultative Committee [UK]

SCC Specialized Common Carrier

SCC Species Survival Commision [of International Union for Conservation of Nature and Natural Resources]

SCC Speed Control Circuit

SCC Split-Conductor Cable

SCC Squamous Cell Carcinoma [Medicine]

SCC Standard Cubic Centimeters [also scc]

SCC Standards Coordinating Committee(s) [of Institute of Electrical and Electronics Engineers, US]

SCC Standards Council of Canada

SCC State Consultative Council

SCC St. Clair College [Canada]

SCC Storage Connecting Circuit

SCC Stress-Corrosion Cracking

SCC Structural Concrete Consortium [UK]

SCC Subcarrier Channel

SCC Sudan Chamber of Commerce

SCC Super Cloud Cluster [also scc]

SCC Superconducting Characterization Cryostat

SCC Superconducting Coil

SCC Supreme Court of Canada

SCC Switching Control Center

SCC Synchronous Channel Check

SCC Synchronous Communications Controller

ScC Scandium Carbide

scc standard cubic centimeters [also SCC]

SCCA Society of Company and Commercial Accountants [UK]

SCCA Sports Car Club of America

SCCB Site Configuration Control Board

SCCC Self-Consistent Charge and Configuration (Molecular Orbital Calculation) [Physical Chemistry]

SCCC Single Channel Communications Controller

SCCC Single-Colored Crystal Class

SCCD Surface Charge-Coupled Device

SCCDEST Steering Committee on Crossborder Data Exchange in Science and Technology [US]

SCCEH Self-Consistent Charge-Extended Hueckel (Procedure) [Physical Chemistry]

SCCF Satellite Communication Control Facility

SCCG Subcritical Crack Growth [Metallurgy]

SCCH Standard Cubic Centimeters-per Hour [also scch]

SCCHLL Standards Coordinating Committee on High-Level Languages [of Institute of Electrical and Electronics Engineers, US]

SCCIG Stress Corrosion Crack Initiation and Growth [Metallurgy]

SCCM Standard Cubic Centimeter per Minute [also sccm]

Sccm Succinimide

SCCO Samarium Cerium Copper Oxide (Superconductor)

SCCO Strontium Calcium Copper Oxide (Superconductor) [also SrCCO]

Sc(Cp)₃ Tris(cyclopentadienyl)scandium

SCCS Ship Command and Control System

SCCS Sodium Chemistry Control System

SCCS Source Code Control System

SCCS Special Consultative Commission on Security [of Organization of American States]

SCCS Standard Cubic Centimeter per Second [also sccs]

SCCS Standards Council Customer Services [of Standards Council of Canada]

SCCS Straight-Cut (Numerical) Control System

SCCS Switching Control Center System

SCCT Specialist in Community College Teaching

SCCU Single Channel Control Unit

SC Cu Steel-Cored Copper (Electric Wire)

SCD Satellite Control Department

SCD Scintillation Detector

SCD Screening Completion Date

SCD Segmented-Array CCD (Charge-Coupled Device) Detector

SCD Semiconductor Device

SCD Sickle-Cell Disease [Medicine]

SCD Single-Crystal Diffraction

SCD Single-Crystal Diffractometer

SCD Source Control Document

SCD Source Control Drawing

SCD Space Control Document

SCD Specification Control Document

SCD Specification Control Drawing

SCD Standard Color Display

SCD Structures and Controls Division [of US Air Force]

SCD Subcarrier Discriminator

SCD Superconducting Device

SCD Supercritical Carbon Dioxide

ScD *(Scientiae Doctor)* – Doctor of Science

ScD Scintillation Detector

scd screwed

SCDA Safing, Cooldown, and Decontamination Area

SCDC Single Commutation Direct Current (Signaling)

SCDC Steel Castings Development Center [US]

SCDF Scanning Electron Diffraction

ScDHyg Doctor of Science in Hygiene

SCDI Serious Chemical Distribution Incident

SCDM Self-Consistent Diagrammatic Method

S-CDMA Synchronous Code-Division Multiple Access

ScDMed Doctor of Medical Science

SCDP Simulation Control Data Package

SCDP Society for Certified Data Processors

SCDR Seller Critical Design Review

SCDR Shuttle Critical Design Review [of NASA]

SCDR Software Critical Design Review

SCDR Subcontractor Critical Design Review

SCDSB Suppressed-Carrier Double Sideband

SCDU Signal Conditioning and Display Unit

SCDW Surface Charge Density Wave [Solid-State Physics]

SCE Saturated Calomel Electrode [Physical Chemistry]

SCE School of Chemical Engineering

SCE Secondary Chemical Equilibrium

SCE Semiconductor Epitaxy

SCE Service Contre-Espionnage [Counter-Intelligence Service, France]

SCE Service in Conservation Education [of International Union for Conservation of Nature and Natural Resources]

SCE Signal Conditioning Electronics

SCE Signal Conditioning Equipment

SCE Signal Conversion Electronics

SCE Signal Conversion Equivalent

SCE Single-Center Expansion

SCE Single Charge Exchange

SCE Single-Cotton Enameled (Electric Wire) [also sce]

SCE Single Cycle Execute

SCE Sister Chromatid Exchange [Cytology]

SCE Situation Caused Error

SCE Society of Carbide Engineers [now Society of Carbide and Tool Engineers, US]

SCE Society of Christian Engineers [US]

SCE Society of Cuban Engineers

SCE Space-Charge Effect [Electronics]

SCE Standard Calomel Electrode [Physical Chemistry]

SCE Stratified Charge Engine

SCE Strongly Correlated-Electron (System)

SCE System Control Element

SCEA Shipping Conference Exemption Act

SCEA Signal Conditioning Electronic(s) Assembly

SCEAND Standing Committee on External Affairs and National Defense [Canada]

SCEAIT Standing Committee on External Affairs and International Trade [Canada]

SCEAR Scientific Committee on the Effects of Atomic Radiation [of United States]

SCEC Sunshine Coast Environment Council [Australia]

SCEDR Special Committee on Electronic Data Retrieval

SCEEE Southeastern Center for Electrical Engineering Education [US]

SCEI Swedish Council of Environmental Information

SCEL Signal Corps Engineering Laboratory [US]

SCEL Small Components Evaluation Loop

SCEL Standing Committee on Education in Librarianship [UK]

SCEM Self-Consistent Eikonal Method [Physics]

SCEM Single-Channel Electron Multiplier

SCEM Superconducting Cryo-Electron Microscope; Superconducting Cryo-Electron Microscopy

SCEME Society of Chief Electrical and Mechanical Engineers [UK]

SCEO Syndicats de Cadres, d'Employés, d'Ouvriers [Trade Union of Executives and Employees, France]

SCEO System Civil Engineering Office

SCEP Self-Consistent Electron Pairs [Physics]

SCEPC Senior Civil Emergency Planning Committee [of NATO]

SCEP-CEPA Self-Consistent Electron-Pair Coupled Electron Pair Approximation [Physics]

SCEPTRE Systems for Circuit Evaluation and Prediction of Transient Radiation Effects

SCEPTRON Spectral Comparative Pattern Recognizer

SCER Sheffield Center for Environmental Research [UK]

SCERA Senate Committee on Environment, Recreation and the Arts

ScErAs Scandium Erbium Arsenide

SCERP Southwest Center for Environmental Research and Policy [Texas, US]

SCERT Systems and Computers Evaluation and Review Technique

SCES Storrs Cooperative Extension Service [of University of Connecticut, US]

SCET Society of Civil Engineering Technicians [UK]

SCETV South Carolina Educational Television [US]

SCEU Selector Channel Emulator Unit

SCF Saskatchewan Cancer Foundation [Canada]

SCF Safing and Deservicing Facility

SCF Satellite Control Facility

SCF Science Computing Facility

SCF Scientific Computing Feature

SCF Self-Consistent Field [Quantum Mechanics]

SCF Self-Consistent Fluctuation [Physics]

SCF Semicircular Focusing [Spectroscopy]

SCF Sequenced Compatibility Firing; Sequential Compatibility Firing

SCF Short-Chain Fat

SCF Single-Crystal Ferrite

SCF SNAP (Space Nuclear Auxiliary Power) Critical Facility

SCF Sociedad Chilena de Física [Chilean Physical Society]

SCF Sociedad Colombiana de Física [Colombian Physical Society]

SCF Société Chimique de France [French Chemical Society]

SCF Sodium Cleaning Facility

SCF Spacecraft Control Facility

SCF Standard Charge Factor

SCF Standard Cubic Feet [also scf]

SCF Standing Committee on Fishing

SCF Standing Committee on Forestry

SCF Staphylococcal Clumping Factor [Medicine]

SCF Statistical Collection File

SCF Stem Cell Factor [Immunochemistry]

SCF Stress Concentration Factor [Mechanics]

SCF Subcritical Flow [Fluid Mechanics]

SCF Subtractive Color Formation

SCF Sunnyvale Control Facility [California, US]

SCF Supercritical Field [Physics]

SCF Supercritical Flow

SCF Supercritical Fluid [Thermodynamics]

SCF Switched-Capacitor Filter

SCF System Control Facility

scf standard cubic feet [also SCF]

SCFA Swiss Commodities and Futures Association

SCFBR Steam-Cooled Fast Breeder Reactor

SCFD Standard Cubic Feet per Day [also scfd]

SCF-EHT Self-Consistent Field Extended Hueckel Theory [Physics]

SCFEL Standard COSMIC (Computer Software Management and Information Center) Facility Equipment List

SCFH Standard Cubic Feet per Hour [also scfh]

SCF-KKR-CPA Self-Consistent Field Korringa-Kohn-Rostoker Coherent Potential Approximation [Physics]

SCFL Source-Coupled FET (Field-Effect Transistor) Logic

SCF-LMTO-ASA Self-Consistent Field Linear Muffin Tin Orbital Atomic Sphere Approximation [Physics]

SCFM Self-Consistent Field Method [Quantum Mechanics]

SCFM Standard Cubic Feet per Minute [also scfm]

SCFM Subcarrier Frequency Modulation

SCFM Subcritical Fracture Mechanics

SCFMO Self-Consistent Field Molecular Orbital (Calculation) [also SCF-MO, or SCF MO] [Physics]

SCFP Self-Consistent Fluctuation Phonon (Approach) [Physics]

SCFPA Structural Cement-Fiber Products Association [US]

SCFS Standard Cubic Feet per Second [also scfs]

SCFT Self-Consistent Field Theory [Quantum Mechanics]

scf/ton standard cubic foot (or feet) per ton [Unit]

SCF UHF Self-Consistent Field Ultrahigh Frequency (Level) [Physics]

SCF-Xα Self-Consistent Field Xα (Quantum Mechanics Model) [Physics]

SCF-Xα-SW Self-Consistent Field Xα Scattered Wave (Model) [Physics]

SCFZ Slow Cooling Float Zone

SCG Scan Generator

SCG Silicon Carbide Graphite

SCG Slow Crack Growth [Mechanics]

SCG Solution Crystal Growth [Metallurgy]

SCG Sphygmocardiograph(y)

SCG Steel Carriers Group [US]

SCG South Central Gyre [Oceanography]

SCG Subcritical Crack Growth [Mechanics]

SCGA Service to Canadians Graduating Abroad

SCGA Sodium-Cooled Graphite Assembly [Nuclear Engineering]

SCGA Southern Cotton Ginners Association [US]

SC GLOBAL Annual International Superconductor Applications Convention

SCGM Société Canadienne de Genie Mécanique [Canadian Society for Mechanical Engineering]

SCGO(p) Strontium Chromium Gallium Oxide (Compound) [also $SrCr_{9p}Ga_{12-9p}O$]

SCGSS Supercritical Gas Storage System

SCH Seizures per Circuit per Hour [Telecommunications]

SCH Separate Confinement Heterostructure

ScH Socket Head (Screw)

Sch Schedule [also sch]

Sch Scheelite [Mineral]

Sch Schilling [Currency of Austria]

Sch Schneiderite [Mineral]

Sch Scholar [also sch]

Sch School [also sch]

Sch Schooner [also sch]

SCHAZ Subcritical Heat-Affected Zone [Metallurgy]

SCHC Semicarbazide [Formula]

Schdlr Scheduler [also schdlr]

Sched Schedule [also sched]

Schem Schematic(s) [also sched]

SCHFA Self-Consistent Hartree-Fock Approximation [Quantum Mechanics]

Sc(HFAC)₃ Scandium Hexafluoroacetylacetonate [also Sc(AcAc)₃, Sc(acac)₃]

SChg Supercharge(r) [also S Chg, or S-Chg]

SCHIRP Salal/Cedar Hemlock Interagency Research Project [Canada]

Schm Schematic(s) [also schm]

SCHMOO Space Cargo Handler and Manipulator for Orbital Operations

Schol Scholar [also schol]

School Bull School Bulletin [of National Geographic Society, US]

Schr-R Rhein-Westfal TÜV Schriftenreihe des Rheinisch-Westfalischen Technischen Überwachungsvereins [Publications of the Technical Inspection Association of North Rhine-Westphalia, Germany]

SCHS Small Component Handling System

SCHS South Carolina Historical Society [US]

Schweissen Schneiden Schweissen und Schneiden [German Publication on Welding and Cutting]

Schweiz Alum Rundsch Schweizer Aluminum Rundschau [Swiss Aluminum Review]

Schweiz Ing Archit Schweizer Ingenieur und Architekt [Swiss Journal for Engineers and Architects]

Schweiz Maschinenmarkt Schweizer Maschinenmarkt [Swiss Journal on Machinery and Equipment]

Schweiz Mineral Petrogr Mitt Schweizer Mineralogische und Petrographische Mitteilungen [Swiss Mineralogical and Petrographical Communications]

Schweiz Tech Z Schweizerische Technische Zeitschrift [Swiss Technical Journal; published by Schweizerischer Technischer Verband]

SCI Scalable Coherent Interface

SCI Science Citation Index [of Institute for Scientific Information, US]

SCI Security Container Institute [US]

SCI Serial Communications Interface

SCI Servicio Central de Informatica [Central Informatics Service, Madrid Spain]

SCI Setor de Circuito Impresso [Printed Circuit Division; of Brazilian National Institute for Space Research]

SCI Ship-Controlled Interception

SCI Società Ceramica Italiana [Italian Ceramic Society]

SCI Società Chimica Italiana [Italian Chemical Society]

SCI Société de Chimie Industrielle [Society for Industrial Chemistry, France]

SCI Society of the Chemical Industry [UK]

SCI Society of Computer Intelligence

SCI Soft Cast Iron

SCI State College of Iowa [Cedar Falls, US]

SCI Steel Construction Institute [UK]

SCI Summer Computer Institute [of Boston Computer Society, US]

SCI Swedish Corrosion Institute

SCI Switch Closure In

SCI Switched Collector Impedance

SCI System Control Interface

Sci Science; Scientific; Scientist [also sci]

Sci Science [US Magazine]

SCIA Smart Card Industry Association [US]

Sci Abstr Science Abstracts [of INSPEC–Information Services Physics, Electrical and Electronics, and Computers and Control, UK]

Sci Abstr A Science Abstracts, Section A: Physics Abstracts [of INSPEC–Information Services Physics, Electrical and Electronics, and Computers and Control, UK]

Sci Abstr B Science Abstracts, Section B: Electrical Engineering Abstracts [of INSPEC–Information Services Physics, Electrical and Electronics, and Computers and Control, UK]

Sci Am Scientific American [US Publication]

SCIAS Society of Chemical Industry, American Section [US]

Sci Atmos Sin Scientia Atmospherica Sinica [Journal on Atmospheric Sciences published in PR China]

SCIBP Special Committee for the International Biological Program

SCIBS Submerged Combustion In-Bath Smelting [Metallurgy]

Sci Bull Science Bulletin

SCIC Saskatchewan Council for International Cooperation [Canada]

SCIC Semiconductor Integrated Circuit

SCIC Single-Column Ion Chromatography

SCIC Southern Corn Improvement Conference [US]

SCIC Shropshire Chamber of Commerce and Industry [UK]

SCIC Suppressed-Conductivity Ion Chromatography

Sci Ceram Science of Ceramics [Journal published in the US]

SCICEX Submarine Science Ice Experiment [of Office of Naval Research, US]

SCICFNDT Standing Committee for International Cooperation within the Field of Nondestructive Testing [Netherlands]

Sci China A Science in China, Series A [Published in PR China]

Sci China B Science in China, Series B [Published in PR China]

SCICOLL Scientific Collections Permit Database [of US Geological Survey]

Sci Comp Scientific Computer; Scientific Computing [also sci comp]

Sci Comput Program Science of Computer Programming [Journal published in the Netherlands]

Sci Comput World Scientific Computing World [Magazine of Institute of Physics, UK]

SCICON Scientific Control

SCID Severe Combined Immunodeficiency [Medicine]

Sci Dig Science Digest [US]

Sci Dimens Science Dimension [Publication of the National Research Council of Canada]

SCIE Scientific Computing Information Exchange Council

Sci Electr Scientia Electrica [Publication on Electrical Sciences; published in Switzerland]

Sci Eng Science and Engineering [Indian Publication]

Sci Eng Rep Natl Def Acad Scientific and Engineering Reports of the National Defense Academy [Japan]

Sci Eng Rep Saitama Univ A Science and Engineering Reports of Saitama University, Series A [Japan]

Sci Eng Rep Saitama Univ B Science and Engineering Reports of Saitama University, Series B [Japan]

Sci Eng Rep Saitama Univ C Science and Engineering Reports of Saitama University, Series C [Japan]

Sci Eng Rep Tohoku Gakuin Univ Science and Engineering Reports of Tohoku Gakuin University [Japan]

Sci Eng Rev Doshisha Univ Science and Engineering Review of Doshisha University [Japan]

SCI Expanded Science Citation Index Expanded [of Institute for Scientific Information, US]

SCIF Sound and Communications Industries Federation [UK]

SCIF Static Column Isoelectric Focusing [Physical Chemistry]

SCIFER Sounding of the Cleft Ion Fountain Energization Region

SciFi Conference on Scintillating and Fiber Detectors [US]

Sci-Fi Science Fiction [also sci-fi]

SC IGBP Scientific Committee for International Geosphere/Biosphere Program [also SC-IGBP]

SCIGR Standing Committee on International Geoscientific Relations

Sci Hort Scientia Horticulturae [Journal on Horticultural Science]

Sci Ind (Philips) Science and Industry (Philips) [Publication of Philips Laboratories, Eindhoven, Netherlands]

scil *(scilicet)* – namely

SCIM Scanned Transmission Electron Image Microscopy

SCIM Selected Categories in Microfiche

SCIM Speech Communication Index Meter

SCIM Standard Cubic Inches per Minute [also scim]

Sci Mon Scientific Monthly [US]

SCI Monogr Society of the Chemical Industry Monographs [UK]

SCIMR Sullivan Center for In-Situ Mining Research [of New Mexico Institute of Mining and Technology, Socorro, US]

SCIO Subsystems Computer Input/Output

SCIOS Scottish Collaborative Initiative in Optical Sciences

SCIP Scanning for Information Parameters

SCIP Self-Contained Instrument Package

SCIP Society of Competitor Intelligence Professionals [US]

Sci Pap Coll Arts Sci, Univ Tokyo Scientific Papers of the College of Arts and Sciences, University of Tokyo [Japan]

Sci Pap Inst Phys Chem Res Scientific Papers of the Institute of Physical and Chemical Research [Japan]

Sci Prog Science Progress [Journal published by Oxford University Press, UK] [also Sci Prog Oxf]

Sci Publ Affairs Science and Public Affairs [Publication of The Royal Society, UK]

SCIR Standing Committee for Installation Rebuilding [UK]

Sci Rep Science Report; Scientific Report

Sci Rep Kanazawa Univ Science Reports of Kanazawa University [Japan]

Sci Rep Osaka Univ Science Reports of Osaka University [Japan]

Sci Rep Res Inst, Tohoku Univ Science Reports of the Research Institutes, Tohoku University [Japan]

Sci Rep Res Inst, Tohoku Univ A, Phys Chem Metall Science Reports of the Research Institutes, Tohoku University, Series A, Physics, Chemistry and Metallurgy [Japan]

Sci Rep Saitama Univ A Science Reports of the Saitama University, Series A [Japan]

Sci Rep Saitama Univ B Science Reports of the Saitama University, Series B [Japan]

Sci Rep Tohoku Univ I Science Reports of the Tohoku University, First Series [Japan]

Sci Rep Tohoku Univ II Science Reports of the Tohoku University, Second Series [Japan]

Sci Rep Tohoku Univ III Science Reports of the Tohoku University, Third Series [Japan]

Sci Rep Tohoku Univ IV Science Reports of the Tohoku University, Fourth Series [Japan]

Sci Rep Tohoku Univ V Science Reports of the Tohoku University, Fifth Series [Japan]

Sci Rep Tohoku Univ VI Science Reports of the Tohoku University, Sixth Series [Japan]

Sci Rep Tohoku Univ VII Science Reports of the Tohoku University, Seventh Series [Japan]

Sci Rep Tohoku Univ VIII Science Reports of the Tohoku University, Eighth Series [Japan]

Sci Rep Yokohama Natl Univ I Science Reports of the Yokohama National University, Section I [Japan]

Sci Rep Yokohama Natl Univ II Science Reports of the Yokohama National University, Section II [Japan]

Sci Res Scientific Research(er)

Sci Rev Scienca Revuo [Science Review; formerly published in Belgrade, Yugoslavia]

Sci Rev Setsunan Univ A Scientific Review of Setsunan University, A [Japan]

Sci Rev Setsunan Univ B Scientific Review of Setsunan University, B [Japan]

SCI/RT (Development of) Scalable Coherent Interface for Real-Time

SCIS Safety Containment Isolation System [Nuclear Engineering]

SCIS Standard Cubic Inches per Second [also scis]

SCISEARCH Science Citation Index (Database) [also Scisearch]

Sci Sin A, Math Phys Astron Tech Sci Scientia Sinica, Series A, Mathematical, Physical, Astronomical and Technical Sciences [PR China]

Sci Sinter Science of Sintering [Publication of the International Institute for the Science of Sintering]

SCIT South China Institute of Technology [Canton, PR China]

SCIT Standard Change Integration and Tracking

SCIT/CMA Standard Change Integration and Tracking/Configuration Management Accounting

Sci Tech Science and Technology; Scientific and Technical [also sci tech]

Sci Tech Sciences et Techniques [French Publication on Science and Technology]

Sci Tech Inf Process Scientific and Technical Information Processing [Translation of: *Nauchno-Tekhnicheskaya Informatisaya, Seriya 1 (USSR)*; published in the US]

Sci Technol Sciences et Technologies [French Publication on Science and Technology]

Sci Technol Dimens Science and Technology Dimensions [Canadian Publication]

Sci Technol Libr Science and Technology Libraries [Journal published in the US]

Sci Technol Rev Science and Technology Review [of Lawrence Livermore National Laboratory, University of California, US]

SCITF Spacecraft Integration and Test Facility

SCIU Selector Control Interface Unit

SCIU Spacecraft Interface Unit

Sci Vie Science et Vie [French Journal on Science and Life]

SCJ Science Council of Japan

SCL Secondary Coolant Line [Nuclear Reactors]

SCL Secondary Coolant Loop [Nuclear Reactors]

SCL Security Log

SCL Self-Checking Logic

SCL Sequential Control Logic

SCL Single Channel Monitoring

SCL Society for Computers and Law [UK]

SCL Society for Construction Law [UK]

SCL Society of County Librarians [UK]

SCL Solar Calorimetry Laboratory [of Queen's University, Kingston, Ontario, Canada]

SCL South Caroliniana Library [of University of South Carolina, Columbia, US]

SCL Space Charge Layer

SCL Space Charge Limited; Space Charge Limitation

SCL Specification Change Log

SCL Strong Coupling Limit

SCL Structural Ceramics Laboratory [of Korea Institute for Science and Technology, Seoul]

SCL Superconducting Lens

SCL Systems Control Language

SCLC Sequence Centrifugal Layer Chromatography

SCLC Space-Charge-Limited Conduction

SCLC Space-Charge-Limited Current

SCLCP Side-Chain Liquid-Crystalline Polymer

SCLDF Sierra Club Legal Defense Fund [US]

SCLF Self-Contained Liquid-Filled (Cable)

SCLM Scanning Confocal Laser Microscopy

SCLM Software Configuration and Library Management

SCLM Stability, Control, and Load Maneuvers

SCLO Self-Consistent Linear Orbital [Physics]

SCLO Self-Consistent Local Orbit(al Method) [Physics]

SCLog Security Log

SCLR Single-Configuration Linear Response (Theory) [Physics]

SCLS Surface Core-Level Shift

Sclt Scarlet [also sclt]

SCM Safety Council of Maryland [US]

SCM Samarium Calcium Manganate

SCM Saturable-Core Magnetometer

SCM Scientific Calculator Machine

SCM Screen-Cam Format

SCM Self-Center Molecules

SCM Self-Consistent Multipolar

SCM Semiconductor Material

SCM Senarmont Compensation Method

SCM Series-Characteristic Motor

SCM Service Command Module

SCM Session Control Module

SCM Shunt-Characteristic Motor

SCM Signal Conditioning Module

SCM Simulated Core Mockup

SCM Simulated Core Model

SCM Single-Channel Modem

SCM Single Column Model

SCM Site Configuration Message

SCM Small Core Memory

SCM Software Configuration Management

SCM Special Council of Ministers [of European Community]

SCM Spinal Changes in Microgravity

SCM Standard Cubic Meter [also scm]

SCM STARAN (Stellar Attitude Reference and Navigation) Control Module

SCM State Certified Midwife

SCM Station Class Mark [Telecommunications]

SCM Studiecentrum voor Kernenergie [Nuclear Energy Research Center; Mol, Belgium]

SCM Subdivision of County Municipality

SCM Subscribers Concentration Module

SCM Subsystem Configuration Management

SCM Subsystem Configuration Monitoring

SCM Superconducting Cryoelectron Microscopy

SCM Superconducting Cryogenic Microscope

SCM Superconducting Magnet

SCM Superconducting Material

SCM Supervision Control Module

SCM Symmetrically Cyclically Magnetized (Condition) [Physics]

SCM System Control Module

S&CM Surfaced on One, or Two Sides and Center Matched [Lumber]

ScM *(Scientiae Magister)* – Master of Science

S/cm Siemens per centimeter [also S cm^{-1}]

s^{-1}cm^{-3} one per second per cubic centimeter [also 1/s·cm^3, or 1/s/cm^3]

SCMA Southern Cypress Manufacurers Association [US]

SCMA Systems Communications Management Association

SCMCC State Catchment Management Coordinating Committee [of New South Wales, Australia]

SCMF Self-Consistent Mean Field (Method) [Physics]

SCMHC South Carolina Mental Health Commision [US]

ScMHyg Master of Science in Hygiene

SCMM Semiconductor Memory Module

SCMO Samarium Calcium Manganate

SCMO Subsidiary Communication Multiplex Operation

SC-MOSFET Surface-Channel Metal-Oxide Semiconductor Field-Effect Transistor

SCMP Self-Consistent Madelung Potential [Physics]

SCMP Software Configuration Management Plan

SCMP System Contractor Management Plan

SCMP Systems Cooperative Marketing Program

SCMR Stanford Center for Materials Research [Stanford University, California, US]

SCMS Small Cumulus Microphysical Study

SCMU Species Conservation Monitoring Unit [of World Conservation Monitoring Center]

SCN Safety Council of Nebraska [US]

SCN Satellite Conference Network

SCN Satellite Control Network

SCN Scan

SCN Scientific Computing Network

SCN Self-Compensating Network

SCN Sensitive Command Network

SCN Shortest Connected Network

SCN (Space) Shuttle Contractor Node [NASA]

SCN Sociedad de Ciencias Naturales [Society of Natural Sciences, Venezuela]

SCN Soybean Cyst Nematode [Zoology]

SCN Specification Change Notice

SCN Strontium Cobalt Nitride

SCN Succinonitrile

SCN Switched Communication Network

SCN Thiocyano (Radical) [Symbol]

SCN$^-$ Thiocyanate (Ion) [Symbol]

Scn Scan(ner) [also scn]

ScN Scandium Nitride

SCNA Sudden Cosmic-Noise Absorption [Communications]

SCNAc Thiocyanato Acetic Acid [also SCN-Ac, or SCN-ac]

SCNAcac Thiocyano-Acetylacetonate [also SCN-Acac, or SCN-acac]

SCNCO Strontium Calcium Neodymium Copper Oxide (Superconductor) [also SrCNCO]

SCNM Southwest College of Naturopathic Medicine [Tempe, Arizona, US]

SCNN Simple Cubic (Lattice) with Nearest Neigbor Interactions (Model) [Crystallography]

SCNPWC Standing Committee for Nobel Prize Winners Congresses [Germany]

SCNR Scanner [also Scnr]

SCNR Scientific Committee of National Representatives [of NATO Supreme Headquarters, Allied Powers Europe]

SCO Samarium Chromium Oxide (or Samarium Chromite)

SCO Samarium Copper Oxide (Superconductor)

SCO Santa Cruz Operating System [Unix]

SCO Santa Cruz Operation [US Company]

SCO Seasonal Climate Outlook

SCO Self-Consistent Orbital [Physics]

SCO Southern College of Optometry [Memphis, Tennessee, US]

SCO Start Checkout

SCO Stiper Critical Oxygen

SCO Strontium Cobalt Oxide

SCO Strontium Copper Oxide

SCO Subcarrier Oscillator

SCO Switch Closure Out

Sco X-1 Scorpio X-1 [Astronomy]

SCOB Scattered Clouds, or Better

SCOBOL Structured Common Business-Oriented Language

SCOCl Strontium Copper Oxygen Chloride

SCOCLIS Standing Conference of Cooperative Library and Information Services [UK]

SCODA Scan Coherent Doppler Attachment

SCOE Sematech Center for Excellence [of Rensselaer Polytechnic Institute, Troy, New York, US]

SCOE Special Checkout Equipment

SCOF Self-Contained Oil-Filled (Cable)

SCOFF Sharp Cut-Off Fringing Field

SCOLCAP Scottish Libraries Cooperative Automation Project

SCOM Site Cutover Manager

ScoMIA Scottish Marine Industries Association

SCOMO Satellite Collection of Meteorological Observations

SCONUL Standing Conference of National and University Libraries [UK]

SCOOP Scientific Computation of Optimal Programs

SCOOP Self-Coupled Optical Pickup

SCOOPS Scheme Object-Oriented Programming System

SCOPCAS Standing Committee on Pollution Clearance at Sea

SCOPE Schedule-Cost-Performance

SCOPE Scientific Committee on Problems of the Environment [of International Council of Scientific Unions]

SCOPE Sequential Customer Order Processing Electronically

SCOPE Simple Communications Programming Environment

SCOPE Special Committee on Paperless Entries

SCOPE Specifiable Coordinating Positioning Equipment

SCOPE Standing Committee on Professional Education

SCOPE Standing Committee on Professional Exchange [Denmark]

SCOPE Systematic Computerized Processing in Cataloguing

SCOPE System for Coordination of Peripheral Equipment

Scope Oscilloscope [also scope]

Scope Radarscope [also scope]

SCOPEP Steering Committee on the Performance of Electrical Products

ScOPr^i Scandium Isopropoxide

SCOPT Subcommittee on ProgrammingTechnology [of Association for Computing Machinery, US]

SCOR Scientific Computer On-Line Resource

SCOR Scientific Committee on Oceanic Research [of International Council of Scientific Unions]

SCOR Scientific Committee on Oceanographic Research [Canada]

SCOR Special Committee on Oceanic Research [of International Council of Scientific Unions]

SCOR Self-Calibrating Omnirange

SCORE Satellite Computer-Operated Readiness Equipment

SCORE Selection Copy and Reporting

SCORE Signal Communications by Orbiting Relay Equipment

SCORE System for Computerized Olympic Results and Events

SCORPI Subcritical Carbon-Moderated Reactor Assembly for Plutonium Investigations

SCORPIO Single Crystal Orientation Rapid Processing and Interpretation Operation

SCORPIO Subcritical, Carbon-Moderated Reactor Assembly for Plutonium Investigations [UK]

SCORPIO Subject-Content-Oriented Retriever for Processing Information On-Line [US]

SCORTEC Shear Controlled Orientation Technology

SCOS Subsystem Computer Operating System

SCOST Special Committee on Space Technology

SCOSTEP Scientific Committee on Solar Terrestrial Physics [US]

S-COSY Scaled Correlation Spectroscopy [also SCOSY, or S/COSY]

SCOT Shippers for Competitive Ocean Transportation [US]

Scot Scotland; Scottish

SCOTA Scottish Offshore Training Association

SCOTICE Scotland-Iceland Submarine Cable [also SCOT-ICE]

SCOTS Surveillance and Control of Transmission Systems

SCOTT Supercritical, Once-Through Tube (Experiment) [Nuclear Reactors]

SCOTT-R Supercritical, Once-Through Tube Reactor Experiment

SCOUG Southern California On-Line User Group [US]

SCOUT Signal Computer, Oscilloscope and Universal Tester

SCOUT Surface-Controlled Oxide Unipolar Transistor

SCOWR Special Committee on Water Research [of International Council of Scientific Unions]

Sco X-1 Scorpio X-1 [Astronomy]

SCP Safety Control Program

SCP Save Cursor Position

SCP Scanner Control Power

SCP Screen Color Photography

SCP Secondary Cross-Connection Point

SCP Seismic Cone Penetrometer

SCP Semiconductor Physics

SCP Serial Character Printer

SCP Service Control Point [Telecommunications]

SCP Sheffield City Polytechnic [UK]

SCP Single-Cell Protein [Biochemistry]

SCP Site Characterization Plan

SCP Small Cardioactive Peptide [Biochemistry]

SCP Sociedad Cientifica Paraguaya [Paraguayan Scientific Society]

SCP Société de Chimie Physique [Society of Physical Chemistry, France]

SCP Soil Conservation Program [of State of Queensland, Australia]

SCP Sodium Cellulose Phosphate

SCP Software Control Procedure

SCP South Celestial Pole [Astronomy]

SCP Space-Charge Polarization

SCP Species Conservation Program [of Species Survival Commision–International Union for Conservation of Nature and Natural Resources]

SCP Specific Candle Power

SCP Spherical Candlepower [also scp]

SCP Spontaneous Crack Propagation

SCP Spray Conversion Processing

SCP Station Call Processor

SCP Stored Command Program

SCP Stromberg-Carlson Practices

SCP Student Conservation Program [US]

SCP Subscribers Call Processing

SCP Subsystem Control Port

SCP Supervisory Control Program

SCP Surface Charge Profiler; Surface Charge Profiling

SCP Surveillance Communication Processor

SCP Symbol(ic) Conversion Program

SCP System Control Processor

SCP System Control Program(ming)

ScP Scandium Phosphide

SCPA Small Cardioactive Peptide A [also SCPa]

SCPBE Species Chemical Potential/Bond Energy (Model) [Physical Chemistry]

SCPC Single Carrier per Channel; Single Channel per Carrier

SCPD Scratch Pad

SCPI Scientists Committee for Public Information [US]

SCPI Standard Commands for Programmable Instrument(ation)s

SCPI Structural Clay Products Institute [Washington, DC, US]

SC PMOSFET Surface Channel P-Type Metal-Oxide Semiconductor Field-Effect Transistor

SCPO Senior Chief Petty Officer

SCPO Special Committee on Peacekeeping Operations [of United Nations] [also known as Committee of 33]

ScPO Scandiophosphate

SCPP Self-Consistent Phase-Phonon Approach [Solid-State Physics]

SCPRF Structural Clay Products Research Foundation [US]

SCPS Single-Chain Polystyrene

SCPS Subscribers Call Processing Subsystem

SCPS Sulfochlorophenol S

SCPS Synchronous Composite Packet Switching

SCPT Seismic Cone Penetrometer Testing

SCPT Self-Consistent Perturbation Theory [Physics]

SCQ Super-Clean Quality (Stainless Steel)

SCQS Society of Chief Quantity Surveyors [UK]

SCR Saturable-Core Reactor [Electrical Engineering]

SCR Scan Control Register

SCR Scanning Control Register

SCR Schedule Change Request

SCR Selective Catalytic Reduction (Process) (for Flue Gas)

SCR Selective Catalyzer Reaction

SCR Selective Chopper Radiometer

SCR Selenium-Controlled Rectifier

SCR Self-Consistent Renormalization (Theory) [Physics]

SCR Semiconductor-Controlled Rectifier

SCR Sequence-Control Register

SCR Servo-Controlled Robot

SCR Seychelles Rupee [Currency]

SCR Short-Circuit Radio

SCR Signal Conversion Relay

SCR Silicon-Controlled Rectifier [also scr]

SCR Simulation Control Room

SCR Single Character Recognition

SCR Single-Cycle (Nuclear) Reactor

SCR Single-Stage Cold-Rolled; Single-Staged Cold Rolling [Metallurgy]

SCR Site Characterization Report

SCR Sneak Circuit Report

SCR Sodium-Cooled Reactor

SCR Software Change Request

SCR Southern Council of Research [US]

SCR Strip-Chart Recorder

SCR Styrene Chloroprene Copolymer

SCR Surface Contour Radar

SCR System Change Request

.SCR Script [File Name Extension]

Scr Screen [also scr]

Scr Screw [also scr]

Scr Streptococcus Cremoris [Microbiology]

SCRA South Carolina Research Authority [US]

SC&RA Specialized Carriers and Rigging Association [US]

SCRAM Scottish Campaign to Resist the Atomic Menace

SCRAM Selective Combat Range Artillery Missile

SCRAM Space Capsule Regulator and Monitor

SCRAM Static Column (Dynamic) Random-Access Memory

Scramjet Supersonic Combustion Ramjet [also scramjet] [Aeronautics]

SCRAP Super-Caliber Rocket-Assisted Projectile

Scrap Process Recycl Mag Scrap Processing and Recycling Magazine [of Institute of Scrap Recycling Industries, US]

SCRATA Steel Castings Research and Trade Association [UK]

SCRC Scientific Computing Resource Center

SCRCC Soil Conservation and Rivers Control Council [New Zealand]

SC/RDL Signal Corps, Research and Development Laboratories [of US Army]

SC Re Sri Lanka Rupee [Currency]

SCRF Scripps Clinic and Research Foundation [La Jolla, California, US]

ScrF Streptococcus cremoris F [Microbiology]

Scr Fac Sci Nat Univ Purkyn Brun Phys Scripta Facultatis Scientiarum Naturalium Universitatis Purkynianae Brunensis, Physica [Journal of the University of Brno–Faculty of Natural Sciences, Physics Section, Czech Republic]

SCRG System Change Review Group [of NASA Marshall Space Flight Center, Greenbelt, Maryland, US]

SCRI Science Court and Research Institute [US]

SCRI Scottish Crop Research Institute

SCRI Steel Can Recycling Institute [US]

SCRI Supercomputer Computational Research Institute

SCRI Superconductor Computations Research Institute [of Florida State University, Tallahassee, US]

SCRIBE Sub-Nanometric Cutting and Ruling by an Intense Beam of Electrons

SCRIPT Scientific and Commercial Interpreter and Program Translator

SCRL Station Configuration Requirement List

Scr Mater Scripta Materialia [Published in the US]

Scr Metall Scripta Metallurgica [later Scr Metall Mater; now Scr Mater]

Scr Metall Mater Scripta Metallurgica et Materialia [Published in the US] [now Scr Mater]

SCRN Screen (Function) [Computer Programming]

Scrn Screen [also scrn]

Scrng Screening [also scrng]

SCRO Strontium Calcium Rubidium Oxide (Superconductor) [also SrCRO]

SCRS Scalable Cluster of RISC (Reduced Instruction Set Computer) Systems

SCRS Serialized Control and Reporting System

SCRS Society of Collision Repair Specialists [US]

SCRS Stokes Coherent Raman Spectroscopy

SCRS Strip-Chart Recording System

SCRSY Serialized Control and Record System

SCRT Serial Choice Reaction Timer

SCRT Subscribers Circuit Routine Tester

SCRTD South California Rapid Transit District [US]

SCS Satellite Communication System

SCS Satellite Control System

SCS Scattering Cross-Section

SCS School of Chemical Sciences

SCS School of Computer Science

SCS Scientific Computer System

SCS Scientific Control System

SCS Secondary Coolant System [Nuclear Reactors]

SCS Second Critical Speed

SCS Section/Chapter Support Committee [of Institute of Electrical and Electronics Engineers– Regional Activities Committee/Technical Activities Committee, US]

SCS Security Control System

SCS Selected Classification Service [of Organization for Economic Cooperation and Development–Information Policy Group, France]

SCS Self-Consistent (Differential) Scheme [Materials Science]

SCS SEM (Scanning Electron Microscope) Cold Stage

SCS Semi-Closed System

SCS Sensor Coordinate System

SCS Separate Channel Signaling

SCS Sequence Coding System [US]

SCS Serbian Chemical Society

SCS Silicon Carbide Spraying

SCS Silicon Chip Stack

SCS Silicon-Control(led) Switch

SCS Simulated Compton Scattering

SCS Simulation Control Subsystem

SCS Singapore Computer Systems [Singapore]

SCS Single-Channel Simplex

SCS Small Computer System

SCS SNA (System Network Architecture) Character String

SCS Society for Computer Simulation [now Society for Computer Simulation International, US]

SCS Society of Cosmetic Scientists [UK]

SCS Sodium Characterization System

SCS Soil Conservation Service [of US Department of Agriculture]

SCS South Carolina State [US]

SCS South China Sea

SCS Southern Computer Service [US]

SCS Space Cabin Simulator

SCS Spacecraft Checkout Station

SCS Spacecraft Control System

SCS Speed Control Switch

SCS Stabilization and Control Subsystem; Stabilization and Control System [of NASA Apollo Program]

SCS Stabilization Cabin Simulator

SCS Standard Coordinate System

SCS State Conservation Strategy

SCS Static Coupling Scheme

SCS Sterically Controlled Substitution [Chemistry]

SCS Strontium Chromium Antimonate

SCS Substituent Chemical Shift

SCS Superconductor Science

SCS Surface Contamination Standard

SCS System Configuration Switch

ScS Scandium Sulfide

ScS Specialist in Science

SC&S Strapped, Corded and Sealed

SCSA Signal Computing System Architecture

SCSA Soil and Water Conservation Society of America [US]

SCSA Steering Committee for Sustainable Agriculture [US]

SCSAG Small-Curvature Semiclassical Adiabatic Ground State Approximation [Physics]

SCSC Soil and Crop Sciences Center [of Texas A&M University, College Station, US]

SCSC South Carolina State College [Orangeburg, US]

SCSC Southern Connecticut State College [New Haven, US]

SCSC Standing Committee on Soil Conservation [Australia]

SCSC St. Cloud State College [Minnesota, US]

SCSDB South Carolina State Development Board [US]

SCSE South Carolina Society of Engineering [US]

SCSEP Summer Canada Student Employment Program

SCSI Small Computer Serial Interface

SCSI Small Computer Systems Interface

SCSI Society for Computer Simulation International [US]

SCSI/IEEE Small Computer System Interface/Institute of Electrical and Electronic Engineers

SCSP Supervisory Computer Software Package

SCSPLS South Carolina Society of Professional Land Surveyors [US]

SCSR Ship Construction Subsidy Regulations

SCSSD Surface-Constrained Soft Sphere Dipole (Model) [Physics]

SCSSS Standard Canadian Shield Saline Solution

SCST Society of Commercial Seed Technologists [US]

SCST State Council of Science and Technology [PR China]

SCSTP Scientific Committee on Solar Terrestrial Physics [US]

SC STRUCT Spacecraft Structure

SCSU South Carolina State University [Orangeburg, US]

SCSU Southern Connecticut State University [New Haven, US]

SCSU System Control Signal Unit

SCT Scanning Telescope

SCT Schottky Clamped Transistor

SCT Short Contact Time

SCT Sector [also Sct, or sct]

SCT Single Cassette Tape Reader

SCT Société des Ceramiques Téchniques [French Ceramic Technology Company]

SCT Society of Cleaning Technicians [US]

SCT Special Characters Table

SCT Statistical Communication Theory

SCT Step Control Table

SCT Stratus-to-Cumulus Transition [Meteorology]

SCT Strong-Coupling Theory [Physics]

SCT Strontium Chromium Tantalate

SCT Subroutine Call Table

SCT Subscriber Carrier Terminal

SCT Sugar-Coated Tablet(s) [Pharmacology]

SCT Superconductor Technology

SCT Surface and Coating Technology

SCT Surface and Coatings Technology [US Journal]

SCT Surface-Charge Transistor

SCT Surface-Crack Tension Specimen [Mechanics]

Sct Scatter(ed) [also sct]

Sct Sector [also sct]

SCTA Society of Calorimetry and Thermal Analysis [Japan]

sctd scattered [also sct'd]

SCTE Society of Cable Television Engineers [UK]

SCTE Society of Carbide and Tool Engineers [US]

SCTF Slab Core Test Facility

SCTF Sodium Chemical Technology Facility

Sc(THD)$_3$ Tris(2,2,6,6-Tetramethyl-3,5-Heptanedionato)-scandium [also Sc(thd)$_3$]

SCTI Sodium Components Test Installation [US]

SCTL Short-Circuited Transmission Line

SCTL Small Components Test Loop

Sc(THD)$_3$ ScandiumTris(2,2,6,6-Tetramethyl-3,5-Heptanedionate) [also Sc(thd)$_3$]

Sc(TMHD)$_3$ Scandium Tris(2,2,6,6-Tetramethyl-3,5-Heptanedionate) [also Sc(tmhd)$_3$]

SCTO Stalled Call Timed Out

SCTOC Satellite Communication Test Operations Center

SCTP Straight Channel Tape Print

SCTP Surface Cleaning Technology Program [of Research Triangle Institute, Research Triangle Park, North Carolina, US]

SCTPP Straight Channel Tape Print Program

Sctr Sector [also sctr]

(S)CTS Secondary Clear to Send

SCTS Simulation Control and Training System

Scts Securities [also scts]

SCTSAG Small-Curvature (Tunneling) Semiclassical Adiabatic Ground State Approximation [Physics]

SCTTL Schottky Clamped Transistor-Transistor Logic

SCTY Security [also Scty, or scty]

SCU Santa Clara University [California, US]

SCU Scanner Control Unit

SCU Secondary Control Unit

SCU Sensor Control Unit

SCU Sequence Control Unit

SCU Serial Communication Unit

SCU Service and Cooling Umbilical

SCU Servicing Control Unit

SCU Signal Conditioner Unit; Signal Conditioning Unit

SCU Signal Control Unit

SCU Southern Cross University [Western Australia]

SCU Station Control Unit

SCU Storage Control Unit

SCU Subscribers Concentrator Unit

SCU Sulfur Coated Urea

SCU System Control Unit

Scu Stratocumulus (Cloud) [Meteorology]

Scuba Self-Contained Underwater Breathing Apparatus [also scuba]

SCUC Satellite Communications Users Conference

SCUC Sunshine Coast University College [Australia]

SCUL Simulation of the Columbia University Libraries [US]

SCULL Serial Communication Unit for Long Links

SCUMRA Société Centrale de l'Uranium et des Minérals et Métaux Radioactifs [Uranium and Radioactive Minerals and Metals Company, France]

SCUP School Computer Use Plan

SCUP Scupper [also Scup, or scup]

SCUREF South Carolina University Research and Education Foundation [US]

SCV Seville Composite Vehicle

SCV Subclutter Visibility [Radar]

SCVD Short-Circuit Voltage Depression

SCW Silicon Carbide Whisker

SCW Space-Charge Wave [Physics]

SCW Special Case Waste

SCW Standard for Clinical Work

SCW State College of Washington [US]

SCW Static Concentration Wave (Method) [Metallurgy]

SCW Supercritical Water [Nuclear Engineering]

SCW Supercritical Wing [Aeronautics]

SCWC Sierra Club of Western Canada

SCWG Satellite Communications Working Group [of NATO]

SCWIST Society for Canadian Women in Science and Technology

SCWM Special Commission on Weather Modification

SCWO Supercritical Water Oxidation (Process)

SCWPLR Special Committee for Workplace Product Liability Reform [US]

SCWST Society of Canadian Women in Science and Technology

SCX Strong Cation Exchange(r) [Chemistry]

SCXI Signal Conditional Extension for Instrumentation

Scy Secretary [also scy]

SCYO Strontium Cerium Yttrium Oxide (Superconductor) [also SrCYO]

SCZ Suez Canal Zone

SD Doctor of Science

SD Salivary Duct [Anatomy]

SD Salt Dome [Geology]

SD Salvage Depot

SD Sampled Data [Statistics]

SD Sample Delay

SD Sample Designation

SD Sample Distribution [Statistics]

SD San Diego [California, US]

SD Santo Domingo [Dominican Republic]

SD Sawdust

SD Scaling and Display

SD Schematic Drawing

SD Schottky Defect [Solid-State Physics]

SD Schottky Diode [Electronics]

SD Science Department

SD *(Scientiae Doctor)* – Doctor of Science

SD Scintillator Dosimeter

SD Scratch Distance

SD Screw Dislocation [Materials Science]

SD Screw-Down (Mechanism)

SD Seasonal Derating

SD Seasoned Dry (Lumber)

SD Second Difference

SD Sedimentary Deposition

SD Selective Decoupling

SD Selenium Diode

SD Self-Aligning Roller Bearing, Double-Row, Outer-Ring Raceway Spherical, Inner Ring with Three Integral Ribs [Symbol]

SD Self-Destroying

SD Self-Diffusion [Solid-State Physics]

SD Semiconductor Device

SD Send Data

SD Send Digits

SD Senior Dean

SD Serial Data

SD Serializer/Deserializer

SD Service Depot

SD Service Door

SD Shape Discrimination

SD Shear Direction [Mechanics]

SD Shell Dressing

SD Shockley Diode

SD Short Delay

SD Sidereal Day [Astronomy]

SD Sight Draft

SD Signal Department [UK Navy]

SD Signal Display

SD Signal Distributor

SD Signal-to-Distortion (Ratio)

SD Significant Digit

SD Silicon Detector

SD Singapore Dollar [Currency]

SD Single Density (Disk)

SD Single Domain [Solid-State Physics]

SD Size Distribution

SD Skin Decontamination

SD Skin Depth

SD Skin Dose [Radiology]

SD Slope Discontinuities (on Stress-Strain Diagram) [Mechanical Testing]

SD Slow-Down

SD Smoke Detector

SD Soft Drawing; Soft-Drawn [Metallurgy]

SD Software Development

SD Soil Dynamics [Geology]

SD Solar Day [Astronomy]

SD Solid-State (Electron) Detector

SD Somigliana Dislocation [Materials Science]

SD Source Data

SD Source Document

SD South Dakota [US]

SD Space Division [of US Air Force Systems Command]

SD Space Division [of Rockwell International Corporation, US]

SD Spare Disposition

SD Spark Discharge

SD Special Delivery

SD Special Directive

SD Specialist Degree

SD Specially Denatured (Alcohol)

SD Specification Document

SD Spectral Diffusion

SD Speech Detector

SD Spin Density [Solid-State Physics]

SD Spin-Dependent (Recombination) [Physics]

SD Spin Doublets [Physics]

SD Spinodal Decomposition [Metallurgy]

SD Spray Direction

SD Spray Drying

SD Sputter-Deposited; Sputter Deposition

SD Square-Law Detector

SD Stage Duration (Time)

SD Standard Data

SD Standard Deviation [also sd]

SD Standard Displacement

SD Standardization Dictionary

SD Standards Development

SD Start Date

SD State Department

SD Statistical Design (of Experiments)

SD Statistical Division [of Australian Bureau of Statistics]

SD Steam Distillation [Chemical Engineering]

SD Steepest Descent [also sd]

SD Stellar Distance

SD Straight Dynamite

SD Strength Differential

SD Structural Design

SD Structural Dynamics

SD Submarine Department

SD Sudan [ISO Code]

SD Super Density

SD Superdislocation [Materials Science]

SD Supply Duct

SD Surface Dose [also sd]

SD Sustainable Development

SD Sweep Driver

SD Symmetric Device

SD Synchrodyne Demodulation

SD Synchronous Detector

SD Synthetic Drug

SD System Dynamics

SD Systems Design(er)

SD System(s) Development

S/D Sea Damaged

S/D Sight Draft [also SD]

S/D Statement of Differences

S-D Signal-to-Distortion Ratio [also S/D]

S-D Single-Domain [Solid-State Physics]

S+D Speech plus Duplex [also S+DX] [Telegraphy]

Sd Sand [also sd]

Sd Sardine

Sd Sound [Geography]

Sd Spiral Galaxy, Transitional Type [Astronomy]

S$ Singapore Dollar [Currency]

sd seasoned [also sd]

sd *(sine die)* – without day; without appointing (or setting) a day

.sd Sudan [Country Code/Domain Name]

σ_y/d yield stress per relative density [Symbol]

6D Six-Dimensional [also 6-D]

7D Seven Digit (Number)

SDA Schweizerische Depeschenagentur [Swiss News Agency]

SDA Screen Design Aid

SDA Semicarbazide Diacetic Acid

SDA Service Delivery Arrangement

SDA Shaft Drive Axis

SDA Share Distribution Agency

SDA Smoothed Density Approach

SDA Soap and Detergent Association [US]

SDA Software Development Association

SDA Software Disk Array

SDA Source-Data Acquisition

SDA Source-Data Automation

SDA Specially Denatured Alcohol

SDA Specific Dynamic Action

SDA Spherical Deflection Analyzer

SDA Stearidonic Acid

SDA Supplier Data Approval

SDA Surface Design Association [US]

SDA Symbolic Device Address

SDA System Display Architecture

SDAA Skein Dyers Association of America [US]

SDAD Satellite Digital and Analog Display

SDADS Satellite Digital and Analog Display System

SDAF Solid Rocket Booster Disassembly Facility [of NASA]

SDAID System Debugging Aid

S Dak South Dakota [US]

SDAL Switched Data Access Line

SDAM Single DOS (Disk-Operating System) Application Mode

SDAP Systems Development Analysis Program

SDAS Scientific Data Automation System

SDAS Secondary Dendrite Arm Spacing [Metallurgy]

SDAT Spacecraft Data Analysis Team [of NASA]

SDAT Symbol Device Allocation Table

S-DAT Stationary-Head Digital Audio Tape

SDAU Safety Data and Analysis Unit [UK]

SDB Safe Deposit Box

SDB Segment Descriptor Block

SDB Shallow Draft Barge

SDB Silicon Direct Bonding

SDB Society for Developmental Biology [US]

SDB Source Database

SDB Standard Device Byte

SDB State Description Block

SDB Storage Data Bus

SDB Strength and Dynamics Branch [of US Air Force]

SDB Symbolic Debugger

SDBL Sight Draft, Bill of Lading Attached

S/D-B/L Sight Draft with Bill of Lading

Sd Bl Sand Blast(ing)

SDBM Monothiodibenzoylmethane

SDBMS Spatial Database Management System

SDBP Small Database Package

SDBP Small Database Project

SDBS Samson Database Services [Netherlands]

SDBS Sodium Dodecylbenzene Sulfonate

SDBY Standby [also Sdby, or sdby]

SDC Scientific Documentation Center

SDC Semiconductor Devices Council [of Joint Electron Device Engineering Council, US]

SDC Shell Development Company

SDC Shutdown Cooling

SDC Signal Data Converter

SDC Simulation Director Console

SDC Society of Dyers and Colourists [UK]

SDC Software Development Center

SDC Software Development Computer

SDC Spares Disposition Code

SDC Special Devices Center

SDC Specific Damping Capacity

SDC Stabilization Data Computer

SDC Strategic Defense Command [US]

SDC Structural Design Criteria

SDC Submersible Decompression Chamber

SDC Submersible Diving Chamber

SDC Switched Digital Capability

SDC Synchro-to-Digital Converter

SDC System Development Corporation [US]

SDCA Society of Dyers and Colorists of Australia

S-DCB Short-Double Cantilever Beam (Specimen) [also s-DCB] [Mechanical Testing]

SDCC San Diego Computer Center [California, US]

SDCC Small-Diameter Component Cask

(S)DCD Secondary Data Carrier Detect

SDCE Society of Die Casting Engineers [US]

SDCL Supplier Documentation Checklist

SDCP Summary Development Cost Plan

SDCR Source Data Communication Retrieval

SDCS SAIL (Shuttle Avionics Integration Laboratory) Data Communications System [of NASA]

SDCS Single Differential Cross-Section [Physics]

SDD Selective Dissemination of Documentation

SDD Shuttle Design Directive [of NASA]

SDD Software Description Database [Internet]

SDD Software Description Document

SDD Software Design Description

SDD Software Design Document

SDD Source-Detector Distance

SDD Speech Direction Detector

SDD Stored Data Description

SDD Synthetic Dynamic Display

SDD System Design Description

SDD System Design Document

SDDC Sodium Dimethyldithiocarbamate

SDDF Salmonid (Fish) Demonstration and Development Farm [Canada]

SDDH South Dakota Department of Highways [US]

SDDL Stored Data Definition Language

SDDS Scientific Digitial Documentation System

SDDS Spin Decoupling Difference Spectroscopy

SDDTTG Stored Data Definition and Translation Task Group

SDE Sigma Delta Epsilon [Ithaca, New York, US]

SDE Small Debond Energy (of Composite Materials)

SDE Society of Data Educators [now Association for Computer Educators, US]

SDE Source Data Entry

SDE Space Division Evaluator

SDE Spin Density Excitation [Solid-State Physics]

SDE Statistical Design of Experiments

SDE Students for Data Education

SDE Submission and Delivery Entity

SDE Syntax Directed Editor

SDE System Development Engine

SDEA Shop and Display Equipment Association [UK]

SDEC Sodium Diethyl Dithiocarbamate

SDEP Staff Diversity Enhancement Program [of US Department of Energy]

Sdelovaci Tech Sdelovaci Technika [Czech Technical Journal]

SDES San Diego Engineering Society [California, US]

SDEV Systems Development

SDM Shape Deposition Manufacturing

SDM Single Domain Model [Solid-State Physics]

SDMC Sodium Dimethyl Dithiocarbamate

SDF Safing and Deservicing Facility

SDF Saskatchewan Development Fund [Canada]

SDF Satellite Distribution Frame

SDF Scottish Decorators Federation

SDF Seasonal Derating Factor

SDF Ship Design File

SDF Simplified Directional Facility

SDF Single Degree of Freedom [also 1DF]

SDF Software Development Facility

SDF Source Development Fund

SDF Space Delimited File

SDF Space Delimited Format

SDF Spin-Density Fluctuation [Solid-State Physics]

SDF Spin Density Functional (Formalism) [Solid-State Physics]

SDF Standard Data Format

SDF State Defense Forces [US]

SDF Statistical Distribution Function

SDF Stepdown Fix

SDF Stored Data Fine

SDF Student Data Form [US]

SDF Supergroup Distribution Frame [Telecommunications]

SDF Superlattice Dark Field (Image) [Microscopy]

SDF Synthetic Discriminant Function

SDF System Development Facility

.SDF Standard Data Format [File Name Extension]

SDFA State Defense Forces Association [US]

SDFAUS State Defense Forces Association of the United States

SDFC Saskatchewan Development Fund Corporation [Canada]

SDFC Space Disturbance Forecast Center [of Environmental Science Services Administration, US]

SDFF Spectroscopically Determined Force Field; Spectroscopically Defined Force Field

SDFL Schottky Diode FET (Field-Effect Transistor) Logic

SDFS Spacelab Data Flow System

SDFS Standard Disk Filing System

SD/FS Smoke Detector/Fire Suppression

SD/FS Supplier Documentation

SDG Simulated Data Generator

SDG Snow Depth Gauge

SDG Supplier Documentation Group

Sdg Siding [also sdg]

SDGE San Diego Gas and Oil Company [California, US]

SDGS South Dakota Geological Survey [US]

SDH Seasonal Derating Hours

SDH Self-Dumping Hopper

SDH Software Development Handbook

SDH Sorbitol Dehydrogenase [Biochemistry]

SDH Succinic Dehydrogenase [Biochemistry]

SDH Synchronous Digital Hierarchy

SDH System Development Handbook

SdH Shubnikov-de Haas (Effect) [Solid-State Physics]

SDHE Spacecraft Data-Handling Equipment

SdHO Shubrikov-de Haas Oscillations [Solid-State Physics]

SDHT Selectively-Doped Heterojunction Transistor

SDHW Solar Domestic Hot Water

SDI Selective Dissemination of Information

SDI Silt Density Index

SDI Single Document Interface

SDI Sludge Density Index [Sewage Treatment]

SDI Software Development Interface

SDI Source Data Information

SDI Space Defense Initiative (System)

SDI Standard Digital Interface

SDI Standard Disk Interconnect(ion)

SDI Standard Disk Interface

SDI Steel Deck Institute [US]

SDI Steel Door Institute [US]

SDI Strategic Defense Initiative [US]

SDIA Soap and Detergent Industry Association [UK]

SDIF Software Development and Integration Facility

SDIF Standard Document Interchange Format

SDILINE Selective Dissemination of Information using MEDLINE (MEDLARS (Medical Literature Analysis and Retrieval System) On-Line System) [US]

SDIM System for Documentation and Information in Metallurgy (Database) [of Bundesanstalt für Materialforschung und −prüfung, Germany]

SDIO Serial Digital Input/Output [also SDI/O]

SDIO Strategic Defense Initiative Organization [now Ballistic Missile Defense Organization, US]

SDIO/IST Strategic Defense Initiative Organization Innovative Science and Technology (Program) [of US Office of Naval Research] [also SDIO-IST]

SDIO/IST-ONR Strategic Defense Initiative Organization Innovative Science and Technology (Program) of the US Office of Naval Research]

SDIO/ISTP Strategic Defense Initiative Organization Innovative Science and Technology Program [of US Office of Naval Research] [also SDIO-ISTP]

SDIT Service de Documentation et d'Information Techniques de l'Aéronautique [French Technical Documentation and Information Service for Aeronautics] [also SDITA]

SDK Software Developer('s) Kit

SDK System Design Kit

SDL Semiconductor Diode Lasers, Inc. [US]

SDL Software Design Language

SDL Software Development Laboratory

SDL Software Development Language

SDL Space Disturbances Laboratory [of Environmental Science Services Administration, US]

SDL Space Dynamics Laboratory [of Utah State University, Logan, US]

SDL Specification and Description Language

SDL Standard Distribution List

SDL Surface Display Library

SDL Submersible Diver Lock-Out

SDL System Descriptive Language

SDL System Design Language

SDL System Directory List

S-DL S-Shaped Pattern, Magnetization Vector Direction Down Left (↙) [Solid-State Physics]

Sdl Saddle [also sdl]

SDLC Software Development Life Cycle

SDLC Synchronous Data Link Communication

SDLC Synchronous Data Link Control

SDLC/BSC Synchronous Data Link Communication/Binary Synchronous Communication

SDLE Solute Drag-Like Effect [Metallurgy]

SD(ln V) Standard Deviation of Volume Logarithm [also SD(ln v)]

SDLTS Scanning Deep Level Transient Spectroscopy

SDM Schwarz Differential Medium

SDM Selective Dissemination on Microfiche

SDM Semiconductor Disk Memory

SDM Sequence Division Multiplex(ing)

SDM Shutdown Mode

SDM Shuttle Data Management [NASA]

SDM Slovensko Društvo za Materiale [Slovenian Society for Materials]

SDM Space-Division Multiplex(ing)

SDM Standardization Design Memorandum

SDM STARAN (Stellar Attitude Reference and Navigation) Debug Module

SDM Statistical Delta Modulation

SDM Subsystem Design Manuals [of NASA]

SDM Symmetric Dimer Model [Chemistry]

SDM Synchronous Digital Machine

SDM System Definition Manual

SDM System Development Multitasking

SDMA Shared Direct Memory Access

SDMA Space-Division Multiple Access

SDMA Surgical Dressings Manufacturers Association [UK]

SDMH Symmetric Dimethylhydrazine

SDMM Scanning Desorption Molecule Microscopy

SDMS SCSI (Small Computer Systems Interface) Device Management System

SDMS Software Development and Maintenance System

SDMS Spatial Data Management System

SDN Société des Nations [League of Nations]

SDN Software Defined Network

SDN Software Development Note

SDN Subscriber's Directory Number

SDN Succinodinitrile

SDN Synchronous Digital Network

SDNO Strontium Dysprosium Niobium Oxide (Superconductor) [also SrDNO]

SDNS Secure Data Network Service

SDNS Software Defined Network Service

SDO Shielded Diatomic Orbitals [Physics]

SDO Standards Development Organization

SDO Standards-Developing Organization

SDO Synthetic Drying Oil

SDOF Single Degree of Freedom [also 1DOF]

6DOF Six Degrees-of-Freedom

SDOS Surface Density of State [Physics]

SDP Sedoheptulose-1,7-Diphosphate [Biochemistry]

SDP Shut-Down Procedure [Nuclear Reactors]

SDP Shuttle Data Processor [of NASA]

SDP Site Data Processor

SDP Signal Data Processor

SDP Signal Dispatch Point

SDP Single Domain Particle (Processing) [Solid-State Physics]

SDP Site Data Processor

SDP Slowing-Down Power [Nuclear Engineering]

SDP Software Development Plan

SDP Software Development Project

SDP Source Data Processing

SDP Special Development Program

SDP Spin Density Product [Solid-State Physics]

SDP Sputter Depth Profiling

SDP Standard Depth of Penetration

SDP Stationary Diffraction Pattern

SDP Steepest Descent Path

SDP Sudanese Pound [Currency]

SDP 4,4'-Sulfonyldiphenol

SDP Supervisors' Development Program [of National Safety Council, US]

SDP Supplier Data Package

SDP Supplier Development Program [of Industrial Research Assistance Program, Canada]

SDPA Styrenated Diphenyl Amine

SDPC Shuttle Data Processing Complex [of NASA]

SDPC Spin-Dependent Photoconductivity

SDPL Servomechanisms and Data Processing Laboratory [US]

SDPL Shut-Down Power Level [Nuclear Reactors]

SDR Search Decision Rule

SDR Service Difficulty Report

SDR Signal-Dependent Response

SDR Signal-to-Distortion Ratio

SDR Significant Deficiency Report

SDR Silicon Diode Rectifier

SDR Single-Drift Region

SDR Small Development Requirement

SDR SNAP (Space Nuclear Auxiliary Power) Developmental Reactor

SDR Sodium-Cooled Deuterium-Moderated Reactor [US]

SDR Software Design Requirements

SDR Software Design Review

SDR Somalian Democratic Republic

SDR Space Division Regulation [of US Air Force Systems Command]

SDR Spacelab Disposition Record [of European Space Agency]

SDR Special Drawing Rights

SDR Spin-Dependent Recombination [Physics]

SDR Spin-Dependent Resonance [Physics]

SDR Staff Development Review

SDR Standard Dimension Ratio

SDR Statistical Data Recorder

SDR Storage Data Recorder

SDR Storage Data Register

SDR Streaming Data Request

SDR Süddeutscher Rundfunk [Broadcasting Station of Southern Germany]

SDR System Design Review

S-DR S-Shaped Pattern, Magnetization Vector Direction Down Right (↘) [Solid-State Physics]

Sdr Sender [also sdr]

SDRAM Synchronous Dynamic Random-Access Memory

SDRB Software Design Review Board

SDRB Supplier Documentation Review Board

SDRD Supplier Data Requirements Description

SDRD Supplier Documentation Review Data

SDRI Sealed Double-Ring Infiltrometer

SDRL Supplier Data Requirements List

SD-ROM Super Density Read-Only Memory

SDRT Slot Dipole Ranging Test

SDS Safe (Gas) Delivery System

SDS Safety Data Sheet

SDS Salzgitter-Dolomit-Schlacke (Process) [Metallurgy]

SDS Sampled Data System

SDS Satellite Data System

SDS Scientific Data Systems

SDS Secure Digital Switch

SDS Shared Data Set

SDS Short-Time Smearing Distortion

SDS Shuttle Dynamic Simulation; Shuttle Dynamic Simulator [of NASA]

SDS Signal-Dependent Stereo

SDS Simulating Digital System

SDS Simulation Data Subsystem

SDS Site Data System

SDS Sodium Dodecyl Sulfate

SDS Software Design Specification

SDS Software Development Specification

SDS Software Development System

SDS South Dakota State [US]

SDS Soxhlet/Dean-Stark (Extractor) [Chemical Engineering]

SDS Spacecraft Data Storage

SDS Space Data System

SDS Space Documentation Service [of European Joint Documentation Service]

SDS Spin-Dependent Scattering [Physics]

SDS Staff Development Summary

SDS Steering Damping System

SDS Stored Data Smooth

SDS Stripe Domain Structure

SDS Sulzer Dainippon Sumitomo (Process)

SDS Supplier Delivery Schedule

SDS Sysops Distribution System

SDS System Data Synthesizer

SDS System Dynamics Society [US]

SDSC San Diego State College [California, US]

SDSC San Diego Supercomputer Center [US]

SDSC South Dakota Safety Council [US]

SDSE Self-Deployable Structure Element

SDSHC South Dakota State Highway Commission [US]

SDSI Shared Data Set Integrity

SDSL Single-Line Digital Subscriber Line

SDSL Symmetric Digital Subscriber Line

SDSMT South Dakota School of Mines and Technology [Rapid City, US]

SDS-PAGE Sodium Dodecylsulfate–Polyacrylamide Gel Electrophoresis

SDSS Sloan Digital Sky Survey (Experiment) [of Fermilab, Batavia, Illinois, US]

SDSS Software Development Support System

SDSS Space Division Shuttle Simulator [of US Air Force Systems Command]

SDSS Spatial Decision Support System

SDSS Super-Duplex Stainless Steel

SDS/SDR Signal-Dependent Stereo/Signal-Dependent Response

SDST Science Data Support Team

SD-STB Streaming-Data Strobe

SDSU San Diego State University [California, US]

SDSU South Dakota State University [Brookings, US]

SDSW Sense Device Status Word

SDT Scaling and Display Task

SDT Schematic Design Technology

SDT Serial Data Transfer

SDT Self-Destroying Tracer

SDT Serial Data Transmitter

SDT Shaft Drilling Technician [Mining]

SDT Shuttle Data Tape [NASA]

SDT Simulated Data Tape

SDT Single Delayed Trapping

SDT Smart Damping Treatment

SDT Society of Dairy Technology [UK]

SDT Soft Drink Technologist; Soft Drink Technology

SDT Spin-Dependent Tunneling (Effect) [Solid-State Physics]

SDT Start-Data-Traffic

SDT Step-Down Transformer

SDT Structural Dynamic Test

SDT System Development Tool

SDTA Stress-Dependent Thermal Activation

SDTA Structural Dynamic Test Article

SDTF Scottish Dairy Trade Federation [UK]

SDTR Serial Data Transmitter/Receiver

SDTS Spatial Data Transfer Standard

SDTV Standard Definition Television

SDU Satellite Data Unit

SDU Signal Distribution Unit

SDU Source Data Utility

SDU Station Display Unit

SDU Subsurface Disposal Unit

SDU System Data Unit

SDV Scram Discharge Volume [Nuclear Reactors]

SDV Single Dimer Vacancy [Chemistry]

SDV Slowed-Down Video

SDV Switched Digital Video

SD(V) Standard Deviation of Volume [also SD(v)]

SDVF Software Development and Verification Facilities

SDVS Slotted Disk Velocity Selector

SDW Segment Descriptor Word

SDW Solvent De-Waxing (Process)

SDW Spin-Density Wave [Solid-State Physics]

SDW Standing Detonation Wave

SDW Static Distortion Wave

SDWA Safe Drinking Water Act [US]

SDW-P Spin-Density Wave-to-Paramagnetic (Transition) [Solid-State Physics]

SDX Satellite Data Exchange

SDX Selective Deposition by Ex-Diffusion

SDX Storage Data Acceleration

SD(X) Standard Deviation of X [Symbol]

S+DX Speech plus Duplex [also S+D] [Telegraphy]

SDZ San Diego Zoo [California, US]

SE Safety Engineer(ing)

SE Safety Evaluation

SE Saint Etienne [France]

SE Sahara Español [Spanish Sahara]

SE Sample Exercise

SE Sanitary Engineer(ing)

SE Santiago del Estero [Argentina]

SE Saponification Equivalent

SE Satellite Engineer(ing)

SE School of Engineering

SE Schottky Effect [Solid-State Physics]

SE Schrödinger Equation [Quantum Mechanics]

SE Science and Engineering [also S&E]

SE Secondary Education

SE Secondary Electrode

SE Secondary Electron [also S(E)]

SE Secondary Emission

SE Secretary of Energy [US]

SE Seebeck Effect [Pysics]

SE Seignette-Electric

SE Self-Aligning Roller Bearing, Double-Row, Outer-Ring Raceway Spherical, Rollers Guided by Separate Center Guide Ring in Outer Ring [Symbol]

SE Self-Energy [Physics]

SE Self-Excitation

SE Sensor Element

SE Series Expansion [Mathematics]

SE Service Entrance

SE Service Expansion

SE Shielding Effect

SE Shielding Effectiveness [Nuclear Reactors]

SE Shifting Equilibrium

SE Shot Effect [Electronics]

SE Shubnikov Effect [Solid-State Physics]

SE Side Entry

SE Sign Extend

SE Silver Enhancement

SE Simultaneous Engineering

SE Single-Engine(d)

SE Single-Error (Correction)
SE Site Engineer
SE Size Effect
SE Size Exclusion (Chromatography)
SE Skin Effect
SE Slip End
SE Small End
SE Smoke Eliminator
SE Society for Endocrinology [US]
SE Society of Ethnobiology [US]
SE Society of Engineers [UK]
SE Socio-Economic(s)
SE Software Engineer(ing)
SE Soil Ecologist; Soil Ecology
SE Soil Engineer(ing)
SE Soil Erosion [Geology]
SE Solar Eclipse
SE Solar Energy
SE Solar Engine
SE Solid Electrolyte
SE Solvent Extraction
SE Somatic Effect [Biology]
SE Sonar Engineer(ing)
SE Soret Effect [Physics]
SE Sound Energy
SE South-East
SE Spark Erosion
SE Spectroellipsometric; Spectroellipsometer; Spectroellipsometry
SE Spectroscopic Ellipsometer; Spectroscopic Ellipsometric; Spectroscopic Ellipsometry
SE Spike Energy (Method) [Physics]
SE Spin Echo [Physics]
SE Spontaneous Emission
SE Stabilization Energy
SE Standard Electrode
SE Standard Error
SE Staphylococcal Enterotoxin [Microbiology]
SE Starch Equivalent
SE Stark Effect [Spectroscopy]
SE State Estimation; State Estimator
SE Static Electrification
SE Static Electricity
SE Stationary Engineer(ing)
SE Statistical Equilibrium
SE Steam Emulsion
SE Steam Engine
SE Stellar Energy
SE Stimulated Echo
SE Stimulated Emission (Spectroscopy)
SE Stock Exchange [also S/E]
SE Stokes-Einstein (Relation) [Physics]
SE Stop Element
SE Stored Energy

SE Strain Energy
SE Street Elbow (Pipe Fitting)
SE Stretched Exponential
SE Strong Electrolyte
SE Structural Element
SE Structural Engineer(ing)
SE Subacute Endocarditis [Medicine]
SE Subcritical Experiment [Atomic Physics]
SE Suction Effect
SE Sulfoethyl
SE Superelastic(ity)
SE Supported End
SE Support Equipment
SE Surface Energy
SE Surface Engineer(ing)
SE Surface-Enhanced; Surface Enhancement
SE Sweden [ISO Code]
SE System Element
SE System(s) Engineer(ing)
SE System-Enhanced; System Enhancement
SE- Sweden [Civil Aircraft Marking]
SE2 Scientists and Engineers for Secure Energy [also SESE, US]
S&E Safety and Environment [also SE]
S&E Science and Engineering [also SE]
S&E Science and Engineering [NASA Marshall Space Flight Center Directorate, Huntsville, Alabama, US]
S&E Surfaced on One Side and One Edge [Lumber]
S&E-1 Science and Engineering Data Stream 1 [NASA]
S&E-2 Science and Engineering Data Stream 2 [NASA]
S/E Stock Exchange [also SE]
Se Search [also se]
Se Selenium [Symbol]
Se^{2-} Divalent Selenium Ion [also Se] [Symbol]
Se^{4-} Tetravalent Selenium Ion [Symbol]
Se^{6-} Hexavalent Selenium Ion [Symbol]
Se$_8$ Molecular Selenium [Symbol]
Se-70 Selenium-70 [also ^{70}Se, or Se^{70}]
Se-72 Selenium-72 [also ^{72}Se, or Se^{72}]
Se-74 Selenium-74 [also ^{74}Se, or Se^{74}]
Se-75 Selenium-75 [also ^{75}Se, or Se^{75}]
Se-76 Selenium-76 [also ^{76}Se, or Se^{76}]
Se-77 Selenium-77 [also ^{77}Se, or Se^{77}]
Se-78 Selenium-78 [also ^{78}Se, or Se^{78}]
Se-79 Selenium-79 [also ^{79}Se, or Se^{79}, or Se]
Se-80 Selenium-80 [also ^{80}Se, or Se^{80}]
Se-81 Selenium-81 [also ^{81}Se, or Se^{81}]
Se-82 Selenium-82 [also ^{82}Se, or Se^{82}]
Se-83 Selenium-83 [also ^{83}Se, or Se^{83}]
Se-84 Selenium-84 [also ^{84}Se, or Se^{84}]
se *(salvo errore)* – errors excepted
/se (maximum number of) segments option [MS-DOS Linker]
.se Sweden [Country Code/Domain Name]
Σ(E) Self-Energy Operator [Symbol]

$\sigma\text{-}\epsilon$ stress-strain diagram [also $\sigma\text{-}\varepsilon$]

SEA Scandium Erbium Arsenide

SEA Scanning Electrostatic Analysis

SEA Science and Education Administration [of US Department of Agriculture]

SEA Sea Education Association [US]

SEA Self-Extracting Archive (File)

SEA Side Entry Anode

SEA Silicon Elastimeter Ablator

SEA Simultaneous Engineering Automation

SEA Single European Act [European Community]

SEA Société d'Electronique et d'Automatisme [Society of Electronics and Automation, France]

SEA Software Engineering Architecture

SEA Software Engineering Associates [US]

SEA Software Engineering of America [US]

SEA Sound Emission Analysis

SEA Southeast Asia

SEA Spherical Electrostatic Analyzer

SEA Standard Extended Attribute

SEA Standby Emergency Assistance

SEA Staphylococcal Enterotoxin A [Microbiology]

SEA Static Error Analysis

SEA Static Exchange Approximation

SEA Statistical Energy Analysis

SEA Structural Engineers Association [US]

SEA Sudden Enhancement of Atmospherics

SEA Systems Effectiveness Analyzer

.SEA Self-Extracting Archive [Macintosh File Name Extension]

SEAB Secretary of Energy Advisory Board [US]

SEAB Serial Bus

SEAC Safety and Environmental Advisory Council [of Westinghouse Hanford Company, Washington, US]

SEAC Solar Energy Advisory Committee [Western Australia]

SEAC Standards Eastern Automatic Computer (or Calculator) [of National Bureau of Standards, US]

SEAC State Energy Advisory Council [of New South Wales, Australia]

SEAC Structural Engineers Association of Colorado [US]

SeaCat Sea Category

SEACC Southeast Alaska Conservation Council [US]

SEACC Structural Engineers Association of Central California [also SEAOCC, US]

SEACF Support Equipment Assembly and Checkout Facility

SEACOM Southeast Asia Commonwealth Submarine Cable [Australia/New Guinea/ Papua/Hong Kong/ Malaysia/Singapore]

SEADAG Southeast Asia Development Advisory Group

SeaDAS SeaWiFS (Sea-viewing Wide Field-of View Sensor) Data Analysis System

SeaDataTPR Sea Data Temperature and Pressure Recorder

SEADS Shuttle Entry Air Data Sensor [NASA]

SEADS Shuttle Entry Air Data Subsystem [NASA]

SEADS Shuttle Entry Air Data System [NASA]

SEAFDC Southeast Asian Fisheries Development Center [Philippines]

SEAFIRE South-East Asia Fire Research Experiment

SEAFP Safety and Environmental Aspects of Fusion Power (Study) [Europe]

Seagas South Eastern Gas Board (Process) [UK]

SEAID Support Equipment Abbreviated Items Description

SEAIMP Solar Eclipse Atmospheric and Ionospheric Measurements Project

SEAISI Southeast Asia Iron and Steel Institute [Philippines]

SEAISI Newsl SEAISI Newsletter [of Southeast Asia Iron and Steel Institute, Philippines]

SEAISI Q SEAISI Quarterly [of Southeast Asia Iron and Steel Institute, Philippines]

SEAL Screening External Access Link

SEAL Sea, Air and Land (Personnel)

SEAL Segmentation and Reassembly Layer

SEAL Signal Evaluation Airborne Laboratory [of Federal Aviation Administration, US]

SEAL Software Engineering and Analysis Laboratory

SEAL Southeast Area Libraries [UK]

SEAL Standard Electronic Accounting Language

SEALAB Undersea Laboratory (Program) [of US Navy] [also Sealab]

SEALION Sea Level Instrumentation and Observation Network

SEALS Severe Environmental Air Launch Study

SEALS Stored Energy Actuated Lift System

SEAM Scanning Electron Acoustic Microscope; Scanning Electron Acoustic Microscopy

SEAM Scanning Electroacoustic Microscopy

SEAM (Conference on the) Scientific and Engineering Applications of the Macintosh (Computer) [US]

SEAM Software Engineering and Management

SEAM Surface-Embedded Atom Method

SEAM Surface Environment and Mining Program

SEAM Symposium on the Engineering Aspects of Magnetohydrodynamics

SEAMEO Southeast Asian Ministers of Education Organization [Thailand]

SEAMS Southeast Asia Mathematical Society [Singapore]

SEAN Scientific Event Alert Network [of Smithsonian Institute, Washington, DC, US]

SEANC Structural Engineers Association of Northern California [also SEAONC, US]

Seance Acad Sci Seance de l'Académie de Sciences [Session of the Academy of Sciences, Paris, France]

SEAO Structural Engineers Association of Oregon [US]

SEAOC Structural Engineers Association of California [US]

SEAP Save the Environment from Atomic Pollution

SEAP Service Element Access Point

SEAPEX Southeast Asia Petroleum Exploration Society

SEAPG Support Equipment Acquisition Planning Group

SEAQ Stock Exchange Automated Quotations

SEARCC Southeast Asia Regional Computer Confederation [Singapore]

SEARCH Scientific Exploratory Archeological Research

SEARCH Systemized Excerpts Abstracts and Reviews of Chemical Headlines [US]

Search Health Search for Health [Journal published in the US]

SEARCHS Shuttle Engineering Approach/Rollout Control Hybrid Simulation [of NASA]

SEAS (Committee for the) Scientific Exploration of the Atlantic Shelf

SEAS School of Engineering and Applied Science [also SE&AS]

SEAS Shipboard Environmental Automated System

SEAS Strategic Environmental Assessment System [of US Environmental Protection Agency]

SEAS System Engineering and Analysis Support

SE&AS School of Engineering and Applied Science [also SEAS]

SEASC Structural Engineers Association of Southern California [also SEAOSC, US]

SEASCO Southeast Asia Science Cooperation Office [India]

SEAT Sociedad Espanola de Automóviles de Turismo [Spanish Society for Tourist Vehicles]

SEATAC Seattle-Tacoma (International Airport) [US]

SEATAC Southeast Asian Agency for Regional Transport and Communications Development [Malaysia]

Seatag Bull Seatag Bulletin [Published in the UK]

SEATO Southeast Asia Treaty Organization [Disbanded]

SeaWiFS Sea-View(ing) Wide-Field-of-View Sensor [also SeaWIFS]

SEB Scanning Electron Beam

SEB Scientific Equipment Bay

SEB Secondary Electron Bremsstrahlung

SEB Sociedade Entomológica do Brasil [Brazilian Entomological Society]

SEB Society for Economic Botany [US]

SEB Society for Experimental Biology [UK]

SEB Source Evaluation Board

SEB Staphylococcal Enterotoxin B [Microbiology]

SEB Support Equipment Building

SEB Systems Engineering Branch [of NASA]

SE(B) Single-Edge Notch Bend [Mechanical Testing]

SEBA Styrene and Ethylbenzene Association [of Synthetic Organic Chemical Manufacturers Association, US]

SEBCO Samarium Erbium Barium Copper Oxide (Superconductor)

SEBM Society for Experimental Biology and Medicine [US]

SEBS Styrene-Ethylene-Butadiene-Styrene (Block Copolymer)

SE by E Southeast by East

SE by S Southeast by South

SEC Office of the Secretary [of US Geological Survey]

SEC Sacramento Engineers Club [California, US]

SEC Safeguards Equipment Cabinet

SEC Sanitary Engineering Center [US]

SEC Science Executive Committee

SEC Secondary Electron Conduction

SEC Secondary Electron-Coupled (Vidicon)

SEC Secondary Emission and Conduction

SEC Secondary Emission Control

SEC Securities and Exchange Commission [US]

SEC Sequential Events Controller

SEC Shuttle Events Control [of NASA]

SEC Simple Electronic Computer

SEC Single-Edge-Cracked (Test Specimen) [Mechanical Testing]

SEC Single Electron Capture

SEC Single Error Correction

SEC Size-Exclusion Chromatography

SEC Solar Energy Collection; Solar Energy Collector

SEC Solar Energy Conversion (System)

SEC Soluble Elastic Capsule(s) [Pharmacology]

SEC Sonoelectrochemical; Sonoelectrochemistry

SEC Source Evaluation Committee

SEC South Equatorial Current [Oceanography]

SEC Southern European Country

SEC Space Engineering Center [of NASA]

SEC Space Environment Center [of National Oceanic and Atmospheric Administration, US]

SEC Special-Event Charter Flight

SEC Spectroelectrochemical; Spectroelectrochemistry

SEC Standard Engineering Control Room

SEC Standard Error of Calibration [Spectroscopy]

SEC Standard Evaluation Circuit

SEC Steam-Engine Cycle

SEC Switching Equipment Congestion

.SEC Data Security [File Name Extension]

Sec Secretary [also sec]

Sec Section; Sector [also sec]

Sec Security [also sec]

sec *(salvo errore calculi)* – arithmetical errors excepted

sec secant [Symbol]

sec second [also s, or ʺ] [Unit]

sec secondary

sec- secondary [Chemicals]

sec *(secundum)* – according to

sec⁻¹ inverse secant [also arcsec] — sec^{-1}

sec⁻¹ one per second [also s⁻¹, 1/s, or 1/sec] — sec^{-1} [also s^{-1}, 1/s, or 1/sec]

sec⁻² one per second squared [also s⁻², 1/s², or 1/sec²] — sec^{-2} [also s^{-2}, $1/s^2$, or $1/sec^2$]

SECA Solar Energy Construction Association

SECAL Selective Calling

SECAL Separate Engineering Control Air Limits

SECAM Séquentiel Couleur à Mémoire [Sequential Color and Memory (Television System)]

SECAP System Experience Correlation and Analysis Program

SECAR Secondary Radar [also Secar, or secar]

SECAR Structural Efficiency Cones with Arbitrary Rings

SECB Square Edge Close Butt (Joint) [Welding]

SECC Single Edge Contact Cartridge

SECCAN Science and Engineering Clubs of Canada

SECD Saskatchewan Economic and Cooperative Development [Canada]

SECD Spacelab Experiment Channel Data

secd secondary [also sec'd]

SECDED Single-Bit Error Correction and Double Bit Error Detection (Logic)

SECDED Single Error Correction and Double Error Detection (Logic)

SECE School of Electrical and Computer Engineering

SECED Society of Earthquake and Civil Engineering Dynamics [UK]

sec-ft second-foot [also sec ft]

sech hyperbolic secant [Symbol]

sech^{-1} inverse hyperbolic secant [also arsech]

SEC-HPLC Size Exclusion Chromatography–High-Performance Liquid Chromatography [also SEC HPLC]

SECI Société d'Etudes Chimiques pour l'Industrie [Society for Chemical Studies for Industry, France]

SEC-LALLS Size Exclusion Chromatography–Low-Angle Laser-Light Scattering [also SEC LALLS]

SECM Scanning Electrochemical Microscopy

Secn Section [also secn]

SECNAV Secretary of the Navy [US]

SECO Self-Regulating Error-Correct Coder-Decoder

SECO Sequential Coding

SECO Sequential Control

SECO Station Engineering Control Office

SECO Sustainer-Engine Cutoff

SECOFI Secretariado de Comercio y Fomento Industrial [Department of Commerce and Industry, Mexico] [also Secofi]

SECON Secondary Electron Conduction

SECOR Sequential Collation of Range (of Geodetic Satellite) [also Secor]

SECOR Sequential Correlation Range

SECORD Secure Voice Cord Board

SECOR/DME Sequential Collation of Range/DistanceMeasuring Equipment [also Secor/DME]

SECOSS Spin-Echo Correlation (with) Shift Scaling [Physics]

SEC Solar Energy Conversion Project [of Massachusetts Institute of Technology, Cambridge, US]

SECPS Secondary Propulsion System

SECR Shuttle Engineering Control Room [of NASA]

Secr Secretary [also secr]

Secs Sections [also secs]

SECS Selective Electron Capture Sensitization

SECS Sequential Events Control System

SECS Shuttle Events Control System [of NASA]

SECS Single Electron Capacitance Spectroscopy

SECS Solar Energy Conversion Systems

SECS Standard Cubic Centimeters per Second [also secs]

SECSY Spin-Echo Correlated (or Correlation) Spectroscopy

SECT Spacelab Experiment Channel Tape

SECT Single-Edge Cracked Tension (Specimen) [Mechanical Testing]

Sect Section [also sect]

sectl sectional [also sect'l]

SECTAM Southeastern Conference on Theoretical and Applied Mechanics [US]

Sec Treas Secretary-Treasurer [also Sec-Treas]

SECU Slave Emulator Control Unit

Secur Security [also secur]

SECURE Systems Evaluation Code Under Radiation Environment

SECURITY International Trade Fair for Security with Congress [Essen, Germany]

Secur Manage Security Management [Publication of American Society for Industrial Security]

SECV Sociedad Espanola de Cerámica y Vidrio [Spanish Society of Cermics and Glass]

SECV State Electricity Commission of Victoria [Australia]

Secy Secretary [also secy]

SECYT Secretario de Ciencia y Tecnologia [Secretariat of Science and Technology, Argentina] [also SECyT]

SED Safety Evaluation Document

SED Sanitary Engineering Division

SED Science and Engineering Directorate [of NASA]

SED Secondary Electron Detector

SED Self-Electrooptic Effect Device

SED Semiconductor Electronics Division [of National Institute for Standards and Technology, Gaithersburg, Maryland, US]

SED Semiconductor Equipment Division [of Eaton Corporation, US]

SED Semiequilibrium Dialysis

SED Skin Erythema Dose [also sed]

SED Slow Electron Diffraction

SED Society for Electrical Development [US]

SED Solar Electrodynamic(s)

SED Space Environment Division [of NASA]

SED Spectral Energy Distribution

SED Spherical Equivalent Diameter

SED Spin Echo Difference [Physics]

SED Status Entry Devices

SED Stimulated Emission Device

SED Stochastic Electrodynamics

SED Stokes-Einstein-Debye (Law) [Physics]

SED Stream(-Oriented) Editor

SED Strong Exchange Degeneracy

SED Surface Engineering Division [of ASM International, US]

SE&D Software Engineering and Development

SEd Specialist in Education

SEDA Safety Equipment Distributors Association [US]

SEDAC Socioeconomic Data and Applications Center [of Consortium for International Earth Science Information Networks]

SEDCO Saskatchewan Economic Development Corporation [Canada]

SEDD Special Extra Deep Drawing

SEDD Systems Evaluation and Development Division [of NASA]

s-EDDA Symmetrical N,N' Ethylenediaminediacetic Acid

SEDFB Surface-Emitting Distributed Feedback

sediment sedimentary

Sediment Geol Sedimentary Geology [Journal published in the US]

Sedimentol Sedimentological; Sedimentologist; Sedimentology

Sedimentol Sedimentology [International Journal]

SEDIT Sophisticated String Editor

SEDIX Selected Dissemination of Indexes

SEDM Selective Excitation Double Moessbauer (Spectroscopy)

SEDM Society for Experimental and Descriptive Malacology [US]

SEDM Status Entry Device Multiplexer

SeDMC Selenium Dimethyl Dithiocarbamate

SEDOR Spin Echo (Quadrupole-Quadrupole) Double Resonance [Physics]

SEDP Solar Energy Development Program [Canada]

SEDR Systems Engineering Department Report

SEDRIS Synthetic Environment Data Representation and Interchange Specification

SEDS Space Electronic Detection System

SEDS Society for Educational Data Systems

SEDS Society for the Exploration and Development of Space [US]

SEDS Stress Enhanced Diffusion/Solubility

SEE Sabotage, Espionage and Embezzlement

SEE Sample Exchange Evaluation

SEE School of Electrical Engineering

SEE Secondary Electron Emission

SEE Secondary Emission Enhancement

SEE Selected Energy Epitaxy

SEE Self-Electrooptic Effect

SEE Single-Event Effect

SEE Society of Electronic Engineers [India]

SEE Society of Environmental Engineers [UK]

SEE Society of Explosives Engineers [US]

SEE Solar Energy Engineer(ing)

SEE Southeastern Electric Exchange

SEE Space Environments and Effects

SEE Special-Purpose End-Effector

SEE Standard End-Effector

SEE Systems Equipment Engineer

Se(E) Electronic Stopping Cross-Section [Symbol]

SEEA Software Error Effects Analysis

SEEB Southeastern Electricity Board [UK]

Seebeck EMF Seebeck Electromotive Force [also Seebeck emf]

SEECCIASDI Standing European Economic Community Committee of the International Association of the Soap and Detergent Industry [Belgium]

SEED Science/Engineering Education Division [of Oak Ridge Institute for Science and Education, Tennessee, US] [also S/EED]

SEED Self-Electrooptic-Effect Device

SEED Sustainable Energy and Environment Division [of the United Nations Development Program]

SEED Newsl Sustainable Energy and Environment Division Newsletter [of the United Nations Development Program]

SEEDS Society, Environment and Energy Development Studies

SEEH Super Energy-Efficient Home

SEEK State of the Environment Education Kit [Canada]

SEEK Systems Evaluation and Exchange of Knowledge

SeekEof Seek End-of-File Status [Pascal Programming Function]

SeekEoln Seek End-of-Line Status [Pascal Programming Function]

SEELFS Surface Electron Energy Loss Fine Structure

SEELFS Surface Extended Energy Loss Fine Structure

SEEN Syndicat d'Etudes de l'Energie Nucléaire [Syndicate for the Study of Nuclear Energy, Belgium]

SEENY Society of Engineers of Eastern New York [US]

seeo *(salvo errore et omissione)* – errors and omission excepted

SEEP Students to Explore and Experience Physics (Project) [of American Physical Society, US]

SEEPRO Self-Employed Professional

SEER Société des Electriciens, des Electroniciens et des Radioélectriciens [Society of Electrical, Electronic and Radio Engineers, France]

SEER Student Exposition on Energy Resources [of National Energy Foundation, US]

SEER Surveillance, Epidemology and End Result

SEERE Secretary for Energy Efficiency and Renewable Energy [of US Department of Energy]

SEERS Simultaneous Electrochemical Electron Spin Resonance Spectroscopy

SEES School of Electronic Engineering and Computer Systems [of University College of North Wales, UK]

SEES Secondary Electron Emission Sensing

SEES Secondary Electron Emission Spectroscopy

SEES Secondary Electron Emission Spectrum

SEES Support Equipment and Engineering Services

se et o *(salvo errore et omissione)* – errors and omissions excepted

SEF Scanning Electron Fractograph(y)

SEF Secondary Emission Factor

SEF Shared Equipment and Facilities (Grant) [of Natural Sciences Engineering and Research Council, Canada]

SEF Small End Forward

SEF Société Electrométallurgique Française [French Electrometallurgical Society]

SEF Software Engineering Facility

SEF Space Education and Foundation [US]

SEF Standard External File

SEF Storage Extension Frame

SEF Stress Enhancement Factor

SEFA South East Forest Alliance

SE/FAC Support Equipment/Facility

SEFAR Sonic End Fire for Azimuth and Range

SEFE Secondary Electron Field Emission

SEFEL Secrétariat Européen des Fabricants d'Emballages Métalliques Légers [European Secretariat of Manufacturers of Light Metal Packaging, Belgium]

SEFI Société Européenne pour la Formation des Ingénieurs [European Society for Engineering Education, Belgium]

SEFIT Single Electron Fit

SEFOR Southwest Experimental Fast Oxide Reactor [US]

SEFR Shielding Experimental Facility Reactor [of Idaho National Engineering Laboratory, Idaho Falls, US]

SEFT Spin Echo Fourier Transform

SEG Scientific Ecology Group, Inc.

SEG Segment (Function) [also Seg] [Programming]

SEG Selective Epitaxial Growth

SEG Society of Economic Geologists [US]

SEG Society of Exploration Geophysicists [US]

SEG Software Engineering Group

SEG Special Effects Generator

SEG Standardization Evaluation Group

SEG Systems Engineering Group

Seg Segment(al) [also seg]

SEGAS Southeastern Gas (Process)

SEGB Southeastern Gas Board [UK]

SEGD Society of Environmental Graphic Designers [US]

Se-Ge Selenium-Germanium (Interface)

SEGF Surface Embedded Green Function (Method)

SEGJ Society of Exploration Geophysicists of Japan

Segm Segment [also segm]

SEG Si Selective Epitaxial Grown Silicon

SEH Structured Exception Handling

SEHF Spin Extended Hartree-Fock [Physics]

SEI Safety Equipment Institute [US]

SEI Secondary Electron Image; Secondary Electron Imaging

SEI Society of Engineering Illustrators [US]

SEI Software Engineering Institute [of US Department of Defense at Carnegie Mellon University, Pittsburgh, Pennsylvania, US]

SEI Space Exploration Initiative (Program) [of NASA]

SEI Statute for Encouragement of Investment [Taiwan]

SEI Stockholm Environment Institute [Sweden]

SEI Support Equipment Installation

SE&I Systems Engineering and Integration [also SEI]

SEIA Security Industry Equipment Association [now Security Industry Association, US]

SEIA Solar Energy Industries Association [US]

SEIAC Science Education Information Analysis Center [of Educational Resources Information Center, US]

SEIC Simposio Europeo sulle Inibitori do Corrozione [European Symposium on Corrosion Inhibitors, Italy]

SEI/CMM Software Engineering Institute/Capability Maturity Model

SEICO Support Equipment Installation and Checkout

SEICO System Engineering Instrumentation and Checkout

SEIDAM System of Experts for Intelligent Data Management

SEIFSA Steel and Engineering Industries Federation of South Africa

SEI Int Environ Bull SEI International Environmental Bulletin [of the Stockholm Environment Institute, Sweden]

SEINA Solar Energy Institute of North America [US]

SEIP Smelter Environmental Improvement Project [of Falconbridge Limited., Sudbury, Ontario, Canada] [Metallurgy]

SEIP Space Exploration Initiative Program [of NASA]

SEIP Strategy for Exploration of the Inner Planets

SEIP Systems Engineering Implementation Plan

SEIRS Suppliers and Equipment Information Retrieval System [of International Civil Aviation Organization]

seism seismic [also seism]

Seism Instrum Seismic Instruments [Translation of: *Seismicheskie Pribory: Instrumental'naye Sredstva Seismicheskikh Nablyudenii* (USSR); Russian Journal published in the US]

Seismol Seismological; Seismologist; Seismology [also seismol]

Seismol Geol Seismology and Geology [Chinese Publication]

Seismol Res Lett Seismological Research Letters [of Seismological Society of America, US]

Seismol Ser Earth Phys Branch Seismological Series of the Earth Physics Branch [of Natural Resources Canada]

Seism Prib Instrum Sreds Seism Nabl Seismicheskie Pribory: Instrumental'naye Sredstva Seismicheskikh Nablyudenii [Russian Journal on Seismic Instruments]

SEIT Satellite Educational and Informational Television

SEITA Société d'Exploitation Industrielle des Tabacs et des Allumettes [Society for the Industrial Exploitation of Tabak and Matches, France]

SEIU Service Employees International Union [US]

SEJ Society of Environmental Journalists [US]

SEJF Sri Lankan Environmental Journalists Forum

SEJJ Step-Edge Josephson Junction [Electronics]

SEJ J Society of Environmental Journalists Journal

SEK Swedish Krona [Currency]

SEL Single-Engine Land(ing)

SEL Single-Event Latchup

SEL Software Engineering Laboratory

SEL Solar Energy Laboratory [of University of Wisconsin, Madison, US]

SEL Space Environment Laboratory [of National Oceanic and Atmospheric Administration, US]

SEL Staff Engineer Loop

SEL Standard Engineering Laboratory

SEL Stanford Electronics Laboratories [of Stanford University, California, US]

SEL Static Equivalent Load(ing)

SEL Surface Emitting Laser

SEL Systems Engineering Laboratories [US]

Sel Select(ion); Selectivity; Selector [also sel]

sel select(ed)

sel select (function) [MS-DOS Command]

SELBUS Systems Engineering Laboratories Data Bus

SELC Southeastern Louisiana College [Hammond, US]

SELCAL Selective Calling

SELDOM Selective Dissemination of MARC (Machine-Readable Catalogue) [of University of Saskatchewan, Canada]

SELECTAVISION Selected Television Video Disk System [also Selectavision]

SELEX Segmented Large X-Baryon Spectrometer (Experiment) [of Fermilab, Batavia, Illinois, US]

SELEX Selective (Spin) Exchange [Physics]

SELFOC Self-Focusing Optical (Instrument)

SELNI Società Elettronucleare Italiana [Italian Electronuclear Society]

SELPER Sociedad de Especialistas Latinoamericanos en Percepción Remota [Society of Latin-American Remote Sensing Specialists]

SELR Saturn Engineering Liaison Request

SELRIP Selected Release Improvement Program

SELS Selsyn (Motor) [also Sels]

SELSPOT Selective Spot (Recognition)

SEM Scanning Electron Micrograph; Scanning Electron Microscope; Scanning Electron Microscopy

SEM Secondary Electron Multiplier

SEM Secondary Emission Microscopy

SEM Seller's Engineering Memo(randum)

SEM Silylethoxymethyl

SEM Singularity Expansion Method

SEM Society for Experimental Mechanics [US]

SEM Solar Energy Material

SEM Solar Equipment Manufacturers

SEM Space-Environment Monitor(ing)

SEM Standard Electron(ic) Module

SEM Standard Error of Mean

SEM Standard Estimating Module

SEM Station Engineering Manual

SEM Station d'Essais de Montluçon [Montluçon Experimental Station, France]

SEM Strategic Enterprise Management

SEM Surface-Engineered Material

SEM System Engineering Management

Sem Seminar(y) [also sem]

SEMA Secretaria Especial do Meio Ambiente [Special Secretariat for the Environment, Brazil]

SEMA Société d'Economie et de Mathematiques Appliqués [Society for Economy and Applied Mathematics, France]

SEMA Specialty Equipment Manufacturers Association

SEMA Specialty Equipment Market Association [US]

SEMA Spray Equipment Manufacturers Association [UK]

SEMA Storage Equipment Manufacturers Association [UK]

SEMAA Safety Equipment Manufacturers Agents Association [US]

SEMATECH Sematech Center of Excellence [of Rensselaer Polytechnic Institute, Troy, New York, US] [also Sematech]

Sem Arthrop Seminars in Arthropathy [Journal]

SEM-BEI Scanning Electron Microscopy–Backscattered Electron Image

SEM-BS Backscattered Scanning Electron Micrograph; Backscattered Scanning Electron Microscopy [also BS-SEM]

SEM-BSE Scanning Electron Microscopy/Backscattered Electron (Analysis) [also SEM/BSE]

SEM-C Scanning ElectronMicroscopy/Channeling

Sem Cancer Biol Seminars in Cancer Biology [Journal published in the US]

Sem Cell Biol Seminars in Cell Biology [Journal published in the US]

Sem Cell Devel Biol Seminars in Cell and Developmental Biology [Journal published in the US]

SEM-Cl 2-(Trimethylsilyl)ethoxymethyl Chloride

SEMCOR Semantic Correlation

SEME School of Electrical and Mechanical Engineering

SEME Sociedad Espanola de Microscopía Electrónica [Spanish Society for Electron Microscopy]

SEM-EBIC Scanning Electron Microscopy Electron Beam-Induced Current [also SEMEBIC or SEM/EBIC]

SEM-EBSP Scanning Electron Microscopy–Electron Backscattered Spectroscopy (Technique) [also SEM/EBSP]

SEM-ECP Scanning Electron Microscopy Electron Channeling Pattern (Technique) [also SEM/ECP]

SEM-EDAX Scanning Electron Microscopy Energy Dispersive X-Ray (Fluorescence) Analysis [also SEM/EDAX]

SEM-EDS Scanning Electron Microscopy Energy Dispersive (X-Ray) Spectroscopy [also SEM/EDS, or SEMEDS]

SEM-EDX Scanning Electron Microscopy/Energy Dispersive X-Ray Analysis [also SEM/EDX, SEM-EDXA, or SEM/EDXA]

SEM-EDXA Scanning Electron Microscopy/Energy Dispersive X-Ray Analysis [also SEM/EDX]

SEM-EMPA Scanning Electron Microscope–Electron Microbeam Analyzer [also SEM/EMPA]

SEMI Semiconductor Equipment and Materials Institute [US]

SEMI Semiconductor Equipment and Materials International

SEMI Sports and Exercise Medicine Institute [Canada]

SEMICON Semiconductor [also Semicon or semicon]

SEMICON Semiconductor Equipment and Materials Institute Conference

Semicond Semiconductor [also semicond]

Semicond Semiconductors [English Translation of Russian Journal Jointly Published by Maik Nauka, Russia, and the American Institute of Physics, US]

Semicond Int Semiconductor International [US Publication]

Semicond Sci Technol Semiconductor Science and Technology [Journal of Institute of Physics, UK]

Semicond Technol Semiconductor Technology

SEMICON Tech Proc SEMICON Technical Proceedings [of Semiconductor Equipment and Materials Institute, US]

Semi-EV Semi-Efficient Vulcanization

Sem Immunol Seminars in Immunology [Journal published in the US]

Semi-IPN Semi-Interpenetrating Polymer Network [also semi-IPN]

SEMIMAG Conference on Application of High Magnetic Fields in Semiconductor Physics

SEMIRAD Secondary-Electron Mixed Radiation Dosimeter

SEMLAM Semiconductor Laser Amplifier

SEMLAT Semiconductor Laser Array Technique

SEMP SERDP (Strategic Environmental Research and Development Program) Ecosystem Management Program [Joint US Program of the Department of Defense, the Department of Energy and the Environmental Protection Agency]

SEMPA Scanning Electron Microscopy with Polarization Analysis [also SEM-PA]

SEMPA Secondary Electron Microscopy with Polarization Analysis [also SEM-PA]

SEMPE Socio-Economic Model of the Planet Earth

Sem Psych Seminars in Psychiatry [Journal]

SEMS Severe Environment Memory System

SEMS Solar Environment Monitor Subsystem

SEMS Space Environment Monitor System

SEMSC Southeast Missouri State College [Cape Girardeau, US]

SEM-STEM Scanning Electron Microscopy/Scanning Transmission Electron Microscopy [also SEM/STEM]

SEMUT Suspectral Editing (using a) Multiple Quantum Trap [Physics]

SEM-VC Scanning Electron Microscope–Voltage Contrast

Semy Seminary [also semy]

SEN Sense Command

SEN Single-Edge Notched (Specimen) [Mechanical Testing]

SEN Society of European Nematologists [now European Society for Nematologists, Scotland]

SEN Software Error Notification

SEN State Enrolled Nurse [UK]

SEN Steam Emulsion Number

SEN Submerged Entry Nozzle

Sen Senate; Senator [also sen]

Sen Senegal

Sen Senior [also sen]

Sen Sensation [also sen]

Sen Sensor [also sen]

SENA Société d'Energie Nucléaire des Ardennes [Society for Nuclear Energy in the Ardennes, Belgium]

SENB Single-Edge Notched Beam (Bending) [Mechanical Testing]

SEND Shared Equipment Need Date

SENECA Semantic Networks for Conceptual Analysis

SENER Strain Energy Density [Mechanics]

SENL Standard Equipment Nomenclature List

SENN Società Elettronucleare Nazionale [National Electronuclear Society, Italy]

Se-77 NMR Selenium-77 Nuclear Magnetic Resonance [also ^{77}Se NMR, or ^{77}Se-NMR]

Senr Senior [also senr]

Sens Sensitive; Sensitivity [also sens]

Sens Sensor [also sens]

Sens Act A Sensors and Actuators A [Swiss Publication]

Sens Act B Sensors and Actuators B [Swiss Publication]

sensi sensible

Sens Mater Sensors and Materials [International Journal]

Sensor International Exhibition and Congress for Sensor and Systems Technology [Germany]

Sens Rev Sensor Review [Published in the UK]

Sensy Sensibility; Sensitivity [also sensy]

SENTA Société d'Etudes Nucléaire et de Techniques Advances [Society for Nuclear Studies and Advanced Techniques, France]

SENTOS Sentinel Operating System

SEO Satellite for Earth Observation [India]

SEO Senior Education Officer

SEO Socio-Economic Objective

SEO Special Engineering Order

SeO Seleninyl (Radical) [Symbol]

SeO$_2$ Selenonyl (Radical) [Symbol]

SeO$_3{}^{2-}$ Selenite [Symbol]

SEOCS Sun-Earth Observatory and Climatology Satellite

SE ODF Series Expansion Orientation-Distribution Function

SEOS Synchronous Earth Observation Satellite

S/EOS Standard Earth Observation Satellite

SEP Polysiloxan Treated EPDM (Ethylene Propylene Diene Monomer)

SEP Salmonid Enhancement Program [of Department of Fisheries and Oceans, Canada]

SEP Schichten-Erfassungs-Programm [(Geological) Strata (Data) Acquisition Program, of NLfB–Niedersächsisches Landesamt für Bodenforschung, Hannover, Germany]

SEP Search Effectiveness Probability

SEP Secco Etch Pits [Metallurgy]

SEP Self-Elevating Platform

SEP Separation Parameter

SEP Set Endpoint

SEP Skeletal Electron Pair

SEP Société Européenne de Physique [European Physical Society, Switzerland]

SEP Société Européenne de Propulsion [European Propulsion Society, France]

SEP Society of Engineering Psychologists

SEP Solar Energy Program

SEP Somatosensory Evoked Potential [Medicine]

SEP Source Evaluation Panel

SEP Space Electronic Package

SEP Standard Electrode Potential

SEP Standard Electronic Package

SEP Standard Error of Performance [Spectroscopy]

SEP Star Epitaxial Planar

SEP Steam Ejector Pump

SEP Steam-Electric Plant

SEP Stimulated Emission Pumping (Spectroscopy)

SEP Strain Energized Powder

SEP Strain Energizing Process

SEP Support Equipment Package

SEP Systematic Evaluation Program

Sep Separate; Separation; Separator [also sep]

Sep September

sep separate(d)

sep septet [Spectroscopy]

SEPA Science Education Programme for Africa [Kenya]

SEPA Southeastern Power Administration [US]

SEPA State Environmental Policy Act [Washington State, US]

SEPAP Shuttle Electrical Power Analysis Program [of NASA]

SEPAR Shuttle Electrical Power Analysis Report [of NASA]

SEPB Single-Edge-Precracked-Beam (Test) [Mechanical Testing]

SEPBOP Southeastern Pacific Biological Oceanographic Program

SEPC Società Esplodenti et Prodotti Chimiche [Society for Explosives and Chemical Products, Italy]

SEPC Space Exploration Program Council [of NASA]

SEPD Special Environmental Powder Diffractometer [of Argonne National Laboratory's Intense Pulsed Neutron Source, US]

sepd separated [also sep'd]

SEPDUMAG Separate (Magnetic) Sound and Picture, Dual Track [Sound Recording]

SEPE Single Escape Peak Efficiency

SEPE Société Européenne de Presse et d'Edition [European Society for Press and Edition, France]

SEPES Synchrotron-Radiation-Excited Photoelectron Spectroscopy

SEPFA South-East Professional Fishermen's Association [US]

Sepg Separating [also sepg]

SEPG Software Engineering Process Group

SEPGA Southeastern Pecan Growers Association [US]

SePh Phenylselenide

SEPI Société d'Exploitation de Produits Industriels [Society for the Exploitation of Industrial Products, France]

SEPIL Selectivity Exciting Probe Ion Luminescence

SEPM Society of Economic Paleontologists and Mineralogists [US]

SEPMAG Separate (Magnetic) Sound and Picture (Recording)

Sepn Separation [also sepn]

SEPOL Settlement Problem-Oriented Language

SEPOL Soil-Engineering Problem-Oriented Language

SEPOX Selective Polycrystalline Silicon Oxidation

SEPP Secure Encryption Payment Protocol

SEPP Software Exchange Pilot Project

SEPP Southeastern Power Pool

SEPP Symposium on Electrometallurgical Plant Practice

Sep Purif Methods Separation and Purification Methods [Journal published in the US]

SEPR Société d'Etude de la Propulsion par Réaction [Society for the Study of Propulsion by Reaction Technique, France]

SEPS Poly(styrene-b-(Ethylene-co-Propylene)-b-Styrene)

SEPS Service Environment Power System

SEPS Service-Module Electrical Power System

SEPS Severe Environment Power System

SEPS Solar Electric Propulsion Stage

SEPS Solar Electric Propulsion System

Sep Sci Technol Separation Science and Technology [Journal published in the US]

SEPT Special Environment Powder Diffractometer

Sept September

SEPTA Southeastern Pennsylvania Transportation Authority [US]

SEPUP Science Education for Public Understanding Program [of Lawrence Hall of Science, Berkeley, California, US]

SEPWin Schichten-Erfassungs-Programm unter Windows (Betriebssystem) [Windows (Geological) Strata (Data) Acquisition Program, of NLfB–Niedersächsisches Landesamt für Bodenforschung, Hanover, Germany]

Seq Sequence(r); Sequential [also seq]

seq (sequens) – the following (one)

SEQI Sociedad Espanola de Quimíca Industrial [Spanish Society for Industrial Chemistry]

seqq (sequentia) – the following (ones)

SEQUEL Structured English Query Language

SER Safety Evaluation Report

SER Sandia Engineering Reactor [of Sandia Laboratories, Albuquerque, New Mexico, US]

SER Satellite Equipment Room

SER Selective Electron Reflection

SER Sequence-of-Events Recorder

SER Sequential Events Recorder

SER Seychelles Rupee [Currency]

SER Significant Event Report

SER Single Electron Response

SER Single-Ended Radiant Tube

SER Single-End Recuperative (Burner)

SER Slip-Energy Recovery

SER SNAP (Space Nuclear Auxiliary Power) Experimental Reactor

SER Smooth Endoplasmic Reticulum [also sER] [Cytology]

SER Stable Element Reference

SER State of the Environment Report

SER Surface-Enhanced Raman (Scattering)

SER Symbol Error Rate

SER System Environment Recording

sER Smooth Endoplasmic Reticulum [also SER] [Cytology]

Ser Series [also ser]

Ser (−)-Serine; Seryl [Biochemistry]

Ser Service [also ser]

ser serial

ser series-wound (dynamo)

SERA Socialist Environment and Resources Association [UK]

SERAI Société d'Applications pour l'Industrie [Society for Industrial Applications, Belgium]

SERAPE Simulator Equipment Requirements for Accelerating Procedural Evolution

SERB Shuttle Engineering Review Board [NASA]

SERB Study of the Enhanced Radiation Belt [of NASA]

SERB Systems Engineering Review Board

Serb Serbia(n)

Serb Acad Sci Arts, Bull Serbian Academy of Sciences and Arts, Bulletin

Serb Acad Sci Arts, Glass Serbian Academy of Sciences and Arts, Glass

Serb Acad Sci Arts, Monogr Serbian Academy of Sciences and Arts, Monographs

Serb Acad Sci Arts, Monogr Dep Tech Sci Serbian Academy of Sciences and Arts, Monographs, Department of Technical Sciences

SERC Science and Engineering Research Council [now Engineering and Physical Sciences Research Council, UK]

SERC Solar Energy Research Center [Baghdad, Iraq]

SERC Southeastern Electric Reliability Council

SERC Space Engineering Research Center [of NASA at the University of Cincinnati, Ohio, US]

SERC State Emergency Response Commission [US]

SERC Bull SERC Bulletin [of Science and Engineering Research Council, UK]

SERCNET Science and Engineering Research Council Network [UK]

SERDES Serializer/Deserializer

SERDP Strategic Environmental Research and Development Program [Joint US Program of the Department of Defense, the Department of Energy and the Environmental Protection Agency]

SERDP Inf Bull Strategic Environmental Research and Development Program Information Bulletin [US]

SERENDIP Search for Extraterrestrial Radio Emission from Nearby Developed Intelligent Populations [of University of California at Berkeley, US]

SEREP System Environment Recording and Edit(ing) Program

SEREP System Error Recording Editing Program

SERF Sandia Engineering Reactor Facility [of Sandia National Laboratories, Albuquerque, New Mexico, US]

SERF Special Environmental Research Facility

SERF Special Extensive Routine Functions

SERF Subsurface Environmental Research Facility

SERG Solar Energy Research Group

Serg Sergeant [also Sergt, or Sgt]

SERGE Socially and Ecologically Responsible Geographers [US]

Serg Maj Sergeant Major

Sergt Sergeant [also Serg, or Sgt]

SERI Science and Engineering Research Institute [of Doshisha University, Kyoto, Japan]

SERI Society for the Encouragement of Research and Invention [US]

SERI Solar Energy Research Institute [now National Renewable Energy Laboratory]

SERI TP Solar Energy Research Institute Technical Protocol [also SERI Tech Prot]

SERI TR Solar Energy Research Institute Technical Report [also SERI Tech Rep; now NREL TR]

SERIS State of the Environment Reporting Information System

SERIWA Solar Energy Research Institute of Western Australia

SERIX Swedish Environmental Research Index [of Swedish Council of Environmental Information]

SERJ Supercharged Ejector Ramjet [Aeronautics]

SERL Science and Engineering Research Laboratory [Waseda University, Tokyo, Japan]

SERL Services Electronics Research Laboratory [UK]

SERL Signals Engineering Research Laboratory

SERl Silicon Electronics Research Laboratory [of AT&T Bell Laboratories, Murray Hill, New Jersey, US]

SERMAG Société d'Etudes et de Recherches Magnetiques [French Manufacturer of Magnetic Materials and Products] [also Sermag]

Serol Serological; Serologist; Serology

SERPS Service Propulsion System

SERPS State Earnings-Related Pension [also Serps, UK]

Serr Serration; Serrator [also serr]

SERRS Surface-Enhanced Resonance Raman Scattering

SERS Science and Engineering Research Semester

SERS Selective Equipment Removal System

SERS Shuttle Equipment Record System [of NASA]

SERS Special Education and Rehabilitation Services (Division) [of US Department of Education]

SERS Surface-Enhanced Raman Scattering

SERS Surface-Enhanced Raman Spectroscopy

SERT Single-Electron Rise Time

SERT Society of Electronic and Radio Technicians [UK]

SERT Space Electric Rocket (Engine) Test(s)

SERUG SII (Systems Integrators Inc.) Eastern Regional Users Group [US]

SERV Surface-Effect Rescue Vehicle

Serv Service [also serv]

Servo Servomechanism [also servo]

Servo Servomotor [also servo]

Servs Services

SES School of Engineering and Science

SES Science Ethics Society [US]

SES Scientific Exploration Society [UK]

SES Secondary Electron Spectroscopy

SES Senior Executive Service [of Office of Personnel Management, US]

SES Service Evaluation System

SES Sesone [Herbicide]

SES Shuttle Engineering Simulation [of NASA]

SES Society of Engineering Science [US]

SES Socio-Economic Status

SES Sodium 2,4-Dichlorophenoxyethyl Sulfate

SES Software Engineering Standard

SES Solar Energy Society [US and Canada]

SES Southeastern Section Meeting [of American Physical Society, US]

SES Special Emphasis Study

SES Special Exchange Service

SES Standards Engineering Society [US]

SES Static Excitation System

SES Strategic Engineering Survey

SES Suffield Experimental Station [now Defense Research Establishment Suffield, Canada]

SES Superexcited Electronic States

SES Surface-Effect Ship

SES Surface Environmental Surveillance

SES Swedish Engineers Society [US]

SES System External Storage

SE&S Square Edge and Sound

SESA Society for Environmental Stress Analysis

SESA Society for Experimental Stress Analysis [Name Changed to Society for Experimental Mechanics, US]

SESA Solvent Extraction Spherical Agglomeration

SESAAS Sustaining Engineering Support for Agency-wide Administrative Support

SESC SGN Eurisys Services Corporation [US]

SESC Shuttle Events Sequential Control [of NASA]

SESC Solar Energy Society of Canada

SESC Southeastern State College [Durant, Oklahoma, US]

SESC Space Environment Services Center [of National Oceanic and Atmospheric Administration at Boulder, Colorado, US]

SESCA Scanning Electron Spectroscopy for Chemical Applications

SESCI Solar Energy Society of Canada, Inc.

SESD Scanning Electron Stimulated Desorption

SESdg Square-Edge Siding

SESDA Small Engine Servicing Dealers Association [US]

SESE Secure Echo-Sounding Equipment

SESE Single Entry Single Exit

SESEF Shipboard Electronic Systems Evaluation Facility

SESET Semi-Selective Excitation

SESF Superlattice Extrinsic Stacking Fault [also S-ESF] [Solid-State Physics]

SESI Solar Energy Society of Ireland

SESIBA European Workshop on Surgery-Engineering: Synergy in Biomaterials Applications

SESL Space Environment(al) Simulation Laboratory [of NASA]

SESLP Sequential Explicit Stochastic Linear Programming

SESO Single End Shutoff

SESOC Single End Shutoff Connector

SESOC Surface-Effect Ship for Ocean Commerce

SESOME Service, Sort and Merge

SESP Science and Engineering Student Program [US]

SESP Servicão Especialais para Saúde Pública [Special Public Health Service, Brazil]

SESP Surface Environmental Surveillance Project

SESPA Scientists and Engineers for Social and Political Action [now Science for the People (Group); of American Physical Society, US]

sesqui- prefix meaning "one-and-a-half," often used to indicate a proportion of 2:3 in chemical compounds, e.g., manganese sesquioxide (Mn_2O_3)

SESS Saskatoon Engineering Students Society [Saskatchewan, Canada]

Sess Session [also sess]

SE(S)TMA Sheffield Engineers (Small) Tools Manufacturers Association [UK]

SE SUPP Safety Evaluation Supplement [also SE Supp]

SET Secure Electronic Transaction

SET Self-Extending Translator

SET Set End-Point Titration

SET Single Electron Transfer

SET Single Electron Transistor

SET Single Electron Tunneling

SET Single Exposure Technique

SET Société d'Explosifs Titanite [Titanite Explosives Company, France]

SET Softwave Engineering Technologist; Software Engineering Technology

SET Software Engineering Terminology

SET Software Engineering Toolkit

SET Solar Energy Thermionic Program

SET Solar Energy Thermionics

SET Space Electronics and Telemetry

SET (Wirtschaftsverband) Stahlbau und Energietechnik [Energy Technology and Steel Construction Association, Germany]

SET Standard d'Echange et de Transfer [Information Exchange Standard used in France]

SET Standard Enskog Theory [Physics]

SET Stepped Electrode Transistor

SET Straining Electrode Test

SET Système d'Echange et de Transfer [Information Exchange and Transfer System]

SET Systems Environment Team

SE(T) Single Edge Notch Tension (Specimen) [Mechanical Testing]

SE2 Scientists and Engineers for Secure Energy [also SESE, US]

.SET Driver Set [Lotus 1-2-3 File Name Extension]

.SET Image Settings [Paradox File Name Extension]

.SET Setup File(s) [WordPerfect File Name Extension]

Set Settlement [also set]

Set Settling [also set]

SETA Scottish Egg Trade Association [UK]

SETA Set Arithmetic (Instruction)

SETA Simplified Electronic Tracking Apparatus

SETA Southeastern Telecommunications Association [US]

SETAB Set Tabular Material

SETAC Society of Environmental Toxicology and Chemistry [Canada]

SETAF Southern European Task Force

SETAR Serial Event Time and Recorder

SETB Set Binary (Instruction)

SETBC Society of Engineering Technologists of British Columbia [Canada]

SETC Set Character (Instruction)

SETC Society of Environmental Toxicology and Chemistry [Canada]

SETD Scheduled Estimated Time of Departure

SETE Secretariat for Electronic Test Equipment [of NASA and US Department of Defense]

SETE Société d'Expansion Technique et Economique [Society for Technical and Economic Expansion, France]

SETEC Semiconductor Equipment Technology Center [US]

SETEC Société d'Etudes du Trafic et des Communications [Society for Traffic and Communication Studies, France]

SETEXT Structure Enhanced Text [Internet]

SETF South East Trawl Fishery

SETF STARAN (Stellar Attitude Reference and Navigation) Evaluation and Training Facility

SETG Salmon Enhancement Task Group [Canada]

SETFIA South East Trawl Fishery Industry Association

SETI Search for Extraterrestrial Intelligence (Program) [of NASA]

SeTl Selenium Tellurium (Spin Glass)

SETIM Service des Equipements et des Techniques Instrumentales [Instrumental Techniques and Equipment Service, France]

SETMAC South East Trawl Ministerial Advisory Committee

SETP Society of Experimental Test Pilots [US]

SETR Specific Equipment Type Rating [of National Air Traffic Services, UK]

SETS Solar Energy Thermionic Conversion System

SETU Société d'Etudes et de Travaux pour l'Uranium [Uranium Research Society, France]

SEU Selective Excitation Unit

SEU Single Event Upset

SEU Small End Up

SEU Smallest Executable Unit

SEU Socio-Ecological Union [Russia]

SEU Source Entry Utility

SEU Structure and Evolution of the Universe (Program) [of NASA]

SEUG Screaming Eagles User Group [US] [*Note:* The "Eagle" is a Computer System]

SEUL Support Equipment Utilization List

SE USA South Eastern United States of America [also SE-USA]

SEV Schweizerischer Elektrotechnischer Verein [Swiss Electrotechnical Association]

SEV Special Equipment Vehicle

Sev Severe; Severity [also sev]

SEVAS Secure Voice Access System

SEVEC Society for Educational Visits and Exchanges in Canada

SEW Satellite Early Warning

SEW Sonar Early Warning

SEW Stahl-Eisen-Werkstoffblatt [Materials Data Sheets on Iron and Steel published by the Verein Deutscher Eisenhüttenleute, Germany]

SEW Steady Expansion Wave

SEW Sulzer-Escher Wyss [Swiss Manufacturer of Turbines, Pumps, Ship Propellers, Compressors, Papermaking Machinery, Refrigeration Equipment, Wastewater Treatment Equipment, Air Purification Equipment, etc.]

SEW Surface Electromagnetic Wave (Spectroscopy)

Sew Sewer [also sew]

Sew Sewing [also sew]

SEWM Strong and Electroweak Matter (Conference) [Particle Physics]

SEWP Science and Engineering Workstation Procurement (Contract)

SEWS Satellite Early Warning System

SEWS Surface Electromagnetic Wave Spectroscopy

SEX 1-Acetyloctahydro-3,5,7-Trinitro-1,3,5,7-Tetrazocine [Symbol]

SEXAFS Surface Extended X-Ray Absorption Fine Structure (Spectroscopy)

SEXAFS Surface-Extended X-Ray Absorption Spectroscopy

SEXAFS Synchrotron Extended X-Ray Absorption Fine Structure (Spectroscopy)

SEXAFS UHV Surface Extended X-Ray Absorption Fine Structure–Ultra-High Vacuum (Chamber)

SEXAFS $\chi(k)$ Surface Extended X-Ray Absorption Fine Structure Photoelectron Wavevector Dependent Oscillation

SEY Secondary Electron Yield

Seybold Rep Desktop Publ Seybold Report on Desktop Publishing [US]

Seybold Rep Publ Syst Seybold Report on Publishing Systems [US]

SF Safe

SF Safe, R&QA (Reliability and Quality Assurance) and Protective Services [NASA Kennedy Space Center Directorate, Florida, US]

SF Safety Factor [also S/F]

SF Safety Fuse

SF Sample Function [Statistics]

SF Sandfly Fever [Medicine]

SF San Fernando [California, US]

SF San Fernando [Spain]

SF San Francisco [California, US]

SF San Francisco Operations Office [of US Department of Energy, California]

SF Santa Fe [New Mexico, US]

SF Santa Fé [Argentina]

SF Salt Film (Treatment)

SF Sampled Filter

SF Saturated Flow

SF Scalar Field
SF Scale Factor
SF Scarlet Fever [Medicine]
SF Scattering Factor
SF Schmid Factor [Materials Science]
SF Science Fiction [also sf]
SF Science Frontiers [US]
SF Screw Feed(er)
SF Sea Floor
SF Secondary Flow
SF Secondary Fuel
SF Seed Fund(ing)
SF Select Frequency
SF Selection Filter
SF Self-Feeding
SF Semifinished; Semifinishing [Metallurgy]
SF Semi-Fixed
SF Semifriable
SF Separated Flow
SF Separate Function
SF Sequential File
SF Service Factor
SF Set Flag (Function) [Computers]
SF Shape Factor
SF Shear Fault
SF Shear(ing) Flow
SF Shear Force
SF Shift Forward
SF Shock Front [Physics]
SF Shop Foreman
SF Short Format
SF Shrink Forming
SF Shift and Function (Keys)
SF Shuttle Facility [of NASA]
SF Side Frequency
SF Signal Flare
SF Signal(ing) Frequency [also sf]
SF Sign Flag [Computers]
SF Significant Figure
SF Silage Film [Agriculture]
SF Silica Fiber
SF Single Feeder
SF Single Frequency [also sf]
SF Sink-Float (Separation) [Metallurgy]
SF Sinking Fund [Industrial Engineering]
SF Sintering Furnace [Metallurgy]
SF Sioux Falls [South Dakota, US]
SF Site Factor [Hydropower Plants]
SF Size Factor
SF Size Fraction
SF Skin Friction [Fluid Mechanics]
SF Skip Flag [Computers]
SF Slave Fermion [Particle Physics]

SF Sliding Filter
SF Sliding Friction
SF Slip Flow
SF Slow-Fast (Wave)
SF Soap Film
SF Sodium Fusion
SF Solar Flare [Astrophysics]
SF Solar Furnace
SF Solid Film
SF Solid Fuel
SF Source Function [Astrophysics]
SF Space Flight
SF Space Foundation [Houston, Texas, US]
SF Spatter-Free [Metallurgy]
SF Spatter-Free Applications (of Welding Electrodes)
SF Spectral Factorization [Mathematics]
SF Spectrofluorometer
SF Spent (Nuclear) Fuel
SF Spindle Fiber [Cytology]
SF Spin Flip [Solid-State Physics]
SF Spin Fluctuation [Solid-State Physics]
SF Spontaneous Fission [Nuclear Physics]
SF Spotface(d); Spotfacing
SF Spotted Fever [Medicine]
SF Spotting Fluid
SF Spray Forming
SF Square Foot [Unit]
SF Stacking Fault [Crystallography]
SF Stagg Field (Reactor) [Chicago, Illinois, US]
SF Standard Form
SF Star Field
SF State Forest [US]
SF Static Field
SF Static Firing [Aeronautics]
SF Static Friction
SF Steady Flow
SF Steam Flow
SF Stereoscopic Fluoroscopy
SF Steric Factor [Chemistry]
SF Stochastic Flow
SF Stopping Factor
SF Store and Forward [also S/F]
SF Straightforward
SF Strain Figure [Metallurgy]
SF Stream Function
SF Stress Field
SF Stress-Free
SF Strong-Focusing
SF Strong Force
SF Strontium Ferrite
SF Structure–Fine
SF Structural Foam
SF Structural Formula [Chemistry]

SF Structure Factor [Solid-State Physics]

SF Subcontractor Furnished

SF Subfile

SF Subject Field

SF Sun Furnace

SF Supercritical Flow

SF Supercritical Fluid

SF Superfluid(ity)

SF Supersonic Flight

SF Supersonic Frequency

SF Suprafacial (Process)

SF Surface Finisher; Surface Finishing

SF Surface Foot (or Feet) [Unit Equal to One Square Foot] [also sf]

SF Surface Force [Mechanics]

SF Swinging Field

SF Swiss-French [also sf] [Keyboard Code]

SF$_6$ Sodium Hexafluoride

S/F Safety Factor [also SF]

S/F Single Flow

S/F Store and Forward [also SF]

S-F Superconductor-Ferromagnet (Heterostructure)

Sf Spodoptera frugiperda [Biochemistry]

Sf Suriname Florin (or Guilder) [also Sf; Currency]

S(f) Fiber-Reinforced S-Glass [Symbol]

S(f) Spectrum [Symbol]

S(f) Frequency Domain [Symbol]

S(f) Spectral Density [Symbol]

SFA Scientific Film Association [UK]

SFA Segment Frequency Algorithm

SFA Semifriable Abrasive

SFA Semifriable Alumina

SFA Short-Field Aircraft

SFA Single Failure Analysis

SFA Société Française d'Acoustique [French Acoustical Society]

SFA Source Function Analysis

SFA Stress Field-Induced (Magnetic) Anisotropy [Solid-State Physics]

SFA Sunfinder Assembly

SFA Surface Force Apparatus

SFA Swedish Foundry Association

S&FA Shipping and Forwarding Agent

SFACA Solid Fuel Advisory Council of America [US]

SFACI Software Flight Article Configuration Inspection

SFAI Steel Furnace Association of India

SFAPS Spaceflight Acceleration Profile Simulator [of NASA]

SFAR Special Federal Aviation Regulation [US]

SFAR System Failure Analysis Report

SF$_6$-Ar Sulfur Hexafluoride-Argon (Gas Mixture)

SFAS Safety Feature Activation System

SFASC Stephen F. Austin State College [Nacogdoches, Texas, US]

SFAX Secure Facsimile

SFB San Francisco Bay [California, US]

SFB Semiconductor Functional Block

SFB Sender Freies Berlin [Radio Free Berlin, Germany]

SFB Sensor Feedback

SFB *(Sonderforschungsbereich)* – German for "Special Research Field"

SFB State Feedback

SFBAEC San Francisco Bay Area Engineering Council [California, US]

SFBC Single Flexible Boundary Condition

SFC Sectored File Controller

SFC Sector-Focused Cyclotron

SFC Secure Flight Communication

SFC Selection Filter Control

SFC Selector File Channel

SFC Sequential Function Chart

SFC Serial File Copy

SFC Sergeant, First Class [also Sfc]

SFC Single-Fiber-Composite (Test) [Materials Science]

SFC Single Filament Composite

SFC Slow Furnace Cooling [Metallurgy]

SFC Société Française de Céramique [French Ceramic Society]

SFC Société Française de Chimie [French Chemical Society]

SFC Society of Flavour Chemists [US]

SFC Solar Forecast Center [of US Air Force]

SFC Special Facilities Contract

SFC Special Forces Commander

SFC Specific Fuel Consumption

SFC State Forestry Commission [South Carolina, US]

SFC Supercritical Fluid Chromatography

SFC System File Checker

Sfc Surface [also sfc]

Sfc Sergeant, first class [also SFC]

SF-CD Surface Finishing CD-ROM (Compact-Disk Read-Only Memory) [of ASM International, US and Insitute of Materials, UK]

SFC/FT-IR Supercritical Fluid Chromatography/Fourier Transform Infrared Spectroscopy [also SFC/FTIR]

SFCG Spectrum Fatigue Crack Growth [Mechanics]

SFC-GC Supercritical Fluid Chromatography/Gas Chromatography [also SFC/GC]

SFCI Spirit of the Future Creative Institute [US]

SFC-IR Supercritical Fluid Chromatography–Infrared

SFCL Superconducting Fault Current Limiter

SFC-LC Supercritical Fluid Chromatography/Liquid Chromatography [also SFC/LC]

SFC-MS Supercritical Fluid Chromatography/Mass Spectroscopy [also SFC/MS]

SF$_6$/CO$_2$ Sodium Hexafluoride/Carbon Dioxide (Gas Mixture)

SFCS Secondary Flow Control System

SFCS Slats and Flaps Control System [Aerospace]

SFCS Survival Flight Control System

SFC-SFE Supercritical Fluid Chromatography/Supercritical Fluid Extraction [also SFC/SFE]

SFCVB San Francisco Convention and Visitors Bureau [California, US]

SFD Start Frame Delimiter

SFD Sudden Frequency Deviation

SFD Switching Field Distribution [Recording Media]

SFD System Function Description

SFDA Sulfofluorescein Diacetate, Sodium Salt

SFDR Spurious-Free Dynamic Range

SFDS System Functional Design Specification

SFDT Site Format Dump Tape

SFDU Standard Formatted Data Unit

SFE Smart Front End

SFE Société Financière Europénne [European Financial Society]

SFE Société Française d'Energie Nucléaire [French Society for Nuclear Energy]

SFE Société Française des Explosifs [French Society for Explosives]

SFE Society of Fire Engineers

SFE Solar-Flare Effect [Astrophysics]

SFE Stacking-Fault Energy [Crystallography]

SFE Supercritical Fluid Extraction (Process) [Chemical Engineering]

S/Fe Sulfur-to-Iron (Ratio)

SFEA Space and Flight Equipment Association [US]

SFEC San Francisco Engineering Council [California, US]

SFE-GC Supercritical Fluid Extraction/Gas Chromatography [also GFE/GC]

SFE-IR Supercritical Fluid Extraction/Infrared (Spectroscopy) [also GFE/IR]

SFEM Southern Farm Equipment Manufacturers [US]

SFEN Société Française d'Energie Nucléaire [French Society for Nuclear Energy]

SFEP Self-Consistent Electron Pairs

SFE-RC Supercritical Fluid Extraction and Retrograde Condensation [also GFE/RC]

SFERT Spinning Satellite for Electric Rocket Test

SFES Southern Forest Experiment Station [of US Department of Agriculture in Florida]

SFE-SFC Supercritical Fluid Extraction/Supercritical Fluid Chromatography [also SFE/SFC]

SFET Standard Free Energy of Transfer

SFEX Solar Flare X-Ray Polarimeter

SFF Self-Forming Fragment

SFF Skin-Friction Factor [Fluid Mechanics]

SFF Solar Forecast Facility [of US Air Force]

SFF Solid Freeform Fabrication

SFF Spirit of the Future Foundation [now Spirit of the Future Creative Institute, US]

SFF Standard File Format

SFFT Superconducting Flux Flow Transistor

SFG Signal Frequency Generator

SFG Sulfurless Fine Grain (Powder)

SFG Sum Frequency Generation

SFHA Scottish Federation of Housing Associations

SFHEA Scottish Further and Higher Education Association

SFHEPC Société Franco-Hellénique d'Explosifs et Produits Chimique [French-Hellenic Society for Explosives and Chemical Products, Greece]

SFI Sector Facility Indexing (Project) [of US Environmental Protection Agency]

SFI Specification Foundation International [US]

SFI Stress-Free Interface [Materials Science]

SFI Support for Innovation

Sfi Streptomyces fimbriatus [Microbiology]

Sfi Sulfide

SFIA Surplus Facilities Inventory and Assessment Plan

SFIEC Société Franco-Italienne d'Explosifs Cheddite [French-Italian Society for Cheddite Explosives]

SFIO Strontium Iron Indium Oxide

SFIR Specific Force Integrating Receiver

SFIT Swiss Federal Institute of Technology [also Ecole Polytechnique Fédérale, or Eidgenössische Technische Hochschule]

SFITV Société Française des Ingénieurs et Techniciens du Vide [French Society of Vacuum Engineers and Technologists]

SF Keys Shift and Function Keys

SFK Step Fixture Key

SFL Safe Fatigue Life

SFL Saskatchewan Federation of Labour [Canada]

SFL Secondary Freon Loop

SFL Solid-Film Lubrication

SFL Substrate-Fed Logic

SFL Suriname Florin (or Guilder) [Currency]

SFLRP Society of Federal Labor Relations Professionals [US]

SFM Scanning Force Microscope; Scanning Force Microscopy

SFM Société Française de Métallurgie [French Society of Metallurgy] [now Société Française de Métallurgie et de Matériaux, France]

SFM Special Functions Module

SFM Split-Field Motor

SFM Sum Frequency Mixing

SFM Superferromagnet(ic); Superferromagnetics; Superferromagnetism

SFM Surface Feet per Minute [also SFPM, sfm or sfpm]

SFM Svenska Föreningen för Materialteknik [Swedish Society for Materials Technology]

SFM Swinging Field Magnetization

SFM Switching-Mode Frequency Multiplier

SFMA Scottish Furniture Manufacturers Association

SFMC Société Française de Minéralogie et de Crystallographie [French Society of Mineralogy and Crystallography]

SFMCTG Société Française de des Munitions de Chasse, de Tir et de Guerre [French Society for Hunting, Shooting and Military Ammunition]

SFME Société Française de Microscopie Electronique [French Electron Microscopy Society]

SFME Storable Fluid Management Experiment

SFMM Société Française de Métallurgie et de Matériaux [French Society for Metallurgy and Materials] [also SF2M]

SFMR Stepped Frequency Microwave Radiometer

SFMRB Surface Finishing Market Research Board [of American Electroplaters and Surface Finishers Society]

SFM/STM Scanning Force Microscopy/Scanning Tunnneling Microscopy

SFMT Shuttle Facility Modification Team [of NASA]

SFMT Svenska Föreningen för Materialteknik [Swedish Society for Materials Technology]

SFN Stereotactic and Functional Neurosurgery [International Journal]

SF$_6$-N$_2$ Sulfur Hexafluoride-Nitrogen (Gas Mixture)

SF$_6$-N$_2$O Sulfur Hexafluoride-Nitrous Oxide (Gas Mixture)

SFNA Stabilized Fuming Nitric Acid

SFNL Step-Function Nonradiative Lifetime

SFO San Francisco International Airport [California, US]

SFO Service Fuel Oil

SFO Space Flight Operations

SFO Stratospheric Fall-Out

SFO Strontium Ferrite [SrFe$_{12}$O$_{19}$]

SF$_6$-O$_2$ Sulfur Hexafluoride-Oxygen (Gas Mixture)

Sfo Sulfone

SF/O Screen Formatter/Optimizer

SFOC Space Flight Operations Center

SFOC Space Flight Operations Complex

SFOD San Francisco Ordnance District [Oakland, California, US]

SFOF Spaceflight Operations Facility [of NASA]

SFOM Segregation Figures of Merit

SFOM Shuttle Flight Operations Manual [of NASA]

SFOP Safety Operating Procedure [of NASA Kennedy Space Center, Florida, US]

SFORD Single-Frequency Off Resonance Decoupling

SFP San Francisco Press [California, US]

SFP Science Foundation of Physics [Australia]

SFP Screen Filtration Pressure

SFP Second Focal Point

SFP Security Filter Processor

SFP Shop Floor Programming

SFP Single Failure Point

SFP Slack Frame Program

SFP Société Française de Physique [French Physical Society]

SFP Société Française de Production [French Production Society]

SFP Solar Flare Proton

SFP Spent (Nuclear) Fuel Pit

SFP Spent (Nuclear) Fuel Pool

SFP Standard Front Page

SFP Summary Flight Plan

S4P Symmetric Four-Point (Bending) Loading [Mechanical Testing]

SFPA Single Failure Point Analysis

SFPA Southern Forest Products Association [US]

SFPD San Franscisco Police Department [California, US]

SFPE Society of Fire Protection Engineers [US]

SFPE Bull SFPE Bulletin [of Society of Fire Protection Engineers, US]

SFPL San Francisco Public Library [California, US]

SFPM Surface Feet per Minute [also SFM, sfm, or sfpm]

SFPPL Short Form Provisioning Parts List

SFPS Single Failure Point Summary

SFPS Society of Fire Protection Specialists

SFPT Société Française de Photogrammétrie et des Télédétection [French Society of Photogrammetry and Remote Sensing]

SFPT Society of Fire Protection Technicians [US]

SFQ Single Flux Quantum [Electronics]

SFQG Single Flux Quantum Gate [Electronics]

SFQL Structured Full-Text Query Language

SFR Shunt Field Rheostat

SFR Signal Frequency Receive

SFr Swiss Franc [Currency of Switzerland and Liechtenstein] [also SFR]

SFRA Science Fiction Research Association [US]

SFRC School of Forest Resources and Conservation [of University of Florida, Gainesville, US]

SFRC Steel Fiber Reinforced Concrete

SFRCS Steam and Feedwater Rupture Control System

SFRT Short-Fiber Reinforced Thermoplastic(s) [also SFRTP]

SFS Saybold Furol Seconds [Unit]

SFS Scandium Iron Silicide (Superconductor)

SFS Sequential Forward Selection

SFS Shuttle Flight Status [NASA]

SFS Singapore Finishing Society

SFS Society of Food Service Systems [US]

SFS Sodium Formaldehyde Sulfoxylate

SFS Software Facilities and Standards

SFS Sorption Filter System

SFS Specimen Feeder System

SFS Striction Free Silder [Magnetic Disk Drives]

SFS Strong-Focusing Synchrotron

SFS Strontium Iron Antimonate

SFS Suomen Fyysikkoseura [Finnish Physical Society]

SFS Svenska Fysikersamfundet [Swedish Physical Society]

SFS Symbolic File Support

SFS System File Server

S4S Surfaced on Four Sides [Lumber]

SFSA Scottish Field Studies Association

SFSA Steel Founders Society of America [US]

SFSC San Francisco State College [California, US]

SFSC Shunt Feedback Schottky Clamped

SFSD Star Field Scanning Device

SFSJ Surface Finishing Society of Japan

SF-SOO Kennedy Space Center–Safety Operations Office [of NASA]

SFSP Spent (Nuclear) Fuel Storage Pool

SFSR Soviet Federated Socialist Republic

SFSS Satellite Field Service Station

SFST Society of Fiber Science and Technology [Japan]

SF Stud Science Fiction Studies [Journal of the Science Fiction Research Association, US]

SFSU San Francisco State University [California, US]

SFSU Signal Frequency Signaling Unit

SFSV Sandfly Fever Sicilian Virus

SFT Saab Friction Tester

SFT Simulated Flight Test

SFT Société Française des Télécommunications [French Telecommunications Society]

SFT Société Française des Thermiciens [French Society of Heating Engineers]

SFT Stacking-Fault Tetrahedron [Crystallography]

SFT Static Firing Test [Aeronautics]

SFT Strontium Iron Tantalate

SFT Super-Fast Train

SFT Symposium on Fusion Technology

SFT System Fault Tolerance

Sft Shaft [also sft]

Sft Shift [also sft]

Sft Software [also sft]

sft soft

SFTA Structural Fatigue Test Article

SFTE Society of Flight Test Engineers [US]

SFTF Static Firing Test Facility [Aeronautics]

SF2M Société Française de Métallurgie et de Matériaux [French Society for Metallurgy and Materials] [also SFMM; formerly Société Française de Métallurgie]

SF2M-SMT Société Française de Métallurgie et de Matériaux–Surface Modication Technologies [International Conference of the French Society for Metallurgy and Materials]

SFTP Science for the People (Group) [of American Physical Society, US]

SFTR Shift Register [Computers]

Sftr Shipfitter

Sftr Shopfitter

SFTS Standard Frequency and Time Signals

SFTU Somali Federation of Trade Unions

Sftwd Softwood [also sftwd]

SFTWE Software [also Sftwe]

SFU Simon Fraser University [Burnaby, British Columbia, Canada]

SFU Space Flyer Unit

SFU Special Function Unit

SFU Standard Firing Unit

SFU Status Fill-in Unit

SFU-R Space Flyer Unit–Retrieval

SFV Sandfly Fever Virus

SFV San Fernando Valley [California, US]

SFV Saybolt Furol Viscosimeter

SFV Semliki Forest Virus

SFV Société Française du Vide [French Vacuum Society]

SFVSC San Fernando Valley State College [Northridge, California, US]

SFW Software [also Sfw]

SFWE Static Feed Water Electrolysis

SFWEM Static Feed Water Electrolysis Module

SFWMD South Florida Water Management District [US]

SFX Sound Effects

SFXD Semi-Fixed [also Sfxd, or sfxd]

SFXTL (The Great) San Francisco Crystal Fair [California, US]

SFXU St. Francis Xavier University [Antigonish, Nova Scotia, Canada]

SG Safety Glass

SG Safety Guide

SG Salivary Gland [Anatomy]

SG Samarium Gallate

SG San German [Puerto Rico]

SG Saw Gummer; Saw Gumming

SG Sawtooth Generator

SG Scanning Gate

SG Screen Grid [Electronics]

SG Screw Gauge

SG Seabird Group [of Royal Society for the Protection of Birds, UK]

SG Sebaceous Gland [Anatomy]

SG Secretary General

SG Seeded-Gel (Abrasive Grain)

SG Semigroup [Mathematics]

SG Senior Grade [also sg] [Military]

SG Set Gate

SG Signal Generation; Signal Generator

SG Signal Ground

SG Silica Gel

SG Silicate Glass

SG Silicon-Dioxide Glazing

SG Silicon Germanate

SG Singapore [ISO Code]

SG Silver Glass

SG Single Girder (Crane)

SG Single Groove (Insulator)

SG Skin Graft(ing) [Medicine]

SG Slide Gauge

SG Slip Gauge

SG Smoke Generator

SG Snow Grains [Meteorology]

SG Society of Genealogists [UK]

SG Sol-Gel (Process) [Physical Chemistry]

SG Solicitor General

SG Solution Growth

SG Solution of Glucose

SG Space Group [Crystallography]

SG Space-to-Ground

SG Spanish Guinea [now Equatorial Guinea]

SG Spark Gap

SG Specialist in Gunnery

SG Specific Gravity [also sg]

SG Spheroidal Graphite [Metallurgy]

SG Spinel Ganglion [Anatomy]

SG Spin Glass [Solid-State Physics]

SG Spray Gun

SG Spun Glass

SG Spur Gear(ing)

SG Standard Gauge

SG Standing Group

SG Steel Girder

SG Steam Generator

SG Stereoscopic Graphics

SG Stern-Gerlach (Experiment) [Atomic Physics]

SG Strain Gauge [also SG]

SG Strategic Grant

SG Stratigraphic; Stratigrapher; Stratigraphy

SG Structural Geologist; Structural Geology

SG Study Group

SG Study Guide

SG Subgroup [Mathematics]

SG Subsurface Geology

SG Supergiant (Star) [Astronomy]

SG Supergranule

SG Supergroup [Telecommunications]

SG Suppressor Grid

SG Surface Grinder; Surface Grinding

SG Surgeon General (of the United States)

SG Swiss-German [Keyboard Code] [also sg]

SG Symbol Generator

SG Synchronous Generator

SG System Gain

SG System Generation

SG Water Soluble Granules [Chemistry]

s^{-1}G^{-1} one per second per Gauss [also $1/(s \cdot G)$]

2G Second Generation (Computer)

S/G Solid/Gas Ratio

S/G Strain Gage [also SG]

S-G Sonken-Galamba (Corporation) [US]

S-G Symington-Gould (Corporation) [US]

.sg Singapore [Country Code/Domain Name]

SGA Schweizerische Gesellschaft für Automatik [Swiss Society for Automation]

SGA Scottish Glass Association

SGA Sea Grant Association [US]

SGA Shared Global Area

SGA Society for Geology Applied to Mineral Deposits [Germany]

SGA Sod Growers Association [US]

SGA Sol-Gel by Acid Catalyst [Physical Chemistry]

SGA Soluble-Gas Atomization [Metallurgy]

SGA Specific Gravity Analysis; Specific Gravity Analyzer

SGA Systems Global Area

SG&A Selling, General and Administrative (Expenses)

SGAA Stained Glass Association of America [US]

SGAE Studiengesellschaft für Atomenergie [Nuclear Energy Society, Austria]

SGA of M-A Sod Growers Association of Mid-America [US]

SGAP Society for Growing Australian Plants

SGAT Study Group on Accounting Terminology

SGAUA Scitex Graphic Arts Users Group

SGB Schweizerischer Gewerkschaftsbund [Federation of Swiss Trade Unions]

SGB Sol-Gel by Base-Catalyst [Physical Chemistry]

SGBD Secondary Grain Boundary Dislocation [Materials Science]

SGC Schweizer Gesellschaft für Chronometrie [Swiss Society of Chronometry]

SGC Second Generation Computer [also 2GC]

SGC Soft Gelatin Capsule(s) [Pharmacology]

SGC Sol-Gel Coating

SGC Solicitor General of Canada

SGC Stabilized Ground Cloud

SGC Standard Gas Cycle

SGC Standard Geographical Classification

SGC Superior Geocentric Conjunction [Astronomy]

SGC Supergroup Connector [Telecommunications]

SGC Swept Gain Control

2GC Second Generation Computer [also SGC]

SGCAS Study Group on Certification of Automatic Systems

SGCC Safety Glazing Certification Council [US]

SGCD Society of Glass and Ceramic Decorators [US]

SGCF SNAP (Space Nuclear Auxiliary Power) Generalized Critical Facility

SGCI Schweizerische Gesellschaft für die Chemische Industrie [Swiss Society for the Chemical Industry]

S-GCOS Space-Based Global Change Observation System

SGCS Second Generation Computer System [also 2GCS]

SGD Seafloor Geosciences Division [of Naval Ocean Research and Development Activity, US]

SGD Self-Generating Dictionary

SGD Signal Ground

SGD Singapore Dollar [also SG$; Currency]

SGD Society of Glass Decorators [now Society of Glass and Ceramic Decorators, US]

SGD Straight Gelatin Dynamite

sgd signed

SGDE Sustainable Gross Domestic Expenditure

SGDF Supergroup Distribution Frame [Telecommunications]

SGDP Sustainable Gross Domestic Product

SGDT Store Global Descriptor Table

SGE Secondary Group Equipment

SGE Starch Gel Electrophoresis

SGEC Société Générale d'Explosifs Cheddite [Cheddite Explosives Society]

SGEM Società Generale di Esplosivi e Munizione [Explosives and Ammunition Society, Italy]

SGEN Signal Generator [also Sgen]

SGEN System Generator [also Sgen]

SGER Small Grants for Exploratory Research (Program) [of National Science Foundation, US]

S Ger Southern Germany; South German

SGEMP System Generated Electromagnetic Pulse

SGF Sequence Generating Function

SGF Société du Gaz de France [French Gas Society]

SGF Société Générale de Financement [General Financing Society, Canada]

SGF Solution Growth Facility

SGF Standard Geopotential Feet

SGF Strong Geometrically Frustrated

SGF Supplementary Ground Field

SGFM Surface Green's Function Matching [Physics]

SGFP Steam Generator Feed Pump

SGG Small Galaxy Groups (Meeting)

SGGG Strontium Gadolinium Gallium Garnet

SGGG Strontium Gadolinium Gallium Germanate

SGGG:YIG Strontium Gadolinium Gallium Garnet Doped Yttrium Iron Garnet

SGHW Steam-Generating Heavy-Water (Reactor) [of UK Atomic Energy Authority]

SGHWR Steam-Generating Heavy-Water (Moderated) Reactor [also SGHR]

SGI Silicon Graphics, Inc. [US]

SGI Swedish Geotechnical Institute

SGID State Geographic Information Database [US]

SGIMC Società Generale per l'Industria Mineraria e Chimica [Mineral and Chemical Industries Society, Italy]

SG Iron Spheroidal-Graphite (Cast) Iron [also SG iron]

SGIS Steam Generator Isolation Signal

SGIS Student Guidance Information Center

SGIT Special Group Inclusive Tour

SGJP Satellite Graphic Job Processor

SGL Society of Gas Lighting [US]

SGL Surface Gridding Library

2GL Second Generation Language

Sgl Signal [also sgl]

sgl single

S-Glass High-Strength, High-Modulus, High-Temperature Fiberglass (for Demanding Applications) [also S-glass]

S Gld Suriname Guilder [Currency]

SGLS Space-Ground Link Station

SGLS Space-to-Ground Link System; Space-to-Ground Link Subsystem

SGLSY Space-Ground Link System

SGM Schweizerische Gesellschaft für Mikrotechnik [Swiss Society for Microtechnology]

SGM Schweizerischer Verband der Grosshändler und Importeure der Motorfahrzeugbranche [Swiss Association of Automotive Wholesalers and Importers]

SGM Shaded Graphics Modeling

SGM Shape Group Method

SGM Ship-to-Ground Missile

SGM Société Générale des Minérais [General Mineral Company, Zaire]

SGM Society for General Microbiology [UK]

SGM Spark Gap Modulation

SGMC Standing Group Meteorological Committee [now NATO Military Committee Meteorological Group]

SGML Standard(ized) Generalized Mark-Up Language [Internet]

SGML Study Group for Mathematical Learning [US]

SGMP Simple Gateway Monitoring Protocol

SGMT Simulated Greenwich Mean Time [Astronomy]

SGN Scan Gate Number

SGN Sign (of Argument) (Function) [Programming]

Sgn Sign [also sgn]

sgn signum function (in mathematics) [Symbol]

SGNMOS Screen-Grid N-Channel Metal-Oxide Semiconductor

Sgnr Signature [also sgnr]

SGO Samarium Gallium Oxide (or Samarium Gallate)

SGO Schweizerische Gesellschaft für Oberflächentechnologie [Swiss Society for Surface Technology]

SGO Semigroup of Operations [Mathematics]

SGO Silicon Germanium Oxide

SGOBE Self-Consistent Group Orbital and Bond Electronegativity

SGOR Solution Gas-Oil Ratio

SGOS Shuttle Ground Operations Simulation; Shuttle Ground Operations Simulator [of NASA]

SGOS Silicon Gate Oxide Semiconductor

SGOT Serum Glutamine (or Glutamic) Oxalacetic Transaminase [Biochemistry]

SGP Shell Gasification Process

SGP Single Ground Point

SGP Society of General Physiologists [US]

SGP Southern Great Plains [US]

SGP Strain Gauge Package

SGP Strategic Grants Program [Canada]

SGP Supergroup [also Sgp] [Telecommunications]

SGPA Stained Glass Professionals Association [US]

SGPF Spencer's Gulf Prawn Fishery [South Australia]

SGPT Serum Glutamate Pyruvate Transaminase; Serum Glutamic Pyruvic Transaminase [Biochemistry]

SGR Self-Generation Reactor

SGR Self-Generation Recycle

SGR Set Graphics Rendition

SGR Short Growth Rate

SGR Sodium Graphite Reactor

SGR Steam Gas Recycling

SGRAM Synchronous Graphics Random-Access Memory

SGRCA Sodium Graphite Reactor Critical Assembly [US]

SGRP Special Guard-Ring Position

SGRPA Spherical Grid Retarding Potential Analyzer

SGS Saskatchewan Geological Society [Canada]

SGS Scene Generation System

SGS Self-Governing System

SGS Silver Gallium Diselenide (Semiconductor)

SGS Single-Green-Silk-Covered (Electric Wire)

SGS Small Grain Size [Metallurgy]

SGS Society for General Systems

SGS Steep Glide Slope

SGS Strain-Gauge System

SGS Strontium Gallium Antimonate

SGS Sveriges Geofysika Selskap [Swedish Geophysical Society]

SGS Symbol Generator and Storage

SGSC Strain Gage Signal Conditioner

SGSE Steady-Gradient Spin-Echo [Physics]

SGSE Supercritical Gas Solvent Extraction

SGSI Symposium on Gas-Surface Interactions

SGSMP Schweizerische Gesellschaft für Strahlenbiologie und Medizinische Physik [Swiss Society for Radiobiology and Medical Physics]

SGSP Single-Groove, Single-Petticoat (Insulator)

SGSP Society for Glass Science and Practices [US]

SGSP Sol-Gel Microsphere Pelletization

SGSR School of Graduate Studies and Research

SGSR Scientific Group for Space Research [Greece]

SGSR Society for General Systems Research [now International Society for the Systems Sciences, US]

SGSTS Space-to-Ground Subsystem Test Set

SGSV/DV Steam Generator Stop Valve Dump Valve

SGT Segment Table

SGT Silicon Gate Transistor

SGT Society of Glass Technology [UK]

SGT Special Gas Taper Thread [Symbol]

SGT Strain Gauge Technology

SGT Strain Gauge Transducer

SGT Strontium Gallium Tantalate

Sgt Sergeant [also SGT]

SGTE Scientific Group Thermodata Europe (Solution and Pure Substance Database)

SGTIM Study Group on Topoclimatological Investigation and Mapping [Poland]

Sgt Maj Sergeant Major

SGTR Steam Generator Test Rig

SGTS Standby Gas Treatment System

SGU Schweizerische Gesellschaft für Umweltschutz [Swiss Society for Environmental Protection]

SGU Sveriges Geologiska Undersökning [Swedish Geological Survey]

SGUR Steering Group on Uranium Resources

SGW Silanized Glass Wool

SGW Simulated Groundwater

SGWCPFA Spencer's Gulf and West Coast Prawn Fishermen's Association [South Australia]

SGX Selector Group Matrix

SGZ Surface Ground Zero

SH Safety Helmet

SH Saint Helena [South Atlantic Island]

SH Saint Helena [ISO Code]

SH Sample and Hold [also S&H, or S/H]

SH Sample Holder

SH Scheme of Hueckel [Physical Chemistry]

SH Scleroscope Hardness

SH Scratch Hardness

SH Screw Head

SH Secondary Heating [Metallurgy]

SH Second Harmonic [also 2H]

SH Section Header; Section Heading

SH Self-Heating

SH Send Hub

SH Sensible Heat

SH Serum Hepatitis [Medicine]

SH Service Hours

SH Session Handler

SH Shock Hardening [Metallurgy]

SH Shut Height (of Presses)

SH Siemens-Hell (Printer)

SH Single Heterostructure

SH Simulation of Hypothetical Nucleus (Method) [Metallurgy]

SH Socket Head (Screw)

SH Södertörns Högskola [Södertörns Institute of Technology, Sweden]

SH Solid Height (of Springs) [Symbol]

SH Southern Hemisphere

SH Space Hardware

SH Specific Head [Fluid Mechanics]

SH Specific Heat

SH Specific Humidity [Meteorology]

SH Specimen Holder

SH Spectroheliograph(y)

SH Spectroheliometer; Spectroheliometry

SH Spherical Harmonics

SH Spitz-Holter (Valve)

SH Squashhead [Explosive]

SH Stack Height (of Write Head) [Magnetic Storage Media]

SH Staphylococcal Hemolysin [Immunology]

SH Start of Heading

SH Static Head [Fluid Mechanics]

SH Steinhart and Hart (Thermistor)

SH Steric Hindrance [Chemistry]

SH Strain Harden(ing) [Metallurgy]

SH Stripe Height [Magnetic Recording Media]

SH Subject Heading

SH Suction Head

SH Sulfhydryl (Group) [Symbol]

SH Superheat(er); Superheating [Thermodynamics]

SH Surface Harden(ing) [Metallurgy]

SH Swift-Hohenberg (Equation)

SH Switch Handler

2H Second Harmonic [also SH]

S&H Safety and Health

S&H Sample and Hold [also S/H, or SH]

S/H Sample/Hold

S(H) Magnetothermopower [Symbol]

SH$_2$ Supercritical Hydrogen

.SH Bourne Shell Script File [File Name Extension] [also .sh]

Sh Share [also sh]

Sh Sheet [also sh]

Sh Shell [also sh]

Sh Sherwood Number (of Mass Transfer) [Symbol]

Sh Shield [also sh]

Sh Shift [also sh]

Sh Shigella [Genus of Enterobacteria]

Sh Shilling [Currency of Kenya]

Sh Shock [also sh]

Sh Shore [also sh]

Sh Shot [also sh]

Sh Shoulder [also sh]

Sh Shower [also sh]

Sh Shunt [also sh]

S7/h6 Medium Drive Fit; Basic Shaft System [ISO Symbol]

sh hyperbolic sine [Symbol]

sh sheep [Immunochemistry]

σ(H) magnetic field dependent conductivity [Symbol]

SHA Salvia Horminum Agglutinin [Immunology]

SHA Secure Hash Algorithm [of National Security Agency, US]

SHA Sidereal Hour Angle [Astronomy]

SHA Society for Historical Archeology [US]

SHA Sodium Hydroxide Addition

SHA Software Houses Association [UK]

SHA Solid Homogeneous Assembly

SHA Special Handling Area

SHA Spherical-Harmonic Analysis

SHA System Hazard Analysis

SHAB Soft and Hard Acids and Bases

Sh Abs Shock Absorber

SHAC Society for the History of Alchemy and Chemistry [UK]

SHAC Superhigh Activity Catalyst

SHADCOM Shipping Advisory Committee

SHADW Shadow Printing (Code) [also Shadw, or shadw]

SHAEF Supreme Headquarters, Allied Expeditionary Forces

SHAG Simplified High Accuracy Guidance [of Honeywell Inc., US]

SHALTA Skin, Hide and Leather Traders Association [UK]

Shanghai Iron Steel Res Inst Tech Rep Shanghai Iron and Steel Research Institute Technical Report [PR China]

SHAP Sintered Hydroxyapatite

SHAPA Solids Handling and Processing Association [UK]

SHAPE Supreme Headquarters (of the) Allied Powers (in) Europe [of NATO]

.SHAR Self-Expanding Shell Archive [UNIX File Name Extension] [also .shar]

SHARE System for Heat and Radiation Energy

SHARES Shared Acquisition and Retention System

SHARP Sensitive, Homogeneous and Resolved Peaks

SHARP SHIPS (Bureau of Ships) Analysis and Retrieval Project [US Navy]

SHARP Small Hydro Analysis and Reporting Program [of Canada Center for Mineral and Energy Technology, Natural Resources Canada]

SHARP Stationary High-Altitude Relay Platform

Sharp Tech J Sharp Technical Journal [Published by Sharp Co. Ltd., Japan]

SHAS Shrapnel Hole Analysis System

SHAU Subject Heading Authority Unit

SHB Spectral Hole Burning [Physics]

Sh Bl Shot Blasting

SHC Sacrificial Hyperconjugation [Chemistry]

SHC Silicones Health Council [US]

SHC Southern Horizon Circle [Astronomy]

SHC Strain Hardening Coefficient [Metallurgy]

SHC Superior Heliocentric Conjunction

SHC Superposition of Harmonic Currents (Method) [Metallurgy]

SHCA Safety Helmet Council of America [US]

ShCh Shaped Charge [Explosives]

Sh Con Shore Connection

SHC-RAM Stacked-High-Capacity Random-Access Memory

SHCS Sand Hill Crane Survey [US]

SHCS Socket Head Cap Screw

SHD Single High Density

SHD Slant Hole Distance

SHD Society for the History of Discoveries [US]

SHD Storage, Handling and Distribution

SHD Storage–Handling–Distribution [Journal published in the UK]

SHD Super Heavy Duty

SHD Superheterodyne Demodulation

SHD Super High Duty

Sh D Shipping Dry [also sh d]

Shdn Shutdown [also Shdn]

SHE Semi-Homogeneous Experiment

SHE Sodium Heat Engine

SHE Spacecraft Handling Equipment

SHE Spontaneous Hall-Effect [Electronics]

SHE Standard Hydrogen Electrode

SHE Strain Hardening Exponent [Metallurgy]

SHE Subject Headings for Engineering [US]

S&HE Sundays and Holidays Excepted [also s&he]

SHe Supercritical Helium

SHEBA Surface Heat Budget in the Arctic [of US Department of Energy]

SHEBA Surface High-Energy Electron Beam Apparatus

SHED Sealed Housing for Evaporative Determination (Test) [Fuel Evaporation Derived Pollution Measurement Test used by US Environmental Protection Agency]

SHED Segmented Hypergraphic Editor

SHED Solar Heat Exchanger Drive

SHEDA Storage and Handling Equipment Distributors Association [UK]

SHEED Scanning High-Energy Electron Diffraction

SHEED Secondary High-Energy Electron Diffraction

Sheet Met Ind Sheet Metal Industry [International Journal]

Sheet Met Tubes Sect Sheet Metal Tubes Sections [Journal published in Germany]

SHEFA Shetland Islands-Faero Islands Submarine Cable

SHELL99 50 Years of the Nuclear Shell Model [1949-1999]

Shell Petrochem Shell Petrochemicals [Publication of Shell Company, UK]

Shelv Shelving [also shelv]

SHEP Solar High-Energy Particles [Physics]

SHERB Sandia Human Error Rate Bank [of Sandia National Laboratory, Albuquerque, New Mexico, US]

SHERI System for Hierarchical Experts of Resource Inventories (Program) [Canada]

SHERP Social Housing Energy Retrofit Program [Joint Program of Ontario Ministry of Energy and Ontario Ministry of Housing, Canada]

SHESI Safety, Health and Environmental Services International

SHF Sensible-Heat Factor [Air Conditioning]

SHF Storage Handling Facility

SHF Super-High Feed (Milling)

SHF Super-High Frequency [Frequency: 3 to 30 Gigahertz; Wavelength: 1 to 10 Centimeters] [also shf]

SHF Superhyperfine (Splitting) [Spectroscopy]

SHFG Spherical Harmonic Spatial Grid Method [of Colorado State University, Fort Collins, US]

SHFS Superhyperfine Splitting [Spectroscopy]

SHFS Superhyperfine Structure [Spectroscopy]

SHFT Shift [also Shft, or shft]

SHFTR Shift Register [Computers]

SHG Second Harmonic Generation [also 2HG]

SHG Segmented Hypergraphics

SHG Simulation of Hypothetical-Nucleus Method in Generalized Form [Metallurgy]

SHG Special High Grade (of Slab Zinc) [Metallurgy]

2HG Second Harmonic Generation [also SHG]

SHGF Scottish Hang Gliding Federation

SHGSS Subharmonic Giant Shapiro Step [Physics]

S-HHFE Symmetric Heavy-Hole Free Electron [Solid-State Physics]

S-HHFE Symmetric Heavy-Hole Free Exciton [Solid-State Physics]

SHHOFI Safety and Health Hall of Fame International

SHI Sumitomo Heavy Industries, Limited [Japan]

SHI Surface Hazard Index

S-HI System-Human Interaction

Sh I Sheet Iron [also ShI]

SHIA Southern Hardwood Traffic Association [now Forest Products Traffic Association, US]

SHIEF Shared Information Elicitation Facility

SHIELD Sylvania High Intelligence Electronic Defense

Shimadzu Rev Shimadzu Review [Published by Shimadzu Corporation, Kyoto, Japan]

Shimizu Tech Res Bull Shimizu Technical Research Bulletin [Published by Shimizu Corporation, Japan]

SHIMWTS Shipboard Integrated Membrane Wastewater Treatment System

SHINCOM Ship-Integrated Communication (System)

SHINE State-wide Health Information Network [US]

Shinko Electr J Shinko Electric Journal [Published by Shinko Electric Company Limited., Tokyo, Japan]

SHINNADS Shipboard Integrated Navigation and Design System; Shipboard Integration Navigation and Display System; Shipboard Integrated Processing and Display System

SHINS Shipping Instruction

SHIOER Statistical Historical Input/Output Error Rate Utility

Ship Shiplap [also ship]

SHIPDES Ship Descriptions (Database) [of Maritime Information Center, Netherlands]

Shipg Shipping [also shipg]

SHIPIR Ship Infrared (Software)

SHIPS Bureau of Ships [US Navy]

SHIRAN S-Band High-Accuracy Ranging and Navigation [also Shiran, or shiran]

Shirley Inst Bull Shirley Institute Bulletin [UK]

Shirley Inst Mem Shirley Institute Memoirs [UK]

SHIRTDIF Storage, Handling and Retrieval of Technical Data in Image Formation

SHIV Chimeric Virus Model [Microbiology]

SHIV Schweizerischer Handels- und Industrieverein [Swiss Association of Industry and Commerce]

SHJ Single Heterojunction [Electronics]

SHK *(Sanitär–, Heizungs– und Klimatechnik)* – German for "Sanitary, Heating and Air-Conditioning Engineering"

SHL Shift Left [also shl] [Logical Operator]

SHL Studio to Head End Link

Shl Shell [also shl]

Shl Shellac [also shl]

SHLB Simulation Hardware Load Boxes

SH/LD Shift/Load (Input) [Digital Logic]

Shld Shield [also shld]

Shld Shoulder [also shld]

Shldr Shareholder [also shldr]

Shldr Shoulder [also shldr]

SHLMA Southern Hardwood Lumber Manufacturers Association [now Hardwood Manufacturers Association, US]

SHLW Solidified High-Level (Radioactive) Waste

shlw shallow

SHLWS Simulated High-Level (Radioactive) Waste Slurry

SHLWS T/S Simulated High-Level (Radioactive) Waste Slurry Treatment/Storage

SHM Sensible Heat Meter

SHM Servohydraulic Machine

SHM Simple Harmonic Motion [Mechanics]

SHM Society for Hybrid Microelectronics [US]

SHM Superhard Material

.SHM Library Shell Macro (File) [WordPerfect File Name Extension]

SH(M) Servo-Hydraulic(s)

SHMOKE Second-Harmonic Magneto-Optic Kerr Effect [also SH-MOKE]

SHN Scleroscope Hardness Number

SHO Simple Harmonic Oscillator [Mechanics]

SHO Spin-Hamilton Operator [Physics]

SHO Super-High Output

SHOALS Scanning Hydrographic Operational Airborne Lidar Survey (System) [of US Army Corps of Engineers]

SHOC Software Hardware Operational Control

Shock Vib Dig Shock and Vibration Digest [of Vibration Institute, US]

SHODOP Short-Range Doppler [also Shodop, or shodop]

SHOM Service Hydrographique et Océanographique de la Marine [Naval Hydrographic and Oceanographic Service, France]

SHOP Shell Higher Olefins Process

SHORAN Short-Range Navigation [also Shoran, or shoran]

Short Wave Mag Short Wave Magazine [UK]

SHOT Society for the History of Technology [US]

SHOW Scripps Institute/University of Hawaii/ Oregon State University/University of Wisconsin [US]

Showa Wire Cable Rev Showa Wire and Cable Review [Japan]

SHP Secondary Hydrogen Precipitation [Metallurgy]

SHP Shaft Horsepower [also shp]

SHP Shock Heated Pinch [Physics]

SHP St. Helena Pound [also SH£; Currency]

SHP Strain Hardening Parameter [Metallurgy]

SH£ St. Helena Pound [also SHP; Currency]

Shp Ship [also shp]

Shp Shipping [also shp]

SHPB Split Hopkinson Pressure Bar [Mechanical Testing]

SHPDT Shipping Date

SHPE Society of Hispanic Professional Engineers [US]

Shpg Shipping [also shpg]

SHPO State Historical Preservation Office [Washington State, US]

SHPS Sodium Hydroxide Purge System

Shpt Shipment [also shpt]

SHPNM Ship-to Name

SHQEA Supreme Headquarters of the European Army

SHR Safety Hoist Ring

SHR Semihomogeneous (Nuclear) Fuel Reactor

SHR Shift Right [also shr] [Logical Operator]

Shr Share [also shr]

Shrap Shrapnel [also shrap]

SHRD Supplemental Heat Rejection Devices

Shrd Shroud [also Shrd]

SHRI Scottish Horticultural Research Institute

SHRIMP Scalar Heteronuclear Recoupled Interactions by Multiple Pulse [Physics]

Shrop Shropshire [UK]

SHRP Strategic Highway Research Program [of National Research Council, US]

SHRP-IDEA Strategic Highway Research Program–Ideas Deserving Exploratory Analysis [of National Research Council, US]

SHRS Supplementary Heat Removal System

Shrtg Shortage [also shrtg]

SHS Sample-Hold-Sample (Circuit)

SHS Scandinavian Herpetological Society [Denmark]

SHS Scottish History Society [UK]

SHS Self-Propagating High-Temperature Synthesis

SHS SEM (Scanning Electron Microscope) Hot Stage

SHS Simulation Hardware System

SHS Small Hydro Society [US]

SHS Solar Heating System

SHS Steam Heating System

SHS Surveyors Historical Society [US]

SHS Swivel Head Screw

Sh S Sheet Steel [also ShS]

SHSG Spherical Harmonic Spatial Grid

SHSMB Safety and Health Standards Management Board

SHSU Sam Houston State University [Huntsville, Texas, US]

SHT Society for the History of Technology [US]

SHT Solution Heat Treated; Solution Heat Treatment [Metallurgy]

SHT Standard Heat Treatment [Metallurgy]

S(H,T) Field and Temperature Dependent Magnetic Viscosity Coefficient [Symbol]

Sht Sheet [also sht]

sht short

sh t short ton [also sh tn, or sh ton]

SHTB Saskatchewan Highway Traffic Board [Canada]

SHTC Short-Time Constant

Shtg Shortage [also shtg]

Shth Sheath [also shth]

Shthg Sheathing [also shthg]

SHTL Small Heat-Transfer Loop [Nuclear Reactors]

sh tn short ton [also sh t, or sh ton]

sh tn wt short ton-weight [Unit]

sh tn wt/in² short ton-weight per square inch [also sh tn wt/sq in]

sh tn wt/cu in short ton-weight per cubic inch [also sh tn wt/in³]

sh ton short ton [also sh t, or sh tn]

Shtr Shutter

SHTTP Secure Hypertext Transfer Protocol [also S-HTTP]

SHTU Short Ton Unit [also shtu]

SHU Saarberg-Holter-Lurgi (Process) [Metallurgy]

SHU Seong Hwa University [South Korea]

SHU Seton Hall University [South Orange, New Jersey, US]

SHU Sheffield Hallam University [Sheffield, UK]

Shutdn Shutdown [also shutdn]

SHV Serum Hepatitis Virus

SHV Superheated Vapor

SHVG Stabilized High-Voltage Generator

.SHW Show [File Name Extension] [also .shw]

SH W Short Wave [also SH/W, Sh W, sh w, or sh/w]

Shwr Shower [also shwr]

SHWY Super Highway [also Shwy, or shwy]

SHY Syllable Hyphen (Character)

SHYDRO Small Hydro Costing Program [of Panel on Energy Research and Development, Canada]

SI Center Support Operations [NASA Kennedy Space Center Directorate, Florida, US]

SI Sacro-Illiac [Anatomy]

SI Safety Injection

SI Safety Inspection

SI Salt Institute [US]

SI Sample Input

SI Sample Interval

SI Sandwich Islands

SI Scientific Information

SI Scientific Instrument

SI Screen-Grid Input

SI Secondary Ion

SI Seismic Instrument(ation)

SI Selectivity Index

SI Self-Inductance; Self-Induction

SI Self-Interstitial [Crystallography]

SI Semi-Insulating; Semi-Insulator

SI Serial In

SI Serial Input

SI Serial Interface

SI Service Indicator

SI Service Interruption

SI Shadow Image

SI Shetland Islands [Scotland]

SI Shift-In (Character)

SI Shirley Institute [UK]

SI Shrinkage Index (of Soil) [Geology]

SI Signal-to-Interference (Ratio)

SI Signal-to-Intermodulation (Ratio)

SI Silicone (Polymer)

SI Silicon Iron

SI Simultaneous Iteration

SI Simple Icosahedral (Lattice) [also si] [Crystallography]

SI Simple Integration

SI Single Instruction

SI Silver Institute [US]

SI Slovenia [ISO Code]

SI Small Intestine [Anatomy]

SI Smithsonian Institution [Washington, DC, US]

SI Social Impact

SI Society of Indexers [UK]

SI Soft Impingement

SI Soft Iron

SI Solar Inertial

SI Solar Influences

SI Solomon Islands

SI Sound Insulation

SI Source Index

SI Southern Indiana [US]

SI Space Industrialization

SI Space Institute [of University of Tennessee, Tullahoma, US]

SI Spark Ignition (Engine)

SI Special Instruction

SI Specific Impulse [Aerospace]

SI Specific Inventory

SI Speech Interpolation

SI Speed Indicator

SI Stability Index (of Intermetallic Compounds) [Metallurgy]

SI Standard Interval

SI Start of Ignition

SI Staten Island [New York, US]

SI Static Induction

SI Station Identification

SI Stationary Interface (Approximation) [Materials Science]

SI Stellar Interferometer; Stellar Interferometry

SI Stereoisomer(ism)

SI Sterol-Inhibiting; Sterol Inhibitor

SI Stochastically Independent (Events); Stochastic Independence

SI Storage Immediate

SI Statement of Intention

SI Statutory Instrument

SI Straight-In Approach

SI Strategic Intelligence

SI Strong Interaction [Physics]

SI Structural Integrity

SI Structure Index

SI Styrene-Isoprene (Copolymer) [also S-I]

SI Sulfur Institute [US]

SI Sustainable Indicators

SI Superexchange Interaction [Solid-State Physics]

SI Superimposability; Superimposition

SI Supplementary Information

SI Sustainable Industry (Project) [of US Environmental Protection Agency]

SI Surface-Interface (Problem) [Composite Engineering]

SI Surface Ionization

SI Surveyors Institute [UK]

SI Swap-In

SI Syndicat d'Indicative [Tourist Association, France]

SI System Information

SI *(Système International d'Unités)* – International System of Units

SI System-Integrated; System(s) Integration [also S-I]

S/I Surface/Interface (Interaction) [Physical Chemistry]

S-I Signal-to-Intermodulation (Ratio) [also S/I]

S-I Styrene-Isoprene (Copolymer) [also SI]

Si Sidereal (Time) [Astronomy]

Si Silicon [Symbol]

Si *(Sinus)* – Spanish for "Gulf"

Si^{4+} Silicon Ion [Symbol]

Si^{4-} Silicide [Symbol]

Si II Silicon II [Symbol]

Si III Silicon III [Symbol]

Si-27 Silicon-27 [also ^{27}Si, or Si^{27}]

Si-28 Silicon-28 [also ^{28}Si, Si^{28}, or Si]

Si-29 Silicon-29 [also ^{29}Si, or Si^{29}]

Si-30 Silicon-30 [also ^{30}Si, or Si^{30}]

Si-31 Silicon-31 [also ^{31}Si, or Si^{31}]

Si-32 Silicon-32 [also ^{32}Si, or Si^{32}]

3C-Si Cubic Silicon [Ramsdell Notation: 3 Refers to Number of Silicon Layers, and C Refers to the Cubic Symmetry]

2H-Si Hexagonal Silicon [Ramsdell Notation: 2 Refers to Number of Silicon Layers, and H Refers to the Hexagonal Symmetry]

4H-Si Hexagonal Silicon [Ramsdell Notation: 4 Refers to Number of Silicon Layers, and H Refers to the Hexagonal Symmetry]

9R-Si Rhombohedral Silicon [Ramsdell Notation: 15 Refers to Number of Silicon Layers, and R Refers to the Rhombohedral Symmetry]

dc-Si Diamond-Cubic Silicon (Phase)

dh-Si Diamond-Hexagonal Silicon (Phase)

si simple icosahedral (lattice) [also SI] [Crystallography]

.si Slovenia [Country Code/Domain Name]

SIA Saskatchewan Institute of Agrologists [Canada]

SIA Sasquatch Investigations of Mid-America [US]

SIA Satellite Intensity Analysis

SIA Scaffold Industry Association [US]

SIA Schweizerischer Ingenieur– und Architektenverein [Swiss Association of Engineers and Architects]

SIA Screened Interaction Approximation [Physics]

SIA Security Industry Association [US]

SIA Self-Interstitial Atom [Crystallography]

SIA Semiconductor Industry Association [US]

SIA Sequence of Initial Actions

SIA Service in Information and Analysis [UK]

SIA (Space) Shuttle Induced Atmosphere

SIA Singapore Airlines

SIA Social Impact Assessment

SIA Society for Industrial Accountants

SIA Society for Industrial Archeology [US]

SIA Society of Insurance Accountants [US]

SIA Software Impact Assessment

SIA Solar Inertial Attitude

SIA Solvents Industry Association [UK]

SIA Sprinkler Irrigation Association [US]

SIA Standard Instrument Approach [Aeronautics]

SIA Standard Interface Adapter

SIA Stereo-Image Alternator

SIA Subminiature Integrated Antenna

SIA System-Initiated Archival

SIA Systems Integration Area

SIAA Singapore Industrial Automation Association

SIAB N-Succinimidyl(4-Iodoacetyl)aminobenzoate

SIAC Specialized Information Analysis Center

Si(Ac)$_4$ Silicon(IV) Acetate

SIAD Society of Industrial Artists and Designers [UK]

SIAE Scottish Institute of Agricultural Engineering [UK]

SIAF Service Indicator Associated Field

SIAL Sigma-Aldrich Family (of Chemical Companies)

Sial Silicon-Aluminum (Layer) [also sial] [Geology]

SiAl Silicon Aluminide (Ceramics)

Si-Al Silicon-Aluminum (Alloy)

Si:Al Aluminum-Doped Silicon

Si(Al):H Hydrogenated Aluminum-Doped Silicon

SIALON Silicon Aluminum Oxynitride (Ceramic) [also SiAlON, or Sialon]

SIAM Shipborne Ice Alert and Monitoring

SIAM Signal Information and Monitoring

SIAM Society for Industrial and Applied Mathematics [US]

SIAM J Appl Math Society for Industrial and Applied Mathematics, Journal of Applied Mathematics [US]

SIAM J Comput Society for Industrial and Applied Mathematics, Journal of Computing [US]

SIAM J Control Optim Society for Industrial and Applied Mathematics, Journal of Control and Optimization [US]

SIAM J Discr Math Society for Industrial and Applied Mathematics, Journal of Discrete Mathematics [US]

SIAM J Math Anal Society for Industrial and Applied Mathematics, Journal of Mathematical Analysis [US]

SIAM J Matrix Anal Appl Society for Industrial and Applied Mathematics, Journal on Matrix Analysis and Applications [US]

SIAM J Numer Anal Society for Industrial and Applied Mathematics, Journal on Numerical Analysis [US]

SIAM J Optim Society for Industrial and Applied Mathematics, Journal of Optimization [US]

SIAM J Sci Comput Society for Industrial and Applied Mathematics, Journal on Scientific Computing [US]

SIAM J Sci Stat Comput Society for Industrial and Applied Mathematics, Journal on Scientific and Statistical Computing [US]

SIAM Rev Society for Industrial and Applied Mathematics, Review [US]

SIAO Smithsonian Institute Astrophysical Observatory [US]

SIAP Shipbuilding Industry Assistance Program

SIAP Statistical Institute for Asia and the Pacific

SIAP Systems Integration Assurance Program

SIAPE Société Industrielle d'Acide Phosphorique et Engrais (Process) [Chemical Engineering]

SIAPS Standard Instrument Approach Procedures [Aeronautics]

SIAS Safety Injection Actuation Signal

SIAS Sussex Industrial Archeological Society [UK]

Si:As Arsenic-Doped Silicon

Si-Au Silicon-Gold (Alloy)

SIAT Single Integrated Attack Team

SIB Satellite Ionospheric Beacon

SIB Screen Image Buffer

SIB Securities and Investments Board [UK]

SIB Serial Interface Board

SIB Shipbuilding Industry Board [UK]

SIB Special Intelligence Bureau

SIB Special Investigation Board

SIB Strategic Information Branch [of Industry Canada]

SIB System Integration Board

SiB Silicon Monoboride (Ceramic)

Si:B Boron-Doped Silicon

Sib Siberia(n)

SiBC Silicon Borocarbide

Si-B-C Silicon-Boron-Carbon (Compound)

Si-B-C-N Silicon-Boron-Carbon-Nitrogen (Compound)

Siber J Chem Siberian Journal of Chemistry [Russia]

SIBGRAPI Simpósios Brasileiros de Computação Gráfica e Processamento de Imagens [Brazilian Symposium on Computer Graphics and Image Processing]

SIB-IC Strategic Information Branch–Industry Canada

SIBL Scanning Ion-Beam Lithography

SIBM Strain-Induced (Grain) Boundary Migration [Materials Science]

SiBN Silicon Boronitride

Si-B-N Silicon-Boron-Nitrogen (Compound)

SIBOR Singapore Interbank Offered Rate [also Sibor]

Si-Bronze Silicon Bronze

SIBS Società Italiana Biologia Sperimentale [Italian Society of Experimental Biology]

SIBS Stellar-Inertial Bombing System

SibSSR Siberian Soviet Socialist Republic [USSR]

SIBT Symposium on Ion-Beam Technology

SIC Science Information Council [of National Science Foundation, US]

SIC Scientific Information Center

SIC Self-Interaction Correction [Physics]

SIC Self-Interaction Correlation [Physics]

SIC Semiconductor Integrated Circuit

SIC Service Indicator Code

SIC Shanghai Institute of Ceramics [of Chinese Academy of Science, PR China]

SIC Shuttered Image Converter

SIC Silicon Integrated Circuit

SIC Special Interest Committee

SIC Specific Inductive Capacitance

SIC Standard Industrial Classification

SIC Standard Industrial Code

SIC Structural Integrity Center [of UK Atomic Energy Authority]

SIC Sulfur-Impregnated Concrete

SIC Switching Integrated Circuit

SIC Synthetic Interstitial Claywater

SiC Silicon Carbide

SiC$_f$ Silicon Carbide Fibers [also SiCf, or SiC(f)]

SiC$_p$ Silicon Carbide Particulate [also SiCp, or SiC(p)]

SiC$_w$ Silicon Carbide Whiskers [also SiCw, or SiC(w)]

3C-SiC Cubic Silicon Carbide [Ramsdell Notation: 3 Refers to Number of Silicon and Carbon Bilayers Necessary to Produce a Unit Cell, and C Refers to the Cubic Symmetry] [also β-SiC]

2H-SiC Hexagonal Silicon Carbide [Ramsdell Notation: 2 Refers to Number of Silicon and Carbon Bilayers Necessary to Produce a Unit Cell, and H Refers to the Hexagonal Symmetry]

4H-SiC Hexagonal Silicon Carbide [Ramsdell Notation: 4 Refers to Number of Silicon and Carbon Bilayers Necessary to Produce a Unit Cell, and H Refers to the Hexagonal Symmetry]

6H-SiC Hexagonal Silicon Carbide [Ramsdell Notation: 6 Refers to Number of Silicon and Carbon Bilayers Necessary to Produce a Unit Cell, and H Refers to the Hexagonal Symmetry]

8H-SiC Hexagonal Silicon Carbide [Ramsdell Notation: 8 Refers to Number of Silicon and Carbon Bilayers Necessary to Produce a Unit Cell, and H Refers to the Hexagonal Symmetry]

10H-SiC Hexagonal Silicon Carbide [Ramsdell Notation: 10 Refers to Number of Silicon and Carbon Bilayers Necessary to Produce a Unit Cell, and H Refers to the Hexagonal Symmetry]

15R-SiC Rhombohedral Silicon Carbide [Ramsdell Notation: 15 Refers to Number of Silicon and Carbon Bilayers Necessary to Produce a Unit Cell, and R Refers to the Rhombohedral Symmetry]

21R-SiC Rhombohedral Silicon Carbide [Ramsdell Notation: 21 Refers to Number of Silicon and Carbon Bilayers Necessary to Produce a Unit Cell, and R Refers to the Rhombohedral Symmetry]

Sic Sicilian; Sicily

SICA Society of Industrial and Cost Accountants [Canada]

SiC/Al Silicon-Carbide Reinforced Aluminum (Composite)

SiC$_f$/Al Silicon-Carbide Fiber Reinforced Aluminum (Composite) [also SiC(f)/Al]

SiC$_p$/Al Silicon-Carbide Particulate Reinforced Aluminum (Composite) [also SiC(p)/Al]

SiC$_w$/Al Silicon-Carbide Whisker Reinforced Aluminum (Composite) [also SiC(w)/Al]

SiC-Al-B-C Silicon Carbide with Aluminum, Boron and Carbon Additions

SiC/AlN Silicon Carbide/Aluminum Nitride (Heterostructure)

SiC/Al$_2$O$_3$ Silicon-Carbide Reinforced Alumina (Composite)

SiC$_f$/Al$_2$O$_3$ Silicon-Carbide Fiber Reinforced Alumina (Composite) [also SiC(f)/Al$_2$O$_3$]

SiC$_p$/Al$_2$O$_3$ Silicon-Carbide Whisker Reinforced Alumina (Composite) [also SiC(p)/Al$_2$O$_3$]

SiC$_w$/Al$_2$O$_3$ Silicon-Carbide Whisker Reinforced Alumina (Composite) [also SiC(w)/Al$_2$O$_3$]

SiC-Al$_2$O$_3$ Silicon-Carbide with Alumina (or Aluminum Oxide) Additions

SICAN Silizium-Anwendung und CAD/CAT Niedersachsen [Silicon Application and CAD/CAT of Lower Saxony, Germany]

SICASP SSR (Secondary Sureveillance Radar) Improvements and Collision Avoidance System Panel [of International Civil Aviation Organization]

SiC-B-C Silicon Carbide with Boron and Carbon Additions

SICC Strain-Induced Corrosion Cracking

SICCAPH Special Interest Committee for Computers and the Physically Handicapped [now Special Interest Group for Computers and the Physically Handicapped, US]

SiC/CAS Silicon Carbide Reinforced Calcium Aluminosilicate [also SiC-CAS]

SiC CVD Silicon Carbide Chemical Vapor Deposition [also SiC-CVD]

SiC CVC Silicon Carbide Chemical Vapor Condensation [also SiC-CVC]

SICD Spacelab I/O (Input/Output) Channel Data

SICE Society of Instrument and Control Engineers [Japan]

SICEJ Society of Instrument and Control Engineers of Japan

SiC(f) Silicon Carbide Fiber [also SiCf, or SiC_f]

SiC(f)/Al Silicon-Carbide Fiber Reinforced Aluminum (Composite) [also SiCf/Al, or SiC_f/Al]

SiC(f)/Al$_2$O$_3$ Silicon-Carbide Fiber Reinforced Alumina (Composite) [also SiCf/Al$_2$O$_3$, or SiC_f/Al$_2$O$_3$]

SiC(f)/SiC Silicon-Carbide Fiber Reinforced Silicon Carbide (Composite) [also SiCf/SiC, or SiC_f/SiC]

SiC(f)/Si$_3$N$_4$ Silicon-Carbide Fiber Reinforced Silicon Nitride (Composite) [also SiC(f)/Si$_3$N$_4$, or SiC_f/Si$_3$N$_4$]

SiC(f)/Ti Silicon-Carbide Fiber Reinforced Titanium (Composite) [also SiC(f)/Ti, or SiC_f/Ti]

SiC/GaN Silicon Carbide/Gallium Nitride (Heterostructure)

SiC(i) Intrinsic Silicon-Carbide

SiC(i)-Si(p) Intrinsic Silicon-Carbide–P-Type Silicon (Heterojunction)

SiC/LAS Silicon Carbide (Fiber) Reinforced Lithium Aluminosilicate (Glass) [also SiC-LAS]

SiCl$_2$H$_2$ Dichlorosilane

SiCl$_2$H$_2$/HCl/H$_2$ Dichlorosilane/Hydrogen Chloride/Hydrogen (Gas Mixture)

SiC/LMAS Lithium-Magnesium Aluminosilicate Glass Coated Silicon Carbide (Ceramics)

SiClPh$_3$ Triphenylsilyl Chloride

SICLSD Self-Interaction-Corrected Local Spin Density [also SIC-LSD] [Physics]

SICLSD Self-Interaction Correction (or Correlation) Local Spin Density (Calculation) [also SIC-LSD] [Physics]

SiC/MAS Silicon Carbide Reinforced Magnesium Aluminosilicate [also SiC-MAS]

SiC MMC Silicon-Carbide Reinforced Metal-Matrix Composite

Si CMOS IC Silicon Complementary Metal-Oxide Semiconductor Integrated Circuit

SICN Société Industrielle de Combustibles Nucléaire [Nuclear Fuels Company, France]

Si-C-N Silicon-Carbon-Nitrogen

SICO Special Interest Committee

SICO Switched-In for Checkout

SICODCPT Special Interest Committee on Digital Computer Programmer Training

SICOM Securities Industry Communications [US]

SICOMP Siemens Industrial Computer

SICOS Sidewall Base Contact Structure

SICP Selected Ion Current Profile

SiC(p) Silicon Carbide Particulate [also SiCp, or SiC_p]

SiC(p)/Al Silicon-Carbide Particulate Reinforced Aluminum (Composite) [also SiCp/Al, or SiC_p/Al]

SiC(p)/Al$_2$O$_3$ Silicon-Carbide Particulate Reinforced Alumina (Composite) [also SiC_p/Al$_2$O$_3$]

Si-29 CP MAS-NMR Silicon-29 Cross-Polarization Magic-Angle Spinning-Nuclear Magnetic Resonance [also ^{29}Si MAS-NMR]

SiC(p/)SiC Silicon-Carbide Particulate Reinforced Silicon Carbide (Composite) [also SiCp/SiC, or SiC_p/SiC]

SiC(p)/Si$_3$N$_4$ Silicon-Carbide Particulate Reinforced Silicon Nitride (Composite) [also SiC(p)/Si$_3$N$_4$, or SiC_p/Si$_3$N$_4$]

SiC(p)/Ti Silicon-Carbide Particulate Reinforced Silicon Nitride (Composite) [also SiC(p)/Ti, or SiC_p/Ti]

Si-Cr-Ni Silicon-Chromium-Nickel (Coating)

Si-Cr-Ti Silicon-Chromium-Titanium (Coating)

SICS Safety Injection System

SICS Semiconductor Integrated Circuit

SICS Standard Industrial Classification System

SiC/Si Silicon Carbide on Silicon (Substrate)

3C-SiC/Si Cubic Silicon Carbide on Silicon [Ramsdell Notation: 3 Refers to Number of Silicon and Carbon Bilayers Necessary to Produce a Unit Cell, and C refers to the Cubic Symmetry]

6H-SiC/Si Hexagonal Silicon Carbide on Silicon [Ramsdell Notation: 6 Refers to Number of Silicon and Carbon Bilayers Necessary to Produce a Unit Cell, and H refers to the Hexagonal Symmetry]

SiC/SiC Silicon-Carbide Reinforced Silicon Carbide (Composite)

SiC$_f$/SiC Silicon-Carbide Fiber Reinforced Silicon Carbide (Composite) [also SiC(f)/SiC]

SiC$_p$/SiC Silicon-Carbide Particulate Reinforced Silicon Carbide (Composite) [also SiC(p)/SiC]

SiC$_w$/SiC Silicon-Carbide Whisker Reinforced Silicon Carbide (Composite) [also SiC(w)/SiC]

SiC/Si$_3$N$_4$ Silicon-Carbide Reinforced Silicon Nitride (Composite)

SiC$_f$/Si$_3$N$_4$ Silicon-Carbide Fiber Reinforced Silicon Nitride (Composite) [also SiC(f)/Si$_3$N$_4$]

SiC$_p$/Si$_3$N$_4$ Silicon-Carbide Particulate Reinforced Silicon Nitride (Composite) [also SiC(p)/Si$_3$N$_4$]

SiC$_w$/Si$_3$N$_4$ Silicon-Carbide Whisker Reinforced Silicon Nitride (Composite) [also SiC(w)/Si$_3$N$_4$]

SiC/Ti Silicon-Carbide Reinforced Titanium (Composite)

SiC$_f$/Ti Silicon-Carbide Fiber Reinforced Titanium (Composite) [also SiC(f)/Ti]

SiC$_p$/Ti Silicon-Carbide Particulate Reinforced Silicon Nitride (Composite) [also SiC(p)/Ti]

SiC$_w$/Ti Silicon-Carbide Whisker Reinforced Silicon Nitride (Composite) [also SiC(w)/Ti]

SiC UV Silicon Carbide Ultraviolet (Photodiode)

SiC(w) Silicon Carbide Whiskers [also SiCw, or SiC_w]

SiC(w)/Al Silicon-Carbide Whisker Reinforced Aluminum (Composite) [also SiCw/Al, or SiC$_w$/Al]

SiC(w)/Al$_2$O$_3$ Silicon-Carbide Whisker Reinforced Alumina (Composite) [also SiC$_w$/Al$_2$O$_3$, or SiC$_w$/Al$_2$O$_3$]

SiC(w)/SiC Silicon-Carbide Whisker Reinforced Silicon Carbide (Composite) [also SiC(w)/SiC, or SiC$_w$/SiC]

SiC(w)/Si$_3$N$_4$ Silicon-Carbide Whisker Reinforced Silicon Nitride (Composite) [also SiC(w)/Si$_3$N$_4$]

SiC(w)/Ti Silicon-Carbide Whisker Reinforced Silicon Nitride (Composite) [also SiC(w)/Ti]

SID Scanning Imaging Device

SID Scheduled Issue Date

SID Scientific Intelligence Digest

SID Security Identifier

SID Seismic Intrusion Detector

SID Selected Ion Detection

SID Serial Input Data

SID Silicon Imaging Device

SID Simulation Interface Device

SID Single Ion Detection

SID Serial Interface Device

SID Sheep Industry Development (Program) [of American Sheep Producers Council, US]

SID Shuttle Integration Device [of NASA]

SID Signal Identification

SID Silicon Imaging Device

SID Simulator Interface Devices

SID Singapore Dollar [Currency]

SID Society for Information Display [US]

SID Society for International Development [Italy]

SID Society of Industrial Designers [US]

SID Sodium Ionization Detector

SID Solomon Island Dollar [Currency]

SID Solubilization by Incipient Development

SID Sound Interface Device

SID Speed-Increasing Diode

SID Standard Instrument Departure

SID Standard Interface Document

SID Station Identification

SID Structural Inspection Database

SID Substitutional-Interstitial Diffusion [Metallurgy]

SID Sudden Ionospheric Disturbance [Geophysics]

SID Surface-Induced Dissociation

SID Swift Interface Device

SID Symbolic Instruction Debugger

SID Synchronous Identification

SID System Identification [Telecommunications]

SID System Interface Document

SID Syntax Improving Device

SiD Sidereal Day [Astronomy]

SI$ Solomon Island Dollar [Currency]

Sid Siderite [Mineral]

SIDA Saskatchewan Implement Dealers Association [Canada]

SIDA Software Industry Development Association [Canada]

SIDA Swedish International Development Agency

SIDAC Silicon Diode Alternating Current [also Sidac, or sidac]

SIDASE Significant Data Selection

SIDC Strategy for International Development and Cooperation

SID Dig Society for Information Display Digest [US]

SIDE Suprathermal Ion Detector Experiment

SIDE (International Interdisciplinary Meeting on) Symmetries and Integrability of Difference Equations

Sider Latinoam Siderurgia Latinoamericana [Iron and Steelmaking Journal of Instituto Latinoamericana del Fierro y el Acero, Chile]

SIDES Source Input Date Edit System

SIDF System Independent Data Format

SIDH System Identification for Home Systems

Si-DLC Silicon-Containing Diamond Like Carbon (Coating)

Si-29 DNP Silicon-29 Dynamic Nuclear Polarization (Spectroscopy) [also ^{29}Si DNP]

Si-29 DNP-DPMAS Silicon-29 Dynamic Nuclear Polarization Direct-Polarization Magic-Angle Spinning (Spectroscopy) [also ^{29}Si DNP-DPMAS]

Si-29 DNP-MAS Silicon-29 Dynamic Nuclear Polarization Magic-Angle Spinning (Spectroscopy) [also ^{29}Si DNP-MAS]

Si-29 DNP-NMR Silicon-29 Dynamic Nuclear Polarization Nuclear Magnetic Resonance [also ^{29}Si DNP-NMR]

SIDR Standard Inside Diameter Ratio

SIDS Screening Information Data Set [of Organization for Economic Cooperation and Development, France]

SIDS Simulated Instrument Datasets

SIDS Small Island Developing States

SIDS Speech Identification System

SIDS Static Inspirable Dust Spectrometer

SIDS Stellar Inertial Doppler System

SIDS Sudden Infant Death Syndrome [Medicine]

SIDT Store Interrupt Descriptor Table

SIE School of Industrial Engineering

SIE Science Information Exchange

SIE Secondary Ion Emission

SIE Selective Ion Electrode

SIE Selective Ion Exchange

SIE Shuttle Interface Equipment [of NASA]

SIE Single Instruction Execute

SIE Society of Industrial Engineers [UK]

SIE Spark-Ignition Engine

SIEAPT Siemens-APT (Automatic Programmed Tools)

SIEM Shadow Image Electron Microscopy

Siemens-Albis Ber Siemens-Albis Berichte [Technical Reports of Siemens-Albis, Zurich, Switzerland]

Siemens Compon Siemens Components [German Publication]

Siemens Compon (Engl Ed) Siemens Components (English Edition)

Siemens Energietech Siemens Energietechnik [German Publication of Siemens AG on Energy Technology]

Siemens Forsch Entwickl-Ber Siemens Forschungs- und Entwicklungsberichte [German Research and Development Reports of Siemens AG]

Siemens Rev Siemens Review [German Publication of Siemens AG]

Siemens-Z. Siemens-Zeitschrift [German Journal published by Siemens AG]

SIEP Società Italiana dell'Esplosivo Prometheus [Italian Society for Promethius Explosives]

SIEPR Sociedad du Ingenieros Estructurolos de Puerto Rico [Society of Structural Engineers of Puerto Rico]

SIER Selective Ion Exchange Resin

Si:Er Erbium-Doped Silicon

SIES Soils and Irrigation Extension Service [Australia]

SIES Supervision, Inspection, Engineering and Services

Si(EtO)$_4$ Silicon Tetraethoxide (or Tetraethylorthosilicate, Ethyl Silicate, or Tetraethyoxysilane)

SIF Security and Intelligence Foundation [US]

SIF Selective Identification Feature

SIF Senior Industrial Fellowship [of Natural Sciences and Engineering Research Council, Canada]

SIF Signaling Information Field

SIF Società Italiana di Fisica [Italian Physical Society]

SIF Sound Intermediate Frequency

SIF Space Industrial Fellowship [of Space Foundation, US]

SIF Standard Interchange Format

SIF Storage Interface Facility

SIF Stress Intensity Factor [Mechanics]

S-I-F Superconductor-Insulator-Ferromagnet (Heterostructure) [Solid-State Physics]

SiF$_x$ Fluorosilyl [General Formula]

SiF$_x$ Silicon Fluoride [General Formula]

SIFAC Space Industry Forum in Atlantic Canada

SIFDT Selected Ion Flow Drift Tube

Si FIB Silicon Implation by Focused Ion Beam (Technique)

SiF$_4$/SiH$_4$ Tetrafluorosilane/Silane (Gas Mixture)

SiF$_4$/SiH$_4$/H$_2$ Tetrafluorosilane/Silane/Hydrogen (Gas Mixture)

SIFT Selected Ion Flow Tube (Apparatus)

SIFT Share Internal FORTRAN Translator

SIFT Software-Implemented Fault Tolerance

SIFT Stanford Information Filtering Tool [of Stanford University, California, US]

SIG Schweizerische Industriegesellschaft [Swiss Industrial Society]

SIG Service d'Information Géologique [Geological Information Service, France]

SIG Special Interest Group

SIG Structural Integrity Group

SIG Sub-Interface Generator

Sig Sigmoid [Biology]

Sig Signal [also sig]

Sig Signature [also sig]

Sig Significance; Significant [also sig]

sig *(signa)* – label [Medical Prescriptions]

SIGA Schweizerische Interessengemeinschaft für Abfall(ver)minderung [Swiss Waste Reduction Association]

Si:Ga Gallium-Doped Silicon

SI-GaAs Semi-Insulating Gallium Arsenide

Si:GaAs Galium Arsenide-Doped Silicon

SIGACT Special Interest Group on Automata and Compatibility Theory [of Association for Computing Machinery, US]

SIGACT News SIGACT News [of Association for Computing Machinery, US]

SIGADA Special Interest Group on Ada (Programming Language) [US]

SIGAPL Special Interest Group on APL (A Programming Language) [US]

SIGARCH Special Interest Group for Architecture of Computer Systems [of Association for Computing Machinery, US]

SIGARCH Comput Archit News SIGARCH Computer Architecture News [of Association for Computing Machinery, US]

SIGART Special Interest Group on Artificial Intelligence [of Association for Computing Machinery, US]

SIGBDP Special Interest Group on Business Data Processing [of Association for Computing Machinery, US]

SIGBIO Special Interest Group on Biomedical Computing [of Association for Computing Machinery, US]

SIGBIO Newsl SIGBIO Newsletter [of Association for Computing Machinery, US]

SigC Signal Corps [of US Army]

SIGCAPH Special Interest Group for Computers and the Physically Handicapped [US]

SIGCAPH Newsl SIGCAPH Newsletter [of Association for Computing Machinery, US]

SIGCAS Special Interest Group for Computers and Society [of Association for Computing Machinery, US]

SIGCAS Comput Soc SIGCAS Computers and Society [of Association for Computing Machinery, US]

SIGCAT Special Interest Group on CD-ROM Applications and Technology [of Association for Computing Machinery, US]

SIGCHI Special Interest Group on Computer Hierarchy and Interfaces [of Association for Computing Machinery, US]

SIGCHI Bull SIGCHI Bulletin [of Association for Computing Machinery, US]

SIGCOMM Special Interest Group on Data Communications [of Association for Computing Machinery, US]

SIGCOMM Comput Commun Rev SIGCOMM Computer Communication Review [of Association for Computing Machinery, US]

SIG CONDR Signal Conditioner [also Sig Condr]

SIGCOSIM Special Interest Group on Computer Systems Installation Management [of Association for Computing Machinery, US]

SIGCPR Special Interest Group for Computer Personnel Research [of Association for Computing Machinery, US]

SIGCSE Special Interest Group on Computer Science Education [of Association for Computing Machinery, US]

SIGCSE Bull SIGCSE Bulletin [of Association for Computing Machinery, US]

SIGCUE Special Interest Group for Computer Uses in Education [of Association for Computing Machinery, US]

SIGCUE Bull SIGCUE Bulletin [of Association for Computing Machinery, US]

SIGDA Special Interest Group for Design Automation [of Association for Computing Machinery, US]

SIGDA Newsl SIGDA Newsletter [of Association for Computing Machinery, US]

SIGDOC Special Interest Group for Systems Documentation [of Association for Computing Machinery, US]

SiGe Silicon Germanide (Semiconductor)

Si/Ge Silicon on Germanium (Substrate)

Si-Ge Silicon-Germanium (Superlattice)

SiGeC Silicon Germanium Carbide (Semiconductor)

$Si_{1-x}Ge_x$:H Hydrogenated Silicon Germanide [General Formula]

$Si-GeH_4-B_2H_6$ MBE Silicon-Germanium Tetrahydride–Diborane Molecular-Beam Epitaxy

SiGe HBT Silicon Germanide Heterojunction Bipolar Transistor

SiGe/Si Silicon Germanide/Silicon (Heterostructure)

SIGForth Special Interest Group on Forth (Programming Language) [of Association for Computing Machinery, US]

SIGForth Newsl SIGForth Newsletter [of Association for Computing Machinery, US]

SIGGEN Signal Generator [also Sig Gen]

SIGGRAPH Special Interest Group on Computer Graphics [of Association for Computing Machinery, US]

SIGGRAPH Comput Graph SIGGRAPH Computer Graphics [of Association for Computing Machinery, US]

SIGHAN Signal Handling (Program)

SIGI System for Interactive Guidance and Information

SIGIR Special Interest Group on Information Retrieval [of Association for Computing Machinery, US]

SIGIR Forum SIGIR Forum [of Association for Computing Machinery, US]

SIGINT Signal Intelligence

SIGLE System for Information on Gray Literature in Europe

SIGMA Sealed Insulating Glass Manufacturers Association [US]

SIGMA Shielded Inert Gas Metal-Arc Welding

SIGMA Society of Independent Gasoline Marketers of America [US]

SIGMA Supersonic Inert Gas Metal Atomization; Supersonic Inert Gas Metal Atomizer

SiGMA Supersonic Inert Gas Metal Atomizer [of National Institute of Standards and Technology, Gaithersburg, Maryland, US]

SIGMA System for Integrated Genome Map Assembly

SIGMET Significant Meteorological Information

SIGMETRICS Special Interest Group on Measurement and Evaluation [of Association for Computing Machinery, US]

SIGMETRICS Perf Eval Rev SIGMETRICS Performance Evaluation Review [of Association for Computing Machinery, US]

SIGMICRO Special Interest Group on Microprogramming [of Association for Computing Machinery, US]

SIGMICRO Newsl SIGMICRO Newsletter [of Association for Computing Machinery, US]

SIGMICRO TC-MICRO Newsl SIGMICRO TC-MICRO Newsletter [of Association for Computing Machinery, US]

SIGMOD Special Interest Group on Management of Data [of Association for Computing Machinery, US]

SIGMOD Rec SIGMOD Record [of Association for Computing Machinery, US]

SIGN Special Interest Group Network

Sign Signature [also sign]

sign *(signatum)* – signed

sign signify

Signal Draht Signal und Draht [German Publication on Signals and Wires]

Signal Process Signal Processing [Journal published in the Netherlands]

Signal Process, Image Commun Signal Processing: Image Communication [Journal published in the Netherlands]

SIGNE Solar Interplanetary Gamma Neutron Experiment [French Satellite]

SIGNUM Special Interest Group on Numerical Mathematics [of Association for Computing Machinery, US]

SIGNUM Newsl SIGNUM Newsletter [of Association for Computing Machinery, US]

SIGOA Special Interest Group on Office Automation [now SIGOIS, US]

SIGOIS Special Interest Group on Office Information Systems [of Association for Computing Machinery, US]

SIGOIS Bull SIGOIS Bulletin [of Association for Computing Machinery, US]

SIGOP Signal Operation

SIGOPS Special Interest Group on Operating Systems [of Association for Computing Machinery, US]

SIGOPS Oper Syst Rev SIGOPS Operating Systems Review [of Association for Computing Machinery, US]

SIGPC Special Interest Group on Personal Computing Systems [of Association for Computing Machinery, US]

SIGPC Notes SIGPC Notes [of Association for Computing Machinery, US]

SIGPIP Signal Peripheral Interchange Program

SIGPLAN Special Interest Group on Programming Languages [of Association for Computing Machinery, US]

SIGPLAN Not SIGPLAN Notices [of Association for Computing Machinery, US]

SIGSAC Special Interest Group on Security, Audit and Control [of Association for Computing Machinery, US]

SIGSAC Rev SIGSAC Review [of Association for Computing Machinery, US]

SIGSAM Special Interest Group for Symbolic and Algebraic Manipulation [of Association for Computing Machinery, US]

SIGSAM Bull SIGSAM Bulletin [of Association for Computing Machinery, US]

SIGSCSA Special Interest Group on Small Computing Systems and Applications [of Association for Computing Machinery, US]

SIGSD Special Interest Group for System Documentation [of Association for Computing Machinery, US]

SIG/SIC Special Interest Group/Special Interest Committee

SIGSIM Special Interest Group on Simulation [of Association for Computing Machinery, US]

SIGSIM Newsl SIGSIM Newsletter [of Association for Computing Machinery, US]

SIGSMALL Special Interest Group on Small and Personal Computing Systems and Applications [of University of Western Ontario, Canada]

SIGSMALL Special Interest Group on Symbolic and Algebraic Manipulation [of Association for Computing Machinery, US]

SIGSMALL Newsl SIGSMALL Newsletter [of Association for Computing Machinery, US]

SIGSMALL/PC Special Interest Group on Symbolic and Algebraic Manipulation/Personal Computers [of Association for Computing Machinery, US]

SIGSMALL/PC Notes SIGSMALL/PC Notes [of Association for Computing Machinery, US]

SIGSOFT Special Interest Group on Software Engineering [of Association for Computing Machinery, US]

SIGSOFT Softw Eng Notes SIGSOFT Software Engineering Notes [of Association for Computing Machinery, US]

SIGTTO Society of International Gas Tanker and Terminal Operators [UK]

SIGUCC Special Interest Group on University Computer Centers [of Association for Computing Machinery, US]

SIGUCCS Special Interest Group for University and College Computing Services [of Association for Computing Machinery US]

SIGUCCS Newsl SIGSUCCS Newsletter [of Association for Computing Machinery, US]

SIGWX Significant Weather [Meteorology]

SiH Silicon Monohydride

Si:H Hydrogenated Silicon

SiH$_3$ Silyl [Symbol]

Si-HET Silicon Heterojunction Bipolar Transistor

SiH$_4$/CH$_4$ Silane/Methane (Gas Mixture)

SiH$_4$/DMA Silane/Dimethylamine (Gas Mixture)

SiH$_4$/H$_2$ Silane/Hydrogen (Gas Mixture)

SiH$_4$/He Silane/Helium (Gas Mixture)

Si$_2$H$_6$/He Disilane/Helium (Gas Mixture)

SII Standards Institution of Israel

Si:I Iodized Silicon

SIIAS Staten Island Institute of Arts and Sciences [New York, US]

Si IC Silicon Integrated Circuit (Technology)

SIIE Secondary Ion-Ion Emission

SiIMPATT Silicon Impact Avalanche Transit Time

SIIMS Secondary Ion Imaging Mass Spectroscopy

SIIN Small Islands Information Network [Maintained by the University of Prince Edward Island, Charlottetown, Canada]

SIIO Senior Installation IRM (Information Resource Management) Official

SIIRS Smithsonian Institution Information Retrieval System [US]

SiKα Silicon K-Alpha (Radiation) [also SiK$_\alpha$]

SIKM Senter for Internasjonal Klima- og Miljöforskning [International Center for Climatic and Environmental Research, University of Oslo, Norway]

Si-K XAFS Chromium-Potassium X-Ray Absorption Fine Structure

Si-K XANES Chromium-Potassium X-Ray Absorption Near-Edge Structure

SIL Safety Information Letter

SIL Scanner Input Language

SIL Single In-Line

SIL Solid Immersion Lens

SIL Sound Interference Level

SIL Special Instruments Laboratory [of National Chemical Laboratory, Pune, India]

SIL Speech Interference Level

SIL Steam Isolation Line

SIL Store Interface Link

SIL Systems Implementation Language

SIL Systems Integration Laboratory

Sil Silence [also sil]

Sil Silicone

Sil Sillimanite [Mineral]

Sil Silver [Abbreviation; Symbol: Ag]

Si/La-214 Silicon on Lanthanum Barium Copper Oxide Superconductor (La$_2$BaCu$_4$O$_5$)

SILAFAE Simposia Latinoamericana par Fisica de Alta Energia [Latin American Symposium on High Energy Physics, San Juan, Puerto Rico, US]

Si(Li) Lithium-Drifted Silicon (Detector)

Silic Silicate

Silica Ind Silica Industries [International Journal]

Silic Ind Silicates Industriels [Belgian Journal on Industrial Silicates]

SIL MO Solid Immersion Lens (Near-Field) Magneto-Optical (Recording Technique)

SILO Sealed Interface Local Oxidation [Materials Science]

SILS Shipboard Impact Locator System

Sil S Silver Solder [also SilS]

SILT Stored Information Loss Tree

SILTS Shuttle Infrared Leeside Temperature-Sensing System [of NASA]

Silvae Genet Silvae Genetica [Journal published in Germany]

Silver Inst Lett Silver Institute Letter [Published in the US]

SILW Solidified Intermediate-Level Waste

Silwr Silverware [also silwr]

SIM Salon International Medical

SIM Scanning Ion Microscope; Scanning Ion Microscopy

SIM Scientific Instrument(ation) Module

SIM Secondary Ion Microscope; Secondary Ion Microscopy

SIM Selected Ion Monitoring; Selective Ion Monitoring [also sim]

SIM Service Instructions Message

SIM Signal Isolator Module

SIM Single-In-Line Memory

SIM Single-In-Line Module [Electronics]

SIM Sistema Interamericano de Metrología [Interamerican Metrology Institute]

SIM Society for Industrial Microbiology [US]

SIM Society for Information Management [US]

SIM Software Interface Module

SIM Strain-Induced Martensite [Metallurgy]

SIM Stress-Induced Martensite [Metallurgy]

SIM Student/Industry Message Board [of American Iron and Steel Institute–Iron and Steel Society, US]

SIM Structure Inversion Method [Materials Science]

SIM Subscriber Interface Module [Telecommunications]

SIM Superconductor–Insulator–Metal

SIM Surface-Induced Mineralization (Process)

SIM Synchronous Induction Motor

SIM Synchronous Interface Module

5-SIM Dimethyl-5-Sulfo Isophthalate

SiM Sidereal Month [Astronomy]

Sim Simulate(d); Simulation; Simulator [also sim]

sim *(similis)* – of similar composition [Medical Prescriptions]

SIMA Scientific Instrument Manufacturers Association [UK]

SIMA Secondary Ion Mass Analysis; Secondary Ion Mass Analyzer

SIMA Secondary Ion Microanalysis; Secondary Ion Microanalyzer

SIMA Special Import Measures Act

SIMA Steel and Industrial Managers Association [UK]

SIMA Strain-Induced Melt Activation [Metallurgy]

Sima Silicon-Magnesium (Layer) [also sial] [Geology]

SIMAJ Scientific Instrument Manufacturers Association of Japan

SIMAS Shuttle Information Management Accountability System [of NASA]

Si-29 MAS-NMR Silicon-29 Magic-Angle Spinning-Nuclear Magnetic Resonance [also ^{29}Si MAS-NMR]

Simazine 2-Chloro-4,6-Bis(ethylamino)-s-Triazine [also simazine]

SIMBAD Set of Identifications, Measurements and Bibliography for Astronomical Data [Maintained by Centre de Données Astronomiques de Strasbourg, France]

Si-MBE Silicon Molecular Beam Epitaxy

SIMCHE Simulation and Checkout Equipment

SIMCOM Simulation Complex

SIMCOM Simulator Compiler

SIMCOM Simulator Computer

SIMCON Scientific Inventory and Management Control

SIMCON Software for Integrated Manufacturing Consortium [of National Research Council of Canada]

SIM COORD Simulation Coordinator

SIMD Single-Input, Multiple-Data (Stream)

SIMD Single-Instruction (Stream), Multiple-Data (Stream)

SIMDIS Simulated Distillation Software

SIME Sian Institute of Metallurgical Engineering [Sian, PR China]

SIME Società Italiana delle Microscopia Elettronica [Italian Society for Electron Microscopy]

Si-Me Silicon-Metal (Bond)

SIMEA Società Italiana Meridionale per l'Energia Atomica [Society for Nuclear Energy in Southern Italy]

Si(MeO)$_4$ Silicon Tetraethoxide

SIMEON Simplified Control

SIMF Svenska Institut för Metalforskning [Swedish Institute for Metals Research, Stockholm]

SIMFAC Simulation Facility

SIMFUEL Simulated Used CANDU (Canadian Deutreium Uranium) Reactor Fuel [of Atomic of Energy of Canada Limited, Whiteshell Laboratories, Pinawa, Manitoba]

SIMICOR Simultaneous Multiple Image Correlation

SIMILE Simulation of Immediate Memory in Learning Experiments

SIMIT Size-Induced Metal-Insulator Transition

SiMITATT Silicon Mixed Tunneling Avalanche Transit Time

SIMLAB (Computer) Simulation Laboratory

SIMM Single-In-Line Memory Module [Electronics]

SIMM STN International, Internet and Multimedia (Project) [of Fachinformationszentrum Karlsruhe, Germany]

SIMM Symbolic Integrated Maintenance Manual

SIMMS Secondary Ion Microscope Mass Spectroscopy

SIMNAV Simulation and Forecasting System for Marine Traffic

SIMNS Simulated Navigation System

SIMO Simultaneous(ly) [also simo]

SIMO Simultaneous Motion [also simo]

SIMO Dump Simultaneous Supply Water and Waste Water Dump [Aerospace]

SIMON Software Interface Monitor

SIMOS Stacked-Gate Injection Metal-Oxide Semiconductor

Si-MOSFET Silicon Metal-Oxide-Semiconductor Field-Effect Transistor

SIMOX Separation by Implantation of Oxygen (Technique); Separation by Implanted Oxygen (Technique); Silicon Implanted with Oxygen (Technique) [also Simox] [Electronics]

SIMOX SOI Separation by Implanted Oxygen Siliconon-Insulator (Technique) [Electronics]

SIMP Satellite Information Message Protocol

SIMP Società Italiana di Mineralogia e Petrologia [Italian Society for Mineralogy and Petrology]

SIMP Specific Impulse [Aerospace]

SIMPAC Simplified Programming for Acquisition and Control

SIMPAC Simulation Package

SIMPLE Secondary Isotope Multiplets of Partially Labeled Entities [Physics]

SIMPLE System for Integrated Maintenance and Program Language Extension

SIMPP Simple Image Processing Package

SI-MQW Semi-Insulating Multiple Quantum Well [Electronics]

SIMR Swedish Institute of Metals Research

SIMR Systems Integration Management Review

SIMRA Scientific Instruments Manufacturers Research Association [UK]

SIMS Scandinavian Simulation Society [Finland]

SIMS Sea Ice Monitoring Site

SIMS International Conference on Secondary Ion Mass Spectrometry

SIMS Secondary Ion Mass Spectroscopy

SIMS Secondary Ion Mass Spectrum

SIMS Security Information Management System

SIMS Shuttle Imaging Microwave System [NASA]

SIMS Shuttle Inventory Management System [NASA]

SIMS Simulated Mission Support [NASA]

SIMS Single-Item, Multi-Source

SIMS Supply Inventory Management System

SIMS Symbolic Integrated Maintenance System

SIMS Symposium for Innovation in Measurement Science

SIMSIDP Secondary Ion Mass Spectroscopy Image Depth Profiling [also SIMS-IDP]

SIM/SIMS Scanning Ion Microscopy/Secondary Ion Mass Spectroscopy

SIMT Stress-Induced Martensitic Transformation [Metallurgy]

SIMTEL Simulation and Teleprocessing

SIMTOP Silicon Nitride Masked Thermally Oxidized Postdiffused Mesa Process [Electronics]

SIMU Simulated Inertial Measurement Unit

Simul Simulation [also simul]

simul simultaneous

SIMULA Simulation Language

SIMULA Newsl SIMULA Newsletter [UK Publication on Computer Simulation]

Simulcast Simultaneous Broadcast [also simulcast]

Simul Dig Simulation Digest [of Association for Computer Machinery, US]

Simul Games Simulation and Games [Journal published in the US]

SIN Schweizerisches Institut für Nuklearforschung [Swiss Institute for Nuclear Research; Now Part of Paul Scherrer Institute, Villigen]

SIN Sensitive Information Network

SIN Simultaneous Interpenetrating (Polymer) Network

SIN Sine (Function) [Programming] [also Sin, or sin]

SIN Single Interconnection Network

SIN Social Insurance Number

SIN Subject Indicator Number

SIN Support Information Network

SIN Symbolic Integrator

SIN-1 3-Morpholinosyndnonimine

Si_xN_y Silicon Nitrides [General Formula]

$Si_3N_4(f)$ Silicon Nitride Fiber

$Si_3N_4(w)$ Silicon Nitride Whisker

Sin Salmonella infantis [Microbiology]

sin sine [Symbol]

sin^{-1} inverse sine [also arcsin]

SINA Service für Industrie und Nukleartechnische Anlagen GmbH [German Service Company for Industry and Nuclear Engineering Plants; located at Pforzheim]

SINAD Signal-to-Noise (Ratio) and Distortion

SINADS Shipboard Integrated Navigation and Display System

SINAP Satellite Input to Numerical Analysis and Prediction

SINB Southern Interstate Nuclear Board [US]

SINCGARS Single Channel Ground/Automatic Recovery System

SINDA Systems Improved Numerical Differencing Analyzer

SINDO Scaled Intermediate Neglect of Differential Overlap [Physics]

Sing Singapore

sing singular

Singap J Phys Singapore Journal of Physics

$[Si(NCN)_2]_n$ Silylcarbodiimide [General Formula]

Sing\$ Singapore Dollar [Currency]

SINGNET Singapore Information Network [also Singnet]

SiN:H Hydrogenated Silicon Nitride [also $SiN_x:H$]

sinh hyperbolic sine [Symbol]

$sinh^{-1}$ inverse hyperbolic sine [also arsinh]

Si-Ni Silicon-Nickel (Alloy System)

Si/Ni Silicon/Palladium (Heterojunction)

$Si(NMe_2)_4$ Tetrakis(dimethylamido)silicon

Si-29 NMR Silicon-29 Nuclear Magnetic Resonance [also ^{29}Si NMR]

SIN Newsl SIN Newsletter [of Schweizerisches Institut für Nuklearforschung, Switzerland] [now PSI Nucl Part Phys Newsl]

SINPFEMO Radio Signal Quality Code

SINPO Radio Transmission Signal Quality Code

SINQ Schweizer Intensitäts-Neutronenquelle [Swiss High-Intensity Neutron Source; of Paul-Scherrer Institute, Villigen]

SINQ-NR Schweizer Intensitäts-Neutronenquelle–Neutronenradiographie [Swiss High-Intensity Neutron Source–Neutron Radiography; of Paul-Scherrer Institute, Villingen]

SINR Swiss Institute for Nuclear Research [Now Part of Paul Scherrer Institute, Villigen]

SINS Ship's Inertial Navigation System

SINS Superconductor-Insulator-Normal (Metal)-Superconductor

$Si_3N_4(Sc_2O_3)$ Scandia-Doped Silicon Nitride

Si_3N_4/Si Silicon Nitride on Silicon (Substrate)

Si_3N_4-SiC Silicon Nitride-Silicon Carbide (Composite)

SINSS Shipboard Ice Navigation Support System

SINTERING International Conference on the Science, Technology and Applications of Sintering

SINWT Science Institute of the Northwest Territories [Canada]

$Si_3N_4(Yb_2O_3)$ Ytterbia-Doped Silicon Nitride

$Si_3N_4(Y_2O_3)$ Yttria-Doped Silicon Nitride

$Si_3N_4(ZrO_2)$ Zirconia-Doped Silicon Nitride

SIO Scripps Institute of Oceanography [of University of California at San Diego, US]

SIO Serial Input/Output

SIO Skidaway Institute of Oceanography [Savannah, Georgia, US]

SIO Social Investment Organization [US]

SIO Staged in Orbit

SIO Start Input/Output

SIO Step Input/Output

SIO Systems Integration Office

Si-O Silicon-Oxygen (Bond)

SiO Silicon Monoxide (Coating)

SiO_x Native Silicon Oxide [also SiOx]

SiO^- Orthosilicate (Ion) [Symbol]

SiO_2-Al_2O_3 Silicon Dioxide–Aluminum Oxide (System)

SIOC Serial Input/Output Channel

SiOC Silicon Oxycarbide (Ceramic) [also SIOC, or Sioc]

SiOD Deuterated Silanol

$Si(OEt)_4$ Silicon Tetraethoxide (or Tetraethylorthosilicate, or Tetraethoxysilane)

SiO:F Fluorine-Doped Silicon Oxide

SiO_2:F Fluorine-Doped Silicon Dioxide

SiOH Silanol

SiO_x:H Hydrogenated (Native) Silicon Oxide

SIOL Stress-Induced Order Locking [Materials Science]

SiO_2-MnO_2 Silicon Dioxide-Manganese Dioxide (System)

SiON Silicon Oxynitride (Ceramic)

SIOP Selector Input/Output Processor

SIOP Single Integrated Operations Plan

SiO_2-PMMA Silicon Dioxide–Poly(methylmethacrylate)

SiO_2/Si Silicon Dioxide on Silicon (Substrate)

SiO_2/SiC Silicon Dioxide Reinforced Silicon Carbide (Composite)

SiO_2/SiO_2 Silicon Dioxide Reinforced Silicon Dioxide (Composite)

SiO_2/Ti Silicon Dioxide/Titanium (Interface)

SIOUX Sequential Iterative Operation Unit X

SiOx Native Silicon Oxide [also SiO_x]

SiO_2-Y_2O_3 Silicon Dioxide-Yttrium Oxide (System)

SIP Safety Injection Pump

SIP Saskatchewan Institute of Petrology [of University of Saskatchewan, Canada]

SIP Scientific Information Processing; Scientific Information Processor

SIP Scientific Instrument Package

SIP Separation Instrument Package

SIP Sheetmetal Insert Process

SIP Short Irregular Pulse

SIP Single-In-Line Package [Electronics]

SIP Sintering/Hot Isostatic Pressing [Powder Metallurgy]

SIP Sleep Inducing Peptide [Biochemistry]

SIP Smithsonian Institution Press [Washington, DC, US, and London, UK]

SIP Société d'Industrie Pétrolifère [Petroleum Industry Society, France]

SIP Society for Invertebrate Pathology [US]

SIP Software Instrumentation Package

SIP Sputter Ion Plating

SIP Sputter Ion Pump

SIP Standard Information Package

SIP Standard Interface Panel

SIP State Implementation Plan [of US Environmental Protection Agency]

SIP Stay-In-Place

SIP Strain Isolator (or Isolation) Pad [of NASA Space Shuttle]

SIP Submerged Injection Process

SIP Symbolic Input Program

SIP System Implementation Plan

Si:P Phosphorus-Doped Silicon

SIPA Science-Based Industrial Park Administration [Taiwan]

SIPA Single Intensity Profile Analysis

SIPA Stress-Induced Preferential Absorption

SIPB Safety Injection Permissive Blocks

SIPC Securities Investor Protection Corporation [US]

SIPC Simply Interactive Personal Computer

Si-P-C-N Silicon-Phosphorus-Carbon-Nitrogen

Si/Pd Silicon on Palladium (Substrate)

SIPE Società Italiana Prodotti Esplodenti [Italian Society for Explosives and Related Products]

SIPER Swedish Institute for Production Engineering Research

SIPES Society of Independent Professional Earth Scientists [US]

Si(P):H Hydrogenated Phosphorus-Doped Silicon

SIPI Scientists Institute for Public Information [US]

SIPN Semi-Interpenetrating Polymer Network [also S-IPN]

SIPN Simultaneous Interpenetrating (Polymer) Network

SIPN Stress-Induced Preferential Nucleation [Metallurgy]

SIPO Serial-In, Parallel-Out (Shift Register) [Computers]

SIPO Serial Input/Parallel Output [Computers]

SIPOP Satellite Information Processor Operational Program

SIPOS Semi-Insulating Polycrystalline Oxygen-Doped Silicon

SIPP Single In-line Pin Package [Electronics]

SIPP Society of Irish Plant Pathologists

SIPP Sodium Iron Pyrophosphate

SIPR Servico de Informaciónes de la Puerto Rico [Puerto Rico Information Service]

SIPRE Snow, Ice and Permafrost Research Establishment [US Army]

SIPRI Stockholm International Peace Research Institute [Sweden]

SIPROS Simultaneous Processing Operating System

SIPS SAC (Strategic Air Command) Intelligence Data Processing System [of US Air Force]

SIPS Simulated Input Preparation System

SIPS Small Instrument Pointing System

SIPS Società Italiana per il Progresso delle Scienze [Italian Society for the Advancement of Science]

SIPS Sonar Image Processing System

SIPS Spacelab Input Processing System

SIPS Sputter-Induced Photon Spectroscopy

SIPS State Implementation Plan System [of US Environmental Protection Agency]

SIPS Statistical Interactive Programming System

SIP/SIMM Single-In-Line Package/Single-In-Line Memory Module [Electronics]

SIPSS Science, Informatics and Professional Services Sector [of Public Works and Government Services Canada]

SIPT Shock-Induced Phase Transformation [Materials Science]

SIPT Slurry Integrated Performance Test

Si(Pt):H Hydrogenated Platinum-Doped Silicon

SIPWR System Integrated Pressurized Water Reactor

SIR Safe Integral (Nuclear) Reactor

SIR Safety Investigation Regulations

SIR Savings-to-Investment Ratio

SIR Scientific and Industrial Research

SIR Scientific Information Retrieval

SIR Segment Identification Register

SIR Selected Ion Recording

SIR Selective Information Retrieval

SIR Semantic Information Retrieval

SIR Serial Infrared (Port)

SIR Shuttle Imaging Radar [NASA]

SIR Signal-to-Interference Ratio

SIR Simultaneous Impact Rate

SIR Service International des Recherches [International Research Service]

SIR Software Initiated Restart

SIR Spaceborne Imaging Radar

SIR Special Information Retrieval

SIR Specification Information Retrieval

SIR Standardized Indonesian Rubber

SIR Statistical Information Retrieval

SIR Stepped Interference Reflection (Method)

SIR Styrene Isoprene Rubber

SIR Submarine Intermediate Reactor

SIR Supersonic Infantry Rocket

SIR Surface Insulation Resistance

SIR Sustained Information Rate

SIR Symbolic Input Routine

SIR Systems Integration Review

SIR System Interface Requirements

SIRA Scientific Instruments Research Association [UK]

SIR-A Shuttle Imaging Radar A (System) [of NASA]

SIRAID Scientific Instruments Research Association Information and Data Service [UK]

SIRC Special Investment Research Contract

SIR-C Shuttle Imaging Radar C (System) [of NASA]

SIR-C Spaceborne Imaging Radar C

SIRDS Single-Image Random Dot Stereogram

SIRE Satellite Infrared Experiment [also Sire]

SIRE Society for the Investigation of Recurring Events [US]

SIRE Syracuse Information Retrieval Experiments [of Syracuse University, New York State, US]

SIRIM Standards and Industrial Research Institute of Malaysia

Siriraj Hosp Gaz Siriraj Hospital Gazette

SIRIS Stratospheric Infrared Interferometric Spectrometer

SIRL Structured Interactive Robot Language

SIRM Saturation Isothermal Remanent Magnetization

SIRM Scanning Infrared Microprobe

SIRR Software Integration Readiness Review

SIRRS Surface-Induced Resonant Scattering

SIRS Salary Information Retrieval System

SIRS Satellite Infrared Spectrometer

SIRS Security Information Retrieval System

SIRS Specification Information Retrieval System

SIRS Surface Infrared Spectroscopy

SIRSA Special Industrial Radio Service Association [US]

SIRT Simultaneous Iterative Reconstruction Technique

SIRT Status and Implementation of Recommendations on Technical Questions [of European Civil Aviation Conference, France]

SIRTF Space Infrared Telescope Facility [of Infrared Processing and Analysis Center, NASA-Jet Propulsion Laboratory, Pasadena, California, US]

SIRTF FIR Space Infrared Telescope Facility Far Infrared (Technology)

SIRU Strapdown Inertial Reference Unit

SIRWT Safety Injection Reserve Water Tank

SIS Safety Injection Signal

SIS Safety Injection Symbol

SIS SAIL (Shuttle Avionics Integration Laboratory) Interface System [of NASA]

SIS Satellite Interceptor System

SIS Saturated Interference Spectroscopy

SIS Scanning Imaging Spectrometer

SIS School of Industrial Science [of Cranfield Institute of Technology, UK]

SIS Schwerionensynchrotron [Heavy-Ion Synchrotron of Gesellschaft für Schwerionenforschung mbH in Darmstadt, Germany]

SIS Science Information Service

SIS Science Information System [of US Geological Survey/Biological Resources Division]

SIS Scientific Information System

SIS Scientific Instruction Set

SIS Scientific Instrument Services [US]

SIS Scientific Instrument Simulation

SIS Scientific Instrument Society [UK]

SIS Secret Intelligence Service [UK]

SIS Selected Inventor Service [of Organization for Economic Cooperation and Development–Information Policy Group, France]

SIS Semiconductor-Insulator-Semiconductor [also S-I-S, or S/I/S]

SIS Serial Input Special

SIS Short-Interval Scheduling

SIS Shuttle Information System [of NASA]

SIS Shuttle Interface Simulator [of NASA]

SIS Signalling Interworking Subsystem

SIS Simulation (or Simulator) Interface Subsystem

SIS Simulation (or Simulator) Interface System

SIS Single Integrated Simulation

SIS Software Implementation Specification

SIS Software Integrated Schedule

SIS Sound-in-Sync (Transmission)

SIS Spatial Information System

SIS Special Industrial Services [of United Nations Industrial Development Organization]

SIS Standard Interface Specification

SIS Standardiseringskommissionen i Sverige [Swedish Standards Institute]

SIS Standards Information Service

SIS Standards Information System [of National Institute of Standards and Technology, Gaithersburg, Maryland, US]

SIS Statistical Information System

SIS Strategic Information System

SIS Stress-Induced Segregation [Metallurgy]

SIS Structure-Related Internal Stress [Materials Science]

SIS Styrene-Isoprene-Styrene (Block Copolymer) [also S-I-S]

SIS Superconductor-Insulator-Superconductor [also S-I-S, or S/I/S]

SIS Supersymmetry and Integrable Systems (Conference)

SIS Supplier Identification System

SIS Support Information System

SIS Sveriges Standardiseringskommission [Swedish Commission for Standardization]

SIS System Identification System

SIS System Interface Specification [of NASA Planetary Data System]

SIS System Interrupt Supervisor

SIS Systems Integration Schedule

S-I-S Styrene-Isoprene-Styrene (Block Copolymer) [also SIS]

S/I/S Semiconductor/Insulator/Semiconductor [also S-I-S, or SIS]

S/I/S Superconductor/Insulator/Superconductor [also S-I-S, or SIS]

Sis Siltstone [also sis]

SISA Strontium Isotope Sample Analysis

SISAM Spectrometer with Interference Selective Amplitude Modulation

Si:Sb Antimony-Doped Silicon

SISD Scientific Information Systems Division

SISD Single-Instruction (Stream), Single-Data (Stream)

SISDATA Statistical Information System Databank [Italy]

SISEX Shuttle Imaging Spectrometer Experiment [of NASA]

SISF Superlattice Intrinsic Stacking Fault [also S-ISF] [Solid-State Physics]

SISFET Semiconductor-Insulator-Semiconductor Field-Effect Transistor

Si SG Spheroidal Graphite Silicon (Cast Iron) [Metallurgy]

Si SG Iron Spheroidal Graphite Silicon (Cast) Iron [also Si SG iron]

SISI Southern Institute of Science and Industry [US]

SISI Surveillance and In-Service Inspection

Si-Si Silicon-Silicon (Reaction)

Si/SiB Silicon/Silicon Boride (Multilayer)

SiSiC Siliconized Silicon Carbide

Si-SiC Silicon-Silicon Carbide (Ceramics) [also SiSiC]

Si/SiGe Silicon/Silicon Germanide (Heterostructure)

Si/SiO₂ Silicon/Silicon Dioxide (Heterojunction)

Si-SiO₂ Silicon-Reinforced Silicon Dioxide (Composite)

SISIR Singapore Institute of Standards and Industrial Research

SIS-JJ Superconductor-Insulator-Superconductor Josephson Junction

SiSn Silicon Stannate (Semiconductor)

SISO Serial-In, Serial-Out (Shift Register)

SISO Serial Input/Serial Output

SISO Single Input, Single Output

SI/SO Shift In/Shift Out

Si/SOI Silicon/Silicon on Insulator (Substrate)

SISRI Shanghai Iron and Steel Research Institute [PR China]

SISS Single-Item, Single-Source

SISS Submarine Integrated Sonar System [also SISSY]

SISSA Scuola Internazionale Superiore di Studi Avanzati [International School for Advanced Studies, Trieste, Italy]

SIST Society for Imaging Science and Technology [US]

Sist Autom Sistemi e Automazione [now Sist Impresa]

Sist Impresa Sistemi e Impresi [Italian Publication on Systems and Automation]

SISTM Simulation Incremental Stochastic Transition Matrices

SISWG STS (Space Transportation System) Integrated Schedule Working Group [of NASA]

SIT Safety Injection Tank

SIT Self-Induced Transparency

SIT Separation-Initiated Timer

SIT Service Improvement Team

SIT Shandong Institute of Technology [Jinan, PR China]

SIT Shibaura Institute of Technology [Tokyo, Japan]

SIT Shonan Institute of Technology [Kanagawa, Japan]

SIT Shuttle Integrated Test [of NASA]

SIT Shuttle Interface Test [of NASA]

SIT Silicon-Intensified Target; Silicon-Intensifier Target

SIT Silicon-Intensifier Tube

SIT Social Implications of Technology (Society) [of Institute of Electrical and Electronics Engineers, US]

SIT Society of Instrument Technology [UK]

SIT Software Integration Test

SIT Special Information Tone(s)

SIT Specific Interaction Theory [Physics]

SIT Spontaneous Ignition Temperature

SIT Static Induction (or Inductance) Transistor

SIT Stevens Institute of Technology [Hoboken, New Jersey, US]

SIT Strain-Induced Transformation [Metallurgy]

SIT Strontium Indium Tantalate

SIT Structural Integrity Test

SIT Subarea Index Table

SIT Sugar Industry Technologist

SIT System Initialization Table

SIT System Integration Test

.SIT Stuff-It (Compressed) [Macintosh File Name Extension)

SiT Sidereal Time [Astronomy]

Sit Situate; Situation [also sit]

SITA Société Internationale des Télécommunications Aéronautiques [International Society for Aeronautical Telecommunications, France]

SITB Symmetric(al) Incoherent Twin (Grain) Boundary [Materials Science]

SITC Salford Information Technology Center [of Salford University, UK]

SITC Satellite International Television Center

SITC Standard International Trade Classification [of United Nations]

SITC Standard International Trade Commodity

sitd situated [also sit'd]

SITE Satellite Instructional Television Experiment [India]

SITE Sculpture in the Environment (Group) [US]

SITE Search Information Tape Equipment

SITE Sodium Selenite–(Bovine) Insulin–(Human) Transferrin–Ethanolamine (Mixture) [Biochemistry]

SITE Spacecraft Instrumentation Test Equipment

SITE Superfund Innovative Technology Evaluation (Project) [US]

SITEL Société des Ingénieurs des Télécommunications et d'Electronique [Society of Electronic and Telecommunication Engineers, Belgium]

SITES Smithsonian Institution Traveling Exhibition Service [US]

SITH Static Induction Thyristor

SITI Swiss Institute for Technical Information

Si/TiSi₂ Silicon Reinforced Titanium Disilicide

SITL Static Induction Transistor Logic

Sitn Situation [also sitn]

SITOY Seoul International Toy Fair [South Korea]

SITPRO (Committee for the) Simplification of International Trade Procedures Board [UK]

SITRA South India Textile Research Association

SITREP Situation Report [also SIT-REP]

SITRO Swedish Industries Trade Representative Office

SITS 4-Acetamido-4'-Isothiocyanatostilbene-2,2'-Disulfonic Acid

SITS SAGE (Semiautomatic Ground Environment) Intercept Target Simulation

SITS Social Implications of Technology Society [of Institute of Electrical and Electronics Engineers, US]

SITS Société Internationale de la Traitement Superficiel [International Society for Surface Treatment]

SITT Systems Integration of Triad Technology

SITU Society for the Investigation of the Unexplained [US]

SITVC Secondary Injection Thrust Vector Control

Sitz ber Akad Wiss Wien Sitzungsberichte der Akademie der Wissenschaften, Wien [Transactions of the Academy of Sciences, Vienna, Austria]

Sitz ber, Österr Akad Wiss Math-Natwiss Kl I Sitzungsberichte, Österreichische Akademie der Wissenschaften, Mathematisch-Naturwissenschaftliche Klasse, Abteilung I [Transactions of the Austrian Academy of Sciences, Mathematical Sciences Class, Section II] [also Sitz ber, Oesterr Akad Wiss Math-Nat wiss Kl I]

Sitz ber, Österr Akad Wiss Math-Natwiss Kl II Sitzungsberichte, Österreichische Akademie der Wissenschaften, Mathematisch-Naturwissenschaftliche Klasse, Abteilung II [Transactions of the Austrian Academy of Sciences, Mathematical Sciences Class, Section II [Sitz ber, Oesterr Akad Wiss Math-Nat wiss Kl II]

SIU Seamen's (or Seafarers) International Union

SIU Serial Interface Unit

SIU Side-In Unit

SIU Signal Input Units

SIU Signal Interface Unit

SIU Significant Industrial Uses

SIU Southern Illinois University [US]

SIU Static Input Unit

SIU System Input Unit

SIU System Interface Unit

SIU *(Système International d'Unités)* – International System of Units

SIUC Southern Illinois University, Carbondale [US]

S/IUCRC State/Industry University Cooperative Research Center [US]

SIUE Southern Illinois University, Edwardsville [US]

SIUS Southern Illinois University, Springfield [US]

SIUSM Southern Illinois University School of Medicine [Springfield, US]

SIV Simian Immunodeficiency Virus

Siv Sieve [also siv]

SIVE Shuttle Interface Verification Equipment [of NASA]

SI/VLSI Semi-Insulating Very Large-Scale Integration

SIWA Scottish Inland Waterways Association

SIW Smithsonian Institute, Washington [Washington, DC, US]

SIWL Single Isolated Wheel Load

SIX Sodium Isopropyl Xanthogenate

Si:X Extrinsic Silicon

SIXPAC System for Inertial Experiment Priority and Attitude Control

SiY Sidereal Year [Astronomy]

Si/Y-123 Iron on Yttrium Barium Copper Oxide Superconductor ($Yba_2Cu_3O_x$)

SIZ Security Identification Zone

SJ Saint John [New Brunswick, Canada]

SJ Saint John's [Newfoundland, Canada]

SJ Sample Job

SJ San Jose [California, US]

SJ San Jose [Luzon, Philippines]

SJ San José [Guatemala, Costa Rica, Uruguay and Chile]

SJ San Juan [Texas, US]

SJ San Juan [Puerto Rico and Argentina]

SJ Screw Jack

SJ Steel Jacket

SJ Svalbard and Jan Mayen [ISO Code]

S/J Signal-to-Jamming (Ratio) [also S-J or SJ]

SJA Sophora Japonica Agglutinin [Immunology]

SJAC Society of Japanese Aerospace Companies

SJAE Steam Jet Air Ejector

SJC Saint John's College [Cambridge, UK]

SJCC Spring Joint Computer Conference [US]

SJCM Standing Joint Committee on Metrication

SJD *(Scientiae Juridicae Doctor)* – Doctor of Juridical Science

SJE Supersonic Jet Epitaxy

SJF Shortest Job First

SJF Syndicat des Journalistes Français [Union of French Journalists]

SJFC St. John Fisher College [US]

SJG Scottish Journal of Geology [of Geological Society, UK]

SJGS San Joaquin Geological Society [California, US]

SJI Steel Joist Institute [US]

SJOD San Jacinto Ordnance Depot [Channelview, Texas, US]

SJP Safe Job Procedures

SJP Serialized Job Processor

SJRWMD St. Johns River Water Management District [US]

SJSC San Jose State College [now San Jose State University]

SJSU San Jose State University [California, US]

SJU Shanghai Jiao-Tong University [Shanghai, PR China] [also SJTU]

SJU St. John's University [New York City, US]

SK Saito and Krumbhaar (Model) [Solid-State Physics]

SK Saskatchewan [Canada]

SK Seek Command [Computers]

SK Sherrington-Kirkpatrick (Theory) [Physics]

SK Slater-Koster (Method) [also S-K] [Physics]

SK Slovakia [ISO Code]

SK Snoek-Köster (Relaxation) [Metallurgy]

SK South Kensington (Porosity Test)

SK South Korea(n)

S-K Slater-Koster (Method) [also SK] [Physics]

S-K Stranski-Krastanov (Film Growth) [also SK] [Solid-State Physics]

S(K) Static Structure Factor [Symbol]

Sk Sack [also sk]

Sk Sink [also sk]

Sk Sketch [also sk]

Sk Skip [also sk]

Sk Stark Number (of Heat Transfer) [Symbol]

S(k) Planar Structure Factor (in Solid-State Physics) [Symbol]

S(k) Static Spin Correlation Function (in Physics) [Symbol]

S(κ) Single-Particle Scattering Function [Symbol]

sk skip [Woven Fabrics]

.sk Slovakia [Country Code/Domain Name]

SKA Studienkommission für Atomenergie [Advisory Committee on Nuclear Energy, Switzerland]

SKB Schweizerischer Koordinationsausschuß für Biotechnologie [Swiss Coordinating Committee for Biotechnology]

SKB Skew Buffer

SKD Semi-Knocked Down

Sked Schedule [also sked]

SKEM Scanning Kerr Electron Microscope; Scanning Kerr Electron Microscopy

SKEW Square Kinetic Energy Well [Physics]

Skewphos 2,4-Bis(diphenylphosphino)pentane [also skewphos]

SKF Schweinfurter Kugellagerfabrik GmbH [German Manufacturer of Bearings and Other Machine Elements and Accessories; located at Schweinfurt]

SK&F Smith Kline and French Inter-American Corporation [Pharmaceutical Manufacturer]

SKIL Scanner Keyed Input Language

SKILL Satellite Kill

Skillings' Min Rev Skillings' Mining Review [Journal published in the US]

SkinnyDIP Thin-Packaged Dual In-Line Package [Electronics]

Skin Pharmacol Appl Skin Physiol Skin Pharmacology and Applied Skin Physiology [International Journal]

SKIP Simple Key-Management for Internet Protocols

SKIP Skeletal Isomerization Process [Chemical Engineering]

SKIZ Skeleton by Influence Zone

SKK Schmiedewerke Krupp-Kloeckner GmbH [Krupp-Kloeckner Forging Plant at Bochum, Germany]

SKKR Screened Korringa-Kohn-Rostoker (Model) [Atomic Bond-Calculation Method]

SKL Skip Lister

SKM Strategic Knowledge Management

SKN Skin Effects (of Hazardous Materials)

SKN Strontium Potassium Niobate

SKO Soft Kill Option [Biomedicine]

Skoda Rev Skoda Review [Czech Republic]

Skogind Skogindustri [Norwegian Publication on the Forestry Industry]

SKOR Sperry-Kalman Optimum Reset

SKP Skip-Line Printer

SKP Snoek-Koester Peak [Physics]

SKR Skip Record Processor

SKr Swedish Krona [also Skr] [Currency]

Skr Nor Vidensk-Akad Oslo I Skrifter utgitt av det Norske Videnskaps-Akademi i Oslo I, Matematisk-Naturvidenskaplige Klasse [Transactions of the Norwegian Academy of Sciences at Oslo I]

Skr Nor Vidensk-Akad Oslo II Skrifter utgitt av det Norske Videnskaps-Akademi i Oslo II [Transactions of the Norwegian Academy of Sciences at Oslo II]

SKS Scanning Kinetic Spectroscopy

SKS (Earthquake) Shear Wave that has Passed through the Earth's Core [Symbol] [also S']

Skt Socket [also skt]

SKU Stockkeeping Unit

SKU# Stockkeeping Unit Number

SKW South Koreran Won [also SK W] [Currency]

SKWOC Structured Keyword Out of Context

Skylab Sky Laboratory [of NASA]

Sky Telesc Sky and Telescope [Journal published in the US]

SL Saberliner (Airplane)

SL Safety Lamp

SL Safety Limit

SL Salvage Loss

SL San Luis [Argentina]

SL Scandinavian Lancers (Process) [Metallurgy]

SL Schinke-Luntz (Potential)

SL Schmid's Law [Materials Science]

SL School Library

SL Scientific Laboratory

SL Scientific Liaison

SL Sea Level [also sl]

SL Searchlight

SL Section Line

SL Self-Aligning Roller Bearing, Double Row, Outer-Ring Raceway Spherical, Two Integral Inner-Ring Ribs, Rollers Guided by Cage [Symbol]

SL Selective Leaching

SL Semiconductor Laser

SL Semi-Liquid

SL Sensation Level

SL Send Left

SL Separate-Lead (Type Cable)

SL Separately-Leaded (Cable)

SL Separate-Loading

SL Sequential Logic

SL Service Life

SL Shared-Logic (System)

SL Shear Lag (Limit) [Mechanics]

SL Shelf Life

SL Shift Left

SL Shock Load(ing)

SL Shrinkage Limit (of Soil) [Geology]

SL Side-Looking (Radar)

SL Sierra Leone(an)

SL Sierra Leone [ISO Code]

SL Signal Light

SL Significance Level [Statistics]

SL Simulation Language

SL Single Layer

SL Single-Lead

SL Single-Loop

SL Skylight

SL Slip Line [Crystallography]

SL Sodium Lamp

SL Soil Loss

SL Sonic Log(ging) [Oil Exploration]

SL Sonoluminescence; Sonoluminescent

SL Sound Level

SL Sound Locator [also S-L]

SL Source Language

SL Source Level [Acoustics]

SL Spacelab [European Space Agency/NASA]

SL Space Lattice [Crystallography]

SL Spectral Line

SL Sphingolipid [Biochemistry]

SL Spin Label(ling) [Physical Chemistry]

SL Spin Liquid [Solid-State Physics]

SL Spirit Level

SL Split Lens [Optics]

SL SPS (Super Proton Synchrotron) and LEP (Large Electron Positron Collider) (Division) [of CERN–European Laboratory for Particle Physics, Geneva, Switzerland]

SL Squadron Leader [also S/L]

SL Sri Lanka(n)

SL Standard Label

SL Standard Lamp

SL Standard Load (Rating) [Automobiles]

SL Staphylococcal Leukocidin [Biochemistry]

SL Static Load(ing)

SL Steam Locomotive

SL Stereolithography

SL Stefan's Law [Statistical Mechanics]

SL St. Lawrence (River) [US/Canada]

SL St. Louis [Missouri, US]

SL Stokes Law

SL Straight Line

SL Strained Layer [Solid-State Physics]

SL Streamline(d); Streamlining [also S/L]

SL Stretcher Levelling [Metallurgy]

SL Striated Layer

SL Structural Ledge (Theory) [Crystallography]

SL Sublaminar

SL Subscriber Line

SL Subscriber Loop

SL Summary Language

SL Superlattice [Solid-State Physics]

SL Surge Line [Meteorology]

SL Suspended Load [Geology]

SL Switched Line

SL Symbolic Language

SL Symmetry Line

SL System Librarian; System Library

SL Systems Language

S&L Savings and Loan (Association) [also S and L]

S/L Shop/Laboratory

S/L Solid-in-Liquid (Solution)

S/L Solid/Liquid (Interface) [also S-L]

S/L Spacelab [European Space Agency/NASA]

S/L Squadron Leader [also SL]

S/L Streamline(d); Streamlining [also SL]

S-L Solid-Liquid (Interface) [also S/L]

S-L Sound Locator [also SL]

Sl Seal [also sl]

Sl Slate [also sl]

Sl Slide [also sl]

S(λ) Relative Spectral Energy (Power) Distribution [Symbol]

s/L second(s) per liter [also s/l, s L^{-1}, sec/L, sec/l, s/ltr, or s ltr^{-1}]

sl *(sine loco)* – without place

sl slight(ly)

sl slip [Woven Fabrics]

s(λ) spectral responsivity (of a detector) [Symbol]

σ-L flow stress versus edge-to-edge particle spacing (curve) [Metallurgy]

SLA Saskatchewan Library Association [Canada]

SLA School Library Association [UK]

SLA Scottish Library Association

SLA Semiconductor Laser Amplifier

SLA Service Lane Attendant

SLA Shared Line Adapter

SLA Source Location Analysis

SLA Spacecraft Lunar Module Adapter [of NASA]

SLA Spacelab Product Assurance Department [of ERNO, Germany]

SLA Special Libraries Association [US]

SLA Static Lattice Approximation [Physics]

SLA Statistical Local Area

SLA Stereolithography Apparatus

SLA Stored Logic Array

SLA Strontium Lanthanum Aluminiate

SLA Support and Logistics Areas

SLA Swine Leucocyte Antigen [Immunology]

SLA Synchronous Line Adapter

SLAA Sierra Leone Airport Authority

SLAAS Sri Lanka Association for the Advancement of Science

SLAB Small Amount of Bits [also slab]

Slabop Obz Slaboproudy Obzor [Czech Publication]

SLAC Spacelab Action Center

SLAC Stanford Linear Accelerator [California, US]

SLAC Stanford Linear Accelerator Center [of US Department of Energy at Stanford University, California, US]

SLAC Subscriber Line Audio Processing Circuit

SLACSS SLAC (Stanford Linear Accelerator) Summer School (on Particle Physics) [at Stanford University, California, US]

SLAC/SSRL Stanford Linear Accelerator Center/Stanford Synchrotron Radiation Laboratory [of Stanford University, California, US]

SLACTC SLAC (Stanford Linear Accelerator) Topical Conference [Stanford University, California, US]

SLADE Society of Lithographic Artists, Designers and Engravers [UK]

SLAET Society of Licenced Aircraft Engineers and Technologists [UK]

SLAG Safe Launch Angle Gate

SLAHTS Stowage List and Hardware Tracking System

SLAK Spacelab Late Access Kit

SLAM Scanning Laser Acoustic Microscope; Scanning Laser Acoustic Microscopy

SLAM Side Load Arrest Mechanism

SLAM Simulation Language for Alternative Modeling

SLAM Single-Layer Aluminum Metallized; Single-Layer Aluminum Metallization

SLAM Space-Launched Air Missile

SLAM Standoff Land Attack Missile

SLAM Strategic Low-Altitude Missile

SLAM Supersonic Low-Altitude Missile

SLAM Symbolic Language Adapted for Microcomputers

SLAMS Simplified Language for Abstract Mathematical Structures

SLAMS State and Local Air Monitoring System [US]

SLANG Systems Language

SLANT Simulator Landing Attachment for Night Landing Training

SLAP Sign-Labeled Polarization Transfer

SLAP Slot Allocation Procedures

SLAP St. Lawrence (River) Action Plan [of Environment Canada]

SLAP Symbolic Language Assembler Program

S/LAP Shiplap [also S/Lap]

SLAR Select ADC (Analog-to-Digital Conversion) Register

SLAR Side-Looking Airborne Radar

SLAR Slant Range [Navigation]

SLASH Seiler Laboratory ALGOL Simulated Hybrid

SLAST Submarine-Launched Anti-Surface Ship Torpedo

S Lat South Latitude [also Slat]

SLATE Small Lightweight Altitude Transmission Equipment

SLATE Stimulated Learning by Automated Typewriter Environment

Slav Slavic; Slavonian; Slavonic

SLB Scanning Laser Beam

SLB Setor de Lançamento de Balões [Balloon Launching Facility; of Brazilian National Institute for Space Research]

SLB Side-Lobe Blanking [Radar]

SLB Spectral Line Broadening

SLBM Ship-Launched Ballistic Missile

SLBM Submarine-Launched Ballistic Missile

SLBMDWS Submarine-Launched Ballistic Missile Detection and Warning System

SLC Salt Lake City [Utah, US]

SLC Scanning Laser Caliper

SLC Searchlight Control [also slc]

SLC Selector Channel

SLC Self-Lubricating Compound

SLC Shift Left and Count Instructions

SLC Shuttle Launch Control [NASA]

SLC Side-Lobe Cancellation; Side-Lobe Canceller

SLC Side-Lobe Clutter

SLC Simulated Linguistic Computer

SLC Single Layer Ceramic

SLC Single Line Control

SLC Small Logic Controller

SLC Spacelab Project Control [of ERNO, Germany]

SLC Space Launch Complex

SLC Specific Line Capacity [also slc]

SLC Stanford Linear Collider [of Stanford University, California, US]

SLC St. Lawrence College [Canada]

SLC Storz Lauener Control (of Roll Gap) [Metallurgy]

SLC Straight-Line Capacitance (Capacitor) [also slc]

SLC Strategic Laboratory Council

SLC Subscriber Line Concentrator

SLC Subscriber Loop Carrier

SLC Surface Laminar Circuitry

SLC Sustained-Load Cracking [Mechanics]

SLC Systems Life Cycle

SL&C Shipper's Load and Count

SLCA Single Line Communications Adapter

SLCA Spacelab Contract Administration [of ERNO, Germany]

SLCB Single-Line Color Bar

SLCC Saturn Launch Computer Complex [of NASA]

SLCC Spacelab Configuration Management [of ERNO, Germany]

SLCD Spring-Loaded Camming Device

SLCL Shop/Lab(oratory) Configuration Layout

SLCM Scanning Laser Confocal Microscope; Scanning Laser Confocal Microscopy

SLCM Sea-Launched Cruise Missile

SLCM Software Life Cycle Management

SLCM Submarine Launched Cruise Missile

SLCO Strontium Lanthanum Copper Oxide (Superconductor) [also SrLCO]

SLCP Saturn Launch Computer Program [of NASA]

SLCRS Supplementary Leak Collection and Release System

SLCS Standby Liquid Control System

SLCSAT Submarine Laser Communications Satellite

SLCU Synchronous Line Control Unit

SLCWA Spacelab Conference Work Area

SLD Scientific Liaison Division [of National Hydrology Research Institute, Canada]

SLD Simulated Launch Demonstration

SLD Slim Line Diffuser

SLD Stiff-Leg Derrick

SLD Straight-Line Depreciation

SLD Superluminescent Diode

SLD Source Language Debug

SLD Synchronous Line Driver

Sld Slide [also sld]

Sld Solder [also sld]

sld sailed

sld sealed

sld solid

SLDC Single-Loop Digital Controller

SLDF Sierra-Club Legal Defense Fund [US]

SLDPF Spacelab Data Processing Facility

SLDPF Spacelab Data Processing Function

SLDR System Loader

Sldr Solder [also sldr]

SLDS Swedish Land Data Information System

Slds Slides [also slds]

SLDT Store Local Descriptor Table

SLDTSS Single Language Dedicated Time-Sharing System

SLE Segment Limits End

SLE Sierra Leonean Leone [Currency]

SLE Small Local Exchange

SLE Society of Logistics Engineers [US]

SLE Spacelab Engineering [of ERNO, Germany]

SLE St. Louis (Equine) Encephalitis [Medicine]

SLE Stochastic Liouville Equation [Statistical Mechanics]

SLE Superheat Limit Explosion

SLE Systemic Lupus Erythematosis [Medicine]

slea *(sine loco et anno)* – without place and year

SLEAF Scanning Laser Environmental Airborne Fluorosensor

SLEAT Society of Laundry Engineers and Allied Trades

SLECR Spacelab Engineering Console Room

SLECR Spacelab Engineering Control Room

SLED Single Large Expensive Disk

SLED Surface Light-Emitting Diode

SLEDGE Simulating Large Explosive Detonable Gas Experiment

SLEEP Scanning Low-Energy Electron Probe

SLEEP Swedish Low-Energy Experimental (Nuclear) Pile

SLEJF Sri Lankan Environmental Journalists Forum

SLEMU Spacelab Engineering Model Unit

SLEP Second Large ESRO (European Space Research Organization) Project

SLEP Service Life Extension Program

SLES Sodium Laureth Sulfate

SLEW Static Load Error Washout (System)

SLF Shuttle Landing Facility [formerly Orbiter Landing Facility] [of NASA Kennedy Space Center, Florida, US]

SLF Single-Layer Film

SLF Skandinaviska Lackteknikers Forbund [Federation of Scandinavian Paint and Varnish Technologists, Finland]

SLF Straight-Line Frequency (Capacitor) [also slf]

SLF Streamline Flow

SLF Stud Leveling Foot

SLF Surface Loss Function

SLF System Library File

SLFA Stress and Longitudinal Field-Induced (Magnetic) Anisotropy [Solid-State Physics]

SL&FD Stimulation, Logging and Formation Damage [Oil and Gas Exploration]

SLFE Slim Ferroelectric

SLFK Sure Lock Fixture Key

SLG Scanning Laser Gauge

SLG Strontium Lanthanum Gallate

SLGC Splitless Gas Chromatography Technique

SLGO Strontium Lanthanum Gallium Oxide

slgt slight

SLH Speech, Language and Hearing

SLH St. Luke Hospital [New York City, US]

SLHA Society of Lincolnshire History and Archeology [UK]

SLHC St. Luke Hospital Center [New York City, US]

SLI Sea Level Indicator

SLI Solid-Liquid Interface

SLI Spacelab Integration and Test [of ERNO, Germany]

SLI Starter, Lighting and Igniter (Battery)

SLI Starting Lighting Ignition

SLI Steam Line Isolation

SLI Suppress Length Indicator

SLI Synchronous Line Interface

SL&I System Load and Initialization

SLIB Source Library

SLIB Subsystem Library

SLIC Selective Listing in Combination

SLIC Semiconductor Laser International Corporation [US]

SLIC Shear Longitudinal Inspection Characterization

SLIC Simulation Linear Integrated Circuit

SLIC Subscriber Line Interface Circuit

SLIC Subscriber Loop Interface Circuit

SLIC State Land Information Council [of New South Wales, Australia]

SLIC System Level Integration Circuit

SLICE Selective Line Insertion Communication Equipment

SLICE St. Louis Institute of Consulting Engineers [US]

SLICE Surrey Library Interactive Circulation Experiment [of University of Surrey, UK]

SLICE System Life-Cycle Estimation

SLICP State Land Information Capture Programs [Western Australia]

SLIC/SLAC Subscriber Line Interface Circuit/Subscriber Line Audio Processing Circuit

SLIES Stratospheric Limb Infrared Emission Spectrometer

SLIH Second Level Interrupt Handler

SLILS Superconductor-Nonsuperconductor Layer-Insulator-Nonsuperconductor Layer-Superconductor [also S-L-I-L-S, or S/L/I/L/S]

SLIM Subscriber Line Interface Module

Slim Software Life-Cycle Management (System)

SLIP Serial-Line Internet Protocol

SLIP Symmetric List Processor

SLIPR Source Language Input Program

SLIS School of Library and Information Science

SLIS Shared Laboratory Information System

SLISIR Sri Lanka Institute of Scientific and Industrial Research [Colombo]

SLISP Symbolic List Processing

SLIV Steam Line Isolation Valve

SLL Satellite Line Link

SLL Shelf Life Limit

SLL Sierra Leonean Leone [Currency]

SLLA Sri Lanka Library Association

SLLS Surface Laser Light Scattering

SLM Scanning Laser Microscope; Scanning Laser Microscopy

SLM Scientific Laboratory–Medical

SLM Semi-Liquid Material

SLM Single-in-Line Memory

SLM Sound Level Meter

SLM Spatial Light Modulator

SLM Standard Laboratory Module

SLM Standard Liter per Minute [also slm]

SLM Statistical Learning Model

SLM Subscriber Loop Multiplex

SLM Synchronous Line Module

slm standard liter per minute [also SLM]

SLMA Southeastern Lumber Manufacturers Association [US]

SLMA Steel Lintel Manufacturers Association [UK]

SLMG Self-Launching Motor Glider

SLMP Self-Loading Memory Print

SLN Spacelab Payload [of ERNO, Germany]

Sln Selection [also sln]

SLNP Swiss League for Nature Protection

SLO San Luis Obispo [California, US]

SLO Segment Limits Origin

SLO Shift Left Out

SLO Spacelab Operations [of ERNO, Germany]

SLO Swept Local Oscillator

SLOC Source Lines of Code

SLOCOP Specific Linear Optimal Control Program

SLOD St. Louis Ordnance District [Missouri, US]

SLOMAR Space Logistics Maintenance and Rescue

SLONET San Luis Obispo Network [California, US] [also Slonet]

SLOP Ship, Load-on Top (Tank) [also Slop]

SLOP St. Louis Ordnance Plant [Missouri, US]

SLOR Swept Local Oscillator Receiver

SLOS Star Line of Sight

SLOSH Sea, Lake and Overland Surge from Hurricane (Program) [US]

SLOSRI Shift Left Out/Shift Right In [also SLO/SRI]

SLOTH Suppressing Line Operands to Hexadecimal

Slov Slovenia(n)

SLP Sales Price

SLP San Luis Potosi [Mexico]

SLP Searchlight Probing

SLP Segmented Level Programming

SLP Seismic Line Program

SLP Silk-Like Peptide [Biochemistry]

SLP Skip-Lot Plan

SLP Source Language Processor

SLP Spacelab Program Office [of ERNO Spacelab]

SLP Straight Line Programming

SLP Super Long Play

α-**SLP** Alpha Silk-Like Peptide [Biochemistry]

Slp Slope [also slp]

SLPB Spacelab Program Board

SLPH Solid Photography

SLPM Standard Liter(s) per Minute [also slpm]

SLPP Serum Lipophosphoprotein [Biochemistry]

SLPP Sea Level Pilot Project

SLPP St. Lawrence Power Project [Canada and USA]

SLPP-SO Sea Level Pilot Project–Southern Ocean

SLPS Supersolidus Liquid Phase Sintering [Metallurgy]

SLPSR Spacelab Planning Support Room

SLQTM Sales Quantity This Month

SLR Side-Looking Radar

SLR Simple Left-to-Right

SLR Single Lens Reflex (Camera)

SLR Slush on Runway

SLR Solid/Liquid Phase Reaction (Process) [Materials Science]

SLR Spectral Line Reversal

SLR Spin-Lattice Relaxation [Solid-State Physics]

SLR Standardized Sri Lanka Rubber

SLR St. Lawrence River [US/Canada]

SLR Storage Limits Register

SLR Superlattice Reflection [Crystallography]

SLRAP Standard Low-Frequency Range Approach

SLRC Salt Lake Research Center [of US Bureau of Mines at Salt Lake City, Utah]

SLRE Self-Loading Random-Access Edit

SL Re Sri Lanka Rupee [Currency]

SLRN Select Read Numerically

SLRS Selective Laser Reactive Sintering

SLRV Shuttle Launched Research Vehicle

SLRV Surveyor Lunar Roving Vehicle [of NASA]

SL/RN Stelco-Lurgi/Republic Steel-National Lead Process [Metallurgy]

SLS Saskatchewan Land Surveyors [Canada]

SLS Secondary Landing Site [Aerospace]

SLS Segment Long Spacing

SLS Selective Laser Scanning

SLS Selective Laser Sintering

SLS Sequential Lateral Solidification

SLS Side-Lobe Suppression

SLS Side-Looking Sonar

SLS Signaling Link Selection

SLS Smart Line Scanner

SLS Soda-Lime-Silicate (Glass)

SLS Sodium Lauryl Sulfate

SLS Solid-Liquid Separation

SLS Sortie Lab Simulator

SLS Source Library System

SLS Spacelab Simulator

SLS Space Life Sciences

SLS Specialist in Library Science

SLS Statement Level Simulator

SLS Static Light Scattering

SLS Strained-Layer Superlattice [Solid-State Physics]

SLS St. Lawrence Seaway [Canada/USA]

SLS Superlarge Scale (Integrated Circuit)

SLS Superlattice Structure [Solid-State Physics]

SLSA Saskatchewan Land Surveyors Association [Canada]

SLSA Secondary Lead Smelters Association [US]

SLSA Shuttle Logistics Support Aircraft [NASA]

SLSA St. Lawrence Seaway Authority [Canada]

SLSSC St. Lawrence Seaway Development Corporation [US]

SL SER Shift Left Serial Input [Digital Logic]

SLSF Sodium Loop Safety Facility [Nuclear Reactors]

SLSFC Severe Local Storms Forecasting Center [Kansas City, Missouri, US]

SLSI Super-Large-Scale Integration

SLSL Strained Layer Superlattice [Solid-State Physics]

SLSMS Spacelab Support Module Simulator

SLSO Self-Limiting Sinusoidal Oscillator

sl sol slightly soluble

SLSS Systems Library Subscription Service [of IBM Corporation, US]

SL-SS Spacelab Subsystems

SL-SSS Spacelab Subsystem(s) Segment

sl st slip stitch [Woven Fabrics]

SLT Second Law of Thermodynamics

SLT Simulated Launch Test

SLT Solid-Logic Technique; Solid-Logic Technology

SLT Spacelab Technology [ERNO, Germany]

SLT Standard(ized) Light Source

SLt Sub-Lieutenant [also Sub Lt]

S/Lt Second Lieutenant

Slt Searchlight [also slt]

Slt Sleet [also slt]

slt select

SLTB Society for Low-Temperature Biology [UK]

SLTC Sri Lanka Technical College [Colombo]

SLTC Society of Leather Technologists and Chemists [UK]

SLTE Self-Loading Tape Edit

SLTF Shortest Latency Time First

SLTMAS Structural Loads Test Measurement Acquisition System

SLU Secondary Logic Unit

SLU Serial Line Unit

SLU Source Library Update

SLU Special Line Unit

SLU Spoken Language Understanding

SLU St. Lawrence University [Canton, New York, US]

SLU St. Louis University [Missouri, US]

SLU Subscriber Line Usage

SLUC Standard Level User Charge

slug slug [Unit]

slug/cu ft slug(s) per cubic foot [also slugs/ft^3]

slug-ft^2 slug-feet squared [Unit]

slug/ft^3 slug(s) per cubic foot [also slugs/cu ft]

slug/ft·s slug(s) per foot-second [also slug/ft·sec]

slug-ft/s slug-feet per second [also slug-ft/sec]

slug-ft/s^2 slug-feet per second squared [also slugft/sec^2]

slug-ft^2/s slug-feet squared per second [also slugft^2/sec]

slug-ft^2/s^2 slug-feet squared per second squared [also slug-ft^2/sec^2]

slug-ft^2/s^3 slug-feet squared per second cubed [also slug-ft^2/sec^3]

slug/ft·sec slug(s) per foot-second [also slug/ft·s]

slug-ft/sec slug-feet per second [also slug-ft/s]

slug-ft/sec^2 slug-feet per second squared [also slugft/s^2]

slug-ft^2/sec slug-feet squared per second [also slugft^2/s]

slug-ft^2/sec^2 slug-feet squared per second squared [also slug-ft^2/s^2]

slug-ft^2/sec^3 slug-feet squared per second cubed [also slug-ft^2/s^3]

SLUK Spacelab Utility Kit

SLUMO Second Lowest Unoccupied Molecular Orbital [Physical Chemistry]

SLUR Share Library User Report

SLURP Spiny Lobster Undersea Research Project [of Florida State University, Tallahassee, US]

SLUS Subscriber Line Usage System

SLUSM St. Louis University School of Medicine [Missouri, US]

SLV Satellite Launch Vehicle [of NASA]

SLV Schweißtechnische Lehr− und Versuchsanstalt [Institute for Welding Education and Research, Hanover, Germany]

SLV Schweizerischer Landwirtschaftlicher Verein [Swiss Agricultural Association]

SLV Space Launch Vehicle

SLV Sulfolobus Virus

SLV-1 Sulfolobus Virus, Type 1

Slv Sleeve [also slv]

SLVC Super-Linear Variable Capacitor

SLW Specific Leaf Weight

SLW Spectral Line Width

SLW Straight-Line Wavelength (Capacitor) [also slw]

SLW Supercooled Liquid Water

SLW Super Light-Weight

SLWD Super-Long Working Distance (Objective)

SLWL Straight-Line Wavelength

SLWT Super Light -Weight Tank

sly slowly [also sly]

SLZA Scandinavian Lead Zinc Association [Sweden]

SM Master of Science

SM Safety Margin

SM Salt Mixture

SM Sales Manager

SM Sample Management (Package)

SM Sample Mean [Quality Control]

SM San Marino [California, US]

SM San Marino [Europe]

SM San Marino [ISO Code]

SM San Marcos [Texas, US]

SM San Martin [Peru]

SM Santa Maria [California, US]

SM Santa Marta [Colombia]

SM Santa Monica [California, US]

SM Saturation Magnetization

SM Scanning Microscope; Scanning Microscopy

SM Scheduled Maintenance [also S/M]

SM School of Medicine

SM School of Metallurgy

SM School of Mines

SM Science Museum

SM *(Scientiae Magister)* – Master of Science

SM Scientific Memorandum

SM Search Menu

SM Secondary Mineral

SM Security Mechanism

SM Search Manual [of Micro Powder Diffraction Search/Match (System]

SM Self-Mass

SM Semantic Model

SM Semiconductor Material

SM Semiconductor Memory

SM Semi-Mat

SM Semimetal

SM Sensor Material

SM Sequence and Monitor

SM Sergeant Major

SM Service Management; Service Manager

SM Service Module [NASA Apollo Program]

SM Service Monitor(ing)

SM Servomechanism

SM Servomotor [also S/M]

SM Set Mode

SM Shape Memory [Materials Science]

SM Shared Memory

SM Shear Modulus

SM Sheet Metal

SM Shell Membrane [Biology]

SM Shell Model [Nuclear Physics]

SM Short Module

SM Shuttle Management [NASA Kennedy Space Center, Florida, US]

SM Shuttle Mission [of NASA]

SM Sidereal Month [Astronomy]

SM Siemens-Martin (Steelmaking) [also S-M] [Metallurgy]

SM Signaling Mode

SM Simulator
SM Skull Melting [Metallurgy]
SM Slow Motion
SM Smart Material
SM Smooth Muscle [Anatomy]
SM Soft Magnet(ic)
SM Soft Manual
SM Soft Mode
SM Software Management
SM Solar Mirror
SM Solder Mask
SM Soldier's Medal
SM Solid Mechanics
SM Solid Modeling
SM Source Module
SM Spatially Multiplexed (Hologram)
SM Special Memorandum
SM Specialty Metal
SM Spin Magnetism [Solid-State Physics]
SM Stable Member
SM Standard Matched (Joint)
SM Standard Model
SM Starting Material
SM Static Memory
SM Statistical Mechanics
SM Statistical Modeling
SM Statistical Multiplexer
SM Statute Mile [also sm]
SM Steady Motion
SM Steckel (Rolling) Mill [Metallurgy]
SM Stereomicroscope; Stereomicroscopy
SM Stiffness Matrix [Mechanics]
SM Storage Mark
SM Strategic Missile
SM Strength of Materials
SM Stress-Directed Migration; Stress-Induced Migration
   [Metallurgy]
SM Striated Muscle [Anatomy]
SM Strip Mine; Strip Mining
SM Strong Matter
SM Strontium Molybdate
SM Structural Material
SM Structure Memory
SM Student Manual
SM Submarine [also S/M]
SM Suhl Magnon [Solid-State Physics]
SM Super-Mirror (Guide) [Physics]
SM Supply Management; Supply Manager
SM Support Module
SM Standard Matched (Joint)
SM Surface Measure [also sm]
SM Surface Metallurgy
SM Surface Mining
SM Surface Missile

SM Surface Mobility
SM Surface Modification
SM Surface Mount(ing)
SM Switched Multipoint
SM Switching Multiplexer
SM Symmetric Matrix [Mathematics]
SM Synchronous Modem
SM Synchronous Motor
SM Synodic Month [Astronomy]
SM Synthetic Material
SM Synthetic Metal
SM System(s) Management; System(s) Manager [also S/M]
SM System Mechanics
SM2 Switched Multipoint 2-Wire
SM4 Switched Multipoint 4-Wire
S&M Surfaced and Matched [Lumber]
S&M Surveillance and Maintenance
S/M Scheduled Maintenance [also S/M]
S/M Service/Maintenance
S/M Servo-Motor [also SM]
S/M Silane/Methane (Mixture)
S/M Sort/Merge
S/M Sort/Message
S/M Structural/Mechanical
S/M Submarine [also SM]
S/M System(s) Management; System(s) Manager [also SM]
S-M Siemens-Martin (Steelmaking) [also SM] [Metallurgy]
Sm Samarium [Symbol]
Sm Smith (Antigen) [Immunology]
Sm Smithsonite [Mineral]
Sm Spiral Galaxy, Transitional Type, Magellanic
   [Astronomy]
Sm^{2+} Divalent Samarium Ion [also Sm^{++}] [Symbol]
Sm^{3+} Trivalent Samarium Ion [also Sm^{+++}] [Symbol]
Sm-I Smectic I (Phase of Liquid Crystal) [also SmI]
Sm II Samarium II [Symbol]
Sm-123 Samarium Barium Copper Oxide [$SmBa_2Cu_3O_{7-x}$]
Sm-144 Samarium-144 [also ^{144}Sm, or Sm^{144}]
Sm-145 Samarium-145 [also ^{145}Sm, or Sm^{145}]
Sm-147 Samarium-147 [also ^{147}Sm, or Sm^{147}]
Sm-148 Samarium-148 [also ^{148}Sm, or Sm^{148}]
Sm-149 Samarium-149 [also ^{149}Sm, or Sm^{149}]
Sm-150 Samarium-150 [also ^{150}Sm, or Sm^{150}]
Sm-151 Samarium-151 [also ^{151}Sm, or Sm^{151}]
Sm-152 Samarium-152 [also ^{152}Sm, or Sm^{152}]
Sm-153 Samarium-153 [also ^{153}Sm, or Sm^{153}]
Sm-154 Samarium-154 [also ^{154}Sm, or Sm^{154}]
Sm-155 Samarium-155 [also ^{155}Sm, or Sm^{155}]
Sm-156 Samarium-156 [also ^{156}Sm, or Sm^{156}]
S/m Siemens per meter [also S m^{-1}]
sm small
sm statute mile [also SM]
sm surface measure [also SM]
s/m second(s) per meter [also s m^{-1}]

s/m^3 second(s) per cubic meter [also s m^{-3}]

$s^{-1}m^{-3}$ one per second per cubic meter [also 1/s·m^3, or 1/s/m^3]

SMA Santa Maria, Azores [Meteorological Station Identifier]

SMA Saskatchewan Mining Association [Canada]

SMA Scale Manufacturers Association [US]

SMA Scandium Magnesium Aluminate

SMA Science Masters Association

SMA Screen Manufacturers Association [US]

SMA Seafood Marketing Association [US]

SMA Second-Moment Approximation

SMA Semimajor Axis [Mathematics]

SMA Sequential Multiple Analyzer

SMA Shape-Memory Alloy [Metallurgy]

SMA Sheffield Metallurgical Association [UK]

SMA Shelving Manufacturers Association [US]

SMA Shielded Metal-Arc (Cutting)

SMA Silicomolybdic Acid

SMA Simultaneous Multi-Wavelength Acquisition

SMA Singapore Manufacturers Association

SMA Single-Mode Approximation [Physics]

SMA Society of Management Accountants [Canada]

SMA Society of Manufacturers Agents [now Society of Manufacturers Representatives, US]

SMA Society of Maritime Arbitrators [US]

SMA Society of Medieval Archeology [UK]

SMA Society of Mineral Analysts [US]

SMA Society of Municipal Arborists [US]

SMA Society of Museum Archeologists [UK]

SMA Soft Magnetic Alloy

SMA Software Maintenance Association [US]

SMA Solid Motor Assembly [Aerospace]

SMA Spherical Mirror Analyzer

SMA Spring Manufacturers Association [now Spring Manufacturers Institute, US]

SMA Standard Methods Agar

SMA Steel Merchants Association [UK]

SMA Stucco Manufacturers Association [US]

SMA Styrene-Maleic Anhydride (Copolymer) [also S-MA or S/MA]

SMA Surface Modeling and Analysis

SMA Surface Mounted Assembly

SMA Surface Mounting Applicator

SMA Survey and Mapping Alliance [UK]

S&MA Safety and Mission Assurance [Aerospace]

SmA Symmetrical Anhydride

Sm-A Smectic A (Phase of Liquid Crystal) [also SmA]

Sm-A$_1$ Smectic A$_1$ (Phase of Liquid Crystal) [also SmA$_1$]

Sm-A$_2$ Smectic A$_2$ (Phase of Liquid Crystal) [also SmA$_2$]

Sma Serratia Marcescens [Microbiology]

SMAA Society of Management Accountants of Alberta [Canada]

SMA-ABS Styrene-Maleic Anhydride/Acrylonitrile Butadiene-Styrene (Blend) [also SMA/ABS]

SMAB Solid Motor Assembly Building [Aerospace]

Sm(Ac)$_3$ Samarium(III) Acetate

SMAC Senate Military Affairs Committee

SMAC Shielded Metal-Arc Cutting

SMAC Society of Management Accountants of Canada

SMAC Special Mission Attack Computer

SMAC State Minerals Advisory Council [of New South Wales, Australia]

SMAC Storage Multiple Access Control

SMAC Surface Modification and Coatings Technology [of The Metallurgical Society, US]

SMAC System Monitor and Control

Sm(ACAC)$_3$ Samarium Acetylacetonate [also Sm(AcAc)$_3$, or Sm(acac)$_3$]

SMACNA Sheetmetal and Air Conditioning Contractors National Association [US]

SMACR Society for Measurement, Automatic Control and Robotics [Poland]

SMACS Simulated Message Analysis and Conversion Subsystem

SMACT Surface Modification and Coating Technology (Committee) [of The Minerals, Metals and Materials Society Materials Design and Manufacturing Division, US]

SMAE Society of Model Aeronautical Engineers [UK]

SMAF Specific Microphage Arming Factor [Biochemistry]

SMAG Star Magnitude

SMaj Sergeant Major

SMAL Structural Macroassembly Language

SMAL Surface and Microstructural Analysis Laboratory [US]

SMALC Sacramento Air Logistics Command [of US Air Force in California]

SMALGOL Small Computer Algorithmic Language

Small Bus Comput News Small Business Computer News [US]

Small Comput Libr Small Computer Library [US]

SMAO Society of Management Accountants of Ontario [Canada]

SMAP Small Manufacturers Assistance Program

SMA/PC Styrene Maleic Anhydride (Copolymer)/ Polycarbonate (Blend) [also SMA/PC]

SMArchS Master of Science in Architectural Science

SMART Salton's Magical Automatic Retrieval of Text

SMART Satellite Maintenance and Repair Technique(s)

SMART Self-Healing Multinodal Alternate Route Topology

SMART Self-Monitoring Analysis and Reporting Technology

SMART Sequential Mechanism for Automatic Recording and Testing

SMART Short Memory Augmented Rate Theory [also SM-ART]

SMART Shuttle Meeting Action-Item Review Tracking [NASA]

SMART Sorption Method using Adaptive Rate Technology

SMART Steckel Mill Advanced Rolling Technology [Metallurgy]

SMART Styrene Monomer Advanced Reheat Technology (Process)

SMART System for the Mechanical Analysis and Retrieval of Text

SMART System Monitor Analysis and Response Technique

SMART Systems Management Analysis, Research and Test

SMARTIE Simple Minded Artificial Intelligence

SMARTIE Standards and Methods Assessment using Rigorous Techniques in Industrial Environments

Smart Mater Struct Smart Materials and Structures [of Institute of Physics, UK]

SMARTRing Self-Healing Multinodal Alternate Route Topology Ring

SMARTS Software Management and Resource Tracking System

SMAS Switched Maintenance Access System

SMASCH Specimen Management System for California Herbaria [US]

SMATCH Simultaneous Mass and Temperature Change

SMATS Source Module Alignment Test Site

SMATV Satellite Master Antenna Television

SMAW Shielded Metal-Arc Welding

SMAW Short-Range Man Portable Antitank Weapon

SMAWT Short-Range Man Portable Antitank Weapon Technology

Smaze Smoke and Haze [also smaze]

SMB Server Message Block [Joint Microsoft/IBM/Intel Protocol]

SMB Société Mathématique de Belgique [Belgian Mathematical Society]

SMB Space Meteorology Branch

SMB Superconducting Magnetic Bearing

SMB Supersonic Molecular Beam

SMB Surface Modification Branch [of Naval Research Laboratory, US Navy, Washington, DC, US]

SMB Surveys and Mapping Branch [of Natural Resources Canada]

SMB System Monitor Board

SMB Systems Management Branch

Sm-B Smectic B (Phase of Liquid Crystal) [also SmB]

Sm-B$_{cryst}$ Smectic B$_{cryst}$ (Phase of Liquid Crystal) [also SmB$_{cryst}$]

Sm-B$_{hex}$ Smectic A$_{hex}$ (Phase of Liquid Crystal) [also Sm$_{hex}$]

SMBA Scottish Marine Biological Association [UK]

SmBCO Samarium Barium Copper Oxide (Superconductor)

Sm-BC Smectic BC (Phase of Liquid Crystal) [also SmBC]

smbl semi-mobile

SMBT Master of Science in Building Technology

SMBT Sodium-2-Mercaptobenzothiazole

SMC Sapporo Medical College [Japan]

SMC Scientific Manpower Commission [now Commission on Professionals in Science and Technology, US]

SMC Segmented Maintenance Cask

SMC Sheet Molding Composite

SMC Sheet Molding Compound

SMC Shieldalloy Metallurgical Corporation [US]

SMC Small Magellanic Cloud [Astronomy]

SMC Smooth Muscle Cell [Anatomy]

SMC Société Mathématique du Congo [Mathematical Society of Congo]

SMC Space and Missile Systems Center [of US Air Force Materiel Command]

SMC Spin Muon Collaboration (Experiment) [of CERN–European Laboratory for Nuclear Research, Geneva, Switzerland]

SMC Sten Machine Carbine

SMC Storage Module Controller

SMC Supermicrocomputer

SMC Supply and Maintenance Command [US Army]

SMC Surface Mount(ed) Component

SMC Surface Movement Control

SMC Switch Maintenance Center

SMC System Management Center

SMC System Monitor Controller

SMC Systems, Man and Cybernetics (Society) [of Institute of Electrical and Electronics Engineers, US]

SMC X-1 Small Magellanic Cloud X-1 [Astronomy]

Sm-C Smectic C (Phase of Liquid Crystal) [also SmC]

SMCAA Sheet Molding Compound Automotive Alliance [US]

SM CAP(S) Small Capital Letter(s) [also Sm Cap(s), or sm cap(s)]

SMCBA Surface Mount and Circuit Board Association [Australia]

SMC/BMC Sheet Molding Compound/Bulk Molding Compound

SMCC 4-(N-Maleimidomethyl)cyclohexane-1-Carboxylic Acid N-Hydroxysuccinimide

SMCC Shuttle Mission Control Center [of NASA]

SMCC Succinimidyl 4-(N-Maleimidomethyl)cyclohexane-1-Carboxylate)

SMCC Simulation Monitor and Control Console

SMCC Subsurface Microbial Culture Collection

SMC-C Sheet Molding Compound–Continuous

SmCCO Samarium Cerium Copper Oxide (Superconductor)

SMC-C/R Sheet Molding Compound–Continuous/Random

SMCE School of Materials Science and Engineering [of University of New South Wales, Kensington, Australia]

SMCG Slow-Moving Consumer Goods [also smcg]

SMCH Standard Mixed Cargo Harness

SMCNA Sheetmetal Contractors National Association [now Sheetmetal and Air Conditioning Contractors National Association, US]

SmCo Samarium-Cobalt (Magnet Material)

Sm(CoFeCu) Samarium-Cobalt-Iron-Copper (Magnet Material)

Sm(CoFeMnCu) Samarium-Cobalt-Iron-Manganese-Copper (Magnet Material)

Sm(CoFeCuZr)$_z$ Samarium-Cobalt-Iron-Copper-Zirconium (Magnet Material)

Sm(Cp)$_3$ Tricyclopentadienylsamarium

SMC-R Sheet Molding Compound–Random

SMCRA Surface Mining Control and Reclamation Act [Canada]

SMCS Separation Monitor and Control System

SMCS Systems, Man and Cybernetics Society [of Institute of Electrical and Electronics Engineers, US]

SMCU Separation Monitoring Control Unit

SMC X-1 Small Magellanic Cloud X-1 [Astronomy]

SMD Sauter Mean Diameter

SMD Scientific and Medical Division

SMD Semiconductor Magnetic-Field Detector

SMD Singular Multinomial Distribution [Mathematics]

SMD Spacelab Mission Development

SMD Special Measuring Device

SMD Spectral Momentum Density

SMD Standardized Military Drawing

SMD Storage Module Drive

SMD Structural Materials Division [of Minerals, Metals and Materials Society of American Institute of Mining, Metallurgical and Petroleum Engineers, US]

SMD Surface-Mount(ed) Device

SMD Systems Manufacturing Division

SMD Systems Measuring Device

SM³/d Standard Cubic Meters per Day [Unit]

SMDA Safe Medical Device Act [US]

SMDA Saskatchewan Mineral Development Agreement [Canada]

SMDC Saskatchewan Mining and Development Corporation [Canada]

SMDC Shielded Mild Detonating Cord

SMDC Sodium Methyl Dithiocarbamate

SMDC Superconductive Materials Data Center

SMDE Static Mercury Drop Electrode

SMDE Stationary Mercury Drop Electrode

SMD/EMPMD Structural Materials Division/Electronic, Magnetic and Photonic Materials Division (Joint Committee) [of Minerals, Metals and Materials Society of American Institute of Mining, Metallurgical and Petroleum Engineers, US]

SMDF SAGE (Semi-Automatic Ground Environment) Main Distributing Frame

SMD/MSD Structural Materials Division/Materials Science Division (Joint Committee) [of Minerals, Metals and Materials Society of American Institute of Mining, Metallurgical and Petroleum Engineers, US]

SMDR Station Message Detail(ed) Recording

SMDR Surface-Mode Dispersion Relation

SMDS Shell Middle Distillate Synthesis (Process)

SMDS Switched Multimegabit Data Service

SMDS Switched Multimegabit Digital Service

SME School of Materials Engineering

SME School of Military Engineering

SME Semiconductor-Metal Eutectic

SME Shape-Memory Effect [Materials Science]

SME Small and Medium-Sized Enterprises; Small-to-Medium Sized Enterprises

SME Sociedad Matemática Española [Spanish Mathematical Society]

SME Society for Mining, Metallurgy and Exploration, Inc. [US]

SME Society of Manufacturing Engineers [US]

SME Society of Mechanical Engineers

SME Society of Military Engineers [US]

SME Society of Mining Engineers [of American Institute of Mining, Metallurgical and Petroleum Engineers, US]

SME Solar Mesosphere Explorer

SME Solid-Metal Embrittlement [Metallurgy]

SME Surface-Modified Electrode

SM&E Semiconductor Materials and Equipment

Sm-E Smectic E (Phase of Liquid Crystal) [also SmE]

SME-AIME Society of Mining Engineers of the American Institute of Mining, Metallurgical and Petroleum Engineers [US]

SMEAR Span/Mission Evaluation Action Request

SMEAT Skylab Medical Experiments Altitude Test

SMEC Single Module (Automotive) Engine Controller

SMED Single-Minute Exchange of Die

SMEDP Standard Methods for the Examination of Dairy Products

SMEF Special Microelectronics Fund [of Natural Sciences and Engineering Research Council, Canada]

SMEK Summary Message Enable Keyboard

SMEM Serial Memory

S-MEM Minimum Essential Medium (Eagle) for Spinner Modification [Biochemistry]

S-MEM Minimum Essential Medium (Eagle) for Suspension Cultures [Biochemistry]

SMEMA Surface-Mount Equipment Manufacturers Association [US]

SMENET Society for Mining, Metallurgy and Exploration Network [US]

SMEP Sociedade Microscopia Electronica do Portugal [Electron Microscopy Society of Portugal]

SMEPR Strain-Modulated Electron Paramagnetic Resonance

SMES Shuttle Mission Engineering Simulator [of NASA]

SMES Shuttle Mission Evaluation Simulation; Shuttle Mission Evaluation Simulator [of NASA]

SMES Superconducting Magnetic Energy Storage

SMES Superconductive Magnetic Energy Storage Initiative [of US Pentagon, Arlington, Virginia]

SMET Simulated Mission Endurance Testing

SME&T Science, Mathematics, Engineering and Technology

SMETDS Standard Message Trunk Design System

SMEX Small Explorer [Aerospace]

SMF Sheetmetal Formability; Sheetmetal Forming

SMF Shielded Materials Facility [of US Department of Energy at Hanford, Washington, US]

SMF Single Mode Fiber

SMF Sociedad Mexicana de Fisica [Mexican Society of Physics]

SMF Société Mathématique de France [French Mathematical Society]

SMF Software Maintenance Function

SMF Solar Magnetic Field [Astrophysics]

SMF Standard Message Format

SMF Static Magnetic Field

SMF System Management Facility; System Manager Facility

SMF System Measurement Facility

Sm-F Smectic F (Phase of Liquid Crystal) [also SmF]

SMFA Simplified Modular Frame Assignment

Sm-Fe Samarium-Iron (Alloy System)

SmFe Samarium Ferrite

Sm-Fe-T Samarium-Iron-Transition Metal (Alloy) [also Sm-Fe-TM]

Sm-Fe-TM Samarium-Iron-Transition Metal (Alloy) [also Sm-Fe-T]

SMFP Scattering Mean Free Path [Physics]

SMG Schweizerische Mathematische Gesellschaft [Swiss Mathematical Society]

SMG Schweizerische Metallurgische Gesellschaft [Swiss Metallurgical Society]

SMG Software Message Generator

SMG Sort-Merge Generator

SMG Spacecraft Meteorology Group; Spaceflight Meteorology Group

SMG Sub-Machine Gun

SMG Submicrograin; Submicrometer Grained; Submicron Grained [Metallurgy]

SMG Superconductive Magnetic-Energy Storage

SMG Surfactant-Mediated Growth

SMG System Management Group

Sm-G Smectic G (Phase of Liquid Crystal) [also SmG]

SMGCS (Study Group on) Surface Movement Guidance and Control Systems

SMGE Selective Multisolvent Gradient Elution [Chemistry]

SMGM School of Mining, Geology and Metallurgy

SMgO Sintered Magnesia

SMGP Strategic Missile Group

SMH Shifted Morse Hybrid (Potential) [Physical Chemistry]

Sm-H Smectic H (Phase of Liquid Crystal) [also SmH]

SMHEA Snowy Mountains Hydroelectric Authority [Australia]

SMHI Swedish Meteorological and Hydrological Institute

SMI Santa Maria Island [Azores]

SMI Scanning Microscopy International [Chicago, Illinois, US]

SMI Simulated Machine Indexing

SMI Society for Machine Intelligence [US]

SMI Sorptive Minerals Institute [US]

SMI Spring Manufacturers Institute [US]

SMI Start Manual Input

SMI Static Memory Interface

SMI Structural Materials Industries, Inc. [Hoboken, New Jersey, US]

SMI Structure of Management Information

SMI Supplemental Medical Insurance

SMI Swiss Meteorological Institute

SMI System Management Interrupt

SMI System Memory Interface

SMI Systems Management and Integration

SMIA Serial Multiplexer Interface Adapter

SMIC Study of Man's Impact on the Climate

SMIC Surveying and Mapping Industry Council

SMIDA Saskatchewan/Manitoba Implement Dealers Association [Canada]

SMIE School of Metallurgical Industrial Engineering

SMIE Solid-Metal-Induced Embrittlement [Metallurgy]

SMIEEE Senior Member of the Institute of Electrical and Electronics Engineers [US]

SMIF Standard Mechanical Interface

Smil Sawmill [also smil]

SMIL Statistics and Market Intelligence Library [of Department of Trade and Industry, UK]

SMIL Synchronized Multimedia Integration Language

SMILE Significant Milestone Integration Lateral Evaluation

SMILE Sub-Micron Laser Experiment

SMIM Selective Metastable Ion Monitoring

S-MIME Secure Multipurpose Internet Mail Extension

SMIO Saskatchewan Meteorological Inspection Office [of Atmospheric Environment Service, Canada]

SMIO Systems Management and Integration Office

SMIP Small Manufacturers Incentive Program

SMIP Structure Memory Information Processor

SMIPP Sheet Metal Industry Promotion Plan [US]

SMIR Shuttle Multispectral Infrared Radiometer [NASA]

SMIRE Senior Member of the Institution of Radio Engineers [UK]

SMIRR Shuttle Multispectral Infrared Radiometer [NASA]

SMiRT (International Conference on) Structural Mechanics in Reactor Technology

SMIS Safety Management Information System

SMIS Society for Management of Information Systems [now Society for Information Management, US]

SMIS Symbolic Matrix Interpretation System

SM Isl Santa Maria Island [Azores]

SMIT Simulated Mid-Course Instruction Test [of NASA]

SMIT Superconducting Multilayer Interconnect Technology

SMIT System Management Interface Tool [Advanced Interactive Executive Unix Administration Tool]

SMITE Simulation Model of Interceptor Terminal Effectiveness

Smith Inst Smithsonian Institution [Washington, DC, US]

Smith Mag Smithsonian Magazine [of Smithsonian Institution, Washington, DC, US]

Smith Misc Coll Smithsonian Miscellaneous Collections [of Smithsonian Institution, Washington, DC, US]

Sm-J Smectic J (Phase of Liquid Crystal) [also SmJ]

SMK Software Migration Kit

Sm-K Smectic K (Phase of Liquid Crystal) [also SmK]

Smk Smoke [also smk]

smkls smokeless

Smk Sh Smoke Shell [also smk sh]

Smk Sig Smoke Signal [also smk sig]

SML Science Museum of London [UK]

SML Shoals Marine Laboratory [of National Oceanic and Atmospheric Administration, US]

SML Software Master Library

SML Standard Meta Language

SML Structure Mold Line

SML Structures and Materials Laboratories [of National Research Council of Canada, Ottawa]

SML Submonolayer [Physical Chemistry]

SML Surface Mixed Layer

SML Symbolic Machine Language

S&ML Sampling and Mobile Laboratories

Sm-L Smectic L (Phase of Liquid Crystal) [also SmL]

sml small

SMLCC Synchronous Multiline Communications Coupler

SMLE Short Magazine Lee-Enfield (Rifle)

SMLE Short Model Lee-Enfield (Rifle)

SMLE Small-Medium Local Exchange

SMLM Simple-Minded Learning Machine

Smls Seamless (Pipe, or Tube) [also smls]

Smlt Smelting [also smlt]

SMM (International) Shipbuilding, Machinery and Marine Technology Exhibition/Conference [Hamburg, Germany]

SMM School of Mines and Metallurgy; School of Mining and Metallurgy

SMM Shared Main Memory

SMM Semiconductor Memory Module

SMM Shared Multiport Memory

SMM Sociedad Matemática Mexicana [Mexican Mathematical Society]

SMM Soft Magnetic Material

SMM (International) Soft Magnetic Materials Conference

SMM Solar Maximum Mission (Satellite) [also SolarMax]

SMM Standard Method of Measurement

SMM Start of Manual Message

SMM Storage Media Module

SMM Subsystem Measurement Management

SMM System Management Mode

SMM System Management Monitor

SMMA Small Motor Manufacturers Association [US]

SMMA Styrene Methyl Methacrylate

SMMC Standard Monthly Maintenance Charge

SMMC System Maintenance Monitor Console

SMMD Specimen Mass Measurement Device

SMMDA Smaller Manufacturers Medical Device Association [US]

SMME Society for Mining, Metallurgy and Exploration [US]

SMMIB International Conference on Surface Modification of Metals by Ion Beams

SMMIB Surface Modification of Metals by Ion Beams

SMMP Standard Methods of Measuring Performance

SMMR Scanning Multichannel Microwave Radiometer

SMMT Society of Motor Manufacturers and Traders [UK]

SMN Servicio Meteorologico Nacional [National Meteorological Service, Argentina]

SmN Samarium Nitride

Sm/RNP Smith/Ribonucleoprotein (Antigen) [Immunology]

SMO Small Magnetospheric Observatory

SMO Stabilized Master Oscillator

SMOA Second Moment of Area

SMOBC Solder Mask Over Bare Copper

SMOC Simulation Mission Operation Computer

SMOC Spacelab Mission Operations Center

SMODOS Self-Modulating Derivative Optical Spectrometer

SMOH Since Major Overhaul

SMOG Special Monitor Output Generator

Smog Smoke and Fog [also smog]

SMOI Second Moment of Inertia

SMOKE Surface Magneto-Optic Kerr Effect

SmOPr^i Samarium Isopropoxide

SMOS Sagnac Magneto-Optic Sensor

SMOW Standard Mean Ocean Water

SMWDA Sydney Metropolitan Waste Disposal Authority [Australia]

SMP Scanning Microscope Photometer

SMP Shape Memory Polymer

SMP Simple Management Protocol

SMP Smith Miller and Patch Limited [Canadian Pharmaceutical Manufacturer]

SMP Software Management Plan

SMP Soil Management Program [of Tasmania, Australia]

SMP Solid Metallic Precursor

SMP Sort/Merge Program

SMP South Magnetic Pole

SMP Spatial Mobilized Planes (Theory) [Physics]

SMP Structures and Materials Panel [of NATO Advisory Group for Aerospace Research and Development]

SMP Sucrose Monopalmitate

SMP Surface-Mount(ed) Package

SMP Symbolic Manipulation Program

SMP Symmetric(al) Multiprocessing; Symmetric(al) Multiprocessor

SMP System Modification Program

.SMP Sample [File Name Extension] [also .smp]

SmP Samarium Phosphide

Smp Sampler [also smp]

SMPA Scottish Master Plasterers Association [UK]

SMPAM Symposium on Magnetic Propeties of Amorphous Materials

SMPB 4-(p-Maleimidophenyl)butyric Acid N-Hydroxysuccinimide Ester

SMPC Shared Memory Parallel Computer

SMPC Specialty Metals Processing Consortium [US]

SMPE Society of Marine Port Engineers [US]

SMPE Society of Motion Picture Engineers [now Society of Motion Picture and Television Engineers, US]

SMPGA Surface-Mountable Pin Grid Array

SMPL Structures, Materials and Propulsion Laboratory [of National Research Council of Canada, Ottawa]

Smpl Sample [also smpl]

SMPM Structural Materials Property Manual

SM/PM System Management/Performance Monitor

SMPP Sintered Metal Powder Process

SM PRO Steelmaking Proceedings [Publication of AIME Iron and Steel Society, US]

SMPS Simplified Message Processing Simulation

SMPS Single-Machine Power System

SMPS Switch(ed)-Mode Power Supply

SMPTE Society of Motion Picture and Television Engineers [US]

SMPTE J SMPTE Journal [of Society of Motion Picture and Television Engineers, US]

SMPTRB Shuttle Main Propulsion Test Requirement Board [of NASA]

SMQ Structure Module Qualification (Test)

SMQORC Staggered Modified Quadrature Overlapped Raised Cosine

SMQW Strained Multi(ple)-Quantum Well [Electronics]

SMR Series Mode Rejection

SMR Shield Mockup (Nuclear) Reactor [US]

SMR Society of Manufacturers Representatives [US]

SMR Solid Moderator (Nuclear) Reactor

SMR Source, Maintenance and Recoverability (Code)

SMR Special Mobile Radio (Equipment)

SMR Standardized Malaysian Rubber

SMR Standard Mortality Ratio

SMR State and Message Register

SMR Static Maintenance Reactor

SMR Super Metal Rich

SMR Support Management Room

SMR Surface Movement Radar

SMR Systems Management Request

SMRA Senior Medical Research Association [US]

SMRA Spring Manufacturers Research Association [UK]

SMRA Surface Mining and Reclamation Act [US]

SMRAM System Management Random Access Memory

SMRB Safety in Mines Research Board [UK]

SMRC Steel Machinability Research Consortium [US]

SMRC Swedish Maritime Research Center

SMRD Spin Motor Rate Detector

SMRD Spin Motor Rotational Detector

SMRD Spin Motor Run Discrete

SMRE Safety in Mines Research Establishment [UK]

SMRE Surface Mining Reclamation and Enforcement Office [US]

SMRF Sydney Magnetic Resonance Foundation [Australia]

SMRI Safety in Mines Research Institute [US]

SMRI Solution Mining Research Institute [US]

SMRI Sugar Milling Research Institute [South Africa]

SMRI Superconducting Magnetic Resonance Imaging

SMRL Seitz Materials Research Laboratory [of University of Illinois-Urbana, US]

SMRL Sudan Medical Research Laboratory [Khartoum]

SMRM Scanning Magnetoresistance (or Magnetoresistive) Microscopy; Scanning Magnetoresistance (or Magnetoresistive) Microscopy

SMRM Solar Maximum (Satellite) Repair Mission

SMRS Specialized Mobile Radio System

SMRS Surveys, Mapping and Remote Sensing

SMRSS Surveys, Mapping and Remote Sensing Sector [of Natural Resources Canada]

SMRT Single Message Rate Timing

SMRU Sea Mammals Research Unit [UK]

SMRVS Small Modular Recovery Vehicle System [of US Atomic Energy Commission]

SMS Schloemann-Siemag AG [Major German Manufacturer of Metallurgical Plant and Equipment]

SMS School of Materials Science

SMS School of Mathematical Sciences

SMS Senior Master Sergeant

SMS Separation Mechanism Subsystem

SMS Service Management System [Telecommunications]

SMS Shared Mass Storage

SMS Shuttle Mission Simulator [of NASA]

SMS Single-Domain Magnetic Spike [Solid-State Physics]

SMS Site Management System

SMS Site-Mixed Slurry

SMS Slope Monitoring System

SMS Slovak Metallurgical Society

SMS Small Magnetospheric Satellite [of NASA]

SMS Small Messaging System

SMS Society of Materials Science [Japan]

SMS Solar Maximum Satellite

SMS Speech Mail System

SMS Specific Magnetic Saturation

SMS Standard Modular System

SMS (NetWare) Storage Management Services

SMS Storage Management Subsystem

SMS Storage Management System

SMS Strategic Missile Squadron

SMS Styrene Methylstyrene

SMS Submicrocrystalline Structure [Metallurgy]

SMS Supra-Thermal Ion Mass Spectrometer

SMS Surface Mass Sensitivity

SMS Surface Missile System

SMS Synchronous Meteorological Satellites [Series of NASA Satellites]

SMS System-Managed Storage

SMS Systems Maintenance Service

SMS Systems Management Server

SMS Tin Manganese Selenide (Semiconductor)

SMS Tin Molybdenum Sulfide [$SnMo_6S_8$]

SMS-1 First Synchronous Meteorological Satellite [of NASA]

SM&S Systems Management and Sequencing

SmS Samarium Sulfide

SMSA Silica and Molding Sands Association [US]

SMSA Standard Metropolitan Statistical Area [US] [now Metropolitan Statistical Area]

SMSAE Surface Missile System Availability Evaluation

SMSC Semimagnetic Semiconductor

SMSC Space and Missile Systems Center [of US Air Force Materiel Command]

SMSCC Shuttle Mission Simulator Computer Complex [of NASA]

SmSCNO Samarium Strontium Copper Niobium Oxide (Superconductor)

SMSE School of Materials Science and Engineering [of University of New South Wales, Kensington, Australia]

SmSe Samarium Selenide

SMSF Southwest Marine Support Facility [of National Oceanic and Atmospheric Administration at San Diego, California, US]

SMS/GEOS Synchronous Meteorological Satellite/ Geostationary Operational Environmental Satellite [of NASA]

SMSgt Senior Master Sergeant

SMSI Standard Manned Spaceflight Initiator

SMSI Strong Metal-Support Interaction

SMSP School of Materials Science and Physics [of Thames Polytechnic, London, UK]

SMSR Systems Management Service Request

SMSS Ship Motion Sensor System

SMSS Software Manufacturing Support System

SMSS System Management Support Service

SMSST Satellite-Measured Skin Surface Temperature

SMST School of Materials Science and Technology

SMST (Conference on) Shape Memory and Superelastic Technologies [Monterrey, California, US]

SMSW Store Machine Status Word

SMT School of Mines and Technology

SMT Seeded-Melt Texturing [Solid-State Physics]

SMT Selective Message Transaction

SMT Semiconductor Manufacturing Technology

SMT Service Module Technician

SMT Société de Micro-Informatique et de Telecommunications [Society for Micro-Informatics and Telecommunications, France]

SMT Spontaneous Morphology Transition

SMT Square Mesh Tracking

SMT Station Management (Protocol)

SMT Storage Management Task

SMT Strontium Manganese Tantalate

SMT Sulfamethazine

SMT (International Conference on) Surface Modification Technologies

SMT Surface-Mount Technology

SMT Survey of Manufacturing Technology

SMT Tin Manganese Telluride (Semiconductor)

SMTA Scottish Motor Trade Association [UK]

SMTA Sewing Machinery Trade Association [now American Apparel Machinery Trade Association, US]

SMTA Sewing Machine Trade Association [UK]

SMTA Surface-Mount Technology Association [US]

SMTAS Shuttle Model Test and Analysis System [of NASA]

SMT Association Surface Mount Technology Association [Name Changed to Surface Mount and Circuit Board Association]

SmTe Samarium Telluride

SMTF Standardless Thin Film Procedure

SMTF Standard Metallurgical Thin Film (Software) [of Tracor Northern, US]

Sm(TFAC)$_3$ Samarium Trifluoroacetylacetonate [also Sm(tfac)$_3$]

smth smooth

Sm(THD)$_3$ Samarium Tris(2,2,6,6-Tetramethylheptanedionate) [also Sm(thd)$_3$]

Sm(TMHD)$_3$ Samarium Tris(2,2,6,6-Tetramethylheptane-dionate) [also Sm(tmhd)$_3$]

SMTI Selective Moving Target Indicator

SMTI Sodium Mechanisms Test Installation [Nuclear Engineering]

SMTI Southeastern Massachusetts Technological Institute [US]

SMTP Simple Mail-Transfer Protocol [Internet]

SMTS Scottish Machinery Testing Station

SMTS Senior Member of the Technical Staff

SMU Saint Mary's University [Halifax, Nova Scotia, Canada]

SMU Saint Mary's University [San Antonio, Texas, US]

SMU Secondary Multiplexing Unit

MU Self-Maneuvering Unit

SMU Shape Memory Unit

SMU Soft Mockup

SMU Southern Methodist University [Dallas, Texas, US]

SMU Space Maneuvering Unit [of NASA]

SMU Speaker/Microphone Unit

SMU Store Monitor Unit

SMU Sun-Moon University [Chung-nam, South Korea]

SMU Super-Module Unit

SMU System Maintenance Unit

SMU System Management Utility

SMU System Monitoring Unit

SMUD Sacramento Municipal Utility District [California, US]

SMUF Simulated Milk Ultrafiltrate

Smust Smoke and Dust [also smust]

SMUT System for Musical Transcription [of Indiana University, Bloomington, US]

SMV Science Museum of Victoria [Melbourne, Australia]

SMV Surveying and Mapping Victoria (State) [Australia]

SMVisS Master of Science in Visual Science

SMVP Shuttle Master Verification Plan [of NASA]

SMVRD Shuttle Master Verification Requirements Document [of NASA]

SMVU Survey of Motor Vehicle Use

SMW School of Mine Warfare [Yorktown, Virginia, US]

SMW Standard Materials Worksheet

SMWG Strategic Missile Wing

SMWIA Sheet Metal Workers International Association

SMWBA Scottish Master Wrights and Builders Association

SMX Semimicro Xerography

SMX Submultiplexer Unit

SMX Sulfamethoxazole

SmxDP (Workshop on) Small-x and Diffractive Physics

SMYS Specified Minimum Yield Strength; Specified Minimum Yield Stress [Mechanics]

SMZ Spontaneous Magnetization

SN Sabouraud-Noiré (Unit) [also S-N] [Microbiology]

SN Samarium Nickelate

SN San Nicolas [Argentina]

SN Saponification Number [Chemistry]

SN Schweizerische Norm [Swiss Standard]

SN School of Nursing

SN Schottky Noise [Electronics]

SN Scientific Note

SN Semiconductor Network

SN Senegal [ISO Code]

SN Sensory Nerve [Anatomy]

SN Separation Nozzle (Process)

SN Serial Number [also S/N]

SN Service Node [Telecommunications]

SN Ship Noise

SN Ship Nut

SN Shipping Note [also S/N]

SN Shoji-Nishiyama (Relationship) [also S-N] [Crystallography]

SN Shot Noise [Electronics]

SN Sierra Nevada [California, US]

SN Signal Node

SN Signal-to-Noise (Ratio) [also S-N, or S/N]

SN Silicon Nitride

SN Skid Number (of Pavement Surface)

SN Slow Neutron

SN Sodium Nitrate

SN Society for Neuroscience [US]

SN Solar Neutrino [Astrophysics]

SN Soon [Amateur Radio]

SN Space Network

SN Spinal Nerve [Anatomy]

SN Spontaneous Nucleation [Metallurgy]

SN Spray Nozzle

SN Standardization News [Publication of American Society for Testing and Materials, US]

SN Stock Number

SN Stress-Number (Curve) [also S-N] [Mechanics]

SN Subject Name

SN Suhl-Nakamura (Interaction) [Solid-State Physics]

SN Superconductor–Normal; Superconducting–Normal [also S-N] [Solid-State Physics]

SN Supernova [Astronomy]

SN Switching Network

SN Swivel Nut

SN Synaptic Neuron [Anatomy]

SN Synchronizer

(SN)$_n$ Polythiazyl

S/N Serial Number [also SN]

S/N Shipping Note [also SN]

S/N Signal-to-Noise (Ratio) [also SN, or S-N]

S-N Sabouraud-Noiré (Unit) [also SN] [Microbiology]

S-N Shoji-Nishiyama (Relationship) [Crystallography]

S-N Signal-to-Noise (Ratio) [also SN, or S/N]

S-N Stress-Number (Curve) [also SN] [Mechanics]

S-N Superconductivity-to Normal (Transition) [Solid-State Physics]

Sn Sign [also sn]

Sn Siren [also sn]

Sn *(Stannum)* – Tin [Symbol]

Sn^{2+} Divalent Tin Ion [also Sn] [Symbol]

Sn^{4+} Tetravalent Tin Ion [Symbol]

Sn II Tin II [Symbol]

Sn III Tin III [Symbol]

Sn-108 Tin-108 [also ^{108}Sn, or Sn^{108}]

Sn-111 Tin-111 [also ^{111}Sn, or Sn^{111}]

Sn-112 Tin-112 [also ^{112}Sn, or Sn^{112}]

Sn-113 Tin-113 [also ^{113}Sn, or Sn^{113}]

Sn-114 Tin-114 [also ^{114}Sn, or Sn^{114}]

Sn-115 Tin-115 [also ^{115}Sn, or Sn^{115}]

Sn-116 Tin-116 [also ^{116}Sn, or Sn^{116}]

Sn-117 Tin-117 [also ^{117}Sn, or Sn^{117}]

Sn-118 Tin-118 [also ^{118}Sn, or Sn^{118}]

Sn-119 Tin-119 [also ^{119}Sn, or Sn^{119}]

Sn-120 Tin-120 [also ^{120}Sn, or Sn^{120}]

Sn-121 Tin-121 [also ^{121}Sn, or Sn^{121}]

Sn-122 Tin-122 [also ^{122}Sn, or Sn^{122}]

Sn-123 Tin-123 [also ^{123}Sn, or Sn^{123}]

Sn-124 Tin-124 [also ^{124}Sn, or Sn^{124}]

Sn-125 Tin-125 [also ^{125}Sn, or Sn^{125}]

Sn-126 Tin-126 [also ^{126}Sn, or Sn^{126}]

Sn-127 Tin-127 [also ^{127}Sn, or Sn^{127}]

sn sine of the amplitude (in mathematics) [Symbol]

s.n. *(suo nomen)* – by its own name [Medical Prescriptions]

SNA Sambucus Nigra Agglutinin [Immunology]

SNA Sociedade Nacional de Agricultura [National Agricultural Society, Brazil]

SNA Société National d'Agriculture [National Agricultural Society, France]

SNA Société National de l'Amiante [National Asbestos Society, Quebec, Canada]

SNA Society of Naval Architects [Japan]

SNA System of National Accounts

SNA Systems Network Architecture

SnaB Sphaerotilus natans [Microbiology]

Sn(Ac)$_2$ Tin(II) Acetate (or Stannous Acetate) [also Sn(ac)$_2$]

Sn(Ac)$_4$ Tin(IV) Acetate (or Stannic Acetate) [also Sn(ac)$_4$]

Sn(ACAC)$_4$ Tin(IV) Acetylacetonate [also Sn(AcAc)$_4$, or Sn(acac)$_4$]

SNACS Share News on Automatic Coding Systems

SNACS Single Nuclear Attack Case Study [of US Department of Defense]

SNADS Systems Network Architecture Distribution Services

SNAE Society of Norwegian American Engineers

SNAF Société Nationale des Architectes de France [French Society of Architects]

SNAFU Situation Normal, All Fouled Up [also snafu]

SNAG Society of North American Goldsmiths [US]

Sn-Ag Tin-Silver (Alloy)

Sn-Ag-Cu Tin-Silver-Copper (Alloy)

SNAJ Society of Naval Architects of Japan

Sn-Al Tin-Aluminum (Alloy)

SNAM School of Naval Aviation Medicine [of US Navy]

SNAM Società Nazionale Metandotti (Process) [Chemical Engineering]

SNAME Society of Naval Architects and Marine Engineers [also SNA&ME, Jersey City, New Jersey, US]

SNAP Selective Niobium Anodization Process [Metallurgy]

SNAP Significant New Alternatives Policy

SNAP Significant New Alternatives Program [of US Environmental Protection Agency]

SNAP Simplified Numerical Automatic Programmer

SNAP Single Number Access Plan

SNAP S-Nitroso-N-Acetylpenicillamine [Biochemistry]

SNAP Space Nuclear Auxiliary Power (Reactor) [of US Atomic Energy Commission]

SNAP Standard Network-Access Protocol

SNAP Steerable Null Antenna Processor

SNAP Strategic Network Architecture Plan

SNAP Streamlined Noncompeting Application Program [of National Institutes of Health, Bethesda, Maryland, US]

SNAP Streamlined Noncompeting Awards Process

SNAP Structural Network Analysis Program

SNAP Sub-Network Access Protocol

SNAP Succinimidyl Nitroazidophenyl Aminoethyldithiopropionate; N-Succinimidyl-3-[(2-Nitro-4-Azidophenyl)-2-Aminoethyldithio]propionate

SNAP Synchronous Nuclear Array Processor

SNAP System Net Activity Program

SNAP System for Nuclear Auxiliary Power (Program)

SNAPI SNA (Systems Network Architecture) Parallel Interface

SNAPS Standard Notes and Parts Selection

SNAPS Switching Node and Processing Sites

SNA/RJE Systems Network Architecture/Remote Job Entry (Workstation)

SNAS Singapore National Academy of Sciences

SNAZOXS 8-Hydroxy-7-(4-Sulfo-1-Naphthylazo)-5-Quinolinesulfonic Acid; 7-(4-Sulfo-1-Naphthylazo)-8-Hydroxyquinoline- 5-Sulfonic Acid [also Snazoxs, or snazoxs]

SNB 2-(Phenylmethylthio)ethanamine

SNBHW Swedish National Board of Health and Welfare

Sn(Bi) Bismuth-Doped Tin

Sn-Bi-Ag Tin-Bismuth-Silver (Alloy)

Snblw Snowblower [also snblw]

SNBU Switched Network Backup

Sn(BuO)$_4$ Tin IV Butoxide (or Tetrabutoxytin)

SNBzl p-Nitrobenzylthio-

SNC 2-(Phenylthio)ethanamine

SNC Significant Noncompliance

SNC Stored-Program Numerical Control

SNC Synchronous Network Clock

SNCA Société Nationale de Construction Aéronautique [National Aeronautical Construction Company, France]

SNCB Société Nationale des Chemins de Fer Belges [Belgian National Railways Company]

Sn-Cd Tin-Cadmium (Alloy)

SNCF Société Nationale des Chemins de Fer Français [French National Railways Company]

SNCIAWPRC Singapore National Committee of the International Association on Water Pollution Research and Control

SNCIAWPRC Swiss National Committee of the International Association on Water Pollution Research and Control

SNCIAWPRC Swedish National Committee of the International Association on Water Pollution Research and Control

SNCMP Service National des Champs Magnetique Pulsés [National Office for Pulsed Magnetic Fields, Toulouse, Institut National des Sciences Appliquées, France]

SNCO Strontium Neodymium Copper Oxide (Superconductor) [also SrNCO]

SNCR Selective Non-Catalytic Reduction (Process)

Sn-Cu Tin-Copper (Alloy)

SNCUNESCO Swedish National Commission for the United Nations Educational, Scientific and Cultural Organization

SND 8-(Methylthio)quinoline

SND Sap No Defect [also snd] [Lumber]

SND Scientific Numeric Database

Snd Send [also snd]

Snd Sound [also snd]

2nd second

SNDC S-Type Negative Differential Conductance

SNDCF Sub-Network Dependent Convergence Function

SNDP Sustainable National Domestic Product

SNDRCV Send/Receive [also SR, or S/R]

SNDT Society for Nondestructive Testing [US]

SNE 8-Quinolinethiol

SNE Single-Nylon Enameled (Electric Wire)

Sn(E) Nuclear Stopping Cross-Section [Symbol]

SNECMA Société Nationale d'Etudes et de Construction de Moteurs d'Aviation [National Aircraft Engine Company, France]

SNEMSA Southern New England Marine Sciences Association [US]

SNENG New England Section [of the American Physical Society, US]

SNEP Sierra Nevada Ecosystem Program [US]

Sn(Et)$_4$ Tetraethyltin

Sn(EtO)$_2$ Tin(II) Ethoxide

SNEWS Secure News Server [Internet]

SNF Second Normal Form

SNF Skilled Nursing Facility

SNF Société Nobel Française [French Nobel Society]

SNF Spent Nuclear Fuel

SNF System Noise Figure

Snflk Snowflake(s) [also snflk]

SNF WP Spent Nuclear Fuel Waste Package

SNG Schweizerische Naturforschende Gesellschaft [Swiss Society of Natural Sciences]

SNG Single-Precision Number [Computers]

SNG Substitute Natural Gas

SNG Supplemental Natural Gas

SNG Synthetic Natural Gas

sngl single

SNH Société Nationale de Hydrocarbures [National Hydrocarbon Society, France]

SNH Sub-Net Handlers

SnH₃ Stannyl (Radical) [Symbol]

SNHS Saskatchewan National History Society

SNI San Nicolas Island [Philippines]

SNI Scanned Nanoprobe Instrument

SNI Selective Notification of Information

SNI Sequence-Number Indicator

SNI SNA (Systems Network Architecture) Network Interconnection

SNI Syndicat National des Instituteurs [National Union of (Primary School) Teachers, France]

SNI Signal-to-Noise Improvement

SNI Standard Network Interface

SNI Strontium Nickel Iridate

SNIAS Société Nationale Industrielle Aérospatiale [National Industrial Society for Space Research, France]

SNICP Sub-Network Independent Convergence Protocol

SNICS Sputtered Negative Ion Cesium Source

SNICT Sistema Nacional de Informação Cientifica e Tecnologica [National System for Scientific and Technological Information, Brazil]

SNID Studies in National and International Development

SNIF Short-Term Note Issuance Facility

SNIFTIRS Subtractively Normalized Interfacial Fourier Transform Infrared Spectroscopy

Sn(In) Indium-Doped Tin

SNIPEF Scotland and Northern Ireland Plumbing Employers Federation

SNIPER Saddle-Node Infinite Period

SNIRI Shikoku National Industrial Research Institute [of Agency of Industrial Science and Technology, Japan]

SNJ Switching Network Junction

SNL Sandia National Laboratories [Albuquerque, New Mexico, US]

SNL Selected Nodes List

SNL Standard Nomenclature List

SNLA Sandia National Laboratories, Albuquerque [New Mexico, US]

SNLC Senior NATO Logisticians Conference [Belgium]

SNM Shielded Nonmetallic (Cable)

SNM Society of Nuclear Medicine [US]

SNM Special Nuclear Material(s)

SNM Spent Nuclear Material

SNM Sun Net Manager

SNMA Swedish National Maritime Administration

SNMC Shielded Nonmetallic Cable

SNME Society of Naval Architects and Marine Engineers

Sn(MeO)₂ Tin(II) Methoxide

SNMMS Standard Navy Maintenance and Material Management System [US Navy]

SnMnSe Tin Manganese Selenide (Semiconductor)

SnMnTe Tin Manganese Telluride (Semiconductor)

SNMP Simple Network-Management Protocol [Internet]

SNMP Spent Nuclear Material Pool

SNMS Secondary Neutral Mass Spectrometry

SNMS Ship Noise Management System

SNMS Sputtered Neutral Mass Spectroscopy

SNMT Society for New Materials and Technologies [Czech Republic]

SNMTS Slovak New Materials and Technology Society [Slovak Republic]

SNN Second Nearest Neighbor (Approximation) [Crystallography]

SNN Strontium Nickel Nitride

S+N/N Signal plus Noise-to-Noise (Ratio)

SNND Small Neural Network with Delay

SNNE Sustainable Net National Expenditure

SNO 2-(Methylsulfonyl)ethanamine

SNO Serial Number [also S NO, or S No]

SNO Strontium Niobium Oxide (Superconductor) [also SrNO]

SNO Sudbury Neutrino Observatory [Sudbury, Ontario, Canada]

SnO Stannous Oxide

SNOBOL String-Oriented Symbolic (Programming) Language

SnO₂:F Fluorine-Doped Stannic Oxide

SNOI Sudbury Neutrino Observatory Institute [of Queen's University, Kingston, Ontario, Canada]

SNOM Scanning Near-Field Optical Microscope; Scanning Near-Field Optical Microscopy

SnO₂:Pd Palladium-Doped Stannic Oxide

SNORT Supersonic Naval Ordnance Research Track [US Navy]

SNOS Silicon-Nitride-Oxide-Silicon

SnO-SiO₂ Stannous Oxide-Silicon Dioxide (System)

SNOTEL Snow Pack Telemetry (Radio)

SNOW Service d'Hydrographie at d'Océanologie de la Marine [Naval Hydrographic and Oceanographic Service, France]

SNOWTAM Snow Notice to Airmen

SNP Serial Number/Password

SNP Single Nucleotide Polymorphism [Biochemistry]

SNP Society for Natural Philosophy [US]

SNP Sodium Nitroprusside

SNP Soluble Nucleoprotein [Biochemistry]

SNP Standard Network Package

SNP Statistical Network Processor

SNP Strategic Network Plan

SNP Strontium Nickel Platinate

SNP Synchronous Network Processor

SNp p-Nitrophenylthio-

SnP Stannic Phosphide

SNPA Sub-Network Point of Attachment

SNPA N-Succinimidyl-p-Nitrophenyl Acetate

Sn-Pb Tin-Lead (Alloy)

Sn-Pb-Ag Tin-Lead-Silver (Alloy)

Sn-Pb-Cd Tin-Lead-Cadmium (Alloy)

SNPM Standard and Nuclear Propulsion Module

SNPN Société Nationale de Protection de la Nature [National Society for the Protection of Nature, France]

SNPO Space Nuclear Propulsion Office [of NASA]

SNPPS Standard NASA Personnel and Payroll System

SNPS Satellite Nuclear Power Station

SNR School of Natural Resources [of University of Michigan, US]

SNR Selective Noncatalytic Reduction (of Nitrogen Oxide)

SNR Signal-to-Noise Ratio [also snr]

SNR Society for Nautical Research [UK]

SNR Strontium Nickel Rhodate

SNR Supernova Remnant [Astronomy]

SNR Supplier Nonconformance Report

SNR$_0$ Signal-to-Noise Ratio without Interference

SNRC Soreq Nuclear Research Center [Yavne, Israel]

SNR FOM Signal-to-Noise Ratio Figure of Merit

snRNP Small Nuclear Ribonucleoprotein Particles [Biochemistry]

SNR Strontium Sodium Ruthenate

SNS School of Natural Sciences

SNS (Scuola Normale Superiore) – Italian for "Higher Normal School"

SNS Selected Numeric Service [of Organization for Economic Cooperation and Development–Information Policy Group, France]

SNS Silicon/Nitride/Silicon

SNS Simulated Network Simulations

SNS Software Notification Service

SNS Space Navigation System

SNS Spallation Neutron Source [of Oak Ridge National Laboratory, Tennessee, US]

SNS Spanish Nuclear Society

SNS Static Nonlinear System

SNS Sulfonyl Nucleophilic Substitution [Physical Chemistry]

SNS Superconducting-Normal (Conducting)-Superconducting (Junction); Superconductor-Normal (Material)-Superconductor (Junction)

SNS Sympathetic Nervous System [Anatomy]

SnS Stannous Sulfide

SNSB Swedish National Space Board

Sn-Sb Tin-Antimony (Alloy) (or Antimonial Tin)

Sn(Sb) Antimony-Doped Tin

Sn-Sb-Cu Tin-Antimony-Copper (Alloy)

SNSC Silicon Nitride Solar Cell

SNSE Society of Nuclear Scientists and Engineers [US]

SnSe Stannous Selenide (Compound)

SNSF Swiss National Science Foundation

SNSF Sur Nedbórs Virkning på Skog og Fisk [Acid Precipitation Effects on Forests and Fish]

SNSH Snow Showers [also Sn Sh]

Sn/Si Tin on Silicon (Substrate)

SNS-JJ Superconductor–Normal (Material)–Superconductor Johsephson Junction

SNSO Space Nuclear Systems Office [US]

Snsr Sensor [also snsr]

SNSS School of Natural Science Society

SNT Sign-on Table

SNT Society for Nondestructive Testing [now American Society for Nondestructive Testing, US]

SNT Society of Nontraditional Technology [Japan]

SNT Strontium Niobate Tantalate

SNTA Sodium Nitrilotriacetate

SnTe Tin(II) (or Stannous) Telluride (Semiconductor)

SNTEMP Stream Network/Stream Segment Temperature Model [US]

SNU Seoul National University [South Korea]

SNU Solar Neutrino Unit [Astrophysics]

SNU Somalia National University [Mogadisho]

SNUPPS Standardized Nuclear Unit Power Plant System

SNUR Significant New Use Rule

SNV Schweizerische Normenvereinigung [Swiss Standards Association]

SNVB Society for Northwestern Vertebrate Biology [US]

SNW Silicon Nitride Whisker

Snw Snow [also snw]

Snwfl Snowfall [also snwfl]

SNWSC Space and Naval Warfare Systems Command [US Navy]

SNX Succinonitrile

SNY New York State Section [of American Physical Society, US]

SN-5Yb Silicon Nitride containing 5% Ytterbium

Sn-Zn Tin-Zinc (Alloy)

SO Center of Symmetry at both, the Closed-Packed and the Octahedral Site [Crystallographic Symbol Appended to Space Group Denotation]

SO Satellite Orbit

SO Saytzeff Orientation [Crystallography]

SO Schöniger Oxidation

SO Second-Order

SO Sectional Office [Data Communications]

SO Seismic Observatory

SO Seller's Option [also so]

SO Semi-Open (System)

SO Send Only [also S/O]

SO Sequential Organization [Computers]

SO Serial Out
SO Service Order
SO Servo-Operation
SO Shale Oil
SO Shared Object (File)
SO Shift-Out (Character)
SO Shop Order
SO Short-Range Order(ing) [Solid-State Physics]
SO Shut-Off [also S/O]
SO Silicon Oxide
SO Signal Oscillator
SO Slow Operating; Slow Operate; Slow to Operate
SO Small Outline
SO Small Outline Package [Electronics]
SO Small Overshoot [Magnetic Recording]
SO Smithsonian Observatory [Cambridge, Massachusetts, US]
SO Solar Observatory
SO Solid Oxide
SO Somalia [ISO Code]
SO Southern Oscillation [Oceanography]
SO Special Orthogonal Group (in Mathematics) [Symbol]
SO Spin Orbit (Interaction) [Solid-State Physics]
SO Stand-Off
SO Stationary Orbit
SO Stationery Office [also Her Majesty's Stationery Office, UK]
SO Station Operation; Station Operator
SO Statistical Orientation
SO Sub-Office [Telecommunications]
SO Sulfinyl (Radical) [Symbol]
SO Support Operations [NASA Kennedy Space Center Directorate, Florida, US]
SO Surface Oxidation; Surface Oxide
SO Switch-Over [also S/O]
SO Symmetry Operations [Crystallography]
SO System Optimization
.SO Shared Object File [File Name Extension] [also .so]
SO_2 Sulfonyl (Radical) [Symbol]
SO_2 Sulfur Dioxide
SO_3 Sulfur Trioxide
SO^- Sulfite (Ion) [Symbol] [also So^-]
SO^- Sulfate (Ion) [Symbol] [also So^-]
SO_x Sulfur Oxides (i.e., SO_2, SO_3, S_2O_3, and S_2O_7) [also SOx]
S_2O^- Thiosulfate (Ion) [Symbol]
S_2O_5 Sulfuryl (Radical) [Symbol]
S/O Send Only [also SO]
S/O Shipper's Option
S/O Shut-Off [also SO]
S/O Sulfur-to-Oxygen (Ratio)
S/O Switch-Over [also SO]
So Socket [also so]
So South(ern) [also so]
sO_2 Oxygen Saturation [Symbol]
$\sigma(\omega)$ emission/absorption cross-section [Symbol]
SOA Saccharose Octaacetate [Biochemistry]

SOA Safe Operating Area
SOA School of Artillery
SOA Science Office Area
SOA Scientific Offline Area
SOA Secure Science Operations Area
SOA Semiconductor Optical Amplifier
SOA Start of Address
SOA Start of Authority
SOA State-of-the-Art
SOAC System On A Chip
So Am South America(n) [also So Amer]
SOAP Self-Optimizing Automatic Pilot
SOAP Society of Office Automation Professionals
SOAP Spectrometric Oil Analysis Program
SOAP Symbolic Optimizer and Assembly Program; Symbolic Optimum Assembly Programming
SOAR Safe Operating Area [also Soar] [Electronics]
SOAR Service Oriented Accounts Receivable
SOAR Shuttle Orbital Applications and Requirements [NASA]
SOARS Shuttle Operation Automated Reporting System [of NASA]
SOATS Support Operation Automated Training System
SOB Shipped on Board
SOB Shortness of Breath
SOB Start of Block
SOBLIN Self-Organizing Binary Logical Network
Sobre Deriv Cana Azucar Sobre los Derivados de la Cana de Azucar [Cuban Publication on Sugar Cane Derivatives]
SOBS Society of Bookbinders [UK]
SOC Satellite Operations Center [US]
SOC Science Operations Center
SOC Scottish Ornithologists Club
SOC Self-Organized Criticality [Nuclear Reactors]
SOC Self-Organizing Control
SOC Semivolatile Organic Compounds
SOC Separated Orbit Cyclotron
SOC Set Override Clear
SOC Silicon-on-Ceramic
SOC Simulation Operations Center
SOC Simulation Operation Computer
SOC Single Occupancy Cell (Method)
SOC Single Orbit Computation
SOC Southern Oregon College [Ashland, US]
SOC Space Operations Center
SOC Specialized Oceanographic Center [UK]
SOC Specific Optimal Controller
SOC Spin-Orbit Coupling [Quantum Mechanics]
SOC Start of Climb [Aeronautics]
SOC State of Charge
SOC Superposition of Configuration
SOC Synthetic Organic Chemical
SOC Synthetic Organic Compound
SOC System On a Chip

SOC System Option Controller

SOC Systems Operation Center

SoC Spacifficity-of-Complexity

Soc Society [also soc]

Soc Socket [also soc]

soc social

SOCA Studies of the Ocular Complications of AIDS (Acquired Immunodeficiency Syndrome)

Soc An *(Société Anonyme)* – French for "Incorporated Company"

SOCAR Shuttle Operational Capability Assessment Report [of NASA]

SOCATA Société de Construction d'Avions de Tourisme et d'Affaires [Commercial and Business Aircraft Company, France]

SOCBRO Society of Chief Building Regulation Officer [UK]

SOCC Salvage Operational Control Center

SOCC Satellite Operation Command and Control

SOCC Satellite Operations Control Center

SOCCER SMART's Own Concordance Constructor Extremely Rapid [of Cornell University, Ithaca, New York, US]

Soc Chem Ind Monogr Society of the Chemical Industry Monographs [UK]

SOCCS Study of Computer Cataloguing Software [UK]

SOCCS Summary of Component Control Status

SOCDET Sydney Oil Company Drilling and Exploration [Australia]

SOCEX Southern Ocean Cloud Experiment(s) [Oceanography]

SOCEX I First Southern Ocean Cloud Experiment [Oceanography]

SOCEX II Second Southern Ocean Cloud Experiment [Oceanography]

Soc Geogr Social Geographer; Social Geography

SOCH Spacelab Orbiter Common Hardware

SOCI Second-Order Configuration Interaction

Soc Sci Med Social Science and Medicine (Journal)

SOCKO Systems Operational Checkout

SOCKS Socket Secure (Server) [Internet]

SOCl$_2$ Thionyl Chloride

SOCM Standoff Cluster Munitions

SOCMA Synthetic Organic Chemical Manufacturers Association [US]

SOCO International ICSC (International Computer Science Convention) Symposium on Soft Computing

SOCO Saskatchewan Opportunities Corporation [Canada]

SOCO Source Code

SOCO Switched-Out for Checkout

SOCOFIDE Société Congolaise de Financement du Développement [Congo Development and Financing Corporation]

SOCOM Solar Optical Communications System

SOCOTEC Société de Contrôles Techniques [Technical Controls Company, France]

SOCR Sustained Operations Control Room

SOCRATES Study of Complementary Research and Teaching in Engineering Science

SOCRATES System for Organizing Content to Review and Teach Educational Subjects [of University of Illinois, US]

SOCRATES System for Organizing Current Reports to Aid Technologists and Scientists [of Department of Scientific Information Services, Department of National Defense, Ottawa, Canada]

SOCS Spacecraft Orientation Control System

SOCS Stranded Oil in Coarse Sediment (Model)

SOCS Subsystem Operation and Checkout System

Soc Work Health Care Social Work in Health Care (Journal)

Socy Society [also socy]

SOD Sample Orientation Distribution [Crystallography]

SOD Savannah Ordnance Depot [Illinois, US]

SOD Second Order Differencing (Method)

SOD Seneca Ordnance Depot [New York, US]

SOD Serial Output Data

SOD Short-Time Overshoot Distortion

SOD Sierra Ordnance Depot [Herlong, California, US]

SOD Sioux Ordnance Depot [Sidney, Nebraska, US]

SOD Small Object Detector

SOD Small Outline Diode

SOD Société d'Etudes pour l'Obtention du Deuterium [Society for the Study of Deuterium Production, France]

SOD Source-Object Distance

SOD Springfield Ordnance District [Massachusetts, US]

SOD Statement of Direction

SOD Superintendent of Documents [of Government Printing Office, US]

SOD Superoxide Dismutase [Biochemistry]

SOD System Operational Design

SODA Source-Oriented Data Acquisition

SODAC Society of Dyers and Colorists

SODAR Sonic Radar [also Sodar, or sodar]

SODAR Sound Detection and Ranging [also Sodar, or sodar]

SODAR Sound Radar [also Sodar, or sodar]

SODAS Structure-Oriented Description and Simulation

SODB Science Organization Development Board [of National Academy of Sciences, US]

SODB Shuttle Operational Data Book [of NASA]

SODB Start of Data Block

SODC Standard Oil Development Company

SODERN Société d'Etudes et Réalisations Nucléaires [Society for the Study and Fabrication of Nuclear Products, France]

SO-DIMM Small-Outline Dual In-Line Memory Module

Sodium MBT Sodium Mercaptobenzothiazole [also sodium MBT]

Sodium TCA Sodium Trichloroacetate [also sodium TCA]

SO$_3$·DMF Sulfur Trioxide Dimethylformamide (Complex)

SODS Shuttle Operational Data System [of NASA]

SODS Skylab Orbit-Deorbit System [of NASA]

SODT Swiss Office for the Development of Trade

SOE Second-Order Effect

SOE Significant Operating Experience

SOE Shipbuilding and Offshore Engineering

SOE Society of Ontario Electrologists [Canada]

SOE Standard Operating Environment

SOE State of the Environment [also SoE]

S1E Surfaced on One Edge [Lumber]

So E Asia Southeast Asia(n)

SOEC Second-Order Elastic Constant [Mechanics]

SOEC Statistical Office of the European Communities

SOEEA Saskatchewan Outdoor and Environmental Education Association [Canada]

SOEKOR Suidelike Olie-Eksplorasiek-Orporasie [Southern Oil Exploration Corporation, South Africa] [also Soekor]

SOEP Solar-Oriented Experimental Package [of NASA]

SOER Significant Operating Event Report

SOER State of the Environment Report

SOERO Small Orbiting Earth Resources Observatory

SOEST School of Ocean and Earth Science and Technology [of University of Hawaii, Honolulu, US]

SOF Safety of Flight

SOF Satisfactory Operation Factor

SOF Shape and Orientation Factor

SOF Shorts, Open and Fixture

SOF Spreading Ocean Floor (Concept) [Geology]

SOF Start of Format

SOF Start of Frame

SOF Storage Oscilloscope Fragment

SOFA Save Our Fisheries Association [Newfoundland, Canada]

SOFAR Sound Finding and Ranging [also Sofar, or sofar]

SOFAR Sound Fixing and Ranging [also Sofar, or sofar]

SOFC Solid-Oxide Fuel Cell

SOFCS Self-Organizing Flight Control System

SOFE Symposium on Fusion Engineering [of Institute of Electrical and Electronics Engineers, US]

S/OFF Sign-Off [also S-OFF]

SOFFEX Swiss Options and Financial Futures Exchange

SOFHT Society of Food Hygiene Technology [UK]

SOFI Soziologisches Forschungsinstitut [Sociological Research Institute, Germany]

SOFI Spray-on Foam Insulation

SOFIA Southern Forest Inventory and Analysis [US]

SOFIA Surface of the Ocean Fluxes and Interaction with the Atmosphere

SOFNET Solar Observing and Forecasting Network [US Air Force]

SOFR State of the Forests Report

S of R Society of Rheology [also SoR, US]

SOFRECOM Société Française d'Etudes et de Réalisations d'Equipements de Télécommunications [French Society for the Study and Utilization of Telecommunications Equipment]

SOFRELEC Société Française d'Etudes et de Réalisations d'Equipements Electriques [French Society for the Study and Utilization of Electric Equipment]

SOFREM Société Française d'Electrométallurgie [French Electrometallurgical Company]

SOFT Simple Output Format Translator

SOFT Society of Forensic Toxicologists [US]

SOFT Software [also Soft]

SOFT Space Operations and Flight Techniques

SOFT Symposium on Fusion Technology

SOFTA Shippers Oil Field Traffic Association [US]

Soft PZT Lead Zirconate Titanate Doped for Low-Coercive-Field Production

Softw Software [also softw]

Softw Eng J Software Engineering Journal [of British Computer Society and Institution of Electrical Engineers]

Softw Eng Notes Software Engineering News [of Association for Computing Machinery, US]

Softw Eng Workstn Software for Engineering Workstations [Journal published in the UK]

Softw Maint News Software Maintenance News [Published in the US]

Softw–Pract Exp Software–Practice and Experience [UK]

Softw Prot Software Protection [Journal published in the US]

Softw World Software World [Journal published in the UK]

SOG Silicon-on-Glass (Substrate)

SOG Small-Outline Integrated Circuit Package with Gull-Wing Leads

SOG Spin-on-Glass

SOG Strongly Orthogonal Geminal

SOGAT Society of Graphical and Allied Trades [UK]

SOGC Saskatchewan Oil and Gas Corporation [Canada]

So Gr Soft Grind [also so gr]

SOGREAH Société Grenobloise d'Etudes et d'Applications Hydrauliques [Grenoble Society for Hydraulic Studies and Applications, France]

SOH Ohio Section [of American Physical Society, US]

SOH Sense Of Humor [Internet Jargon]

SOH Start of Header; Start of Heading

SOHC Single Overhead Camshaft

SOHIC Stress-Oriented Hydrogen-Induced Cracking

SOHL Stillwater Outdoor Hydro(power) Laboratory [Oklahoma, US]

SOHO Small Office/Home Office

SOHO Solar and Heliospheric Observatory [of NASA and European Space Agency]

SOHR Solar Hydrogen Rocket Engine

SOI Semiconductor-on-Insulator

SOI Silicon-on-Insulator

SOI Southern Oscillation Index [Oceanography]

SOI Specific Operating Instruction

SOI Standard Operating Instruction

SOI Stand-Off Insulator

SOI Statement of Intention

SOI Statement of Interest

SOI Survey of India

SOIC Small-Outline Integrated Circuit (Package)

SOI CMOS Silicon-on-Insulator Complementary Metal-Oxide Semiconductor [also SOI/CMOS]

SOI CMOS LSI Silicon-on-Insulator Complementary Metal-Oxide Semiconductor Large-Scale Integration

SOICS Summary of Installation Control Status

SOIE Subcommittee on Oceans and International Environment [of US Senate Foreign Relations Committee]

Soil Dyn Earthq Eng Soil Dynamics and Earthquake Engineering [Journal published in the UK]

Soil Geogr Soil Geography; Soil Geographer

Soil Sci Soil Science [Journal published in the US]

Soil Sci Soc Am J Soil Science Society of America Journal [US]

Soil Surv Horiz Soil Survey Horizons [of Soil Science Society of America, US]

Soil Tillage Res Soil and Tillage Research [US Publication]

SOIR (Study Group on) Simultaneous Operations on Parallel (or Near-Parallel) Instrument Runways

SOIS Shipping Operations Information System

SOI/SIMOX Silicon-on-Insulator/Silicon Implanted with Oxygen

SOJ Small-Outlet, J-Leaded Package [Electronics]

SOJ/TSOP Small-Outlet, J-Leaded Package/Thin, Small Outline Package [Electronics]

SOJ Stand-Off Jammer; Stand-Off Jamming

SOK Second-Order Kinetics

SOL Office of the Solicitor [US]

SOL Safety of Life

SOL School of Librarianship

SOL Simulation-Oriented Language

SOL Solid Waste Database [US/Canada]

SOL Source/Object Library

SOL System-Oriented Language

SOL Systems Optimization Laboratory [of Stanford University, California, US]

Sol Solar [also sol]

Sol Soldier [also sol]

Sol Solenoid [also sol]

Sol Solicitor [also sol]

Sol Solubility; Solubility; Solution [also sol]

sol soluble [Chemistry]

SOLACE School of Librarianship Automatic Cataloguing Experiment [of University of New South Wales, Australia]

SOLAGRAL Solidarités Agricoles et Alimentaires [Agricultural and Food Solidarities, France]

Solan N-(3-Chloro-4-Methylphenyl)-2-Methylpentanamide [also solan]

SOLAR Serialized On-Line Automatic Recording

SOLAR Storage On-Line Automatic Retrieval [of Washington State University, Pullman, US]

SOLARIS Submerged Object Locating and Retrieving Identification System

SolarMax Solar Maximum Mission (Satellite) [also SMM]

SOLAS (International Convention for the) Safety of Life at Sea

Sol Bull Solar Bulletin [of American Association of Variable Star Observers]

SOLCEC Symposium on Localized Corrosion and Environmental Cracking

Sol Cells Solar Cells [Journal published in Switzerland]

SOLCON Solar Constant [also Sol Con]

Soldadura Construcao Met Soldadura Construção Metalica [Portuguese Publication on Soldering in Metal Construction]

Solder Surf Mt Technol Soldering and Surface Mount Technology [UK Publication]

SOLE Society of Logistics Engineers [US]

SOLE Systems Operator Loading Evaluation

Sol En Solar Energy [Journal of the International Solar Energy Society] [also Sol Energy]

Sol En J Solar Energy Journal [of International Solar Energy Society] [also Sol Energy J]

Sol En Mater Solar Energy Materials [Journal published in the Netherlands] [also Sol Energy Mat]

Sol-Gel Sci Technol Sol-Gel Science and Technology [Journal published in the US]

SOLID Self-Organizing Large Information Dissemination System

Solid Fuel Chem Solid Fuel Chemistry [Journal published in the US]

Solid Mech Arch Solid Mechanics Archives [Journal published in the UK]

Solid-State Abstr Solid State Abstracts

Solid-State Commun Solid State Communications [Published in the US]

Solid-State Electron Solid-State Electronics [Published in the UK]

Solid-State Ion Solid State Ionics [Published in the Netherlands]

Solid-State Ion Diffus React Solid State Ionics–Diffusion and Reactions [Published in the Netherlands]

Solid-State Mater Sci Solid-State Materials Science [Journal published in the US]

Solid-State Phenom Solid-State Phenomena [Journal published in the US]

Solid-State Phys Solid-State Physics [Journal published in Japan]

Solid-State Technol Solid-State Technology [Journal published in the US]

SOLINET Southeastern Library Network [US]

SOLINOX SO_x (Sulfuric Oxide) Linde NO_x (Nitric Oxide) (Process) [Chemical Engineering]

Solion Solution Ion [also solion]

SOLIS Symbionics On-Line Information System

Solketal 2,2-Dimethyl-1,3-Dioxolane-4-Methanol, Acetone Ketal of Glycerine

SOLL Second-Order Local Linearization

SOLLAC Société Métallurgique de Normandie [Major French Metallurgical Company] [also Sollac]

Soln Solution [also soln]

SOLO Selective Optical Lock-On

SOLOGS Standardization of Operations and Logistics

SOLOMON Simultaneous Operation Limited Ordinal Modular Network; Simultaneous Operation Linked Ordinal Modular Network

Sol Phys Solar Physics; Solar Physicist

Sol Phys Solar Physics [Journal published in the Netherlands]

Sol Prob Solutions to Problems

SOLR Sidetone Objective Loudness Rating [Telecommunications]

SOLRAD Solar Radiation Satellite [Series of US Satellites] [also Solrad]

SOLS Spacelab/Orbiter Interface Simulator

SOLSPEC Solar Spectrum [also Sol Spec]

SOLSTICE Solar Stellar Irradiance Comparison Experiment

Sol Syst Res Solar System Research [Translation of: *Astronomicheskii Vestnik (USSR)*; published in the US]

Sol Terr Environ Res Jpn Solar Terrestrial Environmental Research in Japan [of Institute of Space and Aeronautical Science, Japan]

Sol Today Mag Solar Today Magazine

Sol Trt Solution Treatment [Metallurgy]

SOLUG San Antonio On-Line User Group [Texas, US]

SOLV Solenoid Valve

SOLV Super-Open-Frame Low Voltage

Solv Solvent [also solv]

Solv Extr Ion Exch Solvent Extraction and Ion Exchange [Journal published in the US]

Sol Wind Technol Solar and Wind Technology [Journal published in the UK]

Soly Solubility [also soly]

Solys Solubilities [also solys]

SOLZ Second Order Laue Zone [Crystallography]

SOM Scanning Optical Microscope; Scanning Optical Microscopy

SOM School of Mines

SOM Self-Organizing Machine

SOM Shift Operations Manager

SOM Ship Operations Manager [of NASA]

SOM Small Office Microfilm System

SOM Society of Occupational Medicine [UK]

SOM Somali Shilling [Currency]

SOM Space Oblique Mercator

SOM Spares Optimization Model

SOM Spin-Orbit Multiplet [Physics]

SOM Standard On-Line Module

SOM Standard Operating Manual

SOM Start of Message

SOM Strength of Materials

SOM System Object Model

Som Somali(a)

Som Somersetshire [UK]

SoMa South of Market [San Francisco, California, US]

SOMADA Self-Organizing Multiple-Access Discrete Address

Soman Methylphosphonofluoridic Acid-1,2,2-Trimethylpropylester [also soman]

SOMED School of Mines and Energy Development [of University of Alabama at Tuscaloosa, US]

SOMED Bibliothek der Sozialmedizin [Social Medicine Library, University of Bielefeld, Germany]

SOMER State of the Marine Environment Reporting [Australia]

SOMF Start of Minor Frame

SOMIEX Société Malienne d'Impôts et d'Expôts [Malian Import-Export Company]

SOMIS Security Office Management Information System

SOMO Semi-Occupied Molecular Orbital [Physical Chemistry]

SOMO Singly Occupied Molecular Orbital [Physical Chemistry]

SOMP Start-of-Message Priority

SOMS Shuttle Orbiter Medical System [of NASA]

Soms Somersetshire [UK]

SOMSEM Scanning Optical Microscopy in Scanning Electron Microscopy

Som Sh Somali Shilling [Currency]

SON Scientific Organization of the Netherlands

SON Self-Organizing Network

SON Society of Nematologists [US]

SON Statement of Need

SON Supra-Optic Nucleus [Anatomy]

S/ON Sign-On [also S-ON]

SONAC Sonar Nacelle

Sonacelle Sonar Nacelle [also sonacelle]

SONAD Speech-Operated Noise Adjusting Device

SONAR Sound Navigation and Ranging [also Sonar, or sonar]

SONAS Société Nationale d'Assurance [National Insurance Company, Zaire]

SONATRACH Société Nationale pour la Recherche, la Production, le Transport, la Transformation et la Commercialisation des Hydrocarbures [Algerian National Hydrocarbon Research, Production, Transportation, Transformation and Commercialization Company]

SONCM Sonar Countermeasures

SONCR Sonar Control Room

SONET Synchronous Optical Network (Technology) [also Sonet]

SONGS San Onofre Nuclear Generating Station [San Clemente, California, US]

SONIC System-Wide On-Line Network for Informational Control

SONOAN Sonic Noise Analyzer

SONORA Société Nigérienne de Commercialisation de l'Arachide [Niger Peanut Commercialization Organization]

SONOS Semiconductor–Oxide–Nitride–Oxide–Silicon

SONRES Saturated Optical Nonresonant Emission Spectroscopy

Soobsh AN GSSR Soobshcheniya Akademii Nauk Gruzinskoi SSR [Publication on Materials of the Academy of Sciences of the Georgian Soviet Socialist Republic] [also Soobsh Akad Nauk Gruz SSR]

SOOP Single-Operator Operational Precision

SOP Scranton Ordnance Plant [Pennsylvania, US]

SOP Secondary Oxygen Pack

SOP Second-Order Perturbation (Theory) [Physics]

SOP Simulation Operations Plan

SOP Small-Outline Package [Electronics]

SOP Spacelab Opportunity Payload

SOP Special Operating Procedure

SOP Standard Operating Procedure

SOP Standing Operating Procedure

SOP Strategic Orbit Point

SOP Student Outreach Program [of ASM International, US]

SOP Subsystem(s) Operating Procedure

SOP Subsystem Operating Program

SOP Sum-of-Products (Form)

SOP Super Olefin Polymer

SOP Surface Optical Phonon [Solid-State Physics]

SOP Systems Operation Plan

SOPA Society of Professional Archeologists [US]

SOPAC South Pacific Applied Geoscience Commission

SOPAC South Pacific Countries

SOPC Shuttle Operations and Planning Center [of NASA]

SOPC Shuttle Operations Planning Complex [of NASA]

SOPG Science Operations Planning Group

Soph Sophomore [also soph]

SOPHYA Supervisor of Physics Analysis

SOPI Serial-Out, Parallel-In

SOPM Standard Orbital Parameter Message

SOPP Sodium Orthophenylphenate

SOPPA Second-Order Polarization Propagator Approximation [Physics]

SOPR Spanish Open-Pool (Nuclear) Reactor [Spain]

SOPS Servo-Optical Projection System

SO₃·Pyr Sulfur Trioxide Pyridine (Complex)

SOQUEM Société Québécoise d'Exploration Minière [Quebec Mining Exploration Society, Canada]

SOR Self-Regulating Organization

SOR Shut-Off Rod(s) [Nuclear Reactors]

SOR Simultaneous Overelaxation (Method)

SOR Small Outrigger

SOR Society for Occupational Research [US]

SOR Specification Operational Requirement

SOR Specific Operating Requirement

SOR Stable Orbit Rendezvous

SOR Start of Record

SOR Statement of Requirement

SOR Statutory Orders and Regulations

SOR Steam/Oil Ratio

SOR Successive Overrelaxation [Physics]

SOR Surface Oxidation-Reduction (Process)

SOR Synchronous Orbital Resonance

SOR Synchrotron Orbital Radiation

SOR Synthetic Oxidation Reduction

SoR Society of Rheology [also SofR, US]

SoR Society of Roadcraft [UK]

SORC Site Operations Review Committee

SORCE Software Resource Center

SORD Separation of Radar Data

Soreq NRC Soreq Nuclear Research Center [Yavne, Israel]

SOR/F Sterling Orbiter Refrigerator/Freezer [of NASA Space Shuttle] [also SORF]

SORG Stratospheric Ozone Review Group [UK]

SORIN Societa Ricerche ed Impianti Nucleare [Society for Nuclear Research and Equipment, Italy]

SORM Set-Oriented Retrieval Module

SORO Special Operations Research Office [US]

SORP Surface Oxidation-Reduction Process

SORPTR South Repeater

SORSA Spatially Oriented Referencing Systems Association

SORTE Summary of Radiation Tolerant Electronics

SORTI Satellite Orbital Track and Intercept

SORTIE Supercircular Orbital Reentry Test Integrated Environment

SOS Save Our Souls [International Distress Signal]

SOS Scheduled Oil Sampling

SOS Schedule Optimization Study

SOS Semi-Open System

SOS Serial Output Special

SOS Service Order System

SOS Share Operating System

SOS Shop Out of Stock

SOS Silicon-on-Sapphire

SOS *(Si opus sit)* – if necessary [also sos] [Medical Prescriptions]

SOS Singlet-Only (Electron) Scattering

SOS Smoke Obscuring Screen

SOS Sniping, Observation and Scouting Distress

SOS Solid-on-Solid (Model) [Solid-State Physics]

SOS Somali Shilling [Currency]

SOS Sophisticated Operating System

SOS Source of Supply

SOS Space-Oriented Satellite

SOS Speed of Service

SOS Spin-Orbit Scattering [Physics]

SOS Stabilized Optical Sight

SOS Standards and Open Systems

SOS Start of Significance [Data Communications]

SOS Station Operating Supervisor

SOS Sum of States; Sum over States [Statistical Mechanics]

SOS Superlattice of Superlattices [Solid-State Physics]

SOS Symbolic Operating System

sos *(si opus sit)* – if necessary [also SOS] [Medical Prescriptions]

S1S Surfaced on One Side [Lumber]

SOSA Spin-Orbit Split Array [Physics]

SOSC Smithsonian Oceanographic Sorting Center [US]

SOSC Southern Oregon State College [US]

SOSC Space Operations Support Center

SOSC Special Offshore Symposium

SOS-CI Sum over States Configuration Interaction [Physics]

So Sh Somali Shilling [Currency]

SOSI Serial-Out, Serial-In

SOSI Shift-In, Shift-Out

SOSI Southern Oilfield Supply and Manufacturing, Inc. [US]

S1S1E Surfaced on One Side and One Edge [Lumber]

SOSP Silicon-on-Spinel

SOSS Shipboard Oceanographic Survey System

SOS/SOI Silicon-on-Sapphire/Silicon-on-Insulator (Technology)

SOST Special Operator Service Traffic

S1S2E Surfaced on One Side and Two Edges [Lumber]

SOSTEL Solid-State Electronic Logic

SOSU Southwestern Oklahoma State University [Weatherford, US]

SOSUS Sound Surveillance System

SOT Sacro-Occipital (Chiropractic) Technique [Medicine]

SOT Second-Order Twin [Metallurgy]

SOT Small Outline Transistor

SOT Society of Toxicology [US]

SOT Start of Text

SOT State of Termination

SOT Station Operator Terminal

SOT Strap-on Tank

SOT Subscriber Originating Trunk

SOT Syntax-Oriented Translator

SOTA State of the Art

SOTAS Stand-Off Target Acquisition System

SOT DAT Source Test Data System [of US Environmental Protection Agency]

SOTEAG Shetland Oil Terminal Environment Advisory Group [formerly Sullom Voe Environmental Advisory Group, UK]

SOTELEC Société Française des Télécommunications [French Telecommunications Society]

SOTER Soil and Terrain Database [US]

SOTIM Sonic Observation of the Trajectory and Impact of Missiles

SO$_3$·TMA Sulfur Trioxide Trimethylamine (Complex)

SOTS Sub-Orbital Tank Separation

SOTUS Sequentially-Operated Teletypewriter Universal Selector

SOU Scandinavian Ornithological Union [Sweden]

Soudage Tech Connexes Soudage et Techniques Connexes [French Journal on Soldering and Joining Techniques]

Sound Vib Sound and Vibration [Journal published in the US]

SOUP Software Utility Package

SOUP Submarine Operational Update Program

SOUT Swap-Out

SOUTC Satellite Operators and Users Technical Committee [US]

South Afr Comput J South African Computer Journal [of Computer Society of South Africa]

South Afr J Antarct Res South African Journal of Antarctic Research

South Afr J Chem South African Journal of Chemistry [of South African Chemical Institute]

South Afr J Libr Inf Sci South African Journal of Library and Information Science [of South African Institute for Librarianship and Information Science]

South Afr Med J South African Medical Journal

South Afr J Phys South African Journal of Physics

South Afr J Sci South African Journal of Science

South Afr Mach Tool Rev South African Machine Tool Review

South Afr Mech Eng South African Mechanical Engineer [of South African Institute of Mechanical Engineers]

SOUTHCOM Southern Command [of US Army in Panama] [Disbanded]

SOUTHEASTCON Southeastern Convention [of Institute of Electrical and Electronics Engineers, US]

Southern J Appl Forestry Southern Journal of Applied Forestry [of Society of American Foresters]

Southern Med J Southern Medical Journal [US Publication]

Southwestern J Anthropol Southwestern Journal of Anthropology [US] [also Southwestern J Anthrop]

SOV Separation of Variables [Mathematics]

SOV Shutoff Valve

SOV Solenoid-Operated Valve

SOV Sound on Vision

Sov Sovereign(ty) [also sov]

Sov Soviet

Sov Aeronaut Soviet Aeronautics [Translation of: *Izvestiya Vysshikh Uchebnykh Zavedenii, Aviatsionnaya Tekhnika*; published in the US]

Sov Appl Mech Soviet Applied Mechanics [Translation of: *Prikladnaya Mekhanika (Ukrainian SSR)*; published in the US]

Sov Arch Intern Med Soviet Archives of Internal Medicine

Sov Astron Soviet Astronomy [Translation of: *Astronomicheskii Zhurnal*; published in the US]

Sov Astron Lett Soviet Astronomy Letters [Translation of: *Pis'ma v Astronomicheskie Zhurnal*; published in the US]

Sov At Energy Soviet Atomic Energy [Translation of: *Atomnaya Energiya*; published in the US]

Sov Cast Technol Soviet Casting Technology [Translation of: *Liteinoe Proizvodstvo*; published in the US]

Sov Chem Ind Soviet Chemical Industry [Published in the US]

Sov Electr Eng Soviet Electrical Engineering [Translation of: *Elektrotekhnika*; published in the US]

Sov Electrochem Soviet Electrochemistry [Translation of: *Elektrokhimiya*; published in the US]

Sov Energy Technol Soviet Energy Technology [Published in the US]

Sov Eng Res Soviet Engineering Research [Selective translation of: *Vestnik Mashinostroeniya* and *Stanki i Instrument*; published in the US]

Sov Geol Phys Soviet Geology and Geophysics [Translation of: *Geologiya i Geofizika*; published in the US]

Sov J Appl Phys Soviet Journal of Applied Physics [Translation of: *Izvestiya Sibirskogo Otdeleniya Akademii Nauk SSSR, Tekhnicheskikh*; published in the US]

Sov J At Energy Soviet Journal of Atomic Energy [Translation of: *Atomnaya Energiya*]

Sov J Autom Inf Sci Soviet Journal of Automation and Information Science [Translation of: *Avtomatika (Ukrainian SSR)*; published in the US]

Sov J Chem Phys Soviet Journal of Chemical Physics [Translation of: *Khimicheskaya Fizika*; published in the UK]

Sov J Commun Technol Electron Soviet Journal of Communications Technology and Electronics [Translation of: *Radiotekhnika i Elektronika*; published in the US]

Sov J Comput Syst Sci Soviet Journal of Computer and Systems Sciences [Translation of: *Tekhnicheskaya Kibernetika*; published in the US]

Sov J Contemp Phys Soviet Journal of Contemporary Physics [Translation of: *Izvestiya Akademia Nauk Armyanskoi SSR Fizika*; published in the US]

Sov J Frict Wear Soviet Journal of Friction and Wear [Translation of: *TrenieIznos (Byelorussian SSR)*; published in the US]

Sov J Glass Phys Chem Soviet Journal of Glass Physics and Chemistry [Translation of: *Fizika i Khimiya Stekla*; published in the US]

Sov J Low Temp Phys Soviet Journal of Low Temperature Physics [Translation of: *Fizika Nizkikh Temperatur*; published in the US]

Sov J NDT Soviet Journal of Nondestructive Testing [Translation of: *Defektoskopiya*; published in the US] [also Sov J Nondestr Test]

Sov J Non-Ferrous Met Soviet Journal of Non-Ferrous Metals [Published in the US]

Sov J Nucl Phys Soviet Journal of Nuclear Physics [Translation of: *Yadernaya Fizika*; published in the US]

Sov J Opt Technol Soviet Journal of Optical Technology [Translation of: *Optiko-Mekhanicheskaya Promyshlennost*; published in the US]

Sov J Part Nucl Soviet Journal of Particles and Nuclei [Translation of: *Fizika Elementarnykh Chastits i Atomnoya Yadra*; published in the US]

Sov J Plasma Phys Soviet Journal of Plasma Physics [Translation of: *Fizika Plazmy*; published in the US]

Sov J Quantum Electron Soviet Journal of Quantum Electronics [Translation of: *Kvantovaya Elektronikaya, Moskva*; published in the US]

Sov J Remote Sens Soviet Journal of Remote Sensing [Translation of: *Issledovanie Zemli iz Kosmosa*; published in Switzerland]

Sov J Superhard Mater Soviet Journal of Superhard Materials [Translation of: *Sverkhtverdyne Materialy*; published in the US]

Sov J Water Chem Technol Soviet Journal of Water Chemistry and Technology [Published in the US]

Sov Mach Sci Soviet Machine Science [Translation of: *Mashinovedenie*; published in the US]

Sov Mater Sci Soviet Materials Science [Translation of: *Fiziko-Khimicheskaya Mekhanika Materialov*; published in the US]

Sov Mater Sci Rev Soviet Materials Science Reviews [English Translation published in the US]

Sov Math Soviet Mathematics [Translation of: *Izvestiya Vysshikh Uchebnykh Zavedenii, Matematika*; published in the US]

Sov Meteorol Hydrol Soviet Meteorology and Hydrology [Translation of: *Meteorologiya i Gidrologiya*; published in the US]

Sov Microelectron Soviet Microelectronics [Translation of: *Mikroelektronika*; published in the US]

Sov Min Sci Soviet Mining Science [Published in the US]

Sov Non-Ferrous Met Res Soviet Non-Ferrous Metals Research [Published in the UK]

Sov Phys-Acoust Soviet Physics–Acoustics [Translation of: *Akusticheskii Zhurnal*; published in the US]

Sov Phys-Collect Soviet Physics–Collection [Translation of: *Litovskii Fizicheskii Sbornik*; published in the US]

Sov Phys-Crystallogr Soviet Physics–Crystallography [Translation of: *Kristallografiya*; published in the US]

Sov Phys-Dokl Soviet Physics–Doklady [Translation of: *Doklady Akademii Nauk SSSR*; published in the US]

Sov Phys J Soviet Physics Journal [Translation of: *Izvestiya Vysshikh Uchebnykh Zavedenii, Fizika*; published in the US]

Sov Phys-JETP Soviet Physics–JETP (Journal of Experimental Theoretical Physics) [Translation of: *Zhurnal Eksperimental'noi i Teoreticheskoi Fiziki*; published in the US]

Sov Phys-JETP Lett Soviet Physics–JETP (Journal of Experimental Theoretical Physics) Letters [Translation of: *Zhurnal Eksperimental'noi i Teoreticheskoi Fiziki*; published in the US]

Sov Phys-Leb Inst Rep Soviet Physics–Lebedev Institute Reports [Translation of: *Sbornik Kratkie Soobshcheniya po Fizike, AN SSSR, Fizikcheskii Institut im PN Lebedeva*; published in the US]

Sov Phys-Semicond Soviet Physics–Semiconductors [Translation of: *Fizika i Tekhnika Poluprovodnikov*; published in the US]

Sov Phys-Solid St Soviet Physics–Solid State [Translation of: *Fizika Tverdogo Tela*; published in the US] [also Sov Phys Solid State]

Sov Phys-Tech Phys Soviet Physics–Technical Physics [Translation of: *Zhurnal Tekhnicheskoi Fiziki*; published in the US]

Sov Phys-Usp Soviet Physics–Uspekhi [Translation of: *Uspekhi Fizicheskii Nauk*; published in the US]

Sov Powder Metall Met Ceram Soviet Powder Metallurgy and Metal Ceramics [Translation of: *Poroshkovaya Metallurgiya*; published in the US]

Sov Prog Chem Soviet Progress in Chemistry [Translation of: *Ukrainskii Khimischeskii Zhurnal*; published in the US]

Sov Surf Eng Appl Electrochem Soviet Surface Engineering and Applied Electrochemistry [Translation of: *Elektronnaya Obrabotka Materialov*; published in the US]

Sov Surf Sci Soviet Surface Science [Published in the US]

Sov Tech Phys Lett Soviet Technical Physics Letters [Translation of: *Pis'ma v Zhurnal Tekhnicheskoi Fiziki*; published in the US]

SOW Stand-Off Weapon

SOW Start-of-Word

SOW Statement of Work

SOW Subdivision of Work

SOW Sunflower Ordnance Works [Lawrence, Kansas, US]

SOWP Screen-Oriented Word Processor

SOWP Society of Wireless Pioneers [US]

SOX Solid Oxygen [also SOx]

SOX Sound Exchange

SOX Supercritical Oxygen

SOx Sulfur Oxides (i.e., SO_2, SO_3, S_2O_3, and S_2O_7) [also SOX, or SO_x]

SP Reflected Earthquake Wave Having Travelled One Leg as a Shear Wave and the Other as a Push Wave [Symbol]

SP Sacrificial Protection [Physical Chemistry]

SP Saddle Point

SP Sampling Period

SP Sampling Plan

SP Sampling Probe [Leak Testing]

SP San Pablo [Luzon, Philippines]

SP San Pedro [California, US]

SP San Pedro [Paraguay]

SP São Paulo [Brazil]

SP Satellite Processor

SP Saturation Point [also sp]

SP Saturation Polarization

SP Saturation Potential

SP Saturation Pressure [Physical Chemistry]

SP Scalable Parallel (System)

SP Scalable Power-Parallel (Server)

SP Scalar Product [Mathematics]

SP Scanning Probe

SP Scratch Pad

SP Screw Pitch

SP Screw Press

SP Screw Propeller

SP Screw Pump

SP Seaplane

SP Secular Perturbation [Astrophysics]

SP Sedimentation Potential

SP Seismic Prospecting

SP Self-Potential [Physics]

SP Self-Propelled

SP Semipermeability; Semipermeable

SP Semipolar

SP Semi-Public

SP Send Processor

SP Separation Process

SP Sequential Phase

SP Sequential Processing; Sequential Processor

SP Serial Printer

SP Serial-to-Parallel

SP Service Pack(age)

SP Service Processor

SP Servo Power

SP Setting Point [Chemistry]

SP Sewer Pipe

SP Shadow Photograph(y)

SP Shear Pin

SP Shear Plane [Mechanics]

SP Shear Plate

SP Shift Pulse

SP Shipping Port

SP Ship Propeller

SP Shockley Partial(s) [Materials Science]

SP Shop Planning [Industrial Engineering]

SP Shore Patrol

SP Shore Police

SP Shot-Peened; Shot Peening [Metallurgy]

SP Shoulder Pin

SP Shoulder Pitch

SP Shubnikov Phase [Solid-State Physics]

SP Shuttle Projects Office [of NASA Kennedy Space Center, Florida, US]

SP Sidereal Period [Astronomy]

SP Sigma Pile [Nuclear Engineering]

SP Signal Plate [Television]

SP Signal Processing; Signal Processor [also S/P]

SP Silicon Carbide Whisker/Calcium Phosphate (Composite)

SP Silicon Processing

SP Silicophosphate

SP Silver Point [Chemistry]

SP Simple Pendulum

SP Simple Pit (in Wood)

SP Singapore Polytechnic (Institute)

SP Singing Point [Control Systems]

SP Single-Perforated (Propellant)

SP Single-Phase

SP Single Plaquette

SP Single-Pole (Switch) [also sp]

SP Single Precision

SP Single Programmer

SP Skeletal Point (Stress) [Mechanics]

SP Slightly Polished; Slight Polishing

SP Slip Plane

SP Small Particle

SP Small Pica (Font)

SP Small Punch (Test) [Mechanical Testing]

SP Smokeless Propellant

SP Society of Protozoologists [US]

SP Softening Point

SP Soil Pipe

SP Soil Profile [Geology]

SP Solar Panel

SP Solar Physicist; Solar Physics

SP Solar Plasma
SP Solar Plexus [Anatomy]
SP Solar Power
SP Solar Proton
SP Solidification Point [also sp]
SP Solidification Processing
SP Solid Phase
SP Solid Propellant
SP Solubility Parameter [Physical Chemistry]
SP Solubility Product [Physical Chemistry]
SP Soluble Powder
SP Sorption Pump(ing)
SP Sort(ing) Program
SP Sound-Powered (Telephone)
SP Sound Pressure
SP Soundproof(ing)
SP Sound Propagation
SP Source Program [Data Processing]
SP Southern Pacific (Railroad) [US]
SP Southern Pine
SP South Pacific
SP South Pole
SP Space (Character)
SP Space Plasma
SP Space Propulsion
SP Spain [ISO Code]
SP Spare Part
SP Spark Plug
SP Spatial Domain
SP Specialist [US Army Grade] [also Sp]
SP Special Paper
SP Special Project
SP Special Publication
SP Special Purpose
SP Special-Thermal Processing
SP Specialty Polymer
SP Spectrophotometer; Spectrophotometry
SP Spectrum Projector
SP Speech Processing
SP Speed Plug
SP Spherical Powder [Metallurgy]
SP Spill Prevention
SP (Neél) Spin Pair (Model) [Solid-State Physics]
SP Spin Packet [Physics]
SP Spin Paramagnetism [Solid-State Physics]
SP Spin Polarization [Physics]
SP Spin-Polarized (Function) [Physics]
SP Splash-Proof
SP Splenopentin [Biochemistry]
SP Spontaneous Potential
SP Spool Piece
SP Square Planar [also sp]
SP Square Punch

SP Square Pyramidal; Square Pyramid
SP Stabilized Pinch
SP Stable Particle
SP Stable Phase [Physical Chemistry]
SP Stack Pointer
SP Standard, or Peculiar
SP Standard Play
SP Standard Potential [Physical Chemistry]
SP Standpipe
SP Starting Point
SP Starting Potential [Electronics]
SP Starting Price
SP State Park
SP Static Pressure
SP Stationary Phase
SP Station Pointer
SP Statistical Physicist; Statistical Physics
SP Statistical Planning
SP Statistical Probability
SP Steam Point
SP Steam Pressure
SP Stem Protector
SP Stereo-Power; Stereoscopic Power
SP Sticking Potential
SP Sticking Probability
SP Stirling Process [Thermodynamics]
SP Stochastic Process
SP Stopping Power
SP Storage Pump
SP Stored Program
SP St. Paul [Minnesota, US]
SP Streak Photography
SP Streak Plate [Microbiology]
SP Strictly Private (Communication)
SP Strike Plate [Building Trades]
SP Structural Plastic(s)
SP Structured Programming
SP Structure Parallelism
SP Sublimation Pump
SP Substance P (Protein) [Immunochemistry]
SP Suction Pump
SP Sugar Pine
SP Sulfopropyl
SP Sulfur Point
SP Summary Punch(ing)
SP Sum-of-Products (Form)
SP Sunderland Polytechnic [Sunderland, UK]
SP Superparamagnet(ic); Superparamagnetism
SP Superplastic(ity)
SP Super-Precision
SP Supervisory Package
SP Supervisory Panel
SP Supervisory Printer

SP Supervisory Process
SP Supervisory Program
SP Supply Post
SP Support Processor
SP Surface Physicist; Surface Physics
SP Surface Plasma [Plasma Physics]
SP Surface Plasmon [Solid-State Physics]
SP Surface Plate [Machining]
SP Surface Potential
SP Surface Pressure
SP Surface Profile
SP Surge Protector
SP Surveillance Procedure
SP Survival Probability
SP Switching Point [Telecommunications]
SP Switch Panel
SP Symbol Programmer
SP Synthetic Polymer
SP Synthetic Potentiostat
SP System Performance
SP System Processor
SP System Product
SP System(s) Programmer; System(s) Programming
SP Thiophenol
SP Water Soluble Powder
SP- Poland [Civil Aircraft Marking]
SP-1 Stored Program 1
S&P Standards and Protocols
S&P Structures and Propulsion Laboratory [of NASA]
S&P Systems and Programming
S/P Serial-to-Parallel [also S-P]
S/P Signal Processor [also SP]
S-P Serial-to-Parallel [also S/P]
S-P (Earthquake) Shear-Push (Interval)
S-P Superparamagnetic(s)
Sp Sapwood [also sp]
Sp Semiprecious (Stones)
Sp Spain; Spaniard; Spanish
Sp Spare [also sp]
Sp Specialty; Specialist [also sp]
Sp Species [also sp]
Sp Specimen [also sp]
Sp Speed [also sp]
Sp Spelling [also sp]
Sp Sphincter [Anatomy]
Sp Spike [also sp]
Sp Spin [also sp]
Sp Spinning [also sp]
Sp Spinel [Mineral]
Sp Spiral [also sp]
Sp Spirit [also sp]
Sp Spur (of a Matrix) [Symbol]
Sp Sputter(ing) [also sp]

Sp Square Planar [also sp]
S£ Syrian Pound [Currency]
sp hybrid orbital of one s-orbital and one p-orbital [Symbol]
sp sapwood [also Sp]
sp (sine prole) – without issue
sp spare [also Sp]
sp special; specialty; specialist [also Sp]
sp species [also Sp]
sp specific
sp specimen [also Sp]
sp speed [also Sp]
sp spelling [also Sp]
sp spike [also Sp]
sp spin [also Sp]
sp spinning [also Sp]
sp spiral [also Sp]
sp spirit [also Sp]
sp square planar [also Sp]
sp² hybrid orbital of one s-orbital and two p-orbitals [Symbol]
sp³ hybrid orbital of one s-orbital and three p-orbitals [Symbol]
.sp Spain [Country Code/Domain Name]
SPA Salt Producers Association [now Salt Institute, US]
SPA S-Band Power Amplifier
SPA Scandinavian Packaging Association [Norway]
SPA Schedules Planning and Analysis
SPA School of Pacific Administration [Sydney, Australia]
SPA Scottish Publishers Association
SPA Screen Printing Association [UK]
SPA Seaplane Pilots Association [US]
SPA Self-Publishing Association [UK]
SPA Semi-Permanently Associated
SPA Separated Pair Approximation [Physics]
SPA Servo Power Amplifier
SPA Servo Power Assembly
SPA Servo Preamplifier
SPA Shared Peripheral Area
SPA Signal Processor Assembly
SPA Single-Pole Approximation
SPA Small Part Analysis
SPA Société Protectrice des Animaux [Society for the Protection of Animals, France]
SPA Software Process Assessment
SPA Software Producers Association [UK]
SPA Software Publishers Association [US]
SPA Solid-Phase Amplification
SPA Solid Phosphoric Acid
SPA Southeastern Peanut Association [US]
SPA Southern Pine Association [now Southern Forest Products Association, US]
SPA Southwestern Power Administration [of US Department of the Interior]

SPA Space Processing Application

SPA Specialist in Public Administration

SPA Specially Protected Area

SPA Special Policy Adjustment

SPA Special-Purpose Automation

SPA Spectrum Analyzer

SPA Spin-Polarization Analysis; Spin-Polarization Analyzer [Physics]

SPA Spot Profile Analysis; Spot Profile Analyzer

SPA Steering Position Amplifier

SPA Stereo Power Amplifier

SPA Stress-Pattern Analysis

SPA Subject to Particular Average

SPA Sudden Phase Anomaly [Physics]

SPA (International Workshop on Critical Currents in) Superconductors for Practical Applications

SPA Surface Peak Area

SPA Surface Photoabsorption

SPA Surface Plasmon Absorption [Solid-State Physics]

SPA Systems and Procedures Association [now Association for Systems Management, US]

SP-A Slightly Polishing-Abrasive

SP-A Specific Protein A [Biochemistry]

S&PA Safety and Product Assurance

SpA *(Société par Actions)* – French for "Limited Partnership"

SpA *(Società per Azioni)* – Italian for "Public Limited Company"

SPAC Salinity Program Advisory Council [Australia]

SPAC Spacecraft Performance Analysis and Command [of NASA]

SPAC Spatial Computer

SPAC Surface Protection Against Corrosion

S-PAC Student-Professional Awareness Committee [of Institute of Electrical and Electronics Engineers, US]

S Pac South Pacific

SPACDAR Specialist Panel on Automatic Conflict Detector and Resolution

SPACE Satellite, Precipitation and Cloud Experiment

SPACE Self-Programming Automatic Circuit Evaluator

SPACE Sequential Position and Covariance Estimation

SPACE Sidereal Polar Axis Celestial Equipment

SPACE Spacecraft Prelaunch Automatic Checkout Equipment

SPACE Spatial and Chemical Shift-Enclosed Excitation

SPACE Symbolic Programming Anyone Can Enjoy

SPACECOM Space Communications [also Spacecom]

Space Commun Space Communications [Published in the Netherlands]

SPACECOMPS Space Components (Database) [of European Space Agency]

Space J Space Journal [US]

Spacelab Space Laboratory [of Europen Space Agency]

Space Sci Rev Space Science Reviews [Journal published in the Netherlands]

Space Technol Space Technology [Journal published in the UK]

SPACHEE South Pacific Action Committee for Human Ecology and Environment

SPACON Space Control [also Spacon]

SPACS Sodium Purification and Characterization System

SPAD Satellite Position Predictor and Display

SPAD Satellite Protection for Area Defense

SPAD SDLC (Synchronous Data Link Control) Packet Assembly/Disassembly

SPAD Shuttle Payload Accommodation Document [of NASA]

SPAD Simplified Procedures for Analysis of Data

SPAD Subsystem Positioning Aid Device

SPADATS Space Detection and Tracking System [also Spadats]

SPADE SCPC (Single Channel per Carrier)/PCM (Pulse-Code Modulation), Multiple Access Demand-Assigned Equipment [also Spade]

SPADE Spare Parts Analysis, Documentation and Evaluation

SPADE SPARTA (Spatial Antimissile Research Tests in Australia) Acquisition Digital Equipment

SPADE Sperry Air Data Equipment

SPADES Solar Perturbation and Atmospheric Density Measurement Satellite

SPADETS Space Detection System

SPADNS 2-(p-Sulfophenylazo)-1,8-Dihydroxy-3,6-Naphthalenedisulfonic Acid (Trisodium Salt); 3-(4-Sulfophenylazo)-4,5-Dihydroxy-2,7-Naphthalene-disulfonic Acid

SPADS Shuttle Problem Action (or Analysis) Data System [of NASA]

SPAEF Société des Pétroles de l'Afrique Equatoriale Française [French Equatorial Africa Petroleum Company]

SPAES Spin-Polarized Auger Electron Spectroscopy

SPAF Simulation Processor and Formatter

SP-AF Shuttle Projects Office–Air Force Support Office [at NASA Kennedy Space Center, Florida, US]

SPAG Standard Promotion Application Group [Europe]

SPAH Spacelab Payload Accommodations Handbook

SPAI Screen Printing Association International [US]

SPAID Society for the Prevention of Asbestosis and Industrial Diseases [UK]

SPAIRS Single Potential Alteration Infrared Spectroscopy

SPALDA Scottish Peat and Land Development Association

SPA-LEED Spot Profile Analysis–Low-Energy Electron Diffraction; Spot Profile Analyzer–Low-Energy Electron Diffraction

SPAM Satellite Processor Access Method

SPAM Scanning Photoacoustic Microscopy

SPAM Ship Position and Attitude Measurement

SPAM Stupid Person's AdvertiseMent [Internet Jargon]

Sp Am Spanish America(n) [Sp Amer]

SPAMS Ship Position and Attitude Measurement System

SPAN Self-Doped Sulfonated Polyaniline

SPAN Solar Particle Alert Network [of NASA]

SPAN Spacecraft Analysis

SPAN Space Physics Analysis Network

SPAN Span (Text) Analysis

SPAN Statistical Processing and Analysis

SPAN Stored Program Alphanumerics

SPAN Sulfonated Polyaniline

SPAN Sulfophenylazonaphthalene

Span Spaniard; Spanish

SPANA Structural Panel Association of North America

SPANDAR Space and Range Radar

SPANRAD Superposed Panoramic Radar Display

SPANS Spatial Analysis System

SPAQUA Sealed Package Quality Assurance

SPAR Seagoing Platform for Acoustic Research

SPAR Society of Photographers and Artists Representatives [US]

SPAR Space Processing Applications Rocket

SPAR Special Products and Applied Research

SPAR Structural Performance Analysis and Redesign

SPAR Super-Precision Approach Radar

SPAR Symbolic Program Assembly Routine

SPAR Synchronous Position Altitude Recorder

SP-AR Super-Precision Anti-Resonance

SPARC Scalable Processor Architecture [also Sparc]

SPARC Secreted Protein Acidic and Rich in Cysteine [Protein Made from Mouse Parietal Yolk Sac Cells]

SPARC Space Air Relay Communications

SPARC Space Program Analysis and Review Council [US Air Force]

SPARC Space Research Conic [of NASA]

SPARC Stratospheric Processes and their Role in Climate

SPARCS Solar Pointing Aerobee Rocket Control System

SPARPES Spin-Polarized Angle-Resolved Photoelectron Spectroscopy

SPARS Space Precision Attitude Reference System

SPARS Space Prediction Attitude Reference System

SPARS Spatially Resolved Spectroscopy

SPARS Society of Professional Audio Recording Services [US]

SPARSA Sferics Pulse, Azimuth, Rate and Spectrum Analyzer

SPARSA Sferics, Position, Azimuth, Range Spectrum Analyzer

SPART Space Research and Technology

SPARTA Spatial Antimissile Research Tests in Australia; Special Antimissile Research Tests in Australia

SPARTECA South Pacific Regional Trade and Economic Cooperation Agreement

SPAS Shuttle Pallet Satellite [Germany]

SPAS Shuttle Planning and Analysis System

SPAS Slow Position Annihilation Spectroscopy

SPASM System Performance and Activity Software Monitor

SPASUR Space Surveillance (System) (or Space Detection and Tracking System) [of US Navy]

SPAT Self-Propelled Antitank

SPAT Silicon Precision Alloy Transistor

SPATE Stress-Pattern Analysis by Thermal Emission

Spatial Vis Spatial Vision [Journal published in the Netherlands]

SPAU Signal Processing Arithmetic Unit

SPAYZ Spatial Property Analyzer

SPB Seaplane Base

SPB Short Packed Bed (Reactor) [Physical Chemistry]

SPB Simple Payback (Period)

SPB Stored Program Buffer

SPBC Snap Pack Battery Cartridge

SPBEC South Pacific Bureau for Economic Cooperation [Fiji]

sPBI Derivatized Polybenzimidazole

SPBM Single Point Buoy Mooring

SPBN Sulfophenylbutylnitrone

2-SPBN 2-Sulfophenylbutylnitrone; N-tert-Butyl-α-(2-Sulfophenyl)nitrone, Sodium Salt

SPC Saskatchewan Power Commission [Canada]

SPC Saskatchewan Power Corporation [Canada]

SPC School of Physics and Chemistry

SPC Screen-Monitored Programmed Control

SPC Secretariat of the Pacific Community [New Caledonia]

SPC Self-Propelled Caterpillar

SPC Serial-to-Parallel (Data) Converter

SPC Series-Parallel Control

SPC Shipping and Packing Cost

SPC Shuttle Processing Contract [NASA]

SPC Shuttle Processing Contractor [NASA]

SPC Simple Point Charge [Physics]

SPC Single-Paper-Covered (Electric Wire)

SPC Single Point Connection

SPC Skip Spaces (Function) [Programming]

SPC Small Pancake-Shaped Cell(s)

SPC Small Peripheral Controller

SPC Software Productivity Consortium [Reston, Virginia, US]

SPC Software Publishing Corporation [US]

SPC Solid-Phase Crystallization

SPC Solid Proton Conductor

SPC South Pacific Commission [of Secretariat of the Pacific Community, New Caledonia]

SPC Soy Protein Council

SPC Special Political Committee

SPC Special Premiers Conference [Canada]

SPC Special-Purpose Computer

SPC Starting Point Code

SPC Static Power Converter

SPC Statistical Process Control

SPC Statistical Product Control

SPC Steel Processing Center [of Colorado School of Mines, Golden, US]

SPC Stones per Carat [Diamond Drill Bits]

SPC Stored-Program Command

SPC Stored-Program Computer

SPC Stored-Program Control [also spc]

SPC Strategic Planning Committee [of Institute of Electrical and Electronics Engineers, US]

SPC Sulfur Polymer Cement

SPC Synoptic Properties Code

SPC Syrian Petroleum Company

SP2C Thiophene-2-Carboxylic Acid

SP3C Thiophene-3-Carboxylic Acid

SPCA Society for the Prevention of Cruelty to Animals

SPCAN Space Plasma Computer Analysis Network

SPCC Spill Prevention Control and Countermeasure (Plan) [of US Environmental Protection Agency]

SPCC State Pollution Control Commission [of New South Wales, Australia]

SPCC STS (Space Transportation System) Processing Control Center [of NASA]

SPC-FP Simple Point Charge (Model with) Flexible Bonds and Polarization [Physics]

SPCGM Self-Propelled Caterpillar Gun-Mount

SPCL Spectrum Cellular Corporation [US]

spcl special

SPCN Sciences Physiques, Chimiques, Naturelles [Physical, Chemical and Natural Sciences]

SPCO Strontium Praseodymium Copper Oxide (Superconductor) [also SrPCO]

SPCP Single Parity Check Product (Code)

SPCS Secondary Pyramidal Cross-Slip System [Crystallography]

SPCS Storage and Processing Control System

SPCS Stored Program Controlled Switch [Telecommunications]

SPC/SPS Statistical Product Control/Statistical Problem Solving

SPC/SQC Statistical Product Control/Statistical Quality Control

Spct Specialist [also spct]

SPCU Simulation Process Control Unit

SPCUS Sweet Potato Council of the United States

SPCW Stored Program Command Word

SPCZ Southern Pacific Convergence Zone

SPD Serial Presence Detect

SPD Severe Plastic Deformation [Mechanics]

SPD Shockley Partial Dislocation [Materials Science]

SPD Singular Point Detection (Technique) [Solid-State Physics]

SPD Society of Publication Designers [US]

SPD Software Product Description

SPD Solar Physics Division [of American Astronomical Society, US]

SPD Space Development and Commercial Research Division [of NASA]

SPD Spray Deposited; Spray Deposition

SPD Standard Practice Directive

SPD Standard Program Device

SPD Steamer Pays Dues [also spd]

SPD Structural Plastics Division [of Society of the Plastics Industry, US]

SPD Surface Potential Difference

SPD Surge Protective Device

SPD Surge-Protective Devices [of Institute of Electrical and Electronics Engineers–Power Engineering Society Committee, US]

SPD Synchronous Phased Detector

SPD Synergetics Paradigm and Dichotomy

Spd Speed [also spd]

Spd Spread [also spd]

SPDB Subsystem Power Distribution Box

Spd Bk Speed Brake

SPDC Severe Plastic Deformation Consolidation [Mechanics]

SPDC Specialized Products Distribution Center

SPDCI Standard Payload Display and Control Interface

s,p,d,f (electron) energy sublevel [Symbol]

SPDL Standard Page Description Language

Spdl Spindle [also spdl]

SPDM Special-Purpose Dexterous Manipulator

SPDM Subprocessor with Dynamic Microprocessing

SPD MMC Spray Deposited Metal-Matrix Composite

SPDMS Shuttle Processing Data Management System [of NASA]

SPDP Society of Professional Data Processors

SPDP 3-(2-Pyridyldithio)propionic Acid N-Hydroxysuccinimide Ester

SPDP N-Succinimidyl 3-(2-Pyridyldithio)propionate

SPDR Software Preliminary Design Review

SPDS Safe-Practice Data Sheet

SPDSC Saskatchewan Professional Drivers Safety Council [Canada]

Spdsh Spreadsheet [also spdsh]

SPDT Single-Pole, Double-Throw (Switch)

SPDTDB Single-Pole, Double-Throw, Double-Break [also SPDT DB]

SPDTNC Single-Pole, Double-Throw, Normally-Closed [also SPDT NC]

SPDTNCDB Single-Pole, Double-Throw, Normally-Closed, Double-Break [also SPDT NCDB]

SPDTNO Single-Pole, Double-Throw, Normally-Open [also SPDT NO]

SPDTNODB Single-Pole, Double-Throw, Normally-Open, Double-Break [also SPDT NODB]

SPDT SW Single-Pole, Double-Throw Switch

SPDU Session Protocol Data Unit

SPE Sampler Performance Evaluation

SPE Service Propulsion Engine [of NASA Apollo Program]

SPE Shuttle Project Engineering [NASA]

SPE Signal Processor Element

SPE Smith-Purcell Effect [Physics]

SPE Sociedade Portuguesa dos Engenheiros [Portuguese Society of Engineers]

SPE Society for Photographic Education [US]

SPE Society of Petroleum Engineers [of American Institute of Mining, Metallurgical and Petroleum Engineers, US]

SPE Society of Plastics Engineers [US]

SPE Solid Particle Erosion

SPE Solid-Phase Epitaxial; Solid-Phase Epitaxy

SPE Solid-Phase Extraction

SPE Solid Polymer Electrode; Solid Polymer Electrolysis; Solid Polymer Electrolyte

SPE Spherical Probable Error

SPE Spontaneous Emission (Spectroscopy)

SPE Static Phase Error

SPE Stimulated Photon Echo

SPE Stored Program Element

SPE Systems Performance Effectiveness

Spe Sphaerotilus secies [Microbiology]

SPEA Southeastern Poultry and Egg Association [US]

SPE-AIME Society of Petroleum Engineers of the American Institute of Mining, Metallurgical and Petroleum Engineers [US]

SPEAM Sun-Photometer Earth Atmospheric Measurements

SPE ANTEC Society of Petroleum Engineers Annual Technical Conference [of American Institute of Mining, Metallurgical and Petroleum Engineers, US]

SPEAR Stanford Positron-Electron Assymetric Ring [of Stanford Linear Accelerator Center, of Stanford University, California, US] [also Stanford Positron-Electron Accelerator Ring]

SPEARS Satellite Photoelectric Analog Rectification System

SPEARS Satellite Photo-Electronic Analog Rectification System

SPEC Service de Physique de l'Etat Condensé [Condensed State Physics Division; of Centre d'Etudes de Saclay, France]

SPEC Society for the Promotion of Environment Conservation

SPEC Society of Pollution and Environmental Control

SPEC South Pacific Bureau for Economic Cooperation

SPEC Speech Predictive Encoding Communication

SPEC Speech Predictive Encoding System

SPEC Standard Performance Evaluation Council [US]

SPEC Stock Precision Engineered Components

SPEC Stored Program Educational Computer

SPEC Stratton Park Engineering Company, Inc. [US]

SPEC System Performance Evaluation Cooperative

Spec Specia(ist); Specialty [also spec]

Spec Specification [also spec]

Spec Spectrographer; Spectrography; Spectrographic

spec special

spec specific(ally)

SPEC CE Saclay Service de Physique de l'Etat Condensé de la Centre d'Etudes de Saclay [Condensed State Physics Division of the Saclay Nuclear Research Center, France]

Spec Chem Specialty Chemicals [Journal published in the UK]

Spec Hum Specific Humidity [Meteorology]

Spec Gr Spectrographic Grade [Chemistry]

Specif Specification [also specif]

specif specific(ally)

SPECLAB Spectroscopy Laboratory [of US Geological Survey]

Spec Lib Special Library [also Spec Libr]

Spec Libr Special Libraries [Journal of the Special Libraries Association, US]

SPE Comput Users Newsl SPE Computer Users Newsletter [of Society of Petroleum Engineers, US]

SPECON System Performance Effectiveness Conference

Spec Publ Special Publication [also spec publ]

Spec Rep Inst Technol, Shimizu Corp Special Report of Institute of Technology, Shimizu Corporation [Japan]

Specs Specifications [also specs]

Spec Ships Special Ships [Journal published in the UK]

Spec Steels Rev Special Steels Review [Published in the UK]

SPECT Single-Photon Emission Computed Tomography

Spect Spectrometer; Spectrometry; Spectroscopy

Spec Techn Publ Special Technical Publication

SPECTRE Spectral Radiation Experiment

Spectrochim Acta Spectrochimica Acta [UK]

Spectrochim Acta A, Mol Spectrosc Spectrochimica Acta A, Molecular Spectroscopy [Journal published in the UK]

Spectrochim Acta B, At Spectrosc Spectrochimica Acta B, Atomic Spectroscopy [Journal published in the UK]

Spectrom Spectrometer; Spectrometry

Spectrosc Spectroscope; Spectroscopy

Spectrosc (Canada) Spectroscopy (Canada) [Journal published in Canada]

Spectrosc Int Spectroscopy International [US Publication]

Spectrosc Lett Spectroscopy Letters [Journal published in the US]

Spectrosc (Oregon) Spectroscopy (Oregon) [Journal published in Eugene, Oregon, US]

Spectry Spectroscopy

Specul Speculation [also specul]

Specul Sci Technol Speculations in Science and Technology [Journal published in the UK]

SPECVER Specification Verification [Spec Ver]

Spec Vol Specific Volume [also spec vol]

SPED Supersonic Planetary Entry Decelerator

SpEd Specialist in Education

SPEDAC Solid-State, Parallel, Expandable, Differential Analysis Computer

SPEDE System for Processing Educational Data Electronically

SPEDI Standard Process for Electronic Data Interchange

SPE Drill Eng SPE Drilling Engineering [of Society of Petroleum Engineers, US]

SPEDTAC Stored Program Educational Transistorized Automatic Computer

SPEE Society for the Promotion of Engineering Education

SPEE Society of Petroleum Evaluation Engineers [US]

SPEE Special-Purpose End-Effector [Robotics]

Speech Commun Speech Communication [Dutch Publication]

Speech Monogr Speech Monographs

Speech Technol Speech Technology [Journal published in the US]

SPEED Selective Potentiostatic Etching by Electrolytic Dissolution

SPEED Self-Programmed Electronic Equation Delineator

SPEED Signal Processing in Evacuated Electronic Device

SPEED Spring-Loaded, Precision, Edgewise and Energy Delivery (System)

SPEED Subsistence Preparation by Electronic Energy Diffusion

SPEELS Spin-Polarized Electron Energy Loss Spectroscopy

SPEF Scottish Print Employers Federation

SPEFC Solid Polymer Electrolyte Fuel Cell

SPE Format Eval SPE Formation Evaluation [of Society of Petroleum Engineers, US]

SPEG Solid-Phase Epitaxial Growth [Electronics]

SPEG Statistical Polynomial-Exponential Gap (Law)

SPEPD Space Power and Electric Propulsion Division [of NASA]

SPEPC South Pacific Environment Protection Convention

SPE-PEM Solid Polymer Electrode–Proton Exchange Membrane

SPE Prod Eng SPE Production Engineering [of Society of Petroleum Engineers, US]

SPERAD Spectrally-Scanning Radiometer

SPERC Society of Petroleum Engineers Reservoir Characterization [of American Institute of Mining, Metallurgical and Petroleum Engineers, US]

SPE Reserv Eng SPE Reservoir Engineering [of Society of Petroleum Engineers, US]

SPERT Schedule Performance Evaluation and Review Technique; Schedule Program Evaluation and Review Technique

SPERT Special Power Excursion (Nuclear) Reactor Test

SPES Soft Particle Energy Spectrometer

SPES Stored Program Element System

SPE/SPI Society of Plastics Engineers/Society of the Plastics Industry [US]

SPESS Stored Program Electronic Switching System

SPET Solid Propellant Electric Thruster

SPEX-M SPEX Mill [*Note:* SPEX is a Tradename]

SPEXAFS Spin-Polarized Extended X-Ray Absorption Fine Structure

SPF Scottish Pharmaceutical Federation

SPF Shortest Path First

SPF Single Point Failure

SPF Site Population Factor

SPF Sociedade Portuguesa de Fisica [Portuguese Physical Society]

SPF Software Production Facility

SPF Solartechnik Prüfung Forschung [Solar Technology–Testing and Research]

SPF South Pacific Forum

SPF Spacelab Processing Facility

SPF Spectrofluorometer; Spectrofluorometry

SPF Spectrophotofluorometer; Spectrophotofluorometry

SPF Spruce, Pine and Fir (Lumber)

SPF Standard Project Flood

SPF Standard Program Facility

SPF Structured Programming Facility

SPF Subscriber Plant Factor

SPF Sun Protection Factor [Medicine]

SPF Superplastic Formability; Superplastically Formed; Superplastic Forming [Metallurgy]

SPF System Performance Factor

SPF System Programming Facility

SPFA Single Point Failure Analysis

SPFA Steel Plate Fabricators Association [US]

SPF/DB Superplastic Forming/Diffusion Bonding [also SPF-DB]

SPFC Solid Polymer Fuel Cell

SPFed Superplastically Formed [Metallurgy]

SPFFA South Pacific Forum Fisheries Agency [Solomon Islands]

SP-FGS Shuttle Projects Office–Flight and Ground Systems Office [at NASA Kennedy Space Center, Florida, US]

SPFI Single-Point Fuel Injection

SPFP Single-Point Failure Potential

SPFR Spectral Flux Radiometer

SPFS Small Payload Flight System

SP4T Single-Pole Quadruple Throw

SP4T SW Single-Pole Quadruple Throw Switch

SPFW Single-Phase Full-Wave

SPG Scan Pattern Generator

SPG Schweizerische Physikalische Gesellschaft [Swiss Physical Society]

SPG Screw-Pitch Gauge

SPG Self-Propelled Gun

SPG Single-Point Ground(ing)

SPG Soft Page [also Spg]

SPG Sort Program Generator

SPG Statistical Power Gap (Law)

SPG Superior Parathyroid Gland [Anatomy]

SPG Sync(hronizing) Pulse Generator

Spg Spring [also spg]

SPGA Staggered Pin-Grid Array

SPGK Shell Polygasoline and Kerosine (Process)

Sp Gr Specific Gravity [also sp gr]

SPGS Secondary Power Generating Subsystem

SPGT Serum Pyruvic Glutamic Transaminase [Biochemistry]

Sp Guin Spanish Guinea [now Eq Guin]

SPH Styrenated Phenol

SPh Phenyl Thiolester; Phenylthio Ester

Sph Spectrohelioscope [also sph]

Sph Sphere; Spherical [also sph]

Sph Streptomyces phaeochromogenes [Microbiology]

SPHCT Simplified Perturbed Hard-Chain Theory [Chemistry]

SP/HD Spool Piece Head [also SP/Hd]

SPHE Society of Packaging and Handling Engineers [US]

S-PHE Sulfated Poly(β-Hydroxyether)

SPHE Newsl SPHE Newsletter [of Society of Packaging and Handling Engineers, US]

SPHER Shell Pellet Heat Exchange Retorting (Process)

spher spherical

SPHE Tech J SPHE Technical Journal [of Society of Packaging and Handling Engineers, US]

SPHF Spin-Polarized Hartree-Fock [Physics]

Sp Ht Specific Heat [also sp ht]

SPHW Single-Phase, Half-Wave

Sphyg Sphygmomanometer [also sphyg]

SPI Schedule Performance Index

SPI Science Publishers Inc. [US]

SPI SCSI (Small Computer System Interface) Parallel Interface

SPI Security Parameters Index

SPI Selective Population Inversion [Physics]

SPI Self-Paced Instruction

SPI Serial Peripheral Interface

SPI Service Provider Interface

SPI Shared Peripheral Interface

SPI Single Particle Impact

SPI Single-Point Injection

SPI Single Processor Interface

SPI Single Program Indicator

SPI Site Population Index

SPI Slagging Pyrolysis Incinerator

SPI Society of Professional Investigators [US]

SPI Society of the Plastics Industry [US]

SPI Sonic Pulse-Echo Instrument

SPI Space Power Institute [of Auburn University, Normal, Alabama, US]

SPI Special Plot Identification

SPI Special Position Identification Pulse

SPI Specific Productivity Index

SPI Standard Power Interface

SPI Station Program Identification

SPI Surface Penning Ionization [Physics]

SPI Surface Position Indicator

SPI Swiss Performance Index

SPI Sydney Pollution Index

SPI Symbolic Pictorial Indicator

spi spots per inch [Scanners]

SPIA Single Peak Intensity Analysis

SPIA Solid Propellant Information Agency [of Johns Hopkins University, Baltimore, Maryland, US]

SPIA-LPIA Solid (Propellant Information Agency) and Liquid Propellant Information Agency [of Johns Hopkins University, Baltimore, Maryland, US]

SPIAM Sodium Purity In-Line Analytical Module

SPIAP Shuttle/Payload Integration Activities Plan [of NASA]

SPIB Southern Pine Inspection Bureau [US]

SPIC Society of Plastics Industry of Canada

SPIC Spacelab Payload Integration and Coordination

SPICE Spacelab Payload Integration and Coordination in Europe

SPICE Simulation Program with Integrated-Circuit Emphasis [also Spice]

SPICE Spacecraft, Planetary and Probe Ephemeris, Instrument, C-Matrix, Event File (Information Storage System) [of NASA Planetary Data System]

SPID Service Profile/Provider Identifier

SPID Submersible Portable Inflatable Dwelling

SPIDAC Specimen Input to Digital Automatic Computer

SPIDEL Société pour l'Information et Documentation Electronique [Society for Electronic Information and Documentation, France]

SPIDER Sonic Pulse-Echo Instrument Designed for Extreme Resolution

SPIDF Support Planning Identification File

SPIDPO Shuttle Payload Integration and Development Program Office [at NASA Johnson Space Center, Houston, Texas, US]

SPIE Scavenging-Precipitation-Ion Exchange

SPIE Self-Programmed Individualized Education

SPIE Simulated Problem, Input Evaluation

SPIE Society of Photo-Optical Instrumentation Engineers [now SPIE-International Society for Optical Engineering, US]

SPIE International Society for Optical Engineering [US]

SPIE/IS&T International Society for Optical Engineering/ Imaging Science and Technology (Symposium)

SPIE Proc Society of Photo-Optical Instrumentation Engineers Proceedings

SPIES Surface Penning Ionization Electron Spectroscopy

SPIF Shuttle Payload Integration Facility [of NASA]

SPIF Standard Payload Interface Facility [of NASA]

SPIG Symposium on the Physics of Ionized Gases

SPII Shuttle Program Implementation Instruction [of NASA]

SPIKE Science Planning Intelligent Knowledge-Based Environment [of Space Telescope Science Institute, US]

SP-ILS Shuttle Projects Office–Integrated Logistics Support Office [of NASA Kennedy Space Center, Florida, US]

SPIMS Shuttle Program Information Management System [of NASA]

SPIN International Symposium on High-Energy Spin Physics

SPIN Science Procurement Information Network

SPIN Searchable Physics Information Notices (Database) [of American Institute of Physics, US]

SPIN Stichting Studenten Physica in Nederland [Netherlands Physics Students Foundation]

SPIO Server Program IRM (Information Resource Management) Official

SPIPES Spin-Polarized Inverse Photoemission Spectroscopy

SPIR Search Program for Infrared Spectra [of Canada Institute for Scientific and Technical Information, Canada]

SPIR Single Pilot Instrument Rating

SPIR Space-Based Infrared (Surveillance System)

SPIRAT Strategic Program for Innovative Research on AIDS Therapies [US]

SPIRES Stanford Public Information (and) Retrieval System [of Stanford University, California, US]

SPIREX South Pole Infrared Explorer [of University of Chicago, Illinois, US]

SPIRS SERDP (Strategic Environmental Research and Development Program) Principal Investigator Reporting System [of US Department of Defense, US Department of Energy and US Environmental Protection Agency]

SPIS Sound Powered Intercom System

SPIS Standard Production Information System

SPIS Surface Penning Ionization Spectroscopy

Spis Bulg Akad Nauk Spisanie na Bulgarskata Akademiya na Naukite [Publication of the Bulgarian Academy of Sciences]

SPIT Selective Printing of Items from Tape

SPIT Sodium Selenite–Sodium Pyruvate–(Bovine) Insulin–(Human) Transferrin (Mixture) [Biochemistry]

SPITE Sodium Selenite–Sodium Pyruvate–(Bovine) Insulin–(Human) Transferrin–Ethanolamine (Mixture) [Biochemistry]

SPIW Shipping, Ports and Inland Waterways [of Economic and Social Commission for Asia and the Pacific, United Nations]

SPIW Special-Purpose Individual Weapon

SPIXE Scanning Proton-Induced X-Ray Emission

SPK Single-Point Keying

Spk Spectro-Heliokinematograph [also spk]

SP-KLH Substance P–Keyhole Limpet Hemocyanin [Immunology]

Spkr Speaker [also spkr]

SPL Scratch Pad Line

SPL Seattle Public Library [Washington, US]

SPL Serialized Parts List

SPL Signal Processing Language

SPL Simple Programming Language

SPL Simulation Programming Language

SPL Software Parts List

SPL Software Programming Language

SPL Sound Pressure Level

SPL Sound Pulse Level

SPL Source Program Library

SPL Space Physics Laboratory [US]

SPL Space(borne) Programming Language

SPL Special-Purpose Logic

SPL Spent Potlining

SPL Spooler

SPL Supplementary Flight Plan

SPL Supply, Procurement and Logistics (Division) [of CERN–European Laboratory for Particle Physics, Geneva, Switzerland]

SPL System Programming Language

.SPL Spell Checker [File Name Extension]

.SPL Print Spooling [Windows File Name Extension] [also .spl]

Spl Special(ty) [also spl]

Spl Splice [also spl]

SPLC Sequential Programmable Logic Controller

SPLC Standard Point Location Code

SPLD Simple Programmable Logic Device

SPLEED Spin-Polarized Low-Energy Electron Diffraction

SPLEEM Spin-Polarized Low-Energy Electron Microscopy

SPLIT Sunstrand Processing Language Internally Translated

SPLL Self-Propelled Launcher/Loader

SP-LMO Shuttle Projects Office–Logistics Management Office [at NASA Kennedy Space Center, Florida, US]

SPL-PTF Dryseal Special Taper Pipe Thread [Symbol]

SP LSDF Spin-Polarized Local Spin Density Function(al) [alsp SP-LSDF]

Splty Specialty [also splty]

SPLX Simplex [also Splx]

Sply Supply [also sply]

SPM Samarium Lead Manganate

SPM Scanning Probe Microscope; Scanning Probe Microscopy

SPM Scratch Pad Memory

SPM Self-Propelled Mount

SPM Semipermeable Membrane

SPM Sequential Processing Machine

SPM Shock Pulse Method

SPM Signal Processing Modem

SPM Single Point Mooring

SPM Six Point Mooring

SPM Small Perturbation Method

SPM Sociedade Portuguesa de Materiais [Portuguese Society for Materials]

SPM Sociedade Portuguesa de Matemática [Portuguese Society for Mathematics]

SPM Société Polonaise de Mathématique [Polish Mathematical Society]

SPM Software Performance Monitor

SPM Solar Proton Monitor

SPM Solar Power Module

SPM Source Program Maintenance

SPM Split-Phase Motor

SPM Standard Prototype Microcomputer

SPM Strokes per Minute [also spm]

SPM Subscriber's Private Meter

SPM Subsystem Project Manager

SPM Sun Probe-Mars [of NASA]

SPM Superparamagnet(ic); Superparamagnetics; Superparamagnetism

SPM Surface Plasmon Microscopy

SPM Symbol Processing Machine

SPM Synaptosomal Plasma Membrane [Cytology]

SPM System Performance Monitor

SpM Specialist in Microbiology

SPMA Society of Post-Medieval Archeology [UK]

SPMA Southwest Parks and Monuments Association [US]

SPM/AFM Scanning Probe Microscope/Atomic Force Microscope

SPMC Solid Polyester Molding Compound

SPMDS Spin-Polarized Metastable Atom Deexcitation Spectroscopy

SPME Solid-Phase Microextraction

SPME Spectroscopic Phase Modulated Ellipsometer; Spectroscopic Phase Modulated Ellipsometry

SPMI Southeastern Pine Marketing Institute [now Southeastern Lumber Manufacturers Association, US]

S-PMMA Sulfur-Derivatized Polymethylmethacrylate

SPMO Samarium Lead Manganate (or Samarium Lead Manganese Oxide)

SPMOL Source Program Maintenance On-Line

SPMS Software Production Monitor System

SPMS Solar Particle Monitoring System

SPMS Special Purpose Manipulator System

SPMS System Program Management Survey

SPMT Strategic Planning and Management Team

SPN Service Protection Network

SPN Shuttle Project Notice [of NASA Kennedy Space Center, Florida, US]

SPN Single Packet Network

SPN Synthetic Polymer Network

SPNC Society for the Promotion of Nature Conservation [UK]

spnd suspend

SPNFZ South Pacific Nuclear Free Zone

SPNFZT South Pacific Nuclear Free Zone Treaty

SPNI Society for the Protection of Nature in Israel

SPNM Scanning Plasmon Near-Field Microscope

SPNR Society for the Promotion of Nature Reserves [now Society for the Promotion of Nature Conservation, UK]

SPNR Spin-Polarized Neutron Reflectivity

SP-NR Superior-Processing Natural Rubber

SPNS Switched Private Network Service

SPO Sacramento Peak Observatory [of US Air Force/ Geophysical Research Directorate]

SPO Separate Partition Option

SPO Shuttle Projects Office [of NASA Kennedy Space Center, Florida, US]

SPO Solar Programs Office

SPO South Pacific Ocean

SPO Spacelab Projects Office

SPO Spare Parts Order

SPO Surface Plasmon Oscillation [Solid-State Physics]

SPO System Program Office

SP&O System Planning and Operation

SPOC Shuttle Payload Operations Contractor [of NASA]

SPOC Shuttle Portable On-Board Computer [of NASA]

SPOC Simulated Processing of Ore and Coal

SPOC Single-Point Orbit Calculator

SPOC Single SAR (Search and Rescue) Points of Contact [of Committee on Space Research Search and Rescue Satellite]

SPOCC South Pacific Organizations Coordinating Committee

SPOCK Simulated Procedure for Obtaining Common Knowledge

SPOCK Standard Performance-Oriented Computer Keyboard

SPOF Single Point of Failure

S-Pol S-Polarization [also s-pol]

SPOM STS (Space Transportation System) Planning and Operations Management [NASA]

spont spontaneous

SPOOF Society for the Protection of Old Fishes [US]

SPOOF Structure and Parity Observing Output Function

SPOOL Simultaneous Peripheral Operations On-Line (File)

SPOOL Simultaneous Peripheral Output On-Line

S&POP Student and Professional Outreach Program [of ASM International, US]

SP-OPI Shuttle Projects Office–Operations Planning and Integration Office [of NASA Kennedy Space Center, Florida, US]

SPORA Special Operational Research and Analysis Group [of Eurocontrol]

SPOREs Specialized Programs of Research Excellence [US]

SP-OSO Shuttle Projects Office–Off-Site Offices [at NASA Kennedy Space Center, Florida, US]

SPOSS Society for the Promotion of Science and Scholarship [US]

SPOT Satellite Positioning and Tracking

SPOT Shared Product Object Tree [Computers]

SPOT Smithsonian Precision Optical Tracking

SPOT Système Probatoire pour l'Observation de la Terre [Earth Observation System–Series of French Satellites]

SPOUT System Peripheral Output Utility

SPP Science Power Platform

SPP Second Principal Plane

SPP Seleno-bis-Succinimidyl Propionate

SPP Sequenced (Data) Packet Protocol

SPP Signal Processing Peripheral

SPP Simulation Planning Panel

SPP Small Periodic Perturbations (Method)

SPP Society of Professional Pilots [US]

SPP Solar Physics Payload

SPP Soluble Reactive Phosphate

SPP Space Payloads and Projects

SPP Space Power Plant

SPP Special-Purpose Processor

SPP Specialty Powder Products [of Inco Limited, Canada]

SPP Species Plantarum Project [of International Organization for Plant Information]

SPP Speed/Power Product (of Semiconductors)

SPP Spin-Polarized Photoemission

SPP Surface Plasmon Polariton [Solid-State Physics]

SPP Syndiotactic Polypropylene

SpP Spanish Patent [also Sp Pat]

s-PP Syndiotactic Polypropylene [also sPP]

spp species (plural) [Chemistry; Biology]

SPPAC Salinity Pilot Program Advisory Council [now Salinity Program Advisory Council]

SP-PAI Shuttle Projects Office–Project Assessment and Integration Staff [at NASA Kennedy Space Center, Florida, US]

Sp Pat Spanish Patent [also SpP]

SP-PCO Shuttle Projects Office–Program Control Office [at NASA Kennedy Space Center, Florida, US]

SPPD Spin-Polarized Photoelectron Diffraction

SPPES Spin-Polarized Photoelectron Spectroscopy

SPPAY Semipost-Pay (Pay-Station)

Sp Pd Specific Productivity

SPPF Solid-Phase Pressure Forming

SPPF Syracuse Pulp and Paper Foundation [New York State, US]

Sp Ph Split Phase [Electrical Engineering]

SPPIL Shuttle Preferred Pyrotechnic Items List [of NASA]

SPPL Spare Parts Provisioning List

SP-PMS Shuttle Projects Office–Performance Management Systems Office [at NASA Kennedy Space Center, Florida, US]

SPPO Spacelab Payload Project Office

SPPP Spacelab Payloads Processing Project

SPPS Scalable Power Parallel System

SPPS Semipost-Pay Pay-Station

SPPS Single-Phase Power Supply

SPPU Signaling Preprocessing Unit

SPQ 6-Methoxy-N-(3-Sulfopropyl)quinolinium (or 3-(6-Methoxyquinolino)propanesulfonate

SPQ Synthetic Pipeline Quality Gas

SPQR Small Profits and Quick Returns

SPQW Stepped-Potential Quantum Well [Electronics]

SPR Sandia Pulsed Reactor [of Sandia National Laboratories, Albuquerque, New Mexico, US]

SPR Send Priority and Route

SPR Silicon Power Rectifier

SPR Simplified Practice Recommendation

SPR Society of Psychical Research [US]

SPR Software Problem Report

SPR Solid Phase Regrowth [Materials Science]

SPR Special Purpose Register

SPR Spreading Resistance Method [Solid-State Physics]

SPR Statistical Pattern Recognition

SPR Storage Protection Register

SPR Strategic Petroleum Reserve [US]

SPR Structure-Property Relationship [Materials Science]

SPR Subcontractor Performance Review

SPR Superparamagnetic Resonance

SPR Supervisory Printer Read

SPR Surface Plasmon Resonance (Spectroscopy)

SPR Swimming Pool (Nuclear) Reactor

Spr Spare [also spr]

Spr Spring [also spr]

Spr Sprinkler [also spr]

SPRA Space Probe Radar Altimeter

Sprache Daterverarb Sprache und Datenverarbeitung [German Publication on Language and Data Processing]

SPRAG STS (Space Transportation System) Payloads Requirements and Analysis Group [of NASA]

SPRAT Small Portable Radar Torch

SPRC Self-Propelled Robot Craft

SPRD Special Pesticide Review Division

SPRD Surplus Production (Nuclear) Reactor Decommissioning

Sprd Spread [also sprd]

SPRDA Solid Pipeline Research and Development Association

SPRDS Steam Pipe Rupture Detector System

SPREAD Support Program for Remote Entry of Alphanumeric Displays [UK]

SPREAD Systems Programming, Research, Engineering and Development [of IBM Corporation, US]

Spreadware Spreadsheet Software [also spreadware]

Sprecher Energ Rev Sprecher Energie Review [Publication of Sprecher Energie AG, Switzerland]

Sprechsaal Int Ceram Glass Mag Der Sprechsaal–International Ceramics and Glass Magazine [Germany]

SPREE Shuttle Potential and Return Electron Experiment [of NASA]

SPREP South Pacific Regional Environment Program [Part of United Nations Environment(al) Program]

S Pres Star Present

SPRI Scott Polar Research Institute [of University of Cambridge, UK]

SPRI Single-Ply Roofing Institute [US]

SPRING Sistema de Processamento de Informações Geográficas [Geographic Information Processing System (Project); of Brazilian National Institute of Space Research]

SPRING Switched Private Networking

Springer Proc Phys Springer Proceedings in Physics [International Publication]

SPRINT Solid-Propellant Rocket Intercept Missile

SPRINT Special Police Radio Inquiry Network

SPRITE Solid-Propellant Rocket Ignition Test and Evaluation

Sprn Suppression

SPROM Switched Programmable Read-Only Memory

SPRPAE Spin-Polarized Random-Phase Approximation with Exchange

SPRS Signal Processing Router/Scheduler

Sprs Springs [Part of City Names; e.g. Silver Springs, etc.]

SPRT Sequential Probability Ratio Test

SPRT Small Purchases Reengineering Team

SPRT Standard Platinum Resistance Thermometer

SPRU Science Policy Research Unit [of Sussex University, UK]

SPS Saddle Point Singularity [Mathematics]

SPS Samples per Second [also S/S]

SPS School of Practical Science

SPS Secondary Power System

SPS Secondary Propulsion System

SPS Serial-Parallel-Serial (Mode)

SPS Service Propulsion Subsystem

SPS Service Propulsion System [of NASA Apollo Program]

SPS Ship Predictor System

SPS Shock Protection System

SPS Short Period Superlattice [Solid-State Physics]

SPS Shuttle Procedures Simulator [of NASA]

SPS Shrouded Plasma Spray

SPS Single Particle Scattering

SPS Society of Physics Students [US]

SPS Society of Polymer Science [Japan]

SPS Software Products Scheme

SPS Solar Power Satellite

SPS Solar Power System

SPS Solar Probe Spacecraft

SPS Solid-Phase Sintering [Metallurgy]

SPS Solution Phase Synthesis [Physical Chemistry]

SPS Spark Plug Socket

SPS Spectrum Planning Subcommittee

SPS Speech Processor Set

SPS Speed Switch

SPS (Fachmesse für) Speicherprogrammierbare Steuerungen [Exhibition on Programmable Logic Control Systems, Germany]

SPS Standard Pipe Size

SPS Standard Pressed Steel Industries Limited [US]

SPS Standby Power Supply

SPS Standby Power System

SPS Statement of Prior Submission

SPS Statistical Problem Solving

SPS String Process System

SPS Submerged Production System

SPS Subsea Production System

SPS Sulfonated Polystyrene

SPS Super Proton Synchrotron [of CERN–European Laboratory for Particle Physics, Geneva, Switzerland]

SPS Surface Photovoltage Spectroscopy

SPS Swiss Physical Society

SPS Switching Power Supply

SPS Symbolic Program(ming) System

SPS (NATO Advanced Research Workshop on) Symmetry and Pairing in Superconductors

SPS System Performance Score

SPS Syndiotactic Polystyrene [also sPS]

SPS Systolic Processing Superchip

SP&S Special Processes and Sequencing

sPS Syndiotactic Polystyrene [also SPS]

SPSA Single Phase Statistical Analyzer

SP-SA Slightly Polishing, Slightly Abrasive

SPSC Super Proton Synchrotron (Experiments) Committee [of CERN–European Laboratory for Particle Physics, Geneva, Switzerland]

SP/SC Speed Plug/Support Cylinder

SPSD Science and Professional Services Directorate [of Supply and Services Canada]

SPSE Society of Photographic Scientists and Engineers [US]

SPSE Spin-Polarized Secondary Electrons [Solid-State Physics]

SPSH Six-Phase Soil Heating

SPSL Society for the Protection of Science and Learning [UK]

SPSLS Short-Period Strain-Layer Superlattice [Solid-State Physics]

SP-SMO Shuttle Projects Office–Site Management Office [at NASA Kennedy Space Center, Florida, US]

SPSO Science Processing Support Office

SP-SOP Surface Plasmon–Surface Optical Phonon (Mode) [Solid-State Physics]

Sp Speak Am Spanish Speaking America

SPSR Service Planning Segment Replacement

SPSR Society for the Promotion of Scientific Research [Austria]

SPSS Science and Professional Services Sector [of Public Works and Government Services Canada]

SPSS Shield Plug Storage Station

SPSS Statistical Package for the Social Sciences

SPST Single-Pole, Single-Throw (Switch)

SPST Society of Photographic Science and Technology [Japan]

SPSTJ Society of Photographic Science and Technology of Japan

SPSTM Spin-Polarized Scanning Tunneling Microscopy [SP-STM]

SPSTNC Single-Pole, Single-Throw, Normally-Closed (Relay) [also SPST NC]

SPSTNO Single-Pole, Single-Throw, Normally-Open (Relay) [also SPST NO]

SPSTNODM Single-Pole, Single-Throw, Normally-Open, Double-Make [also SPST NODM]

SPST SW Single-Pole Single-Throw Switch

SP SW Single-Pole Switch [also SPSW]

SPT Scaled Particle Theory [Physics]

SPT School of Plastics Technology [Israel]

SPT Sectors per Track

SPT Selective Population Transfer

SPT Shared Page Table

SPT Silicon Planar Thyristor

SPT Simultaneous Pouring Technique [Metallurgy]

SPT Single Particle Theory [Physics]

SPT Single Photon Timing [Physics]

SPT Society for Philosophy and Technology [US]

SPT Society of Photo-Technologists [US]

SPT Special Purpose Test

SPT Star Point Transfer [Telecommunications]

SPT Statistical Perturbation Theory [Physics]

SPT Steering Position Transducer

SPT Strength-Probability-Time (Diagram) [Physics]

SPT Strontium Plutonium Titanate

SPT Structural Phase Transition [Materials Science]

SPT Structured Programming Technique

SPT Symbolic Program Tape

SPT Symbolic Program Translator

SPT Symmetry and Perturbation Theory [Physics]

SPT System Page Table

Spt Support [also spt]

SPTA Southern Pressure Treaters Association [US]

SPTC Share Purchase Tax Credit

SPTD Supplemental Provisioning Technical Documentation

SPTDT Short Path Thermal Desorption Tube

SPTF Screen Printing Technical Foundation [US]

SPTF Shortest Programming Time First

SPTF Sodium Pump Test Facility [Nuclear Engineering]

SPTR Current Value of SP (Stack Pointer) Register [also SPtr] [Pascal Function]

SP3T Single-Pole, Triple-Throw [also SPTT]

SP3T SW Single-Pole, Triple-Throw Switch [also SPTT SW]

SPU Scientific Programs Unit [of Australian Nature Conservancy Agency]

SPU Seattle Pacific University [Washington, US]

SPU Signal Processing Unit

SPU Slave Processing Unit

SPU St. Paul University [Alberta, Canada]

SPU St. Petersburg University [Russian Federation]

SPUC Serial Peripheral Unit Controller

SPUC/DL Serial Peripheral Unit Controller/Data Link

SPUNFIT Space Union Fitting (Wrench)

SPUPS Spin-Polarized Ultraviolet Photoelectron Spectroscopy

SPUR Source Program Utility Routine

SPUR Space Power Unit Reactor

SPURM Special Purpose Unilateral Repetitive Modulation

SPURT Spinning Unguided Rocket Trajectory

SPV Sampler Performance Validation

SPV Sealpox Virus

SPV Sheeppox Virus

SPV Surface Photo-Voltage

SPV Swinepox Virus

S-PVC Suspension Polyvinyl Chloride

Sp Vol Specific Volume [also sp vol]

SPVPF Shuttle Payload Vertical Processing Facility [of NASA]

SPVT Spin-Polarized Vacuum Tunneling

SPW Self-Penalty Walk

SPW Single Project Wafer

SPW Spring Plunger Wrench

SpW Special Weapon

SPWLA Society of Professional Well Log Analysts [US]

SPWM Single-Sided Pulsewidth Modulation

SPX Sequenced (or Sequential) (Data) Packet Exchange

SPX Simplex [also Spx]

SPX Superheat Power Experiment [US]

SPXEC Special Executive

SPXPS Spin-Polarized X-Ray Photoelectron Spectroscopy

SPY Secondary Particle Yields (Experiment) [of CERN–European Laboratory for Particle Physics, Geneva, Switzerland]

SPY Second Polar Year [1932–33]

spzd specialized

SQ Self-Quenched; Self-Quenching [Counters, Dectectors, or Oscillators]

SQ Semi-Quantitative

SQ Signal Quality (Detector)

SQ Solid-Quenched; Solid Quenching [Metallurgy]

SQ Space Quantization [Quantum Mechanics]

SQ Square (Lattice Model) [Physics]

SQ Squareness Ratio (i.e., Ratio of Remanent Magnetization M_R to Saturation Magnetization M_S) [Symbol]

SQ Squeezed (Files) [Computers]

SQ Standard Quality

SQ Stationary Quantum

SQ Structural Quality

SQ Superquick

SQ Surface Quality

S$_{mag}$(±Q) Magnetic Structure Factor [Symbol]

S$_{nuc}$(Q) Nuclear Structure Factor [Symbol]

2SQ 2-Mercaptoquinoline

3SQ 3-Mercaptoquinoline

4SQ 4-Mercaptoquinoline

7Q- Malawi [Civil Aircraft Marking]

Sq Sequence [also sq]

Sq Solar Quiet Daily Magnetic Variation [Symbol]

Sq Squadron [also sq]

Sq Square [also sq]

Sq Squall [also sq]

S(q) Spin Structure Factor; (Static) Structure Factor [Symbol]

sq sequence [also Sq]

sq squadron [also Sq]

sq square [also Sq]

sq squall [also Sq]

sq *(sequens)* – the following

SQA Semiquantiative Analysis

SQA Software Quality Assurance

SQA Supplier Qualification Alliance [Canada]

SQA Supplier Quality Assurance

SQA System Queue Area

Sqa Solar-Flare Effect (or Sudden Augmentation of Solar Quiet Daily Magnetic Variation) [Symbol]

SQAM Superposed Quadrature Amplitude Modulation

SQAP Software Quality Assurance Plan

SQAP Swedish Question Answering Project [of Research Institute of National Defense]

SQAPS Supplier Quality Assurance Provisions

SQC Soldier's Qualification

SQC Statistical Quality Control

Sq Cg Squirrel Cage

sq ch square chain [Unit]

sq cm square centimeter [also cm^2]

SQC Rep Statistical Quality Control Report

SQC/SPC Statistical Quality Control/Statistical Process Control

SQD Signal Quality Detection; Signal Quality Detector

Sq(D) Dive Squadron

Sqd Squadron [also Sqd]

sq dkm square dekameter [also dkm^2]

Sqd Ldr Squadron Leader

sq dm square decimeter [also dm^2]

Sqdn Squadron

SQE Signal Quality Error (Test)

S(Q,ε) Neutron Scattering Law for Unpolarized Neutrons [also S(Q,ε)] [Nuclear Physics]

Sq E Square Edge [Lumber]

Sq E&S Square Edge and Sound [Lumber]

SQF Subjective Quality Factor

sq ft square foot [also ft^2]

sq ft/h square feet per hour [also sq ft/hr, ft²/h, or ft²/hr]

sq ft/lb square feet per pound [also ft²/lb]

SQG Small-Quantity Generator

sq hm square hectometer [also hm²]

sq in square inch [also in²]

sq in/min square inches per minute [also in²/min, or in min^{-2}]

sq in/s square inches per second [also sq in/sec, in²/s, or in²/sec]

sq km square kilometer [also km²]

SQL Squelch [also sql]

SQL Standard Query Language

SQL Structured Query Language

SQL/DS Structured Query Language/Data System

SQL/LIMS Structured Query Language/Laboratory Information Management System

SQLN Squall Line [Meteorology]

SQM Scaled Quantum Mechanical Method

SQM Surface Quality Monitor(ing System)

sq m square meter [also m²]

sq m/g square meter(s) per gram [also sq m/gm, m²/g, or m² g^{-1}]

sq mi square mile [also mi²]

sq mil square mil [also mil²]

SQMOFF Scaled Quantum Mechanical Oligomer Force Field (Method)

sq mm square millimeter [also mm²]

sq mu square micron [also sq μ, mu², or μ²]

SQN Spin Quantum Number [Quantum Mechanics]

Sqncr Sequencer [also sqncer]

sq nm square nanometer [also nm²]

SQO Seminar on Quantum Optics

S(q,ω) Structure Factor (in Solid-State Physics) [also S(Q,ω)] [Symbol]

SQPR Staggered Quadrature Partial Response

SQPSK Staggered Quadrature Phase Shift Keying

sqq *(sequentia)* – the following ones

SQR Sequence Relay

SQR Service Request

SQR Square Root (Function) [Programming]

SQR Supplier Quality Representative

Sqr Square (of Argument) [Pascal Function]

SQRL Subpicosecond and Quantum Radiation Laboratory [US]

SQRD Square Root of Multicomponent Diffusivity Matrix (Analysis) [Materials Science]

sq rd square rod [Unit]

Sqrs Squares [also sqrs]

Sq Rt Square Root [also sq rt]

SQRT Square Root [Computers]

Sqrt Square Root [Pascal Function]

SQS Schweizerisches Qualitätssicherungszertifikat [Swiss Quality Assurance Certificate]

SQU Squaric Acid

Squad Squadron [also squad]

S-Quark Strange Quark [also s-quark] [Particle Physics]

SQUID Sperry Quick Updating of Internal Documentation

SQUID Superconducting Quantum Interference Device

SQUIN Sequential Quadrature Inband (Video) System [UK]

SQUIRE Submarine Quickened Response

SQW Single Quantum Well [Electronics]

SQW Square Wave [Electrical Engineering]

SQW Strained Quantum Well [Electronics]

SQW LED Single Quantum Well Light-Emitting Device

sq yd square yard [also yd²]

SR Microsegregation Ratio [Metallurgy]

SR Saarländischer Rundfunk [Broadcasting Station of the Saarland, Germany]

SR Safety Requirements

SR Safety Rod [Nuclear Reactors]

SR Sampling Rate

SR Sandmeyer Reaction [Chemistry]

SR San Remo [Italy]

SR Sarcoplasmic Reticulum [Cytology]

SR Saturable Reactor

SR Saturation Ratio [Meteorology]

SR Saturation Resonance

SR Saytzeff's Rule [Chemistry]

SR Savannah River [South Carolina, US]

SR Scanning Radiometer

SR Scanning Raster

SR Scan Ratio

SR Schematic Representation

SR Schottky Rectifier

SR Scientific Report

SR Scientific Research

SR Scientific Result(s)

SR Scientific Review

SR Scram Rod [Nuclear Reactors]

SR Seaman Recruit

SR Search Radar

SR Secondary Radar

SR Secondary Radiation

SR Secondary Recrystallization [Metallurgy]

SR Sedimentary Rock

SR Sedimentation Rate [Medicine]

SR Segregation Roasting [Metallurgy]

SR Seismic Ray

SR Selection Rules [Physics]

SR Selective Reflection [Physics]

SR Selective Ringing

SR Selenium Rectifier

SR Self-Aligning Roller Bearing, Single-Row, Radial-Contact, Outer-Ring Raceway Spherical, Inner Ring with Ribs [Symbol]

SR Self-Reconfiguring

SR Self-Rectifying

SR Self-Repulsion

SR Self-Reversal

SR Semantic Routine
SR Semiregular
SR Send and/or Receive [also S/R]
SR Send-Receive (Method)
SR Sequence Rule
SR Series Relay
SR Series Resonant (Circuit)
SR Sex Ratio [Biology]
SR Seychelles Rupee [Currency]
SR Sheet and Roll
SR Shift Register [also S/R]
SR Shift Reverse
SR Shift Right
SR Shim Rod [Nuclear Reactors]
SR Shipping Receipt
SR Shipping Route
SR Shock Resistance; Shock-Resistant
SR Short Range
SR Short Rifle
SR Sigma (Nuclear) Reactor
SR Sigmatropic Reaction
SR Signal Regulation
SR Silicone Resin
SR Silicone Rubber
SR Simpson Rule [Mathematics]
SR Slew Rate
SR Slide Rule
SR Slip Range
SR Slip Ring (Motor)
SR Slow Release (Relay)
SR Slow Running
SR Society of Radiographers [UK]
SR Society of Rheology [also S of R, or SOR, US]
SR Silicone Rubber
SR Snap Ring
SR Solar Radiation
SR Solar Research
SR Solid Rocket
SR Sorter-Reader
SR Sound Ranging
SR Sound Receiver
SR Sound Recorder; Sound Recording
SR Sound Rating
SR Space Research
SR Spallation Reaction [Nuclear Physics]
SR Special Register
SR Special Regulation
SR Special Report
SR Specific Resistance [also sr]
SR Specific Rotation [Optics]
SR Spectroradiometer; Spectroradiometry
SR Spectroscopic Research
SR Specular Reflectance; Specular Reflection; Specular Reflector

SR Speech Recognition
SR Speed Rating
SR Speed Ratio
SR Speed Recorder
SR Speed Regulation; Speed Regulator
SR Spherical Radius
SR Spin Resonance [Physics]
SR Spin Rotation [Physics]
SR Split Ring
SR Spray Reacted; Spray Reaction
SR Spring-Return (Valve)
SR Sputter Rate
SR Square Root
SR Stability Ratio
SR Standard Repair
SR Statement of Requirements
SR State Road [US]
SR Stateroom
SR Statistical Research
SR Status Register
SR Status Report
SR Status Review
SR Steam-Refined; Steam Refining [Chemical Engineering]
SR Steel Research
SR Stellar Rotation [Astronomy]
SR Step Reaction [Chemistry]
SR Stepped Reflector
SR Stereo Record(ing)
SR Stereoscopic Radiography
SR Steric Repulsion
SR Stirred Reactor [Chemical Engineering]
SR Stochastic Resonance
SR Storage Register
SR Storage Rings [Nuclear Engineering]
SR Storage and Retrieval
SR Straight-Run (Gasoline)
SR Strain Rate [Mechanics]
SR Strain Relaxation [Metallurgy]
SR Strain Rod
SR Strategic Reconnaissance (Plane)
SR Stress Relief; Stress Relieving [Metallurgy]
SR Stress Relaxation [Metallurgy]
SR Strontium Ruthenate
SR Structural Relaxation
SR Structure Resonance [Spectroscopy]
SR Study Requirement
SR Styrene Rubber
SR Subroutine [also S/R]
SR Summary Report
SR Sunrise
SR Support Reaction (Load) [Mechanics]
SR Support Request
SR Support Room

SR Surface Recombination [Solid-State Physics]

SR Surface Region

SR Surface Rheology

SR Suriname [ISO Code]

SR Surveillance Radar

SR Sweep Rate [Radar]

SR Swissair–Swiss Air Transport Company

SR Switch Register

SR Switching Regulator

SR Synaptic Ribbon [Medicine]

SR Synchrotron Radiation

SR Synthetic Rubber

SR Systems Research

SR$_1$ (Earthquake) Shear Wave Reflected once from the Earth's Surface [also SS] [Symbol]

SR$_2$ (Earthquake) Shear Wave Reflected twice from the Earth's Surface [also SSS] [Symbol]

SR-4233 3-Amino-1,2,4-Benzotriazine-1,4-Dioxine

S&R Search and Rescue

S/R Send and/or Receive [also SR]

S/R Shift Register [also SR]

S/R Stimulus/Response

S/R Subroutine

S-R Set-Reset (Memory)

Sr Sarcomer(ic) [Anatomy]

Sr Senior [also sr]

Sr Sir

Sr Sister [also sr]

Sr Strontium [Symbol]

Sr^{2+} Strontium Ion [also Sr^{++}] [Symbol]

Sr II Strontium II [Symbol]

Sr-80 Strontium-80 [also ^{80}Sr, or Sr^{80}]

Sr-81 Strontium-81 [also ^{81}Sr, or Sr^{81}]

Sr-82 Strontium-82 [also ^{82}Sr, or Sr^{82}]

Sr-83 Strontium-83 [also ^{83}Sr, or Sr^{83}]

Sr-84 Strontium-84 [also ^{84}Sr, or Sr^{84}]

Sr-85 Strontium-85 [also ^{85}Sr, or Sr^{85}]

Sr-86 Strontium-86 [also ^{86}Sr, or Sr^{86}]

Sr-87 Strontium-87 [also ^{87}Sr, or Sr^{87}]

Sr-88 Strontium-88 [also ^{88}Sr, or Sr^{88}]

Sr-89 Strontium-89 [also ^{89}Sr, or Sr^{89}]

Sr-90 Strontium-90 [also ^{90}Sr, or Sr^{90}]

Sr-91 Strontium-91 [also ^{91}Sr, or Sr^{91}]

Sr-92 Strontium-92 [also ^{92}Sr, or Sr^{92}]

Sr-93 Strontium-93 [also ^{93}Sr, or Sr^{93}]

Sr-94 Strontium-94 [also ^{94}Sr, or Sr^{94}]

Sr-95 Strontium-95 [also ^{95}Sr, or Sr^{95}]

Sr-97 Strontium-97 [also ^{97}Sr, or Sr^{97}]

S(r) Mean Rotational Entropy [Symbol]

S/ρ Total Mass Stopping Power [Symbol]

sR (earthquake) shear wave reflected at surface on observing station [Symbol]

sr Senior [also Sr]

sr Sister [also Sr]

sr specific resistance [also SR]

sr steradian [Unit]

.sr Suriname [Country Code/Domain Name]

Σ_t/ρ Mass Absorption Coefficient [Symbol]

Σ_a/ρ True Mass Absorption Coefficient [Symbol]

σ_y/ρ specific strength (in mechanics) [Symbol]

SRA Science Research Associates [US]

SRA Scientific Review Administrator

SRA Shop Replaceable Assembly

SRA Snubber Reducing Adapter

SRA Society of Research Administrators [US]

SRA Special Rules Airspace

SRA Special Rules Area

SRA Spin Reference Axis

SRA Spontaneous Resistive Anisotropy

SRA Spring Research Association [now Spring Research and Manufacturers Association, UK]

SRA State Recreation Area [US]

SRA Stress-Relief Annealed; Stress-Relief Annealing [Metallurgy]

SRA Sulfo-Ricinoleic Acid [Biochemistry]

SRA Support Requirements Analysis

SRA Surveillance Radar Approach

SRAA Singapore Robot and Automation Association [now Singapore Industrial Automation Association]

SRAAW Short-Range Anti-Armor Weapon

SRAAWH Short-Range Anti-Armor Weapon, Heavy [also SRAAW(H)]

SRAAWL Short-Range Anti-Armor Weapon, Light [also SRAAW(L)]

Sr(Ac)$_2$ Strontium(II) Acetate [also Sr(ac)$_2$]

Sr(ACAC)$_2$ Strontium(II) Acetylacetonate [also Sr(AcAc)$_2$, or Sr(acac)$_2$]

SRAG Space Radiation Analysis Group

SRAG Sydney Rainforest Action Group [Australia]

SRAIS Statewide Resource Information and Accounting System [of New South Wales, Australia]

SRAM Semi Random-Access Memory

SRAM Short-Range Attack Missile

SRAM Static Random-Access Memory [also S-RAM, or sRAM]

SRAMA Spring Research and Manufacturers Association [UK]

SRAPI Speech Recognition Application Program Interface

SRARQ Selective Repeat-Automatic Repeat Request

SRAS State Rural Assistance Scheme [of New South Wales, Australia]

SRATS Solar Radiation and Thermospheric Satellite [Japan]

SRAU Simulated Remote Acquisition Unit

SRAW Short-Range Antitank Weapon

SRB Screw Rest Button

SRB Société Royale de Belgique [Royal Society of Belgium]

SRB Solid Rocket Booster [of NASA]

SRB Sorter-Reader Buffered

SRB Source Review Board

SRB Source-Route Bridge [Computers]

SRB State Reserve Board [US]

SRB Sulfate-Reducing Bacteria

SRB Surface Radiation Budget

SRBAB SRB (Solid Rocket Booster) Assembly Building [of NASA]

SRB-ARF Solid Rocket Booster Assembly and Refurbishment Facility [of NASA]

SrBCO Strontium Barium Copper Oxide (Superconductor) [also SBCO]

SRBDF Solid Rocket Booster Disassembly Facility [of NASA]

SRBDM Short-Range Bomber Defense Missile

SRBE Société Royale Belge des Electriciens [Royal Belgian Society of Electrical Engineers]

SRBG Société Royale Belge de Géographie [Royal Belgian Society of Geography]

SRBII Société Royale Belge des Ingénieurs et des Industriels [Royal Belgian Society of Engineers and Manufacturers]

SRBM Short-Range Ballistic Missile

SRBP Synthetic Resin Bonded Paper

SRBPF Solid Rocket Booster Processing Facility [of NASA]

SRBPP Saskatchewan River Basin Planning Program [Canada]

SRBSA Saskatchewan Road Builders Safety Association [Canada]

SRBSN Sintered Reaction-Bonded Silicon Nitride

SRC Sample Return Container

SRC Sanitary Refuse Collector

SRC Saskatchewan Research Council [Canada]

SRC Saturable Reactor Control

SRC Scandinavian Rubber Conference

SRC Science Research Center (of Excellence Program) [of Korea Science and Engineering Foundation, South Korea]

SRC Science Research Council [later Science and Engineering Research Council; now Engineering and Physical Sciences Research Council, UK]

SRC Scientific Research Council [Baghdad, Iraq]

SRC Second Radiation Constant

SRC Semiconductor Research Cooperative

SRC Semiconductor Research Corporation [of Cornell University, Ithaca, New York, US]

SRC Sequencer/Ring Counter

SRC Short-Range Cluster

SRC Solvent-Refined Coal (Process)

SRC Sound Ranging Control

SRC Source [also Src]

SRC Source Range Channel

SRC Spares Receiving Checklist

SRC Spokane Research Center [Washington State, US]

SRC Standard Reporting Conditions

SRC Standard Requirements Code

SRC Stanford Research Center [California, US]

SRC Static Recoverable Creep [Mechanics]

SRC Steel Research Center

SRC Stimulus/Response Compare

SRC Stored Response Chain

SRC Stress-Relief Cracking [Metallurgy]

SRC Synchronous Remote Control

SRC Synchrotron Radiation Center [Madison, Wisconsin, US]

SRC Systems Research Center [US]

SRC Source [File Name Extension] [also .src]

SRCB Software Requirements Change Board

SRCB Software Requirements Control Board

SRCBD Software Requirements Change Board Directive

SRCBD Software Requirements Control Board Directive

SRCC Strikes, Riots and Civil Commotion [also SR&CC, srcc, or sr&cc]

SrCCO Strontium Cerium Copper Oxide (Superconductor) [also SCCO]

SrCCO Strontium Calcium Copper Oxide (Superconductor) [also SCCO]

SRCD Society for Research in Child Development [US]

Srch Search [also srch]

SRCL Spectrally-Resolved Cathodoluminescence

SRCNET Science Research Council Network [now Science and Engineering Research Council Network, UK]

SRC Newsl Semiconductor Research Corporation Newsletter [Published by Cornell University, Ithaca, New York, US]

SrCNCO Strontium Calcium Neodymium Copper Oxide (Superconductor) [also SCNCO]

SRCP Successive Reaction Counterpoise (Method)

SRCR System Run Control Record

SRCRA Shipowners Refrigerated Cargo Research Association [UK]

SrCRO Strontium Calcium Rubidium Oxide (Superconductor) [also SCRO]

SRCT Static Recrystallization Critical Temperature [Metallurgy]

SrCYO Strontium Cerium Yttrium Oxide (Superconductor) [also SCYO]

SRD Screen Reader System

SRD Secret Restricted Data

SRD Self-Reading Dosimeter

SRD Shuttle Requirements Definition [of NASA]

SRD Shuttle Requirements Document [of NASA]

SRD Space Research Division [of NASA Office of Aeronautics and Space Technology]

SRD Standard Reference Data

SRD Step Recovery Diode

SRD Subdivision of Regional District

SRD System Residence Device

SRD Systems Requirements Document

S-RD Shipper-Receiver Difference

(S)RD Secondary Received Data

SRDA Sodium Removal Development Apparatus [Nuclear Engineering]

SRDA Strategic Regional Diversification Agreement [Canada]

SRDAS Service Recording and Data Analysis System

SRDC Sugar Research and Development Corporation [US]

SRDC Semiconductor Research and Development Center [of IBM Corporation at Fishkill, New York, US]

SRDE Signals Research and Development Establishment [UK]

SRDH Subsystems Requirement Definition Handbook

SRDH System Requirements Definition Handbook

SRDL Signal Research and Development Laboratory [of US Army at Fort Monmouth, New Jersey, US]

SRDM Subrate Data Multiplexer

SrDNO Strontium Dysprosium Niobium Oxide (Superconductor) [also SDNO]

Sr(DPM)$_2$ Strontium (II) Dipivaloylmethanoate [also Sr(dpm)$_2$]

SRDRAM Self-Refreshed Dynamic Random-Access Memory

SRDS Standard Reference Data System

SRDS Systems Research and Development Service [of Federal Aviation Administration, US]

SRE Scanning Reference Electrode

SRE Sending Reference Equivalent

SRE Senior Research Engineer

SRE Signaling Range Extender

SRE Single Region Execution

SRE Site Resident Engineer

SRE Society of Reliability Engineers [US]

SRE Sodium Reactor Experiment [of US Atomic Energy Commission]

SRE Speech Recognition Equipment

SRE STDN (Space Tracking and Data Network) Ranging Equipment

SRE Stimulated Raman Effect

SRE Surveillance Radar Element

SRE Surveillance Radar Equipment

SRe Seychelles Rupee [Currency]

SREA Street Rod Equipment Association [US]

SRED Scientific Research and Experiments Department [of US Navy]

SR&ED Scientific Research and Experimental Development

SR-EDD Synchrotron-Radiation Energy-Dispersive Diffraction

SREELS Spatially Resolved Electron Energy Loss Spectroscopy

SREL Savannah River Ecology Laboratory [US]

SREM School of Resource and Environmental Management [of Australian National University, Canberra]

SREM Scanning Reflection Electron Microscope; Scanning Reflection Electron Microscopy

SREMP Source Region Electromagnetic Pulse

SREP State Rivers and Estuaries Policy [of New South Wales, Australia]

SRESS Solid Rocket (Booster) Electrical Subsystem

SRET Scanning Reference Electrochemical Technique

SREX Strontium Extraction (Process)

SRF Self-Resonant Frequency

SRF Semiconductor Research Foundation [Miyagi, Japan]

SRF Semi-Reinforcing Furnace (Black)

SRF Service Request Flag

SRF Short Rotation Forestry

SRF Shuttle Refurbishment Facility [of NASA]

SRF Smithsonian Research Foundation [US]

SRF Software Recording Facility

SRF Software Recovery Facility

SRF Sorter-Reader Flow

SRF Spacecraft Research Foundation [US]

SRF State Revolving Funds [US]

SRF Stereoscopic Rangefinder

SRF Sugar Research Foundation [US]

SRF Surface Roughness Factor

SRFB Space Research Facilities Branch [of National Research Council of Canada]

SRF Black Self-Reinforcing Furnace Black [also SRF black]

SRFDD Semiconductor Research Faculty Database and Directory

SrFe Strontium Ferrite

Sr(FOD)$_2$ Strontium Bis(1,1,1,2,2,3,3-Heptafluoro-7,7-Dimethyl-4,6-Octanedionate) [also Sr(fod)$_2$]

SRFOV Stray Radiation Field of View

SRFT Saturation Recovery Fourier Transform

SRG Safety Regulation Group [of Civil Aviation Authority, UK]

SRG Schweizerische Radio− und Fernsehgesellschaft [Swiss Radio and Television Broadcasting Corporation]

SRG Shift Register

SRG Shift-Register Generator

SRG Short Range

SRG Statistical Research Group

SRG Steel Research Group [of Northwestern University, Evanston, Illinois, US]

SRG Stock Removal Grinding

SRG Suriname Guilder (or Florin) [Currency]

SRGB Sustained RGB (Red-Green-Blue) (Color Values)

SRGHAZ Subcritically Reheated Grain-Coarsened Heat-Affected Zone [Metallurgy]

SRH Subsystems Requirements Handbook [Aerospace]

Sr(HFAC)$_2$ Strontium Hexafluoroacetylacetonate [also Sr(hfac)$_2$]

SRI Saudi Riyal [Saudi Arabia]

SRI Shift Right In

SRI Society of the Rubber Industry [Japan]

SRI Southern Research Institute

SRI Southwest Research Institute [San Antonio, Texas, US] [also SWRI, or SwRI]

SRI Space Research Institute [Canada]

SRI Spalling Resistance Index

SRI Spring Research Institute [US]

SRI Stanford Research Institute [California, US]

SRI Statistical Reference Index [US]

SRI Steel Recycling Institute [US]

SRI Steel-Related Industries

SRI Synchrotron Radiation Instrumentation

SRIAER Scientific Research Institute for Atomic Energy Reactors [Russian Federation]

SRID Single Radial Immunodiffusion

SRIC Short Rotation Intensive Cultures [Microbiology]

S/RID Standards/Identification Document

SRIF Somatostatin Release Inhibiting Factor [Biochemistry]

SRIF Special Recovery Investment Fund

SRIM Selected Research in Microfiche [of National Technical Information Service, US Department of Commerce]

SRIM Structural Reaction Injection Molding

SRIN Ship Research Institute of Norway

SRIOC Shanghai Research Institute of Organic Chemistry [PR China]

SRIP Society for the Research and Investigation of Phenomena [Malta]

SRIS Spatial Reference Information System

SRISSP Scientific Research Institute of Solid-State Physics [Russian Federation]

SRJ Society of Rheology of Japan

SRL Satellite Research Laboratory [of Office of Research and Applications, National Oceanic and Atmospheric Administration in Suitland, Maryland, US]

SRL Savannah River Laboratory [of US Department of Energy at Aiken, South Carolina]

SRL Science Radar Laboratory

SRL Science Reference Library [of British Library]

SRL Scientific Research Laboratory

SRL Shift Register Label

SRL Shift Register Latch

SRL Shimadzu Research Laboratory [Manchester, UK]

SRL Siemens Research Laboratories

SRL (*Société à Responsabilité Limitée*) – French for "Limited (Liability) Company"

SRL Stability Return-Loss [Telecomminications]

SRL Steel Research Laboratory [of Nippon Steel Corporation, Chiba, Japan]

SRL Structural Research Laboratory

SRL Structural Return-Loss [Telecommunications]

SRL Superconductivity Research Laboratory [of International Superconductivity Technology Center, Japan]

SRL Surface Reaction Layer

SRL System Reference Library

SRL Systems Research Laboratories, Inc. [US]

SrLα Strontium L-Alpha (Radiation) [also SrL$_\alpha$]

SrLCO Strontium Lanthanum Copper Oxide (Superconductor) [also SLCO]

SRLCP Solute-Rich Liquid-Crystal Pulling [also SR-LCP]

SRL-ISTEC Superconductivity Research Laboratory of the International Superconductivity Technology Center [Japan]

Sr-LM Strontium-Substituted Lanthanum Manganite

SRLS Source Reactance Lossless Switch

SRLY Series Relay

SRLZ Southern Rock Lobster Zone [Australia]

SRM Safety, Reliability, Maintainability

SRM Scrim-Reinforced Material

SRM Security Reference Monitor

SRM Sediment Routing Model

SRM Service Reference Model

SRM Shock Remanent Magnetization

SRM Short Range Modem

SRM Short Rifle Military

SRM Simple Reservoir Model

SRM Society for Range Management [US]

SRM Solid Rocket Motor

SRM Source-Range Monitor

SRM Southern Rocky Mountains [US]

SRM Specification Requirements Manual

SRM Standard Reference Material

SRM Strategic Reconnaissance Missile

SRM Surface Reflection Microscope; Surface Reflection Microscopy

SRM System Resources Manager

SrM M-Type Strontium Ferrite

SRMA Service de Recherche de Métallurgie Appliquée [Applied Metallurgy Research Service, of Centre d'Etudes Nucléaires de Saclay, France]

SRMA Split-Channel Reservation Multiple-Access

SRMA/CE Saclay Service de Recherche de Métallurgie Appliquée/Centre d'Etudes Nucléaires de Saclay [Applied Metallurgy Research Service of the Saclay Nuclear Research Center, France]

SRMC Solid Rocket Motor Case

SRMC Stimulus/Response Measurements Catalog

SRMCASE Symmetry-Restricted-Multiconfiguration Annihilation of Single Excitations [Physics]

SRME Submerged Repeater Monitoring Equipment

SRMEM Symposium on Recycling of Metals and Engineering Materials

SRMLE Short Rifle Military, Lee-Enfield [UK]

SRM&QA Safety, Reliability, Maintainability and Quality Assurance

SRMS Shuttle Remote Manipulator System [NASA]

SRMS (Space) Station Remote Manipulator System

SRMS Structure Resonance Modulation Spectroscopy

SRMU Space Research Management Unit [US]

SRN Software Release Notice

SRN State Registered Nurse [UK]

SRNA Saskatchewan Registered Nurses Association [Canada]

sRNA Soluble Ribonucleic Acid [Biochemistry]

SrNCO Strontium Neodymium Copper Oxide (Superconductor) [also SNCO]

SRNFC Source Range Neutron Flux Channel

SRNFM Source Range Neutron Flux Monitor

SRNH Service Request Non-Honored

SrNO Strontium Niobium Oxide (Superconductor) [also SNO]

SRO Savannah River Operation [of US Atomic Energy Commission at Aiken, South Carolina]

SRO Senior Reactor Operator

SRO Sharable and Read Only

SRO Short-Range Order(ed); Short-Range Ordering [also sro] [Solid-State Physics]

SRO Silicon-Rich Oxide

SRO Single Room Occupancy

SRO Singly Resonant Oscillator

SRO Special Road Oil

SRO Specification Release Order

SRO Standing Room Only

SRO Strontium Ruthenium Oxide (or Strontium Ruthenate)

SRO Supervisor Range Operations

SrO Strontium Oxide

SROA Swedish Railway Officers Association

SROB Short-Range Omnidirectional Beacon

Sr(OEt)$_2$ Strontium (II) Ethoxide

SrO-Fe$_2$O$_3$ Strontium Oxide-Ferric Oxide (System)

SRON Space Research Organization Netherlands [Utrecht, Netherlands]

SROP Short-Range Order Parameter [Solid-State Physics]

SrOPr^i Strontium Isopropoxide

SRP Savannah River Plant [of US Department of Energy at Aiken, South Carolina, US]

SRP Scaling and Root Planing [Dentistry]

SRP Science Research Program [Singapore]

SRP Self-Reinforced Plastics

SRP Shared Resources Programming

SRP Signal Reference Point

SRP Short-Range-Order Parameter [Solid-State Physics]

SRP Slot Reference Point

SRP Society for Radiological Protection [UK]

SRP Solute Retarding Parameter

SRP Sound Radiation Pressure

SRP Spreading Resistance Probe (Technique) [Solid-State Physics]

SRP Spreading Resistance Profiling [Solid-State Physics]

SRP Standard Review Plan

SRP Static Reservoir Pressure [Geology]

SRP Step-Reaction Polymerization

SRP Strain Range Partitioning [Mechanics]

SRP Suggested Retail Price

SRP Summer Research Program

SRP Surveillance Radar Processor

.SRP Script [File Name Extension] [also .srp]

SRPA Scientific Research on Priority Areas (Program) [Japan]

SRPA Sintering, Re-Pressing-Annealing (Process) [Materials Science]

SRPA Spherical Retarding Potential Analyzer

SRPBA Scottish River Purification Boards Association [UK]

SrPCO Strontium Praseodymium Copper Oxide (Superconductor) [also SPCO]

SRPI Server-Requester Programming Interface

SRPO Science Resources Planning Office [US]

SRPES Synchrotron Radiation Photoemission Spectroscopy

SRPES Synchrotron Radiation Photoelectron Spectroscopy

SRPS Scottish Railway Preservation Society [UK]

SRPT Spin-Reorientation Phase Transistion [Physics]

SRP-UDP Standard Review Plan Update and Development Program

SRQ Service Request [Data Processing]

SR&Q Safety, Reliability and Quality

SR&QA Safety, Reliability and Quality Assurance

SRR Search and Rescue Radio

SRR Search and Rescue Region

SRR Serially Reusable Resource

SRR Shift Register Recognizer

SRR Short Range Recovery

SRR Shuttle Requirements Review [of NASA]

SRR Site Readiness Review

SRR Software Requirements Review

SRR Sound Recorder Reproducer

SRR Sri Lankan Rupee [Currency]

SRR Super-Regenerative Reception

SRR System Requirements Review

SrR Strontium Recovery

SRRB Search and Rescue Radio Beacon

SRRC Scottish Research Reactor Center

SRRC Southern Regional Research Center [of US Department of Agriculture in New Orleans, Louisiana]

SRRC Synchrotron Radiation Research Center [Hsinchu, Taiwan]

SRRL Southern Regional Research Laboratory [US]

SRRP Source Reduction Review Program

SRRS Surface Resonance Raman Spectroscopy

SRRT Schweizerische Gesellschaft für Reinraumtechnik [Swiss Society for Clean Room Technology]

SRS Safety and Reliability Society [UK]

SRS Safety Reporting System

SRS Savannah River Site [of US Department of Energy at Aiken, South Carolina, US]

SRS Selective Record Service

SRS Selenium Rectifier System

SRS Self-Sustained Reaction Sintering [Metallurgy]

SRS Shock-Induced Reaction Synthesis

SRS Silent Running Society [US]

SRS Simulated Remote Site

SRS Simulated Remote Station

SRS Slave Register Set

SRS Slow-Reacting Substance [Medicine]

SRS Société Royale des Sciences [Royal Society of Sciences, Belgium]

SRS Sodium Removal Station [Nuclear Engineering]

SRS Software Requirements Specification

SRS Soil Research Station [New Zealand]

SRS Sonobuoy Reference System

SRS Sound Retrieval System

SRS Space Research Scientist

SRS Spatial Reference System

SRS Specialized Rework System

SRS Specification Revision Sheet

SRS Specular Reflection Spectrometry

SRS Statistical Reporting Service [of US Department of Agriculture]

SRS Stimulated Raman Scattering

SRS Stimulated Raman Spectroscopy

SRS Strain-Rate Sensitivity [Mechanics]

SRS Student Records System

SRS Subscriber Response System [Cable Television]

SRS Supplemental Restraint System [Automobiles]

SRS Support Requirements System

SRS Surface Raman Spectroscopy

SRS Surface Reflectance Spectroscopy

SRS Surtsey Research Society [Iceland]

SRS Swiss Railways Society [UK]

SRS Synchrotron Radiation Source

SRS Synchrotron Radiation Source [Daresbury, UK]

SRS Synchronous Relay Satellite

SRS System Requirement Specification

SrS Strontium Sulfide

SRSA Scientific Research Society of America

SRSA Society of Radiographers of South Africa

SRS-A Slow-Reacting Substance of Anaphylaxis [Medicine]

SR-SAXS Synchrotron Radiation Small-Angle X-Ray Scattering

SRSC Slippery Rock State College [Pennsylvania, US]

SRSC Sul Ross State College [Alpine, Texas, US]

SRSCC Simulated Remote Station Control Center

SrSe Strontium Selenide (Semiconductor)

SR SER Shift Right Serial Input [Digital Logic]

SRSF SRB (Solid Rocket Booster) Refurbishment and Subassembly Facility [of NASA]

SRSF SRB (Solid Rocket Booster) Receiving and Subassembly Facility [of NASA]

SRSK Short-Range Station Keeping

SRSL Société Royale des Sciences de Liège [Royal Scientific Society of Liege, Belgium]

SRSO Silicon-Rich Silicon Oxide

SrSO Strontiosulfate

SRSR Schedule and Resources Status Report

SRSS Shuttle Range Safety System [of NASA]

SRSS Simulated Remote Sites Subsystem

SR-SS Sunrise–Sunset

SRSS Shuttle Range Safety System [of NASA]

SRST Spent Resin Storage Tank

SRT Search Radar Terminal

SRT Secondary Ranging Test

SRT Self-Reinforcing Thermoplastic(s)

SRT Shuttle Requirements Traceability [NASA]

SRT Slow Rise Time

SRT Society of Radiological Technologists [US]

SRT Soft Return [also Srt]

SRT Solids Retention Time

SRT Sorting [also Srt, or srt]

SRT Space-Station Redesign Team [of NASA]

SRT Special Rated Thrust

SRT Special Relativity Theory [Physics]

SRT Specification Requirements Table

SRT Speech Reception Threshold

SRT Spin Reorientation Transition [Physics]

SRT Standard Radio Telegraph

SRT Station Readiness Test

SRT Strontium Rhenium Tantalate

SRT Submarine Thermal Reactor

SRT Supply Response Time

SRT Supporting Research and Technology [also SR&T]

SRT Systems Readiness Test

SRT Systems Reliability Team

SR&T Supporting Research and Technology [also SRT]

SRTA Stationary Reflector, Tracking Absorber [Solar Collector]

SRTC Savannah River Technology Center [of US Department of Energy]

SRTC Scientific Research Tax Credit [Canada]

SRTF Shortest Remaining Time First

Sr(THD)$_2$ Strontium Bis(2,2,6,6-Tetramethyl-3,5-Heptanedionate) [also Sr(thd)$_2$]

SrTiO$_3$/Nb Niobium-Doped Strontium Titanate

SrTiO$_3$/YBCO Strontium Titanate/Yttrium Barium Copper Oxide

SrTl Strontium Thallium (Compound)

Sr(TMHD)$_2$ Strontium Bis(2,2,6,6-Tetramethyl-3,5-Heptanedionate) [also Sr(tmhd)$_2$]

SRTS Self-Regulating/Temperature Source [also SR/TS]

(S)RTS Secondary Request to Send

SRU Ship Repair Unit

SRU Shop Replaceable Unit; Shop Replacement Unit

SRU Société de Raffinage d'Uranium [French Uranium Refining Company, France]

SRU Space Replaceable Unit

SRU Sulfur Recovery Unit

SRV Safety Release Valve

SRV Schweizerische Raumfahrt-Vereinigung [Swiss Astronautics Association]

SRV Simian Rotavirus

SRV Smelting Reduction Vessel

SRV Sooty Ringspot Virus

SRV Space Rescue Vehicle

SRV Statens Räddningsverk [National Civil Defense Administration, Sweden]

SRV Static Recovery [Metallurgy]

SRV Styling Research Vessel

SRV Submarine Rescue Vessel [also srv]

SRV Surface Recombination Velocity [Solid-State Physics]

SRV SA11 Simian Rotavirus SA11

Srv Surveillance [also srv]

Srvl Survival [also srvl]

SRX Static Recrystallization [Metallurgy]

SRXPS Spin-Resolved X-Ray Photoelectron Spectroscopy

SRXRF Synchrotron Radiation X-Ray Fluorescence

SrYNO Strontium Yttrium Niobium Oxide (Superconductor) [also SYNO]

SrYRCO Strontium Yttrium Ruthenium Copper Oxide (Superconductor) [also SYRCO]

SRZ Special Rules Zone

SRZ Surveillance Radar Zone

SS Saddle Surface [Mathematics]

SS Sample Size [Statistics]

SS Sample Stabilizer

SS Sandstone

SS San Salvador [El Salvador]

SS San Sebastian [Spain]

SS Satellite Switching

SS Saturated Steam

SS Saturation Spectroscopy

SS Saturation State

SS Scatter Slit

SS Schlieren System [Optics]

SS School of Science

SS Schwann Sheath [Anatomy]

SS Science Service [Washington, DC, US]

SS Scintillation Spectrometer

SS Secondary Spectrum

SS Secondary Storage

SS Seconds [also ss] [Time]

SS Secretary of State

SS Secret Service

SS Self-Screening [Radar Detection]

SS Self-Shielding [Nuclear Engineering]

SS Self-Starter; Self-Starting

SS Selector Switch

SS Select Standby

SS Semet-Solvay (Oil Gasification Process) [Chemical Engineering]

SS Semiconductor Science

SS Semifinal Splice

SS Semi-Solid

SS Semi-Steel

SS Semi-Submersible

SS Sempervivium Society [UK]

SS Separation System

SS Sequential Storage

SS Serum Sickness [Medicine]

SS Servo System

SS Set Screw

SS Shear Strength [Mechanics]

SS Shear Stress [Mechanics]

SS (Earthquake) Shear Wave Reflected once from the Earth's Surface [also SR_1] [Symbol]

SS Shift Supervisor

SS Ship Service

SS Shop Supervisor

SS Shoulder Screw

SS Shubnikov Group [Solid-State Physics]

SS Shuck Split (Application) [Pesticides]

SS Shuttle System

SS Sieve Shaker

SS Signaling System

SS Signal Selector

SS Signal Splitter; Signal Splitting

SS Signal Strength

SS Silicon Steel

SS Silver Steel

SS Single Scan

SS Single-Shot (Firearm)

SS Single Sideband

SS Single-Sided

SS Single-Signal (Receiver)

SS Single-Silk-Covered (Electric Wire)

SS Single-Space

SS Single-Strength (Glass)

SS Single String

SS Singlet State [Spectroscopy]

SS Sky Survey

SS Sky-Wave Synchronized (Long-Range Navigational System)

SS Slave Station

SS Sliding Scale [Aeronautics]

SS Slipstream

SS Slope Stability [Geology]

SS Slow Setting

SS Slow-Slow (Wave)

SS Small-Scale

SS Sodium Salicylate

SS Soft Sector(ing)

SS Soil Science; Soil Scientist

SS Soil Solution [Geology]

SS Solar Still [Chemical Engineering]

SS Solar Spectrum

SS Solar System

SS Solid/Solid

SS Solid Solution

SS Solid Sphere

SS Solid-State [also S/S, or S-S]

SS Sonic Speed (or Speed of Sound)

SS Sound Scattering

SS Sound Spectrograph(y)

SS Sound Spectrogram

SS Sound Suppression; Sound Suppressor

SS Space Science

SS Space Shuttle [also S/S]

SS Space Simulator

SS Space Station

SS Space Suit

SS Space Switch

SS Spanish Sahara

SS Spark Source

SS Spark Spectrometer

SS Spark Spectrum

SS Spatial Statistics

SS Special Service

SS Special Source Materials

SS Specialty Steel

SS Specific Speed

SS Spectral Series

SS Speech Synthesis

SS Spherical Symmetry

SS Spin Spiral (Calculation) [Solid-State Physics]

SS Spin-Stabilized; Spin Stabilization [Aeronautics]

SS Spread Spectrum

SS Springer and Schwink (Strain-Aging Experiment) [Metallurgy]

SS Stack Segment [Computers]

SS Stainless Steel

SS Standard Specification

SS Star Shell

SS Start-Stop (Character)

SS State Space [Control Engineering]

SS Stationary Satellite

SS Stationary State(s) [Quantum Mechanics]

SS Station Set

SS Station-to-Station

SS Statistical Standards

SS Steady-State

SS Steamship [also S/S]

SS Stellar Spectroscopy

SS Stellar Spectrum

SS Step Scanning

SS Step Signal

SS Stereoscopic Society [UK]

SS Stereoselective; Stereoselectivity [Chemistry]

SS Steric Strain

SS Stokes Scattering [Physics]

SS Storage-to-Storage (Operation) [also S-S]

SS Streaky Sulfides [Metallurgy]

SS Structural Safety

SS Structural Stability [Materials Science]

SS Structure Sensitivity

SS Sturm Sequence [Mathematics]

SS Submarine [US Navy Symbol]

SS Subscriber Switching

SS Subsonic(s)

SS Subsystem [also S/S]

SS Suhl-Schuller (Exchange Bias Field Calculation) [Solid-State Physics]

SS Summing Selector

SS Sunday School

SS Sunset

SS Superconductor Science

SS Supersaturated; Supersaturation

SS Supersaturated Solution [Physical Chemistry]

SS Supersensitive; Supersensitivity

SS Supersonic(s)

SS Surface Science; Surface Scientist

SS Surveyor Spacecraft [of NASA]

SS Suspended Solids [Chemistry]

SS Swedish Standard

SS Sworn Statement

SS System Safety

SS Systems Science; Systems Scientist

SS System(s) Software [also SSW]

SS System Specification

SS System Summary

SS System Supervisor

SS Water Soluble Powder for Seed Treatment [Agriculture]

201SS AISI-SAE 201 Austenitic Stainless Steel

301SS AISI-SAE 301 Austenitic Stainless Steel

302SS AISI-SAE 302 Austenitic Stainless Steel

304SS AISI-SAE 304 Austenitic Stainless Steel

309SS AISI-SAE 309 Austenitic Stainless Steel

310SS AISI-SAE 310 Austenitic Stainless Steel

316SS AISI-SAE 316 Austenitic Stainless Steel

321SS AISI-SAE 321 Austenitic Stainless Steel

329SS AISI-SAE 329 Austenitic Stainless Steel

403SS AISI-SAE 403 Martensitic Stainless Steel

410SS AISI-SAE 410 Martensitic Stainless Steel

430SS AISI-SAE 430 Ferritic Stainless Steel

431SS AISI-SAE 431 Ferritic Stainless Steel

434SS AISI-SAE 434 Ferritic Stainless Steel

436SS AISI-SAE 436 Ferritic Stainless Steel

442SS AISI-SAE 442 Ferritic Stainless Steel

446SS AISI-SAE 446 Ferritic Stainless Steel

501SS AISI-SAE 501 Heat-Resisting Stainless Steel

502SS AISI-SAE 502 Heat-Resisting Stainless Steel

SS_α Supersaturated Alpha Ferrite [Metallurgy]

S_cS (Earthquake) Shear Wave Partially Reflected from Core [Symbol]

S&S Safeguards and Security; Security and Safeguards

S&S Substation and Switchyard

S+S Speech plus Simplex [also S+SX] [Telecommunications]

S/S Samples per Second [also SPS]

S/S Single Sideband

S/S Solidification/Stabilization (Mechanism)

S/S Solid-in-Solid (Solution)

S/S Solid/Solid (Interface) [also S-S]

S/S Solid State [also SS, or S-S]

S/S Source/Sink

S/S Space Shuttle [also SS]

S/S Start/Stop (Character)

S/S Steamship [also SS]

S/S Subsystem [also SS]

S-S Simmons-Smith (Reaction) [Chemistry]

S-S Solid-Solid (Interface) [also S/S]

S-S Storage-to-Storage (Operation) [also SS]

S-S Succinonitrile-Salol

S_1-S_2 Series Field [Controllers]

S_3-S_4 Series Field [Controllers]

6S- Somalia [Civil Aircraft Marking]

ss *(scilicet)* – namely

ss single-stranded (DNA, or RNA Molecules)

ss *(semis)* – one half [also s̄s̄] [Medical Prescriptions]

SSA Safe Sector Altitude

SSA Salvia Sclarea Agglutinin [Immunology]

SSA S-Band Single Access

SSA Scandinavian Shipowners Association

SSA Seismological Society of America [US]

SSA Selected Surface Aging (Treatment) [Metallurgy]

SSA Selective Service Act

SSA Semiconductor Safety Association [US]

SSA Semiotic Society of America [US]

SSA Sequential Spectrometer Accessory

SSA Serial Storage Architecture

SSA Serial Systems Analyzer

SSA Shuttle Simulation Aircraft [of NASA]

SSA Single-Line Synchronous Adapter

SSA Single-Stage Annealing [Metallurgy]

SSA Smoke Suppressant Additive

SSA Soaring Society of America, Inc. [US]

SSA Social Security Administration [of US Department of Health and Human Services]

SSA Solid-State Abstracts

SSA Solid-State Amorphization [Metallurgy]

SSA Solid-State Amplifier

SSA Space Suit Assembly

SSA Specific Service Agreement

SSA Specific Surface Area

SSA Spherical Sector Analyzer

SSA Spontaneous Self-Assembly [Physical Chemistry]

SSA Spring Service Association [US]

SSA State Space Analysis

SSA Static Strain Aging [Metallurgy]

SSA Steady-State Analysis

SSA Substructure Analysis

SSA 5-Sulfosalicylic Acid

SSA Swedish Steel Producers Association

S-Sa Succinonitrile-Salol

SS&A Space Systems and Applications

SSAA Swedish Society of Aeronautics and Astronautics

SSAB Svenska Stal AB [Swedish Steel Corporation, Domnarvet]

SSAC Signaling System Alternating Current

SSAC Society for the Study of Architecture in Canada

SSAC Space Science Advisory Committee [Europe]

SSAC Space Science Analysis and Command [of NASA]

SS-Acid 4-Amino-5-Hydroxy-1,3-Naphthalenedisulfonic Acid; 1-Amino-8-Naphthol-2,4-Disulfonic Acid [also SS-acid, 2S-Acid, or 2S-acid]

SSADM Structured Systems Analysis and Design Methodology

SSAE Society of Senior Aerospace Executives [US]

SSAEC Society for the Study of Alchemy and Early Chemistry [now Society for the History of Alchemy and Chemistry, UK]

SSAI Sternberg State Astronomical Institute [of Moscow University, Russian Federation]

SSA IA SSA (Serial Storage Architecture) Industry Association [US]

SSAL Simplified Short Approach Light

SSAld Sulfosalicylaldehyde [also Ssald]

SSALF Simplified Short Approach Light System with Squenced Flashing Lights

SSALR Simplified Short Approach Light System with Runway Alignment Lights

SSALS Simplified Short Approach Light System

SSALT Solid-State Radar Altimeter

SSAP Session Service Access Point

SSAP Source Service Access Point

SSAP Statistical Signal and Array Processing Committee [of Institute of Electrical and Electronic Engineers, US]

SSAR Site Safety Analysis Report

SSAR Society for the Study of Amphibians and Reptiles [US]

SSAR Solid-State Amorphization Reaction [Metallurgy]

SSAR Spin-Stabilized Aircraft Rocket

SSAR Spotlight Synthetic Aperture Radar

SSAR Standard Safety Analysis Report

SSAS Solid-State Activator Society [Japan]

SSAS Station Signaling and Announcement Subsystem

SSAT Solid-State Amorphization Transformation [Metallurgy]

SSAT Space Shuttle Access Tower [now Flight Support Station] [of NASA]

SSAT Shuttle Service and Access Tower [of NASA]

SSB 1-Pyrrolidinecarbodithioate

SSB Serial Systems Bus

SSB Single Sideband [also ssb]

SSB Single-Strand Binding (Protein) [Biochemistry]

SSB Small-Scale Bridging [Materials Science]

SSB Société Scientifique de Bretagne [Scientific Society of Brittany, University of Rennes, France]

SSB Société Scientifique de Bruxelles [Scientific Society of Brussels, Belgium]

SSB Source Selection Board

SSB Space Science Board [of National Research Council, US]

SSB Spinning Side Band (Pattern) [Nuclear Magnetic Resonance]

SSB Spring Stop Button

SSB Steady-State Bridging [Materials Science]

SSB Stockholms Stadsbibliotek [Stockholm City Library, Sweden]

SSB Subscriber Busy

SSB Subsurface Barrier

SSBAM Single Sideband Amplitude Modulation [also SSB-AM, or SSB AM]

SSBC Summary Sheet Bar Chart

SSBD Single Sideboard

SSBFM Single Sideband Frequency Modulation [also SSB-FM, or SSB FM]

SSBM Single Sideband Modulation

SSBN Fleet Ballistic Missile Submarine [US Navy Symbol]

SSBO Single Swing Blocking Oscillator

SSBSC Single Sideband Suppressed Carrier [also SSB-SC, or SSB SC]

SSBSCAM Single Sideband-Suppressed Carrier Amplitude Modulation [also SSB-SC-AM or SSB-SC/AM]

SSBSCASK Single Sideband-Suppressed Carrier Amplitude Shift Keyed [also SSB-SC-ASK, or SSB-SC/ASK]

SSBSCOM Single Sideband Suppressed-Carrier Optical Modulation [also SSB-SC-OM or SSB-SC/OM]

SSBSCPAM Single Sideband-Suppressed Carrier Pulse-Amplitude Modulation [also SSB-SC-PAM, or SSB-SC/PAM]

SSBUV Shuttle Solar Backscatter Ultraviolet (Radiometer) [NASA]

SSBW Surface Skimming Bulk Wave

SSC Sacramento State College [California, US]

SSC Saline Sodium Citrate (Buffer)

SSC Saskatchewan Safety Council [Canada]

SSC Saskatchewan Science Center [Canada]

SSC Savannah State College [Georgia, US]

SSC SCANNET (Scandinavian Information Retrieval Network) Service Center [Sweden]

SSC Schizophrenia Society of Canada

SSC Screen-to-Specimen Distance [Physics]

SSC Sealed Storage Cask [Nuclear Engineering]

SSC Seattle Specialty Ceramics Inc. [Washington, US]

SSC Secondary Schools Certificate

SSC Sectoral Skills Council [of Ontario Ministry of Education and Training, Canada]

SSC Sector Switching Center

SSC Selector Switching Center

SSC Sensor Signal Conditioner

SSC Settled Sludge Concentration

SSC Shape-Selective Catalyst

SSC Ship Structure Committee [US]

SSC Short Segmented Cask [Nuclear Engineering]

SSC (Space) Shuttle System Contractor [NASA]

SSC Signalling and Supervisory Control

SSC Single-Silk-Covered (Electric Wire) [also ssc]

SSC Single Specimen Compliance

SSC Single-Stage Command

SSC Singapore Science Center

SSC Sintered Silicon Carbide [also SSiC]

SSC Slow-Scan Camera

SSC Slow-Scan CCD (Charge-Coupled Device)

SSC Société Suisse de Chronométrie [Swiss Society of Chronometry]

SSC Solid-Solution Cermet

SSC Solid-State Chemistry

SSC Solid-State Circuit

SSC Solid-State Circuits (Council) [of Institute of Electrical and Electronics Engineers, US]

SSC Solid-State Communications

SSC Solid-State Control

SSC Solid-State Frequency Converter

SSC Southern State College [Magnolia, Arkansas, US]

SSC Space Science Committee [France]

SSC Special Service Campaign [of Toyota Corporation]

SSC Special Service Center

SSC Species Survival Commission [of International Union for Conservation of Nature and Natural Resources]

SSC Spectroscopy Society of Canada

SSC Spin-Spin Contact [Physics]

SSC Spin-Spin Coupling [Physics]

SSC Standard Saline Citrate

SSC Station Selection Code

SSC Statistical Society of Canada

SSC Stout State College [Menomonie, Wisconsin, US]

SSC Stellar Simulation Complex

SSC (John C.) Stennis Space Center [of US Naval Research Laboratory and NASA in Mississippi]

SSC Stored Sequence Command(s)

SSC Stress Shielding Coefficient [Mechanics]

SSC Subsurface Contamination

SSC Subsystem Computer

SSC Subsystem Sequence Controller

SSC Sudden (Magnetic) Storm Commencement

SSC Sulfide Stress Cracking

SSC Superconducting Supercollider

SSC Supersonic Car

SSC Supply and Services Canada [Now Part of Public Works and Government Services Canada]

SSC Surface Science Center [of State University of New York at Buffalo, US]

SSC Surveillance System Control

SSC Swedish Space Corporation

SSC Swivel Screw Clamp

SSC System Support Controller

(SSC)$_{db}$ Stress Shielding Coefficient for Crack-Front Debonding [Mechanics]

(SSC)$_{el}$ Elastic Stress Shielding Coefficient [Mechanics]

SSCA Sports Car Club of America [US]

SSCA Strobed Single Channel Analyzer

SSCA Surface Sampler Control Assembly

SSCB State Soil Conservation Board [US]

SSCC Scandinavian Society for Clinical Chemistry [Finland]

SSCC Scottish Sporting Car Club

SSCC Signaling System, Common Channel

SSCC Solid State Circuits Conference

SSCC Solid State Circuits Council [of Institute of Electrical and Electronics Engineers, US]

SSCC Spin Scan Cloud Camera

SSCC Sulfide Stress Corrosion Cracking

SSCC Superconductity Super-Collider Center [of US Department of Energy]

SSCC Support Services Control Center

SSCC Swiss Scientific Computing Center

SSCD Society of Small Craft Designers [US]

SScD Doctor of Social Science

SSCE Sodium Saturated Calomel Electrode

SSCE Swedish Society of College Engineers

SS/CF Signal Strength/Center Frequency

SSCH Solid-State Chemistry Conference

SSCHS Space Shuttle Cargo Handling System [of NASA]

SSCI Sanitation Suppliers and Contractors Institute [US]

SSCI Steel Service Center Institute [Cleveland, Ohio, US]

SSCI Steel Shipping Container Institute [US]

SSCI Swiss Society of Chemical Industry

SSCL Shuttle System Commonality List [of NASA]

SSCNO Samarium Strontium Copper Niobium Oxide (Superconductor)

SSCNS Ships Self-Contained Navigation System

SSCP Small Self-Contained Payload

SSCP Space Shuttle Contingency Plan [of NASA]

SSCP System Services Control Point

SSCPA Single Site Coherent Potential Approximation [Physics]

SSCP-LU System Services Control Point to Logical Unit (Session)

SSCP-PU System Services Control Point to Physical Unit (Session)

SSCP-SSCP System Services Control Point to System Services Control Point (Session)

SSCR Scottish Society of Crop Research

SSCR Set Screw

SSCR Solid-State Chemical Reaction

SSCR Spectral Shift Control Reactor; Spectrum Shift Controlled Reactor

SSCR Steady-State Creep Rate [Mechanics]

SSCS Safety and Security Communications System

SSCS Solid-State Chemical Sensor

S&SCS Scintillation and Semiconductor Counter Symposium

SSCSP Space Shuttle Crew Safety Panel [of NASA]

SSCU Special Signal Conditioning Unit

SSCV Semisubmersible Crane Vessel

SSCW Single-Silk-Covered (Electric) Wire

SSD Scientific Systems Development, Inc. [US]

SSD Sector Studies Directorate [of Human Resources and Labour Canada]

SSD Shared Secret Distribution [Telecommunications]

SSD Single Scattering Distribution [Physics]

SSD Single-Sided Disk

SSD Single Station Doppler

SSD Small Signal Diode

SSD Sodium Sulfate Decahydrate

SSD Solid-State Decomposition

SSD Solid-State Detector

SSD Solid-State Device

SSD Solid-State Division

SSD Solid-State Division [of Oak Ridge National Laboratory, Tennessee, US]

SSD Solid-State Disk

SSD Solid-State Drive

SSD Space Science Division [of NASA Ames Research Center, Moffett Field, California, US]

SSD Spacecraft Software Division

SSD Space Systems Division [of US Air Force]

SSD Spin-Spin Dipole [Physics]

SSD Static Sensitive Device

SSD Statistically Stored Dislocation [Materials Science]

SSD Statistical Subdivision

SSD Steady-State Diffusion [Physics]

SSD Steady-State Distribution [Chemical Engineering]

SSD Structural Stability Diagram [Materials Science]

SSD Subsurface Distribution

SSD System Summary Display

SS&D Synchronization Separator and Digitizer

SSDA Service Station Dealers of America [US]

SSDA Synchronous Serial Data Adapter

SSDB Space Shuttle Data Base [of NASA]

SSDC Semi-Submersible Drilling Caisson

SSDC Signalling System, Direct Current

SSDC Single Steel Drilling Caisson

SSDC System Safety Development Center

SSDD Single-Sided, Double-Density (Disk)

SSDD Software System Design Document

SSDH Subsystem Data Handbook

SSDM (International Conference on) Solid-State Devices and Materials

ssDNA Single-Stranded DNA (Deoxyribonucleic Acid) (Molecule) [Biochemistry]

SSDOO Space Science Data Operations Office [of NASA]

SSDR Steady-State Determining Routine

SSDS Space Science Data Services [of NASA Office of Space Science]

SSDS Space Station Data System

SSDS System of Social and Demographic Statistics

SSDT Society of Soft Drink Technologists [US]

SSE N,N-Diethyl Dithiocarbamate

SSE Safe(ty) Shutdown Earthquake

SSE School of Science and Engineering

SSE Secondary Salt Effect

SSE Separate Statistical Ensembles [Statistical Mechanics]

SSE Single-Silk-Covered Enameled (Electric Wire)

SSE Sliding Stereo Effect

SSE Society for Scientific Exploration [US]

SSE Society for the Study of Evolution [US]

SSE Software Support Environment

SSE Software System Engineering

SSE Solid Solubility Extension

SSE Solid-State Electrochemistry

SSE Solid-State Electrolyte

SSE Solid-State Electronics

SSE Solid-State Engineer(ing)

SSE Southeastern Section [of American Physical Society, US]

SSE South-Southeast; South by Southeast

SSE Space Science Enterprise [of NASA]

SSE Space Shuttle Engine(s) [of NASA]

SSE Special Support Equipment

SSE Stockholm School of Economics [Sweden]

SSE Subsystem Element

SSE Subsystem Support Equipment

SSE Sulfate Saturated Electrode

SSE Sum of Squares Error

SSEAT Surveyor Scientific Evaluation Advisory Team [of NASA]

SSEB South of Scotland Electricity Board

SSEC Selective Sequence Electronic Calculator

SSEC Solid-State Electrochemist(ry)

SSEC Solid-State Electronics Center [of Honeywell Inc., Plymouth, Minnesota, US]

SSEC Solid-State Equipment Corporation [US]

SSEC Space Science and Engineering Center [of University of Wisconsin at Madison, US]

SSED Solid-State Electronic Device

SSED Solid-State Electronics Directorate [of Wright-Patterson Air Force Base, US]

SSED Space Systems Engineering Division [of US Air Force]

SS-EDDIV S,S-Ethylenediamine-N,N'-di-α-Isovalerate [also SS-eddiv]

SS-EDDP S,S-Ethylenediamine-N,N'-di-α-Propionate [also SS-eddp]

S-SEED Symmetric Self-Electrooptic-Effect Device

SSEG Scottish Solar Energy Group

SSEG Current Value of SS (Stack Segment) Register [also SSeg] [Pascal Function]

SSEGB Société Suisse des Explosifs [Swiss Society for Explosives]

SSEOS Space Shuttle Engineering and Operations Support [of NASA]

SSEP Self-Stimulated Emission Pumping

SSEP System Safety Engineering Plan

SSEPA Society of Spanish Engineers, Planners and Architects [US]

SSERC Social Sciences and Engineering Research Council [Canada]

SSerg Staff Sergeant [also S Serg]

SSES Small Smart Electronic Switch

SSESCO Supercomputer Systems Engineering and Services Company [US]

SSESM Spent Stage Experimental Support Module

SSET Space Shuttle External Tank [of NASA]

SSET SRM (Solid Rocket Motor) Stacking Enhancement Tool

S-SET Superconducting Single Electron Transistor

SSEXP Steady-State Excitation Photoselection

SSF Safe Shutdown Facility

SSF Saybolt Seconds Furol [Unit]

SSF Sample Size and Frequency [Statistics]

SSF Scottish Software Federation

SSF Self-Similar Flow [Fluid Mechanics]

SSF Semi-Solid Forged; Semi-Solid Forging (Technology)

SSF Service Storage Facility

SSF Simultaneous Saccharification and Fermentation (Process)

SSF Single-Sided Frame

SSF Societas Scientiarium Fenneca [Finnish Scientific Society]

SSF Société Spéléologique de France [Speleological Society of France]

SSF Solid-State Fermentation

SSF Southern Shark Fishery [Australia]

SSF Space Station Freedom [Joint Project of the Canadian Space Agency, the European Space Agency and NASA]

SSF Special Service Force

SSF SRB (Solid Rocket Booster) Storage Facility [of NASA]

SSF Steady-State Film

SSF Steady-State Flow [Chemical Engineering]

SSF Subsonic Flow [Fluid Mechanics]

SSF Superlattice Stacking Fault [Crystallography]

SSF Supersonic Flight

SSF Supersonic Flow [Fluid Mechanics]

SSF Supersonic Frequency

SSF Surface-Sensitive Feature

SSF Symmetrical Switching Function

SSF System Support Facility

SSFA Stainless Steel Fabricators Association [UK]

SSFC Sequential Single Frequency Code

SSFC Sir Sanford Fleming College [Peterborough, Ontario, Canada]

SSFC Site-Site Function Counterpoise (Method)

SSFC Solid-State Frequency Converter

SSFD Suggested Standard for Future Design

SSFE Scandinavian Society of Forest Economics [Sweden]

SSFF Solid Smokeless Fuels Federation [UK]

SSFF Space Station Furnace Facility [of NASA]

SSFGSS Space Shuttle Flight and Ground System Specification [NASA]

SSFI Scaffolding, Shoring and Forming Institute [US]

SSFL Santa Susana Field Laboratory [Los Angeles, California, US]

SSFL Schottky-Barrier Coupled Schottky-Barrier Gate FET (Field-Effect Transistor) Logic

SSFL Steady-State Fermi Level [Physics]

SSFM Single Sideband Frequency Modulation

SSFMAC Southern Shark Fishery Management Advisory Committee [Australia]

SSFPO Space Station Freedom Program Office

SSFR Solar Spectral Flux Radiometer

SSFS Space Shuttle Functional Simulator [of NASA]

SSFS Special Services Forecasting System

SSG Guided-Missile Submarine [US Navy Symbol]

SSG Scientific Steering Group

SSG Search Signal Generator

SSG Second Stage Graphitization [Metallurgy]

SSG Semiconductor Strain Gauge

SSG Small Signal Gain

SSG Staff Sergeant [also S Sgt, SSGt, or S/Sgt]

SSG Standard Signal Generator

SSG Stonehenge Study Group [US]

SSG Submarine Guided Missile

SSG Superconducting Superheated Granules

SSG Super Spin-Glass [Solid-State Physics]

SSG Supervising Scientist Group [Australia]

SSG Symbolic Stream Generator

SSGA Sterling Silversmiths Guild of America [US]

SSGA System Suppport Gate Array

SSGD Secondary School Graduation Diploma [Canada]

SSGF Self-Consistent Surface Green Function (Model) [Physics]

SSGLCS Supersonic Gas-Liquid Cleaning System

SSGS Standard Space Guidance System

S Sgt Staff Sergeant [also SSgt, SSG, or S/Sgt]

SSH Schwartz-Slawsky-Herzfeld (Theory) [Physics]

SSH Small-Scale Hydropower (Project)

SSH Snowshoe Hare [Zoology]

SSH Solid Solution Hardening [Metallurgy]

SSH Su-Schrieffer-Heeger (Model) [Physics]

SSh Somalia Shilling [Currency]

SSHB Society for the Study of Human Biology [UK]

SSHB Station Set Handbook

SSHG Surface Second Harmonic Generation

SSHQ Space Station Headquarters [Canada]

SSHRC Social Sciences and Humanities Research Council [Canada]

SSHTM Small-Scale, High-Temperature Melter

SSHW Symmetric Shear Horizontal Wave

SSI Sector Scan Indicator

SSI Server Side Includes [Internet]

SSI Setchkin Self-Ignition (Cell)

SSI Significant Structural Item

SSI Single-Source Image

SSI Single System Image

SSI Sky Survey Instrument

SSI Small-Scale Integrated; Small-Scale Integration

SSI Solid-State Ionics

SSI Space Studies Institute [US]

SSI Special Secretariat for Informatics [Brazil]

SSI Special Scientific Interest

SSI Specialty Steel Industry

SSI Spin-Spin Interaction [Physics]

SSI Stainless Steel Industry

SSI Start Signal Indicator

SSI Steady-State Irradiation

SSI Storage-to-Storage Instruction

SSI Supplemental Security Income (Program) [US]

SSI Surface Solutions Inc. [US]

SSI Surface Survey Information

SSI Svenska Silikatforskningsinstitut [Swedish Institute for Silicate Research]

SSI Synchronous Systems Interface

SSIBD Shuttle System Interface Block Diagram [of NASA]

SSiC Sintered Silicon Carbide [also SSC]

SSID Shuttle Stowage Installation Drawing [of NASA]

SSIDA Steel Sheet Information and Development Association [UK]

SSIE Smithsonian Science Information Exchange, Inc. [of Smithsonian Institution, Washington, DC, US]

SSIG Single Signal

SSIM Stabilization of Stress-Induced Martensite; Stabilized Stress-Induced Martensite [Metallurgy]

SSIM Standard Schedules Information Manual

SSIMS Scanning Secondary Ion Mass Spectroscope; Scanning Secondary Ion Mass Spectroscopy

SSIMS Static Secondary Ion Mass Spectroscopy

SSIMT Suppressed Sidewall Injection Magnetotransistor

SSINA Specialty Steel Industry of North America

SSIO Space Shuttle Input/Output

SSIP Scotian Shelf Ichthyoplankton [Canada]

SSIP Shuttle Student Involvement Program [of NASA]

SSIP Systems Software Interface Processing

SSIS Scanning Surface Inspection System

SSIS Space Station Information System

SSISE Solid-State Ion-Selective Electrode

SSITP Shuttle System Integrated Test Plan [of NASA]

SSIUS Specialty Steel Industry of the United States; Stainless Steel Industry of the United States

SSJ Seismological Society of Japan

SSJ Self-Screening Jammer

SSJ Sole Source Justification

SSJ Solid-State Joint [Metallurgy]

SSJ Spectroscopical Society of Japan

SSK Soil Stack [Building Trades]

SSK Solid-State Kinetics

SSK Submarine–Hunter Killer [US Navy Symbol]

SSL Scientific Subroutine Library

SSL Secure Socket Layer [Internet]

SSL Self-Aligned Super-Injection Logic (Technology) [also S^2L]

SSL Serpentine Superlattice [Solid-State Physics]

SSL Shift and Select

SSL Sodium Stearoyl Lactylate

SSL Software Slave Library

SSL Software Specification Language

SSL Software Support Laboratory (for Space and Earth Scientists) [of University of Colorado, US]

SSL Solid-State Lamp

SSL Solid-State Laser

SSL Source Statement Library

SSL Space Sciences Laboratory [of University of California at Berkeley, US]

SSL Space Sciences Laboratory [of NASA at Marshall Spaceflight Center in Huntsville, Alabama, US]

SSL Storage Structure Language

SSL Strong Segregation Limit [Metallurgy]

SSL Surface Science Laboratories [US]

SSL System Software Loader

SSL System Specification Language

SSLC Super-Speed Liquid Chromatography

SSLC Synchronous Single-Line Controller

S Sleep Synchronized (or Slow-Wave) Sleep [also S sleep]

SSLO Secondary School Liaison Office

SSLO Solid-State Local Oscillator

SS LORAN Sky-Wave-Synchronized Long-Range Navigation(al System) [also SS Loran]

SSLS Standard Space Launch System

SSLV Standard Space Launch Vehicle

SSM N,N,-Dimethyl Dithiocarbamate

SSM Samarium Strontium Manganate

SSM Satellite System Monitor

SSM Sault Ste. Marie [Michigan, US and Ontario, Canada]

SSM Sault Ste. Marie, Michigan [Meteorological Station Identifier]

SSM Scanning SQUID (Superconducting Quantum Interference Device) Field

SSM Schonsted Magnetometer

SSM Semiconductor Storage Model

SSM Semi-Solid Material

SSM Semi-Solid Metal (Casting Process) [Metallurgy]

SSM Semi-Solid Metal Forming (Process) [Metallurgy]

SSM Semi-Solid Metalworking (Casting Process) [Metallurgy]

SSM Sensor Microwave/Imager

SSM Ship-to-Ship Missile

SSM Short Static Mixer

SSM Shuttle Support Manager [NASA]

SSM Single-Sideband Multiplier

SSM Small-Scale Melter

SSM Small Semiconductor Memory

SSM Solid-State Maser

SSM Solid-State Material

SSM Solid-State Microbattery

SSM Spacecraft Systems Monitor

SSM Special Safeguarding Measures

SSM Special Sensor Magnetometer [USDOD Defense Meteorological Satellite Program]

SSM Special Sensor Microwave

SSM Spectral Scanning Microwave

SSM Spread Spectrum Modulation

SSM Stacking Stability Map [Materials Science]

SSM Staff Sergeant Major

SSM Standard Schedule Message

SSM State Space Method [Control Engineering]

SSM State Space Model [Control Engineering]

SSM Stress-Strain Measurement

SSM Stress-Strain Microprobe

SSM Subsystem Manager

SSM Support System Module

SSM Surface-to-Surface Missile

SSM Systems Support Module

S&SM Surfaced on One, or Two Sides and Standad Matched [Lumber]

SSMA School Science and Mathematics Association [US]

SSMA Scientific Society of Measurement and Automation [Hungary]

SSMA Scottish Steel Makers Association [UK]

SSMA Spread-Spectrum Multiple Access

SSMB Space Shuttle Maintenance Baseline [NASA]

SSMB Special Services Management Bureau

SSMBE Surface Segregation Molecular Beam Epitaxy

SSMC Single Scattering Monte Carlo [Physics]

SSMC Software Servomotor Control

S2MC (Laboratoire de) Science des Surfaces et Matériaux Carbones [Surface Science and Carbonaceous Materials (Laboratory); of Institut National Polytechnique de Grenoble, France]

SSMD Silicon Stud-Mounted Diode

SSME Scientific Society for Mechanical Engineering [Hungary]

SSME Space Shuttle Main Engine [NASA]

SSMEC Space Shuttle Main Engine Controller [NASA]

SSMECA Space Shuttle Main Engine Controller Assembly [NASA]

SSME CONF Space Shuttle Main Engine Conference [NASA]

SSMF Signaling System, Multifrequency

SSM/I Special Sensor Microwave/Imager

SSM/I Spectral Scanning Microwave/Imager

SSMO Samarium Strontium Manganate

SSMS Spark-Source Mass Spectrometry; Spark-Source Mass Spectroscopy

SSMSR Spring School on Muon Spin Research [Physics]

SSMT Stress Survival Matrix Test

SSM/T Special Sensor Microwave/Temperature Sounder

SSM/T2 Special Sensor Microwave/Water Vapor Sounder

SSMTG Solid-State and Molecular Theory Group

SSMUTA Sheet and Strip Metal Users Technical Association [UK]

SSMWD Solid-State Microwave Device

SSN Segment Stack Number

SSN Shuttle Support Node [NASA]

SSN Sintered Silicon Nitride

SSN Social Security Number [US]

SSN Specification Serial Number

SSN Submarine, Nuclear-Powered [US Navy Symbol]

SSN Switched Services Network

SSNB Stainless Steel News Bureau [US]

SSNM Strategic Special Nuclear Material(s)

SSNMH Scipio Society of Naval and Military History [US]

SSNNP Sustainable Social Net National Product

SSNOEDS Steady-State Nuclear Overhauser Effect Difference Spectroscopy

SSNPP Small-Size Nuclear Power Plant

SSO Ethylxanthate

SSO Safety Signification Operation

SSO Sanitary Sewer Overflow

SSO Schizophrenia Society of Ontario [Canada]

SSO Shi, Seinfeld and Okuyama (Nucleation Theory) [Metallurgy]

SSO Siding Spring Observatory [at Coonabarabran, Australia]

SSO Single Stage to Orbit

SSO Solar Synchronous Orbit

SSO Source Selection Official

SSO Space Shuttle Orbiter [of NASA]

SSO Steady-State Oscillation

SSO Subsystem Operation (in Spacelab)

SSO Sun-Synchronous Orbit

SSO Support System for OEX (Orbiter Experiments) [NASA]

SSO Symmetrically Split Operator

SSOC Solid-State Optoelectronics Consortium [of National Research Council of Canada, Ottawa]

SSO-FFT Symmetrically Split Operator Fast Fourier Transform (Method)

SSOG Satellite System Operations Guide

SSOP Space Systems Operating Procedure(s)

SSOPZ Satellite System Operations Plan

SSOU Space Station Operation and Utilization

SSOUT System Output Unit

SSOZ Site-Site Ornstein-Zernike (Equation) [Physics]

SSP 1,2-Distearoyl Palmitin

SSP Satellite Sub-Point

SSP Scientific Subroutine Package

SSP Semi-Solid Processing

SSP Service Switching Point

SSP Signaling and Switching Processor

SSP Signal Switching Point

SSP Silarylene-Siloxane Polymer

SSP Silicon Switch Processor

SSP Silk-Screen Printing

SSP Small Sortie Payload

SSP Société Suisse de Physique [Swiss Physical Society]

SSP Society of Satellite Professionals [now Society of Satellite Professionals International, US]

SSP Society of Scholarly Publishing [US]

SSP Sodium Sampling Package

SSP Solid-State Physics

SSP Solid-State Plasma

SSP Space Shuttle Program

SSP Space Station Program

SSP Space Summary Program

SSP Special Services Protection

SSP Spectroscopy Society of Pittsburgh [Pennsylvania, US]

SSP Standard Switch Panel

SSP Start of Sustained Pressure [Aerospace]

SSP Startup Service Package

SSP Static Sodium Pot

SSP Static Spontaneous Potential [Oil Exploration]

SSP Steady State Process

SSP Steam Service Pressure

SSP Stock Savings Plan [Canada]

SSP Stream Sedimentation Problems

SSP Sub-Satellite Point

SSP Subsurface Science Program

SSP Sum of Squares and Products [Statistics]

SSP Super-Shockley Partials (or Partial Dislocations) [Materials Science]

SSP Suspended Solid Phase (Separation)

SSP Sustained Shock-Wave Plasma

SSP System Status Panel

SSP System Support Program

ssp subspecies [Chemistry; Biology]

SSPA Solid-State Power Amplifier

SSPA Swedish Steel Producers Association

SSPC Spacelab Stored Program Command

SSPC Steel Structures Painting Council [US]

SSPCM Slovenian Society for Process Control and Measurements

SSPD Shuttle System Payload Data [NASA]

SSPD Shuttle System Payload Definition (Study) [NASA]

SSPD Shuttle System Payload Description [NASA]

SSPD Sine Squared Phi Plot

SSPDA Space Shuttle Payload Data Activity [NASA]

SSPDB Subsystem Power Distribution Box

SSPDS Space Shuttle Payload Data Study [NASA]

SSPE Saline Sodium Phosphate EDTA (Ethylenediaminetetraacetic Acid) (Buffer)

SSPE Subacute Sclerosing Planencephalitis [Medicine]

SS-PEG Mono-Methoxypolyethylene Glycol Succinimidyl Succinate

SSPF Signal Structure Parametric Filter

SSPF Space Station Processing Facility [of NASA]

SSPGSE Space Shuttle Program Ground Support Equipment [of NASA]

SSPI Security Service Provider Interface

SSPI Society of Satellite Professionals International [US]

SSPI Solid-State Physics Institute [Moscow, Russia]

SSPL Steady-State Power Level

SSPM Space Shuttle Program Manager [NASA]

SSPMA Sump and Sewage Pump Manufacturers Association [US]

SSPO Space Shuttle Program Office [of NASA]

SSPO Space Station Program Office

SSPO Strategic Systems Project Office [of US Navy]

SSPP Scandinavian Society for Plant Physiology [Sweden]

SSPP System Safety Program Plan

SSPPSG Space Shuttle Payload Planning Steering Group [of NASA]

SSPRO Space Shuttle Program Resident Office [of NASA]

SSPS Satellite Solar Power Station

SSPS Satellite Solar Power System

SSPS Solid-State Protection System

SSPS Space Shuttle Program Schedule [NASA]

SSPS Suspended Solid-Phase Separation

SSPTB Space Shuttle Propulsion Test Bed [of NASA]

SSPTF Santa Susana Propulsion Test Facility [Los Angeles, California, US]

SSPWR Small-Size Pressurized-Water (Nuclear) Reactor

SSQ Smallest Sum of Squares (of Errors) [Statistics]

SSQ Standardless Semi-Quantitative (Analysis)

SSQD Single-Sided, Quad-Density (Disk)

SSQS Soil and Sediment Quality Section [of Environment Canada]

SSR Radar Picket Submarine [US Navy Symbol]

SSR Secondary Surveillance Radar

SSR Senior Site Representative

SSR Separate Superheater Reactor

SSR Shift Staging Room

SSR Shop Support Request

SSR Single-Signal Receiver

SSR Site Suitability Report

SSR Slow Strain Rate [Mechanics]

SSR Société Suisse de Radiodiffusion [Swiss Society of Broadcasting]

SSR Society for the Study of Reproduction [US]

SSR Solar System Research

SSR Solid-State Reaction

SSR Solid-State Refining

SSR Solid-State Relay

SSR Soviet Socialist Republic

SSR Specification Status Report

SSR Spin-Spin Relaxation [Solid-State Physics]

SSR Spin-Stabilized Rocket

SSR Staff Support Room

SSR Station Set Requirement

SSR Steady-State Reactor

SSR Stereoselective Reaction [Organic Chemistry]

SSR Stereospecific Reaction [Organic Chemistry]

SSR Stretched Surface Recording

SSR Subsynchronous Resonance

SSR Sum of the Squared Residuals

SSR Supply Support Request

SSR Switching Selector Repeater

SSR Synchronous Stable Relaying

SSR System Status Report

SSRA Scottish Seaweed Research Association

SSRA Solid-State Reaction Amorphization [Metallurgy]

SSRA Spread-Spectrum Random Access

SSR-A Secondary Surveillance Radar Providing Coded Aircraft Identity

SSRC Structural Stability Research Council [US]

SSRC Surface Science Research Center [of University of Liverpool, UK]

SSRC Swedish Space Research Committee

SSR-C Secondary Surveillance Radar Providing Coded Aircraft Altitude

SS&RC Standards Screening and Review Committee [US]

SSRD Solid-State Radiation Detector

SSRD Station Set Requirements Documents

SSRE Society for Social Responsibility in Engineering

SSRI Swedish Silicate Research Institute

SSRL Stanford Synchrotron Radiation Laboratory [of Stanford University, California, US]

SSRM Space Station Remote Manipulator

SSRMS Space Station Remote Manipulator System

SSRN System Software Reference Number

ssRNA Single-Stranded RNA (Ribonucleic Acid) (Molecule) [Biochemistry]

SSRP Secondary Surveillance Radar Processor

SSRP Siam Society under Royal Patronage [Thailand]

SSRP Simple Server Redundancy Protocol

SSRP Space Science Radioastronomy Program [of Canadian Space Agency]

SSRP Spectral Surface Radiation Program

SSRP Stanford Synchrotron Radiation Project [of Stanford University, California, US]

SSRP Symmetry Selection Rule Procedure

SSRR Station Set Requirements Review

SSR/RPG Secondary Surveillance Radar/Regional Planning Group [of International Civil Aviation Organization]

SSRS Self-Sustained Reaction Sintering [Metallurgy]

SSRS Society for Social Responsibility in Science [US]

SSRS Source Storage and Retrieval System

SSRS Start-Stop-Restart System

SSR-S Secondary Surveillance Radar with Selective Interrogation Capability

SSRSJC Saudi-Sudanese Red Sea Joint Commission [Saudi Arabia]

SSRT Slow Strain Rate Technique [Mechanics]

SSRT Slow Strain Rate Tensile (Test) [Mechanics]

SSRT Slow Strain Rate Test(ing) [Mechanics]

SSRT Subsystem Readiness Test

SSRT Stainless Steel Research Team [of Posco Steel Company, Pohang, South Korea]

SSRTP Solid-Surface Room-Temperature Phosphorescence

SSS Scientific Subroutine System

SSS Sea-Surface Salinity

SSS Selective Service System

SSS Self-Similar Systems (Conference) [of Joint Institute for Nuclear Research, Dubna, Russia]

SSS Sequential Slow Scanning

SSS (Earthquake) Shear Wave Reflected twice from the Earth's Surface [Symbol] [also SR_2]

SSS Shipboard Survey Subsystem

SSS Simulation Study Series

SSS Single-Signal Superheterodyne (Receiver)

SSS Small-Scale System

SSS Small Scientific Satellite

SSS Small Solar Satellite

SSS Sodium 1,4-Styrene Sulfonate

SSS Solid-State Sensor

SSS Solid-State Science

SSS Solid-State Sintering

SSS Solid-State Switching

SSS Sound Suppression System

SSS Space Shuttle Support

SSS Space Shuttle System

SSS Special Safety Safeguards

SSS Special-Special-Special (Threefold Grain Boundary Edge) [Materials Science]

SSS Spin-Spin Splitting [Physics]

SSS Stage Separation Subsystem

SSS Station Set Specification

SSS Statistical Self-Similarity [Physics]

SSS Statistical Support Staff [of Division of Computer Research and Technology, US]

SSS Steering and Suspension System

SSS Strategic Satellite System

SSS Structual Space Station

SSS Subscribers Switching Subsystem

SSS Subsystem Segment

SSS Sun-Synchronous Satellite

SSS Superlattice Surface States [Materials Science]

SSS Supersaturated Solution [Chemistry]

SSS Supersaturated Solid Solution [Materials Science]

SSS Swedish Scientific Satellite

SSS System Safety Society [US]

SSSA Society of Solid-State Actuators [US]

SSSA Soil Science Society of America [US]

SSSA J Soil Science Society of America Journal [US]

SSSB Society for the Study of Social Biology [US]

SSSC Single-Sideband Suppressed Carrier

SSSC Solid-State Science Committee [of National Research Council, US]

SSSC Surface-Subsurface Surveillance Center

SSSD Single-Sided, Single-Density (Disk)

SSSD Solid-State Science Division [Later Part of Materials Science and Technology Division, Argonne National Laboratory, US]

SSSERC Scottish Schools Science Equipment Research Center

SSSF Self-Scanner Stop Failure

SSSF Surface and Soundings Systems Facility [of National Center for Atmospheric Research, Boulder Colorado, US]

SSSI Site of Special Scientific Interest

SSSI Soil Science Society of Ireland

SSSM Self-Starting Synchronous Motor

SSSP Solid-State Shear Pulverization

SSSP Space Settlement Studies Program [US]

SSSP Station-to-Station Send Paid

SSSP Structured Sound Synthesis Project [Canada]

SSSP System Startup Service Package

SSSR Sojus Sowjetskckh Sozialisticheskich Republik [Union of Soviet Socialist Republics]

SSSS Society for Social Studies in Science [US]

SSSS Space Shuttle System Specification [NASA]

SSSS Standard Submarine Sonar System [also S4, or S^4]

SSSS Staphylcoccal Scalded Skin Syndrome [Medicine]

SSSS Supersaturated Solid Solution [Materials Science]

SSST Solid-State Science and Technology

SSSSA Soil Science Society of South Africa

SSSV Semi-Submersible Support Vessel

SSSV Subsurface Safety Valve

SSSWP Seismological Society of the Southwest Pacific [New Zealand]

SST Salt-Spray Test [Metallurgy]

SST Saskatchewan Science and Technology [Canada]

SST School of Science and Technology

SST Science Society of Thailand

SST Sea Surface Temperature (Chart)

SST Secondary Surge Tank [Nuclear Reactor]

SST Semiconductor Science and Technology

SST Semiconductor-Semimetal Transition

SST Serum Separation Tube

SST Shuttle Support Team [NASA]

SST Simulated Structural Test

SST Simultaneous Self-Test

SST Single Sheet Tester

SST Single-Shell Tank

SST Single-Sideband Transmission

SST Single Strip Tester

SST Single System Trainer

SST Society of Surveying Technicians [UK]

SST Solid-State Technology

SST Solid-State Theory

SST Solid-State Transformation [Materials Science]

SST Solvent-Suppression Techniques

SST Spacecraft Systems Test

SST Spread-Spectrum Technology

SST Stainless Steel [also SSt]

SST Stationary-State Treatment [Physics]

SST Standard Stereographic Triangle

SST Stress-Stiffening Technique

SST Strong-Segregation Theory [Metallurgy]

SST Structural Static Test

SST Subscriber Transferred

SST Subsystems Test

SST Subsystem Terminal (on Spacelab)

SST Superconductor Science and Technology [Journal of Institute of Physics, UK]

SST Super Sharp Tube

SST Supersonic Telegraphy

SST Supersonic Transport

SST Symmetric Single Trapping

SST Synchronous System Trap

SST System Segment Table

SST Systems Services and Technology

SS/T Steady-State/Transient (Analysis)

Sst Streptomyces stanford [Microbiology]

Sst I Streptomyces stanford I (One) [Microbiology]

Sst II Streptomyces stanford II (Two) [Microbiology]

S(σ,T) (Uniaxial) Stress and Temperature Dependent Thermopower [Symbol]

S Staff South Staffordshire [UK]
SSTC Space Shuttle Test Conductor [NASA]
SSTC Single Sideband Transmitted Carrier
SSTC Solid-State Track Recorder
SSTDMA Satellite-Switched Time-Division Multiple Access [also SS/TDMA, or SS-TDMA]
SSTF Shortest Seek Time First
SSTI Secure Services Technology, Inc. [US]
SSTI Subsynchronous Torsional Interaction
SSTO Single Stage to Orbit
SSTP Subsystems Test Procedure
SSTR Solid-State Track Recorder
SSTS Space Shuttle Transportation System [of NASA]
SSTTS Space Shuttle Television Transmission Service [of NASA]
SSTV Sea Skimming Test Vehicle [US Navy]
SSTV Slow-Scan Television
SSU Saybolt Seconds Universal [Unit]
SSU Secondary Sampling Unit
SSU Sequential Shunt Unit
SSU Session Support Utility
SSU Single Signaling Unit
SSU Species Services Unit [of Bureau of Meteorology, Australia]
SSU Stout State University [Menomonie, Wisconsin, US]
SSU Stratospheric Sounding Unit
SSU Subscriber Switching Unit
SSU Subsequent Signal Unit
SSU Supply and Services Union
SSUD Secret Service Uniformed Division [of US White House, Washington, DC]
SSUDACI Space Station User Development Atlantic Canada Initiative [of University of Prince Edward Island, Charlottetown]
SSUS Spinning Solid Upper Stage [Aerospace]
SSUS-A SSUS for Atlas-Centaur Class Spacecraft
SSUS-D SSUS for Delta Class Spacecraft
SSUSP SSUS Project
SSV Settled Sludge Volume
SSV Ship-to-Surface Vessel (Radar)
SSV Space Shuttle Vehicle
SSV Standby Safety Vessel
SSV Supersaturated Vapor [Physical Chemistry]
SSV Supersonic Vehicle
SSVW Symmetric Shear Vertical Wave
SSW Simple Spin Wave (Theory) [Solid-State Physics]
SSW Solid-State Welding
SSW South-Southwest; South by Southwest
SSW Space Switch [Aerospace]
SSW Standing Spin Wave [Solid-State Physics]
SSW Steady Shock Wave [Physics]
SSW Substitute Seawater
SSW Sudden Stratospheric Warming [Meteorology]
SSW Surface Science Western [of University of Western Ontario, Canada]

SSW Synchro Switch
SSW Systems Software [also SS]
SSWA Sanitary Supply Wholesalers Association [US]
SSWG Supplementary Strategies Working Group [of Australian and New Zealand Environment and Conservation Council]
SSWIA Stainless Steel Wire Industry Association [UK]
SSWO Special Service Work Order
SSWS Static Safe Work Station
SSWWS Seismic Sea-Wave Warning System
SSX Small System Executive
S+SX Speech plus Simplex [also S+S] [Telecommunications]
SSY Small-Scale Yielding [Mechanics]
S/SYS Subsystem [also S/Sys]
SSZ Society of Systematic Zoology [US]
ST Saffman-Taylor Effect (in Metallurgy) [also S-T]
ST Sample Treatment
ST Sao Tome and Principe [ISO Code]
ST Saturation Temperature
ST Saturation Transfer
ST Sawtooth
ST Schiff Test [Analytical Chemistry]
ST Schmitt Trigger [Electronics]
ST Science and Technology
ST Science Team
ST Scientific and Technical
ST Screw Thread
ST Secondary Twin [Metallurgy]
ST Secretary Treasurer
ST Seed Technologist; Seed Technology
ST Segment between QRS Complex and T-Wave (in Electrocardiogram) [also S-T] [Medicine]
ST Segment Table
ST Selective (Gear) Transmission
ST Selectric Typewriter
ST Select Time
ST Self-Test
ST Semiconductor Technology
ST Separation Technology
ST Sequential Timer
ST Serial (Data) Transmission
ST Series Transformer
ST Serous Tissue [Anatomy]
ST Shear Transformation [Metallurgy]
ST Shiatsu Therapist; Shiatsu Therapy [*Note:* "Shiatsu" is a Japanese Massage Technique]
ST Ship Track (Mission)
ST Shock Test(ing)
ST Shock Tube
ST Short Term
ST Short-Transverse (Direction) [also S-T]
ST Sidetone
ST Signaling Tone [Telecommunications]
ST Similarity Transformation [Mathematics]

ST Similitude Theory [Physics]
ST Simplification Task
ST Single-Throw (Switch)
ST Single-Tone (Amplitude Model) [Telecommunications]
ST Sintered Thoria
ST Skin Temperature
ST Slip Trace (Analysis) [Metallurgy]
ST Solar Temperature
ST Solar-Terrestrial
ST Solar Time [Astronomy]
ST Solar Transit
ST Solution Treated; Solution Treatment [Metallurgy]
ST Sorption Trap
ST Sounding Tube
ST Sound Telegraphy
ST Sound Track
ST Sound Transmission; Sound Transmitter
ST Sound Trap [Television]
ST Southern Tablelands [New South Wales, Australia]
ST Spacelab Technology
ST Space Technology
ST Space Telescope
ST Space-Time [Relativity]
ST Space Travel
ST Special Tooling
ST Speech Technologist; Speech Technology
ST Spin–Pseudospin (Model) [Solid-State Physics]
ST Spin Temperature [Solid-State Physics]
ST Spring Tide
ST Stabilizer Tube
ST Standard Time
ST Standard Ton [Unit]
ST Standard Treatment
ST Star Time
ST Starting Time
ST Starting Transient
ST Star Tracker [Aerospace]
ST Start Signal
ST Static Thrust
ST Steam Tables
ST Steam Turbine
ST Steel Technologist; Steel Technology
ST Stereo-Telescope
ST Stishovite [Mineral]
ST Storage Tank
ST Storage Tube
ST Straflo Turbine
ST Stratigraphic Trap [Geology]
ST Stream Tube [Fluid Mechanics]
ST Street Tee (Pipe Fitting)
ST Strontium Titanate
ST Structural Tee (Cut from S-Type Steel Shape) [AISI/AISC Designation]
ST Structural Textile

ST Studio-to-Transmitter (Radio Station)
ST Subscriber Terminal
ST Substrate Transfer
ST Sucrose Tallowate
ST Suomalainen Tideakatemia [Finnish Academy of Sciences, Helsinki]
ST Suomen Tasavalta [Republic of Finland]
ST Suomen Tiedeseuro [Finnish Scientific Society]
ST Superconductor Technology
ST Super-Tough
ST Supplementary Terms
ST Surface Technology
ST Surface Tension
ST Surge Tank
ST Surveying Technician
ST Survey Technology
ST Switching Technique
ST Symmetrical Tilt [Physics]
ST Symbol Table
ST Synchronous Transmission
ST Syntax Tree
ST Systems Technology
ST System Table
ST- Sudan [Civil Aircraft Marking]
S(T) Temperature Dependent Long-Range Order Parameter [Symbol]
S(T) Thermopower [Symbol]
S$_{mag}$(T) Magnetic Thermopower [Symbol]
S&T Science and Technology; Scientific and Technical
S-T Saffman-Taylor Effect (in Metallurgy) [also ST]
S/T Search/Track
S/T Set, or Toggle
S/T Specific Heat/Temperature (Curve)
S-T Axisymmetric Stress State [Mechanics]
S-T Segment between QRS Complex and T-Wave (in Electrocardiogram) [also ST] [Medicine]
S-T Short-Transverse (Direction) [also ST]
7T- Algeria [Civil Aircraft Marking]
St Saint [Geography]
St Start(ing); Starter [also st]
St State
St Static(s) [also st]
St Station [also st]
St Statue [also st]
St Status [also st]
St Statute [also st]
St Steam [also st]
St Steel
St Stefan Number (of Heat Transfer) [Symbol]
St Stimulus
St Stitch
St Stirling (Formula) [Mathematics]
St Stoke [Unit]
St Stokes Number 1 (of Fluid Dynamics) [Symbol]

St Stone

St Storage

St Store

St Strait

St Stratus (Cloud) [Meteorology]

St Street

St Structure; Structural [also st]

St Styrene

St$_2$ Stokes Number 2 (for Rotameters) [Symbol]

S(t) ICDTS (Isothermal Capacitance Transient Spectroscopy) Spectrum [Symbol]

S(t) Mean Translational Entropy [Symbol]

S(t) ST PAM (Single-Tone Pulse-Amplitude Modulation) Signal [Symbol]

S$_N$(t) Survival Probability [Symbol]

s^{-1}T^{-1} one per second tesla [also 1/s·T or 1/(s·T)]

s(T) single-step free energy excess [Symbol]

st stitch [Woven Fabrics]

st stone [Unit]

st store

/st stack option [MS-DOS Linker]

s t short ton [Unit]

s(t) (voltage) signal [Symbol]

Σ3^n Low-Sigma (Grain Boundary) (in Materials Science) [Symbol]

ΣT Sigma Tau (Student Society)

Σ(θ) Tilt Angle Dependent Surface Free Energy [Symbol]

σ(T) temperature dependent (electrical) conductivity [Symbol]

σ(T) yield stress [Symbol]

σ$_0$(T) temperature dependent critical shear stress [Symbol]

σ(t) time-dependent stress [Symbol]

σ(Θ) scattering cross section (in Rutherford backscattering spectrometry) [Symbol]

σ(θ) angle-dependent surface tension [Symbol]

6T Sexithiophene [Biochemistry]

STA Atlas of Crystal Structure Types [Crystallography]

STA Meteorological Station

STA Sail Training Association [US]

STA Saskatchewan Trucking Association [Canada]

STA Science and Technology Advisor

STA Science and Technology Agency [Japan]

STA Shiatsu Therapy Association [*Note:* "Shiatsu" is a Japanese Massage Technique]

STA Short-Time Annealing [Metallurgy]

STA Shuttle Training Aircraft [of NASA]

STA Silicotungstic Acid

STA Simultaneous Thermal Analysis; Simultaneous Thermal Analyzer

STA Slurry Transportation Association

STA Society of Typographic Arts [US]

STA Solanum Tuberosum Agglutinin [Immunology]

STA Solar Trade Association [UK]

STA Solution (Heat) Treated and Aged (Condition) [also ST&A] [Metallurgy]

STA Spanning Tree Algorithm

STA Static Test Article

STA Steel-Tape-Armored (Cable)

STA Store Answer

STA Straight-In Approach

STA Structural Test Article

STA Supersonic Tunnel Association [US]

ST&A Solution (Heat) Treated and Aged [also STA] [Metallurgy]

Sta Statine (or 3S,4S-4-Amino-3-Hydroxy-6-Methylheptanoic Acid)

Sta Station(ary) [also sta]

Sta Statinyl

STAAS Surveillance and Target Acquisition Aircraft System

Stab Stabilization; Stabilize(r) [also stab]

STAB Surface Treatment for Aluminum Bonding

STABEX Stabilization of Export Earnings

STABLE Suppression of Transient Acceleration by Levitation Evaluation

STAC Science and Technology Advisory Committee [US]

STAC Software Timing and Control

STAC Space Telescope Action Center

STACI Space Technology Atlantic Canada Initiative [Halifax, Nova Scotia, Canada]

STACO Standing Committee

STDAS Structural Test Data Acquisition System

STAD Start Address

STADAC Station Data Acquisition and Control

STADAN Satellite Tracking and Data Acquisition Network; Space Tracking and Data Acquisition Network [of NASA] [also Stadan]

STADN Space Tracking and Acquisition Network

STADU System Termination and Display Unit

STAE Second-Time-Around Echo [Radar]

STAE Specify Task Asynchronous Exit

STAEDT Simulation Tool for Analysis and Evaluation of Data-Signal Transmission

STAESA Society of Turkish Architects, Engineers and Scientists in America [US]

STAF Scientific and Technological Applications Forecast

STAF Strategic Air Force Staffs [Staffordshire, UK]

STAFDA Specialty Tools and Fasteners Distributors Association [US]

STAFF Stafford (Student) Loan [US]

STAFF Stellar Acquisition Flight Feasibility

Staff Staffordshire [also Staffs, UK]

STAG Science and Technology Advisory Group [Taiwan]

STAG Shuttle Turnaround Analysis Group [of NASA]

STAG Soils, Trees and Grass Program [of Commonwealth Scientific and Industrial Research Organization, Australia]

STAG Steam and Gas Turbine

STAGG Small Turbine Advanced Gas Generator

STAGS Structural Analysis of General Shells

Stahl Eisen Stahl und Eisen [Publication on Iron and Steelmaking of Verein Deutscher Eisenhüttenleute, Germany]

STAI Subtask ABEND (Abnormal Ending) Intercept

Stainl Steel High Perform Alloys Stainless Steel High Performance Alloys [South Africa]

Stainl Steel Ind Stainless Steel Industry [Journal published in the UK]

Stainl Steels Dig Stainless Steels Digest [of ASM International, US]

Stain Technol Stain Technology [Journal published in the US]

STAIR Structural Analysis Interpretive Routine

STAIRS Storage and Information Retrieval System

STAIRS/VS Storage and Information Retrieval System/Virtual Storage

STAJ Science and Technology Agency of Japan

Stal' Izv Vyssh Uchebn Zaved Chern Metall Izvestiya Vysshikh Uchebnykh Zavedenii, Chernaya Metallurgiya [Journal on Steel in the USSR]

STALO Stable Local Oscillator [also Stalo or stalo] [Radar]

STALO/COHO Stable Local Oscillator/Coherent Oscillator [also Stalo/Coho, or stalo/coho] [Radar]

Stalpeth Grooved Steel Strip Sheath–Grooved Aluminum Strip Sheath–Sprayed PET (Polyethylene Terephthalate) Coating (Cable) [also stalpeth]

STAM Shared Tape Allocation Manager

ST AM Single-Tone Amplitude Modulation [Telecommunications]

Sta Mi Statute Mile [0.86842 Nautical Miles] [also sta mi]

STAMO Stabilized Master Oscillator

STAMOS Sortie Turnaround Maintenance Operations Simulation

STAMP Systems Tape Addition and Maintenance Program

Stamp Stamping [also stamp]

Stamp Q Stamping Quarterly [US]

STAMPS Scientific and Technological Aspects of Materials Processing in Space

Stan Stanchion [also stan]

Stan Standard(ization) [also stan]

STANAG Standardization Agreement [of NATO]

STANAVFORLANT Standing Naval Force Atlantic [of NATO]

Stand Standard [also stand]

Stand Eng Standards Engineering [of Standards Engineering Society, US]

Stand News Standardization News [of American Society for Testing Materials, US]

Stanki Instrum Stanki i Instrumenty [Russian Journal on Engineering Research–Instruments and Equipment]

STAO Science Teachers Association of Ontario [Canada]

STAP Science and Technology Action Plan

STAP Statistics Panel [of International Civil Aviation Organization]

Staph Staphylococcus [also staph] [Microbiology]

STAPPA State and Territorial Air Pollution Program Administrators [US]

STAR Satellite Telecommunications with Automatic Routing

STAR Scientific and Technical Aerospace Reports [of NASA]

STAR Self-Defining Text Archival

STAR Self-Testing and Repairing (Computer)

STAR Serial Transmitter and ROM (Read-Only Memory)

STAR Shield Test Air Reactor

STAR Ship-Tended Acoustic Radar

STAR Shuttle Turnaround Analysis Report [of NASA]

STAR Silicon TARget (Tracker) [of CERN–European Laboratory for Particle Physics, Geneva, Switzerland]

STAR Simulator Training Acceptance Review

STAR Solenoidal Tracker At RHIC (Relativistic Heavy Ion Collider) (Experiment) [of Brookhaven National Laboratory, Upton, New York, US]

STAR Space Thermionic Auxiliary Reactor

STAR Speed through Air Resupply

STAR Standard Terminal Approach Route

STAR Standard Terminal Arrival Route

STAR Star and Stellar Systems Advisory Committee [of European Space Research Organization]

STAR Steam Active Reforming (Process)

STAR Steerable Array Radar

STAR Stellar Attitude Reference

STAR Student Team on Alumni Relations

STAR Studies, Tests and Applied Research

STAR String Array Processor

STAR Submarine Test and Research

STAR System to Automate Records

STARAD Starfish Radiation Satellite [of NASA]

STARAN Stellar Attitude Reference and Navigation [also Staran, or staran]

STARE Steerable Telemetry Antenna Receiving Equipment

STARFIRE System to Accumulate and Retrieve Financial Information with Random Extraction

Star Metall Splav Starenie Metallicheskik Splavov [Russian Journal on Metallic Alloys]

STARP Strategic Technologies in Automation and Robotics Program

STARR Schedule, Technical and Resources Report

StaRRCar Self-Transit Rail and Road Car [also StaRRcar]

STARS Satellite Telemetry Automatic Reduction System [of NASA]

STARS Satellite Transmission and Reception Specialist

STARS Shuttle Traceability and Reporting System [of NASA]

STARS Software Technology for Adaptable, Reliable Systems (Project) [of US Department of Defense] [also Stars]

STARS Stellar Seismology Mission [of European Space Agency]

STARS Supplying Trends and Rating System

START Selections to Active Random Testing

START Short Term Aid for Research and Technology

START Spacecraft Technology and Advanced Reentry Test

START Special Treatment and Rehabilitative Training (Program) [US]

START Strategic Arms Reduction Talks [Between the US and the USSR]

START Success Through Assisted Reproductive Technologies [Toronto, Canada]

START Summit Technology and Research Transfer Center

START Systematic Tabular Analysis of Requirements Technique

START System for Analysis, Research and Training [of International Geosphere/Biosphere Program]

STARTS Software Tools for Application to Large Real-Time Systems

STAT N-(4-Methylphenyl)-2-[(4-Methylphenyl)sulfonyl amino]benzenecarbothiamide, Ion (1−)

STAT Slotted Tube Atom Trap

STAT Shipping Transit Analysis Tabulation

Stat Static(s) [also stat]

Stat Station [also stat]

Stat Stationery [also stat]

Stat Statistical; Statistician; Statistics [also stat]

Stat Statue [also stat]

Stat Statute [also stat]

Stat Status [also stat]

stat *(statim)* – immediately, at once, or without delay [Medical Prescriptions]

stat stationary

stat- prefix of an electrical unit in the centimeter, gram and second (cgs) system of units (e.g. statA, statC, etc.)

statA statampere [Unit]

statC statcoulomb [Unit]

StatCan Statistics Canada

Stat Comp Statistical Computing

STATE Simplified Tactical Approach and Terminal Equipment

State Geol J State Geological Journal [of Association of American State Geologists, US]

State World Popul State of the World's Population [Annual Report of United Nations Fund for Population Activities]

statF statfarad [Unit]

STATFOR Specialist Panel on Air Traffic Statistics and Forecasts [of Eurocontrol]

STAT-H N-(4-Methylphenyl)-2-[(4-Methylphenyl)sulfonyl amino]benzenecarbothiamide

statH stathenry [Unit]

Statis Statistical; Statistician; Statistics [also statis]

Stat J UN Econ Comm Eur Statistical Journal of the United Nations Economic Commission for Europe [Netherlands]

STATLAB Statistics Laboratory [also Statlab]

STATLIB Statistical Library [also Stablib]

stat℧ statmho [Unit]

Stat Mech Statistical Mechanics

STAT MUX Statistical Multiplexer [also Stat Mux]

statΩ statohm [also stat-ohm] [Unit]

statΩ⁻1 statmho [also stat-mho] [Unit]

STATPAC Statistics Package

Stat Phys Statistical Physicist; Statistical Physics

Stat Plan Statistical Planning [also stat plan]

Stat Prob Statistical Probability [also stat prob]

Stat Res Statistical Research(er) [also stat res]

statS statsiemens [Unit]

StatsCan Statistics Canada

Stat Sci Statistical Science [also stat sci]

StatSci Statistical Science [Publication of the Institute of Mathematical Sciences, US]

Stat Softw Newsl Statistical Software Newsletter [of Gesellschaft für Strahlen- und Umweltforschung, Germany]

statT stattesla [Unit]

STAT-USA Statistics-USA [Statistical Database of the US Department of Commerce on Economics, Business and Trade]

statV statvolt [Unit]

statWb statweber [Unit]

Staub Reinhalt Luft Staub–Reinhaltung der Luft [German Journal on Dust and Air Pollution Abatement published by VDI Berufsgenossenschaftliches Institut für Arbeitssicherheit]

STB Save the Bush Project [of Australian Nature Conservancy Agency]

STB Scheldahl Thermal Blanket

STB Segment Tag Bit

STB Simple Two-Band Model [Solid-State Physics]

STB Sodium Trimethyl Borohydride

STB Solid Target Boronization

STB Stock Tank Barrel [also stb]

STB Stop Bar

STB Strobe

STB Subsystems Test Bed

STB Sumitomo Top and Bottom Blowing (Process) [Metallurgy]

STB Symmetrical Tilt (Grain) Boundary) [Materials Science]

Stbd Starboard [also stbd]

STB/D Stock Tank Barrels per Day [also STB D⁻¹, stb/d, or stb d⁻¹]

STBL Stratus-Topped Boundary Layer [Meteorology]

stbl stable

STBR Stirred Tank Bioreactor

STBS Stirred Tank Biological Reactor

STBY Standby [also Stby, or stby]

STC Sample Treatment Center

STC Sample Treatment Chamber

STC Satellite Test Center [of US Air Force]

STC Satellite Telecommunications Center

STC Science and Technology Center

STC Science and Technology Centers (Program) [of National Science Foundation, US]

STC Scientific and Technical Committee [of European Space Research Organization, France]

STC Self-Tuning Controller

STC Sensitivity Time Control [Radar]

STC Set Carry Flag [Computers]

STC SHAPE (Supreme Headquarters, Allied Command Europe) Technical Center [of NATO]

STC Short-Term Cycle (Annealing Furnace) [Metallurgy]

STC Short Time Constant

STC Sliding Twin-Crossbar

STC Society for Technical Communication [US]

STC Society of Telecommunications Consultants [US]

STC Sodium Hydroxy Tricarboxylate

STC Sound Transmission Class

STC South Texas College [Houston, Texas, US]

STC Spacecraft Test Conductor

STC Space Technology Center

STC Specimen Transfer Chamber

STC Stacked Capacitor

STC Staff Training Center

STC Station Technical Control

STC Standard Telephone Cable

STC Standard Telephones and Cables

STC Standard Test Configuration

STC Standard Transmission Code

STC St. Cloud, Minnesota [Meteorological Station Identifier]

STC Structural Thermoplastic Composite

STC Superconductivity Technology Center [of Los Alamos National Laboratory, New Mexico, US]

STC Supplemental Type Certificate

STC Surgical Textiles Conference [UK]

STC System(s) Test Complex

S&TC Science and Technology Center [of Westinghouse Corporation, Pittsburgh, Pennsylvania, US]

STCA Short-Term Conflict Alert System

STCB Subtask Control Block

STCC Spacecraft Technical Control Center

STCC Standards Council of Canada

STCL Source-Term Control Loop [of Hanford Engineering Development Laboratory, Washington, US]

STCP Short-Term Cost Plan

STCQES Science and Technology Center for Quantized Electronic Structures

STC RAM Stacked Capacitor Random-Access Memory

STCS Science and Technology Center for Superconductivity [of National Science Foundation at Northwestern University, Evanston, Illinois, US]

STCW (International Convention on) Standards of Training, Certification and Watchkeeping for Seafarers

STCW System Time Code Word [Computers]

STD Salinity-Temperature-Depth (Recorder) [Oceanography]

STD Separation Technology Division

STD Set Direction Flag [Computers]

STD Sexually Transmitted Disease [Medicine]

STD Shear Trough Diffuser

STD Shuttle Test Director [of NASA]

STD Small Test Detector

STD Society of Typographic Designers [UK]

STD Statistical Time Division

STD Stepwise Thermal Desorption

STD Subscriber Trunk Dialing [UK]

STD Surface Thermodynamics

(S)TD Secondary Transmitted Data

Std Standard [also std]

STDA Selenium-Tellurium Development Association [also S-TDA, US]

STDA Solution Treated and Double Aged; Solution Treating and Double Aging [Metallurgy]

STDA StreetTalk Directory Assistance

std atm standard atmosphere [Unit]

STDAUX Standard Auxillary [also StdAux]

STDB Singapore Trade Development Board

STDC Standards Council of Canada

STDCE Surface Tension Driven Convection Experiment

std cm³/s standard cubic centimeters per second [Unit]

std cu ft standard cubic foot [Unit]

Std Dev Standard Deviation [also std dev]

Std Err Standard Error [also std err]

STDI Selenium and Tellurium Development Institute [US]

STDIN Standard Input [also stdin]

STDIOH Standard Input/Output Header [C Programming Language]

STDL Standard Distribution List [also STD L]

STDM Statistical Time-Division Multiplex(ing); Statistical Time-Division Multiplexer

STDM Synchronous Time-Division Multiplexing

std m³/hr standard cubic meters per hour [Unit]

STDN Space(flight) Tracking and Data Network [of NASA]

STDOUT Standard Output [also stdout]

STDP Single-Throw Double-Pole (Switch)

STDPRN Standard Printer

Std Ref Standard Reference

Std Res Standard Resolution

STDS System for Thermal Decomposition Studies

Std Spec Standard Specification

STDTT Strain Transient Dip Test Technique [Metallurgy]

Stdt Student [also stdt]

stdy steady

stdz standardize

Stdzn Standardization [also stdzn]

STE Sediment Transport Equation [Hydrology]

STE Self-Trapped Exciton(s) [Solid-State Physics]

STE Shield Test Experiment [US]

STE Society of Telecom Executives [UK]

STE Society of Test Engineers [UK]

STE Society of Tractor Engineers [US]

STE Spacecraft Test Equipment

STE Span Terminating Equipment

STE Special Test Equipment

STE Standard Telephone Equipment

STE Stearate (Crystal)

STE Sterolester

STE Stimulated Spin-Echo [Physics]

STE Supergroup Translating Equipment [Telecommunications]

STE System Test Engineer

St+E Stahl und Eisen [Publication on Iron and Steelmaking of Verein Deutscher Eisenhüttenleute, Germany]

Ste *(Sainte)* – French for "Saint" [Geography]

S2E Surfaced on Two Edges [Lumber]

S(θ,E) Neutron Scattered Intensity [Symbol]

STEAG Steinkohlen Elektrizität AG (Prozess) [Electricity from Coal (Process)]

STEAM Stimulated Echo Acquisition Mode

STEAM Streptonigrin, Thioguanine, Endoxan, Actinomycin D, Mitomycin C [Anti-Cancer Medication]

STEAR Strategic Technologies in Automation and Robotics (Program) [of Canadian Space Agency]

STEB Standard Test and Evaluation Bottle [of Air Force Materials Laboratory, US]

STEBIC Scanning Transmission Electron Beam Induced Current (Method)

STEC Short-Term Exposure Criteria

STEC Surface Treatment Enhancement Council

STEC Solar-(to-)Thermal Energy Conversion

ST-ECF Space Telescope–European Coordinating Facility [of European Southern Observatory, Garching/Munich, Germany]

STECR Space Telescope Engineering Console Room

STED Scanning Transmission Electron Diffraction

STED Solar Turboelectric Drive

STEDCON Conference on Science and Technology of Electron Devices

Steel Constr Steel Construction [South African Publication]

Steel Constr Today Steel Construction Today [UK Publication]

Steel Dig Steel Digest [Published in the US]

Steel Founders Res J Steel Founders Research Journal [of Steel Founders Society of America]

Steel Furn Mon Steel Furnace Monthly [Indian Publication]

Steel Ind Jpn Ann Steel Industry of Japan Annual

Steelmaking Proc Steelmaking Proceedings [Publication of AIME Iron and Steel Society, US]

Steel Mark News Steel Market News [Published in Sweden]

Steel Res Steel Research [Journal published in Germany]

Steel Technol Int Steel Technology International [Journal published in UK]

Steel Times Int Steel Times International [Journal published in the UK]

Steel Today Tomorrow Steel Today and Tomorrow [Journal published in the Japan]

Steel USSR Steel in the USSR [Translation of: *Stal': Izvestiya Vysshikh Uchebnykh Zavedenii, Chernaya Metallurgiya*; published in the UK]

Steer Steering [also steer]

STEG Société Tunisienne de l'Electricité et du Gaz [Tunesian Electricity and Gas Company]

TEG Solar Thermoelectric Generator

Steklo Keram Steklo i Keramika [Russian Journal on Glass and Ceramics]

STEL Short-Term Exposure Limit(s)

STELLA Satellite Transmission Experiment Linking Laboratories

STEM Scanning Transmission Electron Microscope; Scanning Transmission Electron Microscopy

STEM Soluble Trace Element(s)

STEM Stay Time Extension Module [of NASA]

STEM Storable (or Stored) Tubular Extendable Member [Solar Cells]

STEM-BF Scanning Transmission Electron Microscopy–Bright Field [also STEM/BF]

STEM-DF Scanning Transmission Electron Microscopy–Dark Field [also STEM/DF]

STEM-EDS Scanning Transmission Electron Microscopy–Energy-Dispersive (X-Ray) Spectrometry [also STEM/EDS]

STEM-EDX Scanning Transmission Electron Microscopy–Energy-Dispersive X-Ray Analysis [also STEM/EDX]

STEM-EELS Scanning Transmission Electron Microscopy–Electron Energy Loss Spectros-copy [also STEM/EELS]

STEM-TEM Scanning Transmission Electron Microscopy–Transmission Electron Microscopy [also STEM/TEM]

ST Eng Space Telescope Engineer

STEP Safety Test Engineering Program [of US Atomic Energy Commission]

STEP Saskatchewan Trade and Export Partnerhip, Inc. [Canada]

STEP Satellite Test of the Equivalence Principle [of European Space Agency]

STEP Scientific and Technical Exploitation Program

STEP SEMI (Semiconductor Equipment and Materials Institute) Technical Education Program [US]

STEP Simple Transition to Electronic Processing

STEP Space Technology Experiments Platform

STEP Specialized Technique for Efficient Typesetting

STEP Standard Exchange of Product Model Data

STEP Standard for the Exchange of Product Data

STEP Standard Tape Executive System

STEP Standard Terminal Program

STEP Stratosphere-Troposphere Exchange Project

STEP Summer Temporary Employment Program [Canada]

STEP Supervisory Tape Executive Program

STEP Support for Technology-Enhanced Productivity

STEPP Society of Teachers in Education of Professional Photography [US]

STEPR Saturation Transfer Electron Paramagnetic Resonance

STEPS Solar Thermionic Electric Power System

STEPS Superconducting Technology for Electric Power System (Program) [of US Department of Energy]

STER Solar Terrestrial Environmental Research

Ster Sterilize(r) [also ster]

Ster Sterling [also ster]

sterad steradian [also sr]

STERAO Stratospheric-Tropospheric Experiments: Radiation, Aerosols and Ozone

Stereo Stereophonic(s); Stereophony [also stereo]

Stereo Stereo Sound System [also stereo]

Stereo Stereoscope; Stereoscopic; Stereoscopy [also stereo]

Stereo Stereotype [also stereo]

Stereot Funct Neurosurg Stereotactic and Functional Neurosurgery [International Journal]

STESR Saturation Transfer Electron Spin Resonance [Physics]

STET Specialized Technique for Efficient Typesetting

STEV Short-Term Exposure Value

St Ex Stock Exchange [also St Exch]

STF Safety Test Facility

STF Saskatchewan Teachers Federation [Canada]

STF Satellite Tracking Facility [US Air Force]

STF Sectoral Training Fund [of Ontario Ministry of Education and Training, Canada]

STF Serum Thymic Factor [Biochemistry]

STF Shield Test Facility (Reactor) [US]

STF Short Title File

STF Slater-Type Function [Physics]

STF Some Test Failed

STF Spin Test Facility [Physics]

STF Stacking Fault [Crystallography]

STF Strain-to-Failure [Fracture Mechanics]

STF Stratiform

STF Structural Fatigue Test [Mechanics]

STF Superconducting Thin Film

STF System Transfer Function [Control Engineering]

.STF Structured File [Lotus Agenda File Name Extension]

Stf Staff [also stf]

St(f) Steel Fiber

STFA Secondary Transverse Field Anneal(ing) [Solid-State Physics]

STFA Stress and Transverse Field-Induced (Magnetic) Anisotropy [Solid-State Physics]

STFF Sodium Tripolyphosphate

STFI Sächsisches Textilforschungsinstitut [Saxon Institute of Textile Research, Germany]

STFI Skogsindustrins Tekniska Forskningsinsitut [Swedish Pulp and Paper Research Institut]

STFI Strategic and Tactical Fire Initiation

STFI Medd Skogsindustrins Tekniska Forskningsinsitut Meddelande [Publication of the Swedish Pulp and Paper Research Institute]

STFLT Shuttle Telescope Flight [NASA Widefield Planetary Camera Mission]

STG Sawtooth Generator

STG Space Task Group [of NASA]

Stg Stage [also Stg, or stg]

Stg Starting [also stg]

Stg Sterling [also stg]

Stg Storage [also stg]

stg strong

STGB Symmetrical Tilt Grain Boundary [Materials Science]

Stge Storage [also stge]

S-2-Glass High-Strength Glass (for Commercial Applications)

STGS South Texas Geological Society [US]

STGWG State and Tribal Government Working Group [US]

STH Self-Trapped Holes [Solid-State Physics]

STH Somatotropic Hormone [Biochemistry]

S(θ,H) Angle of Incidence and Magnetic Field-Dependent Signal [Symbol]

6th sixth

7th seventh

STHF Supertransferred Hyperfine Field [Solid-State Physics]

SThM Scanning Thermal Microscopy

sthrn southern

STHO Strontium Titanyl Hydroxy Oxalate

STI Safety Training Institute [US]

STI Scientific and Technical Information [also ST&I]

STI Screw Thread Insert

STI Service Tools Institute [now Hand Tools Institute, US]

STI Spatiotemporal Intermittency [Physics]

STI Special Thread for Helical Coil Wire Screw Thread Inserts [Symbol]

STI Speech Transmission Index

STI Steel Tank Institute [US]

STI Steel Tube Institute [US]

STI Strategic Transition Initiatives Division [of US Department of Energy]

STI Superconductor Technologies Inc. [Santa Barbara, California, US]

STI Swiss Tropical Institute [Basle]

STI Synergistic Technologies Inc. [of Research Triangle Park, Durham, North Carolina, US]

S&TI Scientific and Technical Information [also STI]

STIA (Directorate for) Scientific, Technological and International Affairs [of National Science Foundation, US]

STIB Stratosphere-Troposphere Interactions and the Biosphere

STIBC Society of Translators and Interpreters of British Columbia [Canada]

STI Bull Scientific and Technical Information Bulletin [of NASA]

STIC Scientific and Technical Intelligence Center [of US Department of Defense]

STICS Scanning Thermal Ion Composition Spectrometer

STID Scientific and Technical Information Division [of NASA]

Stien Stilbenediamine [also stien]

STIF Scientific and Technical Information Facility

STIF Short-Term Irradiation Facility

Stiff Stiffener [also stiff]

STIKSCAT Stick Scatterometer

STIL Short-Term Inhalation Limit

STIL Software Test and Integration Laboratory

STIM Scanning Transmission Ion Microscopy

STIMS Sputtering (Technique) with Thermo-Ionization Mass Spectrometry

STIN Science and Technology Information Network

ST-IN Straight-In (Approach)

STINA Steel Tube Institute of North America [US]

STINFO Scientific and Technical Information

STINFO Scientific and Technical Information Officers [US]

STINGS Stellar Inertial Guidance System

STIP Science and Technology Internship Program [of Natural Resources Canada]

STIP Science Teaching Improvement Program [US]

STIP Scientific and Technical Information Processing

Stip Stipend(iary)

STIR Shield Test Irradiation Reactor [US]

STIR Statistics Indexing and Retrieval Project [of Loughborough University of Technology, UK]

Stir Stirring [also stir]

STIRAP Stimulated Raman Adiabatic Passage

STIRD SAIL (Shuttle Avionics Integration Laboratory) Test Implementation Requirements Document [of NASA]

STIS Space Telescope Imaging Spectrograph [Hubble Space Telescope]

STIS Specialized Textile Information Service [UK]

S/TK Sectors per Track [also S/tk]

Stk Stack [also stk]

Stk Stock [also stk]

Stk Strake [also stk]

Stk Exch Stock Exchange

Stk No Stock Number

STKOUT Strikeout [also Stkout, or stkout]

Stkyd Stockyard [also Stkyd, or stkyd]

STL Schottky Transistor Logic

STL Sound Transmission Loss

STL Space Technology Laboratory [of US Air Force at Inglewood, California, US]

STL Stacked Triangular Lattice [Crystallography]

STL Standard Telecommunication Laboratories [Harlow, Essex, UK]

STL Standard Telegraph Level

STL Standard Template Library

STL Statistical Time Lag

STL Strip Line [Electrical Engineering]

STL Studio-to-Transmitter Link

STL Studio Transmitter Link

STL System Test Loop

Stl Steel

Stl Stilbene [also stl]

stl structural

STLB Schottky TTL (Transistor-Transistor Logic) Bipolar

STLD Society of Television Lighting Directors [UK]

STLE Society of Tribologists and Lubrication Engineers [US]

STLE Spec Publ STLE Special Publication [of Society of Tribologist and Lubrication Engineers, US]

STLE Tribol Trans STLE Tribology Transactions [of Society of Tribologist and Lubrication Engineers, US]

STLHE Society for Teaching and Learning in Higher Education

STLO Scientific and Technical Liaison Office

STLOS Star Line of Sight

STLV Simian T-Lymphotropic Virus

Stl WG Steel Wire Gauge

STM Scanning Tunneling Microscope; Scanning Tunning Microscopy

STM Science, Technology and Medicine; Scientific, Technical and Medical [also ST&M]

STM (International Group of) Scientific, Technical and Medical Publishers [Netherlands]

STM Service Test Module

STM Short-Term Memory

STM Signal Termination Module

STM Standard Thermal Model

STM Standard Trapping Model [Materials Science]

STM Statement [also Stm, or stm]

STM Static Test Model

STM Structural Test Model

STM Structural Thermal Model

STM Supersonic Tactical Missile

STM Support Test Manager

STM Synchronous Transfer Mode

ST&M Science, Technology and Medicine; Scientific, Technical and Medical [also STM]

Stm Steam [also stm]

Stm Storm [also stm]

STM/AFM Scanning Tunneling Microscope/Atomic Force Microscope; Scanning Tunneling Microscopy/Atomic Force Microscopy

S2MC (Laboratoire de) Science des Surfaces et Matériaux Carbones [Surface Science and Carbonaceous Materials (Laboratory); of Institut National Polytechnique de Grenoble, France]

STMCGMW Subcommittee for Tectonic Maps of the Commission for the Geological Map of the World [Russian Federation]

STM-CVD Scanning-Tunneling-Microscopy-Assisted Chemical Vapor Deposition

STME Space Transportation Main Engine [of NASA and US Air Force]

St Mgr Station Manager

STMIS Space Telescope Management Information System

STMIS System Test Manufacturing Information System

STMRA Sparsely Truncated Multiresonance Approximation

STMS Short Test of Mental Status

STM-S Scanning Tunneling Microscopy and Spectroscopy [also STM&S, or STM/S]

STM/STS Scanning Tunneling Microscopy/Scanning Tunneling Spectroscopy

Stmt Statement [also stmt]

STM/TS Scanning Tunneling Microscopy/Tunneling Spectroscopy

STMU Special Test and Maintenance Unit
STN Satellite Tracking Network [of NASA]
STN Scientific and Technical Network
STN (The) Scientific and Technical Information Network
STN Software Trouble Note
STN Strontium Titanium Niobate
STN Super-Twisted Nematic (Liquid Crystal Display)
STN Switched Telecommunications Network
Stn Station [also stn]
Stn Stone [also stn]
stn stainless
STN International (The) Scientific and Technical Information Network International
STN/MPD Scientific and Technical Information Network/Materials Properties Databases
STNMR Stochastic Nuclear Magnetic Resonance [Physics]
stnr stationary
STNV Satellite Tobacco Necrosis Virus
STO Sawtooth Oscillation
STO Segment Table Origin
STO Slater-Type (Atomic) Orbitals [Physics]
STO Stock-Tank Oil
STO Store Number (Function)
STO Strontium Titanyl Oxalate
STO System Test Objectives
Sto Stow(age)
STOA Solution (Heat) Treated and Overaged [also ST&OA] [Metallurgy]
STOC Selected Table of Contents
STOCC Space Telescope Operations Control Center
Stoch Process Appl Stochastic Processes and Their Applications [Netherlands]
Stoch Stoch Rep Stochastics and Stochastics Reports [UK]
STOIAC Static Technology Office Information Analysis Center [of Battelle Memorial Institute, Columbus, Ohio, US]
STOL Short Takeoff and Landing
STOL Short-Time Overload(ed); Short-Time Overloading
STOL Systems Test and Operation Language
STOLed Short-Time Overloaded
STOM Scanning Tunneling Optical Microscope; Scanning Tunneling Optical Microscopy
STOM SPADE Terminal Operator's Manual
STOMP Subsurface Transportation Over Multiple Phases
Stone+tec International Trade Fair for Stone Winning and Dressing Engineering [Nuremburg, Germany]
STO-nG Slater-Type Orbital at the n-Gaussian Level [Physics]
STOP Stable Ocean Platform
STOP Statistical Operations Processor
STOP Supersonic Transport Optimization Program [of NASA]
STOPS Shipboard Toxicological Operational Protective System
STOQ Storage Queue
Stor Storage [also stor]
Storage–Handl–Distrib Storage–Handling–Distribution [Journal published in the UK]

STORC Self-Ferrying Trans-Ocean Rotary-Wing Crane
.STORE Businesses Offering Goods [Internet Domain Name]
STORES Syntactic Tracer Organized Retrospective Enquiry System
STORET Storage and Retrieval
STORI International Conference on Nuclear Physics at Storage Rings
STORLAB Space Technology Operations and Research Laboratory [US]
STORM Statistically-Oriented Matrix Program
STORM Stormscale Operational and Research Meteorology (Program)
STOR-M Saskatchewan Tokamak Reactor Machine [of University of Saskatchewan, Saskatoon, Canada]
STOS Store String
STOVL Short Takeoff and Vertical Landing
STOW System Takeoff Weight
Stow Stowage [also stow]
STOX Speech and Telegraphy in Voice Channel
STO/YBCO Strontium Titanate/Yttrium Barium Copper Oxide (Superconductor)
STP 2,5-Dimethoxy-4,α-Dimethyl Benzenethianamine [Pharmacology]
STP School of Theoretical Physics
STP Scientifically Treated Petroleum
STP Secure Transfer Protocol
STP Selective Tape Print
STP Self-Test Program
STP Serenity, Tranquility, Peace
STP Sewage Treatment Plant
STP Shielded Twisted Pair [Electric Cables]
STP Shuttle Technology Panel [of NASA]
STP Signal Transfer Point [Telecommunications]
STP Silver-Bearing Tough Pitch (Copper)
STP Simultaneous Test Procedure
STP Simultaneous Track Processor
STP Situation Target Proposal
STP Slater-Transfer Preuss [Physics]
STP Solar Terrestrial Physics
STP Southampton Transducer Protocol
STP Space Technology Payload
STP Space Test Program
STP Special Technical Publication
STP Standard Technical Publication
STP Standard Temperature and Pressure
STP Static Pressure
STP Stop (Character)
STP Streaming Potential [Chemistry]
STP Subsystem Test Plan
STP Synchronized Transaction Processing
STP System Test Plan
STP System Test Procedure
Stp Stamp [also stp]
StP Strain Point (of Glass) [also STP, ST P or St P]

STPA Service Technique des Programmes Aéronautiques [Aeronautical Program Service; of French Ministry of Defense]

ST PAM Single-Tone Pulse-Amplitude Modulation [Telecommunications]

STP Copper Silver-Bearing Tough Pitch Copper [also STP copper]

STPF Shield Test Pool Facility [Nuclear Engineering]

STPF Stabilized Temperature Platform Furnace

STPH Static Phase Error

STPI Science and Technology Policy Institute [US]

STPL Sidetone Path Loss

STPL Standard Test Processing Language

ST PM Single-Tone Pulse Modulation [Telecommunications]

STP Newsl Solar Terrestrial Physics Newsletter [of Scientific Committee on Solar Terrestrial Physics, US]

STPO Science and Technology Policy Office [US]

STPP Sodium Tripolyphosphate

ST PPM Single-Tone Pulse-Position (or Phase) Modulation [Telecommunications]

STPS Single Triangular Potential Sweep [Radar]

STPSR Space Telescope Planning Support Room

STPSS Science and Technology Program Support Section [Canada]

STPST Stop-Start

STPT Society of Town Planning Technicians [UK]

STPTC Standard of Tar Products Testing Committee [UK]

STQ Solution (Heat) Treated and Quenched; Solution (Heat) Treatment and Quenching [also ST&Q, or ST & Q] [Metallurgy]

STR Segment Table Register

STR Self-Tuning Regulator

STR Short Tandem Repeat

STR Short-Term Revitalization

STR Side-Tone Reduction

STR Speed-Tolerant Recording

STR Standardized Thai Rubber

STR Status Register

STR Stirred Tank Reactor [Chemical Engineering]

STR Store Task Register

STR String (Procedure) [Pascal Programming]

STR Submarine Thermal Reactor [Nautilus Prototype Basic Design – 1951]

STR Symbol Timing Recovery

STR Synchronous Transmitter/Receiver

STR System Test Review

Str Steamer [also str]

Str Strainer [also str]

Str Strait [Geography]

Str Streptococcus [Microbiology]

Str String [also str]

Str Strip [also str]

Str Strobe [also str]

Str Stroke [also str]

Str Structure [also str]

STR$ String (Function) [Programming]

STRAC Strategic Army Corps [US]

STRAD Signal Transmission, Reception and Distribution

STRAD Switching, Transmitting, Receiving and Distributing

STRADAP Storm Radar Data Processor

STRAF Strategic Army Forces [US]

Strahlentherap Strahlentherapie [German Journal on Radiation Therapy]

STRAIN Structural Analytical Interpreter

STRANGE The Physics of Strangeness (Conference) [Particle Physics]

STRAP Star Tracking Rocket Attitude Positioning; Stellar Tracking Rocket Attitude Positioning

STRASACOL Joint Polymer Physics Action of the University of Strasbourg, Saclay Nuclear Research Center and Collège de France [France]

Strat Strategic; Strategy [also strat]

STRATA Sulfide Stress Cracking Test Assistant (Software)

Stratigr Stratigrapher; Stratigraphic; Stratigraphy [also Stratigr]

Strato cu Stratocumulus (Cloud) [Meteorology]

STRATCOM Strategic Communications Command [US Army]

Stratosph Stratosphere; Stratospheric [also Statosph]

STRATSAT Strategic Satellite [US Air Force]

STRATWARM Stratospheric Warming [also Stratwarm]

STRAW Simultaneous Tape Read and Write

Strb Strobe

STRC Science and Technology Research Center [of University of North Carolina, US]

STRC Scientific and Technical Research Commission

STRC Scientific and Technical Research Council

STRC Spacelink Teacher Resource Center [US]

STRCH Stretch [Strch]

STRCT Scientific and Technical Research Council of Turkey [Ankara]

Strd Strand [also strd]

STRE Side-Tone Reference Equivalent

Strength Mater Strength of Materials [Translation of: *Problemy Prochnosti (USSR)*; published in the US]

STREP Standard Repair Price

Strep Streptococcus [Microbiology]

Strep A Streptococcus A [Microbiology]

Strep B Streptococcus B [Microbiology]

STRESS Structural Engineering Systems Solver [Computer-Programming Language]

STRFLD Star Field

Strg Steering [also strg]

Strg Strength [also strg]

Strg String [also strg]

strg strong

STRI Sequential Tracking, Registration and Information

STRI Smithsonian Tropical Research Institute [Barro Colorado Island in Gatun Lake, Panama Canal Zone]

STRIFE Stressed Life (Testing)

STRIG Status of Trigger (Statement) [Programming]

STRIG OFF Check Status of Trigger Off (Statement) [Programming]

STRIG ON Check Status of Trigger On (Statement) [Programming]

STRINGS Introductory School on String Theory [of International Center for Theoretical Physics, Trieste, Italy]

STRIP Standard Taped Routines for Image Processing

STRIPS Standard Research Institute Problem Solver

STRIVE Standard Techniques for Reporting Information on Value Engineering

S Trk Star Tracker [Aerospace]

STRL Science and Technical Research Laboratory

STRL Solar Thermal Research Laboratory [of University of Waterloo, Ontario, Canada]

strl structural

STRLEN Character String Length

STRN Standard Technical Report Number

STRNUM Character String to Numeric String (Conversion)

STRO Scandinavian Tire and Rim Organization [Sweden]

STROBE Satellite Tracking of Balloons and Emergencies

STROBES Shared Time Repair of Big Electronic Systems

Strojír Výroba Strojírenska Výroba [Czech Scientific Publication]

Strojnícky Čas Strojnícky Časopis [Czech Scientific Publication]

Strojniški Vestn Strojniški Vestnik [Yugoslavian Scientific Publication]

STRP Strategic Transit Research Program [of Canadian Urban Transit Association]

STRPOS Character String Position

STRR Statistical Treatment of Radar Returns

STR$ String Representation of Argument Value (Function) [Programming]

StrSchV *(Strahlenschutzverordnung)* – German Radiation Protection Ordinance [also StrlSchV]

Struct Structural; Structure [also struct]

Struct Chem Structural Chemist(ry)

Struct Chem Structural Chemistry [International Journal]

StructE Structural Engineer [also StructEng]

Struct Eng Structural Engineering

Struct Eng A Structural Engineer, Part A [Journal of Institution of Structural Engineers, UK]

Struct Eng B Structural Engineer, Part B [Journal of Institution of Structural Engineers, UK]

Struct Eng Pract Structural Engineer Practice [Journal published in the US]

Struct Geol Structural Geology; Structural Geologist

Struct Mat Structural Material

Struct Optim Structural Optimization [Journal published in the US]

Struct Program Structured Programming [Journal published in the US]

Struct Saf Structural Safety [Journal published in the Netherlands]

Strukt Sverplast Met Strukturnaya Sverplastichnost' Metallov [Russian Journal on the Structure of Superplastic Metals published in Moscow]

STRUDL Structural Design Language [Computer Programming Language]

STRV Space Technology Research Vehicle

STS Satellite Tracking Station

STS Satellite Transmission System

STS Scanning Tunneling Spectroscopy

STS Science and Technology Studies

STS Science and Technology Society [US]

STS Science, Technology and Society

STS Scientific and Technical Services

STS Scientific and Technological Services

STS Scientific Terminal System

STS Secure TTY (Teletypewriter) System

STS Sequence Tagged Site [Genetics]

STS Serological Test for Syphilis [Medicine]

STS Shared Tenant Services [of Federal Communications Commission, US]

STS Ship-to-Shore (Communication) [also sts]

STS Short-Term Store; Short-Term Storage

STS Shuttle Test Station [of NASA]

STS Shuttle Transportation System [of NASA]

STS Silver Thiosulfate Solution

STS Slater Transition State [Physics]

STS Sodium Tetradecyl Sulfate

STS Solid-State Software

STS Space-Time-Space [Relativity]

STS Space-Time Structure [Relativity]

STS Space Transportation System [of NASA]

STS Special Treatment Steel

STS Speedway Transport System

STS Standard Technical Specification

STS Standard Test for Syphilis [Medicine]

STS Standard Test Signal

STS Static Test Stand

STS Status [also Sts, or sts]

STS Steam Turbine Ship

STS Stimulated Thermal Scattering

STS Strontium Titanium Silicate

STS Structural Transition Section

STS Subscriber Transferred Signal

S2S Surfaced on Two Sides [Lumber]

sts stitches [Woven Fabrics]

STSC Southwest Texas State College [San Marcos, US]

S2S&CM Surfaced on Two Sides and Center Matched [Lumber]

STScI Space Telescope Science Institute [Baltimore, Maryland, US]

STSF Secured Tank Storage Facility

STS/FCC Shared Tenant Services/Federal Communication Commission [US]

STSG Scottish Transport Studies Group

STSG Shuttle Test Group [of NASA]

STSGT (Spring School on Non-Perturbative Aspects of) String Theory and Supersymmetric Gauge Theories [at International Center for Theoretical Physics, Trieste, Italy]

STSI Space Telescope Science Institute [Baltimore, Maryland, US]

ST SIM Space Telescope Simulation

S2S&M Surfaced on Two Sides and (Standard, or Center) Matched [Lumber]

STSMC Statistical Thermodynamic Supermolecule Continuum [Physics]

S2S1E Surfaced on Two Sides and One Edge [Lumber]

STSOPO Shuttle Transportation Systems Operations Program Office [at NASA Johnson Space Center, Houston, Texas, US]

STSR System Test Summary Report

STSS Strowger Telephone Switching System [US]

S2S&SM Surfaced on Two Sides and Standard Matched [Lumber]

STSU Southwest Texas State University [San Marcos, US]

STT Secure Transaction Technology [Internet]

STT Seek Time per Track

STT Single Transmission Time

STT Spacelab Transfer Tunnel

STTE Spectroscopy Time-Tagged

STT Standards Technology Training [of American Society for Testing and Materials, US]

STT Standard Tube Test

STT Strontium Titanium Tantalate

STT Studies, Tests and Trials

STT Surface Tension Transfer (Welding)

$\sigma(\theta,\mathbf{T})$ Angle and Temperature Dependent Surface Tension [Symbol]

STTA Scottish Timber Trade Association

STTA 1,1,1-Trifluoro-4-Mercapto-4(2-Thienyl)3-Butene-2-One

STTC Secondary Teachers Technical Certificate

STTE Spectroscopy Time-Tagged Event

STTL Schottky Transistor-Transistor Logic [also ST²L]

STTMA Screw Thread Tool Manufacturers Association [UK]

STTR Small Business Technology Transfer (Program) [US]

STTR Stochastic Time Resolved (Process) [Physics]

STTR-CIDNP Stochastic Time-Resolved Chemically-Induced Dynamic Nuclear (Spin) Polarization [Physics]

STTS Scottish Tramway and Transport Society

STU Santo Tomas University [Manila, Philippines]

STU Secure Telephone Unit

STU Shanghai Technical University [PR China]

STU Signal Transmission Unit

STU Slovak Technical University

STU Special Test Unit

STU Star Tracker Unit [Aerospace]

STU Story Understander [Artificial Intelligence]

STU St. Thomas University [Fredericton, New Brunswick, Canada]

STU Submersible Test Unit

STU Subscribers Trunk Unit

STU Supertransuranic(s) [Chemistry]

STU Swiss Federation of Trade Unions

STU Systems Test Unit

STU System Timing Unit

STU System Transmission Unit

Stu Streptomyces tubercidicus [Microbiology]

Stu Student [also stu]

Stub Acme Stub Acme (Screw) Thread [Symbol]

STUC Scottish Trade Union Congress

STUD Safety Training Update

Stud Student; Studies; Study [also stud]

Stud Appl Math Studies in Applied Mathematics [Journal of the Society for Industrial and Applied Mathematics, US]

Stud Cercet Calc Econ Cibern Econ Studiişi Cercetări de Calcul Economic si Cibernetica Economica [Romanian Journal on Economical Computation and Cybernetics]

Stud Cercet Doc Studiişi Cercetări de Documentare [Romanian Journal on Documentation]

Stud Cercet Fiz Studiişi Cercetări de Fizica [Romanian Journal on Physics]

Stud Cercet Geol Geofiz Studiişi Cercetări de Geologie, Geofizica, Geografie [Romanian Journal on Earth Sciences]

Stud Cercet Mat Studiişi Cercetări Matematice [Romanian Journal on Mathematics]

Stud Cercet Mec Apl Studiişi Cercetări de Mecanica Aplicata [Romanian Journal on Applied Mechanics]

Stud Environ Sci Studies in Environmental Science [Journal published in the US]

Stud Geophys Geod Studia Geophysica et Geodaetica [Czech Journal on Geophysics and Geodesy]

Studia Forest Suecica Studia Forestalia Suecica [Journal on Forestry; published in Sweden]

Stud J Inst Electron Telecommun Eng Students Journal of the Institution of Electronics and Telecommunication Engineers [India]

Stud Math Studia Mathematica [Mathematical Journal published in Wroclaw, Poland]

Stud Math Geol Studies in Mathematical Geology [of International Association for Mathematical Geology, US]

Stud Rep Hydrol Studies and Reports in Hydrology [of International Hydrological Program, France]

Stud Sci Math Hung Studia Scientiarum Mathematicarum Hungarica [Journal on Mathematical Sciences published in Hungary]

Stud Surf Sci Catal Studies in Surface Science and Catalysis [Journal published in the US]

STUF Sudan Trade Union Federation

STUFA Studiengesellschaft für Automobilstraßenbau [Road Research Association, Germany]

STUFF Systems to Uncover Facts Fast

STUMP Story Understanding and Memory Program

STURP Shroud of Turin Research Project [US]

STUVA Studiengesellschaft für Unterirdische Verkehrsanlagen [Underground Transportation Research Association, Germany]

STUVA-Forschungsber STUVA-Forschungsberichte [Research Reports of Studiengesellschaft für Unterirdische Verkehrsanlagen, Germany]

STV Sawtooth Voltage

STV Schweizerischer Technischer Verband [Swiss Technical Association]

STV Separation Test Vehicle

STV Small Test Vessel

STV Standard Tar Visco(si)meter

STV Standard Test Vehicle

STV Subscription Television

STV Suction Throttling Valve

STV Surveillance Television

STV/BPO Suction Throttling Valve/Bypass Orifice

STVW Symmetrical Triangle Voltage Waveform

STW Stifting voor Technische Wetenschappen [Netherlands Foundation for Technical Sciences, Utrecht]

STW Standard Weight (Crucible) [Metallurgy]

ST W Storm Water [also St W]

STWP Society of Technical Writers and Publishers [now Society for Technical Communication, US]

St WP Steam Working Pressure

STW-Utrecht Stifting voor Technische Wetenschappen–Utrecht [Netherlands Foundation for Technical Sciences–Utrecht]

Stwy Stairway [also stwy]

STX Saxitoxin [Biochemistry]

STX Start-of-Text (Character)

neo-STX neo-Saxitoxin [Biochemistry]

STXM Scanning Transmission X-Ray Microscope; Scanning Transmission X-Ray Microscopy

(S)TxD Secondary Transmit Data

STXM Scanning Transmission X-Ray Microscopy

STY Space Time Yield [Physics]

.STY Style [Ventura/Word/WordPerfect File Name Extensions]

Sty Styrene

sty stationary

SU Saga University [Japan]

SU Saitama University [Japan]

SU Salford University [UK]

SU Search Unit [Ultrasonic Testing]

SU Seattle University [Washington, US]

SU Seikei University [Tokyo, Japan]

SU Selectable Unit

SU Serial Unit

SU Service Unit [Industrial Engineering]

SU Setsunan University [Japan]

SU Set-Up [also S/U]

SU Shandong University [Jinan, PR China]

SU Sheffield University [UK]

SU Shinshu University [Nagano, Japan]

SU Shizuoka University [Hamamatsu, Japan]

SU Signaling Unit

SU Single User

SU Soochow University [Taipei, Taiwan]

SU Solventless Urethane [British Propellant]

SU Sonics and Ultrasonics [Institute of Electrical and Electronics Engineers Group, US]

SU Soviet Union

SU Soviet Union [ISO Code]

SU Speech Understanding [Artificial Intelligence]

SU Sport-Utility (Car)

SU Stanford University [California, US]

SU Start Up

SU State University

SU Station Unit

SU Stockholm University [Sweden]

SU Storage Unit

SU Streamline Upwind (Method) [Mechanics]

SU Strontium Unit

SU Structural Unit (Model) [Materials Science]

SU Subunit

SU Suffolk University [Boston, Massachusetts, US]

SU Sunyatseni University [PR China]

SU Support Unit

SU Suppressor

SU Sussex University [UK]

SU Sydney University [Australia]

SU Synchronization Unit

SU Synchronization Utility

SU Syracuse University [New York State, US]

SU- United Arab Republic (Egypt) [Civil Aircraft Marking]

SU(2) Special Unitary Group in Two Dimensions [also SU_2] [Particle Physics]

SU(3) Unitary Symmetry Theory [also SU_3] [Particle Physics]

SU(6) Combination of SU(3) and SU(2) [also SU_6] [Particle Physics]

S/U Set-Up [also SU]

Su Submittal [also su]

Su Succinimide

Su Sulfur [Abbreviation; Symbol: S]

Su Suspension

Su Sucre [Currency of Ecuador]

Su Sunday

su solute

.su Soviet Union [Country Code/Domain Name]

SUA Silver Users Association [US]

SUA State Universities Association [US]

SUAMC Southern University A&M (Agricultural and Mechanical) College [Baton Rouge, Louisiana, US]

SUAS System for Upper Atmospheric Sounding

SUB Stockholm Universitet Bibliotek [Stockholm University Library, Sweden]

SUB Substitute (Character) [Data Communications]

Sub Subaltern [UK]

Sub Sublimation [also sub]

Sub Submarine [also sub]

Sub Subroutine [also sub]

Sub Subscriber [also sub]

Sub Subscript [also sub]

Sub Subscription [also sub]

Sub Substitute; Substitution [also sub]

Sub Subtract(ion) [also sub]

Sub Suburb(an) [also sub]

Sub Subway [also sub]

sub submerge(d)

SUBA Subbituminous A (Coal) [Geology]

SUBB Subbituminous B (Coal) [Geology]

SUBC Subbituminous C (Coal) [Geology]

SUBCAL Subcaliber [also SUB-CAL, subcal or sub-cal]

SUBCHAR Substring from Character String (Extraction)

SUBCON Subcontracting Industries Exhibition

Subd Subdivision [also subd]

SUBDIZ Submarine Defense Identification Zone

SUBDOC Subdocument [also Subdoc, or subdoc]

SUBIC Submarine Integrated Control System

Subj Subject [also subj]

Subl Sublime; Sublimation [also subl]

Subln Sublimation [also subln]

Sub Lt Sub Lieutenant [also SLt] [Canadian Forces]

Subm Submarine [also subm]

SUBMO Submarine Motion

subn suburban

Subord Subordinate [also subord]

Subprogram Subordinate Program [also subprogram]

SUBROC Submarine Rocket [US Navy Submarine-to-Air-to-Underwater Missile] [also Subroc]

Subroutine Subordinate Routine [also subroutine]

Subs Subsidiary [also subs]

Subs Substance [also subs]

Subs Substitute [also subs]

Subscpt Subscript [also subscpt]

subseq subsequent

Subsid Subsidiary [also Subsid, or subsid]

SUBST Substitute String for Path (Command) [also subst]

Subst Substitute [also subst]

SUBSTA Subscriber's Station [Telecommunications]

Substa Substation [also substa]

Substr Substring [also substring]

Substr Substructure [also substr]

Substring Subordinate String [also substring]

SUBTASS Submarine Towed Array Sonar System

SUBTIL Synthesized User-Based Terminology Index Language

SUC Saskatchewan Universities Commission [Canada]

SUC Secretary of the University Council

SUC Sucre [Currency of Ecuador]

SUC Sulfanilamide–Urea–Chlorobutol (Compound) [Biochemistry]

Suc Succinic Acid

Suc Sucrose [Biochemistry]

Succ Return Successor of Argument [Pascal Function]

SUCCEED Southeastern University and College Coalition for Engineering Education [US National Science Foundation Program Including Clemson University, Cooper Union University, Drexel University, Florida A&M University, Florida International University, Florida State University, New Jersey Institute of Technology, Ohio State University, Polytechnic University, University of Pennsylvania, and University of South Carolina]

SUCCESS Subsonic Aircraft: Contrails and Cloud Effects Special Study [of NASA]

Succr Successor [also succr]

SucON Sucrose Octanitrate [Biochemistry]

Suct Suction [also suct]

SUD SSPF (Space Station Processing Facility) Utilization Document

Sud Sudan(ese)

SUDAA Stanford University, Division of Aeronautics and Astronautics [California, US]

SUDIC Sulfur Development Institute of Canada

SUDOSAT Sudan Domestic Satellite

Sud£ Sudanese Pound [Currency of Sudan]

Sud Promihlen Metal Sudostroitelnaja Promihlenost, Metallurgia [Russian Journal on Metallurgy]

SUDT Silicon Unilateral Diffused Transistor

Sudura Încercari Mater Sudura si Încercari de Materiale [Romanian Journal on Soldering and Materials Joining]

SUE Société Universelle des Explosifs [A French Explosives Company]

SUEB Scottish Universities Entrance Board

SUEDE Surface Evaluation and Definition

Suf Suffix [also suf]

SuFET Superconducting Field-Effect Transistor

Suff Suffix [also suff]

Suff Suffolk [UK]

suff sufficient

suff suffocating

SUFFER Save Us From Formaldehyde Environmental Repercussions [now Citizens United to Reduce Emissions of Formaldehyde Poisoning Association, US]

suffoc suffocating

SUFIR Superfast Inversion Recovery

SUFOI Scandinavian Unidentified Flying Object Information [Denmark]

SUG Smartmac (Computer) User Group [US]

SuG Scheidhauer-und-Geissing (Fireclay Shaping Method) [Metallurgy]

Sug Sugar [also sug]

SU/GBD Structural Unit/Grain Boundary Dislocation [Materials Science]

SUGI SAS Inc. Users Group International

Sugg Suggestion [also sugg]

SUH Stanford University Hospital [California, US]

SUHF Spin-Unrestricted Hartree-Fock [Physics]

SUHL Sylvania Ultrahigh Level Logic; Sylvania Universal High-Speed Logic

SUI Standard Universal Identifier

SUI State University of Iowa [Iowa City, US]

SUIPR Stanford University, Institute for Plasma Research [California, US]

SUL Small University Library

SUL Sudanese Pound [Currency]

S-UL S-Shaped Pattern, Magnetization Vector Direction Up Left (↖) [Solid-State Physics]

SULC Super-Ultralow Carbon (Steel)

SULC Steel Super-Ultralow Carbon Steel [also SULC steel]

SULFAN (Stabilized) Sulfuric Anhydride [also Sulfan]

Sulfd Sulfide

Sulfo-C-Acid 7-Amino-1,3,5-Naphthalenetrisulfonic Acid

Sulfo-NHS Sulfo-N-Hydroxysuccinimide

Sulfo-SIAB Sulfosuccinimidyl(4-Iodoacetyl)aminobenzoate

Sulfo-SMCC 4-(N-Maleimidomethyl)cyclohexane-carboxylic Acid 3-Sulfo-N-Hydroxysuccinimide Ester

Sulfotepp Tetraethyl Dithiopyrophosphate [also sulfotepp]

SULFRED Sulfur Reduction (Process)

SULF-X Sulfur Extraction (Process)

SULIRS Syracuse University Libraries Information Retrieval System [US]

SULIS Sulzer Literaturverteilung und −sortierung [Sulzer Literature Dissemination and Sorting System of Sulzer Brothers Ltd., Winterthur, Switzerland]

SULPEL Sulfur Pelletization (Process)

Sult Sultan [also sult]

Sulzer Tech Rev Sulzer Technical Review [of Sulzer Brothers Ltd., Winterthur, Switzerland]

SUM Set-Up Module

SUM Shallow Underwater Mobile

SUM Sport-Utility Minivan

SUM Surface-to-Underwater Missile

SUM System Utilization Monitor

Sum Sumatra

Sum Summary [also sum]

SUMAC Sheffield University Metals Advisory Center [UK]

SUMC Space Ultrareliable Modular Computer

SUMC Stanford University Medical Center [California, US]

SUMEX Stanford University Medical Experiment [of SUMC]

SUMEXAIM Stanford University Medical Experiment–Applications of Artificial Intelligence to Medical Research [US]

Sumitomo Chem Rev Sumitomo Chemical Review [Published by Sumitomo Metal Industries Ltd., Osaka, Japan]

Sumitomo Electr Tech Rev Sumitomo Electric Technical Review [Published by Sumitomo Electric Industries Ltd., Osaka, Japan]

Sumitomo Light Met Tech Rep Sumitomo Light Metal Technical Reports [Published by Sumitomo Metal Industries Ltd., Nagoya, Japan]

Sumitomo Met Sumitomo Metals [Published by Sumitomo Metal Industries Ltd., Tokyo, Japan]

Sumitomo Met News Sumitomo Metals News [Published by Sumitomo Metal Industries Ltd., Tokyo, Japan]

Sumitomo Spec Met Tech Rep Sumitomo Special Metals Technical Report [Published by Sumitomo Metal Industries Ltd., Osaka, Japan]

Summ Summary [also summ]

Summa Bras Math Summa Brasiliensis Mathematicae [Brazilian Publication on Mathematics; published in Rio de Janeiro, Brazil]

Summer Mtg Am Cryst Assoc Summer Meeting of the American Crystallographic Association [US]

SUMMIT Sperry Univac Minicomputer Management of Interactive Terminals

SUMMIT Supervisor of Multiprogramming, Multiprocessing, Interactive Time-Sharing

Summ Rep Electrotech Lab Summary Reports of the Electrotechnical Laboratory [Japan]

Sum Pch Summary Punch [also Sum Pch]

SUMS Sheffield University Metallurgical Society [UK]

SUMS Shuttle Upper-Atmosphere Mass Spectrometer [of NASA]

SUMS Sperry Univac Material System

SUMT Sequential Unconstrained Minimization Technique

Sun Sunday

SUNET Swedish University Network

Sun Fol Sun Follower; Sun Following [also sun fol]

SUN Symbols, Units, Nomenclature Commission [of International Union of Pure and Applied Physics]

SUNA Switchmen's Union of North America [Now Part of United Transportation Union]

Sund Sunday

SUNFED Special United Nations Fund for Economic Development

SUNI Southern Universities Nuclear Institute [US]

SUNJ State University of New Jersey [New Brunswick, US]

SUNMOS Sandia/University of New Mexico Operating System [US] [also Sunmos]

Sunoco Sun Oil Company [US]

Sun OS Sun Microsystems Operating System (for Unix)

SUNY State University of New York [US]

SUNY Albany State University of New York at Albany [US] [also SUNYA]

SUNY Binghamton State University of New York at Binghamton [US]

SUNY Buffalo State University of New York at Buffalo [US]

SUNY-CF State University of New York–College of Forestry [Syracuse, US]

SUNY-DMC State University of New York–Downstate Medical Center [Brooklyn, US]

SUNY-ESC State University of New York–Empire State College [Saratoga Spings, US]

SUNY-MC State University of New York–Maritime College [Bronx, US]

SUNY Stony Brook State University of New York at Stony Brook [US]

SUNY Syracuse State University of New York at Syracuse [US]

SUNY-UMC State University of New York–Upstate Medical Center [US]

suo nom *(suo nomen)* – by its own name [Medical Prescriptions]

SUOTL Stanford University Office of Technology Licensing [California, US]

SUP Sintering Under Pressure [Metallurgy]

SUP Stanford University Press [California, US]

SUP Start-Up Procedure

SUP Suppressor (Grid) [also Sup, or sup]

SUP Surrey University Press [London, UK]

SUP Syracuse University Press [New York State, US]

.SUP Supplemental Dictionary [WordPerfect File Name Extension]

Sup Supercript [also sup]

Sup Support [also sup]

Sup Supplement(al) [also sup]

Sup Supply [also sup]

Sup Suppress [also sup]

sup *(supra)* – above

sup superior

sup supremum [Mathematics]

SUPA School of Urban and Public Affairs [of Carnegie-Mellon University, Pittsburgh, Pennsylvania, US]

SUPA Society of University Patent Administrators [US]

SUPAERO Ecole Nationale Supérieure de l'Aéronautique et de l'Espace [National Higher School of Aeronautics and Aerospace, France]

SUPARCO Space and Upper Atmosphere Research Committee [Pakistan]

Supcrit Supercritical(ity) [also supcrit] [Nuclear Reactors]

Supe Super(numerary) [also supe]

SUPER SUbmicron Positive Dry Etch Resist

Super Superintendent [also super]

super superfine

super superior

Supercond Superconductivity; Superconductor [also supercond]

Supercond Ind Superconductor Industry [Journal published in the US]

Supercond News Superconductor News [Newsletter of Superconductor Applications Association, US]

Supercond Rev Superconductivity Review [International Journal]

Supercond Sci Technol Superconductor Science and Technology [Journal of the Institute of Physics, UK]

Superhet Superheterodyne (Receiver) [also superhet]

SuperJANET Super Joint Academic Network [UK]

Superlattices Microstruct Superlattices and Microstructures [Journal published in the UK]

supers supersaturated

Superstr Superstructure [also superstr]

SupHqs Supreme Headquarters [Military]

Suphtr Superheater [also suphtr]

sup,lub supremum, least upper bound [Mathematics]

SUPO Super-Power Water Boiler (Nuclear Reactor) [US]

Supp Supplement [also supp]

Supp Suppository [also supp] [Medical Prescriptions]

supp supplemental examination

supp supplementary

Suppl Supplement [also suppl]

Suppl Supplier [also suppl]

Suppl Br Telecommun Eng Supplement to British Telecommunications Engineering [Journal published in the UK]

Suppl Educ Monogr Supplementary Educational Monographs [US]

Support Care Cancer Supportive Care in Cancer (Journal)

Suppr Suppression [also suppr]

SUPPS Supplementary Procedures

Supr Supervisor [also supr]

supr supreme

SUPRA Submersible Underwater Pipeline Repair Apparatus

Supramol Sci Supramolecular Science [International Journal]

SUPROX Successive Approximation [Computers]

Suprscpt Superscript [also suprscpt]

Suprt Support [also suprt]

SUPSALV Supervisor of Salvage [of US Navy]

Supt Superintendent [also supt]

Supt Support [also supt]

SuptDoc Superintendent of Documents [Military]

Supv Supervisor [also supv]

supv supervisory

Sup VC Superior Vena Cava [also sup vc] [Anatomy]

Supvr Supervisor [also supvr, Supv'r, or supv'r]

supy supervisory [also supy]

SUR Société Université et Recherche [Society for Universities and Research, Switzerland]

SUR Society for Universities and Research [Switzerland]

SUR Speech Understanding Research

SUR State University of Rutgers [Piscataway, New Jersey, US]

S-UR S-Shaped Pattern, Magnetization Vector Direction Up Right (↗) [Solid-State Physics]

Sur Surface [also sur]

Sur Surcharge [also sur]

Sur Suriname

Sur Surplus [also sur]

SURA Southeastern Universities Research Association [of Oak Ridge National Laboratory, Tennessee, US]

SURA/ORNL Southeastern Universities Research Association/Oak Ridge National Laboratory [Tennessee, US]

SURANO Surface Radar and Navigation Operation

SURCAL Surveillance Calibration (Satellite)

SURDD Southern Utilization Research and Development Division [of US Department of Agriculture]

SURE Shuttle Users Review and Evaluation [NASA]

SURE Symbolic Utilities Revenue Environment

SURE Systems for Underwriting Risk Evaluation

SURE Sulfur Recovery (Process)

SURF Support of User Records and Files

SURF Surface Storage Facility

Surf Surface [also surf]

Sur f Suriname Florin [Currency]

Surfactant Sci Ser Surfactant Science Series [Journal published in the US]

Surf Coat Technol Surface and Coatings Technology [Journal published in Switzerland]

Surf Eng Surface Engineer(ing)

Surf Eng Surface Engineering [Publication of the Institute of Metals, UK]

SUR/FIN American Electroplaters and Surface Finishers Society Annual Conference and Exhibit(ion) [now Conference and Exhibition for the Surface Finishing Industry]

SUR/FIN Conference and Exhibition for the Surface Finishing Industry

Surf Interf Anal Surface and Interface Analysis [Journal published in the UK]

Surf Phys Chem Mech Surface Physics, Chemistry and Mechanics [Translation of Russian Journal published in the US]

Surf Sci Surface Science; Surface Scientist

Surf Sci Surface Science [Journal published in the Netherlands]

Surf Sci Lett Surface Science Letters [Journal published in the Netherlands]

Surf Sci Rep Surface Science Reports [Journal published in the Netherlands]

Surf Sci Spectra Surface Science Spectra [Journal of the American Vacuum Society, US]

Surf Technol Surface Technology; Surface Technologist

Surf Technol Surface Technology [Journal published in Switzerland]

Surfacing J Surfacing Journal [UK]

Surfacing J Int Surfacing Journal International [UK]

Surg Surgeon; Surgery; Surgical [also surg]

SURGE Sorting, Updating, Report Generating Equipment

Surg Forum Surgery Forum [Journal]

Surg Gen Surgeon General

Surg Gynec Obstet Surgical Gynecology and Obstetrics (Journal) [US]

SURI Syracuse University Research Institute [New York State, US]

SURIC Surface Ship Integrated Control (System)

SURL Smart Uniform Resource Locator [Internet]

SURMAC Surface Magnetic Confinement [Physics]

Surr Surrey [UK]

SURRC Scottish Universities Research and Reactor Center

SURS Standard Umbilical Retraction System

SURSAT Surveillance Satellite [also Sursat]

SURSULF Surface Hardening Sulfur Catalyst

SURTEC International Congress and Exhibition for Surface Technology [also Surtec]

Surv Survey(ing) [also surv]

Surv Surveyor [also surv]

Surv Geophys Surveys in Geophysics [Journal published in the UK]

Surv High Energy Phys Survey in High Energy Physics [Journal published in the Switzerland]

SUS Saybolt Universal Seconds [Unit]

SUS Setting Up Samples

SUS Silicon Unilateral Switch

SUS Single Underwater Sound

SUS Single-User System

SUS Small Ultimate Size

SUS Speech Understanding System [Artificial Intelligence]

SUSD State University of South Dakota [Vermillion, US]

SUSDP Standard for the Uniform Scheduling of Drugs and Poisons

SU/SGBD Structural Unit/Secondary Grain Boundary Dislocation [Materials Science]

SUSI (Conference on) Structures Under Shock and Impact

SUSIE Sequential Unmanned Scanning and Indicating Equipment

SUSIE Stock Updating Sales Invoicing Electronically

SUSIM Solar Ultraviolet Spectral Irradiance Monitor

SUSOPS Sustained Operations Programm [of Department of National Defense, Canada]

Susp Suspension [also susp]

suspd suspended [also susp'd]

Suspn Suspension [also suspn]

Suss Sussex [UK]

SUSSP Scottish University Summer School in Physics [UK]

SUSY International Conference on Supersymmetries in Physics

SUSY Supersymmetric (Model) [Particle Physics]

SUSY GUT Supersymmetric Grand Unified Theory [Particle Physics]

SUT Sample under Test

SUT Science University of Tokyo [Chiba, Japan]

SUT Shanghai University of Technology [PR China]

SUT Society for Underwater Technology [UK]

SUT State University of Tirana [Albania]

SUT Step-Up Transformer

SUT Swinburne University of Technology [Australia]

SUT System Under Test

SUTAGS Shuttle Uplink Text and Graphics Scanner [of NASA]

SUTW Super-Ultra Thin Window

SUV Saybolt Universal Viscosimeter

SUV Saybolt Universal Viscosity [Unit]

SUV Solar Ultraviolet

SUV Sport-Utility Vehicle

SV Effective Circular Tooth Thickness (of Splines) [Symbol]

SV El Salvador [ISO Code]

SV Safety Valve

SV Sailing Vessel

SV Saint Vernant (Principle) [Metallurgy]

SV Salvage Value

SV Sample Variance [Statistics]

SV Saponification Value

SV Satellite Vehicle

SV Saturated Vapor

SV Saturation Voltage

SV Sawtooth Voltage
SV Secular Variations [Astronomy]
SV Self-Verification
SV Seminal Vesicle [Anatomy]
SV Service Voltage
SV Servovalve
SV Shuttle Vehicle
SV Signature Verification
SV Silicon Valley [California, US]
SV Simulated Video
SV (Single) Silk Varnish (Insulation)
SV Single Value
SV Sinus Venosus [Biology]
SV Sleeve Valve
SV Slide Valve
SV Snake Venom
SV Sodani-Vitole (Change of Paidar-Pope-Vitek Hardening Model) [Metallurgy]
SV Solenoid Valve
SV Sound Velocity
SV Space Vehicle [also S/V]
SV Spatial Vision
SV Specific Volume
SV Speed Variation
SV Spherical Valve
SV Spin Valve [Solid-State Physics]
SV Standard Value
SV State Vector
SV Status Valid
SV Stereoscopic Video
SV Stereo Vision
SV Stop Valve
SV Striking Velocity
SV Stripping Voltammetry
SV Subclavian Vein [Anatomy]
SV Supply Valve [also S/V]
SV Surface Volume
SV Surge Voltage
SV Sweden [ISO Code]
SV Sympathetic Vibration [Physics]
SV Synaptic Vesicle [Biology]
SV Synchronous Vibrator
SV Synchronous Voltage
6V- Senegal [Civil Aircraft Marking]
SV+ StrataView Plus
S/V Solid/Vapor (Interface) [also S-V]
S/V Space Vehicle [also SV]
S/V Surface Area per Volume
S/V Surface-to-Volume (Ratio)
S/V Supply Valve [also SV]
S/V$_p$ Surface-to-Pore Volume (Ratio) [Powder Metallurgy]
S$_v$/V^2 Normalized Noise Power Spectral Density [Symbol]
Sv Sievert [Unit]
S(v) Mean Vibrational Entropy [Symbol]

sv solvent
sv *(sub verbo)* – under the following word or heading
.sv El Salvador [Country Code/Domain Name]
$\sigma(v_0)$ 2,200 meters per second cross section (in radiography) [Symbol]
SVA Shared Virtual Area
SVA Snake Venom Agglutinin [Immunology]
SVA Styrene-Vinyl-Acrylonitrile (Thermoplastics)
SVAB Shuttle Vehicle Assembly Building [of NASA]
SVAC Shuttle Vehicle Assembly and Checkout [NASA] [also SVA&C]
SVAFB South Vandenberg Air Force Base [California, US]
SVAK Snake Venom Agglutinin from Naja Naja Kaouthia [Immunology]
SVAM Snake Venom Agglutinin from Naja Mossambica Mosambica [Immunology]
Svar Proizvod Svarochnoe Proizvodstvo [Russian Journal on Welding Production]
SVAT Soil Vegetation Atmosphere Transfer
SVB Shuttle Vehicle Booster [of NASA]
SVB Space Vehicle Booster
SVC El Salvador Colon [Currency]
SVC Secure Voice Communication
SVC Service Message [Radio Engineering]
SVC Simultaneous Voltage Control
SVC Society of Vacuum Coaters [US]
SVC Static Var Compensator
SVC Superior Vena Cava [Anatomy]
SVC Supervisor Call (Instruction)
SVC Supervisory Cell
SVC Switched Virtual Call
SVC Switched Virtual Circuit
Svc Service [also svc]
svcbl serviceable
SVCC Schweizer Verein der Chemiker-Coloristen [Swiss Association of Chemists and Colorists]
SVCC Submerged Vertical Continuous Casting [Metallurgy]
Svce Service [also svce]
SVCS Star Vector Calibration Sensor
SVCT Schweizer Vereinigung diplomierter Chemiker HTL (Höhere Technische Lehranstalt) [Swiss Association of Certified Chemists (of the Higher Technical Institution)]
SVD Saturated Vapor Density
SVD Schweizerische Vereinigung für Dokumentation [Swiss Documentation Association]
SVD Simultaneous Voice and Data
SVD Singular(-Matrix) Value Decomposition
SVD System Verification Diagram
SVDF Segmented Virtual Display File
SVDN State-Variable-Dependent Noise
SVDS Space Vehicle Dynamic Simulator
SVE Schweizerische Vereinigung für Elektrotechniker [Swiss Association of Electrical Engineers]
SVE Senior Video Engineer
SVE Smoluchowski-Vlasov Equation [Physics]

SVE Society for Vector Ecology [US]

SVE Society for Visual Education [US]

SVE Soil Vapor Extraction

SVEAG Sullom Voe Environmental Advisory Group [now Shetland Oil Terminal Environment Advisory Group, UK]

SVEC Space Vacuum Epitaxy Center [of University of Houston, Texas, US]

Svensk Papperstid Svensk Papperstidning [Swedish Publication on Papermaking]

SVER Spatial Visual Evoked Response

Sverkhprovod Fiz Khim Tekn Sverkhprovodnie Fiziki i Khimiya Tekhnika [Russian Journal on Superconductor Physics and Technology]

Sverkhtverd Mater Sverkhtverdye Materialy [Russian Journal on Materials]

SVES Satellite Video Exchange Society [Canada]

SVF Schweizer Vereinigung von Färbereifachleuten [Swiss Association of Dyers]

SVF Simple Vector Format

SVFF Simplified Valence Force Field [Physics]

SVFR Special Visual Flight Rules

SVG Silicon Valley Group, Inc. [San Jose, California, US]

SVG Super-Velocity Gas

SVGA Standard Virtual Graphics Adapter

SVGA Super Video Graphics Array

Svgs Savings [also svgs]

SVH Solar Vacuum Head

S-VHS Separate Y-C Video Home System

S-VHS Super Video Home System

SVI Schweizerisches Verpackungsinstitut [Swiss Packaging Institute, Zurich]

SVI Secondary Virus Infection [Medicine]

SVI Service Interception

SVIA Specialty Vehicles Institute of America [US]

SVIC Shock and Vibration Information Center

SVIC Sociedad Venezolana de Ingenieros Consultores [Venezuelan Society of Consulting Engineers]

SVIH Sociedad Venezolana de Ingenieros Hidraulica [Venezuelan Society of Hydraulic Engineers]

SVIN State-Variable-Independent Noise

S/VISSR Stretched Visible and Infrared Spin-Scan Radiometer

SVL Servicing Log [also SVLog]

SVL Single Vibronic Level

SVL Sodium Vapor Lamp

SV-L-M Saint Venand-Levy-Mises (Equation) [Metallurgy]

SVLF Single Vibronic Level Fluorescence

SVLog Servicing Log [also SVL]

SVM Scanning Viscoelasticity Microscope; Scanning Viscoelasticity Microscopy

SVM School of Veterinary Medicine

SVM Silicon Video Memory

SVM Slant Visibility Meter

SVM System Virtual Machine

SVMT Schweizerischer Verband für Materialtechnik [Swiss Association for Materials Technology]

SVN Switched Virtual Network

SVO Space Vehicle Operations [NASA Kennedy Space Center Directorate, Florida, US]

SVOC Semi-Volatile Organic Compound

SVP Saturated Vapor Pressure; Saturation Vapor Pressure [Thermodynamics]

SVP Schwinger's Variational Principle [Physics]

SVP Service Processor

SVP Society of Vertebrate Paleontology [US]

SVP Software Verification Plan

SVPG Schwerizer Verein der Petroleumgeologen [Swiss Association of Petroleum Geologists]

SVR Alcohol (90%) [Pharmacology]

SVR Slant Visual Range

SVR Software Verification Report

SVR Super Video Recorder

SVR Supply-Voltage Rejection

Svr Server [also svr]

svr severe

SVR# System V Release Number [of AT&T Corporation, US]

svrl several

SVRR Software Verification Readiness Review

SVRS Solochrome Violet RS [Biochemistry]

SVS Single Virtual Storage

SVS Society of Visting Scientists [UK]

SVS Space Vision System

SVS Suit Ventilation System

SVS Supervisory Signal

SVS Switched Voice Service

SVT Superior Vacuum Technology Inc.) [Eden Prairie, Minnesota, US]

SVT System Validation Testing

SVTL Service Valve Test Laboratory

SVTN Secure Voice Teleconferencing Network

SVTP Sound Velocity, Temperature and Pressure; Sound, Velocity, Temperature and Pressure

SVW Schweizerischer Verband für die Wärmebehandlung (der Werkstoffe) [Swiss Heat Treatment Association]

SW Salt Water

SW Sandwich Winding; Sandwich-Wound [Composite Materials]

SW Scattered Wave

SW Scattering Wave [Physics]

SW Seawater

SW Seiberg-Witten (Equation) [Physics]

SW Seismic Wave

SW Self-Aligning Roller Bearing, Double-Row, Inner-Ring Raceway Spherical [Symbol]

SW Series Winding; Series-Wound [Electrical Engineering]

SW Shear Wave

SW Shipper's Weight

SW Shock Wave

SW Short Wave [also S-W, s-w, or sw]

SW Shunt Winding; Shunt-Wound [Electrical Engineering]

SW Sine Wave

SW Single Wheel
SW Single Weight
SW Sky Wave
SW Slow Wave
SW Software [also S/W]
SW Soft Water
SW Softwood
SW Solar Wind [Geophysics]
SW Solid Waste
SW Sound Wave
SW Sound Wormy [Lumber]
SW South Wales [Australia]
SW Southwest(ern) [also sw]
SW Space Wave
SW Specific Weight
SW Spherical Washer
SW Spherical Wave
SW Spin Wave [Solid-State Physics]
SW Spiral Wave
SW Spot Welding
SW Square Wave
SW Staebler-Wronski (Effect) [Solid-State Physics]
SW Standing Wave [also S-W]
SW Stationary Wave
SW Station Wagon
SW Statistical Weight
SW Status Word
SW Stillinger-Weber (Potential) [Physics]
SW Stoner-Wohlfarth (Model) [also S-W] [Solid-State Physics]
SW Stop Watch
SW Storm Water
SW Stress Wave
SW Stud-Arc Welding
SW Sum-of-Weights (Method)
SW Surface Wave
SW Surface Wind [also S-W]
SW S-Wave [Geophysics]
SW Switch(ing)
SW Switchband-Wound
6W- Senegal [Civil Aircraft Marking]
S&W Smith and Wesson (Revolver)
S/W Salt-in-Water (Solution)
S/W Software [also s/w]
S/W Span-to-Width (Ratio)
S/W Surface Wind [also SW]
S-W Short-Wave [also SW, sw, or s-w]
S-W Standing Wave [also SW]
S-W Stoner-Wohlfarth (Model) [also SW] [Solid-State Physics]
Sw Swash [also sw]
Sw Swede(n); Swedish
Sw Swiss; Switzerland
Sw Switch(er) [also sw]

SWA Scheduler Work Area
SWA Single Wire Armor
SWA Sound Wave Analyzer
SWA Southern Wholesalers Association [US]
SWA Southern Woodwork Association [US]
SWA Southwest Asia(n)
SWA South West Africa(n)
SWA Steel Window Association [UK]
SWA Support Work Authorization
SWA System Work Area
SWAA Spacelab Window Adapter Assembly
SWAC Special Weapons Ammunition Command
SWAC (National Bureau of) Standards Western Automatic Computer
SWACS Space Warning and Control System
SWAD Special Warfare Aviation Detachment [of US Army]
SWAD Subdivision of Work Authorization Documents
SWADE Surface Wave Dynamics Experiment
SWADS Scheduler Work Area Data Set
SWAFEC Southwest Atlantic Fisheries Advisory Commission
SW Afr South West Africa(n)
SWAIS Simple Wide Area Information Server [Internet]
SWALCAP Southwest Academic Libraries Cooperative Automation Project [UK]
SWALK Sealed With A Loving Kiss [Internet Jargon]
SWAM Standing Wave Area Monitor
SWAMI Sidewall Masked Isolation
SWAMI Software-Aided Multifont Input
SWAMWU South West Africa Mine Workers Union [Namibia]
SWAN Service Wide Area Network
SWANA Solid Waste Association of North America
SWAP Shared Wireless Access Protocol
SWAP Standard Wafer Array Programming
SWAP Stewart-Warner Array Program
SWAP Stress Wave Analyzing Program
SWASS Screwworm Adult Suppression System
SWAT Sidewinder Acquisition Track
SWAT Special Warfare Armored Transporter
SWAT Special Weapons and Tactics
SWAT Stress Wave Analysis Technique
SWATH Small-Waterplane-Area Twin-Hull Ship
S Wave Negative Downward Deflection Following R Wave of QRS Complex (of Electrocardiogram) [also S wave] [Medicine]
SWB Short Wheelbase
SWB Single Weight Baryta
SWB Solar-Wasserstoff-Bayern (Process) [Chemical Engineering]
SWB Southwestern Bell (Telephone Company) [US]
SWB Summary of World Broadcasts (Database) [UK]
Swbd Switchboard [also swbd]
Sw Bhd Swash Bulkhead [Oil Tankers]
SWBP Service Water Booster Pump

SWBS Ship Work Breakdown Structure

SWBXT Synchrotron White Beam X-Ray Topography

SW by S Southwest by South

SW by W Southwest by West

SWC Shock Wave Consolidation

SWC Solid Waste Cask

SWC Special Weapons Center

SWC Stepwise Cracking [Metallurgy]

SWC Surge Withstand Capability

SWCAA Soil and Water Conservation Association of Australia

SWCC Second World Climate Conference

Swch Switch [also swch]

SWCL Sea Water Conversion Laboratory

SWCRD Soil and Water Conservation Research Division [US]

SWCS Soil and Water Conservation Society [US]

SWD Self-Wiring Data

SWD Sliding Watertight Door

SWD Smaller Word

SWD Standing-Wave Detector

SWD Surface Wave Device

SWDA Solid Waste Disposal Act [US]

SWDC Sleep-Wake Disorders Center [at Montefiore Medical Center, New York City, US]

SWDL Safe Winter Driving League [US]

SWDR Single Way Dynamic Range

SWE Secondary Work Embrittlement [Metallurgy]

SWE Seiberg-Witten Equation

SWE Society of Women Engineers [US]

SWE Software Engineer(ing)

SWE Spherical Wave Expansion

SWE Status Word Enable

SWEAT Student Work Experience and Training

S-Web Supported Web [Solar Cell Manufacture]

SWEC Stone and Webster Engineering Corporation [US]

Swed Swede(n); Swedish

Swed Dent J Swedish Dental Journal

Swed Dent J Suppl Swedish Dental Journal Supplement

SwedP Swedish Patent

SWEEP Small Window Energy Extension Process

SWEEP Soil and Water Environmental Enhancement Program [Canada]

Swep Methyl-n-3,4-Dichlorophenylcarbamate [also swep]

SWEPP Stored Waste Examination Pilot Plant

SWET Simulated Water Entry Test

SWF Service Workers Federation [San Marino]

SWF Shortwave Fade-Out

SWF Statistical Weight Factor

SWF Stress Wave Factor

SWF Structural Weight Fraction

SWF Sudden Wave Fade-Out

SWF Südwestfunk [Broadcasting Station in Southwestern Germany]

SWFG Secondary Waveform Generator

SWF/ISS Stress Wave Factor/Interlaminar Shear Strength [Mechanics]

SWFR Slow Write/Fast Read [also SW/FR]

Sw Fr Swiss Franc [Currency of Switzerland and Liechtenstein] [also SwFr]

SWG Science Working Group

SWG Society of Women Geographers [US]

SWG Software Working Group

SWG Special Working Group [of NATO]

SWG Standard Wire Gauge

SWG Steel Wire Gauge

SWG Stubs Wire Gauge

SWG Switchgear [also Swg]

Swg Swinging [also swg]

SWGB South Western Gas Board [UK]

Swg Bkt Swinging Bracket

Swgr Switchgear [also swgr]

SWH Solar Water Heater

SWI Sealant and Waterproofers Institute [US]

SWI Short-Wave Interference

SWI Sidewall Interface [Laser Technology]

SWI Software Interrupt

SWI Special World Interval

SWI Steel Window Institute [US]

SWIE South Wales Institute of Engineers [Australia]

SWIEEECO Southwestern Institute of Electrical and Electronics Engineers Conference and Exhibition [US]

SWIFT Selected Words in Full Title

SWIFT Signal Word Index of Field and Title

SWIFT Significant Word in Full Title

SWIFT Society for Worldwide Interbank Financial Telecommunications

SWIFT Society for Worldwide Interbank Financial Transactions

SWIFT Software Implemented Friden Translator

SWIFT Stored Waveform Inverse Fourier Transform

SWIFT Strength of Wings Including Flutter [Aeronautics]

SWIFTLASS Signal Word Index of Field and Title Literature Abstracts Specialized Search

SWIFTSIR Signal Word Index of Field and Title Science Information Retrieval

SWINPC Southwest Institute of Nuclear Physics and Chemistry [PR China]

SWIP Seal Worm Intervention Program [Canada]

SWIP Southwest Institute of Physics [PR China]

SWIP Standing Wave Impedance Probe

SWIR Short-Wave(length) Infrared

SWIRA Swedish Industrial Robot Association

SWIRECO Southwestern Institute of Radio Engineers Conference and Electronics Show [US]

SWIRLS Stratospheric Wind Infrared Limb Sounder

SWIRS Solid Waste Information Retrieval System [of US Environmental Protection Agency]

SWIS Shrinkage/Warpage Interface-to-Stress Software

SWISH Simple Web Indexing System for Humans [Internet]

Swiss Biotech Swiss Biotech–Swiss Review for Biotechnology

Swiss Bull Mineral Petrol Swiss Bulletin of Mineralogy and Petrology [of University of Basle, Switzerland]

Swiss Chem Swiss Chem–Swiss Review for the Chemical Industry

Swiss Contam Contr Swiss Contamination Control–Swiss Review for Clean Room Technology

Swiss Food Swiss Food–Swiss Review for the Food Industry

Swiss Materials Swiss Materials–Swiss Review for Materials Technology

Swiss Med Swiss Med–Swiss Review for Medicine and Medical Technology

SwissP Swiss Patent

Swiss Pharma Swiss Pharma–Swiss Review for the Pharmaceutical Industry

Swiss Plastics Swiss Plastics–Swiss Review for the Plastics Industry

SWISSPRO Swiss Association for the Simplification of the Procedures in International Trade

Swiss Vet Swiss Vet–Swiss Review for Veterinary Medicine

SWITS Solid Waste Information and Tracking System

SWITT Surface Wave Independent-Tap Transducer

Switz Switzerland

SwKn Swedish Krona [Currency]

SWL Safe Working Load

SWL Shortwave Listener

SWL Soil Washing Laboratory [US]

SWL Specific Work Load

SWL Standard Wavelength

SWL Sulfite Waste Liquor

SWL Surface Wave Line

SWL Swaziland Lilangeni [Currency]

SWLA Southwestern Library Association [US]

SWLI Southwestern Louisiana Institute [Lafayette, US]

SWM Solid Waste Management

SWM Standing-Wave Meter

SWMA Scottish Wirework Manufacturers Association

SWMA Steel Wool Manufacturers Association [UK]

SWMAS Shropshire and West Midlands Agricultural Society [UK]

SWMSU Southwest Missouri State University [Springfield, US]

SWMU Solid Waste Management Unit

SWMV Soil-Borne Wheat Mosaic Virus

SW-NE Southwest-Northeast

SWNT Single-Walled Nanotube

SWOB Salaries, Wages, Overhead and Benefits

SWOC Steelworkers Organization Committee [US]

SWOF Switchover Operation Failure

SWOP Structural Weight Optimization Program

SWOP Switchable Input Operation

SWOPAMP Switchable Input Operational Amplifier [also SWOP AMP]

SWOPS Single Well Offshore Production System

S-Word Spindle-Speed Word [Numerical Control]

SWOV Stichting Wetenschappelijk Onderzoek Verkeersveiligheid [Foundation for Scientific Road Safety Research, Netherlands]

SWP Safe Working Pressure

SWP Saskatchewan Wheat Pool [Canada]

SWP Service Water Pump

SWP Simple Web Printing [Internet]

SWP Society for Women in Plastics [US]

SWP Software Pack

SWP Sound Wave Photograph(y)

SWP South-West Pacific

SWP Soviet Warsaw Pact

SWP Special Work Permit

SWP Square-Wave Polarography

SWP Systematic Withdrawal Plan

S-W-P Squires-Weiner-Phillips (Creep) [also SWP] [Metallurgy]

.SWP Swap [File Name Extension] [also .swp]

SWPA Southwestern Power Administration [US]

SWPA Steel Works Plant Association [US]

SWPA Submersible Wastewater Pump Association [US]

SWPP Southwest Power Pool [US]

SWPPP Storm Water Pollution Prevention Plan

SWQI South West Queensland Initiative [Australia]

SWR Service Water Reservoir

SWR Short-Wavelength Radar

SWR Sine Wave Response

SWR Single Wafer Reactor

SWR Sodium Water (Nuclear) Reactor

SWR Spin Wave Resonance [Solid-State Physics]

SWR Square Wave Recording [Electronics]

SWR Standing-Wave Ratio

SWR State Wildlife Reserve [US]

SWR Steel Wire Rope

SWRA Selected Water Research Abstracts [of Water Resources Scientific Information Center, US]

SwRI Southwest Research Institute [San Antonio, Texas, US] [also SWRI, or SRI]

SWRT Short-Wave Radio Telephony

SWS Safe Working Stress [Mechanics]

SWS Service Water System

SWS Shallow Water Submersible

SWS Shift Word, Substituting

SWS Single-White-Silk-Covered (Electric Wire)

SWS Slow-Wave Sleep [Psychology]

SWS Software System

SWS Switch Scan

SWSA Southern Wood Seasoning Association [US]

SWSAC Statistical Weights of Stable Atomic Configurations

SWSC Southwestern State College [Weatherford, Oklahoma, US]

SWSI Single Width, Single Inlet

SWST Service Water Storage Tank

SWST Society of Wood Science and Technology [US]

SWT Scottish Wildlife Trust [UK]

SWT Short-Wave Telegraphy

SWT Southwest Texas

SWT Structured Walkthrough [Computers]

SWT Supersonic Wind Tunnel

SWTL Surface-Wave Transmission Line

Swtr Seawater

SWTS Solid Waste Technology Support

SWTSU Southwest Texas State University [San Marcos, US]

Swtz Switzerland

SWU Selective Work Unit

SWU Separate (or Separative) Work Unit

SWU Southwestern University [Georgetown, Texas, US]

SWUCNET Southwest Universities Computer Network [UK]

SWULSCP Southwest University Libraries Systems Cooperative Project [now Southwest Academic Libraries Cooperative Automation Project, UK]

SW USA South Western United States of America [also SW-USA]

SWVR Standing-Wave Voltage Ratio

SWW Severe Weather Warning [Meteorology]

Swy Stopway [also swy]

SX Simplex [also sx]

SX Single Crystal

SX Solvent Extraction

SX- Greece [Civil Aircraft Marking]

ΣΞ Sigma Xi (Student Society)

SXA Soft X-Ray Absorption

SXANES Surface X-Ray Absorption Near-Edge Structures

SXAPS Soft X-Ray Appearance Potential Spectroscopy

SXAPS Soft X-Ray Appearance Potential Spectrum

SXAS Soft X-Ray Absorption Spectroscopy

SXB Spring Extension Bar

SXE Soft X-Ray Emission

SXES Soft X-Ray Emission Spectroscopy

SX/EW Solvent Extraction/Electrowinning (Process) [also SX-EW, or SXEW]

SXGM Spectrum X-Gamma Mission

SXI Solar X-Ray Imager

SxLFD (Workshop on) Small-x Physics and Light-Front Dynamics in Quantum Chromodynamics

SXM Scanning X-Ray Microscopy

SXM Soft X-Ray Microscopy

SXM (Conference on Development and Technological Application of) Scanning Probe Methods

SXMCD Soft X-Ray Magnetic Circular Dichroism

SXN Section [also Sxn, or sxn]

SXPES Soft X-Ray Photoelectron Spectroscopy [also SXPS]

SXPM Scanning X-Ray Photoelectron Microscopy

SXPS Soft X-Ray Photoelectron Spectroscopy [also SXPES]

SXPS Soft X-Ray Photoemission Spectroscopy

SXR Scanning X-Ray Radiography

SXRA Synchrotron X-Ray Fluorescence Analysis

SXRD Surface X-Ray Diffraction

SXRG Supervoltage X-Ray Generator

SXRM Soft X-Ray Microscopy

SXRT Society of X-Ray Technology [UK]

SXS Step by Step

SXS Soft X-Ray (Emission) Spectroscopy

SXT Society of X-Ray Technology [UK]

SXT Space(craft) Sextant [of NASA]

Sxt Sextant

SXW Standing X-Ray Wavefield

SY Serrated Yield(ing) [Materials Science]

SY Shoulder Yaw

SY Sidereal Year [Astronomy]

SY South Yemen

SY Stripping Yield

SY Syria [ISO Code]

SY System [also Sy, or sy]

6Y- Jamaica [Civil Aircraft Marking]

Sy Symbol [also sy]

Sy Synthetic(s)

Sy Syli [Currency of Guinea]

sy square yard [also sq yd, or yd²]

sy synchronize

σ/Y (Residual) Stress to Yield Stress [Mechanics]

SyAS Surrey Archeological Society [UK]

SYBCO Samarium Yttrium Barium Copper Oxide (Superconductor)

SYCLOPS SYFA (System for Access) Current Logic Operating System

SYCOM Synchronous Communications

Sycomom Syndicat des Constructeurs Belges de Machine-Outils pour le Travail des Métaux [Association of Belgian Manufacturers of Machine-Tools for Metalworking Trades]

SYD Sum-of-the-Years Digits

SYD Yemeni Dinar [Currency of People's Republic of Yemen]

SYDAS System Data Acquisition System

SYDEC Selective Yield Delayed Coking (Process) [Chemical Engineering]

SYDOX Sydney Oxidation (Process)

SYEP Summer Youth Employment Program

SYFA System for Access

Syl Syllabus [also syl]

.SYL Syllabus [File Name Extension] [also .syl]

SYLCU Synchronous Line Control Unit

SYLK Symbolic Link (Format)

.SYM Symbols [File Name Extension] [also .sym]

Sym Symbol(ic) [also sym]

Sym Symmetry [also sym]

Sym Symphysis [Anatomy]

sym symmetrical

sym- symmetrical (structure of an organic compound) [Symbol]

SYMAN Symbol Manipulation

Symb Symbol(ic) [also symb]

Symb Comput Symbolic Computation

SYMBIOSIS System for Medical and Biological Information Searching [of New York State University, US]

Symb Log Symbolic Logic

SYMDEB (Microsoft) Symbolic Debug Utility [also symdeb]

Symm Symmetrical; Symmetry [also symm]

Symp Symposium [also symp]

Symp Soc Exper Biol Symposium of the Society for Experimental Biology [UK]

Symp Zool Soc Lond Symposium of the Zoological Society of London [UK]

SYMPAC Symbolic Program for Automatic Control

Sym Set Symbol Set

SYN Synchronous (Idle Character)

.SYN Synonym [File Name Extension] [also .syn]

Syn Synchronization; Synchronous [also syn]

Syn Synonym(ous) [also syn]

Syn Synthesis; Synthesizer; Synthetic [also syn]

Sync Synchronization; Synchronize; Synchronous [also sync]

snch synchronize; synchronous

Synchromesh Synchronous Mesh [also synchromesh] [Automobiles]

SYNCOM Synchronous-Orbit Communications

SYNCOM Synchronous-Orbit(ing) Communications Satellite [Series of 3 Communications Satellites Launched by NASA between February 1963 and August 1964] [also Syncom]

Syncrude Synthetic Crude Oil [also syncrude]

Syncrude Synthetic Crude Oil Consortium

Synd Syndicate [also synd]

synd syndicate(d)

Syndet Synthetic Detergent [also syndet]

Synfuel Synthetic Fuel [also synfuel]

Syngas Synthetic Gas [also syngas]

SYNO Strontium Yttrium Niobium Oxide (Superconductor) [also SrYNO]

Syn Pyr Synthetic Pyrethroid

SYNROC Synthetic Rock (Process) [Nuclear Waste Management]

SYNROC-C Synthetic Rock C (Process) [Nuclear Waste Management]

SYNSAT Synergetic Saturization (Process)

Synscp Synchroscope [also synscp]

SYNSEM Syntax and Semantics

synt synthetic

Syntan Synthetic Tannin; Synthetic Tanning Material [also syntan] [Leather Industry]

Synth Synthesis

Synth Synthetic(s) [also synth]

synth synthetic

Synthehol Synthetic Alcohol [also synthehol]

Synth Met Synthetic Metals [Journal published in Switzerland]

SYNTHOIL Synthetic Oil (Process) [also Synthoil or synthoil]

SYNTOL Syntagmatic Organization of Language

SYNTRAN Synchronous Transmission

SYP Syrian Pound [also SY£; Currency]

Syr Syria(n)

Syr Syrup [also syr]

SYRCO Strontium Yttrium Ruthenium Copper Oxide (Superconductor) [also SrYRCO]

SYROCO Symposium on Robotic Control [of International Federation of Automatic Control, Austria]

syry syrupy [also syry]

SYS System File(s) Transfer (Command) [also sys]

SYS Systemic Effects (of Hazardous Materials)

.SYS System Configuration [File Name Extension]

.SYS System Device Driver [File Name Extension]

Sys System [also sys]

sys systemic [Medicine]

SYSADMIN System Administration [also SysAdmin]

SYSADMSH System Administration Shell [Santa Cruz Operating System Unix] [also sysadmsh]

SYSBACK System Backup and Recovery

SYSCAP System of Circuit Analysis Programs

Sysco Sydney Steel Corporation [Sydney, Nova Scotia, Canada]

SYSCOM System Command [also Syscom]

SYSCON System Configuration [also Syscon]

SYSCTLG System Catalog [also Sysctlg]

SYSDES System Design [also Sysdes]

SYSDEV System Development [also Sysdev]

SYSDOC System Documentation [also Sysdoc]

SYSDYN System Dynamics [also Sysdyn]

SysE Systems Engineer [also SysEng]

Sys Eng Systems Engineer(ing)

SYSEV System Evaluation [also SYSEVA]

SYSGEN System Generation [also SysGen, or sysgen]

SYSIN System Input

SYSINT System Integration [also Sysint]

SYSLIB System Library [or Syslib]

SYSLO System Loader

SYSLOG System Log

SYSMOD System Modification

SYSOP System(s) Operation; System(s) Operator [also Sysop]

SYSOP System Optimization [also Sysop]

SYSOUT System Output

SYSPOP System Programmed Operator

SYSRC System Reference Count

SYSREQ System Request [also SysReq]

SYSRES System Residence Volume

SYSRES Systems Research(er) [also SysRes, or sysres]

Sys Sci Systems Science; Systems Scientist [also Syst Sci]

Syst System

Syst Systematic(s) [also syst]

Syst Anal–Model Simul Systems Analysis–Modelling–Simulation [Journal published in Germany]

Syst Comput Jpn Systems and Computers in Japan [Journal published in the US]

Syst Control Systems and Control [Journal published in Japan]

Syst Control Inf Systems, Control and Information [Publication of the Institute of Systems, Control and Information Engineers, Japan]

Syst Control Lett Systems and Control Letters [Journal published in the Netherlands]

Syst Dev System Development [Journal published in the US]

Syst Dyn Rev System Dynamics Review [of System Dynamics Society, US]

SYSTEC International Trade Fair and Congress for System Integration, Automation, Technology and Quality Assurance [also Systec, Germany]

Sys Tech System Technology; System Technologist [also Syst Tech]

System Systematic(s) [also system]

System Bot Systematic Botany [of American Society of Plant Taxonomists, US]

System Bot Monogr Systematic Botany Monographs [of American Society of Plant Taxonomists, US]

System Ent Systematic Entomology [Publication of the Royal Entomological Society of London, UK]

SYSTEMS Computers, Communications, Applications–International Trade Fair and Congress [Germany]

Syst Int Systems International [Journal published in the UK]

Syst Integr Systems Integration [Journal published in the US]

Syst Log Systèmes Logiques [Swiss Journal on Logic Systems]

SYSTRAN Systems Analysis Translator

Syst Res Systems Research(er)

Syst Res Systems Research [Journal published in the UK]

Syst Res Inf Sci Systems Research and Information Science [Journal published in the UK]

Syst Sci Systems Science; Systems Scientist [also Sys Sci]

Syst Sci Systems Science [Polish Journal]

Sys Tech System Technology; System Technologist [also Syst Tech]

Syst Technol Systems Technology [UK Journal]

Syst Tech Rep Systems Technical Report [Journal published in Japan]

Syst 3X World Systems 3X World [Journal published in the US]

Syst User Systems User [Journal published in the US]

SYSUP System Supervision; System Supervisor

SYSVER System Verification [also Sysver, or sysver]

SYU Synchronization Signal Unit

SYU Synchronization Signal Utility

SYVAC Systems Variability Analysis Code

SZ Santiago-Zamora [Ecuador]

SZ Send Z-Modem [Unix Operating System]

SZ Separation Zone

SZ Sintered Zirconia

SZ Swaziland [ISO Code]

Sz Seizure [also sz]

Sz Size [also sz]

SZA Solar Zenith Angle [also sza] [Astronomy]

SZI Strontium Zinc Iridate

SZL Swaziland Lilangeni [Currency]

SZM Sensing-Zone Method [Chemistry]

SZM Structure (or Structural) Zone Model [Materials Science]

SZP Strontium Zinc Platinate

SZR Strontium Zinc Rhodate

SZVR Silicon Zener Voltage Regulator

SZW Stretch Zone Width [Fracture Mechanics]

SZW_c Critical Stretch Zone Width [Fracture Mechanics]

T   Absolute Temperature [Symbol]

T   Analyzer Response Function (in Auger Spectroscopy) [Symbol]

T   Axisymmetric Loading (in Mechanics) [Symbol]

T   Breaking Toughness (of Geotextiles) [Symbol]

T   Burning Time (of Rockets) [Symbol]

T   Circular Tooth Thickness (of Gears) [Symbol]

T   Critical Shear Stress (for Soil, or Rock) [Symbol]

T   Cutting Time (in Machining) [Symbol]

T   Depth of Focus [Symbol]

T   Englehardt Value (in Draw Fracture Test) [Symbol]

T   Equivalent Blackbody Temperature [Symbol]

T   Fractional Transmission of Analyzer and Detector (in Ion Spectroscopy Scattering) [Symbol]

T   Heat Treated to Produce Stable Tempers other than F, H, or O [Basic Temper Designation for Aluminum Alloys]

T   Isotopic Spin (in Nuclear Physics) [Symbol]

T   Kinetic Energy [Symbol]

T   Long Transverse Direction (in Fracture Testing of Rectangular Specimens) [Symbol]

T   Metallurgical and Materials Engineering (Technology) [Discipline Category Abbreviation]

T   Number of Teeth (of Gears, etc.) [Symbol]

T   Period (of Vibrations, Oscillations, etc.) [Symbol]

T   Pseudospin (in Physics) [Symbol]

T   Reverberation Time (in Acoustics) [Symbol]

T   D-Ribosylthymine (in Biochemistry) [Symbol]

T   Student's T-Statistic [Symbol]

T   Surface Tension [Symbol]

T   Tab [Aerospace]

T   Tab(ulation); Tabulator

T   Table

T   Tablespoon(ful)

T   Tablet

T   Tabriz [Iran]

T   Tachometer

T   Tack(ing)

T   Taenia [Genus of Tapeworms]

T   Tag(ging)

T   Tahiti

T   Tail

T   Tailings

T   Taipei [Taiwan]

T   Taiwan(ese)

T   Talk(er); Talking

T   Talk/Monitor

T   Tallahassee [Florida, US]

T   (+)-Talose [Biochemistry]

T   Tampa [Florida, US]

T   Tamp(er); Tamping

T   Tangent(ial)

T   Tangier [Morocco]

T   Tango [Phonetic Alphabet]

T   Tank

T   Tanker; Tanking

T   Tan(ning)

T   Tanzania(n)

T   Tap

T   Tapping

T   Tape

T   Taper

T   Taper Chamfer (of Taps) [Symbol]

T   Target(ing)

T   Tarnish(ing)

T   Tarsal; Tarsus [Anatomy]

T   Tashkent [Uzbekistan]

T   Tasmania(n) [Australia]

T   (Uppercase) Tau [Greek Alphabet]

T   Tautomer(ism) [Chemistry]

T   Taxi(ing) [Aeronautics]

T   Taxus [Genus of Coniferous Tees Including the Yews]

T   Tbilisi [Georgia]

T   Teach(er); Teaching

T   Team

T   Tear(ing)

T   Tearing

T   Technical; Technician; Technique

T   Technological; Technologist; Technology

T   Tee (Joint)

T   Teem(ing) [Metallurgy]

T   Tegucigalpa [Honduras]

T   Teheran [Iran]

T   Telegraph(y)

T   Telemeter(ing); Telemetry

T   Telephone; Telephony

T   Telescope; Telescopic

T   Tempe [Arizona, US]

T   Temperature

T   Temper(ed); Tempering

T   Tempo
T   Tenacity
T   Tennessee [US]
T   Tensile; Tensional
T   Tension
T   Tensor
T   Tent
T   Tentacle [Zoology]
T   Tera- [SI Prefix]
T   Term
T   Terminal
T   Terminal (in Electrical Engineering) [Often Followed by a Number, e.g., T_1, or T1 Denotes Terminal 1]
T   Termination; Terminator
T   Terrace
T   Terrain
T   Territorial; Territory
T   Tertiary [Geological Period]
T   Tesla [Unit]
T   Test(ing); Tester
T   Testacella [Genus of Slug]
T   Testis [Anatomy]
T   Testosterone [Biochemistry]
T   Tetanus [Medicine]
T   Tether(ed)
T   Tetragon(al)
T   Tetragonal (Crystal) System [Symbol]
T   Tetrahedral; Tetrahedron
T   Tetrahedral Site (in Crystallography) [Symbol]
T   Tetraspore [Botany]
T   Tetrode
T   Tetrose [Biochemistry]
T   Texan
T   Texas [US]
T   Text
T   Textile
T   Textural; Texture
T   Thai(land)
T   Thallus [Botany]
T   Thames [US]
T   Theorem
T   Theoretical; Theoreticist; Theory
T   Therapeutic; Therapist; Therapy
T   Thermal(ism)
T   Thermodynamic(s)
T   Thermodynamic Temperature [Symbol]
T   Thermometer
T   Thermopile
T   Thermoplastic(s)
T   Thermoplastic (Conductor Insulation) [Symbol]
T   Thermoset(s)
T   Thermostabilization; Thermostabilize(d); Thermostabilizer
T   Thickness

T   Thomson-Friedenreich (Antigen) [Immunology]
T   Thoracic; Thorax [Anatomy]
T   Thread
T   (−)-Threonine; Threonyl [Biochemistry]
T   Threshold
T   Throat
T   Thrombin(ogen) [Biochemistry]
T   Thromboplastin [Biochemistry]
T   Thrombus [Medicine]
T   Throttle
T   Throughput [Symbol]
T   Throw
T   Throwing
T   Thrust
T   Thuja [Genus of Trees Including the Arborvitae]
T   Thunder(storm)
T   Thymine [Biochemistry]
T   Thymidine [Biochemistry]
T   Thymus [Anatomy]
T   Thyroid [Anatomy]
T   Thyroxine [Biochemistry]
T   Tiber [Italy]
T   Tibet(an)
T   Tibia [Anatomy]
T   Tick
T   Tidal; Tide(s)
T   Tientsin [PR China]
T   Tight(ness)
T   Tigris (River) [Southwest Asia]
T   Tijuana [Mexico]
T   Tilia [Genus of Trees Including the Basswoods]
T   Tilt(ing)
T   Timber
T   Time
T   Timer; Timing
T   Time Factor [Symbol]
T   Time Inversion (in Particle Physics)
T   Time Period [Symbol]
T   Time Reversal (in Physics) [Symbol]
T   Tinea [Medicine]
T   Tint(ing)
T   Tip(ping)
T   Tipper [Mining]
T   Tirana [Albania]
T   Tire
T   (Maximum) Tire Speed of 118 mph (or 190 km/h) [Rating Symbol]
T   Tissue
T   Titania (or Titanium Dioxide)
T   Titanium [Abbreviation; Symbol: Ti]
T   Titer; Titrant; Titration
T   Titrimeter; Titrimetry
T   Tobacco
T   Toggle

T	Togo		T	Tract
T	Tokyo [Japan]		T	Traction(al)
T	Toll		T	Tractor
T	Tolerant; Tolerance		T	Trade(r); Trading
T	Tolerance (in Machining) [Symbol]		T	Traffic
T	Toledo [Ohio, US]		T	Trail
T	Toluene		T	Trailer
T	Tomography		T	Train(ing)
T	Ton, Metric [Unit]		T	Trainer (Aircraft) [US Air Force and US Navy Symbol]
T	Tonga [South Pacific Islands]		T	Training Mission (Vehicle) [USDOD Symbol]
T	Tone		T	Trajectory
T	Toner		T	Trans (Configuration) [Chemistry]
T	Tongs		T	Transaction
T	Tongue		T	Transduce(r)
T	Tonnage		T	Transduction [Microbiology]
T	Tonus [Biology]		T	Transfer(ence)
T	Tool(ing)		T	Transform(ation); Transformer
T	Tool Life [Symbol]		T	Transformation Matrix [Waveguides]
T	Tooth		T	Transient
T	Top		T	Transistor
T	Topeka [Kansas, US]		T	Transit(ion)
T	Topic(al)		T	Transition Metal
T	Topographic; Topography		T	Translation(al)
T	Topological; Topology		T	Translucency; Translucent
T	Torch		T	Transmissibility
T	Tornado		T	Transmission
T	Toroid(al)		T	Transmission Coefficient [Symbol]
T	Toronto [Canada]		T	Transmissivity [Symbol]
T	Torpedo		T	Transmit(ted); Transmitter
T	Torque [Symbol]		T	Transmittance [Symbol]
T	Torr [Unit]		T	Transparency; Transparent
T	Torsion(al)		T	Transpiration; Transpire [Biology]
T	Torsion [Symbol]		T	Transplant(ation)
T	Torsional Modulus [Symbol]		T	Transport(ation)
T	Torsional Moment [Symbol]		T	Transpose Matrix [Mathematics]
T	Tortoise		T	Transversal; Transverse
T	Torus		T	Transverse Direction
T	Total(ity)		T	Trapezoid(al)
T	Touch		T	Trap
T	Tough(ness)		T	Trapping
T	Toulouse [France]		T	Trauma(tic)
T	Tower		T	Travel
T	Town(ship)		T	Travel (of Machines, or Tools) [Symbol]
T	Toxic(ity)		T	Traverse
T	Toxicological; Toxicologist; Toxicology		T	Tray
T	Toxigenicity [Microbiology]		T	Treat(ment)
T	Toxin [Biochemistry]		T	Treasurer; Treasury
T	Trace; Tracing		T	Tree
T	Trace Command [Computers]		T	Trematode [Genus of Flatworms]
T	Tracer		T	Trend
T	Trachea [Anatomy]		T	Trenton [New Jersey, US]
T	Track		T	Treponema [Genus of Microorganisms]
T	Tracker; Tracking		T	Trial

T	Triangle; Triangular
T	Triassic [Geological Period]
T	Triaxial(ity)
T	Triceps [Anatomy]
T	Trichinella [Genus of Roundworms]
T	Trichomonas [Genus of Protozoans]
T	Trichuris [Genus of Whipworms]
T	Trigger
T	Trigonal (Phase) [Crystallography]
T	Trillion
T	Trillium [Botany]
T	Trim(mer); Trimming
T	Trinitrotoluene
T	Triode [Semiconductor Symbol]
T	Triose [Biochemistry]
T	Tripoli [Libya]
T	Triplet
T	Tritium [also t, ^{3}H, or H-3]
T	Trolley
T	Trough
T	Trowel
T	Troy (Weight) [Unit]
T	Truck(er); Trucking
T	True(ness)
T	Trunk
T	Truss
T	Truth [Particle Physics]
T	Truth [Computers]
T	Trypanosoma [Genus of Flagellated Parasites]
T	Trypsin(ogen) [Biochemistry]
T	Tsingtao [PR China]
T	Tsuga [Genus of Coniferous Trees Including the Hemlocks]
T	Tube; Tubing; Tubular
T	Tubular Propellant [Symbol]
T	Tucson [Arizona, US]
T	Tuesday
T	Tulsa [Oklahoma, US]
T	Tumble(r); Tumbling
T	Tumor(ous)
T	Tuna
T	Tune(r); Tuning
T	Tungsten-Type High-Speed Tool Steels [AISI-SAE Symbol]
T	Tunis [Tunisia]
T	Tunisia(n)
T	Tunnel(ing)
T	Turbid(ity)
T	Turbine
T	Turbocharge(d); Turbocharger
T	Turbulence; Turbulent
T	Turin [Italy]
T	Turk(ish); Turkey
T	Turn(ing)
T	Turret
T	Turtle
T	Twin [Biology]
T	Twin(ning) [Crystallography]
T	Twine
T	Twist(ing)
T	Twisting Moment [Symbol]
T	Tympanum [Anatomy]
T	Type
T	Typha [Botany]
T	Yarn Number [Symbol]
T	Wu's Tensor [Symbol]
α-T	2-Methyl-1,3,5-Trinitrobenzene
-T	Training Aircraft [US Navy Suffix]
/T	Title Name Option [MS-DOS Shell]
/T	Turbo Directory Option [Turbo Pascal]
T+	Time Postintegration [Aerospace]
T−	Time Prior to Launch [Aerospace]
$\dot{T}$	Cooling Rate [Symbol]
T*	Characteristic Temperature [Symbol]
T*	Curie Temperature [Symbol]
T*	(Electron) Hole-Superconductor [Symbol]
T′	Electron-Superconductor [Symbol]
°T	True Degrees [Symbol]
T$^+$	Positive Translation Operation (in Crystallography) [Symbol]
T^{++}	Bloch Wave Transmission Probability (in Solid-State Physics) [Symbol]
T$^-$	Negative Translation Operation (in Crystallography) [Symbol]
T	Bloch Wave Transmission Probability (in Solid-State Physics) [Symbol]
T$_\parallel$	Tension Parallel to Surface [Symbol]
T$_\perp$	Tension Perpendicular to Surface [Symbol]
T$_0$	Absolute Temperature [Symbol]
T$_0$	Ambient Temperature [Symbol]
T$_0$	Curie-Weiss Temperature (in Solid-State Physics) [Symbol]
T$_0$	Film Thickness at Time T$_0$ [Symbol]
T$_0$	Initial Temperature [Symbol]
T$_0$	Matrix Fracture Toughness (of Composites) [Symbol]
T$_0$	Melting Temperature [Symbol]
T$_0$	Normal Thickness (of Surface Rocks) [Symbol]
T$_0$	(Unconfined, or Uniaxial) Tensile Strength (of a Cylindrical Rock, or Soil Specimen) [Symbol]
T$_0$	Time Out [Air Navigation]
T$_0$	Unbalanced Torque [Symbol]
T$_{1/2}$	Half-Life (of Radioactive Substances) [Symbol]
T^{-1}	Bandwidth [also 1/T] [Symbol]
T^{-1}	Reciprocal of Temperature [also 1/T] [Symbol]
T^{-1}	One per Tesla [also 1/T] [Symbol]
T$_1$	Initial Temperature [Symbol]
T$_1$	Temperature of Source (for Reversible Engines) [Symbol]
T^{-2}	One per Tesla Squared [also 1/T^2] [Symbol]

T$_2$ Final Temperature [Symbol]

T$_2$ Temperature of Condenser (for Reversible Engine) [Symbol]

T$_3$ 3,3',5-Triiodo-L-Thyronine [Biochemistry]

T$_4$ L-Thyroxine (or Tetraiodothyronine) [Biochemistry]

T$_A$ Ambient Temperature (or Free-Air Temperature) [Semiconductor Symbol]

T$_A$ Annealing Temperature [Symbol]

T$_A$ Austempering Temperature (in Metallurgy) [Symbol]

T$_a$ Aging Temperature (in Metallurgy) [Symbol]

T$_{ad}$ Adiabatic Temperature [Symbol]

T$_{AF}$ Antiferromagnetic Temperature [Symbol]

T$_B$ Blocking Temperature (for Superparamagnetism) [also T$_b$] [Symbol]

T$_b$ Boiling Temperature [Symbol]

T$_b$ Time Back [Air Navigation]

T$_{BDT}$ Brittle-to-Ductile Transition Temperature (in Metallurgy) [Symbol]

T$_C$ Case Temperature [Semiconductor Symbol]

T$_C$ Curie Temperature (in Solid-State Physics) [also T$_c$] [Symbol]

T$_C$ Temperature in Degrees Celsius [Symbol]

T$_c$ Corrected Temperature [Symbol]

T$_c$ Critical Ordering Temperature (in Solid-State Physics) [Symbol]

T$_c$ Critical Temperature [Symbol]

T$_c$ Critical Transition Temperature (for Superconductivity) [Symbol]

T$_c$ Crystallization Temperature [Symbol]

T$_0^c$ Bardeen-Cooper-Schrieffer Superconductor Transition Temperature [Symbol]

T$_0^c$ Zero-Resistance Temperature [Symbol]

T$_{c0}$ Mean-Field (Transition) Temperature (for Superconductivity) [Symbol]

T$_{c1}$ Lower-Field (Transition) Temperature (for Superconductivity) [Symbol]

T$_{c2}$ Upper-Field (Transition) Temperature (for Superconductivity) [Symbol]

T$_{CAF}$ Charge-Exchange Antiferromagnetic Temperature [Symbol]

T$_{CO}$ Charge-Ordering Antiferromagnetic Temperature [Symbol]

T$_{cr}$ Crystallization Temperature [Symbol]

T$_d$ Decomposition Temperature (or Point) [Symbol]

T$_d$ Deposition Temperature [Symbol]

T$_d$ Period of Natural Vibrations [Symbol]

T$_d$ Thermodynamic Dewpoint Temperature [Symbol]

T$_{db}$ Ductile-to-Brittle Transition Temperature (in Metallurgy) [Symbol]

T$_{DBT}$ Ductile-Brittle Transition Temperature [also T$_{dbt}$] [Symbol]

T$_{DT}$ Ductility-Transition Temperature (in Metallurgy) [also T$_{dt}$] [Symbol]

T$_E$ Equilibrium Transformation Temperature (in Metallurgy) [Symbol]

T$_E$ Eutectic Temperature [Symbol]

T$_e$ Blackbody Temperature per Electron [Symbol]

T$_e$ Effective Temperature [Symbol]

T$_e$ Equilibrium (Melting) Temperature [Symbol]

T$_e$ Eutectoid Temperature [Symbol]

T$_e$ Extrusion Temperature [Symbol]

T$_F$ Ferrite Formation Temperature (in Metallurgy) [Symbol]

T$_F$ Ferry Temperature (in Physics) [Symbol]

T$_F$ Fractional Tolerance [Symbol]

T$_F$ Temperature in Degrees Fahrenheit [Symbol]

T$_f$ Depth of Field [Symbol]

T$_f$ Film Thickness at Time T$_f$ [Symbol]

T$_f$ Finish(ing) Temperature (in Metallurgy) [Symbol]

T$_f$ Formation Temperature [Symbol]

T$_f$ (Spin) Freezing Temperature (in Physics) [Symbol]

T$_f$ Final Temperature [Symbol]

T$_f$ Shear Strength [Symbol]

T$_f$ Thermodynamic Frost-Point Temperature [Symbol]

T$_g$ (Material) Growth Temperature [Symbol]

T$_g$ Gas Temperature [Symbol]

T$_g$ Glass-Transition Temperature [Symbol]

T$_g$ Growth Temperature [Symbol]

T$_H$ Homologous Temperature [Symbol]

T$_{HT}$ Heat Treatment Temperature [Symbol]

T$_i$ Inflection Temperature [Symbol]

T$_i$ Initial Temperature [Symbol]

T$_i$ Internal Shear Stress [Symbol]

T$_i$ Internal Stress Field [Symbol]

T$_i$ Thermodynamic Ice-Bulb Temperature [Symbol]

T$_i$ Time of Integration (in Yarn Unevenness Testing) [Symbol]

T$_{irr}$ Irreversible Temperature (in Thermodynamics) [Symbol]

T$_j$ Junction Temperature [Semiconductor Symbol]

T$_{J(max)}$ Critical Junction Temperature [Semiconductor Symbol]

T$_K$ Kauzmann Catastrophe Temperature (in Physics) [Symbol]

T$_K$ Kondo Temperature (in Metallurgy) [also T$_k$] [Symbol]

T$_{KT}$ Kosterlitz-Thouless (Zero-Resistance) Transition Temperature (of Superconductors) [Symbol]

T$_L$ Complex Reflection Coefficient [Symbol]

T$_L$ Liquidus Temperature (in Thermodynamics) [Symbol]

T$_L$ Lot Traceability [Symbol]

T$_l$ Lower Temperature [Symbol]

T$_\ell$ Liquidus Temperature (in Thermodynamics) [Symbol]

T$_\lambda$ Bulk Transition Temperature (of Superconductors) [Symbol]

T$_M$ Melting Transition Temperature [Symbol]

T$_M$ Member Traceability [Symbol]

T$_M$ Mercury-Vapor Turbine [Symbol]

T$_M$ Morin Temperature (of Magnetic Transition) [Symbol]

T$_m$ Mean Temperature [Symbol]

T$_m$ (Absolute) Melting Temperature [Symbol]

T$_m$ Transformation Temperature (in Metallurgy) [Symbol]

T$_{ma}$ Actual Mean Temperature [Symbol]

T_{MI} Metal-Insulator Phase Transition [Symbol]

T_{mp} Melting Temperature [Symbol]

T_{ms} Standard Mean Temperature [Symbol]

T_N Néel (Transition) Temperature (in Solid-State Physics) [Symbol]

$T_{N\parallel}$ Longitudinal Néel Temperature (in Solid-State Physics) [Symbol]

$T_{N\perp}$ Basal Plane Néel Temperature (in Solid-State Physics) [Symbol]

T_{N1} Lower Néel Temperature (in Solid-State Physics) [Symbol]

T_{N2} Upper Néel Temperature (in Solid-State Physics) [Symbol]

T_{np} Peak-to-Peak Distance between Negative and Positive Magnetic Transition [Symbol]

T_{nr} No Recrystallization Temperature (in Metallurgy) [Symbol]

T_o Minimum Ordering Temperature (in Solid-State Physics) [Symbol]

T_{ox} Oxidation Temperature [Symbol]

T_P Pearlite Formation Temperature (in Metallurgy) [Symbol]

T_p Peak Temperature [Symbol]

T_p Period (of Oscillation) [Symbol]

T_p Polarization Temperature [Symbol]

T_p Preform Temperature (in Metallurgy) [Symbol]

T_{pn} Peak-to-Peak Distance between Positive and Negative Magnetic Transition [Symbol]

T_{PR} Preroughening Temperature [Symbol]

T_{PS} Phase Separation Temperature [Symbol]

T_q Quenching Temperature (in Metallurgy) [Symbol]

T_R Recrystallization Temperature [Symbol]

T_R Reversible Spin-Orientation Transition (in Physics) [Symbol]

T_R Roughening Temperature [Symbol]

T_r Reaction Temperature [Symbol]

T_r Reduced Temperature (in Thermodynamics) [also T_{red}] [Symbol]

T_{rg} Reduced Glass Temperature [Symbol]

T_{RT} Room Temperature [also T_{rt}] [Symbol]

T_S Serial Traceability [Symbol]

T_S Steam Turbine [Symbol]

T_s Minimum Segregation Temperature (in Metallurgy) [Symbol]

T_s Softening Temperature (of Glass, Plastics, etc.) [Symbol]

T_s Spinodal Temperature (in Materials Science) [Symbol]

T_s Start Temperature (in Metallurgy) [Symbol]

T_s Temperature of Solid [Symbol]

T_s Spindle Torque (in Machining) [Symbol]

T_s Substrate Temperature [Symbol]

T_{SF} Spin-Flip Temperature (in Physics) [Symbol]

T_{sp} Spin Rotation Temperature (in Physics) [Symbol]

T_{stg} Storage Temperature [Semiconductor Symbol]

T_t Total Time [Symbol]

T_u Upper Temperature [Symbol]

T_V Verwey Transition Temperature [Symbol]

T_v Time Factor (for Soil Consolidation) [Symbol]

T_W Wetting Transition Temperature [Symbol]

T_w Sticky Limit (of Soil) [Symbol]

T_w Thermodynamic Wet-Bulb Temperature [Symbol]

T_w Toughness Index (of Soil) [Symbol]

T_x Crystallization Temperature [Symbol]

T_x Fractional Tolerance [Symbol]

T_y Pure Axisymmetric Loading [Symbol]

T_z Damage Depth (of a Single-Crystal Silicon Specimen) [Symbol]

T-0 Time Zero [Aerospace]

T1 Cooled from an Elevated-Temperature Shaping Process and Naturally Aged to a Substantially Stable Condition [Temper Designation for Aluminum Alloys]

T1 Carrier Facility for DS-1 Formatted Digital Signal Transmission at a Rate 1.544 mbps [Data Communications]

T1 Terrestrial Digital Circuit Class One [Data Communications]

T2 Cooled from an Elevated-Temperature Shaping Process, Cold-Worked and Naturally Aged to a Substantially Stable Condition [Temper Designation for Aluminum Alloys]

T2 Diiodothyronine [Biochemistry]

T3 Carrier Facility for DS-3 Formatted Digital Signal Transmission at a Rate 44.746 mbps [Data Communications]

T3 Solution Heat-Treated, Cold-Worked and Naturally Aged to a Substantially Stable Condition [Temper Designation for Aluminum Alloys]

T3 Triiodothyronine [also T-3] [Biochemistry]

T4 Hexahydro-1,3,5-Trinitro-1,3,5-Triazine

T4 Solution Heat-Treated and Naturally Aged to a Substantially Stable Condition [Temper Designation for Aluminum and Magnesium Alloys]

T4 3,3',5,5'-Tetraiodothyronine [Biochemistry]

T5 Cooled from an Elevated-Temperature Shaping Process and Artificially Aged [Temper Designation for Aluminum and Magnesium Alloys]

T6 Solution Heat-Treated and Artificially Aged [Temper Designation for Aluminum and Magnesium Alloys]

T7 Solution Heat-Treated and Stabilized [Temper Designation for Aluminum Alloys]

T8 Solution Heat-Treated, Cold Worked, and Artificially Aged [Temper Designation for Aluminum and Magnesium Alloys]

T9 Solution Heat-Treated, Artificially Aged and Cold Worked [Temper Designation for Aluminum Alloys]

T10 Cooled from an Elevated-Temperature Shaping Process, Cold Worked and Artificially Aged [Temper Designation for Aluminum Alloys]

2,4,5-T 2,4,5-Trichlorophenoxyacetic Acid

2,4,6-T 2,4,6-Trichlorophenol

2D Two-Dimensional [also 2-D]

3D Three-Dimensional [also 3-D]

6T Sexithiuophene [Biochemistry]

t actual tooth thickness (of involute splines) [Symbol]

t   central thickness (of concentric lenses) [Symbol]

t   circular tooth thickness (of pinions, etc.) [Symbol]

t   cutting time (in machining) [Symbol]

t   depth [Symbol]

t   general variable, or parameter [Symbol]

t   number of teeth (on gear pinions) [Symbol]

t   number of threads (on worm gears) [Symbol]

t   object-to-image distance (in radiography) [Symbol]

t   shear flow [Symbol]

t   Student t value (in statistics) [Symbol]

t   tau [English Equivlalent]

t   teaspoon(ful)

t   technical

t   temperature (usually in °C) [Symbol]

t   temporary

t   tense

t   tension(al)

t   territorial

t   tertiary [Chemistry]

t   tetragonal (crystal) system [Symbol]

t   thick(ness)

t   time [Symbol]

t   times

t   metric ton [Unit]

t   short ton [Unit]

t   top quark (in particle physics) [30-50 gigaelectronvolts] [Symbol]

t   total

t   trans- [Chemistry]

t   transfer (command) [Edlin MS-DOS Line Editor]

t   transitional

t   transpose (of a matrix) [Mathematics]

t   complex transmission coefficient [Symbol]

t   trace

t   transmit

t   triplet [Spectroscopy]

t   triton [Symbol]

t   troy (weight) [Unit]

t   truth [also T]

t   truth quark (in particle physics) [Symbol]

t   wet-bulb depression (in meteorology) [Symbol]

t-   tertiary [Chemistry]

t^+ lifetime of positive particle (in physics) [Symbol]

t_+ cationic transport number (in physical chemistry) [Symbol]

$t-$ lifetime of negative particle (in physics) [Symbol]

t_- anionic transport number (in physical chemistry) [Symbol]

t' Fresnel transmission coefficient (in optics) [Symbol]

t' time coordinate in the second system [Symbol]

t_0 incubation time (in chemistry) [Symbol]

t_0 reference temperature t_0 (in °C) [Symbol]

t_0 temperature (non-absolute) [Symbol]

t_0 temperature of the inferred zero resistance point [Symbol]

t_0 time for elution of unretained solvent (in chromatography) [Symbol]

$t_{1/2}$ half-life (of radioactive substances) [Symbol]

t_1 initial temperature [Symbol]

t_1 start(ing) time (in metallurgy) [Symbol]

t_2 completion (or finish) time (in metallurgy) [Symbol]

t_2 final temperature [Symbol]

t_{50} mean time to failure [Symbol]

t_A austempering time (in metallurgy) [Symbol]

t_a access time [Semiconductor Symbol]

t_a anodic pulse length (in pulse reversal plating) [Symbol]

$t_{a(A)}$ access time from address [Semiconductor Symbol]

$t_{a(G)}$ access time from output enable low [Semiconductor Symbol]

$t_{a(S)}$ access time, chip select to data out [Semiconductor Symbol]

$t_{a(R/W)}$ access time, chip read/write to data out [Semiconductor Symbol]

t_{av} average thickness [Symbol]

t_B brittle fracture transition temperature (in metallurgy) [Symbol]

t_b permissible (steel) stress in helical reinforcement [Symbol]

t_c lower shelf fracture toughness (in mechanics) [Symbol]

t_c temperature in degrees celsius [Symbol]

t_c cathodic pulse length (in pulse reversal plating) [Symbol]

t_c completion time (in metallurgy) [Symbol]

t_c critical temperature [Symbol]

t_{cal} calculated thickness [Symbol]

$t_{c(rd)}$ read cycle time [Semiconductor Symbol]

$t_{c(wr)}$ write cycle time [Semiconductor Symbol]

t_d delay time [Semiconductor Symbol]

t_{DBL} ductile-to-brittle lower transition temperature (in metallurgy) [Symbol]

t_{DBU} ductile-to-brittle upper transition temperature (in metallurgy) [Symbol]

$t_{dis(G)}$ output disable time after output enable high [Semiconductor Symbol]

$t_{dis(S)}$ output disable time after chip select high [Semiconductor Symbol]

$t_{dis(W)}$ output disable time after write enable low [Semiconductor Symbol]

$t_{d(off)}$ turn-off delay time [Semiconductor Symbol]

$t_{d(on)}$ turn-on delay time [Semiconductor Symbol]

$t_{en(G)}$ output enable time after output enable low [Semiconductor Symbol]

$t_{en(S)}$ output enable time after chip select low [Semiconductor Symbol]

$t_{en(W)}$ output enable time after write enable high [Semiconductor Symbol]

t_F temperature in degrees Fahrenheit [Symbol]

t_f fall time [also t_f] [Semiconductor Symbol]

t_f finish(ing) time (in metallurgy) [also t_f] [Symbol] t_fr   forward recovery time [Semiconductor Symbol]

t_H   HIGH state output time [Semiconductor Symbol]

t_h   hold time [Semiconductor Symbol]

t_h(A)   address hold time [Semiconductor Symbol]

t_h(D)   data hold time [Semiconductor Symbol]

t_i   ice-bulb temperature [Symbol]

t_i   transport number of i-th ion (in physical chemistry) [Symbol]

t_K   temperature in Kelvin [Symbol]

t_L   LOW state output time [Semiconductor Symbol]

t_n   nucleation time [Symbol]

t_off   turn-off time (in electronics) [Symbol]

t_on   turn-on time (in electronics) [Symbol]

t_p   complex transmission coefficient for p-polarization [Symbol]

t_p   polarization time [Symbol]

t_p   temperature (in degree Celsius) at constant pressure [Symbol]

t_p   temperature on platinum resistance scale [Symbol]

t_p   propagation delay (for Schottky TTL logic) [Semiconductor Symbol]

t_p   pulse time [Semiconductor Symbol]

t_pd   power dissipation time [Semiconductor Symbol]

t_PHL   propagation delay time, HIGH-to-LOW (Gate) output [Semiconductor Symbol]

t_PLH   propagation delay time, LOW-to-HIGH (gate) output [Semiconductor Symbol]

t_R   retention time (in chromatography) [Symbol]

t_R   temperature in degrees Rankine [Symbol]

t_R'   adjusted retention time (in chromatography) [Symbol]

t_r   rise time [Semiconductor Symbol]

t_r   rupture lifetime (or time to rupture) [Symbol]

t_rms   root-mean-square thickness [Symbol]

t_rr   reverse recovery time [Semiconductor Symbol]

t_s   complex transmission coefficient for s-polarization [Symbol]

t_s   set-up time [Semiconductor Symbol]

t_s   start(ing) time (in metallurgy) [Symbol]

t_s   storage time [Semiconductor Symbol]

t_s   temperature of a substance, substrate, etc [Symbol]

t_su   set-up time [Symbol]

t_su(A)   address set-up time [Semiconductor Symbol]

t_su(D)   data set-up time [Semiconductor Symbol]

t_su(S)   chip select set-up time [Semiconductor Symbol]

t_v   temperature (in °C) at constant volume [Symbol]

t_v(A)   output data valid after address change [Semiconductor Symbol]

t_w   peak width (in liquid chromatography) [Symbol]

t_w   pulse average time [Semiconductor Symbol]

t_w   temperature of water [Symbol]

t_w   wafer thickness [Semiconductor Symbol]

t_w   wet-bulb temperature (in meteorology) [Symbol]

t_w(W)   write pulse width [Semiconductor Symbol]

τ   applied stress [Symbol]

τ   average (corrosion) pit initiation time [Symbol]

τ   band index [Symbol]

τ   charged lepton (in particle physics) [Symbol]

τ   decay time [Symbol]

τ   density [Symbol]

τ   friction angle (in machining) [Symbol]

τ   flow stress [Symbol]

τ   half-life [Symbol]

τ   incubation time (in solidification) [Symbol]

τ   initiation rate (in chemistry) [Symbol]

τ   interfacial sliding resistance (in materials science) [Symbol]

τ   Kirchhoff stress [Symbol]

τ   lifetime [Symbol]

τ   line tension of dislocations (in materials science) [Symbol]

τ   magnetic wave vector [Symbol]

τ   monolayer time [Symbol]

τ   period of vibration [Symbol]

τ   photoelectric attenuation coefficient [Symbol]

τ   power factor of dielectrics [Symbol]

τ   pseudospin (in solid-state physics) [Symbol]

τ   pulse duration [Symbol]

τ   relaxation time (in physics) [Symbol]

τ   resolved shear stress [Symbol]

τ   Saybolt Universal viscosity [Symbol]

τ   scattering time [Symbol]

τ   simple shear stress [Symbol]

τ   sound transmission coefficient [Symbol]

τ   spin echo time (in physics) [Symbol]

τ   superparamagnetic relaxation [Symbol]

τ   surface free energy [Symbol]

τ   (lower-case) tau [Greek Alphabet]

τ   tau phase (of a material) [Symbol]

τ   Thomson heat (in physics) [Symbol]

τ   time (interval) [Symbol]

τ   time constant [Symbol]

τ   time-phase displacement (in physics) [Symbol]

τ   torque (on a current loop) [Symbol]

τ   torque (in rotary motion) [Symbol]

τ   tractive force (on a wetted surface per unit area) [Symbol]

τ   translation vector (in crystallography) [Symbol]

τ   transmission factor (in physics) [Symbol]

τ   transmittance [Symbol]

τ   turbidity coefficient [Symbol]

τ   volume resistivity [Symbol]

$\bar{\tau}$ mean (or average) lifetime [Symbol]

$\bar{\tau}$ mean relaxation time (in physics) [Symbol]

$\vec{\tau}$ shear stress vector [Symbol]

τ∓   combined shear stress [Symbol]

τ_∞   non-zero lower bound of interfacial shear stress (in materials science) [Symbol]

τ_∞   time constant at infinite temperature [Symbol]

τ_I   resolved shear stress, stage I [Symbol]

τ_{II} resolved shear stress, stage II [Symbol]

τ_{III} resolved shear stress, stage III [Symbol]

τ_0 critical (or initial) interfacial shear stress (in materials science) [Symbol]

τ_0 frictional stress [Symbol]

τ_0 Peierls-Nabarro stress (in solid-state physics) [Symbol]

τ_0 resolved shear stress [Symbol]

τ_{aa} actual stress amplitude [Symbol]

τ_{al} fatigue limit amplitude [Symbol]

τ_{bas} basal plane shear stress [Symbol]

τ_c concentrated shear stress [Symbol]

τ_c fiber-matrix bond strength (of composite) [Symbol]

τ_{CRSS} critical resolved shear stress [also τ_{crss}] [Symbol]

τ_{cy} cyclic shear stress [Symbol]

τ_d diffuse transmittance [Symbol]

τ_d interface strength (in materials science) [Symbol]

τ_e transmittance [Symbol]

τ^F interfacial shear stress related to sliding (in materials science) [Symbol]

τ_f shear strength [Symbol]

τ_{flow} characteristic time of fluid flow [Symbol]

τ_{fr} frame time (in electronics) [Symbol]

τ_{gb} grain boundary dislocation stress (in materials science) [Symbol]

τ_i internal transmittance [Symbol]

τ_i induction time (in ellipsometry) [Symbol]

τ_{id} ideal shear strength [Symbol]

τ_{LT} shear stress in longitudinal plane (of composites) [Symbol]

τ_{LYS} lower yield stress [also τ_{lys}] [Symbol]

τ_m matrix shear strength [Symbol]

τ_m matrix yield stress [Symbol]

τ_m Orowan stress (in metallurgy) [Symbol]

τ_{max} maximum shear stress [Symbol]

τ_{min} minimum shear stress [Symbol]

τ_{nr} radiative transition rate (in physics) [Symbol]

τ_P Peierls stress (in solid-state physics) [Symbol]

τ_p particle by-passing yield stress [Symbol]

τ_p value of time characterizing speed of transition of fluid from a given state to equilibrium state [Symbol]

τ_{py} particle yield strength [Symbol]

τ_R resolved shear stress [Symbol]

τ_R response time [Symbol]

τ_r radiative transition rate (in physics) [Symbol]

τ_r regular transmittance [Symbol]

τ_s frictional shear stress [Symbol]

τ_s resolved shear stress, saturation level [Symbol]

τ_t equivalent shear stress [Symbol]

τ_{th} theoretical yield strength [Symbol]

τ_{TT} shear stress in transverse plane (of a composite) [Symbol]

τ_v diffuse transmittance [Symbol]

τ_v viscous time [Symbol]

τ_{xy} shear stress in Y-Z plane with x-axis perpendicular and y-axis parallel [Symbol]

τ_{xz} shear stress in Y-Z plane with x-axis perpendicular and z-axis parallel [Symbol]

τ_y shear yield strength [Symbol]

τ_{ys} yield stress [Symbol]

τ_{yx} shear stress in X-Z plane with y-axis perpendicular and x-axis parallel [Symbol]

τ_{yz} shear stress in X-Z plane with y-axis perpendicular and z-axis parallel [Symbol]

τ_{zx} shear stress in X-Y plane with z-axis perpendicular and x-axis parallel [Symbol]

τ_{zy} shear stress in X-Y plane with z-axis perpendicular and y-axis parallel [Symbol]

Θ Angle of Reflection [Symbol]

Θ Azimuth [Symbol]

Θ Curie-Weiss Temperature (in Solid-State Physics) [Symbol]

Θ Image Transfer Constant (in Electronics) [Symbol]

Θ Magnetic Interaction Strength [Symbol]

Θ Misorientation Angle (in Crystallography) [Symbol]

Θ Modulus of Canonical Distribution (in Statistical Mechanics) [Symbol]

Θ Scattering Angle [Symbol]

Θ (Upper-case) Theta [Greek Alphabet]

Θ_B Planar Bond Angle [Symbol]

Θ_D Debye Temperature (in Solid-State Physics) [Symbol]

Θ_E Einstein Temperature (in Physics) [Symbol]

Θ_F Faraday Rotation (in Optics) [Symbol]

Θ_H Hall Angle (in Electrical Engineering) [Symbol]

Θ_N Néel Temperature (in Solid-State Physics) [Symbol]

Θ_p Transmission Coefficient (in Physics) [Symbol]

θ absolute temperature [Symbol]

θ addendum angle (of milled bevel gears) [Symbol]

θ (plane) angle [Symbol]

θ angular displacement [Symbol]

θ angular phase difference (in electrical circuits) [Symbol]

θ anneal(ing) temperature [Symbol]

θ argument of a complex number (e.g., $\theta = \text{arc } z$; $z = a + jb = re^{j\theta}$) [Symbol]

θ attitude angle (of sleeve bearings) [Symbol]

θ azimuthal angle (in polar coordinate systems) [Symbol]

θ ballistic throw (of ballistic galvanometers) [Symbol]

θ banking angle (of highway curves) [Symbol]

θ Bragg (diffraction) angle (in solid-state physics) [Symbol]

θ center angle (of shoe brakes) [Symbol]

θ Compton electron recoil angle (in quantum mechanics) [Symbol]

θ contact angle (in physical chemistry) [Symbol]

θ coverage (in materials science) [Symbol]

θ Curie Temperature (in solid-state physics) [Symbol]

θ Debye temperature (in solid-state physics) [Symbol]

θ (angular) deflection [Symbol]

θ direction of a complex number [Symbol]

θ draft (of casting patterns in metallurgy) [Symbol]

θ duty cycle [Symbol]

θ galvanometer deflection [Symbol]

θ groove angle (of V-belts) [Symbol]

θ hydraulic transmissivity (of geotextiles, etc.) [Symbol]

θ image transfer constant (in electrical engineering) [Symbol]

θ magnetooptical rotation [Symbol]

θ notch angle (in fluid flow) [Symbol]

θ orientation angle (in crystallography) [Symbol]

θ phase angle (in physics) [Symbol]

θ polar angle [Symbol]

θ porosity [Symbol]

θ rotation angle [Symbol]

θ second coordinate (usually) in cylindrical, polar and spherical coordinate systems [Symbol]

θ sidereal time (in astronomy) [Symbol]

θ temperature (above ice-point in Celsius) [Symbol]

θ (lower-case) theta [Greek Alphabet]

θ theta phase (of a material) [Symbol]

θ tilt angle [Symbol]

θ time [Symbol]

θ_I work-hardening rate in stage I (critical resolved shear stress) [Symbol]

θ_{II} work-hardening rate in stage II (critical resolved shear stress) [Symbol]

θ_0 Curie-law constant (in solid-state physics) [Symbol]

θ_1 initial temperature [Symbol]

θ_2 final temperature [Symbol]

θ_a advancing contact angle (in physical chemistry) [Symbol]

θ_B Bragg angle (in solid-state physics) [Symbol]

θ_B Brewster angle (in optics) [Symbol]

θ_C Cassie contact angle (in physical chemistry)[Symbol]

θ_c critical angle (in total internal reflection, spectroscopy, etc.) [Symbol]

θ_c critical coverage (in spectroscopy) [Symbol]

θ_c critical temperature (for superconductivity) [Symbol]

θ_D Debye temperature (in solid-state physics) [Symbol]

θ_e equilibrium contact angle (in physical chemistry) [Symbol]

θ_F Faraday rotation (in optics) [Symbol]

θ_{Frem} remanent Faraday rotation (in physics) [Symbol]

θ_{Fsat} saturation Faraday rotation (in physics) [Symbol]

θ_H Hall angle (in electrical engineering) [Symbol]

θ_H orientation of applied field [Symbol]

θ_{hkl} Bragg angle of lattice plane set (hkl) (in crystallography) [Symbol]

θ_i angle of incidence [Symbol]

θ_i fraction of surface sites occupied (in materials science) [Symbol]

θ_K Kerr rotation (in physics) [Symbol]

θ_M macroscopic contact angle (in physical chemistry) [Symbol]

θ_p paramagnetic (Curie) temperature [Symbol]

θ_r angle of refraction [Symbol]

θ_W Wenzel contact angle (in physical chemistry) [Symbol]

θ_x angle of force "F" with x-axis [Symbol]

θ_Y Young contact angle (in physical chemistry) [Symbol]

θ_y angle of force "F" with y-axis [Symbol]

θ_z angle of force "F" with z-axis [Symbol]

ϑ angle [Symbol]

ϑ angular phase displacement [Symbol]

ϑ (crystal) lattice orientation [Symbol]

ϑ reluctance [Symbol]

ϑ (lower-case) theta (variant) [Greek Alphabet]

ϑ temperature (usually in degrees Celsius) [Symbol]

ϑ time constant [Symbol]

3° Tertiary (Alcohol, Amine, etc.) [Symbol]

TA Tail Antenna [Aeronautics]

TA Tail Assembly (or Empennage) [Aeronautics]

TA Tailhook Association [US]

TA Taiwan Airlines

TA Tannic Acid

TA Tape Address

TA Tape Armored (Electric Cable)

TA Target

TA Tartaric Acid

TA Task Agreement

TA Task Analysis

TA Teaching Assistant

TA Technical Advisor

TA Technical Agriculture

TA Technonet Asia [Singapore]

TA Tel Aviv [Israel]

TA Telecom Australia

TA Telegraphic Address

TA Telluric Acid

TA Tensile Axis; Tension Axis [Mechanics]

TA Tensor Analysis [Mathematics]

TA Terephthalic Acid

TA Terminal Adapter

TA Terminal Address

TA Terminal Area

TA Territorial Army [UK]

TA Tertiary Air

TA Test Access

TA Test Article

TA Thenoylacetone

TA Theoretical Air

TA Thermal Analysis; Thermal Analyzer

TA Thermionic Arc

TA Thermoacoustic(s)

TA Thermoanalysis; Thermoanalytical

TA Thermoplastic and Asbestos Type (Conductor Insulation) [Symbol]

TA Thioamide

TA Thrust Ball Bearing, Single-Direction, Flat Seats, Grooved Raceways [Symbol]

TA Ticket Agent

TA Time, Actual

TA   Time Analysis; Time Analyzer
TA   Titrimetric Analysis
TA   o-Toluic Acid
TA   Total Air
TA   Trace Analysis [Analytical Chemistry]
TA   Tractor-Drawn Artillery
TA   Training Agency
TA   Transactional Analysis [Psychotherapy]
TA   Transatlantic
TA   Transient Absorption
TA   Transition Altitude [Aeronautics]
TA   Transmission Authenticator
TA   Transport Association [UK]
TA   Transportation Alternatives [US]
TA   Transverse Acoustic(al)(Mode)
TA   Travel Authorization
TA   Triacetic; Triacetate
TA   Triacetin
TA   Triple-Action,
TA   Triplex Annealing [Metallurgy]
TA   Tropical Air
TA   Truncus Arteriosus [Embryology]
TA   Trunnion Angle
TA   Tungstic Acid
TA   Turbo-Alternator
TA   Turn Altitude [Aeronautics]
TA   Turbocharged and Aftercooled (Engine)
TA   Turbulence Amplifier
T&A   Time and Attendance
T&A   Tonsillectomy and Adenoidectomy [also T and A]
T/A   Turnaround
Ta   Tantalum [Symbol]
Ta-176   Tantalum-176 [also ^{176}Ta, or Ta^{176}]
Ta-177   Tantalum-177 [also ^{177}Ta, or Ta^{177}]
Ta-178   Tantalum-178 [also ^{178}Ta, or Ta^{178}]
Ta-179   Tantalum-179 [also ^{179}Ta, or Ta^{179}]
Ta-180   Tantalum-180 [also ^{180}Ta, or Ta^{180}]
Ta-181   Tantalum-181 [also ^{181}Ta, Ta^{181}, or Ta]
Ta-182   Tantalum-182 [also ^{182}Ta, or Ta^{182}]
Ta-183   Tantalum-183 [also ^{183}Ta, or Ta^{183}]
Ta-184   Tantalum-184 [also ^{184}Ta, or Ta^{184}]
Ta-185   Tantalum-185 [also ^{185}Ta, or Ta^{185}]
tA   transpose of matrix "A" [Mathematics]
t(A)   exposure time of (metallic) A atoms [Symbol]
3A-   Monaco [Civil Aircraft Marking]
TAA   Tallowamine Acetate
TAA   Technical Assistance Agreement
TAA   Technical Assistance Administration [of United Nations]
TAA   Temporary Access Authorization
TAA   Thioacetamide
TAA   Thrust Ball Bearing, Single-Row, Angular Contact [Symbol]
TAA   Time Averaged Amplitude
TAA   Titanium Diisopropoxide Bis(2,4-Pentanedionate)

TAA   Track Average Amplitude [Magnetic Recording Media]
TAA   Transportation Association of America [US]
TAA   Trialkylamine
TAA   Triamylamine
TAA   Triansyl Amine
TAA   Turbine-Alternator Assembly
TAAB   Tetrabenzo[b,f,j,n][1,5,9,13]tetraazacyclohexadecine
TAAC   Technical Assessment Advisory Council
TAAC   Tetraalkyl Ammonium Cation
TAALS   Tactical Army Aircraft Landing System
Ta-Al   Tantalum-Aluminum (Alloy System)
TAAM   Tomahawk Airfield Attack Missile
TAAR   Target Area Analysis Radar
TAAS   Three-Axis Attitude Sensor
TAB   Büro für Technikfolgenabschätzung (beim Deutschen Bundestag) [Technological Impact Assessment Bureau (of the German Bundestag)]
TAB   Tabular Language
TAB   Tabulate; Tabulation [also Tab, or tab]
TAB   Tab(ulation) Character
TAB   Tape-Automated Bond(ing)
TAB   Technical Abstract Bulletin
TAB   Technical Activities Board [of Institute of Electrical and Electronics Engineers, US]
TAB   Technical Advisory Bureau
TAB   Technical Analysis Branch
TAB   Technical Assistance Board [of United Nations]
TAB   Technical Assistance Bureau [of International Civil Aviation Organization]
TAB   Technical Assessment Board
TAB   Teflonized Acetylene Black
TAB   Testing, Adjusting and Balancing
TAB   2-Thiolacetoxybenzanilide
TAB   Typhoid Fever and Paratyphoid A and B (Vaccine) [Immunology]
TAB   Tyres, Accessories, Batteries [UK Publication]
TA&B   Testing, Adjusting and Balancing [also TAB]
TaB   Tantalum Boride
Tab   Table [also tab]
Tab   Tablet(s) [also tab] [Medical Prescriptions]
TABC   Tab(ulator) Character
TABD   2,5,7,10-Tetraazabicyclo[4.4.0]decane
TABESIM   Tabulating Equipment Simulator
Tabl   Tablet [also tabl]
TABN   1,3,5,7-Tetraazabicyclo[3.3.1]nonane
TABR   Tab(ulator) [also Tabr, or tabr]
TABS   Tailored Abstracts [of Information Services Physics, Electrical and Electronics, and Computers and Control, UK]
TABS   Telephone Automated Briefing Service
TABS   Terminal Access to Batch Service
TABS   Total Automated Broker System
TABS   Total Aviation Briefing Service
TABS   Transparent Acrylonitrile-Butadiene-Styrene (Plastics)
TABSAC   Targets and Backgrounds Signature Analysis Center [US]

TABSIM Tab(ulating) Simulator

TABSOL Tabular Systems-Oriented Language

TABSTONE Target and Background Signal-to-Noise Evaluation

TABT Typhoid–Paratyphoid A and B–Tetanus (Vaccine) [Immunology]

TABTD Typhoid–Paratyphoid A and B–Tetanus–Diphtheria (Vaccine) [Immunology]

Tabun Dimethyl Phosphoramidocyanidic Acid, Ethyl Ester [also tabun]

Ta(BuO)₅ Tantalum (V) Butoxide

TAC Kennedy Space Center Landing Site [of NASA in Florida, US]

TAC Tables Annuelles Internationales de Constantes et Données Numériques [Annual International Tables of Constants and Numerical Data]

TAC Tactical Air Command [of US Air Force]

TAC Tangential-Momentum Accomodation Coefficient

TAC Tanners Association of Canada

TAC Technical Advisory Committee

TAC Technical Advisory Council

TAC Technical Assistance Center

TAC Technical Assistance Committee [of United Nations]

TAC Technical Assistance Contract

TAC Technology Assessment Conference [of National Institutes of Health, Bethesda, Maryland, US]

TAC Telemetry and Command

TAC TELENET (Telecommunication Network) Access Controller

TAC Television Advisory Committee

TAC Telex Data Acquisition and Control

TAC Terminal Access Controller [Internet]

TAC Terminal Area Charts

TAC Test Access Control

TAC Tested Additive Chemicals

TAC Texas Agricultural Commission [US]

TAC Thermostatically-Controlled Air Cleaner

TAC Timber Association of California [US]

TAC Time Amplitude Converter

TAC Time-to-Amplitude Converter

TAC Titanium Aluminum Carbide

TAC Total Allowable Catch

TAC Total Available Carbohydrates

TAC Total Average Cost

TAC Toxic Air Contaminant(s)

TAC Trade Advisory Council [UK]

TAC TRANSAC (Transistorized Automatic Computer) Assembler Compiler

TAC Trans-Anti-Cis (Configuration) [Chemistry]

TAC Transformer Analog Computer

TAC Transistorized Automatic Control

TAC Translation Accomodation Coefficient

TAC Translator–Assembler–Compiler

TAC Transport Advisory Council

TAC Transportation Association of Canada

TAC Trapped Air Cushion

TAC Triallyl Cyanurate

TAC Triallyl-1,3,5-Triazine-2,4,6-Trione

TAC Tribunal Anti-Dumping Canada

TAC TST (Thermal Spray Technology) Applications Center [of Drexel University, Philadelphia, Pennsylvania, US]

TAC Tunneling Association of Canada

TaC Tantalum Carbide (Abrasive)

Ta-C Tantalum-Carbon (System)

TACA Transportes Aéreas Centroamericana [Central American Air Transport Company, Honduras]

TACAMO Take Charge and Move Out

TACAN Tactical Air Command and Navigation System [of US Air Force]

TACAN Tactical Air Navigation (System) [also Tacan, or tacan]

TACAN-DME TACAN Distance Measuring Equipment

TACC Tactical Air Control Center

TACCAR Time-Averaged Clutter Coherent Airborne Radar

TACCO Tactical Coordinator [also Tacco]

TACD 1,4,7-Triazacyclodecane

TACDA The American Civil Defense Association [US]

TACDACS Target Acquisition and Data Collection System

TACDEN Tactical Data Entry Device [US Army]

TACE tris-Anisyl-Chloroethylene

TACE Turbine Automatic Control Equipment

TACEX Tactical Excercise

TACF Temporary Alteration Control Form

TACF Time-Autocorrelation Function

TACFU Things Are Completely Fouled Up

TACH 1,3,5-Triazacyclohexane

Ta-C:H Tantalum-Bearing Hydrogenated Carbon

Tach Tachometer [also tach]

Tach cis,cis-1,3,5-Triaminocyclohexane [also tach]

TACI Test Access Control Interface

TACL Telecommunications Analysis Center Library

TACL Time and Cycle Log

TACM Temperature Dependence of Adiabatic Compressibility Minimum

TAcM Triacetylmethane

TACMA The Association of Control Manufacturers [UK]

TACMAR Tactical Multifunction Array Radar

TACN 1,4,7-Triazacyclononane [also Tacn, or tacn]

TaCN Tantalum Carbonitride

TACNAV Tactical Navigation [also Tacnav]

TACO Test and Checkout Operations

TACODA Target Coordinate Data

TACOL Thinned Aperture Computed Lens

TACOM Tank-Automotive Command [of US Army]

TACOM-ARDEC Tank-Automotive Command–Army Research, Development and Engineering Center [of US Army at Picatinny Arsenal, New Jersey]

TACOMSAT Tactical Military Communications Satellite [of US Air Force] [also Tacomsat]

TACOS Tool for Automatic Conversion of Operational Software

TACP Titanium Bis(acetylacetonate) Dipropanolate

TACPD 1,4,8,12-Tetraazacyclopentadecane

TACPOL Tactical Procedure-Oriented Language

TACR Time and Cycle Record

TACRINE 9-Amino-1,2,3,4-Tetrahydroacridine [also Tacrine]

TAC/RA TACAN (Tactical Air Navigation) Radar Altimeter

TACRAHD Tactical Routing-Indicator-Lookup and Header-Preparation Device

TACRV Tracked Air Cushion Research Vehicle

TACS Tactical Air Control System

TACS Technical Assignment Control System

TACS Test Assembly Conditioning Station

TACS Total Access Communications System [UK]

TACSAT Tactical Communications Satellite [also Tacsat, US]

TACSATCOM Tactical Satellite Communications

TACSR/ACS Thermal-Resistant, Aluminum Alloy Conductor, Aluminum-Clad, Steel-Reinforced

TACT Technological Aids to Creative Thought

TACT Terminal Activated Channel Test

TACT Transistor and Component Tester

TACT Transonic Aircraft Technology

tact tactical

TACTAS Tactical Towed Array Sensor

TACTD 1,4,8,11-Tetraazacyclotetradecane

TACTRI Taiwan Agricultural Chemicals and Toxic Substances Research Institute

TACV Tracked Air Cushion Vehicle

Ta-Cu Tantalum-Copper (Alloy System)

TAD Target Acquisition Data

TAD Technical Approach Document

TAD Telephone Answering Device

TAD Temperature and Dewpoint ,

TAD Tensile Axis Direction [Mechanics]

TAD Terminal Address Designator

TAD Thrust-Augmented Delta [of NASA]

TAD Tobyhanna Army Depot [of US Army at Betts Army Air Field, Tobyhanna, Pennsylvania, US]

TAD Top Assembly Drawing

TAD Transaction Application Drive

tAD (Metric) Tons, Air Dried [Unit]

1,3-TADAB 4-(2-Thiazolylazo)-1,3-Diaminobenzene

TADC Technical Application Development Center

TADC Theory of the Average Dielectric Constant

TADDOL 1,1,4,4-Tetraphenyl-2,3-O-Isopropylidene-L-Threitol

TADIC Telemetry Analog-to-Digital Information Converter

TADIL Tactical Digital Information Link

TADOG Towed Array Deep Operating Gear

TADP Terminal Area Distribution Processing [also TAD/P]

TADS Tactical Automatic Digital Switching

TADS Teletypewriter Automatic Dispatch System

TADS Temperature Analysis and Detection System

TADS Thermal Analysis Data Station

TADS Time, Attendance, and Distribution System

TADS/PNVS Tactical Automatic Digital Switching/Pilot Night Vision System

TADSS Tactical Automatic Digital Switching System

TAE Technical Aeronautical Engineering [Israel]

TAE Technische Akademie Esslingen [Esslingen Academy of Engineering, Germany]

TAE Transportable Applications Executive [Computers]

TAE Tris-Acetate EDTA (Ethylenediaminetetraacetic Acid) (Buffer)

TAEC Thailand Atomic Energy Commission

TAEC Thiolated Aminoethylcellulose

TAEM Terminal Area Energy Management

TAERS Transportation Army Equipment Record System

TAETACN 1,4,7-tris(2-Aminoethyl)-1,4,5-Triazacyclononane [also taetacn]

Ta(EtO)$_5$ Tantalum(V) Ethoxide

TAF TATA Associated Factor [*Note:* TATA is a DNA Sequence]

TAF Terminal Aerodrome Forecast

TAF Top of Active Fuel

TAF Torque Amplification Factor

TAF Transaction Facility

TAF Triangular Ferromagnet [Solid-State Physics]

TAFIC Tasmanian Fishing Industry Council [Australia]

TAFO Trialkyl Phosphine Oxide

TAFOR Terminal Airport Forecast

TAFUBAR Things Are Fouled Up Beyond All Recognition

TAG Tactical Air Group

TAG Technical Activities Guide [of Institute of Electrical and Electronics Engineers, US]

TAG Technical Advisory Group

TAG Technical Air-to-Ground

TAG Technical Advisory Group

TAG Technical Assistance Grant

TAG Technical Assistance Group

TAG The Acrylonitrile Group [US]

TAG The Adjutant General

TAG Time Automated Grid

TAG Time-Dependent, Autoregressive, Gaussian (Model) [Solar Engineering]

TAG Trans-Atlantic Geotraverse (Expedition) [of National Oceanic and Atmospheric Admninistration, US]

TAG Transient Analysis Generator

TAG Triacylglycerol

Tag Tagliabue (Closed Cup Flash Point Tester)

TAGA Technical Association of the Graphic Arts [US]

TAGA/NSTF Technical Association of the Graphic Arts/National Scholarship Trust Fund (Fellowship) [US]

TAGA Proc TAGA Proceedings [of Technical Association of the Graphic Arts, US]

TAGN Triaminoguanidine Nitrate [Biochemistry]

TAGS Text and Graphics System

TAGS The Atlantic Groundfish Strategy (Program) [Newfoundland, Canada]

TAH Total Artificial Heart [Medicine]

TA/H Turn Altitude Height [Aeronautics]

TaH Tantalum Hydride

Ta-Hf Tantalum-Hafnium (Alloy System)

TAI Temps Atomique International [International Atomic Time]

TAI Thai Airways International

TAI Time to Auto-Ignition [Mechanical Engineering]

TAIC Tokyo Atomic Industrial Consortium [Japan]

TAIC Triallyl Isocyanurate

TAID Thrust Augmented Improved Delta [of NASA]

TAIDC Taiwan Agricultural and Industrial Development Corporation [Taiwan]

TAIMACO Taiwan Machinery Manufacturing Corporation

TAINS TERCOM (Terrain Contour Matching) Aided Inertial Navigation System

TAIO Technology Applications Information Organization [of Strategic Defense Initiative Organization, US]

TAIR Test Assembly Inspection Record

Ta-Ir Tantalum-Iridium (Alloy System)

TAISU Tennessee Agricultural and Industrial State University [Nashville, US]

Tajik Tajikstan

Tajik SSR Tajik Soviet Socialist Republic [USSR]

Takaoka Rev Takaoka Review [Published by Takaoka Electric Manufacturing Co. Ltd., Tokyo, Japan]

Takenaka Tech Res Rep Takenaka Technical Research Report [Published by Takenaka Co. Ltd., Osaka, Japan]

TAKIS Tutmonda Asocio pri Kibernetiko, Informatiko, kaj Sistemiko [World Association of Cybernetics, Computer Science and System Theory, Germany]

TAL Tandem Accelerating Laboratories [of McMaster University, Hamilton, Canada]

TAL Terminal Application Language

TAL Transaction Application Language

TAL Trans-Alpine

TAL Transatlantic Abort Landing; Trans-Atlantic Landing

TAL Tungsten-Arc Lamp

TAL Turkish Air Lines

tal *(talis)* – such [Medical Prescriptions]

TALA Tactical Landing Approach

TALAR Tactical Approach and Landing Radar; Tactical Landing Approach Radar

TALC Time, Attendance and Labor Collection

TALES The Association of Library Equipment Suppliers [UK]

Ta(Lig)$_2$ Tantalum (II) Ligand [*Note:* "Lig" Represents Ethyl, Methyl, or Benzoylacetate]

Ta(Lig)$_3$ Tantalum (III) Ligand [*Note:* "Lig" Represents Ethyl, Methyl, or Benzoylacetate]

θ-Al$_2$O$_3$ Theta Aluminum Oxide (or Theta Alumina)

TALON Texas, Arkansas, Louisiana, Oklahoma and New Mexico [US]

TALS Transatlantic Abort Landing Sites

TALS Transfer Airlock Section

TALTC Test Access Line Termination Circuit

TA Luft *(Technische Anleitung zur Reinhaltung der Luft)* – German Technical Instructions on Air Pollution Prevention

TAM Technical Association of Malaysia

TAM Technische Akademie Mannheim [Mannheim Academy of Engineering, Germany]

TAM Telecommunications Access Method

TAM Telephone Answering Machine

TAM Terminal Access Method

TAM Test Access Multiplexer

TAM Theoretical and Applied Mechanics

TAM Thermal Analytical Model

TAM 2-(Thiazolyl-2-Azo)-4-Methoxyphenol

TAM Total Available Market

TAM Triallyl Trimellithate

T&AM Technical and Administrative Management Company [US]

TAMA N-Methylanilinium Trifluoroacetate

TAMAC Trimellitic Anhydride Monoacid

TAMCO Training Aid for MOBIDIC (Mobile Digital Computer) Console Operations

TAMDA Timber and Allied Materials Development Association [South Africa]

TAMDP Tetraamyl Methylenediphosphonate

TAME Thermal Arrest Memory Effect [Metallurgy]

TAME Nα-p-Tosyl-L-Arginine Methyl Ester [Biochemistry]

TAME 1,1,1-Tris-(aminomethyl)ethane

Ta(MeO)$_5$ Tantalum(V) Methoxide

TAMIS Telemetric Automated Microbial Identification System

TAMM Tetrakis(acetylmercuri)methane

TAMM Tetrakis(acetoxymercuri)methane

Ta-Mo Tantalum-Molybdenum (Alloy System)

TAMOS Terminal Automatic Monitoring System

TAMOS Terminal Auto-Operator and Monitor System

TAMPL Thermal Analysis of Materials Processing Laboratory [of Tufts University, Medford, Massachusetts, US]

TAMRF Texas A&M Research Foundation [US]

TAMS Telenet Access Management System

TAMSA Tubos de Acero de Mexico SA [Mexican Steel Tubing Manufacturer]

TAMS Tunnel Air Monitoring System

TAMU Texas A&M (Agricultural and Mechanical) University [College Station, US]

TAMVEC Texas A&M (University) Variable Energy Cyclotron [US]

TAN Tananarive, Madagascar [Space Tracking and Data Network Site]

TAN Tangent (Function) [Programming] [also Tan, or tan]

TAN Textile Association of the Netherlands

TAN 1-(2-Thiazolylazo)-2-Naphthol

TAN Titanium Aluminum Nitride

TAN Total Acid Number [Chemistry]

TaN Tantalum Nitride (Compound)

Tan Tandem [also tan]

tan tangent [Symbol]

tan^{-1} inverse tangent [also arctan] [Symbol]

Ta-Nb Tantalum-Niobium (Alloy System)

TANCA Technical Assistance to Non-Commonwealth Countries [UK]

tan δ dissipation factor (in electrical engineering) [Symbol]

tan δ tangent delta (or dielectric loss factor) [Symbol]

tan δ tangent delta (i.e. ratio of loss modulus to storage modulus) [Symbol]

T and A Tonsillectomy and Adenoidectomy [also T&A]

TANDEM Tactical Naval Decision Making (System)

TANE Telephone Association of New England [US]

TANGO Testing for Adjacent Nuclei with a Gyration Operator [Physics]

tanh hyperbolic tangent [Symbol]

tanh^{-1} inverse hyperbolic tangent [also artanh] [Symbol]

Ta-Ni Tantalum-Nickel (Alloy System)

tan φ loss tangent (in electronics) [Symbol]

TANS Tactical Air Navigation System

Tantalum Prod Int Study Cent Q Bull Tantalum Producers International Study Center Quarterly Bulletin [Belgium]

Tanulm M Tud Akad Szam tech Autom Kut Intez Tanulmanyok Magyar Tudomanyos Akademia Szamitastechnikaiu es Automatizalasi Kutato Intezet [Publication on Computer Engineering and Automation of the Hungarian Academy of Sciences]

TAO Technical Assistance Operation [of United Nations]

TAO Technical Assistance Order

TAO 4-(2-Thiazolylazo)orcinol

TAO Tokyo Astronomical Observatory [Japan]

TAO Tropical Atmosphere Ocean (Array) [of National Oceanic and Atmospheric Administration–Pacific Marine Environmental Laboratory, US]

TaO$_x$ Tantalum Oxides (i.e., TaO$_2$, Ta$_2$O$_4$, and Ta$_2$O$_5$) [also TaOx]

Ta(OBu)$_5$ Tantalum(V) Butoxide

Ta(OEt)$_5$ Tantalum(V) Ethoxide

Ta(OMe)$_5$ Tantalum(V) Methoxide

Ta(OR)$_5$ Tantalum(V) Alkoxide [General Formula]

Ta-Os Tantalum-Osmium (Alloy System)

Ta$_2$O$_5$-TiO$_2$ Tantalum Pentoxide-Titanium Dioxide (System)

TaOx Tantalum Oxides (i.e., TaO$_2$, Ta$_2$O$_4$, and Ta$_2$O$_5$) [also TaO$_x$]

TAP ortho-Phthalate Tantalum Hydrogen

TAP Tactical Area Positioning

TAP T-Angle Plate

TAP Tank Advisory Panel [Nuclear Engineering]

TAP Tape Automatic Positioning

TAP Tape Automatic Preparation

TAP TARGET (Technology Accessible Resource Gives Employment Today) Access Program

TAP Technical Achievement Plan

TAP Technical Advisory Panel

TAP Technological Assessment and Planning [Canada]

TAP Technological Awareness Program

TAP Technology Assessment Program [Canada]

TAP Technology Apprenticeship Program [US]

TAP Technology Assistance Program [of Science Council of British Columbia, Canada]

TAP Telecommunications Access Partnerships [of (Ontario) Ministry of Economic Development, Trade and Tourism, Canada]

TAP Telelocator Alphanumeric Protocol

TAP Telemetry Acceptance Pattern

TAP Terminal Access Processor

TAP Terminal Applications Package

TAP Test Access Path

TAP Test Access Port

TAP Test Administration Plan

TAP Test Analysis Package

TAP Test Article Protector

TAP Test Assistance Program

TAP 1,4,5,8-Tetraazaphenanthrene

TAP Tetraazaporphinate

TAP Tetrainosylpyrrole

TAP 2,3,5,6-Tetraminopyridine

TAP Thermal Analyzer Program

TAP Thermodynamics and Physical Properties Package [of University of Houston, Texas, US]

TAP Time and Percussion (Fuse)

TAP Time-Sharing Assembly Program

TAP Tomographic Atom Probe

TAP Total Air Pressure

TAP Trade Assistance Program [Canada]

TAP Trans-Alaska Pipeline

TAP Transportes Aéreos Portugueses [Portuguese Air Transport Company]

TAP Tresun Address Processing

TAP Triallyl Phosphate

TAP 2,3,6-Triaminopyridine

TAP Triamylphosphate

TAP Triglycol Acetate Phthalate

TAP Triisoamyl Phosphate

T-AP Auxiliary Transport Vessel [US Navy Symbol]

Tap Tapping [also tap]

TAPA 2-(2,4,5,7-Tetranitro-9-Fluorenylideneaminooxy) propionic Acid

TAPAC Tape Automatic Positioning and Control

TAPCIS The Access Program for the CompuServe Information Service [US]

tapCuPc Copper Tetra-4(2,4-Di-tert-Amylphenoxyphthalo-cyanine)

TAPE Tape Automatic Preparation Equipment

TAPE Technical Advisory Panel for Electronics [of US Air Force]

TAPF Time-Averaged Precession Frequency

TAPH2 Tetraazaporphine

TAPI Telephony Applications Programming Interface

Tapline Trans-Arabian Pipeline

Ta(PMCp)Cl$_4$ Pentamethylcyclopentadienyltantalum Tetrachloride

TAPO Trialkyl Phosphine Oxide

TAPP Tarapur Atomic Power Project [India]

TAPP Thermochemical and Physical Properties (Database)

TAPP Two-Axis Pneumatic Pickup

TAPPI Technical Association of the Pulp and Paper Industry [US]

TAPPI Eng Conf TAPPI Engineering Conference [of Technical Association of the Pulp and Paper Industry, US]

TAPPI J TAPPI Journal [of Technical Association of the Pulp and Paper Industry, US]

TAPPIK Technical Association of the Pulp and Paper Industry of Korea [South Korea]

TAPRE Tracking in an Active and Passive Radar Environment

TAPS Tactical Area Positioning System

TAPS Terminal Area Positive Separation [of Federal Aviation Administration, US]

TAPS Triple Action Process STB [of Sumitomo Corporation, Japan]

TAPS Tropical Assimilation and Prognosis System

TAPS 3-{[Tris(hydroxymethyl)methyl]amino}-1-Propane-sulfonic Acid

TAPS N-Tris(hydroxymethyl)methyl-3-Aminopropanesulfonic Acid

TAPS Turbo-Alternator Power System

TAPSO 3-[N-Tris(hydroxymethyl)methylamino]-2-Hydroxypropanesulfonic Acid

TAPT 1,4,5,8-Tetraazaphenanthrene [also tapt]

TAPTACN 1,4,7-Tris(3-Aminopropyl)-1,4,7-Triazacyclononane [also taptacn]

Taq Thermus aquaticus [Microbiology]

TAR Tactical Air Reconnaissance

TAR Tactical Attack Radar

TAR Tartrate

TAR Technical Action Request

TAR Technical Analysis Request

TAR Technical Assistance Request

TAR Technical Association of Refractories [Japan]

TAR Terminal Address Register

TAR Terminal Area (Surveillance) Radar

TAR Terrain-Avoidance Radar

TAR Test Action Requirement

TAR Test Agency Report

TAR 4-(2-Thiazolylazo)resorcine; 4-(2-Thiazolyazo)resorcinol

TAR Thrust-Augmented Rocket

TAR Track Address Register

TAR Trajectory Analysis Room

.TAR Tape Archive [UNIX File Name Extension] [also .tar]

TARA Truck-Frame and Axle Repair Association [US]

TARABS Tactical Air Reconnaissance and Aerial Battlefield Surveillance System

TARAD Tracking Asynchronous Radar Data

TARAN Tactical Attack Radar and Navigation

TARC The Army Research Council [US]

TARC Trace Analysis Research Center [of Dalhousie University, Halifax, Canada]

TARC Through Axis Rotational Control

TARDIS Time and Attendance Recording Analysis

TARDIS Titles Automated Register and Document Information System [Canberra, Australian Capital Territory]

TARE Telegraph Automatic Routing Equipment

TARE Telemetry Automatic Reduction Equipment

TARE Transistor Analysis Recording Equipment

Ta-Re Tantalum-Rhenium (Alloy System)

TARF Thermally Activated Radio-Frequency (Method)

TARFOX Tropospheric Aerosol Radiative Forcing Observational Experiment [of National Oceanic and Atmospheric Administration–Pacific Marine Environmental Laboratory, US]

TARFU Things Are Really Fouled Up

TARGA TrueVision Advanced Raster Graphics Adapter

TARGET Team to Advance Research for Gas Energy Transformation [US]

TARGET Technology Accessible Resource Gives Employment Today

TARGET Thermal Advanced Reactor Gas-Cooled Exploiting Thorium [Nuclear Engineering]

TARGIT Three Axis Route Gyro Inertial Tracker

TARI Taiwan Agricultural Research Institute

TARIF Technical Apparatus for Rectification of Indifferent Films

TARL Telecom Australia Research Laboratory [Clayton, Victoria, Australia]

TARMAC Terminal Area Radar Moving Aircraft

Tarn Tarnish [also tarn]

TARO Tanzania Agricultural Research Institute

TARP Tactical/Acoustic Replay Package

TARP Technical Assistance Research Program

TARP Test and Repair Processor

TARP Texas Advanced Research Program [US]

Tarp Tarpaulin [also tarp]

TARPOL Tariff Policy [of European Civil Aviation Conference, France]

TARPS Tactical Aerial Reconnaissance Pod System

TARS Technical Assistance Recruitment Service [of United Nations]

TARS Test Analysis Retrieval System

TARS Terrain Analog Radar Simulator

TARS Three-Axes Reference System

TARS Turn-Around Ranging Station

TART Tip-Angle Reduced T1-Imaging

TART Twin Accelerator Ring Transfer [Nuclear Engineering]

Tart Tartaric Acid

tart tartrate

Tart A Tartaric Acid [also tart a]

Ta-Ru Tantalum-Ruthenium (Alloy System)

.tar.Z Compressed Archived (Files) [UNIX File Name Extension]

TAS Tactical Analysis System

TAS Tanzanian Shilling [Currency]

TAS Target Acquisition System

TAS Teacher at Sea (Program) [of National Oceanic and Atmospheric Administration, US]

TAS Telecommunications Authority Singapore
TAS Telemetry Antenna Subsystem
TAS Telephone Access Server
TAS Telephone Answering Service
TAS Telephone Answering System
TAS Temperature Actuated Switch
TAS Tennessee Academy of Science [Nashville, US]
TAS Tenure Administration System
TAS Terminal Address Selector
TAS Test Access Selector
TAS Test and Set
TAS Test Article Specification
TAS The Aviation Society [UK]
TAS Thermal Aerobic Stabilization (Process)
TAS Time and Attendance System
TAS Triple-Axis Spectrometer
TAS Tris(dimethylamino)sulfonium
TAS Tris(dimethylamino)sulfur
TAS True Air Speed
TAS Turkish Academy of Sciences [Ankara]
TaS Tantalum Monosulfide
Tas Tasmania(n)
TASC Tabular Sequence Control
TASC Center for Technology and Social Change
TASC Tactical Articulated Swimmable Carrier
TASC Technical Assistance and Support Center
TASC Telecommunications Alarm Surveillance and Control
TASC Terminal Area Sequence and Control
TASC The Analytical Sciences Corporation [San Antonio, Texas, US]
TASC Tracor-Northern Acquisition System Chassis
TASCC Tandem Accelerator Superconducting Cyclotron [Chalk River, Ontario, Canada]
TASCC Test Access Signaling Conversion Circuit
TASCD Texas Association of Soil Conservation Districts [US]
TASCON Television Automatic Sequence Control
TASE Tenacity-as-Specified Elongation [Textile Materials]
TaSe$_2$ Tantalum Diselenide
2H-TaSe$_2$ Hexagonal Tantalum Diselenide [Ramsdell Notation: 2 Refers to Number of Tantalum and Selenium Bilayers Necessary to Produce a Unit Cell, and H Refers to the Hexagonal Symmetry]
TASES Tactical Airborne Signal Exploitation System
TASF Tris(dimethylamino)sulfonium Fluoride
TAS-F Tris(dimethylamino)sulfur Difluoride
TAS Fluoride Tris(dimethylamino)sulfonium Fluoride [also TAS fluoride]
TASI Time Assignment Speech Interpolation
TASK Temporary Assembled Skeleton [Information Technology]
TASM (Borland) Turbo Assembler [Computers]
Tasm Tasmania(n) [also Tasman]
TASMAN Tasmania-Australia Submarine Cable
TASO Television Allocation(s) Study Organization [US]
TASP Teamed-Architectural Signal Processor

TASP Toll Alternatives Studies Program
TASP Tropical Area Special Product
TAS-PAC Tactical Analysis System for Production, Accounting and Control
TASPAWS Tasmanian Parks and Wildlife Service [Australia]
TASPR Technical and Schedule Performance Report
TASQUE University of Tasmania Consultative Unit [Australia]
TASR Terminal Area Surveillance Radar
TASR Thermal Activation Strain Rate [Materials Science]
TASRA Thermal Activation Strain Rate Analysis [Materials Science]
TASS Tactical Avionics System Simulator
TASS Tactical Signal Simulator
TASS Technical Assembly
TASS Telegrafnoje Agenstwo Sowjetskowo Sojusa [News Agency of the USSR]
TASS Teleprinter Automatic Switching System
TASS Trouble Analysis System
TASSA Thermally Assisted Solid-State Amorphization
3-A SSC 3-A Sanitary Standards Committee [US]
TAST Thermoacoustic Sensing Technique
TASTW Third ASEAN (Association of Southeast Asian Nations) Science and Technology Week
TAT Target Aircraft Transmitter
TAT Technical Acceptance Team
TAT Tetraallyltin
TAT Thematic Apperception Test [Psychology]
TAT Thermally Assisted Tunneling [Electronics]
TAT Thrust-Augmented Thor [of US Air Force]
TAT Total Air Temperature
TAT Trans-Anti-Trans (Configuration) [Chemistry]
TAT Trans-Atlantic Telecommunications
TAT Transatlantic Telephone Cable
TAT Turnaround Time
TAT Tyrosine Aminotransferase [Biochemistry]
TATB Tetraphenylarsonium Tetraphenylborate
TATB Triamino Trinitrobenzene
TATC Terminal Air Traffic Control
TATC Transatlantic Telephone Cable
TATCS Terminal Air Traffic Control System
TATDI Deiced Total Air Temperature
TATE Triaminotriethylamine [also Tate]
Ta(TFE)$_5$ Tantalum Trifluoroethoxide
TATHS Tool and Trades History Society [UK]
TATI Total Air Temperature Indicator
Ta-Ti Tantalum-Titanium (Alloy System)
TATM Triallyl Trimellitate
TATNB Triaminotrinitrobenzene
TATOA Toronto Area Transit Operating Authority [Canada]
TATr Tyrosine Aminotransferase Regulator [Biochemistry]
TATRP Texas Advanced Technology Research Program [US]
TAU International Workshop on Tau Lepton Physics

TAU Tel Aviv University [Israel]

TAU Test Access Unit

TAU Thesaurus Alphabetical Up-to-Date

TAU Trunk Access Unit

TAUP Topics in Astroparticle and Underground Physics (Conference)

TAURUS Transfer and Automated Registration of Uncertified Stock [UK]

TAUN Technical Assistance of the United Nations

TAV Transatmospheric Vehicle

TAV Transverse Acoustoelectronic Voltage

TAVE Thor-Agena Vibration Experiment [of NASA]

TAVG Temperature, Average

TAW Technische Akademie Wuppertal [Wuppertal Academy of Engineering, Germany]

Ta-W Tantalum-Tungsten (Alloy System)

TAWAR Tactical All-Weather Attack Requirements

TAWCS Tactical Air Weapons Control System

TAWDS Target Acquisition Weapon Delivery System

TAX 1-Acetylhexahydro-3,5-Dinitro-s-Triazine [Symbol]

TAX Taxiway Lights [Aeronautics]

TAX Taxi(ing) [Aeronautics]

Ta-X Tantalum-Nonmetal

TAXI Transparent Asynchronous Transceiver(/Receiver) Interface

TAXIR Taxonomic Information Retrieval

Ta-Zr Tantalum-Zirconium (Alloy System)

TB Talk Back

TB Tariff Bureau

TB Taste Bud [Anatomy]

TB T-Bolt

TB Technical Bulletin

TB Temper Brittle(ness) [Metallurgy]

TB Terabyte [also Tbyte]

TB Terminal Base

TB Terminal Board

TB Terminal Box

TB Terminal Bud [Botany]

TB Theoretical Biology

TB Thermal Barrier [Aeronautics]

TB Thermal Biology

TB Thermobalance

TB Thrust Bearing

TB Thrust Brake

TB Thymol Blue [Organic Chemistry]

TB Tidal Bore [Oceanography]

TB Tight-Binding (Approximation) [Solid-State Physics]

TB Tightly Bound

TB Tilt (Grain) Boundary [Materials Science]

TB Time Base [Electronics]

TB Time (Duration) of Burn [Aeronautics]

TB Tip Blunting [Cutting Tools]

TB Title Block (on Drawings) [also T/B]

TB Toggle Buffer

TB Tone Burst

TB Top Blowing; Top Blown [Metallurgy]

TB Top Boundary

TB Torch Brazing

TB Torpedo Boat

TB Torsion Balance

TB Tracer Bullet

TB Tracheobronchial; Tacheobronchitis [Anatomy]

TB Track Ball [Computers]

TB Train Bridge [Horology]

TB Transborder

TB Transition Band [Metallurgy]

TB Transmitter-Blocker (Cell) [also T-B]

TB Trial Balance [also tb]

TB Tributyl Phosphate

TB Triple-Braid(ed) (Electric Wire)

TB Tubercle Bacillis [Medicine]

TB Tuberculosis [also Tb, or tb]

TB Tungsten Bronze

TB Twin (Grain) Boundary [Materials Science]

TB Twist (Grain) Boundary [Materials Science]

2,4,5-TB 4-(2,4,5-Trichlorophenoxy)butanoic Acid

T/B Title Block (on Drawings) [also TB]

T/B Top and Bottom

T-B Transmitter-Blocker (Cell) [also TB]

$T_c(B)$ Magnetic Induction Dependent Critical Temperature [Symbol]

Tb Terabyte [Unit]

Tb Terbium [Symbol]

Tb^{3+} Trivalent Terbium Ion [also Tb^{+++}] [Symbol]

Tb^{4+} Tetravalent Terbium Ion [Symbol]

Tb-147 Terbium-147 [also ^{147}Tb, or Tb^{147}]

Tb-148 Terbium-148 [also ^{148}Tb, or Tb^{148}]

Tb-149 Terbium-149 [also ^{149}Tb, or Tb^{149}]

Tb-150 Terbium-150 [also ^{150}Tb, or Tb^{150}]

Tb-151 Terbium-151 [also ^{151}Tb, or Tb^{151}]

Tb-152 Terbium-152 [also ^{152}Tb, or Tb^{152}]

Tb-153 Terbium-153 [also ^{153}Tb, or Tb^{153}]

Tb-154 Terbium-154 [also ^{154}Tb, or Tb^{154}]

Tb-155 Terbium-155 [also ^{155}Tb, or Tb^{155}]

Tb-156 Terbium-156 [also ^{156}Tb, or Tb^{156}]

Tb-157 Terbium-157 [also ^{157}Tb, or Tb^{157}]

Tb-158 Terbium-158 [also ^{158}Tb, or Tb^{158}]

Tb-159 Terbium-159 [also ^{159}Tb, Tb^{159}, or Tb]

Tb-160 Terbium-160 [also ^{160}Tb, or Tb^{160}]

Tb-161 Terbium-161 [also ^{161}Tb, or Tb^{161}]

Tb-162 Terbium-162 [also ^{162}Tb, or Tb^{162}]

Tb-163 Terbium-163 [also ^{163}Tb, or Tb^{163}]

Tb-164 Terbium-164 [also ^{164}Tb, or Tb^{164}]

t(B) exposure time of (nonmetallic) B atoms [Symbol]

tb tablespoon(ful) [Unit]

tb tuberculosis [also TB, or Tb]

TBA N-tert-Butyl Acrylamide

TBA tert-Butyl Alcohol (or tert-Butanol)

TBA tert-Butyl Arsine

TBA Testbed Aircraft

TBA Test Boring Association [US]

TBA Tetra-n-Butyl Ammonium

TBA Thiobarbituric Acid [Biochemistry]

TBA Thiobenzanilide

TBA Tight-Binding Approximation [Solid-State Physics]

TBA Tires, Batteries and Accessories

TBA To Be Added [also tba]

TBA To Be Announced [also tba]

TBA o-(p-Toluyl)benzoic Acid

TBA Torsional Braid Analysis

TBA Tribenzyl Amine

TBA Tribromoaniline

TBA Tribromoanisole

TBA Tributylamine

TBA 2,3,6-Trichlorobenzoic Acid

2,3,6-TBA 2,3,6-Trichlorobenzoic Acid

tBA tert-Butylacrylate

tBa tert-Butylaniline

tba to be added [also TBA]

tba to be announced [also TBA]

2tBa 2-tert-Butylaniline

TBAA Tetrabutyl Ammonium Acetate

TBAB Tetrabutyl Ammonium Bromide; Tetra-n-Butyl Ammonium Bromide [also TBABr]

TBABF$_4$ Tetrabutyl Ammonium Tetrafluoroborate [also TBATFB]

TBABH$_4$ Tetrabutyl Ammonium Borohydride

TBABr Tetrabutyl Ammonium Bromide; Tetra-n-Butyl Ammonium Bromide [also TBAB]

TBABr$_3$ Tetrabutyl Ammonium Tribromide

TBAC Tetrabutyl Ammonium Chloride [also TBACr]

TBAC Tetrabutyl Ammonium Cyanoborohydride

TBAC Treasury Board Advisory Committee [US]

Tb(Ac)$_3$ Terbium(III) Acetate

Tb(ACAC)$_3$ Terbium(III) Acetylacetonate [also Tb(AcAc)$_3$, or Tb(acac)$_3$]

TBACC Tetrabutyl Ammonium Chlorochromate

TBA-CN Tetrabutyl Ammonium Cyanide [also TBACN]

TBAC/FLM Treasury Board Advisory Committee on Federal Land Management [US]

TBACl Tetrabutyl Ammonium Chloride [also TBAC]

TBADC Tetrabutyl Ammonium Dichromate

TBAF Tetrabutyl Ammonium Fluoride

TBAH Tetrabutyl Ammonium Hydroxide [also TBA-OH]

TBAHFP Tetrabutyl Ammonium Hexafluorophosphate [also TBAPF$_6$]

TBAHP Tetrabutyl Ammonium Hypophosphite

TBAHS Tetrabutyl Ammonium Hydrogen Sulfate

TBAI Tetrabutyl Ammonium Iodide

TBAl Tri-i-Butyl Aluminum

TBAN Tetrabutyl Ammonium Nitrate

TBAN Tight-Binding Anderson-Newns (Model) [Solid-State Physics]

TBA-OH Tetrabutyl Ammonium Hydroxide [also TBAH]

TBAP Tetrabutyl Ammonium Perchlorate

TBAPF$_6$ Tetrabutyl Ammonium Hexafluorophosphate

TBAS Tetrabutyl Ammonium Succinimide

TBATD Torsion in a Bridgman Anvil-Type Device [Mechanical Testing]

TBATFB Tetrabutyl Ammonium Tetrafluoroborate [also TBABF$_4$]

TBATPB Tetrabutyl Ammonium Tetraphenylborate

TBAX Tube Axial

TBB tert-Butylbenzene

TBB Tetrabromobenzene

TBB Tight-Binding Bond [Chemistry]

TBB Tri-i-Butylboron

TBBA Terephthal-bis-Butylaniline; Terephthalbutylaniline; Terephthalylidenebisbutylaniline

TBBA tert-Butylbenzylamine

TBBA Tetrabromobisphenol A

TBBA Thai-Britain Business Association

TBBA-AE Tetrabromobisphenol A Allyl Ether

TBBA-EO Tetrabromobisphenol A Bis(hydroxyethyl Ether)

TBBDP Tetrabutyl Butylenediphosphonate

TBBF Top Baseband Frequency

TBBPA Tetrabromobisphenol A

TB12BQ Tetrabromo-1,2-Benzoquinone

TB14BQ Tetrabromo-1,4-Benzoquinone

TBBS N-tert-Butyl-2-Benzothiazole Sulfenamide

TBBS The Bread Board System [Bulletin Board System]

TBC Tensile Bolting Cloth

TBC Terminal Buffer Controller

TBC 4-tert-Butylcatechol

TBC Thermal Barrier Coating

TBC Time-Base Corrector

TBC Token-Bus Controller

TBC Toss Bomb Computer

TBC Treasury Board of Canada

TBC Tributyl Citrate

TBCCO Thallium Barium Calcium Copper Oxide (Superconductor)

TBCCO/LAO Thallium Barium Calcium Copper Oxide on Lanthanum Aluminate

TB Cell Transmitter-Blocker (Cell) [also T-B Cell, T-B cell, or TB cell]

TBCO Thallium Barium Copper Oxide (Superconductor) [Tl$_2$Ba$_2$Cu$_2$O$_x$]

TbCo Terbium Cobalt (Alloy)

Tb-Co CMF Terbium-Cobalt Composition(al) Modulated Film

TbCoFe Terbium Cobalt Iron (Alloy)

TB-CPA-GPM Tight-Binding Coherent-Potential Approximation Generalized-Perturbation Method [Physics]

TBCTO Terbium Barium Copper Titanium Oxide (Superconductor)

TBD Target-Bearing Designator

TBD To Be Defined [also tbd]

TBD To Be Determined [also tbd]

TBD To Be Developed [also tbd]

TBD Torpedo-Boat Destroyer

TBD 1,5,7-Triazabicyclo[4.4.0]dec-5-ene

TB-d Tight-Binding d (Band)

tbd to be defined [also TBD]

tbd to be determined [also TBD]

tbd to be developed [also TBD]

TBDA tert-Hexylborane-N,N-Diethylaniline

TBDB18C6 Tetra-t-Butyldibenzo-18-Crown-6

TBDC Bis(tetrabutyl Ammonium) Dichromate

TBDF Transborder Data Flow

TBDG Triaxial Borehole Deformation Gauge [Geology]

TBDMS tert-Butyldimethylsilyl

TBDMSCl tert-Butyldimethylsilyl Chloride

TBDMSI 1-tert-Butyldimethylsilyl)imidazole

TBDMSIM 1-(tert-Butyl Dimethyl Silyl) Imidazole

TBDMSTF tert-Butyl Dimethyl Silyl Trifluoromethane Sulfonate

TBDMS-Triflat tert-Butyl Dimethyl Silyl Trifluoromethane Sulfonate [also TBDMS-triflat]

TBDP Transborder Data Processing

TBDPE Tetrabromodipentaerythritol

TbDy Terbium Dysprosium (Alloy)

TbDyFe Terbium Dysprosium Iron (Alloy)

TbDyHoFe Terbium Dysprosium Holmium Iron (Alloy)

TbDyZn Terbium Dysprosium Zinc (Alloy)

TBE Teledyne Brown Engineering [US]

TBE Tetrabromoethane

TBE Tile and Brick Manufacturers (Association) of Europe [Switzerland]

TBE 1,1,2,2-Tetrabromoethane

TBE To Be Evaluated [also tbe]

TBE Total Beam Energy

TBE Total Binding Energy [Physics]

TBE Tris-Borate EDTA (Ethylenediaminetetraacetic Acid) (Buffer)

TBEDP Tetrabutyl Ethylenediphosphonate

TBEP Tri(butyletherethyl)phosphate

TBEM Terminal-Based Electronic Mail

TBET Truncated Brunauer-Emmett-Teller (Method) [Physical Chemistry]

TBF Butylformamidine

TBF Phosphoric Acid Tributyl Ester

TBF Tail Bomb Fuse

TBF TATA Binding Protein [*Note:* TATA is a DNA Sequence]

TBF N-tert-Butylformamide

TBF N-tert-Butyl Formamidine

TBF Test de Bon Fonçtionnement [Spacelab Functional Test]

TBF Transmitted Brightfield [Microscopy]

TbFe Terbium-Iron (Alloy)

Tb-Fe CMF Terbium-Iron Composition(al) Modulated Film

TbFeCo Terbium-Iron Cobalt (Alloy)

TbFeCoZr Terbium-Iron Cobalt Zirconium (Alloy)

TBG Thyroid-Binding Globulin; Thyroxine-Binding Globulin [Biochemistry]

TBG Time-Base Generator

TBG Tribologie-Beratungs-Gesellschaft e.V. [German Tribology Consulting Society]

TBGA Tape Ball Grid Array

TBH Taylor, Bishop and Hill (Model) [Metallurgy]

TBH Technical Benzene Hexachloride

TBHC tert-Butyl Hypochlorite

TbHoFe Terbium Holmium Iron (Alloy)

TBHP tert-Butyl Hydroperoxide

TBHPA Tetrabutyl Hypophosphoric Acid

TBHQ tert-Butyl Hydroquinone

TBI Tape and Buffer Index

TBI Target-Bearing Indicator

TBI The Brookings Institution [US]

TBI Throttle Body Injection

TBI Through-Bulkhead Initiator

TBI Toronto Biotechnology Initiative [Canada]

TBI Total Body Irradiation [Radiology]

TBI Turn and Bank Indicator [Aeronautics]

TbIG Terbium Iron Garnet

TBIS Transverse Biased Initial Susceptibility

TBITS Treasury Board Information Technology Standards

Tbit/in² Terabit(s) per square inch [also Tbit in^{-2}, Tb/in², or Tb in^{-2}]

.TBK Toolbook [also .tbk] [File Name Extension]

TBL Terminal Ballistics Laboratory [US Army]

TBL Thermal Boundary Layer

TBL Through Bill of Lading

.TBL Table [also .tbl] [File Name Extension]

Tbl Table [also tbl]

Tbl Trouble [also tbl]

TBL DEF Table Definition [also Tbl Def]

TB-LMTO Tight-Binding Linear Muffin-Tin Orbital [also TBLMTO] [Physics]

TB-LMTO-ASA Tight-Binding Linear Muffin-Tin Orbital Atomic-Sphere Approximation [Physics]

TBL NUM Table Number [also Tbl Num]

TBL OPT Table Option [also Tbl Opt]

TBM Temporary Bench Mark

TBM Terabit Memory (System)

TBM tert-Butylmethylether

TBM Thyssen Basic-Oxygen Metallurgy

TBM Tone Burst Modulation

TBM Tunnel Boring Machine

TBMA Textile Bag Manufacturers Association [US]

TBMAC Tributylmethyl Ammonium Chloride

T/B MAR Top and Bottom Margin [also T/B Mar]

TBMD Tight-Binding Molecular Dynamics [Physics]

TBMDP Tetrabutyl Methylenediphosphonate

TBME Tactical Ballistic Missile Experiment [also TBMX, or TBX]

TBME tert-Butyl Methyl Ether

Tb(MeBB)$_3$ Terbium(III) Methylenebenzoylbenzoate

Tb(MeOBB)$_3$ Terbium(III) Methoxybenzoylbenzoate

TBMP 2-tert-Butyl-4-Methyl Phenol

TBMPSiBr tert-Butyl-Methoxyphenylbromosilane

TBMT Transmitter Buffer Empty

TBMX Tactical Ballistic Missile Experiment [also TBME, or TBX]

TBN Turbostratic Boron Nitride [also tBN, or t-BN]

TBN To Be Negotiated [also tbn]

TBN Total Base Number [Chemistry]

TBN 2,4,6-Tri-tert-Butylnitrobenzene

TBNPA Tribromoneopentyl Alcohol

TBO Time before (or between) Overhauls

TBO 3-(Trimethylsilyloxy)-3-Butene-2-one

TBOAA Tuna Boat Owners Association of Australia

TBOS Tetrabutyl Orthosilicate

TBP TATA-Box Binding Protein [Note: TATA is a DNA Sequence]

TBP tert-Butyl Phenol

TBP tert-Butyl Phosphine

TBP Tetrabenzoporphine; 29H,31H-Tetrabenzo[b,g,l,q]porphine

TBP 2,2'-Thiobis(4,6-Dichlorophenol)

TBP To Be Provided [also tbp]

TBP Transferrin-Binding Protein [Biochemistry]

TBP Tribromophenol

TBP Tributyl Phosphate [also tbp]

TBP Tri-n-Butyl Phosphate [also tbp]

TBP Tributylphosphine

TBP Tributylphosphoric Acid

TBP Trigonal Bipyramidal [Crystallography]

TBP True Boiling Point [also tbp]

2,4,6-TBP 2,4,6-Tribromophenyl Ester

ТВП Tau Beta Pi (Student Society)

TBPA Tetrabromophthalic Anhydride

TBPA Textile Bag and Packaging Association [US]

TBPAE Tribromophenyl Allyl Ether

TBPBr Tetrabutylphosphonium Bromide

TBPCl Tetrabutylphosphonium Chloride

TBPCO Thallium Barium Praseodymium Copper Oxide (Superconductor)

TBPFe Iron Tetrabutylphosphonium

TBPO Tributyl Phosphine Oxide

TBPEHS Tau Beta Pi Engineering Honour Society [US]

TPBG Tetra tert-Butyl Phosphinogallane

TBPMI Tribromophenylmaleimide

TBPO Tertiary Butylperoctoate

TBPO Tributyl Phosphine Oxide

TBPS Terabits per second [also Tbps]

TBPy 2-tert-Butylpyridine [also tBPy]

TBq Terabecquerel [Unit]

TBQMD Tight-Binding Quantum Molecular Dynamics [Physics]

TBq/mmol Terabecquerel per millimole [also TBq mmol^{-1}]

TBR Table Base Register

TBR To Be Received [also tbr]

TBR To Be Revised [also tbr]

Tbr Timber [also tbr]

TBRC Top-Blown Rotary (Steelmaking) Converter [Metallurgy]

TBRI Technical Book Review Index

Tbrs Timber [also tbrs]

TBS Talk between Ships

TBS Tape-Bearing Surface [Magnetic Media]

TBS Task Breakdown Structure

TBS tert-Butyldimethylsilyl

TBS 4-tert-Butylphenol Salicylate

TBS Tethered Balloon System

TBS Tetrapropylenebenzenesulfonate

TBS The Biodeterioration Society [US]

TBS Tight Building Syndrome [Medicine]

TBS To Be Specified [also tbs]

TBS To Be Superseded [also tbs]

TBS To Be Supplied [also tbs]

TBS Treasury Board Secretariat

TBS Tribromosalan

TBS Tris Buffered Saline

TB&S Top, Bottom and Sides

tbs tablespoon(ful) [Unit]

T$_3$-BSA 3,3',5-Triiodo-L-Thyronine–Bovine Serum Albumin [Immunology]

T$_4$-BSA L-Thyroxine–Bovine Serum Albumin [Immunology]

TBS-BSA Tris Buffered Saline with Bovine Serum Albumin [Immunology]

TBSCl tert-Butyldimethylsilyl Chloride

TBSn Tetra-n-Butyltin [also TBT]

tbsp tablespoon(ful) [Unit]

TBS-T Tris Buffered Saline with Tween-20 [Note: TWEEN-20 is a Trademark for Polyoxyethylene (20) Sorbitan Monolaurate]

TBSV Tomato Bushy Stunt Virus

TBT Technical Barriers to Trade [of General Agreement on Tariffs and Trade]

TBT Tetra-n-Butyltin [also TBSn]

TBT Tetrabutyl Titanate

TBT Tight-Binding Theory [Physics]

TBT Tri-n-Butyltin

TBTA t-Butyl Trichloroacetimidate

TBTAc Tri-n-Butyltin Acetate

TBTCl Tri-n-Butyltin Chloride

TBTD N,N,N',N'-Tetrabutyl Thiuram Disulfide

TBTD Tri-n-Butyltin Deuteride

TBTE Tight-Binding Total-Energy (Model) [Physics]

TBTF Tri-n-Butyltin Fluoride

TBTH Tri-n-Butyltin Hydride

Tb(THD)$_3$ Terbium Tris(2,2,6,6-Tetramethyl-3,5-Heptanedionate) [also Tb(thd)$_3$, Tb(TMHD)$_3$, or Tb(tmhd)$_3$]

TBTO bis(Tri-n-butyltin) Oxide

TBTO Tributyltin Oxide

TBTO Tributyl Trioxide

TBTS Bis(tri-n-butyltin)sulfide

TBTU O-Benzotriazol-1-yl-N-Tetramethyluronium Tetrafluoroborate

TBTU N,N,N'-Tributyl Thiourea

TBTUP Tight-Binding Theory with Universal Parameters [Physics]

TBU Tape Backup Unit

TBU Terminal Buffer Unit

TBU Transmit Baseband Unit

TBUP Tributylphosphine

TBV Total Burgers Vector [Crystallography]

TBWP Triple-Braid(ed) Weatherproof (Electric Wire)

TBX Tactical Ballistic-Missile Experiment [also TBME, TBMX, or TBX]

T2BX Toluene-to-Benzene and Xylene (Process)

Tb-YSZ Terbia-Doped Yttria-Stabilized Zirconia

Tbyte Terabyte [also TB]

TBZ Thiabendazole

TbZn Terbium-Zinc (Alloy)

TC Tab Card

TC Tactical Computer

TC Tag Code

TC Tail Cone

TC Take Credit (Memorandum)

TC Tantalum Capacitor

TC Tape Command

TC Tariff Commission

TC Task Control

TC Taxiway Centerline [Aeronautics]

TC Taylor-Couette (Apparatus) [Fluid-Mechanics]

TC T-Carrier

TC Teachers College

TC Technical Classification

TC Technical College

TC Technical Commission

TC Technical Committee

TC Technical Communication

TC Technical Conference

TC Technical Consultant

TC Technical Control

TC Technical Cooperation

TC Technical Council [of Institute of Electrical and Electronics Engineers–Power Engineering, US]

TC Telecommunications [also T/C]

TC Telecommuting

TC Teleconference; Teleconferencing

TC Telemetry and Command

TC Telephone Channel

TC Telephone Company

TC Telescoping Collar

TC Television Control

TC Temperature Coefficient

TC Temperature Compensating; Temperature Compensation

TC Temperature Conductivity

TC Temperature Control(ler)

TC Temper Color [Metallurgy]

TC Temporary Council

TC Term Coordination

TC Terminal Computer

TC Terminal Concentrator

TC Terminal Congestion

TC Terminal Control(ler)

TC Terminal Count

TC Terra Cotta

TC Tertiary Compound

TC Tertiary Consumer [Ecology]

TC Tesla Coil

TC Test Case

TC Test Certificate

TC Test Chamber

TC Test Chart

TC Test Code

TC Test Coil

TC Test Conductor

TC Test Console

TC Test Control

TC Test Controller

TC Tetracycline [Microbiology]

TC Tetrahedral Cubic (Crystal)

TC Texture Coefficient [Crystallography]

TC Theoretical Chemist(ry)

TC Theoretical Climatologist; Theoretical Climatology

TC Thermal Conditioning

TC Thermal Conduction; Thermal Conductivity; Thermal Conductor

TC Thermal Continuum

TC Thermal Control

TC Thermal Conversion; Thermal Converter

TC Thermal Crack [Metallurgy]

TC Thermal Cracking [Chemical Engineering]

TC Thermal Creep [Mechanics]

TC Thermal Cutting [Metallurgy]

TC Thermal Cycle

TC Thermal Cycling [Atomic Physics]

TC Thermionic Cathode

TC Thermionic Converter

TC Thermocell

TC Thermochemical; Thermochemist(ry)

TC Thermocolor

TC Thermocompression; Thermocompressor

TC Thermocouple

TC Thermocurrent

TC Thermoplastic Composite

TC Thiocarbamyl

TC Thiokol Corporation [US]

TC Third Class

TC Thread Count [Fabrics]

TC Threshold Circuit

TC Thrust Chamber [also T/C] [Aerospace]

TC Thrust Coefficient [Aerospace]

TC Thrust Control [Aerospace]

TC Time Clock

TC Time Code

TC Time Constant

TC Time Control(led)

TC Time Correlation

TC Timed Closing

TC Time to Circular

TC Time to Computation

TC Timing Channel

TC Tin Cry [Metallurgy]

TC Tissue Culture

TC TOGA/COARE (Tropical Ocean and Global Atmosphere/ Coupled Ocean Atmosphere Response Experiment) [also T/C]

TC Toggle Clamp

TC To Contain

TC Tokyo Convention [of International Civil Aviation Organization in 1963]

TC Toll Center

TC Toll Completing

TC Toroidal Coil

TC Toroidal Combustion

TC Torque Converter

TC Torsion Couple

TC Total Carbon [Metallurgy]

TC Total Cholesterol [Biochemistry]

TC Total (Number of Spring) Coils [Symbol]

TC Total Count

TC Total Counter

TC Town Council

TC Toxic Concentration

TC Traceability Code

TC Tracer Composition

TC Tracking Camera

TC Traffic Control

TC Traffic-Signal Control(ler)

TC Training Course

TC Trans-Canada

TC Trans-Cis (Configuration) [Chemistry]

TC Transconductance; Transconductor [Electronics]

TC Transfer Channel

TC Transfer Coefficient

TC Transistorized Carrier

TC Transmission Channel(ing)

TC Transmission Coefficient [Physics]

TC Transmission Control [Telecommunications]

TC Transmitter Clock

TC Transverse Contraction

TC Transmission Control (Character)

TC Transmission Controller

TC Transmitting Circuit

TC Transnational Committee [of Institute of Electrical and Electronics Engineers–Regional Activities Board/ Technical Activities Board, US]

TC Transient Converter

TC Transparent Conductor

TC Transportation Commodity

TC Transportation Corps [of US Army]

TC Transport Canada

TC Transverse Colon [Anatomy]

TC Trash Coordinator

TC Tray Cable

TC Treatment Code

TC Trichloroacetate; Trichloroacetic Acid

TC Tri-Cities [Madison–Venice–Granite City, Illinois, US]

TC Tri-Cities [Richland–Kennewick–Pasco, Washington State, US]

TC Tricrystal [also TX]

TC Trinity College [Cambridge, UK]

TC Trinity College [Dublin, Ireland]

TC Trip Coil

TC Tropical Cyclone

TC True Complement [Mathematics]

TC Truncated Cylinder

TC Trunk Control

TC Trusteeship Council [of United Nations]

TC Tuned Circuit

TC Tungsten Cathode

TC Tunnel Car

TC Turbine Cylinder

TC Turbocompound (Engine)

TC Turbocompressor

TC Turbostratic Carbon

TC Türkiye Cumhuriyeti [Republic of Turkey]

TC Turks and Caicos Islands [ISO Code]

TC Twin Cities [Auburn–Lewiston, Maine, US]

TC Twin Cities [Minneapolis–St. Paul, Minnesota, US]

TC Twiss Coefficient

(TC) External Transmitter Clock

TC- Turkey [Civil Aircraft Marking]

TC_{Hi} Toxic Concentration High [also TC_{HI}]

TC_{LO} Toxic Concentration Low [also TC_{Lo}]

3TC Lamivudine [AIDS Drug]

T&C Threads and Couplings

T&C Time and Charges

T/C Telecommunications [also TC]

T/C Temperature Compensating

T/C Termination Check

T/C Thermal Cycle (Test)

T/C Thrust Chamber [also TC] [Aerospace]

T/C TOGA/COARE (Tropical Ocean and Global Atmosphere/Coupled Ocean Atmosphere Response Experiment) [also TC]

T-C Tetragonal-to-Cubic (Transformation) [also t-c] [Metallurgy]

Tc Talc [Mineral]

Tc Technetium [Symbol]

Tc Teracycle [Symbol]

Tc-93 Technetium-93 [also ^{93}Tc, or Tc^{93}]

Tc-94 Technetium-94 [also ^{94}Tc, or Tc^{94}]

Tc-95 Technetium-95 [also ^{95}Tc, or Tc^{95}]

Tc-96 Technetium-96 [also ^{96}Tc, or Tc^{96}]

Tc-97 Technetium-97 [also ^{97}Tc, or Tc^{97}]

Tc-99 Technetium-99 [also ^{99}Tc, or Tc^{99}]

Tc-100 Technetium-100 [also ^{100}Tc, or Tc^{100}]

Tc-101 Technetium-101 [also ^{101}Tc, or Tc^{101}]

Tc-105 Technetium-105 [also ^{105}Tc, or Tc^{105}]

t-c tetragonal-to-cubic (transformation) [also T-C] [Metallurgy]

$\tau_0(c)$ concentration dependent critical shear stress [Symbol]

TCA Sodium Trichloroacetate

TCA Tactical Combat Aircraft

TCA Tanners Council of America [US]

TCA Teach Cable Assembly

TCA Technical Cooperation Administration [US]

TCA Telecommunications Association [US]

TCA Telemetering Control Assembly

TCA Television Companies Association Ltd. [UK]

TCA Temperature Control Assembly

TCA Temporary Change Authorization

TCA Terminal Communications Adapter

TCA Terminal Control Area [Aeronautics]

TCA Textile Converters Association [UK]

TCA The Commonwealth of Australia

TCA The Technical Center for Agricultural and Rural Cooperation [of European Commission]

TCA Thermal Critical Assembly [US]

TCA Thermal Cycle Annealed; Thermal Cycle Annealing [Metallurgy]

TCA Thermocentrifugometric Analysis

TCA Thermostatically Controlled Air-Cleaner

TCA Thrust Chamber Assembly [Aerospace]

TCA Tile Council of America [US]

TCA Tilt-Up Concrete Association [US]

TCA Time Correlation Analysis

TCA Time of Closest Approach [Astronomy]

TCA Tissue Culture Association [US]

TCA TOGA/COARE (Tropical Ocean and Global Atmosphere/ Coupled Ocean Atmosphere Response Experiment) Central Archive

TCA Total Cost Assessment

TCA Trace Contamination Analysis

TCA Trans-Canada Airlines

TCA Transcontinental Control Area

TCA Translation Controller Assembly

TCA Transmission Control Area

TCA Tricarboxylic Acid (Cycle)

TCA Trichloroacetate; Trichloroacetic Acid

TCA Trichloroaniline

TCA 2,4,6-Trichloroanisole

TCA Tricyclic Antidepressant [Pharmacology]

TCA Turbulent Contact Absorber

TCA Twin Cities Arsenal [Minneapolis/St.Paul, Minnesota, US]

TCA Two-Channel Analyzer

TCAA Tile Contractors Association of America [US]

TCAA Trichloroacetic Aldehyde

TCAAN Trans-Canada Advertising Agency Network

TCAB Temperature of Cabin

TCAB Tetrachloroazobenzene

TCAB Twin-Carbon Arc Brazing

TC$_5$ABr Tetrapentylammonium Bromide

TC$_6$ABr Tetrahexylammonium Bromide

TC$_7$ABr Tetraheptylammonium Bromide

TC$_8$ABr Tetraoctylammonium Bromide

TC$_{18}$ABr Tetraoctadecylammonium Bromide

TCAC Technical Committee on Agricultural Chemicals

TCA/CFC Trichloroacetic Acid/Chlorofluorocarbon (Solvent)

TC$_5$ACl Tetrapentylammonium Iodide

TC$_6$ACl Tetrahexylammonium Iodide

TC$_7$ACl Tetraheptylammonium Iodide

TCAD Traffic Alert and Collision Avoidance Device

TCAE Thermal Coefficient of Area(l) Expansion

TC$_8$AF Tetraoctylammonium Fluoride

TCAI Texas College of Arts and Industries [Kingsville, US]

TCAI Tutorial Computer-Assisted Instruction

TCAM Telecommunications Access Method

TCAM-IMS/VS Telecommunications Access Method– Information Management System/Virtual Storage

TCAN Trichloroacetonitrile

TCAP Tank Characterization Advisory Panel [Nuclear Engineering]

T/CAP Thermal Capacitor

TCA Rep TCA Report [of Tissue Culture Association, US]

TCART Toronto Center for Advanced Reproductive Technology [Canada]

TCAS T-Carrier Administration System

TCAS Traffic Alert and Collision Avoidance System

TCAT Test Coverage Analysis Tool

TCAW Twin-Carbon Arc Welding

TCB Task Control Block

TCB Technical Coordinator Bulletin

TCB Terminal Control Block

TCB Tetracarboxybutane

TCB Textile and Clothing Board

TCB Transfer Control Block

TCB Trichlorobenzene

TCB Trichloroborazine

TCB Trusted Computing Base

TCBA Tesla Coil Builders Association [US]

TCBA Trichlorobutyl Alcohol

TCBC The Conference Board of Canada

TCBC Trichlorobenzylchloride

TCBCO Thallium Calcium Barium Copper Oxide (Superconductor)

TCBH Time-Consistent Busy Hour

TCBI N-2,6-Trichloro-4-Benzoquinone Imine

TCBL Tradewind Cumulus Boundary Layer [Meteorology]

TCBM Transcontinental Ballistic Missile

TCBO Trichlorobutylene Oxide

TCBOC Trichlorobutoxycarbonyl

TCBOC 2,2,2-Trichloro-1,1-Dimethylethyl Chloroformate

TCBOC-Cl Trichloro-tert-Butoxycarbonyl Chloride [also TCBOC-Chloride]

TCBQ p-Tetrachlorobenzoquinone

TC12BQ Tetrachloro-1,2-Benzoquinone

TC14BQ Tetrachloro-1,4-Benzoquinone

TCBS Tesla Coil Builders Society [US]

TCBV Temperature Coefficient of Breakdown Voltage [Electronics]

TCC Tag(liabue) Closed Cup (Flash Point Tester) [Chemistry]

TCC Technical Change Center [UK]

TCC Technical Computing Center

TCC Technical Control Center

TCC Technical Coordination Committee

TCC Technological Change Committee

TCC Telecommunications Coordinating Committee

TCC Television Control Center

TCC Temperature Coefficient of Capacitance [Electronics]

TCC Temperature Coefficient of Conductivity [Electronics]

TCC Temporary Council Committee [of NATO]

TCC Test Control Center

TCC Test Controller Console

TCC Texas Chiropractic College [Pasadena, Texas, US]

TCC Thermal Conduction Control(ling)

TCC Thermal Conductivity Cell

TCC Thermal Control(led) Coating

TCC Thermofor Catalytic Cracking (Process) [Chemical Engineering]

TCC Through-Connected Circuit

TCC Time Compression Coding

TCC Toll Center Code

TCC Tons of Clean Coal [also tcc]

TCC Torque Converter Clutch (Solenoid)

TCC Toxic Chemicals Committee [US]

TCC Tracking and Control Center

TCC Traffic Control Center

TCC Trans-Cis-Cis (Geometry) [Chemistry]

TCC Transfer Channel Control

TCC Transmission Control Character

TCC Transmission Cross Coefficient [Physics]

TCC Transport and Communications Commission [of United Nations]

TCC Transportation Commodity Classification

TCC Trichlorocarbane

TCC 3,4,4'-Trichlorocarbanilide

TC-C Tien and Copley–Compression (Behavior) [Metallurgy]

TCCA Textile Color Card Association [US]

TCCA Trichloroisocyanuric Acid

TCCAUS Textile Color Card Association of the United States

TCCB Technical College of Cape Breton [Nova Scotia, Canada]

TCCH Tetrachlorocyclohexane

TCCP Technical Cooperation Council of the Philippines

TCCP DB Technical Council on Computer Practices, Database Committee

TCCSR Telephone Channel Combination and Separation Racks

TCCTB Technical Center for Clay, Tiles and Bricks [France]

TCD Tank Characterization Database [Nuclear Engineering]

TCD Telemetry and Command Data

TCD Temperature Coefficient of Decay [Electronics]

TCD Test Completion Date

TCD Test Control Document

TCD The College Board [US]

TCD Thermal Conductivity Detector

TCD Thyratron Core Driver

TCD Time Code Division

TCD Trans-Canada Dermapeutics Limited

TCD Transistor-Controlled Delay

TCD Trinity College Dublin [Ireland]

T&CD Timing and Countdown

TCDD Tetrachlorinated Dibenzo-p-Dioxin; 2,3,7,8-Tetrachlorodibenzo-p-Dioxin; Tetrachlorodibenzo-p-Dioxin

2,3,7,8-TCDD 2,3,7,8-Tetrachlorodibenzo-p-Dioxin

TCDF Tetrachlorodibenzofuran

TCDMS Telecommunications/Data Management System

TCDNB Trichlorodinitrobenzene

TCDS Tandem Cylindrical Deflector Spectrometry

TCDS Thyristor Controlled Drive System

TCDT Terminal Countdown Demonstration Test

TCE Telecommunication Engineering

TCE Telemetry and Command Equipment

TCE Telemetry Checkout Equipment

TCE Telephone Company Engineered

TCE Temperature Coefficient of Expansion

TCE Tetrachloroethane

TCE Thermal Canister Experiment

TCE Thermal Coefficient of Expansion

TCE Thermal Conversion Efficiency

TCE Tons of Coal Equivalent

TCE Total Composite Error

TCE Transmission Control Element

TCE 1,1,1-Trichloroethane

TCE 2,2,2-Trichloroethanol

TCE Trichloroethene

TCE 2,2,2-Trichloroethyl

TCE Trichloroethylene

TCε Temperature Coefficient of Dielectric Constant [Electronics]

Tce Terrace [also tce]

TCEA Training Center for Experimental Aerodynamics [of NATO]

TCEBU Technical Center of the European Broadcasting Union [Brussels, Belgium]

TCED Thrust Control Exploratory Development [Aerospace]

TCEF Phosphoric Acid Mono(2,2,2-Trichloroethyl)ester

T Cell Thymus (Lymophocyte) Cell [also T cell] [Immunology]

TCENM Triscyanoethylnitromethane; Tris(2-Cyanoethyl) nitromethane

TCEP tris(Carboxyethyl)phosphine

TCEP Trichloroethyl Phosphate

TCEP Triscyanoethoxypropane; 1,2,3-Tris(2-Cyanoethoxy) propane

TCETACN 1,4,7-tris(2-Cyanoethyl)-1,4,7-Triazacyclononane [also tcetacn]

TCF Phosphoric Acid Tris(methylphenyl) Ester

TCF Tank Checkout Facility [Nuclear Engineering]

TCF Technical Control Facility

TCF Technical Cooperation Fund

TCF Terminal Configuration Facility

TCF Time Correlation Function

TCF Toho (Polyacrylonitrile-Based) Carbon Fiber

TCF Transparent Conducting Film

TCF Trillion Cubic Feet [also tcf]

TCFTD Triple Constant Fraction Time Discriminator

TCG Test Call Generator

TCG Test Control Group

TCG Time Code Generator

TCG Time-Controlled Gain

TCG Transponder Control Group

TCG Tune-Controlled Gain

TCG Twinned Columnar Growth (Casting Defect) [Metallurgy]

TCGF T-Cell Growth Factor [Biochemistry]

TCH Thiocarbohydrazide

TCH Threshold Crossing Height

TCH Trigger Chamber

TC (hkl) Texture Coefficient for (hkl) Plane [Crystallography]

TCHC Thermally Conductive Hydrocarbon

TCHP Tris(chlorohexyl)phosphate

TcHPO Tris(cyclohexylphosphione Oxide [also TCHPO]

TCHQ-DE Tetrachlorohydroquinone Dialkyl Ether

TChT Thermochemical Treatment

TCI Technical Critical Item

TCI Technology Conformance Inspection

TCI Technology Commercialization Initiative [of US Department of Energy]

TCI Telemetry Components Information

TCI Terrain-Clearance Indicator [Aeronautics]

TCI Textual Command Interface

TCI Theoretical Chemistry Institute

TCI True/Complement I

TCID Test Configuration Identifier Document

TCID Tissue Culture Infectious Dose

TCID$_{50}$ Tissue Culture Median Infectious Dose

TCIL Transmission Control Indicator Light [Automobiles]

TCIM Tactical Communication Interface Module

TCIPO TOGA/COARE (Tropical Ocean and Global Atmosphere/Coupled Ocean Atmosphere Response Experiment) International Project Office

TCIS Telex Computer Inquiry Service

TCJ Tanners Council of Japan

TCL Target Compound List [of US Environmental Protection Agency]

TCL Terminal Command Language

TCL Terminal Control Language

TCL ω-Thiocaprolactam [also tcl]

TCL Time and Cycle Log

TCL Toll Circuit Layout

TCL Tool Command Language

TCL Toxicity Characteristic Leaching (Procedure)

TCL Toxicity Chemical Leaching (Procedure)

TCL Traction Control

TCL Transcrystalline Layer [Metallurgy]

TCL Transfer Chemical Laser

TCL Transistor-Coupled Logic

TCl Terephthaloyl Chloride

tcl ω-thiocaprolactam [also TCL]

TClAc Trichloroacetic Acid

TC/LD Thermocouple/Lead Detector

TCLE Thermal Coefficient of Linear Expansion

TCLP Toxicity Characteristic Leach(ing) Procedure

TCLP Toxicity Chemical Leach(ing) Procedure

TCLR Toll Circuit Layout Record

TCM Telecommunications Monitor

TCM Telemetry Code Modulation

TCM Temperature Control Model

TCM Terminal Capacity Matrix

TCM Terminal Charge Management

TCM Terminal-to-Computer Multiplexer

TCM Test Call Module

TCM Tetrachloromercurate

TCM Thermal Conduction Module

TCM Thermochemical Machining

TCM Thermoplastic Cellular Molding

TCM Time Compression Multiplex(ing)

TCM Tool Condition Monitoring

TCM Torque-Coil Magnetometer

TCM Total Catchment Management

TCM Trajectory Correction Maneuver

TCM Traditional Chinese Medicine

TCM Transient Conductance Measurement (Technique) [Physics]

TCM Transmission Control Module [Automobiles]

TCM Trellis Code(d) Modulation

TCM Tricarbonyl Methylcyclopentadienyl Manganese

TCM Trichloromethane

T/cm² Tesla(s) per square centimeter [also T cm^{-1}]

TCMA Tooling Component Manufacturing Association [US]

TCMD Transportation Control and Movement Document

TCMF Touch-Calling Multifrequency

TC$_{14}$MgCl Tetradecylmagnesium Chloride

TCMP Toxic Chemicals Management Program

TCMS Telephone Cost Management System

TCMS Test, Checkout and Monitor(ing) System

TCMS Toll Centering and Metropolitan Sectoring

TCMTB Thicyanomethylthio Benzothiazole

TCN Tantalum-Copper-Niobium (Superconductor)

TCN Telecommunications Cooperative Network

TCN Telecommunications Network

TCN Test Change Notice

TCN Tetrachloronaphthalene

TCN Transportation Control Number

TCN Transport Canada (Data Processing) Network

TCNA 2,3,5,6-Tetrachloro-4-Nitroanisole

TCNA Tube Council of North America [US]

tCnA trans-Cinnamic Acid

TCNB 1,2,4,5-Tetrachloro-3-Nitrobenzene

TCNB Tetracyanobenzene

TCNBT Thiocarbamyl Nitro Blue Tetrazolium [also TC-NBT]

TCNE Tetracyanoethanide

TCNE Tetracyanoethenide

TCNE Tetracyanoethylene

TCNEO Tetracyanoethylene Oxide

TCNP Tetracyanopyridine

TCNQ Tetracyanoquinodimethane; 7,7,8,8-Tetracyano-p-Quinodimethane; Tetracyanoquinodimethanide

TC-NR Technical Classified Natural Rubber

TCO Taken Care of

TCO Telenet Central Office

TCO Temperature Coefficient of Offset

TCO Termination Contracting Officer

TCO Test and Checkout

TCO The Carnegie Observatories [US]

TCO Thermal Cutoff

TCO Total Cost of Ownership

TCO Transparent Conducting (or Conductive) Oxide [Solar Cells]

TCO Trunk Cutoff

TCOE Temperature Coefficient of Expansion

3COM Computer, Communications and Compatibility Corporation [US]

TCOP Test Checkout Plan(s)

TCOS Trunk Class of Service [Telecommunications]

TCP Tank Characterization Plan [Nuclear Engineering]

TCP Tape-Carrier Package

TCP Tape Conversion Program

TCP Task Control Packet

TCP Task Control Program

TCP Technical Cooperation Program [of Australia, Canada, UK and US]

TCP Technology Commercialization Program [Canada]

TCP Terminal Control Program

TCP Test Checkout Procedure

TCP Test Control Package

TCP Thermochemical Process(ing)

TCP Thrust Chamber Pressure

TCP Tool Center Point [Robotics]

TCP Topologically Close-Packed [also tcp] [Crystallography]

TCP Total Carbon Pool [Ecology]

TCP Transmission Control Protocol

TCP Traffic Control Post

TCP Transmission Channelling Pattern

TCP Transmission Control Program

TCP Transmission Control Protocol

TCP Transmitter Control Pulse

TCP Transport Control Protocol

TCP Tribasic Calcium Phosphate; Tricalcium Phosphate

TCP Trichlorophenol

TCP Trichlorophenoxyacetic Acid

TCP Tricresyl Phosphate

TCP Tricritical Point

α-TCP Alpha Tricalcium Phosphate

β-TCP Beta Tricalcium Phosphate

T&CP Test and Checkout Procedure

TCPA Tantawangalo Catchment Protection Association

TCPA Town and Country Planning Association [UK]

TCPC Tab Card Punch Control

TCPC Trichloropropylcarbinol

TCPE 2-(2,4,5-Trichlorophenoxy)ethanol

TCPIP TOGA COARE TOGA/COARE (Tropical Ocean and Global Atmosphere/Coupled Ocean Atmosphere Response Experiment) Program International Panel

TCP/IP Transmission Control Protocol /Internet Protocol

TCPL Trans-Canada Pipeline

TCPO TOGA/COARE (Tropical Ocean and Global Atmosphere/Coupled Ocean Atmosphere Response Experiment) Project Office

TCPO Bis(2,4,6-Trichlorophenyl)oxalate

TCPO 1,1,1-Trichloropropene 2,3-Oxide

TCPO 3,3,3-Trichloropropylene Oxide

TCPP Tetra-p-Chlorophenylpyrrole

TCPP Trischloropropyl Phosphate

TCPPA 2-(2,4,5-Trichlorophenoxy)propanoic Acid

2,4,5-TCPPA 2-(2,4,5-Trichlorophenoxy)propanoic Acid

TCPS Tissue Culture Polystyrene

TCPSA Thermally Coupled Pressure Swing Absorption (Cycle)

TCPSPC Time-Correlated Pulsed Single Photon Counting [Physics]

TCQ The Centre for Quality [of Ngee Ann Polytechnic, Singapore]

TCQD Tetracyanoquinodimethane

TCR Tank Characterization Reporting [Nuclear Engineering]

TCR Tape Cassette Recorder

TCR T-Cell Receptor [Immunology]

TCR Teleconference Room

TCR Telemetry Compression Routine

TCR Temperature Coefficient of Resistivity

TCR Test Compare Results

TCR Test Constraints Review

TCR Thermal Concept Review

TCR Thermochemical Reduction

TCR Thermofor Catalytic Reforming (Process) [Chemical Engineering]

TCR Transfer Control Register

TCR Transient Call Register

TC&R Telemetry Command and Ranging

Tcr Tracer [also tcr]

TCRA Telegraphy Channel Reliability Analyzer

TCRC Toshiba Cambridge Research Center [UK]

TCRC Treatment Charges and Refining Charges

TCRC Twin Cities Research Center [Minneapolis/St. Paul, Minnesota, US]

TCRDL Toyota Central Research and Development Laboratories, Inc. [Aichi, Japan]

TCRF Toxic Chemical Release Form

TCRI Toyota Central Research Institute [Japan]

TCRSD Test and Checkout Requirements Specification Documentation

TCS Teaching Company Scheme [UK]

TCS Technical Change Summary

TCS Technical Concurrence Sheets

TCS Technology Club of Syracuse [New York, US]

TCS Telecentric System

TCS Telecommunications Control System

TCS Telecommunications System

TCS Telephone Conference Summary

TCS Telex Communications Service [US]

TCS Temperature Coefficient of Sensitivity

TCS Temperature Control System

TCS Terminal Communications Subsystem

TCS Terminal Communications System

TCS Terminal Control System

TCS Terminal Countdown Sequencer

TCS Terminal Count Sequence

TCS Terne-Coated Steel

TCS Test Control Supervisor

TCS Test Control System

TCS Texas Center for Superconductivity [of University of Houston, US]

TCS The Classification Society [now Classification Society of North America, US]

TCS The Coastal Society [US]

TCS The Constant Society [US]

TCS The Cousteau Society [US]

TCS The Crustacean Society [US]

TCS Theoretical Computer Science

TCS Thermal Conditioning Service

TCS Thermal Control Subsystem; Thermal Control System

TCS Thermal Conversion Spectroscopy

TCS Thermal Cycle Stimulation; Thermal Cycle Stimulator

TCS Thomson Cross-Section [Physics]

TCS Time Collection System

TCS Tool Coordinate System

TCS Total Communications System

TCS Total Concept Sales

TCS Total Control System

TCS Total Current Spectroscopy

TCS Total (Integrated) Cross-Section [Atomic Physics]

TCS Traffic Control Station

TCS Transaction Control System

TCS Transmission-Controlled Spark

TCS Transportable Communication System

TCS Transportation and Communication Service

TCS Tribasic Calcium Silicate; Tricalcium Silicate

TCS True Chemical Shift [Physical Chemistry]

Tc/s Teracycles per second [Unit]

TCSA 3,3',4',5-Tetrachlorosalicylanilide

TCSAR Telenet Communications Service Authorization Request

TCSBP The Capability Shall Be Provided

TCSC Trainer Control and Simulation Computer

TC/SC Technical Committee/Subcommittee

TCSCF Two-Configuration Self-Consistent Field [Physics]

TCSES Trusted Computer System Evaluation System

TCSP Tandem Cross Section Program

TCSP Tourism Council of the South Pacific

TCSPC Time-Correlated Single Photon Counting [Physics]

TCSR Ammunition Type Classification Summary Reports

TCSS Teleconferencing Software System

TCSSS Thermal Control Subsystem Segment

TCST TOGA/COARE (Tropical Ocean and Global Atmosphere/Coupled Ocean Atmosphere Response Experiment) Science Team

TCST Trichlorosilanated Tallow

TCSUH Texas Center for Superconductivity, University of Houston [US] [also TcSUH]

TCSUH HTS Texas Center for Superconductivity, University of Houston High-Temperature Superconductor (Workshop) [US] [also TcSUH HTS]

TCT Tag(liabue) Closed-Cup (Flash Point) Tester [Chemistry]

TCT Tasmanian Conservation Trust [Australia]

TCT Telecommunication Technology

TCT Terminal Control Table

TCT Thermochemical Treatment [Metallurgy]

TCT Thermocycling Treatment [Metallurgy]

TCT Thrombin Clotting Time [Medicine]

TCT Toll Connecting Trunks

TCT Trans-Cis-Trans (Geometry) [Chemistry]

TCT Translator and Code Treatment Frame

TCT Transmission Computed Tomography

TCT Trichromacy Theory

TCT Tricrotonylidene Tetramine

TC-T Tien and Copley–Tension (Behavior) [Metallurgy]

TCTA 1,4,7-Triazacyclononane-1,4,7-Triyl-2,2′,2″-Tris(acetate) [also tcta]

TCTFB 1,1,2,2-Tetrachloro-3,3,4,4-Tetrafluorocyclobutane

TCTFC 1,1,2-Trichloro-2,3,3-Trifluorocyclobutane

TCTFE Trichlorotrifluoroethane

TC-TG-CVI Time Control(led), Temperature Gradient Chemical Vapor Infiltration

TCTI Time Compliance Technical Instruction

TCTNB 1,3,5-Trichloro-2,4,6-Trinitrobenzene

TCTO Time Compliance Technical Order

TCTP Tetrachlorothiophene

TCTS Trans-Canada Telephone System [now Telecom Canada]

TCTSB Transport Canada Technical Services Branch [of International Civil Aviation Organization]

TCTSC Transport Canada Technical Systems Center [of International Civil Aviation Organization]

TCTTE Terminal Control Table Terminal Entry

TCTTMA Tank Conference of the Truck Trailer Manufacturers Association [US]

TCU Tape Control Unit

TCU Telemetry Control Unit

TCU Teletypewriter Control Unit

TCU Terminal Control Unit

TCU Test Control Unit

TCU Texas Christian University [Fort Worth, US]

TCU Timing Control Unit

TCU Towering Cumulus [Meteorology]

TCU Transmission Control Unit

T-Curve Toughness Curve [also T-curve] [Mechanics]

TCV Technology Characterization Vehicle

TCV Temperature Control Valve

TCV Temperature Control Voltage

TCV Terminal Configured Vehicle

TCVA Trichlorovinylamine

TCVC Tape Control Via Console

TCVD Technical Committee on Veterinary Drugs

TCVE Thermal Coefficient of Volume Expansion

TCW Time Code Word

TCWB Tinned Copper Wire Braid [Electrical Engineering]

TCWG Telecommunications Working Group

TCWI Terrain Clearance Warning Indicator

TCX Tetrachloroxylene

TCXO Temperature-Compensated Crystal Oscillator

TCZD Temperature-Compensated Zener Diode

TD Chad [ISO Code]

TD Tabular Data

TD Tank Destroyer

TD Tap Density [Powder Metallurgy]

TD Tape Drive

TD Tardive Dyskinesia [Medicine]

TD Technical Data

TD Technical Directive

TD Technical Director

TD Technical Division

TD Technical Document

TD Technology Department

TD Technology Development

TD Technology Division

TD Telemetry Data

TD Telephone Directory

TD Temperature Differential

TD Temporarily Disconnected

TD Term Deposit

TD Term Diagram

TD Terminal Digit

TD Terminal Distributor

TD Termination Date

TD Territorial Decoration

TD Test Design (Specification)

TD Test Director [of NASA]

TD Test Distributor

TD Testing Device

TD Text and Data (Exchange)

TD Theoretical Density

TD Thermal Deburring [Metallurgy]

TD Thermal Desorption

TD Thermal Diffusion

TD Thermal Donor

TD Thermodiffusion

TD Thermodynamic(s)

TD Thermoluminescent Dating [Archeology]

TD Thoracic Duct [Anatomy]

TD Thor-Delta Satellite

TD Thoria-Dispersed; Thoria Dispersion [Metallurgy]

TD Thorium Dioxide

TD Threading Dislocation [Materials Science]

TD Timed Disintegration (Capsule) [Pharmacology]

TD Time Delay [also T/D]

TD Time Dependence; Time-Dependent

TD Time Deposit [also T/D]

TD Time Difference

TD Time Domain [Electronics]

TD Timing Device

TD Tissue Dose [Radiology]

TD To Deliver

TD Tolerance Dose [Radiology]

TD Toluene Diamine

TD Tolylenediamine

TD Tool Design

TD Top Down

TD Torpedo Detonating

TD Total Depth

TD Total Distillate

TD Total Dust Sampler

TD Touchdown [also T/D]

TD Townsend Discharge [Electronics]

TD Toxic Dose

TD Toyota Diffusion (Process) [Metallurgy]

TD Trace Direction

TD Track Data

TD Track Display

TD Transducer

TD Transfer Dolly

TD Transformation Dislocation [Materials Science]

TD Transmission and Distribution

TD Transmission Detector

TD Transmission Distributor

TD Transmit(ted) Data

TD Transmitter-Detector

TD Transmitter-Distributor

TD Transverse Direction

TD Treasury Department [US]

TD Triple-Defect (Jump)

TD Tunesian Dinar [Currency]

TD Tunnel Diode

TD Turbulent Diffusion

TD Turbine Drive

TD Turbo-Diesel (Engine)

TD Twist Drill

TD$_{HI}$ Toxic Dose High

TD$_{LO}$ Toxic Dose Low

T&D Transmission and Distribution [Institute of Electrical and Electronic Engineers–Power Engineering Society Committee, US]

T&D Transportation and Docking

T/D Time Delay [also TD]

T/D Time Deposit

T/D Touchdown [also TD]

T-4 Thyroxine [Biochemistry]

2D Two Dimensional [also 2-D]

2½D 2½-Dimensional [also 2½-D]

2.5D 2.5 Dimensional (Image) [also 2.5-D]

2.75D 2.75 Dimensional (Image) [also 2.75-D]

3D Three Dimensional [also 3-D]

3-D Dangerous, Difficult and Dirty (Tundish Maintenance) [also DDD] [Metallurgy]

3-D Dull, Dirty and Dangerous (Work Environment)

T$ New Taiwan Dollar [Currency of Taiwan]

td *(ter in die)* – three times a day [Medical Prescriptions]

t/d tons per day [Unit]

TDA Tamm-Dancoff Approximation [Quantum Mechanics]

TDA Target Docking Adapter

TDA Technology Development Corporation

TDA Telecommunications Dealers Association [UK]

TDA Temporary Danger Area

TDA Terminal Diode Amplifier

TDA Tetracarboxylic Dianhydride

TDA Tetradecenyl Acetate

TDA Textile Distributors Association [US]

TDA 2,2'-Thiodiacetic Acid

TDA Timber Development Association [US]

TDA Time-Deposit Accounting

TDA Time-Domain Analysis

TDA Titanium Development Association [US]

TDA Toll Dial Assistance

TDA Tracking and Data Acquisition

TDA Tridecyl Alcohol

TDA Tridecyl Amine

TDA Trisdecylamine

TDA Tris(dioxaheptyl)amine

TDA Transportation Development Center [Canada]

TDA Transport Distribution Analysis

TDA Tunnel Diode Amplifier

TDA-1 Tris[2-(2-Methoxyethoxy)ethyl]amine

TDA-1 Tris(3,6-Dioxaheptyl)amine

7-TDA 7-Tetradecen-1-yl Acetate

11-TDA 11-Tetradecen-1-yl Acetate

E-11-TDA trans-11-Tetradecenyl Acetate

Z-7-TDA cis-7-Tetradecenyl Acetate

Z-9-TDA cis-9-Tetradecenyl Acetate

Z-11-TDA cis-11-Tetradecenyl Acetate

2D ACAR Two-Dimensional Angular Correlation of Annihilation Radiation [also 2D-ACAR or 2D/ACAR]

2D ACPAR Two-Dimensional Angular Correlation (of) Positron Annihilation Radiation [also 2D-ACPAR, or 2D ACPAR]

TDAE Tetrakis(dimethylamino)ethylene

TDAE-C$_{60}$ Tetrakis(dimethylamino)ethylene–Fullerene

TDAL Tetradecanal

Z-7-TDAL cis-7-Tetradecanal

2D ALE Two-Dimensional Arbitrary Lagrangian-Eulerian (Code) [also 2D-ALE]

TDAS Tris(dimethylamino)arsine

TDAFP Turbine Driven Auxiliary Feedwater Pump [Nuclear Reactors]

TDAL Tetradecenal

TDAS Traffic Data Administration System

T-DAY Termination of War Day [also T-Day or T-day]

TDB Technical Development Bureau [of Nippon Steel Corporation, Chiba, Japan]

TDB Technical Divisions Board [of ASM International, US]

TDB Terminology Database

TDB Toxicology Data Bank [of National Institutes of Health, Bethesda, Maryland, US]

TDB Trade and Development Board [of United Nations]

TDB Traffic Database

TDB Transition Description Block

3DB Three-Dimensional Database

TDBA Tetradecyl Dimethyl Benzyl Ammonium

tDBA 2-tert-Butyl-9,10-Dibromoanthracene

3D BEG Three-Dimensional Blume-Emery-Griffiths (Model) [also 3D-BEG]

TDBI Test During Burn-In [Electronics]

TDBP 5,10,15,20-Tetrakis(3,5-Di-t-Butylphenyl) Porphinate

TDBTU N,N,N',N'-Tetramethyl-O-(3,4-Dihydro-4-Oxo-1,2,3-Benzotriazin-3-yl)uronium

TDC Tabular Data Control

TDC Target Data Collection

TDC Telegraphy Data Channel

TDC Tennessee Department of Conservation [US]

TDC Thermal Diffusion Coefficient

TDC Time Delay Closing

TDC Time-to-Digital Converter

TDC Tone-Digital Command (Technique)

TDC Top Dead Center

TDC Total Density Control

TDC Track Detection Circuit

TDC Trade Development Council

TDC Transistor Digital Circuit

TDC Transport Development Center [of Transport Canada]

TDC Trendata Computer System

TDC Two-Dimensional Chromatography [also 2DC]

TDC Type Directors Club [US]

2DC Two-Dimensional Chromatography [also TDC]

TDCA Technical Documentation Center for the Army [Netherlands]

3D-CAD Three-Dimensional Computer-Aided Design [also 3D CAD]

TDCB Tapered Double-Cantilever Beam (Test Specimen) [Mechanical Testing]

TDCC Transportation Data Coordinating Committee [US]

2D C/C Two-Dimensional Carbon-Carbon Composite

3D C/C Three-Dimensional Carbon-Carbon Composite

TDC-DF Tennessee Department of Conservation–Division of Forestry [US]

TDC-GD Tennessee Department of Conservation–Geology Division [US]

TDCHF Time-Dependent Coupled Hartree-Fock (Method) [Quantum Mechanics]

2D-CIDNP Two-Dimensional Chemically Induced Dynamic Nuclear Polarization

TDCM Technology Development for the Communications Market

TDCM Transistor Driver Core Memory

2-D C-13 NMR Two-Dimensional Carbon-13 Nuclear Magnetic Resonance [also 2-D ^{13}C-NMR, 2D C-13 NMR, or 2D ^{13}C NMR]

TDCO Torpedo Data Computer Operator

TDCP Trisdichloropropyl Phosphate

2-D CRAMPS Two-Dimensional Combined Rotation and Multiple Pulse Spectroscopy [also 2D CRAMPS]

TDCS Time Division Circuit Switching

TDCS Total Desorption Cross-Section

TDCS Triple Differential Cross-Section

2-D CSL Two-Dimensional Coincidence Site Lattice [also 2D CSL] [Materials Science]

3-D CSL Three-Dimensional Coincidence Site Lattice [also 3D CSL] [Materials Science]

TDCTL Tunnel-Diode Charge-Transformer Logic

TDD Task Description Document

TDD Target Detection Device

TDD Technical Data Department

TDD Technical Data Digest

TDD Telecommunication Device for the Deaf

TDD Telemetry Data Digitalizer

TDD Telephonic Device for the Deaf

TDD Teletype Device for the Deaf

TDD Test Design Description

TDD Thermodynamic Drag

TDD Threshold Damage Density

TDD Time Division Duplexing

TDD Total Difference Density

TDD Two Detector Delay

2DD Double-Density Disk [also DDD]

TDDA Tetradecadienol Acetate; Tetradecadienyl Acetate

(Z,E)-9,11-TDDA cis-9, 11-trans-11-Tetradecadienyl Acetate

(Z,E)-9,12-TDDA cis-9, 12-trans-11-Tetradecadienyl Acetate

TDDB Time-Dependent Dielectric Breakdown

TDDL Time-Division Data Link

TDDOL Tetradecadienol

(Z,E)-9,11-TDDOL cis-9, trans-11-Tetradecadienol

2D DOS Two-Dimensional Density of States [Solid-State Physics]

TDDR Technical Data Department Report

TDDR Total-Direct-Diffuse Radiometer

TDDS Topographic Digital Data System

TDE Technical Development Establishment [India]

TDE Terminal Display Editor

TDE Tetrachlorodiphenylethane

TDE 2,2'-(2,5,8,11-Tetraoxa-1,12-Dodecanediyl) bisethylenoxide

TDE Thermodynamic Equilibrium

TDE Time-Dependent Embrittlement [Metallurgy]

TDE Total Data Entry

TDE Transverse Doppler Effect [Physics]

TDE Turbo-Diesel Engine

2DE Two-Dimensional Electrophoresis [also TDE]

3D8 Three-Dimensional Eight-Grain Junction [Metallurgy]

TDEC Technical Division and Engineering Center

TDEC Telephone Line Digital Error Checking

TDEFWP Turbine Driven Emergency Feedwater Pump [Nuclear Reactors]

2DEG Two-Dimensional Electron Gas [also 2-DEG, or TDEG] [Solid-State Physics]

2DEG-FET Two-Dimensional Electron Gas Field-Effect Transistor [Solid-State Physics]

TDEP Tracking Data Editing Program [of NASA]

TDEPR Thermally Detected Electron Paramagnetic Resonance

TDES Tennessee Department of Employment Security [US]

2DES Two-Dimensional Electron System [also TDES]

TDF Task Deletion Form

TDF Technology Demonstration Facility

TDF 1,1,1,2,2,3,3,7,7,8,8,9,9,9-Tetradecafluorononanel-4,6-Dione [also tdf]

TDF Tetradecenyl Formate

TDF Total Dietary Fiber [Medicine]

TDF Trace Definition File [IBM OS/2]

TDF Transborder Data Flow

TDF Transmitted Dark-Field

TDF Trunk Distribution Frame

TDF Three Degrees of Freedom

TDF Two Degrees of Freedom

TDF Two-Dimensional Flow [also 2DF]

.TDF Trace Definition File [IBM OS/2 File Name Extension] [also .tdf]

.TDF Typeface Definition File [File Name Extension]

2DF Two-Dimensional Flow [also TDF]

3DF Three-Dimensional Flow

Z-9-TDF cis-9-Tetradecenyl Formate

3D4 Three-Dimensional Four-Grain Junction [Metallurgy]

3D4a Three-Dimensional Four-Grain Junction, axisymmetric [Metallurgy]

TdeV Tokamak de Varennes [Quebec, Canada]

TDF Transmitted Darkfield [Microscopy]

TDFA Total Dietary Fiber Assay [Medicine]

TDFAB Total Dietary Fiber Assay Bulletin [US]

3D-FEM Three-Dimensional Finite-Element Analysis (Model)

TDFI Turbo Diesel Fuel Injection

2DFI Two-Dimensional Fourier Imaging

2D FIR Two-Dimensional Far Infrared [also 2-D FIR]

TDFS Temperature Dependence of Flow Stress [Mechanics]

TDFSS Time-Dependent Fluorescence Stokes Shift [Spectroscopy]

2D-FT-ELDOR Two-Dimensional Fourier Transform Electron-Electron Double Resonance

2D-FT-ESR Two-Dimensional Fourier Transform Electron-Spin Resonance

TDG Tap Density Gauge [Powder Metallurgy]

TDG Test Data Generator

TDG Thiodiglycol

TDG Transmit Data Gate

TDG Transport of Dangerous Goods

TDGA Transport of Dangerous Goods Act [Canada]

TDGL Time-Dependent Ginzburg-Landau (Theory) [Solid-State Physics]

TDH Temperature-Dependent Hall [Electronics]

TDH Total Dynamic Head

2D-HAF Two-Dimensional Heissenberg Antiferromagnet [Solid-State Physics]

3D-HAF Three-Dimensional Heissenberg Antiferromagnet [Solid-State Physics]

2D-HAHA Two-Dimensional Hartmann-Hahu (Spectroscopy)

TDHF Time-Dependent Hartree-Fock [Quantum Mechanics]

TDHG Time-Dependent Hartree Grid (Method) [Quantum Mechanics]

TDHP Trans-2,6-Dimethyl-3,6-Dihydro-2H-Pyran

2DHS Two-Dimensional Hole System

TDI Technical Development Institute [Japan]

TDI Telecommunications Data Interface

TDI Textile Dye Institute [US]

TDI Toluene Diisocyanate; Toluene-2,4-Diisocyanate; Toluene-2,6-Diisocyanate

TDI Tolylene Diisocyanate

TDI Tool and Die Institute

TDI Total Dielectric Isolation

TDI Trade Data Interchange

TDI Transport Device Interface

2,4-TDI Toluene-2,4-Diisocyanate

2,6-TDI Toluene-2,6-Diisocyanate

TDIA Transient Data Input Area

2-D IIR Two-Dimensional Imaging Infrared

TDIO Tuning Data Input/Output [also TDI/O]

TDIS Travel Documents and Issuance System

3DIX Three-Dimensional Interaction Accelerator

3D J-NMR Three-Dimensional J-Resolved Nuclear Magnetic Resonance [also 3D-J-NMR]

TDKIE Temperature Dependence of the Kinetic Isotope Effect [Physics]

TDL Task Description Language

TDL Toxic Dose Level

TDL Transaction Definition Language

TDL Tunnel-Diode Logic

TDLAS Tunable Diode Laser Absorption Spectrometry

TDLDA Time-Dependent Local Density Approximation [Physics]

2D LJ Two-Dimensional Lennard-Jones (System) [also 2D-LJ] [Physical Chemistry]

2D LRO Two-Dimensional Long-Range Order(ing) [also 2D-LRO] [Solid-State Physics]

3D LRO Three-Dimensional Long-Range Order(ing) [also 3D-LRO] [Solid-State Physics]

TDLU Terminal Ductal Lobular Unit

TDM Tandem

TDM Technical Document Management

TDM Telemetric Data Monitor

TDM Telephone Directory Memory

TDM Template Descriptor Memory

TDM Tertiary Dodecyl Mercaptan

TDM Time-Division Multiplex(ing); Time-Division Multiplexer

TDM Torpedo Detection Modification

TDM Trehalose Dicorynomycolate [Biochemistry]

TDMA Tape Direct Memory Address

TDMA Time-Division Multiple Access

3DMA Three-Dimensional Microgravity Accelerometer

TDMAAs tris-Dimethylamino Arsenic

TDMAC Tridodecylmethylammonium Chloride

TDMAC-Heparin Tridodecylmethylammonium Heparinate

TDMAE Tetrakis(dimethylamino)ethylene

TDMAMP Trisdimethylaminomethyl Phenol

2D-MAS-NMR Two-Dimensional Magic-Angle Spinning Nuclear Magnetic Resonance

TDMAT Tetrakisdimethylamino Titanium

TDMAT-TiN Tetrakisdimethylamino Titanium Titanium Nitride

3D-MEMS Three-Dimensional Microelectromechanical Device

TDMF Time Dependent Mean Field Theory [Physics]

TDMS Technical Document Management System

TDMS Telegraph(ic) Distortion Measurement Set

TDMS Terminal Data Management System

TDMS Terminal Display Management System

TDMS Thermal Desorption Mass Spectrometry;Thermal Desorption Mass Spectroscopy

TDMS Time-Shared Data Management System

TDMS Toxicology Data Management System

TDMS Transmission Distortion Measuring Set

TDM/SDM Time-Division Multiplex/Space-Division Multiplex (Switching Stage)

TDM-VDMA Time-Division Multiplex–Variable Destination Multiple Access

TDN Target Doppler Nullifier

TDN Thoria-Dispersed Nickel [also TD-Nickel]

TDN Total Digestible Nutrient [Medicine]

T-DNA Tumor-Desoxyribonucleic Acid [Biochemistry]

3DNEPH Three-Dimensional Nephanalysis [Oceanography]

TD-Nichrome Thoria-Dispersed Nichrome (Alloy)

TD-Nickel Thoria-Dispersed Nickel [also TDN]

2D-NMR Two-Dimensional Nuclear Magnetic Resonance Spectroscopy [also 3DNMR]

3D-NMR Three-Dimensional Nuclear Magnetic Resonance Spectroscopy [also 2DNMR]

2D-NOE Two-Dimensional Nuclear Overhauser Effect [also 2DNOE] [Nuclear Physics]

TDNS Total Data Network System

TDO Tallow Diaminopropanedioleate

TDO Tapered Roller Bearing, Two-Row, Double-Cup Single-Cone Adjustable [Symbol]

TDO Technology Development Officer

TDO Technology Development Organization (Program) [of New York State Science and Technology Foundation, US]

TDO Time Delay Opening

TDO Training Development Officer

TDO Treasury Department Office

Tdo Tornado [also tdo]

TDOA Thermally Detected Optical Absorption

TDO-Carbonate 4,6-Diphenylthieno[3,4-d]-1,3-Dioxol-2-one 5,5-Dioxide

2DOF Two Degrees Of Freedom

3DOF Three Degrees Of Freedom

TDOL Tetradecenol

7-TDOL 7-Tetradecen-1-ol

9-TDOL 9-Tetradecen-1-ol

11-TDOL 11-Tetradecen-1-ol

Z-7-TDOL cis-7-Tetradecanol

Z-9-TDOL cis-9-Tetradecanol

Z-11-TDOL cis-7-Tetradecanol

TDOS Tape-Disk Operating System

TDOS Total Density of States [Solid-State Physics]

TDOS Tunneling Density of States [Solid-State Physics]

TDP Table-Driven Programming

TDP Tag Distribution Protocol

TDP Technical Data Package

TDP Technical Development Plan

TDP Teledata Processing

TDP Telelocator Data Protocol

TDP Temperature and Dewpoint

TDP Tetradecylphosphonium

TDP Thermodynamic Properties

TDP 4,4'-Thiodiphenol

TDP Thymidine Diphosphate [Biochemistry]

TDP Toluene Disproportionation Process

TDP Top-Down Parsing [Computers]

TDP Tracking Data Processor

TDP Trade and Development Program [of International Development Cooperation Agency, US]

TDP Traffic Data Processor

TDP Tris(dimethylamino)phosphine

3DP Three-Dimensional Printing

TDPA Titanium Di(dioctylpyrrophosphate Oxyacetate)

TDPA Tridecyl Phosphoric Acid

TDPAC Time Differential Perturbed Angular Correlation

2-d PBC 2-d Periodic Boundary Conditions [Materials Science]

3-d PBC 3-d Periodic Boundary Conditions [Materials Science]

TDPI Tasmanian Department of Primary Industry [Australia]

TDPL Top Down Parsing Language

TDPO Technology Development Program Office [of US Department of Energy]

TDPO Tris(dimethylamino)phosphine Oxide [also tdpo]

2DPS Two-Dimensional Periodic Structure [also TPS]

TDPSK Time-Differential Phase Shift Keying [also TD-PSK]

TDPT Time-Dependent Perturbation Theory [Physics]

TDQ Trimethyl Dihydroquinoline

TDQ 2,2,4-Trimethyl-1,2-Dihydroquinoline

2D-QE Two-Dimensional Quadrupolar Echo [also 2DQE] [Physics]

TDQM Time-Dependent Quantum-Mechanical (Calculation)

TDQP Trimethyldihydroquinoline Polymer

TDR Tape Data Register

TDR Target Discrimination Radar

TDR Technical Data Relay

TDR Technical Data Report

TDR Technical Design Review

TDR Technical Documentary Report; Technical Documentation Report

TDR Temperature-Dependent Resistor

TDR Temporarily Disconnected at (Subscribers) Request

TDR Test Discrepancy Report

TDR Terminal Digit Requested

TDR Time-Delay Relay

TDR Time Distribution Record

TDR Time-Domain Reflectometer; Time-Domain Reflectometry

TDR Tone Dial Receiver

TDR Torque Differential Receiver (or Repeater)

TDR Traffic Data Record

TDR Transactional Document Recorder

TDR Transmit Data Register

TD&RA Twist Drill and Reamer Association [UK]

TDRE Tracking and Data Relay Experiment

TDRI Tropical Development and Research Institute [now Overseas Development Natural Resources Institute, UK]

TDRM Time Domain Reflectometry Microcomputer

TDRP Treatment Development Research Project

TDRR Test Data Recording and Retrieval

TDRS Telemetry Downlist Receiving Site

TDRS Text Data Retrieval System

TDRS Three-Dimensional Reinforced Solid

TDRS Tracking and Data Relay Satellite [of NASA]

2D-RSM Two-Dimensional Reciprocal Space Mapping [Physics]

TDRSS Tracking and Data Relay Satellite System [of NASA]

TDS Tactical Dosimetry System

TDS Tape Data Selector

TDS Target Designation System

TDS Technical Data System

TDS Technology Demonstration Satellite

TDS Temperature Diffuse Scattering

TDS Tertiary Data Set

TDS Test Data Sheet

TDS Test Data System

TDS Test Development Series

TDS Test Development Station

TDS Testing Data System

TDS Thermal Desorption Spectroscopy

TDS Thermal Desorption Spectrum

TDS Thermal Diffuse Scattering

TDS Thermodynamic Density of States [Physics]

TDS Three-Dimensional Structure [Materials Science]

TDS Time-Distance-Speed

TDS Time-Division Switching

TDS Time-Domain Spectroscopy

TDS Titanium Descaling Salt [Metallurgy]

TDS Total Density of States [Physics]

TDS Total(ly) Dissolved Solids

TDS Total Distribution Solution

TDS Track Data Simulator

TDS Track Data Storage

TDS Tracking and Data (Acquisition) System

TDS Transaction Distribution System

TDS Transaction-Driven System

TDS Transistor-Display and Data-Handling System

TDS Translation and Docking Simulator

TDS Tunneling Density of States [Physics]

%TDS Percent Total Dissolved Solids

tds *(ter die sumendum)* – to be taken three times daily [Medical Prescriptions]

3D6 Three-Dimensional Six-Grain Junction [Metallurgy]

TD&SA Telephone, Data and Special Audio

2D-SAS Two-Dimensional Small-Angle Scattering [also 2-D SAS]

3D-SAX Three-Dimensional Small-Angle Scattering of X-Rays [also 3-D SAX]

TDSCF Time-Dependent Self-Consistent Field [Physics]

TDS-Cl Thexyldimethylsilyl Chloride

TDSE Time-Dependent Schroedinger Equation [Quantum Mechanics]

3DSE Three-Dimensional Schroedinger Equation [Quantum Mechanics]

TDSF Tasmanian Department of Sea Fisheries [Australia]

TDSF Triangle-Dimer Stacking Fault [Crystallography]

3D SIMS Three-Dimensional Secondary Ion Mass Spectrometry [also 3-D-SIMS]

TDSR Transmitter Data Service Request

TDSS Time-Dependent (Fluorescence) Stokes Shift [Spectroscopy]

TDSS Time Dividing Spectrum Stabilization

TDS-Triflate Thexyldimethylsilyl Trifluoromethanesulfonate [also TDS-Triflate]

TDT Target Designation Transmitter

TDT Task Dispatch Table

TDT Thermodynamic Tailing

TDT Thermodynamic Topping

TDT Time Domain Transmission [Electronics]

TDT Trigger Discharge Tube

TdT Terminal Deoxynucleotidyl Transferase [Biochemistry]

2D2 Two-Dimensional Two-Grain Junction [Metallurgy]

2D3 Two-Dimensional Three-Grain Junction [Metallurgy]

3D2 Three-Dimensional Two-Grain Junction [Metallurgy]

3D3 Three-Dimensional Three-Grain Junction [Metallurgy]

3D3a Three-Dimensional Three-Grain Junction, axisymmetric [Metallurgy]

TDTA Di-p-Toluoyltartaric Acid

TDTA Tetramethylenediaminetetraacetate [also tdta]

TDTA Toluoyl-D-Tartaric Acid

2D t-J Two-Dimensional t-J Model [Physics]

TDTL Time-Dependent Thermal Lensing

TDTL Tunnel-Diode Transistor Logic

2DTLC Two-Dimensional Thin Layer Chromatography

TDTMABr Tetradecyltrimethyl Ammonium Bromide

TDU Target Detection Unit

TDU Time Display Unit

TDU Tokyo Denki University [Japan]

TDV Technology Development Vehicle

TDV Test Data Van

TDV Total Discontinuity Vector

TDVC Transition Dipole Vector Coupling (Model) [Physics]

TDVM Time-Dependent Variational Method [Physics]

TDVP Time-Dependent Variational Principle [Physics]

2D VRH Two-Dimensional Variable Range Hopping [also 2-D VRH] [Physics]

3D VRH Three-Dimensional Variable Range Hopping [also 3-D VRH] [Physics]

1D VRHC One-Dimensional Variable Range Hopping Conduction [also 1-D VRHC] [Physics]

2D VRHC Two-Dimensional Variable Range Hopping Conduction [also 2-D VRHC] [Physics]

3D VRHC Three-Dimensional Variable Range Hopping Conduction [also 3-D VRHC] [Physics]

TDW Ton(s) Deadweight [also tdw] [Unit]

TDWG Taxonomic Databases Working Group for Plant Sciences [of International Union for Conservation of Nature and Natural Resources]

TDWR Terminal Doppler Weather Radar

TDX Thermal Demand Transmitter

TDX Time Division Exchange

TDX Torque Differential Transmitter

TDX Text and Data Exchange

3D-XY Three-Dimensional X-Y (Model)

TDY Task Dictionary

TDY Temporary Duty

TDY Tour of Duty

TDZ Touchdown Zone

TDZE Touchdown Zone Elevation

TDZL Touchdown Zone Lights

TE Early Transition (Metal)

TE Echo Time

TE Taxiway Edge-Lighting

TE Technical Education

TE Technical Engineer

TE Technical Exchange

TE Telecommunications Engineer(ing)

TE Telephone Engineer(ing)

TE Television Electronics

TE Television Engineer(ing)

TE Temperature Effect

TE Temperature Efficiency

TE Temper Embrittlement [Metallurgy]

TE Tensile Elongation [Mechanics]

TE Terminal Emulation; Terminal Emulator

TE Terminal Enable

TE Terminal Equipment

TE Terminal Exchange

TE Terrestrial Ecology

TE Test Engineer(ing)

TE Test Equipment

TE Textile Engineer(ing)

TE Thermal Efficiency

TE Thermal Electron

TE Thermal Element

TE Thermal Energy

TE Thermal Engineer(ing)

TE Thermal Etching

TE Thermal Evolution

TE Thermal Expansion

TE Thermionic Emission

TE Thermoelastic(ity)

TE Thermoelectric(ity)

TE Thermoemission

TE Thermoplastic Elastomer

TE Threefold Edge (for Grain Boundaries) [Materials Science]

TE Threshold Extension

TE Tidal Energy

TE Time Earliest

TE Time Effect

TE Time Expected

TE Timing Electronics

TE Tokamak Experiment [Plasma Physics]

TE Tool Engineer(ing)

TE Top Edge

TE Topographical Engineers

TE Torsional Energy

TE Total Elongation (of Ropes)

TE Total Energy

TE Totally Enclosed

TE Toxicity Equivalent

TE Trace(r) Element

TE Traffic Enforcement

TE Traffic Engineer(ing)

TE Trailing Edge

TE Transcendental Equation [Mathematics]

TE Transesterification

TE Transfer Equilibrium

TE Transient Equilibrium

TE Transition Element

TE Transition Editor

TE Transportation Engineer(ing)

TE Transverse Electric (Mode)

TE Trapped Electron

TE Trend Error

TE Triboelectric(ity)

TE Trunk Equalizer

TE Trunk Expansion

TE Turbine-Electric; Turbo-Electric (Drive)

TE Tunnel Effect [Quantum Mechanics]

TE Turbine Efficiency

TE Turbine Engine

TE Tyndall Effect [Optics]

TE$_{10}$ Transverse Electric Dominant Field Configuration in Rectangular Waveguide (Mode)

TE/2 (Oberon) Terminal Emulator/2

T(E) Transmission Coefficient (of Semiconductor)

T&E Test and Evaluation [also T/E]

T/E Transporter/Erector

T-E Temperature-Efficiency (Index); Thermal-Efficiency (Index) [Ecology]

T(ε) Track Scan Profile (of Magnetic Recording Media) [Symbol]

3-E Three-Element (Body) [Mechanics]

Te Tellurium [Symbol]

Te Tetra [Chemistry]

Te Tetrahedral Site (in Crystallography) [Symbol]

Te^{2-} Divalent Tellurium Ion [also Te] [Symbol]

Te^{4-} Tetravalent Tellurium Ion [Symbol]

Te^{6-} Hexavalent Tellurium Ion [Symbol]

Te$_8$ Molecular Tellurium [Symbol]

Te II Tellurium II [Symbol]

Te III Tellurium III [Symbol]

Te-118 Tellurium-118 [also ^{118}Te, or Te^{118}]

Te-119 Tellurium-119 [also ^{119}Te, or Te^{119}]

Te-120 Tellurium-120 [also ^{120}Te, or Te^{120}]

Te-121 Tellurium-121 [also ^{121}Te, or Te^{121}]

Te-122 Tellurium-122 [also ^{122}Te, or Te^{122}]

Te-123 Tellurium-123 [also ^{123}Te, or Te^{123}]

Te-124 Tellurium-124 [also ^{124}Te, or Te^{124}]

Te-125 Tellurium-125 [also ^{125}Te, or Te^{125}]

Te-126 Tellurium-126 [also ^{126}Te, or Te^{126}]

Te-127 Tellurium-127 [also ^{127}Te, or Te^{127}]

Te-128 Tellurium-128 [also ^{128}Te, or Te^{128}]

Te-129 Tellurium-129 [also ^{129}Te, or Te^{129}]

Te-130 Tellurium-130 [also ^{130}Te, or Te^{130}]

Te-131 Tellurium-131 [also ^{131}Te, or Te^{131}]

Te-132 Tellurium-132 [also ^{132}Te, or Te^{132}]

Te-133 Tellurium-133 [also ^{133}Te, or Te^{133}]

Te-134 Tellurium-134 [also ^{134}Te, or Te^{134}]

Te-135 Tellurium-135 [also ^{135}Te, or Te^{135}]

$\tau(E_F)$ time between interactions of electrons with phonons, impurities, imperfections, etc. in lattice of a solid [Symbol]

τ-ϵ shear stress/shear strain (curve) [Mechanics]

TEA Technical Engineers Association

TEA Technical Exchange Agreement

TEA Tensile Energy Absorption

TEA Tetraethylammonium

TEA Thermal Energy Analyzer

TEA Thermal Expansion Analysis

TEA Thermal Expansion Anisotropy

TEA Total Exposure Assessment

TEA Transferred-Electron Amplifier

TEA Transversely Excited Atmosphere

TEA Transversely Excited Atmospheric-Pressure (Laser)

TEA Triethanolamine

TEA Triethyl Aluminum

TEA Triethylamine [also Tea]

TEA Triethylammonium

TEA Turbine Engine Alloy(s)

TEA Tunnel-Emission Amplifier

TEAA Triethyl Ammonium Acetate

TeAA Tetraazylazide

TEAABS Trieethanolamine Alkylbenzenesulfonate [also TEA-ABS]

TEAB Tetraethyl Ammonium Bicarbonate

TEAB Tetraethyl Ammonium Borohydride

TEAB Tetraethyl Ammonium Bromide [also TEABr, or TEA-Br]

TEAB Triethylamine Borane (Complex)

TEABF$_4$ Tetraethyl Ammonium Tetrafluoroborate [also TEATFB]

TEABr Tetraethyl Ammonium Bromide [also TEAB, or TEA-Br]

TEAC Technology Education Advisory Council

TEAC Tetraethyl Ammonium Chloride [also TEACl, or TEA-Cl]

TEAC Tetraethyl Ammonium Cyanide [also TEA-CN, or TEACN]

TEACl Tetraethyl Ammonium Chloride [also TEAC, or TEA-Cl]

TEA-CN Tetraethyl Ammonium Cyanide [also TEACN]

TEA CO$_2$ Transversely-Excited Atmospheric Pressure Carbon Dioxide (Laser) [also TEA-CO$_2$]

TEAE Triethylaminoethyl

TEAF Triethyl Ammonium Formate

TEAH Tetraethyl Ammonium Hydroxide [also TEA-OH]

TEAHA Trans-East African Highway Authority [Ethiopia]

TEAHS Tetraethyl Ammonium Hydrogen Sulfate

TEAI Tetraethyl Ammonium Iodide

TEAl Triethyl Aluminum

TEA Laser Transversely Excited Atmospheric Pressure Laser [also TEA laser]

TEAM Technique for Evaluation and Analysis of Maintainability

TEAM Techniques for Effective Alcohol Management

TEAM Teleterminals Expandable Added Memory

TEAM Terminology Evaluation and Acquisition Method

TEAM Test and Evaluation of Air Mobility

TEAM The European-Atlantic Movement [UK]

TEAM Total Environment Analysis and Management

TEAM Training Equipment and Maintenance

TEAMS Test, Evaluation and Monitoring System

TEA-OH Tetraethyl Ammonium Hydroxide [also TEAH]

TEAP Tetraethyl Ammonium Perchlorate

TEAP Transportation Emergency Assistance Plan

TEAP Triethyl Ammonium Phosphate

TEAS Tetraethyl Ammonium Succinimide

TEAS Thermal Energy Atom Scattering

TEAS Triethanolamine Stearate

TEAs Triethylarsine

TEASER Tunable Electron Amplifier for the Stimulated Emission of Radiation [also Teaser, or teaser]

TEAT Triethanolamine Titanate

TEATFA Tetraethyl Ammonium Trifluoroacetate

TEATFB Tetraethyl Ammonium Tetrafluoroborate [also TEABF$_4$]

TEATS Tetraethyl Ammonium Toluenesulfonate

TEAZ Tip-Emission Adjusted Zone [Fracture Mechanics]

TEB Thread Environment Block

TEB Transient Electric Birefringence

TEB Triethoxybutane

TEB Triethylbenzene

TEB Triethylboron

TEBA Triethylbenzyl Ammonium

TEBC Total Equivalent Boron Content [Metallurgy]

TEBIMA Tris-[2-(N-Ethylbenzimidazolyl)methane]amine [also tebima]

TEBOL Terminal Business-Oriented Language

TEC Tactical Electromagnetic Coordinator

TEC Tantalum Electrolytic Capacitor

TEC Tasmanian Environment Center [Australia]

TEC Technical Education Consultant

TEC Telephone Engineering Center

TEC Test Equipment Center

TEC Texas Employment Commission [US]

TEC Texas Environmental Center [Houston, US]

TEC The Electrification Council [US]

TEC Thermal Elongation Coefficient

TEC Thermal Expansion Coefficient

TEC Thermionic Energy Converter

TEC Thermoelectric Cooler

TEC Tokyo Electronics Corporation [Japan]

TEC Total Electron Content

TEC Total Energy Distribution

TEC Total Environment Center

TEC Total Estimated Cost

TEC Transearth Coast

TEC Transient Electron Current

TEC Transmission Electron Diffraction

TEC Triethyl Citrate

TEC Triple Erasure Connection

TECC Technology Education for Children Council [US]

TECDA Thai Environmental and Community Development Association

TECH Division of Chemical Technicians [of American Chemical Society, US]

Tech Technician; Technics; Technique [also tech]

Tech Technological; Technologist; Technology [also tech]

tech technical(ly)

tech technical grade (chemical)

Tech Assoc Refract Technical Association of Refractories [Journal published in Japan]

Tech Bull Technical Bulletin [also tech bull]

Tech Bull Vevey Technical Bulletin Vevey [Published by Vevey Engineering Works Ltd., Vevey, Switzerland]

Tech Ceram Bull Technical Ceramics Bulletin [UK]

Tech Ceram Int Technical Ceramics International

Tech Commun Technical Communication [Publication of Society for Technical Communication, US]

Tech Dept Technical Department

Tech Dig Int Electron Dev Mtg Technical Digest of the International Electron Devices Meeting

Tech Dig PVSEC Technical Digest of the (International) Photovoltaics Science and Engineering Conference [of Institute of Electrical and Electronics Engineers, US]

Tech Div Technical Division

Tech Doc Technical Document(ation)

Tech Doc Hydrol Technical Documents in Hydrology [Publication of of International Hydrological Program, France]

Tech gesch Technikgeschichte [Publication on the History of Technology of Verein Deutscher Ingenieure, Germany]

Tech Gr Technical Grade (Chemical)

Tech Inf Technical Information

Tech Inf, Process Autom – Electr Power Install Technical Information, Process Automation–Electrical Power Installations [German Publication]

Tech Lit Technical Literature

Tech Mar Environ Sci Technique in Marine Environmental Science [Publication of International Council for the Exploration of the Sea, Denmark]

Tech Meet Technical Meeting

Tech Meet Magn Technical Meeting on Magnetics [of Institute of Electrical and Electronics Institute of Japan]

Tech Mech Technische Mechanik [German Publication on Engineering Mechanics]

Tech Mess Technisches Messen [German Publication Measurement Engineering]

Tech Mitt Technische Mitteilungen [German Technical Communications]

Tech Mitt Haus Tech Technische Mitteilungen Haus der Technik [German Technical Communications of Haus der Technik]

Tech Mitt Krupp Technische Mitteilungen Krupp [Technical Communications of Krupp GmbH, Essen, Germany]

Tech Mitt Krupp (Engl Ed) Technische Mitteilungen Krupp (English Edition) [Technical Communications of Krupp GmbH–English Edition]

Tech Mitt Krupp, Werksber Technische Mitteilungen Krupp, Werksberichte [Technical Communications and Works Reports of Krupp GmbH, Essen, Germany]

Tech Mitt PTT Technische Mitteilungen Post–Telefon–Telegraf [Technical Communications of Swiss Federal Post Telephone and Telegraph Office, Berne, Switzerland]

Tech Mod La Technique Moderne [French Publication on Modern Techniques]

Techn Technical; Technician; Technics [also techn]

Tech News Technical News [US]

Tech News Govt Ind Res Inst, Nagoya Technical News of the Government Industrial Research Institute, Nagoya [Japan]

Technion-IIT Technion-Israel Institute of Technology [Haifa]

Techno-COM Trade Fair for Sensorics, Actorics, Networks and Telecommunications in Buildings, Administration and Production [Germany]

Techno Jpn Techno Japan [Journal published in Japan]

Technol Technological; Technologist; Technology [also technol]

Technol Conserv Technology and Conservation [Journal published in the US]

Technol Cult Technology and Culture [Journal published in the US]

Technol Forecast Soc Change Technological Forecasting and Social Change [Published in the US]

Technol Inf, Soc Technologies de l'Information et Société [Publication of Université de Québec, Canada]

Technol Prod Technology in Production [US Publication]

Technol Rep Iwate Univ Technology Reports of the Iwate University [Japan]

Technol Rep Kansai Univ Technology Reports of the Kansai University [Japan]

Technol Rep Kyushu Univ Technology Reports of the Kyushu University [Japan]

Technol Rep Osaka Univ Technology Reports of the Osaka University [Japan]

Technol Rep Seikei Univ Technology Reports of the Seikei University [Japan]

Technol Rep Tohoku Univ Technology Reports of the Tohoku University [Japan]

Technol Rep Yamaguchi Univ Technology Reports of the Yamaguchi University [Japan]

Technol Rev Technology Review [of Massachusetts Institute of Technology, Cambridge, US]

Technol Sci Inf Technology and Science of Informatics [Journal published in the UK]

Technol Today Technology Today [Published in the US]

Technol Train Technology and Training [Journal published in Taiwan]

Technol Util Technology Utilization [Journal published in the US]

TECHNONET Technical Information Network [of National Research Council of Canada]

Tech Phys Technical Physics; Technical Physicist

Tech Phys Technical Physics [Translation of: *Zhurnal Tekhnicheskoi Fiziki*; Published by the American Institute of Physics, US and MAIK Nauka, Russia]

Tech Phys Lett Technical Physics Letters [Translation of: *Pis'ma v Zhurnal Tekhnicheskoi Fiziki*; Published by the American Institute of Physics, US and MAIK Nauka, Russia]

Tech Publ Prepr Technical Publications and Preprints [of Fermilab, Batavia, Illinois, US]

Tech Rep Technical Report

Tech Rep Autom Res Lab, Kyoto Univ Technical Reports of Automation Research Laboratory, Kyoto University [Japan]

Tech Rep Hydrol Technical Reports in Hydrology [of International Hydrological Program, France]

Tech Rep Inst At Energy, Kyoto Univ Technical Reports of the Institute of Atomic Energy, Kyoto University [Japan]

Tech Rep Kumamoto Univ Technical Reports of the Kumamoto University [Japan]

Tech Rep Sumitomo Light Metals Technical Reports of Sumitomo Light Metals Company [Japan]

Tech Res Cent Finl Electr Nucl Technol Publ Technical Research Center of Finland Electrical and Nuclear Technology Publication [Finland]

Tech Res Cent Finl Res Rep Technical Research Center of Finland Research Report [Finland]

Tech Res Rep Shimizu Corp Technical Research Report of Shimizu Corporation [Japan]

Tech Rev Technical Review

TECHS Thermal Election Capped Hemisphere Spectrometer

Tech Sci Inf Technique et Science Informatique [Publication on Information Science and Technology of Association Française pour le Cybernetique, Economique et Technique, France]

Tech Serv Technical Service

Tech Serv Q Technical Services Quarterly [Published in the US]

Tech Spec Technical Specification

Tech Überwachung Technische Überwachung [German Publication on Technical Inspection of Vereinigung der Technischen Überwachungsvereine]

Tech Zpr Technicke Zpravy [Czech Technical Periodical]

Tec Ital Tecnica Italiana [Italian Technical Periodical]

TECMA Technical Ceramics Manufacturers Association [US]

Tec Metal Tecnica Metalurgica [Spanish Publication on Metallurgy]

Tecnol Elettr Tecnologie Elettriche [Italian Publication on Electrical Engineering]

Tecnol Filo Tecnologie del Filo [Italian Publication on Wire Technology]

Tecnol Mecc Tecnologie Maccaniche [Italian Publication on Mechanical Engineering]

Tecnol Quim Tecnologia Quimica [Cuban Publication Chemical Engineering]

TECOM Test and Evaluation Command [US Army]

TECR Technical Reason

Tect Tectonic; Tectonics [also tect]

Tect Tectonics [Publication of the American Geological Union, US]

Tectonophys Tectonophysics [International Journal]

TED Tiny Editor [Computers]

TED Technical Evaluation and Development

TED Technology Evaluation and Development

TED Television Disk

TED Test Engineering Division [US Navy]

TED Thermal Expansion Difference

TED Thermoelastic Damping

TED Threshold Extension Demodulator

TED Tiny Editor [Computers]

TED Total Energy Detector

TED Trailing Edge Down

TED Transferred Electron Device

TED Transient Enhanced Diffusion

TED Translation Error Detector

TED Transmission Electron Diffraction

TED Triethylenediamine

TED Triple Extra D [French High-Speed Steel]

TED Trunk Encryption Device

TED Turbine Engine Division [US Air Force]

TEDA Triethylenediamine

TE/DC Traffic Enforcement/Driver Control

TEDE Temperature-Enhanced Displacement Effect

TeDEC Tellurium Diethyl Dithiocarbamate

TEDF Treated Effluent Disposal Facility

TEDGA Triethylene Glycol Diacetate

TEDMA Triethylene Glycol Dimethacrylate

TeDMC Tellurium Dimethyl Dithiocarbamate

TEDP Tensor of Effective Dielectric Permittivity

TEDP O,O,O,O-Tetraethyl Dithiophosphate

TEDP Tetraethyl Dithiopyrophosphate

TEDS Tactical Expendable Drone System

TEDS Twin Exchangeable Disk Storage

TEDTA Thiobis(ethylene)diamimetetraacetate [also tedta]

TEDTA Thiobis(ethylene)diamimetetraacetic Acid

TEE Telecommunications Engineering Establishment [UK]

TEE Thermionic Electron Emission

TEE Thermoelastic Effect [Physics]

TEE Thermoelectric Effect [Physics]

TEE Torpedo Experimental Establishment [UK]

TEE Trans Europe Express

TEEC Total Energy and Environmental Conditioning

TEED N,N,N',N'-Tetraethyl 1,2-Ethanediamine

TEELS Transmission Electron Energy Loss Spectroscopy

TEEM Techno-Economic-Environmental Model

TEEM Thermionic Electron Emission Microscope; Thermionic Electron Emission Microscopy

TEEN N,N,N',N'-Tetraethyl 1,2-Ethanediamine

TEEOC Turret Electronics and Electrooptical Console

TEES Texas Engineering Experiment Station [of NASA]

TEES Thermochemical Environmental Energy System

TEF Thorne Ecological Foundation [now Thorne Ecological Institute, US]

TEF Transverse Electrical Field

TEFA Total Esterified Fatty Acid

TEFC Totally Enclosed Fan-Cooled (Motor)

TEFL Teaching (of) English as a Foreign Language

TEG Tactical Electronics Group

TEG Tetraethylene Glycol

TEG Thermoelectric Generator

TEG Triethylgallium [also TEGa]

TEG Training and Education Group [of Library Association, UK]

TEG Triethylene Glycol

TEGa Triethylgallium [also TEG]

TEGAS Test Generation and Simulation

TEGDA Triethylene Glycol Diacetate

TEGDN Triethylene Glycol Dinitrate

TEGeCl Triethylgermanium Chloride

TEGFET Two-Dimensional Electron Gas Field-Effect Transistor

TEGMA Terminal Elevator Grain Merchants Association [US]

TEGDA Triethylene Glycol Dimethacrylate

TeGe Tellurium Germanide

TEGMN Triethylene Glycol Mononitrate

TEGN Triethylene Glycol Dinitrate

2e/h Josephson Frequency-Voltage Ratio (4.835977×10^{14} Hz/V) [Symbol]

Teh Fiz Tehnika Fizika [Yugoslav Publication on Technical Physics]

TEHO Tail-End-Hop-Off

TEHP Tris(2-Ethylhexyl)phosphate

TEHPO Tris(2-Ethylhexyl)phosphine Oxide

Teh Rud Geol Metal Tehnika Rudarstvo Geologiji i Metalurgija [Yugoslav Publication on Geology and Metallurgy]

TEI Technologika Ekpaideutika Idrimata [Technological Educational Institution of Higher Learning, Greece]

TEI Terminal Endpoint Identifier

TEI Text Encoding Initiative [Text Information Interchange Standard]

TEI Thorne Ecological Institute [US]

TEI Total Employee Involvement

TEI Total Environmental Impact

TEIn Triethyl Indium [also TEI]

TEIC Tissue Equivalent Ionization Chamber

TEIMS The Executive Information Management System

TE Int TE (Tecnologie Elettriche) International [Journal on Electrical Engineering published in Italy]

TEJ Transverse Expansion Joint

TEK Traditional Ecological Knowledge

TEKES Technology Development Center of Tampere University of Technology [Finland]

Tekh Elektrodin Tekhnicheskaya Elektrodinamika [Russian Journal of Electrodynamics]

Tekh Kibern Tekhnicheskaya Kibernetika [Russian Journal of Computer and Systems Sciences]

Tekh Kino Telev Tekhnika Kino i Televideniya [Russian Journal on Kinematography and Television Engineering]

Tekh Misul Tekhnicheska Misul [Bulgarian Engineering Publication]

Tekh Legkik Splav Tekhnologiya Legkik Splavov [Russian Journal on Light Metal Technology]

Tekstil Prom Tekstilnaya Promyshlennost [Russian Publication on Textile Technology]

Tek Tidskr Teknisk Tidskrift [Swedish Technical Journal]

Tek Ukebl Teknisk Ukeblad [Norwegian Technical Communication]

TEL Task Execution Language

TEL (Aeronautical) Telecommunications Service

TEL Tetraethyl Lead

TEL Tritium Engineering Laboratory [Japan]

Tel Telegram [also tel]

Tel Telegraph(y) [also tel]

Tel Telephone; Telephony [also tel]

TELA Triethanolamine

TE-LAB Tether Laboratory Demonstration System

TELCO Telephone Company

TELCOM Telecommunications

TELCON Teletypewriter Conference

Telebanking Television/Telephone Banking

Telec Telecommunication [also telec]

Telecom Telecommunication(s) [also Telecomm, Telecommun, telecom, telecomm, or telecommun]

Telecommun J Telecommunication Journal [of International Telecommunication Union, Switzerland]

Telecommun J Aust Telecommunication Journal, Australia [of Telecommunication Society of Australia]

Telecommun Policy Telecommunications Policy [UK Publication]

Telecommun Radio Eng 1, Telecommun Telecommunications and Radio Engineering, Part 1, Telecommunications [Translation of: *Elektrosvyaz (USSR)*; published in the US]

Telecommun Radio Eng 2, Radio Eng Telecommunications and Radio Engineering, Part 2, Radio Engineering [Translation of: *Radiotekhnika, Moskva (USSR)*; published in the US]

Telecom Rep Telecom Report [Published by Siemens AG, Munich, Germany]

TELEDAC Telemetric Data Converter

TELEDOC Telecommunications Documentation

TELEFAX Facsimile Transmission Service [also Telefax] [Europe]

Telefunken Z Telefunken Zeitschrift [Journal published by Telefunken GmbH, Heilbronn, Germany]

Teleg Telegram [also teleg]

Teleg Telegraph(y) [also teleg]

TELEMAIL Electronic Mail Service [also Telemail]

TELEMAT Telematic(s) [also Telemat, or telemat]

Telemat India Telematics India [Published in India]

Telemat Inf Telematics and Informatics [Published in the UK]

TELENET Telecommunication Network [also Telenet; now SprintNet]

Telenorma Nachr Telenorma Nachrichten [Technical Newsletter published by Telefonbau und Normalzeit, Frankfurt, Germany]

TELEPAC Packet-Switching Telecommunications Network

TELEPAC Telemetering Package

TELEPAC Telephone Package

Teleph Telephone [also teleph]

Teleph Eng Manage Telephone Engineer and Management [US]

Teleputer Television/Computer System

Teleran Television and Radar Air Navigation (System)

Telesat Telecommunications Satellite [Canada]

Teletex (International) Super-Telex Service

Telettra Rev Telettra Review [Published in Italy]

Telev Television [also telev]

Telev, J R Telev Soc Television, Journal of the Royal Television Society [UK]

Telev/Radio Age Television/Radio Age [Published in the US]

Telex Teleprinter Exchange Service [also telex, TEX, TLX, or TX]

Telg Telegram [also telg]

Telg Telegraph(y) [also telg]

TELINET Telefax Library Network [US]

Tellus A, Dyn Meteorol Oceanogr Tellus, Series A, Dynamic Meteorology and Oceanography [Publication of the Swedish Geophysical Society]

Tellus B, Chem Phys Meteorol Tellus, Series B, Chemical and Physical Meteorology [Publication of the Swedish Geophysical Society]

TELOPS Telemetry On-Line Processing System

TELPAL Tel Aviv-Palo Submarine Cable [Israel/Italy]

TELRY Telegraph Reply

TELS Transmission Energy Loss Spectroscopy

TELS Turbine Engine Load Simulator

TELSAM Telephone Service Attitude Measurement

TELSCA Transmission Energy Loss Electron Spectroscopy for Chemical Analysis

TELSCOM Telemetry-Surveillance-Communications

TELSIM Teletypewriter Simulator

TELTIPS Technical Effort Locator and Technical Interest Profile System [US Army]

TELUS Telemetric Universal Sensor

TEM Thermal Expansion Mismatch

TEM Thermal Expansion Molding

TEM Thermoelectric Electromagnetic (Pump)

TEM Total Energy Management

TEM Transmission Electron Micrograph; Transmission Electron Microscope; Transmission Electron Microscopy

TEM Transmittance Electron Microscopy

TEM Transverse Electromagnetic (Mode)

TEM Triethoxymethane

TEM Triethylenemelamide

TEM Triethylenemelamine [Pharmacology]

TEM Triethyl Methanetricarboxylate

TEM Triple Extra M [French High-Speed Steel]

TE/m³ Toxicity Equivalent per Cubic Meter [TE m^{-3}]

TEMA Tank Equipment Manufacturers Association

TEMA Telecommunication Engineering and Manufacturing Association [UK]

TEMA Tubular Exchanger Manufacturers Association [US]

TEM-AEM Transmission Electron Microscopy–Analytical Electron Microscopy

TEMAG Tether Magnetic Field Experiment [of NASA]

TEM-BF Transmission Electron Microscopy–Bright Field (Image) [also TEM BF]

TEM-CL Transmission Electron Microscopy–Cathodoluminescence (Technique) [also TEM-CL, or TEM/CL]

TEM-DF Transmission Electron Microscopy–Dark Field (Image) [also TEM DF]

Te5MC Tellurium Pentamethylene Dithiocarbamate

TeMeAN Tetramethylammonium Nitrate

TEM-EBIC Transmission Electron Microscopy–Electron Beam Induced Current

TEMED N,N,N',N'-Tetramethylethylenediamine

TEM-EDS Transmission Electron Microscopy–Energy-Dispersion Spectrum [also TEM EDS]

TEM-EDS Transmission Electron Microscopy–Energy-Dispersive (X-Ray) Spectroscopy [also TEM EDS]

TEM-EELS Transmission Electron Microscopy–Electron Energy Loss Spectroscopy [TEM EELS]

Temephos O,O,O'-Tetramethyl-O,O'-Thio-Di-p-Phenylene Phosphorothionate [also temephos]

TEMF Thermo-Electromotive Force [also temf]

TEM FFT Transmission Electron Microscopy Fast-Fourier Transform

TEM-IP Transmission Electron Microscope Imaging Plate

TEML (Borland) Turbo Editor Macro Language

TEM Mode Transverse Electromagnetic Mode [also TEM mode]

TEMO Topological Effect on Molecular Orbitals [Physical Chemistry]

TE Mode Transverse Electric Mode [also TE mode]

TEMP Test and Evaluation Master Plan

TEMP Thermal Energy Management Processes

TEMP Total Energy Management Professional

TEMP Transportation Energy Management Program

Temp Temperature [also temp]

Temp Template [also temp]

temp temporary

temp *(tempore)* – in the time of

TEMPEST Transient Energy Momentum and Pressure Equation Solutions

TEMPO 2,2,6,6-Tetramethylpiperidine-N-Oxide (Radical)

TEMPO Tetramethylpiperidinyloxy-

TEMPO 2,2,6,6-Tetramethyl-1-Piperidinyloxy

TEMPO Total Evaluation of Management and Production Output

tempo temporary

TEMPOS Timed Environment Multipartioned Operating System

TEMS Test Equipment Maintenance Set

TEMS Toyota Electronic(ally) Modulated Suspension [Automobiles]

TEMS Transport Environment Monitoring System

TEM-SAED Transmission Electron Microscope/Selected Area Electron Diffraction [also TEM/SAED]

TEM-SAD Transmission Electron Microscope/Selected Area Diffraction [also TEM/SAD]

TEM-STEM Transmission Electron Microscope/Scanning Transmission Electron Microscope; Transmission Electron Microscopy/Scanning Transmission Electron Microscopy [also TEM/STEM]

TEM-TED Transmission Electron Microscopy/Transmission Electron Diffraction [also TEM/TED]

TEMW Transverse Electromagnetic Wave

TEN Trisepoxy Novolac (Resin)

TeNA Tetranitroaniline

TeNAns Tetranitroanisole

TeNAzxB Tetranitroazoxybenzene

TeNB Tetranitrobenzene

TeNBPh Tetranitrobiphenyl

TeNCbl Tetranitrocarbanilide

TeNCbz Tetranitrocarbazole

TeNDG Tetranitrodiglycerin

TeNDMBDNA Tetranitrodimethylbenzidinedinitramine

TeNDPhA Tetranitrodiphenylamine

TeNDPhEta Tetranitrodiphenylethane

TeNDPhEth Tetranitrodiphenylether

TeNDPhEtlA Tetranitrodiphenylethanolamine

TENE Total Estimated Net Energy

TeNHzB Tetranitrohydrazobenzene

TeNMA Tetranitromethylaniline [also TeNMe]

TeNMe Tetranitromethane

Te-125 NMR Tellurium-125 Nuclear Magnetic Resonance [also ^{125}Te NMR]

TeNN Tetranitronaphthalene

Tenn Tennessee [US]

Tenn Hist Quart Tennessee Historical Quarterly [of Tennessee Historical Society, US]

TeNOx Tetranitrooxanilide

TeNPhMNA Tetranitrophenylmethylnitramine [also TeNPhMeNA]

TENS Transcutaneous Electronic Nerve Stimulator (or Stimulation) [Medicine]

Tens Tension [also tens]

TENSIR Tendon Supported Inspection Robot

TeNT Tetranitrotoluene

tent tentative

TEN-TELECOM Trans-European Telecommunications Network [of European Union]

TeNTMB 3,5,3',5'-Tetranitro-4,4'-Tetramethyldiamino-biphenyl [also TeNTMeB]

TENV Totally Enclosed Naturally Ventilated (Motor)

TEO Telephone Equipment Order

TEO Thermoplastic Elastic Olefin

TEO Transferred Electron Oscillator

TEOA Tetraoctyl Ammonium

TEOA Triethanolamine

TEOC-ONp 2-Trimethylsilylethyl Oxycarbonyl-p-Nitrophenolate

TEOM Transformer Environment Overcurrent Monitor

Teor Eksp Khim Teoreticheskaya i Eksperimental'naya Khimiya [Russian Publication on Theoretical and Experimental Chemistry]

Teor Mat Fiz Teoreticheskaya i Matematischeskaya Fizika [Russian Publication on Theoretical and Mathematical Physics]

Teor Veroyatn Prim Teoriya Veroyatnostei i ee Primeneniya [Russian Journal on the Theory of Probability and Its Applications]

TEOS Tetraethylorthosilane (or Tetraethyoxysilane)

TEOS Tetraethyl Orthosilicate

TEP Technical Evaluation Panel

TEP Technology Enhancement Project [of Canadian Passport Office]

TEP Terminal Error Program

TEP Test Evaluation Plan

TEP Test Executive Processor

TEP Tetraethyl Pyrophosphate

TEP Tetraethyl Pyrophosphite

TEP Thermoelectric Power

TEP Transmitter Experimental Package

TEP Transportation Energy Panel

TEP Triethyl Phosphate

TEP Triethylphosphene

TEP Tritium Extraction Plant

TEPA Tetraethylenepentamine

TEPA Triethylenephosphoramide

Tepa Tris(1-Aziridinyl)phosphine Oxide [also tepa]

TEPAC Tube Engineering Panel Advisory Council [of Electronic Industries Association, US]

TEPb Tetraethyl Lead

TEPD Total End-Point Dose

TEPG Thermionic Electrical Power Generator

TePh Phenyltelluride

TePhUr Tetraphenylurea

TEPI Triethyl Phosphite

TEPIAC Thermophysical and Electronic Properties Information Analysis Center [US]

TEPICO Threshold Electron Photo-Ion Coincidence

Te5MC Tellurium Pentamethylene Dithiocarbamate

Teploenerg Teploenergetika [Russian Publication on Thermal Engineering]

Teplofiz Vys Temp Teplofizika Vysokikh Temperatur [Russian Publication on High-Temperature Physics]

TEPOS Test Program Operating System

TEPP Tetraethyl Pyrophosphate

TEPRSSC Technical Electronic Product Radiation Safety Standards Committee

TEPZ Taichung Export Processing Zone [Taiwan]

TER Tergitol

TER Test Equipment Readiness

TER Thermal Eclipse Reading

TER Thermal Expansion Rubber

TER Thyssen-Extrem-Rechtkant (Process) [Metallurgy]

TER Time Estimating Relationship

TER Torpedo Effective Range

TER Total External Reflection

TER Townsend Energy Ratio [Electronics]

TER Transmission Electron Radiography

TER Transmission Equivalent Resistance

TER Transportation Engineering Research

TER Travel Expense Report

TER Triple Ejector Rack

TER Triple Extra R [French High-Speed Steel]

Ter Teratogen(ic) [also ter]

Ter Terazzo [also ter]

Ter Terrace [also ter]

Ter Territorial; Territory [also ter]

Ter Tertiary [Geology]

ter tertiary

TERA Total Energy Resource Analysis

TERA Transient Error Reconstruction Approach (Algorithm)

tera- SI prefix representing 10^{12}

TERAC Tactical Electromagnetic Readiness Advisory Council

Teraflops Tera Floating-Point Operations per Second [Unit]

Terat Teratology (Journal)

Terbufos O,O-Diethyl-S-(tert-Butyl)methyl Phosphorodithioate [also terbufos]

TERC Technical Education Research Center [Cambridge, Massachusetts, US]

TERCOM Terrain Contour Matching [also Tercom]

TERD Tritium Effects Research Division [of Sandia National Laboratories, Albuquerque, New Mexico, US]

TEREC Tactical Electromagnetic Reconnaissance

TERENA Trans-European Research and Education Networking Association

TERI Tesla Electronics Research Institute [Prague, Czech Republic]

TERIB Test and Evaluation, Reliance and Investment Board

TERL Test Engineer Readiness List

TERL Test Equipment Readiness List

TERLS Thumba Equatorial Rocket Launching System [India]

Term Terminal [also Term, or term]

Term Terminate; Termination [also term]

Term Terminology [also term]

TermNet International Network for Terminology

Termotec Termotecnica [Italian Publication on Thermal Engineering]

TERMPWR Terminator Power

TERN Terrestrial Ecosystem Research Network [Germany]

TERP Terrain Elevation Retrieval Program

TERPS Terminal Instrument Approach Procedures

TERPS Terminal Planning System

TERPY 2,2',6',2"-Terpyridine [also terpy]

Terr Terrace [also terr]

Terr Territorial; Territory [also terr]

TERS Tactical Electronic Reconnaissance System

TERSE Tunable Etalon Remote Sounder of Earth

TERSS Tasmanian Earth Resources Satellite Station [Australia]

tert- tertiary [Chemistry]

tert-BOC tert-Butyloxycarbonyl

TES 2-([2-Hydroxy-1,1-bis(hydroxymethyl)ethyl]amino) ethanesulfonic Acid

TES Technical Enforcement Support

TES Tetraethylsilane

TES N,N,N',N'-Tetraethyl Sulfamide

TES Text Editing System

TES Thermal Emission Spectrometer

TES Thermal Energy Storage

TES Thermal Evaluation Spectroscopy

TES Thermoelectric Series

TES Time Encoded Speech

TES Total Energy System

TES Translational Energy Spectroscopy

TES Transportable Earth Station

TES Triethylsilane; Triethylsilyl

TES Trisethanesulfonic Acid

TES N-tris(Hydroxymethyl)methyl-2-Aminoethanesulfonic Acid

TES Tropical Emission Spectrometer

TES Tropospheric Emission Spectrometer

TES Troubleshooting Expert System

TESA Television Electronics Service Association [Canada]

TESb Triethyl Antimony

TESCl Triethylsilyl Chloride

TESH Technical Shop

TESi Tetraethylsilane

TESICO Threshold Electron-Secondary Ion Coincidence (Technique)

TESL Teaching English as a Second Language

TESLA Technical Standards for Library Automation Committee [of Library and Information Technology Association, US]

Tesla Electron Tesla Electronics [Publication of Tesla Electronics Research Institute, Prague, Czech Republic]

TESn Tetraethyltin

TeSn Tellurium Stannide

TESS Tactical Electromagnetic Systems Study

TESS Thermocouple Emergency Shipment Service

TEST Thesaurus of Engineering and Science Terms [of Engineers Joint Council, US]

TESTAS Turkish Electronics and Trade Association

Testg Testing [also testg]

Test Meas World Test and Measurement World [US Publication]

TES-Triflat Triethylsilyl-Trifluoromethanesulfonate [also TES-triflat]

TET Telemedicine and Educational Technology

TET Test Evaluation Team

TET Tetrachloride

TET Tetraethyltin

TET Tetraethylthiuram Disulfide

TET Thermoelectric Thermometer

TET Thermometric Enthalpy Titration [Chemistry]

TET Total Elapsed Time

TET Transient Energy Transfer

TET Transportation Energy Technology

TET Transportation Engineering Technology

TET Triethylene Tetramine

TET Turbine Entry Temperature

2,3,2-TET 1,4,8,11-Tetraazaundecane

Tet Tetrahedron [also tet]

TETA 1,4,8,11-Tetraazacyclotetradecane-N,N′,N″,N‴-Tetraacetate; Tetraazacyclotetradecane-N,N′,N″,N‴-Tetraacetic Acid

TETA Tetraethanol Propane

TETA Triethylenetetramine [also Teta]

TETA Triethylenetriamine

TETBr Triethyltin Bromide

TETCl Triethyltin Chloride

TetCP Tetracalcium Phosphate

TETD Tetraethylthiuram Disulfide

TETeA Triethylenetetramine

θ-(ET)$_2$I$_3$ Theta Bis(ethylenedithio)tetrathiafulvalene Triiodine

Tetrahed Asym Tetrahedron Asymmetry [Now Part of Tetrahedron Letters]

Tetrahed Lett Tetrahedron Letters [Journal on Biochemistry published in the US]

TETM Tetraethyl Thiuram Monosulfide

TE-TM Transverse Electric–Transverse Magnetic (Mode)

TETN Triethylamine

TETOC (Council for) Technical Education and Training for Overseas Countries [UK]

TETR Test and Training Satellite [US]

tetr tetragonal

TETRA Telemedicine and Educational Technology Resources Agency [of Memorial University of Newfoundland, St. John's, Canada]

TETRA Terminal Tracking Telescope

TETRA Tetrachloromethane also Tetra, or tetra]

TETRA Tetraethyleneglycol [also Tetra, or tetra]

Tetra-Di-Salt Tetramethylammonium Dinitrate

TETRAEN Tetraethylenepentamine

Tetrag Tetragon(al) [also tetrag]

Tetrah Tetrahedral; Tetrahedron [also Tetrahed, tetrahed, or tetrah]

Tetrahed Asym Tetrahedron Asymmetry [Now Part of Tetrahedron Letters]

Tetrahed Lett Tetrahedron Letters [Journal on Biochemistry published in the US]

Tetrahydro-DOC Tetrahydrodeoxycorticosterone [Biochemistry]

TETRAPHOS-1 1,1,4,7,10,10-Hexaphenyl-1,4,7,10-Tetra-phosphadecane [also TETRAPHOS I]

TETRAPHOS-2 Tris(2-Diphenylphosphinoethyl)phosphine [also TETRAPHOS II]

Tetra-Salt Tetramethylammonium Nitrate

Tetra-Tetryl Tetra(2,4,6-Trinitro)phenylnitraminomethyl) methane [also Tetra-tetryl]

Tetrazene 4-Amidino-1-(Nitrosamino-Amidino)-1-Tetrazene [also tetrazene]

TETRENE Tetraethylene Pentamine [also Tetrene]

Tetrg Tetragon(al) [also Tetrg, or tetrg]

Tetrh Tetrahedral; Tetrahedron [also tetrh]

TETROON Tethered Meteorological Balloon [also Tetroon, or tetroon]

Tetroxyl 2,4,6-Trinitrophenylmethoxynitramine

Tetryl N-Methyl-N,2,4,6-Tetranitrobenzenamide

Tetryl N,2,4,6-Tetranitromethylaniline

Tetrytol N,2,4,6-Tetranitromethylaniline plus Trinitrotoluene [Explosive Mixture]

TEU Tetraethylurea

TEU Total Equivalent Units

TEU Twenty-Foot Equivalent Unit

Teut Teuton(ic) [also Teut, or teut]

TEV (Workshop on) Physics at a High Luminosity Tevatron Collider [at Fermilab, Batavia, Illinois, US]

TEV Thermoelectric Voltage

TeV Teraelectronvolt [Unit]

TeV Tevatron (Collider) [of Fermilab, Batavia, Illinois, US]

TEVA Total Equivalent Volt-Ampere

TEVROC Tailored Exhaust Velocity Rocket

TEW Tactical Electronic Warfare

TEW Thyssen Edelstahlwerke AG [Major German Steel Manufacturer]

TEW Transverse Elastic Wave

TEW Transverse Electric Wave

TEWA Threat Evaluation and Weapon Assignment [of Department of National Defense, Canada]

TEWD Tactical Electronic Warfare Division [of Naval Research Laboratory, Washington, DC, US]

TEWG Transitional Environmental Working Group

TEWI Total Equivalent Warming Impact

TEWS Tactical Electronic Warfare System

TEW Tech Ber TEW Technische Berichte [Technical Reports of Thyssen Edelstahlwerke AG, Krefeld, Germany]

TEX Target Excitation

TEX Tau-Epsilon-Chi System

TEX Teleprinter Exchange Service [also Telex, telex, TEX, TLX, or TX]

TEX Test Executive

Tex Texan; Texas [US]

Tex Textile(s) [also tex]

Tex Trivedi (Microsegregation Model) [Metallurgy]

Texaco Texas Oil Company [US]

Tex J Sci Texas Journal of Science [US]

TEXPLOT Texas Instruments Plotter

Texpo American Fashion Textiles Exposition

TEXT Texas Experimental Tokamak [Plasma Physics]

Text Textile [also text]

Textbk Textbook [also textbk]

TextE Textile Engineer [also TextEng]

Text Eng Textile Engineer(ing) [also Textile Eng]

Text Horiz Textile Horizons [Publication of The Textile Institute, UK]

Textile Chem Color Textile Chemists and Colorist [Publication of American Association of Textile Chemists and Colorists]

Textile Horiz Textile Horizons [Publication of The Textile Institute, UK]

Textile Res J Textile Research Journal [Publication of Textile Research Institute, US]

Textile Technol Dig Textile Technology Digest [Publication of Institute of Textile Technology, US]

TEXTIR Text Indexing and Retrieval

TEXTLINE Text On-Line (Database) [UK]

TEXTOR Tokamak Experiment for Technology-Oriented Research

Texture Cryst Solids Texture of Crystalline Solids [UK Publication]

Textures Microstruct Textures and Microstructures [UK Publication]

TEY Total Electron Yield

Tez Tetrazole

TEZG Tribological Experiments in Zero Gravity

TF French Southern Territories [ISO Code]

TF Tail Fin [Aerospace]

TF Tail Fin [Biology]

TF Tangential Force

TF Tape Feed

TF Task Force

TF Technological Forecasting

TF Temperature Factor

TF Tensile Fracture [Mechanics]

TF Ternary Fission

TF Territorial Forces

TF Test Facility

TF Test Fixture

TF Test Frame

TF Test Frequency

TF Textile Foundation [US]

TF Thermal Fatigue

TF Thermal Flow

TF Thermos Flask

TF Thick Film

TF Thin Film

TF Thomas-Fermi (Equation) [Atomic Physics]

TF Thomson-Freundlich (Equation) [Physics]

TF Threshold Factor

TF Throttled Flow; Throttling Flow [Fluid Mechanics]

TF Tierra de Fuego [Chile/Argentina]

TF Time Fuse

TF Tool Foundation [Netherlands]

TF Toroidal Field

TF Toxicology Forum [US]

TF Tractor Feed (Printer)

TF Transaction File

TF Transcendental Function [Mathematics]

TF Transcription Factor [Biochemistry]

TF Transformer

TF Transfer Function [Controllers]

TF Transgranular Fracture [Metallurgy]

TF Transitional Field

TF Transitional Flow

TF Transonic Flow

TF Trim and Form

TF Tritium Fluoride

TF Trunk Frame

TF Tuning Fork

TF Turbofan

TF Turbulent Flow

TF Twin Fraction [Metallurgy]

TF Twist Factor (of Yarn)

TF IIA Transcription Factor II A [Biochemistry]

TF IIB Transcription Factor II B [Biochemistry]

TF- Iceland [Civil Aircraft Marking]

Tf 2-Thenoyltrifluoroacetone

Tf Trifyl (or Trifluoromethanesulfonyl)

TFA Taiwan Foodstuffs Association, Inc.

TFA Taiwan Forest Administration

TFA Tank Focus Area [Nuclear Engineering]

TFA Target Factor Analysis

TFA The Ferroalloys Association [US]

TFA Technology Forecasting and Assessment

TFA Texas Forestry Association [US]

TFA Textile Finishers Association [UK]

TFA Thin Film Analysis

TFA Thin-Film Approximation

TFA Thomas-Fermi Approximation [Atomic Physics]

TFA Timing Filter Amplifier

TFA Trifluoroacetate; Trifluoroacetic Acid

TFA Trifluoroacetamide

TFA Trifluoroacetyl [also Tfa]

TFA Trifluoroacetylacetone [also tfa]

TFA 1,1,1-Trifluoro 2,4-Pentanedionate [also tfa]

TFAA Trifluoroacetic Anhydride

TFAC Trifluoroacetylacetonate; 1,1,1-Trifluoroacetylacetonate; Trifluoroacetylacetone; Trifluoroacetylacetonato- [also tfac]

TFAc Trifluoroacetic Acid

TFACAC 1,1,1-Trifluoroacetylacetone

(TFAC)Cu(PMe₃) 1,1,1,5,5,5-Tetrafluoroacetylacetonate Copper (I) Trimethylphosphine Copper [also (tfac)Cu(PMe₃)]

TFAI 1-(Trifluoroacetyl)imidazole

TFAME Trifluoroacetatic Acid Methylester

TFA-ME Trifluoroacetate Methylester

TFAP Tropical Forest Action Plan [of Food and Agricultural Organization]

TFB Towed Flexible Barge

TFBPA Textile Fibers and Byproducts Association [US]

TFBQ p-Tetrafluorobenzoquinone

TF14BQ Tetrafluoro-1,4-Benzoquinone

TFC International Conference on Transparent Ferroelectric Ceramics

TFC Tasmanian Forestry Commission [Australia]

TFC Telefilm Canada

TFC Thin-Film Capacitor

TFC Thin-Film Chemistry

TFC Time from Cutoff

TFC Transmission Fault Control

TFC Transparent Ferroelectric Ceramic(s)

TFC 3-(Trifluoromethylhydroxymethylene)-d-Camphorate [also tfc]

TFC Turbulent Forced Convection

Tfc Traffic

TFCA The Federation of Commodity Associations [UK]

TFCF Turbulent Free Convective Flow

TFCG Thin Film Crystal Growth

TFCS Triplex Flight Control Subsystem; Triplex Flight Control System

TFCX Tokamak Fusion Core Experiment [Plasma Physics]

TFCRI Tropical Fish Culture Research Institute [Malaysia]

TFD Television Feasibility Demonstration

TFD Thermofield Dynamics

TFD Thin Film Deposition

TFD Thin Film Detector

TFD Time Frequency Division [of National Institute of Standards and Technology, Gaithersburg, Maryland, US]

TFD Transcription Factor Database [of United Nations International Center for Genetic Engineering and Biotechnology]

TFD Transflective Device

TF/D Time-Frequency Dissemination

TFDD Text File Device Driver

TFDL Technisch Fysische Dienst voor de Landboov [Technical and Physical Engineering Service for Agriculture, Netherlands]

TFDU Thin Film Deposition Unit

TFDW Thomas-Fermi-Dirac-Weizaecker (Model) [Physics]

TFE Tetrafluoroethane

TFE Tetrafluoroethylene

TFE The Fertilizer Institute [US]

TFE Thermionic Field Emission

TFE Thermionic Fuel Element

TFE Thermo-Fluid Engineering

TFE Thin-Film Evolution

TFE Time from Event

TFE Toronto Futures Exchange [Canada]

TFE 2,2,2-Trifluoroethanol

TFE 2,2,2-Trifluoroethyl

TFE Turbofan Engine

TFE/HFP Tetrafluoroethylene/Hexafluoropropylene (Copolymer)

TFEL Thin-Film Electroluminescence; Thin-Film Electroluminescent

TFEM Thin-Film Electron Microscopy

TFEO Tetrafluoroethylene Epoxide

TFEP tris(2,2,2-Trifluoroethyl)phosphate

TFE/P Tetrafluoroethylene and Propylene Copolymer

TFE/P/T Tetrafluoroethylene, Propylene and Vinylidene Fluoride Terpolymer

TFER Threshold for Full Epitaxial Regrowth [Crystallography]

TFET Tetrode Field-Effect Transistor

TFET Triode Field-Effect Transistor

TFF Time to Free Fall

TFFE Thin-Film Ferroelectrics [also TFF]

TFFET Thin-Film Field-Effect Transistor [also TFFET]

TFG Thin-Film Growth

TFG Transmit Format Generator

TFH Thin-Film Head

TFHQ Tetrafluorohydroquinone

TFI Teppich-Forschungsinstitut [German Carpet Research Institute]

TFI The Fertilizer Institute [US]

TFI Thick-Film Ignition

TFI Thick-Film-Integrated (Module) [Automobiles]

TFI Threaded Full Length

TFI Timed Fuel Injection

TFI Time from Ignition

TFI Total Forest Industries

TFI Tropical Forest Initiative

TFIC Thin-Film Integrated Circuit

TFIH Transverse Flux Induction Heating

TFK Transportforskningskommission [Commission for Transport Research, Sweden]

TFL Telemetry Format Load

TFL Through Flowline

TFL Time from Launch

TFLED Thin-Film Light-Emitting Diode [also TFLED]

TFLOPS Tera-Floating-Point Operations per Second [also Tflops]

TFM Tagged Font Metric

TFM Tamed Frequency Modulation

TFM Tape File Management

TFM Theoretical Fracture Mechanics

TFM Thin-Film Memory

TFM Time-Quantized Frequency Modulation

TFM Total Flowmeter

TFM Trifluoromethane

TFM 3-Trifluoromethyl-4-Nitrophenol

.TFM Tagged Font Metric [File Name Extension]

TFMA Trifluoromethylamine

TFMC Thin-Film Memory Chip

TFMC Tris(fluoromethylhydroxymethylene)-d-Camphorato

TFMC-Eu Tris[3-(Trifluoromethylhydroxymethylene)-d-Camphorato], Europium

TFMC-Pr Tris[3-(Trifluoromethylhydroxymethylene)-d-Camphorato], Praseodymium

TFME Thin-Film Mercury Electrode

TFMH Thin-Film Magnetic Head

TFMIC Trifluoromethyl Isocyanate

TFML Thin-Film Multilayer

TFMP Thin-Film Melt Polymerization

TFMRC Thermo-Fluid Mechanics Research Center [UK]

TFMS Text and File Management System

TFMS Trifluoromethanesulfonate; Trifluoromethanesulfonyl

TFMS Trunk and Facilities Maintenance System

TFMSA Trifluoromethane Sulfonic Acid

TFN Transfer Function [Controllers]

TFN Tungavik Federation of Nunavut [Canada]

TFO Trifluoromethanesulfonic Acid Anhydride [also Tf2O]

TFO Tropospheric Fall-Out

TFO Tuning-Fork Oscillator

TfO Trifluoromethanesulfonate

TFOBr Trifluoroacetyl Hypobromide

TFOH Trifluoroacetic Acid

TFOI Trifluoroacetyl Hypoiodide

TFOP 2-(Trifluoromethylsulfonyloxy)pyridine

TFOT Thin-Film Oven Test

TFOV Telescope Field of View

TFOV Total Field of View

TFP Thermoplastic Fiber Placement

TFP Thin-Film Physics

TFP Thin-Film Polarizer

TFP Transfiber Photography

TFP Trees on Farms Program [US]

TFP Trifluo(ro)piperazine

TFP Trifluoropropene

TFP Trifluoropropylene

TFPIA Textile Fiber Products Identification Act [US]

TFR Terrain Following Radar

TFR Theoretical Final Route

TFR Thick-Film Resistor

TFR Tightly Folded Resonator

TFR Total Fertility Rate

TFR Transaction Formatting Routine

TFR Trouble and Failure Report

Tfr Transfer [also tfr]

TFRI Taiwan Fisheries Research Institute

TFRI Taiwan Forestry Research Institute

TFRS Tungsten Fiber-Reinforced Superalloy

TFS Tape File Supervisor

TFS Telemetry Format Selection

TFS Telephone Feature System

TFS Tetrafluorosilane

TFS Thin-Film Science

TFS Thin-Film Solid

TFS Thulium Iron Silicide

TFS Time-of-Flight SIMS (Secondary-Ion Mass Spectrometry)

TFS Tin-Free Steel

TFS Traffic Flow Security [Telecommunications]

TFS Traffic Forecasting System [Telecommunications]

TFS Transfer Standard

TFS Triplet Fine Structure [Spectroscopy]

TFS Tritium Filling Station

TFS True Fracture Stress [Mechanics]

TFSA Thin-Film Spreading Agent

TFSC Thin-Film Semiconductor

TFSI Toronto Fertility Sterility Institute [Canada]

TFSM Terminal Flow Simulation Model

TFS&T Thin Film Science and Technology

TFSUS Task Force on the Scientific Uses of Space Stations

TFT Thermal Field Theory (Workshop)

TFT Thin-Film Technology

TFT Thin-Film Transistor

TFT Threshold Failure Temperature

TFT Time-to-Frequency Transformation

TFT Trees for Tomorrow [US]

TFT Trifluorotoluene

TFTA Tetrafluoroterephthalic Acid

TFT LCD Thin-Film Transistor Liquid-Crystal Display [also TFT-LCD, or TFT/LCD]

TFTP Television Facility Test Position

TFTP Tetrafluorothiophenol

TFTP Trivial File Transfer Protocol

TFTP2 2,3,4,5-Tetrafluorothiophenol

TFTP4 2,3,5,6-Tetrafluorothiophenol

TFTR Tokamak Fusion Test Reactor [of Princeton University, New Jersey, US]

TFTR Toroidal Fusion Test Reactor

TFTS Target Facing Type Sputtering [Materials Science]

TFU Test Facility Utilization

TFU Timing and Frequency Unit

TFV Triple-Force Vector (Process)

TFVA Training Film and Video Association [UK]

TFX Tactical Fighter Experimental

TFX Toxic Effects (of Hazardous Materials)

TFX Transverse Flux

TFXRD Thin-Film X-Ray Diffraction

TFZ Transfer Zone

TG Ground Systems [NASA Kennedy Space Center Directorate, Florida, US]

TG Tacho-Generator

TG Taper Gauge

TG Telegraph(y)

TG Temperature Gradient

TG Tempered Graphite [Metallurgy]

TG Terminal Guidance

TG Terminator Group

TG Thermal Gradient

TG Thermal Gravimeter

TG Thermogravimeter; Thermogravimetric; Thermogravimetry

TG Thickness Gauge

TG Third Generation (Computer) [also 3G]

TG Thyroid Gland [Anatomy]

TG Togo [ISO Code]

TG Tone Generator

TG Top Grille

TG Torpedo Group

TG Total Graph

TG Touch and Go (Landing)

TG Town Gas

TG Transgranular [Metallurgy]

TG Triethyleneglycol

TG Triglyceride

TG Trunk Group

TG Tuned Grid

TG Turbo-Generator

TG Twin Grain [Metallurgy]

TG- Guatemala [Civil Aircraft Marking]

T&G Tongue(d) and Groove(d)

3G Third Generation (Computer) [also TG]

Tg Target [also tg]

Tg Teragram [Unit]

Tg Tugrik [Currency of Mongolia]

tg treppe + geländer [German Journal on Stairs and Handrails]

tg trans-gauche [also t-g] [Chemistry]

.tg Togo [Country Code/Domain Name]

2g twice normal gravity (or twice normal weight) [Symbol]

TGA Gesellschaft Technische Gebäudeausrüstung [Building Equipment Society; of Verein Deutscher Ingenieure, Germany]

TGA Internationale Fachmesse Technische Gebäudeausrüstung [International Trade Fair for Building Equipment, Leipzig, Germany]

TGA The Glutamate Association [US]

TGA Therapeutic Goods Administration

TGA Thermal Gravimetric Analysis; Thermal Gravimetric Analysis; Thermogravimetric Analysis; Thermogravimetric Analyzer

TGA Thioglycolic Acid

TGA Trace Gas Analysis; Trace Gas Analyzer

TGA Triglycollamic Acid

TGa Thioglycolic Acid

TGAC Technical Grade Active Constituent

TGA/FTIR Thermogravimetric Analysis/Fourier-Transform Infrared (Spectroscopy [also TGA-FTIR]

TGA-IR Thermogravimetric Analysis–Infrared [also TGA/IR]

TGAP Triglycidyl P-Aminophenol

TGA/DTA Thermogravimetric Analysis/Differential Thermal Analysis [also TGA-DTA]

TGA/EGA Thermogravimetric Analysis/Evolved Gas Analysis [TGA-EGA]

TGA/FTIR Thermogravimetric Analysis/Fourier Transform Infrared (Spectroscopy) [also TGA-FTIR, or TGA/FT-IR]

TGB Thermogravity Balance

TGB Tongued, Grooved and Beaded

TGB Torpedo Gun Boat

TGB Twist Grain Boundary [Materials Science]

TGC Teleglobe Canada

TGC Third Generation Computer [also 3GC]

TGC Titanium Germanium Carbide

TGC Tomato Genetics Cooperative [US]

TGC Transmitter Gain Control [also tgc]

TGC Traffic Guidance Computer

TGC Travel Group Charter

3GC Third Generation Computer [also TGC]

TGCA Transportable Ground Control Approach

TGCS Third Generation Computer System [also 3GCS]

TGD Temperature Gas Diverter [also T-GD]

TG-DTA Thermogravimeter–Differential Thermal Analyzer; Thermogravimetric Differential Thermal Analysis; Thermogravimetry–Differential Thermal Analysis [also TG/DTA]

TG-DTA-DTG Thermogravimeter–Differential Thermal Analyzer–Differential Thermogravimeter; Thermogravimetric and Differential Thermogravimetric Differential Thermal Analysis; Thermogravimetry–Differential Thermal Analysis–Differential Thermogravimetry [also TG/DTA/DTG]

TG-DTA-MS Thermogravimeter–Differential Thermal Analyzer–Mass Spectrometer; Thermogravimetry–Differential Thermal Analysis–Mass Spectrometry [also TG/DTA/MS]

TGE Triglycidyl Ether

TGETPM Triglycidyl Ether of Triphenyl Methane

TGF Through Group Filter

TGF Timed Gel Formation

TGF Transforming Growth Factor [Biochemistry]

TGF Transgranular Fracture [Metallurgy]

TGF-α Transforming Growth Factor-α [Biochemistry]

TGF-β Transforming Growth Factor-β [Biochemistry]

TGF-β1 Transforming Growth Factor-β1 [Biochemistry]

TGF-β2 Transforming Growth Factor-β2 [Biochemistry]

TGF-β3 Transforming Growth Factor-β3 [Biochemistry]

TGF-β5 Transforming Growth Factor-β5 [Biochemistry]

TGF-βRI Transforming Growth Factor-β Receptor I [Biochemistry]

TGFA Triglyceride Fatty Acid

TGFB Triglycine Fluo(ro)beryllate

TGFE Total Gibbs Free Energy [Thermodynamics]

TGFO Trihexyl Phosphine Oxide

TGG Third Generation Gyroscope

tg±g∓ trans-gauche plus, or minus, gauche minus, or plus [Chemistry]

TGH Toronto General Hospital [Canada]

TGI Target Group Index [US]

TGI Target Intensifier

TGIC Triglycidyl Isocyanurate

TGID Transmission Group Identifier

TGID Trunk Group Identification

TGL Thermal Gradient Lamp

TGL Touch and Go Landing

3GL Third Generation Language

Tgl Toggle [also tgl]

TGM Toroidal Gate Monochromator

TGM Trunk Group Multiplexer

Tgm Telegram [also tgm]

TGMDA Tetraglycidyl-4,4'-Methylene Dianiline

TGMDA/DDS Tetraglycidyl Methylene Dianiline modified with Diaminodiphenylsulfone

TG/MS Thermal Gravimetric/Mass Spectrometer; Thermal Gravimetric/Mass Spectrometry; Thermogravimeter/Mass Spectrometer; Thermogravimetry/Mass Spectrometry

TGMV Technical Group for Machine Vision [now Machine Vision Association, US]

TGN Trunk Group Number

TGO Time to Go

TGO Toxic Gas Ordinance

TGOWG Teleoperator Ground Operations Working Group

TGPAP Triglycidyl-p-Aminophenol

TGS Tactical Ground Support

TGS Taguchi Gas Sensor

TGS Taxying Guidance System

TGS Telemetry Ground Station

TGS Telemetry Ground System

TGS Texas Geographic Society [US]

TGS Translator Generator System

TGS Triglycerine Sulfate

TGS Triglycine Sulfate

TGSSC Transgranular Stress-Corrosion Cracking [Metallurgy]

TGSE Tactical Ground Support Equipment

TGSE Telemetry Ground Support Equipment

TGSe Triglycine Selenate

TGSO Tertiary Groups Shunt Operation

TGSS Triglycine Sulfate Selenate

TGSSC Transgranular Stress-Corrosion Cracking

Tg S/yr Teragram(s) of Sulfur per Year [Unit of Measurement for Global Sulfur Emission]

TGT TDRSS (Tracking and Data Relay Satellite System) Ground Terminal

TGT Thromboplastin Generation Test [Medicine]

TGT Tree in Graph Theory

Tgt Target [also tgt]

t-g-t trans-gauche-trans [Chemistry]

TGTP Tuned-Grid Tuned-Plate (Circuit)

TGU Tezukayama Gakuin University [Japan]

TGU Tokyo Gakugei University [Japan]

TGU Turbine-Generator Unit

TGV Train à Grande Vitesse [French High-Speed Train Operating between Paris and Lyon]

TGW Teatum-Gschneidner-Waber (Atomic Radius) [Physical Chemistry]

TGW Thyssen Getriebe- und Kupplungswerke GmbH [German Transmission and Clutch Manufacturer located at Cassel]

TGWU Transport and General Workers Union [UK]

Tg/yr Teragram(s) per year [Unit]

TG-ZC Thermal Grooving–Zero Creep (Theory) [Mechanics]

TGZM Temperature-Gradient Zone Melting

TH *Technische Hochschule* – German for "Institute of Technology," or "University of Technology"

TH Texture Hardening [Metallurgy]

TH Thailand [ISO Code]

TH Thermal Head

TH Thermal Heating

TH Third Harmonic [also 3H]

TH Through-Hole (Mount)

TH Thionine [Biochemistry]

TH Thyristor

TH Tool Holder

TH Total Hardness (of Water)

TH Total Head [Water Turbines]

TH Transformation Hardened; Transformation Hardening [Metallurgy]

TH (Data) Transmission Header

TH Triode-Hexode

TH True Heading

TH Turbulent Heating

TH Tyrosine Hydroxylase [Biochemistry]

T*(H) Irreversible Temperature (of Superconductors) [Symbol]

T$_c$(H) Magnetic Field Dependent Critical Transition Temperature [Symbol]

T&H Toloui and Hellawell (Solidification) [Metallurgy]

3H Third Harmonic [also TH]

3-H Hot, Heavy and Hazardous (Work Environment)

Th Thorium [Symbol]

Th (−)-Threose [Biochemistry]

Th Thickness

Th Thursday

Th-223 Thorium-223 [also ^{223}Th, or Th^{223}]

Th-224 Thorium-224 [also ^{224}Th, or Th^{224}]

Th-225 Thorium-225 [also ^{225}Th, or Th^{225}]

Th-226 Thorium-226 [also ^{226}Th, or Th^{226}]

Th-227 Thorium-227 [also ^{227}Th, Th^{227}, or RdAc]

Th-228 Thorium-228 [also ^{228}Th, Th^{228}, or RdTh]

Th-229 Thorium-229 [also ^{229}Th, or Th^{229}]

Th-230 Thorium-230 [also ^{230}Th, Th^{230}, or Io]

Th-231 Thorium-231 [also ^{231}Th, Th^{231}, or UY]

Th-232 Thorium-232 [also ^{232}Th, Th^{232}, or Th]

Th-233 Thorium-233 [also ^{233}Th, or Th^{233}]

Th-234 Thorium-234 [also ^{234}Th, Th^{234}, UX$_1$ or UX$_l$]

Th-235 Thorium-235 [also ^{235}Th, or Th^{235}]

th thermie [Unit]

th theta [English Equivalent]

th thousand

.th Thailand [Country Code/Domain Name]

t/h ton(s) per hour [also t/hr, or t h^{-1}]

THA Tetrahydroaminacrine

THA Tetrahydroaminoacridine

THA Tetrahydro Kendall's Compound A

THA Total Hip Athroplasty [Medicine]

THA Triheptyl Amine

ThA Thorium A [also Po-216, ^{216}Po, or Po^{216}]

t/ha ton(s) per hectare [Unit]

THAB Tetrahexyl Ammonium Benzoate

THABr Tetrahexyl Ammonium Bromide

Th(ACAC)$_4$ Thorium Acetylacetonate [also Th(AcAc)$_4$, Th(acac)$_4$]

THACl Tetrahexyl Ammonium Chloride

THAHS Tetrahexyl Ammonium Hydrogen Sulfate

Thai Thailand [also Thail]

THAM Trimethylol Aminomethane

THAM Tris(hydroxymethyl)aminomethane

THAN Tetraheptyl Ammonium Nitrate

THB Temperature-Humidity Bias [Meteorology]

THB Thai Baht [Currency of Thailand]

ThB Thorium B [also Pb-212, ^{212}Pb, or Pb^{212}]

tHb Total Hemoglobin [Biochemistry]

THBP Trihydroxy Butyrophenone

THBP 2,4,5-Trihydroxybutyrophenone

THC Tentative Human Consensus

THC Δ-9-Tetrahydrocannabinol [Biochemistry]

THC Thermal Converter

THC Thrust Hand Controller

THC Toronto Harbor Commission [Canada]

THC Total Hydrocarbon

THC Thiosemicarbazone

THC Translation(al) Hand Controller

ThC Thorium C [also Bi-212, ^{212}Bi, or Bi^{212}]

ThC′ Thorium C′ [also ThC$_1$, Po-212, ^{212}Po, or Po^{212}]

ThC″ Thorium C″ [also ThC$_2$, Tl-208, ^{208}Tl, or Tl^{208}]

THCA Tetrahydrocannabinolcarbonic Acid

THCE Total Hydrocarbon Emission(s)

TH Chemnitz Technische Hochschule Chemnitz [Chemitz University of Technology, Germany]

THCS (Outlet) Temperature, Hot-Channel Sodium [Nuclear Reactors]

THD Technische Hochschule Darmstadt [Darmstadt University of Technology, Germany]

THD Technische Hogeschool Delft [Delft University of Technology, Netherlands]

THD 2,2,6,6-Tetramethyl-3,5-Heptanedione [also thd]

THD 1,1,6,6-Tetraphenyl Hexadiynediamine (Polydiacetylene)

THD Texas Highway Department [US]

THD Total Harmonic Distortion

THD Total Head [Water Turbines]

.THD Thread [File Name Extension] [also .thd]

2HD Double High Density (Disk) [also DHD]

ThD Thorium D [also Pb-208, ^{208}Pb, or Pb^{208}]

Thd Thread [also thd]

Thd Ribosyl-Thymine [Biochemistry]

Thd 1,1,1-Trifluoro-2,4-Hexanedionate [also thd]

THDA Toluene Hydrodealkylation (Process)

TH Darmstadt Technische Hochschule Darmstadt [Darmstadt University of Technology, Germany]

THDM Translucent Humic Degradation Matter [Geology]

Thdr Thunder [also thdr]

THDS Thermal Helium Desorption Spectrometry

THDS Time Homogeneous Data Set

THE Tetrahydrocortisone [Biochemistry]

THE Thunderstorm Event [Meteorology]

THE 3α,17α,21-Trihydroxypregnane-11,20-Dione

THECB Texas Higher Education Coordinating Board [US]

THECB ATP Texas Higher Education Coordinating Board Advanced Technology Program [US]

THEED Tetrahydroxyethyl Ethylene Diamine

THEED Transmission High-Energy (Scanning) Electron Diffraction

THEIC Tris(hydroxyethyl)isocyanurate

Th-Em Thorium Emanation (or Thoron)

THEN Tetrahydroxyethyl Ethylenediamine Tetranitrate

theo theoretical

Theor Theoretical; Theory [also theor]

Theor Theorem [also theor]

Theor Appl Climatol Theoretical and Applied Climatology [Journal published in Austria]

Theor Appl Fract Mech Theoretical and Applied Fracture Mechanics [Published in the Netherlands]

Theor Appl Genetics Theoretical and Applied Genetics [Published in Germany and the US]

Theor Appl Mech Theoretical and Applied Mechanics [Published in the US]

Theor Chem Theoretical Chemist(ry)

Theor Chim Acta Theoretica Chimica Acta [German Publication]

Theor Comput Fluid Dyn Theoretical and Computational Fluid Dynamics [Journal published in Germany]

Theor Comput Sci Theoretical Computer Science [Published in the Netherlands]

Theor Exp Chem Theoretical and Experimental Chemistry [Translation of: *Teoreticheskaya i Eksperimental'naya Khimiya (USSR)*; published in the US]

Theor Found Chem Eng Theoretical Foundation of Chemical Engineering [Journal published in the US]

Theor Math Phys Theoretical and Mathematical Physics [Translation of: *Teoreticheskaya i Matematicheskaya Fizika (USSR)*; published in the US]

Theor Mech Theoretical Mechanics

Theor Metrol Theoretical Metrologist; Theoretical Metrology

Theor Phys Theoretical Physicist; Theoretical Physics

Theor Popul Biol Thereoretical Population Biology [Journal published in the US]

Theory Probab Appl Theory of Probability and Its Applications [Translation of: *Teoriya Veroyatnostei i ee Primeneniya (USSR)*; published by the Society for Industrial and Applied Mathematics, US]

THEP Theoretical High-Energy Physics

THERDAS Thermochemical Databank System [of Institute for Theoretical Metallurgy at RWTH Aachen, Germany]

Therm Thermistor [also therm]

Therm Thermometer [also therm]

therm thermal

Thermal EMF Thermal Electromotive Force [also thermal emf]

Therm Biol Thermal Biology

Therm Cap Thermal Capacity

Thermec International Conference on Thermomechanical Processing of Steels and Other Materials

Therm Eng Thermal Engineering [Translation of: *Teploenergetika (USSR)*; published in the UK]

THERMINIC Thermal Investigations of Integrated Circuits and Microstructures (Meeting) [Europe]

Thermistor Thermally Sensitive Resistor [also thermistor]

Thermo Thermostat [also thermo]

Thermochem Thermochemical; Thermochemist(ry) [also thermochem]

Thermochim Acta Thermochimica Acta [Published in the Netherlands]

Thermodyn Thermodynamic(s) [also Thermod, thermod, or thermodyn]

THERMOPROCESS International Exhibition for Industrial Furnaces and Thermic Production Processes, Dusseldorf, Germany] [also thermoprocess]

Thermophys Thermophysics

Thermosp Thermosphere [Meteorology]

Therm Prop Thermal Properties

THERP Technique for Human Error Rate Prediction

THERSYST Databank System for Thermophysical Properties

THF Tetrahydrocompound F

THF Tetrahydrofolic Acid

THF Tetrahydrofuran

THF $3\alpha,11\beta,17\alpha,21$-Tetrahydroxypregnane-20-One

THF Tremendously High Frequency

THFA Tetrahydrofolic Acid

THFA Tetrahydrofurfuryl Alcohol

THFC Tris(heptafluoropropylhydroxymethylene)camphorato

THFC-Eu Tris[3-(Heptafluoropropylhydroxymethylene)-d-Camphorato], Europium

THF-DMF Tetrahydrofuran Dimethyl Formamide

THFTDA Tetrahydrofuran Tetracarboxylic Dianhydride

THG Telecommunications Heritage Group [UK]

3HG Third-Harmonic Generation [also THG]

THHF Tetrahydrohomofolate

Th(HFAC)$_4$ Thorium Hexafluoroacetylacetonate [also Th(hfac)$_4$]

THHP Tetrahydrohomopteroic Acid

THI 1-Methyl-3,4,5-Trihydroxy 1H-Indole

THI Technische Hochschule Ilmenau [Ilmenau Institute of Technology, Germany]

THI Temperature-Humidity Index [Meteorology]

THI Total Height Index

Thia Thiadiazole

THIB α,β,β'-Trihydroxyisobutyrate

TH Ilmenau Technische Hochschule Ilmenau [Ilmenau University of Technology, Germany]

ThinDIP Thin-Packaged Dual In-Line Package [Electronics]

Thin-Walled Struct Thin-Walled Structures [Journal published in the UK]

Thiotepa Triethylenethiophosphoramide [also thiotepa]

4-Thio UMP 4-Thiouridine 5'-Monophosphate [Biochemistry]

THIOX Monothiooxalate

THIP 4,5,6,7-Tetrahydroisoxazolo[5,4-d]-2H-3-Pyrimidone

THIR Temperature Humidity Infrared Radiometer [of Nimbus-7–NASA Meteorological Research Satellite]

Thiram bis-(Dimethylthiocarbamyl)disulfide [also thiram]

33 Met Prod Thirty-Three (33) Metal Producing [Journal published in the US]

33 Met Prod Nonferrous Ed Thirty-Three (33) Metal Producing Nonferrous Edition [Journal published in the US]

thk thick

TH Karlsruhe Technische Hochschule Karlsruhe [Karlsruhe University of Technology, Germany]

THL True Heavy Liquid

THL Trans-Hybrid Loss [Telecommunications]

THL Tungsten-Halogen Lamp

Thld Threshold [also thld]

THM Tetradecylhydrogenmaleate (Crystal)

THM Tons of Hot Metal [Metallurgy]

THM Trihalo(geno)methane

thm therm

THMA Trailer Hitch Manufacturers Association [US]

TH Magdeburg Technische Hochschule Magdeburg [Magdeburg University of Technology, Germany]

TH Merseburg Technische Hochschule Merseburg [Merseburg University of Technology, Germany]

Th-Mn Thorium-Manganese (Alloy System)

THN 1,2,3,4-Tetrahydronaphthalene

thn thin

Th-Ni Thorium-Nickel (Alloy System)

Tho D-Ribosylthymine [Biochemistry]

THOD Terre Haute Ordnance Depot [Indiana, US]

ThOD Theoretical Oxygen Demand

Thold Threshold

THOMAS The House (of Representatives) Open Multimedia Access System [US]

THOMIS Total Hospital Operating and Medical Information System

Thomson-CSF Thomson-Compagnie Générale de Télégraphie sans Fil [Major French Telecommunications Company]

Thomson-CSF/LCR Thomson-Compagnie Générale de Télégraphie sans Fil/Laboratoire Central de Recherche [Central Research Laboratory of Thomson-Compagnie Générale de Télégraphie sans Fil, Orsay, France]

THOPS Tape Handling Operational System

THOR Tandy High-Intensity Optical Recording; Tandy High-Performance Optical Recording

THOR Tape-Handling Option Routine

THOR Transistorized High-Speed Operations Recorder

THOREX Thorium Extraction (Process) [Nuclear Engineering]

THORP Thermal-Oxide Reprocessing Plant

THORS Thermal-Hydraulic Out-of-Reactor Safety Facility

thous Thousand

THP Terminal Handling Processor

THP Tetrathydropapaveroline

THP Tetrahydropyran(yl)

THP Theoretical Horsepower [also thp]

THP Thermohydrogen Processing

THP Thrust Horsepower

THP Trihexyl Phosphate

THP Through-the-Hole Plating

Thp Tetrahydropyranyl

THPA Tetrahydrophthalic Anhydride; cis-1,2,3,6-Tetrahydrophthalic Anhydride

THPA Triheptyl Amine

THP(B) (s)-2-Methyl-1,4,5,6-Tetrahydropyrimidine-4-Carboxylic Acid

THPC Tetrakis(hydroxymethyl)phosphonium Chloride

THPE 1,1,1-Tris(4-Hydroxyphenyl)ethane

THPED N,N'-Tetrakis(2-Hydroxypropyl)-1,2-Ethylenediamine

THPFB Treated Hard Pressed Fiberboard

THPO Trihexylphosphine Oxide

THPS Tetrakis(hydroxymethyl)phosphonium Sulfide

THPy 2-(2-Thienyl) Pyridine [also thpy]

THQ Telecommunications Headquarters

THQ 2,3,5,6-Tetrahydroxy-1,4-Benzoquinone

THQ Thermionic Quadrupole

THR Total Hip Replacement [Medicine]

THR Transmit(ter) Holding Register

Thr (−)-Threonine; Threonyl [Biochemistry]

Thr Threshold [also thr]

t/hr (imperial) tons per hour [also t/h] [Unit]

THRA Tasmanian Historical Research Association

thrftr thereafter

thrm thermal

ThroB/L Through Bill of Lading [also Thro B/L]

Thromb Haem Thrombosis and Haemostasis (Journal)

Thromb Res Thrombosis Research (Journal)

Throt Throttle [also throt]

T-HR-T Temperature versus Heating-Rate Transformation (Diagram)

thru through

thrut throughout

THS Temkin-Hillert and Staffanson (Treatment) [Materials Science]

THS Tennessee Historical Society [US]

THS Tetrabutylammonium Hydrogen Sulfate

THS The Hovercraft Society [UK]

THS The Hydrographic Society [UK]

THS Titanic Historical Society [US]

.THS Thesaurus [File Name Extension]

thsd thousand

THSP Thermal Spraying [Metallurgy]

THT Technische Hogeschool Twente [Twente University of Technology, Netherlands]

THT Tetrahydrothiophene

THT Thermohydrogen Treatment

THTF Thermal Hydraulic Test Facility

THTFAC 2-Thenoyltrifluoroacetone

THTMP 2,7,12,17-Tetrahexyl-3,8,13,18-Tetramethyl Porphine

THTR Thorium High-Temperature Reactor

THTRA Thorium High-Temperature Reactor Association [US]

THU Tsing-Hua University [Hsin Chu, Taiwan]

TH/U (Technische Hochschule/Universität) – German for "Institute of Technology/Technical University"

Th-U-D Thorium-Uranium-Deuterium (Oxides)

Thur Thursday [also Thurs]

THW (Technisches Hilfswerk) – German Technical Relief Organization

THWM Trinity High-Water Mark

ThX Thorium X [also Ra-224, ^{224}Ra, or Ra^{224}]

THXA Trihexylamine

THY Thymoma [Medicine]

Thy Thyroid [Anatomy]

Th:YAG Thorium-Doped Yttrium Aluminum Garnet (Laser)

THYMOTRO Thyratron Motor Control

THYNET Thyssen Network [of Thyssen Stahl AG, Germany] [also Thynet]

Thyssen Edelst Tech Ber Thyssen Edelstahl Technische Berichte [Technical Reports of Thyssen Edelstahl, Germany]

Thyssenforsch Thyssenforschung [Research Reports published by Thyssen AG, Germany]

Thyssen Tech Ber Thyssen Technische Berichte [Technical Reports published by Thyssen AG, Germany]

Thyssen Tech Inf Thyssen Technische Information [Technical Information Series published by Thyssen AG, Germany]

Thyssen Tech Inf Offshore-Tech Thyssen Technische Information, Offshore-Technik [Technical Information Series on Offshore Engineering published by Thyssen AG, Germany]

THz Terahertz [Unit]

Thz Thiozolidine [also thz]

Thzl Thiazol

TI Information Systems [NASA Kennedy Space Center Directorate, Florida, US]

TI Tactical Information

TI Tape Inverter

TI Tamarind Institute [US]

TI Target Identification

TI Technical Information

TI Technical Integration

TI Technical Interchange

TI Technical Investigation

TI Technological Institute

TI Telechelic Ionomer [Organic Chemistry]

TI Temperature Index

TI Temperature Indicator

TI Temperature Inversion [Meteorology]

TI Terminal Interface

TI Terrestrial Interference

TI Test Instruction

TI Texas Instruments, Inc. [Dallas, US]

TI Textile Institute [UK]

TI Thermal Image

TI Thermal Imagery; Thermal Imaging [Electronics]

TI Thermally Induced; Thermal Induction

TI Thermal Inspection

TI Thermal Insulation

TI Thermal Ionization

TI Thousand Islands [St. Lawrence River, US/Canada]

TI Threaded Insert

TI Thread Institute [US]

TI Time Independence; Time Independent

TI Time Index

TI Time Integral

TI Time Integrated; Time Integration

TI Time Interpolation

TI Time Interval

TI Tohoku Institute [of Japan Fisheries Agency]

TI Topological Information

TI Track Identifier

TI Track Initiator [Aeronautics]

TI Training Integration; Training Integrator

TI Transfer Impedance [Electrical Engineering]

TI Transfer Ionization

TI Transfrigoroute International [Switzerland]

TI Transportation Institute [US]

TI Trypsin Inhibitor [Biochemistry]

TI Tunnel-Thin Insulator

TI Turbine Inlet

TI Turing Instability [Linear Analysis]

TI Turn Indicator

TI- Costa Rica [Civil Aircraft Marking]

Ti Titanium [Symbol]

Ti Tumor-Induced (Plasmid) [Genetics]

Ti²⁺ Divalent Titanium Ion [also Ti⁺⁺] [Symbol]

Ti³⁺ Trivalent Titanium Ion [also Ti⁺⁺⁺] [Symbol]

Ti⁴⁺ Tetravalent Titanium Ion [Symbol]

Ti II Titanium II [Symbol]

Ti-43 Titanium-43 [also ⁴³Ti, or Ti⁴³]

Ti-45 Titanium-45 [also ⁴⁵Ti, or Ti⁴⁵]

Ti-46 Titanium-46 [also ⁴⁶Ti, or Ti⁴⁶]

Ti-47 Titanium-47 [also ⁴⁷Ti, or Ti⁴⁷]

Ti-48 Titanium-48 [also ⁴⁸Ti, or Ti⁴⁸]

Ti-49 Titanium-49 [also ⁴⁹Ti, or Ti⁴⁹]

Ti-50 Titanium-50 [also ⁵⁰Ti, or Ti⁵⁰]

Ti-51 Titanium-51 [also ⁵¹Ti, or Ti⁵¹]

tI Pearson symbol for body-centered space lattice in tetragonal crystal system (this symbol is followed by the number of atoms per unit cell, e.g. tI6, tI16, etc.)

ti ionic transport number (in physical chemistry) [Symbol]

TIA Taxation Institute of Australia

TIA Telecommunications Industry Association [US]

TIA Territorial and International Affairs

TIA Thanks In Advance [Internet]

TIA The Internet Adapter

TIA Time-Interval Analysis; Time Interval Analyzer

TIA Transient Ischemic Attack [Medicine]

TIA Typographers International Association [US]

TIAA Timber Importers Association of America [US]

TIAC Thermal Insulation Association of Canada

Ti(ACAC)₂ Titanylacetylacetonate (or Titanium Acetylacetonate) [also Ti(AcAc)₂, or Ti(acac)₂]

TIAG Telecommunications Industry Advisory Group

TIAL Titanium-Aluminum (Composite) [also Tial, or tial]

TiAl Titanium Aluminide (or Titanium Monoaluminide)

Ti-Al Titanium-Aluminum (Alloy System)

TiAlCr Titanium Aluminum Chromium (Alloy)

TiAlN Titanium Aluminum Nitride (Compound)

Ti/Al₂O₃ Alumina-Reinforced Titanium (Composite)

Ti:Al$_2$O$_3$ Titanium-Doped Alumina (or Aluminum Oxide) (Laser)

TIALON Titanium Aluminum Oxynitride (Ceramic) [also TiAlON, or Tialon]

Ti(AmO)$_4$ Titanium Amyl Oxide

TIARA Target Illumination and Recovery Aid

TIAS Target Identification and Acquisition System

TIAS True Indicated Air Speed

TIB Target Identification Bomb

TIB Technical Information Bureau [UK]

TIB Technische Informationsbibliothek (und Universitätsbibliothek Hannover) [Technical Library and Document Delivery Center, University Library of Hanover, Germany]

TIB Through Ice Bathymetry

TIB Transferred Interaction Block

TIB Triisopropylbenzene

TiB Titanium-Boron (Chemical Vapor Deposition Process)

TiB Titanium Monoboride

TIBA Triiodobenzoic Acid

TIBA 2,3,5-Triiodobenzoic Acid

TIBA Triisobutyl Aluminum

TIBA Traffic Information Broadcast by Aircraft

TIBA Turkish Industrialists and Businessmen's Association

TIBAL Triisobutyl Aluminum

TIBBS Trace Integrated Bare Board System

TIBC Total Iron-Binding Capacity [Biochemistry]

TiB CVD Titanium-Boron Chemical Vapor Deposition (Process)

Ti-Be-Cr Titanium-Beryllium-Chromium (Alloy System)

TIBER Tokamak Ignition/Burn Experimental Reactor [US]

TIBER Tokamak Ignition/Burn Experimental Research [Plasma Physics]

TIBER/ITER Tokamak Ignition/Burn Experimental Reactor/International Thermonuclear Experimental Reactor [US]

TIBEX Tidewater Bermuda Experiment [Oceanography]

TIBI Training in Business and Industry (Program) [Canada]

Ti(bipy)$_3$ Titanium Tris(2,2'-Bipyridine)

TIBOE Transmitting Information by Optical Electronics

TIBP Triisobutylphosphate [also TiBP]

Ti-BR Butadiene Rubber Based on Titanium Catalyst

TIBS Through Ice Bathymetry System

TIBS Towed In-Flight Bathymetry System

TiB$_2$/TiSi$_2$ Titanium Diboride/Titanium Disilicide (Bilayer)

TiB$_2$/TiSi$_2$ p-Si Titanium Diboride/Titanium Disilicide P-Type Silicon (Schottky Diode)

Ti(BuO)$_4$ Titanium(IV) n-Butoxide (or Tetrabutyl Titanate)

TIC Tantalum-Niobium International Study Center [US]

TIC TAB (Tape-Automated Bond) in Cap

TIC Tape Intersystem Connection

TIC Target Intercept Computer

TIC Task Interrupt Control

TIC Technical Information Center

TIC Technical Institute Council

TIC Telecommunications Information Center

TIC Telemetry Instruction Conference

TIC Temperature-Indicating Controller; Temperature Indication and Control

TIC Temperature of Initial Combustion

TIC Tentatively Identified Compound(s)

TIC Terminal Identification Code

TIC Thermo-Induced Chemistry

TIC Time-Integrating Correlator

TIC Total Inorganic Carbon

TIC Total Installed Cost

TIC Total Ion Chromatography

TIC Total Ion Current

TIC Total Item Change

TIC Transducer Information Center [of Battelle Memorial Institute, Columbus, Ohio, US]

TIC Transfer in Channel

TIC Transformations during Interrupted Coolings [Metallurgy]

TIC Turbulence Instrument Cluster

TiC Titanium Carbide

TICA Technical Information Center Administration

TICA Thermal Insulation Contractors Association [UK]

TiC/Al Titanium-Carbide (Particles) in Aluminum (Matrix)

TICAS Taxonometric Intra-Cellular Analytic System

TICCI Technical Information Center for Chemical Industry [India]

TICCIT Time-Shared Interactive Computer-Controlled Information(al) Television

TiC-Co Cobalt-Bonded Titanium Carbide

TICE Time Integral Cost Effectiveness

TICE Training Information and Communication Enhancemement

Ti-C:H Titanium-Bearing Hydrogenated Carbon

Tickle Titanium Tetrachloride [TiCl$_4$]

TICM Test Interface and Control Module

TiCN Titanium Carbonitride

Ti-Co Titanium-Cobalt (Alloy System)

Ticoss Time-Compressed Single-Sideband System [also ticoss]

TICP Total Ion Current Profile

TiCpCl$_3$ Cyclopentadienyltitanium Trichloride

Ti(Cp)$_2$Cl$_2$ Bis(cyclopentadienyl)titanium Dichloride (or Titanocene Dichloride)

Ti-Cr Titanium-Chromium (Alloy System)

TICS Teacher Interactive Computer System

TICS Telecommunication Information Control System

TICS Tetraisocyanatesilane

TICS Tidal Current System

TICS Time-Reversal Invariant Closed Shell

TICS Total Integral Cross-Section

TICT Twisted Intramolecular Charge Transfer

TICTAC Time Compression Tactical Communications

TiC-TaC-WC Titanium Carbide–Tantalum Carbide–Tungsten Carbide

TIC/TOC Total Inorganic Carbon/Total Organic Carbon

TIC-TOC Telecommunications Information Center/Technical Office for Consumers [UK]

TiCu Titanium Copper (Compound)

Ti-Cu Titanium-Copper (Alloy System)

TID Target Identification

TID Technical Information Division

TID Test Instrument Division

TID Thermal Ionization Detector

TID Thermionic Detector

TID Time Independent Data

TID Total Ion Detector

TID Total Ionizing Dose

TID Transient Ion Drift

TID Traveling Ionospheric Disturbance

TID 3-(Trifluoromethyl)-3-(m-Iodophenyl)diazirine

tid *(ter in die)* – three times a day [Medical Prescriptions] [also TID]

TIDA Travel and Industrial Development Association [UK]

tidac *(ter in die ante cibum)* – three times daily before meals [Medical Prescriptions]

TIDAR Time Delay Array Radar

TIDB Tester-Independent Database

TIDDAC Time in Deadband Digital Attitude Control

TIDE Thermal Ion Dynamics Experiment

TIDE Transponder Interrogation and Decoding Equipment

TIDES Time Division Electronic Switching System

TIDF Temperature-Independent Delayed Fluorescence

TIDF Trunk Intermediate Distribution Frame

TIDI Time-Division (Sound)

TIDMA Tape Interface Direct Memory Access

Ti(DPM)$_2$(i-OPr)$_2$ Titanium(II) Bis(diphenylphosphino)-methane Isopropoxide

Tidskr Dok Tidskrift for Dokumentation [Swedish Publication on Technical Literature and Documentation]

TIDU Technical Information and Documentation Unit [of Department of Scientific and Industrial Research, UK]

TIDU Technical Intelligence Documents Unit [US]

TIDY Track Identity

TIE Technical Information Exchange

TIE Technical Integration and Evaluation

TIE Terminal Interface Equipment

TIE Tetraiodoethylene

TIE Time Interval Error

TIE Total Ion Electropherogram

TIE Toxicity Identification Evaluation

TIEM The Innovation and Entrepreneurial Management

TIES Technical Investigations and Engineering Services [Canada]

TIES Technical Investigations and Engineering Studies

TIES Technical Investigations and Engineering Support [of Department of National Defense, Canada]

TIES Telecom Information Exchange Services [of International Telecommunication Union]

TIES Time-Independent Escape Sequence

TIES Total Integrated Engineering System

TIES Translators and Interpreters Educational Society

TIES Transmission and Information Exchange System

Ti(EtO)$_4$ Titanium(IV) Ethoxide (or Tetraethyl Titanate)

TIF Tag(ged) Image File

TIF Tape Inventory File

TIF Task Initiation Form

TIF Technical Integration Forum

TIF Telephone Interference (or Influence) Factor

TIF Temperature-Independent Factor

TIF Terminal Independent Format

TIF Text Interchange Format

TIF True Involute Form

.TIF Tagged Image File [File Name Extension] [also .tif]

TiF Time Fuse

TIFAST Technical Integration for Army Simulation and Training [US]

Ti-Fe Titanium-Iron (Alloy System)

TIFF Tag(ged) Image File Format

TIFO Text Intercharge Format

TIFP Tactical Information Fusion Prototype

TIFR Tata Institute of Fundamental Research [Bombay, India]

TIFS Total In-Flight Simulator

TiFs Time Fuse [also TiFz]

TIG Technical Information Group

TIG Telegram Identification Group

TIG Time of Ignition

TIG Tungsten Inert-Gas (Welding)

TIGA Texas Instruments Graphics Adapter

TIGA Transport Issues Group Australia

TIGAS Topsoe Integrated Gasoline Synthesis (Process) [Chemical Engineering]

TiGB Tilt Grain Boundary [Materials Science]

TIGER Telephone Information Gathering, Evaluation and Review

TIGER Topologically Integrated Geographic Encoding and Referencing (System) [US Census Database]

TIGER Total Information Gathering and Reporting

TIGS Terminal-Independent Graphic System

TIH Trunk Interface Handler

TII (European Association for the Transfer of) Technologies, Innovation and Industrial Information [Luxembourg]

TII Terminal IGES (Initial Graphics Exchange Specification) Interface

TII Tooling Inspection Instrumentation

TIIAL The International Institute of Applied Linguistics

TIIAP Telecommunications and Information Infrastructure Assistance Program [of US National Information Infrastructure]

TIIF Tactical Image Interpretation Facility

Ti(i-OPr)$_2$ Titanium(II) Isopropoxide

Ti(i-OPr)$_4$ Titanium(IV) Isopropoxide

TIIPS Technically Improved Interference Prediction System

TIIT Technion Israel Institute of Technology [Haifa, Israel]

Tijdschr Ned Elektron- & Radiogenoot Tijdschrift van het Nederlands Elektronica- en Radiogenootschap [Journal of the Netherlands Electronics and Radio Engineering Association]

TiKα Titanium K-Alpha (Radiation) [also TiK$_\alpha$]

TiKβ Titanium K-Beta (Radiation) [also TiK$_\beta$]

TIL Tumor-Infiltrating Lymphocytes [Medicine]

TIL Two Interdigitation Level

Ti/La-214 Titanium on Lanthanum Barium Copper Oxide Superconductor [Ti/La$_2$BaCu$_4$O$_5$]

Ti:LiNbO₃ Titanium-Doped Lithium Niobate

TILS Technical Information and Library Services [UK]

TILS Tumor-Infiltrating Lymphocytes [Medicine]

TIM Table Input to Memory

TIM Technical Information Manager

TIM Technical Information Memo(randum)

TIM Technical Interchange Meeting

TIM Teeth in Mesh [Gear Trains]

TIM Test Instrumented Missile

TIM 2,3,9,10-Tetramethyl-1,4,8,11-Tetraazacyclotetradecane-1,3,8,10-Tetraene

TIM Thermoset Injection Molding

TIM Time Interval Meter

TIM Time Interval Monitor

TIM Time Meter [Telecommunications]

TIM Toll Interface Module

TIM Total Information Management

TIM Total Ion Monitoring

TIM Transistor Information Microfile

TIM Transmitter Intermodulation

TIM Triiodomethane

TIMA Thermal Insulation Manufacturers Association [US]

Timber Bull Timber Bulletin [of Food and Agricultural Organization]

Timber Harv Timber Harvesting [Published in the US]

TIMCON Timber Packaging and Pallet Conference [UK]

TIME Test, Inspection, Measurement and Evaluation

TIME Total Industry Marketing Effort

TIME Transferred Ionized Molten Energy

TIMEC Technology Institute for Medical Devices Canada

Ti(MeO)₄ Titanium Methoxide (or Tetramethyl Titanate)

TIMET Titanium Metals Corporation [US] [also Timet]

TIMI Technology Independent Machine Interface

TIMIS Train Management Information System

TIMIX Texas Instruments Mini/Microcomputer Information Exchange [Name Changed to The International Microcomputer Information Exchange, US]

TIMIX The International Microcomputer Information Exchange [also TiMix, US]

TIMIXE Texas Instruments Mini/Microcomputer Information Exchange Europe [also TiMixE, Netherlands]

TIMM Thermionic Integrated Micromodules

Ti MMC Titanium Metal-Matrix Composite

Timminco Timmins Mining Company [Toronto, Canada]

Ti-Mn Titanium-Manganese (Alloy System)

Ti-Mo Titanium-Molybdenum (Alloy System)

TIMS Telephone Information Management System

TIMS Text Information Management System

TIMS The Institute of Management Sciences [of Operations Research Society of America, US]

TIMS Thermal Ionization Mass Spectrometry; Thermal Ionization Mass Spectroscopy

TIMS Transmission Impairment Measuring Set

TIMSA Thermal Insulation Manufacturers and Suppliers Association [UK]

TIM/TOM Table Input to Memory/Table Output from Memory

TIN Tax(payer) Identification Number [US]

TIN Temperature Independent

TIN Triangulated Irregular Network

TIN Transportation Information Network

TiN Titanium Nitride

TINA Telecommunication Information Networking Architecture

TINA Tris(isononyl) Amine

TINA Truth in Negotiations Act [US]

Ti-Nb-Zr Titanium-Niobium-Zirconium (Alloy System)

Ti-Nb-Al Titanium-Niobium-Aluminum (Alloy System)

Tinct Tincture [also Tr] [Medical Prescriptions]

TINET Transparent Intelligent Network

Ti(NEt₂)₄ Tetrakis(diethylamido)titanium

TINFO Tieteellisen Informoinnin Neuvosto [Council for Scientific Information and Research Libraries, Finland]

Ti/n-GaAs Titanium on N-Type Gallium Arsenide (Substrate)

TiNi Titanium Nickelide

Ti-Ni Titanium-Nickel (Alloy System)

Tin Int Tin International [Journal published in the UK]

TINP Telenet Intranetworking Protocol

TINS Thermal Imaging Navigation Set

TINS Thermal Inelastic Neutron Scattering

TiN/Si Titanium Nitride on Silicon (Substrate)

TINST Turbo Pacal Installation Program

TINSTAAFL There is no such Thing as a Free Lunch [Internet]

TINT Track in Track

TiN/Ti Titanium Nitride–Titanium (Bilayer)

Tin Uses Tin and Its Uses [of International Tin Research Council, UK]

TiN/VN Titanium Nitride/Vanadium Carbide (Coating)

TIO Target Insertion Orbit

TIO Technology Innovation Office [of US Environmental Protection Agency]

TIO Terminal Input/Output

TIO Test Input/Output

TIO Time Interval Optimization

TIO Time Inversion Operation

TiO Titanium Monoxide

TIOA Terminal Input/Output Area

TIOA Tri(isooctyl) Amine [also TiOA]

Ti(OAc)₄ Titanium(IV) Acetate

Ti(OBu)$_4$ Titanium(IV) n-Butoxide (or Tetrabutyl Titanate)

TIOC Terminal Input/Output Controller

TiO(DPM)$_2$ Titanium Oxide Bis(diphenylphosphino)-methane

Ti(OEt)$_4$ Titanium(IV) Ethoxide (or Tetraethyl Titanate)

TIOM Triisooctyl Trimellitate

Ti(OMe)$_4$ Titanium Methoxide (or Tetramethyl Titanate)

TiO$_2$-MnO$_2$ Titanium Dioxide-Manganese Dioxide (System)

TIOPO Tri(isooctyl)phosphine Oxide

Ti(OPr)$_4$ Titanium Propoxide (or Tetrapropyl Titanate)

Ti(O-i-Pr)$_2$ Titanium(II) Isopropoxide [also Ti(OPr^i)$_2$, or Ti(O^{i}Pr)$_2$]

Ti(O-i-Pr)$_4$ Titanium(IV) Isopropoxide [also Ti(OPr^i)$_4$, or Ti(O^{i}Pr)$_4$]

Ti(OR)$_2$(ACAC)$_2$ Titanium (II) Alkoxide Bis(acetylacetonate) [also Ti(OR)$_2$(acac)$_2$]

Ti(OR)$_3$ Titanium (III) Alkoxide [General Formula]

Ti(OR)$_4$ Titanium (IV) Alkoxide [General Formula]

TiO$_2$-SiO$_2$ Titanium Dioxide-Silicon Dioxide (System)

TIOT Task Input/Output Table

TiO(THD)$_2$ Titanium Oxide Bis(2,2,6,6-Tetramethyl-3,5-Heptanedionate) [also TiO(thd)$_2$]

TIOTM Tri(isooctyl) Trimellitate

TIOWQ Terminal Input/Output Wait Queue

TiOx Titanium Oxides (i.e., TiO, TiO$_2$, TiO$_3$ and Ti$_2$O$_3$) [also TiO$_x$]

TIP Tape Input

TIP Technical Information Panel [of NATO Advisory Group for Aerospace Research and Development]

TIP Technical Information Processing; Technical Information Processor

TIP Technical Information Program

TIP Technical Information Project

TIP Technology Inflow Program [of Department of Foreign Affairs and International Trade, Canada]

TIP Telefiche Image Processor

TIP Teletype Input Processing

TIP Temperature-Independent Paramagnetism

TIP Temperature-Indicating Paint

TIP Terminal Interface Package

TIP Terminal Interface Processor

TIP Terminal Interface Program

TIP Texture Independent Path

TIP Thermal Image Processing; Thermal Image Processor

TIP Thermally-Induced Porosity

TIP Thermodynamics of Irreversible Processes

TIP Thrust Inlet Pressure [Aerospace]

TIP Time-Independent Perturbation (Theory) [Physics]

TIP Titanium(IV) Isopropoxide

TIP Tokyo Institute of Polytechnics [Japan]

TIP Tool Interface Place

TIP Total Isomerization Process

TIP Transaction Interface Package

TIP Transaction Interface Processor

TIP Transaction Internet Protocol

TIP Transient In-Core Probe

TIP Traveling In-Core Probe

TIPA Tank and Industrial Plant Association

TIPA Tetraisopropylpyrophosphoramide

TIPA Triisopropanolamine

TIPACS Texas Instruments Planning and Control System

Ti-Pd Titanium-Palladium (Alloy System)

TIPDSiCl$_2$ 1,3-Dichloro-1,1,3,3-Tetraisopropyldisiloxane

TIPI Tactical Information Processing and Interpretation

TIPL Teach Information Processing Language

TIPMA Tri(isopentyl)methyl Amine

Ti(PMCp)Cl$_3$ Pentamethylcyclopentadienyltitanium Trichloride

Ti(PMCp)$_2$Cl$_2$ Bis(pentamethylcyclopentadienyl)titanium Dichloride

TIPP Time-Phasing Program

TIPP Tri(isopropyl)phosphate

TIPPC Texas Instruments Personal Programmable Calculator Club [US]

TIPPO Tri(isopentyl)phosphine Oxide

Ti(Pr^i)$_4$ Titanium Isopropoxide

TIPRO Texas Independent Producers and Royalty Owners Association [US]

Ti(PrO)$_4$ Titanium Propoxide (or Tetrapropyl Titanate)

Ti(Pr^{i}O)$_4$ Titanium Tetraisopropoxide

TIPS Technical Information Processing System

TIPS Technical Information Programming System

TIPS Telemetry Impact Prediction System [of US Air Force]

TIPS Terminal Interface Processor

TIPS Text Information Processing System

TIPS The Image Processing Software

TIPS Thermal Impulse Printer System

TIPS Thermally Induced Phase Separation [Materials Science]

TIPS Thousands of Instructions per Second

TIPS Triisopropylsilyl

TIPS TrueVision Image Processing Software

TIPS-Cl Triisopropylsilylchloride

TIPS/SV The Image Processing Software/Signature Verification

TIPS Triflate Triisopropylsilyl Trifluoromethanesulfonate [also TIPS triflate]

TIPT Toronto Institute of Pharmaceutical Technology [Canada]

TIPT 2,4,6-Tri(isopropyl) Thiophenol

TIPTOM Towards Improved Performance of Tool Materials

TIPTOP Tape Input/Tape Output [also TIP/TOP or TIP TOP]

TIPTOP The Internet Pilot TO Physics

TIQ Task Input Queue

TIQ Tetrahydroisoquinone

TIQM Time-Independent Quantum-Mechanical (Calculation) [Physics]

TIR Target Illuminating Radar

TIR Target Instruction Register

TIR Technical Information Release

TIR Technical Information Report

TIR Technical Intelligence Report

TIR Test Incident Report

TIR Thermal Infrared

TIR Total Indicator Reading

TIR Total Indicator Runout

TIR Total Internal Reflection

TIR Transient Impulse Resonance

TIR Transient Impulse Response

TIR Transport International de Marchandises par la Route [International Road Transportation Organization, France]

TIR Transport International Routier [International Road Transportation] [*Note:* License Plate Designation on European Lond-Distance Trucks]

TIRAM Taper Insulated Random-Access Memory

TIRC Toxicology Information Response Center [US]

TIRDO Tanzania Industrial Research and Development Organization

Tire Bus Tire Business [Published in the US]

TIREC TIROS (Television Infrared Observation Satellite) Ice Reconnaissance

Tire Sci Technol Tire Science and Technology [Journal published in the US]

TIRF Traffic Injury Research Foundation [Canada]

TIRH Theoretical Indoor Relative Humidity

TIRKS Trunk Integrated Record Keeping System

TIROS Topographical Infrared Operations Satellite

TIROS Television Infrared Observation Satellite [Series of 10 Weather Satellites Launched by NASA between April 1960 and July 1965] [also Tiros]

TIROS N Television Infrared Observation Satellite N [Third Generation of Meteorological Satellites launched by National Oceanic and Atmospheric Administration, US]

TIRP Textile Information Retrieval Program [US]

TIRP Total Internal Reflection Prism

TIRPF Total Integrated Radial Peaking Factor

TIRS Thermal Infrared Scanner

TIS Target Information Sheet

TIS Target Information System

TIS Technical Information Service

TIS Technical Information Service [of Canada Center for Remote Sensing]

TIS Technical Inspection and Safety (Commission) [of CERN–European Laboratory for Particle Physics, Geneva, Switzerland]

TIS Technology Information System

TIS Terminal Interface Subsystem

TIS Termination Inventory Schedule

TIS Test Interface Subsystem

TIS The Information System

TIS Timber Industry Strategy

TIS Time Integration Spectroscopy

TIS Total Information System

TIS Total Integrated Scattering

TIS Total Ionic Strength [Physical Chemistry]

TIS Tracking Instrumentation Subsystem

TIS Transponder Interrogation Sonar

TIS Trench-Isolated (Transistor)

Ti:S Titanium-Doped Sapphire (Laser)

TiS Titanium(I) Sulfide (or Titanium Monosulfide)

TISA Time Interpolation and Spatial Averaging

TISAB Total Ionic Strength Adjustment Buffer

Ti:Sapphire Titanium-Doped Sapphire (Laser)

TISB Technical Information Service Bureau [US]

TISC Tire Industry Safety Council [US]

Ti-Si Titanium-Silicon (Alloy System)

TiSiC Titanium Silicomonocarbide

Ti/SiC Titanium on Silicon Carbide

TISO Tropical Intra-Seasonal Oscillation

TiSiON Titanium Silicon Oxynitride (Compound)

TiSi$_2$/Si Titanium Silicide on Silicon (Substrate)

TISR Temperature-Induced Spin Reorientation [Physics]

Tiss Tissue [also tiss]

TISSS Tester-Independent Support Software System

TISTR Thailand Institute of Scientific and Technological Research

TiSULC Titanium-Stabilized Ultralow Carbon

TiSULC Steel Titanium-Stabilized Ultralow Carbon Steel [also TiSULC steel]

TIT Tatung Institute of Technology [Taipei, Taiwan]

TIT Test Item Taker

TIT Tohoku Institute of Technology [Sendai, Japan]

TIT Tokyo Institute of Technology [Japan]

TIT Turbine Inlet Temperature [Aerospace]

Tit Titer [also tit]

Tit Title [also tit]

tit ternary digit [also Tit, or TIT]

Ti-Ta Titanium-Tantalum (Alloy System)

Titanium Zirconium Titanium and Zirconium [Japanese Publication]

Ti-Ta-V Titanium-Tantalum-Vanadium (Alloy System)

TiTech Tokyo Institute of Technology [Japan]

TITF Test Item Transmittal Form

Ti-TMS Titanium Transition-Metal-Oxide Molecular Sieve

TITOS Television Infrared Orbital Satellite

Titr Titrate; Titration [also titr]

Titrn Titration [also titr]

TITS Test Instrument Tracking System

TITUS Traitement de l'Information Textile Universelle et Sélective [Processing of General and Selected Textile Information (Database); of Institut Textile de France]

TIU Tape Identification Unit

TIU Terminal Interface Unit

TIU Terrestrial Interface Unit

TIU Toxicologically Insignificant Usage

TIU Trypsin Inhibitor Unit [Biochemistry]

TIU Typical Information Use

TIUC Textile Information Users Council [US]

TIU/mg Trypsin Inhibitor Unit per milligram [Unit]

TIU/mL Trypsin Inhibitor Unit per milliliter [also TIU/ml]

TIUPIL Typical Information Use per Individual

TIV Total Indicator Variation

Ti-V Titanium-Vanadium (Alloy System)

TIWE Tropical Instability Wave Experiment [Oceanography]

TiWN Titanium Tungsten Nitride

TIWU Transport and Industrial Workers Union [Trinidad]

Ti/Y-123 Titanium on Yttrium Barium Copper Oxide (Superconductor) [Ti/YBa$_2$Cu$_3$O$_x$]

Ti-Y-TZP Titanium/Yttrium Tetragonal Zirconia Polycrystal

TIZ Technisches Informations-Zentrum [Technical Information Center, Germany]

TIZ Traffic Information Zone

TIZ Int Powder Mag TIZ International Powder Magazine

Ti-Zr Titanium-Zirconium (Alloy System)

TJ Tadjikistan [ISO Code]

TJ Temperature Jump (Method)

TJ Terajoule [Unit]

TJ Thermojunction

TJ Triple Junction (of Grain Boundaries) [Materials Science]

TJ Turbojet (Engine)

TJ- Cameroon [Civil Aircraft Marking]

t-J Time-Antiferromagnetic Exchange (Model) [Solid-State Physics]

t-J Time versus Current Density (Model)

TJC Trajectory Chart

TJD Trajectory Diagram

TJE Turbojet Engine

TJF Test Jack Field

TJID Terminal Job Identification

TJMTS Teleman-Joensson Multiple Time Step [Physics]

TJNAF Thomas Jefferson National Accelerator Facility [Newport News, Virginia, US] [also Jefferson Laboratory]

TJR Trunk and Junction Routing

TJS Transverse Junction Stripe

TJT Tetrode Junction Transistor

TJU Thomas Jefferson University [Philadelphia, Pennsylvania, US]

TK Taka [Currency of Bangladesh]

TK Takeuchi-Kuramoto (Theory) [Metallurgy]

TK Technologies of Kobe Steel [Publication of Kobe Steel Ltd., Tokyo, Japan]

TK Thymidine Kinase [Biochemistry]

TK Tokelau [ISO Code]

TK Tool Kit

TK Transmission Kossel (Technique) [Physics]

Tk Taka [Currency of Bangladesh]

Tk Tank [also tk]

Tk Track [also tk]

Tk Truck [also tk]

Tk Trunk [also tk]

T$_d$(κ) Conductivity-Dependent (Material) Deposition Temperature [Symbol]

TKA Thermokinetic Analysis

T/kbar Tesla(s) per kilobar [also T kbar^{-1}]

TKE Transverse Kerr Effect [Optics]

TKK Tanaka Kininzoku Kogyo Corporation [Kanagawa, Japan]

TKL Thermodynamics and Kinetics Laboratory [of University of Toronto, Canada]

TKN Total Kjeldahl Nitrogen (Concentration) [Chemistry]

TKO Trunk Offer

TKOFF Takeoff [also TKOF]

TKP Tribasic Potassium Phosphate; Tripotassium Phosphate

TKPP Tetrapotassium Pyrophosphate

TKS Tessman-Kahn-Shockley (Theory) [Materials Science]

TKSC Tsukuba Space Center [of National Space Development Agency of Japan]

TKT Transkaryotic Therapies, Inc. [US]

Tkt Ticket [also tkt]

TK Technol Kobe Steel TK Technologies of Kobe Steel [Publication of Kobe Steel Ltd., Tokyo, Japan]

TK/TK Track to Track [also Tk/Tk]

TkV Tracked Vehicle

TKW Thousand Kernel Weights [Cereal Crops]

TL Late Transition (Metal)

TL Parallel to Reinforcement Fiber Direction

TL Tantalum Lamp

TL Tape Library

TL Target Label

TL Target Language

TL Target Loss

TL Task Leader

TL Taxiway Light(s)

TL Taxonomic Literature [Index]

TL Taylor Lattice [Metallurgy]

TL Technical Liaison

TL Team Leader

TL Temporal Lobe (of the Cerebral Cortex) [Anatomy]

TL Testing Laboratory

TL Test Link

TL Test Log

TL Thermal Life

TL Thermal Load

TL Thermoluminescence; Thermoluminescent

TL Thin Layer

TL Thin-Layer Leaching

TL Threshold Lights [Aeronautics]

TL Thrust Level

TL Tie Line

TL Time Lag

TL Time Lapse

TL Time Limit(er)

TL Tolerance Limit

TL Tool Life

TL Top Layer

TL Torr-Liters [Unit]

TL Total Lipid [Biochemistry]

TL Total Loss [also T/L, or tl]

TL Transaction Language

TL Transient Load

TL Transition Level [Aeronautics]

TL Transmission Level

TL Transmission Line

TL (Sound) Transmission Loss

TL (Data) Transmit Level

TL Transverse to Longitudinal (Direction) [also T-L]

TL Triboluminescence; Triboluminescent

TL Trilayer; Triple Layer

TL (Grain Boundary) Triple Line [also tl] [Materials Science]

TL Truck Load [also tl]

TL Tubal Ligation [Medicine]

TL Tube-Line (Bale Wrapper) [Agriculture]

TL Tungsten Lamp

TL Turbine Liquefier

TL Turkish Lira [Currency]

TL- Central African Republic [Civil Aircraft Marking]

TL1 Taxonomic Literature, First Edition [Index]

TL2 Taxonomic Literature, Second Edition [Index]

T/L Talk and Listen

T/L Time Line

T/L Total Loss [alst TL, or tl]

T/L Transformer Load

T-L Transverse to Longitudinal (Direction) [also TL]

Tl Thallium [Symbol]

Tl Tool [also tl]

Tl$^+$ Monovalent Thallium Ion [Symbol]

Tl^{3+} Trivalent Thallium Ion [also Tl^{+++}] [Symbol]

Tl-191 Thallium-191 [also ^{191}Tl, or Tl^{191}]

Tl-192 Thallium-192 [also ^{192}Tl, or Tl^{192}]

Tl-193 Thallium-193 [also ^{193}Tl, or Tl^{193}]

Tl-194 Thallium-194 [also ^{194}Tl, or Tl^{194}]

Tl-195 Thallium-195 [also ^{195}Tl, or Tl^{195}]

Tl-196 Thallium-196 [also ^{196}Tl, or Tl^{196}]

Tl-197 Thallium-197 [also ^{197}Tl, or Tl^{197}]

Tl-198 Thallium-198 [also ^{198}Tl, or Tl^{198}]

Tl-199 Thallium-199 [also ^{199}Tl, or Tl^{199}]

Tl-200 Thallium-200 [also ^{200}Tl, or Tl^{200}]

Tl-201 Thallium-201 [also ^{201}Tl, or Tl^{201}]

Tl-202 Thallium-202 [also ^{202}Tl, or Tl^{202}]

Tl-203 Thallium-203 [also ^{203}Tl, or Tl^{203}]

Tl-204 Thallium-204 [also ^{204}Tl, or Tl^{204}]

Tl-205 Thallium-205 [also ^{205}Tl, or Tl^{205}]

Tl-206 Thallium-206 [also ^{206}Tl, Tl^{206}, RaE″, or RaE$_2$]

Tl-207 Thallium-207 [also ^{207}Tl, Tl^{207}, AcC″, or AcC$_2$]

Tl-208 Thallium-208 [also ^{208}Tl, Tl^{208}, ThC″, or ThC$_2$]

Tl-209 Thallium-209 [also ^{209}Tl, or Tl^{209}]

Tl-210 Thallium-210 [also ^{210}Tl, Tl^{210}, RaC″, or RaC$_2$]

Tl-221 Tl$_2$Ba$_2$CuO$_x$ [A Thallium Barium Copper Oxide Superconductor]

Tl-222 Tl$_2$Ba$_2$Cu$_2$O$_x$ [A Thallium Barium Copper Oxide Superconductor]

Tl-1122 TlBaCa$_2$Cu$_2$O$_x$ [A Thallium Barium Calcium Copper Oxide Superconductor]

Tl-1212 TlCa$_2$BaCu$_2$O$_x$ [A Thallium Barium Calcium Copper Oxide Superconductor]

Tl-1223 TlBa$_2$Ca$_2$Cu$_3$O$_x$ [A Thallium Barium Calcium Copper Oxide Superconductor]

Tl-1234 TlBa$_2$Ca$_3$Cu$_4$O$_x$ [A Thallium Barium Calcium Copper Oxide Superconductor]

Tl-1245 TlBa$_2$Ca$_4$Cu$_5$O$_x$ [A Thallium Barium Calcium Copper Oxide Superconductor]

Tl-1324 TlCa$_3$Ba$_2$Cu$_4$O$_x$ [A Thallium Barium Calcium Copper Oxide Superconductor]

Tl-2122 Tl$_2$CaBa$_2$Cu$_2$O$_x$ [A Thallium Barium Calcium Copper Oxide Superconductor]

Tl-2201 Tl$_2$Ba$_2$Cu$_1$O$_x$ [A Thallium Barium Copper Oxide Superconductor]

Tl-2212 Tl$_2$Ba$_2$CaCu$_2$O$_x$ [A Thallium Barium Calcium Copper Oxide Superconductor]

Tl-2223 Tl$_2$Ba$_2$Ca$_2$Cu$_3$O$_x$ [A Thallium Barium Calcium ˙ Copper Oxide Superconductor]

Tl-2234 Tl$_2$Ba$_2$Ca$_3$Cu$_4$O$_x$ [A Thallium Barium Calcium Copper Oxide Superconductor]

Tl-2324 Tl$_2$Ca$_3$Ba$_2$Cu$_4$O$_x$ [A Thallium Barium Calcium Copper Oxide Superconductor]

Tl-2623 Tl$_2$Ba$_6$Ca$_2$Cu$_3$O$_{10}$ [A Thallium Barium Calcium Copper Oxide Superconductor]

$\tau(\lambda)$ spectral internal transmittance [Symbol]

TLA Thiolactic Acid [Biochemistry]

TLA Three Letter Acronym [Computers]

TLA Trilauryl Amine

TLA Truck Loggers Association

TlAc Thallium(I) Acetate

Tl(Ac)$_3$ Thallium(III) Acetate

Tl(ACAC) Thallium(I) Acetylacetonate [also Tl(AcAc), or Tl(acac)]

Tl(ACAC)$_3$ Thallium(III) Acetylacetonate [also Tl(AcAc)$_3$, or Tl(acac)$_3$]

TLAHBr Trilauryl Ammonium Bromide

TLAHCl Trilauryl Ammonium Chloride

TLAHI Trilauryl Ammonium Iodide

Tl:Al$_2$O$_3$ Thallium-Doped Alumina (Laser)

TLAN Trilauryl Ammonium Nitrate

TlAP Thallium Acid Phthalate

TLAS Trilauryl Ammonium Sulfate

TLASCN Trilauryl Ammonium Thiocyanate

TLAH$_2$SO$_4$ Trilauryl Ammonium Sulfate

TLAS Tactical, Logical and Air Simulation

TLB Translation Lookaside Buffer

TLBE Thin-Layer Bulk Electrolysis

TLBCCO Thallium Lead Barium Calcium Copper (Superconductor)

TlBr Thallium(I) Bromide (or Thallous Bromide)

TLC Tank Landing Craft

TLC Task Level Control

TLC Technical LAN (Local Area Network) Coordinator

TLC Telecommand

TLC Tender Loving Care

TLC Thick Molding Compound

TLC Thin-Layer Chromatography [also tlc]

TLC Total Loss Control

TLC Trilateral Labeling Committee

TLC Typed Lambda Calculi

TLCC Trades and Labour Congress of Canada

TLCE Transmission Line Conditioning Equipment

TLCF Team Leader Computing Facility

TLCF Thin-Layer Chromatography Combined with Fluorimetry

TLC-FID Thin-Layer Chromatography–Flame Ionization Detector

TLCK Nα-Tosyl-L-Lysine-Chloromethyl Ketone [Biochemistry]

TlCl Thallium(I) Chloride (or Thallous Chloride)

TLCP Thermotropic Liquid Crystalline Polymer

TlCp Cyclopentadienylthallium

TLCT Total Life-Cycle Time

TLCTI Total Loss Control Training Institute

TLCV Thin-Layer Cyclic Voltammetry

TLD Thermoluminescence Dosimeter; Thermoluminescent Dosimeter

TLD Thermoluminescent Detector

TLD Top Level Domain

TLD Trapped Lattice Dislocation [Materials Science]

TLDPC Tidal Land Development Planning Commission [Taiwan]

TLE Temperature-Limited Emission

TLE Theoretical Line of Escape

TLE Thin-Layer Electrochemical (Cell)

TLE Timeline Engineer(ing)

TLE Tracking Light Electronics

TlEtO Thallium(I) Ethoxide

TLEV Transitional Low Emission Vehicle

TLF Trunk Line Frame

TLF Two-Level Fluctuator

TlF Thallium(I) Fluoride

Tlg Telegraph [also tlg]

t(lg) ton, long [Unit]

LGPC Thin-Layer Gel Permeation Chromatography

Tl(HFAC) Thallium(I) Hexafluoroacetylacetone [also Tl(hfac)]

Tl-HTSC Thallium High-Temperature Superconductor

TLI Telephone Line Interface

TLI The Laser Institute [Edmonton, Alberta, Canada]

TLI Thermionics Laboratory, Inc. [Hayward, California, US]

TLI Transport Layer Interface

TlI Thallium(I) Iodide (or Thallous Iodide)

TLIB The Teaching Library [of University of California at Berkeley, US]

TLIU Transmission Line Interface Unit

TLK Terrace-Ledge-Kink (Model) [Metallurgy]

TLK Test Link

TLK Tritium Laboratory Karlsruhe [of Kernforschungszentrum Karlsruhe, Germany]

Tlk Talking [also tlk]

TLL Teflon Luer Lock

TLL Thin-Layer Leaching

TLLM Temperature and Liquid Level Monitor

TLM Tandem Link Module

TLM Tape-Laying Machine

TLM Thin Lipid Membrane [Biochemistry]

TLM Transition Line Model

TLM Transmission Light Microscope; Transmission Light Microscopy

TLM Transmission Line Model [Metallurgy]

TLM Transmission Line Method [Metallurgy]

TLM Transmission Lorentz Microscopy

TLM Trouble Locating Manual

TLm Median Tolerance Limit [Lethal Temperature]

Tlm Telemeter(ing); Telemetry [also tlm]

TLMA Tunnel Lining Manufacturers Association [UK]

TLMA(NO$_3$) Trilauryl Methyl Ammonium Nitrate

TLMAN Trilauryl Methyl Ammonium Nitrate

TLMI Tag and Label Manufacturers Institute [US]

TLMS Tape Library Management System

TLN Trunk Line Network

Tl:NaI Thallium-Doped Sodium Iodide (Crystal)

Tl NMR Thallium Nuclear Magnetic Resonance

TlNO$_3$ Thallium(I) Nitrate (or Thallous Nitrate)

TLO Total Loss Only [also tlo]

TLO Tracking Local Oscillator

TlOEt Thallium(I) Ethoxide

TlOH Thallium Hydroxide (or Thallous Hydroxide)

TLP Telephone Line Patch

TLP Tension Leg Platform

TLP Threshold Learning Process

TLP Top Load Pad

TLP Torpedo Land Plane [US Navy/Coast Guard]

TLP Total Language Processor

TLP Toxicity Leaching Procedure

TLP Transient Liquid Phase [Metallurgy]

TLP Transient Lunar Phenomena [Astronomy]

TLP Translation Lookaside Buffer

TLP Transmission Level Point

TLPB Transient Liquid Phase Bonding [Metallurgy]

TLR Temperature Lapse Rate [Meteorology]

TLR Toll Line Release

TLR Top Luminaire Retainer

TLRI Taiwan Livestock Research Institute

TLRS Tramway and Light Railway Society [UK]

TLRV Tracked Levitated Research Vehicle

TLS Tactical Landing System

TLS Tape Librarian System

TLS Target Level of Safety

TLS Telecommunication Liaison Staff

TLS Telemetry Listing Schedule

TLS Terminal Landing System

TLS Thermal Line Softening

TLS Three-Stage Least Squares

TLS Time Line Sheet

TLS Top Left Side

TLS Total Library System

TLS Transformed Least Squares

TLS Translunar Space [Astronomy]

TLS Transport Level Security

TLS Two-Level (Tunneling) System [Electronics]

TL/s Torr-Liters per Second [Unit]

TLSA Torso Limb Suit Assembly

TLSA Transparent Line Sharing Adapter

TlSb Thallium Antimonide (Semiconductor)

TLSCCO Thallium Lead Strontium Calcium Copper Oxide (Superconductor)

Tlscp Telescope

TLSM Trail Lead Smelter Modernization [Canada]

TL/SX/EW Thin-Layer Leaching followed by Solvent Extraction and Electrowinning

TLT Terminal List Table

TLT Thermal Laser Treatment

TLT Third Law of Thermodynamics

TLT Transportable Link Terminal

TLTA Toluoyl-L-Tartaric Acid

TLTA Two-Loop Test Apparatus

TLTLM Tri-Layer Transmission Line Model

Tl(TMHD) Thallium(I) (2,2,6,6-Tetramethyl-3,5-Heptanedionate) [also Tl(tmhd)]

TLTP Trunk Line Test Panel

TLTR Translator [also Tltr]

TLU Table Lookup

TLU Terminal Logic Unit

TLV Threshold Limit Value [of American Conference of Governmental Industrial Hygienists]

TLV Tracked Levitated Vehicle

TLV Transporter-Loader Vehicle

TLV Troop Landing Vessel

TLV-C Threshold Limit Value–Ceiling Limit

TLV-TWA Threshold Limit Value–Time-Weighted Average

TLV-Skin Threshold Limit Value–(Human) Skin

TLV-STEL Threshold Limit Value–Short-Term Exposure Limit

TLV-TWA Threshold Limit Value–Time-Weighted Average

TLW Titanium-Lead-Tungsten (Compound)

Tlwd Tailwind

TLWS Trunk and Line Workstation

TLX Teleprinter Exchange Service [also TELEX, Telex, telex, TEX, or TX]

.TLX Telex [File Name Extension] [also .tlx]

TlX Thallium Halide [Formula]

TLZ Titanium-Lead-Zinc (Compound)

TLZ Transfer on Less than Zero

TM Table Maintenance

TM Tactical Missile

TM Tape Mark [Data Communications]

TM Tape Module

TM Teacher's Manual

TM Team Member

TM Technical Magazine

TM Technical Management; Technical Manager

TM Technical Manual

TM Technical Memo(randum)

TM Technical Monograph

TM Teeth per Minute [Machining]

TM Telemeter; Telemetry

TM Telemonitor(ing)

TM Temperature Measurement

TM Temperature Meter

TM Temperature Monitor

TM Tempered Martensite [Metallurgy]

TM Tensile Modulus (of Elasticity)

TM Terrestrial Magnetism

TM Test(ing) Machine

TM Test Method

TM Test Mode

TM Text Management; Text Manager

TM Thematic Mapper [NASA Landsat Project]

TM Theoretical Mathematician; Theoretical Mathematics

TM Theoretical Mechanics

TM Thermal Margin

TM Thermal Modeling

TM Thermal Motion

TM Thermomagnetic(s)

TM Thermomechanic(al); Thermomechanics

TM Thermomigration [Electronics]

TM Thermoplastic Molding

TM Thermoremanent Magnetism [Geophysics]

TM Thickness Measurement

TM Threshold Marker

TM Time and Materials [also T&M, or T-M]

TM Time and Motion (Study) [Industrial Engineering]

TM Time Management

TM Time, Mechanization (Fuse)

TM Time Modulation; Time Modulator

TM Timing Module

TM Tissue Mechanics

TM Titrimetric; Titrimetry

TM Tone Modulation

TM Tool Material

TM Topical Meeting

TM Topographical Map

TM Torque Magnetometry

TM Torsion Modulus [Mechanics]

TM Town Manager

TM Trademark

TM Trade-Manufacturer

TM Traffic Management

TM Traffic Mix

TM Traffic Model

TM Training Manual

TM Transcendental Meditation

TM Transfer Matrix [Controllers]

TM Transfer Mode

TM Transformation Matrix [Waveguides]

TM Transition Metal

TM Translational Motion [Mechanics]

TM Transmission Matrix

TM Transportation Management

TM Transverse Magnetic (Mode)

TM Traveling Microscope

TM Trench Mortar

TM Trimolecular (Reaction) [Chemistry]

TM Tropical Month [Astronomy]

TM Tube Mill(ing)

TM Tunneling Model [Electronics]

TM Turbomachine(ry)

TM Turing Machine [Computers]

TM Turkmenistan [ISO Code]

TM Twist Multiplier (of Yarn)

TM Tympanic Membrane [Anatomy]

TM Type Metal

.TM Appointment Scheduler [File Name Extension]

3M Microsensors, Microactuators and Microsystems [also MMM]

3M Minnesota Mining and Manufacturing, Inc. [St. Paul, US]

T&M Time and Materials [also TM, or T-M]

T&M Training and Mentoring

T-M Tetragonal-to-Monoclinic (Transformation) [also t-m] [Metallurgy]

Tm (Polymer) Melting Temperature

Tm Thulium [Symbol]

Tm Tumor [Medicine]

Tm² Tesla-meter squared [Unit]

Tm^{3+} Thulium Ion [also Tm^{+++}] [Symbol]

Tm-161 Thulium-161 [also ^{161}Tm, or Tm^{161}]

Tm-162 Thulium-162 [also ^{162}Tm, or Tm^{162}]

Tm-163 Thulium-163 [also ^{163}Tm, or Tm^{163}]

Tm-164 Thulium-164 [also ^{164}Tm, or Tm^{164}]

Tm-165 Thulium-165 [also ^{165}Tm, or Tm^{165}]

Tm-166 Thulium-166 [also ^{166}Tm, or Tm^{166}]

Tm-167 Thulium-167 [also ^{167}Tm, or Tm^{167}]

Tm-168 Thulium-168 [also ^{168}Tm, or Tm^{168}]

Tm-169 Thulium-169 [also ^{169}Tm, or Tm^{169}, or Tm]

Tm-170 Thulium-170 [also ^{170}Tm, or Tm^{170}]

Tm-171 Thulium-171 [also ^{171}Tm, or Tm^{171}]

Tm-172 Thulium-172 [also ^{172}Tm, or Tm^{172}]

Tm-173 Thulium-173 [also ^{173}Tm, or Tm^{173}]

Tm-174 Thulium-174 [also ^{174}Tm, or Tm^{174}]

Tm-175 Thulium-175 [also ^{175}Tm, or Tm^{175}]

Tm-176 Thulium-176 [also ^{176}Tm, or Tm^{176}]

T²/m Square Tesla(s) per meter [also T² m^{-1}]

T/μ_B Tesla(s) per Bohr magnetron [also T μ_B^{-1}]

.tm Turkmenistan [Country Code/Domain Name]

t-m tetragonal-to-monoclinic (transformation) [also T-M] [Metallurgy]

t/m² tons per square meter [also t m^{-2}]

TMA Tape Motion Analysis; Tape Motion Analyzer

TMA Telecommunications Managers Association [UK]

TMA Telemetry Manufacturers Association

TMA Terminal Control Area [Aeronautics]

TMA Terminal Maneuvering Area

TMA Testability Measure Analyzer

TMA Test Module Adapter

TMA Tetrahydroaminacrine

TMA Tetramethyl Ammonium

TMA Thermal Mechanical Analyzer

TMA Thermomagnetic Analysis; Thermomagnetic Analyzer

TMA Thermomechanical Analysis; Thermomechanical Analyzer

TMA Titanium-Molybdenum-Aluminum (Alloy)

TMA Titrimetric Analysis

TMA Toy Manufacturers of America

TMA Transition Metal Atom

TMA Transport Museum Association [US]

TMA Trimellitic Acid

TMA Trimellitic Anhydride

TMA 3,4,5-Trimethoxyamphetamine

TMA Trimethyladipic Acid; 2,2,4-Trimethyladipic Acid; 2,4,4-Trimethyladipic Acid

TMA Trimethyl Aluminum [also TMAl]

TMA Trimethylamine

TMA Trimethylammonium

TMA Trimethylanilinium

TMA-d9 Perdeuterotrimethylamine

TMA-3 Trimethyl Amphetamine

Tma Trimethylamine

TMAA n-Trimethylamine-Alane [Biochemistry]

TMAB Tetramethyl Ammonium Bromide

TMAB Tetramethyl Ammonium Borohydride

TMABF$_4$ Tetramethyl Ammonium Tetrafluoroborate

TMABH$_4$ Tetramethyl Ammonium Borohydride

TMAC Tangential Momentum Accomodation Coefficient

TMAC Tetramethylammonium Carbonate

TMAC Tetramethylammonium Chloride

TMAC Trimellitic Anhydride Chloride

TMAc Trimethylacetic Acid

Tm(Ac)$_3$ Thulium(III) Acetate

Tm(ACAC)$_3$ Thulium(III) Acetylacetonate [also Tm(AcAc)$_3$, or Tm(acac)$_3$]

TMACl Trimellitic Anhydride Chloride

TMA-d9 Perdeuterotrimethylamine

TMA-DPH 1-(4-Trimethylammoniumphenyl)-6-Phenylhexa-1,3,5-Hexatriene (p-Toluene Sulfonate)

TMAEMC Trimethylammonium Ethyl Methacrylate Chloride [also TMA-EMC]

TMAH Tetramethylammonium Hydroxide

TMAH Trimethylammonium Hydroxide

TMAH Trimethylanilinium Hydroxide

TMAHF$_2$ Tetramethyl Ammonium Hydrogen Difluoride

TMAHFP Tetramethyl Ammonium Hexafluorophosphate

TMAHS Tetramethyl Ammonium Hydrogen Sulfide

TMAI Tetramethyl Ammonium Iodide

TMAl Trimethylaluminum [also TMA]

TMAl Transition-Metal Aluminide [General Formula]

TMAl$_3$ Transition-Metal Trialuminide [General Formula]

TMAMA Textile Machinery and Accessory Manufacturers Association

T-Man Treasury Department Man

TMAO Trimethylamine N-Oxide

TMA-OH Tetramethyl Ammonium Hydroxide [also TMAOH]

TMAP Thermal Modeling and Analysis Project [of National Oceanic and Atmospheric Administration–Pacific Marine Environmental Laboratory, US]

TMAPF$_6$ Tetramethyl Ammonium Hexafluorophosphate

TMAs Trimethylarsine

2MASS Two Micron All Sky Survey [of Infrared Processing and Analysis Center, NASA Jet Propulsion Laboratory, Pasadena, California, US]

TMAT Tetramethyl Ammonium Tribromide

TMAT 2,4,6-tris[2-Methylaziridine-1-yl]-1,3,5-Triazine

TMATFB Tetramethyl Ammonium Tetrafluoroborate

TMAV$_2$O$_5$ Tetramethyl Ammonium Vanadium Pentoxide

TMAX Maximum Time

TMB Tetramethyl Benzidine; 3,3',5,5'-Tetramethylbenzidine; N,N,N',N'-Tetramethylbenzidine

TMB Transportation Management Bulletin

TMB Trench Mortar Bomb

TMB Trimesitylborane

TMB Trimethoxybenzoate

TMB Trimethoxybenzoic Acid

TMB 2,4,6-Trimethoxyboroxin

TMB Trimethylbenzene

TMB Trimethylboron

TMB Trimethylbromosilane

TMB 2,2,3-Trimethylbutane

TMB-4 1,1'-Trimethylene-Bis(4-Formylpyridinium Bromide) Dioxime

TMB-8 3,4,5-Trimethoxybenzoic Acid 8-(Diethylamino)octyl Ester

TMBA Thickness Measurement by Beam Alignment

TMBA 3,4,5-Trimethylbenzaldehyde

TMBA Trimethylborate/Alcohol Azeotrope

TmBCO Thulium Barium Copper Oxide (Superconductor)

TMBMA Tris(1-Methyl-Benzimidazole-2-ylmethyl)amine [also Tmbma, or tmbma]

TMBO 2,4,6-Trimethoxyboroxine

TMBQ Tetramethylbenzoquinone

Tmbr Timber [also tmbr]

TMBS Trimethylbromosilane

TMBU Table Maintenance Block Update

TMBZPS 3,3',5,5'-Tetramethylbenzidinepropanesulfonic Acid

TMC 7,8-Dihydro-5,10,15,20-Tetramethylporphinate; 7,8-Dihydro-5,10,15,20-Tetramethylporphine

TMC Tape Management Catalog

TMC Terminal Control

TMC Ternary and Multinary Compounds

TMC Test Module Coil

TMC Test Monitoring Console

TMC 1,4,8,11-Tetramethyl-1,4,8,11-Tetraazacyclotetradecane

TMC Texas Medical Center [Houston, US]

TMC The Maintenance Council [of American Trucking Association, US]

TMC Thompson Machine Gun

TMC Thermomechanical Control

TMC Thick Molding Compound

TMC Titanium Matrix Composite

TMC Total Molding Concept

TMC Tourism Ministers Council

TMC Traffic Message Channel

TMC Transition Metal Carbide

TMC Transition Metal Chemistry

TMC Transition Metal Complex

TMC Transition Metal Compound

TMC Transmission Maintainance Center [Telecommunications]

TMC Trimethyl Cyclohexanol

TM12C4 Tetramethyl-12-Crown-4

TM18C6 Tetramethyl-18-Crown-6

14TMC 1,4,8,11-Tetramethyl-1,4,8,11-Tetraazacyclotetradecane

15TMC 1,4,8,12-Tetramethyl-1,4,8,12-Tetraazacyclopentadecane

16TMC 1,5,9,13-Tetramethyl-1,5,9,13-Tetraazacyclohexadecane

TMC-Amine 3,3,5-Trimethylcyclohexylamine

TMCB Tetramethoxycarbonyl Benzophenone

TMCC Time-Multiplexed Communication Channel

TmCd Thulium-Cadmium (Compound)

TMCDT Trimethylcyclododecatriene

TMC-ol 3,3,5-Trimethyl Cyclohexanol

TMCOMP Telemetry Computation

TMC-one 3,3,5-Trimethylcyclohexanone

TMCP Thermomechanical Control(led) Process(ing); Thermomechanical(ly) Controlled Processing

TMCP Tris(monochloroisopropyl) Phosphate

TMCS Toshiba Minicomputer Complex System

TMCS Trimethylchlorosilane

TMD Tactical Munitions Dispenser

TMD Technical Management Division

TMD Technology Marketing Division [of Canada Center for Mineral and Energy Technology, Natural Resources Canada]

TMD Tensor Meson Dominance [Particle Physics]

TMD Tetramethylene Diamine [also tmd]

TMD Theoretical Maximum Density

TMD Tilt (Type) Misfit Dislocation [Materials Science]

TMD Torpedo and Mine Department

TMD Transient Mass Distribution Code

TMD Trimethylhexamethylene Diamine; 2,2,4-Trimethylhexamethylene Diamine; 2,4,4-Trimethylhexamethylene Diamine

TMDA Tetramethylenediamine

3M/DARPA/ONR 3M Company/Defense Advanced Research Projects Agency/Office of Naval Research (Joint Program) [US]

TMDAU Trimethylolglycoluril

TMDDA Trimethylenediamine-N,N'-Diacetate [also tmdda]

TMDE Test, Measurement and Diagnostic Equipment [also TM&DE]

TMDFP Trimethyldifluorophosphine

TMDHQ Trimethyl Dihydroquinoline; 2,2,4-Trimethyl-1,2-Dihydroquinoline

TMDI Trimethylhexamethylene Diisocyanate; 2,2,4-Trimethylhexamethylene Diisocyanate; 2,4,4-Trimethylhexamethylene Diisocyanate

TMDL Total Maximum Daily Load (of Pollutants) (Program) [of US Environmental Protection Agency]

TMDP 4,4'-Trimethylenedipyridine

TMDS 1,1,3,3-Tetramethyldisilazane

TMDTA Trimethylenediaminetetraacetic Acid

TMDU Tokyo Medical and Dental University [Japan]

TMDXI Tetramethyl Xylene Diisocyanate

Tmdz Hexahydro-5,5,7-Trimethyl 1H-1,4-Diazepine [also tmdz]

TME Technical Unit of Mass, Metric

TME Telemetric Equipment

TME Tempered Martensite Embrittlement [Metallurgy]

TME Tensile Modulus of Elasticity

TME Tetramethylethylene

TME Thermomagnetic Effect

TME Thermomechanical Effect

TME Thick Molding Compound

TME Trimethylolethane

TMECO Time of Main Engine Cutoff [Aerospace]

TMED N,N,N',N'-Tetramethyl Ethylene Diamine

TMEDA N,N,N',N'-Tetramethyl 1,2-Ethanediamine; Tetramethylethylenediamine

TMEMT Trimethylolethylmethane Trinitrate

TMEN N,N,N',N'-Tetramethyl Ethylene Diamine

Tmen Trimethylene Diamine

TMETD Trimethyl Ethyl Thiuram Disulfide

TMF Thermal Mechanical Fatigue; Thermomechanical Fatigue

TMF Transmission Monitoring Facility

TMF Transporter Maintenance Facility [of NASA]

TMF Trunk Maintenance File

TMF-IP In-Phase Thermomechanical Fatigue [Metallurgy]

TMF-OP Out-of-Phase Thermomechanical Fatigue [Metallurgy]

TMG Methyl-1-Thio-β-D-Galactopyranoside [Biochemistry]

TMG Methyl β-D-Thiogalactoside [Biochemistry]

TMG Tape Manufacturers Group [UK]

TMG Technical Management Group

TMG Tetramethyleneglutaric Acid [Biochemistry]

TMG Tetramethylguanidine [Biochemistry]

TMG Thermal Meteoroid Garment

TMG Thermomagnetic Gravimeter; Thermomagnetic Gravimetry

TMG Track Made Good

TMG Trimethylgallium [also TMGa]

TMG Turret Machine Gun

TMGa Trimethylgallium [also TMG]

TMGE Thermo-Magnetic-Galvanic Effect

TMGeBr Trimethylgermanium Bromide

TMGeCl Trimethylgermanium Chloride

TMH Tasmania Museum (and Art Gallery), Hobart [Australia]

TMH Tons per Man-Hour [Unit]

TMHD 2,2,6,6-Tetramethylheptane-3,5-Dione; 2,2,6,6-Tetramethyl-3,5-Heptanedionate; 2,2,6,6-Tetramethyl-3,5-Heptanedionato-; 2,2,6,6-Tetramethyl-3,5-Heptanedione [also tmhd]

TMHD 1,1,1-Trifluoro-5-Methyl-2,4-Hexanedionate; Trifluoromethylhexanedionato- [also tmhd]

Tm:Ho:YLF Thulium- and Holmium-Doped Yttrium Lithium Fluoride

TMHPD 1,1,1-Trifluoro-6-Methyl-2,4-Heptanedionate [also tmhpd]

TMHR Tandem Mirror Hybrid Reactor

TMI Technical Management Item(s)

TMI Telesat Mobile International [Canada]

TMI Test, Measurement and Inspection (Conference and Exhibition)

TMI Thermomechanical Instability [Metallurgy]

TMI Three Mile Island [near Harrisburg, Pennsylvania, US]

TMI Timing Measurement Instrument

TMI Trade Media International Corporation [New York City, US]

TMI Trimethyl Indium [also TMIn]

TMI TRMM (Tropical Rainfall Measuring Mission) Microwave Imager [Japan]

TM&I Test, Measurement and Inspection

TMIC Tosylmethylisocyanide

TMIC Toxic Materials Information Center [Joint US Center National Science Foundation and Department of Energy]

TMIN Minimum Time

TMIn Trimethyl Indium [also TMI]

TMIS Technical Management Information System

TMIS Technical and Management Information System [of NASA]

TMIS Technical Meetings Information Service [US]

TMIS Television Measurement Information System

TMIS Television Metering Information System

TMIS Trimethyliodosilane

TMITEC Tokyo Metropolitan Industrial Technology Center [also TMITC, Japan]

TMJ Temporo-Mandibular Joint (Syndrome) [Dentistry]

TMK Thio-Michler's-Ketone [Organic Chemistry]

TMK Trivedi, Magnin and Kurz (Dentritic Growth Model) [Metallurgy]

TML Tandem Matching Loss

TML Tetramethyl Lead

TML Total Mass Loss

.TML Template [File Name Extension] [also .tml]

Tml Terminal [also tml]

TM-Ln Transition-Metal/Lanthanide Metal (Alloy)

TM/LP Thermal Margin/Low Pressure (Trip)

TMM Temperature-Stable Microwave Material

TMM Test Message Monitor

TMM Transfer Matrix Method

TMM Transfer Matrix Modelling

TMM Transverse Magnetic

TMM Transverse Magnetic Mode

TMM Trimethylenemethane

TM-M Transition Metal–Metalloid (Alloy)

TMMA N,N,N',N'-Tetramethylmalonamide

TMMB Truck Mixer Manufacturers Bureau [US]

TMMC Tetramethyl Ammonium Manganese Chloride

TMMC Transition Metal Monocarbide

TMMD Tactical Moving Map Display

TMMMT Trimethylolmethylmethanetrinitrate

TM Mode Transverse Magnetic Mode [also TM mode]

TMN Technical and Management Note

TMN Transition Metal Nitride

TMN Trimethyladipic Dinitrile

TMN 2,2,4-Trimethyladipic Dinitrile; 2,4,4-Trimethyladipic Dinitrile

TMN Trimethylamine Nitrogen

TMN 1,4,6-Trimethylnaphthalene

TMNIN Nickel Tetramethylammonium Trinitrite

TM-NM Transition Metal/Nonmagnetic Metal

TMNO Trimethylamine-N-Oxide

TMO Telegraph Money Order

TMO Terbium Molybdenum Oxide

TMO Time Out

TMO Test Manufacturing Order

TMO Thallium Manganese Oxide

TMO Tool Manufacturing Order

TMO Transition Metal Oxide

TMO Trimethylamine N-Oxide

3MO Trimethylene Oxide

TMOD 1,1,1-Trifluoro-7-Methyl-2,4-Octanedionate [also tmod]

T-MOKE Transverse Magnetooptical Kerr Effect

T-MOR Trench Mortar [also T-Mor]

TMORPO Tris(4-Morpholinyl)phosphine Oxide [also tmorpo]

TMOS Tetramethoxysilane

TMOS T-Type Metal Oxide Semiconductor

TMP Temporary (File)

TMP Terminal Monitor Program

TMP Terminal Panel

TMP 5,10,15,20-Tetramesityl Porphinate

TMP 2,2,6,6-Tetramethylpiperidine

TMP The Madison Project [US]

TMP Theory, Modeling and Practice

TMP Thermomechanically Processed; Thermomechanical Process(ing)

TMP Thermomechanical Processing Conference

TMP Thermomechanical Pulp(ed); Thermomechanical Pulping [Papermaking]

TMP Thermomolecular Pressure

TMP Thymidine Monophosphate; Thymidine-5'-Monophosphate [Biochemistry]

TMP Time Management Processor

TMP Toa (Steel Company) Mist Patenting (System)

TMP Transparent Multiprocessing

TMP Transmembrane Potential

TMP Trimetaphosphate

TMP Trimethoprim [Pharmacology]

TMP Trimethylammonium Propanediol Iodide

TMP Trimethylol Propane

TMP 2,2,4-Trimethyl Pentane

TMP Trimethylphenol

TMP Trimethyl Phosphate

TMP Trimethyl Phosphine

TMP Turbomolecular Pump

.TMP Temporary [File Name Extension] [also .tmp]

[TMP]² Thermomechanical Processing Theory, Modeling and Practice

Tmp Temperature [also tmp]

Tmp 2,2,6,6-Tetramethyl-1-Piperidinyl [also tmp]

tmp temporary

TMPA Thermal Melt Polyamide

TMPA N,N,N',N'-Tetramethyl-1,3-Propanediamine

TMPA Transocean Marine Paint Association [Netherlands]

TMPAH Trimethylphenylammonium Hydroxide

TMPD N,N,N',N'-Tetramethylphenylenediamine

TMPD 2,2,4-Trimethyl 1,3-Pentanediol

TMPDA N,N,N',N'-Tetramethyl-p-Phenylenediamine

TMPDE Trimethylolpropanediallylether

TMPDF Trade Marks, Patents and Designs Federation [UK]

TMPEG Methoxypolyethylene Glycol Tresylate

TMPG N,N,N',N'-Tetramethyl-N''-Phenyl Guanidine [Biochemistry]

TMPI Trimethyl Phosphite

TMPIP Tetramethylpiperidine

TMPO Trimethylphosphine Oxide

TMP-OLAC Thermomechanically Processed On-Line Accelerated Cooling [Metallurgy]

TMPP 5,10,15,20-Tetrakis(methoxyphenyl)porphine

tmpry temporary

TMPyP Tetrakis(1-Methyl-4-Pyridinium)porphine Tetratosylate

TMPS Poly(tetramethyl-p-Silphenylene Siloxane

TMPS Tetramethyl-P-Silphenylene Siloxane

TMP-SMX Trimethoprim–Sulfamethoxazole

TMPT Trimethylolpropane Trimethacrylate

TMPTA Trimethylolpropane Triacrylate

TMPTETA Trimethylolpropanetriethoxytriacrylate

TMPTMA Trimethylolpropane Trimethacrylate

TMPV Torquemotor Pilot Valve

TMQ Trimethyl Dihydroquinoline; 2,2,4-Trimethyl-1,2-Dihydroquinoline

TMR Tandem Mirror Reactor

TMR Teledyne Materials Research

TMR TOPEX (Ocean Topography Experiment) Microwave Radiometer

TMR Topical Magnetic Resonance

TMR Transverse Magnetoresistance [Solid-State Physics]

TMR Triple Modular Redundancy [Computers]

TMR Tunneling Magnetoresistance [Solid-State Physics]

TMR Tunneling Magnetoresistance Program [of European Community]

Tmr Timer

TMRBM Transportable Medium Range Ballistic Missile

TMRS Traffic Measuring and Recording System

TMS Tactical Missile Squadron

TMS Tape Management Software

TMS Tape Management System

TMS Tape Music Search

TMS Telecommunications Message Switch

TMS Telecommunications Mission Services

TMS Telemotor System

TMS Telephone Management System

TMS Telex Management System

TMS Tetramethylene Sulfone

TMS Tetramethylsilane

TMS Temperature Measurements Society [US]

TMS Tesla Memorial Society [US]

TMS Test Management System

TMS Tethered Manned Submersible

TMS Tetramethylsilane

TMS Texas Microprocessor Software [of Texas Instruments, Dallas, US]

TMS The Magnolia Society [US]

TMS The Masonry Society [US]

TMS Thematic Mapper Simulator [of NASA]

TMS The Metallurgical Society [now The Minerals, Metals and Materials Society, US]

TMS The Minerals, Metals and Materials Society [formerly The Metallurgical Society]

TMS The Metals Society [UK]

TMS Theoretical Materials Science

TMS Thermal Mechanical Simulator [Metallurgy]

TMS Thermionic Mass Spectrometry

TMS Thermomechanical Stability [Metallurgy]

TMS Time and Motion Study [Industrial Engineering]

TMS Time-Shared Monitor System

TMS Time-Multiplexed System

TMS Titan Missile Site [of US Air Force]

TMS Trademark Society [US]

TMS Traffic Management System

TMS Tramway Museum Society [UK]

TMS TMO (Transition-Metal Oxide) Molecular Sieve

TMS Transition Metal Sulfide

TMS Transmission Measuring Set

TMS Transmission Moessbauer Spectroscopy

TMS Transportation Management System

TMS Trimethylenesulfone

TMS 2,2,4-Trimethyl Hexanedioic Acid

TMS Trimethylsilane

TMS Trimethylsiloxane

TMS Trimethylsilyl-

TMS Trochoidal Mass Spectrometer

TMSA Technical Marketing Society of America [US]

TMSA Trimethylsilyl Amine

TMS-AIME The Minerals, Metals and Materials Society of the American Institute of Mining, Metallurgical and Petroleum Engineers [formerly The Metallurgical Society of American Institute of Mining, Metallurgical and Petroleum Engineers, US]

TMSAN Trimethylsilyl Acetonitrile

TMSAQ 8-Trimethylsilylaminoquinoline [also Tmsaq, or tmsaq]

TMS-L(+)-Arabinose Trimethylsilyl-L-(+)-Arabinose (Sugar) [Biochemistry]

TMS/ASM The Minerals, Metals and Materials Society/American Society for Materials International (Joint Conference) [US]

TMSb Trimethyl Antimony

TMSC Telecommunications Mission Services Contractor

TMSCl Trimethylsilyl Chloride

TMS-CMS The Minerals, Metals and Materials Society–Conference Management System [of the American Institute of Mining, Metallurgical and Petroleum Engineers, US]

TMSCN Trimethyl Silanecarbonitrile

TMSCN Trimethylsilyl Cyanide

TMSDEA N-(Trimethylsilyl)diethylamine

TMSDMA N-(Trimethylsilyl)dimethylamine

TMS-Dulcitol Trimethylsilyldulcitol (Sugar) [Biochemistry]

TMS-EMC The Minerals, Metals and Materials Society–Electronic Materials Committee [US]

TMS EMPMD The Minerals, Metals and Materials Society Electronic, Magnetic and Photonic Materials Division [also TMS EMPMD, US]

TMS/FEMS The Minerals, Metals and Materials Society/Federation of European Materials Societies (Joint Symposium)

TMS-D(−)-Fructose Trimethylsilyl-D(−)-Fructose (Sugar) [Biochemistry]

TMS-D(+)-Galactose Trimethylsilyl-D(+)-Galactose (Sugar) [Biochemistry]

TMS-α-D(+)-Glucose Trimethylsilyl-α-D-(+)-Glucose (Sugar) [Biochemistry]

TMS-β-D(+)-Glucose Trimethylsilyl-α-D-(+)-Glucose (Sugar) [Biochemistry]

TMS-Glycerol Trimethylsilylglycerol (Sugar) [Biochemistry]

TMSI 1-(Trimethylsilyl)imidazole

TMSI Trimethylsilyl Iodide

TMSi Tetramethylsilane

TMSIM Trimethylsilyl Imidazole

TMS-meso-Inositol Trimethylsilyl-meso-Inositol (Sugar) [Biochemistry]

TMS-D(+)-Mannitol Trimethylsilyl-D(+)-Mannitol (Sugar) [Biochemistry]

TMS-D(+)-Mannose Trimethylsilyl-D(+)-Mannose (Sugar) [Biochemistry]

TMS MATH Times Mathematics (Font) [also Tms Math]

TMS MDMD The Minerals, Metals and Materials Society Materials Design and Manufacturing Division [also TMS-MDMD]

TMSn Tetramethyltin

TMSO 3-Trimethylsilyl-2-Oxazolidinone

TMSO Tetramethylene Sulfoxide

TMSOTf Trimethylsilyl Trifluoromethanesulfonate

TMS PRICM The Minerals, Metals and Materials Society Pacific Rim International Conference on Advanced Materials and Processing

TMS-L(+)-Rhamnose Trimethylsilyl-L(+)-Rhamnose (Sugar) [Biochemistry]

TMS-D(−)-Ribose Trimethylsilyl-α-D-(+)-Ribose (Sugar) [Biochemistry]

TMSRD Telecommunications Mission Services Requirements Document

TMS RMN Times Roman (Font) [also Tms Rmn]

TMSS Tetrakis(trimethylsilyl)silane

TMSS Total Mine Simulation System

TMS SMD The Metallurgical Society Structural Materials Division [also TMS-SMD]

TMS-D(−)-Sorbitol Trimethylsilyl-D(−)-Sorbitol (Sugar) [Biochemistry]

TMS-Sucrose Trimethylsilyl-Sucrose (Sugar) [Biochemistry]

TMS-D(+)-Trehalose Trimethylsilyl-D(+)-Trehalose (Sugar) [Biochemistry]

TMS-Xylitol Trimethylsilylxylitol (Sugar) [Biochemistry]

TMS-D(+)-Xylose Trimethylsilyl-D(+)-Xylose (Sugar) [Biochemistry]

TMT Telemation

TMT Testing Methods and Techniques

TMT Tetramethyltin

TMT Thermoelastic Martensitic Transformation [Metallurgy]

TMT Thermomechanical Test(ing)

TMT Thermomechanically Treated; Thermomechanical Treatment [Metallurgy]

TMT Thermomechanicallly Mill Treated; Thermomechanical Mill Treatment [Metallurgy]

TMT Transmit [also Tmt, or tmt]

TMT Trimercaptotriazine

TMT Turbine Motor Train

TMTA Trimethylenediamine-N,N,N',N'-Tetraacetic Acid [also Tmta]

TMTBr Trimethyltin Bromide

TMTC Through-Mode Tape Converter

TMTCl Triethyltin Chloride

TMTD Tetramethylthiuram Disulfide

TMTFTH Trimethyl-(α-Trifluorotolyl)ammonium Hydroxide

TMTH Triethyltin Hydroxide

TMTM Tetramethylthiuram Monosulfide

Tm(TMHD)$_3$ Thulium(III) Tris(2,2,6,6-Tetramethyl-3,5-Heptanedionate) [also Tm(tmhd)$_3$]

TMTP Tetra-m-Tolylporphinate

TMTSF Tetramethyltetraselenafulvalene

(TMTSF)$_2$ClO$_4$ Tetramethyltetraselenafulvalene Perchlorate

TMTSF-DMTCNQ Tetramethyltetraselenafulvalene Dimethylenetetracyanoquinoline

(TMTSF)$_2$PF$_6$ Tetramethyltetraselenafulvalene Phosphorus Hexafluoride

(TMTSF)$_2$X Tetramethyltetraselenafulvalene Complex [X Represents Perchlorate, Phosphorus Hexafluoride, etc.]

TMTSV Triangular Modulated Triangular Sweep Voltammetry

TMTT Tetramethylthiuram Tetrasulfide

TMTTF Tetramethyltetrathiafulvalene

TMTU Tetramethythiourea

TMTU N,N,N',N'-Tetramethythiourea

TMU Technical Monitor Unit

TMU Temperature Measurement Unit

TMU Test Maintenance Unit

TMU Tetramethyl Urea

TMU Time Measurement Unit

TMU Tokyo Metropolitan University [Japan]

TMU Transmission Message Unit

T-MUN Trench Munition [also T-Mun]

TMV Technical Maintenance Vehicle

TMV Tobacco Mosaic Virus

TMVS Trimethylvinylsilane [also tmvs]

TMVS Transition Metal Vapor Source

TMW Transverse Magnetic Wave

tmw tomorrow

TMWS Terrestrial Microwave Station

TMWS Tunable Millimeter Wave Source

TMX Tamoxifen [Anti-Cancer Drug]

TMX Tandem Mirror Experiment

TMX Telemeter Transmitter

TMX Terminal Multiplexer

TMXDI Tetramethyl Xylene Diisocyanate

TMXO Tactical Miniature Crystal Oscillator

TMX-U Tandem Mirror Experiment-Upgrade

Tm:YAG Thulium-Doped Yttrium Aluminum Garnet (Laser) [also Tm^{3+}:YAG]

TmZn Thulium-Zinc (Compound)

TN Task Number

TN Taylor-Nabarro (Lattice) [Materials Science]

TN Technical Note

TN Telnet (Protocol) [Internet]

TN Tenascin [Cytology]

TN Tennessee [US]

TN Terminal Node

TN Thermal Neutron

TN Thermal Noise [Electronics]

TN Thermonuclear

TN Thionalide

TN Thyssen-Niederrhein AG [German Steel Manufacturer and Forging Plant]

TN Thyssen-Niederrhein (Process) [Metallurgy]

TN Total Nitrogen (Concentration)

TN Total Normality [Chemistry]

TN Toughened Nylon

TN Track Number [Aeronautics]

TN Tracor Northern, Inc. [US]

TN Trade Name

TN Transcendental Number [Mathematics]

TN Transnational

TN Transport Number [Physical Chemistry]

TN Trivial Name [Chemistry]

TN True Negative

TN Tunesia [ISO Code]

TN Twisted Nematic (Liquid Crystal Display)

TN- Republic of Congo [Civil Aircraft Marking]

Tn Thoron [also Rn-220, ^{220}Rn, or Rn^{220}]

Tn Tone [also tn]

Tn Town [also tn]

Tn Train [also tn]

Tn Trimethylene Diamine

Tn Troponin [Biochemistry]

Tn-I Troponin I [Biochemistry]

tn ton [Unit]

tn trimethylene diamine (or 1,3-diaminopropane) [Abbreviation used in Coordination Compound Formulas, e.g., Co(tn)$_3$]

.tn Tunesia [Country Code/Domain Name]

τ(N) Shear Stress at N Cycles [Symbol]

TNA Tapered Roller Bearing, Two-Row, Double-Cup Single Cone, Nonadjustable [Symbol]

TNA Technology Needs Assessment

TNA Telex Network Adapter

TNA Tetranitroaniline

TNA Thermal-Neutron Activation

TNA Thermal-Neutron Analysis

TNA Transient Network Analyzer

TNA 2,4,6-Trinitroaniline

TNAA Thermal Neutron Activation Analysis

TNAA Thermal Neutrons Atomic Absorption

TNAL Texas National Accelerator Laboratory [US]

TNALC Texas National Accelerator Laboratory Commission [US]

TNAmPh Trinitroaminophenol

TNAns Trinitroanisole

TNAZ 1,3,3-Trinitroazetidine

TNB Trinitrobenzene

TNBA Tri-n-Butyl Aluminum

TNBA Tri-n-Butylamine

TNBA Trinitrobenzaldehyde

TNBA Trinitrobenzoic Acid

TNBP Tri-n-Butylphosphate [also TnBP]

TNBS Trinitrobenzenesulfonate

TNBS 2,4,6-Trinitrobenzenesulfonic Acid

TNBT Tetranitroblue Tetrazolium

TNBT Tetra-N-Butyltitanate

TNBzN Trinitrobenzyl Nitrate

TNC TERENA (Trans-European Research and Education Networking Association) Networking Conference

TNC Terminal Network Controller

TNC Terminal Node Controller

TNC Tetranitrocarbazole

TNC The Nature Conservancy [US]

TNC Threaded-Nut Coupling

TNC Total Nonstructural Carbohydrates

TNC Trade Name Classification

TNC Trade Negotiations Committee [General Agreement on Tariffs and Trade]

TNC Transnational Corporation

TNC Transport Network Controller

TNC Trinitrocellulose

TN-C Tenascin-C [Cytology]

TNCB 2,4,6-Trinitrochlorobenzene

TNCIAWPRC Thai National Committee of the International Association on Water Pollution Research and Control

TNCIAWPRC Turkish National Committee of the International Association on Water Pollution Research and Control

TNClB Trinitrochlorobenzene

TNCrs Trinitrocresol

TND Trace Narcotics Detector

TND Tunesian Dinar [Currency]

TNDC Thai National Documentation Center [Bangkok, Thailand]

Tndcy Tendency [also tndcy]

TNDMA Trinitrodimethylaniline [also TNDMeA]

TNDPhA Trinitrodiphenylamine

TNDS Total Network Data System

TNEB Trinitroethylbenzene

TNEDV Trinitroethyldinitrovalerate [also TNEtDNV]

TNETB Trinitroethyltrinitrobutyrate [also TNEtTNBu]

TNEtB Trinitroethylbenzene

TNEtDNV Trinitroethyldinitrovalerate [also TNEDV]

TNEtTNBu Trinitroethyltrinitrobutyrate [also TNETB]

TNEL Total Noise Exposure Level

TNEP Total Noise Equivalent Power [Electronics]

TNF Teaching, Nontenure Track, Full-Time Faculty (Position)

TNF Theater Nuclear Forces [of NATO]

TNF Third Normal Form

TNF Transfer on No Overflow

TNF 2,4,7-Trinitro-9-Fluorenone

TNF Tumor Necrosis Factor [Biochemistry]

TNF-α Tumor Necrosis Factor-α [Biochemistry]

TNF-β Tumor Necrosis Factor-β [Biochemistry]

TNFE Twisted Nematic Field Effect [Electronics]

TNG Technology Networking Group [of National Research Council of Canada]

TNG Trinitroglycerin

Tng Training [also Tng]

tngt tonight

TNHP Texas Natural Heritage Program [US]

TNI The Networking Institute [US]

TNIP Terrestrial Network Interface Processor

TNIRI Tohoku National Industrial Research Institute [of Agency of Industrial Science and Technology, Japan]

TNIS Thermal Neutron Inelastic Scattering

TNL Terminal Net Loss

Tnl Tunnel [also tnl]

TNLDIO Tunnel Diode [also TnlDio]

TNM Tetranitromethane

TNM Tokyo National Museum [Japan]

TNM Trinitromethane [also TNMe]

TNMA Trinitromethylaniline [also TNMeA]

TNMe Trinitromethane [also TNM]

TNMeA Trinitromethylaniline [also TNMA]

TNMel Trinitromelamine

TNMes Trinitromesitylene

TNMIP Turbo Network Management Information Processor

TNMOC Total Non-Methane Organic Carbon

TNMR Topical Nuclear Magnetic Resonance

TNN Trinitronaphthalene

2NN Second Nearest Neighbor (Distance) [Crystallography]

TNO Tetranitrooxanilide

TNO The Netherlands Organization

TNO Thyssen Niederrhein AG, Werk Oberhausen [Oberhausen Plant of Thyssen Niederrhein, Germany]

TNO (Organisatie voor) Toegepast-Naturwetenschappelijk Onderzoek [Organization for Applied Scientific Research, Netherlands]

TNO Trade Negotiations Office

TNO Transnational Operation

TNOA Tri-n-Octylamine

TNOBA 2,4,6-Trinitrobenzoic Acid

TNO Mag TNO Magazine [Publication of Organisatie voor Toegepast-Naturwetenschappelijk Onderzoek, Netherlands]

TNOP Total Network Operations Plan

TNP 2,3-Tetranaphthoporphyrine

TNP Theoretical Nozzle Performance

TNP Trinitrophenol [Biochemistry]

TNP Trinitrophenyl [Biochemistry]

β-TNP β-Trinaphthylphosphate

TNPA Tri-n-Propyl Aluminum

TNPAL Trinitrophenylaminolauryl-

TNPAL-OH N-(2,4,6-Trinitrophenyl)-12-Aminolauric Acid

2,3-TNPFe Iron 2,3-Tetranaphthoporphyrine

TNPh Trinitrophenol

TNPhBuNA Trinitrophenylbutylnitramine [also TNPhBNA]

TNPhDA Trinitrophenylenediamine

TNPhEtNA Trinitrophenylethylnitramine [also TNPhENA]

TNPhMeNA Trinitrophenylmethylnitramine [also TNPhMNA]

TNPhMeNAPh Trinitrophenylmethylnitraminophenol [also TNPhMNAPh]

TNPht Trinitrophenetole

Tnpk Turnpike [also tnpk]

TNPP Trisnonylphenyl Phosphite

TNR Thermal-Neutron Radiograph(y)

TNR Thermonuclear Reaction; Thermonuclear Reactor

TNR Trinitroresorcinol [Biopchemistry]

TNRCC Texas Natural Resource Conservation Commission [US]

TNRF Thermal Neutron Research Facility

TNRF-1 Thermal Neutron Radiation Facility 1 [of Japan Atomic Energy Research Institute, Tokai, Japan]

TNRF-2 Thermal Neutron Radiation Facility 2 [of Japan Atomic Energy Research Institute, Tokai, Japan]

TNRLC Texas National Research Laboratory Commission [US]

TNS The Next Step

TNS Toluidinylnaphthylenesulfonate

TNS 6-(p-Toluidino)-2-Naphthalenesulfonic Acid

TNS Transaction Network Service [US]

2,6-TNS 6-(p-Toluidino)-2-Naphthalenesulfonic Acid

TNSC Tanegashima Space Center [of National Space Development Agency of Japan]

TNStl Trinitrostilbene

TNT Terminal Name Table

TNT Thallium Nitrate Trihydrate

TNT Total Network Test (System)

TNT Transient Nuclear Test

TNT Trinitrotoluene; 2,4,6-Trinitrotoluene

Tn-T Troponin T [Biochemistry]

TNTAB Trinitrotriazidobenzene

TNTBP Trinitrotoluene plus Black Powder [Explosive Mixture]

TNTCB Trinitrotrichlorobenzene [also TNTClB]

TNTMNA Trinitrotolylmethylnitramine

TNTN [1]Benzothieno[2,3-b][1]benzothiophene

TNTU O-(5-Norbornene-2,3-Dicarboximido)-N,N,N'N'-Tetramethyluronium Tetrafluoroborate

TNV Tobacco Necrosis Virus

TNX Thanks [Amateur Radio]

TNX Trinitroxylene

TNXCD Transaction Type Code

TNZ Transfer to Non-Zero

TO Operations Management [NASA Kennedy Space Center Directorate, Florida, US]

TO Takeoff [also T-O, or T/O]

TO Tantalum Oxide

TO Task Order

TO Technical Officer

TO Technical Order

TO *(Telegrafenordnung)* – German Telegraph Regulations

TO Telegraph Office

TO Telegraph Offices [of International Telecommunications Union]

TO Tensor Operation [Mathematics]

TO Test Operation

TO Thermo-Osmosis

TO Third-Order

TO Tilt Order

TO Time Out

TO Tin Oxide

TO Toll Office [Data Communications]

TO Tonga [ISO Code]

TO Traffic Order

TO Transfer Orbit [Aerospace]

TO Transformer Oil

TO Transistor Outline (Package)

TO Transverse Optic(al) (Mode)

TO Turbine Outlet

TO Turnover [also to]

TO$_1$ Transverse Optic(al), Rocking Vibration (Mode)

TO$_2$ Transverse Optic(al), Symmetric Stretching Vibration (Mode)

TO$_3$ Transverse Optic(al), Asymmetric Stretching Vibration (Mode)

T$_2$O Tritium Oxide [also TOD]

T&O Test and Operation(s)

T/O Takeoff [also T-O, or TO]

T/O Takeover

T/O Tetragonal-to-Orthorhombic (Transformation) [also T-O] [Metallurgy]

T/O Transfer Order

To Ton, Long [Unit]

TΩ Teraohm [Unit]

T(ω) Transmittance [Symbol]

θ-ω phase-modulation frequency characteristics [Electronics]

TOA Table of Authorities [also ToA]

TOA Takeoff Angle

TOA Telephone Office Automation

TOA tert-Octyl Acrylamide

TOA Time of Arrival

TOA Time-Off Award

TOA Tools-Oriented Approach

TOA Top of the Atmosphere

TOA Total Obligationary Authority [UK]

TOA Tri-n-Octylamine

TOAA (Study Group on) Takeoff Obstacle Accountability Areas

TOAA Trawler Owner's Association of Australia

TOABr Tetraoctyl Ammonium Bromide

TOACl Tetraoctyl Ammonium Chloride

TOADM Tele-Operated Area Defense Mine

TOADS Terminal-Oriented Administrative Data System

TOAHBr Trioctyl Ammonium Bromide

TOAHCl Trioctyl Ammonium Chloride

TOAHI Trioctyl Ammonium Iodide

TOAHSCN Trioctyl Ammonium Thiocyanate

TOAH$_2$SO$_4$ Trioctyl Ammonium Sulfate

TOAS Transmission Optical Absorption Spectroscopy

TOB TAB (Tape-Automated Bond) on Board

TOB Technical Operations Board

TOB Translation Operations Branch [of Department of the Secretary of State, Canada]

Tob Brz Tobin Bronze

TOBO 2-(p-Tolyl)benzoxazole

TOC Table of Coincidences

TOC Table of Contents [also ToC]

TOC Tag(liabue) Open-Cup (Flash Point Test)

TOC Task-Oriented Costing

TOC Technical Office for Consumers [UK]

TOC Television Operating Center

TOC Test Operations Center

TOC Test Operations Change

TOC Threshold Odor Concentration

TOC Timber Operators Council [US]

TOC Time of Contact

TOC Top of Climb [Aeronautics]

TOC Total Optical Color

TOC Total Organic Carbon

TOC Total Organic Compounds

ToC Table of Contents [also TOC]

TOCC Technical and Operational Control Center

TOCC Technical and Operational Coordination Center

TOCl Total Organic Halogen(s)

TOC-NDIR Total Organic Carbon/Nondispersive Infrared

TOCP Tri-o-Cresyl Phosphate

TOCS Terminal-Oriented Computer System

TOCS Testing Open Communications Systems (Project) [Canada]

TOCSY Total Correlation Spectroscopy

TOD Technical Objective Directive

TOD Technical Objective Document

TOD Theoretical Oxygen Demand

TOD Time of Day

TOD Time of Delivery

TOD Time of Departure

TOD Tooele Ordnance Depot [Utah, US]

TOD Top of Duct

TOD Total Oxygen Demand

TOD Tritium Oxide [also T_2O]

TODA Takeoff Distance Available

Today's Educ Today's Education [Publication of the National Education Association, US]

Today's Off Today's Office [Published in the US]

Today Technol Today Technology [Published in the US]

TODS Test-Oriented Disk System

TODS Transaction on Database System

TOE Tons of Oil Equivalent

TOE Total Operating Expenses

TOEC Third-Order Elastic Constant [Mechanics]

TOEFL Test of English as a Foreign Language

Toegep Wet TNO Toegepaste Wetenschap TNO [Dutch Publication of (Organisatie voor) Toegepast-Naturwetenschappelijk Onderzoek]

TOES The Other Economic Summit

TOES Trade-Off Evaluation System

TOF Phosphoric Acid Trioctyl Ester

TOF Test Operations Facility

TOF Time of Filing

TOF Time of Flight

TOF Tone Off

TOF To Order From

TOF Top of File

TOF Top of Form

TOF Tri(isooctyl)phosphate

TOFA Tall Oil Fatty Acid

TOFA Time-of-Flight Analysis

TOFC Trailer on Flatcar

TOF-ERDA Time-of-Flight Energy Recoil Detection Analysis

TOFF Thin Overlay for Friction

TOFI Time of Flight Isochronous (Spectrometer)

TOF-ICISS Time-of-Flight Impact Collision Ion Scattering Spectroscopy

TOFISS Time-of-Flight Ion Scattering Spectroscopy [also TOF ISS, or TOF-ISS]

TOFLEIS Time-of-Flight Low-Energy Ion Scattering [also TOF LEIS, or TOF-LEIS]

TOFMS Time-of-Flight Mass Spectrometer; Time-of-Flight Mass Spectrometry [also TOF-MS]

TOFO Trioctylphosphine Oxide

TOFPD Time-of-Flight Powder Diffractometer [also TOF-PD]

TOFS Time-of-Flight Spectrograph

TOF-SIMS Time-of-Flight Secondary-Ion Mass Spectrometry [also TOF SIMS, or TOFSIMS]

Tog Toggle

tog together

TOGA Tropical Ocean and Global Atmosphere (Experiment) [of World Meteorological Organization]

TOGA COARE Tropical Ocean and Global Atmosphere/Coupled Ocean Atmosphere Response Experiment [also TOGA/COARE]

TOGA NEG Tropical Ocean and Global Atmosphere Numerical Experiment Group

TOGA/TAO Tropical Ocean and Global Atmosphere/Tropical Atmosphere Ocean (Experiment)

TOH Total Organic Halogen

TOH Tyrosine Hydroxylase [Biochemistry]

TOHM Terohmmeter

Tohoku Geophys J, Sci Rep Tohoku Univ Tohoku Geophysical Journal, Science Reports of the Tohoku University [Japan]

Tohoku Geophys J, Sci Rep Tohoku Univ, Fifth Ser Tohoku Geophysical Journal, Science Reports of the Tohoku University Fifth Series [Japan]

Tohoku Math J Tohoku Mathematical Journal [of Tohoku University, Japan]

TOHP Takeoff Horsepower

TOI Technical Operation Instruction

TOI Transfer Orbit Insertion [Aerospace]

T1ICL Type-I Interband Cascade Laser

TOIRS Transfer Orbit Infrared Earth Sensor

TOJ Track on Jam(ming)

TOK Türk Otomatik Kontrol Milli Komitesi [Turkish Committee for Automation and Control]

Tokyo Astron Bull Tokyo Astronomical Bulletin [Japan]

Tokyo Astron Obs, Kiso Inf Bull Tokyo Astronomical Observatory, Kiso Information Bulletin [Japan]

Tokyo Astron Obs Rep Tokyo Astronomical Observatory Report [Japan]

Tokyo Astron Obs Time Latit Bull Tokyo Astronomical Observatory Time and Latitude Bulletins [Japan]

TOL Temporary Occupation Licence

TOL Test-Oriented Language

Tol Tolerance [also tol]

Tol Toluene

TOLAR Terminal On-Line Availability Reporting

TOLD Telecommunications On-Line Data System

TOLED Transparent Organic Light Emitting Device

TOLIP Trajectory Optimization and Linearized Pitch

TO-LO Transverse Optic(al)–Longitudinal Optic(al) (Mode Splitting)

TOLR Transmitting Objective Loudness Rating

TOLT Teleprocessing On-Line Test

TOLTEP Teleprocessing On-Line Test Executive Program

TOLTS Total On-Line Testing System

TOLZ Third-Order Laue Zone [Crystallography]

TOM Table Output from Memory

TOM Teleprinter-on-Multiplex

TOM Tool Material

TOM Translation On-Line Marketplace [of Public Works and Government Services of Canada]

TOM Translator Octal Mnemonic

TOM Trioctyl Mellitate

TOM Typical Ocean Model

TOMAL Task-Oriented Microprocessor Application(s) Language

TOMCAT Telemetry On-Line Monitoring Compression and Transmission

TOMP Tris(hydroxymethyl)phosphine

TOMS Torus Oxygen Monitoring System [Nuclear Engineering]

TOMS Total Ozone Mapper (or Mapping) System [of NASA]

TOMS Total Ozone Mapping Spectrometer [of NASA]

TOMUIS Three-Dimensional Ozone Mapping with Ultraviolet Imaging Spectrometer

ToMV Tomato Mosaic Virus

TON Tone On

Ton Ton-force [Unit]

ton ton, long [Unit]

ton/cu yd (long) ton per cubic yard [also ton/yd^3] [Unit]

tonf ton-force [Unit]

tonf/in² ton(-force) per square inch [also TSI or tsi]

tonf/mm² ton(-force) per square millimeter [Unit]

ton/h (metric) ton(s) per hour [also ton h^{-1}]

ton/hr ton(s) per hour [also ton hr^{-1}, or t/hr]

Ton/in³ Ton-force per square inch [also Ton/sq in]

Tonind Ztg Tonindustrie-Zeitung [German Journal of the Brick and Clay Industries]

Tonind Ztg keram Rdsch Tonindustrie-Zeitung und keramische Rundschau [German Journal and Review of Ceramics and the Brick and Clay Industries]

TONLAR Tone-Operated Net Loss Adjuster [also Tonlar, or tonlar]

Tonn Tonnage [also tonn]

Ton/sq in Ton-force per square inch [also Ton/in³]

ton/yd³ (long) ton per cubic yard [also ton/cu yd]

TOO Target of Opportunity

TOO Test Operations Order

TOO Threshold of Odor

TOOL Test-Oriented Operation Language

Tool Tooling [also tool]

Tool Alloy Steels Tool and Alloy Steels [Journal published in India]

Tool Prod Tooling and Production [Journal published in the US]

Tool Prog Tooling Progress [Published in the US]

TOOLS Total Operating On-Line System

Tool Trends Tooling Trends [Published in the US]

TOOS N-Ethyl-N-(2-Hydroxy-3-Sulfopropyl)-m-Toluidine; 3-(N-Ethyl-m-Toluidino)-2-Hydroxypropanesulfonic Acid

TOP Tape Output

TOP Technical and Office Protocol

TOP Technical Office Protocol [Communications]

TOP Technical Operating Procedure

TOP Temporary Operating Permit

TOP Terrestrial Observation Panel [Global Climate Observation and Supervision/Global Terrestrial Observing System]

TOP Tether Optical Phenomena

TOP Tongan Pa'anga [Currency of Tonga]

TOP Topeka, Kansas [Meteorological Station Designator]

TOP Total Obscuring Power

TOP Transaction-Oriented Programming

TOP Transient Overpower Accident [Nuclear Reactors]

TOP Tri(isooctyl)phosphate

TOP Trioctyl Phosphate

Top Topic [also top]

TOPCAT Texas On-Board Program of Computer-Assisted Training [US]

TOPCOPS The Ottawa Police Computerized On-Line Processing System [Canada]

TOPES Telephone Office Planning and Engineering System

TOPEX (Ocean) Topography Experiment [of NASA]

TOPEX Typhoon Operation Experiment

TOPEX/Poseidon (Ocean) Topography Experiment–Poseidon [A Joint US-France Project]

TOPIC Time Ordered Programmer Integrated Circuit

TOPICS Transport Operations Program for Increasing Capacity and Safety

TOPIS Topographic Information Sytstem [Digital Topographic Database of South Australia]

TOPM Takeoff Performance Monitor

TOPM Tetra(isooctyl)pyromellitate

TOPO Tri-n-Octylphosphine Oxide

TOPO-250K 1:250,000 Topographic Map Series [of Australian Surveying and Land Information Group]

Topog Topographical; Topography [also Topogr, topogr, or topog]

TOPP Task-Oriented Plant Practice

TOPP Terminal-Operated Production Program

TOPR Taiwan Open Pool Reactor

TOPS 3-[Ethyl-(3-Methylphenyl)amino]-1-Propanesulfonic Acid

TOPS N-Ethyl-N-Sulfopropyl-m-Toluidine

TOPS Telemetry Operations

TOPS Telephone Order Processing System

TOPS Teletype Optical Projection System

TOPS Test and Operating System

TOPS The Operational PERT (Production Evaluation and Review Technique) System

TOPS Thermoelectric Outer Plant Spacecraft

TOPS Time-Sharing Operating System

TOPS Total Operations Processing System

TOPS Traffic Operator Position System [Telecommunications]

TOPS Transducer Operated Pressure System

TOPS Transistorized Operational Phone System

TOPS Trends in Optics and Photonics Series [Publication of the Optical Society of America]

TOPSe Trioctylphosphine Selenide

TopSec Top Secret

TOPSI Topside Sounder, Ionosphere

TOPSY Territorial Operational System

TOPSY Test Operations and Planning System

TOPSY Thermally Operated Plasma System

TOPSYS Tools for Parallel Systems

Top Times Topical Times [of American Topical Association, US]

TOPTS Test-Oriented Paper-Tape System

TOPV Trivalent Oral Polio Vaccine

T1QWL Type-I Quantum-Well Laser

TOR Take-Off Run [Aeronautics]

TOR Technical Operations Research

TOR Technology-Oriented Research

TOR Telegraph-on-Radio

TOR Teleprinter-on-Radio

TOR Teleprinter over Radio

TOR Time of Receipt

TOR Transpolyoctenamer

TOR Tropospheric Ozone Research (Project) [Part of EUREKA]

Tor Toronto [Canada]

Tor Torpedo [also tor]

Tor Torque [also tor]

TORA Takeoff Run Available [Aeronautics]

TORC Traffic Overflow Reroute Control; Traffic Overload Reroute Control

TORO TOCSY (Total Correlation Spectroscopy)–ROESY (Rotating Frame Overhauser Effect Spectroscopy)

Torp Torpedo [also torp]

TORPCM Torpedo Counter-Measures

torr cm torr-centimeter(s) [also torr-cm]

torr·L/s torr-liter(s) per second [also torr·L/sec]

torr·L/s-He torr liter(s) per second of helium [also torr·L/sec-He]

TORRO Tornado and Storm Research Organization [UK]

torr s torr-second [also torr-s]

TorT Torpedo Tube

TOS Tactical Operations System

TOS Taken Out of Service

TOS Tape-on-Substrate (Technology)

TOS Tape-Operating System

TOS Temporarily Out of Service

TOS Terminal-Oriented Software

TOS Terminal-Oriented System

TOS Test Operating System

TOS Thermal Oxidative Stability

TOS Time-Ordered System

TOS TIROS (Television Infrared Observation Satellite) Operational Satellite (System) [of NASA]

TOS Top of Stack

TOS Top of Steel

TOS Traffic Orientation Scheme

TOS Transfer Orbit Stage

Tos Toluenesulfonyl

Tos Tosyl

TOSA Takeoff Space Available [Aeronautics]

TOSBAC Toshiba Scientific and Business Automatic Computer

TOSC p-Toluenesulfonyl Chloride

TOSCA Topological Oscillation Search with kinematiCal Analysis (Experiment) [of CERN–European Laboratory for Particle Physics, Geneva, Switzerland]

TOSCA Toxic Substances Control Act [US]

TOSCW Top of Stack Control Word

TOSD Telephone Operations and Standards Division [of Rural Electrification Association, US]

Toshiba Rev Toshiba Review [Japan]

TOSI Trade Office of Swiss Industries

TOSL Terminal-Oriented Service Language

TOSMIC p-Toluenesulfonyl Methylisocyanide

TosMIC Tosylmethylisocyanide [also TOSMIC]

TOSP Top of Stack Pointer

TOSPIX-i Toshiba (Company) Image Analysis System [also TosPIX-i]

TOSS Terminal-Oriented Support System

TOSS Test Operation Support Segment

TOSS TIROS (Television Infrared Observation Satellite) Operational Satellite System [of NASA]

TOSS Total Suppression of Sidebands

TOSS Transient and/or Steady State

TOSS Two-Step Oxidation of Sidewall Surface

TOSSA Transfer Orbit Sun Sensor Assembly [Aerospace]

TOST Turbine Oxidation Stability Test

TOT Time of Tape

TOT Time of Transmission

TOT Transfer of Technology

TOT Tritium Oxide [also T_2O]

tot total(ly)

TOTE Terminal On-Line Test Executive

TOTEM Total Cross Section, Elastic Scattering and Diffraction Dissociation (Experiment) at the LHC Large Hadron Collider [of CERN–Laboratory for Particle Physics, Geneva, Switzerland]

TOTM Tri(isooctyl)trimellitate; Trioctyl Trimellitate

TOTU O-([Carbethoxy]cyanomethylenamino)-N,N,N',N'-Tetramethyluronium Tetrafluoroborate

TOU Technologieorientierte Unternehmensgründungen (Modellversuch) [Establishment of Technology-Oriented Companies (Experiment); of Bundesministerium für Forschung und Technology, Germany]

TOUGH Transport of Unsaturated Groundwater and Heat (Code)

Toute Electron Toute l'Electronique [French Electronics Magazine]

TOVALOP Tanker Owners Voluntary Agreement concerning Liability for Oil Pollution [of International Tanker Owners Pollution Federation, UK]

TOVC Top of Overcast

TOVS TIROS (Television Infrared Observation Satellite) Operational Vertical Sounder [of NASA]

TOW Takeoff Weight

TOW Tank and Orbiter Weight [NASA Space Shuttle]

TOW Transport on Water Association [UK]

TOW Tube-Launched, Optically-Tracked, Wire-Guided (Antitank Missile) [also tow]

TOWA Terrain and Obstacle Warning and Avoidance

Townsend Let Doct Townsend Letter for Doctors [Published at Port Townsend, Washington, US]

TOX Thermal Oxide [also Tox]

TOX Total Organic Halide(s)

TOX Toxaphene

TOXBACK Toxicology Information Service [of National Library of Medicine, Bethesda, Maryland, US]

TOXBIB Toxicology Bibliography (Database) [of National Library of Medicine, Bethesda, Maryland, US]

TOXCON Toxicology Information Conversation On-Line Network [now TOXLINE]

ToxFAQs Toxicology Frequently Asked Questions [of Agency for Toxic Substances and Disease Registry, US]

Toxicol Toxicological; Toxicologist; Toxicology [also toxicol]

Toxicol Appl Pharmacol Toxicology and Applied Pharmacology [Journal of the Society of Toxicology, US]

TOXLINE Toxicology Information On-Line (Database) [of National Library of Medicine, Bethesda, Maryland, US]

TOXNET Toxicology Network [US]

TOXTIPS Toxicology Test in Progress (Database) [of National Library of Medicine, Bethesda, Maryland, US]

Toxy Toxicity [also toxy]

Toyota Gosei Tech Rev Toyota Gosei Technical Review [Published by Toyoda Gosei Co. Ltd., Aichi, Japan]

Toyo Kohan Tech Rep Toyo Kohan Technical Report [Published by Toyo Kohan Co, Kudamatsu, Japan]

Toyo's Tech Bull Toyo's Technical Bulletin [Published by Toyo Communication Equipment Co. Ltd., Kanagawa, Japan]

t oz troy ounce [Unit]

TP East Timor [ISO Code]

TP Tail Pipe

TP Tandem Propellers [Shipbuilding]

TP Tank Pressure

TP Tape [also Tp, or tp]

TP Target Practice

TP Technical Pamphlet

TP Technical Paper

TP Technical Physics

TP Technical Program

TP Technical Publication

TP Technological Properties

TP Tectonophysical; Tectonophysicist; Tectonophysics

TP Teesside Polytechnic [Middlesbrough, Cleveland, UK]

TP Telemetry Processing

TP Telephone

TP Telephotography

TP Telephotometer; Telephotometry

TP Teleprinter

TP Teleprocessing; Teleprocessor

TP Teletype Printer

TP Telenet Processor

TP Temperature Pod

TP Temperature Profile

TP Temperature Program(mer)

TP Tender Panel

TP Terminal Pole

TP Terminal Processor

TP Terphenyl

TP Terrestrial Planets

TP Testing Pressure

TP Testosterone Propionate [Biochemistry]

TP Test Paper [Chemistry]

TP Test Piece

TP Test Plate

TP Test Point

TP Test Port

TP Test Position

TP Test Procedure

TP Test Program

TP Test-Stop Pulse

TP Tetracritical Point

TP Text Processing

TP Thames Polytechnic [London, UK]

TP Theoretical Physicist; Theoretical Physics

TP Theoretical Plate

TP Thermal Paint

TP Thermal Power

TP Thermal Printer

TP Thermal Process(ing)

TP Thermal Properties

TP Thermal Protection

TP Thermophysic(al); Thermophysicist; Thermophysics

TP Thermopile

TP Thermoplastic(s)

TP Thermoplastic Preform

TP Theta(tron) Pinch [Plasma Physics]

TP Thiophosphamide

TP Thromboplastin [Biochemistry]

TP Throttle Positioner

TP Throughput

TP Throwing Power [Electroplating]

TP Thrust Plate [Geology]

TP Thrust Roller Bearing, Single-Direction, Flat Seats, Cylindrical Rollers [Symbol]

TP Thymidine Phosphate [Biochemistry]

TP Thymolphthalein

TP Thymopoietin [Biochemistry]

TP Tidal Power

TP Time Period

TP Time Pulse

TP Time to Perigee

TP Timing Pulse

TP Title Page [also tp]

TP Toll Point

TP Toll Prefix

TP Torpedo Part of Beam

TP Torque Pressure

TP Torsion(al) Pendulum

TP Total Pressure

TP Total Protein [Biochemistry]

TP Township

TP Tracking Powder

TP Track Pitch [also tp] [Magnetic Recording Media]

TP Tracks per Inch [also tpi]

TP Trading Partner

TP Tractor Propeller

TP Training Plan

TP Trajectory Parabola

TP Transaction Paper

TP Transaction Processing (System)

TP Transform Processing

TP Transient Program

TP Transition Period

TP Transition Point

TP Transition Probability

TP Transmission Planar

TP Transportation Planning

TP Transport Phenomena

TP Transport Protocol [of International Organization for Standardization]

TP Tree Preservation

TP Tribophysic(al); Tribophysicist; Tribophysics

TP Triphenyl Phosphate

TP Triphosphate

TP Triple-Play (Tape)

TP Triple Point [also tp]

TP Triple Point (for Grain Boundaries) [Materials Science]

TP Triple Pole [also 3P]

TP Triple Play

TP Tris[2-(Diphenylphosphino)ethyl]phosphine

TP True Position

TP True Positive

TP Tryptophan Pyrrolase [Biochemistry]

TP Tuned Plate

TP Turbine Power

TP Turbine Pump

TP Turbo Pascal (Programming Software)

TP Turboprop(eller)

TP Turbopump

TP Turning Point

TP Tutorial Program

TP Twin Paradox [Relativity]

TP Twin(ning) Plane [Crystallography]

TP Two Photon (Transition) [Solid-State Physics]

TP Type

.TP Turbo Pascal Binary Configuration File [File Name Extension]

TP0 Transport Protocol, Class 0 [of International Organization for Standardization]

TP1 Transport Protocol, Class 1 [of International Organization for Standardization]

TP2 Transport Protocol, Class 2 [of International Organization for Standardization]

TP3 Transport Protocol, Class 3 [of International Organization for Standardization]

TP4 Transport Protocol, Class 4 [of International Organization for Standardization]

2,4,5-TP 2-(2,4,5-Trichlorophenoxy)propionic Acid

T/P Tank Piercing

T-P Temperature–Pressure (Diagram)

T-5'-P Thymidine 5'-Monophosphate [Biochemistry]

Tp Township [also tp]

Tp Troop [also tp]

T(p) Optical Transfer Function [Symbol]

T$ Tongan Pa'anga [Currency of Tonga]

tP Pearson symbol for primitive (simple) space lattice in tetragonal crystal system (this symbol is followed by the number of atoms per unit cell, e.g. tI6, tI16, etc.)

tp title page [also TP]

3P Three-Pole

3P Triple-Pole [also TP]

3Ψ Three-Fold Symmetry Distribution of Second Harmonic Field [Symbol]

TPA Tantalum Producers of America [US]

TPA Tape Pulse Amplifier

TPA Technical Publications Association

TPA Telephone Pioneers of America [US]

TPA Tennessee Pharmaceutical Association [US]

TPA Terephthalic Acid

TPA Test Preparation Area

TPA 12-O-Tetradecanoylphorbol 13-Acetate

TPA Tetragonolobus Purpureas Agglutinin [Immunology]

TPA Tetraphenyl Arsonium

TPA Tetrapropyl Ammonium

TPA Texas Pharmaceutical Association [US]

TPA The Plasminogen Activator [Biochemistry]

TPA Thermal Plasma Analyzer

TPA Timber Producers Association [US]

TPA Tissue Plasminogen Activator [Biochemistry]

TPA Toll Pulse Accepter

TPA Traffic Pattern Altitude

TPA Transient Program Area

TPA Transmission Products Association [US]

TPA Transpolypentenamer

TPA Tri-Party Agreement

TPA Triphenylamine

TPA Triphenylarsine

TPA Tri-n-Propylamine

TPA Two-Photon Absorption

TPA Two-Point Approximation

TPa Terapascal [Unit]

T/Pa Tersla(s) per Pascal [also Tpa^{-1}]

Tpa Tris(2-Picolinyl)amine [also tpa]

tPA Tissue Plasminogen Activator [also t-PA] [Biochemistry]

t/pa ton(ne)s per annum

TPABF$_4$ Tetrapropyl Ammonium Tetrafluoroborate

TPABr Tetrapropyl Ammonium Bromide

TPACl Tetrapropyl Ammonium Chloride

TPAD Temperature-Programmed Ammonia Desorption

TPAD Time Differential Perturbed Angular Distribution [Physics]

TPAD Trunnion Pin Acquisition (or Attachment) Device [Aerospace]

T-PAD Terminal Packet Assembler/Disassembler

TPAHS Tetrapropyl Ammonium Hydrogen Sulfate

TPAI Tetrapropyl Ammonium Iodide

TPAM Teleprocessing Access Method

Tp[10]aneN3 N,N′,N″-Tris(2-Pyridylmethyl)-1,4,7-Triazacyclodecane [also tp[10]aneN3]

TPAO Triphenyl Arsine Oxide

TPAO Turkish Petroleum Corporation

TPA-OH Tetrapropyl Ammonium Hydroxide [also TPAOH]

TPAP Tetrapropylammonium Perruthenate

TPAPF₆ Tetrapropyl Ammonium Hexafluorophosphate

2PAPMM Two-Pulse Amplitude and Phase Modulation Modem

TPAs Triphenylarsine

TPAsCl Tetraphenylarsonium Chloride

TPAsO Triphenylarsenic Oxide

TPAsTPB Tetraphenylarsonium Tetraphenylborate

TPB Tetradecylpyridinium Bromide

TPB Tetraphenylbenzidine

TPB Tetraphenylborate

TPB 1,1,4,4-Tetraphenyl-1,3-Butadiene

TPB Three-Point Bending [also 3PB]

TPB Triphenylboron

2PB Two-Point Bending

3PB Three-Point Bending [also TPB]

TPBF Three-Point Bending Fatigue

TPBi Triphenyl Bismuth

TPBM Triphenylbromomethane

Tpbn N,N,N′,N′-Tetrakis(2-Pyridylmethyl)-1,4-Butanediamine [also tbpn]

TPBVP Two-Point Boundary Value Problem

TPC 7,8-Dihydro-5,10,15,20-Tetraphenylporphinate

TPC Tactical Pilotage Chart

TPC Taiwan Power Company

TPC Technical Practices Committee [of NACE (National Association of Corrosion Engineers) International, US]

TPC Technology Partnerships Canada [of Industry Canada]

TPC Telecommunications Planning Committee

TPC Telemetry Preprocessor Computer

TPC Thermofor Pyrolytic Cracking (Process)

TPC Thermoplastic Composite

TPC Three-Phase Catalysis

TPC Thymolphthalein Complexone [Biochemistry]

TPC Time Pickoff Control

TPC Time Projection Chamber [Nuclear Engineering]

TPC Time-to-Pulse-Height Converter

TPC Tire Performance Criteria

TPC Totally Pyrolytic Carbon

TPC Totally Pyrolytic Cuvette

TPC Total Process Control

TPC Total Project Costs

TPC Trade Practices Commission

TPC Transaction-Processing Performance Council [US]

TPC Transient Photoconductivity

TPC Trans-Pacific Cable

TPC Trifluoroacetylprolyl Chloride

TPC Triphenyl Carbinol

TPC Triple-Paper-Covered (Cable)

TPC True Power Control

TP18C6 Tetraphenyl-18-Crown-6

TP&C Thermal Protection and Control

t/pc ton(s) per piece [Unit]

TPCA Tobacco Products Control Act [Canada]

TPCD Tetraphenylcyclopentadienone

Tpchxn N,N,N′,N′-Tetrakis(2-Pyridylmethyl)-1,2-Cyclohexanediamine [also tpchxn]

TPCIPP 5,10,15,20-Tetrakis(p-Chlorophenyl)porphinate

TPCK Tosylamino-2-Phenylethyl Chloromethyl Ketone; N-Tosyl-L-Phenylalanine Chloromethyl Ketone

TPCM Triphenylchloromethane

tpcm turns per centimeter (of yarn) [Unit]

TPCOMP Tape Compare

TPC RPC Technical Practices Committee/Reference Publications Committee [of NACE (National Association of Corrosion Engineers) International, US]

TPCU Test Power Control Unit

TPCU Thermal Preconditioned Unit

TPCV Turbine Power Control Valve

TPD Temperature-Programmed Decomposition

TPD Temperature-Programmed Desorption

TPD 1,3,4,5-Tetrathiapentalene-2,5-Dione

TPD Theoretical Physics Department

TPD Thermoplastic Photoconductor Device

TPD Third Party Database

TPD Time Pulse Distributor

TPD Tons per Day [also tpd]

TPD Transverse Photothermal Deflection

TPD Two-Photon Dissociation

TPDB Third Party Database

TPDC Tanzania Petroleum Development Corporation

TP-DDI Twisted Pair Distributed Data Interface

TPDICS Two-Photon Differential Cross-Section

TPDMDS 1,1,3,3-Tetraphenyl-1,3-Dimethyl Disilazane

TPDS Test Procedures Development System

3PDT Triple-Pole, Double-Throw [also TPDT]

3PDT SW Triple-Pole, Double-Throw Switch [also TPDT SW]

TPDU Transport Protocol Data Unit

TPDUP Tape Duplicate

TPE Technical Plastics Extruder

TPE Teleprocessing Executive

TPE Test Project Engineer

TPE Tetraphenyl Ethene

TPE Tetraphenylethylene

TPE Thermoplastic Elastomer

TPE Thermoplastic Polyethylene

TPE Three-Point Extension

TPE Transmission Parity Error

TPE Transplutonium Elements

TPE Tris-Phosphate EDTA (Ethylenediaminetetraacetic Acid)

TPE Turboprop Engine

TPE Twisted Pair Ethernet

TPE Two-Photon Excitation [Physics]

TPE Two Pion Exchange [Particle Physics]

TPE Typhoid-Paratyphoid-Enteric (Pathogen)

TPEA Tetrapentyl Ammonium Halide

TPE-A Thermoplastic Polyetheramide Elastomer

TPE-E Thermoplastic Polyetherester Elastomer

TPEF Two-Photon Excited Fluorescence [Physics]

TPE-FKM Thermoplastic Fluorelastomer

TPE HNC/MS Three-Point-Extension Hypernetted Chain Mean Spherical (Equation) [Physics]

TPEN N,N,N'N'-tetrakis(2-Pyridylmethyl)ethylenediamine

TPEN N,N,N',N'-tetrakis(2-Picolinyl)ethylenediamine [also Tpen, or tpen]

TPE-NBR Thermoplastic Elastomer Based on Nitrile Butadiene Rubber

TPE-NR Thermoplastic Elastomer Based on Natural Rubber

TPE-O Thermoplastic Elastomer Based on Polyolefins; Thermoplastic Elastomer–Olefinic [also TPO]

TPEON Tripentaerythritol Octanitrate

TPEP 1-(3-Trifluoromethylphenyl)-2-Ethylaminopropane

TPEPICO Threshold Photoelectron Photo-Ion Coincidence [Physics]

TPES Threshold Photoelectron Spectrum [Physics]

TPES Two-Photon Excitation Spectrum [Physics]

TPE-S Thermoplastic Elastomer Based on Styrene Butadiene Styrene; Thermoplastic Elastomer– Styrenic

T-PES Terminated Polyethersulfone

TPE-U Thermoplastic Elastomer Based on Polyurethanes

TPF Phosphoric Acid Triphenyl Ester

TPF Taper per Foot

TPF Terminal Phase Finalization; Terminal Phase Finish

TPF Tetraphenylfuran

TPF Time Prism Filter

TPF Track Pick Fragments

TPF Transaction Processing Facility

TPF Transfer Phase Final

TPF Tug Processing Facility

TPF Two-Phase Flow

TPF Two Photon Fluorescence

TPFA Taiwan Provincial Farmers Association

TPFI Terminal Pin Fault Insertion

TPFP Thermoplastic Fluoropolymer

TPG Telecommunications Program Generator

TPG Time Pulse Generator

TPG Tin Plate Gauge

TPG Total Pressure Gage

TPG Transmission Project Group [of Central Electricity Generating Board, UK]

TPG Triphenylguanidine [Biochemistry]

TPG Tripropylene Glycol

TPG Trypticase Peptone Glucose [Biochemistry]

Tpg Topping [also tpg]

TPGDA Tripropylene Glycol Diacrylate

TPGC Temperature-Programmed Gas Chromatography

TPGeCl Triphenylgermanium Chloride

TPH Tocco (Inc.) Profile Hardening [Metallurgy]

TPH Ton(s) per Hour [also tph]

TPH Total Petroleum Hydrocarbon(s)

2PH Two-Phase

TPHA Tetraethylene Pentamine Heptaacetic Acid

TPHA Texas Public Health Association [US]

TPhA Terephthalic Acid

TPHC Time-to-Pulse-Height Converter

TPHCWG Total Petroleum Hydrocarbon Compounds Working Group [of US Environmental Protection Agency]

TPhP Triphenylphosphate

TPhPO Triphenyl Phosphine Oxide

TPHT Two-Phase Heat Transfer

2ph-TDA Two-Particle One-Hole Tamm-Dancoff Approximation [Physics]

TPI Tape Phase Inverter

TPI Taper per Inch

TPI Target Position Indicator

TPI Tax and Price Index

TPI Teeth per Inch

TPI Tennessee Polytechnic Institute [Cookeville, US]

TPI Terminal Phase Initiate; Terminal Phase Initiation

TPI Theoretical Physics Institute [of University of Minnesota, Minneapolis, US]

TPI Thermoplastic Imide

TPI Thermoplastic Polyimide

TPI Threads per Inch

TPI Town Planning Institute [UK]

TPI Tracks per Inch [also tpi]

TPI Transmission Performance Index

TPI Treponema Pallidum Immobilization (Venereal Disease Test)

TPI Triosephosphate Isomerase [Biochemistry]

TPI Tropical Products Institute [UK]

TPI Truss Plate Institute [US]

TPI Turns per Inch

tpi tracks per inch [also TPI]

TPIBC 2,3,7,8-Tetrahydro-5,10,15,20-Tetraphenyl-porphinate

TPIC Town Planning Institute of Canada

TPIS Tropical Pesticides Information Service [UK]

TPISC Tantalum Producers International Study Center [Belgium]

TPIVPP 5,10,15,20-Tetrakis[o-(Pivaloamido)phenyl]porphine

Tpk Turnpike [also Tpke, tpke, or tpk]

TPL Table Producing Language

TPL Telephoto Lens

TPL Tetraphenyl Lead

TPL Tempol

TPL Terminal Per Line

TPL Terminal Processing Language

TPL Test Parts List

TPL Test Processing Language

TPL Toll Pole Line

TPL Toronto Public Library [Canada]

TPL Total Phospholipid [Biochemistry]

TPL Transaction Processing Language

TPL Tritium Process(ing) Laboratory [of Japan Atomic Energy Research Institute]

TPL Turbo Pascal Library (File)

TPL Two-Photon Luminescence

.TPL Turbo Pascal Library [File Name Extension]

TPLA The Product Liability Alliance [US]

TPLAB Tape Label (Information)

TPM Tape Preventive Maintenance

TPM Technical Performance Measure(ment)

TPM Technical Program Manager

TPM Telemetry Processor Module

TPM Terminal Phase Maneuver

TPM Terminal Phase Midcourse

TPM Tetraphenylmethane

TPM Tilting Plate Micrometer

TPM Timber Products Manufacturers [US]

TPM Total Population Management [Entomology]

TPM Total Preventive Maintenance

TPM Transactions Per Minute

TPM Transfer Phase Midcourse

TPM Transmission and Processing Model

TPM Transport Planning Mobilization

TPM Triphenylmethane

TPM Triphenylmethyl

.TPM Turbo Pascal Map [File Name Extension]

TPM1 Terminal Phase Midcourse 1

TPM2 Terminal Phase Midcourse 2

Tpm 1,1,1-Trifluoro-5,5-Dimethyl-2,4-Hexanedionate [also tpm]

TPMA Thermodynamic Properties of Metals and Alloys

TPMA Timber Products Manufacturers Association [now Timber Products Manufacturers, US]

Tpmbn N,N,N',N'-Tetrakis(2-Pyridylmethyl)butane-2,3-Diamine

TPMM Teleprocessing Multiplexer Module

TP/mm Taper per Millimeter

TPMP Triphenylmethyl Phosphonium

TPMS Technical Performance Measurement System

TPMS Transaction Processing Management System

TPMS Triply Periodic Minimal Surfaces [Crystallography]

TPN α-Tetrahydrofuranyloxy Phenylnitroxide

TPN β-Triphosphopyridine Nucleotide [also β-TPN] [Biochemistry]

TPN Two-Port Network

α-TPN α-Nicotinamide Adenine Dinucleotide Phosphate [Biochemistry]

β-TPN β-Triphosphopyridine Nucleotide [also TPN] [Biochemistry]

3'-TPN β-Nicotinamide Adenine Dinucleotide 3'-Phosphate [Biochemistry]

TPNH Triphosphopyridine Nucleotide, Reduced Form [also β-TPNH] [Biochemistry]

α-TPNH α-Nicotinamide Adenine Dinucleotide Phosphate, Reduced Form [Biochemistry]

β-TPNH β-Triphosphopyridine Nucleotide [also TPNH] [Biochemistry]

3'-TPNH β-Nicotinamide Adenine Dinucleotide 3'-Phosphate, Reduced Form [Biochemistry]

TPNS Teleprocessing Network Simulator

TPO Technical Program Office(r)

TPO Technical Project Office(r)

TPO Telecommunications Program Objective

TPO Temperature-Programmed Oxidation

TPO Test Program Outline

TPO Thermoplastic Olefin; Thermoplastic Elastomer Based on Polyolefins; Thermoplastic Elastomer–Olefinic [also TPE-O]

TPO Tree Preservation Order

TPO Tropical Pacific Ocean

TPO Tryptophan Pyrrolase [Biochemistry]

TPO/EPDM Thermoplastic Olefin/Ethylene-Propylene Diene Monomer

TPOP Time-Phased Order Point

TPODS Two-Photon Oscillator Strength Density

TPORT Twisted Pair Port Transceiver [of AT&T Company, US]

TPOT Tetrapropyl Orthotitanate

TPP Technical Program Plan

TPP Telephony Preprocessor

TPP Test Plan and Procedures

TPP Test Point Pace

TPP 5,10,15,20-Tetraphenyl Porphinate

TPP Tetraphenylporphine

TPP Tetraphenylporphyrine

TPP Tetraphenylpyrrole

TPP Theoretical Pairwise Potential [Physics]

TPP The Pharmaceutical Press [London, UK]

TPP Thermal Power Plant

TPP Thermal Protective Performance (Rating)

TPP Thiamine Pyrophosphate

TPP Tidal Power Plant

TPP Transuranium Processing Plant [of Oak Ridge National Laboratory, Tennessee, US]

TPP Triphenyl Phosphate

TPP Triphenyl Phosphine

TPP Triphenyl Phosphite

θ,φ,ψ (Three) Eulerian Angles [Symbol]

TPPb Tetraphenyl Lead

TPPA N,N',N'',N'''-Tetraphenylpyromellitamide

TPPC Total Package Procurement Concept

TP-PCB Teleprocessing Program Communication Block

TPPCF Thousand Particles per Cubic Foot [also tppcf]

TPPD Technical Program Planning Division

TPPH2 5,10,15,20-Tetraphenylporphine

TPPI Time-Proportional Phase Incrementation

TPPME 1,1,1-Tris(diphenylphosphinomethyl)ethane [also Tppme, or tppme]

TPPN N,N,N',N'-Tetrakis(2-Pyridylmethyl)propane-1,2-Diamine [also Tppn, or tppn]

TPPO Thermoplastic Polyolefin

TPPO Triphenyl Phosphine Oxide

TPPP Transient Plastic-Phase Processing

TPPS 5,10,15,20-Tetrakis-(p-Sulfonatophenyl)porphinate

TPPT Triphenyl Phosphorothionate

TPQ Threshold Planning Quantities

TPQ Two-Piston Quenching [Metallurgy]

TPR Tape Programmed Row

TPR Teleprinter [also Tpr]

TPR Telescopic Photographic Recorder

TPR Temperature-Programmed Reaction

TPR Temperature-Programmed Reduction

TPR Temperature-Programmed Reoxidation

TPR Temperature, Pulse, Respiration [Medicine]

TPR Terrain Profile Recorder

TPR Test Problem Report

TPR Thermoplastic Recording [Electronics]

TPR Thermoplastic Resin(s)

TPR Thermoplastic Rubber

TPR Total Peripheral Resistance

TPR Total Pitting Rate [Corrosion Engineering]

TPR T-Pulse Response

Tpr Taper [also tpr]

Tpr Teleprinter [also tpr]

Tpr Trooper [also tpr]

TPRA Tape-to-Random Access

TPRC Thermophysical Properties Research Center [of Purdue University, West Lafayette, Indiana, US]

TPRC Trade Policy Research Center [UK]

TPRC Data Ser Thermophysical Properties Research Center Data Series [of Purdue University, West Lafayette, Indiana, US]

TPRE Twin Plane Re-Entrant Edge (Mechanism) [Crystallography]

TPRE Two-Photon Raman Excitation

T PROC Test Procedure [also T Proc]

T PROF Temperature Profile [also T Prof]

TPRS Temperature-Programmed Reflection Spectroscopy

TPRU Tropical Pesticides Research Unit [UK]

TPS Tape Programming System

TPS Task Parameter Synthesizer

TPS Technical Problem Summary

TPS Technical Publishing Software

TPS Telecommunications Programming System

TPS Telemetry Processing Station

TPS Temperature-Programmed Sulfation

TPS Terminal Polling System

TPS Terminals per Station

TPS Term Preferred Shares

TPS Test Preparation Sheet

TPS Test Program Set

TPS Tetrapropylenebenzenesulfonate

TPS Text Processing Service

TPS Text Processing System

TPS Thermal Power System

TPS Thermal Protection Subsystem; Thermal Protection System

TPS Thermoprocessing System

TPS Throttle Position Sensor [Automobiles]

TPS Tool Plant Solution

TPS Transaction Processing System

TPS Transactions per Second

TPS Transient Plane Source (Technique)

TPS Translator Processing System

TPS 2,4,6-Triisopropylbenzenesulfonyl

TPS Triphenylsilane

TPS Triphenyl Sulfonium Chloride

TPS Twisted Pair Shielded (Cable)

TPS Two-Dimensional Periodic Structure [also 2DPS] [Materials Science]

TPSA 1,1,1-Triphenyl Silanamine

TPSb Triphenyl Antimony

TPSbBr$_2$ Triphenyl Antimony Dibromide

TPSbCl$_2$ Triphenyl Antimony Dichloride

TPSbS Triphenyl Antimony Sulfide

TPSCl 2,4,6-Triisopropylbenzenesulfonyl Chloride

TPSCl Triphenylsilyl Chloride

TPSD Tons per Stream Day [Chemical Engineering]

TPSE Thermal Protection Subsystem Experiments

TPSF Turbulent Pulse Sheet Former [Papermaking]

TPSH 2,4,6-Triisopropyl Benzenesulfonic Acid Hydrazide

TPSI Torque-Pressure in Pounds per Square Inch [also tpsi]

TPSI N-(2,4,6-Triisopropylbenzenesulfonyl) Imidazole

TPSn Tetraphenyltin

TPSNI 1-(2,4,6-Triisopropylbenzenesulfonyl)-4-Nitroimidazole

TPSNT 1-(Triisopropylbenzenesulfonyl)-3-Nitro-1H-1,2,4-Triazole

T-PSR Temperature Pre-Soak Rail

TPSRS Terminal Primary and Secondary Radar System

TPSSIMS Temperature-Programmed Static Secondary Ion Mass Spectrometry (or Spectroscopy)

TPST 1-(Triisopropylbenzenesulfonyl)-1H-1,2,4-Triazole

3PST Triple-Pole, Single-Throw [also TPST]

3PST SW Triple-Pole Single-Throw Switch [also TPST SW]

3P SW Triple-Pole Switch [also TP SW]

TPT Temporary Power Tap(s)

TPT Tetraisopropyl Titanate

TPT Tetraphenylthiophene

TPT Titanium (IV) Isopropoxide

TPT Thermal Penetration Theory

TPT Transmission Path Translator

TPT Triphenyltetrazolium Chloride

TPT 2,4,6-Tris(2-Pyridyl)-1,3,5-Triazine

TP-T Target Practice with Tracer

TpT Thymidylyl Thymidine [Biochemistry]

TPTA Transient Pressure Test Article

TPTA Trimethylolpropane Trimethacrylate

TPTA Triphenyltin Acetate [also TPTAc]

Tptan N,N',N"-Tris(2-Pyrodylmethyl)-2,5,8-Triazanonane [also tptan]

TPTB The Powers That Be

TPTC Triphenyltin Chloride [also TPTCl]

Tptcd N,N',N"-Tris(2-Pyrodylmethyl)-1,5,9-Triazacyclododecane [also tptcd]

Tptcn N,N',N"-Tris(2-Pyrodylmethyl)-1,4,7-Triazacyclononane [also tptcn]

TPTD Tetraisopropylthiuram Disulfide

TPTF Triphenyltin Fluoride

TPTF Two Phase Flow Test Facility

TPTG Tuned-Plate Tuned-Grid (Oscillator)

TPTH Triphenyltin Hydroxide

TPTH N-p-Toluenesulfonyl-N'-Phenylthiobenzhydrazide [also TPTH-H]

Tptn N,N,N',N'-Tetrakis(2-Pyridylmethyl)-1,3-Propanediamine [also tptn]

TPTP Tape-to-Tape

TPTP Tetra-p-Tolyl Porphinate

TpTpT Thymidylyl Thymidylyl Thymidine [Biochemistry]

TPTU O-(1,2-Dihydro-2-Oxo-1-Pyridyl)-N,N,N',N'-Tetramethyl Uronium Tetrafluoroborate

TPTZ 2,3,5-Triphenyl-2H-Tetrazolium Chloride

TPTZ 2,4,6-(Tripyridyl)-s-Triazine

TPTZ 2,4,6-Tris(2-Pyridyl)-1,3,5-Triazine

TPU Tape Preparation Unit

TPU Task Processing Unit

TPU Telecommunications Processing Unit

TPU Thermal Processing Unit

TPU Thermoplastic Polyurethane; Thermoplastic Urethane

TPU Threatened Plants Unit [of International Union for Conservation of Nature and Natural Resources]

TPU Time Pickoff Unit

TPU Turbo Pascal (Precompiled) Unit (File)

.TPU Turbo Pascal (Precompiled) Unit File [File Name Extension]

TPU/ABS Thermoplastic Polyurethane/Acrylonitrile-Butadiene-Styrene

TPUMOVER Turbo Pascal Unit Mover (File)

TPUN Test Procedure Update Notice

TPUR Thermoplastic Polyurethane

TPUR-ABS Thermoplastic Polyurethane–Acrylonitrile-Butadiene-Styrene

TPUR-PA Thermoplastic Polyurethane–Polyamide

TPUR-PC Thermoplastic Polyurethane–Polycarbonate

TPUR-PVC Thermoplastic Polyurethane–Polyvinyl Chloride

TPUS Transportation and Public Utilities Service [of General Services Administration, US]

TPWB Three Program Wire Broadcasting

TPV Temperature–Pressure–Volume (Diagram) [also T-P-V]

TPV Thermophotovoltaic(s)

TPV Thermoplastic Vulcanizate

TPV Thermoproteus Virus

TPV-1 Thermoproteus Virus, Type 1

TPV-2 Thermoproteus Virus, Type 2

TPW Turbo Pascal for Windows

TPX Polymethylpentene [also PMP]

TPX Tokamak Physics Experiment [Nuclear Engineering]

TPXPP Tetrakis(p-substituted Phenyl) Porphinate

TPYEA Tris[2-(1H-Pyrazol-1-yl)ethyl]amine

TQ Total Quality

2Q 2-Q Spin State [Solid-State Physics]

3Q 3-Q Spin State [Solid-State Physics]

TQA Total Quality Assurance

TQC Technical Quality Control

TQC Total Quality Commitment

TQC Total Quality Control

TQCA Textile Quality Control Association [US]

TQCM Temperature-Controlled Quartz Crystal Microbalance

TQD Tapered Roller Bearing, Four-Row, Cup Adjusted, Contact Angle Vertex Inside [Symbol]

TQE Time Queue Element

TQE Total Quality Excellence

TQF Triple-Quantum-Filtered

TQF-COSY Triple-Quantum-Filtered Correlation Spectroscopy

TQFP Tape Quad Flat Pack [Electronics]

TQFP Thin Quad Flat Pack [Electronics]

TQI Tapered Roller Bearing, Four-Row, Cup Adjusted, Contact Angle Vertex Outside [Symbol]

TQL Total Quality Leadership

TQM Total Quality Management

TQMS Triple Quadrupole Mass Spectrometry

$\tau(q,n)$ multifractal spectrum [Symbol]

2QT Double Quantum Transition

2QT-ENDOR Double Quantum Transitions (in) Electron Nuclear Double Resonance

T-Quark Truth (or Top) Quark [also t-quark] [Particle Physics]

TQW Triple Quantum Well [Electronics]

TR Table Run [Leather Processing]

TR Tail Rotor

TR Tamper Resistance; Tamper-Resistant

TR Tannery Run [Leather Processing]

TR Tape Recorder; Tape Recording [also T/R]

TR Tape Resident

TR Technical Reference

TR Technical Report [also T/R]

TR Technical Research
TR Technical Review
TR Technology Report [also T/R]
TR Teleradiograph(y)
TR Temperature Rate
TR Temperature Recorder
TR Temperature Regulator
TR Temper Rolled; Temper Rolling [Metallurgy]
TR Temporary Register
TR Tennessee River [US]
TR Terminal Ready
TR Terminal Room
TR Terrestrial Radiation
TR Test Reactor
TR Test Report
TR Test Request
TR Test Result
TR Thermal Radiation; Thermal Radiator
TR Thermal Reactor
TR Thermal Regulation
TR Thermal Resistance
TR Thermit Reaction [Metallurgy]
TR Thermoregulator
TR Thiokol Rubber
TR Thrust Reverser [Aerospace]
TR Tidal Range; Tide Range
TR Tie Rod
TR Tile Red
TR Time-Delay Relay
TR Time-Release (Capsule) [Pharmacology]
TR Time Resolution; Time-Resolved
TR Times Roman (Font)
TR Time to Retrofire [Aerospace]
TR Tollens' Reagent
TR Tons Registered
TR Top Register
TR Top Running (Crane)
TR Torque Receiver [Control Engineering]
TR Torque Repeater [Control Engineering]
TR Torsional Rigidity
TR Total Radiation
TR Total Reaction
TR Total Reflection [Optics]
TR Tracheid (of Wood)
TR Tracking Regulator
TR Traffic Route [Telecommunications]
TR Transaction
TR Transformer
TR Transformer-Rectifier
TR Transient Reflectance
TR Transient Response
TR Transistor Radio
TR Transition Region

TR Translation Register
TR Transmissibility Ratio
TR Transmission Report
TR Transmit-Receive [also T/R]
TR Transmitter-Receiver [also T/R]
TR Transparencies
TR Transportation Request [also T/R]
TR Transportation Research
TR Transverse Resonance
TR Trapezoidal Rule [Mathematics]
TR Tropical Rainforest
TR Trouble Report
TR Truck Ramp
TR True Range
TR Turbine Rotor
TR Turkey [ISO Code]
TR Turkish Reactor
TR Turnover Rate [Chemical Engineering]
TR Tyndall-Roentgen (Effect) [Physics]
TR- Gabon [Civil Aircraft Marking]
TR³ Time-Resolved Resonant Raman Spectroscopy [also TRRR]
TR³ Time-Resolved Resonant Raman Spectrum [also TRRR]
T&R Transmit and Receive
T/R Tape Recorder; Tape Recording [also TR]
T/R Technical Report [also TR]
T/R Tip/Ring
T/R Transformer Rectifier
T/R Transmit/Receive [also TR]
T/R Transmitter/Receiver [also TR]
T/R Transportation Request [also TR]
T/R Turnaround Requirements
3R Reclaim, Reuse, Reduce [also RRR, R^3 or R3]
3R Reduce, Reuse, Recycle [also RRR, R^3 or R3]
Tr Tincture [also Tinct] [Medical Prescriptions]
Tr Trace(r) [also tr]
Tr Trace (of a Matrix) [Mathematics]
Tr Track [also tr]
Tr Train [also tr]
Tr Transfer [also tr]
Tr Transformation [also tr]
Tr Transistor [also tr]
Tr Transition [also tr]
Tr Translation; Translator [also tr]
Tr Transmitter [also tr]
Tr Transport(ation) [also tr]
Tr Transposition [also tr]
Tr Treasure(r) [also tr]
Tr Troop [also tr]
Tr Trustee [also tr]
T-ρ Temperature-Resistivity (Curve)
tr translate(d)
tr transpose (of a matrix) [Mathematics]
.tr Turkey [Country Code/Domain Name]

TRA Technical Requirement Analysis
TRA Temporary Reserved Airspace
TRA Test and Repair Analysis
TRA The Reclamation Association [UK]
TRA Thermal Radiation Analysis
TRA Tire and Rim Association [US]
TRA Training Requirements Analysis
TRA Tubular Reactor Assembly
TRA Turnaround Requirements Analysis
Tr A Trace of Matrix "A" [Mathematics]
Tra Transfer [also tra]
TRAA Towing and Recovery Association of America [US]
TRAAC Transit Research and Attitude Control
TRAACS Transit Research and Attitude Control Satellite
TRAB Transient Absorption (Technology)
TRABTECH Transient Absorption Technology
TRAC Teacher Reseach Associate (Program) [of US Department of Energy]
TRAC Test Report and Certification (System) [also TraC]
TRAC Texas Reconfigurable Array Computer
TRAC Text Reckoning and Compiling
TRAC Thermally Regenerative Alloy Cell
TRAC Tracks Radioactive Components
TRAC Traction Control System
TRAC Transient Radiation Analysis by Computer
TRAC Transmit Control
TRAC Trends in Analytical Chemistry [Dutch Publication]
Trac Tractor [also trac]
TRACALS Traffic Control and Landing System
TRACALS Traffic Control, Approach and Landing System
TRACE Tactical Readiness and Checkout Equipment
TRACE Tape-Controlled Reckoning and Checkout Equipment
TRACE Tape-Controlled Recording and Checkout Equipment; Tape-Controlled Recording Automatic Checkout Equipment
TRACE Task Reporting and Current Evaluation
TRACE Taxiing and Routing of Aircraft Coordination Equipment
TRACE Teleprocessing Recording for Analysis by Customer Engineers
TRACE Time-Shared Routines for Analysis, Classification and Evaluation
TRACE Tolls Recording and Computing Equipment
TRACE Toronto Region Association of Computer Enthusiasts [Canada]
TRACE Tracking and Communications, Extraterrestrial
TRACE Transistor Radio Automatic Circuit Evaluator
TRACE Transportable Automated Control Environment
Trace Anal Trace Analysis [Journal published in PR China]
Trace Subst Environ Health Trace Substances and Environmental Health [International Journal]
TRACOM Tracking Comparison [also Tracom or tracom]
TRACON Terminal Radar Approach Control (Facility)
TRACS Terminal Radar and Control System
TRACS Test and Repair Analysis/Control System

TRACS Tool-Record Accountability System
TRACS Traffic-Actuated Computerized Signal(s) [also Tracs]
TRACS Transport and Road Abstracting and Cataloguing System [of Transport and Road Research Laboratory, UK]
Tracts Mod Phys Tracts in Modern Physics [US]
TRADA Timber Research and Development Association [UK]
Trade Marks J Trade Marks Journal [Published in the UK]
TRADEX Target Resolution and Discrimination Experiment
TRADIC Transistor Digital Computer
TRADIC Transistorized Airborne Digital Computer
TRAFFIC Trade Records Analysis of Flora and Fauna in Commerce (Program) [of World Wide Fund for Nature]
Traffic Safety Mag Traffic Safety Magazine [of National Safety Council, US]
TRAIN Telerail Automated Information Network [of Association of American Railroads, US]
Train Training [also train]
Train Dev J Training and Development Journal [of American Society for Training and Development, US]
Train Off Training Officer [Journal published in the UK]
Trait Signal Traitement du Signal [French Publication on Signal Processing]
Trait Therm Traitement Thermique [French Publication on Heat Treatment]
Traj Trajectory [of traj]
TRALA Truck Renting and Leasing Association [US]
TRAM Target Recognition (and) Attack Multisensor
TRAMAR Tropical Rain Mapping Radar
TRAMEX Tertiary Amine Extraction (Method)
TRAMEX Transuranic Metal Extraction (Process)
TRAMP Time-Shared Relational Associative Memory Program
TRAMPS Temperature Regulator and Missile Power Supply
TRAMPS Traffic Measurement and Path Search
Tran Transaction [also tran]
Tran Transit [also tran]
Tran Transmit [also tran]
Tranciatura Stampaggio Tranciatura e Stampaggio [Italian Publication on Stamping and Punching]
TRANDIR Translation Director
TRANET Tracking Network
TRANPRO Transaction Processor
Trans Transaction [also trans]
Trans Transfer(ence) [also trans]
Trans Transformer [also trans]
Trans Transistor [also trans]
Trans Transition [also trans]
Trans Translation; Translator [also trans]
Trans Transmission
Trans Transmitter [also trans]
Trans Transparency; Transparent [also trans]
Trans Transport(ation) [also trans]
Trans Transposition [also trans]
trans translated (by)
trans transverse

trans- prefix denoting a stereoisomer that has atoms, or atom groups attached to opposite sides of a molecular chain [Chemistry]

TRANSAC Transistorized Automatic Computer

Trans ACA Transactions of the American Crystallographic Association [US]

Trans AGU Transactions of the American Geophysical Union [US]

Trans AIME Transactions of the American Institute of Mining, Metallurgical and Petroleum Engineers

TRANSALT Transition Altitude [Aerospace]

Trans Am Cryst Assoc Transactions of the American Crystallographic Association [US]

Trans Am Geophys Union Transactions of the American Geophysical Union [US]

Trans Am Inst Min Eng Transactions of the American Institute of Mining, Metallurgical and Petroleum Engineers [US]

Trans Am Math Soc Transactions of the American Mathematical Society [US]

Trans Am Microsc Soc Transactions of the American Microscopical Society [US]

Trans Am Nucl Soc Transactions of the American Nuclear Society [US]

Trans Am Phil Soc Transactions of the American Philosophical Society [Philadelphia, Pennsylvania, US]

Trans AMS Transactions of the American Mathematical Society [US]

Trans AMS Transactions of the American Microscopical Society [US]

Trans Am Soc Agric Eng Transactions of the American Society of Agricultural Engineers [US]

Trans ANS Transactions of the American Nuclear Society [US]

Trans ASAE Transactions of the American Society of Agricultural Engineers [US]

Trans ASHRAE Transactions of the American Society of Heating, Refrigeration and Air Conditioning Engineers [US]

Trans ASM Transactions of the American Society of Metals [now Transactions of ASM International, US]

Trans ASME Transactions of the American Society of Mechanical Engineers [US]

Trans ASME, J Appl Mech Transactions of the American Society of Mechanical Engineers, Journal of Applied Mechanics [US]

Trans ASME, J Basic Eng Transactions of the American Society of Mechanical Engineers, Journal of Basic Engineering [US]

Trans ASME, J Bioeng Transactions of the American Society of Mechanical Engineers, Journal of Bioengineering [US]

Trans ASME, J Biomech Eng Transactions of the American Society of Mechanical Engineers, Journal of Biomechanical Engineering [US]

Trans ASME, J Dyn Syst Transactions of the American Society of Mechanical Engineers, Journal of Dynamic Systems [US]

Trans ASME, J Dyn Syst Meas Control Transactions of the American Society of Mechanical Engineers, Journal of Dynamic Systems, Measurement and Control [US]

Trans ASME, J Electron Packag Transactions of the American Society of Mechanical Engineers, Journal of Electronic Packaging [US]

Trans ASME, J Energy Resour Technol Transactions of the American Society of Mechanical Engineers, Journal of Energy Resources Technology

Trans ASME, J Eng Gas Turbines Power Transactions of the American Society of Mechanical Engineers, Journal of Engineering for Gas Turbines and Power [US]

Trans ASME, J Eng Ind Transactions of the American Society of Mechanical Engineers, Journal of Engineering for Industry [US]

Trans ASME, J Eng Mater Technol Transactions of the American Society of Mechanical Engineers, Journal of Engineering Materials and Technology [US]

Trans ASME, J Eng Power Transactions of the American Society of Mechanical Engineers, Journal of Engineering for Power [US]

Trans ASME, J Fluids Eng Transactions of the American Society of Mechanical Engineers, Journal of Fluids Engineering [US]

Trans ASME, J Heat Transf Transactions of the American Society of Mechanical Engineers, Journal of Heat Transfer [US]

Trans ASME, J Ind Transactions of the American Society of Mechanical Engineers, Journal of Industry [US]

Trans ASME, J Ind Eng Transactions of the American Society of Mechanical Engineers, Journal of Industrial Engineering [US]

Trans ASME, J Lubr Technol Transactions of the American Society of Mechanical Engineers, Journal of Lubrication Technology [US]

Trans ASME, J Mech Des Transactions of the American Society of Mechanical Engineers, Journal of Mechanical Design [US]

Trans ASME, J Mech Transm Autom Des Transactions of the American Society of Mechanical Engineers, Journal of Mechanisms, Transmissions, Automation in Design [US]

Trans ASME, J Power Transactions of the American Society of Mechanical Engineers, Journal of Power [US]

Trans ASME, J Press Vessel Technol Transactions of the American Society of Mechanical Engineers, Journal of Pressure Vessel Technology [US]

Trans ASME, J Sol Energy Eng Transactions of the American Society of Mechanical Engineers, Journal of Solar Energy Engineering [US]

Trans ASME, J Tribol Transactions of the American Society of Mechanical Engineers, Journal of Tribology [US]

Trans ASME, J Turbomach Transactions of the American Society of Mechanical Engineers, Journal of Turbomachinery [US]

Trans ASME, J Vib Acoust Stress Reliab Des Transactions of the American Society of Mechanical Engineers, Journal of Vibration, Acoustics, Stress and Reliability in Design [US]

Trans ASME Ser A Transactions of the American Society of Mechanical Engineers, Series A [US]

Trans ASME Ser B Transactions of the American Society of Mechanical Engineers, Series B [US]

Trans ASME Ser C Transactions of the American Society of Mechanical Engineers, Series C [US]

Trans ASME Ser D Transactions of the American Society of Mechanical Engineers, Series D [US]

Trans ASME Ser E Transactions of the American Society of Mechanical Engineers, Series E [US]

Trans ASM Q Transactions of the American Society of Metals Quarterly [US]

Trans BCS Transactions of the British Ceramic Society [also Trans Br Ceram Soc, or Trans Brit Ceram Soc]

Trans Biomed Eng Transactions on Biomedical Engineering [of Institute of Electrical and Electronics Engineers, US]

Trans Bose Res Inst Transactions of the Bose Research Institute [India]

Trans Br Ceram Soc Transactions of the British Ceramic Society [also Trans Brit Ceram Soc, or Trans BCS]

TRANSCAER Transportation Community Awareness and Emergency Response

TRANSCAN Canary Islands Submarine Cable

Transceiver Transmitter/Receiver [also transceiver, or TRANSCEIVER]

Trans Cerv V Transverse Cervical Vein (of Heart) [also trans cerv v] [Anatomy]

Trans Chin Weld Inst Transactions of the China Welding Institute [PR China]

Trans CHMT Transactions on Components, Hybrids and Manufacturing Technology [of Institute of Electrical and Electronics Engineers, US]

TRANSCRIPT Financial (Student) Aid Transcript [US]

trans-Cypenphos trans-1,3-Cyclopentadienediylbis-(diphenylphosphine)

TRANSDOC Transportation Documents (Database) [of European Conference of Ministers of Transport, France]

TRANSDOC Transport Documentation (Database) [of Organization for Economic Cooperation and Development, France]

Transducer Technol Transducer Technology [UK Publication]

TRANSEC Transmission Security

Trans Elec Dev Transactions on Electron Devices [of Insitute of Electrical and Electronics Engineers–Electron Devices Society, US]

Trans Electr Supply Auth Eng Inst NZ Transactions of the Electric Supply Authority Engineers Institute of New Zealand

Transf Transfer [also transf]

Transf Transformer [also transf]

Trans Faraday Soc Transactions of the Faraday Society [of Royal Society of Chemistry, UK]

Trans GCAGS Transactions of the Gulf Coast Association of Geological Societies [US]

Trans GRC Transaction of the Geothermal Resources Council [US]

Trans ICS Transactions of the Indian Ceramic Society [also Trans Indian Ceram Soc]

TRANSIF Transient State Isoelectric Focusing [Physical Chemistry]

Trans IIM Transactions of the Indian Institute of Metals [also Trans Indian Inst Met]

TRANSIM Transportation Simulator

Trans Indian Ceram Soc Transactions of the Indian Ceramic Society [also Trans ICS]

Trans Indian Inst Met Transactions of the Indian Institute of Metals [also Trans IIM]

Trans Inf Process Soc Jpn Transactions of the Information Processing Society of Japan

Trans Inst Chem Eng Transactions of the Institution of Chemical Engineers [UK]

Trans Inst Electr Eng Jpn A Transactions of the Institute of Electrical Engineers of Japan, Part A

Trans Inst Electr Eng Jpn B Transactions of the Institute of Electrical Engineers of Japan, Part B

Trans Inst Electr Eng Jpn C Transactions of the Institute of Electrical Engineers of Japan, Part C

Trans Inst Electr Eng Jpn D Transactions of the Institute of Electrical Engineers of Japan, Part D

Trans Inst Electr Eng Jpn E Transactions of the Institute of Electrical Engineers of Japan, Part E

Trans Inst Electron Inf Commun Eng A Transactions of the Institute of Electronics, Information and Communication Engineers A [Japan]

Trans Inst Electron Inf Commun Eng B Transactions of the Institute of Electronics, Information and Communication Engineers B [Japan]

Trans Inst Electron Inf Commun Eng B-I Transactions of the Institute of Electronics, Information and Communication Engineers B-I [Japan]

Trans Inst Electron Inf Commun Eng B-II Transactions of the Institute of Electronics, Information and Communication Engineers A [Japan]

Trans Inst Electron Inf Commun Eng C Transactions of the Institute of Electronics, Information and Communication Engineers C [Japan]

Trans Inst Electron Inf Commun Eng C-I Transactions of the Institute of Electronics, Information and Communication Engineers C-I [Japan]

Trans Inst Electron Inf Commun Eng C-II Transactions of the Institute of Electronics, Information and Communication Engineers C-II [Japan]

Trans Inst Electron Inf Commun Eng D Transactions of the Institute of Electronics, Information and Communication Engineers D [Japan]

Trans Inst Electron Inf Commun Eng D-I Transactions of the Institute of Electronics, Information and Communication Engineers D-I [Japan]

Trans Inst Electron Inf Commun Eng D-II Transactions of the Institute of Electronics, Information and Communication Engineers D-II [Japan]

Trans Inst Electron Inf Commun Eng E Transactions of the Institute of Electronics, Information and Communication Engineers E [Japan]

Trans Inst Eng Aust Transactions of the Institution of Engineers of Australia

Trans Inst Eng Aust, Civ Eng Transactions of the Institution of Engineers of Australia, Civil Engineering Transactions

Trans Inst Eng Aust, Mech Eng Transactions of the Institution of Engineers of Australia, Mechanical Engineering Transactions

Trans Inst Mar Eng Transactions of the Institute of Marine Engineers [UK]

Trans Inst Meas Control Transactions of the Institute of Measurement and Control [UK]

Trans Inst Met Finish Transactions of the Institute of Metal Finishing [UK]

Trans Inst Min Metall A Transactions of the Institution of Mining and Metallurgy, Section A [UK]

Trans Inst Min Metall B Transactions of the Institution of Mining and Metallurgy, Section B [UK]

Trans Inst Min Metall C Transactions of the Institution of Mining and Metallurgy, Section C [UK]

Trans Inst Prof Eng NZ, Civ Eng Sect Transactions of the Institution of Professional Engineers New Zealand, Civil Engineering Section

Trans Inst Prof Eng NZ, Electr/Mech/Chem Eng Sect Transactions of the Institution of Professional Engineers New Zealand, Electrical/Mechanical/Chemical Engineering Section

Trans Inst Syst Control Inf Eng Transactions of the Institute of Systems, Control and Information Engineers [Japan]

Trans Iron Steel Inst Jpn Transactions of the Iron and Steel Institute of Japan

Trans Iron Steel Soc Transactions of the Iron and Steel Society [of American Institute of Mining, Metallurgical and Petroleum Engineers, US]

Trans J Br Ceram Soc Transactions and Journal of the British Ceramic Society

Trans JIM Transactions of the Japan Institute of Metals

Trans Jpn Assoc Refrig Transactions of the Japanese Association of Refrigeration

Trans Jpn Foundrymen's Soc Transactions of the Japan Foundrymen's Society

Trans Jpn Inst Met Transactions of the Japan Institute of Metals

Trans Jpn Inst Met Suppl Transactions of the Japan Institute of Metals Supplement

Trans Jpn Soc Aeronaut Space Sci Transactions of the Japan Society for Aeronautical and Space Sciences

Trans Jpn Soc Compos Mater Transactions of the Japan Society for Composite Materials

Trans Jpn Soc Mech Eng Transactions of the Japan Society of Mechanical Engineering

Trans Jpn Soc Mech Eng A Transactions of the Japan Society of Mechanical Engineering A

Trans Jpn Soc Mech Eng B Transactions of the Japan Society of Mechanical Engineering B

Trans Jpn Soc Strength Fract Mater Transactions of the Japan Society for Strength and Fracture of Materials

Trans Jpn Weld Res Inst Transactions of the Japan Welding Research Institute

Trans Jpn Weld Soc Transactions of the Japan Welding Society

Trans JSNDI Transactions of the Japanese Society of Nondestructive Inspection

Trans JWRI Transactions of the Japan Welding Research Institute

Trans Kokushikan Univ Fac Eng Transactions of the Kokushikan University Faculty of Engineering [Japan]

Trans Korean Inst Electr Eng Transactions of the Korean Institute of Electrical Engineers [South Korea]

Transl Translation; Translator [also transl]

transl translated (by)

Trans Leningrad Electrotech Inst Transactions of the Leningrad Electrotechnical Institute [USSR]

TRANSLEV Transition Level [Aeronautics]

translu translucent

Transm Transmission [also transm]

transm transmit(ted)

Transm Distrib Transmission and Distribution [US Publication]

Trans Met Finish Assoc India Transactions of the Metal Finishing Society of India

Trans Met Heat Treat Transactions of the Metal Heat Treatment [PR China]

Trans Met Soc Transactions of the Metallurgical Society [of American Institute of Mining, Metallurgical and Petroleum Engineers]

Trans Met Soc AIME Transactions of the Metallurgical Society of the American Institute of Mining, Metallurgical and Petroleum Engineers]

Trans Min Metall Assoc, Kyoto Transactions of the Mining and Metallurgical Association, Kyoto [Japan]

TRANSMUX Transmission Multiplexer

Transnatl Data Commun Rep Transnational Data and Communications Report [US]

Trans Natl Res Inst Met Transactions of the National Research Institute for Metals [Japan]

Trans NY Acad Sci Transactions of the New York Academy of Sciences [also Trans NYAS]

Trans Orth Res Soc Transactions of the Orthopedic Research Society [US]

Transp Transportation [also transp]

TRANSPAC Transpacific Submarine Cables [Hawaii–Japan–Philippines]

TRANSPAC Data Transmission via Packet-Switching Network [French Public Data Communication Network] [also Transpac]

Transp Eng Transport Engineer [Publication of the Institute of Road Transport Engineers, UK]

Transp J Transportation Journal [of American Society of Transportation and Logistics, US]

TRANSPLAN Transaction Network Service Planning Model

Transpo International Transport Exhibition

TRANSPONDER Transmitter/Responder [also Transponder, or transponder]

TRANSPORT Database on National and International Transportation

Transp Plan Technol Transportation Planning and Technology [Journal published in the UK]

Trans Powder Metall Assoc India Transactions of the Powder Metallurgy Association of India

Transp Porous Media Transport in Porous Media [Journal published in the Netherlands and the US]

Transpt Transport [also transpt]

Transp Res A, Gen Transportation Research, Part A, General [Published in the UK]

Transp Res B, Methodol Transportation Research, Part B, Methodological [Published in the UK]

Transp Res C, Emer Technol Transportation Research, Part B, Emerging Technologies [Published in the UK]

Transp Res Circ Transportation Research Circular [of Transportation Research Board, US]

Transp Res Rec Transportation Research Record [of Transportation Research Board, US]

Transp Sci Transportation Science [Publication of the Operations Research Society of America]

Transp Theory Stat Phys Transport Theory and Statistical Physics [Published in the US]

Trans RAS Transactions of the Russian Academy of Sciences

Trans R Soc Edinb Transactions of the Royal Society of Edinburgh [UK]

Trans R Soc Can Transactions of the Royal Society of Canada

Trans R Soc South Afr Transactions of the Royal Society of South Africa

Trans S Afr Inst Electr Eng Transactions of the South African Institute of Electrical Engineers

TRANSSIB Transsiberian Railway [also Transsib]

Trans SME Transactions of the Society of Mining Engineers [of the American Institute of Mining, Metallurgical and Petroleum Engineers, US]

Trans Soc Biomater Transactions of the Society for Biomaterials [US]

Trans Soc Comput Simul Transactions of the Society for Computer Simulation [US]

Trans Soc Instrum Control Eng Transactions of the Society of Instrument and Control Engineers [Japan]

Trans Soc Min Eng Transactions of the Society of Mining Engineers [of the American Institute of Mining, Metallurgical and Petroleum Engineers, US]

Trans Soc Pet Eng Transactions of the Society of Petroleum Engineers [of the American Institute of Mining, Metallurgical and Petroleum Engineers, US]

Trans Soc Rheol Transaction of the Society of Rheology [US]

Trans South Afr Inst Electr Eng Transactions of the South African Institute of Electrical Engineers

Trans SPE Transactions of the Society of Petroleum Engineers [of the American Institute of Mining, Metallurgical and Petroleum Engineers, US]

Trans Tech Publ Ltd Trans Tech Publications Limited [Aedermannsdorf, Switzerland]

Trans TMS-AIME Transactions of The Metallurgical Society of the American Institute of Mining, Metallurgical and Petroleum Engineers [US]

Trans TMS-AIME A Transactions of The Metallurgical Society of the American Institute of Mining, Metallurgical and Petroleum Engineers, Part A [US]

Trans TMS-AIME B Transactions of The Metallurgical Society of the American Institute of Mining, Metallurgical and Petroleum Engineers, Part B [US]

transv transverse

Transyl Transylvania(n)

trans-ZEATIN 6-[4-Hydroxy-3-Methylbut-2-eneamino]purine) cytokinin [Biochemistry]

Trans Zimb Sci Assoc Transactions of the Zimbabwe Scientific Association

TRAP Teledyne Research Assistance Program [US]

TRAP Terminal Radiation Airborne Program

TRAP Tracker Analysis Program

TRAPATT Trapped Plasma Avalanche Transit Time (Diode)

TRAPATT Trapped Plasma Avalanche Triggered Transit (Diode)

TRAPR Tethered Remote Automatic Pipeline Repairer

TRASYS Thermal Radiation Analysis System

Tratt Finit Trattamenti & Finiture [Italian Publication on Metal Coating and Finishing]

Trav Travel [also trav]

Trav Traverse [also trav]

Trav Com Int Etude Bauxites, Alumine Alum Travaux du Comité International pour l'Etude des Bauxites, d'Alumine et d'Aluminium [Papers of the International Committee for Bauxites, Alumina and Aluminum Studies]

TRAVIS Traffic Retrieval Analysis Validation and Information System

TRAW Tape Read and Write

TRAWL Tape Read and Write Library

TRAX Total Reflection-Angle X-Ray Spectrometer

TRB Technical Review Board

TRB *(Technische Regeln für Druckbehälter)* – German Technical Regulations for Pressure Vessels

TRB Technology Review Board

TRB Test Review Board

TRB Transportation Research Board [US]

TRB Transport Research Board [Canada]

TRBDD Tribromo-p-Dibenzodioxin

TRBDF Tribromodibenzofuran

TRbF *(Technische Regeln für brennbare Flüssigkeiten)* – German Technical Regulations for Flammable Liquids

TRBL Troubleshooting [also Trbl]

TRC Tape Record Coordinator

TRC Technical Research Center

TRC Technical Resources Center [US]

TRC Technical Review Committee

TRC Technology Reports Center [of Department of Trade and Industry, UK]

TRC Telemetry and Remote Control

TRC Temperature-Ratio Criterion (for Amorphization) [Metallurgy]

TRC Texas Railroad Commission [US]

TRC Textile Research Council [UK]

TRC Time Reporting Center [of Ontario Hydro, Toronto, Canada]

TRC Total Residual Chlorine

TRC Tracking, Radar-Input and Correlation

TRC Trade Relations Council [US]

TRC Traffic Counts and Listings

TRC Transmit Receive Control

TRC Transverse Redundancy Check

TRC Tukuba Research Center [Japan]

TRC Twin-Reflex Camera

TRCARS Time-Resolved Coherent Anti-Stokes Raman Scattering

TRC-AS Transmit Receive Control–Asynchronous Start/Stop

TRCBZ Trichlorobenzene [also Trcbz, or TrcBz]

TRCC T-Carrier Restoration Control Center

TRCCC Tracking Radar Central Control Console

TRCDF Trichlorodibenzofuran

TRCDO Trichloro-p-Dibenzodioxin

TRCDS Time Resolved Circular Dichroism Spectroscopy

TRCF Technical Research Center of Finland [at Espoo, Finland]

Trckg Trucking [also trckg]

Tr Colon Transverse Colon [also tr colon] [Anatomy]

TRCP Tape Recorder Control Panel

TRCP Trichlorophenol

TRC-SC Transmit Receive Control–Synchronous Character

TRC-SF Transmit Receive Control–Synchronous Framing

Trctr Tractor [also trctr]

TRD Tape Read

TRD Technical Resource Document

TRD Test Requirements Document

TRD Timed Release Disconnect

TR&D Transport Research and Development

3rd third

TRDC Tourism Research and Data Center

TRDF Time-Resolved Dispersed Fluorescence

TRDS Time-Resolved Differential Scanner

Trdta Trimethyldiaminetetraacetate [also trdta]

TRDTO Tracking Radar (Data) Takeoff

Trdtra Trimethylenediamine-N,N,N'-Triacetate [also trdtra]

TRE Telecommunications Research Establishment [UK]

TRE Thermochemical Resonance Energy

TRE Thyssen Roheisen-Entschwefelung (Prozess) [Thyssen Pig-Iron Desulfurization Process] [Metallurgy]

TRE Tokai Research Establishment [of Japan Atomic Energy Research Institute, Oarai, Japan]

TRE Topological Resonance Energy [Physics]

TRE Torso Rotation Experiment [International Space Station]

TRE Toxic Reduction Evaluation

TRE Transmit Reference Equivalent

TRE Type Rate Examiner

.TRE Tree List File [File Name Extension]

Treas Treasurer; Treasury [also treas]

TREAT Transient Radiation Effects Automated Tabulation

TREAT Transient Reactor Test Facility [of Argonne National Laboratory-West, Idaho Falls, US]

TREB Treble [also Treb, or treb]

TREC Tethered Remote Camera

TRED Taxonomic Resources Expertise Directory (for North America) [of US Geological Survey/ Biological Resources Division]

TREDAT Tree Database [Forestry Database of the Commonwealth Scientific and Industrial Research Organization, Australia]

TREE Transient Radiation Effects on Electronics

TREE Training, Research, Environment and Education

TREECD International Forestry Database on Compact Disk

TREELS Time-Resolved Electron Energy Loss Spectroscopy

TREES Tree Seed Center Database [of Commonwealth Scientific and Industrial Research Organization, Australia]

TREES Turbine Rotors Examination and Evaluation System

TREM Tape Reader Emulator Module

TREN Triaminotriethylamine

TREN Tris(2-Aminoethyl) Amine [also Tren, or tren]

TREND Tropical Environment Data

Trends Anal Chem Trends in Analytical Chemistry [Published in the Netherlands]

Trends Biochem Sci Trends in Biochemical Sciences [Publication of the International Union of Biochemistry]

Trends Biotechnol Trends in Biotechnology [International Publication]

Trends End-Use Mkts Plast Trends in End-Use Markets for Plastics [Published in the US]

Trends Neurosci Trends in Neuroscience [International Publication]

Trends Pharmacol Sci Trends in Pharmacological Sciences [International Publication]

Trends Polym Sci Trends in Polymer Science [Published in the Netherlands]

Trends Telecommun Trends in Telecommunications [Published in the Netherlands]

Trenie Iznos Trenie i Iznos [Byelorussian Journal of Friction and Wear]

TREPR Time-Resolved Electron Paramagnetic Resonance

TRES Time-Resolved Emission Spectrum

TRES Thermally Regenerative Electrochemical System

TREX Tube Reduction Extrusion

TREXP Time-Resolved Excitation Photoselection

TRF T-Cell Replacing Factor [Biochemistry]

TRF Temperature Recovery Factor

TRF Thyrotropin-Releasing Factor [Biochemistry]

TRF Time-Resolved Fluorescence

TRF Transportation Research Forum [US]

TRF Tripoli Rocketry Association [US]

TRF Tritium Removal Facility

TRF Tropical Forests Program [of International Union for Conservation of Nature and Natural Resources]

TRF Tropical Rainforest

TRF TSH (Thyroid-Stimulating Hormone) Releasing Factor [Biochemistry]

TRF Tuned Radio-Frequency (Receiver) [also trf]

TRF I T-Cell Replacing Factor I [Biochemistry]

TRF II T-Cell Replacing Factor II [Biochemistry]

TRF III T-Cell Replacing Factor III [also TRF$_M$] [Biochemistry]

Trf Transfer [also trf]

TRFA Total-Reflection Radiographic Fluorescence Analysis

TRFB Tariff Board

Trfc Traffic [also trfc]

TRFCS Temperature Rate Flight Control System

TRFD Time-Resolved Fluorescence Depletion

TR-FIA Time-Resolved Fluorescence Immunoassay

TRF Newsl TRF Newsletter [of Transportation Research Forum, US]

Trfr Transfer [also trfr]

TRFS Time-Resolved Fluorescence Spectroscopy

TRG Technical Research Group

TRG Technical Review Group

TRG Tertiary Research Group [UK]

TRG Tip/Ring to Ground

Trg Training [also trg]

TRGB Tail Rotor Gearbox

TRH Thyrotropin-Releasing Hormone [Biochemistry]

TRH-Gly Thyrotropin-Releasing Hormone Glycine [Biochemistry]

TRI Technical Research Institute

TRI Textile Research Institute [US]

TRI Theoretical Research Institute [of General Magnaplate Corporation, US]

TRI The Refractories Institute [US]

TRI Time Reversal Invariance [Physics]

TRI Tin Research Institute, Inc. [Columbus, Ohio, US]

TRI Toxic (Chemical) Release Inventory [Section 313 of Superfund Amendments and Reauthorization Act (SARA) of 1986 of the US Environmental Protection Agency]

TRI Triode [also Tri, or tri]

TRI Truck Research Institute [US]

Tri Trichloroethylene [also tri]

tri tribenzo[b,f,j][1,5,9]triazacyclododecine

tri trigonal

TRIA Temperature Removable Instrument Assembly

TRIAC Triode Alternating-Current (Switch) [also Triac, or triac]

TRIAD Technical Research, Investigations and Developments

TRIAD Triode Ion-Assisted Deposition

TRIAL Technique for Retrieving Information from Abstracts of Literature

TRI/Austin Texas Research Institute Austin, Inc. [US]

TRIAX Triaxial (Cable) [also triax]

TRIB Tire Retread Information Bureau [US]

TRIB Transfer Rate of Information Bits

Trib Tributary [also trib]

tribas tribasic

Tribol Tribological; Tribologist; Tribology [also tribol]

Tribol Int Tribology International [Published in the UK] [also Trib Int]

Tribol Trans Tribology Transactions [Published in the US] [also Trib Trans]

TRIC Tank Riser Interface Containment [Nuclear Reactors]

TRIC Television and Radio Industries Club [UK]

TRIC Time-Resolved Ion Correlation

tric triclinic [Crystallography]

TRICE Transistorized Real-Time Incremental Computer Equipment

TRICINE N-tris(Hydroxymethyl)methylglycine; Trismethylglycine [Biochemistry]

tricl triclinic [Crystallography]

TRID Track Identity

TRIDEC Tri-Cities Industrial Development Council [Richland–Kennewick–Pasco, Washington State, US]

TRIDOP Tri-Doppler

Trien Triethylene Tetramine [also trien]

TRIF Technical Requirements Industry Forum

Triflate Trifluoromethanesulfonate [also triflate]

TRIG Tank Riser Interface Glovebox [Nuclear Reactors]

trig trigonal [Crystallography]

Trig Trigonometric; Trigonometry [also trig]

TRIGA Training Reactor, Isotope-Production, General Atomics

Trigon Trigonometric; Trigonometry [also trigon]

Triglym Triethylene Glycol Dimethyl Ether

TRILD Time-Resolved Intracavity Laser Detection

TRILO Nitrilotriacetic Acid

TRIM Tailored Retrieval and Information Management

TRIM Test, Rework and Inspection Management

TRIM Test Rules for Inventory Management

TRIM Transport of Ion in Metal (Code)

TRIM Transport of Ions in Matter

Trim Trimmer [also trim]

trim trimetric

TRIMCAM 1,3,5-tris[{(2,3-Dihydroxyphenyl)methyl}-carbamoyl]benzene

TRIMCSR Transport of Ions in Matter including Collision Cascade, Sputtering and Replacement Events (Code)

TRIMIS Tri-Service Medical Information Systems [of US Department of Defense]

Trimpsi Tris(dimethylphosphinomethyl)butylsilane [also trimpsi]

TRIMS Time-Resolved Ion Momentum Spectrometry

Tr Inst Teor Astron Trudy Instituta Teoreticheskoi Astronomii [Russian Journal on Theoretical Astronomy]

3R Int Three R (Reduce, Reuse, Recycle) International [Journal published in Germany]

TRIO Telecommunications Research Institute of Ontario [Canada]

TRIP The Road Information Program [Canada]

TRIP Thunderstorm Research International Program

TRIP Transformation-Induced Plasticity [Metallurgy]

TRIPHOS Tris(diphenylphosphinoethyl)phenylphosphine

TRIPHOS 1,1,1-Tris(diphenylphosphinomethyl)ethane

TRIPS Trade Related Intellectual Property Rights [General Agreement on Tariffs and Trade]

TRIPS Travel Information Processing System

TRIR Time-Resolved Infrared

TRIREGVER Tri-Regional Perspective on the Verification Synergies among the NPT (Nonproliferation Treaty), CTBT (Comprehensive Test Ban Treaty) and Fissile Material Cutoff [Canada]

TRIRS Time-Resolved Infrared Spectroscopy

TRIS Time-Resolved Infrared Spectroscopy

TRIS Transportation Research Information Services [of US Department of Transportation]

TRIS Trisamine

TRIS Tris(hydroxymethyl)aminomethane [also Tris]

tris- prefix denoting that a grouping, or radical occurs three times in a molecule [Chemistry]

TRIS·ADP Tris(hydroxymethyl)aminomethane Salt of Adenosine 5'-Diphosphate [also Tris-ADP] [Biochemistry]

Tri-Salt Trimethylammoniumnitrate

TRIS·ATP Tris(hydroxymethyl)aminomethane Salt of Adenosine 5'-Triphosphate [also Tris-ATP] [Biochemistry]

TRIS-EDTA Tris(hydroxymethyl)aminomethane–Ethylenediaminetetraacetic Acid (Buffer Solu-tion) [also Tris-EDTA]

TRIS-HCl Hydrochloric Acid, Tris(hydroxymethyl)aminomethane Salt [also Tris-HCl] [Biochemistry]

TRISNET Transport Research Information Services Network [of US Department of Transportation]

Tritan Triphenylmethane

Trit Triturate [also trit]

TRITC Tetramethylrhodamine Isothiocyanate [Biochemistry]

TRITC-Dextran Tetramethylrhodamine Isothiocyanate Dextran [Biochemistry]

TRIU Tri-Universities [Canada]

TRIUL Tri-Universities Libraries (of British Columbia) [Canada]

TRIUMF Tri-Universities Meson Facility [of University of British Columbia, Vancouver, Canada]

TRIUMF-NRC-VPI Tri-Universities Meson Facility–National Research Council-Virginia Polytechnic Institute (Collaboration)

TRIUMF-TPC Tri-Universities Meson Facility–Time Projection Chamber

TRIX Total Rate Imaging of X-Rays

Trk Track [also trk]

Trk Trunk [also trk]

Trkr Tracker [also trkr]

TRL Technical Research Laboratory

TRL Test Readiness List

TRL Thermodynamics Research Laboratory

TRL Transistor-Resistor Logic

TRL Transuranium Research Laboratory [of Oak Ridge National Laboratory, Tennessee, US]

TRL Trunk Register Link

TRL Turkish Lira [Cuurrency of Turkey]

Trlr Trailer [also trlr]

TRLS Time-Resolved Laser Spectroscopy

Trls Trials [also trls]

TRM Tamper-Resistant Module

TRM Terminal Response Monitor

TRM Test Request Message

TRM Thermal Remanent Magnetization; Thermoremanent Magnetization

TRM Transient Reflectance Measurement (Technique) [Physics]

TRM Transmit-Receive Module

.TRM Terminal [File Name Extension] [also .trm]

TrM Track Magnetic

TRMA Time Random Multiple Access

TRMDT Transmission Date

TRMG Tread Rubber Manufacturers Group [US]

TRMI Tubular Rivet and Machine Institute [US]

Trml Terminal [also trml]

TRMM Tropical Rainfall Measuring Mission [A Joint US-Japan Mission]

TRMP Target Radiation Measurement Program

TRMS TDMA (Time-Division Multiple Access) Reference and Monitor Station

TRMS Total Report Management Solutions

TRMS True Root Mean Square

TRN Technical Research Note

TRN Television Relay Network

TRN Threaded Read News [Internet]

TRN Token Ring Network

Trn Transfer [also trn]

Trn Trunnion [also trn]

tRNA Transfer (or Transport) Ribonucleic Acid [also t-RNA, or T-RNA] [Biochemistry]

tRNA-AA Transfer (or Transport) Ribonucleic Acid–Amino Acid (Complex) [also t-RNA-AA, or T-RNA-AA] [Biochemistry]

Trnbkl Turnbuckle [also trnbkl]

TR News Transport Research News [of Transportation Research Board, US]

Trng Training [also trng]

Trnsln Translation [also trnsln]

Trnst Transit [also trnst]

trnt transient

TRO Temporary Restraining Order

TROA Time-Resolved Optical Absorption

TROC Trichloroethoxycarbonyl-

TROCA Tangible Reinforcement Operant Conditioning Audiometry

TROFF Trace Off [Programming]

troff text run-off [Unix Processor]

TROLL Technion Robotics Laboratory Language

TROM Teletext Read-Only Memory

TROMEX Tropical Meteorological Experiment

TRON The Real-Time Operating-System Nucleus (Project) [Japan]

TRON Trace On (Command) [Programming]

TRON Assoc TRON (The Real-Time Operating-System Nucleus) Association [Japan]

TROO Transponder On-Off

TROP Time-Phased Order Point

Trop Tropical; Tropics [also trop]

Trop Tropolone [Biochemistry]

Trop Tropopause [also trop] [Meteorology]

TROPAG Tropical Agriculture (Database) [of Royal Tropical Institute, Netherlands]

Trop Biol Tropical Biologist; Tropical Biology

TROPO Tropospheric Scatter Communication [also Tropo, or tropo]

Tropone 2,4,6-Cycloheptatriene-1-one

Tropop Tropopause [also tropop]

Troposp Troposphere [also troposp]

TROS Tape Resident Operating System

TROS Transformer Read-Only Storage

Trotyl 2-Methyl-1,3,5-Trinitrobenzene

TROV Tethered Remotely Operated Vehicle

tr oz troy ounces [Unit]

TRP Technical Review Panel

TRP Technology Reinvestment Program (Grant) [Joint US Grant of the Department of Defense/Advanced Research Projects Agency, the Department of Commerce, the Department of Energy, NASA and the National Science Foundation International]

TRP Technology Research Partnerships [US]

TRP Television Remote Pickup

TRP Thermal Ribbon Printer

TRP Total Radiation Pyrometer

TRP Toxicology Review Panel

TRP Tuition Reimbursement Plan

Trp Trap [also trp]

Trp (−)-Tryptophan; Tryptophyl [Biochemistry]

TRPA Tahoe Regional Planning Agency [Nevada, US]

TRPF Transmit/Receive Parity Failure

TRPGDA Tripropylene Glycol Diacrylate

TRPL Time-Resolved Photoluminescence

TRPP Training Plan and Procedure

Trps Transpose [also trps]

Tr Pt Transition Point [also Tr Pt, or tr pt]

Trpy 2,2',2"-Tripyridyl [also trpy]

TRQ Task Ready Queue

Trq Torque

TRR Tape Read Register

TRR Target Ranging Radar

TRR Teaching and Research Reactor

TRR Technical Research Report

TRR Test Readiness Review

TRR Thailand Research Reactor [of Office of Atomic Energy for Peace, Bangkok]

TRR Thermal Radio Radiation

TRR Time-Resolved Reflectivity (Method)

TRR Topical Report Request

TRR Topical Report Review

TRR-1 Thailand Research Reactor 1 [of Office of Atomic Energy for Peace, Bangkok]

TRRB Test Readiness Review Board

TRRB Trade Relations Research Bureau [UK]

TRRF The Refrigeration Research Foundation [US]

TRRFA Time-Resolved X-Ray Fluorescence Analysis

TRRL Transport and Road Research Laboratory [UK]

TRRL Tree-Ring Research Laboratory [of University of Arizona, Tucson, US]

TRMM Tropical Rainfall Measuring Mission

TRRN Terrain [also Trrn]

TRRR Time-Resolved Resonant Raman Spectroscopy [also TR^3]

TRRR Time-Resolved Resonant Raman Spectrum [also TR^3]

TRRRS Time-Resolved Resonance Raman Spectroscopy [also TR^{3}S]

TRRS Time-Resolved Raman Spectroscopy

TRRS Total Reflection Raman Scattering

TRRS Two-Photon Resonant Raman Scattering

TRS Tandy Radio Shack (Corporation) [Radio Shack Division of Tandy Corporation, Fort Worth Texas, US]

TRS Technical Reports Server [of NASA Center for AeroSpace Information]

TRS Teleoperator Retrieval System

TRS Telephone Referral Service

TRS Telephone Repeater Station

TRS Terrestrial Remote Sensing

TRS Terrestrial Repeater Station

TRS Test Response Spectrum

TRS Tetrahedral Research Satellite

TRS Thermal Residual Stress [Mechanics]

TRS Time Reference System

TRS Time-Resolved Spectroscopy

TRS Toll Room Switch

TRS Top Right Side

TRS Total Reducing Sugar [Biochemistry]

TRS Tough Rubber Sheath

TRS Transmission Regulated Spark

TRS Transmit-Receive Switch [also trs]

TRS Transverse Rupture Strength [Mechanics]

TRS Tree-Ring Society [US]

TRS Troubleshooting Record Sheet

TRS Trustees [also Trs, or trs]

TRS Tug Rotational System

Trs Transposition [also trs]

Trs Tritylsulfenyl

TRSA Terminal Radar Service Area

TRSB Telecommunications Regulatory Service Branch

TRSB Time Reference Scanning Beam (System)

TRSB Transcriber [also Trsb, or trsb]

TR/SBS Teleoperator Retrieval/Skylab Boost System

TRSC Thailand Remote Sensing Center

TRSD Test Requirements Specification Document

TRSDOS Tandy Radio Shack Disk-Operating System [also TRS-DOS]

TRSL Test Requirements and Specification Language

trsl translate

Trsn Transaction [also trsn]

Trsp Transport [also trsp]

TRSR Taxi and Runway Surveillance Radar

TRSSGM Tactical-Range Surface-to-Surface Guided Missile

TRSSM Tactical Range Surface-to-Surface Missile

TRSSSV Tubing-Retrievable Subsurface Safety Valve

TRST Tunneling Real-Space Transfer

Trst Transit [also trst]

TRT Technology, Research and Telecommunications

TRT Tensile Recoil Test

TRT Traffic Route Testing [Telecommunications]

TRT Turnaround Time [Computers]

Trt Trityl

TrT Track True [Magnetic Media]

TRTA Traders Road Transport Association [UK]

TRTL Transistor-Resistor-Transistor Logic

TRTPC Time-Resolved Transient Photoconductivity

TRU Tire Recycling Unit

TRU Total Recycle Unit

TRU Transmit-Receive Unit

TRU Transportable Radio Unit

TRU Transuranic (Waste)

TRU Transuranium Processing Facility [Nuclear Engineering]

Tru9 Thermus ruber 9 [Microbiology]

Trudy Mat Inst VA Seklova Trudy Matematicheskogo Instituta imeni V. A. Steklova [Mathematical Publication of the Academy of Sciences of the USSR]

TRUEX Transuranium Extraction (Process)

TRUF Transferable Revolving Underwriting Facility [Business Finance]

TRUMP Teller Register Unit Monitoring Program

TRUMP Total Revision and Upgrading of Maintenance Procedures

TRUMP-S Transuranic Management Through Pyropartitioning Separation

Trun Trunnion

Trun Ang Trunnion Angle

Trunc Truncate (Real-Type to Integer Value) [Pascal Function]

TRUPP Transuranium Processing Plant

TRUSAF Transuranic Waste Storage and Assay Facility

TRV Tobacco Rattle Virus

TRV Tobacco Ringspot Virus

TRV Transient Recovery Voltage

TRW Thompson, Ramo and Woolridge (Process) [Physics]

TRX Transaction [also Trx, or trx]

TRX Two-Region Physics Critical Experiment

TRXF Total-Reflection X-Ray Fluorescence

TRXRD Time-Resolved X-Ray Diffraction [also TRXD]

TRXRF Time-Resolved X-Ray Fluorescence

TRXRF Total Reflectance X-Ray Fluorescence

Try Tryptophan [Biochemistry]

TS Absolute Temperature versus Entropy (Diagram) [also T-S]

TS Structural Steel Tubing [AISI/AISC Designation]

TS Tapered Roller Bearing, Single-Row [Symbol]

TS Tape Status

TS Taper Shank

TS Tape Storage

TS Tape System

TS Target Strength [Acoustics]

TS Target System

TS Tar Sand

TS Taylor Series [Mathematics]

TS Technical Section

TS Technical Service

TS Technical Specification

TS Technical Support

TS Technical Support [NASA Kennedy Space Center Directorate, Florida, US]

TS *Technikerschule* – German for "School of Technology"

TS Telegraph Service

TS Telemetry Simulator

TS Temperature Scale

TS Temperature Sensitivity

TS Temperature Sensor

TS Temporary Storage

TS Tensile Stage

TS (Ultimate) Tensile Strength [also ts]

TS Tensile Stress

TS Tension Spring

TS Teratology Society [US]

TS Terrestrial Space

TS Testing Stage

TS Test Set

TS Test Site

TS Test Solution

TS Test Switch

TS Test Specification

TS Test Specimen

TS Test Stand

TS Test Station

TS Tetrahedral Site [also ts] [Crystallography]

TS Theoretical Science; Theoretical Scientist

TS Theoretical Statistics; Theoretical Statistician

TS Thermal Shield [Nuclear Reactors]

TS Thermal Shock

TS Thermal Spike

TS Thermal Spray(ing)

TS Thermal Storage

TS Thermal Stress

TS Thermal System

TS Thermal Switch

TS Thermoset

TS Thermosonic(s)

TS Thermostatic(s)

TS Thermosyphon (or Thermosiphon)

TS Thorium Series

TS Thrust Roller Bearing, Self-Aligning, Single-Direction, Flat Seats, Symmetrical Barrel-Shaped Rollers [Symbol]

TS Thunderstorm

TS Time Schedule

TS Time Series [Statistics]

TS Time Sharing

TS Time Signal

TS Time Slot

TS Time Stamp

TS Time Standard

TS Time Switch

TS Timing System

TS Titanium Silicate

TS Toll Switching

TS Tool Steel

TS Topographic Survey(ing)

TS Top Secret

TS Torch Soldering

TS Torpedo Shell

TS Torsional Strain

TS Total Solids [Chemistry]

TS Touch-Sensitive; Touch Sensitivity

TS (Peak) Toughness Structure

TS Tourette's Syndrome

TS Toxic Shock

TS Toxic Substance

TS Trace Specification

TS Trade Secret

TS Transfer Station

TS Transformer Station

TS Transient Stability

TS Transient State

TS Transition Set

TS Transition State

TS Translation Stage

TS Transmission Service

TS Transmission System

TS Transmitting Station

TS Transonic(s)

TS Transportation Science

TS Transportation System

TS Transverse Section

TS Triple Scattering

TS Triple Space [Computers]

TS Triplet State [Physics]

TS Trivial Solution [Mathematics]

TS Tunneling Spectroscopy

TS Tunnel(l)ing State [Electronics]

TS Turbine Shaft

TS Turboshaft Engine

TS Twin System [Metallurgy]

TS Typesetting

TS Unavailable Energy (i.e., Product of the Absolute Temperature T and the Entropy S) [Symbol]

TS- Tunesia [Civil Aircraft Marking]

2S 2-Sulfate

T&S Turn and Slip Indicator [Aeronautics]

T/S Treatment/Storage

T-S Absolute Temperature versus Entropy (Diagram) [also TS]

T-S Target-to-Substrate (Distance) [Physics]

T-S Temperature-Salinity (Diagram) [also TS]

Ts Tosyl (or p-Toluenesulfonate)

ts tetrahedral site [Crystallography]

τ-σ shear stress–tensile stress (diagram)

TSA Tactical Situation Analysis

TSA Target Service Agent

TSA Technical Support Alliance

TSA Technology Student Association [US]

TSA Telecommunications Society of Australia

TSA Telephony Services Architecture

TSA Temperature-Swing Absorption

TSA Temperature-Swing Adsorption

TSA Test Start Approval

TSA Texas Safety Association [US]

TSA Textile Services Association

TSA Thermal-Swing Adsorption

TSA Thrust Roller Bearing, Self-Aligning, Single-Direction, Flat Seats, Asymmetrical Barrel-Shaped Rollers [Symbol]

TSA Time Series Analysis [Mathematics]

TSA Time Slot Access

TSA Toluenesulfonamide

TSA Toluenesulfonic Acid

TSA Total Scan Area

TSA Total Surface Area

TSA Transportation Standardization Agency [of US Department of Transportation]

TSA Transverse Spherical Aberration

TSA Tube Support Assemblies

TSA Tourette's Syndrome Association [US]

TSA Turnstile Antenna

TSA Two-Sample Analysis (Test) [Materials Science]

TSA Two-Step Absorption

TSA Two-Step Annealing [Metallurgy]

TSAC Time Slot Assignment Circuit

TSAC Title, Subtitle and Caption

TSAC Tracking System Analytical Calibration

TSAIM Time Dependence of Sound Acoustic Impedance

TSALEN N,N'-Bis(thiosalicylidene)ethylene Diamine

TSAM Time Series Analysis and Modelling

Ts3[9]aneN3 Tris(tosyl)-1,4,7-Triazacyclononane

TSAO Transportation Safety Association of Ontario [Canada]

TSAP Technical and Scientific Advisory Panel

TSAP Transport Service Access Point

TSAP GOOS Technical and Scientific Advisory Panel for Global Ozone Observing System

TSAPI Telephony Services Application Program Interface

TSAPS Texas Section of the American Physical Society [US]

TSAU Time Slot Access Unit

TSAZ Target Seeker-Azimuth

TSB Taiwan Supply Bureau

TSB Technical Service Bulletin

TSB Technical Support Building

TSB Temporary Stowage Bag

TSB Termination Status Block

TSB Terminal Status Block

TSB Thermal Systems Branch [of National Renewable Energy Laboratory, Golden, Colorado, US]

TSB Toronto School of Business [Canada]

TSB Trade Show Bureau [US]

TSB Twin Sideband

TSBP Tri-sec-Butylphosphate

TSBU 2-Trimethylsiloxy-1,3-Butadiene

TSC Tape System Calibrator

TSC Technical and Scientific Center

TSC Technical Service Center

TSC Technical Service Contractor

TSC Technical Services Council

TSC Technical Subcommittee

TSC Technical Support Center

TSC Test Set Connection

TSC Test Setup Complete

TSC Test Shipping Cask [Nuclear Engineering]

TSC Test Support Coordinator

TSC Test System Controller

TSC Thermally Stimulated Current

TSC Thermal Stress Cracking

TSC Thiosemicarbazide

TSC Three-State Control

TSC Time-Sharing Control

TSC Titanium Silicon Carbide

TSC Totally Self-Checking [Computers]

TSC Towson State College [Baltimore, Maryland, US]

TSC Trans-Syn-Cis (Geometry) [Chemistry]

TSC Transit Switching Center

TSC Transmitter Start Code

TSC Transportation Safety Committee [of Society of Automotive Engineers, US]

TSC Transportation Systems Center

TSC Transport Support Component

TSC Tree Seed Center (Canberra) [of Commonwealth Scientific and Industrial Research Organization, Australia]

TSC Trenton State College [New Jersey, US]

TSC Troy State College [Alabama, US]

TSC Tsukuba Space Center [Japan]

TSC Turbosupercharger

TSC Two-Stage Command

TSC Two-Stroke Cycle

TSC Two-Subcarrier System [Television]

TSCA Timing Single Channel Analyzer

TSCA Toxic Substances Control Act [of US Environmental Protection Agency]

TSCAC Acetone Thiosemicarbazone

TSCAINV Toxic Substances Control Act Chemical Substances Inventory [of US Environmental Protection Agency]

TSCAP Thermally Stimulated Capacitance

TSCC Telemetry Standards Coordination Committee

TSCC Tilt-Scan CCD (Charge-Coupled Device) Camera

TSCCO Thallium Strontium Calcium Copper Oxide (Superconductor)

TSC-2,5-D Hexane-2,5-Dione Bis(4-Methyl Thiosemicarbazone)

TSCF Task Schedule Change Form

TSCF Textronix Standard Codes and Formats

TSCION TSC I/O (Input/Output) Number

TsCl Tosyl Chloride (or p-Toluenesulfonyl Chloride)

TSCLT Transportable Satellite Communications Link Terminal

TSCM Tardy and Senarmont Compensation Method

TSCO Test Support Coordination Office

TSCO Test Support Coordinator

TSCP Training Simulator Control Panel

TSCPHAL Phthalaldehyde-Bis(thiosemicarbazide)

TSC-PSC Transport-Support-Component to Protocol-Support-Component (Interface) [also TSC-to-PSC]

TSCr Thermally Stimulated Creep (Spectroscopy)

TSCS Thermally Stimulated Current Spectroscopy

TSCT Time-Sharing Control Task

TSCTES Total Spin Coherence Transfer Echo Filtering Spectroscopy

TSC-to-PSC Transport-Support-Component to Protocol-Support-Component (Interface) [also TSC-PSC]

TSD Technical Services Division

TSD Test Start Date

TSD Thermally Stable (Drill Bit) Diamond

TSD Thermally-Stimulated Depolarization

TSD Thermally-Stimulated Desorption

TSD Thermally-Stimulated Discharge

TSD Thermal Shock Damage

TSD Thermal Spray Division [of ASM International, US]

TSD Thermionic Specific Detector

TSD Time Slice Detection [Computers]

TSD Touch Sensitive Digitizer

TSD Traffic Situation Display

TSD Transportation System Design

TSD Treatment, Storage and Disposal

TSD Triode Sputtering Deposition

TSD Two-Stage Demagnetization

TSDA Tandem Spherical Deflector Analyzer

TSDA Thiosemicarbazide-N,N-Diacetic Acid

TSDC Thermally Stimulated Depolarization Current

TSDC Thermally Stimulated Discharge Current

TSDD Temperature-Salinity-Density-Depth (Relationship)

TSDF Target System Data File

TSDF Treatment, Storage and Disposal Facility

TSDIS TRMM (Tropical Rainfall Measuring Mission) Science and Data Information System [Japan]

TSDM Time-Shared Data Management (System)

TSDOS Time-Shared Disk Operating System

TSDR Treatment, Storage, Disposal and Recycling

TSDU Target System Data Update

TSDU Transport Service Data Unit

TSDW Transverse Spin Density Wave [Solid-State Physics]

TSE Tactical Support Equipment

TSE Temperature-Sensitive Element

TSE Terminal Source Editor

TSE Test of Spoken English

TSE Test Support Equipment

TSE The Semware Editor

TSE Time Slice End [Computers]

TSE Tokyo Stock Exchange [Japan]

TSE Toronto Stock Exchange [Canada]

TSE Translation Stage Equipment

TSE Transportation Support Equipment

TSE Türk Standardlari Enstitüsü [Turkish Standards Institute]

TSE Twist Setting Efficiency

TSE Two-Stroke Engine

TSEC Threatened Species and Ecological Communities [Section of Environment Australia]

TSEC Translinear Scanning Electro-Optical Camera [ER-2 NASA Ames High-Altitude Aircraft]

T/sec Tesla(s) per second [also T sec^{-1}, T/s, or T s^{-1}]

TSED Transmission Scanning Electron Diffraction

TSEE Thermally-Stimulated Emission of Exoelectrons; Thermally-Stimulated Exoelectron Emission

TSeF Tetraselenafulvalene [also TSF]

TSEF Total Structural Echo Factor

TSEI Transportation Safety Equipment Institute [US]

TSEI Truck Safety Equipment Institute [now Transportation Safety Equipment Institute, US]

TSEM Transmission Scanning Electron Microscopy

TSEM Transmission Secondary Electron Multiplier [also tsem]

TSEQ Time Sequence

TSeT Tetraselenatetracene

TSF Telegraphie, Téléphonie Sans Fil [Wireless Telegraphy and Telephony Service, France]

TSF Ten Statement FORTRAN

TSF Tetraselenafulvalene [also TSeF]

TSF Texaco Selective Finishing (Process)

TSF Texas Section Fall Meeting [of American Physical Society, US]

TSF Thermal Stress Fracture

TSF Thermoplastic Structural Foam

TSF Thin-Solid Film

TSF Thin-Solid Film (Journal) [US]

TSF Through Supergroup Filter [Telecommunications]

TSF Ton(s) per Square Foot [also tsf]

TSF Tower Shielding Facility

TSF Transverse Shielding Factor

TSF Treasury Security Force [of US Department of the Treasury]

TSF Twin (Growth) Stacking Fault [Materials Science]

TSF Typical Shape Function (Analysis)

TSFC Thrust-Specific Fuel Consumption [Aerospace]

TSFO Transportation Support Field Office [US]

TSF/O Teklogix Screen Formatter/Optimizer

TSFP Transportation Safety and Field Program [Canada]

TSFS Trunk Servicing Forecasting System

TSFZ Travelling Solvent Floating Zone (Method)

TSG Technical Service Group

TSG Technical Steering Group

TSG Time Signal Generator

TSG Time Slot Generator

TSG Transversely Adjusted Gap

TSG Tri-Service Group [of NATO]

TSGA Tanzania Sisal Growers Association

TSGCEE Tri-Service Group on Communications and Electronic Equipment [of NATO]

TSGP Test Sequence Generator Program [of ESTEC–European Space Agency Technical Center]

TSgt Technical Sergeant [also T Sgt, or T/Sgt]

TSH Tensor Surface Harmonic (Theory) [Physics]

TSH Through-Surface Hardening [Metallurgy]

TSH Thyroid-Stimulating Hormone [Biochemistry]

TSH Tolylene Sulfohydrazide

TSH Toluenesulfonyl Hydrazide

TSH Trajectory-Surface-Hopping (Calculation)

TSH Triaxial Sensor Head

TSh Tanzanian Shilling [Currency]

TSHA Texas State Historical Association [US]

TSHEO Tensor Surface Harmonic Equivalent Orbital [Physics]

TSHR Thyrotropin Receptor [Biochemistry]

TSHS Tennessee State Horticultural Society [US]

Tshwr Thundershower [also tshwr]

TSI Task Status Index

TSI Telesensory Systems, Inc. [US]

TSI Terminal Specific Interface

TSI Test Structure Input

TSI Thermal Systems, Inc. [US]

TSI Threshold Signal-to-Interference Ratio

TSI Time Slot Interchanger

TSI Ton-Force per Square Inch; Ton(s) per Square Inch [also tsi]

TSI Transmitting Station Identification

TSI Transmitting Subscriber's Identification

TSI Transportation Safety Institute [US]

TSI N-Trimethylsilyl Imidazole

tsi ton-force per square inch; ton(s) per square inch [also TSI]

TSIC Thermally Stimulated Ionic Current

TSID Track Section Identification

TSIE Transformed Spectral Index of the External Index

TSI J Part Instrum Transportation Safety Institute Journal of Particle Instrumentation [US]

TSIM 1-(Trimethylsilyl)imidazole; N-Trimethylsilyl Imidazole

Tsim Trivedi (Microsegregation Model) Simplified by Tewari and Laxmanan [Metallurgy]

TSIMS Telemetry Simulation Submodule

TSIOA Temporary Storage Input/Output Area

TSIS Turbine Shaft Inspection System

TSIU Telephone System Interface Unit

TSJP Tanaka Solid Junction Project [of Exploratory Research for Advanced Technology Organization–Japan Research Development Corporation, Yokohama]

Tsk Task [also tsk]

TSL Test Source Library

TSL Texas State Library [Austin, US]

TSL Thermal Superlattice [Materials Science]

TSL Tilted Superlattice [Materials Science]

TSL Titanium-Doped Sapphire Laser

TSL Total Sediment Load

TSL Total Service Life

TSL Tri-State Logic

TSL Tungsten Strip Lamp

TSL Two-Stage Liquefaction

TSLD Troubleshooting Logic Diagram

TSLE Two-Step Laser Excitation

TSLS Two-Stage Least Squares

TSM Tail Service Mast [Aerospace]

TSM Tandem Scanning Microscope

TSM Telephony Signaling Module

TSM Terminal Server Manager

TSM Terminal Support Module

TSM Time-Shared Monitor

TSM Trade Study Management

TSMA Traffic Safety Markings Association [UK]

TSMDA Test-Section Meltdown Accident [Nuclear Reactors]

TSMFN Tunneling-Stabilized Magnetic Force Microscopy

TSMG Thompson Sub-Machine Gun

TSMT Transmit [also TSMIT, TSMT, or Tsmt]

TSMTR Transmitter [also Tsmtr]

TSN Task Sequence Number

TSN Tryptone Sulfite Neomycin [Biochemistry]

TSNI 1-(p-Toluenesulfonyl)-4-Nitro Imidazole

TSO Technical Service Order [US]

TSO Technical Standard Order

TSO Technical Standing Order

TSO Telecommunications Service Order

TSO Telephone Service Observations

TSO Terminating Screening Office [Telecommunications]

TSO Time-Sharing Option

TSO Time Since Overhaul

TSO trans-Stilbene Oxide

TSO Tropical Studies Organization [of University of San José, Costa Rica]

TSO Two Stages to Orbit [Aerospace]

TSO Titanium Silicon Oxide

TSODB Time Series Oriented Database

TsOEt Ethyl Tosylate (or Ethyl p-Toluenesulfonate)

TsOH Tosyl Hydroxide

T-Solvent Tetrahydro-2-[(Tetrahydro-2-Furanyl)methoxy]-2H-Pyran

TSOP The Society for Organic Petrology [US]

TSOP Thin, Small Outline Package [Electronics]

TsOR Tosylate [Chemistry]

TSORT Transmission System Optimum Relief Tool

TSOS Time-Sharing Operating System

TSO/VTAM Time-Sharing Option/Virtual Telecommunications Access Method

TSP Technical Steering Panel

TSP Technical Support Package

TSP Test Software Program

TSP Tippins Strip Processing

TSP Thermosetting Plastics

TSP Thermosetting Polymer

TSP Thermostable Polymer

TSP Time and Space Processing

TSP Time-Shared Processing

TSP Titanium Sublimation Pump

TSP Torpedo Seaplane

TSP Total Suspended Particulates

TSP Total Systems Performance

TSP Traffic Service Position [Telecommunications]

TSP Triple Superphosphate

TSP Tribasic Sodium Phosphate

TSP 3-Trimethylsilyl 1-Propanesulfonic Acid, Sodium Salt

TSP Trisodium Phosphate

TSP Twisted Shielded Pairs [Electric Cables]

Tsp Trichophyton Species [Immunology]

tsp teaspoon(ful) [Unit]

TSPA Total System Performance Assessment

TSPA Triethylenethiophosphoramide

TSPC Thermally Stimulated Polarization Current

TSPC Tropical Stored Products Center [of Ministry for Overseas Development, UK]

TSPC/DC Thermally-Stimulated Polarization (Current)/Depolarization Current

TSPE Texas Society of Professional Engineers [US]

TSPG Tomographic (Steel) Sheet Profile Gauge

tspn teaspoon(ful) [Unit]

TSPP Tetrasodium Pyrophosphate

TSPS Time Sharing Programming System

TSPS Traffic Service Position System [Telecommunications]

TSPSCAP Traffic Service Position System Capacity Program

TSPZ Torres Strait Protected Zone [Australia]

TSPZA Torres Strait Protected Zone Authority [Australia]

TSQ Time and Superquick

TSQ Triple State Quadrupole (Spectrometer)

TSQAP Tasmanian Shellfish Quality Assurance Program [Australia]

Tsqls Thundersqualls [also Tsqls]

TSQMS Triple State Quadrupole Mass Spectrometry

TSQS Triple State Quadrupole Spectrometer

TSR Test Schedule Request

TSR Technical Service Request

TSR Technical Status Review

TSR Technical Summary Report

TSR Telecommunications Service Request

TSR Temporary Storage Register

TSR Terminate-and-Stay-Resident (Program)

TSR Test Status Report

TSR Test Summary Report

TSR Thermal Shock Resistance

TSR Thermal Shock Rig

TSR Thermochemical Sulfate Reduction

TSR Thermosetting Resin(s)

TSR Thyroid Secretion Rate [Medicine]

TSR Torpedo-Spotter Reconnaissance (Plane)

TSR Total Stress Range [Mechanics]

TSR Tower Shielding Reactor [of Oak Ridge National Laboratory, Tennessee, US]

TSR Trace Sulfur Removal (Process)

TSR Tube-Shift Radiography

TSR Tunnel Stress Relaxation [Solid-State Physics]

TSR Turbine Steam Rate

TSRA Total System Requirements Analysis

TSRC Tubular and Split Rivet Council [now Tubular Rivet and Machine Institute, US]

TSRO Topological Short-Range Order(ing) [Solid-State Physics]

TSRP Training and Simulation Requirements Plan

TSRST Thermal Stress Restrained Specimen Test

TSS Tactical Strike System

TSS Task State Segment

TSS Tangential Signal Sensitivity [Electronics]

TSS Technical Staff Surveillance

TSS Ternary Solid Solution [Materials Science]

TSS Telecommunication Standards Section [of International Telecommunication Union]

TSS Telecommunication Switching System

TSS Telephone Support System

TSS Teletype Switching Subsystem

TSS Terminal Security System

TSS Terminal Solid Solubility; Terminal Solid Solution

TSS Terminal Support Subsystem

TSS Tethered Satellite System

TSS Thermal Spray Society [of ASM International, US]

TSS Thermal Synthesizer System (Software) [of NASA]

TSS Threatened Species Strategy

TSS Time-Sharing System

TSS Toll Switching System

TSS Total Suspended Solids

TSS Toxic Shock Syndrome [Medicine]

TSS Transmission Surveillance System

TSS Tromsø Satellite Station [of Swedish Space Corporation]

TSS Trunk Servicing System

TSS Tug Structural Support

TSS-1R Tethered Satellite System–Reflight Mission

TSSA Technical Standards and Safety Authority [Canada]

TSSA Test Scorer and Statistical Analyzer

TSSA Thunderstorm with Sandstorm

TSSC Technical and Scientific Societies Council [US]

TSSC Telecommunications System Status Center

TSS-C Transmission Surveillance System–Cable

TSSD Terminal Solid Solubility for Dissolution

TSSG Top-Seeded Solution Growth (of Superconductors)

TS/SI Top Secret/Sensitive Information

TSSMCP Time-Sharing System Message and Control Program

TSSST Time-Space-Space-Space-Time [also TS^{3}T]

TSST Toxic Shock Syndrome Toxin

TSS/TSO Time-Sharing System/Time-Sharing Option

TSSU Test Signal Switching Unit

TSSX Tape Subsystem Extender

TST Technical Support Team

TST Telecommunications Services Tax [Canada]

TST Thermal Spray Technology

TST Time-Space-Time (Network) [also T-S-T]

TST Tokamak Steady-State Technology (Project) [of Oak Ridge National Laboratory, Tennessee, US]

TST Torres Strait Treaty [Australia]

TST Transaction Step Task

TST Transient State Theory

TST Transition-State Theory [Physics]

TST Trans-Syn-Trans (Geometry) [Chemistry]

TST Trimethylsilyl Thiazole

TST Triple-Sorbant Trap

TST Two Square Theorem

.TST Test [File Name Extension] [also .tst]

1-TST 1-(Trimethylsilyl) Thiazole

2-TST 2-(Trimethylsilyl) Thiazole

T-S-T Time-Space-Time (Network) [also TST]

Tst Test [also tst]

TSTA Transmission, Signaling and Test Access

TSTA Tritium Systems Test Assembly [of Los Alamos National Laboratory, New Mexico, US]

TSTC Texas State Technical College [Waco, US]

TSTCM Toronto School of Traditional Chinese Medicine [Canada]

TSTE Texas Society of Telephone Engineers [US]

TSTFA Tasmania Sashimi Tuna Fishermens's Association [Australia]

TSTI Texas State Technical Institute [US]

Tstm Thunderstorm [also tstm]

TSTN Triple Supertwisted Nematic

T Storm Thunderstorm [also T storm]

TSTR Transition Strain Rate [Mechanics]

TSTPAC Transmission and Signalling Test Plan and Analysis Concept

TSTU N-[(Dimethylamino){(2,5-Dioxo-1-Pyrrolidinyl)oxy}-methylene]-N-Methyl MethanaminiumTetrafluoroborate

TSU Tandem Signal Unit

TSU Tape Search Unit

TSU Technical Service Unit

TSU Test Signal Unit

TSU Time Standard Unit

TSU Texas Southern University [Houston, US]

TSU Tolstoi State University [Russian Federation]

TSU Troy State University [Alabama, US]

TSU Trunk Switching Unit

TSUN Turbo Sun [Computers]

TSUS Tariff Schedules of the United States (of America)

TSV Tab Separated Values

TSV Through Sight Video

TSV Tobacco Streak Virus

TSV Toronto School of Video [Canada]

TSV Townsville, Australia [Meteorological Location Designator]

TSVD Thermally-Stimulated Voltage Decay

Tsvetn Met Tsvetnye Metally [Russian Journal on Nonferrous Metals published in Moscow]

Tsvetn Metall Tsvetnaya Metallurgia [Russian Journal on Metallurgical Engineering]

Tsvetn Metall Nauchno-Tekh Sb Tsvetnaya Metallurgia Nauchno-Tekhnicheskii Sbornik [now Tsvetn Metall]

TSVM Temperature Dependence of Sound Velocity Maximum

TSVP *(Tournez s'il vous plaît)* – French for "Please Turn Over"

TSW Task Status Word

TSW Tele-Software

TSW Test Software

TSW Test Switch

TSW Time Switch

TSW The Searchers Workbench

TSW Tube Socket Weld

TSWB Tinned Steel Wire Braid

TSWV Tomato Spotted Wilt Virus

TSX Time-Sharing Executive

TT Tachiki and Teramoto (Method) [also T-T] [Metallurgy]

TT Tail-to-Tail (Linkage)

TT Target Transformation

TT Technical Translation

TT Technology Transfer

TT Telegraphic Transfer [also T/T]

TT Teletype(writer)

TT Teletypewriter Terminal

TT Template Technique

TT Temporarily Transferred

TT Tensile Test(ing)

TT Terminal Timing

TT 2,2',5',2"-Terthiophene

TT Test(ing) Temperature

TT Test Tube

TT Tetrode Transistor

TT Textile Technologist; Textile Technology

TT Thermal Tempering

TT Thermal Transformation

TT Thermal Treatment

TT Thermionic Tube

TT Thoriated Tungsten

TT Thrust Tapered Roller Bearing [Symbol]

TT Thrust Termination; Thrust Terminator [Aerospace]

TT Thymol Turbidity

TT Tide Tables [of US Coast Guard]

TT Timing and Telemetry [also T/T]

TT Torpedo Tube

TT Torque Tube

TT Total Temperature

TT Total Time

TT Touch-Tone (Telephone)

TT Towed Target [Military]

TT Tracking and Telemetry

TT Tracking Telescope

TT Traffic Tester

TT Transaction Terminal

TT Transducer Technology

TT Transfer Tube

TT Transformation Toughened; Transformation Toughening

TT Transformation Twin [Crystallography]

TT Transportation Technologist; Transportation Technology

TT Transport Theory [Atomic Physics]

TT Transport Trust [UK]

TT Trans-Trans (Configuration) [Chemistry]

TT Transverse Plane (of Composite) [Symbol]

TT Treinto y Tres [Uruguay]

TT Trinidad and Tobago

TT Trinidad and Tobago [ISO Code]

TT Trinity Test [First Atomic Explosion on July 16, 1945 at Jornada de Muerto, New Mexico, US]

TT 2,3,5-Triphenyl Tetrazolium

TT Triple Thermoplastic (Electric Wire)

TT True Track

TT Truncated Tetrahedron [Crystallography]

TT Truth Table

TT Tunnel Triode

TT Turbine Trip

TT Tuberculin-Tested [Agriculture]

TT Typewriter Text

TT- Chad [Civil Aircraft Marking]

TT30 30-Foot-Pound Indexing Temperature [Charpy Impact Test]

35TT 35-Mil Lateral Expansion Indexing Temperature [Charpy Impact Test]

T&T Travel and Tourism

T&T Turbine Trip and Throttle (Valve)

T/T Telegraphic Transfer

T/T Terminal Timing

T/T Timing/Telemetry [also TT]

T/T$_m$ Homologous Temperature [Symbol]

T-T Tachiki and Teramoto (Method) [also TT] [Metallurgy]

T-T Twin-Twin (Grain Boundary) [Materials Science]

T(t) Time Rate of Change of Temperature [Symbol]

T/Θ Reduced Temperature (i.e. Temperature to Paramagnetic Temperature Ratio) [Symbol]

T$_c$/Θ$_D$ Critical Temperature to Debye Temperature (Ratio) [Symbol]

tt triplet of triplets [Spectroscopy]

.tt Trinidad and Tobago [Country Code/Domain Name]

τ(T) surface free energy [Symbol]

τ(T) temperature-dependent life time [Symbol]

τ(t) time-dependent (local) shear stress [Symbol]

2×2 Two-by-Two (Truck)

2θ Diffraction (or Scattering) Angle (i.e., Twice the Bragg Angle) [Symbol]

TTA Thenoyltrifluoroacetone

TTA Thermomechanical Test Area

TTA Thenoyltrifluoroacetone

TTA Time-Temperature Austenitization [Metallurgy]

TTA Time to Apogee

TTA Traditional Theoretical Approach

TTA Traffic Trunk Administration

TTA Transformation-Toughened Alumina

TTA Transport-Triggered Architecture

TTA Travel and Tourism Administration [US]

TTA Triethylene Tetramine

TTA Tri-p-Tolylamine

TTA Triplet-Triplet Absorption [Spectroscopy]

2TA Zirconia-Toughened Alumina

TTAC Telemetry Tracking and Command

TTAH Thenoyltrifluoroacetone

T/TAL Tandem Computers Transaction Application Language

TTAWG Technology Thrust Area Working Group [US]

TTB Technology Test Bed

TTB Tetragonal Tungsten Bronze

TTB Toll Testboard

TTB Troop Transport Boat

TTB Trunk Test Buffer

TTBBP 3,3',5,5'-Tetrakis(t-Butyl)biphenyl-4,4'-Diole

TTBC Thick Thermal Barrier Coating

TTBS Trinidad and Tobago Bureau of Standards

TTBT Threshold Test Ban Treaty [Between the US and the USSR]

TTBWR Twisted Tape Boiling Water Reactor

T2BX Toluene-to-Benzene and Xylene (Process)

TTC Tape-to-Card

TTC Technical Transfer Council [Australia]

TTC Technology Transfer Center

TTC Technology Transfer Committee

TTC Telecommunications Techniques Corporation [US]

TTC Telecommunication Training Centre [Jabalpur, India]

TTC Telemetry, Tracking and Command

TTC Teletypewriter Center

TTC Television Training Center [UK]

TTC Terminating Toll Center

TTC Test Transfer Cask [Nuclear Engineering]

TTC n-Tetratetracontane

TTC Tetrazolium Chloride

TTC Texas Technological College [Lubbock, Texas, US]

TTC Thermal Transfer Equipment

TTC Thermal Transient Equipment

TTC Thiatricarboxyanine

TTC Time to Circularize (Orbit) [Aerospace]

TTC Toronto Transit Commission [Canada]

TTC Tracking, Telemetry and Command

TTC Transformation-Toughened Ceramics

TTC Transportation Technology Center [of Sandia National Laboratory, Albuquerque, New Mexico, US]

TTC Trans-Trans-Cis (Geometry) [Chemistry]

TTC 2,3,5-Triphenyl-2H-Tetrazolium Chloride

TTC Trithiocarbonate

TTC Tritium Technology Conference

TTC Tunnel Thermal Control

TT&C Tracking, Telemetry and Command

TT&C Tracking, Telemetry and Control

TTCA Thrust Translation Control(ler) Assembly [also T/TCA] [Aerospace]

TTCM Tracking, Telemetry, Command and Monitoring Station

TTCN Tree (and) Tabular Combined Notation

TTCP Test Transmission Control Protocol

TTCP The Technical Cooperation Program [Australia/Canada/UK/US]

TTCP The Technical Cooperation Program [of Department of National Defense, Canada]

TTCV Tracking Telemetry, Command and Voice

TTD Technical Training Division [of Air Training Command, US Air Force]

TTD Temporary Text Delay

TTD Textile Technology Digest [Publication of the Institute of Textile Technology, US]

TTD Tetraethylthiuram Disulfide

TTD Thickened Tailings Discharge [Mining]

TT$ Trinidad and Tobago Dollar [also TTD] [Currency]

TTDA Triethylenetetraminediacetate [also Ttda, or ttda]

TTDF Tariff and Trade Data File [of General Agreement on Tariffs and Trade]

TTDL Terminal Transparent Delay Language

TTDU Technology Transfer Diffusion Unit

TTE Telephone Terminal Equipment

TTE Time to Event

TTE Tropical Testing Establishment [UK]

TTEB Time-to-Transmit the Encyclopedia Brittanica

TTEC Thai Technical and Economic Cooperation Office

TTeF Tetratellurafulvalene

TTeF-TCNQ Tetratellurafulvalene Tetracyanoquinoline

TTEGDA Tetraethylene Glycol Diacrylate

TTEL Tool and Test Equipment List

TTF Taiwan Textile Federation

TTF Tetrathiofulvalene; 1,1',3,3'-Tetrathiofulvalene

TTF Teaching, Tenure Track, Full-Time Faculty (Position)

TTF 2-Thenoyltrifluoroacetone

TTF Time-to-Failure [also ttf]

TTF Timber Trade Federation [UK]

TTF Torsional-Mode Tuning Fork

TTF Transient Time Flowmeter

TTF Transmission Test Facility

TTF TrueType Font

.TTF TrueType Font [File Name Extension]

TTFA Thallic Trifluoroacetate

TTFA Thallium(III) Trifluoroacetate

TTFBr Tetrathiafulvalenium Bromide

TTFC Textile Technical Federation of Canada

TTFN Ta-Ta For Now [Internet Jargon]

TTF-TCNQ Tetrathiafulvalene Tetracyanoquinodimethane

TTFTT 1,1',3,3'-Tetrathiafulvalene Tetrathiolate

TTG Technical Translation Group

TTG Time to Go

TTH Tieftemperaturhydrierung (Process) [Low-Temperature Hydration (Process)]

10th tenth

20th twentieth

1,000th thousandth

TTHA Triethylenetetraminehexaacetic Acid

TTHDCM Thin and Thick Film High-Density Ceramic Module

TTHE Thermal Transient Histogram Equivalent

TTHMK Ta Tschung-Hua Min-Kuo [Republic of China–Taiwan]

TTI Teletype Test Instruction

TTI The Technological Institute [US]

TTI Time-Temperature Indicator

TTI Toyota Technological Institute [Nagoya, Japan]

TTI Transparent Thermal Insulation

T2000I Transport 2000 International [UK]

T2ICL Type-II Interband Cascade Laser

TTIP Titanium Tetraisopropoxide

TTK Tampereen Teknillinen Korkeakoulu [Tampere University of Technology, Finland]

TTK Tie Trunk

TTL Taxi-Track Light

TTL Time to Live [Internet]

TTL Through-the-Lens (Camera)

TTL Transistor-to-Transistor Link

TTL Transistor-Transistor Logic [also T²L]

TTL/LS Transistor-Transistor Logic/Large Scale

TTLM Through-the-Lens Light Metering

TTLS Transistor-Transistor Logic Schottky

TTM Two-Tone Modulation

TTMA Truck Trailer Manufacturers Association [US]

TTMAPP Tetrakis(4-N-Trimethylaminophenyl)porphine

TTμL Transistor-Transistor-Micrologic [also TTML]

TTMP Transit Time Magnetic Pumping

TTMPP Tris(2,4,6-Trimethoxyphenyl)phosphine

TTMS Telephoto Transmission Measuring Set

TTN Technology Transfer Network [of US Environmental Protection Agency–Office of Air Quality Planning and Standards]

TTN Technology Transfer Nodes

TTN Thallium Nitrate Trihydrate

TTN Thallium Trinitrate

TT/N Test Tone to Noise (Ratio)

TTN BBS Technology Transfer Network Bulletin Board System [of US Environmental Protection Agency–Office of Air Quality Planning and Standards]

TTO Technology Transfer Office

TTO Total Toxic Organics

TTO Traffic Trunk Order

TTO Transmitter Turn-Off

TTOF Tandom Time-of-Flight (Mass Spectrometry)

TTP Tabular Tape Processor

TTP Tape-to-Print

TTP Technical Task Plan

TTP Test Transfer Port

TTP 1,4,8,11-Tetrathiacyclotetradecane

TTP Tetratolylporphinate

TTP Tetratolylpyrrole

TTP Thermal-Transfer Printer; Thermal-Transfer Printing

TTP The Technology Partnership [UK]

TTP Thymidine Triphosphate; Thymidine 5'-Triphosphate [Biochemistry]

TTP Time-Temperature-Precipitation (Diagram)

TTP Time to Perigee

TTP Transient Thermal Processing

TTP Trunk Test Panel

TTP Tris(4-Tolyl)phosphate

TTPB Trunk Test Panel Buffer

TTPC Time-Resolved Transient Photoconductivity

TTPD Threshold Temperature Programmed Desorption

T2QWL Type-II Quantum-Well Laser

TTR Target Tracking Radar

TTR Thermal Test Reactor

TTR Time-Transfer Receiver

TTR Tonopah Test Range [of US Atomic Energy Commission in Nevada]

TTR Toshiba Training Reactor [Japan]

TTR Transition Temperature Range [Metallurgy]

TTR Transmission Test Rack

TTR Trunk Test Rack

TTR Type-Token Ratio

TTRAN Tape Transfer

TTRANCWB Tape Transfer Ceramic Wiring Board

TTRC Thrust Travel Reduction Curve [Aerospace]

TTRM Transition Thermoremanent Magnetization

TTS TDMA (Time-Division Multiple Access) Terminal Simulation

TTS Tearing Topography Surface

TTS Technology Transfer Society [US; also T2S, or T²S]

TTS Telecommunications Terminal System

TTS Teletypesetter

TTS Temperature-Time-Sensitization (Diagram)

TTS Temporary Threshold Shift [Acoustics]

TTS Test and Training Satellite

TTS Tetrathiosquarate

TTS Text-To-Speech

TTS Time-Temperature Superposition

TTS Time to Station

TTS Tire Transformation System

TTS Transaction Terminal System

TTS Transaction Tracking System

TTS Transit Time Speed

TTS Transit Time Spread

TTS Transmission Test Set

TTS Transportable Transformer Substation

TTS Transputer Technology Solutions [UK]

TTS Transverse Tensile Stress

TTS Trouble Tracking System

TTS Tubular-Tin-Source (Process)

T2S Second-Stage Ternary Structure [Metallurgy]

TTSA Trinidad and Tobago Scientific Association

TTSMSP Time-Temperature-Stress-Moisture Superposition Principle

TTSP Time-Temperature Superposition Principle

TTSPN Two-Terminal Series Parallel Networks

TTSSP Time-Temperature-Stress Superposition Principle

TTT Template Tracing Technique

TTT Thermal-Temporal Treatment [Metallurgy]

TTT Tetrathiatetracene

TTT Time-Temperature-Transformation (Diagram) [also T-T-T]

TTT Trans-Trans-Trans (Geometry) [Chemistry]

TTT Triethyl Trimethylene Triamine

TTTF Thermal Treatment Technology Test Facility

TTTL Transistor-Transistor-Transistor Logic [also T³L]

T_1-T_2-T_3-T_4 Alternating-Current Primary [Controllers]

TTU Tennessee Tech(nological University) [Cookeville, US]

TTU Terminal Time Unit

TTU Texas Tech(nological) University [Lubbock, US]

TTU Through-Transmission Ultrasonics

TTU Timing Terminal Unit

TTU Transportable Treatment Unit

TTV Termination, Test and Verification

TTV Total Thickness Variation [Electronics]

TTVD Trap-to-Trap Distillation

TTW Teletype(writer)

TTX Tetrodotoxin [Biochemistry]

TTY Teletype(writer)

TTYBS Teletype(writer), Backspace

TTYC Teletype(writer) Controller

TTYL Talk to You Later [Internet Jargon]

TTYPP Teletype Point-to-Point

TTYS Teletypesetter

TTY STN Teletype Station

TTZ Transformation-Toughened Zirconia

TU Take-Up (Mechanism) [Mechanical Engineering]

TU Tamagawa University [Tokyo, Japan]

TU Tamkang University [Taiwan]

TU Tape Unit

TU Technical University

TU Technical Utilization

TU *(Technische Universität)* – German for "Technical University"

TU Technology Utilization

TU Teikyo University [Utsunomiya, Japan]

TU Temple University [Philadelphia, Pennsylvania, US]

TU Test Unit

TU Thank You [Amateur Radio]

TU Thermal Unit

TU Thiourea [also tu]

TU Timing Unit

TU Tohoku University [Sendai, Japan]

TU Tokai University [Kanagawa, Japan]

TU Tokushima University [Japan]

TU Tokyo University [Japan]

TU Tongji University [PR China]

TU Top Up

TU Tottori University [Japan]

TU Toxic Unit

TU Toyama University [Japan]

TU Trade Union

TU Traffic Unit [Telecommunications]

TU Transfer Unit [Chemical Engineering]

TU Transmission Unit

TU Transport Unit [Computers]

TU Transuranic(s); Transuranium

TU Transurethral [Medicine]

TU Trent University [Peterborough, Ontario, Canada]

TU Triangle Universities [North Carolina State University, University of North Carolina, and Duke University, US]

TU Tritium Unit

TU Tsinghua University [Peking, PR China]

TU Tsukuba University [Japan]

TU Tufts University [Medford, Massachusetts, US]

TU Tulane University [New Orleans, Louisiana, US]

TU Tulsa University [Oklahoma, US]

TU Turbopump Unit

TÜ Technische Überwachung [Journal on Technical Inspection published by VdTÜV, Germany]

TU- Ivory Coast [Civil Aircraft Marking]

Tu Tuesday [also Tue, or Tues]

Tu Tungsten [Abbreviation; Symbol: W]

Tu Tupolev [Series of Russian Airplanes]

tu thiourea

TUA Technische Universität Aachen [Technical University of Aachen, Germany]

TUA Telecommunications Users Association [UK]

TUA TOW (Tube-Launched, Optically-Tracked, Wire-Guided) under Armor

TUA Tractor Users Association [UK]

TUA Trade Union Act [US]

TU Aachen Technische Universität Aachen [Technical University of Aachen, Germany]

TUAC Trade Union Advisory Committee [of Organization for Economic Cooperation and Development, France]

TUAT Tokyo University of Agriculture and Technology [Japan]

TUB Technical University of Budapest [Hungary]

TUB Technische Universität Berlin [Technical University of Berlin, Germany]

TUB Technische Universität Braunschweig [Technical University of Braunschweig, Germany]

Tub Tubing [also tub]

Tube Int Tube International [Journal published in the UK]

Tube Pipe Technol Tube and Pipe Technology [Journal published in the UK]

TU Berlin Technische Universität Berlin [Technical University of Berlin, Germany]

TU Braunschweig Technische Universität Braunschweig [Technical University of Braunschweig, Germany]

Tubul Struct Tubular Structures [Journal published in the US]

TUC Trade Union Congress [UK]

TUCA Transient Undercooling Accident [Nuclear Reactors]

TUCC Triangle Universities Computing Center [of North Carolina State University, University of North Carolina, and Duke University, US]

TU Chemnitz Technische Universität Chemnitz [Technical University of Chemnitz, Germany]

TU Chemnitz-Zwickau Technische Universität Chemnitz-Zwickau [Technical University of Chemnitz-Zwickau, Germany]

TU Clausthal Technische Universität Clausthal [Technical University of Clausthal, Clausthal-Zellerfeld, Germany]

TUCM Trade Union Congress of Malawi

TUCOWS The Ultimate Collection of Winsock Software [Internet]

TUCP Trade Union Congress of the Philippines

TUCOP Transient Undercooled Overpower Accident [Nuclear Reactors]

TUCSA Trade Union Council of South Africa

TUD Technical University of Dresden [Germany]

TUD Technical University of Denmark [at Copenhagen and Lyngby]

TUD Technische Universiteit Delft [Delft University of Technology, Netherlands]

TUD Technology Utilization Division [of NASA]

TUD Tunesian Dinar [Currency]

TU Dresden Technische Universität Dresden [Technical University of Dresden, Germany]

TUE Technische Universiteit Eindhoven [Eindhoven University of Technology, Netherlands]

TUE Thiourea Extraction

TUE Transuranic (or Transuranium) Element(s)

Tues Tuesday [also Tu, or Tue]

TUF Thermal Utilization Factor [Nuclear Engineering]

TUF Transmitter Underflow

TUFCDF Thorium-Uranium Fuel Cycle Development Facility [of Brookhaven National Laboratory, Upton, New York, US]

TUFI Toughened Uni-Piece Fibrous Insulation

TUG Tape Unit Group

TUG Towed Universal Glider

TUG TRANSAC (Transistorized Automatic Computer) Users Group [US]

Tug Tugrik [Currency of the Mongolian Republic]

TU Graz Technische Universität Graz [Technical University of Graz, Austria]

TUH Temple University Hospital [Philadelphia, Pennsylvania, US]

TU Hamburg-Harburg Technische Universität Hamburg-Harburg [Technical University of Hamburg-Harburg, Germany] [also TUHH]

TUHI Technical University for Heavy Industry [Miskolc, Hungary]

TUHSC Temple University Health Sciences Center [Philadelphia, Pennsylvania, US]

TUI Technical University of Istanbul [Turkey]

TUI (WordPerfect) Text-Based User Interface

TUI Transurethral Incision (of the Prostate) [Medicine]

TUIAFPW Trade Unions International of Agricultural, Forestry and Plantation Workers [Italy]

TUIMWE Trade Unions International of Miners and Workers in Energy

TUIP Transurethral Incision of the Prostate [Medicine]

TUK Technical University Karlsruhe [Germany]

TUK Technical University of Krakow [Poland]

TU Karlsruhe Technische Universität Karlsruhe [Technical University of Karlsruhe, Germany]

TUL Technical University of Lulea [Sweden]

TUL Tula Peak, New Mexico [Space Tracking and Data Network Site]

TUL Tulane University of Louisiana [New Orleans, US]

TUL Turkish Lira [Currency]

TUM Technical University of Munich [Germany]

TUM Technological University of Malaysia [Kuala Lumpur]

TUM Tuning Unit Member

TU Magdeburg Technische Universität Magdeburg [Technical University of Magdeburg, Germany]

TU Merseburg Technische Universität Merseburg [Technical University of Merseburg, Germany]

TUMC Temple University Medical Center [Philadelphia, Pennsylvania, US]

TUMM Tokyo University of Mercantile Marine [Japan]

TUMS Table Update and Management System [of Stanford University, California, US]

TUMS Temporary Usage Measured Service

TU München Technische Universität München [Technical University of Munich, Germany]

TuMV Turnip Mosaic Virus

Tun Tunesia(n)

Tun Tuning [also tun]

TUNA Transurethral Needle Ablation (of the Prostate) [Medicine]

TUNGAR Tungsten-Argon (Rectifier) [also Tungar, or tungar]

TUNIS Toronto University (Computer) System [Canada]

TUNNETT Tunneling Transit Time (Diode)

TUNS Technical University of Nova Scotia [Halifax, Canada]

TUP Technical University of Poznan [Poland]

TUP Technology Utilization Program

TUP Telephony User Part

TUPS Technical User Performance Specifications [of United States Telephone Association]

TUR Test Uncertainty Ratio

TUR Traffic Usage Recorder

TUR Transurethral Resection (of the Prostate) [Medicine]

TUr Thiourea

Tur Turbine [also tur]

Tur Turret [also tur]

Turb Turbine [also turb]

Turb Turbulence [also Turbc, turbc, or turb]

Turbo Gen Turbo-Generator [also Turbo-Gen]

turbt turbulent

TURDOK Turkish Scientific and Technical Documentation Center

TURI Toxics Use Reduction Institute [US]

Turk Turk(ey); Turkish

Turkist Turkistan(i)

Turkm Turkmen(istan)

TurkmSSR Turkmen Soviet Socialist Republic [USSR]

TURP Transurethral Resection of the Prostate [Medicine]

Turp Turpentine [also turp]

TURPS Terrestrial Unattended Reactor Power System

TUS Technische Universität Stuttgart [Technical University of Stuttgart, Germany]

TUS Tucson, Arizona [Meteorological Station Designator]

TUSM Tufts University School of Medicine [Boston, Massachusetts, US]

TUSM Tulane University School of Medicine [New Orleans, Louisiana, US]

tuss *(tussis)* – cough

TUST Teikyo University of Science and Technology [Yamanashi, Japan]

TU Stuttgart Technische Universität Stuttgart [Technical University of Stuttgart, Germany]

TUT Tampere University of Technology [Finland]

TUT Toyohashi University of Technology [Aichi, Japan]

TUT Transistor Under Test

TUT Twente University of Technology [Netherlands]

.TUT Tutorial [File Name Extension] [also .tut]

TUV Thermal Ultraviolet

TÜV Technischer Überwachungsverein [Technical Inspection Association, Germany]

TUVR Total Ultraviolet Radiation

TU Wien Technische Universität Wien [Technical University of Vienna, Austria]

TV Tape Velocity

TV Television

TV Temperature-Volume (Diagram) [also T-V]

TV Tenth Value

TV Terminal Velocity

TV Test Vector

TV Test Vehicle

TV Test Voltage

TV Thermal Vacuum

TV Thermal Vibration

TV Thermal Voltage

TV Thermionic Valve

TV Thoracic Vertebrae [Anatomy]

TV Threshold Value

TV Throttle Valve

TV Thrust Vector [Aerospace]

TV Torsional Vibration [Mechanics]

TV Total Vehicle

TV Track Velocity

TV Transfer Vector

TV Transport Vehicle

TV Tricuspid Valve [Anatomy]

TV Tube Valve

TV Tube Voltmeter

TV Tuvalu [ISO Code]

$2V_m$ Crack-Mouth Opening Displacement [Symbol]

T-V Temperature-Volume (Diagram) [also TV]

Tv Traverse [also tv]

TVA Taiwan Visitor's

TVA Tax on Value Added

TVA Tennessee Valley Authority [US]

TVA Technical Valuation Society [US]

TVA Terminal Volume Addendum

TVA Thermovaporimetric Analysis

TVA Thrust Vector Alignment [Aerospace]

TVAEL Tennessee Valley Authority Engineering Laboratory [US]

TVAR Test Variance [also TVar]

T-VASI Tee Visual Approach Slope Indicator

TVB Tennessee Valley Basin [US]

TvB Television Bureau of Advertising [US]

TVC Tag Vector Control

TVC Television Camera

TVC Thermal-to-Voltage Converter

TVC Thermal Vacuum Chamber

TVC The Vice-Chancellor

TVC Thrust Vector Control [Aerospace]

TVCA Thrust Vector Control Actuator [Aerospace]

TVCD Thrust Vector Control Driver [Aerospace]

TV CONF Television Conference

TVCS Thrust Vector Control System [Aerospace]

TVD Tank-Vapor Database

TVDA Terminal Volume Discount Agreement

TVDC Test Volts, Direct Current

TVDP Terminal Vector Display Unit

TVDR Tag Vector Display Register

TVDS Towed Vertically Directive Source

TVE Television Enginee(ring)

TVE Total Vertical Error

TVE Town and Village Enterprise

TVEC Tangent Vector Error Criterion

TVEL Tape Velocity

TVEL Track Velocity

TVEXPIS Television Experiment Interconnecting Station

TVF Tape Velocity Fluctuation

TVF Television Film

TVF TOC (Table of Contents) Verbosely from File [Unix Operating System]

TVFS Toronto Virtual File System

TVG Time-Varied Gain

TVG Triggered Vacuum Gap

TVI Television Interference

TVI Tomasetti Volatile Indicator

TVIST Television Information Storage Tube

TVL Television Line

TVL Tenth-Value Layer

TVM Tachometer Voltmeter

TVM Television Monitor

TVM Track Via Missile

TVM Transistor Voltmeter

TVN Test Verification Network

TVN Total Volatile Nitrogen

TVO Thermoplastic Vulcanite(s)

TVO Tractor Vaporizing Oil

TVOC Television Operating (or Operations) Center

TVOL Television On-Line

TVOR Terminal VHF (Very High Frequency) Omnirange

TVP Temperature–Volume–Pressure [also T-V-P]

TVP Teoriya Veroyatnostei i ee Primeneniya [Russian Journal on the Theory of Probability and Its Applications]

TVP Test Verification Program

TVP Textured Vegetable Protein [Biochemistry]

TVP True Vapor Phase (Process)

TVP True Vapor Pressure

TVPCS Triton Vector Processing Computer System

TVPPA Tennessee Valley Public Power Association [US]

TVPRS Transient-Viscous Phase Reaction Sintering [Metallurgy]

TVR Tag Vector Response

TVR Television Recorder; Television Recording

TVR Tennessee Valley Region [US]

TVRO Television Receive-Only

TVS Thermostatic Vacuum Switch

TVS Toxic Vapor Suit

TVS Transient Voltage Suppressor

TVS Triangle Voltage Sweep

TVSA Thrust Vector Control Servoamplifier [Aerospace]

TV-SAT Television Satellite [France/Germany]

TVSi Tetravinylsilane

TVSS Transient Voltage Surge Suppressor

TVSSIS Television Subsystem Interconnecting Station

TV SYS Television System [also TV Sys]

TVT Television Terminal

TVT Television Typewriter

TVT Tenth-Value Thickness

TVT Thermal Vacuum Test

TVT Tetravinyltin

TVTA Thermal Vacuum Test Article

TVW Tag Vector Word

TW Tail Warning (Radar)

TW Tail Water

TW Tail Wave [Aeronautics]

TW Tail Wind

TW Taiwan [ISO Code]

TW Tape Word

TW Terawatt [Unit]

TW Text Word

TW Thermal Wave [Physics]

TW Thermal Wire

TW Thermit Welding

TW Thermoplastic Type (Conductor Insulation) [Symbol]

TW Thumb Wheel

TW Tidal Wave

TW Time Words

TW Topical Workshop

TW Torsional Wave

TW Toxic Waste

TW Track Width [also tw] [Magnetic Recording Media]

TW Trade Wind

TW Transit Working

TW Translational Wave

TW Transverse Wave

TW Travelling Wave

TW Truncated Wave

TW Twaddell (Scale of Specific Gravity) [also Tw]

TW Two-Way

TW Typewriter

T/W Thrust Weight [Aerospace]

T/W Thrust-to-Weight (Ratio) [Aerospace]

Tw Twaddell (Scale of Specific Gravity) [also TW]

Tw Twist(ed) [also tw]

°Tw Degree(s) Twaddell [Symbol]

.tw Taiwan [Country Code/Domain Name]

3W World Wide Web [also WWW, or W3]

TWA Task Work Area

TWA The Waferboard Association [Canada]

TWA Time-Weighted Average

TWA Transaction Work Area

TWA Trans-World Airlines [US]

TWA Travelling-Wave Accelerator

TWA Travelling-Wave Amplifier

TWA Two-Way Alternate

TWAEC Time-Weighted Average Exposure Criteria

TWAEV Time-Weighted Average Exposure Value

TWAIT Terminal Wait

TWARO Textile Workers Asian Regional Organization

TWAS Third World Academy of Sciences [Italy]

T Wave Ventricular Relaxation (in Electrocardiogram) [also T wave] [Medicine]

TWB Tailor-Welded Blanket

TWB Typewriter Buffer

TWC Technician Working Circuit

TWC Texas Western College [El Paso, US]

TWC The Williams Companies [US]

TWC Three-Way (Automotive) Catalyst

TWC Three-Way Converter

TWC Total Water Content

TWC Trinity Western College [Canada]

TWC (American) Truncated Whitworth Coarse Thread

TW-C Truncated Wave with Compressive Dwell

TWCA Texas Water Conservation Association [US]

TWCA Tank Waste Characterization Plan [Nuclear Engineering]

TWCP Topical Workshop on Collider Physics

TWCRT Travelling-Wave Cathode-Ray Tube

TWD (New) Taiwanese Dollar [Currency]

TWD Technical Work Document

TWD Toxic Waste Disposal

TWD Toxic Waste Dump

TWD Twisted Double Shielded (Cable)

2WD Two Wheel Drive [also TWD]

twd towards

TWDD Two-Way/Delay Dial

TWEAT Ternary Waste Envelope Assessment Tool [Nuclear Engineering]

TWEB Transcribed Weather Broadcast [US]

TWEC Tidal Wave Energy Conversion (System)

TWERLE Tropical Wind Energy Conservation and Reference Level Experiment

TWF Third World Forum [Egypt]

TWF Third World Foundation [UK]

TWF (American) Truncated Whitworth Fine Thread

TWG Technical Working Group

TWG Telemetry Working Group

TWHA (Western) Tasmanian (Wilderness National Parks) World Heritage Area [Australia]

TWI Thermal Wave Imaging

TWI The Welding Institute [now World Center for Information and Technology in Welding, Cambridge, UK]

TWI Trade Weighted Index

TWI Bull The Welding Institute Bulletin [UK]

TWID Two-Way/Immediate Dial

TWIG Tracked Wing In Ground-Effect

TWINAX Twinaxial (Cable) [also twinax]

TWINS Tank-Waste Information Network System [Nuclear Engineering]

Twist Twisting [also twist]

TWK Typewriter Keyboard

TWL Total Weight Loss

TWM Toxic Waste Management

TWM Traveling Wedge Micrometer

TWM Traveling-Wave Maser

TWMBK Traveling-Wave Multiple-Beam Klystron

TWME Two-Way Memory Effect [Metallurgy]

TWMR Tungsten Water-Moderated Reactor

TWN Third World Network

TWOM Traveling-Wave Optical Maser

T-Word Tool Function Word [Numerical Control]

TWP Technical Wordprocessing (System)

TWP Tropical Western Pacific

TWP Twisted Wire Pair [Electric Cables]

Twp Township [also twp]

TWPS Technical Wordprocessing System

Twps Townships [also twps]

TWR Tail-Warning Radar (Set)

TWR Tape Write

TWR Tape Write Register

TWR Teletypewriter Exchange Service

TWR (Control) Tower [Aeronautics]

TWR Trans-World Radio

Twr Tower [also twr]

Twrg Towering [also twrg]

TWRI Texas Water Resources Institute [US]

TWRS Tank Waste Remediation System [Nuclear Engineering]

TWS Tail-Warning Radar Set

TWS Tasmanian Wilderness Society [Australia]

TWS Thermal Weapon Sight [of US Army]

TWS The Wilderness Society [Australia]

TWS The Wildlife Society [US]

TWS Track-While-Scan [Electronics]

TWS (American) Truncated Whitworth Special Thread

TWS Two-Way Simultaneous

TWSB Twin Sideband

TWSM Two-Way Shape Memory [Metallurgy]

TWSME Two-Way Shape Memory Effect [Metallurgy]

TWSO Transuranic Waste Systems Office [of US Department of Energy at Rocky Flats]

TWST Torus Water Storage Tank [Nuclear Engineering]

3W SW Three-Way Switch [also TW SW]

TWT Thin-Wire Thermometer

TWT Traffic Work Table

TWT Traveling-Wave Tube

TWT Trisonic Wind Tunnel [of Rockwell International Corporation, US]

TW-T Truncated Wave with Tensile Dwell

TWTA Traveling-Wave Tube Amplifier

TWTC Taipei World Trade Center [Taiwan]

TW-TC Truncated Wave with Tensile and Compressive Dwell

TWTF Tank Waste Task Force [Nuclear Engineering]

TWU Telecommunication Workers Union

TWU Transport Workers Union [US]

TWU Trinity Western University [Langley, British Columbia, Canada]

TWUA Textile Workers Union of America [US]

TWWS Two-Way/Wink Start

TWX Teletype Message

TWX Teletype Wire Transmission

TWX Teletypewriter Exchange (Service) [Canada and US]

TWX Time Wire Transmission

TWY Taxiway [also Twy, or twy]

TWYL Taxiway Link

TX Task Extension

TX Telephone Exchange

TX Teleprinter Exchange (Service) [also TELEX, Telex, telex, TEX, or TLX]

TX Television Receiver

TX Temperature–Concentration (Diagram) [also T-X]

TX Texas [US]

TX Text [also Tx, or tx]

TX Thromboxane [Biochemistry]

TX Time to Equipment Reset

TX Torque Synchro Transmitter

TX Torque Transmitter

TX Translation Hand Controller X-Axis Direction [Aerospace]

TX Transmit (Mode)

TX Transmitter [also Tx]

TX Tricrystal [also TC]

3X- Guinea [Civil Aircraft Marking]

T-X Temperature–Concentration (Diagram) [also TX]

T-X Travel-Time (Graph) [Geophysics]

Tx Toxin

$T_x(x)$ Crystallization Temperature as a Function of Concentration [Symbol]

TXA Task Extension Area

TXB2 Thromboxane B2 [Biochemistry]

TXC Telephone Exchange–Crossbar

TxC Transmitter Clock [Data Processing]

TXD Telephone Exchange–Digital

TXD Transmit(ted) Data

TxD Transmit Data to Equipment [Data Processing]

TXDS Qualifying Toxic Dose

TXE Telephone Exchange–Electronic

.TXF Tax Exchange Format [File Name Extension]

TXIB Texanol Isobutyrate

TXN Transaction

TXRF Total-Reflection X-Ray Fluorescence

TX/RX Transmit/Receive; Transmitter/Receiver [also TX-RX]

TXS Telephone Exchange–Strowger

TXT Text [also Txt, or txt]

TXT Text File [File Name Extension] [also .txt]

TXT NUM Text Box Number [also Txt Num]

TXT OPT Text Box Options [also Txt Opt]

TXT2STF Text To Structured File

TY Translation Hand Controller Y-Axis Direction [Aerospace]

TY Trinkhaus and Yoo (Nucleation Rate Computation) [Metallurgy]

TY Tropical Year [Astronomy]

TY- Dahomey [Civil Aircraft Marking]

3Y Three Times Yield Strength [Mechanics]

Ty Territory [also ty]

t/y tons per year [also t/yr]

TYAA Textured Yarn Association of America [US]

TYDAC Typical Digital Automatic Computer

TYMV Turnip Yellow Mosaic Virus

TYP Taiwan Yellow Pages

Typ Typewriter [also typ]

Typ Type

Typ Typographical; Typographer; Typography [also typ]

typ typical

Typh Typhoon [also typh]

Typo Typographical; Typographer; Typographer [also Typogr, typogr, or typo]

Typo Typographical Error [also typo]

TYPOUT Typewriter Output Routine

Typsg Typesetting [also typsg]

Typw Typewriter [also typw]

Tyr (−)-Tyrosine; Thyrosyl [Biochemistry]

t/yr tons per year [also t/y]

Tyramine 4-Hydroxyphenethyl Amine

Tyres Access Tyres and Accessories [UK Publication]

TYS Tensile Yield Strength [Mechanics]

TZ Tanzania [ISO Code]

TZ *(Technische Zeitschrift)* – German for "Technical Journal"

TZ Test Zone

TZ Time Zone

TZ Trade Zone

TZ Translation Hand Controller Z-Axis Direction [Aerospace]

TZ Transmitter Zone

TZ Trapezoidal Model [Metallurgy]

TZ Triple-Zeta Basis

TZ- Mali [Civil Aircraft Marking]

Tz Tetrazine [also tz]

Tz Triazole [also tz]

tz trapezoidal

TZD True Zenith Distance [Astronomy]

TZ2P Triple-Zeta plus Double Polarization [Physics]

TZDB Time-Zero-Dielectric Breakdown

TZE Transverse Zeeman Effect [Physics]

TZM Molybdenum-Base Titanium-Zirconium (Alloy)

TZ Met bearb Technische Zeitschrift für Metallbearbeitung [German Journal on Metalworking]

TZM-Mo TZM Molybdenum (Alloy System)

TZP Tetragonal Zirconia Polycrystal [also t-ZrO$_2$]

TZP Tetragonal Zirconia Precipitate

TZP Tribasic Zinc Phosphate

t-ZrO$_2$ Tetragonal Zirconia (Polycrystal) [also TZP]

TZS Tanzanian Shilling [Currency of Tanzania]

TZT Tetrazoline-5-Thione

U

U   Accelerating Voltage (or Potential) [Symbol]
U   Activation Energy [Symbol]
U   Average Velocity of Fluid Flow [Symbol]
U   Bulk (Substrate) [Semiconductor Symbol]
U   Charging Energy [Symbol]
U   Coulomb Interaction [Symbol]
U   Coulomb Repulsion [Symbol]
U   Degree of Consolidation (or Percent Consolidation of Soil) [Symbol]
U   Dislocation Self-Energy (in Materials Science) [Symbol]
U   Group Velocity (of Waves) [Symbol]
U   Internal Energy [Symbol]
U   Non-Ohmic Resistance [Symbol]
U   North-South Wind Component [Symbol]
U   Number of Revolutions (of Springs) [Symbol]
U   Ouguiya [Currency of Mauritania]
U   (Vortex) Pinning Energy [Symbol]
U   Pitch Line Velocity (of Thrust Bearings) [Symbol]
U   Potential Difference [Symbol]
U   Radiotelephony [4,000 to 23,000 kc/s Range]
U   Radius of Worm Gear Throat [Symbol]
U   Scattering Potential [Symbol]
U   (Total) Strain Energy [Symbol]
U   Substitutional Solute [Symbol]
U   Total Uplift (on a Structure) [Symbol]
U   Uganda(n)
U   Ulcer(ous)
U   Ulmus [Genus of Deciduous Trees Including the Elms]
U   Ulna [Anatomy]
U   Ulothrix [Genus of Algae]
U   Ultrasonic(s)
U   Unary
U   Unassemble (Command) [Pascal Programming]
U   Unclassified [also u]
U   Underfloor

U   Underground
U   Underwater Attack (Vehicle) [USDOD Symbol]
U   Underwater Launched (Vehicle) [USDOD Symbol]
U   Ungulate [Zoology]
U   Uniform(ity)
U   Uniform [Phonetic Alphabet]
U   Union
U   Unipolar(ity)
U   Unit(ary) [also u]
U   Unit Energy [Symbol]
U   Universal(ity)
U   Universe
U   University
U   Unknown
U   Unlicensed
U   Unpredictable
U   Unsatisfactory [also u]
U   Unsharpness
U   Unstability; Unstable; Unstabilize(d)
U   Update
U   Upper [also u]
U   Uppsala [Sweden]
U   Upright
U   Upset(ter)
U   Uranium [Symbol]
U   Uranyl
U   Uracil [Biochemistry]
U   Urea
U   Uridine [Biochemistry]
U   Urether [Anatomy]
U   Urethra [Anatomy]
U   Urinate; Urination
U   Urine
U   Uruguay(an)
U   Usability; Usable
U   Use(r)
U   Useful(ness)
U   Utah [US]
U   Uterine; Uterus [Anatomy]
U   Utilize(d); Utilization
U   Utility
U   Utility (Aircraft) [US Army and US Navy Symbol]
U   Utrecht [Netherlands]
U   (Linear) Velocity [Symbol]
U   Work Performed on a Body [Symbol]
/U   Unit Directories Option [Turbo Pascal]
U%   Mean Deviation Unevenness (in Textiles) [Symbol]
-U   Utility (Aircraft) [US Navy Suffix]
U*   Friction Velocity (in Meteorology) [Symbol]
U_0 Cooper-Pair Condensate Energy [Symbol]
U_0 Pinning Energy (in Solid-State Physics) [Symbol]
U_0 Polar-State Formation Energy [Symbol]
U_0 Strain Energy per Unit Volume [Symbol]

U_0 Thermal Activation Energy [Symbol]

U_1 Uranium 1 [also U, U_I, U-238, ^{238}U, or U^{238}]

U_2 Uranium 2 [also U_{II}, U-234, ^{234}U, or U^{234}]

U^{4+} Tetravalent Uranium Ion [Symbol]

U_e Electric Energy Density [Symbol]

U_e Electron Interaction Energy [Symbol]

U_{eff} Effective Energy [Symbol]

U_g Geometric Unsharpness (in Radiography) [Symbol]

U_i Relative Humidity with Respect to Ice [Symbol]

U_L Ultimate Load [Symbol]

U_λ Energy Density of Radiation in Wavelength Range $\lambda \to \lambda + d\lambda$ (for Rayleigh Scattering) [Symbol]

U_m Magnetic Energy Density [Symbol]

U_m Molecular Interaction Energy [Symbol]

U_r Modulus of Resilience [Symbol]

U^s Surface Energy [Symbol]

U_s Consolidation Ratio (for Soil) [Symbol]

U_w Relative Humidity with Respect to Water [Symbol]

U-100   100 Units of Insulin per Milliliter, or Cubic Centimeter of Solution [Unit]

U-227   Uranium-227 [also ^{227}U, or U^{227}]

U-228   Uranium-228 [also ^{228}U, or U^{228}]

U-229   Uranium-229 [also ^{229}U, or U^{229}]

U-230   Uranium-230 [also ^{230}U, or U^{230}]

U-231   Uranium-231 [also ^{231}U, or U^{231}]

U-232   Uranium-232 [also ^{232}U, or U^{232}]

U-233   Uranium-233 [also ^{233}U, or U^{233}]

U-234   Uranium-234 [also ^{234}U, U^{234}, U_2, or U_{II}]

U-235   Uranium-235 [also ^{235}U, U^{235}, or AcU]

U-236   Uranium-236 [also ^{236}U, or U^{236}]

U-237   Uranium-237 [also ^{237}U, or U^{237}]

U-238   Uranium-238 [also ^{238}U, U^{238}, U, U_1, or U_I]

U-239   Uranium-239 [also ^{239}U, or U^{239}]

U-240   Uranium-240 [also ^{240}U, or U^{240}]

u   (unified) atomic mass unit (1.6605×10^{-27} kg) [Symbol]

u   bulk (substrate) [Semiconductor Symbol]

u   cloud total optical depth [Symbol]

u   crack opening displacement [Symbol]

u   displacement [Symbol]

u   excess hydrostatic pressure [Symbol]

u   group velocity (of waves) [Symbol]

u   internal energy [Symbol]

u   (crystal) lattice energy [Symbol]

u   (dislocation) line unit vector (in materials science) [Symbol]

u   mobility (of electric charges) [Symbol]

u   neutral stress (or pore water pressure) (of soil) [Symbol]

u   normalized variable (or normalized observed value) (in statistics) [Symbol]

u   number of defects per unit (in statistics) [Symbol]

u   particle velocity [Symbol]

u   real object distance (in optics) [Symbol]

u   specific internal energy [Symbol]

u   thermal velocity (of electrons) [Symbol]

u   unclassified [U]

u   unidirectional

u   unit(ary) [also U]

u   unit uplift (on a structure) [Symbol]

u   unit vector [Symbol]

u   unsatisfactory [also U]

u   update

u   upper [also U]

u   upsilon [English Equivlalent]

u   up(-type) quark (in particle physics) [Symbol]

u   velocity [Symbol]

u   velocity component (in Cartesian system) [Symbol]

u   wind speed [Symbol]

$\bar{u}$ average number of defects per unit (in statistics) [Symbol]

$\bar{u}$ excess hydrostatic pressure [Symbol]

$\bar{u}$ temporal mean of velocity component "u" [Symbol]

$\dot{u}$ dislocation line direction (in materials science) [Symbol]

u' ion mobility [Symbol]

u_+ mobility of positive ions [Symbol]

u_- mobility of negative ions [Symbol]

u' total load per unit area of concrete slab or per unit length of beam [Symbol]

u_0 hydrostatic pressure [Symbol]

$\bar{u}^2$ mean (or average) square velocity (in kinetic molecular theory) [Symbol]

u_o velocity of observer (e.g., Doppler equation) [Symbol]

u_p cube crushing strength of concrete (preliminary test) [Symbol]

u_r displacement in radial direction [Symbol]

u_s velocity of source (of sound, or light) (e.g., Doppler equation) [Symbol]

u_θ displacement with respect to azimuthal angle [Symbol]

u_w cube crushing strength of concrete (works test) [Symbol]

u_w neutral stress (or pore water pressure) (of soil) [Symbol]

u_x displacement (or velocity component) in x-direction [Symbol]

u_y displacement (or velocity component) in y-direction [Symbol]

u_z displacement (or velocity component) in z-direction [Symbol]

v discharge velocity (of water through soil or rock) [Symbol]

v (mean) flow velocity (of a fluid) [Symbol]

v phase velocity (of waves) [Symbol]

v specific volume [Symbol]

v (lower-case) upsilon [Greek Alphabet]

v (linear) velocity [Symbol]

v velocity component (in Cartesian system) [Symbol]

v volume [Symbol]

v volume fraction [Symbol]

$\bar{v}$ drift velocity (of electrons) [Symbol]

$\bar{v}$ mean particle velocity [Symbol]

$\bar{v}$ temporal mean of velocity component "v" [Symbol]

v_+ velocity of cations [Symbol]

v_- velocity of anions [Symbol]

v_0 initial velocity [Symbol]

v_0 volume occupied by a gas at the ice point (0°C) [Symbol]

v_1 initial volume [Symbol]

v_1 specific volume of phase 1 [Symbol]

v_1 velocity (or speed) of light in medium 1 [Symbol]

v_2 final volume [Symbol]

$\bar{v^2}$ mean (or average) square velocity (in kinetic molecular theory) [Symbol]

v_4 wind movement four meters above the surface [Symbol]

v_{100} volume occupied by a gas at the steam point (100°C) [Symbol]

v_α volume of the alpha phase (of an alloy) [Symbol]

v_{av} average speed (or velocity) [Symbol]

v_β volume of the beta phase (of an alloy) [Symbol]

v_c average velocity of dislocation climb (in materials science) [Symbol]

v_c critical velocity [Symbol]

v_c critical (specific) volume [Symbol]

v_d drift velocity (of electrons) [also v_D] [Symbol]

v_e velocity of extraordinary ray [Symbol]

v_F Fermi velocity (of electrons) [Symbol]

v_f final velocity [also v_f] [Symbol]

v_g group velocity (of waves) [Symbol]

v_g molar volume [Symbol]

v_i initial velocity [Symbol]

v_i mobility of i-th ion [Symbol]

v_k kink velocity [Symbol]

v_L flux line velocity [Symbol]

v_l molar volume of liquid at (absolute) temperature T [Symbol]

v_o velocity of observer (e.g., Doppler equation) [Symbol]

v_o velocity of ordinary ray [Symbol]

v_p particle velocity [Symbol]

v_R Rayleigh wave velocity [Symbol]

v_r reduced volume [Symbol]

v_r resonance velocity [Symbol]

v_s molar volume of solid at melting temperature T [Symbol]

v_s spin-wave velocity (in solid-state physics) [Symbol]

v_s velocity of source (e.g., Doppler equation) [Symbol]

v_x velocity component in x-direction [Symbol]

v_y velocity component in y-direction [Symbol]

v_z velocity component in z-direction [Symbol]

UA Ukraine [ISO Code]

UA Ultimate Analysis [Chemistry]

UA Ultra-Accelerator

UA Ultra-Audible

UA Underaged; Underaging [Metallurgy]

UA Underwater Association [UK]

UA Unified Agenda (of Federal Regulatory and Deregulatory Actions) [Federal Register, US]

UA United Airlines [US]

UA United Atom (Model) [Physics]

UA Unit of Account

UA Universidad de las Americas [University of the Americas, Santa Catarina Martir, Mexico]

UA Universidad de los Andes [University of the Andes, Bogota, Colombia]

UA Universidade de Aveiro [University of Aveiro, Portugal]

UA University of Aberdeen [Scotland]

UA University of Adelaide [South Australia]

UA University of Akron [Ohio, US]

UA University of Alabama [US]

UA University of Alaska [US]

UA University of Alberta [Edmonton, Canada]

UA University of Algiers [Algeria]

UA University of Arkansas [US]

UA University of Arizona [Tucson, US]

UA University of Aston [UK]

UA University of Athens [Greece]

UA University of Auckland [New Zealand]

UA Unnumbered Acknowledge [Data Communications]

UA Until Advised [also ua]

UA Up Aiming

UA Upper Atmosphere

UA User Address [Computers]

UA User Agent

UA User Area

UAA United Arab Airlines [Egypt]

UAA University Aviation Association [US]

UAA University of Alaska at Anchorage [US]

UAA Upper Advisory Area [Aeronautics]

UAA Utility Arborist Association [US]

UAAEE United Arab Atomic Energy Establishment

UAAS Ukranian Academy of Arts and Sciences

UAB Unemployment Assistance Board

UAB Unidirectional Address Bus

UAB Universidad Autónoma de Barcelona [Autonomous University of Barcelona, Spain]

UAB University of Alabama at Birmingham [US]

UAB University of Aston in Birmingham [UK]

UAB Until Advised By [also uab]

UABC Universidad de la Baie de California [University of the Gulf of California, Tijuana, Mexico]

UABS Union of American Biological Societies [US]

UAC Ulster Automobile Club [Northern Ireland]

UAC Unified Agricultural Cooperative

UAC Uninterrupted Automatic Control

UAC Universidad Autónoma de Chihuahua [Autonomous University of Chihuahua, Mexico]

UAC Universidad Autónoma de Coahuila [Autonomous University of Coahuila, Saltillo, Mexico]

UAC Upper Airspace Center

UAC Upper Airspace Control

UAC Urban Affairs Council [US]

UACA United American Contractors Association [US]

UACC Upper Area Control Center [Aeronautics]

UACM University of Arizona College of Medicine [Tucson, US]

UACN Unified Automated Communications Network

UACN University of Alaska Computer Network [US]

UAD Upper Advisory Route [Aeronautics]

UADPS Uniform Automatic Data Processing System

UADS User Attribute Data Set

UADW Universal Alliance of Diamond Workers [Belgium]

UAE United Arab Emirates

UAE Unrecoverable Application Error [Computers]

UAE Unsteady Adiabatic Expansion

UAE Dh UAE Dirham [Currency of the United Arab Emirates]

UAEM Union of Associations of European Meatmeal Producers [Netherlands]

UAEU United Arab Emirates University [Al-Ain]

UAF Union des Artisans de France [French Union of Craftsmen]

UAF University of Alaska at Fairbanks [US]

UAF University of Arkansas at Fayetteville [US]

UAFMMEEC Union of Associations of Fish Meal Manufacturers in the European Economic Community [Germany]

UAFSME University of Alaska at Fairbanks School of Mineral Engineering [US]

UAG Université des Antilles et de la Guyane [University of the Antilles and Guyana, Guadeloupe, French West Indies]

UAG User Advisory Group

UAG USSR Academy of Science Proceedings: Geographical Series

UAGI University of Alaska Geophysical Institute [US]

UAH University of Alabama at Huntsville [US]

UAHS Ulster Archeological Heritage Society [Northern Ireland]

UAIDE Users of Automatic Information Display Equipment

UAIS Universal Aircraft Information System

UAJ University of Alaska at Juneau [US]

UAL United Airlines [US]

UAL University of Arkansas Library [Fayetteville, US]

UAL User Agent Layer

U/Al Uranium-Aluminum (Alloy)

UALR University of Arkansas at Little Rock [US]

UAM Ultrasonically Assisted Machining

UAM Underwater-to-Air Missile

UAM Union Africaine et Malgache [African and Malagasy Union] [now UAMCE]

UAM Universidad Autónoma de Madrid [Autonomous University of Madrid, Spain]

UAM Universidad Autónoma Metropolitana [Autonomous Metropolitan University, Mexico]

UAM Université Aix-Marseille [University of Aix-Marseille, France]

UAM University of Arkansas at Monticello [US]

UAM Urban Airshed Model

UAM User Authentication Method

UAMCE Union Afro-Malagasie pour la Coopération Economique [Afro-Malagasy Union for Economic Cooperation, Ivory Coast]

UAMI Universidad Autónoma Metropolitana–Ixtapalapa [Autonomous Metropolitan University of Ixtapalapa, Federal District of Mexico]

UAMPT Union Africaine et Malgache de Poste et de Télécommunications [African and Malagasy Postal and Telecommunications Union]

UAMR United Association Manufacturers Representatives [US]

UAMUA Universidad Autónonma Metropolitana Unidad Azcapotzalco [Free University of Metropolitan Azcapotzalco, Mexico City]

UAM-V Urban Airshed Model, Version-V

UAN Unified Automatic Network

UANA Union of African News Agencies

UANL Universidad Autónoma de Nuevo León [Autonomous University of Nuevo Leon, Monterrey, Mexico]

UAO Unexplained Aerial Object

UAOS Ulster Agricultural Organization Society [Northern Ireland]

UAP Unidentified Atmospheric Phenomena

UAP Universal Availability of Publications

UAP Universidad Autónoma de Puebla [Autonomous University of Puebla, Mexico]

UAP University of Alberta Press [Edmonton, Canada]

UAP Upper Air Project

UAP User Area Profile

UAP User Application Processing [also uap]

UAPB University of Arkansas at Pine Bluff [US]

UAPT United Association for the Protection of Trade [UK]

UAQ Universidad Autónoma de Queretaro [Autonomous University of Queretaro, Mexico]

UAR Unit Address Register

UAR United Arab Republic

UAR Upper Air Route

UAR Upper Atmosphere Research

UARAEE United Arab Republic Atomic Energy Establishment

UARC Upper Atmosphere Research Corporation

UARI Union of Applied Research Institutes [Burma]

UARC University Archives [Division of Bancroft Library, University of California at Berkeley, US]

UARI University of Alabama Research Institute [US]

UARS Upper Atmosphere Research Satellite [of NASA]

UART Universal Asynchronous Receiver/Transmitter [also UAR/T]

UARTO United Arab Republic Telecommunication Organization

UAS Ukrainian Academy of Sciences

UAS Ulster Archeological Society [Northern Ireland]

UAS Union of African States

UAS Unit Approval System [of Federal Aviation Administration, US]

UAS Universidad Autónoma de Sudeste [Autonomous University of Sudeste, Campeche, Mexico]

UAS USSR Academy of Sciences [Moskow]

UAS Uzbek Academy of Sciences [Tashkent]

UAs Uranium Arsenide

UASAL Utah Academy of Sciences, Arts and Letters [Salt Lake City, US]

UASB Upflow Anaerobic Sludge Blanket

UASC United Agencies Shipping Company, Limited

UASIF Union des Associations Scientifiques et Industrielles Françaises [Union of French Scientific and Industrial Associations]

UASLP Universidad Autónoma de San Luis Potosí [Autonomous University of San Luis Potosi, Mexico]

UASRC University of Arizona Space Research Center [Tucson, US]

UAT University of Alabama at Tuscaloosa [US]

UAT University of Arizona at Tucson [US]

UAUM Underwater-to-Air-to-Underwater Missile

UAV Unmanned Aerospace Vehicle

UAV Unmanned Air (or Aerial) Vehicle

UAV Upper Atmosphere Vehicle

U-AVLIS Uranium-Atomic Vapor Laser Isotope Separation (Program) [of US Department of Energy]

UAW United Auto(mobile) Workers

UAW-CIO United Auto(mobile) Workers of the Congress of Industrial Organizations [US]

UAX Unit Automatic Exchange

UAZ Universidad Autónoma de Zacatecas [Autonomous University of Zacatecas, Mexico]

UAZ University of Arizona [Tucson, US]

UB Ulan Bator [Mongolian People's Republic]

UB Unassemble Both (Command) [Pascal Programming]

UB Unbalance [also U/B]

UB Ungermann-Bass

UB Unibus (Adapter)

UB Union of Burma

UB Universidad de Barcelona [University of Barcelona, Spain]

UB Universidade de Brasília [University of Brasilia, Brazil]

UB Universität Basel [University of Basel, Switzerland]

UB *(Universitätsbibliothek)* – German for "University Library"

UB Universitatii Bucuresti [University of Bucharest, Romania]

UB Universitatea din Braşov [University of Brasov, Romania]

UB Université de Besançon University of Besançon, France]

UB Universitet u Beograd [University of Belgrade, Serbia]

UB University of Ballarat [Victoria State, Australia]

UB University of Bath [UK]

UB University of Belgrade [Serbia]

UB University of Bihar [Muzaffarpur, India]

UB University of Birmingham [UK]

UB University of Bradford [UK]

UB University of Brandon [Manitoba, Canada]

UB University of Bratislava [Slovakia]

UB University of Bridgeport [Connecticut, US]

UB University of Bristol [UK]

UB University of Budapest [Hungary]

UB University of Buffalo [New York State, US]

UB Upper Bound [Mathematics]

UB Upper Brace

UB Urinary Bladder [Anatomy]

UB Utility Bridge

UB I Université de Bordeaux I [University of Bordeaux I, France]

UB II Université de Bordeaux II [University of Bordeaux II, France]

UB III Université de Bordeaux III [University of Bordeaux III, France]

U/B Unbalance [also UB]

UBA Umweltbundesamt [Federal Office of the Environment, Germany]

UBA Unblocking Acknowledgement Signal

UBA Unibus Adapter

UBA Universidad de Buenos Aires [University of Buenos Aires, Argentina]

UBAEC Union of Burma Atomic Energy Center [Rangoon, Burma]

UBARI Union of Burma Applied Research Institute [Rangoon, Burma]

UBB Universal Black Box

UBC Unbalanced Cable

UBC Uniform Building Code [US]

UBC Universal Bibliographical Control

UBC Universal Block Channel

UBC Universal Book Code

UBC Universal Buffer Controller

UBC Universidad Bolivia Católica [Catholic University of Bolivia, La Paz]

UBC University of British Columbia [Vancouver, Canada]

UBC Used Beverage Can

UBC Used Beverage Container

UBCW United Brick and Clay Workers of America

UBD Unique But Duplicative

UBD Utility Binary Dump

UBDT Unter-Bad-Düsen-Technik [Submerged Bath Nozzle Technology] [Metallurgy]

UBE Universal Bus Exercisor

UBEC Union of Banana-Exporting Countries [Panama]

UBF Underground Baggage Facility

UBFF Urey-Bradley Force Field

UBHR User Block Handling Routine

UBi Uranium Bismuthide

UBIC Universal Bus Interface Controller

UBIS Ultrasonic Boiler Inspection System

Ubitron Undulating Beam Interaction Electron Tube [also ubitron]

UBJ Union Ball Joint

UBK Unblock

UBL Unbalanced Load [Electrical Engineering]

UBL Unblocking (Signal)

UBLS University of Botswana, Lesotho and Swaziland

UBM Unbalanced Magnetron (Sputtering)

UBM Under Bump Metallization

UBM Uniform Boundary Model [Metallurgy]

UBM Unit Bill of Materials [also UB/M]

UBP Université Blaise Pascal [Blaise Pascal University, Clermont, France]

UBPVLS Uniform Boiler and Pressure Vessel Laws Society [US]

UBR Unspecified Bit Rate

UBS Union Bank of Switzerland

UBS Unit Backspace (Character)

UBS United States Biological Survey

UBSI Unbleached Sulfite (Pulp) [Papermaking]

UBSS Unbleached Sulfate Softwood (Pulp) [Papermaking]

UBT Universal Book Tester

UB/TIB Universitätsbibliothek/Technische Informationsbibliothek Hannover [University Library/Technical Library and Document Delivery Center, University of Hanover, Germany]

UBTL Utah Biomedical Test Laboratory [at University of Utah Research Park, Salt Lake City, US]

UC Uddeholm Corporation

UC Ultracentifugation; Ultracentrifuge

UC Ultraclean

UC Umbilical Cord [Anatomy]

UC Unclassified

UC Uncoated (Paper)

UC Uncontrolled

UC Under Construction [also U/C]

UC Undercool(ed); Undercooling

UC Undercorrected

UC Undercurrent

UC Unemployment Compensation

UC Unichannel

UC Unit Call

UC Unit Cell [also uc] [Crystallography]

UC Unit Cooler

UC Unit Cost

UC Unit Cylinder

UC Universal Constant [Physics]

UC Universal Cooperative [US]

UC Universal Crown (Rolling Mill) [Metallurgy]

UC Universidad de Chile [University of Chile, Santiago]

UC Universidad de Colima [University of Colima, Mexico]

UC Universidad de Concepción [University of Concepcion, Chile]

UC Universidad de Córdoba [University of Cordoba, Spain and Argentina]

UC Università di Cagliari [University of Cagliari, Italy]

UC Université de Caen [University of Caen, France]

UC University College

UC University of Cairo [Egypt]

UC University of Calcutta [India]

UC University of California [US]

UC University of Cambridge [UK]

UC University of Canberra [Australian Capital Territory]

UC University of Canterbury [UK]

UC University of Canterbury [Christchurch, New Zealand]

UC University of Chattanooga [Tennessee, US]

UC University of Chicago [Illinois, US]

UC University of Cincinnati [Ohio, US]

UC University of Cologne [Germany]

UC University of Colorado [Colorado Springs, US]

UC University of Connecticut [Storrs, US]

UC University of Copenhagen [Denmark]

UC Unrestricted Ceiling [Aeronautics]

UC Unsatisfactory Condition

UC Up-Converter [also U-C, or U/C]

UC Upper Case [also uc]

UC Uranium Monocarbide

UC Usual Conditions [Commerce]

UC Utilization Coefficient

.UC2 Compressed File [UltraCompressor File Name Extension] [also .uc2]

U/C Undercharge [also u/c]

U/C Under Construction [also UC, or u/c]

U/C Under Current [also UC]

U-C Up-Converter [also UC, or U/C]

U-C Uranium-Carbon (Alloy)

uc upper case [also UC]

u/c undercharge [also U/C]

uc usual conditions [Commerce]

UCA Uddeholm Corporation Alloy

UCA Ulster Chemists Association [Northern Ireland]

UCA Uncommitted Component Array

UCA Universal Cellular Automaton

UCA Universidad Católica del Argentina [Catholic University of Argentina, Buenos Aires]

UCA Universidad Centroamericano [Central American University, Managua, Nicaragua]

UCA University of Central Arkansas [Conway, US]

UCAA University College of Addis Ababa [Ethiopia]

UCADIA Union Centroamericana de Asociaciónes de Ingenieros y Arquitectos [Central American Union of Engineers and Architects, Costa Rica]

U Cal University of California [US]

U Cal-SD University of California at San Diego [US]

UCAM Uncoupled Angular Momentum (Representation)

UCAP Utilization Center for Agricultural Products [of Iowa State University, Ames, US]

UCAP University Committee on Academic Personnel [of University of California at Berkeley, US]

UCAR Union Carbide Carbon Dioxide (Process) [also Ucar] [Chemical Engineering]

UCAR University Corporation for Atmospheric Research [US]

UC-Ar Ultraclean Argon (Gas)

UCARCOL Union Carbide Solvent (Process) [Chemical Engineering]

UCAR Newsl UCAR Newsletter [of University Corporation for Atmospheric Research, US]

UCAR/NSF University Corporation for Atmospheric Research/National Science Foundation [US]

UCAR/OFPS University Corporation for Atmospheric Research/Office of Field Project Support [US]

UCATS Uniform Configuration and Technical Support [NASA]

UCATT Union of Construction, Allied Trades and Technicians [UK]

UCB Union Chemique Belge [Belgian Chemical Union]

UCB Unit Control Block

UCB Universal Character Buffer

UCB Universidad Central de Barcelona [Barcelona Central University, Spain]

UCB Universidad Central de Bayamón [Bayamon Central University, Puerto Rico]

UCBL Université Claude Bernard [Claude Bernard University, Lyons, France]

UCB University College of Botswana [Gaborone]

UCB University of California at Berkeley [US]

UCB University of Colorado at Boulder [US]

UCBL Université Claude Bernard, Lyon [Claude Bernard University, Lyons, France]

UCB-LBL University of California at Berkeley–Lawrence Berkeley Laboratory [US]

UCC Undecreciated Capital Cost

UCC Unified Classification Code

UCC Uniformation Classification Committee [now National Railroad Freight Committee, US]

UCC Uniform Code Council [US]

UCC Uniform Commercial Code [US]

UCC Universal Checkout Console

UCC Universal Classification Code

UCC Universal Conference Circuit

UCC Universal Copyright Convention

UCC Universidad Católica de Chile [Catholic University of Chile, Santiago]

UCC Universidad Católica de Cuyo [Catholic University of Cuyo, San Juan, Argentina]

UCC University Center Corporation [US]

UCC University College of Cardiff [of University of Wales, UK]

UCC University Computing Center

UCC University Computing Company [US]

UCC University of Cape Coast [Ghana]

UCCA Universities Central Council on Admissions [UK]

UCC-APT University Computing Company Automatic Programmed Tool

UCCB University College of Cape Breton [Nova Scotia, Canada]

UCCM University of Cincinnati College of Medicine [Ohio, US]

UCCRS Underwater Coded Command Release System

UCCS Universal Camera Control System

UCCS University of Colorado at Colorado Springs [US]

UCD Uniform Call Distributor

UCD Universal Control Drive

UCD University of California at Davis [US]

UCD University of Colorado at Denver [US]

UCD Urine Collection Device

UCDA University and College Designers Association [US]

UCDP Uncorrected Data Processor

UCDWR University of California, Department of War Research [US]

UCE Universidad Central del Ecuador [Central University of Ecuador, Quito]

UCE Unsolicited Commercial E-Mail

UCEA University Council for Educational Administration [US]

UCEPCEE Union du Commerce des Engrais des Pays de la Communauté Economique Européenne [Union of the Fertilizer Producing Countries of the European Economic Community, Belgium]

UCF Universal Conductance Fluctuations [Electronics]

UCF Université de Clermont-Ferrand [University of Clermont-Ferrand, France]

UCF University of Central Florida [Orlando, US]

UCF Utility Control Facility

UCG Ultrasound Cardiogram [Medicine]

UCG Underground Coal Gasification [Mining]

UCG United Communications Group [Bethesda, Maryland, US]

UCG Urine Chorionic Gonadotropin (Test) [Medicine]

UCHF Uncoupled Hartree-Fock [Quantum Physics]

UCHSC University of Colorado Health Sciences Center [Denver, US]

UCI Ultrasonic Contact Impedance (Process) [Mechanical Testing]

UCI University of California at Irvine [US] [also UC-I]

UCI User Class Identifier

UCI Utility Card Input

UCIDT University Consortium for Instructional Development and Technology [US]

UCIS Uprange Computer Input System

UCITS Undertaking for Collective Investment in Transferable Securities

UCJ Ulmann, Chalmers and Jackson (Theory) [Metallurgy]

UCK Unit Check

UCL University of Central Lancashire [Preston, UK]

UCL Undercooled Liquid

UCL Undercorrected Lens

UCL Universal Communications Language

UCL Université Catholique de Louvain [Catholic University of Leuven, Belgium]

UCL University Center of Luxembourg [Luxembourg City]

UCL University College of London [UK]

UCL University of California at Livermore [US]

UCL University of Chicago Library [US]

UCL Upper Catastrophic Limit

UCL Upper Confidence Limit [Statistics]

UCL Upper Control Limit [Statistics]

UCL$_c$ Upper Control Limit for c Control Chart (in Statistics) [Symbol]

UCL$_p$ Upper Control Limit for p Control Chart (in Statistics) [Symbol]

UCL$_R$ Upper Control Limit for R Control Chart (in Statistics) [Symbol]

UCL$_{Rm}$ Upper Control Limit for Moving-Average (R_m) Control Chart (in Statistics) [Symbol]

UCL$_u$ Upper Control Limit for u Control Chart (in Statistics) [Symbol]

UCL Upper Control Limit for Control Chart (in Statistics) [Symbol]

UCLA Uncommitted Logic Array

UCLA University of California at Los Angeles [US]

UCLEA University and College Labor Education Association [US]

UCLJ University of California at La Jolla [US]

UCLP Universidad Católica de La Plata [Catholic University of La Plata, Argentina]

UCLR University Center for Laser Research [of Oklahoma State University, Stillwater, US]

UCLRL University of California Lawrence Radiation Laboratory [US]

UCM Universal Canister Mount

UCM Universal Communications Monitor

UCM User Communications Manager

UCM Universidad Complutense de Madrid [Complutense University of Madrid, Spain]

UCMC University of California Medical Center [San Francisco, US]

UC-Mill Universal Crown (Rolling) Mill [Metallurgy]

UCML University of California Microwave Laboratory [US]

UCMP University of California Museum of Paleontology [Berkeley, US]

UCMT Upper Critical Mixture Temperature

UCN Ultra-Cold Neutron [Physics]

UCN Uniform Control Number

UCNI Unclassified Controlled Nuclear Information

UCNI Unified Communications, Navigation and Identification [also U/CNI]

UCNSA Universidad Católica de Nuestra Señora de la Asunción [Catholic University of Our Lady of Asuncion, Paraguay]

UCNT University College of the Northern Territory [Australia]

UCNW University College of North Wales [of University of Wales, Bangor, UK]

UCO United Cooperatives of Ontario [Canada]

UCO Utility Compiler [also UCOM]

UCO Utility Compiler [also UCOM]

UCOL University Committee on Libraries [of University of California at Berkeley, US]

UCON Utility Control

UCONN University of Connecticut [Storrs, US] [also UConn]

UCORC University of California Operations Research Center [also UC/ORC, US]

UCOS Uprange Computer Output System

UCOS User Class of Service

UCOWR Universities Council on Water Resources [US]

UCP Ubiquitous Crystallization Process

UCP Ultra-Carbon Powder

UCP Uninterruptible Computer Power

UCP Unit Construction Principle

UCP Universidade Catolica do Pernambuco [Pernambuco Catholic University, Brazil]

UCP University of California Press [US]

UCP University of Chicago Press [US]

UCP Update Control Process

UCP Upper Collector Plate

UCP Utility Control Program

UCPA United Cerebral Palsy Association [US]

UCPL Union Centrale des Producteurs de Lait [Central Union of Milk Producers, Switzerland]

UCPR Universidad Católica de Puerto Rico [Catholic University of Puerto Rico, Ponce]

UCPSS (International Symposium on) Ultra-Clean Processing on Silicon Surfaces

UCPTE Union pour la Coordination de la Production et du Transport de l'Electricité [Union for the Coordination of the Production and Transport of Electricity, Switzerland]

UCPTE Union for the Coordination of the Production and Transport of Electric Power [Switzerland]

UCPU Urine Collection and Pretreatment Unit

UCR Unconditioned Response

UCR Universidad de Costa Rica [University of Costa Rica, San José]

UCR University of California at Richmond [US]

UCR University of California at Riverside [US]

UCR University of Costa Rica [San José]

UCR Unsatisfactory Condition Report

UCRI Union Carbide Research Institute [US]

UCRL University of California Radiation Laboratory [of US Department of Energy at Lawrence Livermore National Laboratory]

UCRL University of California Research Laboratory [US]

UCRP University Coal Research Program [of US Department of Energy]

UCRS Uniform Contractor Reporting System

UCS Ultimate Compressive Strength

UCS Unbalance Control System

UCS Unicode Conversion Support

UCS Unconditioned Stimulus

UCS Uniform Chromaticity Scale

UCS Universal Call Sequence

UCS Universal Camera Site

UCS Universal Character Set

UCS Universal Classification System

UCS Universal Communications Subsystem

UCS Universal Control System

UCS Universidad Católica de Santiago [Catholic University of Santiago, Chile]

UCS University College of Swansea [of University of Wales, UK]

UCS University College of Swaziland [Kwaluseni]

UCS User Control Store

UCS User Coordinate System

UCS Utility Conservation System

UCS Utility Control Services

UCS Utility Control System

UCSB Universal Character Set Buffer

UCSB University of California at Santa Barbara [US]

UCSC University of California at Santa Cruz [US]

UCSD Universal Communications Switching Device

UCSD University of California at San Diego [US]

UCSDSM University of California at San Diego School of Medicine [US]

UCSEL University of California Structural Engineering Laboratory [US]

UCSF University of California at San Francisco [US]

UCSF Universidad Católica de Santa Fé [Catholic University of Santa Fe, Argentina]

UC SIM University of Chicago Scanning Ion Probe

UCSM University of Connecticut School of Medicine [Farmington, US]

UCSSL University of California Space Sciences Laboratory [also UC/SSL, US]

UCST Upper Critical Solution Temperature [Physical Chemistry]

UCST Upper Critical Solution Type (Phase Diagram) [Physical Chemistry]

UCSTR Universal Code Synchronous Transmitter Receiver

UCT Universal Continuity Tester

UCT Universal Coordinated Time

UCT University of Cape Town [South Africa]

UCTA Urine Collection Transfer Assembly

UCTE Union of Canadian Transport Employees

UCU Union College and University [Schenectady and Albany, New York State, US]

UCV Universidad Católica de Valparaíso [Catholic University of Valparaiso, Chile]

UCV Universidad Central de Venezuela [Central University of Venezuela, Caracas]

UCV Upper Calorific Value

UCW Union of Communications Workers [UK]

UCW Unit Control Word

UCW University College of Wales [Aberystwyth, Cardiff and Bangor, UK]

UCWE Underwater Countermeasures and Weapons Establishment

UCWI University College of the West Indies [Kingston, Jamaica]

UCX Ultrix Connection

UD UAE Dirham [Currency of the United Arab Emirates]

UD Ultradisperse(d)

UD Underground Distribution [Electrical Engineering]

UD Unidirectional

UD Uniform Distribution

UD Universal Donor (of Type-O Blood Group) [Immunology]

UD Universität Dortmund [University of Dortmund, Germany]

UD University of Dacca [Bangladesh]

UD University of Dayton [Ohio, US]

UD University of Delaware [Newark, US]

UD University of Denver [Colorado, US]

UD University of Detroit [Michigan, US]

UD University of Dublin [Ireland]

UD University of Dundee [UK]

UD University of Durham [UK]

UD Unplanned Derating

UD Update [also U/D]

UD User Data

U/D Update [also UD]

UDA Uramil-N,N-Diacetic Acid

Uda Undecylamine

UDAC User Digital Analog Controller

UDAR Universal Digital Adaptive Recognizer

UDAS Underground Data Acquisition System

UDAS Unified Direct Access Standard

UDAS Unified Direct Access System

UDB Update Buffer

UDC Uganda Development Corporation

UDC Ultrasonic Doppler Cardioscope [Medicine]

UDC Unidirectional Composite

UDC Unidirectional Current

UDC Universal Decimal Classification

UDC Universal Digital Control

UDC Universal Digital Controller

UDC University of the District of Columbia [Washington, US]

UDC Upper Dead Center

UDC Urban Development Corporation [US]

UDC Urban District Council

UDC User Defined Code

UDC User Defined Command

UDCA Ursodeoxycholic Acid [Biochemistry]

UDD Ultradisperse Diamonds (Powder)

UDDF Up and Down Drafts

UDE Universal Data Entry

UDE Universal Data Exchange

UDE University of Denver [Colorado, US]

UDEAC Union de Douane et Economique de l'Afrique Centrale [Central African Economic and Customs Union]

UDEAO Union de Douane et Economique de l'Afrique Occidental [West African Economic and Customs Union]

UDEC Unitized Digital Electronic Calculator

UDEC Universal Digital Electronic Computer

UDET Unsymmetrical Diethylenetriamine

UDEX Universal Dow Extraction (Process) [Chemical Engineering]

UDF UHF (Ultrahigh Frequency) Direction Finding (Station)

UDF Unducted Fan

UDF Unidirectional Freezing

UDF Unit Derating Factor

UDF Universal Disk Format

UDF Unkinkable Domestic Flex

UDF User Defined Function

UDF Utility and Data Flow

UDG Unit Derating Generation

UDG User Defined Gateway

UDH Unplanned Derating Hours

UDH1 Unplanned Derating Hours, Class 1

UDI Urban Development Institute [Canada]

UDK User-Defined Key

UDL Unidirectional Laminate

UDL Uniform Data Language

UDL Uniform Data Link

UDL Universal Data Language

UDL Update Link

UDLC Universal Data Link Control

UDM Unified Defect Model [Materials Science]

UDM Union of Democratic Mineworkers [UK]

UDMH Unsymmetrical Dimethylhydrazine

UDMH/H Unsymmetrical Dimethylhydrazine/Hydrazine Blend

UDOFT Universal Digital Operational Flight Trainer

UDOP UHF (Ultrahigh Frequency) Doppler; Universal Doppler

UDP Ultradispersed Diamond Powder

UDP Ultradispersed Powder

UDP Uniform Distribution Pattern

UDP Unitary Development Plan

UDP United Data Processing

UDP University Distinguished Professor

UDP Update and Development Program

UDP Uridine Diphosphate; Uridine-5'-Diphosphate [Biochemistry]

UDP User Datagram Protocol

UDP User-Developed Program

UDP User Development Program

UDP User Datagram Protocol [Internet]

UDP User Development Program [of Canadian Space Agency]

5'-UDP Uridine-5'-Diphosphate

UDPAG Uridine 5'-Diphospho-N-Acetylglucosamine [Biochemistry]

UDPG Uridine Diphosphate Glucose; Uridine 5'-Diphosphoglucose [Biochemistry]

UDPGA Uridine Diphosphoglucuronic Acid; Uridine 5'-Diphosphoglucuronic Acid [Biochemistry]

UDP-GAL Uridine 5'-Diphosphogalactose [also UDPGal, or UDP-gal] [Biochemistry]

UDPGDH Uridine 5'-Diphosphoglucose Dehydrogenase [also UDPGD] [Biochemistry]

UDR Universal Document Reader

UDRI University of Dayton Research Institute [Ohio, US]

UDRPS Ultrasonic Data Recording and Processing System

UDS Ultrasonic Distance Sensor

UDS Unidirectionally Solidified; Unidirectional Solidification [Metallurgy]

UDS Uniform-Droplet Spray (Process) [Metallurgy]

UDS Uniscope Display System

UDS Universal Data Systems

UDS Universal Distributed System

UDS Universal Documentation System

UDS Unscheduled DNA (Deoxyribonucleic Acid) Synthesis [Biochemistry]

UDS Uranium Demonstration System [of US Department of Energy at Lawrence Livermore National Laboratory, University of California]

UDSR United Duroc Swine Registry [*Note:* Duroc Refers to a Breed of Swine]

UDT Underwater Demolition Team [of US Navy]

UDT Unified Dislocation Theory [Materials Science]

UDT Uniform Data Transfer

UDT Universal Data Transcriber

UDT Universal Document Transport

UDTC University of Dublin, Trinity College [Ireland]

UDTI Universal Digital Transducer Indicator

UDTMA Unsymmetrical N-Diethylenetriamine Acetic Acid

UDTS Universal Data Transfer Service

UDU Underwater Demolition Unit [of US Navy]

UE Uniform Elongation

UE Union Européenne [European Union]

UE University of Edinburgh [Scotland]

UE University of Essex [UK]

UE University of the East [Manila, Philippines]

UE Uranium Enrichment [Nuclear Engineering]

UE User Equipment

UEA Ulex Europaeus Agglutinin [Immunology]

UEA Union Européenne de l'Ameublement [European Furniture Manufacturers Federation, Belgium]

UEA United Epilepsy Association [US]

UEA Uranium Enrichment Associates

UEA Utah Education Association [US]

UEA I Ulex Europaeus Agglutinin I (with L-Fructose Affinity) [Immunology]

UEA II Ulex Europaeus Agglutinin II (with N,N'-Diacetylchitobiose Affinity) [Immunology]

UEAC Union des Etats de l'Afrique Centrale [Union of Central African States] [Defunct]

UEATC Union Européenne pour l'Agrément Technique dans la Construction [European Union for Technical Aprroval in Building Construction, France]

UEC United Engineering Center [New York City, US]

UEC Universidade Estadual de Campinas [University of Campinas, Brazil]

UEC University of Electro-Communications [Tokyo, Japan]

UEC Utah Engineers Council [US]

UECU Union for Experimenting Colleges [US]

UEDDA Unsymmetrical Ethylenenediamine Diacetic Acid [also u-EDDA]

UEE Union of Electrical Engineers [Estonia]

UEEB Union des Exploitations Electrique en Belgique [Union of Electricity Users in Belgium, Brussels]

UEF Union des Etudiants de France [Union of French Students]

UEF Union Europäischer Forstberufsverbände [Union of European Forestry Associations, Germany]

UEG Underwater Engineering Group

UEG Universidad de Estado de Guayaquil [State University of Guayaquil, Ecuador]

UEI User Equipment Information

UEIS United Engineering Information Service [US]

UEL University Engineering Laboratory [of University of Cambridge, UK]

UEL Upper Explosion (or Explosive) Limit

UEl Uniform Elongation

UEM Universidade Estadual de Maringa [Maringa State University, Brazil]

UEO Union de l'Europe Occidentale [Western European Union]

UEP Underwater Electric Potential

UEP Unequal Error Protection

UEP Union Européenne des Paiements [European Payment Union, France]

UEP Universidade Estadual Paulista [Paulista State University, Sao Paulo, Brazil]

UER Union Européenne de Radiodiffusion [European Broadcasting Union, Switzerland]

UER Unique Equipment Register

UER Unité d'Enseignement et de Recherche [Teaching and Research Unit, France]

UER Unsatisfactory Equipment Report

UERD Underwater Explosives Research Division [US Navy]

UERJ Universidade do Estado de Rio de Janeiro [Rio de Janeiro State University, Brazil]

UERL Underwater Explosives Research Laboratory [Woods Hole, Massachusetts, US]

UERPIC Underground Excavation and Rock Properties Information Center [US]

UES Uniform Emission Standard

UES United Engineering Societies

UES Universal Energy Systems, Inc. [Dayton, Ohio, US]

UES Universidad de El Salvador [University of El Salvador, San Salvador]

UESA Ukrainian Engineers Society of America [US]

UET United Engineering Trustees [US]

UET Universal Emulating Terminal

UET Universal Engineered Tractor

UET University of Engineering and Technology [Lahore, Pakistan]

UEW United Electrical Workers [US]

UEW Unsteady Expansion Wave [Fluid Mechanics]

UEX Unit Exception

UF Ultrafilter; Ultrafiltration

UF Ultrafine

UF Ultrasonic Frequency

UF Unavailability Factor

UF Underflow [also U/F]

UF Underground Feeder

UF Unified Field [Relativity Theory]

UF Universal Format

UF Università di Ferrara [University of Ferrara, Italy]

UF Universität Frankfurt [University of Frankfurt, Germany]

UF Université de Fribourg [University of Fribourg, Switzerland]

UF University of Florida [Gainesville, US]

UF Urea-Formaldehyde [also U-F]

UF Urethane Foam,

UF User Friendliness; User-Friendly

UF Utility File

UF Utility Function

U/F Underflow [also UF]

U-F Urea-Formaldehyde [also UF]

UFA Ultra Fine-Grained

UFA Uncorrelated Factors Approximation

UFA Unesterified Fatty Acids

UFA Unsaturated Flow Apparatus

UFA Until Further Advised [also ufa]

UFAA United Food Animal Association [US]

UFAC Upholstered Furniture Action Council [High Point, North Carolina, US]

UFAM Universal File Access Method [Computers]

UFAW United Fishermen and Allied Workers Union

UFAW Universities Federation for Animal Welfare [UK]

UFB Unfit for Broadcast

UFB Universidade Federal do Bahia [Federal University of Bahia, Brazil]

UFC Uniform Fire Code [US]

UFC Uniform Freight Classification

UFC Universal Frequency Counter

UFC Universidade Federal do Ceará [Federal University of Ceara, Brazil]

UFC Universities Funding Council [UK]

UFCM University of Florida College of Medicine [Gainesville, US]

UFCS Underwater Fire Control System

UFCW United Food and Commercial Workers

UFD User File Directory

UFDC Used-Fuel Disposal Center [Canada]

UFE Undergraduate Faculty Enhancement (Workshop) [of National Science Foundation, US]

UFEDC Unified Fixed Exchangeable Disk Coupler

UFESA United Fire Equipment Service Association [US]

UFF Universidade Federal Fluminense [Niterói, Brazil]

UFFC Ultrasonics, Ferroelectrics and Frequency Control (Society) [of Institute of Electrical and Electronics Engineers, US]

UFFCS Ultrasonics, Ferroelectrics and Frequency Control Society [of Institute of Electrical and Electronics Engineers, US]

UFFI Urea-Formaldehyde Foam Insulation

UFG Ultrafine Grain(ed)

UFG Ultrafine Grinding; Ultrafine Ground

UFG Universidade Federal de Goiás [Federal University of Goias, Goiânia, Brazil]

UFHSC University of Florida Health Science Center [Gainesville, US]

UFI Union des Foires Internationales [Union of International Fairs, France]

UFI Upstream Failure Indication

UFI Usage Frequency Indicator

UFI User-Friendly Interface

UFIPTI Union Franco-Iberique pour la Coordination de la Production et du Transport de l'Electricité [Franco-Iberian Union for Coordinating the Production and Transmission of Electricity, France]

UFL Upper Flammability (or Flammable) Limit

UFla University of Florida [Gainesville, Florida, US]

UFM Ultrafine Magnetic (Particles)

UFM Ultrafine Medium

UFM Ultrasonic Force Mode [Atomic Force Microscopy]

UFM Universidade Federal do Maranhão [Federal University of Maranhao, Brazil]

UFM Upper Felsic Metapyroclastics [Geology]

UFM User to File Manager

UFMG Universidade Federal de Minas Gerais [Federal University of Minas Gerais, Belo Horizonte, Brazil]

UFN Until Further Notice

UFO Unidentified Flying Object [also Ufo]

UFOD Union Française des Organismes de Documentation [French Union of Documentation Organizations]

UFOIRC UFO (Unidentified Flying Object) Information Retrieval Center [US]

UFOKAT Umweltforschungskatalog [Environmental Research Catalogue published by the Umweltbundesamt (German Federal Department of the Environment)]

UFOP Universidade Federal de Ouro Preto [Federal University of Ouro Preto, Brazil]

UFORDAT Umweltforschungsdatenbank [Environmental Research Database for Austria, Germany and Switzerland; Maintained by STN International]

UFP Ultrafine Particle

UFP Ultrafine Powder

UFP Universidade Federal de Paraíba [Federal University of Paraiba, Brazil]

UFP Universidade Federal de Pernambuco [Federal University of Pernambuco, Recife, Brazil]

UFP Utility Facilities Program

UFP-NMR Ultrafine Particle Nuclear Magnetic Resonance

UFPT Ultrafine Particle Technology

UFR Underfrequency Relay

UFRGN Universidade Federal do Rio Grande do Norte [Federal University of Rio Grande do Norto, Natal, Brazil]

UFRGS Universidade Federal do Rio Grande do Sul [Federal University of Rio Grande do Sul, Porto Alegre, Brazil]

UFRJ Universidade Federal do Rio de Janeiro [Federal University of Rio de Janeiro, Brazil]

UFRO (International) Union of Forest Research Organizations

UFRP Universidade Federal Rural do Pernambuco [Federal Rural University of Pernambuco, Brazil]

UFS Ultimate Factor of Safety

UFS United Farmers and Stockowners

UFS Unix (Operating System) File System

UFSC Universidade Federal de São Carlos [Federal University of Sao Carlos, Brazil]

UFSC Universidade Federal de Santa Catarina [Federal University of Santa Catarina, Florianopolis, Brazil]

UFSCar Universidade Federal de São Carlos [Federal University of Sao Carlos, Brazil]

UFSM Universidade Federal de Santa Maria [Federal University of Santa Maria, Brazil]

UFSM Universidad Federico Santa Maria [University of Federico Santa Maria, Chile]

UFT Unified Field Theory [Relativity Theory]

UFT Union des Fédérations des Transports [Union of Transport Federations, France]

UFT United Federation of Teachers [US]

UFTP Ultrafine Tungsten Powder

UFTR University of Florida Teaching Reactor [US]

UFU Ulster Farmers Union [Northern Ireland]

UFVA University Film and Video Association [US]

UFVF University Film and Video Foundation [US]

UFW United Farm Workers (Union) [US]

UFWOC United Farm Workers Organizing Committee [US]

UFZ Umweltforschungszentrum [Environmental Research Center, Leipzig-Halle, Germany]

UG Uganda [ISO Code]

UG Ultrasonic Generator

UG Undergraduate

UG Underground

UG Uniaxial Gauge

UG Universidad de Guadalajara [University of Guadalajara, Mexico]

UG Universitatii din Galati [University of Galati, Romania]

UG Université de Genève [University of Geneva, Switzerland]

UG University of Georgia [Athens, US]

UG University of Glasgow [UK]

UG University of Guelph [Ontario, Canada]

UG Urban Geography

UG User Group

UG User Guide

.ug Uganda [Country Code/Domain Name]

u(α) orientation difference distribution function (in crystallography) [Symbol]

UGA Ultra-Graphics Accelerator

UGA Uncommitted Gate Array

UGA Unitary Group Approach [Mathematics]

UGA University of Georgia at Athens [US]

Ugan Uganda(n)

U-GAS Utility Gas (Process) [Chemical Engineering]

UGC University Grants Commission [Canada]

UGC University Grants Committee [UK]

UGCW United Glass and Ceramic Workers (of North America)

UGET Union Générale des Etudiants Tunésiens [General Union of Tunesian Students]

UGF Universidade de Gama Filho [Gama Filho University, Rio de Janeiro, Brazil]

UGG United Grain Growers [Canada]

U/g Hb Units per gram of Hemoglobin [Unit]

U-GHS Essen Universität-Gesamthochschule Essen [University and Polytechnic Institute of Essen, Germany]

U-GHS Siegen Universität-Gesamthochschule Siegen [University and Polytechnic Institute of Siegen, Germany]

UGI United Gas Improvement Company (Process) [Chemical Engineering]

UGIC Union Générale des Ingénieurs et Cadres [General Union of Engineers and Higher Technical Staffs]

UGLIAC United Gas Laboratory Internally Programmed Automatic Computer

UGM Underground Mining

UGPP Uridine Diphosphoglucose Pyrophosphorylase [Biochemistry]

UGS Ugandan Shilling [Currency of Uganda]

UGS Undergraduate Student

UGS Uplink Gateway Services [also ugs]

UGS Upper Guide Structure

UGS Utah Geological Survey [US]

UGT Unión General de Trabajadores [General Union of Workers, Spain]

UGT Urogenital Tract [Anatomy]

ugt urgent

UGTA Union Générale des Travailleurs Algériens [General Union of Algerian Workers]

UGTC Union Générale des Travailleurs Centrafrique [General Union of Centrafrican Workers]

UGTT Union Générale des Travailleurs Tunésiens [General Union of Tunesian Workers]

UH Ultrahigh

UH Unavailable Hours

UH Unit Heater

UH Universität Hannover [University of Hannover, Germany]

UH Universidad de Havana [University of Havana, Cuba]

UH University of Hartford [West Hartford, Connecticut, US]

UH University of Hawaii [Honolulu, US]

UH University of Heidelberg [Germany]

UH University of Helsinki [Finland]

UH University of Houston [Texas, US]

UH University of Hull [UK]

UH Upper Half [also uh]

Uh Utah [US]

U7/h6 Force Fit; Basic Shaft System [ISO Symbol]

UHA University of Hawaii [Honolulu, US]

UHB Ultrahigh Bypass (Jet Engine)

UHB University of Hamburg [Germany]

UHC Ultrahigh Carbon (Steel)

UHC Unsaturated Hydrocarbon

UHCH Unit Handling Conveyor Association [now Conveyors Section of Material Handling Institute, US]

UHCI Universal Host Controller Interface

UHCS Ultrahigh Carbon Steel

UHD Ultrahigh Density

UHDMRH Ultrahigh Density Magnetic Recording Head (Project)

UHDPE Ultrahigh Density Polyethylene

UHE Ultra-High Energy

UHF Ultrahigh Frequency [Frequency: 300 to 3,000 Megahertz; Wavelength: 10 Centimeters to 1 Meter] [also uhf]

UHF Uniform Heat Flux

UHF Unrestricted Hartree-Fock (Theory) [Quantum Mechanics]

UHF Upstream Heat Flow [Metallurgy]

UHFDF Ultrahigh Frequency Direction Finder

UHFPCD Ultrahigh Frequency Photoconductive Decay [also UHF-PCD]

UHF/VHF Ultrahigh Frequency/Very High Frequency [also uhf/vhf]

UHI Upper Head Injection

UHINFO University of Hawaii Information [World Wide Web Site]

UHIP University Health Institute Plan

UHK University of Hong Kong

UHL Universitäre Hochschule Luzern [University College of Luzern, Switzerland]

UHL User Header Label

UHM Ultrahigh Modulus

UHM University of Hawaii at Manoa [Honolulu, US]

UHM(f) Ultrahigh Modulus (Carbon) Fiber

UHML Upper-Half Mean-Length

UHMW Ultrahigh Molecular Weight

UHMWPE Ultrahigh Molecular Weight Polyethylene [also UHMW PE, or UHMW-PE]

UHP Ultrahigh Performance

UHP Ultrahigh Power(ed)

UHP Ultrahigh Pressure

UHP Ultrahigh Purity

UHP Unit Horsepower [also uhp]

UHP Upper Half Plane

UHPEAF Ultrahigh-Power Electric-Arc (Steelmaking) Furnace [also UHP-EAF] [Metallurgy]

UHPEM Ultrahigh Purity Electronic Material

UHPFB Untreated Hard Pressed Fiberboard

UHPM (International) Ultrahigh-Purity Base Metals Conference

UHPS Underground Hydroelectric Pumped Storage

UHR Ultrahigh Reduction

UHR Ultrahigh Resistance

UHR Ultrahigh Resolution

UHRF Ultrahigh Resolution Facsimile

UHR-HVEM Ultrahigh Resolution High-Voltage Electron Microscope; Ultrahigh Resolution High-Voltage Electron Microscopy

UHRM Ultrahigh Resolution Microscope; Ultrahigh Resolution Microscopy

UHRNMR Ultrahigh-Resolution Nuclear Magnetic Resonance [also UHR-NMR]

UHR-TEM Ultrahigh Resolution Transmission Electron Microscope; Ultrahigh Resolution Transmission Electron Microscopy

UHS Ultra-High Strength

UHS Undoped Heterostructure [Electronics]

UHS University Hospital Society [Canada]

UHS(f) Ultrahigh Strength (Carbon) Fiber

UHSM Ultrahigh Strength Material

UHSR Ultra-High-Speed Radiography

UHSS Ultrahigh Strength Steel

UHT Ultimate High Temperature

UHT Ultra-Heat Treated (Milk)

UHT Ultrahigh Temperature

UHT Universal Hardness Tester

UHTREX Ultrahigh Temperature Reactor Experiment [US]

UHV Ultrahigh Vacuum

UHV Ultrahigh Voltage

UHV-AES Ultrahigh Vacuum Auger Electron Spectroscopy [also UHV AES]

UHV-CVD Ultrahigh Vacuum Chemical Vapor Deposition [also UHV/CVD, or UHV CVD]

UHV-ECRCVD Ultrahigh Vacuum Electron Cyclotron Resonance Chemical Vapor Deposition

UHV-EM Ultrahigh Vacuum Electron Microscope; Ultrahigh Vacuum Electron Microscopy [also UHV EM, or UHVEM]

UHV-EM Ultrahigh Voltage Electron Microscope; Ultrahigh Voltage Electron Microscopy [also UHV EM, or UHVEM]

UHVEM Center Ultrahigh Vacuum Electron Microscopy Center [Osaka, Japan]

UHV FI-STM Ultrahigh-Vacuum Field-Ion Scanning Tunneling Microscopy

UHV-HREM Ultrahigh Vacuum High-Resolution Electron Microscopy [also UHV HREM]

UHV-HRTEM Ultrahigh Vacuum High-Resolution Transmission Electron Microscopy [also UHV HRTEM]

UHV-RTCVD Ultrahigh-Vacuum Rapid Thermal Chemical Vapor Deposition [also UHV/RTCVD, or UHV RTCVD]

UHV-SKEM Ultrahigh Vacuum Scanning Kerr Electron Microscope; Ultrahigh Vacuum Scanning Kerr Electron Microscopy [also UHV SKEM]

UHV-SPM Ultrahigh Vacuum Scanning Probe Microscope; Ultrahigh Vacuum Scanning Probe Microscopy [also UHV SPM]

UHV-STC Ultrahigh Vacuum Scanning-Specimen Transfer Chamber [also UHV STC]

UHV-STEM Ultrahigh Vacuum Scanning Transmission Electron Microscopy; Ultrahigh Vacuum Scanning Transmission Electron Microscopy [also UHV STEM]

UHV-STM Ultrahigh-Vacuum Scanning Tunneling Microscope; Ultrahigh-Vacuum Scanning Tunneling Microscopy [also UHV/STM, or UHV STM]

UHV-TEM Ultrahigh-Vacuum Transmission Electron Microscopy [also UHV/TEM, or UHV TEM]

UHV-TGA Ultrahigh-Vacuum Thermogravimetric Analysis [also UHV/TGA, or UHV TGA]

UHV-XPS Ultrahigh-Vacuum X-Ray Photoelectron Spectroscopy [also UHV/XPS, or UHV XPS]

UHY University of Han Yang [Kyongju, South Korea]

UI Ultraionization

UI Ultrasonic Imaging

UI Ultrasonic Inspection

UI Ultrasonic Interferometer; Ultrasonic Interferometry

UI Unemployment Insurance

UI Unit of Issue

UI Unix International

UI University of Iceland [Reykjavik, Iceland]

UI University of Idaho [Moscow, US]

UI University of Illinois [US]

UI University of Indonesia [Djakarta, Indonesia]

UI University of Iowa [US]

UI University of Islamabad [Pakistan]

UI University of Istanbul [Turkey]

UI Unix International

UI Uranium Institute [UK]

UI Urban Initiatives [US]

UI User Interface

UIA Ultrasonic Industry Association [US]

UIA Union Internationale des Architectes [International Union of Architects, France]

UIA Union Internationale Astronomique [International Astronomical Union]

UIA Union of International Associations

UIB Unemployment Insurance Benefits

UIBC Unsaturated Iron-Binding Capacity

UIBPA Union Internationale de Biophysique Pure et Appliquée [International Union of Pure and Applied Biophysics, Hungary]

UIC Underground Injection Control

UIC Unemployment Insurance Commission

UIC Union Internationale des Chemins de Fer [International Union of Railways, France]

UIC University of Illinois at Chicago [US]

UIC Upper Information Center,

UIC Upper Iowa College [Fayette, Iowa, US]

UIC User Identification Code

UIC User Interface Circuit

UICB Union Internationale des Centres du Bâtiment [International Union of Building Centers, Netherlands]

UICC Union Internationale Contre le Cancer [International Union for Cancer Prevention]

UICM University of Iowa College of Medicine [Iowa City, US]

UICP Union Internationale de la Couverture et Plomberie [International Union of Roofing and Plumbing, France]

UID User Identifier

UIE UNESCO Institute for Education

UIE Union Internationale d'Electrothermie [International Union for Electroheat, France]

UIE Union Internationale des Etudiants [International Union of Students]

UIEO Union of International Engineering Organizations

UIESP Union Internationale pour l'Etude Scientifique de Population [International Union for Scientific Population Studies]

UIF Unrestricted Industrial Funds

UIG User Instruction Group

UIGGM United Institute of Geology, Geophysics and Mineralogy [Russia]

UIIG Union International de l'Industrie du Gaz [International Gas Union, Switzerland]

UIL Unione Italiana del Lavoro [Italian Labor Union]

UIL Univac Interactive (Computer) Language

UILA Union Internationale des Laboratoires Independants [International Union of Independent Laboratories, UK]

UILERMS Union Internationale des Laboratoires d'Essai et de Recherche pour les Matériels et les Structures [International Union of Testing and Research Laboratories for Materials and Structures, France]

UIM Ultra-Intelligent Machine

UIM Union of International Motorboating

UIMS User-Interface Management System

UIO Union of International Organizations

UIO Universal Input/Output

UIOC Universal Input/Output Controller

UIP Union Internationale d'Assus de Propriétaires de Wagon [International Union of Private Railway Truck Owners Association, Switzerland]

UIP University Interactions Program

UIP University of Illinois Press [Urbana, US]

UIP User-Interface Platform

UIP User-Interface Presentation

UIPC Underground Injection Practices Council [US]

UIPPA Union Internationale de Physique Pure et Appliquée [International Union of Pure and Applied Physics, Sweden]

UIPR Universidad Interamericana de Puerto Rico [Inter-American University of Puerto Rico, San Germán]

UIPRE Union Internationale de la Presse Radiotechnique et Electronique [International Union of the Radiotechnical and Electronics Press, Germany]

UIR Union Internationale de Radiodiffusion [International Broadcasting Union, Switzerland]

UIR Upper Flight Information Region

UIR Upwelling Infrared

UIR User Instruction Register

UIS Unemployment Insurance Service [of Employment and Training Administration, US]

UIS Uniform Interevent Steps

UIS United Inventors and Scientists

UIS Universal Illumination System

UIS Universal Infinity System (Optics) [Microscopy]

UIS Universidad Industrial de Santander [Santander Industrial University, Bucaramanga, Colombia]

UIS Upper Information Service

UIS Uranium Institute Symposium [UK]

UIS Urban (Geographic) Information System

UIS User Interface Service

UISPI Urethane Institute, Society of the Plastics Industry [US]

UISRC University-Industry Steel Research Center [of Colorado School of Mines, Golden, US]

UIT Ultraviolet Imaging Telescope

UIT Union Internationale de Télécommunication [International Telecommunications Union, Switzerland]

UIT University of Industrial Technology [Kanagawa, Japan]

UITA Union of International Technical Associations [France]

UITA Bull UITA Bulletin [of Union of International Technical Associations, France]

UITP Union Internationale des Transports Publics [International Union of Public Transport, Belgium]

UITP University-Industry Training Partnership

UIU University of Illinois at Urbana [US]

UIU Upper Iowa University [Fayette, US]

UIUC University of Illinois at Urbana-Champaign [US]

UJ Universal Joint

UJ University of Jilin [Changchun, PR China]

UJCL Universal Job Control Language

UJF/CNRS/INPG Université Joseph Fourier/Conseil National de la Recherche Scientifique/Institut National Polytechnique de Grenoble [Joseph Fourier University/ National Scientifique Research Council/National Polytechnic Institute of Grenoble, France]

UFJ-LSP Université Joseph Fourier–Laboratoire des Spectrométrie Physique [Physical Spectrometry Laboratory of Joseph Fourier University, St. Martin d'Hères, France]

UJFG Université Joseph Fourier de Grenoble [Joseph Fourier University of Grenoble, France]

UJM Uncorrected Jet Model

UJSE Union of Japanese Scientists and Engineers [Tokyo, Japan]

UJT Unijunction Transistor

$U_{eff}(J,T)$ Effective Vortex Pinning Energy (in Physics) [Symbol]

UK United Kingdom

UK United Kingdom [ISO Code]

UK Universität Karlsruhe [University of Karlsruhe, Germany]

UK University of Kansas [Lawrence, US]

UK University of Karachi [Pakistan]

UK University of Kansas [Lawrence, US]

UK University of Keele [UK]

UK University of Kentucky [Lexington, US]

UK University of Kumamoto [Japan]

UK University of Kuwait [Kuwait City]

.uk United Kingdom [Country Code/Domain Name]

UKα Uranium K-Beta (Radiation) [also UK$_\alpha$]

UKACC United Kingdom Automatic Control Council [of Institution of Electrical Engineers]

UKAFFP United Kingdom Association of Frozen Food Producers

UKAATS United Kingdom Advanced Air Traffic System

UKAC United Kingdom Automation Council

UKAEA United Kingdom Atomic Energy Authority

UKAEA-R United Kingdom Atomic Energy Authority–Risley

UKAEA Rep United Kingdom Atomic Energy Authority Report

UKAIIM United Kingdom Association for Information and Images Management [UK]

UKAIP United Kingdom Aeronautical Information Publication

UKAPE United Kingdom Association of Professional Engineers

UKASA United Kingdom Agricultural Students Association

UKASE University of Kansas Automated Serials System [US]

UKASTA United Kingdom Agricultural Supply Trade Association

UKB Universal Keyboard

UKβ Uranium K-Beta (Radiation) [also Uk$_\beta$]

UKC University of Kansas City [Missouri, US]

UKCC United Kingdom Central Council

UKCC United Kingdom Countryside Commission

UKCEED United Kingdom Center for Economic and Environmental Development [also UK CEED, or UK-CEED]

UKCCD United Kingdom Council Computing Development

UKCIS United Kingdom Chemical Information Service

UKCM University of Kentucky College of Medicine [Lexington, US]

UKCSMA United Kingdom Cutlery and Silverware Manufacturers Association

UKCTRAIN United Kingdom Catalogue Training [for United Kingdom Machine-Readable Catalogue]

UKDA United Kingdom Dairy Association

UK gal United Kingdom Gallon [also UKgal, or UKG]

UK gal/min United Kingdom gallon per minute [also UKGPM, or UK GPM]

UKELA United Kingdom Environmental Law Association

UK EPSRC United Kingdom Engineering and Physical Sciences Research Council [formerly Science and Engineering Research Council]

UK ESCA United Kingdom Electron Spectroscopy for Chemical Analysis (Users Group) [also UKESCA]

UKGPA United Kingdom Glycerin Producers Association

UKGPM United Kingdom Gallon per Minute [also UK gpm, or UK gal/min]

UKH Umformtechnisches Kolloquium Hannover [Hanover Metalworking Colloquium, Germany]

UKH University King Hassan II [Casablanca, Morocco]

UKIC United Kingdom Institute of Conservation

UKIPA United Kingdom and Ireland Particleboard Association

UKIRT United Kingdom Infrared Telescope

UKISC United Kingdom Industrial Space Committee

UKIO United Kingdom Information Office

UKITO United Kingdom Information Technology Organization

UKJGA United Kingdom Jute Goods Association

UKLDS United Kingdom Library Database System

UKLEO United Kingdom Laser and Electro-Optics Trade Association [UK]

UKMARC United Kingdom Machine-Readable Catalogue [of British Library Automated Information Service]

UKMC University of Kentucky College of Medicine [Lexington, US]

UKMCA United Kingdom Module Constructors Association [UK]

UKMRF United Kingdom Meteorological Research Facility

UKMO United Kingdom Meteorological Office

UKNCIAWPRC United Kingdom National Committee of the International Association on Water Pollution Research and Control

UKNR University of Kansas Nuclear Reactor [Lawrence, US]

UKOBA United Kingdom Outboard Boating Association

UKOLUG United Kingdom On-Line Users Group

UKOOA United Kingdom Offshore Operators Association

UKOOG United Kingdom Onshore Operators Group

UK£ United Kingdom Pound Sterling [Currency of Northern Ireland and the UK]

UKPA United Kingdom Pilots Association

UKPIA United Kingdom Petroleum Industry Association [also UK-PIA]

UKPMA United Kingdom Preserves Manufacturers Association

UKPO United Kingdom Post Office

UK qt United Kingdom quart [Unit]

Ukr Ukraine; Ukrainian

UKRAS United Kingdom Railway Advisory Service

Ukr Fiz Zh Ukrainskii Fizichnii Zhurnal [Ukrainian Journal of Physics]

Ukr Khim Zh Ukrainskii Khimicheskii Zhurnal [Ukrainian Journal of Chemistry]

Ukr Mat Zh Ukrainskii Matematicheskii Zhurnal [Ukrainian Journal of Mathematics]

UkrSSR Ukrainian Soviet Socialist Republic [USSR]

UKSATA United Kingdom-South Africa Trade Association

UKSIA United Kingdom Sugar Industry Association

UKSM United Kingdom Scientific Mission (in the USA)

UKSPA United Kingdom Science Park Association

UKTS United Kingdom Treaty Series

UL Ultralarge

UL Ultralight

UL Ultralow

UL Uncontrolled Language

UL Underload

UL Underwriters' Laboratories [US]

UL Unit Load

UL Universidade de Lisboa [University of Lisbon, Portugal]

UL Université de Lausanne [University of Lausanne, Switzerland]

UL Université de Liège [University of Liège, Belgium]

UL Université de Lille [University of Lille, France]

UL Université de Limoges [University of Limoges, France]

UL Université Laval [Laval University, Sainte Foy, Quebec, Canada]

UL Universiteit Leiden [Leyden University, Netherlands]

UL University Librarian; University Library

UL University of Lancaster [UK]

UL University of Lanzhow [PR China]

UL University of Leeds [UK]

UL University of Leicester [UK]

UL University of Leoben [Austria]

UL University of Lethbridge [Alberta, Canada]

UL University of Liverpool [UK]

UL University of London [UK]

UL University of Louisville [Kentucky, US]

UL University of Lowell [Massachusetts, US]

UL Univerza v Ljubljana [University of Ljubljana, Slovenia]

UL Unlubricated

UL Unordered List

UL Uplink [also U/L]

UL Upload

UL Upper Left

UL Upper Level

UL Upper Limit

UL Upper List

UL User Language

UL I Université de Lyon I [University of Lyons I, France]

UL II Université de Lyon II [University of Lyons II, France]

U/L Units per liter [also U L^{-1}]

U/L Unlimited

U/L Uplink [also U/L]

ULA Fairbanks, Alaska [Space Tracking and Data Network Site]

ULA Uncommitted Logic Array

ULA Uniform Line Array

ULAP University Libraries Automation Program [US]

ULAU Union of Latin American Universities [Mexico]

ULB Universal Logic Block

ULB Université Libre de Bruxelles [Free University of Brussels, Belgium]

ULC Ultralow Carbon (Steel)

ULC Underwriters Laboratories of Canada

ULC Uniform Loop Clock

ULC Universal Logic Circuit

ULC Upper and Lower Case

ULCB Ultralow Carbon Bainitic; Ultralow Carbon Bainite [Metallurgy]

ULCB Steel Ultralow Carbon Bainitic Steel [also ULCB steel]

ULCC Ultralarge Crude Carrier

ULC IF Ultralow Carbon/Interstitial-Free (Steel) [also ULC/IF, or ULC-IF]

ULCL University of London Central Library [UK]

ULCRF University of London Central Research Fund [UK]

UL/CSA Underwriters Laboratory/Canadian Standards Association

ULD Ultralarge Diameter

ULD Ultralow Density

ULD Unit Load Device

ULDLPE Ultralow-Density Linear Polyethylene [also ULD LPE, or ULD-LPE]

ULDPE Ultralow-Density Polyethylene [also ULD PE, or ULD-PE]

ULE Ultralow Emission

ULE Ultralow Expansion

ULEA University Labor Education Association [now University and College Labor Education Association, US]

ULEV Ultralow Emission Vehicle

ULF Ultralow Frequency [also ulf]

ULG Upholstery Leather Guild [now Autoleather Guild, US]

ULI University of Lisbon [Portugal]

ULI Urban Land Institute [US]

ULICP Universal Log Interpretation Computer Program

ULIE University of Leeds Institute of Education [UK]

ULISYS University Library System

ULJ Updated Lagrangian-Janmann (Formulation)

ULL Universal Leveling Loop

Ull Ullage [also ull]

ULLN Upper Limit Log Normal Distribution

ULLV Unmanned Lunar Logistics Vehicle [of NASA]

ULM Ultralight Material(s)

ULM Ultrasonic Light Modulator

ULM Universal Line Multiplexer

ULM Universal Logic Module

ULMA University Laboratory Managers Association [now Analytical Laboratory Managers Association, US]

ULMM Universal Length-Measuring Machine

ULMS Undersea Long-Range Missile System

ULMS Underwater-Launched Missile System

UL/MSHA Underwriters Laboratory/Mine Safety and Health Administration [US]

ULN Ultralow Noise

ULN Universal Link Negotiation

ULO University of London Observatory [Mill Hill, UK]

ULO Unmanned Launch Operations

ULOW Unmanned Launch Operations Western Test Range

ULP Ultralow Pressure

ULP Universidad de La Plata [University of La Plata, Argentina]

ULP Universidad de Las Palmas [University of Las Palmas, Gran Canaria, Spain]

ULP Université Libre de Paris [Free University of Paris, France]

ULPA United Lightning Protection Association [US]

ULPGC Universidad de Las Palmas de Gran Canaria [Spain]

ULPR Ultralow Pressure Rocket

ULR Ultra-Long Range

ULS Ultra-Light Steel

ULS Upward Looking Sonar

ULS Universidad Literaria de Salamanca [University of Salamanca, Spain]

ULSAB Ultra-Light Steel Auto Body

ULSAS Upward Looking Sonar Array System

ULSCS University of London Shared Cataloguing System [UK]

ULSD University of Louisville School of Dentistry [Kentucky, US]

ULSI Ultra-Large-Scale Integrated; Ultra-Large-Scale Integration

ULSIC Ultra-Large Scale Integrated Circuit

ULSI MOS Ultra-Large-Scale Integrated Metal-Oxide Semiconductor (Device) [also ULSI-MOS]

ULSV Unmanned Launch Space Vehicle

ULT Unique Last Term

ULT Untyped Lambda Calculi [Mathematics]

Ult Ultimatum [also ult]

ult *(ultimo)* – last day of month

Ult ultimate(ly)

ULTI Ultralow-Temperature Isotropic (Carbon)

ulto *(ultimo)* – last day of month; in, or of last month

ULTRA Universal Language for Typographic Reproduction Applications

ULTRA Special Symposium on Ultrafast Quantum Optics [of Optical Society of America]

ULTRA LSI Ultra Large Scale Integration [also Ultra LSI]

Ultrason Imaging Ultrasonic Imaging [Published in the US]

Ultrason Symp Proc Ultrasonic Symposium Proceedings [of Institute of Electrical and Electronics Engineers, US]

Ultrason Technol Ultrasonic Technology (Journal)

Ultrasound Med Biol Ultrasound in Medicine and Biology [Published in the UK]

ULU Union of Latin-American Universities

ULV Ultra-Low Volume

ULW Ultra Lightweight

UM Ultramicroscope; Ultramicroscopy

UM Ultrasonic Machining

UM Unbalanced Magnetron

UM Unimolecular

UM Unitary Matrix [Mathematics]

UM Unitas Malacologica [Netherlands]

UM Unit of Measure [also U/M]

UM Universal (Rolling) Mill [Metallurgy]

UM Universal Motor

UM Universidad de Madrid [University of Madrid, Spain]

UM Universidad de Monterrey [University of Monterrey, Mexico]

UM Universidade MacKenzie [MacKenzie University, Sao Paulo, Brazil]

UM Universität München [of University of Munich, Germany]

UM Università di Milano [University of Milan, Italy]

UM Université de Montpellier [University of Montpellier, France]

UM Université de Montréal [University of Montreal, Quebec, Canada]

UM University of Madras [India]

UM University of Maine [Orono, US]

UM University of Malaysia [Kuala Lumpur]

UM University of Manchester [UK]

UM University of Manitoba [Winnipeg, Canada]

UM University of Maryland [College Park, US]

UM University of Massachusetts [US]

UM University of Melbourne [Australia]

UM University of Miami [Coral Gables, Florida, US]

UM University of Michigan [US] [also U-M, or U Mich]

UM University of Minnesota [Minneapolis, US]

UM University of Mississippi [University, US]

UM University of Missouri [US]

UM University of Moncton [New Brunswick, Canada]

UM University of Montana [Missoula, US]

UM University of Montreal [Quebec, Canada]

UM University of Munich [Germany]

UM Unscheduled Maintenance [also U/M]

UM Upper Mantle (of the Earth)

UM Ural Mountains [Russia]

UM US Minor (Outlying Islands) [ISO Code]

U/M Unit of Measure [also UM]

U/M Unmanned

U/M Unscheduled Maintenance [also UM]

u/m undermentioned

UM I Université de Montpellier I [University of Montpellier I, France]

UM II Université de Montpellier II [University of Montpellier II, France]

UM III Université de Montpellier III [University of Montpellier III, France]

UMA Ultrasonic Manufacturers Association [US]

UMA Unified Memory Architecture

UMA Unión Mathemática Argentina [Argentinian Mathematical Union]

UMA Universal Measuring Amplifier

UMA University of Massachusetts at Amherst [US]

UMAA University of Michigan at Ann Arbor [US] [also UM-AA]

UMARP Upper Mantaro Archeological Research Project [Peru]

UMASS Unlimited Machine Access from Scattered Sites

U Mass University of Massachusetts [also U MASS]

UMB University of Massachusetts at Boston [US]

UMB University of Minnesota at Bemidji [US]

UMB Upper Memory Block [Lotus/Intel/Microsoft]

Umb Umbilical; Umbilicus

Umb Umbilicus

UMBC Umbilical Cord Cable

UMBC University of Maryland Baltimore County [US]

UMC Underwater Manifold Center

UMC Unibus Micro Channel

UMC Unidirectional Molding Compound

UMC Uniform Mechanical Code [US]

UMC Universidade de Mogi das Cruzes [University of Mogi das Cruzes, Brazil]

UMC University Materials Council

UMC University Medical Center

UMC University of Missouri at Columbia [US]

UMC Upper Mantle Commission [UK]

UMC Upstate Medical Center [of State University of New York, Syracuse, US]

UMCA United Mining Council of America [US]

UMCA Universities Mission to Central Africa

UMCC United Maritimes Consultative Committee

UMD Unitized Microwave Device

UMD University of Maryland [College Park, US]

UMD University of Michigan at Dearborn [US]

UMD University of Minnesota at Duluth [US]

UMDNJ University of Medicine and Dentistry of New Jersey [Newark, US]

UMES University of Michigan Executive System

UMF Ultra-Microfiche

UMF University of Michigan at Flint [US]

UMF Urea Melamine Formaldehyde

UMFA United Mineworker's Federation of Australia

Umform Tech Umform-Technik [German Publication on Metalworking]

U/mg Units per milligram [Unit]

UMGAP Upper Midwest Gap Analysis Program [US]

UMH Ultra-Microhardness (Test) [also Hum]

UMI Underwater Mining Institute [US]

UMI Unione Matematica Italiana [Italian Mathematical Union]

UMI Universal Measuring Instrument

UMI University Microfilms International (Database) [US]

U Mich University of Michigan [US] [also UM, or U-M]

UMIDD University of Michigan–Industrial Development Division [of University of Michigan, Detroit, US]

U Minn University of Minnesota [Minneapolis, US] [also U MINN]

UMIS Ultra-Micro-Indentation System [Mechanical Testing]

UMIST University of Manchester, Institute of Science and Technology [UK]

UMKC University of Missouri at Kansas City [US]

UML Ultra-Short, Medium and Long Waves

UML Unified (or Universal) Modeling Language

UML Universal Machine Language

UML University of Massachusetts at Lowell [US]

U/mL Units per milliliter [also U/ml]

UMLC Universal Multiline Controller

UMLER Universal Machine Language Equipment Register

UMLS Unified Medical Language System [of National Library of Medicine, Bethesda, Maryland, US]

UMM Universal Measuring Microscope

UMM University of Minnesota at Minneapolis [US]

UMM University of Minnesota at Moorhead [US]

UMMC University of Michigan Medical Center [Ann Harbor, US]

UMMC University of Miami Medical Center [Coral Gables, US]

UMMC University of Mississippi Medical Center [Jackson, US]

UMMS University of Massachusetts Medical School [Worcester, US]

UMMS University of Michigan Medical School [Ann Arbor, US]

UMNE University of Maryland Nuclear Experiment [US]

UMO Unmanned Orbital

U-Mo Uranium-Molybdenum (Alloy)

UMODE University of Manitoba Optokinetic Decay Experiment [Canada]

UMOS U-Type Metal-Oxide Semiconductor

UMP University of Manitoba Press [Winnipeg, Canada]

UMP University of Minnesota Press [Minneapolis, US]

UMP Unrestricted Moeller-Plesset (Theory) [Physics]

UMP Upper Mantle Program [Geophysics]

UMP Upper Mantle Project [of International Council of Scientific Unions]

UMP Uridine Monophosphate; Uridine-5'-Monophosphate

5'-UMP Uridine-5'-Monophosphate

UMPA Universal Microprobe Mass Analyzer

UMPLIS Umweltplanungsinformationssystem [Information System on Environmental Planning, Germany]

UMR Unimolecular Reaction [Physical Chemistry]

UMR Unipolar Magnetic Region

UMR University of Missouri-Rolla [US]

UMR Upper Maximum Range

UMR Upper Mississippi River [US]

UMRB University of Missouri Research Board [US]

UMRC University of Missouri Research Council [US]

UMRCC Upper Mississippi River Conservation Committee [US]

UMRR University of Missouri Research Reactor [US]

UMRRC University of Missouri Research Reactor Center [US]

UMS Underwater Monitoring System

UMS United Mexican States

UMS Universal Maintenance Standards

UMS Universal Measuring Standard

UMS Universal Memory System

UMS Universal Multiprogramming System

UMS University Mailing Service

UMSC University of Minnesota at Saint Cloud [US]

UMSC Upper Mississippi Science Center [of US Geological Survey/Biological Resources Division at LaCrosse, Wisconsin]

UMSL University of Missouri at St. Louis [US]

UMSM University of Maryland School of Medicine [Baltimore, US]

UMSM University of Miami School of Medicine [Florida, US]

UMSP University of Minnesota at Saint Paul [US]

UMSNH Universidad Michoacana de San Nicolas de Hidalgo [Michoacana University of San Nicolas de Hidalgo, Morelia, Mexico]

UMS/VS Universal Multiprogramming System/Virtual Storage

UMT Ultrasonic Machine Tool

UMT Union Marocain des Travailleurs [Moroccan Labor Union]

UMT Universal Military Training

UMTA Urban Mass Transit Authority

UMTA Urban Mass Transportation Administration [of US Department of Transportation]

UMTRA Uranium Mill Tailings Radiation Control Act [US]

UMTRAP Uranium Mill Tailings Remedial Action Program [US]

UMTRI University of Michigan Transportation Research Institute [Ann Arbor, US]

UMTRIS Urban Mass Transportation Research Information Service [of US Department of Transportation]

UMTS Universal Military Training Service [US]

UMVF Unmanned Vertical Flight

UMW United Mine Workers (International)

UMWA United Mineworkers of America

Umwelt-Mag Umwelt-Magazin [German Environmental Magazine]

UN Unassigned [also Un, or un]

UN Unified Constant Pitch Thread [Symbol]

UN United Nations

UN Universidad de Navarra [University of Navarra, San Sebastian, Spain]

UN Université de Nancy [University of Nancy, France]

UN Université de Nantes [University of Nantes, France]

UN University of Nanking [Nanking, PR China]

UN Université de Neuchâtel [University of Neuchatel, Switzerland]

UN University of Nagoya [Japan]

UN University of Nebraska [Lincoln, US]

UN University of Nevada [Reno, US]

UN University of Newcastle [New South Wales, Australia]

UN University of Nottingham [UK]

UN Upper Facet of Normal-Position Orientation [Mechanical Testing]

UN Uranium Mononitride

UN Uranyl Nitrate

UN Urban Network

UN Urinary Nitrogen

UN I Université de Nancy I [University of Nancy I, France]

UN II Université de Nancy II [University of Nancy II, France]

Un Union [also un]

UNA Unibus Network Adapter

UNA United Nations Association [UK]

UNA Universal Night Answering

UNA Universidad Nacional Agraria [National Agrarian University, Lima, Peru]

UNA Universidad Nacional de Asunción [National University of Asuncion, Paraguay]

UNA University of North Alabama [Florence, US]

UNA Uranium Nickel Aluminide

Unabr unabridged

UNAC United Nations Association of Canada

UNACAST United Nations Advisory Committee on the Application of Science and Technology

Unacc unaccompanied

UNADA United Nations Atomic Development Authority

UNADS Univac Automated Documentation System

UNAEC United Nations Atomic Energy Commission

UNAH Universidad Nacional Autónoma de Honduras [National Autonomous University of Honduras, Tegucigalpa]

UNAIDS United Nations Joint Programme on AIDS (Acquired Immunodeficiency Syndrome)

UNAIS United Nations Association International Service [UK]

UNALC User Network Access Link Control

UNAM Universidad Nacional Autónoma de Mexico [National Autonomous University of Mexico, Mexico City]

UNAMACE Universal Automatic Map Compilation Equipment

UNAN Universidad Nacional Autónoma de Nicaragua [National Autonomous University of Nicaragua, Léon]

UNATAC Union d'Assistance Technique pour l'Automobile et la Circulation Routière [Union of Technical Assistance for Motor Vehicle and Road Traffic, Switzerland]

unavbl unavailable

UNB Ultra-Narrow Bandwidth [also UNBW]

Unb Unbound [also unb]

UNB Universal Navigation Beacon

UNB University of New Brunswick [Fredericton, Canada]

U-Nb Uranium-Niobium (Alloy)

Unbal Unbalance(d) [also unbal]

UNBC University of Northern British Columbia [Prince George, Canada]

UNBM University of New Brunswick Museum [Moncton, Canada]

UNBSJ University of New Brunswick, St. John [Canada]

UNBTAO United Nations Bureau of Technical Assistance Operations

UNBW Ultra-Narrow Bandwidth [also UNB]

U-Nb-Zr Uranium-Niobium-Zirconium (Alloy)

UNC UUEncoded (Unix-to Unix Encoded) Netnews Collator [Unix Operating System]

UNC Unified National Coarse (Screw Thread) [Symbol]

UNC United Nations Command

UNC United Nations Convention

UNC Universal Naming Convention

UNC Universidad Nacional de Colombia [National University of Colombia, Bogota]

UNC Universidad Nacional de Córdoba [National University of Cordoba, Argentina]

UNC Universidad Nacional de Cuyo [Cuyo National University, Mendoza, Argentina]

UNC University of North Carolina [US]

UNC University of Northern Colorado [Greeley, US]

unc unconditional

UNCA United Nations Correspondents Association

UNCA University of North Carolina at Asheville [US]

UNCAITNDPS United Nations Convention Against Illicit Traffic in Narcotic Drugs and Psychotropic Substances

UNCAST United Nations Conference on the Applications of Science and Technology

UNCC University of North Carolina at Charlotte [US]

UNCCH University of North Carolina at Chapel Hill [US] [also UNC-CH]

UNCD United Nations Conference on Disarmament

UNCDF United Nations Capital Development Fund

UNCED United Nations Conference on Environment and Development

UNCentro Universidad Nacional del Centro de la Provincia de Buenos Aires [Central National University of the Province of Buenos Aires, Tandil, Argentina]

UNCF United Negro College Fund [US]

UNCFI United Nations Commission for Indonesia

UNCG University of North Carolina at Greensboro [US]

UNCHE United Nations Conference on the Human Environment

UNCHR United Nations Commission for Human Rights

UNCHS United Nations Center for Human Settlements [Kenya]

UNCI Université Nationale de la Côte d'Ivoire [National University of the Ivory Coast, Abidjan]

UNCIAWPRC Uruguayan National Committee of the International Association on Water Pollution Research and Control [also UNC-IAWPRC]

UNCIO United Nations Conference on International Organization

UNCIP United Nations Commission for India and Pakistan

UNCITRAL United Nations Commission on International Trade Laws

UNCJIN United Nations Crime and Justice Information Network

UN Clim Change Bull The United Nations Climate Change Bulletin

UNCLOS United Nations Conference(s) on the Law of the Sea

UNCLOS United Nations Convention on the Law of the Sea [also UNLOSC]

UNCM University of Nebraska College of Medicine [Lincoln, US]

UNCM User Network Control Machine

UNCMAC United Nations Command Military Armistice Commission

UNCOD United Nations Conference on Desertification

UNCOK United Nations Commission on Korea

UNCOL Universal Computer-Oriented Language

UNCOPUOS United Nations Committee on the Peaceful Uses of Outer Space

UNCPBA Universidad Nacional de Centro de la Provincia de Buenos Aires [Central National University of the Province of Buenos Aires, Argentina]

UNCPICPUNE United Nations Conference for the Promotion of International Cooperation in the Peaceful Uses of Nuclear Energy

UNCPUOS United Nations Committee on the Peaceful Uses of Outer Space

UNCPUS United Nations Committee on the Peaceful Uses of the Seabed

UNCR University of North Carolina at Raleigh [US]

UNCRD United Nations Centre for Regional Development

UNCS Unified Network Command Structure

UNCSC United National Continental Shelf Convention

UNCSTD United Nations Center for Science and Technology for Development [US]

UNCTAD United Nations Conference on Trade and Development

UNCTAD/GATT United Nations Conference on Trade and Development/General Agreement on Tariffs and Trade

UNCTC United Nations Center for Transnational Corporations

UNCW University of North Carolina at Wilmington [US]

unctld uncontrolled

Und Underline; Underlining [also und]

UND Unit Derating

UND University of New Delhi [India]

UND University of North Dakota [Grand Forks, US]

UND University of Notre Dame [Indiana, US]

UNDA University of Notre Dame, Australia

UNDCC United Nations Development Cooperation Cycle

UNDCP United Nations (International) Drug Control Programme

UNDD United Nations Development Decade

UNDD United Nations Demographic Division

UNDE Union of National Defense Employees

UN Devel Update United Nations Development Update [Journal]

UNDEX United Nations Documents Index

Undgrd Underground [also undgrd]

UNDH Unit Derating Hours

UNDIS United Nations Documentation Information System

UNDK Undock [also Undk]

UNDLN Underline [also Undln]

Und-oder-Nor Steuertech Und-oder-Nor und Steuerungstechnik [German Publication on And-or-Nor Logic and Control Engineering]

UNDOF United Nations Disengagement Observer (or Observation) Force

UNDP United Nations Development Program

UNDP Universitaires Notre-Dame de la Paix [Universities of Notre-Dame de la Paix, Namur, Belgium]

UNDP-SEED United Nations Development Program–Sustainable Energy and Environment Division

UNDRLN Underline; Underlining [also Undrln]

UNDRO United Nations Disaster Relief Organization

UNDTCD United Nations Department of Technical Cooperation for Development

UNDV Under Voltage

Undw Underwater [also undw]

UNE University of New England [Armidale, New South Wales, Australia]

Une Unnilennium [Element 109]

UNEC United Nations Economic Commission

UNECA United Nations Economic Commission for Africa [also UN-ECA]

UNECE United Nations Economic Commission for Europe [also UN-ECE]

UNECI Union National des Etudiants de la Côte d'Ivoire [National Union of Ivory Coast Students]

UNECLA United Nations Economic Commission for Latin America [also UN-ECLA, Chile]

UNECLAC United Nations Economic Commission for Latin America and the Caribbean [also UN-ECLAC, or UN/ECLAC]

UNEDA United Nations Economic Development Administration

UNEF Unified National Extra Fine (Screw Thread) [Symbol]

UNEF Union Nationale des Etudiants Françaises [National Union of French Students]

UNEF United Nations Emergency Force

UNEGA Union Européenne des Fondeurs et Fabricants de Corps Gras Animaux [European Union of Animal Fat Producers, France]

UNEM Union Nationale des Etdudiants Marocains [National Union of Moroccan Students]

UNEP United Nations Environment Programme

UNEP-C2E2 United Nations Environment Program–Collaborating Center on Energy and Environment

UNEPCOM Commission for the United Nations Environment Program [Russia]

UNEP-IETC United Nations Environment Program–International Environmental Technology Centre [Japan]

UNEPIE United Nations Environment Program, Industry and Environment

UNEPNET-LAC United Nations Environment Program Network for Latin America and the Caribbean

UNEP/UNESCO/ICRO United Nations Environmental Program/United Nations Educational, Scientific and Cultural Organization International Cell Research Organization (Panel on Microbiology) [Sweden]

UNES Universidad Nacional del Salvador [National University of El Salvador, San Salvador]

UNESCAP United Nations Economic and Social Commission for Asia and the Pacific [also UN-ESCAP]

UNESCO United Nations Educational, Scientific and Cultural Organization [also Unesco]

UNESCO-ROSTA United Nations Educational, Scientific and Cultural Organization/Regional Office for Science and Technology in Africa [Kenya]

UNESCO Stat Yrb UNESCO Statistical Yearbook

Unesid Union de Empresas Siderurgicas [Union of the Iron and Steel Industry, Spain]

UNESOB United Nations Economic and Social Office in Beirut

UNESP Universidade Nacional de Engenharia de São Paulo [Sao Paulo University of Engineering, Araraquara, Brazil]

UNET Universidad Nacional Experimental de Táchira [National Experimental University of Tachira, San Cristóbal, Venezuela]

UNETAS United Nations Emergency Technical Aid Service

UNETPSA United Nations Educational and Training Program for Southern Africa

Unexc unexcavated

Unexpl unexplained

UNF Unified National Fine (Screw Thread) [Symbol]

unf unfinished

Unf uniform(ly)

UNFAO United Nations Food and Agricultural Organization

UNFB United Nations Film Board

UNFDAC United Nations Fund for Drug Abuse and Control

UNFICYP United Nations Force in Cyprus

UNFN United Nations Fund for Namibia

UNFPA United Nations Fund for Population Activities

UNFSTD United Nations Fund for Science and Technology Development

Ung *(unguentum)* – ointment [Medical Prescriptions]

UNGA United Nations General Assembly

UNH University of New Hampshire [Durham, US]

UNH Uranium Nitrate Hexahydrate (or Uranyl Nitrate Hexahydrate)

Unh Unnilhexium [Element 106]

UNHCR United Nations High Commissioner for Refugees

UNHHSF United Nations Habitat and Human Settlements Foundation

UNI Ente Nazionale Italiano di Unificazione [Italian National Standards Association]

UNI UNC Nuclear Industries, Inc. [US]

UNI Universidad Nacional de Ingeniera [National University of Engineering, Lima, Peru]

UNI University of Northern Iowa [Cedar Falls, US]

UNI User-Network Interface

UNIBUS Universal Bus [also Unibus]

UNIC United Nations Information Center

UNICA Association of Caribbean Universities and Research Institutes [Puerto Rico]

UNICAMP Universidade Estadual Campinas [Campinas State University, Brazil] [also Unicamp]

UNICAT Union Catalog(ue)

UNICAT/TELECAT Union Catalog(ue)/Telecommunications Catalog(ue)

UNICC United Nations International Computing Center

UNICCAP Universal Cable Circuit Analysis Program

UNICEF United Nations International Children's Emergency Fund [Name changed to United Nations Children's Fund]

UNICHAL Union Internationale des Distributeurs de Chaleur [International Union of Heat Distributors, Switzerland]

UNICIS University of Calgary Information System [Canada]

UNICOM Universal Integrated Communications; Universal Integrated Communication System

UNICOMP Universal Compiler

UNICON Unidensity Coherent Light Recording

UNICOS Universal Compiler FORTRAN Compatible

UNIDIR United Nations Institute for Disarmament Research

UNIDO United Nations Industrial Development Organization

UNIDO Newsl UNIDO Newsletter [of United Nations Industrial Development Organization]

unif uniform

UNIFE Union des Industriels Ferroviaires Européennes [Union of European Railway Industries, France]

UNIFEM United Nations Development Fund for Women

UNIFET Unipolar Field-Effect Transistor [also Unifet, or unifet]

UNIFREDI Universal Flight Range and Endurance Data Indicator

UNIGES Union des Groupes Economiques de Sénégal [Union of Economic Groups of Senegal]

UNII Unlicensed National Information Infrastructure

UNIMARC Universal Machine-Readable Catalog

Union Burma J Sci Technol Union of Burma Journal of Science and Technology [of Union of Applied Research Institute]

UNIPAUL Universidade Estadual Paulista [Paulista State University, Sao Paulo, Brazil]

UNIPEDE Union Internationale des Producteurs et Distributeurs d'Energie Electrique [International Union of Producers and Distributors of Electric Power, France]

UNIPOL Universal Procedure-Oriented Language

UNIQUE Uniform Inquiry Update Element

Unirr unirradiated

UNIS United Nations Information Service

UNIS (Sperry) Univac Industrial System

UNISA University of South Africa [Pretoria]

UNISAP Univac Share Assembly Program

UNISIST United Nations Information System in Science and Technology [of UNESCO]

UNISIST/ICSU United Nations Information System in Science and Technology/International Council of Scientific Unions

UNISOR University Isotope Separator at Oak Ridge [of Oak Ridge Associated Universities, Tennessee, US]

UNISTAR User Network for Information Storage, Transfer, Acquisition and Retrieval

UNISULF Unocal Sulfur Removal (Process) [Chemical Engineering]

UNISYS United Information System

UNIT Instituto Uruguayo de Normas Técnicas [Uruguay Standards Institute]

UNITAC United Nations Information, Training and Analysis Center

UNITAP United Nations Intermunicipal Technical Assistance Program

UNITAR United Nations Institute for Training and Research

UNITEC University Information Technology Corporation [US]

UNITEL Universal Teleservice

Unité PCPM Unité de Physico-Chimie et de Physique des Matériaux [Physical Chemistry and Materials Physics Unit, Université Catholique de Louvain, Louvain-de-Neuve, Belgium]

UNITRAC Universal Trajector Compiler

Univ University [also univ]

univ universal

UNIVAC Universal Automatic Computer [also Univac]

Univac Technol Rev Univac Technology Review [Japan]

Univ Cal Publ Am Archeol Ethnol University of California Publications in American Archeology and Ethnology [US]

Univ Comput University Computing [Published in the UK]

UNIVERSE Universities Expanded Ring and Satellite Experiment [UK]

Univ Iowa Stud education University of Iowa Studies in Education [US Publication]

Univ Mag University Magazine [US]

Univ Leeds Inst Educ Bull University of Leeds Institute of Education Bulletin [UK]

Univ Oxford, Dept Eng Sci Rep University of Oxford, Department of Engineering Science Reports [UK]

UNIVSERV United Nations International Voluntary Service Fund

Univ Tripoli Bull Fac Eng University of Tripoli Bulletin of the Faculty of Engineering [Libya]

Univ Wales Rev University of Wales Reviews [UK]

UNIX Universal Executive (Operating System) [of AT&T Bell Laboratories, US] [also Unix]

Unix Rev Unix Review [AT&T Bell Laboratories, US]

UNJ Unified Thread with a 0.15011p to 0.18042p Controlled Root Radius [Symbol]

UNJC Unified Coarse Thread with a 0.15011p to 0.18042p Controlled Root Radius [Symbol]

UNJF Unified Fine Thread with a 0.15011p to 0.18042p Controlled Root Radius [Symbol]

unkn unknown

UNL United Nations Library

UNL Universidad Nacional del Litoral [Litoral National University, Santa Fe, Argentina]

UNL University of Nebraska at Lincoln [US]

UNL University of Nebraska Library [Lincoln, US]

UNL University of Nevada Library [Reno, US]

unl unlimited

UNLC United Nations Liaison Committee

Unlch Unlatch(ing) [also unlch]

Unld Unload(ing) [also unld]

unlgtd unlighted

Unlk Unlock(ing) [also unlk]

UNLOSC United Nations Law-of-the-Sea Convention [also UNCLOS]

UNLP Universidad Nacional de La Plata [National University of La Plata, Argentina] unltd unlimited

UNLV University of Nevada at Las Vegas [US]

UNM Unified National Miniature (Screw Thread) [Symbol]

UNM University of New Mexico [Albuquerque, US]

UNMA Unified Network Management Architecture

UNMA United Nations Model Assembly [US]

UNMA University of New Mexico at Albuquerque [US]

Unmanned Syst Unmanned Systems [Publication of Association for Unmanned Vehicle Systems, US]

UNMC University of Nebraska Medical Center [Omaha]

UNML University of New Mexico Library [Albuquerque, US]

UNMOGIP United Nations Military Observer Group in India and Pakistan

UNM/SNL University of New Mexico/Sandia National Laboratories [Albuquerque, US]

UNN Universidad Nacional de Nicaragua [National University of Nicaragua, Léon]

UN/NA United Nations/North America (Number) [On Cylinder Labels of USDOT]

UNNSAD Unit Neutral Normalized Spectral Analytical Density

UN Number United Nations Number [On Cylinder Labels]

UNO United Nations Organization

UNO University of Nebraska at Omaha [US]

UNO University of New Orleans [Louisiana, US]

Uno Unniloctium [Element 108]

UNODS United Nations Optical Disk System

UNOLS University National Oceanographic Laboratory System

UNOOSA United Nations Office of Outer Space Affairs

UNOS United Network for Organ Sharing [US]

UNOSAD United Nations Outer Space Affairs Division

UNOQ Unit of Quantity

UNP Uluru National Park [Ayers Rock/Mount Olga, Northern Territory, Australia]

UNP University of Nebraska Press [Lincoln, US]

Unp Unnilpentium [Element 105]

UNPA United Nations Postal Administration

UNPAAERD United Nations Program of Action of African Economic Recovery and Development

Unpkd unpacked

UNPOC United Nations Peace Observation Commission

UNPS Universal Power Supply

Unpub unpublished

Unq Unnilquadium [Element 104]

UNR Unified Constant Pitch Thread with a 0.108p to 0.144p Controlled Root Radius [Symbol]

UNR Universidad Nacional de Rosario [National University of Rosario, Argentina]

UNR University of Nevada at Reno [US]

UNRC Unified National Coarse Screw Thread with a 0.108p to 0.144p Controlled Root Radius [Symbol]

unrel unreliable

UNRF Unified National Fine Screw Thread with a 0.108p to 0.144p Controlled Root Radius [Symbol]

UNRFNRE United Nations Revolving Fund for Natural Resources Exploration

UNRIPS United Nations Regional Institute for Population Studies [Ghana]

UNRISD United Nations Research Institute for Social Development

UNRR Unidirectional Non-Reversing Relay

UNRRA United Nations Relief and Rehabilitation Administration

Unrstd unrestricted

UNRWA United Nations Relief and Works Agency

UNS Unified National Special (Screw Thread) [Symbol]

UNS United Nations System

UNS Universidad Nacional del Sur [National University of the South, Bahia Blanca, Argentina]

UNS University of Novi Sad [Serbia]

UNS Unified Numbering System

Uns Unnilseptium [Element 107]

uns unsymmetrical(ly)

uns- unsymmetrical (structure of an organic compound) [also unsym]

UNSA Universidad Nacional de San Agustín [San Agustin National University, Arequipa, Peru]

UNSAA Universidad Nacional de San Antonio Abad [San Antonio Abad National University, Cusco, Peru]

unsat unsatisfactory

Unsatd unsaturated [also unsat'd]

unsatur unsaturated

UNSC United Nations Security Council

UNSC United Nations Special Commission

UNSCC United Nations Standards Coordinating Committee

UNSCCUR United Nations Scientific Conference on the Conservation and Utilization of Resources

UNSCEAR United Nations Scientific Committee on the Effects of Atomic Radiation [United Nations Environment Programme]

UNSF United Nations Special Fund

UNSM Universidad Nacional de San Marcos [National University of San Marcos, Lima, Peru]

UNSM University of Nevada School of Medicine [US]

UNSO United Nations Statistical Office

UNSO United Nations Sudano-Sahelian Office

UNSR Unclassified National Security Related

UNSSOD United Nations Special Session on Disarmament

Unst unstable [also unstbl]

unstdy unsteady

UNSW University of New South Wales [Sydney, Australia]

unsym unsymmetrical(ly)

unsym- unsymmetrical (structure of an organic compound) [also uns-]

UNT Universidad Nacional Técnologia [National Technological University, Buenos Aires, Argentina]

UNT Universidad Nacional de Trujillo [Trujillo National University, Peru]

UNT Universidad Nacional do Tucumán [Tucuman National University, San Miguel de Tucumán, Argentina]

UNT University of Newcastle-upon-Tyne [UK]

UNT University of North Texas [Denton, US]

u(ν,T) Energy Density for Blackbody Radiation (ν = Frequency; T = Absolute Temperature) [Symbol]

UNTAA United Nations Technical Assistance Administration

UNTAM United Nations Technical Assistance Mission

UNTE Upper-Nose Temper Embrittlement [Metallurgy]

UNTRA Union of National Television and Radio Organizations of Africa

UNTS Union Nationale des Travailleurs Sénégalaises [National Union of Senegalese Workers]

UNTSFA United Nations Trust Fund for Southern Africa

UNTSO United Nations Truce Supervisory Organization

UNU United Nations University [Japan]

UNUT University of Newcastle-upon-Tyne [UK]

UNV United Nations Volunteers

UNVC United Nations Vienna Conference

UN/VER/CHAIR United Nations Verification: Chairmanship

UNWC University of Northwestern California [US]

UNWMG Utility Nuclear Waste Management Group

UNZ University of New Zealand [Auckland]

UO Unit Operations [Chemical Engineering]

UO Unit Operator [Mathematics]

UO Universidad de Oviedo [University of Oviedo, Spain]

UO Universitetet i Oslo [University of Oslo, Norway]

UO Université d'Orleans [University of Orleans, France]

UO University of Oklahoma [Norman, US]

UO University of Oregon [Eugene, US]

UO University of Osaka [Japan]

UO University of Oslo [Norway]

UO University of Otago [Dunedin, New Zealand]

UO University of Ottawa [Ontario, Canada]

UO University of Oulo [Finland]

UO University of Oxford [UK]

U/O Used On

UO$_x$ Uranium Oxides [e.g., UO_2, UO_3, U_3O_4, etc.] [also Uox]

UO$_2$ Uranyl (Radical) [Symbol]

UOB Union of Burma

UOB Université Omar Bongo [Omar Bongo University, Gabon Republic]

UOC Ultimate Operating Capability

UOC University of Calgary [Alberta, Canada]

UOD Umatilla Ordnance Depot [Oregon, US]

UOD Units of Optical Density

UOF Unplanned Outage Factor

UOF Uranium Oxide Fuel

U of A University of Alberta [Edmonton, Canada]

U of C University of Calgary [Alberta, Canada]

U of I University of Idaho [Moscow, US]

U of M University of Michigan [Ann Arbor, US]

U of M University of Mississippi [Jackson, US]

U of MN University of Minnesota [Minneapolis, US]

U of Nfld University of Newfoundland [St. John's, Canada]

U of O University of Ottawa [Canada]

U of Penn University of Pennsylvania [Philadelphia, US]

U of S University of Saskatchewan [Saskatoon and Regina, Canada]

U of S University of Sydney [New South Wales, Australia]

U of SAf Union of South Africa [also U of SAfr]

U of T University of Toronto [Canada]

U of U University of Utah [Salt Lake City, US]

U of V University of Virginia [Charlottesville, US]

U of W University of Waterloo [Ontario, Canada]

UOG User and Operations Guide

UOH Unplanned Outage Hours

UOHSC University of Oklahoma Health Sciences Center [Oklahoma City, US]

UOHSC University of Oregon Health Sciences University [Portland, US]

UOMHS University of Osteopathic Medicine and Health Sciences [Des Moines, Iowa, US]

UON Unless Otherwise Noted [also uon]

UOP Undistorted Output Power [also uop]

UOP Universal Oil Products Company [US]

UOP University of Osaka Prefecture [Japan]

UOP University of Ottawa Press [Canada]

UOP University of the Pacific [Stockton, California, US]

UOPWA United Office and Professional Workers of America

UOR Unplanned Outage Rate

UOR Unusual Occurrence Report

UOS Underwater Ordnance Station [Newport, Rhode Island, US]

UOS Universal Operator Station

UOSD University of Oklahoma School of Dentistry [US]

UOSAT University of Surrey Satellite [UK]

UOV Unit of Variance

UOW University of Waterloo [Ontario, Canada]

UOW University of Wisconsin [US]

UOx Uranium Oxides [e.g., UO_2, UO_3, U_3O_4, etc.] [also Uo$_x$]

UP Increment/Flag Direction Up [Computers]

UP Ulster Polytechnic [Newtownabbey, Northern Ireland]

UP Ultrapure; Ultrapurification; Ultrapurifier

UP Ultrasonic (Spray) Pyrolysis (Method) [Solid-State Physics]

UP Ultraspherical Polynomials [Mathematics]

UP Union Pacific (Railroad) [US]

UP Uniprocessor

UP United Press [Now Part of United Press International]

UP Unit Process

UP Universal Programmer

UP Università di Padova [University of Padua, Italy]

UP Universidad Pontifica [Pontifical University, Madrid, Spain]

UP Universidade de Porto [University of Porto, Portugal]

UP University of Panama [Panama City]

UP University of Panjab [Lahore, Pakistan]

UP University of Pennsylvania [Philadelphia, US]

UP University of Pittsburgh [Pennsylvania, US]

UP University of Poona [India]

UP University of Portland [Oregon, US]

UP University of the Pacific [Stockton, California, US]

UP University of the Philippines [Quezon City]

UP University Professor

UP Unrotating Projectiles

UP Unsaturated Polyester (Thermoset) [also UPE]

UP Unsolicited Proposal

UP Unstable Particle [Physics]

UP Upper Part

UP Uranium Phosphide

UP Urea Phosphate

UP Uridine Phosphate

UP User-Programmable

UP Utility Path

UP Utility Program

UP I Université de Paris I [University of Paris I, France]

UP II Université de Paris II [University of Paris II, France]

UP III Université de Paris III [University of Paris III, France]

UP IV Université de Paris IV [University of Paris IV, France]

UP V Université de Paris V [University of Paris V, France]

UP VI Université de Paris VI [University of Paris VI, France]

UP VII Université de Paris VII [University of Paris VII, France]

UP IX Université de Paris IX [University of Paris IX, France]

U-5'-P Uridine 5'-Monophosphate

Up Upper [also up]

U$_0(\Phi)$ Fluence Dependent Pinning Energy [Symbol]

UPA Ultrafine-Particle Analyzer

UPA United Printers Association [UK]

UPA University Photographers Association (of America) [US]

UPA University Press of America, Inc. [Lanham, Maryland, US]

UPA Uranium Palladium Aluminide

UPA Uranium Producers of America [US]

UPA Uridylyl(3'-5')adenosine [Biochemistry]

UPA Utah Pharmaceutical Association [US]

UpA Uridylyladenosine [Biochemistry]

UPAA University Photographers Association of America [US]

UPACS Universal Performance Assessment and Control System

UPADI Union Panamericana de Asociaciones de Ingenieros [Pan-American Federation of Engineering Societies, Venezuela]

UPAO University Professors for Academic Order [US]

UpApA Uridylyladenylyladenosine [Biochemistry]

UPAU Uttar Pradesh Agricultural University [India]

UPB University of Pittsburgh at Bradford [US]

UPB Upper Bound [Mathematics]

U-Pb Uranium-Lead (Geochronometry)

UPBS Unsolicited Proposals Brokerage Service [of Public Works and Government Services Canada]

UPC Ultrapure Carrier

UPC Uniform Product Code

UPC United Press Canada

UPC Unit of Processing Capacity

UPC Universal Peripheral Controller

UPC Universal Product Code

UPC Universidad Politécnico de Cataluñya [Polytechnic University of Catalunya, Barcelona, Spain]

UPC Uridylyl(3'-5')cytidine [Biochemistry]

UPC-E Universal Product Code–Europe

UpC Uridylylcytidine [Biochemistry]

UpCase Convert Character to Uppercase [Pascal Function]

UPCS Universal Process Control Software

UPD Underpotential (Metal) Deposition; Underpotentially Deposited (Metal)

UPD Uniaxial Plastic Deformation [Mechanics]

Upd Update [also upd]

UPDATE Unlimited Potential Data through Automation Technology in Education

UPDEA Union of Producers and Distributors of Electric Power in Africa [Ivory Coast]

Updfts Updrafts [also updfts]

UP/DOC Universal Programmer/Documentor

Updt Update [also updt]

Updt Supp Update Supplement

UPE Unsaturated Polyester (Thermoset) [also UP]

UPE University of Port Elizabeth [South Africa]

UPE University Professor Emeritus

UPE Unnatural Parity Exchange

UPEI Union Pétrole Européenne Indépendante [Independent European Petroleum Union, France]

UPEI University of Prince Edward Island [Charlottetown, Canada]

UPenn University of Pennsylvania [Philadelphia, US]

UPES Ultraviolet Photoelectron Spectroscopy

UPES Ultraviolet Photoemission Spectroscopy

UPF Ultraspherical Polynomial Filter

UPFDA United Products Formulators and Distributors Association [US]

UpG Uridylylguanosine [Biochemistry]

Upg Upgrade

UPH Unit(s) per Hour [also uph]

UPHA Utah Public Health Association [US]

UPHS Underground Pumped Hydro Storage

UPI Unione Paraguaya Industrial [Paraguayan Industrial Union]

UPI Union Postale Internationale [International Postal Union]

UPI United Press International [US]

UPI Universal Personal Identifier

UPIICSA Unidad Profesional Interdisciplinaria de Ingenieria de Ciencias Sociales y Administrativas [Interdisciplinary Professional Union of Engineering, Social and Administrative Sciences; of Instituto Politecnico Nacional, Mexico]

UPIR Uttar Prasesh Irrigation Research Institute [India]

UPitt University of Pittsburgh [Pennsylvania, US]

UPIU United Paperworkers International Union

UPJ University of Pittsburgh at Johnstown [Pennsylvania, US]

UPL Universal Programming Language

UPL Up-Conversion Luminescence

UPL User Program(ming) Language

UPLIFTS University of Pittsburgh Linear File Tandem System [US]

UPLK Uplink [also Uplk]

UPLR Uniform Packaging and Labeling Regulation [US]

UPM Universidad Politecnico de Madrid [Madrid Polytechnic University, Spain]

UPM Universal Permissive Module

UPM University of Petroleum and Minerals [Dhahran, Saudi Arabia]

UPM Unix Programmer's Manual

UPM User Profile Management

UPMC Université Pierre et Marie Curie [University of Pierre and Marie Curie, Paris, France]

UPMC Urban Planning Ministers Conference

UPN Unique Project Number

UPO Undistorted Power Output [also upo]

UPO Unified Proposal Outline [US]

UPOS Utility Program Operating System

UPOV Union Internationale pour la Protection des Obtentions Végétales [International Union for the Protection of New Varieties of Plants]

UPP United Papermakers and Paperworkers

UPP Universal PROM (Programmable Read-Only Memory) Programmer

UPP University of Pennsylvania Press [Philadelphia, US]

UPP Unsolicited Proposals Program [Canada]

UPP User Parameter Processing

UPPE Ultraviolet Photometric and Polarimetric Explorer

UPR Ultrasonic Paramagnetic Resonance

UPR Universidad de Puerto Rico [University of Puerto Rico]

UPR Unsaturated Polyester Resin

UPR Uranium Production Reactor

Upr upper

Uprav Syst Mash Upravlyanyschie Systemi i Mashini [Russian Journal on Systems and Machinery]

UPRB Usines des Poudreries Réunies de Belgique [Belgian Gunpowder Manufacturer]

UPRC Universidad de Puerto Rico, Cayey [University of Puerto Rico at Cayey]

UPRH Universidad de Puerto Rico, Humacao [University of Puerto Rico at Humacao]

UPRM Universidad de Puerto Rico, Mayagüez [University of Puerto Rico at Mayaguez]

UPRRP Universidad de Puerto Rico, Río Piedras [University of Puerto Rico at Rio Piedras]

UPRSJ Universidad de Puerto Rico, San Juan [University of Puerto Rico at San Juan]

UPS Ultraviolet Photoelectron Spectrometry; Ultraviolet Photoelectron Spectroscopy; Ultraviolet Photoelectron Spectrum

UPS Ultraviolet Photoemission Spectroscopy

UPS Underwater Photography Society [US]

UPS Uninterrupted Power Supply; Uninterruptible Power Supply; Uninterruptible Power System

UPS Unitary Pricing System [of Organization of Petroleum Exporting Countries]

UPS United Parcel Service [US]

UPS Universal Processing System

UPS Université de Paris-Sud [University of Paris-South, Orsay, France]

UPS Université Paul Sabatier [Paul Sabatier University, Toulouse, France]

UPS University of Puget Sound [Tacoma, Washington, US]

UPS Upper Sideband

UPS Upright Perigee Stage

UPSC United Parcel Service Canada

UPSD University of the Pacific School of Dentistry [San Francisco, California, US]

UPSE Universal Power Service Equipment

UPSI User Program Sense Indicator

Upslp Upslope [also upslp]

UPSM University of Pennsylvania, School of Medicine [Philadelphia, US]

UPSM University of Pittsburgh School of Medicine [Pennsylvania, US]

UPT Unipolar Transistor

UPT Universal Polarization Transfer

UPT Universal Portable Telephone

URZ Unreacted Zone [Metallurgy]

UPT User Process Table

UPTLM Uplink Telemetry

UPTS Undergraduate Pilot Training System

UPU Universal Postal Union [of United Nations]

UpU Uridylyluridine [Biochemistry]

UpUpG Uridylyluridylylguanosine [Biochemistry]

U-Pu-Zr Uranium-Plutonium-Zirconium (Alloy)

UPV Universidad País Vasco [Pais Vasco University, Spain]

UPVC Unplasticized Polyvinyl Chloride

UPW Union of Post Office Workers [UK]

UQ University of Quebec [Quebec City, Canada]

UQ University of Queensland [St. Lucia, Australia]

U-Quark Up Quark [also u-quark] [Particle Physics]

UQAC University of Quebec at Chicoutimi [Canada]

UQAH University of Quebec at Hull [Canada]

UQAM University of Quebec at Montreal [Canada]

UQAR University of Quebec at Rimouski [Canada]

UQATR University of Quebec at Trois-Rivières [Canada]

UQB University of Queensland at Brisbane [Australia]

UQC Université de Québec à Chicoutimi [University of Quebec at Chicoutimi, Canada]

UQH Université de Québec à Hull [University of Quebec at Hull, Canada]

UQM Université de Québec à Montréal [University of Quebec at Montreal, Canada]

UQP University of Queensland Press [St. Lucia, Australia] [also UQ Press]

UQR Université de Québec à Rimouski [University of Quebec at Rimouski, Canada]

UQSL University of Queensland at St. Lucia [Australia]

UQT User Queue Table

UQTR Université de Québec à Trois-Rivières [University of Quebec at Trois-Rivières, Canada]

u Quark up quark [also U quark] [Particle Physics]

UR Unattended Repeater

UR Under Running (Crane)

UR Uniform Resonance

UR Unit Record

UR Unit Register

UR Universidad de la República [University of the Republic, Montevideo, Uruguay]

UR Universidad Regiomontana [Mining and Materials University, Monterrey, Mexico]

UR Universal Recipient (of Type AB Blood Group) [Immunology]

UR Università di Roma [University of Rome, Italy]

UR Université de Reims [University of Reims, France]

UR Université de Rennes [University of Rennes, France]

UR Université de Rouen [University of Rouen, France]

UR University of Rajshahi [Bangladesh]

UR University of Rangoon [Burma]

UR University of Reading [UK]

UR University of Regensburg [Germany]

UR University of Regina [Canada]

UR University of Richmond [Virginia, US]

UR University of Rochester [New York State, US]

UR University of Roorkee [India]

UR University of Ryukyus [Okinawa, Japan]

UR University Research

UR Unsatisfactory Report

UR Unsulfonated Residue

UR Upper Right

UR Ural River [Russia/Kazakhstan]

U/R Under-Range

U/R Up Range

U/R Uranium/Radium (Ratio)

Ur Urea

U(r) (London) interaction potential (in solid-state physics) [Symbol]

URA Ultrared Absorption

URA Ultrasonic Resin Analyzer

URA Universities Research Association [Washington, DC, US]

URA Urine Receptacle Assembly [Medicine]

URA User Requirements Analysis

U/Ra Uranium-Radium (Ratio)

URAEP University of Rochester Atomic Energy Project [US]

URAG Uranium Resources Appraisal Group [of Natural Resources Canada, Canada]

Urban Aff Q Urban Affairs Quarterly [International Journal]

Urban Des Int Urban Design International [US Journal]

URBANICOM Association Internationale Urbanisme et Commerce [International Association for Town Planning and Distribution, Belgium]

URBEMISS Urban Emissions Model [of California Air Resources Board, US]

URBM Ultimate Range Ballistic Missile

URC Uniform Resource Characteristics

URC Uniform Resource Citation

URC Unit Record Control

URC Universal Resource Characteristic [Internet]

URC User Resource Center [of National Institutes of Health, Bethesda, Maryland, US]

URC Utilities Research Commission

URD Underground Residential Distribution (Cable)

URD (Department of) Urban and Regional Development [Australia]

URD User Requirements Document

Urd Uridine [Biochemistry]

Urdox Uranium Dioxide [also urdox]

URDU Urban Regional Development Unit [of Australian Council of Social Services]

URE Unintentional Radiation Exploitation

URE Unit Record Equipment

UREP Unix RSCS (Remote Spooling Communications System) Emulation Protocol (Protocol)

Urethane Plast Prod Urethane Plastics and Products [Published in the US]

Urethanes Technol Urethanes Technology [Published in the UK]

URF Unit Risk Factor

URF University Research Fellow

URG Universal Radio Group

URG University Research Grant

URGC University Research Grant Committee

URHN Université de Rouen–Haute Normandie [University of Rouen–High Normandy, France]

URI Underground Research, Inc. [US]

URI Uniform Resource Identifier

URI University of Rhode Island [Providence, US]

URI University Research Initiative (Program) [of Defense Advanced Research Projects Agency, US]

URI University Research Instrumentation (Program) [of US Department of Energy]

URI Upper Respiratory Infection [Medicine]

URIF University Research Incentive Fund [Canada]

URI-GSO University of Rhode Island–Graduate School of Oceanography [Providence, US]

URIP University Research Initiative Program [of University of California at Santa Barbara, US]

URIPS Undersea Radioisotope Power Supply

URIR Unified Radioactive Isodromic Regulator

URISA Urban and Regional Information Systems Association [US]

URL Underground Research Laboratory [of Whiteshell Nuclear Research Establishment, Pinawa, Manitoba, Canada]

URL Uniform Resource Locator [Internet]

URL User Requirements Language

URLL University of Rochester Laser Laboratory [New York State, US]

URMUR User Requested Minor User Registration (Pilot Project) [Canada]

URN Uniform Resource Name [Internet]

URN Uniform Resource Number [Internet]

UrN Urea Nitrate

Urn Urine

Urol Urologist; Urology

Urol Int Urologia Internationalis [International Journal]

Urol Res Urological Research (Journal)

UROP Undergraduate Research Opportunities

URP Unit Record Processor

UrP Urea Picrate

URPA University of Rochester, Department of Physics and Astronomy [US]

URPIS Urban and Regional Planning Information Systems [now Australasian Urban and Regional Information Systems Association, Inc.]

URR Ultrared Reflection

URR Unidirectional Reversing Relay

URRI Urban Regional Research Institute [of Michigan State University, East Lansing, US]

URS Ultra-Rapid Solidification

URS Unate Ringe Sum

URS Uniform Reporting System

URS United Research Service

URS Unit Record System

URS Universal Reference System

URS Universal Regulating System

URS Unmanned Repeater Station

URS Update Report System

URS Uranium Ruthenium Silicide

URSG Uprava Republike Slovenije za Geofiziko [Geological Survey of Solvenia]

URSI Union Radio-Scientifique Internationale [International Scientific Radio Union, Belgium]

URSMD University of Rochester School of Medicine and Dentistry [New York State, US]

URST Unrecrystallized Solution Treatment [Metallurgy]

URT United Republic of Tansania

URT Upper Respiratory Tract [Anatomy]

URT Uranium Research Technology

URT Urotropin [Organic Chemistry]

URTNA Union of (National) Radio and Television Organizations of Africa

URTU United Road Transport Union [UK]

Uru Uruguay(an)

URV Undersea Rescue Vehicle

URV Undersea Research Vehicle

URV Upper Range Value

URW United Rubber Workers

US Ultimate Strength

US Ultrasonic(s)

US Ultrasonic Society

US Ultrasound [also U/S]

US Ultrastructure [Molecular Biology]

US Unassemble Source (Command) [Pascal Programming]

US Underground Storage

US Under-Secretary

US Underside

US Undersize(d)

US Understressed; Understressing

US Unified Statistical (Theory) [Physics]

US United Services [Armed Forces]

US United States (of America)

US United States (of America) [ISO Code]

US Unit Separator (Character) [Data Communications]

US Unit Switch

US Unit System [Physics]

US Universal System (of Lens Apertures)

US Universidad de Santiago [University of Santiago, Chile]

US Universidad de Sarragossa [University of Sarragossa, Spain]

US Universidad de Sonora [University of Sonora, Hermosillo, Mexico]

US Università di Siena [University of Siena, Italy]

US Université de Sherbrooke [University of Sherbrooke, Quebec, Canada]

US University of Salford [UK]

US University of Sarajevo [Bosnia]

US University of Saskatchewan [Regina and Saskatoon, Canada]

US Universidad de Sevilla [University of Sevilla, Spain]

US University of Sheffield [UK]

US University of Sofia [Bulgaria]

US University of Southampton [UK]

US University of Stockholm [Sweden]

US University of Strathclyde [UK]

US University of Stuttgart [Germany]

US University of Sudbury [Ontario, Canada]

US University of Surrey [Guildford, Surrey, UK]

US University of Sussex [UK]

US University of Sydney [New South Wales, Australia]

US University of Syracuse [New York State, US]

US Unmanned System

US Unstabilized

US Upper Stage (of a Rocket)

US Upstream

US Upwelling Sonar

US Uranium(I) Sulfide (or Uranium Monosulfide)

US Uranium Series [Nuclear Physics]

US Utah State [US]

US I Université de Strasbourg I [University of Strasbourg I, France]

US II Université de Strasbourg II [University of Strasbourg II, France]

US III Université de Strasbourg III [University of Strasbourg III, France]

U/S Ultrasound [also US]

U/S Unserviceable

U(s) Disturbance Input [Automatic Control Symbol]

us *(ubi supra)* – where mentioned above

Us *(ut supra)* – as above

.us United States [Country Code/Domain Name]

USA Ultra-Small Angle

USA Undercar Specialists Association [of National Exhaust Distributors Association, US]

USA Union of South Africa [now Republic of South Africa]

USA United Space Alliance

USA United States of America

USA United States Army

USA Universidad de San Andrés [University of San Andres, Bolivia]

USA University of South Africa [Pretoria]

USA University of South Australia [The Levels]

USA University of St. Andrews [UK]

USAA United Service Automobile Association [US]

USAA United States Armor Association [US]

USAA United States Army, Atlantic [Virginia]

USAAC United States Army Air Corps

USAADC United States Army Air Defense Command [Ent Air Force Base, Colorado, US]

USAAML United States Army Aviation Materiel Laboratories

USAASO United States Army Aeronautical Services Office

USAAVLABS United States Army Aviation Materiel Laboratories

USAAVNS United States Army Aviation School

USAAVNTA United States Army Aviation Test Activity

USAB United States Activities Board [of Institute of Electrical and Electronics Engineers, US]

USABAAR United States Army Board for Aviation Accident Research

USABC United States Advanced Battery Consortium

USABESRL United States Army Behavioral Science Research Laboratory

USABL United States Army Biological Laboratories [Fort Detrick, Frederick, Maryland, US]

USABRL United States Army Ballistics Research Laboratory

USAC United States Activities Committee [now United States Activities Board]

USAC United States Automobile Club

USACA United States Advanced Ceramics Association

USACA Rep USACA Report [of United States Advanced Ceramics Association]

USACDA United States Arms Control and Disarmament Agency

USACDC United States Army Combat Developments Command [Fort Belvoir, Virginia, US]

USACDCAVNA United States Army Combat Developments Command Aviation Agency

USACDCCBRA United States Army Combat Developments Command Chemical-Biological-Radiological Agency

USACE United States Army Corps of Engineers [also USACOE]

USA-CRREL United States Army Cold Regions Research and Engineering Laboratory [of Hanover, New Hampshire]

USACSC United States Army Computer Systems Command

USACSSC United States Army Computer Systems Support (and Evaluation) Command

USADATCOM United States Army Data Support Command [also USADC]

USADSC United States Army Data Services Command

USAE United States Army, Europe [Germany]

USAEC United States Atomic Energy Commission [Abolished]

USAEC United States Army Environmental Center

USAEC Rep United States Atomic Energy Commission Report

USAECOM United States Army Electronics Command

USAEPG United States Army Electronic Proving Ground

USAERDA United States Army Electronic Research and Development Agency

USAER&DC United States Army Engineering Research and Development Center [Vicksburg, Mississippi] [also USAERDC]

USAERDL United States Army Engineering Research and Development Laboratories

USAERDL United States Army Electronics Research and Development Laboratory

USAEWES United States Army Engineer Waterways Experiment Station [Vicksburg, Mississippi]

USAF Undersecretary of the Air Force [US]

USAF United States Air Force

USAF United States Army Force

USAF United States of America Foundation of Research and Education

USAFECI United States Air Force Extension Course Institute

USAFA United States Air Force Academy [at Colorado Springs, Colorado]

USAFB United States Air Force Base

USAFCRL United States Air Force Cambridge Research Laboratories

USAFE United States Air Force in Europe [Germany]

USAFETAC United States Air Force Environmental Technical Applications Center

USAFI United States Armed Forces Institute [of US Department of Defense]

USAFIC United States Association of Firearm Instructors and Coaches (International)

USAFIT United States Air Force Institute of Technology

USAFMC United States Air Force Materiel Command

USAFO United States Air Force Office [Washington, DC]

USAFSC United States Air Force Systems Command

USAFSO United States Air Force Southern Command [at Albrook Air Force Base, Panama Canal Zone] [also USAFSC]

USAFSS United States Air Force Security Service

USA Funds United Students Aid Fund [also USA Funds, US]

USAHQ United States Army Headquarters [Pentagon, Washington, DC]

USAICC United States Army Intelligence Corps Command [at Fort Holabird, Maryland, US]

USAID United States Agency for International Development [of International Development Cooperation Agency, US]

USAIDSC United States Army Information and Data Systems Command

USAID/WEC United States Agency for International Development/World Environment Center

USAir United States Airlines

US Air Force Wright Aeronaut Lab US Air Force Wright Aeronautical Laboratories [Publication of US Air Force]

USALMC United States Army Logistics Management Center

USAM Ultrasonically Assisted Machining (Process)

USAM Unified Space Applications Mission [of NASA]

USAM Unique Sequential Access Method

USAM User Spool Access Method

USAMC United States Aerospace Medical Center [at Brooks Air Force Base, San Antonio, Texas]

USAMC United States Army Materiel Command [Washington, DC]

USAMC United States Army Missile Command [at Redstone Arsenal in Huntsville, Alabama]

USAMMCS United States Army Missile and Munitions Center and School [at Redstone Arsenal, Huntsville, Alabama]

USAMP United States Automotive Materials Partnership [of United States Council for Automotive Research]

USAMRA United States Army Materials Research Agency [US]

USAMTL United States Army Materials Technology Laboratory [Watertown, Massachusetts]

USA/MULTI-VER United States of America Perspective on Multilateral Verification of Arms Control Agreements

USAN United States Adopted Name [Pharmacology]

USANC United States Army Nurse Corps [Brooke Army Hospital, Fort Sam Houston, Texas]

USANDL United States Army Nuclear Defense Laboratory

USANS UHR (Ultrahigh Resolution) Small-Angle Neutron Scattering

USANWSG United States Army Nuclear Weapon Surety Group

USAO University of Science and Arts of Oklahoma [Chickasha, US]

USAOMC United States Army Ordnance Missile Command [Redstone Arsenal, Alabama, US]

USAP United States Antarctic Program

USAP United States Army, Pacific [Fort Shafter, Hawaii, US]

USAPC United States Army Petroleum Center

USAPHS United States Army Primary Helicopter School

USAPO United States Antarctic Projects Office

USARAL United States Army, Alaska [Fort Richardson]

USARATL United States Army, Atlantic [Virginia]

USARCAR United States Army, Caribbean [Panama]

USAREPG United States Army Electronic Proving Ground

USAREUR United States Army, Europe [Germany]

USARIEM United States Army Research Institute of Environmental Medicine [Natick, Massachusetts]

USARL United States Army Research Laboratory [Fort Monmouth, New Jersey, US]

USARO United States Army Research Office [of Department of Defense in Research Triangle Park, North Carolina, US] [also US ARO]

USARP United States Antarctic Research Program [of National Science Foundation International]

USARP United States Atlantic Research Program

USARPA United States Army Radio Propagation Agency

USARPAC United States Army, Pacific [Hawaii]

USARSOUTHCOM United States Army Southern Command [Panama] [Disbanded]

USART Universal Synchronous/Asynchronous Receiver/Transmitter [also USAR/T]

USARV United States Army, Vietnam

USAS United States of America Standards

USASA United States Army Security Agency [Arlington, Virginia, US]

USASCAF United States Army Service Center for Armed Forces

USASCC United States Army Strategic Communications Command [Washington, DC, US]

USASCII United States of America Standard Code for Information Interchange [also US ASCII, or ASCII]

USASCSOCR United States of America Standard Character Set for Optical Character Recognition

USASDC United States Army Strategic Defense Command

USASF United States Army Special Forces

USASI United States of America Standards Institute [now American National Standards Institute]

USASMSA United States Army Signal Missile Support Agency

USATACOM United States Army Tank-Automotive Command

USATC United States Army Topographic Command

USATEA United States Army Transportation Engineering Agency

USATECOM United States Army Test and Evaluation Command

USATHAMA United States Army Toxic and Hazardous Material Agency [now Army Environmental Center]

USATIA United States Army Transportation Intelligence Agency

USAXS Ultra-Small-Angle X-Ray Scattering

USB Unified S-Band [also U-SB]

USB Universal Serial Bus

USB Upper Sideband

USB Upper Surface Blowing

U-SB Unified S-Band [also USB]

USb Uranium Antimonide

USBAP University Small Business Assistance Program

USBC United States Bureau of Census

USBE Unified S-Band Equipment

USBE Universal Serials and Book Exchange [US]

USBG United States Botanical Garden [Washington, DC, US]

USBGN United States Board on Geographic Names

USBI United Space Booster, Inc. [US]

USBIA United States Bureau of Indian Affairs [of US Department of the Interior]

USBLS United States Bureau of Labor Statistics

USBM-P1S United States Bureau of Mines High-Purity Sulfur (99.999%) (Standard)

USBoM United States Bureau of Mines [of US Department of the Interior, Washington, DC] [also USBoM, USBM, BoM, or BOM]

USBR United States Bureau of Reclamation

USBS Unified S-Band Subsystem; Unified S-Band System

USBS United States Bureau of Standards [Now Part of NIST, US]

USBSM United States Bureau of Standards and Measures [Now Part of National Institute of Standards and Technology, Gaithersburg, Maryland, US]

USBT Upper Surface Blowing Technique

USBTC University-Small Business Technology Consortium [US]

US Bur Mines Bull United States Bureau of Mines Bulletin

US Bur Mines Inf Circ United States Bureau of Mines Information Circular

US Bur Mines Rep Invest United States Bureau of Mines, Report Investigation

USBYD United States Bureau of Yards and Docks

USC Ultra-Selective Conversion (Process)

USC Ultrasonic Cleaning; Ultrasonically Cleaned

USC Ultrasupercritical(ity)

USC United States Code

USC United States Congress

USC Universal Service Circuit

USC Universidad de San Carlos [San Carlos University, Guatemala City, Guatemala]

USC Universidad de Santiago de Chile [University of Santiago de Chile]

USC Universidad de Santiago de Compostela [University of Santiago de Compostela, Spain]

USC University of San Carlos [Philippines]

USC University of Santa Clara [California, US]

USC University of Santa Cruz [California, US]

USC University of South Carolina [Columbia, US]

USC University of Southern California [Los Angeles, US]

USC Urban Studies Center [of University of Louisville, Kentucky, US]

USC User Support Component

USC Utah Safety Council [US]

USCA United States Contract Awards (Database)

USCA Universal Signal Conditioning Amplifier

USCAC United States Continental Army Command [Fort Monroe, Virginia, US]

USCAL University of Southern California Aeronautical Laboratory [Los Angeles, US]

USCAR United States Council for Automotive Research

USCB United States Census Bureau

USCB United States Customs Bureau

USCEA United States Council for Energy Awareness

USCEC University of Southern California Engineering Center [Los Angeles, US]

USCEE University of Southern California, Department of Electrical Engineering [Los Angeles, US]

USCEF United States–China Education Fund

USCF United States Churchill Fund

USCG United States Coast Guard

USCGA United States Coast Guard Academy [New London, Connecticut]

USC&GS United States Coast and Geodetic Survey [of US Department of Commerce] [also UCCGS]

USCI United Satellite Communications, Inc. [US]

USCIB United States Council for International Business [also USC IB]

USCID United States Committee on Irrigation and Drainage [US]

USCIDF United States Committee on Irrigation, Drainage and Flood Control [now USCID]

USCIGW Union of Salt, Chemical and Industrial General Workers [UK]

USCLASS United States Classifications (Database)

USCMC University of Southern California Medical Center [Los Angeles, US]

USCMI United States Commission on Mathematical Instruction

US CMS United States Compact Muon Solenoid (Collaboration Experiment) [at Fermilab, Batavia, Illinois, US]

USCOE United States Corps of Engineers

USCOLD United States Committee on Large Dams

USCP University of South Carolina Press [US]

US-CRS United States Working Group on Computer Reservation Systems [of European Civil Aviation Conference]

USCS United States Commercial Standard

USCS United States Customs Service

USCSC United States Civil Service Commission [Abolished]

USCSCV United States Committee for Scientific Cooperation with Vietnam

USCTI United States Cutting Tool Institute

USCV Union Scientifique Continentale du Verre [Continental Scientific Glass Union, Charleroi, Belgium]

USCWT United States Hundredweight [also US cwt]

USD Ultimate Strength Design

USD Ultrasonic Detector

USD (Office of the) Under-Secretary of Defense [US]

USD Uniform Symbol Description

USD United States Dollars [Currency of Guam, Panama Canal Zone, Puerto Rico, USA and Western Samoa] [also $US, or US$]

USD Universidad de Santo Domingo [University of Santo Domingo, Dominican Republic]

USD University of San Diego [California, US]

USD University of South Dakota [Vermillion, US]

USD Upstream Detector

USD Uranium Series Disequilibria [Nuclear Physics]

US$ United States Dollar [Currency of Guam, Panama Canal Zone, Puerto Rico, USA and Western Samoa] [also USD, or $US]

USDA United States Department of Agriculture

USDA United States Department of the Army

USDA-ARS United States Department of Agriculture–Agricultural Research Service

USDA/CRIS United States Department of Agriculture/ Current Research Information System

USDA Forest Serv Agr Hdbk USDA Forest Service, Agricultural Handbooks [of United States Department of Agriculture]

USDA Forest Serv Gen Tech Rep USDA Forest Service, General Technical Reports [of United States Department of Agriculture]

USDA List Publ USDA List of Publications [of United States Department of Agriculture]

USDA-NRCS United States Department of Agriculture–Natural Resource Conservation Service [Lincoln, Nebraska]

USDA/NSF United States Department of Agriculture/ National Science Foundation (Program) [US]

USDC United States Department of Commerce

USDC United States Display Consortium [Government-Industry Partnership]

USDDM United States Department of Data Management

USDDMS United States Department of Defense Manned Spaceflight

USDEA United States Department of External Affairs

US Dept Agric For Serv United States Department of Agriculture Forest Service

USDGA United States Durum (Wheat) Growers Association [US]

USDHEW United States Department of Health, Education and Welfare [now USDHHS]

USDHHS United States Department of Health and Human Services

USDI United States Department of the Interior

USDL United States Department of Labor

US$/gal United States Dollar(s) per Gallon [Unit]

US$/kW United States Dollar(s) per Kilowatt (of Electricity) [Unit]

USDMV United States Department of Motor Vehicles

USDN United States Department of the Navy

USDNA United States Defense Nuclear Agency

USDNE United States Department of Nuclear Engineering

USDOA United Stated Department of Agriculture [also USDoA, or US-DOA]

USDOA/CRIS United States Department of Agriculture/ Current Research Information System

USDOA-FS United States Department of Agriculture–Forest Service

USDOC United States Department of Commerce [also USDoC, or US-DOC]

USDOD United States Department of Defense [also USDoD, or US-DOD]

USDOE United States Department of Energy [also USDoE, or US-DOE]

USDOE-BES USDOE Basic Energy Sciences (Division)

USDOE/PETC USDOE–Pittsburgh Energy Technology Center [Pennsylvania]

USDOE Rep United States Department of Energy Report

USDOF United States Department of Fisheries [also USDoF, or US-DOF]

USDOI United States Department of the Interior [also USDoI, or US-DOI]

USDOJ United States Department of Justice [also USDoJ, or US-DOJ]

USDOL United States Department of Labor [also UsDoL, or US-DOL]

USDOM United States Department of Mines [also USDoM, or US-DOM]

USDON United States Department of the Navy [also USDoN, or US-DON]

USDOS United States Department of State [USDoS, or US-DOS]

USDOT United States Department of Transportation [also USDoT, or US-DOT]

USDOT/UMTA United States Department of Transportation/ Urban Mass Transportation Administration [at Washington, DC]

US-DPC United States–Demonstration Poloidal Coil

USDR&E Undersecretary of Defense for Research and Engineering [US]

USDT United States Department of the Treasury

US$/t United States Dollars per ton [Unit]

US$/Wp United States Dollar(s) per peak watt [Unit]

USE Underground Service Entrance

USE Unified S-Band Equipment

USE Union des Syndicats d'Electricité [Union of Electricity Associations]

USE United States Engineers

USE Unit Support Equipment

USE Univac Scientific Exchange [US]

USE Université de Saint-Etienne [University of Saint-Etienne, France]

USE Upper-Shelf Energy (in Charpy Impact Test) [Mechanical Testing]

USe Uranium(I) Selenide (or Uranium Monoselenide)

USEA United States Energy Association

USEA Q USEA Quarterly [Publication of United States Energy Association]

USEC United System of Electronic Computers

USEF United States Expeditionary Forces

USEIA United States Energy Information Administration [of US Department of Energy]

USEMA United States Electronic Mail Association

USENET User's Network [Internet]

Usenix Unix Users Association [US]

USEPA United States Environmental Protection Agency [also US-EPA, or US EPA]

USEPA, Publ Bibl United States Environmental Protection Agency, Publications Bibiliography

USER User System Evaluator

USERC United States Environment and Resources Council

USERDA United States Energy Research and Development Administration [formerly United States Atomic Energy Commission; Now Part of US Department of Energy] [also US ERDA]

US ERDA Tech Inf Center United States Energy Research and Development Administration Technical Information Center [Oak Ridge, Tennessee]

USERID User Identification

USERP United Scientists for Environmental Responsibility and Protection

USES United States Electronic Service

USES United States Employment Service [of Employment and Training Administration, US]

USETI United States Environmental Training Institute

USF United States Form Thread

USF United States Frigate

USF Università degli Studi Ferrara [Technical University of Ferrara, Italy]

USF University of San Francisco [California, US]

USF University of South Florida [Tampa, US]

USF Unsaturated Flow

USF Unsteady Flow

USFA United States Fire Administration [of US Department of Commerce] [Now Part of Federal Emergency Management Agency]

USFAA United States Field Artillery Association [now Association of the United States Army]

USFCC United States Federation for Culture Collections

USFCF United States Frigate Constellation Foundation

USFDA United States Food and Drug Administration [of US Department of Health and Human Services]

USFED United States Federal Specifications Board [also USFed]

USFGC United States Feed Grains Council

USFM Ultrasonic Flowmeter

USFMC University of South Florida Medical Center [Tampa, US]

USFPL United States Forest Products Laboratory [Madison, Wisconsin]

USFS United States Foreign Service [of US Department of State]

USFS United States Forest(ry) Service [of US Department of Agriculture]

USFS United States Frequency Standard

USFSS United States Federation of Scientists and Scholars

USFSS United States Fleet Sonar School

USFWS United States Fish and Wildlife Service [of US Department of the Interior, Washington]

USFX Universidad do San Francisco Xavier [St. Francis Xavier University, Sucre, Bolivia]

USG Ultrasonic Generator

USG Undoped Silica(te) Glass

USG United States (Standard Plate) Gauge

USG United States Gallon [also USGAL, USgal, or US gal]

USG United States Government

USG University of Surrey, Guildford [UK]

USGA Ultrasonic Gas Atomization [Powder Metallurgy]

US gal United States Gallon [also USgal, or USG]

US gal/min United States gallon per minute [also USGPM, or US GPM]

USGAO United States General Accounting Office [also US GAO, or GAO]

USGC United States Geodynamics Committee [of National Academy of Science]

USGCA United States Government Contract Awards [now United States Contract Awards]

USGCRP United States Global Change Research Program

US GeoData United States Digital Cartographic Data Sets [of US Geological Survey]

US Geol Surv Bull United States Geological Survey Bulletin [also USGS Bull]

USGM United States Government Manual [of United States General Services Administration]

USGovt United States Government

USGovtPtgOff United States Government Printing Office [Washington, DC]

USGPM United States Gallons per Minute [also US GPM, or US gal/min]

USGPO United States Government Printing Office [Washington, DC] [also US GPO or GPO]

USGR United States Government Report

USGRDR United States Government Research and Development Report [of National Institute of Standards and Technology, Gaithersburg, Maryland]

USGRR United States Government Research Reports

USGS United States Geological Society

USGS United States Geological Survey [of US Department of the Interior]

USGSA United States General Services Administration

USGS/BRD United States Geological Survey/Biological Resources Division

USGS/BRD HQ United States Geological Survey/Biological Resources Division National Headquarters [Reston, Virginia]

USGS/GD United States Geological Survey–Geologic Division

USGS MRP United States Geological Survey/Mineral Resources Program

USGS/NMD United States Geological Survey/National Mapping Division

USGS NMP United States Geological Survey/National Mapping Program

USGS/OPS United States Geological Survey/Office of Program Support

USGSB United States Geological Survey Bulletin [also US Geol Surv Bull]

USGS/WRD United States Geological Survey/Water Resources Division

USGW Undersea Guided Weapon

USGW Underwater-to-Surface Guided Weapon

USH Ultrasonic Holography

USH University of Sag Hor [PR China]

USh Ugandan Shilling [Currency of Uganda]

USHIGEO United States National Committee on the History of Geology

USI Ultrasonic Imaging

USI Ultrasonic Inspection

USI Ultrasonic Interferometer; Ultrasonic Interferometry

USI Ultrasonic Society of India

USI Union of Students in Ireland

USI United Schools International

USI United Services Institution [UK]

USI United States Industrial Chemicals Company

USI United States Industry

USI Universal Software Interface

USI University of Southern Indiana [US]

USI Update Software Identity

USI User System Interface

USi Uranium Silicide

U-Si Uranium-Silicon (Alloy)

USIA United States Information Agency

USIB Unsaturated Iron-Binding Capacity [Biochemistry]

USIC United States Information Center

USICA United States International Communications Agency

USICU United States International Standard Payload Rack Checkout Unit

USICF United States International Communication Facility

USIDCA United States International Development Cooperation Agency

USIG Ultrasonic Impact Grinding

US IGES/PDES United States Initial Graphics Exchange Specification/Product Design Exchange Specification Organization [also US IGES-PDES]

USIMC United States International Marketing Center [UK]

Usine Nouv Usine Nouvelle [French Publication on Modern Factories]

USIO Unlimited Sequential Input/Output

USIS United States Information Service

USISC United States International Service Carriers

USITA United States Independent Telephone Association [now United States Telephone Association]

USITC United States International Trade Commission

USIU United States International University [San Diego, California]

USJJA Ultra-Small Josephson Junction Array

USJPRS United States Joint Publication Research Service

USL Underwater Sound Laboratory [US Navy]

USL United States Laboratories

USL United States (Steamship) Lines

USL Universal Sign Language

USL University of Sierra Leone [Freetown]

USL University of Sri Lanka [Peradeniya]

USL Upper Specification Limit

USL Upper Specified Limits

USL$_x$ Upper Specification Limit for Value "x" [Symbol]

USLDMA United States Lanolin and Derivatives Manufacturers Association

USLE Universal Soil Loss Equation

USLP Universidad de San Luis Potosi [University of San Luis Potosi, Mexico]

USLS United Lake Survey [of US Army Corps of Engineers]

USLSA United States Livestock Sanitary Association

USM Ultrasonically Machined; Ultrasonic Machining

USM Ultrasound Microscopy

USM Underwater-to-Surface Missile

USM Unlisted Securities Market

USM United States Mail

USM United States Marines

USM United States Microcomputer (Corporation)

USM United States Mint [of US Department of Treasury]

USM Università degli Studi Milano [Technical University of Milan, Italy]

USM Université Scientifique et Médicale [Scientific and Medical University, Grenoble, France]

USM University of Southern Maine [Portland, US]

USM University of Southern Mississippi [Hattiesburg, US]

USM Unlisted Securities Market

USMA Underfeed Stoker Makers Association [UK]

USMA United States Maritime Administration [of US Department of Transportation]

USMA United States Metric Association

USMA United States Military Academy [West Point, New York State]

USMAC United States Management Advisory Committee

USMACV United States Military Assistance Command, Vietnam [Disbanded]

USMA Newsl USMA Newsletter [of United States Metric Association]

USMC United States Marine Corps

USMC United States Maritime Commission

USMCA United States Marine Corps, Aviation

USMCCCA United States Marine Corps Combat Correspondents Association

USMCEB United States Marine Corps Equipment Board [Quantico, Virginia]

USMCOC United States/Mexico Chamber of Commerce

USMDR United States Minimum Daily Requirements [of US Food and Drug Administration]

USMH United States Marine Hospital [New York City]

USMHS United States Marine Hospital Service

USML United States Marine Laboratory [Sandy Hook, New York]

USML United States Microgravity Laboratory

USMM United States Merchant Marine

USMMA United States Merchant Marine Academy [at Kings Point, New York State]

USMMC United States Marine Mammal Commission

USMP United States Microgravity Payload

USMP United States Military Police

USMSEC United States Materials Science and Engineering Council

USMT United States Military Transport

USMUN United States Mission to the United Nations

USN Ultrasonic Nebulization; Ultrasonic Nebulizer

USN United States Navy

USNA United States National Army

USNA United States Naval Academy [Annapolis, Maryland]

USNASC United States Naval Aviation Safety Center [Norfolk, Virginia]

USNAD United States Naval Ammunition Depot [Crane, Indiana]

USNAOTS United States Naval Aviation Ordnance Test Station [Chincoteague, Virginia]

USNAS United States Naval Air Service

USNATC United States Naval Air Test Center [Patuxant River, Maryland]

USNCAM National Congress on Applied Mechanics [of American Society of Mechanical Engineers, US] [also US Nat Congr Appl Mech]

USNBS United States National Bureau of Standards [now National Institute of Standards and Technology]

USNC United States National Committee

USNC/IEC United States National Committee of the International Electrotechnical Commission

USNC/CIE United States National Committee/CIE (Commission Internationale de l'Eclairage (International Commission on Illumination)) [also USNCCIE]

USNCCr United States National Committee for Crystallography

USNC/IAH United States National Committee of the International Association of Hydrogeologists

USNC/IPS United States National Committee of the International Peat Society [also USNCIPS]

USNC/FID United States National Committee of FID [Federation Internationale de Documentation (International Federation of Documentation)] [also USNCFID]

USNC/SCOR United States National Committee for the Scientific Committee on Oceanic Research [also USNCSCOR]

USNC/TAM United States National Committee on Theoretical and Applied Mechanics [also USNCTAM]

USNC/URSI United States National Committee for URSI [Union Radio-Scientifique Internationale (International Union of Radio Science)] [also USNCURSI]

USNC/WEC United States National Committee of the World Energy Conference [also USNCWEC]

USNEL United States Navy Electronics Laboratory [Now Subdivided into Naval Undersea Warfare Center and Naval Command, Control and Communications Center]

USNFEC United States Naval Facilities Engineering Command [Alexandria, Virginia]

USNG United States National Guard

USNH United States National Herbarium [Washington, DC]

USNHS United States National Health Survey

USNI United States Naval Institute

USNIS United States Naval Investigation Service

USNM United States National Museum

USNMFS United States National Marine Fisheries Service

USNMRL United States Navy Medical Research Laboratory

USNO United States Naval Observatory [Washington, DC]

USNOO United States Naval Oceanographic Office

USNORDA United States Naval Ocean Research and Development Activity

USNPC United States Naval Photographic Center

USNPS United States National Parks Service

USNPS United States Naval Postgraduate School [Monterey, California]

USNR United States Naval Research

USNR United States Naval Reserve

USNRC United States National Research Council [also US NRC]

USNRC United States Nuclear Regulatory Commission [Washington, DC]

USNRDL United States Naval Radiological Defense Laboratory

USNRRC United States Nuclear Reactor Regulatory Committee

USNS United States National Society

USNS United States Naval Ship

USNS United States Naval Station

USNS/ISSMFE United States National Society for the International Society of Soil Mechanics and Foundation Engineering

USNUSL United States Navy Undersea Laboratory

USNUWL United States Navy Underwater Laboratory

USO Ultrastable Oscillator

USO United Service Organizations

USO Unit Security Officer [Canada]

USO Universal Service Order

USO Unmanned Seismic Observatory [of US Department of Defense]

USOA Uniform System of Accounts

USOAR Uniform System of Accounts Revision

USOC Uniform Service Order Code

USOC Universal Service Ordering Code

USOE United States Office of Education

USOI United States Office of Information

USONR United States Office of Naval Research [of US Navy at Arlington, Virginia] [also US-ONR]

USOOG United States Office of Oil and Gas [of US Department of the Interior, Washington, DC]

USOPA United States Ordnance Producers Association

USORC United States Organized Reserve Corps

USOS United States Occupational Health Standards

USOS United States Operations Center

USOSF Union of Shipowners for Overseas Shrimp Fisheries [Greece]

USOTP United States Office of Telecommunications Policy

USOW Underground Sources of Drinking Water

USP Unique Sales Proposition; Unique Selling Proposition

USP United States (of America) Patent [also US Pat]

USP United States Pharmacopoeia

USP Universidade do São Paulo [University of Sao Paulo, Brazil]

USP University of St. Petersburg [Russian Federation] [formerly Leningrad University]

USP University of the South Pacific

USP Usage Sensitive Pricing

USPA United States Parachute Association

USPA United States Patents (Database)

USPA United States Pilots Association

USPA United States Potters Association

USPA United States Psychotronics Association

USPAS United States Particle Accelerator School [Fermilab, Batavia, Illinois]

US Pat United States Patent [also USP]

USPC Ulster Society for the Preservation of the Countryside [Northern Ireland]

USPC United States Pharmacopoeia Convention

USPCA Ulster Society for the Prevention of Cruelty to Animals [Northern Ireland]

USPD Ultrasonic Position Decoder

USPE Utah Society of Professional Engineers [US]

USPEC United States Paper Exporters Council

US Pharm United States Pharmacopoeia

USPHS United States Public Health Service

Usp Khim Uspekhi Khimii [Uzbek Chemical Reviews]

Usp Fiz Nauk Uspekhi Fizicheskikh Nauk [Uzbek Journal on Physical Sciences]

Usp Mat Nauk Uspekhi Matematicheskikh Nauk [Uzbek Journal on Mathematics]

USPHS United States Public Health Service

USPIRG United States Public Interest Research Group

USPMF United States Patent Model Foundation

USP-NF United States Pharmacopoeia–National Formulary [also USP/NF]

USPO United States Patent Office

USPO United States Post Office

USPS United States Postal Service

USP-SC Universidade de São Paulo–São Carlos [University of Sao Paulo at Sao Carlos, Brazil]

USPTO United States Patent and Trademark Office [of US Department of Commerce]

USQ University of Southern Queensland [Australia]

USQ Unreviewed Safety Question

USQ Unsqueezed (Files)

USQD Unreviewed Safety Question Determination

US qt United States quart [Unit]

USQMC United States Quartermaster Corps

USR United States ROSAT (Roentgen Satellite) [of NASA]

USR User (Function) [Programming]

USR User Service Routine

USR US Robotics, Inc. [Skokie, Illinois]

USRA Undergraduate Students Research Awards

USRA United States Railway Association

USRA Universities Space Research Association [US]

USR BBS US Robotics, Inc. Bulletin Board System

USR BOX User-Defined Box [also Usr Box]

USRC United States Rubber Company

USRDA United States Recommended Daily Allowance [of US Food and Drug Administration]

USRDA United States Recommended Dietary Allowance [of US Food and Drug Administration]

USRFP United States Request for Proposals

USRL Underwater Sound Reference Laboratory [US Navy]

USR NUM User-Defined Box Number [also Usr Num]

USR OPT User-Defined Box Options [also Usr Opt]

USRP Undergraduate Science Research Programme [of National University of Singapore]

USRS United States Rocket Society

USRSDC United States Roentgen Satellite Science Data Center [of NASA Astrophysics Data Facility, Goddard Space Flight Center, Greenbelt, Maryland]

USRT Universal Synchronous Receiver/Transmitter

USS Ultimate Shear Strength [Mechanics]

USS Ultrasonic Society

USS Ultrasonic Spectroscopy

USS Unformatted System Service

USS Uniform Symbol(ogy) Specification

USS Unique Support Structure

USS United States Senate

USS United States Ship

USS United States Standard

USS United States Steamer

USS United States Steel Corporation [Gary, Indiana]

USS Utility Support Structure

USSA Underground Security Storage Association

USSA Uniaxial Split Sphere Apparatus

USSA United States Student Association

USSA User Supported Software Association [UK]

USSAF United States Strategic Air Force

USSC United Solar Systems Corporation [US]

USSC United States Supreme Court

USSEA United States Space Education Association

USSF United States Space Foundation

USSG United States Standard Gauge

USSI Ultrasonic Society of India

USSIA United States Shellac Importers Association

USSR Union of Soviet Socialist Republics [Dissolved]

USSR Comput Math Math Phys USSR Computational Mathematics and Mathematical Physics [Translation of: *Zhurnal Vychislitel'noi Matematiki i Matematicheskoi Fiziki;* published in the UK]

USSS United States Secret Service [of US Department of the Treasury]

USStd United States Standard

USStd Sieve United States Standard Sieve

US Steel United States Steel Corporation

USStl WG United States Steel Wire Gauge [also USSTL WG]

USSTS United States Student Travel Service

USSWG United States Steel Wire Gauge

UST Ultrasonic Testing

UST Ultrasonic Transducer
UST Ultrasonic Transmission; Ultrasonic Transmitter
UST Ultrasonic Treatment
UST Ultrasonotomography
UST Unblock Spade Terminal (Command)
UST Underground Storage Tank
UST Underground Storage Tank (Regulation) [of US Environmental Protection Agency]
UST United States Testing Company
UST Universal Servicing Tool
UST University of Santo Tomás [Manila, Philippines]
UST University of Science and Technology
USTA Union des Syndicates des Travailleurs Algériens [Algerian Workers Trade Union]
USTA United States Telephone Association
USTA United States Trademark Association
USTAG United States Technical Advisory Group [of International Standards Organization] [also US TAG]
USTAN Universidad do San Thomás Aquinas de la Norte [University of San Thomas Aquinas de la Norte, San Miguel de Tucumán, Argentina]
USTB University of Science and Technology, Beijing [PR China]
USTC United States Tariff Commission
USTD United States Treasury Department
USTD University Science and Technology Division [of Science and Engineering Research Council, UK]
USTID Underground Storage Tank Integrated Demonstration
USTL Université des Sciences et Technique du Languedoc [Languedoc University of Science and Technology, France]
USTR United States Trade Representative
USTS United States Travel Service [now US Travel and Tourism Administration]
USTSA United States Telephone Suppliers Association
USTTA United States Travel and Tourism Administration
USU Ural State University [Ekaterinburg, Russia]
USU Utah State University (of Agriculture and Applied Science) [Logan, US]
usu usual(ly)
USV Ultrasonic Vibration
USV Unsaturated Vapor
USVA United States Veterans Administration
USVI United States Virgin Islands
USW Ultra-Short Wave
USW Ultrasonic Wave
USW Ultrasonic Welding
USW Undersea Warfare
USW United States Wheat Associates
USW United Steelworkers of America
USW Unsteady Shock Wave [Physics]
USWA United Steelworkers of America
USWB United States Weather Bureau
USWES United States Waterways Experiment Station [of US Army at Vicksburg, Mississippi]

USWL University of Southwestern Louisiana [Lafayette, US]
USWL Upstream Water Level
USZ Unstabilized Zirconia
UT Ultrasonic Test(ing)
UT Ultrathin
UT Umbilical Tower [Aerospace]
UT Uncontrolled Term
UT Undertage Maschinenfabrik GmbH [German Manufacturer of Metallurgical and Mining Plant and Equipment; located at Saarbruecken-Dudweiler]
UT Uniaxial Tension
UT Unified Theory [Physics]
UT Unipolar Transistor
UT Unit Tester
UT Universal Telescope
UT Universal Thread
UT Universal Time [Astronomy]
UT Universidad de Táchira [Tachira State University, San Cristóbal, Venezuela]
UT Università di Torino [University of Turin, Italy]
UT Università di Trento [University of Trent, Italy]
UT Università di Trieste [University of Trieste, Italy]
UT Universiteit Twente [University of Twente, Netherlands]
UT Universitetet i Trondheim [University of Trondheim, Norway]
UT University of Tampa [Florida, US]
UT University of Tasmania [Hobart]
UT University of Teheran [Teheran, Iran]
UT University of Tennessee [Tullahoma, US]
UT University of Texas [US]
UT University of Tokyo [Japan]
UT University of Toledo [Ohio, US]
UT University of Tripoli [Libya]
UT University of Trondheim [Norway]
UT University of Tsukuba [Japan]
UT University of Tulsa [Oklahoma, US]
UT University of Turku [Finland]
UT Upper Troposphere
UT Up Time [Computers]
UT Urban Transportation
UT Urethane Technology
UT Urinary Tract [Anatomy]
UT User Terminal
UT Usual Terms [Commerce]
UT Utah [US]
UT 0 Universal Time 0 (Zero) [also UT_0] [Astronomy]
UT 1 Universal Time 1 (i.e., Universal Time Corrected for Variations in Longitude) [also UT_1] [Astronomy]
UT 2 Universal Time 2 (i.e., Universal Time Corrected for Seasonal Variations) [also UT_2] [Astronomy]
Ut Uterus [Anatomy]
ut usual terms [Commerce]
u(t) transducer input [Symbol]
UTA Uniform Time Act [US]

UTA Union de Transport Aérien [Air Transport Union, France]

UTA University of Texas at Arlington [US]

UTA University of Texas at Austin [US]

UTA Unresolved Transition Array

UTA Upper Control Area

UTA Urban Transportation Administration

UTA User Transfer Address

UTAA University of Toronto Alumni Association [Canada]

UTACA Unión dos Trabajadores Aviaciones, Communicaciones y Aliadas [Aviation, Communication and Allied Workers Union, Venezuela]

UTAD Universidade de Trás-os-Montes e Alto Douro [University of Tras-os-Montes and Alto Douro, Vila Real, Portugal]

Utah Int Rev Utah International Review [US]

UTANG University of Toronto Antinuclear Group [Canada]

UTAP Unified Transportation Assistance Program [US]

UTAP Urban Transport Assistance Program

UTas University of Tasmania [Hobart]

U-Tb Uranium-Terbium (Ferromagnet)

UTC Unione dos Trabajadores de Colombia [Union of Colombian Workers]

UTC United Technologies Corporation [East Hartford, Connecticut, US]

UTC United Technology Center

UTC Unit Test Case(s)

UTC Universal Test Communicator

UTC Universal Test Console

UTC Universal Time Code

UTC Universal Time Coordinated [i.e., Greenwich Mean Time] [Astronomy]

UTC Université de Technologie de Compiegne [Compiegne University of Technology, France]

UTC University of Tennessee at Chattanooga [US]

UTC Utilities Telecommunications Council [US]

UTCC University of Tennessee Computing Center [US]

UTCD University of Tennessee College of Dentistry [Memphis, US]

UT Chart Universal Time Chart [Astronomy]

UTCHS University of Tennessee Center for Health Sciences [Memphis, US]

UTCS Urban Traffic Control System

UTD Universal Tone Decoder

UTD Universal Transfer Device

UTD University of Texas at Dallas [US]

UTD Utilization to Date

Utd united

UTDC Urban Transportation Development Corporation [Canada]

ut dict *(ut dictum)* – as directed [Medical Prescriptions]

U2D2 Low-Field Antiferromagnetic Phase [Solid-State Physics]

UTE Universal Test Equipment

UTE Universidad Tecnica del Estado [State Technical University, Chile]

UTe Uranium(I) Telluride (or Uranium Monotelluride)

UTEC Kongress-Messe für Umwelttechnik [Congress/Fair for Environmental Engineering, Austria]

UTEC Utah University College of Engineering [US]

UTECH Umwelttechnologieforum [Environmental Technology Forum, Berlin, Germany]

UTenn University of Tennessee [Knoxville, US]

UTEOS Undoped Tetraethylorthosilane (or Undoped Tetraethyoxysilane)

UTEP University of Texas at El Paso [US]

UTF Ultrathin Film

UTF Underground Test Facility

UTFSM Universidad Técnico do Federico Santa Maria [Federico Santa Maria Technical University, Valparaiso, Chile]

UTG Universal Tone Generator

UT&GS Uplink Text and Graphics System

UTH Universal Test Head [Mechanics]

UTH University of Texas at Houston [US]

UTH Upper Troposphere Humidity

UTHSC University of Texas Health Science Center [San Antonio, US]

UTI Union des Télécommunications Internationales [International Telecommunication Union, United Nations]

UTI Universal Text Interchange

UTI Universal Text Interface

UTI Universal Time, International

UTI Urinary Tract Infection [Medicine]

U-Ti Uranium-Titanium (Alloy)

UTIA University of Toronto Institute of Aerophysics [Canada]

UTIAS University of Toronto Institute of Aerospace Studies [also Utias, Canada]

UTICS University of Texas Institute for Computer Science [US]

UTIG University of Texas Institute for Geophysics [US]

Util Utility [also util]

UTK Ultrasonic Trim Knife

UTK University of Tennessee at Knoxville [US]

UTL Universidade Tecnica de Lisboa [Technical University of Lisbon, Portugal]

UTL University of Toronto Library [Canada]

UTL Upper Threshold Limit

UTL Upper Tolerance Limit

UTL User Trailer Label

UTLAS University of Toronto Library Automation System [Canada]

UTLM Up Telemetry

UTM Universal Test(ing) Machine

UTM Universal Test Message

UTM Universal Transverse Mercator (Projecton) [Cartography]

UTMS University of Texas Medical School [Galveston, US]

UTNL University of Tokyo Nuclear (Engineering Research) Laboratory [Japan]

UTO Ultrathin Oxide

UTO Universidad Tecnico de Oruro [Oruro University of Technology, Bolivia]

UTO University of Texas at Odessa [US]

UTOA United TVRO (Television Receive Only) Owners Association [US]

UTOL Universal Translator-Oriented Language

UTOP United Technological Organizations of the Philippines

UTP Ultrafine Tungsten Powder

UTP United Trade Press [London, UK]

UTP Universal Tape Processor

UTP University of Tokyo Press [Japan]

UTP University of Toronto Press [Canada]

UTP Unshielded Twisted Pair (Cable)

UTP Upper Threshold Point

UTP Upper Turning Point

UTP Uridine Triphosphate [Biochemistry]

5'-UTP Uridine-5'-Triphosphate [Biochemistry]

UTPC Utah Tourist and Publicity Council [US]

UTPL Ultrathin Polymer Layer

UTQGS Uniform Tire Quality Grading System

UTR Universal Teaching Reactor

UTR Universal Training Reactor [US]

UTR University of Texas at Richardson [US]

UTRA Underground Transportation Research Association [Germany]

UTRC United Technologies Research Center [East Hartford, Connecticut, US]

UTRCA Upper Thames River Conservation Authority [London, Ontario, Canada]

UTRR University of Teheran Research Reactor [Iran]

UTS Ultimate Tensile Strength [Mechanics]

UTS Ultimate Tensile Stress [Mechanics]

UTS n-Undecyltrichlorosilane

UTS Underwater Technology School

UTS Underwater Telephone System

UTS Unified Transfer System

UTS United Transfer System

UTS Unit Test Station

UTS Universal Terminal System

UTS Universal Test Station

UTS Universal Test(ing) System

UTS Universal Time-Sharing System

UTS University of Technology, Sydney [Australia]

UTS Update Transaction System

UTS Urban Transportation System

UTS Urine Transfer System

UTSA University of Texas at San Antonio [US]

UTSI University of Tennessee Space Institute [Tullahoma, US]

UTSL Ultrathin Superlattice [Solid-State Physics]

UTSN Used Truck Sales Network [US]

UTSP Urban Transportation System Planning

Ut sup *(ut supra)* – as above

UTS/VS Universal Time-Sharing System/Virtual Storage

UTS/YS Ultimate Tensile Strength to Yield Strength (Ratio) [Mechanics]

UTT Utility Tactical Transport

UTTAS Utility Tactical Transport Aircraft System

UTTC Universal Tape-to-Tape Converter

UTU United Transportation Union [US]

UTU Universidad del Trabajo del Uruguay [Labour University of Uruguay]

UTV Uncompensated Temperature Variation

UTW Ultrathin Window

UTW United Telegraph Workers

UTWA United Textile Workers of America

UU Ultimate User

UU Unencode/Undecode

UU University of Utah [Salt Lake City, US]

UU Universiteit Utrecht [University of Utrecht, Netherlands]

UU UUE (Unix-to Unix Encode)/UUD (Unix-to-Unix Decode)

UU Uppsala University [Sweden]

UUA Univac Users Association [US]

UUCP Unix-to-Unix Copy Protocol [Internet]

UUCPnet Unix-to-Unix Copy Protocol Network

UUD Unix-to-Unix Decoding [also UUDECODE]

UUE University of the United (Arab) Emirates [Al-Ain]

UUE Unix-to-Unix Encoding [also UUENCODE]

UUI User-To-User Information [AT&T Feature]

UUID Universal Unique Identifier

UUM Underwater-to-Underwater Missile

UUMP Unification of Units of Measurement Panel [of International Civil Aviation Organization]

Uun Ununnilium [Element 110]

UUNET Unix-to-Unix Network [also Uunet]

UUO Unimplemented User Operation

UUP University of Utah Press [Salt Lake City, US]

UUT Unit Under Test

UUT University of Utah [Salt Lake City, US] [also UUt]

UUTR University of Utrecht [Netherlands]

U-U-U-U Polyuridylic Acid [Biochemistry]

UUV Unmanned Underwater Vehicle

UV Ultraviolet [also uv]

UV Ultraviolet Spectroscopy

UV Ultravisible

UV Undervoltage

UV Under Voltage [also U/V]

UV Unit Vector [Mathematics]

UV UniVerse (System) [Database Management]

UV Universidad de Valencia [University of Valencia, Spain]

UV Università di Venezia [University of Venice, Italy]

UV University of Vermont [Burlington, US]

UV University of Victoria [British Columbia, Canada]

UV University of Vienna [Austria]

UV University of Virginia [Charlottesville, US]

UV Unmanned Vehicle

UV Upper Volta

U/V Under Voltage [also UV]

U-V Uranium-Vanadium (Alloy System)

UVA Ultraviolet Absorber; Ultraviolet Absorption

UV-A Ultraviolet, Range A [from 315 to 400 nm]

UVa University of Virginia [Charlottesville, US]

UvA Universiteit van Amsterdam [University of Amsterdam, Netherlands]

U-VAC Interstitial-Vacancy Interaction [Materials Science]

UVAI Ultraviolet Auroral Imager

UVAR University of Virginia Reactor [at Nuclear Research Facility, Charlottesville, US]

UVAR NRF University of Virginia Reactor–Nuclear Research Facility [of University of Virginia, Charlottesville, US]

UVAS Ultraviolet Absorption Spectrophotometry

UVASER Ultraviolet Amplification by Stimulated Emission of Radiation [also Uvaser, or uvaser]

UV-B Ultraviolet, Range B [from 280 to 315 nm]

UVBIS Ultraviolet Bremsstrahlung Isochromat Spectroscopy

UVBY Ultraviolet, Blue, and Yellow (System) [also uvby]

UVC Unidirectional Voltage Converter

UVC Uniform Vehicle Code [US]

UV-C Ultraviolet, Range C [from 100 to 280 nm]

UVCB Under-Voltage Circuit Breaker

UVCE Unconfined Vapour Cloud Explosion

UVCM University of Virginia College of Medicine [Charlottesville, Virginia, US]

UVD Under-Voltage Device

UVD Universal Velocity Distribution

UVDM Ultraviolet Data Manager

UVE United Verde Extension [Arizona, US]

UV/EB Ultraviolet Electron Beam (Curing)

UV EPROM Ultraviolet-Erasable Programmable Read-Only Memory [also UVEPROM, or UV-EPROM]

UV Eraser Ultraviolet Eraser

UVF Unmanned Vertical Flight

UV/F$_2$/Ar Ultraviolet Argon-Diluted Fluorine Gas (Cleaning)

UV/F$_2$/H$_2$ Ultraviolet Hydrogen-Diluted Fluorine Gas (Cleaning)

UVFL Ultraviolet Fluorescence

UVHSC University of Virginia Health Sciences Center [Charlottesville, US]

UVI Ultraviolet Imager

UVic University of Victoria [British Columbia, Canada]

UVIP Used Vehicle Information Package [of Ministry of Consumer and Commercial Relations, Ontario, Canada]

UV-IR Ultraviolet-Visible Infrared (Spectroscopy) [also UV/IR]

UVIS Ultraviolet Imaging Spectrograph

UVIS University of Virginia Information Services [Charlottesville, US]

UVL Ultraviolet Lamp

UVL Ultraviolet Light

UV-LED Ultraviolet Light-Emitting Diode [also UV LED]

UVM Ultraviolet Microscope; Ultraviolet Microscopy

UVM Universitas Viridis Montis [University of the Green Mountains, i.e., University of Vermont, Burlington, US]

UVMC University of Vermont Medical Center [Burlington, US]

UVMC University of Virginia Medical Center [Charlottesville, US]

UV/O$_3$ Ultraviolet Ozone (Cleaning) [also UV-O$_3$]

UVOC Ultraviolet Ozone Cleaning

UV-ODMR Ultaviolet-Excited Optically Detected Magnetic Resonance

UVOM Ultraviolet Optical Microscope; Ultraviolet Optical Microscopy

UVOS Ultraviolet Optical Spectroscopy

UVOX Ultraviolet Oxidation (Process)

UVP Ultraviolet Photography

UVP Undervoltage Protection

UVPD Ultraviolet Photodesorption

UVPES Ultraviolet Photoelectron Spectroscopy [also UVPS]

UV PROM Ultraviolet Programmable Read-Only Memory [also UVPROM, or UV-PROM]

UVPS Ultraviolet Photoelectron Spectroscopy [also UVPES]

UVR Ultraviolet Radiation

UVR Ultraviolet Reflectance

UVR Undervoltage Relay

UVRR Ultraviolet Resonance Raman Spectroscopy [also UVR^2]

UVS Ultraviolet Spectrometer; Ultraviolet Spectrometry

UVS Ultraviolet Spectroscopy

UVS Ultraviolet Spectrum

UVS Ultraviolet Stabilizer

UVS Unmanned Vehicle System

UVSG Ultraviolet Spectrometry Group [UK]

UVSM University of Virginia School of Medicine [Charlottesville, US]

UVSOR Ultraviolet Synchrotron Orbital Radiation

[uvtw] Miller-Bravais indices for crystallographic direction in hexagonal systems [Symbol]

UVV *(Unfallverhütungsvorschrift)* – German for "Accident Prevention Regulation"

UV-VIS Ultraviolet-Visible (Spectroscopy) [also UVVIS, UV/VIS, UV/Vis, or UV-Vis]

UV-VIS-IR Ultraviolet-Visible Infrared (Spectroscopy) [also UV-Vis-IR]

UV-VIS-NIR Ultraviolet-Visible Near-Infrared (Spectroscopy) [also UV-Vis-NIR]

UV-VIS-NMR Ultraviolet-Visible Nuclear Magnetic Resonance [also UV-Vis-NMR]

(uvw) zone indices (of a crystal) [Symbol]

[uvw] indices for a crystallographic direction [Symbol]

<uvw> family of (crystallographic) directions [Symbol]

UW Ultrasonic Wave

UW Underwater [also U/W]

UW Underwriter [also U/W]

UW Unique Word

UW Universität Wien [Vienna University, Austria]

UW University of Wales [UK]

UW University of Warsaw [Poland]

UW University of Warwick [UK]

UW University of Washington [Seattle, US]

UW University of Waterloo [Ontario, Canada]

UW University of Windsor [Ontario, Canada]

UW University of Winnipeg [Manitoba, Canada]

UW University of Wisconsin [US]

UW University of Witwatersrand [Johannesburg, South Africa]

UW University of Wollongong [New South Wales, Australia]

UW University of Wyoming [Laramie, US]

UW Upset Welding

U/W Underwater [also UW]

U/W Underwriter [also UW]

U/W Used With

UWA United Weighers Association [US]

UWA University of Washington [Seattle, US]

UWA University of Western Australia [Perth]

UWA User Working Area

UWAL University of Washington Aeronautics Laboratory [Seattle, US]

UWash University of Washington [Seattle, US]

UWB Ultra-Wideband (Wave)

UWBR Ultra-Wideband Radar

UWC University of the Western Cape [Cape Town/Bellville, South Africa]

UWC University of Wisconsin Center [US]

UWCC University of Wales College of Cardiff [UK]

UWCHS University of Wyoming College of Health Sciences [Larimie, US]

UWCM University of Wales College of Medicine [Cardiff, UK]

UWE Underwater Equipment

UWE Underwater Explosive

UWE University of the West of England [Bristol, UK]

UWEC University of Wisconsin at Eau Claire [US]

UWEX University of Wisconsin-Extension [US]

UWH Underwater Habitat

UWI University of the West Indies [Campuses at Kingston, Jamaica, Bridgetown, Barbados and Port of Spain, Trinidad and Tobago]

UWis University of Wisconsin [US]

UWIST University of Wales Institute of Science and Technology [UK]

UWM University of Wisconsin at Madison [US]

UWM University of Wisconsin at Milwaukee [US]

UWME University of Wyoming/Mechanical Engineering Department [US]

UW-MEMS-SRI University of Wisconsin–Microelectromechanical System–Synchrotron Radiation Instrumentation (Meeting) [Madison, US]

UWMS University of Washington Medical School [Seattle, US]

UWNDS Upper Winds [Meteorology]

UWO University of Western Ontario [London, Canada]

UWP Underwater Photography

UWP University of Washington Press [Seattle, US]

UWP University of Waterloo Press [Ontario, Canada]

UWP University of Wisconsin Press [Madison, US]

UWP Utility Nuclear Waste and Transportation Program [US]

UWRA Urban Water Research Association [Australia]

UWRC University of Wisconsin Rehabilitation Center [Milwaukee, US]

UWRC Urban Wildlife Research Center [now National Institute for Urban Wildlife, US]

UWRR University of Wyoming Research Reactor [US]

UWS University of Western Sydney [New South Wales, Australia]

UWS User Work Station

UWSM University of Washington School of Medicine [Seattle, US]

UWSP University of Wisconsin at Stevens Point [US]

UWT Uniform Wall Temperature

UWTR University of Washington Test Reactor [Seattle, US]

UWY Upper Airway [Aeronautics]

UX Unexploded (Ordnance)

UX Uranium Monochalcogenides [General Formula; X Represents Sulfur, Selenium, or Tellurium]

UX Uranium Monopnictides [General Formula; X Represents Bismuth, Antimony, Arsenic, or Phosphorus]

UX_1 Uranium X_1 [also UX_I, Th-234, ^{234}Th, or Th^{234}]

UX_2 Uranium X_2 [also Pa-234, ^{234}Pa, or Pa^{234}]

U(x) Interaction Energy [Symbol]

ux *(uxor)* – wife

UXAA Unexploded Anti-Aircraft (Shell)

UXB Unexploded Bomb

UXD Ultimate X-Ray Detector

UXIB Unexploded Incendiary Bomb

UXO Unexploded Ordnance

UXPM Unexploded Parachute Mine

UXTGM Unexploded Gas-Type Mine

U(x,y,z) Potential (in Schrödinger Equation) [Symbol]

UY Unit Years

UY University of York [UK]

UY University of Yunnan [Kunming, PR China]

UY Uranium Y [also Th-231, ^{231}Th, or Th^{231}]

UY Uruguay [ISO Code]

.uy Uruguay [Country Code/Domain Name]

UYP Upper Yield Point [Mechanics]

UYP Uruguayan Nuevo Peso [Currency]

UYS Upper Yield Stress [Mechanics]

UYVDRA Upper Yarra Valley and Dandenong Regional Authority [of Victoria State, Australia]

UZ Univerzitet u Zagreb [University of Zagreb, Croatia]

UZ University of Zimbabwe

UZ University of Zurich [Switzerland]

UZ Uranium Z [also Pa-234, ^{234}Pa, or Pa^{234}]

UZ Uzbekistan [ISO Code]

Uz Uzbek(istan) [also Uzb]

uz Uzbekistan [Country Code/Domain Name]

Uzb Uzbek(istan) [also Uz]

Uzb Khim Zh Uzbekskii Khimicheskii Zhurnal [Uzbek Journal of Chemistry]

UzSSR Uzbek Soviet Socialist Republic [USSR]

V   Acoustic-Emission Amplitude [Symbol]

V   Activation Volume [Symbol]

V   Civil Engineering (Technology) [Discipline Category Abbreviation]

V   Coefficient of Variance (or Coefficient of Variation) [Symbol]

V   Convertiplane [US Air Force Symbol]

V   Cutting Speed [Symbol]

V   Effective Value of AC Voltage [Symbol]

V   Efflux Velocity [Symbol]

V   Electron-Electron Interaction Parameter [Symbol]

V   Five [Roman Numeral]

V   (Fluid) Flow Velocity [Symbol]

V   Growth Rate [Symbol]

V   Maximum Tire Speed of 149 mph (or 240 km/h) [Rating Symbol]

V   Mean Particle Volume (in Quantitative Metallography) [Symbol]

V   Pitch Line Velocity (of Spur Gears) [Symbol]

V   (Electrical) Potential (in Free Space) [Symbol]

V   (Electrical) Potential Difference [Symbol]

V   Potential Energy [Symbol]

V   Present Value (of an Amount) [Symbol]

V   Radiotelephony [156 to 174 Mc/s Range]

V   Rate of Traverse (in Welding, etc.) [Symbol]

V   Reflected Acoustic Signal Variation [Symbol]

V   Rotation Factor (of Roller Bearings) [Symbol]

V   Rubbing Speed of a Worm [Symbol]

V   Shear(ing) Force [Symbol]

V   Speed [Symbol]

V   Staff Personnel (Aircraft) [USDOD Symbol]

V   Surface Speed (in Machining) [Symbol]

V   Transverse Shear Load [Symbol]

V   Vacancy [also v]

V   Vaccination; Vaccine

V   Vaccinia [Medicine]

V   Vacuum

V   Vaduz [Liechtenstein]

V   Vagus [Anatomy]

V   Valence (or Valency)

V   Valencia [Spain]

V   Valerian(a) [Botany]

V   Valid(ation)

V   (+)-Valine; Valyl [Biochemistry]

V   Valley

V   Vallisneria [Genus of Edible Plants Including Wild Celery]

V   Valparaiso [Chile]

V   Value

V   Valve

V   Van

V   Vancouver [British Columbia, Canada]

V   Vane

V   Vanadium [Symbol]

V   Vanuatu

V   Vapor(ization); Vaporizer

V   Varactor

V   Variability; Variable

V   Variance; Variant

V   Variation(al)

V   Varistor

V   Varnish

V   Varnished Cambric Type (Conductor Insulation) [Symbol]

V   Vascular

V   Vat

V   Vatu [Currency of Vanuatu]

V   V-Coefficient (in Quantum Mechanics)

V   Vector [also v]

V   Vector Velocity (of an Element of Fluid) [Symbol]

V   Vegetable

V   Vegetation

V   Vehicle

V   Vein [also v]

V   Velocity [also v]

V   Velocity Vector [Symbol]

V   Vendor

V   Vene; Venous

V   Veneer

V   Venezuela(n)

V   Venice [Italy]

V   Venom(ous)

V   Vent(ing)

V   Ventilation; Ventilator

V   Ventricle [Anatomy]

V   Venule [Anatomy]

V   Verb(al)

V   Verification; Verifier; Verify

V   Verdet Constant (in Optics) [Symbol]

V   Vermont [US]

V   Vernier

V   Version [also v]

V   Vertebra(l) [Anatomy]

V   Vertebrata; Vertebrate [Zoology]

V   Vertex [also v]

V   Vertical [also v]

V   Vertical Shear Load [Symbol]

V   Vessel

V   Vesuvius [Italian Volcano]

V   Vial

V   Vibrate; Vibration; Vibrator

V   Vibrio [Genus of Bacteria]

V   Victor [Phonetic Alphabet]

V   Victoria [Canada and Hong Kong]

V   Victoria [Australian State]

V   Victory

V   Video

V   Vienna [Austria]

V   Vietnam(ese)

V   View(ing)

V   View (Command) [Programming]

V   Vignetting Factor [Symbol]

V   Villus [Anatomy]

V   Vine

V   Vinifera [Botany]

V   Vinyl

V   Violet

V   Viper

V   Vipera [Genus Comprising Vipers]

V   Viral; Virus

V   Virgin

V   Virginia [US]

V   Virtual

V   Virtue

V   Virological; Virologist; Virology

V   Virulence; Virulent

V   Viscometer; Viscometry

V   Viscosity; Viscous [also v]

V   Vise

V   Visibility; Visible

V   Vision

V   Visual(ization)

V   Visual Telegraphy [Symbol]

V   Visit(or)

V   Vitamin

V   Vitis (or Vine Family) [Botany]

V   Vitreous; Vitrification

V   Vitrified Type (Grinding Wheel) Bond

V   Vocal

V   Void

V   Voice

V   Volatile; Volatility; Volatilization

V   Volcano

V   Volga [Russian Federation]

V   Vollonia [Genus of Snails]

V   Volt [Unit]

V   Voltage [Symbol]

V   Voltaic

V   Voltmeter

V   Volume

V   Volume of Three-Dimensional Structural Elements, or Test Volume (in Quantitative Metallography) [Symbol]

V   Volvox [Botany]

V   Volumetric; Volumetry

V   Vortex

V   Vorticity

V   Vowel

V   (60-Degree) V-Thread with Truncated Crest and Root [Symbol]

V   Vertical and Short Takeoff and Landing Aircraft [USDOD Symbol]

V   Vulcanizate; Vulcanize(d); Vulcanization

V   Vulnerability; Vulnerable

$-V$ Negative Direct-Current Test Voltage

$+V$ Positive Direct-Current Test Voltage

V^* Activation Volume [Symbol]

V^* Volume Flow [Symbol]

$\bar{V}$ Average Grain Volume (in Metallography) [Symbol]

$\bar{V}$ Mean Particle Volume (in Quantitative Metallography) [Symbol]

$\bar{V}$ Mean (Atomic) Volume [Symbol]

$\bar{V}$ Molar Volume [Symbol]

$\dot{V}$ Volume Flux [Symbol]

V^0 Standard (Electrode) Potential [Symbol]

V_0 Initial Vacancy Concentration (in Physical Metallurgy) [Symbol]

V_0 Initial Velocity [Symbol]

V_0 Initial (or Original) Volume [Symbol]

V_0 Maximum Voltage [Symbol]

V_0 Undisturbed Liquid-Stream Velocity [Symbol]

V_0 Void Volume [Symbol]

V_0 Volume at Absolute Zero (0K) [Symbol]

V_0 Volume at Reference Temperature (usually, $t = 0°C$) [Symbol]

V_0 Volume at Temperature $t = 0°C$ [Symbol]

V_1 Velocity in Medium 1 [Symbol]

V^{2+} Divalent Vanadium Ion [also V^{++}] [Symbol]

V^{3+} Trivalent Vanadium Ion [also V^{+++}] [Symbol]

V^{4+} Tetravalent Vanadium Ion [Symbol]

V^{5+} Pentavalent Vanadium Ion [Symbol]

V_{12} Seebeck Potential (in electronics) [Symbol]

V_{100} Volume at Temperature $t = 100°C$ [Symbol]

V_A Anode Voltage [Semiconductor Symbol]

V_A Available Potential [Symbol]

V_A Volume of Component, Phase, Solid, etc. "A" [Symbol]

V_a Accelerating Voltage (or Potential) [Symbol]

V_a Anode (or Plate) Voltage [Semiconductor Symbol]

V$_a$ Seepage Velocity (for Soil, or Rock) [Symbol]

V$_\alpha$ Volume Fraction of Alpha Phase (of an Alloy) [Symbol]

V$_{ac}$ Alternating-Current Voltage [Semiconductor Symbol]

V$_{AK}$ Anode-Cathode Voltage [Semiconductor Symbol]

V$_B$ Breakdown Voltage [Semiconductor Symbol]

V$_b$ Substrate Bias [Semiconductor Symbol]

V$_\beta$ Volume Fraction of Beta Phase (of an Alloy) [Symbol]

V$_{BB}$ Base Supply Voltage (Direct-Current) [Semiconductor Symbol]

V$_{BC}$ Average, or Direct-Current Voltage, Base to Collector [Semiconductor Symbol]

V$_{BE}$ Average, or Direct-Current Voltage, Base to Emitter [Semiconductor Symbol]

V$_{bp}$ Backplane Voltage [Liquid-Crystal Display Symbol]

V$_{br}$ Breakdown Voltage [Semiconductor Symbol]

V$_{(BR)}$ Breakdown Voltage, Direct Current [Semiconductor Symbol]

V$_{(BR)CBO}$ Collector-Base Breakdown Voltage, Emitter Open [Semiconductor Symbol]

V$_{(BR)CEO}$ Collector-Emitter Breakdown Voltage, Base Open [Semiconductor Symbol]

V$_{(BR)CER}$ Collector-Emitter Breakdown Voltage, Resistance between Base and Emitter [Semiconductor Symbol]

V$_{(BR)CES}$ Collector-Emitter Breakdown Voltage, Base Short-Circuited to Emitter [Semiconductor Symbol]

V$_{(BR)CEV}$ Collector-Emitter Breakdown Voltage, Specified Voltage between Base to Emitter [Semiconductor Symbol]

V$_{(BR)CEX}$ Collector-Emitter Breakdown Voltage, Specified Circuit between Base to Emitter [Semiconductor Symbol]

V$_{(BR)EBO}$ Emitter-Base Breakdown Voltage, Collector Open [Semiconductor Symbol]

V$_{(BR)ECO}$ Emitter-Collector Breakdown Voltage, Base Open [Semiconductor Symbol]

V$_{(BR)E1E2}$ Emitter-Emitter Breakdown Voltage [Semiconductor Symbol]

V$_{(BR)GSS}$ Gate-Source Breakdown Voltage [Semiconductor Symbol]

V$_{(BR)GSSF}$ Forward Gate-Source Breakdown Voltage [Semiconductor Symbol]

V$_{(BR)GSSR}$ Reverse Gate-Source Breakdown Voltage [Semiconductor Symbol]

V$_{B2B1}$ Interbase Voltage [Semiconductor Symbol]

V^C Coulomb Repulsion [Symbol]

V$_c$ Corrosion Potential [Symbol]

V$_c$ Output Voltage Across Capacitor [Semiconductor Symbol]

V$_c$ Unit Cell Volume [Symbol]

V$_c$ Collector Voltage (of Transistor) [Semiconductor Symbol]

V$_c$ Critical Voltage [Symbol]

V$_c$ Critical (Molar) Volume [Symbol]

V$_c$ Cutting Speed (in Machining) [Symbol]

V$_c$ Unit-Cell Volume [Symbol]

V$_{C1}$ Output Voltage Across Capacitor 1 [Semiconductor Symbol]

V$_{C2}$ Output Voltage Across Capacitor 2 [Semiconductor Symbol]

V$_{CB}$ Average, or Direct-Current Voltage, Collector to Base [Semiconductor Symbol]

V$_{CB(fl)}$ Collector-Base Direct-Current Open-Circuit Voltage (Floating Potential) [Semiconductor Symbol]

V$_{CBO}$ Collector-Base Voltage, Direct-Current, Emitter Open [Semiconductor Symbol]

V$_{CC}$ Collector Supply Voltage, Direct Current [Semiconductor Symbol]

V$_{CE}$ Average, or Direct-Current Voltage, Collector to Emitter [Semiconductor Symbol]

V$_{CE(fl)}$ Collector-Emitter Direct-Current Open-Circuit Voltage (Floating Potential) [Semiconductor Symbol]

V$_{CEO}$ Collector-Emitter Voltage, Direct-Current, Base Open [Semiconductor Symbol]

V$_{CE(ofs)}$ Collector-Emitter Offset Voltage [Semiconductor Symbol]

V$_{CER}$ Collector-Emitter Voltage, Direct-Current [Semiconductor Symbol]

V$_{CES}$ Collector-Emitter Voltage, Direct-Current, Base Short-Circuited to Emitter [Semiconductor Symbol]

V$_{CE(sat)}$ Collector-Emitter Direct-Current Saturation Voltage [Semiconductor Symbol]

V$_{CEV}$ Collector-Emitter Voltage, Direct-Current, Specified Voltage between Base to Emitter [Semiconductor Symbol]

V$_{CEX}$ Collector-Emitter Voltage, Direct-Current, Specified Circuit between Base to Emitter [Semiconductor Symbol]

V$_{DD}$ Drain-Drain (Supply) Voltage [Semiconductor Symbol]

V$_{DG}$ Drain-Gate Voltage [Semiconductor Symbol]

V$_{do}$ Diffusion Voltage [Symbol]

V$_{dp}$ Hopping Parameter (in Physics) [Symbol]

V$_{DRM}$ Forward Breakover Voltage for Thyristors and Trigger Devices [Semiconductor Symbol]

V$_{DS}$ Drain-Source Voltage [also V$_{ds}$] [Semiconductor Symbol]

V$_{DS(on)}$ Drain-Source On-State Voltage [Semiconductor Symbol]

V$_{DU}$ Drain-Substrate Voltage [Semiconductor Symbol]

V$_E$ Emitter Voltage [Semiconductor Symbol]

V$_e$ Volume of Elementary Cell [Symbol]

V$_e$ Emitter Voltage (of Transistor) [Semiconductor Symbol]

V$_e$ Exciting Voltage (of X-Ray Tube) [Symbol]

V$_{EB}$ Average, or Direct-Current Voltage, Emitter to Base [Semiconductor Symbol]

V$_{EB(fl)}$ Emitter-Base Direct-Current Open-Circuit Voltage (Floating Potential) [Semiconductor Symbol]

V$_{EBO}$ Emitter-Base Voltage, Direct-Current, Collector Open [Semiconductor Symbol]

V$_{EB1(sat)}$ Emitter Saturation Voltage [Semiconductor Symbol]

V$_{EC}$ Average, or Direct-Current Voltage, Emitter to Collector [Semiconductor Symbol]

V$_{EC(fl)}$ Emitter-Collector Direct-Current Open-Circuit Voltage (Floating Potential) [Semiconductor Symbol]

V$_{EC(ofs)}$ Emitter-Collector Offset Voltage [Semiconductor Symbol]

V$_{EE}$ Emitter Supply Voltage [Semiconductor Symbol]

V$_{eff}$ Effective Voltage [Symbol]

V_F Direct-Current Forward Voltage (Without Alternating Component) for Rectifier and Signal Diodes [Semiconductor Symbol]

V_F Direct-Current Forward Voltage for Voltage-Reference and Voltage-Regulator Diodes [Semiconductor Symbol]

V_f Alternating Component (Root-Mean-Square Value) of Forward Voltage [Semiconductor Symbol]

V_f Chip Velocity (in Machining) [Symbol]

V_f Filament (or Heater) Voltage [Symbol]

V_f Free (Molar) Volume [Symbol]

V_f Volume Fraction [Symbol]

V_f Volume Fraction of Composite Fibers [Symbol]

V_f Final Velocity [Symbol]

V_f Free (Molar) Volume [Symbol]

V_f Volume Fraction of Composite Fibers [Symbol]

$V_{F(AV)}$ Forward Voltage, Direct-Current (With Alternating Component) [Semiconductor Symbol]

V_{FB} Flatband Voltage [Semiconductor Symbol]

V_{fb} Flatband voltage [Semiconductor Symbol]

V_{fb} Flatband Voltage [also V_{fb}] [Semiconductor Symbol]

V_{fk} Voltage between Heater and Cathode [Semiconductor Symbol]

V_{FM} Maximum (Peak) Total Forward Voltage [Semiconductor Symbol]

$(V_f)_{0K}$ Free Volume at 0K [Symbol]

$V_{F(RMS)}$ Total Root-Mean-Square Forward Voltage [Semiconductor Symbol]

$(V_f)_T$ Free Volume at Temperature T [Symbol]

V_G Gate Voltage [also V_g] [Semiconductor Symbol]

V_G Growth Rate [also V_g] [Semiconductor Symbol]

V_G Isothermal Potential [Symbol]

V_G Volume of Gas [also V_g] [Symbol]

V_g Grid Voltage [Semiconductor Symbol]

V_g Grid Bias [Semiconductor Symbol]

V_{g1} Grid Bias [Semiconductor Symbol]

V_{g2} Screen-Grid Bias [Semiconductor Symbol]

V_{GB} Gate-to-Bulk Potential [Semiconductor Symbol]

V_{GG} Gate Supply Voltage [Semiconductor Symbol]

V_{GS} Gate-Source Voltage [Semiconductor Symbol]

V_{gs} Gate-Source Voltage [Semiconductor Symbol]

V_{GSF} Forward Gate-Source Voltage [Semiconductor Symbol]

$V_{GS(off)}$ Gate-Source Cutoff Voltage [Semiconductor Symbol]

V_{GSR} Reverse Gate-Source Voltage [Semiconductor Symbol]

$V_{GS(th)}$ Gate-Source Threshold Voltage [Semiconductor Symbol]

V_{GU} Gate-Substrate Voltage [Semiconductor Symbol]

V_H Hall Voltage [Semiconductor Symbol]

V_H Hartree Potential (in Physics) [Symbol]

V_I Inflection-Point Voltage [Semiconductor Symbol]

V_I Input Voltage [Semiconductor Symbol]

V_i Indicated Airspeed [Symbol]

V_i Induced Voltage [Symbol]

V_i Input Potential (or Voltage) [Symbol]

V_i Instantaneous Applied Test Voltage at Time of Partial-Discharge (Corona) [Symbol]

V_i Ionization Potential [Symbol]

V_i Voltage Drop across Interface [Symbol]

V_i Volume Fraction of Phase i [Symbol]

$\overline{V}_i$ Partial Molar Free Energy [Symbol]

$V_{ij}^{(s)}$ Interaction Energy Between the i-th and j-th Atom which are Neighbors of the s-Order (in Physics) [Symbol]

V_{IH} HIGH Level Input Voltage [Semiconductor Symbol]

$V_{IH(min)}$ Minimum HIGH Level Input Voltage [Semiconductor Symbol]

V_{in} Input Voltage [Symbol]

V_{IL} LOW Level Input Voltage [Semiconductor Symbol]

$V_{IL(max)}$ Maximum LOW Level Input Voltage [Semiconductor Symbol]

V_{IO} Input Offset Voltage for Operational Amplifier [Semiconductor Symbol]

V_j Instantaneous Value of Applied Voltage at which j-th Partial-Discharge (Corona) Pulse Occurs [Symbol]

V_L Output Voltage Across Inductor [Symbol]

V_L Voltage Across Load Resistance [Symbol]

V_L Volume of Liquid [also V_l] [Symbol]

V_l Velocity of Longitudinal Wave [Symbol]

V_M Millman's Voltage [Symbol]

V_m Matrix Volume (or Volume Fraction of Matrix) [Symbol]

V_m Molar Volume (of Ideal Gas) at Standard Temperature and Pressure [22.414×10^{-3} m^3/mol)] [Symbol]

V_m Molecular Volume [Symbol]

V_m Volume Fraction of (Composite) Matrix [Symbol]

V_m Volume of Mobile Phase [Symbol]

V_{MC} Minimum Control Speed (in Aeronautics) [Symbol]

V_{mn} Standard Molar Volume (22.414×10^{-3} m^3/mol) [Symbol]

V_n Volume (Standard Temperature and Pressure) [Symbol]

V_{NE} Never Exceed Speed (in Aeronautics) [Symbol]

V_{NH} HIGH Level Direct-Current Noise Margin [Semiconductor Symbol]

V_{NHE} Normal Hydrogen Electrode Potential [Symbol]

V_{NL} LOW Level Direct-Current Noise Margin [Semiconductor Symbol]

V_{NO} Normal Operating Limit Speed (in Aeronautics) [Symbol]

V_{NOR} NOR Gate Output Voltage [Semiconductor Symbol]

V_o Output Voltage [also V_o]

V_{OBI} Base-1 Peak Voltage [Semiconductor Symbol]

V_{off} Off-State Voltage for Thyristor [Semiconductor Symbol]

V_{OH} HIGH Level Output Voltage [Semiconductor Symbol]

$V_{OH(min)}$ Minimum HIGH Level Output Voltage [Semiconductor Symbol]

V_{OL} LOW Level Output Voltage [Semiconductor Symbol]

$V_{OL(max)}$ Maximum LOW Level Output Voltage [Semiconductor Symbol]

V_{on} On-State Voltage for Thyristor [Semiconductor Symbol]

V_{OR} OR Gate Output Voltage [Semiconductor Symbol]

V_{OUT} Output Voltage [also V_{out}]

V_P Peak-Point Voltage [Semiconductor Symbol]

V_P Pinch-Off Voltage (of FET Transistor) [Semiconductor Symbol]

V_p Particulate Volume Fraction (or Volume Fraction of Particles) [Symbol]

V_p Plate (or Anode) Voltage [Semiconductor Symbol]

V_p Pore Volume [Symbol]

V_p Practicable Cutting Speed [Symbol]

V_p Primary Voltage (i.e., Voltage in Primary Coil of Transformer, etc.) [Symbol]

V_{PP} (Projected) Peak-Point Voltage [also V_{pp}, or V_{P-P}] [Semiconductor Symbol]

V_R Direct-Current Reverse Voltage (Without Alternating Component) for Rectifier and Signal Diodes [Semiconductor Symbol]

V_R Direct-Current Reverse Voltage for Voltage-Reference and Voltage-Regulator Diodes [Semiconductor Symbol]

V_R Voltage across Resistor (in Alternating-Current Circuit) [Symbol]

V_r Alternating Component (Root-Mean-Square Value) of Reverse Voltage [Semiconductor Symbol]

V_r Reverse Voltage [Semiconductor Symbol]

$V_{R(AV)}$ Reverse Voltage, Direct-Current (With Alternating Component) [Semiconductor Symbol]

V_{REL} Relative Velocity [also V_{rel}]

V_{RL} Voltage Across Load Resistance [Semiconductor Symbol]

V_{RLC} Voltage across a Resistor-Inductor-Capacitor Combination [Semiconductor Symbol]

V_{RM} Maximum (Peak) Total Reverse Voltage [Semiconductor Symbol]

V_{RRM} Peak Reverse Voltage, Repetitive [Semiconductor Symbol]

$V_{R(RMS)}$ Total Root-Mean-Square Reverse Voltage [Semiconductor Symbol]

V_{RSM} Peak Reverse Voltage, Nonrepetitive [Semiconductor Symbol]

V_{RT} Reach-Trough Voltage [Semiconductor Symbol]

V_{RWM} Working Peak Reverse Voltage [Semiconductor Symbol]

V_s Source Voltage, Direct Current [Semiconductor Symbol]

V_s Volume of Solid [also V_s] [Symbol]

V_s Volume of Sphere [Symbol]

V_s Applied Root-Mean-Square Test Voltage [Symbol]

V_s Sample Volume [Symbol]

V_s Secondary Voltage (i.e., Voltage in Secondary Coil of Transformer, etc.) [Symbol]

V_s Shear Velocity (in Metal Cutting) [Symbol]

V_s Source Voltage [Symbol]

V_s Stopping Potential (or Voltage) [Symbol]

V_s Surface Potential [Symbol]

V_{SD} Source-Drain Voltage [Semiconductor Symbol]

V_{SCE} Saturated Calomel Electrode Potential [Symbol]

V_{SS} Source Supply Voltage [Semiconductor Symbol]

V_{SU} Source-Substrate Voltage [Semiconductor Symbol]

V_T Threshold Voltage [Semiconductor Symbol]

V_T Volume at Temperature T [Symbol]

V_t Partial-Discharge Pulse Voltage [Symbol]

V_t True Airspeed [Symbol]

V_t Volume at Temperature $t \neq 0°C$ [Symbol]

V_{th} Threshold Voltage [also $V_{(TO)}$] [Semiconductor Symbol]

V_{to} Turnoff Voltage [Semiconductor Symbol]

V_{tr} Volume at Triple Point [Symbol]

V_v Valley-Point Voltage [Semiconductor Symbol]

V_v Volume Fraction (i.e., Sum of Volumes of Structural Features Divided by Total Test Volume) (in Quantitative Metallography) [Symbol]

V_v Volume of Vapor [Symbol]

V_v Volume Fraction of Void Content [Symbol]

V_w Volume Fraction of Whiskers [Symbol]

V_w Welding Speed [Symbol]

V_x Vacancy (x Refers to a Chemical Element, e.g., V_{Ga} is a gallium vacancy) [Symbol]

V_x Velocity Component in X-Direction [Symbol]

V_x Voltage in x Direction [Symbol]

V_x Volume of Body which is Rotationally Symmetrical about x Axis [Symbol]

V_{xx} Magnetoresistance [Symbol]

V_{xy} Hall Voltage (in Electronics) [Symbol]

V_y Velocity Component in Y-Direction [Symbol]

V_y Voltage in y Direction [Symbol]

V_y Volume of Body which is Rotationally Symmetrical about y Axis [Symbol]

V_z Reference Voltage, Regulator Voltage, Direct Current [Semiconductor Symbol]

V_z Zener (Breakdown) Voltage [Semiconductor Symbol]

V_z Velocity Component in Z-Direction [Symbol]

V_z Voltage in z Direction [Symbol]

V_{ZM} Reference Voltage, Regulator Voltage, Direct Current at Maximum Rated Current [Semiconductor Symbol]

V-4 Four-Cylinder V-Type (Automotive) Engine

V-6 Six-Cylinder V-Type (Automotive) Engine

V-8 Eight-Cylinder V-Type (Automotive) Engine

V-10 Ten-Cylinder V-Type (Automotive) Engine

V-12 Twelve-Cylinder V-Type (Automotive) Engine

V-16 Sixteen-Cylinder V-Type (Automotive) Engine

V-46 Vanadium-46 [also ^{46}V and V^{46}]

V-47 Vanadium-47 [also ^{47}V and V^{47}]

V-48 Vanadium-48 [also ^{48}V and V^{48}]

V-49 Vanadium-49 [also ^{49}V and V^{49}]

V-50 Vanadium-50 [also ^{50}V and V^{50}]

V-51 Vanadium-51 [also ^{51}V, V^{51}, or V]

V-52 Vanadium-52 [also ^{52}V and V^{52}]

V-53 Vanadium-53 [also ^{53}V and V^{53}]

16 V Sixteen Valve (Automotive) Engine

v degree of freedom (in statistics) [Symbol]

v discharge velocity (of water through soil or rock) [Symbol]

v displacement [Symbol]

v electrical potential [Symbol]

v eddy diffusivity [Symbol]

v flow velocity (of a fluid) [Symbol]

v mean reflectivity [Symbol]

v phase velocity (of waves) [Symbol]

v   real image distance (in optics) [Symbol]

v   specific volume [Symbol]

v   thermopile voltage [Symbol]

v   true airspeed [Symbol]

v   vacancy [also V]

v   vacuum

v   valve

v   vapor [also V]

v   variable; variation

v   vector [also v]

v   vein [also V]

v   (linear) velocity [Symbol]

v   velocity component (in Cartesian system) [Symbol]

v   vent

v   verify

v   Verdet constant (in optics) [Symbol]

v   version [also V]

v   *(versus)* – against

v   vertex [also V]

v   vertical [V]

v   very

v   *(vide)* – see

v   violet

v   virtual

v   viscous; viscosity [also V]

v   visibility

v   voice

v   volt [Unit]

v   (instantaneous) voltage [Symbol]

v   volume [also V]

v   volume fraction [Symbol]

v   volumetric rainfall [Symbol]

v   vowel

v%   volume percent [also vol%]

$\bar{v}$ drift velocity (of electrons) [Symbol]

$\bar{v}$ mean particle velocity [Symbol]

$\bar{v}$ temporal mean of velocity component "v" [Symbol]

$\overline{v^2}$ mean (or average) square velocity (in kinetic molecular theory) [Symbol]

v$_+$ velocity of cations [Symbol

v$_-$ velocity of anions [Symbol]

v$_{||}$ velocity perpendicular to axis [Symbol]

v$_{\perp}$ velocity along axis [Symbol]

v$_0$ crosshead speed (of mechanical testing system) [Symbol]

v$_0$ initial velocity [Symbol]

v$_0$ muzzle velocity [Symbol]

v$_0$ volume occupied by a gas at the ice point (0°C) [Symbol]

v$_1$ initial volume [Symbol]

v$_1$ specific volume of phase 1 [Symbol]

v$_1$ velocity of light in medium 1 [Symbol]

v$_2$ final volume [Symbol]

v$_{100}$ volume occupied by a gas at the steam point (100°C) [Symbol]

v$_a$ axial speed [Symbol]

v$_{\alpha}$ volume of the alpha phase (of an alloy) [Symbol]

v$_{av}$ average speed (or velocity) [Symbol]

v$_{\beta}$ volume of the beta phase (of an alloy) [Symbol]

v$_{bc}$ instantaneous value of alternating component of base-collector voltage [Semiconductor Symbol]

v$_{be}$ instantaneous value of alternating component of base-emitter voltage [Semiconductor Symbol]

v$_{(BR)}$ breakdown voltage (instantaneous total) [Semiconductor Symbol]

v$_C$ capacitance voltage for RC transient circuit [Semiconductor Symbol]

v$_C$ junction critical voltage [Semiconductor Symbol]

v$_c$ average velocity of dislocation climb (in materials science) [Symbol]

v$_c$ circular velocity (of comets) [Symbol]

v$_c$ critical velocity [Symbol]

v$_c$ critical (specific) volume [Symbol]

v$_{cb}$ instantaneous value of alternating component of collector-base voltage [Semiconductor Symbol]

v$_{ce}$ instantaneous value of alternating component of collector-emitter voltage [Semiconductor Symbol]

v$_D$ drift velocity (of electrons) [also v$_d$] [Symbol]

v$_e$ elliptical velocity (of comets) [Symbol]

v$_e$ exhaust velocity [Symbol]

v$_e$ velocity of extraordinary ray [Symbol]

v$_{eb}$ instantaneous value of alternating component of emitter-base voltage [Semiconductor Symbol]

v$_{ec}$ instantaneous value of alternating component of emitter-collector voltage [Semiconductor Symbol]

v$_F$ Fermi velocity (of electrons) [Symbol]

v$_F$ instantaneous total forward voltage [Semiconductor Symbol]

v$_f$ final velocity [also v$_f$] [Symbol]

v$_f$ volume fraction of fibers of fiber-reinforced composite [Symbol]

v$_g$ group velocity (of waves) [Symbol]

v$_g$ molar volume [Symbol]

v$_h$ hyperbolic velocity (of comets) [Symbol]

v$_h$ volume fraction of high-modulus phase of a uniformly dispersed aggregate composite [Symbol]

v$_i$ initial velocity [Symbol]

v$_i$ internal velocity [Aerospace Symbol]

v$_i$ velocity of *i*-th ion [Symbol]

v$_{id}$ ideal velocity [Symbol]

v$_k$ kink velocity [Symbol]

v$_L$ flux line velocity [Symbol]

v$_l$ molar volume of liquid at (absolute) temperature *T* [Symbol]

v$_l$ volume fraction of low-modulus phase of a uniformly dispersed aggregate composite [Symbol]

v$_m$ volume fraction of matrix of fiber-reinforced composite [Symbol]

v$_{max}$ maximum (value of) instantaneous voltage [Symbol]

v_o velocity of observer (e.g., Doppler equation) [Symbol]

v_o velocity of ordinary ray [Symbol]

v_p escape (or parabolic) velocity (of comets) [Symbol]

v_p particle velocity [Symbol]

v_p phase velocity [Symbol]

v_R instantaneous total reverse voltage [Semiconductor Symbol]

v_R Rayleigh wave [Symbol]

v_R resistance voltage for RC transient circuit [Semiconductor Symbol]

v_r reduced volume [Symbol]

v_r resonance velocity [Symbol]

v_s molar volume of solid at melting temperature T [Symbol]

v_s spin-wave velocity (in solid-state physics) [Symbol]

v_s staging velocity [Aerospace Symbol]

v_s velocity of source (e.g., Doppler equation) [Symbol]

v_{st} molar volume of steam at (absolute) temperature T [Symbol]

v_t volume occupied by a gas at temperature $t \neq 0°C$ [Symbol]

v_x velocity along x-axis [Symbol]

v_x velocity component in x-direction [Symbol]

v_y velocity component in y-direction [Symbol]

v_z velocity along y-axis [Symbol]

v_z velocity-component in z-direction [Symbol]

VA   Vacuum Arc

VA   Vacuum Anneal(ed); Vacuum Annealing

VA   n-Valeric Acid

VA   Value Added

VA   Value Analysis

VA   Vanadium-Aluminum (Alloy)

VA   Vane Anemometer

VA   Vanillic Acid

VA   Variance Analysis

VA   Variometer

VA   Vatican [ISO Code]

VA   Vector Addition

VA   Vector Analysis

VA   Vehicle Assembly

VA   Velocity Analysis; Velocity Analyzer

VA   Velocity at Apogee

VA   Ventral Aorta [Zoology]

VA   Vermiculite Association [US]

VA   Vertical Alignment

VA   Veterans Administration [US]

VA   Vibrational Analysis

VA   Vibro-Acoustic(s) [also V-A]

VA   Vice-Admiral [also VAdm, or Vice-Adm]

VA   Vickers-Armstrong (Gun)

VA   Victor Alfa [Radiotelephony]

VA   Video Amplifier

VA   Vinyl Acetate

VA   Vinyl Alcohol

VA   Virginia (State) [US]

VA   Virtual Address

VA   Visual Acuity

VA   Visual Aid

VA   Visual Age

VA   Visual Approach

VA   Visual Axis

VA   Vital Area

VA   Voest Alpine [Major Austrian Steel and Steel Plant Manufacturer]

VA   Vocational Agriculture

VA   Voice-Activated (System)

VA   Volt-Ammeter

VA   Volt-Ampere [also V-A, va, or v-a]

VA   Volumetric Analysis

V/A   Video/Analog

V/A   Volt(s) per Ampere [also V A^{-1}]

V/A   Volume-to-Area (Ratio)

V-A   Vibro-Acoustic(s) [also VA]

V-A   Volt-Ampere [also v-a, VA, or va]

V/Å   Volt(s) per Angstrom [also V Å^{-1}]

Va   Virginia (State) [US]

.va   Vatican [Country Code/Domain Name]

VAA   Vanadium Acetylacetonate

VAA   Vehicle Assembly Area

VAA   Viewpoint Adapter Assembly

VAA   Vinyl Acetic Acid

VAA   Viscum Album Agglutinin [Immunology]

VAA   Voice Access Arrangement

VAAC   Vanadyl Acetylacetonate

VAAC   Vectored Thrust Advanced Aircraft

VAAE   Vaccine Associated Adverse Events [Immunology]

VAAP   Volunteer Army Ammunition Plant [of US Army in Tennessee]

VAB   Vehicle Assembly Building [of NASA at Kennedy Space Center, Florida, US]

VAB   Vertical Assembly Building

VAB   Voice Answerback

VAC   Vacuum-Arc Cast(ing)

VAC   Value-Added Carrier

VAC   Variable Amplitude Correction

VAC   Variance at Completion

VAC   Vector Analog Computer

VAC   Vehicle Assembly and Checkout

VAC   Vertical Assembly Component

VAC   Video Amplifier Chain

VAC   Vinyl Acetate [also Vac]

VAC   Volts of Alternating Current [also Vac, vac, V AC, V/AC, v ac, V/ac, V A-C, or v a-c]

$V(Ac)_3$ Vanadium(III) Acetate

Vac   Vacancy [also vac]

Vac   Vacuum [also vac]

% vac   percent of vacuum [also pct vac] ◆

V(ACAC)$_3$ Vanadium(III) Acetylacetonate [also V(AcAc)$_3$, or V(acac)$_3$]

VACASULF Vacuum Desulfurization (Process)

VACC Value-Added Common Carrier

VACF Velocity Autocorrelation Function

VACM Vector Averaging Current Meter

Vacmetal Gesellschaft für Vakuummetallurgie mbH [German Vacuum Metallurgy Company]

VACP Visual, Auditory, Cognitive and Psychomotor

VACR Variable Amplitude Correction Rack

VAC(RMS) Volt, Alternating Current (Root-Mean Square)

Vac Sci Vacuum Science; Vacuum Scientist

Vac Sci Technol Vacuum Science and Technology [Published in PR China]

VACTL Vertical Assembly Component Test Laboratory

VAD Vacuum Arc Decarburization (Process) [Metallurgy]

VAD Vacuum Arc Degassing (Process) [Metallurgy]

VAD Vacuum Arc Deposition

VAD Value-Added Dealer

VAD Value-Added Distributor

VAD Vandenberg (Air Force Base) Addendum Document [US]

VAD Vapor Axial Deposition; Vapor-Phase Axial Deposition

VAD Velocity-Azimuth Display

VAD Ventricular Assist Device

VAD Voltmeter Analog-to-Digital (Converter)

VAD Volunteer Aid Detachment

VADA Voluntary Agricultural Development Aid (Program) [of Canadian International Development Agency, Canada]

VADAC Voice Analyzer Data Converter

VADC Video Analog to Digital Converter

VADC Vacuum Arc Decarburization Converter [Metallurgy]

VADC Voltmeter Analog-to-Digital Converter

VADD Value Added Disk Drive

VADE Versatile Auto Data Exchange

VADER Vacuum Arc Double Electrode Remelting [Metallurgy]

VADIS Voice and Data Integrated System

V Adm Vice-Admiral [also VA, or ViceAdm]

VADMS Visualization, Analysis and Design in the Molecular Sciences Laboratory [of Washington State University, Pullman, US]

VADS Velocity-Aligned Doppler Spectroscopy

VADS Verdix Ada Development System

VADSL Very-High-Rate Asymmetric Digital Subscriber Line

VAE V-Activating Enzyme [Biochemistry]

VAE Vinyl Acetate Ethylene

VAEP Variable Attributes Error Propagation

VAF Vacancy Availability Factor [Solid-State Physics]

VAF Vacuum-Arc Furnace

VAF Velocity Autocorrelation Function

VAF Vernacular Architecture Forum [US]

VAF Volume Advantage Factor (of Nuclear Reactors)

VAFB Vandenberg Air Force Base [California, US]

VAFC Volume Average Fuel Consumption

VAFI Victorian Association of Forest Industries [Australia]

VAG Vernacular Architecture Group [UK]

Vag Vagina [also vag] [Anatomy]

VAH Veterans Administration Hospital [Iowa City, US]

VAI Video-Assisted Instruction

VAI Voest-Alpine Industrieanlagenbau [Voest-Alpine Plant Engineering Division, Austria]

VAIL Variable Axis Immersion Lens

VAK Vertical Access Kit

VAK Vertical Assembly Kit

VAK Versuchs-Atomkraftwerk Kahl [Nuclear Power Pilot Plant at Kahl, Germany]

VA/kg Volt-Ampere per kilogram [also VA kg^{-1}]

Vak-Tech Vakuum-Technik [German Publication on Vacuum Technology]

VAL (Numerical) Value (Function) [also Val] [Computers]

VAL Vehicle Authorization List

VAL Vinyl Alcohol [also VAl]

VAL Visual Approach and Landing

VAL Voice Application Language

.VAL Validity Checks [Paradox File Name Extension]

V-Al Vanadium-Aluminum (Alloy)

Val Valid(ation) [also val]

Val (+)-Valine; Valyl

Val Valley [also val]

Val Value; Valuation [also val]

val valinate

Val valued

VALCO Volta Aluminum Company [also Valco]

VALD Research Vessel Valdivia [Germany]

Valid Validation [also valid]

VALL Variable Angle Load Lock (System)

VALOR Variable Locale and Resolution

VALSAS Variable Length Word Symbolic Assembly System

VALU Videotex Application Link Utilities

VAM Vasicular-Arbuscular Mycorrhizae [Botany]

VAM Vector Airborne Magnetometer

VAM Vinyl Acetate Monomer

VAM Virtual Access Method

VAM Virtual Adjunct Method

VAM Vogel's Approximation Method [Physics]

VAM Voltammeter

VAMAS Versailles Project on Advanced Materials and Standards [Joint Project of the Commission of the European Communities and the G-7 Nations]

VAMC Veterans Administration Medical Center [US]

VAMP Variable Anamorphic Motion Picture(s)

VAMP Vector Arithmetic Multiprocessor

VAMP Vincristine Amethopterin [Pharmacology]

VAMP Visual-Acoustic-Magnetic Pressure

VAN United States Navy Ship "Vanguard" [Space Tracking and Data Network]

VAN Value-Added Network

VANOS *(Verstellbares Nockenwellensystem)* – German for "Variable Camshaft Control"

VANS Value-Added Network Services

VANSGL Value-Added Network Services General License [UK]

VAOR VHF (Very High Frequency) Aural Omnirange

VAP Valence Alteration Pair

VAP Value Added Process

VAP Vinylacetylene Polymer

VAP Visual Aids Panel [of International Civil Aviation Organization]

VaP Vanadium Permandur

Vap Vapor [also vap]

VAPEX Vapor Extraction

VAPI Video Application Programming Interface

VAPI Visual Approach Path Indicator

VAPP Vector and Parallel Processor

Vap P Vapor Pressure [also Vap Press, vap press, or vap p]

Vap Prf Vapor-Proof [also Vap-Prf, vap prf, or vap prf]

VAPS Virtual Applications Prototyping System

VAPS Virtual Avionics Prototyping System

VAPS VSTOL (Vertical and/or Short Takeoff and Landing) Approach System

VA-PVP Vinyl Acetate–Vinylpyrrolidone Copolymer

Vapzn Vaporization [also vapzn]

VAR Vacuum Arc Refined; Vacuum Arc Refining [Metallurgy]

VAR Vacuum Arc Remelted; Vacuum Arc Remelting [Metallurgy]

VAR Value-Added Reseller

VAR Value-Added Resource

VAR Value-Added Retailer

VAR Variable [Computers]

VAR Variable Parameters [Pascal Programming]

VAR (Population) Variance [also Var, var, or σ^2] [Statistics]

VAR Varian Data Machine

VAR Variometer [also Var, or var]

VAR Varmeter [also Var, or var]

VAR Verification Analysis Report

VAR Vertical Aircraft Rocket

VAR VHF (Very High Frequency) (Visual-) Aural Range

VAR Video-Audio Range

VAR Visual and Aural Range; Visual-Aural (Radio) Range

VAR Volt-Ampere, Reactive [also var]

VAR Volunteer Air Reserve

Var Variable [also var]

Var (Population) Variance [also VAR, var, or σ^2] [Statistics]

Var Variant; Variation; Variety [also var]

Var Variometer [also var]

Var Varmeter [also var]

Var Varnish [also var]

Varactor Variable Reactor [also varactor]

VARAD Varying Radiation

VARC Variable Axis Rotor Control

Var-Cam Varnished-Cambric (Conductor Insulation)

VAR/ESR Vacuum Arc Remelted/Electroslag Refined; Vacuum Arc Remelting/Electroslag Refining [also VAR-ESR]

VARHM Var-Hour Meter [also varhm]

VARI Vacuum-Assisted Resin Injection

Varicap Variable-Capacitance Diode [also Varicap, or varicap]

Varindor Variable Inductor [also varindor]

Varistor Variable Resistor [also varistor]

Varistor Variable Transistor [also varistor]

Varitran Variable-Voltage Transformer [also varitran]

Varn Varnish [also varn]

VAROC Variable-Geometry Radial Outflow Compressor

Var Osc Variable Oscillator [also var osc]

VARPT Variational Perturbation Theory [Physics]

VARPTR Variable Pointer (Function) [Programming]

VARR Variable Range Reflector

VARR Visual Aural Radio Range

VARS Vertical Azimuth Reference System

Var(t) variance for time "t" [Symbol]

VAR(V) Volume Variance [also Var(V)]

VAR/VAD Vacuum-Arc Remelting/Vacuum-Arc Decarburization

Var(X) Variance of general function "X" [Symbol]

Var(x) Variance of variable "x" [Symbol]

VAS Value-Added Service

VAS Variable-Angle Spinning

VAS Vector Addition System

VAS Venomological Artifact Society [US]

VAS Vibration Absorption System

VAS Victorian Agricultural Strategy [Australia]

VAS Videodisk Authorizing System

VAS Virtual Address Space

VAS Visible Absorption Spectrophotometry

VAS VISSR (Visible/Infrared Spin-Scan Radiometer) Atmospheric Sounder [of Geostationary Operational Environmental Satellite, National Oceanic and Atmospheric Administration, US]

VAS Vortex Advisory System

VAs Volt-Ampere-second [Unit]

Vas Vaseline

VASCA Valve and Semiconductor Manufacturers Association [US]

VASCAR Visual Average Speed Computer and Recorder

VASG Vapor-Air Specific Gravity

VASE Variable-Angle Spectroscopic Ellipsometer; Variable-Angle Spectroscopic Ellipsometry

VASE FTIR Variable-Angle Spectroscopic Ellipsometry Fourier Transform Infrared [also VASE-FTIR]

VASI Visual Approach Slope Indicator

VASIS Visual Approach Slope Indicator System

VAs/m Volt-Ampere-second per meter [also VAs m^{-1}]

VASP Value-Added Service Producer

VASP Value-Added Service Provider

VASP Visible Absorption Spectrophotometry

VASS Variable-Angle Sample Spinning

VASS ViTS (Video Teleconferencing System) Automated Scheduling System

VAST Variable Array Storage Technology

VAST Vector Alignment Search Tool [of National Center for Biological Information, US]

VAST Versatile Automatic Specification Tester

VAST Versatile Avionic Shop Test (System)

VAST Versatile Avionics System Tester

VAST Vibration and Strength Analysis (Computer Program)

VAT Value-Added Tax [also vat]

VAT Vatu [Currency of Vanuatu]

VAT Vibrationally Assisted Tunnel(l)ing

VAT Vibro-Acoustic Test(ing)

VAT Virtual Address Translator

VAT Voice Activation Technology

VATA Vibro-Acoustic Test Article

VATE Versatile Automatic Test Equipment

VATF Vibration and Acoustic Test Facility

VATI Vermont Agricultural and Technical Institute [Randolph Center, US]

VATLS Visual Airborne Target Location System

V-ATPase Vacuolar Type H+-Adenosine 5'-Triphosphatase [Biochemistry]

VA/T/VTA Vibro-Acoustic/Thermal/Vacuum Test Article

VAU Vertical Arithmetic Unit

VAUS Value Added Utilization System

VAV Variable Air Volume

VAV Ventricular Assist Valve [Medicine]

VA/VE Value Analysis/Value Engineering

VAW Vereinigte Aluminiumwerke AG [German Manufacturer of Aluminum and Aluminum Products]

VAW Visual Analyst Workbench

VAW Volcanic Ash Warning

VAWR Vector Averaging Wind Recorder

VAWT Vertical Axis Wind Tunnel

VAWT Vertical Axis Wind Turbine

VAWTG Vertical Axis Wind Turbine Generator

VAX Virtual Address Extension [Digital Equipment Corporation Mainframe Computer]

VAX VTOL (Vertical Takeoff and Landing) Attack-Plane Experimental

VAXCOM VAX (Virtual Address Extension) Committee

VAX GKS Virtual Address Extension Graphics Kernel System

VAX PCA Virtual Address Extension Performance and Coverage Analyzer

VAX Prof VAX Professional [Publication of Digital Equipment Corporation, US]

VAX SPM Virtual Address Extension Software Performance Monitor

VAX/VMS Virtual Address Extension/VAX (Virtual Address Extension) Management System [of Digital Equipment Corporation, US]

VB Vacuum Bottoms

VB Valence Band [Solid-State Physics]

VB Valence Bond [Physical Chemistry]

VB Valve Box

VB Vanadium Bismuthate

VB Vanadium Monoboride

VB Vapor Barrier [Civil Engineering]

VB Variable Block

VB Variable Bomb

VB Vascular Bundle [Botany]

VB Vector Boson [Particle Physics]

VB Vertical Beam

VB Vibrator

VB (Microsoft) Visual Basic

VB Voice Band

VB Voice Box [Anatomy]

vβ Vee-Beta-Technology (Laser)

VBA Vibrating Beam Accelerometer

VBA Vinyl Benzoic Acid

VBA Visual Basic for Applications [Computers]

V band Microwave Frequency Band of 46.00 to 56.00 gigahertz

V-BAR Velocity Vectory Axis [Aerospace]

VBAS Von Braun Astronomical Society [US]

VBB Verein der Bibliothekare an öffentlichen Bibliotheken [Association of Librarians in Public Libraries, Germany]

VBBA Vietnam-Britain Business Association

VBBT Vanadium Bismuthate–Barium Titanate (Compound)

VBC Vacuum Barrier Corporation [Woburn, Massachusetts, US]

VBCDC Victorian Brown Coal Development Council [Australia]

VBD Veterinary Biologics Division [Now Part of Animal and Plant Health Inspection Service, US Department of Agriculture]

VBD Voice Band Data

VBDOS Valance Band Density of States [Solid-State Physics]

VBE Valinolbutylether

VBE Valinol t-Butyl Ether

VBE VESA (Video Electronics Standards Association) BIOS (Basic Input/Output System) Extension

VBE Volumetric Balance Equation

VBE/AI VESA (Video Electronics Standards Association) BIOS (Basic Input/Output System) Extension/Audio Interface

VBE/PM VESA (Video Electronics Standards Association) BIOS (Basic Input/Output System) Extension for Power Management

VBF Variable Block Format

VBFEh Verein zur Betriebsfestigkeitsforschung in der Eisenhüttenindustrie [Association for Reseach on Operational Strength of Materials in the Iron and Steel Industries]

VBG Verband der Berufsgenossenschaften [Federation of Professional/Trade Associations, Germany]

VBI Vertical Blanking Interval

VBI Vital Bus Inverter

VBL Vertical Bloch Line [Solid-State Physics]
VBL Voyager Biological Laboratory [of NASA]
VBM Valence Band Maximum [Solid-State Physics]
VBM Valence Bond Maximum [Physical Chemistry]
VBM Valence-Bond Method [Physical Chemistry]
VBM Vibratory Ball Mill(ing)
VBMA Vacuum Bag Manufacturers Association [US]
VBNS Very-High-Speed Backbone Network Service [of MCI Communications and National Science Foundation, US]
VBO Valence Band Offset [Solid-State Physics]
VBO Venezuelan Bolivar [Currency]
VBOMF Virtual Base Organization and Maintenance Processor
VBP Virtual Block Processor
VBR Variable Bit Rate
VBR Vinyl Bromide [also VBr]
VBR Virtual Bit Rate
VBRA Vehicle Builders and Repairers Association [UK]
VBRUN Visual Basic Runtime
VBS Valence Band Spectrum [Solid-State Physics]
VBS Vapor Barrier Sorbent (Blanket)
VBWR Vallecitos Boiling-Water Reactor [at Pleasanton, California, US]
.VBX Visual Basic Extension [File Name Extension]
VBXES Valence Band X-Ray Emission Spectroscopy
VC Vacuum Calorimeter
VC Vacuum Chamber
VC Vacuum Crystallizer
VC Valence Compound
VC Valence Crystal
VC Vanadium Carbide
VC Vane Control
VC Vapor Compression
VC Variable Capacitor
VC Variant Combination
VC Variant-Variant Correspondence [Metallurgy]
VC Varnished-Cambric (Insulated Electric Wire)
VC Vascular Cell [Botany]
VC Vascular Cylinder [Botany]
VC Vatican City
VC Vector Character
VC Vector Colorimeter
VC Vector Control [also V/C]
VC Vector Current [Particle Physics]
VC Velocity Coefficient
VC Velocity Counter [also V/C]
VC Vena Cava (of Heart) [also vc] [Anatomy]
VC Vena Contracta [also vc] [Fluid Mechanics]
VC Venture Capital(ist)
VC Verification Condition
VC Vernier Caliper
VC Versatility Code
VC Vertical Circle [Astronomy]
VC Vertical Coverage [Navigation]

VC Veterinary Corps [US]
VC Vice-Chancellor
VC Vice-Consul
VC Victoria Cross [UK]
VC Video Correlator
VC Vietcong
VC Vinyl Chloride
VC Vinylidene Chloride
VC Virtual Call
VC Virtual Cathode
VC Virtual Channel
VC Virtual Circuit
VC Virus Code [of International Committee on Taxonomy of Viruses Database]
VC Viscosity Cup
VC Visible Capacity
VC Visual Capacity
VC Visual Computer
VC Vitamin Chemist(ry)
VC Vitreous Carbon
VC Vocal Cord [Anatomy]
VC Voice Card
VC Voice Coil (of Loudspeaker)
VC Void Content
VC Voltage Circuit
VC Voltage Comparator
VC Voltaic Cell
VC Volt-Coulomb [also vc]
VC Volume Concentration
VC St. Vincent and Grenadines [ISO Code]
V$_2$C Divanadium Carbide
V/C Vector Control [also VC]
V/C Velocity Counter [also VC]
vc vena cava (of Heart) [also VC] [Anatomy]
VCA Value Control Amplifier
VCA Vanadium-Chromium-Aluminum-Type (Titanium Alloy)
VCA Viral Capsid Antigen [Immunology]
VCA Virtual-Crystal Approximation
VCA Voice Connecting Arrangement
VCA Voltage-Controlled Amplifier
VCAC Vice Chancellor's Advisory Council [of University of California at Berkeley, US]
VCAD Video Control and Drawing Chip
VCAM Vascular Cell Adhesion Molecule
VCAP Vehicle Charging and Potential Experiment
VCAP Vinyl-Capped Addition Polyimide
VCASS Visually Coupled Airborne Systems Simulator
VCAT Veterinary College Admission Test
VCB Vertical Location of the Center of Buoyancy
VCBA Variable Control Block Area
VCC VAXcluster Control Center (Unit)
VCC Venture Capital Corporation
VCC Verification Code Counter

VCC Vertical Continuous Casting

VCC Video Compact Cassette

VCC Virtual Channel Connection

VCC Visual Communications Congress

VCC Voice Control Center

VCC Voice-Controlled Carrier

VCC Voltage-Controlled Capacitor

VCCR Vienna Convention on Consular Relations

VCCS Video and Cable Communications Section

VCCS Voltage-Controlled Current Source

VCD Vacuum Carbon Deoxidation [Metallurgy]

VCD Valve Coil Driver

VCD Vapor Compression Distillation

VCD Variable-Capacitance Diode

VCD Variable Center Distance

VCD Verification Control Document

VCD Vibrational Circular Dichroism

VCD Vinylcyclohexene Dioxide

VCD Virtual Communications Driver

VCD Voltage Charge Device

VCD/HXSA Vinyl Cyclohexene Dioxide/Hexenyl Succinic Anhydride (Resin)

VCDO Vinyl Cyclohexene Diepoxide

VCDS Vapor Compression Distillation Subsystem

VCE Collective Emitter Voltage

VCE Variational Cumulant Expansion

Vce Voice [also vce]

VCESE Virginia Center for Electrochemical Science and Engineering [of University of Virginia, Charlottesville, US]

VCF Velocity Correlation Function

VCF Vinyl Chloroformate

VCF Visual Coordination Facility

VCF Voltage-Controlled Filter

VCF Voltage-Controlled Frequency

VCF&L Victoria Department of Conservation, Forests and Lands [Australia]

VCG Vapor Crystal Growth

VCG Vectorcardiogram; Vectorcardiography [Medicine]

VCG Verification Condition Generator

VCG Vertical Line through Center of Gravity

VCG Vertical Location of the Center of Gravity

VCGS Vapor Crystal Growth System

VCH Vinylcyclohexane

VCHD Vinylcyclohexene Dioxide

VCH 4-Vinyl Cyclohexane

VCI Vehicle Cone Index

VCI Velocity Change Indicator

VCI Verband der Chemischen Industrie [Chemical Industry Federation, Germany]

VCI Virtual Channel Indicator

VCI Virtual Circuit Identifier

VCI Volatile Corrosion Inhibitor

VC Inf Vena Cava Inferior [also vc inf] [Anatomy]

VCIS Voice Communications Instrument System

V-CITE Vertical-Cargo Integration Test Equipment

VCL Vectorized Link Cell (Method) [Molecular Dynamics]

VCL Vertical Center Line

VCL N-Vinyl Caprolactam

VCL Visual Component Library

VCLF Vertical Cask-Lifting Fixture

VCM Vacuum Condensible Material

VCM Variable-Capacitance Micromotor

VCM Variational Cellular Method

VCM Vehicle Condition Monitor

VCM Vibrating-Coil Magnetometer

VCM Victorian Chamber of Mines [Australia]

VCM Vinyl Chloride Monomer

VCM Volatile Combustible Material; Volatile Condensible Material

VCM Voltage-Controlled Multivibrator

V cm^{-2} Volt(s) per square centimeter kelvin [also Vcm^{-2}, or V/cm^2]

V/cm Volt(s) per centimeter [also V cm^{-1}]

V/cm^2 Volt(s) per square centimeter [also V m^{-2}]

VCMA Vacuum Cleaner Manufacturers Association [US]

V cm^{-2}K^{-1} Volt(s) per square centimeter kelvin [also Vcm^{-2}K^{-1}, V/cm^2·K, or V/cm^2-K]

VCMO Volatile Condensible Materials–Optical

VCN Vanadium Carbonitride

VCN Verification Completion Notice

VCN Visual Communication Network

VCNO Vice Chief of Naval Operations [of US Department of the Navy]

VCNR Voltage-Controlled Negative (Differential) Resistance

VCO Variable Crystal Oscillator

VCO Voice-Controlled Oscillator

VCO Voltage-Controlled Oscillation; Voltage-Controlled Oscillator

VCO Voluntary Conservation Organization

VCOAD Voluntary Committee on Overseas Aid and Development [UK]

VCOC Venture-Capital Operating Company

VCONFIG Video Configuration

VCONFIG VGA (Visual Graphics Adapter) Configuration

VCOS Visual Caching Operating System

VCP Vacuum Condensing Point

VCP Vacuum Cup Pencil

VCP Vandenberg (Air Force Base) Contract Report

VCP Vinylcyclopentane

VCP Virtual Control Panel

V(Cp)$_2$ Bis(cyclopentadienyl)vanadium (or Vanadocene)

V(Cp)$_2$Cl$_2$ Bis(cyclopentadienyl)vanadium Dichloride (or Vanadocene Dichloride)

VCp(CO)$_4$ Cyclopentadienylvanadium Tetracarbonyl

VCPI Virtual Control Program Interface

VCR Variable Compression Ratio

VCR Variable-Configuration Resonator

VCR Vehicle Condition Report

VCR Vertical Crater Retreat

VCR Video Cartridge Recorder

VCR Video Cassette Recorder; Video Cassette Recording

VCR Vincristine [Pharmacology]

VCR Visual Control Room

VCR Voltage-Controlled Resistor

VCR Volumetric Compression Ratio

V-Cr Vanadium-Chromium (Alloy)

VCRD V-22 Composite Repair Development (Program)

VCRG VHF (Very-High Frequency) Channel Requirements Group [of International Civil Aviation Organization]

VCRO Validity Check and Readout

V-Cr-Ti Vanadium-Chromium-Titanium (Alloy System)

VCS Vacuum Science Workshop [of Battelle National Laboratory, US]

VCS Validation Control System

VCS Vehicle Control Station [of Defense Research Establishment Suffield, Canada]

VCS Ventilation Control System

VCS Verification Control Sheet

VCS Vice Chief of Staff

VCS Victorian Computer Society [Australia]

VCS Video Cassette System

VCS Video Communications System

VCS Video Computer System

VCS Vibration Control System

VCS Visually Coupled System

VCS VMS (Voice Message System) Controller Software

VCS Voice Command System

VCS Voltage Calibration Set

VCSEL Vertical-Cavity Surface-Emitting Laser

VCSR Vehicle Control Station Research [of Defense Research Establishment Suffield, Canada]

VCSR Voltage-Controlled Shift Register

VCSUSA Chief of Staff, US Army

VCSUSAF Vice Chief of Staff, US Air Force

VCT Variable Capacitance Transducer

VCT Victoria, Texas [Meteorological Station Designator]

VCT Voltage Control Transfer

VCT Volume Control Tank

VCTCA Virtual Channel-to-Channel Adapter

Vctr Vector [also vctr]

VCU Variable Correction Unit

VCU Video Combiner Unit

VCU Video Control Unit

VCU Virginia Commonwealth University [Richmond, US]

VC-VA Vinyl Chloride/Vinyl Acetate (Copolymer)

VC-VAC Vinyl Chloride/Vinyl Acetate (Copolymer)

VC-VDC Vinyl Chloride/Vinylidene Chloride (Copolymer)

VCVS Voltage-Controlled Voltage Source

VCXO Voltage-Controlled Crystal Oscillator

VCZ N-Vinyl Carbazole [also Vcz]

VD Vacuum Decarburization

VD Vacuum-Degassed; Vacuum Degassing [also V-D]

VD Vacuum Deposited; Vacuum Deposition

VD Vacuum Destillation

VD Vacuum Diffusion

VD Vacuum Diode

VD Vacuum Dried; Vacuum Drying

VD Vandyke

VD Vapor Density [also vd]

VD Varactor Diode

VD Vas Deferens [Anatomy]

VD Vat Dye

VD Vector Diagram

VD Velocity Distribution

VD Velocity of Detonation

VD Venereal Disease [Medicine]

VD Vibration Damper; Vibration Damping

VD Vibration Detector

VD Video Disk

VD Vinylidene

VD Virtual Data

VD Virtual Device

VD Virtual Displacement

VD Viscous Damping [Mechanical Engineering]

VD Viscous Drag [Fluid Mechanics]

VD Voice Data

VD Void [also Vd, or vd]

VD Voltage-Dependence; Voltage-Dependent

VD Voltage Detector

VD Voltage Divider

VD Voltage Doubler

VD Voltage Drop

VD Volume Damper

VD Volume Diffusion

VD Volume Dose

VD Voronoi Diagram

VD Vortex Dynamics

V/D Voice Data

V-D Vacuum Degassing [also VD]

vd various dates

VDA Valve Driver Assembly

VDA Variable Data Area

VDA Verbal Delay Announcement

VDA Verband der Deutschen Automobilindustrie [German Autombile Manufacturers Federation]

VDA Verband Deutscher Adreßbuchverleger [German Association of Directory Publishers]

VDA Vertical Danger Angle

VDA Video Distribution Amplifier

V&DA Video and Data Acquisition

V&DA Video and Data (Processing) Assembly

VDA-FS Verband der Deutschen Automobilindustrie/Freeform Surfaces [German Autombile Industry Association/Freeform Surfaces]

VDAM Virtual Data Access Method

V-Day Victory Day [also V-day]

VDAS Vibration Data Acquisition System

VDB Vector Data Buffer

VDB Verein Deutscher Bibliothekare [German Library Association]

VDB Video Display Board

VDC Vaccine-Damaged Children

VDC VAX (Virtual Address Extension) Data Console

VDC Versatile Digital Controller

VDC Video Data Controller

VDC Video Display Controller

VDC Video-Documentary Clearinghouse [US]

VDC Vinylidene Chloride

VDC Virginia Department of Conservation [US]

VDC Virtual Device Coordinate

VDC Viscosity-Density Constant

VDC Visual Display Controller

VDC Volts of Direct Current [also Vdc, vdc, V DC, V/DC, v dc, V/dc, V D-C, or v d-c]

VDCE Victoria (State) Department of Conservation and Environment [Australia]

VDCT Direct Current Test Volts [also vdct]

VDCU Videograph Display Control Unit

VDCW Direct Current Working Volts [vdcw]

VDD Verification Description Document

VDD Version Description Document

VDD Vietnamese Dong [Currency]

VDD Virtual Device Driver

VDD Visual Display Data

VDD Voice Digital Display

VdDB Verein der Diplom-Bibliothekare an Wissenschaftlichen Bibliotheken [Association of Graduate Librarians in Academic/Research Libraries, Germany]

VDDI Voyager Data Detailed Index [NASA]

VDDL Voyager Data Distribution List [NASA]

VDDM Virtual Device Driver Manager

VDDS Voyager Data Description Standards [NASA]

VDE Variable Displacement Engine

VDE Variable Display Equipment

VDE Verband Deutscher Elektrotechniker [German Association of Electrical Engineers]

VDE Vertical Detachment Energy

VDE Video Display Editor

VDE Visual Development Environment

VDE Voice-Data Entry

VDE Fachber VDE Fachberichte [Technical Reports of Verband Deutscher Elektrotechniker, Germany]

VDEh Verein Deutscher Eisenhüttenleute [Association of German Iron and Steel Engineers]

VDES Voice-Data Entry System

VDET Voltage Detector

VDETS Voice Data Entry Terminal System

VDE/VDI Verband Deutscher Elektrotechniker/Verein Deutscher Ingenieure [Association of German Electrical Engineers/Association of German Engineers]

VDEW Vereinigung Deutscher Elektrizitätswerke [German Association of German Electric Power Companies]

VDF Vapor-Deposited Carbon Fiber

VDF VHF (Very High Frequency) Direction Finder; VHF (Very High Frequency) Direction Finding

VDF Video Frequency

VDF Vinylidene Fluoride

VDF Virginia Division of Forestry [US]

VDFG Variable Diode Function Generator

VdFuOI Verband der Deutschen Feinmechanischen und Optischen Industrie [German Association for Precision Mechanics and Optics]

VDG Van de Graaff (Accelerator) [also vdG]

VDG Variable Drive Group

VDG Verein Deutscher Giessereifachleute [Association of German Foundry Engineers and Technologists]

VDG Vereinigung Deutscher Gewässerschutz [German Water Pollution Control Association]

VDG Video Data Generator

VDG Video Display Generator

VdG Van de Graaf (Accelerator) [also VDG]

VDGH Verband der Diagnostica- und Diagnosticageräte-Hersteller [German Diagnostics and Diagnostic Equipment Manufacturer]

VDGS Visual Docking Guidance System

VDHF Variational Dirac-Hartree-Fock [Quantum Mechanics]

VDI Vendor Documentation Inventory

VDI Verein Deutscher Ingenieure [Association of German Engineers]

VDI Vertical Display Indicator

VDI Video Device Interface

VDI Video Display Input

VDI Video Display Interface

VDI Virtual Device Interface

VDI Visual Display Input

VDI Visual Doppler Indicator

VDI/ADB Verein Deutscher Ingenieure–Gesellschaft Produktionstechnik [Association of German Ingenieure–Production Engineering Society]

VDI-BDW Verein Deutscher Ingenieure–Bundesverband der Deutschen Wirtschaft e.V. [Association of German Engineers–Federation of German Trade and Commerce]

VDI-Ber VDI-Berichte [Technical Reports of the Verein Deutscher Ingenieure]

VDI-EKV Verein Deutscher Ingenieure–Gesellschaft Entwicklung Konstruktion Vertrieb [Association of German Engineers–Development, Design and Distribution Society]

VDI Forschungsh VDI Forschungshefte [Research Journal of the Verein Deutscher Ingenieure, Germany]

VDI-GKE Verein Deutscher Ingenieure–Gesellschaft Konstruktion und Entwicklung [Association of German Engineers–Design and Development Society]

VDI-KLM Verein Deutscher Ingenieure–Kommission Lärmminderung [Association of German Engineers–Commission on Noise Abatement, Germany]

VDI Nachr Verein Deutscher Ingenieure Nachrichten [News Magazine of the Verein Deutscher Ingenieure, Germany]

VDIR View Direction

VDISK Virtual Disk [also VDisk]

VDI-TZ Verein Deutscher Ingenieure–Technologiezentrum [Technology Center of the Association of German Engineers]

V/div Volt(s) per (oscilloscope) division [Unit]

VDI/VDE Verein Deutscher Ingenieure/Verband Deutscher Elektrotechniker [Association of German Engineers/ Association of German Electrical Engineers, Germany]

VDI/VDE-GMA Verein Deutscher Ingenieure/Verband Deutscher Elektrotechniker–Gesellschaft für Meß– und Automatisierungstechnik [Association of German Engineers/Association of German Electrical Engineers– Society for Measurement and Automation Technology]

VDI-W Verein Deutscher Ingenieure/Werkstoffgesellschaft [Materials Society of the Verein Deutscher Ingenieure, Germany]

VDI-Z Verein Deutscher Ingenieure Zeitschrift [Journal of the Verein Deutscher Ingenieure, Germany]

VdL Verband der Lackindustrie e.V. [Association of the Lacquer and Varnish Industries, Germany]

VDK Vicinal Diketone

VDKSPI V. D. Kuznetsov Siberian Physiotechnical Institute [Russian Federation]

VDL Van Diemen's Land [Former Name of "Tasmania"]

VDL Verband Deutscher Diplomlandwirte [Association of German Agricultural Engineers]

VDL Verband Deutscher Lokomotivindustrie [German Locomotive Manufacturers Federation]

VDL Vienna Definition Language [Computer Language]

VDL Vision Development Language

VDLUFA Verband Deutscher Landwirtschaftlicher Untersuchungs– und Forschungsanstalten [Federation of German Institutes for Agricultural Research]

VDM Varian Data Machines

VDM Vector Dominance Model

VDM Verein Deutscher Metallhändler [German Association of Metal Mongers, Germany]

VDM Vereinigte Deutsche Metallwerke AG [A German Manufacturer of Metallic Materials]

VDM Video Display Module

VDM Virtual Device Metafile

VDM Virtual DOS (Disk-Operating System) Machine [of IBM Corporation, US]

VDM Visual Display Module

VDMA Variable Destination Multiple Access

VDMA Verband Deutscher Maschinen– und Anlagenbau [German Machinery and Plant Manufacturers Association]

VDMAD Virtual Direct Memory Access Device

VDOS Vacuum Distillation/Overflow Sampler

VDOS Vibrational Density of States [Solid-State Physics]

VDP Vacuum Diffusion Pump

VDP Vacuum Distillation Process

VDP Van der Pauw Structure

VDP Vapor Deposition Polymerization

VDP Vertical Data Processing

VDP Video Data Processor

VDP Visual Descent Point

VdP Van de Pauw (Method) [Physics]

vdP-H Van de Pauw-Hall (Analysis) [Physics]

VDPI Verein Deutscher Postingenieure [Association of German Postal Engineers]

VDPI Voyager Data Processing Instruction [NASA]

VDR Video Disk Recorder; Video Disk Recording

VDR Voice and Data Recording

VDR Voice Digitization Rate

VDR Voltage-Dependent Resistor

VDRA Voice and Data Recording Auxiliary

VDRG Verband Deutscher Rundfunk– und Fernsehfachgroßhändler [Association of German Radio and Television Wholesalers]

VDRIVER Video Driver

VDRIVER VGA (Visual Graphics Adapter) Driver

VDRL Venereal Disease Research Laboratory (Test) [of US Public Health Service]

VDS Valence Defect Structure

VDS Vandenberg (Air Force Base) Data Systems

VDS Vapor-Deposited Silica (Process)

VDS Variable-Depth Sonar

VDS Vehicle Dynamics Simulator

VDS Verband Deutscher Studentenschaft [German Student Association]

VDS Verein Deutscher Schleifmittelwerke e.V. [German Abrasives Manufacturer Association]

VDS Vereinigung Deutscher Schmelzhütten [German Association of Smelting Works]

VDS Very Difficultly Soluble

VDS Video Distribution System

VDS Virtual DMA (Direct Memory Access) Services

VDS Visual Docking Simulator

VDS Voice Data Switch

VDSI Verein Deutscher Sicherheitsingenieure [Association of German Safety Engineers]

VDSL Very-High-Data-Rate Digital Subscriber Line

VDT Variable Density Tunnel (Aerodynamics)

VDT Vehicle Data Table

VDT Video Data Terminal

VDT Video Dial Tone

VDT Video Display Terminal

VDT Visual Display Terminal

Vdt Validate [also vdt]

VDTA Vacuum Dealers Trade Association [US]

VdTÜV Vereinigung der Technischen Überwachungsvereine e.V. [Federation of Technical Inspection Associations, Germany]

VDU Video Display Unit

VDU Visual Display Unit

VDUC Video Display Unit Controller

VDUC Visual Display Unit Controller

VD/VOD/VID Vacuum Degassing/Vacuum-Oxygen Decarburization/Vacuum Injection Degassing (Process) [Metallurgy]

VDW Valley Density Wave [Physics]

VDW Van der Waals (Force) [also vdW] [Physical Chemistry]

VDW Verein Deutscher Werkzeugmaschinenfabriken e.V. [German Machine-Tool Manufacturers Association]

vdW Van der Waals (Force) [also VDW] [Physical Chemistry]

VDWG Van der Waals Gap [also VdWG, or VWG] [Physical Chemistry]

VDWS Van der Waals Surface [also VdWS] [Thermodynamics]

VDX Videotex

V/dyn/cm² Volt per dyne per square centimeter [also $V \ dyn^{-1} \ cm^{-2}$]

VE Vacuum Engineer(ing)

VE Valance Electron

VE Value Engineer(ing)

VE Vane Engine

VE Vapor Ejector

VE Variable Element

VE Vehicle Emission

VE Vehicle Engineer(ing)

VE Vehicle Engineering [NASA Kennedy Space Center Directorate, Florida, US]

VE Venezuela [ISO Code]

VE Vernal Equinox [Astronomy]

VE Vernier Engine [Aerospace]

VE Vibrational Energy [Physical Chemistry]

VE Vibration Eliminator

VE Vinyl Ester

VE Vinyl Ether

VE Virtual Endoscopy [Medicine]

VE Viscoelastic(ity)

VE Visual Examination

VE Visual Exempted

VE Vitreous Enamel

VE Vocational Education

VE Volume Expander; Volume Expansion

V(E_F) average drift velocity [Symbol]

.ve Venezuela [Country Code/Domain Name]

−ve negative [Electrical Engineering]

+ve positive [Electrical Engineering]

VEA Value Engineering Association

VEA Virtual Effective Address

VEB Variable Elevation Beam

VEB Venezuelan Bolivar [Currency]

VEB *(Volkseigener Betrieb)* – State Owned Company [*Note:* Former East German Corporate Form]

VEC Valance-Electron Concentration

VEC Variable Energy Cyclotron

VECI Vehicle Emission Control Information

VECI Vehicular Equipment Complement Index

VECIB Vehicle Engineering Change Implementation Board [of NASA Kennedy Space Center]

VECO Vernier Engine Cutoff [Aerospace]

VECOS Vehicle Checkout Set

VECP Value Engineering Change Proposal

VECP Value Engineering Cost Proposal

VECU Vacuum Pump Exhaust Cleanup System

VED Vapor-Phase Epitaxial Deposition

Vedar Visible Energy Detection and Ranging [also vedar]

VE-Day Victory in Europe Day [also VE day, V-E Day, or V-E day]

VEDC Vitreous Enamel Development Council [UK]

VEDS Vehicle Emergency Detection System [of NASA]

VEdS Vocational Education Specialist

VEE Venezuelan Equine Encephalomyelitis [Medicine]

VEEC Victorian Environmental Education Council [Australia]

VEEI Vehicle Electrical Engine Interface

VEELS Vibrational Energy-Loss Electron Spectroscopy

VEEV Venezuelan Equine Encephalomyelitis Virus

VEFCO Vertical Functional Checkout

VEG Verband des Elektro-Großhandels e.V. [Electrical Wholesalers Association, Germany]

Veg Vegetable [also veg]

VEGA Vegetable Growers of America [US]

VEGA Video-7 Enhanced Graphics Adapter [of Video-7, Inc., US]

VEGF Vascular Endothelial Growth Factor [Biochemistry]

VEGG Vernier Engine/Gas Generator

Veg J Vegetarian Journal [US]

Veg Times Vegetarian Times [US]

Veg Voice Vegetarian Voice [US]

VEH Valence Effective Hamiltonian

Veh Vehicle [also Veh, or veh]

VEH ID Vehicle Identification [also Veh ID]

Veh Syst Dyn Vehicle Systems Dynamics [Publication of International Association for Vehicle Systems Dynamics, Netherlands]

VEI Vehicle End Item

Vel Vellum [also vel]

Vel Velocity [also vel]

VELCOR Velocity Correction [also Velcor]

VELD Valence Electron Localization Degree (Model) [Solid-State Physics]

VELF Velocity Filter

VELF Verwaltung für Ernährung, Landwirtschaft und Forsten [Food, Agriculture and Forestry Administration, Germany]

Veloc Velocity [also veloc]

VEM Vasoexcitor Material

VEM Virtual Electrode Model

VEMM Virtual Expanded Memory Manager

VEMMI Versatile Multimedia Interface

VEN Virtual Equipment Number

Ven Venice [Italy]

ven venerable

VenAmCham Venezuelan-American Chamber of Commerce and Industry [Venezuela]

Venez Venezuela(n)

Vent Ventilate(d); Ventilation; Ventilator [also vent]

VENUS Valuable and Efficient Network Utility Service

VEO Voluntary Environmental Organization

VEOMP Valence Electron-Only Model Potential [Solid-State Physics]

VEP Visual Evoked Potential

VEPB Variable Energy Positron Beam,

VEPCO Virginia Electric and Power Company [US]

VEPE Vehicle Experimental and Proving Establishment [Canada]

VEPIS Vocational Education Program Information System

VER Verify [Computers]

VER Version [Data Communications]

VER Visual Evoked Response

VER Voluntary Export Restraint

Ver (Computer) Program Version [also ver]

Ver Verification; Verifier; Verify [also ver]

Ver Vernier [also ver]

Ver Version [also ver]

Ver Vertical [also ver]

VERA Verbundstudie Ernährungserhebung und Risikofaktorenanalytik [Cooperative Study: Nutritional Survey and Risk Factor Analysis, Germany]

VERA Verglasungsanlage für Radioaktive Abfälle (Prozess) [Radioactive Waste Vitrification Plant (Process)]

VERA Versatile Experimental Reactor Assembly

VERA Vision Electronic Recording Apparatus

VERDAN Versatile Differential Analyzer

Verif Verification [also verif]

VERL Vacuum Evaporation on Running Liquids (Process)

VERLORT Very Long-Range Tracking (Radar)

VERN Vernier [also Vern]

VERNITRAC Vernier Tracking by Automatic Correlation

VERONICA Very Easy Rodent-Oriented Net-Wide (or Network) Index to Computer(ized) Archives [also Veronica] [Internet]

VEROS Vacuum Evaporation on Running Oil Surface (Process)

Verpack-Rundsch Verpackungs-Rundschau [German Packaging Review]

VERPROT Verification Prototype

VERR Verify Read Access

VERRES Verification Resumé

Verres Réfract Verres et Réfractaires [French Publication on Glass and Refractories]

vers versed sine [also versine]

VERSCOM Verification for Special Commission

Versine versed sine [also vers]

VER SYN Verification Synergies

Vert Vertebra [also vert] [Anatomy]

Vert Vertebrate [also vert]

vert vertical

Ver Tag Verification Tagging [also ver tag]

VERTIC Verification Technology Information Center [UK]

VERVIS Vertical Visibility [Meteorology]

VERW Verify Write Access

VES Vapor Extraction System

VES Variable Elasticity of Substitution

VES Vibrational Excitation Spectroscopy

VES Video Encoding Standard

Ves Vessel [also ves]

VESA Video Electronics Standards Association

VESC Vehicle Equipment Safety Commission [US]

VESIAC Vela Seismic Information Analysis Center [US]

VESR Vellecitos Experimental Superheat Reactor [US]

VEST Volunteer Engineers, Scientists and Technicians

Vestn Akad Nauk BSSR Vestnik Akademii Nauk BSSR [Publication of the Academy of Sciences of the Byelorussian Soviet Socialist Republic]

Vestn Akad Nauk Kazakh SSR Vestnik Akademii Nauk Kazakhskoi SSR [Publication of the Academy of Sciences of the Kazakh SSR]

Vestn Akad Nauk SSSR Vestnik Akademii Nauk SSSR [Publication of the Academy of Sciences of the USSR]

Vestn Leningr Univ Fiz Khim Vestnik Leningradskogo Universiteta Fizika i Khimiya [Russian Publication of Leningrad University on Physics and Chemistry]

Vestn Mashinostr Vestnik Mishinostroeniya [Russian Publication on Engineering Research]

Vestn Mosk Univ, Fiz-Astron Vestnik Moskovskogo Universiteta, Fizika-Astronomiya [Russian Publication of Moscow University on Astronomical Physics]

Vestn Mosk Univ, Khim Vestnik Moskovskogo Universiteta, Khimiya [Russian Publication of Moscow University on Chemistry]

Vestn Mosk Univ, Vychisl Mat Kibern Vestnik Moskovskogo Universiteta, Vychislitel'naya Matematika i Kibernetika Russian Publication of Moscow University on Computational Mathematics and Cybernetics]

Vestsi Akad Navuk BSSR, Biyal Navuk Vestsi Akademii Navuk BSSR, Seriya Biyalogichnykh Navuk [Publication of the Academy of Sciences of the Byelorussian SSR on Biological Sciences]

Vestsi Akad Navuk BSSR, Fiz Energ Navuk Vestsi Akademii Navuk BSSR, Seriya Fizika Energetychnykh Navuk [Publication of the Academy of Sciences of the Byeolorussian SSR on Physics–Energy Sciences]

Vestsi Akad Navuk BSSR, Fiz-Mat Vestsi Akademii Navuk BSSR, Seriya Fizika-Matematichnykh Navuk [Publication of the Academy of Sciences of the Byelorussian SSR on Physics–Mathematical Sciences]

Vestsi Akad Navuk BSSR, Fiz-Tekh Vestsi Akademii Navuk BSSR, Seriya Fizika-Technichnykh Navuk [Publication of the Academy of Sciences of the Byelorussian SSR on Physics–Technical Sciences]

Vestsi Akad Navuk BSSR, Khim Navuk Vestsi Akademii Navuk BSSR, Seriya Khimichnykh Navuk [Publication of the Academy of Sciences of the Byelorussian SSR on Chemical Sciences]

Vestsi Akad Navuk BSSR, Ser Biyal Navuk Vestsi Akademii Navuk BSSR, Seriya Biyalogichnykh Navuk [Publication of the Academy of Sciences of the Byelorussian SSR on Biological Sciences]

Vestsi Akad Navuk BSSR, Ser Fiz Energ Navuk Vestsi Akademii Navuk BSSR, Seriya Fizika Energetychnykh Navuk [Publication of the Academy of Sciences of the Byelorussian SSR on Physics–Energy Sciences]

Vestsi Akad Navuk BSSR, Ser Fiz-Mat Vestsi Akademii Navuk BSSR, Seriya Fizika-Matematichnykh Navuk [Publication of the Academy of Sciences of the Byelorussian SSR on Physics–Mathematical Sciences]

Vestsi Akad Navuk BSSR, Ser Fiz-Tekh Vestsi Akademii Navuk BSSR, Seriya Fizika-Technichnykh Navuk [Publication of the Academy of Sciences of the Byelorussian SSR on Physics–Technical Sciences]

Vestsi Akad Navuk BSSR, Ser Khim Navuk Vestsi Akademii Navuk BSSR, Seriya Khimichnykh Navuk [Publication of the Academy of Sciences of the Byelorussian SSR on Chemical Sciences]

VESV Vesicular Exanthema of Swine Virus

VET Vibrational Energy Transfer [Physical Chemistry]

VET Visual Editing Terminal

Vet Veteran [also vet]

Vet Veterinary; Veterinarian [also vet]

VETA Verification Test Article

VETDOC Veterinary Documentation (Database) [UK]

Veter Veterinarian; Veterinary [also veter]

V(EthO)$_4$ Vanadium Ethoxide (or Vanadium Ethylate)

VETL Vehicle Emission Testing Laboratory [Canada]

Vet Med Veterinary Medicine [US Journal]

Vet Med Small Animal Clin Veterinarian Medicine and Small Animal Clinician [US Journal]

VETS Veteran Employment and Training Service [of US Department of Labor]

Vets Veterans [also vets]

VEV Voice-Excited Vocoder

VEW Vereinigte Edelstahlwerke AG [Major Austrian Steelmaker]

VE-WO$_3$ Vacuum Evaporated Tungsten Trioxide

VEWS Very Early Warning System

VF Vacuum Feedthrough

VF Vacuum Filter; Vacuum Filtration

VF Vacuum Flask

VF Vaccum Freezing

VF Vacuum Furnace

VF Valence Force [Physics]

VF Valet and Fert (Theory) [Materials Science]

VF Value Foundation [now Miles Value Foundation, US]

VF Variable Feedforward

VF Variable Frequency

VF Vector Field

VF Velocity Filter

VF Vertical Flight

VF Very Fine

VF Vibrational Frequency

VF Vibration Foundation [now Vibration Institute, US]

VF Video Frequency

VF Vinyl Fluoride

VF Vinylidene Fluoride

VF Virtual Floppy

VF Viscous Flow

VF Visibility Factor [Electronics]

VF Visual Field

VF Vogel-Fulcher (Law) [Physics]

VF Voice Foundation [US]

VF Voice Frequency [also vf]

VF Voltage-to-Frequency [also V-F, or V/F]

VF Vulcanized Fiber

V/F (Alternate) Voice/Facsimile (Transmission)

VF$_2$ Vinylidene Difluoride

VF$_3$ Vinylidene Trifluoride

VFA Vicia Faba Agglutinin [Immunology]

VFA Volatile Fatty Acid

VFA Volunteer Fire Alarm

VFAT Virtual File Allocation Table

VFB Velocity Feedback

VFB Vertical Format Buffer

VFB Voltage, Flatband

VFB Voltage Feedback

VFC Variable File Channel

VFC Vertical Format Control

VFC Vertical Forms Control

VFC Video Film Converter

VFC Video Frequency Carrier

VFC Voice-Frequency Carrier

VFC Voltage-to-Frequency Converter

V.FC Version.Fast Class [Data Communications Standard]

VFCGL Visiting Fellowships in Canadian Government Laboratories [of Natural Sciences and Engineering Council, Canada]

VFCT Voice-Frequency Carrier Telegraph(y)

VFCT Voice-Frequency Carrier Teletype

VFD Vacuum Fluorescent Diode

VFD Vacuum Fluorescent Display

VFD Variable Frequency Drive

VFDB Vereinigung zur Förderung des Deutschen Brandschutzes [German Association for the Advancement of Fire Protection]

VFE Vendor Furnished Equipment

VFE Verband der Führungskräfte der Eisen- und Stahlerzeugung und –verarbeitung e.V. [Association of Iron and Steel Manufacturing and Processing Executives, Germany]

V-Fe-Mn Vanadium-Iron-Manganese (Alloy System)

VFET Variable Field-Effect Transistor

VFET Vertical Field-Effect Transistor

VFF Valence-Force Field [Physics]

VFF Vertical Flight Foundation [Alexandria, Virginia, US]

VFF Victorian Farmers Federation [Australia]

VFFT Voice Frequency Facility Terminal

VFH Vertical Flow Horizontal

VFI Vehicle Flight Instrumentation

VFI Verification Flight Instrumentation

VFL Variable Field Length

VFL Variable Focus Lens

VFL Vertex Focal Length

VFL Voice Frequency Line

VFM View Finder Monitor

VFM Vinyl Fluoride Monomer

VFMED Variable Format Message Entry Device

VFO Variable-Frequency Oscillator

VFO Voice-Frequency Oscillator

VFOP VFR (Visual Flight Rules) Operations Panel [of International Civil Aviation Organization]

VFOX Visual FoxPro [also VFP]

VFP Variable Frequency Pulse

VFP Visual FoxPro [also VFOX]

VFPYP Visual FoxPro Yellow Pages

VFR Visual Flight Rules

VFRG Visual Flight Rules Group [of International Civil Aviation Organization]

VFS Vacuum Furnace System

VFS Velocity-Focusing Spectrograph

VFS Vertical Full Scale

VFSS Variable Frequency Selection System

VFSTC Valley Forge Space Technology Center [of General Electric Company, Valley Forge, Pennsylvania, US]

VFT Verification Flight Test

VFT Very Fast Train

VFT Vibrational Frame Transformation

VFT Voice-Frequency Telegraph(y) [also vft]

VFT Voice-Frequency (Carrier) Telegraphy [also vft]

VFT Voice-Frequency Telephone; Voice-Frequency Telephony [also vft]

VFT Volume Fraction Transfer (Scheme) [Metallurgy]

VfT Verkaufsgesellschaft für Teererzeugnisse mbH [German Tar Products Sales Company]

VFT-3 Three-Channel Voice Frequency Telegraph(y)

VFT-6 Six-Channel Voice Frequency Telegraph(y)

VFTEh Verein zur Förderung der Forschung im Transportwesen der Eisenhüttenindustrie [Association for the Promotion of Transportation Research in the Iron and Steel Industries Transport]

VFTG Voice Frequency Telegraph

VFTH Vogel-Fulcher-Tamman-Hesse (Equation) [Physics]

VFU Vertical Format Unit

VFU Vocabulary File Utility

VFVC Vacuum Freezing Vapor Compression

VF$_2$-VF$_3$ Vinylidene Difluoride–Vinylidene Trifluoride (Copolymer)

VFW Video For (Microsoft) Windows

VFW Vereinte Flugtechnische Werke-Fokker [A German Aircraft and Aerospace Manufacturer]

VFWF Verein zur Walzwerksforschung [Rolling Mill Research Society, Germany]

VFX Variable-Frequency Crystal Oscillator

VG Vacuum Gauge

VG Vacuum Generator

VG Vapor Growing; Vapor-Grown

VG Vector Generator

VG Vector Geometry

VG Velocity Gradient

VG Venturi Ga(u)ge

VG Vertical Grain [Lumber]

VG Very Good [also vg]

VG (Automatic) Vickers Gun

VG Virgin Islands (British) [ISO Code]

VG Viscosity Grade

VG Voice Grade

VG Voltage Gain

VG Voltage Gradient

VG Voorhees and Glicksman (Theory) [Metallurgy]

v(G) Crystal Potential [Symbol]

vg Very Good [also VG]

VGA Vacuum Gas Atomization [Powder Metallurgy]

VGA Vapor Generation Accessory

VGA Variable Gain Amplifier

VGA Vicia Graminea Agglutinin [Immunology]

VGA Video Graphics Array

VGA Video Graphics Adapter

VGAA Vegetable Growers Association of America

VGADDS Vertical General Area Detector Diffraction System

VGAHi Visual Graphics Adapter, High (Mode)

VGALo Visual Graphics Adapter, Low (Mode)

VGAM Vector Graphics Access Method

VGAMed Visual Graphics Adapter, Medium (Mode)

VGATEST Visual Graphics Adapter Test

VGB Vereinigung der Großkraftwerksbetreiber e.V. [Association of Large Power Plant Operators, Germany]

VGB Kraftwerkstech VGB-Kraftwerkstechnik [German Publication on Power Plant Engineering of Vereinigung der Grosskraftwerksbetreiber]

VGB Kraftwerkstech (Ger Ed) VGB-Kraftwerkstechnik (German Edition) [German Publication of Vereinigung der Grosskraftwerksbetreiber]

VGC Vacuum Gauge Calibration

VGC Vacuum Gauge Control(ler)

VGC Video Graphics Controller

VGC Viscosity-Gravity Constant

VGCF Vapor-Grown Carbon Fiber

VGCS Vacuum Gage Calibration System

VGDC Victorian Geographic Data Committee [Australia]

v-GeO$_2$ Vitreous Germanium Dioxide

VGF Vertical Gradient Freeze (Crystal Growing Technique)

VGF Vertical Gradient Furnace (Method) [Metallurgy]

VGI Voice Group Interface

VGKL Verband der Deutschen Groß– und Außenhandels für Krankenpflege- und Laborbedarf [German Laboratory and Hospital Supplies Wholesalers and Exporters Association]

VGL Virus Control Laboratory [of Oak Ridge National Laboratory, Tennessee, US]

VGM Virtual Graphics Machine

VGIMU Velocity to be Gained as Related to IMU (Inertial Measurement Unit) Orientation [Aerospace]

VGO Vacuum Gas Oil (Process) [Chemical Engineering]

VGOR Vandenberg (Air Force Base) Ground Operations Requirement

VGOR Vehicle Ground Operation Requirements

VGP Vehicle Ground Point

VGP Virtual Geomagnetic Pole

VGP Void Growth Parameter [Metallurgy]

VGPI Visual Glide Path Indicator

VGR Video Graphics Recorder

VGR Voyager (Program) [of NASA]

V-Gr Vanadium-Graphite (System)

VG/RIA Vision Group of Robotics Industries Association [now Automated Vision Association, US]

VGS Vermont Geological Survey [US]

VGS Vision Guidance System

VGS Visual Guidance System

VGSI Visual Glide Slope Indicator

VGT Variable Geometry Truss

VGT Vehicle Ground Test

VGU Video Generation Unit

VGVT Vertical Ground Vibration Test

VGX Velocity To Be Gained (Body X-Axis) [Aerospace]

VGY Velocity To Be Gained (Body Y-Axis) [Aerospace]

VGZ Velocity To Be Gained (Body Z-Axis) [Aerospace]

VH Vanadium Hydride

VH Varhour Meter

VH Velocity Head [Fluid Mechanics]

VH Vickers Hardness [Metallurgy]

VH Vilsmeier-Haack (Synthesis) [Chemistry]

VH Vinylic Hydrogen

VH Virtual Height [Geophysics]

VH Vitreous Humor [Anatomy]

VH- Australia [Civil Aircraft Marking]

V&H Vertical and Horizontal (Resolution)

V/H Velocity-to-Height (Ratio)

V/H Vertical/Horizontal (Adjustment)

VHA Very-High Accuracy

VHA Very-High Altitude

VHAA Very-High-Altitude Abort

VHAP Volatile Hazardous Air Pollutants

VHB Very-High Brightness

VHD Very-High Density

VHD Video High Density

VHDL Very-High-Density Lipoprotein [Biochemistry]

VHDL Very-High-Density Logic

VHDL VHSIC (Very-High-Speed Integrated Circuit) Hardware Description Language

VHE Very-High Energy

VHES Vitro Hanford Engineering Services [Washington State, US]

VHF Very-High Frequency [Frequency: 30 to 300 megahertz; wavelength: 1 to 10 meters] [also vhf]

VHF/AM Very-High Frequency Amplitude Modulator

VHF Commun VHF Communications [Published in Germany]

VHFDF Very-High Frequency Direction Finding [also VHF/DF]

VHF-GD Very-High Frequency Glow Discharge

VHF/PTN VHF Radio Connection to Public Telephone Network

VHF/UHF Very-High Frequency/Ultrahigh Frequency [also vhf/uhf]

VHI Vapor Hazard Index

VHIPS Very-High Impact Polystyrene

VHL Vacuum Hydraulic Lamination

VHL Von Hippel-Lindau (Disease) [Medicine]

VHLL Very-High-Level (Computer) Language

VHM Vector Helium Magnetometer

VHM Very-High Modulus

VHM Virtual Hardware Monitor

VHN Vickers Hardness Number [Metallurgy]

VHO Very-High Order

VHO Very-High Output

VHOK Vertical Hatch Operations Kit [NASA]

VHOL Very-High-Order (Computer) Language

VHP Vacuum Hot Pressed; Vacuum Hot Pressing

VHP Very-High Performance

VHP Very-High Pressure

VHPBH Vacuum Hot Pressing (Separate) Brazed-on-Hub

VHPed Vacuum Hot Pressed

VHPIC Very-High-Performance Integrated Circuit

VHPing Vacuum Hot Pressing

VHR Very-High Reduction

VHR Very-High Resolution

VHRR Very-High Resolution Radiometer

VHS Vancouver Historical Society [Canada]

VHS Van Hove Singularity [also vHS]

VHS Very-High Speed

VHS Video Home System

VHS Virtual Host Storage

VHS *(Volkshochschule)* – German for "Adult Education Center"

VHS-C Video Home System–Compact

VHSI Very-High-Scale Integrated; Very-High-Scale Integration

VHSI Very-High-Speed Integrated; Very-High-Speed Integration

VHSIC Very-High-Scale Integrated Circuit; Very-High-Speed Integrated Circuit

VHT Vacuum Heated Treated; Vacuum Heat Treatment [Metallurgy]

VHT Vapor Hydration Test

VHT Very-High Temperature

VHTR Very-High-Temperature Reactor
VHV Very-High Vacuum
VHV Very-High Voltage
VI Vancouver Island [British Columbia, Canada]
VI Variable Integral
VI Variable Intensity
VI Vertical Integration
VI Vibration Institute [US]
VI Vibration Isolation; Vibration Isolator
VI Vickers Indentation [Mechanical Testing]
VI Video Integrator
VI Virgin Islands [US]
VI Virgin Islands (US) [ISO Code]
VI Virtual Image
VI Virtual Instrument(ation)
VI Viscosity Index [Chemical Engineering]
VI Visual Information
VI Visual Inspection
VI Visual Intelligence Corporation [US]
VI Visual Interactive (Editor) [Unix Operating System]
VI Vocational Institute
VI Voice Input
VI Volume Indicator
V-I Voltage-Current (Characteristics)
Vi Virginium [Obsolete; now Francium]
Vi Virulent Surface Antigen [Immunology]
vi (vide infra) – see below
Vi Virgin Islands (US) [Country Code/Domain Name]
VIA Value-Impact Assessment
VIA VAX (Virtual Address Extension) Information Architecture
VIA VectorNet Interface Adapter
VIA Versatile Interface Adapter
VIA Videotex Industry Association [US]
VIA Visual Interactive Access
VIAS Voice Intelligibility Analysis Set
VIAS Voice Interference Analysis Set
VIB Variationally Induced Breathing
VIB Vertical Integration Building [Aerospace]
VIB Vibrationally-Induced Breathing (Model)
Vib Vibrate; Vibration; Vibrator [also vib]
VIBL Variable Intensity Back Lighting
VIC Vapor Injection Cure (Process)
VIC Variable Instruction Computer
VIC Victoria (State) [Australia] [also Vic]
VIC Video Interface Chip
VIC Vienna International Center [Austria]
VIC Virtual Interaction Controller
VIC Visitor Information Center
Vic Vicinity [also vic]
vic- vicinal (compound) [Chemistry]
VICA Vocational Industrial Clubs of America [US]
VICAM Virtual Integrated Communication Access Method
VICAR Video Image Communication and Retrieval

VICARS Visual Integrated Crime Analysis and Reporting System
Vicastrong Vickers-Armstrong (Gun)
VICC Vehicle Information Center of Canada [Markham, Ontario, Canada]
Vice-Adm Vice-Admiral [also VA, or VAdm]
VicEPA Victoria (State) Environmental Protection Authority [Australia]
VICI Velocity Indicating Coherent Integrator
VICOM Visual Communication
VID Vacuum Injection Degassed; Vacuum Injection Degassing [Metallurgy]
VID Video [also Vid, or vid]
VID Vietnamese Dong [Currency]
VID Virtual Image Distance
Vid (vide) – see
VIDA VAX/IBM (Virtual Address Extension/IBM Corporation) Data Access
VIDA Ventricular Impulse Detection and Alarm (Device)
VIDAC Virtual Data Acquisition and Control
VIDAC Visual Data Acquisition
VIDAC Visual Information Display and Control
VIDAMP Video Amplifier
VIDAS Video Digital Analysis System [also Vidas]
VIDAT Visual Data Acquisition
VIDD Vertical Interval Data Detector
VIDE Virus Identification Data Exchange [of Research School for Biological Sciences, Australian National University, Canberra]
Vide Couches Minces Vide, les Couches Minces [French Journal on Vacuum and Thin Films published by Societé Française du Vide]
VIDEdB VIDE (Virus Identification Data Exchange) Database [of Research School for Biological Sciences, Australian National University]
VIDEO VORTEX (Versatile Omnitask Real-Time Executive) Interactive Data Entry Operation
Videotex Viewp Videotex Viewpoint [Published in the UK]
VIDF Vertical Side of Intermediate Distribution Frame
VIDF Video Frequency
Vid i (vide infra) – see below
VIDIAC Video Input-to-Automatic Computer
VIDIAC Vidicon Input-to-Automatic Computer
VIDIAC Visual Information Display and Control
VIDIST Vacuum Injection Distillation
VIDO Veterinary Infections Disease Organization [Canada]
VIDP Vacuum Injection Degassing and Pouring [Metallurgy]
VIDS Video Interactive Display System
VIDS Visual Information Display System
Vid s (vide supra) – see above
VIE Victorian Institute of Engineers [Australia]
VIE Visual-Indicator Equipment
VIE Virtual Information Environment
Viet Vietnam(ese)
VIEW Virtual Instrument Engineering Workbench

VIEW Visual Information Enhanced Workstation

VIF Vacuum Induction Furnace

VIF Visiting Industrial Fellowship

VIF Virtual Interface

VIF Virtual Interrupt Flag

VIFC VTOL (Vertical Takeoff and Landing) Integrated Flight Control

VIFCS VTOL (Vertical Takeoff and Landing) Integrated Flight Control System

VIFI Voyager Information Flow Instruction [NASA]

VIG Video Integrating Group

VIG Visual Integrating Group

VIGS Video Interactive Gunnery Simulator

VIGS Virtual-Induced Gap States

VIGS Visual Glide Slope

VIH Voltage-Input High

VIK Verein für Internationale Krankentransporte [German Association for International Ambulance Services]

VIK Vereinigtes Institut für Kernforschung [Nuclear Research Institute, Russia]

VIK Vereinigung Industrielle Kraftwirtschaft [Association of the Power-Supply Industry, Germany]

VIL Vertical In-Line [Electronics]

VIL Voltage-Input Low

VIL Volume Imaging Lidar

Vil Village [also vil]

VILP Vector Impedance Locus Plotter

VILSU V. I. Lenin State University [Russian Federation]

VIM Vacuum Induction Melted; Vacuum Induction Melting [Metallurgy]

VIM Vendor Independent Mail

VIM Vendor Independent Messaging (Interface)

VIM Vibrational Microlamination

VIM Video Interface Module

VIM Vision Input Module

VIM Vocational Instructional Materials

VIM/ESR/VAR Vacuum Induction Melted/Electroslag Refined/Vacuum Arc Remelted; Vacuum Induction Melting/Electroslag Refining/Vacuum Arc Remelting [also VIM-ESR-VAR] [Metallurgy]

VIMIS Vertically-Integrated Metal-Insulator Semiconductor

VIMP Vancouver Island Mainland Pipeline [Canada]

VIMS Victorian Institute of Marine Science [Australia]

VIMS Virginia Institute of Marine Science [US]

VIMS Visual and Infrared Mapping Spectrometer

VIMS Visualization in Materials Science (Program) [of National Science Foundation, US]

VIMSIS Victorian Institute of Marine Sciences Information System [Australia]

VIMTPG Virtual Interactive Machine Test Program Generator

VIM/VAR Vacuum Induction Melted/Vacuum Arc Remelted; Vacuum Induction Melting/Vacuum Arc Remelting [also VIM-VAR] [Metallurgy]

VIN Vehicle Identification Number

VIN Voltage Input

VIND Vicarious Interpolations Not Desired

VINES Virtual Networking (Operating) System

VINS Velocity Inertial Navigation System

VINS Vermont Institute of Natural Science [US]

VINS Very Intense Neutron Source

VINT Video Integration

VIO Video Input/Output

VIO Virtual Input/Output

VIOC Variable Input/Output Code

VIOL Violuric Acid

Viol violet

VIOLET Voice Input/Output Lexically Endowed Terminal

VIOS Versatile Instrument Operating Software

VIP Vacuum Injection Pouring [Metallurgy]

VIP Variable Information Processing

VIP Variable Input Phototypesetting

VIP Vasoactive Intestinal Peptide (or Polypeptide) [Biochemistry]

VIP Vector Instruction Processor

VIP VectorNet Interface Processor

VIP Vee-Shaped Isolation Regions filled with Polycrystalline Silicon [Electronics]

VIP Verifying Interpreting Punch

VIP Verkehrs-Informations-Programm [Computer-Controlled Radio Traffic Service, Germany]

VIP Versatile Information Processor

VIP Vertical Ionization Potential

VIP Very Important Person

VIP Videodisk Innovation Project [of Utah State University, Logan, US]

VIP Video Information Provider

VIP Virtual Interrupt Pending

VIP Vision Integrated with Positioning

VIP Visual Image Processor

VIP Visual Information Processing; Visual Information Processor

VIP Visual Information Projection

VIP Visual Information Protocol

ViP Visual Programming

VIPER Verifiable Integrated Processor for Enhanced Reliability

VIPER Video Processing and Electronic Reduction

VIPER Visualization in Phase Equilibrium Relationships (Program) [US]

VIPP Variable Information Processing Package

VIPS Variable Information Processing System

VIPS Video Ice Particle Sampler

VIPS Voice Interruption Priority System

VIR Vacuum Infrared

VIR Valves-in-Receiver System

VIR Vertical Interval Reference (Signal)

VIR Vertical Interval Retrace

VIR Visible and Infrared Spin-Scan Radiometer

VIR Vulcanized India Rubber

Vir Virginia [US]

Vir Virgo (or Virgin) [Astronomy]

Virginia Tech Virginia Polytechnic Institute and State University [Blacksburg, US]

VIRL Visible and Near Infrared Lidar

VIRNS Velocity Inertia Radar Navigation System

VIROC Visible System of Information Retrieval by Optical Coordination

Virol Virological; Virologist; Virology [also virol]

Virol Virology [Journal published in the US]

VIRR Visible and Infrared Radiometer

VIRS Vertical Interval Reference Signal

VIRS Visible and Infrared Scanner

VIRSR Visible Infrared Scanning Radiometer

VIRSS Visual and Infrared Screening Smoke (Grenade)

Virtual Rev The Virtual Review [of Brown University, Providence, Rhode Island, US]

VIS Verification Information System

VIS Veterinary Investigation Service [UK]

VIS Video Information System

VIS Videotex Information System

VIS Visible Spectroscopy

VIS Visual Instruction Set

VIS Visual Instrumentation Subsystem; Visual Instrumentation System

VIS Voice Information System

Vis Viscosity [also vis]

Vis Viscount

Vis Visibility; Visible; Vision; Visual [also vis]

VISAM Virtual Indexed Sequential Access Method

VISC Video Disc

ViSC Visualization in Scientific Computing

Visc Viscosity [also visc]

visc viscous

Vis Comput Visual Computer [Published in Germany]

Visible Lang Visible Language [Published in the US]

VisiCalc Visible Calculator

VISIT Vehicle Internal Systems Investigative Team (Project) [US]

VISITT Vendor Information System for Innovative Treatment Technologies

VISLAP Viscoelastic Lamination Theory Program

VIS-NIR Visible/Near Infrared

VISQI Visual Quality Indicator

Vis Res Vision Research(er)

Vis Res Vision Research [Journal published in Germany]

Vis Res Visiting Researcher

Vis Sci Vision Science; Vision Scientist

VISSR Visible and Infrared Spin-Scan Radiometer [of Geostationary Operational Environmental Satellite; of National Oceanic and Atmospheric Administration, US]

VISTA Validation Intelligent Software for Thermal Applications

VISTA Verbal Information Storage and Text Analysis

VISTA Vertical Integration of Science, Technology and Applications

VISTA Visual Interpretation System for Technical Applications

VISTA *(Visuelles Interpretationssystem für Technische Anwendungen)* – German for "Visual Interpretation System for Technical Applications"

VISTA Volunteers in Service to America

Vistas Astron Vistas in Astronomy [Published in the UK]

VIS-UV Visible/Ultraviolet Spectrometer

VIT Variable Inductance Transducer

VIT Variable-Impedance Transformer

VIT Vehicle Integration Test

VIT Vertical Interval Test (Signal)

VIT Vertriebsingenieurtagung [Conference of Sales Engineers; of Verein Deutscher Ingenieure, Germany]

Vit Vitamin [also vit]

vit vitreous

VITA VME (Versa Module Europe) Bus International Trade Association

VITA Volunteers for International Technical Assistance [US]

Vit A Vitamin A (or Retinol)

Vit A_1 Vitamin A_1 (or Retinol)

Vit A_2 Vitamin A_2

VITAL Variably Initialized Translator for Algorithmic Languages

VITAL VAST (Versatile Automatic Specification Tester) Interface Test Application Language

Vit B Vitamin B (or Thiamine)

Vit B_1 Vitamin B_1 (or Thiamine)

Vit B_2 Vitamin B_2 (or Riboflavin)

Vit B_3 Vitamin B_3 (or Panthothenic Acid, or Niacin)

Vit B_5 Vitamin B_5

Vit B_6 Vitamin B_6

Vit B_{12} Vitamin B_{12} (or Cobalamin)

Vit B_{17} Vitamin B_{17} (or Laetrile)

Vit B_c Vitamin B_c (or Folic Acid)

Vit B_T Vitamin B_T (or Carnitine)

Vit C Vitamin C (or Ascorbic Acid)

Vit D Vitamin D (or Antirachitic Vitamin, or Sunshine Vitamin)

Vit D_2 Vitamin D_2 (or Ergocalciferol)

Vit D_3 Vitamin D_3 (or Cholecalciferol)

Vit E Vitamin E

VITEAC Video Transmission Engineering Advisory Committee [US]

Vit F Vitamin F

Vit G Vitamin G (or Riboflavin)

Vit H Vitamin H (or Biotin)

Vit K Vitamin K (or Phylloquinone, or Antihemorrhagic Vitamin)

Vit K_1 Vitamin K_1 (or Phytonadione)

Vit K_2 Vitamin K_2 (or Menaquinone)

Vit K_3 Vitamin K_3 (or Menadione)

Vit M Vitamin M (or Folic Acid)

VITO Vlaamse Institut voor Technologische Onderzoek [Flemish Institute for Technological Research, Belgium]

Vit P Vitamin P

Vit PP Vitamin PP (or Niacin)

vitr vitreous

Vitr Enameller Vitreous Enameller [Published in the UK]

Vit Ret Vitamin Retailer [US Journal]

VITS Vertical Insertion Test Signal

VITS Vertical Interval Test Signal

ViTS Video Teleconferencing System

VITT (Space) Vehicle Integration Test Team

Vit U Vitamin U

VIU Video Interface Unit

VIU Voice Interface Unit

VIURAM Video Interface Unit Random-Access Memory [also VIU RAM]

VIV Variable Inlet Vane

Viz *(videlicet)* – namely; that is

VJ Vacuojunction [Electronics]

VJ Vacuum Jacket(ed) [Electrical Engineering]

VJ Vacuum Jet

VJ-Day Victory in Japan Day [also VJday, V-J Day, or V-J day]

VJFET Vertical Junction Field-Effect Transistor

VJ HEADER Van Jacobsen Header (Compression) [Unix Operating System]

VK Volterra Kernel

VKα Vanadium K-Alpha (Radiation) [also Vk$_\alpha$]

VKβ Vanadium K-Beta (Radiation) [also Vk$_\beta$]

VKE Verband Kunststofferzeugende Industrie [Plastics Producers Association, Germany and Switzerland]

VKI Verband Kunststoff-Industrie [Plastics Industry Association, Switzerland]

VKI Verein für Konsumenten-Information [Consumer Information Association, Austria]

VKIFD Von Kármán Institute for Fluid Dynamics [Brussels, Belgium]

VKW Vereinigte Kesselwerke AG [German Manufacturer of Steam Generators, Mechanical Stokers, Fluidized-Bed Equipment and Steam and Nuclear Plant Accessories located at Duesseldorf]

VL δ-Valerolactam

VL Valerolactone

VL Vapor Lamp

VL Vapor-Liquid (Interface) [also V/L, or V-L]

VL Vapor Lock [Gasoline Engines]

VL Variable-Length [Computers]

VL Vegard's Law [Metallurgy]

VL Vertical Ladder

VL VESA (Video Electronics Standards Association) Local

VL Vibrational Luminescence

VL Video Logic

VL Virtual Library

VL Visible Language

VL Visible Light

VL Visual Learning

VL Visible Line

VL Vortex Line [Fluid Mechanics]

Vℓ Interfacial Interaction Potential [Symbol]

V-L Vapor-Liquid (Interface) [also V/L, or VL]

V(λ) Luminosity Function (or Spectral Luminous Efficiency Function) [Symbol]

V(λ) Balancing Voltage [Symbol]

VLA Very Large Array [of National Radio Astronomy Observatory, Socorro, New Mexico, US]

VLA Very-Low Altitude

VLAN Virtual Local-Area Network

VLB Very-Long Baseline [Electronics]

VLB Very-Low Brightness

VLB Vincaleucoblastine [Biochemistry]

VLBA Very-Long-Baseline Array

VLBI Very-Long-Baseline Interferometer; Very-Long-Baseline Interferometry

VL-BUS VESA (Video Electronics Standards Association) Local-Bus

VLC Video Level Controller

VLC Visual Logic Controller

VLCC Very-Large Cargo Carrier

VLCC Very-Large Crude Carrier

VLCR Variable Length Cavity Resonance

VLCS Voltage-Logic Current-Switching

VLD Variable-Length Decoder

VLD Very-Low Density

VLD Visible Laser Diode

Vld valid

VLDL Very-Low-Density Lipoprotein [Biochemistry]

VLDPE Very-Low-Density Polyethylene [also VLD-PE]

VLE Vapor Levitation Epitaxy

VLE Vapor/Liquid Equilibrium

VLE Voice Line Expansion

VLED Visible Light-Emitting Diode

VLF Variable-Length Field [Computers]

VLF Vertical Launch Facility [Aerospace]

VLF Very-Low Frequency [Frequency: 3 to 30 kilohertz; wavelength: 10,000 to 100,000 meters] [also vlf]

VLFEM Very-Low Frequency Electromagnetic(s) [also VLF-EM]

VLHC Very-Large Hadron Collider (Workshop) [of Fermilab, Batavia, Illinois, US]

VLFS Variable Low Frequency Standard

VLINE Vertical Line [also VLine, or Vline]

VLINK Visited Link [Internet]

VLIW Very-Large Instruction Word

VLIW Very-Long Instruction Word

VLM Vereinigte Leichtmetallwerke GmbH [German Manufacturer of Light Metal Alloys and Products]

VLM Virtual Loadable Module

VLM Visible Laser Module

VLM VLX Large Memory [*Note: VLX* is a Trade Name]

VLMR Very-Large (Positive) Magnetoresistance [Solid-State Physics]

VLN Very-Low Nitrogen (Steelmaking Process) [Metallurgy]

vlnt violent

VLO Varnish Linseed Oil

VLP Variable Leakages Path (Sensor)

VLP Very-Low Pressure

VLP Video Long Play(er); Video Long-Playing

VLP Virus-Like Particle [Microbiology]

VLPCVD Very-Low-Pressure Chemical-Vapor Deposition

VLPE Very-Long-Period Experiment

VLPP Very-Low-Pressure Pyrolysis

VLPR Very-Low-Pressure Reactor

VLPS Vandenberg (Air Force Base) Launch Processing System

VLR Variable Length Record

VLR Variable Linear Resistor

VLR Very-Long Range

VLR Very-Low Range

VLR Visitor Location Register [Telecommunications]

VLS Vacancy Lattice Site [Solid-State Physics]

VLS Vacuum Loading-Assist System

VLS Vacuum Loading System

VLS Valence Level Spectroscopy

VLS Vandenberg Launch and Landing Site [of Vandenberg Air Force Base, California, US]

VLS Vapor-Feed-Gases Liquid-Catalyst Solid-Crystalline (Whisker Growth); Vapor-Liquid-Solid (Whisker Growing Process) [Materials Science]

VLS Vehicle Locator System

VLS Very-Large Scale

VLS Virtual Linkage System

VLS Volume Loadability Speed

VLS/CVD Vapor-Liquid-Solid (Transformation)/Chemical-Vapor Deposition

VLSI Very-Large-Scale Integrated; Very-Large-Scale Integration

VLSIC Very-Large-Scale Integrated Circuit

VLSI/CAD Very-Large-Scale Integration/Computer-Aided Design

VLSIIC Very-Large-Scale Integration Implementation Center [Canada]

VLSIPS Very Large Scale Immobilized Polymer Synthesis

VLSI Syst Des VLSI (Very-Large-Scale-Integrated) Systems Design [Published in the US]

VLSI Tech Symp VLSI (Very-Large-Scale-Integrated) Technical Symposium [US]

VLSI/ULSI Very-Large-Scale Integration/Ultra-Large-Scale Integration (Technology)

VLS-LCVD Vapor-Liquid-Solid Laser-Assisted Chemical Vapor Deposition

VLSW Virtual Line Switch

VLT Variable List Table

VLT Very-Low Temperature

VLT Video Lotto Terminal

Vlt Violet

V(l,t) Velocity Difference Between Two Points Separated by Distance l [Symbol]

Vlv Valve [also vlv]

VLVS Voltage-Logic Voltage Switching

VM Vacuum Melting

VM Vacuum Metallizing

VM Vacuum Metallurgy

VM Vacuum Mixer; Vacuum Mixing

VM Vacuum Monochromator

VM Varnishmaker; Varnishmaking

VM Vector Meson [Particle Physics]

VM Vector Message

VM Vector Method

VM Vector Model (of Atoms)

VM Velocity Meter [also V/M]

VM Velocity-Modulated (Tube) [also vm]

VM Velocity Modulation

VM Vening Meinesz (System) [Geodesy]

VM Venturi Meter

VM Vertical Measurement

VM Vibratory Mill

VM Victorian Museum [Australia]

VM Videomicroscope; Videomicroscopy

VM Video Measurement (System)

VM Video Monitor

VM Viomycin [Microbiology]

VM Virgin Material

VM Virgin Metal

VM Virtual Machine

VM Virtual Mass

VM Virtual Memory

VM Volatile Matter

VM Voltage Multiplication; Voltage Multiplier

VM Voltage Measurement

VM Volume Measurement

VM Voltmeter

VM Volumetric Measure

V/M Velocity Meter [also VM]

V/m Volt(s) per meter [also v/m, VPM, Vpm, vpm, or V m^{-1}]

VMA Vacuum Metallizers Association [now Association of Industrial Metallizers, Coaters and Laminators, US]

VMA Valid Memory Address

VMA Valve Manufacturers Association [US]

VMA DL-Vanillomandelic Acid; Vanillylmandelic Acid [Biochemistry]

VMA (Space) Vehicle Maintenance Area

VMA Virtual Machine Assist

VMA Virtual Memory Address

VMA Virtual Memory Allocation

VMACS Video Marshall Access Control System [of NASA Marshall Space Flight Center, Huntsville, Alabama, US]

VMAI Veterinary Medicine Association of Ireland

VMAPS Virtual Memory Array Processing System

V-51 MAS Vanadium-51 Magic Angle Spinning [also ^{51}V MAS]

VMB Velocity-Modulated (Electron) Beam

VMB Virtual Machine Boot

V/μbar volt per microbar [also V μbar^{-1}]

VM/BSE Virtual Machine/Basic System Extension

VMC Variable Message Cycle

VMC Variable Modulus Counter

VMC Variational Monte Carlo (Method)

VMC Vermont Monitoring Cooperative [of University of Vermont, Burlington, US]

VMC Vertical Machining Center

VMC Vertical Motion Carriage

VMC Video Matrix Control

VMC Virtual Machine Control

VMC Visual Meteorological Conditions

VMCB Virtual Machine Control Block

VMCF Virtual Machine Communication Facility

VMCM Vector Measuring Current Meter

VM/CMS Video Monitor/Cambridge Monitoring System [of IBM Corporation, US]

VMD Vacuum Metallurgy Division [of American Vacuum Society, US]

VMD Vector Meson Dominance [Particle Physics]

VMD Vertical Magnetic Dipole

VMD *(Veterinariae Medicinae Doctor)* – Doctor of Veterinary Medicine

VMD Virtual Manufacturing Device

VMDF Vertical Side of Main Distribution Frame

VME Versa Module Europe (Bus) [Computers]

VME Virtual Machine Environment

VME Virtual Memory Environment

VMEbus Versa Module Europe Bus

VMET Velocity-Modulated Electron Tube

VMF Vertical Maintenance Facility

VMFP Volumetric Mean Free Path

VMG Vickers Machine Gun

VMgBr Vinylmagnesium Bromide

VMGSE (Space) Vehicle Measuring Ground Support Equipment

VMH Vickers Microhardness [Metallurgy]

VMH Visual Maneuvering Height

VMI Vertical Motion Index

VMI Virginia Military Institute [Lexington, US]

VMIC VLSI (Very-Large-Scale Integration) Multi-Interconnection Conference

VMID Virtual Machine Identifier

V/mil Volt(s) per mil [also v/mil, VPM, vpm, or V mil^{-1}]

VMJ Vertical Multijunction

VML Veterinarian Medical Laboratory

VMM Variable Mission Manufacturing

VMM Virtual Machine Monitor

VMM Virtual Memory Manager

V/mm Volt(s) per millimeter [also V mm^{-1}]

V/μm Volt(s) per micrometer [also V μm^{-1}]

V-Mo Vanadium-Molybdenum (Alloy System)

VMOS Vertical-Channel Metal-Oxide Semiconductor

VMOS Vertical Metal-Oxide Semiconductor (Technology)

VMOS Virtual Memory Operating System

VMOS V-Type Metal-Oxide Semiconductor

VMOSFET V-Type Metal-Oxide Semiconductor Field-Effect Transistor

VMP Vegetation Management Program [Northern Territory, Australia]

VMP Vertical Microprogramming

VMP Virtual Modem Protocol

VM&P Varnishmakers and Painters (Naphtha) [also VMP]

VMPA Vancouver Museums and Planetarium Association [Canada]

VMPA Verband der Materialprüfämter [Federation of Materials Testing Institutes, Germany]

VMRI Veterinary Medicine Research Institute [of Iowa State University, Ames, US]

VMRS (Space) Vehicle Maintenance Reporting Standards

VMS Valve Monitoring System

VMS Variable Magnetic Shunt

VMS VAX (Virtual Address Extension) Management System

VMS Velocity Measuring System

VMS Victorian Military Society [UK]

VMS Virtual Machine Storage

VMS Virtual Memory System

VMS Visual Management System

VMS Voice Mailbox Service

VMS Voice Message System

VMS Voice Messaging Service

VMS Volcanogenic Massive Sulfide

V/ms Volt(s) per millisecond [also V/msec]

V/μs Volt(s) per microsecond [also V/μsec]

VM/SE Virtual Machine/System Extension

V/msec Volt(s) per millisecond [also V/ms]

VM/SP Virtual Machine/System Product

VMT Variable Microcycle Timing

VMT Variable Mu Tube

VMT Velocity-Modulated Tube

VMT Vertical Magnetotransistor

VMT Video Matrix Terminal

VMT Virtual Memory Technique

VMTAB Virtual Machine Table

VMTSS Virtual Machine Time-Sharing System

VMU Voice Management Unit

VMX Voice Mail Exchange

VMX Voice Message Exchange

VN 1-Phenyl-2-Chloro Ethanone

VN Valence Number

VN Valeronitrile

VN Vanadium Nitride

VN Verify Number

VN Vertical Needle

VN Vietnam [ISO Code]

VN β-Vinylnaphthalene

VN Volatile Nitrogen

V/N Volt(s) per Newton [also V N^{-1}]

V(N$_d$) Volume of Average Number of d-Electrons per Atom [Symbol]

VNA Virtual Network Architecture

VNA Visiting Nurse Association [US]

VNA Very Narrow Aisle (Truck)

VNAV Vertical Navigation

VNb Vanadium-Niobium (Alloy)

VNbN Vanadium-Niobium-Nitrogen (Alloy)

VNbTa Vanadium-Niobium-Tantalum (Alloy)

VNC Vallecitos Nuclear Center [of General Electric Company, Pleasanton, California, US]

VNC Victorian Naturalists' Club [Australia]

VNC Voice Numerical Control

VNCIAWPRC Venezuelan National Committee of the International Association on Water Pollution Research and Control

VND Vietnamese Dong [Currency]

V(NEt$_2$)$_4$ Tetrakis(diethylamido)vanadium

VNG Virginia National Guard [US]

VNIR Visible and Near Infrared

VNL Via Net Loss

VNLF Via Net Loss Factor

V/nm Volt(s) per nanometer [also V nm^{-1}]

V-51 NMR Vanadium-51 Nuclear Magnetic Resonance [also ^{51}V NMR]

VNN Vacant National Number

VNP Vinylnitrate Polymer

VNS Visiting Nurse Service

VNSP Vacant Nozzle Shield Plug

VNTR Variable Number Tandem Repeats [Genetics]

VNTT Bis(vanillin)triethylenetetramine

VO Space Vehicle (or Shuttle) Operations [NASA Kennedy Space Center Directorate]

VO Vacuum-Tube Oscillator

VO Vanadium Oxide

VO Vanadyl (Radical) [Symbol]

VO Vaporizing Oil

VO Vatican Observatory [Italy]

VO Vegetable Oil

VO (Space) Vehicle Operations

VO Verbal Orders

VO Vinyl-Oxi-Etoxi-Methyloxirane

VO Virtual Object

VO Voice Output

VO Voice-Over

VO Void

VO$_x$ Vanadium Oxides (i.e., VO, V$_2$O$_2$, V$_2$O$_3$, V$_2$O$_4$, or V$_2$O$_5$) [also VOx]

V/O Volume-to-Outflow (Ratio) [Ecology]

Vo Voice

V-ω Signal Amplitude versus Modulation Frequency [Symbol]

v/o volume percent

V$_c$(ω) (angular) frequency dependent critical velocity [Symbol]

VOA Voice of America [US]

VOA Volatile Organic Analysis; Volatile Organic Analyte

VOC Vacancy-Ordered Cesium Chloride [Solid-State Physics]

VOC Variable Output Circuit

VOC Vinyloxycarbonyl

VOC Voice (Format), Creative

VOC Voice of the Customer

VOC Voice-Operated Coder

VOC Volatile Organic Carbon

VOC Volatile Organic Chemical(s)

VOC Volatile Organic Compound(s)

VOC Volatile Organic Component(s)

Voc Vocabulary [also voc, Vocab, or vocab]

VOCAL Voluntary Organizations in Communication and Language [UK]

VOC-Arid ID Volatile Organic Compounds–Arid Integrated Demonstration (Network Database)

VOCCl Vinyloxycarbonyl Chloride

VOCD Vacuum Oxygen Carbon Deoxidation [Metallurgy]

VOCED Vocational Education and Training Database

VOCl Volatile Organic Chlorinated Compound

VOCl$_3$ Vanadium (V) Oxytrichloride (or Vanadyl Trichloride)

Vocoder Voice Coder; Voice-Operated Coder [also Vocoder, or vocoder]

VOCOM Voice Communications [also Vocom]

VOD Vacuum-Oxygen Decarburization (Process) [Metallurgy]

VOD Velocity of Detonation

VOD Vertical Onboard Delivery

VOD Video On Demand

VOD Voice-Operated Device

VODACOM Voice Data Communications

VODAS Voice-Operated Device, Anti-Singing [also Vodas, or vodas]

VODAT Voice-Operated Device for Automatic Transmission [also Vodat, or vodat]

VODC Vacuum-Oxygen Decarburization Converter (Process) [Metallurgy]

Voder Voice Decoder; Voice Encoder [also voder]

Voder Voice-Operation Demonstrator [also voder]

VOEST Vereinte Oesterreichische Eisen– und Stahlwerke [Major Austrian Iron and Steel Works; located at Linz]

VOF Volume of Fluid

VoFu *(Vollzugsordnung für den Funkdienst)* – German Radio Regulations

Vogad Voice-Operated Gain-Adjusted (or Adjusting) Device [also vogad]

VOH Verification Off Hook [Telecommunications]

VOH Voltage-Output High

VOHI Vermont Oil Heat Institute [US]

VOI Volume of Interest

VOIP Valence Orbital Ionization Potential

VOIR Venus Orbiter Imaging Radar; Venus Orbiting Imaging Radar

VOIS Visual Observation Instrumentation Subsystem; Visual Observation Instrumentation System

VOK Voice of Kenya

VOL Volatile Organic Liquid

VOL Voltage-Output Low

VOL (Disk) Volume Label (Command) [also vol]

Vol Volcano [also vol]

Vol Volume [also vol]

Vol Voluntary; Volunteer [also vol]

vol% volume percent

Volat volatile; volatilize(d)

Volaty Volatility [also volaty]

Volcas Voice-Operated Loss Control and Suppressor [also volcas]

VOLERE Voluntary/Legal/Regulatory

Vol Pct Volume Percent [also vol pct]

VOLS Voluntary Overseas Library Service [UK]

Vols Volumes [Publishing]

VOLSCAN Volume Scanning

VOLSER Volume/Serial Number

Volt volatilize

Voltage TIF Voltage Telephone Interference (or Influence) Factor [also voltage TIF]

VOLTAN Voltage Amperage Normalizer

VOM Volt-Ohm-Meter

VOM Volt-Ohm-Milliammeter

VOMD VAFB (Vandenberg Air Force Base) Operations and Maintenance Documentation

VON Victorian Order of Nurses

VO(OEt)$_3$ Vanadium Triethoxide Oxide

VO(OPr^i)$_3$ Vanadyl (III) Isopropoxide

VOPc Vanadyl Phthalocyanine

Vopr At Nauki Tekh Ser, Fiz Radiats Povredhdenii Radiats Mater Voprosy Atomnoi Nauki i Tekhniki, Seriya: Fizika Radiatsionnyk Povredhdenii i Radiatsionnoe Materialovedenie [Ukrainian Publication on Radiation Physics and Radiation and Materials]

Vopr At Nauki Tekh Ser, Obshch Yad Fiz Voprosy Atomnoi Nauki i Tekhniki, Seriya: Obshchaya i Yadernaya Fizika [Ukrainian Publication on Atomic Sciences and Nuclear Physics]

Vopr Psikhol Voprosy Psikhologii [Russian Journal Psychology]

VOPS ViTS (Video Teleconferencing System) Operations

VOPS VoTS (Voice Teleconferencing System) Operations

VOR Vaughan Orientation Relationship [Materials Science]

VOR VHF (Very High Frequency) Omnidirectional Radio Range; VHF (Very High Frequency) Omnidirectional Range; VHF (Very High Frequency) Omnirange

VOR Visions of Reality

VOR Voice-Operated Recording

VORD Vibrational Optical Rotatory Dispersion

VORDAC VOR/DME (VHF Omnirange/Distance Measuring Equipment) for Area Coverage; VOR/DME (VHF Omnirange/Distance Measuring Equipment) for Average Coverage [also Vordac, or vordac]

VOR/DME VHF Omnirange/Distance Measuring Equipment [also VOR-DME]

VOR/DMET VOR/DME (VOR/Distance Measuring Equipment) Compatible with TACAN (Tactical Air Navigation) [also VOR-DMET]

VORG Vatican Observatory Research Group

VORG Victorian Ornithological Research Group [Australian]

VORI Vasilov Optical Research Institute [Moskow, Russian Federation]

VORI Viticultural and Oenological Research Institute [South Africa]

VOR/ILS VOR (VHF Omnirange)/Instrument Landing System [also VOR-ILS]

VORTAC Variable Omni Range, Tactical; VHF Omnirange Co-located with TACAN (Tactical Air Navigation) [also Vortac, or vortac]

VORTEX Versatile Omnitask Real-Time Executive

VOS Verbal Operating System

VOS Vertical Obstacle Sonar

VOS Virtual Operating System

VOS Vision on Sound

VOS Vision Optical System

VOS Voice-Operated Switch

VOS Voice Operating System

VOS Volunteer Observer (or Observing) Ship

VOSC VAST (Versatile Automatic Specification Tester) Operating System Code

VORG Vatican Observatory Summer School [at Castel Gandolfo, Rome, Italy]

VOST Volatile Organic Sampling Train

VOT VHF (Very-High Frequency) Omni Test

VOT Voice-Operated Control

VOT Voice-Operated Device

VOT Voice-Operated Regulator

VOT Voice-Operated Transmission; Voice-Operated Transmit(ter)

VOT VOR (VHF Omnirange) Test Signal

VoTel *(Vollzugsordnung für den Telegraphendienst)* – German Telegraphy Regulations

VoTS Voice Teleconferencing System

Vou Voucher [also vou]

VOW Volunteer Ordnance Works [Chattanooga, Tennessee, US]

VOTA Vibration Open Test Assembly

VOTERM Voice Terminal

VOTS VAX (Virtual Address Extension)/OSI (Open System Interconnect) Transport Services (Software)

VOX Voice-Activated

VOX Voice-Operated Circuit

VOX Voice Operated Transmission; Voice Operated Transmitter [also vox]

VOx Vanadium Oxides (i.e., VO, V_2O_2, V_2O_3, V_2O_4, or V_2O_5) [also VO_x]

Vox Sang Vox Sanguinis [International Journal on Blood]

Voy Voyage [also voy]

VP Vacuum Press(ing)

VP Vacuum Product

VP Vacuum Pump(ing)

VP Validation Parameter [Computers]

VP Valve Port

VP Vanadium Phosphide

VP Vanishing Point

VP Vapor Pressure [also vp]

VP Vapor-Releasing Product

VP Variable Parameter

VP Variable Pressure

VP Variable Proportional

VP Vector-Processing; Vector Processor [Computers]

VP Vector Product [Mathematics]

VP Velocity Potential [Fluid Mechanics]

VP Velocity Pressure [Mechanics]

VP Vent Pipe

VP Vertex Power [Optics]

VP Vertical Polarization (of Antennas)

VP Verification Polarization

VP Verifying Punch

VP Verify Position

VP Vice-President [also V-P]

VP Vice-Principal [also V-P]

VP Video Port

VP View Plane

VP Viewpoint

VP Viewport

VP Vinylpropionate

VP Vinyl Pyridine (Copolymer)

VP Virtual Path

VP Virtual Processor

VP Virtual Program

VP Virus Protection

VP Viscoplastic(ity)

VP Visiting Professor

VP Voiceprint

VP Vulnerable Point

V/P Vanadium/Phosphorus (Ratio)

V-P Vice-President [also VP]

V-P Vice-Principal [also VP]

Vp Vapor [also vp]

vp vapor pressure [also VP]

VPA Valproic Acid

VPA VAX (Virtual Address Extension) Performance Advisor

VPA Vegetable Protein Association [UK]

VPA Véry Pistol Ammunition

VPA Vinylphosphonic Acid

VPA Virginia Port Authority [US]

VPAD Vapor-Phase Axial Deposition

VPAD Videotex (Data) Packet Assembly/Disassembly

VPAM Virtual Partitioned Access Method

VPAR Virtual Partition

VPB Virtually Pivoted Beam (Laser)

VP-B Bahamas [Civil Aircraft Marking]

VPC Vacuum Photocell

VPC Vapor-Phase Chromatography

VPC Vector Producer Cells [Medicine]

VPC Vertebrate Pests Committee

VORG Victorian Ornithological Research Group [Australia]

VPC Vinylpyridine Copolymer

VPC Virtual Path Circuit

VPCA Video Prelaunch Command Amplifier

VPCS Vector Processing Computer System

VP/CSS Virtual Program/Conversation Software System

VPD Vacuum Products Division

VPD Vapor-Phase Deacidification

VPD Vapor-Phase Decomposition

VPD Virtual Printer Device

VPD Vital Product Data

VPD AAS Vapor-Phase Decomposition/Atomic Absorption Spectroscopy

VPDN Virtual Private Data Network

VPDS Virtual Private Data Service [of MCI Communications, US]

VPE Vapor-Phase Epitaxial; Vapor-Phase Epitaxy [Solid-State Physics]

VPE Video Port Extensions

VPE Visual Programming Environment

VPE Vortex Pinning Energy

VPED Vapor-Phase Epitaxial Deposition

VPF Vector Point Function [Physics]

VPF Vector Product Format

VPF Vehicle Protection Factor

VPF Vertical Processing Facility

VP-F Falkland Islands [Civil Aircraft Marking]

VPFIN Vice-President of Finance

VPG Vapor Growth

VP-G British Guiana [Civil Aircraft Marking]

VPH Vickers Plate Hardness [Metallurgy]

VPH Vickers Pyramid Hardness [Metallurgy]

VP-H British Honduras [Civil Aircraft Marking]

VPHD Vertical Payload Handling Device

VPI Vacuum Pressure Impregnation

VPI Valve Position Indicator

VPI Vapor-Phase Inhibitor

VPI Virginia Polytechnic Institute [Blacksburg, Virginia, US]

VPI Virtual Path Identifier

VPI Voiceprint Identification

VPi Vinyl Pivalate

VPIS Vice President of Information Systems

VPISU Virginia Polytechnic Institute and State University [Blacksburg, Virginia, US] [also VPI&SU]

VPk Volts, Peak

2004 VPL ◆ VR

VPL Vancouver Public Library [Canada]
VPL Vanishing Point Left
VPL Virtual Programming Language
VP-L Leeward Islands [Civil Aircraft Marking]
VPLCC (Space) Vehicle Propellant Loading Control Center
VPM Vector Proton Magnetometer
VPM (Space) Vehicle Project Manager
VPM (Motor) Vehicles per Mile
VPM Vibrations per Minute [also vpm]
VPM Video Port Manager
VPM Visco-Plastic Model
VPM Volts per Meter [also Vpm, vpm, V/m, or v/m]
VPM Volts per Mil [also vpm, V/mil, or v/mil]
VP-M Malta [Civil Aircraft Marking]
VPN Virtual Page Number
VPN Virtual Private Network
VPO Valve Position Option
VPO Vanadium-Phosphorus Oxides
VPO Vapor-Pressure Osmometer; Vapor-Pressure Osmometry
VPO Vegetation Protection Ordinance [Brisbane, Queensland, Australia]
VPP Vertebrate Pest Program
VP-P Western Pacific Islands [Civil Aircraft Marking]
VPPA Variable Polarity Plasma-Arc (Welding Process)
vppm volume parts per million [Unit]
VPPO Vice-President of Plant Operations
VPR Vanishing Point Right
VPR Vapor-Phase Reactor
VPR Vapor-Phase Reflow
VPR Virtual PPI (Plan-Position Indication) Reflectoscope (Chart)
V Pres Vice-President [also V-Pres]
VPRF Variable Pulse Repetition Frequency
VPRI Victoria (State) Plant Research Institute [Burnley, Australia]
VPS Vacuum Plasma Spray(ed); Vacuum Plasma Spray(ing)
VPS Vainstein-Presnyakov-Sobelman (Approximation) [Physics]
VPS Vapor Phase (Reflow) Soldering
VPS Vapor-Phase Synthesis
VPS Vector Processing System
VPS Vibrations per Second [also vps]
VPS Video Playback System
VPS Video Printing System
VPS Video Program System
VPS Virtual Print System
VPS Virtual Programming System
VPS Voice Position Reports
VPS Voice Processing System
VPSA Vacuum Pressure-Swing Adsorption (Process)
VPSC Vault, Process, Structure, Configuration
VPSC Visco-Plastic Self-Consistent (Model)
VPSD Vacuum Plasma Structural Deposition (Process)

VP SEM Variable-Pressure Scanning Electron Microscope [also VPSEM, or VP-SEM]
VPSP Vinylpolystyrylpyridine
VPSS Vector Processing Support Subsystem
VPSW Virtual Program Status Word
VPT Vapor-Pressure Thermometer
VPT Variational Perturbation Theory
VPT Virtual Print Technology
VPT Viscosity–Pressure–Temperature (Diagram) [also V-P-T]
VPT Voice plus Telegraph
VPT Volume–Pressure–Temperature (Diagram) [also V-P-T]
VPU Variable Pitch Unit
VP-V St. Vincent [Civil Aircraft Marking]
VP&VLE Vapor Pressures and Vapor Liquid Equilibria (Database) [of National Physics Laboratory, University of Illinois at Urbana-Champaign, US]
VPW Visiting Professorships for Women Program [of National Science Foundation, US]
VP-X Gambia [Civil Aircraft Marking]
VP-Y Zimbabwe [Civil Aircraft Marking]
VPy 4-Vinylpyridine [also vpy]
VPZ Virtual Processing Zero
VQ Vapor-Quenched; Vapor Quenching
VQ-B Barbados [Civil Aircraft Marking]
VQ-F Fiji Islands [Civil Aircraft Marking]
VQFP Very-Small Quad Flat Pack [Electronics]
VQFP Very-Thin Quad Flat Pack [Electronics]
VQ-G Grenada [Civil Aircraft Marking]
VQ-H St. Helena [Civil Aircraft Marking]
VQ-L St. Lucia [Civil Aircraft Marking]
VQ-M Mauritius [Civil Aircraft Marking]
VQMC Variational Quantum Monte Carlo (Method) [Physics]
VQMG Vice-Quartermaster-General
VQMNC Vertical Quasi-Multilayered Nanocomposite
VQN Vibrational Quantum Number [Physical Chemistry]
VQ-S Seychelle Islands [Civil Aircraft Marking]
VQ-ZA Lesotho [Civil Aircraft Marking] [also VQ-ZD]
VQ-ZE Botswana [Civil Aircraft Marking] [also VQ-ZH]
VQ-ZI Swaziland [Civil Aircraft Marking]
VR Vacuum Residual Oil
VR Validation and Recovery
VR Vane Radiometer
VR Vapor Recovery
VR Variance Ratio (i.e., Ratio of Variances s_1^2 and s_2^2) [Statistics]
VR Variable Reluctance
VR Variety Reduction
VR Velocity Ratio [Mechanical Engineering]
VR Velocity, Relative
VR Vertex Renormalization
VR Vibrational Relaxation
VR Vibration Reduction
VR Victoria River [Australia]

VR Video Recorder
VR Viewing Room
VR Vinyl Resin
VR Vinyl Roll
VR Virtual Reality
VR Virtual Route
VR Visible Radiation
VR Vision Research(er)
VR Voice Recognition
VR Voltage-Regulated; Voltage-Regulating; Voltage Regulation; Voltage Regulator [also V-R]
VR Voltage Relay
VR Volta River [Africa]
VR Volume Resistivity [Electrical Engineering]
VR Vulcanized Rubber
$V(\mathbf{r})$ Atomic Electrostatic Potential [Symbol]
$V(\mathbf{r})$ Coulomb Potential for Point Charge [Symbol]
$V(\mathbf{r})$ Pair Interaction Potential (in Physics) [Symbol]
$V(\mathbf{r})$ Spherical Volume of Fractal Material with Radius r [Symbol]
$V_{att}(\mathbf{r})$ Attractive Interaction Potential [Symbol]
$V_{rep}(\mathbf{r})$ Repulsive Interaction Potential [Symbol]
$v(\mathbf{r})$ electronic potential [Symbol]
$v_x(\mathbf{r})$ exchange contribution to the potential [Symbol]
VRA Volta River Authority
VRA Voluntary Restraint Agreement
VRAM Variable Rate Adaptive Multiplexing
VRAM Video Random-Access Memory
VRB VHF (Very-High Frequency) Recovery Beacon
VRB Voice Rotating Beacon
VR-B Bermuda [Civil Aircraft Marking]
vrbl variable
VRC Variable Resistive Component
VRC Varian Research Center [US]
VRC Vertical Reciprocating Conveyor
VRC Vertical Redundancy Check [Computers]
VRC Very Rapid Calorimetry
VRC Visual Record Computer
VRC Voice-Recognition Chip
VRCI Variable Resistive Components Institute [US]
VRC/LRC Vertical Redundancy Check/Longitudinal Redundancy Check [Computers]
VRCS Vernier Reaction Control System
VRD Vacuum-Tube Relay Driver
VRD Variable Ratio Divider
VRD Vertical Resin Duct (in Wood)
VRD Victoria River District [Australia]
VRDCA Victoria River District Conservation Association [Australia]
VRDDO Variable Retention of Diatomic Differential Overlap
VRE Voice Recognition Equipment
VRE Voltage-Regulated Extended
VRECL Variable Length Record Size [BASIC Programming]
VREW Vegetative Rehabilitation and Equipment Workshop

VRF Vertical Random Format
VRF Vertical Removal Fixture
VRFI Voice Reporting Fault Indicator
Vrfn Verification [also vrfn]
VRFWS Vehicle Rapid Fire Weapon System
VRG Vertical Reference Gyro
VR-G Gibraltar [Civil Aircraft Marking]
VRH Variable-Range-Hopping (Mott Law) [Physics]
VR-H Hong Kong [Civil Aircraft Marking]
VRHC Variable Range Hopping Conduction [Physics]
VRID Virtual Route Identifier
VRIP Volume Related Incentive Pricing
VRL Vertical Recovery Line
VRL Vertical Reference Line
VRL Vibration Research Laboratory [US]
VRM Variable Range Marker
VRM Vertical Retreat Mining
VRM Viscous Remanent Magnetization
VRM Visible Record Machine
VRM Voltage Regulator Module
VRM Virtual Reality Markup Language [now Virtual Reality Modeling Language]
VRML Virtual Reality Modeling Language [Internet]
V RMS Volts, Root-Mean-Square [also VRMS, V rms, or v rms]
vRNA Viral Ribonucleic acid [also V-RNA, or vRNA]
Vrnr Vernier
VRO Vendor Repair Order
VROM Video Read-Only Memory
VROOMM (Borlund) Virtual Real-Time Object-Oriented Memory Manager
VROT Victorian Rare, or Threatened Plants [Australia]
VRP Vehicle Recycling Partnership [US Consortium of General Motors, Ford Motor Company and Chrysler Corporation]
VRP Visual Record Printer
VRP Visual Reporting Point
VRP Volta River Project
VRPS Voltage-Regulated Power Supply
VRR Verification Readiness Review
VRR Visual Radio Range
VRS Vehicle Replacement System
VRS Vertical Raster Scanning
VRS Vibration Reduction System
VRS Video Recall System
VRS Voice Recognition System
VRS Volatile Reducing Substances
.VRS Graphics Driver [WordPerfect File Name Extension]
VRSA Voice Reporting Signal Assembly
VRSS Vehicular Radar Safety System
VRSS Voice Reporting Signal System
VRT Variable Reactance Transformer
VRT Vessel Residence Time; Vessel Resident Time
VRT Vibration-Rotation Transition

VRT Vibration-Rotation Tunnel(l)ing
VRT Voltage Regulation Technology
VRT Volume-Rendering Technique
V(r,t) Hydrodynamic Velocity
V-RTIF Vandenberg (Air Force Base) Real-Time Interface
VRU Voice Response Unit
VRU Voltage Readout Unit
VR-U Brunei [Civil Aircraft Marking]
VR-W Sarawak [Civil Aircraft Marking]
VRX Virtual Resource Executive
VRX-MP Virtual Resource Executive–Multiprocessor
vry very
VS Vacuum Science; Vacuum Scientist
VS Vacuum Spectrometer; Vacuum Spectrometry
VS Vacuum Switch
VS Vacuum System
VS Valence Shell
VS Valence State
VS Valve Seat
VS Vanadium Steel
VS Vanadium(I) Sulfide
VS Vapor-Solid (Interface) [Metallurgy]
VS Variable Speed
VS Variable Star [Astronomy]
VS Variable Store [Computers]
VS Variable Sweep
VS Vector Sum
VS Velocity Spectrograph(y)
VS Vent Sealing
VS Vent Stack
VS Venturi Scrubber [Chemical Engineering]
VS Vermont State [US]
VS Vernier Scale
VS Vertical Sounding
VS Vertical Speed [Aeronautics]
VS Vertical Stabilizer [Aeronautics]
VS Very Soluble [also vs]
VS Very Strong [also vs]
VS Very Susceptible [also vs]
VS Vesicular Stomatitis [Medicine]
VS Vestigial Sideband
VS Veterinary Surgeon; Veterinary Surgery
VS Vibrating Sieve(r)
VS Vibrational Spectroscopy
VS Vibrational Spectrum
VS Vinyl Stabilizer
VS Virginia Semiconductor Inc. [Fredericksburg, US]
VS Virtual Storage
VS Virtual System
VS Visible Spectrum
VS Visible Speech (Process)
VS Vision Sensor
VS Visualization System

VS Visual Signaling (Equipment)
VS Vital Sign(s) [Medicine]
VS Vital Statistics
VS Vocal Synthesis; Vocal Synthesizer
VS Voice Synthesis; Voice Synthesizer
VS Voice Switch
VS Volatile Storage
VS Voltmeter Switch
VS Volume Shadowing
VS Volumetric Solution
VS Vortex Street [Fluid Mechanics]
VS Vortex Structure
V/S Volume-to-Surface (Area) (Ratio)
V-S Vapor-Solid (Interface) [Metallurgy]
Vs Volt-second [also V·s]
V/s Volt(s) per second [also V/sec, or $V\,s^{-1}$]
V(s) Command [Automatic Control Symbol]
vs *(versus)* – against
Vs very soluble [also VS]
vs very strong [also VS]
vs very susceptible [also VS]
vs *(vide supra)* – see above
VSA Vacuum-Swing Adsorption (Process)
VSA Variable Stability Aircraft
VSA Vehicle Security Association [US]
VSA Verification Site Approval
VSA Very Small Array [Joint UK-Spain Microwave Telescope Project Erected in the Canaries]
VSA Vessel Sharing Agreement [Ocean Shipping]
VSA Vibrating-String Accelerometer
VSA Vicia Sativa Agglutinin [Immunology]
VSA Virginia Safety Association [US]
VSA Vocal Server Adapter
Vs/A Volt-second(s) per Ampere [also V·s/A, or $Vs\,A^{-1}$]
VSAM Virtual Sequential Access Method
VSAM Virtual Storage Access Method
VSAM Virtual System Access Method
Vs/Am Volt-second(s) per Ampere-meter [also $Vs\,A^{-1}\,m^{-1}$]
VSAT Very-Small-Aperture Terminal
VSB Variable Shaped Beam [Integrated Circuit Manufacture]
VSB Vestigial Sideband
VSB VME (Versa Module Europe) Subsystem Bus [Electronics]
VSb Vanadium Antimonide
VSBS Very Small Business System
VSBS Voluntary Standard Bodies
VSB-SC Vestigial Sideband–Suppressed Carrier [also VSBSC, VS-SC, or VSSC]
VSBO Vysoké Skola Banská Ostrava [Technical University of Ostrava, Czech Republic]
VSBY Visibility [also Vsby, or vsby]
VSC Vacuum Suction (Slag) Cleaner [Metallurgy]
VSC Valdosta State College [Georgia, US]

VSC Variable Speed Control

VSC Variable Speech Control

VSC Vibration Safety Cutoff

VSC Video System Controller

VSC Virginia State College [Petersburg, US]

VSC Virtual Subscriber Computer

VSC Voltage-Saturated Capacitor

VSC Voltage-Stabilizing Circuit

V/SCE Voltage per Saturated Calomel Electrode

VSCF Variable-Speed Constant-Frequency

VSD Vacuum Stream Degassing [Metallurgy]

VSD Variable Speed Drive

VSD Vehicle Systems Dynamics

VSD Ventricular Septal Defect [Medicine]

VSD Vertical Situation Display [Aeronautics]

VSD Voltage-Stabilizing Device

VSDA Video Software Dealers Association [US]

VSDM Variable Scope Delta Modulation

VSE Vancouver Stock Exchange [Canada]

VSE Vermont Society of Engineers [US]

VSE Vessel Steam Explosion

VSE Vibrational Stark Effect [Physics]

VSE Virtual Storage Equipment

VSE Virtual Storage Extended

VSE Volume-Selective Excitation

VSE/AF Virtual Storage Extended/Advanced Function

V/sec Volt(s) per second [also V/s, or V s^{-1}]

V-sec/rad volt-second per radian [also V-s/rad]

VSEPR Valence-Shell Electron Pair Repulsion

VSETUP Video Setup

VSETUP VGA (Visual Graphics Adapter) Setup

VSEW Verband Schweizerischer Elektrizitätswerke [Swiss Electric Power Association]

VSF Vertical Scanning Frequency

VSF Voice Store and Forward

VSF Volume-Size Factor

VSFR Vertical Seismic Floor Response

VSG Variable Speed Gear

VSG Verband Schweizerischer Graphiker [Swiss Federation of Graphic Art Designers/Illustrators]

VSG Vereinigte Schmiedewerke GmbH [Major German Forging Plant located at Essen]

VSG Vertical Sweep Generator

VSGT Verband Schweizerischer Gummi— und Thermoplast-Hersteller [Swiss Associaton of Rubber and Thermoplastic Manufacturers]

VSI Vacuum-Super-Insulation

VSI Vacuum Super Isolation

VSI Vertical-Speed Indicator

VSI Video-Conference Systems, Inc. [US]

VSI Video Simulation Interface

VSI Video Sweep Integrator

VSI Vinyl Siding Institute [US]

VSI Virginia Semiconductor Inc. [Fredericksburg, US]

VSI Virtual Socket Interface

VSI Virtual Storage Interrupt

VSI Voltage Source Inverter

VSIE Valence State Ionization Energy

VSIG Vereinigung des Schweizerischen Import— und Großhandels [Swiss Association of Importers and Wholesalers]

VSIO Virtual Serial Input/Output

V-SiO$_2$ Vitreous Silica

VSIP Valence State Ionization Potential

VSIV Vesicular Stomatitis Virus

VSJ Vacuum Society of Japan

VSKPS Verband Schweizerischer Kunststoff-Press— und Spritzwerke [Swiss Association of Plastics Molding and Extrusion Companies]

VSL Variable Specification List

VSL Ventilation Sampling Line

VSL Very Small Laser

VSL Victoria State Library [Melbourne, Australia]

VS-LCVD Vapor-Solid Laser-Assisted Chemical Vapor Deposition

VSLE Very Small Local Exchange

VSM Value Servomotor

VSM Variable-Speed Mixer

VSM Variable-Speed Motor

VSM Variational Stiffness Method

VSM Vereinigte Schmirgel— und Maschinenfabriken AG [German Manufacturer of Grinding and Sanding Supplies and Equipment located at Hanover]

VSM Verein Schweizerischer Maschinenindustrieller [Swiss Machinery Manufacturers Association]

VSM Vertical Scan Machine (Induction Machine)

VSM Vertical Section of Multilayers

VSM Vertical Separation Minimum [Aeronautics]

VSM Vestigial Sideband Modulation

VSM Vibrating-Sample Magnetometer; Vibrating-Sample Magnetometry

VSM Vibrating Space Modulator

VSM Virtual Shared Memory

VSM Virtual Storage Management

VSM Visual System Management

Vs m Volt-second meter [also Vs-m, or Vs·m]

Vs/m^2 Volt-second per square meter [also Vs m^2]

VSMA Vibrating Screen Manufacturers Association [US]

VSMF Virtual Search Microfilm File

VSMF Visual Search Microfilm File

VSMF Visual Search on Microfilm

VSMPO Verkhnaya Salda Metallurgical Production Association [Commonwealth of Independent States]

VSM RTM Vibrating-Sample Magnetometer–Rotational Transverse Magnetometer; Vibrating-Sample Magnetometry–Rotational Transverse Magnetometry

VSMS Vermont State Medical Society [US]

VSMS Video Switching Matrix System

VSN Video Switching Network

VSN Volume Serial Number

VSNU Vereniging van Nederlandse Universiteiten [Association of Universities in the Netherlands]

VSO Vertically Self-Organized

VSO Very Small Outline (Package) [Electronics]

VSO Very Stable Oscillator

VSO Voltage-Sensitive Oscillator

VSO Voluntary Services Overseas

VSOP Vector Signal Operations Package [Software]

VSOP Very-Small-Outline Package [Electronics]

VSOP Very Superior Old Product [On Wine Bottles, etc.]

VSOP VLBI (Very-Long Baseline Interferometry) Space Observatory Project

VSOS Virtual Storage Operating System

VSP Verein der Schweizer Presse [Swiss Press Association]

VSP Vertical Seismic Profile

VSP Vertical Speed [Aeronautics]

VSP Virtual Switching Point

VSP Vision Statistical Processor

VSP Voith-Schneider Propeller [Naval Architecture]

VSP Voltage Set Point

VSP Voluntary Savings Plan

VSPC Virtual Storage Personal Computer

VSQG Very Small Quantity Generator

VSQW Variable-Strain Quantum Well [Electronics]

VSR Validation Summary Report

VSR Vallecitos Superheat Reactor [at Pleasanton, California, US]

VSR Variable Speed Range

VSR Vertical Storage and Retrieval

VSR Very Short Range (Radar)

V-s/rad volt-second per radian [also V-sec/rad]

VSRBM Very Short Range Ballistic Missile

VSRT Vertical Spindle Rotary Table

VSS Vacuum Slag Suction [Metallurgy]

VSS Vapor Sampling System

VSS Vapor Suppression System [Nuclear Engineering]

VSS Variable Stability System

VSS Vehicle Sampling System

VSS Vehicle Speed Sensor

VSS Video Storage System

VSS Virtual Storage System

VSS Voice Signaling System

VSS Volatile Suspended Solids

VSS Voltage for Substrate and Sources; Voltage to Substrate and Sources

VSSC Vikram Sarabhai Space Center [Trivandrum, India]

VS-SC Vestigial Sideband–Suppressed Carrier [also VSBSC, VSB-SC, or VSSC]

V/SSE Voltage per Sulfate Saturated Electrode

VSSG Vertical Separation Study Group [(Specialist Panel on) Navigation and Separation of Aircrafts]

VST Vacuum Science and Technology

VST Variable Speed (Friction) Tester

VST Variable Speed Transmission

VST Verband Schweizerischer Transportanstalten [Swiss Transportation Federation]

VST Volume Sensitive Tariff

VSTAG Vandenberg (Air Force Base) Shuttle Turnaround Analysis Group [US]

VSTK Vysoka Skola Technicka v Kosiciach [Technical University of Kosice, Kosice, Slovakia]

VSTL Viking Science Test Lander [of NASA]

VSTO Virginia State Travel Office [US]

VSTOL Vertical and/or Short Takeoff and Landing (Aircraft) [also V/STOL]

VSTS Virginia State Travel Service [US]

VSV Vapor-Spray-Vapor (System)

VSV Vesicular Stomatitis Virus

VSV-G Vesicular Stomatitis Virus Glycoprotein [Biochemistry]

VSW Very Short Wave (Radar)

VSW Voltage Standing Wave

VSWR Voltage Standing Wave Ratio

VSYNC Vertical Synchronization

VT Vacuum Technologist; Vacuum Technology

VT Vacuum Test(ing)

VT Vacuum-Tight

VT Vacuum Tube

VT Valid Test

VT Vapor Trail

VT Vapor Transport

VT Variable Temperature

VT Variable Time

VT Vehicular Technology (Society) [of Institute of Electrical and Electronics Engineers, US]

VT Venturi Tube

VT Vermont [US]

VT Vertical Tab(ulation Character) [Data Communications]

VT Vertical Tail [Aeronautics]

VT Vibration Test(ing)

VT Video Tape

VT Video (Display) Terminal

VT Video Transmit

VT Vidicon Tube

VT Vinyl Toluene

VT Virtual Technology

VT Virtual Temperature [Meteorology]

VT Virtual Terminal

VT Visual Telegraphy

VT Visual Test(ing)

VT Voice Tube

VT Voltage Transformer

VT- India [Civil Aircraft Marking]

V(T) Temperature Dependent Velocity [Symbol]

V(θ) Angle-Dependent Voltage Change [Symbol]

Vt Vent [also vt]

Vt Vermont [US]

V/t Volumetric flow rate [Symbol]

v(T) temperature dependent velocity [Symbol]

VTA Vacuum Tube Amplifier

VTA Vertical Transition Approximation

VTA Video Trade Association [UK]

VTA Voice Terrain Advisory [Aeronautics]

VTAB Vertical Tab(ulation Character)

VTAC Video Timing and Control

VTAM Virtual Telecommunications Access Method

VTAM Virtual Terminal Access Method

VTB Voltage Time to (Cable) Breakdown

VTBS Video Tape/Book Set

VTC Vocational Training Center

VTC Vinyl Trichlorosilane

VTCC Video Teleconferencing Control Center

VTCS Vehicular Traffic Control System

VTD Vacuum Tube Detector

VTD Vertical Tape Display

VTDC Vacuum Tube Development Committee

V/TDD Video/Teletype Device for the Deaf

VTDI Variable Threshold Digital Input

VTE Variable Time Expansion

VTE Vasileion Tis Ellados [Republic of Greece]

VTE Vertical Tube Evaporator

VTEC Variable Valve Timing Electronic Control (Engine)

VTEC Verotoxic E. Coli [Biotechnology]

VTEO Vinyl Trioxysilane

VTEU Vaccine and Treatment Evaluation Unit

VTF Variable Time Fuse

VTF Vertical Test Fixture

VTF Vertical Test Flight

VTF Via Terrestrial Facilities

VTF Vogel-Tammann-Fulcher (Parameter) [Physics]

VTI Video Terminal Interface

V-Ti Vanadium-Titanium (Alloy)

V-Ti-Ta Vanadium-Titanium-Tantalum (Alloy)

VTL Variable Threshold Logic

VTLA Video Tape Live Audio

VTLS Virginia Technical Library System [of Virginia Polytechnic Institute, Blacksburg, US]

VTM Vacuum Tube Modulator

VTM Vibration Test Module

VTM Voltage-Tunable Magnetron

VTMO Vinyl Trimethoxysilane

VTMOEO Vinyl Tris(2-Methoxyethoxy)silane

VTMS Vessel Traffic Management System

VTMS Vinyltrimethylsilane [also vtms]

VTN Verification Test Network

VTNS Virtual Telecommunications Network Service

VTO Vacuum Tube Oscillator

VTO Voltage-Tuned Oscillator

VTO Volumetric Top-Off

VTOC Volume Table of Contents

VTP Vehicle Test Plan

VTP Verification Test Plan

VTP Verification Test Program

VTR Verificatión Text Report

VTR Video Tape Recorder; Video Tape Recording

VTOC Volume Table of Contents

VTOHL Vertical Takeoff and Horizontal Landing

VTOL Vertical Takeoff and Landing

VTOVL Vertical Takeoff and Vertical Landing

VTP Viewdata Terminal Program

VTP Virtual Terminal Protocol

VTP Volume–Temperature–Pressure [also VT-P]

VTPR Vertical Temperature Profile Radiometer

VTR Video Tape Recorder; Video Tape Recording

VTR Voltage Transformation Ratio

VTRS Video Tape Response System

VTS Vandenberg (Air Force Base) Tracking Station

VTS Variable Transition State

VTS Vehicular Technology Society [of Institute of Electrical and Electronics Engineers, US]

VTS Vertical Test Site

VTS Vertical Test Stand

VTS Vertical Test System

VTS Vessel Traffic Services

VTS Viewscan Text System

VT-SIFDT Variable-Temperature Selected Ion Flow Drift Tube

VTST Variational Transition State Theory

VT&T Verification Test and Training

VTTC Video Tape Time-Code,

VTU Vienna Technical University [Austria]

V+TU Voice plus Teleprinter Unit

VTÜV Vereinigung der Technischen Überwachungsvereine [Federation of German Inspection Associations]

VTVM Vacuum-Tube Voltmeter

VTX Vertex [also Vtx, or vtx]

VTX Videotex

VU Valence Unit [also vu]

VU Vanderbilt University [Nashville, Tennessee, US]

VU Vanuatu [ISO Code]

VU Vehicle Unit

VU Vehicle Utility

VU Voice Unit

VU Volume Unit [also vu]

vu valence unit [also VU]

VUA Virtual Unit Address

VUA Vrije Universiteit Amsterdam [Free University Amsterdam, Netherlands]

VUC Victoria University College [New Zealand]

VUCDT Ventilation Unit Condensate Drain Tank

VUE Visual User Environment

VUI Video User Interface

Vulc Vulcanization; Vulcanizing [also vulc]

VUM Victoria University of Manchester [UK]

Vuoto Sci Tecnol Vuoto Scienza e Tecnologia [Italian Publication on Vacuum Science and Technology]

VUP VAX (Virtual Address Extension) Unit of Performance

VUPC Vyskumneho Ustavu Papieru a Celuloy [Paper and Cellulose Research Institute, Bratislava, Slovakia]

VUSEIAR Vyskumny Ustav Socialno-Ekonomickych Informacii a Automatizacie v Riadeni [Riadeni Research Institute for Social-Economic Information and Automation, Bratislava, Slovakia]

VUT Victoria University of Technology [Australia]

VUTB Vysoké Uceni Technické v Brno [Technical University of Brno, Czech Republic]

VUTI Variations of Upper Tropospheric Ionization

VUTS Verification Unit Test Set

VUU Virginia Union University [Richmond, US]

VUV Vacuum Ultraviolet

VUV Vatu [Currency of Vanuatu]

VUVR Vacuum Ultraviolet Radiation

VUVS Vacuum Ultraviolet Spectroscopy

VUW Victoria University of Wellington [New Zealand]

VUZ Vysshikh Uchebnykh Zavedenii [Higher Education Institute, Russia]

VV Vaccinia Virus

VV Vacuum Valve

VV Varicella Virus

VV Velocity Vector

VV Vent Valve [also V/V]

VV Vertical Velocity [also V/V]

VV Vertical Visibility [Meteorology]

VV Void Volume

VV Volt Velocity

V&V Validation and Verification; Verification and Validation [also V/V]

V/V Vent Valve [also VV]

V/V Vertical Velocity [also VV]

V/V Volume by (in, per, or to) Volume [also v/v]

V-V Variant-Variant (Interaction)

V-V Velocity–Volume

V_{1D}/V_{3D} First Harmonic/Third Harmonic (Magnetic Erasure Ratio)

vv *(vice versa)* – conversely

V/v volume by (in, per, to) volume [also V/V]

%v/v percent volume in volume

VVA Vapor Vacuum Arc (Source)

VVA Vicia Villosa Agglutinin [Immunology]

VVA-A_4 Vicia Villosa Agglutinin Isolectin A_4 [Immunology]

VVA-A_2B_2 Vicia Villosa Agglutinin Isolectin A_2B_2 [Immunology]

VVA-B_4 Vicia Villosa Agglutinin Isolectin B_4 [Immunology]

VVC Virtual ViTS (Video Teleconferencing System) Connection

VVC Voltage-Variable Capacitance; Voltage-Variable Capacitor

VVD Verkehrsverein Düsseldorf [Duesseldorf Travel Association, Germany]

VVDS Video Vertex Decision Storage

VVE Vertical Vertex Error

VVI Vancouver Vocational Institute [Canada]

VVI Vertical Velocity Indicator

VVM Valve Voltmeter

VVP Voluntary Protection Program

VVR Variable Voltage Rectifier

VVR Void Volume Ratio

VvTP Vereniging voor Technische Physica [(Student) Association for Technical Physics, of Delft University of Technology, Netherlands]

VVVF Variable Voltage Variable Frequency

VW Valley Wave

VW Very Weak [also vw]

VW Virtual Work

VW Visualization Workstation

VW Volkswagen AG [Wolfsburg, Germany]

VW Volmer-Weber (Growth Mode) [also V-W] [Metallurgy]

V-W Vanadium-Tungsten (Alloy System)

vw very weak [also VW]

VWA Verwaltungs– und Wirtschaftsakademie Wuppertal [Wuppertal Academy of Administration and Economics, Germany]

VWB Visual Workbench

VWD Variable Wavelength

VWDI Vinyl Window and Door Institute [US]

vWF von Willebrand Factor [Electroimmunodiffusion Assay]

VWG Van der Waals Gap

VWI Vertical Wafer Integration

VWL Variable Word Length

VWOA Veteran Wireless Operators Association [US]

VWP Variable Width Pulse

VWPI Vacuum Wood Preservers Institute [US]

VWPT Vibrating-Wire Pressure Transducer

VWQMN Victorian Water Quality Monitoring Network [Australia]

VWRRC Virginia Water Resources Research Center [US]

VWS Variable Word Size

VWS VMS (VAX (Virtual Address Extension) Management System) Workstation Software

VWS Vortex Wake System

VWSS Vertical Wire Sky Screen

VX Volume–Concentration (Diagram) [alsoV-X]

.VXD Virtual Device [Microsoft File Name Extension]

VxD (Microsoft) Virtual Extended Driver

VXO Variable(-Frequency) Crystal Oscillator

Vy very

Vyber Inf Organ Vypocet Tech Vyber Informaci z Organizacni a Vypocetni Techniky [Czech Publication]

Vychisl Metody Program Vychislitel'naya Metody i Programmirovanie [Russian Publication on Computational Methods and Computer Programming]

Vychisl Seismol Vychislitel'naya Seismologiya [Russian Publication on Computational Seismology]

Vychisl Tekh Vopr Kibern Vychislitel'naya Tekhnika i Voprosy Kibernetiki [Russian Publication on Computer Technology and Cybernetics]

Vysokomol Soed A Vysokomolekulyarnye Soedineniya, Seriya A [Russian Polymer Science Journal, Series A]

Vysokomol Soed B Vysokomolekulyarnye Soedineniya, Seriya B [Russian Polymer Science Journal, Series B]

Vysokomol Soed, Kratk Soobshcheniya Vysokomolekulyarnye Soedineniya, Kratkie Soobshcheniya [Russian Polymer Science Journal, Physics Section]

VZ Varicella Zoster (Virus)

V(z) Acoustic Material Signature [Symbol]

V(z) Electric Potential (Function) [Symbol]

V(z) Interference Pinning Potential [Symbol]

VZA Viewing Zenith Angle

VZM Video Zoom Microscope

VZV Varicella Zoster Virus

W

W   Absorbed Energy [Symbol]

W   Atomic Interchange Energy [Symbol]

W   (Energy) Band-Width [Symbol]

W   Bond-Bond Repulsion [Symbol]

W   Deviation from Spherical Form (of Bearing Balls) [Symbol]

W   Dislocation Line Energy (in Materials Science) [Symbol]

W   Disorder [Symbol]

W   Flow Resistance (in Fluid Mechanics) [Symbol]

W   Interface Roughness (in Materials Science) [Symbol]

W   (Total) Load [Symbol]

W   Long Wave [Symbol]

W   Macroscopic Growth Rate [Symbol]

W   Mass Fraction [Symbol],

W   Maximum Tire Speed of 168 mph (or 270 km/h) [Rating Symbol]

W   Partial-Discharge (Corona) Energy (in Electronics) [Symbol]

W   Potential Energy [Symbol]

W   Quer Wave [Symbol]

W   Radiant Flux [Symbol]

W   Radiotelegraphy [110 to 150 kc/s Range]

W   Reflection Coefficient [Symbol]

W   Rotational Energy [Symbol]

W   Solution Heat-Treated [Basic Temper Designation for Aluminum Alloys]

W   Sound Power [Symbol]

W   Special Search (Aircraft) [US Navy Symbol]

W   Specimen Width (in Testing) [Symbol]

W   Spectral Radiant Emittance [Symbol]

W   Stability Factor (in Electronics) [Symbol]

W   Statistical Probability (of Thermodynamic System) [Symbol]

W   Stored Energy [Symbol]

W   Strain Energy (Density) Function [Symbol]

W   Tool Wear Factor [Symbol]

W   (−)-Tryptophan; Tryptophyl [Biochemistry]

W   Tungsten [Symbol]

W   Velocity of a Medium (in Doppler Equation) [Symbol]

W   Wabasca Bitumen [Alberta Oil Sands Technology and Research Authority]

W   Waco [Texas, US]

W   Wait(ing)

W   Wales [UK]

W   War(fare)

W   Ward

W   Ware(house)

W   Warm

W   Warsaw [Poland]

W   Wash(ing)

W   Washington [US Capital]

W   Washington [US State]

W   Waste

W   Watch

W   Water

W   Water Equivalent (in Meteorology) [Symbol]

W   Water-Hardening Tool Steels [AISI-SAE Symbol]

W   Watt [Unit]

W   Wattmeter

W   Wave

W   Wax

W   Weak

W   Weak [Particle Physics]

W   Wear(ing)

W   Weather

W   Weatherability; Weathering

W   Weather Mission (Aircraft, Missile or Rocket) [USDOD Symbol]

W   Weave

W   Web(bing)

W   Wedge

W   Wednesday

W   Week

W   Weigh(ing); Weight

W   Weld(ed); Welder; Welding

W   Welfare

W   Well

W   Wellington [New Zealand]

W   Welsh

W   West(ern)

W   Wet(tability); Wetting

W   Whale; Whaling

W   Wheat

W   Wheel

W   Whisker [also w]

W   Whiskey [Phonetic Alphabet]

W   Whitehorse [Yukon Territory, Canada]

W   White(ness)

W   Whole Tooth Depth (of Gears) [Symbol]

W   Wichita [Kansas, US]

W   Wick

W   Wide [also w]

W   Width [Symbol]

W   Wild

W   Wilderness

W   Wildlife

W   Win(ning)

W   Winch

W   Wind

W   Windhoek [Namibia]

W   Window

W   Windsor [Ontario, Canada]

W   Wing

W   Winnipeg [Manitoba, Canada]

W   Winter

W   Wipe(r)

W   Wire

W   Wisconsin [US]

W   With

W   Withdraw(al)

W   Wobble(r)

W   Wolf

W   *(Wolfram)* – Tungsten [Symbol]

W   Womb

W   Won [Currency of South Korea]

W   Wood

W   Wool

W   Worchester [Massachusetts, US]

W   Word

W   Work [Symbol]

W   Workable; Workability

W   Worker; Working

W   Workfunction (in Physics)

W   Working Strength (of Bolts) [Symbol]

W   Work of Formation [Symbol]

W   Worksheet

W   Workstation

W   Work-to-Break (in Tensile Testing) [Symbol]

W   World

W   Worm

W   Wrap(ping)

W   Wrench

W   Wring(ing)

W   Wrist

W   Write; Writing

W   Wroclaw [Poland]

W   W-Type Structural Steel Shape [AISI/AISC Designation]

W   Wuchereria [Genus of Parasitic Worms]

W   Wuhan [PR China]

W   Wulff Shape [Crystallography]

W   Wyoming [US]

-W   Special Search (Aircraft) [US Navy Suffix]

$\hat{W}$ Wigner Transform (in Physics) [Symbol]

$W^{\pm}$ Charged Weak Boson (with Mass of 78 GeV) (in Particle Physics) [Symbol]

W^+ W-Plus Meson (in Particle Physics) [Symbol]

W^- W-Minus Meson (in Particle Physics) [Symbol]

W_0 Specific Weight [Symbol]

W^{2+} Divalent Tungsten Ion [also W^{++}] [Symbol]

W^{4+} Tetravalent Tungsten Ion [Symbol]

W^{5+} Pentavalent Tungsten Ion [Symbol]

W^{6+} Hexavalent Tungsten Ion [Symbol]

W_A (Thermodynamic) Work of Adhesion [Symbol]

W_α Mass Fraction of Alpha Phase (of an Alloy) [Symbol]

$W_{\alpha'}$ Weight Fraction of Proeutectoid Ferrite (for Hypoeutectoid Iron-Carbon Alloy) (in Metallurgy) [Symbol]

W_{ad} Work of Adhesion (in Thermodynamics) [Symbol]

W_β Mass Fraction of Beta Phase (of an Alloy) [Symbol]

W_C Cohesional Work (in Thermodynamics) [Symbol]

W_c Density-Corrected Wear Rate [Symbol]

W_e Centrifuge Moisture Equivalent (of Soil) [Symbol]

W_e Weight Fraction of Eutectic Microconstituent [Symbol]

W_{elec} Electrical Work [Symbol]

$W_{Fe3C'}$ Weight Fraction of Proeutectoid Cementite (or Iron Carbide) (for Hypereutectoid Iron-Carbon Alloy) [Symbol]

W_h Hysteresis Loss [Symbol]

W_I Wulff Shape (of Interface) (in Crystallography) [Symbol]

W_i Mass Fraction of Phase i [Symbol]

W_i Partial-Discharge (Corona) Energy (in Electronics) [Symbol]

W_i Work of Immersion (in Physics) [Symbol]

$W_{ij}^{(s)}$ Pair Interchange Energy (in Physics) [Symbol]

W_{in} Input Energy [Symbol]

W_k Fluorescence Yield [Symbol]

W_k Kink Formation Energy (in Solid-State Physics) [Symbol]

W_L Mass Fraction of Liquid Phase (of an Alloy) [Symbol]

W_{mech} Mechanical Work [Symbol]

W_{out} Output Energy [Symbol]

W_P Peierls Energy (in Physics) [Symbol]

W_p Plastic Work [Symbol]

W_p Weight Fraction of Pearlite (for Iron-Carbon Alloys) [Symbol]

$W_{p,T}^E$ Water-Electrostatic Gibbs Potential [Symbol]

W1   Warrant Officer, First Class [Canada]

W1   Water-Hardening Plain-Carbon Tool Steels (with 0.6-1.4% Carbon) [AISI-SAE Symbol]

W2   Warrant Officer, Second Class [Canada]

W2   Water-Hardening Tool Steels (with 1.41% Carbon and 0.25% Vanadium) [AISI-SAE Symbol]

W3   World Wide Web [also WWW or 3W]

W4   What-Works-With-What [also WWWW]

W5   Water-Hardening Tool Steels (with 1.1% Carbon and 0.5% Chromium) [AISI-SAE Symbol]

W-1   Warrant Officer [US Army, Air Force, Navy, Marine Corps and Coast Guard Grade] [also W1]

W-2 Chief Warrant Officer, First Class [US Army, Air Force, Navy, Marine Corps and Coast Guard Grade] [also W2]

W-3 Chief Warrant Officer, Second Class [US Army, Air Force, Navy, Marine Corps and Coast Guard Grade] [also W3]

W-4 Chief Warrant Officer, Third Class [US Army, Air Force, Navy, Marine Corps and Coast Guard Grade] [also W4]

W-176 Tungsten-176 [also ^{176}W, or W^{176}]

W-177 Tungsten-177 [also ^{177}W, or W^{177}]

W-178 Tungsten-178 [also ^{178}W, or W^{178}]

W-179 Tungsten-179 [also ^{179}W, or W^{179}]

W-180 Tungsten-180 [also ^{180}W, or W^{180}]

W-181 Tungsten-181 [also ^{181}W, or W^{181}]

W-182 Tungsten-182 [also ^{182}W, or W^{182}]

W-183 Tungsten-183 [also ^{183}W, or W^{183}]

W-184 Tungsten-184 [also ^{184}W, or W^{184}]

W-185 Tungsten-185 [also ^{185}W, or W^{185}]

W-186 Tungsten-186 [also ^{186}W, or W^{186}]

W-187 Tungsten-187 [also ^{187}W, or W^{187}]

W-188 Tungsten-188 [also ^{188}W, or W^{188}]

2W Two Weeks

w air velocity [Symbol]

w complex number [Symbol]

w displacement [Symbol]

w dry season in winter [Subtype of Climate Region, e.g., in Aw, Caw, Cbw, etc.]

w energy density [Symbol]

w flow rate [Symbol]

w fluorescence yield [Symbol]

w (magnetic core) lamination width [Symbol]

w load per unit distance [Symbol]

w mass flux velocity [Symbol]

w mixing ratio [Symbol]

w molar weight [Symbol]

w normalized excitation vector (in electron microscopy) [Symbol]

w rate of evaporation [Symbol]

w specific weight [Symbol]

w vacancy jump ratio (in solid-state physics) [Symbol]

w velocity component (in Cartesian system) [Symbol]

w vertical velocity [Symbol]

w warm

w waste

w water

w water content (of a soil or rock mass) [Symbol]

w water-to-air ratio [Symbol]

w weak

w weather

w web thickness (of a girder or rail) [Symbol]

w weekly

w weight

w wet

w whisker [also W]

w white

w wide [also W]

w width [Symbol]

w width of cut (in machining) [Symbol]

w width of welding bead [Symbol]

w Wien's displacement constant (in statistical mechanics) [Symbol]

w with [also w/]

w wind

w write

w write (command) [Edlin MS-DOS Line Editor]

w/ with [also w]

$\bar{w}$ temporal mean of velocity component "w" [Symbol]

w* convective velocity [Symbol]

w_0 characteristic velocity of the fluid [Symbol]

w_0 gauge width [Symbol]

w_c critical crack width [Symbol]

w_e energy density of electric field [Symbol]

w_H hygroscopic water content (of soil, or rock) [Symbol]

w_i weight fraction of molecules within polymer-molecule size range "i" [Symbol]

w_L liquid limit (of soil) [Symbol]

w_m energy density of magnetic fields [Symbol]

w_o optimum moisture content (of soil) [Symbol]

w_p plastic limit (for rock or soil) [Symbol]

w_s shrinkage ratio (of soil) [Symbol]

WA Warren-Averbach (Grain Size Analysis) [also W-A] [Metallurgy]

WA Washington (State) [US]

WA Waste Acid

WA Water Absorption

WA Water Analysis; Water Analyzer

WA Water Atomized; Water Atomization [Powder Metallurgy]

WA Watertown Arsenal [of US Army in Massachusetts, US]

WA Watervliet Arsenal [of US Army in New York, US]

WA Wave Analysis; Wave Analyzer

WA Wave Antenna

WA Waveform Analysis; Waveform Analyzer

WA Weak Acid

WA Wear Analysis

WA Weighted Average

WA Welding Abstracts [of the World Center for Information and Technology in Welding]

WA Well-Annealed [Metallurgy]

WA West Africa(n)

WA Western Airlines [US]

WA Western Australia

WA Wet Analysis; Wet Analytical

WA Wetting Angle [Physical Chemistry]

WA Wide-Angle (Lens)

WA Wildlife Area

WA Wire Armored (Cable)

WA Wire Association [now Wire Association International, US]

WA *(Wissenschaftlicher Ausschuss)* – German for "Scientific Committee"

WA Women's Auxiliary

WA Woolwich Armstrong (Gun)

WA Woolwich Arsenal [UK]

WA Word After [Radio Communications]

WA Work Accomplished

WA Work Area

WA Work Authorization

WA Wrong Answer

WA Wrought Alloy

W/A Water/Air (Ratio) [also w/a]

W/A Watt(s) per Ampere [also W A^{-1}]

W-A Warren-Averbach (Grain Size Analysis) [also WA] [Metallurgy]

W-A Wilbur-Anderson (Unit)

W3A World Wide Web Applets [also W^{3}A]

WAA War Assets Administration

WAA Washington Area Astronomers (Meeting) [at United States Naval Observatory, Washington, DC]

WAA Water Authorities Association [UK]

WAA Western Awning Association [US]

WAA World Aluminum Abstracts [of Aluminum Association, US]

WAAC Women's Army Auxiliary Corps [US]

WAAF Women's Auxiliary Air Force [UK]

WAAG Waveform Acquisition and Arbitrary Generator; Waveform Analysis and Arbitrary Generator

WAAG Waveguide Acquisition and Arbitrary Generator (Board)

WAAIME Women's Auxiliary of the American Institute of Mining, Metallurgical and Petroleum Engineers [US]

WAAM Wide Area Antiarmor Munitions

WAAP World Association of Animal Production [Italy]

WAAS Wide Area Augmentation System

WAAS Women's Auxiliary Army Service [UK]

WAAS World Academy of Arts and Sciences [Sweden]

WAASC Women's Army Auxiliary Service Corps [UK]

WAAVP World Association for the Advancement of Veterinary Parasitology

WABCO Westinghouse Automotive Brake Company [US]

WABE Western Association of Broadcast Engineers [US]

WABI Windows Application Binary Interface

WABI Windows Applications Basic Interpreter

WAC Washington Administrative Code [US]

WAC Waste Acceptance Criteria [Nuclear Engineering]

WAC Watts, Alternating Current

WAC Weapon Aiming Computer

WAC Weighted Average Cost

WAC West Africa Committee [UK]

WAC Wet Analytical Chemistry

WAC Women's Army Corps [US]

WAC Work Accomplished Code

WAC Worked-all-Continents (Certificate) [International Amateur Radio]

WAC World Aeronautical Chart

WAC Write Address Counter

WACA Western Agricultural Chemicals Association [US]

WACA World Airline Clubs Association [Canada]

WACBDP Wide-Angle Convergent Beam Diffraction Pattern [Physics]

WACES Wyoming Association of Consulting Engineers and Surveyors [US]

WACHO Western Association of Canadian Highway Officials

WACISS Wide Area Coverage Infrared Search System

WACK Wait before Transmitting Positive Acknowledgement

WACM Western Association of Circuit Manufacturers [US]

WACMR West African Council for Medical Research

WACS Wide Angle Collimated Display System

WACS Workshop Attitude Control System

WACU West African Customs Union

WAD Weak Acid Dissociables

WAD Work Authorization Document

WAD Worst Area Difference

WADB West African Development Bank

WADC Wright Air Development Center [US Air Force]

WADC-AML Wright Air Development Center–Aeromedical Laboratory [of US Air Force at Dayton, Ohio, US]

WADD Wright Air Development Division [of US Air Force at Wright-Patterson Air Force Base, Dayton, Ohio, US]

WADE World Association of Document Examiners [US]

WADEX Word and Author Index

WADF Western Air Defense Force

WADR Waste Acid Detoxification and Reclamation [Nuclear Engineering]

WADS Wide Area Data Service

WAD/NU Work Authorization Document Number

WAE Wheel Abrasion Experiment [for NASA Mars Pathfinder Mission]

WAE Worked All Europe (Certificate) [International Amateur Radio]

WAEC West African Economic Community

WAEC West African Examinations Council [Ghana]

WAEI Western Association of Electrical Inspectors [US]

WAEP World Association for Element Building and Prefabrication [Germany]

WAEPA Western Australian Environment Protection Authority

Waermeu Stoffuebertrag Warme– und Stoffuebertragung [German Journal on Heat and Matter Transfer]

WAES Workshop on Alternative Energy Strategies [UK]

W-AESD WEC (Westinghouse Electric Corporation) Advanced Energy Systems Division [US]

WAF Width across Flats

WAF Wiring around Frame

WAF Women in the Air Force [US]

W Af West Africa(n) [also W Afr]

WAFC World Area Forecast Center

WAFF Western Australian Farmers Federation

WAFIC Western Australian Fishing Industry Council

W Afr West Africa(n) [also W Af]

WAFRI West-African Fisheries Research Institute

WAFRU West-African Fungicide Research Unit

WAFS Women's Auxiliary Ferrying Squadron [US Army]

WAFS World Area Forecast System

WAFWA Western Association of the Fish and Wildlife Association [US]

WAG Water-Alternating Gas

WAg Chemical Etchant of Hydrofluoric Acid, Nitric Acid and a 5% Aqueous Solution of Silver Nitrate; Mixture Ratio 2:1:2

W-Ag Tungsten-Silver (Alloy System)

WAGR Western Australian Government Railways

WAGR Windscale Advanced Gas-Cooled Reactor [UK]

WAHERB Western Australian Herbarium Plant Specimen Database

WAHS Warm-Air Heating System

WAHT Weighted Average Holding Time

WAI Web Application Interface [Internet]

WAI William Andrews Inc. [Norwich, New York, US]

WAI Wire Association International [US]

WAIS Wechsler Adult Intelligence Scale [Psychology]

WAIS Wide-Area Information Server [Internet]

WAIS Wide-Area Information Service

WAIS Wide-Area Information System

WAIS-R Wechsler Adult Intelligence Scale–Revised [Psychology]

WAIT Western Australian Institute of Technology

WAIT What Alloy Is That (Test) [Metallurgy]

WAITRO World Association of Industrial and Technological Research Organizations [Denmark]

WAITS Wide-Area Information Transfer System

WAK Wait Acknowledge

WAK Wiederaufbereitungsanlage Karlsruhe [Karlsruhe (Nuclear Fuel) Reprocessing Plant, Germany]

WAK Write Access Key

WAL Watertown Arsenal Laboratory [of US Army in Massachusetts, US]

WAL Wide-Angle Lens

WAL Wider, All Length [also wal] [Construction]

WAL World Association of Lawyers [US]

WAL Wright Aeronautical Laboratory [of US Air Force at Dayton, Ohio]

W-AL Westinghouse-Astronuclear Laboratory [US]

W-Al Aluminum-Doped Tungsten

WALA West African Library Association

WALDO Wichita Automatic Linear Data Output [US]

WALIP Western Australian Land Information Program

WALIS Western Australian Land Information System

Wall St Comput Rev Wall Street Computer Review [US]

W/Al$_2$O$_3$ Alumina Fiber Reinforced Tungsten (Composite)

WALOPT Weapons Allocation and Desired Ground Zero Optimizer [US Military]

WALTZ Wideband Alternating-Phase Low-Power Technique for Zero Residue Splitting

WAM Words A Minute

WAM Workshop on Advanced Materials [of International Union of Pure and Applied Chemistry]

WAM Worth Analysis Model

WAMCE Western Association of Minority Consulting Engineers

WAMDII Wide Angle Michelson Doppler Imaging Interferometer

WAMI World Association of Medical Informatics

WAMIC Women's Auxiliary of the Mining Industry of Canada

WAML Western Association of Map Libraries [US]

WAML Wright Aero-Medical Laboratory [US]

Wamoscope Wave-Modulated Oscilloscope

WAMPRI Western Australian Mining and Petroleum Research Institute

WAMRL Western Australian Marine Research Laboratories

WAMRU West African Maize Research Unit

WAMS World Association of Military Surgeons

WAMU West African Monetary Union

WAN Wide-Area Network

WAN Women's Aquatic Network [US]

WANC Western Australian Naturalists Club

WAND Working Party on Access to the National Database [UK]

WANEF Westinghouse Astronuclear Experimental Facility [US]

WANHS Wiltshire Archeological and Natural History Society [UK]

WANL Westinghouse Astronuclear Laboratory [US]

WAN/LAN Wide Area Network/Local Area Network

WANS Wide-Area Network System

WAOI Wide-Angle Acoustooptic Interaction

WAOS Welsh Agricultural Organization Society [UK]

WAP Waste Analysis Plan

WAP Work Assignment Procedure

WAPA Western Area Power Administration [US]

WAPD Westinghouse Atomic Power Division [US]

WAPDA Water and Power Development Authority [Pakistan]

WAPET Western Australia Petroleum Proprietary

WAPF West African Pharmaceutical Federation [Nigeria]

WAPS Waste Acceptance Product Specifications [of US Department of Energy]

WAR Warehouse Action Request

WAR Work Authorization Report

War Warning [also war]

War Warrant(y) [also war]

War Warwickshire [UK]

WARC World Administrative Radio Conference

WARC-BS World Administrative Radio Conference for Broadcasting Satellites [also WARC/BS]

WARC-MOB World Administrative Radio Conference for Mobile Services [also WARC/MOB]

WARC-MR World Administrative Radio Conference on Maritime Radio [also WARC/MR]

WARC-MT World Administrative Radio Conference on Maritime Telecommunications [also WARC/MT]

WARC-ORB World Administrative Radio Conference for Geostationary Satellite Orbit [also WARC/ORB]

WARC-ST World Administrative Radio Conference on Space Telecommunications [also WARC/ST]

WARDA West African Rice Development Association [Bouake, Ivory Coast]

Ward's Automot Rep Ward's Automotive Reports [US]

WARF Wisconsin Alumni Research Foundation [US]

WARFAC War Game Facility

Warhd Warhead [also warhd]

WARI Waite Agricultural Research Institute [South Australia]

WARI Western Australian Agricultural Research Institute

WARIS Western Arid Resource Information System [Queensland, Australia]

WARIS Water Resources Information System [New South Wales, Australia]

WARLA Wide Aperture Radio Location Array

WARM Wood and Solid Fuel Association of Retailers and Manufacturers [UK]

WARMER World Action for Recycling Materials and Energy from Rubbish

W/Armt With Armament [also W/armt]

WARN Worker Adjustment and Retraining Notification

Warn Warning [also warn]

WARR Waste Acid Release Reduction

WARRS West African Rice Research Station

Warw Warwickshire [UK] [also Warws]

WAS Washington Academy of Sciences [US]

WAS West Australian Standard (Time) [Greenwich Mean Time −7:00]

WAS Wisconsin Archeological Society [US]

WAS Worcestershire Archeological Society [UK]

WAS Worked-all-States (Certificate) [US Amateur Radio]

WAS World Aquaculture Society [US]

WAS World Archeological Society [US]

WASA West African Science Association

WASA West African Shippers Association [UK]

WASA Western Australian Specialty Alloys [Perth, Australia]

WASAL Wisconsin Academy of Sciences, Arts and Letters [US]

WASAR Wide Application System Adapter

WASAW Western Australian Sewage and Waste Quality Infrastructure Program

WASC West African School Certificate

WASC Western Administrative Support Center [of National Oceanic and Atmospheric Administration, US]

WASCO Water and Soil Conservation Organization [New Zealand]

Wash Washer [also wash]

Wash Washington (State) [US]

Wash, DC Washington, District of Columbia [US]

WASHO Western Association of State Highway Officials [US]

WASID Water and Soil Investigation Department [Pakistan]

WASMAC Western Australian Survey and Mapping Advisory Council

WASME World Assembly of Small and Medium Enterprises [India]

WASNA Western Agricultural Society of North America [US]

WASP Weightless Analysis Sounding Probe [of NASA]

WASP Williams Aerial Systems Platform

WASP Workshop Analysis and Scheduling Program

WASP Women's Air Force Service Pilots [US]

WASP World Association of Seaweed Processors [France]

WASSP Wire Arc Seismic Section Profiler

WASTAC Western Australian Satellite Technology Applications Consortium

Waste Environ Today Waste and Environment Today

Waste Manage Waste Management [Published in the UK]

WASU West African Students Union

WASWC World Association of Soil and Water Conservation [US]

WAT Web Action Time

WAT Weight, Altitude and Temperature [also W-A-T]

WAT West Africa Time [Greenwich Mean Time +1:00]

WAT Wide Area Telecommunications (Service)

WAT Work Adjustment Training (Program)

WATBRU West African Timber Borer Research Unit

WATCH Watch Trust for Environmental Education [UK]

WATCON Waterloo Concordance [of University of Waterloo, Ontario, Canada]

WATDOC Water Resources Document Reference System [of Environment Canada]

WATER Water Analysis to Evaluate and Recommend (Program)

Water Air Soil Pollut Water, Air and Soil Pollution [Published in the US]

Water Eng Mgmt Water Engineering and Management [Published in the US]

Water Int Water International [Publication of International Water Resources Association, US]

WATERLIT Water Literature (Database) [of South African Water Information Center]

Water Poll Control Water Pollution and Control [Published in Canada]

Water Qual Int Water Quality International [Publication of International Association on Water Pollution Research and Control, UK]

Water Res Water Research [Publication of International Association on Water Pollution Research and Control, UK]

Water Resour Bull Water Resources Bulletin [of American Water Resources Association]

Water Resour Congr Rep Water Resources Congress Report [US]

Water Resour Monogr Ser Water Resources Monograph Series [Publication of American Geophysical Union, US]

Water Resour Res Water Resources Research [Publication of American Geophysical Union, US]

Water Res Technol Water Research and Technology [Publication of International Association on Water Pollution Research and Control]

Water SA Water South Africa [Publication of Water Research Commission of South Africa]

Water Sci Technol Water Science and Technology [Published in the US]

Water Serv Water Services [Published in the US]

Water Wastes Dig Water and Wastes Digest [Published in the US]

WATFIV University of Waterloo FORTRAN V [Canada]

WATFOR University of Waterloo FORTRAN IV [Canada]

WATRS West Atlantic Route System [US]

WATS Wide-Area Telecommunications (Access) Service

WATS Wide-Area Telephone Service [US]

WATS Wide-Area Telephone System

WATS Work Authorization Tracking System

WATSTOR Water Data Storage and Retrieval System [of US Geological Survey]

WATTE West African Tropical Testing Establishment

WATTec Welding and Testing Technology Exhibition and Conference [US]

WAU Wausau, Wisconsin [Meteorological Station Designator]

W Austr West Australia(n) [also W Austral]

.WAV Waveform Audio [Microsoft File Name Extension]

WAVE Water Alliances for Voluntary Efficiency (Program) [of US Environmental Protetion Agency]

WAVES Waveform And Vector Exchange Program

WAVES Women Accepted for Volunteer Emergency Service [now Women in the United States Navy]

Waves Random Media Waves in Random Media [Journal of the Institute of Physics, UK]

WAVFH World Association of Veterinary Food Hygienists

WAWRC Western Australian Water Resources Council

Wax Waxing [also wax] [Paper Products]

WAXD Wide-Angle X-Ray Diffraction

WAXS Wide-Angle X-Ray Scattering

WAXS Wide-Angle X-Ray Diffraction Scattering

WB Tungsten Boride

WB Warm-Blooded (Animal)

WB Water Base(d)

WB Water Bath

WB Water Board

WB Water Boiler

WB Waybill [also W/B]

WB Weak Base

WB Weak Beam

WB Weather Bureau [Australia]

WB Weigh Batcher

WB Weld Brazing

WB Westbound

WB Wet Bulb

WB Wheel Balance(r)

WB Wheel Base

WB White Band

WB White Star, Blinker, Parachute

WB Wideband [also W/B]

WB Wire-Bonded; Wire Bonding

WB Wire Brush

WB Women's Bureau [of US Department of Labor]

WB Wool Bureau [US]

WB Word Before

WB The World Bank [Washington, DC, US]

WB Write Boundary [Electronics]

WB Write Bubble [Electronics]

WB-4101 2-([2,6-Dimethoxyphenoxyethyl)aminomethyl)-1,4-Benzodioxane

W&B Warrington and Boon (Method) [Materials Science]

W&B Weight and Balance

W/B Waybill [also WB]

W/B Wideband [also WB]

Wb Weber [also wb]

WBA World Buffalo Association [US]

Wb/A Weber(s) per Ampere [also Wb A^{-1}]

Wb/A/m Weber(s) per Ampere-meter [also Wb A^{-1} m^{-1}, or Wb/(A·m)]

WBAN Weather Bureau, Air Force and Navy [US]

W band Microwave Frequency Band of 56.00 to 100.00 gigahertz

WBAR Wing Bar [also Wbar]

WBBA Western Bird Banding Association [US]

WBBCC Wide Bay Burnett Conservation Council [Australia]

WBC Western Boundary Currents [Oceanography]

WBC White Blood Cell

WBC Whole-Body Count [Medicine]

WBC World Business Council [US]

WBCO Waveguide below Cutoff

WBCT Wideband Current Transformer

WBCV Wideband Coherent Video

WBD Wideband Data

WBD Wire Bound

WBDC Winnipeg Business Development Corporation [Manitoba, Canada]

WBDCS Wide-Band Data Collection System

WBDF Weak Beam Dark Field (Method) [Electron Microscopy]

WBDI Wideband Data Interleaver

WBDL Wideband Data Link

WBEC World Bank Environment Community

WBEM Web-Based Enterprise Management [Internet]

w-BEO Wurtzite-Type Beryllium Oxide

WBFM Wideband Frequency Modulation

WBGT Wet Bulb Globe Temperature

WBGT Wet Bulb Globe Thermometer

WBGU Wissenschaftlicher Beirat der Bundesregierung Globale Umweltveränderungen [Scientific Advisory Council of the German Government on Global Changes]

WBIF Wideband Intermediate Frequency

WBIS Wechsler-Bellevue Intelligence Scale [Psychology]

WBL Wideband Limiting

WBLC Waterborne Logistics Craft

WBM Wheeler, Boettinger and McFadden (Formulation) [Metallurgy]

Wb/m² Weber(s) per square meter [also Wb m^{-2}]

WBMA Western Building Material Association [US]

WBMA Wirebound Box Manufacturers Association [US]

WBMS World Bureau of Metal Statistics [UK]

WBN Wurtzite-Type Boron Nitride [also w-BN or wBN]

WBNL Wideband Noise Limiting

WBNS Water-Boiler Neutron Source (Reactor) [US]

WB/NWRC Weather Bureau/National Weather Records Center [US]

WBO Weather Bureau Office

WBO Wien-Bridge Oscillator

WBP Weather and Boil-Proof

WBPA Wideband Power Amplifier

WBR Water Boiler Reactor

WBR Work Bench Rack

WBRA Wagon Building and Repairing Association [UK]

WBRR Weather Bureau Radar Remote System

WBRS Wideband Remote Switch

WBS Washington/Baltimore Section [of Materials Research Society, US]

WBS Weight and Balance Measuring System

WBS Wellington Botanical Society [New Zealand]

WBS Wideband System

WBS Wide-Body STOL (Short Takeoff and Landing)

WBS Work Breakdown Structure

Wb/s Weber(s) per second [also Wb s^{-1}, or Wb/sec]

WBSARC World Broadcasting Satellite Administrative Radio Conference

WBSC Wide-Band Signal Conditioner

W/β-SiC Tungsten-Reinforced Beta Silicon Carbide

WBSW Wideband Switch

WBSXRT White-Beam Synchrotron X-Ray Topography

WBT Wet-Bulb Temperature

WBT Wet-Bulb Thermometer

WBT White Beam Topography

WBT Wide-Band Terminal

WBT Women in Broadcast Technology [US]

WB-TEM Weak-Beam Transmission Electron Microscopy

WBTS Wideband Transmission System

Wb turn Weber turn [Symbol]

WBVTR Wideband Video Tape Recorder

W by N West by North

W by S West by South

WC Tungsten Carbide

WC Walnut Council [US]

WC War Cabinet [UK]

WC War College

WC Warren-Cowley (Parameter) [Materials Science]

WC Waste Collection; Waste Collector

WC Water Closet [also wc]

WC Water Column

WC Water Content

WC Water Conservation(ist)

WC Water-Cooled; Water Cooling

WC Water Current

WC Water Cycle [Ecology]

WC Weapon Carrier

WC Weather Center

WC Weierstrass Condition [Mathematics]

WC Weld Crack(ing)

WC Wellington College [Canada]

WC Wenner Configuration

WC West Coast

WC Western Cedar

WC Weston (Standard) Cell

WC Whale Center [US]

WC Wigner Crystal

WC Wilkinson's Catalyst [Chemistry]

WC Wilson Chamber [Atomic Physics]

WC Wind Chill (Factor) [Meteorology]

WC Wing Commander [also W/C]

WC Wiper Crown (Cutter Insert) [Machining]

WC Wire Cable

WC Wire Chief

WC Without Charge [also wc]

WC Wood Chemistry

WC Wood Cover

WC Word Count

WC Work Cell

WC Work Center

WC Workers' (or Workmen's) Compensation

WC World Calendar

WC World Coordinate(s)

WC Worsted Count [Textiles]

WC Woven Composite

WC Write and Compute

W&C Wire and Cable

W/C Water-to-Cement Ratio (of Concrete) [also w/c]

W/C Wing Commander [also WC]

W-C Wright-Crossman (Model for Composite Materials)

W₂C Ditungsten Carbide

W3C World Wide Web Consortium

W/c Watts per Candle [also W c^{-1}]

wc with costs [Commerce]

w/c water-to-cement ratio (of concrete) [also W/C]

WCA Water Conservation Area

WCA Weeks-Chandler-Andersen (Theory)

WCA Western College Association [US]

WCA Wind Correction Angle

WCA Wireless Cable Association [US]

WCA Workmen's Compensation Act [Canada]

WCA World Ceramic Abstracts

WCA World Communication Association [US]

WCACP World Congress of Anatomic and Clinical Pathology

WCAM Western Canada Aviation Museum

WCAP Westinghouse Commercial Atomic Power [US]

WCAP World Climate Applications Program

WCASP World Climate Applications and Services Program

WCB Weatherproof Circuit Breaker

WCB Way Control Block

WCB Wildlife Conservation Board [California, US]

WCB Workers' Compensation Board [Canada]

WCB World Council for the Biosphere

WCBBC Workers' Compensation Board of British Columbia [Canada]

WCB-ISEE World Council for the Biosphere–International Society for Environmental Education

WCBO Workers' Compensation Board of Ontario [Canada]

WCC Wall Colmonoy Corporation [Madison Heights, Michigan, US]

WCC War Claims Commission

WCC War Crimes Commission

WCC Western Carolina College [Cullowhee, North Carolina, US]

WCC Wildfire Coordinating Committee [of National Association of State Foresters, US]

WCC Winnipeg Convention Center [Manitoba, Canada]

WCC Wire-Line Common Carrier

WCC Wisconsin Conservation Commission [US]

WCC Work Cell Control

WCC Workmen's Compensation Commission [Canada]

WCC World-Class Competitiveness

WCC World Climate Conference

WCC World Computer Conference

WCC World Congress Center [Atlanta, Georgia, US]

WCC World Crafts Council [Denmark]

WCC Write Control Character

WCCES World Council of Comparative Education Societies [France]

WCCF West Coast Computer Fair [San Francisco, California, US]

WCCI World Council for Curriculum and Instruction [US]

WC(Co) Cobalt Tungsten Carbide

WC-Co Cobalt-Bonded Tungsten Carbide

WC/Co Tungsten Carbide (Particles) in Cobalt (Matrix)

WCCPPS Waster Channel and Containment Pressurization and Penetration System

WCCU Wireless Communications Control Unit

WCCU Wireless Crew Communication Unit

WCCV-1 White Clover Cryptic Virus, Type 1

WCCV-2 White Clover Cryptic Virus, Type 2

WCD Work Control Desk

WCDB Wing Control During Boost

WCDMP World Climate Data and Monitoring Program

WCDP World Climate Data Program

WCDT Wet Countdown Demonstration Test

WCE Watt Committee on Energy [of Institution of Mechanical Engineers, UK]

WCE Winnipeg Commodity Exchange [Manitoba, Canada]

WCED World Commission on Environment and Development

WCEE Women's Council on Energy and the Environment [US]

WCEE World Conference on Earthquake Engineering

WCEMA West Coast Electronic Manufacturers Association [US]

W-CeO₂ Ceria-Doped Tungsten

WCES Wolfson Center for International Affairs [UK]

WCF Waste Calcining Facility [of Idaho National Engineering Laboratory, Idaho Falls, US]

WCF Water Concentration Factor

WCF White Cathode Follower

WCF Wind Chill Factor

WCF Winston Churchill Fund [US]

WCF Work Control File

WCF Workload Control File

WCFA Wildlife Conservation Fund of America [US]

WCG Weakly Compact Generated

WCGA World Computer Graphics Association [US]

WCGM Writable Character Generation Memory

WCGM Writable Character Generation Module

W-C:H Tungsten-Bearing Amorphous Hydrogenated Carbon

WCHS Western Colorado Horticultural Society [US]

WCI Waiting for Calls Indicator

WCI White Cast Iron

WCI Wildlife Conservation International [of New York Zoological Society, US]

WCI With Controlled Impurities [also wci] [Solid-State Physics]

WCI Wood Conservation Inc. [Canada]

WCI Workmen's Compensation Insurance

WCIP World Climate Impact Studies Program

WCIRP World Climate Impact Assesment and Response Strategies Program

WCISP World Climate Impact Studies Program

WCK Wildlife Clubs of Kenya (Association)

WCL Water Coolant Loop [Nuclear Engineering]

WCL Word Control Logic

WCL World Confederation of Labour [Belgium]

WCLC Women's Computer Literacy Center [US]

WCLIB West Coast Lumber Inspection Bureau [US]

WCLP Women's Computer Literacy Project [now Women's Computer Literacy Center, US]

WCLS Water Coolant Loop System [Nuclear Engineering]

WCM Wired Core Matrix

WCM Word Combine and Multiplex

WCM Writable Control Memory

W/cm Watt(s) per centimeter [also $W\,cm^{-1}$]

W/cm² Watt(s) per square centimeter [also $W\,cm^{-2}$]

W/cm³ Watt(s) per cubic centimeter [also $W\,cm^{-3}$]

WCMC World Conservation Monitoring Center [of International Union for Conservation of Nature and Natural Resources]

WCMIA West Coast Metal Importers Association [US]

W/cm·K Watt(s) per centimeter kelvin [also W/cm/K, or W cm^{-1} K^{-1}]

WCMMF World Congress on Man-Made Fibers

WCMT Wolfson Centre for Magnetic Technology [Cardiff, Wales, UK]

WCNDT World Conference on Nondestructive Testing

WCNURC West Coast (and Polar Regions) National Undersea Research Center [at University of Alaska Fairbanks, US]

WCO World Customs Organization

W-Co Tungsten-Cobalt (Alloy System)

WCOS Work Cell Operator Station

WCOTP World Confederation of Organizations of the Teaching Profession

WCP Waste Compliance Plan [Nuclear Engineering]

WCP Waste Collector Pump

WCP Water Circulating Pump,

WCP World Climate Program [of World Meteorological Organization]

WCP World Council of Peace

WC(p)/Co Tungsten Carbide (Particles) in Cobalt (Matrix) [also WC$_p$/Co]

WCPS World Confederation of Productive Science [Norway]

WCPSC Western Conference of Public Services Commissioners

WCR Water-Cement Ratio

WCR Water-Cooled Reactor

WCR Water Cooler

WCR Word Control Register

WCR Word Count Register

W-Cr Tungsten-Chromium (Alloy System)

WCRA Weather Control Research Association [now Weather Modification Association, US]

WCRA West Coast Railway Association [US]

W-Cr-N Tungsten-Chromium-Nitrogen (Alloy System)

WCRP World Climate Research Program [of World Meteorological Organization]

WCS Warehouse Control System

WCS Waste Collection System

WCS Whatman Compression Screw

WCS Wildlife Conservation Society [US]

WCS Work Center Supervisor

WCS Work Control System

WCS World Congress on Superconductivity

WCS World Conservation Strategy

WCS World Coordinate System

WCS Writable Control Store

WCSAC War Cabinet Scientific Advisory Committee [UK]

WCSC Western Connecticut State College [Danbury, US]

WCSI World Center for Scientific Information

WCSICEC Working Committee of the Scientific Institutes for Crafts in the European Community [Germany]

WCSS Waste Characterization and Sampling System

WCT Water-Cooled (Electron) Tube

WCT Weak-Coupling Theory [Particle Physics]

WCTOC World Congress of Theoretical Organic Chemists

WCTP Wire Chief Test Panel

WCU West Chester University [Pennsylvania, US]

WCU Western Carolina University [Cullowhee, North Carolina, US]

W/Cu Tungsten (Particles) in Copper (Matrix)

W/Cu Tungsten Fiber-Reinforced Copper (Composite)

W-Cu Tungsten-Copper (Alloy System)

WCVA Washington Convention and Visitors Association [Washington, DC, US]

WCVM Western College of Veterinary Medicine [Canada]

WCX Weak Cation Exchange

WCY World Communication Year [of United Nations]

WD War Department [Now Split into Department of the Army and Department of the Air Force]

WD Water-Dispersible; Water-Dispersed; Water Dispersion

WD Watt Demand-Meter

WD Wave Diffraction

WD Waveform Distortion

WD Wavelength Dispersion; Wavelength Dispersive

WD Weapons Division

WD Weld Decay

WD Welding Design(er)

WD White Dwarf [Astronomy]

WD Width [also Wd or wd]

WD Wigner-Dyson (Distribution) [Physics]

WD Williams Domain [Solid-State Physics]

WD Winchester Disk

WD Wind Direction

WD Wire Drawing

WD Wiring Diagram

WD Withdrawal; Withdrawn [also Wd]

WD Work Design

WD Work Distance

WD Working Directory

WD Working Distance [Microscopy]

WD Working Draft

WD Working Drawing

WD Works Department

WD Worm Drive

WD Write Data

WD Write Direct

W/D Write Down

Wd Wind [also wd]

Wd Withdrawal; Withdrawn [also WD]

Wd Wood [also wd]

Wd Word [also wd]

wd well-done

WDA Weighted Density Approximation

WDA Well Drillers Association [UK]

WDA Wildlife Disease Association [US]

WDAF Works Directorate of the Air Force

WDAX Wavelength Dispersive X-Ray Diffraction Analysis

WDB Werkstoffdatenbank [Materials Database, of Verein Deutscher Eisenhüttenleute, Germany]

WDB Wideband

WDB World Databank [of Harvard University, Cambridge, Massachusetts, US]

WDC Waste Disposal Cask

WDC Watts, Direct Current

WDC Western Digital Corporation [San Jose, California, US]

WDC Western Defense Command

WDC World Data Centers

WD&C Waste Dislodging and Conveyance

WDC-A World Data Center–A [US]

WDCM World Data Center on Microorganisms [also WDC-M]

WDCMGG World Data Center for Marine Geology and Geophysics [of NOAA National Geophysical Data Center, US]

WDCS Whale and Dolphin Conservation Society [UK]

WDCS Writable Diagnostic Control Store

WDD Well Drawdown

WDDES World Digital Data for the Environmental Sciences

WDE Weak Diffusion Expansion

WD EPMA Wavelength-Dispersive Electron Probe Microanalysis

WDF Washington Department of Fisheries [US]

WDF Waste-Derived Fuel

WDF Wave Digital Filter

WDF Wigner Distribution Function [Physics]

WDF Woodruff [also Wdf, or wdf]

WDG Water Dispersible Granular [Agriculture]

Wdg Winding [also wdg]

WDI Wilhelm-Dyckerhoff-Institut [Wilhelm Dyckerhoff Institute, Wiesbaden, Germany]

WDI Wind Direction Indicator

WD-IGI Woodall-Duckham Il Gas Internazionale (Process)

WDL Wien Displacement Law [Statistical Mechanics]

WDL (Microsoft) Windows Driver Library

Wdly widely

WDM Wavelength Division Multiplex(ing); Wavelength Division Multiplexer

WDM (Microsoft) Windows Driver Model

WDOE Washington (State) Department of Ecology [US]

WDOH Washington (State) Department of Health [US]

WDP Women in Data Processing [San Diego, California, US]

WDPC Western Data Processing Center [of University of California at Los Angeles, US]

WDR Westdeutscher Rundfunk [Broadcasting Station in Western Germany],

wdr wider

WDRAM Windows Dynamic Random Access Memory

WDQD Word Queue Directory

WDS Wavelength-Dispersive Spectrometer; Wavelength-Dispersive Spectrometry

WDS Wavelength-Dispersive (X-Ray) Spectroscopy

WDS Width Distribution Skew [Physics]

WDS Word Digital Sum

Wdsprd widespread

WDST Western Daylight Saving Time

WDT Weight Data Transmitter

WDTRS Westinghouse Development Test Requirement Specification

WDW Wirtschaftsverband Deutscher Werbeagenturen e.V. [German Association of Advertising Agencies, Germany]

Wdwk Woodwork [also wdwk]

Wdwkg Woodworking [also wdwkg]

WDX Wavelength Dispersive X-Ray Analysis [also WDXA]

WDX/EPMA Wavelength Dispersive X-Ray Analysis/Electron Probe Microanalysis

WDXRF Wavelength Dispersive X-Ray Fluorescence (Spectrometer) [also WD-XRF]

WE E-Plane Half-Power Width [Antennas]

WE Wankel (Combustion) Engine

WE Water Engineer(ing)

WE Water Equivalent [Meteorology]

WE Wave Energy

WE Wave Equation [Physics]

WE Weak Electrolyte

WE Weight Engineer(ing)

WE Welding Electrode

WE Welding Engineer(ing)

WE Wigner-Eckart (Theorem) [Quantum Mechanics]

WE Whitham Equation

WE Wind Energy

WE Wind Engineer(ing)

WE Wind Erosion

WE Women in Energy [US]

WE Work(ing) Envelope [Mechanical Engineering]

WE Working Electrode

WE Working Elongation (of a Rope)

WE Working Equation

WE Write Enable

W&E Wildlife and Ecology Division [of Commonwealth Scientific and Industrial Research Organization, Australia]

W/E Weekend [also w/e]

W/E Week Ending [also w/e]

W/E Wing Elevon [Aeronautics]

We Weber Number 1 (of Surface Tension) [Symbol]

W(ε) Statistical Probability for Energy Fluctuation of Magnitude ε [Symbol]

WEA Westinghouse Engineers Association [US]

WEA White Etching Areas [Metallurgy]

WEA Workers Education Association [Sweden]

WEA Workers Educational Association [UK]

Wea Weather [also wea]

WEAA Western European Airports Association

WEA-N Westinghouse Engineers Association National [US]

WEAP World Environment Action Plan

WEARCON Weather Observation and Forecasting Control System

Web World Wide Web

.WEB Entities Emphasizing the World Wide Web [Internet Domain Name] [also .web]

WEBA World Educational Broadcasting Assembly

WEBB Water, Energy and Biogeochemical Budgets

WebNFS Web Network File System [Internet]

WEC Washington Energy Conference [US]

WEC Western European Country

WEC Westinghouse Electric Corporation [Pittsburgh, Pennsylvania, US]

WEC Wind Energy Conversion (System)

WEC World Energy Conference [UK]

WEC World Engineering Conference

WEC World Environment Center [US]

WECAFC Western Central Atlantic Fisheries Commission [Italy]

WECC Western Engineering Conference and Competition [Canada]

WECC Western European Calibration Cooperation

WECM Wirecut Electrochemical Machining

WECO Western Electric Company [US]

WECON Western Electronics Show and Convention [of Institute of Electrical and Electronics Engineers, US]

WECOM Weapons Command [US Army]

WECPNL Weight Equivalent Continuous Perceived Noise Level

WECS Wind Energy Conservation System

WECS Wind Energy Conversion System

WED Weak Exchange Degeneracy [Particle Physics]

WED Western Economic Diversification [Canada]

WED Western Europe Daylight (Time) [Greenwich Mean Time + 1:00]

WED World Energy Day [of United Nations]

WED World Environment Day

Wed Wednesday

WEDA Western Dredging Association [US]

WEDA Wholesale Engineering Distributors Association [UK]

WEDC Western Economic Diversification Canada

WEDC Western Engineering Design Competition [Canada]

WEDSS Whole Earth Decision Support System

WEE Western Equine Encephalitis [Medicine]

WEE Wind Erosion Equation [Geology]

WEEC World Energy Engineering Congress

WEETAG Women's Employment, Education and Training Advisory Group

WEEV Western Equine Encephalitis Virus

WEF Waste Environment Federation [of Purdue University, West Lafayette, Indiana, US]

WEF With Effect From [also wef]

WEF World Education Fellowship [UK]

WE&FA Welsh Engineers and Founders Association [UK]

WEFAX Weather Facsimile (via Geosynchronous Weather Satellites) [of Environmental Science Services Administration, US]

WEFLD Wide-Extended Fault-Like Defect [Crystallography]

WEG Wind Energy Group [UK]

WEGS Western European Geological Surveys

WEI Western European Institute (for Wood Preservation) [Belgium]

WEI Work Efficiency Institute [Finland]

WEIA Washington Environmental Industries Association [US]

Weight Eng J Weight Engineering Journal [of Society of Allied Weight Engineers, US]

WEIN (International Symposium on) Weak and Electromagnetic Interactions in Nuclei

WELC Western Environmental Law Center [Eugene, Oregon, US]

WELC World Electrotechnical Congress

Weld Welding [also weld]

Weld Abstr Welding Abstracts [of the World Center for Information and Technology in Welding]

Weldasearch Comprehensive Database on Welding, Brazing, Soldering and Related Topics

Weld Des Fabr Welding Design and Fabrication [Published in the US]

WeldE Welding Engineer [also WeldEng]

Weld Eng Welding Engineer(ing)

WELD EXPO Canadian Welding Exposition

Weld Inf Newsl Welding Information Newsletter [of American Welding Institute, US]

Weld Innov Q Welding Innovation Quarterly [Published in the US]

Weld Int Welding International [Publication of Welding Institute, UK]

Weld J Welding Journal [of American Welding Society, US]

Weld J Res Suppl Welding Journal Research Supplement [of American Welding Society, US]

Weld Met Welding Metallurgist; Welding Metallurgy

Weld Met Fabr Welding and Metal Fabrication [Published in the UK]

Weld Prod Welding Production [Russian Journal published in the UK]

Weld Res Abroad Welding Research Abroad [Publication of Welding Research Council, US]

Weld Res Bull Welding Research Bulletin

Weld Res Counc Bull Welding Research Council Bulletin [US]

Weld Res Counc Prog Rep Welding Research Council Progress Report [US]

Weld Res Int Welding Research International [Published in the UK]

Weld Res News Welding Research News [Publication of Welding Research Council, US]

Weld Res Suppl Welding Research Supplement

Weld Res Brit Welding Research in Britain [UK Publication]

Weld Rev Welding Review [Published in the UK]

Weld Source Welding Source [Publication of Welding Institute of Canada]

Weld Tech Welding Technology

Weld World Welding in the World [Publication of International Institute of Welding, UK]

Weld Technol Can Welding Technology for Canada [Publication of Welding Institute of Canada]

Weld World Welding in the World [Journal of the International Institute of Welding]

WELGAS Western Leg Gas [of Far North Liquids and Associated Gas System]

WELL Whole Earth Lectronic Link [Bulletin Board System]

WEM Western European Metal Trades Employers Organization [Germany]

WEMA Western Electronic Manufacturers Association [now American Electronics Association, US]

WEMA Winding Engine Manufacturers Association

WEMC Western European Metrology Club

WEOG Western European and Others Group [of G-77–Group of Seventy Seven (Nations)]

WEP Water-Extended Polyester (Resin)

WEP World Economic Prospects [UK]

WEPSD World Engineering Partnership for Sustainable Development

WEPZA World Export Processing Zones Association [US]

WER Worth Estimating Relationship

WERC Waste Management Education and Research Consortium [of New Mexico State University, University of New Mexico, New Mexico Institute of Mining and Technology, Navajo Community College, Sandia National Laboratory and Los Alamos National Laboratories] [also Waste and Energy Reduction Consortium]

WERC Western Ecological Research Center [of US Geological Survey/Biological Resources Division at Davis, California]

WERC World Environment and Resources Council [Belgium]

Werkst Konstr Werkstoffe und Konstruktion [German Journal on Materials and Design]

Werkst Korros Werkstoffe und Korrosion [German Journal on Materials and Corrosion]

Werkstatt Betr Werkstatt und Betrieb [German Journal on Workshop and Factory]

Werk Wir Das Werk und Wir [German Publication on Industrial Plants and People]

WES Waterways Experiment Station [of US Army at Vicksburg, Mississippi]

WES Western Europe Standard (Time) [Greenwich Mean Time +1:00]

WES Wind Electric System

WES Women's Engineering Society

WES Writing Equipment Society [UK]

WES Wyoming Engineering Society [US]

WESAR Westinghouse Safety Analysis Report

WESA SAT Western Sahara Satellite Survey

WESCON Western Electronics Show and Convention [of Institute of Electrical and Electronics Engineers, US]

WESF (Nuclear) Waste Encapsulation and Storage Facility

WESRAC Western Research Application Center [US]

W/E&SP With Equipment and Spare Parts

WEST Western Energy Supply and Transmission (Associates) [US]

WESTAC Western Transportation Advisory Council

WESTAR Waterways Experiment Station Terrain Analyzer Radar

Westars Western Union Satellite System [US]

WESTEC Western Metal and Tool Exposition and Conference [of ASM International, US]

West Hem Western Hemisphere

Westinghouse STC Westinghouse Science and Technology Center [Pittsburgh, Pennsylvania, US]

WESTPAC Program Group for the Western Pacific

WET Waste Extraction Test

WET Western European Time

WETAC Westinghouse Electronic Tubeless Analog Computer [of Westinghouse Electric Corporation, Pittsburgh, Pennsylvania, US]

WETF Weightless Environment Training Facility [of NASA] [also WET-F]

WETNET Distribution and Joint Analysis Network for SSM/I (Special Sensor Microwave/ Imager) Data [of NASA Marshall Space Flight Center, Greenbelt, Maryland, US]

W(EtO)$_5$ Tungsten (V) Ethoxide (or Pentaethyl Tungsten)

W(EtO)$_6$ Tungsten (VI) Ethoxide (or Hexaethyl Tungsten)

WEU Western European Union

WEU Wide End Up (Mold) [Metallurgy]

WEX Word Expander

WF Wall Friction [Fluid Mechanics]

WF Wallis and Futuna Islands [ISO Code]

WF Wannier Function [Solid-State Physics]

WF Warm Front [Meteorology]

WF Wave Filter

WF Waveform

WF Wave Frequency [also W/F]

WF Wave Front [Physics]

WF Wave Function [Quantum Mechanics]

WF Weak Ferromagnetic (Phase) [Solid-State Physics]

WF Wedge Filter

WF Weight Fraction

WF Weighting Factor [Hydropower Plant]

WF Weighting Function

WF Weiss Field [Solid-State Physics]

WF Wheat Flour

WF Widefield

WF Wiedemann-Franz (Ratio) [also W-F] [Solid-State Physics]

WF Wigner Force [Particle Physics]

WF Wind Finding (Radar)

WF Wing Flap

WF Wiper Flat (Cutter Insert) [Machining]

WF Wire Foundation [US]

WF Wire Frame

WF Wood Fiber

WF Wood Flour

WF Wool Fat

WF Work Factor

WF Work Function [Solid-State Physics]

WF Working Fluid

WF Write Forward

WF Wrong Font [also wf]

W/F Wave Frequency [also WF]

W/F Wow and Flutter

W-F Wiedemann-Franz (Ratio) [also WF] [Solid-State Physics]

WFA Waveform Analysis; Waveform Analyzer

WFA White Fish Authority [UK]

WFA Wire Fabricators Association [US]

WFA Wisteria Floribunda Agglutinin [Immunology]

WFA World Federation of Advertisers [Belgium]

WFAIT Western Foundation for Advanced Industrial Technology

WFBSC World Federation of Building Service Contractors [UK]

WFC Wallops Flight Center [of NASA in Virginia, US]

WFC World Food Conference [of United Nations Food and Agricultural Organization]

WFC World Food Council [of United Nations Food and Agricultural Organization]

WFCA Western Forestry and Conservation Association [US]

WFCC World Federation for Culture Collections [UK]

WFCMV Wheeled Fuel-Consuming Motor Vehicle

WFD Waveform Digitizer

WFD Waveform Distortion

WFDA Wholesalers and Floorcovering Distributors Association [UK]

WFE Wiped-Film Evaporation

WFEA World Federation of Education Associations

WFEB Worchester Foundation for Experimental Biology [Maryland, US]

WFEO World Federation of Engineering Organizations [France]

WF&Eq Wave Filters and Equalizers [Institute of Electrical and Electronic Engineers–Parts, Hybrids, and Packaging Committee]

WFF Wallops Flight Facility [of NASA at Wallops Island, Virginia, US]

WFF Well-Formed Formula [also wff]

WFG Waveform Generator

WFG Westfälische Ferngas AG [Major German Gas Supplier located in Dortmund]

WFG Wet Film Gauge

WFI Wirtschaftsförderungsinstitut [Institute for Economic Advancement, Austria]

WFI Wood Foundation Institute [US]

WFI World Federation of Investors [Belgium]

WFIA Western Forest Industries Association [US]

WFL Wiedemann-Frank-Lorenz (Law) [Solid-State Physics]

WFL Work-Flow Language

WFM Wired For Management

WFMC Welding Filler Material Control

WFMI World Federation for the Metallurgical Industry

WFMT Wet Fluorescent Magnetic-Particle Testing

WFN White Fuming Nitric (Acid)

WFNA White Fuming Nitric Acid

WF$_6$-NH$_3$-H$_2$ Tungsten Hexafluoride/Ammonia/Hydrogen (Gas)

WFO Work for Others

WFOV Wide Field-of-View

WFP Warm Front Passage

WFP World Federation of Parasitologists [Netherlands]

WFP World Food Program [of United Nations]

WFPA Washington Forest Protection Association [US]

WFPA World Federation for the Protection of Animals

WFPC Widefield Planetary Camera [also WF/PC]

WFPIS Whole Farm Plan Incentives Scheme [of Victoria State, Australia]

WFPL Western Forest Products Laboratory [Canada]

WFPLCA World Federation of Pipeline Contractors Associations [US]

WFPMA World Federation of Personnel Management Associations [US]

WFPP Whole Farm Planning Program [of Tasmania, Australia]

WFR Weil-Felix Reaction [Medicine]

WFR Wiedemann-Franz Ratio [Solid-State Physics]

Wfr Wafer [also wfr]

WFRC Western Fisheries Research Center [of US Geological Survey/Biological Resources Division at Seattle, Washington]

WFS Wet Flexural Strength

WFS World Fertility Survey

WFS World Future Society [US]

WFSF World Futures Studies Federation [Hawaii, US]

WFSW World Federation of Scientific Workers [UK]

WFT Winograd Fourier Transform

WFTU World Federation of Trade Unions

WFU Wake Forest University [North Carolina, US]

WFUMB World Federation for Ultrasound in Medicine and Biology

WFUNA World Federation of United Nations Associations

WFW (Microsoft) Windows For Workgroups

WG Water-Dispersible Granules [Agriculture]

WG Water Gas

WG Water Gauge

WG Waveguide [also W/G]

WG Weapons Grade

WG Weighing

WG Weight Guaranteed [also wg]

WG Welding Generator

WG West Germany

WG Wire Gauge

WG Working Gauge

WG Working Group

WG Worm Gear(ing)

WG Woven Glass

W-G Water-Glycol (Mixture) [also W/G]

W/G Waveguide [also WG]

W/g Watt(s) per gram [also W/gm or W g^{-1}]

WGA Western Governors Association [US]

WGA Wheatgerm (*Triticum vulgaris*) Agglutinin [Immunology]

WGA Wild Goose Association [US]

WGA Wiskundig Genootschap te Amsterdam [Mathematical Society of Amsterdam, Netherlands]

WGA Working Group on Agriculture

WGA Wyoming Geological Association [US]

W-Ga Tungsten-Gallium (Alloy System)

WGBC Waveguide Operating Below Cutoff

WGC West Georgia College [now West Georgia University, US]

WGD Waste Generation Database

WG/DM Working Group on Data Management [International Satellite Cloud Climatology Project]

WGDT Waste Gas Decay Tank

WGE Western Gold Exposition [US]

W Ger West German(y)

WGES World's Greatest Environment Statement

WGF Women's Gas Federation [UK]

WGFC Wyoming Game and Fish Commission [US]

WG-GGI Working Group on Geodesy and Geographic Information

WGI World Glacier Inventory

WGL Waveguide Laser

WGL Wissenschaftsgemeinschaft Gottfried Wilhelm Leibniz [Gottfried Wilhelm Leibnitz Scientific Society, Germany]

WGLR Wissenschaftliche Gesellschaft für Luft– und Raumfahrt [Scientific Society for Aeronautics and Astronautics, Germany]

WGM Waveguide Mode

W/gm Watt(s) per gram [also W/g or W g^{-1}]

WGMA Work Glove Manufacturers [US]

WGMP Workshop on Geometric Methods in Physics

WGMS World Glacier Monitoring Service [Switzerland]

WGN White Gaussian Noise [Physics]

WGNE Working Group on Numerical Experimentation

WGNHS Wisconsin Geological and Natural History Survey [US]

WGPD Waveguide Photodetector

WGS Waveguide Glide Slope

WGS Wideband Ground Station

WGS Wiltshire Geological Services [Australia]

WGS Work Group System

WGS World Geodetic Spheroid

WGS World Geodetic System

WGS84 World Geodetic Spheroid 1984

WGSC Wide Gap Spark Chamber

WGST Waste Gas Storage Tank

WGT Western Greenwich Time

Wgt Weight [also wgt]

WGTA Western Grain Transportation Act

WGU West Georgia University [Carrollton, US]

WGW Wind-Generated Wave

WH H-Plane Half-Power Width [Antennas]

WH Waste Heat

WH Water Hammer [Fluid Mechanics]

WH Water-Hardened; Water-Hardening [Metallurgy]

WH Water Heater

WH Weber-Hermite (Equation) [Mathematics]

WH Western Hemlock

WH West Hartford [Connecticut, US]

WH Williamson-Hall (Method)

WH Woodward-Hoffmann (Rules) [Organic Chemistry]

WH Work Harden(ed); Work Hardening [Metallurgy]

WH World Heritage

WH Write Head [Electronics]

W&H Wage and Hour Division

Wh Watthour [also wh, Whr, or whr]

wh which

Wh white

WHA Tungsten Heavy Alloy

WHA Western Hardwood Association [US]

WHA Wood Heating Association [US]

WHA World Heritage Area [Australia]

WHAM Waveform Hold and Modify

What's New Comput What's New in Computing [UK Publication]

WHB Waste-Heat Boiler

WHC Washington Hospital Center [Washington, DC, US]

WHC Watthour Meter with Contact Device

WHC Westinghouse Hanford Company [Washington State, US]

WHC Workstation Host Connection

WhC White Compound [1,9-Dicarboxy-2,4,6,8-Tetranitro-phenazine -N-Oxide]

WHCA White House Communications Agency [Washington, DC, US]

WHCES Wave Hill Center for Environmental Studies [Bronx, US]

WHCLIS White House Conference on Library and Information Services [US]

WHD Wage and Hour Division [of US Employment Standards Administration]

WHDM Watthour Demand Meter

WHEC Wildlife Habitat Enhancement Council [US]

WHEC World Hydrogen Energy Conference

WHERF Wood Heating Education and Research Foundation [US]

WHETS Washington Higher Education Telecommunication System

Whf Wharf [also whf]

WHH Werthamer-Helfand-Hohenberg (Formula) [Materials Science]

WhH White House [Washington, DC, US]

WHI Western Highway Institute [US]

Which Comput Which Computer? [UK Publication]

WHIMS Wet High-Intensity Magnetic Separation

WHIMS-CF Wet High-Intensity Magnetic Separation and Cationic Flotation

WINE Windows Emulator

WHIPS Windows Highly Integrated Program

Wh/kg Watthour per kilogram [also Wh kg^{-1}]

WHL Watthour Meter with Loss Compensator

Wh/L Watthour per liter [also Wh L^{-1}]

Whl Wholesale [also whl]

Whlr Wholesaler [also whlr]

Wh Lt White Light [also wh lt] [Optics]

WHM Watthour Meter

WHM Watthour Motor

WHMBL Woods Hole Marine Biological Laboratory [Massachusetts, US]

WHMI Workplace Hazardous Materials Information

WHMIS Workplace Hazardous Materials Information System

WHO World Health Organization [of United Nations]

WHOI Woods Hole Oceanographic Institute [Woods Hole, Massachusetts, US]

WHO STD World Health Organization Standard [also WHO Std]

WHO Tech Rep Ser World Health Organization, Technical Report Series

WHP Water Horsepower [also W HP, or w hp]

WHP World Hydrocarbon Program [of Stanford Research Institute, California, US]

Wh P White Phosphorus

WHPH Width at Half-Peak Height [X-Ray Diffraction Analysis]

WHPS Watt, High-Pressure Sodium

WHR Waste Heat Recovery

WHR Work Hardening Rate

Whr Watthour [also whr, Wh, or wh]

WHRA Welwyn Hall Research Association [UK]

WHRM Wildlife/Habitat Relationship Model

WHS Wisconsin Historical Society [US]

WHS Work-Hardened Zone [Metallurgy]

Whse Warehouse [also whse]

Whsg Warehousing [also whsg]

Whsle Wholesale [also whsle]

WHSV Weight Hourly Space Velocity

WHT Watthour Meter, Thermal Type

WHT Welded and Heat-Treated; Welding and Heat Treatment [Metallurgy]

wht white

WHTC Wolfson Heat Treatment Center [University of Aston in Birmingham, UK]

W/Hz Watt(s) per Hertz [also W Hz^{-1}]

WI Wascana Institute [Saskatchewan, Canada]

WI Water Injection

WI Weak Interaction [Particle Physics]

WI Wedge Interferogram

WI Welding Institute [now The Welding Institute, Cambridge, UK]

WI Wentworth Institute [US]

WI West India(n)

WI West Indies

WI Wetland Inventory [US]

WI Windward Islands

WI Wisconsin [US]

WI Wrought Iron

W&I Weighing and Inspection

W/I Within [also w/i]

WIA Women in Aerospace [US]

WIAS West Indies Associated States

WIAB Wistar Institute of Anatomy and Biology [US]

WIACO World Insulation and Acoustic Congress Organization [France]

Wiad Elektrotech Wiadomosci Elektrotechniczne [Polish Journal on Electrical Engineering]

Wiad Hutn Wiadomosci Hutnicze [Polish Journal on Metallurgy]

Wiad Telekomum Wiadomosci Telekomunikacyjne [Polish Journal on Telecommunication]

WIAS Weierstraß-Institut für Angewandte Analysis und Stochastik [Weierstrass Institute for Applied Analysis and Stochastics, Berlin, Germany]

WIAS West Indies Associated States

WIBS Wool Industry Bureau of Statistics [UK]

WIC Water Information Center [US]

WIC Welding Institute of Canada

WIC Wildlife Information Center [US]

WICAT World Institute for Computer-Assisted Teaching [US]

WICB Women in Cell Biology [US]

WICE World Industry Council for the Environment

WICEM World Industry Conference on Environmental Management [of International Committee for Conservation, France]

WICP Wacker Ingot Casting Process [Metallurgy]

WICS Westinghouse Integrated Compiling System

Wid Width [also wid]

WIDE Wiring Integration Design

WID Women in Development [of Development Assistance Committee of the Organization for Economic Cooperation and Development]

WIDJET Waterloo Interactive Direct Job Entry Terminal (System) [of University of Waterloo, Ontario, Canada]

WIE With Immediate Effect

WIE Women in Engineering

WIF Water Immersion Facility

WIF Wrought Iron Front (Boiler) [also wif]

WIFT (Stored) Waveform Inverse Fourier Transform

WIG WWW (World Wide Web) Interest Group

WIGSYM International Wigner Symposium [Physics]

WIGUT West Indies Group of University Teachers [Jamaica]

WIHS Women's Interagency HIV (Human Immunodeficiency Virus) Study [US]

WIIU Workers International Industrial Union

WIL White Indicating Lamp

WILCO Will Comply [Radio Communication]

Wildl Soc Bull Wildlife Society Bulletin

Wilts Wiltshire [UK]

WIM Weigh-in-Motion (Sorting System)

WIM Women in Mining [US]

WIMA Writing Instrument Manufacturers Association [US]

WIMP Weakly Interacting Massive Particle

WIMP Window, Icon, Mouse, Pointer

WIMP Windows, Icons, Mouse and Pulldown Menus

WIMP Wireless Implanted Magnetic Resonance Probes

WIMP Wisconsin Interactive Molecule Processor (Software) [also Wimp]

WIMR West Indies Marine Laboratory [Jamaica]

WIMSE Women in Materials Science and Engineering [of Materials Research Society, US]

WIN (International Workshop on) Weak Interactions and Neutrinos [Particle Physics]

WIN Welding Information Network [of American Welding Institute, US]

WIN Wissenschaftsnetz [German Research Network]

WIN Windows (Operating System) [of Microsoft Corporation, US] [also Win]

WIN Work Incentive Program [US]

Win Window [also win]

Win Windows (Operating System) [of Microsoft Corporation, US] [also WIN]

W/in² Watt(s) per square inch [Unit]

WIN3.1 Windows 3.1 (Operating System) [Microsoft Corporation Operating System] [also Win3.1]

WIN95 Windows 95 (Operating System) [Microsoft Corporation Software published in 1995] [also Win95]

WIN98 Windows 98 (Operating System) [Microsoft Corporation Software published in 1998] [also Win98]

WIN2000 Windows 2000 (Operating System) [Microsoft Corporation Software published in 2000] [also Win2000]

WINA Webb Institute of Naval Architecture [Glen Cove, New York, US]

WINB Western Interstate Nuclear Board [US]

WINC Watts, Incandescent

WINCO Westinghouse Idaho Nuclear Company [US]

WINCO Winter College on Optics [at International Center for Theoretical Physics, Trieste, Italy]

WINCON Winter Conference on Aerospace and Electronic Systems [US]

WINCON Winter Convention [of Institute of Electrical and Electronics Engineers, US]

WIND Global Geospace Science Wind Spacecraft [of NASA]

WIND Weather Information Network and Display

W Ind West Indies

Wind Eng Wind Engineering [Publication of European Wind Energy Association, UK]

WINDII Wind Imaging Interferometer

WINDI RAC Wind Imaging Interferometer Remote Analysis Computer

Wind Is Windward Islands

Window Ind Window Industries [Published in the UK]

Windpower Mon Newsmag Windpower Monthly Newsmagazine [Published in Denmark]

WINDS Weather Information Network and Display System

WINForum Wireless Information Networks Forum

WINMIC Windsor Metric Information Center [Ontario, Canada]

WINS Weight Information Networking System

WINS (Microsoft) Windows-Internet Naming Service

WINSOCK (Microsoft) Windows Open Systems Architecture

WINSOCK (Microsoft) Windows Sockets

WINTEM Wind and Temperature Forecast

WINWORD Word For (Microsoft) Windows

WINWOX Winfrith Wet Oxidation (Process)

WIP Waste Immobilization Plant

WIPE Waste Immobilization Process

WIP Wissenschafts-Integrations-Programm [Science Integration Program, Germany]

WIP Women in Information Processing [US]

WIP Work-in Process

WIP Work-in-Progress

WIPE Waste Immobilization Process Experiment

WIPO World Intellectual Property Organization [of United Nations]

WIPP Waste Isolation Pilot Plant [near Carlsbad, New Mexico, US]

WIPP-MIIT Waste Isolation Pilot Plant–Materials Interface Interactions Test (Program) [US Nuclear Waste Management]

WIPS Weather Image Processing System

W-Ir Tungsten-Rhodium (Alloy System)

Wir Wiring [also wir]

WIRA Wax Importers and Refiners Association [US]

WIRA Wool Industries Research Association [UK]

WIRCCWS WHO (World Health Organization) International Reference Center for Community Water Supply [now International Reference Center of UNESCO and WHO]

WIRDS Weather Information Remoting and Display System

WIRE International Wire and Cable Trade Fair [Duesseldorf, Germany]

WIRE Web Information Repository for the Enterprise [of European Strategic Program for Research (and Development) in Information]

WIRE Wide-Field Infrared Explorer [of Infrared Processing and Analysis Center, NASA Jet Propulsion Laboratory, Pasadena, California, US]

WIRECOM Wire Communication [also Wirecom]

Wire Ind Wire Industry [Published in the UK]

Wire J Wire Journal [of Wire Association International, US]

Wire J Int Wire Journal International [of Wire Association International, US]

Wireless Eng Wireless Engineer [Published in the UK]

Wire Technol Wire Technology [Published in the US]

Wire Technol Int Wire Technology International [Published in the US]

Wire World Int Wire World International [Published in Germany]

WIRTC Western Industrial Research and Training Center

WIS WATS (Wide Area Telephone Service) Information System

WIS Weizmann Institute of Science [Rehovot, Israel]

WIS Wheat Information Service [Japan]

WIS Women in Science

WIS Workforce Information System

WIS World Information Systems

Wis Wisconsin [US]

WISA West Indies Sugar Association

WISA Wire Industry Suppliers Association [US]

WISARD Web Interface for Searching Archival Research Data [of NASA Astrophysics Data Facility, Goddard Space Flight Center, Greenbelt, Maryland, US]

WISC Writable-Instruction-Set Computer; Writable-Instruction-Set Computing

Wisc Wisconsin [US]

WISCLAND Wisconsin Initiative for Statewide Cooperation on Land Cover Analysis and Data [US]

WISE Wang (Computer) Intersystem Exchange

WISE Washington Internships for Students of Engineering [of American Society for Engineering Education, US]

WISE Wolfson Institute for Surface Engineering [of Leeds University, UK]

WISE Women in Science and Engineering

WISE WordPerfect Information System Environment

WISE World Information Service on Energy [Netherlands]

WISE World Information Systems Exchange

WISHA Washington (State) Industrial Safety and Health Act [US]

WISI World Information System on Informatics

WISICA West Indies Sea Island Cotton Association

WISITEX World Instrumentation Symposium and International Trade Exposition

WISP Waves in Space Plasma Program [of National Research Council of Canada]

WISP Winter Icing and Storms Project

WISP-HF Waves in Space Plasma–High-Frequency [of National Research Council of Canada]

WISP-OMV Waves in Space Plasma–Orbital Maneuvering Vehicle

Wiss Z Friedrich-Schiller Univ Jena Natwiss Reihe Wissenschaftliche Zeitschrift der Friedrich-Schiller-Universität Jena Naturwissenschaftliche Reihe [Scientific Journal of Friedrich Schiller University at Jena, Natural Sciences Series, Germany]

Wiss Z Karl-Marx-Univ Leipz Math-Natwiss Reihe Wissenschaftliche Zeitschrift der Karl-Marx-Universität Leipzig Naturwissenschaftliche Reihe [Scientific Journal of Karl-Marx University at Leipzig, Natural Science Series, Germany]

Wiss Z Tech Hochsch Chemnitz Wissenschaftliche Zeitschrift der Technischen Hochschule Chemnitz [Scientific Journal of the Chemnitz University of Technology, Germany]

Wiss Z Tech Hochsch Ilmenau Wissenschaftliche Zeitschrift der Technischen Hochschule Ilmenau [Scientific Journal of the Ilmenau University of Technology, Germany]

Wiss Z Tech Hochsch Otto von Guericke Wissenschaftliche Zeitschrift der Technischen Hochschule Otto von Guericke [Scientific Journal of the Otto von Guericke University of Technology, Magdeburg, Germany]

Wiss Z Tech Univ Dresd Wissenschaftliche Zeitschrift der Technischen Universität Dresden [Scientific Journal of the Dresden University of Technology, Germany]

Wiss Z Tech Univ Chemnitz Wissenschaftliche Zeitschrift der Technischen Universität Chemnitz [Scientific Journal of the Chemnitz University of Technology, Germany]

Wiss Z Tech Univ Otto von Guericke Magde Wissenschaftliche Zeitschrift der Technischen Universität Otto von Guericke, Magdeburg [Scientific Journal of the Otto von Guericke University of Technology, Magdeburg, Germany]

WIST (Study Group on) Wind Shear and Turbulence [of International Civil Aviation Organization]

WISTCI Wisconsin University–Theoretical Chemistry Institute [US]

WIT Washington Institute of Technology [US]

WIT Web Interactive Talk [Internet]

WIT Wentworth Institute of Technology [Boston, Massachusetts, US]

WIT Wessex Institute of Technology [Ashurst, Southampton, UK]

WITB Wood Industry Training Board [UK]

WITG Western International Trade Group [US]

WITS Washington Integrated Telephone System [US]

WITS (Hazardous) Waste Inventory Tracking System

WITS Waterloo Interactive Terminal System [of University of Waterloo, Ontario, Canada]

WITS Welliste Information Transfer Specification [of University of Tulsa, Oklahoma, US]

Wits Witwatersrand [South Africa]

WIU Western Illinois University [Macomb, US]

WJ Water Jet

WJ Wigner-Jordan (Method) [Quantum Mechanics]

WJC Water-Jet Cutting

WJCC Western Joint Computer Conference [US]

WJP Water-Jet Pump

WJTA Water Jet Technology Association [US]

WK Wiener Kernel [Physics]

WK Wiener-Khinchine (Relation) [Mathematics]

WK Wolff-Kishner (Reduction) [Organic Chemistry]

.WK1 Worksheet [Lotus 1-2-3 File Name Extension]

W/K Watt(s) per Kelvin [also W K^{-1}]

W-K Potassium-Doped Tungsten

Wk Week [also wk]

Wk Work [also wk]

wk weak

WKB Wentzel-Kramers-Brillouin (Theory) [Quantum Mechanics]

.WKB Workbook [WordPerfect File Name Extension]

WKBJ Wentzel-Kramers-Brillouin-Jordan (Theory) [Quantum Mechanics]

Wkbk Workbook [also wkbk]

Wkd Weekdays [also wkd]

.WKE Worksheet [Lotus 1-2-3 File Name Extension]

Wkg Working [also wkg]

W/kg Watt(s) per kilogram [also W kg^{-1}]

Wkg Ppr Working Paper(s) [also wkg ppr]

wkly weekly [also wkly]

wkn weaken

WKNL Walter Kidde Nuclear Laboratories [US]

.WKQ Spreadsheet [File Name Extension]

WKQDR Work Queue Directory

WKR Wolff-Kishner Reduction [Organic Chemistry]

.WKS Worksheet [Lotus 1-2-3 File Name Extension]

WKSC Western Kentucky State College [now Western Kentucky University, US]

Wksp Workshop [also wksp]

WKT Wong-Koplik and Tomanic [Porous Media]

WKU Western Kentucky University [Bowling Green, US]

.WKZ Compressed Spreadsheet [File Name Extension] [also .wkz]

WL Wafer Loading [Electronics]

WL Wagner-Lifshitz (Model) [Metallurgy]

WL Warm Layer

WL Wash Load [Geology]

WL Water Line

WL Waved (Slip) Line

WL Wavelength [also W/L]

WL Weak Localization

WL Weeping Lubrication

WL Weight Loss

WL Western Larch

WL Wet Location

WL Wetting Layer

WL Wheel Load

WL White Light [Optics]

WL Willis and Lake (Grain Size Method) [also W&L] [Metallurgy]

WL Wind Load(ing)

WL Wired Logic

WL Wire List

WL Word Line

WL Working Level

WL Working Life

WL Working Limit

WL Work Life [Chemical Engineering]

WL Work Load

WL Wright Laboratory [of Wright-Patterson Air Force Base, Dayton, Ohio, US]

W&L Willis and Lake (Grain Size Method) [also WL] [Metallurgy]

W/L Dislocation Line Energy per Unit of Length (in Materials Science) [Symbol]

W/L Wavelength [also WL]

W/L Width-to-Length Ratio [also w/l]

Wl Wheel [also wh]

WLA Wagner-Lifshits-Ardell (Theory) [Metallurgy]

W-La$_2$O$_3$ Lanthania-Doped Tungsten

WL/ASC Wright Laboratory/Aeronautical Systems Center [of US Air Force Materiel Command]

WLC Wavelength Comparator

WLC Wisconsin Library Consortium [US]

Wldg Welding [also wldg]

WLCP (International) Workshop on Liquid Crystalline Polymers

WLF Williams, Landel and Ferry (Equation) [Physics]

WLFA Wildlife Legislative Fund of America [US]

WLI Wavelength Interval

WLM Working-Level Minute

WLM Working Level Month

WLN Washington Library Network [US]

WLN Western Library Network [US]

WLN Wiswesser Line Notation

W Lon West Longitude [also W Long]

WLP Wasserstoff-Lichtbogen-Pyrolyse (Prozess) [Hydrogen-(Electric) Arc-Pyrolysis (Process)]

WLPSA Wildlife Preservation Society of Australia

WLR Wechsler-Lieberman-Read (Formalism) [also W-L-R] [Metallurgy]

WL-RDCB Wedge-Loaded Rectangular Double Cantilever Beam [Mechanical Testing]

WLSC West Liberty State College [West Virginia, US]

WLT Wissenschaftlicher Arbeitskreis Lasertechnik [Study Group on Laser Technology, Germany]

WL-TR Wright Laboratory Technical Report [of Wright Patterson Air Force Base, Dayton, Ohio, US]

WLU Wilfred Laurier University [Waterloo, Ontario, Canada]

WLUS World Land Use Survey

WLZS World Lead-Zinc Symposium

WM Warranty Manager

WM Waste Management

WM Waste Minimization

WM Wastewater Management

WM Watermark

WM Water Miscible

WM Wave Mechanics [Quantum Mechanics]

WM Wavemeter

WM Wave Motion

WM Wattmeter

WM Weather Modification

WM Welding Machine

WM Welding Metallurgy

WM Weld Metal

WM Wet-Milled; Wet Milling

WM Whistler Mode

WM White Matter [Anatomy]

WM White Metal

WM White Mountains [New Hampshire, US]

WM Whole Molecule [also wm]

WM Windmill

WM Wire Mesh

WM Wobble Motor

WM Women Marines

WM Wood Meal

WM Wood's Metal

WM Word Mark

WM Working Memory

W&M Washburn and Moen (Wire Gauge)

W/M Weight and/or Measurement

Wm² Watt(s) meter squared [Unit]

W/m² Watt(s) per square meter [also W m^{-2}]

W/m³ Watt(s) per cubic meter [also W m^{-3}]

wm whole molecule [also WM]

Wmk Watermark [also wmk]

W/mK Watt(s) per millikelvin [also W mK^{-1}]

W/m·K Watt(s) per meter kelvin [also W/m-K, W/m/K, or W m^{-1} K^{-1}]

W/m/K Watt(s) per meter per kelvin [also W m^{-1}·K^{-1}]

W/m²·K Watt(s) per square meter kelvin [also W/m²-K, W/m²/K, or W m^{-2} K^{-1}]

W/m²·K⁴ Watt(s) per square meter kelvin to the fourth [also W/m²/K⁴, or W m^{-2} K^{-4}]

W/m²·sr Watt(s) per square meter steradian [also W/m²-sr, or W m^{-2} sr^{-1}]

WMA Wallcovering Manufacturers Association [UK]

WMA Waste Management Authority

WMA Waste Management Authority [of New South Wales, Australia]

WMA Waterbed Manufacturing Association [US]

WMA Water-Heater Manufacturers Association [UK]

WMA Weather Modification Association [US]

WMA Welding Manufacturers Association [UK]

WMA Wellington Mathematical Society [New Zealand]

WMA Wildlife Management Area

WMA World Medical Association

WMATA Washington Metropolitan Area Transit Authority [Washington, DC, US]

WMB Walnut Marketing Board [US]

WMB Waste Management Branch

WMBA Wire Machinery Builders Association [now Wire Industry Suppliers Association, US]

WMC Waste Management Compartment

WMC Western Mapping Center [of US Geological Survey]

WMC Western Mining Corporation [Australia]

WMC Wool Manufacturers Council [of Northern Textile Association, US]

WMC Workflow Management Coalition

WMC World Manufacturing Congress

WMC World Materials Congress

WMC World Meteorological Center

WMC World Meteorological Congress

WMC World Mining Congress (and Exhibition)

W/m/°C Watt(s) per meter per degree Celsius [also W m^{-1} °C^{-1}]

W/m² °C Watt(s) per square meter degree Celsius [also W m^{-2} °C^{-1}]

WMCE Western Montana College of Education [Dillon, Montana, US]

WMD Water Management District

WMDA Woodworking Machinery Distributors Association [US]

WME Women and Mathematics Education [US]

WMEC Western Military Electronics Center [US]

W(MeO)₆ Tungsten(VI) Methoxide (or Hexamethyl Tungstate)

WMF Washington Materials Forum [Consortium of the American Ceramics Society, American Chemical Society, American Institute of Chemical Engineers, American Physical Society, American Society of Mechanical Engineers, American Vacuum Society, ASM International, Electrochemical Society, Inc., Federation of Materials Societies, Materials Research Society, Mineralogical Society of America, Society for Hybrid Microelectronics, Society of Photo-Optical Instrumentation Engineers, and The Minerals, Metals and Materials Society]

WMF Windows Metafile(s) [Computers]

.WMF Windows Metafile Format [Microsoft File Name Extension] [also .wmf]

W-MgO Magnesia-Doped Tungsten

WMHS Wall-Mounted Handling System

WMI Walther-Meissner-Institut [Walther Meissner Institute, Garching, Germany]

WMI Wildlife Management Institute [of US Department of the Interior]

WMI Wolfson Microelectronics Institute [UK]

WMIA Woodworking Machinery Importers Association [US]

WMIB Waste Management Information Bureau [UK]

W/m/K Watt(s) per Meter per Kelvin [also W m^{-1} K^{-1}]

WMKLEM W. M. Keck Laboratory of Engineering Materials [of California Institute of Technology, Pasadena, California, US]

W/mm² Watt(s) per Square Millimeter [also W mm^{-2}]

WMMA Wood(working) Machinery Manufacturers of America [US]

WMMI Western Museum of Mining and Industry [Colorado Springs, US]

WMMP Waste Management Master Plan

WMMPA Wood Moulding and Millwork Producers Association [US]

Wmn Woman [also wmn]

WMO World Meteorological Organization [of United Nations]

W-Mo Tungsten-Molybdenum (Alloy System)

WMOA Waste Minimization Opportunity Assessment

WMP Waste Management Plan

WMR Wave-Making Resistance

Wmr warmer

WMS Warehouse Management System

WMS Waste Management System

WMS Washington Medical School [US]

WMS Welsh Mines Society [UK]

WMS Whaling Museum Society [US]

WMS Wheeler, Murray and Schaefer (Dendritic Solidification Approach) [Metallurgy]

WMS Wieland Multiplex System

WMS Wire Mesh Screen

WMS World Magnetic Survey [of International Union of Geodesy and Geophysics]

WMS World Mariculture Society [now World Aquaculture Society, US]

WMSA Woodworking Machinery Suppliers Association [UK]

WMSB World Magnetic Survey Board [of International Union of Geodesy and Geophysics]

WMSC Waste Management Steering Committee

WMSC Weather Message Switching Center

WMSI Weak Metal-Support Interaction

WMSI Western Management Science Institute [US]

WMSO Wichita Mountains Seismological Observatory [US]

WMT Waste Monitor Tank

WMTDD Waste Management and Transportation Development Division [of US Department of Energy]

WMU Western Michigan University [Kalamazoo, US]

W&M Wire G Washburn & Moen Wire Gauge

WMY World Mathematical Year

WN Tungsten Nitride

WN Wave Normal [Physics]

WN Wave Number [Physics]

WN White Noise [Physics]

WN Whole Number

WN Wobbe Number [Thermodynamics]

WN Wrong Number

W$_2$N Ditungsten Nitride

W-N Tungsten-Nitrogen (Alloy System)

Wn Won [Currency of North Korea]

WNA World Nature Association [US]

W-Na Tungsten-Sodium (Alloy System)

WNAAA Women of the National Agricultural Aviation Association [US]

WNAMA Women's Network in Aquatic and Marine Affairs [now World Aquaculture Society, US]

WNAR Western North American Region [US]

WNC West North Central [US Geographic Region]

Wnd Wind [also wnd]

WNE (Microsoft) Windows Network Environment

WNEC Western New England College [Springfield, Massachusetts, US]

WNGA Wholesale Nursery Growers of America [US]

WNIC Wide-Area Network Interface Coprocessor

W/NiFe Tungsten (Particles) in Nickel-Iron (Matrix)

WNIM Wide-Area Network Interface Module

WNMU Western New Mexico University [Silver City, US]

WNNLSA White Noise Non-Linear System Analysis

WNO Wrong Number

WNRB Wyoming National Resources Board [US]

WNRC Washington National Records Center [US]

WNRE Whiteshell Nuclear Research Establishment [of Atomic Energy of Canada Limited at Pinawa, Manitoba, Canada]

WNRM Women in Natural Resources Management (Program) [of International Union for Conservation of Nature and Natural Resources]

WNSM Welsh National School of Medicine [Cardiff, UK]

WNT Waste Neutralizer Tank

WNW West-Northwest; West by Northwest

WNYNRCR Western New York Nuclear Research Center Reactor [US]

WNYS Western New York Section [of Materials Research Society, US]

WO War Office [UK]

WO Warnier-Orr (Diagram)

WO Warrant Officer

WO Wave Optics

WO Welding Operator

WO White Oak

WO Wireless Officer

WO Word Pointer

WO Work Order

WO Write Only

WO Write Out

W/O Water-in-Oil (Emulsion)

W/O Widow/Orphan (Protection) [Word Processing]

W/O Without [also w/o]

W/O Write-Off [also w/o]

W-O Oxygen-Doped Tungsten

W(ω) Noise Spectral Density [Symbol]

w/o weight percent

W/o without [also W/O]

WOA Warrant Officers Association [US]

.WOA Windows Swap Files [File Name Extension]

WOAS When on Active Duty

WOAUS Warrant Officers Association of the United States

WOB Weight on Bit

WOBO World Organization of Building Officials [Canada]

WOC Waiting on Cement (Time)

WOC Water-Oil Concentration

WOC Water/Oil Contact

WOC Welding Operator Certificate

WOCA World Outside Communist Areas

WOCE World Ocean Circulation Experiment (Program) [of World Meteorological Organization]

Wochenbl Papierfabr Wochenblatt für Papierfabrikation [German Weekly on Papermaking]

WOD Wingate Ordnance Depot [Gallup, New Mexico, US]

WODA World Organization of Dredging Associations [US]

WODC World Ozone Data Center [of US/Canada Atmospheric Environment Service]

W/O E&SP Without Equipment and Spare Parts

W(OEt)$_5$ Tungsten(V) Ethoxide (or Pentaethyl Tungstate)

W(OEt)$_6$ Tungsten(VI) Ethoxide (or Hexaethyl Tungstate)

WOF Work of Fracture

WOG Water, Oil and Gas

WOGCC Wyoming Oil and Gas Conservation Commission [US]

WOHMA Waste Oil Heating Manufacturers Association [US]

WΩ/K² Watts-ohm per kelvin squared [also WΩ K⁻²]

WOL Wedge-Opening Loaded (Specimen) [Mechanical Testing]

WOL White Oak Laboratory [of US Naval Surface Warfare Center, Silver Spring, Maryland]

WOM Wear of Materials (Conference)

WOM Write-Only Memory

WOM Write Optional Memory

W(OMe)₅ Tungsten(V) Methoxide (or Pentamethyl Tungstate)

W(OMe)₆ Tungsten(VI) Methoxide (or Hexamethyl Tungstate)

WOMI Wireless Operation and Maintenance Instructions

WONG Weight on Nose Gear

WOO Waiting on Orders

WOO Western Operation Office [of NASA]

Wood Fiber Sci Wood and Fiber Science [Publication of Society of Wood Science and Technology, US]

Wood Res Wood Research [Publication of Wood Research Institute, Japan]

Wood Sci Technol Wood Science and Technology [Published in Germany and the US]

Wood Sci Technol (NY) Wood Science and Technology (New York) [Published in the US]

Wood Tech Wood Technologist; Wood Technology [also Wood Technol]

W/O OFF Widow/Orphan Protection Off [also W/O Off] [Word Processing]

W/O ON Widow/Orphan Protection On [also W/O On] [Word Processing]

WOR Water-Oil Ratio

WORB Work Roll Bending [Metallurgy]

WORC Washington Operations Research Council [US]

WORCRA Worner/Conzinc Riotinto of Australia (Process) [Metallurgy]

Worcs Worcestershire [UK]

Word Inf Process Word and Information Processing [Published in the UK]

WORD PTR Word Pointer [also Word Ptr]

WORF (Laboratory) Window Observational Research Facility [of NASA]

Workstn Workstation

World Alum Abstr World Aluminum Abstracts [of Aluminum Association, US]

World Alum Dyn World Aluminum Dynamics [Published in the US],

World Arch World Archeology [International Journal]

World Cem World Cement [Published in the US]

World Ceram World Ceramics [Published in Germany]

World Clim Rep World Climate Report

WORLDDIDAC World Association of Manufacturers and Distributors of Educational Material [Switzerland]

World Drug Rep World Drug Report [of United Nations International Drug Control Programme]

World Forest Res World Forest Resources [of Food and Agricultural Organization]

World Min World Mining [Published in the US]

World Packag News World Packaging News [Publication of World Packaging Organization, France]

World Pat Inf World Patent Information [Published in the US]

World Rev Nutr Diet World Review of Nutrition and Dietetics

World Rivers Rev World Rivers Review [Quarterly Newsletter of the International Rivers Network, US]

World Sci World Scientific [Published in Singapore]

World Steel World of Steel [Published in Japan]

World Steel Metalwork Export Man World Steel and Metalworking, Export Manual [Published in Germany]

World Textile Abstr World Textile Abstracts [Publication of Shirley Institute, UK]

WorldWIDE World Women in Defense of the Environment

Work People Work and People [Australian Publication]

Workstn Mag Workstation Magazine [UK Publication]

WORM Write Once, Read Many Times (Memory)

WORM Write Once, Read Mostly

WORM CD Write Once, Read Many Times Compact Disk

W or W/O With or Without [also w or w/o]

WOS Web Offset Section [of Printing Industries Association, US]

WOS Workstation Operating System

WOS Wilson Ornithological Society [US]

W-Os Tungsten-Osmium (Alloy System)

WOSA (Microsoft) Windows Open Systems Architecture

WOSAC Worldwide Synchronization of Atomic Clocks

WOSC World Organization of Systems and Cybernetics [France]

WOSS Work Order Status System

WOT Wide-Open Throttle (Switch)

WOTA World Organization of Tourism and Automobiles

WOTS Work Order Tracking System

WOTS Wallops Orbital Tracking System [of NASA Wallops Flight Facility, Wallops Island, Virginia, US]

WOW Waiting on Weather

WOW Weight on Wheels

WO/W Without Weapons

WOWS Wire Obstacle Warning System

WOWS Women Ordnance Workers

WOX Wet Oxidation (Process)

WP Tungsten Phosphide

WP Warm Processing

WP Warsaw Pact

WP Waste Package

WP Waste Processing

WP Waste Prevention

WP Water Pollution

WP Water Power

WP Waterproof(ing) [also wp]

WP Water Pump

WP Water Purification

WP Wave Packet [Physics]

WP Wave Propagation

WP Way Point [also W/P]

WP Wear Plate

WP Weather Permitting

WP Weatherproof(ing)

WP Weatherproof Type (Conductor Insulation) [Symbol]

WP Welding Procedure

WP Welding Processing

WP Weld Pool [Metallurgy]

WP Western Pine

WP West Point [New York State, US]

WP Wettable Powder

WP Wheel Pulser

WP White Phosphorus

WP White Pine

WP White Plains [New York State, US]

WP Whole Pattern

WP Wildlife Protection

WP Wind Power

WP Wind Profiler

WP Wing Profile [Aeronautics]

WP *(Wirtschaftspatent)* – German for "Industrial Patent"

WP Wolf-Palissa (Chart) [Astronomy]

WP Wood Physicist; Wood Physics

WP Wood Pulp

WP WordPerfect (Software)

WP Word Processing; Word Processor [also W/P]

WP Working Party

WP Working Plane

WP Working Point

WP Working Pressure

WP Work Package

WP Workpiece

WP Workplace

WP Workspace Pointer

WP Workspace-Register Pointer

WP Write Permit

WP Write Protect(ed); Write Protection

WP4.2 WordPerfect, Version 4.2 (Software)

WP5.0 WordPerfect Version 5.0 (Software)

WP6.0 WordPerfect Version 6.0 (Software)

WP6.1 WordPerfect Version 6.1 (Software)

WP8.0 WordPerfect Version 8.0 (Software)

W/P Way-Point [also WP]

W/P Word Processing; Word Processor [also WP]

Wp Peak Watt

WPA Western Pine Association [US]

WPA Whale Protection Act [Australia]

WPA Wire Products Association [UK]

WPA Wisconsin Pharmaceutical Association [US]

WPA With Particular Average [also wpa]

WPA Works Project Administration [US]

WPA Wyoming Pharmaceutical Association [US]

WPAFB Wright Patterson Air Force Base [Dayton, Ohio, US]

WPALC Wright Patterson Air Logistics Command [of US Air Force]

WPAP World Population Action Plan [of United Nations]

WPB War Production Board

WPB Waste Paper Basket

WPB Write Printer Binary

WPBW Wide Parabolic Well

WPBS Welsh Plant Breeding Station [UK]

WPC War Problems Commission

WPC Water Pillow Cooling

WPC Water Pollution Control

WPC Water Pollution Council [US]

WPC Watts per Candle [also Wpc, wpc, or W/c]

WPC Wheat Protein Concentrate [Biochemistry]

WPC William Paterson College [US]

WPC Woman Police Constable

WPC Wood-Plastic Combination

WPC Wood-Plastic Composite

WPC WordPerfect (Software) Corporation [US]

WPC World Petroleum Congress [UK]

WPC World Population Conference [of United Nations]

WPC World Power Conference

WPC World Print Council [US]

WPCA Water Pollution Control Act [US]

WPCA Water Pollution Control Administration [US]

WPCAA Water Pollution Control Act Amendment [US]

WPCF Water Pollution Control Federation [US]

WPCP Water Pollution Control Plant

WPD Write Printer Decimal

.WPD Windows Printer Description [File Name Extension] [also .wpd]

WPDA Writing Push Down Acceptor

WPDES Waste Pollution Discharge Elimination System

WPDN Wind Profiler Demonstration Network

WPDP Waste Processing and Disposal Program [US]

WPDraw WordPerfect Draw [Word Processing]

WPE Weight per Epoxide

WPESS Width-Pulse Electronic Sector Scanning

WPF World Productivity Forum

Wpf Waterproof [also wpf]

Wpfg Waterproofing [also wpfg]

WPG Frigate-Type Patrol Gunboat [US Navy Symbol]

WPG Water Pipe Ground

WPG Weight per Grain [Forestry]

.WPG WordPerfect Graphics [File Name Extension]

WPHD Write-Protect Hard Disk

WPI Waste Policy Institute [of Virginia Polytechnic Institute and State University, Blacksburg, US]

WPI Wholesale Price Index

WPI Worcester Polytechnic Institute [Massachusetts, US]

WPI World Patents Index [UK]

.WPK WordPerfect Keyboard [File Name Extension]

WPL Waste Pickle Liquor

WPL Wave Propagation Laboratory [of National Oceanic and Atmospheric Administration, US]

WPL Wright-Patterson Laboratories [of Wright-Patterson Air Force Base, Dayton, Ohio, US]

WPLA Wool Products Labeling Act [US]

WPM Wobbles per Minute

WPM Words per Minute [also wpm]

WPM Write Protect Memory

.WPM WordPerfect Macro [File Name Extension]

WPMA Windows/Presentation Manager Association [US]

WPMA Wood Products Manufacturers Association [US]

WPN Wind Profiler Network [US]

Wpn Weapon [also wpn]

WPNS Weapons Division [of US Army, Naval Air Warfare Center, China Lake, California]

Wpns Weapons [also wpns]

WPO Weak Phase Object

WPO World Packaging Organization [France]

WPO World Ploughing Organization [UK]

WPOS Workplace Operating System

WPP Waterproof Paper Packing

WPP Water-Purification Process

WPPA World Population Plan of Action

Wppb weight parts per billion [Unit]

wppm weight parts per million [Unit]

WPPSS Washington Public Power Supply System [US]

wppt weight parts per trillion [Unit]

WPQW Wide Parabolic Quantum Well [Electronics]

WPR Water Production Rate

WPR Write Permit Ring

WPRD Water Produced

WP(REI) Wildlife Protection (Regulations and Exports and Imports) Act [Australia]

WPRL Water Pollution Research Laboratory [UK]

WPRT Write Protect

WPS Warm Pre-Stressing [also wps]

WPS Water Phase Salt

WPS Welding Procedure Specification

WPS Wet Powder Spraying

WPS (Microsoft) Windows Printing System

WPS Wireless Preservation Society [UK]

WPS Word Processing Software

WPS Word Processing Specialist

WPS Word Processing System

WPS Words per Second [also wps]

WPS Workplace Shell [IBM Operating System 2]

WPS Workstation Presentation Services

WPSA Wildlife Preservation Society of Australia

WPSA World Poultry Science Association [Germany]

WPSG WordPerfect (Software) Support Group [US]

WPSMA Welsh Plate and Steel Makers Association [UK]

WPSQ Wildlife Preservation Society of Queensland [Australia]

WPT Way-Point

WPT Word Processing Technician

WPTI Wildlife Preservation Trust International [US]

WPUET West Pakistan University of Engineering and Technology [Lahore]

WPVM (Microsoft) Windows Parallel Virtual Machine

WPW Wide Parabolic (Quantum) Well [Electronics]

WPWin WordPerfect for Windows

WPWRA Wallpaper, Paint and Wallcovering Research Association [UK]

WPY World Population Year

WQ Water Quality

WQ Water-Quenched; Water Quenching

.WQ Compressed Spreadsheet [File Name Extension]

.WQ1 Spreadsheet [File Name Extension]

WQA Water Quality Act [US]

WQA Water Quality Association [US]

WQASD Water Quality and Aquatic Science Division [of National Hydrology Research Institute, Canada]

WQB Water Quality Based

WQB Water Quality Branch [of Environment Canada]

WQC Water Quality Certification

WQC Wheat Quality Council [US]

WQM Weld Quality Monitor

WQMP Water Quality Management Project

WQP Water Quality Project [now National Water Center, US]

WQPA Western Quick Printers Association

WQR Wasteform Qualification Report

WQRC Water Quality Research Council [US]

WR Walden's Rule [Physical Chemistry]

WR Wall Receptacle

WR Wand Reader

WR Warehouse Receipt [also W/R]

WR Wassermann Reaction [Medicine]

WR Waste Reduction

WR Water Reactive (Material)

WR Water Repellency; Water Repellent

WR Water Research

WR Water Resistance; Water-Resistant

WR Water Resources

WR Wave Resistance

WR Wear Resistance; Wear-Resistant

WR Weight Reduction

WR Welding Rectifier

WR Welding Research

WR Welding Rod

WR Weather Radar

WR Whiteshell Reactor [of Whiteshell Nuclear Research Establishment, Pinawa, Manitoba, Canada]

WR Wildlife Reserve

WR Wilson Repeater

WR Wind Rose

WR Wire Rod

WR Wire Rope

WR Wittig Reaction [Organic Chemistry]

WR Wolf-Rayet (Star) [Astronomy]

WR Wood Ray

WR Working Register

WR Woven Roving

WR Wurtz Reaction [Organic Chemistry]

W/R Warehouse Receipt [also WR]

Wr Write [also wr]

WR-1 Whiteshell Reactor No. 1 [of Whiteshell Nuclear Research Establishment, Pinawa, Manitoba, Canada]

WRA Water Research Association [UK]

WRA Water Resources Abstracts

WRA Water Resources Administration

WRA Weapon Replaceable Assembly

WRA Western Railroad Association [US]

WRA Whiteshell Research Area [Canada]

WRA Wild Rivers Act [US]

WRAC Willow Run Aeronautical Center [US]

WRAC Waste and Recycling Advisory Committee [of Australian and New Zealand Environment and Conservation Council]

WRAC Water Resources Advisory Committee [of Australian Environment Council]

WRAC Women's Royal Army Corps [UK]

WRADAC Water Research Association Distribution Analog Center [UK]

WRAF Women's Royal Air Force [UK]

WRAIR Walter Reed Army Institute of Research [US]

WRAIS Wide Range Analog Input Subsystem

WRAM Windows Random Access Memory

WRAMA Warner-Robins Air Materiel Area [US]

WRAP Waste Receiving and Packaging

WRAP Waste Reduction Audit Protocol

WRAP Waste Reduction Always Pays (Program)

WRAP Weighted Regression Analysis Program

WRB Wire, Rod and Bar

WRB Wissenschaftliches Rechenzentrum Berlin [Berlin Supercomputer Center, Germany]

WRBC Weather Relay Broadcast Center

WRC War Resources Council

WRC Wastewater Research Center [of University of British Columbia, Vancouver, Canada]

WRC Water Research Center [UK]

WRC Water Reseach Commission

WRC Water Resources Commission

WRC Water Resources Congress [US]

WRC Water Resources Council

WRC Watson Research Center [of IBM Corporation at Yorktown Heights, New York, US]

WRC Weather Relay Center [US]

WRC Webster Research Center [of Xerox Corporation at Webster, New York, US]

WRC Welding Research Council [US]

WRC Western Red Cedar

WRC Western Research Center [of Canada Center for Mineral and Energy Technology/Natural Resources Canada in Devon, Alberta]

WRC Western Reserve College [of Case Western Reserve University, Cleveland, Ohio, US]

WRC Weston Research Center [Ontario, Canada]

WRC Westwater Research Center [of University of British Columbia, Vancouver, Canada]

WRC World Radiation Center

WRCA Water Resources Center Archive [of University of California at Berkeley, US]

WRCA Western Red Cedar Association [US]

WRCB Water Resources Control Board [of California Environmental Protection Agency, US]

WRCC Wildlife Research Coordinating Committee [West Africa]

WR CHK Write Check [also Wr Chk]

Wrckg Wrecking [also wrckg]

WRCLA Western Red Cedar Lumber Association [US]

WRCSA Water Research Commission of South Africa

WRD Water Resources Division [of US Geological Survey]

WRD Woolwhich Research Department [UK]

Wrd Word

WRDC Wool Research and Development Corporation [Australia]

WRDC Wright Research and Development Center [of Wright-Patterson Air Force Base, Dayton, Ohio, US]

WRDC-TR Wright Research and Development Center Technical Report [of Wright-Patterson Air Force Base, Dayton, Ohio, US]

WRD/ES Woolwhich Research Department, Explosives Section [UK]

WRE Weapons Research Establishment [Australia]

W-Re Tungsten-Rhenium (Alloy System)

WREAFS Waste Reduction Evaluation at Federal Sites [US]

WREC Westinghouse Reactor Evaluation Center [US]

W Rec Warehouse Receipt [also w rec]

W-Re-Mo Tungsten-Rhenium-Molybdenum (Alloy System)

WRENDA World Request for Neutron Data Measurements (Database) [of International Atomic Energy Agency, Austria]

WRESAT Weapons Research Establishment Satellite [Australia]

WRF World Research Foundation [Woodland Hills, California, US]

WRG Waterways Recovery Group [UK]

wrg wrong

WRH Walter Reed Hospital [Washington, DC, US]

W-Rh Tungsten-Rhodium (Alloy System)

WRI Walter Reed Institute [US]

WRI War Risk Insurance [of US Maritime Administration]

WRI Western Research Institute [US]

WRI Wire Reinforcement Institute [US]

WRI Wood Research Institute [of Kyoto University, Japan]

WRI World Resources Institute [Washington, DC, US]

.WRI Write [File Name Extension]

WRISE Waste Reduction Institute for Scientists and Engineers

WRITE Waste Reduction Innovative Technology Evaluation

WRITELN Write Line [Pascal Programming] [also Writeln]

WRIU Write Interface Unit

WRK (Microsoft) Windows Resource Kit

WRK Woodward's Reagent K [Organic Chemistry]

WRL Westinghouse Research Laboratories [Pittsburgh, Pennsylvania, US]

WRL Willow Run Laboratory [US]

WRL Woodward's Reagent L [Organic Chemistry]

WRLA Western Retail Lumbermen's Association [now Western Building Material Association, US]

WRM Water Routing Model

WRM Western Reserve Manufacturing Company, Inc. [US]

WRM World Rainforest Movement [Malaysia]

wrm warm

WRME Wood Raw Material Equivalent

WRMFNT Warm Front [Meterorology]

WRMI We Really Mean It! [Internet Jargon]

WRMS Watts, Root-Mean-Square [also Wrms]

Wrng Warning [also wrng]

WRNI Wide Range Neutron Indicator

WRNI Wide Range Nuclear Instrument

Wrnt Warrant [also wrnt]

WRNS Women's Royal Naval Service [UK]

WROW Wabash River Ordnance Works [Newport, Indiana, US]

w(r,φ) magnetic interaction energy [Symbol]

WRPC Water Resources Planning Commission [Taiwan]

WRPC Weather Records Processing Center

WRP/CI Worldwide Reinforced Plastics/Composites Institute

WRPDB Western Regional Production Development Board [Nigeria]

WRQ Westinghouse Resolver/Quantizer

WRR Woomera Rocket Range [Adelaide, Australia]

WRRC Willow Run Research Center [US]

WRRI Water Resources Research Institute [of University of Wyoming, Laramie, US]

WRRL Western Regional Research Laboratory [US]

WRRR Walter Reed Research Reactor [US]

WRRS Wire Relay Radio System

WRS Waveguide Raman Spectroscopy

WRS Working Reference System

WRSA World Rabbit Science Association [UK]

WRSIC Water Resources Scientific Information Center [of US Department of the Interior]

WRSSSV Wireline-Retrievable Subsurface Safety Valve

WRT With Respect To [also wrt]

wrt wrought

WRTA Western Railroad Traffic Association [US]

WRTB Wire Rope Technical Board [US]

WRTC Working Reference Telephone Center

WRU Western Reserve University [Now Part of Case Western Reserve University, Cleveland, Ohio, US]

WRU Who Are You? [Telecommunications and Computers]

W-Ru Tungsten-Ruthenium (Alloy System)

WRVS Women's Royal Voluntary Service [UK]

WS Samoa [ISO Code]

WS Wait State [Computers]

WS Wallops Station [now Wallops Flight Facility]

WS Wannier-Stark (Quantization) [Solid-State Physics]

WS Washington State [US]

WS Watch Supervisor

WS Water-Dispersible Powder for Slurry Treatment (of Seeds) [Agriculture]

WS Water Softener; Water Softening

WS Water Solid

WS Water Solubility; Water-Soluble

WS Water Storage

WS Water Supply

WS Wave Soldering [Electronics]

WS Wave Spectrum

WS Weather Satellite

WS Weather Service [US]

WS Weather Ship

WS Weatherstrip(ping)

WS Weapon System

WS Wedge Spectrogram; Wedge Spectrograph

WS Weigh Scale

WS Welding Solidification

WS Western Samoa(n)

WS West Siberia(n)

WS Wet Strength

WS Wetted Surface

WS Whole Antiserum [Immunology]

WS Wigner-Seitz (Cell) [Crystallography]

WS Williamson Synthesis [Organic Chemistry]

WS Wind Shear [Meteorology]

WS Wind Sock

WS Wind Speed

WS Winter School

WS Winter Semester

WS Wireless Station

WS Wire Send

WS Wire Shear

WS Woodcock Survey [US]

WS Wood Science; Wood Scientist

WS WordStar (Software)

WS Working Space [Computers]

WS Working Storage

WS Working Stress

WS Work-Softened; Work Softening [Metallurgy]

WS Work Space [Computers]

WS Workstation [also W/S]

WS Work Support

WS Wright Stain [Biology]

WS Wyoming State [US]

WS The Wilderness Society [US]

W/S Water to-Solution Ratio [also w/s]

W/S Work Station [also WS]

Ws Wattsecond [Unit]

WSA Web Sling Association [now Web Sling and Tiedown Association, US]

WSA Weed Society of Australia

WSA Wet-Gas Sulfuric Acid (Process) [Chemical Engineering]

WSA Wholesalers Stationers Association [US]

WSA Work Sciences Association [UK]

WSA World Sign Associates [US]

WSAD Weapon Systems Analysis Division [US Navy]

WSAPI Web Site Application Program Interface

WSAS Weather Service Airport Station [US]

WSB Water Spray Boiler

WSB Wheat Soy Blend [Food Science]

WSC Wave Spectrum Concept

WSC Water-Soluble Crystal

WSC Water Systems Council [US]

WSC Wayne State College [Nebraska, US]

WSC Western Snow Conference [US]

WSC West South Central [US Geographic Region]

WSC White Sands Complex [of NASA in New Mexico, US]

WSC White Star Cluster [Astronomy]

WSC Winona State College [Minnesota, US]

WSC Wire Strand Core

WSC Wisconsin State College [Platteville, US]

W/Sc Tungsten/Scandium (Multilayer)

WSCA World Surface Coatings Abstracts (Database) [of Paint Research Association, UK]

WSCC Western State College of Colorado [Gunnison, US]

WSCC Western States Chiropractic College [Portland, Oregon, US]

WSCC Western Systems Coordinating Council

WSCD Water Survey Canada Division [of Environment Canada]

WSCF (Nuclear) Waste Sampling and Characterization Facility

WSCIT Wisconsin State College and Institute of Technology [Platteville, US]

WSCMO Weather Service Contract Meteorological Observatory [US]

WSCS Western Society of Crop Science [US]

WSCS Wide Sense Cyclo-Stationary

WSD Water Supply and Destination

WSD Working Stress Design

WSD World Systems Division

WS$ Western Samoan Dollar [Currency]

WSDA Water and Sewer Distributors of America [US]

WSE Washington Society of Engineers [US]

WSE Water Saline Extract

WSE Weapon System Efficiency

WSE Western Society of Engineers [US]

WSE Winnipeg Stock Exchange [Manitoba, Canada]

WSE Women in Science and Engineering

WSE World Society for Ekistics [Greece]

Wsec Wattsecond [Unit]

WSED Weapon Systems Evaluation Division [of Institute for Defense Analysis, US]

WSEG Weapons System Evaluation Group

WSEIAC Weapon Systems Effectiveness Industry Advisory Committee [of US Department of Defense]

WSEP (Nuclear) Waste Solidification Engineering Prototype Plant [of US Atomic Energy Commission]

WSF Wire Seal Flange

WSF Workstation Facility

WSFA Water-Soluble Fatty Acid

WSFI Wood and Synthetic Flooring Institute [US]

WSfI Whole Surface Imager

WSFO Weather Service Forecast Office [US]

WSG WAN (Wide Area Network) Security Gateway

WSG Wired Shelf Group

WSG (International) Wool Study Group

WSGCP Washington Sea Grant College Program [US]

WSGS Woodcock Singing Ground Survey [US]

WSGS Wyoming State Geological Survey [US]

WSGT White Sands Ground Terminal [of NASA in New Mexico, US]

WSH (Microsoft) Windows Scripting Host

WSHA Washington State Horticultural Association [US]

WSHFT Wind Shift [Meteorology]

Wshg Washing [also wshg]

WSHS Wyoming State Historical Society [US]

WSI Wafer-Scale Integration [Electronics]

WSI Walter-Schottky-Institut [Walter Schottky Institute, Technical University of Munich, Germany]

WSI Water Solidity Index

WSI Wave-Sediment Interaction

WSI Weather Service, Inc. [US]

WSI Whole Sky Imager

W-Si Silicon-Doped Tungsten

WSi$_2$/Si Tungsten Disilicide/Silicon (Multilayer)

WSIA Water Supply Improvement Association

WSIB Workplace Safety and Insurance Board [Canada]

W Sib West Siberia(n)

WSiC Tungsten Silicocarbide

WSI-MSTE Washington Systemic Initiative in Mathematics, Science and Technology Education

W-SiO$_2$ Silica Doped Tungsten

WSIT Washington State Institute of Technology [US]

WSK Water Servicing Kit

WSL Wannier-Stark Ladder [Solid-State Physics]

WSL Warren Spring Laboratory [UK]

WSL Washington State Library [US]

WSL Weak Segregation Limit

WSM Western Society of Malacologists [US]

WSM Wyoming State Museum [Cheyenne, US]

WSMA Washington State Medical Association [US]

WSMA Wyoming State Medical Association [US]

WSMC Western Space and Missile Center [US]

WSMO Weather Service Meteorological Observatory [US]

WSMO Weather Service Meteorological Offices [US]

WSMR White Sands Missile Range [of US Army at Las Cruces, New Mexico, US] [formerly WSPG]

WSN Western Society of Naturalists [US]

WSO Weapons Systems Officer

WSO Weather Service Office [Australia]

WSO World Safety Organization

WSOW Weldon Springs Ordnance Works [Missouri, US]

WSP Water-Soluble Polymer

WSP Water Supply Point

WSP Weizmann Science Press [Jerusalem, Israel]

WSP White Star-Parachute

WSPA Washington State Pharmaceutical Association [US]

WSPA Western States Petroleum Association [US]

WSPA World Society for the Protection of Animals

WSPACS Weapon System Programming and Control System

WSPC World Scientific Publishing Company, Inc. [US]

WSPE Wisconsin Society of Professional Engineers [US]

WSPG White Sands Proving Ground [now White Sands Missile Range, US]

WSQW Wide Single Quantum Well [Electronics]

WSR Weather Search Radar

WSR Weather Surveillance Radar

W/sr Watt(s) per steradian [also W sr^{-1}]

WSRA Wild and Scenic Rivers Act [US]

WSRC Westinghouse Savannah River Company [US]

WSRC Wisconsin Synchrotron Radiation Center [US]

W/sr·m² Watt(s) per steradian square meter [also W sr^{-1} m^{-2}]

WSRO World Sugar Research Organization [UK]

WSRT Westerbork Synthesis Radio Telescope

WSRTC Westinghouse Savannah River Technology Center [Aiken, South Carolina, US]

WSS War Savings Stamp

WSS Waste Sampling System

WSS Winterbon-Sigmund-Sanders (Irradiation Damage Model)

WSS World Salt Symposium

WSSA Weed Science Society of America [US]

WSSC Wetlands of Special State Concern [US]

WSSH White Sands Space Harbor [at Las Cruces, New Mexico, US]

WSSS Western Society of Soil Science [US]

WST Western Samoan Tala [Currency of Western Samoa]

WST Wholesale Sales Tax [Australia]

WST Word Synchronizing Track

WST Water Storage Tank

WSTC Westinghouse Science and Technology Center [Pittsburgh, Pennsylvania, US]

WSTCC Washington State Trade and Convention Center [Seattle, US]

WSTDA Web Sling and Tiedown Association [US]

Wstd Ct Worsted Count [Woven Fabrics]

WSTF White Sands Test Facility [of NASA in New Mexico, US]

WSTI Wood Science and Technology Institute [Corvallis, Oregon, US]

WSTS Westinghouse Science Talent Search [of Science Service, US]

WSU Washington State University [Pullman, US]

WSU Wayne State University [Detroit, Michigan, US]

WSU Wichita State University [Wichita, Kansas, US]

WSU Wisconsin State University [US]

WSU Wright State University [Dayton, Ohio, US]

WSUEC Wisconsin State University at Eau Claire [US]

WSULC Wisconsin State University at La Crosse [US]

WSUM Wisconsin State University at Menomonie [US]

WSUO Wisconsin State University at Oshkosh [US]

WSUOPR Washington State University Open-Pool Reactor [Pullman, US]

WSUP Wayne State University Press [Detroit, Michigan, US]

WSUP Wisconsin State University at Platteville [US]

WSU Press Wayne State University Press [Detroit, Michigan, US]

WSURF Wisconsin State University at River Falls [US]

WSUS Wisconsin State University at Superior [US]

WSUSP Wisconsin State University at Stevens Point [US]

WSU TC Washington State University–Tri-Cities [Richland, US]

WSUW Wisconsin State University at Whitewater [US]

WSW West-Southwest; West by Southwest

WSW Wind Shear Warning [Meteorology]

WSWA Weed Society of Western Australia

WSWPC Washington State Water Power Company [US]

WSW/RGS Wind Shear Warning/Recovery Guidance System

WSW/RS Wind Shear Warning/Recovery System

WSX Word-Select Line

WSX Word-Select Signal

WSZ Wood Supply Zone

WSZ Word Size (Function)

WSZ Wrong Signature Zero

WT Structural Tee (Cut from W-Type Steel Shape) [AISI/AISC Designation]

WT Wait Time

WT Walkthrough

WT Wastewater Treatment

WT Water Tank

WT Water Technologist; Water Technology

WT Watertight

WT Water Tower

WT Water Treatment

WT Water Tube

WT Water Tunnel

WT Water Turbine

WT Wave Theory

WT Wave Train [Physics]

WT Wave Tube

WT Weapon Training

WT Wear Test(ing)

WT Welding Technologist; Welding Technology

WT Welding Transformer

WT Wetting Transition [Metallurgy]

WT Wide Track (Tractor)

WT Wild Type [Genetics]

WT Will Talk

WT Wilms' Tumor [Medicine]

WT Wilson's Theorem [Mathematics]

WT Wind Technologist; Wind Technology

WT Wind Tee

WT Wind Tunnel [also W/T]

WT Wireless Telegraphy [also W/T]

WT Wire Technology

WT Wood Technologist; Wood Technology

WT Word Terminal

WT Work Task

WT Work Truck

WT World Trade

WT Write Through

W/T Wind Tunnel [also WT]

W/T Wireless Telegraphy [also WT]

Wt Weight [also wt]

W(t) Weight Change per Unit Area [Symbol]

wt% weight percent [Unit]

w(t) weight gain or loss as a function of time [Symbol]

WTA Water Transport Association [US]

WTA West Test Area

WTA Willingness to Avoid

WTA World Textile Abstracts (Database) [of the Shirley Institute, UK]

W3A World Wide Web Applets

WTAA World Trade Alliance Association

WTB Water Tube Boiler

WTBA Water Tube and Boilermakers Association [UK]

WTC Washington Technology Center [US]

WTC Wastewater Technology Center [Canada]

WTC Water Transport Committee

WTC World Trade Center

WTC World Trade Conference [also United Nations Conference on Trade and Development]

W3C World Wide Web Consortium

WTCA Wood Truss Council of America [US]

WTCA World Trade Centers Association [US]

WTCC Wet Tropics Consultative Committee

WTCI Western Telecommunications Institute [US]

WTCP Wynzial Technologii Chemicznej Politechniki [Chemical Technology Polytechnic, Poznan, Poland]

wtd weighted

Wtd Av Weighted Average [also Wtd Avg]

WTE Waste-to-Energy

WTE World Telecommunications Exhibit(ion) [Geneva, Switzerland]

WTF Waste Treatment Facility

WTF World Time-Capsule Fund [US]

wt frac Weight Fraction

Wt/gal weight per gallon [Unit]

WTGS West Texas Geological Society [US]

wth width

WTHA West Texas Historical Association [US]

W-ThO$_2$ Tungsten-Thoria (Alloy System) (or Thoriated Tungsten)

WTHS West Tennessee Historical Society [US]

WTI World Trade Index

WTI World Translations Index

WTIA Welding Technology Institute of Australia

WTIC World Trade Information Center [US]

WTM Wind Tunnel Model

WTM Write Tape Mark

WTM Write Tape Mask

WTMA Wood Tank Manufacturers Association [US]

WTMS World Trade in Minerals (Database) System

WTMU Wildlife Trade Monitoring Unit [of World Conservation Monitoring Center]

WTNP Works Technical News Policy

WTO Warsaw Treaty Organization

WTO World Tourism Organization

WTO World Trade Organization [of United Nations]

WTO Write to Operator

WTOR Write to Operator with Reply

WTP Width Table Pointer

WTP Willingness to Pay

Wt pct weight percent [Unit]

wt ppb weight parts per billion [Unit]

wt ppm weight parts per million [Unit]

wt ppt weight parts per trillion [Unit]

WTR Western Test Range [of NASA]

WTR Westinghouse Test Reactor [US]

WTR Working Group on the Toxicology of Rubber Auxiliaries

Wtr Water [also wtr]

WTRA Wool Textile Research Council [UK]

WTRD Water Treatment Research Division [South Africa]

Wtr Sys Water System

WTRXLPE Water Tree Retardant Crosslinked Polyethylene

WTS Waste Tank Safety [Nuclear Engineering]

WTS Wastewater Treatment System

WTS World Terminal Synchronous

WTSA Wood Turners and Shapers Association [US]

WTSAP Wet Tropics Structural Adjustment Package

WTSC West Texas State College [Canyon, US]

WTSC Wet Tantalum Slug Capacitor

Wtspt Waterspout [also wtspt]

WTSU West Texas State University [Canyon, US]

WTT Wind Tunnel Test

WTT Weapons Tactical Trainer

WTTF Wastewater Treatment Test Facility

WTU Warsaw Technical University [Poland]

WTU Workers Trade Union [Spain]

WTU Wroclaw Technical University [Poland]

WTV Wound Tumor Virus

WTW Wissenschaftlich-Technische Werkstätten GmbH [German Manufacturer of Scientific and Technical Equipment located at Weilheim]

WTWA World Trade Writers Association [US]

WU Waikato University [Hamilton, New Zealand]

WU Warwick University [UK]

WU Waseda University [Tokyo, Japan]

WU Washington University [St. Louis, Missouri, US]

WU Western Union (Company) [US]

WU Willamette University [Salem, Oregon, US]

WU Willstätter Unit(s) [Pharmacology]

WU Write-Up

WUA Workers Union Association [Sudan]

WUC Work Unit Code

WUCF Work Unit Code File

WUDO Western Union Defense Organization

WUE Water Use Efficiency

WUG Word-Processing Users Group

WUI Western Union International [US]

WUIS Work Unit Information System

WUJS World Union of Jewish Students [Israel]

WULTUO World Union of Liberal Trade Union Organizations [Switzerland]

WUPPE Wisconsin Ultraviolet Photo-Polarimeter Experiment

WUR Where-Used Report

WUR World University Roundtable

WURDD Western Utilization Research and Development Division [US]

WUS Word Underscore (Character)

WUS World University Service [Switzerland]

WUSC World University Service of Canada

WUSCI Western Union Space Communications, Inc. [US]

WUS-US World University Service–United States

WUTC Western Union Telegraph Company [US]

WUX Western Union Telegram

WV Water Vapor

WV Wave Vector [Physics]

WV West Virginia [US]

WV Wind Vane

WV Wind Velocity [also W/V, wv, or w/v]

WV Wirtschaftsvereinigung Eisen− und Stahlindustrie [Iron and Steel Industries Association, Germany]

WV Women's Volunteers

WV Working Voltage

W/V Weight by (or per) Volume [also w/v]

W/V Wind Velocity [also WV, W/V, or w/v]

W-V Tungsten-Vanadium (Alloy System)

w/v weight by (or per) volume [also W/V]

WVA Wetlands Value Assessment

WVA World Veterinary Association

W Va West Virginia [US]

WVADSM Wetlands Value Assessment Decision Support Model [US]

W Va Tpk West Virginia Turnpike [US]

WVAS Wake Vortex Avoidance System

WVC World Veterinary Congress

WVDC Working Voltage, Direct Current

WVDP West Valley Demonstration Project [US]

WVE Water Vapor Electrolysis

WVE Water Vapor Embrittlement [Metallurgy]

WV EPSCoR West Virginia Experimental Program to Stimulate Competitive Research [National Science Foundation Program at West Virginia University, Morgantown, US]

WVGS West Virginia Geological and Economic Survey [US]

WVHS West Virginia Historical Society [US]

WVHS West Virginia Horticultural Society [US]

WVIPC West Virginia Industrial and Publicity Commission [US]

WVIT West Virginia Institute of Technology [Montgomery, US]

WVL Water-Vapor Laser

WVL West Vancouver Laboratory [British Columbia, Canada]

WVLC West Virginia Library Commission [US]

WVNS West Valley Nuclear Services Company [US]

WVOW West Virginia Ordnance Works [Point Pleasant, US]

WVP Waste Vitrification Plant

WVRG West Valley Reference Glass [US]

WVS West Virginia State [US]

WVS Women's Voluntary Service [UK]

WVSC West Virginia Safety Council [US]

WVSC West Virginia State College [US]

WVSG Valley Glass Sludge Glass [US]

WVSMA West Virginia State Medical Association [US]

WVSP West Valley Support Program [US]

WVSPA West Virginia State Pharmaceutical Association [US]

WVT Water Vapor Transmission

WVTR Water Vapor Transmission Rate

WVU West Virginia University [Morgantown, US]

WVUB West Virginia University Board [Morgantown, US]

WVUL West Virginia University Library [Morgantown, US]

WVUMC West Virginia University Medical Center [Morgantown, US]

WVUSP West Virginia University School of Pharmacy [Morgantown, US]

WVV Wissenschaftlicher Verein für Verkehrswesen [Scientific Association for Traffic and Transportation, Germany]

WW Wall-to-Wall

WW Warehouse Warrant [also W/W]

WW Waste Water

WW Water-Washable

WW Water Waste

WW Water Wheel

WW Water Wave

WW Water White

WW Whirlwind

WW Wide Well

WW Wide Woods

WW Wigner-Weisskopf (Approximation) [Physics]

WW Wilderness Watch [US]

WW Wire Way

WW Wire-Wound (Resistor)

WW Wire Wrap

WW Worldwide

WWI World War I [also WW I]

WWII World War II [also WW II]

ww with warrants [Securities Exchange]

W/W Warehouse Warrant [also WW]

W/W Weight to (or in) Weight [also w/w]

W/W Wheel Well

W/W With Warrants [Securities Exchange]

W/W With Weapons

W/w weight to (or in) weight [also w/w]

WWB Westerly Wind Bursts [Meteorology]

WWBA Western Wooden Box Association [US]

WWC Wastewater Coalition [US]

WWCC Western Weed Control Conference [US]

WWCC World Wide Cost Comparison

WWCCS World-Wide Command and Control System

WWCE Western Washington College of Education [Bellingham, US]

WWD Weather Working Days [also wwd]

WWD Weltwirtschaftsdatenbank [Database on Global Economy and Business, Germany]

WWD West Wind Drift [Oceanography]

WWEMA Water and Wastewater Equipment Manufacturers Association [US]

WWER Water-Moderated Water-Cooled (Nuclear) Reactor

WWF Whole Wheat Flour

WWF World Wide Fund for Nature

WWF World Wildlife Fund [now Worldwide Fund for Nature]

WWFA WWF–Australia

WWFC Water-Washed Filler Clay

WWF-Canada WWF–Canada

WWF-France WWF–France

WWF-US WWF–United States

WWG Warrington (Iron) Wire Gauge

WWG Worldwide Guide

WWHS Western Washington Historical Society [US]

WWI World Watch Institute

WWIS World Wide Information System [Internet]

WWM Wastewater Management

WWMCCS Worldwide Military Command Control System [also WW MCCS]

WWMS Waste Water Management System

WWMP Western Wood Molding Producers

W/WMS Water/Waste Management Subsystem; Water/Waste Management System

WWP Working Water Pressure

WWP World Weather Program

WWPA Western Word Processing Association

WWPA Western Wood Products Association [US]

WWR Wire-Wound Resistor

WWREA Wire and Wire Rope Employers Association [UK]

WWSC Western Washington State College [Bellingham, US]

WWSMA Wood Wool Slab Manufacturers Association [UK]

WWSN Worldwide Seismology Network [of National Institute of Standards and Technology, US]

WWSSN World Wide Standard Seismograph Network

WWTP Waste Water Treatment Plant

WWU Water/Wastewater Utilities

WWU Western Washington University [Bellingham, US]

WWU Westfälische Wilhelms-Universität [University of Muenster, Germany]

WWW Wooten, Winer, Weaire (Theory)

WWW World Weather Watch [of World Meteorological Organization]

WWW World Wide Web [also W3, or 3W]

WWWW What-Works-With-What [also W4]

WWWW World Wide Web Worm [also W4]

WX Weather (Message)

W(x) Strain Energy Function [Symbol]

WXTRN Weak External Reference

WY Wet Year

WY Wyoming [US]

WY-14643 4-Chloro-6-(2,3-Xylidino)-2-Pyrimidinylthioacetic Acid

W/Y Wrist Yaw [Mechanical Engineering]

W-Y$_2$O$_3$ Yttria-Doped Tungsten

Wyo Wyoming [US]

WyoUniv University of Wyoming [Laramie, US]

WYOR Wettest Year of Record [also WYR]

WYSBYGI What You See Before You Get It [Computers]

WYSIAWYG What You See is Almost What You Get [Computers]

WYSIAYG What You See is All You Get [Computers]

WYSIWYG What you See is What You Get [Computers]

WYWO While You Were Out [Internet Jargon]

WZ *(Weltzeit)* – German for "World Time"

WZ Wurtzite (Structure) [also Wz]

W(Z) Water Vapor Profile [Symbol]

WZB Wissenschaftszentrum Berlin (für Sozialforschung GmbH) [Berlin Science Center (for Social Research), Germany]

W-ZrO$_2$ Zirconia-Doped Tungsten

WZW Weiss-Zumino-Witten (Theory) [Physics]

X Amplitude of Displacement [Symbol]

X Arithmetic Average [Symbol]

X Atmospherics (in Radio Engineering) [Symbol]

X Atomic Fraction (e.g. X_{Ni} is the Atomic Fraction of Nickel in an Alloy) [Symbol]

X Bond-Site Repulsion [Symbol]

X (Upper-case) Chi [Greek Alphabet]

X Christ(ian)

X Computer

X Concentration [Symbol]

X Contact (between Electrical Lines) [Symbol]

X Controlled Condition [Control Systems]

X Cross

X Crystal [also x]

X Dimension (e.g. Length, Thickness, etc.) [Symbol]

X Electronegativity [Symbol]

X Experimental Stage (Aircraft, Missile, or Rocket) [USDOD Symbol]

X Explicit Sexual Content [Motion-Picture Rating]

X Exposure (in Radiometry) [Symbol]

X Exposure Rate [Symbol]

X Extension

X Experiment(al) [also x]

X Explosive

X Extension [also x]

X Extra [also x]

X Extraordinary

X Facsimile Equipment [Symbol]

X Force Component [Symbol]

X Halogen [Symbol]

X Index

X Internal Stress Tensor [Symbol]

X Kienboeck [Radiology]

X Loading Hole Offset (in Tension and Bending Testing of Arc or Similar Shaped Specimens) [Symbol]

X Mean (or Average Value) [Symbol]

X Mining Engineering (Technology) [Discipline Category Abbreviation]

X No Priority

X No Protest [Negotiable Instruments]

X Parameter (or Value) "X" [Symbol]

X Phase Distortion [Symbol]

X Quantity of Component Adsorbed [Symbol]

X Radial Load Factor (of Roller Bearings) [Symbol]

X Phase Distribution [Symbol]

X Radiotelegraphy [405 to 535 kc/s Range]

X Reactance (or Reactive Impedance) [Symbol]

X Reactor

X Repellent

X Research Aircraft [USDOD Symbol]

X Saint Andrew's Cross [Symbol Used in Packaging for Transport]

X Sample Ions (in Ion Chromatography) [Symbol]

X Solubility [Symbol]

X Ten [Roman Numeral]

X Transistor

X Transmission; Transmitter

X Volume Fraction [Symbol]

X Xanthosine

X X (Sex) Chromosome [Genetics]

X Xenon [Abbreviation; Symbol: Xe]

X Xerographic; Xerography

X Xylan

X X-Ray [Phonetic Alphabet]

X X-Ray Tube [Symbol]

X X-Ray Unit [Symbol]

X X-Unit

X (+)-Xylose [Biochemistry]

/X Execute Option [Turbo Pascal]

$\overline{X}$ Average Value [Symbol]

$\overline{X}$ (Sample) Mean [Symbol]

$\overline{X}$ Mean Displacement [Symbol]

$\overline{X}$ Median [Symbol]

$\overline{X}$ Process Average (in Statistics) [Symbol]

$\overline{X}$ Ten Thousand [Roman Numeral]

X^- Halogen Ion (i.e., X = Bromine, Chlorine, Fluorine or Iodine) [Symbol]

X_0 Original Value of Quantity X [Symbol]

X_0 Reference Point of Controlled Condition (of a Control System) [Symbol]

X_0 Zero Frequency Deflection (of Spring-Mass System) [Symbol]

X_1 Reactance of Electric Circuit 1 [Symbol]

X_2 Molecular Halogen [Symbol]

$\overline{X}^2$ Mean Square Displacement [Symbol]

X_A Mole Fraction of "A" Atoms [Symbol]

X_b Grain Boundary Concentration (in Materials Science) [Symbol]

X_C Capacitive Reactance [also X_c] [Symbol]

X_c Bulk Solute Concentration [Symbol]

X_c Limiting Compressive Stress in Fiber Direction (of Composite) [Symbol]

X_D Deformation Depth [Symbol]

X_d d-Band Electronegativity [Symbol]

X_e Spectral Concentration (of Radiometric Quantity) [Symbol]

X_{eq} Reactive Component of a Series Equivalent Circuit [Symbol]

X_h Running Range (of Control Systems) [Symbol]

X_i Mole Fraction of Component "i" in a Mixture [Symbol]

X_K Set Point (of Control Systems) [Symbol]

X_L Inductive Reactance [Symbol]

X_m Mutual Reactance (of Two Electric Circuits) [Symbol]

X_{MB} Electronegativity in the Martynov-Batsanov Scale (in Physics) [Symbol]

X_{max} Largest Observed Value in a Sample (in Statistics) [Symbol]

X_{min} Smallest Observed Value in a Sample (in Statistics) [Symbol]

X_n Nominal Value of Parameter "X" [Symbol]

X Polyhalide [Formula]

X_p Equivalent Parallel Reactance [Symbol]

X_p Proportional Band (of Control Systems) [Symbol]

X_s Scratch Depth [Symbol]

X_s Interfacial Concentration [Symbol]

X_s Surface Reactance [Symbol]

X_t Limiting Tensile Stress in Fiber Direction (of Composite) [Symbol]

X_v Volume Fraction [Symbol]

X.21 Level One Standard for X.25 [Data Communications]

X.25 CCITT-ISO International Standard Packet-Switching Protocol [Data Communications]

X.28 Asynchronous Terminal Interface Standard [Data Communications]

X.400 CCITT-ISO Electronic Mail Standard [Data Communications]

X.500 CCITT-ISO Standard for Electronic Directory Services

x amount of impurity additions (in metallic conductors) [Symbol]

x (overall) composition [Symbol]

x concentration [Symbol]

x deviation of controlled condition (in control systems) [Symbol]

x displacement (in x-direction) [Symbol]

x distance [Symbol]

x elevation [Symbol]

x experimental [also X]

x extension

x extra [also X]

x first rectangular coordinate (of a point) [Symbol]

x haploid number of chromosomes [Genetics]

x length [Symbol]

x mean result (in statistics) [Symbol]

x measured value (in statistics) [Symbol]

x mole fraction [Symbol]

x observed value (in statistics) [Symbol]

x one of the axes in a rectangular coordinate system [Symbol]

x power of magnification

x random variate [Symbol]

x real part of a complex variable (e.g., $x + yi$) [Symbol]

x repellent

x space coordinate [Symbol]

x unknown (or variable) [Symbol]

x xi [English Equivalent]

$\bar{x}$ estimate for population mean (in statistics) [Symbol]

$\bar{x}$ sample mean (in statistics) [Symbol]

x' space coordinate in the second system [Symbol]

$\dot{x}$ first derivative of x with respect to time [Symbol]

$\ddot{x}$ second derivative of x with respect to time [Symbol]

x_0 amplitude (of a wave) [Symbol]

x_1 mole fraction of solvent [Symbol]

x_2 mole fraction of solute [Symbol]

x_a atomic fraction of element α [Symbol]

x_A mole fraction of component A [Symbol]

x_α composition of alpha phase [Symbol]

x_β composition of beta phase [Symbol]

x_c percolation limit (in solid-state physics) [Symbol]

x camera length (in photography) [Symbol]

x_γ atomic fraction of element γ [Symbol]

x_γ composition of gamma phase [Symbol]

$\bar{x}_I$ I-th class mark (in statistics) [Symbol]

x_i impurity concentration [Symbol]

x_i individual result (in statistics) [Symbol]

x_i mole fraction of component i [Symbol]

x_m matrix composition [Symbol]

x_m maximum overshoot (of control systems) [Symbol]

x_p percolation threshold (in solid-state physics) [Symbol]

x_v mole fraction of water vapor in moist air [Symbol]

x_v volume fraction [Symbol]

x_{vi} mole fraction of ice-saturated with water vapor [Symbol]

x_{vw} mole fraction of water-saturated with water vapor [Symbol]

x_w deviation (of a control system) [Symbol]

Ξ (Upper-case) Xi [Greek Alphabet]

Ξ Xi Particle (or Xi Hyperon) (in Particle Physics) [Symbol]

Ξ^- Xi-Minus Particle (i.e., Negatively Charged Xi Hyperon) [Symbol]

$\overline{\Xi}^-$ Xi-Minus Antiparticle (i.e., Negatively Charged Xi Antihyperon) [Symbol]

Ξ^0 Xi-Zero Particle (i.e., Neutral Xi Hyperon) [Symbol]

$\overline{\Xi}^0$ Xi-Zero Antiparticle (i.e., Neutral Xi Antihyperon) [Symbol]

ξ chain charge density (of polymers) [Symbol]

ξ characteristic length (in mechanics) [Symbol]

ξ coherence length (in physics) [Symbol]

ξ coordinate [Symbol]

ξ correlation length (of superconductors) [Symbol]

ξ damping ratio (in physics) [Symbol]

ξ displacement (of sound) [Symbol]

ξ efficiency (of defects) [Symbol]

ξ electric susceptibility [Symbol]

ξ extinction length (in electron microscopy) [Symbol]

ξ kinetic energy correction factor (of thrust bearings) [Symbol]

ξ mass accommodation coefficient [Symbol]

ξ metal ratio (on phase diagrams) [Symbol]

ξ nucleation coefficient [Symbol]

ξ resolution broadening (in spectroscopy) [Symbol]

ξ root-mean-square roughness [Symbol]

ξ sound particle velocity [Symbol]

ξ stability function [Symbol]

ξ (dislocation) unit vector (in materials science) [Symbol]

ξ (lower-case) xi [Greek Alphabet]

$\xi_\parallel$ longitudinal (or parallel) correlation length (of superconductors) [Symbol]

$\xi_\perp$ transverse (or perpendicular) correlation length (of superconductors) [Symbol]

ξ_0 coherence length (in physics) [Symbol]

ξ_0 electron-hole completion factor [Symbol]

ξ_c critical displacement [Symbol]

ξ_d disorder correlation length (in physics) [Symbol]

ξ_g extinction length [Symbol]

ξ_{hkl} extinction length of hkl plane (in microstructural analysis) [Symbol]

ξ_i equilibrium normal mode amplitude [Symbol]

ξ_k kinematical extinction length [Symbol]

ξ_{max} maximum sound particle velocity [Symbol]

XA Allocate Expanded Memory

XA Cross-Arm

XA Extended Architecture

XA Extended Attribute

XA Transmission Adapter

XA Ximenic Acid

XA Xyloascorbic Acid

XA X-Ray Absorption

XA- Mexico [Civil Aircraft Marking]

X-A X-Axis

Xa Activated Factor X [Biochemistry]

xa ex all [Securities Exchange]

XAA X-Ray Absorption Analysis,

XAAP Exploratory Airborne Acoustic Processor

XACS X-Ray Absorption Cross-Section

XACT X (Computer) Automatic Code Translation

XAD Transmit Address

XAD X-Ray Anomalous Dispersion

XADS X-Ray Anomalous Dispersion Microscopy

XAE X-Ray Absorption Edge

XAES X-Ray Absorption Edge Spectrometry

XAES X-Ray Excited (or Induced) Auger Electron Spectroscopy

XAFO Xenon Arc Fade-O-meter

XAFS X-Ray Absorption Fine Structure

XAFS $\chi(k)$ X-Ray Absorption Fine Structure Photoelectron Wavevector Dependent Oscillation

XAI X-Ray Absorption Index

XAJ X-Ray Absorption Jump

XAL Xenon Arc Lamp

XAM X-Ray Analytical Method

XAN Xanthate [also Xan]

XANES X-Ray Absorption Near-Edge Spectroscopy

XANES X-Ray Absorption Near-Edge Structure

Xao Xanthosine [Biochemistry]

XAPIA X.400 Application Program Interface Association

XAPS X-Ray Appearance Potential Spectroscopy

XARM Cross-Arm [also X-ARM, or x-arm]

XAS X-Ray Absorption Spectroscopy

XAS X-Ray Absorption Spectrum

XAS X-Ray Excited Auger Spectroscopy

XAWD Xenon Arc Weathering Device

XB Crossbar

XB Experimental Bomber

XB- Mexico [Civil Aircraft Marking]

X2B Hexadecimal to Binary [Restructured Extended Executor (Language)]

Xba Xanthomonas badrii [Microbiology]

X band Microwave Frequency Band of 5.20 to 10.90 Gigahertz

XBAR Crossbar [also Xbar]

XBD Xanthene Basic Dye

XBC External Block Controller

XBF Experimental Boundary File

XBIS X-Ray Bremsstrahlung Isochromat Spectroscopy

XBL Extension Bell

XBM Extended BASIC Mode [Computers]

.XBM X-Windows Bitmaps [File Name Extension]

XBO Xenon Burner [Arc Lamp]

XBT Crossbar Tandem

XBT Expendable Bathythermograph

XC Cross Channel

XC Cross Country

XC Exchange Correlation

XC Extremely Clean (Sputtering Process)

XC- Mexico [Civil Aircraft Marking]

X2C Hexadecimal to Character [Restructured Extended Executor (Language)]

X-C Ex coupon [Securities]

XCD East Caribbean Dollar [Currency of Antigua-Barbuda, Grenada, St. Christopher Nevis, St. Kitts, St. Lucia, St. Vincent and the Grenadines]

XCD Exceed [also Xcd]

XCH Exchange [also XCHG, Xchg, or Xch]

XCL Exclusive [also Xcl, or xcl]

XCLB X-Ray Compositional Line Broadening [Spectroscopy]

XCMD External Command

XCO Crystal-Controlled Oscillator

XCON Expert Configuration; Expert Configurer

XCONN Cross Connection

XCOPY Extended Copy (Command) [also xcopy]

XCor Current X-Coordinate of Turtle [Turbo Pascal Turtlegraphics]

X-CORE Cross Core [also X-Core or X-core]

XCORFE X-Nucleus Correlation with Fixed Evolution Time [Two-Dimensional Nuclear Magnetic Resonance]

XCP Xenon Critical Point

x cp ex coupon [Securities Exchange]

X-CR Chloroprene Rubber with Reactive Groups

XCS X (Ten) Call Seconds

XCS Xerox Computer Services

XCSRA Cross Channel Special Rules Area [UK]

XCTD Expendable Current Temperature Density (Profiler)

XCU Crosspoint Control Unit

X-CUT Crystal Cut, Orientation I [also X CUT, X- cut, or X cut]

XCV Experimental Composite Vehicle

XCVR Transceiver [also Xcvr]

XD Crossed [also Xd]

XD Crystal Detector

XD Deallocate Expanded Memory

XD Ex-Directory

XD Exothermic Dispersion

XD Exploratory Development

XD Xanthene Dye

XD Xenon Dioxide

XD X-Ray Density

x-d ex dividend [also xd, XD, or X-D]

X2D Hexadecimal to Decimal [Restructured Extended Executor (Language)]

XDAAP Exploratory Development of an Airborne Acoustic Processor

XDC X-Ray Double Crystal Diffraction

XDCR Transducer [also Xdcr]

XDF Exchange Data Format

XDF Extended Density Format

XDH Xanthine Dehydrogenase [Biochemistry]

XDI Xylylene Diisocyanate

x div *(ex dividendum)* – without dividend

XDM Experimental Development Model

XDM Exploratory Development Model

XDM Extended Dipolar Modulation

XDP Crosslinked Fibrin Degradation [Biochemistry]

XDP Xanthosine Diphosphate; Xanthosine-5'-Diphosphate [Biochemistry]

XDP X-Ray Density Probe

XDP Xylenyl Diphenyl Phosphate

XDR Extended Data Representation

XDR External Data Representation (Protocol) [Internet]

XDS Xerox Data System

XDTA Xylylenediaminetetraacetate [also Xdta, or xdta]

XDTF Xenon Dioxide Tetrafluoride

XDUP Extended Disk Utilities Program

XE Experimental Engine

XE Xenon Effect [Atomic Physics]

Xe Xenon [Symbol]

Xe-121 Xenon-121 [also ^{121}Xe, or Xe^{121}]

Xe-122 Xenon-122 [also ^{122}Xe, or Xe^{122}]

Xe-123 Xenon-123 [also ^{123}Xe, or Xe^{123}]

Xe-124 Xenon-124 [also ^{124}Xe, or Xe^{124}]

Xe-125 Xenon-125 [also ^{125}Xe, or Xe^{125}]

Xe-126 Xenon-126 [also ^{126}Xe, or Xe^{126}]

Xe-127 Xenon-127 [also ^{127}Xe, or Xe^{127}]

Xe-128 Xenon-128 [also ^{128}Xe, or Xe^{128}]

Xe-129 Xenon-129 [also ^{129}Xe, or Xe^{129}]

Xe-130 Xenon-130 [also ^{130}Xe, or Xe^{130}]

Xe-131 Xenon-131 [also ^{131}Xe, or Xe^{131}]

Xe-132 Xenon-132 [also ^{132}Xe, or Xe^{132}]

Xe-133 Xenon-133 [also ^{133}Xe, or Xe^{133}]

Xe-134 Xenon-134 [also ^{134}Xe, or Xe^{134}]

Xe-135 Xenon-135 [also ^{135}Xe, or Xe^{135}]

Xe-136 Xenon-136 [also ^{136}Xe, or Xe^{136}]

Xe-137 Xenon-137 [also ^{137}Xe, or Xe^{137}]

Xe-138 Xenon-138 [also ^{138}Xe, or Xe^{138}]

Xe-139 Xenon-139 [also ^{139}Xe, or Xe^{139}]

Xe-140 Xenon-140 [also ^{140}Xe, or Xe^{140}]

Xe-141 Xenon-141 [also ^{141}Xe, or Xe^{141}]

Xe-142 Xenon-142 [also ^{142}Xe, or Xe^{142}]

Xe-143 Xenon-143 [also ^{143}Xe, or Xe^{143}]

Xe-144 Xenon-144 [also ^{144}Xe, or Xe^{144}]

Xe-145 Xenon-145 [also ^{145}Xe, or Xe^{145}]

XEAES X-Ray Excited Auger Electron Spectroscopy

XEAPS X-Ray Excited Electron Appearance Potential Spectroscopy

XEB X-Ray Emission Band

XEC Execute [also Xec]

XEC X-Ray Elastic Compliances

XEC X-Ray Elastic Constants

XECF Experimental Engine–Cold Flow

XeCl Xenon Chloride (Excimer Laser)

XED X-Ray Energy Dispersive Diffractometer

XEDS X-Ray Energy Dispersive Spectroscopy [also X-EDS]

XFCN External Function

XEG X-Ray Emission Gauge

XEM Exoelectron Microscopy

XEOF X-Ray Excited Optical Fluorescence Spectroscopy

XEQ Execute

Xerox PARC Xerox Palo Alto Research Center [California, US]

XES Exoelectron Spectroscopy

XES Xerox Engineering System

XES X-Ray Emission Spectroscopy

XES X-Ray Emission Spectrum

XES X-Ray Energy Spectrometry

XESD X-Ray Induced Electron-Stimulated (Ion) Desorption

XET Transparent End-of-Transmission

XET X-Ray Topography

XETB Transparent End-of-Transmission Block
XETX Transparent End-of-Text
XEU European Currency Unit [Currency of the European Community]
XF Extra Fine
XF X-Ray Fluorescence
XFA X-Ray Fluorescence Analysis
XFC Extended Function Code
XFC Transfer Charge
XFER Transfer [also Xfer]
XFH X-Ray Fluorescence Holography
XFL Xenon Flash Lamp
XFLU X-Ray Fluorescence
XFM "X" Force Microscope [*Note:* "X" Represents the Force Being Measured, e.g., Atomic Force, Magnetic Force, Friction, etc.]
X-FORMER Transformer [also X-former]
XFMR Transformer [also Xfmr]
XFR X-Ray (Quasi-)Forbidden Reflection
XFS X-Ray Fluorescence Spectroscopy
XG Xanthan Gum
XGA Extended Graphics Array
X-Gal 5-Bromo-4-Chloro-3-Indolyl β-D-Galactopyranoside [also X-gal] [Biochemistry]
X-GalNAc 5-Bromo-4-Chloro-3-Indolyl N-Acetyl-β-D-Galactosaminide [Biochemistry]
X-GlcNAc 5-Bromo-4-Chloro-3-Indolyl N-Acetyl-β-D-Glucosaminide [Biochemistry]
XGAM Experimental Guided Aircraft Missile
X-GIC Halogen-Graphite Intercalation Compound
X-Glc 5-Bromo-4-Chloro-3-Indolyl β-D-Glucopyranoside [also X-gal] [Biochemistry]
X-GlcA 5-Bromo-4-Chloro-3-Indolyl β-D-Glucuronide [Biochemistry]
XGP Xerox Graphic Printer
XGPRT Xanthine-Guanine Phosphoribosyltransferase [Biochemistry]
XH Experimental Helicopter
XHAIR Cross Hair [also Xhair]
XHB Extra-Hard Bolt (Temper) [Metallurgy]
XHC Xylometazoline Hydrochloride
XHFR Experimental High-Frequency Radar
XHMO Extended Hueckel Molecular Orbital [Physical Chemistry]
Xho Xanthomonas holcicola [Microbiology]
XhoI Xanthomonas holcicola I [Microbiology]
XHTB Extended Hueckel Tight-Binding [Physical Chemistry]
XHV Extreme(ly) High Vacuum
XI Xylitone Isomer
x-i ex interest; without interest [also X-I, XI, or xi]
XIA Xenon Ion Accelerator
XIAES X-Ray Induced Auger Emission Spectroscopy
XIC Transmission Interface Converter
XICS Xerox Integrated Composition System

XID Exchange Identification; Exchange Identifier
XIE X-Ray Imaging Experiment
XIF X-Ray Induced Fluorescence
XIIR Halogenated Isobutylene-Isoprene Rubber
XIL Xenon-Ion Laser
XIL Xenon Isotope Lamp
XIM X-Ray Inspection Module
XIMCE Xian Institute of Metallurgy and Construction Engineering [PR China]
XIN X-Ray Interferometer; X-Ray Interferometry
x in ex interest [also x int]
XIO Execute Input/Output
XIOP Block-Multiplexer Input/Output
XIOS Extended Input/Output System
XIS X-Ray Isochromat Spectroscopy
XIT Extra Input Terminal
XITB Transparent Intermediate Text Block
XJU Xi'an Jiaotong University [Sian, PR China]
XL Crystal [also Xl, XTAL, Xtal, X-TAL, Xtal, XTL, or Xtl]
XL Excellence and Leadership
XL Excess Loss
XL Execution Language
XL Extended Logic (Interface)
XL Extra Large
XL Extra Load (Rating) [Automobiles]
XL Extra Long
XL Xenon Lamp
$\bar{x}(\lambda)$ tristimulus value for spectral component "x" in CIE System [Symbol]
.XLA Add-In [Microsoft-Excel File Name Extension]
XLAT Translate [also Xlat]
XLBIB Extra-Large Burn-In Bath
XLC (Project) Excellence and Leadership for Communities [US Community-Based Environmental Project]
XLDP Xylose-Lysine-Deoxycholatepeplon [Biochemistry]
X-LENS Lens Multibeam Antenna
XLHDPE Cross-Linked High-Density Polyethylene
X-LINKED Cross-Linked [also X-linked, or X-Linked]
.XLK Backup [Microsoft-Excel File Name Extension]
XLL Extensible Link Language
XLL Extra Lightly Loaded (Line) [also Xll]
XLM Crossed Lamellar Structure [Metallurgy]
XLM Excel Macro Language
XLPE Cross-Linked Polyethylene [also X-LPE, XLP, or X-LP]
XLPI Cross-Linked Polyimide
XLPS Cross-Linked Polystyrene
XLR Experimental Liquid Rocket
.XLS Microsoft Excel Document File [also .xls] [File Name Extension]
.XLT Template [Microsoft-Excel File Name Extension]
XLY Xylene Light Yellow
XM Expanded Memory
XM Experimental Missile
XM Extra Marker

XM Map Expanded Memory Pages

XM Xylene Musk

XM Xylyl Mercaptan

XM- New Zealand [Civil Aircraft Marking]

XMA Extended Memory Allocation (Emulator) [also xma]

XMA X-Ray Microanalysis; X-Ray Microanalyzer

Xma Xanthomonas malvacaerum [Microbiology]

X-Man 5-Bromo-4-Chloro-3-Indolyl α-D-Mannopyranoside [Biochemistry]

XMAS Extended Mission Apollo Simulation [of NASA]

Xmas Christmas

XMA2EMS Expanded Memory Allocation to Expanded Memory Specification (Mapping) [also xma2ems]

XMBA Executive Master of Business Administration

XMC Directionally-Reinforced Molding Compound

XMC Extra-High-Strength Molding Compound

XMC Xylyl Methyl Carbamate

XMCD X-Ray Magnetic Circular Dichroism [Physics]

XMFR Transformer [also Xmfr]

XMIT Transmit [also X-MIT, Xmit, or X-mit]

XMITTER Transmitter [also X-MITTER, Xmitter, or X-mitter]

XMITTING Transmitter [also X-MITTING, Xmitting, or X-mitting]

XML Extensible Markup Language [Internet]

XMM Expanded Memory Manager

XMM Extended Memory Manager [Lotus/Intel/Microsoft]

XMOS Cross Metal-Oxide Semiconductor [also X MOS, or X-MOS]

XMOS High-Speed Metal-Oxide Semiconductor [also X MOS, or X-MOS]

XMP Xanthosine Monophosphate; Xanthosine-5'-Monophosphate [Biochemistry]

XMPA X-Ray Electron Microprobe Analysis

XMR X-Ray Micro-Radiography

XMS Expanded Memory Status

XMS Extended Memory Specification [Lotus/Intel/Microsoft]

XMS Xerox Memory System

XMSN Transmission [also Xmsn]

XMT Exempt [also Xmt]

XMT Exempted Addressee

XMT Transmit [also Xmt]

XMT X-Ray Microtomography

XMTD Transmitted [also Xmtd]

XMTG Transmitting [also Xmtg]

XMTL Transmittal [also Xmtl]

XMTR Transmitter [also Xmtr]

XN Execution Node

XN Intersection

XN Xanthinol Niacinate [Biochemistry]

Xn Christian

XNBR Nitrile-Butadiene Rubber with Reactive Groups

XND X-Ray Neutron Diffraction

X-NeuNAc 5-Bromo-4-Chloro-3-Indolyl α-D-N-Acetylneuraminic Acid [Biochemistry]

XNOR Exclusive Not Or

XNOS Experimental Network Operating System

XNS Xerox Network Services

XNS Xerox Network System (Protocol)

Xnty Christianity [also Xty]

XO Cross-Office

XO Crystal Oscillator

XO Xanthine Oxidase [Biochemistry]

XO Xylenol Orange

XOAE Execution Orbital Analysis Engineer

XOF CFA (Communauté Financière Africaine) Franc [Currency of Benin, Congo, Gabon, Ivory Coast, Niger, Senegal, Togo and Upper Volta]

XOFF Transmit(ter) Off [also X-OFF]

XON Transmit(ter) On [also X-ON]

XOP Extended Operation

XOR Exclusive Or [also EOR]

XOS Cross-Office Slot

XOS Xerox Operating System

XOW Express Order Wire

XP Cross-Polarization

XP Explosionproof

XP Xanthate Pyrolysis

XP Xanthation Process

XP Xeroderma Pigmentosum [Medicine]

XP Xerographic Paper

XP Xerographic Printer

XP Xerographic Process

XP X-Ray Photoelectron

XPAN Cross-Linked Polyaniline

XPD Cross-Polarization Discrimination

XPD X-Ray Photoelectron Diffraction

XPD X-Ray Powder Diffraction

XPDA X-Ray Powder Diffraction Analysis

XPD/AED X-Ray Photoelectron and Auger Electron Diffraction

XPDR Transponder

XPE Cross-Linked Polyethylene

XPED X-Ray Photoelectron Diffraction [also X- PED]

XPES X-Ray Photoelectron Spectroscopy [also X-PES]

XPES X-Ray Photoemission Spectroscopy [also X-PES]

XPF CFP (Communauté Financière Pacifique) Franc [Currency of French Polynesia and Tahiti]

XPFC Explosionproof Fan-Cooled (Motor)

XPG X/Open Portability Guide

XPI Cross-Linked Polyimide

XPI Cross-Polarization Interference

XPL Explosive [also Xpl]

XPM Expanded Metal

.XPM X-Windows Pixelmaps [File Name Extension]

XPN External Priority Number

X-POL Cross-Polarization [also XPOL, X-Pol, or Xpol]

XPP Xylenes Plus Process
XPRM Xerox Print Resources Manager
XPS Expandable Polystyrene
XPS Extruded Polystyrene
XPS X-Ray Photoelectron Spectroscopy
XPS X-Ray Photoelectron Spectrum
XPS X-Ray Photoemission Spectroscopy
XPSW External Processor Status Word
XPT Crosspoint [also Xpt]
XPT External Page Table
XPTP Xylenes-Plus Transalkylation Process
X/Q Relative Concentration [Symbol]
XR External Relations Service [of UNESCO]
XR External Reset
XR Index Register
XR Xeroradiograph(y)
XR X-Ray(s)
XR Xylene Red
x-r ex rights [also xr]
XRA X-Ray Analysis
XRA X-Ray Astronomy
XRAC X-Ray Analysis Computer
X-Ray Sci Technol X-Ray Science and Technology (Journal) [also X-Ray Sci Tech]
X-Ray Spectrum X-Ray Spectrometry [Published in the UK]
XRB X-Ray Beam
XRBR X-Ray Beam Resist
XRC Xerox Research Center
XRC X-Ray Calorimetry
XRC X-Ray Cathode
XRC X-Ray Counter
XRC X-Ray Crystal
XRC X-Ray Crystallography
XRC X-Ray Rocking Curve
XRCC Xerox Research Center of Canada
XRCF X-Ray Calibration Facility
XRCL X-Ray Crystallography Laboratory [of University of Kansas, Lawrence, US]
XRCS X-Ray Crystal Spectrometer
XRCT X-Ray Cascade Tube
XRD X-Ray Detector
XRD X-Ray Diffraction
XRD X-Ray Diffractometer; X-Ray Diffractometry
XRD X-Ray Dispersion
XRD X-Ray Dosage
XRD X-Ray Radiation Dose
XRDA X-Ray Diffraction Analysis
XRDM X-Ray Diffraction Microscopy
XRDP X-Ray Diffraction Pattern
XRDRSA X-Ray Diffraction Residual Stress Analysis
XRDT X-Ray Diffraction Topography
XRE X-Ray Effect
XRE X-Ray Emission
X-Ref Cross-Reference [also x-ref]

XRES X-Ray Emission Spectrum
XREP Auxiliary Report
XRF X-Ray Film
XRF X-Ray Filter
XRF X-Ray Fluorescence (Spectroscopy)
XRF X-Ray Fluorometry
XRFA X-Ray Fluorescence Analysis
XRF/ICP X-Ray Fluorescence/Inductively Coupled Plasma (Technique)
XRFS X-Ray Fluorescence Spectroscopy
XRI X-Ray Intensity
XRI X-Ray Interference
XRI X-Ray Ionization
XRIC X-Ray Intensities Computation
XRITC Bis(quinolizino)rhodamineisothiocyanate
XRITC Quinolizino-Substituted Fluorescein Isothiocyanate
XRL X-Ray Laser
XRL X-Ray Lithography
XRLB X-Ray Line Broadening
XRLBA X-Ray Line Broadening Analysis
XRLM Extended Range Lance Missile
XRLS X-Ray Line Spectrum
XRM External Relational Memory
XRM X-Ray Mapping
XRM X-Ray Microanalysis
XRM X-Ray Microscope; X-Ray Microscopy
XRM X-Ray Monochromator
XRMF X-Ray Micro-Fluorescence
XRMM X-Ray Multi-Mirror Mission [of European Space Agency]
XRND X-Ray Neutron Diffraction
XRP X-Ray Photon
XRPD X-Ray Powder Diffraction
XRPD X-Ray Powder Diffractometer
XRPDF X-Ray Powder Diffraction File
XRPM X-Ray Projection Microscope; X-Ray Projection Microscopy
XRR X-Ray Radiograph(y)
XRR X-Ray Reflection; X-Ray Reflectivity
XRR X-Ray Reflectometer; X-Ray Reflectometry
XRR X-Ray Refraction
XRR X-Ray Resin
XRS X-Ray Scattering
XRS X-Ray Scintillography
XRS X-Ray Spectrograph(y)
XRS X-Ray Spectrometer; X-Ray Spectrometry
XRS X-Ray Spectroscopy
XRS X-Ray Spectrum
XRSAS X-Ray Small-Angle Scattering
XRT Extensions for Real-Time
XRT X-Ray Technologist; X-Ray Technology
XRT X-Ray Telescope
XRT X-Ray Therapy
XRT X-Ray Transition

XRT X-Ray Tube

X-Rts Ex rights [Securities Exchange]

XRV X-Ray Vidicon

XRWL X-Ray Wavelength

XS Extra Strong

XS Get Expanded Memory Status

XS Xerces Society [US]

X-S Cross-Section

XS-3 Excess-Three (Code) [Computers]

XSA X-Ray Stress Analysis

XSAD X-Ray Small-Angle Diffraction

XSADP Cross-Sectional Selected Area Electron Diffraction (Pattern)

X-SAR Extended Synthetic Aperture Radar; X-Band Synthetic Aperture Radar

XSCE X-Ray Spectroscopy from Channeled Electrons

XSDC XTE (X-Ray Timing Explorer) Science Data Center [of NASA Astrophysics Data Facility, Goddard Space Flight Center, Greenbelt, Maryland, US]

XSE X-Ray Stress Evaluation

XSECT Cross Section [also X-SECT, X-sect, or Xsect]

XSFA, X-Ray Surface Forces Apparatus

XSL Extensible Style Language [Internet]

XSM Experimental Strategic Missile

XSMD Extended Storage Module Drive (Interface)

XSONAD Experimental Sonic Azimuth Detector

XSP Cross-Sectional Point

XSPT External Shared Page Table

XSR X-Ray Synchrotron Radiation Source

XSSI Extended Server Side Includes [Internet]

XSST X-Band Satellite Suitcase Terminal

XSTM Cross-Sectional Scanning Tunneling Microscopy

XSTR Transistor [also Xstr]

XSTS Cross-Sectional Scanning Tunneling Spectroscopy

XSTX Transparent Start-of-Text

XSW X-Ray Standing Wave

XSWIS X-Ray Standing Wave Interference Spectroscopy

XSWT X-Ray Standing Wave Technique

XSYN Transparent Synchronous

XT Crosstalk [also X-TALK, X-Talk, Xtalk, or CT]

XT Expanded Technology

XT Xenon Trioxide

XT X-Ray Tomography

XT- Upper Volta [Civil Aircraft Marking]

X(t) Input [Symbol]

x(t) time-dependent distance [Symbol]

ξ(T) temperature-dependent coherence length (in physics) [Symbol]

XTA Modified Terephthalic Acid

XTA X-Ray Texture Analysis

XTACACS Extended Terminal Access Controller Access Control System

XTAL Crystal [also Xtal, X-TAL, X-tal, XL, Xl, XTL, or Xtl]

X-TALK Crosstalk [also X-Talk, Xtalk, or CT]

XTASI Exchange of Technical Apollo Simulation Information [NASA]

X2B Hexadecimal to Binary [Restructured Extended Executor (Computer Language)]

XTC External Transmit Clock [also XTCLK]

X2C Hexadecimal to Character [Restructured Extended Executor (Computer Language)]

XTCLK External Transmit Clock [also XTC]

X2D Hexadecimal to Decimal [Restructured Extended Executor (Computer Language)]

XTE X-Ray Timing Explorer [of NASA]

XTED Cross-Sectional Transmission Electron Diffraction

XTEL Cross Tell

XTEM Cross-Sectional Transmission Electron Microscope; Cross-Sectional Transmission Electron Microscopy

XTEN Xerox Telecommunications Network [US]

Xtian Christian [also Xn]

XTK Cross-Track

XTL Crystal(line) [also Xtl, XTAL, X-TAL, X-tal, Xtal, XL, or Xl]

XTLO Crystal Oscillator

XTM Cross-Sectional Transmission Electron Microscope; Cross-Sectional Transmission Electron Microscopy

XTM X-Ray Tomographic Microscopy

XTP Xanthosine Triphosphate; Xanthosine-5'-Triphosphate [Biochemistry]

XTPA Extended Transaction Processing Architecture

XTR Extra [also Xtr]

XTS Cross-Tell Simulator

XTS X-Windows Tracking System

XTSI Extended Task Status Index

XTT 2,3-bis(2-Methoxy-4-Nitro-5-Sulfophenyl)-2H-Tetrazolium-5-Carboxanilide

XTTD Transparent Temporary Text Delay

Xty Christianity [also Xnty]

XU Xiamen University [Beijing, PR China]

XU X-Ray Unit [also Xu, xu, or X]

XU X-Unit [also Xu, xu, or X]

XU- Cambodia [Civil Aircraft Marking]

xu(CuKα₁) copper x-unit; 1537.400 xu (1.002078×10^{-13} m) [Symbol]

XUG Xyvision (Typesetting System) Users Group [US]

xu(MoKα₁) molybdenum x-unit; 707.831 xu (1.002099×10^{-13} m) [Symbol]

XUPS Photoelectron Spectroscopy between Ultraviolet and X-Ray [also XUCPES]

XUV Extreme Ultraviolet

XUV X-Ray Ultraviolet

XUVPES Photoelectron Spectroscopy between Ultraviolet and X-Ray [also XUPS]

XV- Vietnam [Civil Aircraft Marking]

XVR Exchange Voltage Regulator

XVR Transceiver [also Xvr]

XW Xenon Weatherometer

XW- Laos [Civil Aircraft Marking]

xw ex warrants [Securities Exchange]

X-Warr Ex warrants [Securities Exchange]

XWAVE Extraordinary Wave [Xwave]

XWIND Cross Wind (Direction) [also Xwind]

XWO Xenotest Weather-O-Meter

XWRC Xerox Webster Research Center [Webster, New York, US]

XX Double Extra

XX Double Strength

XX Two X-Chromosomes (i.e., Number of X Chromosomes in Normal Human Females) [Genetics]

$\overline{\text{XX}}$ Twenty Thousand [Roman Numeral]

X-X Axes (Through any Point) [Symbol]

XXL Double Extra Long

XXS Double Extra Strong

XXX Triple Extra

XXX Triple Strength

X_1-X_2-X_3 Transformer Secondary [Controllers]

XXXL Triple Extra Long

XXXS Triple Extra Strong

XY One X-Chromosome and One Y-Chromosome (i.e., Number of Chromosomes in Normal Human Males) [Genetics]

XY Scalar Product of X and Y [Mathematics]

X-Y X-Y Plane

XY- Burma [Civil Aircraft Marking]

(XY) Scalar Product of X and Y [Symbol]

[XY] Vector Product of X and Y [Symbol]

X·Y Dot Product of X and Y [Symbol]

X×Y Cross Product of X and Y [Symbol]

X∧Y Cross Product of X and Y [Symbol]

$X^H Y$ Normal Male with Antihemophilic Factor [Symbol]

$X^h Y$ Hemophilic Male [Symbol]

$X^h Y^H$ Hemophilic Female [Symbol]

$X^h Y^h$ Homozygous Female with Overt Hemophilia (in Medicine) [Symbol]

Xyl Xylose

XYO_{3-x} Perovskite-Type Oxide (X Represents a Divalent Alkaline Earth Metal, such as Barium, Calcium or Strontium, and Y represents a Tetravalent Transition Metal such as Cobalt, Iron, Manganese, Nickel, etc.) [Symbol]

XYR X-Y Recorder

XY TAF X-Y Triangular Ferromagnet [Solid-State Physics]

XYY One X and Two Y-Chromosomes (i.e., Genetic Condition in Males with Extra Y Chromosome) [Genetics]

(x,y,z) Cartesian coordinates [Symbol]

x,y,z distance coordinates [Symbol]

XZ- Burma [Civil Aircraft Marking]

Y Admittance [Symbol]

Y 2-Amino-3-(4-Hydroxyphenyl)propanoic Acid

Y Axial Load (or Thrust) Factor (of a Rolling Bearing) [Symbol]

Y Biaxial Elastic Modulus [Symbol]

Y Force Component [Symbol]

Y Hypercharge [Symbol]

Y Hyperon (in Particle Physics) [Symbol]

Y Luminance (in Television Engineering) [Symbol]

Y Manipulated Variable (of Control Systems) [Symbol]

Y Mobile Phase Ions (in Ion Chromatography) [Symbol]

Y Mole Fraction [Symbol]

Y Photoelectric Yield [Symbol]

Y Prototype (Aircraft, Missile or Rocket) [USDOD Symbol]

Y Radiotelegraphy [1,605 to 3,800 kc/s Range]

Y Response of Interest (in Statistics) [Symbol]

Y Steady-State Amplitude [Symbol]

Y Strength Factor (of Gears) [Symbol]

Y Surface Displacement Yield (in Physics) [Symbol]

Y Transfer Function (of Control Systems) [Symbol]

Y (–)-Tyrosine; Tyrosyl [Biochemistry]

Y (Upper-case) Upsilon [Greek Alphabet]

Y Vertical Deflection (in Electronics) [Symbol]

Y Yangtse (River) [Tibet/China]

Y Yankee [Phonetic Alphabet]

Y Yard [also yd, or y]

Y Yaw [Aerospace]

Y Y (Sex) Chromosome [Genetics]

Y Y Connection (in Electrical Engineering) [Symbol]

Y Year [also y, Yr, or yr]

Y Yeast

Y Yellow [also y]

Y Yellowknife [Northwest Territories, Canada]

Y Yen [Currency of Japan]

Y Yield(ing)

Y Yoke

Y Yokohama [Japan]

Y Yolk

Y Young's Modulus [Symbol]

Y Youngstown [Ohio, US]

Y Yttria (or Yttrium Oxide) [Y_2O_3]

Y Yttrium [Symbol]

Y Yuan [Currency of PR China]

Y Yucca [Botany]

Y Yucatán [Mexican Peninsula]

Y Yugoslavia(n)

Y_0 Reference Point of Manipulated Condition (of Control Systems) [Symbol]

Y^{3+} Yttrium Ion [also Y^{+++}] [Symbol]

Y_c Limiting Compressive Stress Transverse to Fibers (of Composites) [Symbol]

Y_g Regulation Ratio (in Control Engineering) [Symbol]

Y^H Young's Modulus at Constant Magnetic Field "H" [Symbol]

Y_h Total Range of Manipulated Variable (of Control Systems) [Symbol]

$\hat{Y}_i$ Calibration Equation (in Molecular Spectroscopy) [Symbol]

Y_i Accepted Reference Value (in Molecular Spectroscopy) [Symbol]

Y_n^m Spherical Harmonics [Symbol]

Y_T Total Admittance [Symbol]

Y_t Limiting Tensile Stress Transverse to Fibers (of Composite) [Symbol]

Y_v Admittance of Vacuum [Symbol]

Y-82 Yttrium-82 [also ^{82}Y, or Y^{82}]

Y-83 Yttrium-83 [also ^{83}Y, or Y^{83}]

Y-84 Yttrium-84 [also ^{84}Y, or Y^{84}]

Y-85 Yttrium-85 [also ^{85}Y, or Y^{85}]

Y-86 Yttrium-86 [also ^{86}Y, or Y^{86}]

Y-87 Yttrium-87 [also ^{87}Y, or Y^{87}]

Y-88 Yttrium-88 [also ^{88}Y, or Y^{88}]

Y-89 Yttrium-89 [also ^{89}Y, or Y^{89}, or Y]

Y-90 Yttrium-90 [also ^{90}Y, or Y^{90}]

Y-91 Yttrium-91 [also ^{91}Y, or Y^{91}]

Y-92 Yttrium-92 [also ^{92}Y, or Y^{92}]

Y-93 Yttrium-93 [also ^{93}Y, or Y^{93}]

Y-94 Yttrium-94 [also ^{94}Y, or Y^{94}]

Y-95 Yttrium-95 [also ^{95}Y, or Y^{95}]

Y-96 Yttrium-96 [also ^{96}Y, or Y^{96}]

Y-123 $YBa_2Cu_3O_x$ [A Yttrium Barium Copper Oxide Superconductor] [also Y123]

Y-123.5 $YBa_2Cu_{3.5}O_{7.5-z}$ [A Yttrium Barium Copper Oxide Superconductor] [also Y123.5]

Y-124 $YBa_2Cu_4O_x$ [A Yttrium Barium Copper Oxide Superconductor] [also Y124]

Y-125 $YBa_2Cu_5O_x$ [A Yttrium Barium Copper Oxide Superconductor] [also Y125]

Y-132 $YBa_3Cu_2O_x$ [A Yttrium Barium Copper Oxide Superconductor] [also Y132]

Y-137 $YBa_3Cu_7O_x$ [A Yttrium Barium Copper Oxide Superconductor] [also Y137]

Y-142 $YBa_4Cu_2O_x$ [A Yttrium Barium Copper Oxide Superconductor] [also Y142]

Y-143 $YBa_4Cu_3O_x$ [A Yttrium Barium Copper Oxide Superconductor] [also Y143]

Y-152 $YBa_5Cu_2O_x$ [A Yttrium Barium Copper Oxide Superconductor] [also Y152]

Y-202 $Y_2Cu_2O_5$ [A Yttrium Barium Copper Oxide Superconductor] [also Y202]

Y-211 Y_2BaCuO_5 [A Yttrium Barium Copper Oxide Superconductor] [also Y211]

Y-247 $Y_2Ba_4Cu_7O_5$ [A Yttrium Barium Copper Oxide Superconductor] [also Y247]

Y-385 $Y_3Ba_8Cu_5O_x$ [A Yttrium Barium Copper Oxide Superconductor] [also Y385]

y deflection (of springs) [Symbol]

y deviation of manipulated variable (of control systems)

y displacement (of a wave) [Symbol]

y form factor (for bevel and helical gears) [Symbol]

y object size (in optics) [Symbol]

y one of the axes in a rectangular coordinate system [Symbol]

y parallax (in electron stereomicroscopy) [Symbol]

y pure imaginary coefficient of a complex variable (e.g., x + yi) [Symbol]

y response (in statistics) [Symbol]

y second rectangular coordinate (of a point) [Symbol]

y space coordinate [Symbol]

y specific (normal) acoustic admittance [Symbol]

y unknown (or variable) [Symbol]

y yard [also yd, or Y]

y year [also Y, Yr, or yr]

y yellow

y yield factor (or minimum design seating stress) of gaskets [Symbol]

$\bar{y}$ average responses (in statistics) [Symbol]

y' image size (in optics) [Symbol]

$\dot{y}$ first derivative of y with respect to time [Symbol]

$\ddot{y}$ second derivative of y with respect to time [Symbol]

y_0 amplitude (of a wave) [Symbol]

y_{fb} common-base small-signal short-circuit forward transfer admittance [Semiconductor Symbol]

y_{fc} common-collector small-signal short- circuit forward transfer admittance [Semiconductor Symbol]

y_{fe} common-emitter small-signal short-circuit forward transfer admittance [Semiconductor Symbol]

y_{fs} common-source small-signal short-circuit forward transfer admittance [Semiconductor Symbol]

$y_{fs(imag)}$ common-source small-signal forward transfer susceptance [Semiconductor Symbol]

$y_{fs(real)}$ common-source small-signal forward transfer conductance [Semiconductor Symbol]

y_g regulating ratio (of control systems) [Symbol]

y_h regulating range (of control systems) [Symbol]

y_{ib} common-base small-signal short-circuit input admittance [Semiconductor Symbol]

y_{ic} common-collector small-signal short- circuit input admittance [Semiconductor Symbol]

y_{ie} common-emitter small-signal short-circuit input admittance [Semiconductor Symbol]

$y_{ie(imag)}$ imaginary part of small-signal short-circuit input admittance, common emitter [Semiconductor Symbol]

$y_{ie(real)}$ real part of small-signal short-circuit input admittance, common emitter [Semiconductor Symbol]

y_{is} common-source small-signal short-circuit input admittance [Semiconductor Symbol]

$y_{is(imag)}$ common-source small-signal input susceptance [Semiconductor Symbol]

$y_{is(real)}$ common-source small-signal input conductance [Semiconductor Symbol]

y_l longitudinal wave [Symbol]

y_{ob} common-base small-signal short-circuit output admittance [Semiconductor Symbol]

y_{oc} common-collector small-signal short- circuit output admittance [Semiconductor Symbol]

y_{oe} common-emitter small-signal short-circuit output admittance [Semiconductor Symbol]

$y_{oe(imag)}$ imaginary part of small-signal short-circuit output admittance, common emitter [Semiconductor Symbol]

$y_{oe(real)}$ real part of small-signal short-circuit output admittance, common emitter [Semiconductor Symbol]

y_{os} common-source small-signal short-circuit output admittance [Semiconductor Symbol]

$y_{os(imag)}$ common-source small-signal output susceptance [Semiconductor Symbol]

$y_{os(real)}$ common-source small-signal output conductance [Semiconductor Symbol]

y_{rb} common-base small-signal short-circuit reverse transfer admittance [Semiconductor Symbol]

y_{rc} common-collector small-signal short- circuit reverse transfer admittance [Semiconductor Symbol]

y_{re} common-emitter small-signal short-circuit reverse transfer admittance [Semiconductor Symbol]

y_{rs} common-source small-signal short-circuit reverse transfer admittance [Semiconductor Symbol]

$y_{rs(imag)}$ common-source small-signal reverse transfer susceptance [Semiconductor Symbol]

$y_{rs(real)}$ common-source small-signal reverse transfer conductance [Semiconductor Symbol]

y_T thermal resistivity coefficient [Symbol]

¥ Yen [Currency of Japan]

YA Yacht Architect(ure)

YA Yagi Antenna

YA Yet Another

YA Yosemite Association [US]

YA Yttrium Aluminate

YA Yttrium Aluminide

Y-A Y-Axis

YA- Afghanistan [Civil Aircraft Marking]

Y(a) Stress Intensity Factor [Fracture Mechanics]

YABA Yacht Architects and Brokers Association [US]

YAC Yeast Artificial Chromosome (Cloning System) [Genetics]

YAC Young Astronaut Council [US]

Y(Ac)$_3$ Yttrium(III) Acetate

Y(ACAC)$_3$ Yttrium(III) Acetylacetonate [also Y(AcAc)$_3$, or Y(acac)$_3$]

YACC Yet Another Compiler Compiler [Unix Computer Tool]

Yad Energ Yadrena Energiya [Bulgarian Journal on Nuclear Energy]

Yad Fiz Yadernaya Fizika [Russian Journal of Nuclear Physics]

YAG Yttrium Aluminum Garnet [also Yag, or yag]

YAg Yttrium Silver (Compound)

YAG(Cr) Chromium-Doped Yttrium Aluminum Garnet (Laser) [also YAG:Cr]

YAG(Er) Erbium-Doped Yttrium Aluminum Garnet (Laser) [also YAG:Er]

YAG(Ho) Holmium-Doped Yttrium Aluminum Garnet (Laser) [also YAG:Ho]

YAG(Nd) Neodymium-Doped Yttrium Aluminum Garnet (Laser) [also YAG:Nd]

YAG(Th) Thorium-Doped Yttrium Aluminum Garnet (Laser) [also YAG:Th]

YAG(Tm) Thulium-Doped Yttrium Aluminum Garnet (Laser) [also YAG:Tm]

YAG(Yb) Ytterbium-Doped Yttrium Aluminum Garnet (Laser) [also YAG:Yb]

YAG:YIG Yttrium Aluminum Garnet Doped Yttrium Iron Garnet

YAHOO Yet Another Hierarchically Officious Oracle [Internet]

Yale J Biol Med Yale Journal of Biology and Medicine [of Yale University, New Haven, Connecticut, US]

YAM Yet Another Modem (Device)

YAM Yttrium-Aluminum Monoclinic [Metallurgy]

YAO Ytterbium Aluminum Oxide

YAO Yttrium Aluminum Oxide

YAP Yttrium Aluminum Perovskite Oxide

YAP(Nd) Neodymium-Doped Yttrium Aluminum Perovskite (Laser)

YAPP Yankee Atomic Power Plant [Haddam Neck, Connecticut, US]

YAS Yorkshire Archeological Society [UK]

YAs Yttrium Arsenide (Semiconductor)

YB Yearbook

YB Yellow Brass

Y/B Yellow/Blue (Ratio)

Yb Ytterbium [Symbol]

Yb^{2+} Divalent Ytterbium Ion [also Yb^{++}] [Symbol]

Yb^{3+} Trivalent Ytterbium Ion [also Yb^{+++}] [Symbol]

Yb-164 Ytterbium-164 [also ^{164}Yb, or Yb^{164}]

Yb-165 Ytterbium-165 [also ^{165}Yb, or Yb^{165}]

Yb-166 Ytterbium-166 [also ^{166}Yb, or Yb^{166}]

Yb-167 Ytterbium-167 [also ^{167}Yb, or Yb^{167}]

Yb-168 Ytterbium-168 [also ^{168}Yb, or Yb^{168}]

Yb-169 Ytterbium-169 [also ^{169}Yb, or Yb^{169}]

Yb-170 Ytterbium-170 [also ^{170}Yb, or Yb^{170}]

Yb-171 Ytterbium-171 [also ^{171}Yb, or Yb^{171}]

Yb-172 Ytterbium-172 [also ^{172}Yb, or Yb^{172}]

Yb-173 Ytterbium-173 [also ^{173}Yb, or Yb^{173}]

Yb-174 Ytterbium-173 [also ^{174}Yb, or Yb^{174}]

Yb-175 Ytterbium-174 [also ^{175}Yb, or Yb^{175}]

Yb-176 Ytterbium-175 [also ^{176}Yb, or Yb^{176}]

Yb-177 Ytterbium-176 [also ^{177}Yb, or Yb^{177}]

Yb(Ac)$_3$ Ytterbium(III) Acetate

Yb:BaY$_2$F$_8$ Ytterbium-Doped Barium-Yttrium Fluoride [also Yb^{3+}:BaY$_2$F$_8$]

YbBCO Ytterbium Barium Copper Oxide (Superconductor)

YBC Yttrium Barium Cuprate

YBCAO Yttrium Barium Copper Aluminum Oxide (Superconductor)

YBCCO Yttrium Barium Copper Cobalt Oxide (Superconductor)

YBCFO Yttrium Barium Copper Iron Oxide (Superconductor)

YBCGO Yttrium Barium Copper Gallium Oxide (Superconductor)

YBCHC Yttrium Barium Copper Hydroxy Carbonate (Superconductor)

YBCNO Yttrium Barium Copper Nickel Oxide (Superconductor)

YBCNO Yttrium Barium Copper Niobium Oxide (Superconductor)

YBCO Yttrium Barium Copper Oxide (Superconductor)

YBCO-Ag Yttrium Barium Copper Oxide–Silver (Composite Superconductor)

YBCOC Yttrium Barium Copper Oxide Carbonate (Superconductor)

YBCO(H) Proton Implanted Yttrium Barium Copper Oxide (Superconductor)

YBCO/LAO Yttrium Barium Copper Oxide on Lanthanum Aluminate

YBCO/PBCO Yttrium Barium Copper Oxide on Praseodymium Barium Copper Oxide [also YBCO/PrBCO]

YBCO-SiO$_2$ Yttrium Barium Copper Oxide-Silicon Dioxide (Composite Superconductor)

YBCO/STO Yttrium Barium Copper Oxide on Strontium Titanate

Yb(Cp)$_3$ Tris(cyclopentadienyl)ytterbium

YBCO/LAO Yttrium Barium Copper Oxide on Lanthanum Aluminate

YBCO/PBCO Yttrium Barium Copper Oxide on Praseodymium Barium Copper Oxide

Y:BCSCO Yttrium-Doped Bismuth Calcium Strontium Copper Oxide (Superconductor)

YBCZO Yttrium Barium Copper Zinc Oxide (Superconductor)

YBDSA Yacht Brokers, Designers and Surveyors Association [UK]

Yb(FOD)$_3$ Ytterbium Tris[3-(Heptafluorodimethyloctane-dionate)] [also Yb(fod)$_3$]

Yb(HFC)$_3$ Ytterbium Tris[3-(Heptafluoropropylhydroxy-methylene)camphorate] [also Yb(hfc)$_3$]

YbIG Ytterbium-Iron Garnet

Yb:KYF$_4$ Ytterbium-Doped Potassium-Yttrium Fluoride [also Yb^{3+}:KYF$_4$]

YBL Yellow with Black

Yb:LiYF$_4$ Ytterbium-Doped Lithium-Yttrium Fluoride [also Yb^{3+}:LiYF$_4$]

YBMO Yttrium Barium Magnesium Oxide (Superconductor)

YbN Ytterbium Nitride

YBNO Yttrium Barium Nickel Oxide (Superconductor)

YBNO Yttrium Barium Niobium Oxide (Superconductor)

YbOPr^i Ytterbium(III) Isopropoxide [also Yb(O-I-Pr)]

YBP Years before Present

YBPC Young Black Programmers Coalition [US]

YBRA Yellowstone-Bighorn Research Association [US]

YbSe Ytterbium Selenide (Semiconductor)

YbSiON Yttrium Silicon Oxynitride

YbTe Ytterbium Telluride (Semiconductor)

Yb(TFC)$_3$ Ytterbium Tris[3-(Trifluoromethylhydroxy-methylene)camphorate] [also Yb(tfc)$_3$]

Yb(THD)$_3$ Ytterbium(III) Tris(2,2,6,6-Tetramethyl-3,5-Heptanedionate) [also Yb(thd)$_3$]

Yb(TMHD)$_3$ Ytterbium(III) Tris(2,2,6,6-Tetramethyl-3,5-Heptanedionate) [also Yb(tmhd)$_3$]

Yb:YAG Ytterbium-Doped Yttrium Aluminum Garnet (Laser)

Yb:YAO Ytterbium-Doped Yttrium Aluminum Oxide

YBT Yoshida Buckling Test

YBZO Yttrium Barium Zinc Oxide (Superconductor)

YC Yellow Cake

YC Yield Condition [Mechanics]

YC Yield Criterion [Mechanics]

YC York College [of City University of New York, US]

YC Yttrium Chromite

YCA Yttrium Calcium Aluminate

YCBCO Yttrium Calcium Barium Copper Oxide (Superconductor) [also YCaBCO]

YCCNO Yttrium Calcium Copper Niobium Oxide (Superconductor)

YCeBCO Yttrium Cerium Barium Copper Oxide (Superconductor)

YCF Yacht-Club de France [Yacht Club of France]

YCIT York College of Information Technology [Toronto, Canada]

YCM Yeast Cell Mutation [Genetics]

YCM Yttrium Calcium Manganite

YCM Yttrium-Cobalt-Molybdenum (Compound)

YCmBCO Yttrium Curium Barium Copper Oxide (Superconductor)

YCO Yttrium Chromium Oxide

Y-Cop Quantitative Analysis of Yttrium Barium Copper Oxide by Coprecipitation Method

YCor Current Y-Coordinate of Turtle [Turbo Pascal Turtlegraphics]

Y(Cp)$_3$ Tris(cyclopentadienyl)yttrium

YCPF Yellow Creek Processing Facility [US]

YCS Yeast Chromium Source

YCS Yukon Conservation Society [Canada]

YCSCGO Yttrium Calcium Strontium Copper Gallium Oxide

YCU Yokohama City University [Japan]

YCu Yttrium Copper (Compound)

Y-Cu Yttrium-Copper (Alloy System)

Y-CUT Crystal Cut, Orientation II [also Y CUT, or Y cut]

YCZ Yellow Caution Zone

YD South Yemen Dinar [Currency]

YD Yacht Design(er)

YD Yankee Dryer

YD Young-Dupré (Equation) [Thermodynamics]

yd yard [also Y, or y]

yd^2 square yard [also sq yd]

yd^3 cubic yard [also cu yd]

Yda Yesterday [also yda]

YDD Yemeni Dinar [Currency of the People's Republic of Yemen]

yd/lb yard(s) per pound [Unit]

Y(DPM)$_2$ Yttrium (II) Dipivaloylmethanoate (or Tris(dipivaloylmethanato)yttrium(III)) [also Y(dpm)$_2$]

yds yards

YDSA Yacht Designers and Surveyors Association [UK]

YDSP Young Defence Scientists Program [Singapore]

YDT Yukon Daylight Time [Greenwich Mean Time +10:00]

YE Yemen [ISO Code]

YE Young Equation [Physics]

YE- Yemen [Civil Aircraft Marking]

.ye Yemen [Country Code/Domain Name]

YEA Yale Engineering Association [of Yale University, New Haven, Connecticut, US]

YEC Youngest Empty Cell

YECIP Yukon Energy Conservation Incentive Program [Canada]

YEE Youth and Environment Europe [Denmark]

yel yellow [also Yel]

YEP Young Eucalyptus Program [Australia]

YER Yemeni Riyal (or Rial) [Currency of the Arab Republic of Yemen]

YES Yeast Extract Sucrose [Biochemistry]

YES Youth Engineering and Science (Camps of Canada)

YEU Yugoslav Electricity Union [Belgrade]

YF Yellow Fever [Medicine]

YF Yield Factor [Industrial Engineering]

YF Yttrium Ferrite

YF Yukawa Force [Nuclear Physics]

YFC Yakima Firing Center [of US Army in Washington State, US]

YFCM Yttrium-Iron-Cobalt-Molybdenum (Compound)

Y-Fe Yttrium-Iron (Alloy System)

YFO Yttrium Ferrite [YFe$_5$O$_{12}$]

YFL Yukon Federation of Labour [Canada]

YFN Yttrium-Iron Nitride

YFS Yttrium Iron Silicide

YFU Youth for Understanding [US]

YFV Yellow Fever Virus

YG Yeast Glycogen [Biochemistry]

YG Yellow-Green (Algae) [Botany]

YG Yellowish Green

YGaIG Yttrium Gallium Iron Garnet

YGO Yttrium Germanium Oxide (Superconductor)

YGS Yorkshire Geological Society [UK]

YH Young-Helmholtz (Theory) [Color Vision]

Y(HFAC)$_3$ Yttrium Hexafluoroacetylacetonate (also Y(hfac)$_3$]

Y-HTSC Yttrium High-Temperature Superconductor

YI Yeast Intervase [Biochemistry]

YI Yellowness Index

YI Young's Interferometer [Optics]

YI- Iraq [Civil Aircraft Marking]

YIC Yukawa Interaction Constant [Nuclear Physics]

YIES Yamanashi Institute of Environmental Sciences [Woodridge, Illinois, US]

YIG Yttrium Iron Garnet [also Yig, or yig]

YIG(Ga) Gallium-Doped Yttrium-Iron Garnet

YIG/GGG/YIG Yttrium Iron Garnet/Gadolinium Garnet/Yttrium Iron Garnet

Y i-Bu Yttrium Isobutyrate

YIL Yellow Indicating Lamp

YIO Yellow Iron Oxide

YIP Young Investigator Program [of Office of Naval Research, Washington, DC, US]

YIP Youth Internship Canada [Human Resources Development Canada]

Y-IR Thermoplastic Isoprene Rubber

YJ- New Hebrides [Civil Aircraft Marking]

YK Yttrium K (Absorption Edge) [X-Ray Analysis]

YK- Syria [Civil Aircraft Marking]

YKB Yukon Bibliography (Database) [of University of Alberta, Edmonton, Canada]

YL Yield Line

YL Yoke Length (of Write Head) [Electronics]

YL Young-Laplace (Equation)

$\bar{y}(\lambda)$ tristimulus value for spectral component "y" in CIE System [Symbol]

YLα Yttrium L-Alpha (Radiation) [also Yl$_\alpha$]

YLβ Yttrium L-Beta (Radiation) [also Yl$_\beta$]

Yld Yield [also yld]

YLF Yttrium Lithium Fluoride (Laser)

YLF(Er) Erbium-Doped Yttrium Lithium Fluoride (Laser) [also YLF:Er]

YLF(Ho) Holmium-Doped Yttrium Lithium Fluoride (Laser) [also YLF:Ho]

YLF(Nd) Neodymium-Doped Yttrium Lithium Fluoride (Laser) [also:YLF:Nd]

ylsh yellowish

Y/L/Y YBCO (Yttrium Barium Copper Oxide/LSCO (Lanthanum Strontium Copper Oxide)/YBCO (Yttrium Barium Copper Oxide) (Trilayer)

YM Yamakawa and Maeta (Method) [Metallurgy]

YM Yamamoto Method [Physics]

YM Yankee Machine

YM Yellow Metal

YM Young's Modulus [Mechanics]

YM Yttrium Manganite

YM Yvon's Method

Y/M Yard/Meter (Ratio) [also y/m]

YMBA Yacht and Motor Boat Association

YMC Yeast Mold Count [Biochemistry]

YMC Yttrium-Oxide Microconcrete

Y(Me$_3$Ac)$_3$ Yttrium Trimethylacetate

YMMV Your Mileage May Vary

YMO Yellow Mercury Oxide

YMO Yttrium Manganate

YMO Yttrium Molybdenum Oxide

YMP Yucca Mountain Site Characterization Project [of US Department of Energy, Lawrence Livermore National Laboratory] [*Note:* Yucca Mountain is in Nevada]

YMP/LLNL Yucca Mountain Site Characterization Project/Lawrence Livermore National Laboratory [of US Department of Energy]

YMPO Yucca Mountain Project Office [of US Department of Energy in Nevada]

YMTV Yaba Monkey Tumor Virus

YN Yttrium Nitride

Y/N Yes/No (Prompt)

YNA Young Naturalists Association [Canada]

YNB Yale News Bureau [of Yale University, New Haven, Connecticut, US]

Y-NBR Thermoplastic Nitrile-Butadiene Rubber

YNCIAWPRC Yugoslavian National Committee of the International Association on Water Pollution Research and Control

YNHA Yosemite Natural History Association [now Yosemite Association]

Y-Ni Yttrium-Nickel (Alloy)

Y:NiO Yttrium-Doped Nickel Oxide

Y-89-NMR Yttrium-89 Nuclear Magnetic Resonance [also ^{89}Y NMR]

YNP Yellowstone National Park [US]

YNU Yeung Nam University [Kyungpook, South Korea]

YNU Yokohama National University [Japan]

YNWT Yukon and Northwest Territories [Canada]

YO Yerkes Observatory [of University of Chicago; located at Williams Bay, Wisconsin, US]

YO Yerzley Oscillograph

YO Yttrium Monoxide

Y(OAc)$_3$ Yttrium (III) Acetate

Y_2O_3-Al_2O_3 Yttria-Alumina (Alloy System)

YOB Year of Birth [also yob]

Y_2O_3(Ce) Cerium-Doped Yttria [also Y_2O_3:Ce]

Y_2O_3-Cr_2O_3 Yttria-Chromia (Alloy System)

Y_2O_3(Er) Erbium-Doped Yttria [also Y_2O_3:Er]

Y(OEt)$_3$ Yttrium(III) Ethoxide [General Formula]

Y_2O_3(Eu) Europium-Doped Yttria [also Y_2O_3:Eu]

YOF Year-of-the-Ocean Foundation [now Ocean Outlook, US]

Yokogawa Tech Rep Yokogawa Technical Report [Published by Yokogawa Electric Works Ltd., Tokyo, Japan]

Yokohama Med Bull Yokohama Medical Bulletin [Japan]

Y(OMe)$_3$ Yttrium(III) Methoxide

Y(OPr^i) Yttrium Isopropoxide [also Y(O-i-Pr)]

Y(OR)$_3$ Yttrium Alkoxide [General Formula]

Yorks Yorkshire [UK]

YOTO (International) Year of the Ocean

Y$_2$O$_3$(Tb) Terbium-Doped Yttria [also Y$_2$O$_3$:Th]

Y$_2$O$_3$(ThO$_2$) Yttria-Stabilized Thoria [also Y$_2$O$_3$:ThO$_2$]

Y$_2$O$_3$(ZrO$_2$) Yttria-Stabilized Zirconia [also Y$_2$O$_3$:ZrO$_2$]

YP Yamazaki Process

YP Year of Publication

YP Yellow Pages

YP Yellow Pine

YP Yield Phenomenon [Mechanics]

YP Yield Point [Mechanics]

YP Young Professionals

YP Yttrium Phosphide (Semiconductor)

YP Yukawa Particle [Particle Physics]

YP Yvon Photometer

YPBCO Yttrium Praseodymium Barium Copper Oxide (Superconductor) [also YprBCO]

YPCBCO Yttrium Praseodymium Calcium Barium Copper Oxide (Superconductor) [also YprCaBCO]

YPd Yttrium Palladium (Compound)

YPE Yield Point Elongation [Mechanics]

YPF Yacimientos Petrolíferos Fiscales S.A. [Argentinian State Oil Company]

YPFB Yacimientos Petrolíferos Fiscales Bolivianos [Bolivian State Petroleum Company]

YPG Yuma Proving Ground [of US Army in Arizona, US]

YPP Yield Point Phenomenon [Mechanics]

YPRC Yerkes Primate Research Center [Atlanta, Georgia, US]

YPrBCO Yttrium Praseodymium Barium Copper Oxide (Superconductor) [also YPBCO]

YPrCaBCO Yttrium Praseodymium Calcium Barium Copper Oxide (Superconductor) [also YPCBCO]

YPS Yellow Pages Service

Y-PSZ Yttria Partially-Stabilized Zirconia

YPV Yatapoxvirus

YR Yellowness Reduction

YR Yellow–Red

YR Yellow River [PR China]

YR Yemeni Riyal [Currency of Northern Yemen]

YR Yield Region [Mechanics]

YR Yield Requirement

YR Yukon River [Alaska/Canada]

YR- Rumania [Civil Aircraft Marking]

Yr Year [also yr]

yr your

Yrb Yearbook [also yrb]

YRB Yellowstone River Basin [US]

Yrb Am Phil Soc Yearbook of the American Philosophical Society [Philadelphia, Pennsylvania, US]

Yrb Forest Prod Yearbook of Forest Products [of Food and Agricultural Organization]

Yrb Forest Prod Stat Yearbook of Forest Products Statistics [of Food and Agricultural Organization]

Yrb Nat Soc Study Educ Yearbook of the National Society for the Study of Education [Published by University of Chicago Press, US]

Yrb Phys Anthrop Yearbook of Physical Anthropology [US]

Yrbk Yearbook [also yrbk]

YRI Yano Research Institute [Tokyo, Japan]

YRI Yemeni Riyal (or Rial) [Currency of the Arab Republic of Yemen]

YRL Yokosuka Research Laboratory [Kanagawa, Japan]

Yrs Years [also yrs]

yrs yours

YS Yacht Surveyor

YS Yellow Sea

YS Yellow Spot

YS Yield Spectroscopy

YS Yield Strain [Mechanics]

YS Yield Strength [Mechanics]

YS Yield Stress [Mechanics]

YS Yield Surface [Mechanics]

YS Youhao Sulfur

YS$_l$ Lower Yield Stress (in Mechanics) [Symbol]

YS$_u$ Upper Yield Stress (in Mechanics) [Symbol]

YS$_2$ Yttrium Disilicate [Y$_2$O$_3$·2SiO$_2$]

Y$_2$S$_3$ Diyttrium Trisilicate [2Y$_2$O$_3$·3SiO$_2$]

YS- El Salvador [Civil Aircraft Marking]

Y-S Yttrium-Sulfur (Alloy System)

YSB Yacht Safety Bureau

YSb Yttrium Antimonide (Semiconductor)

Y-SBR Thermoplastic Styrene-Butadiene Rubber

YSC Yttria-Stabilized Chromia

YSCGO Yttrium Strontium Copper Gallium Oxide (Superconductor)

YSCMO Yttrium Strontium Copper Molybdenum Oxide (Superconductor)

YSCNO Yttrium Strontium Copper Niobium Oxide (Superconductor)

YSCRO Yttrium Strontium Copper Rhenium Oxide (Superconductor)

YSCWO Yttrium Strontium Copper Tungsten Oxide (Superconductor)

YSF Yield Safety Factor [Mechanics]

YSF Youth Science Foundation [Canada]

YSGG Yttrium Scandium Gadolinium Garnet

YSGG(Er) Erbium-Doped Yttrium Scandium Gadolinium Garnet (Laser)

YSI Yellow Springs Instrumentation, Inc. [US]

YSI Yukon Science Institute [Canada]

YSiAlON Yttrium Silicon Aluminum Oxynitride

Y-Sin Quantitative Analysis of Yttrium Barium Copper Oxide Single Crystal

YSLF Yield Strength Load Factor [Mechanics]

Y-Sol Quantitative Analysis of Yttrium Barium Copper Oxide by Solid-Phase Reaction Method

YSSP Young Scientist Summer Program [of International Institute of Applied Systems Analysis, Austria]

YST Yttria-Stabilized Thoria

YST Yttrium Stannate Titanate

YST Yukon Standard Time [Greenwich Mean Time +9:00]

YSTZ Yttria-Stabilized Tetragonal Zirconia

YSU Youngstown State University [Youngstown, Ohio, US]

Y/S/Y YBCO (Yttrium Barium Copper Oxide)/Strontium Titanate/YBCO (Yttrium Barium Copper Oxide)(Trilayer)

YSZ Yttria-Stabilized Zirconia

YT Mayotte [ISO Code]

YT Yarn Texturing

YT Yield Temperature

YT Yttrium Tantalate

YT Yukawa Theory [Nuclear Physics]

YT Yukon Territory [Canada]

Y(t) Output [Symbol]

YTbBCO Yttrium Terbium Barium Copper Oxide (Superconductor)

YTD Year to Date

Y(THD)$_3$ Yttrium Tris(2,2,6,6-Tetramethyl-3,5-Heptanedionate [also Y(thd)$_3$]

Y2K Year 2000 (Virus)

Y(TMHD)$_3$ Yttrium Tris(2,2,6,6-Tetramethyl-3,5-Heptanedionate [also Y(tmhd)$_3$]

YTO Yttrium Tantalum Oxide

Ytterbium Triflate Ytterbium(III) Trifluoromethanesulfonate [also ytterbium triflate]

YTO YIG (Yttrium Iron Garnet) Tuned Oscillator

Y(θ,ϕ) Spherical Harmonics [Symbol]

YTS Youth Training Scheme [UK]

YTS Yuma Test Station [Arizona, US]

YTT Yield Time Test

YTZ Yttria-Toughened Zirconia

Y-TZP Yttria-Stabilized Tetragonal Zirconia Polycrystal

1Y-TZP Tetragonal Zirconia Polycrystal Stabilized with 1 mol% Yttria

YU Yale University [New Haven, Connecticut, US]

YU Yamagata University [Japan]

YU Yamaguchi University [Japan]

YU Yamanashi University [Japan]

YU Yonsei University [South Korea]

YU York University [Toronto, Ontario, Canada]

YU Youngstown University [Ohio, US]

YU Yugoslavia [ISO Code]

YU Yukon (Territory) [Canada]

YU- Yugoslavia [Civil Aircraft Marking]

Yu Yuan [Currency of PR China]

Yu Yukon (Territory) [Canada]

.yu Yugoslavia [Country Code/Domain Name]

YUD Yugoslavian Dinar [Currency]

YUDC York University Development Center [Toronto, Ontario, Canada]

Yugo Yugoslavia(n) [also Yugos]

Yuk Yucatan [Mexico]

YUO Yale University Observatory [New Haven, Connecticut, US]

YUP Yale University Press [New Haven, Connecticut, US] [also YU Press]

YV Yield Value

YV- Venezuela [Civil Aircraft Marking]

YVO$_4$(Nd) Neodymium-Doped Yttrium Vanadate (Laser)

YVO$_4$/YSZ Yttrium Vanadate/Yttrium -Stabilized Zirconia

YW York Whiting

YWIA Your Welcome In Advance [Internet Jargon]

YX One Y and One X Chromosome (i.e., Number of X and Y Chromosomes in Normal Human Males) [Genetics]

Y(x) Integral Function [Symbol]

YY Year [also yy] [Date]

YY Yellow-Yellow [Double Star Rocket]

Y-Y Axes (Through any Point) [Symbol]

y-y yellow-yellow [Double Star Rocket]

YYEF Yttrium-Ytterbium-Erbium Fluoride

YY/MM/DD Year/Month/Day [also yy/mm/dd]

YYTF Yttrium-Ytterbium-Thulium Fluoride

Y-Z X-Z Plane

YZn Yttrium Zincide (Compound)

3Y-ZrO$_2$ Zirconia Stabilized with 3 mol% Yttria

YZT Yttrium Zirconate Titanate

Z

Z   Absolute Viscosity [Symbol]

Z   Acoustic Impedance [Symbol]

Z   Airship [USDOD Symbol]

Z   Altitude [Symbol]

Z   Attenuation Length [Symbol]

Z   Atomic Number [Symbol]

Z   Avogadro's Number [Symbol]

Z   Azimuth [Symbol]

Z   cis- [Chemistry]

Z   Compressibility Factor (in Thermodynamics) [Symbol]

Z   Coordination Number (in Physics) [Symbol]

Z   Disturbance Variable (in Control Systems) [Symbol]

Z   Figure of Merit (in Electronics) [Symbol]

Z   Force Component [Symbol]

Z   Glutamic Acid [Biochemistry]

Z   Glutamine [Biochemistry]

Z   Greenwich Mean Time (or Zulu Time) [Symbol]

Z   Heating Rate [Symbol]

Z   Impedance [Symbol]

Z   Impingement Rate [Symbol]

Z   Lidar Zenith Pointing [Symbol]

Z   Modified Knudsen Number (in Fluid Mechanics) [Symbol]

Z   Number of Atomic Monolayers [Symbol]

Z   Number of Atoms per Unit of Structure (for Chemical Elements) [Symbol]

Z   Number of Bearing Balls, or Rollers [Symbol]

Z   Number of Chemical Formula Units per Unit of Structure [Symbol]

Z   Number of Teeth (of Gears, Splines, etc.) [Symbol]

Z   Partition Function (of Molecular System) [Symbol]

Z   Phenylmethoxycarbonyl

Z   Planning Stage (Aircraft, Missile or Rocket) [USDOD Symbol]

Z   Radiotelegraphy [4,000 to 25,110 kc/s Range]

Z   Reflection Coefficient Ratio (i.e., R_p/R_s) [Symbol]

Z   Section Modulus [Symbol]

Z   Self-Inductance [Symbol]

Z   Standard Gaussian Random Variable [Symbol]

Z   Standard Normal Distribution [Symbol]

Z   Supply Function (in Field-Ion Microscopy) [Symbol]

Z   Surface Collision Factor [Symbol]

Z   Thermoelectric Figure of Merit [Symbol]

Z   Valence [Symbol]

Z   Vertical (Geomagnetic) Intensity [Symbol]

Z   Viscosity Index (in Chemical Engineering) [Symbol]

Z   Zagreb [Croatia]

Z   Zaire

Z   Zambesi (River) [Southern Africa]

Z   Zambia(n)

Z   Zeldovich Factor (in Physics) [Symbol]

Z   Zener-Hollomon Parameter [Symbol]

Z   Zenith (Distance) [Astronomy]

Z   Zenith Angle (in Astronomy) [Symbol]

Z   Zeolite [Mineral]

Z   (Upper-case) Zeta [Greek Alphabet]

Z   Zero [also z]

Z   Zero Zone Time [Astronomy]

Z   Zimbabwe

Z   Zinc [Abbreviation; Symbol: Zn]

Z   Zinc Blende [Mineral]

Z   Zirconia

Z   Zirconolite [Mineral]

Z   Zizania [Genus of Edible Plants Including Wild Rice]

Z   Z-Marker [Navigation]

Z   Zone [also z]

Z   Zone Factor (of Gears) [Symbol]

Z   Zone Marker [Navigation]

Z   Zoom(ing)

Z   Zoospore [Biology]

Z   Zulu [Phonetic Alphabet]

Z   Zulu Time (or Greenwich Mean Time)

Z   Zurich [Switzerland]

Z   Zygote; Zygotic [Embryology]

Z-   Benzyloxycarbonyl

-Z   Administraive (Aircraft) [US Navy Suffix]

.Z   Compressed File (UNIX File Name Extension in Compress/Uncompress Program]

(Z)- Z-Isomer (on Same Sides of Molecule) [Chemistry]

Z*   Effective Valence [Symbol]

Z*   Electromigration Parameter [Symbol]

Z′   Real Component of Electrical Impedance [Symbol]

Z′   Imaginary Component of Electrical Impedance [Symbol]

Z_+ Charge on Positive Ions (in a Solution) [Symbol]

Z_- Charge on Negative Ions (in a Solution) [Symbol]

Ż   Sputtering Rate [Symbol]

Z^0 Z-Zero Boson (i.e., Neutral Weak Boson with Mass of 89 GeV) (in Particle Physics) [Symbol]

Z_0 Characteristic Impedance (or Surge Impedance) [Symbol]

Z_0 Impedance of Vacuum [Symbol]

Z_1 Collision Number (i.e., Number of Collisions per Second per Molecule) [Symbol]

Z_1 Load Impedance [Symbol]

Z_1 Number of Collisions per Unit Time [Symbol]

Z_{11} Bimolecular Collision Rate for Unlike Molecules [Symbol]

Z_{12} Bimolecular Collision Rate for Pure Substance [Symbol]

Z_A (Area-Related) Impingement Rate [Symbol]

Z^a Refractive Index of Species "a" [Symbol]

Z_a Atomic Partition Function [Symbol]

Z_{AB} Collision Number for Dissimilar Molecules "A" and "B" [Symbol]

Z_b Section Modulus for Bending [Symbol]

Z_c Section Modulus for Compression [Symbol]

Z_α Susceptibility Renormalization Constant [Symbol]

Z_d Dynamic Impedance (for a Rejector Circuit) [Symbol]

Z_f Feedback Impedance [Symbol]

Z_i Charge on the i-th Ion (in a Solution) [Symbol]

Z_i Image Impedance [Symbol]

Z_i Input Impedance [Symbol]

Z_i Ionic Partition Function [Symbol]

Z_i Valence of Ion Species "i" [Symbol]

Z_L Load Impedance [Symbol]

Z_l Acoustic Impedance for Longitudinal Wave [Symbol]

Z_l Load Impedance [Symbol]

Z_{LSL} Standard Normal Distribution for Lower Specification Limit (in Statistics) [Symbol]

Z^m Magnetic Valence [Symbol]

Z_{out} Output Impedance [Symbol]

Z_p Polar Section Modulus [Symbol]

Z_p Impedance in Primary (Transformer) Coil [Symbol]

Z_R Terminating Impedance [Symbol]

Z_S Source Impedance [also Z_s] [Symbol]

Z_s Impedance in Secondary (Transformer) Coil [Symbol]

Z_s Surface Impedance [Symbol]

Z_T Total Impedance [Symbol]

Z_t Section Modulus for Torsion [Symbol]

$Z_{\theta JA(t)}$ Junction-to-Ambient Transient Thermal Impedance [Semiconductor Symbol]

$Z_{\theta JC(t)}$ Junction-to-Case Transient Thermal Impedance [Semiconductor Symbol]

$Z_{\theta(t)}$ Transient Thermal Impedance [Semiconductor Symbol]

Z_{USL} Standard Normal Distribution for Upper Specification Limit (in Statistics) [Symbol]

Z_V Volume Collision Rate (for a Gas) [Symbol]

Z_Z Zener Impedance [Symbol]

z altitude [Symbol]

z axial distance (in cylindrical polar coordinate system) [Symbol]

z cloud thickness [Symbol]

z (molecular) collision rate (for a gas) [Symbol]

z complex number (i.e., $z = a + bj = re^{j\varphi}$) [Symbol]

z dependent variable (e.g., $z = f(x,y)$) [Symbol]

z displacement (in z-direction) [Symbol]

z distance from load axis to axis coinciding with cross-sectional center of gravity of column [Symbol]

z ionic charge [Symbol]

z ionic valence [Symbol]

z number of faradays required for a reaction [Symbol]

z one of the axes in a space rectangular coordinate system [Symbol]

z order of interference [Symbol]

z positive integer (e.g., 0, 1, 2, 3...) [Symbol]

z space coordinate [Symbol]

z specific (normal) acoustic impedance [Symbol]

z system length [Symbol]

z third (or space) rectangular coordinate (of a point) [Symbol]

z unit vector in z-direction [Symbol]

z unknown or variable [Symbol]

z valence [Symbol]

z vertical resolution [Symbol]

z zero

z zeta [English Equivalent]

z zone

z "z" transform operator [Symbol]

.z packed file [UNIX File Name Extension in Pack/Unpack Program]

z^* complex conjugate $z = a - jb$ of $z = a + jb$ [Symbol]

z_i charge on the i-th ion in a solution [Symbol]

z_{if} intermediate-frequency impedance [Semiconductor Symbol]

z_m modulator-frequency load impedance [Semiconductor Symbol]

z_p coherent depth of penetration [Symbol]

z_{rf} radio-frequency impedance [Semiconductor Symbol]

z_v video impedance [Semiconductor Symbol]

z_v volume collision rate [Symbol]

z_z reference impedance, regulator impedance (small signal at I_Z) [Semiconductor Symbol]

z_{zk} reference impedance, regulator impedance (small signal at I_{ZK}) [Semiconductor Symbol]

z_{zm} reference impedance, regulator impedance (small signal at I_{ZM}) [Semiconductor Symbol]

ζ coefficient [Symbol]

ζ coordinate [Symbol]

ζ cutting angle (of milled bevel gears) [Symbol]

ζ damping ratio (or factor) [Symbol]

ζ electrokinetic potential (or zeta potential) [Symbol]

ζ extinction coefficient [Symbol]

ζ (local) Fermi energy [Symbol]

ζ Gibbs function (or zeta function) [Symbol]

ζ (thermal) roughness exponent [Symbol]

ζ stress intensity factor [Symbol]

ζ (lower-case) zeta [Greek Alphabet]

ζ zeta phase (of a material) [Symbol]

Z39.50 Z39.50 Interchange Protocol [Data Communications]

ZA South Africa [ISO Code]

ZA Zenith Angle [also za] [Astronomy]

ZA Zenith Attraction

ZA Zeranin Alloy

ZA Zinc Aluminate

ZA Zero Absolute

ZA Zero Access

ZA Zero Adjuster

ZA Zero and Add

ZA Zinc-Aluminum (Foundry Alloys)

ZA Zirconia-Alumina

ZA Zirconium Aluminate

ZA Zone Axis [Crystallography]

Z-A Z-Axis

ZA- Albania [Civil Aircraft Marking]

Z/a² Charge Density [Symbol]

.za South Africa [Country Code/Domain Name]

ZAA Zeeman Atomic Absorption [Physics]

ZAAS Zeeman Atomic Absorption Spectroscopy

ZABF Zone-Axis Brightfield (Image) [Microscopy]

ZABF TEM Zone-Axis Brightfield Transmission Electron Microscope (Image) [also ZABF-TEM]

ZAC Zinc Alkyl Catalyst

ZAC Zinc Aluminum Coater

ZAC Zinc-Aluminum Coaters (Association) [US]

Zac Zacatecas [Mexico]

ZADCA Zinc Alloy Die Casters Association [UK]

ZADI Zentralstelle für Agrardokumentation und –information [Agricultural Documentation and Information Center, Germany]

ZADP Zone Axis Diffraction Pattern [Crystallography]

ZAED Zentralstelle für Atomenergiedokumentation [Nuclear Energy Documentation Center, Germany]

ZAF Atomic Number, Absorption and Fluorescence (Correction) [Physics]

ZAHD Zerewitinoff Active Hydrogen Determination [Chemistry]

ZAI Zaire [Currency of Zaire]

ZAI Zero-Access Instruction

ZAI Zero-Address Instruction

ZAK Zero Administration Kit

ZAL Long-Fiber-Reinforced ZAM (Zinc-Aluminum Matrix Alloy)

ZAL Zirconium-Arc Lamp

ZALF Zentrum für Agrarlandschafts– und Landnutzungsforschung e.V. [Agricultural Land and Land- Use Research Center, Muncheberg, Germany]

Z Allg Wiss theor Zeitschrift für Allgemeine Wissenschaftstheorie [German Journal of General Theoretical Sciences]

ZAM Zinc-Aluminum Matrix Alloy

ZAMEFA Metal Fabricators of Zambia

ZAMM Zeitschrift für Angewandte Mathematik und Mechanik [German Journal of Applied Mathematics and Mechanics]

ZAMP Zeitschrift für Angwandte Mathematik und Physics [Journal of Applied Mathematics and Physics, Switzerland]

ZAMS Zero Age Main Sequence

Zan Zanzibar

Z Anal Chem Zeitschrift für Analytische Chemie [German Journal of Analytical Chemistry]

Z Angew Math Mech Zeitschrift für Angewandte Mathematik und Mechanik [German Journal of Applied Mathematics and Mechanics]

Z Angew Math Phys Zeitschrift für Angewandte Mathematik und Physik [Swiss Journal of Applied Mathematics and Physics]

Z Angew Phys Zeitschrift für Angewandte Physik [German Journal of Applied Physics]

Z Anorg Allg Chem Zeitschrift für Anorganische und Allgemeine Chemie [German Journal of Inorganic and General Chemistry]

Z Anorg Chem Zeitschrift für Anorganische Chemie [German Journal of Inorganic Chemistry]

Zanz Zanzibar

ZAP Republic of South Africa Patent

ZAP Zeitschrift für Angewandte Physik [German Journal of Applied Physics]

ZAP Zinc Ammonium Phosphate

ZAP Zone Axis Pattern [Crystallography]

ZAPM Zone Axis Pattern Map [Crystallography]

ZAPP Zero Assignment Parallel Processor

ZAPS Zonal Air Pollution System

ZAR Southern African Rand [Currency of South Africa and Namibia]

ZAR Zeus Acquisition Radar

ZARC (Symposium on) Zirconium Alloys for Reactor Components

ZARM Zentrum für Angewandte Raumfahrttechnologie und Mikrogravitation [Center for Applied Space Technology and Microgravitation at University of Bremen, Germany]

Zarub Elektron Tekh Zarubezhnaya Elektronnaya Tekhnika [Russian Journal on Electrical Engineering and Electronics]

ZAS Zero-Acess Storage

ZAS Zirconia-Alumina-Silica (System)

ZAS Short-Fiber-Reinforced ZAM (Zinc-Aluminum-Matrix Alloy)

Zashch Met Zashchita Metallov [Russian Journal on Protection of Metals]

Zashch Pokrytiya Met Zashchita Pokrytiya na Metallakh [Russian Journal on Protection of Metallic Materials]

ZAT Zirconia-Alumina-Titania (System)

ZAT Zirconium Aluminum Titanate

ZAV Zentralstelle für Arbeitsvermittlung [Central Employment Bureau, Germany]

ZAV Zero Air Voids

Zavod Lab Zavodskaya Laboratoriya [Russian Journal on Industrial Laboratory Technology]

ZAW Zentralausschuß der Deutschen Wirtschaft [Central Committee of the German Industry]

ZAW Zero Administration for (Microsoft) Windows

ZAW Whisker-Reinforced ZAM (Zinc-Aluminum Matrix Alloy)

ZB Zero Balance

ZB Zero-Based

ZB Zinc-Blende [Mineral]

ZB Zinc Borate

ZB Zone Boundary

ZBA Zero Bias Annealed; Zero Bias Annealing

ZBA Zero Bias Anomaly

ZBA/RBA Zero Bias Annealing/Reverse Bias Annealing (Effect)

ZBB Zero-Based (Imaginary) Budget(ing) (Process)

ZBCP Zero-Bias Conductance Peak [Physics]

ZBD Zener Breakdown [Electronics]

ZBDC Zinc Bis(dibenzyldithiocarbamate)

z-BEO Zinc Blende-Type Beryllium Oxide

ZBF Zirconium Barium Fluoride

ZBID Zero Bit Insertion/Deletion

ZBL Deutsche Zentralbibliothek für Landbauwissenschaften [German Central Library for Agricultural Sciences, Bonn]

ZBLA Zirconium Tetrafluoride–Barium Difluoride–Lanthanum Trifluoride–Aluminum Trifluoride (Glass) [ZrF_4–BaF_2–LaF_3–AlF_3]

ZBLAL Zirconium Tetrafluoride–Barium Difluoride–Lanthanum Trifluoride–Aluminum Trifluoride–Lithium Fluoride (Glass) [ZrF_4–BaF_2–LaF_3–AlF_3–LiF]

ZBLAN Zirconium Tetrafluoride–Barium Difluoride–Lanthanum Trifluoride–Aluminum Trifluoride–Sodium Fluoride (Glass) [ZrF_4–BaF_2–LaF_3–AlF_3–NaF]

ZBLANTh$_2$ Zirconium Tetrafluoride–Barium Difluoride–Lanthanum Trifluoride–Aluminum Trifluoride–Sodium Fluoride (Glass) [ZrF_4–BaF_2–LaF_3–AlF_3–NaF–ThF_4]

ZBM Deutsche Zentralbibliothek für Medizin [German Library of Medicine, Bonn]

ZBR Zone-Bit Recording

Z(Br) p-Bromobenzyloxycarbonyl

Zb Rad Prir-Mat Fak Ser Fiz Zbornik Radova Prirodno-Matematickog Fakulteta, Seija za Fiziku [Yugoslav Journal on Mathematics and Physics]

ZBS Zero Bias Schottky Diode

Zb Ved Pr Vys Sk Tech Kosiciach Zbornik Vedeckych Prace Vysokej Skoly Technickej v Kosiciach [Journal of the Technical University of Kosice, Slovakia]

ZBW Deutsche Zentralbibliothek für Wirtschaftswissenschaften [German Library of Economics, Kiel]

ZBX Zinc Bis(butylxanthate)

ZBX Zinc Butyl Xanthogenate

ZC Z Center

ZC Zener Current [Electronics]

ZC Zenithal Chart [Astronomy]

ZC Zenith Camera

ZC Zeolite Catalyst

ZC Zernicke Contrast

ZC Ziegler Catalyst [Chemistry]

ZC Zinc Chromate

ZC Zinc Cobaltate

ZC Zinc Cuprate

ZC Zirconia-Ceria (System)

ZC Zirconium Catalyst

ZC Zonal Centrifuge [Biology]

ZC Zone Control

ZC Zone Crushing

Z(c) Boundary Profile [Symbol]

Z-CAV Zoned Constant Angular Velocity

ZCG Zinc Chloride Glass

Z-Chloride Carbonochloridic Acid Phenylmethyl Ester [also Z-chloride]

ZCG Zinc Crown Glass

ZCI Zyderm Collagen Implant [Medicine]

Z(2-Cl)-ONSu N-(2-Chlorobenzyloycarbonyloxy)succinimide

ZCO Zinc Cobalt Oxide

ZCO Zinc Copper Oxide

Z-Contrast STEM Z-Contrast Scanning Tunneling Electron Microscope

ZCR Zero Crossing Rate

ZCS Zemansky Chemical System

ZCS Zero Carbon Steel

ZCS Zinc Cadmium Selenide

ZCS Zone Communication Station

ZCT Zero Creep Technique [Metallurgy]

ZCTU Zambia Congress of Trade Unions

ZCTU Zimbabwe Congress of Trade Unions

ZCX Zinc Cellulose Xanthate

ZD Zener Diode

ZD Zenith Distance [Astronomy]

ZD Zeolithic Diffusion

ZD Zero Defects [Industrial Engineering]

ZD Zero Depression

ZD Zero Drift

ZD Zoned Decimal

Z$ Zimbabwe Dollar [Currency]

0D Zero-Dimensional [also 0-D]

ZDA Zero Displacement Adiabatic (Approximation)

ZDA Zinc Development Association [UK]

ZDBC Zinc Dibutyl Dithiocarbamate [also Zinc Bis(N,N-Dibutyldithiocarbamate), or Zinc Di-n-Butyldithiocarbamate)

ZDBP Zinc Dibutyl Dithiophosphate [also Zinc Di-n-Butyldithiophosphate]

ZDBP Zirconium 1-10-Decanediylbis(phosphonate)

ZDC Zeus Defense Center

ZDCTBS Zeus Defense Center Tape and Buffer System

ZDDP Zinc Didecyldithiophosphate

ZDDP Zinc Didodecyl Dithiophosphate

ZDDS Zentrale Deutsche Datenstation [German Central Data Station] [Aerospace]

ZDDS Zero-Dispersion Double Spectrometer

ZDE Zentralstelle für Dokumentation Elektrotechnik [Documentation Center for Electrical Engineering, Germany]

ZDEC Zinc Diethyl Dithiocarbamate [also Zinc Bis(N,N-Diethyldithiocarbamate)

ZDF Zweites Deutsches Fernsehen [Second German Television–A Television Station]

ZDGG Zeitschrift der Deutschen Geologischen Gesellschaft [Journal of the German Geological Society]

ZDH Zentralverband des Deutschen Handwerks [German Trade Federation]

ZDL Zero Delay Lockout [Computers]

ZDMC Zinc Bis(N,N-Dimethyldithiocarbamate); Zinc Dimethyl Dithiocarbamate

Z-DNA Left-Spiraling DNA (Deoxyribonucleic Acid) Section [Genetics]

ZDO Zero-Differential Overlap [Physics]

ZDOA Zero Differential Overlap Approximation [Physics]

ZDP 5-Aminoimidazole-4-Carboxamide-1-β-D-Ribofuranosyl-5'-Diphosphate [Biochemistry]

ZDP Zero Delivery Pressure

ZDP Zinc Dithiophosphate

ZDR Zeus Discrimination Radar

ZDS Zenith Data System

ZDS Zilog Development System

ZDT Zero-Ductility Transition [Metallurgy]

Z Dtsch Geol Ges Zeitschrift der Deutschen Geologischen Gesellschaft [Journal of the German Geological Society]

ZDV Zidovuline [Azidothymidine Based AIDS Drug]

ZDVR Zener Diode Voltage Regulator

ZE Zeeman Effect [Spectroscopy]

ZE Zeeman Energy [Atomic Physics]

ZE Zener Effect [Electronics]

ZE Zero Emission

ZE Zero Energy

ZE Zimm Equation [Chemistry]

ZE Zone Electrophoresis

ZEA Zero-Energy Assembly

ZEB Zenith-Enhanced Backscatter

ZEB Zero-Energy Band [Solid-State Physics]

ZEBRA Zero-Energy Breeder Reactor Assembly [UK]

ZEBS Zentrale Erfassungs– und Bewertungsstelle für Umweltchemikalien [Environmental Chemicals Registration and Evaluation Center, Berlin, Germany]

ZEDC Zinc Ethyldithiocarbamate

ZEEP Zero-Energy Experimental Pile [of Chalk River Nuclear Laboratories, Ontario, Canada]

ZEF Zero Extraction Force

ZEG Zero Economic Growth

ZEG Zero-Energy Growth

ZEH Zinc Bis(2-Ethyl Hexanoate)

Z Eisenb wes Verk tech Zeitschrift für Eisenbahnwesen und Verkehrstechnik [German Journal on Railroad and Traffic Engineering]

Zeiss Inf Zeiss Information [Publication of Carl Zeiss Optische Werke, Oberkochen, Germany]

ZEKE Zero Electron Kinetic Energy

ZEKES Zero Electron Kinetic Energy Spectrometry

ZEL Zeeman Energy Level [Atomic Physics]

Z Elektrochem Zeitschrift für Elektrochemie [German Journal of Electrochemistry]

Zellstoff Papier Zellstoff und Papier [German Publication on Pulp and Paper]

ZEM Zeiss Elrepho Meter

Zem Kalk Gips Zement-Kalk-Gips (International) [German Periodical for the Cement, Lime and and Gypsum Industries]

Zen Zenith Angle [Astronomy]

Z Energiewirtsch Zeitschrift für Energiewirtschaft [German Journal on the Power-Supply Industry]

ZENIT Zentrum für Innovation und Technologie [Innovation and Technology Center in Mülheim/Ruhr, Germany]

ZENITH Zero-Energy Nitrogen-Heated Thermal Reactor [UK]

ZENR Zero-Energy Nuclear Reactor

Zentralbl Geol Paleont Zentralblatt für Geologie und Paleontologie [German Journal of Geology and Paleontology]

Zentralbl Mineral Zentralblatt für Mineralogie [German Journal of Mineralogy]

Zeo Zeolite [Mineral]

Zeo Zeosphere [Chemistry]

ZEP Zymogen Enzyme Precursor [Biochemistry]

ZEPC Zinc Ethyl Phenyl Dithiocarbamate

ZEPP Zeppelin (Antenna) [also Zepp, or zepp]

ZEPR Zero-Energy Power Reactor

ZER Zero-Energy Reactor

ZER Zero-Energy Resonance

ZERLINA Zero-Energy Reactor for Lattice Investigations and Study of New Assemblies

ZERT Zero Reaction Tool

ZES Zero-Energy System

Zesz Nauk AGH, Metal Odlew Zeszyty Naukowe AGH, Metalurgia i Odlewnictwo [Polish Journal on Mining and Metallurgy]

Zesz Nauk Politech Czestochow, Tech Zeszyty Naukowe Politechniki Czestochowskiej, Nauki Techniczne [Polish Journal on Science and Technology]

Zesz Nauk Politech Lodz, Elektr Zeszyty Naukowe Politechniki Lodzkiej, Elektryka [Polish Journal on Electrical Engineering]

Zesz Nauk Politech Lodz, Fiz Zeszyty Naukowe Politechniki Lodzkiej, Fizyka [Polish Journal on Physics]

Zesz Nauk Politech Lodz, Mech Zeszyty Naukowe Politechniki Lodzkiej, Mechanika [Polish Journal on Mechanics]

Zesz Nauk Politech Slask, Chem Zeszyty Naukowe Politechniki Slaskiej, Chemia [Polish Journal on Chemistry]

Zesz Nauk Politech Slask, Hutn Zeszyty Naukowe Politechniki Slaskiej, Hutnictwo [Polish Journal on Metallurgy and Metalworking]

Zesz Nauk Politech Slask, Mech Zeszyty Naukowe Politechniki Slaskiej, Mechanika [Polish Journal on Mechanics]

ZETA Zero-Energy Thermonuclear Apparatus [of Atomic Energy Research Establishment , UK]

ZETR Zero-Energy Thermal Reactor [of Atomic Energy Research Establishment, UK]

ZEUS Zero-Energy Uranium System [of Atomic Energy Research Establishment, UK]

ZEV Zero-Emission Vehicle

ZEVIS Zentrales Verkehrsinformationssystem [Central Traffic Information System, Germany]

ZF Zero Field

ZF Zero Frequency

ZF Zinc Ferrite

ZF Zone Finder

ZF Zone of Fire [also Z/F]

ZFA Zeitschrift für Allgemeinmedizin [German Journal for General Medicine] [also ZfA]

Z-Factor Zeldovich Factor [also Z-factor] [Symbol]

ZfBR Zeitschrift für deutsches und internationales Baurecht [Journal for German and International Building Law]

ZFC Zero Failure Criterion

ZFC Zero-Field Cooled; Zero-Field Cooling [Solid-State Physics]

ZFC-FC Zero-Field Cooled–Field-Cooled; Zero-Field Cooling–Field Cooling [Solid-State Physics]

ZFCW Zero Field Cooling and Warming [Vibrating Sample Magnetometer]

ZFE Zero Field Emission [Electronics]

ZFF Zinc Formalin Fixture

ZFF Zinc Fume Fever [Medicine]

ZfF Zentrale für Funkberatung [Radio Information Center, Germany]

ZFHGSS Zero-Field Half-Integer Giant Shapiro Step [Physics]

ZFI Zentralinstitut für Isotopen– und Strahlenforschung [Institute for Isotope and Radiation Research, Leipzig, Germany]

ZfK Zentralstelle für Korrosionsschutz [Corrosion Control Center, Germany]

Z Flugwiss Weltraumforsch Zeitschrift für Flugwissenschaften und Weltraumforschung [German Journal of Aeronautics and Space Research]

ZFM Zeiss Fluorescence Microscope

ZFMA Zip Fastener Manufacturers Association [UK]

ZFMK Zoologisches Forschungsinstitut und Museum Alexander Koenig [Alexander Koenig Zoological Research Institute and Museum, Bonn, Germany]

ZFO Zinc Ferrite [$ZnFe_2O_4$]

ZFP Zero Flux Plane

ZFRF Zircon Flower Resin Filler

ZFS Zapon Fast Scarlet

ZFS Zero-Field Splitting

ZFS Zinc Fluosilicate

ZFS Zinc Iron Selenide (Semiconductor)

ZfS Zentralstelle für Solartechnik [Solar Technology Center at Hilden, Germany]

ZFW Zentralinstitut für Festkörperphysik und Werkstofforschung [Institute for Solid-State Physics and Materials Research, Germany]

ZFW Zero Fuel Weight

ZG Zero Gravity (or Zero Weight) [also OG]

ZG Zoological Garden [also Zoo, or zoo]

ZGA Zambia Geographical Association

ZGB Ziff-Gulari-Barshad (Model)

ZGE Zero-Gravity Effect

ZGP Zero-Gravity Processing

ZGP Zinc Germanium Diphosphide (Semiconductor)

ZGPS Zero Gradient Proton Synchrotron

ZGS Zero-Gradient Synchrotron [of Argonne National Laboratory, Illinois, US]

ZGS Zirconia Grain Stabilized; Zirconia Grain Stabilization

ZGV Zentrale für Gußverwendung [Casting Applications Center, Germany]

ZH Zener-Hillert (Method) [Metallurgy]

ZH Zonal Harmonics [Mathematics]

ZH Zone Heating

Z(H) Magnetic Field Dependent Impedance [Symbol]

Zh Zhukovsky Number (of Fluids) [Symbol]

ZHA Zone-Hardened Alloy

Zh Anal Khim Zhurnal Analiticheskoi Khimii [Russian Journal of Analytical Chemistry]

Zh Eksp Teor Fiz Zhurnal Eksperimental'noi i Teoreticheskoi Fiziki [Russian Journal of Experimental and Theoretical Physics]

Zh Eksp Teor Fiz Pis'ma Zhurnal Eksperimental'noi i Teoreticheskoi Fiziki, Pis'ma [Russian Journal of Experimental and Theoretical Physics, Letters]

ZhETF Zhurnal Eksperimental'noi i Teoreticheskoi Fiziki [Russian Journal of Experimental and Theoretical Physics]

Zh Fiz Khim Zhurnal Fizicheskoi Khimii [Russian Journal of Physical Chemistry]

Zh Nauchn Prikl Fotogr Kinematogr Zhurnal Nauchnoi i Prikladnoi Fotografii i Kinematografii [Russian Journal of Photography and Kinematography]

Zh Neorg Khim Zhurnal Neorganicheskoi Khimii [Russian Journal of Inorganic Chemistry]

Zh Nevropatol Psikhiat Imeni S.S. Korsak Zhurnal Nevropatologii i Psikhiatrii-Imeni S.S. Korsakova [Russian Journal on Neuropathology and Psychiatry]

Z-HOUR Zero Hour [also Z-Hour or Z-hour]

Zh Obshchei Khim Zhurnal Obshchei Khimii [Russian Chemical Journal]

Zh Prikl Khim Zhurnal Prikladnoi Khimii [Russian Journal of Applied Chemistry]

Zh Prikl Mekh Tekh Fiz Zhurnal Prikladnoi Mekhaniki i Tekhnicheskoi Fiziki [Russian Journal of Applied Mechanics and Technical Physics]

Zh Prikl Spektrosk Zhurnal Prikladnoi Spektroskopii [Russian Journal of Applied Spectroscopy]

Zh Prikl Spektrosk BSSR Zhurnal Prikladnoi Spektroskopii BSSR [Byelorussian Journal of Applied Spectroscopy]

ZHR Zenithal Hourly Rate [Astronomy]

ZHS Zero Hoop Stress

Zh Strukt Khim Zhurnal Strukturnoi Khimii [Russian Journal of Structural Chemistry]

ZHT Zero Heat Transfer

Zh Tekh Fiz Zhurnal Tekhnicheskoi Fiziki [Russian Journal on Technical Physics]

ZhTF (Pis'ma v) Zhurnal Tekhnicheskoi Fiziki [Russian Journal on Technical Physics, Letters]

Zh Vychisl Mat Mat Fiz Zhurnal Vychislitel'noi Matematiki i Matematicheskoi Fiziki [Russian Journal of Computational Mathematics and Mathematical Physics]

ZI Zeeman Interaction [Physics]

ZI Ziegelindustrie International [German Periodical on the International Bricklaying and Brickmaking Industries]

ZI Zinc Institute, Inc. [New York, US]

ZI Zone of the Interior

ZI Zoom In

ZI Zwitter Ion [Chemistry]

ZIA Zone Immunoassay

ZIAC Zentralinstitut für Anorganische Chemie [Institute for Inorganic Chemistry, Berlin, Germany]

ZIB (Konrad) Zuse Institut Berlin [(Konrad) Zuse Institute at Berlin, Germany]

ZIC Zero Integrated Curvature (Grain) [Materials Science]

ZID Zimbabwe Dollar [Currency]

ZIE Zimbabwe Institution of Engineers

ZIF Zero Insertion Force (Socket)

ZIFT Zigote Intra-Fallopian Transfer [Infertility Treatment]

ZIF/ZEF Zero Insertion Force/Zero Extraction Force

ZIL Zigzag In-Line (Package) [Electronics]

Zimb Zimbabwe

Zimb Eng Zimbabwe Engineer [of Zimbabwe Institution of Engineers]

Zinc/Cadmium Res Dig Zinc/Cadmium Research Digest [US Publication]

ZINCOM Zambia Industrial and Commercial Association

Zinc Res Dig Zinc Research Digest [US Publication]

Zinc TPP 5,10,15,20-Tetraphenyl-21H,23H- Porphine Zinc [also zinc TPP]

Zinc Triflate Zinc Trifluoromethanesulfonate [also zinc triflate]

ZINEB Zinc Ethylenebis(dithiocarbamate) [also Zineb or zineb]

ZIP Zero Inventory Purchasing

ZIP Zigzag In-Line Package [Electronics]

ZIP Zintl-Ion Phase [Chemistry]

ZIP Zinc Impurity Photodetector

ZIP Zone Improvement Plan (Code) [US] [also Zip]

.ZIP Compressed File [PKWare File Name Extension] [also .zip]

ZIPCD Zip Code [also ZIP CD, ZIP Cd or Zip Cd]

Zir Zircon [Mineral]

ZIRAM Zinc Bis(N,N-Dimethyldithiocarbamate [also Ziram, or ziram]

ZIS Zentralinstitut für Schweißtechnik [German Welding Institute]

ZISCH Zirkulation und Schadstoffeintrag in der Nordsee [Circulation and Pollution of the North Sea (Project), Germany]

Zisin, J Seismol Soc Jpn Zisin, Journal of the Seismological Society of Japan

ZIS Mitt ZIS (Zentralinstitut für Schweißtechnik) Mitteilungen [Communications of the German Welding Institute]

ZIS Rep ZIS (Zentralinstitut für Schweißtechnik) Report [Report of the German Welding Institute]

ZIT Zhengzhou Institute of Technology [Zhengzhou, PR China]

ZIX Zinc Isopropylxanthate

ZIX Zinc Isopropyl Xanthogenate

ZK- New Zealand [Civil Aircraft Marking]

ZKG Zement-Kalk-Gips (International) [German Periodical for the Cement, Lime and and Gypsum Industries]

ZKIEM Zeo-Karb Ion Exchange Membrane [Chemical Engineering]

Z Kristallogr Zeitschrift für Kristallographie [German Journal of Crystallography]

Z Kristallogr Kristallgeom Kristallphys Kristallchem Zeitschrift für Kristallographie, Kristallgeometrie, Kristallphysik und Kristallchemie [German Journal of Crystallography, Crystal Geometry, Crystal Physics and Crystal Chemistry]

Zkw Zambian Kwacha [Currency of Zambia]

ZkW Zero Kilowatt

ZL Zero Level [Acoustics]

ZL Zero Lift [Aeronautics]

ZL Zero Line

ZL Zeroth Law (of Thermodynamics)

ZL Zhurkow's Law

ZL Zipf's Law

ZL Zip Length

ZL Zirconium Lamp

ZL Zodiac(al) Light [Geophysics]

ZL Zone Law [Crystallography]

ZL Zone Leveling [Electronics]

ZL Zoom Lens

ZL- New Zealand [Civil Aircraft Marking]

Zl Zloty [Currency of Poland]

$\bar{z}(\lambda)$ tristimulus value for spectral component "z" in CIE System [Symbol]

ZLA Zero-Lift Angle [Aeronautics]

ZLA Zymate Laboratory Automation

Z-L-Arg-MCA N(α)-Carbobenzoxy-L-Arginine-4-Methylcoumaryl-7-Amide [Biochemistry]

Z-Laser Zone Laser [also Z-laser]

ZLC Zinc, Lead and Cadmium Abstracts [of Zinc Development Association, UK]

ZLF Zirconium Lanthanum Fluoride

ZLM Zero-Lift Moment [Aeronautics]

ZLO Zloty [Currency of Poland]

Z Logistik Zeitschrift für Logistics [German Journal of Logistics published by Deutsche Gesellschaft für Logistik]

ZM Zambia [ISO Code]

ZM Zanstra Method [Physics]

ZM Zeisel Method [Chemistry]

ZM Zeldovitch Mechanism [Physics]

ZM Zerewitinoff Method [Chemistry]

ZM Zinc Manganate

ZM Z (Impedance) Meter

ZM Z-Marker; Zone Marker [Aeronautics]

ZM Zone Melted; Zone Melting [Chemical Engineering]

ZM Zone Meridian [Astronomy]

ZM Zwitterion Mechanism [Chemistry]

.zm Zambia [Country Code/Domain Name]

ZMA Zinc Manganese Arsenide (Semiconductor)

ZMA Zinc Metaarsenite

ZMA Zirconia–Magnesia–Yttria (Ceramics) [ZrO_2–MgO–Y_2O_3]

ZMAR Zeus Multifunction Array Radar

Z-Marker Zero Marker (Beacon) [also Z-marker] [Navigation]

ZMAR/MAR Zeus Multifunction Array Radar/Multifunction Array Radar

Z Math Log Grundl Math Zeitschrift für Mathematische Logik und Grundlagen der Mathematik [German Journal of Mathematical Logic and Fundamental Mathematics]

Z Mat kd Zeitschrift für Materialkunde [German Journal of Materials Science; published by Deutsche Gesellschaft für Materialkunde]

ZMB Zero Moisture Basis

ZMBI Zinc-2-Mercaptobenzimidazole

ZMBT Zinc-2-Mercaptobenzothiazole

Z5MC Zinc Pentamethylene Dithiocarbamate

ZMCM Zimm Main Chain Motion [Chemistry]

ZMD Zeisel Methoxy Determination [Chemistry]

Z Metallkd Zeitschrift für Metallkunde [German Journal of Metallurgy; published by Deutsche Gesellschaft für Metallkunde] [now Zeitschrift für Materialkunde]

Z Meteorol Zeitschrift für Meteorologie [German Journal of Meteorology]

Z Met kd Zeitschrift für Metallkunde [German Journal of Metallurgy; published by Deutsche Gesellschaft für Metallkunde] [now Zeitschrift für Materialkunde]

ZMI Zehnder-Mach Interferometer

ZMK Zambian Kwacha [Currency of Zambia]

ZMMBI Zinc-Methyl-2-Mercaptobenzimidazole

ZMO Zinc Manganese Oxide

ZMPC (International Symposium on) Zeolites and Microporous Crystals

ZMR Zone-Melting Recrystallization (Process)

ZMS Zinc Manganese Sulfide (Semiconductor)

ZMSe Zinc Manganese Selenide (Semiconductor)

ZMT Zentralvereinigung Medizin-Technik e.V. [German Medical Technology Association]

ZMT Zinc Manganese Telluride (Semiconductor) [also ZMTe]

ZN Ziegler-Natta (Process) [also Z-N] [Chemical Engineering]

ZN Ziehl-Neelsen (Stain) [Microbiology]

ZN Zinc Niobate

ZN Zsigmondy Number [Chemistry]

Zn Zinc [Symbol]

Zn Zone [also zn]

Zn^{2+} Zinc Ion [also Zn^{++}] [Symbol]

Zn II Zinc II [Symbol]

Zn-62 Zinc-62 [also ^{65}Zn, or Zn^{62}]

Zn-63 Zinc-63 [also ^{63}Zn, or Zn^{63}]

Zn-64 Zinc-64 [also ^{64}Zn, or Zn^{64}]

Zn-65 Zinc-65 [also ^{65}Zn, or Zn^{65}]

Zn-66 Zinc-66 [also ^{66}Zn, or Zn^{66}]

Zn-67 Zinc-67 [also ^{67}Zn, or Zn^{67}]

Zn-68 Zinc-68 [also ^{68}Zn, or Zn^{68}]

Zn-69 Zinc-69 [also ^{69}Zn, or Zn^{69}]

Zn-70 Zinc-70 [also ^{70}Zn, or Zn^{70}]

Zn-71 Zinc-71 [also ^{71}Zn, or Zn^{71}]

Zn-72 Zinc-72 [also ^{72}Zn, or Zn^{72}]

ZnAA Zinc Acetylacetonate

$Zn(Ac)_2$ Zinc(II) Acetate

$Zn(ACAC)_2$ Zinc(II) Acetylacetonate [also $Zn(AcAc)_2$, or $Zn(acac)_2$]

Zn-Ag Zinc-Silver (Alloy System)

Zn-Al Zinc-Aluminum (Alloy System)

Z Nat forsch Zeitschrift für Naturforschung [German Journal of Scientific Research] [also Z Naturf]

Z Nat forsch A Zeitschrift für Naturforschung A [German Journal of Scientific Research A] [also Z Naturf A]

Z Natforsch B Zeitschrift für Naturforschung B [German Journal of Scientific Research B] [also Z Naturf B]

Z Nat heilk Zeitschrift für Naturheilkunde [Journal of Naturopathy; published in Switzerland] [also Z Naturheilk]

ZNC Ziegler-Natta Catalyst [Chemistry]

Zn-Cd Zinc-Cadmium (Alloy System)

ZnCdS Zinc Cadmium Sulfide (Phosphor)

ZnCdSe Zinc Cadmium Selenide (Semiconductor)

ZnCdSe/InP Zinc Cadmium Selenide on Indium Phosphide

Zn_3Cit_2 Zinc Citrate

$ZnCO_3$:Fe Iron-Doped Zinc Carbonate

ZnCoSe Zinc Cobalt Selenide (Semiconductor)

Zn-Cu Zinc-Copper (Alloy System)

ZnDTP Zinc Dialkyl Dithiophosphate

ZNE Ziegler-Nichols Experiment [Chemistry]

ZNF Zirconium Sodium Fluoride

Zn-Fe Zinc-Iron (Alloy System)

ZnFeS Zinc Iron Sulfide (Semiconductor)

ZnFeSe Zinc Iron Selenide (Semiconductor)

ZnFeTe Zinc Iron Telluride (Semiconductor)

Zn/GaAs Zinc/Gallium Arsenide (Electrical Contact)

ZnGeP Zinc Germanium Phosphide (Semiconductor)

Zn(HFAC)$_2$ Zinc Hexafluoroacetylacetonate [also Zn(hfac)$_2$]

ZNI Ziegler-Natta Initiator [Chemistry]

Zn:InGaAs Zinc-Doped Indium Gallium Arsenide (Semiconductor)

ZnL Zinc Ligand [Chemistry]

ZnMgSSe Zinc Magnesium Sulfur Selenide

ZnMnS Zinc Manganese Sulfide (Semiconductor)

ZnMnSe Zinc Manganese Selenide (Semiconductor)

ZnMnTe Zinc Manganese Telluride (Semiconductor)

Zn-Ni Zinc-Nickel (Alloy System)

ZnO Zinc Oxide

Z(NO$_2$) p-Nitrobenzyloxycarbonyl

ZnO:Al Aluminum-Doped Zinc Oxide

ZnO-Al$_2$O$_3$ Zinc Oxide-Aluminum Oxide (System)

ZnO:B Boron-Doped Zinc Oxide

ZnO-Bi$_2$O$_3$ Zinc Oxide-Bismuth Trioxide (System)

ZnOEP 2,3,7,8,12,13,17,18-Octaethylporphine Zinc(II)

ZnO:Fe^{2+} Divalent Iron-Doped Zinc Oxide

ZnO:Ga Gallium-Doped Zinc Oxide

ZnO:Ge Germanium-Doped Zinc Oxide

ZnO:In Indium-Doped Zinc Oxide

ZnO/PPA Zinc Oxide-to-Polyacrylic Acid (for Polyelectrolite Cements]

ZnO/Si Zinc Oxide on Silicon (Substrate)

ZnO:Si Silicon-Doped Zinc Oxide

ZnO-SiO$_2$ Zinc Oxide-Silicon Dioxide (System)

ZnO:Zn Zinc-Doped Zinc Oxide

ZNP Ziegler-Natta Polymer(ization) [Chemistry]

ZNP Zirconium n-Propoxide

ZnPc Zinc(II) Phthalocyanine (Organic Semiconductor)

ZnPCTP Zinc Pentachlorothiophenol

Zn-Pd Zinc-Palladium (Alloy System)

ZNR Zinc-Oxide Nonlinear Resistance

ZnS Zinc Sulfide (Structure)

ZnSb Zinc Antimonide

Zn-Sb-Mg Zinc-Antimony-Magnesium (Alloy System)

ZnS:Co^{2+} Cobalt-Doped Zinc Sulfide

ZnSe Zinc Selenide (Semiconductor)

ZnSe:Cr Chromium-Doped Zinc Selenide

ZnSe/GaAs Zinc Selenide on Gallium Arsenide

ZnSe:K Potassium-Doped Zinc Selenide (Semiconductor)

ZnSe:Li Lithium-Doped Zinc Selenide (Semiconductor)

ZnSe:N Nitrogen-Doped Zinc Selenide (Semiconductor)

ZnSe:Na Sodium-Doped Zinc Selenide (Semiconductor)

ZnSe:P Phosphorus-Doped Zinc Selenide (Semiconductor)

ZnSe/Si Zinc Selenide on Silicon (Substrate)

ZnSe/ZnTe Zinc Selenide/Zinc Telluride (Superlattice)

ZnSe/ZnTe SL Zinc Selenide/Zinc Telluride Superlattice

ZnS/GaN Zinc Sulfide/Gallium Nitride (Heterostructure)

ZnSO Zincosulfate

ZnSSe Zinc Sulfur Selenide (Semiconductor)

ZnSSe:Cl Chlorine-Doped Zinc Sulfur Selenide (Semiconductor)

ZnS/Si Zinc Sulfide on Silicon (Substrate)

ZnS(Mn) Manganese-Doped Zinc Sulfide (Film)

Zn-SPS Sulfonated Polystyrene, Zinc Salt

ZNT Ziegler-Natta Telomerization

ZNTB Zambia National Tourist Bureau

ZnTe Zinc Telluride (Semiconductor)

ZnTe:Cl Chlorine-Doped Zinc Telluride (Semiconductor)

Zn$_x$Te$_y$O$_z$:Er Erbium-Doped Zinc Tellurium Oxide

ZnTe-SiO$_2$ Zinc Telluride-Silicon Dioxide (System)

Znth Zenith [also znth]

Zn(TMHD)$_2$ Zinc Bis(2,2,6,6-Tetramethyl-3,5-Heptanedionate) [also Zn(tmhd)$_2$]

ZnTPP 5,10,15,20-Zinc Tetraphenylporphyrin

ZnTsPP Zinc Tosyl Pyrophosphate

ZnWO Zincotungstate

ZnWO$_4$:Nb Niobium-Doped Zinc Tungstate

ZnWO$_4$:Sb Antimony-Doped Zinc Tungstate

ZO Zero Order

ZO Zinc Oxide

ZO Zirconium Oxide

Z$_2$O (Phenylmethoxycarbonyl)$_2$O

ZOC Zinc Oxide Coating

ZOC Zone of Convergence

ZODIAC Zone Defense Integrated Active Capability

ZOE Zero Energy

ZOE Zinc Oxide-Eugenol (Cement)

ZOE Zone of Exclusion

ZOF Zirconium Oxide Fiber

ZOF Zone of Flow

ZOH Zero Order Hold [Control Systems]

ZOI Zero-Order Interpolator

ZOLZ Zero-Order Laue Zone [Crystallography]

ZOLZ-CBEDP Zero-Order Laue Zone–Convergent-Beam Electron Diffraction Pattern [Crystallography]

ZOM Zone of Mixing

Z(OMe) p-Methoxybenzyloxycarbonyl

Z-ONSu N-(Benzyloxycarbonyloxy)succinimide

Zoo Zoological Garden [also zoo]

.ZOO Compressed File [File Name Extension] [also .zoo]

Zoogeogr Zoogeographical; Zoogeographer; Zoogeography [also zoogeogr]

Zool Zoological; Zoologer; Zoology [also zool]

ZOP Zero-Order Predictor

ZOP Ziegler Oligomerization Process

ZOPA Zinc Oxide Producers Association [of Conseil Européen des Fédérations de l'Industrie Chimique, Belgium]

ZOPFAN Zone of Peace, Freedom and Neutrality (Declaration)

ZOR Zeitschrift für Operations Research [German Journal of Operations Research]

ZOR Zero-Order Reaction [Physical Chemistry]

ZOR Zero-Order Release

ZOR Zone of Radiation

ZOR, Methods Models Oper Res Zeitschrift für Operations Research, Methods and Models of Operations Research [German Journal of Operations Research]

ZOS Zone of Silence [Acoustics]

ZP Zeeman Pattern [Physics]

ZP Zeiss Planetarium [Jena, Germany]

ZP Zener Product [Physics]

ZP Zenith Point [Astronomy]

ZP Zenith Process

ZP Zernicke-Prins (Equation) [Physics]

ZP Zero Page

ZP Zero Point

ZP Zeta Potential [Physics]

ZP Zimm Plot [Chemistry]

ZP Zinc Phosphate

ZP Zisman Plot

ZP Zone Plate [Optics]

ZP Zone Punch

ZP Zone Purification [Metallurgy]

ZP Zulauf Process

ZP Zwitterionic Polymerization

ZP- Paraguay [Civil Aircraft Marking]

ZPA Zero-Page Addressing

ZPA Zero Period Acceleration

ZPA Zeus Program Analysis

ZPC Zinc Phosphate Cement

ZPC Zirconium-Potassium Chloride

ZPCK N-Benzyloxycarbonyl-L-Phenylalanine Chloromethyl Ketone [Biochemistry]

ZPCK N-Carbobenzyloxy-L-Phenylalanyl Chloromethyl Ketone [Biochemistry]

ZPD Zinc Pentamethylenedithiocarbamate

ZPDA Zinc Pigment Development Association [UK]

ZPE Zero-Point Energy [Statistical Mechanics]

ZPE Zero-Point Entropy [Statistical Mechanics]

ZPEN Zeus Project Engineer Network

ZPF Zeitschrift für das Post– und Fernmeldewesen [German Journal for Postal and Telecommunication Engineering]

ZPF Zernicke-Prins Formula [Physics]

ZPF Zirconium-Potassium Fluoride

ZPG Zero Population Growth

Z Phys A Zeitschrift für Physik A [German Journal of Physics]

Z Phys A, At Nuclei Zeitschrift für Physik A, Atomic Nuclei [German Journal of Physics A, Atomic Nuclei]

Z Phys B Zeitschrift für Physik B [German Journal of Physics]

Z Phys C Zeitschrift für Physik C [German Journal of Physics C]

Z Phys Chem Zeitschrift für Physikalische Chemie [German Journal of Physical Chemistry]

Z Phys Chem A Zeitschrift für Physikalische Chemie A [German Journal of Physical Chemistry A]

Z Phys Chem A Zeitschrift für Physikalische Chemie A [German Journal of Physical Chemistry A]

Z Phys Chem B Zeitschrift für Physikalische Chemie B [German Journal of Physical Chemistry B]

Z Phys Chem, Leipz Zeitschrift für Physikalische Chemie, Leipzig [German Journal of Physical Chemistry, Leipzig]

Z Phys Chem, Neue Folge Zeitschrift für Physikalische Chemie, Neue Folge [German Journal of Physical Chemistry, New Series]

Z Phys C, Part Fields Zeitschrift für Physik C, Particles and Fields [German Journal of Physics C, Particles and Fields]

Z Phys D Zeitschrift für Physik D [German Journal of Physics D]

Z Phys D, At Mol Clusters Zeitschrift für Physik D, Atoms, Molecules and Clusters [German Journal of Physics D, Atoms, Molecules and Clusters]

Z Physiol Chem Zeitschrift für Physiologische Chemie [German Journal of Physiological Chemistry]

ZPI Zeiss Planetarium Instrument

ZPI Zone of Priority Investigation

ZPI Zone Position Indicator

ZPID Zentralstelle für Psychologische Information und Dokumentation [Psychological Information and Documentation Center, Trier, Germany]

ZPL Zero Phonon Line

ZPL Zero-Potential Layer

ZPNR Zero-Power Nuclear Reactor

ZPO Zeus Project Office

ZPP Zero-Point Pressure [Statistical Mechanics]

ZPPR Zero Power Physics Reactor [of Argonne National Laboratory-West, Idaho Falls, US]

ZPPR Zero-Power Plutonium Reactor [of US Atomic Energy Commission]

ZPR Zero-Power Reactor

ZPR-I Zero-Power Reactor I [Submarine Thermal Reactor, Mark I, 1950] [also ZPR-1]

ZPR-II Zero-Power Reactor II [Savannah River, 1952] [also ZPR-2]

ZPR-III Zero-Power Reactor III [Fast Assembly, 1955] [also ZPR-3]

ZPR-IV Zero-Power Reactor IV [Neutron Source, 1953] [also ZPR-4]

ZPR-IV' Zero-Power Reactor IV' [Neutron Source, 1957] [also ZPR-4']

ZPR-V Zero-Power Reactor V [Fast Thermal Assembly, 1956] [also ZPR-5]

ZPR-VI Zero-Power Reactor VI [Fast Assembly, 1959] [also ZPR-6]

ZPR-VII Zero-Power Reactor VII [Thorium- Uranium-Deuterium Oxides, 1957] [also ZPR-7]

ZPR-VIII Zero-Power Reactor VIII [Rocket Reactor Criticals, 1963] [also ZPR-8]

ZPR-IX Zero-Power Reactor IX [Fast Source Reactor, 1959] [also ZPR-9]

ZPR(ANL) Zero-Power Reactor at Argonne National Laboratory [Illinois, US]

ZPRF Zero-Power Reactor Facility [of NASA]

ZPR(NASA) Zero-Power Reactor of NASA [US]

ZPS 3-(2-Benzothiazolylthio) 1-Propanesulfonic Acid

ZPT Zero-Power Test

ZPV Zero-Point Vibration [Statistical Mechanics]

ZPV Zoomed Port Video

ZPVE Zero-Point Vibrational Energy [Statistical Mechanics]

ZQC Zero-Quantum Coherence

ZQL Zippered Quilted Liner

ZQS Zero-Quantum Spectroscopy

ZR Maximum Tire Speed of over 149 mph (or 240 km/h) [Rating Symbol]

ZR Zachariasen Rule [Physics]

ZR Zaire [ISO Code]

ZR Zero Reset (Flag)

ZR Zero Resistance

ZR Zhang-Rice (Singlet) [also Z-R] [Spectroscopy]

ZR Zone-Refined; Zone Refining [Metallurgy]

ZR Zone Row

ZR Zoom Ratio

Z-R Zhang-Rice (Singlet) [also ZR] [Spectroscopy]

Zr Zirconium [Symbol]

Zr^{4+} Zirconium Ion [Symbol]

Zr II Zirconium II [Symbol]

Zr-86 Zirconium-86 [also ^{86}Zr, or Zr^{86}]

Zr-87 Zirconium-87 [also ^{87}Zr, or Zr^{87}]

Zr-88 Zirconium-88 [also ^{88}Zr, or Zr^{88}]

Zr-89 Zirconium-89 [also ^{89}Zr, or Zr^{89}]

Zr-90 Zirconium-90 [also ^{90}Zr, or Zr^{90}]

Zr-91 Zirconium-91 [also ^{91}Zr, or Zr^{91}]

Zr-92 Zirconium-92 [also ^{92}Zr, or Zr^{92}]

Zr-93 Zirconium-93 [also ^{93}Zr, or Zr^{93}]

Zr-94 Zirconium-94 [also ^{94}Zr, or Zr^{94}]

Zr-95 Zirconium-95 [also ^{95}Zr, or Zr^{95}]

Zr-96 Zirconium-96 [also ^{96}Zr, or Zr^{96}]

Zr-97 Zirconium-97 [also ^{97}Zr, or Zr^{97}]

ZRA Zeolex Reinforcing Agent

ZRA Zero Range Approximation

ZRA Zero-Resistance Ammeter; Zero-Resistance Ammetry

ZrAA Zirconium Acetylacetonate

Zr(Ac)$_4$ Zirconium(IV) Acetate

Zr(ACAC)$_4$ Zirconium(IV) Acetylacetonate [also Zr(AcAc)$_4$, or Zr(acac)$_4$]

ZRADAS Zero-Resistance Ammeter Data Acquisition System [of Defense Research Establishment Pacific, Canada]

Zr/Al Zirconium/Aluminum (Getter)

Zr-Al Zirconium-Aluminum (Alloy System)

Zr(BuO)$_4$ Zirconium (IV) Butoxide

ZrC Zirconium Carbide (Ceramics)

Zr(Cp)Cl$_3$ Cyclopentadienylzirconium Trichloride

Zr(Cp)$_2$Cl$_2$ Bis(cyclopentadienyl)zirconium Dichloride

Zr(Cp)$_2$ClH Bis(cyclopentadienyl)zirconium Chloride Hydride

Zr-Cu Zirconium-Copper (Alloy System)

Zr(EtO)$_4$ Zirconium (IV) Ethoxide

Zr-Fe Zirconium-Iron (Alloy System)

ZRGG Zhonghua Renmin Gonghe Guo [People's Republic of China]

Zr-H Zirconium-Hydrogen (System)

Zr(HFAC)$_4$ Zirconium Hexafluoroacetylacetonate [also Zr(hfac)$_4$]

ZRIME Zhengzhou Research Institute of Mechanical Engineering [PR China]

ZrK Zirconium K (Absorption Edge) [X-Ray Analysis]

ZRL Zurich Research Laboratory [of IBM Research Division, Rüschlikon, Switzerland]

ZrLβ Zirconium L-Beta (Radiation) [also ZrL$_\beta$]

ZRM Zone Reserved for Memory

ZrN Zirconium Nitride

Zr(NEt$_2$)$_4$ Tetrakis(diethylamido)zirconium

Zr-Nb Zirconium-Niobium (Alloy System)

Zr-Ni Zirconium-Nickel (Alloy System)

c-ZrO$_2$ Cubic Zirconia Polycrystal [also CZP]

m-ZrO$_2$ Monoclinic Zirconia Polycrystal [also MZP]

t-ZrO$_2$ Tetragonal Zirconia Polycrystal [also TZP]

Zr(OAc)$_4$ Zirconium (IV) Acetate

ZrO$_2$-Al$_2$O$_3$ Zirconium Oxide-Aluminum Oxide (System)

ZrO$_2$-Al$_2$O3-SiO$_2$ Zirconium Oxide–Aluminum Oxide–Silicon Dioxide (System)

Zr(OBu)$_4$ Zirconium (IV) Butoxide

Zr(OBu^n)$_4$ Zirconium (IV) n-Butoxide

Zr(OBu^t)$_4$ Zirconium (IV) tert-Butoxide

ZrO$_2$-CaO Zirconium Oxide–Calcium Oxide (System)

ZrO$_2$-CeO$_2$ Zirconium Oxide–Ceric Oxide (System)

Zr(OEt)$_4$ Zirconium (IV) Ethoxide

Zr(OH)$_4$ Zirconium (IV) Hydroxide

ZrO(Me$_3$Ac)$_2$ Zirconyl Trimethylacetate

Zr(OPr^i) Zirconium (IV) Isopropoxide [also ZrOPr^i]

Zr(OPr^n)$_4$ Zirconium (IV) n-Propoxide [also ZrOPr^n]

Zr(OR)$_4$ Zirconium(IV) Alkoxide [General Formula; R represents an Alkyl Group]

ZrOS Zirconyl Sulfide

ZrO$_2$:Si Silicon-Doped Zirconia

ZrO$_2$-SiO$_2$ Zirconium Oxide–Silicon Dioxide (System)

ZrO$_2$-Y$_2$O$_3$ Zirconium Oxide-Yttrium Oxide (System)

ZRP Zinc-Rich Paint

Zr(PMCp)Cl$_3$ Pentamethylcyclopentadienylzirconium Trichloride

Zr(PMCp)$_2$Cl$_2$ Bis(pentamethylcyclopentadienyl)zirconium Dichloride

Zr(Pr^n) Zirconium n-Propoxide

Zr(PrO)$_4$ Zirconium (IV) n-Propoxide

Zr$_3$Rh(H) Hydrogenated Zirconium Rhenium

Zr-Sn Zirconium-Tin (Alloy System)

Zr-TFA Zirconium Trifluoroacetylacetonate [also Zr(tfa)]

Zr(TFAC)$_4$ Zirconium (IV) Trifluoroacetylacetonate [also Zr(tfac)$_4$]

Zr(THD)$_4$ Zirconium(IV) (2,2,6,6-Tetramethyl-3,5-Heptanedionate) [also Zr(thd)$_4$]

Zr-Ti Zirconium-Titanium (Alloy System)

Zr(TMHD)₄ Zirconium Tetrakis(2,2,6,6-Tetramethyl-3,5-Heptanedionate) [also Zr(tmhd)₄]

Zr-V Zirconium-Vanadium (Alloy System)

Zr-W Zirconium-Tungsten (Alloy System)

ZRZ Zaire [Currency of Zaire]

ZS Cycloserine [Microbiology]

ZS Zeise's Salt [Chemistry]

ZS Zero and Subtract

ZS Zero Slope

ZS Zero Sound

ZS Zero Suppression

ZS Zinc Silicate

ZS Zirconia-Silica

ZS Zirconium Silicate

ZS Zonal Soil [Geology]

ZS Zoological Society

ZS- South Africa [Civil Aircraft Marking]

ZSA Zimbabwe Scientific Association

ZSBF Zimm-Stockmayer Branching Factor [Chemistry]

ZSF Zero Skip Frequency

ZSL Zero Slot LAN (Local-Area Network)

ZSL Zero Suppression Logic

ZSL Zoological Society of London [UK]

ZSMV Zero Shear Melt Viscosity [Chemical Engineering]

ZSP Zinc Silicate Primer

ZSSA Zoological Society of Southern Africa

ZST Zero Strength Time

ZST Zwicky Strength Theory

ZSW Zentrum für Sonnenenergie– und Wasserstoff-Forschung [Solar Energy and Hydrogen Research Center, Germany]

ZT Zenith Telescope [Optics]

ZT Zenith Tube

ZT Zero Time

ZT Zimm Theory [Chemistry]

ZT Zinc Telluride

ZT Zone Theory [Color Vision]

ZT Zone Time [Astronomy]

ZT- South Africa [Civil Aircraft Marking]

ZTA Zirconia-Toughened Alumina

ZTC Zirconia-Toughened Ceramic(s)

ZTDI Zentralstelle für Textildokumentation und Information [Textile Documentation and Information Center, Germany]

ZTE Zero Time Exchange

Z Tierpsychol Zeitschrift für Tierpsychologie [German Journal of Animal Psychology]

ZTM Zirconia-Toughened Mullite

ZTO Zero Time Outage

ZTO Zinc Tellurium Oxide

ZTO-Chromate Zinc Tetraoxychromate

ZTP 5-Aminoimidazole-4-Carboxamide-1-β-D-Ribofuranosyl-5'-Triphosphate [Biochemistry]

ZTP Ziegler Tetraethyllead Process [Chemical Engineering]

ZTP Z Toroidal Pinch

Z(θ,ϕ) Normalized Tesseral Harmonic [Symbol]

ZTS Zoom Transfer Scope

ZTS Z-Transition State

ZU Zhejiang University [Hangchow, PR China]

ZU Zhongshan University [Guangzhou, PR China]

ZU- South Africa [Civil Aircraft Marking]

ZUMA Zentrum für Umfragen, Methoden und Analysen e.V. [Association for Surveys, Methods and Analyses, Germany]

ZURF Zeus Up-Range Facility

ZV Zahn Viscometer

ZV Zeitfuchs Viscometer

ZV Zener Voltage [Electronics]

ZV Zwitterionic Viologen [Chemistry]

ZVD Zinc Vapor Deposition

ZVEI Zentralverband Elektrotechnik- und Elektronikindustrie [Electrical and Electronics Manufacturers Association, Germany]

Z Vermess wes Zeitschrift für Vermessungswesen [German Journal of Surveying]

ZVGGD Zentralverband der Genossenschaftlichen Großhandels– und Dienstleistungsunternehmen e.V. [Association of Cooperative Wholesale and Service Businesses, Germany]

ZVS Zentralstelle für die Vergabe von Studienplätzen [Universities Central Admission Council, Germany]

ZVS Zero Voltage Switching

.zw Zimbabwe [Country Code/Domain Name]

ZWD Zimbabwe Dollar [Currency]

Z Werkstofftech Zeitschrift für Werkstofftechnik [German Journal of Materials Engineering]

ZWF Zeitschrift für Wirtschaftliche Fertigung und Automatisierung [Journal of Manufacturing and Automation, Germany]

ZWF CIM Zeitschrift für wirtschaftliche Fertigung und Automatisierung–CIM [Journal of Computer-Integrated Manufacturing and Automation, Germany]

Z Wirtsch Fert Autom Zeitschrift für Wirtschaftliche Fertigung und Automatisierung [Journal of Manufacturing and Automation, Germany]

ZWL Zentralstelle für wissenschaftliche Literatur [Scientific Literature Center, Berlin, Germany]

ZWOK Zirconium-Water Oxidation Kinetics

ZWS Zeolite Water Softening

Z-X Z-X Plane

ZY Zinc Yellow

ZZ Zero Zone [Navigation]

Z-Z Axes (Through any Point) [Symbol]

Appendix

Greek Alphabet*

A α a Alpha	Z ζ z Zeta	Λ λ l Lambda	Π π p Pi	Φ ϕ φ ph Phi
B β b Beta	H η ē Eta	M μ m Mu	P ρ r Rho	X χ ch Chi
Γ γ g Gamma	Θ θ ϑ th Theta	N ν n Nu	Σ σ s Sigma	Ψ ψ ps Psi
Δ δ d Delta	I ι I Iota	Ξ ξ x Xi	T τ t Tau	Ω ω ō Omega
E ϵ ε e Epsilon	K κ k Kappa	O o o Omicron	Y υ u Upsilon	

* The lowercase letter following each set of uppercase and lowercase Greek letters indicates the English equivalent.

Roman Numerals

I = 1	II = 2	III = 3	IV = 4	V = 5	VI = 6	VII = 7	VIII = 8	IX = 9
X = 10	XX = 20	XXX = 30	XL = 40	L = 50	LX = 60	LXX = 70	LXXX = 80	XC = 90
C = 100	CC = 200	CCC = 300	CD = 400	D = 500	DC = 600	DCC = 700	DCCC = 800	CM = 900
M = 1000	MM = 2000							

Examples:	99 = XCIX	990 = CMXC	999 = CMXCIX	1003 = MIII	2001 = MMI

Prefixes for the SI System

Prefix	Symbol	Multiplication Factor	Exponential Expression	Meaning
exa	E	1 000 000 000 000 000 000	10^{18}	one quintillion
peta	P	1 000 000 000 000 000	10^{15}	one quadrillion
tera	T	1 000 000 000 000	10^{12}	one trillion
giga	G	1 000 000 000	10^{9}	one billion
mega	M	1 000 000	10^{6}	one million
kilo	k	1 000	10^{3}	one thousand
hecto*	h	100	10^{2}	one hundred
deka*	da	10	10^{1}	ten
Base Unit		1	10^{0}	one
deci*	d	0.1	10^{-1}	one-tenth
centi*	c	0.01	10^{-2}	one-hundredth
milli	m	0.001	10^{-3}	one-thousandth
micro	μ	0.000 001	10^{-6}	one-millionth
nano	n	0.000 000 0001	10^{-9}	one-billionth
pico	p	0.000 000 000 0001	10^{-12}	one-trillionth
femto	f	0.000 000 000 000 0001	10^{-15}	one quadrillionth
atto	a	0.000 000 000 000 000 001	10^{-18}	one quintillionth

Examples:

teraohm:	$1\ T\Omega = 10^{12}\ \Omega$	kilogram:	$1\ kg = 10^{3}\ g$	milliliter:	$1\ mL = 10^{-3}\ L$
gigajoule:	$1\ GJ = 10^{9}\ J$	kilometer:	$1\ km = 10^{3}\ m$	microhm:	$1\ m\Omega = 10^{-6}\ \Omega$
gigawatt:	$1\ GW = 10^{9}\ W$	hectoliter:	$1\ hL = 10^{2}\ L$	nanosecond:	$1\ ns = 10^{-9}\ s$
megahertz:	$1\ MHz = 10^{6}\ Hz$	decigram:	$1\ dg = 0.1\ g$	picowatt:	$1\ pW = 10^{-12}\ W$
meganewton:	$1MN = 10^{6}\ N$	centimeter:	$1\ cm = 10^{-2}\ m$	femtovolt:	$1\ fV = 10^{-15}\ V$

* Nonpreferred

Base Quantities and Units of the SI System

Physical Quantity	Quantity Symbol	Unit Name	Unit Symbol	Definition
BASE UNITS				
Length	l	meter	m	1,650 763.73 times the wavelength of the orange-red spectral line (in a vacuum) of the krypton-86 (^{86}Kr) isotope.
Mass	m	kilogram	kg	The standard mass of a cylinder made of platinum-iridium alloy and kept by the Bureau International des Poids et Mésures (BIPM) in Sèvres, France.
Time	t	second	s	The duration of 9,192,631,770 periods of the radiation resulting from the cesium-133 (^{133}Cs) isotope shifting between two hyperfine ground-state energy levels.
Electric current	I	ampere	A	The amount of constant current required to produce 2×10^{-7} newtons of force per meter of length (N/m) in two infinitely long parallel conductors of negligible cross section in a vacuum separated from each other by one meter.
Temperature (thermodynamic)	T	kelvin	K	The fraction 1/273.16 of the absolute temperature of the triple point of water.
Luminous intensity	I	candela	cd	The luminous intensity, in perpendicular direction, of a surface of 1/600,000 square meter (m²) of a black body radiator at the temperature of freezing platinum under a pressure of 101,325 newtons per square meter (N/m²).
Amount of substance	n	mole	mol	The amount of pure substance of a system containing the same number of elementary entities as there are atoms in 0.012 kilogram of the nuclide carbon-12 (^{12}C).
SUPPLEMENTARY UNITS				
Plane angle	θ	radian	rad	The measure of the central angle of a circle determined by two radii and an adjoining arc, all one meter in length.
Solid angle	Ω	steradian	sr	The solid angle subtended at the center of a sphere by a portion of the surface of the sphere whose area is equal to the square of the radius.

Derived SI Units with Special Names and Symbols

Physical Quantity	Quantity Symbol	Unit Name	Unit Symbol	Relationship
COHERENT UNITS				
Absorbed dose (radiation)	D	gray	Gy	$1\ Gy = 1\ m^2/s^2$
Acceleration	a	meter per second squared	m/s^2	$1\ m/s^2 = 1\ m/s/s$
Acceleration of free fall	g	meter per second squared	m/s^2	$1\ m/s^2 = 1\ m/s/s$
Activity (of radionuclides)	A	becquerel	Bq	$1\ Bq = 1/s$
Angular acceleration	α	radian per second squared	rad/s^2	$1\ rad/s/s$
Angular momentum	L	Joule-second	Js	$1\ Js = 1\ Nm \cdot 1\ s$
Angular velocity	ω	radian per second	rad/s	$1\ rad/s = 180°/\pi/s$
Area	A	square meter	m^2	$1\ m^2 = 10^4\ cm^2 = 10^6\ mm^2$
Celsius temperature	t, ϑ	degree celsius	°C	$0°C = 273\ K\ (kelvin)$
Capacity	V	liter	L, l, ℓ	$1\ L = 1\ dm^3 = 10^{-3}\ m^3$
Capacity rate	—	watt per kelvin	W/K	$1\ W/K = 1\ J/(s \cdot K)$
Concentration (of amount of substance)	C	mole per cubic meter	mol/m^3	$1\ mol/m^3 = 10^{-3}\ mol/dm^3 = 10^{-3}\ mol/L$
Current density	J	ampere per square meter	A/m^2	$1\ A/m^2 = 10^{-6}\ A/mm^2$
Density; mass density	ρ, d	kilogram per cubic meter	kg/m^3	$1\ kg/m^3 = 10^{-3}\ g/cm^3$
Diffusion coefficient	D	square meter per second	m^2/s	$1\ m^2/s = 10^4\ cm^2/s$
Dose equivalent	—	sievert	Sv	$1\ Sv = 1\ m^2/s^2$
Dynamic viscosity	η	pascal-second	Pa·s	$1\ Pa \cdot s = 1\ N \cdot s/m^2$
Electric capacitance	C	farad	F	$1\ F = 1\ A \cdot s/V = 1\ C/V$
Electric charge; quantity of electricity	Q	coulomb	C	$1\ C = 1\ A \cdot s$ $1\ Ah = 3,600\ A \cdot s$
Electric charge density	D	coulomb per cubic meter	C/m^3	$1\ C/m^3 = 10^{-6}\ C/cm^3$
Electric conductance	G	siemens	S	$1\ S = 1V/A = 1/\Omega$
Electric conductivity	κ	ampere per volt-meter	A/V·m	—
Electric field strength	E	volt per meter	V/m	$1\ V/m = 1\ W/(A \cdot m)$
Electric flux density	D	coulomb per square meter	C/m^2	$1\ C/m^2 = 10^{-4}\ C/cm^2$
Electric resistance	R	ohm	Ω	$1\ \Omega = 1\ V/A$
Electric resistivity	ρ	ohm-meter	$\Omega m, \Omega \cdot m$	—
Electromotive force; electric potential; potential difference	E, $\mathscr{E}$, V	volt	V	$1\ V = 1\ W/A$
Energy	E	joule	J	$1\ J = 1\ Nm = 1\ Ws = 1\ kgm^2/s^2$
Energy density	E	joule per cubic meter	J/m^3	$1\ J/m^3 = 10^{-6}\ J/cm^3$
Entropy	S	joule per kelvin	J/K	—
Force	F	newton	N	$1\ N = 1\ kgm/s^2$
Frequency	f, ν	hertz	Hz	$1\ Hz = 1/s$
Heat flow (rate)	Q	watt	W	$1\ W = 1\ J/s$
Heat flux (density)	q	watt per square meter	W/m^2	$1\ W/m^2 = 10^{-6}\ W/mm^2$
Heat quantity	Q	joule	J	$1\ J = 1\ Nm = 1\ Ws$

Physical Quantity	Quantity Symbol	Unit Name	Unit Symbol	Relationship
Heat transfer coefficient	α	watt per meter squared kelvin	W/(m²·K)	—
Inductance	L	henry	H	1 H = 1 V·s/A
Illumination	I	lux	lx	1 lx = 1 lm/m² = 1 cd · sr/m²
Irradiance	E	watt per square meter	W/m²	1 W/m² = 10^{-6} W/mm²
Kinematic viscosity	ν	square meter per second	m²/s	1 m²/s = 10^{6} mm²/s
Luminance	L	candela per square meter	cd/m²	—
Luminous flux	F	lumen	lm	1 lm = cd · sr
Magnetic field strength; magnetizing force	H	ampere per meter	A/m	—
Magnetic flux	ϕ	weber	Wb	1 Wb = 1 V·s
Magnetic flux density; magnetic induction	B	tesla	T	1 T = 1 Wb/m²
Magnetic permeability	μ	henry per meter	H/m	1 H/m = 1 Vs /(A·m)
Magnetomotive force	$\mathscr{F}$	ampere	A	—
Modulus of elasticity; Young's modulus	E	pascal	Pa	1 Pa = 1 N/m²
Molar heat capacity	C_m	joule per mole kelvin	J/(mol·K)	—
Moment (of force)	M	newton-meter	Nm, N·m	1 Nm = 1 J = 1 Ws
Moment of inertia (of mass)	J	kilogram-meter squared	kg m², kg·m²	—
Momentum	p	newton-second	Ns	1 Ns = 1 kgm/s
Permittivity	ϵ, ε	farad per meter	F/m	1 F/m = 1 A·s/V·m
Plane angle	θ	radian	rad	1 rad = 1 m/m = 180°/π
Power	P	watt	W	1 W = 1 J/s = 1 Nm/s = 1 VA
Pressure	p	pascal, newton per meter squared	Pa N/m²	1 Pa = 1 N/m²
Radiance	L	watt per square meter steradian	W/m² · sr	—
Radiant flux	Φ, Φ_e	watt	W	1 W = 1 J/s
Radiant intensity	I, I_e	watt per steradian	W/sr	—
Modulus of rigidity; shear modulus	G	pascal	Pa	1 Pa = 1N/m²
Solid angle	Ω	steradian	sr	1 sr = 1 m²/m²
Specific energy	u	joule per kilogram	J/kg	—
Specific entropy	s	joule per kilogram kelvin	J/(kg·K)	—
Specific heat capacity	c	joule per kilogram kelvin	J/(kg·K)	—
Specific volume	v, v	cubic meter per kilogram	m³/kg	m³/kg = 10^{6} mm³/g
Speed of rotation; number of revolutions	n	one per second	1/s	1/s = 60/min
Stress	σ, τ	pascal	Pa	1 Pa = 1 N/m²
Surface tension	σ	newton per meter, joule per square meter	N/m J/m²	1 N/m = 1 kg/s² 1 J/m² = 1 kg/s²
Thermal conductivity	λ	watt per meter kelvin	W/(m·K)	1 W/(m·K) = 1 kgm/(s³·K)

Physical Quantity	Quantity Symbol	Unit Name	Unit Symbol	Relationship
Thermal diffusivity	a	square meter per second	m^2/s	—
Torque	T	newton-meter	Nm, N·m	1 Nm = 1 J = 1 Ws
Velocity	v, v	meter per second	m/s	1 m/s = 3.6 km/h
Volume	V	cubic meter	m^3	$1\ m^3 = 1000\ dm^3 = 1000\ L$
Volumetric flow rate	q_v	cubic meter per second	m^3/s	$1\ m^3/s = 1000\ L/s$
Volumetric heat release rate	—	watt per cubic meter	W/m^3	$1\ W/m^3 = 1\ J/(s·m^3)$
Wavenumber	$\bar{v}$	one per meter	1/m	—
Work	W	joule	J	$1\ J = 1\ Nm = 1\ Ws = 1\ kgm^2/s^2$
NONCOHERENT UNITS				
Time; time interval	t	year	y	1 y ≈ 365.25 d ≈ 8,766 h
Time; time interval	t	day	d	1 d = 24 h = 1,440 min
Time; time interval	t	hour	h	1 h = 60 min = 3,600 s
Time; time interval	t	minute	min	1 min = 60 s
Plane angle	θ	degree	°	$1° = (\pi/180)·rad = 60'$
Plane angle	θ	minute	'	1' = 1°/60 = 60"
Plane angle	θ	second	"	1" = 1'/60 = 1°/3,600

Selected Quantities and Units of the English System

Physical Quantity	Unit Name	Unit Abbreviation or Symbol*	Equivalent in Metric Units
Acceleration; linear velocity	foot per second squared	ft/s²	3.0480×10^{-1} m/s²
	inch per second squared	in/s²	2.5400×10^{-2} m/s²
	mile (US statute) per hour	mph	1.609344 km/h
	foot per hour	ft/h	0.3048×10^{-3} m/h
	foot per minute	ft/min	0.3048×10^{-3} m/min
	foot per second	ft/s	0.3048×10^{-3} m/s
	inch per minute	in/min	2.54 cm/min
	inch per second	in/s	2.54 cm/s
Angular velocity	revolution per minute	rpm, rev/min	1.0472×10^{-1} rad/s
	revolution per second	rps, rev/s	6.2832 rad/s
Area	acre	—	4.0469×10^{3} m²
	square mile	mi²	2.5900 km²
	square yard	yd²	8.3613×10^{-1} m²
	square foot	ft²	9.2903×10^{-2} m²
	square inch	in²	6.4516×10^{-4} m²
	circular mil	cmil	5.0671×10^{-10} m²
	square mil	sq mil	6.4516×10^{-10} m²
Bending moment; torque	pound-foot	lbf-ft	1.3558 Nm
	pound-inch	lbf-in	1.1298×10^{-1} Nm
	ounce-inch	ozf-in	7.0616×10^{-3} Nm
Current density	ampere per square inch	A/in²	1.5500×10^{-1} A/cm²
	ampere per square foot	A/ft²	10.7640 A/m²
Electric field strength	volt per mil	V/mil	39.3701 kV/m
Electrical resistivity	ohm-circular mil per foot	Ω-cmil/ft	1.6624×10^{-3} $\mu\Omega \cdot$m
Energy, work	British thermal unit	Btu	1.0545×10^{3} J
	British thermal unit (IT)	Btu$_{IT}$	1.0551×10^{3} J
	British thermal unit (mean)	Btu$_{mean}$	1.0558×10^{3} J
	foot-pound force	ft-lbf	1.355818 J
	foot-poundal	ft-pdl	4.2140×10^{-2} J
Flow rate	gallon (UK liquid) per hour	UK gal/h	7.5768×10^{-2} L/min
	gallon (US liquid) per hour	US gal/h	6.3090×10^{-2} L/min
	gallon (UK liquid) per minute	UK gal/min	4.5461 L/min
	gallon (US liquid) per minute	US gal/min	3.7854 L/min
	cubic foot per hour	ft³/h	4.7195×10^{-1} L/min
	cubic foot per minute	ft³/min	28.31 L/min

Physical Quantity	Unit Name	Unit Abbreviation or Symbol*	Equivalent in Metric Units
Force and force/length	ton-force	tonf	8.8964 kN
	pound-force	lbf	4.4482 N
	ounce-force	ozf	2.7801×10^{-1} N
	kip (1000 lbf)	—	4.4482×10^{3} N
	poundal	pdl	1.3826×10^{-1} N
	pound-force per foot	lbf/ft	14.5939 N/m
	pound-force per inch	lbf/in	1.7513 N/m
	kip per inch	kip/in	1.7513×10^{5} N/m
Fracture toughness	kilopound square root of inch	ksi $\sqrt{\text{in}}$	1.0988 MPa $\sqrt{\text{m}}$
Heat content	British thermal unit per cubic foot	Btu/ft^3	37.2589 kJ/m^3
	British thermal unit per pound	Btu/lb	2.3260 kJ/kg
Heat flow intensity; heat input	British thermal unit per square foot hour	Btu/ft^2· h	3.1546 W/m^2
	joule per inch	J/in	39.3701 J/m
Impact strength	foot-pound force per foot	ft-lbf/ft	4.4482 J/m
	foot-pound force per inch	ft-lbf/in	53.3787 J/m
	foot-pound force per square foot	ft-lbf/ft^2	14.5900 J/m^2
	foot-pound force per square inch	ft-lbf/in^2	2.1020×10^{3} J/m^2
Length	mile (US statute)	mi	1.6093 km
	rod	—	5.0292 m
	yard	yd	9.144×10^{-1} m
	fathom (ocean depth)	—	1.8288 m
	foot	ft, '	3.0480×10^{-1} m
	inch	in, "	2.540×10^{-2} m = 2.540 cm
	microinch	μin	2.5400×10^{-2} μm
	mil	—	25.400 μm
Mass; density	hundredweight (long)	cwt	50.8024 kg
	hundredweight (short)	cwt	45.3592 kg
	ton (long 2240 pounds)	—	1.0160×10^{3} kg
	ton (short 2000 pounds)	—	9.0718×10^{2} kg
	pound	lb	4.5359×10^{-1} kg
	ounce (avoirdupois)	oz, oz av	2.8350×10^{-2} kg
	ounce (troy)	oz t	3.1103×10^{-2} kg
	slug	—	14.5939 kg
	grain	gr	6.4799×10^{-5} kg

Physical Quantity	Unit Name	Unit Abbreviation or Symbol*	Equivalent in Metric Units
Mass per unit length	pound per foot	lb/ft	1.4882 kg/m
	pound per inch	lb/in	17.8580 kg/m
Mass per unit area	pound per square foot	lb/ft^2	4.8824 kg/m^2
	ounce per square yard	oz/yd^2	3.3906×10^{-2} kg/m^2
	ounce per square foot	oz/ft^2	3.0515×10^{-2} kg/m^2
	ounce per square inch	oz/in^2	43.9500 kg/m^2
Mass per unit volume	pound per gallon (UK liquid)	lb/UK gal	99.7763 kg/m^3
	pound per gallon (US liquid)	lb/US gal	119.8264 kg/m^3
	pound per cubic foot	lb/ft^3	16.01846 kg/m^3
	pound per cubic inch	lb/in^3	2.7690×10^4 kg/m^3
	ounce per cubic inch	oz/in^3	1.7300×10^3 kg/m^3
Mass per unit time	pound per hour	lb/h	1.2600×10^{-4} kg/s
	pound per minute	lb/min	7.5599×10^{-3} kg/s
	pound per second	lb/s	4.5359×10^{-1} kg/s
Moment of inertia	pound-square foot	lb-ft^2	4.2140×10^{-2} kg· m^2
	pound-square inch	lb-in^2	2.9264×10^{-4} kg· m^2
Momentum	pound-foot per second	lb-ft/s	1.3826×10^{-1} kg-m/s
	pound-inch per second	lb-in/s	1.1521×10^{-2} kg-m/s
Power	horsepower (550-ft-lbf/s)	hp	7.4570×10^2 W
	horsepower (electric)	hp	7.4600×10^2 W
	horsepower (UK)	hp	7.4570×10^2 W
	British thermal unit per hour	Btu/h	2.9307×10^{-1} W
	British thermal unit per minute	Btu/min	17.585 W
	British thermal unit per second	Btu/s	1.0551×10^3 W
	foot-pound force per hour	ft-lbf/h	3.7662×10^{-4} W
	foot-pound force per minute	ft-lbf/min	2.2597×10^{-2} W
	foot-pound force per second	ft-lbf/s	1.3558 W
Power density	watt per square inch	W/in^2	1.5500×10^3 W/m^2
Pressure	gallon (US liquid) of water (60°F)	US gal H$_2$O	3.782 kg H$_2$O
	pound of water (60°F)	lb H$_2$O	4.54×10^{-4} m^3
	inch of mercury (32°F)	in Hg	3.3864×10^3 Pa
	inch of mercury (60°F)	in Hg	3.3771×10^3
	pound per square foot	psf	4.8824 kg/m^2
	pound per square inch	psi	6.8948×10^3 Pa
	pound per square inch absolute	psia	—
	pound per square inch differential	psid	—
	pound per square inch gage	psig	—
Section modulus	foot-cubed	ft^3	2.8317×10^{-2} m^3
	inch-cubed	in^3	1.6387×10^{-5} m^3
Specific heat	British thermal unit per pound degree fahrenheit	Btu/lb·°F	4.1868×10^3 J/kg·K
Stress	ton-force per square in	tsi	13.7895 MPa
	kilopound per square inch	ksi	6.8948 MPa
	pound per square inch	psi	6.8948×10^{-3} MPa

Physical Quantity	Unit Name	Unit Abbreviation or Symbol*	Equivalent in Metric Units
Temperature	Fahrenheit	°F	5/9 (°F − 32) °C
	Rankine	°R	9/5 (°C + 273.15)
Thermal conductivity	British thermal unit per foot hour degree fahrenheit	Btu/ft·h·°F	1.7307 W/m·K
	British thermal unit inch per hour square foot degree fahrenheit	Btu·in/h·ft²·°F	1.4423×10^{-1} Wm·K
	British thermal unit inch per second square foot degree fahrenheit	Btu·in/s·ft²·°F	5.1922×10^{2} Wm·K
Viscosity	centipoise	cP	0.001 Pa·s
	centistoke	cSt	10^{-6} m²/s
	poise	P	0.1 Pa· s
	stoke	St	0.0001 m²/s
	square foot per hour	ft²/h	9.2903×10^{-2} m²/h
	square foot per second	ft²/s	9.2903×10^{-2} m²/s
Volume; capacity	gallon (UK liquid)	UK gal	4.5461×10^{-3} m³
	gallon (US liquid)	US gal	3.7854×10^{-3} m³
	cubic yard	yd³	7.6455×10^{-1} m³
	cubic foot	ft³	2.8317×10^{-2} m³
	cubic inch	in³	1.6387×10^{-5} m³
	quart (UK liquid)	qrt	1.1365×10^{-3} m³
	quart (US liquid)	qrt	9.4635×10^{-4} m³
	pint (UK liquid)	pt	5.6826×10^{-4} m³
	pint (US liquid)	pt	4.7318×10^{-3} m³
	fluid ounce (UK liquid)	fl oz	2.8413×10^{-5} m³
	fluid ounce (US liquid)	fl oz	2.9574×10^{-5} m³
	barrel (US liquid)	bbl	119 L = 0.119 m³
	barrel (oil)	bbl	158 L = 0.158 m³
Volume per unit time	cubic foot per minute	ft³/min	4.7195×10^{-4} m³/s
	cubic foot per second	ft³/s	2.8317×10^{-2} m³/s
	cubic inch perminute	in³/min	2.7312×10^{-2} m³/s

* For some of the English units there are several possible abbreviations or symbols, these can be found throughout the main section of this publication.

Mathematical Symbols and Signs

Symbol	Explanation	Symbol	Explanation	Symbol	Explanation
...	Baseline ellipses (e.g., a...)	$\ngtr$	Neither greater than, nor equal to (e.g., $a \ngeq b$)	∟	Right angle (i.e., 90° angle)
⋯	Axis ellipses (e.g., $a \cdots b$)	$\nless$	Not less than (e.g., $a \nless b$)	∠	Measured angle
=	Equal to (e.g., $a = b$)	$\nleq$	Neither less than, nor equal to (e.g., $a \nleq b$)	∢	Spherical angle
≡	Equivalent to (e.g., $a \equiv b$)			°	Degrees (e.g., $\alpha = 45°$)
≠	Not equal to (e.g., $a \neq b$)	∞	Infinity (e.g., $x \to \infty$)	′	Minutes (e.g., $\alpha = 45°32′$)
$\not\equiv$	Not equivalent to (e.g., $a \not\equiv b$)	→, ≐	Approaches or tends to (e.g., $x \to a$)	″	Seconds (e.g., $\alpha = 45°32′17″$)
∝, ~	Proportional to, or vary as (e.g., $a \propto b$ or $a \sim b$)	+	Plus (e.g., $a + b$)	∇	Nabla (or del operator) (i.e., vector differential operator whose components in x, y, and z are: directions are: $\nabla_x = \partial/\partial x$; $\nabla_y = \partial/\partial y$, and $\nabla_z = \partial/\partial z$)
~	Similar to (e.g., $\triangle ABC \sim \triangle DEF$)	−	Minus (e.g., $a - b$)	
~	Asymptotically equal to (e.g., $f(x) \sim g(x)$)	±	Plus or minus (e.g., $\pm a$)	
		∓	Minus or plus (e.g., $\mp b$)	
$\not\sim$	Not similar to (e.g., $\triangle ABC \not\sim \triangle DEF$)	×, ·, *	Multiplied by (e.g., $a \times b$, or $a \cdot b$)	∇f, grad f	Gradient of f (e.g., gradient of (scalar function) $f(x_1, x_2, x_3)$ is a vector whose components are: $(\nabla f)_{xi} = \partial f/\partial x_i$, $i = 1, 2, 3$)
$\not\sim$	Not asymptotically equal to (e.g., $f(x) \not\sim g(x)$)	−, /, ÷, :	Divided by (e.g., a/b, $a \div b$, etc.)	
≍	Asymptotically equivalent to	/	Per, or Divided by (e.g., 10 meters per second is: 10 m/s)	$\tilde{N} \times a$, ∇a, div a	Divergence of a (e.g., the divergence of vector function $a(x,y,z)$ [vector field] is: $\nabla \cdot a = \partial a_x/\partial x + \partial a_y/\partial y + \partial a_z/\partial z$)
$\not\asymp$	Not asymptotically equivalent	↑	North(ern direction)	
≈	Approximately equal to (e.g., $a \approx b$)	→	East(ern direction)	$\tilde{N} \times a$, rot a	Curl (or rotation) of a (e.g., curl a curl (or rotation) of vector function $a(x,y,z)$ is the cross product of the del operator and the vector: $(\nabla \times a)_{ik} = \nabla_i a_k - \nabla_k a_i$)
$\not\approx$	Not approximately equal to (e.g., $a \not\approx b$)	↓	South(ern direction)	
		←	West(ern direction)	
≃	Similar or equal to (e.g., $\triangle ABC \simeq \triangle DEF$)	↗	Northeast(ern direction)	
		↙	Southwest(ern direction)	
$\not\simeq$	Not similar or equal to (e.g., $\triangle ABC \not\simeq \triangle DEF$)	↖	Northwest(ern direction)	
		↘	Southeast(ern direction)	Δ, ∇^2, div grad	Laplacian (operator) (i.e., $\Delta = \nabla \cdot \nabla = \nabla^2 = \partial^2/\partial x^2 + \partial^2/\partial y^2 + \partial^2/\partial z^2$)
≅	Congruent to (e.g., $\triangle ABC \cong \triangle DEF$)	∵	Because	
		∴	Therefore	
≅	Asymptotically identical to (e.g., $f \cong g(x)$)	∷	Identical (or equals)	$a \times b$, $a \wedge b$	Cross product (or vector product, or outer product) of a and b (e.g., $(a \times b)_{ik} = a_i b_k - b_i a_k$)
		%	Percent (e.g., 5% = 5/100)	
$\not\cong$	Not congruent to (e.g., $\triangle ABC \not\cong \triangle DEF$)	‰	Per thousand (e.g., 8‰ = 8/1000)	$a \cdot b$, $\vec{a} \cdot \vec{b}$	Dot product (or scalar product, or inner product) of a and b (e.g., $a \cdot b = a_x b_x + a_y b_y = a_z b_z$)
$\not\cong$	Asymptotically identical to (e.g., $f \not\cong g(x)$)	( )	Parentheses (e.g., $(a + b)^2$)	
$\triangleq$	Defined as (e.g., $a \triangleq b$)	[ ]	Brackets (e.g., $A = [x(a + b)^3 + c])$	A, a, $\bar{A}$, $\vec{a}$	Vector A
$\triangleq$	Corresponds to (e.g., $30 \triangleq 15$ kg)	{ }	Braces (e.g., $\{[(3a^2 b + b) - 4ab] + b^2\}$)	AB. ab	Vector from A to B
>	Greater than (e.g., $a > b$)	⟨ ⟩	Angle brackets (e.g., $\langle ab \rangle$)	A_x, A_y, A_z	Coordinates of vector A in x, y and z direction
<	Less than (e.g., $a < b$)	[ ]	Double brackets	$\overline{AB}$	Length of Line from A to B
≥	Greater than or equal to (e.g., $a \geq b$)	∥	Parallel to (e.g., $b \parallel h$)	$\widehat{AB}$	Arc from A to B
≤	Less than or equal to (e.g., $a \leq b$)	$\nparallel$	Not parallel to (e.g., $b \nparallel h$)	sgn (x)	Signum x (e.g., sgn $(x) = 1$ if $x > 0$; sgn (x) 0 if $x = 0$; sgn $(x) = -1$, if $x < 0$)
≫	Much greater than (e.g., $a \gg b$)	⇈	Parallel and equidirectional to (e.g., a ⇈ b)	
≪	Much less than (e.g., $a \ll b$)			
⋙	Much much greater than (e.g., $a \ggg b$)	⇅	Parallel, but not equidirectional to (e.g., a ⇅ b)	$|x|$	Absolute Value of x (e.g., $|6| = 6$, $|-7| = 7$)
		⊥	Perpendicular to (e.g., $x \perp y$)	$x|y$	Divide Symbol (i.e., x divides y)
⋘	Much much less than (e.g., $a \lll b$)	△	Triangle (e.g, $\triangle ABC$)	arg z	Argument of z (e.g., in $z = |z| e^{j\theta}$, angle θ is the argument of complex number z)
		□	Square (e.g., $\square ABCD$)	
$\not>$	Not greater than (e.g., $a \not> b$)	∠	Angle (e.g., $\angle ABC$)	$n!$	n factorial (e.g., $3! = 1 \times 2 \times 3 = 6$)

Symbol	Explanation	Symbol	Explanation	Symbol	Explanation
$\binom{n}{k}$	Binomial coefficient (i.e., $n!/k!(n-k)!)$	$[a,\infty)$, $[a,\to[$	Interval infinite to the right, closed at the left (i.e., open at the right)	sin	Sine (function) (e.g., $\sin\theta = y/r$)
^nC_k, $_nC_k$	Binomial coefficient	$f(x)$, $F(x)$	Function of x	csc	Cosecant (function) (i.e., the reciprocal of the sine function), (e.g., $\csc\theta = 1/\sin\theta = r/y$)
$\left(\dfrac{n}{p}\right)$	Legendre symbol	$g \circ f$	Composite function (e.g., $(g \circ f)(x) = g(f(x)))$	cos	Cosine (function) (e.g., $\cos\theta = x/r$)
$\sum$	Summation (e.g., $\sum\limits_{i=1}^{n} x_i$)	lim	Limit (e.g., $\lim_{x\to a} f(x) = b$, i.e., b is the limit of $f(x)$ as x approaches a)	sec	Secant (function) (i.e., the reciprocal of the cosine (function), e.g., $\sec\theta = 1/\cos\theta = r/x$)
$\prod$	Product (e.g., $\prod\limits_{i=1}^{n} x_i$)	Δ	Delta (or finite difference)		
$\coprod$	Coproduct	Δx, Δy	Increment of x, increment of y	tan	Tangent (function) (e.g., $\tan\theta = y/x$)
$[x]$	Integral part of x	Δf	Delta f (i.e., difference between two functions, e.g., $\Delta f(x) = f(x_1) - f(x_2))$	cot	Cotangent (function) (i.e., the reciprocal of the tangent (function), e.g., $\cot\theta = x/y$)
$\sqrt{x}$	Square root of x (e.g., $\sqrt{x} = x^{1/2}$)				
$\sqrt[n]{x}$	nth root of x (e.g., $\sqrt[n]{x} = x^{1/n}$)	$f'(x)$, $f''(x)$,..., $f^{(n)}(x)$	First, second, ..., nth derivative of function $f(x)$	ctn	Cotangent (function) (i.e., the reciprocal of the tangent (function), e.g., $\cot\theta = x/y$)
a', a'', a'''	a prime, a double prime, a triple prime				
a_1, a_2, ...	a sub(script) one; a sub(script) two, etc.	$\dot{x}$	First derivative of x with respect to time	arcsin	Arc sine (function) (i.e., arc-sin x is the inverse sine of x) [also $\sin^{-1}$]
$a.b$	a point b (e.g., 3.1)	$\ddot{x}$	Second derivative of x with respect to time		
a,b	a comma b (e.g., 6,3)	$df(x)$	Differential of function $f(x)$ (e.g., $df(x) = f'(x)\,dx$)	arccsc	Arc cosecant (function) (i.e., arccsc x is the inverse cosecant of x) [also $\csc^{-1}$]
a^x	xth Power of a (e.g., a^2, a^3, etc.)				
e	Naperian log base ($e = 2.718281828459045235...$)	$df(x,y)$	Total differential of function $f(x,y)$ (i.e., $df(x,y) = (\partial f/\partial f)\,dx + (\partial f\,\partial y)\,dy$)	arccos	Arc cosine (function) (e.g., arccos x is the inverse cosine of x) [also $\cos^{-1}$]
$\exp x$	Exponential Function of x (i.e., $\exp x = e^x$; $e = 2.718...$)	dx, dy	Differential of x, differential of y, etc.	arcsec	Arc secant (function) (e.g., arcsec x is the inverse secant of x) [also $\sec^{-1}$]
$\log_a$	Logarithm to the base a (i.e., $\log_a x$ is the reciprocal of a^x	dy/dx	First derivative of y with respect to x		
$\log_e$, ln	Naperian (or natural) logarithm (i.e., to the Base e; $e = 2.718...$)	d^2y/dx^2	Second derivative of y with respect to x	arctan	Arc tangent (function) (e.g., arctan x is the inverse tangent of x) [also $\tan^{-1}$]
π	Pi ($\pi = 3.141592653589...$)	d^ny/dx^n	nth derivative of y with respect to x	arccot	Arc cotangent (function) (e.g., arccot x is the reverse cotangent of x) [also $\cot^{-1}$]
i,j	Imaginary part of complex number (e.g., $j^2 = -1$)	∂u	Partial derivative of u		
z^*	Complex conjugate	$\partial u/\partial x$	(First) partial derivative of u with respect to x	arcctn	Arc cotangent (function) (e.g., arcctn x is the inverse cotangent of x) [also ctn^{-1}]
Re, $\Re$	Real part (of complex number)	$\partial^2 u/\partial x^2$	Second partial derivative of u with respect to x		
Im, $\Im$	Imaginary part (of complex number)			sinh	Hyperbolic sine (i.e., $\sinh x = \frac{1}{2}(e^x - e^{-x})$)
$(\mathbf{a}_{ik})$	Matrix a ik	$\int$	Integral of (e.g., $\int f(x)\,dx = F(x) + C$)		
$|\mathbf{a}_{ik}|$	Determinant of the matrix a ik	$\int_b^a$	(Definite) Integral of... between limits a and b (e.g., $\int_b^a f(x)\,dx$	csch	Hyperbolic cosecant (function) (i.e., csch $x = 1/\sinh x$)
$\|\mathbf{a}_{ik}\|$	Determinant of the matrix a ik				
det $(\mathbf{a}_{ik})$	Determinant of the matrix a ik	$\iint$	Double integral	cosh	Hyperbolic cosine (function) (i.e., $\cosh x = \frac{1}{2}(e^x + e^{-x})$)
(a,b)	Open interval (i.e., (a,b) means: $a < x < b$)	$\iiint$	Triple integral	sech	Hyperbolic secant (function) (i.e., sech $x = 1/\cosh x$)
$[a,b)$, $[a,b[$	Half-open interval (i.e. open at the right)	$\oint$	Contour integral (or line integral)	tanh	Hyperbolic tangent (function) (i.e., $\sinh x/\cosh x$)
$[a,b]$, $\langle a,b\rangle$	Closed interval (i.e., $[a,b]$ or $\langle a,b\rangle$ mean: $a \le x \le b$)	mod	Modulo (e.g., $x \equiv y \bmod p$ means: x congruent to y modulo p)	coth	Hyperbolic cotangent (function) (i.e., $\coth x = \cosh x/\sinh x$)

Symbol	Explanation	Symbol	Explanation	Symbol	Explanation
$\sinh^{-1}$	Inverse hyperbolic sine (i.e., $\sinh^{-1} x = 1/\sinh x$) [also arsinh]	$\subseteq$	Reflex subset of (or contained in or equals) (e.g., $A \subseteq X$)	$\mathbb{C}$	Complex number system (i.e., $z = a + bj$; z is the complex number, a and b are the respective real and imaginary parts of z; a and b are real numbers, j is an imaginary unit)
$\cosh^{-1}$	Inverse hyperbolic cosine (i.e., $\cosh^{-1} x = 1/\cosh x$) [also arcosh]	$\nsubseteq$	Not reflex subset of (or not contained in or equals) (e.g., $A \nsubseteq Y$)		
$\tanh^{-1}$	Inverse hyperbolic tangent (i.e., $\tanh^{-1} x = 1/\tanh x$) [also artanh]	$\subsetneq$	Subset of ... but does not equal		
		$\Subset$	Double subset of	$\mathbb{R}$	Real number system (i.e., integer, rational and irrational numbers)
$\coth^{-1}$	Inverse hyperbolic cotangent (i.e., $\coth^{-1} x = 1/\cot x$) [also arcoth]	$\supset$	Proper superset of (or contains) (e.g., B $\supset$ A)		
		$\not\supset$	Not superset of (or does not contain) (e.g., A $\not\supset$ B)	$R\{x,y\}$, xRy	Relation (between mem-bers (or elements) x and y)
$\operatorname{sn} x$, $\operatorname{cn} x$, $\operatorname{dn} x$	Jacobian elliptic functions	$\supseteq$	Reflex superset of (or contains or equals) (e.g., $B \supseteq A$)	ω	Order type of the set of positive integers
$\delta(x)$	Dirac delta function	$\nsupseteq$	Not reflex superset of (or does not contain or equal)	c	power of continuum
$\alpha(x)$	Gamma function	$\supsetneq$	Superset of ... but does not equal		
$J_\nu(x)$	Bessel function	$\Supset$	Double Superset of		**MATHEMATICAL LOGIC**
$\wp(x)$	Weierstrass elliptic function	$\ni$	Contains as a member (or element) (e.g., $P \ni Q$)	$\neg, \sim, /$	Negation (e.g., $\neg p$ or $\sim p$ mean: not p)
$\mathscr{L}$, L	Laplace transform (i.e., $L(f) = \mathscr{L} = \int_0^\infty e^{-st} f(t)\, dt$)	$\cap$	Intersection of (e.g., $X \cap Y$)	$\wedge$	Conjunction (e.g., $p \wedge q$ means: p and q)
$L(f)$	Laplace transform of function $f(t)$	$\Cap$	Double intersection of	$\vee$	Disjunction (e.g., $p \vee q$ means: p or q)
$L(f')$	Laplace transform of the first derivative of function f	$\cup$	Union of (e.g., $X \cup Y$)	$\underline{\vee}$	Exclusive or (e.g., $p \underline{\vee} q$ means: p exlusive or q)
$L(f'')$	Laplace transform of the second derivative of function f	$\Cup$	Double union of		
		$\uplus$	Multiset Union of (or U plus)	$\Rightarrow, \rightarrow$	Implication (e.g., $p \Rightarrow q$ or $p \rightarrow q$ mean: p implies q or if p then q)
$L^{-1}(F)$	Inverse Laplace transform of function $f(t)$	$\emptyset, \Lambda, 0,$ $\{\ \}$	Empty set (or null set)		
$\mathscr{F}$, F	Fourier transform (i.e., $F(x) = 1/\sqrt{2\pi} \int_{-\infty}^{+\infty} f(t)\, dt$)	$\circ$	Operator or operation (e.g., $a \circ b = b \circ a$)	$\Leftarrow, \leftarrow$	Implication (e.g., $p \Leftarrow q$ or $p \leftarrow q$ mean: p implied by q or if q then p)
		(x,y,z), $\langle x,y,z\rangle$	Ordered set of members (or elements) x, y, and z (e.g., $(x,y,z) \neq (x,z,y)$)	$\Leftrightarrow, \leftrightarrow, \equiv$	Equivalence (e.g., $p \Leftrightarrow q$, $p \leftrightarrow q$ or $p \equiv q$ mean: p is equivalent to q or p if and only if q)
	SET THEORY; RELATIONS	$\{x,y,z\}$	Unordered set of members (or elements) x, y and z		
X, Y	Set X, Set Y	$\{x_i\}_{i\in I}$	Set whose members (or elements) are x_i, where $i \in I$)	$=$	Identity (e.g., $p = q$ means: p is identical to q)
x, y	Member (or element) x, member (or element) y	$X \times Y$	Cartesian product (i.e., the set of all (x,y) such that $x \in X$ and $y \in Y$)	$\vdash$	Assertion (e.g., $\vdash p$ means: p is a tautology from; $p \vdash q$ means: q follows from p)
$\{x\|P(x)\}$, $\{x{:}P(x)\}$	Set of all elements x having property P				
$\in$	Member (or element) of (e.g., $x \in X$, $y \in Y$)	card X, $\|X\|$	Cardinal of set X	$\dashv$	Mirrored assertion (e.g., $\dashv p$ means: p is a mirrored tautology from; $p \dashv q$ means: p follows from q)
$\notin$	Not a member (or element) of (e.g., $x \notin Y$, $y \notin X$)	$\mathbb{N}$	Natural number system (i.e., 1,2,3,...)		
$\setminus, -$	Difference of (or set minus) (e.g., $x \in (A\setminus B)$ means: $x \in A$ and $x \notin B$)	$\mathbb{N}_0$	Positive integer number system (i.e., 0,1,2,3, ...)	$\succ$	Succession (e.g., $p \succ q$ means: p follows (or succedes or equals q)
$\subset$	Proper subset of (or contained in) (e.g., $A \subset X$)	$\mathbb{I}$	Integer (number) system (i.e.,...,−2,−1,0,1,2, ...)	$\nsucc$	Does not succeed (or follow)
$\not\subset$	Not subset of (or not contained in) (e.g., $A \not\subset Y$)	$\mathbb{I}^-$	Negative integer (number) system (i.e., −1,−2,−3, ...)	$\prec$	Precedence (e.g., $p \prec q$ means: p precedes q)

Symbol	Explanation	Symbol	Explanation	Symbol	Explanation
$\nprec$	Does nor precede	R	Range value (i.e., $R = X_{\max} - X_{\min}$)	X', X^t, ${}^tX$, tX	Transpose of matrix X
$\nsucceq$	Neither succeeds (or follows) nor equals	ρ	(Linear) correlation coefficient (i.e., $-1 \leq \rho \leq 1$)	$\mathrm{Tr}\,X$, $\mathrm{Sp}\,X$	Trace (or spur) of matrix X
$\succeq$	Succession or Identity (e.g., $p \succeq q$ means: p follows (or succeeds) q)			E, e, P	Idempotent
$\preceq$	Precedence or Identity (e.g., $p \preceq q$ means: p precedes or equals q)	**TOPOLOGY**		G/H	Factor group (or quotient group) (i.e., group whose members (or elements) are the co-sets gH of normal subgroup H of a group G)
$\npreceq$	Neither precedes nor equals	E^n	Euclidian n space		
$\exists$, Π	Existential quantifier (i.e., there exists a …, or there is a …)	T_2	Hausdorff space		
		H	Hilbert space	GL(n,R)	(Full) linear group of degree n over field R
$\forall$, Σ	Universal quantifier (i.e., for all …, or for every …)	S^n	n sphere	O(n,R)	(Full) orthogonal group of degree n over field R
$\ni$	Such that …	d(p,q)	distance between points p and q in metric space	SO(n,R), $O^+(n,R)$	Special orthogonal group of degree n over field R
0, 1	Truth, falsity values	FrX, fr$X$	Frontier (or boundary) of X	Hom(P,Q)	Group of all homomor-phisms of P and Q
		dim X	Dimension of X		
STATISTICS AND PROBABILITY		int X	Interior of X	(C,α), (C,P)	Cesàro summability
		$\pi_1(X)$	Fundamental group of space X		
X, Y	Random variables X and Y	$\pi_n(X)$, $H^*(X)$	Homotopy group	C_k, C^k	Class of functions with continuous kth derivative
E(X), $E(X)$	Expectation of X (e.g., E(X) = $\mu = \sum x_i p_i$)	$H_n(X)$, $H_*(X)$	Homology group	C^0, $C(X)$	Space of continuous functions
$E(X\backslash Y)$	Conditional expectation (or expectation of X, if Y is …; e.g., $E(X\backslash Y \leq 7)$ means: expectation of X if $Y \leq 7$)	$H^n(X)$, $H^*(X)$	Cohomotopy group	L^p, L_p	Space of functions with integrable absolute pth power
		cl X, $\bar{X}$, X	Closure of set X	Lip α, Lip$_\alpha$	Lipschitz function
$P(X)$, Pr(X), $\mathcal{P}(X)$	Probability of X (e.g., $P(X \geq 5)$ means: probability that $X \geq 5$)	**OTHER ALGEBRAIC SYMBOLS, NUMERICAL FUNCTIONS, TENSORS AND OPERATORS**		e, e	Neutral (or identity) element (of an algebraic system) (e.g., for every element a of group G: $a \circ e = e \circ a = a$)
$P(X\backslash Y)$, Pr$(X\backslash Y)$, $\mathcal{P}(X\backslash Y)$	Conditional probability (e.g., P of X, if Y is …; $P X < 5\backslash Y > 2.5$) means: probability of $X < 5$, if $Y > 2.5$)	$[X,Y]$	Lie product (or Poisson bracket)	a^{-1}, a^*	Inverse element (of an algebraic system) (e.g., for every element a of group G: $a \circ a^{-1} = a^{-1} \circ a = e$)
		$[P{:}p]$	Dimension of P over p		
e	(Random) Error	$\otimes$	Kronecker product (or tensor product)	$\otimes$	Addition (or plus) operator (e.g., $a \otimes c < b \otimes c$)
$[\varepsilon]$	Error sum	$\oplus$	Direct sum	$\ominus$	Subtraction (or minus operator)
$\bar{X}$	Arithmetic average (i.e., measure of central tendency) (Sample) mean (e.g., = $1/n \sum x_i$)	$\wedge$	Grassmann product (or exterior product)	$\otimes$, $\odot$	Multiplication operator (e.g., $a \odot c < b \odot c$)
σ	Standard deviation (i.e., positive square root of variance σ^2)	Ax, xA	Image of x under transformation A	$\oplus$	Division operator (e.g., $a \oplus c < b \oplus c$)
σ^2, Var, var	Variance (e.g., $\sigma^2 = 1/(n-1) \sum (x_i -)^2$)	δ_{ij}	Kronecker delta (i.e., $\delta_{ii} = 1$, while $\delta_{ij} = 0$, for $i \neq j$)		
		A^*, $\tilde{A}$	Hermitian conjugate (or adjoint) of A		

Symbols and Values for Selected Fundamental Constants*

Name	Symbol	Numerical Value SI Units	Numerical Value CGS Units
Acceleration of gravity (or free fall)	g	9.806 m/s^2	9.806×10^2 cm/s^2
Atomic mass unit (amu)	u	1.6605×10^{-27} kg	1.6605×10^{-24} g
Avogadro constant	N_A	6.0221×10^{23} mol^{-1}	6.0221×10^{23} mol^{-1}
Boltzmann constant	k	1.3807×10^{-23} J/K	1.3807×10^{-16} erg//K
Bohr magneton	μ_B	9.2740×10^{-24} J/T	9.2740×10^{-21} erg/G
Bohr radius	a_0	5.2918×10^{-11} m	5.2918×10^{-9} cm
Electron charge; elementary charge	e	1.6028×10^{-19} C	1.6028×10^{-20} emu
Electron rest mass	m_e	9.1094×10^{-31} kg	9.1094×10^{-23} g
Electron volt	eV	1.6022×10^{-19} J	3.8293×10^{-20} cal
Faraday constant	$\mathscr{F}$, F	9.6485×10^4 C/mol	9.6485×10^3 emu/mol
Fine structure constant	α	7.2974×10^{-3}	7.2974×10^{-3}
Gas constant; molar gas constant	R	8.3145 J/m·K	8.3145×10^7 erg/mol·K
Gravitational constant	G	6.6726×10^{-11} m^3/s^2·kg	6.6726×10^{-3} cm^3/s^2·g
Magnetic flux quantum	Φ_0	2.0673×10^{-15} Wb	2.0673×10^{-7} G·cm^2
Molar volume of ideal gas (at STP)	V_m	22.4141×10^{-3} m^3/mol	22.4141×10^3 cm^3/mol
Neutron rest mass	m_n	1.6749×10^{-27} kg	1.6749×10^{-24} g
Nuclear magneton	μ_N	5.0508×10^{-27} J/T	5.0508×10^{-24} erg/G
Permeability of vacuum	μ_0	1.2566×10^{-6} H/m	Unity (in cgs-emu system)
Permittivity of vacuum	ϵ_0, ε_0	8.8542×10^{-12} F/m	Unity
Planck's constant	h	6.6261×10^{-34} J·s	6.6261×10^{-37} erg·s
Planck's constant ($\hbar = h/2\pi$)	$\hbar$	1.0546×10^{-34} J·s	1.0546×10^{-37} erg·s
Proton rest mass	m_p	1.6726×10^{-27} kg	1.6726×10^{-24} g
Rydberg constant	R_∞	1.0974×10^7 m^{-1}	1.0974×10^5 cm^{-1}
Specific electron charge	e/m_e	1.7588×10^{11} C/kg	1.7588×10^7 emu/g
Stefan-Boltzmann constant	σ	5.6705×10^{-8} W/m^2·K^4	5.6705×10^{-5} erg/s·cm^2·K^4
Velocity (or speed) of light (in vacuum)	c	2.9979×10^8 m/s	2.9979×10^{10} cm/s

*Symbols and values for other fundamental constants can be found throughout the main section of this book.

Symbols and Conversion Factors of Electric and Magnetic Units (SI and CGS-ESU System)

Quantity	Symbol	SI Unit and Symbol	CGS-ESU Unit and Symbol	Conversion
Electric capacitance	C	farad (F)	statfarad (statF)	$1\ F \approx 8.9876 \times 10^{11}$ statF
Charge; quantity of electricity	Q	coulomb (C)	statcoulomb (statC); Franklin (Fr)	$1\ C \approx 2.9979 \times 10^{9}$ statC
Electric conductance	G	siemens (S), mho (℧)	statsiemens (statS), statmho (stat℧)	$1\ S \approx 8.9876 \times 10^{11}$ statS
Electric current	I	ampere (A)	statampere (statA)	$1\ A \approx 2.9979 \times 10^{9}$ statA
Electromotive force; potential difference	$\mathscr{E}$, V	volt (V)	statvolt (statV)	$1\ V \approx 3.3356 \times 10^{-3}$ statV
Inductance	L	henry (H)	stathenry (statH)	$1\ H \approx 1.1126 \times 10^{-12}$ statH
Magnetic flux	ϕ	weber (Wb)	statweber (statWb)	$1\ Wb \approx 3.3356 \times 10^{-3}$ statWb
Magnetic flux density; magnetic induction	B	tesla (T)	stattesla (statT)	$1\ T \approx 3.3356 \times 10^{-7}$ statT
Resistance	R	ohm (Ω)	statohm (statΩ)	$1\ \Omega \approx 1.1126 \times 10^{-12}$ statΩ

Symbols and Conversion Factors of Magnetic Units (SI and CGS-EMU System)

Quantity	Symbol	SI Unit and Symbol	CGS-EMU Unit and Symbol	Conversion
Field strength	H	Ampere-turn per meter (A/m, or A-t/m)	Oersted (Oe)	$1\ A/m = 1.2566 \times 10^{-2}$ Oe
Flux	Φ	Weber (Wb)	Maxwell (Mx); abweber (abWb)	$1\ Wb = 10^{8}$ Mx
Induction	B	Tesla (T)	Gauss (G); abtesla (abt)	$1\ T = 1\ Wb/m^2 = 10^4$ G
Magnetization	M * I **	Ampere-turn per meter (A/m or A-t/m)	Maxwell per square centimeter (Mx/cm²)	$1\ A/m = 10^{-3}\ Mx/cm^2$
Magnetomotive force	$\mathscr{F}$	Ampere (A) or Ampere-turn (A-t)	Gilbert (Gb)	$1\ A = 1\ A\text{-}t = 1.2566$ Gb
Permeability (vacuum)	μ_0	Henry per meter (H/m)	Unitless **	$4\pi \times 10^{-7}$ H/m = 1 emu
Relative pemeability	μ_r * μ' **	Unitless	Unitless	$\mu_r = \mu'$
Susceptibility	χ_m * χ **	Unitless	Unitless	$\chi_m = 4\pi \cdot \chi$

*In SI System   **In CGS-EMU System

Symbols and Conversion Factors of Electric Units (SI and CGS-EMU System)

Quantity	Symbol	SI Unit and Symbol	CGS-EMU Unit and Symbol	Conversion
Capacitance	C	farad (F)	abfarad (aF)	$1 \text{ F} = 10^{-9} \text{ aF}$
Charge, quantity of electricity	Q	coulomb (C)	abcoulomb (aC)	$1 \text{ C} = 0.1 \text{ aC}$
(Space) charge density	ρ	coulomb per cubic meter (C/m³)	abcoulomb per cubic centimeter (aC/cm³)	$1 \text{ C/m}^3 = 10^{-7} \text{ aC/cm}^3$
Conductance	G	siemens (S), mho (℧)	absiemens (aS), abmho (a℧)	$1 \text{ S} = 1 \text{ ℧} = 10^{-9} \text{ aS}$
Current	I	ampere (A)	abampere (aA)	$1 \text{ A} = 0.1 \text{ aA}$
Current density	J	ampere per square meter (A/m²)	abampere per square centimeter (aA/cm²)	$1 \text{ A/m}^2 = 10^{-5} \text{ aA/cm}^2$
Dipole moment	p	coulomb-meter (Cm)	abcoulomb centimeter (aCcm)	$1 \text{ Cm} = 10 \text{ aCcm}$
Electromotive force; potential difference	$\mathscr{E}$, V	volt (V)	abvolt (aV)	$1 \text{ V} = 10^8 \text{ aV}$
Field strength	E	volt per meter (V/m)	abvolt per centimeter (aV/cm)	$1 \text{ V/m} = 10^6 \text{ aV/cm}$
Inductance	L	henry (H)	abhenry (aH)	$1 \text{ H} = 10^{-9} \text{ aH}$
Power	P	watt (W)	abwatt (aW)	$1 \text{ W} = 1 \text{ aW}$
Resistance	R	ohm (Ω)	abohm (aΩ)	$1 \text{ Ω} = 10^9 \text{ aΩ}$
Resistivity	ρ	ohm-meter (Ωm)	abohm-centimeter (aΩcm)	$1 \text{ Ωm} = 10^{11} \text{ aΩcm}$
Polarization, charge density (of surface)	P	coulomb per square meter (C/m²)	abcoulomb per square centimeter (aC/cm²)	$1 \text{ C/m}^2 = 10^{-5} \text{ aC/cm}^2$

Symbols, Atomic Numbers and Atomic Weights for the Chemical Elements*

Element	Symbol	Atomic Number	Atomic Weight	Element	Symbol	Atomic Number	Atomic Weight
Actinium	Ac	89	(227)	Mercury	Hg	80	200.59
Aluminum	Al	13	26.98	Molybdenum	Mo	42	95.94
Americium	Am	95	(243)	Neodymium	Nd	60	144.24
Antimony	Sb	51	121.75	Neon	Ne	10	20.18
Argon	Ar	18	39.95	Neptunium	Np	93	237.05
Arsenic	As	33	74.92	Nickel	Ni	28	58.71
Astatine	At	85	(210)	Niobium	Nb	41	92.91
Barium	Ba	56	137.33	Nitrogen	N	7	14.01
Berkelium	Bk	97	(247)	Nobelium	No	102	(259)
Beryllium	Be	4	9.012	Osmium	Os	76	190.2
Bismuth	Bi	83	208.98	Oxygen	O	8	16.00
Boron	B	5	10.81	Palladium	Pd	46	106.42
Bromine	Br	35	79.90	Phosphorus	P	15	30.97
Cadmium	Cd	48	112.41	Platinum	Pt	78	195.09
Calcium	Ca	20	40.08	Plutonium	Pu	94	(244)
Californium	Cf	98	(251)	Polonium	Po	84	(~210)
Carbon	C	6	12.01	Potassium	K	19	39.10
Cerium	Ce	58	140.12	Praseodymium	Pr	59	140.91
Cesium	Cs	55	132.91	Promethium	Pm	61	(145)
Chlorine	Cl	17	35.45	Protactinium	Pa	91	231.04
Chromium	Cr	24	52.00	Radium	Ra	88	226.03
Cobalt	Co	27	58.93	Radon	Rn	86	(222)
Copper	Cu	29	63.55	Rhenium	Re	75	186.21
Curium	Cm	96	(247)	Rhodium	Rh	45	102.91
Dysprosium	Dy	66	162.50	Rubidium	Rb	37	85.47
Einsteinium	Es	99	(254)	Ruthenium	Ru	44	101.07
Erbium	Er	68	167.26	Samarium	Sm	62	150.36
Europium	Eu	63	151.96	Scandium	Sc	21	44.96
Fermium	Fm	100	(257)	Selenium	Se	34	78.96
Fluorine	F	9	19.00	Silicon	Si	14	28.09
Francium	Fr	87	(223)	Silver	Ag	47	107.87
Gadolinium	Gd	64	157.25	Sodium	Na	11	22.99
Gallium	Ga	31	69.72	Strontium	Sr	38	87.62
Germanium	Ge	32	72.59	Sulfur	S	16	32.06
Gold	Au	79	196.97	Tantalum	Ta	73	180.95
Hafnium	Hf	72	178.49	Technetium	Tc	43	98.91
Helium	He	2	4.003	Tellurium	Te	52	127.60
Holmium	Ho	67	164.93	Terbium	Tb	65	158.93
Hydrogen	H	1	1.008	Thallium	Tl	81	204.38
Indium	In	49	114.82	Thorium	Th	90	232.04
Iodine	I	53	126.90	Thulium	Tm	69	168.93
Iridium	Ir	77	192.22	Tin	Sn	50	118.70
Iron	Fe	26	55.85	Titanium	Ti	22	47.90
Krypton	Kr	36	83.80	Tungsten	W	74	183.85
Lanthanum	La	57	138.91	Uranium	U	92	238.03
Lawrencium	Lr	103	(260)	Vanadium	V	23	50.94
Lead	Pb	82	207.21	Xenon	Xe	54	131.30
Lithium	Li	3	6.941	Ytterbium	Yb	70	173.04
Lutetium	Lu	71	174.97	Yttrium	Y	39	88.91
Magnesium	Mg	12	24.31	Zinc	Zn	30	65.38
Manganese	Mn	25	54.94	Zirconium	Zr	40	91.22
Mendelevium	Md	101	(258)				

*Elements 104 through 110 have been discovered, but lack official names and symbols. The unofficial names and symbols are given below:

Element 104: Unnilquadium (Unq); Kurchatovium (Ku); Rutherfordium (Rf)

Element 105: unnilpentium (Unp); Hahnium (Ha); Nielsbohrium

Element 106: unnilhexium (Unh)

Element 107: unnilseptium (Uns)

Element 108: unniloctium (Uno)

Element 109: unnilennium (Une)

Element 110: ununnilium (Uun)

Periodic Table of the Elements *

* For each element, the atomic number, symbol and name as well as the number of electrons per shell (in italic) have been listed. Elements of Subgroups A and B are indicated by a heavy border

Period	Shells filled	Group I A	Group I B	Group II A	Group II B	Group III B	Group III A	Group IV B	Group IV A	Group V B	Group V A	Group VI B	Group VI A	Group VII B	Group VII A	Group VIII (Group VIIIB)	Group VIII	Group VIII	Group 0 (Group VIIIa)
1	1. (K)	1 H Hydrogen *1*																	2 He Helium *2*
2	2. (L)	3 Li Lithium *1*		4 Be Beryllium *2*			5 B Boron *3*		6 C Carbon *4*		7 N Nitrogen *5*		8 O Oxygen *6*		9 F Fluorine *7*				10 Ne Neon *8*
3	3. (M)	11 Na Sodium *1*		12 Mg Magnesium *2*			13 Al Aluminum *3*		14 Si Silicon *4*		15 P Phosphorus *5*		16 S Sulfur *6*		17 Cl Chlorine *7*				18 Ar Argon *8*
4	3. (M) / 4. (N)	19 K Potassium *1*	29 Cu Copper *10, 1*	20 Ca Calcium *2*	30 Zn Zinc *10, 2*	21 Sc Scandium *1, 2*	31 Ga Gallium *10, 3*	22 Ti Titanium *2, 2*	32 Ge Germanium *10, 4*	23 V Vanadium *3, 2*	33 As Arsenic *10, 5*	24 Cr Chromium *5, 1*	34 Se Selenium *10, 6*	25 Mn Manganese *5, 2*	35 Br Bromine *10, 7*	26 Fe Iron *6, 2*	27 Co Cobalt *7, 2*	28 Ni Nickel *8, 2*	36 Kr Krypton *10, 8*
5	4. (N) / 5. (O)	37 Rb Rubidium *1*	47 Ag Silver *10, 1*	38 Sr Strontium *2*	48 Cd Cadmium *10, 2*	39 Y Yttrium *1, 2*	49 In Indium *10, 3*	40 Zr Zirconium *2, 2*	50 Sn Tin *10, 4*	41 Nb Niobium *4, 1*	51 Sb Antimony *10, 5*	42 Mo Molybdenum *5, 1*	52 Te Tellurium *10, 6*	43 Tc Technetium *6, 1*	53 I Iodine *10, 7*	44 Ru Ruthenium *7, 1*	45 Rh Rhodium *8, 1*	46 Pd Palladium *10*	54 Xe Xenon *10, 8*
6	5. (O) / 6. (P)	55 Cs Cesium *1*	79 Au Gold *10, 1*	56 Ba Barium *2*	80 Hg Mercury *10, 2*	57 La Lanthanum *1, 2*	81 Tl Thallium *10, 3*	72 Hf Hafnium *2, 2*	82 Pb Lead *10, 4*	73 Ta Tantalum *3, 2*	83 Bi Bismuth *10, 5*	74 W Tungsten *4, 2*	84 Po Polonium *10, 6*	75 Re Rhenium *5, 2*	85 At Astatine *10, 7*	76 Os Osmium *6, 2*	77 Ir Iridium *7, 2*	78 Pt Platinum *9, 1*	86 Rn Radon *10, 8*
7	6. (P) / 7. (Q)	87 Fr Francium *1*		88 Ra Radium *2*		89 Ac Actinium *1, 2*		104 Unq Unnilquadium *2, 2*		105 Unp Unnilpentium *3, 2*		106 Unh Unnilhexium *4, 2*		107 Uns Unnilseptium *5, 2*		108 Uno Unniloctium *6, 2*	109 Une Unnilennium *7, 2*	110 Uun Ununnilium	

Lanthanide Series

Element	O	P	N
58 Ce Cerium		*2*	*2*
59 Pr Praseodymium		*2*	*3*
60 Nd Neodymium		*2*	*4*
61 Pm Promethium		*2*	*5*
62 Sm Samarium		*2*	*6*
63 Eu Europium		*2*	*7*
64 Gd Gadolinium	*1*	*2*	*7*
65 Tb Terbium		*2*	*9*
66 Dy Dysprosium		*2*	*10*
67 Ho Holmium		*2*	*11*
68 Er Erbium		*2*	*12*
69 Tm Thulium		*2*	*13*
70 Yb Ytterbium		*2*	*14*
71 Lu Lutetium	*1*	*2*	*14*

Actinide Series

Element	P	Q	O
90 Th Thorium	*2*	*2*	
91 Pa Protactinium	*1*	*2*	*2*
92 U Uranium	*1*	*2*	*3*
93 Np Neptunium	*1*	*2*	*4*
94 Pu Plutonium	*1*	*2*	*5*
95 Am Americium	*1*	*2*	*6*
96 Cm Curium	*1*	*2*	*7*
97 Bk Berkelium	*1*	*2*	*8*
98 Cf Californium	*1*	*2*	*9*
99 Es Einsteinium	*1*	*2*	*10*
100 Fm Fermium	*1*	*2*	*11*
101 Md Mendelevium	*1*	*2*	*12*
102 No Nobelium	*1*	*2*	*13*
103 Lr Lawrencium	*1*	*2*	*14*

Major Elementary Particles*

Type	Group	Particle	Antiparticle	Symbol	Charge	Mass (in MeV)	Mass (m_e)	Mean life (in s)	Spin	Iso-spin	Strange-ness	Decay type
Photon		Photon	(identical)	γ	0	0	0	stable	1	—	0	—
Leptons		Neutrino		$\nu_e,\ \nu_\mu,\ \bar\nu_e,\ \bar\nu_\mu$	0	0	0	stable	1/2	—	—	—
Leptons		Electron	Positron	e^- / e^+	−1 / +1	0.511	1	stable	1/2	—	—	—
Mesons	Muons	μ^- meson	μ^+ meson	μ^- / μ^+	−1 / +1	105.65	206.77	2.212×10^{-6}	1/2	—	—	$\mu^- \to e + \nu + \bar\nu$
Mesons	Pions	π^0 meson	π^0 mesons	π^0	0	135	264.20	1.8×10^{-16}	0	1	0	$\pi^0 \to \gamma + \gamma$
Mesons	Pions	π^+ meson	π^- meson	π^+ / π^-	+1 / −1	139.54	273.1	2.55×10^{-8}	0	1	0	$\pi^+ \to \mu^+ + \nu$
Mesons	Kaons	K^+ mesons	K^- mesons	K^+ / K^-	+1 / −1	494	966.7	1.23×10^{-8}	0	1/2	+1	$K^+ \to \pi^- + \pi^0$
Mesons	Kaons	K^0 mesons	K^0 mesons	K^0 / $\bar K^0$	0	497.8	974.2	$K_1^0\!:\ 1 \times 10^{-10}$ / $K_2^0\!:\ 6 \times 10^{-8}$	0	1/2	+1	$K_1^0 \to \pi^+ + \pi^-$; $K_2^0 \to \pi^+ + \pi^- + \pi^0$
Mesons	η mesons	η^+ mesons	η^- mesons	η^+ / η^-	0	548.6	1074	ca. 10^{-20}	0	0	0	$\eta \to \pi^+ + \pi^- + \pi^0$
Baryons	Nucleons	Proton	Antiproton	p / $\bar p$	+1 / −1	938.256	1836.12	stable	1/2	1/2	0	
Baryons	Nucleons	Neutron	Antineutron	n / $\bar n$	0	939.550	1838.65	1×10^{3}	1/2	1/2	0	$n \to p + e^- + \nu$
Baryons	Hyperons	Λ^0 hyperon	$\bar\Lambda^0$ antihyperon	Λ^0	0	1115.4	2182.8	2.6×10^{-10}	1/2	0	−1	$\Lambda^0 \to p + \pi^-$
Baryons	Hyperons	Σ^+ hyperon	$\bar\Sigma^+$ antihyperon	Σ^+	+1	1189.4	2327.7	0.79×10^{-10}	1/2	1	−1	$\Sigma^+ \to p + \pi^0$
Baryons	Hyperons	Σ^0 hyperon	$\bar\Sigma^0$ antihyperon	Σ^0	0	1192.4	2332	1×10^{-14}	1/2	0	−1	$\Sigma^0 \to \Lambda^0 + \gamma$
Baryons	Hyperons	Σ^- hyperon	$\bar\Sigma^-$ antihyperon	Σ^-	−1	1197	2342	1.58×10^{-10}	1/2	0	−1	$\Sigma^- \to n + \pi^-$
Baryons	Hyperons	Ξ^0 hyperon	$\bar\Xi^0$ antihyperon	Ξ^0	0	1314	2571	3×10^{-10}	1/2	−1	−2	$\Xi^0 \to \Lambda^0 + \pi^0$
Baryons	Hyperons	Ξ^- hyperon	$\bar\Xi^-$ antihyperon	Ξ^-	−1	1321.4	2586	1.8×10^{-10}	1/2	−1/2	−2	$\Xi^- \to \Lambda^0 + \pi^-$
Baryons	Hyperons	Ω^- hyperon	$\bar\Omega^-$ antihyperon	Ω^-	−1	1672	3278	$\approx 10^{-10}$	3/2	−1/2	−3	$\Omega^- \to \Xi^0 + \pi^-$

* Mesons and baryons have strong interactions and are collectively known as hadrons. The collective name for protons and neutrons is nucleons. The neutrinos are: e neutrino (ν_e) and e antineutrino ($\bar\nu_e$) from beta decay (β decay), and μ neutrino (ν_μ) and μ antineutrino ($\bar\nu_\mu$) from mu decay (μ decay). The elementary charge (or electron charge) is 1.6028×10^{-19} C. The unit of the spin of an elementary particle is $\hbar$ (eta). 1 electronvolt (eV) = 1.6022×10^{-19} V; 1 megaelectronvolt (MeV) = 10^6 eV. The electron rest mass, m_e, is the electron rest mass of 9.1094×10^{-31} kg. The hyperons are: lambda hyperon Λ^0 and lambda antihyperon $\bar\Lambda^0$; sigma hyperon (Σ^+, Σ^0, Σ^-) and sigma antihyperon ($\bar\Sigma^+$, $\bar\Sigma^0$, $\bar\Sigma^-$), xi hyperon (Ξ^0, Ξ^-) and xi antihyperon ($\bar\Xi^0$, $\bar\Xi^-$) and omega hyperon Ω^- and omega antihyperon $\bar\Omega^-$.

Elementary Particles: Major Gauge Bosons and Quarks

Group	Name	Symbol	Charge	Mass (in MeV)	Strangeness
Gauge bosons	Photon	γ	0	0	0
	Gluon	g	0	0	—
	Weak Bosons: Charged	W±	±1	8×10^4	
	Neutral	Z^0	0	9×10^4	
Quarks	Up quark	u	$+\frac{2}{3}$	5	0
	Down quark	d	$-\frac{1}{3}$	10	0
	Charmed quark	c	$+\frac{2}{3}$	1.4×10^3	0
	Strange quark	s	$-\frac{1}{3}$	1.5×10^2	−1
	Bottom quark (or b quark)	b	$-\frac{1}{3}$	4.8×10^3	
	Truth quark (or top quark)	t	$+\frac{2}{3}$	$>1.8 \times 10^4$	

Bosons are elementary particles with integral spin (e.g., photons, pions, etc.) that obey Bose-Einstein statistics. Fermions are elementary particles with semi-integral spin (e.g., electrons, protons, neutrons, etc.) that obey Fermi statistics. Gauge bosons, as defined by gauge theory, are elementary particles with a mass of 0 and a spin of 1.

Selected Biological and Botanical Symbols and Signs

◯	Individual cell or organism, specifically a female [General Biology]	|w	Plant useful to wildlife [Botany]
☐	Individual cell or organism, specifically a male [General Biology]	♂	Staminate flower or plant [Botany]
♂	Male cell or organism [General Biology]	♀	Pistillate flower or plant [Botany]
♀	Female cell or organism [General Biology]	×	Hybrid [Genetics]
⚲	Neuter cell or organism [General Biology]	+	Wild type [Genetics]
◯, ⊙, ①	Annual plant [Botany]	F	Filial generation (i.e., offspring of parental generation) [Genetics]
⊙, ⊙, ♂	Biennial plant [Botany]	$F_1, F_2, F_3, ..., F_n$	First, second, third, ... n-th filial generation (i.e., offspring of first, second, third, n-th parental generation) [Genetics]
△	Evergreen plant [Botany]	P	Parental generation [Genetics]
⊙	Monocrapic plant [Biology]		

Selected Chemical Symbols and Signs

CHEMICAL REACTIONS				
$\rightarrow$	Direction of reaction (i.e., activity goes to the right)	$\uparrow$	Gas (written after compound, e.g., $H_2\uparrow$)
$\leftarrow$	Direction of reaction (i.e., activity goes to the left)	$\downarrow$	Precipitate (written after compound, e.g., $BaSO_4\downarrow$)
$\rightleftharpoons$, $\leftrightarrow$	Reversible reaction	Δ	Presence of heat
$+$	Addition (e.g., $2H + O_2$)	$=$, $\rightleftharpoons$	Equivalence of amounts in quantitative equations
CHEMICAL COMPOUNDS				
$\alpha, \beta, \gamma, \ldots$	Position of substituting atoms or groups in organic compounds	$-$	Single bond in structural formulas
$\alpha, \beta, \gamma, \ldots$	Attachment of side chain of ring compounds	$=$	Double bond in structural formulas
$+$	Dextrorotation	$\equiv$	Triple bond in structural formulas
$-$	Levorotation	$R-$	Alkyl radical
[ ]	Radical containing another radical (e.g., $Fe_3[Fe(CN)_6]_2$)	$\bigcirc$	Benzene ring
Ions				
$^-, ^{--}, ^{---}, \ldots$	Single, double, triple... negative charge	$^{+1}, ^{+2}, ^{+3}, \ldots$	Single, double, triple... positive charge
$^{-1}, ^{-2}, ^{-3}, \ldots$	Single, double, triple... negative charge	$', '', ''', \ldots$	Single, double, triple...charge, or valence
$^+, ^{++}, ^{+++}, \ldots$	Single, double, triple... positive charge		

ASCII Character Codes and Symbols*

Character/ Symbol	Decimal Code	Hexadecimal Code	Description	Character/ Symbol	Decimal Code	Hexadecimal Code	Description
	000	00	Null (NUL)	Ç	128	80	Uppercase C cedilla
☺	001	01	Start of heading (SOH)	ü	129	81	Lowercase u umlaut or diaeresis
☻	002	02	Start of text (STX)	é	130	82	Lowercase e acute accent
♥	003	03	End of text (EXTX)	â	131	83	Lowercase a circumflex
♦	004	04	End of transmission (EOT)	ä	132	84	Lowercase a umlaut or diaeresis
♣	005	05	Enquiry (ENQ)	à	133	85	Lowercase a grave accent
♠	006	06	Acknowledge (ACK)	å	134	86	Lowercase a degree
●	007	07	Bell (BEL)	ç	135	87	Lowercase c cedilla
◘	008	08	Backspace (BS)	ê	136	88	Lowercase e circumflex
○	009	09	Horizontal tabulation (HT)	ë	137	89	Lowercase e umlaut or diaeresis
◎	010	0A	Line feed (LF)	è	138	8A	Lowercase e grave accent
♂	011	0B	Vertical tabulation (VT)	Ï	139	8B	Uppercase I umlaut or diaeresis
♀	012	0C	Form feed (FF)	Î	140	8C	Uppercase I circumflex
♪	013	0D	Carriage return (CR)	Ì	141	8D	Uppercase I grave accent
♫	014	0E	Shift out (SO)	Ä	142	8E	Uppercase A umlaut or diaeresis
▯	015	0F	Shift in (SI)	Å	143	8F	Uppercase A degree
►	016	10	Data link escape (DLE)	É	144	90	Uppercase E acute accent
◄	017	11	Device control 1 (DC 1; X-ON)	æ	145	91	Lowercase ae ligature
↕	018	12	Device control 2 (DC 2)	Æ	146	92	Uppercase AE ligature
‼	019	13	Device control 3 (DC-3; X-OFF)	ô	147	93	Lowercase o circumflex
¶	020	14	Device control 4 (DC4)	ö	148	94	Lowercase o umlaut or diaeresis
§	021	15	Negative acknowledge (NAK)	ò	149	95	Lowercase o grave accent
▬	022	16	Synchronous idle (SYN)	û	150	96	Lowercase u circumflex
↨	023	17	End of transmission block (ETB)	ù	151	97	Lowercase u grave accent
↑	024	18	Cancel (CAN)	ÿ	152	98	Lowercase y umlaut or diaeresis
↓	025	19	End of medium (EM)	Ö	153	99	Uppercase O umlaut or diaeresis
→	026	1A	Substitute (SUB)	Ü	154	9A	Uppercase U umlaut or diaeresis
←	027	1B	Escape (ESC)	¢	155	9B	US cent symbol
∟	028	1C	File separator (FS)	£	156	9C	British pound symbol
↔	029	1D	Group separator (GS)	¥	157	9D	Japanese yen symbol
▲	030	1E	Record separator (ES)	₧	158	9E	Pesetas symbol
▼	031	1F	Unit separator (US)	ƒ	159	9F	Dutch guilder symbol
	032	20	Space	à	160	A0	Lowercase a grave accent
!	033	21	Exclamation point	í	161	A1	Lowercase i acute accent
"	034	22	Double quote	ó	162	A2	Lowercase o acute accent
#	035	23	Number sign (pound sign)	ú	163	A3	Lowercase u acute accent
$	036	24	Dollar sign	ñ	164	A4	Lowercase n tilde
%	037	25	Percent sign	Ñ	165	A5	Uppercase N tilde

Character/ Symbol	Decimal Code	Hexadecimal Code	Description	Character/ Symbol	Decimal Code	Hexadecimal Code	Description
&	038	26	Ampersand	a	166	A6	Feminine ordinal indicator
'	039	27	Apostrophe (closing single quote)	o	167	A7	Masculine ordinal indicator
(	040	28	Opening parenthesis	¿	168	A8	Inverted question mark
)	041	29	Closing parenthesis	⌐	169	A9	Beginning of line
*	042	2A	Asterisk	¬	170	AA	End of line
+	043	2B	Plus	½	171	AB	One half
,	044	2C	Comma	¼	172	AC	One forth (one quarter)
-	045	2D	Hyphen (minus)	¡	173	AD	Inverted exclamation point
.	046	2E	Period (point)	«	174	AE	Opening double guillemet
/	047	2F	Slant (solidus); forward slash	»	175	AF	Closing double guillemet
0	048	30	Zero	░	176	B0	Box [shade 1]
1	049	31	One	▒	177	B1	Box [shade 2]
2	050	32	Two	▓	178	B2	Box [shade 3]
3	051	33	Three	│	179	B3	Box [top bottom]
4	052	34	Four	┤	180	B4	Box [left top bottom]
5	053	35	Five	╡	181	B5	Box [left top bottom]
6	054	36	Six	╢	182	B6	Box [left top bottom]
7	055	37	Seven	╖	183	B7	Box [left bottom]
8	056	38	Eight	╕	184	B8	Box [left bottom]
9	057	39	Nine	╣	185	B9	Box [left top bottom]
:	058	3A	Colon	║	186	BA	Box [top bottom]
;	059	3B	Semicolon	╗	187	BB	Box [left bottom]
<	060	3C	Less than sign	╝	188	BC	Box [left top]
=	061	3D	Equal sign	╜	189	BD	Box [left top]
>	062	3E	Greater than sign	╛	190	BE	Box [left top]
?	063	3F	Question mark	┐	191	BF	Box [left bottom]
@	064	40	Commercial at	└	192	C0	Box [top right]
A	065	41	Uppercase A	┴	193	C1	Box [left top right]
B	066	42	Uppercase B	┬	194	C2	Box [left right bottom]
C	067	43	Uppercase C	├	195	C3	Box [top right bottom]
D	068	44	Uppercase D	─	196	C4	Box [left right]
E	069	45	Uppercase E	┼	197	C5	Box [left top right bottom]
F	070	46	Uppercase F	╞	198	C6	Box [top right bottom]
G	071	47	Uppercase G	╟	199	C7	Box [top right bottom]
H	072	48	Uppercase H	╚	200	C8	Box [top right]
I	073	49	Uppercase I	╔	201	C9	Box [right bottom]
J	074	4A	Uppercase J	╩	202	CA	Box [left top right]
K	075	4B	Uppercase K	╦	203	CB	Box [left right bottom]
L	076	4C	Uppercase L	╠	204	CC	Box [top right bottom]

Character/ Symbol	Decimal Code	Hexadecimal Code	Description	Character/ Symbol	Decimal Code	Hexadecimal Code	Description
M	077	4D	Uppercase M	=	205	CD	Box [left right]
N	078	4E	Uppercase N	╬	206	CE	Box [left top right bottom]
O	079	4F	Uppercase O	⊥	207	CF	Box [left top right]
P	080	50	Uppercase P	⊥⊥	208	D0	Box [left top right]
Q	081	51	Uppercase Q	╤	209	D1	Box [left right bottom]
R	082	52	Uppercase R	╥	210	D2	Box [left right bottom]
S	083	53	Uppercase S	╙	211	D3	Box [top right]
T	084	54	Uppercase T	╘	212	D4	Box [top right]
U	085	55	Uppercase U	╒	213	D5	Box [right bottom]
V	086	56	Uppercase V	╓	214	D6	Box [right bottom]
W	087	57	Uppercase W	╫	215	D7	Box [left top right bottom]
X	088	58	Uppercase X	╪	216	D8	Box [left top right bottom]
Y	089	59	Uppercase Y	┘	217	D9	Box [left top]
Z	090	5A	Uppercase Z	┌	218	DA	Box [right bottom]
[	091	5B	Opening square bracket	█	219	DB	Box [shade 4]
\	092	5C	Reverse slant; backslash	▄	220	DC	Box [left shade]
]	093	5D	Closing square bracket	▌	221	DD	Box [top shade]
^	094	5E	Caret (circumflex)	▐	222	DE	Box [right shade]
_	095	5F	Underscore (low line)	▀	223	DF	Box [bottom shade]
`	096	60	Grave accent	α	224	E0	Lowercase Greek alpha
a	097	61	Lowercase a	β	225	E1	Lowercase Greek beta
b	098	62	Lowercase b	Γ	226	E2	Uppercase Greek gamma
c	099	63	Lowercase c	π	227	E3	Lowercase Greek pi
d	100	64	Lowercase d	Σ	228	E4	Uppercase Greek sigma
e	101	65	Lowercase e	σ	229	E5	Lowercase Greek sigma
f	102	66	Lowercase f	μ	230	E6	Lowercase Greek mu
g	103	67	Lowercase g	τ	231	E7	Lowercase Greek tau
h	104	68	Lowercase h	Φ	232	E8	Uppercase Greek phi
i	105	69	Lowercase i	Θ	233	E9	Uppercase Greek theta
j	106	6A	Lowercase j	Ω	234	EA	Uppercase Greek Omega
k	107	6B	Lowercase k	δ	235	EB	Lowercase Greek delta
l	108	6C	Lowercase l	∞	236	EC	Infinity symbol
m	109	6D	Lowercase m	$\varnothing$	237	ED	Empty set (null set) symbol
n	110	6E	Lowercase n	$\in$	238	EE	Member (element) symbol
o	111	6F	Lowercase o	$\cap$	239	EF	Intersection symbol
p	112	70	Lowercase p	$\equiv$	240	F0	Equivalent sign
q	113	71	Lowercase q	$\pm$	241	F1	Plus or minus sign
r	114	72	Lowercase r	$\geq$	242	F2	Greater than or equal sign
s	115	73	Lowercase s	$\leq$	243	F3	Less than or equal sign

Character/ Symbol	Decimal Code	Hexadecimal Code	Description	Character/ Symbol	Decimal Code	Hexadecimal Code	Description
t	116	74	Lowercase t	⌠	244	F4	Integral (top)
u	117	75	Lowercase u	⌡	245	F5	Integral (bottom)
v	118	76	Lowercase v	÷	246	F6	Divide
w	119	77	Lowercase w	≈	247	F7	Approximately equal sign
x	120	78	Lowercase x	°	248	F8	Degree (ring)
y	121	79	Lowercase y	•	249	F9	Circle (solid)
z	122	7A	Lowercase z	∘	250	FA	Small centre dot
{	123	7B	Opening brace	√	251	FB	Root
|	124	7C	Vertical bar	^{n}	252	FC	Power of n
}	125	7D	Closing brace	^{2}	253	FD	Power of 2
~	126	7E	Tilde	■	254	FE	Square (solid)
	127	7F	Delete		255	FF	Undefined

*ASCII (American Standard Code for Information Interchange) is a set of binary codes that represent the most commonly used letters, numbers, and symbols.

Selected Letter and Nonletter Symbols in Science Technology

Symbol	Description	Symbol	Description
®	Registered trademark	$_{L}F$	Line feed [Data Communication]
©	Copyright	$_{N}L$	New line [Data Communication]
™	Trademark	$_{V}T$	Vertical tab [Data Communication]
SM	Servicemark	S_O	(Character) Shift out [Data Communication]
₵	Centerline	S_I	(Character) Shift in [Data Communication]
B_S	Backspace [Data Communication]	▲	Open/Eject [Electronics]
S_P	Space [Data Communication]	▶	Play [Electronics]
$_{H}T$	Horizontal tab [Data Communication]	▶▶	Fast Forward [Electronics]
$_{F}F$	Form feed [Data Communication]	◻	Rewind [Electronics]
$_{C}R$	Carriage return [Data Communication]	■	Stop [Electronics]
E_C	(Printer) Escape character [Data Communication]	●	Record [Electronics]